2021 22nd International Conference on Electronic Packaging Technology (ICEPT 2021)

Xiamen, China
14 – 17 September 2021

Pages 767-1532

IEEE Catalog Number: CFP21553-POD
ISBN: 978-1-6654-1392-3

Copyright © 2021 by the Institute of Electrical and Electronics Engineers, Inc.
All Rights Reserved

Copyright and Reprint Permissions: Abstracting is permitted with credit to the source. Libraries are permitted to photocopy beyond the limit of U.S. copyright law for private use of patrons those articles in this volume that carry a code at the bottom of the first page, provided the per-copy fee indicated in the code is paid through Copyright Clearance Center, 222 Rosewood Drive, Danvers, MA 01923.

For other copying, reprint or republication permission, write to IEEE Copyrights Manager, IEEE Service Center, 445 Hoes Lane, Piscataway, NJ 08854. All rights reserved.

*** This is a print representation of what appears in the IEEE Digital Library. Some format issues inherent in the e-media version may also appear in this print version.

IEEE Catalog Number: CFP21553-POD
ISBN (Print-On-Demand):   978-1-6654-1392-3
ISBN (Online):   978-1-6654-1391-6

Additional Copies of This Publication Are Available From:

Curran Associates, Inc
57 Morehouse Lane
Red Hook, NY 12571 USA
Phone: (845) 758-0400
Fax: (845) 758-2633
E-mail: curran@proceedings.com
Web: www.proceedings.com

TABLE OF CONTENTS

EVALUATION OF AGING PERFORMANCE OF THERMAL GEL SUBJECTED TO LASER FLASH TESTS.. 1

Yimin Yao, Yonglun Xu, Xue Bai, Yunsong Pang, Linlin Ren, Xiaoliang Zeng, Fei Deng, Jian-Bin Xu, Rong Sun

RESEARCH ON CONFORMAL PHASED ARRAY T/ R MODULE BASED ON LCP SUBSTRATE 7

Yan Luo, Kai Liu, Qiu Gao, Lei Ding, Yi Zhou, Lichun Wang

MECHANICAL RESPONSE OF BNNS-REINFORCED ALUMINUM COMPOSITES UNDER UNIAXIAL COMPRESSION.. 12

Jinming Li, Yuhua Huang, Baoshan Zeng, Chenzefang Feng, Fulong Zhu

FABRICATION OF SOFT CONDUCTIVE MICROSPHERES AND THEIR APPLICATION IN ELECTROMAGNETIC INTERFERENCE SHIELDING SHEETS 17

Luhui Zhang, Zuomin Lei, Jinming He, Deliang Zhu, Yougen Hu, Rong Sun

FATIGUE LIFE EVALUATION AND TEST METHOD FOR REPRESENTATIVE PRINTED CIRCUIT BOARD.. 21

Jun Tong, Hao Chen, Hefeng Liu

METHOD FOR PREPARING SILICON PHOTONIC CHIP EDGE PACKAGING STRUCTURE BASED ON INCLINED DEEP ETCHING PROCESS... 25

Heng Zhao, Laisheng He, Junbo Feng, Bangtong Ge

LIFE PREDICTION OF GOLD-ALUMINUM BONDING SYSTEM BASED ON FAILURE PHYSICS UNDER MULTI-STRESS COUPLING... 29

Menglin Li, Longfei Chen, Xianshun Zhang, Dongfei Zheng, Zeping Xiao

LOW-TEMPERATURE BONDING OF HIGH-POWER DEVICE USING CU-AG COMPOSITE NANOPARTICLE PASTE 34

Jiaxin Liu, Qing Wang, Yun Mou, Mingxiang Chen

RESEARCH ON THE UNIFORM TEMPERATURE OF HEAT DISSIPATION FOR THE REVERSE OBLIQUE MICROCHANNEL.. 38

Qinglin Tang, Dongcheng Liu, Yanping Zeng, Xin Lan, Lihua Zheng

A METHOD OF RESEARCH FOR THE RELIABILITY OF SOLDER JOINT SHAPE 44

Wenchao Tian, Xuewei Hou, Hao Cui, Xuegui Feng

INVESTIGATION OF THE RDL RELIABILITY BASED ON RF CHARACTERIZATION 48

Hongyue Wang, Weijie Zhang, Yijun Shi, Si Chen, Zhiwei Fu, Xiaofeng Yang, Bin Zhou

EFFECT OF NI_3SN_4 NANOPARTICLES ON GRAIN REFINEMENT IN SAC305 FREESTANDING SOLDER BALLS AND SAC305/CU BGA JOINTS 51

Xiaolei Ren, Xiaoying Liu, Longjiang Zou, Yunpeng Wang, Ning Zhao

RESEARCH ON EFFECT OF ANNEALING ON COPPER DEPOSITED BY ELECTROPLATING IN HIGH DENSITY TSV 55

Wei Wang, Fei Geng, Peng Sun, Yulong Ren, Huan Liu, Kai Zhang

DESIGN AND OPTIMIZATION OF A PNEUMATIC DOD SOLDER BALL 3D PRINTING SYSTEM 58

Zhixian Min, Huiming Pan, Dinglei Zhao, Sheng Liu, Zhiqin Wang, Zhiwen Chen

STRAIN RATE AND TEMPERATURE EFFECTS ON TENSILE PROPERTIES OF MONOCRYSTALINE CU6SN5 BY MOLECULE DYNAMIC SIMULATION 63

Jian Zhang, Wei Huang, Kai-Lin Pan

EFFECT OF ELECTROMIGRATION ON INTERFACIAL REACTION IN NI/SN63PB37/CU BGA SOLDER JOINTS 67

Fei Zhang, Shuai Chen, Zhidan Liu, Wenlong Wang

OPTIMIZATION OF SOLDER HEIGHT FOR STENCIL PRINTING PROCESS PERFORMANCE ON LENGTH-WIDTH RATIO 71

Dezhi Su, Cen Wang, Hongkun Wang, Junxiang Zhao, Hao Cheng, Lejun Zhang

HIGH STRENGTH AND DENSITY CU-CU JOINTS FORMATION BY LOW TEMPERATURE AND PRESSURE SINTERING OF DIFFERENT MASS RATIO OF CU MICRON-NANOPARTICLES PASTE 75

Zhongwei Huang, Jian Wen, Yu Zhang, Qiang Liu, Huacong Li, Jin Tong, Peilin Liang, Guannan Yang, Chengqiang Cui

RELIABILITY ASSESSMENT IN WELDING PROCESS OF SIP WITH DUAL-CHAMBER BY FINITE ELEMENT ANALYSIS 80

Dezhi Su, Fuxin Wang, Lejun Zhang, Cen Wang, Huihui Yang, Wenyu Jiang

ON-DIE CLOCK TREE LOW PSIJ THROUGH PDN OPTIMIZATION 85

Vinod Arjun Huddar

THE SHAPE CONTROL PROCESS OF A CU/SNAG SOLDER JOINT WITH A NI INSERTION USING THERMO-COMPRESSION BONDING 88

Mingang Fang, Zhuo Chen, Fuliang Wang, Chu Tang, Wenhui Zhu

DESIGN AND FABRICATION OF MULTI-LAYER SILICONE MICROCHANNEL COOLER FOR HIGH-POWER CHIP ARRAY 94

Tao Wei, Haojie Huang, Yupa Ma, Jiyu Qian

LASER RAPID SYNTHESIS OF ULTRA-SMALL NI NANOPARTICLES EMBEDDED GRAPHENE FOR HIGH-PERFORMANCE SUPERCAPACITORS 99

Fangcheng Wang, Zhuo Zhang, Guangyao Zhao, Mingjie Liu, Hongjin Fan, Cheng Yang

SIGNAL INTEGRITY DESIGN AND ANALYSIS OF HIGH BANDWIDTH MEMORY ON SILICON INTERPOSER* 103

Jin Hu, Tao Li, Yuqing Fan

STUDY ON THE TRANSPORT PERFORMANCE OF MICROSTRIP CIRCUIT BOARD WITH VOIDS IN SOLDER LAYER 107

Zhidan Liu, Zhiping Zhao, Fei Zhang, Shuai Chen

112G HIGH SPEED INTERFACE PACKAGE DESIGN AND SIMULATION 111

Jiangtao Zhang, Li Zhang, Jian Pang, Tuobei Sun, Keqing Ouyang

MICRO-VISION IMAGE STITCHING SYSTEM FOR LARGE-SCALE AND FINE-FEATURED CIRCUIT SUBSTRATES 115

Yuanyang Wei, Jian Gao, Yongbin Zhong, Lanyu Zhang

STUDY IN MULTILAYER WIRING TECHNOLOGY ON HIGH-HEAT-CONDUCTION SUBSTRATES 121

Lei Ding, Jing Chen, Kai Liu, Yan Luo, Yue Zhao, Lichun Wang

METHOD OF PREDICTING THE MAXIMUM STRESS OF BGA SOLDER JOINTS BASED ON BP NEURAL NETWORK 126

Huaiquan Zhang, Chunyue Huang, Shuaidong Liao, Shoufu Liu

THERMAL STRESS STUDY OF 3D IC BASED ON TSV AND VERIFICATION OF THERMAL DISSIPATION OF STI 131

Shuaidong Liao, Chunyue Huang, Huaiquan Zhang, Shoufu Liu

A HUMIDITY-SENSITIVE CAPACITOR BASED ON FAN-OUT PANEL LEVEL PACKAGE TECHNOLOGY 136

Shuhan Hou, Tingyu Lin

MICROSTRUCTURES PROPERTIES OF BARIUM-STRONTIUM TITANATE (BST) CERAMICS DOPED WITH B-LI GLASSES FOR LTCC TECHNOLOGY APPLICATIONS 140

Linjiang Tang, Xiaofeng Sun, Minghua Zhang, Chengan Wan

SPECTRUM ANALYSIS AND APPLICATION OF XY PLATFORM SERVO SYSTEM OF THE HIGH-PRECISION PACKAGING EQUIPMENT 144

Shnjin Liu, Yunbo He, Zesheng Li, Qihao Oian, Huilong Liao

DESTRUCTIVE PHYSICAL ANALYSIS METHODS OF FLIP CHIP PACKAGING DEVICES FOR HIGH RELIABILITY 148

Zhou Shuai, Weng Zhangzhao, Qiu Baojun, Luo Daojun, Wang Xiaoqiang, Ma Kaixue

SIMULATION ANALYSIS OF COUPLING COIL OF 13.56MHZ MAGNETIC COUPLING RESONANT WIRELESS ENERGY TRANSMISSION SYSTEM 153

Bihong Zhan, Wei Xia, Chunshui Xiong, Sheng Liu

STRESS ANALYSIS OF CU/SN BUMP EUTECTIC BONDING INTERFACE 159

Xinpeng Chen, Ruixia Huo, Daowei Wu, Wansheng Liu

CHALLENGEABLE MECHANICAL ISSUES IN MICROELECTRONIC PACKAGES FOR DEVELOPMENTS 163

Xiangdong Xue, Jianghao Wei, Yuming Wang, Jianxin Yang

THE INTERFACIAL REACTION OF CU/RNULTILAYER SN -CU -SN / CU JOINT IN SOLDERING 169

Min Shang, Chong Dong, Xiangxu Chen, Shaocheng Wu, Haoran Ma, Haitao Ma

STUDY ON THE INFLUENCE OF DIFFERENT FILLER FRACTIONS ON THE PROPERTIES OF THERMAL INTERFACE MATERIALS 173

Wenbo Ye, Zhenyu Wang, Xiangliang Zeng, Linlin Ren, Rong Sun, Zhibin Wen, Xiaoliang Zeng

PACKAGING OF A MEMS SENSOR IN AN ACTIVE INTERVENTIONAL BLOOD PRESSURE MONITORING CATHETER 177

Guanzhe Xu, Junshu Lin, Rongying Yu, Zebo Zhang, Meng Gao, Le Ye

HEAT TRANSFER ANALYSIS OF PHASE CHANGE MATERIALS WITH METAL FOAMS 180

Yan Zhang, Huihui Wang, Pei Lu, Jingyu Fan, Qixuan Tu, Johan Liu

DELAMINATION REDUCTION BY MATERIAL AND PROCESS OPTIMIZATION 185

Tina Li, Aaron He, Phoebe Chen, Aaron Lai, Yuan Hang, Colin Feng

ELECTROMECHANICAL CO-DESIGN AND EXPERIMENTAL TESTING OF PACKAGE LAYER IN STRUCTURALLY EMBEDDED PHASED ARRAY ANTENNA 189
Jinzhu Zhou, Zhenyu Gu, Yu Si, Mei Wang, Ping'An Wang

THE FORMATION OF CN-SN IMC INTERCONNECTION BY SOLID-LIQUID INTERDIFFUSION BONDING FOR 3D GLASS WAFER STACKING 193
Yangquan Su, Kuili Ren, Yiyong Huang, Mingchuan Zhang, Daquan Yu

RESEARCH ON THE RELIABILITY OF BOARD LEVEL INTERCONNECT SOLDER JOINTS UNDER THERMAL-MECHANICAL COUPLING 197
Xin Liu, Yongjian Yu, Weikun Xie, Kaihong Zhang, Kai Zhu

APPROACH TOWARDS ACCURATE MODELING OF THERMAL RESISTANCE IN THERMAL MANAGEMENT OF PCB 202
Jianghao Wei, Tao Wan, Xiangdong Xue, Yuming Wang

STUDY ON WARPAGE AND PEELING MITIGATION OF WAFER LEVEL DURING METAL PLATING PROCESS 208
Dayang Li, Ming Xiao, Zhiqi Wang

THE STUDY OF EFFECTS TO THE THERMO-MECHANICAL PERFORMANCE OF THE FIRST LEVEL THERMAL INTERFACE MATERIALS 213
Zhenyu Wang, Linlin Ren, Xiangliang Zeng, Wenbo Ye, Yonglun Xu, Xiaoliang Zeng, Yunsong Pang, Sun Rong

FACILE PREPARATION OF COBALT HYDROXIDE BASED SUPERCAPACITOR WITH HIGH VOLUMETRIC ENERGY DENSITY AT HIGH VOLUMETRIC POWER DENSITY 219
Peng Liu, Fangcheng Wang, Zhuo Zhang, Jing Li, Hongjin Fan, Cheng Yang

DIE CHIPPING FDC DEVELOPMENT AT WAFER SAW PROCESS 224
Dongpeng Xue, Caiden Zhong, Elley Zhang, Weiting Jiang, Cong Zhang

HYBRID-EMBEDDED SIP PACKAGE DESIGN 226
Louise Tan, Chender Chen, Cc Liao

A NOVEL CU@SN@AG CORE-SHELL PARTICLES FOR DIE ATTACHMENT IN POWER DEVICE PACKAGING 229
Jiahao Liu, Hui Xiao, Xiaotong Guo, Xinjie Wang, Zhijun Yao, Xingchao Mao, Hao Liu, Hongtao Chen

EFFECT OF DIMENSION OF BOARD AND MICRO-BUMPS ON INTERCONNECTION STRESS UNDER DROP TEST 233
Mingtao Lv, Shimei Liu, Taotao Chen, Yunpeng Liu, Junfu Liu, Hu He

THERMOMECHANICAL AND ELECTRICAL PROPERTIES OF THE $SIO_2/ZRW_2O_8/EPOXY$ COMPOSITE 239
Chaofan Li, Suibin Luo, Shuhui Yu, Baojin Chu, Rong Sun

HIGH EFFICIENCY TESTING SYSTEM FOR 5G POWER AMPLIFIER 244
Zongqi Cai, Sha Tang, Jun Luo, Xing Li, Xiaoqiang Wang, Daojun Luo

RESEARCH ON HIGH-SPEED SERDES INTERFACE TESTING TECHNOLOGY 248
Weikun Xie, Guangqiang Cao, Weiwei Ji

HIGH-PERFORMANCE THERMAL GREASE WITH THE ADDITION OF SILVER PARTICLES 253

Xiangliang Zeng, Zhenyu Wang, Wenbo Ye, Linlin Ren, Xiaoliang Zeng, Xinnian Xia, Rong Sun

SOP WELDING JOINT BENDING STRESS FINITE ELEMENT ANALYSIS AND OPTIMIZATION 257

Gong Jinfeng, Huang Chunyue, Li Maolin, Liu Shoufu

COMPARATIVE ANALYSIS OF TEMPERATURE-INDUCED MICRO-SCALE DEFORMATION OF PACKAGE BY EXPERIMENT AND FINITE ELEMENT ANALYSIS 261

Cheng Zhong, Chenglong Li, Tao Peng, Yunxia Wang, Gang Li, Pengli Zhu, Jibao Lu, Rong Sun, Ching-Ping Wong

ACTIVE HEAT DISSIPATION BY CHIP ON THERMOELECTRIC COOLER FOR HIGH-POWER LED 265

Shuang Li, Jinglong Liu, Yang Peng, Mingxiang Chen

TUNING THE CURING TEMPERATURE OF POLYIMIDE PRECURSOR: PLOY AMIDE ESTER 270

Kuangyu Wang, Liang Shan, Guoping Zhang, Rong Sun

VISCOELASTIC CHARACTERIZATION AND SIMULATION OF THERMAL INTERFACE MATERIALS 274

Cheng Zhong, Chenglong Li, Yunxia Wang, Jibao Lu, Linlin Ren, Rong Sun, Ching-Ping Wong

CHARACTERIZATION AND VERIFICATION OF VISCOELASTIC CONSTITUTIVE PARAMETERS OF UNDERFILL MATERIAL 278

Cheng Zhong, Chenglong Li, Lu Lu, Yunxia Wang, Gang Li, Pengli Zhu, Jibao Lu, Rong Sun, Ching-Ping Wong

SYNTHESIS AND PROPERTIES STUDY OF A THERMOPLASTIC POLYIMIDE WITH HIGH GLASS TRANSITION TEMPERATURE FOR WAFER LEVEL PACKAGE 283

Wen Liu, Jinhui Li, Jinshan Liu, Tao Wang, Ao Zhong, Guoping Zhang, Qiang Liu, Rong Sun, Daquan Yu

THE EFFECT OF THERMAL-INDUCED WARPAGE AND DEGENERATION OF THERMAL INTERFACE MATERIALS ON THE THERMAL PERFORMANCE OF A FLIP-CHIP PACKAGE 287

Ruoyu Jiang, Cheng Zhonz, Haozhe Wang, Chenglong Li, Yi Zheng, Linlin Ren, Jibao Lu, Rong Sun, Ching-Ping Wong

ENHANCED DISCHARGED ENERGY DENSITY IN POLYETHERIMIDE COMPOSITES BY BORON NITRIDE/ALUMINUM NITRIDE HYBRID FILLERS 292

Xudong Wu, Shiyi Gao, Xin Wu, Shuo Zhang, Zhijun Cao, Daniel Q. Tan

THE STUDY ON THERMAL AGING MECHANISM OF SILICONE MATERIALS FOR LED ENCAPSULATION 295

Jiabao Gu, Huanxiang Xu, Bo Peng, Zilian Liu, Gang Zhu

DEFECT LOCALIZATION AND OPTIMIZATION OF PIND FOR LARGE SIZE CQFP DEVICES 298

Shinan Wang, Yong Ma, Kaihong Zhang, Yongjian Yu, Yongkang Wan, Weikun Xie

EXPLORATION OF THE SYNTHESIS METHOD OF QUATERNARY COPOLYMERIZED THERMOPLASTIC POLYIMIDE 301
Jinshan Liu, Jinhui Li, Fangfang Niu, Tao Wang, Wen Liu, Guopinz Zhang, Rong Sun

IMPACT FORCE CONTROL OF HIGH-SPEED WIRE BONDING MACHINE BASED ON FUZZY ACTIVE DISTURBANCE REJECTION CONTROLLER 305
Yachao Liu, Jian Gao, Boyu Zhan, Lanyu Zhang

FAILURE ANALYSIS OF ANISOTROPIC CONDUCTIVE ADHESIVE PACKAGES IN NARROW-PITCH FLIP CHIP PACKAGING 311
Gui Chen, Yan Wang, Xiaoyu Xiao, Yamei Yan, Wenhui Zhu

FAILURE MECHANISM OF NICKEL-CHROMIUM THIN FILM CHIP RESISTORS 316
Zhiyuan Mao, Gaoming Shi, Weili Li, Xianjun Kuang, Fuyao Mo

FUZZY TUNING ALGORITHM FOR FEEDFORWARD PARAMETER BASED ON IC PACKAGE FOR MASS TRANSFER OF MICRO-LED EQUIPMENT XY MOTION PLATFORM 320
Wenbin Fan, Yunbo He, Guofu Qiu

NUMERICAL ANALYSIS OF THE MICROSCOPIC FACTORS INFLUENCING THE THERMAL CONDUCTIVITY OF AL_2O_3/AIN POLYMER COMPOSITES 326
Nan Cheng, Xiaoxin Lu, Jiabin Huang, Jibao Lu, Shen Xu, Sun Rong, Jianbin Xu, Ching-Ping Wong

THE INFLUENCE ANALYSIS OF GEOMETRY ON VOID IN MOLDED UNDERFILL FOR FLIP CHIP 331
Yamei Yan, Gui Chen, Xiaoyu Xiao, Yan Wang, Wenhui Zhu

A COST-SAVING THERMAL TEST CHIP DESIGN IN A TEST VEHICLE OF LARGE BGA 336
Jianjun Sun, Yuanting Lai, Hao Yang, Jian Pang, Tuobei Sun, Keqing Ouyang

THERMODYNAMIC SIMULATION AND ANALYSIS OF METAL BUMPS IN FLIP-CHIP MICRO-LED PACKAGING 340
Xiaoyu Xiao, Yamei Yan, Gui Chen, Wenhui Zhu

STRESS-STRAIN STUDY OF QFN SOLDER JOINTS WITH DIFFERENT STRUCTURAL PARAMETERS UNDER RANDOM VIBRATION LOADING 346
Maolin Li, Chun-Yue Huang, Zhuo Wang, Wei Wei

COMPARISON BETWEEN TWO NUMERICAL METHODS FOR THE COMPUTATION OF THERMAL CONDUCTIVITIES OF PARTICULATE COMPOSITES: FEM AND GEODICT 351
Xiaoxin Lu, Jiabin Huang, Jianbin Xu, Jibao Lu, Sun Rong, Ching-Ping Wong

RESEARCH ON POINT-TO-POINT MOTION CONTROL OF PACKAGING EQUIPMENT 356
Zesheng Li, Yunbo He, Shujin Liu, Qihao Qian

NUMERICAL ANALYSIS ON THE EFFECT OF MICROSTRUCTURES ON THE THERMAL AND MECHANICAL PROPERTIES OF CARBON FIBER/AL_2O_3 THERMAL PAD 360
Shu Liu, Xiaoxin Lu, Jiabin Huang, Jibao Lu, Shen Xu, Sun Rong, Jianbin Xu, Ching-Ping Wong

QUALITY INSPECTION OF OPTICAL LENS IN IC PACKAGING EQUIPMENT BASED ON MTF 365
Qihao Qian, Yunbo He, Zesheng Li, Shujin Liu, Huilong Liao

STRESS ANALYSIS AND PARAMETER OPTIMIZATION OF FINE-PITCH BGA SOLDER JOINTS UNDER CANTILEVER PLATE TORSION CONDITIONS 371
Zhuo Wang, Chunyue, Jinfeng Gong, Huaiquan Zhang, Shuaidong Liao, Shoufu Liu

THE INFLUENCE AND OPTIMIZATION OF DESIGN PARAMETERS ON INTEGRATED CIRCUITS PACKAGE WARPAGE 376
Qiang Wei, Cao Ting, Weidong Liu, Ning Sun, Qu Fang, Jie Liu, Huan Yang, Xiaojian Ma

THE IN-SITU OBSERVATION OF MICROSTRUCTURE, GRAIN ORIENTATION EVOLUTION AND ITS EFFECT ON CRACK PROPAGATION PATH IN SAC305 UNDER EXTREME TEMPERATURE CHANGES 381
Kexin Xu, Xing Fu, Min Liu, Zhiwei Fu, Si Chen, Yijun Shi, Yun Huang, Hongtao Chen

A VERTICAL TRANSMISSION LEADLESS SURFACE-MOUNTABLE CERAMIC PACKAGE WITH HIGH CORE PROPORTION 385
Zhizhuang Qiao, Linjie Liu, Yangfan Zhou, Ke Wang

RELIABILITY AND THERMAL DEGRADATION OF FIRST-LEVEL THERMAL INTERFACE MATERIALS 388
Yunpeng Su, Junhong Li, Qiangquiang Ma, Linlin Ren, Xiaoliang Zeng, Rong Sun

COUPLING DAMAGE ACCUMULATION OF DIE-ATTACH SOLDER LAYER WITH DISTRIBUTED VOID DEFECTS FOR POWER ELECTRONICS 393
Yidian Shi, Cheng Peng, Wenhui Zhu, Taotao Chen, Yunpeng Liu, Junfu Liu, Hu He

RESEARCH ON WIRE SWEEP OF INTEGRATED CIRCUIT PACKAGING BASED ON THREE-DIMENSIONAL FLOW SIMULATION 398
Fang Qu, Ting Cao, Huan Yang, Ning Sun, Qiang Wei, Xiaojian Ma, Weidong Liu

CHARACTERISTICS OF 10–110GHZ TRANSMISSION LINES ON FUSED SILICA SUBSTRATE FOR MILLIMETER-WAVE MODULES 403
Tian Yu, Kai Xue, Ke Li, Yihang Liang, Daquan Yu

LOW TEMPERATURE BONDING BY SINTERING OF AG NANOPARTICLE PASTE WITH THE ASSISTANCE OF MOD 408
Xun Liu, Yulei Yuan, Junjie Li, Li Liu, Rong Sun

CU-CU JOINT FORMATION BY SINTERING OF SELF-REDUCIBLE CU NANOPARTICLE PASTE ASSISTED BY MOD UNDER AIR CONDITION 412
Yulei Yuan, Xun Liu, Junjie Li, Pengli Zhu, Rong Sun

RESEARCH ON THE BOARD LEVEL RELIABILITY OF CQFJ CERAMIC PACKAGE 416
Zhen-Tao Yang, Fei Yu, Lin-Jie Liu, Ling Gao

THERMAL AND OPTICAL MODELING ON INTELLIGENT LED HEADLIGHTS 420
Yikang Qin, Miao Cai, Xindong Chen, Jinyang Li, Daoguo Yang, Guoqi Zhang

RESEARCH ON THE DESIGN AND PROCESSING TECHNOLOGY OF CQFJ CERAMIC PACKAGE 424
Fei Yu, Zhen-Tao Yang, Lin-Jie Liu, Ling Gao

RESEARCH ON DOUBLE-LAYER NETWORKS-ON-CHIP FOR INTER-CHIPLET DATA SWITCHING ON ACTIVE INTERPOSERS 428
Xiaolong Duan, Min Miao, Zhuanzhuan Zhang, Liang Sun

INVESTIGATION OF THE INFLUENCES OF THERMAL STRESSES AND JOULE HEATING WITHIN A PIEZORESISTIVE MEMS PRESSURE SENSOR USING THE FINITE ELEMENT MODELING.. 434
    Chunming Zhou, Peng Zhou, Yuehua Hu, Hao Zhang

ACCELERATED AGING AND LIFETIME EVALUATION OF POLYURETHANE PACKAGING MATERIAL FOR OPTICAL FIBER HYDROPHONE.. 440
    Wenyuan Liao, Canxiong Lai, Rui Gao, Shaohua Yang, Guoguang Lu, Shuwang Li

CHARACTERIZING THE DIE ATTACH LAYER DELAMINATION EFFECT ON THE HEAT TRANSFERRING PERFORMANCE IN LED PACKAGE WITH ENTROPY GENERATION ANALYSIS 443
    Binjie Ai, Miao Cai, Daoguo Yang, Guangsheng Lu, Kailin Zhang, Guoqi Zhang

SHEAR PROPERTIES AND FRACTURE BEHAVIORS OF CU/SN-37PB/CU SOLDER INTERCONNECTIONS AT CRYOGENIC TEMPERATURES.. 447
    Ruyu Tian, Chunlei Wang, Yanhong Tian

RESEARCH ON 3D INTERPOSER/CHIP STACKING TECHNOLOGY AND RELIABILITY 450
    Ning Zhang, Baoxia Li, Qiucheng Yan, Daowei Wu

HIGHLY CONDUCTIVE SILVER NANOWIRE TRANSPARENT ELECTRODES HYBRIDIZED WITH LAMINATED MULTI-LAYER MXENE 454
    Pengchang Wang, Maoliang Jian, Chi Zhang, Majiaqi Wu, Huaying Hu, Lianqiao Yang

SYNTHESIS OF AIR-SINTERABLE COPPER NANOPARTICLES FOR DIE-ATTACHMENT 458
    Yue Yao, Liang Xu, Pengli Zhu, Tao Zhao, Rong Sun, Yinachao Huo

THE PARTICLE INTERACTION ANALYSIS FOR NANOPARTICLES IN UNDERFILL FOR FLIP-CHIP PACKAGING-.. 462
    Mingyong Du, Ning Wang, Xiaomeng Du, Tao Zhao, Pengli Zhu, Rong Sun

A 3D TSV-MEMS BASED HETEROGENEOUS INTEGRATION TECHNOLOGY FOR RF APPLICATION.. 466
    Min Huang, Tinglei Wang, Fan Hou, Ping Su, Chao Sun, Huakai Luan

THE EFFECT OF ANNEALING TIME ON THE MECHANICAL PROPERTIES OF TSV-CU 470
    Yadong Li, Pei Chen, Fei Qin

STUDY ON CURRENT CARRYING CAPACITY OF A NOVEL INTERCONNECT MATERIAL ZRTE₃ 475
    Xiaokun Wen, Liangyi Ni, Wenyu Lei, Li Yang, Yuan Liu, Pengzhen Zhang, Haixin Chang, Wenfeng Zhang

INTERACTION OF SILANE COUPLING AGENTS WITH NANO-SILICA PROBED BY NANO-IR 479
    Pengli Zhu, Jianjun Ruan, Mingyong Du, Ning Wang, Xiaomeng Du, Tao Zhao, Xiaodong Li, Jiakai Cao, Jianping Zhang, Xiaoyao Sun

LIGHTWEIGHT AND COMPRESSIBLE EXPANDABLE POLYMER MICROSPHERES/SILVER FLAKES COMPOSITES FOR HIGH-EFFICIENCY ELECTROMAGNETIC INTERFERENCE SHIELDING.. 483
    Jianhong Wei, Yadong Xu, Zhiqiang Lin, Zuomin Lei, Xianzhu Fu, Yougen Hu, Rong Sun

STUDY OF EFFICIENT AUTOMATIC DETECTION OF COPLANARITY AND POSITION OF CCGA DEVICES 489
    Qi Zhang, Xiaoyan Liu, Leida Chen, Xujing Nan

MODELING AND SIMULATION OF INTERCONNECTION STRUCTURE COMPENSATION DESIGN IN HIGH-SPEED MODULES 493
Qi Zheng, Huimin He, Lijuan Bai, Yubo Wang, Dejian Li, Shunfeng Han, Bofu Li, Dameng Li, Fengman Liu, Liqiang Cao

STUDY ON GOLD WIRE SWEEP IN CANTILEVER CHIP-STACKED PACKAGE DURING MOLDING PROCESS 497
Sicheng Cao, Daoguo Yang, Wangyun Li, Xiyou Wang, Shirui Xue, Zhanfei Yun

X-SHAPED THROUGH GLASS VIA AND ITS TRANSMISSION PERFORMANCE IN KA BAND 501
Qiangwen Wang, Yuhua Guo, Rui Wang

PROCESS DEVELOPMENT AND FAILURE ANALYSIS OF SUPER-SIZE EMBEDDED SILICON FAN-OUT PACKAGE 506
Dongzhi Fu, Shuying Ma, Jiao Wang

SURFACE MODIFICATION OF GRAPHITE AND ITS EFFECT ON THERMAL AND MECHANICAL PROPERTIES OF GRAPHITE-BASED THERMAL INTERFACE MATERIALS 512
Yuexing Zhang, Hong He, Junwei Li, Chenxu Zhang, Rong Sun, Meng Han, Ping Zhang

DESIGN, FABRICATION, AND TEST OF AN EMBEDDED SI-GLASS MICROCHANNEL HEAT SINK FOR HIGH-POWER RF APPLICATION 517
Jianyu Du, Weihao Li, Xu Gao, Deyin Zheng, Yuchi Yang, Zetian Wang, Haoran Zhao, Jiajie Kang, Wei Wang

A HIGH-Q INDUCTOR BASED ON FAN-OUT PANEL LEVEL PACKAGE TECHNOLOGY 521
Xulei Niu, Zixu Wang, Shaopan Lin, Guochi Huang, Tingyu Lin

EFFECTS OF CETYLTRIMETHYLAMMONIUM BROMIDE (CTAB) ON ELECTROPLATING TWIN-STRUCTURED COPPER INTERCONNECTS 525
Yi Dong, Zhe Li, Li-Ying Gao, Xiao Li, Zhi-Quan Liu, Rong Sun

ANALYSIS FOR THERMAL CONTACT RESISTANCE OF PRESS-PACK IGBTS 530
Yakun Zhang, Tong An, Fei Qin, Yanpeng Gong, Chen Liang

ANALYSIS ON THE THERMAL STRESS OF AL-SI THIN FILM USING DIC METHOD 535
Huiming Pan, Guoliang Xu, Zhiwen Chen, Chongming Zhang, Chao Sun, Sheng Liu, Li Liu

FLEXIBLE THERMAL INTERFACE MATERIALS THROUGH CONTROLLING THE RATIO OF SILICONE OIL FUNCTIONAL GROUPS 539
Wendian Tu, Linlin Ren, Guoping Du, Rong Sun, Xiaoliang Zeng, Junwei Li

SIMULATION AND OPTIMIZATION OF 3D HETEROGENEOUS INTEGRATION OF INERTIAL MICRO-SYSTEM 543
Nanxin Wang, Shenglin Ma, Yufeng Jin, Chaoyang Xing, Nannan Li, Peng Sun

STUDY ON IMAGE ALIGNMENT TECHNOLOGY BASED ON CCD THERMAL REFLECTION METHOD 549
He Yang, Weikang Si, Dazheng Wang, Yingying Gao, Libing Zheng

A THERMAL NETWORK MODEL FOR THERMAL ANALYSIS IN AUTOMOTIVE IGBT MODULES 555
Yanzhong Tian, Tong An, Fei Qin, Yanpeng Gong, Chen Liang

NOVEL WATER-SOLUBLE PROTECTIVE ADHESIVE FOR WAFER'S LASER DICING 560
Deliang Sun, Jinhui Li, Yuxi Yi, Guoping Zhang, Rong Sun, Mingqi Huang

A TGV-BASED ANTENNA IN PACKAGE FOR 5G MM-WAVE APPLICATION 564
Sha Xu, Dianyang Shi, Chunbing Guo

LOADING RATE ON MODE II FRACTURE TOUGHNESS OF SINTERED SILVER 568
Yanning Li, Yanwei Dai, Fei Qin, Shuai Zhao

COMPARATIVE RESEARCH OF INFRARED THERMOGRAPHY AND ELECTRICAL
MEASUREMENT METHOD FOR THE THERMAL CHARACTERISTICS TEST OF GAN
HEMT DEVICES 572
Zhiwei Fu, Bingjie Zheng, Xu Huang, Bin Zhou, Xiaofeng Yang, Huaixin Guo

RESEARCH PROGRESS OF EXTREME LOW TEMPERATURE RELIABILITY OF TYPICAL
ELECTRONIC INTERCONNECTION STRUCTURES 576
Zhaoning Sun, Xiaotong Guo, Zhenbo Zhao, Yiqing Ni, Guanghui He

PROPERTIES OF ROOM TEMPERATURE BONDED AND UV CURED TEMPORARY
BONDING ADHESIVE FOR ULTRA-THIN WAFER'S HANDLING 581
Xujun Li, Qiang Liu, Deliang Sun, Zhipeng Li, Guoping Zhang, Rong Sun

EFFECT OF THERMOMIGRATION ON EVOLUTION OF INTERFACIAL INTERMETALLIC
COMPOUNDS IN CU/NI/SN-3.5AG MICROSOLDER JOINTS FOR 3D INTERCONNECTION 585
Chu Tang, Wenhui Zhu, Zhuo Chen

IMPROVEMENT OF AU ELECTRODE BY GLASS OPTIMIZATION FOR LTCC
APPLICATION 591
Tingnan Yan, Dawei Wang, Yuanyuan Wang, Jinhao Jia

STUDY ON THE HEAT DISSIPATION PERFORMANCE OF SYMMETRICAL BROKEN-
LINE MICROCHANNEL RADIATOR 594
Pengfei Wang, Hongyue Wang, Yijun Shi, Xiangjun Lu, Jile Xu

FCCSP(MUF) MOLD-FLOW VOID RISK PREDICTION WITH DIFFERENT SUBSTRATE
SURFACE AND BUMP HEIGHT DESIGN 598
Freedman Yen, Nicholas Kao, David Lai, Yu Po Wang

RESEARCH ON ELECTROMIGRATION BEHAVIOR OF CU PILLAR BUMPS UNDER
PULSE CURRENT STRESS 603
Jile Xu, Xiangjun Lu, Zhiwei Fu, Chenbing Qu, Xiao Luo, Pengfei Wang

SIMULATION ON TSV PROTRUSION FROM ATOMIC TO MICRON SCALES 607
Xiaoting Luo, Zhiheng Huang, Yuezhong Meng, Shan Ren, Hui Yan, Qizhuo Li

BGA CHIP TORSION FINITE META ANALYSIS AT HIGH TEMPERATURE 613
Yu-Qing Yue, Chao Jiang, Yan-Ting Chen, Jing Wei, Lv Wu

THERMAL ANALYSIS OF HIGH-POWER LIGHT-EMITTING DIODE USING
THERMOREFLECTANCE THERMOGRAPHY 617
Dazheng Wang, Libing Zheng, Weikang Si, He Yang, Yingying Gao

LOW-TEMPERATURE CU/SIO₂ HYBRID BONDING USING A NOVEL TWO-STEP
COOPERATIVE SURFACE ACTIVATION 621
Qiushi Kang, Chenxi Wang, Ge Li, Shicheng Zhou, Yanhong Tian

KEY FACTOR ANALYSIS OF NANO SILICA ON THE DISPERSION IN UNDERFILL 626
*Xiaomeng Du, Ning Wang, Mingyong Du, Leicong Zhang, Tao Zhao, Pengli Zhu, Rong Sun,
Jiakai Cao, Jianjun Ruan*

EFFECT OF FILLER, TOUGHENING AGENT AND COUPLING AGENT ON THE CURING SHRINKAGE OF EPOXY-BASED UNDERFILLS 630

Xiaohui Peng, Jinbao Yang, Tao Peng, Pengli Zhu, Xing Ouyang, Yan Pan, Gang Li, Rong Sun, Yajing Yang, Liang Peng

SOLID-LIQUID MIXING-STATE ORGANIC LENSES FOR DEEP-ULTRAVIOLET LIGHT-EMITTING DIODES TO ENHANCE THE LIGHT-EXTRACTION EFFICIENCY 634

Zihao Deng, Jiexin Li, Jiayong Liang, Jiayi Li, Jiasheng Li, Xinrui Ding, Zongtao Li

NUMERICAL SIMULATION ANALYSIS OF FLEXIBLE PRINTED CIRCUITS UNDER BENDING CONDITIONS 638

Yongchao Liu, Haozhe Wang, Xianqin Hu, Chao Peng, Xu Long, Jibao Lu, Rong Sun, Ching-Ping Wong

TENSILE DEFORMATION MECHANISM OF SN-37PB SOLDER ALLOY AT CRYOGENIC TEMPERATURES 642

Xiaotong Guo, Kun Zhang, Jiahao Liu, Yong Li, Xinlang Zuo, Hui Xiao, Guanghui He

THE RELIABILITY ASSESSMENT OF PULSE-DRIVEN LIGHT EMITTING DIODES 646

Shen Yaoyang, Sun Bo

SEQUENTIAL ANALYSIS OF DROP IMPACT AND THERMAL CYCLING OF ELECTRONIC PACKAGING STRUCTURES 651

Yongchao Liu, Xu Long, Haozhe Wang, Jibao Lu, Rong Sun, Ching-Ping Wong

ENHANCED OPTICAL PERFORMANCE AND THERMAL STABILITY OF QUANTUM DOT CONVERTERS FOR LASER SOURCE 655

Jiayong Liang, Jiexin Li, Zihao Deng, Yihua Qiu, Zongtao Li, Jiasheng Li

MICRON-SCALE SILVER FLAKE PASTE SINTERING WITHOUT PRESSURE FOR POWER ELECTRONIC DIE ATTACHMENT 659

Shijun Huang, Ruidong Luo, Zhen Wu, Cai-Fu Li

WARPAGE MEASUREMENT OF SUBSTRATES AND PRINTED CIRCUIT BOARDS WITH SHADOW MOIRÉ 663

Xingjia Huang, Changping Ou, Shengcong Zhu, Yixiu Huang

RELIABILITY ANALYSIS OF THERMAL INTERFACE MATERIALS (TIMS) IN LARGE SIZE FCBGA PACKAGE 667

Yi Zheng, Cheng Zhong, Haozhe Wang, Jibao Lu, Rong Sun, Ching-Ping Wong

STUDY ON WARPAGE AFTER POST SOLIDIFYING OF ULTRATHIN FINGERPRINT PACKAGE PRODUCTS 671

Ning Sun, Xiaojian Ma, Fang Qu, Qiang Wei, Huan Yang, Weidong Liu, Ting Cao

INFLUENCE OF IMC MORPHOLOGY ON FATIGUE STRESS, STRAIN AND LIFE OF SOLDER LAYER BETWEEN SIC CHIP AND DBC SUBSTRATE IN IGBT UNDER THERMAL CYCLING 674

Guang Yang, Fengshun Wu, Longzao Zhou, Xinghe Luan, Xinrui Zou, Hui Liu, Yang Wan, Xiaowei Zhang, Bin Wang

RESEARCH ON RELIABILITY OF SOLDER LAYER IN IGBT MODULE PACKAGING 679

Panwang Chi, Shengru Lin, Yesu Li, Yiping Liu, Jicun Lu, Ming Li, Liming Gao

THE STUDY ON ELECTROMIGRATION OF SOLDER JOINTS UNDER THERMAL CYCLING LOAD 684

Leyi Niu, Xiaodi Tian, Fei Jia

EVALUATION OF FATIGUE CRACK GROWTH IN SOLDER LAYER OF IGBT MODULE UNDER POWER CYCLE BY USING J-INTEGRAL METHOD 689
Kai Yang, Longzao Zhou, Fengshun Wu, Yi Zhang, Yang Han, Zhou Zhang, Yang Wan, Xiangmiao Huang, Dayong Huang

STUDY ON LEADFRAME OVERFLOW PREVENTION OF SOLDERING PASTE USING FLUID-STRUCTURE COUPLING ANALYSIS 695
Guangsheng Lu, Daoguo Yang, Wangyun Li, Xiyou Wang, Xiangli Wei, Binjie Ai

SIMULATION STUDY ON THERMOMECHANICAL RELIABILITY IN EMBEDDED DIE PACKAGE FABRICATION PROCESS 699
Zhou Zhou, Haibo Fan, Yuning Shi

WETTABILITY IMPROVEMENT OF SOLDER IN FLUXLESS SOLDERING UNDER FORMIC ACID ATMOSPHERE 705
Yuhao Bi, Siliang He, Wangyun Li, Daoguo Yang, Hiroshi Nishikawa

THE EFFECT OF FLUX ON SI-AL WIRE BONDING RELIABILITY 709
Yao Zhang, Wuxing Cao, Jun Zhang, Liu Yang, Pei Zhang, Jiao Yang

FINITE ELEMENT ANALYSIS OF THERMAL CONTACT RESISTANCE IN PRESS-PACK IGBT MODULE 714
Tong An, Rui Zhou, Fei Qin, Yanpeng Gong, Chen Liang

ORTHOGONAL EXPERIMENT FOR ANALYZING THE IMPACT OF THERMAL STRESS ON THE RELIABILITY OF AN EMC PACKAGE 718
Yulong Li, Yi Zheng, Haozhe Wang, Jibao Lu, Rong Sun, Wenhui Zhu, Ching-Ping Wong

OPTICAL PERFORMANCE ANALYSIS OF UVC-LED PACKAGE STRUCTURE BASED ON RAY-TRACING SIMULATION 722
Jinyang Li, Wangyun Li, Daoguo Yang, Yikang Qin, Sicheng Cao, Feixiang Liu

DETERMINATION OF PARAMETERS IN MIXED-MODE COHESIVE ZONE MODELS FOR MODIFIED BUTTON SHEAR TESTS BY PARTICLE SWARM OPTIMIZATION 726
Wenyu Wu, Ke Xue, Weijing Dai, Dashun Liu, Dali Yang

DESIGN AND SIMULATION OF 3D ANTENNA BASED ON CONICAL VIA STRUCTURE 732
Ziyu Liu, Junhao Wang, Zhiyuan Zhu, Lin Chen, Qingqing Sun

COMPARATIVE STUDY ON THE EFFECTS OF FE AND NI ADDITIONS ON THE ELECTROMIGRATION PROPERTIES OF SN58BI SOLDER JOINTS 736
Zhuangzhuang Hou, Yaru Dong, Lingyao Sun, Ying Liu, Yongiun Huo, Yingxia Liu, Xiuchen Zhao

LOW TEMPERATURE AND SHORT TIME AU/SN SOLID-LIQUID DIFFUSION BONDING FOR 3D INTEGRATION 740
Ziyu Liu, Wang Wenchao, Zhu Zhiyuan, Chen Lin, Sun Qingqing

COPPER ADHESION PROMOTERS FOR POLYIMIDE: HETEROCYCLIC COMPOUNDS ADDITIVES CONTAINING AMINO AND HYDROXYL GROUPS 746
Yingying Li, Guoping Zhang, Jinhui Li, Changqing Li, Ao Zhong, Deliang Sun

UNDERFILL FILLER SETTLING EFFECT ON THE ADHESIVE FORCE OF FLIP CHIP PACKAGES 750
Guolin Zhao, Houya Wu, Yuanyuan Yang, Gang Li, Pengli Zhu, Rong Sun, Wenhui Zhu

THE INFLUENCE OF DIFFERENT PHOSPHOR COATING METHODS ON THE TEMPERATURE OF LED 755

Kun Chen, Deming Hu, Liang Yang

AN INFRARED LASER TEMPORARY BONDING MATERIAL USED FOR DEVICE WAFER THINNING AND COMPLETION OF BACKSIDE PROCESSING TECHNOLOGY 759

Zhenwen Ye, Deliang Sun, Mingqi Huang, Guoping Zhang, Jianwen Xia

THE STUDYS OF ADHESION AND CONTACT THERMAL RESISTANCE OF TIM1 763

Yunsong Pang, Meng Han, Ting Liang, Xue Bai, Liang Li, Yonglun Xu, Bin He, Daifeng Ai, Liuxin Wang, Linlin Ren, Xiaoliang Zeng, Rong Sun

DESIGN AND VERIFICATION OF TDDB TEST STRUCTURES FOR TSV 767

Kai Li, Si Chen, Xiaofeng Yang, Guoyuan Li, Bin Zhou

REVIEW ON ERROR COMPENSATION AND COOPERATIVE CONTROL OF MULTI-AXIS MOTION PLATFORM 771

Zhiwei Zhou, Jian Gao, Lanyu Zhang, Jindi Zhang, Yuheng Luo

THE EFFECT OF DUAL ULTRASONIC-ASSISTED SOLDERING PROCESS ON THE PROPERTIES OF CU/40%ZN+60%SAC0307 POWDER/AL JOINT 776

Zhaoqi Jiang, Guisheng Gan, Qianzhu Xu, Shiqi Chen, Peng Ma, Yufeng Tang, Xiangtao Xu

EFFECTS OF SURFACE OXIDATION TREATMENTS ON THE INTERFACIAL ADHESION BETWEEN COPPER AND UNDERFILL 781

Bin Wang, Haoliang Lin, Houya Wu, Yuanyuan Yang, Gang Li, Pengli Zhu

FEASIBILITY ANALYSIS OF CRACK INITIATION IDENTIFICATION OF SINTERED SILVER FOR A FAST LIFETIME PREDICTION 786

Jiuyang Tang, Jing Zhang, Guoqi Zhang, Pan Liu

INVESTIGATION ON THE WARPAGE OF FAN-OUT WAFER-LEVEL PACKAGING USING CURING REACTION KINETICS OF COMPOSITES 791

Wei Li, Jianhong Huang, Jinbo Xiao, Yiyong Huang, Houdun Zhang, Daquan Yu

INORGANIC STABILIZER INTRODUCED FLUX SYSTEM WITH HIGH TACKINESS: AN EFFICIENT AND NOVEL MATERIAL SOLUTION FOR THE MICRO LED MASS TRANSFER 796

Liangzheng Ji, Jing Zhang

STUDIES ON THE EFFECTS OF SOLDERING LAYER STRUCTURES ON TEC MODULE PERFORMANCE AND THERMAL STRESS 800

Dianru Yu, Zhihao Yin, Guanghui Liu, Hongyan Xu, Ju Xu

A FLEXIBLE ULTRASONIC SENSOR BASED ON PIEZOELECTRIC MICROMACHINED ULTRASONIC TRANSDUCERS (PMUTS) 805

Zhihao Tong, Zhipeng Wu, Songsong Zhang, Huicong Liu, Liang Lou

EFFECT OF TEMPERATURE ON THE FATIGUE DAMAGE OF SAC305 SOLDER 809

Xu Long, Ying Guo, Xiaotong Chang, Yutai Su, Hongbin Shi, Tao Huang, Bingyi Tu, Yanpei Wu

ENHANCEMENT OF SILANE COUPLING AGENTS ON THE UNDERFILL ADHESION UNDERGOING HYDROTHERMAL AGING 813

Haoliang Lin, Yuanyuan Yang, Bin Wang, Gang Li, Houya Wu, Pengli Zhu

REVIEW OF COPPER-SILVER CORE-SHELL SINTERING PASTES: TECHNOLOGY AND FUTURE TRENDS 817

Zejun Zeng, Jing Zhang, Pan Liu

MECHANICAL BEHAVIOR OF LEAD-FREE SOLDER AT HIGH TEMPERATURES AND HIGH STRAIN RATES 823

Xu Long, Tianxiong Su, Chao Chang, Jiaqiang Huang, Xiaotong Chang, Yutai Su, Hongbin Shi, Tao Huang, Yanpei Wu

EMPLOYING SINGLE-CRYSTAL COBALT SUBSTRATES TO CONTROL βSN GRAIN ORIENTATIONS IN SOLDER INTERCONNECTIONS 827

Ce Li, Xufeng Chang, Bingguang Wang, Zhaolong Ma, Xingwang Cheng

TOPOLOGY OPTIMIZATION DESIGN OF META-MATERIAL HEAT SPREADER 832

Xue Bai, Qinghua Hu, Xiaoliang Zeng, Rong Sun, Jianbin Xu

EFFECT OF SURFACE STRESS ON INDENTATION RESPONSE OF ELASTIC-PLASTIC MATERIALS 836

Xu Long, Ziyi Shen, Xiaotong Chang, Yutai Su, Hongbin Shi, Tao Huang, Yanpei Wu, Bingyi Tu

SURFACE PASSIVATION OF CU NANOFIBER FILMS FABRICATED BY ELECTROSPINNING FOR TRANSPARENT CONDUCTIVE FILMS 839

Zhefei Sun, Ming-Sheng Wang, Zhihao Zhang

RESEARCH ON THERMAL CHARACTERISTICS OF HIGH POWER 3D MICROCHANNEL MULTICHIP PACKAGE 843

Xiangli Wei, Daoguo Yang, Wangyun Li, Xiyou Wang, Guangsheng Lu, Shirui Xue

A COMPREHENSIVE SIMULATION STUDY OF WARPAGE OF FAN-OUT PANEL-LEVEL PACKAGE USING ELEMENT BIRTH AND DEATH TECHNIQUE 847

Yun-Kai Deng, Bing-Xian Yang, Wei-Lin Hu, Xin-Ping Zhang

STUDY OF PACKAGE RELIABILITY ACCORDING TO THE EPOXY MOLDING COMPOUND 852

Eunsol Jo, Jung-Rae Park, Cheong-Ha Jung, Gu-Sung Kim

SIMULATION ANALYSIS OF THE INFLUENCE OF TEST CONDITIONS ON THE BONDING STRENGTH OF PCB PADS 854

Huanhuan Wang, Miao Cai, Daoguo Yang, Hengjian He, Yanchen Wu, Lei Song

EFFECT OF SAC0307 CONTENT ON PROPERTIES OF CU/ZN POWDER/AL JOINT BY ULTRASONIC EXCITATION AT LOW TEMPERATURE 859

Qianzhu Xu, Guisheng Gan, Zhaoqi Jiang, Shiqi Chen, Tian Huang, Cong Liu, Xiangtao Xu

A BUBBLE-FREE ELECTROOSMOTIC PUMP WITH POLYANILINE-WRAPPED PLATINUM-COATED TITANIUM MESH ELECTRODES 864

Qian Yang, Liang Li, Wen Chen, Meng Gao, Le Ye

FINITE ELEMENT SIMULATION STUDY OF INTERFACIAL CRACK PROPAGATION IN THE UNDERFILLED FC-BGA PACKAGE 870

Hong-Guang Wang, Min-Bo Zhou, Jiu-Bin Fei, Li Sun, Bing-Xian Yang, Wei-Lin Hu, Chang-Bo Ke, Xin-Ping Zhang

ENHANCEMENT OF COPPER NANOPARTICLE PASTE BY PRESSURE-LESS SINTERING ON DIFFERENT SUBSTRATES IN PT-CATALYZED FORMIC ACID ATMOSPHERE 875

Junlong Li, Yang Xu, Panju Shang, Yinghui Wang, Tadatomo Suga

INFLUENCES OF THE SOLDER SIZE ON GROWTH OF INTERFACIAL CU_6SN_5 AND MECHANICAL PERFORMANCE OF SN-3.0AG-0.5CU/(111)CU JOINTS SUBJECTED TO MULTIPLE REFLOW SOLDERING 880
Ming-Qiang Chen, Min-Bo Zhou, Yun-Wei Li, Xin-Ping Zhang

SIMULATION OF ELECTROMAGNETIC WAVE PROPAGATION IN 3D INTEGRATED MODULE BASED ON 3D ADI-FDTD ALGORITHM 885
Yinhui Han, Min Miao, Jin Li

A COMPARATIVE STUDY ON THE INFLUENCES OF VARIOUS NICKEL POWDERS ON THE EMI SHIELDING PERFORMANCE OF CONDUCTIVE POLYMER COMPOSITES 891
Yong Wang, Dingkun Tian, Yadong Xu, Baotan Zhang, Tao Zhao, Yougen Hu, Rong Sun

A X BAND CERAMIC PACKAGE WITH KILOWATT-LEVEL HIGH-POWER AND LOW LOSS 895
Yangfan Zhou, Linjie Liu, Zhizhuang Qiao, Ke Wang, Gai Liu

ANISOTROPIC BN NANOSHEET/POLYMER COMPOSITE BULK MATERIAL: A STUDY ON MECHANICAL PROPERTY 898
Chen Jing, Zeng Xiao Liang, Zhangguo Qi, Hu Qing Hua, Ye Huai Yu, Liu Pan

SOFT DEGRADATION AND RECOVERY UNDER ESD STRESS OF E-MODE GAN HEMTS WITH P-GAN GATE 902
Mei Wang, Zhiyuan He, Yan Ren, Lichao Hao, Zhaohui Wu, Bin Li

FABRICATION OF FLEXIBLE PRINTED CIRCUITS ON POLYIMIDE SUBSTRATE BY USING AG NANOPARTICLE INK THROUGH 3D DIRECT-WRITING AND RELIABILITY OF THE PRINTED CIRCUITS 907
Cheng-Bo Li, Xiao Ma, Hai-Jun Huang, Min-Bo Zhou, Xin-Ping Zhang

STUDY ON THE ELECTROMAGNETIC AND THERMAL CHARACTERISTICS OF AEROSTATIC SPINDLE FOR WAFER GRINDING 912
Chen Zhao, Ye Lezhi, Song Xuanjie, Wang Zhiyue, Liu Guangjie, Zhao Yumin

COPPER FILLING OF HIGH ASPECT RATIO THROUGH CERAMIC HOLES: EFFECT OF CONVECTION ON ELECTROCHEMICAL BEHAVIOR OF ADDITIVES 918
Wang Qing, Liu Jiaxin, Wu Yilin, Chen Mingxiang

THE EFFECT OF TOUGHENING AGENTS ON CAPILLARY UNDERFILL IN THE FLIP CHIP PACKAGE 923
Yuanyuan Yang, Houya Wu, Tao Peng, Jinbao Yang, Bin Wang, Haoliang Lin, Gang Li, Pengli Zhu, Rong Sun

PREPARATION AND PROPERTIES OF LOW MELTING POINT SN-P-F-O-MATRIX PHOSPHOR-IN-GLASS FOR WHITE LED 928
Deming Hu, Liang Yang, An Xie, Chunyan Cao, Xiayun Shu, Chenrui Fan, Jinrong Deng, Kun Chen

ARTIFICIAL NEURAL NETWORKS MODELING TECHNOLOGY FOR SUBSTRATE INTEGRATED SUSPENDED LINE 932
Shuxia Yan, Nana Yang, Zhifeng Chen, Peng Huang, Weiguang Shi

NON-CYANIDE ELECTROPLATING OF GOLD-TIN EUTECTIC ALLOY FOR FLIP-CHIP PACKAGING OF LED 936
Mingliang Huang, Chao Fang, Feifei Huang, Yan Yan

FABRICATION OF SIC NANO-PORE ARRAYS STRUCTURE BY METAL-ASSISTED PHOTOCHEMICAL ETCHING 940
Zijian Li, Dachuang Shi, Yun Chen, Maoxiang Hou, Jian Gao, Xin Chen

SIMULATION AND OPTIMIZATION OF INKJET-PRINTED OUTLINES TO IMPROVE PATTERN FIDELITY 944
Shaowei Hu, Wanchun Yang, Mingyu Li

SYNTHESIS AND CHARACTERIZATION OF AG-37AT.%CU SOLID SOLUTION NANOPARTICLES FOR HIGH RELIABILITY PACKAGING 947
Wanchun Yang, Shaowei Hu, Wei Zheng, Mingyu Li

EXPERIMENTAL ANALYSIS FOR BUMP SHEAR METHOD OF BSOB WIRE BONDING PROCESS 950
Zhu Chenjun, Zhang Pingsheng, Wen Zehai, Li Hui, Wu Yilong, Dong Dong

LOW TEMPERATURE CURING COPOLYIMIDE WITH MONOMER CONTAINING PYRAZINE MOIETY 954
Changqing Li, Guoping Zhang, Jinhui Li, Yingying Li

MICROSTRUCTURE EVOLUTION AND INTERFACIAL REACTION OF CO-P/SAC305/CO-P SOLDER JOINTS AT HIGH CURRENT DENSITY 958
Tao Fan, Donghua Yang, Haotong Qin, Yuqian Chen, Tao Chen, Xiang Zhai, Teng Ran

RESEARCH ON FAILURE MECHANISMS OF LOCALIZED IGBT DRIVE BOARD 961
Tiezhu Chen, Liang Zhou

SUBSTRATE SOLDER RESIST CRACK ROOT CAUSE INVESTIGATION THROUGH FINITE ELEMENT ANALYSIS 963
Ye Zhang, Tan Boowei, Tan Chow-Khong

MICROSTRUCTURE AND MECHANICAL PROPERTIES OF JOINTS BETWEEN GAAS SOLAR CELL ELECTRODE AND AG INTERCONNECTOR UNDER TEMPERATURE THERMAL CYCLE 967
Yuhan Ding, Zhichao Wang, Xueming Hua, Chen Shen, Min Wang, Jusha Ma, Bin Qian

DESIGN METHOD OF TRIAXIAL VIBRATION FIXTURE FOR COMPLEX INTEGRATED CIRCUITS 972
Xiaoqiang Wang, Bin Li, Chuanjin Deng, Rui Deng, Ruolei Wang, Kun Jiang

GROWTH BEHAVIOR OF INTERFACIAL INTERMETALLIC COMPOUNDS OF CO-20%P/SOLDER JOINT UNDER TEMPERATURE GRADIENT 978
Haotong Qin, Donghua Yang, Tao Fan, Tao Chen, Chunhong Zhang, Yuqian Chen

AN ACCURATE SIMULATION METHOD OF PACKAGE WARPAGE EXPERIMENTAL RESULTS BASED ON FEM 982
Liqiang Neng, Tingting Song, Guangping Shao, Jian Pang, Sun Tuobei, Keqing Ouyang

SIMULATION AND EXPERIMENTAL ANALYSIS OF A COST-EFFECTIVE MINIATURIZED TRANSCEIVER FOR X-BAND APPLICATION 986
Yu Ban, Jie Liu

EXTRACTION, OPTIMIZATION AND FAILURE DETECTION APPLICATION OF PARASITIC INDUCTANCE FOR HIGH-FREQUENCY SIC POWER DEVICES 990
Minghui Yun, Kailin Zhang, Miao Cai, Yiren Yang, Changqi Feng, Song Wei, Daoguo Yang, Guoqi Zhang

SIMULATION ANALYSIS OF RESIDUAL STRESS OF SINTERED NANO-SILVER UNDER MULTILAYER STACKED MODULE 995
   Zhentang Liang, Hongyue Wang, Bin Zhou, Guoyuan Li

LOW TEMPERATURE BONDING POLYCRYSTALLINE DIAMOND TO SI BY USING AU THIN-LAYER FOR HIGH-POWER SEMICONDUCTOR DEVICES 999
   Yi Zhong, Shuchao Bao, Ke Li, Mingchuan Zhang, Daquan Yu

RESEARCH ON RELIABILITY LIFE EVALUATION METHOD BASED ON AIRBORNE T/R COMPONENTS 1002
   Rui Deng, Chuanjin Deng, Ruolei Wang, Weixi Gong, Zongqi Cai, Kun Jiang

EFFECT OF GRAIN SIZE ON DIELECTRIC PROPERTIES AND RELIABILITY FOR ULTRA-THIN MLCCS 1006
   Kulun Jiang, Rong Sun, Xiuhua Cao, Lei Zhang, Shuhui Yu, Bo Li, Zhenxiao Fu

A QUICK CORRECTION METHOD FOR THE BOARD-LEVEL FINITE ELEMENT ANALYSIS OF QFP DEVICE UNDER VIBRATION 1012
   Zicheng Sa, Shang Wang, Jiayun Feng, Guangliang Yu, Ning Zhang, Yanhong Tian

PREPARATION AND MICROWAVE PROPERTIES OF TIO$_2$/PTFE COMPOSITES REINFORCED BY MULLITE FIBERS 1015
   Zhangzhao Weng, Jun Luo, Hongfeng Lv, Shuai Zhou, Xiaoqiang Wang, Daojun Luo

A SEGMENTED PLASMA ETCHING METHOD FOR 2.5D/3D THROUGH SILICON VIAS 1019
   Yuanwei Lin

STUDY ON THERMAL CYCLE AND INSULATION CHARACTERISTICS OF HIGH RELIABLE IGBT MODULE 1025
   Tao Chen, Minghua Zhang, Binbin Zhang, Qing Chen, Kun Tian, Yan Li

WARPAGE BEHAVIOR STUDY AND OPTIMIZATION FOR ULTRA-THIN POP MEMORY WITH MULTI-STACKED CHIPS 1029
   Dongmei Xia, Bei Wang, Chengyu Liao, Zuyao Liu, Hongwen He, Lu Liu

RESEARCH ON POWER DEVICE STRUCTURE BASED ON FO PACKAGE METHOD 1034
   Guan Qiang Song, Jia Ren Huo, Juntao Wang, Debo Liu, Chenwei Zhang, Jing Jiang, Huaiyu Ye

A BROADBAND MODEL OF STACKING TSV CHANNELS FOR NONDESTRUCTIVE DEFECT LOCALIZATION IN 3D ICS AND MICROSYSTEM 1040
   Chenbing Qu, Liwei Wang, Si Chen, Chen Sun, Zhiwei Fu, Guan-Lin Feng

A HYBRID DEGRADATION MODELING OF LIGHT-EMITTING DIODE USING PERMUTATION ENTROPY AND DATA-DRIVEN METHODS 1044
   Minzhen Wen, Zhou Jing, Mesfin S. Ibrahim, Jiajie Fan, Guoqi Zhang

RESEARCH ON ADJUSTMENT OF THE ELECTRICAL RESISTIVITY OF ALUMINUM NITRIDE CERAMIC 1050
   Jinhu Fan, Zirong Tang, Jie Wang

SOLDER PREFORMS COMPOSED OF HIGH CU-CONTENT SN-XCU ALLOYS FOR POWER ELECTRONIC PACKAGING AND CHARACTERIZATION OF THE PROCESSING PERFORMANCE AND JOINT'S PROPERTIES 1054
   Ru-Zeng Shi, Ming-Qiang Chen, Hai-Jun Huang, Min-Bo Zhou, Xin-Ping Zhang

THE MECHANICAL AND PHYSICAL PROPERTIES OF THE PHASES IN THE MICROSTRUCTURE OF SN-BI SOLDER ALLOY 1059
Chuyi Lei, Xinghe Luan, Zhigao Liu, Hongbo Qin, Bin Hou, Wangyun Li

SCALABLE MODELING AND ANALYSIS OF TSV USING BUMPLESS INTERCONNECTS TECHNOLOGY 1064
Zewei Li, Zhikuang Cai, Lei Pan, Lu Liu, Binbin Xu, Yufeng Guo

THE INFLUENCE OF MOLDING COMPOUND PROPERTIES ON SYSTEM-IN-PACKAGE RELIABILITY FOR 5G APPLICATION 1069
Dashun Liu, Kai Chen, Yijing Qin, Yong Zhong, Zhaorong Wan, Richeng Liu, Dong Lu, Ke Xue, Jingshen Wu

THERMAL RESISTANCE OF EUTECTIC GA-IN-SN/PARTICLES BINARY THERMAL INTERFACE MATERIALS 1075
Wendong Wang, Meijuan Lv, Jingdong Guo

RESEARCH ON BEOL FAILURES OF THE CHIP-PACKAGE INTERACTION BY SHEAR TESTS OF THE BUMPS 1081
Shizhao Wang, Lianghao Xue, Hongjie Wang, Rui Li, Can Sheng, Yameng Sun, Sheng Liu

EVALUATION OF SOLDER JOINTS RELIABILITY OF BALL GRID ARRAY ASSEMBLY IN ASTRONAVIGATION MODULES 1085
W. L. Qin, Z. P. Yan, M. H. Wu

NOVEL DESIGN OF SIC MOSFET ACTIVE DRIVE CIRCUIT BASED ON IMPROVED AUXILIARY BRANCH METHOD 1088
Li Yuhong, Niu Pingjuan, Mei Yunhui, Ning Pingfan, Zhao Di, Bai Jie

OPTICAL PATH DESIGN OF A HIGH COUPLING EFFICIENCY DFB LASER PACKAGE BASED ON COLLIMATOR 1094
Xiaomeng Lv, Ao Liao, Weihua Xu, Yangzhi Liu, Yilong Wu, Lan Qiao, Yong Zhao, Junli Yang

SIO_2/SIO_2 BONDING TECHNOLOGY RESEARCH ON WAFER-LEVEL 3D STACKING 1097
Zhang Peng, Chengyu Yu, Kai Cen, Jie Pu, Pengcheng Xia, Chengqian Wang

HIGH SPEED CU PLATING TECHNOLOGY FOR WAFER LEVEL PACKAGING 1101
Jian Wang, David Wang, Zhaowei Jia

ATE BOARD DESIGN SOLUTION OF SMALL PIN PITCH DUT 1107
Fei Pan, Qing Zhou, Jie Zhou

A STUDY OF COPPER OXIDIZATION MECHANISM AT METAL INTERFACE 1111
Ruolin Zhang, Ying Tang, Xu Wang, William Cao, Pradeep Rai

IRREGULAR HEAT DISSIPATING COVER BASED ON A1N MATERIAL 1117
Jianhui Liu, Yuxin Guo, Xiaowei Zhao

EVALUATION AND REDUCTION OF OPTICAL CROSSTALK IN QUANTUM DOT COLOR-CONVERTED MINI/MICRO-LED DISPLAYS 1121
Yuanjie Cheng, Jeffery C. C. Lo, Xing Qiu, S. W. Ricky Lee

STUDY ON A KIND OF THERMAL SOURCE CHIP FOR THE PERFORMANCE ANALYSIS OF MICRO CHANNEL HEAT SINK: SIMULATION AND EXPERIMENTAL VALIDATION 1126
Ming Zhao, Jian Zhang, Qian Lu, Weiwei Xiang, Yangyang Li, Miaomiao Jiang, Huijie Ye, Ting Peng

PLANE POSITION MEASUREMENT FOR μLED BASED ON SINGLE CAMERA 1131
Jie Bai, Pingjuan Niu, Erdan Gu

CURE SHRINKAGE CHARACTERIZATION AND WARPAGE SIMULATION
OPTIMIZATION OF EPOXY MOLDING COMPOUND FOR 5G APPLICATION 1134
Kai Chen, Dashun Liu, Yijing Qin, Yong Zhong, Zhaorong Wan, Richeng Liu, Dong Lu, Ke Xue, Jingshen Wu

FAILURE MECHANISM STUDY FOR FLIP CHIP QFN CRACK ISSUE UNDER
TEMPERATURE CYCLING TEST.. 1139
Ke Xue, Dong Lu, Kai Chen, Yijing Qin, Dayang Li, Qi Tang

STUDY ON THE ANTIOXIDATION OF COATED NANA-COPPER... 1144
Weijie Zhang, Quan Zhou, Chenshan Gao, Debo Liu, Jun Li, Huaiyu Ye

THE IMPACT OF PACKAGING MATERIALS ON THERMOMECHANICAL RELIABILITY
OF FC-LGA (FLIP-CHIP LAND GRID ARRAY) PACKAGE FOR 5G APPLICATION 1148
Dayuan Wan, Ke Xue, Weijing Dai, Yijing Qin, Dong Lu, Zhiqi Wang, Yi Chen

FAILURE ANALYSIS AND RELIABILITY IMPROVEMENT OF CRIMPING ASSEMBLY OF
COPPER-CLAD ALUMINUM CONDUCTORS FOR AEROSPACE.. 1153
Rui Cao, Meng Meng, Zhibin Wang, Yarong Chen, Yu Ye, Xueyin Huang

A 3D-MEMS BASED ARCHITECTURE FOR CS-CMOS HETEROGENEOUS INTEGRATION........... 1158
Min Huang, Pengfei Liu, Hongze Zhang, Yan Wu, Jian Zhu

DESIGN AND OPTIMIZATION OF TEMPERATURE SENSOR BASED ON LGA PACKAGE
STRUCTURE.. 1161
Jianhui Liu, Juntao Wang, Jun Li

HIGHER ASPECT RATIO TSV STRUCTURE ECP BOTTOM-UP PLATING PROCESS...................... 1165
Yinuo Jin, Bo Zheng, Jian Wang, David H. Wang, Qixing Yu

MONTE CARLO BASED STOCHASTIC FINITE ELEMENT MODEL FOR UNCERTAINTY
QUANTIFICATION IN FLIP CHIP BGA ELECTRONIC PACKAGING 1169
Liu Chu, Jiajia Shi, Robin Braun

EFFECT OF BUMP SHAPES ON THE ELECTROMIGRATION RELIABILITY OF COPPER
PILLAR SOLDER JOINTS 1173
Zhekun Fan, Zhankun Li, Junhui Li, Jinqing Xiao, Yunpeng Liu, Junfu Liu, Taotao Chen

STUDY ON HERMETIC PACKAGE OF ANTIRADIATION DIRECT-HEAD TRANSMITTER........... 1179
Chuanwei Wang, Yukun Wu, Jiabo Zhang, Kuang Pan, Daochang Wang, Yi Liang

STUDY ON DIE-BONDING KEY TECHNOLOGY OF UV LED PACKAGING................................... 1183
Wei Liu, Chunfang Zi, Qingping Lin

STUDY ON PERFORMANCE OPTIMIZATION OF NANOMETER COPPER PASTE 1187
Qiang Liu, Jian Wen, Yu Zhang, Yu Liu, Guannan Yang, Zhongwei Huang, Chengqiang Cui

STUDY ON THE EFFECT OF ASSEMBLY ERRORS ON THE ELECTROSTATIC TUNING
ABILITY IN MICRO UMBRELLA SHELL RESONATORS.. 1191
Lu Xu, Bin Luo, Jintang Shang, Zhaoxi Su, Shouyu Han, Yinghui Zhang

RESEARCH ON INTEGRATED METASURFACE LENS FOR HIGH GAIN MULTIBEAM
SYSTEM IN PACKAGE APPLICATION 1196
Yuxiang Zheng, Weikang Wan, Qidong Wang, Liqiang Cao

STUDY ON THERMAL STABILITY OF ALL COPPER INTERCONNECT STRUCTURES UNDER THERMAL SHOCK.. 1201

Hao Li, Jun Shen, Jiacheng Xie

MACHINE LEARNING BASED PREDICTION OF WIRE BONDING PROFILE IN 3D STACKED INTEGRATED MICROELECTRONIC PACKAGING 1206

Zhengping Ou, Junyu Long, Shuquan Ding, Yun Chen, Maoxiang Hou, Yunbo He, Xin Chen, Jian Gao

THE DESIGN OF A VOLTAGE CONVERSION SIP BASED ON ELECTROTHERMAL COUPLING METHOD... 1210

Hao-Hang Su, Shuai Fu

RELIABILITY ANALYSIS ON WAFER BONDING PROCESS OF MEMS CIRCULATOR 1215

Dongxue Luo, Yan Wu, Junfeng Sun, Miao Yu

EXPERIMENTAL TESTS AND STRESS ANALYSIS OF SNPB SOLDER JOINTS IN A CERAMIC POP DEVICE 1219

Kaiyu Guo, Honglei Ran, Jun Wang

ANALYSIS OF FACTORS AFFECTING THE OUTGASSING RATE OF MEMS VACUUM PACKAGING MATERIALS.. 1224

Yong Yang, Bin Zhou, Xuanjun Dai, Yun Huang

FATIGUE LIFE PREDICTIONS OF SNPB SOLDER BALL IN A CERAMIC POP DEVICE 1230

Yu Yao, Zirui Cui, Honglei Ran, Jun Wang

DESIGN AND FABRICATION OF A SOFT MICRO-ACTUATOR BASED ON DISTRIBUTED MAGNETIC COMPOSITE 1235

Langkun Wang, Shimei Liu, Hu He

ELECTROLESS COPPER DEPOSITION WITH PYRAMIDAL MICRO-CONES MORPHOLOGY FOR LOW-TEMPERATURE CU-CU BUMP INTERCONNECTIONS....................... 1241

Yiming Chen, Yiqiao Wei, Zhuo Chen, Wenjing Zhang, Fuliang Wang, Wenhui Zhu

THE EFFECT OF SILICON ANISOTROPY ON THE THERMAL STRESS OF TSV STRUCTURE OF 3D PACKAGING CHIP UNDER THERMAL CYCLIC LOADS................................ 1246

Jingyang Liang, Minjie Ning, Chao Ding, Tianhan Liu, Zongbei Dai, Hongbo Qin

HIGH YIELD AND HIGH THROUGHPUT LITHOGRAPHY SOLUTION FOR EMERGING HIGH DENSITY FAN-OUT PANEL LEVEL PACKAGING.. 1250

Junbo Jiang, Kang Zhang, Di He, Chen Xiang, Cheng-Tar Wu, Minghao Shen

AN OPTIMIZATION METHOD OF ULTRA HIGN SPEED DIFFERENTIAL STRUCTURE FOR BGA PACKAGE.. 1255

Nong Jin, Zhizhuang Qiao, Linjie Liu, Ke Wang, Yangfan Zhou, Zan Ren

REFLOW SOLDERING PROCESS OPTIMIZATION BASED ON SURFACE EVOLVER SOLDER JOINT SHAPE SIMULATION AND FINITE ELEMENT ANALYSIS OF PCB ASSEMBLY.. 1258

Xing Jin, Wenlong Wang, Wenzhong Zhao, Yuting Zhang, Xiaopeng Tan, Shuai Chen

A NOVEL BUMPING METHOD FOR FLIP-CHIP INTERCONNECTION............................. 1262

Kun Li, Gaowei Xu, Quan Zhou, Wei Gai, Yanhong Wu, Jie Ren, Zhen Wang

DESIGN AND TEST OF TRANSMISSION LINE IN SFQ CIRCUIT 1265

Quan Zhou, Lingyun Li, Zhen Wang, Gaowei Xu, Le Luo, Xiaoming Xie, Kun Li, Jie Ren

THE INFLUENCE OF SOLDERING VOIDS IN POWER DEVICES.. 1269
Jianming Fang, Min Wang, Xuanlong Chen

DOUBLE-SIDED ELECTROPLATING PROCESS FOR THROUGH GLASS VIAS (TGVS)
FILLING 1273
Ke Li, Heng Wu, Weijian Chen, Daquan Yu

SAWING INVESTIGATION FOR THIN WAFER LAMINATED WITH DIE ATTACH FILM 1277
Haiyan Liu, Qingyu Pan, Sean Xu, Jianhong Wang, Lu Li, Xueting Wu

BLADE DICING ON WAFER SAW STUDY 1282
Qiuchen Zhang, Hongbin Xia, Shwu Miin Tan

3D DISLOCATION MULTI-STACK FAN-OUT PACKAGE OF ULTRA-THIN DIES FOR
HETEROGENEOUS INTEGRATION.. 1288
Lijun Chen, Feng Chen, Fengwei Dai

THE MULTI-DIMENSION CO-DESIGN FOR THE PACKAGE OF THE HIGH-SPEED MULTI-
FUNCTION CHIP... 1292
*Yuan Guan, Yubo Wang, Dejian Li, Shunfeng Han, Bofu Li, Dameng Li, Fengman Liu,
Huimin He*

EFFECT OF PLASMA AND STAGING TIME ON THE UNDERFILL VOIDS IN FINE PITCH
FLIP-CHIP PACKAGE.. 1297
*Saif Wakeel, Dominic Koey Poh Meng, Stella Wong Wun Chin, Jos Philipsen, Pieter
Gommers, Annelies Joosten*

THE STUDY OF FAR-UVC 222-NM EXCILAMP AND GERMICIDAL-UVC 254-NM LOW-
PRESSURE HG LAMP: OPTICAL CHARACTERISTICS AND SERVICE LIFE 1302
*Fanny Zhao, Guoshuai Dong, Hao Wu, Guoming Yang, Brian Shieh, S. W. Ricky Lee,
Ronghua Deng*

EFFECTIVENESS VALIDATION ON EQUIVALENT MODEL FOR WAFER-LEVEL
WARPAGE PREDICTION.. 1305
Guoli Sun, Jiahui Wei, Fei Qin, Yanwei Dai, Kui Li, Baoxia Li

MULTI-CHIPS HIGH-DENSITY INTERCONNECTION DESIGN ON INFO PLATFORM 1311
Jiang Qiang, Zhou Guodan, Wang Zongwei, Pang Jian, Sun Tuobei, Keqing Ouyang

FIRST-PRINCIPLES STUDY ON THE MECHANICAL PROPERTIES OF CU₃SI COMPOUND........... 1315
Jian Wang, Xiaowei Xu, Chao Ding, Tianhan Liu, Zongbei Dai, Hongbo Qin

STUDY ON THE EFFECT OF SILICA SUBMICRON PARTICLE SIZE AND CONTENT ON
FRACTURE TOUGHNESS OF FILLED EPOXY RESIN 1320
Qin Zhou, Xuecheng Yu, Leicong Zhang, Pengli Zhu, Rong Sun

EFFECT OF TEMPERATURE CYCLING, HIGH TEMPERATURE STORAGE AND STEADY-
STATE OPERATION LIFE TEST ON RELIABILITY OF GAN HEMTS...................................... 1323
Mao Mao, Sha Tang, Zhizhe Wang, Rui Deng, Chuanjin Deng, Si Chen, Yan Ren

EFFECT OF IMC MORPHOLOGY ON THE CURRENT DENSITY AND TEMPERATURE
GRADIENT OF LINE-TYPE CU/SN/CU SOLDER JOINT 1327
Jiaqi Tang, Tianhan Liu, Chao Ding, Jian Wang, Hongbo Qin, Wangyun Li

DESIGN AND ANALYSIS OF MOSFET BASED ON FAN-OUT PANEL-LEVEL PACKAGE
TECHNOLOGY... 1332
Zhi Liang, Dongdong Shao, Kunpeng Ding, Chuang Tian

ADSORPTION BEHAVIOR OF HCOOH ON THE CRYSTAL SURFACES OF CU(111) AND (100) 1336

Liheng Jiang, Tianhan Liu, Siliang He, Hongbo Qin, Wangyun Li, Daoguo Yang

INNER MICROCRACK INDUCED BY INTERMETALLIC COMPOUND IN RIGHT-ANGLE AU/SN3.0AG0.5CU/AU SOLDER JOINTS 1340

Wu Yue, Zhenyu Zhang, Bunv Liang, Jing Li, Cheng Xue

A COMPACT CHIP FILTER FOR 5G COMMUNICATION FRONTEND BASED ON GLASS IPD TECHNOLOGY 1343

Zhitao Zhang, Ling Gu, Shanwen Hu, Yuehua Zhang, Xinlei Zhang

ENCAPSULATION TECHNIQUES OF PEROVSKITE SOLAR CELLS.......................... 1346

Qi Wu, Wenfeng Li, Mengyu Chen, Cheng Li

EFFECT OF NI-CNTS ON WETTING PROPERTIES, MICROSTRUCTURE, AND CREEP RESISTANCE OF SN58BI-0.1ER COMPOSITE SOLDER 1352

Qi Li, Fengmei Liu, Yaoyong Yi, Xueying Zhang, Haitao Gao

A PROCESS IMPROVEMENT IN SILVER-INDIUM TRANSIENT LIQUID PHASE BONDING METHOD FOR THE HIGH-POWER ELECTRONICS AND PHOTONICS PACKAGING 1357

Donglin Zhang, Xiuchen Zhao, Yingxia Liu, Ying Liu, Yongjun Huo

RESEARCH ON SIP SIGNAL INTEGRITY BASED ON ANSYS SIWAVE IN WEARABLE MEDICAL SYSTEMS... 1361

Zheng Yang, Yingke Gao, Shenglong Li, Chuanchuan Sun, Yunfu Zhao

LIGHTWEIGHT SILVER NANOWIRE AEROGEL FOR ELECTROMAGNETIC INTERFERENCE SHIELDING 1366

Fei Peng, Yi Fang, Zhehao Han, Wenbo Zhu, Mingyu Li

A FACILE BONDING MATERIAL TO ENABLE INTERCONNECTION AMONG COMPLEX SURFACES THROUGH AGNWS AEROGEL 1370

Zhehao Han, Mingyu Li, Wenbo Zhu, Fei Peng, Yi Fang

HIGH-SENSITIVITY FLEXIBLE SENSOR BASED ON SILVER NANOWIRE AEROGEL.................... 1374

Yi Fang, Mingyu Li, Zhehao Han, Wenbo Zhu, Fei Peng

SOLDERING OF GRAPHENE ASSEMBLED FILMS WITH ULTRASONIC ASSISTANCE AND ITS UTILIZATION POTENTIALITY IN ELECTRONIC DEVICES 1378

Huaqiang Fu, Yong Xiao, Daping He

THEORETICAL CALCULATION AND SIMULATION OF BGA PACKAGE STRESS UNDER TEMPERATURE CYCLING LOAD 1382

Zhang Yueping, Hou Chuantao, Tong Jun, Wang Long

LOSS MODELLING AND ANALYSIS OF A HIGH-EFFICIENCY WIRELESS POWER TRANSFER SYSTEM FOR AUTOMATED GUIDED VEHICLE APPLICATIONS.............................. 1387

Jincheng Yu, Wai Leong Ng, Minglu Xia, Ziyang Gao

THREE DIMENSIONAL WAFER-LEVEL VACUUM PACKAGING OF MEMS RESONANT ACCELEROMETER 1393

Ziji Wang, Chaoyang Xing, Jin Zhang, Zhaoxi Su, Wenqi Li, Bin Luo, Jintang Shang

GLASS REFLOW AND THERMO-MECHANICAL STRESS SIMULATION FOR THROUGH GLASS VIA IN GLASS-SILICON COMPOSITE INTERPOSER.. 1397

Wenqi Li, Chaoyang Xing, Jianfeng Zhang, Ziji Wang, Zhaoxi Su, Bin Luo, Jintang Shang

RESEARCH ON THE APPLICATION OF FEEDFORWARD + HIGH-ORDER ITERATIVE LEARNING OF PERMANENT MAGNET LINEAR MOTOR IN WIRE BONDING MACHINE 1401
Haomiao Wu, Yunbo He, Xiaohui Lin, Haolin Li

MACHINE LEARNING ENABLED OPTIMIZATION OF PICK-UP PROCESS FOR THIN DIE............ 1407
Peilun Yao, Haibin Chen, Jinglei Yang, Jingshen Wu

DESIGN AND ANALYSIS OF A FAST-SPEED FLIP-CHIP BONDING SYSTEM WITH FORCE CONTROL 1412
Zhongyuan Zhu, Hui Tang, Jiedong Li, Sifeng He

DEEP LEARNING PRODUCT CLASSIFICATION FRAMEWORK BASED ON THE MOTIVATION OF TARGET CUSTOMERS 1418
Fei Sun, Ding-Bang Luh, Qidong Wang, Yulin Zhao, Yue Sun

CHARACTERIZATION OF LONGITUDINAL THERMAL CONDUCTIVITY OF GRAPHENE FILM 1422
Jiajia Chen, Xinjian Gong, Sihua Guo, Yong Zhang, Jin Chen, Yuanyuan Wang, Johan Liu

RANDOM VIBRATION RESPONSE ANALYSIS OF SIP POWER SUPPLY MODULE........................ 1426
Tao Lin, Weiyin Wang, Jun Li

THE INFLUENCE MECHANISM OF PROCESS DEFECTS OF PLANE WELDING ON SIGNAL TRANSMISSION IN MICROWAVE COMPONENTS 1429
Ruining Li, Jun Tian, Song Xue, Congsi Wang, Yan Wang, Cheng Zhou, Jing Liu, Zhihai Wang, Kunpeng Yu, Haitao Shi, Daoheng Sun

EPOXY FLUX PREVENT HOT TEAR AT VIPPO SOLDER JOINTS 1435
Elaina Zito, Dave Bedner, Ning-Cheng Lee

PROCESS OPTIMIZATION FOR CCGA SURFACE MOUNT ASSEMBLY BASED ON PHYSICS OF FAILURE... 1441
Hui Xiao, Weiming Li, Yabing Zou, Xiaotong Guo, Jiahao Liu

210°C REFLOW TECHNOLOGY STUDY IN 3D PACKAGING 1445
Lingyao Sun, Yaru Dong, Zhuangzhuang Hou, Xiuchen Zhao, Yongjun Huo, Ying Liu, Yingxia Liu

RESEARCH ON THE HUMIDITY RESISTANCE RELIABILITY OF DIFFERENT PACKAGING STRUCTURES 1449
Hui Xuan, Zheng Yu, Hua Wu, Wanchun Ding, Guohua Gao

ELECTROPLATING NANOTWINNED COPPER FOR ULTRAFINE PITCH REDISTRIBUTION LAYER (RDL) OF ADVANCED PACKAGING TECHNOLOGY 1453
Yu-Bo Zhang, Li-Yin Gao, Xiao Li, Zhe Li, Xu-Liang Ma, Zhi-Quan Liu, Rong Sun

MIXED CU NANOPARTICLES AND CU MICROPARTICLES WITH PROMISING LOW-TEMPERATURE AND LOW-PRESSURE SINTERING PROPERTIES AND INOXIDIZABILITY FOR MICROELECTRONIC PACKAGING APPLICATIONS 1459
Haiqi Lai, Jian Wen, Guannan Yang, Yu Zhang, Chengqiang Cu

EFFECT OF DAISY CHAIN STRUCTURE ON ELECTROMIGRATION RELIABILITY OF MICROBUMPS: A SIMULATION STUDY 1463
Chenkan Yan, Dun Wang, Kaihong Zhang, Huibin Zhang, Yongjian Yu, Weikun Xie, Kai Zhu, Yongkang Wan

NOVEL LOW-DIELECTRIC FLUORINATED CARBON FIBER/POLYIMIDE MATERIALS WITH HIGH ELONGATION 1469

Tao Wang, Jinhui Li, Fangfang Niu, Liang Shan, Guoping Zhang, Rong Sun, Ching-Ping Wong

INFLUENCE OF POROSITY ON THE MECHANICAL PROPERTIES OF HYBRID SILVER SINTERED JOINT 1473

Zunyu Guan, Fred Fuliang Le, Jingshen Wu, Jinglei Yang, Rinse Van Der Meulen, Haibin Chen

INTEGRATION DESIGN OF MM-WAVE RADAR ANTENNAS BASED ON FOWLP 1478

Lihong Liu, Chen Cheng, Jionajiong Gu, Tianhua Shen, Quanbing Li

EXPERIMENTAL AND COMPUTATIONAL STUDY OF SHIELDING EFFECTIVENESS OF METAL GRIDS 1481

Zhiqiang Lin, Xuebin Liu, Xinrong Shi, Yougen Hu, Rong Sun

STUDY ON THE MECHANICAL PROPERTIES OF ULTRA-LOW DIELECTRIC FILM BY TENSILE TEST 1485

Lei Wang, Fei Xiao, Jun Wang

SIMULATION RESEARCH ON ELECTROMIGRATION OF BGA DEVICES 1489

Wenchao Tian, Yiming Zhang, Yong Chen, Si Chen, Hao Cui

MEASUREMENT PROCESS OPTIMIZATION IN USING LOCK-IN THERMOGRAPHY FOR FAULT LOCALIZATION OF COWOS PACKAGES 1494

Shuanshe Chao, Na Mei, Xinyi Lin, Tuo Bei Sun, Dan Yang, Keqing Ouyang

MICROSTRUCTURE AND MECHANICAL CHARACTERISTICS OF A NOVEL CU/CU_3SN JOINT WITH CU/SN PREFORM 1498

Hongyan Xu, Wei Zhang, Xuan Liu, Ju Xu

THICKNESS MEASUREMENT OF MULTI-LAYER THIN FILM ALLOY BY X-RAY FLUORESCENCE SPECTROMETER 1504

Sung-Hua Zhong, Liang-Pin Chen

NANO SILVER PARTICLES PREPARED BY SPARK ABLATION AS PACKAGING INTERCONNECTION MATERIAL 1507

Jin Tong, Yu Zhang, Peilin Liang, Zhongwei Huang, Chengqiang Cui, Keju Zhong

SILVER FILM PREPARED BY SPARK ABLATION FOR CONDUCTIVE PATTERN 1511

Peilin Liang, Yu Zhang, Jin Tong, Zhongwei Huang, Chengqiang Cui, Keju Zhong

SIMULATION AND OPTIMUM DESIGN OF AL-SI ALLOY BY DOUBLE-SIDED LASER WELDING 1514

Liangchen Tan, Shanshan Li, Jianfeng Zhong

SIGNIFICANT EFFECT OF TEMPERATURE AND SOLDERS ON THE GROWTH BEHAVIOR OF CU_6SN_5 ON (110) CU SINGLE CRYSTAL 1518

Chong Dong, Min Shang, Ying Guo, Xiangxu Chen, Jun Zhang, Changlong Dong, Haoran Ma, Haitao Ma

A STRETCHABLE HIGH-SENSITIVE CAPACITIVE SENSOR FOR AERODYNAMIC PRESSURE MEASUREMENT 1521

Xiaofeng Yang, Jian Chen, Si Chen, Bin Zhou, Yijun Shi, Zhiwei Fu

3D WAFER LEVEL PACKAGING FOR SAW FILTER USING THIN GLASS CAPPING WITH THROUGH GLASS VIAS.. 1525
Zuohuan Chen, Jin Zhao, Feng Jiang, Heng Wu, Mingchuan Zhang, Daquan Yu

WARPAGE CHARACTERISTIC OF GLASS INTERPOSER WITH DIFFERENT CTE'S AND THICKNESS 1529
Jin Zhao, Fei Qin, Daquan Yu, Zuohuan Chen, Shuai Zhao

Author Index

Design and Verification of TDDB Test Structures For TSV

Kai Li[1,2]
[1] School of Micro-Electronics, South
China University of Technology
[2] Science and Technology on Reliability
Physics and Application of Electronic
Component Laboratory
Guangzhou, China
likai2020@outlook.com

Si Chen[2*]
Science and Technology on Reliability
Physics and Application of Electronic
Component Laboratory
Guangzhou, China
chensiceprei@yeah.net

XiaoFeng Yang[2]
Science and Technology on Reliability
Physics and Application of Electronic
Component Laboratory
Guangzhou, China
yxf004@hotmail.com

GuoYuan Li[1]
School of Micro-Electronics, South
China University of Technology
Guangzhou, China
phgyli@scut.edu.cn

Bin Zhou[2]
Science and Technology on Reliability
Physics and Application of Electronic
Component Laboratory
Guangzhou, China
Zhoubin722@163.com

Abstract—In this paper, two TDDB test structures of TSV, including Single-TSV and Dual-TSV, have been designed. In order to compare the availability between these two test structures, the electric field simulation and TDDB failure analysis have been carried out. The simulation results show that the Single-TSV is more prone to abnormal breakdown for its higher electric field between surface pads during TDDB test. On the contrary, the Dual-TSV has a greater probability of breakdown in the dielectric liner of TSV, because the maximum of electric field is located at the dielectric liner. The conclusion of TDDB failure analysis support the point of simulation. The Dual-TSV has better availability than the Single-TSV for the TDDB test of TSV. Furthermore, we discussed the guidelines for the design of TSV TDDB test structures.

Keywords—TSV, TDDB Test Structure, electric field simulation, failure analysis.

I. INTRODUCTION (*HEADING 1*)

Through Silicon Via (TSV) is one of the most important interconnection technologies in 3D integration. It realizes the vertical interconnection among the chips by metal vias that penetrate the chips and wafers. TSV plays the role of signal conduction, heat transfer and mechanical support [1]. TSV has been widely used in different fields for its good electrical performance, low power consumption, high interconnection density, small size, etc. Although TSV technology has many advantages, there are still many reliability problems to be solved.

The reliability of the TSV dielectric liner is particularly important to the reliability of TSV. More and more researches draw much attention on the Time Dependent Dielectric Breakdown (TDDB) problem caused by the accumulation of defects in the TSV dielectric liner [2,4]. On the one hand, Coefficient of Thermal Expansion (CTE) mismatch in the TSV interface caused by the thermal load will lead to crack in TSV dielectric liner [5]; On the other hand, as the interconnect material of TSV, Cu is easy to diffuse into TSV dielectric liner in the form of Cu ions, reducing the reliability of TSV dielectric liner [6]. The accumulation of these defects aggravates TDDB effect in TSV dielectric layer.

In order to explore the TDDB effect of the TSV dielectric liner, many targeted test structures have been proposed. However, inappropriate TDDB test structures will induce breakdown in other locations, but not in TSV dielectric liner [7]. Therefore, it is very important to design an appropriate test structure for TSV TDDB test. In this paper, two TDDB test structures of TSV: Single-TSV and Dual-TSV, have been designed. The availability of these two test structures are verified by electric field simulation and TDDB failure analysis.

II. TWO TDDB TEST STRUCTURES FOR TSV

A. Single-TSV

The Single-TSV TDDB test structure is composed of surface pads, a TSV-Cu, and P+ doped region, as shown in fig.2. The doped region is used as an embedded conductive layer to form a Metal-Oxide-Semiconductor (MIS) test structure with TSV-Cu and TSV dielectric liner.

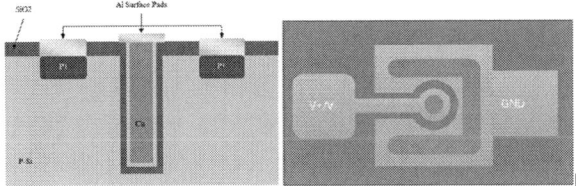

Fig. 1. The structure of Single-TSV

B. Dual-TSV

The Dual-TSV TDDB test structure is composed of surface pads and two TSV-Cu, as shown in fig.3. In this structure, in order to form a MIS test structure, another TSV-Cu is used as embedded conductive layer, rather than the commonly used doping region.

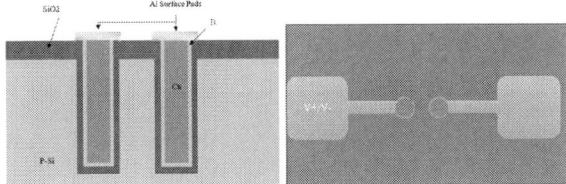

Fig. 2. The structure of Dual-TSV

978-1-6654-1392-3/21 $31.00 © 2021 IEEE

III. Simulation Analysis

A. Simplification of finite element model

For the dielectric, the greater the applied electric field, the greater the probability of TDDB. Therefore, TSV dielectric liner should be the position with the largest electric field in a test structure to ensure that TDDB can occur in the TSV dielectric liner accurately. Based on this point, we can simulate the electric field distribution of these two tests structures with the use of ANSYS to compare the availability for TSV TDDB test. Before simulation, we should have some simplification for the finite element model:

(1) The dielectric liner and barrier are ignored for their extremely small size relative to the whole model.

(2) The problem of non-uniform voltage in metal conductor is ignored.

B. Electric Field Simulation

Finite element models of Single-TSV and Dual-TSV are shown in Fig.3. The voltage load is divided into two parts, one is applied with 10V, the other is applied with 0V. Ansys analysis type is transient analysis.

The electric field distribution of Single-TSV is shown in Fig.4. We can find that the electric field intensity between surface pads $(2.1 \times 10^5 \sim 4.7 \times 10^5 V/M)$ is much higher than that between the P+ doped region and TSV $(4 \times 10^4 \sim 2.1 \times 10^5 V/M)$ in the Single-TSV, which means that the dielectric between surface pads has greater probability to occur TDDB than the TSV dielectric liner. The dielectric between surface pads is a potential location to abnormal breakdown. On the contrary, for the Dual-TSV, the largest electric field $(4.37 \times 10^4 \sim 3.97 \times 10^5 V/M)$ is in the TSV dielectric liner, instead of the dielectric between surface pads $(< 4.37 \times 10^4 V/M)$, as shown in Fig.5.

The results of simulation show that the Dual-TSV has better availability than the Single-TSV for TSV TDDB test.

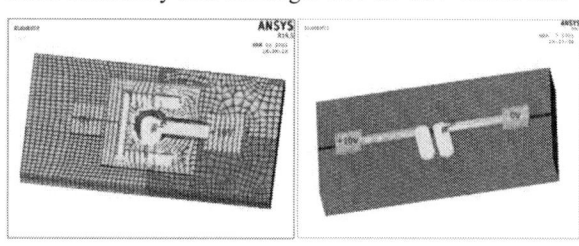

(a) Single-TSV   (b) Dual-TSV

Fig. 3. Finite Element Model (after meshing and loading)

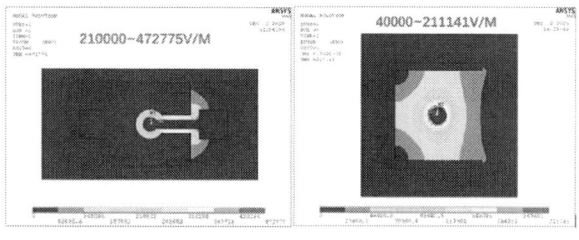

(a) between Surfaces Pads   (b) between P+ and TSV

Fig. 4. Electric Field Distribution of Single-TSV

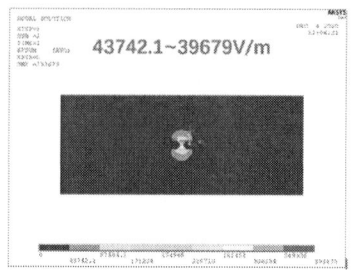

Fig. 5. Electric Field Distribution of Dual-TSV

IV. Failure Analysis

A. TDDB test

In order to verify the conclusion of electric field simulation further, TDDB tests are carried out on these two test structures. The method of TDDB test is Constant Voltage Stress (CVS).

These two test structures are applied to a constant voltage of 70V by using probe stage and semiconductor analyzer (1500A). The ambient temperature is set to 150°C to speed up the TDDB test. The leakage current of test structures and the duration time of CVS is monitored by 1500A. When the leakage current of test structures at a certain time is more than 100 times of that at the initial time, TDDB occurs [8].

As shown in Fig.6, TDDB occurs in both test structures after a period of time, but it occurs first in Single-TSV. This may be due to the abnormal breakdown of the Single-TSV.

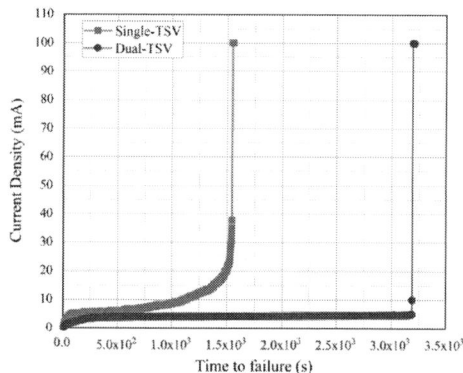

Fig. 6. I-t curve of TSV TDDB test structures during TDDB test

B. Breakdown Location

As shown in Fig.7(a), we can find that the Single-TSV has a breakdown between surface pads obviously. This breakdown position is same as the breakdown position predicted in the electric field simulation. From the Fig.7(b), the surface pads of Dual-TSV doesn't occur TDDB, so we speculate that TDDB occurred in the TSV dielectric liner.

(a) Single-TSV   (b) Dual-TSV

Fig. 7. Surface Pads of TSV TDDB test structrues after TDDB test

In order to confirm whether the breakdown position of Dual-TSV is in the TSV dielectric liner, we used Emission Microscope (EMMI) to locate the breakdown position of Dual-TSV. As shown in Figure 8, A light point represents a breakdown point. There is a breakdown point in the TSV dielectric liner under forward voltage test mode and reverse voltage test mode respectively. This means TDDB occur in the TSV dielectric liner, supporting the point of simulation: The Dual-TSV has better availability than the Single-TSV for TSV TDDB test.

Incidentally, The reason why the breakdown position of Dual-TSV after TDDB test is not the point with maximum electric field, is because the occurrence of TDDB has a randomness and is not only related to the electric field intensity.

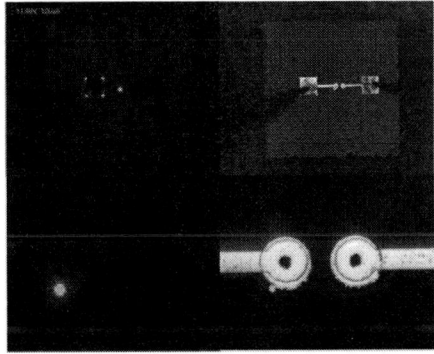

(a) Forward voltage test mode (left is VDD, right is GND)

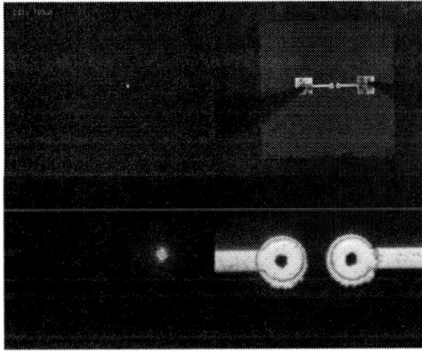

(b) Reverse voltage test mode (left is GND, right is VDD)

Fig. 8.   Emission Microscope Photo

C. Breakdown cross section

In order to explore the depth information and cross section of breakdown location for the Dual-TSV, the breakdown point located by the EMMI above is trenched by the Focused Ion Beam (FIB). After trenching, the cross section of breakdown position is observed by the Scanning Electron Microscope (SEM), as shown in Fig.9.

A serious breakdown within the TSV dielectric liner was observed in Fig.9. Compared with the non-breakdown sample, the breakdown sample shows serious Cu sliding along the interface of TSV, which is caused by the huge impact at the moment of breakdown. Meanwhile, there is a large area of scorch at the interface of TSV, which is due to the huge short-circuit current generated after breakdown.

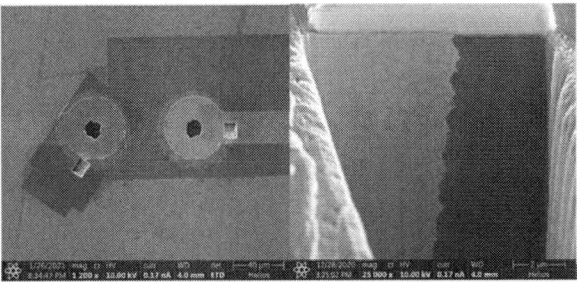

(a) Breakdown-top view   (b) Non-Breakdown-section (As Received)

(c) Forward Breakdown-section   (d) Reverse Breakdown-section

Fig. 9.   Cross section of Dual-TSV after TDDB test and FIB

V. DISSCUTION

One of the difficulties in the design of TSV TDDB test structure is that the TDDB test in the TSV is more difficult to carry out than that in the MOS. During TDDB test, in order to apply electric field to the dielectric, it is necessary to form a MIS test structure. The MIS test structure in TSV is vertical, which lead to the need for an embedded conductive layer in the Si substrate. As a result, the embedded conductive layer should have additional pads to connect itself with test probe for being loaded.

The Single-TSV need a special shape of surface pad to connect P+ doped region, which will increase the complexity of surface pads. The more complex the surface pads is, the more likely abnormal breakdown will occur. The P+ doped region is replaced by another TSV in the Dual-TSV, which reduces the complexity of the structure. Therefore, the potential abnormal breakdown points are reduced. Manufacturing cost has also been reduced for no having doping process.

Based on the results of simulation and failure analysis, we consider that there should be two guidelines for TSV TDDB test structure design:

(1) The test structure should be as simple as possible to avoid potential abnormal breakdown.

(2) The location planned to TDDB should be the position with maximum electric field intensity.

VI. CONCLUSION

In this paper, two TDDB test structures of TSV: Single-TSV and Dual-TSV, have been designed. For the Single-TSV, the largest electric field is located on the interval of surface pads. For the Dual-TSV, the maximum electric field intensity

978-1-6654-1392-3/21 $31.00 © 2021 IEEE   769

is in the TSV dielectric liner. It is mean that the Single-TSV is more prone to occur TDDB in the dielectric between surface pads. From the TDDB failure analysis, TDDB occurred on the interval of surface pads of Single-TSV is observed clearly. For Dual-TSV, the TDDB location is determined in the TSV dielectric liner by the EMMI. The Dual-TSV has better availability than the Single-TSV for TSV TDDB test. From the observation of breakdown cross section, we can find that TDDB will not only cause serious Cu sliding along the interface of TSV, but also make the TSV dielectric liner burned down, which affects the reliability of TSV in deep.

ACKNOWLEDGMENT

This research was supported by the National Natural Science Foundation of China (NSFC) under Grant No. 61804032, the Guangzhou Science and Technology Plan Project under Grant No. 201904010333 and Grant No. 201904010457, the Academician Fund under Grant No.ZHD201806.

REFERENCES

[1] K. Croes et al., "Reliability Challenges Related to TSV Integration and 3-D Stacking," in IEEE Design & Test, vol. 33, no. 3, pp. 37-45, June 2016, doi: 10.1109/MDAT.2015.2501302.

[2] Jiawei Marvin Chan,Chuan Seng Tan,Kheng Chooi Lee.Effects of Copper Migration on the Reliability of Through-Silicon Via (TSV)[J].IEEE transactions on device & materials reliability,2018,18(4):520-528.

[3] Seung-Ho Seo,Joo-Sun Hwang,Jun-Mo Yang, et al.Failure mechanism of copper through-silicon vias under biased thermal stress[J].Thin Solid Films: An International Journal on the Science and Technology of Thin and Thick Films,2013,546(Nov.1):14-17.

[4] Yunlong Li,Yoshiyuki Oba,Chen Wu, et al.Hydrogen outgassing induced liner/barrier reliability degradation in through silicon via's[J].Applied physics letters,2014,104(14):142906-1-142906-3.

[5] Han, Chang-Fu,Lin, Jen-Fin.Thermally-induced failures of copper through-silicon via structures evaluated by the strain energy density model[J].Thin Solid Films: An International Journal on the Science and Technology of Thin and Thick Films,2016,615(Sep.30):281-291.

[6] J. M. Chan, X. Cheng, K. C. Lee, W. Kanert and C. S. Tan, "Reliability Evaluation of Copper (Cu) Through-Silicon Vias (TSV) Barrier and Dielectric Liner by Electrical Characterization and Physical Failure Analysis (PFA)," 2017 IEEE 67th Electronic Components and Technology Conference (ECTC), Orlando, FL, USA, 2017, pp. 73-79, doi: 10.1109/ECTC.2017.77.

[7] Choi, Hyun-Jun,Choi, Seung-Man,Yeo, Myung-Soo, et al.An experimental study on the TSV reliability: Electromigration (EM) and time dependant dielectric breakdown (TDDB)[C].//2012 IEEE International Interconnect Technology Conference.2012:1-3.

[8] K. Hummler,B. Sapp,J.R. Lloyd, et al.TSV and Cu-Cu Direct Bond Wafer and Package-Level Reliability[C].//2013 IEEE 63rd electronic components and technology conference: ECTC 2013, Las Vegas, Nevada, USA, 28-31 May 2013, pages 1-806, [v.1].2013:41-48.

Review on Error Compensation and Cooperative Control of Multi-axis Motion Platform

1st Zhiwei Zhou
State Key Laboratory of Precision Electronic Manufacturing Equipment and Technology, Guangdong University of Technology
School of Electromechanical Engineering, Guangdong University of Technology
Guangzhou, China
810225808@qq.com

2nd Jian Gao*
State Key Laboratory of Precision Electronic Manufacturing Equipment and Technology, Guangdong University of Technology
School of Electromechanical Engineering, Guangdong University of Technology
Guangzhou, China
jian_gao2004@163.com

3rd Lanyu Zhang
State Key Laboratory of Precision Electronic Manufacturing Equipment and Technology, Guangdong University of Technology
School of Electromechanical Engineering, Guangdong University of Technology
Guangzhou, China
lyuzhang@qq.com

4th Jindi Zhang
State Key Laboratory of Precision Electronic Manufacturing Equipment and Technology, Guangdong University of Technology
School of Electromechanical Engineering, Guangdong University of Technology
Guangzhou, China
861491590@qq.com

5th Yuheng Luo
State Key Laboratory of Precision Electronic Manufacturing Equipment and Technology, Guangdong University of Technology
School of Electromechanical Engineering, Guangdong University of Technology
Guangzhou, China
1249664568@qq.com

Abstract—With the continuous improvement of technical indexes of precision electronic packaging equipment, its performance is primarily confirmed by the motion features of the multi-axis motion platform, which plays a key role for the precision packaging equipment. In recent years, the control technologies of multi-axis motion platforms have been rapidly developed to satisfy the demands of high precision and high acceleration. Nevertheless, under the condition of high-speed and large stroke motion, the motion platform usually has very limited positioning accuracy. The positioning accuracy of a multi-axis motion stage has become increasingly prominent for high-end electronic packaging equipment, and will determine their performance. In view of the motion planning, dynamic error compensation, and multi-axis cooperative control involved in the multi-axis motion platform, the research progress on this topic is investigated and described in this paper. The key issues related to the motion platform are summarized and the future development trend is prospected.

Keywords—Packaging equipment, Multi-axis motion platform, Motion planning, Error compensation, cooperative control

I. INTRODUCTION

In recent years, high-speed large-stroke multi-axis motion platform has been increasingly developed and widely used in such advanced and sophisticated fields as micro-electronic system [1], semiconductor package [2], aerospace equipment [3], micro-nano manufacturing [4], and precision imaging [5], which can satisfy the requirements of the faster processing. With the constant improvement of the high-end equipment technical index, precision motion platform covers the types of complex production lines operating onerous, processing complex, single-axis motion platform is difficult to meet the task allocation, multipath operation, high-efficiency machining, and the workpiece positioning and another complicated industrial environment, high accuracy

multi-axis motion platform design demands should be formed. Compared with the structure, performance, and operation mode of the single-axis alignment platform, the multi-axis motion platform has the characteristics of multi-operation applicability, high-performance machining, and strong robustness [6].

At present, the IC chip manufacturing process has reached 5nm node, electronic manufacturing has entered the post-Moore era, and advanced packaging technology has promoted the development of high-density integration of the semiconductor industry. Chip packaging technology is rapidly changing. In the case of physical process limits, system-level packaging of electronic devices to improve system performance and integration has become a parallel packaging trend [7]. Therefore, large stroke, high precision, high-density packaging equipment has become the core of the further development of the semiconductor industry. As a key component of packaging equipment, the precision positioning performance of a multi-axis motion platform determines the quality of the whole equipment. Based on this, this paper introduces the research progress, technical difficulties, and solutions of the motion planning, dynamic error compensation, and multi-axis cooperative control involved in the multi-axis motion platform, and prospects the future development trend.

II. MOTION PLANNING

To enhance the performance of the multi-axis motion platform and solve its motion planning problem, it is necessary to design the optimal control path from the given position to the terminal position. According to the control principle, motion planning includes two aspects: speed control [8] and trajectory planning [9].

The speed control methods are divided into linear acceleration and deceleration [10], exponential acceleration

978-1-6654-1392-3/21 $31.00 © 2021 IEEE

and deceleration [11], S-curve control [12], cubic displacement curve acceleration and deceleration [13], trigonometric acceleration and deceleration [14], and NURBS [15] (non-uniform rational B-splines) methods. To solve the problems such as velocity fluctuation, vibration, and acceleration mutation caused by the system's high-speed start and stop, Wu et al. [16] designed a new speed planning method for NURBS curves, According to the discreteness of velocity step change and the principle of uniform velocity design at interpolation time, the velocity fluctuation is caused by the modified circular algorithm of acceleration coefficient, and the smooth motion control of the curve is realized under dynamic conditions. Zhao et al. [17] designed a quintic polynomial velocity planning method for suspended parallel robots. In this method, the continuity of the curve was optimized by connecting points of the zero instantaneous velocity trajectory segment, and the completeness of the planned velocity curve of the robot was proved by tracking the end-effector with an optical capture system. Gai et al. [18] proposed an adaptive S-type acceleration and deceleration speed planning method. The third-order polynomial and the recursive average filter are introduced to guarantee the continuity of the initial and final velocity of the system. Due to the pause in the processing of small line segments, Zhang et al. [19] designed an automatic adjustment interpolation control method for the two-level interpolation mode. The method introduced velocity planning, real-time interpolation technology, and DDA (Digital Differential Analyzer, digital integral) fine interpolation technology, through the speed planning and forward-looking data real-time optimization of each line of the actual corner speed, and with variable speed DDA interpolation method to realize the micro line segment fine interpolation.

Considering that the lack of information interaction in speed planning affects the quality and efficiency of machining, Sun et al. [20] designed a dynamic programming method for real-time variable speed machining, which adjusted the speed according to the radius of arc transition of adjacent machining paths to reduce the stable transition time Liu et al. [21] designed a dynamic speed control method that fitted discrete data. In this method, many discrete points were fitted by a five-fold spline curve in a smooth transition mode, and the curvature of discrete junctions was output by a differential geometric numerical analysis method. The velocity prospective program segment was determined according to the curvature of the junctions after weighted accumulation, which effectively avoided the problem that too few velocity prospective program segments could not slow down quickly. Leng et al. [22] proposed the curved surface interpolation acceleration and deceleration control method. In this method, the discrete characteristic of sampling interpolation is introduced to output the velocity expression, and the adaptive acceleration and deceleration control approach is adopted to deal with the puzzle of determining deceleration points in advance before interpolation.

It was explored that the speed control method under constraints could improve the connecting speed at the corner of a line segment. Yang et al. [23] designed a dynamic speed adjustment method with quadratic B-spline speed curve, bow height error, and driving performance as constraints. Based on the penalty function and inverse rank 2 quasi-Newton method, the velocity curve is designed with robustness and smoothness. Haddad et al. [24] designed a velocity planning method with curvature control, velocity, and torque constraints. Combined with the trapezoidal velocity profile constraint, this method is helpful to increase computational efficiency and reduce running time.

To solve the technical challenges of trajectory planning, the trajectory of terminal equipment is discretized to obtain the spatial position and direction. The core trouble of trajectory planning is to discretize the movement route of terminal equipment to obtain the spatial position and direction. Trajectory planning methods are mainly divided into optimal control algorithm, visual servo control algorithm, and interpolation algorithm. To solve the problems of position error and attitude error in a dynamic environment, Chen and Liao [25] proposed the optimal trajectory method based on the swarm motion control method. The method determines the trajectory path with the minimum motion time and energy consumption to improve the motion efficiency. Azizi and Khani [26] designed a smooth trajectory planning algorithm for parallel robots. This algorithm introduces the fifth-order B-spline curve to find the geometric constraints of the robot, which is beneficial to approximate the algorithm model and improve the algorithm execution efficiency. Chesi et al. [27] designed a path control method based on dynamic system analysis. In this method, Rodriguez formula parameters are used, Linear Matrix Inequality (LMI) constraint optimization method is introduced to handle the maximization of typical performance, and path tracking is implemented by image controller. Geng et al. [28] proposed a new interpolation method for modifying command points. In this method, the command points were divided into different positions, and the discrete points were compressed into continuous smooth curves with the use of quintuple spline interpolation to reduce the tool machining time, surface roughness, and attitude error.

III. Dynamic Error Compensation

With the continuous development of the micro-nano processing technology，ultra-precision equipment is more and more extensive, and the dynamic error has become the technical evaluation index of equipment components. To ameliorate the dynamic precision of the operating platform, the influence of dynamic error is reduced by the error compensation method. According to the compensation mechanism, the dynamic error compensation methods are divided into mechanism compensation and control method compensation.

The basic principle of the mechanism and device compensation method is to realize system dynamic error compensation by compensating the device. Zhang et al. [29] designed a uniaxial macro-micro motion platform with a workspace(40mm), speed($0.2m/s^2$), acceleration(8g), and accuracy(30nm), as shown in Figure 1. The macro-motion platform is dominated via voice coil motor to achieve large-stroke micron-level motion precision, the micro-motion platform is driven by piezoelectric ceramics and preloaded spring to achieve nanometer motion precision, and the platform uses absolute grating ruler and encoder as closed-loop feedback to ensure the position accuracy. From the perspective of control theory, the micromanipulator plays the role of an error compensator. Shinno et al. [30] proposed a macro and micro dual-drive motion platform with 1M stroke and 20nm resolution. The error real-time dynamic compensation is implemented by utilizing a VCM and air floating guide rail to drive the micromanipulator. Because of the high resolution of the compensation device, the system can obtain high positioning accuracy. However, the high cost of the compensation device and the small drive travel limit the application scope of the compensation method of mechanism and device.

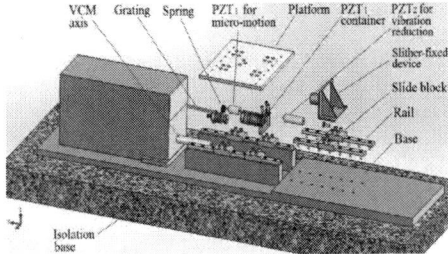

Fig. 1. Macro-Micro Motion Platform [29]

Control compensation error compensation is carried out by using a control algorithm. Chiu et al. designed a synchronization control method. The method takes into account the characteristics of nonlinear contour comprehensively, but it has the problem of relative degree limitation. Therefore, Tomizuka and Chiu [31] proposed a control method to enhance the position coordination ability of mechanical systems. The relative degree limitation is eliminated by the integrator back-extension technology, and the contour tracking performance will be effectively improved. Wang et al. [32] designed a contour error compensation method based on an active compensation structure. In this method, the position loop and velocity loop are partitioning compensated by introducing the coupling compensation of reference instruction, which improves the contour tracking accuracy of large curvature trajectory by 15%, reduces the uniaxial following error, and is beneficial to the suppression of external disturbance. Xiao et al. [33] designed an error compensation method based on a neural network. The system model was identified by the neural network as the feedforward compensator, and the neural network was trained by the hybrid algorithm. The problem of the larger system contour error caused by the difference of dynamic characteristics of each axis was solved. Wu et al. proposed a nonlinear PID control method[34]. The error model and the compensator are established to ameliorate the tracking precision of the platform. Jia et al. proposed a precise position compensation method[35]. The periodic errors were analyzed dynamically by the contour tangential error retroversion method, and the errors were compensated by the nonlinear gain scheduling method to solve the contradiction between rapid feed and operation error. Yuan and Zhao [36] proposed a control tactic on account of the global coordinate system. The global coordinate system was established by the contour error and contour trajectory, and then the iterative learning control method was introduced to compensate for the unmodeled dynamics in the contour tracking process. The complementary sliding mode control strategy was used to suppress the external disturbance and system parameter changes to achieve accurate motion control.

IV. MULTI-AXIS COOPERATIVE CONTROL

The basic principle of multi-axis cooperative control is a control method in which the controller and the driving device transfer data at high frequency and without jitter, so that the output states of the controlled objects in the system remain relatively consistent. The control problem is related to position control and torque control. The multi-axis cooperative control method is divided into the coupling synchronous control method and the uncoupling synchronous control method.

Given coupled synchronous control methods are mainly divided into cross-coupling control and deviation coupling control. The cross-coupling control decomposes the system error to each axis for compensation without modifying the control parameters. Its conception comes from the master-slave cross-coupling controller proposed by Sarachik and Ragazzini. Koren [37] improved the master-slave structure cross-coupling controller and proposed an asymmetric cross-coupling controller. By introducing an integrator and a digital comparator into the dual-axis positioning system, the following error compensation of each axis was realized effectively. however, the dynamic response of the axis was weakened. To improve the feed speed without affecting the system performance, Chuang [38] designed an automatic speed adjustment method based on adaptive control and cross-coupling control, which effectively established a closed-loop feedback loop between the profile error model and the feed speed, and ameliorated the operation precision of the multi-axis equipment. To diminish the machining error of dual-axis CNC equipment, Yeh et al. [39] combined cross-coupling control with fuzzy logic technology, proposed a fuzzy logic cross-coupling controller based on online estimation method, and used the fuzzy rule generation method to reduce the following lag, thus effectively improving the tracking accuracy and contour feed speed of machine tools. Zhong et al. [40] proposed an improved linear coupling controller to reduce the profile machining error of CNC machine tools at a feed speed of 30m/min by analyzing the steady-state error and adjusting the controller parameters adaptably. Yu and Chen[41] proposed a coupling synchronous control method, which reduced the synchronization error of the biaxial system through neural network cross-coupling controller and had strong robustness for uncertain dynamic systems and nonlinear systems. Considering the unmodeled nonlinear system, Zhao et al. devised a motion control method on account of the predictive analysis[42]. The estimated observer was used to suppress external disturbances, and the decoupling dynamic model at the joint end was built to weaken the system synchronization error and boost the contour following accuracy. Chin et al. [43] designed a pre-estimation method based on fuzzy control technology, which used the position errors to output the compensation terms, and made fine pre-compensation for the given motion trajectory to improve the machining grade of parts. Cross-coupling control is beneficial to ameliorate the synchronization of double motors, but it is not applicable to double motors and above systems. Perez-Pinal et al. [44] devised a decoupling control method suitable for the multi-motor system. The control principle was based on the deviation operation of each motor speed to realize the compensation of each motor speed and reduce the difference of moment of inertia between each motor. Li et al. [45] designed a mean deviation coupling synchronous control method based on switching functions. This method can reduce the complexity of the mechanismdesign, and has a strong adjustment effect on optimization and random disturbance suppression. Given the low motor synchronization performance under unknown disturbance, Li and Li [46] designed a dual-motor automatic control system and adopted the deviation coupling strategy of intelligent PI control to control the brushless DC motor, to solve the contradiction between the speed and smoothness of the motion trajectory.

Uncoupled synchronous control methods are mainly divided into master-slave control and electronic virtual spindle control. Cruz-Ortiz et al. [47] adopted the master-slave control structure based on time-varying gain to locate the robot and planned the trajectory path of the robot operator through the boundary constraint method to reduce the risk of collision of the robot with obstacles and improve the robot position tracking capability. The master-slave control has strong intuitiveness and a simple control structure, but poor robustness. Lorenz and Schmidt [48] proposed the electronic virtual spindle control method, which replaced the traditional mechanical long shaft with the virtual shaft to make the control shaft move in real-time with the virtual shaft, and ensured the synchronization and stability of the control system through the shaft coupling torque feedback link. Modak et al. [49] designed a virtual spindle control model based on improved inter-shaft stiffness characteristics to effectively improve inter-shaft synchronization and high-bandwidth servo control of paper machine transmission systems. Geng et al. [50] proposed an improved electronic virtual spindle control method for computing delay between load torque and reference torque and synchronization of speed and position in case of load disturbance. The load torque was predicted by the design state observer, and then the feedback was given to the virtual motor to reduce the system setting time and ameliorate the system anti-interference ability. To weaken the state trajectory near the sliding mode surface through repeatedly, construct a new reaching law by solving differential equations. Under the condition of load velocity and torque saturation, the velocity and position synchronization errors were significantly reduced, and the synchronization between axles was improved. Considering that the load torque is related to speed, external interference, and other factors, the traditional electronic virtual spindle control adopts electromagnetic torque as the feedback signal. When the load torque changes, the speed needs to stabilize for a while, and the actual feedback of the load torque cannot represent the change of the torque. Zhang et al.[51] proposed an electronic virtual spindle control method based on an equivalent load torque observer. Tracking control sliding mode controller was introduced to diminish the impact of low-speed contact on the synchronization precision in the pre-alignment process of the shaftless printing system, reduce the output synchronization error of nonlinear and load disturbance in the printing process, and improve printing equipment control system performance and efficiency.

V. DEVELOPMENT TREND

Based on the investigation of multi-axis motion platform studies at home and abroad and to meet the high-end electronic packaging demands, we summarized the following aspects for further study :

(1) Nanoscale motion platform is susceptible to environmental and vibration factors, and the suppression of external interference by moving platform should be further studied by combining visual algorithm and bionics algorithm.

(2) Research on multi-axis cooperative control is mostly based on continuous-time, and the system needs discrete sampling data in practical application. Therefore, multi-axis cooperative control based on discrete-time should be studied.

(3) The multi-axis motion controller is designed in the multi-input multi-output mode, and the underlying control algorithm is improved to achieve the intelligent control of the multi-axis motion platform under complex working conditions.

ACKNOWLEDGMENT

This work is supported by the National Natural Science Foundation of China under Grant 52075106, and the Guangdong Provincial R&D Key Projects under Grant 2018B090906002, Grant 17ZK0091.

REFERENCES

[1] L. Y. Zhang, J. Gao, and X. Chen, "A rapid vibration reduction method for macro-micro composite precision positioning stage," IEEE Trans. Ind. Electron, vol. 64, pp. 401–411, January 2017.

[2] C. Z.Wang, D. W. Yang, Y. Zhang, and Y. N. Zhang, "Design Methodology of Parallel Mechanisms Based on the Concept of Generalized Moving Platform," J. Mech. Eng, vol. 57, pp. 86–99, January 2021.

[3] P. Yan, Z. Zhang, L. Guo, and P. B. Liu, "Control and applications of ultra high precision mechatronics," Control Theory and Applications, vol. 31, pp. 56–69, October 2014.

[4] C. Lin, J. Yu, Z. H. Wu, and Z. L. Shen, "Decoupling and control of micromotion stage based on hysteresis of piezoelectric actuation," Microsyst. Technol, vol. 25, pp. 3299–3309, May 2019.

[5] Q. Zhang, J. G. Zhao, P.Yan, H. Y. Pu, and Y.Yang, "A novel amplification ratio model of a decoupled XY precision positioning stage combined with elastic beam theory and Castigliano's second theorem considering the exact loading force," Mech. Syst. Signal Process, vol. 136, pp. 1–16, February 2020.

[6] C. A. Nelson, M. A. Laribi, and S. Zeghloul, "Multi-robot system optimization based on redundant serial spherical mechanism for robotic minimally invasive surgery," Robotica, vol. 37, pp. 1202–1213, July 2019.

[7] G. Q. Lu, A. Cereska, G. Augustinavicius, and R. Maskeliunas, "Intelligent control and performance evaluation of a novel precise positioning stage," J. Intell. Fuzzy. Syst, vol. 36, pp. 1205–1214, December 2019.

[8] L.Wang, L. J. Lin, Y. Chang, and D. Song, "Velocity Planning for Astronaut Virtual Training Robot with High-Order Dynamic Constraints," Robotica, vol. 38, pp. 1–17, February 2020.

[9] K. Dejan, V. Bastiaan, and H. Eldert, "Coverage trajectory planning for a bush trimming robot arm," J. Field. Robot, vol. 37, pp. 283–308, February 2020.

[10] H. Wang, W. Heng, J.H.Huang, B. Zhao, and L. Quan, "Smooth point-to-point trajectory planning for industrial robots with kinematical constraints based on high-order polynomial curve," Mech. Mach. Theory, vol. 139, pp. 284–293, September 2019.

[11] Y. Liu, Y. Liu, and X.Tian, "Trajectory and velocity planning of the robot for sphere-pipe intersection hole cutting with single-Y welding groove," Robot. CIM-INT. Manuf, vol. 56, pp. 244–253, April 2019.

[12] Z. N. Li, T. Wang, B. R. Wang, and D. J. Chen, "Trajectory planning for manipulator in Cartesian space based on constrained S-curve velocity," CAAI Transactions on Intelligent Systems, vol. 14, pp. 655–661, July 2019.

[13] S. Q. Zhao, D. Yu, C. Geng, and W. Y. Han, "Research on Lookahead Interpolation Algorithm Based on Cubic Polynomial Displacement Incremental Model," China Mechanical Engineering, vol. 24, pp. 1066–1073, June 2013.

978-1-6654-1392-3/21 $31.00 © 2021 IEEE

[14] H. T. Wang, D. B. Zhao, L. Y. Lu, and K. Liu, "Parametric Curve Look-ahead Interpolatin Algorithm with Flexible Acceleration and Deceleration Method," China Mechanical Engineering, vol. 23, pp. 299–304, May 2012.

[15] V. Sathiya and M. Chinnadurai, "Evolutionary Algorithms-Based Multi-Objective Optimal Mobile Robot Trajectory Planning," Robotica, vol. 37, pp. 1363–1382, March 2019.

[16] C. Wu, J. C. Wu, S. P. Yang, Y. H. Zhou, and Q. C. Ma, "S-type series velocity planning algorithm based on NURBS curves," Comput. Integra. Manuf. Sys, vol. 21, pp. 3249–3255, December 2015.

[17] T. Zhao, B. Zi, S. Qian, and J. Zhao, "Algebraic Method-Based Point-to-Point Trajectory Planning of an Under-Constrained Cable-Suspended Parallel Robot with Variable Angle and Height Cable Mast," Chin. J. Mech. ENG-EN, vol. 33, pp. 45–62, November 2020.

[18] R. L. Gai, H. Lin, L. M. Zheng, and Y. Huang, "Design and Implementation of Velocity Planning Algorithm for High Speed Machining," J. Chin. Mini-Micro. Comput. Syst, vol. 30, pp. 1067–1071, June 2009.

[19] Y. N. Zhang, D. B. Zhao, K. Liu, and S. Chen, "A real-time look-ahead direct interpolation algorithm for small line," J. Southeast. U: Nat. Sci. Ed, vol. 40, pp. 726–730, April 2010.

[20] S. J. Sun, H. Lin, D. Yu, L. M. Zheng, and J. G. Yu, "A Look-ahead Path Planning Algorithm with High Speed and High Precision," J. Mech. Eng, vol. 52, pp. 170–176, November 2016.

[21] Z. Z. Liu, T. Y. Wang, C. Z. Ren, and Q. J. Liu, "Dynamic velocity look-ahead control based on weighted accumulation of curvature," J. Machine Des, pp. 89–92, March 2012.

[22] H. B. Leng, W. U. Yi-Jie, and X. H. Pan, "Research on cubic polynomial acceleration and deceleration control model for high speed NC machining," J. Zhejiang Univ, vol. 9, pp. 358–365, June 2008.

[23] M.Yang, X. C. Zhao, Z. S. Zhong, Y. Yue, and P. Gao, "Adaptive Velocity Planning under Complex Constraints for 5-axis CNC Systems," J. Mech. Eng, vol. 56, pp. 173–183, November 2020.

[24] M. Haddad, W. Khalil, and H. E. Lehtihet, "Trajectory Planning of Unicycle Mobile Robots With a Trapezoidal Velocity Constraint," IEEE Trans. Robot, vol. 26, pp. 954–962, November 2010.

[25] C. T. Chen and T. T. Liao, "A hybrid strategy for the time and energy efficient trajectory planning of parallel platform manipulators," Robot. CIM-INT. Manuf, vol. 27, pp. 72–81, February 2011.

[26] M. R. Azizi and R. Khani, "An algorithm for smooth trajectory planning optimization of isotropic translational parallel manipulators," P. I. Mech. Eng. C-J. Mec, vol. 230, pp. 1987–2002, May 2016.

[27] G. Chesi, "Visual Servoing Path Planning via Homogeneous Forms and LMI Optimizations," IEEE Trans. Robot, vol. 25, pp. 281–291, April 2009.

[28] C. Geng , D. Yu , L. Zheng , H. Zhang, and F. Wang, "A tool path correction and compression algorithm for five-axis CNC machining," J. Syst. Sci. Complex, vol. 26, pp. 799–816, April 2013.

[29] L. Y. Zhang, J. Gao, and X. Chen, "A rapid dynamic positioning method for settling time reduction through a macro-micro composite stage with high positioning accuracy," IEEE Trans. Ind. Electron, vol. 65, pp. 4849–4860, June 2018.

[30] H.Shinno, H.Yoshioka, and H.Sawano, "A newly developed long range positioning table system with a sub-nanometer resolution," CIRP. Ann-Manuf. Techn, vol. 60, pp. 403–406, March 2011.

[31] M.Tomizuka and G.C.Chiu, "Coordinated Position Control of Multi-Axis Mechanical Systems," J. Dyn. Syst-T. ASME, vol. 120, pp. 389–393, September 1998.

[32] R. K. Wang, Z. C. Yu, and J. Wang, "Active compensation of contour error of X-Y linear motor precision motion platform," Optics Precis. Eng, vol. 27, pp. 1536–1543, July 2019.

[33] B. X. Xiao, Q. J. Wang, W. B. Ang, and T. L. Lou, "The Simulation Research of Contour Error Control Methods Based on Neural Networks," J. Simul, vol. 15, pp. 572–574, April 2003.

[34] Y. S. Wu, Y. W. Wang, L. Yu, and J. X. Wang, "NLPID Based Cross-Coupled Control for Multi-axis Motion Control Systems," Control Eng. China, vol. 27, pp. 830–834, May 2020.

[35] Z. Y. Jia, D. N. Song, J. W. Ma, X. X. Zhao, and W. W. Su, "Adaptive estimation and nonlinear variable gain compensation of the contouring error for precise parametric curve following," Scientia Sinica Technologica, vol. 60, pp. 1494–1504, July 2017.

[36] H. Yuan and X. M. Zhao, "Learning Complementary Sliding Mode Contouring Control Based on Global Task Coordinate Frame for Direct Drive XY Table," Transactions of China Electrotechnical Society, vol. 35, pp. 2141–2148, May 2020.

[37] Y. Koren, "Cross-Coupled Biaxial Computer Control for Manufacturing Systems," J. Dyn. Syst-T. ASME, vol. 102, pp. 265–272, January 1980.

[38] H. Y. Chuang and C. H. Liu, "Cross-Coupled Adaptive Feedrate Control for Multiaxis Machine Tools," J. Dyn. Syst-T. ASME, vol. 113, pp. 451–457, September 1991.

[39] Z. M. Yeh, Y. S. Tarng, and Y. S. Lin, "Cross-coupled fuzzy logic control for biaxial servo mechanisms," Proceedings of IEEE 5th International Fuzzy Systems, vol. 2, pp. 1184–1190, October 1996.

[40] Q. Zhong, Y. Shi, J. Mo, and S. Huang, "A Linear Cross-Coupled Control System for High-Speed Machining," Int. J. Adv. Manuf. Tech, vol. 19, pp. 558–563, May 2002.

[41] C. H. Yu and T. C. Chen, "Robust Neural Network Controller Design For A Biaxial Servo System," Asian J. Control, vol. 9, pp. 390–401, December 2007.

[42] C. Zhao, C. Yu and J. Yao, "Dynamic Decoupling Based Robust Synchronous Control for a Hydraulic Parallel Manipulator," IEEE Access, vol. 7, pp. 30548–30562, January 2019.

[43] J. H. Chin, Y. M. Cheng, and J. H. Lin, "Improving contour accuracy by Fuzzy-logic enhanced cross-coupled precompensation method,"Robot. CIM-INT. Manuf, vol. 20, pp. 65–76, February 2004.

[44] F. J. Perez-Pinal, C. Nunez, and R. Alvarez, "Compassion of multi-motor synchronization techniques," The 30th Annual Conference of me IEEE industrial Electronics Society. Korea, vol. 10, pp. 2–6, October 2004.

[45] L. Li, L. Sun, and S. Zhang, "Mean deviation coupling synchronous control for multiple motors via second-order adaptive sliding mode control," Isa. T, vol. 62, pp. 222–235, May 2016.

[46] Z. Li and Z. C. Li, "Double motor synchronous control system based on deviating coupling strategy," Unifying Electrical Engineering and Electronics Engineering, pp. 849–855, June 2014.

[47] D. Cruz-Ortiz, I. Chairez, and A. Poznyak, "Robust control for master-slave manipulator system avoiding obstacle collision under restricted working space," IET Control. Theory Appl, vol. 14, pp. 1375–1386, May 2020.

[48] R. D. Lorenz and P. B. Schmidt, "Synchronized motion for process automation," Conference Record of the IEEE Industry Applications Society Annual Meeting.San-Diego, vol. 2, pp. 1693–1698, November 1989.

[49] J. P. Modak, K. S. Zakiuddin, G. D. Mehta, and M. K. Sonpimple, "Electronic Line Shafting-Control for Paper Machine Drives," International Journal of Scientific and Engineering Research, vol. 3, pp. 1–4, March 2013.

[50] Q. Geng, W. Liu, H. Wang, Z. Zhou, and G. Zhang, "An Improved Electronic Line Shafting Control for Multi-motor Drive System Based on Sliding Mode Observer," Math. Probl. Eng, vol. 8, pp. 1–13, April 2019.

[51] C. F. Zhang, Y. Y. Xiao, J. He, and M. Yan, "Improvement of electronic line-shafting control in multi-axis systems," International Journal of Automation and Computing, vol. 15, pp. 1–8, August 2018.

The Effect of Dual Ultrasonic-assisted Soldering Process on the Properties of Cu/40%Zn+60%SAC0307 Powder/Al Joint

Zhaoqi Jiang
College of Materials Science and Engineering
Chongqing University of Technolog
Chongqing, China
2581284141@qq.com

Guisheng Gan*
College of Materials Science and Engineering
Chongqing University of Technolo
Chongqing, China
ggs@cqut.edu.cn

Qianzhu Xu
College of Materials Science and Engineering
Chongqing University of Technolog
Chongqing, China
351317290@qq.com

Shiqi Chen
College of Materials Science and Engineering
Chongqing University of Technolog
Chongqing, China
793608729@qq.com

Peng Ma
College of Materials Science and Engineering
Chongqing University of Technolo
Chongqing, China
1639440567@qq.com

Yufeng Tang
College of Materials Science and Engineering
Chongqing University of Technolog
Chongqing, China
1186462638@qq.com

Xiangtao Xu
Chongqing Pingwei Volt Integrate Circuit Sealing and Application Industry Research Institute Co., Lt
Chongqing, China
xuxt@perfectway.cn

Abstract: **In this paper, a mixed powder of 40%Zn+60%SAC0307 is used as the solder, and a new type process of dual ultrasonic-assisted soldering is used to achieve the interconnection of copper and aluminum dissimilar materials. By comparing the different positions of the ultrasonic vibrators act, it can be concluded that when ultrasonic waves are applied to the side of the hard-to-soldering base material (Al side) under the same ultrasonic power, the overall interconnection quality of the joint will be improved. At the same time, by comparing the difference in shear strength and microstructure of the joints obtained by dual ultrasonic and traditional single ultrasonic-assisted soldering, it is concluded that the soldering effect of the dual ultrasonic-assisted soldering process is more ideal. The Cu side forms an IMC layer of Cu_5Zn_8, and an Al-Zn solid solution layer is formed on the Al side. The microstructure of the interface on both sides of the joint is uniform and smooth; meanwhile, there are no obvious defects. The shear strength of the joints obtained by the dual ultrasonic process can enhance about 18% compared with the single ultrasound, the average strength can reach 26.33MPa.**

Keywords—dual ultrasonic-assisted soldering; copper and aluminum dissimilar materials; SAC solder

I. INTRODUCTION

Copper metal has good thermoelectric properties and excellent mechanical properties, but its production cost is relatively high. In order to further reduce costs, aluminum is used to replace the original copper material in some sub-important parts, which not only guarantees product quality but also reduces costs.[1] As a result, the problem of Cu-Al dissimilar material interconnection has become a research hotspot. The interconnection of Cu-Al dissimilar materials can adopt interconnection processes such as induction diffusion brazing, friction stir welding, and ultrasonic-assisted brazing.[2-4] Among them, ultrasonic-assisted soldering can obtain good joints without flux. As for the

solder used in the interconnection of Cu-Al dissimilar materials, Zn-based solder is the most commonly used. Although satisfactory soldering results can be obtained, the soldering temperature is too high, and the microstructure composition of the joint is different under different cooling rates. Its mechanical properties will also be affected to a certain extent.[5-7] This article uses SAC0307 low melting point lead-free solder and dual ultrasonic-assisted soldering technology to achieve high-quality interconnection of Cu-Al dissimilar materials at 220°C. At the same time by comparing four different soldering processes, it is concluded that the effect of dual ultrasonic-assisted soldering is better than the traditional single ultrasonic-assisted soldering.

II. EXPERIMENTAL PROCEDURES

T2 copper with a specification of φ=5mm, h=3mm, a purity of 99.98% and 1060 aluminum alloy with a specification of φ=5mm, h=3mm are used as base materilas. Adopt Sn99Ag0.3Cu0.7 (SAC0307) No. 4 powder (particle size 20-38μm) and 99.9% purity Zn particle powder (particle size 1μm) mixed uniformly with the mass fraction of 40%Zn+60%SAC0307 solder. Two ultrasonic generators with different powers, YP5020-6Z-1000W and YPS23B-ZB-2000W are used. The frequencies of the two ultrasounds are both 20KHZ, as the No. 1 callout in Fig. 1 shows that 2000W ultrasound acts on the Cu side (referred to as the upper ultrasound); and the No. 7 callout in Fig. 1 is 1000W ultrasound which acts on the Al side (referred to as the lower ultrasound).

Before soldering, use 400# sandpaper to polish off the oxide film on the copper and aluminum surface and clean the surface with alcohol, then place the aluminum material on the lower side of the fixture, and fill the uniformly mixed solder powder into the fixture, drip an appropriate amount of alcohol to compact the solder. Then the copper material is placed on top of the solder to form a Cu/40%Zn+60%SAC0307 powder/Al structure. Place the fixture on the heating device (measured temperature 220℃), the upper ultrasound acts on Cu side and the lower ultrasound acts on Al side, heat the fixture for 205 seconds, turn on the ultrasonic for 5 seconds, and then keep this temperature for 90 seconds. Remove the fixture and cool it to room temperature. According to the difference of ultrasound power, the experiment was divided into four groups as shown in Table 1. Groups 1- 4 ultrasonic process are referred to S600, X600, S1200, S600X600.

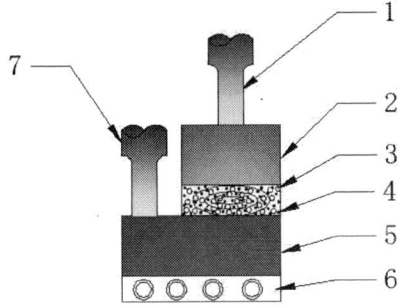

Fig. 1 Dual ultrasonic-assisted soldering system:
1:upper ultrasonic vibrator; 2:Cu base material; 3:SAC0307 solder powder; 4:Zn powder; 5:Al base material; 6:heating device; 7:lower ultrasonic Vibrator

Table 1. The position of ultrasound action

Number of groups/position of ultrasound action	Group 1	Group 2	Group 3	Group 4
Upper ultrasound (position 1 in the picture)	600W	0W	1200W	600W
Lower ultrasound (position 7 in the picture)	0W	600W	0W	600W

The CMT-2503 microcomputer-controlled electronic universal testing machine was used for shear strength test; the LEICA-DM-2500M metallurgical microscope and German Zeiss company's field emission scanning electron microscope (ΣIGMAHD) were used to observe the microstructure of joints; the Oxford EDX analysis accessories was used to determine the composition of joints.

III. RESULTS AND DISCUSSIONS

As shown in Fig. 2 (a-b) and Fig. 3 (a-d), when the total ultrasonic power is few (Group 1 and Group 2), the solder is mostly granular and the Cu/interface IMC is thinner. At the same time, comparing Group 1 with Group 2, it can be seen that when the total ultrasonic power is constant (600W), and the ultrasonic vibrator acts on the Al side, there is a certain amount of Sn in the solder seam. However, while the ultrasonic vibrator acts on the Cu side, most of the Sn in the solder seam is concentrated on the Al side to form a Sn-rich layer with a certain thickness. As shown in Fig. 2 (c-d) and

Fig. 3 (e-h), when the total ultrasonic power is large (Group 3 and Group 4), the melting degree of the solder will increase due to the increase of ultrasonic power intake (1200W). Some of the solder is melted and the other part remains granular, the IMC on the Cu/interface side is thicker (compared to the Group 1 and Group 2). At the same time, comparing the Group 3 with the Group 4, there is a small amount of Sn on the Cu/interface side of the Group 3, but almost all the Sn of the Group 4 is enriched at the Al/interface to form a thicker Sn-rich layer. Moreover, the morphology at the Cu/interface and the Al/interface of the Group 4 is better than the Group 3. It was verified that the dual ultrasonic process is better than single ultrasonic under the same total ultrasonic power, and no obvious phase interference effect is found. It can be seen from the comparison and analysis of the SEM（Fig. 2 and Fig. 3）, by observing and comparing the microstructure at the Cu and Al interface, it can be found that the interface generated by the dual ultrasonic-assisted soldering process （Group 4） compared to the other three processes （Group 1-3） is relatively flat, and the microstructure is relatively smooth, whether it is the Cu/soldering interface or the Al/soldering interface.

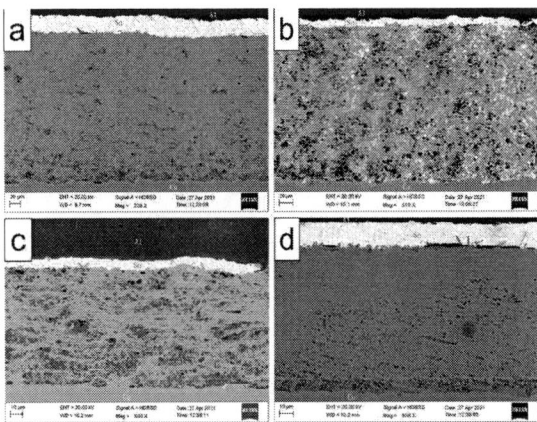

Fig. 2 SEM of solder seam obtained by four different ultrasonic processes: (a) Group 1, (b) Group 2, (C) Group 3, (d) Group 4

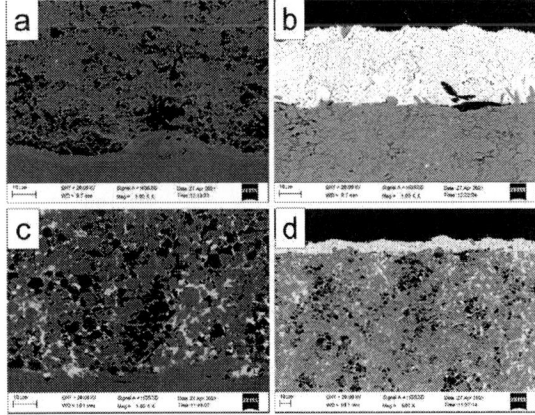

Fig. 3 SEM at Cu/interface and Al/interface of joints obtained by four different ultrasonic processes: (a) copper/interface of Group 1; (b) aluminum/interface of Group 1; (c) copper/interfe of Group 2; (d) aluminum /interface of Group 2

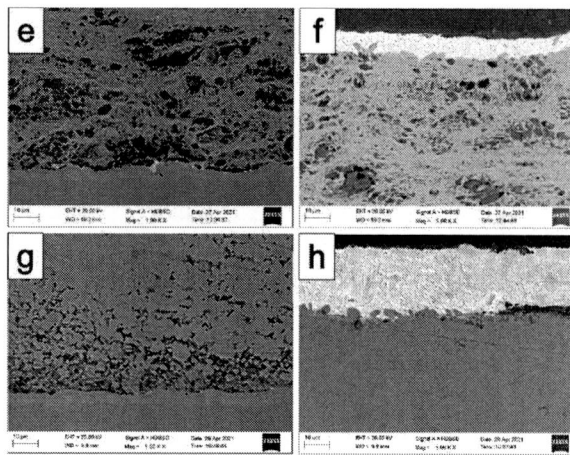

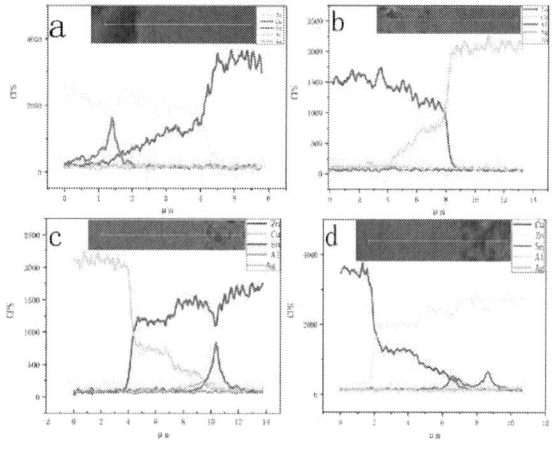

Fig. 3 SEM at Cu/interface and Al/interface of joints obtained by four different ultrasonic processes: (e) copper/interface of Group 3; (f) aluminum /interface of Group 3; (g) copper/interface of Group 4; (h) aluminum /interface of Group 4

Fig. 4 shows the EDX at the Cu/interface of the joints obtained by four different ultrasonic processes. The composition of IMC can be predicted by the platform, as shown in Fig. 4 (a-d), the Cu/Zn atomic ratio is: 42/58, 42/58, 38/62 and 38/62. From the Cu-Zn phase diagram and the Cu/Zn atomic ratio, it can be seen that the IMC on the Cu/interface side of all joints is Cu_5Zn_8. However, the length of the platform does not represent the thickness of the IMC.

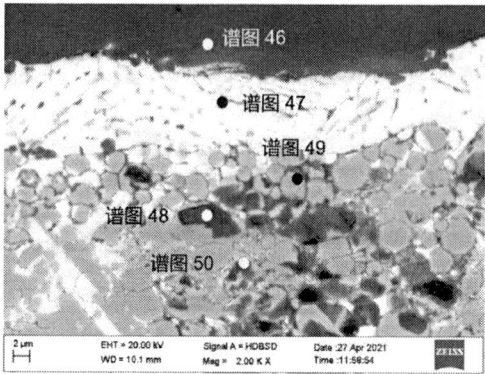

Fig. 5 The microstructure of the Al/solder interface with Group 2 process

Table 2. Atomic percentage of each point in Figure 5

Element	46	47	48	49	50
Al	93.93	2.39	4.39	1.46	4.88
Cu	0.11	0.17	0.48	0.33	0.93
Zn	4.64	3.57	93.78	97.82	84.21
Ag	0.01	0.98	0.00	0.12	0.27
Sn	1.31	92.88	1.35	0.26	9.71
Total	100.00	100.00	100.00	100.00	100.00

As shown in Fig. 6, the process used in Group 1 is S600 and its shear strength is 23.44MPa; the process used in Group 2 is X600 and its shear strength is 23.91MPa; the process used in Group 3 is S1200 and its shear strength is 22.33MPa; the process used in Group 4 is S600X600 and its shear strength is 26.33MPa. Through the data of Group 1-4, the following conclusions can be drawn: ①Comparing Group 1 with Group 3 shows that for the Cu-Al dissimilar material interconnection, it is not that the greater the ultrasonic power, the higher the joint strength; on the contrary, the increase of power may slightly reduce the mechanical properties. ②Comparing Group 1 with Group 2 shows that when the ultrasonic power is the same, the position of the ultrasonic vibrator acting is different, and the soldering effect is also slightly different. For the interconnection of Cu-Al dissimilar materials, the strength is slightly increased when the ultrasonic vibrator acts on the aluminum side, but the improvement rate is less than 2%. It shows that for the interconnection of Cu-Al dissimilar materials, when ultrasonic vibrator act on the side of the hard-to-weld base material, the overall soldering effect will be slightly improved. ③Comparing the Group 4 with the Group 3, it can be seen that the mechanical properties of the joints of the dual ultrasonic-assisted soldering will be significantly improved (about 18%) compared with the single ultrasonic-assisted soldering. It is also proved that the effect of the dual ultrasonic-assisted soldering is not only superposition on the power of two single ultrasonics; ④Comparing the Group 4 with the Group 1, 2, and 3, not only the shear strength of the joints is significantly improved, but also the mechanical properties of the joints obtained by the double ultrasonic action are more stable than the other three processes.

Fig. 4 EDX of IMC at Cu/interface:
(a) Group 1, (b) Group 2, (C) Group 3, (d) Group 4

As shown in Fig. 5 and Table 2, EDX analysis shows that there is a Sn-rich layer near the Al side, and the upper part of the Sn-rich layer near the solder side contains more Zn elements. There is a very thin Al-Zn solid solution layer between the Sn-rich layer and Al. At the same time, as shown in Fig. 2, it can be seen that when use the S600 process, a part of Sn remains in the solder seam, but when the power increases, the Sn in the solder is all concentrated at the Al interface. When the dual ultrasonic-assisted soldering process is adopted, the Sn-rich layer of the solder seam is relatively flat and the microstructure is relatively uniform. At the same time, the Zn in the solder seam will form a needle-like structure, and grow from the Sn-rich layer/solder interface to the Sn-rich layer.

978-1-6654-1392-3/21 $31.00 © 2021 IEEE

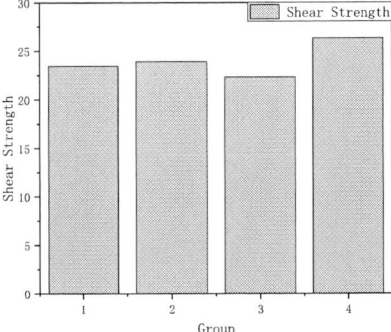

Fig. 6 The shear strength of joints obtained by using four different ultrasonic processes

Fig. 7 shows the fractures of the joints obtained by the four soldering processes after shear strength testing, and the fracture locations are all at the solder. However, the joints obtained by the dual ultrasonic process showed ductile fracture, the dimples are relatively large and the quantity is more. While the joints obtained by the other three processes showed ductile and brittle mixed fracture, the dimples are relatively small and the quantity is less. Therefore, the shear strength of the joints obtained by the dual ultrasonic process was better than the other three processes.

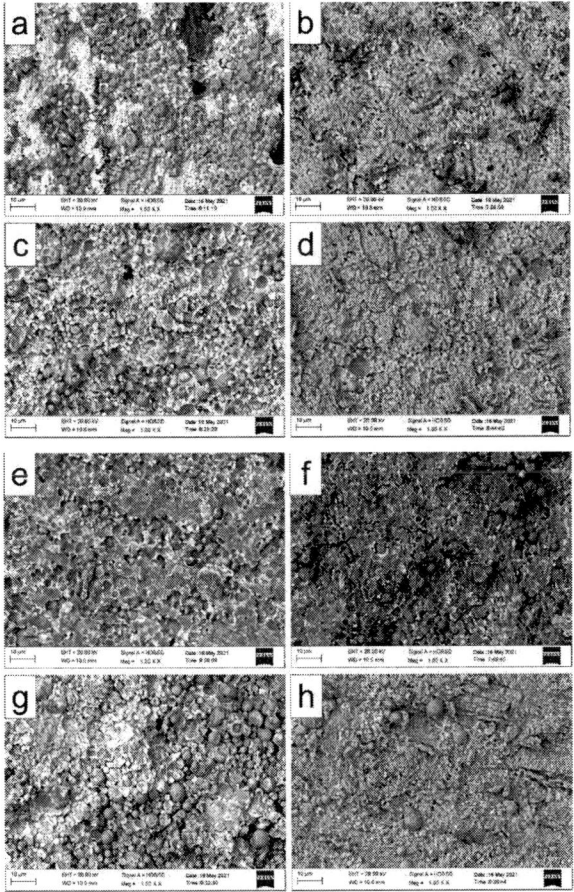

Fig. 7 The fracture of the joints obtained by four different ultrasonic processes: (a) Al side fracture of Group 1, (b) Al side fracture of Group 2, (c) Al side fracture of Group 3, (d) Al side fracture of Group 4, (e) Cu side fracture of Group 1, (f) Cu side fracture of Group 2, (g) Cu side fracture of Group 3, (h) Cu side fracture of Group 4

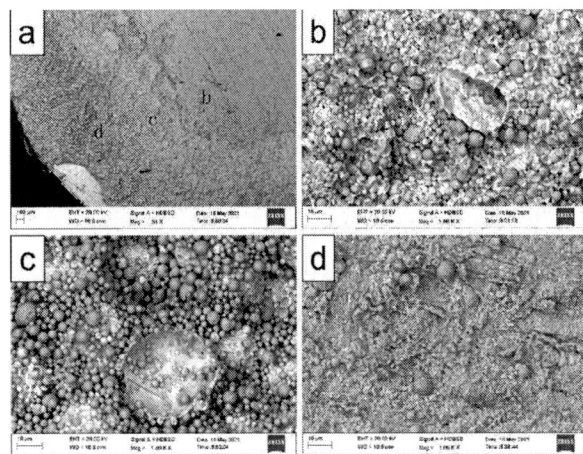

Fig. 8 SEM shows the fracture morphology of the joints obtained by the Group 4 process: (a) SEM of the fracture, (b) a partial enlarged view of b in a, (c) a partial en larged view of c in a, (d) a partial enlarged view of d in a

Fig. 8 shows the microstructure of the fracture after dual ultrasonic action. It can be seen that the entire fracture is divided into three areas from the inside to the outside, namely the melting area, the semi-melting area and the solid area (unmelting area). Zn still remains granular in the melting zone, but its proportion is relatively fewer, most of the composition in this area is molten SAC0307. The semi-melted area is composed of melted SAC0307 solder mixed with unmelted Zn particles. However, the proportion of Zn particles relative to the melting area has increased. In the solid region, both SAC0307 and Zn particles are solid, and the Zn particles are wrapped around the SAC0307 particles. It shows that the temperature in the dual ultrasonic-assisted soldering process gradually decreases from the center to the edge area of the joint. This phenomenon may be caused by the gradual weakening of ultrasonic energy from the center to the edge area of the joint during the soldering.

IV. CONCLUSION

The high-quality interconnection of Cu-Al dissimilar materials was successfully achieved after dual ultrasonic-assisted soldering using 40%Zn+60%SAC0307 solder. Through research, it is found that dual ultrasonic-assisted soldering has a better soldering effect than traditional single ultrasonic-assisted soldering. After dual ultrasonic action, the interface of Cu/solder side and Al/solder side both achieve good bonding, the Cu side forms an IMC layer of Cu_5Zn_8, and an Al-Zn solid solution layer is formed on the Al side. Compared with the joints obtained by single ultrasonic-assisted soldering, the Sn-rich layer and IMC layer is more flat and the microstructure is more uniform. The average shear strength of the joints is significantly improved after dual ultrasonic action, reaching 26.33MPa, which is about 18% higher than that of single ultrasonic, and the fracture mode is ductile fracture. There are more dimples at the fracture and the dimples are larger. In summary, the overall performance of the joints obtained by the dual ultrasonic process is better.

ACKNOWLEDGMENT

This work was supported by the National Natural Science Foundation of China (Grant No.61974013 and

61774066), and the Scientific and Technological Research Program of Chongqing Municipal Education Commission(NO.KJZD-K202101101), Chongqing Research Program Basic Research Frontier Technology (cstc2020jcyj-msxmX0819), University Innovation Research Group of Chongqing (CXQT20023), Graduate Innovation Project of Chongqing University of Technology (ycx20192039, clgycx20201001 and clgycx20201002), respectively.

REFERENCES

[1] B. W. Dong, X. Dong, L. Bao, L. Zhang, and W. M. Long, "Research status of copper-aluminum dissimilar metal brazing materials", Welding, Vol.005, PP.7-12, 2019.

[2] T. Saeid, A. Abdollah-Zadeh, and A. Sazgari, "Weld ability and mechanical properties of dissimilar aluminum−copper lap joints made by friction stir welding", Journal of Alloys & Compounds, Vol.1-2, PP.652-655, 2010.

[3] X. G. Wang, F. J. Yan, X. G. Li, and C. G. Wang, "Induction diffusion brazing of copper to aluminium", Science & Technology of Welding & Joining, 2017.

[4] Y. Xiao, M. Y. Li, and J. Kim, "Ultrasound-assisted brazing of large area Cu/Al dissimilar metals used for package heat dissipation substrate", IEEE, International Conference on Electronic Packaging Technology, 2014.

[5] Y. Xiao, H. J. Ji, M. Y. Li, and J. Kim, "Ultrasound-induced equiaxial flower-like $CuZn5/Al$ composite microstructure formation in $Al/Zn − Al/Cu$ joint", Materials Science & Engineering A, Vol.31, PP.135-139, 2014.

[6] F. Ji, and S. B. Xue, "Growth behaviors of intermetallic compound layers in Cu/Al joints brazed with $Zn−22Al$ and $Zn−22Al−0.05Ce$ filler metals", Materials & Design, 2013.

[7] Z. W. Niu, Y. Zheng, J. H. Huang, H. Yang, J. Yang, and S. H. Chen, "Interfacial structure and properties of Cu/Al joints brazed with Zn-Al filler metals", Materials Characterization, Vol.138, PP.78-88, 2018.8.

Effects of Surface Oxidation Treatments on the Interfacial Adhesion between Copper and Underfill

Bin Wang
Shenzhen Institute of Advanced Electronic Materials, Shenzhen Institute of Advanced Technology, Chinese Academy of Sciences; Nano Science and Technology Institute, University of Science and Technology of China
Shenzhen, China / Suzhou, China
wang.bin@siat.ac.cn

Haoliang Lin
Shenzhen Institute of Advanced Electronic Materials, Shenzhen Institute of Advanced Technology, Chinese Academy of Sciences;
Department of intelligent manufacturing, WuYi University
Shenzhen, China / Jiangmen, China
hl.lin@siat.ac.cn

Houya Wu*
Shenzhen Institute of Advanced Electronic Materials, Shenzhen Institute of Advanced Technology, Chinese Academy of Sciences
Shenzhen, China
hy.wu1@siat.ac.cn

Yuanyuan Yang
Shenzhen Institute of Advanced Electronic Materials, Shenzhen Institute of Advanced Technology, Chinese Academy of Sciences
Shenzhen, China
yy.yang@siat.ac.cn

Gang Li
Shenzhen Institute of Advanced Electronic Materials, Shenzhen Institute of Advanced Technology, Chinese Academy of Sciences
Shenzhen, China
gang.li@siat.ac.cn

Pengli Zhu*
Shenzhen Institute of Advanced Electronic Materials, Shenzhen Institute of Advanced Technology, Chinese Academy of Sciences
Shenzhen, China
pl.zhu@siat.ac.cn

Abstract—Delamination between the underfill and the copper pillar or copper solder pad is one of the major drawbacks of microelectronic packaging, leading to premature failure of the entire device. In this study, the chemical oxidation method was used to improve the adhesive strength between copper and underfill. It was found that the adhesive strength of copper and the underfill was significant increased by oxidation treatment of the copper. The wettability of the copper substrates was improved by the three oxidants ($Na_2S_2O_8$, $CuCl_2$, and $KMnO_4$) and was reduced by the oxidant of H_2O_2. SEM characterization results showed that different microstructures were formed on the oxidized copper surface. Through the analysis of the results of XPS, the surface layer of the oxidized copper changed from Cu metal layer to CuO layer. These results indicated that the wettability, microstructure, and oxide layer affected the adhesive strength of the copper/underfill interface.

Keywords—Underfill, Interface adhesion, Oxidation surface treatment, Advance packaging

I. INTRODUCTION

Nowadays, good adhesion between copper and epoxy resin is the key to ensuring the reliability of many types of electrical interconnections. Copper-epoxy adhesion has become the focus of research in the field of electronic packaging[1, 2]. Delamination failure frequently occurs in IC packages because of the poor adhesive strength between copper and underfill. Due to the thermal expansion coefficient of copper and underfill do not match, additional stress was caused during the solder reflowing process, then delamination or copper pillar cracking would occur. Fig.1 shows a typical delamination and copper pillar cracking failure. In order to prevent these failures, it is necessary to enhance the adhesive strength between the underfill and copper [3, 4].

Lots of research has been conducted to increase the adhesive strength between the copper and underfill[5-7]. Generally, good adhesion between copper and epoxy resin is obtained by changing the surface chemistry and surface morphology of the metal. Surface chemistry changes, obtaining by chemical oxidation, organic inhibitor priming (such as mercaptobenzothiazole), ion implantation, self-assembled monolayer (SAM) covering, vacuum deposition, and UV cleaning, metal plating, have been investigated to obtain better adhesion[8-11]. These surface chemical modification methods improved the adhesion between copper and epoxy resin by changing the physical and chemical properties of the copper surface, e.g., increasing mechanical interlocking, generating new interlayer materials, and forming new chemical bonds[12-15].

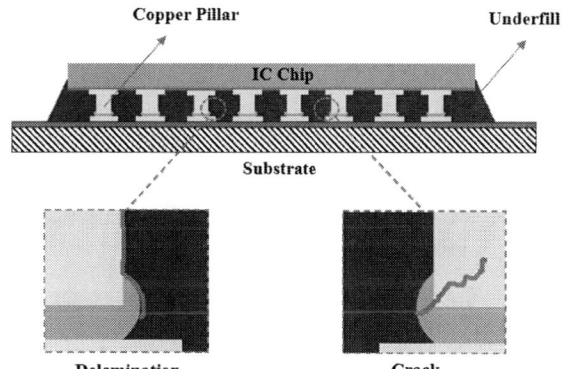

Fig. 1. Failures of copper pillar in flip chip.

Among all the methods, chemical oxidation has become the focus of research due to its low cost, merits of simple operation, and better performance. In this study, the chemical oxidation method was used to improve the adhesive strength between copper and underfill. Four kinds of oxidants ($Na_2S_2O_8$, $CuCl_2$, H_2O_2, and $KMnO_4$) are used for the surface treatment of the copper substrate, generating a unique oxide layer on the copper surface to improve the adhesion between the copper and the underfill. The die shear test was used to investigate the adhesive strength between the oxidized copper and underfill. Then the fracture mode between the underfill and the copper substrate was observed after the die shear test. Finally, the contact angle measurement, SEM, XPS were conducted to characterize the physical and chemical changes on the surface of copper after oxidation.

II. EXPERIMENT

A. Materials

Electroplated copper plate (Electroplating copper, nickel, and gold sequentially), ethyl alcohol (99.9%, aladdin), hydrochloric acid (60%, aladdin). Four kinds of oxidizing agents are purchased from Aladdin Reagent Co. (Shanghai, China). Among them, $Na_2S_2O_8$, $CuCl_2$, and $KMnO_4$ were mixed with pure water to prepare a solution with a concentration of 0.1mol/L, respectively; and the concentration of H_2O_2 solution is 30%. Commercial underfill was used, consisting of epoxy and SiO_2 fillers.

B. Preparations of specimens

Before the oxidation process, the copper substrate was pretreated following several steps: the copper substrate was firstly polished with 2000 and 4000 grit sandpaper in turn; then, the copper was ultrasonically cleaned in absolute ethanol to remove the residues on the surface; the substrate was dried in an oven at 60°C for 10 minutes; after that, the substrate was immersed in 10% HCl (aq) solution for 10 minutes to remove the natural oxides on the copper surface; the treated copper substrate was ultrasonically cleaned in distilled water for 5 minutes; finally, the cleaned substrate was dried under a nitrogen atmosphere. After the pretreatment, the copper substrate was immersed in the four oxidant solutions ($Na_2S_2O_8$, $CuCl_2$, H_2O_2, and $KMnO_4$) for 10 minutes to produce an oxide layer on the copper substrate. After the oxidation, the copper substrate was taken out, ultrasonically cleaned in pure water for 5 minutes, and dried under a nitrogen atmosphere. Finally, a copper substrate covered by a copper oxide layer was obtained, and then the samples were vacuum stored for use.

C. Die shear test

As shown in Fig. 2, a sandwich structure sample was prepared for the die shear, which consists of a silicon die (2×2 mm), oxidized copper substrate (5×5 mm), and a layer of underfill (thickness of around 20 μm) filled the gas of the silicon die and the copper substrate. The sandwich structure is cured at 150 °C for 2 h. Later, the samples are sheared by the test device (DAGE 4000) at room temperature 25°C and high temperature 260°C respectively. It is worth noting that one side of the sample should be parallel to the cutter head of the shear test device to correctly measure the accurate shear strength. The formula for calculating adhesion strength is as follows:

$$P = \frac{F}{s} \tag{1}$$

Where, P = adhesive strength, F = peak shear force that causing damage, s = contact area, respectively.

D. Characterizations

In order to better understand the fracture mode between the underfill and the copper substrate after the die shear test, a parallel coaxial light microscope was used to observe the cross-section of the sample after the experiments. A contact Angle measuring instrument (OCA20) is used to detect whether the wettability of the copper surface after oxidation is enhanced or decreased. The surface structure of the oxidized copper substrate was observed by a scanning electron microscopy (thermoscientific Apreo 2 S). Surface analytical characterization was also performed on the samples to determine the level of oxidation. X-ray photoelectron spectroscopy (XPS) was used to determine the concentrations and the valence structure of the elements.

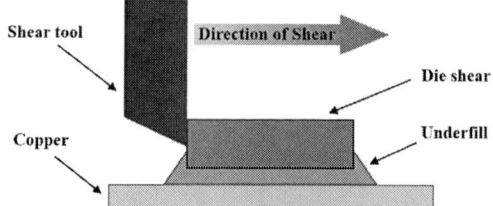

Fig. 2. Scheme of the die shear test

III. RESULTS AND DISCUSSION

A. Die shear strength results

Fig. 3 a) and b) show the die shear strength results of the underfill and the copper substrate at room temperature 25°C and high temperature 260°C, respectively. It can be seen from Figure. 3 a) that the shearing strength of the blank sample (without oxidation treatment) is only 81.2 MPa at room temperature. The shearing strength of the copper substrate treated by $Na_2S_2O_8$ shows no obvious improvement (82.5Mpa). However, the shearing strength has been significantly improved by the other three oxidants, which were 90.4 ($CuCl_2$), 91.1 (H_2O_2), and 96.4Mpa ($KMnO_4$), respectively.

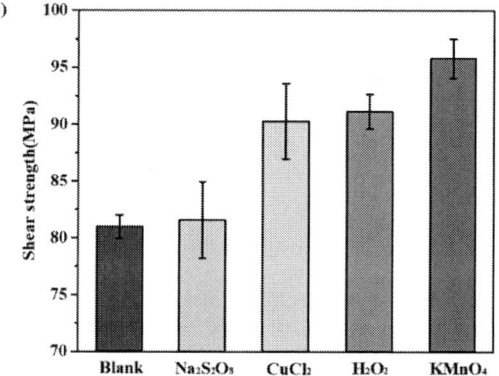

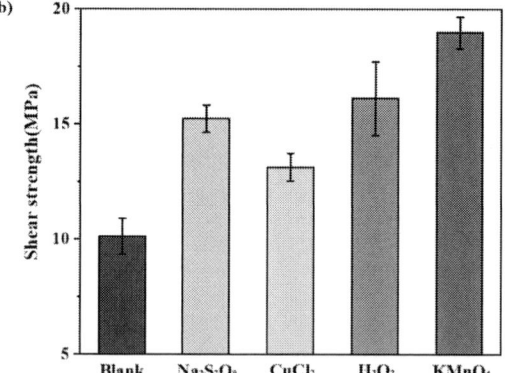

Fig. 3. a) shear strength at 25°C, b) shear strength at 260°C.

As shown in Fig. 3 b), the bonding strength of the copper substrate without oxidation is only 10.3 MPa at high temperature. However, after oxidized by oxidants, the bonding strength of the copper was significantly increased at different degrees. Among them, $KMnO_4$ shows the highest shearing strength, 18.7Mpa, much higher than the blank sample. This indicates that the bonding strength between the copper and the underfill is greatly dependent on the oxidization treatment of the copper substrate.

Fig. 4 a) and b) show the fracture-section of the samples after the die shear test at 25°C and 260°C, respectively. The fracture models of the samples are different. Without oxidation treatment, the fracture path was mainly along the interface between the copper substrate and the underfill, leaving a few underfill on the substrate; in contrast, the fracture path was mainly extending to bulk material of the underfill when the samples were oxidized, leaving more underfill on the substrates. A possible explanation is that the crack path tends to propagate from the interface to the bulk material as the interfacial bonding strength (enhanced by oxidation) is stronger than the bulk material strength.

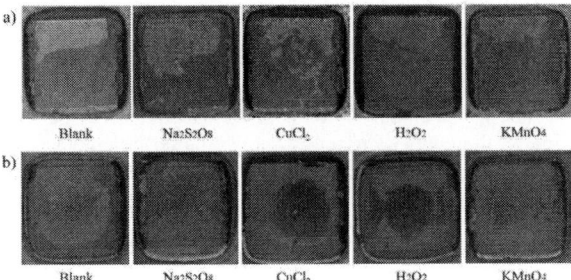

Fig. 4. Fracture-section of the die shear test samples: a) die shear test at 25°C, b) die shear test at 260°C.

B. Surface analysis results

The wettability of the copper surface is determined by the water contact angle. The surface is considered hydrophilic when the contact angle is below 90° and is considered hydrophobic otherwise. The water contact angle results are shown in Fig. 5 and Fig. 6. The contact angle of the copper surface without oxidation was 85.3°. By contrast, the copper substrate after oxidation changed greatly. The contact angle of the copper oxidized by H_2O_2 is the largest, 97.9°, while the copper oxidized by $CuCl_2$ is the smallest, 59.8°. It can be considered that the former is hydrophobic and has poor wettability, while the latter is hydrophilic and has the best wettability.

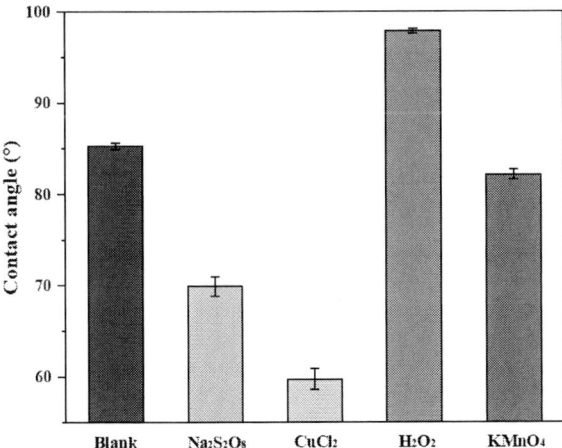

Fig. 5. Contact angle measurement results of copper substrate

However, it is found that the wettability of the copper surface after oxidation has no obvious corresponding relationship with the improvement of the bond strength by comparing the wettability results with the adhesive strength.

As a result, wettability is not the main factor in improving the bond strength of the copper after oxidation.

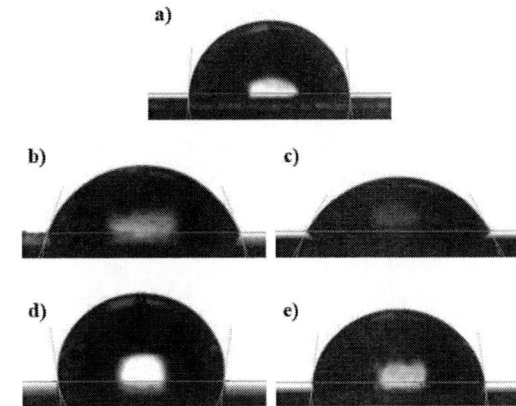

Fig. 6. Contact angle measurement results: a) unoxidized, b) oxidized by $Na_2S_2O_8$, c) oxidized by $CuCl_2$, d) oxidized by H_2O_2, e) oxidized by $KMnO_4$.

The morphologies of the copper substrate before and after oxidation were detected by SEM. As shown in Fig. 7 a) and b), before oxidation, the copper surface is relatively smooth, though many tiny grooves existed on the surface. The grooves are caused by the grinding and polishing processing.

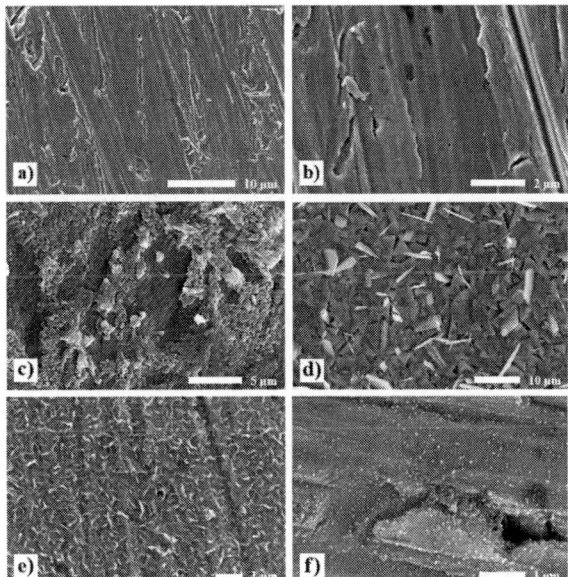

Fig. 7. Micro morphologies of copper surface: a) and b) unoxidized, c) oxidized by $N_2S_2O_8$, d) oxidized by $CuCl_2$, e) oxidized by H_2O_2, f) oxidized by $KMnO_4$.

On the contrary, the oxidized copper surface showed a completely different phenomenon. As shown in Fig. 7 c), many flake-like structures appeared on the $Na_2S_2O_8$ oxidized surface, which were loosely and unevenly distributed. Oxidized by $CuCl_2$, the copper surface appeared a considerable number of crystals, which were in the shape of bars, squares, diamonds, etc., as shown in Fig. 7 d). the crystals were also loosely distributed. Fig. 7 e) shows the oxidized H_2O_2 surface, where numerous dense fish-scale protrusions are tightly formed on the oxidized copper surface. These protrusions can be mechanically interlocked with the underfill, which is of great help in improving the adhesion between the copper and the underfill. Similarly as shown in Fig. 7 f), a great amount of fine particles are formed on the surface of the copper oxidized by $KMnO_4$, and the particle size of which is about 50nm. It can be considered that these particles can also mechanically interlocked with the underfill, and hence enhance the adhesive strength of the underfill and the copper. According to the SEM results, it was found that protrusions of different shapes appeared on the surface of the copper oxidized by these four oxidants. These protrusions were considered to be oxides formed after copper oxidation. Due to the role of these oxides, the adhesive strength between copper and underfill can be enhanced.

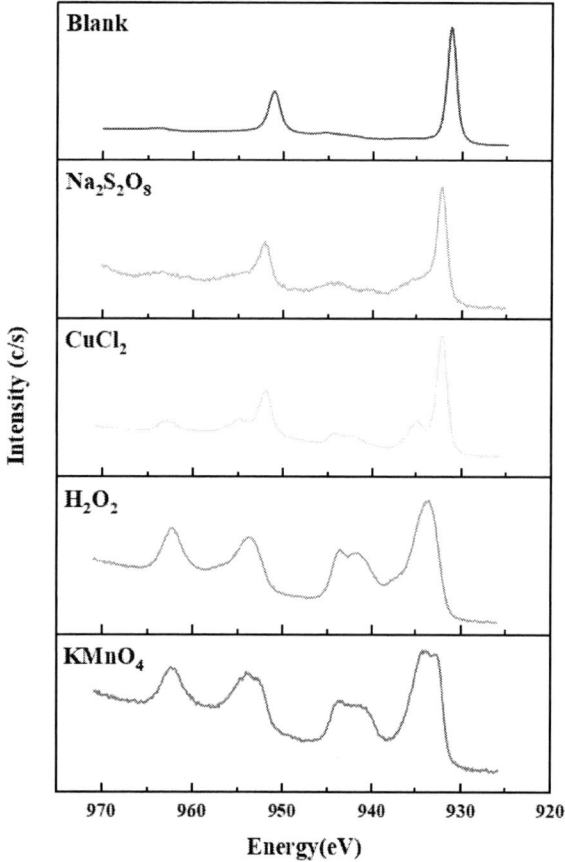

Fig. 8. XPS narrow scans around copper *2p* peaks on unoxidized and oxidized Cu surface.

X-ray photoelectron spectroscopy (XPS), is a popular method for analyzing the chemical properties of the surface of a substance, which can detect the element composition, binding energy, chemical potential, and electronic state of the elements. Fig. 8 shows the high-resolution narrow scans of the five kinds of copper substrates. Examination of XPS narrow scans around copper *2p* peaks on the unoxidized copper surface, there are no other peaks except for two obvious peaks in the spectrum. There is no doubt that that is the characteristic of Cu metal, demonstrating the oxide on the copper surface is removed after treatment with hydrochloric acid solution.

For the oxidized copper surface, the results of the XPS spectra show a great difference. In the spectra, the copper surface that oxidized by $Na_2S_2O_8$ and $CuCl_2$ shows two weak shake-up satellites of Cu^{2+} oxidation state at 945eV and 963eV. The difference is the satellite peak intensity of the latter is stronger than that of the former. Both the positions of the two spectra were also shifted compared with the spectra of the copper without oxidation. Combined with the XPS standard spectral library of Cu 2p, a Cu_2O layer is formed on the copper surface oxidized by $Na_2S_2O_8$, Cu_2O and CuO are present on the copper surface oxidized by $CuCl_2$. And This also indicates the oxidation effect of $Na_2S_2O_8$ and $CuCl_2$ on copper is poor. The XPS spectra of copper oxidized by H_2O_2 and $KMnO_4$ show that there are strong shake-up satellites of Cu^{2+} oxidation state at 945eV and 963eV, and these two spectra were more shifted, indicating a thicker CuO layer is formed on the surface of copper.

The surface analysis of XPS results suggested that the oxide on the copper surface is removed after treatment with hydrochloric acid solution in the experiment. When the copper surface is oxidized by oxidant, Cu_2O layer and CuO layer were generated on the surface, and the thickness of CuO layer was affected by the of type oxidant. Refering the shearing test results, it is found that the adhesive strength of the copper surface treated with hydrochloric acid is much smaller than that of the copper surface treated with oxidant. The adhesive strength of copper surface oxidized by $Na_2S_2O_8$ and $CuCl_2$ was not improved as much as that treated by H_2O_2 and $KMnO_4$. It is comfirming that the oxidation effect of H_2O_2 and $KMnO_4$ on copper is better than that of $Na_2S_2O_8$ and $CuCl_2$. It's quite obvious that the copper surface changes from a Cu metal layer to a CuO oxide layer after oxidation, and the CuO oxide layer can greatly improve the adhesive strength between copper and underfill.

IV. CONCLUSION

In this study, the chemical oxidation method was used to improve the adhesive strength of the copper/underfill interface. Die shear test and surface characterization showed that the shearing strength was increased from 81.0 to 91.8 MPa at 25°C and from 10.1 to 19.0 MPa at 260°C when $KMnO_4$ treated the copper substrate. The surface morphologies of copper before and after oxidation treatment were different, such as dense scale-like protrusions were formed on the surface of copper treated by H_2O_2, and granular protrusions were generated on the surface of copper treated by $KMnO_4$. These microstructures were possibly beneficial to increase the adhesive strength of the copper and the underfill. XPS results showed that the surface layer of the oxidized copper changed from Cu metal layer to CuO layer. It is possibly that the CuO oxide layer is helpful to enhance the improvement of adhesion. Therefore, it was found that the microstructure and CuO oxide layer had a positive influence on the adhesive strength of the copper/underfill interface.

ACKNOWLEDGMENT

This study is supported by the National Key R & D Project from Minister of Science and Technology of China (2020YFB0311800), Shenzhen basic research plan (JCYJ20190807154409372), the National Natural Science Foundation of China (61704182), and the GuangDong Basic and Applied Basic Research Foundation (2020A1515111003).

REFERENCES

[1] C. L. Gan et al., "Effects of Bonding Wires and Epoxy Molding Compound on Gold and Copper Ball Bonds Intermetallic Growth Kinetics in Electronic Packaging," (in English), J Electron Mater, vol. 43, no. 4, pp. 1017-1025, Apr 2014, doi: 10.1007/s11664-014-3011-y.

[2] C. Q. Cui, H. L. Tay, T. C. Chai, R. Gopalakrishnan, and T. B. Lim, "Surface treatment of copper for the adhesion improvement to epoxy mold compounds," (in English), Elec Comp C, pp. 1162-1166, 1998, doi: Doi 10.1109/Ectc.1998.678863.

[3] C. Zhang, J. Hankett, and Z. Chen, "Molecular Level Understanding of Adhesion Mechanisms at the Epoxy/Polymer Interfaces," Acs Applied Materials & Interfaces, vol. 4, no. 7, pp. 3730-3737, Jul 2012, doi: 10.1021/am300854g.

[4] K. Yamanaka, T. Ooyoshi, and T. Nejime, "Effect of underfill on electromigration lifetime in flip chip joints," (in English), J Alloy Compd, vol. 481, no. 1-2, pp. 659-663, Jul 29 2009, doi: 10.1016/j.jallcom.2009.03.063.

[5] S. J. Cho et al., "The effect of the oxidation of Cu-base leadframe on the interface adhesion between Cu metal and epoxy molding compound," (in English), Ieee T Compon Pack B, vol. 20, no. 2, pp. 167-175, May 1997, doi: Doi 10.1109/96.575569.

[6] P. S. Ho et al., "Study on Factors Affecting Underfill Flow and Underfill Voids in a Large-die Flip Chip Ball Grid Array (FCBGA) Package," 2007 2007: IEEE, doi: 10.1109/eptc.2007.4469683.

[7] H. Abe et al., "Cu Wire and Pd-Cu Wire Package Reliability and Molding Compounds," (in English), 2012 Ieee 62nd Electronic Components and Technology Conference (Ectc), pp. 1117-1123, 2012.

[8] F. A. Akgul et al., "Influence of thermal annealing on microstructural, morphological, optical properties and surface electronic structure of copper oxide thin films," (in English), Mater Chem Phys, vol. 147, no. 3, pp. 987-995, Oct 15 2014, doi: 10.1016/j.matchemphys.2014.06.047.

[9] S. Dong Kil, L. Hyo Sug, and J. Im, "Chemical and Mechanical Analysis of PCB Surface Treated by Argon Plasma to Enhance Interfacial Adhesion," Ieee T Electron Pa M, vol. 32, no. 4, pp. 281-290, 2009, doi: 10.1109/tepm.2009.2029700.

[10] C. L. Gan et al., "Reliability assessment and mechanical characterization of Cu and Au ball bonds in BGA package," (in English), J Mater Sci-Mater El, vol. 24, no. 8, pp. 2803-2811, Aug 2013, doi: 10.1007/s10854-013-1174-6.

[11] C. P. Liu et al., "A novel decapsulation technique for failure analysis of epoxy molded IC packages with Cu wire bonds," (in English), Microelectronics Reliability, vol. 52, no. 4, pp. 725-734, Apr 2012, doi: 10.1016/j.microrel.2011.11.005.

[12] T. P. Ang, T. S. A. Wee, and W. S. Chin, "Three-dimensional self-assembled monolayer (3D SAM) of n-alkanethiols on copper nanoclusters," (in English), J Phys Chem B, vol. 108, no. 30, pp. 11001-11010, Jul 29 2004, doi: 10.1021/jp049006r.

[13] M. P. K. Turunen et al., "Pull-off test in the assessment of adhesion at printed wiring board metallisation/epoxy interface," (in English), Microelectronics Reliability, vol. 44, no. 6, pp. 993-1007, Jun 2004, doi: 10.1016/j.microrel.2004.01.001.

[14] S. Gunther, H. Marbach, R. Hoyer, R. Imbihl, L. Gregoratti, and M. Kiskinova, "Spatial variations of the interface composition during surface chemical reactions," (in English), J Electron Spectrosc, vol. 114, pp. 989-996, Mar 2001, doi: Doi 10.1016/S0368-2048(00)00262-0.

[15] Y. Feng et al., "Corrosion Protection of Copper by a Self‐Assembled Monolayer of Alkanethiol," Journal of The Electrochemical Society, vol. 144, no. 1, pp. 55-64, 2019, doi: 10.1149/1.1837365.

Feasibility Analysis of Crack Initiation Identification of Sintered Silver for a Fast Lifetime Prediction

Jiuyang Tang
Academy for Engineering and Technology, Fudan University
Shanghai, China
jiuyangtang20@fudan.edu.cn

Jing Zhang
Heraeus Materials Technology Shanghai Ltd.
Shanghai, China
j.zhang@heraeus.com

Guoqi Zhang
ECTM, Delft University of Technology
Delft, the Netherlands
Shenzhen Institute of Wide-bandgap Semicontuctors
Shenzhen, China
g.q.zhang@tudelft.nl

Pan Liu*
Academy for Engineering and Technology, Fudan University
Shanghai, China
Yiwu Research Institute of Fudan University
Yiwu, China
panliu@fudan.edu.com

Abstract—Traditional packaging materials such as solder paste were studied for decades, which is possible to detect initial cracks for a reliability lifetime estimation. While novel die-attach materials such as sintered silver are developing towards higher working temperature and higher current density, it is not clear whether initial cracks are also helpful for reliability assessment. Therefore, in this work, a series of FEM simulations were established for a response surface model with power electronics chip sizes to predict sintered silver joint reliability. Impact factors for resistance were analyzed and compared such as die size and the thickness of the joint. Sintered silver layer sandwiched by copper substrate and terminals with constant current supply was generated and simulated for resistance fluctuation. Through the high-precision DC resistance measurement setup based on the four-probe method, the resistance over 50 nΩ is possible to detect, thus lead to crack growth. In order to study the geometry sensitivity of cracks, preset arc-shape cracks were modeled to simulate crack generation. The coupling of resistance and crack were analyzed through von Mise strain distribution. With a proper geometric configuration of the die-attach layer, it is possible to minimize the testing time for new joint materials through a high precision electrical resistance measurement and simulation-assisted models.

Keywords—*sintered silver, finite element analysis, cracks evolution, failure prediction*

I. INTRODUCTION

Die-attach materials become increasingly important in power module packages since they are often considered as one of the weakest parts of the system. Solder is gradually replaced by sintered silver material due to the development trend of high working temperature. Sintered silver provides advent-ages of high thermal conductivity, longer lifetime, high electrical conductivity, proper coefficient of thermal expansion (CTE), low-temperature processing, etc.[1, 2]. However, the reliability of such new material still needs investigation, especially under different working conditions[3, 4]. Meanwhile, the demand for rapid product development puts an urgent need to develop new and fast testing methods and protocols for accurate crack initiation and propagation measurements for the sintered silver joint[5-7].

Therefore, in this work, a feasibility analysis of crack initiation under different geometry designs was carried out using FEM simulation methods for a fast lifetime prediction.

Previous work showed that it was possible to detect the viscoplastic deformation prior to crack and the crack propagation in the solder joint with high precision resistance measurement setup. J.Zhang et. al[8, 9] experimentally prove the combination of resistance measurement and the proposed prediction model can detect the initiation of a single crack, which confirm that high-precision electrical measurement can distinguish and monitor in situ both viscoplastic deformation and crack initiation and propagation. G.S. Khinda et. al[10] used the four-probe method to monitor the resistance changes of nano-silver sintering. And the quantified resistance characterizes the reliability of sintering, which can be potentially used in the process set up. Pushparajah Rajaguru et. al[11] used the finite element method to analyze the sintered silver interconnection between the silicon carbide chip and the copper substrate in the power electronic module assembly, and proposes a model for predicting the fatigue damage of the sintered silver layer. Such set up provides a non-destructive, in-situ solution for crack detecting.

In this work, a series of Finite Element Method (FEM) simulations were established for a response surface model with power electronics chip sizes to predict sintered silver joint reliability. The simulation was established using the FEM through COMSOL Multiphysics 5.5. It investigated the resistance change influenced by chip size, joint thickness, and crack geometry. Different crack initiation was set in simulation to compare the influence to cracks propagation, which led to a fast lifetime reliability prediction. It was found that crack propagation was linked to a fast growth of resistance, which made it possible to identify the crack initiation point through a high precision DC electrical resistance measurement (Agilent 34970A Data acquisition unit equipped with Agilent 3490). The Four-Probe Method was designed for on-board electrical resistance monitoring. Preset arc-shape cracks were modeled to simulate crack generation. And the purpose is to analyze whether proper

978-1-6654-1392-3/21 $31.00 © 2021 IEEE

geometric configuration of the die-attach layer is possible to minimize the testing time for new joint materials through an electrical resistance increase rate.

II. SIMULATION MODEL DESCRIPTION

The Finite Element Method was applied for the feasibility analysis of crack initiation identification of sintered silver for a fast lifetime prediction, which required a multi-physics approach using electrical-thermal-mechanical simulations. Since the hardware for resistivity measurement is Agilent 34970A Data acquisition, it is difficult to detect resistivity fluctuation within 50 nΩ. Therefore, the resistivity change for different crack thicknesses and crack radius was investigated and simulated in Fig. 1. It is apparent that the resistance in general follows a linear trend with crack thickness and crack length. At a certain crack length, the crack growth under a certain level is difficult to be detected when the total resistance fluctuation is under 50 nΩ. Therefore, a detailed resistance analysis was carried out from 10-30 μm thickness increase. It is clear that the resistance increase follows a polynomial fitting. At 20 μm, the sharp increase indicated the crack growth of the thickness is possible to be detected. Further calculations will be shown in result discussion part.

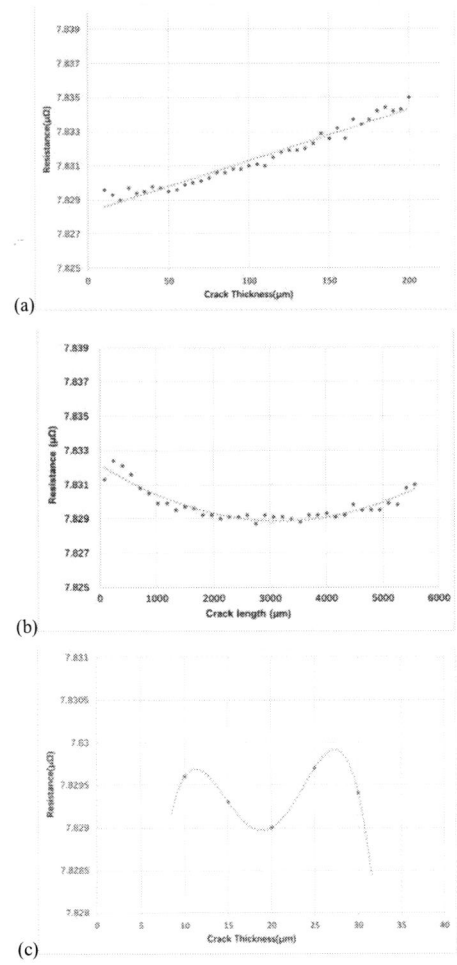

(a)

(b)

(c)

Fig. 1. The resistivity change for different (a) crack thickness (y = 3E-5x + 7.8283) (b) crack length (y = 3E-10x^2 - 2E-06x + 7.8322) (c) thickness range from 10-50 μm (y = -2E-07x^4 + 2E-05x^3 − 4E-4x^2+4.7E-3x+7.8112)

In this work, the main constraints are the geometry of the sintered silver joint and copper pad, the thermoelectric properties of the material, and the stress-strain behavior of the material system. Meanwhile, the mechanical boundary conditions are defined for fixed upper and lower surfaces of the copper pads to simplify the finite element model, and the silver-copper interface is selected as the experimental object.

COMSOL 5.5 version with electric current, thermal transfer in solid and solid mechanics modules were chosen to calculate the resistivity change and the strain-stress analysis. The geometry design of the system is shown in Fig. 2. Components dimensions are listed in Table 1. Copper is selected to be pad material and the material properties of the electric assembly used for the finite element model are listed in Table 2.

(a)

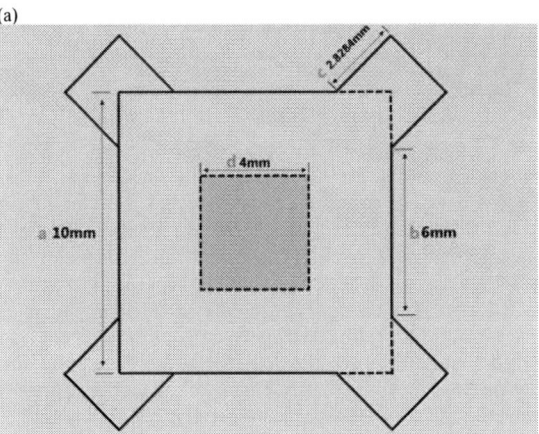

(b)

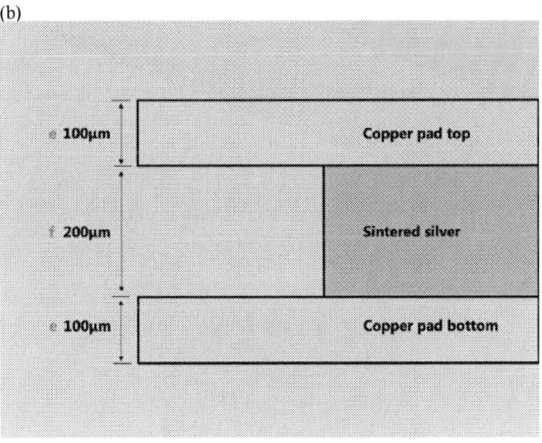

Fig. 2. schematic of geometry design of the system

TABLE I. GEOMETRY DESIGN PARAMETERS

Parameter	Material	Thickness	Dimension
Pad	copper	0.1mm	a=10mm b=6mm c=2.828mm
Joint	Sintered silver	0.2mm	4x4mm²

978-1-6654-1392-3/21 $31.00 © 2021 IEEE

TABLE II.   MATERIAL PROPERTIES

Material properties	Material	
	Sintered Silver	Copper
Thermal conductivity [W/(m·k)]	220	400
Density [kg/m3]	8580	8960
Heat capacity at constant pressure [J/(kg · k)]	234	385
Young's module [Pa]	10E9	110E9
Poisson's radio	0.37	0.35
Coefficient of thermal expansion [1/k]	20.3E-6	17E-6
Relative permittivity	1	1
Electrical conductivity [S/m]	1.25E7	6E7

The copper pad geometry contains one rectangle with two terminals in corners, which is designed for the four-probe measurement. The mesh structure used is shown in Fig. 3. Sintered silver was sandwiched by two copper pad, with terminals connected to current supplier of 1A. The sinter joint was assumed to be rectangular and the size is 4 mm×4 mm×0.2 mm. The size of the employed copper pad is 10 mm×10 mm×0.1 mm. The measurement was generated through four-probe measurement methods. When the four metal probes of A, B, C, and D are in contact and pressed against the material to be measured with a certain pressure, a current I is passed between the two probes of A and D, and a potential difference V is generated between the probes of B and C. Consequently, the resistance of the tested material is obtained and the influence of contact resistance was expelled. Therefore, assuming that the contact resistance at the interface to be negligible, the electrical resistance of interconnect was determined using (1).

$$R = (V_C - V_B) / I \qquad (1)$$

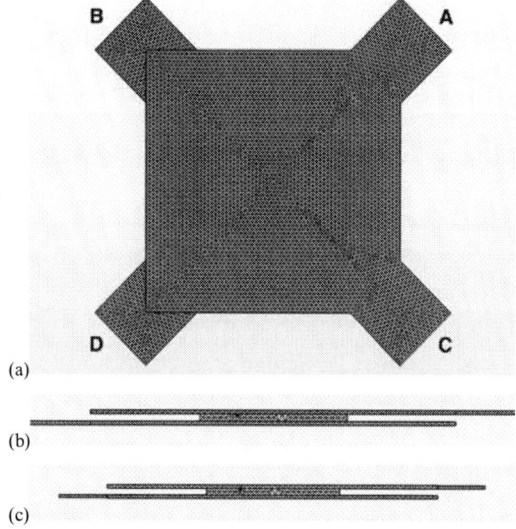

Fig. 3.   (a) top, (b) side and (c) section view of the FE model

In this simulation work, to simplify the simulation and concentrate on the crack initiation, the porosity of sintered silver layer was neglected. Such simplifying assumption is to prevent the intrinsic pores located in sintered silver layer to prevent the deformations of the cracks. The simulation of models with different chip sizes and joint thickness indicated that selecting the proper chip size and the appropriate sintered silver joint thickness prolonged the life of the sintered joint.

III.   RESULT DISCUSSION

A. Resistance vs die attach layer geometry

In this work, a series of FEM simulations were established for a response surface model with power electronics chip sizes to predict sintered silver joint reliability, as shown in Fig.4. Simulation output of resistance set was mapped, while the die size ranging from 4 to 100 mm² and the die attach layer thickness ranging from 20 to 300 μm. It showed the die size and thickness were important geometric factors when monitoring the resistivity of die attach layer. The impact factor of die size is more influential than thickness. When keeping same voltage/current source, the resistance drop of die attach layer following the exponential trend, as shown in Fig. 5.

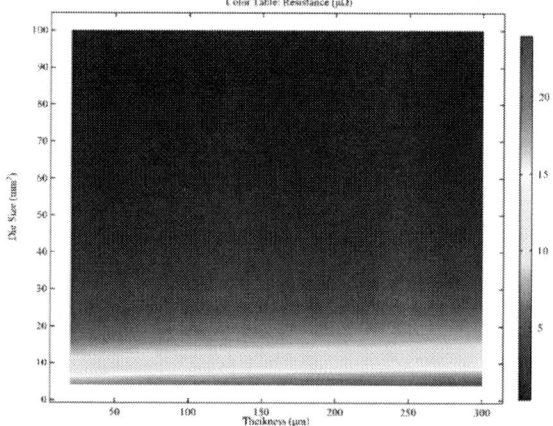

Fig. 4.   Resistance map with axis x: thickness, y: die size

Taken into consideration of the SiC common die sizes, and the pressureless sintering geometry limitation, in the following work, 4x4 mm² and 200 μm thickness of the sintered silver layer were chosen as the die-attach layer geometry.

B. Resistance vs crack geometry

As for crack analysis, preset arc-shaped cracks with different thicknesses and different radius were simulated. It is found that through crack thickness growth from 20 μm to 200 μm, the resistance change was within 50 nΩ, as shown in Fig. 1. It is calculated that resistance of 50 nΩ can be detected by the electrical resistance measurement, when crack thickness grows from 10 μm to 20 μm, with the crack length of 1.571 mm. Therefore, the following crack thickness was preset to be 20 μm.

When keeping the crack thickness as 20 μm, crack radius impact to resistance was also analyzed. As shown in Fig. 6, die attach layer resistance reached peak value with the radius of 3950 μm, which also corresponds to the greatest length of the crack.

978-1-6654-1392-3/21 $31.00 © 2021 IEEE

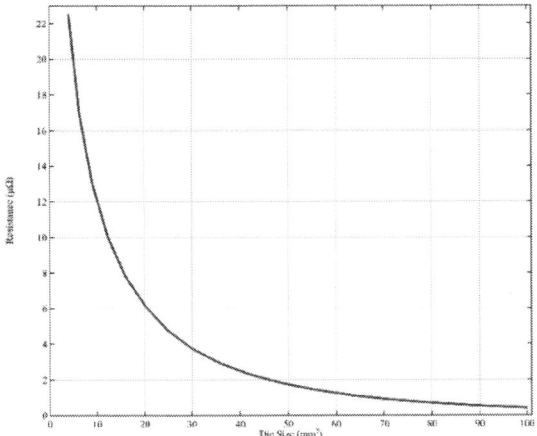

Fig. 5. profile between resistance and die size

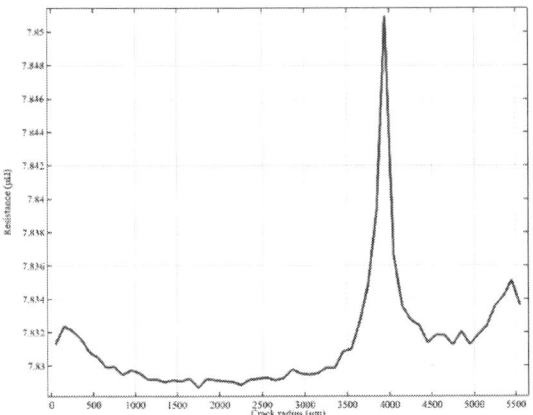

Fig. 6. profile between resistance and crack radius

C. Von Mise strain distribution analysis of cracks

Through von Mise stress distribution of the top surface of die attach layer, top view of Von Mise stress distributed of joint is shown in Fig.7 which apparently indicated stress change brought by the propagation of crack, and it can also infer direction of the crack moving which contribute to deformation analysis. It is clear in Fig. 10 that strain-stress was concentrated at the edge of the crack to the side surface. Before the crack grew through the die attach layer, the stress value reached the top. After the crack went through the total die attach layer, the stress value dropped significantly, as shown in Fig. 8. In order to detail analyze the surface strain-stress conditions, the surface layer as specified in Fig. 11 was chosen. It is clear that through side-surface stress analysis, the top surface suffers the greatest stress, which usually leads to the die attach delamination failures at the side surface connections. Through joint analysis, it is apparent that the max stress concentrated not only at the bottom of the crack, but also at the corners. Therefore, it indicated that the crack growth and corner delamination/cracks are expected failures.

It worth to mention that the von Mise stress has nothing to do with crack initiation. It is an indication of the onset of plastic flow. It gives an indication of the crack initiated failure mechanism analysis.

The feasibility shown by the results of simulation is validation needed, and it will be verified by experiment in the future as shown in Fig.9. We plan to develop a special fixture for 4x4 mm^2 chip size sintered silver connection which has a suitable shape for the four-probe method. The die-attach layer will be pre-set with series cracks, with a cyclic test carrying out under constant current to promote propagation of crack. Furthermore, the resistance change will be delivered from detection load to the control unit and then to the upper computer for data collection. Consequently, the equipment (Agilent 34970A Data acquisition/switch unit equipped with Agilent 3490) detect the resistance change and validate that the crack propagation occurs by the sectional characterization method is expected. Pre-set cracks of sintered silver layer are expected to be fabricated via designed masks during screen printing. Resistance are monitored for crack initiation, thus to verified the simulation model via experimental data.

Fig. 7. Top view of Von Mise Stress distributed of Joint

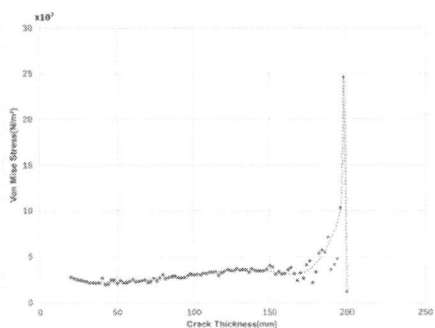

Fig. 8. maxium Von Mise Stress distributed of crack surface with increase of crack thickness

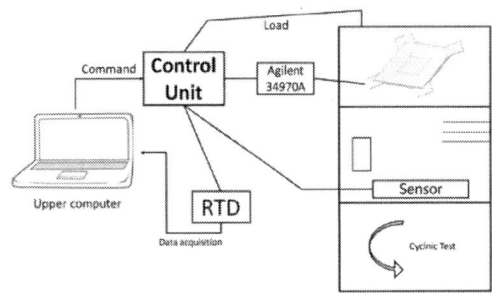

Fig. 9. Schematic of experimental setup

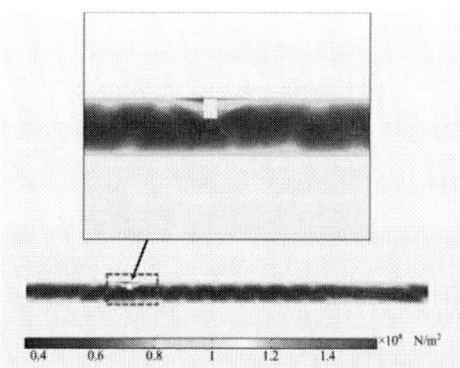

Fig. 10. Side surface of Von Mise Stress distributed of Joint

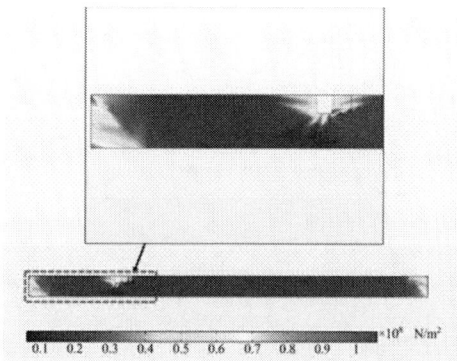

Fig. 11. Section of Von Mise Stress distributed of the joint

IV. CONCLUSION

In this work, a series of FEM simulations were established for a response surface model with power electronics chip sizes to predict sintered silver joint reliability. It showed the die size and the thickness of the joint were the top influential geometric factors. The simulation investigated the influence of chip size, joint thickness, and resistance change. Sintered silver layer sandwiched by copper substrate and terminals with constant current supply was simulated. The simulation of models with different chip sizes and joint thickness indicated that selecting the proper chip size and the appropriate sintered silver joint thickness prolonged the life of the sintered joint.

Furthermore, different crack initiation was set in the simulation to compare the influence to cracks propagation, which led to a fast lifetime reliability prediction. It was found that crack propagation was linked to a fast growth of resistance, which made it possible to identify the crack initiation point through high precision DC electrical resistance measurement. Preset arc-shape cracks were modeled to simulate crack generation. The coupling of resistance and crack were analyzed through von Mise strain distribution. The study showed that with a proper geometric configuration of the die-attach layer, it is possible to minimize the testing time for new joint materials through an electrical resistance increase rate.

Corresponding experiments will be carried out in future work via a setup of sintered silver fixture and resistance measuring system. Pre-set cracks of sintered silver layer are fabricated via designed masks during screen printing. Resistance are monitored for crack initiation, thus to verified the simulation model via experimental data.

ACKNOWLEDGMENT

This project was supported by the Key-Area Research and Development Program of Guangdong Province (Funder ID: 2019B010131001); Science and Technology Innovation Base of Shanghai Municipal Science and Technology Commission, Shanghai SiC Power Device Engineering and Technology Research Center, Fudan University(Funder ID: 19DZ2253400); Fudan University-Heraeus cooperation project(Funder ID:20549).

REFERENCES

[1] J. G. Bai, J. N. Calata, L. Guangyin, and L. Guo-Quan, "Thermomechanical reliability of low-temperature sintered silver die-attachment," *Thermal and Thermomechanical Proceedings 10th Intersociety Conference on Phenomena in Electronics Systems, 2006*, pp. 1126-1130, 30 May-2 June 2006

[2] J. G. Bai, J. N. Calata, and G. Q. Lu, "Processing and characterization of nanosilver pastes for die-attaching SiC devices," *IEEE Transactions on Electronics Packaging Manufacturing,* vol. 30, no. 4, pp. 241-245

[3] C. Weber, M. Hutter, S. Schmitz, and K. Lang, "Dependency of the porosity and the layer thickness on the reliability of Ag sintered joints during active power cycling," *2015 IEEE 65th Electronic Components and Technology Conference (ECTC),* pp. 1866-1873, 26-29 May 2015

[4] X. Li, G. Chen, L. Wang, Y. H. Mei, X. Chen, and G. Q. Lu, "Creep properties of low-temperature sintered nano-silver lap shear joints," *Materials Science and Engineering a-Structural Materials Properties Microstructure and Processing,* vol. 579, pp. 108-113

[5] D. Guang-Cheng, C. Xu, and l. Guo-Quan, "Fatigue lifetime prediction of a novel interface material with plastic failure for power electronics packaging," *2008 11th Intersociety Conference on Thermal and Thermomechanical Phenomena in Electronic Systems,* pp. 793-798, 28-31 May 2008

[6] C. Weber, H. Walter, M. V. Dijk, M. Hutter, O. Wittler, and K. Lang, "Combination of Experimental and Simulation Methods for Analysis of Sintered Ag Joints for High Temperature Applications," *2016 IEEE 66th Electronic Components and Technology Conference (ECTC),* pp. 1335-1341, 31 May-3 June 2016

[7] N. Heuck *et al.,* "Aging of new Interconnect-Technologies of Power-Modules during Power-Cycling," *CIPS 2014; 8th International Conference on Integrated Power Electronics Systems,* pp. 1-6, 25-27 Feb 2014

[8] J. Zhang, S. v. d. Zwaag, H. W. Zeijl, and G. Q. Zhang, "On the use of high precision electrical resistance measurement for analyzing the damage development during accelerated test of Pb-free solder interconnects," *2013 IEEE 63rd Electronic Components and Technology Conference,* pp. 192-199, 28-31 May 2013

[9] J. Zhang, "Fast Qualification of Solder Reliability in Solid-state Lighting System," *Electrical Engineering Mathematics & Computer ence,*

[10] G. S. Khinda *et al.,* "Effects of Oven and Laser Sintering Parameters on the Electrical Resistance of IJP Nano-Silver Traces on Mesoporous PET Before and During Fatigue Cycling," *2019 IEEE 69th Electronic Components and Technology Conference (ECTC),* pp. 1946-1951, 28-31 May 2019

[11] P. Rajaguru, H. Lu, and C. Bailey, "Sintered silver finite element modelling and reliability based design optimisation in power electronic module," *Microelectronics Reliability,* vol. 55, no. 6, pp. 919-930

978-1-6654-1392-3/21 $31.00 © 2021 IEEE

Investigation on the Warpage of Fan-Out Wafer-Level Packaging Using Curing Reaction Kinetics of Composites

Wei Li
Department of Microelectronics and Integrated Circuit
Xiamen University
Xiamen, China
36120200155819@stu.xmu.edu.cn

Jianhong Huang
Product Development Department
Xiamen Sky Semiconductor Technology Co.Ltd.
Xiamen, China
huangjh@sky-semi.com

Jinbo Xiao
Product Development Department
Xiamen Sky Semiconductor Technology Co.Ltd.
Xiamen, China
xiaojb@sky-semi.com

Yiyong Huang
Research and Development Department
Xiamen Sky Semiconductor Technology Co.Ltd.
Xiamen, China
huangyy@sky-semi.com

Houdun Zhang
Engineering Department
Xiamen Sky Semiconductor Technology Co.Ltd.
Xiamen, China
Zhanghd@sky-semi.com

Daquan Yu*
Department of Microelectronics and Integrated Circuit;
Xiamen University; Xiamen Sky Semiconductor Technology Co.Ltd.
Xiamen, China
yudaquan@xmu.edu.cn

Abstract— **As an advanced edition of the standard wafer-level package (WLP), Fan-Out WLP (FOWLP) has played a critical role in the industry of integrated circuits with higher integration levels and condense external contacts. Since TSMC/Apple buzz, Fan-Out packaging is still maintaining its centrality as a popular option for mega-trend driven applications like 5G, HPC (Networking) and SiPs (Consumers). Discrepant with the design defects of conventional WLP schemes, the number of I/O connections is unlimited in FOWLP. Thus, FOWLP takes the single die diced from the whole wafer and coating them with epoxy mold compound (EMC). Then space allocated between each die for additional I/O connection points is design and manufactured by the foundry. Afterward, the redistribution layers (RDL) are classical means to interconnect the die and solder joints instead of the substrate. Nevertheless, the main issue and challenge in the engineering process is warpage, which is fostered by the discrepancy of the coefficient of thermal expansion between the heterogeneous materials. Given the whole real industrial process, the warpage of all stages is prompted by utilizing the continuity simulation. According to the process modeling, the post-processes are exclusively taken into account, such as: post-mold cure (PMC), temporary bonding, after formation passivation one (FP1), after solder mask formation (SMF) and de-bonding. For these five main steps of the rapid change of expansion and contraction in the compression molding process, based on the composite material curing reaction kinetics, the element birth and death methods are utilized. Moreover, to the best of our knowledge, it's the first time to analysis the influence of cavity and cutting channel model on warpage. Hence, a bar-level finite element model rather than a quarter-wafer-level finite element model is utilized. Experimental tests were conducted on 8-inch wafers to verify the simulation results. Compared with the experimental results, both finite element models have proved the effectiveness and accuracy of warpage prediction.**

Keywords—Coefficient of thermal expansion (CTE), element birth and death method, fanout wafer-level packaging (FOWLP), finite-element modeling, warpage

I. INTRODUCTION

In recent years, SAW (surface acoustic wave) filter, as an indispensable semiconductor, has found a vital role in various applications, such as the Internet of Things, Smart Grid, New Energy Vehicles, Industry 4.0, and other areas [1-3]. Moreover, with the rise of the enlarging market such as new energy vehicles, smart wearable devices, intelligent city infrastructure, SAW filter are booming in a new perspective of industrial 4.0.

Owing to Fan-Out WLP (FOWLP) technology can cover both high-end and low-end applications for its stable processs compatibility [4], it is an viable alternative for SAW filter packaging. Conventionally, SAW filter packages are utilizing glass trenching technology followed by micro-vias to realize the electrical interconnection between the glass interposer and the Cu pads [5]. Nevertheless, this scheme is intractable for high I/O density packaging for the challenging reliability problems [6].

SAW (surface acoustic wave) is an elastic wave that is generated and propagated on the surface of piezoelectric substrate material, and its amplitude decreases rapidly as the depth of the substrate material increases. The basic structure of SAW filter is to make two acoustoelectric transducers-interdigital transducers (IDT) on the burnished surface of the substrate material with piezoelectric characteristics, which are used as transmitting transducer and receiving transducer respectively. Traditional dielectric filters generally have the characteristics of low loss, large bandwidth, and high power tolerance. But its fatal weakness is that it is too large to adapt to the trend of miniaturization of mobile phones. The SAW filter has the advantages of small size, suitable for micro-encapsulation, good consistency, and no need to adjust. This article takes SAW filters for mobile phones in wireless communication systems (technical index: T_x end center frequency f_0 is 902.5 MHz, bandwidth is 25 MHz; R_x end f_0 is 947.5 MHz, bandwidth is 25 MHz) as an example to introduce

978-1-6654-1392-3/21 $31.00 © 2021 IEEE

the ladder structure SAW filter packaging process, and focus on the simulation calculation of the warpage of each process.

The thermal expansion mismatch between the constituents and the degree of crosslinking reaction of polymers during packaging curing processes are the main reasons for the warpage during the whole process. To quantitatively characterize these two driving forces and assess the process effect on warpage deformation, experimental and numerical analyses were applied to study the warpage evolution of over-molded ball grid array (BGA) package under post-mold curing (PMC) thermal histories. To quantitatively characterize the mechanism of these effects during the molding stage, the warpage evolution of the over-molded ball grid array (BGA) package under post-mold curing (PMC) thermal histories are meticulously analyzed next by performing the experimental and FEA simulation study. During the isothermal molding stage, shrinkage of EMC can significantly induce the increase of the maximum of warpage. In this paper, a kind of numerical modeling algorithm that the thermochemical cure kinetics are incorporated into the curing and chemical aging-induced shrinkage strains is proposed. Then the experimentally obtained warpage evolutions are setting to verify the simulation molding with cure-dependent viscoelastic relaxation modulus. Above the material glass transition point, two stages are simultaneously superpositioned: on the one hand, thermal expansion mismatch-driven warpage change during non-isothermal stages; on the other hand, the chemical shrinkage-induced warpage evolution during isothermal aging. The maximum warpage of wafers should be minimized to adapt the lithography devices in the process of manufacture. Hence, the study of mechanisms of residual deformation and stress generation is decisive to packaging materials selection and process optimization.

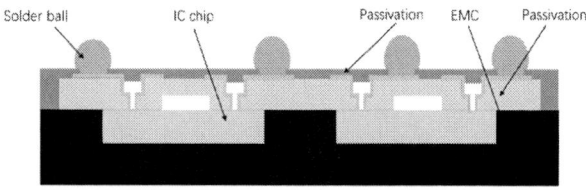

Fig. 1. The schematic view of the SAW filter structure.

TABLE I.   MAIN PARAMETERS OF A SINGLE PACKAGE

Parameters	Design Value(um)
Total package height	414±39
Roof layer thickness	45±3
Wall layer thickness	15±3
Roof opening	60±5
Wall opening	40±5
Chip thickness	200±10
EMC thickness	275±10
Chip edge to EMC edge	50±5
Cu thickness	7±3
Ball diameter	100±20
Ball height	60±5

II. PACKAGE AND PROCESS

A. Package Information

The package size of the ultra-thin SAW filter is 950×750 μm, and the height including the ball grid array is 400 μm, while the size of the SAW filter is 780×570×200 μm. Fig. 1 is a schematic diagram of the SAW filter package. The key structure dimensions are shown in Table 1. This structure can be compatible with TSV technology to achieve denser circuits and BGA. In order to prevent ultra-thin wafers from cracking during operation and transportation, 1000 μm thick glass is used for temporary bonding, and laser stripping is used before cutting. Compared to 8-inch wafers, the area of a single chip is extremely small, and the material of the SAW filter is LiTaO₃ (LT). LiTaO₃ crystal has high stability and high chemical properties (insoluble and water), the Curie point is higher than 600℃, it is not prone to depolarization, low dielectric loss, and high detection rate. It has become a pyroelectric infrared detection. The best choice for the application material of the device. However, its coefficient of thermal expansion (CTE) is significantly higher than that of packaging materials, especially the thermal mismatch with commonly used EMC materials.

The effect of curing agents on the post-warping process, especially PMC, has not been reported. In the literature, there is a large error between the theoretical model of the elastic body and the actual value. This paper uses a calibration factor to correct the warpage results of the finite element simulation of the elastic model, but only based on this the encapsulation results of the different geometric models The unfolding prediction lacks a theoretical basis. We have noticed that the hardening agent content of EMC materials and the intricate influence of composite materials are the pain points that lead to the prediction and optimization of warpage problems in the industry. Often rely too much on experiments to ensure the reliability of the project. In this paper, a theoretical analysis is made on the relationship between the curing agents and temperature and the material properties of composite materials. In addition, considering the high temperature, the thermodynamic properties of the die attach layer material, especially the problem of die offset caused by the viscoelastic-viscous flow state are analyzed.

B. FOWLP Process flow

Fig. 2 shows the process flow of FOWLP. First, a layer of high-strength adhesive glue (to control offset) is spin-coated on the glass (temporary carrier support) surface. Then, aligning the mark points, more than 20,000 dies are laid. Second, The key process in FOWLP is the compression molding process, by which multiple dies are embedded inside the mold compound (Fig. 3). Third, the temporary bonding medium has light-transmitting properties, and the temporary bonding glue is heated by a laser. Heat until the lock loading force disappears, and remove the temporary plate. Fourth, rotate 180° to temporarily bond the glass window to the mobile phone of the LT chip to control the warpage. Fifth, according to the layout, use a photo etching machine to make a wall layer on the front of the LT. Sixth, according to the layout, continue to use the lithography machine to make the roof layer on the front of the LT chip. Seventh, according to the layout, continue to make the SMF layer while making the RDL and complete the BGA. Finally, the backside of EMC is thinned, and laser marking is completed and cuts into a single chip.

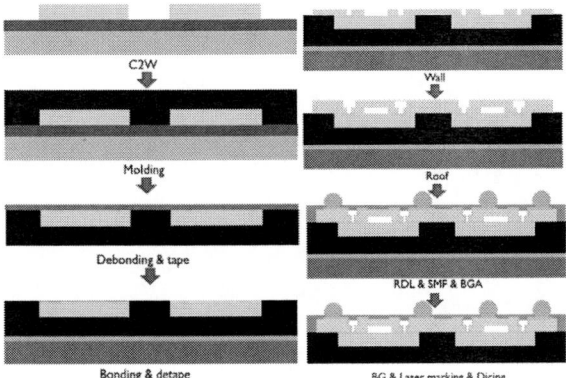

Fig. 2. The process flow of the FOWLP.

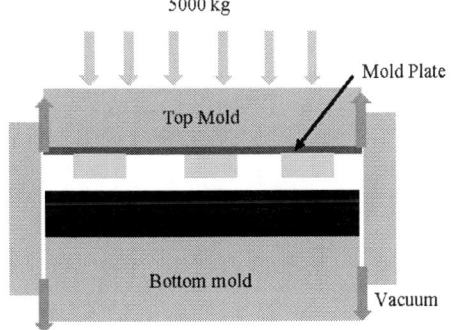

Fig. 3. Schematic cross section for compression molding structure.

C. The theoretical analysis

- Curing Strain of EMC

The total strain of a linear elastic body is the sum of the curing strain (ε_c), thermal strain (ε_T) and elastic strain (ε_e), defined as

$$\boldsymbol{\varepsilon}_e + \boldsymbol{\varepsilon}_T + \boldsymbol{\varepsilon}_c = B\mathbf{u} \qquad (1)$$

- Cure Kinetics of EMC

$$H_t = \int_0^t \frac{dH}{dt} dt \qquad (2)$$

$$\dot{\varphi} = \frac{d\varphi}{dt} = \left(k_1 + k_2\varphi^m\right)\left(1-\varphi\right)^n \qquad (3)$$

- Viscosity Mathematical Model of EMC

$$\eta(\varphi,T) = \eta_0 \left(\frac{\varphi_g}{\varphi_g - \varphi}\right)^{C_1 + C_2\varphi} \qquad (4)$$

$$\eta_0(T) = A_3 \exp\left(\frac{E_3}{RT}\right) \qquad (5)$$

- Cure Shrinkage of EMC During Curing

There are two shrinkage parts during the curing stage: one is the shrinkage in the rubbery/liquid state and the other the shrinkage in the glassy/solid state. Thus, both of the stages have the superposition impact on the maximum wafer warpage in the final room temperature. In the design mission, it is significant to predict and control the warpage.

III. EXPERIMENTAL RESULTS AND DISCUSSION

In Xiamen Sky Semiconductor Technology Co. Ltd., material analysis was conducted for warpage analysis, and experimental research was carried out using three EMC materials provided by the manufacturer with different thicknesses and curing processes. The experimental results are:

- In the case of the same molding thickness, the thinner the chip thickness, the greater the warpage;

- When the chip thickness is the same, the carrier material used in molding has little effect on the warpage absolute value;

- The warpage of bonding and the warpage change of PMC after molding and then de-bonding have not much effect.

Then, Fig. 4 shows the actual situation in the SAW filter packaging process. Fig. 5 shows the time-temperature curve of the key process of molding.

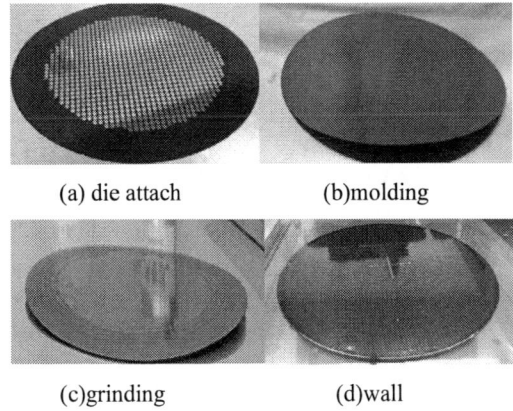

(a) die attach   (b)molding

(c)grinding   (d)wall

Fig. 4. Experimental results at different stages.

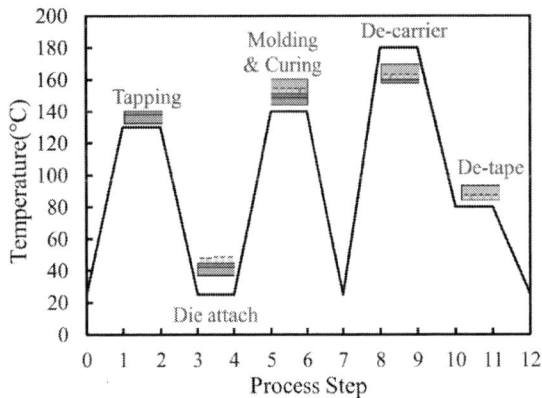

Fig. 5. Temperature time curve of the whole process.

The warping direction of the wafer is usually defined by the active surface as the standard, as shown in Fig. 5. In these cases, we assume the active face is upwards. When the warpage is higher on both peripheral edges than the middle part, it is denoted as concave, and the warpage value is positive (edge height - middle height). If the two sides are lower than the middle partial, it is denoted as convex, the warpage value is negative (edge height - middle height). In this paper, this standard is also used to describe the warpage direction.

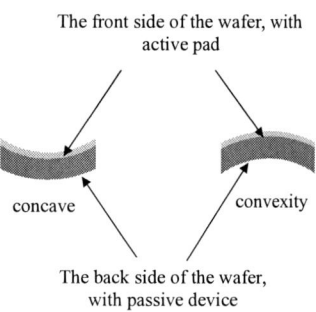

Fig. 6. The concave-convex morphology definition of warpage.

Based on the deformation mechanism determined in the experimental warpage analysis, a solidification-related modeling method was used to simulate the warpage evolution process of the over-molded package in the PMC thermoforming process. The chemical-thermo-mechanical coupling constitutive model is used to consider the influence of polymerization conversion and chemical aging on warpage. First, the EMC curing kinetic model is used to perform numerical analysis by determining the degree of curing conversion of the entire thermal process. The warpage model based on the finite elements is adopted, and the viscoelastic constitutive characteristics of EMC are considered, and the warpage evolution process of the material is simulated.

A. Finite element model

Fig. 7 shows an idealized bar warping model and its boundary conditions. The model cuts a strip from the radius of the wafer. The length of the model is equal to the radius of the wafer, and the width of the model is equal to the average width of a single packaged chip. Set the Cartesian working coordinate system, and the plane settings on the x-axis and y-axis coordinate planes are subject to symmetrical boundary conditions ($U_y = UR_x = UR_z = 0$; ($U_x = UR_y = UR_z = 0$), and the other is parallel to the x-axis, is constrained in the y-direction ($U_y = 0$), and the original nodes are also fixed, and all degrees of freedom are constrained. APDL-based live and death element technology are very suitable for analyzing the process flow in Fig. 2. Assuming that the EMC thickness is 0.25 mm, the warpage after temporary bonding is shown in Fig. 8.

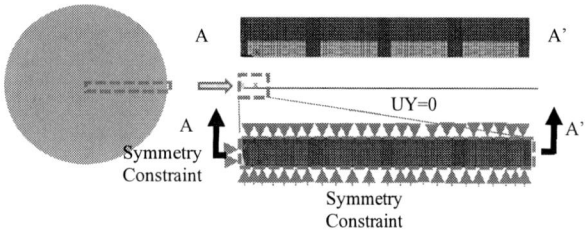

Fig. 7. The strip finite element model was established.

On account of molding, deposition is the most illustrative step, creating a reconstituted wafer/panel and processing it is at the very core of fan-out manufacturing. In addition, the wafer molding process is complex, the temperature and boundary conditions vary in different stages. Moreover, the layers of the whole structure are augmented to realize RDL.

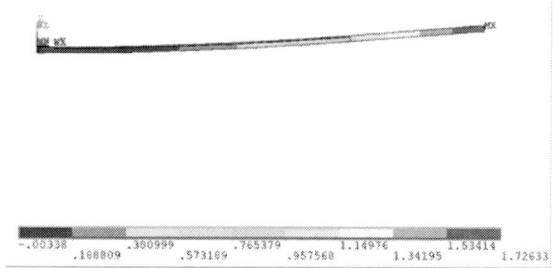

Fig. 8. After temporary bonding.

B. Effective curing shrinkage

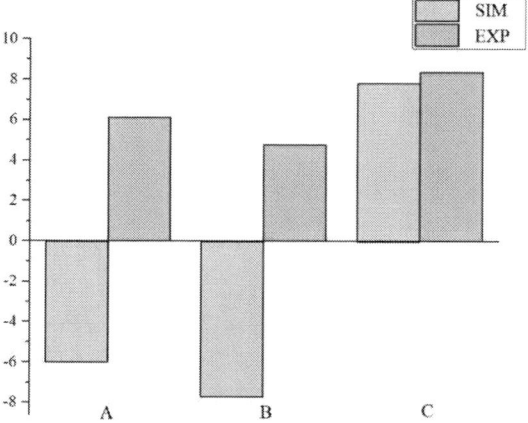

(a) Ignoring the curing shrinkage

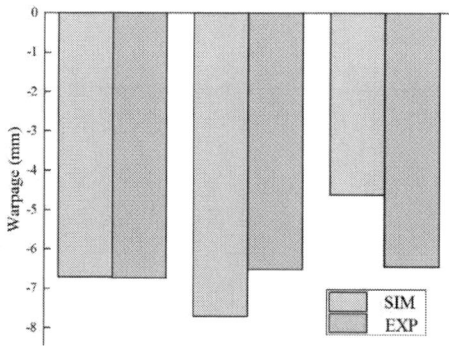

(b) Considering the curing shrinkage

Fig. 9. The effective of curing shrinkage to the warpage

Obviously, due to ignoring the curing shrinkage, the discrepancy of simulation and experimental results are huge in Fig. 9(a). However, by introducing the EMC material supplied by the vendor, and considering the effective curing shrinkage rate of 0.27%, the experimental and simulation results before and after the CTE is corrected to 9 ppm/k are shown in Fig. 9(b). Further, from Fig. 9(b), by comparing with the experimental results, it is found that the warpage value can be well predicted by considering the curing shrinkage. Thereby, the packaging process of saw filter wafer can achieve well consensus between the simulations and experiments. Moreover, the curing shrinkage parameters are convenient to obtain for the industry application, in which the differential scanning calorimetry (DSC) test is conducted. In this paper, the DSC instrument TA Q2000 is utilized.

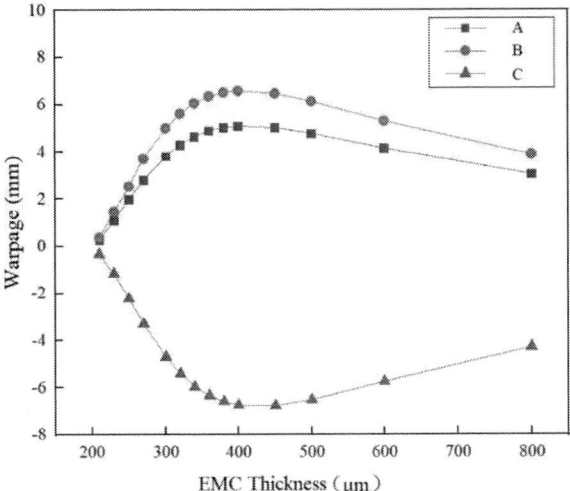

Fig. 10. The strip finite element model was established

Therefore, the three different EMC materials provided by the vendor were analyzed in Fig. 10. It was found that the warpage value of the three EMC materials first increased and then decreased with the thickness of EMC. It is recommended to use a thinner EMC thickness for packaging. Under the same EMC material thickness, A-type EMC is recommended, and this conclusion is basically consistent with the experimental recommendations.

IV. CONCLUSIONS

In this paper, the warpage of FOWLP for SAW filters was studied. Assuming the die and carrier to be elastic and isotropic, the elastic model of the molding compound is adopted to analyze the debonding warpage since the molding compound is fully cured. On the basis of the two finite element models established, the effects of materials, dimensions and processes on the maximum warpage are analyzed and compared, and suggestions for optimization are proposed.

A series of parametric studies are performed by changing the die thickness, die pitch distance and the EMC thickness. Since the mechanical and thermal behavior of the molding compound and adhesive are different when the temperature is below/above the glass transition temperature, the viscoelastic model and elastic model are consequently utilized to mimic the PMC. Meanwhile, an improved FOWLP manufacturing process is proposed, resulting in the maximum warpage value within 1 mm on an 8-inch wafer.

ACKNOWLEDGMENT

The authors would like to appreciate the financial support from the Science and Technology Major Project of Xiamen City No. 3502Z20201004.

REFERENCES

[1] CHEN C, YU D, WANG T, et al. Warpage Prediction and Optimization for Embedded Silicon Fan-Out Wafer-Level Packaging Based on an Extended Theoretical Model [J]. IEEE Transactions on Components, Packaging and Manufacturing Technology, 2019, 9(5): 845-53.

[2] CHE F X, HO D, DING M Z, et al. Modeling and design solutions to overcome warpage challenge for fan-out wafer level packaging (FO-WLP) technology [M]. 2015 IEEE 17th Electronics Packaging and Technology Conference (EPTC). 2015: 1-8.

[3] LAU J H, LI M, TIAN D, et al. Warpage and Thermal Characterization of Fan-Out Wafer-Level Packaging [J]. IEEE Transactions on Components, Packaging and Manufacturing Technology, 2017, 7(10): 1729-38.

[4] LAU J H, LI M, YANG L, et al. Warpage Measurements and Characterizations of Fan-Out Wafer-Level Packaging With Large Chips and Multiple Redistributed Layers [J]. IEEE Transactions on Components, Packaging and Manufacturing Technology, 2018, 8(10): 1729-37.

[5] QIN F, ZHAO S, DAI Y, et al. Study of Warpage Evolution and Control for Six-Side Molded WLCSP in Different Packaging Processes [J]. IEEE Transactions on Components, Packaging and Manufacturing Technology, 2020, 10(4): 730-8.

[6] WU M-L, LAN J-S. Simulation and Experimental Study of the Warpage of Fan-Out Wafer-Level Packaging: The Effect of the Manufacturing Process and Optimal Design [J]. IEEE Transactions on Components, Packaging and Manufacturing Technology, 2019, 9(7): 1396-405.

Inorganic Stabilizer Introduced Flux System with High Tackiness: an Efficient and Novel Material Solution for the Micro LED Mass Transfer

Liangzheng Ji
Electronic deparment
Heraeus Material Technolo
Shanghai Ltd.
Shanghai, China
liangzheng.ji@heraeus.com

Jing Zhang*
Electronic deparment
Heraeus Material Technolo
Shanghai Ltd.
Shanghai, China
j.zhang@heraeus.com

Abstract— As Micro LED technology becomes increasingly matured, several assembly process options were proposed, and seems feasible and reliable to be used in mass production. Usually, these processes require a tacky flux to adhere the die to the substrate when mass transferring, and meanwhile to provide certain activation to facilitate the joining process. However, the work life of such flux is limited, which leads to low printing efficiency and unstable assembly quality. In this work, a novel inorganic compound stabilized tacky flux system was successfully developed for newly emerging carrier assistant Micro LED mass transfer technology. This designation can extend the working life from less than 30 mins to more than 180 mins. Such a significant improvement in working life was realized by introducing an inorganic stabilizer, which reacted with chemicals to strengthen the inner connection, furtherly realized precise and efficient tackiness in-situ controlling. Moreover, this tacky flux system didn't make any compromising on flux's inherent characteristic of promoting soldering. Inorganic compound stabilized tacky flux provided a new approach for material solutions in Micro LED mass transfer.

Keywords— Micro LED, Mass transfer, Tacky flux, Inorganic stabilizer

I. INTRODUCTION

Micro LED, also known as mLED or µLED, is based on inorganic semiconductor technology, which was first invented in 2000 by Prof. Jiang's research group from Kansas State University [1]. It is defined as an LED with a size of a few microns and a few hundred microns [2-3]. With nearly 20 years' development, many potential applications including display, optical communication, indoor positioning, and biomedicine have been identified by numerous researchers [4-6]. Among these applications, the display seems to be the most promising and closest to mass production, due to its unique advantages of self-illumination, high resolution, high brightness, low power consumption, fast response, high integration, high stability, thin thickness, long life, etc. [7-9].

However, there are still technical hurdles to overcome to achieve large scale commercialization, such as LED full color, wavelength consistency, driver technology, and mass transfer [2]. Currently, the largest bottleneck seems to be mass transfer, which refers to the rapid and precise transfer of millions or tens of millions of three color Micro LED dies to a target specific location of the substrate to achieve electrical connection and mechanical support. Therefore, many efforts

have been made by numerous institutes and companies, and multiple solutions were reported, including precise pick and place (van der Waals force, magnetic force and electrostatic force), selective releasing, roll-printing and fluidity self-assembly [10-14]. Recently, the mass transfer solutions: carrier transferring becomes more popular, which is depicted in Fig. 1. There are many factors to determine the transfer quality, including tape (carrier) adhesion, Micro LED die metallization, adhesion and processibility of flux, and so on. Nevertheless, the flux material plays a more crucial role in this special mass transfer and bonding process, since it has much more flexibility to be adjusted and modified based on process needs. However, there is a tendency for traditional flux to degrade fast during printing, since continuous printing breaks most of the inner connection of the chemical system and hard to restore in a short time. This eventually leads to Micro LED low transfer yield and efficiency, as flux has to be replaced on frequent basis. Thus, it is of vital importance to develop a flux system with the right and stable properties to achieve proper yield during mass production.

In Heraeus, we have developed a new type of tacky flux system to facilitate the popular carrier transferring process. This tacky flux system combines the benefits of traditional flux and glues with the specially designed sticky property, which ensures successful die transfer from tape to pre-fluxed substrate and improves the transfer yields. Furthermore, with the addition of some inorganic stabilizer, the new flux system has a significantly longer work time due to inorganic compounds acted as cross linkers to bond isolated chemical

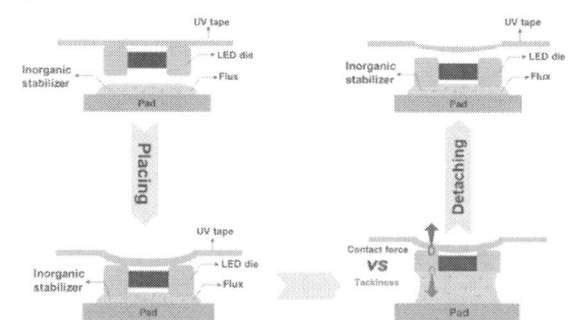

Fig.1. Schematic diagram of Micro LED carrier transfer process

molecular, to rebuild the inner connection. Importantly, the utilization of inorganic stabilizer not only can extend the working life in the printing process but also makes up for some minor defects of hardware including die and substrate, since additional inorganic compounds can offset the less of tin caused by substrate warpage and die defects.

II. EXPERIMENTAL PROCEDURE

To prepare the stabilized tacky flux, there are 2 steps: 1. Made the pure tacky flux, 2. Add in the stabilizer. The tacky flux is prepared by classic heating- stirring- cooling method. In the formulation, organic components such as rosin, thickener, organic acid, amine, organic halogen compound, solvent and additives were mixed up in a certain proportion, and the final mixed solution was heated on a thermostat plate for a certain time at the appropriate reaction temperature, dropping speed and stirring speed. Once all the chemicals dissolved sufficiently, stop heating, and transfer to another container for cooling with appropriate stirring. When cooling to room temperature, tacky flux was made.

The next step is to mix the fresh prepared tacky flux and inorganic stabilizer with a disperser, at a certain rotational speed for a period of time. The inorganic-stabilizer powder used in this study had an average particle size of ~5-15 μm. Finishing the mixing, inorganic compounds stabilized tacky flux was prepared. As to this product, all the functional chemicals guaranteed good particle dispersion, flowability and viscosity for stencil printing, long time working stability, solderability, reliability onto tin pre-coated Micro LED dies.

The tackiness was measured according to IPC-M-650 standard. Malcom tackiness tester was applied to measure the tackiness in the condition of 25 °C ± 2 °C and 50% ± 10% relative humidity (RH), and the equipment should have a stainless steel test probe with a nominal 5.1mm ± 0.13 mm diameter bottom surface, which is smooth, flat, and aligned parallel to the plane of the subject test specimen. Before measuring, using a stencil with the size of 6.3 mm in diameter and 0.25 mm thick to print out the flux. Placed the specimen slide under the test probe once the printing finished. Brought the test probe at a rate of 2.5 mm/min ± 0.5 mm/min, and applied a force of 300 g ± 30 g to the specimen. Within five seconds following the application of this force, withdraw the probe from the specimen at a rate of 2.5 mm/min ± 0.5 mm/min, and recorded the result of the tackiness value. Taken at least five measurements under the same test conditions and averaged all the readings.

After confirmed the tacky flux is suitable for printing, placed the tacky flux placed on the bare stencil, rolled in the printer with preset automatic continuous printing mode. In addition, proceeded to the printing process to coat onto Cu pad of FR4 substrate using 30 μm stencil. Furtherly, placed die onto the pad. This board will be reflowed in a N_2 protected reflow oven with a certain profile.

III. RESULTS AND DISCUSSION

To verify the effects of tackiness on die missing ratio (the proportion of that dies fail to successfully adhere to a specific location on the substrate in the Micro LED transfer process) in tape assistant carrier transfer process, tacky fluxes with different tackiness (112 gf, 121 gf, 129 gf, and 133 gf) were selected to test die missing ratio. The tackiness of flux ranging from 112gf to 138gf through formulation adjustment minorly. Interestingly, an approximately linear correlation was

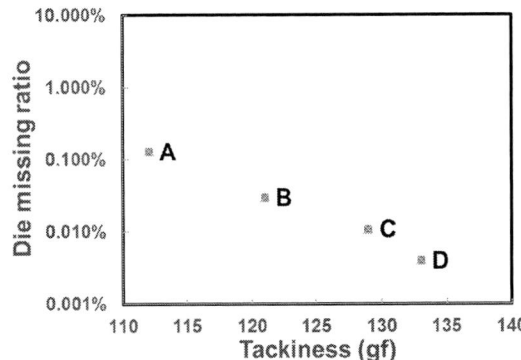

Fig.2. The correlation between tackiness and die missing ratio

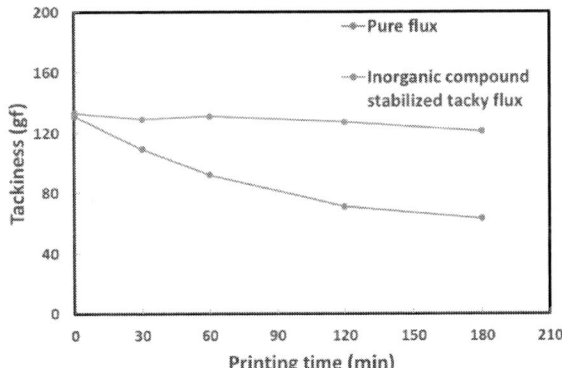

Fig.3. Trend chart of tackiness with printing time

observed between flux tackiness and die missing ratio at the logarithmic scale (Fig.2). When the flux tackiness increased from 112 gf to 138 gf, the die missing ratio would be decreased from 0.13% to 0.04%. As for Micro LED technology, 0.13% LED die missed in the transfer process, means heavily rework workload. Ultimately, leads to unacceptable production cost. Therefore, it seems tackiness stability is the key parameter, which can directly determine the success of the carrier transfer process.

Although with standard pure flux, one can achieve great transfer yield in the carrier transfer process as well, the flux's tackiness decreased dramatically in the printing process, as can be seen in Fig.3, pure flux's tackiness decreased dramatically from 131 gf to 65 gf with the prolonging of rolling time, which may lead to a catastrophic transfer yield. The reason for such degradation is that printing can break the inner connection (mainly hydrogen bond) of chemicals in flux (Fig.4.b). What is more, rolling damaged most hydrogen bonds, and hard to restore in a such short time. What is worse, absorbed water diluted the flux, causing the tackiness to decrease further. With the continuous process of the aforementioned mechanism, tackiness gradually decreases in an irreversible direction.

When we introduce an inorganic stabilizer into the flux system, the tackiness instability had been greatly improved. The tackiness of stabilized flux can maintain for a relatively long period, after 180 min continuous printing, tackiness stable at around 130 gf (Fig.3). There is no doubt that inorganic compounds played a vital role in tackiness stabilization. Inorganic compounds acted as cross linkers to bond isolated chemical molecular through coordination react-

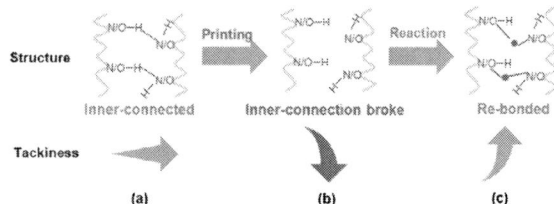

Fig.4. Schematic diagram of flux inner- connection variation (a) before, (b) after printing and (c) inorganic compound stabilization

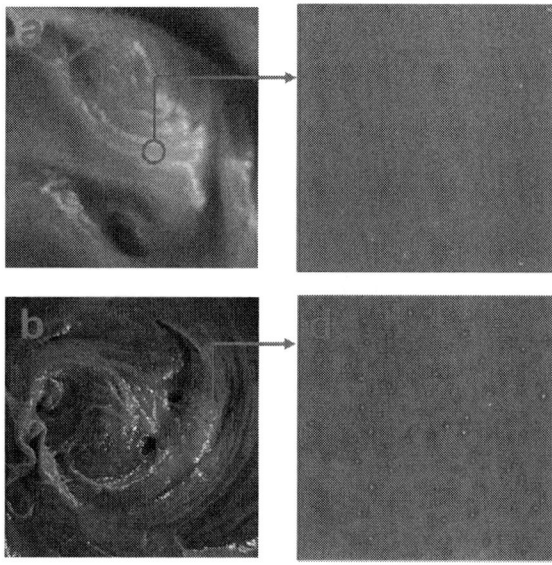

Fig.5. The appearance of pure flux (a) and inorganic compound stabilized tacky flux (b), microscope image of pure flux(c) and inorganic compound stabilized tacky flux (d).

ion, to realize the inner connection rebuilding (Fig.4.c). In this type of reaction, inorganic stabilizer acted as center metal ions providing empty orbits to coordinate with organic molecule ligands, which have lone pair electrons, then form two- or three- dimensional structured framework to achieve the purpose of strengthening the inner connection. This can offset the tackiness declining, which was caused by inner connection breaking in the printing process for pure chemicals. In addition, although absorbed water had some effects on tackiness, water can promote the dissociation of organic acids in flux, which in turn cleaned the surface of some inorganic compounds, making them easier to connect with organic chemicals. Further, strengthen the inner connection. As shown in Fig.5, a comparison of classic pure flux (Fig. 5 a and c) and our inorganic compound stabilized tacky flux (Fig. 5 b and d). Compared to pure flux, the inorganic stabilizer was well dispersed in the tacky flux system, this can greatly improve reaction uniformity of inorganic-stabilizer and organic compounds in tacky flux system, which provide possibilities for in-situ tackiness controlling precisely and effectively.

Furthermore, this inorganic stabilized tacky flux system not only guaranteed the tackiness stability through chemical reaction but also without any compromising on other comprehensive performance, even have improvements on some of the performance. Voiding weas in the same level of around 7%-9% for both standard flux and newly designed tacky flux. SPI detectability significantly improved since the presence of organic particles can improve the recognition of optical devices.

IV. CONCLUSION

In conclusion, a novel inorganic compound stabilized tacky flux system was developed for carrier assistant Micro LED mass transfer and bonding technology. The introduction of an inorganic stabilizer can strengthen the inner connection of flux, ensure the stable controlling of tackiness through in-situ crosslink reaction to bond isolated chemicals appropriately., even in the long period of continuous printing. Additionally, this stabilizer can also lead to better SPI detectability, voiding performance and die shear strength. Base on such improvement, this new stabilized flux system will strengthen the popularity of the carrier assistant Micro LED mass transfer, and greatly accelerate the market adoption of Micro LED technology.

REFERENCES

[1] S. Jin, J. Li, J. Lin, H. Jiang, "GaN microdisk light emitting diodes". Appl. Phys. Lett., vol. 76, no. 5, pp. 631-633, 2000.

[2] X. Zhou, P. Tian, C. Sher, J. Wu, "Growth, transfer printing and color conversion techniques towards full-color micro LED display". Prog. Quan. Electro. vol. 71, pp.100263, 2020.

[3] J. Li, B. Luo, Z. Liu, "Micro-LED Mass Transfer Technologies" 21st International Conference on Electronic Packaging Technology (ICEPT). 2020.

[4] J. Lin, H. Jiang, "Development of micro LED". Appl. Phys. Lett., vol. 116, pp. 116: 100502.1-100502.8, 2020.

[5] P. Tian, X. Liu, S. Yi, Y. Huang, S. Zhang, X. Zhou, L. Hu, L. Zheng, R. Liu, "High-speed underwater optical wireless communication using a blue GaN-based micro-LED". Opt. Express, vol. 25, pp. 1193-1201, 2017.

[6] S. Lee, K. Park, C. Koo, H. Yoo, S. Kim, C. Ah, G. Sung, K. Lee, "Water-resistant flexible GaN LED on a liquid crystal polymer substrate for implantable biomedical applications". Nano Energy, vol. 1, pp. 145-151, 2012.

[7] S. Lee, J. Kim, J. Shin, H. Lee, I. Kang, K. Gwak, D. Kim, D. Kim, K. Lee, "Optogenetic control of body movements via flexible vertical light-emitting diodes on brain surface". Nano Energy, vol. 44 pp. 447-455, 2018.

[8] P. Tian, J. J. D. Makendary, E. Gu, Z. Chen, Y. Sun, G. Zhang, M. Dawson, R. Liu, "Fabrication, characterization and applications of flexible vertical InGaN micro light emitting diode arrays". Opt. Express, vol. 24, no.1, pp.699-707, 2016.

[9] Eric, M, Virey, N. Baron, "Status and prospects of micro LED Displays". SID International Symposium Digest of Technology Papers, vol.49, no.2, pp. 593-596, 2018.

[10] C. Bower, M. Meitl and D. Kneeburg, "Micro-transfer-printing: Heterogeneous integration of microscale semiconductor devices using elastomer stamps". SENSORS, 2014 IEEE, Valencia, pp. 2111-2113, 2014.

[11] M. Meitl, E. Radauscher, S. Bonafede, D. Gomez, T. Moore, C. Prevatte, B. Raymond, B. Fisher, K. Ghosal, and A. Fecioru.. "Invited Paper: Passive matrix displays with transfer-printed microscale inorganic LEDs". SID Symposium Digest of Technical Papers, vol.47, no. 1, pp. 743-746, 2016.

[12] M. Choi, B. Jang, W. Lee, S. Lee, T. Kim, H. Lee, J. Kim and J. Ahn "Stretchable active matrix inorganic Light-Emitting Diode display enabled by overlay-aligned Roll-Transfer printing". Adv. Func. Mater. vol. 1606005, pp.1-10, 2017.

[13] V. R. Marinov, "Laser-enabled extremely-high rate technology for μLED assembly". SID Symposium Digest of Technical Papers vol. 49, no. 1, pp. 692-695, 2018.

[14] B. Andreas, G. Dariusz, et al. "Flexible display and method of formation with sacrificial release layer". US patent: 09318475, 2016.

Studies on the Effects of Soldering Layer Structures on TEC Module Performance and Thermal Stress

Dianru Yu
Institute of Electrical Engineering
Chinese Academy of Sciences
University of Chinese Academy of
Sciences
Beijing, China
yudianru@mail.iee.ac.cn

Zhihao Yin
Institute of Electrical Engineering
Chinese Academy of Sciences
University of Chinese Academy of
Sciences
Beijing, China
yinzhihao20@mails.ucas.ac.cn

Guanghui Liu
School of Materials Science and
Engineering
Beihang University
Beijing, China
liuguanghui@buaa.edu.cn

Hongyan Xu
Institute of Electrical Engineering
Chinese Academy of Sciences
University of Chinese Academy of
Sciences
Beijing, China
hyxu@mail.iee.ac.cn

Ju Xu*
Institute of Electrical Engineering
Chinese Academy of Sciences
University of Chinese Academy of
Sciences
Beijing, China
xuju@mail.iee.ac.cn

Abstract—**Effects of soldering layer structures on the performance and thermal stress of thermoelectric cooling device (TEC) were studied by employing a three-dimensional finite element electric-thermal-mechanical modeling method established via COMSOL simulations. AuSn eutectic alloys and SnSb soldering were applied as joints to connect TE legs and metal bonded ceramic substrate (MBCs) for TEC's hot and cold ends, respectively. The temperature and stress distribution effects of the device based on different soldering layer structures were analyzed. The soldering layer structures include thickness, void size, number of voids, and void distribution. The results showed the maximum heat dissipation rate (Q_{max}), the maximum temperature difference(ΔT_{max}), and the related coefficient-of-performance (COP) of the device are decreased due to the reduction of the thickness of the soldering layer and the number of voids. Also, the more voids in the soldering layer greatly enhanced the maximum stress and temperature gradient. This enhancement will be more obvious when the voids stayed at the edge of the soldering layers.**

Keywords—TEC, finite element analysis, soldering layer structures, thermal stress, the performance of the device

I. INTRODUCTION

Worldwide energy production is mainly based on fossil fuels, and it leads to global warming over the years—research focusing on the production and management of energy from renewable sources. The thermoelectric(TE) phenomenon was observed in 1821 by Thomas Seebeck and in 1834 by Peltier. The TE materials are used for power generation or cooling. The thermoelectric cooler can convert electrical energy directly into thermal energy according to the Peltier effect[1]. It has the advantages of high reliability, no moving parts, no refrigerant, long life, small size, and no environmental pollution. Due to these, it is of particular interest in recent years because of its potential application in temperature control[1].

A typical TEC consists of p-type and n-type thermoelectric elements alternately connected to the conductor and bonded on a substrate by soldering layer. The soldering layer plays the role of electrical, thermal, and mechanical connections in TEC and is one of the weakest parts of the device structure. As one of the primary heat transfer paths, except TE leg materials, the soldering layer is a critical layer that affects the performance of TEC devices, such as coefficiency-of-performance and reliability. When dynamic heat resources are applied, the soldering layer is subjected to variable stresses due to the mismatch of coefficient of thermal expansion (CTE) between different device materials[2]. Structures including defects in the soldering layer resulting in creep effects in long-term service can significantly affect the performance and reliability of the device. Therefore, it is necessary to investigate the effect of soldering layer structures on device performance and thermal stress.

The thickness of the soldering layer may vary due to the process, and regardless of whether leaded or lead-free solder is chosen, the soldering layer will have more or less voids in the middle of soldering joints, for example, the bubbles may be generated by the high-temperature cracking of the organic matter in the brazing material resulting in the voids are created after cooling[3], the voids may also be generated due to the soldering wettability problems either from soldering itself, the defects of soldering pads, or the metal layer electrode surface at the end of TE leg. The presence of voids increased the thermal resistance of the soldering layer. It triggered a current constriction effect, which leads to an increase in the power consumption of the TEC and a decrease in the cooling efficiency.

II. FINITE ELEMENT MODELING AND ASSUMPTIONS

A. Finite element model analysis process

In this paper, the finite element simulation contains electric, thermal, and mechanical fields. The electric field analysis gives the voltage and current distribution in the device, and the temperature distribution can be obtained based on the electric field distribution and boundary conditions. Then mechanical field distribution can be

978-1-6654-1392-3/21 $31.00 © 2021 IEEE

simulated. The correlation between the soldering layer structure including and TEC performance has been explored by conducting the multi-physics simulation process. Under the coupled multiple physical field, the parameters relations of the TEC three-dimensional finite element electric-thermal-mechanical model are shown in Fig. 1.

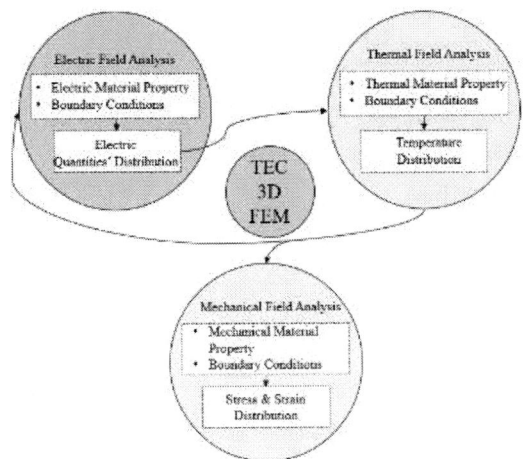

Fig. 1 Multiple physical field coupling and the parameters relations in the electro-thermo-mechanical FEM[11]

B. Finite element model parameters and assumptions

The finite element simulation model of a TEC device(shown in Fig. 2) is constructed using a multi-layer structure. Only two pairs of thermocouples are selected for simplifying and reducing the calculation time due to the symmetry structure of the TEC structure, see Fig. 3.

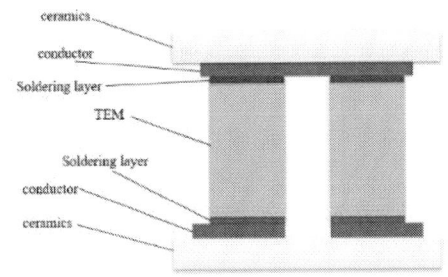

Fig. 2 The schematic diagram of a TEC with a seven-layer structure

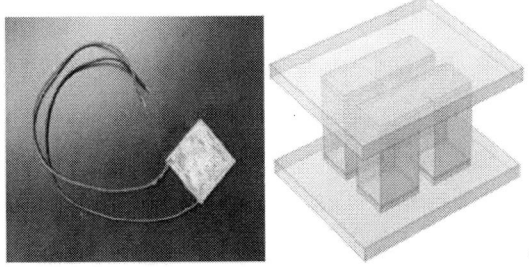

Fig. 3 TEC module and its partially finite element model

COMSOL Multiphysics 5.6 was used to perform the relevant finite element simulation. The following assumptions were made for the simulation the model: 1) The mass diffusion between different material layers during the operation was neglected in this analysis. Each soldering

layer's interface was considered ideal contact, with no contact thermal resistance. 2) Heat was mainly transferred orderly through the 7-layer structure via the direction perpendicular to the substrate. The heat flow between the thermoelectric elements and the substrate's sides was considered thermally insulated, and no convective heat transfer and thermal radiation were considered[4]. 3) The terminal leads were simplified, and no wires were led out. 4) The ambient temperature is 273.15 K. 5) The soldering layer and the TE leg cross-section were equal. The geometric parameters of the model are shown in Table I.

TABLE I. GEOMETRIC PARAMETERS

Parameters and Unit	Parts			
	Ceramic	Conductor	TE elements	Soldering layer
Long (mm)	4	1	1	1
Width (mm)	5	2.9	1.2	1.2
Height (mm)	0.3	0.1	0.17	0.02~0.1

In this simulation, the insulated thermally conductive ceramic substrate is Al_2O_3, the cold-end solder is SnSb, the hot-end solder is AuSn, the conductor is Cu, and the p-type and n-type thermoelectric elements materials are doped Bi_2Te_3. Bi_2Te_3 is a typical thermoelectric material in the low-temperature region, and the temperature variation of the device is not big(normally less than 100ºC) during operation. So, material property parameters with temperature variation were taken as constant values. The parameters of the above materials are shown in Table II.

TABLE II. PARAMETERS OF MATERIAL

Parameters and Unit	Material				
	Al_2O_3	SnPb	AuSn	Bi_2Te_3	Cu
Thermal conductivity [W/(m·K)]	35	60	57	2.4	400
Specific heat capacity [J/(kg·K)]	730	150	150	154	385
Electrical conductivity [S/m]	/	6.214e7	4.587e7	1.4286e5	5.998e7
Relative dielectric constant	5.7	1	1	1	1
Seebeck coefficient [V/K]	/	/	/	168e-6	/
Young's modulus[GPa]	400	70	68	31.5	110
Poisson's ratio	0.22	0.4	0.405	0.33	0.35
CTE[10e-6/K]	6.5	24.7	16	13	17

III. ANALYSIS OF SIMULATION RESULTS ON DEVICE PERFORMANCE

A. Effects of soldering layer thickness

Device performance at different soldering layer thickness is analyzed while keeping others geometric parameters constant. Simulations results for COP, Q_{max} and

ΔT_{max} that can be achieved by varying soldering layer thickness are shown in Figs. 4-6.

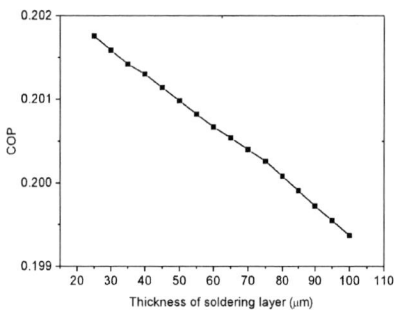

Fig. 4 Coefficient-of-performance variation versus the thickness of soldering layer

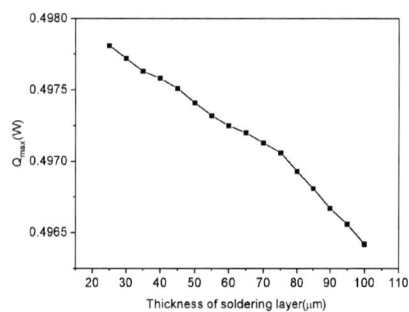

Fig. 5 Maximum heat dissipation rate variation versus the thickness of soldering layer

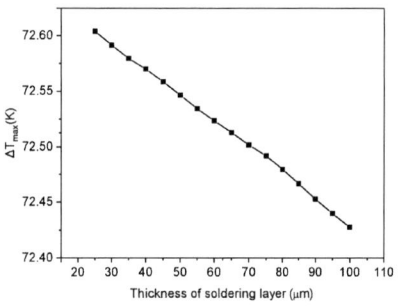

Fig. 6 Maximum temperature difference variation versus the thickness of soldering layer

Figs. 4-6 show COP, Q_{max} and ΔT_{max} decrease with thicker soldering layer. It should because a thicker soldering layer increases thermal and electrical resistance. In summary, to obtain a significant cooling efficiency, the thickness of the soldering layer must be decreased once it achieves an excellent bonding with strong mechanical strength.

B. Effects of soldering layer voids

TEC is susceptible to the formation of voids in the soldering layer due to imcomplete solder materials and process treatments. The actual voids size and formation location are difficult to predict, so spherical voids present in the soldering layer center were introduced to simplify the

model and calculations. In addition, three different void numbers on performance were analyzed[5].

Fig. 7 shows the comparison of the temperature difference between the two ceramics subtrate ends of the device with the different radius of the voids when the presence of one, three and five spherical voids in the soldering layer of the device model. It indicates that the temperature difference shows an parabolic curve type change along the increase of void radius and has no obvious change for different voids number when the void radius is between 4 and 8μm.

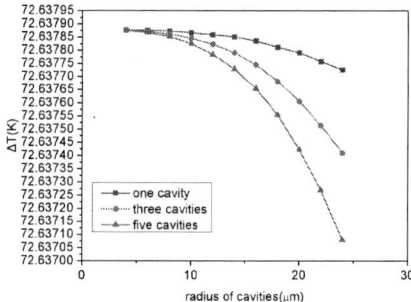

Fig. 7 Temperature difference of TEC versus the size of voids and void numbers

However, the temperature difference gradually decreases as the radius of the void increases, and the situation with five voids is worse than one void and three voids. All the temperature difference appears as an overall exponential variation with the radius of the void. Due to the thermal conductivity of the air in the void is lower than that of the soldering layer material, and the thermal resistance is linearly related to the air volume, bigger void radius and more void number related to larger air volume results in the the thermal resistance of the soldering layer According to one-dimensional Fourier's law of thermal conductivity, when we keep one end of the temperature same, if the temperature of the other end of the device decreases, the temperature difference should also decreases. When the radius of void increase from 4μm to 24μm, the temperature difference with five voids decrease over five times as much as one void.

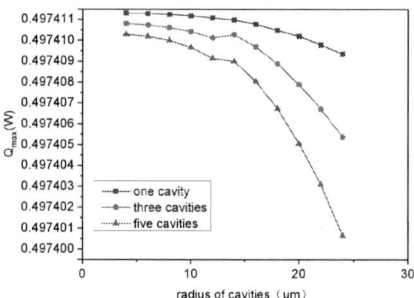

Fig. 8 Maximum heat dissipation rate drop for different radius and number of voids

Fig. 8 shows the maximum heat dissipation rate of the device as a function of the radius of the void for the same case. It indicates that an increase in the number of voids leads to a decrease in the maximum heat dissipation rate of the device. The maximum heat dissipation rate decreases due to the larger radius of the void. In this case, the temperature difference between the two ends of the device is constant. The increase in the number of voids and the increase in the radius of the voids leads to an increase in the thermal resistance of the soldering layer, and the maximum heat dissipation rate decreases when the heat flow of the device becomes smaller, as can be seen from Fourier's law of thermal conductivity. Figs. 7-8 present a square exponential relationship with the radius of the void, which then shows a linear relationship with the volume of the void, following the theory. When the radius of void increase from 4μm to 24μm, the maximum heat dissipation rate with five voids decrease near five times as much as one void.

IV. ANALYSIS OF SIMULATION RESULTS ON DEVICE THERMAL STRESS

The temperature change caused by the power cycle during service makes the cycle of shear strain and spatial temperature gradient between the parts due to the mismatch of thermal expansion coefficients, increasing voids, cracks, and even delamination. The thermal stress in the soldering layer of the middle and edge thermoelectric elements are different. To simplify the calculation, only the thermal stress in the soldering layer of two TE elements are analyzed and the actual situation is more complex.

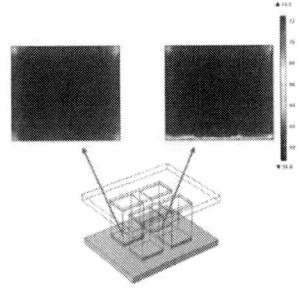

Fig. 9 stress distribution in the ideal soldering layer

Fig. 9 shows that the stress of the left soldering layer with no void connected to the electrode is concentrated in the four corners. In contrast, the stress of the right soldering layer connected to the conductor is mainly concentrated in the edge near the inner side of the conductor surface. The stress value changes rapidly from about 60MPa to about 72.5 MPa in a tiny area, with a variation of 20.83%. Prolonged exposure to this large stress gradient can lead to creep in the soldering layer, and problems such as cracking and its expansion can easily occur at the edge of the soldering layer near the inner side of the conductor surface, leading to failure of the TEC. The thermal expansion coefficients of the thermoelectric elements, solder, deflector, and ceramic substrate do not match. Due to these, the heat is mainly transferred along the z-direction, which makes the whole module have a certain degree of warpage, with higher stress in the center and lower stress in the edge area. Additional attention should be paid to this part of the TEC production process, such as increasing the area of the soldering layer in this part.

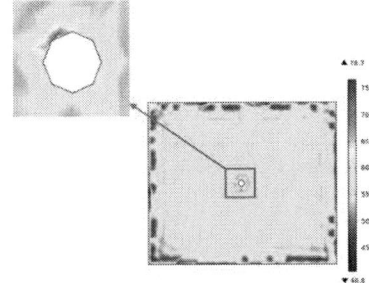

Fig. 10 The stress distribution in the soldering layer containing one void

When a void appears in the soldering layer, the maximum stress value rises from 72.5 MPa to 76.7 MPa. The stress in the central area of the solder layer remains around 60 MPa at most locations, and the stress in the edge area is higher. Still, the maximum value is concentrated around the void, and the crack may start to expand in part with the most elevated stress around the void. As the number of voids in the device increases, the soldering layer's maximum stress value increases accordingly.

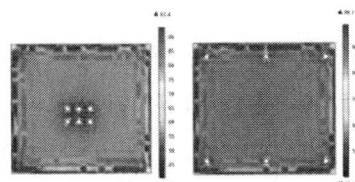

Fig. 11 Stress distribution for different void location cases

Fig. 11 shows the stresses in the soldering layer with different distributions for the same number of voids. When the voids are scattered in the edge area of the soldering layer, the maximum stress increases by 5.67% compared to the concentration distribution in the middle. It indicates that the voids appearing at the edge position have a more significant effect on the stress in the soldering layer and are more likely to generate cracks and expand, causing device failure. Table III shows the maximum Von-Mises stress value of the soldering layer and its distribution law under different numbers and distribution of voids obtained from the simulation. The higher the number of voids, the higher the maximum stress value. The effect of voids at edge locations on device performance is more pronounced.

TABLE III.   VON-MISES STRESS UNDER DIFFERENT FAILURE TYPES

type	Von-Mises stress	
	Maximum stress （MPa）	*Location*
No void	72.5	Edge of soldering layer
One void	76.7	Edge of void
Three voids	91.8	Edge of voids
Six voids	93.4	Edge of voids
Six voids(Edge)	98.7	Edge of voids

Compared to other power electronics, such as Insulated Gate Bipolar Transistor(IGBTs), TECs should focus on the stress conditions of the device rather than the maximum junction temperature. Chen[4] et al. proposed a method to assess the health status of the soldering layer based on the temperature gradient, which was cited in this paper to evaluate the health status of the TEC. The temperature gradient in the heat flow direction (z-direction) of the soldering layer cross-section is obtained by simulation. The energy is high where the temperature gradient is high, as is the thermal stress. Considering the temperature gradient and stress conditions together can provide a good analysis of the health of the device

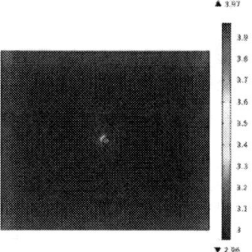

Fig. 12 Temperature gradient distribution of a soldering layer containing a void

The locations of the maximum temperature gradients for different number and distribution of voids are given in Table IV. As can be seen from Fig. 12, the location of the maximum temperature gradient is inside the voids. One void inside the soldering layer increases the maximum temperature gradient by 0.22 K/mm than when there is no void. The greater the number of voids, the greater the maximum temperature gradient.

TABLE IV.   TEMPERATURE GRADIENT UNDER DIFFERENT FAILURE TYPES

Empty number	Temperature gradient	
	Maximum gradient (K/mm)	Location
No void	3.6	Edge of the soldering layer connected to the conductor
One void	3.82	Inside the void
Three voids	4.48	Inside the voids
Six voids	4.77	Inside the voids
Six voids(Edge)	4.89	Inside the voids

Soldering layer defects can change the energy distribution of the soldering layer. The change in energy distribution is reflected in the difference in a temperature gradient. The energy near the void is considerable, manifested by a big temperature gradient, and the soldering layer is prone to creep, cracking, and expansion.

The temperature gradient distribution pattern of the soldering layer voids is almost identical to the stress distribution [4], so the two can be considered together to assess the soldering layer health status of the device.

V. CONCLUSION

In this paper, we presented the effects of different soldering layer thickness, void size, number of cavities, and cavity distribution in the soldering layer between TEC legs and MBC substrate on the performance and reliabilities of TEC devices using a 3D FEM modeling method. The results showed the maximum heat dissipation rate(Q_{max}), the maximum temperature difference(ΔT_{max}), and the related coefficient-of-performance(COP) of the device are decreased due to the reduction of the thickness of the soldering layer and the number of voids. The existence of voids deteriorated the reliability of TE device via the enhancement of the maximum stress and maximum temperature gradient in the soldering layer, especially more voids or voids staying on the edge of soldering layers

The results showed in this paper should be pretty helpful guidance for the experimental researcher to improve the performance and reliability of TEC devices.

REFERENCES

[1] C. Li, D. Jiao, J. Jia, F. Guo and J. Wang, "Thermoelectric cooling for power electronics circuits: Small signal modeling and controller design," 2013 IEEE Energy Conversion Congress and Exposition, 2013, pp. 2201-2207.

[2] YD. Wu et al., "Effect of Solder Voids on IGBT Thermal and Stress Performance," in High Power Converter Technology, vol. 36, no. 10, pp. 17-23.

[3] Songbai Xue, "Microelectronic welding technology".

[4] M Chen et al., "Healthy Evaluation on IGBT Solder Based on Electro-Thermal-Mechanical Analysis," in Transactions of China Electrotechnical Society, vol. 30, no . 020, pp. 252-260.

[5] J Zhang et al., "Effect of Die Attach Void on IGBT Thermal Reliability," in Research & Progress Of SSE, vol. 31, no. 005, pp. 517-521.

[6] XL Zhang et al., "The simulation and die attach analysis on IGBT thermal model," in Journal of Functional Materials and Devices, vol. 17, no. 006, pp. 555-558.

[7] L Xu et al., "Influence of Solder Void to Thermal Distribution of IGBT Module," in Journal of China Academy of Electronics and Information Technology, vol. 9, no. 02, pp. 125-129.

[8] YJ Tian et al., "Thermal Fatigue Effects on IGBT Die Attach Reliability," in Research & Progress OF SSE, vol. 034, no. 003, pp. 288-292.

[9] F Xiao et al., "Influence of Voids in Solder Layer on the Temperature Stability of IGBTs," in High Voltage Engineering, vol. 044, no. 005, pp. 1499-1506.

[10] GT Zheng et al., "Effects of Void Area in Solder Layer on Resistance and Thermal Impedance of Power Devices, " in Semiconductor Technology, vol. 35, no. 011, pp. 1059-1063.

[11] M. Jiang et al., "Finite Element Modeling of IGBT Modules to Explore the Correlation between Electric Parameters and Damage in Bond Wires," 2019 IEEE Energy Conversion Congress and Exposition (ECCE), 2019, pp. 839-844.

[12] Abid M, et al., "Design Optimization of a Thermoelectric Cooling Module Using Finite Element Simulations," in Journal of Electronic Materials, vol. 47, no. 47, pp. 4845-4854.

A Flexible Ultrasonic Sensor Based on Piezoelectric Micromachined Ultrasonic Transducers (pMUTs)

Zhihao Tong[1,2]
[1]The School of Microelectronics
Shanghai University
Shanghai, China
[2]Shanghai Industrial μTechnology
Research Institute
Shanghai, China
zhihaotong0929@foxmail.com

Zhipeng Wu[2]
[2]Shanghai Industrial μTechnology
Research Institute
Shanghai, China
whuwuzhipeng@163.com

Songsong Zhang[1,2]
[1]The School of Microelectronics
Shanghai University
Shanghai, China
[2]Shanghai Industrial μTechnology
Research Institute
Shanghai, China
songsong.zhang@sitrigroup.com

Huicong Liu[3]*
[3]The School of Mechanical and Electric
Engineering
Soochow University
Suzhou, China
hcliu078@suda.edu.cn

Liang Lou[1,2]*
[1]The School of Microelectronics
Shanghai University
Shanghai, China
[2]Shanghai Industrial μTechnology
Research Institute
Shanghai, China
liang.lou@sitrigroup.com

Abstract—As a large number of robots enter the manufacturing industry, the safety issue of robot obstacle avoidance becomes a key challenge. A flexible ultrasonic sensor is proposed to meet the requirement of the robot safety in this paper. The sensor is composed of two piezoelectric micromachined ultrasonic transducers (pMUTs) at the resonant frequency of 100kHz and a flexible printed circuit board (FPCB), which can be easily attached to the curved robot surface through deformation. The size of pMUT and flexible ultrasonic sensor are 3.5mm × 4mm and 10mm × 25mm, respectively. By comparing the experimental results of the sensing ability of the flexible ultrasonic sensor in the bend state and the flat state, it can be known that the sensing range of the sensor in the bend state is consistent with that of the flat state. The distance and angle sensing range are 25cm ~ 65cm and -35° ~ +35°, respectively. The flexible ultrasonic sensor has the potential to be installed on the curved robot surface to sense objects.

Keywords—*pMUTs, flexible ultrasonic sensor, robot obstacle avoidance*

I. INTRODUCTION

As a large number of industrial robots are put into the production of the manufacturing industry, the mode of industrial production has gradually changed from robots performing simple and repetitive tasks to human-robot collaborative work. In this production mode, work efficiency of robots have been improved. But the robots have got rid of the limitation of the protective fence, the risk of collision accidents has increased. In order to reduce the harm caused by collisions, robots should have the ability to perceive obstacles [1]. Therefore, it is necessary to study a sensing system to solve the problem of robot obstacle avoidance. At present, common security sensing systems include tactile, visual, and proximity sensors/sensing systems [2-8]. The tactile sensor detects the sensor resistance change [2], capacitance change [3] or voltage change [4] caused by contact to realize the collision perception of the touching object. However, the tactile sensor can only sense the existence of obstacles after the collision, and the security is not high. The visual sensing system based on depth camera [5] or visual sensor [6] could effectively detect the relative position between the robot and

the obstacle, which improves the security of robot obstacle avoidance. But the cost of the visual system is expensive and the installation is complicated. Another kind of proximity sensors based on capacitive sensing [7] or laser sensing principle [8] could detect the obstacles without contact. However, these sensors either have a short sensing range with a large size, or the structures are rigid so that it is not easy to be installed on the robot through deformation. Therefore, it is of great significance to research a kind of proximity sensor with large sensing range, small size, flexible structure for robot obstacle avoidance.

In the research work of this paper, a flexible ultrasonic sensor based on piezoelectric micromachined ultrasonic transducers (pMUTs) is proposed. The sensor is processed by micro electro mechanical systems (MEMS) technology with a small size and a large sensing range. The flexible printed circuit board (FPCB) is used as the base of the sensor, so that it can be bent like a band-aid and attached to the robot. And the sensing range of flexible ultrasonic sensor is not affected when it is in a certain bend state. This paper verifies the possibility of the flexible ultrasonic sensor to realize obstacle sensing in the bend state, which lays the foundation for its attachment on the surface of the robot to realize the safe robot obstacle avoidance function.

II. SENSING PRINCIPLE AND SENSOR FABRICATION

A. pMUTs

With the gradual maturity of MEMS technology, micromachined ultrasonic transducers (MUTs) have been extensively developed. According to the principle of energy conversion, MUTs can be divided into two types: capacitance micromachined ultrasonic transducers (cMUTs) and piezoelectric micromachined ultrasonic transducers. Among them, cMUTs are a kind of micro capacitor based on the bending vibration of the diaphragm induced by the electrostatic attraction between the suspended diaphragm and the substrate. When the cMUTs is connected with AC voltage, the Coulomb Force attracts the diaphragm to the substrate, which causes the vibration of the diaphragm to produce ultrasonic waves. Such cMUTs generally need to apply high bias voltage (100V-1000V), which increases the risk of device

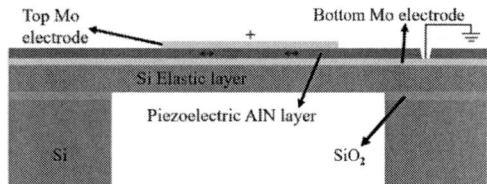

Fig. 1. Cross-section of the pMUTs.

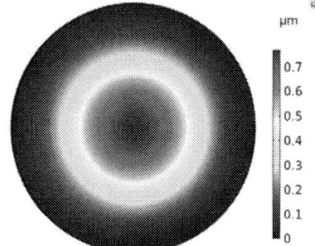

Fig. 2. The mode shape of the circular diaphragm PMUTs at the first resonance frequency.

failure. In contrast, pMUTs have been developed due to the advantages of no DC bias, low voltage drive and easy miniaturization [9]. Aluminum nitride (AlN) [10], zinc oxide (ZnO) [11] and Lead zirconate titanate (PZT) [12] are widely used as piezoelectric materials in pMUTs. Among them, AlN has attracted increasing attention due to excellent electrical insulation and compatibility with complementary metal oxide semiconductor process. In this paper, the piezoelectric material of pMUTs in the proposed flexible ultrasonic sensor is AlN.

As shown in Fig. 1, the diaphragm of pMUTs is composed of a Silicon (Si) layer, a AlN layer, and molybdenum (Mo) layers, which serve as the elastic layer, piezoelectric layer and electrode layers, respectively. Due to the piezoelectric effect, the device can achieve mutual transform of acoustic energy (mechanical energy) and electrical energy. When an AC signal of appropriate frequency is applied to the electrodes, the piezoelectric AlN layer will generate transverse stress and vibration of the diaphragm, thereby emitting ultrasonic waves. On the contrary, when the diaphragm receives ultrasonic waves, the electronic signal will be detected by the electrodes. The pMUTs have multiple resonance frequencies. When the pMUTs work at the 1st resonant frequency, the longitudinal resonance displacement amplitude of the center point and the surface sound pressure generated both are the largest. Therefore, the first resonance frequency is the preferred operating frequency of the pMUTs designed in this article, and the mode shape as shown in Fig. 2. The diaphragm is generally designed to a certain resonance frequency according to the application.

B. Time-of-flight detection method

Time-of-flight (ToF) detection method is also called pulse-echo detection method. The transmitter pMUT emits a burst of ultrasonic wave pulse, and the ultrasonic wave is reflected back to the receiver pMUT after encountering obstacles. When the reflected echo is captured by the receiver pMUT, there is a time interval between it and the transmitting time, which is called the ultrasonic time of flight, t. It can be seen that the time of flight, t, is related to the sound wave speed, c, and the propagation distance, L, i.e. $t = L / c$.

When two pMUTs respectively are used as transmitter and receiver of separate ultrasonic sensor, as shown in Fig. 3(a), there is a certain distance between them. Therefore, the

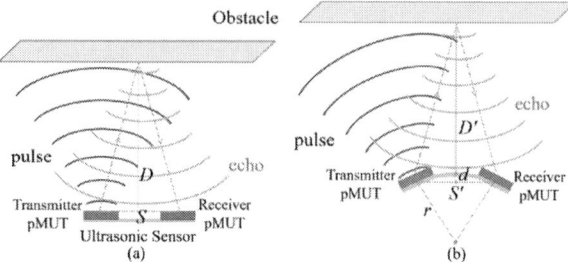

Fig. 3. ToF detection method of ultrasonic range finding as the ultrasonic sensor is in (a) flat state and (b) bend state.

propagation distance of the ultrasonic wave is not twice of the distance between the ultrasonic sensor and the obstacle. The influence of the distance between the pMUTs should be considered. And the propagation distance of ultrasonic wave, L, is:

$$L = ct = \sqrt{S^2 + 4D^2}\,, \tag{1}$$

where, S is the distance between pMUTs, D is the distance between the ultrasonic sensor and the obstacle. Hence, the distance, D, could be obtained as:

$$D = \frac{\sqrt{c^2t^2 - S^2}}{2}\,. \tag{2}$$

When the ultrasonic sensor is in bend state, the distance between the pMUTs will change, as shown in Fig. 3(b), at this time the actual distance S' between the pMUTs is:

$$S' = 2r\sin\frac{90S}{\pi r}\,, \tag{3}$$

where r is the radius of curvature of the ultrasonic sensor in the bend state. Hence, the actual distance between the ultrasonic sensor and the obstacle, D', could be obtained as:

$$\begin{aligned} D' = D - d &= \frac{\sqrt{c^2t^2 - S'^2}}{2} - (r - r\cos\frac{90S}{\pi r}) \\ &= \frac{\sqrt{c^2t^2 - 4r^2\sin^2(90S/\pi r)}}{2} - r + r\cos\frac{90S}{\pi r} \end{aligned} \tag{4}$$

C. Fabrication of the flexible ultrasonic sensor

The designed flexible ultrasonic sensor, as shown in Fig. 4, consists of two pMUTs (size of 3.5mm × 4mm). These devices are fixed on a piece of FPCB with thickness of 0.4mm using the heat curing die attach adhesive (Delo DA3700). The pads between pMUTs and FPCB are connected by gold wire bonding. Two pairs of interconnection lines are soldered on

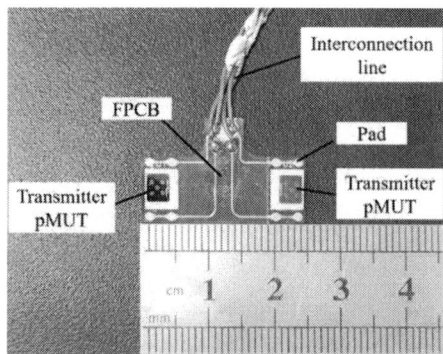

Fig. 4. Flexible ultrasonic sensor.

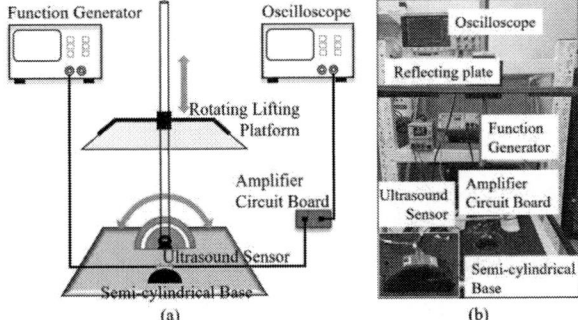

Fig. 5. (a) Schematic diagram and (b) experimental environment of the distance and angle experiment.

the FPCB. The four lines and pads are respectively connected to the electrodes of two pMUTs to transmit voltage signals. The two pMUTs are separated by 20mm, and the overall size of the flexible ultrasonic sensor is about 10mm × 25mm × 1mm.

III. EXPERIMENTAL SETUP AND RESULTS

A. The setup of experimental environment

The echo signal sensing experiment of the flexible ultrasonic sensor is conducted by using the experimental environment shown in Fig. 5. The experimental setup is composed of a rotating lifting platform, a function generator (KEYSIGHT 33621A), a digital storage oscilloscope (KEYSIGHT DSOX2014A), a signal amplifier circuit board and a semi-cylindrical base. The flexible ultrasonic sensor is fixed on the base of the platform, and the distance and angle sensing experiment of the sensor can be measured by moving and rotating the reflection plate. During the experiment, the transmitter pMUT is driven by the function generator, which generates a string of 100kHz, 5Vpp sinusoidal burst signal with a period of 10 cycles. The reflected ultrasonic echo is captured by the receiver pMUT, and recorded and displayed on the oscilloscope after being amplified by the amplifying circuit board. In addition, the semi-cylindrical base can be used to perform the sensing experiment of the flexible ultrasonic sensor in the bend state. The semi-cylindrical base is made by the 3D additive manufacturing process, and its radius of curvature, r of 2.7cm, simulates the radius of the thinnest joint of the UR3 collaborative robot.

B. Experimental Results

Fig. 6 describes the experimental results of the change in the amplitude of the echo signal voltage, when the flexible ultrasonic sensor and the plate are separated by 25cm ~ 60cm. The red curve shows the sensing results of the flexible ultrasonic sensor in the flat state, when the reflection plate is directly above the sensor and moves in 5cm steps. Similarly, the black curve represents the experimental results of the sensor in the bend state. From the Fig. 5, it can be seen that the echo signal gradually weakens as the distance increases. And when the sensor is in the bend state, the signal amplitude is always smaller than that of the flat state at the same position. However, as the sensing distance continues to increase, the signal amplitude difference between the two states gradually decreases until the two amplitudes are equal at the sensing distance of 65cm, which is also the maximum sensing distance of the flexible ultrasonic sensor. The experimental results show that the influence of the bend deformation of the sensor on the echo signal amplitude gradually reduces as the sensing distance increases, and the maximum sensing distance of the

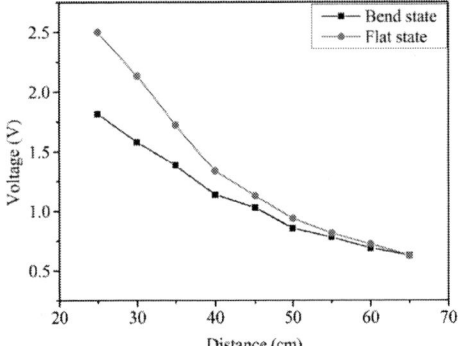

Fig. 6. The distance sensing experimental results of flexible ultrasonic sensor.

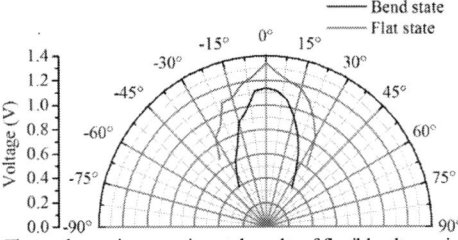

Fig. 7. The angle sensing experimental results of flexible ultrasonic sensor.

sensor is not affected by the bend deformation.

In addition, the angle sensing experiment of the flexible ultrasonic sensor has been conducted. In this experiment, the distance between the reflection plate and the sensor is 40cm, and the reflection plate rotates from -40° to +40° with step of 5°. Due to the sensing angle of ±40°exceeds the maximum angular sensing range of the sensor, the receiver pMUT cannot measure the reflected echo signal. Fig. 7 shows the echo signal voltage amplitude measured by the receiver pMUT when the sensing angle is from -35° to +35°. The experimental results show that the echo signal amplitude in the bend state is always smaller than that amplitude in the flat state. And the influence of the bend deformation of the flexible ultrasonic sensor on the signal amplitude basically does not change with the sensing angle, which is different from the experimental results in the distance sensing experiment. Additionally, it can be known that the angular sensing range of the sensor is not affected in the bend state, which is the same as the range of ±35° in the flat state.

IV. CONCLUSIONS

In this work, a flexible ultrasonic sensor based on pMUTs is proposed, which can be deformed to adapt to curved robot surfaces. The sensor consists of two pMUTs 20mm apart, one of which as transmitter and the other as receiver. Due to the pMUTs are prepared by MEMS technology, the flexible ultrasonic sensor is smaller in volume, size of 10 × 25mm. For compare the sensing ability of the flexible ultrasonic sensor in the bend state and the flat state, the sensing experiments of the sensor on the semi-cylindrical base and the conventional plane were conducted, respectively. The radius of curvature of the semi-cylindrical base is 2.7cm. It can be seen from the experimental results that, although the echo signal amplitude of the flexible ultrasonic sensor in the bend state is weaker, compared to the flat state, the sensing range of the sensor is not reduced. The distance sensing range is both 25cm ~ 60cm, and the angle sensing range is both ±35°. In this paper, the possibility of flexible ultrasonic sensors for sensing obstacles in the bend state is verified, which is helpful to develop its potential for application on curved robot surface.

ACKNOWLEDGMENT

The authors would like to thank the National Key Research and Development Program of China (2018YFB1307700) and Shanghai Industrial μTechnology Research Institute (SITRI) for the device fabrication and financial support of this work.

REFERENCES

[1] V. Villani, F. Pini, F. Leali, C. Secchi, "Survey on human-robot collaboration in industrial settings: Safety, intuitive interfaces and applications," Mechatronics, vol. 55, pp. 248-266, 2018.

[2] X. Wang, Y. Gu, Z. Xiong, Z Cui, T Zhang, "Silk-molded flexible, ultrasensitive, and highly stable electronic skin for monitoring human physiological signals," Advanced Materials, vol. 26, no. 9, pp. 1336-1342, 2014.

[3] Z. Ji, H. Zhu, H. Liu, N. Liu, T. Chen, Z. Yang, et al., "The Design and Characterization of a Flexible Tactile Sensing Array for Robot Skin," Sensors (Basel), vol. 16, no. 12, 2016.

[4] R. S. Dahiya, M. Valle, G. Metta, L. Lorenzelli, "Tactile Sensing Arrays for Humanoid Robots using Piezo-Polymer-FET Devices," 2015, pp. 301-306.

[5] B. Schmidt and L. Wang, "Depth camera based collision avoidance via active robot control," Journal of Manufacturing Systems, vol. 33, no. 4, pp. 711-718, 2014.

[6] S. Robla, JR. Llata, C. Torre-Ferrero, EG. Sarabia, V. Becerra, and J. Perez-Oria, "Visual sensor fusion for active security in robotic industrial environments," Eurasip Journal on Advances in Signal Processing, vol. 2014, no. 1, pp. 1-20, 2014.

[7] Y. Huang, X. Cai, W. Kan, S. Qiu, X. Guo, C. Liu, et al., "A flexible dual-mode proximity sensor based on cooperative sensing for robot skin applications," Rev Sci Instrum, vol. 88, no. 8, pp. 085005, Aug. 2017.

[8] X. Chen, M. Zhao, L. Xiang, F. Sugai, and M. Inaba, "Development of a low-cost ultra-tiny line laser range sensor," in 2016 IEEE/RSJ International Conference on Intelligent Robots and Systems (IROS), Daejeon, South Korea, 2016, pp. 111-116.

[9] W. Liu, L. He, X. Wang, J Zhou, W. Xu, N. Smagin, et al., "3D FEM Analysis of High-Frequency AlN-Based PMUT Arrays on Cavity SOI," Sensors, vol. 19, no. 20, pp. 4450-4464, 2019.

[10] E. Herth, L. Valbin, F. Lardet-Vieudrin, E. Algré, "Modeling and detecting response of micromachining square and circular membranes transducers based on AlN thin film piezoelectric layer," Microsystem Technologies, vol. 23, pp. 3873-3880, 2018.

[11] J. Li, W. Ren, G. Fan, C. Wang, "Design and Fabrication of Piezoelectric Micromachined Ultrasound Transducer (pMUT) with Partially-Etched ZnO Film," Sensors, vol. 17, no. 6, pp. 1381-1393, 2017.

[12] D. Dezest, O. Thomas, F. Mathieu, L. Mazenq, C. Soyer, J. Costecalde, et al., "Wafer-scale fabrication of self-actuated piezoelectric nanoelectromechanical resonators based on lead zirconate titanate (PZT)," Journal of Micromechanics and Microengineering, vol. 25, no. 3, pp. 035002, 2015.

Effect of Temperature on the Fatigue Damage of SAC305 Solder

Xu Long*
School of Mechanics, Civil Engineering and Architecture Northwestern Polytechnical University Xi'an China
xulong@nwpu.edu.cn

Ying Guo
School of Mechanics, Civil Engineering and Architecture Northwestern Polytechnical University Xi'an China
1261024184@qq.com

Xiaotong Chang
School of Mechanics, Civil Engineering and Architecture Northwestern Polytechnical University Xi'an China
xtchang@nwpu. edu. cn

Yutai Su
School of Mechanics, Civil Engineering and Architecture Northwestern Polytechnical University Xi'an China
suyutai@buaa.edu.cn

Hongbin Shi
Xi'an Institute Consumer BG, DFR Department HUAWEI DEVICE CO LTD Xi'an China
shihongbin2@huawei.com

Tao Huang
Space Research Institute of Electronics and Information Technology Aerospace Science and Technology Corporation Xi'an China
huangt89@163.com

Bingyi Tu
Space Research Institute of Electronics and Information Technology Aerospace Science and Technology Corporation Xi'an China
977793390@qq.com

Yanpei Wu
Space Research Institute of Electronics and Information Technology Aerospace Science and Technology Corporation Xi'an China
wuyanpei2007@126.com

Abstract—In recent years, the field of electronic packaging has been rapidly developing. The traditional lead-containing solders have been abandoned because of their pollution to the environment and harm to human body. Therefore, lead-free solders are gradually beginning to be used in the electronic packaging industry, among which, the Sn-3.0Ag-0.5Cu (SAC305) solder is the most frequently used material in lead-free solder. As the fundamental elements, solder joints in electronic components are affected by periodical mechanical stress and strain for the while service life, so it is important to understand the fatigue damage about them. In this paper, SAC305 solder specimens were experimentally investigated to uniaxial cyclic loading at different temperatures from 288.15K to 373.15K under the same strain rate of 10^{-3}/s and the same maximum strain of 0.10. In order to explain the accumulative damage due to fatigue at different temperatures and the mechanical properties, SAC305 solder specimens after cyclical loading were tested through tensile force until fracture at the same temperature. Furthermore, the differences of microstructure were observed by a scanning electron microscope (SEM). It could be known that the fatigue damage of the specimens is different at different temperatures, and the failure morphology of the solder specimens also varies.

Keywords—electronic packaging, SAC305 solder, uniaxial cyclic loading, fatigue damage

I. INTRODUCTION

Electronic products have been developing to be miniaturized and complicated, and the electronic packaging structure has become an important guarantee for ensuring the normal operation of miniaturized and flexible electronic devices. The package structure is a typical laminated structure, which contains a variety of materials but their thermal expansion coefficients are different. In the service process, the thermal deformation mismatch is caused by the temperature change, which results in the local material stress concentration and plastic deformation, and finally the deterioration of fatigue lifetime for the package structure. The quality of packaging structure directly affects the

product function and user experience. Since the Sn-3.0Ag-0.5Cu (SAC305) solder has become the most widely applied lead-free material in the field of electronic packaging, more comprehensive studies should be performed to solve mechanical reliability problems of SAC305 solder under extreme service conditions with considering the effect of thermal expansion coefficient mismatch.

Many existing studies have investigated the mechanical performance of lead-free material under torsion [1], creep [2], and shear [3]. In addition to the mechanical loading, Long et al. [4] reported the uniaxial tensile behavior under different strain rates and current densities, explained the current stress response of electro-mechanical coupling loading, and found that when the current density gradually increased, the fracture mode of SAC305 solders changed. The effects of higher temperature reduced low-cycle fatigue endurance [5]. Pang et al. [6] established a low-cycle fatigue model for lead-free solder, in order to evaluate the endurance under fatigue test, and determined the parameters of the low-cycle fatigue model. The crack area and length in development had been calculated by using a resistance technique [7]. Finally, Lall et al. [8] believed that as the temperature increase, the movement of dislocations increases, resulting in an increase in the range of plastic strain and a decrease in the peak stress,, which ultimately affect the fatigue lifetime for lead-free solders.

The uniaxial low-cycle fatigue behavior for SAC305 material at different temperatures was studied by using a universal testing machine equipped with a temperature controller. The control parameters included test temperatures (288.15K ~ 373.15K), strain rate of 10^{-3}/s and maximum strain of 0.10. Through the analysis of the experimental data, the relationships between the cycles and accumulative damage during fatigue were obtained, and the critical parameters related to the temperatures were found. Finally, the solder specimens after fatigue loading were further tested by uniaxial tensile force until fracture at the same temperature, and the microstructure differences were

---

The Natural Science Foundation of Shaanxi Province (No. 2021KW-25) and the Astronautics Supporting Technology Foundation of China (No. 2021-HT-XG).

obtained and analyzed by the scanning electron microscope (SEM).

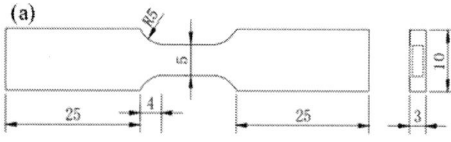

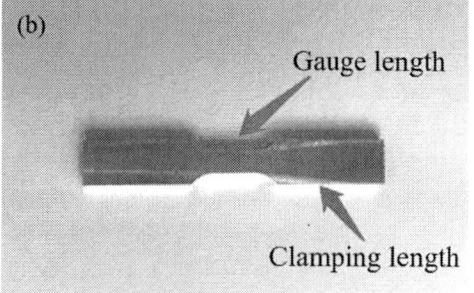

Fig. 1. Description of SAC305 solder specimen. (a) Schematic diagram of specimen size (mm); (b) Image of prepared specimen.

II. EXPERIMENTAL PROCEDURE

The geometric dimension of the specimen can be observed in Fig. 1 (a). These specimens were made by mechanical processing the SAC305 solder rods with a high purity. In order to eliminate the residual stress produced due to machining process, the samples were annealed at 417K for 3h and cooled to room temperature [9]. One of the prepared specimens is shown in Fig. 1 (b), which were stored in the room temperature and dry environment.

A universal testing machine was used with a furnace that can provide temperature, so as to represent the actual working environment of electronic packaging by controlling the strain rate and maximum strain of the mechanical test machine in Figs. 2 and 3. As shown in Fig. 3, the velocity of the clamps is controlled by displacement to be 10^{-3}/s, and thus the strain rate is controlled to be the constant of 10^{-3}/s. It could be considered as a fatigue test at low strain rate [10]. The gauge length of the specimens is 10mm and the applied maximum strain is at 0.10. In the experimental study, three specimens were tested for each loading condition, and the average value was taken for analysis. These data were recorded by the monitor with the mechanical testing machine.

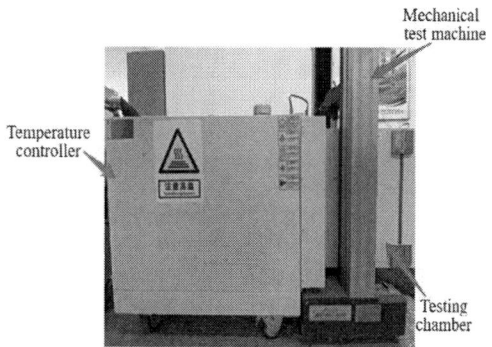

Fig. 2. Universal mechanical testing machine

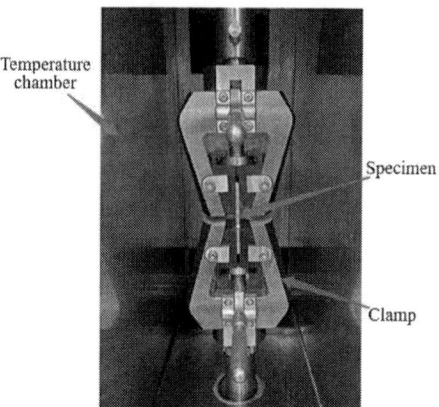

Fig. 3. Testing instrument for fatigue behavior of SAC305 solder specimens.

III. RESULE ANALYSIS

When the SAC305 samples were applied with the strain rate of 10^{-3}/s and the maximum strain of 0.10, the relationship curves between the peak stress and cycle times under different temperatures were shown in Fig. 4. Generally, the peak strength is found to be decreasing with the increasing temperature. The points from experimental data in each curve could be largely fitted to the following formula [8]:

$$\sigma = k_1/1 + k_2 \exp(-k_3 N) + k_4 N \qquad (1)$$

where σ represents the peak stress, N is the cycles, k_1, k_2 are material parameters related to temperature, and k_3, k_4 are constants. The relationships between parameters k_1, k_2 and the test temperatures were given in Table I. In Fig. 4, the decreasing trends of the peak stress were similar at different test temperatures. The proposed empirical formula could well represent the decreasing trend of the peak strength with the increasing temperature. In particular, the decreasing rates were rapid within 50N, and then tended to stabilize. Another point was that the higher temperature, the greater reduction in the peak stress. This is consistent with the temperature softening effect. It could be seen that the damage is different under different test temperatures, that is, the damage and temperature are closely correlated.

In order to further investigate the accumulated damage in fatigue, tensile tests of solder specimens were performed after fatigue loading. Fig. 5 shows that the tensile stress and strain curves of SAC305 solder specimens deteriorated because of temperature effects. It was obvious to observe that the initial modulus and strengths are greatly reduced.

TABLE I. COEFFICIENT VALUES RELATED TO TEMPERATURE

Temperature (K)	Material Parameters	
	k_1	k_2
288.15	20.21335	-0.28642
308.15	13.04941	-0.55829
343.15	7.96135	-0.70802
373.15	3.79982	-0.86259

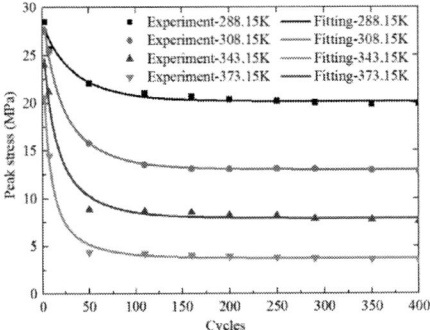

Fig. 4. Effects of cycles and temperature on peak stress

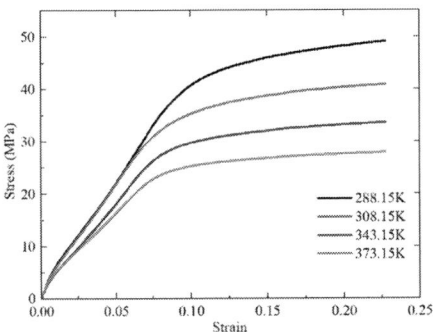

Fig. 5. Tensile test of solder specimens after fatigue loading

To characterize the tensile behavior of the SAC305 solders after the low-cycle fatigue test, the fractured specimens were compared as shown in Fig. 6. The following rules could be found: most of the fracture modes for the specimens were ductile fracture, with some apparent necking phenomenon at the same time. And the stretch length of the SAC305 solder materials increased while the temperature from 288.15K to 373.15K, which can be elucidated by temperature softening. As shown in Fig. 6, the size and depth of the dimples increased with the increase of temperature [11]. It has been proved that temperature affected the size of the dimples by acting on the plasticity and hardening index of the material at same strain rate, so the plasticity of the material was improved.

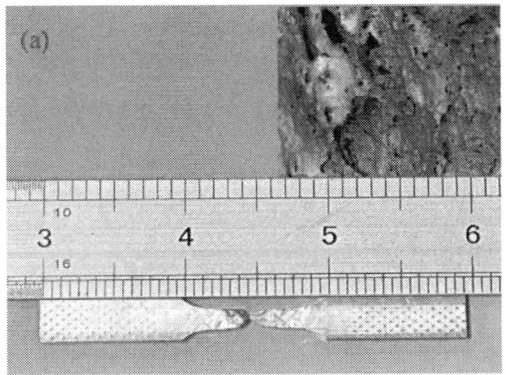

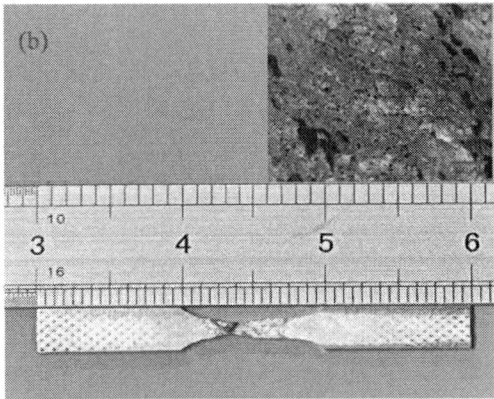

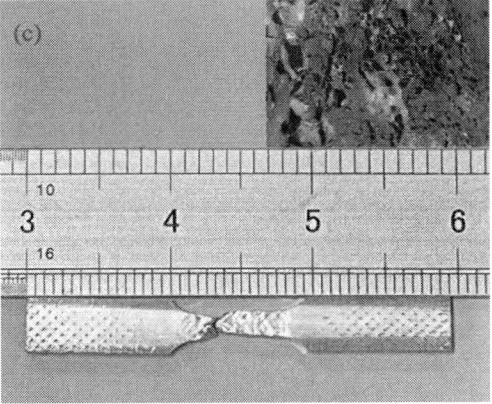

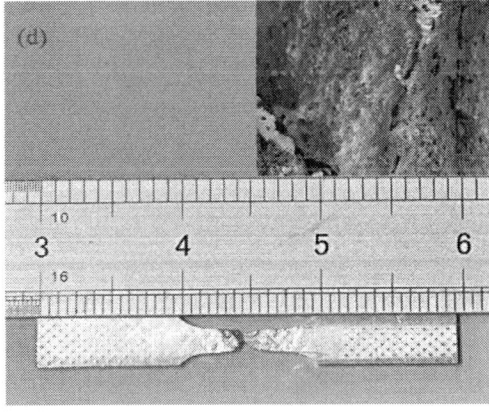

Fig. 6. Description of cross-section at different temperature. (a)288.15K; (b)308.15K; (c)343.15K; (d)373.15K

IV. CONCLUSION

In this paper, the low-cycle fatigue experiments for SAC305 solders were investigated in the temperature range between 288.15K and 373.15K at the same strain rate (10^{-3}/s) up to the same maximum strain of 0.10. And under the combined action of temperatures and strain rates,, the tensile properties of SAC305 materials were tested. It could be found that the strength of materials gradually decreases and the damage increases when the temperature is from 288.15K to 373.15K. Finally, morphology of the SAC305 solder specimens was observed by SEM. When the temperature rise from 288.15K to 373.15K, the plasticity increased and ductile failure occurred due to the softening behavior of the material.

The damage model based on the analysis of experimental data was studied, but it did not give the physical mechanism of fatigue damage. The damage model has been actively studied by the authors based on entropy which combines Newtonian mechanics and the second law of thermodynamics. And the unified creep plasticity (UCP) and damage model will be combined to explain the damage mechanism of lead-free solders fatigue.

ACKNOWLEDGMENT

This work was supported by the Natural Science Foundation of Shaanxi Province (No. 2021KW-25) and the Astronautics Supporting Technology Foundation of China (No. 2021-HT-XG).

REFERENCES

[1] T. Liang, Y. F. Liang, J. Fang, C. Gang, and C. Xu, "Torsional fatigue with axial constant stress for Sn–3Ag–0.5Cu lead-free solder [J]," *International Journal of Fatigue,* vol. 67, no. Oct., pp. 203-211, 2014.

[2] T. H. Gu, V. S. Tong, C. M. Gourlay, and T. B. Britton, "In-situ study of creep in Sn-3Ag-0.5Cu solder [J]," *Acta Materialia,* vol. 196, no. Sep., pp. 31-43, 2020.

[3] A. Bing, G. Q. Gu, W. F. Zhang, and Y. P. Wu, "Shear creep behavior of Sn-3Ag-0.5Cu solder bumps in ball grid array [J]," *IEEE,* 2011.

[4] W. Tang, X. Long, Y. Liu, C. Du, and F. Jia, "Effect of electric current on constitutive behaviour and microstructure of SAC305 solder joint [C]," in *2018 IEEE 20th Electronics Packaging Technology Conference (EPTC),* 2018.

[5] J. H. L. Pang, B. S. Xiong, and T. H. Low, "Low cycle fatigue models for lead-free solders [J]," *Thin Solid Films,* vol. 462, no. Sep., pp. 408-412, 2004.

[6] J. H. L. Pang, B. S. Xiong, and T. H. Low, "Creep and Fatigue Characterization of Lead Free 95.5Sn-3.8Ag-0.7Cu Solder [C]," in *Electronic Components and Technology Conference, 2004. Proceedings. 54th,* 2004, vol. 2, pp. 1333-1337.

[7] M. Duek and C. Hunt, "Low cycle isothermal fatigue properties of lead-free solders [J]," *Soldering and Surface Mount Technology,* vol. 19, no. 4, pp. 25-32, 2007.

[8] P. Lall, N. Fu, and J. Sunlinf, "Cyclic stress-strain behavior of SAC305 lead free solder: effects of aging, temperature, strain rate, and plastic strain range [C]," in *Electronic Components & Technology Conference,* 2016, pp. 927-939.

[9] M. Mustafa, Z. Cai, J. C. Suhling, and P. Lall, "The effects of aging on the cyclic stress-strain behavior and hysteresis loop evolution of lead free solders [J]," *IEEE,* pp. 927-939, 2011.

[10] M. W. Xie, G. Chen, J. Yang, and W. L. Xu, "Temperature- and rate-dependent deformation behaviors of SAC305 solder using crystal plasticity model [J]," *Mechanics of Materials,* vol. 157, p. 103834, 2021.

[11] I. Abdullah, M. N. Zulkifli, A. Jalar, R. Ismail, and M. A. Ambak, "Relationship of mechanical and micromechanical properties with microstructural evolution of Sn-3.0Ag-0.5Cu (SAC305) solder wire under varied tensile strain rates and temperatures [J]," *Journal of Electronic Materials,* vol. 48, pp. 2826-2839, 2019.

Enhancement of Silane Coupling Agents on The Underfill Adhesion Undergoing Hydrothermal Aging

Haoliang Lin
Shenzhen Institute of Advanced Electronic Materials
Shenzhen Institute of Advanced Technology, Chinese Academy of Sciences;
Department of intelligent manufacturing, WuYi University, Shenzhen, China / Jiangmen, 529020, China
Shenzhen 518055, China
hl.Lin@siat.ac.cn

Bin Wang
Shenzhen Institute of Advanced Electronic Materials
Shenzhen Institute of Advanced Technology, Chinese Academy of Sciences;
Nano Science and Technology Institute, University of Science and Technology of China
Shenzhen, China / Suzhou, China;
Shenzhen 518055, China
wang.bin@siat.ac.cn

Houya Wu*
Shenzhen Institute of Advanced Electronic Materials
Shenzhen Institute of Advanced Technology, Chinese Academy of Sciences
Shenzhen 518055, China
hy.wu1@siat.ac.cn

Yuanyuan Yang
Shenzhen Institute of Advanced Electronic Materials
Shenzhen Institute of Advanced Technology, Chinese Academy of Sciences
Shenzhen 518055, China
yy.yang@siat.ac.cn

Gang Li
Shenzhen Institute of Advanced Electronic Materials
Shenzhen Institute of Advanced Technology, Chinese Academy of Sciences
Shenzhen 518055, China
gang.li@siat.ac.cn

Pengli Zhu*
Shenzhen Institute of Advanced Electronic Materials
Shenzhen Institute of Advanced Technology, Chinese Academy of Sciences
Shenzhen 518055, China
pl.zhu@siat.ac.cn

Abstract—**In this study, three kinds of underfill materials were used to investigate the enhancement of silane coupling agents on the adhesive strength of the underfill and the silicon substrate undergoing boiling water soaking. Adhesion properties between the underfill and the silicon were evaluated by shearing test. The fracture interfaces of the shearing tests are observed by SEM. The results show that the adhesive strength between the silicon and the underfill decreases significantly without silanes, and the bonding between the resin and the filler becomes loose by hydrothermal aging. Underfill containing aminosilane has higher water resistance, which can improve the adhesive strength of underfill, and can help to strengthen the bonding between underfill and filler after hydrothermal aging.**

Keywords—Adhesion, hydrothermal aging, underfill, interface

I. INTRODUCTION

Underfill is widely applied in the flip-chip packages as a liquid thermosetting polymer material between silicon dies and substrates. It not only can effectively improve the reliability of flip chip packaging, but also provides a solution for the application of flip chip technology from ceramic substrates to organic substrates[1]. With the development of integrated packaging technology, the size of electronic devices is increasing. At the same time, the spacing of C4 bumps or micro-bumps in the devices is getting smaller and smaller, such as ASIC (Application Specific Integrated Circuit), HBM (High Bandwidth Memory), and other electronic devices. This will gradually increase the area of underfill and reduce the thickness of underfill. Therefore, the underfill reliability of electronic products become increasingly important. The failure modes of the LSI (Large-scale integrated circuit) package are mainly divided into two categories, i.e., internal delamination and fracture underfill. Therefore, the underfill attachment structure of the electronic component is a weak point of the device[2]. However, the packages are generally vulnerable to environmental attack, especially under hydrothermal aging conditions[3]. As a plasticizer of polymer matrix, the water may increase the dielectric constant of the underfill material and lower its glass transition temperature[4]. At the same time, as a polymer material, underfill can easily absorb water in a humid environment. The swelling of the underfill caused by the absorption of water will introduce swelling stress at the interface between the swelling underfill and the passivation[5][6], which weakens the interfacial adhesion strength. The weakened adhesive strength of the underfill often results in reliability problems of the chip, such as delamination and the crack，as shown in Fig. 1. Therefore, it is highly important to enhance the adhesive strength of the underfill against hydrothermal aging.

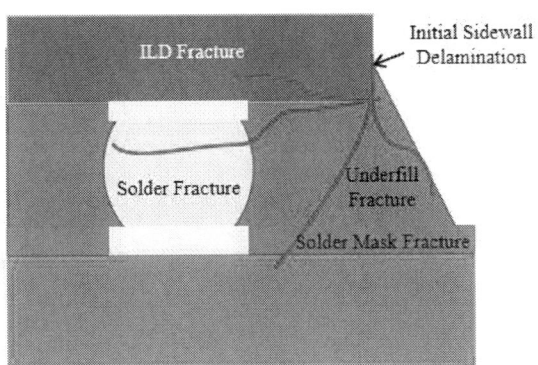

Fig. 1. Shearing test setup

978-1-6654-1392-3/21 $31.00 © 2021 IEEE

Currently, there are three main interfacial adhesion mechanisms，physical adsorption，contact and Chemical adsorption, in which chemisorption is the strongest adsorption mechanism. This is the main covalent and ionic bonds between the two surfaces results[7]. By adding appropriate chemical additives to the underfill material, such as silane coupling agent, chemical adsorption can be enhanced. The silane coupling agents are used to bridge the resins and silicon substrate. When the silane coupling agent links the inorganic and organic interfaces, a bonding layer of organic matrix-silane coupling agent-inorganic matrix is formed, and the adhesive strength is improved. At present, some studies have shown that adding an appropriate silane coupling agent to the underfill through a shear test can improve the adhesion between the underfill material and the substrate. Vincent *et al.* reported that the adhesion of underfill to alumina can be increased by using silane coupling agents containing amine, epoxy resin and vinyl functional groups, in which vinyl functional silane can also increase the adhesion of underfill to solder masks such as FR4[7]. However, there is no systematic study on the effect of silane coupling agent on the bonding and fracture behavior of underfill interface undergoing hydrothermal aging.

In this study, three kinds of underfills are prepared: without coupling agent, adding a methoxysilane, and adding a aminosilane, were selected to prepare Si/underfill/Si adhesion structures. The fracture behavior of the adhesion structure of underfill undergoing different hydrothermal aging time was systematically studied, and the enhancement effect of silane coupling agent on underfill was discussed. This research provides valuable guidance for the selection of silane coupling agent in manufacturing process planning and electronic packaging design.

II. EXPERIMENTAL PROCEDURE

A. Underfill Materials

In this study, three kinds of underfills were prepared: no coupling agent (BLANK), addition of methoxysilane (SCA1), and addition of aminosilane (SCA2). Three kinds of underfill material parameters are shown in Table I. A silicon wafer with a diameter of 8 inches and a thickness of 500 ~ 15 um was produced. After cleaning by plasma treatment, the silicon wafer was cut into square molds with dimensions of 2 mm×2 mm and 5 mm×5 mm, which were used as the die and Si-Substrate of the adhesion structure, respectively.

TABLE I.   UNDERFILL MATERIAL PARAMETERS

Material parameters	BLANK	SCA1	SCA2
Hardener type	Amine	Amine	Amine
Resin type	Epoxy	Epoxy	Epoxy
Silane coupling agent type	No addition	Methoxysilane	Aminosilane
Filler average size /μm	1.0	1.0	1.0
Filler content /wt%	20	20	20
Filler type	Spherical silica	Spherical silica	Spherical silica

B. Adhesion test

Sandwich structures were prepared as shown in Fig. 2, which was used to evaluate the adhesive strength of the underfills by the die shear test. The sandwich structures were cured at 150 ℃ for 2 hours. After curing, the sandwich structures were immersed in boiling water for hydrothermal aging. Three conditions of hydrothermal aging are considered, including aging time of 0 hours, 24 hours, and 48 hours. In order to ensure the validity of the data, 20 identical samples were tested in each group of die shear tests. Finally, optical microscope and scanning electron microscope (SEM) were used to observe the fracture interface after the adhesive test.

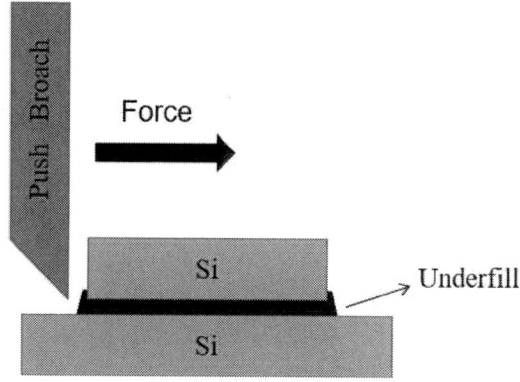

Fig. 2. Shearing test setup

III. RESULTS AND DISCUSSION

A. Results of shear strength

The shearing test results show that the shearing strength of the three kinds of underfills is almost the same before aging, all-around 90 MPa, as presented in Fig. 3. However, after hydrothermal aging, the shearing strength of the BLANK and the SCA1 decreased dramatically, down to less than 5 Mpa under both conditions of 24 h and 48 h soaking. On the contrary, the adhesion of SCA2 is maintained very well. The shearing strength was only reduced from 90.1 Mpa to 76.7 MPa (24 h soaking) and 69.6 MPa (48 h soaking), respectively. After hydrothermal aging, the adhesion strength of the three underfills has decreased to different degrees. It is supposed that water molecules enter the interfaces between the underfill and the substrate during the hydrothermal aging process and destroy the molecular bond on the interface, which results in adhesive strength decline. Although the adhesion strength after aging shows a downward trend, the interface adhesion strength is still higher than the shear strength of underfilled samples.

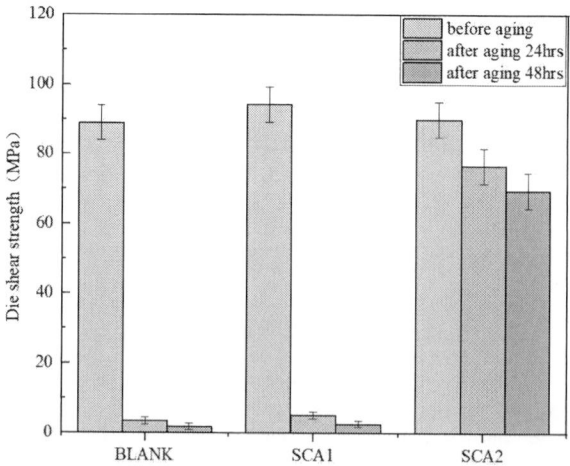

Fig. 3. Shearing strength of underfill under different conditions.

It worth noting that the adhesive strength of SCA2 after aging is significantly higher than that of the other two groups. The explanation is that the amino group of SCA2 has strong reactivity with the epoxy group of the resin, forming C-N bond between the silane and the epoxy resin. Thus, strong chemical bond bridges between the Si substrate and the epoxy underfill is built, which increasing the water resistance of the underfill and well maintain the adhesive strength of the underfill. In contrast, the Methoxy of SCA1 is an inert group, which not very reactive, leading to a weak bond between the Si substrate and the epoxy underfill after hydrothermal aging process.

B. Fracture models

In order to analyze the fracture behavior of the glue at the interface, we observed the interface fracture morphology of the sheared samples by an optical microscope.

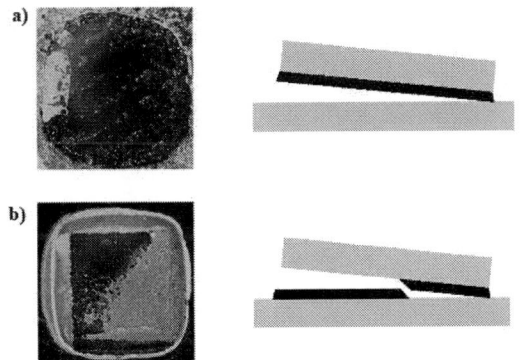

Fig. 4. Fracture models: (a) Interfacial delamination between si and underfill, and (b) Mixed fracture

It is mainly found two fracture models: one is the delamination of the interface between the underfill and the Si substrate, where the underfill and the substrate are completely separates into two parts, as shown in Fig 4(a). In contrast, the other is a mixed fracture mode of interfacial delamination and internal fracture, where the cracks initiates from the interface and then propagates inside the bulk material of the underfill , as shown in Fig. 4(b).

Fig. 5 shows the fracture modes of the three underfills after hydrothermal aging of 0 hours and 24 hours. It can be seen that without hydrothermal aging, all of the three underfills are mainly fractured in a mixed model of interfacial delamination and internal fracture, as shown in Fig. 5(a). However, after 24 hours of hydrothermal aging, the BLANK sample and the SCA1 sample changed to an interfacial delamination model. The debonding interfaces were clean as almost no glue remains on the substrate. It is possible that the water molecules have entered the interfaces between the underfill and the Si substrate during the hydrothermal aging. the water molecules would destroy the molecular bond on the interface. One possible mechanism is that free water molecules hydrolyze the hydroxyl groups bonding the underfill and the Si substrate, thereby breaking the connection. This results in sharp decrement in the shearing strength of the underfill.

On the contrary, with the addition of SCA2 in the underfill, its shear strength has been well maintained after hydrothermal aging. As shown in Fig. 5(b), the fracture model of SCA2 after hydrothermal aging remains the same as before hydrothermal aging. It is possible that the bond of SCA2 the Si substrate is strong enough to prevent the hydrolysis reaction from breaking. In other words, SCA2 has a better water-blocking performance.

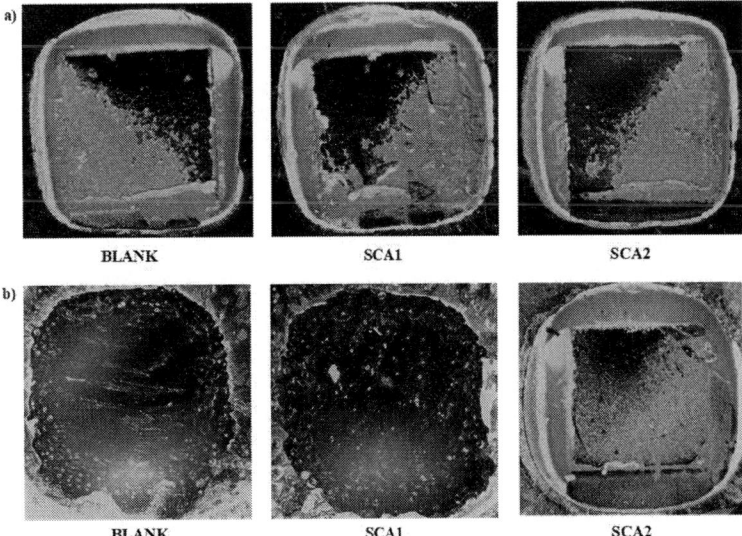

Fig. 5. Fracture modes of the three kinds of underfill: (a)Die shear test before aging, and (b)Die shear test after 24 hours aging.

C. Micromorphology of the fracture section

The micromorphology of the fracture section detected by SEM is presented in Fig. 7. It is found that the silica filler and the epoxy are loosely connected without silane before hydrothermal aging. What is more, the loose connection

becomes worse after hydrothermal aging. This is easy to cause cracks of the bulk material of the underfill. However, with the help of SCA1 and SCA2, the fillers are well wrapped by the epoxy before hydrothermal aging, as shown in Fig. 7 a). After hydrothermal aging, it is found that SCA1

978-1-6654-1392-3/21 $31.00 © 2021 IEEE

sample loosened the bond between filler and epoxy resin. Differently, the filler and the epoxy are still tightly bonded with the help of SCA2 before and after hydrothermal aging, as shown in Fig. 7(b).

A possible explanation is that SCA2 has a better water resistance performance and stronger bonding ability to the inorganic materials than SCA1. First, SCA2 can effectively prevent water from penetrating to the bonding interface, leading to less water introduced to destroy the bonding between the underfill and the substrate; in addition, the bonding between the underfill and the substrate helped by SCA2 is too strong to break by the water molecu.

Fig. 6. Micromorphology of the fracture section : (a) before aging and (b) after aging.

IV. CONCLUSIONS

The effects of silane coupling agent on the adhesive strength between the underfill and the silicon substrate have been investigated. Shearing tests are conducted under the condition of with and without hydrothermal aging process. The main conclusions are as follow:

1) Aminosilane can effectively maintain the adhesion between the underfill and the Si substrate after hydrothermal aging, while the Methoxysilane shows no improvement for the adhesive strength of the underfill.

2) The fracture modes of the shearing samples are relative to their corresponding adhesive strength. Completed separated fracture mode is caused by weaker adhesive strength, while mixture fracture mode indicates strong adhesive strength.

3) The underfill containing aminosilane all showed mixture fracture mode before and after hydrothermal aging, indicating it well water resistance ability and adhesion maintenance performance.

4) Aminosilane can also strongly bond the epoxy resin and the silica filler, preventing the bond breaking from hydrothermal aging.

ACKNOWLEDGMENT

This work was supported by the National Key R & D Project from Minister of Science and Technology of China (2020YFB0311800), Shenzhen basic research plan (JCYJ20190807154409372), the National Natural Science Foundation of China (61704182), and the GuangDong Basic and Applied Basic Research Foundation (2020A1515111003).

REFERENCES

[1] Lejun Wang and C. P. Wong, "Recent advances in underfill technology for flip-chip, ball grid array, and chip scale package applications," International Symposium on Electronic Materials and Packaging (EMAP2000) (Cat. No.00EX458), 2000, pp. 224-231, doi: 10.1109/EMAP.2000.904159.

[2] L. Sun, M. -B. Zhou, T. Sun and X. -P. Zhang, "Effects of underfill thickness on mechanical properties and fracture behavior of Si/underfill/Si adhesion structures," 2020 21st International Conference on Electronic Packaging Technology (ICEPT), 2020, pp. 1-4, doi: 10.1109/ICEPT50128.2020.9202876.

[3] L. Chen, Y. Liu and X. Fan, "Application of water activity-based theory for moisture diffusion in electronic packages using ANSYS," 2018 19th International Conference on Thermal, Mechanical and Multi-Physics Simulation and Experiments in Microelectronics and Microsystems (EuroSimE), 2018, pp. 1-6, doi: 10.1109/EuroSimE.2018.8369947.

[4] J. W. Evans and K. Sinha, "Applications of fracture mechanics to quantitative accelerated life testing of plastic encapsulated microelectronics," Microelectron. Reliab., vol. 80, no. October, pp. 317–327, 2018, doi: 10.1016/j.microrel.2017.10.022.

[5] E. H. Wong, K. C. Chan, R. Rajoo and T. B. Lim, "The mechanics and impact of hygroscopic swelling of polymeric materials in electronic packaging," 2000 Proceedings. 50th Electronic Components and Technology Conference (Cat. No.00CH37070), 2000, pp. 576-580, doi: 10.1109/ECTC.2000.853216.

[6] M. Schneider, U. Gierth, L. Simunkova, P. Gierth, and L. Rebenklau, "Complementary EIS/FTIR study of the degradation of adhesives in electronic packaging," Mater. Corros., vol. 71, no. 11, pp. 1832–1841, 2020, doi: 10.1002/maco.202011772.

[7] M. B. Vincent, L. Meyers and C. P. Wong, "Enhancement of underfill adhesion to die and substrate by use of silane additives," Proceedings. 4th International Symposium on Advanced Packaging Materials Processes, Properties and Interfaces (Cat. No.98EX153), 1998, pp. 49-52, doi: 10.1109/ISAPM.1998.664432.

Review of Copper-Silver Core-Shell Sintering Pastes: Technology and Future Trends

Zejun Zeng
Academy for Engineering and Technology, Fudan University
Shanghai, China
19210860052@fudan.edu.cn

Jing Zhang
Heraeus Materials Technology Shanghai Ltd.
Shanghai, China
j.zhang@heraeus.com

Pan Liu*
Academy for Engineering and Technology, Fudan University
Shanghai, China
Yiwu Research Institute of Fudan University, Yiwu, China
panliu@fudan.edu.cn

Abstract—For power electronics packaging, die attach materials are developing towards higher working temperature and higher current density. Traditional packaging materials such as solder paste and conductive adhesives no longer meet the requirements of advanced packaging, especially for wide-band gap chip packages. Copper-silver core-shell (Cu@Ag) particle is an emerging material that can be used for advanced packaging. It combines the excellent electrical and thermal conductivity of copper and silver particles, while avoiding the shortcomes such as the electromigration of silver and the oxidation of copper. The sintered paste prepared by Cu@Ag has excellent electrical and thermal conductivity and is low in cost compared with commercial silver sintered paste, which has attracted the interest of researchers and companies all over the world. In this work, Cu@Ag particles preparation methods, paste preparation, and sintering process were listed and compared. Furthermore, molecular simulations regarding such core-shell particles were also concluded. The main preparation methods of Cu@Ag particles include physical vapor deposition, spray pyrolysis, direct displacement, liquid-phase chemical reduction, etc. Commonly used sintering methods for Cu@Ag sintering paste include hot pressing sintering, ultrasonic assisted sintering, and pulse current sintering. Copper-silver core-shell nanoparticles have a larger radius of curvature between the grain boundaries than bulk materials due to their smaller size, and the sintering temperature is greatly reduced, so the sintering kinetics is also different from that of bulk materials. The molecular dynamics simulation (MD) software LAMMPS can simulate the particle sintering process and explore the sintering mechanism.

Keywords—*Cu@Ag particles, preparation process, sintering process, molecular dynamics*

I. INTRODUCTION

A. The Development of High Power Devices

Nowadays, modern electronic products are developing in the direction of miniaturization and multi-function. At the same time, electronic products are required to have good electrical-mechanical performance in harsh working environments such as high temperatures, humidity, and high currents, which brings significant challenge to the chip interconnection. The electronic package provides electrical connection or insulation, thermal cooling, mechanical support, and physical protection for power electrical electronics in the circuits. More advanced electronic packaging technologies are needed in areas such as aerospace, automotive electronics, and sensing technology to keep semiconductor devices running more reliably and consistently over the long term. These devices need to maintain switching rate, junction temperature and power density in extreme temperature conditions [1].

Therefore, the development of the wide-band gap semiconductor materials was accelerated especially for silicon carbide (SiC) and hydrogen nitride (GaN). Compared with conventional silicon, SiC has the advantages of high breakdown voltage, large band gap and small heat resistance. In order to ensure the high performance and reliability of the system, when using high-power semiconductor chips such as silicon carbide, the die attach materials must be adapted to severe thermal and current effects. Therefore, die attach materials that connect chips or devices to the entire system play a vital role in maintaining the stability of the system [2]. In order to meet the harsh working environment requirements of high-power devices, new die attach materials must be developed.

B. The Development of Packaging Materials

Interconnect technology in electronic packaging refers to the use of die attach materials to connect the chip and substrate to achieve the mechanical support, heat dissipation and electrical conduction functions of the chip [3]. At present, the main die attach materials are lead-free solder and conductive glue. Lead-free solder is mainly tin-based solder and has added alloying elements such as silver, copper, zinc, indium and bismuth in a certain proportion. Lead-free solder is still the most widely used die attach material, with the advantages of low price and high flexibility. However, there are also obvious defects: insufficient wettability of the solder paste, easy to form curing defects. Moreover, the reliability of the solder paste is low under temperature cycle and mechanical fatigue.

Conductive adhesive is an adhesive that has a certain conductivity after curing or drying. Its conductivity is provided by metal particles (usually silver particles) or flakes, while the adhesion is mainly provided by polymers [4]. Due to the limitations of the conductive adhesive by the aging of the resin substrate, the disadvantages are low thermal conductivity, low temperature resistance, limited current carrying capacity, low impact strength [5]. Therefore, the use of conductive adhesive is limited to low temperature and light-duty applications.

The various defects of lead-free solder and conductive adhesive limit their usage in semiconductor high-power, high-temperature and high-density packages, resulting in the emergence of low-temperature sintering technology. As the particle size of the material decreases, especially when the size drops to the nano/micro scale, the melting point will be significantly lower than that of bulk materials [6]. Therefore, the application of nano-silver/copper particles (Ag/Cu NPs) in electronic packaging greatly reduces the die attach

978-1-6654-1392-3/21 $31.00 © 2021 IEEE

temperature, which result in low temperature sintering. The size effect of the sintered Ag/Cu NPs disappears, which turns it suitable for high temperature applications [7]. Nano-silver paste [8] made of Ag NPs has excellent thermal conductivity, which meet the heat dissipation requirements of high-power devices. However, due to the electromigration phenomenon and high cost of silver, the wide application of nano-silver paste as a die attach material has been limited [9]. Copper is cheaper than silver, has good resistance to electromigration, and its electrical and thermal conductivity are similar silver. Fig. 2 [10] shows the paste made of nano-copper. However, nano-copper particles tend to agglomerate and easily oxidized in the air. The resistivity and melting point of copper oxide produced by the oxidation of nano-copper are much higher than that of copper, which ultimately leads to a decrease in the electrical conductivity of the nano-copper particles and an increase in the sintering temperature, which is very detrimental to the application of sintering.

Fig.1 Silver paste [8].   Fig.2 Copper paste [10].

In order to combine the advantages of copper and silver and avoid the shortcomings, research on composite sintering paste made of a mixture of silver and copper has emerged. On one hand, the addition of copper particles inhibits the electrochemical migration of silver. On the other hand, the two sizes of materials sinter to achieve greater density, while reducing the cost of sintering paste. Li [11] mixed Cu and Ag particles with an average particle size of 61.02 and 69.25 nm to prepare nano-paste, and sintered with Ar-H2 shielding gas at 250 °C for 60 minutes at a pressure of 1.12 MPa. As shown in the Fig. 3 below, the surface melting of copper particles was obvious at 250 °C, while no large-scale melting and condensation occurs. Compared with nano-copper, nano-silver particles form a large-area melting, indicating that the sintering performance of nano-silver particles is higher than that of copper. When the copper-silver molar ratio changed to 3:1, the sintering of the composite paste was better than that of pure copper particles. However, porosity still remained high. After increasing the silver content, due to the sintering

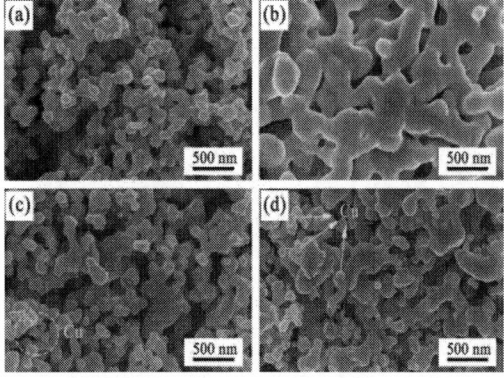

Fig.3 SEM images of sintered (a) Cu, (b) Ag, (c) Cu3-Ag
and (d) Cu2-Ag nanoparticle pastes after sintering at 250 °C [11].

performance of nano-silver, voids were effectively filled up. The shear strength of the joint was 25.41 MPa, which is close to the strength that obtained by pure nano-silver paste (27.03 MPa).

However, mechanical mixing has several disadvantages: only long-term mechanical stirring leads to the two kinds of particles uniformly mixed. If the mixing time is too short, the particles of the two sizes will separate from each other. More importantly, mechanical mixing does not fundamentally solve the oxidation problem of copper: nano-copper will still be exposed to the air and oxidized during the long-term mixing process, which is not conducive to subsequent sintering. In order to avoid the problems of mechanical mixing of copper and silver materials, copper-silver core-shell (Cu@Ag) particles have come into being [12]. Cu@Ag can take into account of the advantages of copper and silver particles (excellent electrical and thermal properties), while overcoming the shortcomings of both (electromigration of silver and oxidation of copper). In binary metal nanoclusters, the interatomic bond length increases as the coordination number of atoms increases. Since the coordination number of internal Cu atoms is greater than that of surface Ag atoms, the bond length between internal Cu atoms and between surface Ag atoms will be closer, thus eliminating stress energy, which leads to the production of copper-silver core-shell structure.

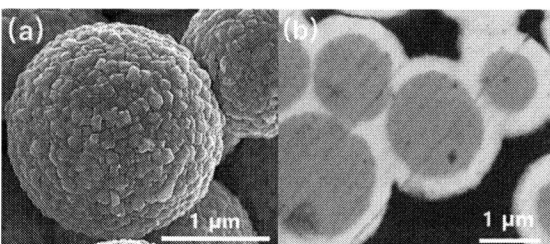

Fig. 4 (a) SEM image of Cu@Ag particles, (b) Cross-sectional BSE
image of Cu@Ag particles [12].

II. PREPARATION OF CU@AG

A. Preparation of particles

Due to the small particle size of copper particles, large surface activity, and easy oxidation, it is necessary to pre-treat the copper powder before coating the silver layer. Generally, it is soaked in dilute sulfuric acid to remove the surface oxide layer, and then washed with distilled water until there is no copper ion residue, and then the silver layer is coated. The preparation of Cu@Ag particles mainly includes physical vapor deposition method, spray pyrolysis method, electrochemical method and chemical synthesis method. Cazayous [13] deposited copper nanoparticles on a thin amorphous carbon film by thermally evaporating copper under ultra-high vacuum, and then deposited a silver layer on its surface, finally obtaining a copper-silver core-shell structure. In the spray pyrolysis reactor system, Jung [14] prepared silver-coated copper particles with different silver contents through a direct liquid-to-particle conversion process, with an average particle size of 0.45 μm. However, the experimental device for preparing Cu@Ag particles by physical method is complicated, which cannot meet the needs of mass preparation in industrial production. At present, the chemical synthesis method uses simple equipments and reaction conditions which shows good development prospects.

Chemical synthesis methods mainly include direct replacement method and liquid phase chemical reduction method.

1) Direct replacement method

The direct displacement method is silver plated by the copper powder particles themselves as a reducing agent for silver ions. The process reaction equation is:

$$Cu + 2[Ag\ (NH_3)_2]^+ = [Cu\ (NH_3)_4]^{2+} + 2Ag\downarrow \quad (1)$$

Luo [15] used silver nitrate as the main salt and prepared a core-shell copper-silver bimetallic powder with an average particle size of 70 nm by a direct replacement method. The powder had good dispersion performance, and the atomic fraction of the silver content on the surface reached 74.28 %. The influence of the process conditions on the coating effect was analyzed. Although the preparation process of direct replacement method was simple and the production cost was low, the silver layer obtained by using this coating method was relatively loose. The silver particles were coated on the surface of the copper particles with a dotted structure. The binding force between them and the copper core was relatively poor, and the antioxidant property of the powder was not desired.

In addition, the composition of the plating solution can be changed by adding complexing agents or additives, which can improve the coating effect of silver on the surface of copper particles. The $[Cu\ (NH_3)_4]^{2+}$ generated during the electroless plating process will be adsorbed on the surface of the copper particles, hindering the continuation of the replacement reaction. Xiao [16] used chemical replacement plating to prepare micron-sized Cu@Ag particles. They used RE-608 copper special extractant as a complexing agent to react with Cu^{2+} to form a complex. The complex was extracted into the organic phase, preventing the formation of $[Cu\ (NH_3)_4]^{2+}$. The oxidation resistance of the final prepared Cu@Ag was improved a lot.

2) Liquid phase chemical reduction method

In order to improve the utilization rate of silver ions in the process of direct replacement and improve the structure of silver coating layer, appropriate reducing agents are often introduced to promote the reduction and deposition of silver particles on the surface of copper powder. The chemical reduction method requires pretreatment of the copper powder particles to remove the surface oxide layer. After sensitization ($SnCl_2$) activation ($PdCl_2$) treatment [17], which can catalyze the reduction of silver ions. Commonly used reducing agents include glucose, ascorbic acid, tartaric acid [18], hydrazine hydrate [19], and formaldehyde. Additives include polyvinyl pyrrolidone (PVP) [19], ethylene diamine tetraacetic acid (EDTA), which play a role in preventing particle agglomeration, adjusting particle size, and promoting uniform precipitation coating. The chemical reduction method is complicated in operation and high in cost. The coating is controllable and stable, which is widely used in the preparation of Cu@Ag.

Yuan [20] used a one-step wet chemical reduction method to prepare Cu@Ag nanoparticles. The reducing agents were glucose and ascorbic acid. The preparation steps were as follows: mix 0.24 mol $CuSO_4 \cdot 5H_2O$, 60 ml ammonia (25 % by weight) and 450 ml of deionized water to obtain solution A. Put solution A into a three-necked flask and heat it to 90°C

in a water bath. Then, 180 ml of glucose solution (1.1 M) was slowly dropped into solution A. Mix 400 ml of ascorbic acid solution (0.375 M) and 16 ml NaOH solution (7 M) to obtain solution B. Slowly drip solution B into A for about 40 minutes under stirring. Finally, 400 ml Ag coating solution containing $AgNO_3$, ammonia and 1 g PVP was slowly dropped into a three-necked flask. Samples with different Ag contents were prepared by changing the Ag/Cu molar ratio of the plating solution. The by-products were removed by rinsing with deionized water and ethanol three times or more, and then the product was filtered and dried in a vacuum oven at 60 °C for 6 hours.

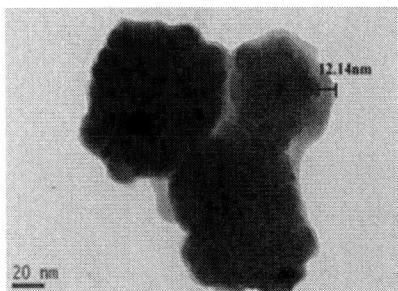

Fig. 5 TEM image of Cu@Ag particles [20].

B. Preparation of paste

After the Cu@Ag particles are perpared, they need to be mixed with organic solvents and dispersants to prepare a paste with a certain viscosity, so as to be coated on the substrate connecting the chip. Hsiao [7] prepared Cu@Ag submicron particles with a particle size of 890 nm by electroless plating, and then mixed with α-terpineol (as a solvent) and ethyl cellulose solution to prepare the paste. The metal particles are 70 wt%, α-terpineol are 20 wt% and ethyl cellulose is 10 wt%. Ji [21] prepared Cu@Ag nanoparticles with a particle size of 70 nm by displacement and chemical reduction methods, and the paste ratio was Cu@Ag NPs: absolute ethanol: deionized water = 90:7:3, ultrasound shake for 5 minutes to improve the dispersibility of the nano powder, and then heat at 60 °C to speed up the evaporation of ethanol. Tian [22] prepared 57.5 nm Cu core and 6.9 nm Ag shell nanoparticles, and used HCl to clean Cu@Ag NPs to remove the PVP layer. Then they were mixed with diethylene glycol (DEG) in a mass ratio of 6:1 and the mixture was stirred ultrasonically for 30 minutes to obtain the expected viscosity and fluidity of the paste. After mixing, it was heated at 60 °C to evaporate the solvent. The shear strength obtained by sintering at 250 °C is 26.5 MPa.

Methods for preparing Cu@Ag paste by mixing organic solvents varies. However, the influence of metal particle proportion, organic solvents and dispersants on the final properties of paste (viscosity, wettability, etc.) has not been thoroughly studied. The viscosity and printability of Cu@Ag paste influenced on the subsequent sintering process. Paste that is too thin or too viscous will cause uneven thickness and poor flatness of the printed layer after printing by the stencil. In addition, the volatilization of organic solvents in the paste of different proportions will directly affect the sintering process. Incomplete volatilization of the organic solvents will leave a large number of pores, making the sintered structure not dense enough thus lead to to low shear strength.

In addition, Kammer [23] selected Cu@Ag particles with an average particle size of 1 μm and an Ag layer thickness of 9 nm, and mixed them with flux to prepare a paste. The flux contains resins, solvents, thickeners and acid activators. Resins were used to control the rheology and surface activity of metal particles and matrixes. Solvents can control the evaporation of fluxes during printing and sintering, and promote contact between particles. Thickeners were used to improve the rheological properties of the paste, and acid activators promote sintering by increasing the activity of the flux. More than 50 kinds of flux formulations have been optimized: flux containing methyl ethyl ketone, isopropanol, methyl pyrrolidone, butyl cellosolve, terpineol, Teckros D85, Troythix XYZ, and malonic acid performed best. The paste exhibited shear thinning behavior and the flux residue after annealing was acceptable, without inhibiting the formation of sintering neck.

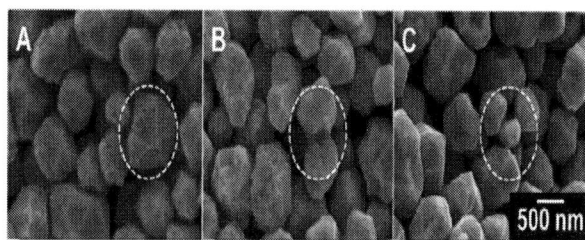

Fig. 6 SEM images of powder samples containing flux after annealing at 205 °C for 5, 30 and 60 min [23].

III. SINTERING PROCESS

A. Sintering mechanism

According to Herring's law of size [24], when the size of metal particles is reduced from micrometers to nanometers, the sintering rate will increase by 12 orders of magnitude, and metal nanoparticles can achieve rapid sintering at lower temperatures. However, this rule no longer applies when nanoparticles appear in a state of agglomeration, so it is important to limit the agglomeration of nanoparticles. In the process of preparing nanoparticles, organic dispersants are often used to prevent agglomeration of nanoparticles during preparation and storage. However, these dispersants hinder the sintering of the nanoparticles during the subsequent sintering process. For nanoparticle sintering, it is important to quickly remove organic dispersants during the sintering process to open the diffusion channels.

B. Sintering methods

Commonly used sintering methods for Cu@Ag paste include temperature assisted pressure sintering, ultrasonic assisted sintering, and pulse current sintering. In temperature assisted pressure sintering, a pressure of more than 10 MPa needs to be applied to the die, which may cause certain damage to the die structures. Most temperature assisted pressure sintering is performed in an inert protective gas atmosphere such as N_2 or Ar. Therefore, pressureless sintering and direct sintering in air have become research hotspots.

1) Temperature assisted pressure sintering

Liu [25] prepared Cu@Ag nanoparticles with an average particle size of 86.15 nm and an Ag layer thickness of 6.8 nm. They were sintered at 225-300 °C for 30 min in argon atmosphere, and the sintering pressure was 5 MPa. The maximum joint strength was 19.7 MPa, which was significantly higher than the joint strength (8.3 MPa) of the sintered copper nano paste without Ag layer. However, the joint structure still contained a lot of pores, as shown in the Fig. 7, indicating that the sintering process is incomplete, resulting in low joint shear strength.

Fig. 7 SEM image of the fracture surface of the joint after sintering at 250 °C [25].

Tian[22] prepared 57.5 nm Cu core and 6.9 nm Ag shell nanoparticles, sintered in ambient air at 250 ℃ for 20 min, the sintering pressure was 5 MPa, and the obtained joint strength was 26.5 MPa.

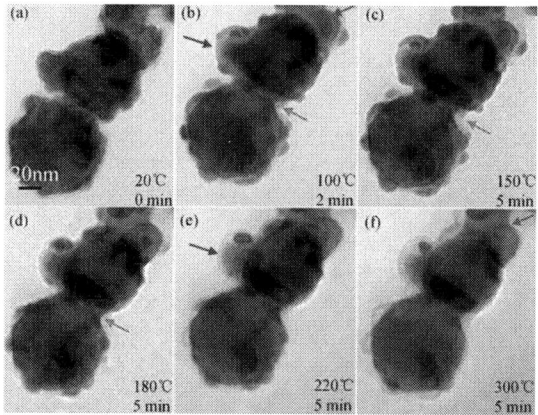

Fig. 8 Sintering process of Cu@Ag particles by TEM heating in-situ [22].

2) Ultrasonic assisted sintering

Ji [20] successfully synthesized Cu@Ag nano paste, and successfully bonded Cu/Cu@Ag NPs paste/Cu in the air through ultrasonic assisted sintering (UAS) at a temperature as low as 160 °C. The sintered microstructure had a dense and crystalline phase, which was completely different from

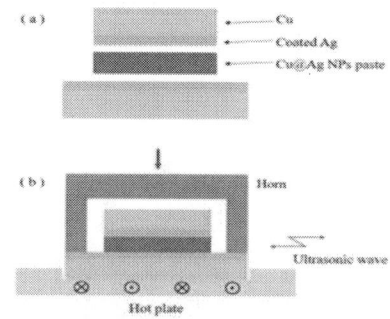

Fig. 9 (a) Cu/ Cu@Ag NPs paste/ Cu sintered sandwich structure diagram, (b) ultrasonic assisted sintering principle [21].

traditional pressure sintering. The shear strength of the joint can reach 54.27 MPa.

In the ultrasonic assisted sintering process, the Cu/Cu@Ag NPs paste/Cu sandwich sample was placed directly at 100 °C for 30 s, then raised to the sintering temperature, and sintered at 150, 160, and 180 °C, respectively. The ultrasonic power was 230 W, the sintering time was 10 s, the ultrasonic frequency was 35 kHz, and the pressure was 0.2 MPa. The experimental schematic diagram is shown in the Fig. 9.

3) Pulsed electric current sintering

Yuan[26] prepared 57.5 nm Cu core and 6.9 nm Ag shell nanoparticles, and prepared a Cu/Cu@Ag NPs paste/Cu sandwich structure using a stencil printing method. Pulsed electric current sintering of Cu@Ag paste to connect Cu-Cu substrate can quickly obtain a high-density and high-strength solder joint structure for metallurgical connection in a short time (less than 200 ms). The shear strength of the sandwich structure increased with the increase of the pulsed electric current. Under a pulsed electric current of 1.0 kA, a shear strength of up to 80 MPa can be obtained.

IV. MOLECULAR DYNAMICS SIMULATION

Due to the size of Cu@Ag nanoparticles, the radius of curvature between the grain boundaries is larger than that of the bulk material, which leads to the reduction of processing temperature. The sintering kinetics is also different from that of the bulk material. The molecular dynamics simulation (MD) software Lammps (Large-scale Atomic/ Molecular Massively Parallel Simulation) is developed by the Sandia Laboratory in the United States for large-scale atomic and molecular dynamics simulations. It performs parallel calculations. The software is free and open source and has high reliability.

A. Two-particle sintering simulation

Wang [27] studied the sintering process of Cu@Ag particles by establishing a double-sphere model. They not only discussed the sintering between two particles of the same size, but also explored the sintering between two unequally large particles. Solid phase sintering is divided into three stages: the first stage is the formation and rapid growth of the sintering neck (independent of the sintering temperature), during which the shrinkage rate rises sharply. The second stage is the growth of the sintering neck related to the sintering temperature, the sintering driving force is reduced after the first stage, so the sintering rate in the second stage is reduced. The third stage is the equilibrium stage, and the shrinkage rate increases slowly. Solid phase sintering is mainly plastic deformation at low temperature, and the surface diffusion becomes the main mechanism of sintering at higher temperature. In addition, two new sintering mechanisms have been discovered in the simulation: (1) Crystallization-amorphization-recrystallization in the solid phase sintering

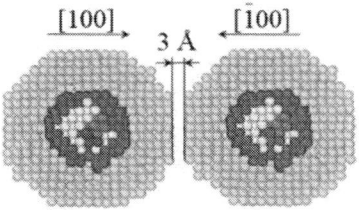

Fig. 10 Double-sphere model of Cu@Ag particles [27].

process; (2) Due to the small size of the two particles and the presence of Cu core, the wetting phenomenon in the sintering process occurs. As the particle size difference increases, the wettability becomes more obvious, so when the particle size difference is large, a denser sintered body can be obtained.

B. Multi-particle sintering simulation

Compared with the double-sphere model, the multi-particle model is closer to the actual sintering process. Wang [28] used a simple cubic (SC) arrangement, as shown in Fig. 11, each simulated box containing 8 nanoparticles. The simulation shows that multi-particle interaction accelerates the sintering process. For smaller core-shell particles, plastic deformation is still the main mechanism of room temperature sintering. At moderate temperature, solid surface diffusion dominates the sintering process. For larger core-shell particles, solid surface diffusion can cause continuous pore shrinkage and elimination.

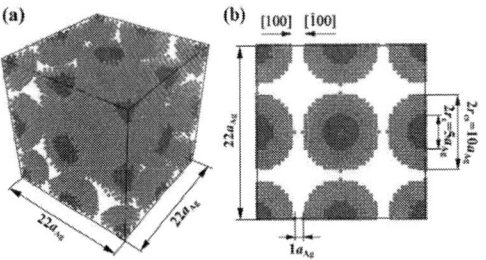

Fig. 11 Multi-particle model of Cu@Ag particles [28].

However, in the actual sintering process, the situation is much more complicated. The close packing of particles (FCC, HCP, etc.) will affect the sintering rate. In addition, the size distribution of Cu@Ag particles is uneven. It is difficult to use molecular dynamics to completely simulate the actual sintering situation. In the existing research, the sintering mechanism can only be explored through simplified models.

V. CONCLUSION

In conclusion, Cu@Ag sintering paste is an excellent material that can be used for advanced packaging which can avoid the shortcomings of pure silver/copper sintering paste. At present, the preparation of Cu@Ag particles in laboratories mostly adopts liquid-phase chemical reduction method, the thickness of the silver layer is controllable. However, due to the limitation of cost in actual production, there is still the problem of incomplete coating. Researches on the organic solvent system of the paste are still lacking. It is extremely important to adjust the organic solvent system for different sintering methods. At present, temperature assisted pressure sintering is the most promising sintering method to be extended to production, and the resulting dense sintered structure provides better reliability. In addition, it is necessary to strengthen the study of molecular dynamics simulation of the sintering process of Cu@Ag particles.

ACKNOWLEDGMENT

The authors would like to thank the funding of Key-Area Research and Development Program of Guangdong Province (Funder ID: 2020B010170002), Science and Technology Innovation Base of Shanghai Municipal Science and Technology Commission, Shanghai SiC Power Device Engineering and Technology Research Center, Fudan

University (Funder ID: 19DZ2253400) and Fudan University-Heraeus cooperation project (Funder ID: 20549).

REFERENCES

[1] V. R. Manikam and K. Y. Cheong, "Die attach materials for high temperature applications: a review," IEEE Trans. Components, Packag. Manuf. Technol., vol. 1, no. 4, pp. 457–478, 2011.

[2] H. S. Chin, K. Y. Cheong, and A. B. Ismail, "A review on die attach materials for SiC-based high-temperature power devices," Metall. Mater. Trans. B-Process Metall. Mater. Process. Sci., vol. 41, no. 4, pp. 824–832, 2010.

[3] I. H. Tavman and T. Evgin, "Thermally conductive polymer nanocomposites for thermal management of electronic packaging," in 2017 IEEE 23rd International Symposium for Design and Technology in Electronic Packaging (SIITME), pp. 64–67, 2017.

[4] R. S. Rorgren and J. Liu, "Reliability assessment of isotropically conductive adhesive joints in surface mount applications," IEEE Trans. Components, Packag. Manuf. Technol. Part B, vol. 18, no. 2, pp. 305–312, 1995.

[5] H. P. Wu et al., "High conductivity of isotropic conductive adhesives filled with silver nanowires," Int. J. Adhes. Adhes., vol. 26, no. 8, pp. 617–621, 2006.

[6] S. A. Paknejad, A. Mansourian, J. Greenberg, K. Khtatba, L. Van Parijs, and S. H. Mannan, "Microstructural evolution of sintered silver at elevated temperatures," Microelectron. Reliab., vol. 63, 2016.

[7] C.-H. Hsiao, W.-T. Kung, J.-M. Song, J.-Y. Chang, and T.-C. Chang, "Development of Cu-Ag pastes for high temperature sustainable bonding," Mater. Sci. Eng. A-STRUCTURAL Mater. Prop. Microstruct. Process., vol. 684, pp. 500–509, 2017.

[8] J. G. Bai, J. N. Calata, and G. Lu, "Processing and characterization of nanosilver pastes for die-attaching SiC devices," vol. 30, no. 4, pp. 241–245, 2007.

[9] H. Alarifi, A. Hu, M. Yavuz, and Y. N. Zhou, "Silver nanoparticle paste for low-temperature bonding of copper," J. Electron. Mater., vol. 40, no. 6, pp. 1394–1402, 2011.

[10] J.-L. Jo, K. Anai, S. Yamauchi, and T. Sakaue, "The properties of Cu sinter paste for pressure sintering at low temperature," in 2019 IEEE 69th Electronic Components and Technology Conference (ECTC), 2019, pp. 76–80, 2019.

[11] J. Li et al., "Depressing of Cu-Cu bonding temperature by composting Cu nanoparticle paste with Ag nanoparticles," J. Alloys Compd., vol. 709, pp. 700–707, 2017.

[12] E. B. Choi and J.-H. Lee, "Dewetting behavior of Ag in Ag-coated Cu particle with thick Ag shell," Appl. Surf. Sci., vol. 480, pp. 839–845, 2019.

[13] M. Cazayous, C. Langlois, T. Oikawa, C. Ricolleau, and A. Sacuto, "Cu-Ag core-shell nanoparticles: A direct correlation between micro-Raman and electron microscopy," Phys. Rev. B, vol. 73, no. 11, 2006.

[14] D. S. Jung, H. M. Lee, Y. C. Kang, and S. B. Park, "Air-stable silver-coated copper particles of sub-micrometer size," J. Colloid Interface Sci., vol. 364, no. 2, pp. 574–581, 2011.

[15] J. S. Luo et al., "Preparation of coated nano-copper-silver bimetal powder by direct displacement method," High Power Laser Part. Beams, vol. 53, no. 9, pp. 1689–1699, 2013.

[16] X. G. Cao and H. Y. Zhang, "Preparation of silver-coated copper powder and its oxidation resistance research," Powder Technol., vol. 226, pp. 53–56, 2012.

[17] D. Zhang et al., "Preparation and Application of Cu-Ag Composite Preforms for Power Electronic Packaging," in 2019 IEEE 69th Electronic Components and Technology Conference (ECTC), pp. 63–68, 2019.

[18] K. Chen, D. Ray, Y. Peng, and Y.-C. Hsu, "Preparation of Cu-Ag core-shell particles with their anti-oxidation and antibacterial properties," Curr. Appl. Phys., vol. 13, no. 7, pp. 1496–1501, 2013.

[19] J. Wen and Y. Tian, "The Synthesis of Cu-Ag core-shell bimetallic nanoparticles for IC bonding," in 2016 17th International Conference on Electronic Packaging Technology (ICEPT), pp. 788–790, 2016.

[20] Y. Yuan, H. Xia, Y. Chen, and D. Xie, "One-step synthesis of oxidation-resistant Cu@Ag core–shell nanoparticles," Micro Nano Lett., vol. 13, no. 2, pp. 171–174, 2018.

[21] H. Ji and M. Li, "Mechanism of ultrasonic-assisted sintering of Cu@Ag NPs paste in air for high-temperature power device packaging," 2018 IEEE 68th Electronic Components and Technology Conference (ECTC). pp. 1270–1275, 2018.

[22] Y. Tian et al., "Sintering mechanism of the Cu-Ag core-shell nanoparticle paste at low temperature in ambient air," RSC Adv., vol. 6, no. 94, pp. 91783–91790, 2016.

[23] M. J. Kammer, A. Muza, J. Snyder, A. Rae, S. J. Kim, and C. A. Handwerker, "Optimization of Cu–Ag core–shell solderless interconnect paste technology," IEEE Trans. Components, Packag. Manuf. Technol., vol. 5, no. 7, pp. 910–920, 2015.

[24] C. Herring, "Diffusional viscosoty of a polycrystalline solid," J. Appl. Phys., vol. 21, no. 5, pp. 437–445, 1950.

[25] J. Liu, Y. Mou, Y. Peng, and M. Chen, "Oxidation-resistant Cu-Ag core-shell nanoparticle paste for high temperature electronic packaging," in 2019 20th International Conference on Electronic Packaging Technology (ICEPT), pp. 1–4, 2019.

[26] Y. Huang, C. Hang, Y. Tian, C. Wang, and H. Zhang, "Rapid sintering of nano copper-silver core-shell paste and interconnection of copper substrates by pulse current," Jixie Gongcheng Xuebao/Journal Mech. Eng., vol. 55, no. 24, pp. 51–56, 2019.

[27] J. Wang, S. Shin, and A. Hu, "Geometrical effects on sintering dynamics of Cu-Ag core-shell nanoparticles," J. Phys. Chem. C, vol. 120, no. 31, pp. 17791–17800, 2016.

[28] J. Wang and S. Shin, "Sintering of multiple Cu-Ag core-shell nanoparticles and properties of nanoparticle-sintered structures," RSC Adv., vol. 7, no. 35, pp. 21607–21617, 2017.

Mechanical Behavior of Lead-free Solder at High Temperatures and High Strain Rates

Xu Long*
School of Mechanics, Civil Engineering and Architecture
Northwestern Polytechnical University
Xi'an, China
email: xulong@nwpu.edu.cn

Tianxiong Su
School of Mechanics, Civil Engineering and Architecture
Northwestern Polytechnical University
Xi'an, China
email:1439199980@qq.com

Chao Chang
School of Applied Science
Taiyuan University of Science and Technology
Taiyuan, China
email: cc@tyust.edu.cn

Jiaqiang Huang
School of Mechanical and Electrical Engineering
Guilin University of Electronic Technology
Guilin, China
email: huangjiaqiang@guet.edu.cn

Xiaotong Chang
School of Mechanics, Civil Engineering and Architecture
Northwestern Polytechnical University
Xi'an, China
email: xtchang@nwpu. edu. cn

Yutai Su
School of Mechanics, Civil Engineering and Architecture
Northwestern Polytechnical University
Xi'an, China
email: suyutai@buaa.edu.cn

Hongbin Shi
Xi'an Institute Consumer BG, DFR Department
HUAWEI DEVICE CO LTD
Xi'an, China
email: shihongbin2@huawei.com

Tao Huang
Space Research Institute of Electronics and Information Technology
Aerospace Science and Technology Corporation
Xi'an, China
email: huangt89@163.com

Yanpei Wu
Space Research Institute of Electronics and Information Technology
Aerospace Science and Technology Corporation
Xi'an, China
email: wuyanpei2007@126.com

Abstract—While moving ahead with science and technology, the service conditions of microelectronic devices are becoming more and more complicated, and the performance requirements of packaging materials are also increasing especially for hash applications. The dependability of solder joints in electronic chips under high temperatures and high strain rates have attracted much attention. In this paper, experimental studies are carried out on the dynamic behavior of Sn-3.0Ag-0.5Cu (SAC305) materials. The dynamic compression experiments are operated by using the Split-Hopkinson Pressure Bar (SHPB) with a heating furnace to achieve the target temperature condition. The true stress- strain curves are received by this system at a temperature of 343K and the strain rates from 959 s^{-1} to 1961s^{-1}. Additionally, in order to investigate the deformation mechanism under impact scenarios, an optical microscopy is utilized to characterize the microstructure of the SAC305 lead-free solder specimen at high temperatures and high strain rates.

Keywords—Dynamic, SHPB, strain rate, temperature, microstructure

I. INTRODUCTION

In the information age, the electronic industry has developed rapidly, and consumer electronic products have been widely used in modern society. The working environment adaptability of electronic products is constantly improving. In the entire packaging structure, solder joints are irreplaceable and have an important effect in mechanical connection. Therefore, a comprehensive understanding of the solder joint materials is essential. When some unexpected harsh situations such as collision, crash and explosion occur,

the strain rate of solder joint structure increases rapidly, which is much higher than that under the normal conditions. In addition, the increase in temperature is also one of the reasons why the stability of the solder joints is reduced. Therefore, in the field of electronic packaging, mechanical behaviors of lead-free solders at high temperatures and high strain rates are getting more concerned, which have great reference values for industrial production applications.

So far, many researchers have employed different methods to carry out intensive studies on the solder material properties in different working environments, and have made outstanding contributions. For understanding the influence of strain rate, Long et al.[1, 2] studied the material properties of SAC305 solder through static tensile experiments and nanoindentation methods. To implement the experimental conditions of high strain rates, SHPB has become a mature way to be utilized to measure the dynamic material properties, especially for metals[3-5]. Moreover, a furnace device is added to the SHPB system in order to increase the specimen temperature to achieve high temperature experiments[6, 7]. Dou et al. also summarized the method of using the SHPB system for high-temperature experiments[8]. Long et al. used the SHPB device to study the dynamic response of SAC305 solder under high strain rate conditions[9]. The high temperature environment has continuously improved the reliability requirements of solder joint materials, so the high temperature experiment of SAC305 is extremely anticipated.

In this paper, the dynamic mechanical properties of SAC305 solder are researched by a split-Hopkinson pressure bar (SHPB) added a furnace device. Through changing the impact speed of the striker bar, the plastic strain rate during the experiment is increased from 959s^{-1} to 1961s^{-1}. Meanwhile,

---

Natural Science Foundation of Shaanxi Province (No. 2021KW-25) and Astronautics Supporting Technology Foundation of China (No. 2021-HT-XG).

978-1-6654-1392-3/21 $31.00 © 2021 IEEE

the target temperature of the experimental specimens is adjusted to 343K by controlling the furnace temperature. The constitutive relationship of lead-free solder specimens under high temperatures and high strain rate conditions is described by the true stress-strain curves. Finally, by using an optical microscope, the deformation mechanism of SAC305 under experimental conditions is further revealed from the perspective of microstructure morphology and metallographic changes.

II. EXPERIMENT MATERIAL AND METHOD

As shown in Fig. 1 (a) and (b), SAC305 lead-free solders in this research are machined to be the shape of cylinders with the height of 6mm and the thickness of 5mm. Then, the cylindrical specimens are annealed at the annealing temperature of 398K for 12 hours[9], the purpose of the operation is to alleviate the residual stress caused in machining process and acquire the homogenous microstructure of solder specimens. Later, the solder specimens are deposited at a room temperature space for twenty days before executing the dynamic compression experiments. Some representative specimens are also shown in the Fig. 1 (c), (d) and (e) after the impact of 7m/s, 11m/s and 14m/s. Apparently, the surface of the material specimens is found to be deformed due to the impact deformation, so the morphology is also investigated in the later section.

In Fig. 2, a striker bar, an incident bar and a transmitted bar constitute the central part of SHPB system. And that these three components are all made of steel and all their diameter is 14.5 mm. Compared with previous ambient testing conditions, a heating furnace is incorporated in the SHPB system to achieve the target temperature condition during the experiments. In the process of experiments, the incident bar is impacted by the striker bar at a speed caused by a gas gun. A one-dimensional elastic compression wave (stress wave) is created in the incident bar and spreads forward along the bar after impact. The experimental specimen is clamped between two long bars, which are the incident bar and the transmitted bar. Both of the end faces of the specimen are in close contact with the cross-sections of the two bars. When the elastic compression wave arrives the right end of the incident bar, the wave will be transferred into two forms of waves. One form of the wave is sent back into the incident bar, and it is named as the reflected wave. The other wave form is called as the transmitted wave and continues to pass the specimen and enter the transmitted bar. The propagation of each wave causes deformation of strain gauge appressed to the middle of the corresponding bar. Subsequently, the strain signals are converted into voltage signals through the connection of strain gauge and Wheatstone bridge. Finally, the voltage signals are captured by the signal receiving system.

As shown in Fig. 3 (a), the experimental specimen is mounted as a matter of convenience to improve the efficiency of grinding and polishing. In this process, SiC abrasive paper is used to grind the specimen roughly. Then, the specimens are further polished by a polishing machine for a smother surface. After the surface preparations, the specimens are etched in the corrosion solution composed with 2%HCl, 3%HNO$_3$ and 95% ethanol absolute for several seconds. An optical microscope is used to observe the microstructures of the specimens under different impact speeds showed in Fig. 3 (b).

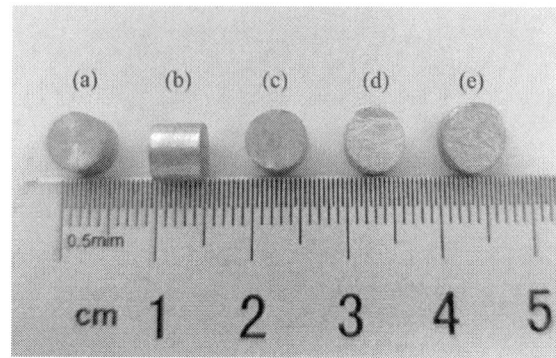

Fig. 1. The specimens of SAC305 solder under different impact speeds. (a) and (b) annealed specimens before tests, (c) specimens after the impact of 7m/s, (d) specimens after the impact of 11m/s, and (e) specimens after the impact of 14m/s.

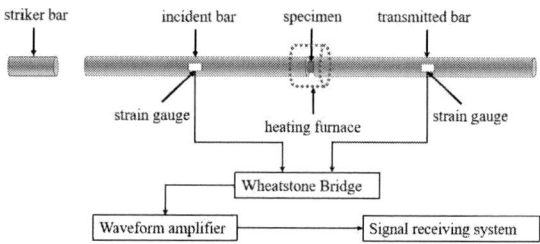

Fig. 2. The schematic of SHPB experimental device.

III. RESULT AND DISCUSSION

Through the Split-Hopkinson Pressure Bar experiments, voltage signals which are the result of the interaction between the strain gauge and the Wheatstone bridge can be obtained directly from the computer of this system. Furthermore, the strains of the incident bar and the transmitted bar can be expressed by Eq. (1).

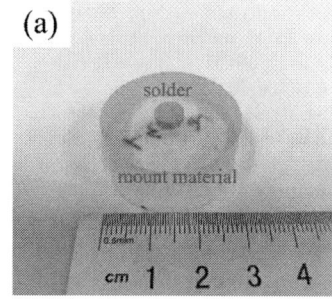

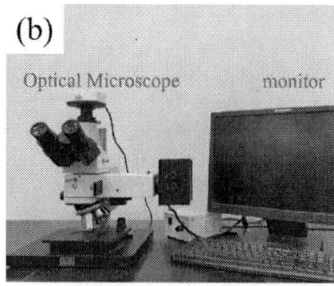

Fig. 3. (a) Mounted experimental specimen, (b) Optical microscope.

$$\varepsilon = \frac{2\Delta U}{k_1 k_2 U_0} \qquad (1)$$

where ε is the strain signal, ΔU is the voltage signal, k_1 is the strain gauge sensitivity factor, k_2 is the dynamic strain gauge magnification factor, U_0 is the initial voltage of the Wheatstone bridge. Afterward, the engineering stress (σ_s) and the engineering strain (ε_s) of specimen can be calculated by Eq. (2) and Eq. (3).

$$\sigma_s = \frac{A_0}{A_s} E \varepsilon_t \qquad (2)$$

$$\varepsilon_s = \frac{2C_0}{L_s} \int_0^t \varepsilon_r \, dt \qquad (3)$$

where A_0 and A_s are the cross-sectional areas of the bars and experimental specimen, respectively. ε_r and ε_t are strains caused by reflected pulses and transmitted pulses, respectively. E is the elastic modulus of the steel. C_0 is the speed of the compression stress wave in bars. L_s is the thickness of the specimen. Subsequently, the true stress and true strain can be obtained based on the relationship of engineering stress, engineering strain and true stress, true strain.

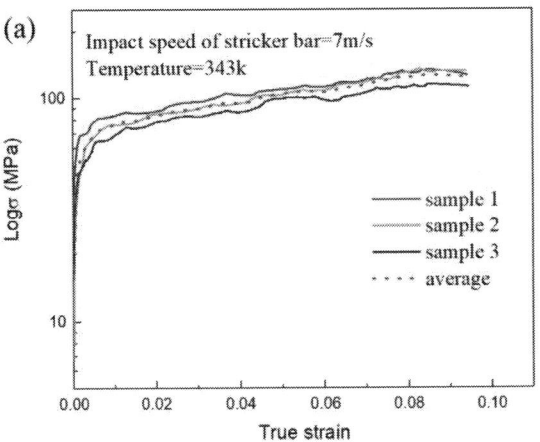

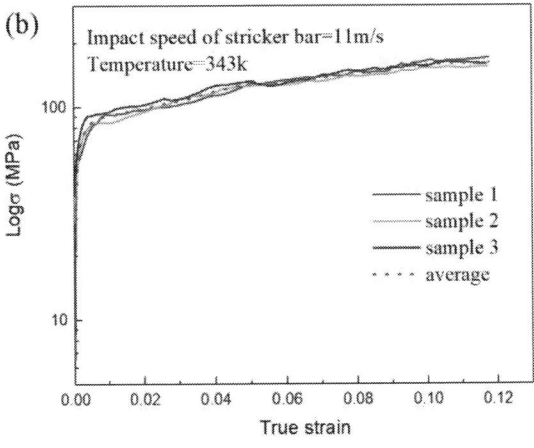

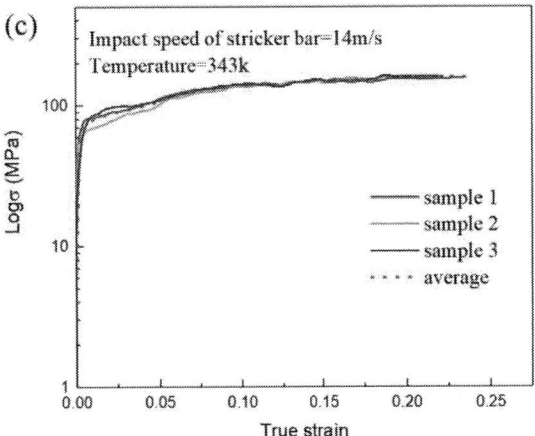

Fig. 4. Constitutive relationship of specimens under different impact speeds of striker bar at the temperature of 343k. (a) 7m/s, (b) 11m/s, (c) 14m/s.

TABLE I.   MECHANICAL PROPERTIES OF SPECIMENS UNDER DIFFERENT IMPACT SPEEDS OF STRIKER BAR AT THE TEMPERATURE OF 343K

Impact speed (m/s)	Maximum true stress (MPa)	Maximum true strain	Plastic strain rate(s^{-1})
7	127.1	0.094	959
11	161.1	0.117	1082
14	159.3	0.235	1961

The constitutive relationship of SAC305 lead-free solder specimens under different impact speeds of striker bar conditions at a high temperature of 343K have been provided in Fig. 4. For each set of impact velocity, it is obvious in Fig. 4 that the experimental results of the three specimens are stable with the difference within the acceptable extent. On the basis of these curves, the maximum true stresses and maximum true strains are obtained conveniently, which have been listed in Table I. At the same time, the corresponding plastic strain rates produced by experiments are also shown in the Table I.

According to Table I, it can be found that the strain rate gradually becomes higher as the impact velocity increases. Nevertheless, the maximum true stress first intensifies and then weakens with the aggravate of impact speed. When the impact speed increases from 7m/s to 11m/s, the maximum true stress increases from 127.1MPa to 161.1MPa. When the impact speed continues to rise to 14m/s, the maximum true stress starts to decrease. In addition, the maximum true strain

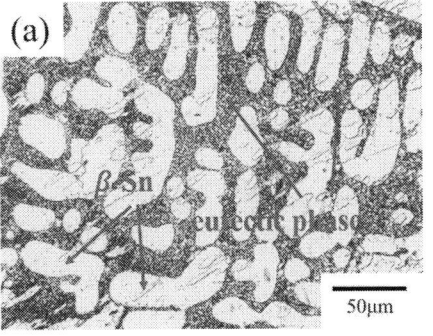

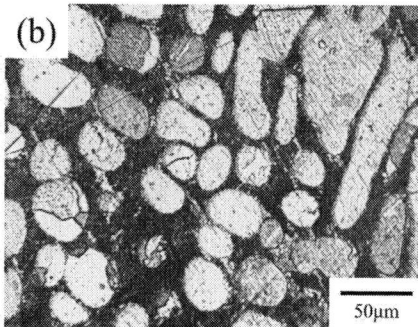

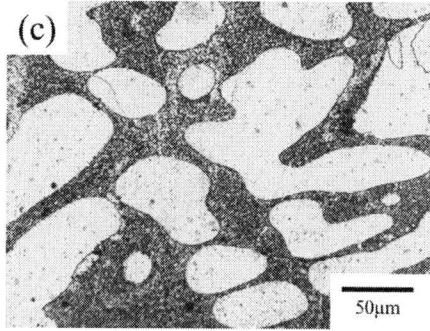

Fig. 5. Microstructure of specimens under different impact speeds of striker bar at the temperature of 343k. (a) 7m/s, (b) 11m/s, (c) 14m/s.

has a positive correlation with impact speed, which increases as the impact speed is enhanced.

So as to characterize the dynamic response of SAC305 solder from the microstructure, the cross-section of the specimen under various working conditions was observed with an optical microscope. It can be seen in Fig. 5 that the microstructures of the specimens are different by various impact speeds. By observing fig. 5 (a), (b) and (c), it is found that the microstructure of solder between β-Sn and eutectic phase is different under changing impact velocity. When the impact speed increases from 7m/s to 11m/s, the microstructure becomes refined when comparing Fig. 5 (a) and (b). This refining phenomenon may be caused by the rapid increase of deformation rate in the specimen due to the instantaneous impact. It is manifest from the fig. 5 (b) and (c), with the impact speed continues to increase to 14m/s, the microstructure becomes coarsened. This coarsening phenomenon may be due to the instant increase of the impact speed that causes the local temperature inside the specimen to exceed the melting point, which intensifies the plastic deformation of the specimen. The specimen local temperature increases rapidly under high strain rate is also a reason for the maximum true stress decreases due to the strength deterioration. By combining the microscopic morphologies with the curves of constitutive relationship, it can be straightforward found that the microstructure changes of SAC305 solder promote the strain rate hardening and ductility enhancement of the solder joint material[9-11]. In order to acquire the microscopic morphology changes at the maximum true stress in more detail, the further researches should to be done near the impact speed of 11m/s.

IV. CONCLUSIONS

In this research, the dynamic mechanical characteristics of SAC305 specimens under different impact speeds at an elevated temperature of 343K are studied, which could promote the understanding of deformation mechanism of SAC305 specimens under the conditions of high temperatures and high strain rates. Different strain rates of the experimental specimen are realized by changing the impact speeds of the striker bar. The experimental results show that at the elevated temperature, the maximum true stress of the specimen first increases and then decreases when the strain rate gradually strengthens, but the maximum true strain could continue to increase. In this process, the material characteristic of strain rate hardening is fully reflected. Simultaneously, the ductility of the material also expand with the growth of the strain rate. What's more, the deformation mechanism is also explained with the observations of microscopic morphology.

ACKNOWLEDGMENT

This work was supported by the Natural Science Foundation of Shaanxi Province (No. 2021KW-25) and the Astronautics Supporting Technology Foundation of China (No. 2021-HT-XG).

REFERENCES

[1] X. Long, X. D. Zhang, W. B. Tang, S. B. Wang, Y. H. Feng and C. Chang "Calibration of a constitutive model from tension and nanoindentation for lead-free solder," Micromachines, vol. 9, no. 11, 2018.

[2] X. Long, W. B. Tang, S. B. Wang, X. He, and Y. Yao, "Annealing effect to constitutive behavior of Sn–3.0Ag–0.5Cu solder," Journal of Materials Science Materials in Electronics, 2018.

[3] S. Acharya, R. K. Gupta, J. Ghosh, S. Bysakh, K. S. Ghosh, D. K. Mondal, and A. K. Mukhopadhyay, "High strain rate dynamic compressive behaviour of Al6061-T6 alloys," Materials Characterization, vol. 127, pp. 185-197, 2017.

[4] D. Chan, N. Xu, D. Bhate, G. Subbarayan, and I. Dutta, "High strain rate behavior of Sn3.8Ag0.7Cu solder alloys and its influence on the fracture location within solder joints," in ASME 2009 InterPACK Conference collocated with the ASME 2009 Summer Heat Transfer Conference and the ASME 2009 3rd International Conference on Energy Sustainability, 2009.

[5] X. Zhang, Y. Zhu, Y. Yang, and C. Xu, "Mechanical behavior of lead at high strain rates," Journal of Materials Engineering and Performance, pp. 1-7, 2019.

[6] M. L. Hu, W. D. Song, D. B. Duan, and Y. Wu, "Dynamic behavior and microstructure characterization of TaNbHfZrTi high-entropy alloy at a wide range of strain rates and temperatures," International Journal of Mechanical Sciences, p. 105738, 2020.

[7] F. Zhang, Z. Liu, Y. Wang, P. L. Mao, X. W. Kuang, Z. L. Zhang, Y. D. Ju, and X. Z. Xu"The modified temperature term on Johnson-Cook constitutive model of AZ31 magnesium alloy with {0002} texture," Journal of Magnesium and Alloys, vol. 8, no. 1, pp. 172-183, 2020.

[8] Q. B. Dou, K. R. Wu, T. Suo, C. Zhang, X. Guo, Y. Z. Guo, W. G. Guo, and Y. L. Li, "Experimental methods for determination of mechanical behaviors of materials at high temperatures via the split Hopkinson bars," Acta Mechanica Sinica, pp. 1-19, 2020.

[9] X. Long, J. M. Xu, S. B. Wang, W. B. Tang, and C. Chang, "Understanding the impact response of lead-free solder at high strain rates," International Journal of Mechanical Sciences, vol. 172, 2020.

[10] X. Long, Y. H. Feng, and Y. Yao, "Cooling and annealing effect on indentation response of lead-free solder," International journal of applied mechanics, 2017.

[11] S. B. Wang, Y. Yao, and X. Long, "Critical review of size effects on microstructure and mechanical properties of solder joints for electronic packaging," Applied Sciences, vol. 9, no. 2, 2019.

Employing Single-Crystal Cobalt Substrates to Control βSn Grain Orientations in Solder Interconnections

Ce Li
Beijing Institute of Technology
School of Materials Science and Engineering
National Key Laboratory of Science and Technology on Materials under Shock and Impact
Beijing, China
920770358@qq.com

Xufeng Chang
Beijing Institute of Technology
School of Materials Science and Engineering
National Key Laboratory of Science and Technology on Materials under Shock and Impact
Beijing, China
3220191010@bit.edu.cn

Bingguang Wang
Beijing Institute of Technology
School of Materials Science and Engineering
National Key Laboratory of Science and Technology on Materials under Shock and Impact
Beijing, China
3220191067@bit.edu.cn

Zhaolong Ma
Beijing Institute of Technology
School of Materials Science and Engineering
National Key Laboratory of Science and Technology on Materials under Shock and Impact
Beijing, China
z.l.ma@bit.edu.cn

Xingwang Cheng
Beijing Institute of Technology
School of Materials Science and Engineering
National Key Laboratory of Science and Technology on Materials under Shock and Impact
Beijing, China
chengxw@bit.edu.cn

Abstract—Lead-free solder joints on the traditional Cu substrates usually contain few βSn grains with random orientations. Due to the strong anisotropy of βSn, some common reliability issues of solder joints such as electromigration are related to βSn grain orientations. In this paper, we proved that βSn grain orientations can be effectively controlled using single-crystal Co substrates through adjusting the interfacial $\alpha CoSn_3$ morphologies. Two single-crystal Co substrates, $(11\bar{2}0)Co$ and $(10\bar{1}0)Co$, were used in this study. The textures of interfacial $\alpha CoSn_3$ were observed by selective etching. The grain orientations of the interfacial $\alpha CoSn_3$ and βSn were examined by electron backscatter diffraction (EBSD). The result indicated that interfacial $\alpha CoSn_3$ presented 2 or 4 dominant orientations related to single Co with fixed orientation relationships (ORs). The interfacial atomic mismatches and crystal growth kinetics represented by the angle between the substrate plane and $(100)CoSn_3$ were analyzed to understand the mechanism of the orientation selection of $\alpha CoSn_3$. On $(11\bar{2}0)Co$, there are only 20 βSn orientations including both single and twinned grains. On $(10\bar{1}0)Co$, there was no [001]Sn perpendicular to the substrate plane, which ought to improve the reliability of solder interconnections

Keywords—Texture; Orientation relationships; EBSD; Interface; Crystal structure

I. INTRODUCTION

Lead-free solder joints on common substrates (e.g. Cu) often contain very few βSn grains with random orientations[1, 2]. Since βSn is highly anisotropic in thermal expansion coefficients, elastic modulus, and diffusivities, each joint is unique on thermomechanical properties and electromigration behaviors[3, 4]. For example, many studies have shown that when the [001]Sn is parallel to the electron flow direction, the solder interconnections will experience the most severe electromigration and tend to fail early[5]. Therefore, it is highly desired to develop a method that can control βSn orientations in solder joints.

The βSn grain orientation is determined by its nucleation in solder joints. When the liquid solder alloy was connected to the substrate, the interfacial intermetallic compounds (IMCs) are generated first, and then with a certain degree of nucleation undercooling, βSn will preferentially nucleate and grow on or near the IMCs. On traditionally used polycrystalline Cu substrates, the interfacial IMCs are Cu_6Sn_5, which have an uneven "scallop-type" morphology that forms conical grooves with different sizes. This geometric structure will reduce the free energy of βSn nucleation and promote βSn to nucleate and grow on or near the interfacial Cu_6Sn_5. However, the interfacial Cu_6Sn_5 grown on the polycrystalline Cu substrate does not have strong textures, and Cu_6Sn_5 is not a good nucleation phase for βSn, which leads to the random orientations of βSn in each solder joint[6].

Compared with the traditional Cu substrates, the solidification of solder alloys on Co, Pd and Pt requires less nucleation undercooling (<4 K), and studies have shown that the IMCs on these substrates are $oS32$-$\alpha CoSn_3$, $oS20$-$PdSn_4$, and $oS20$-$PtSn_4$, which were all shown to be potent nucleants for βSn[7]. For example, studies have shown that when βSn nucleates on an $\alpha CoSn_3$ particle, there will be two reproducible ORs[8]:

$$(100)CoSn_3\|(100)/(010)Sn; \quad [010]/[001]CoSn_3\|[001]Sn \quad (1)$$

Solder joints on polycrystalline Co substrates have textured interfacial $\alpha CoSn_3$ with [100] approximately parallel to the substrate. When βSn in these joints nucleate on $\alpha CoSn_3$, the overall βSn orientations are still widely distributed due to the weak $\alpha CoSn_3$ textures and the twinning of βSn[9]. Studies have shown that when single crystal substrates are used, the interfacial IMCs in solder joints can exhibit strong textures[10]. For example, on the (111)Cu, the interfacial Cu_6Sn_5 forms the roof morphology with the intersection angle of 60°, and on the (001)Cu, 90° crossed Cu_6Sn_5 roofs were also observed. Inspired by these studies, it is expected that strong textured interfacial $\alpha CoSn_3$ could also form if the joints are produced on single Co, and when βSn nucleate and grow on these αCoSn₃, its grain orientations could be effectively controlled.

To test this hypothesis, in this present work, we demonstrated the influence of single Co substrates on βSn structures and grain orientations in BGA solder interconnections. The morphologies of interfacial IMCs were studied first, and the preferred ORs between the IMCs and Co were measured. The interfacial atomic mismatches and crystal growth kinetics, which can be represented by the angle between the substrate plane and $(100)CoSn_3$ were used to explain the orientation selection mechanisms of the interfacial IMCs. Then, the βSn structures and orientations on $(11\bar{2}0)Co$ and $(10\bar{1}0)Co$ were studied and corresponded to the interfacial IMCs textures. Finally, the mechanism governing the orientation symmetry of IMCs and βSn was discussed.

II. MATERIAL AND METHODS

A. Preparation for solder balls

In this work, laboratory-produced Sn3.0wt%Ag0.5wt%Cu (SAC305) and commercial pure tin solder alloys were employed. Compositions were measured by X-Ray Fluorescence spectroscopy (XRF) and given in Table I. The Sn/SAC305 ingot was first rolled into a foil with a thickness of 0.05 mm, and then punched into a disc with a diameter of 1.6 mm, and heated on a quartz plate with ROL1 viscous flux, and finally the solder balls with a diameter of 550 ± 25 μm were produced attribute to the surface tension.

TABLE I.   COMPOSITIONS OF PURE TIN AND SAC305 (WT.%)

Alloys	Sn	Sb	Cu	Fe	Ag	Ni
SAC305	96.19	0.0146	0.54	0.0045	3.07	0.011
Pure Sn	99.89	<0.001	0.00196	0.024	<0.001	0.00254

B. Preparation for solder joints

To prepare solder joints, firstly, the single-crystal Co substrates with their crystal planes $(11\bar{2}0)Co$ and $(10\bar{1}0)Co$ respectively were coated with solder resist, and then exposed to prepare pads with a diameter of 500 μm. The Sn/SAC305 solder balls were soldered on the HARRIS Stay-Clean flux-coated pads in the reflow oven, and the thermal profile is shown in Fig. 1.

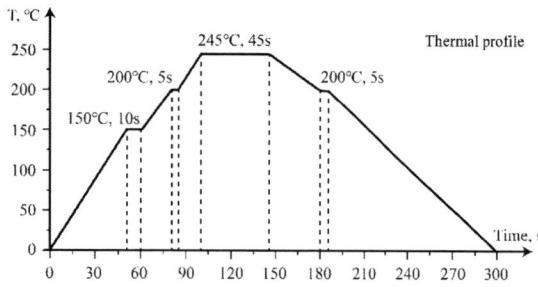

Fig.1 Thermal profile.

C. Microstructure characterization

In order to reveal the interfacial IMCs morphology, the solder joints were etched in 5% NaOH and 3.5% ortho-nitrophenol with deionized water to remove excess βSn. To prepare samples for microstructure analysis and measuring orientations, the solder joints were cold-mounted, wet ground and finally polished carefully sequentially. The solder joints were subjected to cross/ longitudinal section and observation using a thermo Scientific Quanta field emission gun scanning electron microscope (FEG-SEM) equipped with a Nordlys electron backscatter diffraction (EBSD) detector (Oxford Instruments) and an Oxford Instruments INCA x-sight energy dispersive x-ray (EDX) detector. The EBSD patterns were processed by HKL CHANNEL5 software and a series of matlab codes were developed to analyze the orientation data.

III. RESULT AND DISCUSSIONS

A. Morphologies and grain orientations of interfacial IMCs

As we can see in Figs. 2(a) and 3(a), the interfacial IMC plates on two different single-crystal Co substrates were identified as oS32-αCoSn₃ with strong textures, which are significantly different from those on polycrystalline Co substrates[9].

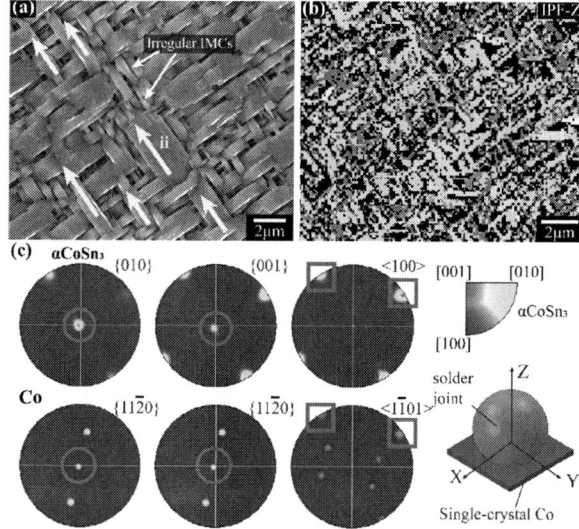

Fig. 2 (a) SEM image of interfacial αCoSn₃ on $(11\bar{2}0)Co$; (b) EBSD-IPFZ map of αCoSn₃; (c) Pole figures of αCoSn₃ and Co.

According to Fig. 2(a), we can find that the interfacial αCoSn₃ on $(11\bar{2}0)Co$ has two dominant orientations, which are marked by red and yellow arrows. The (100) crystal planes (i.e. the largest facet of the plates) of the two kinds of αCoSn₃ have an angle of ~90° and the [010]/[001] (i.e. corners of crystals) of most plates is upward, which is agreed with the pole figures in Fig. 2(c). The cross sections of joints were parallel with the single Co and the interfacial IMCs (i.e. the XY plane) was determined by EBSD. The typical EBSD-IPFZ map and the corresponding Co and αCoSn₃ pole figures are respectively presented in Fig. 2(b) and (c). As shown by the same circles and squares, (010) or (001)CoSn₃ is parallel with the Co substrate surface $(11\bar{2}0)$ and [100]CoSn₃ is parallel with $[1\bar{1}01]Co$. The preferred ORs can be expressed by Equation (2):

$$(11\bar{2}0)Co\|(010)/(001)CoSn_3 \text{ with } [1\bar{1}01]Co\|[100]CoSn_3 \quad (2)$$

The interfacial αCoSn₃ formed on $(10\bar{1}0)Co$ presents 4 dominant orientations (labeled as i, ii, iii, and iv) as shown in Fig. 3(a), which are marked by yellow, blue, red, and green arrows, too. Orientations i-ii, i-iii, ii-iv, and iii-iv are all mirror symmetries. The EBSD map of αCoSn₃ is presented in Fig. 3(b) and the corresponding Co and αCoSn₃ pole figures are listed in Fig. 3(c). The parallel planes and directions are highlighted using same hexagons, triangles, squares and circles, which clearly showed that the $(312)/(321)CoSn_3$ parallel with the single Co substrate and the $[100]CoSn_3$ parallel with $[1\bar{1}01]Co$. The fixed ORs can be written as:

$$(10\bar{1}0)Co\|(3\bar{1}2)/(3\bar{2}1)CoSn_3 \text{ with } [1\bar{1}01]Co\|[100]CoSn_3 \quad (3)$$

These fixed ORs can also be deduced from Equation (2), which means that ORs in Equation (2) and (3) are crystallographically equivalent, though the morphologies of the $\alpha CoSn_3$ grown on $(11\bar{2}0)Co$ and $(10\bar{1}0)Co$ were different.

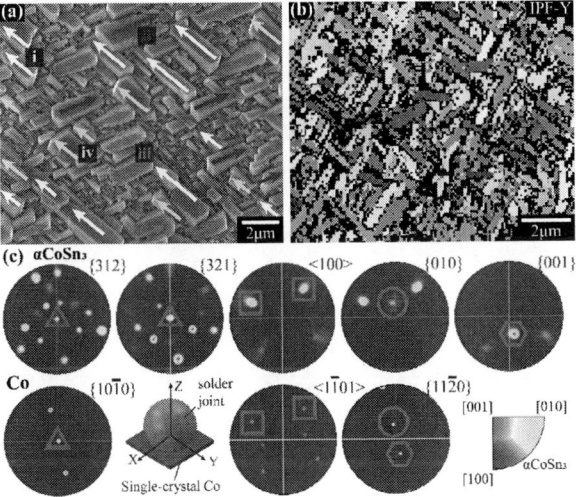

Fig. 3 (a) SEM image of of interfacial $\alpha CoSn_3$ on $(10\bar{1}0)Co$; (b) EBSD-IPFY map of $\alpha CoSn_3$; (c) Pole figures of $\alpha CoSn_3$ and Co.

B. Orientation selections of interfacial aCoSn₃

It should be noted that the crystallographically equivalent $\alpha CoSn_3$ orientations on $(11\bar{2}0)Co$ and $(10\bar{1}0)Co$ represented by ORs in Equation (2) or (3) were not all observed in experiments. For example, 4 $\alpha CoSn_3$ orientations (i.e. the black unit cell wireframes) appeared on $(10\bar{1}0)Co$ and the other two (i.e. the red unit cell wireframes) were not shown in Fig. 4(a). We first analyzed the atomic mismatching of Co and $\alpha CoSn_3$ at the interfaces using the crystal structures in Ref. [11] to clarify the mechanisms of orientation selection. To facilitate the analysis, Co atoms are only analysed here.

As typically presented in Fig. 4(b) and (c), $\alpha CoSn_3$ indicated by the black unit cell wireframes have $(312)/(321)CoSn_3$ parallel with $(10\bar{1}0)Co$. On $(312)/(321)CoSn_3$, the atomic spacing along $[1\bar{3}3]/[13\bar{3}]$ direction is 0.5160 nm and it is 1.2631 nm along $[\bar{1}03]/[\bar{1}30]$. On $(10\bar{1}0)Co$, the corresponding parallel direction $[\bar{1}2\bar{1}0]Co$ has an atomic spacing of 0.5040 nm. The disregistry along the closest-packed directions $[1\bar{3}3]/[13\bar{3}] \| [\bar{1}2\bar{1}0]$ is only 2.4%. For the other two unobserved orientations (red unit cell wireframes), the $(301)/(310)CoSn_3$ plane is parallel with $(10\bar{1}0)Co$. The parallel direction $[010]/[001]CoSn_3 \| [\bar{1}2\bar{1}0]Co$ has an atomic spacing of 0.6268 nm or 0.6270 nm, which features a much bigger disregistry of 24.4% compared with the presented orientations. The analysis on $(11\bar{2}0)Co$ is similar to the $(10\bar{1}0)Co$, and the same conclusions can be drawn.

Based on Kelly and Zhang's Edge-to-Edge theory[12, 13], the largest contribution to the interfacial energy is the disregistries along the closest-packed directions on the heterointerfaces. Therefore, it indicates that the orientation selection of the interfacial $\alpha CoSn_3$ is affected by thermodynamics. Compared with the unobserved orientations, the lattice mismatch between the observed ones and single-crystal Co substrate is smaller, which indicates the lower interfacial energy.

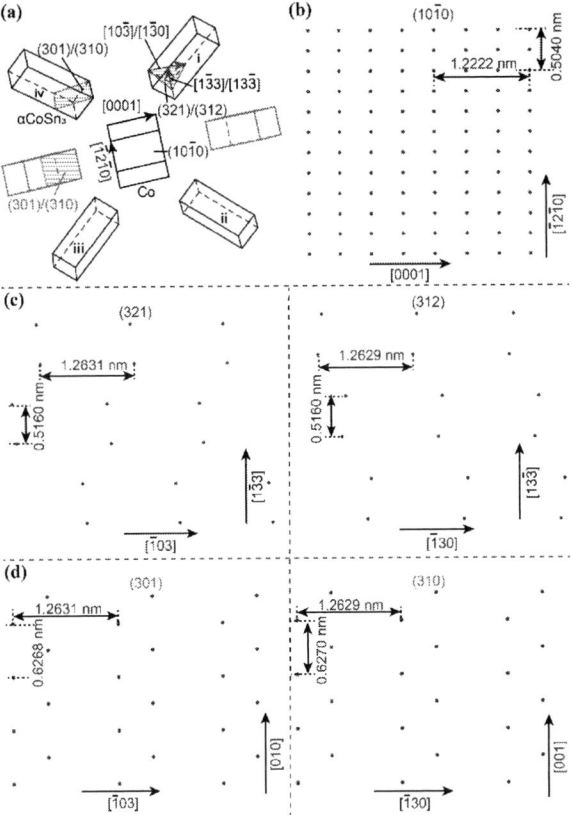

Fig. 4 (a) The fixed ORs between $(10\bar{1}0)Co$ and $\alpha CoSn_3$; (b-d) Distribution of Co atoms on interfaces.

Besides thermodynamics, crystal growth kinetic factors also influence the interfacial $\alpha CoSn_3$ orientations. It's well known that the preferential growth direction of $\alpha CoSn_3$ is [010] or [001] and the growth rate along [100] direction is the smallest and finally produces the largest facet. Therefore, when soldering on Co substrate, the interfacial IMCs having the $(100)CoSn_3$ tilted at larger angles to the Co substrate will have higher growth advantages and will be observed in experiments. Thus, we summarized the angles between the single-crystal Co substrate and $(100)CoSn_3$ including observed and unobserved interfacial $\alpha CoSn_3$ orientations, as shown in Table II.

The angle between (100) of observed $\alpha CoSn_3$ and $(11\bar{2}0)Co$ is $90°$ theoretically and $85.33 \pm 3.74°$ in measurements. However, for unobserved orientations, the angle is only $49.85°$, which is much smaller than the observed ones. The same phenomenon can be found on $(10\bar{1}0)Co$. Compared to unobserved orientations, the observed ones have larger angles between their (100) facets and Co substrate, which satisfied the crystal growth kinetics.

TABLE II.   ANGLES BETWEEN (100)CoSn3 AND THE SUBSTRATE (°)

Co substrate	αCoSn$_3$ orientations	Angles between (100)CoSn$_3$ and the substrate plane(°)	
		Theoretical	Measured
$(11\bar{2}0)$Co	Fig. 2(a)	90	85.33±3.74
	unobserved	49.85	-
$(10\bar{1}0)$Co	Fig. 3(a)	68.14	68.01±8.20
	unobserved	41.88	-

In summary, the orientations of interfacial αCoSn$_3$ are affected by both thermodynamic and kinetic factors, which result in strong textures of αCoSn$_3$ on single-crystal Co.

C. Structures and grain orientations of βSn

βSn structures and grain orientations in SAC305 solder joints on $(11\bar{2}0)$Co and $(10\bar{1}0)$Co are investigated. Structure types of βSn and corresponding frequencies in all tested solder joints are summarized in Table III and typical solder joints are shown in Fig. 5 and Fig. 6.

TABLE III.   BETA-SN STRUCTURES AND GRAIN ORIENTATIONS IN SOLDER JOINTS ON SINGLE CO

Co substrate	$(11\bar{2}0)$Co		$(10\bar{1}0)$Co	
βSn structure	Single grain	Twins	Single grain	Twins
Joints	27		58	
With OR	13	8	34	17
Without OR	2	4	3	4

βSn structures and grain orientations were summarized in Table III. 21/27 joints on $(11\bar{2}0)$Co have βSn orientations meeting the fixed OR given by the unit cell wireframe in Fig. 5(b), including 13 joints have single grain and 8 joints have twinned grain structures. The frequency of solder joints with βSn orientations satisfying the fixed ORs is 77.78%, meaning a better control over βSn structure and grain orientations on $(11\bar{2}0)$Co.

EBSD mapping of the representative joints on $(11\bar{2}0)$Co showed that the joint 1 has a cyclic twined grain and the joint 2 and 3 have a single grain structures as presented in Fig. 5(a). The grain orientations of βSn and single-crystal Co are illustrated by the black and white frames separately. The βSn unit cell wireframe that has the fixed ORs in Fig. 5(b) is highlighted by a pink dashed circle and twinned axes are denoted by black arrows. When the interfacial αCoSn$_3$ grown on $(11\bar{2}0)$Co have morphology as shown in Fig. 2(a), the fixed ORs between βSn, αCoSn$_3$ and Co are presented in Fig. 5(b) according to ORs described in Equation (1) and Equation (2). Parallel planes are identified by the same colors. According to Fig. 5(b) and Fig. 5(c), the direct fixed OR between βSn and Co single crystal can be obtained:

$$(11\bar{2}0)Co\|(100)/(010)/(001)Sn$$

with
$$[1\bar{1}01]Co\|[100]/[010]/[001]Sn \qquad (4)$$

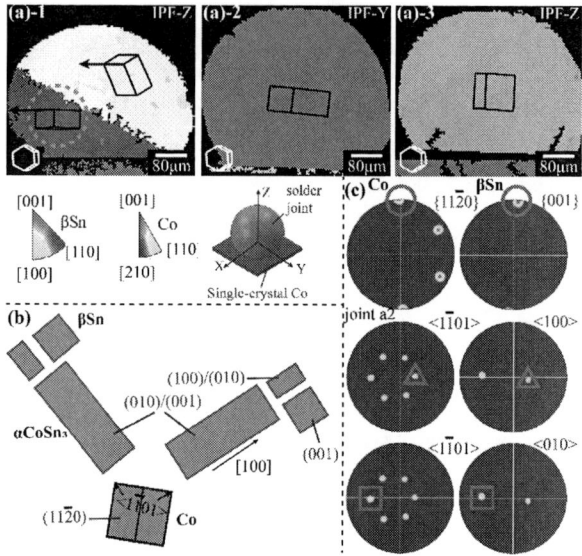

Fig. 5 (a) EBSD maps of 3 of 21 joints meeting the predicted ORs; (b) Fixed ORs shown by the unit cell wireframes; (c) Pole figures of βSn and Co of joint a2.

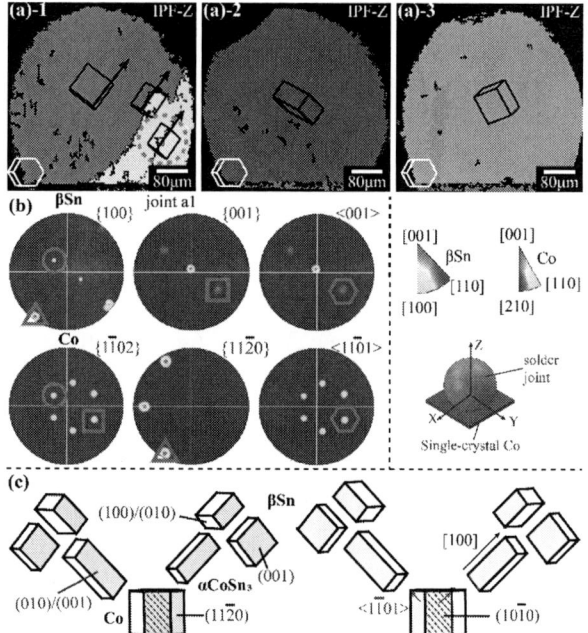

Fig. 6 (a) EBSD maps of 3 of 51 joints meeting the predicted ORs; (b) Pole figures of βSn and Co of joint a1. (c) Fixed ORs presented by the unit cell wireframes

For joints on $(10\bar{1}0)$Co, 51/58 solder joints satisfy the fixed OR presented in Fig. 6(c). As shown in Table III, 34 joints have a single grain and 17 joints have twinned grain microstructures and the frequency of occurrence with βSn orientations satisfying the fixed ORs is 87.93%. It proved that βSn orientations in these joints on $(10\bar{1}0)$Co have been effectively controlled.

EBSD mapping of the representative joints on $(10\bar{1}0)$Co confirmed that the joint 2 and 3 have a single-grain and the joint 1 has a cyclic twined-grain structures. Fig. 6(b) presented the pole figures of βSn and Co of joint 1, in which parallel crystal planes and directions are highlighted by the common circles, squares, triangles, and hexagons. The morphology of interfacial αCoSn₃ grown on (10$\bar{1}$0)Co, which was shown in Fig. 3(a), showed strong textures with four dominant orientations. If βSn nucleates and grows on these αCoSn₃ plates, 8 possible grain orientations will be generated, as presented in Fig. 6(c). Based on pole figures in Fig. 6(b) and the schematic diagram in Fig. 6(c), the fixed OR between Co and βSn can be obtained. This OR can also be expressed by Equation (4), since interfacial αCoSn₃ on (11$\bar{2}$0)Co and (10$\bar{1}$0)Co have the same fixed ORs.

Due to the strong textures of interfacial αCoSn₃, which induces fixed ORs between βSn and the single-crystal Co substrate, both (11$\bar{2}$0)Co and (10$\bar{1}$0)Co are effective in controlling βSn structures and grain orientations. Compared to traditional Cu substrates, the number of βSn orientations in solder interconnections on single Co significantly reduced. For (11$\bar{2}$0)Co, if βSn nucleates and grows on the regular interfacial αCoSn₃, 4 possible orientations will be generated as presented in Fig. 5(b) and 4 twinned grains can be further produced by each grain, whose twinned axis is [010] or [100]Sn, and finally, there will be a total of 20 potential βSn orientations as presented in Fig. 7 (4 original orientations were represented by red dots and the twinned orientations were represented by black dots), which will certainly improve the reliabilities of solder joints. For (10$\bar{1}$0)Co, 40 orientations are expected, whose [001]Sn is not perpendicular to the substrate (i.e. parallel to the current direction), and it is thought to be beneficial for preventing electromigration failure.

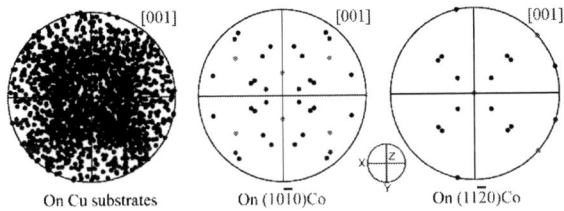

Fig. 7 Pole figures of βSn on the traditional Cu, (10$\bar{1}$0)Co and (11$\bar{2}$0)Co substrate

IV. CONCLUSIONS

In this work, the mechanism of controlling βSn structure and grain orientations in solder interconnections with single-crystal Co substrate was investigated. The morphologies and textures of interfacial αCoSn₃ were studied. The orientation selection of αCoSn₃ was clarified. The nucleation and grain orientations of βSn are determined. The conclusions can be summarized as follows:

1. The interfacial αCoSn₃ on single Co has a plate-like morphology with strong textures. There are 2 dominated orientations on (11$\bar{2}$0)Co and 4 dominated orientations on (10$\bar{1}$0)Co, which can be described as the same reproducible ORs:

(11$\bar{2}$0)Co∥(010)/(001)CoSn₃ and [1$\bar{1}$01]Co∥[100] CoSn₃;

2. On (11$\bar{2}$0)Co and (10$\bar{1}$0)Co, the frequency of occurrence of solder joints with expected βSn orientations is >75%, which reflects a strong controlling effect of single-crystal Co substrate on βSn orientations.

3. Not all crystallographically equivalent orientations of αCoSn₃ have been observed. Compared with the unobserved orientations, the observed ones have higher interfacial thermodynamic stability and growth kinetics.

4. Different from soldering on traditional Cu, the number of βSn orientations on single Co is significantly reduced. The βSn grain orientation on (11$\bar{2}$0)Co is only 20 (including twinned orientations). On (10$\bar{1}$0) Co, there is no [001]Sn perpendicular to the substrate, which was considered to be beneficial for reducing electromigration.

REFERENCES

[1] L. P. Lehman, Y. Xing, T. R. Bieler, et al. "Cyclic twin nucleation in tin-based solder alloys," Acta Materialia, Binghamton, USA, vol. 58, pp. 3546-3556, Jun. 2010.

[2] S. H. Yang, Y. H. Tian, C. Q. Wang. "Investigation on Sn grain number and crystal orientation in the Sn-Ag-Cu/Cu solder joints of different sizes," Journal of Materials Science: Materials in Electronics, Harbin, China, vol. 21, pp. 1174-1180, Dec. 2009.

[3] K. Lee, K. S. Kim, Y. Tsukada, et al. "Effects of the crystallographic orientation of Sn on the electromigration of Cu/Sn-Ag-Cu/Cu ball joints," Journal of Materials Research, Osaka, Japan, vol. 26, pp. 467-474, Feb. 2011.

[4] H. T. Chen, B. B. Yan, M. Yang, et al. "Effect of grain orientation on mechanical properties and thermomechanical response of Sn-based solder interconnects," Materials Characterization, Harbin, China, vol. 85, pp. 64-72, Nov. 2013.

[5] M. H. Lu, D. Y. Shih, P. Lauro, et al. "Effect of Sn grain orientation on electromigration degradation mechanism in high Sn-based Pb-free solders," Applied Physics Letters, New York, USA, vol. 92, 211909, May. 2008.

[6] J. W. Xian, Z. L. Ma, S. A. Belyakov, et al. "Nucleation of tin on the Cu₆Sn₅ layer in electronic interconnections," Acta Materialia, London, UK, vol. 123, pp. 404-415, Oct. 2016.

[7] Z. L. Ma, J. W. Xian, S. A. Belyakov, et al. "Nucleation and twinning in tin droplet solidification on single crystal intermetallic compounds," Acta Materialia, London, UK, vol. 150, pp. 281-294, May. 2018.

[8] Z. L. Ma, S. A. Belyakov, K. Sweatman, et al. "Harnessing hetero-geneous nucleation to control tin orientations in electronic interconnections," Nature Communications, London, UK, vol. 8, 1916, Dec. 2017.

[9] Z. L. Ma, C. M. Gourlay. "Nucleation, grain orientations, and microstructure of Sn-3Ag-0.5Cu soldered on cobalt substrates," Journal of Alloys and Compounds, London, UK, vol. 706, pp. 596-608, Jun. 2017.

[10] Z. H. Zhang, M. Y. Li, Z. Q. Liu, et al. "Growth characteristics and formation mechanisms of Cu₆Sn₅ phase at the liquid-Sn₀.₇Cu/(111)$_{Cu}$ and liquid-Sn₀.₇Cu/(001)$_{Cu}$ joint interfaces," Acta Materialia, Shenzhen, China, vol. 104, pp. 1-8, Dec. 2015.

[11] A. Lang, W. Jeitschko. "Two new phases in the system cobalt-tin: The crystal structures of alpha- and beta-CoSn₃," Zeitschrift Fur Metallkunde, vol. 87, pp. 759-764. 1996.

[12] M. X. Zhang, P. M. Kelly, M. A. Easton, et al. "Crystallographic study of grain refinement in aluminum alloys using the edge-to-edge matching model," Acta Materialia, Brisbane, Australia, vol. 53, pp. 1427-1438, Mar. 2005.

[13] P. M. Kelly, M. X. Zhang. "Edge-to-edge matching-The fundamentals," Metallurgical and Materials Transactions A, Brisbane, Australia, vol. 37A, PP. 833-839, Mar. 2006.

Topology Optimization Design Of Meta-Material Heat Spreader

Xue Bai
Shenzhen Institute of Advanced Electronic Materials, Shenzhen Institutes of Advanced Technology, Chinese Academy of Sciences,
Shenzhen, China
xue.bai@siat.ac.cn

Qinghua Hu
Shenzhen Institute of Advanced Electronic Materials, Shenzhen Institutes of Advanced Technology, Chinese Academy of Sciences,
Shenzhen, China
State Key Laboratory of Power Transmission Equipment & System Security and New Technology School of Electrical Engineering, Chongqing University,
Chongqing, China
qh.hu@siat.ac.cn

Xiaoliang Zeng*
Shenzhen Institute of Advanced Electronic Materials, Shenzhen Institutes of Advanced Technology, Chinese Academy of Sciences,
Shenzhen, China
xl.zeng@siat.ac.cn

Rong Sun
Shenzhen Institute of Advanced Electronic Materials, Shenzhen Institutes of Advanced Technology, Chinese Academy of Sciences,
Shenzhen, China
rong.sun@siat.ac.cn

Jianbin Xu*
Shenzhen Institute of Advanced Electronic Materials, Shenzhen Institutes of Advanced Technology, Chinese Academy of Sciences,
Shenzhen, China
Department of Electronics Engineering, The Chinese University of Hong Kong,
Hong Kong, China
jbxu@ee.cuhk.edu.hk

Abstract—**The increasing power density in electronics calls for novel approaches to manipulate heat flow. Thermal meta-materials, which do not exist naturally but can be designed rationally, have been proved to manipulate heat flow as will recently. The heat flow manipulators constructed by the thermal meta-materials, such as thermal cloak, thermal concentrator and thermal rotator have exhibited the potentials to be applied in novel heat flow guiding in electronics packaging. This work aims to design heat spreader with thermal meta-materials to protect critical device component from heat source and dissipate heat flow in a deterministic way. Commercial finite element software COMSOL Multiphysics has been used in solving topology optimization of heat transfer problem. The physical model has been set as a hot spot heat source with a power density of $1 \times 10^{11}\ W/m^3$ placed in the upper part of a 0.1mm height silicon plate. The size of the plate is $0.6*1\ mm^2$ and the radius of the hot spot is 0.1mm. A critical square component with a size of $0.2*0.2\ mm^2$ is placed in the lower part of the plate. A middle section with a size of $0.2*0.1\ mm^2$ at bottom side of the plate is set to room temperature T0=293.15K to represent that it is connected to a heat sink. Two sections with the size of $0.1*0.1\ mm^2$ at upper side of the plate are also set to room temperature T0 to represent two electrodes connecting to heat sink. The rest edges of the plate are set as convective heat flux to the environment with a heat transfer coefficient h=10 $W/(m^2 \cdot K)$. The simulation results have shown that the average temperature of the critical component is 5.92K lower after the optimization. However, the heat dissipation in the heat source region has been sacrificed under the current the optimization objective. This work has offered an interesting and alternative approach for design heat management devices.**

Keywords—*Thermal meta-materials; Topology optimization; Heat transfer structure*

I. INTRODUCTION

With the rapid development of modern electronics, the traditional heat management has been pushed to reach its limit. The increasing power density in electronics calls for novel approaches to manipulate heat flow. Thermal meta-materials, which do not exist naturally but can be designed rationally, have been proved to manipulate heat flow as will recently. The heat flow manipulators constructed by the thermal meta-materials, such as thermal cloak, thermal concentrator and thermal rotator have been developed based on transformation thermodynamics as well as scattering cancelation method, and have been demonstrated by numerical and experimental studies[1]–[4]. These heat flow manipulators have provided a new approach for critical device component protection and heat management. However, most of studies are restricted to solid heat conduction with rigorous mathematical solved problems and lack of flexibility for general application.

Recently, optimization method has been introduced to the design process of thermal meta-materials in practical electronic heat management. Ercan M. Dede et.al. have proposed thermal metamaterials based on anisotropic thermal-composite design optimization method, and have applied them to the heat flow control on a printed circuit board (PCB) ,which have exhibited the potentials to be applied in novel heat flow guiding in electronics packaging[5], [6]. Topology optimization technique, as one of the most important optimization methods, has been widely used for engineering designs in microchannel heat sink, especially since the fast development of 3D printing[7]. It has been expected to be a promising design tool for heat management in heat conductive, convective and conjugate heat transfer system[8].

978-1-6654-1392-3/21 $31.00 © 2021 IEEE

This work aims to design a heat spreader with thermal meta-materials to protect critical device component from heat source and dissipate heat flow in a deterministic way based on topology optimization method. Commercial finite element software COMSOL Multiphysics have been used in solving topology optimization of heat transfer problem. A heat spreader constructed with high thermal conductive carbon fiber film and paraffin has been proposed, since the phase change material paraffin can extend the application in lithium ion battery heat management.

II. MODEL

A. Physical Model

As the power density continuously increased with the shrinkage of the device size, the hot spots occur in electronic devices have become an urgent issue need to be addressed. In this work, a heat spreader is design to protect critical device component from hot spot and dissipate heat flow in a deterministic way.

For the convenience of simulation, the physical model has been set as a hot spot heat source with a power density of 1×10^{11} W/m^3 placed in the upper part of a 0.1mm height silicon plate. The size of the plate is 0.6*1 mm^2 and the radius of the hot spot is 0.1mm. A critical square component with a size of 0.2*0.2 mm^2 is placed in the lower part of the plate. A middle section with a size of 0.2*0.1 mm^2 at bottom side of the plate is set to room temperature T0=293.15K to represent that it is connected to a heat sink. Two sections with the size of 0.1*0.1 mm^2 at upper side of the plate are also set to room temperature T0 to represent two electrodes connecting to heat sink. The rest edges of the plate are set as convective heat flux to the environment with a heat transfer coefficient h=10 $W/(m^2 \cdot K)$. The basic physical properties of material silicon is used from the build in parameters in COMSOL. Figure 1 (a)shows the general physical model setting of the simulation, and (b) shows the mesh of the model.

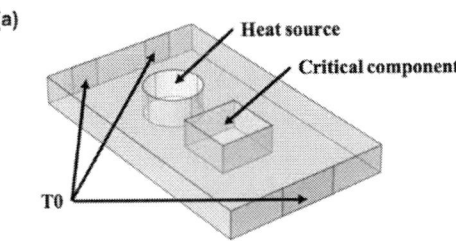

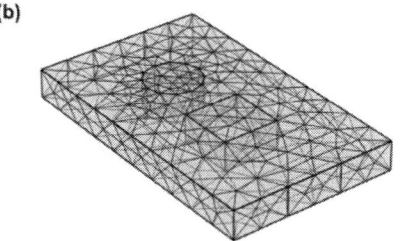

Fig. 1. (a) General physical model setting. (b) The mesh of the model.

The temperature distribution of the plate at steady state has been simulated in 3D under the Heat Transfer in Solid model in COMSOL Multiphysics, as shown in Figure 2.

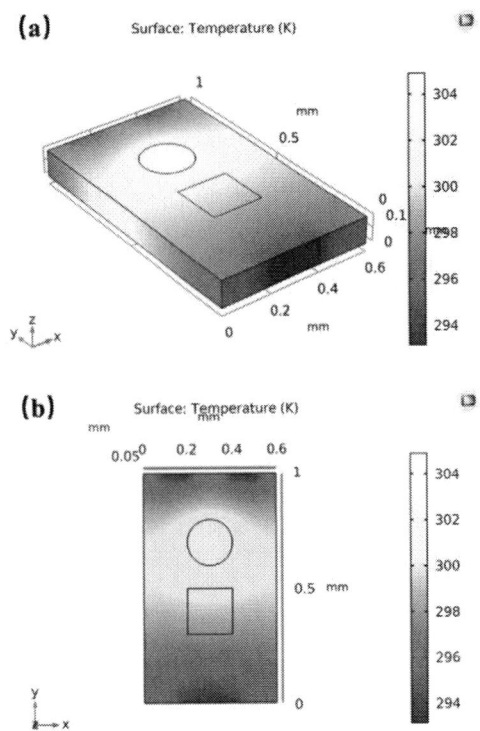

Fig. 2. (a) The 3D temperature distribution of the plate; (b) the top view of the surface temperature distribution of the plate.

B. Topology Optimization

Since the heat conduction in this physical model is symmetry in z-direction, to simplify the optimization problem, we construct the topology optimization simulation in 2D model.

In this work, high thermal conductive carbon fiber film and paraffin have been proposed as two component materials to construct the heat spreader. The density model from the Topology optimization in COMSOL has been employed to predict the optimal material distribution to fulfill the objective function.

The Solid Isotropic Material with Penalization method (SIMP) has been used in the simulation and set as

$$SIMP(\rho) = \rho_{min} + (1 - \rho_{min})\rho_e^p \quad (1)$$

$$0 \leq \rho_{min} \leq 1$$

$$0 \leq \rho_e \leq 1$$

where ρ_{min} is the minimum relative density, ρ_e is the element relative density, p is the penalty factor set as p=3. The thermal conductivity, specific heat capacity and density of the topology optimization are modulated according to

$$\kappa_{SIMP(\rho)} = \sum_{i=1}^{N} SIMP(\rho) \cdot \kappa_i \quad (2)$$

$$C_{p\,SIMP(\rho)} = \sum_{i=1}^{N} SIMP(\rho) \cdot C_{p_i}$$

$$\rho_{SIMP(\rho)} = \sum_{i=1}^{N} SIMP(\rho) \cdot \rho_i$$

Table 1 shows the basic material properties of the carbon fiber film and paraffin.

TABLE I.   MATERIAL PROPERTIES

Material	Carbon fiber film	Paraffin
Thermal Conductivity (W/m K)	400	0.15
Specific Heat capacity (J/kg K)	860	2000
Density（kg/m³）	350	880
Melting temperature （K）	--	313.75
Latent heat (J/kg)	--	205000

In this work, the heat spreader is aim to protect critical device component from hot spot heat source. A domain probe, T_av_csc, for detecting the average temperature of the critical component region has been placed. Therefore, the optimization objective function of the design is:

Minimize: T_av_csc

Considering the balance between maximizing the heat conduction ability and utilizing the phase change latent heat, the constraints have been set as $v_i = 0.2$, where v_i is the average material volume factor.

The initial temperature of the 2D model has been set as the average temperature of the whole domain from the 3D plate model, 298.81K. The optimality tolerance is set as 0.001, and the maximum number of model evaluations is 100.

III. RESULTS AND DISCUSSION

After 13 times optimal iteration, the objective function has reached local minimization. Figure 3 shows the output material volume factor of the topology optimization, which has designed the material distribution to reach the goal of reduce the temperature of the critical component region.

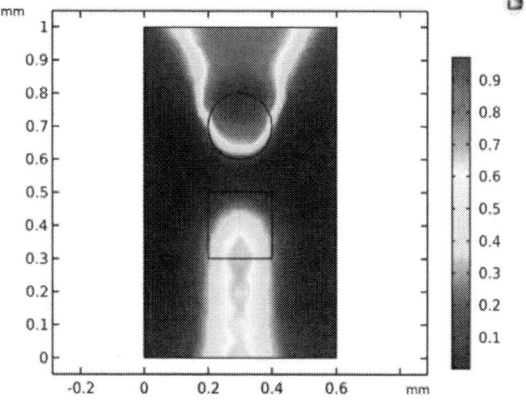

Fig. 3.   Output material volume factor of the topology optimization.

The surface temperature distribution of the plate after topology optimization has been shown in Figure 4 (b)，while Figure 4 (b) shows the top view of the surface temperature distribution of the original plate. Compared with the average temperature of the critical component region from original plate (299.60K), the average temperature of the critical component region from optimized plate is 293.68K, which is lower 5.92K. This result has revealed that the topology optimization method can be applied in design some heat management structures to obtain the deterministic purpose.

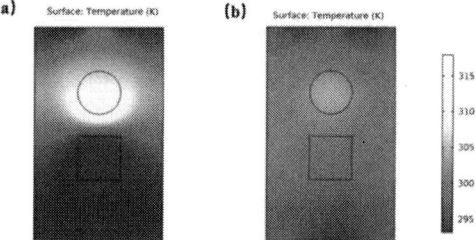

Fig. 4.   (a) The surface temperature distribution of the plate after topology optimization; (b) the top view of the surface temperature distribution of the original plate.

On the other hand, we have noticed that the heat source region in the plate has reached much higher temperature after the optimization. This is due the setting of the objective function. To protect the critical component region, the heat dissipation in the heat source region has been sacrificed. This issue can be addressed by change the optimization objective.

IV. CONCLUSION

A heat spreader with thermal meta-materials based on topology optimization method has been designed under the purpose of protecting critical device component from heat source and dissipating heat flow in a deterministic way. The simulation results have shown that the average temperature of the critical component is 5.92K lower after the optimization. However, the heat dissipation in the heat source region has been sacrificed under the current the optimization objective. This work has offered an interesting and alternative approach for design heat management devices.

ACKNOWLEDGMENT

This work was supported by National Natural Science Foundation of China (Grant No. 62004211)，Guangdong Province Key Field R&D Program Project (No. 2020B010190004), Shanghai Sailing Program in China (No. 19YF1450500), Shenzhen Science and Technology Program (Grant No. RCBS20200714114858221).

REFERENCES

[1]   R. Schittny, M. Kadic, S. Guenneau, and M. Wegener, "Experiments on transformation thermodynamics: Molding the flow of heat," *Phys. Rev. Lett.*, vol. 110, no. 19, p. 195901, 2013.

[2]   S. Narayana and Y. Sato, "Heat flux manipulation with engineered thermal materials," *Phys. Rev. Lett.*, vol. 108, no. 21, p. 214303, 2012.

[3] H. Tiancheng, B. Xue, G. Dongliang, T. John T L, L. Baowen, and Q. Cheng Wei, "Experimental demonstration of a bilayer thermal cloak," *Phys. Rev. Lett.*, vol. 112, no. 5, p. 054302, 2014.

[4] H. Xu, X. Shi, F. Gao, H. Sun, and B. Zhang, "Ultrathin three-dimensional thermal cloak," *Phys. Rev. Lett.*, vol. 112, no. 5, p. 054301, 2014.

[5] E. M. Dede, F. Zhou, P. Schmalenberg, and T. Nomura, "Thermal Metamaterials for Heat Flow Control in Electronics," *J. Electron. Packag.*, vol. 140, no. 1, p. 010904, 2018.

[6] J. C. Kim *et al.*, "Recent Advances in Thermal Metamaterials and Their Future Applications for Electronics Packaging," *J. Electron. Packag. Trans. ASME*, vol. 143, no. 1, pp. 1–15, 2021.

[7] L. Xu, H. Li, X. Ding, and S. Liu, "Design of microchannel heat sink using topology optimization for high power modules cooling," *18th Int. Conf. Electron. Packag. Technol. ICEPT 2017*, no. September, pp. 1092–1097, 2017.

[8] T. Dbouk, "A review about the engineering design of optimal heat transfer systems using topology optimization," *Appl. Therm. Eng.*, vol. 112, pp. 841–854, 2017.

[9] A. Hamed and S. Ndao, "High anisotropy metamaterial heat spreader," *Int. J. Heat Mass Transf.*, vol. 121, pp. 10–14, 2018.

Effect of Surface Stress on Indentation Response of Elastic-plastic Materials

Xu Long*
School of Mechanics, Civil Engineering and Architecture
Northwestern Polytechnical University
Xi'an, China
xulong@nwpu.edu.cn

Ziyi Shen
School of Mechanics, Civil Engineering and Architecture
Northwestern Polytechnical University
Xi'an, China
1154847376@qq.com

Xiaotong Chang
School of Mechanics, Civil Engineering and Architecture
Northwestern Polytechnical University
Xi'an, China
xtchang@nwpu.edu.cn

Yutai Su
School of Mechanics, Civil Engineering and Architecture
Northwestern Polytechnical University
Xi'an, China
suyutai@buaa.edu.cn

Hongbin Shi
Shanghai Institute Consumer BG, DFR Department
HUAWEI DEVICE CO., LTD
Shanghai, China
shihongbin2@huawei.com

Tao Huang
Space Research Institute of Electronics and Information Technology
Aerospace Science and Technology Corporation
Xi'an, People's Republic of China
huangt89@163.com

Yanpei Wu
Space Research Institute of Electronics and Information Technology
Aerospace Science and Technology Corporation
Xi'an, People's Republic of China
wuyanpei2007@126.com

Bingyi Tu
Space Research Institute of Electronics and Information Technology
Aerospace Science and Technology Corporation
Xi'an, People's Republic of China
977793390@qq.com

Abstract—In this study, nanoindentation technology is applied to evaluate the elastic-plastic law of mechanical properties of small-sized in-situ packaging materials. As one of the most appealing advantage, the proposed method requires only a single nanoindentation, which means the nanoindentation experiments can be performed at minor effort. Combined with nanoindentation technology, the mechanical behavior of packaging materials in single-point nanoindentation loading process was simulated by the finite element method. The effect of surface stress on nanoindentation was studied numerically. In the finite element simulations, after applying prestress to the substrate material, the single-point nanoindentation was numerically analyzed up to the same indentation depth. Based on the finite element predictions, the mechanical properties of materials subjected to prestressing in different directions are studied. By comparing the applied load-penetration depth curves, stress contour, plastic zone and pile-up behavior, the influence of surface stress on the elastic-plastic material is elucidated.

Keywords—Nanoindentation, Berkovich indentation, surface stress

I. Introduction

Nowadays, the packaging structure of electronic chips is developing towards miniaturization, and the thickness or area is a sharply reduced trend of the packaging material, so the traditional test method cannot measure the small-size material. The mechanical properties of nanomaterials can be studied by nanoindentation. The hardness of the sample can be obtained by this technique. This method is to determine the contact area by using the penetration depth of the recorded indentation and the measured load applied to the sample, and then determine the hardness. Mechanical properties such as elastic modulus and hardening exponent can also be obtained by this method. Nevertheless, the effect of the indentation response of elastic-plastic materials when considering the residual stress is not well investigated. This is an important factor affecting the mechanical properties of engineering structures.

In the existing literature, a comprehensive computational study has been carried out to the extent that the instrumented sharp indentation can determine the elastoplastic properties of ductility materials, and the sensitivity of this extraction characteristic to the measurement of indentation data variability has been quantified [1]. Through nanoindentation test, Zhan et al. obtained the apparent surface stress and then studied its surface effect. The nanoindentation process is simulated by two-dimensional quasi-continuum method. The results show that the mechanical properties are different when the surface roughness of single crystal copper thin film is different [2]. Long and Wang [3] applied surface tension on the half-space and applied force on it using a rigid body ball. Based on the elastic solution of the obtained force, the contact problem was studied. Clyne performed differential thermal shrinkage on thin copper foil resulting in biaxial residual stresses of the same amount. According to the prediction of the finite element model, it is found that the applied load at a given pressing depth decreases with the increase of residual stress. Because the residual stress change is small, it should be studied by nanoindentation. However, due to changes in residual stress levels, changes in hardness are smaller and more difficult to analyze [4]. Xu and Li studied the nanoindentation unloading behavior of elastic-plastic strain-hardening materials under equi-biaxial residual stress by finite element method [5]. Based on samples at different annealing temperatures and times, Long et al. [6] carried out nano-indentation and constitutive experiments, and the effect of residual stress on SAC305 was investigated. However, the pile-up deformation around indentations caused by residual stress of annealed SAC305 solder has not been studied further. In other metals and alloys, the analytical methods and rate factors presented also apply when the material under study does not have significant residual stresses [7]. In recent years, Long et al. [8] had revealed the relationship between microstructure and constitutive behavior.

The effect of surface stress on strain hardening materials is studied in this paper. The Berkovich indenter was used for finite element simulations of elastic-plastic materials with different surface stresses. The applied load-penetration depth curves, stress contour, plastic zone and pile-up behavior are compared by finite element analysis. The results show that there are significant differences in the properties of materials when the surface stress is present in the substrate material.

II. FINITE ELEMENT MODEL

The indentation problem considered in this paper is shown in Figure 1. Uniformly distributed surface stress is applied in semi-infinite substrate material. Under the action of vertical load P, the Berkovich indenter is pressed into the prestressed substrate material. The resulting indentation depth is denoted by h. It is assumed that elastic-plastic substrates follow the isotropic strain-hardening law of power-law relations. In the elastic deformation stage, the relationship between stress σ and strain ε under uniaxial tension is linear [9].

$$\begin{cases} \sigma = E\varepsilon & , \ for \ \sigma \leq \sigma_y \\ \sigma = R\varepsilon^n & , \ for \ \sigma \geq \sigma_y \end{cases} \quad (1)$$

where σ_y is the initial yield stress, n is the strain hardening exponent, R is a strength coefficient, E is the Young's modulus.

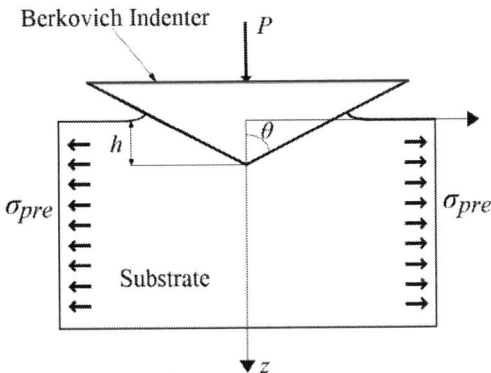

Fig. 1 Schematic of a strain hardening substrate with surface stress compressed by Berkovich indentation

Regarding material, a substrate with Young's modulus of 10GPa, hardening index of 0.1, Poisson's ratio of 0.3 and yield strength of 50 MPa was selected for simulation. The stress-strain curve of the substrate material was shown in Fig. 2.

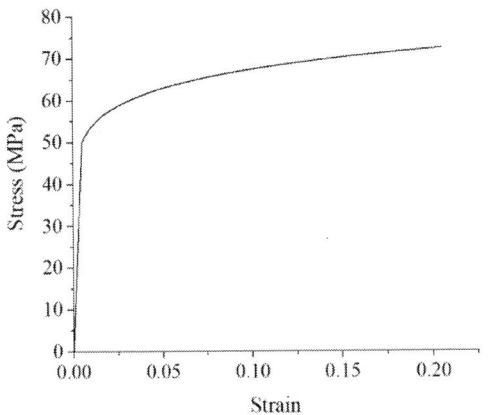

Fig. 2 Stress-strain relationship of material

III. RESULTS AND DISCUSSION

The variation of applied load with penetration depth under different pre-strains is shown in Fig. 3. Penetration depth is 2000nm.The results show that there is a strong correlation

between pressure and pre-strain. At a certain indentation depth, a higher pressure will be generated when the substrate is pressed, and a lower pressure will be generated when the substrate is pulled.

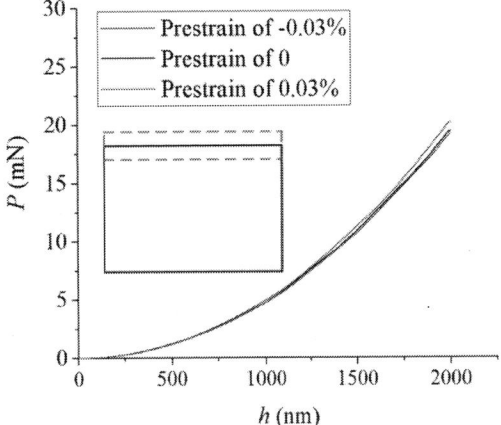

Fig. 3 Applied load-penetration depth curves

As shown in Fig. 4, the distribution of stress is affected by the applied pre-strain. Compared with the situation without pre-strain, the stress distribution develops horizontally when the substrate is under compression, and decreases to a certain extent when the substrate is under tension.

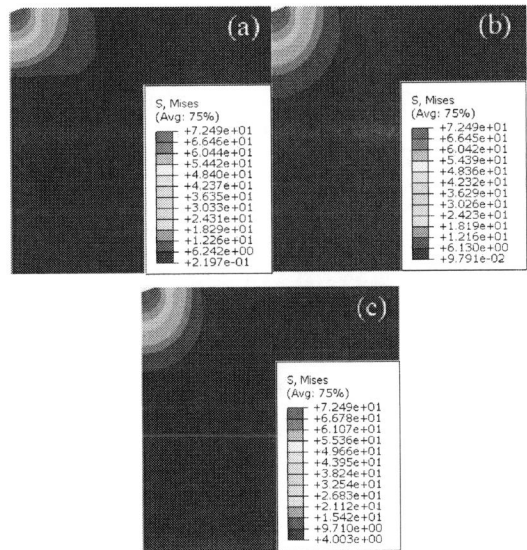

Fig. 4 Stress contour under the indenter. (a) Pre-strain is -0.03%; (b) Pre-strain is 0; (c) Pre-strain is 0.03%

There is a strong correlation between the plastic radius and the pre-strain. When the substrate is compressed, the plastic radius is 4406.80nm. When there is no pre-strain, the plastic radius is 4820.27nm. When the substrate is pulled horizontally, the plastic radius is 5100.07nm. It can be seen that the plastic radius decreases when the substrate is pressed, and increases when the substrate is pulled. This is clearly shown in Fig. 5.

978-1-6654-1392-3/21 $31.00 © 2021 IEEE

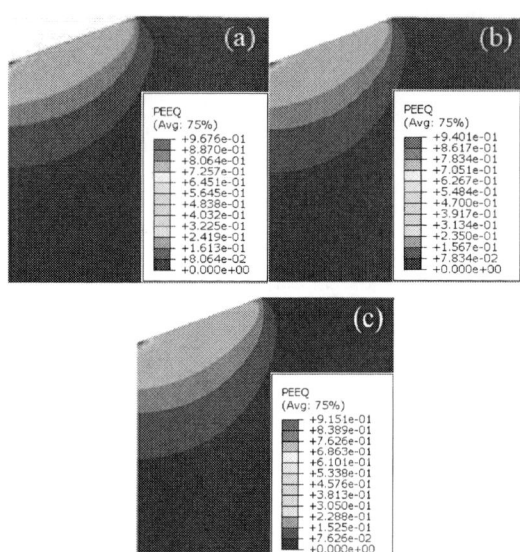

Fig. 5 Equivalent plastic strain contour under the indenter. (a) Pre-strain is -0.03%; (b) Pre-strain is 0; (c) Pre-strain is 0.03%

As shown in Fig. 6, pile-up behavior appears in the substrate under different pre-strains. When the substrate is under compression, the distance from the highest point of accumulation to the reference line is 66.6194 nm. When there is no pre-strain, the distance from the highest point of accumulation to the reference line is 42.2706 nm. When the substrate is under tension, the distance between the highest point of accumulation and the reference line is 36.8833 nm. As shown in Figure 6, the substrate compression will increase the distance, and the substrate tension will decrease the distance.

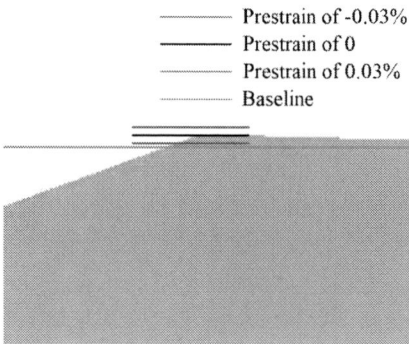

Fig. 6 Pile-up behavior

Residue stress is an important factor affecting the mechanical properties of engineering structures. Surface stress should be taken into account when the *P-h* curve is used

to invert and calculate the value of constitutive parameters. The above numerical analysis of the mechanical properties of materials under prestressing in different directions is of great significance for the subsequent residual stress inversion analysis.

IV. CONCLUSION

By comparing the applied load-penetration depth curves, stress contour, plastic zone and pile-up behavior, the effect of surface stress on the material is discussed. The results show that at a certain indentation depth, the tensile surface stress will produce lower pressure, while the compression surface stress will produce higher pressure. The stress distribution develops horizontally when the substrate is under compression, and decreases somewhat when the substrate is under tension. The plastic radius decreases when the substrate is pressed, and increases when the substrate is pulled. The substrate compression will increase the distance from the highest point to the reference line, and the substrate tension will decrease the distance.

ACKNOWLEDGMENT

This work was supported by the Natural Science Foundation of Shaanxi Province (No. 2021KW-25) and the Astronautics Supporting Technology Foundation of China (No. 2021-HT-XG).

REFERENCES

[1] M. Dao, N. Chollacoop, K. Vliet, T. A. Venkatesh, and S. Suresh, "Computational modeling of the forward and reverse problems in instrumented sharp indentation," Acta Materialia, vol. 49, no. 19, pp. 3899-3918, 2001.

[2] T. Y. Zhang and W. H. Xu, "Surface Effects on Nanoindentation," Journal of Materials Research, vol. 17, no. 7, pp. 1715-1720, 2002.

[3] J. M. Long and G. F. Wang, "Effects of surface tension on axisymmetric Hertzian contact problem," Mechanics of Materials, vol. 56, no. JAN., pp. 65-70, 2013.

[4] A. Clyne, "Use of nanoindentation to measure residual stresses in surface layers," Acta Materialia, 2011.

[5] Zhihui Xu and Xiaodong Li, "Influence of equi-biaxial residual stress on unloading behaviour of nanoindentation," Acta Materialia, 2005.

[6] X. Long, S. Wang, Y. Feng, Y. Yao, and L. M. Keer, "Annealing effect on residual stress of Sn-3.0Ag-0.5Cu solder measured by nanoindentation and constitutive experiments," Materials Science and Engineering: A, vol. 696, no. JUN.1, pp. 90-95, 2017.

[7] X. Long, Xiaodi Zhang, Wenbin Tang, Shaobin Wang, Yihui Feng and Chao Chang, "Calibration of a Constitutive Model from Tension and Nanoindentation for Lead-Free Solder," Micromachines, 2018.

[8] X. Long, B. Hu, Y. Feng, C. Chang, and M. Li, "Correlation of microstructure and constitutive behaviour of sintered silver particles via nanoindentation," International Journal of Mechanical Sciences, vol. 161-162, p. 105020, 2019.

[9] Z. S. Ma, Y. C. Zhou, S. G. Long, X. L. Zhong, and C. Lu, "Characterization of stress-strain relationships of elastoplastic materials: An improved method with conical and pyramidal indenters," Mechanics of Materials, vol. 54, no. NOV., pp. 113–123, 2012.

Surface Passivation of Cu Nanofiber Films Fabricated by Electrospinning for Transparent Conductive Films

Zhefei Sun
Department of Materials Science and Engineering
College of Materials, Xiamen University
Xiamen, People's Republic of China
zhefeisun@stu.xmu.edu.cn

Ming-Sheng Wang*
Department of Materials Science and Engineering
College of Materials, Xiamen University
Xiamen, People's Republic of China
mswang@xmu.edu.cn

Zhihao Zhang*
Department of Materials Science and Engineering
College of Materials, Xiamen University
Xiamen, People's Republic of China
zhzhang@xmu.edu.cn

Abstract—Indium-tin oxide (ITO) transparent conductive films (TCFs) are widely used in electronic displays and solar cells. However, the brittleness and high cost of ITO are critical shortcomings for its applications in the next-generation flexible and low-cost electronics. As a novel substitution of ITO films, copper nanofiber films (CNFs) take the great advantages of low cost, high conductivity and excellent flexibility. Unfortunately, CNFs are sensitive to oxygen, which may cause severe performance deterioration over time. Here, we use an electrospinning (ES) approach to fabricate continuous CuAc₂/PVP nanofibers, followed by their oxidation and reduction to obtain the CNFs. The as-prepared CNFs show transparency of 80.48 % and conductivity of ~60 ohm/sq. Subsequently, a simple solvothermal treatment with sodium formate endows the CNFs with superior anti-oxidation properties. With the assistance of *in-situ* transmission electron microscopy, the thermal response of the CuAc₂/PVP nanofibers and the formation mechanism of Cu nanofibers were investigated. This study may contribute to the application of CNFs in the field of flexible optoelectronics and new display technology.

Keywords—Transparent Conductive Films, Electrospinning, In-situ TEM, Copper Nanofibers

I. INTRODUCTION

Transparent conductive films (TCFs) are key devices in the field of electronic displays and solar cells [1-2]. Nowadays, TCFs are dominantly fabricated by indium tin oxide (ITO), which show excellent transparency and good conductivity. However, its large-scale application is hindered due to the high cost and brittleness [3].

Copper nanofiber films (CNFs) were attracted increasing attention due to their low cost, high conductivity and excellent flexibility. Cui *et al.* fabricated high performance TCFs using continues nanofibers (NFs) fabricated by electrospinning (ES), and the resultant TCFs exhibited high conductivity and transparency [4]. However, the oxidation of Cu nanofibers is inevitable which may cause the deterioration of TCFs during service.

Different methods were developed for the anticorrosion of Cu, *e.g.*, inert material coating on Cu [5-6]. However, these methods were expensive. Recently, Peng *et al.* proposed a novel surface passivation treatment by HCOONa, which can provide superior anti-oxidation for Cu materials without electrical or mechanical property deterioration [7].

In-situ TEM technique is a powerful tool to investigate the mechanism at nanoscale. It has been widely used in the fields

of energy, catalysis, thermoelectric, *etc.* Applying different stimulation (*e.g.*, force, thermal, light, gas, electricity, *etc.*) on the sample may cause different changes, and thus we can analysis the heating process of NFs to directly observe the reaction process by *in-situ* heating TEM [8].

Herein, we use ES to fabricate continues nanofibers containing Cu salts, and then the nanofibers were heated and oxidized at ambient atmosphere to obtain the CuO nanofibers (CuO NFs). The resultant products were then heated at H₂/Ar, and the Cu NFs were finally produced. Using *in-situ* TEM, the reaction process of CuAc₂/PVP NFs was explored and the formation mechanism of Cu NFs was proposed.

II. EXPERIMEMT

A. Materials

N, N-Dimethylformamide (DMF, 99.5%) and Cu acetate (CuAc₂) were purchased from Rhawn. Sodium formate (HCOONa) was purchased from Macklin, polyvinyl pyrrolidone (PVP) was obtained from Sigma-Aldrich and oleylamine from D&B. All the reagents without any purification.

B. Preparation

In a typical synthesis process (Fig. 1), CuAc₂ and PVP were dissolved in DMF. The solution was stirred for overnight to form a homogeneous solution. Then the solution was load in a syringe, which the metal needle was connect to high voltage (the voltage was set at 15 kV) and provide a flow rate of 1 mL/h. An glass substrate or aluminum foil was perpendicularly placed 12 cm away from the needle. A few minutes later, the glass substrate was covered with PVP/CuAc₂ NFs. Next, the NFs were heated up to 500 °C at ambient atmosphere for 2 hours. In order to obtain the CNFs, the CuO NFs were annealed at 300 °C in H₂/Ar atmosphere for an hour. After that, HCOONa was dissolved in H₂O and DMF mixed solution. The as-prepared CNFs and oleylamine were added in aforesaid solution. The mixture was heated to 160 °C for 16 h. The passivated CNFs were noted as CNFs-SF.

C. In-situ Heating TEM

CuAc₂/PVP NFs were load on a microelectromechanical system (MEMS) chip on a heating holder (FEI- NanoEx) and the experiment was realized inside a transmission electron microscope (TEM, FEI Talos-200s). The heating rate was set at 1 °C/s, and maintained 5 min at 100 °C, 200 °C, 300 °C, 400 °C, 500 °C, respectively.

978-1-6654-1392-3/21 $31.00 © 2021 IEEE

D. Characterization

The as-prepared nanofibers were characterized by TEM, X-ray diffraction (XRD, Bruker-axs XRD), scanning electron microscopy (SEM, Zeiss SIGMA), and ultraviolet-visible spectroscopy (UV-vis, Shimadzu UV-2550).

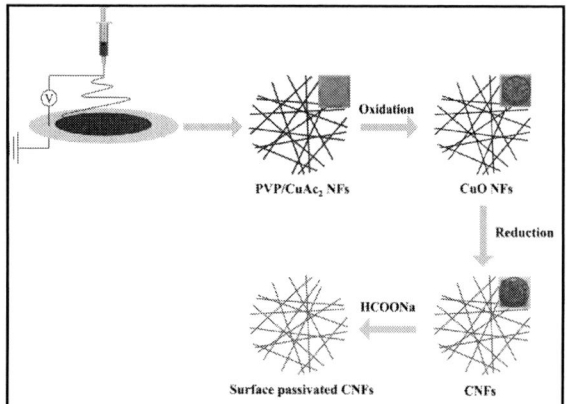

Figure 1 Schematic preparation of CNFs.

III. RESULTS AND DISCUSSION

A. Microstructure characteristics

ES is a promising technique which can produce ultra-long nanofibers with different diameters. Figure 2a shows the as-prepared $CuAc_2$/PVP NFs. The as-prepared $CuAc_2$/PVP NFs had smooth surfaces. However, after annealing, the rough surfaces of NFs were detected in both Fig. 2b and Fig. 2c.

Based on XRD patterns in Fig. 3, the polymer component was removed after heated at air, and $CuAc_2$ was oxidized to CuO simultaneously. Subsequently, Cu NFs were obtained after the reduction of CuO NFs in H_2/Ar atmosphere. The results were basically consistent with previous studies [4,6].

Due to mass loss during heating, the diameters of NFs were sharply reduced. As depicted in Fig. 4, the average diameter of $CuAc_2$/PVP NFs was ~410 nm; however, they reduced to ~209 nm for CuO NFs, and ~166 nm for Cu NFs.

Figure 2 SEM images of (a) $CuAc_2$/PVP NFs (b) CuO NFs (c) Cu NFs.

To fabricate CNFs, the glass substrate was placed 12 cm away from the metal needle. The NFs were collected on the glass substrate. Only 2 min is sufficient to form a network, with the conductivity of ~60 ohm/sq after treatment (measured by four-terminal configuration) and transparency of 80.48 %.

Cu is prone to be oxidized at ambient atmosphere, especially in nanoscale. The as-prepared CNFs were detected to be transformed into Cu_2O in just a few hours, leading to an increase in electrical resistance over time.

A surface passivation method of Cu has been developed [7]. With the help of HCOONa, the high oxidation resistance of Cu can be realized by surface reconstruction. We introduce this method to treat CNFs to enhance the oxidation resistance. But CNFs may be damaged during passivation treatment. While, improving the collecting time during electrospinning procedure can mitigate the damage. Increasing the collecting time leading to the decrease of transparency and increase of conductivity as well.

In order to accelerate the deterioration process. The samples were placed in an oven, and the temperature was set at 60 °C. As illustrated in Fig.5, surface passivation treatment inhibited the oxidation of Cu. Therefore, no peaks of CuO or Cu_2O were found after treatment, while apparent diffraction peaks of Cu_2O were detected in the sample without treatment. Meanwhile, the conductivity of CNFs-SF had no significant change in 24h. However, the performance of CNFs without treatment deteriorated obviously, with 538 % resistance increasement. Fig. 6 showed the resistance changes of CNFs and CNFs-SF, respectively.

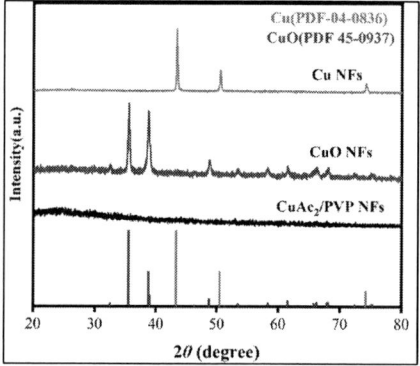

Figure 3 XRD patterns of $CuAc_2$/PVP NFs, CuO NFs and Cu NFs, respectively.

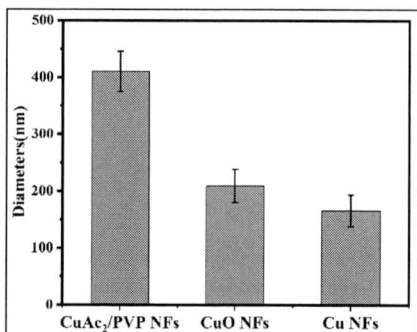

Figure 4 Average diameters of $CuAc_2$/PVP NFs, CuO NFs and Cu NFs, respectively.

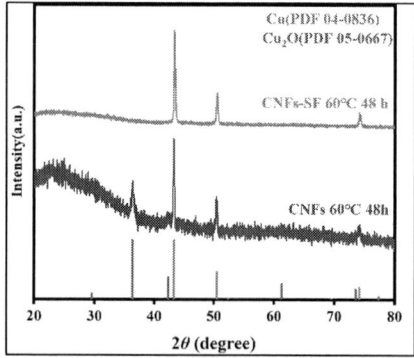

Figure 5 XRD patterns of CNFs and CNFs-SF at 60 °C.

978-1-6654-1392-3/21 $31.00 © 2021 IEEE

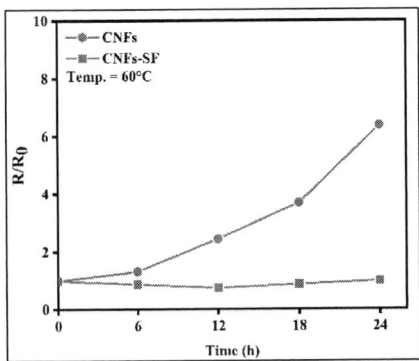

Figure 6 Resistance test of CNFs (red) and CNFs-SF (grey).

B. In situ heating

During heating treatment, the surface morphologies of NFs were changed due to the formation of nanoparticles, and their surfaces aggregated by nanoparticles were not smooth any more. Here, we apply *in-situ* heating TEM to investigate the sintering behavior of $CuAc_2$/PVP NFs. *In-situ* heating experiment was realized on a MEMS chip. First, we directly collected $CuAc_2$/PVP NFs on the chip, then the chip was transferred into a heating holder.

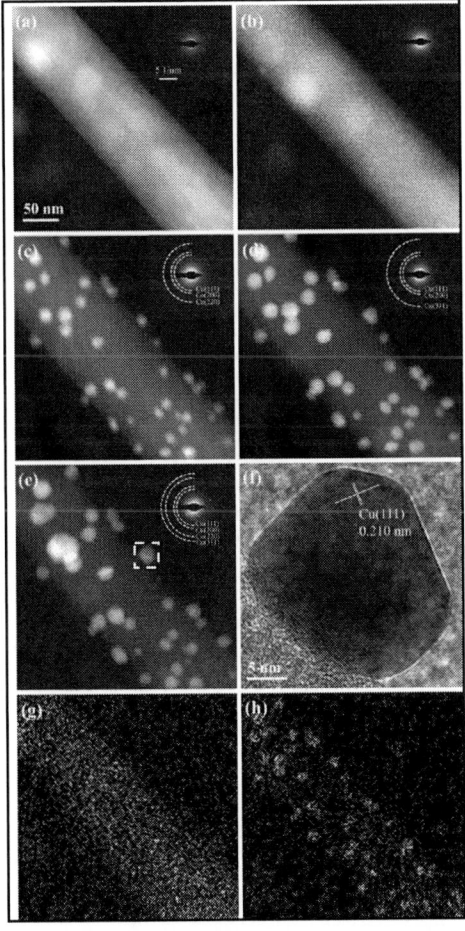

Figure 7 *In-situ* heating of $CuAc_2$/PVP NFs at (a)100 °C, (b) 200 °C, (c) 300 °C, (d) 400 °C (e) 500 °C. (f) HRTEM image of selected area in Figure 7e. (g, h) STEM mapping before and after heating, yellow dots represent Cu.

As we know, TEM has a strict limitation in the thickness of sample. Our nanofibers with diameters of ~410 nm were not able to be characterized by TEM. Accordingly, a small $CuAc_2$/PVP NF with a diameter of 110 nm was sought and characterized.

The Cu atoms of $CuAc_2$/PVP NF were well-distributed, demonstrated by EDS mapping (Fig. 7g), indicating that the copper salt and polymer were mixed homogeneously. The selected area electron diffraction (SAED) demonstrated that no crystalline existed in $CuAc_2$/PVP NFs, corresponding with our XRD result in Fig. 3.

After heating to about 240 °C, close to the decomposition temperature of $CuAc_2$, nanoparticles were gradually generated on the surface of NFs. These nanoparticles were metallic Cu based on the SAED and HRTEM analyses. Moreover, with increasing temperature, the Cu crystal grew up gradually. Interestingly, Cu nanoparticles preferred to nucleate on the surface of NFs rather inside of NFs. So when $CuAc_2$/PVP heated at air, Cu nanoparticles were formed first. Then, nanoparticles were oxidized into CuO. Because of the vacuum environment, we could not observe the transformation of Cu → CuO and decomposition of PVP.

According to the consequence of *in-situ* TEM, we unveil the formation mechanism of CuO NFs. When the $CuAc_2$/PVP NFs were heated to the decomposition temperature of $CuAc_2$, Cu nanoparticles formed first. Then, the Cu nanoparticles were rapidly transformed to CuO in air. Further heating cause the decomposition of PVP on account of existence of oxygen at about 450 °C. Once the PVP was completely decomposed, CuO NFs were formed. Indeed, the CuO NFs were consisted by CuO nanoparticles, formed before PVP decomposition. After annealing at H_2/Ar atmosphere, CuO nanoparticles were reduced to Cu nanoparticles, and the Cu NFs were synthesized. Here, the formation mechanics of Cu NFs was elucidated.

IV. CONCLUSION

In summary, CNFs were successfully fabricated by simple ES, which exhibit high conductivity (~60 ohm/sq) and transparency (80.48 %). HCOONa treatment endowed the as-prepared CNFs with an outstanding oxidation resistance. The electrical resistance had no obvious increase during deterioration experiment (maintain at 60 °C for 24 h) compared with a 583 % resistance increase without treatment. To further understand the thermal response of $CuAc_2$/PVP NFs, *in-situ heating* TEM technique was conducted. The formation of CuO NFs experienced three periods. 1) $CuAc_2$ decomposed to Cu nanoparticles. 2) Cu nanoparticles oxidized to CuO nanoparticles 3) decomposition of PVP and the CuO nanoparticles connect into fibers. During reduction process, the morphology of NFs show no significant change. The CuO NFs transformed into Cu NFs, which consist of Cu nanoparticles. Our work may be helpful for the application of CNFs in flexible displays and further understanding of formation mechanism of metal nanofibers.

ACKNOWLEDGMENT

This work was financially supported by the Fundamental Research Project of Shenzhen under Grant No. JCYJ20190809161213154, the National Natural Science Foundation of China (Grants No. 61471307) and the Fundamental Research Funds for the Central Universities (Grant No. 20720200075)

REFERENCES

[1] Wu H, Kong D S, Ruan Z C, et al. A transparent electrode based on a metal nanotrough network[J]. Nature Nanotechnology, 2013, 8(6): 421-425.

[2] Li D D, Lai W Y, Zhang Y Z, et al. Printable Transparent Conductive Films for Flexible Electronics[J]. Advanced Materials, 2018, 30(10): 24.

[3] Kumar A, Zhou C. The Race to Replace Tin-Doped Indium Oxide: Which Material Will Win? [J]. ACS Nano, 2010, 4(1): 11-14.

[4] Wu H, Hu L, Rowell M W, et al. Electrospun metal nanofiber webs as high-performance transparent electrode[J]. Nano Letters, 2010, 10(10): 4242-8.

[5] Ahn Y, Jeong Y, Lee D, et al. Copper Nanowire–Graphene Core–Shell Nanostructure for Highly Stable Transparent Conducting Electrodes[J]. ACS Nano, 2015, 9(3): 3125-3133.

[6] Jiang D H, Tsai P C, Kuo C C, et al. Facile Preparation of Cu/Ag Core/Shell Electrospun Nanofibers as Highly Stable and Flexible Transparent Conductive Electrodes for Optoelectronic Devices[J]. ACS Applied Materials & Interfaces, 2019, 11(10): 10118-10127.

[7] Peng J, Chen B, Wang Z, et al. Surface coordination layer passivates oxidation of copper[J]. Nature, 2020, 586(7829): 390-394.

[8] Zhang C, Firestein K L, Fernando J F S, et al. Recent Progress of In Situ Transmission Electron Microscopy for Energy Materials[J]. Advanced Materials, 2019, 32(18): 1904094.

Research on Thermal Characteristics of High Power 3D Microchannel Multichip Package

Xiangli Wei
School of Mechanical and Electrical Engineering , Guilin University of Electronic Technology
Guilin, China
weixiangli_guet@163.com

Daoguo Yang*
School of Mechanical and Electrical Engineering , Guilin University of Electronic Technology
Guilin, China
daoguo_yang@163.com

Wangyun Li
School of Mechanical and Electrical Engineering , Guilin University of Electronic Technology
Guilin, China
li.wangyun@guet.edu.cn

Xiyou Wang
School of Mechanical and Electrical Engineering , Guilin University of Electronic Technology
Guilin, China
wxy_07@126.com

Guangsheng Lu
School of Mechanical and Electrical Engineering , Guilin University of Electronic Technology
Guilin, China
guangshenglu@163.com

Shirui Xue
School of Mechanical and Electrical Engineering , Guilin University of Electronic Technology
Guilin, China
shiruixue4268@163.com

Abstract—In order to solve the heat dissipation problem of the high-power three-dimensional System in Package (3D-SiP), the Fractal Theory is applied to optimize the design of microchannel structure to get better thermal characteristics. The effects of microchannel spacing (W), cross-sectional width (L) and length (H) to heat dissipation are optimized with an orthogonal experiment. Then the optimum parameters are obtained by orthogonal test at the W 40 µm, L 40 µm and H 300 µm. Three types of microchannel structures, microchannel heat sink with parallel fin (MPF), microchannel heat sink with staggered fin (MSF) and microchannel heat sink with embedded fin (MEF), are designed and studied on thermal characteristics by ANSYS ICEPAK module. The simulation results show the temperature of MPF, MSF and MEF is lower than that of a conventional microchannel heat sink (CM) by 1.0 °C, 0.8 °C and 1.7 °C. Compared with the others, MEF could get the best heat dissipation which provides a reference for the design of high power 3D-SiP heat dissipation.

Keywords—3D-SiP, Fractal Theory, microchannel, heat sink

I. INTRODUCTION

With the rapid development of 5G communication technology, electronic devices continue to develop in the direction of miniaturization, high power and multi-function. Three-dimensional System in Package (3D-SiP) has become an important development direction for future electronic devices owing to its high flexibility, high integration and low cost. However, the reduction of package size could increase the heat flux density and seriously affect package reliability. Therefore, the heat dissipation of high-power 3D-SiP is an urgent problem. Air cooling is not suitable for 3D-SiP with high heat flux. Liquid has a higher thermal conductivity and specific heat capacity than air. Microchannel liquid cooling could be a suitable choice.

The structure and channel arrangement of the microchannel heat sink can significantly affect heat dissipation. A parallel fin microchannel heat sink with rectangular cross-section (MRCS) has a simple structure and better heat transfer performance[1]. Although microchannel heat sink with micro pin fins (MMPF) can efficiently improve heat dissipation under a slight increase in pressure drop, the structure is relatively complicated[2]. Because of the manufacturing cost and benefits, parallel fin microchannel

heat sink is still a research hotspot. Parallel fin microchannel heat sink can be designed by further optimization to obtain a better heat dissipation, such as large aspect ratio channels. Furthermore, the channel arrangement of the microchannel heat sink is an effective method to improve the cooling effect The conventional microchannel (CM) heat sink, a single inlet/outlet arrangement, has some disadvantages. For example, the pressure drop across a channel is high due to the long flow length[3]. N. Khan et al. [3] reported a flow arrangement (dual-port). There are two fluidic inlets and two outlets. This arrangement is beneficial to reduce the flow length. Self-similarity and iterative generation are proposed in the Fractal Theory. Optimization based on Fractal Theory will make the microchannel geometric structure more reasonable.

In this paper, Fractal Theory is applied to the optimization design of microchannel liquid cooling. Three types of microchannel structures are proposed in this paper. Heat flow coupling analysis is conducted by using the ANSYS ICEPAK module. The effects of three microchannel structures on the heat dissipation characteristics of the multichip hot spots are compared.

II. MODEL PREPROCESSING

A. Model Information

N. Khan et al. [3] presented a 3D-SiP integrated liquid cooling system, as shown in Fig. 1. The 3D-SiP package consists of two silicon carriers and a silicon interposer. Electrical interconnection between the chip and the silicon carrier is achieved by Ball Grid Array (BGA) and Through-Silicon via (TSV). The silicon carrier and silicon interposer are vapor-deposited with Au-Sn solder. The functions of airtightness and flow channel isolation between the silicon carrier and silicon interposer are realized by thermocompression bonding. Finally, the 3D-SIP composition is soldered on a PCB by BGA. In this arrangement, heat is transferred from the chip to the silicon carrier and is rejected to the ambient through the heat exchanger. The 3D-SiP structure is simplified and the microchannel structure is optimized in this paper, as shown in Fig. 2. The structure size parameters are shown in TABLE I.

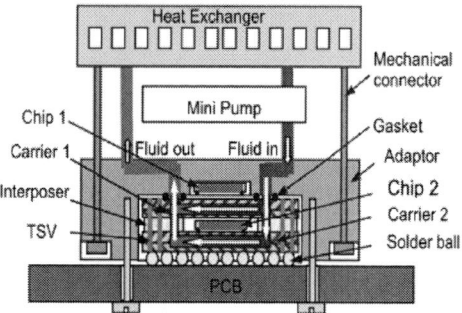

Fig. 1. Scheme of the integrated liquid cooling system for 3D-SiP

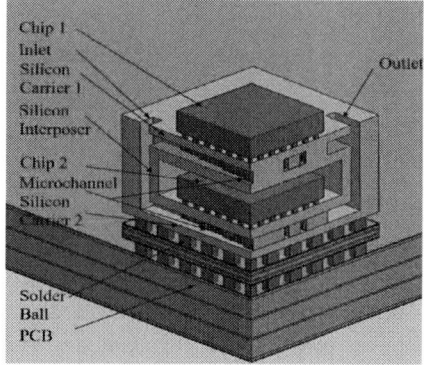

Fig. 2. Schematic simplify model of 3D-SiP

TABLE I.   STRUCTURAL DIMENSION PARAMETERS

Constituent	Parameters (mm)
PCB	30×30×1.6
Chip	4×4×0.5
Silicon Carrier	7×7×0.8
Silicon Interposer	5×5×0.8
Microchannel Cross-section	0.04×0.30
Inlet	0.70×0.25
Outlet	1.40×0.25

B. Modelling and Boundary Conditions

The three-dimensional model of the microchannel is established by Solidworks. Using the ANSYS ICEPAK module heat flow coupling analysis of the three types of microchannel structures is conducted. At room temperature, deionized water is the circulating coolant. The heat flux density of the chip is 100 W/cm². The volume flowrate of the water inlet is 200 ml/min, the fluid temperature at the water inlet is 50 °C and the pressure at the water outlet is 0 Pa.

The fluid dynamics analysis model refers to the flow characteristics and heat transfer characteristics of a single-phase fluid in the microchannel. Hypothesis in model preprocessing[4-6]:

- The flow is a steady-state, incompressible laminar flow.

- The wall of the microchannel is smooth. The Navier-Stokes equation and non-slip boundary conditions are still applicable.

- Ignoring radiation heat dissipation, ignoring the convective heat transfer between outside air and 3D-SiP.

III. MICROCHANNEL STRUCTURE AND OPTIMIZED DESIGN

Considering the 3D-SiP manufacture process and reliability, the design of the microchannel should minimize the large-angle turning of the fluid flow direction and reduce the change of the fluid cross-section. According to the Fractal Theory, three types of fin-type microchannel structures are designed in this paper, as shown in Fig. 3.

The microchannel heat sink with staggered fin (MSF) is made up of the discontinuous microchannel heat sink with parallel fin (MPF) and the interlaced turbulence structure, which is in discontinuous areas. MSF is beneficial to reduce the pressure drop of the fluid and strengthens the convective heat transfer. Fins are embedded in a single microchannel to form the microchannel heat sink with embedded fin (MEF). The fluid presents the vein flow direction. MEF is helpful to improve heat exchange and uniform temperature.

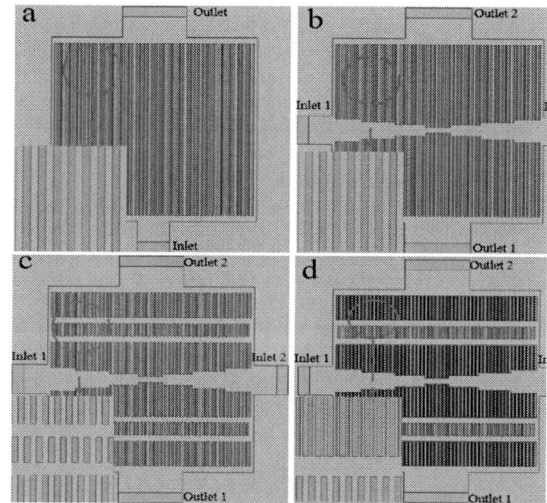

Fig. 3. Schematic diagram of the three types of fin-type microchannel structures. a: conventional microchannel (CM); b: microchannel heat sink with parallel fin (MPF); c: microchannel heat sink with staggered fin (MSF); d: microchannel heat sink with embedded fin (MEF)

The effects of microchannel spacing (W), cross-sectional width (L) and length (H) to heat dissipation are optimized with an orthogonal test. The test level was 4 and the L16(4^3) orthogonal test was designed. The horizontal design of the heat dissipation of microchannel orthogonal test factors is shown in TABLE II. The heat dissipation of microchannel orthogonal test range analysis data is shown in TABLE III.

TABLE II.   THE HEAT DISSIPATION OF MICROCHANNEL L16 (4^3) ORTHOGONAL TEST FACTOR HORIZONTAL DESIGN.

Level	Experimental Factor		
	Microchannel spacing W (μm)	Microchannel cross-sectional width L (μm)	Microchannel cross-sectional length H (μm)
1	40	40	150
2	60	60	200
3	80	80	250
4	100	100	300

TABLE III. THE HEAT DISSIPATION OF MICROCHANNEL ORTHOGONAL TEST DATA

Experiment Number	Experimental Factor			Experimental Result		
	W	L	H	Junction Temperature T1 (°C)	Junction Temperature T2 (°C)	Pressure Drop P (kPa)
1	1	1	1	79.62	79.33	667.96
2	1	2	2	81.52	80.87	326.22
3	1	3	3	82.87	82.04	204.27
4	1	4	4	84.04	82.95	138.25
5	2	1	2	79.54	79.03	391.80
6	2	2	1	82.11	81.46	608.70
7	2	3	4	83.06	82.73	141.71
8	2	4	3	84.94	84.12	196.02
9	3	1	3	79.21	78.97	271.04
10	3	2	4	80.65	79.93	158.22
11	3	3	1	83.94	83.16	563.01
12	3	4	2	84.15	83.26	322.63
13	4	1	4	79.42	78.85	214.87
14	4	2	3	82.32	81.62	230.13
15	4	3	2	82.87	82.12	346.33
16	4	4	1	88.23	87.49	558.36

The range analysis is done according to TABLE III. The influences of each factor on two chip junction temperatures are shown in Fig. 3. It can be seen that the temperature of the lower chip T1 (°C) is slightly lower than the upper T2 (°C). For the junction temperature of high power 3D-SiP chip, the L has the greatest influence, and the W and H have relatively little influence. The optimum parameters are obtained by orthogonal test at the W 40 μm, L 40 μm and H 300 μm. After optimization, the junction temperature of the chip is reduced by about 9 °C, and the pressure drop of the fluid is reduced by 22%.

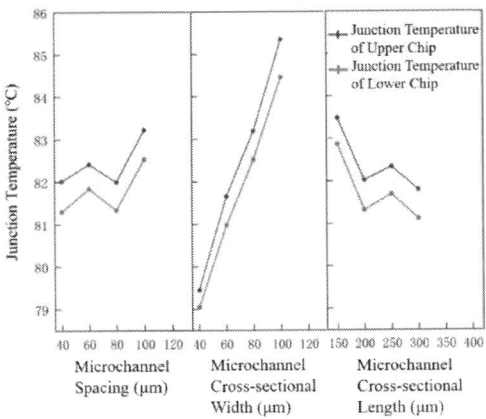

Fig. 4. Influences of each factor on two chip junction temperatures

IV. RESULT AND DISCUSSION

A. Comparison and Analysis

Using the ANSYS ICEPAK module conducts heat flow coupling analysis of the three types of microchannel structures. The closer to the water inlet, the lower the local temperature of the chip, and the closer to the water outlet, the higher the local temperature, as shown in Fig. 5. The maximum temperature of MPF, MSF and MEF is 79.11 °C, 79.26 °C and 78.38 °C, respectively. The upper chip has a slightly higher temperature than the lower. Because the flowrate in the lower carrier is slightly higher than that in the upper carrier and takes away more chip heat per unit time.

The fluid pressure distribution of four microchannel structures is shown in Fig. 6. It can be seen that the fluid pressure gradually decreases from the inlet to the outlet. The maximum pressure drop of MPF, MSF and MEF is 177.09 kPa, 165.59 kPa and 223.66 kPa, respectively.

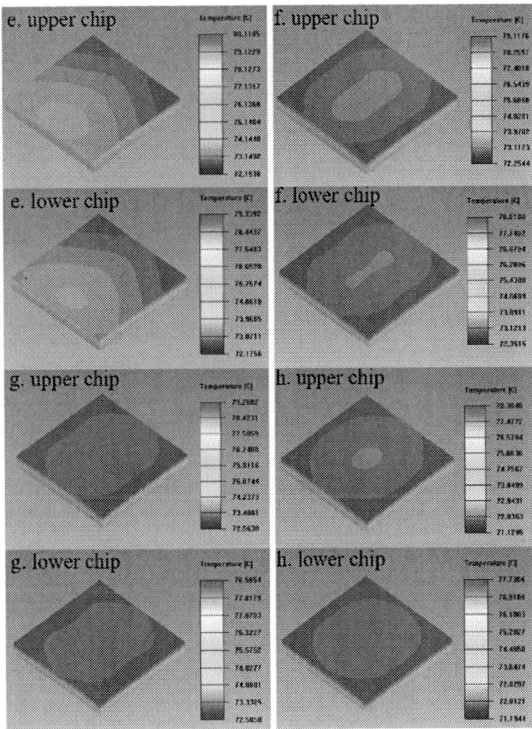

Fig. 5. The upper and lower chips temperature distribution of four microchannel structures. e: conventional microchannel (CM); f: microchannel heat sink with parallel fin (MPF); g: microchannel heat sink with staggered fin (MSF); h: microchannel heat sink with embedded fin (MEF)

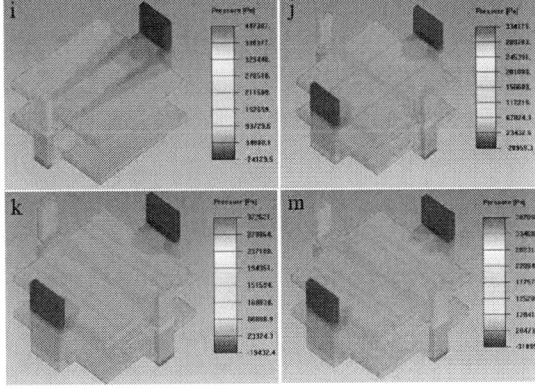

Fig. 6. The fluid pressure distribution of four microchannel structures. i: conventional microchannel (CM); j: microchannel heat sink with parallel fin (MPF); k: microchannel heat sink with staggered fin (MSF); m: microchannel heat sink with embedded fin (MEF)

The package thermal resistance refers to the ratio of the temperature difference between the junction temperature of the device to the outside and the heat dissipation power through the device. Rjc represents the total thermal resistance from the heat source junction of the chip to the package enclosure. The coolant inlet is selected as the lowest temperature reference point of the package enclosure. T1, T2, Rjc and P can be found in TABLE IV.

TABLE IV.   COMPARISONS OF 3D-SIP THERMAL CHARACTERISTICS IN DIFFERENT MICROCHANNEL STRUCTURES

Microchannel heat sink type	T1 (°C)	T2 (°C)	Rjc1(°C/W)	Rjc2(°C/W)	P (kPa)
CM	80.11	79.33	1.88	1.83	299.10
MPF	79.11	78.51	1.81	1.78	177.09
MSF	79.26	78.56	1.82	1.78	165.59
MEF	78.38	77.73	1.77	1.73	223.66

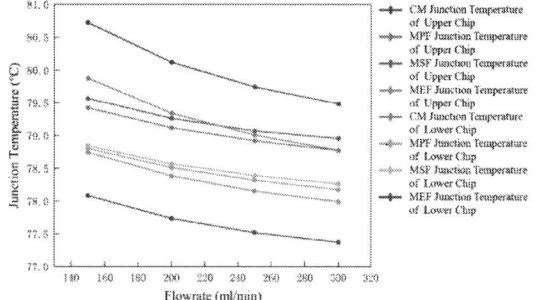

Fig. 7. Junction temperature characterization results

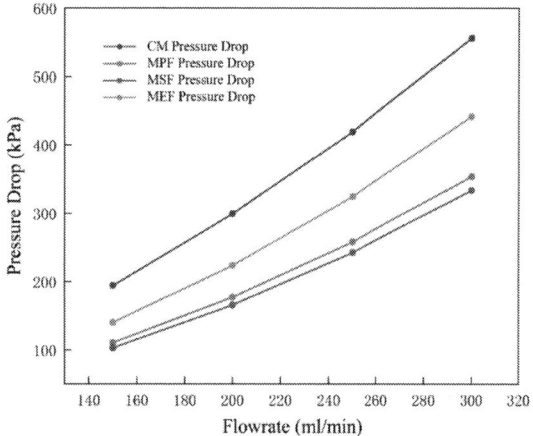

Fig. 8. Pressure dorp characterization results

As shown in Table IV, the chip junction temperature of MPF and MSF is larger. For MEF, the junction temperature is the smallest, the temperature distribution uniformity and heat dissipation is the best. In addition, under different flow rates, Junction temperature characterization results are shown in Fig. 7. Pressure drop characterization results are shown in Fig. 8. With the inlet flow rate increases, the junction temperature of the chip will decrease while the fluid pressure drop will increase. When the water flowrate exceeds 200 ml/min, the effect of flowrate on the decrease of junction temperature slows down and the pressure drop increases obviously. Considering the power loss of the heat exchanger, when the flow rate is 200 ml/min, the thermal characteristics of high power 3D-SiP multichip package are the most reasonable.

V. CONCLUSION

In this paper, Fractal Theory is applied to optimize the design of microchannel structure to get better thermal characteristics. The effects of microchannel spacing (W), cross-sectional width (L) and length (H) on the heat dissipation are optimized with an orthogonal experiment. Three types of microchannel structures, microchannel heat sink with parallel fin (MPF), microchannel heat sink with staggered fin (MSF) and microchannel heat sink with embedded fin (MEF), are designed and studied on thermal characteristics by ANSYS ICEPAK module. The heat flow coupling simulations of microchannel liquid cooling systems are carried out to reveal and contrast thermal and flow characteristics of different types of microchannels. The results can be briefly summarized as follows:

- The optimum parameters are obtained by orthogonal test as the W 40 µm, L 40 µm and H 300 µm. The junction temperature of the chip is reduced by about 9 °C, and the pressure drop of the fluid is reduced by 22%. The temperature of the lower chip is slightly lower than the upper, and the L has the greatest influence on the junction temperature of the chip.

- The temperature of MPF, MSF and MEF is lower than that of CM by 1.0 °C，0.8 °C and 1.7 °C. Compared with the others, MEF could get the best heat dissipation. The highest junction temperature of high power 3D-SiP multichip package is 78.38 °C, which can meet the requirements for the normal engineering application.

ACKNOWLEDGMENT

This research was supported by the Key R & D Plan Project of Guangxi Province (grant No. GuiKe AB20159038) and Science and Technology Planning Project of Guangxi Province under Grant No. GuiKeAD18281021.

REFERENCES

[1] H. Wang, Z. Chen, and J. Gao, "Influence of geometric parameters on flow and heat transfer performance of micro-channel heat sinks," Appl. Therm. Eng., vol. 107, pp. 870-879, 2016.

[2] N. R. Kuppusamy, R. Saidur, N. Ghazali, and H. A. Mohammed, "Numerical study of thermal enhancement in micro channel heat sink with secondary flow," Int. J. Heat Mass Transfer, vol. 78, pp. 216-223, 2014.

[3] N. Khan, L. H. Yu, T. S. Pin, S. W. Ho, V. Kripesh, D. Pinjala, et al., "3-D packaging with through-silicon via (TSV) for electrical and fluidic interconnections," IEEE T. Comp. Pack. Man., vol. 3, pp. 221-228, 2013.

[4] B. Xu, K. T. Ooti, N. T. Wong, and W. K. Choi, "Experimental investigation of flow friction for liquid flow in microchannels," Int. Commun. Heat Mass Transfer, vol. 27, pp. 1165-1176, 2000.

[5] X. F. Peng and G. P. Peterson, "Convective heat transfer and flow friction for water flow in micro channel structures," Int. J. Heat Mass Transfer, vol. 39, pp. 2599-2608, 1996.

[6] M. I. Hasan, A. A. Rageb, M. Yaghoubi, and H. Homayoni, "Influence of channel geometry on the performance of a counter flow microchannel heat exchanger," Int. J. Therm. Sci., vol. 48, pp. 1607-1618, 2009.

A Comprehensive Simulation Study of Warpage of Fan-out Panel-level Package using Element Birth and Death Technique

Yun-Kai Deng*

Lab of Smart Materials and Electronic Packaging in School of Materials Science and Engineering, and Guangdong Provincial Provincial Engineering Technology R&D Center of Electronic Packing Materials and Reliability
South China University of Technology, Guangzhou
510640, China
msykdeng@mail.scut.edu.cn

Bing-Xian Yang

Lab of Smart Materials and Electronic Packaging in School of Materials Science and Engineering, and Guangdong Provincial Provincial Engineering Technology R&D Center of Electronic Packing Materials and Reliability
South China University of Technology, Guangzhou
510640, China
msbingxianyang@mail.scut.edu.cn

Wei-Lin Hu

Lab of Smart Materials and Electronic Packaging in School of Materials Science and Engineering, and Guangdong Provincial Provincial Engineering Technology R&D Center of Electronic Packing Materials and Reliability
South China University of Technology, Guangzhou
510640, China
msweilinhu@mail.scut.edu.cn

Xin-Ping Zhang*

Lab of Smart Materials and Electronic Packaging in School of Materials Science and Engineering, and Guangdong Provincial Provincial Engineering Technology R&D Center of Electronic Packing Materials and Reliability
South China University of Technology, Guangzhou
510640, China
mexzhang@scut.edu.cn

Abstract—In recent years, the fan-out panel-level package (FOPLP) has attracted increasing attention due to its low cost while maintaining the same advantages of the fan-out wafer-level package (FOWLP) such as high I/O density and excellent performance. Compared with FOWLP, FOPLP possesses a higher carrier usage ratio and larger size, thus provides a cost-effective solution. However, FOPLP faces severe warpage problem caused by the mismatch of coefficient of thermal expansion (CTE) between the epoxy molding compound (EMC) and the carrier. In this study, a model of FOPLP featured with mold-first and die face-down approach is built and then used to evaluate and predict the warpage using ANSYS by means of the element birth and death technique based on finite element method (FEM) . The effects of carrier types (glass, steel, FR4 and silicon) on warpage are analyzed. Meanwhile, some key geometric parameters affecting the warpage have also been considered, such as the thickness of the carrier, chips and EMC cap. Simulation results manifest that the difference of CTE between carrier and EMC affects the warping direction. When the CTE difference is small, the modulus is the key factor affecting the warpage direction. Increase in thickness of the carrier and its Young's modulus leads to decrease in warpage. Reducing the mismatch of CTE between carrier and EMC will reduce warpage significantly. The thickness of the EMC cap is also crucial to warpage, a large thickness can result in increase of the rigidity of the package and thus decrease of the warpage.

Keywords—FOPLP; Warpage; Finite element method

I. INTRODUCTION

With the increasing demand for computer, mobile phones and other consumer electronics, most of the products tend to be lighter and thinner to meet the market. Nowadays, more and more attentions have been taken in advanced electronic packaging technology to fulfill the demand. Fan-out panel-level package (FOPLP) is one of the attractive technologies due to its high I/O density, minimal size and great reduction in cost. The panel size of FOPLP can reach up to 24×18 inch or even larger [1–3]. Such a high carrier usage ratio causes severe warpage and die shift issues [4], which hinder the application of FOPLP. It is necessary to predict and control the warpage during packaging processes for securing high reliability packages and wider application of advanced FOPLP technology. The warpage mainly comes from the mismatch of coefficient of thermal expansion (CTE) between the epoxy molding compound (EMC) and the carrier, especially during heating and cooling process. Too large warpage will increase the difficulties in subsequent processing and finally reduce product yield [5].

Due to very small dimension, sophisticated structure and complicated fabrication process of FOPLP, it is relatively difficult yet costly to evaluate accurately the warpage by means of conventional measurement approaches, in particular for the dynamic behavior in the process. In such circumstances, computer modelling method shows distinct advantage and can greatly save manpower and material resources.

In this study, the typical mold-first and face-down manufacturing process of FOPLP is considered, as shown in Fig.1, and a three-dimensional (3D) finite element model of FOPLP will be built and then used to evaluate and predict warpage generated in compression molding and back griding processes. Notably, with the help of the element birth and death technique in ANSYS Workbench 2020R2, it is very efficient and convenient to simulate the process of material reduction manufacturing. The effects of different carrier materials, such as glass, steel, FR4 and silicon, which are usually considered in research and production process, on the package warpage are analyzed. Meanwhile, some key geometric parameters affecting the warpage have also been taken into account, such as the thickness of carrier and chips as well as the thickness of EMC cap.

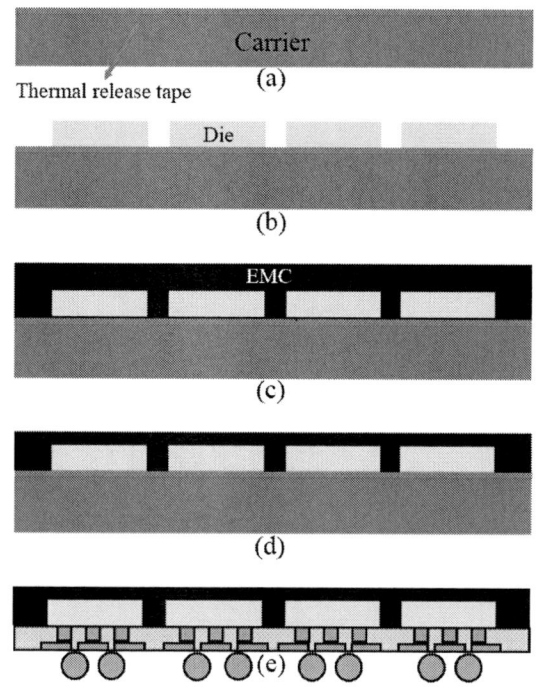

Fig. 1. Schematice of process of mold-first and face-down FOPLP: (a) attach thermal release film, (b) pick and place KGD, (c) compression molding, (d) back grinding, (e) fabricating RDL layer and solder balls.

II. FINITE ELEMENT METHOD

A. Model and material properties

Fig. 2 shows the cross section of the simulation model. Typically, the thickness of thermal release tape is at the micro-scale, so its influence on warpage can be neglected in simulation. In order to improve the calculation efficiency, a 1/4 model of the rectangular symmetric structure is used in the simulation, as shown in Fig. 3 and Fig. 4. The whole model consists of carrier, chip arrays (54×40, with pitch size of 11 mm) and EMC. The geometric parameters of the model are exhibited in Table I, where a, b and t represent length, width and thickness of the carrier, die and EMC, respectively.

Table II lists the material properties of carriers, dies and EMC [3, 6]. All of the materials are considered as linear elastic material. The simulation is conducted using ANSYS 2020R2.

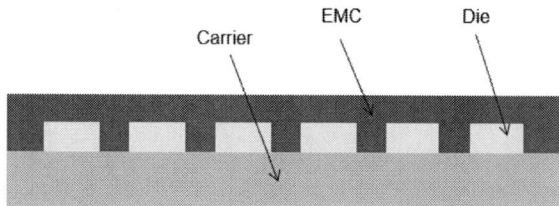

Fig. 2. Cross section diagram of the simulation model

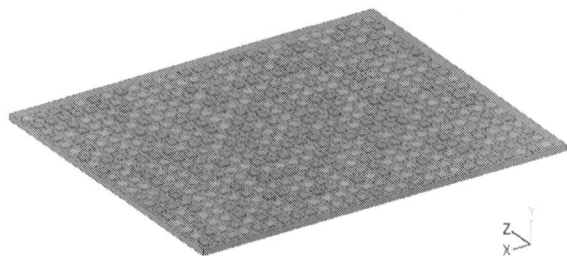

Fig. 3. 1/4 model (in which EMC is hidden to expose chip arrays)

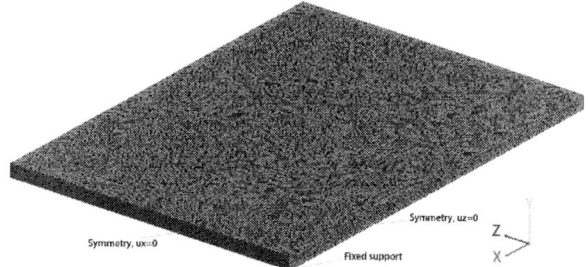

Fig. 4. The mesh shape and boundary conditions of the 1/4 model

TABLE I. GEOMETRIC PARAMETERS OF THE MODEL

Parts	a (mm)	b (mm)	t (mm)
Carrier	610	457	5
Die	7	7	1
EMC	610	457	5

TABLE II. MATERIAL PROPERTIES

Material	Young's modulus (GPa)	Poisson's ratio	CTE (ppm/°C)	Tg (°C)
Steel-1	200	0.33	12	-
Steel-2	200	0.33	15	-
Steel-3*	300	0.33	12	-
FR4	24.6	0.136	15.5	-
Silicon	131	0.28	2.8	-
Glass-1	69	0.25	3	-
Glass-2	70	0.25	5	-
Glass-3	69.3	0.25	7.6	-
EMC	28.5	0.2	8/40	175

* This is supposed to be a high modulus steel (for example, TiB_2 particle reinforced steel) for a comparative study.

B. Boundary conditons

In the simulation, the symmetric boundary condition is applied in two symmetrical faces, meaning that the displacement in the normal direction is zero and the center point is fixed to prevent rigid body displacement, as shown in Fig. 3.

The whole processes include compression molding, post mold curing (PMC) and back grinding. Firstly, the temperature is increased to 170 ℃ for compression molding, which is also the reference temperature (or called stress-free temperature). Then the temperature is further increased to 190 ℃ for PMC process, followed by cooling down to 100 ℃ for back grinding. Finally, it is cooled down to room temperature (25 ℃). Temperature profile of the whole process is depicted in Fig. 5.

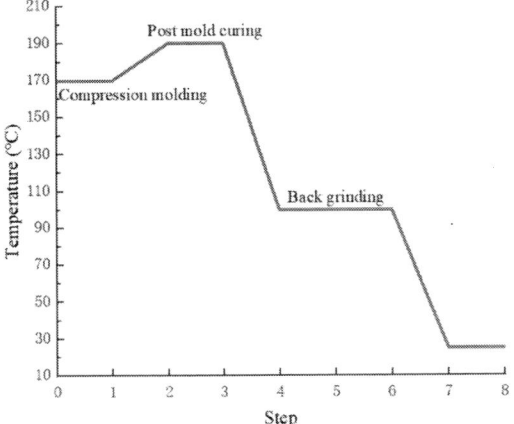

Fig. 5. Temperature profile of the entire process

III. SIMULATION DESIGN AND RESULTS

Effects of different types of carriers and the thickness of carrier, chips as well as EMC cap on the warpage are studied in this work. Simulations under the following four scenarios are performed.

A. Different types of carrier materials

To estimate the influence of different carrier materials on warpage, four different carrier materials are chosen, in which steel and glass have different mechanical and physical properties, as shown in Table I. Thus, eight groups of different carrier types with a constant EMC cap thickness are considered, as presented in Table III.

Fig.6 shows the simulation results of the change in warpage with different carrier types. In general, when the CTE of the carrier is smaller than that of EMC, the warpage value is positive (i.e., smiling face), otherwise it is negative (i.e., crying face), as shown in Fig. 7. However, the CTE of Glass-3 is slightly smaller than that of EMC, then the warpage value is negative, -1.472 mm. This is because the deformation mainly depends on Young's modulus when the carriers have similar CTE values, the material with smaller Young's modulus is more easily able to deform.

For steel carriers, when Young's modulus of the carrier increases from 200 GPa to 300 GPa, i.e., Type 1-1 to 1-3 carriers as shown in Tables II and III, the warpage decreases from -5.689 mm to -4.511 mm. Similar results are also found for Type 2 and 3 carriers, i.e., FR4 and silicon carriers. Increasing the Young's modulus of carrier can lead to decrease of the warpage, but the effect is not obvious.

For glass carriers, i.e., Types 4-1 to 4-3, simulation results indicate that the CTE mismatch between the carrier and the EMC is the major contribution to different warpage values. The mismatch value of CTE changes from 5 ppm/℃ to 0.4 ppm/℃, the warpage is decreased from 16.883 mm to 1.472 mm (absolute value), reduced by 76%. In short,

the smaller the mismatch of CTE between the carrier and the EMC is, the smaller the warpage is.

TABLE III. CARRIER TYPES USED IN SIMULATION

Carrier type	Carrier material	EMC cap thickness (mm)
1-1	Steel-1	2
1-2	Steel-2	2
1-3	Steel-3	2
2	FR4	2
3	Silicon	2
4-1	Glass-1	2
4-2	Glass-2	2
4-3	Glass-3	2

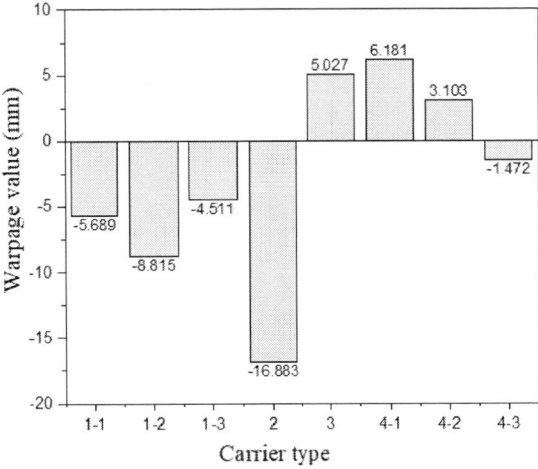

Fig. 6. Change of warpage with different carriers used in simulations

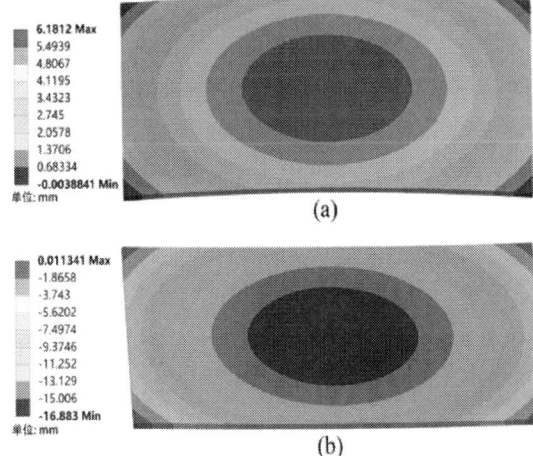

Fig. 7. Post-processing results of FE simulation: (a) negative value and (b) positive value of warpage.

B. Thickness of carrier

To characterize the carrier thickness effect on the panel warpage, five different carrier thicknesses (3, 4, 5, 6 and 7 mm) are considered in simulation, in which Glass-3 is chosen as the carrier, other geometric parameters keep unchanged. Simulation results are presented in Fig. 8. It should be mentioned that Glass-3 is also selected as carrier in subsequent simulations.

It can be seen clearly in Fig. 8 that the absolute value of the panel warpage decreases with the increase of the thickness of the glass carrier, despite a very small difference in warpage between two initial thin carrier plates with the thickness of 3 and 4 mm, respectively. To this end, it is understandable that the warpage is directly related to the stiffness of the carrier plate, and a thicker carrier plate has a higher stiffness than the thinner one, thereby generating a smaller warpage.

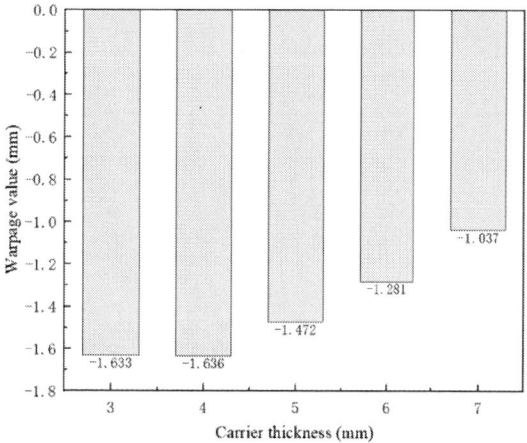

Fig. 8. Simulation results of change in the panel warpage with variation of the carrier thickness

C. Thickness of EMC cap

The thickness of EMC cap refers to the EMC thickness that are thicker than the dies. Obviously, it is very interesting to look at the correlation between the EMC cap thickness and the warpage. Five groups of experimental parameters with the increasing EMC cap thickness (which is actually corresponding to the decreasing EMC grinding thickness) are taken into account in the simulation, as shown in Table IV. Likewise, in the simulation only EMC cap thickness is changing, other geometric parameters are the same as those presented in Table I.

Fig. 9 exhibits a histogram of simulation results of change in panel warpage with the increasing thickness of EMC cap. Apparently, the warpage decreases with increase of the EMC cap thickness; for example, the absolute value of warpage decreases from 2.475 mm to 0.735 mm, about 70% reduction, with increasing the thickness of EMC cap from 1 mm to 3 mm. The results appear reasonable and may be understood in a way that a thicker EMC cap leads to increase of the stiffness (rigidity) of the package and thereby reducing the panel warpage. This actually means a thicker mold brings about lower panel warpage. Nevertheless, a comprehensive consideration should be given to optimal selection of carrier and EMC materials with appropriate parameters, such as CTE and Young's modulus, as well as the glass transition temperature (Tg) of EMC, so as to minimize the panel warpage in FOPLP. Definitely, more in-depth understanding and systematic study are needed.

TABLE IV.   CHANGE OF EMC CAP THICKNESS

Number	EMC cap thickness (mm)	EMC grinding thickness (mm)
2-1	1.0	3.0
2-2	1.5	2.5
2-3	2.0	2.0
2-4	2.5	1.5
2-5	3.0	1.0

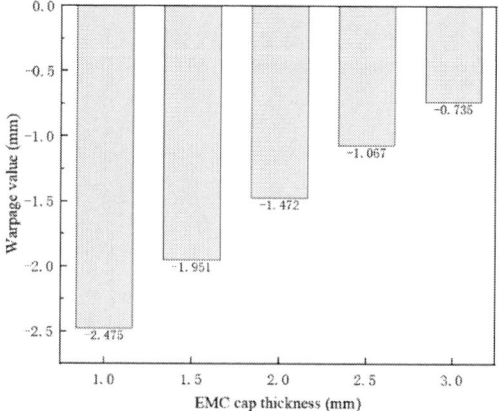

Fig. 9. Simulation results of change in panel warpage with the thickness of EMC cap

D. Thickness of the dies

The simulation results of the influence of die thickness change, in the range of 0.5, 1.0, 1.5, 2.0 and 2.5 mm, on panel warpage are exhibited in Fig. 10, in which the EMC cap thickness is 2 mm. As seen clearly in Fig. 10, the absolute value of the panel warpage increases with the die thickness. This result is similar to that for fan-out wafer-level package (FOWLP) presented in a previous study [8], in which a thicker die exhibited a larger warpage for different EMC materials. Noteworthily, it will be more significant to clarify the influence of change in the ratio of die thickness to EMC thickness on the panel warpage in FOPLP.

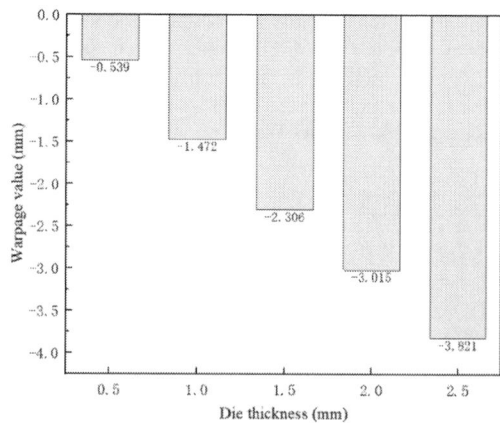

Fig. 10. Simulation results of change in the panel wrpage with the die thickness

978-1-6654-1392-3/21 $31.00 © 2021 IEEE

IV. ANALYSIS AND CONCLUSIONS

According to FEM simulation results obtained in the present study, the following conclusions can be drawn:

1) Increasing the modulus of the carrier can lead to decrease of the panel warpage in fan-out panel-level package (FOPLP), due to the increase of the rigidity (stiffness) of the package.

2) For a slight change in Young's modulus of the carrier, the change of CTE has a significant effect on the warpage. The decrease in the mismatch of CTE between the carrier and the EMC can bring about a significant reduction of the warpage.

3) Thickness of EMC cap is a key parameter influencing the panel warpage. Increasing the thickness of EMC cap can result in increase of the rigidity of the package structure, thereby reducing the warpage. For a constant EMC cap thickness, a thicker die appears to correspond to a larger warpage.

ACKNOWLEDGMENTS

This research is supported by the National Natural Science Foundation of China under grant No. 51775195, and the Key Project of Guangzhou City Science and Technology Plan under grant No. 201807010028.

REFERENCES

[1] S. Chao, Y. Sung and D. Luh, "New wave Fan-out package: For heterogeneous integration," 2017 International Conference on Electronics Packaging (ICEP), Yamagata, Japan, 2017, pp. 321–324.

[2] T. Braun, S. Voges, M. Töpper, M. Wilke, M. Wöhrmann and U. Maaß, " Material and process trends for moving from FOWLP to FOPLP," 2015 IEEE 17th Electronics Packaging and Technology Conference (EPTC), Singapore, 2015, pp. 1–6.

[3] F. X. Che, K. Yamamoto, V. S. Rao and V. N. Sekhar, "Panel Warpage of Fan-Out Panel-Level Packaging Using RDL-First Technology," IEEE Transactions on Components, Packaging and Manufacturing Technology, vol. 10, no. 2, 2020, pp. 304–313.

[4] T. Braun, K. F. Becker, O. Hoelck, S. Voges, L. Boettcher and M. Töpper, "Panel Level Packaging - From Idea to Industrialization -," 2019 IEEE CPMT Symposium Japan (ICSJ), Kyoto, Japan, 2019, pp. 85–87.

[5] T. Braun, K. F. Becker, R. Kahle, R. Kahle and S. Raatz, "Potential and challenges of fan-out panel level packaging," 2016 IEEE CPMT Symposium Japan (ICSJ), Kyoto, Japan, 2016, pp. 132–136.

[6] C. Chen, D. Yu, T. Wang, Z. Xiao and L. Wan, "Warpage Prediction and Optimization for Embedded Silicon Fan-Out Wafer-Level Packaging Based on an Extended Theoretical Model," in IEEE Transactions on Components, Packaging and Manufacturing Technology, vol. 9, no. 5, 2019, pp. 845–853.

[7] Z. Kuang, G. Yang, R. Cui, Y. Zhang and C. Cui, "Warpage simulation and optimization of panel level fan-out package in post molding cure," 2020 21st International Conference on Electronic Packaging Technology (ICEPT), 2020, pp. 1–5.

[8] F.X. Che, D. Ho, M. Z. Ding and X.W. Zhang, "Modeling and design solutions to overcome warpage challenge for fan-out wafer level packaging (FO-WLP) technology," 2015 IEEE 17th Electronics Packaging and Technology Conference (EPTC), Singapore, 2015, pp. 1–8.

Study of Package Reliability according to the Epoxy Molding Compound

1st Eunsol Jo
Electronic Package Research Center
Knagnam University *organization*
Yong-In, Republic of Korea
dmsthf914@gmail.com

2nd Jung-Rae Park
Electronic Package Research Center
Knagnam University *organization*
Yong-In, Republic of Korea
rae9364@daum.net

3rd Cheong-Ha Jung
Electronic Package Research Center
Knagnam University *organization*
Yong-In, Republic of Korea
cjdgk1101@nate.com

Gu-Sung Kim*
lElectronic Package Research Center
Knagnam University *organization*
Yong-In, Republic of Korea
gkim@kangnam.ac.kr

Abstract—**5G technology takes the form of a heterogeneous integration package in which several functions are loaded in one module. However, Heterogeneous Integration technology enables the miniaturization and multifunctionality of the package, but generates more heat on operation compared to a single module. This heat affects the warpage and stress of the package, which affects the performance and reliability of the package. In order to solve this problem, the warpage and stress occurring in the package were confirmed by adjusting the composition ratio of EMC. As a result, as the silica content of EMC increased, warpage decreased, and as the silica content decreased, the stress decreased.**

Keywords—EMC, 5G, Silica, Warpage, Stress, FEM

I. INTRODUCTION

5G technology is an important technology that enables a hyperconnected society beyond simple communication, and is increasingly affecting our living life. Unlike the conventional RF package, the 5G package features Heterogeneous Integration (HI) in which several functions such as antenna, RF, amplifier, and transceiver are mounted in one package. [1] HI technology enables multifunction and miniaturization of the package, but has a problem in that it generates more heat than the conventional single package, which affects chip life and reliability. Therefore, controlling the heat generated when driving the chip is one of the important factors in 5G technology. Among the semiconductor packaging processes, the molding process serves to dissipate heat while protecting the internal chip from the external environment. [2] The material mainly used as a molding material in the semiconductor industry is Epoxy Molding Compound(EMC), a composite material that is a mixture of an epoxy and silica. More than 70% of EMC constituent materials are composed of silica and epoxy resin, and the composition ratio of the two components affects mechanical and thermal properties. If the silica content is higher than epoxy, the thermal conductivity increases but the coefficient of thermal expansion decreases. Conversely, when the content of the epoxy resin is higher than silica, the thermal conductivity decreases while the thermal expansion coefficient increases. [3,4] Since the thermal conductivity and the coefficient of thermal expansion of EMC are in a trade-off relationship, it is difficult between controlling the warpage and increasing the heat dissipation performance at the same time. Warpage caused by the difference in the coefficient of thermal expansion from the substrate makes it difficult to proceed with the subsequent process. And if heat is not released, the reliability of the semiconductor is deteriorated. Therefore, it is essential to properly control the EMC composition ratio and understand the mechanism. In this study, after designing a simple RF package through a 3D modeling program, it was applied with different properties of EMC, and confirmed the warpage and stress caused by heat generated by RF chip operation. The properties of EMC were compared and analyzed by applying the composition of Silica at 80%, 65%, and 50% based on commercial materials that are widely used in the semiconductor industry. [5]

II. FINITE ELEMENTS METHOD

A. Modeling

In this study, ANSYS v.2020 R2 was used to confirm the stress and warpage caused by heat when the chip was operated. Figure 1. (a) shows the overall package model. To make a simple package, it was made only from substrate, solder balls, chip and EMC. In order to increase the efficiency of the analysis, the analysis was performed by cutting the package into quarter considering it is symmetric model. Figure 1. (b) shows quarter of full model. There are approximately 710,000 nodes and 320,000 elements was used in this analysis.

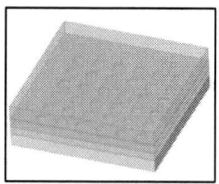 

Fig 1. (a) Full package   (b) Quarter of model

B. Boundary Condition

In order to simulate the situation when operating the chip, the room temperature was defined as a stress free state. 150℃, the temperature generated during operation of the chip, was applied to the chip as a thermal load condition. Warpage and stress behaviors occurring at this time were confirmed. The properties used in the package are specified in Table 1, and the properties of EMC refer to the results of previous studies. [5]

TABLE I.   MATERIAL PROPERITES

Materials	FR-4	Chip	Solder	EMC A	EMC B	EMC C
CTE (ppm/℃)	3.53	2.6	70	14	27	35
Young's Modulus (GPa)	18	163	2.4	20	10	7
Poisson's ratio	0.11	0.27	0.33	0.2	0.23	0.26
Tg(℃)	-	-	-	138	140	145

III. SIMULATION RESULTS

A. Warpage

Figures 2, 3 and 4 show the results with silica content of 80%, 65% and 50% in order. As a result, in the package to which EMC with 80% silica content was applied, the lowest warpage was 1.06um. Warpages of 1.4um and 1.5um were found in 65% and 50%, respectively.

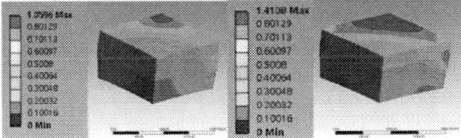

Figure 2. EMC A   Figure 3. EMC B

Figure 4. EMC C

B. Stress

Figures 4,5 and 6 show the results with silica content of 80%, 65%, and 50% in that order. As a result, the lowest stress was found at 69 MPa in the package to which the EMC was applied with a silica content of 50%. 65% and 80% showed stresses of 81 MPa and 99 MPa, respectively.

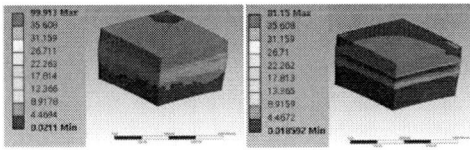

Figure 5. EMC A   Figure 6. EMC B

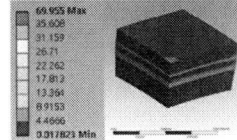

Figure 7. EMC C

IV. DISCUSSION

In this study, warpage and stress were confirmed by using three EMCs with different silica contents in consideration of the chip heating situation. After the simulations, some important result and tendency was found and that are summarize din the following.

1) The higher the silica content in EMC, the smaller the CTE, which is suitable for improving warpage.

2) The higher the silica content, the lower the thermal conductivity, indicating that the stress caused by heat was small.

3) It is considered that a study on heat dissipation according to EMC is need.

4) It is necessary to study the heat dissipation according to the silica size.

5) Considering that there is a trade-off relationship between filling rate and fluidity according to silica content also, it is considered necessary to develop products with high filing rate and fluidity.

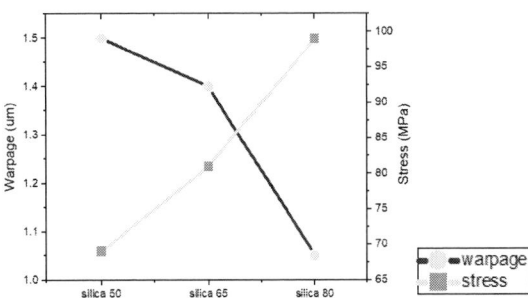

Fig 8. Warpage and stress according to silica content

REFERENCES

[1] Atom O. Watanabe, Sk Yeahia Been Sayeed, Rao R. Tummala (2020), "A Reivew of 5G Front-End Systems package Integration", IEEE

[2] M. Sadeghinia, K.M.b. Jansen, L.J. Ernst (2012), "Characterization and modeling the thermo-mechanical cure-dependent properties of epoxy molding compound", International Journal of Adhesiln & Adhesives, 32, pp. 82-88.

[3] Minseok Seo (2020), "Packages and tests that increase the added value of semiconductors", HANOL, Leecheon.

[4] Joonwoo Lee, Heungsoon Shin, Byeongyel Ko (2003), "Semiconductor Packaging Materials", Korea Institute of Science and Technology Information, Seoul, pp.1-20.

[5] DongKil Shin, JungJoo Lee (1997), "Thermal Stress Analysis of EMC in QFP Electronics Package", The Korean Society of Mechanical Engineers, pp. 771-777.

Simulation analysis of the influence of test conditions on the bonding strength of PCB pads

Huanhuan Wang
College of Mechanical and Electrical Engineering
Guilin University of Electronic Technology
Guangxi, Guilin, China
whh920321342@163.com

Miao Cai*
College of Mechanical and Electrical Engineering
Guilin University of Electronic Technology
Guangxi, Guilin, China
caimiao105@163.com

Daoguo Yang
College of Mechanical and Electrical Engineering
Guilin University of Electronic Technology
Guangxi, Guilin, China
daoguo_yang@163.com

Hengjian He
College of Mechanical and Electrical Engineering
Guilin University of Electronic Technology
Guangxi, Guilin, China
921688895@qq.com

Yanchen Wu
College of Mechanical and Electrical Engineering
Guilin University of Electronic Technology
Guangxi, Guilin, China
wyc1996823@163.com

Lei Song
College of Mechanical and Electrical Engineering
Guilin University of Electronic Technology
Guangxi, Guilin, China
754605552@qq.com

Abstract—**In practice applications, the problem of printed circuit board (PCB) pad cratering is a common in BGA components. Pad cratering starts with the laminate. Cracks may start at the intersection of the copper pad and the laminate and gradually initiate at the prepreg and glass fiber, which can lead to PCB Pad cratering over time. These pads with pits may not fail immediately during early performance testing, but may initiate cracks during later service.**

In this paper, the finite element analysis method is used to study the stress and strain distribution of the pad when a single BGA pad is subjected to a certain tension, and the changes in the stress of each layer node during the process of pulling out the pad. and to simulate the bonding process of pad, adhesive layer and fiber interface. The pin pulling test of PCB material was carried out by 2D simulation with the cohesive method. Finally, it finds that the contact line area between pad edge and adhesive layer is subjected to the greatest pull force. Among the variables, the size of the pad has the greatest influence on the tensile force. The stress change process in the drawing process is the same as [1], which shows that the use of 2D simulation has a certain feasibility. The above conclusions can be used to optimize the design of PCB pads, and have a good reference value to the PCB designer.

Keywords—Bonding strength of PCB pads, FEM, Cohesion, The layers of stress, pad cartering

I. INTRODUCTION

With the progress of the times, people's requirements for the quality of life are getting higher and higher. Microelectronics products are developing towards portability, miniaturization and high performance. Among them, BGA (Ball Grid Array) packaging has become one of the most advanced packaging technologies. However, in practice, pad cratering often occur in BGA components. Pad cratering usually occur under a mechanical stress environment, and pad cratering will appear over time. Pad crater is the initiation and propagation of fine cracks under the BGA Pad in the organic substrate material or PCB laminates. The IV pattern in Fig. 1 [1-2]. It is one of the major failure modes encountered in electronic components, especially in lead-free and halogen-

free processes. Pad cratering begins in the laminate, and cracks may start at the intersection of solder, copper pad, and laminate, which is the stress concentration of crack initiation [3-4].

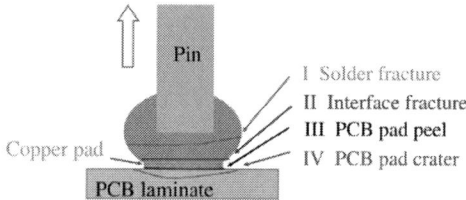

Fig. 1. Classification of failure modes[1].

IPC-9708[5] proposes three Test methods for pad testing, which are the Hot Pin Pull Test, the Ball Pull Test and the Ball Shear Test. Follow-up test pad strength method is most widely used in soldering ball drawing and shear test. Y. Yao et al[6]. studied the effect of drop shock on BGA electronic package through numerical analysis, and established a 3D model to simulate solder joint interconnection and electronic package failure under drop shock on plate surface. The cohesive force method was used to predict the crack initiation and propagation near the IMC/ solder interface. Cai M[1] et al. studied a new test device for pin pulling, and used this device to study the effects of reflux times, different pad types and length of service time on pad reliability. Song F B et al[7]. put forward the influence of different drawing speed, type of solder pad and size of solder pad on the drawing force by using the method of ball shear and pull-out experiment. Effect of different reflow on pad strength under the same pad type. Zhou R et al[8]. combined the Drucker-Prager CAP (DPC) model with the cohesive force model (CZM) to simulate the tensile and shear cracks and the process of crack propagation under external loading, he results show that the exponential CZM model mixed with DPC model can describe the change of the crack damage zone more accurately. Zhang Q M et al., used a ball shear testing machine to shear a single solder joint. Change its shear parameters and then calculate their average tension in failure mode. The validity of the proposed tensile force is verified by a four-point bending experiment with the obtained average tensile force. The purpose of this method is to simulate the loading condition of single solder joint assembly level test.

Sn3Ag05Cu as lead-free solder is harder than SnPb solder, so they can transfer more stress and strain to PCB board. Phenolic cured PCB materials in lead-free solder are more brittle than traditional DICY-cured FR4 materials, and peak temperatures of lead-free components may result in higher strains. Pad failure is more likely to occur in components where stress and strain are concentrated. It is necessary to analyze the force of PCB pads. In this paper, the finite element simulation analysis method is used to evaluate the stress and strain distribution of a single BGA pad under stress, as well as the stress variation in the process of pin pulling. And the method of cohesion force was used to conduct several pin drawing tests on PCB materials to quickly evaluate the influence of pad size, drawing speed and drawing Angle on pad strength.

II. THE ESTABLISHMENT MODEL AND ANALYSIS

A. 3D model building

PCB board is mainly composed of protective film, prepreg layer (semi-cured sheet), epoxy resin (composed of glass fiber cloth and FR-4), which is pressed under high temperature, as shown in Fig. 2. Green oil is mainly high temperature resistance, anti-moisture, reduce the role of copper pollution on the welding tank. Prepreg layer plays the role of bonding core board. It is mainly composed of gel content, flow speed, gel time and volatile matter content. Adhesive content: general resin content of 45%-65%, its content decreases with the increase of the thickness of glass fiber cloth. Resin flow is easy to produce the phenomenon of lack of glue or poor glue, and then produce bubbles, cavities and other phenomena. Volatile content should be less than or equal to 0.3%. In the lamination of volatile materials are prone to form bubbles, resulting in the flow of resin foam. So different PCB board manufacturers of finished product yield is different, PCB board of finished product yield and manufacturing process has a great relationship.

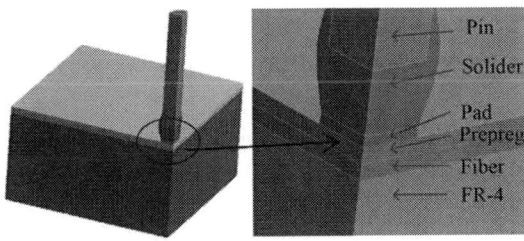

Fig. 2. One quarter of a single BGA pad model.

Due to the small size of the BGA pads in the PCB, failure can occur when subjected to very small stresses, and this failure may not occur immediately. In actual service, the crack will occur along the prepreg layer below the pad, the glass fiber, and then the pad cratering as shown in Fig. 1 IV. The solder pin drawing method is mainly used to evaluate the stress and strain distribution of each layer of a single BGA pad under a single drawing stress. The main methods to evaluate the strength of board PCB pads are drop, bend and stretch. In this paper, PCB board is divided into: adhesive layer, glass fiber layer, FR-4 substrate, as shown in Fig. 2. Material parameters of the model structure are shown in

Table 1. Using ANSYS to build a 3D model for the overall force analysis, in order to improve the speed of operation, the final use of a quarter of the model for analysis. The structural size of the 3D model is shown in Table 2.

TABLE I. MATERIAL PROPERTIES USED IN FEM

Materia l Layer	Type	Young' modulus, (MPa)	Poisson' ratio	CTE, ppm/C,25°C
Pad/Pull Pin	Cu	127000	0.31	17.5
Solder	SAC 305	16700	0.3	20
Prepreg	Epoxy	1610	0.29	14(x, y)
Fiber	Fiber	150000	0.11	10(x, y)
FR-4	FR-4	25000	0.18	16(x, y)

TABLE II. GEOMETRIC PARAMETERS OF THE 3D MODEL STRUCTURE

Item	Dimensions(mm)	
	L*W*H	R*H
Cu pad	-	0.2×0.035
Epoxy	5×5×0.05	-
Fiber	5×5×0.05	-
FR-4	5×5×1.3	-
SAC305	-	0.3×0.5
Cu Pin	-	0.25×1.5

PCB pads can be divided into 4 types[3]: T Pad with lead connection, V Pad with through-hole connection, G Pad with ground connection, O Pad without connection, i.e., single bare Pad. Single bare pad used in 3D simulation.

B. 2D model building

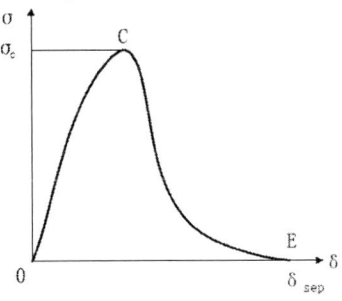

Fig. 3. Stress and opening displacement curves of exponential type CZM.

Analyzed from the atomic level, the cohesive force to be overcome when the surface is separated, The corresponding stress value is the cohesion (for plastic metals, the cohesive stress is the yield stress of the material). The CZM suggested that there exists a small nonlinear region (cohesive zone) due to microcrack or plasticity at the crack tip. The cohesive force σ increases with the increase of external load, and the same is true for the surface relative displacement δ. When the cohesion is the maximum σ_c, a new fracture surface is produced by interface separation. At the same time, it is accompanied by interface damage. The damage degree deepens with the increase of the relative displacement of the interface. Meanwhile, the cohesion force gradually decreases from the peak value to zero (at this time, the corresponding relative displacement reaches the maximum value, also known as the critical relative displacement), as shown in Fig. 3. Cohesive force is divided into exponent type, ladder type, etc. This paper uses exponential cohesion.

Because the strength of BGA pads in PCB is affected by many aspects, such as different pad size, pad type, service time, and the number of reflux, etc. The 2D model established uses the method of exponential cohesion in this paper. Firstly, the model is divided into two parts, one is the pin, tin ball and solder pad, the other is the adhesive layer, glass fiber and FR-4 substrate, using the binding Lagrange self-enhancement algorithm. The parameter properties of the corresponding 2D model material are the same as those of the 3D model. The corresponding parameters of the model on the XY plane established are shown in Table 3. CZM parameters are shown in Table 4. In this paper, there are mainly two methods of pads, O pads and T pads. The two kinds of pads are used for comparative analysis. The size of the T pad on the X-Y plane is 0.6×0.035mm. The thickness is the same as the bare pad. The grid division is shown in Fig 4. the strength of the pad is analyzed by changing different parameters.

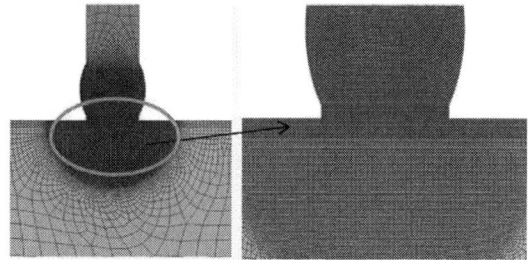

Fig. 4.   Meshing of 2D models

TABLE III.   GEOMETRIC PARAMETERS OF THE 2D MODEL STRUCTURE

type	Size（mm）		
	Plane stress mode		
	X-Y		Z(Thickness)
Cu pin	0.4×1.5		0.4
SAC305	$0.4(H) \times 0.25$（R）$\times 0.2(R)$		0.4
Pad	$0.4 \times 0.0.35$		0.4
Epoxy	5×0.05		5
Fiber	5×0.05		5
FR-4	5×1.3		5

TABLE IV.   PARAMETERS OF THE CZM MODEL

Maximum normal contact stress (MPa)	Contact gap at the completion of debonding (mm)	Maximum equivalent tangential contact stress (MPa)	Tangential slip at the completion of debonding(mm)	Artificial damping coefficien -t(s)
30	1	10	0.33	1E-08

III. THE RESULTS OF THE 3D MODEL ARE DISCUSSED PREPARE

A.   The results of the 3D model

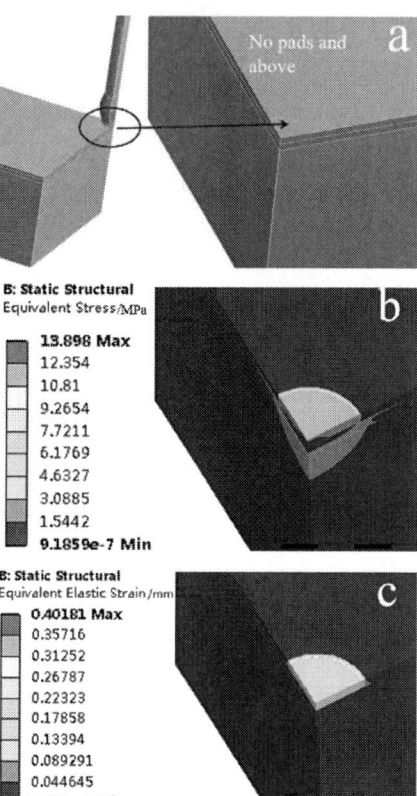

Fig. 5.   Pull pin test under 10N tension. (a) Overall model, (b) Stress distribution, (c) strain distribution.

In the simulation of the model in Fig. 2, O pad was closely used for simulation analysis. The whole simulation method is adopted for the model. The surroundings temperature is 25℃, and the pin is pulled 10N upward. Fig. 4a is a model diagram of no pads, solder balls and solder pins. The stress and strain analysis results of the PCB board are shown in Fig. 5b-c. B1-B2, C1-C2 and D1-D2 are the distance from the midpoint to the edge of the upper surface of the adhesive layer, glass fiber and FR-4 substrate, respectively. A1-A2 is the distance from the midpoint to the edge of the lower surface of the pad, as shown in Fig. 6a. Stress distribution of nodes in each layer is shown in Fig. 6b. As can be seen from the model results. The contact line area between pad edge and adhesive layer is subjected to the greatest stress and strain, followed by glass fiber, adhesive layer and FR-4. This result is the same as in practice, that is, the brittleness fracture or pads cartering is most likely.

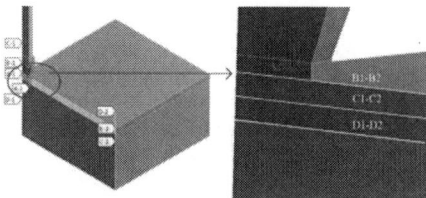

(a).   Finite element model of each layer stress.

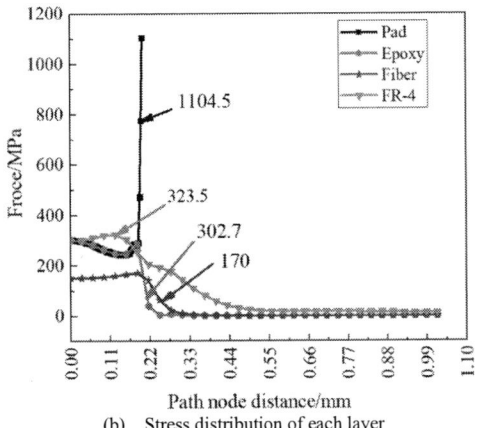

(b). Stress distribution of each layer

Fig. 6. pull pin test under 10N tension.

B. The results of the 2D model

In order to facilitate calculation, 2D model is adopted in this paper. The 2D model is analyzed by changing the drawing Angle and drawing speed respectively. The standard condition is vertical stretching, 25°C ambient temperature, and fixed bottom surface. The 2D model uses separate variable type CZM and Lagrangian binding algorithm. Due to the maximum stress and strain at the interface between the pad and the adhesive layer in the 3D model, brittle fracture or pad cratering is most likely to occur (of course, the interface between the pad and SAC305 will also fail, which is not within the scope discussed in this paper). So think of them as separation layers, the order from top to bottom of a 2D model is the same as that of a 3D model. The only difference is to divide them into two groups. The upper part is divided into: pin, SAC305, pad, the lower part is divided into adhesive layer, glass fiber, FR-4 substrate.

O pad and T pad were used in the 2D model for comparative analysis. The drawing direction is shown in Fig. 7a. Clockwise is negative and counterclockwise is positive. The stress distribution of the drawing Angle is shown in Fig. 7b. It can be seen from the figure that the strength of the pad is the maximum when it is 0°, and the lowest when it is between 30° and 40°. For the same kind of pad, the pad strength at 30°-40° is about 1/2 of that at 0°. The pulling force required for the T pad to be pulled out is greater than that of the O pad, That is, the strength of the T pad is greater than the strength of the O pad. It is mainly caused by the reason that the contact area is proportional to the tensile force under the condition of certain strength. This is why electronic products are most likely to be broken when they encounter obstacles when they fall.

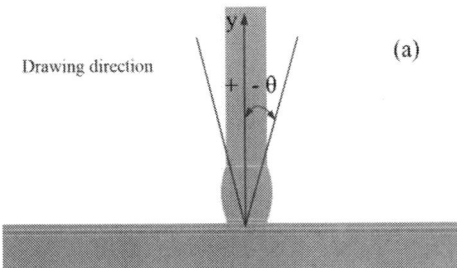

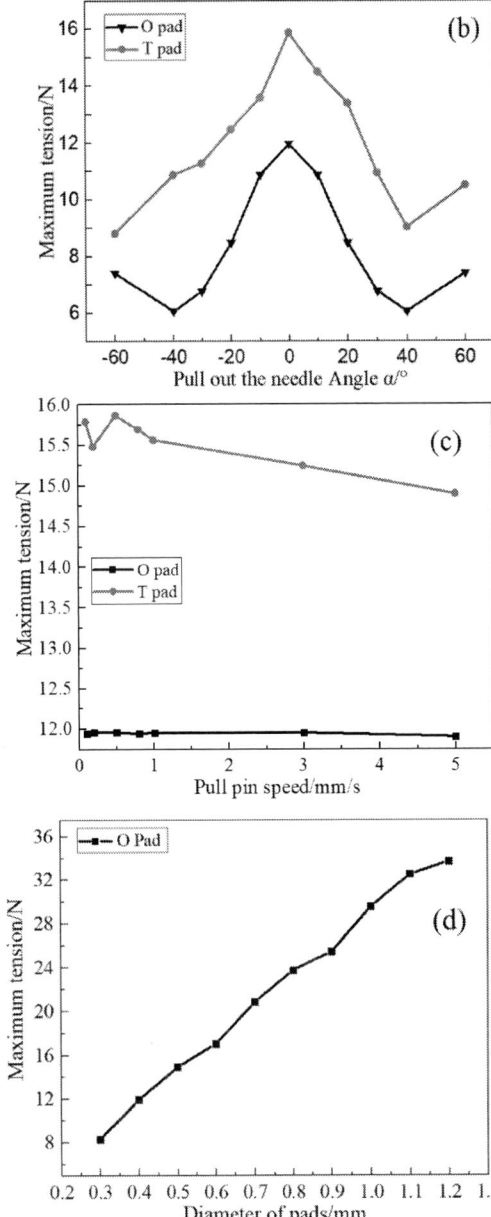

Fig. 7. The influence of various factors on the tension.

Pull the pins at speeds of 0.1mm/s, 0.2mm/s, 0.5mm/s, 0.8mm/s, 1mm/s, 3mm/s, and 5mm/s respectively in the Y direction Fig. 7a. The influence of drawing speed is shown in Fig. 7c on the strength of the pad. It can be seen from the figure that the speed has little effect on the O pad, and the curve almost fluctuates very little. For the T pad, the float is large, and the overall trend is that as the drawing speed increases, the strength of the pad decreases. At the same speed, the pull-off force of T pad is larger. It can be seen from Fig. 7d that the strength of the pad increases approximately linearly as the diameter increases.

978-1-6654-1392-3/21 $31.00 © 2021 IEEE

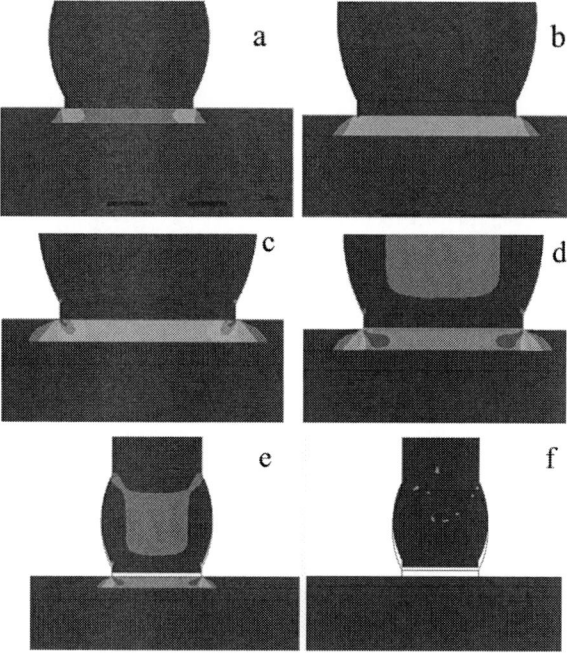

Fig. 8.   Strain change process during drawing

The O pad is pulled out at a speed of 0.5mm/s. The drawing stress distribution is shown in Fig. 9. The strain diagram is shown in Fig. 8 during the drawing process, which is the change from a to f. The strain is concentrated on the edge of the pad and expands to the inside, which is also the extension direction of the crack. It can be seen that the pulling force first rises and then drops to zero. It can be seen from the figure that point A is not zero. It shows that the crack starts to expand when the tensile force is 4.5N. The growth of the AB segment crack is reversible, and its stress is less than that shown in Fig. 3 σ_c. The crack propagation is irreversible in the BCD segment. Because this is the process of cohesion softening. The tension drops to zero at point D, indicating that the pad and the adhesive layer have been completely separated. Among them, point B is the peak value of the tensile force, which determines the bonding strength of the pad.

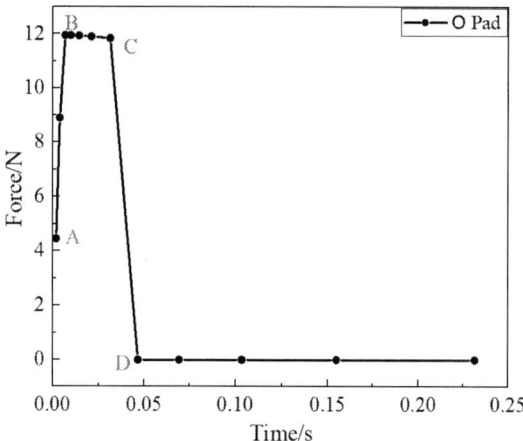

Fig. 9.   Stress change process during drawing.

It can be seen that the size of the pad has the greatest influence on the strength of the pad, followed by the angle of drawing, and the speed has a slight influence on the pad. It is particularly important to choose the size of the pad reasonably.

IV. CONCLUSION

In this paper, a 3D FEM is used to analyze the stress and strain distribution of the pad under stress. Changes in the stress of each layer node during the process of pulling out the pad. It was found that the tensile force of the contact surface between the pad and the adhesive layer was the largest, followed by glass fiber, adhesive layer and substrate. The PCB material was tested for pin-out by using the cohesion method of 2D simulation. It is found that the size of the pad has a greater impact on the strength of the pad in the test conditions, followed by the drawing angle and the drawing speed. This is consistent with the results in the literature[10]. Indicating that the method of using 2D cohesion has certain reference value. Finally these parameters can be used to optimize PCB design. Pad cracking can be improved by choosing pad type or changing PCB material. It has a certain reference value to the practitioners who are engaged in PCB design.

ACKNOWLEDGMENT

This research was by supported the National Natural Science Foundation of China (No. 61865004), the Guangxi Science and Technology Program (No. AB20159007), the Innovation-Driven Development Project of Guangxi Province (No. AA182420), and the Guilin Science Research and Technology Development Program (No. 2020010302).

REFERENCES

[1] Cai M, Xie D J, Chen W B, et al. A novel soldering method to evaluate PCB pad cratering for pin-pull testing[J]. Miroelectronics Reliability, 2013; 53(9-11): 1568-1574.

[2] Xie D J, Geiger D, Shangguan D K, et al. Failure mechanism and mitigation of PCB pad cratering[C]. IEEE Electronic Components & Technology Conference, 2010.

[3] Cai M, Xie DJ, Zhang Z, et al. Investigation on PCB Pad Strength[C]. International Conference on Electronic Packaging Technology & High Density Packaging, 2010.

[4] Xie DJ, Geiger D, Shangguan DK, et al. Failure mechanism and mitigation of PCB pad cratering[C]. 60th electronic components and technology conference; 2010. p. 471-6.

[5] IPC SMT Attachment Reliability Test Methods Task Group. IPC9708: test methods for characterization of PCB pad cratering. IPC standard, 2011.

[6] Yao Y, Keer L M, et al. Cohesive fracture mechanics based numerical analysis to BGA packaging and lead free solders under drop impact[J]. Miroelectronics Reliability, 2013; 53(4): 629-637.

[7] Song F B, Yang C Y, Li S W, et al. Study on test method of Pad Cratering in printed circuit board[J]. China electronic manufacturing and packaging technology annual conference, 2011; 150-155.

[8] .Zhou R, Liu Z W, Zhang J G, et al. Three-dimensional numerical simulation of metal powder compaction crack based on DPC-CZM hybrid model[C]. Material review. 2020; 34(6): 155-159.

[9] Zhang Q M, Lo J, Lee S, et al. Correlation of board and joint level test methods with strain dominant failure criteria for improving the resistance to pad cratering[C]. 2016 17th International Conference on Electronic Packaging Technology (ICEPT), 2016.

[10] Ahmad M, Burlingame J, Guirguis C, et al. Comprehensive methodology to characterize and mitigate BGA pad cratering in printed circuit boards[J]. SMTA, 2009; 22(1): 21-28.

Effect of SAC0307 content on properties of Cu/Zn powder/Al joint by ultrasonic excitation at low temperature

Qianzhu Xu
College of Materials Science and Engineering
Chongqing University of Technology
Chongqing, China
351317290@qq.com

Guisheng Gan*
College of Materials Science and Engineering
Chongqing University of Technology
Chongqing, China
ggs@cqut.edu.cn

Zhaoqi Jiang
College of Materials Science and Engineering
Chongqing University of Technology
Chongqing, China
2581284141@qq.com

Shiqi Chen
College of Materials Science and Engineering
Chongqing University of Technology
Chongqing, China
793608729@qq.com

Tian Huang
College of Materials Science and Engineering
Chongqing University of Technology
Chongqing, China
757704385@qq.com

Cong Liu
College of Materials Science and Engineering
Chongqing University of Technology
Chongqing, China
1010725476@qq.com

Xiangtao Xu
Chongqing Pingwei Volt Integrated Circuit Sealing and Application Industry Research Institute Co., Ltd
Chongqing, China
xuxt@perfectway.cn

Abstract—The ultrasonic-assisted soldering joint of Cu/Zn powder+xSAC0307/Al(x=0%, 5%, 10%, 15%, 20%) at low temperature was studied. In this study, 45μm Zn powder were the main solder, and a small amount of 500nm SAC0307 was mixed into it as the final solder, the soldered was heated to 240℃ in environmental atmosphere under the ultrasonic-assisted, successfully realized bonding of Cu/Al at low temperature. The results show that a complete and continuous Cu_5Zn_8 compound was formed at the copper side interface, a solid solution bonding was formed between the base metal and Zn at the aluminum side. When there was no SAC0307, the bonding between Zn balls in the sorlder was weak, the gap between Zn balls was obvious, and the shear strength of the joint was only 21MPa. When the content of SAC0307 was 15%, tin was melted and extruded to the interface of base metal under pressure, the Zn balls were bonded tightly without defects, and the shear strength of the joint reached the maximum value of 39.51MPa.

Keywords—SAC0307, Ultrasonic-assisted, Shear strength, interface reaction, Zn ball

I. INTRODUCTION

Aluminum is an excellent and cost-effective metal due to its high strength, low density and corrosion resistance, which are widely used in aerospace, military industries and electronic packaging[1]. Due to the widely use of aluminum alloy and composite materials in electronic packaging, people have an urgent need in the industry for direct soldering of aluminum electrode pads without plating silver, nickel and other metals. In addition, due to the high cost performance of aluminum, which is becoming more and more used instead of copper, but because of copper's high electrical conductivity, thermal fatigue resistance and other excellent performance, aluminum can not directly replace copper in most occasions. Therefore, the realization of low temperature and high efficiency interconnection between

copper and aluminum is of great significance in electronic packaging. Compared with the current mainstream lead-free tin based solder, Zn with a lower price and stronger interaction to aluminum is a better copper-aluminum interconnect material. However, the melting point of Zn based solder is high, too high soldering temperature is easy to cause damage to electronic components, aiming at these problems, a lot of research has been carried out[2-3]. These studies have made a great contribution to the improvement of solder properties, but have little effect on changing melting temperature and solderability, so it is difficult to obtain ideal results. So, it is necessary to exploit a new process of soldering to overcome the bottleneck of high melting point of Zn based solder[4]. Zn powder were used as the main solder, in order to strengthen the bonding between Zn balls, a small amount of SAC0307 were mixed into the solder, and ultrasonic assisted bonding was used. Ultrasonic assisted bonding can generate part of bonding energy by vibration, and promote the mutual diffusion between Zn balls and Zn balls, Zn balls and base metal. The Cu/Zn/Al joint under low temperature ultrasonic assisted was successfully realized. In addition, the microstructures of the interface were analyzed, further explained the strengthening mechanism of Zn to Cu interfaces, Al interfaces and the strengthening mechanism of Sn to Zn balls.

II. EXPERIMENTAL PROCEDURES

Cylindrical T2 Cu and 1060 aluminum alloys with a height of 3mm and a diameter of 5mm were used as base materials, polished with 400# sandpaper, and then the surface was cleaned with alcohol, put the cleaned base metal substrate into the fixture of the experimental device (Fig.1), 45μm Zn powder and 500nm SAC0307 powder were used to fill the Cu/Al joint. The whole experimental device was composed of ultrasonic generator(with the power of 600W and the frequency of 20kHz), induction heating unit, and amplitude transformer. During the whole soldering process in the ambient atmosphere, a constant temperature was provided by the heating platform, and a

978-1-6654-1392-3/21 $31.00 © 2021 IEEE

constant pressure of 8MPa was provided by the ultrasonic vibrator at 240°C to the aluminum substrate, until 205s later, start the ultrasound and continue for 5s, then hold the pressure and temperature for 90s to finish the soldering. The microstructure of interface was analyzed by SEM, and the shear strength was tested by universal tensile testing machine.

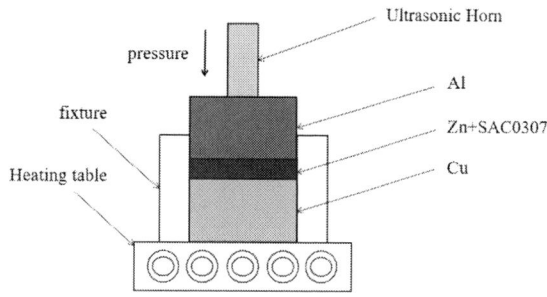

Fig.1 Diagram of sample and experimental device

III. RESULTS AND DISCUSSIONS

The joint of Cu/Zn+SAC0307/Al was successfully obtained under the assistance of ultrasonic at 240 °C. Under the action of ultrasonic, the oxide film on the surface of base metal was easy to be broken, the diffusion between solder and base metal were promoted. The soldering joint with different SAC0307 content was shown in Fig.2 The soldering interface was bonded completely and the bonding formation was good. In Fig.2(a) the Zn balls could be seen in the soldering joint without SAC0307, moreover, there were many gaps between Zn balls. When the content of SAC0307 were 5% from Fig.2(b), white tin phase distributed in the soldering seam, which filled in some defects between Zn balls and strengthen the connection between Zn balls to a certain extent. Due to the increased of SAC0307 content to 10%-15%, the bonding between Zn balls becomes closer, and the distribution of white tin-rich phase was more uniform. When the content of SAC0307 were 20%, the white tin phase between tin balls becomes more and more. Stress concentration resulting in loose soldering, which might caused the similar crack structure(Fig.2c,d).

Fig.3 is the SEM image of interface between base metal and solder, Fig.3(a,c,e,g,i) and Fig.3(b,d,f,h,j) show the aluminum side interface and the copper side interface respectively. Through the action of ultrasonic, the interface undulate to a certain extent, which promoted the bonding between solder and base metal. It could be seen from Fig.3(a-b) that when SAC0307 were not added, incomplete and discontinuous IMC was formed at the interface of copper side, but no obvious IMC and other phases could be seen at the aluminum side. With the addition of SAC0307, the IMC at copper side becames more continuous and complete. Combine with the platform in the EDX in Fig.4(a-e) and the atomic ratio of Cu to Zn in point 4 and 5 of Table 1 all were 35/64, it could be determined that the IMC was Cu_5Zn_8. Due to the Cu element diffused from the Cu substrate to the solder was consumed in a large amount, the excess zinc above the Cu_5Zn_8 layer might form other CuZn compounds with a small amount of copper. It is worth mentioning that there was no Cu_xSn_y compound at the copper side interface, because the Gibbs free energy of

formation of Cu_5Zn_8 is -12.345 kJ/mol, which is lower than for $Cu_3Sn(\Delta G=-7.78KJ/mol)$ and $Cu_6Sn_5(\Delta G=-7.42KJ/mol)$[5-6], in addition, because of a strong affinity exists between Cu and Zn, which lead to the Cu-Zn clusters form easily[7], therefore, tin hardly reacted with copper.

It could be seen from the aluminum side interface in Fig.3(c,e,g,i) that the interface becames undulating due to the action of ultrasonic and the melting of the aluminum substrate into the solder, which strengthened the force between the solder and the aluminum substrate, It could be inferred from the atomic ratios of points 1, 7, 9, 10 and 12 the black particles connected together were an Al-rich phase with solid solution of Zn[8], due to the weak interaction between aluminum and tin, there was no Al-Sn compound or solid solution at the aluminum side interface. The results show that tin had no bonding effect at the base metal interface but effected at bonding between Zn balls in solder seam.

Fig.3(e-g) show that under the action of ultrasonic energy and pressure, molten tin was squeezed to the interface during the soldering process, resulted in a layer of obvious white tin-rich phase gathered at the interface of 10% SAC0307 and 15% SAC0307 in joint, which made the bonding between Zn balls in the solder seam more compact, and the bonding between Zn balls had no defects, so the soldering strength between Zn balls was higher. With the further increased of SAC0307 to 20%, the liquefying or semi-solid degree of the solder increased, and the tin enriched on both sides of the base metal matrix was easy to be extruded out of solder seam. In addition, the solder seam thickness of 0%, 5%, 10%, 15% and 20% was 686.72μm, 515.23μm, 549.1μm, 523μm, 419.89μm respectively, due to the melting or softening of tin, the solder seam was easy to be thinned by pressure, when the SAC content was 20%, the tin rich phases on both sides were extruded out of the solder seam, resulted in further thinning of the solder seam.

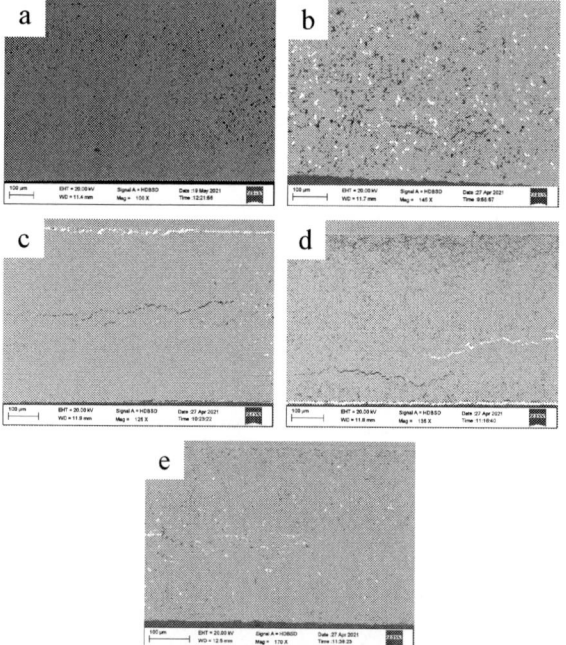

Fig.2 Microstructure of solder with different SAC0307 contents. (a)0%; (b)5%; (c)10%; (d)15%; (e)20%

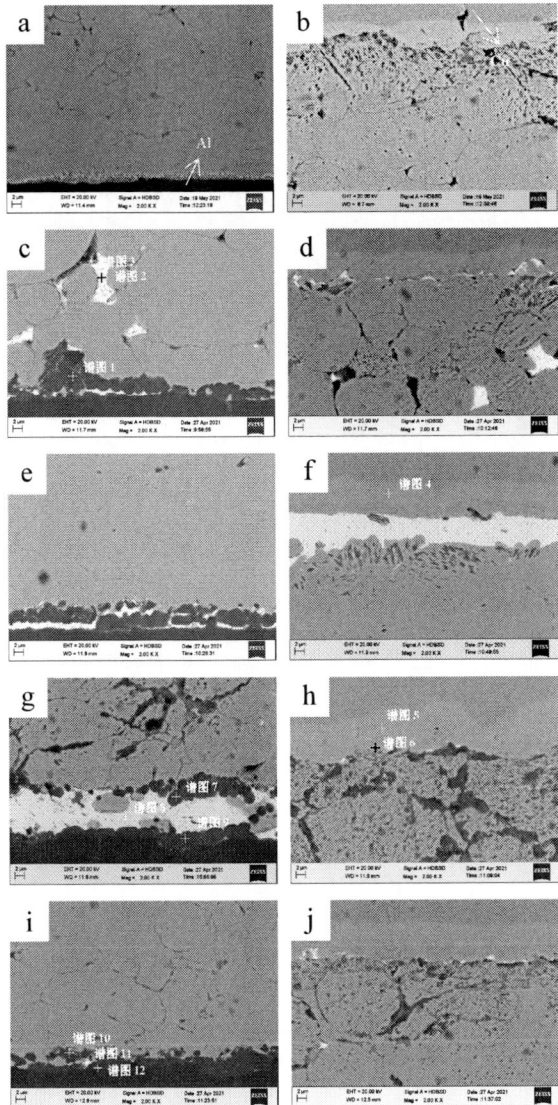

Fig.3 Microstructure of soldering interface with different SAC0307 content.
(ab)0%; (cd)5%; (ef)10%; (gh)15%; (ij)20%

Table 1 Atomic percentage of each point element in Fig.3

element	1	2	3	4	5	6
Al	90.35	1.25	2.69	0.72	0.49	1.29
Zn	8.75	17.87	60.67	63.68	64.42	37.48
Sn	0.95	79.64	29.76	0.15	0.13	54.49
Ag	0	0.14	0.27	0.05	0	0.31
Cu	0.13	0.84	6.62	35.40	34.97	6.43
Total	100	100	100	100	100	100
element	7	8	9	10	11	12
Al	86.97	2.18	91.86	79.89	51.64	93.51
Zn	9.34	7.21	7.25	17.55	7.61	5.58
Sn	3.35	89.86	0.64	1.96	40.16	0.68
Ag	0.01	0.54	0	0.06	0.24	0
Cu	0.33	0.20	0.25	0.53	0.35	0.23
Total	100	100	100	100	100	100

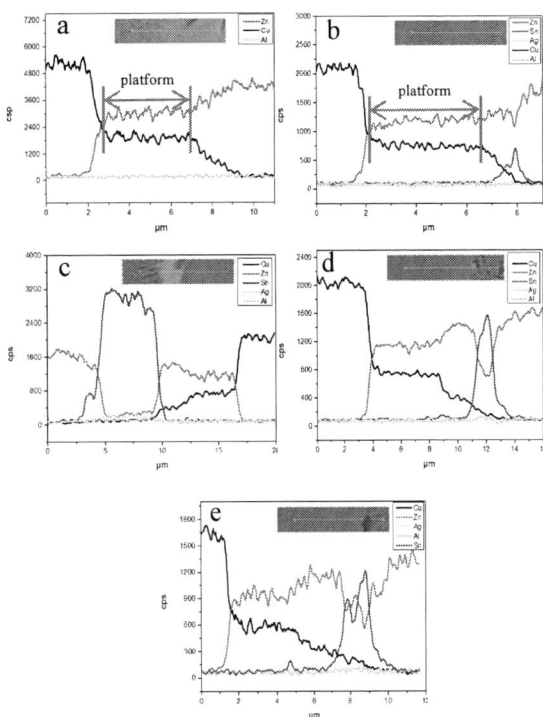

Fig.4 EDX of Cu/Zn interfacial IMC joints with different SAC0307 contents.
(a)0%; (b)5%; (c)10%; (d)15%; (e)20%

Fig.5 shows the bonding between Zn balls of different SAC0307 content. When the content of SAC0307 were 10% and 15%, the bonding between Zn ball and Zn ball was relatively closer, and it was not obvious that the white tin rich phase was between Zn balls. There were obvious boundaries between Zn balls in soldering joints with 10% and 20% SAC0307(Fig.5a,c), while the boundaries between Zn balls with 15% SAC0307 in soldering joints were relatively not obvious, it indicate that the bonding degree between Zn balls was the best, so the bonding strength of 15% SAC0307 was the highest. When the content of SAC0307 were 20%, the obvious white tin phase appears between the Zn balls, which greatly improved the strength of the joint compared with the defects of voids between the Zn balls in the solder joints with 0% and 5% SAC0307 content, but too thick Sn-Zn solid solution or compound between Zn balls might lead to a decline of shear strength of joint.

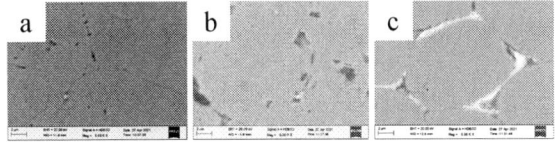

Fig.5 Ball to ball connections in solders with different SAC0307
contents.(a)10%; (b)15%; (c)20%

The shear strength of the joint is as shown in Fig.6 When the content of SAC0307 was 0%, the shear strength of joint was only 21MPa. After adding 5% SAC0307, the shear strength does not change significantly, which was 22.2MPa, this is because when the content of SAC0307 was low, the bonding between Zn balls was weak. When the content of SAC0307 increased to 10%, the defects between the Zn balls were greatly reduced, the tin was evenly distributed, and the shear strength was greatly improved, which reached 35.19MPa. With the further increased of SAC0307 content to 15%, the shear strength of the joint reached the peak of 39.51MPa, and then decreased to 36.45MPa of 20%.

Fig.7 shows the fracture diagram of the joint with different SAC0307 content. When no SAC0307 were added, due to the soldering temperature is 240°C, it is far lower than the melting point of Zn(419.53°C), the Zn balls on the fracture surface was bonding by physical action (ultrasonic friction, pressure, etc.), no obvious metallurgical reaction was occured, this kind of bonding was weak. When 5% SAC0307 were added, this phenomenon had not been significantly changed. It could be seen from Fig.7(c-d) that when the content of SAC0307 were 10% or more, SAC0307 with melting point(217-225°C) lower than soldering temperature reduced the defect between zinc balls, and lamellar tear could be seen on the fracture surface, which was a mainly ductile fracture, this is consistent with the joint shear strength variation. According to Fig.7(a-e), all solder joints with different SAC0307 content were fracture at the soldering seam, but not at the base metal interface, which indicates that the bonding strength at the base metal interface was relatively higher, therefore, it is particularly important to improve the bonding strength between Zn balls in solder. The results show that SAC0307 have a significant effect on the shear strength of joint.

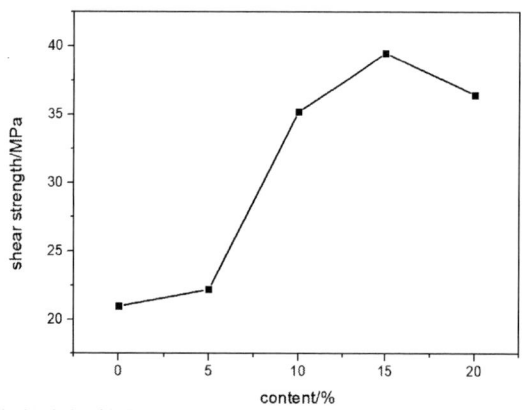

Fig.6 Relationship between different contents of SAC0307 and shear strength

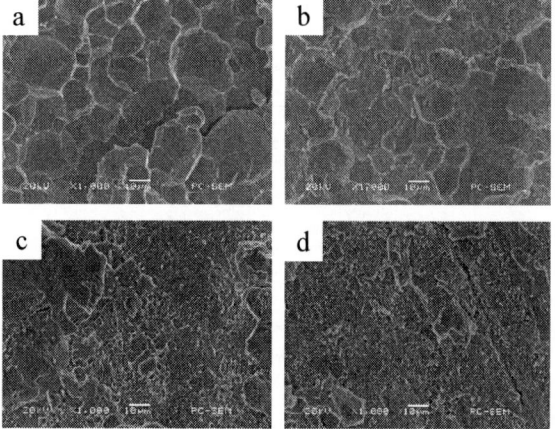

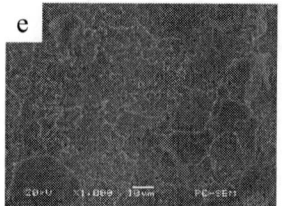

Fig.7 Fracture surface morphology of joint for different SAC0307 contents.
(a)0%; (b)5%; (c)10%; (d)15%; (e)20%

IV. CONCLUSION

500nm SAC0307 powder was added to Zn powder and mixed as solder and successfully realized bonding of Cu/Al at low temperature. According to the microscopic analysis of the soldering interface, Cu_5Zn_8 compound was formed at the copper side interface without Cu_xSn_y compound, and the base metal at the aluminum side diffused to the solder, resulting in the formation of Al-Zn solid solution. When no SAC0307 was added, the bonding between Zn balls in the solder was weak, and there were obvious defects and voids, the shear strength of joint was only 21MPa. When 5% SAC0307 were added, the bonding defects were not significantly improved, and the shear strength was 22.2MPa. Until the content of SAC0307 increased to 10%, the shear strength of the joint reached 35.19MPa, the bonding between the Zn balls in the solder joints was strengthened obviously, and there were no defects. With the further increased of SAC0307 content to 15%, the mechanical properties were greatly improved, the shear strength reached the peak of 39.51MPa, and then decreased. And the corresponding fracture surface morphology with the increased of SAC0307 to more than 10%, lamellar tear appear on the surface of fracture surface and transformed to ductile fracture.

ACKNOWLEDGMENT

This study was supported by the National Natural Science Foundation of China(Grant No.61974013 and 61774066) and the Scientific and Technological Research Program of Chongqing Municipal Education Commission(NO.KJZD-K202101101), Chongqing Research Program Basic Research Frontier Technology(cstc2020jcyj-msxmX0819), University Innovation Research Group of Chongqing (CXQT20023), Graduate Innovation Project of Chongqing University of Technology (ycx20192039, clgycx20201001 and clgycx20201002), respectively.

REFERENCES

[1] Z. W. Niu, J. H. Huang, X. U. Fang-Zhao, K. K.Liu, S. H. Chen, X. K. Zhao, "Current research status and prospect of brazing filler metals for aluminum and aluminum alloys", chinese Journal of Nonferrous Metals, vol. 26, January 2016, pp. 77-87.

[2] C. W. Chang, K. L. Lin, "High-Temperature Mechanical Properties of Zn-Based High-Temperature Lead-Free Solders", Journal of Electronic Materials, vol. 48 November 2019, pp. 135-141.

[3] M. Prach, R. Koleňák, "Soldering of Copper with High-temperature Zn-based Solders", Procedia Engineering, vol. 100, 2015, pp. 1370-1375.

[4] G. S. Gan, L. J. Jiang, S. Q. Chen, Y. Q. Deng, D. H. Yang, Z, Q, Jiang, H. D. Cao, M. Z. Tian, Q. Z. Xu, "Effect of Zn-powder content on the property of Cu/SAC0307 powder/Cu joint under ultrasonic

assisted at low temperature", Soldering & Surface Mount Technology, December 2020, in press.

[5] L. Qu, H.T. Ma, H.J. Zhao, A. Kunwar, N. Zhao, "In situ study on growth behavior of interfacial bubbles and its effffect on interfacial reaction during a soldering process", Appl. Surf. Sci.,vol.305, 2014, pp. 133–138.

[6] D. G. Kim, H. S. Jung, S. B. Jung, "Kinetics of intermetallic layer growth and interfacial reactions between Sn–8Zn–5In solder and bare copper substrate", Materials Science and Technology, Vol. 21, 2005, pp. 381-386.

[7] J. O. Suh, K. N. Tu, G. V. Lutsenko, A. M. Gusak, "Size distribution and morphology of Cu_6Sn_5 scallops in wetting reaction between molten solder and copper", Acta Materialia, Vol. 56, 2008, pp. 1075-1083.

[8] M. L. Huang, Y. Z. Huang, H. T, Ma, J. Zhao, "Mechanical Properties and Electrochemical Corrosion Behavior of Al/Sn-9Zn-xAg/Cu Joints", Journal of Electronic Materials, vol. 40, 2011, pp. 315-323.

A Bubble-free Electroosmotic Pump with Polyaniline-wrapped Platinum-coated Titanium Mesh Electrodes

Qian Yang
Internet of things research center
Advanced institute of information
technology, Peking University
Hangzhou, China
qyang@aiit.org.cn

Liang Li
Internet of things research center
Advanced institute of information
technology, Peking University
Hangzhou, China
lli@aiit.org.cn

Wen Chen
Internet of things research center
Advanced institute of information
technology, Peking University
Hangzhou, China
wchen@aiit.org.cn

Meng Gao*
Internet of things research center
Advanced institute of information
technology, Peking University
Hangzhou, China
mgao@aiit.org.cn

Le Ye*
Institute of Microelectronics
Peking University
Beijing, China
Internet of things research center
Advanced institute of information
technology, Peking University
Hangzhou, China
yele@pku.edu.cn

Abstract—We developed a bubble-free electroosmotic pump (EOP) with polyaniline-wrapped platinum-coated titanium mesh electrodes. These electrodes prepared by the chronopotentiometry (CP) method show a better redox performance, electrochemical activity and stability. The optimal CP parameters are 0.413 mA of anode current (current density, 1.5 mA/cm^2), and 1800 s of anode time (polymerization time). The EOP, assembled with polyaniline-wrapped platinum-coated titanium mesh electrodes, shows a much better performance than that of platinum-coated titanium mesh electrodes. After modification of the electrodes, the flow rate at a voltage of 6 V for purity water increases from 26.9 µL/min to 87.6 µL/min, and the limiting hydrostatic pressure improves from 59.3 Pa/V to 113.9 Pa/V. The polyaniline-wrapped platinum-coated titanium mesh electrodes can also keep undamaged surface morphology and unaltered redox performance after a long time EOP performance test, which can make it suitable for bubble-free uses in microfluidic system, drug delivery, and cooling of microelectronic devices.

Keywords—Electroosmotic pump; Bubble-free; Polyaniline-wrapped electrode

I. INTRODUCTION

The development of bubble-free EOP is very important for the further productization of EOP in its applications, for example microfluidics, drug delivery, bioanalytical systems, and so on. However, the gas bubble generation is easily generated on the surface of the electrodes[1]. Several methods for the development of bubble free EOP have been reported. One is modulating the electric field to control the electrochemical process, such as pulsed direct current and alternating current (AC) mode[2]. Redox electrode is another one, which includes Ag/AgCl[3], Ag/Ag$_2$O[4], and conducting redox polymers, like polyaniline (PANI)[5], poly(3,4-ethylenedioxythiophene) (PEDOT)[1], poly(quinone)[6]. Although these redox electrodes can avoid the gas bubble

generation temporarily, they are not able to obtain high flow rate and long life time.

PANI is an attractive conductive polymer, because of the existence of different oxidation forms and the conductivity of its emeraldine salt. Rudra Kumar[5] et al. proved the feasibility of PANI application in EOPs. They synthesized the NH$_2$-G/PANI composites by chemical polymerization and then prepared the functional electrodes by dip coating NH$_2$-G/PANI paste on nickel foam. Nonetheless, the preparation of this electrode is a complicated process and the adhesion of the NH2-G/PANI composites on the nickel foam also needs to be verified. By contrast, electrochemical polymerization is a simpler method, which can produce polyaniline film in a single step on an inert electrode, for example platinum, gold, carbon, etc.. Moreover, the galvanostatic technique is prominent among various electrochemical polymerization[7] methods, because it is well controllable to form the conductive form of polyaniline.

In this work, we proposed a polyaniline-wrapped platinum-coated titanium mesh (PANI@Pt/Ti) electrode for bubble-free EOPs. The PANI@Pt/Ti electrode synthesized using the galvanostatic method (the chronopotentiometry (CP) technique on a CHI 660E electrochemical workstation), has a much better electroosmosis performance in flow rate, limiting hydrostatic pressure, stability, and bubble-free generation, comparing with platinum-coated titanium(Pt/Ti) mesh electrodes.

II. MATERIALS AND METHODS

A. Materials

Platinum-coated titanium mesh (diameter = 10 mm, diamond hole size = 0.5 mm × 1.0 mm, porosity = 65%) was purchased from Anping County, Hebei Province. Sulfuric acid (H$_2$SO$_4$), nitric acid, acetone were purchased from Sinophram. Aniline, ethanol (95%), sodium phosphate monobasic dihydrate, sodium phosphate dibasic dodecahydrate were obtained from Shanghai Titan Scientific Co., Ltd. Track-

This project is supported by National Key Research and Development Project of China (2019YFB2204900). Yang and Li contributed to this paper equally. *Corresponding author.

etched polycarbonate (PCTE) membranes were purchased from MemberSpace.

B. Methods

Synthesis of Polyaniline-wrapped platinum-coated titanium mesh Electrodes. PANI@Pt/Ti electrodes were synthesized in 0.3 M aniline and 0.5 M H_2SO_4 solution of water using the chronopotentiometry (CP) technique on a CHI 660E electrochemical workstation (CH Instrument, Shanghai). A conventional three-electrode configuration was employed. The platinum-coated titanium (Pt/Ti) mesh electrode (calculated electrode area is $0.275cm^2$) was used as the working electrode, and the saturated calomel electrode (SCE) and platinum wire were used as the reference and the counter electrodes, respectively. Before the polymerization, the platinum-coated titanium mesh electrodes were cleaned and activated in 0.5 M H_2SO_4 aqueous solution, using the cyclic voltammetry (CV) technique by cycling the potential from −0.2 V to 1.3 V (versus SCE) at a scan rate of 50 mV/s for 20 cycles. The CP parameters used to synthesize the PANI@Pt/Ti electrodes were as follows: when the polymeric time was maintained at 1200 s, the anode current was set as 0.138 mA, 0.300 mA, 0.413 mA, 0.550 mA, 0.688 mA, which means the polymeric current densities were 0.5 mA/cm², 1.1 mA/cm², 1.5 mA/cm², 2.0 mA/cm², 2.5 mA/cm², respectively. When the polymeric current density was constant at 1.1 mA/cm² (anode current, 0.300 mA), the polymeric time was set as, 600 s, 1200 s, 1800 s, 2400 s, and 3600 s. After the polymerization, the PANI@Pt/Ti electrodes were washed ultrasonically in purity water.

PCTE Membrane and EOP Assembly. The PCTE membranes with 200 nm pore size and 3×10^8 pore density, were immersed in ethanol (95%) for 10 min and purity water for at least 2 h before the integration. A sandwich electroosmotic unit was consisted of two prepared PANI@Pt/Ti electrodes and one piece of PCTE membrane. The distance between two electrodes was fixed at 0.4 mm and the working area of the PCET membrane was 50.24 mm². The bubble-free EOP was assembled based on the sandwich electroosmotic units.

General Characterization. The surface morphology of electrodes was performed by a Hitachi SU8010 field emission scanning electron microscopy (SEM, Hitachi, Japan). The Electrochemical characteristics of the electrodes was performed in three electrodes system by a CHI 660E electrochemical workstation (CH Instrument, Shanghai). The flow rate was measured using a L-FLOW UNIT flowmeter (Fluigent, France) which was connected to the EOP to form a closed loop. The limiting hydrostatic pressure was measured by reading the liquid level difference in height between the two sides of the EOP when there was no flow observed. During the test, the voltage was applied by a 2450 source

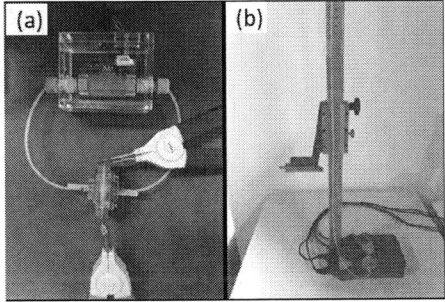

Fig.1. Test devices of: (a) flow rate measurement and (b) limiting hydrostatic pressure measurement.

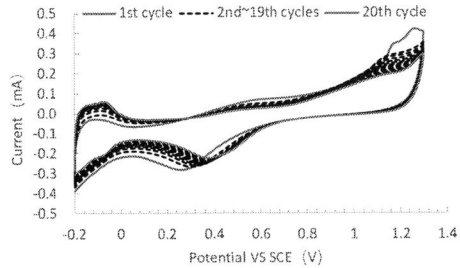

Fig.2. Cyclic voltammograms of platinum-coated titanium mesh electrode from aqueous solution of 0.5 M H_2SO_4, at scan rate of 50 mV/s .

measure unit (SMU) instrument (Keithley, USA). The test devices are shown in Fig. 1.

III. RESULTS AND DISCUSSION

A. Preparation of Polyaniline-wrapped platinum-coated titanium mesh Electrodes

The Pt/Ti mesh electrodes were firstly cleaned and activated in aqueous solution of 0.5 M H_2SO_4. As seen from cyclic voltammograms (CVs) in Fig. 2, two anodic peaks at potential of 1.15 V ~ 1.25 V in the first cycle, are rapidly decreased until disappeared, and another anodic peak at potential of −0.1 V are gradually increasing, within the scanning process. At the end of the scanning with the 20th cycle, the curve is consistent with the reference[8], indicating that the Pt/Ti mesh electrode has been cleaned and activated.

Fig. 3 shows the chronopotentiometric curves of aniline electrochemical polymerization on Pt/Ti mesh electrodes in aqueous solution of 0.3 M aniline and 0.5 M H_2SO_4. The electrodes fabricated at different current densities or polymeric times have the same trend of potential. At the initial 50 s, the potential drops rapidly from 0.8 V to 0.7 V, and then

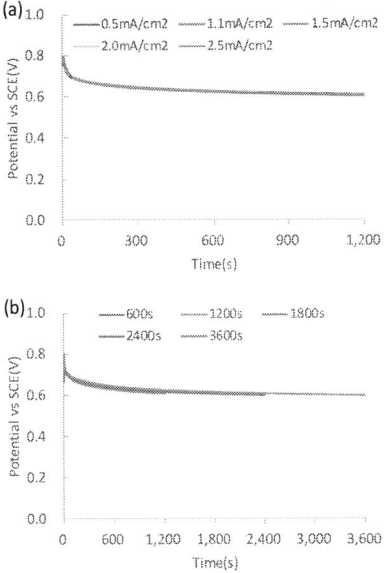

Fig.3. Chronopotentiometric curves of aniline electrochemical polymerization on platinum-coated titanium mesh electrodes in aqueous solution of 0.3 M aniline and 0.5 M H_2SO_4. (a) Different current densities at constant polymeric times of 1200 s. (b) Different polymeric times at constant current densities of 1.1 mA/cm².

tends to 0.6 V at a slow rate. It illustrates that the higher potential is needed in the start of the polymerization of aniline. The reason is that, as generally accepted, the first step of the polymerization of aniline is anodic oxidation on the electrode surface at a higher potential, in order to form the aniline cation radicals. But the higher potential is not required for subsequent polymerization, owing to the autocatalytic nature of aniline electropolymerization.

B. Surface Morphology of Electrodes

Fig. 4 shows SEM images of Pt/Ti mesh electrode and PANI@Pt/Ti electrodes synthesized with different current densities or polymeric time. Before modified, there were Pt nanoparticles of uneven size from 50 nm to 1 μm, distributed on the surface of the Pt/Ti mesh electrode. After the

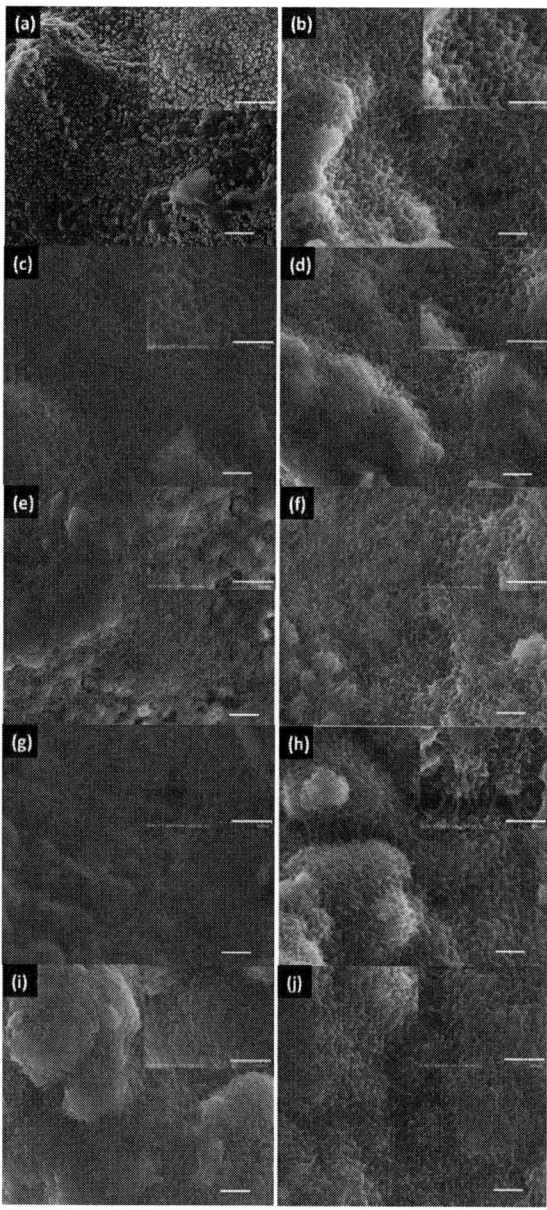

Fig.4. SEM images of (a) Pt/Ti mesh electrodes and PANI@Pt/Ti electrodes by different polymeric parameters, (b) 0.5 mA/cm² for 1200 s, (c) 1.1 mA/cm² for 1200 s, (d) 1.5 mA/cm² for 1200 s, (e) 2.0 mA/cm² for 1200 s,(f) 2.5 mA/cm² for 1200 s, (g) 1.1 mA/cm² for 600 s, (h) 1.1 mA/cm² for 1800 s, (i) 1.1 mA/cm² for 2400 s, (j) 1.1 mA/cm² for 3600s. Scale bar is 4 μm.

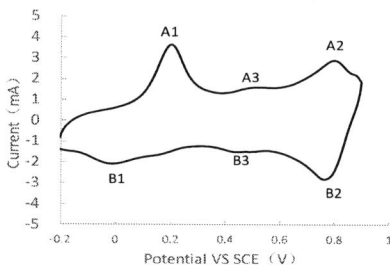

Fig.5. Cyclic voltammograms of PANI@Pt/Ti electrode synthesized at 1.5 mA/cm² for 1200 s, in aqueous solution of 0.5 M H_2SO_4, at scan rate of 50 mV/s.

polymerization of aniline, we can see a polymer film formed by the interweaving and connecting of coral-like structures, indicating the obtained electrodes were completely coated with polyaniline. When the polymeric time was maintained at 1200 s, the polymer film became dense and joined into chunks as the current density increased to 1.5 mA/cm², and then maintained (Fig. 4(b)~Fig. 4(f)). However, there was not a significant difference among the SEM images (Fig. 4(c), Fig. 4(g) ~ Fig. 4(j)) at a current density of 1.1 mA/cm², when the polymeric time was extended. It showed that the surface morphology of PANI@Pt/Ti electrodes was mainly affected by the current density, and had little relation with the polymeric time.

C. Electrochemical characteristics

The cyclic voltammograms of PANI@Pt/Ti electrodes synthesized at 1.5 mA/cm² for 1200 s was recorded in 0.5 M H_2SO_4 solution and presented in Fig. 5. There were two sets of redox peaks A1/B1 and A2/B2, which specifically referred to the redox of polyaniline. Generally, the 3rd inconspicuous redox peak A3/B3 was the characteristic peak for the electrochemical degradation of polyaniline. And the redox peaks were almost negligible. It elucidated that the polyaniline was stable and was hardly degraded.

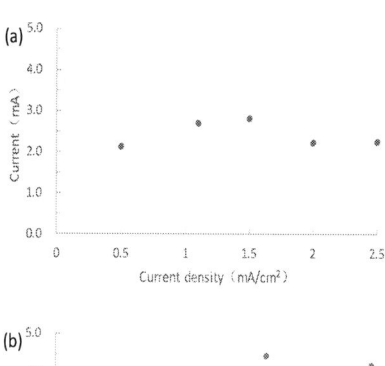

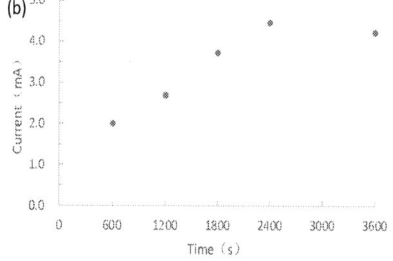

Fig.6. The effect of polymerization parameters on the electrochemical activity of the PANI@Pt/Ti electrodes, (a)Currentdensity, (b) polymeric time.

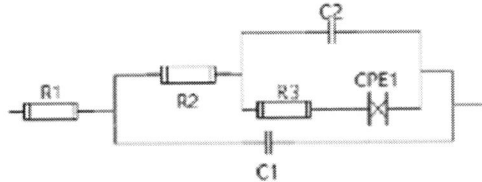

Fig.7. The equivalent circuit for the EIS test.

Compared with the CVs of Pt/Ti mesh electrodes, the response current of PANI@Pt/Ti electrodes was larger. It proved that the electrochemical activity of the modified electrodes was obviously enhanced. The current at peak A1 on the CVs of each PANI@Pt/Ti electrode, were extracted to evaluate the effect of polymerization parameters on the electrochemical activity of the electrodes. Fig. 6 shows that the response current increased less for increasing the current density than extending the polymerization time. As a result, the polymerization time played a more important role to improve the electrochemical activity of the PANI@Pt/Ti electrodes.

The electrochemical impedance spectroscopy (EIS) was recorded in 0.5 M H_2SO_4 solution at open circuit potential. The EIS parameters were used as follows: amplitude of 5 mV, high frequency of 100K Hz, and low frequency of 0.01 Hz. Then the data was simulated through an equivalent circuit (see Fig. 7), in which R1 was the resistance of solution, R2 was the charge transfer resistance, R3 was the membrane resistance, C1 was the membrane capacitance, C2 was the double-layer capacitance, and CPE1 was the constant phase element associated with diffusion processes.

The charge transfer resistance for each PANI@Pt/Ti electrode was shown in Fig. 8. And the Pt/Ti mesh electrodes, of which the obtained charge transfer resistance is 287.9 Ω, was tested for comparison. The charge transfer resistance of PANI@Pt/Ti electrode was about 1 Ω, two orders of magnitude lower than Pt/Ti mesh electrodes. It indicated that

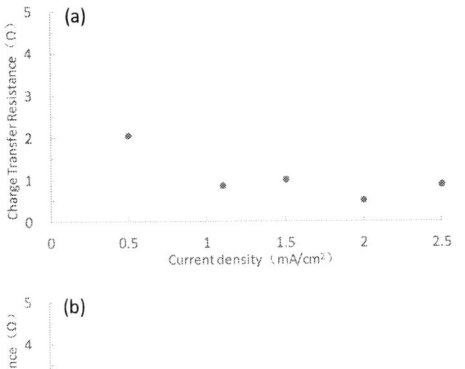

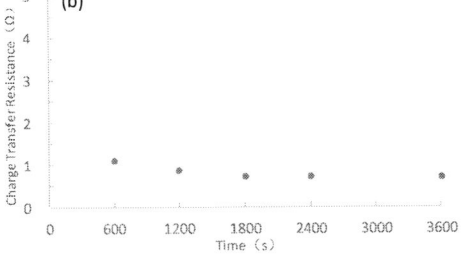

Fig.8. The effect of polymerization parameters on the charge transfer resistance of the PANI@Pt/Ti electrodes, (a) current density, (b) polymeric time.

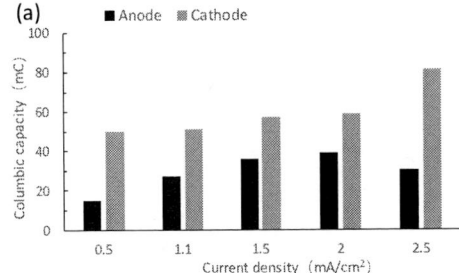

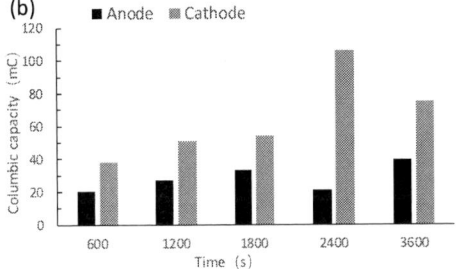

Fig.9. The effect of polymerization parameters on the columbic capacity of the PANI@Pt/Ti electrodes, (a) current density, (b) polymeric time.

charge transfer was more likely to occur on the surface of the PANI@Pt/Ti electrode. The polymerization parameters here did not change the result significantly.

The PANI@Pt/Ti electrode is a consumable electrode. It means that the life time of the electrodes may depend on the columbic capacity both of the anode and cathode. We measured the columbic capacity of PANI@Pt/Ti electrode by applying a constant potential of 1.0 V and −0.2 V (versus SCE) respectively, in 20 mM phosphate buffer at pH 7.0. Fig. 9 shows the effect of polymerization parameters on the columbic capacity of the PANI@Pt/Ti electrodes. The resulting columbic capacity was about 35 mC for anode and 55 mC for cathode, with the polymerization parameter of 1.5 mA/cm^2 and 1800 s.

D. EOP Performance

The EOP were assembled with the sandwich unit, with two PANI@Pt/Ti electrodes prepared by the same polymerization parameter and a PCTE membrane. As a comparison, two Pt/Ti mesh electrodes were used to assemble the reference pump similarly. The EOP performance was tested by applying a constant potential with a SMU instrument, and the operating fluid was purity water.

To measure the flow rate of the EOP, a L-FLOW UNIT flow meter (Fluigent, France), which was connected to the EOP to form a closed loop (see Fig. 1(a)), was employed. Fig. 10 shows the dependence of flow rate on the applied constant voltage for the PANI@Pt/Ti electrodes and Pt/Ti mesh electrodes. The flow rate is linearly dependent on the applied voltage for all of the electrodes. The slopes for PANI@Pt/Ti electrodes with each polymerization parameter were calculated. The highest slope was up to 25.8, though they were mainly distributed in the range of 10~15. Whereas, the slope of Pt/Ti mesh electrodes was 6.3. It is evident that the flow rate is improved by using the PANI@Pt/Ti electrodes. Furthermore, the flow rate increased with the current density and the polymeric time, when the polymeric time or current density was maintained respectively.

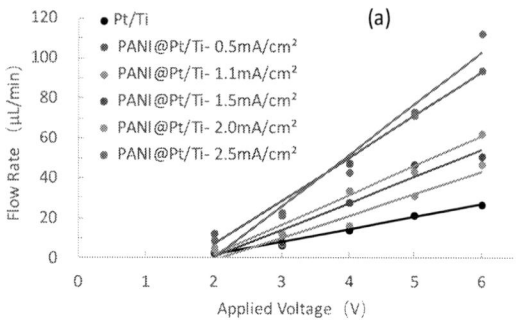

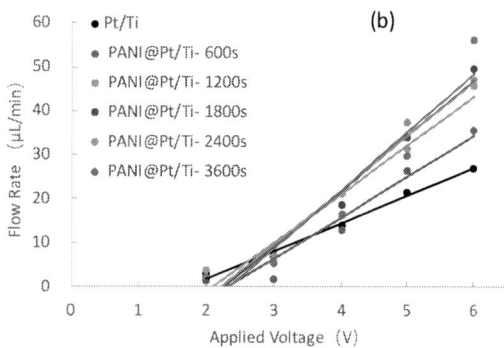

Fig.10. Dependence of flow rate on applied constant voltage for PANI@Pt/Ti electrodes. (a) Maintained the polymeric time at 1200 s, the slopes were 21.5 (0.5 mA/cm^2), 11.1 (1.1 mA/cm^2), 13.3 (1.5 mA/cm^2), 14.7 (2.0 mA/cm^2), 25.8 (2.5 mA/cm^2). (b) Maintained the current density at 1.1 mA/cm^2, the slopes were 9.3 (600 s), =11.1 (1200 s), 12.5 (1800 s), 12.3 (2400 s), 13.2 (3600 s). In comparison, the slope for Pt/Ti mesh electrodes is 6.3.

Completely new PANI@Pt/Ti electrodes with the polymerization parameter of 1.5 mA/cm^2 for 1800 s were prepared and assembled into the EOP. Fig. 11 shows the flow rate-time curve for this pump when the applied external voltage was 6 V. The flow rate of the EOP was initially 126.4 μL/min and then kept 87.6 μL/min after 100 s, which was reduced by 30%. However, the flow rate of the EOP consisted with Pt/Ti mesh electrodes dropped by 36% (from 42.1 μL/min to 26.9 μL/min). The results exhibit that the modification of polyaniline can increase the flow rate and the reliability of the EOP. In addition, compared with the EOP of Pt/Ti mesh electrodes, the EOP with PANI@Pt/Ti electrodes has no gas formation during the experiment.

The EOPs above were then connected to the level differential device (see Fig. 1(b)). The liquid level difference

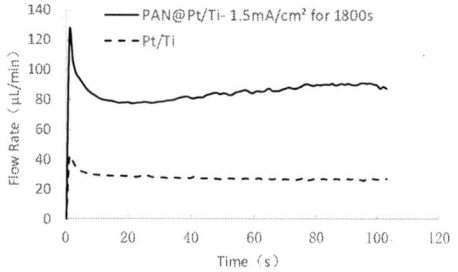

Fig.11. The flow rate of the EOPs with PANI@Pt/Ti electrodes (1.5 mA/cm^2 for 1800 s), and Pt/Ti mesh electrodes, when applying an external voltage of 6 V.

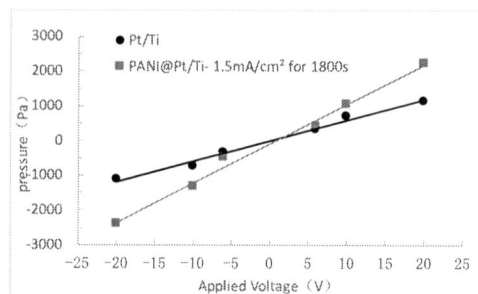

Fig.12. Limiting hydrostatic pressure of the EOPs, the slope for PANI@Pt/Ti electrodes (1.5 mA/cm^2 for 1800 s) is 113.9, and for Pt/Ti mesh electrodes is 59.3.

in height between the two sides of the EOP was recorded until no flow was observed. The recorded differences were further calculated as pressure to report the limiting hydrostatic pressure of each pump. During the test, the voltage was applied by a SMU instrument. The ultimate limiting hydrostatic pressures are shown in Fig. 12. The limiting hydrostatic pressure measured for PANI@Pt/Ti electrodes was about 113.9 Pa/V, while that for the Pt/Ti mesh electrodes was 59.3 Pa/V. The result displayed that a significant improvement in limiting hydrostatic pressure was obtained by using PANI@Pt/Ti electrodes.

After the EOP performance measurement above, the PANI@Pt/Ti electrodes with the polymerization parameter of 1.5 mA/cm^2 for 1200 s were then characterized. As seen in Fig. 13, there was no morphological change after the measurement of EOP performance. Moreover, the cyclic voltammograms displays three sets of redox peaks (A'1/B'1, A'2/B'2, A'3/B'3,),and the position of each peak changed a little. Similarly, the peaks of A'1/B'1 and A'2/B'2 refer to the redox of polyaniline, and the weak peaks A'3/B'3 show small amount of electrochemical degradation for polyaniline. In general, the PANI@Pt/Ti electrodes kept a good surface morphology and redox performance.

IV. CONCLUSION

In summary, we have prepared a PANI@Pt/Ti electrode for bubble-free EOPs. The PANI@Pt/Ti electrode was synthesized in 0.3 M aniline and 0.5 M H$_2$SO$_4$ using the CP method on a CHI 660E electrochemical workstation. We found that the surface morphology of PANI@Pt/Ti electrodes is mainly dominated by the current density, and the polymerization time played a more important role to improve the electrochemical activity. The optimized CP parameters were as follows: anode current was 0.413 mA (current density, 1.5 mA/cm^2), and anode time (polymerization time) was 1800s. Compared with Pt/Ti mesh electrode, the PANI@Pt/Ti

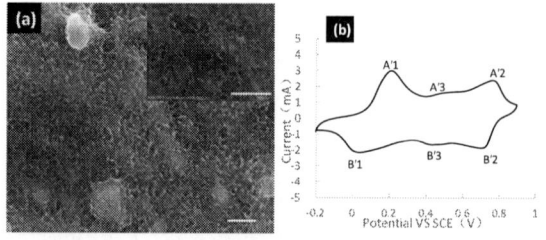

Fig.13. The characterisitics of PANI@Pt/Ti electrodes (1.5 mA/cm^2 for 1200 s). (a) SEM image. (b) Cyclic voltammograms in aqueous solution of 0.5 M H$_2$SO$_4$, at scan rate of 50 mV/s.

electrodes show good redox performance, electrochemical activity and stability. The charge transfer resistance for each PANI@Pt/Ti electrode was about 1 Ω (287.9 Ω, for Pt/Ti mesh electrode). Columbic capacity was measured to indicate the lifetime of PANI@Pt/Ti electrode (about 35 mC for anode and 55 mC for cathode). The EOP assembled with PANI@Pt/Ti electrodes showed better performance than that with Pt/Ti mesh electrodes, which was also bubble-free. The flow rate for purity water increased from 26.9 μL/min to 87.6 μL/min after 100 s, when a 6 V constant voltage was applied. Furthermore, a significant improvement in limiting hydrostatic pressure, from 59.3 Pa/V to 113.9 Pa/V, was obtained by using PANI@Pt/Ti electrodes. The PANI@Pt/Ti electrodes will be widely used in bubble-free EOPs in microfluidic system, drug delivery, and cooling of microelectronic devices.

REFERENCES

[1] Erlandsson Per G. , and Robinson Nathaniel D. "Electrolysis-reducing electrodes for electrokinetic devices," Electrophoresis. vol. 32, March 2011.

[2] Mena E., Tawfik,Francisco, J. Diez, "Maximizing fluid delivered by bubble‐free electroosmotic pump with optimum pulse voltage waveform," Electrophoresis. Vol. 38, 2017.

[3] Luft, G., Kuhl, D., Richter, G. J., "Electroosmotic Pump for Steady Regulated or Controlled Release of Medicaments," Biomed.Tech. Vol. 22, pp.169–173, July 1977

[4] Shin Woonsup, Lee Jong Myung, Nagarale Rajaram Krishna, Shin Samuel Jaeho, and Heller Adam, "A miniature, nongassing electroosmotic pump operating at 0.5 V," Journal of the American Chemical Society. Vol.133, March 2011.

[5] Kumar Rudra, Jahan Kousar, Nagarale Rajaram K., and Sharma Ashutosh, "Nongassing long-lasting electro-osmotic pump with polyaniline-wrapped aminated graphene electrodes," ACS applied materials & interfaces. Vol. 1, January 2015.

[6] Mathi S., Kumar R., Nagarale R. K., and Sharmb A., "Graphitic carbon coupled poly(anthraquinone) for proton shuttle flow in-a-cell application," Physical Chemistry Chemical Physics, Vol.18, pp. 8447–8456, December 2017.

[7] Milica M., Gvozdenović, Branimir Z., Jugović, Jasmina S., Stevanović, Tomislav Lj., Trišović, and Branimir N.Grgur, "Electrochemical Polymerization of Aniline," Electropolymerization. Dr. Ewa Schab-Balcerzak(Ed.), ISBN: 978-953-307-693-5.

[8] Bard A. J., Faulkner L. R., "Electrochemical Methods," New York: John Wiley &Sons, 1980:540.

Finite Element Simulation Study of Interfacial Crack Propagation in the Underfilled FC-BGA Package

Hong-Guang Wang
Lab of Smart Materials and Electronic Packaging in School of Materials Science and Engineering, and Guangdong Provincial Engineering Technology R&D Center of Electronic Packaging Materials and Reliability South China University of Technology
Guangzhou, 510640, China
mswanghongguang@mail.scut.edu.cn

Min-Bo Zhou
Lab of Smart Materials and Electronic Packaging in School of Materials Science and Engineering, and Guangdong Provincial Engineering Technology R&D Center of Electronic Packaging Materials and Reliability South China University of Technology
Guangzhou, 510640, China
msmbzhou@scut.edu.cn

Jiu-Bin Fei
Lab of Smart Materials and Electronic Packaging in School of Materials Science and Engineering, and Guangdong Provincial Engineering Technology R&D Center of Electronic Packaging Materials and Reliability South China University of Technology
Guangzhou, 510640, China
msjiubin@mail.scut.edu.cn

Li Sun
Lab of Smart Materials and Electronic Packaging in School of Materials Science and Engineering, and Guangdong Provincial Engineering Technology R&D Center of Electronic Packaging Materials and Reliability South China University of Technology
Guangzhou, 510640, China
mslisun@mail.scut.edu.cn

Bing-Xian Yang
Lab of Smart Materials and Electronic Packaging in School of Materials Science and Engineering, and Guangdong Provincial Engineering Technology R&D Center of Electronic Packaging Materials and Reliability South China University of Technology
Guangzhou, 510640, China
msbingxianyang@mail.scut.edu.cn

Wei-Lin Hu
Lab of Smart Materials and Electronic Packaging in School of Materials Science and Engineering, and Guangdong Provincial Engineering Technology R&D Center of Electronic Packaging Materials and Reliability South China University of Technology
Guangzhou, 510640, China
msweilinhu@mail.scut.edu.cn

Chang-Bo Ke
Lab of Smart Materials and Electronic Packaging in School of Materials Science and Engineering, and Guangdong Provincial Engineering Technology R&D Center of Electronic Packaging Materials and Reliability South China University of Technology
Guangzhou, 510640, China
mecbke@scut.edu.cn

Xin-Ping Zhang
Lab of Smart Materials and Electronic Packaging in School of Materials Science and Engineering, and Guangdong Provincial Engineering Technology R&D Center of Electronic Packaging Materials and Reliability South China University of Technology
Guangzhou, 510640, China
mexzhang@scut.edu.cn

Abstract—Flip chip ball grid array (FC-BGA) has been used widely since mid-late 1990s as a high-density package technology. The use of underfill can improve the stability and reliability of the package. However, due to the mismatch of coefficient of thermal expansion (CTE), cracking and subsequent propagation are more likely to occur at the interface related to underfill such as underfill/die, underfill/passivation and underfill/solder-mask interfaces, which has become a serious concern for package reliability. In this paper, the propagation behavior of crack at bimaterial interfaces of underfill/die, underfill/passivation and underfill/solder-mask are analyzed by two-dimensional (2D) Virtual Crack Closure Technique (VCCT). The propagation tendency of different interfacial cracks and possible failure forms of the package are predicted according to calculation of fracture mechanics parameters (e.g., energy release rate, G and phase angle, ψ). The effects of temperature on energy release rate are also investigated. Simulation results show that cracks located at different interfaces show distinct propagation tendencies and may lead to different failure forms. The energy release rate increases with temperature when below glass transition temperature (Tg) of underfill. But when temperature reaches Tg, the change of energy release rate depends on the degree of softening effect of underfill and the worsened mismatch of thermal expansion. The risk level of crack located at three different interfaces under thermal load seems to have the order from high to low as follows: underfill/passivation interface, underfill/die interface and underfill/solder-mask interface.

Keywords—FC-BGA; Underfill; Interfacial crack; VCCT; Crack propagation

I. INTRODUCTION

Since early 1990s ball grid array (BGA) package has been widely used for high-density packaging, in particular the subsequently emerged flip chip ball grid array (FC-BGA) has become a mainstream advanced packaging technology due to its high input/output (I/O) density, short interconnect and outstanding electrical performance [1]. The application of underfill can relieve the stress concentration induced by the mismatch of coefficient of thermal expansion (CTE) of various materials used in the package, including silicon chip and organic substrate. Simultaneously, fatigue life of solder joints and reliability of the package can be improved after the usage of underfill [2]. However, the use of underfill introduces many bimaterial interfaces, such as underfill/die, underfill/passivation and underfill/solder-mask interfaces, where cracking is more likely to occur. In the process of device manufacturing (particularly chip packaging) and service, the interfacial crack propagation can lead to interface delamination, solder joint failure and die cracking, which poses great threat to package reliability [3]. Heretofore, however, there is a stark lack of sufficient study and in-depth understanding about the cracking problem and related failure issue in sophisticated package structures containing various materials and multiple interfaces.

The propagation of crack along the bimaterial interface can be studied by Virtual Crack Closure Technique (VCCT). VCCT is based on the theory of linear elastic fracture mechanics, which assumes that the energy needed to separate a surface is the same as the energy needed to close the same surface. In VCCT the energy release rate (G) is used, which can be calculated by taking into account nodal forces and displacements near to the crack-tip [4]. The energy release rate that is composed of mode I and mode II components can be expressed as:

$$G_I = -\frac{1}{2\Delta\alpha} R_Y \Delta v \qquad (1)$$

$$G_{II} = -\frac{1}{2\Delta\alpha} R_X \Delta\mu \qquad (2)$$

where $\Delta\alpha$ is the length of crack extension, R_X and R_Y are reaction forces at the crack-tip node in x and y directions, respectively, $\Delta\mu$ and Δv are the relative displacement between the top and bottom nodes of the crack face in x and y directions, respectively [4], as depicted in Fig. 1.

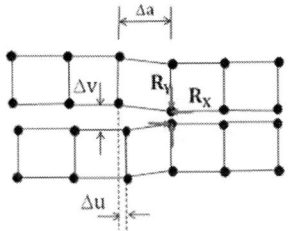

Fig. 1. 2D-VCCT crack geometry schematic [4].

The total energy release rate G can be obtained based on G_I and G_{II} as:

$$G = G_I + G_{II} \qquad (3)$$

Noteworthily, both the energy release rate and the stress intensity factor can be used as the crack driving force, and there is a certain correlation between them. The driving force of crack at the interface is closely related to the mode-mixity phase angle ψ, which is described as follows:

$$\psi = tan^{-1}(K_{II} / K_I) \qquad (4)$$

where K_I and K_{II} are mode I and mode II stress intensity factors respectively.

Generally, the crack propagation condition in the maximum energy release rate criterion is written as:

$$G = G_c \qquad (5)$$

where G_c is the critical value of energy release rate, above which a crack may propagate. For crack propagation along the bimaterial interface, the criterion is expressed as: the crack propagates when G reaches its critical value, i.e., G_c, which is also regarded as the interfacial fracture toughness. Many studies showed that G_c of interfacial crack is not a constant material parameter but depends on the mixed mode

loading [5, 6]. The dependence of the interfacial fracture toughness on the mode-mixity can be expressed as:

$$G_c = G_{Ic} [1 + (1 - \lambda) \tan^2\psi] \qquad (6)$$

where G_{Ic} is the pure mode I interfacial fracture toughness, λ represents the sensitivity of mode II loading to G_c and is in the range of 0–1, ψ is the phase angle used to describe the ratio of mode II to mode I loading [6].

In the present study, based on the two-dimensional (2D) VCCT, finite element analysis (FEA) is performed using ANSYS 19.0 to calculate fracture mechanics parameters of crack at typical interfaces in the underfilled FC-BGA package. The crack propagation direction and tendency at multiple interfaces are predicted. The effect of temperature on the energy release rate of crack is also studied.

II. MODELING AND MATERIAL PROPERTIES

This study mainly takes into account the interfaces related to underfill, BGA solder joints and package substrate were ignored when building the model. Since the flip chip structure is nearly symmetric, only a 2D half model was built to simplify analysis. Fixed support and symmetrical boundary condition were applied to the model. Thermal load was applied to the model and the whole model was assumed in plane strain state. The meshed 2D half model with boundary conditions and local model are shown in Fig. 2. Geometric parameters of the model are shown in Table I.

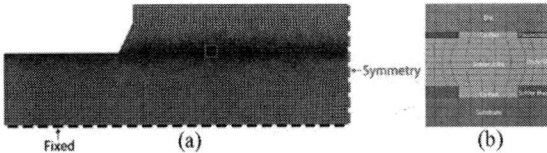

Fig. 2. 2D meshed (a) global and (b) local models in FC-BGA package.

TABLE I. GEOMETRIC PARAMETERS OF THE MODEL

Parameter	Value
Die (mm×mm)	5×0.5
Passivation thickness (μm)	5
Cu pad diameter (μm)	50
Cu pad thickness (μm)	5
Solder joint diameter (μm)	65
Solder joint height (μm)	40
Solder joint pitch (μm)	100
Solder-mask thickness (μm)	10
Substrate (mm×mm)	8×0.8

The crack was preseted at three typical interfaces, i.e., underfill/die, underfill/passivation and underfill/solder-mask interfaces. It should be indicated that these interfaces were studied separately. The crack locations and the mesh around the crack are shown in Fig. 3.

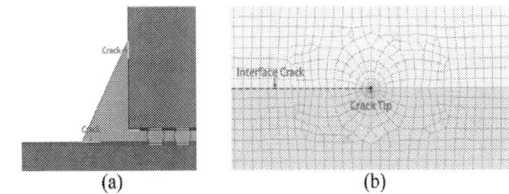

Fig. 3. Crack locations (a) and the mesh around the crack (b).

Table II lists the material properties of the model where Tg represents the glass transition temperature [7]. All of the materials are assumed as linear elastic.

TABLE II. MATERIAL PARAMETERS OF THE MODEL

Material	Young's Modulus (GPa)	Passion's Ratio	CTE (ppm/°C)
Die	131	0.28	2.8
Passivation	5.1	0.25	35
Cu pad	110	0.34	17
Solder joint	$10^{-3} \times [52708 - 67.14 \times T\,(°C) - 0.0587 \times T^2\,(°C)]$	0.40	$21.85 + 0.02039 \times T\,(°C)$
Underfill	9.0/0.23 (Tg: 120 °C)	0.35	25/95 (Tg: 120 °C)
Solder-mask	2.7	0.40	43/145 (Tg: 100 °C)
Substrate	24.5	0.22	16

III. SIMULATION METHOD AND RESULTS

A. Crack propagation direction

During packaging, testing and service, the thermal stress caused by the mismatch of CTE of materials may easily lead to crack initiating at the interface. As is known, the interfacial crack may propagate in different directions and cause different failure forms. To determine the propagation direction of an interfacial crack, an initial crack with different lengths (5, 10 and 20 μm) is preseted at the interface firstly. Then the secondary crack with a length of 3 μm is preseted in a certain direction and the initial crack is assumed to propagate in this direction. Nine different propagation directions, with angle values of 80°, 60°, 40°, 20°, 0°, -20°, -40°, -60° and -80°, are assumed, as depicted in Fig. 4. The total energy release rate of the secondary crack is calculated at 100 °C by VCCT. Crack always tends to propagate in the direction with the largest energy release rate, so the propagation direction of the initial crack can be predicted by comparing energy release rate values of the secondary cracks in different directions.

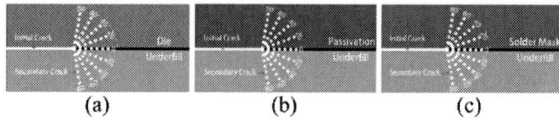

Fig. 4. Models of initial crack at different interfaces in various directions: (a) the underfill/die interface, (b) the underfill/passivation interface, and (c) the underfill/solder-mask interface.

Fig. 5 shows the variation of the total energy release rate of crack at underfill/die, underfill/passivation and underfill/solder-mask interfaces with different propagation directions, in terms of different angles illustrated in Fig. 4. For a short crack at the underfill/die interface (e.g., 5 or 10 μm), the energy release rate increases with the angle on the underfill side (0° to -80°), which means the crack is more likely to propagate along a path perpendicular to the interface when it comes into the underfill; whereas there is an opposite trend for a relatively long initial crack of 20 μm, which shows the largest energy release rate at an angle of 0°, implying that the crack propagates preferentially along the interface. Similarly, when the crack extends into the die, it tends to propagate parallel to the interface as the crack grow driving force, in terms of energy release rate, increases as the angle decreases, and it reaches the maximum at an angle of zero (i.e., at the interface).

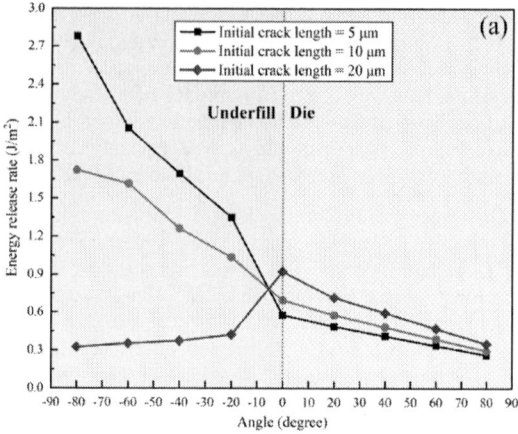

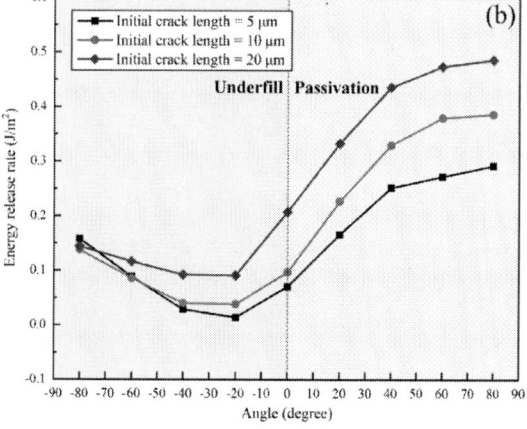

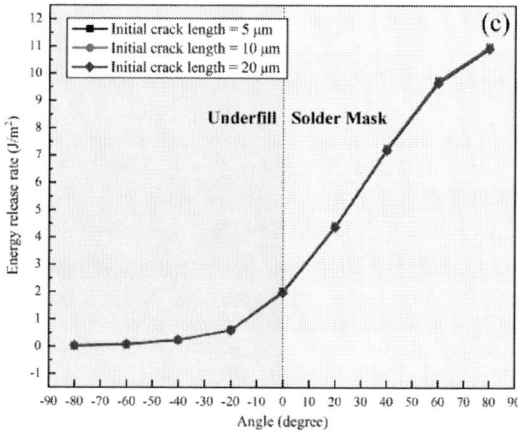

Fig. 5. Energy release rate *vs* angle (direction) for the crack with three initial lengths at different interfaces: (a) underfill/die interface, (b) underfill/passivation interface, and (c) underfill/solder-mask interface.

By using the same analysis method, for a crack at the underfill/passivation interface extending into the underfill or passivation, as shown in Fig. 4(b), the energy release rate increases with the angle away from the interface, as exhibited in Fig. 5(b), which means the crack is more likely to propagate along a path perpendicular to the interface and may penetrate into the passivation and propagate into the die. For a crack at the underfill/solder-mask interface, it is prone to propagate along the direction perpendicular to the interface on the solder-mask side and grow through the solder-mask and then extend into the substrate, as shown in

Fig. 5(c). Seemingly, the energy release rate of a crack at the underfill/solder-mask interface is independent of the initial crack length within a certain range, and the crack appears to preferentially propagate into the solder-mask.

By comparing the energy release rate values of crack at three interfaces along all possible propagation directions, it is clear that the crack at the underfill/solder-mask interface exhibits the highest energy release rate along the direction (angle) away from the interface (i.e., perpendicular to the interface). It appears that the crack is more likely to penetrate into the solder-mask, denoting a high risk for package reliability. Nevertheless, in order to predict accurately at which of the bimaterial interfaces a crack is most likely to propagate, it is imperative to do a comprehensive evaluation of both the crack driving force and the fracture toughness (in terms of critical stress intensity factor, K_c, or critical energy release rate, G_c), of the bimaterial interfaces, so as to determine precisely the interface with the highest risk of cracking and failure.

B. Effect of temperature on energy release rate

In order to study the effect of temperature on the crack growth driving force (i.e., energy release rate) at above three interfaces, cracks with different lengths (2, 5, 8, 10, 15, 20 and 30 μm, respectively) are preseted at three interfaces, respectively. The energy release rate and stress intensity factor of cracks at different temperatures (100 °C, 110 °C, 120 °C and 130 °C) are calculated. The phase angles of different cracks are obtained according to Eq. (4), which can be used to predict crack propagation tendency.

Fig. 6 plots the variation of the total energy release rate at underfill/die, underfill/passivation and underfill/solder-mask interfaces, respectively, under different temperatures. Clearly, the energy release rate of crack at all three interfaces increases with a rise in temperature from 100 °C to 110 °C, both are below Tg of underfill (120 °C) but respectively reaches and exceeds Tg of solder-mask (100 °C). It is worth looking at the energy release rate when temperature reaches Tg of underfill, whose values at the underfill/die interface are lower than that under 100 °C and 110 °C, while being higher than that at the underfill/passivation interface under 100 °C and 110 °C, as depicted in Fig. 6(a) and (b); in particular, the energy release rate at the underfill/solder-mask interface exhibits a complicated change trend with temperature (and with crack length as well), as presented in Fig. 6(c), which is in stark difference from that of Fig. 6(a) and (b). Actually, the change of energy release rate is the result of two factors with opposite effects: the softening effect of underfill and the worsened mismatch of thermal expansion. The former leads to decrease of energy release rate, while the latter results in increase of energy release rate. At the underfill/die interface and underfill/solder-mask interface, the softening effect of underfill is the key factor affecting the change of energy release rate. Understandably, the softened underfill relieves the stress concentration and reduces the driving force of crack propagation. However, the energy release rate at the underfill/passivation interface continues increasing with temperature rising from 100 °C to 130 °C, this is because the softening effect of underfill is not obvious, while the worsened mismatch of thermal expansion takes the lead.

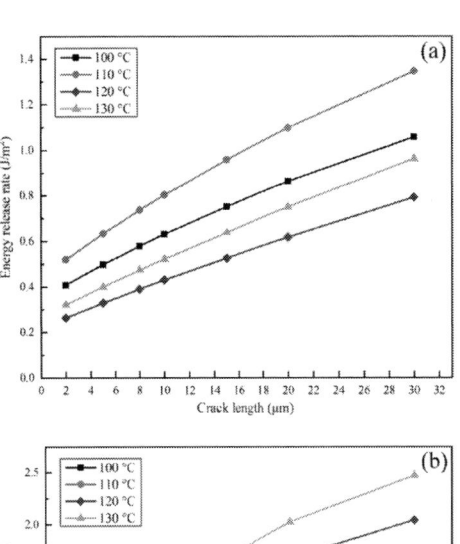

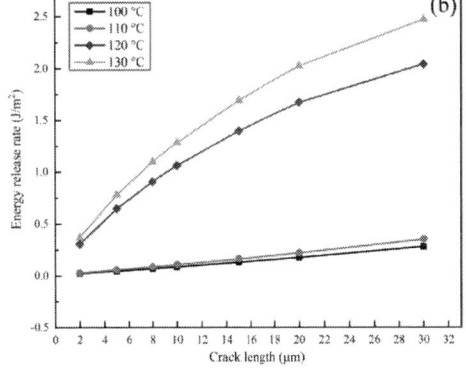

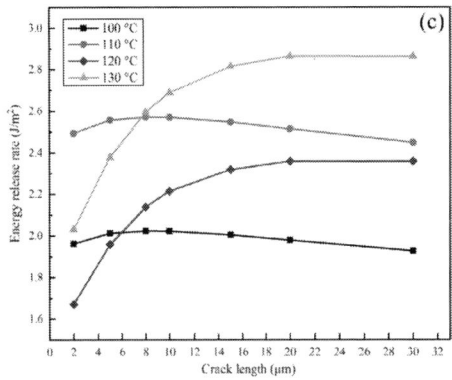

Fig. 6. Energy release rate of the crack at different interfaces of (a) underfill/die, (b) underfill/passivation and (c) underfill/solder-mask at different temperatures.

Further, it can also be seen that the energy release rate increases monotonically with the crack length at underfill/die and underfill/passivation interfaces, as shown in Fig. 6(a) and (b), meaning the increase of the driving force for crack propagation with crack length. On contrast, the energy release rate of crack at the underfill/solder-mask interface exhibits two different trends with crack length at temperatures below and above Tg of underfill, that is, an initial increase and then decrease at 100 °C and 110 °C, a monotonic increase at 120 °C and 130 °C, as presented in Fig. 6(c). This may be caused by crack closure when crack propagates to a certain length. Nevertheless, an in-depth study needs to be done to clarify the mechanism.

C. Prediction of crack propagation tendency

The phase angle changes with loading condition, in terms of the ratio of mode II to mode I loading, and the interfacial fracture toughness (G_c) increases with the increase of proportion of mode II loading according to Eq. (6), thus the change of G_c can be estimated by counting the variation of phase angle. Consequently, the crack propagation tendency can be predicted by taking into account the change of both the energy release rate (G) and its critical value (G_c). When G_c increases and becomes larger than G, the crack may stop propagating. When G_c decreases but G increases, and there is $G > G_c$, then the crack continues propagating. Fig. 7 shows the change of phase angle with the length of crack at underfill/die, underfill/passivation and underfill/solder-mask interfaces under 100 °C and 120 °C. Apparently, phase angle increases with length of crack at underfill/die and underfill/solder-mask interfaces, which means that the proportion of mode II loading increases with crack length, leading to increase of G_c; meanwhile, the energy release rate (G) also increases largely with crack length, as shown in Fig. 6. In this case, whether a crack will propagate depends on the larger one of the two parameters (i.e., G and G_c) according to the energy criterion as expressed in above Eq. (5). However, in contrast, for a crack at the underfill/passivation interface the phase angle decreases but the energy release rate increases with the crack length, as depicted in Fig. 7, which actually implies decrease of the interfacial fracture toughness and increases of the crack driving force, respectively; thus the crack may continue propagating and eventually cause interface delamination. More importantly, as long as crack delamination propagates, it may extend into the solder joint and lead to the solder joint cracking even failure.

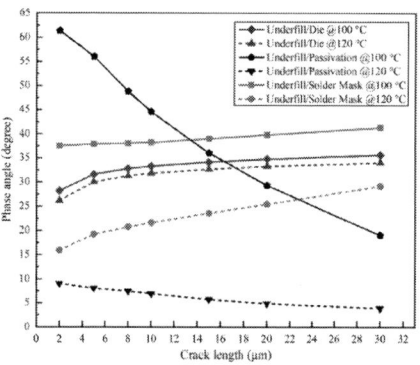

Fig. 7. Changes of the phase angle with the length of crack at underfill/die, underfill/passivation and underfill/solder-mask interfaces.

Based on above results and analysis, the risk level of crack located at three different interfaces under thermal load can be estimated tentatively as follows. For a crack at the underfill/passivation interface, the interfacial fracture toughness decreases when the crack propagates along the interface and may eventually cause interface delamination, which poses a threat to the solder joint. The crack may also penetrate into the passivation and propagate into the die, resulting in die damage, thus the crack at the underfill/passivation interface might be the one with the highest risk level. For crack at underfill/die and underfill/solder-mask interfaces, the interfacial fracture toughness increases as crack propagates. Comparatively, the energy release rate at the underfill/die interface increases monotonically with crack length under different temperatures, while the energy release rate at the underfill/solder-mask interface increases slowly (at 120 °C

and 130 °C) or drops (at 100 °C and 110 °C) when crack exceeds 10 μm. Thus, the crack at the underfill/die interface seems to be at higher risk level for propagation and fracture than at the underfill/solder-mask interface.

IV. CONCLUSIONS

In this study, the propagation behavior and risk of crack at underfill/die, underfill/passivation and underfill/solder-mask interfaces are studied by finite element simulation using two-dimensional (2D) Virtual Crack Closure Technique (VCCT). Based on simulation results, the following conclusions can be drawn:

1) A relatively short crack at the underfill/die interface tends to propagate into the underfill while a longer one is prone to propagate along the interface. Crack at the underfill/passivation interface may propagate through passivation and then extend into the die. Crack at the underfill/solder-mask interface may penetrate the solder-mask and then propagate into substrate.

2) The energy release rate of crack at three interfaces increases with temperature when below Tg of underfill. At temperature above Tg of underfill, the change of energy release rate depends on the degree of softening effect of underfill and the worsened mismatch of thermal expansion.

3) When an interfacial crack propagates along the interfaces, the interfacial fracture toughness increases gradually at underfill/die and underfill/solder-mask interfaces, but decreases at the underfill/passivation interface.

4) The risk level of crack located at three different interfaces under thermal load seems to show the order, from high to low, as follows: underfill/passivation interface, underfill/die interface and underfill/solder-mask interface.

ACKNOWLEDGMENTS

This work was supported by the National Natural Science Foundation of China under Grant No. 51775195, and the Research Fund for the Guangzhou Municipal Science and Technology Program under Grant Nos. 201807010028.

REFERENCES

[1] C. T. Peng, C. M. Liu, J. C. Lin, H. C. Cheng and K. N. Chiang, "Reliability analysis and design for the fine-pitch flip chip BGA packaging," IEEE Transactions on Components and Packaging Technologies, vol. 27, no. 4, pp. 684–693, 2004.

[2] Z. Q. Zhang and C. P. Wong, "Recent advances in flip-chip underfill: materials, process, and reliability," IEEE Transactions on Advanced Packaging, vol. 27, no. 3, pp. 515–524, 2004.

[3] T. Sinha, T. J. Davis, T. E. Lombardi and J. T. Coffin, "A systematic exploration of the failure mechanisms in underfilled flip-chip packages," IEEE 65th Electronic Components and Technology Conference (ECTC), pp. 1509–1517, 2015.

[4] ANSYS, ANSYS Mechanical APDL Theory Reference, Release 19.0, Ansys, Inc., Canonsburg, PA, 2017.

[5] H. C. Cao and A. G. Evans, "An experimental study of the fracture resistance of bimaterial interfaces," Mechanics of Materials, 7.4, 295–304, 1989.

[6] J. W. Hutchinson and Z. Suo, "Mixed mode cracking in layered materials," Advances in Applied Mechanics, 29, 63–191, 1991.

[7] M. Hsieh, C. C. Lee and L. C. Hung, "Comprehensive thermo-mechanical stress analyses and underfill selection of large die flip chip ball grid array," IEEE Transactions on Components, Packaging and Manufacturing Technology, vol. 3, no. 7, pp. 1155–1162, 2013.

Enhancement of Copper Nanoparticle Paste by Pressure-less Sintering on Different Substrates in Pt-catalyzed Formic Acid Atmosphere

Junlong Li
Smart Sensing Research and Design Center
Institute of Microelectronics of Chinese Academy of Sciences
Beijing, China
nuaajunlong@163.com

Yang Xu*
Smart Sensing Research and Design Center
Institute of Microelectronics of Chinese Academy of Sciences
Beijing, China
xuyang@ime.ac.cn

Panju Shang
Material and process technology Department
Central Hardware Engineering Institute of Huawei
Shenzhen, China
shangpanju@huawei.com

Yinghui Wang
University of Chinese Academy of Sciences
Beijing, China
wangyinghui@ime.ac.cn

Tadatomo Suga
Collaborative Research Center
Meisei University
Tokyo, Japan
suga@gakushikai.jp

Abstract—**With the rapid development of wide bandgap semiconductor materials, next generation power devices are attracting increasing attention in electronic application. Due to the high conductivity and electromigration resistance, the bonding technology by copper paste sintering is a promising method to meet the demands for reliability of power devices. However, the oxidation of copper nanoparticle in paste is still the bottleneck in pressure-less sintering process at low temperature. In this paper, a pressure-less sintering process was developed in Pt-catalyzed formic acid atmosphere to improve the sintering property of copper paste at 220°C. The H radicals generated from Pt-catalyzed formic acid atmosphere effectively protected the copper nanoparticles from oxidation through the whole of evaporation of solvent and decomposition of organics, thus facilitated the grains growth. By this process, we achieved denser sintered layers and obtained a 22.5 MPa shear strength with NiPdAu finished substrates. This pressure-less sintering process is of great significance to the high temperature packaging technology.**

Keywords—*Copper nanoparticle paste, Pt-catalyzed formic acid, Powder device, Pressure-less*

I. INTRODUCTION

Due to their excellent physical properties, wide bandgap semiconductor materials have attracted more and more attention. In order to improve the reliability of applications, the requirements for high temperature packaging of power devices have become harsher[1]. Sn-based solders are at a risk of remelting under high temperature operating conditions [2]. Besides, conventional high temperature solders which can form intermetallic compounds with high melting points, such as Au-Sn and Au-Ge [3], Zn-Sn[4] and Zn-Al[5], but the poor thermal conductivity limits their further development.

Owning to the high conductive property and thermal conductivity, the sintering technology using metallic nanoparticles is a promising method for the high temperature packaging of power devices[6]. So far, abundant plentiful studies have shown that Ag sintering technology is an excellent lead-free and reliable solution. However, Ag sintering material has the problem of poor resistance to electromigration. In contrast, copper sintering material exhibits an excellent ability of electromigration resistance and the cost of material is lower[7]. Nevertheless, a challenge for copper sintering technology is that copper nanoparticles tend to be oxidized during the sintering process. Therefore, different methods have been used to improve the antioxidant capacity of copper nanoparticles, such as Cu@Ag core-shell structure[8], formic acid sintering atmosphere[9] and reductive solvent[10]. Meanwhile, most of studies were focus on the pressure assisted sintering process. Considering the demands of practical production, the method of pressure-less sintering has a better potential in the future applications.

In this paper, the pressure-less sintering property of copper paste was enhanced by Pt-catalyzed formic acid atmosphere. The sintering property of copper nanoparticles in different atmospheres were investigated. Moreover, the mechanical property of specimens sintered with different metal layers on bare copper substrates were measured, and the mechanisms were discussed.

II. MATERIALS AND EXPERIMENTAL PROCEDURES

A. Materials

The copper paste used in this study was acquired from Namcis Corporation. The copper paste comprises of large-sized (diameter around 500 nm) and small-sized (diameter around 50 nm) copper nanoparticles. The copper nanoparticles were coated with organics and mixed with low boiling solvents to form the copper paste which needs to be stored at a condition of -40°C. The substrates were bare Cu substrates. Two kinds of metal layers, Ag (4 μm) and NiPdAu (1 μm), finished on the bare Cu substrates by electroplating process were also used in our experiment. The chips of silicon devices were 0.8 mm × 0.8 mm × 0.2 mm, and the surfaces were finished with gold layer (3 μm).

B. Experimental Procedures

The specimens were prepared in following steps: (1) the substrates were ultrasonically cleaned in acetone and ethanol, then citric acid (0.2 mol/L) was used to remove the oxide on the surface; (2) printing the copper paste on the cleaned substrates by dispensing machines (350PCS-SM200 OMEGAX); (3) placing the chips on the paste layer and fabricating sandwich structures. A bonding equipment (HTB-MM & Smart R Box) was used to carry out the sintering process without pressure for copper paste (in Fig. 1 (a)). The prepared specimens were sintered at 220°C in Pt-catalyzed formic acid atmosphere for an hour and the typical profile is shown in Fig.1 (b). Through the whole process, the operating temperature of Pt catalysis was controlled at 160°C.

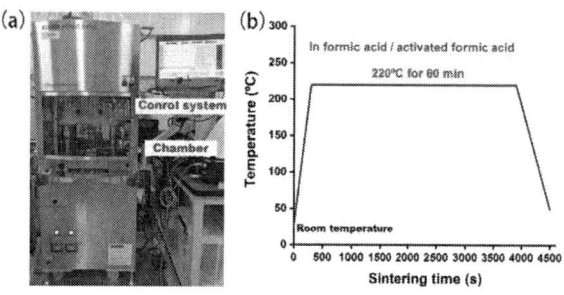

Fig. 1 (a) A bonding equipment for copper paste sintering, (b) a typical temperature-time profile for copper paste sintering process.

To investigate the thermal characteristics of copper paste, the thermogravimetric-differential thermal analysis (TG-DTA) was conducted in a flow of nitrogen atmosphere and the heating rate was 10°C/s. To evaluate the shear strength of specimens, a tester (TRY Precision MFM1200) was used at a speed of 50 μm/s. For different parameters of sintering process, 14 specimens were measured. The SEM (S4800, FE-SEM) was employed to observed the fracture surfaces after shear test. To obtain the characterization of the cross-sectional microstructures of sintered layers, the specimens were firstly mechanically cut and then polished by an ion beam. And the central part of the specimen was employed for every specimen. In order to further verify the effect of the Pt-catalyzed formic acid atmosphere, controlled experiments were also conducted in nitrogen and formic acid atmosphere (without Pt catalysis), respectively.

III. RESULTS AND DISCUSSION

A. Thermal characteristics of the copper paste

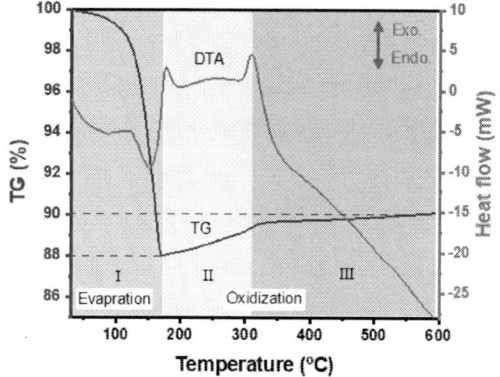

Fig. 2 TG-DTA results of copper paste acquired from Namcis Corporation.

As shown in Fig.2, the results exhibit that the weight of copper was 88% according to the TG curve. At the temperature above 170°C, the weight gradually increased, which indicates copper nanoparticles were oxidized in nitrogen. At 150°C, there was a endothermic peak in DTA curve, which corresponds to the evaporation of solvent and decomposition of organics in the paste. However, there were two exothermic peaks at 170°C and 300°C, respectively.

Owing to the increase of the weight, it indicates that different oxidation reaction occurred when the temperature reach to 170°C and 300°C. Therefore, the sintering process requires to be conducted in a reductive atmosphere to avoid the effects of these reactions.

B. Sintering property in different atmospheres

Fig.3 shows the morphologies of copper paste sintered for 60 min in different atmospheres were observed by SEM. When sintered in N_2 nitrogen atmosphere, no connection formed between copper nanoparticles as shown in Fig. 3 (a) and (d). It infers that the oxidation of copper nanoparticles after the evaporation of solvent, which leads to a failure in the shear test. On the contrary, it is obvious that the neck growth between copper nanoparticles in formic acid (Fig.3 (b) and (e)) and Pt-catalyzed formic acid atmosphere (Fig.3 (c) and (f)). It should be noted that the grain size sintered in Pt-catalyzed formic acid atmosphere was larger than that in formic acid atmosphere. It suggests that the surfaces of copper nanoparticles suffered oxidation followed by reduction in formic acid atmosphere.

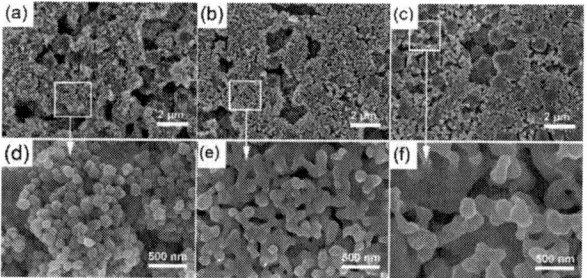

Fig. 3 SEM images of fracture surface sintered at 220°C for 60 min in different atmospheres: (a) and (d) in nitrogen, (b) and (e) in formic acid atmosphere, (c) and (f) in Pt-catalyzed formic acid atmosphere.

Furthermore, the effects of reductive atmospheres on the sintering process of copper paste are explained in Fig.4. For the formic acid atmosphere, the surfaces of copper nanoparticles were oxidized after the evaporation of solvent when the sintering temperature much higher than 170°C. Subsequently, the formic acid molecules reacted with the copper oxide and formed the amorphous layers on the surfaces of copper nanoparticles according to our previous study[11]. As the sintering process proceeded, the decomposition of amorphous layers leaded to the neck growth. In comparison, highly activated H radicals generated from the Pt-catalyzed formic acid atmosphere more effectively protected the surfaces of copper nanoparticles from oxidation in the initial stage of sintering process. Therefore, the neck growth would occur earlier than that in formic acid atmosphere, which facilitated the fast growth of grain size.

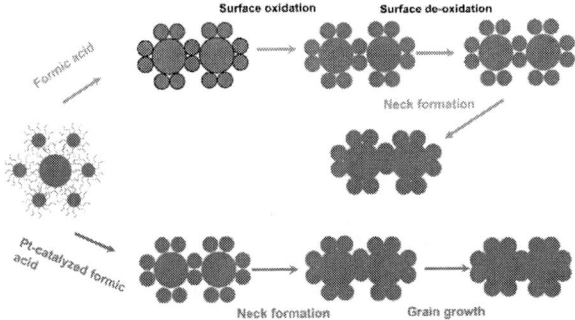

Fig. 4 Schematic diagram of the mechanism of copper nanoparticle sintering process in formic acid atmosphere without and with Pt catalysis.

C. Effects of substrates finished with different metallic layers

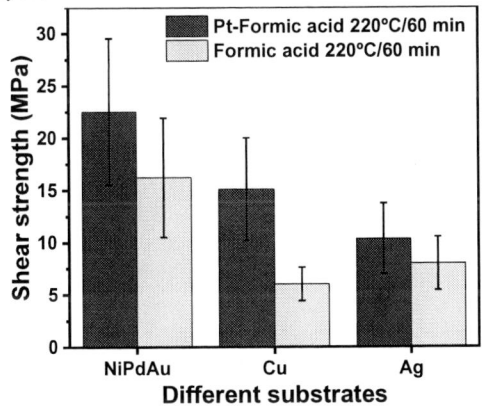

Fig. 5 The shear strength of copper nanoparticle paste sintered on different substrates.

To evaluate the mechanical property of copper paste, the specimens sintered with different substrates in formic acid and Pt-catalyzed formic acid atmosphere. In general, the shear strengths obtained in formic acid atmosphere was much lower than in Pt-catalyzed formic acid atmosphere as shown in Fig.5. This is consistent with the analysis of Fig.3 and Fig.4. Therefore, it is proved that the H radicals can significantly improve the sintering property of copper paste at low temperature. In addition, the shear strengths of 22.5 MPa, 15.1 MPa and 10.5 MPa were obtained in Pt-catalyzed formic acid atmosphere for NiPdAu, copper and Ag substrate respectively. These results demonstrate that the NiPdAu substrate was benefit for the enhancement of sintering property of copper paste. While the effect of Ag substrate on improvement of sintering property was poor. Interestingly, the shear strength obtained with bare copper substrates was the lowest compared with NiPdAu and Ag substrates in formic acid atmosphere, and almost one-third of the shear strength in Pt-catalyzed formic acid. It is deduced that the surfaces of bare copper substrates were at a risk of oxidation along with copper nanoparticles in formic acid atmosphere, according to the significant decrease in shear strength.

Meanwhile, the cross-sectional structures of sintered layer were observed as shown in Fig.6. The thickness of sintered layers were about 20 µm. Obviously, there were large-sized voids in the sintered layers obtained with different substrates in formic acid atmosphere (Fig.6 (b), (d) and (f)). In contrast, there are almost no voids in the sintered layer in Pt-catalyzed formic acid atmosphere (Fig.6 (a) and (c)), except for a few

small-sized voids for NiPdAu substrate (Fig.6 (f)). Therefore, the formation of voids was eliminated effectively by Pt-catalyzed formic acid atmosphere leading to a denser sintered layer. These results were attributed to the fact that H radicals can promote the slow evaporation of solvents at low temperature.

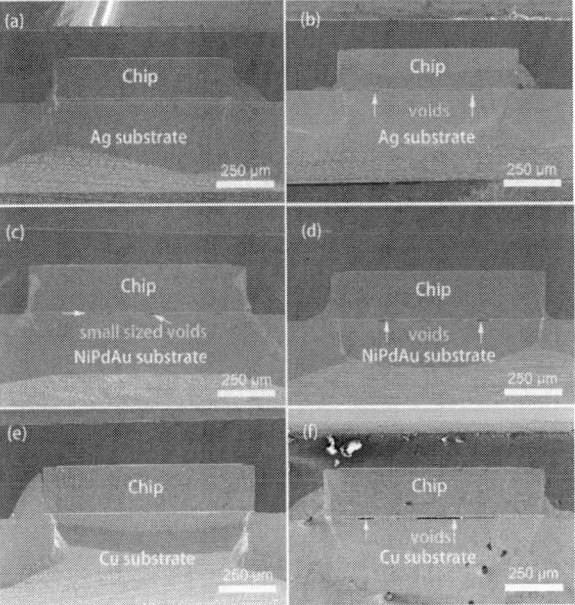

Fig. 6 The SEM images of cross-sections of the copper paste sintered on different substrates: (a), (c) and (e) in Pt-catalyzed formic acid atmosphere; (b), (d) and (f) in formic acid atmosphere.

It is a known fact that the mechanical property is also influenced by the metal layer finished on the bare copper substrates. In order to verify the effects of metal layers on shear strength, SEM imaging was performed to characterize the interfaces of sintered layer/substrates as shown in Fig.7. Fig.7 (a) shows that cracks appeared at the interface of sintered layer/Ag substrate. Meanwhile, there are clearly grain boundaries in Ag layer. Conversely, no obvious grain boundaries could be observed in the metal layers of bare copper and NiPdAu substrate as shown in Fig.7 (b) and (c). Surprisingly, there was a layer of Au-Cu intermetallic compounds at the interface for the NiPdAu substrate and the thickness was around 5 nm as shown in Fig.7 (b).

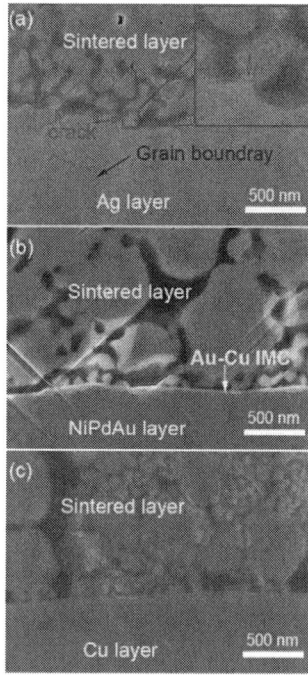

Fig. 7 The SEM images of different interfaces of sintered layer/substrates in Pt-catalyzed formic acid atmosphere.

Based on the above analysis, the sintering mechanisms for copper paste with different metal layer finished on bare copper substrates and are illustrated in Fig.8. Owing to a large surface area for small-sized copper nanoparticles, the diffusion rate of the surface atoms is faster than that of large-sized copper nanoparticles. So the connection between large-sized nanoparticles or sintered layer and substrates is always bridged by the sintering of small-sized nanoparticles. By calculation at 220°C[12], the diffusion rate of copper atoms into Au, Ag and copper metal layers were 1.02×10^{-23} m^2/s, 4.37×10^{-25} m^2/s and 6.26×10^{-26} m^2/s, respectively. However, the fast diffusion of copper atoms to grain boundaries resulted in the density decrease of sintered layer at the interface for Ag substrate. Thus, the cracks occurred at the interface results in an adverse effect on mechanical strength of specimens during the pressure-less sintering process. In contrast, the fast diffusion of copper atoms to gold layer leaded to the formation of Au-Cu intermetallic compounds[13]. Moreover, the IMCs formation does not affect the density of the sintered layer at the interface. Therefore, the NiPdAu substrate is more conductive to enhance the mechanical strength of specimens.

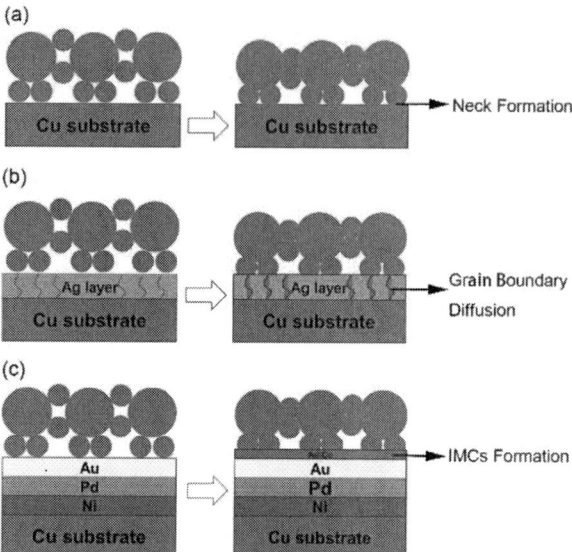

Fig. 8 The sintering mechanism of different substrate: (a) Cu, (b) Ag and (c) NiPdAu.

IV. CONCLUSIONS

The pressure-less sintering process in Pt-catalyzed formic acid atmosphere for copper paste has been successfully demonstrated. The following conclusions in this paper are obtained:

1) The oxidation resistance of copper paste and the grain size were effectively improved in Pt-catalyzed formic acid atmosphere due to the high activity of H radicals.

2) Compare to the formic acid atmosphere, specimens sintered in Pt-catalyzed formic acid atmosphere exhibited higher shear strength and denser sintered layer;

3) A shear strength of 22.5 MPa, 15.1 MPa and 10.5 MPa was obtained at 220°C by sintering on NiPdAu, bare Cu and Ag substrate, respectively. There were no obvious grain boundaries in the Cu and Au layer. But the formation of Au-Cu IMCs at the Cu/Au interface results a higher shear strength due to the faster diffusion rate of copper atoms to the Au layer. However, a large number of copper atoms diffused into the grain boundaries in the Ag layer leading to a decrease in density and even cracks at the Cu/Ag interface.

ACKNOWLEDGMENT

This research was supported by the project "Low temperature bonding technology by copper sintering". All authors are grateful for the facility support from Alpha Design Co. Ltd (HTB-MM & Smart R Box) and Musashi Engineering INC (350PCS-SM200 OMEGAX).

REFERENCES

[1] H. Lee, V. Smet, R. Tummala, "A Review of SiC Power Module Packaging Technologies: Challenges, Advances, and Emerging Issues," IEEE J. Emerg. Sel. Top. Power Electron., vol.8, pp:239-255, 2019.

[2] F. P. McCluskey, M. Dash, Z. Wang, D. Huff, "Reliability of high temperature solder alternatives," Microelectron. Reliab., vol.46, pp:1910-1914, 2006.

[3] J. W. Yoon, B. I. Noh, S. B. Jung, "Interfacial reaction between Au-Sn solder and Au/Ni-metallized Kovar," J. Mater. Sci., vol.22, pp:84-90, 2011.

[4] S. Kim, K. S. Kim, S. S. Kim, K. Suganuma, "Interfacial Reaction and Die Attach Properties of Zn-Sn High-Temperature Solders," J. Electron. Mater., vol.38, pp:266-272, 2009.

[5] Y. Xiao, M. Li, L. Wang, S. Huang, X. Du, Z. Liu, "Interfacial reaction behavior and mechanical properties of ultrasonically brazed Cu/Zn–Al/Cu joints," Mater. Des., vol.73, pp:42-49, 2015.

[6] H. Zhang, H. Bai, Q. Jia, W. Guo, G. Zou, L. Liu, "Stabilizing the sintered nanopore bondline by residual organics for high temperature electronics," Microelectron. Reliab., vol.111, pp:113727, 2020.

[7] H. Ren, F. Mu, S. Shin, L. Liu, G. Zou, T. Suga, "Low temperature Cu bonding with large tolerance of surface oxidation," AIP Adv., vol.9, pp:055127, 2019.

[8] J. Liu, Y. Mou, Y. Peng, M. Chen, "Facile Preparation of Cu-Ag Micro-Nano Composite Paste for High Power Device Packaging," IEEE 70th Electronic Components and Technology Conference, pp:755-761, 2020.

[9] X. Liu, H. Nishikawa, "Low-pressure Cu-Cu bonding using in-situ surface-modified microscale Cu particles for power device packaging," Scr. Mater., vol.120, pp:80-84, 2016.

[10] Z. Yang, S. Carter-Searjeant, M. Green, L. Mills, S. H. J. M. L. Mannan, "High bond strength Cu joints fabricated by rapid and pressureless in situ reduction-sintering of Cu nanoparticles," Mater. Lett., pp:128260, 2020.

[11] J. Li, Y. Xu, X. Zhao, Y. Meng, Z. Yin, Y. Wang, et al., "Enhancement and Mechanism of Copper Nanoparticle Sintering in Activated Formic Acid Atmosphere at Low Temperature," ECS J. Solid State Sci. Technol., vol.10, pp:054004, 2021.

[12] X. Wang, Y. Mei, X. Li, M. Wang, Z. Cui, G. Lu, "Pressureless sintering of nanosilver paste as die attachment on substrates with ENIG finish for semiconductor applications," J. Alloy. Compd., vol.777, pp:578-585, 2018.

[13] C. Chen, Z. Zhang, Q. Wang, B. Zhang, K. Suganuma, "Robust bonding and thermal-stable Ag–Au joint on ENEPIG substrate by micron-scale sinter Ag joining in low temperature pressure-less," J. Alloy. Compd., vol.828, pp:154397, 2020.

Influences of the Solder Size on Growth of Interfacial Cu_6Sn_5 and Mechanical Performance of Sn-3.0Ag-0.5Cu/(111)Cu Joints Subjected to Multiple Reflow Soldering

Ming-Qiang Chen
Lab of Smart Materials and Electronic Packaging in School of Materials Science & Engineering, and Guangdong Provincial Engineering Technology R&D Center of Electronic Packaging Materials and Reliability South China University of Technology
Guangzhou, 510640, China
msmingqiangchen@mail.scut.edu.cn

Min-Bo Zhou*
Lab of Smart Materials and Electronic Packaging in School of Materials Science & Engineering, and Guangdong Provincial Engineering Technology R&D Center of Electronic Packaging Materials and Reliability South China University of Technology
Guangzhou, 510640, China
msmbzhou@scut.edu.cn

Yun-Wei Li
Lab of Smart Materials and Electronic Packaging in School of Materials Science & Engineering, and Guangdong Provincial Engineering Technology R&D Center of Electronic Packaging Materials and Reliability South China University of Technology
Guangzhou, 510640, China
msyunweili@mail.scut.edu.cn

Xin-Ping Zhang*
Lab of Smart Materials and Electronic Packaging in School of Materials Science & Engineering, and Guangdong Provincial Engineering Technology R&D Center of Electronic Packaging Materials and Reliability South China University of Technology
Guangzhou, 510640, China
mexzhang@scut.edu.cn

Abstract—In heterogeneous integration with three-dimensional (3D) packaging, solder joints are used in different levels of packages, which have a wide range of sizes, for instance, from hundreds of microns in BGA joints to a few tens of microns in micro-bump joints. Thus far, it is not yet well understood how the change in joint size affects the growth of interfacial intermetallic compound (IMC) and reliability of joints. In this work, a series of Sn-3.0Ag-0.5Cu solder balls with diameter scaling down from 450 μm to 15 μm were used to fabricate single-sided joints on monocrystalline (111)Cu by reflow soldering at 260 °C for different reflow cycle times of 1, 2 and 4. The influences of solder ball sizes and reflow cycle times on interfacial IMC growth and shear strength of Sn-3.0Ag-0.5Cu/(111)Cu joints were investigated systematically. Results show that interfacial Cu_6Sn_5 grains in Sn-3.0Ag-0.5Cu/(111)Cu joints exhibit mainly scallop-like morphology. The grain size and thickness of the interfacial IMC layer in the joints change non-monotonically with decreasing solder ball diameter, and both of them reach the maximum at a solder ball size of 200 μm. The ripening of IMC grains by grain boundary (GB) migration mechanism during IMC growth occurs as earlier as the size of solder joints reduces and the reflow cycle time increases. In addition, the shear strength of Sn-3.0Ag-0.5Cu/(111)Cu joints increases with decrease of the solder ball diameter.

Keywords—*Heterogeneous integration, Micro-bump solder size effect, Intermetallic compound, Monocrystalline Cu, Multiple reflow*

I. INTRODUCTION

Heterogeneous integration with three-dimensional (3D) packaging enables chips with different functions and sizes to be arranged side by side or stacked through redistribution layers (RDLs) or through-silicon vias (TSVs) so as to achieve higher input/output density and smaller form factor. It has been generally regarded as an effective approach to realizing miniaturization and multifunctionality of electronic products [1]. In 3D heterogeneous integration

packages, solder joints with various sizes are used in different levels of packages, such as BGA joints with the size of hundreds of microns, flip-chip joints with the size of tens of microns and micro-bump joints with ten microns even smaller size [2, 3]. Hitherto, however, the solder joint size effects on the growth behavior of interfacial intermetallic compound (IMC) and reliability of the joints are not sufficiently well understood.

Previous work showed that the growth curves of IMC grain sizes are not monotonic when the solder joint size decreases gradually from hundreds of microns to several tens of microns, and a larger Cu concentration gradient leads to formation of smaller IMC grains [3, 4]. Further, it has been indicated that the decrease of solder joint size can lead to increase in shear strength yet the changes in deformation and fracture behavior of joints [2, 5].

Notably, the package miniaturization has led to the solder joining area of either the solder bump or under bump metallization (UBM) being so small that only a few grains (or even only one grain) exist in the level of micro-bump joints. In such a scenario, the influence of grains with different orientations in the UBM on the interfacial IMC growth has increased significantly, which may cause the anisotropy in the microstructure of interfacial IMC, thereby influencing greatly the performance of solder joints. Many studies pointed out that the IMC nucleation and growth behavior during the interfacial reaction between the solder alloy and the monocrystalline metallization is obviously different from that of the polycrystalline metallization. For example, regular prismatic Cu_6Sn_5 grains can be obtained on (001) Cu substrate and (111) Cu substrate within a short reflow time, but the morphology of Cu_6Sn_5 transforms form prism-shape to scallop-shape with a long reflow time or aging time [6, 7]. Meanwhile, it was indicated that the monocrystalline Cu metallization has no grain boundary defects, which has a beneficial effect on retarding formation of Kirkendall voids in joints, especially for the electromigration problem widely existed in micro-bump solder joints, thereby, less defective or defect-free solder joints can be obtained [8].

So far most of the available studies concentrated on the interfacial IMC growth behavior in solder joints with polycrystalline Cu substrate and the orientation relationship between the IMC and the monocrystalline metallization as well as the IMC growth kinetics in large solder joints (e.g., a few hundred microns or larger), less attention has been given to the interfacial IMC growth and reliability issue of the micro-bump solder joints with constrained dimension of Sn-based solder alloy cap on the monocrystalline Cu metallization. Further, there is few work about the solder volume (size) effects on the interfacial IMC growth and morphological evolution for joints using monocrystalline metallization. Meanwhile, due to multi-layer structures in heterogeneous integration with 3D packaging devices, the solder joints of different sizes may be subjected to several times of reflow soldering. Multiple temperature cycles of reflow soldering may bring about overgrowth of IMC grains, thereby affecting significantly the reliability of the solder joints and packaging structures [9].

In this work, Sn-3.0Ag-0.5Cu solder balls with the decreasing diameter in the range of 450 to 15 μm and the monocrystalline Cu metallization with (111) plane are used to fabricate solder joints with different sizes by reflow soldering for multiple reflow cycles. The effects of monocrystalline Cu metallization (substrate), solder ball sizes and reflow soldering cycle times on growth of interfacial Cu_6Sn_5 in solder joints are systematically investigated. Meanwhile, the shear performance of solder joints with different sizes are also studied.

II. EXPERIMENTAL PROCEDURES

A series of Sn-3.0Ag-0.5Cu solder balls having diameters in the range of 450, 300, 200, 120, 85, 50 and 15 μm were made by re-melting the stencil printed solder powers (Shenzhen Fitech corporation). Then, the solder balls were reflowed on monocrystalline (111) Cu substrate at 260 °C for reflow cycle times of 1, 2 and 4, respectively. The reflowed single-sided joints, i.e., Sn-3.0Ag-0.5Cu/(111)Cu joints, were etched with a 3% $HCl+7\%HNO_3+90\%C_2H_5OH$ solution to remove all the solder above the interfacial IMC layer so as to obverse the microstructural morphologies of the IMC. For observing the cross-sectional microstructure of the joints, the mounted joints were ground using SiC sandpapers and polished in Al_2O_3 suspension, followed by etching using a $2\%HNO_3+3\%HCl+95\%C_2H_5OH$ solution. Morphologies and compositions of the interfacial IMC were analyzed by a scanning electron microscope (SEM, Phenom ProX) equipped with energy-dispersive spectrometer (EDS). Image-Pro Plus 6.0 software of was employed to measure the size of interfacial IMC grains and the thickness of IMC layer. IMC layer thickness was calculated by dividing the IMC layer area by the spread length of the solder joint. A multifunctional mechanical tester (MFM1200, TRY Precision) was used to evaluate the shear strength of joints.

III. RESULTS AND DISCUSSION

Fig. 1 shows the top-view morphologies of interfacial IMC grains in the joints, in which IMC is confirmed to be Cu_6Sn_5 phase thought EDS analysis. After undergoing a single reflow cycle, for solder joints of 450 μm solder ball the interfacial Cu_6Sn_5 grains show prism-like morphology, as presented in Fig. 1(a1); while for solder joints with solder ball diameters less than 450μm, almost all the interfacial Cu_6Sn_5 grains show a typical scallop-like morphology, as exhibited in Fig. 1 (b1)–(f1). For joints subjected to reflow for twice and four times, prismatic IMC grains in the largest joint, as shown in Fig. 1 (a1), gradually change to scallop-shaped, as displayed in Fig. 1 (a2) and (a3). While Cu_6Sn_5 grains in smaller joints still keep scallop-shaped, as presented in Fig. 1(b2)–(f2) and (b3)–(f3). Meanwhile, it can be seen clearly that the boundaries between different Cu_6Sn_5 grains are closer in joints subjected to multiple reflow than those undergoing a single reflow cycle. With increasing reflow cycle time (to 2 and 4), the number of Cu_6Sn_5 grains in the limited soldering region decreases gradually, and Cu_6Sn_5 grains grow rapidly and shrink simultaneously, which is the typical Ostwald ripening behavior. However, the joints with smaller solder balls appear to show the adsorption type growth behavior.

Figs. 2 and 3 shows the size distributions and the change of the average size of Cu_6Sn_5 grains in different solder joints. Apparently, for joints with the same size and undergoing different reflow cycle times, the size distribution interval gradually increases and peaks of curves move to the right direction and go down with increase of reflow cycles, as exhibited in Fig. 2. Noteworthily, the average size of Cu_6Sn_5 grains does not show a monotonic change with decrease of solder ball diameter from 450 to 15 μm under both of single and multiple reflow cycles, as depicted in Fig.3. Firstly, the average size of Cu_6Sn_5 grains increases with solder ball diameter decreasing from 450 to 200 μm and reaches the maximum for joints of 200 μm diameter solder ball, then it shows an opposite trend, i.e., the average grain size of Cu_6Sn_5 decreases with the decreasing solder ball diameter.

During reflow soldering, the formation and growth of IMC are usually accompanied by the dissolution of IMC [10]. In the initial stage of heating to soldering temperature, Cu atoms from Cu substrate react with Sn atoms in the molten solder at the interface to form IMC [3–5,11]. At this stage, the interfacial reaction is the major factor for the IMC growth. As the reflow time proceeds, the IMC grains become larger and gradually touch each other closely, thus the migration rate of Cu atoms to the solder is slowed down and the growth mode of IMC grains changes to the grain ripening, Ostwald ripening and grain adsorption with grain boundary (GB) migration [6, 12, 13]. For Ostwald ripening, atoms released by dissolution of the small one diffuse into the large one to promote the growth of the larger grains, then the small grains gradually disappear. The grain adsorption with grain boundary migration mechanism actually means that the adjacent grains adsorb each other and form GB, then the GB migrates outward and the final ripened IMC grain retains the outline of the original grains. As shown in the red hexagon in Fig. 1 (a3)–(f3), there are boundary lines on the surface of some large grains and protrusions at the grain boundaries.

978-1-6654-1392-3/21 $31.00 © 2021 IEEE

Based on above observation and analysis of morphology evolution and growth of interfacial IMC grains in joints of different solder ball diameters, it is clear that the solder ball size and reflow cycle times have significant influence on the interfacial IMC growth behavior of solder joints. Small joints are more likely to form scallop-like rather than prism-like morphology. With the increase of reflow cycle, the interfacial IMC tends to grow up by grain adsorption mechanism. This is not the same as being reported in previously studies that there is regular arrangement of 60° elongated prism-like IMC grains [5–7] in Sn-based solder alloy/(111)Cu joints with different sizes. This phenomenon may be directly related to the Cu atom diffusion flux and the surface energy of grains [7, 10]. Firstly, the surface area of scallop-like grains is smaller than that of prism-like grains for IMC grains having the same volume and requiring smaller surface energy. Therefore, scallop-like grains can be formed easily.

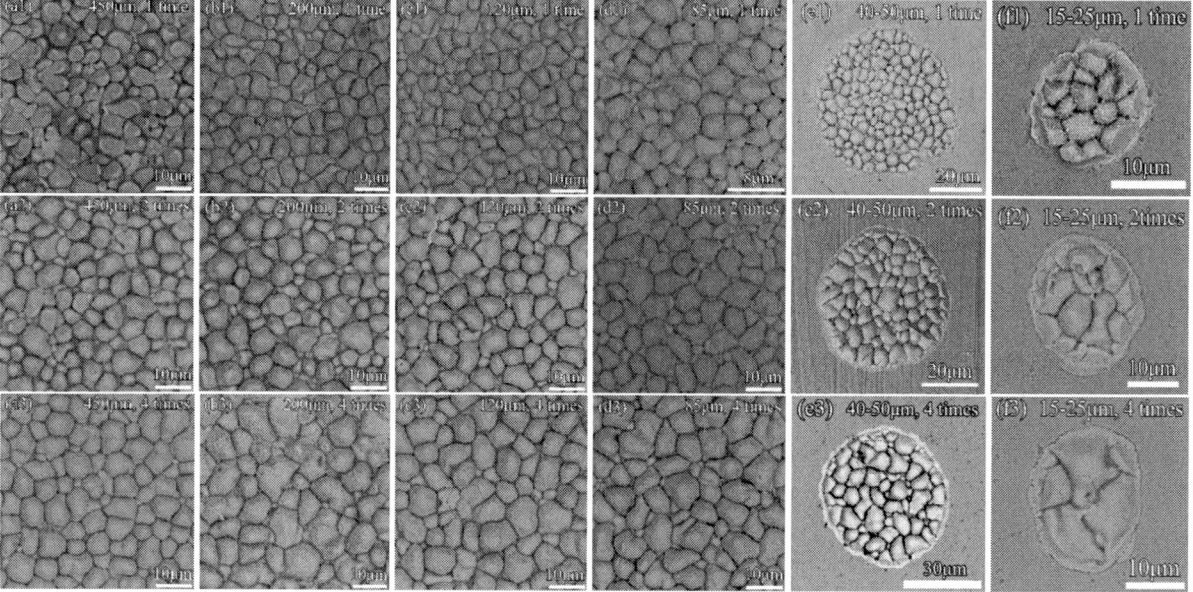

Fig. 1 Morphologies of interfacial IMC grains in Sn-3.0Ag-0.5Cu/(111)Cu joints with different solder ball sizes: (a1–a3) 450 μm, (b1–b3) 200 μm, (c1–c3) 120 μm, (d1–d3) 85 μm, (e1–e3) 40–50 μm and (f1–f3) 15–25 μm, and subjected to reflow soldering at 260 °C for different reflow cycle times: (a1–f1) once, (a2–f2) twice, (a3–f3) four times.

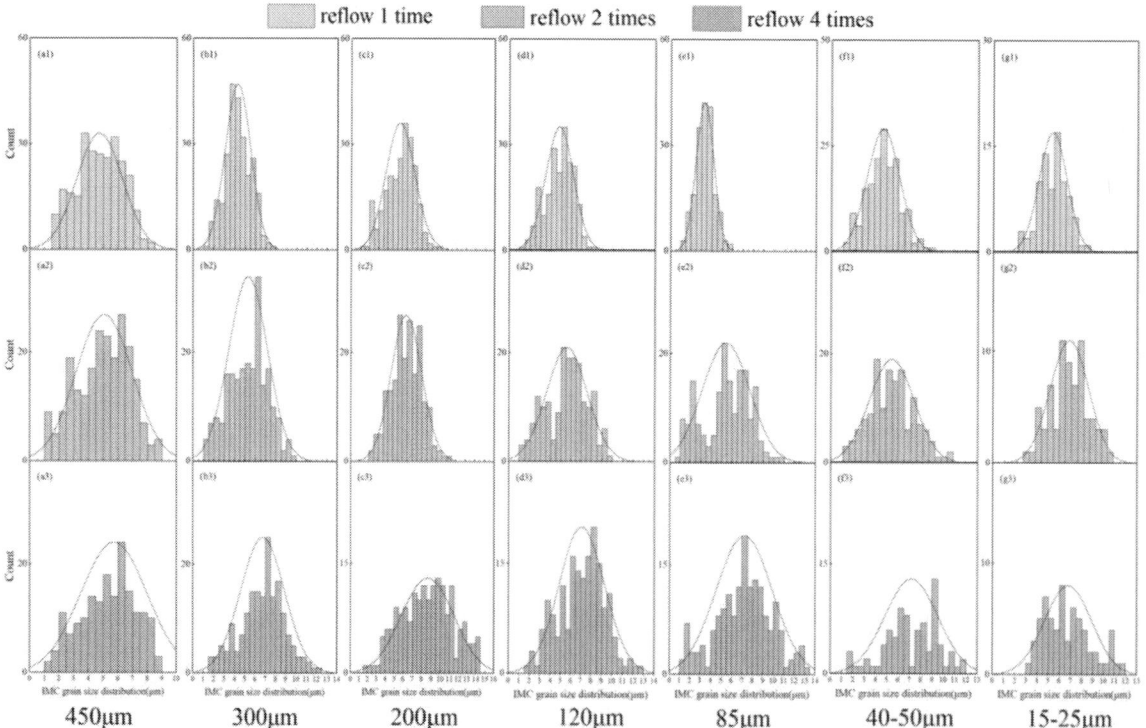

Fig. 2 Size distributions of interfacial IMC grains in Sn-3.0Ag-0.5Cu/(111)Cu joints with different solder ball sizes: (a1–a3) 450 μm, (b1–b3) 300 μm, (c1–c3) 200 μm, (d1–d3) 120 μm, (e1–e3) 85 μm, (f1–f3) 40–50 μm and (g1–g3) 15–25 μm, and subjected to reflow soldering at 260 °C for different reflow cycle times: (a1–g1) once, (a2–g2) twice, (a3–g3) four times.

On the other hand, Cu atoms rapidly diffuse through the channels and grain boundaries between the fine IMC grains under the driving force provided by the Cu concentration in the stage of solid-liquid interfacial reaction. Then, there are sufficient Cu atoms for the growth of prism-like grains. However, the number of IMC grains will decrease with the decreasing solder joint size and the increasing reflow cycle, then the diffusion channels and grains boundaries can also be reduced. These factors cause insufficient Cu atom flux for IMC grain growth. Thus, the interfacial prism-like IMC grains tend to dissolve in the solder to supply Cu atoms and transform to scallop-like grains with a smaller surface area, which is called ripening [6,13]. In sum, Cu atom diffusion flux and surface energy may be the key factor affecting the morphology of IMC.

Fig. 4 shows cross-section microstructure views of joints after reflow for 1, 2 and 4 times. There are no visible voids at interfaces of all joints. Changes of thickness of interfacial IMC in joints with different solder ball diameters subjected to reflow for different cycle times are exhibited in Fig. 5. Obviously, for all joints subjected to reflow for 1, 2 and 4 times, the change in the average IMC layer thickness exhibits a similar trend, and the maximum interfacial IMC layer thickness appears in the joints of 200 μm diameter solder ball. This change tendency is consistent with that of Cu_6Sn_5 grain size shown in Fig.3. Further, as presented in Fig. 5, the IMC layer thickness in the joints with each of five

different solder ball diameters (i.e., 450, 300, 200, 120 and 85 μm, respectively) always increases as the reflow cycle number increases, but the increment thickness of interfacial IMC layer in different size solder joints is distinct. For example, when the reflow cycle time is increased from 1 to 2 and then from 2 to 4, the IMC thickness in joints of 85 μm diameter solder ball is increased by 0.43 μm and 0.37 μm respectively, while being increased by 0.94 μm and 0.91 μm respectively for joints of 450 μm diameter solder ball. This result means that the IMC thickness increases relatively slowly in joints of small solder balls.

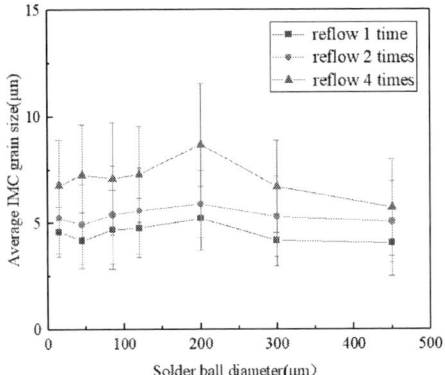

Fig. 3 Relationship between the average size of IMC grains and the diameter of solder balls in Sn-3.0Ag-0.5/(111)Cu joints.

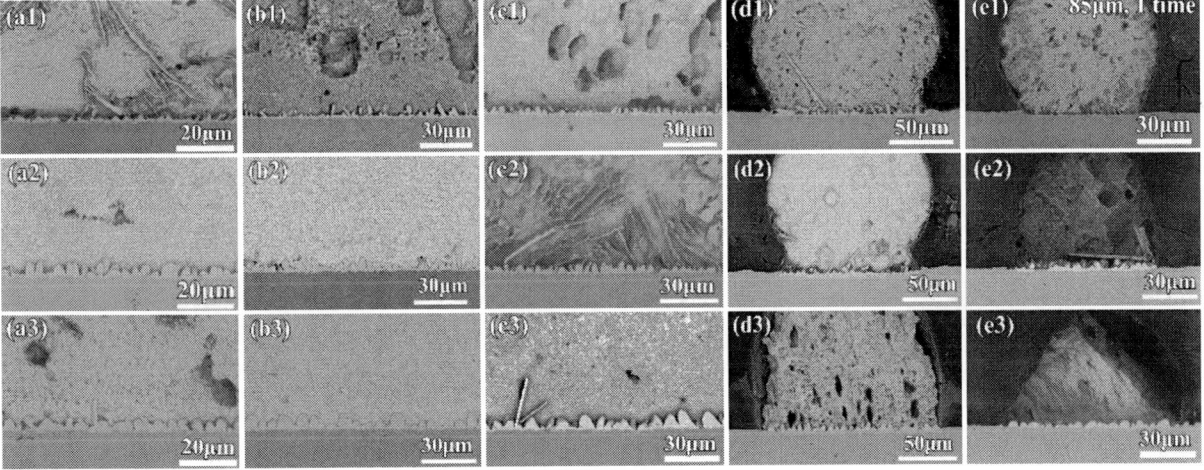

Fig. 4 Morphologies of cross-sectional interfacial IMC layer in Sn-3.0Ag-0.5Cu/(111)Cu joints with different solder ball sizes: (a1–a3) 450 μm, (b1–b3) 300 μm, (c1–c3) 200 μm, (d1–d3) 120 μm and (e1–e3) 85 μm, and subjected to reflow soldering at 260 °C for different reflow cycle times: (a1–e1) once, (a2–2) twice and (a3–e3) four times.

Some studies pointed out that the difference in IMC growth behavior in joints of different sizes is mainly induced by the Cu concentration gradient [3,4,11]. The magnitude of the Cu concentration gradient and the time required to reach saturation should be relevant to the solder volume. The large solder joints have a relatively large Cu concentration gradient and thus need longer time to meet the saturation of Cu. In this case, Cu atoms passing through the interface preferentially dissolve into the solder instead of participating in the interface reaction. Therefore, the grain size and thickness of IMC decrease with increase of the solder volume (size). This can well explain the changing trend of the grain size and thickness of interfacial IMC in solder joints with the solder ball diameter decreasing from 450 to 200 μm in this study. However, this explanation does not apply to the joints formed by solder balls with diameters

below 200 μm because the concentration of Cu atoms at the reflow soldering temperature has already reached saturation in the solder matrix. Further, it is noteworthy that the increment thickness of interfacial IMC layer in joints with solder ball diameters below 200 μm is smaller than that of joints with large solder balls. Under this circumstance, the Cu concentration gradient in small joints with solder ball diameters below 200 μm may not be the major factor affecting the growth of interfacial IMC.

As mentioned above, the Cu atom diffusion takes place mainly via channels between IMC grains in the stage of solid-liquid interfacial reaction [13]. When reflowing for a short time, the concentration gradient of Cu in joints with small solder volume decreases rapidly and reaches saturation. At this time, adjacent IMC grains tend to adsorb each other to grow via grain boundary migration. As the IMC grains grow, the number of the IMC grains gradually decreases, leading to reduction of diffusion channels for Cu atoms. These two factors contribute to weakening of the diffusion of Cu, thereby resulting in small increment thickness of interfacial IMC layer in solder joints with small solder size (volume).

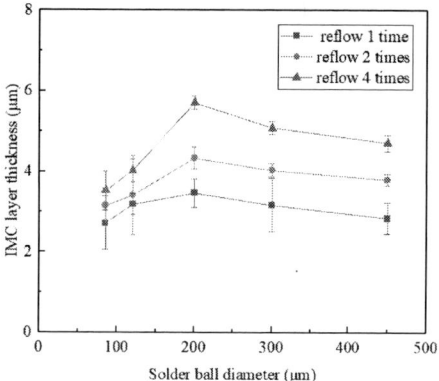

Fig. 5 The average IMC thickness as a function of solder ball diameter of Sn-3.0Ag-0.5Cu/(111)Cu joints subjected to reflow soldering for different cycle times.

Fig. 6 shows the variation of the shear strength of the joints with different sizes subjected to reflow for different reflow cycle times. Clearly, the shear strength increases as the diameter of the solder ball decreases. For the joints with the same size, the shear strength increases with increasing reflow cycle time from 1 to 2 and then to 4.

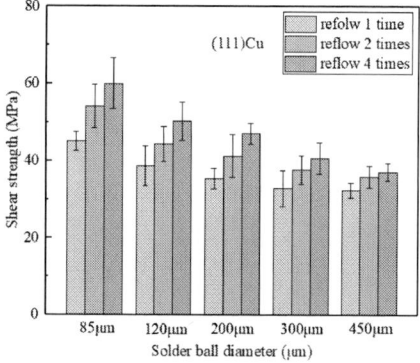

Fig. 6 Variation of shear strength of Sn-3.0Ag-0.5Cu/(111)Cu joints with different solder ball diameters subjected to reflow soldering for different cycle times.

IV. CONCLUSION

The influences of solder ball diameter and reflow cycle time on the growth behavior and morphology evolution of interfacial IMC layer in joints by reflowing Sn-3.0Ag-0.5Cu solder balls on monocrystalline (111)Cu substrate are systematically studied, and the conclusions can be drawn as below:

(1) Cu_6Sn_5 grains formed at interfaces of Sn-3.0Ag-0.5Cu/(111)Cu joints exhibit largely scallop-like morphology. Only small amounts of prism-like Cu_6Sn_5 grains appear at interfaces of the joints formed by the solder ball with diameter above 200 μm.

(2) With the decrease of solder ball diameter from 450 μm to 85 μm, both of the grain size and thickness of interfacial

IMC firstly increase till the solder ball diameter of 200 μm and thereafter decrease. Meanwhile, the increment thickness of the interfacial IMC layer is smaller in small-size (volume) solder joints than in large-size (volume) solder joints.

(3) With the decrease of solder ball diameter and the increase of reflow cycle number, the IMC grains tend to grow dominantly by adsorption and grain boundary migration mechanism.

(4) The shear strength of Sn3.0Ag0.5Cu/(111)Cu joints increases with the decrease of solder ball diameter. For the joints with the same size, those underwent more reflow cycles show higher shear strength.

ACKNOWLEDGMENTS

The work presented in this paper is supported by the National Natural Science Foundation of China (Grant Nos. 51775195 & 51405162) and Guangzhou City Science and Technology Scheme through project No. 201807010028.

REFERENCES

[1] J. H. Lau, Heterogeneous Integrations. Springer, 2019.

[2] Y. D. Wang, Igor M. De Rosa., and K. N. Tu, "Size effect on ductile-to-brittle transition in Cu-solder-Cu micro-joints," 65th Electronic Components and Technology Conference (ECTC), San Diego, CA, USA, pp. 632–639, 2015.

[3] X. F. Zhao, M. B, Zhou, T. Sun, and X. P. Zhang, "Size effect on the interfacial reaction and IMC growth of Sn-3.0Ag-0.5Cu/Cu joints with the decreasing joint size to several tens of microns during reflowing soldering," 19th International Conference on Electronic Packaging Technology (ICEPT), Shanghai, China, pp. 562–565, 2018.

[4] M. L. Huang and F. Yang, "Size effect model on kinetics of interfacial reaction between Sn-xAg-yCu solders and Cu substrate," Sci. Rep., vol. 4, pp. 1–9, May 2014.

[5] Y. H. Tian, C. J. Hang, C. Q. Wang, S. H. Yang, and P. G. Lin, "Effects of bump size on deformation and fracture behavior of Sn3.0Ag0.5Cu/Cu solder joints during shear testing," Mater. Sci. Eng. A, vol. 529, pp. 468–478, Sept 2011.

[6] H. F. Zou, H. J. Yang, and Z. F. Zhang, "Morphologies, orientation relationships and evolution of Cu_6Sn_5 grains formed between molten Sn and Cu single crystals," Acta Mater, vol. 56, pp. 2649–2662, Jun 2008.

[7] Y. H. Tian, R. Zhang, C. J. Hang, L. N. Niu, and C. Q. Wang, "Relationship between morphologies and orientations of Cu_6Sn_5 grains in Sn3.0Ag0.5Cu solder joints on different Cu pads," Mater. Charact., vol 88, pp. 58–68. Feb. 2014

[8] J. Zou, L.P. Mo, F.S. Wu, B. Wang, H. Liu, J. Zhang, and Y.P. Wu, "Effect of Cu substrate and solder alloy on the formation of Kirkendall voids in the solder joints during thermal aging," 11th International Conference on Electronic Packaging Technology & High Density Packaging (ICEPR&HDP), Xi'an, China, pp. 944–948, 2010.

[9] V. Wirth, K. Rendl，and F. Steiner, "Effect of multiple reflow cycles on intermetallic compound creation," 38th International Spring Seminar on Electronics Technology (ISSE), Eger, Hungary, pp. 226–230, 2015.

[10] Y, Zhong, N. Zhao, C. Y. Liu, W. Dong, Y. Y. Qiao, Y. P. Wang, and H.T. Ma, "Continuous epitaxial growth of extremely strong Cu_6Sn_5 textures at liquid-Sn/(111)Cu interface under temperature gradient," Appl. Phys. Lett, vol 111, pp. 22502–22507. Nov, 2017.

[11] S. Annuar, R. Mahmoodian, and K. N. Tu, "Intermetallic compounds in 3D integrated circuits technology: a brief review," Sci. Technol. Adv. Mater., vol. 18, pp. 693–703, Dec. 2017.

[12] Z. H. Zhang, M. Y. Li, and C. Q. Wang, "Fabrication of Cu_6Sn_5 single-crystal layer for under-bump metallization in flip-chip packaging," Intermetallics, vol. 42, pp. 2013, 52–55, Nov. 2013.

[13] H. K. Kim, and K. N. Tu, "Kinetic analysis of the soldering reaction between eutectic SnPb alloy and Cu accompanied by ripening," Phys. Rev. B, vol. 53, pp. 16027–16034, Jun. 1996.

Simulation of Electromagnetic Wave Propagation in 3D Integrated Module Based on 3D ADI-FDTD Algorithm

Yinhui Han
Academy of Smart IC and Networks
Key Laboratory of the Ministry of
Education for Optoelectronic
Measurement Technology and
Instrument, Beijing Information Science
and Technology Universit
Beijing, China
1139853646@qq.vom

Min Miao*
Correspondding author
Academy of Smart IC and Networks
Key Laboratory of the Ministry of
Education for Optoelectronic
Measurement Technology and
Instrument, Beijing Information Science
and Technology Universit
Beijing, China
miaomin@bistu.edu.cn

Jin Li
Academy of Smart IC and Networks
Key Laboratory of the Ministry of
Education for Optoelectronic
Measurement Technology and
Instrument, Beijing Information Science
and Technology University
Beijing, China
candies1026@163.com

Abstract—In order to solve the problem of electromagnetic propagation in three-dimensional (3-D) heterogenous integrated modules. This paper presents an alternating direction implicit finite-difference time-domain (ADI-FDTD) method. This method can be well applied to 3-D electromagnetic problems. The excitation source suitable for this method is given in this paper. In order to effectively analyze electromagnetic problems at infinity, the ADI-FDTD algorithm combines the uniaxial PML(UPML) and the Mur first order absorbing boundary condition; the former is used for wave propagation in direction z, and the latter used for other boundaries. Finally, a demo is given to compare the results calculated by ADI-FDTD with that calculated by conventional FDTD algorithm. Numerical results show that the ADI-FDTD is not constrained by conditional stability, the calculation time can be greatly shortened and the calculation efficiency of FDTD is increased. The introduction of absorption boundary condition makes the numerical simulation more accurate and the calculation more efficient.

Keywords—*ADI-FDTD method; FDTD method; absorbing boundary conditions; computational electromagnetics; three-dimensional heterogenous integration*

I. INTRODUCTION

Along with the microelectronic system gradually evolving toward higher speed and more miniaturization, integration scale is larger and larger, with the port number , the density of interconnect, the chip clock frequency continuously rising, and the electromagnetic interference, signal integrity and power integrity issues are increasingly prominent. And therefore, electromagnetic wave propagation effects for the 3-D integrated heterogenous interconnect must be thoroughly revealed before the design tape off. Due to the high density and complex structure characteristics of 3-D heterogeneous integration, only various numerical calculation methods can be used to accurately analyze electromagnetic wave propagation in interconnects. The FDTD is more widely used to analyze high density problems of 3D heterogeneous integration problems because of its characteristics. As the name implies, the FDTD method can not only obtain the time information related to the electromagnetic problem, but also obtain the required frequency information through a series of simple solution transformations[1]. FDTD has been developing and improving since it has been widely used. In recent years,

many improved algorithms have appeared to meet the different needs of FDTD in different fields to some extent. More and more people begin to use this algorithm, and continue to improve the algorithm level, but also the advantages and disadvantages of this algorithm were summarized, the advantages and disadvantages of the analysis, now the FDTD method has been gradually improved.

The ADI-FDTD method is used here. This method is adopted to solve electromagnetic propagation problems in 3D integrated modules without the constraints of Courant stability conditions. The electromagnetic field components of ADI-FDTD algorithm, which has been initially proposed in Ref. [2], are staggered on the spatial grid, which are the same as those of conventional FDTD algorithm. The difference is that this method is based on the ADI algorithm, which will lead to a three-order coefficient matrix linear equations. Thomas algorithm is very effective in solving this type of linear equations, which makes ADI algorithm hold the advantage of convenient and quick solution compared with other implicit algorithms. This method makes unit time and unit space have no absolute constraint relationship and has unconditional stability.

In order to effectively analyze electromagnetic problems at infinity, the absorption boundary conditions in ADI-FDTD are very important [3]. UPML absorption boundary conditions and Mur first order absorption boundary conditions are often used in ADI-FDTD algorithms. However, for a 3-D electromagnetic propagation problem, if the ADI-FDTD algorithm is applied to PML media, the calculation formula is extremely complex and the calculation cost will increase. The adoption of Mur conditions will lead to large reflection. In this paper, the ADI-FDTD algorithm combines Mur absorption boundary conditions and UPML absorption boundary conditions.

The introduction of absorption boundary condition keeps the unconditional stability of ADI-FDTD algorithm, and balances the calculation cost and reflection effect. The ADI-FDTD algorithm in 3-D space is derived. ADI-FDTD algorithm is written in C++ code. Finally, a demo is given to show that the ADI-FDTD method is suitable for 3D heterogeneous integration problems. Calculations show that, ADI-FDTD takes a longer time to complete an iteration,

978-1-6654-1392-3/21 $31.00 © 2021 IEEE

while FDTD takes a shorter time to complete an iteration, larger time step can be adopted, and the overall numerical calculation can be made more accurate (lower dispersion) and more efficient, with the combination of the two absorption boundary conditions.

II. THREE-DIMENSIONAL ADI-FDTD METHOD

A. $n \rightarrow n+1/2$

The ADI-FDTD method is studied and the algorithm of this process is deduced in detail, the 3-D ADI-FDTD difference calculation is carried out in two steps, and the six partial differential equations in (1) and (2) are discretized by numerical difference in two processes. The electromagnetic field components of ADI-FDTD algorithm are staggered on the spatial grid, which are the same as those of conventional FDTD algorithm.

$$
\left.\begin{aligned}
\varepsilon \frac{\partial E_x}{\partial t} + \sigma E_x &= \frac{\partial H_z}{\partial y} - \frac{\partial H_y}{\partial z} \\
\varepsilon \frac{\partial E_y}{\partial t} + \sigma E_y &= \frac{\partial H_x}{\partial z} - \frac{\partial H_z}{\partial x} \\
\varepsilon \frac{\partial E_z}{\partial t} + \sigma E_z &= \frac{\partial H_y}{\partial x} - \frac{\partial H_x}{\partial y}
\end{aligned}\right\} \tag{1}
$$

$$
\left.\begin{aligned}
-\mu \frac{\partial H_x}{\partial t} - \sigma_m H_x &= \frac{\partial E_z}{\partial y} - \frac{\partial E_y}{\partial z} \\
-\mu \frac{\partial H_y}{\partial t} - \sigma_m H_y &= \frac{\partial E_x}{\partial z} - \frac{\partial E_z}{\partial x} \\
-\mu \frac{\partial H_z}{\partial t} - \sigma_m H_z &= \frac{\partial E_y}{\partial x} - \frac{\partial E_x}{\partial y}
\end{aligned}\right\} \tag{2}
$$

The following labels are used in the derivation of the algorithm discussed in this article.

$$
CA(m) = \frac{\dfrac{2\varepsilon(m)}{\Delta t} - \dfrac{\sigma(m)}{2}}{\dfrac{2\varepsilon(m)}{\Delta t} + \dfrac{\sigma(m)}{2}}
$$

$$
CB(m) = \frac{1}{\dfrac{2\varepsilon(m)}{\Delta t} + \dfrac{\sigma(m)}{2}}
$$

$$
CP(m) = \frac{\dfrac{2\mu(m)}{\Delta t} - \dfrac{\sigma_m(m)}{2}}{\dfrac{2\mu(m)}{\Delta t} + \dfrac{\sigma_m(m)}{2}}
$$

$$
CQ(m) = \frac{1}{\dfrac{2\mu(m)}{\Delta t} + \dfrac{\sigma_m(m)}{2}}
$$

At $n+1/2$ time in the x direction we can get the electric field，At $n+1/2$ time in the y direction we can get the magnetic field , and can be expressed as

$$
E_x^{n+1/2}\left(i+\frac{1}{2}, j, k\right) =
$$
$$
CA\left(i+\frac{1}{2}, j, k\right) \cdot E_x^n\left(i+\frac{1}{2}, j, k\right) + CB\left(i+\frac{1}{2}, j, k\right) \cdot
$$

$$
\left[\frac{H_z^n\left(i+\frac{1}{2}, j+\frac{1}{2}, k\right) - H_z^n\left(i+\frac{1}{2}, j-\frac{1}{2}, k\right)}{\Delta y}\right.
$$

$$
\left.-\frac{H_y^{n+1/2}\left(i+\frac{1}{2}, j, k+\frac{1}{2}\right) - H_y^{n+1/2}\left(i+\frac{1}{2}, j, k-\frac{1}{2}\right)}{\Delta z}\right] \tag{3}
$$

$$
H_y^{n+1/2}\left(i+\frac{1}{2}, j, k+\frac{1}{2}\right) =
$$
$$
CP\left(i+\frac{1}{2}, j, k+\frac{1}{2}\right) \cdot H_y^n\left(i+\frac{1}{2}, j, k+\frac{1}{2}\right)
$$
$$
-CQ\left(i+\frac{1}{2}, j, k+\frac{1}{2}\right) \cdot
$$

$$
\left[\frac{E_x^{n+1/2}\left(i+\frac{1}{2}, j, k+1\right) - E_x^{n+1/2}\left(i+\frac{1}{2}, j, k\right)}{\Delta z}\right.
$$

$$
\left.-\frac{E_z^n\left(i+1, j, k+\frac{1}{2}\right) - E_z^n\left(i, j, k+\frac{1}{2}\right)}{\Delta x}\right] \tag{4}
$$

where Δx, Δy, Δz are the space indexes for Yee cell, the total time is divided into multiple equal periods, each time period is denoted by Δt.

By combining equation (3) and (4), the tridiagonal matrix for solving $E_x^{n+\frac{1}{2}}$ can be given by

$$
a_k E_x^{n+1/2}\left(i+\frac{1}{2}, j, k-1\right) + b_k E_x^{n+1/2}\left(i+\frac{1}{2}, j, k\right)
$$
$$
+c_k E_x^{n+1/2}\left(i+\frac{1}{2}, j, k+1\right) = d_k \tag{5}
$$

where

$$
a_k = -\frac{CB\left(i+\frac{1}{2}, j, k\right) \cdot CQ\left(i+\frac{1}{2}, j, k-\frac{1}{2}\right)}{(\Delta z)^2}
$$

$$
b_k = 1 + \frac{CB\left(i+\frac{1}{2}, j, k\right)}{(\Delta z)^2}\left[CQ\left(i+\frac{1}{2}, j, k+\frac{1}{2}\right)\right.
$$
$$
\left.+CQ\left(i+\frac{1}{2}, j, k-\frac{1}{2}\right)\right]
$$

$$
c_k = -\frac{CB\left(i+\frac{1}{2}, j, k\right) \cdot CQ\left(i+\frac{1}{2}, j, k+\frac{1}{2}\right)}{(\Delta z)^2}
$$

Equation (5) is a special form of equation that can be converted to another form, represented as a matrix

$$
AX = Y \tag{6}
$$

Here

$$X = \begin{bmatrix} E_z^{n+1/2}(1,j) \\ \vdots \\ E_z^{n+1/2}(i,j) \\ \vdots \\ E_z^{n+1/2}(i,j) \end{bmatrix}, Y = \begin{bmatrix} d_1 \\ \vdots \\ d_i \\ \vdots \\ d_{i\max} \end{bmatrix}$$

$$A = \begin{bmatrix} b_1 & c_1 & 0 & 0 & 0 \\ a_2 & b_2 & c_2 & 0 & 0 \\ 0 & \ddots & \ddots & \ddots & 0 \\ 0 & 0 & a_{i\max-1} & b_{i\max-1} & c_{i\max-1} \\ 0 & 0 & 0 & a_{i\max} & b_{i\max} \end{bmatrix}$$

$$(7)$$

The matrix A is a tridiagonal matrix; Gaussian elimination method can be used to solve the tridiagonal matrix to get the electric field $E_z^{n+1/2}$ at time $n+1/2$, and then substituted into (4) to get the $H_y^{n+1/2}$ component. Similarly, the components of $E_y^{n+1/2}$, $E_z^{n+1/2}$, $H_x^{n+1/2}$ and $H_z^{n+1/2}$ can also be solved by solving matrices and plugging in transformations.

B. $n+1/2 \rightarrow n+1$

In the second sub time step, the difference formats of Equation (8) are similar to (3) in the first sub time step, except that the display and implicit difference directions are exchanged.

At $n+1$ time in the x direction we can get the electric field，and can be expressed as

$$E_x^{n+1}\left(i+\frac{1}{2},j,k\right) =$$
$$CA\left(i+\frac{1}{2},j,k\right) \cdot E_x^{n+1/2}\left(i+\frac{1}{2},j,k\right) + CB\left(i+\frac{1}{2},j,k\right) \cdot$$
$$\left[\frac{H_z^{n+1}\left(i+\frac{1}{2},j+\frac{1}{2},k\right) - H_z^{n+1}\left(i+\frac{1}{2},j-\frac{1}{2},k\right)}{\Delta y}\right.$$
$$(8)$$
$$\left. - \frac{H_y^{n+1/2}\left(i+\frac{1}{2},j,k+\frac{1}{2}\right) - H_y^{n+1/2}\left(i+\frac{1}{2},j,k-\frac{1}{2}\right)}{\Delta z}\right]$$

And just like the first sub time step, by solving special equations and matrices, the electromagnetic fields in the x, y, and z directions at time $n+1$ can be obtained.

III. ABSORBING BOUNDARY CONDITIONS AND EXCITATION

Similar to FDFD methods and finite element methods, the FDTD method needs to truncate the open domain into a finite-domain when it is applied to electromagnetic problems in the open domain. In this case, absorption boundary conditions must be introduced to the truncated boundary to simulate the effect of the truncated outer space. The correct selection of absorption boundary conditions is very important in the analysis of electromagnetic problems. Different absorption boundary conditions will have different effects on the final calculation results, which will lead to deviation of

the calculation results, sometimes even completely wrong. In this paper, the two absorption boundary conditions are combined and applied to the electromagnetic propagation simulation problem. On the other hand, adding excitation source is the key to the successful simulation of ADI-FDTD. Using the uniaxial perfect matching layer on the propagation direction z, in order to balance the calculation cost and reflection effect, Mur first order absorption boundary conditions are used for other boundaries. Putting the two boundaries together in an electromagnetic problem has produced significant results in reducing the greater reflectivity and computation required when they exist alone.

The first sub time step iteration in the implicit format direction is

$$E_x^{n+1/2}(i,0,k)\left(1+\frac{v_{\max}dt}{2dy}\right) + E_x^{n+1/2}(i,1,k)\left(1-\frac{v_{\max}dt}{2dy}\right) \quad (9)$$
$$= E_x^n(i,1,k) + E_x^n(i,0,k)$$

$$E_x^{n+1/2}(i,\max y-1,k)\left(1-\frac{v_{\max}dt}{2dy}\right)$$
$$+E_x^{n+1/2}(i,\max y,k)\left(1+\frac{v_{\max}dt}{2dy}\right) \quad (10)$$
$$= E_x^n(i,\max y-1,k) + E_x^n(i,\max y,k)$$

The second sub time step iteration in the explicit format direction is

$$E_x^{n+1}(i,0,k) = E_x^{n+1/2}(i,1,k) + \frac{v_{\max}dt-2dy}{v_{\max}dt+2dy} \cdot$$
$$\left(E_x^{n+1}(i,1,k) + E_x^{n+1/2}(i,0,k)\right) \quad (11)$$

$$E_x^{n+1}(i,\max y,k) = E_x^{n+1/2}(i,\max y-1,k) + \frac{v_{\max}dt-2dy}{v_{\max}dt+2dy} \cdot$$
$$\left(E_x^{n+1}(i,\max y-1,k) - E_x^{n+1/2}(i,\max y,k)\right)$$
$$(12)$$

And the same thing gives us the E_y, E_z component.

IV. NUMERICAL RESULTS

In order to conduct numerical simulation of the interconnection structure in 3-D heterogenous integration, we set up $230 \times 230 \times 230$ cells area as the simulation space, the cell size is $0.5mm \times 0.5mm \times 0.5mm$. Absorption boundary conditions are added to the outside of the simulation space. UPML is used on the wave propagation direction z, and Mur first order absorption boundary conditions are used on other boundaries. The program implementation defines the time step $\Delta t = CFLN\frac{\delta}{2c}$, $\Delta t_0 = \frac{\delta}{2c}$, the time step is controlled by the variable CFLN, $CFLN = \frac{\Delta t}{\Delta t_0}$, when the variable CFLN is 1, the time step just meets the CFL stability condition. The excitation source is set at the computing space center A (115,115,115), and B (45,10,30) and C (45,10,35) are set as observation points. Fig. 1 shows the space required to be solved.

978-1-6654-1392-3/21 $31.00 © 2021 IEEE

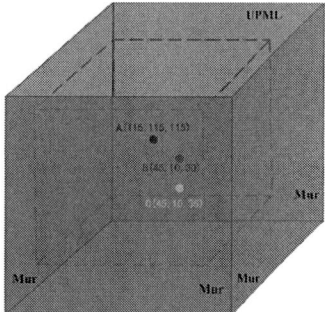

Fig. 1. Solution space.

In the program implementation, the iterative equation is completed by 3-D ADI-FDTD difference equation. Gaussian pulse is used as radiation source in calculation. As shown in Fig.2, a source needs to be placed at the center of the solution space so that the electromagnetic field can be generated in the solution space, and its expression can be expresses as

$$E_{x(i,j,k)}\left(t\right) = \exp\left[-\frac{4\pi\left(t-t_0\right)^2}{\tau^2}\right] \qquad (13)$$

where $t_0 = 30\Delta t$, the pulse peak appears when $t = t_0$; $\tau = 10$ represents the width of the Gaussian pulse, if we want the Gaussian pulse to be approximately zero at $t = 0$, we should choose $t_0 \geq 3\tau$.

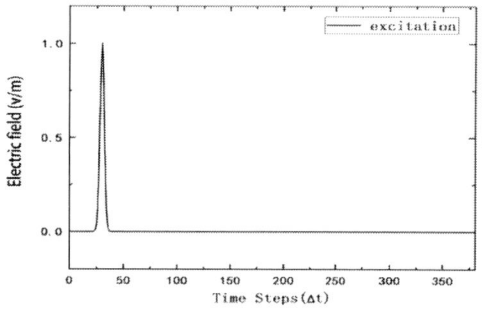

Fig. 2. Gaussian pulse excitation source.

As a comparison, 3D ADI-FDTD and FDTD were simultaneously used for simulation. Firstly, $CFLN = 1.2$ is used for FDTD simulation, taking E_z record of point B in the simulation space. Fig. 3 shows the FDTD diverges quickly as we expected.

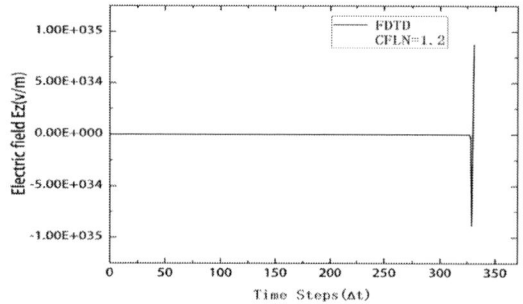

Fig. 3. The variation of E_z with time step (FDTD).

In order to observe this divergence phenomenon more clearly, select a plane in the solution space for observation. As shown in Fig. 4, the radiation field of the excitation source is divergent.

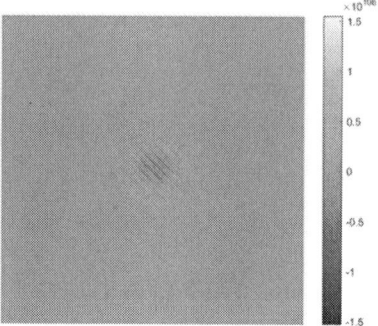

Fig. 4. $z = 30$ plane E_z distribution.

If the ADI-FDTD simulation is used, this situation will not occur, taking $CFLN = 1.2$, record E_z of the same point B in the simulation space. As Fig. 5 shows, the results do not diverge.

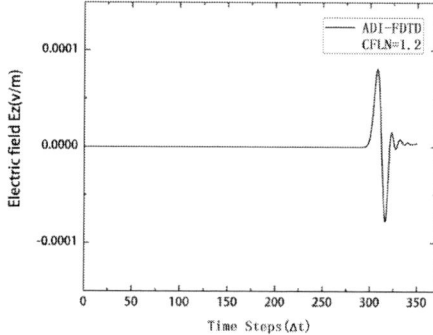

Fig. 5. E_z varies with time step (ADI-FDTD).

And then taking different time steps, ADI-FDTD and FDTD are respectively used to simulate the change of electric field at point C, and the results are compared.

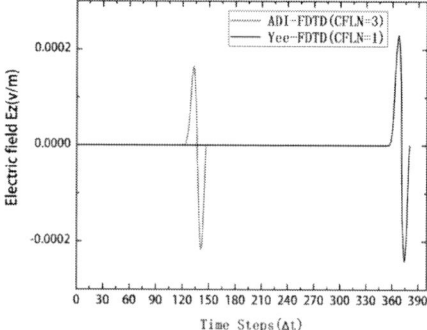

(a) CFLN=3，waveforms of ADI-FDTD and FDTD

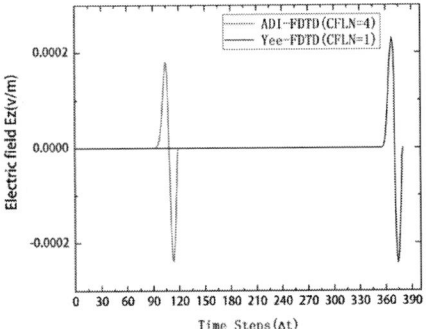

(b) CFLN=4，waveforms of ADI-FDTD and FDTD.

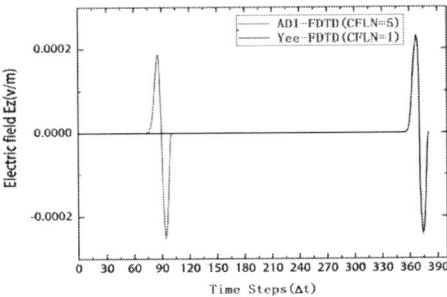

(c) CFLN=5，waveforms of ADI-FDTD and FDTD.

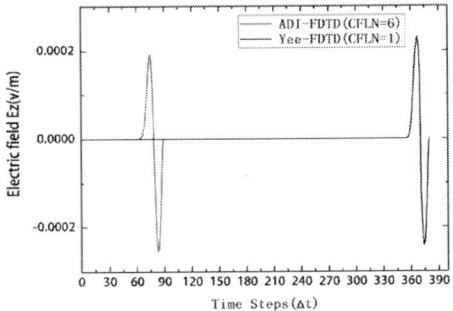

(d) CFLN=6，waveforms of ADI-FDTD and FDTD.

Fig. 6. At the same position in the simulation space, the electric field component waveform changes of the two methods with different CFLN are compared.

By comparing the electric field component waveforms of two different CFLN methods at the same position in the simulation space shown in Fig. 6, it can be found that the electric field waveforms calculated by the ADI-FDTD method are only slightly different from those calculated by the traditional FDTD method, and basically is the same. At the same time, it shows that this method is feasible to apply to 3D heterogeneous integration, and it can solve complex electromagnetic problems in 3D heterogeneous integration, and the results obtained are reliable. The simulation with FDTD method requires 380 time steps; by comparing Fig. 6a, b, c and d, it can be found that when CFLN is 3, 4, 5 and 6 respectively, it takes 147, 119, 100 and 89 time steps respectively to simulate the same electromagnetic physical process by ADI-FDTD method. When different CFLN are taken, the running time required to adopt the two algorithms is shown in Table 1. It takes 167.0s to adopt FDTD method; it can be found that when CFLN is 3, 4, 5 and 6, it takes

154.1s, 115.5s , 93.8s and 79.6s using ADI-FDTD method. Through the research and analysis of the above two methods, we can clearly see that the maximum value that CFLN can get in the traditional FDTD is only 1. Thus, the minimum number of iteration steps is 380. However, ADI-FDTD can take a larger CFLN and reduce the number of iteration steps, the total running time is thus reduced. When CFLN is 3, there is only a small difference between the running time required by the ADI-FDTD and the running time required by the FDTD.

TABLE I.　　THE RUNNING TIME OF TWO ALGORITHMS WITH DIFFERENT CFLN.

	CFLN	Time Steps(Δt)	CPU Time(s)
FDTD	1	380	167.0
ADI-FDTD	3	147	154.1
ADI-FDTD	4	119	115.5
ADI-FDTD	5	100	93.8
ADI-FDTD	6	89	79.6

The influence of absorption boundary conditions in the ADI-FDTD method on the time-domain electromagnetic wave propagation simulation of a vertical interconnect is observed with the equiphase line. The Mur boundary and the binding boundary of Mur and UPML proposed in this paper are added to the outside of the $230\times230\times230$ cell size. Fig. 7 shows using the two kinds of boundary discussed in this paper, the field value distribution of a plane in the simulation space. The following color picture can clearly see the law of field distribution.

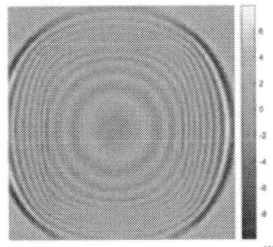

(a) Mur absorption boundary condition.

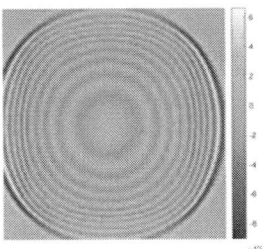

(b) Mur + UPML absorption boundary condition.

Fig. 7. The two kinds of boundary discussed in this paper, the field value distribution of a plane in the simulation space

It can be seen from Fig. 7 that the isophase lines under the two boundary conditions are basically distributed as concentric circles. The binding boundary reflection of Mur and UPML is small. That is, the combination of Mur and

UPML has a good absorption effect. In the future electromagnetic analysis, it is a good choice to use the two boundaries together, which can weaken their disadvantages and enhance their advantages. When the electromagnetic wave propagates to the boundary, there are very few electromagnetic waves reflected back, which are basically absorbed by the absorption boundary, and the calculation time is also accelerated.

V. CONCLUSION

Here, an ADI-FDTD method for interconnection structure in three-dimensional heterogenous integration is proposed. A joint absorption boundary of the joint Mur boundary and the UPML boundary is proposed. In this paper, two methods are used to solve the problem, and the results of the solution are shown in the form of graphs, and these graphs are observed. Numerical results show that the ADI-FDTD method is no longer affected by conditional stability, however, FDTD should be subject to conditional stability. When the time step goes beyond the range specified by CFL, the ADI-FDTD algorithm can still correctly solve the problem we are concerned about. Under the same conditions, ADI-FDTD takes a long time to complete each iteration, while FDTD takes a short time to complete each iteration, and there is a certain proportional relationship in time (For the purposes of this article, it is about 3 times, but in general, it also has to do with the complexity of the computing domain.), as the overall simulation time required by ADI-FDTD algorithm will be affected by CFLN value, the overall simulation time can be reduced by increasing the CFLN value. This enables the simulation to be completed faster and more efficiently. It can be seen that, the introduction of absorbing boundary conditions preserves the unconditional stability of ADI-FDTD algorithm, while implement an balance between the computation cost and the reflection effect. This joint application method can be well used in 3D heterogeneous integration problems.

ACKNOWLEDGMENT

This paper is sponsored by National Natural Science Foundation of China (No. 62074017, 61674016).

REFERENCES

[1] D. Y. Heh and E. L. Tan, "Efficient implementation of 3-D ADI-FDTD method for lossy media," 2009 IEEE MTT-S International Microwave Symposium Digest, 2009, pp. 313-316.

[2] Jiunn-Nan and Fu-Chiarng, "Analysis of stability and numerical dispersion relation of Mur's absorbing boundary condition in the ADI-FDTD method," 2007 European Microwave Conference, 2007, pp. 1385-1388.

[3] W. C. Tay and E. L. Tan, "Implementation of mur first order absorbing boundary condition in efficient 3-D ADI-FDTD," 2009 IEEE Antennas and Propagation Society International Symposium, 2009, pp. 1-4.

A Comparative Study on the Influences of Various Nickel Powders on the EMI Shielding Performance of Conductive Polymer Composites

Yong Wang
Shenzhen Institute of Advanced Electronic Materials, Shenzhen Institute of Advanced Technology, Chinese Academy of Sciences,
Shenzhen, 518055, China
yong.wang1@siat.ac.cn

Dingkun Tian
Shenzhen Institute of Advanced Electronic Materials, Shenzhen Institute of Advanced Technology, Chinese Academy of Sciences,
Shenzhen, 518055, China
dk.tian@siat.ac.cn

Yadong Xu
Shenzhen Institute of Advanced Electronic Materials, Shenzhen Institute of Advanced Technology, Chinese Academy of Sciences,
Shenzhen, 518055, China
yd.xu@siat.ac.cn

Baotan Zhang
Shenzhen Institute of Advanced Electronic Materials, Shenzhen Institute of Advanced Technology, Chinese Academy of Sciences,
Shenzhen, 518055, China
bt.zhang@siat.ac.cn

Tao Zhao
Shenzhen Institute of Advanced Electronic Materials, Shenzhen Institute of Advanced Technology, Chinese Academy of Sciences,
Shenzhen, 518055, China
tao.zhao@siat.ac.cn

Yougen Hu*
Shenzhen Institute of Advanced Electronic Materials, Shenzhen Institute of Advanced Technology, Chinese Academy of Sciences,
Shenzhen, 518055, China
yg.hu@siat.ac.cn

Rong Sun*
Shenzhen Institute of Advanced Electronic Materials, Shenzhen Institute of Advanced Technology, Chinese Academy of Sciences,
Shenzhen, 518055, China
rong.sun@siat.ac.cn

Abstract—The morphology of filler has an important effect on the electrical conductivity and electromagnetic shielding properties of conductive polymer composites (CPCs). In this work, three different nickel powders include spherical nickel (S-Nickel) powder, flaky nickel (F-Nickel) powder, and chain spherical nickel (CS-Nickel) powder were mixed into two-component silica gel and epoxy resin to obtain CPCs, respectively. The electrical conductivity, electromagnetic interference shielding effectiveness (EMI SE), compressibility, and tensile strength of CPCs were investigated. The results show that CPCs with CS-Nickel have the best electrical conductivity and EMI SE, due to the excellent dimensional structure of CS-Nickel result in the denser conductive network. At the same time, compared with silica gel, epoxy resin has a higher cohesion force to fillers, and epoxy resin CPCs show better electromagnetic shielding performance with lower filler content. We believe that our work can provide some experimental support for the selection of filler and matrix for high performance electromagnetic shielding materials.

Keywords—electromagnetic interference shielding, shielding effectiveness, conductive polymer composites, nickel powders, morphology

I. INTRODUCTION

The demand for electromagnetic interference (EMI) shielding materials has increased strongly in the near years, owing to the fast developments of 5G technology. Electromagnetic pollution severely damages the health of people under exposure and normal functioning of surrounding electronics[1]. The materials with high EMI shielding effectiveness (SE) can effectively block the transmission of electromagnetic waves and reduce the electromagnetic pollution between electronic products[2].

As typical EMI shielding materials, conductive polymer composites (CPCs) has excellent versatility, light weight, and corrosion resistance characteristics. CPCs plays an important role in the electrical connection and electromagnetic shielding protection of sensitive electronic components[3]. The electromagnetic interference shielding effectiveness (EMI SE) of CPCs is not only related to the conductivity but also depends on the internal structure of the composite material and the morphology of the filler[4]. A pore structure of carbon nanotube foam that can be customized and reinforced to obtain an efficient and reliable EMI absorber was fabricated by Yu et al[5]. Zou et al. prepared a polypropylene/graphene composite material with high EMI SE through a shrinkage reduction method[6]. Ma et al. invented a novel three-layered sandwich structure of poly(vinylidene fluoride)-based nanocomposites, consisting of graphene nanoplatelets, nickel, and carbon nanotubes[7]. Efficient synthesis of highly aligned laminated pristine graphene films and nacre-like pristine graphene/polymer composites with excellent EMI shielding performance by a scanning centrifugal casting method is reported[8]. However, the effect of fillers with various morphologies on the EMI shielding performance of CPCs has rarely been studied. Therefore, it is extremely meaningful to study the effects of fillers with various morphologies on the EMI shielding performance of CPCs.

Due to the excellent chemical stability, low cost, and relatively high conductivity of nickel powder, which are widely used in EMI shielding materials[9]. Herein, spherical nickel (S-Nickel) powder, flaky nickel (F-Nickel) powder,

978-1-6654-1392-3/21 $31.00 © 2021 IEEE

and chain spherical nickel (CS-Nickel) powder CPCs were fabricated to study the influence on the EMI SE, respectively. Three nickel powders with different morphologies were mixed into addition type two-component silica gel and epoxy resin to obtain CPCs. As a result, with the same filler content, CS-Nickel is easier to construct a continuous conductive network, and the silica gel CPCs with 82 wt% CS-Nickel has higher conductivity (139.27 S/m) and EMI shielding performance (43.46 and 37.61 dB in 10 MHz-3 GHz and 8.2-12.4 GHz, respectively). In addition, the cohesion force of the matrix to the filler also has an important influence on the electrical conductivity of the CPCs. The epoxy resin CPCs with 75 wt% CS-Nickel shows better electrical conductivity (5466.50 S/m) and EMI SE (51.69 and 70.57 dB in 10 MHz-3 GHz and 8.2-12.4 GHz, respectively).

II. MATERIALS AND EXPERIMENTS

A. Materials

S-Nickel powder with a diameter of 2-6 um and CS-Nickel powder with a diameter of 2-5 um and length of 10-15 um were provided by Ningbo Fanuowei Metal Material Co., Ltd. F-Nickel powder with the particle size of 2-8 um was provided by Chengdu Nucleus 857 New Materials Co., Ltd. Silica gel (CX 3561A/B) were purchased from Guangzhou Trancytech Co., Ltd. Epoxy resin was purchased from HEXION Co., Ltd.

B. Preparation

First, two components of silica gel were mixed in a high-speed mixer (FlackTek SpeedMixer Co., Ltd) at 1000 rpm for 1 min, and then nickel powders were filled into the uniformly mixed silica gel at 2000 rpm for 2 min, nickel powders were filled into epoxy resin with the same mixing process. The silica gel filled with 82 wt% nickel powders were hot-pressed in a metal mold at 150 ℃ and 5 MPa for 1 h to obtain S-Nickel, F-Nickel, and CS-Nickel Silica gel CPCs with a thickness of 1 mm. Epoxy resin filled with 75 wt% nickel powders were knife coated in a polytetrafluoroethylene mold and heated in an oven at 165 °C for 2 h to obtain S-Nickel, F-Nickel, and CS-Nickel epoxy resin CPCs with a thickness of 1 mm.

C. Characterization

The microscopic morphology of nickel powders and CPCs sheets were collected by a scanning electron microscope (SEM, Nova Nano SEM 450). The mechanical compression-electrical property of the CPCs was synchronously tested by the automatic load tester (Japan Instrumentation System Co., LTD) connected with a resistance meter (HIOKI RM3545). The conductivity of CPCs was also measured by the resistance meter, and the tensile properties were tested with a precision electronic universal testing machine (AGX-10kN NVD). The EMI SE tests were carried by operating a vector network analyzer (PNA, Keysight N5227B, 10 MHz-67 GHz) connected with an SE tester (Beijing Dingrong, 10 MHz-3 GHz) and waveguide test fixture (Xian Hengda, 8.2-12.4 GHz). The power coefficient of reflectivity (R), transmissivity (T), and absorptivity (A) can be obtained from the measured scattering parameters (S11, S21), and then the total EMI SE (SE$_T$), microwave reflection (SE$_R$), and microwave absorption (SE$_A$) can be calculated as follows[10]:

$$R = |S_{11}|^2$$

$$T = |S_{21}|^2$$

$$1 = A + R + T$$

$$SE_T = -10 \log T$$

$$SE_R = -10 \log (1 - R)$$

$$SE_A = -10 \log \left(\frac{T}{1-R}\right) = SE_T - SE_R - SE_M$$

SE$_M$ is the multiple internal reflection of electromagnetic wave which can be inappreciable when SE$_T$ exceeds 10 dB.

III. RESULTS AND DISCUSS

A. Morphology

Fig. 1(a-c) shows the SEM images of spherical nickel (S-Nickel), flaky nickel (F-Nickel), and chain spherical nickel (CS-Nickel), respectively. The CS-Nickel has a higher aspect ratio compared with S-Nickel. Besides, the specific surface area of F-Nickel and CS-Nickel is also larger than S-Nickel, and it is easier to form conductive networks. Fig. 1(d-f) exhibits the cross-sectional SEM images of silica gel CPCs filled with 82 wt% of S-Nickel, F-Nickel, and CS-Nickel, respectively. Obviously, under the same filler content, the surface distribution density of F-Nickel and CS-Nickel in the silica gel is denser compared with S-Nickel, and it is easier to form conductive networks. Next, we explored the effect of filling nickel powders with different morphologies on the conductivity of epoxy resin CPCs. The cross-sectional SEM images of epoxy resin CPCs filled with 75 wt% different nickels (Fig. 1(g-i)) show the same trend, that is the F-Nickel and CS-Nickel have more conductive paths in the epoxy resin. At the same time, the density of nickel in epoxy resin is denser than that of silica gel. The possible reason is that the cohesive force of the epoxy resin is relatively large, which can make the nickel powder gather together.

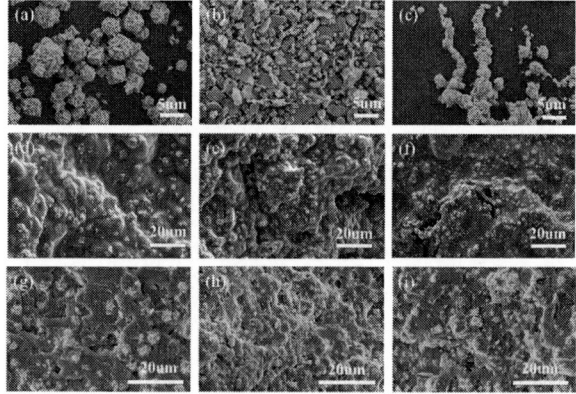

Fig. 1. (a-c) SEM images of nickel powders with different morphologies of spherical, flaky, and chain spherical, respectively; (d-f) cross-sectional SEM images of silica gel CPCs filled with spherical, flaky, and chain spherical nickel, respectively; (d-i) cross-sectional SEM images of epoxy resin CPCs filled with spherical, flaky, and chain spherical nickel, respectively.

B. Electrical properties of CPCs

The influence of filler morphology on the conductive properties of CPC was investigated. Fig. 2 shows the conductivity of silica gel and epoxy resin CPCs with different fillers. Silica gel CPCs with CS-Nickel and F-Nickel have higher electrical conductivity of 139.27 and 80.9 S/m than that of silica gel CPCs with S-Nickel. The reason is that the difference in network construction efficiency caused by the dimensional structure of the filler. Besides, when the matrix is epoxy resin, the overall conductivity of CPCs has been significantly improved, which is mainly due to the different cohesive forces of the matrix to the filler. The epoxy resin has a higher cohesive force to the filler, and the formed conductive network is denser result in lower contact resistance. Epoxy resin CPCs with CS-Nickel and F-Nickel have higher electrical conductivity of 5466.50 and 4877.82 S/m than that of epoxy resin CPCs with S-Nickel.

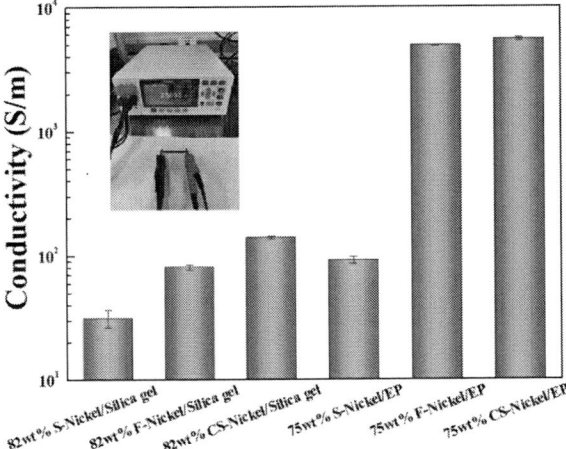

Fig. 2. The conductivity of CPCs with different fillers and matrixes. Illustration shows the test equipment.

C. EMI Shielding properties of CPCs

The electromagnetic interference shielding efficiency (EMI SE) of shielding materials often depends on their electrical conductivity. As shown in Fig. 3a, the average EMI SE value of silica gel CPCs with 82 wt% S-Nickel, F-Nickel and, and CS-Nickel shows 28.15, 35.71, and 43.46 dB, in 10 MHz-3 GHz, respectively. As for epoxy resin CPCs with 75 wt% different nickel fillers, while the average EMI SE value increase to 32.29, 43.83, and 51.69 dB. The CPCs with lower filler content has a higher EMI SE, mainly due to the better cohesion force of epoxy resin to the filler, which improves the formation efficiency of the conductive network and reduces the contact resistance between the fillers. Besides, in the frequency range below 350 MHz, the EMI SE of epoxy resin CPCs shows a significant improvement. The possible reason is that magnetic materials have a better shielding effect in the low frequency band. However, the silica gel CPCs does not show that trait due to insufficient cohesion. As shown in Fig. 3b, the EMI SE of different CPCs were tested in 8.2-12.4 GHz. The average EMI SE value of silica gel CPCs with 82 wt% S-Nickel, F-Nickel, and CS-Nickel reaches 22.73, 35.88, and 37.61 dB, respectively. With the filler are transferred to epoxy resin, the EMI SE of epoxy resin CPCs with 75 wt% filler content increase to 28.17, 58.21, and 70.57 dB, respectively. As shown in Fig. 3c, the SE_R, SE_A, and SE_T of the CPCs were calculated, in 8.2-12.4GHz, and they are sequentially increase from CPCs with S-Nickel to CPCs with F-Nickel and CPCs with CS-Nickel. The CPCs with CS-Nickel and F-Nickel have higher EMI SE compared with CPCs with S-Nickel. This result is consistent with its electrical conductivity. The high aspect ratio and high specific surface area of CS-Nickel and F-Nickel powders facilitate them easier to form conductive paths in the matrix. These results prove that the morphology of nickel powders has an important influence on EMI shielding performance, which offers reference in choosing

various morphology conductive fillers for the EMI shielding materials.

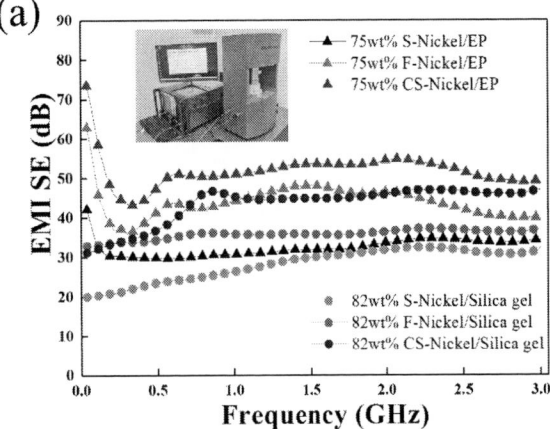

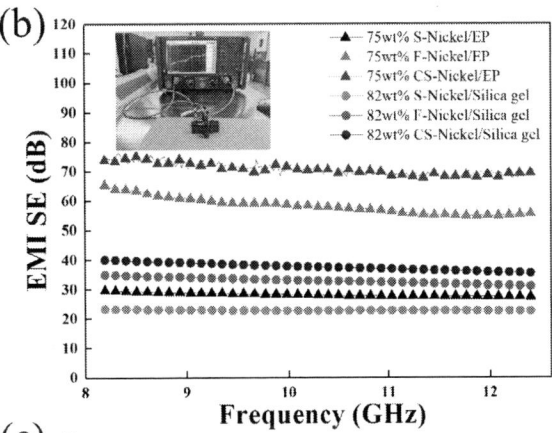

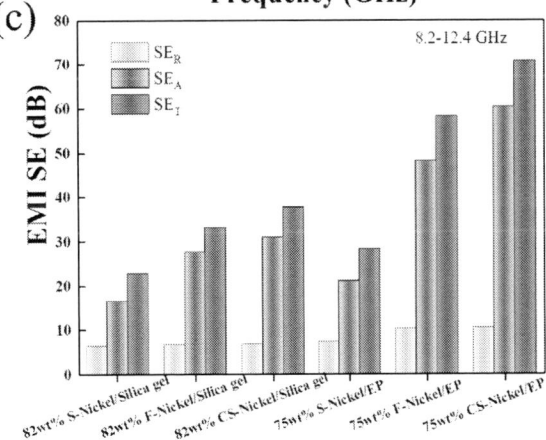

Fig. 3. The EMI SE of CPCs with different fillers and matrixes in 10 MHz-3 GHz and 8.2-12.4 GHz, respectively. Illustration shows test equipment in different frequency bands.

D. Mechanical properties of CPCs

As shown in Fig. 4a, the tensile strength of epoxy resin CPCs with 75 wt% of S-Nickel, F-Nickel, and CS-Nickel decreased from 72.49 MPa to 65.68 MPa, while the tensile fracture strain increased from 3.06% to 5.27%. The S-Nickel with island distribution is used as the reinforcing phase to realize the stress dispersion and improve the tensile strength of the CPCs, and the continuous network formed by CS-

Nickel increases the elongation at break of the material. Besides, the mechanical compression-electrical property of the silica gel CPCs was synchronously tested by the automatic load tester connected with a resistance meter. The results exhibit that the compression rate of the silica gel filled with S-Nickel is about 10% under the pressure of 1 MPa, while that of the F-Nickel, and CS-Nickel filled silica gel are only 5.0% and 5.7%, respectively, as shown in Fig. 4b. At the same time, the resistance of the silica gel filled with S-Nickel, F-Nickel, and CS-Nickel are dramatically decreased from 22.821, 17.812, and 10.694 ohm to 16, 4, and 3 mohm with increasing of the compression strain from the initial state to 5%, respectively.

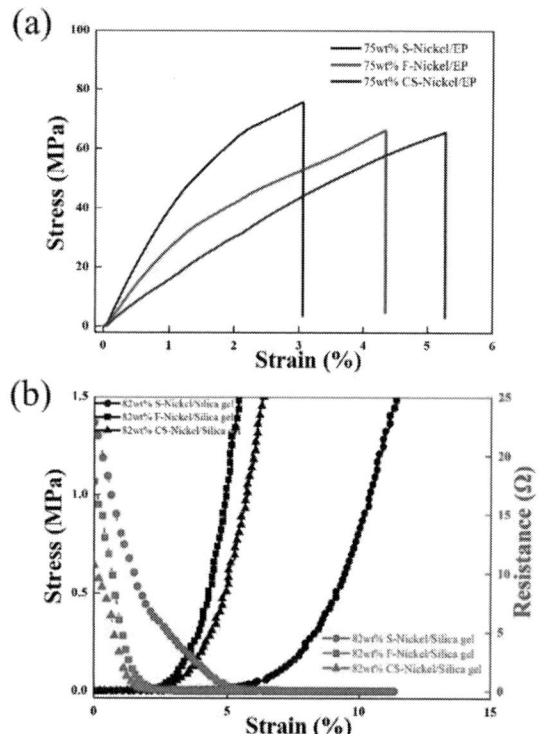

Fig.4. (a) The stress-strain curves of epoxy resin CPCs with different fillers; (b) mechanical compression-electrical property of silica gel CPCs with different fillers.

IV. CONCLUSION

The silica gel and epoxy resin CPCs filled with nickel powders with different morphologies were prepared. The compression rate of the silica gel CPCs filled with S-Nickel powders is higher than that of the F-Nickel and CS-Nickel powders. But the electrical resistance of the CPCs filled with S-Nickel powders is higher than that of the F-Nickel and CS-Nickel powders at the same compression strain. Besides, the epoxy resin CPCs with high cohesion force to filler exhibits higher electrical conductivity at the lower filler content. The silica gel CPCs filled with CS-Nickel powders possess the highest average EMI SE values of about 43.46 and 37.61 dB in 10 MHz-3 GHz and 8.2-12.4 GHz, respectively. And the silica gel CPCs filled with S-Nickel powders exhibit the lowest values of 28.15 and 22.73 dB. The epoxy resin CPCs filled with 75 wt% CS-Nickel powder possess the highest average EMI SE values of 51.69 and 70.57 dB in 10 MHz-3 GHz and 8.2-12.4 GHz, respectively.

ACKNOWLEDGMENT

This work was financially supported by the National Natural Science Foundation of China (62074154), China Postdoctoral Science Foundation (Grant No. 2020M682983), Guangdong Basic and Applied Basic Research Fund (2020A1515110962, 2020A1515110154), and Shenzhen Basic Research Plan (JCYJ20180507182530279).

REFERENCES

[1] W. Zhang, L. Wei, Z. Ma, Q. Fan, and J. Ma, "Advances in waterborne polymer/carbon material composites for electromagnetic interference shielding," *Carbon,* vol. 177, pp. 412-426, 2021.

[2] M. Wang, X.-H. Tang, J.-H. Cai, H. Wu, J.-B. Shen, and S.-Y. Guo, "Construction, mechanism and prospective of conductive polymer composites with multiple interfaces for electromagnetic interference shielding: A review," *Carbon,* vol. 177, pp. 377-402, 2021.

[3] B. Zhao *et al.*, "Dependence of electromagnetic interference shielding ability of conductive polymer composite foams with hydrophobic properties on cellular structure," *Journal of Materials Chemistry C,* vol. 8, no. 22, pp. 7401-7410, 2020.

[4] W.-C. Yu *et al.*, "Selective electromagnetic interference shielding performance and superior mechanical strength of conductive polymer composites with oriented segregated conductive networks," *Chem Eng J,* vol. 373, pp. 556-564, 2019.

[5] Y. Yu *et al.*, "Tailoring hierarchical carbon nanotube cellular structure for electromagnetic interference shielding in extreme conditions," *Mater Design,* vol. 206, 2021.

[6] T. Sun *et al.*, "Self-Reinforced Polypropylene/Graphene Composite with Segregated Structures To Achieve Balanced Electrical and Mechanical Properties," *Ind Eng Chem Res,* vol. 59, no. 24, pp. 11206-11218, 2020.

[7] Q. Qi *et al.*, "An Effective Design Strategy for the Sandwich Structure of PVDF/GNP-Ni-CNT Composites with Remarkable Electromagnetic Interference Shielding Effectiveness," *ACS Applied Materials & Interfaces,* vol. 12, no. 32, pp. 36568-36577, 2020.

[8] Q. Wei *et al.*, "Superhigh Electromagnetic Interference Shielding of Ultrathin Aligned Pristine Graphene Nanosheets Film," *Adv Mater,* vol. 32, no. 14, p. e1907411, Apr 2020.

[9] W. Xiao *et al.*, "Effect of Powder Morphologies on the Property of Conductive Silicone Rubber Filled with Carbonyl Nickel Powder," *J Electron Mater,* vol. 46, no. 11, pp. 6306-6310, 2017.

[10] R. Kumar, S. R. Dhakate, T. Gupta, P. Saini, B. P. Singh, and R. B. Mathur, "Effective improvement of the properties of light weight carbon foam by decoration with multi-wall carbon nanotubes," *J Mater Chem A,* vol. 1, no. 18, 2013.

A X Band Ceramic Package with Kilowatt-level High-power and Low Loss

Yangfan Zhou
The 13th research institute
CETC
Shijiazhuang, China
1017315424@qq.com

Linjie Liu
The 13th research institute
CETC
Shijiazhuang, China
zhouyf19900702@163.com

Zhizhuang Qiao
The 13th research institute
CETC
Shijiazhuang, China
qiaozhizhuang@126.com

Ke Wang
The 13th research institute
CETC
Shijiazhuang, China
Laowu20@126.com

Gai Liu
The 13th research institute
CETC
Shijiazhuang, China
774750445@qq.com

Abstract—With the increase of the power density of microwave power devices, the ceramic packages of microwave power devices are required a high frequency, low loss and high heat dissipation. A X band ceramic package with kilowatt-level high-power and low-loss is designed based on diamond/Cu composites which is a very important heat dissipation material in high-power microwave devices. In order to avoid the air breakdown of arc phenomenon and improve the current carrying capacity of ceramic insulator, the insulator of ceramic package for kilowatt-level microwave high-power device has changed the conventional side metallization structure and adopted the side special-shaped metallization structure. In addition, the RF transmission structure of the package has a low loss by optimizing with HFSS software and is manufactured by multilayer high-temperature co-fired ceramic (HTCC) process. The measured results exhibit the return loss value better than 14 dB and the insertion loss value better than 0.25 dB over a wide frequency range from DC up to 12 GHz.

Keywords—Coplanar waveguide, MMIC, High-temperature co-fired ceramic (HTCC) , Ka band

I. INTRODUCTION

With the rapid development and application of the third-generation semiconductor GaN chip, the development of microwave wireless communication system is accelerated. The output power of the third-generation semiconductor GaN chip is larger than that of the second-generation GaAs chip. As the core devices of wireless communication system and weapon equipment, high-power microwave power devices plays an essential role in the performance of the whole system.

Due to the increasing power of microwave power devices, the power consumption in the circuit system is increasing, and the heat of the chip increases sharply. Excessive heating will cause the chip processing speed to slow down, or even damage and failure. Microwave high-power devices need good heat dissipation performance as a guarantee during their use[1].

In the integrated circuits system, the main functions of the packages are supports, protects, radiates, insulates and connects the power MMIC chip with the external complex circuits system. The packaging of MMIC circuits ensures the integrity of microwave and electrical signals, as well as adequate grounding. And the packaging of MMIC circuits

provide a powerful mechanical structure to protect the circuit from exposure to the external environment, which improve the reliability of the circuit system .

With the wide application of X-band high power microwave devices, the power demand has exceeded kilowatt, and the requirement of the packaging of kilowatt power microwave devices is also increasing. For high-power MMIC circuits, the package not only requires the performance mentioned above, but also requires to have good heat dissipation potential. Traditional power devices packaging technology has been unable to meet the reliability requirements of kilowatt-level power devices, different packaging materials and packaging forms of new technology research and development, to meet the demand of kilowatt-level microwave power devices continuously increasing power[2].

In order to meet the application requirements of kilowatt-level microwave high-power devices, a kilowatt-level high power and low loss ceramic package which is based on high temperature co-fired ceramics (HTCC) technology[3] is proposed in this paper. The X band ceramic package' transmission structure in the MMIC package is very simple and has a low loss. And the ceramic packageis designed based on diamond/Cu composites which is a very important heat dissipation material in high-power microwave devices.

The X band ceramic package' transmission structure were manufactured based on a 90% alumina tape system and tungsten metallization. The tungsten metallization is screen printed on the 90% alumina tape system to transmit radio frequency signals. The relative permittivity value of the X band ceramic package is 9.8 and the dielectric loss tangent is 0.003.

The X band ceramic package is compose of ceramic transmission structure and diamond/Cu composites base. With the 3D numerical simulations soft HFSS, the RF performance of the transmission structure is optimized. By the finite element analysis software, the stress distribution is analyzed. The parallel seam welding sealing process are adopted to ensure the good hermeticity.

II. HIGH-POWER PACKAGE DESIGN

A. RF Design

As seen in Fig.1, the package geometry dimension is 22.00mm*20.00mm. The core area for GaN MMIC chips geometry dimension is 16.00mm*12.00mm. The ceramic radio frequency insulators geometry dimension is 3.00mm*2.00mm.

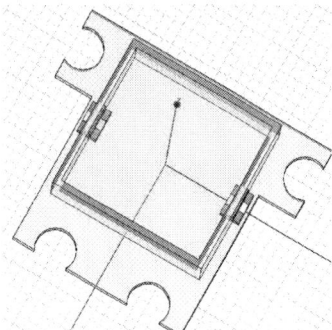

Fig. 1 Schematic diagram of the package

The material of the package wall is selected kovar alloy, which has a close liner thermal expansion coefficient to alumina ceramic. And the radio frequency insulators are inserted between the diamond/Cu composites base and the kovar alloy wall.

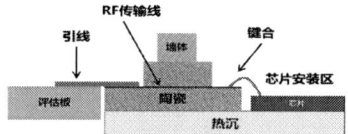

Fig. 2 Schematic diagram of the connection method

By a gold wire bonding method, the package realize the interconnection between the GaN MMIC chip and the package bonding finger. And then by a metal wires and solder, the package realize the interconnection between the printed circuit board and the assembled package. As seen in Fig.2.

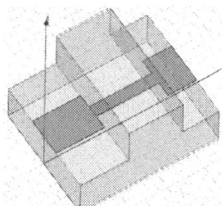

Fig. 3 The simulation model of the transmission structure

The bandwidth of the radio frequency insulators transmission structure is very wide. Range from DC up to X band and the radio frequency insulators transmission structure has a low transmission loss. By using the 3D electromagnetic simulation software HFSS, a simulation model of the radio frequency insulators transmission structure is built. As shown in figure 3.

In order to ensure the good radio frequency insulators transmission performance, the thickness of ceramic medium and the width of the signal line should be optimized with the simulations soft HFSS based on the finite-difference time-domain method algorithm. Because of the actual situation, the conductivity of the electroplated gold film is lesser than that of the bulk gold. the conductivity of the electroplated gold film needs to be reduced, and will be set as 3.96×10^7 S/m[4]. The X band ceramic package simulation model is presented in Fig.1 and the radio frequency insulators transmission structure simulation model is presented in Fig.3.

B. Reliability Design

In the welding process, the thermal stress between metal wall and radio frequency insulator is not balanced. There is solder residue in the corner position. The accumulation of solder in the corner position may cause the ceramic cracking near the ceramic and metal interface, leading to the air tightness failure.The reliability of the welding between the metal and the insulator directly affects the air tightness of the package. The finite element software Ansysworkbench is used to simulate the brazing stress of the package, and the thickness of the metal wall was selected as 0.3mm, 0.5mm and 0.7mm. The stress distribution is shown in the figure 4. The maximum principal stress corresponding to the metal wall thickness of 0.3mm, 0.5mm and 0.7mm is 167Mpa, 171Mpa and 180Mpa respectively. With the increase of the metal wall thickness, the stress is increasing.

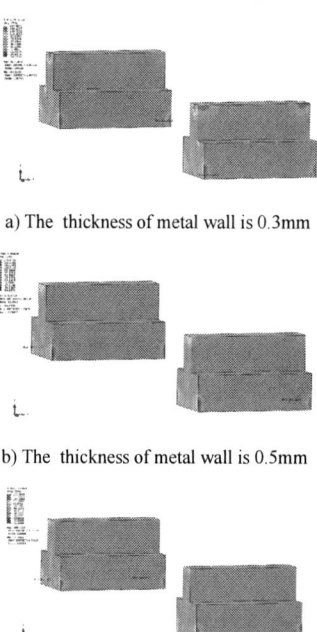

a) The thickness of metal wall is 0.3mm

b) The thickness of metal wall is 0.5mm

c) The thickness of metal wall is 0.7mm

Fig. 4 Under different thickness of metal wall, cloud map of maximum principal stress distribution of the insulator

The sealing method of the ceramic package sample is to adopt parallel seam welding. Combined with the influence of brazing process on the reliability of the package, and comprehensively considering the influence of welding stress and parallel seam welding stress on the reliability, the thickness of the metal wall is preliminarily selected as 0.5mm.

Since the power of the ceramic package sample is more than one kilowatt, the instantaneous current that the output of the package needs to carry greater than 100 amperes.When the power density is greater than the ceramic surface breakdown threshold, the ceramic surface breakdown. This process produces a jet of material which reduce the breakdown threshold of the surrounding air. The electrons generated in this process act as the initial electrons in the air breakdown region, and the air in this region will rapidly ionize, causing the air breakdown. After the air is broken down, there will be an arc phenomenon and the temperature is very high, which will cause the ceramic insulator to burn down.

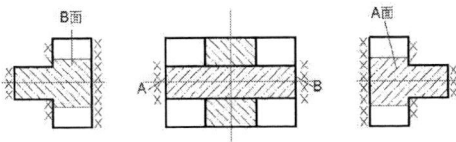

Fig. 5 Diagram of insulator side printing

In order to avoid the air breakdown of arc phenomenon and improve the current carrying capacity of ceramic insulator, the insulator of ceramic package for kilowatt-level microwave high-power device has changed the conventional side metallization structure and adopted the side special-shaped metallization structure. As shown in figure 5.

III. FABRICATION AND MEASURMENTS

The X band ceramic package samples are manufactured based on HTCC technology. To validate the performance of radio frequency insulators structure, The X band ceramic package samples are measured by Agilent vector network analyzer. The X band ceramic package samples is shown in Fig.6.

Fig. 6 The Photograph of package samples

As the radio frequency insulator transmission structure adopts microstrip line structure to transmit radio frequency signals, the X band ceramic package realize the interconnection with the CPW printed circuit board by a metal wires. By using Rogers RO4350B with a PCB thickness of 0.254 mm, the CPW printed circuit board realized 50 ohm impedance matching which is used to connect the input/output ports. The S-parameter of the package are measured by using a 500um pitch air coplanar probes ,the vector network analyzer and matching fixture[5].

The measurement S-parameter of the X band radio frequency ceramic insulator sample is shown in figure7. The measurement results show that the frequency of the ceramic insulator can cover DC ~ 12GHz, The insertion loss is better than 0.25 dB (@DC~12 GHz). The return loss is better than 14 dB (@DC~12 GHz). The insertion loss include the ceramic insulator, test substrate and the wire bonding.

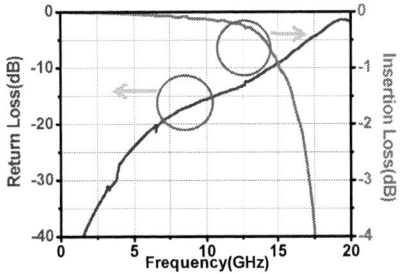

Fig. 7 Measured results of S parameter

IV. CONCLUSION

This paper presents a X band ceramic package with kilowatt-level high-power and low-loss is designed based on diamond/Cu composites which is a very important heat dissipation material in high-power microwave devices. In order to avoid the air breakdown of arc phenomenon and improve the current carrying capacity of ceramic insulator, the insulator of ceramic package for kilowatt-level microwave high-power device has changed the conventional side metallization structure and adopted the side special-shaped metallization structure. In addition, the RF transmission structure of the package has a low loss by optimizing with HFSS software and is manufactured by multilayer high-temperature co-fired ceramic (HTCC) process. The measured results exhibit the return loss value better than 14 dB and the insertion loss value better than 0.25 dB over a wide frequency range from DC up to 12 GHz.

ACKNOWLEDGMENT

This work was supported by the 13th research institute of CETC.

REFERENCES

[1] J.-C. Jeong, I.-B. Yom and D.-P. Jang, "A Ka-Band 6-W High Power MMIC Amplifier with High Linearity for VSAT Applications," in ETRI Jounal, Vol.35, No.3, June 2013, pp. 546-549.

[2] Dong L L,Ahangarkani M,Chen W G, et al. International Journal of Refractory Metals and Hard Materials [J],2018(75):30-42.

[3] William T. Minehan, Kyle Adams and Doug Brown, "High Temperature Co-Fired Ceramic (HTCC) Packages," [J].Advancing microelectronics,2016,43(6):16-19.

[4] Jiang Hu, Yong Zhang and Shanyi Xie, "Micromachined Terahertz Rectangular Waveguide Band pass Filter on Silicon-Substrate", IEEE Microw. Wireless Compon. Lett., vol. 22, no. 12, pp. 636–638, Dec.2012 .

[5] Yangfan Zhou, Linjie Liu, Zhizhuang Qiao,"A Wideband and Low Loss Millimeter-wave MMIC Packaging Based on HTCC Technology " . 2020 21th ICEPT.

Anisotropic BN Nanosheet/polymer Composite Bulk Material：A Study on Mechanical Property

Chen Jing

Academy for Engineering and Technology
Fudan University
Shanghai 200433, Peoples R China
19110860066@fudan.edu.cn

Zeng Xiao Liang
Shenzhen Institute of Advanced Electronic Materials, Shenzhen Institute of Advanced Technology
Chinese Academy of Sciences
Shenzhen 518055, China
xl.zeng@siat.ac.cn

ZhangGuoQi
Department of Microelectronics, Delft
University of Technology
Delft, the Netherlands
G.Q.Zhang@tudelft.nl

Hu Qing Hua
Shenzhen Institute of Advanced Electronic Materials, Shenzhen Institute of Advanced Technology
Chinese Academy of Sciences
Shenzhen 518055, China
qh.hu@siat.ac.cn

Ye Huai Yu

The Key Laboratory of Optoelectronic Technology &Systems, Education Ministry of China, Chongqing University and College of Optoelectronic Engineering, Chongqing University

Chongqing, China

Shenzhen institute of wide-bandgap semiconductors

Shenzhen, China

Liu Pan

Academy for Engineering and Technology
Fudan University
Shanghai 200433, Peoples R China
panliu@fudan.edu.cn

Abstract—With the miniaturization of electronic devices and the development of integrated circuit chips in higher density and higher frequencies, thermal management has become one of the most critical challenges in device performance and reliability. Thermal interface material is bonded between the chip and heat sink. In addition to its thermal conductivity, the mechanical property is essential. Good resilience can effectively fill the bonding gap and reduce thermal resistance, thus improve the overall thermal conductivity. In this paper, three methods are adopted to prepare BN nanosheets/polymer composites. Two polymer composites materials with BN nanosheets oriented were prepared through different suction filtration methods, and the process of homogenization and mixing prepared another sample. Then the microscopic appearance structure of the three materials was analyzed by scanning electron microscope(SEM). The anisotropic compression resilience and tensile properties were studied, respectively. The study showed that BN nanosheet-oriented materials have anisotropic mechanical properties. The better the orientation performance, the greater the difference in the anisotropic mechanical properties of the composite materials. Under the pressure of 0.08MPa, materials with better BN nanosheet orientation have excellent compression and resilience performance, but homogeneous composite materials have the best compression resilience performance. The tensile properties of the three composite materials are quite different. The tensile strength of the homogeneously composite materials can reach 119.59%, while the in-plane and

through-plane tensile strength of the other two materials can reach 102.84%, 38.52%, and 67.65%, 46.67%, respectively.

Keywords — Anisotropy; Mechanical properties ; BN/polymer composite

I. INTRODUCTION

The 5G age introduces high frequencies and doubled networked devices, causing severe heat dissipation problems of electronic devices[1]. Traditional heat-dissipating materials such as aluminum, silver, copper, and ceramics have high density and high processing temperature, limited in electronic devices. Polymers are widely used in electronic devices due to their advantages of light-weight, flexibility, low price, and easy processing and molding[2]. However, due to the random winding of polymer molecular chains, low crystallinity, and the scattering effect of molecular chain vibration on phonons, its thermal conductivity is lower than 0.5 W/mK, which limits its use in high power density electronic devices application[3]. The high thermal conductivity polymer composite material is prepared by adding high thermal conductivity fillers to the polymer matrix. Because of its simple processing technology, easy operation process, low processing cost, it has become the most commonly used material for thermal management of electronic devices[4].

The metals and graphene carbon nanotubes are conductive, and they are not suitable for high-insulation places. Boron nitride (BN) is called "white graphene" and is a crystal composed of nitrogen and boron atoms. Among them, hexagonal boron nitride (h-BN) has a wide bandgap (5.2eV), high thermal conductivity (theoretical calculation is about 2000W/(m·K), empirical 380W/(m·K), high temperature and oxidation resistance. The physical and chemical properties can improve the thermal conductivity of polymers and maintain excellent insulation. In recent years, it has been favored in the field of insulating and thermally conductive composite materials[5-6].

In applying composite materials as thermal interface materials in electronic packaging, high-performance flexible Polymer-based thermal interface materials (TIMs) fill many micro-nano-scale pores caused by incomplete contact between the chip and the heat sink under a certain pressure, thus improving the heat transfer efficiency in the thermal interface and overall heat transfer efficiency[7]. Therefore, the flexibility of the material is also essential. Researchers at Sungkyunkwan University used a pyramid template to assemble BN to form an anisotropic 3D thermally conductive network, which was then transferred to a flexible polymer matrix to form an anisotropic, flexible thermally conductive composite material that can withstand 50% strain[8].

In this paper, two different suction filtration methods are used, and the difference in mechanical properties caused by different orientation methods is compared. The results of the SEM images show that the material orientation rate obtained by method 1(M1) is higher. Its surface shows that BN nanosheets are basically arranged horizontally. In contrast, the BN micro sheets of the composite material that is oriented by suction in the organic silicon are relatively irregular for the viscosity of BN nanosheets in ethanol is lower than that in organic silicon. The DMA tensile properties show that the better-oriented composite materials have better tensile properties and compression resilience.

II. EXPERIMENTAL INVESTIGATIONS

A. Materials

Hexagonal BN powder, size 17-30um, purchased from Merch, Germany; Silicone substrate 1000E purchased from Sigma Chemical Reagent Company; Curing agent and catalyst inhibitor were purchased from Aladdin.

B. Preparation of BN/SiR composites

The preparation process of silicone is as follows: Weigh a certain quality of silicone matrix 1000E, add curing agent and inhibitor, and stir the mixture evenly with a planetary vacuum mixer (Sinomih, VM200s150ML). After the mixture was placed for 20 minutes, the catalyst was added, and the planetary vacuum mixer was used to stir evenly to obtain the organic silicon impregnated with the BN filler in the experiment.

The BN/organosilicon bulk material was prepared by the method of vacuum filtration (M1). The specific steps are as follows: adding a number of BN to the ethanol solution and sonicating it for 30 minutes, then pour the suspension into a suction filter bottle for suction filtration to obtain oriented BN nanoplates. After drying the BN nanosheets, the organic silicon is impregnated, vacuumed, and cured at 80° C. for 5H to obtain a BN/organic silicon bulk material. The specific process is shown in the figure below. This method is simple and easy to operate, suitable for industrial mass production.

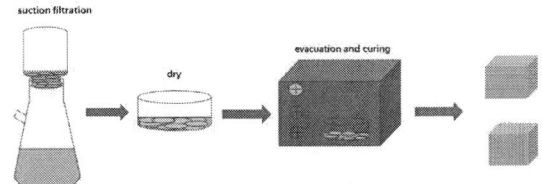

Fig.1 Schematic illustrations of the fabrication procedure of composites material.

In order to compare the advantages of the infiltration method, two sets of comparative samples are added in this experiment. The experimental process of one set of experiments is as follows (M2): Add some BN to the ethanol solution and sonicate it for 30 minutes, then pour the suspension into After suction filtration in a suction filter bottle, drying, mixing with organic silicon to obtain a BN nanoflake/organic silicon solution. The solution was suction filtered to obtain oriented BN nanoplates, which were cured at 80° C. for 5H to obtain BN/organic silicon bulk materials. The mixing experimental process is as follows (M3) : add some BN to the ethanol solution and sonicate it for 30 minutes, then pour the suspension into a suction filter bottle for suction filtration and then dry, mix with silicone to obtain a BN nanoflake/silicone solution, 80°C After curing for 5H, a BN/organic silicon bulk material is obtained.

C. Characterization:

A scanning electron microscope (SEM) study was performed on the FEI NOVA 4500 equipment with an accelerating voltage of 10 kV. Thermal diffusivity (α used laser flash diffuser LFA 467 (Germany NETZSCH) to detect h-BN/PVA composite material). Before testing, cut the composite material into small pieces of 1×1cm, and then spray a thin Graphite powder layer on both sides. Determine the density (ρ) of the composite material by a densitometer. The specific heat capacity is measured using differential scanning calorimetry (DSC, 200 F3, NETZSCH, Germany). The thermal diffusivity, density, and specific heat capacity are characterized at room temperature. The thermal conductivity is calculated according to the following formula: $k = α × ρ × c$. A thermal imaging camera (FLIR E30, FLIR Systems, Inc., USA) was used to record the temperature. In order to study the mechanical properties of composite materials, a dynamic mechanical performance testing machine (DMA Q800, TA Instruments) was used to perform compression cycle tests and viscoelastic performance tests at room temperature. Among them, the maximum compressive force is 18N, and the force area is a small piece of 1×1cm.

III. RESULTS AND DISCUSSION

A. SEM Results

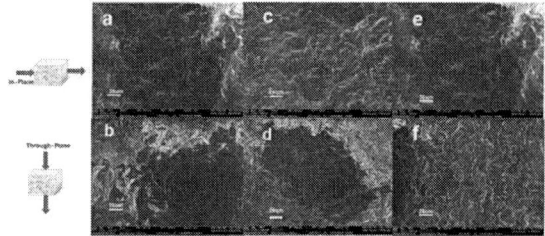

Fig. 2. (ab) In-Plane and Trough-Plane BN/SiR composites image by M1 (cd) In-Plane and Trough-Plane BN/SiR composites image by M2. (ef) In-Plane and Trough-Plane BN/SiR composites image by M3

To observe the filler orientation of hexagonal boron nitride (h-BN) filled composites more intuitively, we use a scanning electron microscope (SEM) to observe the microscopic orientation maps of the transverse and longitudinal sections of the composite. As shown in the figure, in the SEM photo of the cross-section of the composite material, the smooth part represents the polymer matrix. In the transverse section, the orientation of the fillers is basically stacked in the horizontal direction. In the longitudinal section, the orientation of the filler is basically stacked in the vertical direction.

By comparing the composite materials prepared by different methods, it can be seen that the in-plane direction sample surface of M1 in Figure 2-a is basically stacked along the horizontal direction, and the parallel rows can be clearly observed. The vertical grain edges of the BN nanoplates can be seen in the vertical direction. In Figure 2-c. It can be observed that there are more parallel BN nanoplates in the in-plane order. However, we also observed that there are some BN nanoplates arranged in the vertical direction. In the out-of-plane direction, the vertical grain edges of the BN nanoplates and a small number of BN nanoplates arranged in parallel can also be observed. e is the microscopic image obtained by the third method. It can be seen that the vertical and parallel BN nanoplates are more evenly distributed in the in-plane and out-of-plane directions. Through the SEM image, we can observe that the orientation method of M1 is better, and its orientation exhibit a high rate.

B. DMA Compression Test Results

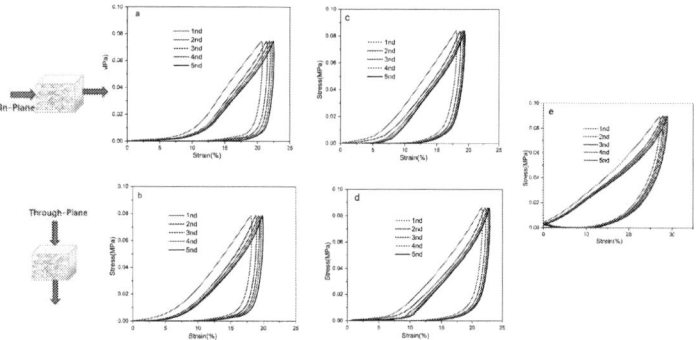

Fig. 3. The typical compression stress–strain curves of BN/SiR composites (ab) In-Plane and Trough-Plane BN/SiR composites by M1 (cd) In-Plane and Trough-Plane BN/SiR composites by M2. (e) BN/SiR composites by M3

Figure 3 shows the true cyclic compressive stress-strain relationship of a polymer material under a stress rate of 0.5N/min. At the beginning of loading, the stress-strain curve is nonlinear, which is a typical feature of soft polymer materials [9]. It does not show an obvious yield point but has a relatively large deformation area. In this area (0-0.08Mpa), the sample has experienced significant strain, the stress increase is slight, and it shows flexibility characteristics. As the stress increases, the stress-strain curve in this area is almost horizontal. The additional stress of the polymer is not enough to deform the polymer that already has a certain amount of deformation, and the polymer shows a specific compressive performance. When the stress gradually releases the pressure at a rate of -0.5N/min, a return path is formed. The strain return path is different from the strain loading path, which is a typical viscoelastic behavior involving elastic and viscous components [10]. The deformation hysteresis in the polymer material during the cyclic compression process may be due to the change of the direction or waviness of the single molecular chain during the loading-unloading cycle. As shown in the Figure 3, the composite material made by M1 shows good cyclic performance, M1 composite's resilience deformation is less than that of M2. Moreover, composites material made by M3 exhibits the best performance.The experimental results show that the M1 material exhibits better resilience performance and better deformability compared with M2 material, which is a required property for a good packaging interface material.

C. DMA Elastic Tests Results

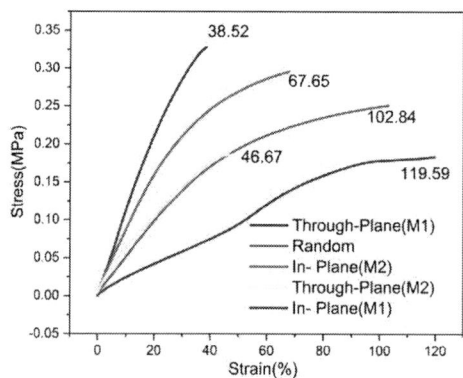

Fig. 4. The typical elastic stress–strain curves of BN/SiR composites

DMA is used to understand the interfacial interactions in the polymer filled with particles[11]. Figure 2 shows the typical stress-strain curves of the different cross-sections of the three composite materials. Studies have shown that the internal organosilicon molecular chain of the uniformly mixed composite material has the stickiest interaction force with the BN nanosheets, and its tensile deformation can reach 119.59%. The composite material prepared by the infiltration method has prominent anisotropic properties. Mechanical properties, its in-plane mechanical properties are relatively excellent, and the tensile deformation exceeds 100%. However, its mechanical properties are the worst for composite materials, and the tensile deformation is only

38.52%. The in-plane and out-of-plane mechanical properties of the mixed first and then oriented composites are weak and only reach 67.65% of the in-plane shape—Variable, 46.67% of the out-of-plane deformation. From the experiment, we found that the mechanical properties of better-oriented composites exhibit better anisotropic properties. The possible reason is that the BN nanoplates arranged in layers have a higher interface strength with the silica molecular chain in the horizontal direction, but in the vertical direction.,the breaking strength is stronger. Compared with composite materials with less obvious orientation, there is no apparent distinction between the interface strength and fracture strength between the horizontal and vertical BN microplate materials, so the anisotropy is less obvious.

D. Conclusions

In this paper, two different suction filtration assited orientation methods are used, one is the method of suction filtration and then infiltration, and the other is the method of mixing and then suction filtration to prepare BN/SiR composite materials. The experimental results show that the composite material prepared by M1 has better anisotropy, its compression resilience and tensile properties are the best, and the in-plane tensile deformation can be achieved. The composite material prepared by M2 has weak anisotropy and in-plane tensile as well as compressive properties. At the same time, a uniformly mixed BN/SiR composite material was prepared as a comparative material in the paper. The experimental results finally show that under the same vacuum filtration pressure, because the viscosity of BN nanosheets in ethanol is relatively tiny compared with that of organic silicon, they can be aligned better, and their orientation performance is better, while the vacuum infiltration method dramatically reduces the preparation time of the composite material. This method is simple, easy to operate, and accessible to industrialized production.

ACKNOWLEDGMENT

This work was supported by the National Key R&D Program of China (2018YFE0204600), the Key-Area Research and Development Program of GuangDong Province (2019B010131001), Guangdong Province Key Field R&D Program Project (No. 2020B010190004).

REFERENCES

[1] A. L. Moore and L. Shi, Materials Today, 2014, 17, 163-174.

[2] H. Chen, V. V. Ginzburg, J. Yang, Y. Yang, W. Liu, Y. Huang, L. Du and B. Chen, Progress in Polymer Science, 2016, 59, 41-85.

[3] N. Burger, A. Laachachi, M. Ferriol, M. Lutz, V. Toniazzo and D. Ruch, Progress in Polymer Science, 2016, 61, 1-28.

[4] Zeng, Xiaoliang, et al. "Fibrous epoxy substrate with high thermal conductivity and low dielectric property for flexible electronics." Advanced Electronic Materials 2.5 (2016): 1500485.

[5] Fu, K., et al. (2021). "Highly Multifunctional and Thermoconductive Performances of Densely Filled Boron Nitride Nanosheets/Epoxy Resin Bulk Composites." ACS Appl Mater Interfaces 13(2): 2853-2867.

[6] Yanming Xue, et al. (2018) "Densely Interconnected Porous BN Frameworks for Multifunctional and Isotropically Thermoconductive Polymer Composites"Adv. Funct. Mater. 2018, 28, 1801205

[7] Han, W., et al. (2020). "Construction of hexagonal boron nitride@polystyrene nanocomposite with high thermal conductivity for thermal management application." Ceramics International 46(6): 7595-7601.

[8] Hong, Haeleen, et al. "Thermal Conductive Composites: Anisotropic Thermal Conductive Composite by the Guided Assembly of Boron Nitride Nanosheets for Flexible and Stretchable Electronics (Adv. Funct. Mater. 37/2019)." Advanced Functional Materials 29.37 (2019): 1970252.

[9] Xue, Yanming, et al. "Densely interconnected porous BN frameworks for multifunctional and isotropically thermoconductive polymer composites." Advanced Functional Materials 28.29 (2018): 1801205.

[10] Bai, Lu, et al. "Effect of PLA crystallization on the thermal conductivity and breakdown strength of PLA/BN composites." ES Materials & Manufacturing 3 (2018): 66-72.

[11] Zhang, Jun, et al. "A facile method to prepare flexible boron nitride/poly (vinyl alcohol) composites with enhanced thermal conductivity." Composites Science and Technology 149 (2017): 41-47.

Soft Degradation and Recovery under ESD stress of E-Mode GaN HEMTs with P-GaN Gate

1st Mei Wang
School of Microelectronics
South China University of Technology
Guangzhou, China
1970170288@qq.com

2nd Zhiyuan He
The 5th Electronics Research Institude of the Minstry of the Industry and Information Technology,
Guangzhou, China
hezhiyuan1988@126.com

3rd Yan Ren
The 5th Electronics Research Institude of the Minstry of the Industry and Information Technology,
Guangzhou, China
184465031@qq.com

4th Lichao Hao
Huada semiconductor CO. LTD
Shanghai, China
haolc@hdsc.com.cn

5th Zhaohui Wu
School of Microelectronics
South China University of Technology
Guangzhou, China
phzhwu@scut.edu.cn

6th Bin Li
School of Microelectronics
South China University of Technology
Guangzhou, China
phlibin@scut.edu.cn

Abstract—The robustness of E-mode AlGaN/GaN high-electron mobility transistors (HEMTs) is investigated with P-GaN gate under ESD testing. In this paper, we further studied the degradation of DC electrical characteristics of the device and its recovery after 24h rest under different level of ESD stress. Output, transfer, on-resistance, gate-leakage characteristics and capacitors were analyzed in detail before and after different level of ESD stress. We mainly found that: i) after ESD experiments, the DC characteristics of the device have undergone some degradation, such as threshold voltage positively shift, gate capacitance decrease, gate leakage current decrease; ii) after resting the device for 24 hours at room temperature, these degradations partially recovered. We infer that electric field redistribution resulted in parameter shift, while further damage caused by ESD is ascribed to degradation formed in the interface between P-GaN layer and metal. Such an investigation can be a significant reference to the popularization of E-mode GaN power devices with field plate.

Keywords—AlGaN/GaN; HEMT; Field plate; ESD; Degradation; Recovery

I. INTRODUCTION

Wide band-gap (WBG) gallium nitride (GaN)-based power electronic devices have superior material properties such as wide bandgap (3.4eV), high-electric breakdown field (3.3MV/cm), high-electron saturation velocity(1.5×10^7cm/s), and high mobility in a readily available heterojunction 2-D electron gas (2DEG) channel, with which density can reach up to 10^{13}cm^{-2} [1-2]. 2DEG is naturally formed between GaN buffer and AlGaN barrier layer due to the piezoelectric and polarization effect. The Gallium nitride high electron mobility transistor (GaN HEMT) is quite popular in the power market in recent years due to its superior electrical performance. With the join of GaN, many prototypes can achieve high efficiency as well as power density. GaN HEMTs have been widely used in many products [3–4]. The enhancement-mode (E-mode) transistors are strongly desired in the market, since it has normally-off behavior, widest available power range [5] and better slew rate control for safe option and reduction of ringing noises [1]. The most commonly used doping material for P-type doping is magnesium. The AlGaN barrier, GaN channel and p-doped gate formed a back-to-back diode [6], which lifts up the potential and contain sufficient negative charges to fully deplete the channel under the gate. When gate is stressed with a positive bias, the p-GaN/AlGaN/GaN diode gradually begins turning on, reforming 2DEG channel and enable the current from drain to source. In order to optimize the electric field inside the device, improve the breakdown voltage of the device and prevent the electric field lines at the gate from being too dense, the field plate is used to extend the length of the gate or source and effectively extends the depletion layer.

GaN HEMT has a wide range of applications, and people will put forward higher requirements for its reliability, and one of them is electrostatic discharge (ESD) robustness. Although GaN has a relatively high breakdown field, ESD could still jeopardize transistors. Device design is changing rapidly, but the research on ESD failure model is still in the early stage of application (AlGaAs/GaAs systems), which has little significance for HEMTs design guidance. ESD characteristics are complex and are not determined by a single factor, but are closely related to design and process parameters. Therefore, it is important to distinguish the effects of different factors on device ESD characteristics and analyze the causes of ESD failure [2]. In the previous studies, transmission line pulse (TLP) testing has been widely used to test the reliability of ESD on P-GaN gate devices, degradation of DC characteristics of devices has been fully monitored in TLP experiment [7-9]. Chen Y Q et.al use low-frequency noise (LFN) technology to measure the defect density, the failure of ESD brought to devices attributed to the increasing number of traps between P-GAN layer and AlGaN buffer layer [7]. The gate of P-GaN HEMT is most prone to failure in ESD events [8]. Stress applied on drain versus source with gate floating or grounded also exhibited different robustness [8]. Further studies indicated that gated device exhibits improved ESD robustness than non-gated structure [9]. The failure caused by ESD stress was thought to be related to trap in P-GaN layer by some researchers [7], while others believe it was due to the hard breakdown of the Schottky structure [8,9]. About ESD stress leads to the threshold voltage shift, several possible mechanisms have been suggested. One theory holds that the threshold voltage shift is the result of ionization of Mg acceptor states in the AlGaN barrier in the gate region. In this case holes from Mg were trapped in the P-GaN layer, which cause fully depletion of 2DEG and make Schotty diode (metal/P-GaN layer) band bending more severe. Thus gate needs a higher V_{th} to open the gate [6]. According to the main characteristics of electrostatic discharge in different occasions, the corresponding electrostatic discharge models can be established to simulate the characteristics of electrostatic discharge, mainly including human model (HBM), machine model (MM), human transmission line pulse model (TLP) and IEC 61000-4-2. Nan-Hung Cheng et al. have applied HBM model to evaluate AlGaN/GaN HEMT's ESD reliability. Device used in this experiment has a metal–insulator–metal (MIM) capacitor structure which is placed on an aluminum nitride (AlN) flip-chip (FC) submount [10]. In the research of recovery mechanisms, X B Xu et.al have observed a good recovery phenomena (rest for 10 days) under repetitive short-circuit stress and attributed to the trapping and releasing process of electron in the P-GaN layer and AlGaN barrier layer [11]. Canato et.al use 365nm UV light to realize recover [6]. Both two letters carried on TLP experiments. It is found that seldom letters studied recovery phenomena under ESD stress.

In this work, ESD behavior of GaN HEMT with P-GaN gate based on human body model (HBM) is fully researched. Four basic DC characteristics were analyzed in detail before and after different level of ESD stress applied on gate in order to observe the soft degradation. Rest the device for 24 hours to confirm whether the degradation of characteristics can recover. Usually the ESD discharge time is only about 200ns, so after 24 hours the distribution of the electric field affected by the ESD stress within the device has disappeared. In sever ESD condition, we aimed at finding soft degradation and recovery mechanism of the P-GaN gate HEMT with field plate.

II. DEVICE DESCRIPTION

Commercially available single-chip, lateral 48A–150V E-mode AlGaN/GaN power HEMTs with p-type gate fabricated on Si substrate are used as device under test (DUT) in this paper. One device was previously used for destructive ESD experiments to obtain the ESD breakdown voltage of the device in preparation for experimental planning. A schematic structure of a p-GaN gated HEMT is shown in Fig. 1. As shown in the Fig. 1, there is a single field plate put on the source electrode and extended to the right side of the gate. In order to test the device conveniently, flip the chip onto the PCB (see in Fig. 2). The threshold voltage (V_{th}) and on-resistance (R_{on}) of the device are 1.49 V and 8.7 mΩ with additional PCB, respectively.

III. EXPERIMENTAL

In this experiment, the Mk2 machine was used to apply ESD stress on the DUT's gate with the human body model (HBM). HBM uses a 100pF capacitor to simulate a human body with a point, and a 1500R resistor to simulate a device with no charge. When the two contact, a discharge is generated. HBM assumes that one end of discharge is in contact with the discharge and the other end is grounded. In this study, the gate terminal is exposed to a discharge while the source terminal is grounded [12]. We destroyed one DUT to obtain a breakdown voltage of 3000V by applying progressively increasing ESD stress continuously. We supposed that continually applying stress to the device without giving stress relief time (e.g., 24 hours) causes premature device breakdown. According to this result, we managed that the ESD voltage level gradually increased from 500V with a step of 500V. Each DC characteristic of the device is measured after each stress test. The recovery characteristics are measured after resting the DUT for 24 hours. Then the ESD voltage will be increased by 500V for

the next round of ESD stress test. Fig. 3 shows the flow chart of the experiment. We use a semiconductor characterization system (Model: Agilent B1505A and B1500) measure DC characteristics at room temperature. Output characteristics are measured at gate-source voltage (V_{gs}) values ranging from 2 to 3.2 V, stepped by 0.2 V, and drain-source voltage (V_{ds}) values ranging from 0 to 6 V. We take the value $R_{ds(ON)}$ when Id is equal to 20A with V_{gs} values ranging from 3 to 5 V. Transfer characteristics are measured at V_{gs} values ranging from 0.5 to 5 V, and set a fixed V_{ds} value of 6V. Gate leakage curves are measured at V_{ds}=0V, V_{gs} varying from 0V to 6V. For capacitances, we mainly consider input capacitance (C_{iss}), output capacitance (C_{oss}), and reverse transfer capacitance (C_{rss}).

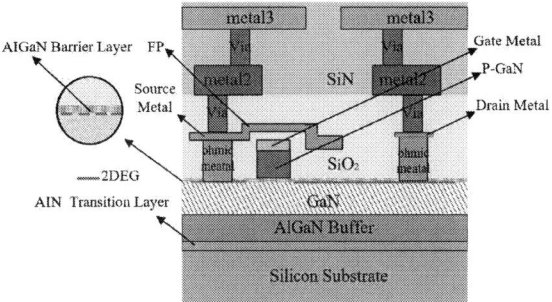

Fig. 1. Schematic view of the AlGaN/GaN HEMT.

Fig. 2. Photograph of the AlGaN/GaN HEMT in PCB package.

IV. RESULTS AND DISCUSSION

After a series of ESD experiments, we finally reached ESD voltage of 5000V, but there was still no breakdown. >2KV HBM gate ESD capability can meet the industrial standard [13], which proved that the DUT has a good ESD reliability. Our primary goal is to explore the degradation and recovery phenomena under ESD, so we need to make sure that the DUT work properly. ESD stress mainly affects the device characteristics related to gate, such as threshold voltage (V_{th}), gate-source capacitance (C_{gs}) and gate leakage current ($I_{g-leakage}$). Fig. 4(a) compares DUT's transfer characteristics before ESD tests with those after ESD tests. From Fig. 4 we can observe that with ESD level increasing, the curve tends to shift positively. However this drift is transverse-shifted without changing the saturation current and G_{m-max}. In particular, we extracted the threshold voltage of each step during the experiment. As shown in the Fig. 4(b), with the increase of ESD voltage, the threshold voltage increases from 1.49V to 1.68V. The shift rate of the threshold voltage can reach up to 12.8%, which indicates that gate electrode may suffer from physical damage. After rest for 24 hours, the V_{th} of the device recovered a little, but it cannot recover to the initial state.

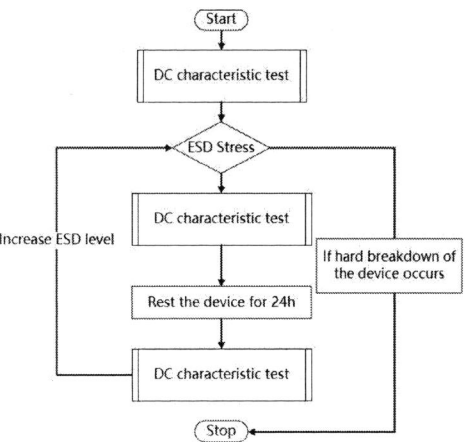

Fig. 3. Measurement procedure in this letter. The ESD level starts from 500V, if there's no breakdown happens to the DUT, the ESD level will increase by 500V.

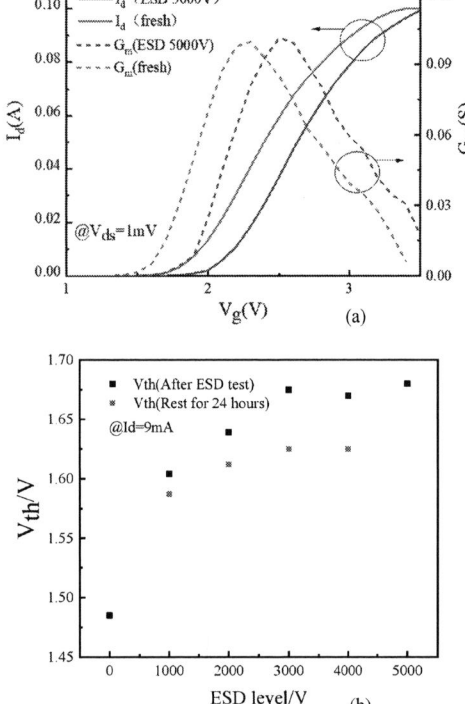

Fig.4. Transfer characteristics curves:(a) measured before and after ESD 5000V; (b) Scatter plot of threshold voltage with ESD voltage

Output characteristic is also a critical information to analysis whether ESD stress has changed drain to source conductive channel. Fig. 5(a) presents the output curves of device before ESD and after 1000V, 3000V, 5000 V ESD stress. As can be seen from the picture, the output current has a significant reduction with the increase of ESD voltage. This is due to the positive shift of V_{th}, the opening status of conductive channel reduced under the same level of V_{gs}. Fig. 5(a) shows that, device rested for 24 hours can recovery the output characteristics to a certain extent. We made a sequence of the output characteristics, and found that it was consistent with the change of threshold voltage with the recovery characteristics of ESD. It proved that the change of output characteristics was mainly caused by the change of

threshold voltage as well. This shows that ESD stress mainly affects the structure characteristics of the gate, the results are consistent with previous study [6].

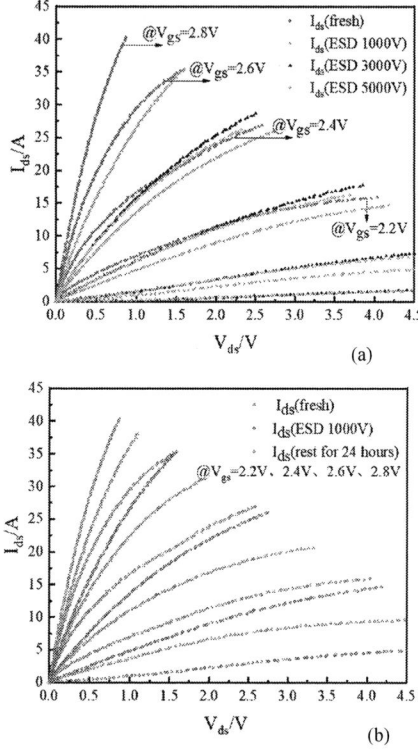

Fig.5. Output characteristics curves:(a) measured before and after ESD stress; (b) measured before and after ESD 1000V, 1000V rest for 24h

Fig. 6(a) shows the conduction resistance fluctuates, the average value, max-min and variance of the conduction resistance are $8.691m\Omega$, $0.63m\Omega$ and $3.54E\text{-}05(m\Omega)^2$. It can be concluded that the conduction resistance unchanged. As can be seen in the Fig. 6(b), the curve move up with the increase of ESD stress. That's because the conduction channel cannot fully open until V_{gs} reaches V_T. When the threshold voltage shifted positively, it suggested that more voltage was needed to fully open the channel and get the smallest R_{on}.

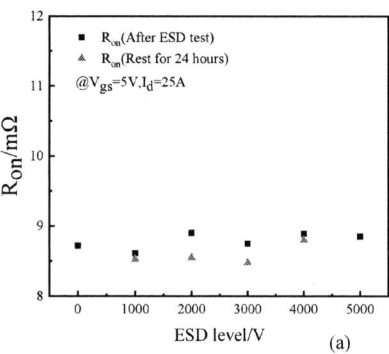

978-1-6654-1392-3/21 $31.00 © 2021 IEEE

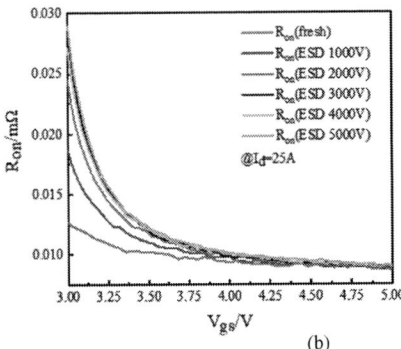

Fig. 6. Conduction resistance characteristics:(a) Scatter diagram of conduction resistance varying with ESD level; (b) Curves of R_{on} measured before and after ESD stress

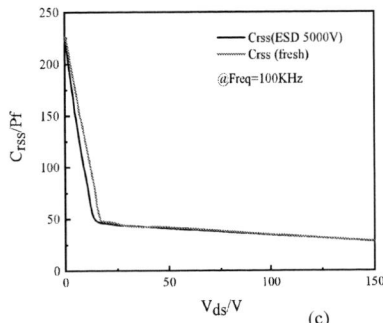

Fig. 7. Capacitances of the DUT:(a) Input capacitance measured before and after ESD stress; (b) Output capacitance measured before and after ESD stress; (c) Reverse transfer capacitance measured before and after ESD stress

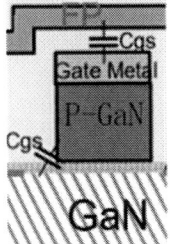

Fig. 8. Schematic cross section of the gate of a device showing locations of C_{gs} capacitances. Some of the metal layers and structures are omitted for simplicity

In order to further analyze gate reliability, we measured C_{iss}, C_{oss} and C_{rss} varied with V_{ds}, their equivalent relationships are shown in equation 1-3. In experiment, these three capacitances need to be measured under drain to source bias. As we can see from Fig. 7, C_{oss} and C_{rss} curves negatively shift a little after ESD stress due to V_{th} shift, while C_{iss} has a significantly decrease after ESD 5000V. When V_{ds} increases, the capacitances drop because the free electrons in GaN are depleted. Higher V_{ds} values extend the depletion region laterally from the field plate edge to the drain, further depleting the 2DEG and eliminating its capacitive component [14]. Combine the experimental results with the equation, we can inference C_{gs} decreased after ESD stress were applied to gate. From Fig. 8, we can know that C_{gs} includes two capacitances: one is between field plate and gate metal, the other is between P-GaN and GaN layer. In order to better understand (and further predict) gate behavior in ESD, it is necessary to characterize the leakage current.

$$C_{iss} = C_{gs} + C_{gd} \quad (1)$$

$$C_{oss} = C_{ds} + C_{gd} \quad (2)$$

$$C_{rss} = C_{gd} \quad (3)$$

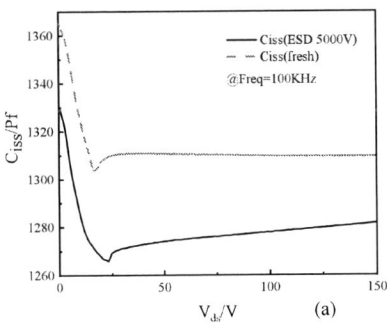

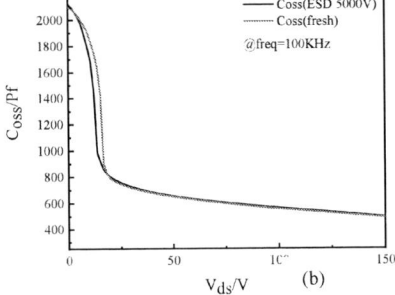

For further analyzing gate, we also measured gate leakage current under $V_{ds}=0V$, as shown in Fig. 9(a) and (b). The gate leakage current mutation is taken as the failure criterion, and it is believed that the increase of leakage current caused by ESD is due to the additional ESD stress-bypassing path formed in MIM structure [10]. We use log scale to evaluate $I_{g-leakage}$, and find that with ESD stress increase, $I_{g-leakage}$ turn to decrease. But when ESD stress reach up to 5000V, $I_{g-leakage}$ is larger than last time. Specifically, we extract Ig-leakage at $V_{gs}=6V$ in Fig. 9(b). It is shown that $I_{g-leakage}$ measured after ESD is smaller than fresh device. Moreover, from Fig. 9(b), resting for 24 hours did not change the damage ESD brought to $I_{g-leakage}$. Field distribution under gate electrode is strongly related to gate leakage. The gate leakage current is generated by the percolation path caused by defects. The percolation path is most likely to be generated in the depletion zone near the metal/p-GaN interface in the p-GaN layer, where the electric field intensity is concentrated [15]. Experimental results also show that, after 5000 ESD stress, $I_{g-leakage}$ turn to increase, which means there was percolation path formed in P-GaN layer close to metal, and these always lead to avalanche multiplication in the space charge region of the Schottky metal/P-GaN layer junction. Meanwhile, we can deduce that C_{gs} decrease is due to the decrease of the capacitance between P-GaN layer (close to metal) and field plate, where defects formed due to the stress. The permittivity (ε) of air is about 1, while the permittivity of SiO2 is about 1.56. The capacitance is proportional to the permittivity, so when defects occur, the capacitance will decrease.

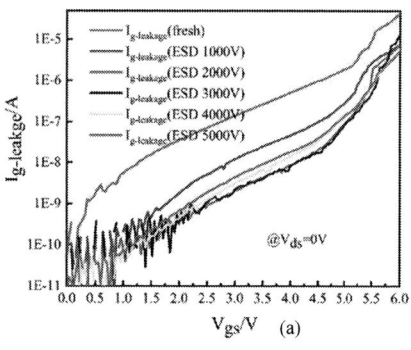

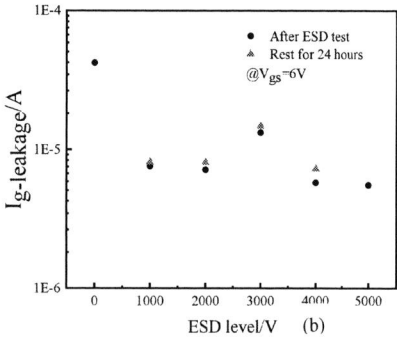

Fig. 9. (a) Gate leakage current measured before and after ESD stress; (b) Scatter diagram of gate leakage current at $V_{gs}=6V$ varying with ESD level

Soft degradation mechanism under ESD stress can be explained as follows: when minor stress applies to the gate, the rapid discharge process results in the variation of electric field distribution near the gate, which result in the positive shift of threshold voltage and decrease of gate leakage. When the ESD level increase, percolation path formed between metal and P-GaN layer. The decrease of capacitance formed between gate and source can be explained in such case. And due to the percolation path, gate leakage current increases after 5000V ESD stress as we have seen in the experiment.

V. CONCLUSION

We have investigated soft degradation and recovery under ESD stress of E-Mode GaN HEMTs with P-GaN Gate and Field Plate. Finally, after 5000V ESD stress, the V_{th} of the device shifted positively and the capacitance C_{gs} decreased. The decrease of $I_{g\text{-}leakage}$ further proved that the characteristic of gate has been damaged by the ESD stress. We suppose the corresponding mechanism could be attributed to degradation of the interface, however further detail needs failure analysis. Our foremost goal is to prove that although the device can sustain >2KV ESD stress, its DC characteristics have been changed. Even the device can still work, it can not function well as before.

ACKNOWLEDGMENT

This work was supported by the Key-Area Research and Development Program of Guangdong Province (2020B010173001,2018B010142001).

REFERENCES

[1] K.J.Chen, O.Häberlen, A.Lidow, C.L.Tsai, T.Ueda, Y.Uemoto, and Y.Wu, "GaN-on-Si power technology: Devices and applications," IEEE Trans. Electronn Devices, vol.64, no.3,pp.779-795,Mar.2017.doi:10.1109/TED.2017.2657579.

[2] Bhawani Shankar, Mayank Shrivastava. Unique ESD behavior and failure modes of AlGaN/GaN HEMTs[C]// Reliability Physics Symposium. IEEE, 2016.

[3] D. Floriot et al., "GH25-10: New qualified power GaN HEMT process from technology to product overview," 2014 9th European Microwave Integrated Circuit Conference, 2014, pp. 225-228, doi: 10.1109/EuMIC.2014.6997833.

[4] S. Nakajima, "GaN HEMTs for 5G Base Station Applications," 2018 IEEE International Electron Devices Meeting (IEDM), 2018, pp. 14.2.1-14.2.4, doi: 10.1109/IEDM.2018.8614588.

[5] Li H, Li X, Wang X D, Lyu X T, Cai H W, Alsmadi Y M, Liu L M, Bala S and Wang J 2019 Robustness of 650 V enhancement-mode GaN HEMTs under various short circuit conditions IEEE Trans. Ind. Appl. 55 1807–16.

[6] Canato E, Meneghini M, Nardo A, et al. ESD-failure of E-mode GaN HEMTs: Role of device geometry and charge trapping[J]. Microelectronics Reliability, 2019, 100-101:113334-.

[7] Chen Y Q, Feng J T, Wang J L, et al. Degradation Behavior and Mechanisms of E-Mode GaN HEMTs With p-GaN Gate Under Reverse Electrostatic Discharge Stress[J]. IEEE Transactions on Electron Devices, 2020, PP(99):1-5.

[8] Y. Xin et al., "Electrostatic Discharge (ESD) Behavior of p-GaN HEMTs," 2020 32nd International Symposium on Power Semiconductor Devices and ICs (ISPSD), Vienna, Austria, 2020, pp. 317-320, doi: 10.1109/ISPSD46842.2020.9170063.

[9] Shankar B, Raghavan S, Shrivastava M. Distinct Failure Modes of AlGaN/GaN HEMTs Under ESD Conditions[J]. IEEE Transactions on Electron Devices, 2020, PP(99):1-8.

[10] Cheng N H, Chen Y F, Chang L B, et al. Improvement in electrostatic discharge robustness of a gallium-nitride-based flip-chip high-electron mobility transistor with a metal–insulator–metal capacitor structure[J]. IEEJ Transactions on Electrical and Electronic Engineering, 2019.

[11] Xu, X.B., Li, B., Chen, Y.Q., Wu, Z.H., He, Z.Y., En, Y.F., Huang, Y., 2020. Analysis of trap and recovery characteristics based on low-frequency noise for E-mode GaN HEMTs with p-GaN gate under repetitive short-circuit stress. Journal of Physics D: Applied Physics.. doi:10.1088/1361-6463/ab713a.

[12] T. S. Speakman, "A Model for the Failure of Bipolar Silicon Integrated Circuits Subjected to Electrostatic Discharge," 1974 Reliability Physics Symposium Proceedings, pp. 60-G9.

[13] M.-D. Ker, J.-J. Peng, and H.-C. Jiang, "ESD test methods on integrated circuits: an overview," in ICECS 2001, Malta, Malta,2001, pp. 1011-1014.

[14] A. Lidow, J. Strydom, M. De Rooij, and D. Reusch, GaN Transistors for Efficient Power Conversion. Hoboken, NJ, USA: Wiley, 2014.

[15] Tallarico A N, Stoffels S, Magnone P, et al. Investigation of the p-GaN Gate Breakdown in Forward-Biased GaN-Based Power HEMTs[J]. IEEE Electron Device Letters, 2016:99-102.

Fabrication of Flexible Printed Circuits on Polyimide Substrate by Using Ag Nanoparticle Ink through 3D Direct-writing and Reliability of the Printed Circuits

Cheng-Bo Li
Lab of Smart Materials and Electronic Packaging in School of Materials Science and Engineering, and Guangdong Provincial Engineering Technology R&D Center of Electronic Packaging Materials and Reliability
South China University of Technology
Guangzhou 510640, China
mschengboli@mail.scut.edu.cn

Xiao Ma*
Lab of Smart Materials and Electronic Packaging in School of Materials Science and Engineering, and Guangdong Provincial Engineering Technology R&D Center of Electronic Packaging Materials and Reliability
South China University of Technology
Guangzhou 510640, China
maxiao@scut.edu.cn

Hai-Jun Huang
Lab of Smart Materials and Electronic Packaging in School of Materials Science and Engineering, and Guangdong Provincial Engineering Technology R&D Center of Electronic Packaging Materials and Reliability
South China University of Technology
Guangzhou 510640, China
mshjhuang@mail.scut.edu.cn

Min-Bo Zhou
Lab of Smart Materials and Electronic Packaging in School of Materials Science and Engineering, and Guangdong Provincial Engineering Technology R&D Center of Electronic Packaging Materials and Reliability
South China University of Technology
Guangzhou 510640, China
msmbzhou@scut.edu.cn

Xin-Ping Zhang*
Lab of Smart Materials and Electronic Packaging in School of Materials Science and Engineering, and Guangdong Provincial Engineering Technology R&D Center of Electronic Packaging Materials and Reliability
South China University of Technology
Guangzhou 510640, China
mexzhang@scut.edu.cn

Abstract—**A 3D printable conductive ink was successfully prepared using polyacrylic acid coated Ag nanoparticles (Ag NPs) as conductive fillers. The Ag NP ink was printed on polyimide (PI) substrate to fabricate circuits by a 3D direct-write printer. The average width and thickness of the printed Ag film circuits are 178 μm and 6 μm, respectively. The resistivity, morphology and microstructure of sintered Ag films subjected sintering at different temperatures for various times were studied systematically. For the Ag film sintered at 300 °C for 60 minutes, the resistivity of the Ag film is greatly reduced to 4.8×10^{-8} Ω·m, which is only 3 times of that of bulk silver. Finally, the cyclic bending reliability of the printed circuits was evaluated by cyclically bending the printed circuit-on-substrate at a radius of 4 mm, and the results show that the relative resistance of the circuits can be maintained at a value smaller than 1.6 after 1000 bending cycles.**

Keywords—Conductive ink; Ag nanoparticles; 3D printing; Flexible electronics; Bending cycle reliability

I. INTRODUCTION

Printed electronics has received extensive attention in a number of fields, such as flexible/stretchable sensors and circuits, electronic skins, wearable electronics and robotics, mainly due to its advantages of facile and cost-effective manufacturing process, as well as environmental benign, as compared with traditional electronic manufacturing technologies like electroless plating processes, vacuum deposition and photolithography [1]. Printed electronic technology has been applied in various applications, such as flexible electrodes, solar cells, flexible sensors and radio frequency identification (RFID) tags [2]. The emergence of this technology enables mass-produce electronic devices on flexible plastics, paper and textile substrates at low cost.

As an essential material for flexible printed electronics, electrically conductive inks with good dispersion stability and high conductivity are crucial to fabrication of highly conductive circuits on flexible substrates. Generally, conductive ink consists of conductive fillers, solvents, dispersants and various additives. The materials used as conductive fillers include metal nanoparticles (NPs), carbon nanotubes, graphene, conductive polymers [3]. Among these materials, carbon nanomaterials and conductive polymers are greatly restricted in their applications due to their poor electrical conductivity (typically 10 to 10^2 S/cm) [4]. Metal

978-1-6654-1392-3/21 $31.00 © 2021 IEEE

NPs are considered as the most suitable material for conductive fillers because of their high electrical conductivity (typically 10^4 to 10^5 S/cm) and low-temperature sintering performance. Currently, for most conductive inks silver nanoparticles (Ag NPs) are used as fillers, because silver is the most electrically conductive metal at room temperature and Ag NPs have good oxidation resistance [5].

It is well known that nanoparticles tend to aggregate due to their extremely high surface energy. Therefore, surface capping agents are needed to modify the nanoparticles to prevent them from agglomerating. So far, the commonly used capping agents are polyacrylic acid (PAA), polyvinylpyrrolidone (PVP) etc [6, 7]. Due to the presence of these capping agents, the sintering process of nanoparticles is hindered at some extent. Nevertheless, sintered Ag NP films exhibit good conductivity after sintering at approximately 200–350 °C [7].

Many previous studies focused on obtaining patterns with excellent conductivity after sintering Ag NPs at low temperature. It has been found that the interaction between chloride ions and silver can desorb the surface capping agent of Ag NPs, thus enables Ag NPs to be sintered at low temperatures [8]. However, limited attention has been paid to the bending fatigue performance and adhesion reliability of printed circuits on flexible substrates, which are very important for the practical application of the conductive ink. Although there were some studies on bending fatigue performance of printed circuits, only some preliminary results were obtained, which were also not satisfactory [9]. In our previous work, conductive inks were successfully prepared by using sodium citrate and PAA capped Ag NP, the printed Ag film circuits showed good conductivity, but no attention was paid to evaluation of the cyclic bending reliability of sintered Ag film circuits [10, 11].

On the other hand, for the mask-free printing techniques of nano silver ink, the most frequently reported one is inkjet printing technology, which has harsh requirements on the viscosity of ink, and is not compatible with high-viscosity inks [12]. Although aerosol printing can be used to print inks with viscosity in the range of 1–1000 cp [1], the equipment is very expensive, thereby being hardly suitable for large-scale applications. 3D direct-write printing is well suited for digital printing of high-viscosity inks with low cost.

In the present study, PAA-capped Ag NPs with an average particle size of about 23 nm were prepared through chemical reduction method. By dispersing silver particles in a mixed solvent of deionized water, propylene glycol and glycerol, a conductive ink with a solid content of 55% was obtained. A 3D direct-write printer was used to print the Ag NP ink on polyimide (PI) substrate to form conductive circuits. Furthermore, the influence of sintering temperature and dwelling time on the resistivity of the sintered Ag film was studied systematically. Finally, the cyclic bending reliability of the printed circuit was evaluated by cyclically bending the circuit-on-substrate specimen with a bending radius of 4 mm for 1000 cycles.

II. EXPERIMENTAL PROCEDURE

Polyacrylic acid (PAA, M_W=5000), 1, 2-propylene glycol, sodium carboxymethyl cellulose (CMC), triethanolamine (TEA) were purchased from Shanghai Macklin Biochemical Co. Ltd. Glycerin was bought from Shanghai Aladdin Biochemical Technology Co. Ltd, and silver nitrate (AgNO₃)

was commercially supplied by Sinopharm Chemical Reagent Co. Ltd.

Ag NPs were synthesized by using $AgNO_3$ as the silver precursor, PAA as the stabilizer, and TEA as the reductant. Typically, 5 g $AgNO_3$ was dissolved into 10 ml deionized water to obtain solution A, 1 g PAA and 12.5 g TEA were dissolved in water to obtain 30 ml solution B. Then the solution A was added into the solution B, followed by a stirring at 600 r/min. The whole reaction was carried out in a water bath, and the mixed solution was heated from 25 °C to 80 °C for a dwelling time of 1 h under continuously stirring. After completion of the reaction, 150 ml ethanol was added to precipitate the Ag NPs, then the concentrated Ag NPs were separated from the mixture by centrifugating at 4000 r/min for 5 min. The solvent of 3D printing Ag ink was prepared by mixing glycerol, propylene glycol, deionized water, and CMC. Finally, the obtained Ag NPs were added into the solvent with a metal loading of 55wt% to prepare 3D printing conductive ink.

The as-prepared Ag NP ink was directly printed on the polyimide (PI) substrate using Voltera V-one 3D direct-write printing platform with a nozzle diameter of 100 μm and a printing speed of 500 mm/min to fabricate conductive circuits. For a comparative study, the Ag films were also fabricated on glass substrates by screen printing and then sintering at various temperatures of 180, 220, 260 and 300 °C for 5–60 minutes.

The morphology and size of Ag NPs were investigated by JEM-1400Plus transmission electron microscope (TEM). Crystal structures of the PAA-Ag NPs powder were characterized by X-ray diffraction (XRD, PANalytical X'pert3) using Cu Kα radiation. Thermogravimetric analysis (TGA) of Ag NP ink was carried out using simultaneous DSC-TG (SDT, Q600, TA) in N_2 with a heating rate of 10 °C/min. A 4-point probe instrument (RTS-9) was used to measure the electrical resistivity of the sintered silver films. The microstructures of sintered Ag films were analyzed using a field emission scanning electron microscope (FE-SEM, NOVA NANOSEM 430). The bending cycle test of the printed circuits was carried out using a Shimadzu EZ-graph materials testing machine. The change of the circuit resistance during bending test was measured in real time through digital multimeter datasheet (DMM6500, Keithley).

III. RESULTS AND DISSCUSSION

The TEM image and XRD patterns of as-synthesized Ag nanoparticles stabilized by PAA are presented in Fig. 1. Apparently, the particles largely exhibit a bimodal size distribution, i.e., large particles of size in the range of 70 to 90 nm surrounded by small particles with sizes ranging from 5 to 20 nm, as shown in Fig. 1(a). This bimodal size distribution is beneficial to forming dense sintered structure during subsequent sintering process [13]. Four strong characteristic peaks can be observed at 38.2°, 44.2°, 64.5° and 77.4°, as depicted in Fig. 1(b). which can be attributed to the (111), (200), (220) and (311) diffraction of crystalline planes of Ag face-centered cubic (FCC) structure, respectively. Noting that Ag oxide peak is not found in the XRD pattern. Above results demonstrate that the Ag nanoparticles are well-crystallized and do not undergo oxidation. The average particle size of PAA-Ag NPs is about 23 nm as calculated by the Scherrer formula.

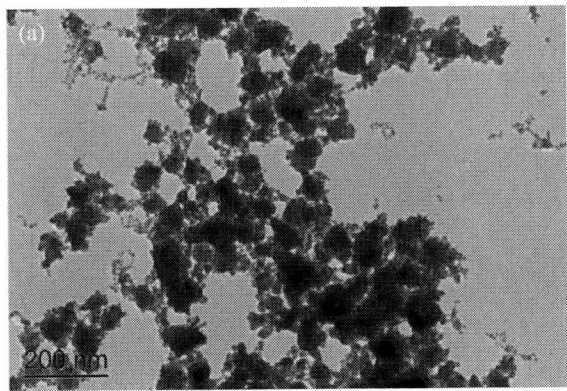

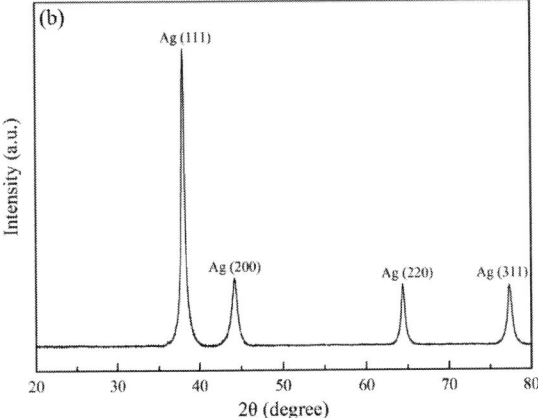

Fig. 1 TEM image (a) and XRD pattern (b) of synthesized Ag NPs capped by polyacrylic acid (PAA)

Thermogravimetric analysis results of as-prepared Ag NP ink are shown in Fig. 2. It can be seen clearly from the TG curve (i.e., blue curve) that the weight percentage of the ink reduces to 56.5% during heating from 25 °C to 450 °C. The slope of the curve gradually decreases as the temperature increases. Notably, when the temperature is lower than 250°, the weight loss is mainly caused by the evaporation of the solvent (e.g., water, propylene glycol and glycerol). As the temperature is further increased, the decomposition of high molecular weight organics (PAA and CMC) in the ink makes the main contribution to the weight loss. In addition, there is an obvious endothermic peak in the DSC curve at 67 °C, implying the evaporation of water in the ink. Two small turning points can be observed at 111 °C and 181 °C, which may be due to the decomposition of propylene glycol and glycerol, respectively, as indicated by a previous study [10].

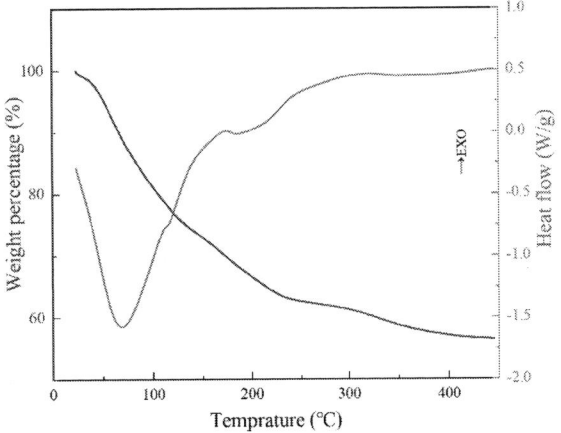

Fig. 2 TG and DSC results of the as-prepared Ag NP ink

The electrical resistivity of the Ag films printed on glass subjected to different sintering temperatures for various periods are presented in Fig. 3. Obviously, the resistivity gradually decreases with increasing both sintering temperature and time. In addition, as the temperature increases, the amplitude of fluctuation of the resistivity decreases gradually at different heating times. Noteworthily, for the Ag film sintered at 180 °C for a short time (5 min and 15 min), the value of resistivity of the sample is too large to be measured by four-probe instrument (RTS-9) used in this work. The resistivity of the Ag films sintered at 200 °C shows a maximum fluctuation in resistivity from 5.8×10^7 to 249 $\mu\Omega \cdot cm$ as the dwelling time is prolonged from 5 to 60 min. When sintering the Ag NP film at 300 °C for 5 and 60 min, respectively, the resistivity of sintered Ag film decreases from 291 to 4.8 $\mu\Omega \cdot cm$. This lowest value of resistivity (4.8 $\mu\Omega \cdot cm$) is approximately three times of that of the bulk Ag, which is 1.59 $\mu\Omega \cdot cm$ at room temperature. Although the sintering temperature is relatively high, the resistivity of the Ag film sample has reached a very low level, meaning that the ink can be used for printing circuits on some heat-resistant substrates such as PI.

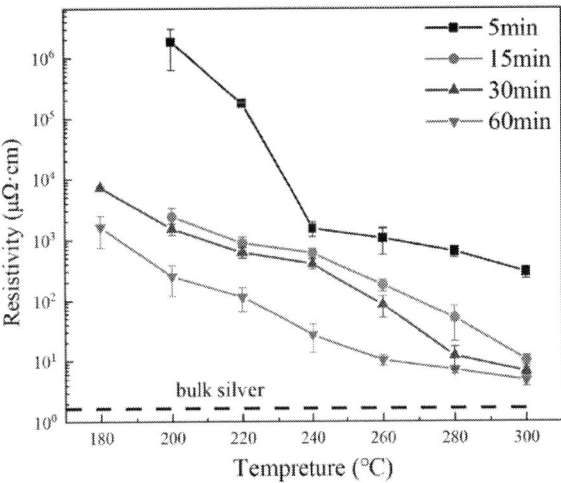

Fig. 3 Resistivity of Ag films after sintering at different temperatures for various dwelling times

In order to clarify the reason for the decrease of resistivity caused by the increase of temperature, SEM observation was carried out on the Ag films sintered at 180–300 °C for different dwelling times. Fig. 4 exhibits SEM images of the Ag films sintered at four different temperatures of 180 °C, 220 °C, 260 °C and 300 °C for 15 minutes. Obviously, as the sintering temperature is increased, Ag NPs gradually coalesce together to form large lumps of Ag, and the density of the Ag film gradually increases. For example, at sintering temperature of 180 °C, there are a large number of small particles in the Ag film, as shown in Fig. 4(a), which means sintering has not started yet. When the temperature is increased to 220 °C, the existence of sintering necks can be found, as exhibited in Fig. 4(b), indicating that the sintering of Ag NPs happens. With further increase of sintering temperature to 260 °C, the surface of Ag NPs seems to melt and particles appear to be connected largely, as demonstrated in Fig. 4(c), manifesting that sintering takes place substantially. After sintering at 300 °C, most boundaries between Ag NPs have disappeared and the surface of the Ag film becomes smoother and denser, as presented in Fig. 4(d), which can result in decrease of the electrical resistivity of sintered Ag film to a lower value.

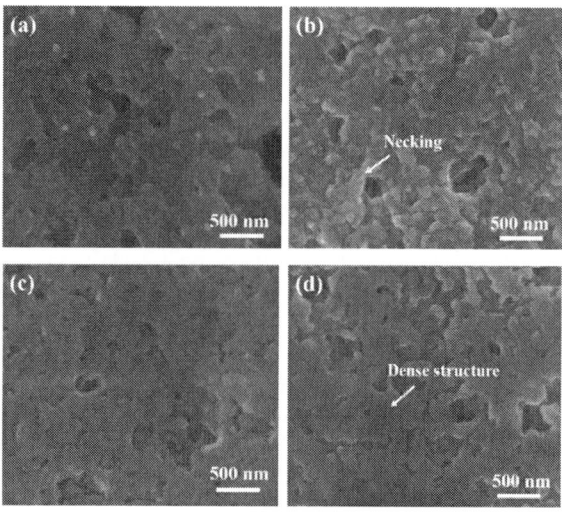

Fig.4 SEM images of Ag films at different sintering temperatures for 15 min: (a) 180 °C, (b) 220 °C, (c) 260 °C, and (d) 300 °C

According to the above thermogravimetric analysis results of the Ag NP ink (Fig. 2), resistivity change (Fig. 3) and morphological characterization (Fig. 4) of the sintered Ag film, it has been demonstrated that when a relatively low sintering temperature is used (e.g., below 240 °C), the presence of high-boiling solvents in the ink hinders the sintering behavior of Ag NPs, resulting in poor sintering quality of the Ag film and thereby higher resistivity. When sintering at relatively high temperature (240–280 °C), the high-boiling solvents quickly evaporate, the surface dispersants decompose, and the sintering of Ag NPs occurs. For sintering at high temperature (above 280 °C), the degree of sintering increases significantly, the sintered Ag film becomes dense and smooth, and the number of conductive paths increases, thereby the resistivity of the Ag film drops sharply. Notably, the lowest resistivity of the sintered Ag film is three times of bulk Ag, which is mainly caused by the voids and a small amount of residual organics in the sintered Ag film.

Fig. 5 displays SEM images of the Ag film circuit printed on PI substrate by using a 3D direct-write printer (Voltera V-one) with a nozzle diameter of 100 μm, followed by sintering at 280 °C for 15 min. It should be indicated that the parameters of the printer during the printing process have a great influence on the quality of the printed Ag film circuits, such as ink pressure (E), printing speed (v) and nozzle height (z). If the ink pressure is too small or the nozzle height is too high, the printed circuits may be discontinuous. Similarly, the printing speed also has a big influence on the shape of the Ag film circuits. The printed Ag film circuit by using the optimized parameters (v=500 mm/min, z=50 μm) is shown in Fig. 5, which has an average line width of 178±10 and an average thickness of 4–6 μm, respectively, and the Ag film tracks are uniform and continuous with sharp edges.

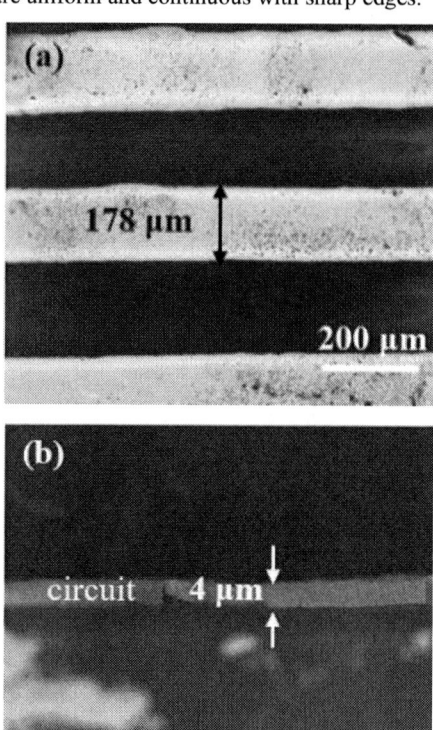

Fig. 5 SEM Images showing (a) line width and (b) thickness of the printed Ag circuit on PI substrate via 3D direct-write printing

When taking into account the application of flexible electronics, the printed conductive circuits on flexible substrate should have adequate mechanical and electrical reliability. In this work, the specimen for the reliability test was prepared by printing Ag NP ink on PI substrate with a line width of 1.0 mm and a length of 50 mm, followed by sintering at 280 ℃ for 15 min. Fig. 6 shows the real-time variation of relative resistance (R/R_0) of the Ag film circuit during bending fatigue test up to 1000 cycles with a bending radius of 4 mm. Obviously, in the initial stage of the cyclic bending test, the relative resistance (R/R_0) of the film increases rapidly, and then enters a steady and slow growth stage. After 1000 bending cycles, the relative resistance value is only about 1.6, indicating that the circuit has good bending fatigue performance. A detailed visual inspection of the film specimen subjected to 1000 cycles shows that there is no visible damage (crack) in the film and no peeling of the film from the PI substrate.

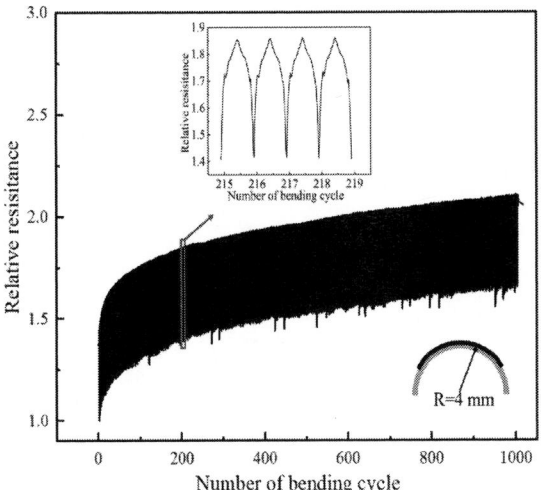

Fig. 6 Change in relative resistance (R/R_0) of the printed Ag NP film circuit on PI substrate during cyclic bending test up to 1000 cycles

IV. CONCLUSION

In summary, by dispersing PAA-capped Ag NPs in a solvent mixed with water, propylene glycol and glycerol, a conductive Ag NP ink with excellent printability is prepared. After printing Ag NP ink on PI substrate using a 3D direct-write printer and sintering the printed Ag film with appropriate sintering parameters (300 °C and 60 min), the sintered Ag film circuit shows a relatively low resistivity of 4.8 $\mu\Omega\cdot$cm, which is only three times of that of bulk silver. The sintered Ag film circuit on PI substrate also exhibits good bending fatigue performance, in terms of very small change of relative resistance even after undergoing 1000 bending cycles.

ACKNOWLEDGMENTS

This work was supported by the National Natural Science Foundation of China under Grant No. 51775195, and the Research Fund for the Guangzhou Municipal Science and Technology Program under Grant Nos. 201807010028 and 201904010316, and the Science and Technology Commissioner Program for Guangdong Enterprises under Grant No. GDKTP2020067700.

REFERENCES

[1] W. Wu, "Inorganic nanomaterials for printed electronics: a review," *Nanoscale*, vol. 9, no. 22, pp. 7342–7372, 2017.

[2] P. S. Karthik and S. P. Singh, "Conductive silver inks and their applications in printed and flexible electronics," *RSC Advances*, vol. 5, no. 95, pp. 77760–77790, 2015.

[3] A. Kamyshny and S. Magdassi, "Conductive nanomaterials for 2D and 3D printed flexible electronics," *Chemical Society Reviews*, vol. 48, no. 6, pp. 1712–1740, 2019.

[4] W. Shen, X. Zhang, Q. Huang, Q. Xu, and W. Song, "Preparation of solid silver nanoparticles for inkjet printed flexible electronics with high conductivity," *Nanoscale*, vol. 6, no. 3, pp. 1622–1628, 2014.

[5] D. Zhu and M. Wu, "Highly conductive nano-silver circuits by inkjet printing," *Journal of Electronic Materials*, vol. 47, no. 9, pp. 5133–5147, Sep. 2018.

[6] Z. Wu, S. Yang, and W. Wu, "Shape control of inorganic nanoparticles from solution," *Nanoscale*, vol. 8, no. 3, pp. 1237–1259, Jan. 2016.

[7] L. Mo, Z. X. Guo, L. Yang et al., "Silver nanoparticles based ink with moderate sintering in flexible and printed electronics," *International Journal of Molecular Sciences*, vol. 20, no. 9, Art. no. 9, Jan. 2019.

[8] M. Grouchko, A. Kamyshny, C. F. Mihailescu, D. F. Anghel, and S. Magdassi, "Conductive inks with a "built-in" mechanism that enables sintering at room temperature," *ACS Nano*, vol. 5, no. 4, pp. 3354–3359, Apr. 2011.

[9] Q. Huang, W. Shen, Q. Xu, R. Tan, and W. Song, "Properties of polyacrylic acid-coated silver nanoparticle ink for inkjet printing conductive tracks on paper with high conductivity," *Materials Chemistry and Physics*, vol. 147, no. 3, pp. 550–556, Oct. 2014.

[10] C. Yin, H. Jin, Z. Zhou, M.B. Zhou, and X.P. Zhang, "Processing and electrical properties of sodium citrate capped silver nanoparticle based inks for flexible electronics," *in 2017 18th International Conference on Electronic Packaging Technology (ICEPT)*, pp. 1572–1576, Aug. 2017.

[11] H. J. Huang, M. B. Zhou, C. Yin, Q. W. Chen, and X. P. Zhang, "Preparation of a low temperature sintering silver nanoparticle ink and fabrication of conductive patterns on PET substrate," in *2018 19th International Conference on Electronic Packaging Technology (ICEPT)*, pp. 482–485, Aug. 2018.

[12] J. Kathirvelan, "Recent developments of inkjet-printed flexible sensing electronics for wearable device applications: a review," *Sensor Review*, vol. ahead-of-print, no. ahead-of-print, Jan. 2020.

[13] S. Ummartyotin, N. Bunnak, J. Juntaro, M. Sain, and H. Manuspiya, "Synthesis of colloidal silver nanoparticles for printed electronics," *Comptes Rendus Chimie*, vol. 15, no. 6, pp. 539–544, Jun. 2012.

Study on the Electromagnetic and Thermal Characteristics of Aerostatic Spindle for Wafer Grinding

1st Chen Zhao
College of Mechanical Engineering and
Applied Electronic Technology
Beijing University of Technology
Beijing, China
1419258965@qq.com

2nd Ye Lezhi*
College of Mechanical Engineering and
Applied Electronic Technology
Beijing University of Technology
Beijing, China
yelezhi@bjut.edu.cn

3rd Song Xuanjie
College of Mechanical Engineering and
Applied Electronic Technology
Beijing University of Technology
Beijing, China
306228246@qq.com

4th Wang Zhiyue
Beijing Electronic Equipment Co.
Ltd of CETC
Beijing, China
wangzhy@45inst.com

5th Liu Guangjie
Beijing Electronic Equipment Co.
Ltd of CETC
Beijing, China
cetcvip@126.com

6th Zhao Yumin
Beijing Electronic Equipment Co.
Ltd of CETC
Beijing, China
1183140477@qq.com

Abstract—**With the development of advanced IC packaging technology, wafers are required to become thinner and thinner. The 12-inch memory wafer has reached within 25um, which requires a higher-performance aerostatic spindle of the wafer grinder. The spindle motor power of the wafer grinder is large, which generates a lot of heat. The heat causes the temperature rise of the aerostatic spindle, which adversely affects the wafer grinding accuracy and the reliability of the spindle motor operation. In this paper, the field-circuit coupling model of the spindle motor in the aerostatic spindle is established. The transient two-dimensional electromagnetic field is analyzed, and the no-load and load characteristics of the motor are obtained, which provides theoretical support for the design of the spindle motor. Using motor loss as the heat source, through the one-way coupling of electromagnetic field and temperature field, establish an aerostatic spindle thermal analysis model, and the steady-state temperature field of the aerostatic spindle and the rotating shaft is simulated and analyzed, and the temperature distribution of the aerostatic spindle and the rotating shaft is obtained, and the thermal deformation is analyzed. This paper has certain reference value for the optimization design of aerostatic spindle.**

Keywords—Wafer grinding; aerostatic spindle; thermal analysis; electromagnetic analysis.

I. INTRODUCTION

The thinning and grinding of the backside of the wafer is a key process step for the back-end packaging of integrated circuits. The excess silicon substrate material on the back of the wafer can be quickly thinned from 775um to less than 100um, thereby meeting the requirements of high-density and small-volume integrated circuit chips [1]. The thinning and grinding method of the back of the wafer mainly includes the rotary table for wafers of 6 inches and below and the rotary type for wafers of 8 inches and above.

Fig.1 shows the basic principle of wafer rotary grinding. The wafer is vacuumized by the wafer supporting stage and the cup grinding wheel is fixed on the grinding disk of the aerostatic spindle rotation shaft. When the silicon wafer is thinned by grinding, the edge of the grinding wheel is adjusted to coincide with the center of the wafer, and the rotation axis of the grinding wheel is at a certain angle with the rotation axis of the wafer, so that the wafer contacts the

semicircle of the grinding wheel, and the grinding wheel feeds along the vertical direction to remove the material[2]. This cutting in grinding method can achieve high efficiency grinding and deep cutting grinding, with stable machining state and high precision of wafer surface.

The aerostatic spindle is composed of a spindle motor and an air bearing. The spindle motor provides the driving force and requires the required grinding torque to be achieved in a certain space, which difficult for the spindle motor design [3]. Electromagnetic loss will generate heat. Although cooling water passes through the spindle, it will also cause uneven heat of the spindle, resulting in thermal deformation. This not only affects the grinding accuracy, but also deteriorates the uniformity of the air film thickness, which can cause damage to the aerostatic spindle in severe cases. Therefore, it is necessary to study the magneto-thermal physical field of the spindle motor [4-6].

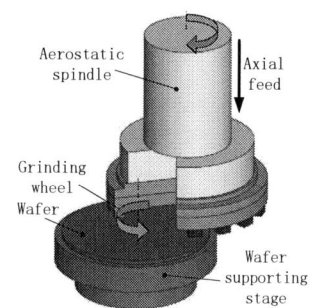

Fig. 1 Schematic diagram of rotary table grinding

II. SPINDLE MOTOR STRUCTURE AND WORKING PRINCIPLE

A. Spindle motor structure

The spindle motor is a surface-mounted three-phase permanent magnet synchronous motor. The winding method of the stator winding is double-layer winding, and the connection method is Y-connection. There are 12 permanent magnets evenly distributed on the spindle motor rotor. The specific motor parameters are shown in Table 1, and the model structure diagram is shown in Fig. 2 [7-8].

978-1-6654-1392-3/21 $31.00 © 2021 IEEE

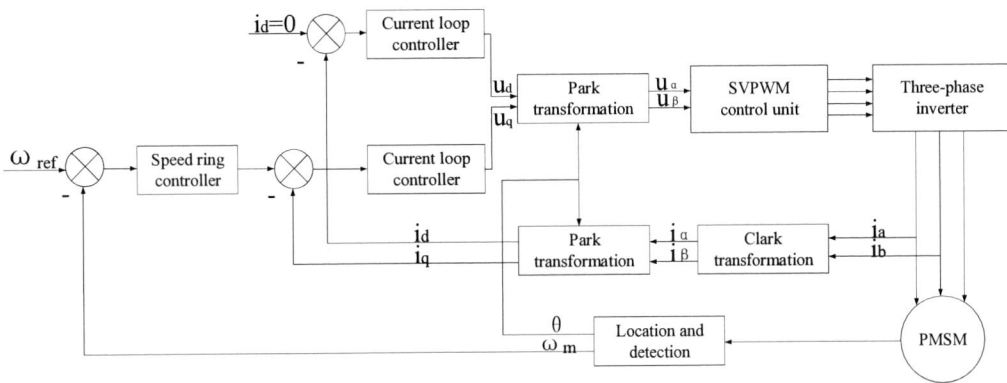

Fig. 3 Structure diagram of motor control system

TABLE 1 MAIN PARAMETERS OF SPINDLE MOTOR

Parameter	Value
Rated Power (P_N/kW)	4.8
Rated Voltage(U_N/V)	340
Rated Speed{n_N/(r·min^{-1})}	2000
Rated Frequency(f_N/Hz)	200
Number of Phase(m)	3
Number of Slots(z)	45
Number of pole pairs(p)	6
Outer Diameter of the Stator(D_1/mm)	117
Inner Diameter of the Stator (D_{i1}/mm)	73.44
Outer Diameter of the Rotor(D_2/mm)	71.5
Inner Diameter of the Rotor(D_{i2}/mm)	34.14

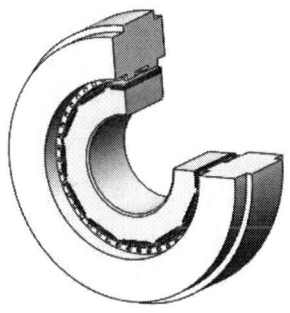

Fig. 2 Model structure diagram of spindle motor

B. Working principle of spindle motor

Spindle motor have three-phase symmetric windings distributed on the stator, permanent magnets are evenly distributed on the rotor, and there is a certain air gap between the stator and the rotor. When the three-phase current flows into the three-phase symmetrical windings of the stator, a circular rotating magnetic field will be formed in the stator windings. At this time, since the magnetic poles of the permanent magnets on the rotor are fixed, according to the principle that the magnetic poles attract and repel each other, the circular rotating magnetic field generated on the stator will drive the permanent magnet poles of the rotor to rotate synchronously.

The rated frequency of the spindle motor is 200 Hz, and a frequency converter is used to implement variable frequency speed control. When adjusting the speed of the spindle motor, it is mainly to change the frequency and voltage of the motor, and adopt vector control methods, including maximum torque/current control (MPTA), $i_d = 0$ control, constant flux linkage control, field weakening control, and maximum output power control and other control strategies. The spindle motor can adopt the $i_d = 0$ control method, and its control system structure diagram is shown in Fig. 3.

III. ELECTROMAGNETIC FIELD MODELING AND ANALYSIS

This paper uses ANSYS Maxwell finite element analysis software to establish a simulation model of the spindle motor, and analyzes the no-load and load characteristics of the motor.

A. Electromagnetic field model

According to the size parameters and working parameters of the spindle motor, set the parameters in the RMxprt module in Maxwell, establish the basic full model of the spindle motor, and create "Maxwell 2D Design", and 1/3 of the finite element analysis will be automatically generated in the Maxwell 2D module Model, the model is shown in Fig. 4.

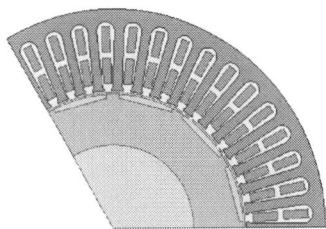

Fig. 4 1/3 spindle motor finite element analysis model diagram

After generating a 1/3 spindle motor finite element analysis model in Maxwell 2D, set the model simulation parameters. Among them, a rotational movement of 2000 rpm is applied to the rotor. The boundary conditions are the Master-slave boundary conditions and the Dirichlet boundary conditions. The excitation source is a voltage source, the amplitude of the voltage source is 160V, and the waveform is a sinusoidal periodic change. The mesh type is surface approximation division, and the triangle mesh element is used by default. The division accuracy of permanent magnets is different from other parts. The division accuracy surface deviation is 0.03575mm, the normal deviation is 15°, and the division accuracy surface deviation of other parts is 0.0585mm, the normal deviation is 15°, and the mesh model is divided as shown in Fig. 5. The solver sets the solution stop time to 50ms and the time step to 0.05ms.

978-1-6654-1392-3/21 $31.00 © 2021 IEEE

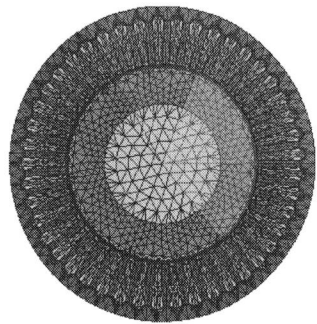

Fig. 5 Result of meshing of spindle motor model

B. Electromagnetic field analysis

(1) Electromagnetic field analysis of the spindle motor under rated power

After solving the calculation in Maxwell 2D, the calculated magnetic flux density distribution cloud map and magnetic field line distribution cloud map of the spindle motor are shown in Fig. 6 and Fig. 7, which shows the state of the motor at 50ms. According to Fig. 6 and Fig. 7, it can be seen that the distribution of the magnetic flux density of the magnetic field of the motor is uniform and symmetric, the distribution of magnetic lines of force is normal, and the phenomenon of magnetic flux leakage is not obvious. The calculated electromagnetic torque curve of the rotor at different speeds is shown in Fig. 8. According to Fig. 8, it can be seen that as the spindle motor speed increases, the electromagnetic torque is continuously reduced. At the rated speed of 2000 rpm, the electromagnetic torque is about 20.2N·m After subtracting the no-load torque of 1.5N·m, it is close to the rated torque of 17.5N·m.

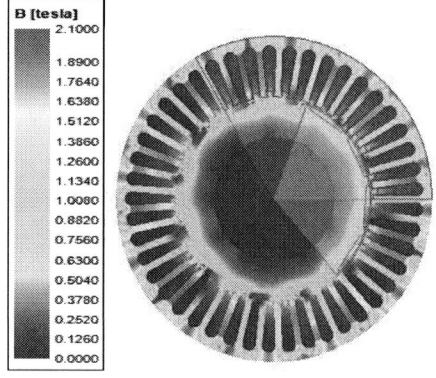

Fig. 6 Two-dimensional magnetic field density distribution cloud map

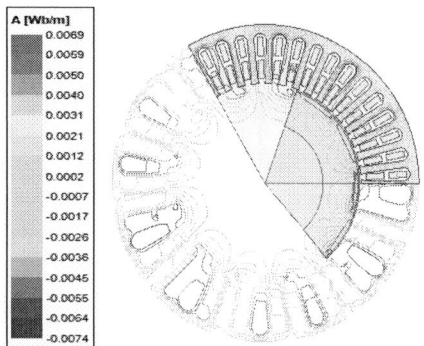

Fig. 7 Cloud map of the distribution of magnetic lines of magnetic field

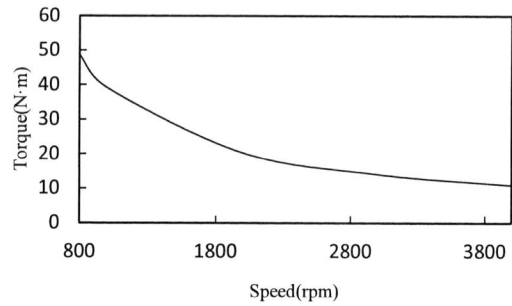

Fig. 8 spindle motor torque curve

(2) No-load characteristics of the spindle motor

When the motor is running with no load, the permanent magnet on the rotor and the three-phase symmetric winding on the stator will generate a no-load back potential due to electromagnetic induction. In the Maxwell 2D solver, the excitation source is set to the open state, and the no-load back potential of the spindle motor described in this paper is obtained as shown in Fig. 9. Fig. 9 shows that the no-load back-potential waveform changes approximately in a sinusoidal cycle, and its effective value is about 171V. The no-load back potential waveform close to the sinusoidal waveform can reduce the stator loss, stator current and the amount of permanent magnet material, and can improve the efficiency of the motor.

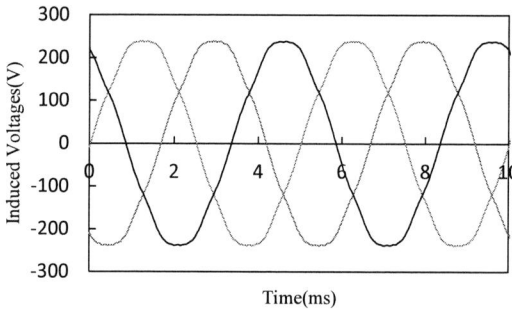

Fig. 9 Waveform diagram of spindle motor no-load back potential

When the motor is running at no load, the interaction between the permanent magnet and the stator core will produce cogging torque. In the Maxwell 2D solver, the excitation source is set to the open state, and the cogging torque of the spindle motor described in this paper is calculated as shown in Fig. 10. According to Fig. 10, it can be seen that the cogging torque waveform changes approximately in a sinusoidal cycle, and the torque peak value is about 0.01N·m, which is much smaller than the rated torque of the spindle motor.

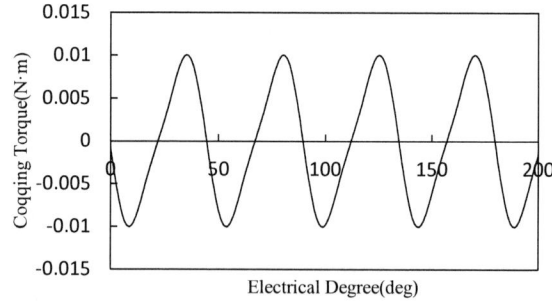

Fig. 10 Spindle motor cogging torque curve

(3) Spindle motor load characteristics

In the Maxwell 2D solver, the delta parameterized analysis can get the motor output power change curve with delta, and the motor efficiency change curve with delta. The output power-delta curve of the spindle motor described in this paper is shown in Fig. 11, and the spindle motor efficiency-delta curve is shown in Fig. 12. According to Fig. 11, as the delta angle increases, the output power of the spindle motor first increases and then decreases. When the delta angle is about 92°, the output power reaches the maximum of about 4.378kW. According to Fig. 12, as the delta angle increases, the spindle motor efficiency first rises straight from 0 and then decreases slowly, and then drops to 0 after a certain angle. When the delta angle is about 25°, the maximum efficiency is about 95%.

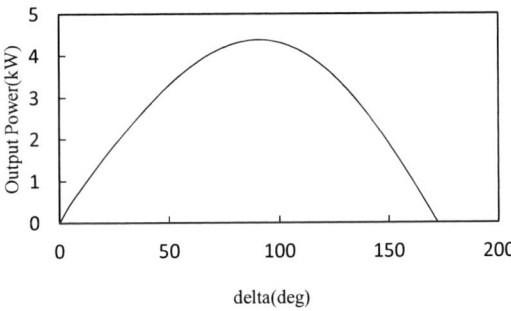

Fig. 11 Spindle motor output power-delta curve

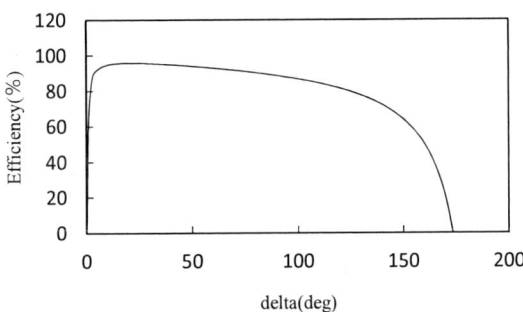

Fig. 12 Spindle motor efficiency-delta curve

IV. MAGNETO-THERMAL COUPLING ANALYSIS

This paper adopts the single-phase coupling method of electromagnetic field and temperature field to analyze the influence of the heat generated by the spindle motor loss on the aerostatic spindle structure. The calculated motor loss is applied as a heat source to the temperature field for temperature analysis, and then the thermal deformation analysis is performed in the structure field.

A. Thermal model

The copper loss and iron loss obtained in the electromagnetic calculation are applied to the thermal analysis model of the aerostatic spindle in the form of heat generation rate. The heat generation rate formula of the motor is:

$$Q = W_q/V \qquad (1)$$

In the formula, Q represents the heat generation rate of the motor, the unit is W/m^3, W_q represents the loss of the motor, the unit is W, and V represents the volume of each part of the motor, the unit is m^3.

The establishment of the aerostatic spindle thermal analysis model is shown in Fig. 13.

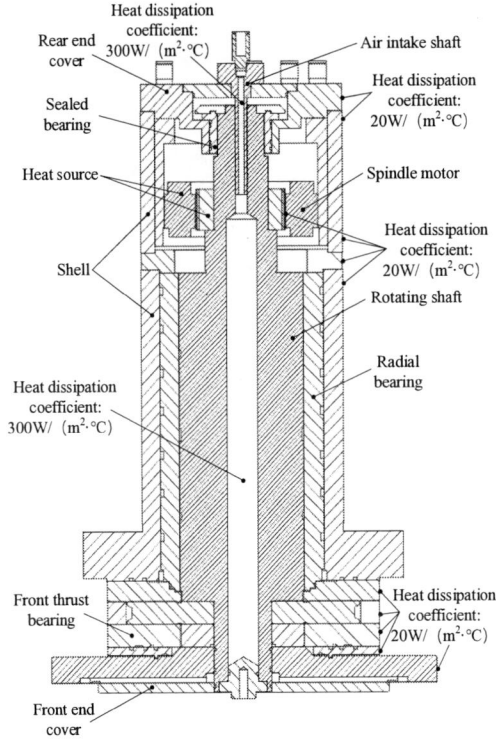

Fig. 13 Aerostatic spindle thermal analysis model

Use the finite element analysis software ANSYS Workbench to analyze the model. First, establish a steady-state thermal analysis module, and import the aerostatic spindle thermal analysis model into this module. Then, the loss calculated by the electromagnetic field is used as the heat source to be imported to the corresponding parts. The mesh type is tetrahedral mesh, and the precision is set to 3mm. The initial temperature is set to 22°C. The boundary conditions mainly set heat convection and heat conduction, and the radiation heat is ignored. The aerostatic spindle shell and the outside air are convective heat exchange, and the heat dissipation coefficient is set to 20 W/(m$^2 \cdot$ °C); thermal convection is generated in the air gap between the stator and rotor of the spindle motor, the equivalent thermal convection coefficient is used to replace the flowing air in the air gap, and the equivalent thermal convection coefficient is set to 20 W/(m$^2 \cdot$ °C); in addition, there is fluid in the rotating shaft and the intake shaft, and the heat dissipation coefficient is set as 300 W/(m$^2 \cdot$ °C). Finally, a simulation analysis is carried out. The cloud map of the simulation calculation results is shown in Fig. 14. The surface temperature curve of the aerostatic spindle shell is shown in Fig. 15. The temperature curve of the inner surface of the rotating shaft is shown in Fig. 16.

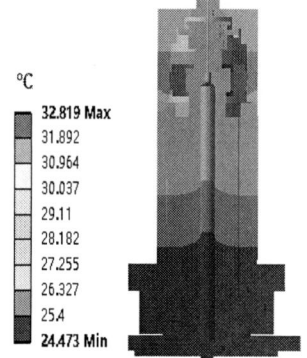

(a) Cloud map of aerostatic spindle temperature distribution

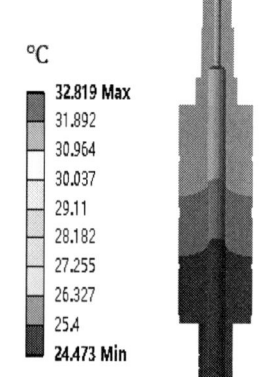

(b) Cloud map of the temperature distribution of the rotating shaft

Fig. 14 Temperature distribution cloud map

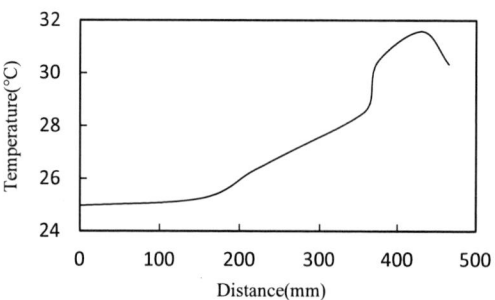

Fig 15 The surface temperature curve of the aerostatic spindle shell

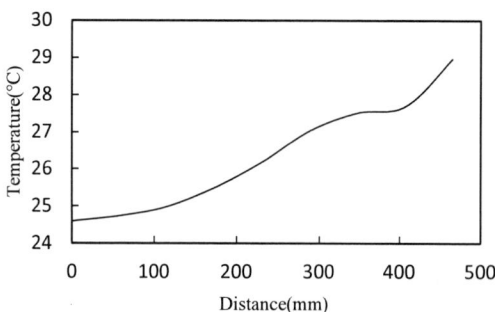

Fig. 16 The temperature curve of the inner surface of the rotating shaft

It can be seen from Fig. 14(a) that the main temperature rise of the aerostatic spindle occurs on the spindle motor windings, and the maximum temperature is 32.819°C. The temperature of the motor winding is transferred to the motor cooling jacket, the aerostatic spindle shell and other components in the form of heat conduction. From Fig. 14(b) and Fig. 16, it can be seen that the highest temperature on the rotating shaft is at its top, reaching 28.988°C, and the temperature is rising from the lower end to the upper end. It can be seen from Fig. 15 that the surface temperature of the aerostatic spindle shell is generally rising from the front cover to the rear cover. The surface temperature of the shell near the spindle motor stator is the highest, reaching 31.581°C.

B. Structural analysis

Because the temperature distribution of the aerostatic spindle is not evenly distributed when it is working, it will produce certain thermal stress and lead to structural deformation. The thickness of the wafer is generally tens of microns. When the wafer is thinned, if the aerostatic spindle is deformed greatly due to thermal stress, the thinning accuracy will be affected.

After the temperature simulation analysis of the aerostatic spindle is completed, the analysis model and simulation results are imported into the steady-state structural analysis module of ANSYS Workbench. The mesh type is tetrahedral mesh with an accuracy of 3mm. Use the front and rear shell surfaces and the outer surface of the bearing as a fixed support. Finally, a simulation analysis is carried out. The cloud map of the simulation calculation results is shown in Fig. 17. The thermal deformation curve of the inner surface of the rotating shaft is shown in Fig. 18. The thermal deformation curve of the lower surface of the front cover is shown in Fig. 19.

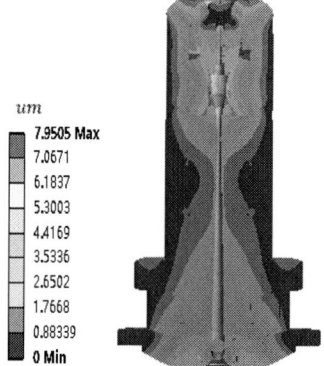

(a) The distribution cloud map of the thermal deformation of the aerostatic spindle

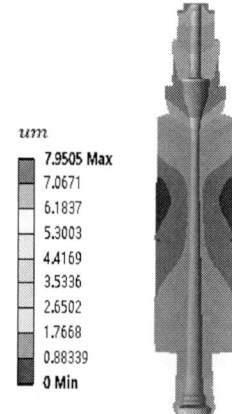

(b) Cloud map of thermal deformation distribution of rotating shaft

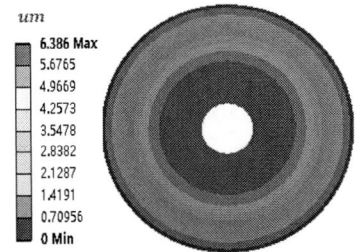

(c) Cloud map of thermal deformation distribution on the lower surface of the front cover

Fig. 17 Thermal deformation distribution cloud map

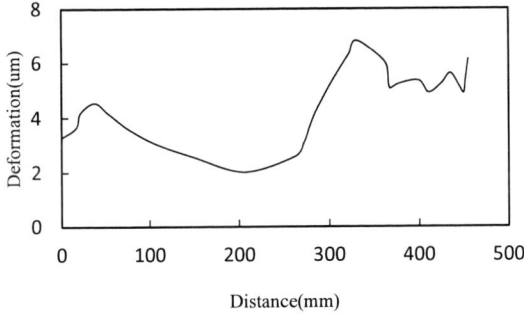

Fig. 18 Thermal deformation curve of the inner surface of the rotating shaft

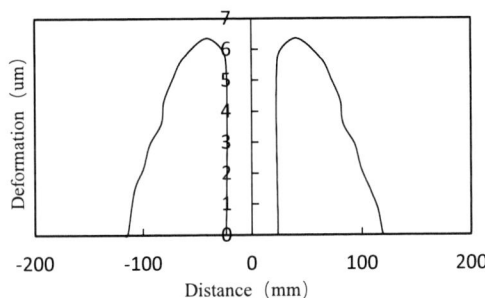

Fig. 19 Thermal deformation curve of the lower surface of the front cover

From Fig. 17(a), it can be seen that the maximum thermal deformation of the aerostatic spindle occurs on the rotor of the spindle motor, which can reach up to 7.9505um. The more uneven the temperature distribution, the greater the thermal stress. It can be seen from Fig. 17(b) and Fig. 18 that the part of the rotating shaft next to the rotor of the spindle motor has the largest thermal deformation, reaching 6.8272um. Since the bottom of the aerostatic spindle front cover is connected to the grinding wheel, it is particularly important to analyze the deformation of the front cover. It can be seen from Fig.

17(c) and Fig. 19 that the thermal deformation of the front cover next to the rotating shaft is the largest, reaching 6.3621um. When the aerostatic spindle is working, its deformation has little effect on grinding accuracy.

V. CONCLUSION

In this paper, the method of magneto-thermal coupling and the finite element analysis method are used to analyze the electromagnetic field of the spindle motor and the temperature field and structure field of the overall structure of the aerostatic spindle. By electromagnetic simulation, the magnetic field characteristics, the no-load characteristics of the motor and the load characteristics of the motor under the rated power are analyzed, and the magnetic density distribution cloud map, magnetic line distribution cloud map, no-load back potential, cogging torque, torque value at different speeds, power and efficiency under different loads and other important parameters. By the simulation of temperature field and structural field, the temperature distribution and thermal deformation distribution of the aerostatic spindle and the rotating shaft are analyzed, and the temperature rise and thermal deformation law, as well as the maximum temperature rise and thermal deformation area are summarized. This paper can give some suggestions for the design and optimization of the spindle motor and aerostatic spindle structure.

REFERENCES

[1] M. Feil, C. Alder, G. Klink,et al. Ultra thin ICs and MEMS elements: techniques for wafer thinning, stress-free separation, assembly and interconnection[J]. Microsystem Technologies, 9(3):176-182.

[2] Gurnett K, Adams T. Ultra-thin semiconductor wafer applications and processes[J]. 2006, 19(4):38-40.

[3] Xinhao Luo,Bin Han,Xuedong Chen,Xueping Li,Wei Jiang. Multi-physics modeling of tunable aerostatic bearing with air gap shape compensation[J]. Tribology International,2021,153.

[4] D. Joo, J. Cho, K. Woo, B. Kim and D. Kim, "Electromagnetic Field and Thermal Linked Analysis of Interior Permanent-Magnet Synchronous Motor for Agricultural Electric Vehicle," in IEEE Transactions on Magnetics, vol. 47, no. 10, pp. 4242-4245, Oct. 2011.

[5] H. Yeo, H. Park, J. Seo, S. Jung, J. Ro and H. Jung, "Electromagnetic and Thermal Analysis of a Surface-Mounted Permanent-Magnet Motor with Overhang Structure," in IEEE Transactions on Magnetics, vol. 53, no. 6, pp. 1-4, June 2017.

[6] J. -H. Woo, T. -K. Bang, H. -K. Lee, K. -H. Kim, S. -H. Shin and J. -Y. Choi, "Electromagnetic Characteristic Analysis of High-Speed Motors With Rare-Earth and Ferrite Permanent Magnets Considering Current Harmonics," in IEEE Transactions on Magnetics, vol. 57, no. 2, pp. 1-5, Feb. 2021.

[7] X. Zhu, X. Wang, C. Zhang, L. Wang and W. Wu, "Design and Analysis of a Spoke-Type Hybrid Permanent Magnet Motor for Electric Vehicles," in IEEE Transactions on Magnetics, vol. 53, no. 11, pp. 1-4, Nov. 2017.

[8] A. Koronides, C. Krasopoulos, D. Tsiakos, M. S. Pechlivanidou and A. Kladas, "Particular Coupled Electromagnetic, Thermal, Mechanical Design of High-Speed Permanent-Magnet Motor," in IEEE Transactions on Magnetics, vol. 56, no. 3, pp. 1-5, March 20

Copper Filling of High Aspect Ratio Through Ceramic Holes: Effect of Convection on Electrochemical Behavior of Additives

Wang Qing
School of Mechanical Science & Technology
Huazhong University of Science and Technology
Wuhan, China
wangq_hust@163.com

Liu Jiaxin
School of Mechanical Science & Technology
Huazhong University of Science and Technology
Wuhan, China
stu_liujx@163.com

Wu Yilin
School of Mechanical Science & Technology
Huazhong University of Science and Technology
Wuhan, China
francis510@yeah.net

Chen Mingxiang
School of Mechanical Science & Technology
Huazhong University of Science and Technology
Wuhan, China
chimish@hust.edu.cn

Abstract—The electrochemical behavior of additives is severely affected by convection, which determines the filling process of through ceramic holes (TCHs). VMS contained a high concentration of cupric and a low concentration of H_2SO_4 was prone to achieve a good leveling effect. The additives including thiazolinyl polydipropyl sulfonate (SH110) as the accelerator, polyethylene glycol (PEG) as the suppressor, and nitrotetrazolium blue chloride (NBT) as the leveler, respectively. The electrochemical measurements such as galvanostatic and cyclic voltammetric measurements were employed to evaluate the electrochemical behaviors of additives. The additive mixture PEG+SH110+NBT significantly promotes copper deposition under weak convection and inhibits copper deposition under strong convection, which is suitable for the defect-free filling of TCHs with an aspect ratio larger than 4.75, and the copper film on the ceramic substrate is smooth and bright.

Keywords—Through ceramic holes; convection; High aspect ratio; electrochemical behavior; Defect-free filling

I. INTRODUCTION

The direct plated copper (DPC) ceramic substrate is usually applied to high power packaging such as deep UV LED[1]. With the development of power devices towards high frequency, miniaturization, and high density, the ceramic substrate is also optimized. DPC ceramic substrate can achieve vertical interconnection by through-holes drilled by laser and filled with copper. Generally, the through ceramic holes (TCHs) have multiple sizes to adjust to various functions. High aspect ratio (AR) TCHs can meet the demand of high-density packaging, and its filling quality directly affects the device's reliability[2]. Therefore, the high-quality filling of high AR TCHs has become a key point of packaging.

Acidic copper sulfate solutions are the predominant virgin make-up solution (VMS) for the copper filling of through-holes. However, only $CuSO_4$ and dilute H_2SO_4 cannot obtain high-quality filling performance, so that additives are usually used to optimize the filling process and improve the filling quality. Additives include accelerator, suppressor, and leveler, where Cl^- is of great important to act as the anchor point to achieve the tight adsorption on the

surface to be electroplated[3][4][5]. The effect of additives is affected by the composition of VMS, current distribution, and convection. The VMS that contains high cupric concentration and low sulfate acid concentration has excellent leveling ability and high limiting current density, which is beneficial to ensure the flatness of the copper film and high filling efficiency in high current density electroplating. For a specific VMS, direct current electroplating results in a strong edge effect, and the current density at the through-hole mouth is high. The leveler plays an important role, mainly adsorbing on the through-hole mouth to prevent the through-hole mouth from preferentially closing, and voids forming in the through-hole center to improve the throwing power (TP). In addition, most additives are also sensitive to convection. Depending on additives' synergistic/antagonistic effect, the filling process can be controlled, and a variety of TP can be formed. By optimizing the concentration of additives, it is helpful to form defect-free filling of the through-holes[6][7].

With the increase of the AR, the convection in the TCHs center weakens, and the convection difference between the TCHs center and the mouth increases gradually. In the Haring-Blum cell, air agitation is customarily used to realize the convection of the electrolyte, which flow rate directly affects the convection around the TCHs mouth, and the convection in the TCHs center also changes[8]. Therefore, there are different adsorption characteristics and synergistic/antagonistic effects between additives. Notably, few papers reported the relationship between electrochemical behaviors of additives and the filling performance of TCHs with multiple ARs under different convection conditions.

Herein, copper filling of various AR TCHs under different convection was investigated. The VMS has a high cupric concentration, and the suppressor, accelerator and leveler were chosen as polyethylene glycol (PEG), thiazolinyl polydipropyl sulfonate (SH110), and nitrotetrazolium blue chloride (NBT), respectively[9][10][11]. Minute studies on the effect of convection on the electrochemical behaviors of the additives were made. The filling quality and the microstructures of the plated copper film were examined.

II. EXPERIMENTAL AND METHODS

A. Electrochemical measurements

The electrochemical behaviors of additives were characterized by galvanostatic measurement (GMs) and voltammetric curve measurement (CV) using a CS2350H (Corretest) potentiostat with a three-electrode system at 25°C. The tested electrolyte had a volume of 100 mL, and it was collocated with a 4mm-diameter platinum rotating disk electrode (Pt-RDE) as a working electrode, a platinum plate electrode with an area of $2\times2cm^2$ as a counter electrode, and a mercurous sulfate electrode (MSE) as a reference electrode. Before each electrochemical test, a thin copper film with a thickness of 500nm was electrodeposited on the Pt-RDE to prepare a copper rotating disk electrode (Cu-RDE) by VMS (0.7 M $CuSO_4$ and 0.61 M H_2SO_4). Additive formulations are shown in Table 1.

The effect of convection on the cathodic polarization of additives was evaluated by GMs, and the convection difference depends on the rotation speed of the Cu-RDE. The Cu-RDE rotation speeds of 10 rpm, 100 rpm, and 500 rpm were performed to simulate the convection condition near the TCHs center with the decreasing ARs, and 1000 rpm was chosen to approximate the convection around the TCHs mouth. The potential difference ($\Delta\eta$) of the Cu-RDE at two rotation speeds was defined by formula (1), which could manifest the influence of additives on copper electrodeposition and the growth trends of the copper film. $\Delta\eta$

$$\Delta\eta=\eta(10/100/500 \text{ rpm})-\eta(1000 \text{ rpm}) \quad (1)$$

The positive $\Delta\eta$ indicated that the copper electrodeposition was promoted with weak convection while inhibited under strong convection. The filling quality was improved with the increasing $\Delta\eta$, and the plating solution was beneficial for the defect-free filling of TCHs. The adsorptive characteristic of additives was characterized by CV measurement, and the parameter range was carried out between 0.2 V vs. MSE and -0.65 V vs. MSE.

B. TCHs filling

The ceramic substrate selected for TCHs filling has a thickness of 380 μm and 500 μm, and the dimension is 6×6 cm². Each kind of ceramic substrate contains 80 μm and 100 μm diameter TCHs. Before TCHs filling, the following pretreatment should be carried out, including clean the substrate with deionized water, magnetron sputtering 2-3μm thick copper seed layer, remove the smear by degreasing, rough the surface by micro etching, and activate surface by pickling.

Two phosphor copper sheets were used as anodes, and they were placed on both sides of the Haring-Blum cell. The ceramic substrate with TCHs was suspended vertically in the middle of the Haring-Blum cell. The additives were diluted from stock solutions and were added to the VMS quantitatively. The 1.5 L homogenized electrolyte was used to fill the TCHs with a current density of 2ASD. Air agitation with a flow rate of 1.6 L/min and 2 L/min were used to achieve different convection and ensure good mass transfer.

C. Characterization

The morphology of the copper electroplating film was observed by SEM (Nova NanoSEM 450). The cross-section of the TCHs was observed by using an optical microscope (VHX-6000).

TABLE I. ADDITIVE FORMULATION OF THE TESTED ELECTROLYTE

Electrolyte No.	Additives		
	PEG ppm	SH110 ppm	NBT ppm
1	0	0	0
2	200	0	0
3	0	8	0
4	0	0	5
5	200	8	5

III. RESULTS AND DISCUSSION

CV measurements were carried out with the electrolytes in Table 1, and a graphical representation of results is shown in Fig. 1. Additives affect the stripping peak area and the starting deposition potential (SDP). VMS has the largest stripping peak area (2.498) and the most minor initial deposition potential, SDP_a is -0.38 V vs. MSE, indicating that the copper deposition is not inhibited and the deposition amount is significant. SH110 results in a slight decrease in the amount of copper deposition, with a stripping peak area of 1.357, and its SDP is close to that of VMS, indicating that SH110 alone has a slight inhibition effect on copper deposition. Both PEG and NBT significantly inhibit copper deposition, showing a significant decrease in the stripping peak area (0.1 and 0.136, respectively), and the SDP moves negatively to SDP_c (-0.52 V vs. MSE) and SDP_b (-0.41V vs. MSE). When PEG,SH110, and NBT work together, the inhibition effect on copper deposition is much more significant than other electrolytes, the stripping peak area is only 0.055, and the SDP is close to that of PEG, proving that the additive mixture has a strong inhibition effect on copper deposition, which is conducive to improving the quality of the copper film.

Based on the CV measurements, the GMs measurements of each additive and their mixture were carried out with the Cu-RDE rotation speeds of 100 rpm and 1000 rpm. The ceramic substrates were electroplated in the corresponding electrolyte with a current density of 2ASD and air agitation flow rate of 1.6 L/min, and the micromorphology of the copper surface was observed. As depict in Fig.2, the overpotential of SH110, NBT, and PEG+ SH110+NBT (the mixture) changes with the convection, their $\Delta\eta$ are all positive and greater than 20 mV, which indicates that they are favorable for TCHs filling. While the overpotential of VMS and PEG is hardly changed since $\Delta\eta$ is 6.9 mV (VMS) and -4.8 mV (PEG), respectively. This phenomenon indicates that VMS and PEG cannot promote the filling of TCHs. Moreover, the order of cathodic potential at 100 rpm is SH110 (-0.434 V vs. MSE), VMS (-0.456 V vs. MSE), NBT (-0.555 V vs.

MSE), the mixture (-0.673 V vs. MSE) and PEG (-0.683 V vs. MSE), which is closely related to the micromorphology of the copper surface. The copper micromorphology has a large grain size and rough surface by VMS, micro-fold by SH110, micro-hollow by SH110, nodulation by NBT, and mirror-bright by the mixture of three additives. It shows that the mixture of three additives can obtain good film quality.

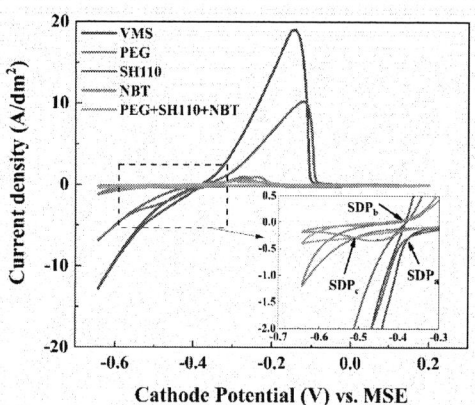

Fig. 1. CV curves of different electrolyte with the Cu-RDE rotation speed of 1000 rpm.

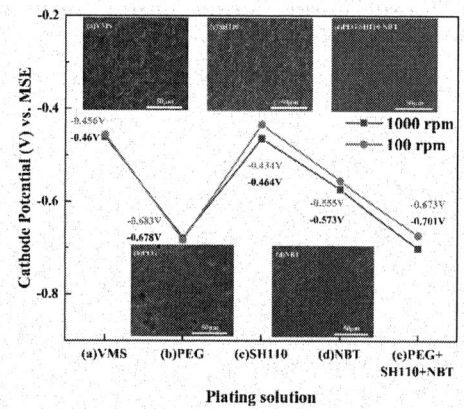

Fig. 2. Cathodic potential and SEM micromorphology of the copper surface treated with different electrolytes.

The CV curves under different convection intensities were used to depict the electrochemical behaviors of the mixture of three additives with different convection. As shown in Fig.3, the faster rotation speed of the Cu-RDE lead to the SDP shift negatively, since SDP_1 is around -0.47 V vs. MSE with Cu-RDE rotation speed of 10 rpm, and SDP_2 is around -0.51 V vs. MSE with Cu-RDE rotation speed between 100 rpm and 1000 rpm. With strong convection, the additives strongly inhibited copper deposition from 0.2 V vs. MSE to SDP_2, which indicates that the additives adsorption is mass transfer controlled. Furthermore, the stripping peak area decreases with convection, and it has the value of 0.017 at 10 rpm with a large amount of copper deposition, which decreases sharply to 0.011 and 0.009 at 100 rpm and 500 rpm, respectively. When the rotation speed further increases to

1000 rpm, the stripping peak area only has the value of 0.06, far less than that at 10 rpm.

In the process of TCHs filling, the air agitation brings strong mass transfer to the electroplating electrolyte. For the ceramic substrate with the same thickness, the convection in the hole center decreases with the AR of the TCHs. According to the above results, the TCHs with large AR will get more copper deposition in the center of the TCHs, so that they will be closed preferentially to reduce the defects such as voids.

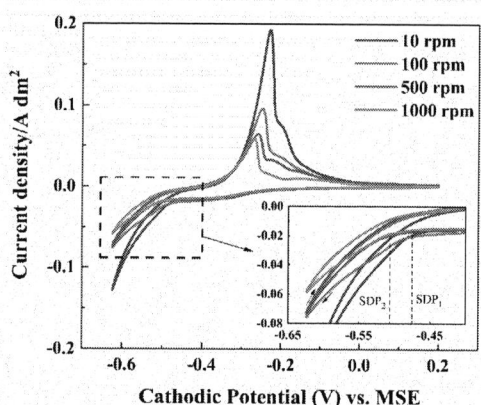

Fig. 3. CV curves of the mixture of three additives. The rotation speeds were 10 rpm, 100 rpm, 500 rpm, and 1000 rpm.

E-t curves in Fig.4 show the interaction among the mixture of three additives; a certain amount of PEG, SH110, and NBT is added to the test vessel every 1000 s of GMs. In the first 1000 s, the adsorption performance of PEG is affected by convection. It is not conducive to the adsorption of PEG on the surface with weak convection, leading to the poor inhibition effect of PEG on copper deposition, and $\Delta\eta$ between 10 rpm and 1000 rpm is up to 124 mV. With the increase of Cu-RDE rotation speed, the adsorption of PEG on the Cu-RDE surface is improved, resulting in the enhancement of cathode polarization, since $\Delta\eta$ between 100 rpm and 1000 rpm sharply decreases to 13 mV. When the convection increases to 500 rpm ~ 1000 rpm, the cathode polarization reaches the limit. Therefore, stronger convection is helpful for PEG adsorption and thus inhibits copper deposition.

SH110 is added to the tested electrolyte at 1000 s. SH110 contains the sulfonic acid group and thiazolyl group, which play an acceleration effect and leveling effect. The cathodic polarization decreases at all rotating speeds, indicating that the significant effect of SH110 at this concentration is the acceleration effect, but not the leveling effect. In addition, the acceleration effect of SH110 is affected by convection. With the increase of convection, the acceleration effect first increases and then decreases, which indicates that weaker convection is conducive to SH110 to play the role of acceleration. When the rotation speed changes from 500 rpm to 1000 rpm, the difference in the potential change process is slight, and the final cathodic potential is the same, which suggests that SH110 has no significant effect on polarization when the rotation speed exceeds 500 rpm. The

978-1-6654-1392-3/21 $31.00 © 2021 IEEE

$\Delta\eta$ is 6 mV (10 rpm ~ 1000 rpm) and 15 mV (100 rpm ~ 1000 rpm).

NBT is a kind of leveling agent based on dyes. When NBT is added to the solution at 2000 s, the cathodic polarization under all convection is significantly enhanced. When the rotation speed is between 10 rpm and 500 rpm, polarization is enhanced with the convection, but with the rotation speed over 500 rpm, the polarization also reaches the limit value. The final $\Delta\eta$ is 46 mV (10 rpm ~ 1000 rpm) and 33mV (100 rpm ~ 1000 rpm), respectively. The results show that strong convection contributes to the cathodic polarization of NBT, the whole filling rate is charge-transfer-controlled with positive $\Delta\eta$, and the plating electrolyte including the three additives is suitable for defect-filling of TCHs with large AR.

Experiments verified the filling performance of the mixture of three additives, the air flow rate of 1.6 L/min and 2 L/min was used, and the TCHs were filled with a current density of 2 ASD in 120 min. The experimental results are shown in Fig. 5. In Fig. 5 (a), stronger air agitation (2 L/min) is not conducive to the filling of TCHs, and a large void is formed in the TCHs center. In Fig. 5 (b-e), using an appropriate air flow rate (1.6 L/min) can achieve defect-free filling of TCHs with AR of 4.75 (380 µm thick ceramic substrate, 80µm diameter TCHs), 5 (500 µm thick ceramic substrate, 100µm diameter TCHs), and 6.25 (500 µm thick ceramic substrate, 80µm diameter TCHs), while voids form in TCHs with AR of 3.8 (380 µm thick ceramic substrate, 100µm diameter TCHs). Therefore, the plating electrolyte with the mixture of three additives is suitable for filling TCHs with AR greater than 4.75. The weak convection in the TCHs center promotes the deposition of copper. Meanwhile, the copper film on the surface of the ceramic substrate is smooth and bright without the dimple. This phenomenon suggests that the filling performance of TCHs is affected by the convection since the convection difference between the TCHs center and the TCHs mouth increases with AR under

the same air agitation conditions. Therefore, the adsorption characteristics of additives are affected, and the growth rate of different parts of the TCHs is adjusted by additives.

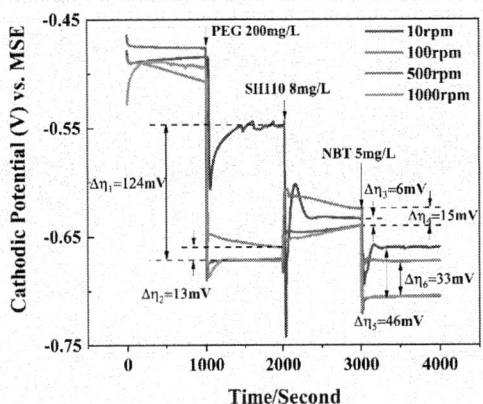

Fig. 4.   E-t curves of additives at multiple convection with the current density of 2 ASD.

IV. CONCLUSION

Effect of convection on the electrochemical behavior of additives for high AR TCHs and the filling performance of additives were investigated.

1） The absorbance of PEG is less on the cathodic surface with weak convection, leading to lower cathodic polarization. Weaker convection is conducive to SH110 to play the role of acceleration, which promotes the deposition of copper. Strong convection contributes to the cathodic polarization of NBT, and the filling rate is charge-transfer-controlled.

2） The filling performance of TCHs is affected by the convection, which significantly promotes copper deposition at the TCHs center (weak convection) and

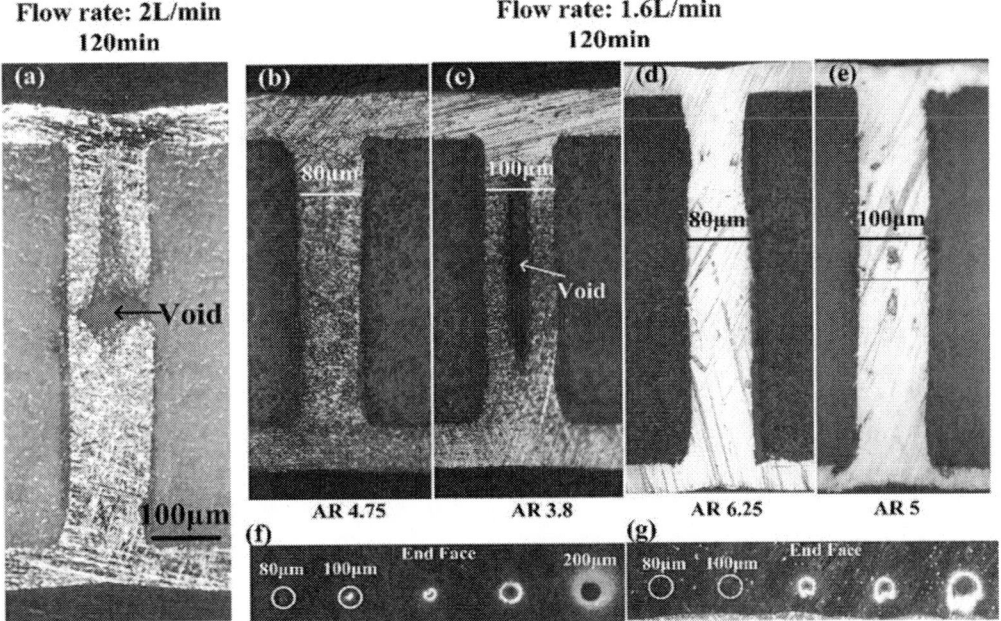

Fig.5.   Cross-section of the TCHs.

inhibits copper deposition around the TCHs mouth (strong convection).

3) The plating electrolyte 0.7M $CuSO_4$, 0.61M H_2SO_4, 55ppm Cl^-, 200ppm PEG, 8ppm SH110, and 5ppm NBT is suitable for the defect-free filling of TCHs with AR larger than 4.75, and the copper film on the ceramic substrate is smooth and bright.

ACKNOWLEDGMENT

This work was supported by the Key Research and Development Projects in Hubei Province (No: 2020BAB068) and National Natural Science Foundation of China (51775219).

REFERENCES

[1] Z. Yang, Q. Sun, H. Cheng, X. Liu, Y. Peng, and M. Chen, "Preparation of three-dimensional ceramic substrate by multiple electroforming for UV-LED hermetic packaging," *Ceram. Int.*, vol. 45, no. 17, pp. 22022–22028, 2019, doi: 10.1016/j.ceramint.2019.07.218.

[2] H. Wu, Y. Wang, Z. Li, and W. Zhu, "Investigations of the electrochemical performance and filling effects of additives on electroplating process of TSV," *Sci. Rep.*, vol. 10, no. 1, pp. 1–12, 2020, doi: 10.1038/s41598-020-66191-7.

[3] F. Wang *et al.*, "Effect of molecular weight and concentration of polyethylene glycol on through-silicon via filling by copper," *Microelectron. Eng.*, vol. 215, no. May, 2019, doi: 10.1016/j.mee.2019.111003.

[4] W. P. Dow, C. W. Lu, J. Y. Lin, and F. C. Hsu, "Highly selective Cu electrodeposition for filling through silicon holes," *Electrochem. Solid-State Lett.*, vol. 14, no. 6, 2011, doi: 10.1149/1.3562278.

[5] X. Wang *et al.*, "Effects of 2,2-Dithiodipyridine as a Leveler for Through-Holes Filling by Copper Electroplating," *J. Electrochem. Soc.*, vol. 166, no. 13, pp. D660–D668, 2019, doi: 10.1149/2.0461913jes.

[6] A. T. Tepzz, "Method of filling through-holes," vol. 1, no. 19, pp. 1–13, 2014.

[7] T. D. A. Jones, A. Bernassau, D. Flynn, D. Price, M. Beadel, and M. P. Y. Desmulliez, "Analysis of throwing power for megasonic assisted electrodeposition of copper inside THVs," *Ultrasonics*, vol. 104, no. January, p. 106111, 2020, doi: 10.1016/j.ultras.2020.106111.

[8] C. Wang and Y. Chen, "Numerical simulation and experiments to improve throwing power for practical PCB through-holes plating," *Circuit World*, no. May 2018, 2019, doi: 10.1108/CW-05-2018-0033.

[9] W.-P. Dow, M.-Y. Yen, W.-B. Lin, and S.-W. Ho, "Influence of Molecular Weight of Polyethylene Glycol on Microvia Filling by Copper Electroplating," *J. Electrochem. Soc.*, vol. 152, no. 11, p. C769, 2005, doi: 10.1149/1.2052019.

[10] W.-P. Dow *et al.*, "Through-Hole Filling by Copper Electroplating," *J. Electrochem. Soc.*, vol. 155, no. 12, p. D750, 2008, doi: 10.1149/1.2988134.

[11] C. Wang, J. Zhang, P. Yang, B. Zhang, and M. An, "Through-Hole Copper Electroplating Using Nitrotetrazolium Blue Chloride as a Leveler," *J. Electrochem. Soc.*, vol. 160, no. 3, pp. D85–D88, 2013, doi: 10.1149/2.035303jes.

The Effect of Toughening Agents on Capillary Underfill in the Flip Chip Package

Yuanyuan Yang
Shenzhen Institute of Advanced Electronic Materials
Shenzhen Institute of Advanced Technology, Chinese Academy of Sciences
Shenzhen, China
yy.yang@siat.ac.cn

Houya Wu
Shenzhen Institute of Advanced Electronic Materials
Shenzhen Institute of Advanced Technology, Chinese Academy of Sciences
Shenzhen, China
hy.wu1@siat.ac.cn

Tao Peng
Shenzhen Institute of Advanced Electronic Materials
Shenzhen Institute of Advanced Technology, Chinese Academy of Sciences
Shenzhen, China
tao.peng1@siat.ac.cn

Jinbao Yang
School of Materials Science and Engineering
Shandong University of Science and Technology
Qingdao, China
jb.yang@siat.ac.cn

Bin Wang
Nano Science and Technology Institute
University of Science and Technology of China
Suzhou, China
wang.bin@siat.ac.cn

Haoliang Lin
Department of intelligent manufacturing
WuYi University
Jiangmen, China
hl.Lin@siat.ac.cn

Gang Li*
Shenzhen Institute of Advanced Electronic Materials
Shenzhen Institute of Advanced Technology, Chinese Academy of Sciences
Shenzhen, China
gang.li@siat.ac.cn

Pengli Zhu*
Shenzhen Institute of Advanced Electronic Materials
Shenzhen Institute of Advanced Technology, Chinese Academy of Sciences
Shenzhen, China
pl.zhu@siat.ac.cn

Rong Sun
Shenzhen Institute of Advanced Electronic Materials
Shenzhen Institute of Advanced Technology, Chinese Academy of Sciences
Shenzhen, China
rong.sun@siat.ac.cn

Abstract— **Toughening agents were introduced in capillary underfill to prevent fillet cracking and promoting reliability in flip-chip package. The effects of toughening agents with typical structures on the performance of capillary underfill such as storage modulus, coefficient of thermal expansion (CTE), glass transition temperature (Tg), adhesion strength, rheological and curing behavior were investigated. It is found that the capillary underfill containing carboxyl terminated rubber showed the highest CTE, lowest adhesion strength, lowest moisture absorption, and highest modulus. The capillary underfill containing silicone elastomer showed low viscosity, fast flow time and proper moisture absorption, fast gel time, CTE, storage modulus and Tg, as well as good adhesion strength in five substrates. Compared with capillary underfill without toughening agent, the viscosity of capillary underfill with toughening agent increases with increasing flow time. The influence of toughening agents with different structures on capillary underfill is varios. Our work provides guidance for selecting the toughening agent for the underfill of the capillary.**

Keywords—Flip-chip; Capillary underfill; Toughening agents

I. INTRODUCTION

Flip chip technology is an effective method to improve packaging density and reliability of electronic products, and to reduce packaging costs due to its outstanding merits of high I/O density, short interconnects, self-alignment, better heat dissipation through the back of the die, smaller footprint, lower profile, high throughput, etc. In a flip-chip package, the active side of a silicon chip is faced down towards and mounted onto a organic substrate,[1-4] as schematically illustrated in Fig. 1.[5] Underfill, usually liquid epoxy resins filled with large amount of SiO_2 filler, is applied in flip-chip

interconnection after cured to minimize the thermal expansion coefficient mismatch between the silicon chip and the substrate on fatigue life. [6-8]

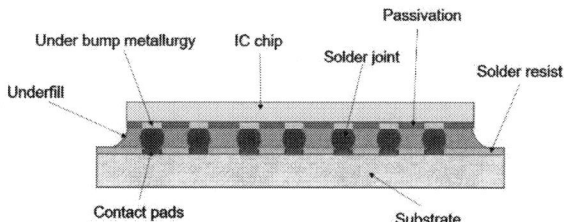

Fig. 1. Generic construction of flip chip package containing capillary underfill.

Capillary underfill is one kind of underfill which mainly injected between the silicon chip and the organic substrate via capillary force.[9,10] One of the critical requirements for the capillary underfill is to provide enough fracture toughness to prevent fillet cracking in flip chip technology.[11] However, the brittle and rigid in nature of cured epoxy resin lead to crack initiation and growth, as well as low impact strength, which limited the applications of epoxy resin in underfill.[12] Nevertheless, the incorporation of a second microphase has been well established into the epoxy polymer to increase its toughness. There are many investigations having been dedicated to increase the fracture toughness of epoxy resin with different mechanisms to adapt its application in underfill. Thermoplastic polymers such as polyimide and polyethersulfone have been introduced into epoxy resin compound to enhance its fracture toughness by forming continuous or co-continuous phase morphology.[13,14] Also, the epoxy resin added poly(acrylonitrile-co-butadiene-co-

978-1-6654-1392-3/21 $31.00 © 2021 IEEE

styrene) (ABS) was found having an enhanced mechanical properties at low ABS contents.[15,16] In addition, inorganic nano-fillers were also added to the epoxy resin to improve its mechanical properties via forming a two-phase microstructure of randomly dispersed silica particles in a continuous epoxy matrix. The toughening mechanism can be explained by the crack pinning/bowing, crack deflection, microcracking and crack bridging.[17,18]

Toughening agents can improve the toughness of the underfill. However, it also affects the physical and chemical properties, processing performance, and interface adhesion strength of the underfill.[19] Five toughening agents with typical structures were selected to systematically investigate the effects of toughening agents on capillary underfill. The flow time, gel time, viscosity, curing curve, CTE and storage modulus before and after the Tg, interfacial adhesion strength with five substrates, and moisture absorption of the capillary underfill with various toughening agents were measured and analyzed. An suitable category of the toughening agent for capillary underfill is obtained, and a universal procedure of evaluating the effect of the additives on capillary underfill in flip chip is established in this paper.

II. EXPERIMENTAL PROCEDURE

A. Preparation of Materials

Bisphenol-F, cycloaliphatic epoxy, and their mixture were used as base epoxy resin. Diethyltoluenediamine was used as the curing agent. Amino-terminated block copolymer, silicone elastomer, silicone block copolymer, carboxyl terminated rubber, acrylic elastomer were added to the capillary underfill as toughening agents. Furthermore, the spherical silica was introduced capillary underfill to reduce its CTE. The capillary underfills with or without toughening agents are shown in Table 1.

Table 1. The capillary underfill and toughening agents used in this study

Sample name	Toughening agent
UF-0	Without toughening agent
UF-1	amino-terminated block copolymer
UF-2	silicone elastomer
UF-3	silicone block copolymer
UF-4	carboxyl terminated rubber
UF-5	acrylic elastomer

B. Preparation of test samples

The CTE test sample was prepared using a PTFE mold with a cuboid groove with the dimension of 8 mm×8 mm×6 mm (height). Similarly, the storage and loss modulus sample was molded into a block of 32 mm×11 mm×3 mm. The capillary underfill was prepared into a sandwich structure on different substrates to test the adhesion strength between the capillary underfill and different substrates. The curing procedure of the abovementioned sample is as follows: the temperature is increased from 25 ℃ to 165 ℃ with 5 ℃/min, keep it at 165 ℃ for 2 h, and then naturally cool to room temperature to obtain a solidified sample.

C. Measurement of Materials Properties

The curing profile was measured by a differential scanning calorimeter (DSC, TA Instrument model Q2000), in which the capillary underfill was heated from 25 ℃ to around 250 ℃ with 5 ℃/min in the sample cell. CTE was investigated in thermo-mechanical analyzer (TMA, Netzsch Instrument model TMA 402 F1) by heating the sample from room temperature to about 250 ℃ at 5 ℃/min, in which the inflection point of thermal expansion was defined as Tg (TMA). Tg (DMA), storage and loss modulus were obtained in a dynamic mechanical analyzer (DMA, TA Instrument model Q80) at dual cantilever mode with 1 Hz sinusoidal strain loading. The adhesion strength on different substrates was measured using a die shear tester (Nordson Instrument system DAGE 4000 PXY). Shear rate dependent viscosity was tested by the modular compact rheometer (Anton Instrument model MCR302), and flow time at 110 ℃ and gel time at 165 ℃ were measured in a microcomputer heating platform (Instrument model JF966-1015). The moisture absorption of specimens was calculated from the change of sample quality after 48h boiling.

III. RESULTS AND DISCUSSION

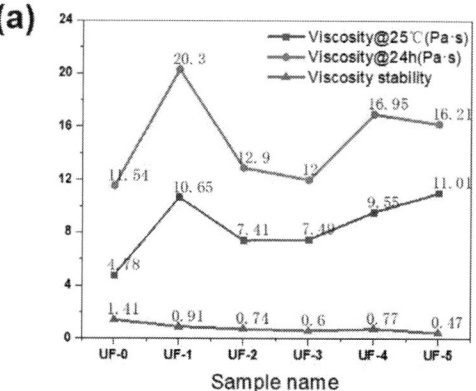

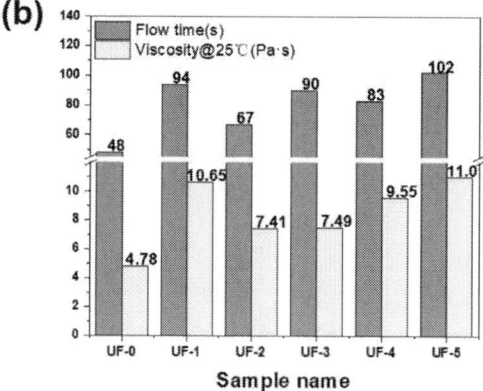

Fig. 2 (a) the viscosity and stability curves of the capillary underfills, (b) the flow time and viscosity comparison chart of capillary underfill with different toughening agents.

The viscosity at 25℃ and 24 hours later as well as the flow time of capillary underfill were tested to study the influence of different structure toughening agents on the rheological properties of capillary underfill. Fig. 2 (a) presents the viscosity and the viscosity stability of the capillary underfill. Since the natural viscosity of the toughening agents is higher than that of epoxy resin, adding toughening agents leads to the increase of initial viscosity of capillary underfill. However, the viscosity stability of the capillary underfill improved after adding the toughening agents with different structures. As the flow time of capillary underfill mainly affects the required dispensing process, a 50 μm gap similar to that in the dispensing process was employed to test the flow time of capillary underfill. The flow time of capillary underfill increases with the addition of toughening agent, as shown in Fig. 2 (b), which is mainly caused by the increased viscosity of the capillary underfill.

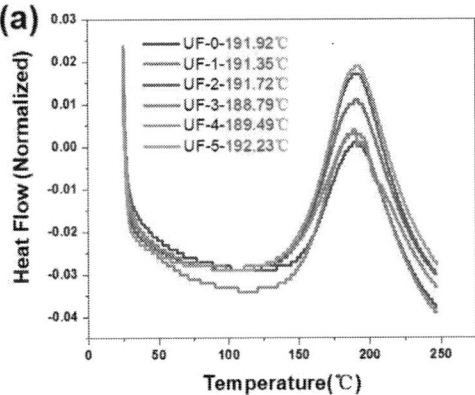

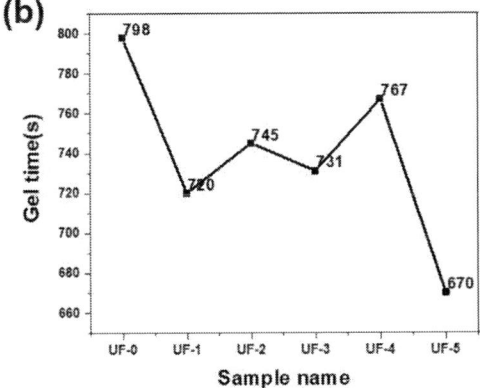

Fig. 3 (a) the DSC curve and (b) the gel time of capillary underfill with different toughening agents.

Fig. 3 (a) presents the curing profile of the capillary underfill without and with various toughening agents. The toughening agents have a negligible effect on the curing peak temperature of the capillary underfills. Fig. 3 (b) displays the relationship between the gel time and the toughening agent at 165 °C. There is a significant decrease in gel time with acrylic elastomer toughening agent, indicating that acrylic elastomer (UF-5) can promote the crosslinking reaction of epoxy resin. Furthermore, the gel time of UF-1 is also significantly lower than that UF-0, which is mainly due to the amino group of the toughening agent promotes the crosslinking reaction of epoxy resin.

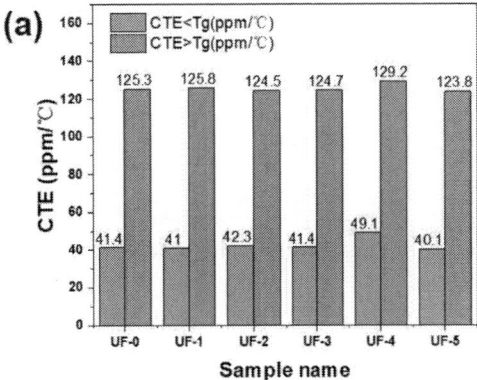

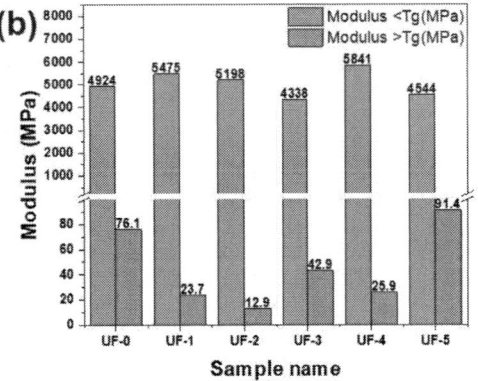

Fig. 4 (a) the CTE before and after Tg, (b) the storage modulus before and after Tg of capillary underfill with different toughening agents.

Fig. 4 (a) shows that all the capillary underfill containing toughening agents almost had no change of CTE (at lower or higher than Tg), except the carboxyl terminated rubber (UF-4) which significantly increased the CTE of the capillary underfill (at lower or higher than Tg). The possible explanation is that the large size of phase separation of traditional carboxyl terminated rubber in epoxy resin changed the crosslinking density of the cured resin. High modulus is necessary for capillary underfill materials to effectively redistribute the stress in the solder joints to the chip and substrate, reducing package warpage. Fig. 4 (b) shows that the different toughening agent has various effects on the storage modulus of the cured underfill before and after Tg. Amino-terminated block copolymer, silicone elastomer and carboxyl terminated rubber toughening agent are preferred for increasing the modulus of the undefill at room temperature, and the acrylic elastomer is suitable for improving the modulus of the undefill after Tg. However, the modulus of capillary underfill before and after Tg decreases with the addition of silicone block copolymer toughening agent.

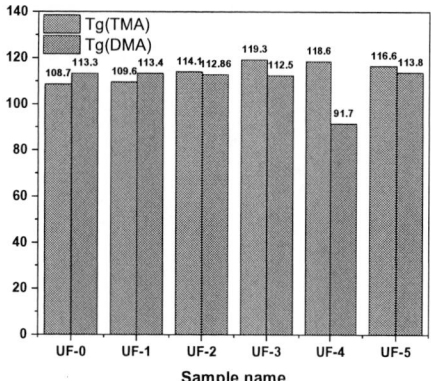

Fig. 5 the Tg of capillary underfill with different toughening agent tested by TMA and DMA.

Tg (measured by TMA and DMA) of capillary underfill is one of the crucial parameters in determining the reliability of the flip chip in thermal cycling or thermo-shock test. The point where the thermal expansion suddenly increases is defined as Tg in TMA, and the peak of the damping coefficient (tan δ) is the Tg in DMA, in which the internal structure of the material changed. Fig. 5 indicates the Tg (measured by TMA and DMA) of capillary underfill. It shows that there is a toughening agent (silicone block copolymer) at which the Tg (TMA) of the cured underfill has the highest value. Furthermore, the cured underfill with toughening agents have higher Tg (TMA) compared with cured underfill without toughening agent. There is a cured underfill with carboxyl terminated rubber toughening agent that shows the lowest Tg (DMA). The cured underfill with toughening agents except carboxyl terminated rubber have little effect on the Tg (DMA).

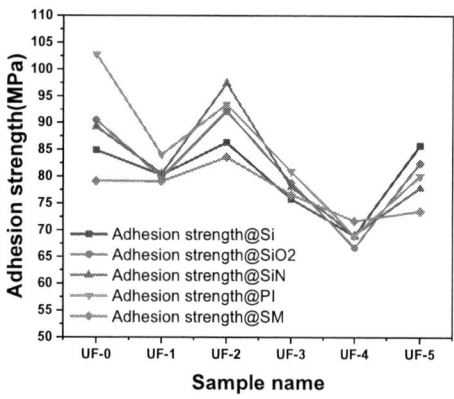

Fig. 6 the adhesion strength in five substrates of capillary underfill with different toughening agents.

Fig. 6 indicates that adding toughening agents can effectively affect the adhesion strength of capillary underfill in five substrates. The study on the effects of toughening agents on the adhesion strength of capillary undefill in Si substrate has shown that there is a trade-off in selecting the species of the toughening agent. On the one hand, silicone elastomer and acrylic elastomer toughening agents are suitable in terms of excellent adhesion strength of capillary

underfill in Si substrate. On the other hand, amino-terminated block copolymer, silicone block copolymer and carboxyl terminated rubber are unfriendly in high adhesion strength of capillary underfill in Si substrate. Among them, the carboxyl terminated rubber has the most obvious damage to adhesion strength of underfill in Si substrate. Considering the silicon oxide, silicon nitride, and polyimide are common passivation layers in flip chip packages, the adhesion strength of capillary underfill with typical structures toughening agents on silicon oxide, silicon nitride and polyimide substrates were investigated. The adhesion strength of capillary underfill with silicone elastomer on SiO_2 substrate is higher than that capillary underfill with the other four toughening agents and UF-0. Similarly, the influence of the toughening agents on the adhesive strength of capillary underfill on SiN substrate is same to SiO_2 substrate. UF-0 has the highest adhesive strength on PI substrates, but the adhesive strength of the capillary underfill decreased with the addition of toughening agent, and the UF-4 with carboxyl terminated rubber has the most obvious decreased in adhesive strength in PI substrates. Solder mask (SM) is directly in contact with underfill in flip chip, so the adhesion strength on SM substrate is a critical parameter to evaluate the anti-cracking of capillary underfill. As shown in Fig. 6, the silicone elastomer is reasonable to improve the adhesion strength of capillary underfill on SM substrate, and the other toughening agent is not suitable for improving the adhesion strength of capillary underfill on SM substrate. As described above, carboxyl terminated rubber is not friendly to the adhesion strength of capillary underfill on all interfaces. However, UF-2 displays better adhesion strength in Si, SiO_2 and SiN substrate because of the good interface compatibility of silicone elastomer with capillary underfill and Si, SiO_2, and SiN substrate.

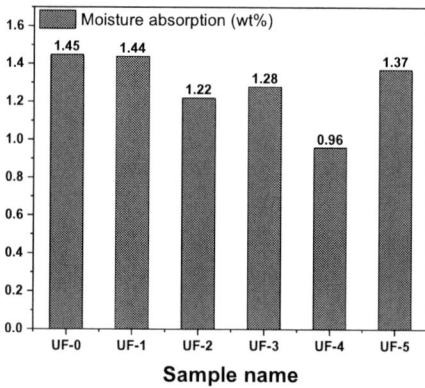

Fig. 7 Percent moisture absorption with toughening agent for the capillary underfill.

The pop-corning delamination or corrosion caused by the moisture absorbed in capillary underfill can result in early failure of the flip chip packaged IC devices. Therefore low moisture absorption is always ideal for the capillary underfill in flip chip package. The results of the moisture absorption percent with toughening agents for the capillary underfill are shown in Fig. 7 where UF-4 with carboxyl terminated rubber is the best in low moisture absorption among these kinds of toughening agents. This result corresponds to the high storage modulus of UF-4.

IV. CONCLUSIONS

The capillary underfill with no toughening agent and with five typical structure toughening agents were prepared and their flow time, gel time, viscosity, curing curve, CTE, storage modulus before and after Tg, interfacial adhesion with five substrates, and moisture absorption were tested. The viscosity of the capillary underfill with toughening agents are greater than that of capillary underfill without toughening agent, and the flow time of capillary underfill is also increased after adding toughening agents, which is mainly caused by the higher viscosity of the toughening agents. However, the viscosity stability of capillary underfill with toughening agents is improved. Consistent with the decrease in the curing peak temperature of capillary underfill with toughening agents, the gel time of capillary underfill is also reduced. The CTE of UF-4 is significantly higher than that of other capillary underfill, indicating that carboxyl terminated rubber toughening agent has an obvious effect on the thermal expansion coefficient of capillary underfill. There have different influences in modulus of capillary underfill by five toughening agents, the carboxyl terminated rubber toughening agent has the most obvious effect on modulus before Tg, and the silicone elastomer has the greatest impact on modulus after Tg. The five toughening agents have dissimilar influences on the interfacial adhesion of capillary underfill, of which carboxyl terminated rubber has the most obvious effect on all substrates. Furthermore, the moisture absorption with toughening agent except carboxyl terminated rubber has higher moisture absorption compared with the underfill wthout toughening agent.

These five typical structure toughening agents have various effects on the property of underfill. UF-2 has low viscosity, flow time and moisture absorption, suitable gel time, CTE, storage modulus and Tg, as well as high adhesion strength in five substrates compared with other underfill with toughening agents, indicating that silicone elastomer is an excellent candidate toughening agent for underfill. However, the addition of carboxyl terminated rubber has a negative effect on the performance of the underfill, indicates that this toughening agent should be used with caution. An suitable category of the toughening agent for capillary underfill is obtained, and a universal procedure of evaluating the effect of the additives on capillary underfill in flip chip is established in this paper.

ACKNOWLEDGMENT

This work was financially supported by the National Natural Science Foundation of China (61704182), Shenzhen basic research plan (JCYJ20190807154409372), National Key R & D Project from Minister of Science and Technology of China (2020YFB0311800).

REFERENCES

[1] C. Wong, S. Lou, Z. Zhang, Flip the chip. Science 2000; 290: 2269.

[2] Z. Zhang, C. Wong, Recent Advances in Flip-Chip Underfill: Materials, Process, and Reliability. IEEE TRANSACTIONS ON ADVANCED PACKAGING 2004; 27: 515-524.

[3] Q. Tong et al., "Recent advances on a wafer-level flip chip packaging process," 2000 Proceedings. 50th Electronic Components and Technology Conference (Cat. No.00CH37070), 2000, pp. 101-106.

[4] R. Thorpe, D. F. Baldwin, B. Smith and L. McGovern, "Yield analysis and process modeling of low cost, high throughput flip chip assembly based on no-flow underfill materials," in IEEE Transactions on Electronics Packaging Manufacturing, vol. 24, no. 2, pp. 123-135, April 2001.

[5] S. Machuga, S. Lindsey, K. Moore and A. Skipor, "Encapsulation Of Flip Chip Structures," Thirteenth IEEE/CHMT International Electronics Manufacturing Technology Symposium, 1992, pp. 53-58.

[6] K. Moon, L. Fan, C. Wong, Study on the Effect of Toughening of No-Flow Underfill on Fillet Cracking. Electronic Components and Technology Conference 2001; 167-173.

[7] M. Paquet, J. Sylvestre, E. Gros and N. Boyer, "Underfill delamination to chip sidewall in advanced flip chip packages," 2009 59th Electronic Components and Technology Conference, 2009, pp. 960-965,.

[8] P. Su, S. Rzepka, M. Korhonen, et al. The effects of underfill on the reliability of flip chip solder joints. Journal of Elec Materi 28, 1017–1022 (1999).

[9] K. Chai and L. Wu, "The underfill processing technologies for flip chip packaging," First International IEEE Conference on Polymers and Adhesives in Microelectronics and Photonics. Incorporating POLY, PEP & Adhesives in Electronics. Proceedings (Cat. No.01TH8592), 2001, pp. 119-123.

[10] P. Miao, Y. Chew, T. Wang and L. Foo, "Flip-chip assembly development via modified reflowable underfill process," 2001 Proceedings. 51st Electronic Components and Technology Conference (Cat. No.01CH37220), 2001, pp. 174-180.

[11] S. Park, C. Feger, Thermal fracture toughness measurement for underfill during temperature change, Microelectronics Reliability, Volume 51, Issue 3, 2011, Pages 685-691.

[12] B. Wetzel, P. Rosso, F. Haupert, K. Friedrich, Epoxy nanocomposites-fracture and toughening mechanisms, Engineering Fracture Mechanics, Volume 73, Issue 16, 2006, Pages 2375-2398.

[13] M. Kimoto, K. Mizutani, Blends of thermoplastic polyimide with epoxy resin: Part II Mechanical studies. Journal of Materials Science 32, 2479–2483 (1997).

[14] K. Mimura, H. Ito, H. Fujioka, Improvement of thermal and mechanical properties by control of morphologies in PES-modified epoxy resins, Polymer, Volume 41, Issue 12, 2000, Pages 4451-4459

[15] A. Torres, I. López-de-Ullibarri, M. Abad, L. Barral, J. Cano, S. García-Garabal, F. Díez, J. López, and C. Ramírez, Study of the effect of poly(acrylonitrile-co-butadiene-co-styrene) on the mechanical properties of an epoxy system. J. Appl. Polym. Sci., 2004, 92: 461-467.

[16] H. Ramakrishna, S. Priya, and S. Rai, Flexural, compression, chemical resistance, and morphology studies on granite powder-filled epoxy and acrylonitrile butadiene styrene-toughened epoxy matrices. J. Appl. Polym. Sci., 2007, 104: 171-177.

[17] Y. Zheng, Y. Zheng, R. Ning, Effects of nanoparticles SiO2 on the performance of nanocomposites, Materials Letters, Volume 57, Issue 19, 2003, Pages 2940-2944, ISSN 0167-577X.

[18] J. Ma, M. Mo, X. Du, P. Rosso, K. Friedrich, H. Kuan, Effect of inorganic nanoparticles on mechanical property, fracture toughness and toughening mechanism of two epoxy systems, Polymer, Volume 49, Issue 16, 2008, Pages 3510-3523

[19] H. L. Tay and C. Q. Cui, "Underfill material requirements for reliable flip chip assemblies," Proceedings of 2nd Electronics Packaging Technology Conference (Cat. No.98EX235), 1998, pp. 345-348.

Preparation and Properties of Low Melting Point Sn-P-F-O-Matrix Phosphor-in-Glass for white LED

Deming Hu[1]
School of Materials Science and Engineering
Xiamen University of Technology
Xiamen, China
hdm9229@163.com

Liang Yang[1*]
School of Materials Science and Engineering
Xiamen University of Technology
Xiamen, China
yangliang86@xmut.edu.cn

An Xie[1]
School of Materials Science and Engineering
Xiamen University of Technology
Xiamen, China
anxie@xmut.edu.cn

Chunyan Cao[1]
School of Materials Science and Engineering
Xiamen University of Technology
Xiamen, China
caoyan_80@xmut.edu.cn

Xiayun Shu[2]
School of Mechanical Engineering
Xiamen University of Technology
CountryXiamen, China
shuxiayun@xmut.edu.cn

Chenrui Fan[1]
School of Materials Science and Engineering
Xiamen University of Technology
Xiamen, China
18065867661@189.cn

Jinrong Deng[1]
School of Materials Science and Engineering
Xiamen University of Technology
Xiamen, China
17859706075@139.com

Kun Chen[1]
School of Materials Science and Engineering
Xiamen University of Technology
Xiamen, China
chenkun19980917@163.com

Abstract—The Phosphor-in-Glass(PIG) for white LED has the characteristics of high physicochemical stability, good heat dissipation and excellent performance, which can effectively avoid the problems of low luminous efficiency, poor heat dissipation and color coordinate shift caused by organic resin cracking, yellowing and phosphor degradation in traditional white LED packaging. In order to prepare PIG with even better properties, the low melting point PIG samples were prepared by melting sintering of Sn-P-F-O glass matrixs and red $CaAlSiN_3:Eu^{2+}$ and green $(BaSr)_2SiO_4:Eu^{2+}$ phosphors. The properties of low melting point PIG were analyzed by PL, XRD, SEM, EDS and LED photoelectric test system. The results show that the phosphor is regularly embedded in the precursor glass, the phosphor morphology is not damaged, and the luminescence characteristics of different phosphors are effectively protected as Sn-P-F-O matrix has low melting point. At the same time, by changing the content of different phosphors and the thickness of the glass, the luminescent performance of the PIG can be flexibly adjusted, and the white LED shows adjustable chromaticity. Finally, the packaged light emitting diodes were obtained approprite CCT of 3000-6000 K and good CRI of 89-91.

Keywords—PIG; white Light emitting diodes; Sn-P-F-O precursor glass

I. INTRODUCTION

Over the last two decades, there has been great deal of development in the widly use of blue chip-excited phosphors for the conversion of white light-emitting diodes (pc-WLEDs). Compared to previous LEDs, these pc-WLEDs have many outstanding features in terms of lighting applications and performance, Such as Long service life and relatively high energy efficiency, environmental friendliness, small size, performance aspects including high luminous efficiency (LE), high color rendering index (CRI)[1-4]. Currently, most of the pc-WLEDs use high light transmission and low manufacturing temperature silicone as the packaging material, followed by a uniform dispersion of

phosphor on the LED chip. High power pc-WLEDs operate at junction temperatures of up to 150-200°C. At such high temperatures, conventional encapsulation materials such as silicones or resins, due to poor heat resistance, non-flame-retardant, intolerant to UV irradiation and susceptible to discolouration at high temperatures and short-wave light, long-term use will cause ageing and yellowing of the organic encapsulation material leading to light fading and colour drift of the light source, thus significantly reducing the quality of the white light. Hence, we need to develop thermally stable alternative materials[5-6]. To date, there have been many reports of glass matrices that have been used as substitutes for silicone to create PIG[7-11]. In this study, low temperature glasses were used instead of epoxy resin or silicon. The LEDs can be produced on a large scale with small size, high current and lighting grade.

II. EXPERIMENTAL

A. Synthesis

Glass plates with a glass composition of $(70-x)SnO-xSnF_2-30P_2O_5$ (x = 30, 40, 50, 55, in mol%,) were prepared by melt cooling method. To prepare the precursor glass, the precursors (SnO, SnF_2 and P_2O_5) were homogeneously mixed and then ground in a grinding bowl for 30 minutes, then the muffle was heated to 800°C, and the mixture was placed in the muffle for one hour, then these precursors are annealed and ground to a powder. The powder was passed through a sieve to the same particle size. Different ratios of $(BaSr)_2SiO_4:Eu^{2+}$ green and $CaAlSiN_3:Eu^{2+}$ red phosphor were then formulated at 5.00, 6.00, 7.00, 8.00 and 9.00 and these blended phosphors were subsequently added to the glass powder to achieve a weight ratio of 5%[12].

B. Characterization

The purity of the phases was determined by X-ray powder diffraction. The microstructure was observed and studied by means of a scanning electron microscope instrument. The glass transition temperature is analysed by differential scanning calorimetry. The excitation emission spectra of the fluorescent glass samples were obtained by testing with a fluorescence spectrometer using a 150 W DC powered sealed xenon lamp operating at 400 V. To ensure comparability between the test results, the In order to ensure comparable results between the samples, the same parameters were used for all tests, the measurement slit was kept at 5 nm for both excitation and emission spectra. The optical parameters of the white LEDs encapsulated in PIG were measured by the LED optoelectronic test system.

III. RESULTS AND DISCUSSION

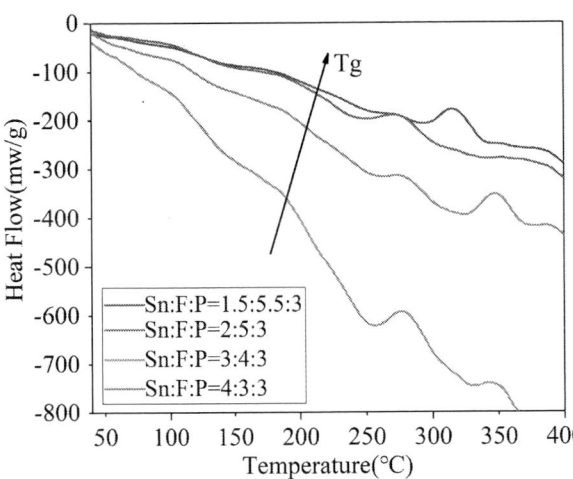

Fig. 1. DSC curves of (70–x)SnO–xSnF$_2$–30P$_2$O$_5$ glasses (x=30, 40, 50, 55).

The composition and content of the glass is very important for many properties of the glass. In fluorophosphate glasses, the skeletal structure of the glass network consists mainly of [PO$_4$] and [PO$_{4-x}$F$_x$] tetrahedra. Therefore, we first optimized the molar content of P$_2$O$_5$ during our experiments. It was shown that when its content is greater than 40 mol%, the precursor glass becomes highly absorbent. when its content is less than 25 mol%, the structure of the glass changes[13]. Therefore we set 30 mol% of P$_2$O$_5$ as the proportioning value.

The measured differential thermal analysis curve of (70-x)SnO-xSnF$_2$-30P$_2$O$_5$ glass is shown in the figure. 1, from which we can obtain the glass transition temperature(TG) of the precursor glass. The TG is an important parameter in determining if an encapsulation material is suitable for high power applications. Based on the TG, the heat treatment temperature of the glass can be accurately formulated. The TG of the studied tin-phosphorus-oxy-fluoride glasses is much lower, between 150°C and 200°C, compared to the previously reported silicate or telluride glasses. Clearly, tin-phosphorus-oxy-fluoride glass cannot be used for higher power applications, but it can still be used for less powerful residential lighting. When tin phosphorus oxyfluoride glass is used in residential lighting, the operating temperature of the chip core does not exceed 150°C, and in some cases it can even be below 100°C, especially if they are fitted with a heat sink. It is worth mentioning that when samples with x=40, 30 were heat treated about 200°C, no more significant performance gaps were observed, proving they can be used in some high power environments.

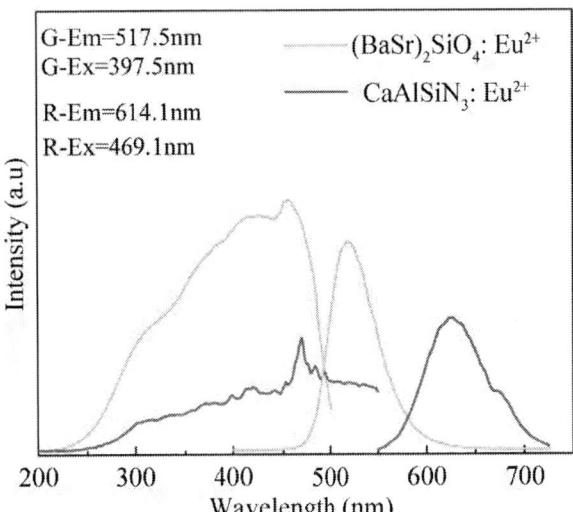

Fig. 2. Emission and excitation spectra of (BaSr)$_2$SiO$_4$:Eu^{2+} and CaAlSiN$_3$:Eu^{2+} phosphors.

Figure. 2 shows that the emission and excitation spectra of the commercially available (BaSr)$_2$SiO$_4$:Eu^{2+} and CaAlSiN$_3$:Eu^{2+} phosphors. It can be seen that both phosphors have relatively wide range of emission peaks and a higher intensity, the broad emission peaks of the G phosphors are concentrated at 517 nm under excitation at 397 nm, due to the 4f→5d energy level jump of Eu^{2+} in the silicate. Under 469 nm excitation, the broad emission peaks of the R phosphors are concentrated at 614 nm, which is due to the 4f→5d energy level jump of Eu^{2+} in nitrides [14-15]. In addition, (BaSr)$_2$SiO$_4$:Eu^{2+} and CaAlSiN$_3$:Eu^{2+} phosphors have relatively high intensity excitation peaks between 360 and 480 nm, and since the blue wavelength is in the range of this excitation peak, it is possible to use these phosphors as WLEDs conversion materials to obtain WLEDs with different color temperatures.

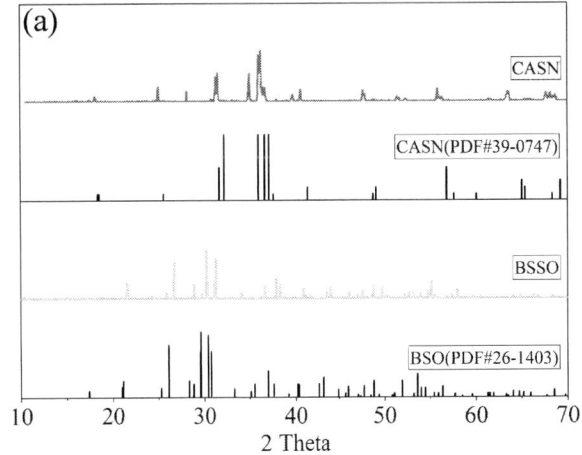

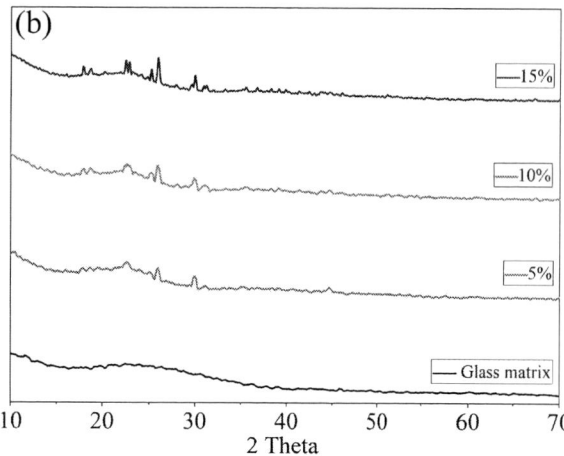

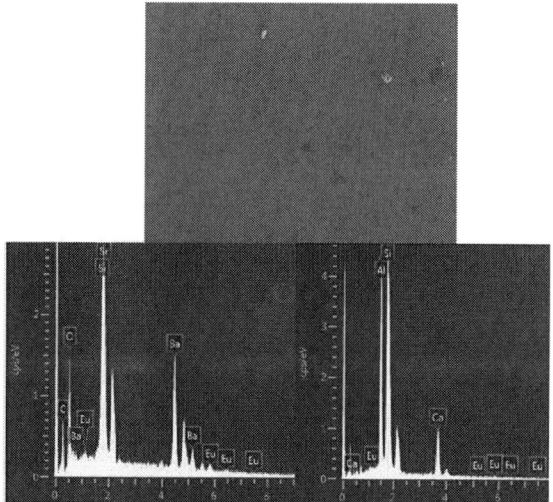

Fig. 3. XRD patterns of (a) $(BaSr)_2SiO_4:Eu^{2+}$ and $CaAlSiN_3:Eu^{2+}$ phosphors, (b) glass matrix and different weight of green/red phosphors in PIG

Fig. 4. SEM image of the phosphor mixed PIG sample and EDS energy spectrum analysis of PIG.

XRD analysis of the phosphors and the corresponding gas chromatography were carried out. As there is no reference standard for commercial green phosphors, a reference standard most similar to it has been chosen for comparison. The XRD pattern of the glass-ceramics changed after sintering, as shown in Figure. 3a and 3b, a broad peak appears at around 25°, which can be seen to be the amorphous region of the glass crystal by comparison with bare glass, and then by comparison with a standard card, it can be concluded that several of the sharper peaks appearing in the PIG are the XRD peaks of the $(BaSr)_2SiO_4:Eu^{2+}$ phosphor. The red $CaAlSiN_3:Eu^{2+}$ phosphor peak does not appear, probably the red phosphor has a very small proportion and its XRD peak may have overlapped with the green phosphor or glass matrix. The micromorphology and elements of the PIG was studied using SEM, as shown in Figure. 4, we can clearly see that the phosphors are regularly distributed on the surface of the precursor glass and the G and R powder particles are relatively uniform in size. This showed that a relatively good sample was obtained and indicates that there is no relatively serious interfacial reaction between the glass matrix and the phosphors. Energy spectral

analysis clearly shows tin, phosphorus, oxygen and fluorine signals in the glass matrix, barium, strontium, silicon and oxygen signals appeared in the $(BaSr)_2SiO_4:Eu^{2+}$ particles, and calcium, silicon, aluminum and nitrogen signals appeared in the $CaAlSiN_3:Eu^{2+}$ particles, this is consistent with the composition of the doped phosphor.

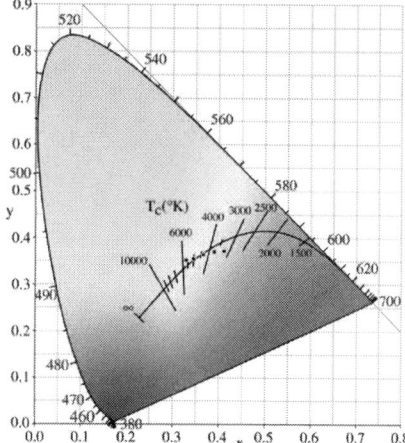

Fig. 5. Chromaticity color coordinates of different ratios of G/R phosphor

In order to investigate the possibility of applying PIG plates with a mixture of G/R phosphors to get WLED in the continuous CCT range from 3000 to 6000 K without changing the luminescence characteristics of the R phosphor. We used different ratios of G/R phosphors and precursor glass. The phosphors and precursor glass were first mixed and ground, then the proportioned powder mixture was placed in a mould and pressed into cylindrical under 20MP, then held at 285°C for 1 min, then annealed to relieve internal stress, polished and packaged in white light emitting diodes. As shown in Figure. 5, the chromaticity color coordinated with the WLED CCT (ratio of G/R phosphor). As the total weight of the phosphor increases and the phosphor G/R ratio decreases, the CCT gradually decreases and the color coordinate gradually shifts from the blue region to the red region. As previously reported, we can obtain white light emitting diodes with different CCT by varying the ratio of G/R phosphor and tuning the specific weight of the phosphor [12].

CONCLUSION

Low melting point PIG samples were prepared by grinding, mixing, sintering, pressing and polishing of certain mass ratio of low melting point glass powder and $(BaSr)_2SiO_4:Eu^{2+}$ and $CaAlSiN_3:Eu^{2+}$ phosphors. The SEM analysis shows that the phosphor is uniformly distributed on the precursor glass and that no strong interfacial reactions between phosphor and precursor glass. WLEDs with different CCT by varying the ratio of G/R phosphor and tuning the specific weight of the phosphor. Finally, we eatimated the various optical properties of the PIG encapsulated in WLEDs. The packaged WLEDs were obtained appropriate CCT of 3000-6000 K and good CRI of 89-91.

Funding : Nature Science Foundation Project of China (NSFC61904156). NSF of Fujian General Program (2020J01293); NSFC to Promote Cross-Strait Science and Technology Cooperation Joint Foundation (U2005212); NSF of Fujian Key Project (2020J02049); Xiamen Major Science and Technology Project (3502Z20201003, 3502ZCQ20201001).

REFERENCES

[1] Chen, D. & Chen, Y. Transparent Ce^{3+}. Y$_3$Al$_5$O$_{12}$ Glass Ceramic for Organic-Resin-Free White-Light-Emitting Diodes. Cerant. 4m. l0, 15325–15329 (2014).

[2] Zhang, R. et al. A New-Generation Color Converter for High-Power White LED: Transparent Ce^{3+}.YAG Phosphor-in-Glass. Laser Photonics Rev. 8, 158–164 (2014).

[3] Oh, J. H., Oh, J. R., Park, H. K., Sung, Y. G. & Do, Y. R. New paradigm of multi-chip white LEDs; combination of an InGaN blue LED and full down-converted phosphor-converted LEDs. Opt. Express 19, A270 (2011).

[4] Mueller-Mach, R., Mueller, G. O. Krames, M. R. & Trottier, T. High-Power Phosphor-Converted Light-Emitting Diodes Based on III-Nitrides. IEEE J. Sel. Top Quant. 8, 339 (2002).

[5] L. Yang, M.X. Chen, Z.C. Lv, S.M. Wang, X.G. Liu, S. Liu, Preparation of a YAG:Ce phosphor glass by screen-printing technology and its application in LED packaging, Opt. Lett. 38 (2013) 2240-2243.

[6] T. Nakanishi, S. Tanabe, Novel Eu^{2+} Activated Glass Ceramics Precipitated with Green and Red Phosphors for High-Power White LED, IEEE J. Sel. Top. Quantum Electron. 15 (2009) 1171-1176.

[7] R. Zhang, H. Lin, Y.L. Yu, D.Q. Chen, J. Xu, Y.S. Wang, A new-generation color converter for high-power white LED: transparent Ce^{3+}: YAG phosphor-in-glass, Laser Photonics Rev. 8 (2014) 158-164.

[8] Y.K. Lee, J.S. Lee, J. Heo, W.B. Im, W.J. Chung, Phosphor in glasses with Pb-free silicate glass powders as robust color-converting materials for white LED applications, Opt. Lett. 37 (2012) 3276-3278.

[9] H. Yoo, Y. Kouhara. H.C. Yoon, S.J. Park, J.H. Oh, Y.R. Do, Sn-P-F containing glass matrix for the fabrication of phosphor-in-glass for use in high power LEDs, RSC Adv. 6 (2016) 111640-111647.

[10] D.Q. Chen, W.D. Xiang, X.J. Liang, J.S. Zhong, H. Yu, M.Y. Ding, H.W. Lu, A.G. Ji, Advances in transparent glass-ceramic phosphors for white light-emitting diodes-A review, J. Eur. Ceram. Soc. 35 (2015) 859-869.

[11] Chung W J, Nam Y H. Review—A Review on Phosphor in Glass as a High Power LED Color Converter[J]. ECSJournal of Solid State Science and Technology, 2020, 9(1):016010.

[12] Yoon H C, Yoshihiro K, Yoo H, et al. Low-Yellowing Phosphor-in-Glass for High-Power Chip-on-board White LEDs by Optimizing a Low-Melting Sn-P-F-O Glass Matrix[J]. Scientific Reports, 2018, 8(1).

[13] Towards long-lifetime high-performance warm w-LEDs: Fabricating chromaticity-tunable glass ceramic using an ultra-low melting Sn-P-F-O glass - ScienceDirect[J]. Journal of the European Ceramic Society, 2017, 38(4):1990-1997.

[14] Zhang, X., Tang, X., Zhang, J. & Gong, M. An efficient and stable green phosphor SrBaSiO$_4$:Eu^{2+} for light-emitting diodes. J. Lumin. 130, 2288–2292 (2010).

[15] Watanabe, H., Wada, H., Seki, K., Itou, M. & Kijima, N. Synthetic Method and Luminescence Properties of SrxCa1−xAlSiN$_3$: Eu^{2+} Mixed Nitride Phosphors. J. Electrochem. Soc. 155, F31 (2008).

978-1-6654-1392-3/21 $31.00 © 2021 IEEE

Artificial Neural Networks Modeling Technology for Substrate Integrated Suspended Line

1st Shuxia Yan
School of Electrical and Electronic Engineering
Tiangong University
Tianjin, China
tjuysx@163.com

2nd Nana Yang
School of Electrical and Electronic Engineering
Tiangong University
Tianjin, China
tjpuynn@163.com

3rd Zhifeng Chen
School of Electrical and Electronic Engineering
Tiangong University
Tianjin, China
czfyywd@163.com

4th Peng Huang
School of Electrical and Electronic Engineering
Tiangong University
Tianjin, China
huangpeng@tjpu.edu.cn

5th Weiguang Shi
School of Electrical and Electronic Engineering
Tiangong University
Tianjin, China
shiweiguang12345@126.com

Abstract—A new method for parameter modeling of self-packaged Substrate Integrated Suspended Line (SISL) devices based on artificial neural network is proposed. The method only uses the input and output signals of self-packaged SISL devices, and does not need to establish the internal structure information of the model. The cost of modeling is greatly reduced and the design cycle of microwave passive devices is shortened. In addition, an improved training method is proposed based on the neural network to learn the relationship between electromagnetic response and geometric parameters, so that the proposed model can efficiently and accurately match the characteristics of SISL devices. The model takes geometric parameters as variables. It can provide the exact and rapid prediction for the electromagnetic response of SISL devices. The validity and accuracy of the parameter modeling technique based on an artificial neural network are verified by a modeling example of a wideband Yagi antenna.

Keywords—artificial neural network, SISL, modeling

I. INTRODUCTION

The Substrate Integrated Suspended Line (SISL) device is a novel self-packaged transmission structure with low loss, high performance and high reliability. It effectively overcomes the disadvantages of the traditional circuit such as large volume and additional metal cavity processing. The patent of SISL was proposed by Professor Ma Kaixue et al in 2007[1]. Accurate modeling of SISL devices can facilitate their development. The special structure of a SISL device is composed of a multilayer dielectric plate, which is partially cut off to form a cavity. This structure makes some existing models ineffective. At present, the main modeling method for the SISL device is the electromagnetic simulation modeling method. The electromagnetic simulation modeling method can solve the problem of low precision, but it takes very long time. In [2], the inductance modeling of the dielectric integrated mount line is carried out, and the established model can meet most of the microwave circuit design. However, the parameters of SISL devices with more and more complex structures are numerous.The process of obtaining parameters by equivalent circuit method takes a lot of time to obtain a high precision model.Therefore, it is necessary to propose a efficient modeling method for SISL devices.

In the field of RF/microwave design, artificial neural network (ANN) has strong learning ability and generalization ability, it is suitable for creating new models and improving the accuracy of existing models. ANN can quickly and accurately establish the corresponding model by training and learning the data. These models can learn nonlinear input-output relationship from geometric parameter to electromagnetic response. Generalized approximation theory is the theoretical basis of neural networks, which shows that neural networks with at the lowest one hidden layer can similar with any nonlinear continuous multidimensional function to any desired accuracy. The neural network model has the advantages of simple model structure, fast simulation speed, and high precision, and has been used in various microwave design applications.

In this paper, a new method of parameter modeling of self-packaged SISL devices based on artificial neural networks is proposed. The data needed for modeling were generated by HFSS software, and then imported into Neuromodelplus software for neural network model training. The established ANN model can provide an accurate and rapid predictions for the electromagnetic response of SISL devices with geometric parameters as variables. In addition, an improved training method is proposed to adjust the electromagnetic responses of self-packaged SISL devices under different parameters, so that the proposed model can efficiently and accurately match the device data. In this paper, the feasibility of neural network parameter modeling is proved by using a wideband Yagi antenna.

II. PROPOSED MLP MODEL

This paper puts forward artificial neural network model of SISL device. This model can study the geometrical parameters and the relationship between the electromagnetic response. Since the artificial neural network can learn nonlinear input-output relationship from the training data, we use the artificial neural network as the structure of the model. The most familiar artificial neural network is a multi-layer perceptual neural network (MLP). It belongs to feed-forward neural network and can approximate general functions, including continuous function and integrable function. MLP can effectively and accurately obtain the non-linear relationship. Presently, MLP neural network has been used to various microwave modeling and optimization problems[4]. Based on these, a parameter modeling method of encapsulated SISL device based on four-layer MLP structure is proposed in this paper. The MLP model structure is shown in Figure 1. In this model, $\mathbf{X}$ is the input variable including geometric parameters X_s and frequency f, and $\mathbf{Y}$ is the output response, namely the electromagnetic response features.

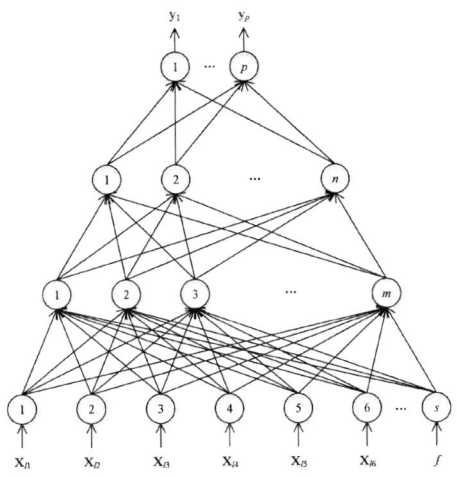

Fig. 1. Parametric model based on MLP.

Where s and p deputy the number of input and output neurons respectively, m and n express the number of first and second hidden neurons respectively. The model is divided into four layers: The first layer is the input layer, where the input variables are geometric parameters and frequency. The second and third layers are the hidden layers of the model, using the sigmoid function as the activation function. The fourth layer is the output layer. The output formula of our model is as follows:

$$ y_i^l = \begin{cases} x_i & l = 1 \\ \sigma\left(\sum_{j=1}^{s} w_{ij}^l x_j + w_j^l\right) & l = 2 \\ \sigma\left(\sum_{j=1}^{m} w_{ij}^l y_j^{l-1} + w_j^l\right) & l = 3 \\ \sum_{j=1}^{n} w_{ij}^l y_j^{l-1} + w_j^l & l = 4 \end{cases} \tag{1} $$

In this function, y_i^l represents the output of the ith hidden neuron in the lth layer. w_{ij}^l is the weight value of the ith hidden neuron in the lth layer and the jth neuron in the next layer. The w_j^l expresses the bias of the ith neuron in the lth layer. These weights and offsets decide the nonlinear relation between the input and output variables. Only by using the training data to train the ANN model well, the model can accurately predict the electromagnetic characteristics of the device. Therefore, the training of the model is an important step to establish the model. In the training process, we first select number of neurons in hidden

layer as a matter of the experience, and then continuously optimize and adjust the weight value in the ANN model, therefore the output of the ANN model can accurately meet with the electromagnetic simulation data. We use quasi Newton training method to change the weight value, so that to reduce the error between the output of neural network model and the training sample. To check up the accuracy of the model, we used the training error to assessment the learning level of the model. And used the test error to verify the prediction level. When the training error and test error are all meet the precision requirement, stop training. The error function is defined as:

$$ E(\mathbf{w}) = \frac{1}{2}\sum_{t=1}^{T}\sum_{j=1}^{q} \| y_j^t(x,\mathbf{w}) - y_{jD}^t(x) \|^2 \tag{2} $$

In the formula, $y_j^t(x,\mathbf{w})$ and $y_{jD}^t(x)$ are the output response of the established ANN model and the simulation data respectively. t is the index of training and testing data, and T is the sum number of training or testing data. q is the number of output variables.

The artificial neural network model is established by the steps shown in Figure 2. Two sets of geometric parameters were generated by using DOE[3] method. This two groups of parameters are simulated by HFSS software, and two groups of data are obtained: a set of training data of neural network model, a set of test data. The test data is used to verify the feasibility of the neural network model. The whole neural network model is trained by training data, so that the internal weight values are optimized continuously. If both training and test errors satisfy the accuracy requirements, then training will be stopped. Otherwise, the training will continue after adjustment.

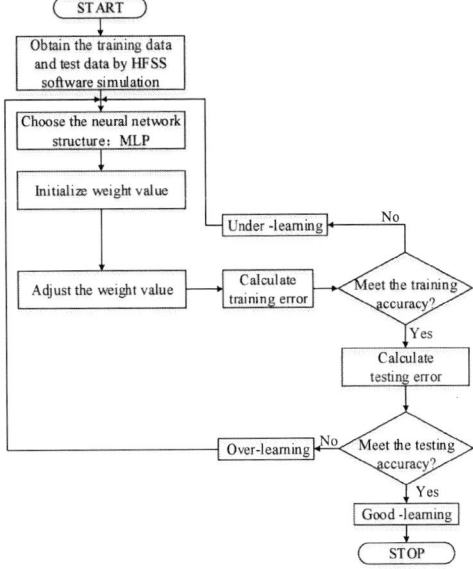

Fig. 2. Training and testing processes for models.

978-1-6654-1392-3/21 $31.00 © 2021 IEEE

III. EXAMPLES

Taking broadband Yagi antenna[6] as an example, the feasibility of the proposed model is verified. Figure 3 shows the structure of the antenna.

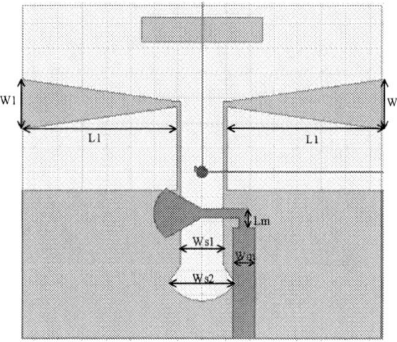

Fig. 3. Structure and parameters of yagi antenna.

As shown in the Figure. 3, six major geometric parameters are selected at Bow-tie driver and Sector slot and sector stub[7]. The geometric parameters are defined as: $X_i=[Wm1,Lm,W1,L1,Ws2,Ws1]^T$. In addition, the MLP model has another input: frequency f. In summary, the input of the MLP model is $X=[Wm1,Lm,W1,L1,Ws2,Ws1,f]^T$. The output value of the model is the dB value of the return loss characteristic S_{11} of the wideband Yagi antenna. Firstly, 81 groups of training data and 25 groups of test data were collected by DOE method. Their frequency ranges from 2 to 10 GHz with intervals of 0.1 GHz. The data range is shown in the table I. Secondly, the 6561 training samples and 2025 test samples are obtained by HFSS software for subsequent MLP modeling and verification of model accuracy.

TABLE I. GEOMETRIC PARAMETERS AND FREQUENCY RANGE

Input variables	Training data			Test data		
	Min	*Max*	*Step*	*Min*	*Max*	*Step*
$Wm1$(mm)	0.693	0.707	0.007	0.69	0.7	0.007
Lm(mm)	1.485	1.515	0.01	1.48	1.5	0.01
$W1$(mm)	3.96	4.04	0.02	3.96	4	0.02
$L1$(mm)	12.87	13.13	0.03	12.88	12.92	0.03
$Ws2$(mm)	0.99	1.01	0.002	0.99	1	0.002
$Ws1$(mm)	3.465	3.535	0.008	3.46	3.51	0.008
f (GHZ)	2	10	0.1	2	10	0.1

There are four layers in our model: the first layer is the input layer, which has 7 input neurons. The second layer and the third layer are hidden layers. The fourth layer is the output layer, which has one output neuron. Table II shows the training error and testing error of the model when the number of hidden layer neurons is different. It can be seen from the Table II that when the number of hidden neurons in the first layer and the second layer is 5 and 3 respectively

respectively, the training error and testing error does not meet the accuracy requirements. With the increase of the number of hidden neurons, the error decreases gradually. The minimum error (training error 0.99%, test error 0.97%) was obtained when the number of hidden neurons was 15 and 16 respectively. The number of neurons continued to increase until there were 35 and 40 neurons respectively, in which case overlearning occurred. The number of neurons selected was 15 and 16. The training sample data and test sample data were introduced into Neuromodelplus to complete the training and testing process of the model. To reveal detailed results, two groups of untrained parameters were used to verify the accurateness of the model. Figure 4 gives the output responses comparison between the electromagnetic simulation and the proposed model under two different sets of input variables. It can be seen that the MLP model is consistent with the electromagnetic simulation data. The HFSS simulation time was 3 hours, while the MLP model only needed 0.1 hours to get the output response. Compared with electromagnetic simulation, the modeling method proposed in this paper consumes less time to predict the electromagnetic response of SISL devices.

TABLE II. TRAINING AND TEST ERRORS UNDER DIFFERENT HIDDEN NEURONS

First layer hidden neurons	Second layer hidden neurons	Training error (%)	Test error (%)
5	3	1.3	1.29
7	5	1.16	1.11
15	16	0.99	0.97
30	30	1.2	1.2
35	40	0.9	9.21

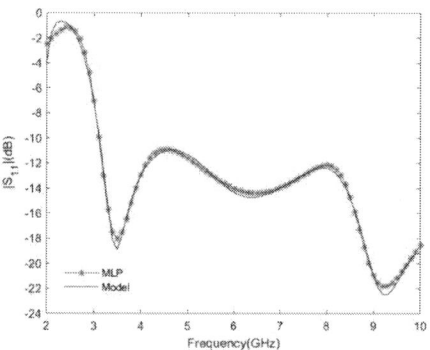

(a) $X_i=[13.0162,0.99125,0.7008,1.501,4.005,3.504]^T$

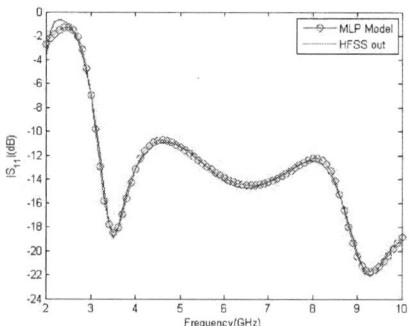

(b) $X_i=[12.95,0.993,0.699,1.487,4.00,3.48]^T$

Fig. 4. Compare the decibel level of S11 between MLP model and HFSS simulation data.

IV. CONCLUSIONS

In this paper, a parameter modeling method of dielectric integrated suspension line based on an artificial neural network is proposed. Compared with the electromagnetic simulation method, the proposed model can get similar accuracy in minor time.The example shows that the proposed model can forecast the electromagnetic response of SISL device rapidly and accurately.

ACKNOWLEDGMENT

This work was supported in part by the National Natural Science Foundation of China (Grant No.61601323), the Scientific Research Project of Tianjin Education Commission (Grant No.2017KJ088), the China Postdoctoral Science Foundation (Project No.2020M680883) and the Natural Science Foundation of Tianjin (Grant No.19JCQNJC03300).

REFERENCES

[1] K. Ma and KT. Chan, "Quasi-planar circuits with air cavities," WO, WO2007149046 A1, 2007.

[2] L. Li, K. Ma and S. Mou, "Modeling of New Spiral Inductor Based on Substrate Integrated Suspended Line Technology," in IEEE Transactions on Microwave Theory and Techniques, vol. 65, no. 8, pp. 2672-2680, Aug. 2017, doi: 10.1109/TMTT.2017.2701374.

[3] W. Zhang et al., "Space Mapping Approach to Electromagnetic Centric Multiphysics Parametric Modeling of Microwave Components," in IEEE Transactions on Microwave Theory and Techniques, vol. 66, no. 7, pp. 3169-3185, July 2018, doi: 10.1109/TMTT.2018.2832120.

[4] F. Feng et al. "Multifeature-Assisted Neuro-transfer Function Surrogate-Based EM Optimization Exploiting Trust-Region Algorithms for Microwave Filter Design," in IEEE Transactions on Microwave Theory and Techniques, vol. 68, no. 2, pp. 531-542, Feb. 2020, doi: 10.1109/TMTT.2019.2952101.

[5] H Zhang, K. Ma, H Fu and N. Yan, "Design of wide stopband lowpass filter using transformed radial stub based on substrate integrated suspended line technology," Microwave and Optical Technology Letters,vol. 63, no. 3, pp. 798-804(7), March 2021, doi: https://doi.org/10.1002/mop.32677

[6] N. Yan, K. Ma and H. Zhang, "A Novel Self-Packaged Substrate Integrated Suspended Line Quasi-Yagi Antenna," in IEEE Transactions on Components, Packaging and Manufacturing Technology, vol. 6, no. 8, pp. 1261-1267, Aug 2016, doi: 10.1109/TCPMT.2016.2585349.

[7] L. Li, K. Ma, N. Yan, Y. Wang and S. Mou, "A Novel Transition From Substrate Integrated Suspended Line to Conductor Backed CPW," in IEEE Microwave and Wireless Components Letters, vol. 26, no. 6, pp. 389-391, June 2016, doi: 10.1109/LMWC.2016.2562631.

[8] S Yan, Y Zhang, X Jin, W Zhang, and W Shi, "Multi-Physics Parametric Modeling of Microwave Passive Components Using Artififficial Neural Networks," Progress In Electromagnetics Research M, vol. 72, pp. 79-88, 2018, doi:10.2528/PIERM18070403

Non-cyanide Electroplating of Gold-Tin Eutectic Alloy for Flip-Chip Packaging of LED

Mingliang Huang*
Packaging Materials
Laboratory, School of
Materials Science &
Engineering
Dalian University of
Technology
Dalian, China
huang@dlut.edu.cn

Chao Fang
School of Chemical
Engineering
Dalian University of
Technology
Dalian, China
fangchao2019@mail.dlut.edu.cn

Feifei Huang
School of Chemical
Engineering
Dalian University of
Technology
Dalian, China
huangff@dlut.edu.cn

Yan Yan
Packaging Materials
Laboratory, School of
Materials Science &
Engineering
Dalian University of
Technology
Dalian, China
21805021yanyan@mail.dlut.edu.cn

Abstract—A high stability non-cyanide gold-tin alloy electroplating solution was developed to prepare Au-Sn bumps in flip-chip packaging of light-emitting diode. 5,5-Dimethyl hydantoin (DMH) was used as the main complexing agent for Au(III) and Sn(IV) was used as the main salt of tin. The additions of complexing agents stabilized the Sn(IV) and improved the stability of electroplating solution. To obtain eutectic composition of Au-Sn coating, the ratio of Au(III) to Sn(IV) was determined to be 1:10. The polarization potential of tin ion was positively shifted, which was approaching to the deposition potential of gold ion. The deposition potential of Au-Sn coating was -0.93 V. The stirring rate had an effect on the content of tin atom in Au-Sn coating. Energy dispersive X-ray spectrometer results showed that the tin content decreased with increasing stirring rate. Scanning electron microscope images showed that the Au-Sn coating became denser with increasing stirring rate. The optimal stirring rate was 100 rpm under a current density of 40 mA/cm2. The content of tin atom in the Au-Sn coating increased when the current density was increased in the range of 20~55 mA/cm2. Voids and larger grains of Au-Sn coating formed at high current density.

Keywords—flip-chip; Au-Sn eutectic alloy; electrodeposition

I. INTRODUCTION

With the miniaturization and the high-power demand of light-emitting diode (LED), the heat dissipation capacity and the luminous efficiency of LED are of great concern. The traditional packaging structure of LED needs to make pads and gold wires on the light emitting chip, which greatly reduces the luminous efficiency. Moreover, sapphire used as the substrate is not conducive to the heat dissipation of LED, which has a low thermal conductivity. Flip-chip (FC) packaging structure has been extensively used in high-power LED because of the high luminous efficiency and excellent heat dissipation performance [1,2], In which, the sapphire is placed on the top of the chip and the solder bumps are used to conduct electricity and heat. Therefore, there is a high requirement for the material of solder bumps. Au-Sn eutectic alloy is one of candidates for FC packaging structure of LED, owing to its excellent conductivity and heat conduction , and excellent corrosion and creep resistances as well [3].

The most effective preparation method for solder bumps is electro-deposition. Cyanide-containing solution has been banned due to its high toxicity. The research on stable and environmentally friendly Au-Sn alloy electroplating solution has attracted great interest. Ivey [4] developed a electroplating solution with neutral pH, in which Au-Sn alloy were deposited using potassium aurate chloride (KAuCl₄) and stannous chloride ($SnCl_2$) as metal ions. Sodium sulfite (Na_2SO_3) was used to complex with Au ions, and L-ascorbic acid was used to prevent the hydrolysis of stannous ion. However, the main drawback of the electroplating solution is the low stability [5]. Huang [6] used 5,5-dimethyl hydantoin (DMH) as the complexing agent of Au(III) and Sn(II) as main tin salt and found that the non-cyanide Au-Sn alloy electroplating solution had a high stability. Walsh [7] found that Sn(II) was easily oxidized to Sn(IV) in the electroplating solution, and consequently Sn(IV) hydrolyzed into stannic acid, resulting in poor stability of the electroplating solution. The chemical reactions were expressed by the equations (1) and (2).

$$O_2 + 4H^+ + 2Sn^{2+} = 2Sn^{4+} + 2H_2O \qquad (1)$$

$$Sn^{4+} + 3H_2O = SnO_2 \cdot H_2O + 4H^+ \qquad (2)$$

Katsunori [8] found that thioglycolic acid was mainly used to stabilize Sn(IV), which can effectively solve the problem of solution stability .

In the present work, a high stability non-cyanide gold-tin alloy solution using Sn(IV) as the main tin salt was investigated. The operation life of the electroplating solution was as long as one week and the conservation time was more than three months. Polarization curve shows that the deposition potential of Au-Sn coating from the electroplating solution. The morphology and composition of Au-Sn coating were optimized to adjust the stirring rate and the current density.

II. EXPERIMENTAL

The cathode of electroplating solution is a silicon wafer (4 mm×5 mm) sputtered with 20 nm Cr as the adhesive layer and 200 nm Au as the seed layer. The anode is a platinum-plated titanium mesh (20 mm×30 mm). The solution consists of HAuCl₄, Na₂SnO₃·3H₂O, DMH, Na₂SO₄, Sn(IV) complexing agent A, accessory complexing agent sorbitol and solution stabilizer B. All the chemical reagents were of analytical grade. The pH of gold-tin alloy solution was adjusted to 7.2 using 1 M KH₂PO₄ and 1 M K₂HPO₄.

The electroplating parameters were controlled by Princeton VersaSTAT 4 electrochemical workstation and the turn on and turn off times were 2 ms and 8 ms, respectively. Polarization curves were measured using glassy carbon electrode as cathode and platinum-titanium mesh as anode. The reference electrode is the saturated calomel electrode(SCE).

This work is supported by the National Natural Science Foundation of China (Grant Nos. 51671046 and U1837208) and the Fundamental Research Funds for the Central Universities (Grant No. DUT20LAB122).

978-1-6654-1392-3/21 $31.00 © 2021 IEEE

The grain size and surface voids of the coating were observed by scanning electron microscope (Zeiss Supra 55). The composition of the coating was analyzed by Energy dispersive X-ray spectrometer.

III. RESULTS AND DISCUSSION

A. Component and deposition potential of Au-Sn alloy electroplating solution

In order to reduce the solution component changes due to the oxidation of Sn(II), Sn(IV) was used as the main salt in the Au-Sn electroplating solution. Table 1 shows the stability of Sn electroplating solution with different complexing agents. Without the complexing agent, the white precipitation appeared immediately. It was happened for the formation of stannic acid by hydrolysis of Sn(IV). The solution usually produced precipitation after one week storage when only one of the complexing agents A and sorbitol was used. When both of the two complexing agents were used, the electroplating solution had a storage life as long as more than three months without any turbidity and precipitation. Therefore, two complexing agents were used to stabilize Sn(IV).

TABLE I. SOLUTION COMPOSITION AND PHENOMENON

Num	Na$_2$SnO$_3$(M)	A(M)	sorbitol(M)	pH	stability
1	0.06	-	-	-	instable
2	0.06	0.72	-	2.16	instable
3	0.06	-	0.72	-	instable
4	0.06	0.36	0.36	2.36	stable
5	0.06	0.36	0.36	7.2	stable
6	0.1	-	-	-	instable
7	0.1	1.2	-	2.16	instable
8	0.1	-	1.2	-	instable
9	0.1	0.6	0.6	2.36	stable
10	0.1	0.6	0.6	7.2	stable

It is clear that the deposition potential of Sn (IV) shifted positively with the addition of complexing agent A and sorbitol (Fig.1). The polarization curves of No. 4 and No. 9 solutions (pH=2.36) show that the onset deposition potentials of Sn(IV) were at around -0.668 V, the hydrogen evolution potential were at about -1.110 V. Polarization curves of No. 5 and No. 10 solutions (pH=7.2) show that the onset deposition potentials of Sn(IV) were about -0.980 V and the hydrogen evolution potentials were at about -1.446 V. With increasing alkalinity, the onset deposition potential of Sn(IV) and the potential of hydrogen evolution shifted negatively. It was determined that tin solution of No. 5 and No. 10 were used to mix with gold solution because the neutral electroplating solution application was more extensive.

Fig. 2 shows the polarization curves of Au-Sn alloy with different concentration ratios of Au(III) to Sn(IV). The beginning deposition potential of Au-Sn alloy was at -0.930 V, and the initial hydrogen evolution potential was at about -1.291 V. The change of onset deposition potential was not obvious with two different ratios of Au(III) to Sn(IV).

Fig. 3 shows the variation on content of tin atom in Au-Sn coating electroplated on 1st, 3rd, 5th and 7th days with two different ratios of Au(III) to Sn(IV). Fig. 3(a) shows the content of tin atom in Au-Sn coating changed a little when the ratio of Au(III) to Sn(IV) is 1:6. The tin contents were as low as 5 at.%, which was significantly lower than the content of tin

atom in Au-Sn eutectic alloy. Fig. 3(b) shows the contents of atom were about 40 at.%, however, when electroplated on the 7th day, the tin content decreased to 27.76 at.%. Therefore, the ratio of Au(III) to Sn(IV) in the electroplating solution was chosen as 1:10, since the composition of electroplated Au-Sn coating was approximate to Au-Sn eutectic alloy. Moreover, the process parameters need to be further optimized to prepare Au-Sn eutectic coating.

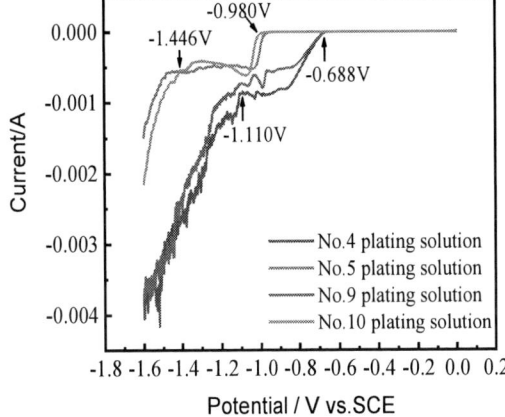

Fig. 1. Polarization curves of Sn(IV) in different electroplating solutions.

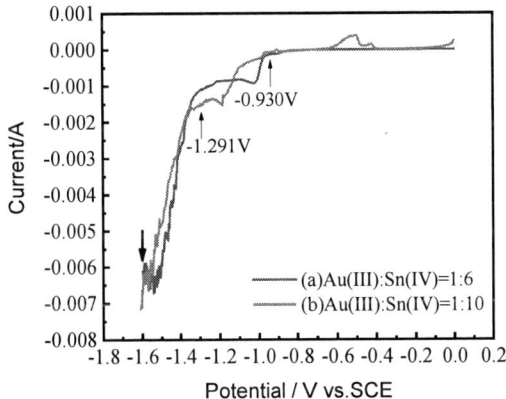

Fig. 2. Polarization curves of Au-Sn in electroplating solutions with two different ratios of Au(III) to Sn(IV): (a) 1:6 and (b) 1:10.

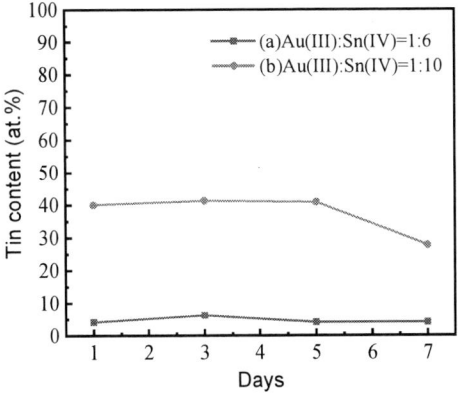

Fig. 3. The tin content of Au-Sn coating electroplated on 1st, 3rd, 5th and 7th days with two different ratios of Au(III) to Sn(IV): (a) 1:6 and (b) 1:10.

B. effect of the stirring rates on morphology and composition of Au-Sn coating

Fig. 4 shows the SEM images of Au-Sn coating at different stirring rates. It is clearly showed that there were obvious voids and many gaps in the Au-Sn coating at stirring rates of 0 rmp (Fig. 4a) and 50 rmp (Fig. 4b). If the stirring rate increased to 150 rpm (Fig. 4d), the grains of Au-Sn coating became smaller and the gaps or voids on the surface became less. The surface of Au-Sn coating was more compact at the stirring rate of 250 rpm (Fig. 4f). The morphology of Au-Sn coating was optimized by increasing stirring speed.

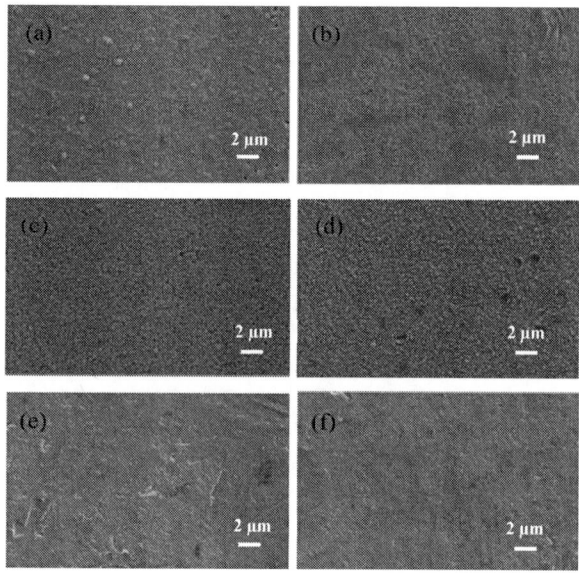

Fig. 4. SEM images of Au-Sn coating at different stirring rates: (a) 0, (b) 50 rpm, (c) 100 rpm, (d) 150 rpm, (e) 200 rpm and (f) 250 rpm.

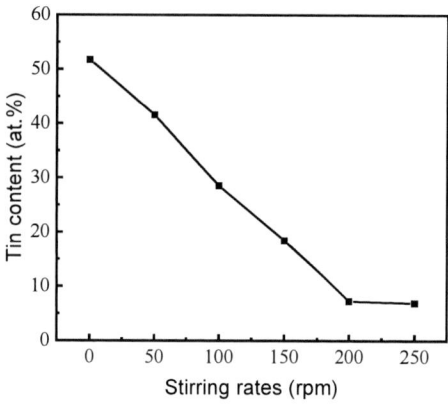

Fig. 5. The tin content of Au-Sn coating as a fuction of the stirring rate.

Fig. 5 shows the variation of tin atom percent in in the Au-Sn coating at different stirring rates. When the solution was not stirred (0 rpm), the content of tin atom in the Au-Sn coating was 51.74 at.%. When the stirring rate was 100 rpm, the tin content decreased to 28.53 at.%. When the stirring rate was 200 rpm, the content of tin atom further decreased to 7.26 at.%. As a result, the content of tin atom in Au-Sn coating decreased with increasing stirring rate. The deposition potential of Au(III)

was more positive than that of Sn(IV), and the increase of stirring rate was beneficial to the deposition of Au(III).

C. effect of the current density on morphology and composition of Au-Sn coating

The current density had an effect on the grain size, surface void and content of tin atom in Au-Sn coating. Fig. 6 shows the SEM images under different current densities at a stirring rate of 100 rpm. Fig. 6(a) shows the surface morphology of the Au-Sn coating under a current density of 20 mA/cm². The grains of Au-Sn coating were fine and compactly arranged. Under a current density of 45 mA/cm², the grains of Au-Sn coating increased, and the surface voids were more and larger. Under a current density of 55 mA/cm², the grain size and the surface voids significantly increased. The grains of Au-Sn coating became larger and the surface voids in the Au-Sn coating increased because of the increase of current density. This phenomenon might be attributed to the concentration polarization in the electroplating process. The concentration of ions in different parts of the cathode surface was inconsistent and the partial discharge was serious. Metal ions discharged in the protruding part resulted in the formation of voids and larger grains.

Fig. 7 shows the content of tin atom in Au-Sn coating as a function of the current density. When the peak current density increased from 20 mA/cm² to 55 mA/cm², the content of tin atom in Au-Sn coating increased from 6.71 % to 38.83 %. The cathodic overpotential increased with increasing current density, which leads to the increase of metal ions with negative potential in the coating. The content of tin atom in Au-Sn coating increased because of the increase of current density.

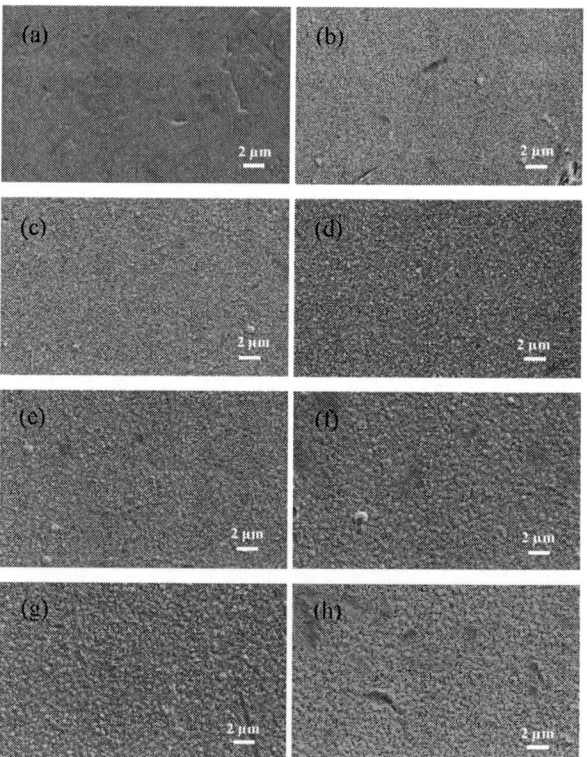

Fig. 6. SEM images of Au-Sn coating under different current densities: (a) 20 mA/cm², (b) 25 mA/cm², (c) 30 mA/cm², (d) 35 mA/cm², (e) 40 mA/cm², (f) 45 mA/cm², (g) 50 mA/cm² and (h) 55 mA/cm².

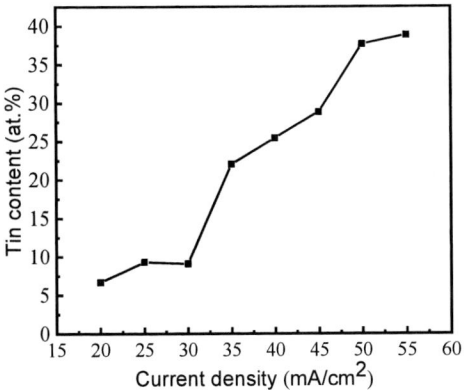

Fig. 7. The tin content of Au-Sn coating as a fuction of the current density.

IV. CONCLUSIONS

1) The high stability non-cyanide Au-Sn eutectic alloy electroplating solution was developed with the additions of two complexing agents of Sn(IV). The onset deposition potential of Sn(IV) was at around -0.98 V and the deposition potential of Au-Sn alloy was at around -0.93 V in the neutral tetravalent tin solution. The ratio of Au(III) to Sn(IV) was determined as 1:10.

2) The composition and morphology of the Au-Sn coating were optimized by adjusting the stirring rate. The content of tin atom in Au-Sn coating decreased and the surface of the Au-Sn coating became more tighter with increasing stirring rate. The optimal stirring rate was 100 rpm under a current density of 40 mA/cm^2.

3) The content of tin atom of Au-Sn coating increased with increasing current density in the range of 20~55 mA/cm^2. At a higher current density, larger grains and more voids formed in the Au-Sn coating.

ACKNOWLEDGMENT

This study is supported by the National Natural Science Foundation of China (Grant Nos. 51671046 and U1837208) and the Fundamental Research Funds for the Central Universities (Grant No. DUT20LAB122).

REFERENCES

[1] S.Y. Huang, R. H. Horng, et al. "Characteristics of flip-chip InGaN-based light-emitting diodes on patterned sapphire substrates," Jpn. J. Appl. Phys, vol. 45, pp. 3430–3432, April 2006.

[2] L B Wang, Y Chen, et al. "Thermal simulation and analysis of high power flip-chip light-emitting diodes," Semiconductor Optoelectronics, vol. 28, pp. 769-773, 2007.

[3] B. S. Lee, C.W. Lee and J.W. Yoon. "Comparative study of Au-Sn and Sn-Ag-Cu as die-attach materials for power electronics applications," Surf. Interface Anal, vol. 48, pp. 493-497, July 2016.

[4] J. Doesburg, and D.G. Ivey. "Microstructure and preferred orientation of Au-Sn alloy plated deposit," Mat. Sci. Eng. B-Solid, vol. 78, pp. 44-52, October 2000.

[5] F.F. Huang, Y.W. Liu, M.L. Huang, "Development of a stable non-cyanide gold-tin electroplating solution for optoelectronic applications," in Proc. 17th International Conference on Electronic Packaging Technology (ICEPT), Wuhan, 2016, pp. 538-541.

[6] F. F. Huang, M. L. Huang, "Complexation behavior and co-electrodeposition mechanism of Au-Sn alloy in highly stable non-cyanide bath," J. Electrochem. Soc, vol. 165, pp. D152-D159, 2018.

[7] F. C. Walsh, C. T. J. Low, "A review of developments in the electrodeposition of tin," Surf. Coat. Tech, vol. 288, pp. 79-94, February 2016.

[8] K. HAYASHI. "Non-Cyanide based Au-Sn alloy plating solution," U.S. Patent 10301734 B2, May 28 2019.

Fabrication of SiC Nano-pore arrays Structure by Metal-assisted Photochemical Etching

Zijian Li
School of Electromechanical Engineering
Guangdong University of Technology
Guangzhou, China
li.zijian2021@foxmail.com

Dachuang Shi
School of Electromechanical Engineering
Guangdong University of Technology
Guangzhou, China
dachuang.shi@gmail.com

Yun Chen*
School of Electromechanical Engineering
Guangdong University of Technology
Guangzhou, China
chenyun@gdut.edu.cn

Maoxiang Hou
School of Electromechanical Engineering
Guangdong University of Technology
Guangzhou, China
maoxiangh@gdut.edu.cn

Jian Gao
School of Electromechanical Engineering
Guangdong University of Technology
Guangzhou, China
gaojian@gdut.edu.cn

Xin Chen
School of Electromechanical Engineering
Guangdong University of Technology
Guangzhou, China
Chenx@gdut.edu.cn

Abstract—Silicon carbide (SiC) is widely used for electronic devices in high temperature and harsh conditions because of its excellent properties. However, SiC is difficult to processed effectively by the traditional wet etching method due to its excellent chemical stability. In this paper, a metal-assisted photochemical etching method was proposed and controllable nano-pore arrays structure in SiC was fabricated. By sputtering a noble metal layer on the bottom surface of the SiC wafer and illuminating UV light on the top surface, SiC could be effectively etched in etchant of hydrofluoric acid and hydrogen dioxide. The effective etching of SiC is attributed to the synergistic effect of abundant photogenerated holes excited by UV light and rapid electron transmission enhanced by the Pt layer, thereby, enhancing the etching reaction rate. Further, it was demonstrated that the diameter of nano-pores could be controlled by adjusting the time of plasma etching polystyrene spheres or changing the etching time. This work will provide an effective means to process wide band gap semiconductor.

Keywords—Metal-assisted photochemical etching; Silicon carbide; Polystyrene spheres; Nano-pore arrays structure

I. INTRODUCTION

Silicon carbide (SiC) belongs to wide band gap semiconductors and is one of the third generation semiconductors. It is widely used in high temperature[1, 2], high frequency[3] and high power[4] electronic devices because of its excellent physical and chemical properties. Especially in the new-energy vehicles field, SiC is the first choice for power chip material. However, SiC is difficult to processed effectively by the traditional wet etching method due to its excellent chemical stability. Therefore, dry etching is mostly used to process SiC wafers in industry at present. However, because of its low etching selectivity, it can cause subsurface damage of substrates. Moreover, dry etching requires vacuum and high temperatures conditions, which makes the equipment cost increase. On the contrary, wet etching has the following advantages: high etching selectivity, simple operation and low cost. Hence, the study of novel and effective wet etching has positive significance for the development of processing semiconductor. In recent years, photochemical etching and metal-assisted chemical etching have aroused a lot of attention as the effective wet etching methods for wet etching semiconductor materials, such as processing Si[5], GaAs[6], and GaN[7]. Nevertheless, using only one of above methods is inefficient for processing SiC duo to its wide band gap and insufficient holes . In previous reports, researchers needed to additionally apply an external electric field while using photochemical etching or metal-assisted chemical etching to process SiC[8-12]. Whereas, Leitgeb et al.[13] and Rittenhouse et al.[14] used the metal-assisted photochemical etching to effectively process SiC without an external electric filed and achieved the fabrication of porous SiC. However, in the above reports, the size of SiC pore structures are uncontrolled.

In this study, we propose a metal-assisted photochemical etching (MAPCE) method to fabricate the SiC nano-pore arrays structure. By sputtering a noble metal layer on the bottom surface of the SiC wafer and illuminating UV light on the top surface, SiC could be effectively etched. Moreover, it was demonstrated that the diameter of the nano-pores could be controlled by regulating the plasmas etching treatment time. In additionally, the mechanism of MAPCE was described. Furthermore, the effect of the etching time on nano-pores was studied. This study provides an effective approach to fabricating SiC array structure.

II. EXPERIMENT

A. Experiment material

A n-type monocrystalline 6H-SiC wafer (0.02-0.1 Ω cm, <0001>, TankeBlue Semiconductor CO., Ltd.) was applied in the experiment. Polystyrene (PS) spheres colloid (510 nm in diameter, 2.5% by weight) used for self-assembly was purchased from PolySciences. Inc., US. The etchant used for etching SiC wafers was made of 21 mL H_2O (produced by Milli-Q), 6 mL H_2O_2 (30 wt.%) and 5 mL HF (49 wt.%).

B. Experiment method

The specific process of the MAPCE is shown in Fig. 1. First of all, the SiC wafer was cut into square samples with dimensions of 5 mm × 5 mm and cleaned in ethanol by ultrasonic treatment for 5 min. Then the SiC wafer was washed by deionized water for 2~3 times and then blew dry by nitrogen. The last cleaning step was performed in a plasma cleaner with oxygen (purity: 99.999%) for 5 min. Subsequently, a self-assembled monolayer of the PS spheres was deposited on the top surface of SiC wafer by the spinning coating method. The SiC wafer dripping with the PS spheres colloidal of 5 μL on the top surface was fixed on the spin-coater (WS--650MZ-23NPPB) whose parameters were set as shown in Table Ⅰ. After spin coating process, the size of PS spheres were shrunk by the plasma etching treatment (power: 100W, no any assisting gas) and the single-layer non-densely packed PS spheres array was obtained. And then, a noble metal Ti layer of ~10 nm thickness and Pt layer of ~45 nm thickness were deposited on the top and bottom surfaces of SiC wafers by the magnetic sputtering (Desk V TSV, Denton Vacuum), respectively. After that, the shrunken PS spheres on the top surface of the SiC wafer were ultrasonically dissolved in acetone for 5 min. Finally, the SiC wafer was immersed in the etchant and irradiated by the UV-light (365 nm) at a distance of 10mm from the top surface of the SiC wafer. The UV-light used in the experiment was produced by UV-LED irradiation head (ZUV-H20MB, OMRON) whose spot diameter was 8 mm and the power was adjusted to 100% by the controller (ZUV-C20H, OMRON).

TABLE I.   THE PARAMETERS OF THE SPIN-COATOR

Step	Parameters	
	Rotate speed / rpm	*Time / min*
1	100	10
2	1000	1
3	2500	2

C. Charateristic

the SiC wafer split manually was characterized by the scanning electron microscopy (SEM, SU8220, HITACHI).

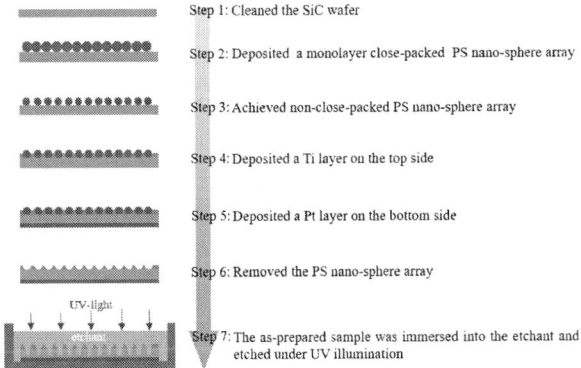

Fig. 1. Scheme diagram of the MAPCE. Step 1, the SiC wafer was clean by ethanol, deionized water and plasma, successively. Step 2, the SiC wafer was deposited with a monolayer close-packed PS sphere array by the colloidal self-assembly. Step 3, the PS spheres were shrunk uniformly under the plasma etching treatment and a non-closed-packed array was obtained. Step 4, assisted by the mask of the non-closed-packed array, a patterned Ti layer was fabricated on the SiC wafer by the magnetic sputtering. Step 5, the SiC wafer was placed upside-down in the magnetic sputter and its bottom side was deposited with a Pt layer. Step 6, residual PS spheres on the top side were removed by the ultrasonic bath with acetone solution. Step 7, the as-prepared sample was immersed into the etchant and its top side was illuminated by a UV light.

III. RESULT AND DISCUSSION

A. Formation mechanism of nano-pore arrays

Fig. 2 shows the PS sphere arrays and nano-pore arrays on the top surface of the SiC wafer. Before the plasma etching treatment, the PS spheres were closely packed into a monolayer (Fig. 2a). Once the plasma etching treatment was implemented, the monolayer closely packed PS spheres were chemically dissociated[15], which reduced gradually the size of the PS spheres. This indicates that the size of the PS spheres be capable of being adjusted by controlling the time of the plasma etching treatment. In Fig. 2b, the PS spheres were etched by the plasma for 20 min, the PS spheres with an average diameter of 154 nm were obtained. Because the PS sphere array served as a template, the sputtered Ti layer only covered the area without the PS spheres. Therefore, the Ti layer with the nano-pore arrays was formed on the top surface of SiC wafer. When the SiC wafer was etched, the Ti layer acted as a mask, so that the area exposed to the etchant on the SiC wafer was corroded and dissolved. Then, the nano-pore array with the average diameter and average depth of 100 nm and 191 nm was formed (as shown in Fig. 2c and 2d). However, from Fig. 2b and 2c, the diameter of the nano-pores was smaller than the diameter of the shrunken PS spheres. This attributed to the fact that the contact zone between the SiC and the PS spheres wafer was smaller than the vertical projection area of PS spheres. Hence, the diameter of the nano-pores was incompletely equal to that of the shrunken PS spheres. According to the above discussion, it was revealed that the diameter of nano-pores be capable of being controlled by adjusting the plasma etching treatment time.

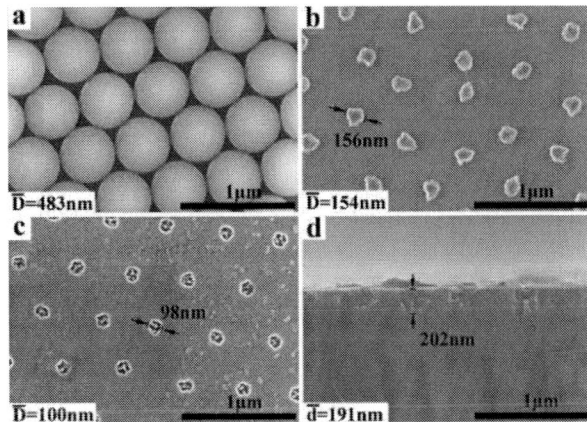

Fig. 2. SEM images of the difference process stages of the SiC wafer. (a) Top view of the monolayer close-packed PS sphere array on the top suface of the SiC wafer. The mean diameter of the PS spheres was 483 nm. (b) Top view of the non-densely packed single layer PS sphere array. The average diameter of the shrunken PS spheres was 154 nm. (c) Top view of the nano-pore array, which was fabricated by the MAPCE and it had an average diameter of 100 nm. (d) Cross-section view of the corresponding nano-pore array, which had an average depth of 191 nm.

According to previous studies[8, 16], the total etching process of SiC is divided into two steps. In the first step, SiC is oxidized to silicon dioxide and carbon dioxide by holes (h^+), and carbon dioxide escapes (eq 1). In the second step, silicon dioxide is dissolved by HF (eq 2).

$$SiC + 4H_2O + 8h^+ \rightarrow SiO_2 + CO_2 \uparrow + 8H^+ \qquad (1)$$

$$SiO_2 + 6HF \rightarrow H_2SiF_6 + 2H_2O \qquad (2)$$

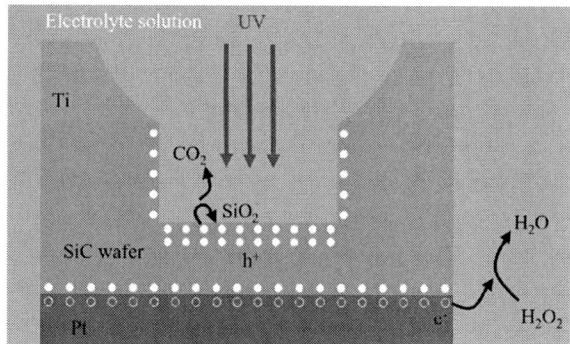

Fig. 3. Mechanism of the MAPCE. The irradiated side of the SiC wafer serves an anode and has the oxidation reaction. The Pt layer serves a cathode and has the reduction reaction.

The mechanism of the MAPCE is shown in Fig. 3. On account of the inequality of the Fermi level between the electrolyte solution and the SiC wafer, when the electrolyte solution is in contact with the SiC wafer, for n-type SiC, the holes in the electrolyte solution flow to the SiC wafer, resulting in the formation of an internal electric field pointing to the electrolyte solution at the electrolyte solution / SiC wafer interface. Under the irradiation of the UV light, electron-hole pairs are excited in the surface of SiC wafer. A part of the photogenerated holes are recombined with electrons. A considerable part of the photogenerated holes migrate to the bottom and sidewall of the nano-pores under the action of the internal electric field, participating in the oxidation reaction of SiC (eq 1). Similar to the electrolyte solution / the SiC wafer interface, the interface between the Pt layer and the SiC wafer forms an internal electric filed

pointing to the Pt layer. The Pt layer at the bottom of SiC wafer acts a cathode[17]. Electrons (e^-) in the SiC wafer are injected into the Pt layer under the action of the internal electric filed and flow to the electrolyte solution to participate in the reduction reaction (eq 3).

$$H_2O_2 + 2H^+ + 2e^- \rightarrow 2H_2O \qquad (3)$$

In summary, the UV light and Pt layer enhance the oxidation reaction and reduction reaction rate, respectively, leading to increase the total etching reaction rate.

B. Effect of etching time on nano-pores

Fig. 4 shows the morphology of nano-pores under different etching times. When the SiC wafer was etched for 5 min (Fig. 4a), the diameter of the nano-pores was 50 nm. When the etching time was extended to 10 min (Fig. 4b), the diameter of the nano-pores slightly increased by 7 nm. As increasing etching time to 15 min (Fig. 4c), the diameter of the nano-pores increased about twice, reaching 119 nm. When the etch treatment to 30 min (Fig. 4d), the diameter of the nano-pores slightly increased again, the increase amplitude was 8 nm. When the etching time was extended to the maximum 60 min (Fig. 4e), the diameter of the nano-pores significantly increased to 249 nm. The relationship between the diameter of the nano-pores and the etching time is shown in Fig. 3f. The diameter of nano-pores increases with the extension of etching time, which indicates that the diameter of the nano-pores also can be controlled by changing the etching time. The lateral etching is ascribed to the excessive holes in nano-pores sidewall.

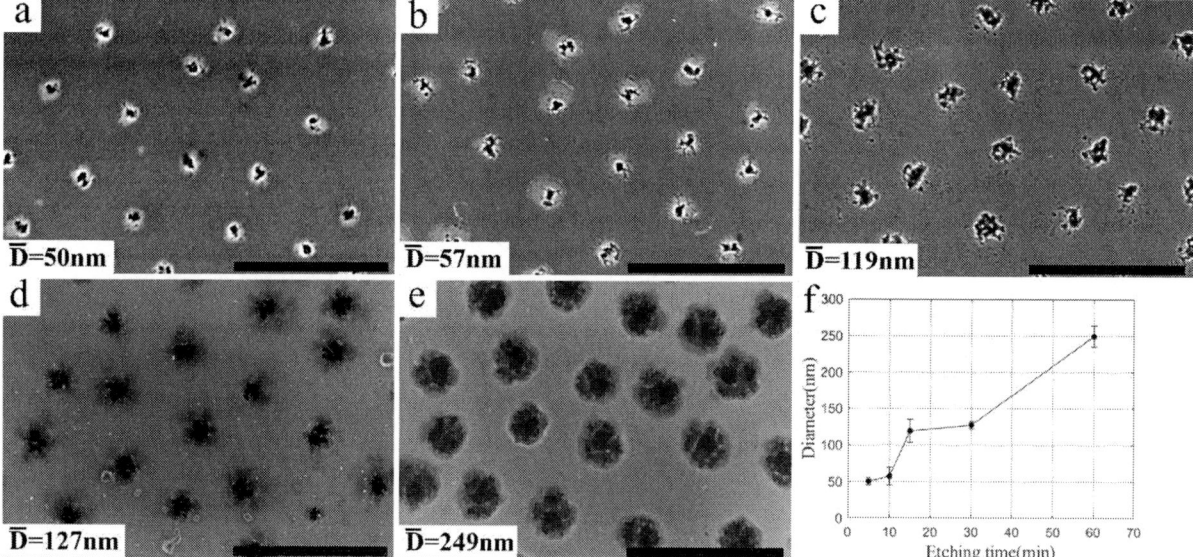

Fig. 4. The top view SEM images of SiC wafer etched for (a) 5 min, (b) 10 min, (c) 15 min, (d) 30 min, (e) 60 min, respectively. The relationship between diameters of nano-pores and etching time was showed as (f). The average diameter of (a)-(e) are 50 nm, 57 nm, 119 nm, 127 nm, 248 nm, respectively. The average diameter represented the diameter of nano-pores. Scale bars in (a)-(e) are 1 μm.

IV. CONCLUSION

In this paper, a metal-assisted photochemical etching method was proposed and SiC nano-pore arrays structure was achieved by exploring this method. The mechanism of nano-pores formation were disclosed as the synergistic effect of UV light excited holes and Pt layer enhanced rapid electron

transmission. Increasing the plasma etching time can reduce the diameter of PS spheres, indicating the diameter of nano-pores can be controlled by tuning the plasma etching treatment time. By exploring the influence of the etching time, it was found that the diameter of nano-pores increased with the increase of the etching time, which indicating that the diameter of nano-pores can be controlled by changing the etching time.. This study provides an effective approach to fabricating SiC array structure.

ACKNOWLEDGMENT

The work presented in the paper is supported the National Natural Science Foundation of China (51975127, U20A6004), the Fund of Research and Development Program of Guangdong Province (2020A0505140008) and Key-Area Research and Development Program of Guangdong Province (2018B090906002).

REFERENCES

[1] Willander, M., et al., Silicon carbide and diamond for high temperature device applications. Journal of Materials Science: Materials in Electronics, 2006. 17(1): p. 1-25.

[2] Guo, X., et al., Silicon Carbide Converters and MEMS Devices for High-temperature Power Electronics: A Critical Review. Micromachines (Basel), 2019. 10(6): p. 406.

[3] Chennu, J.V.P.S., R. Maheshwari and H. Li, New Resonant Gate Driver Circuit for High-Frequency Application of Silicon Carbide MOSFETs. IEEE transactions on industrial electronics (1982), 2017. 64(10): p. 8277-8287.

[4] She, X., et al., Review of Silicon Carbide Power Devices and Their Applications. IEEE transactions on industrial electronics (1982), 2017. 64(10): p. 8193-8205.

[5] Huang, Z., et al., Metal-Assisted Chemical Etching of Silicon: A Review. Advanced Materials, 2011. 23(2): p. 285-308.

[6] Yasukawa, Y., H. Asoh and S. Ono, Site-Selective Metal Patterning/Metal-Assisted Chemical Etching on GaAs Substrate Through Colloidal Crystal Templating. Journal of The Electrochemical Society, 2009. 156(10): p. H777.

[7] Wang, Q., et al., Metal-assisted photochemical etching of GaN nanowires: The role of metal distribution. Electrochemistry Communications, 2019. 103: p. 66-71.

[8] SHOR, J.S. and A.D. KURTZ, Photoelectrochemical etching of 6H-SiC. Journal of the Electrochemical Society, 1994. 141(3): p. 778-781.

[9] Ke, Y., R.P. Devaty and W.J. Choyke, Self-Ordered Nanocolumnar Pore Formation in the Photoelectrochemical Etching of 6H SiC. Electrochemical and solid-state letters, 2007. 10(7): p. K24.

[10] Lihuan, W.H.S.X., Hierarchical Porous Patterns of n-type 6H-SiC Crystals via Photo- electrochemical Etching. Journal of materials science & technology, 2013. 29(7): p. 655-661.

[11] Naderi, N., et al., Enhanced optical performance of electrochemically etched porous silicon carbide. Semiconductor science and technology, 2013. 28(2): p. 25011.

[12] Chen, Y., et al., Hybrid Anodic and Metal-Assisted Chemical Etching Method Enabling Fabrication of Silicon Carbide Nanowires. Small, 2019. 15(7): p. 1803898

[13] Leitgeb, M., et al., Metal assisted photochemical etching of 4H silicon carbide. Journal of physics. D, Applied physics, 2017. 50(43): p. 435301.

[14] Rittenhouse, T.L., P.W. Bohn and I. Adesida, Structural and spectroscopic characterization of porous silicon carbide formed by Pt-assisted electroless chemical etching. Solid state communications, 2003. 126(5): p. 245-250.

[15] Chen, Y., et al., A Facile, Low-Cost Plasma Etching Method for Achieving Size Controlled Non-Close-Packed Monolayer Arrays of Polystyrene Nano-Spheres. Nanomaterials (Basel, Switzerland), 2019. 9(4): p. 605.

[16] Tan, J., et al., Fabrication of uniform 4H-SiC mesopores by pulsed electrochemical etching. Nanoscale research letters, 2014. 9(1): p. 1-5.

[17] Leitgeb, M., et al., Communication-The Role of the Metal-Semiconductor Junction in Pt-Assisted Photochemical Etching of Silicon Carbide. ECS journal of solid state science and technology, 2015. 5(3): p. P148-P150.

Simulation and Optimization of Inkjet-printed Outlines to Improve Pattern Fidelity

Shaowei Hu
Sauvage Laboratory for Smart Materials, School of Materials Science and Engineering Harbin Institute of Technology (Shenzhen)
Shenzhen, China
hswwhhit@163.com

Wanchun Yang
Sauvage Laboratory for Smart Materials, School of Materials Science and Engineering Harbin Institute of Technology (Shenzhen)
Shenzhen, China
yangwanchunvip@126.com

Mingyu Li*
Sauvage Laboratory for Smart Materials, School of Materials Science and Engineering Harbin Institute of Technology (Shenzhen)
Shenzhen, China
myli@hit.edu.cn

Abstract—Inkjet-printing has become a effective method for fabricating flexible electronics. This paper aims to improve the form fidelity via a numerical method. The properties of ink fluid and parameters of printing process are taken into consideration, and some rules for improve pattern fidelity are found. Finally, strategies are proposed according to the results of simulation.

Keywords—printed electronics; fluid simulation; selective treatment.

I. INTRODUCTION

The development of materials and technologies has made it possible to directly fabricate electronics on flexible substrates by additive manufacturing process [1,2]. Inkjet printing is a mature technology and is fit for fabricating prototype products of printed electronics [3]. Different from traditional subtractive manufacturing process and other printing technologies, the pattern fabricated by inkjet printing consists of an array of ink droplets so the shape accuracy depends on the radius of droplets. As a result, the outlines of inkjet-printed pattern are wavy in most conditions. Because ink droplets are deposited line by line, the outlines are classified into two types, horizontal lines and oblique lines. For oblique lines, because the position of droplets are not parallel to them, the pattern fidelity and properties are always worse than horizontal lines [4].

To improve the shape quality of patterns, many methods have been studied. A direct strategy is decreasing the droplet radius to increase the resolution, in this way where some advanced printers have been developed like aerosol jet printing and electrohydrodynamic jet printing [5,6]. But these printers has high price and low printing rate, which run counter to the economic and efficient feature of printing electronics. FUJI DMP printer series are most welcome production this field for its relative high resolution and low price. Some studies tried to optimize the shape accuracy, proper droplet spacing lead to smooth outlines by controlling the effect of surface tension [7,8]. But the parameters are often decided by experimental methods. Selective treatment methods have been studied already [9], but these methods are always too complex to accord with the flexibility of inkjet printing technology.

In this study, a series of simulations was done using the software Fluent, and the shape outlines were derived from the simulation results. The inkjet printing parameters like drop spacing, frequency, and physical properties of ink like surface tension, viscosity and contact angle are all concerned in this numerical model. In this way, we have successfully obtained the outlines of a sequence of droplets, and the results will help to select proper printing parameters for a certain ink and then to get straight and smooth outlines. Next, we add a wetting region into the model to find out the self-assembly effect of the region. Finally, two strategies were designed to improve the shape accuracy by using the simulation results. In the first method, every outline was transformed into horizontal lines and printed first. In the second method, selective treatment is achieved by forming wetting regions with laser marking.

II. EXPERIMENTAL

A. Simulations of droplets

The geometry model and mesh are built in software ICEM. The computational domain is cuboid, and every side of the cube is defined as an individual part, as Fig.1 (a) shows. Then read the mesh file in Fluent, and set right scale. In model settings, multiphase type is Volume of Fluid and viscous model type is Laminar. The viscosity, surface tension, contact angle are also inputted into the model. Several sphere regions are defined as the initial position of droplets, as Fig.1 (b) shows. The time step size is 5×10^{-7} s and the number of time steps is constant between two droplets. Finally derive the outline of the patterns.

B. Selective treatment and printing

Use laser marking to treat the surface selectively. The laser pulse power is about 2.5w, frequency is 20 kHz and moving speed of laser beam is 1000 mm/s. Printer DMP 2850 is utilized to fabricate patterns. First, print a reference line on paper, and two laser marking crosses defining the position are placed at the endpoint of the reference line so that ink droplets can exactly deposit on the treated regions. The printing voltage is 16 V, frequency is 2 kHz and the drop spacing is 20 μm.

III. RESULTS AND DISCUSSION

In order to get smooth outlines, proper printing frequency and drop spacing are crucial parameters. But for different inks, the parameters are usually determined by experimental methods. To replace the experiment by numerical simulations will help to summarize some rules from a wide range of properties of ink liquid. Meanwhile, simulations will help to design and verify strategies to improve form accuracy.

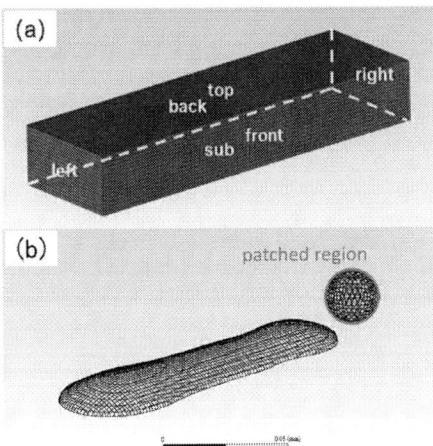

Fig.1. (a) Geometry model and meshes built by ICEM. (b) Simulated droplets and their patched regions in FLUENT

The influence of number of time steps between two droplets (interval time) is studied first.. Results in Fig.2 are a series of simulations with same parameters but different in number of time steps. It shows that the number of time steps has influence on the simulated patterns, and 500 is the most appropriate number to obtain patterns similar to the actual lines, thus in this study all simulations use 500 as the number of time steps. Besides, it should be noticed that the time in simulations is different from that in practice. The former is 500 time steps (2.5×10^{-4}s). The lack of friction in simulations has made the outlines of droplets move faster and higher frequency is thus needed to obtain patterns similar to the actual condition.

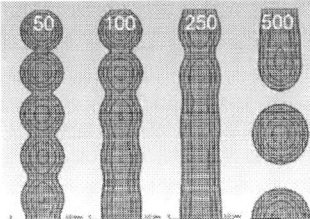

Fig.2. Simulation results with different number of time steps.

To avoid that drop spacing is too large to form a continuous line, the radius of droplets at a different contact angle is calculated. As droplets is spherical cap, its volume is determined by equation:

$$V = 1/3[\pi(3R-h)h^2] \quad (1)$$

Where R, h are radius and height of spherical cap respectively. For a constant volume of droplets with a different contact angle, the radius of spherical cap can be derived from (1), herein the relationship between the radius of droplet cap on the substrate and radius of droplet ball in the air are given by equation:

$$R/r = \{4/[(2/sin\,\theta+cot\,\theta)(1/sin\,\theta-cot\,\theta)^2]\}^{1/3} \quad (2)$$

Where r is the radius of droplet ball in the air, θ is the contact angle of spherical cap. r is constant for ink liquid and printer, and R varies with the contact angle. $2R$ is also the maximum of droplet spacing, thus the drop spacing ranges at a different contact angle can be confirmed.

In parameter settings, the contact angle ranges from 20° to 50°, which covers a common scope for inkjet printing. The surface tension of ink liquid ranges from 0.02 N/m to 0.04 N/m, and the viscosity ranges from 0.01 Pa·s to 0.02 Pa·s, which accord with the requirement of DMP printers.

After all droplets is landed on substrate, the number of time steps still should be taken into account. By printing actual lines and after referring to printing results from other studies, 1000 time steps is enough for proper settings to form a stable smooth line, so taking more calculations makes no sense.

Surface tension (γ) and viscosity (η) are the most important variables in all simulations, as the former is motivation while the latter is the resistance for droplets' merging. Different from γ, η, a series of simulations are done in this study in order to find crucial factors. Simulation results prove the value of γ/η is the factor to determine printing parameters.

Fig.3 shows the simulated maximum spacing for smooth outlines at different γ/η and contact angles, and the black curve is the theoretical maximum droplet spacing. When the contact angle is low (20°-30°), the spacing of higher surface tension droplets is larger than that of low surface tension droplets. Surface tension is the motivation for merging of droplets, therefore, droplets in the low contact angle tend to spread out and it requires relatively high surface tension for droplets to form smooth outlines. When the contact angle is about 35°, all spacing values are similar and close to its maximum, which means that 35° is the best contact angle to form smooth outlines. When the contact angle is relatively high (40°-50°), results differ from that with low contact angles, in these conditions higher viscosity helps to stabilize and balance the droplets forming smooth lines. In total, when the best value of γ/η is 2, it is best for improving form fidelity.

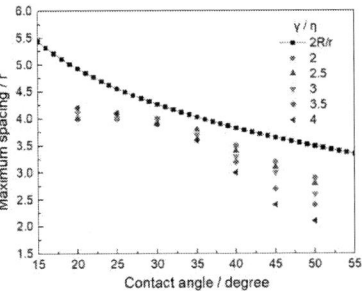

Fig.3.Simulated maximum spacing for droplets with different γ/η to form smooth outlines

Suppose the section shape of ink liquid is circular segment, and the width of printed lines can be calculated via mathematical calculation, shown in Fig.4. For a constant contact angle, the bigger the spacing is, the smaller the width become. As fine lines are needed in a wide range of applications, simulations performed in this study exactly can provide a guidance for fabricating finer lines.

There are two reasons for the jagged outline of oblique lines. First, the array distribution of droplets is not parallel to the line direction. Second, the interval time between two droplets is too long to merge with each other via liquid surface tension. Considering the second reason, short interval time between droplets is required accordingly so as to form smooth outlines, therefore, the outlines should be printed quickly. On the one hand, printing outlines is a strategy for improving shape accuracy. If the oblique lines were transferred into horizontal lines, the problems related with oblique lines would be solved. Hence, a rotatable platform is designed to make this thought tangible, as Fig.5 (b) shows, the rotation angle accuracy of 0.1°. The whole procedures are shown in Fig.5 (c). With assistance of a calculation program, the rotation angle and coordinates of the outlines can be calculated easily.

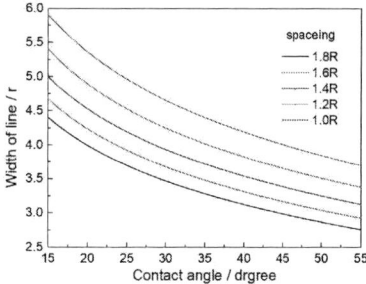

Fig.4.Calculated width of lines with different spacings.

The strategy mentioned above has some disadvantages. First, only straight outlines can be transferred into horizontal lines. Second, it is hard to achieve an accurate alignment of outlines after several rotation turns. Hence the other strategy — selective treatment — is proposed. The aim of selective treatment is to form regions with a different contact angle. The difference in contact angles will lead to a self-assembly effect and it also has been verified in this study. Adding a wetting region in the simulation model, and the width of this region is determined by the calculated results in Fig.4. Simulation results are shown in Fig.6 (a), indicating that wetting regions helps to form continuous and smoother outlines. What's more, droplets in wetting regions can form continuous lines with theoretical maximum spacing, and, as a result, width becomes smaller.

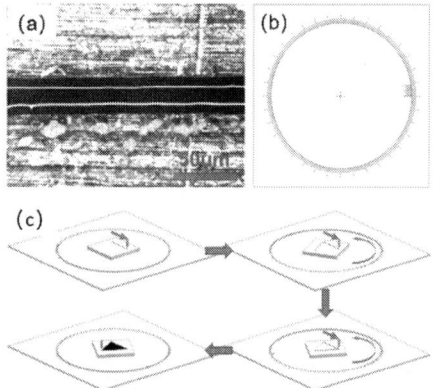

Fig.5. (a) Actual printed smooth horizontal line. (b) Schematic diagram of rotation platform. (c) Process of printing a triangle by outline-first strategy.

Laser marking technology is used for achieving selective treatment, because this technology is as flexible as inkjet printing. According to the Wenzel equation, surface roughness has influence on contact angles:

$$\cos\theta^* = a(\gamma_{SG} - \gamma_{SL})/\gamma_{LG} = a\cos\theta \quad (3)$$

Where θ^* is the contact angle on rough surface, γ_{SG}, γ_{SL} and γ_{LG} are the interface tension between solid and gas, solid and liquid, liquid and gas respectively, and a is a roughness factor equal to the ratio of actual contact area to projection area. For a contact angle below 90°, higher roughness will lead to a lower contact angle. Procedures of the selective treatment is shown in Fig.6 (b). For opaque substrates, low power laser can roughen the surface directly.

However, laser energy can't be absorbed by transparent substrates. To solve it, a transparent substrate is closely placed on a metal panel., Then laser marking will stir up tiny metal particles and these particles will roughen the surface of transparent substrate.

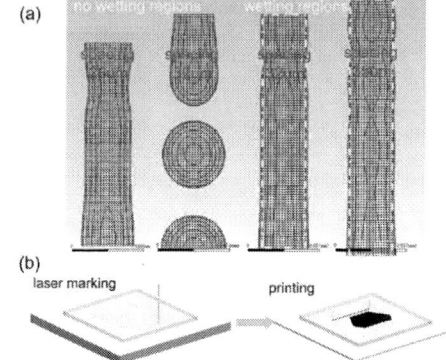

Fig.5. (a) Lines simulated with or without wetting regions. (b) Schematic diagram of selective laser marking treatment on the surface of a transparent substrate.

IV. CONCLUSION

A simulation model has been preliminarily built to study the pattern formation of printing. Simulation results show that the surface tension of ink is the main factor determining the form fidelity within the range of parameters required by printers. For different contact angles, forming smooth outlines needs different surface tension. As approximate simulation results, more factors should be involved such as dynamic contact angles. Finally, two strategies are proposed. Compared to outlines-first printing, selective laser treatment is more flexible and more suitable for improve the fidelity of patterns.

REFERENCES

1. H. W. Choi, Z. Tianlei, S. Madhusudan, "Recent Developments and Directions in Printed Nanomaterials". Nanoscale, 2015, 7, pp.3338–3355.

2. S. Magdassi, K. Alexander, "Conductive Nanomaterials for Printed Electronics" Small, 2014,10(17), pp.3515-3535.

3. P. Calvert, "Inkjet Printing for Materials and Devices". Chemistry of Materials, 2001,13(10), pp.3299-3305.

4. S. Enrico, P. Maxim, B. Reinhard, "The Design Challenge in Printing Devices and Circuits: Influence of the Orientation of Print Patterns in Inkjet-printed Electronics". Organic Electronics, 2016, 37(), pp.428–438.

5. A. Mahajan, C.D. Frisbie, L. F. Francis, "Optimization of Aerosol Jet Printing for High-Resolution, High-Aspect Ratio Silver Lines". ACS Applied Materials & Interfaces, 2013,5(11) pp.4856-4864.

6. M.S. Onses, E. Sutanto, P. M. Ferreira, "Mechanisms, Capabilities, and Applications of High-Resolution Electrohydrodynamic Jet Printing" Small, 2015,11(34), pp.4237-4266.

7. D. Kim, S. Jeong, B. K. Park, "Direct Writing of Silver Conductive Patterns: Improvement of Film Morphology and Conductance by Controlling Solvent Compositions". Applied Physics Letters, 2006, 89(26), pp.264101 1-3.

8. D. Soltman, B. Smith, H. Kang, "Methodology for Inkjet Printing of Partially Wetting Films". Langmuir the Acs Journal of Surfaces & Colloids, 2010, 26(19), pp.15686-15693.

9. P. Nguyen, L. P. Yeo, B. K. Lok, "Patterned surface with controllable wettability for inkjet printing of flexible printed electronics". Acs Applied Materials & Interfaces, 2014, pp.4011-4016.

Synthesis and characterization of Ag-37at.%Cu solid solution nanoparticles for high reliability packaging

Wanchun Yang
Sauvage Laboratory for Smart
Materials, School of Materials Scien
and Engineering
Harbin Institute of Technology
(Shenzhen)
Shenzhen 518055, China
yangwanchunvip126.com

Shaowei Hu
Sauvage Laboratory for Smart
Materials, School of Materials Scienc
and Engineering
Harbin Institute of Technology
(Shenzhen)
Shenzhen 518055, China
hswwhhit@163.com

Wei Zheng
Sauvage Laboratory for Smart
Materials, School of Materials Scienc
and Engineering
Harbin Institute of Technology
(Shenzhen)
Shenzhen 518055, China
754560018@qq.com

Mingyu Li
Sauvage Laboratory for Smart
Materials, School of Materials Scien
and Engineering
Harbin Institute of Technology
(Shenzhen)
Shenzhen 518055, China
myli@hit.edu.cn

Abstract—In this paper, Ag-Cu solid solution nanoparticles (ACSS-NPs) were prepared by liquid phase chemical reduction for high reliability packaging. On the one hand, the ACSS-NPs only with the lattice structure of Ag show excellent oxidation resistance. On the other hand, solute of Cu in Ag phase is expected to inhibit the electrochemical migration of Ag. The pressure sintering was used to prepare the Ag-Ag joints. The shear strength of the joints sintered at 295 ℃ for 30 min under 20MPa, reaches 104.5 MPa, which satisfies the use requirements.

Keywords—Ag-Cu solid solution, nanoparticles, high reliability

I. INTRODUCTION

With the emergence of hybrid and electronic vehicles, higher service temperature and electric current density puts forward higher requirements for interconnection materials[1]. Soldering may be the most popular method, however, Sn-based soft solers are difficult to withstand above 150 °C because of thermomechanical stress. At present, Au80Sn20 eutectic solder is widely used for high reliability power devices packaging[2]. However, high process temperature (above 320 °C) is easy to cause chip cracking due to thermal mismatch, and the formation of AuSn$_4$ greatly increases the brittleness of the joints[3]. At the same time, the price of Au is too high. Sintering bonding is a new technology for interconnection. Different from soldering, sintering is a solid-state material transport process. During sintering, the decrease of total energy drives the atomic diffusion. Since excellent electrical, thermal and antioxidant properties, Ag particles has been used for interconnection by sintering in power electronic since 1987[4]. In the early stage of the development, Ag particles were micron-sized. During sintering process, ultra-high pressure was necessary to assist bonding. So large pressure may result in cracking of die.

With the development of the nano technology, Ag nanoparticles are used for low pressure or pressureless sintering[5,6]. However, Ag is prone to take place electromigration and thermal migration under high electric current density and temperature, which is a major failure mode in electronic interconnections. Guo-Quan Lu et al[7] reported that the sintered silver films failed under electromigration at 150 ℃. Alloying has been proved to be effectively enhanced the EM-resistance. Yu-chen Liu et al proved that the alloy with single-phase (Ag-2.26 at.%) has better EM-resistance than the alloy with multi-phase (Ag-27.98 at.%)[8].

Therefore, supersaturated ACSS-NPs were prepared with single-phase. And ultra-high shear strength Ag-Ag joints were obtained by pressure sintering.

II. EXPRIMENTAL

In this paper, ACSS-NPs were prepared by liquid phase chemical reduction. Cu(NO$_3$)$_2$ ·3H$_2$O and AgNO$_3$ was the precursor. NaBH4 acted as reductant, reducing the Ag atoms and Cu atoms at the same time. Citric acid acted as stabilizer, decomposition occurring above 175 ℃. The original atomic ratio of Ag to Cu was 3:2, and preparation process took place in an ice bath. The phase analysis of ACSS-NPs particles was detected by X-ray diffraction. Ag-Ag joints were obtained by pressure sintering. The joints was measured by shear tester to get the value of shear strength. And transmission electron microscopy (TEM, F30, FEI).

III. RESULTS AND DISCUSSION

Fig. 1 is the TEM image of ACSS-NPs prepared by liquid phase chemical reduction. The mean size of the particles is 15 nm. NaBH$_4$ is highly reductive, when it was added into the reaction system, a mass of Ag$^+$ and Cu^{2+} was reduced simultaneously, and realized multi-locus nucleation under the action of magnetic stirring. At the same time, low temperature conditions made it difficult for the particles to grow, and proned to form small nanoparticles. The EDS result show that the atom ratio of the particles is Ag-37at.%Cu.

Fig. 1 The TEM image of ACSS-NPs

Fig. 2 is the XRD pattern of the ACCS-NPs prepared. The five peaks tested all belong to Ag. No peaks belong to Cu were detected. Based on the results of EDS and XRD, the ASCC-NPs were proved to be single-phase of solid solution. Although, Ag and Cu partly satisfy the Hume-Rothery rules. But they have a solubility limit well below 1 at.% at room temperature in eutectic phase diagram. Because of the lattice mismatch of about 11.7%, the mixing enthalpy is about 6kJ/mol. In addition, nanocrystalline materials can also improve the solid solubility of atoms, and even the immiscible materials in the binary phase diagram can achieve a certain solid solubility in the nanostructure. The reason why

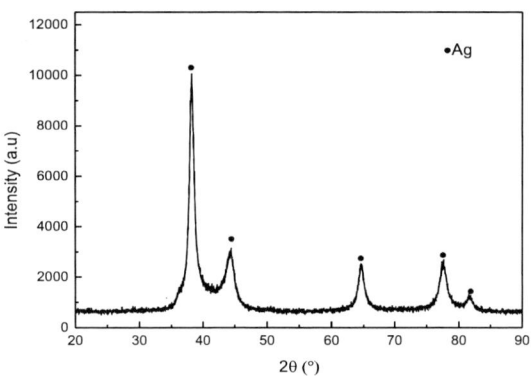

Fig.2 The XRD pattern of ACSS-NPs

Fig.3 is the shear strength of the joints obtained at 295 °C for 30 min under different pressure. With the pressure increasing, the shear strength shows an obvious upward trend. When pressure is 10MPa, the shear strength reached 43.6 MPa, which have satisfied the general operating requirement. When the sintering pressure is increased to 15MPa and 20MPa , the shear strength reached

to 76.4 MPa and 104.5 MPa respectively, which well meets the requirements of high reliability packaging.

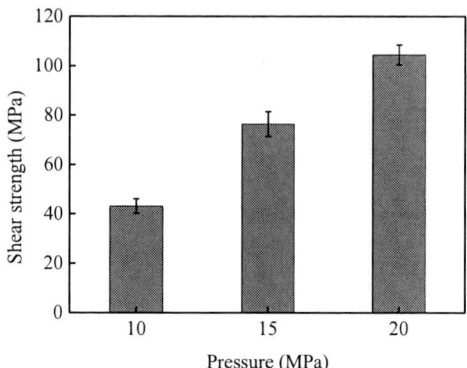

Fig.3 The Shear strength of the joints

In a fact, ACSS-NPs needs to sintered at higher temperature and larger pressure compared to the nano-Ag packaging materials. This phenomenon is because solute Cu atoms cause large lattice distortion in the Ag crystal, which is an obstacle to atomic diffusion and migration. Therefore, the diffusion of atoms in the sintering process requires a greater sintering driving force, that is, higher temperature and pressure. At the same time, the microstructure of the joint evolves slowly in the service process, so the service reliability and high temperature stability can be effectively improved.

IV. CONCLUSSION

Novel ACSS-NPs were prepared by liquid phase chemical reduction for high reliability packaging. the ACSS-NPs only with the lattice structure of Ag show excellent oxidation resistance because that Cu atoms are solute atoms in the crystal lattice of silver. The pressure sintering was used to prepare the Ag-Ag joints. The shear strength of the joints sintered at 295 ℃ for 30 min under 10 MPa, reache 43.6MPa, which satisfy the use requirements. When the sintering pressure is increased to 15MPa and 20MPa , the shear strength reached to 76.4 MPa and 104.5 MPa respectively, which well meets the requirements of high reliability packaging.

ACKNOWLEDGMENT

The authors acknowledge the support from the Science and Technology Innovation Committee of Shenzhen (JCYJ20180306172006392) and the support from National Nature Science Foundation of China (No.52075125).

REFERENCES

[1] J. Watson, G. Castro, A review of high-temperature electronics technology and applications[J], Journal of Materials Science: Materials in Electronics 26 (2015) 9226-9235.

[2] J. Yoon, B. Noh, S. Jung, Interfacial reaction between Au-Sn solder and Au/Ni-metallized Kovar[J], Journal of Materials Science: Materials in Electronics 22 (2011) 84-90.

[3] X. Wei, Y. Zhang, R. Wang, Y. Feng, Microstructural evolution and shear strength of AuSn20/Ni single lap solder joints[J], Microelectron. Reliab. 53 (2013) 748-754.

[4] Novel Large Area Joining Technique for Improved Power Device Performance[J].

[5] J. Yan, G. Zou, A. Wu, J. Ren, J. Yan, A. Hu, Y. Zhou, Pressureless bonding process using Ag nanoparticle paste for flexible electronics packaging[J], Scripta Mater. 66 (2012) 582-585.

[6] Yang W , Zheng W , Hu S , et al. Synthesis of highly antioxidant and low-temperature sintering Cu-Ag core-shell submicro-particles for high-power density electronic packaging[J]. Materials Letters, 2021(12):129781.

[7] J. N. Calata, G. Lu, K. Ngo, L. Nguyen, Electromigration in Sintered Nanoscale Silver Films at Elevated Temperature[J], J. Electron. Mater. 43 (2014) 109-116.

[8] Y. Liu, Y. Yu, S. Lin, S. Chiu, Electromigration effect upon single- and two-phase Ag-Cu alloy strips: An in situ study[J], Scripta Mater. 173 (2019) 134-138.J. Clerk Maxwell, A Treatise on Electricity and Magnetism, 3rd ed., vol. 2. Oxford: Clarendon, 1892, pp.68–73.

Experimental Analysis for Bump Shear Method of BSOB Wire Bonding Process

Zhu Chenjun*
The 29th Research Institute of China Electronics Technology Group Corporation
Chengdu, China
270154750@qq.com

Wen Zehai
The 29th Research Institute of China Electronics Technology Group Corporation
Chengdu, China
187332095@qq.com

Wu Yilong
The 29th Research Institute of China Electronics Technology Group Corporation
Chengdu, China
yilong_wu@126.com

Zhang pingsheng
The 29th Research Institute of China Electronics Technology Group Corporation
Chengdu, China
zhangps168@yeah.net

Li hui
The 29th Research Institute of China Electronics Technology Group Corporation
Chengdu, China
15928647104@163.com

Dong dong
The 29th Research Institute of China Electronics Technology Group Corporation
Chengdu, China
26879241@qq.com

Abstract—The ball stitch on ball(BSOB) bonding mode is very common in wire bond, which greatly enhances the strength of the 2nd bonding and provides well support for the wire heel. The bump that is directly torn off by the wedge will leave a tail wire which will reduce the bonding strength. Seriously, it will lead to bonding failure or flying wire. Therefore, the wire tearing off method of bump needs to be improved. There are four bump shear methods: (a) horizontal bump shear; (b) horizontal reverse bump shear; (c) horizontal reciprocating bump shear; (d) diagonal bump shear.

This paper gets well consistency bumps through the parameter optimization test of different shear modes about 25um gold wire bonding. Secondly, the paper analyzes the bond strength about four bump shear modes combining the relevant parameters obtained from the above experiments.

Keywords—Wire bond; Ball stitch on ball; Bump; Shear mode;

I. INTRODUCTION

Wire ball bonding technology has the advantages of fast bonding speed, non-directionality, and high bonding strength, which is widely used in micro-assembly processes. In the traditional ball bonding process, the first bonding point is ball type, and the second bonding point is wedge type, which has extremely high requirements on the surface state of the pad. Due to the small effective bonding area of wedge bonding point and substrate pads, the flatness and pollution of the substrate pad surface will have a great impact on the bonding strength. Therefore, a special bonding mode—BSOB method is usually used in wire ball bonding to enhance the strength of the wedge bonding[1-3]. The mainstream automatic gold wire ball bonding machine is equipped with a pre-planted gold ball function, and the gold ball can still form a reliable connection on the substrate pad with a poor surface condition[4]. The BSOB wire bonding mode is through ball bond on the chip, the dome wire is on the pre-plated gold wire on the substrate and the wedge bonding is performed, as shown in Fig. 1.

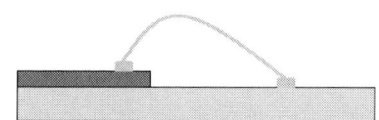

Fig.1. BSOB wire bonding process

In the BSOB mode, the flat surface of the bump is formed by the shear movement of the capillary. Excessive shear distance will result in short tail wire or flying wire, small shear distance will result in a wire tail on the surface of the bump, which greatly reduces the bonding strength of the wedge bond[5]. The shear height parameter of the capillary will also affect the state of the surface of the bump. Therefore, the selection of the bump shear method and the setting of shear parameters have a great influence on the reliability of bonding wires.

This paper analyzes the reliability differences of different bump shear modes by studying the four shear modes of horizontal bump shear, horizontal reverse bump shear, horizontal reciprocating bump shear and diagonal bump shear in 25um gold wire ball bonding. And shear parameters of the diagonal bump shear mode are tested and analyzed, the bump shear parameters are obtained which are conductive to the well bonding strength of the wedge bonding point.

II. BUMP SHEAR METHOD

A. Horizontal Bump Shear Mode

Horizontal shear mode means that the direction of the gold wire is used as the shear direction of the capillary to move horizontally , as shown in Fig. 2a. Unidirectional shear movement requires more accurate control of the shear distance and bump consistency. Generally speaking, the shear distance of the horizontal shear mode is related to the shear height. The greater the shear height, the smaller the diameter of the bump, and the smaller the required shear distance.

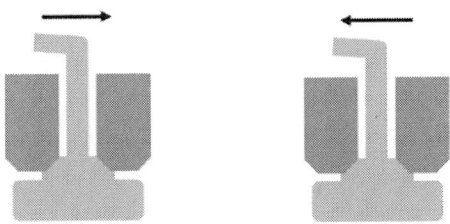

(a) horizontal bump shear   (b) reverse horizontal bump shear

Fig.2. Bump shear modes

B. Horizontal Reverse Bump Shear Mode

Horizontal reverse bump shear mode means that the opposite direction of the gold wire is used as the shear direction of the horizontal movement for the capillary, as shown in Fig. 2b. Reverse shear mode also requires accurate control of the shear distance and consistent bump. Compared to the horizontal shear mode, this mode can provide better support for the wire heel.

C. Horizontal Reciprocating Bump Shear Mode

The horizontal reciprocating bump shear mode combines the above two modes. The capillary performs the shear movement along the front and back of the gold wire direction, as shown in Fig. 3a. This mode is conductive to the formation of bump with a flat surface, but the bidirectional shear movement of the capillary can easily lead to the shortage of flying tail wires. The mode is usually suitable for wires with larger elongation rate.

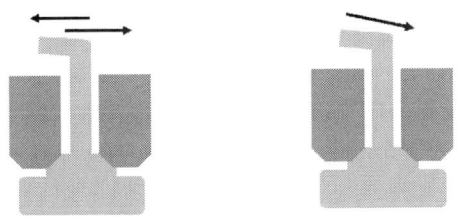

(a) horizontal reciprocating bump shear   (b) diagonal bump shear

Fig.3. Bump shear modes

D. Diagonal Bump Shear Mode

Diagonal shear movement refers to the diagonally downward movement of the capillary along the path of the gold wire, as shown in Fig. 3b. The bump surface of this shear mode is an inclined surface to improve the support of the wire heel, as shown in Fig. 4. It requires extra setting of the start shear height and end shear height, and a larger shear distance is required. Compared with the previous three modes, when the capillary performs diagonal shear, its shearing force on the bumps is greater, so higher bonding strength of bumps and pads is required.

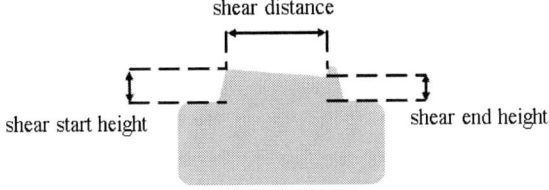

Fig.4. Diagonal bump shear shape

III. EXPERIMENT ANALYSIS

In order to further analyze the differences of the four bump shear modes, firstly the paper used the initial paramters(shear height 20um, shear distance 30um) to pre-plant gold bumps of horizontal bump shear mode and horizontal reverse bump shear mode. The horizontal reciprocating bump shear mode has a small shear distance, pre-planting gold bumps with 20um shear height and 15um shear distance. For diagonal bump shear mode, 25um start shear height, 20um end shear height and 35um shear distance are taken for gold bump pre-planting.

Four bump shear modes used the same bonding parameters to pre-planted thirty gold bumps through 25um gold wire thermoacoustic bonding. The length of the pre-planted gold bump tail wire was measured under a measuring microscope and the average value was taken, the results are shown in Table I.

TABLE I.   THE LENGTH OF GOLD BUMP TAIL

Bump Shear Mode	Tail Length(um)
Horizontal Bump Shear	18.15
Horizontal Reverse Bump Shear	18.08
Horizontal Reciprocating Bump Shear	15.42
Diagonal Bump Shear	6.54

It can be seen from Table I that the first two unidirectional shear modes have long tails, the length of the tail in the horizontal reciprocating bump shear mode is slightly shorter than the former, and diagonal bump shear mode with the shortest tai. Fig. 5 shows the four shear modes of gold bump planted observed under the microscope, the surface of the bumps in the Fig. 5d is the flattest one which are pre-planted by diagonal bump shear mode.

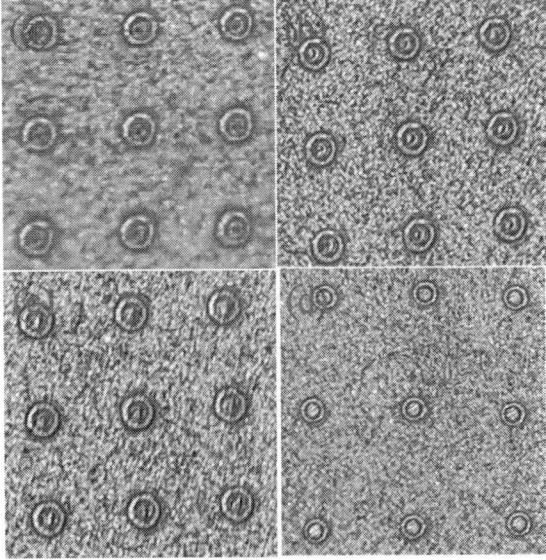

Fig.5. Four shear modes of gold bump

Secondly, according to the above experiments, the shear parameters of the four modes are optimized, the length of the tail wire of bump surface is as short as possible under ensuring the stability of the bonding process. The selection of bump shear parameters is closely related to Free air ball size and bump size. In the experiment, the FAB size is 60um and the bump diameter is 70um. After multiple orthogonal experiment analysis, the shear parameters suitable for the four bump shear modes are obtained, as shown in Table II.

TABLE II.   BUMP SHEAR PARAMETER

Shear mode	Bump Shear Parameter(um)			
	Shear height	Shear distance	Start shear height	End shear height
Horizontal Bump Shear	5	32	/	/
Horizontal Reverse Bump Shear	5	32	/	/
Horizontal Reciprocating Bump Shear	5	18	/	/
Diagonal Bump Shear	/	42	25	18

The optimized parameters were used to pre-plant gold ball, and the length of the tail wire on the bump surface was measured by a measuring microscope. The results are shown in Table III. Compared with the initial shear parameters, the tail wire length in the optimized parameters has been greatly reduced, and the tail wire length for diagonal shear mode is the shortest.

TABLE III.   THE LENGTH OF GOLD BUMP TAIL

Bump Shear Mode	Tail Length(um)
Horizontal Bump Shear	8.21
Horizontal Reverse Bump Shear	8.75
Horizontal Reciprocating Bump Shear	5.41
Diagonal Bump Shear	1.63

Thirdly, using the same bonding parameters to bond the pre-planted gold bump with 25um gold wires, the bonding situation of wedge bond and the support of the bumps for the four modes to the wire heel are obtained.

Although the length of the optimized bump tail wire has been reduced, the remaining tail wire will still cause abnormalities in the wedge bond. The unidirectional shear modes differ in the direction of the shear, the orientation of the remaining tails is also different. When the capillary is bonding, the tail will be squeezed to the front side of the wedge bond or behind the gold wire heel, resulting in abnormal bonding point as shown in Fig. 6 and Fig.7. The tail wire of the reciprocating shear mode is shorter than the former, bur there are still abnormal bonding point, as shown in Fig. 8. The bump bonding effect obtained by the diagonal bump shear mode is the best, and the effective bonding area of the wedge bonding point is the largest, as shown in Fig. 9.

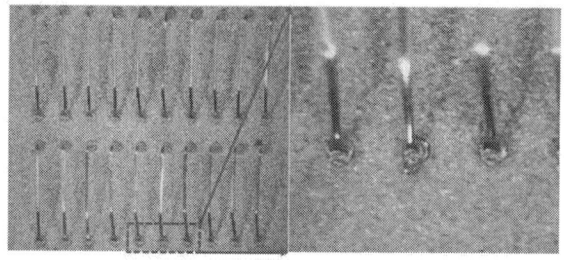

Fig.6. Gold wires of horizontal shear bump

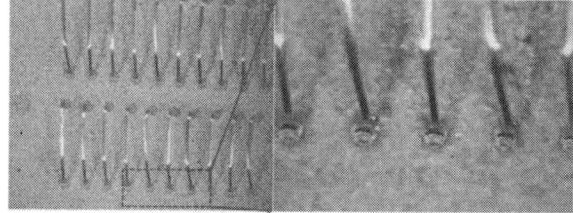

Fig.7. Gold wires of horizontal reverse shear bump

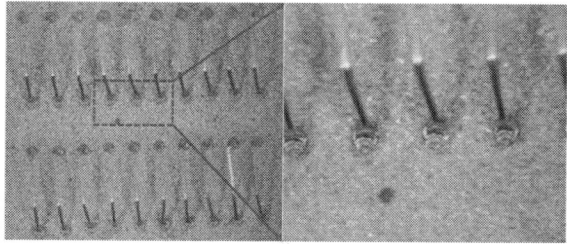

Fig.8. Gold wires of horizontal reciprocating shear bump

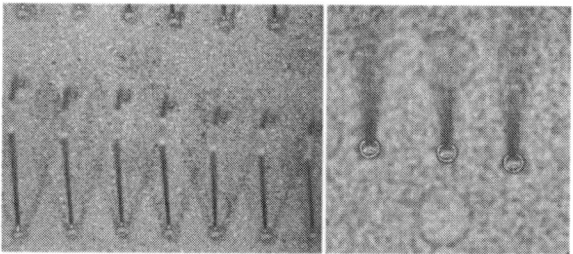

Fig.9. Gold wires of diagonal shear bump

The tensile test was performed on the gold wire obtained in the above experiment, and the average value of the gold wire tensile force are shown in Table IV. Each bump shear modes are bonded 30 gold wires, the wire tension in the diagonal shear mode is much greater than the first three types. Most of the first three modes of failure are the failure of the gold wire heel of wedge bonding point. In the diagonal shear mode, 12 wires were broken at the ball bonding point and 18 wires were broken at the wedge bonding point.

TABLE IV.   THE AVERAGE VALUE OF THE WIRE TENSILE FORCE

Bump Shear Mode	Average Value of Tensile Force(g)
Horizontal Bump Shear	8.11
Horizontal Reverse Bump Shear	7.82
Horizontal Reciprocating Bump Shear	8.40

Bump Shear Mode	Average Value of Tensile Force(g)
Diagonal Bump Shear	11.24

IV. CONCLUSION

The paper is studied the bump shear method of BSOB wire bonding process, analyzed the differences between the four bump shear modes. Pre-planting of bumps on 25um gold wire, by optimized bump shear parameters in the four modes, to ensure the flatness of the bump surface. Through the gold wire bonding experiment and the shear force test on the pre-planted gold bumps, the bond strength of the wedge bonding point under the four bump shear modes was compared. The results show that the diagonal bump shear mode is conductive to the formation of high-reliability gold wire interconnection.

REFERENCES

[1] HARMAN G. Wire bonding in microelectronics[M]. New York: McGraw-Hill Companies, 2010: 36-37..

[2] Yong L , Allen H , Luk T , Irving S. "Simulation and experimental analysis for a ball stitch on bump wire bonding process above a laminate substrate." Electronic Components and Technology Conference, 2006. Proceedings. 56th IEEE, 2006.

[3] REN Z. Silver bonding wire for BSOB (Bond-Stitch-on-Ball)/ BBOS (Bonding-Ball-in-Stitch) [C]. China Semicond Technol Int Conf. Shanghai, China. 2016: 1-3.

[4] TANK C, LIONG J Y. Bond stitch on ball for bare copper wire[C]. 35th IEEE/CPMT Int Elec Manufac Technol Conf. Perak, Malaysia. 2012: 1-7.

[5] Takahashi, Y. , and M. Inoue . "Numerical Study of Wire Bonding-Analysis of Interfacial Deformation Between Wire and Pad." ASME J. of Electronic Packaging, vol.121, 2002.

Low Temperature Curing Copolyimide with Monomer Containing Pyrazine Moiety

1st Changqing Li
Shenzhen Institute of Advanced Electronic Materials, Shenzhen Institute of Advanced Technology, Chinese Academy of Sciences,
Shenzhen 518055, China
cq.li@siat.ac.cn

2nd Guoping Zhang*
Shenzhen Institute of Advanced Electronic Materials, Shenzhen Institute of Advanced Technology, Chinese Academy of Sciences,
Shenzhen 518055, China
gp.zhang@siat.ac.cn

3rd Jinhui Li*
Shenzhen Institute of Advanced Electronic Materials, Shenzhen Institute of Advanced Technology, Chinese Academy of Sciences, Shenzhen 518055, China
jh.li@siat.ac.cn

4th Yingying Li
Shenzhen Institute of Advanced Electronic Materials, Shenzhen Institute of Advanced Technology, Chinese Academy of Sciences,
Shenzhen 518055, China
yy.li2@siat.ac.cn

Abstract—Polyimides (PIs) are a specific kind of polymer that extensively applied in microelectronics for their outstanding thermal stability, thermo-mechanical property and electrical property. Imidization of traditional PIs were commonly completed at about 350 °C, which limits their applicable scope in the process of semiconductor manufacturing. Therefore, it is highly required to achieve superior properties of films when annealed below 250 °C. This study comprehensively investigated thermal and mechanical properties of pyrazine moieties incorporated copolyimide. The thermo-mechanical properties of copolyimide based on 4,4'-(pyrazine-2,5-diyl)dianiline monomer exhibit excellent mechanical properties and low coefficient of thermal expansion when cured at 200 °C (tensile strength of 138 MPa, modulus of 3.1 GPa and CTE of 9.31 ppm/°C) and 250 °C (tensile strength of 191 MPa, modulus of 3.6 GPa and CTE of 9.32 ppm/°C). The aggregation structure as well as molecular orientation was investigated by WAXRD and polarized FTIR spectrum. This research gives an investigation of pyrazine-containing copolyimide for low-temperature curable polyimide and provides opportunities of application in the microelectronics.

Keywords—*Low temperature curing, Polyimide, Microelectronics*

I. INTRODUCTION

Polyimide has been demonstrated with outstanding comprehensive performance, such as heat resistance, chemical resistance, electric characteristics and mechanical strength, and extensively used as excellent polymer materials in flexible circuit boards, organic light emitting diodes and microelectronics.[1] Recently, in multichip packages, silicon substrates are stacked in a memory chip to achieve high-density packaging, and those are thinner than the usual ones due to fitting them in the traditional size.[2] Photosensitive polyimide (PSPI) is used for dielectric materials in re-distribution layers (RDLs) of Fan-out Wafer Level Packages (FOWLPs), which can simplify processing and avoid the use of photoresist in the microelectronic industry.[3] Traditionally, the prepolymers of polyamide acid (PAA) or poly(amic ester) (PAE) are imidized at high temperature (>350 °C).[4] However, such high imidization temperature has limited their application in FOWLPs for the coefficient of thermal expansion (CTE) mismatch of PI, Si and other package materials. In this case, low-temperature curable PI is highly required which can reduce the warpages for the large wafer during the electronic package process in FOWLP.[4],[5]

Several ways for low-temperature cured polyimide has been studied including one step polymerization[6] and employing low-temperature accelerants[7]. However, these reported methods suffered from more or less thorny issues. For instance, system using curing accelerators may lead to severe degradation of mechanical properties in PI films.[8] In addition, curing reagents were usually added separately with a large amount releasing gas during thermal annealing, which should be avoided in package materials and process. To address these problems, we believe that incorporation of alkaline functional group in precursor of polyimide may augment imidization rate of PI films below 250 °C.

In this study, we aimed to improve mechanical property of polyimide, consisting of pyrazine unit, by copolymerization strategy. 4,4'-oxydianiline (ODA) with flexible ether linker may improve the entanglement of each polymer chain to enhance the elongation of PI film. As a result, the obtained low-temperature cured polyimide film gives excellent comprehensively properties.

II. EXPERIMENTS

A. Preparation of Polyimides

The molar ratios of PRZ and ODA were 6:4, 5:5, 4:6, 3:7 and 2:8. For instance, for a molar of PRZ: ODA = 6:4, PRZ (3 mmol) and ODA (2 mmol) were dissolved in DMAc. The resulting solution was stirred until all starting materials had dissolved. And then, BPDA (5 mmol) was added in batch to bottle and kept stirring under N_2 at 0 °C. A viscous solution with a solid content of 12 wt% was obtained after 16 h. PI film could be obtained under stepped heating program with heating rate of 5 °/min.

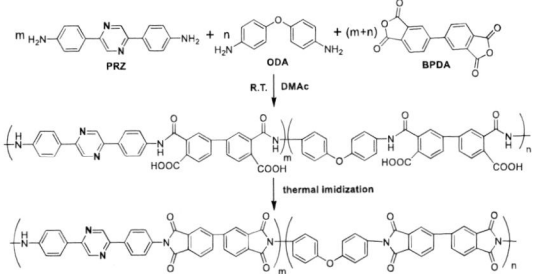

Fig. 1. Synthesis of copolyimide (PRZ/ODA/BPDA).

Table I. Molecular Weights and PDIs of PAAs

Sample	M_n ($\times 10^4$)	M_w ($\times 10^4$)	DPI (M_w/M_n)
coPI-1 (PRZ/ODA/BPDA=6/4/10)	2.96	4.24	1.43
coPI-2 (PRZ/ODA/BPDA=5/5/10)	3.12	4.42	1.41
coPI-3 (PRZ/ODA/BPDA=4/6/10)	3.07	4.20	1.36
coPI-4 (PRZ/ODA/BPDA=3/7/10)	2.76	3.86	1.40
coPI-5 (PRZ/ODA/BPDA=2/8/10)	3.11	4.47	1.43

III. RESULTS AND DISCUSSIONS

A various of PAAs were prepared from various molar ratios of PRZ/ODA with BPDA. The chemical structures of PAAs and PIs were identified with ^{1}H-NMR spectra and FT-IR spectra. The peak of H_a at 9.29 ppm and the peak of H_b at 7.04 ppm are assigned to the protons of pyrazine ring by comparing with proton NMR spectra of PRZ and ODA, respectively. The molar ratios (m:n) of PRZ/ODA in co-PAAs with calculated by integrating the area of two equimolar of H_a and one equimolar of H_b peaks. (Fig. 2) As 4:6 of PRZ/ODA were copolymerized with BPDA, the molar ratio of PRZ/ODA in co-PAA reached nearly 1:1 in backbone of polyamic acid. As exhibited in Fig. 3, imide absorption of C=O is shown at 1772 cm^{-1} and 1714 cm^{-1} combined with C−N−C absorption of imide ring at 1363 cm^{-1}, which also validated the successful construction of imide ring. In this paper, we selected a ratio of 5:5:10 (PRZ: ODA: BPDA) of co-polymer (coPI-3) as an example to investigate thermo-mechanical properties dependence on the temperature.

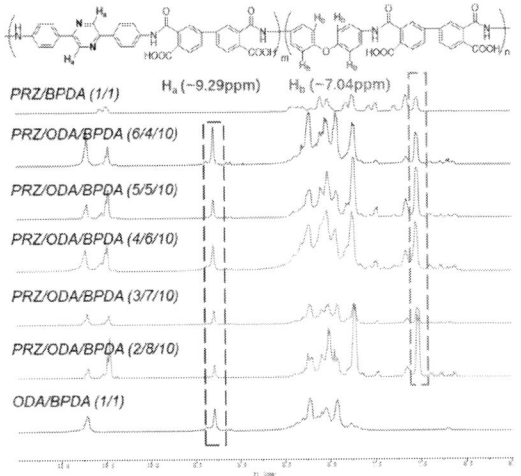

Fig. 2. ^{1}H-NMR spectra of co-PAAs with different molar ratios of PRZ/ODA/BPDA. The chemical shift of hydrogens from pyrazine unit was noted as blue at 9.29 ppm and that of hydrogens from ODA backbone was noted as red at 7.04 ppm.

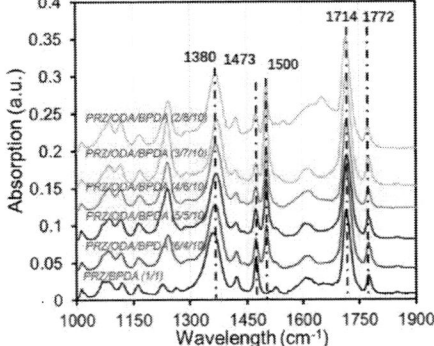

Fig. 3. ATR-FTIR spectra of PI films prepared from copolymers with different molar ratios of PRZ/ODA/BPDA.

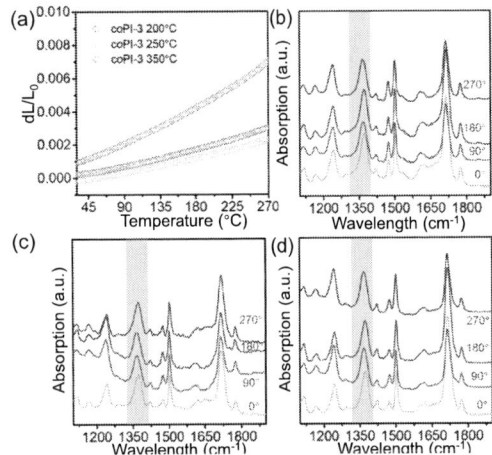

Fig. 4. (a) dimensional change curves versus temperature for coPI-3 under different temperatures; polarized infrared spectra from 1900 cm^{-1} to 1100 cm^{-1} for polyimide films (coPI-3) annealed at different temperatures with different polarized angles: (b)200 °C; (c)250 °C; and (d)350°C. (Green region is noted as stretching vibration of C-N-C at 1363 cm^{-1} of imide ring).

Copolymerization of PRZ, ODA and BPDA bring disorder into the PI chain. CTE values of coPI-3 were increased from 9.31 to 23.2 ppm/°C when annealing the films by improving temperature from 200 °C to 350 °C. (Fig. 4a) Further investigation of chain orientation is done with polarized FTIR to understand the intuitive factor on CTE values. Polarized FTIR was a powerful tool to study the orientation of the groups in polyimide films. Dichroic ratio (R), which represent in-plane orientation, is shown as below:

$$R = \frac{A_{0°} + A_{180°}}{A_{90°}} \qquad (1)$$

$A_{0°}$, $A_{90°}$ and $A_{180°}$ stand for absorption area of IR spectrum at polarization angle of 0°, 90°and 180°, respectively. The vibration of C-N-C at 1363 cm^{-1} of imide ring is parallel to the main chain of polyimide, therefore R of 1363 cm^{-1} is regarded as a signal of in-plane orientation. Fig. 4b-d indicates that absorbances at 1363 cm^{-1} of coPI-3 decrease from 0° to 90°

and increase to 180°. What's more, the absorbances of 1363 cm^{-1} decreases from 180° to 270° and increases from 270° to 0°. The value of R is 3.57, 3.24 and 2.46, which is in one-to-one correspondence with the annealing temperature of copolyimide films at 200 °C, 250 °C and 350 °C. In another word, with increasement of annealing temperature, the PI chains of coPI-3 becomes unfavorable for in-plane orientation.

The resulting PAAs were cured as PI films and the mechanical properties were performed with DMA analysis. As shown in Fig. 5a, coPI-3 has a tensile strength of 150 MPa with tensile moduli of 3.1 GPa at 200 °C and tensile strengths of 186 MPa with tensile moduli of 3.5 GPa at 250 °C. Elongation at break of film cured under 200 °C is 14.3% and that of film cured at 250 °C is 18.3%. When coPI-3 was cured at 350 °C, the film has a tensile strength of 206 MPa with tensile moduli of 4.1 GPa and elongation at break of 46.1%. From a viewpoint of polymer chain structure, the ODA part increases the helical-shaped segments conformation, thereby occupying large elongation under a small tension. Interestingly, coPI-3 cured 350 °C has a comparable tensile strength however a three times higher elongation break when comparing with that cured at 200 °C, which means curing temperature of PI films has a definitive effect on molecular aggregation structure, chain orientation and chain mobility.

The storage modulus curves are summarized in Fig. 5b. The T_gs of coPI-3 annealed at 200 °C, 250 °C and 350 °C were 292 °C, 293 °C and 300 °C, respectively. The TGA of coPI-3s is exhibited in Fig. 5c. The $T_d5\%$ in N_2 were depicted in Table III. XRD patterns of copolyimide (coPI-3) treated at 200 °C and 250 °C show diffraction peaks appeared at 13.9° (d-spacing = 6.4 Å) and 30.0° (d-spacing = 3.0 Å). The secondary diffraction peak at 30.0° of coPI-3 annealed at 250°C seems like a sign of relatively higher intensity compared with annealed at 200°C. By curing film of coPI-3 at 350°C, the first diffraction peak in Fig. 5d at 14.7° (d-spacing = 6.0 Å) becomes sharper, which illustrates a signal of crystal-like order in partial copolyimide.

coPI-1 when annealed at 200 °C, 250°C and 350°C, respectively.

Table II. Mechanical properties of polyimide films.

Sample name	Final curing temperature (°C)	Tensile Strength (MPa)	Elongation (%)	Modulus (GPa)
coPI-3	200	150	14.3	3.1
coPI-3	250	186	18.3	3.5
coPI-3	350	206	46.9	4.1

Table III. T_g, CTE, $T_d5\%$ results of PI films.

Sample Name	Final curing temperature (°C)	T_g (°C)	CTE (ppm/°C)	$T_d5\%$ (°C)
coPI-3	200	292	9.31	558
coPI-3	250	293	9.32	546
coPI-3	350	300	23.2	548

IV. CONCLUSION

In this work, copolymerization of PRZ, ODA and BPDA can achieve comparable thermo-mechanical properties of PI films with traditional high-temperature annealed polyimide cured at 200 °C (tensile strength of 138 MPa, elongation of 14.3%, modulus of 3.1 GPa and CTE of 9.31 ppm/°C).

ACKNOWLEDGMENT

This work was supported by Guangdong Jointed Funding 2020A1515110934, National Natural Science Foundation of China (61904191), Youth Innovation Promotion Association of Chinese Academy of Sciences (2017410), Key R&D Project of Guangdong Province (2020B010180001) and National Key R&D Project from Minister of Science and Technology of China (2017ZX02519).

REFERENCES

[1] D.-J. Liaw, K.-L. Wang, Y.-C. Huang, K.-R. Lee, J.-Y. Lai, C.-S. Ha, "Advanced polyimide materials: syntheses, physical properties and applications," *Prog. Polym.Sci.,* vol. 37, no. 19, pp. 907-974, March 2012.

[2] T. J. Tsukui, "3D High Performance Packaging of Electronic Equipments and Guaranteed Reliability, " *Jpn. Inst. Electron. Packag.*, vol. 11, no. 5, pp. 317-325, June 2008.

[3] Y. Saito, K. Mizoguchi, T. Higashihara, M. Ueda, "Alkaline-developable, chemically amplified, negative-type photosensitive polyimide based on polyhydroxyimide, a crosslinker, and a photoacid generator, " *J. Appl. Polym. Sci.,* vol. 113, no. 6, pp. 3605-3611, September 2009.

[4] T. Enomoto, S. Abe, D. Matsukawa, T. Nakamura, N. Yamazaki, N. Saito, M. Ohe,T. Motobe, Ieee, "Recent Progress in Low Temperature Curable Photosensitive Dielectrics, " International Conference on Electronics Packaging (ICEP), Yamagata, Japan, 2017, pp. 498-501, doi: 10.23919/ICEP.2017.7939431.

[5] R.J. Iredale, C. Ward, I. Hamerton, "Modern advances in bismaleimide resin technology: A 21st century perspective on the chemistry of addition polyimides, " *Progress in Polymer Science,* vol. 69, pp. 1-21, June 2017.

[6] P.K. Tapaswi, C.-S. Ha, "Recent Trends on Transparent Colorless Polyimides with Balanced Thermal and Optical Properties: Design and Synthesis, "*Macromolecular Chemistry and Physics,* vol. 220, no. 3, 1800313, January 2019.

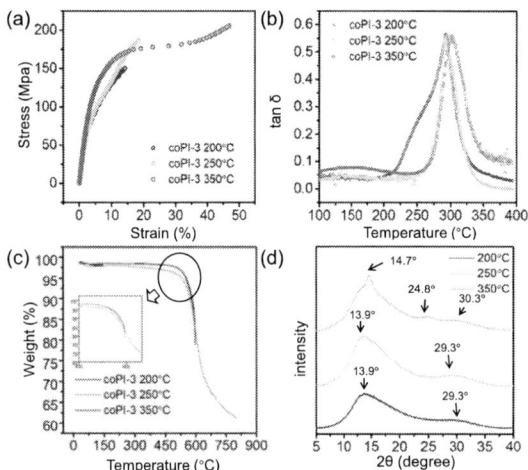

Fig. 5. (a) Stress–strain curves of coPI-3 films measured cured at 200 °C, 250 °C and 350 °C by TMA; (b) DMA curves of coPI-3 cured at different temperatures; (c) TGA curves of coPI-3 cured at different temperatures; (d) XRD patterns of

[7] M. Oba, "Effect of curing accelerators on thermal imidization of polyamic acids at low temperature, " *Journal of Polymer Science Part A: Polymer Chemistry*, vol. 34, no. 4, pp. 651-658, March 1996.

[8] Y. Xu, A. Zhao, X. Wang, H. Xue, F. Liu, "Influence of curing accelerators on the imidization of polyamic acids and properties of polyimide films, " *Journal of Wuhan University of Technology-Mater. Sci. Ed.,* vol. 31, no. 5, pp. 1137-1143, October 2016.

Microstructure evolution and interfacial reaction of Co-P/SAC305/Co-P solder joints at high current density

Tao Fan
College of Materials Science and Engineering
Chongqing University of Technology
Chongqing, China
827356317@qq.com

Donghua Yang*
Chongqing Municipal Engineering
Research Center of Institutions of
Higher Education for Special Welding,
College of Materials and Technology
Chongqing University of technology
Chongqing, China
yangdonghua@cqut.edu.cn

Haotong Qin
College of Materials Science and Engineering
Chongqing University of Technology
Chongqing, China
1278120444@qq.com

Yuqian Chen
College of Materials Science and Engineering
Chongqing University of Technology
Chongqing, China
627168390@qq.com

Tao Chen
College of Materials Science and Engineering
Chongqing University of Technology
Chongqing, China
1916546139@qq.com

Xiang Zhai
College of Materials Science and Engineering
Chongqing University of Technology
Chongqing, China
920067707@qq.com

Teng Ran
College of Materials Science and Engineering
Chongqing University of Technology
Chongqing, China
939392663@qq.com

Abstract—The interfacial reaction of Co-P/SAC305/Co-P solder at current density of 1.0×10^4 A/cm^2 was studied. In Co-P/SAC305/Co-P solder joints, Sn-3.0wt.%Ag-0.5%Cu (SAC305) BGA solder ball as a connection between Co-P UBM and Co-P UBM. Ultrasonic assisted electro-deposition process was used to manufacture the Co-P UBM. The P (phosphorus) concentrations were 5%. Conducted EM test on the solder joints after soldering with the DC stabilized power supply, the solder joints were tested at current density of 1.0×10^4 A/cm^2.The interface intermetallic compound (IMC) of Co-P/SAC/Co-P solder joint was mainly CoSn$_3$. After 200h of EM test，the average IMC thickness at the cathode of the Co-P/SAC/Co-P solder joint increased from 2.10 μm to 5.12 μm, while at the anode increased from 2.15 μm to 12.05 μm. During the EM process, the cathode dissolution phenomenon was found in Co-P/SAC/Co-P solder joints. With the increase of electromigration (EM) time for Co-P/SAC/Co-P solder joints, the IMC morphology at the anode became more regular and the surface smoothness improved. Due to the presence of Cu atoms in the solder, a local (Cu, Co)$_6$Sn$_5$ compound was formed at the anode. Therefore, this provides an important theoretical basis for the resistance of EM performance of the new Co-P UBM.

Keywords—Electromigration; CoSn$_3$; Interfacial reaction; microstructure evolution

I. INTRODUCTION

With the development of the artificial intelligence and the Internet, for the purpose of meeting the needs of light weight, portability, fast computing and multi-function, electronic products require higher integration density while miniaturizing [1-3]. Due to the development of integrated circuit silicon chip technology, the packaging technology has transitioned from two-dimensional packaging to three-dimensional packaging, and the impact of EM on solder joints has become more and more obvious [4-5]. In lead-free solder joints, EM will also cause the redistribution of IMC. IMC decomposes at the cathode and aggregates and grows at the anode, both of which will seriously affect the mechanical and electrical properties of the solder joints.

For the purpose of improving the performance of solder joints, meeting the miniaturization and integration of devices, and predicting the service life and performance of power devices under extreme conditions, it's necessary to find new interconnect materials and technology [6-7]. The study found that filling UBM in the packaged chip can alter the EM performance of the solder joints, and Ni UBM can effectively improve the EM performance of the solder joints. However, the EM resistance of solder joints using Co and Co-P UBM have not been studied. During the soldering process of electronic packaging, the interface IMC formed by Co and Sn is mainly CoSn$_3$. CoSn$_3$ has better mechanical properties than Cu$_6$Sn$_5$. However, the growth kinetic behavior of CoSn$_3$ under the action of EM has not yet been reported. The study of the EM resistance of Co-P UBM and the growth kinetics of CoSn$_3$ under EM test is of great significance to the research in the field of electronic packaging.

II. EXPERIMENTAL

Co-5at.%P UBM was prepared on Cu lines (0.5mm in diameter) by ultrasound-assisted electrodeposition process. The thickness of Co-P UBM was 5μm(±0.4μm). Co-P/SAC305/Co-P line-type solders were prepared by microcomputer electric heating plate at 270 °C for 30 s, the height of the solder joint was 400 μm (±30 μm). The diameter of the SAC305 solder ball is 0.6 mm.

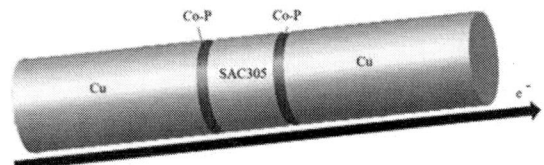

Fig.1 Schematic of the Co-P/SAC305/Co-P line-type solder joints

In this experiment, a regulated DC power supply was used for the experiment, and the current density was $1.0 \times 10^4 A/cm^2$. The temperature of solder joint was 30 ± 3 °C, and the EM test time were 0 h, 50 h, 100 h, and 200 h. The sample structure was shown in Fig.1. The solder joints need to be carefully polished to make them cylindrical in shape before the EM test. The cross-section of solder joints and morphology of IMC under different EM test times were observed by FE scanning electron microscope. The growth curve was obtained, ImageJ was used to detect the thickness of $CoSn_3$.

III. RESULT AND DISCUSSION

A. Microstructure and morphology of solder joints after soldering

As shown in Fig.2, the thickness of Co-P UBM was 2.10 ± 0.5 μm. There was no IMC dispersed in the solder. As a diffusion barrier, Co-P UBM reduced the diffusion rate of Cu atoms in the matrix. There were no cracks and defects at the interface. EDS analysis technology was used to detect the elements in Co-P UBM. The composition in Co-P UBM is Co-5 at.% P. There was no large temperature difference between the positions of the solder joints during soldering, and no difference in the thickness of the IMC at the two interfaces.

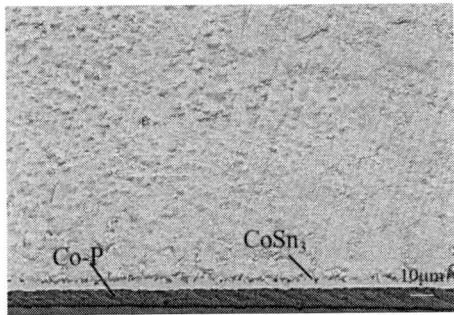

Fig.2 The cross-section SEM morphology of the Co-P /SAC305/Co-P solder joint

B. Interfacial reaction of Co-P/SAC305/Co-P solder joints under EM test

Fig.3 shows the morphology of IMCs and evolution of the microstructure at the interface under EM test with 1.0×10^4 A/cm². The EM test time was 0 h, 50 h, 100 h and 200 h. With the increase of EM time, there is no big difference in IMC morphology at the anode and cathode, but the thickness changes were different. When the EM test time increased from 0 h to 200 h, the thickness of $CoSn_3$ increased from 2.15 to 12.05 μm at the anode, and $CoSn_3$ increased from 2.10 to 5.12 μm at the cathode. No cracks were formed between the interface and the solder. At the cathode of the solder joints, as the EM test time increased, although the interface between the Co-P layer and the IMC was tightly bonded, the Co-P layer still found a little consumption. This was because EM promoted the diffusion of Co atoms from cathode to anode. As a result, Co-P UBM at the cathode was consumed, and the thickness of $CoSn_3$ increased significantly at the anode.

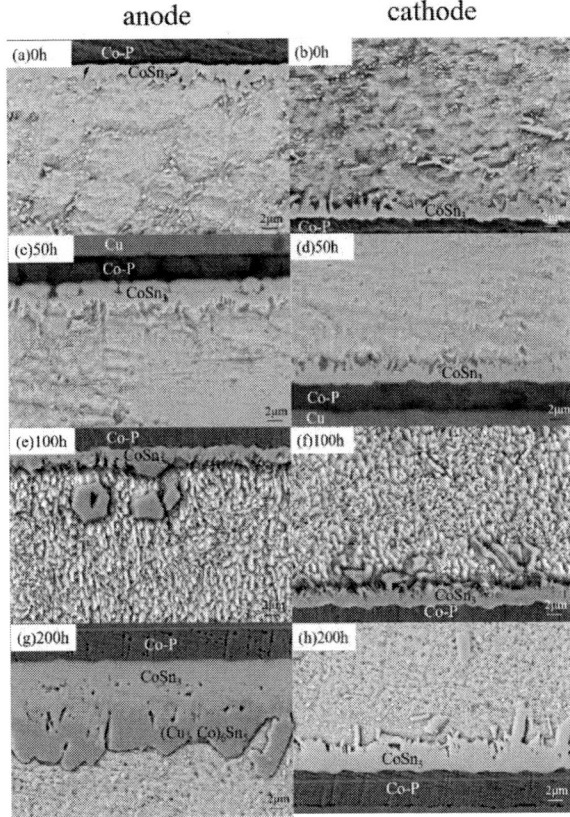

Fig.3 Interfacial reaction of Co-P/SAC305/Co-P under the EM test

The evolution of the interfacial IMC depends on "three abilities" at the interface: (1) The ability of atoms to diffuse in the solder (the ability of atoms to leave the emergency interface in the solder or the ability to diffuse to the anode interface); (2) The reaction ability between atoms at the interface (reaction rate); (3) The dissolving ability of the interface IMC. In the EM process, when the atomic flux flowing into the interface IMC is greater than the atomic flux flowing out of the interface IMC, the thickness of the interface IMC will increase with the extension of the EM test time. Conversely, the thickness of the interface IMC will decrease with the extension of the EM time. We know that the growth of anode IMC during EM test is controlled by diffusion, that is, the ability of elements to diffuse to the interface is less than the reaction ability between atoms at the interface.

For the Sn/Co-P interface, Co atoms are the main diffusion element in the interface reaction process, so only the diffusion flux of Co atoms is considered. During the EM process, the solder near the cathode interface IMC is always in an unsaturated state, so the cathode interface IMC gradually becomes thinner. For the anode interface IMC, the Cu atoms continue to move toward the anode under the electron wind, the Cu atoms in the solder near the anode interface are always in a saturated state, so there is no dissolution phenomenon of the anode interface IMC. The thickness of the interface IMC increased with the extension of the EM test time.

Because of the existence of Cu atoms in solder, during the process of EM test, due to the influence of the electronic wind, the Cu atoms in the solder formed $(Cu, Co)_6Sn_5$ with $CoSn_3$ at the anode as showed in Fig.3(g).

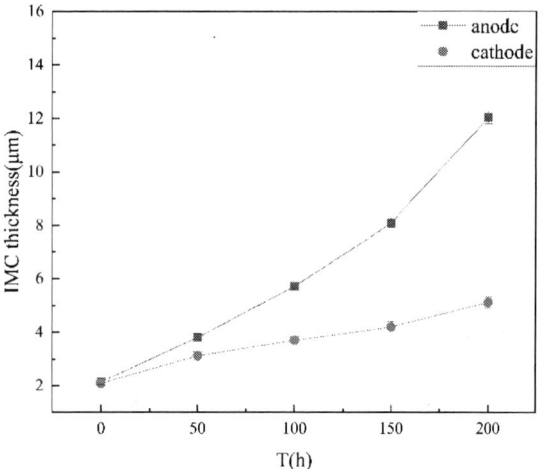

Fig.4 The variation of $CoSn_3$ thickness at the anode and cathode under the EM test over loading time

Fig.4 shows the thickness variation of $CoSn_3$ of solder joints under EM test with 1.0×10^4 A/cm^2 at the anode and cathode. When the EM test time increased from 0 h to 200 h, the thickness of $CoSn_3$ increased from 2.15 to 12.05 μm at the anode, and $CoSn_3$ increased from 2.10 to 5.12 μm at the cathode. The thickness of the IMC at the anode increased by 9.90 μm, while the cathode increased by only 3.02 μm.

C. IMCs in soldering after EM test

Fig.4 shows the IMC inside the Co-P/SAC305/Co-P solder joint. After the 100h and 200h EM test, it was found that scattered $CoSn_3$ was formed inside the solder.

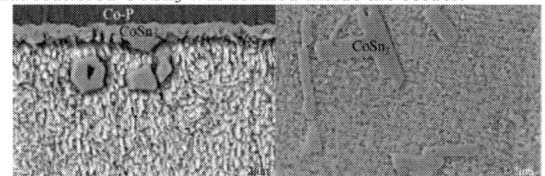

Fig.5 IMC inside the Co-P/SAC305/Co-P solder joint after 200h EM test

This is because the EM caused Co atoms to diffuse from the cathode to the solder, and the Co atoms diffused into the solder met Sn atoms to form the compound CoSn3. It may also be due to the electronic wind that caused the IMC at the interface to diffuse into the solder and formed this phenomenon.

IV. CONCLUSIONS

In this paper, the microstructure evolution of solder joints with time under high current density were studied. The research results under EM test are as follows:

(1) The EM test of Co-P/SAC305/Co-P solder joints were carried out with the current of 19.625 A. As the EM test time increased from 0 h to 200 h, the thickness of $CoSn_3$ increased from 2.10 to 5.12 μm at the cathode, the thickness of $CoSn_3$ increased from 2.15 to 12.05 μm at the anode.

(2) The obvious compound $(Cu, Co)_6Sn_5$ was found in the Co-P/SAC305/Co-P solder joints with the electrification time of 100 h and 200 h at the anode, and there were more $(Cu, Co)_6Sn_5$ at the anode of the solder joint under EM test for 200h.

(3) In the EM experiment, scattered $CoSn_3$ was found in the solder. This is because the EM caused Co atoms to diffuse from the cathode to the solder, and the Co atoms diffused into the solder met Sn atoms to form the compound $CoSn_3$.

ACKNOWLEDGMENT

This research was supported by National Natural Science Foundation of China (Grant No. 61804018), the Scientific and Technological Research Program of Chongqing Municipal Education Commission (KJQN202001124), the Chongqing Research Program of Basic Research and Frontier Technology (cstc2016jcyjA0226) and the Research and Innovation Project of Graduate Students of Chongqing Municipal Education Committee (CLGYCX20203009) respectively.

REFERENCES

[1] M. Abdelaziz, D. E. Xu, G. Wang, Electromigration in solder joints: A cross-sectioned model system for real-time observation. Microelectronics Reliability, 2021, vol. 119, pp. 126-131.

[2] S. Kim, W. Hong, H Nam, M. Mayer, Growth Behavior of Intermetallic Compounds in Various Solder Joints Induced by Electromigration. Journal of Welding and Joining, 2021, vol. 39, pp. 53-59.

[3] H. Qiu, H. Xu, C. Zhang, X. Hu, Q. Li, Influence of Co addition on microstructure evolution and mechanical strength of solder joints bonded with solid–liquid electromigration, Journal of Materials Science: Materials in Electronics, 2021, vol. 23, pp. 1-13.

[4] S. Liang, A. Kunwar, C. Wei, C. Ke, Insight into the preferential grain growth of intermetallics under electric current stressing − A phase field modeling. Scripta Materialia, 2021, vol. 203, pp. 102-109.

[5] X. Mao, R. Zhang, X. Hu. Influence of Ni foam/Sn composite solder foil on IMC growth and mechanical properties of solder joints bonded with solid-liquid electromigration. Intermetallics, 2021, vol. 131, pp. 56-61.

[6] W. Yue, C. Ding, H. B. Qin, C. G. Gong, J. X, Zhang, Crystallographic Characteristic Effect of Cu Substrate on Serrated Cathode Dissolution in Cu/Sn–3.0Ag–0.5Cu/Cu Solder Joints during Electromigration. Materials, 2021, vol. 14, pp. 71-76.

[7] Z. J. Zhang, M. L. Huang. In Situ Observation of Electromigration-Induced Anomalous Precipitation of Ag3Sn Phase in Ag-Containing Solder Joints. Journal of Electronic Materials, 2021, vol. 91, pp. 25-31.

Research on Failure Mechanisms of Localized IGBT drive board

1st Tiezhu Chen
Reliability failure analysis center
The Fifth Electronics Research
Institute of Ministry of Industry an
Information Technology
Guangzhou, China
chentiezhulunwen@126.com

2nd Liang Zhou
Reliability failure analysis center
The Fifth Electronics Research
Institute of Ministry of Industry and
Information Technology
Guangzhou, China
chentiezhulunwen@126.com

Abstract—**In order to confirm the failure mechanism for the failure Localized IGBT drive board, this paper analyzes the key components on the Localized IGBT drive board by using optical microscope observation, electrical characteristic test, X-ray inspection, scanning electron microscope and energy spectrum analysis methods. Experimental resultes show that mechanical stress was the major factor for the failure of Localized IGBT drive board.**

Keywords—localized IGBT drive board, solder joint, cracking, mechanical stress, failure analysis

I. INTRODUCTION

The localized IGBT drive board are generally integrated on various control and protection circuit. Therefore, various functional modules are integrated on the localized IGBT drive board. so that the various components on the drive circuit board are densely and there are certain differences between them. influences. In the process of soldering the device to the PCB board, if the operation is improper, it may cause the PCB board to warp and generate mechanical stress at the solder joint. and the mechanical stress can easily cause the solder joint of device cracks at the solder joint and affect the long-term reliability of PCB. In severe cases, the entire PCB board will fail.

The field installation of localized IGBT drive board was shown in Fig.1. The drive board was fixed to the other PCB with 6 screws. The failure sample returned by customer and the Pin E solder joint of M6 IGBT on the drive board was burned. After some M6 IGBT were replaced, the IGBT drive board can return to normal.

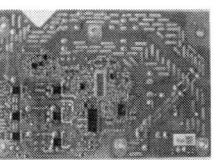

Fig. 1. (a) view of IGBT drive board, (b) Bottom view of IGBT drive board

II. ANALYSIS PROCEDURE

A. External Visual Inspection

The Pin E solder joint of the M6 IGBT on the drive board was cracked. Melting and carbonization morphology were found on the surface of the Pin E. Significant crack morphology was found on the surface of the solder joint nearby M6.

Fig. 2. (a) view of view of IGBT M6, (b) cracks of Pin E IGBT M6, (c) cracks of solder joint nearby IGBT M6

B. Electrical Characteristic Test

In order to determine the failure characteristics and failure modes, characteristic test was performed on 6 IGBTs on the drive board. Test result of electrical parameters of 6 IGBTs on the drive board was shown in Table 1. The test results show that the electrical parameters were in compliance with the specifications.

parameter	$V(BR)_{CES}(V)$	$V_{CEsat}(V)$	$V_{GE (th)} (V)$	$I_{GES}(nA)$
Test Conditions	$V_{GE}=0V, I_C=0.2mA$	$V_{GE}=15V, I_C=6A$	$I_C=0.11mA, V_{CE}=V_{GE}$	$V_{CE}=0V, V_{GE}=20V$
Specification standard	$\geq 600V$	$\leq 2.5V$	$4.3V \leq VGE (th) \leq 5.7V$	$\leq 100nA$
M6	680	2.32	4.90	0.58
M7	674	2.42	4.89	0.60
M5	674	2.29	4.87	0.58
M4	678	2.31	4.89	0.23
M2	674	2.26	4.88	0.24
M3	672	2.34	4.89	0.60

Table. 1 Test results of Electric parameter measurement of 6 IGBTs

C. X-RAY ANALYSIS

In order to detect the internal structure of the sample, X-ray analysis was performed on failure IGBT drive board. The Pin E solder joint of the IGBT M6 was cracked. But no obvious anomaly was found on the die of IGBT M6. Significant crack morphology was found in the solder joint nearby M6.

2021 22nd International Conference on Electronic Packaging Technology (ICEPT)
This work is financially supported by the National High Quality Development Program of China under Grant No. 2020-0093-2-1, the Key R&D Program of Guangong under Grant No. 2019B010143002

978-1-6654-1392-3/21 $31.00 © 2021 IEEE

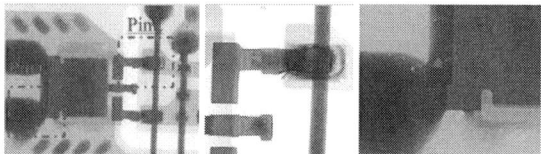

Fig. 3. (a) X-ray view of M6, (b) X-ray view of Pin E of M6, (c) X-ray view of Pin C of M6

D. IGBT M6 Analysis

In order to check whether the solder joints of IGBTs meet the requirements, there were internal structural anomalies or defects related to the failure mode. Metallographic sections of The Pin E solder joint of M6 were obtained after solidification, grind and polish.

The Pin E solder joint of the M6 IGBT was cracked. Sawtooth crack morphology was found on the edge of the solder joint. Significant crack morphology was also found in the solder joint. Loose gray foreign matter was found in the middle of the pad. The EDS test results shown that the main element of the white matter was Sn; The main element of the gray foreign matter were C, O and Sn. That means the white matter was soldering, and the gray matter was SnO2.

Fig. 4.(a)Metallographic section of M6, (b) Pin E of M6

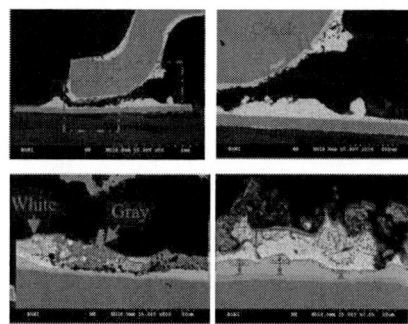

Fig. 5. (a) SEM view of Pin E of sample M6, (b) Magnifying view of Pin E of M6, (c) Magnifying view of Pin E of M6, (d) SEM view of the IMC of Pin E of M6

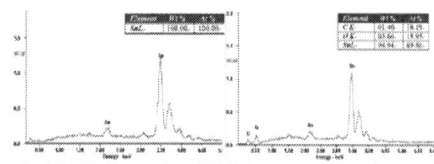

Fig. 6. (a) EDS result of White matter as shown in Fig.5, (b) EDS result of Gray matter as shown in Fig.5

Fig. 7. (a) View of M6 (after chemical decapsulation), (b) Die of M6

E. Overall Analysis

Electrical characteristic tests confirmed that 6 IGBTs on drive board were functioning normally, and the electrical parameters were in compliance with the specifications.

The Pin E solder joint of the M6 IGBT on drive board was cracked.Sawtooth crack morphology affected by mechanical stress were found on the edge of the solder joint . Significant crack morphology was also found in the solder joint. Melting morphology was found on the surface of the Pin E and the pad, with the characteristics of arc-over.

No electrical burnout or mechanical damage was found on the die of IGBT M6, which shown that the overheating and melting of the Pin E was not caused by the overheating of the die, but by the arc-over between the Pin E and the pad.

According to the above analysis, it was confirmed that the solder joint was cracked affected by mechanical stress, and the crack extends and grows toward the center of the solder joint. Then, it was completely disconnected, which resulted in arc-over between the Pin E and the pad, finally causes the full-bridge circuit to burn down.

[1] Xuedong Kong, Yunfei En, "Failure analysis and typical cases of electronic components," National Defense industry Press, pp. 215–218, September 2006.

[2] X Yang, Z Lin, J Ding, Lifetime Prediction of IGBT Modules in Suspension Choppers of Medium/Low-Speed Maglev Train Using an Energy-Based Approach. IEEE Transactions on Power Electronics. 2018; 28127-28132.

[3] Hui Li, Haiyang Long, Ran Yao, A Study on the Failure Evolution to Short Circuit of Nanosilver Sintered Press-Pack IGBT. IEEE Transactions on Components, Packaging and Manufacturing Technology. 2019; 184-187.

[4] Mohamed Halick Mohamed Sathik , Prasanth Sundararajan, Comparative Analysis of IGBT Parameters Variation Under Different Accelerated. IEEE Transactions on Electron Devices. 2020; 1098-1105.

[5] Mohamed Halick Mohamed Sathik , Prasanth Sundararajan, Reliability Assessment of IGBT Through Modelling and Experimental Testing. IEEE Transactions on Electron Devices. 2020; 1-14.

[6] Xu Chuannuo, Cheng Xuezhen, An Analysis Model and Test Monitoring Device for IGBT Intermittent Fault. 2020 the 5th International Conference on Control and Robotics Engineering. 2020; 83-88.

[7] Erik E. Kostandyan, John D. Sørensen, Reliability assessment of IGBT modules modeled as systems with correlated components. Reliability and Maintainability Symposium (RAMS). 2013; 1-6.

[8] Hua Lu, Chris Bailey, Approximate methods for IGBT solder joint stress and fatigue prediction. Electronic System-Integration Technology Conference (ESTC). 2016; 675-680.

Substrate Solder Resist Crack Root Cause Investigation Through Finite Element Analysis

1st Ye Zhang
Suzhou TF-AMD semiconductors co., ltd.
Suzhou, China
steve.zhang@tf-amd.com

2nd Tan BooWei
Suzhou TF-AMD semiconductors co., ltd.
Suzhou, China
boowei.tan@tf-amd.com

3rd Tan Chow-Khong
Suzhou TF-AMD semiconductors co., ltd.
Suzhou, China
chow-khong.tan@tf-amd.com

Abstract—The solder resist crack root cause and methods to prevent it were studied. Solder resist crack with solder extrusion was observed in the sample package after solder ball mount. A finite element model was developed to simulate the solder resist stress. A high stress area at the bump pad edge was observed at the reflow peak temperature, which is the solder resist crack root cause. The factors affecting the solder resist stress were also studied via the finite element method. It was found that reflow temperature, underfill Young's modulus, substrate dielectrics CTE, and die-to-substrate ratio are the key factors to the solder resist stress. Strategies of using low Young's modulus underfill and small die-to-substrate ratio to reduce solder resist crack length were demonstrated through process DOE and design optimization in the test vehicle package.

Keywords—flip chip assembly, solder resist, finite element analysis

I. INTRODUCTION

Integrated circuit (IC) packaging is the last process that turns the IC design into product, hence it is the utmost importance to keep the package structurally safe through design and assembly process development. The recently emerged flip chip package designs such as fine bump pitch (<130um), large die/package size, high stack count substrate laminates has led to a series of challenges in the assembly process development for defect-free product manufacture. [1] Among these challenges, solder resist crack can cause solder extrusion and bridging in between adjacent conductors hence is one of the top concern in the industry.

A solder resist is either a UV cured epoxy or a laminated film, of which the role is to form a barrier between adjacent solder bump to prevent solder bridging during die attach. [2-3] In a flip-chip package, the solder resist locates in the area of solder bump, underfill and substrate dielectric layer, hence its structural stress is also affected by these adjacent component materials.[4] This creates a complex stress environment which leads to many difficulties when in package design and process development. The finite element analysis (FEA) method offers a powerful tool to address this issue, which generates visible results according to different material and design parameter inputs. [5] Through FEA, it is possible to detect high stress area in the solder resist and validate the strategies to reduce such stress.

In this paper, the root cause of solder resist crack and methods to prevent it is studied via the finite element analysis

(FEA) method. To begin with, a flip chip ball grid array (FCBGA) package was used as an example to study the process-failure relation. The solder resist was found to be structurally safe until solder ball mount, after which crack and solder extrusion was observed. This indicates that the crack might happen around the solder ball mount reflow peak temperature. A finite element model was then developed. It was found that at reflow peak, the solder bump melts and the solder resist is squeezed by the bump pad. This forms a high stress area at the bump pad edge which pointing to the IC chip center. The simulated high stress locates at the same area where solder resist happens in the real product, which validates the finite element model. The process, material selection and package design strategies to reduce the solder resist crack were studied through the FEA method. The findings from the simulation were then evaluated in the test vehicle package through process DOE and design optimization.

II. METHODS

A. Test Vehicle

A 35mm×35mm lidded flip chip ball grid array (FCBGA) package with 11mm×13mm silicon die was used as test vehicle. A 4-2-4 layered organic substrate with maximum 3 layers of stack via was used. A UV cured solder resist was deployed on the top and bottom sides of the substrates. The package was then assembled through a standard flip chip process.

B. Finite element model

Ansys Mechanical 20 software was used for FEA modeling. A 1/4 model was developed for package warpage simulation. The finite element model contains substrate, underfill, silicon die, thermal interface, heatsink and heatsink adhesive. The sub modeling method was used to simulate micron area stress, which is shown in Figure 1(b). The sub model contains dielectric layer, solder resist, stack via, solder bump, underfill and silicon die. There are two steps for solder resist stress simulation. First, the warpage of the 1/4 model is simulated, which was then applied as the boundary condition to the sub model. The sub model was then calculated with this boundary condition, and the equivalent stress calculator was used for stress analysis considering the complex stress condition of the solder resist.

978-1-6654-1392-3/21 $31.00 © 2021 IEEE

III. RESULTS

A. Solder Resist Crack Failure Analysis

Figure 1(a) shows the solder resist crack phenomenon detected in an as-assembled 35mm*35mm flip chip package. The crack area observed was located in the die corner area, which pointing to the die center direction. This indicates that the crack is activated by substrate warpage during assembly process. Figure 1(b) shows the x-section image of the crack, which shows the solder extrusion phenomenon. To evaluate the crack behavior, the test vehicle package solder resist was assessed separately after die attach, underfill cure, heatsink attach, and solder ball mount. Table II shows the peak temperature of each condition and the measured crack length. As seen, the solder resist remains structurally safe until after solder ball mount, and the average crack length of 4 randomly chosen samples was measured to be 54.63% of the distance between adjacent conductive components. This result indicate that the solder resist crack phenomenon is correlated to the solder ball mount process. Considering the solder extrusion phenomenon, it is very likely to assume that the crack happens around the ball mount reflow peak temperature.

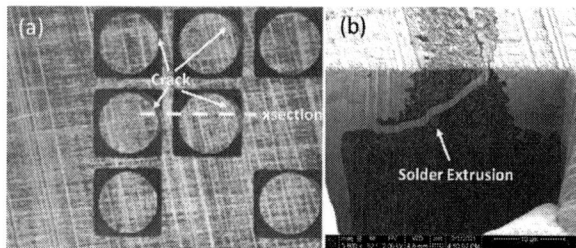

Fig. 1. (a) solder resist phenomenon in the 35mm*35mm test vehicel and (b) xsection of the crack showing the solder extrusion that bridges the adjacent conductive component

TABLE I. SOLDER RESIST CRACK LENGTH AFTER DIFFERENT ASSEMBLY PROCESS CONDITIONS

Process	Temperature Condition	Avg. Crack Length (Percentage of Distance between Adjacent Conductive Components)
Silicon Die Mount	Peak 230°C	0%
Underfill Cure	Peak 160°C	0%
Heat Sink Attach	Peak 150°C	0%
Solder Ball Mount	Peak 240°C	54.63%
3x Reflow	Peak 240°C	64.50%

B. Effect of Solder Ball Mount Temperature to the Solder Resist Stress

Table III shows the material parameters used in the FEA. As discussed above, the solder resist cracks at ball mount reflow peak temperature, hence reducing this temperature could be effective in reducing the solder resist stress. Figure 2(a) shows the simulated warpage of the 35mm*35mm package at 230°C and 240°C. As seen, the package warpage was reduced from 12.64μm to 11.04μm when reducing the reflow temperature to 230°C. Hence the solder resist could be less stressed. Figure 2(b) shows the simulated solder resist stress. The maximum equivalent stress appears in the area adjacent to the bump pad, of which the area is pointing to the

die center. This is in good agreement with the experimental observations, which validate the simulation model. When the reflow temperature was reduced to 230°C, the maximum solder resist stress reduced from 11.95MPa to 10.86MPa. Based on the result, the crack root cause can be summarized as bump pad squeezing the solder resist at high temperature so that the crack originates from the bump pad corner area. Based on the root cause, it can be concluded that methods which reducing the package warpage could be effective in reducing solder resist stress. This explains why reducing the reflow peak temperature can reduce solder resist stress.

TABLE II. TEST VIHECLE MATERIAL PROPERTIES (AT REFLOW PEAK TEMPERATURE)

Material	Mechanical Properties		
	E (GPa)	*CTE*	*Poisson Ratio*
Dielectric	4.90	117	0.31
Core	24	15	0.21
Underfill	0.12	100	0.30
Silicon Die	130	2.80	0.30
TIM	0.50×10^{-3}	280	0.4
Heat Sink	110	18	0;34
Adhesive	0.50×10^{-3}	280	0.4
Solder Resist	2.6	115	0.30

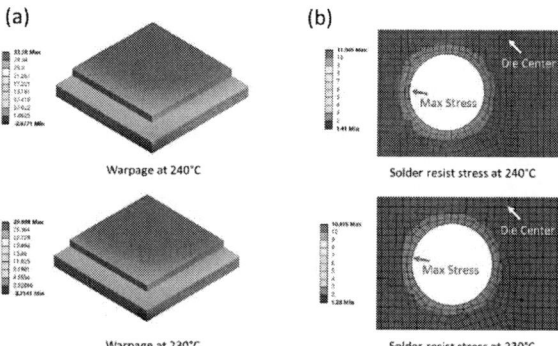

Fig. 2. (a) Simulated package warpage and (b) solder resist stress at reflow temperature of 240°C and 230°C, respectively

C. Effect of Underfill Material Selection to the Solder Resist Stress

Deploy underfill could also affect the solder resist stress, hence it is interesting to know what is the most important factor in underfill selection to solve solder resist crack issue. Table IV shows the underfill parameters used in the FEA. Two factors was evaluated in this elastic model, the Young's modulus (E) and coefficient of thermal expansion (CTE). Three underfill materials were selected, and the package warpage results are shown in Figure 3(a). The package with underfill C shows the maximum warpage of 16.17um, which is significantly larger than that with underfill A (12.64um) and underfill B (12.92um). Figure 3(b) shows the solder resist stress with 3 different underfill types. As seen, underfill B shows the lowest solder resist stress of 11.37MPa compared to underfill A of 11.95MPa and C of 15.99MPa. In comparison, underfill B possesses the lowest CTE mismatch ΔCTE of 10 (above Tg) as well as the lowest E of 0.07GPa (above Tg). To study if the ΔCTE or E is the more important factor, their effects on the solder resist stress was compared. Compared to underfill A, the ΔCTE of underfill B decreased by 33.33% and the E decreased by 41.67%, this leads to a solder resist stress decrease of 4.77%. Compared to underfill C, the ΔCTE of underfill B decreased by 28.57% and the E decreased by 76.67%, the solder resist stress decreased by 28.89%. This comparison illustrates the underfill Young's modulus is the more important factor affecting the solder resist stress. Hence, to solve the solder resist crack issue, it is recommended to deploy underfill with low Young's modulus.

TABLE III.   UNDERFILL MATERIAL PROPERTIES (AT REFLOW PEAK TEMPERATURE)

Material	Mechanical Properties		
	E (GPa)	CTE	Poisson Ratio
Underfill A	0.12	100	0.30
Underfill B	0.07	125	0.30
Underfill C	0.30	101	0.30

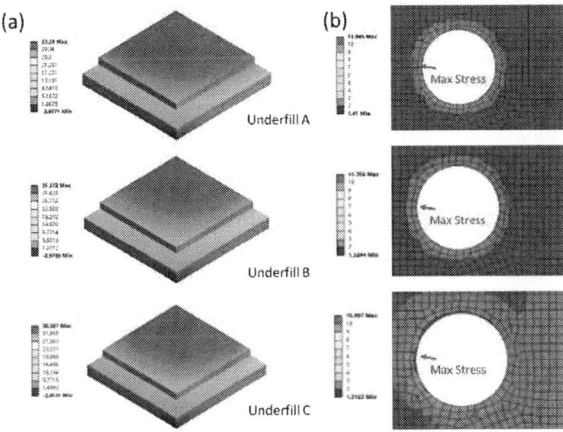

Fig. 3.   (a) Simulated package warpage and (b) solder resist stress at reflow peak of 240°C with different underfill selection

D. Effect of Substrate Dielectric Material Selection to the Solder Resist Stress

The substrate dielectric layer is another component that is adjacent to the solder resist. Herein, the effect substrate dielectric layer properties on the solder resist stress is studied. Still, two factors was studied in this elastic finite element model, the Young's modulus (E) and coefficient of thermal expansion (CTE). Table IV shows the material parameters of three types of dielectric layers. Figure 4(a) shows the package warpage results. As seen, type B dielectric material leads to the lowest package warpage of 12.09um. As for the solder resist stress, type C dielectric material also leads to the lowest stress as shown in Figure 4(b). Comparing to type A and type B, type C dielectric material shows the lowest CTE, but the highest E. Different to the underfill, the type C also shows the highest ΔCTE between dielectric and solder resist. Therefore, it can be summarized that the dielectric material with low CTE can lead to low solder resist stress due to reduced package warpage. Comparing to the underfill with a total thickness of 75um, the dielectric layers are interconnected in the substrate with a total thickness of um. This means the dielectric material

is active rather than passive in controlling the package warpage, hence a low CTE is preferred.

TABLE IV.   DIELECTRIC MATERIAL PROPERTIES (AT REFLOW PEAK TEMPERATURE)

Material	Mechanical Properties		
	E (GPa)	CTE	Poisson Ratio
Dielectric A	4.90	117	0.31
Dielectric B	9.00	67	0.28

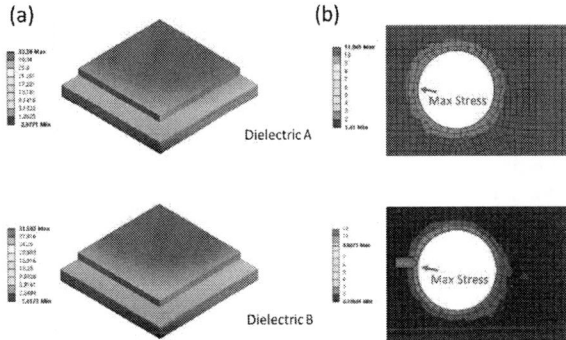

Fig. 4.   (a) Simulated package warpage and (b) solder resist stress at reflow peak of 240°C with different substrate dielectric material

E. Effect of Package Design to the Solder Resist Stress

Beyond process parameter and material selection, it is yet unknow how package design would affect the solder resist crack. As all the solder resist cracks were observed in the die corner area, it is reasonable to assume that a smaller die-to-substrate ratio would be effective in reducing solder resist stress. Figure 5(a) shows the warpage simulation result of two test vehicle package, the 35mm*35mm package with 0.11 die-to-substrate ratio and a 35mm*35mm package with 0.09 die-to-substrate ratio. As seen, when reducing the die-to-substrate ratio to 0.09, the package warpage increased to 14.85um. This is due to smaller die having weaker control to the substrate warpage. Figure 5(b) shows the simulated result of solder resist stress. The maximum solder resist stress of the 0.09 die-to-substrate ratio package is 9.67MPa, compared to that of 11.95MPa in a 0.11 die-to-substrate ratio package. This results indicates that a smaller die is effective to reduce solder resist crack, even though the overall package warpage might be larger.

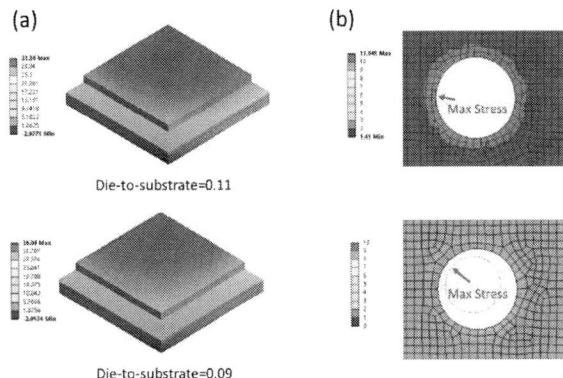

Fig. 5.   (a) Simulated package warpage and (b) solder resist stress at reflow peak of 240°C with different die-to-substrate ratio

IV. DISCUSSION

As discussed above, the key factors affecting solder resist stress in a flip chip package has been summarized. In terms of assembly process, a low ball mount reflow peak temperature is effective in reducing solder resist stress, this is due to a smaller package warpage that leads to the solder resist to be less squeezed by bump pad. In terms of materials, both underfill and substrate dielectric material selection was found to be critical. As for the underfill, a smaller Young's modulus is preferred. This is because the underfill in a flip chip package is more of a passive component due to its thin thickness, hence a low modulus underfill would be better to release the package stress. As for the substrate dielectric material, a low CTE is preferred as this can lead to smaller package warpage. In terms of the package design, a smaller die to substrate ratio is preferred also due to a smaller package warpage.

Validation of these findings from the finite element analysis was made though process DOE and design optimization in the 35mm*35mm test vehicle. First, the effect of ball mount reflow temperature was evaluated and shown in Figure 6(a). As seen, Reducing the ball mount temperature from 240°C to 230°C, the solder resist crack length measurements does not show a clear reduction trend. In practice, the temperature is not homogeneous in the package. Hence a 10°C process margin might not be enough to induce an obvious temperature difference in the solder resist area. This means that although temperature have a straightforward effect to solder resist stress, it is hard in practice to solve the crack issue through precise temperature control as this is engineering-vise not practical. Figure 6(b) shows the solder resist crack length measurements of packages with different underfill. As seen, underfill B leads to the smallest crack length of 37.99% comparing to 54.63% of underfill A and 71.11% of underfill C. A smaller crack length could lead to lower risk of solder bridging, hence choosing underfill with low Young's modulus is demonstrated to be effective. The solder resist crack length of two test vehicles with die-to-substrate ratio of 0.11 and 0.09 was compared in Figure 6(c), and a crack length of 64.50% and 39.90%, respectively, was observed after 3x reflow. No solder resist crack was observed in the test vehicle with 0.09 die-to-substrate ratio after solder ball mount reflow. This validates the small die-to-substrate ratio is effective in reducing the solder resist stress. In summary, although lowering the solder ball mount temperature was found to be not effective, it was demonstrated in practice that low modulus underfill and small die-to-substrate ratio is effective in reducing the solder resist crack. The substrate dielectric material selection was not evaluated due to the lack of proper test vehicle, but this strategy will be continuously studied in our future work.

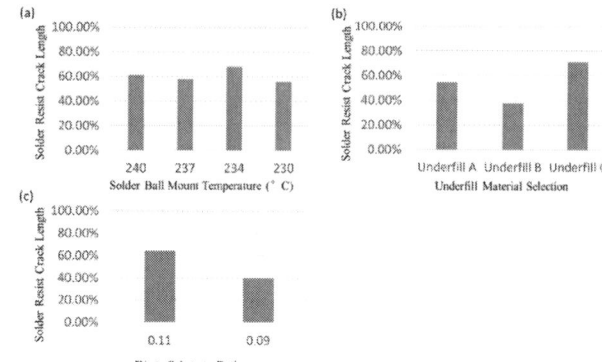

Fig. 6. (a) Effet of solder ball mount temperature on the solder resist crack length (b) effect of underfill material modulus to the solder resist crack length, and (c) effect of die-to-substrate ratio to the solder resist crack length

V. CONCLUSION

In this paper, the solder resist crack root cause is investigated through finite element analysis. The results indicate that the solder resist is squeezed by the bump pad at solder ball mount reflow peak temperature, which causes the crack. The reflow temperature, underfill Young's modulus, substrate dielectric material CTE, and package die-to-substrate ratio are found to be important factors to reduce solder resist stress through finite element analysis. Except ball mount temperature and dielectric material selection, other factors were then assessed and proved to be effective via process DOE and design optimization in the 35mm*35mm test vehicle. In flip chip package design and assembly, it is recommended to use low Young's modulus underfill, low CTE dielectric, and low die-to-substrate ratio to avoid severe solder resist crack.

ACKNOWLEDGEMENT

The author is grateful to Suzhou TF-AMD semiconductors co., ltd. for the support of this research.

REFERENCES

[1] P. C. Kuo, C. H. Wang, K. K. Ho, K. M. Chen, C. Y. Wu and C. L. Yang, "14 nm chip package interaction development with Cu pillar bump flip chip package," 2015 IEEE 65th Electronic Components and Technology Conference (ECTC), 2015, pp. 30-34, May 2015

[2] H. Zhu, Y. Guo, W. Y. Li, A. A. Tseng, and B. Martin, "Micro-mechanical characterizations of solder mask materials," Proceedings of 3rd Electronics Packaging Technology Conference (EPTC 2000), pp. 148-153, Dec 2000

[3] S. Siau, A. Vervaet, S. Degrande, E. Schacht, and A. Van Calster, "Dip coating of dielectric and solder mask epoxy polymer layers for build-up purposes," Applied surface science, 245(1-4), 353-368, 2005

[4] T. Ferguson, and J. Qu, "Effect of moisture on the interfacial adhesion of the underfill/solder mask interface," J. Electron. Packag., 124(2), 106-110, 2002

[5] A. Q. Xu, and H. F. Nied, "Finite element analysis of stress singularities in attached flip chip packages," J. Electron. Packag., 122(4), 301-305 , 2002

Microstructure and mechanical properties of joints between GaAs solar cell electrode and Ag interconnector under temperature thermal cycle

Yuhan Ding
Shanghai Jiao Tong University
School of Materials Science and
Engineering
Shanghai, China
yuhanding@sjtu.edu.cn

Zhichao Wang
Shanghai Institute of Space Power
Sources
Shanghai, China
1124455738@qq.com

Xueming Hua*
Shanghai Jiao Tong University
School of Materials Science and
Engineering
Shanghai, China
xmhua@sjtu.edu.cn

Chen Shen*
Shanghai Jiao Tong University
School of Materials Science and
Engineering
Shanghai, China
cs395@uowmail.edu.au

Min Wang
Shanghai Jiao Tong University
School of Materials Science and
Engineering
Shanghai, China
Wang-ellen@sjtu.edu.cn

Jusha Ma
Shanghai Institute of Space Power
Sources
Shanghai, China
jushama@163.com

Bin Qian
Shanghai Institute of Space Power
Sources
Shanghai, China
qianliyan1967@163.com

Abstract—**Spacecraft play an important role in the national economy and other fields. Solar cell array is the most critical power supply component on spacecraft, and its joining quality determines the success and failure of flight missions. The joints operate in high and low temperature alternating environment when the spacecraft is in orbit. Therefore, the thermal reliability of joints is an important factor to a life of spacecraft. In this paper, the joints between GaAs solar cell electrode and Ag interconnector have been joined using a parallel gap resistance welding (PGRW) process. The joints are conducted a thermal cycling test in a temperature range of −160 to 120 °C. The heating and cooling rate are 15 °C/min and maintains 10 min at −160 °C and 120 °C, respectively. The joints are taken out after different thermal cycle number and pull test are performed. To characterize the microstructure evolution and failure mode at the interface before and after thermal cycle, scanning electron microscope (SEM), transmission electron microscopy (TEM) are conducted to investigate the joining interface. According to the obtained results, the PGRW interface between the Ag interconnector and the Au surface of the GaAs solar cell electrode forms a solid-solution layer in low voltage process. The joining mechanism is solid phase diffusion welding. Melting connection occurs in the central region of two electrodes in high voltage process. The performance of joints with a lower welding voltage shows a downward trend after thermal cycles. And failure occurs at interface after 5000 cycles. The performance of joints with a high welding voltage still keeps invariability. This is owing to the boding zone increases and the joining mechanism varies. The fracture surface also varies form characteristics of dimples to characteristics of cleavage fracture with a low voltage process after 5000 cycles.**

Keywords—*Parallel gap resistance welding; Ag interconnector, GaAs solar cell electrode, Thermal cycle, Microstructure evolution, Mechanical properties.*

I. INTRODUCTION

With the development of space technology, human has landed on the moon, flown around the earth and built a manned space station [1]. The solar cell array is prime power unit to the spacecraft [2]. Welding process is a crucial procedure in the manufacture of solar cell array. The electrodes of solar cell are connected with Ag interconnectors to achieve the series-parallel connection between individual solar cell. Therefore, The joining quality of solar cell is important to the service life of spacecraft.

The process of welding the solar cell mainly are soldering and resistance welding. Traditionally, the joining methods for the solar cell array connections are mainly brazing [3]. But the process of brazing must use the solder to be as an intermediate layer. Besides, the joining the materials need to keep a specified period of time at a higher temperature, the intermetallic compounds formed and grew. Therefore, the development of fast joining technology for solar cell and interconnector is necessary. Compared to the brazing method, parallel gap resistance welding (PGRW) is capable of providing direct joining without heat preservation thus has higher working efficiency [4]. Also, since no solder is required in PGRW. At present, PGRW is generally used for joining between solar cell and interconnector. This method has high welding efficiency and less damage to solar cell, which can meet the welding requirements of space solar cell, it is difficult to control the welding quality of parallel gap resistance welding of battery interconnects because there are many factors affecting welding in the welding process, and the composite interconnects material for space environment is multilayer heterogeneous structure. And large thermal stresses are generated during operation in space environmental, The crack would initiate and propagate at the bonding interface because of thermal-fatigue stress [5, 6].

978-1-6654-1392-3/21 $31.00 © 2021 IEEE

The joining quality of GaAs solar cell and Ag interconnector material is one of pivotal factors of solar array performance, so it is necessary to investigate and analyze the joints in order to improve the service life of spacecraft. In this study, the microstructure of the joints between GaAs solar cell electrode and Ag interconnector are studied. The mechanical properties of joints before and after thermal cycling also are tested. Subsequently, the fracture morphology of specimen are observed.

II. EXPERIMENTAL

A. The welding process

Commercial pure Ag interconnector and GaAs solar cell were used in this study. A schematic diagram of the PGRW process as shown in Fig. 1. The molybdenum electrode arranged side by side. The molybdenum electrodes were on the surface Ag interconnector to ensure adequate contact between GaAs solar cell electrode and Ag interconnector when the welding began. A constant voltage was applied between the two welding electrodes to form a circuit between the welding electrodes. In this study, we carried out two processes to join the Ag interconnector and GaAs solar cell electrode. One process is high voltage mode, and the other is low voltage mode.

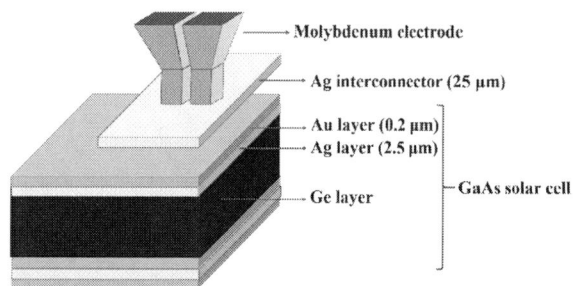

Fig. 1. Schematic diagram of PGRW of GaAs solar cell and Ag interconnector

B. The thermal cycling test

To investigate the stability and reliability of joints at an extreme temperature thermal cycling environment. A temperature cycle box (CTN1701) was applied to realize a temperature variation environment from -160 ℃ to 120 ℃. Fig. 2 shows the schematic illustration of temperature profile of thermal cycling tests The hearting rate and cooling rate were 15 ℃/min. And the stage of maximum temperature and minimum temperature hold 10 minutes. 9 joints were took out after 200 cycles, 500 cycles, 850 cycles, 1500 cycles, 2000 cycles,3000 cycles and 5000 cycles. The eight joints were used to perform pull test and the another were used for microstructure analysis.

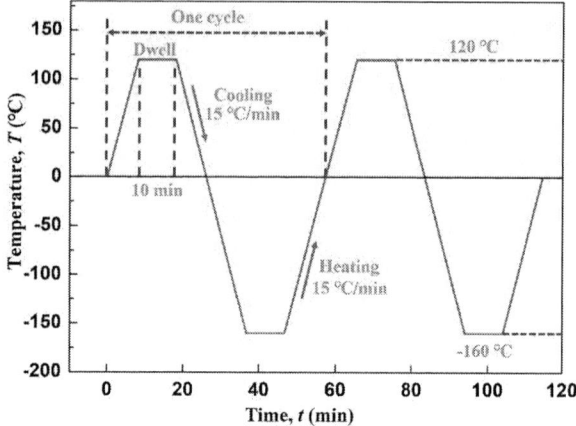

Fig. 2. Schematic illustration of temperature profile of thermal cycling tests

C. Pull test

For mechanical property test, a 45° pull test (JIS Z 3198-6) was conducted at room temperature by using an electronic tension tester. Fig. 2 shows the schematic illustrations of pull test. The rate of tensile was 10 mm/min. For different thermals cycling number, 8 joints were tested and recorded the value of pull force. The characteristic of fracture surfaces of the joints were analyzed by SEM after pull test.

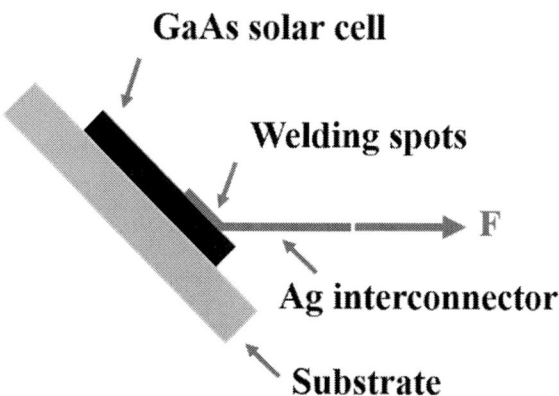

Fig. 3. Schematic illustrations of pull test

D. Characterization

The samples were embedded in epoxy resin after different cycles of thermal cycle. These samples were ground and polished to the central area of joints by an elaborate the all-in-one machine (EM TXP). Then, an ion polishing machine (EM TIC 3X) was used to perform Ion polishing on the cross section of joints. The microstructure and chemical compositions of selected areas were observed by a scanning electron microscope (SEM, MIRA3 LHM) equipped with an energy dispersive spectroscopy (EDS). In addition, Focused Ion beam (FIB) technology was used to prepare transmission electron microscope (TEM) samples. To investigate the initial joining interface and the evolution after thermal cycle, a transmission electron microscope (TEM, Talos F200X G2) was used to analyze the elements diffusion and microstructure evolution.

III. RERULTS AND DISCUSSION

A. Macro-morphology

Fig. 4 shows the macro-morphology of the joint. The Ag interconnector have three legs. The legs were attached to the surface of the electrode of GaAs solar cell. The GaAs solar cell electrode and Ag interconnector are successfully joined via PGRW process. Each PGRW joint possesses three welding spots. It can be observed that the GaAs solar cell side has no macroscopic defects after welding. And there is no spatter near the welding zones.

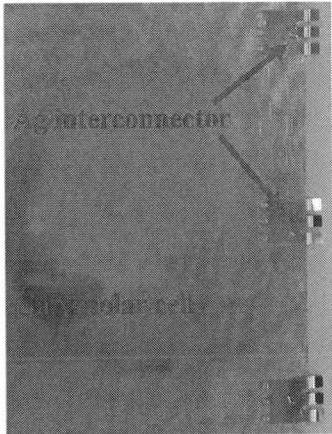

Fig. 4. Macro-morphology of the joints

B. Mechancal procperties

The pulling test is carried to investigate the effects of thermal cycle on the tensile strength of PGRW joints. Fig. 5 displays the tensile strength changes of PGRW joints after thermal cycles ranging from -160~ 120 °C. Fig. 5a shows the tensile strength changes of PGRW joints with low voltage mode. The tensile strength gradually declines with the increase of thermal cycles. Within the 200 cycles, the decline is relatively fast. The value of force is 2.06 N after 200 cycles. But with further thermal cycling number, the strength of joints holds 1.75 N until 3000cycles. After 5000 cycles, the tensile strength decreases to 1.62 N. Because of the coefficient of thermal expansion (CTE) mismatch of the Ag interconnector and GaAs solar cell, the joining interface of Ag interconnector and GaAs solar cell withstood the alternative tensile and compressive stresses during the thermal cycling test. For the joint with a low welding voltage process, thermal stress was generated at the central and edge of the bonding interface during heating-to-cooling and cooling-to-heating conversions, the weak interface bonding line subjected to a large tensile stress during the thermal cycling test, thereby resulting in interface delamination. Compared to the low voltage mode, the tensile strength of joints with the high voltage mode has a little change, the value is approximately 3.0 N, as shown in Fig. 5b. The melting zone occurred at the boing interface, the solar cell electrode and Ag interconnector achieved metallurgical bonding. Therefore, the joint performance is almost not affected by the thermal cycle.

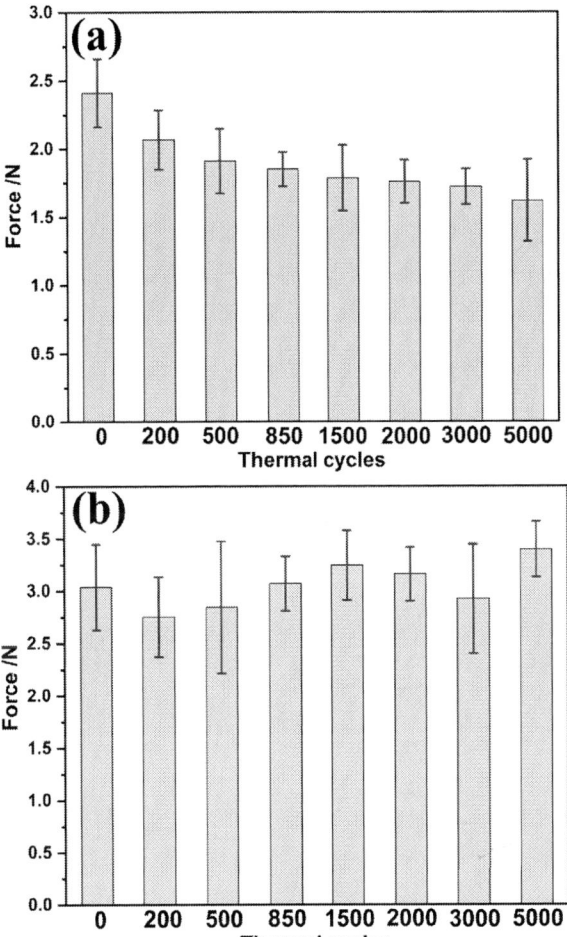

Fig. 5. The tensile strength changes of PGRW joints after thermal cycles ranging from −160 °C to 120 °C.

C. Fracture morphology

In order to study the fracture morphology of Ag/GaAs solar cell PGRW joints after different thermal cycles. SEM is used to observe the fracture morphologies. Fig. 6 shows the fracture surfaces of joints after different thermal cycles with low voltage mode. It can be seen from Fig. 6a, the location of the fracture occurs at the Ag interconnector near the welding spot. A large amount of dimples distribute on the fracture surface, as shown in Fig. 6b. This means that the fracture occurs at the joining interface in a ductile mode. After 1500 cycles, the residual Ag reduces. There are no dimples distributes on the fracture surface, as shown in Fig. 6c and d. For joints underwent thermal cycling for 5000 cycles, there is no dimples distributes on the fracture surface, as shown in Fig. 6f. Because of the CTE mismatch of Ag and solar cell, the stress produced at the joining interface. Under repeated stresses, the fracture displays shear and rheological morphology. Cleavage cracks in the joints initiates and propagates more easily after 5000 thermal cycles than at 0 thermal cycles, and this may be the reason for the reduction of tensile strength.

Fig. 7 shows the fracture surfaces of the joints with high voltage mode. The fracture of as bonded and after 5000 cycles joints the fracture occurs at the Ag interconnector near the welding spot. The welding spot area lager than the joints with low voltage mode. The bonding strength of joints also higher. It can explain that the bonding zone increases with the increases of heat input.

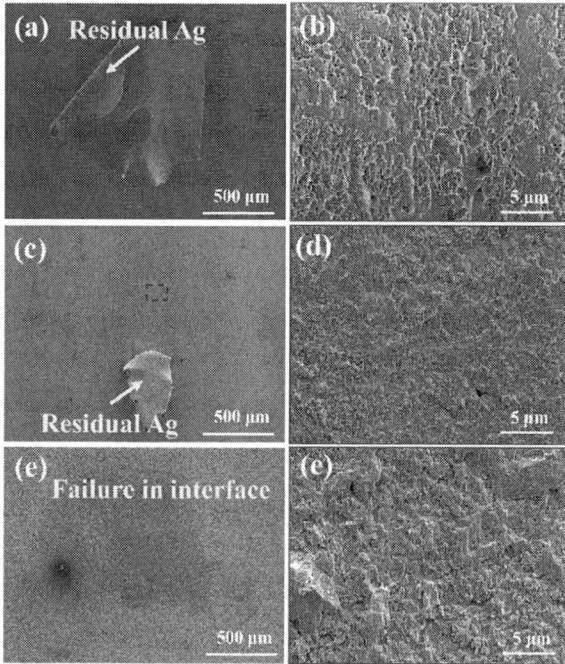

Fig. 6. Fracture surfaces of the joints with low voltage mode: (a) as bonded, (b) magnified view of red rectangle zone, (c) after 1500 cycles, (d) magnified view of purple rectangle zone, (e) after 5000 cycles, (f) magnified view of blue rectangle zone.

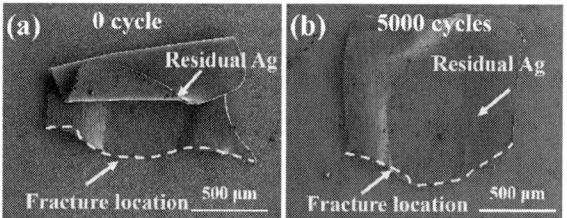

Fig. 7. Fracture surfaces of the joints with high voltage mode, (a) as bonded, (c) after 5000 cycles.

D. Microstruture

The microstructure of the Ag interconnector and GaAs solar cell electrode is investigated to analyze the reason of performance variation. Fig. 8 shows cross section images of the initial joint with a low voltage process. As shown in Fig.8a, the Ag interconnector is successfully joined on the GaAs solar cell electrode. There are two welding spots at the joining interface, and a gap is observed between the two welding spots. Fig. 8b presents the magnified view of red rectangle in Fig. 8a, the joining interface contains some defects. And it can weaken the performance of joints after thermal cycle. To further investigate the interface between Ag interconnector and GaAs solar cell electrode, the Ag-Au bonding interface of the initial joint is investigated. Fig. 9 shows the STEM-EDS results of the foil specimen derived at the region marked by the red rectangle shown in Fig. 8b. The Ag and Au form an inter-diffusion layer. But the voids occur

at the location near the joining interface, which can be the source of cracks and the crack propagation path.

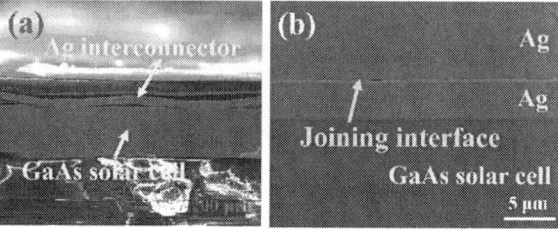

Fig. 8. (a) SEM of the cross-section of joint, (b) magnification of the region marked by red rectangle shown in a.

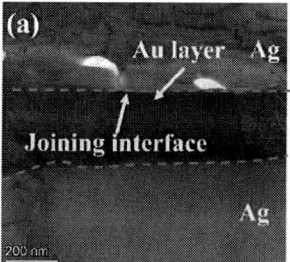

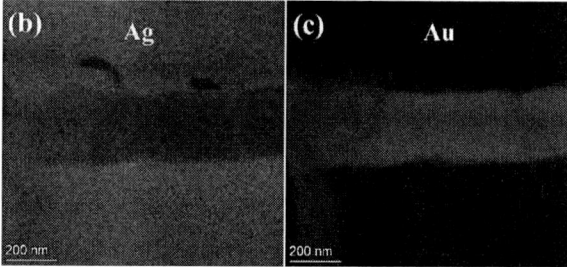

Fig. 9. STEM-EDS results of the foil specimen derived at the region b marked by the red rectangle shown in Fig. 7 b: (a) STEM bright field image, (b, c) corresponding Ag, Au elemental mappings, respectively.

Fig. 10 shows cross section images of the initial joint with a high voltage process. There is no gap at the bonding interface. To study the joining mechanism of the process, the border zone and central zone of the interface are magnified in Fig. 11 and Fig. 12, respectively. Fig. 11a shows the magnification of the region marked by blue rectangle shown in Fig. 10a. The diffusion and melting exist in the bonding interface. It can be observed that the fusion begins at the interface of the Ge substrate/Ag. This is owing to the Ag and Ge can form Ag-Ge binary eutectic under the current and heat during welding process. The melting point of Ag-Ge lower than Au, Ag and Ge [7]. Therefore, the Ag, Au and Ge uniformly distribute in the melting zone. The interface of Ag interconnector and GaAs solar cell achieves metallurgical bonding. The strength of the joints also higher than the joints with low voltage process. In addition, the melting zone increases in the central zone, as shown in Fig. 12. Because the heat concentrates at the location between the electrodes, the Ag interconnector melts more than other zone. The element of Ge, Ag, and Au homogeneously distribute in melting zone.

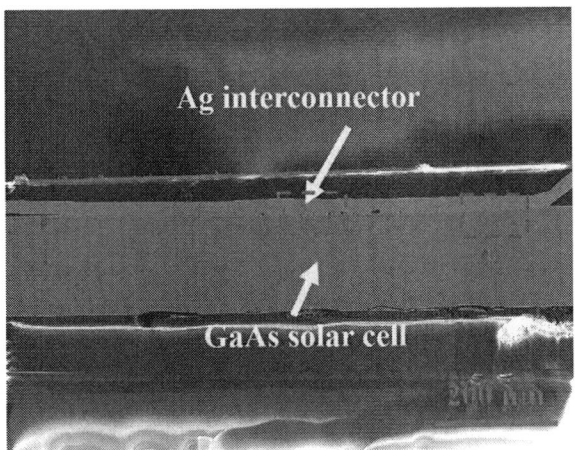

Fig. 10 SEM of the cross-section of joint with high voltage process

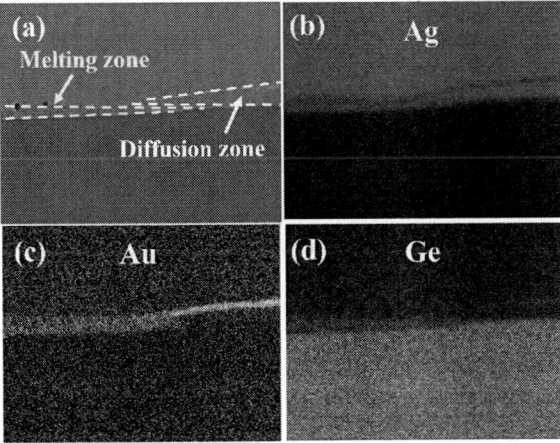

Fig. 11   (a) magnification of the region marked by blue rectangle shown in Fig. 9 a. (b~d) The corresponding Ag, Au, Ge elemental mappings in a, respectively.

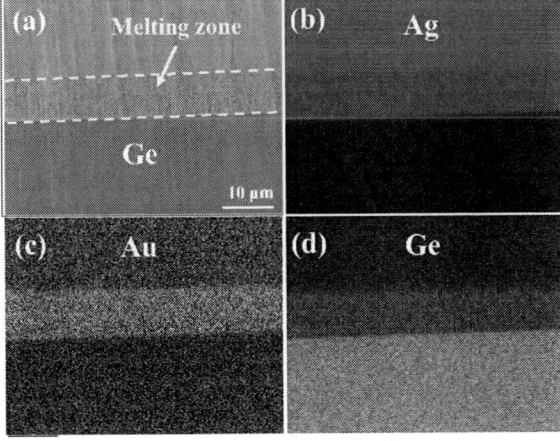

Fig. .12 (a) magnification of the region marked by red rectangle shown in Fig. 9 a, (b~ d) the corresponding Ag, Au, Ge elemental mappings in a, respectively.

IV. CONCLUSION

In summary, the Ag interconnector and the GaAs solar cell electrode are successfully joined by PGRW. The main conclusions are draw as follows:

1)   After 5000 cycles, the performance of joints with a lower welding voltage shows a downward trend, and failure occurs at interface after 5000 cycles. The performance of joints with a high welding voltage still keep invariability after thermal cycles. This is due to the melting area at the joint interface increases with the increase of welding voltage.

2)   For low voltage process, the fracture surface varies form characteristics of dimples to characteristics of cleavage fracture after 5000 cycles. The fracture occurs at Ag interconnector and the interface of Ag/GaAs before and after thermal cycling. For high voltage process, the fracture occurs at Ag interconnector, it also can indicate that the performance of joints can keep unchanged after thermal cycles.

3)   The PGRW interface between the Ag interconnector and the GaAs solar cell electrode is a solid-solution layer in the low voltage process. It forms through elements diffuse. The melting connection occurs in the central region of two electrodes in the high voltage process. With the increases of welding voltage, the melting area of the interface increases and the performance of joints improves.

ACKNOWLEDGMENT

This research is supported by National Natural Science Foundation of China (NSFC, Funding No. U1937601).

REFERENCES

[1]   Polakis T . Images of International Space Station Moon Transit[J]. Blood, 2006, 107(7):2643-52.

[2]   Zou Y , Lin L I , Liu G , et al. Research progress of GaAs Solar cell[J]. Journal of Changchun University of Science and Technology(Natural Science Edition), 2010.

[3]   Schmitt P , Kaiser P , Savio C , et al. Intermetallic Phase Growth and Reliability of Sn-Ag-Soldered Solar Cell Joints[J]. Energy Procedia, 2012, 27:664-669.

[4]   Chu, C. L. , and P. Iles . "Control of parallel gap welding for solar cells." Electronics Components Conference IEEE Xplore, 1988.

[5]   Chen C , Nagao S , Hao Z , et al. Mechanical Deformation of Sintered Porous Ag Die Attach at High Temperature and Its Size Effect for Wide-Bandgap Power Device Design[J]. Journal of Electronic Materials, 2016, 46(3):1-11.

[6]   Sugiura K , Iwashige T , Tsuruta K , et al. Reliability Evaluation of SiC Power Module With Sintered Ag Die Attach and Stress-Relaxation Structure[J]. IEEE Transactions on Components, Packaging and Manufacturing Technology, 2019:609-615. il 1955.

[7]   Manasijevic D , Balanovic L , Markovic I , et al. Study of thermal properties and microstructure of the Ag–Ge alloys[J]. Journal of Thermal Analysis and Calorimetry, 2021(3).

Design method of triaxial vibration fixture for complex integrated circuits

Xiaoqiang Wang
South China University of Technology;
China electronic product reliability and
environmental testing research institute
Guangzhou, China
ps_800@126.com

Bin Li
South China University of Technology
Guangzhou, China
phlibin@scut.edu.cn

Chuanjin Deng*
China electronic product reliability and
environmental testing research institute
Guangzhou, China
tjudeng@126.com

Rui Deng
China electronic product reliability and
environmental testing research institute
Guangzhou, China
dengrui@ceprei.com

Ruolei Wang
China electronic product reliability and
environmental testing research institute
Guangzhou, China
wrl19931011@163.com

Kun Jiang
China electronic product reliability and
environmental testing research institute
Guangzhou, China
jiangkun2086@126.com

Abstract—**In view of the harsh conditions of vibration test of complex integrated circuits in aerospace and the complex characteristics of chip packaging, the vibration fixture optimization design method based on topology optimization and multi-objective genetic algorithm is adopted to optimize the fixture structure design for a complex integrated circuit in PGA273 for aerospace. The sweep frequency vibration and random vibration response of the fixture structure are simulated and analyzed. Finally, the validity and rationality of the fixture design was proved by experiments, which effectively solve the problems of test resonance, poor dynamic response and excessive fixture mass.**

Keywords—Integrated circuit; Fixture design; triaxial vibration fixture; Optimal design; Vibration response simulation

I. INTRODUCTION

The function of vibration test fixture is to transfer the energy and motion of vibration table to the test piece, and its vibration transfer characteristics will directly affect the accuracy and credibility of vibration test. According to MIL-STD-883K [1]and GB/T 2423[2,3], the vibration frequency of integrated circuits is as high as 2000Hz, the sinusoidal acceleration is as high as 70g or the root mean square value of random vibration acceleration is as high as 29.2g .The number of pins of complex integrated circuit chips reaches more than 1,000 levels, and the package is complex and hard to be rigidly fixed, which leads to the structural rigidity of the fixture system decreasing after the fixture is installed, thus it is very easy to cause the vibration fixture to resonate in the test frequency range or transmit vibration magnitude exceeding the allowable error range. Structural characteristics of complex integrated circuit products and harsher test conditions challenge the design method of vibration test fixture.

In the engineering practice of IC vibration fixture design in China, the design is mainly based on technical experience, which requires repeated modification and optimization for complex IC products, with long design time and low efficiency. In recent years, many scholars have studied the structural design of vibration fixture, and obtained the structural modal parameters which can improve the dynamic characteristics of the fixture by finite element simulation. Some scholars' research on optimization design method of vibration fixture based on finite element software mostly adopts the combination of empirical optimization and software simulation[4~8], which only optimizes one aspect of fixture material, structure and size by simulation, but fails to carry out comprehensive and standardized optimization design of fixture design and lacks practical reference value. In addition, the integrated circuit vibration needs to be vibrated in three axes, and because of the large number of samples, in order to improve the test efficiency, the triaxial vibration fixture can realize simultaneous loading in three orthogonal directions. At present, the related research on vibration fixture mainly focuses on uniaxial vibration fixture, while the research on triaxial vibration fixture is relatively few.

On the basis of this, aiming at the aerospace application environment and the high standard technical requirements of vibration fixture design for complex integrated circuit products, this paper puts forward a design method of triaxial vibration fixture for complex integrated circuits, taking the typical PGA273 packaged complex integrated circuit chips for aerospace as an example, and puts forward a design flow of triaxial vibration fixture for complex integrated circuits based on various finite element simulation methods. The tri-axial vibration fixture of integrated circuit is designed reasonably, and the finite element model is established. The integrated optimization design method based on topology optimization and dimensional parameter optimization is adopted to optimize the vibration fixture of integrated circuit. The simulation analysis and test loading verification of the three-axial swept frequency vibration and random vibration response of the fixture are carried out. The results show that the natural frequency, mass, swept frequency vibration and random vibration response of the fixture meet the design and use requirements. This method can provide reference for the optimization design of vibration test fixture of complex integrated circuits.

II. DESIGN REQUIREMENTS AND PROCESSES

A. Fixture design requirements

Complex integrated circuit Vibration test fixtures function to transmit vibrational motion and energy to the test piece. In fact, the vibration table on the connection points and fixtures on the point movement is not exactly the same, at present, China still does not have the integrated circuit vibration test fixture detailed requirements of the standard, so it is not possible to find a basis to determine whether the fixture is qualified. Referring to the common vibration test standard of integrated circuits, the fixture standard proposed by Sandia Corporation of the United States[9], and the literature on vibration fixture design[10], this paper puts forward the basic requirements for the mechanical test fixture of complex integrated circuits:

1. Convenient connection with the vibration table can also be easily connected to the test piece, and should be convenient to change the test direction and install more test samples.

2. The first-order inherent frequency is at least 1.5 times higher than the test maximum frequency to ensure no resonance within the test frequency range.

3. The weight of the fixture should be as light as possible, as the weight of the fixture directly affects the total weight of the vibratory system and the maximum acceleration value that can be achieved when the vibratory table is fitted with the test piece.

4. Sweep frequency vibration 2000Hz internal acceleration response deviation ±20%, random vibration within 1000Hz power spectral density response value deviation is less than ± 1.5dB, 1000 to 2000Hz power spectral density response value deviation is less than ±3dB.

B. Fixture design process

The flow of complex integrated circuit vibration fixture design is shown in Figure 1. First according to the fixture design objectives for the initial shape and function design, and then based on finite element simulation analysis to evaluate and modify the initial scheme, and then by using topology optimization and structural size parameters optimization design methods to optimize the fixture, and the optimized fixture model for sweeping vibration and random vibration response simulation analysis, and finally make the design fixtures meet the modal characteristics, mass requirements, transmission performance and other indicators, if all meet the requirements, the production and processing fixture test verification. For fixtures that do not meet the design requirements, return to continue one or more interactions using topology optimization and structural size optimization design methods, resulting in the design fixtures meeting the complex requirements of vibration testing.

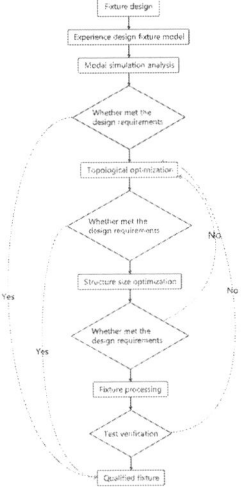

Fig. 1. Design flow of integrated circuit vibration fixture

III. Structural Design

A. Design objects and vibration loads

In the case of a certain type of PGA273 package integrated circuit for aerospace, the sample diagram shows Figure 2, which has 273 Pin-gate array pins. The vibration resistance requirements of its application environment Sine vibration test frequency is 20 to 2000Hz, acceleration 20g, and acceleration response requirements are not more than ±20%, random vibration test frequency is 50 to 2000Hz, power spectral density $30(m/s^2)^2$ /Hz, The total mean square root value is 24.06g, power spectral density response requires a deviation of less than ±1.5dB in 1000Hz and a response value deviation of less than ±3 dB from 1000 to 2000Hz.

Fig. 2. PGA273 packaged integrated circuit

With 30kN thrust of DONGLING vibrator ES-30D-270 for test design, according to the requirements of the fixture system of the first-order inherent frequency is not less than 3000Hz, in order to ensure that the fixture in the test frequency range without resonance, where in the Vibration table ring and horizontal slide mass M_1 is 148kg, sample mass M_3 is 0.05kg, F_0 is 30kN, a is 24.06g*3=72.18g, then the mass of the fixture should meet the $M_2 \leqslant \dfrac{F_0}{1.5a} - (M_1 + M_3) \approx 35\text{kg}$, and the mass of the fixture should be as small as possible to meet the requirements of the modal and transfer characteristics, so that the vibrator has a greater margin for more advanced testing.

B. Vibration test fixture structure design

According to the design process of Figure 1, according to the design requirements of the preceding fixture, combined with the use and characteristics of common centralized fixture combination structure types [11], fixture preliminary solution using the hexagonal mother fixture combination, using 6061-T6 aluminum alloy material. The master fixture of the hexagon can be connected to the vibratory table, and the other faces in addition to the vibration table connection surface can be connected with the sub-fixture, which can be more efficient to realize the requirements of the test method of sample conversion. The sub-fixture is based on the appearance of the sample size, design the corresponding grooves to install the specimen, with a cover or a thin pressure strip to press the specimen, to ensure that the specimen and fixture rigid contact. The

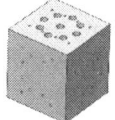 

(a). Master Fixture Structure   (b). Sub-Fixture Structure

structure of the fixture is shown in Figure 3.

Fig. 3. Fixture Structure Diagram

The initial design fixture structure modal simulation results are shown in Table 1 below:

TABLE I.   INITIALLY DESIGNS MODAL SIMULATION RESULTS

Fixture	1st order frequency (Hz)	2st order frequency (Hz).	3rd order frequency (Hz).	4th order frequency (Hz).	Mass (kg)
Mother fixture	2729.7	2731.5	3853.5	5776.7	21.02
Sub-fixtures	8480	10323	12483	16597	0.4
Fixture assembly	2083.7	2088.9	3596.3	5500.8	21.22

The simulation results show that the inherent frequency of the sub-fixture section is much higher than 2000Hz, and the mass is light, so no optimization is required. The first two orders of the master fixture and fixture assembly are close to 2000Hz and of excessive mass, and the design improvement needs to be made using the above optimization method to ensure that the modal characteristics of the fixture meet the requirements during the test and minimize the mass of the fixture.

IV.   VIBRATION FIXTURE DESIGN OPTIMIZATION

A. Layout optimization

Topological optimization has more design freedom than dimensional optimization and shape optimization, and the finite element model used by topology optimization does not need to be parametric, so the first step in fixture optimization should be topology optimization.

The fixtures in this optimization are continuous structure, and the continuous topology optimization method mainly has the SIMP variable density method, the homogenization method, and the horizontal set method[12]. Based on isotropic material, the SIMP variable density method uses the relative density within the interval of 0,1 as the topological design variable, and by defining the empirical formula, the optimal route of transmission of the structure is sought by artificially assuming the nonlinear relationship between the relative density and the elastic modulus of the material, in order to optimize the distribution of the material within the design area[13,14]. SIMP variable density method can get the structure of clear boundaries, rules, and SIMP variable density method is more mature and integrated into most commercial finite element software, with the advantages of easy program implementation, fast computational efficiency, high computational accuracy, this topology optimization design is specifically implemented by imp variable density method.

In order to reduce the amount of computation of the analysis and the possibility of unreasonable solution, the model needs to be simplified as necessary when establishing a topology optimization clip with a limit element model. The principle of simplification is to remove detail features such as small components, small holes, rounded corner and fillets that have little effect on structural stiffness but have a great impact on the complexity of the model. In this study, remove the threads and rounded corner from the screw holes on the fixture.

When selecting the optimization area, because the bottom of the female vibration fixture is bolted with the vibratory table, the middle part of the mother fixture has a fixed hole, and five face design screw holes in the hexagon are used to secure the sub-fixture with bolts, so in addition to the fixed hole and screw hole two parts of the other parts need to be optimized. In order to improve the optimization efficiency, the optimization process is constrained by the optimal distribution of materials, the 1st order frequency is greater than 2000Hz as

the goal, and the optimization results meet the convergence conditions within a limited number of iterations of 100 times.

As can be seen from Figure 4 of the iterative process, the system finds the optimal solution on the 55th iteration..

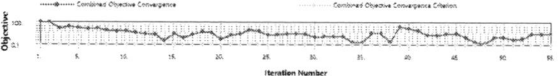

Fig. 4.   iterative process diagram

Fig. 5.   Topology Optimization Results

The approximate shape of the optimization can be obtained by optimizing the results of the topology in Figure 5. Two improved shapes can be obtained based on the results: a. Fill the missing unit with discontinuous boundaries; b. Cut the excess units so that the boundaries are continuous. Structure shape improvements are made based on topological optimization results, as shown in Figure 6.

Fig. 6.   Improves the rear mother fixture structure diagram

The first 4 order frequencies of topologically optimized fixture modal simulation calculations are 2960.2 Hz, 2963.6Hz, 3764.5 Hz, 6297.4Hz, and the optimized model mass is 10.58 kg.

B. Size optimization

After topology optimization, the basic shape of the fixture has been determined, the first-order natural frequency of the master fixture is close to 3000Hz, but the mass of the fixture is still relatively large, in order to ensure that the fixture first-order inherent frequency, the structure size optimization to further improve its inherent frequency and reduce the mass of the fixture.

Structural size optimization is a multi-goal optimization to find the optimal solution to the problem[15~17]. The optimal solution to a multi-goal optimization problem is generally referred to as the Pareto optimal solution[18] . The basic feature of the multi-target genetic algorithm is that it is an effective means of solving Pareto optimal solution sets by using populations with potential solutions from generation to generation to search for multi-directionality and globality[19]. In order to achieve optimal first-order intrinsic frequency and lightest mass of the vibration fixture, the dimension parameter optimization method based on multi-target genetic algorithm(MOGA) is used in this optimization.

The mother fixture is long, wide and high, with the width, depth and diameter and depth of the middle circular groove as the optimization parameters. The size of the vibratory ring and the size of the sub-fixture form factor are used as the constraints of the parameters of the mother fixture length, width and height of 3 dimensions, and the dimension constraints of the side grooves and the middle circular grooves are set according to the size of the fixture structure, and the optimization design is carried out with the first-order inherent frequency of the fixture and the minimization of the fixture mass as the optimization goal. The main steps in the design of structural dimension optimization are as follows:

First, the design parameter sample point selection is carried out by Monte Carlo sampling technique in the experimental design, secondly, the response results of each sample point are solved, the design spatial response surface is established by means of secondary interpolation function, and finally, the multi-target genetic algorithm (MOGA) optimization analysis is achieved by using the response surface analysis results, a different set of Pareto optimal solutions is obtained, and one of the optimal solutions is selected as the optimization scheme according to the requirements.

The structural dimension parameter optimization test design uses the default center conforms to the design method, automatically generating 79 sets of design sample points. The 79 sample points are calculated to form the response face results, and the response face fitting discrete diagram in Figure 7 shows that the design points of the two target parameters are very close to the diagonal, indicating a good fit. Using a multi-target genetic algorithm to implement optimization on the response surface, the convergence criteria are as follows: During this iteration, when 70% of the samples are distributed at the forefront of Pareto optimization, the iteration ends. Where the total population is defined as 1000, and the maximum number of iterations is defined as 100.

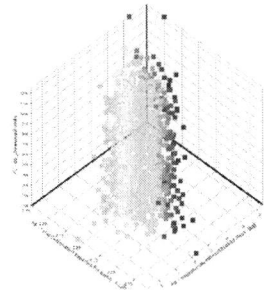

Fig. 7. Response Face Fitting Discrete Plot

In this optimization, the first-order inherent frequency of the fixture is as important as the mass, so the importance of the two design goals is set to default in the objective function, and the weight values of the two target values are 0.5 each.

Structural dimension parameter optimization results in three optimization schemes, the first-order frequency of the three optimization schemes is much higher than the test maximum frequency of 2000Hz, so select the least mass group of optimization scheme as the optimal solution.

Considering the convenience of machining, rounding the dimensions of optimization scheme 2, re-modeling in 3D and calculating the mode of the master fixture, optimizing the comparison of the dimension parameters before and after Table 2, optimizing the 1st order inherent frequency of the post-master fixture is 5908.1Hz, the frequency is much higher than 2,000Hz, 2947.9Hz higher than before topology optimization, an increase of 99.6%. The optimized mass is 2.84kg, which is 7.74kg lower and 73.2% lower than before optimization.

TABLE II. COMPARISON OF DIMENSIONAL PARAMETERS BEFORE AND AFTER OPTIMIZATION

Optimize the parameters	Before topology optimization	After topology optimization	After the size optimized	Round
Fixture length (mm).	200	180	129.73	130
Fixture width (mm).	200	180	126.64	127
Fixture height (mm).	200	140	79.68	80
Side groove width (mm).	/	60	51.55	52
Side groove depth (mm).	/	20	13.72	14
Middle groove diameter (mm).	/	40	46.44	47
Middle groove depth (mm).	/	20	35.72	36
1st order frequency(Hz)	2729.7	2960.2	6160.8	5908.1
Fixture mass(kg).	21.02	10.58	2.64	2.84

The optimized fixture assembly is modal simulation, as shown in Figure 8, with the first four-order modal analysis results being 4141 Hz, 4200.6 Hz, 5578.1 Hz, 8074.4Hz. The simulation results show that the inherent frequency and mass of the optimized fixture meet the design requirements.

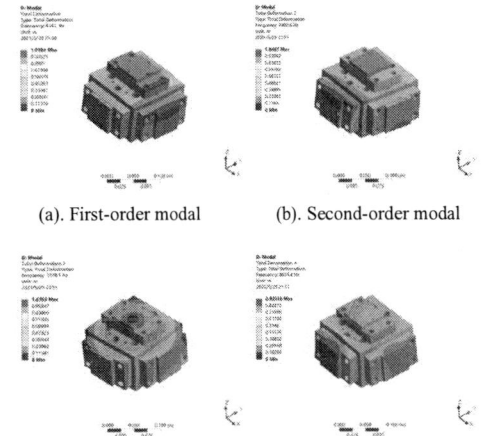

(a). First-order modal   (b). Second-order modal

Fig. 8. Optimizes the first four-order mode of the rear fixture assembly

V. OPTIMIZE THE RESULT VALIDATION

A. Response simulation analysis

Modal analysis can only be extracted to the modal frequency and formation of the fixture, can initially determine the reasonableness of the fixture design, but cannot know the specific response of the fixture in the vibration condition, so it is also necessary to carry out random vibration of the fixture and sweep vibration response analysis, to obtain the response of the fixture under vibration, vibration response analysis steps are as follows:

1. Random vibration analysis steps: Modeling → obtain modal solution→ converted into spectral analysis type → define and apply power spectral density excitation→ solve → look at the results.

2. Sweep vibration analysis steps: Modeling → obtain modal solution→ convert to harmonic response analysis type

→ define and apply acceleration excitation→ solve → see the results.

The preferred choice in this simulation is the fixture mounting state when the excitation load direction is perpendicular to the direction of the vibrating countertop, and the vibration response of the middle point of the jacket cover in the fixture is analyzed.

Applying random vibration excitation is shown in Table 3.

TABLE III.   RANDOM VIBRATION INPUT EXCITATION

Frequency (Hz)	Power spectrum density $(g)^2$/Hz
50	0.1503
100	0.30
1000	0.30
2000	0.30

The random vibration acceleration response results are shown in Figure 9, and the random vibration power spectrum density response is up to $0.3774 g^2$/Hz, within put excitation deviation of 0.99dB, which meets the fixture design requirements.

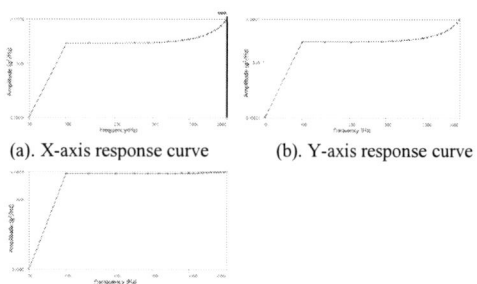

(a). X-axis response curve   (b). Y-axis response curve

(c). Z-axis response curve

Fig. 9.   Random Vibration Response Analysis Results

Applying the sweep vibration excitation is shown in Figure 10.

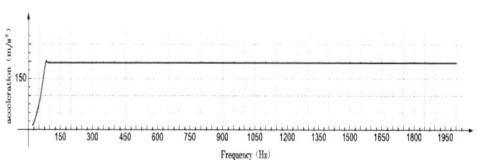

Fig. 10. Random Vibration Response Analysis Results

The results of the sweep vibration acceleration response are shown in Figure 11, the response acceleration and input excitation signal amplitude in the fixture sweep vibration analysis results are not significantly changed, and the sweep acceleration response value is stable at $196 m/s^2$, which meets the response range requirement of ±20%.

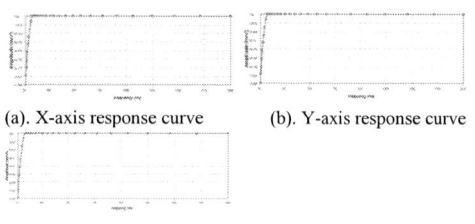

(a). X-axis response curve   (b). Y-axis response curve

(c). Z-axis response curve

Fig. 11. Sweeping Frequency Vibration Response Analysis Results

B. Test verification

After ensuring that the first-order natural frequency, mass and delivery response of the fixture meet the requirements, the fixture is machined and tested for verification, as shown in Figure 12,the mass of the fixture is smaller than the simulation results, 1 acceleration sensor is installed on the vibratory station horizontal slide, the sensor on the horizontal slide acts as the control sensor (black curve), and the sub-clamp cover plate sensor is used as the monitoring sensor (purple/blue curve). As shown in Figure 13, the integrated circuit vibration fixture has no resonance in the sweep frequency of three axes in the 2000Hz test range, and the acceleration response deviation is within ±5%, which meets the requirements of the fixture design. As shown in Figure 14, the random vibration power spectral density response deviation of the three axes of the integrated circuit vibration fixture meets the requirements of the fixture design.

Fig. 12. diagram of fixture system Verification test

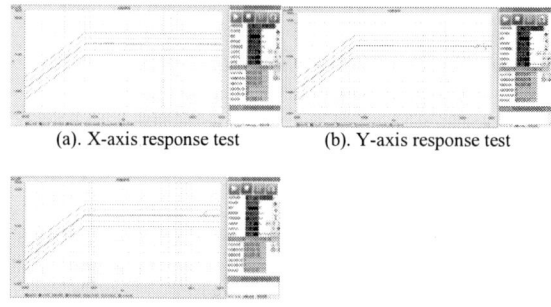

(a). X-axis response test   (b). Y-axis response test

(c). Z-axis response test

Fig. 13. diagram of fixture sinusoidal vibration response test

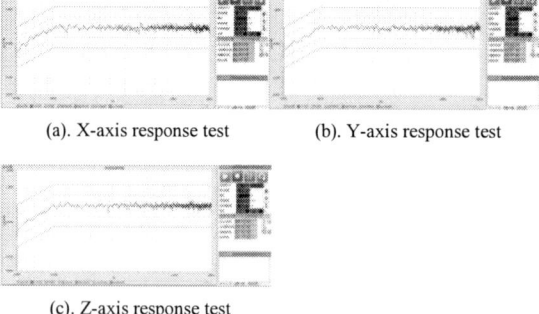

(a). X-axis response test   (b). Y-axis response test

(c). Z-axis response test

Fig. 14. diagram of fixture random vibration response test

VI.   CONCLUSIONS

For the harsh vibration test conditions of complex integrated circuits for aerospace and the difficulty of rigid fixing complex package structures, the corresponding triaxial vibration fixture is designed for aerospace PGA273 package complex integrated circuit, based on the fixture design and test verification results，the following conclusions are obtained:

1) The dimensional parameter optimization method based on topology optimization and multi-target genetic algorithm is used to optimize the inherent frequency improvement and mass reduction of the fixture structure are optimized, and the clamp model is simulated for sweeping vibration and random vibration test response, and finally the cleaning frequency vibration and random vibration response of the fixture are tested, and the test results verify the rationality and effectiveness of the fixture design and optimization method.

2) The design and optimization method of integrated circuit triaxial vibration fixture structure is different from previous optimization method , The first-order inherent frequency of vibration fixture is improved by 116.4 percent and the mass is 86.5 percent lower with optimization, The sweep vibration and random vibration response are in line with the standard requirements.

3) Since the use of empirical design optimization or only single-size optimization of the design of complex integrated circuits vibration fixtures may have low natural frequency, poor vibration acceleration response uniformity, and other common problems, it may provide a certain reference value for the aerospace complex integrated circuit vibration fixture design and improvement

ACKNOWLEDGMENT

This research was supported by the Guangdong Science and Technology Department Key Research and Development Program Project (2020B0404030005), Guangdong Provincial Development and Reform Commission Project (Guangdong Province High-end Chip Reliability Engineering Research Center).

REFERENCES

[1] MIL-STD-883K w/CHANGE2, TEST METHOD STANDARD MICROCIRCUITS [S]. Military and Government Specs & Standards (Naval Publications and Form Center) (NPFC),2017.

[2] GB/T 2423.10-2019, Environmental Test Part 2: Test Method Test Fc: Vibration (Sine)[S] . Beijing: China Standard Press, 2019.

[3] GB/T 2423. 56-2018, Environmental Test Part 2: Test Method Test Fh: Broadband Random Vibration and Guidelines [S]. Beijing: China Standard Press, 2018.

[4] SHANG X Z,XUE J W,MO F, et al. Analysis of Fixture Vibration Characteristics in Test for IGBT Power Module Anti-vibration Performance[J].Mechanical Manufacturing and Automation, 2020,49(6):52-55.

[5] WANG H D, LI G L, WANG Z X, et al. Structural optimization design and analysis of triaxial vibration fixture based on multi-objective genetic algorithm[J].Spacecraft Environment Engineering, 2020,37(2):154-160.

[6] LIU X C, CHEN J, CUI W, et al. Design and Analysis of Dynamic Characteristics of Vibration Fixture with Solid Propellant Rocket Engine[J]. Structure&Environment Engineering, 2020, 47(2): 56-63.

[7] XU H. The Design Optimization by FEA for Vehicle Media System Vibration Fixture[J].Value Engineering, 2019, 38(34):117-119.

[8] MAO L, HU J S. An Optimization Method of Airborne Equipment Vibration Fixture[J].Mechanical Research and Application, 2016,29(2):138-140.

[9] WANG Y Q, LEI P S, FENG R, et al. Mechanical environment test technology[M].Xi'an: Northwest University of Technology Press, 2003.

[10] Wang D S, Ren W F, Liu Q L, et al. The design requirements of the resonance frequency of vibration test fixtures [J].Spacecraft Environment Engineering, 2014, 31(1): 37-41.

[11] DENG C J, DENG R.Summary of design and test methods of integrated circuit vibration test fixture[J]. Electronic Product Reliability and Environmental Testing, 2017, 35(03): 64-68.

[12] LIANG J, LI X J. The application of topology optimization in product design[J].Design, 2021, 34(05):137-139.

[13] QIU F S, JI W Q, XU H C, et al. The vertical tail of topology optimization design study based on improved variable density method[J].Journal of Shenyang Aerospace University, 2013, 30(1): 26-29.

[14] Bendsoe M P, Kikuchi N. Generating optimal topologies in structural design using a homogenization method[J].Computer Methods in Applied Mechanics and Engineering, 1988, 71(I):197-224.

[15] Esebenauer H A, Kobelev H A, Sehumaeher A. Bubble method for topology and shape optimization of structures[J].Structural Optimization, 1994, 8:14 5-149.

[16] GU D D, ZHU H P, CHEN X Q, et al. Position optimization design of passive control device between adjacent structures[J].Journal of Vibration, Measurement&Diagnosis, 2010, 30(1); 11-15.

[17] WANG K, SUN Y Y, MAO Z Y, et al. Comprehensive optimization method for dynamic design of vibration test fixture[J].Journal of Vibration, Measurement&Diagnosis, 2013, 33(3): 483-487.

[18] HAO X H, HU Z G, HOU Q, et al. Optimization of serpentine channel heat sink based on multi-objectivegenetic algorithm[J].Journal of Mechanical E.

[19] CHEN Shun-chang. Frequency Response Analysis and Strutural Optimization Design of a Cabin Triaxial Vibration Fixture[J]. Xi'an: XiDian University, 2015: 50-61.

Growth Behavior of Interfacial Intermetallic Compounds of Co-20%P/Solder Joint under Temperature Gradient

Haotong Qin
College of Materials Science and Engineering
Chongqing University of Technology
Chongqing, China
1278120444@qq.com

Donghua Yang*
Chongqing Municipal Engineering Research Center of Institutions of Higher Education for Special Welding Materials and Technology(Chongqing University of technology)
Chongqing, China
yangdonghua@cqut.edu.cn

Tao Fan
College of Materials Science and Engineering
Chongqing University of Technology
Chongqing, China
827356317@qq.com

Tao Chen
College of Materials Science and Engineering
Chongqing University of Technology
Chongqing, China
1916546139@qq.com

Chunhong Zhang
College of Materials Science and Engineering Chongqing University of Technology
Chongqing, China
zhangchunhong@cqut.edu.cn

YuQian Chen
College of Materials Science and Engineering
Chongqing University of Technology
Chongqing, China
627168390@qq.com

Abstract—In this paper, the growth behavior of interfacial IMC of Co-P/SAC105/Co-P solder joints under the temperature gradients of 2215°C/cm and 3260°C/cm were studied. Co-P UBMs with 20% P content were electroplated on the copper substrates. Co-p/SAC105/Co-P solder joints were prepared by reflow welding process. BGA solder balls were Sn-1.0wt.%Ag-0.5%Cu solder balls with diameters of 400µm. The obtained solder joints were used to two above temperature gradients test, and the loading times were 0h, 100h, 200h and 400h, respectively. It found that the Co-Sn-P layer appeared between the Co-P film and the $CoSn_3$ layer. As the loading time of the temperature gradient increased, the fine needle-shaped $CoSn_3$ near the Sn solder became plate-shaped $CoSn_3$, and finally formed a bulk shape. Compared with 2215°C/cm, the interfacial $CoSn_3$ was relatively continuous and dense, because of the better combination of fine needle-shaped $CoSn_3$ and bulk-shaped $CoSn_3$ under 3260°C/cm. With the loading time increased, the growth of $CoSn_3$ at the hot and cold ends of the solder joints interface showed obvious differences. Intuitively, the thickness of the cold end increases more rapidly than that of the hot end. After loading time of 400h under 2215°C/cm, the total thickness growth of $CoSn_3$ was 18.10µm. After loading time of 400h under 3260°C/cm, the total thickness growth of $CoSn_3$ was 46.74µm. It showed that the growth of $CoSn_3$ at the cold and hot ends shows the more significant asymmetric growth trend under a larger temperature gradient, revealing that the growth of interfacial $CoSn_3$ could be induced by the larger temperature gradient.

Keywords—Growth Behavior; CoSn3; Temperature gradient; IMC thickness

I. INTRODUCTION

With the development of microelectronic products towards multi-functional, high-density and miniaturization, the solder joints are becoming smaller and the percentage of IMC (Intermetallic Compound) in solder joints is becoming larger. In the end, solder joints will be fully made up of IMCs. When the Sn in the solder reacts with Cu on the substrate, Cu-Sn IMC will be formed, which are Cu_6Sn_5 and Cu_3Sn, respectively. However, the Kirkendall voids [1] formed at the Cu_3Sn interface significantly reduce the toughness of the solder joint. Compared with Cu_3Sn and Cu_6Sn_5, $CoSn_3$ has the advantages of higher melting temperature [2-3] and fracture toughness. It might solve the problem of brittleness and hardness of full IMC solder joints. Consequently, Inducing the formation of full $CoSn_3$ IMC solder joint is an important way to make the reliability of solder joints become higher [4]. However, the way to prepare the full $CoSn_3$ IMC solder joints have not been investigated in the literature.

The study found that thermal migration can induce the growth of IMC by enhancing the diffusion of metal atom [5]. However, Co-P UBMs with low P content have good diffusion-barrier properties, which can inhibit the thermal migration to some extent. When the P content reaches 20at%, the diffusion rate of Sn into Co-P film with amorphous structure was relatively fast, and the growth rate of $CoSn_3$ was increase [6]. Therefore, under extreme temperature gradient, the full $CoSn_3$ solder joint is probably to be get when the P content in Co-P films is 20%. In this paper, growth behavior and morphology of $CoSn_3$ at the interface of Co-20%P/solder joints were studied by scanning electron microscopy. And a way to induce the growth of $CoSn_3$ at the interface of the solder joint was also suggested.

II. EXPERIMENT

In this paper, Co-P UBMs with P content of 20% were electroplated on copper substrates. The BGA copper substrate is composed of 99.5 percent pure copper in this experiment, and the size of the substrate is 6mm x 6mm x 0.6mm. The electroplated Co-P UBM had a thickness of 8µm(±1µm). The BGA substrates were carefully polished and ultrasonic cleaned before soldering. After the rosin flux was applied, the Sn-1.0wt.%Ag-0.5%Cu (SAC105) BGA solder balls with the diameter of 400µm were reflowed on the Co-P films, and the peak temperature is 275±3°C.The schematic diagram of the solder joint model is shown in Figure 1.

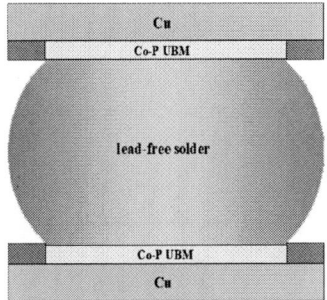

Fig.1 Co-P/SAC105/Co-P Solder joint model diagram

In this experiment, two sets of heat transfer devices with different temperature gradients are designed and built. The schematic diagram of the structure is shown in Figure2. The hot end is the constant temperature heating platform, and the cold end is chilled by semiconductor chilling plates and the coolants respectively.

The surface temperature of the cold and hot ends of the sample under the cooling system of the refrigeration fin is stabilized at 0°C and 270°C respectively, and the surface temperature of the sample at the two ends of the circulating refrigerant system is stabilized at -40°C and 290°C respectively.

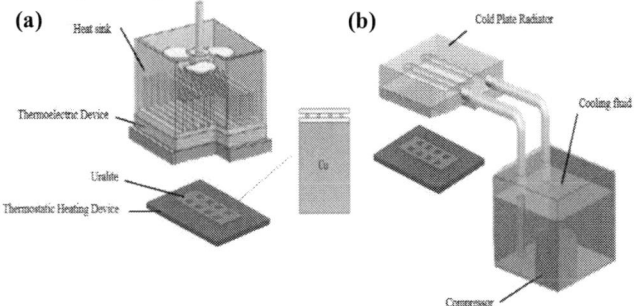

Fig.2 The schematic diagram of temperature gradient devices

Real-time monitoring of the temperature of the hot and cold solder joints is performed to ensure that the center temperature of the solder joints remains consistent during the experiment. The temperature field at the solder joint was simulated by ANSYS. And the actual temperature of the solder joints at the hot and cold ends was obtained，as Figure 3.

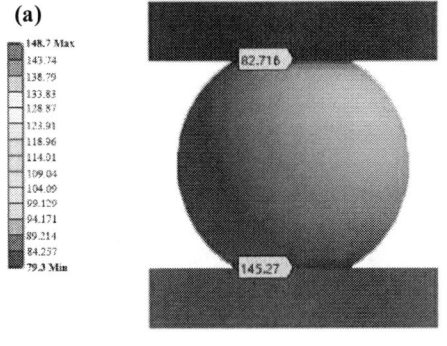

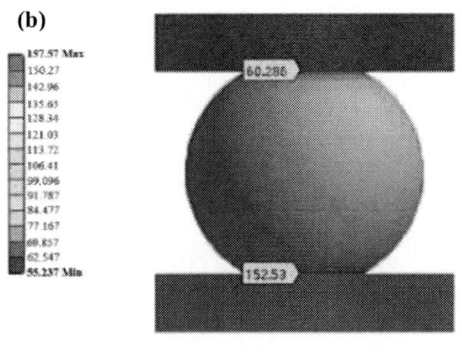

Fig.3 Simulation diagram of the temperature field at the solder joint. (a)148.7~79.3°C; (b)157.57~55.237°C

The actual height of Co-P/SAC105/Co-P solder joint is 280μm(±5μm) through a lot of experiments. Combining the two actual temperatures of the solder joints at the hot and cold ends, the two different actual temperature gradients in the solder joints can be obtained from the following formula:

$$G = \frac{\Delta T}{h} \qquad (1)$$
$$G1 = \frac{\Delta T}{h} \approx 2215°C/cm$$
$$G2 = \frac{\Delta T}{h} \approx 3260°C/cm$$

Two temperature gradients were finally got through ANSYS simulation and calculation. In the cooling system of semiconductor chilling plate, the temperature gradient is 2215°C/cm, and the temperature gradient is 3260°C/cm in the circulating freezing liquid cooling system. The samples were placed in the above two devices which provide two different temperature gradients, and the loading times were 0 hour, 100 hours, 200 hours, and 400 hours, respectively. The FE Scanning electron microscope was used to observe the growth behavior and morphology evolution of IMC at the interface of the solder joints under different temperature gradients with different loading time. The thickness of IMC could be got with the help of Image J software.

III. RESULT AND DISCUSSION

0A. Growth behavior of interfacial $CoSn_3$ under the temperature gradients of 2215°C/cm

The morphology and evolution of the microstructure of the interface IMC of the Co-P/SAC105/Co-P solder joints under 2215°C/cm were showed by Figure 4. The loading time were 0h, 100h, 200h, and 400h, respectively. It indicated that the IMCs at the cold and hot ends are $CoSn_3$ at 20% P content, and the dark gray Co-Sn-P layer is formed between the black Co-P layer and the light white $CoSn_3$ layer.

The $CoSn_3$ near the Co-Sn-P layer at both cold and hot ends of 0h presents a relatively continuous and dense flake structure. There are fine needle-like and short rod-like $CoSn_3$ above the dense layer. As the loading time of the temperature gradient increases, the bulk shaped $CoSn_3$ thickened, but the IMC layer became less continuous compared to the initial sample. And the short rod-shaped $CoSn_3$ on the top gradually disappears. This is mainly due to the combination of the fast-growing bulk $CoSn_3$ and the short rod-shaped $CoSn_3$ under the temperature gradient.

With the increase of the loading time of the temperature gradient, the thickness of $CoSn_3$ at the cold and hot ends of the solder joints increased in various degree. The thickness of the initial sample is close, and the hot end is significantly thinner than the cold end after 400h, revealing that the IMC of the cold and hot ends of the solder joints exhibit asymmetric growth under the temperature gradient.

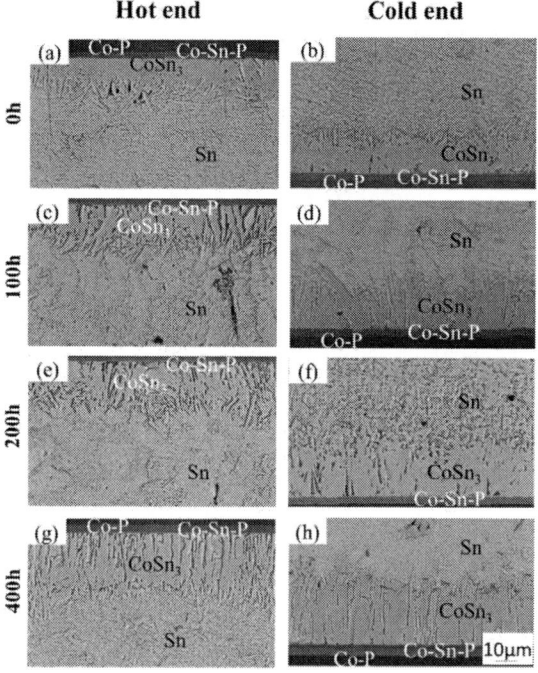

Fig.4 Morphology evolution and microstructure of the interface IMC under 2215°C/cm

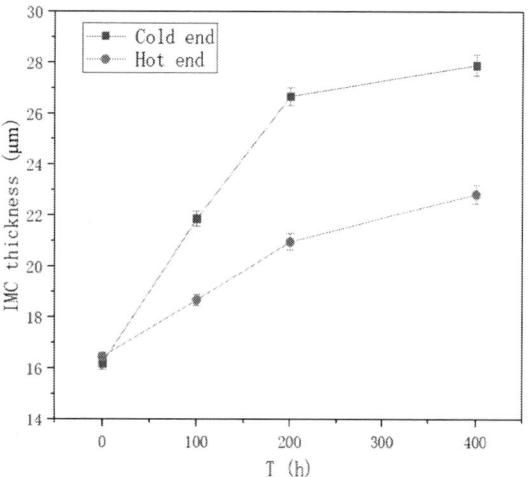

Fig.5 Thickness of interface CoSn₃ with different loading time of the temperature gradient of 2215°C/cm

Figure 5 summarized the thickness of interface $CoSn_3$ at two ends under the temperature gradient of 2215°C/cm. The average thickness of $CoSn_3$ at the hot end increased from 16.19μm to 22.37μm after 400h. At the same time, the average thickness of $CoSn_3$ at the cold end increased from 16.46μm to 27.91μm. It is found that the thickness of $CoSn_3$ of the hot end and the cold end at the initial stage is similar by comparing the $CoSn_3$ IMC thickness of the hot and cold ends. Under the temperature gradient, the thickness of the both ends gradually increased because of heat transfer, but the growth rate of $CoSn_3$ at the hot end was slower than that of cold end. The specific performance is that the interfacial $CoSn_3$ at cold end is thicker, showing the trend of asymmetric growth.

As the loading time of the temperature gradient increased, the growth rate of $CoSn_3$ decreases slightly at two ends. Compared with the hot end, the thickness of $CoSn_3$ at the cold end increased faster. This is because the thickness of the interface $CoSn_3$ increased as the loading time of the temperature gradient increases. It is more difficult for Co atoms to pass through the $CoSn_3$ layer and react with Sn to form $CoSn_3$, resulting in a decrease in the synthesis speed of $CoSn_3$ at the cold and hot ends. With the thickening of the interfacial IMC, the diffusion rate of Co atoms into the Sn in the solder decreases, so the growth rate of $CoSn_3$ is inhibited.

B. Growth behavior of interfacial CoSn₃ under the temperature gradients of 3260°C/cm

The morphology evolution and microstructure of interfacial $CoSn_3$ of the solder joints under the temperature gradient of 3260 °C/cm were showed by Figure 6, and the loading time were respectively 0h, 100h, 200h and 400h.

It is found that the interfacial $CoSn_3$ IMC near the position of Co-Sn-P layer is dense and smooth with the loading time from 0h to 400h under the temperature gradient of 3260°C/cm. The IMC in contact with the Sn solder is dendritic, and the longer the loading time of the temperature gradient, the thicker the dense layer of the IMC, and the smoother the morphology of the IMC. It can be found that the shape of the interface IMC near the position of Co-Sn-P layer has barely changed with the temperature gradient time from 0h to 400h under 3260°C/cm, and the overall IMC presented a relatively dense stage.

The $CoSn_3$ near the solder side had a fine needle-like morphology. As the increase of the loading time of the temperature gradient, the fine needle-shaped $CoSn_3$ gradually evolved into the plate-shaped $CoSn_3$, and finally slowly formed bulk shapes. Similar to the temperature gradient of 2215°C/cm, this is due to the rapid growth of $CoSn_3$ under the action of thermal migration, resulting in the combination of thick massive $CoSn_3$ and fine needle-shaped $CoSn_3$ to form the more complete, dense and continuous $CoSn_3$ IMC layer.

Fig.7 shows that the thickness of the interfacial $CoSn_3$ changes with time at the temperature gradient of 3260°C/cm. At 3260°C/cm for 400h, the average thickness of $CoSn_3$ at the cold end increased from 21.71μm to 49.77μm and that of the hot end increased from 20.61μm to 39.29μm. The initial $CoSn_3$ thicknesses at two ends were similar under 3260°C/cm. Compared with the thickness of interfacial $CoSn_3$ at the hot and cold end with the different loading time of temperature gradient, it can be seen that the IMC at the cold end increased more rapidly than the hot end at the initial stage. The cold end was obviously thicker than the hot end under the temperature gradient. When the loading time of temperature gradient reach 400h, the difference of the thickness at both ends became the largest. Therefore, the difference of the $CoSn_3$ thickness at both ends at the interface of the solder joints is more conspicuous with the increase of the loading time of the temperature gradient.

Hot end   **Cold end**

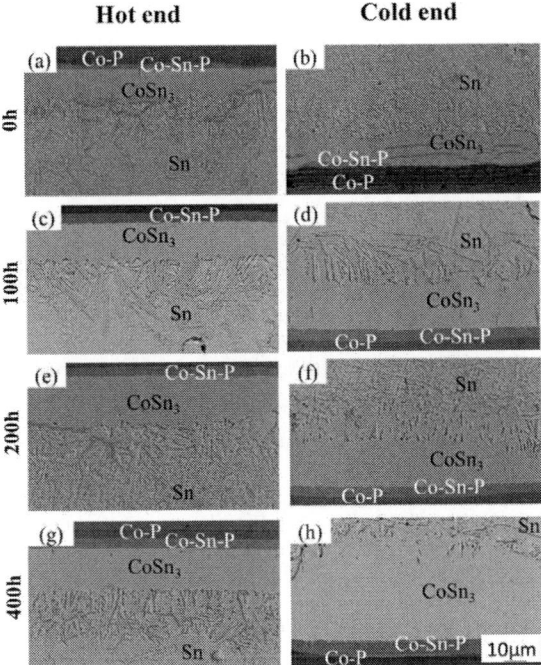

Fig.6 Morphology evolution and microstructure of the interface IMC under 3260°C/cm

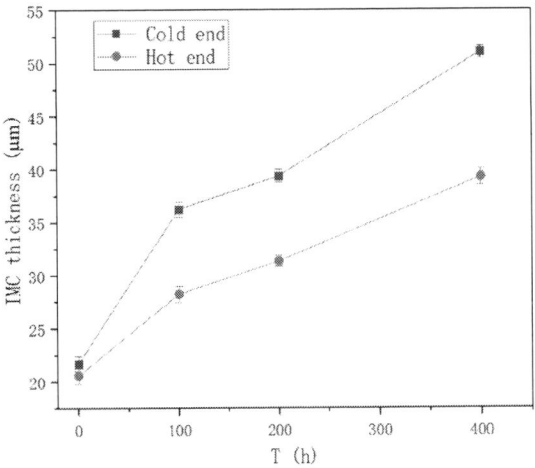

Fig.7 Thickness of interface CoSn₃ with different durations of the temperature gradient of 3260°C/cm

Table1 The average thickness of interfacial IMC

		$h/\mu m$ (0h)	$h/\mu m$ (400h)	$\triangle h/\mu m$	$\triangle d(total)/\mu m$
2215 °C/cm	Cold end	16.19	27.91	11.72	18.10
	Hot end	16.46	22.84	6.38	
3260 °C/cm	Cold end	21.71	49.77	28.06	46.74
	Hot end	20.61	39.29	18.68	

Table1 reveals the thickness of interfacial CoSn₃ at initial stage and loading time of 400h under two different temperature gradients. Under the temperature gradient of 2215°C/cm, the total thickness growth of the interface CoSn₃ at the cold and hot ends of the Co-P /SAC105/ Co-P solder joints are 11.72μm and 6.38μm, respectively. And the total thickness growth of the interfacial CoSn₃ at the cold and hot

ends of the solder joints at 3260°C/cm are 28.06μm and 18.68μm, respectively. It can be seen that the increase of temperature gradient will lead to more obvious asymmetrical growth phenomenon at both ends of the cold and hot. At the same time, as the numerical value of the temperature gradient increase, total thickness growth of IMC at both hot and cold ends also increases, and the proportion of IMC in solder joint increases. Therefore, it presents a way to induce the growth of CoSn₃ at the interface of the solder joint.

IV. CONCLUSION

In this paper, growth behavior of interfacial CoSn₃ of Co-20%P/solder joint under temperature gradient was studied. The conclusions can be made as follows:

(1) The components of the interfacial IMC were CoSn₃ under temperature gradients of 2215°C/cm and 3260°C/cm. When the P content in the coating reached 20%, an obvious Co-Sn-P layer appeared between the Co-P film and the IMC layer. As the increase of loading time of temperature gradient, the fine needle-shaped CoSn₃ near the Sn solder combined with the bulk-shaped CoSn₃, and the interfacial CoSn₃ became smooth and dense.

(2) The average thickness of the interfacial CoSn₃ at two ends both increased. However, the growth of CoSn₃ at the interface of the cold and hot ends of the Co-P/SAC105/Co-P solder joint showed obvious differences. Intuitively, the thickness of the cold end increased more rapidly comparing with the hot end.

(3) The total growth of interfacial CoSn₃ increased from 18.10μm to 46.74μm with the temperature gradient from 2215°C/cm to 3260°C/cm, revealing that the growth of interfacial CoSn₃ could be induced by the larger temperature gradient.

ACKNOWLEDGMENT

This research was supported by National Natural Science Foundation of China (Grant No. 61804018), the Scientific and Technological Research Program of Chongqing Municipal Education Commission (KJQN202001124), the Chongqing Research Program of Basic Research and Frontier Technology (cstc2016jcyjA0226) and the Research and Innovation Project of Graduate Students of Chongqing Municipal Education Committee (clgycx20203009) respectively.

REFERENCES

[1] C.C. Lee, P.J. Wang, J.S. Kim, Are intermetallics in solder joints really brittle 57th Electronic Components and Technology Conference (ECTC) 2007, pp. 648–652.
[2] W. Liu. "Study on Interface Reaction Mechanism and Micromechanical Behavior of Full IMC Micro-interconnect Solder Joints", Mechanical Manufacturing Abstracts-Welding Volume, Vol.1, PP.17-18, 2013.
[3] K. N. Tu, and Y. Liu, "Recent advances on kinetic analysis of solder joint reactions in 3D IC packaging technology", Materials ence & Engineering, vol. 136, pp.1-12, 2019.
[4] Yang D, Cai J, Wang Q, et al. IMC growth and shear strength of Sn–Ag–Cu/Co–P ball grid array solder joints under thermal cycling[J]. Journal of Materials Science: Materials in Electronics, vol. 26, pp.962-969, 2015.
[5] Zhang P, Xue S, Wang J. New challenges of miniaturization of electronic devices: Electromigration and thermomigration in lead-free solder joints[J]. Materials & Design, vol. 192, 2020.
[6] Yang G, D Yang, Li L. Microstructure and morphology of interfacial intermetallic compound CoSn₃ in Sn–Pb/Co–P solder joints[J]. Microelectronics Reliability, vol.55, pp. 2403-2411, 2015.

978-1-6654-1392-3/21 $31.00 © 2021 IEEE

An Accurate Simulation Method of Package Warpage Experimental Results Based on FEM

Liqiang Neng[2*]
Department of Packaging and Testing
ZTE Corporation
Shen zhen, China
neng.liqiang@sanechips.com.cn

Tingting Song[2*]
Department of Packaging and Testing
ZTE Corporation
Shen zhen, China
song.tingting1@sanechips.com.cn

Guangping Shao[2]
Department of Packaging and Testing
ZTE Corporation
Shen zhen, China
shao.guangping@sanechips.com.cn

Jian Pang[2]
Department of Packaging and Testing
ZTE Corporation
Shen zhen, China
pang.jian@sanechips.com.cn

Sun Tuobei[2]
Department of Packaging and Testing
ZTE Corporation
Shen zhen, China
sun.tuobei@sanechips.com.cn

Keqing Ouyang[1]
State Key Laboratory of Mobile
Network and Mobile Multimedia
Technology
Shen zhen, China
ouyangkeqing@sanechips.com.cn

Abstract—In order to realize the diversification of electronic products, it is necessary to integrate more electronic devices, which leads to more and more substrate circuit density, resulting in the increase of package size and the number of substrate layers. At the same time, the mismatch of thermal expansion coefficient of packaging materials is becoming more and more serious, which is bound to lead to larger package warpage. Large warpage may lead to substrate delamination, solder joint cracking and bump electromigration failure, etc. At the same time, it will also cause problems in the placement of electronic devices, which will affect the reliability of the whole package. Therefore, how to have an accurate prediction of package warpage in the early stage of packaging is a difficult problem that needs to be solved urgently. In this paper, based on ANSYS Workbench software, using the method of equivalent packaging materials and equivalent substrate, the warpage values of different packaging forms and sizes are simulated, and compared with the average value of shadow moire results, and the error is less than 13%, which greatly improves the simulation accuracy.

Keywords—Package warpage, Equivalent substrate, shadow moire

I. INTRODUCTION

In recent years, due to the diversification of package functions and the continuous increase in the number of devices that need to be placed inside the package, the package size becomes larger and larger[1]. At the same time, more routing is needed inside the packaging substrate, which leads to the increase of the number of layers of the packaging substrate and the complexity of the packaging structure. This leads to more types of packaging materials and significant differences in the thermal expansion coefficient of packaging materials. In the process of packaging reflow, the large temperature gradient in the process of heating and cooling, and the difference in the heating and cooling rate will cause greater warpage during the reflow process[2,3]. However, a large warpage will also be generated during the processing of the package substrate, which will further increase the value of the package warpage, and the large package warpage will affect the reliability of the entire packaging process.

Large package warpage will affect the solder ball implantation and cause the loss of yield. At the same time, when the warpage is too large, it will not only cause delamination of the packaging substrate, but also affect the installation and yield problems of electronic devices[4,5]. The warpage is caused by multiple parts, mainly by the substrate processing and reflow process. There are many kinds of warpage reasons, and the shadow moire result can only be carried out after the entire package is completed. At this time, if there is warpage problem in the product, it will not only take time to modify, but also need to pay more manpower and material resources. At the same time, the shadow moire results of package warpage are affected by the initial warpage of the substrate processing, and the warpage of the same batch of products also changes greatly, which increases the difficulty of the accuracy of the simulation results. Therefore, it is very difficult to accurately simulate the package warpage by using simulation software.

Therefore, this paper adopts the method of material equivalence and substrate equivalence to simulate the packaging warpage. This method is used to simulate the package models of three different package sizes and different package forms, and the three different simulation results are compared with the shadow moire results. Considering that the shadow moire results have multiple test samples for each package form, the simulation results are compared with the average of the shadow moire results. The results show that all errors can be controlled within 13%, which shows that this method has important guiding significance for the simulation of package warpage in the future.

II. EQUIVALENT METHOD

Many properties of materials will change near the glass transition temperature of the molding compound, which will cause the material's thermal expansion coefficient and Young's modulus to change suddenly at this temperature, resulting in two large changes in the coefficient of thermal expansion and Young's modulus of materials. When we simulate, the selected temperature range will contain two values, which is inconvenient to handle and inaccurate to the results. It is true that it is not easy to converge. After many attempts, it is found that no matter what the glass transition temperature of the material is, it is directly using formula (1) and formula (2) for calculation. At this time, the thermal expansion coefficient and Young's modulus of the material can be equivalent to one value, and the input mode of the material does not need to be considered, which not only saves the calculation time, but also improves the simulation accuracy.

As shown in Fig. 1, the substrate is made up of many thin

layers stacked, and it is composed of a variety of materials. A variety of different material input methods have been tried for this stack. It is found that the results calculated by formula (6) and formula (7) are compared with the results of shadow moire. It is found that the simulation results under this condition are basically the same as those of shadow moire. At the same time, by comparing the calculated thermal expansion coefficient and Young's modulus with the actual measurement results, it is found that the equivalent material properties are basically consistent with the test results.

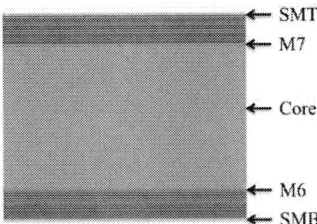

Fig. 1. Composition of substrate stack

$$CTE = \frac{\alpha_1(T_g - T_{final}) + \alpha_2(T_{ref} + T_g)}{T_{ref} - T_{final}} \quad (1)$$

$$E = \frac{E_1(T_g - T_{final}) + E_2(T_{ref} + T_g)}{T_{ref} - T_{final}} \quad (2)$$

$$v_1 = \frac{H_{SMT} + H_{SMB}}{H_{sub}} \quad (3)$$

$$v_2 = \frac{H_{ABF}}{H_{sub}} \quad (4)$$

$$v_3 = \frac{H_M}{H_{sub}} \quad (5)$$

$$v_4 = \frac{H_{Core}}{H_{sub}} \quad (6)$$

$$CTE_{sub} = \frac{CTE_{SM}*E_{SM}*v_1 + CTE_{ABF}*E_{ABF}*v_3 + CTE_M*E_M*v_3 + CTE_{Core}*E_{Core}*v_4}{E_{SM}*v_1 + E_{ABF}*v_2 + E_M*v_3 + E_{Core}*v_4} \quad (7)$$

$$E_{sub} = E_{SM} * v_1 + E_{ABF} * v_2 + E_M * v_3 + E_{Core} * v_4 \quad (8)$$

Symbol description：

CTE:Coefficient of thermal expansion

α_1:Coefficient of thermal expansion less than T_g

α_2:Coefficient of thermal expansion greater than T_g

T_g:Glass transition temperature

T_{final}:Maximum temperature during reflow

T_{ref}:Reference temperature

E:Young's modulus

E_1:Young's modulus less than T_g

E_2:Young's modulus greater than T_g

v_1:Ratio of total thickness of SM layer to total thickness of substrate

v_2:Ratio of total thickness of ABF layer to total thickness of substrate

v_3:Ratio of total thickness of copper layer to total thickness of substrate

v_4 :Ratio of thickness of core layer to total thickness of substrate

H_{SMT}:SMT layer thickness

H_{SMB}:SMB layer thickness

H_{sub}:substrate thickness

H_{ABF}:Thickness of all ABF layers

H_M:Thickness of all copper layers

H_{Core}:Core layer thickness

CTE_{sub}:Thermal expansion coefficient of equivalent substrate

CTE_{SM}:Thermal expansion coefficient of SM layer

E_{SM}:Young's modulus of SM layer

CTE_{ABF}:Thermal expansion coefficient of ABF layer

E_{ABF}:Young's modulus of ABF layer

CTE_{SM}:Thermal expansion coefficient of SM layer

E_{SM}:Young's modulus of SM layer

CTE_{Core}:Thermal expansion coefficient of core layer

E_{Core}:Young's modulus of core layer

E_{sub}:Young's modulus of equivalent substrate

III. SIMULATION METHOD VALIDATION

The verification of this new simulation method adopts three different packaging forms, namely ring and lid packaging forms, of which lid packaging has two different packaging forms, hat and flat. Based on the pod diagram and package information, a three-dimensional finite element simulation model is established. The ring packaging structure is shown in Fig. 2, the flat packaging structure is shown in Fig. 3, and the hat packaging structure is shown in Fig. 4.

Fig. 2. Ring package structure

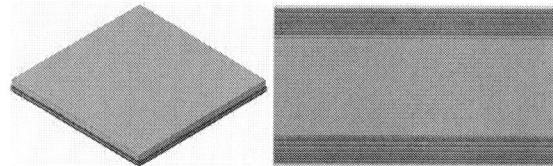

Fig. 3. Flat package structure

Fig. 4. Hat package structure

In the simulation, it is assumed that all materials are isotropic, and the material properties are shown in TABLE I. The material equivalent method is used to calculate the material in TABLE I., and the substrate equivalent method is used to equivalent the substrate material to a coefficient of thermal expansion and Young's modulus. The data obtained are shown in TABLE II. Although the finite element simulation method can predict the package warpage, but because the substrate itself has warpage after production, and this part of the warpage value is difficult to determine, Lin et al[6]. Carried out a lot of experiments, based on the shadow moire test data, determined the zero warpage temperature, which is about 150℃, and we also observe from the shadow moire test data that the package structure warpage is closer to 0 at about 150℃, so the simulation high temperature range in this paper is 150-260℃, the room temperature range is 150-25℃.

TABLE I.   PACKAGING MATERIAL PROPERTIES

Material properties										
Item		Unit	SM	Trace	Core	ABF	Die	Adhesive	Ring	Bump
			S	Copper	C	A	Si	Ad	R	B
Tg		℃	133	/	/	153	/	/	/	95
CTE	α1	ppm/℃	38	17	7	20	2.8	45	17.8	25
	α2	ppm/℃	115			49				95
Young's modulus	E1	GPa	3.5	120.66	23	13	131	0.234	120.66	11
	E2	GPa								0.45

TABLE II.   PACKAGING MATERIAL PROPERTIES AFTER EQUIVALENCE

Item	Unit	Substrate	Die	Ring	Adhesive	Bump
Material	/	/	D	R	Ad	B
CTE	ppm/℃	12.25	2.8	17.8	45	10
Young's modulus	GPa	31.71	131	120.66	0.234	16.43

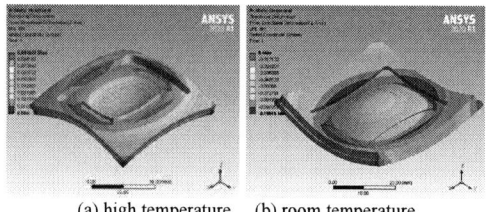

(a) high temperature   (b) room temperature

Fig. 5. Ring package structure simulation results (a) high temperature (b) room temperature

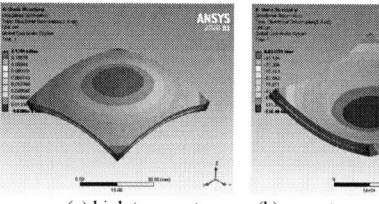

(a) high temperature   (b) room temperature

Fig. 7. Simulation results of flat package structure(a) high temperature (b) room temperature

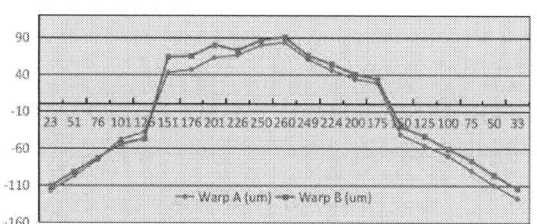

Fig. 6. Shadow moire results of ring packaging structure

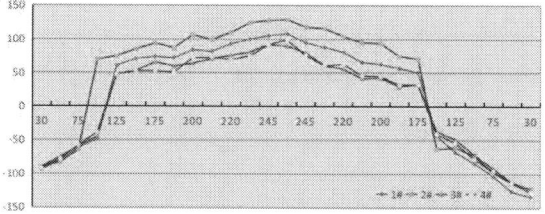

Fig. 8. Shadow moire results of flat packaging structure

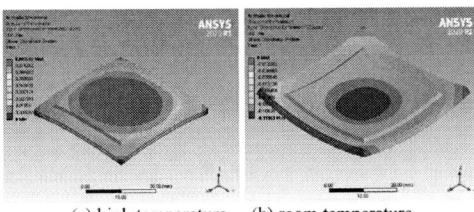

(a) high temperature   (b) room temperature

Fig. 9. Simulation results of hat packaging structure (a) high temperature (b) room temperature

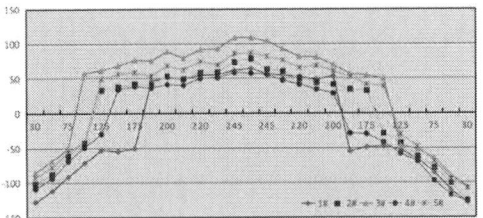

Fig. 10. Shadow moire results of hat packaging structure

TABLE III.   COMPARISON OF SIMULATION AND SHADOW MOIRE TEST RESULTS

package type		simulation result(μm)	test result(μm)					Test mean(μm)	error(%)
Ring package	high temperature	94.68	83.50			90.90		87.20	8.6
	room temperature	109.19	127.30			114.00		120.70	9.5
Flat package	high temperature	120.14	108.00	129.00	90.00	99.00		106.50	12.8
	room temperature	136.49	134.00	122.00	123.00	127.00		126.50	7.9
Hat package	high temperature	83.52	64.00	78.00	108.00	57.00	86.00	78.60	6.2
	room temperature	119.53	129.00	126.00	107.00	128.00	108.00	119.60	0.06

Using ANSYS Workbench software and material equivalent method, the warpage values of ring package, flat package and hat package at high temperature and room temperature are simulated. The simulation results are shown in Fig. 5, Fig. 7 and Fig. 9 respectively, and the shadow moire results are shown in Fig. 6, Fig. 8 and Fig. 10 respectively. Shadow moire results each different packaging form has multiple groups of test sample data, and the shadow moire results are affected by human and the initial warpage results of the substrate, resulting in large differences between different samples of the same packaging. In order to reduce the impact of test data, the comparison of the test data in this time is all carried out by the average value. The comparison result of the simulation result and the average value of the shadow moire is shown in TABLE III. It can be seen from the error percentage in the table that the error between the simulation results of the three different package structures and the average of the shadow moire results can be strictly controlled within 13%, regardless of whether it is high temperature or room temperature. This shows that this method can be effective carry out the preliminary simulation of the project to determine whether the package structure has warpage risk.

IV. CONCLUSION

In this paper, the substrate equivalent method is used to simulate the warpage of the multilayer substrate package structure using ANSYS Workbench. The simulation results are compared

with the average value of the shadow moire results. It is found that the error can be controlled within 13%, and most errors can be controlled within 10%,which shows that this kind of simulation method can effectively carry out the warpage simulation of different package structures, so as to judge the warpage risk in advance, and find the method of optimizing and controlling the warpage.

REFERENCES

[1] Kim. D-H, Joo. S-J, Kwak. D-O, and Kim. H-S, "Warpage Simulation of a Multilayer Printed Circuit Board and Microelectronic Package Using the Anisotropic Viscoelastic Shell Modeling Technique That Considers the Initial Warpage," IEEE T. COMP. PACK. MAN. Vol. 06, No. 11, pp. 1667-1676.

[2] Lien C Y, Chuang Y C, Yao Y, Charn E, and Chen E, "Block-Based Finite Element Modeling, Simulation, and Optimization of the Warpage of Embedded Trace Substrate" 2018 IEEE 20th Elec. Pack. Tech. Conf. (EPTC), pp. 802-806.

[3] Zhang Q, Lo J C C, Lee S W R, and Xu W, "Determination of a Meaningful Warpage Acceptance Criterion for Large PBGA Components Through the Correlation with Scattering in Material Properties" IEEE 68th Elec. Comp. & Tech. Conf. 2018, pp. 718-723.

[4] Y. Bin, W. Xiaofeng and Z. Yabing, "The Study of Thermally Induced Warpage of BGA Package during Reflow Soldering," 2018 19th International Conference on Electronic Packaging Technology (ICEPT), Shanghai, China, 2018, pp. 1411-1414.

[5] Loh W K, Kulterman R, Fu H, and Tsuriya M, "Recent trends of package warpage and measurement metrologies," International Conference on Electronics Packaging. IEEE, 2016, pp. 89-93.

[6] Lin W, "A feasible method to predict thin package actual warpage based on an FEM model integrated with empirical data," IEEE 65th Electronic Components and Technology Conference (ECTC). IEEE, 2015. pp. 1985-1990.

Simulation and experimental analysis of a cost-effective miniaturized transceiver for X-band application

Yu Ban
Information Science Academy of CETC
Beijing, China
email address: yu.ban@outlook.com

Jie Liu
Information Science Academy of CETC
Beijing, China
email address: ljragdoll@126.com

Abstract—In this paper, the design, development and realization of an X-band transceiver in a system-level solution with high performance and a compact size is investigated and verified by the experimental results. Multi-layer low loss substrate based system-in-package (SiP) technology, which integrates digital/analog integrated circuits, monolithic microwave integrated circuits (MMICs) and passive devices, is one of the best candidates for RF system-level integration due to its advantages of low loss, high integration capability and low cost. The electrical model of MMICs, such as: power amplifier (PA), low noise amplifier (LNA), have been respectively created and combined, in order to achieve an overall test-bench for the performance investigation of the X-band transceiver. The output amplitude and phase of the transmitter are simulated and characterized, respectively, in order to validate the design and to prove the accuracy of the MMIC models. In addition, the essential passive components and chip-chip interconnections such as wire-bonding, transmission lines are individually modeled and integrated in the SiP. The interconnection loss between MMICs should be carefully analyzed in the co-integration, and the choice of the packaging method and the in-package transmission line structure is crucial to ensure a good RF performance. A 3-dimensional (3D) system-level package is implemented on a low dk/df substrate which ensures a good radio-frequency (RF) performance with reasonable fabrication cost. Though wire-bonding technique is still feasible for the microwave and millimeter-wave application, the performance of flip-chip technology is more suitable for the high frequency application. Therefore, a combination of wire-bonding and flip-chip has been applied in the final prototype. Finally, both simulation and measurement results of the output amplitude and phase are analyzed and compared, in order to validate the package design flow and electrical modeling.

Keywords—system-in-package; monolithic microwave integrated circuits; electrical modeling

I. INTRODUCTION

Nowadays, people are expecting ever faster wired and wireless data communication or signal processing, which pushes a strict requirement for the performance of high bandwidth (BW) signal transmitters / receivers. On the other hand, consumer electronics are highly competitive and are evolving rapidly to meet the updated requirement of the consumers. A fast research and development cycle, decreasing price and the associated fabrication cost are the most primary factors driving the growth widely and globally. Therefore, a cost-effective hardware solution of a miniaturized and high BW microwave / millimeter-wave transceiver is required [1]. It has been previously approved that system-in-package (SiP) modules have various advantages of shorter chip-chip interconnection length, a broader BW, a more compact chip size as well as lower fabrication cost, which have been widely applied in wired and wireless telecommunication systems, personal mobile devices and consumer electronics, especially in high frequency applications [2].

In this work, we demonstrate a compact X-band transceiver front-end which integrates a transmitter signal chain and a receiver signal chain, together with passive devices, monolithic microwave integrated circuits (MMICs) as well as some relative interconnections and bonding pads, into one package. Meanwhile, an effective methodology and design flow with a system-level integration of SiP technology has been researched and analyzed. SiP technology has a prominent advantage of small package size, which integrates variety of digital / analog and radio frequency (RF) integrated circuits into one package, together with passive components such as capacitors and resistors, which are mounted on the same substrate.

II. BLOCK DIAGRAM OF THE X-BAND TRANSCEIVER

High speed and high BW transceivers have been widely applied both in civil and military application, which consist of transmitter and receiver signal chains. In order to save the chip area and to improve the integration level, a transmitter and a receiver are commonly integrated into one package, using input/output switches or a three-port circulator, which transmit signal only with single direction. However, a high BW circulator is hard to design and realize with the silicon-based CMOS or III-V compound semiconductor technology, thus increasing the difficulty to achieve a highly integrated transceiver module. In this paper, a switchable power amplifier (PA) with GaN technology has been applied in the transceiver, which integrates an X-band PA with high power gain and a selective signal switch between the input and output signal channels.

The block diagram of the complete X-band transceiver is depicted in Fig. 1.

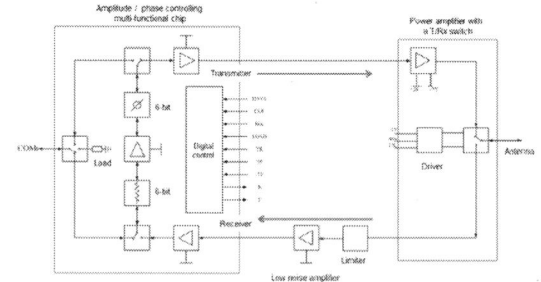

Figure 1. Block diagram of the X-band transceiver with switchable PA

978-1-6654-1392-3/21 $31.00 © 2021 IEEE

As shown in Fig. 1, the upper part of the block diagram depicts the transmitter, which consists of a multi-functional amplitude / phase controlling chip (MFC) and a switchable PA. The input signal (COM port) is firstly amplified, followed by a phase shifting module of the MFC. The amplitude / phase modulated signal is transmitted to the PA MMIC, which is located at last stage of the transmitter, to achieve a high power level sufficient to trigger the antenna. The receiver signal channel is demonstrated in the lower part of Fig. 1. Similar to the transmitter, the receiver signal from the antenna is firstly limited in signal amplitude, before sending to a low noise amplifier (LNA). Finally, the amplitude and phase of the received signal are modulated respectively according to the requirement of following parts. Both of amplitude and phase are digitally controlled with 6-bit digital signals. As both PA and MFC have integrated switches in the chip, the transceiver is applied either as a transmitter or a receiver, depending on the embedded input and output switches controller.

III. CHIP-TO CHIP INTERCONNECTION AND SYSTEM-LEVEL PACKAGE

A. Chip-to-chip interonnection

Integrating a variety of RF MMICs into one package and achieving a high working frequency up to X-band with a broad BW beyond 4 GHz is not easy due to various aspects across the end-to-end signal communication channel. In the transceiver package design, it is highly challenging due to the very tough high speed requirements. In order to minimize the frequency-dependent insertion loss of the chip-to-chip interconnection, the transmission lines and RF vias are carefully designed and optimized [3].

In order to reduce the overall chip size and to realize a cost-effective integrated transceiver in a compact package, both active and passive electronics are integrated within one package, which however creates a challenge as different circuit topologies and their relative performance degradation caused by signal reflection, crosstalk, electromagnetism (EM) and so on. Therefore, the block diagram of the transceiver circuit has specific characteristics and the layout requires a careful consideration and optimization with EM perspective, together with a trade-offs among power dissipation, chip area and performance. Investigation of different types of interconnection between MMICs will also be necessary in the circuit-package co-design to improve the BW and to minimize the signal degradation along the travelling path from input of MFC to the output of the PA. Wire-bonding techniques achieve a high RF and microwave performance and are commonly applied in the X-band application. Meanwhile, flip-chip technology outperforms wire-bonding especially at higher frequencies duo to its shorter interconnection length and lower frequency-dependent insertion loss. Therefore, a combination of wire-bonding and flip-chip has been applied in the final prototype of this design. The choice of SiP package and structure is also crucial to ensure a good overall RF performance. Therefore, during the circuit design and post-layout analysis, a 3-dimensional (3D) system-level package is modeled and optimized with a low dk/df substrate which ensures a good RF performance with reasonable fabrication cost and minimized

device under test (DUT) difference from measurement board to guarantee a more accurate RF measurement [4].

B. System-level package

The quad flat no-leads (QFN) and land grid array (LGA) packages are typically applied as the cost-effective, system-level package technologies, especially in the RF and microwave applications. Both of them have the advantages of high BW of interconnection with a short length and high level of integration with a compact chip size. In the LGA package, multi-layer laminates are used as the low-loss substrate, which help the signal routing and provide low profile and high count of chip I/O pins. Unlikely, the QFN technology has a disadvantage of high density chip integration due to routing difficulties [5]. As it utilizes a large size of copper frame under the chip, limited area can be used as signal routing and only wire-bonding and flip-chip technologies are allowed in the package as the chip-chip and chip-package interconnection. Although, it has been widely utilized and treated as one of the most promising package solutions for the high BW and high power applications, due to its excellent heat release capability, which is able to directly transfer the heat of the power devices from the package to the outside [6].

In this work, the LGA package with an embedded copper block under the PA is used as the SiP technology, in order to achieve a higher signal routing density than QFN, and to improve thermal performance using the copper to transfer the heat from the backside of the high output PA directly to outside of the package. Fig. 2 depicts the cross-section of LGA package with embedded coppers.

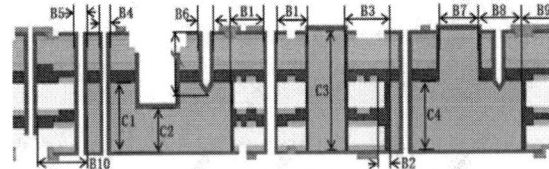

Figure 2. High density LGA technology with an embedded copper block

IV. CIRCUIT DESIGN AND SYSTEM-LEVEL IMPLEMENTATION

A. X-band transmitter

In this work, we demonstrate a compact X-band transceiver front-end which integrates both a transmitter and a receiver into one single package.

The RF transmitter front-end consists of a three-stage PA and an amplitude/ phase controlling module, which is shown in Fig. 3.

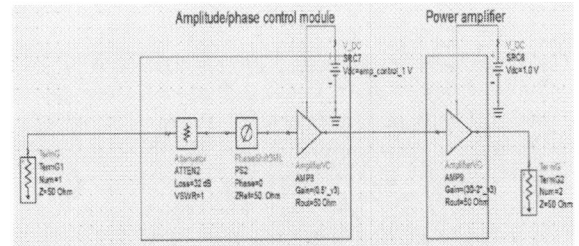

Figure 3. X-band transmitter circuit architecture

978-1-6654-1392-3/21 $31.00 © 2021 IEEE

As shown in Fig. 3, the amplitude / phase controlling module consists of three parts: an attenuator, a phase controller as well as a variable gain amplifier. A solid-state PA (SSPA) MMIC with GaN technology is applied as the last stage of the transmitter, using power combining topologies to achieve a saturated output power over 5 W and a BW beyond 4 GHz. The PA works with a frequency range from 8 to 12 GHz, representing a significant increase in output power in order to drive the antenna.

Apart from active MMICs, some essential passive components are integrated in the transmitter SiP, such as high speed transmission lines, RF vias, bonding wires and pads. The passive devices and chip-chip interconnections are individually modeled and integrated in the miniaturized transceiver, ensuring a good RF performance with reasonable fabrication cost. The parasitics of the passive devices and the influences on the overall system performance have been analyzed and demonstrated [4]. The complete circuit structure and test-bench of the X-band transmitter is depicted in Fig. 4.

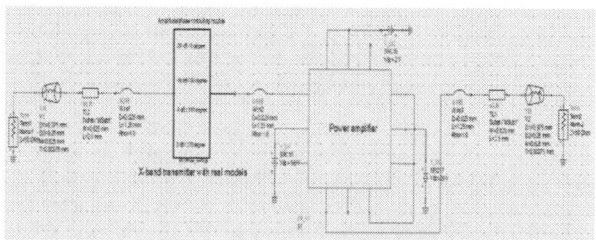

Figure 4. X-band transmitter with real models

B. X-band receiver

Similar to the transmitter, the RF receiver front-end consists of an X-band limiter, a RF LNA and an amplitude/ phase controlling module, which is shown in Fig. 5.

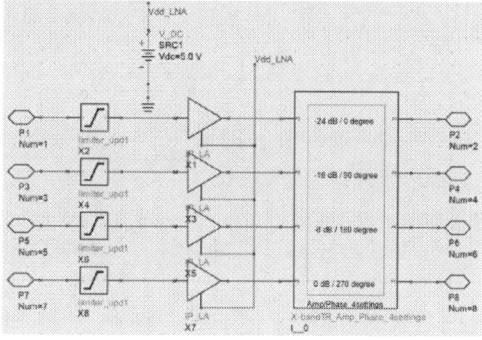

Figure 5. X-band receiver circuit architecture

C. Electrical analysis and system-level simulation

The simulated channel gains of the receiver with different amplitude controllers of 0 dB (red), 8 dB (blue), 16 dB (purple) and 24 dB (light blue) are depicted in the top left of Fig. 6, respectively. As shown in Fig. 6, the BW of the receiver is over 4 GHz and the flatness is good throughout the frequency from 8 GHz to 12 GHz. The phase of the receiver channels are demonstrated in the top right, and the phase shift between neighboring channels is around 90 degree, as expected. The simulated noise figure of the receiver is lower than 2 throughout the frequency range from 8 to 12 GHz, and noise figure of the four channels is 1.652, 1.441, 1.407 and 1.402 at the central frequency of 10 GHz, respectively, as shown in the bottom of Fig. 6.

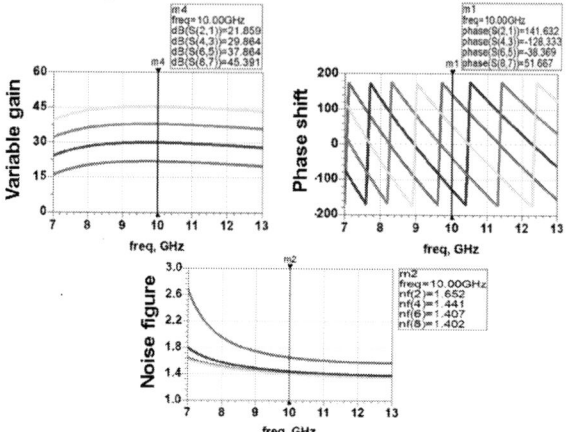

Figure 6. Simulated variable gain, phase shift and noise figure of the receiver

V. SiP LAYOUT AND EXPERIMENTAL VERIFICATION

A. SiP layout

Considering the very high speed and high frequencies application, one should always be careful and verify with measurements results. As shown in the bottom right of Fig. 7, the X-band transceiver with LGA package technology is implemented and fabricated. As the total power consumption of the transceiver including the GaAs / GaN MMICs is around 7.8 W, which generates a high level of heat, a copper frame is placed under the dies in order to achieve better thermal dissipation by transferring the heat of the chips directly to the outside of the package.

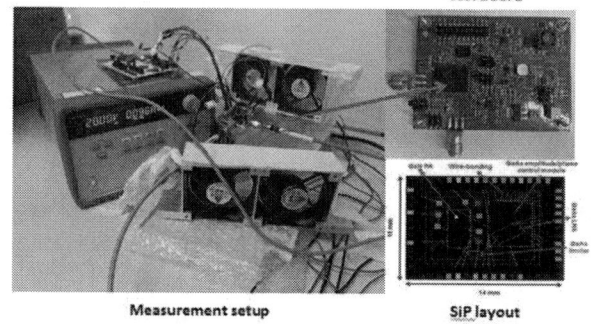

Figure 7. SiP layout and measurement setup

B. Measurement setup

In order to validate the performance, a system-level experimental verification is operated. As shown in Fig. 7, the measurement setup of the X-band transceiver is performed using two SMA connectors and RF cables connecting the

DUT with an Agilent vector network analyzer (VNA), allowing for both amplitude and phase measurement using S-parameters. As the preceding thermal simulation result demonstrates that the maximal in-package temperature may reach as high as 200 °C, exceeding the requirement of 125 °C, it implied that extra power limiting methodology and external cooling systems are required [4]. Therefore, a heat sink is placed under the test board, together with two fans placed around the DUT during the experiments, for a better heat release. Besides, as shown in Fig. 7, a FPGA board is connected with the DUT, providing the digital settings for the amplitude / phase controlling module.

C. Analysis

Fig. 8 illustrates the measurement results of the modulated amplitude / phase with different settings. High integration level of the SiP increases the power density and thus brings a huge challenge of thermal release. This resulted in measurement challenges and restricts the experimental duration and time cycle. A number of valuable conclusions are derived. First of all, we find that the phase difference between the two neighboring phase settings of the controller contributes to a minimal phase resolution of 5 degree, with regard to that of the MFC chip. Meanwhile, on the plot of the amplitude, one can clearly see that the amplitude difference of the neighboring amplitude settings is around 0.48 dB, which is close to the minimal amplitude resolution of 0.5 dB degree of the variable gain amplifier in the MFC chip. Therefore, it concludes that the phase and amplitude resolution of the transceiver are derived from the MFC chip. Besides, as shown in the right of Fig. 8, the BW of the X-band transceiver is limited, mainly due to the frequency-dependent insertion loss of the chip interconnections as well as the parasitic capacitance and inductance of the package.

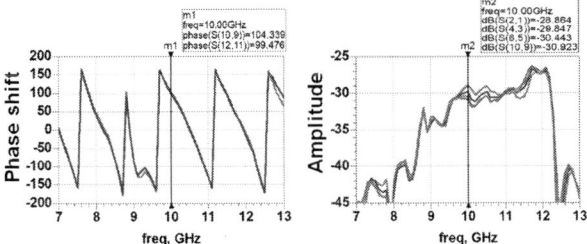

Figure 8. Measurement results of modulated amplitude / phase

VI. CONCLUSION

This paper demonstrates a compact X-band transceiver front-end which integrates a transmitter signal chain and a receiver signal chain into one package. The active MMICs are individually modeled, in order to investigate the transceiver performance. The influence of passive devices, bonding wires and some relative interconnections are also optimized to minimize the frequency-dependent interconnection loss. In order to validate the SiP design, the X-band transceiver front-end is implemented and fabricated in a cost-effective LGA process with a 60-lead package, occupying a core chip size of around 154 mm*mm. Both simulation and measurement results of the amplitude and phase are analyzed and compared, in order to validate the electrical modeling of active MMICs, passive components and interconnections.

ACKNOWLEDGMENT

This work is supported by a project funding of fundamental research and technology from CETC. Besides, the authors would like to thank Mr. Peng Su and Mr. Quan Zeng from SCI/SCC, for the technical support during this work.

REFERENCES

[1] S. Masuda, M. Yamada, Y. Kamada, T. Ohki, K. Makiyama, N. Okamoto, K. Imanishi, T. Kikkawa and H. Shigematsu, "GaN single-chip transceiver frontend MMIC for X-band applications", IEEE/MTT-S International Microwave Symposium Digest, Montreal Canada, pp. 1-3, Jun. 2012.

[2] Y. Ban and J. Liu, "Investigation on two realizations of miniaturized integrated passive devices in a SiP transceiver front-end", 19th International Conference on Electronic Packaging Technology (ICEPT), Shanghai China, pp. 1-4, Aug. 2018.

[3] M. Dittrich and A. Heinig, "Investigation of chip-to-chip interconnection structures for high data rates on a low cost silicon interposer", 19th IEEE Workshop on Signal and Power Integrity (SPI), Berlin Germany, pp. 1-4, May 2015.

[4] Y. Ban and J. Liu, "Investigation on a miniaturized X-band transceiver front-end in a cost-effective SiP solution", 21st International Conference on Electronic Packaging Technology (ICEPT), Guangzhou China, pp. 1-4, Aug. 2020.

[5] T. Zwick and S. Beer, "QFN based Packaging Concepts for Millimeter-Wave Transceivers", IEEE International Workshop on Antenna Technology (iWAT), Tucson US, pp. 335-338, Mar. 2012.

[6] J. Laskar et al., "A SOC/SOP Co-design approach for mmW CMOS in QFN Technology", IEEE Custom Intergrated Circuits Conference (CICC), San Jose USA, pp. 1-8, Sep. 2008.

Extraction, Optimization and Failure Detection Application of Parasitic Inductance for High-Frequency SiC Power Devices

Minghui Yun
School of Mechanical and Electrical Engineering
Guilin University of Electronic Technology
Guangxi, Guilin, China
yunminghui_01@163.com

Kailin Zhang
School of Mechanical and Electrical Engineering
Guilin University of Electronic Technology
Guangxi, Guilin, China
1198583386@qq.com

Miao Cai
School of Mechanical and Electrical Engineering
Guilin University of Electronic Technology
Guangxi, Guilin, China
caimiao105@163.com

Yiren Yang
School of Mechanical and Electrical Engineering
Guilin University of Electronic Technology
Guangxi, Guilin, China
yangyirentrain@126.com

Changqi Feng
School of Mechanical and Electrical Engineering
Guilin University of Electronic Technology
Guangxi, Guilin, China
1643043861@qq.com

Song Wei*
School of Mechanical and Electrical Engineering
Guilin University of Electronic Technology
Guangxi, Guilin, China
swei2020@126.com

Daoguo Yang*
School of Mechanical and Electrical Engineering
Guilin University of Electronic Technology
Guangxi, Guilin, China
daoguo_yang@163.com

Guoqi Zhang
Delft Institute of Microsystems and Nanoelectronics
Delft University of Technology
Delft, CD, The Netherlands

Abstract—Silicon carbide (SiC) is a third-generation semiconductor material with many advantages, such as high thermal conductivity, high critical breakdown voltage, and high saturated electron drift velocity, which can increase the operating frequency of the power conversion system to more than 100kHz. In the high-frequency, the parasitic effect will significantly reduce the switching speed of the power devices, increase power consumption and influence the uniformity of current distribution. In this paper, we established the calculation nodes of each part of the SiC-MOSFET Half-bridge power module and used ANSYS Q3D software to extract the parasitic parameters. Die-Die, Die-DBC-1, Die-DBC-2 and hybrid interconnect package structures were designed to optimize the parasitic inductance. Simulation results indicated that the chip-DBC-1 structure can reduce the parasitic inductance about 30% compared with Chip-Chip structure and effectively control the uniformity of current density on two parallel diode chips ($\Delta L_{diode} < 0.1\%$). In the meantime, an 3D model of partial bond wires broken were designed to get further insight into the variation of parasitic inductances. The correlation mechanism between the partial bond wires broken and inductance change of the D-S terminals current path was studied. The results showed that as the number of broken bond wires increases, the inductance of D-S terminals current path was increased gradually. Finally, by using two-port S-parameters measurement method to extract the inductances of the discrete power device, the experimental results indicated that when one or two bond wires were broken, the inductance of D-S terminals current path increased by 4.69% and 15.69%, respectively. Overview, a risk evaluation method for SiC power devices based on the variation of the parasitic parameters of the bond wire was established.

Keywords—*SiC power devices, Bond wires, Parasitic*

National Natural Science Foundation of China (No. 61865004)

inductance, Packaging

I. INTRODUCTION

Power semiconductor devices are the key electronic components used in power electronic systems, which have been widely used in smart power grid, high-speed railway, frequency conversion system, and new energy power generation. As a potential WBG semiconductor material[1], SiC has several superior performances compared with Si. 1) The intrinsic carrier concentration is smaller, SiC devices can operate at higher temperature (>200℃). 2) The 10× critical electric field makes ultra-high-voltage (> 10KV) power devices. 3) The resistance of a SiC devices can be smaller than that of its Si counterpart, leading to lower parasitic capacitance and higher switching speed, allowed higher operation frequency (>100kHz) with less switching loss. With the increase of working frequency, the switch durations of SiC MOSFET is usually only in a few tens of a nanoseconds, which is easily affected by parasitic parameters. In particular, during the on-off process of power devices, parasitic parameters would inevitably cause the voltage and current oscillations, aggravate EMI, produce unnecessary voltage spikes, and lead to additional switching losses, these unnecessary harm constantly affect the switching characteristics of the power devices.

Using the planar interconnect packaging technology can reduce the parasitic inductance of power modules effectively[2-3], but when large areas of different metal layers are connected, a layer of intermetallic compounds (IMC) will be formed at the interconnection interface, and the interconnection process is easy to produce holes and microcracks. The reliability problem will be particularly prominent when power devices are subjected to electro-thermal-vibration shock for a long time, so reliability is more important than other performance. At the present stage, commercial SiC devices still use the traditional bonding wire package technology and need to use multiple aluminum bonding wires in parallel to carry large voltage and current in actual application. In high-power modules with aluminum bonding packaging technology, module failures caused by bonding wire faults account for about 70% of the total failures[4], so the reliability of the bond wire is particularly important.

In previous studies, three-dimensional (3D) numerical modeling[5-6] and different measurements techniques[7-8] have been employed in engineering practice to extract the parasitic of power modules and hence, characterize their high frequency performance. However, the research is mainly based on the power module, topological model and equivalent circuit model to characterize the internal distribution of parasitic inductance. Without a clear analysis of its physical structure, the research results cannot be directly corresponding to the real object and cannot guide the optimization of parasitic inductance more intuitively. A large number of research papers have been published, proposing different detection methods for bond wire lift-off[4, 9-12], techniques proposed is mainly through measurement voltages and currents at available terminals: the collector, the emitter, the gate, and the auxiliary emitter, which provides the return path for the power device. But these detection methods have a common shortcoming: the transmission characteristics of MOSFET/IGBT chips are very sensitive to temperature changes, so it is difficult to accurately detect the changes of voltage and current at the exposed terminals in actual working condition. Therefore, a bonding wire failure detection method independent of temperature changes is needed.

In this paper, the inductance calculation nodes were established based on 3D physical model of a SiC half bridge module. By using Q3D software, we characterized the parasitic inductance in detail and completed the optimization design of the packaging structure. Based on the optimization model, the simulation analysis revealed a phenomenon that the break of the bond wires can significantly affect the inductance of the G-S terminals current path, which is verified by experiment. Then, a risk evaluation method for bond wire is proposed, which does not depend on the working condition and can avoid the influence of temperature on the test.

II. MODELING AND PARASITIC EXTRACTION

Fig. 1 shows a schematic diagram of a typical half-bridge SiC-MOSFET power device structure. From top to bottom, the structure of each layer is as follows: aluminum bonding wire, DBC substrate, upper solder layer, copper substrate, bottom solder layer, thermal grease and heatsink. A typical half-bridge structure is composed of two Phase-Leg modules (High Phase-Leg and Low Phase-Leg), each of which is consisted in parallel by one SiC-MOSFET chip and two SiC-Diode chips. By using the external drive system to control power device periodic on-off to complete the inverting process. Fig. 2 shows the operating principle of the High Phase-Leg switching process of the half-bridge inverter

circuit, where T_1, T_2 are the SiC-MOSFET chips and D_1, D_2 are the diode chips. The external driver subsystem generates different square wave signals (U_{g1} and U_{g2}) to the MOSFET (T_1 and T_2) gate, and controls the alternating work of T_1 and T_2 chips. So in half of a switching cycle, the T_1 chip of High Phase-Leg bridge and the D_2 chip of Low Phase-Leg bridge constitute a commutating path. However, the increase of the parasitic inductances can obviously delay the commutation time, which will affect the switching frequency of the device. Therefore, the parasitic inductance of commutator path is selected as the main evaluation index of power module optimization.

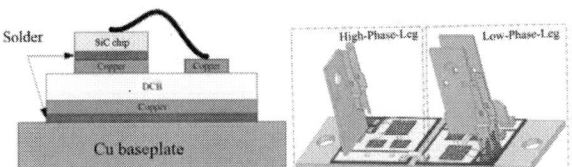

Fig1. SiC Power device structure and half bridge model

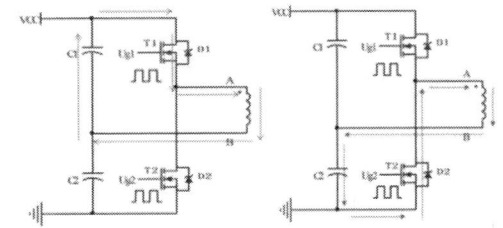

Fig2. Operating principle of the High Phase-Leg switching process

Table 1 shows the main parameters of the simulation model. The commonly used switching frequency of SiC MOSFET is between 40kHz and 300kHz, so 200kHz frequency is determined as the initial condition for parasitic parameter extraction. As shown in Fig. 3, the contact positions of terminal, source, drain and gate are set as calculation nodes respectively, and the input and the output sink of the excitation source are set on each node as required. Base on the above settings, the inductances of the SiC power module can be divided into several parts. 1) L_{tot}: Inductance of the commutation path. 2) $L_{D-terminal}$: Inductance of the drain-terminal. 3) L_D: Inductance of the drain path of MOSFET. 4) L_S: Inductance of the source path of MOSFET. 5) L_{DBC}: Inductance of the DBC interconnection path. 6) L_A: Inductance of the anode path of diode. 7) $L_{S-terminal}$: Inductance of the source-terminal. 8) L_G: Inductance of the gate path of MOSFET. Among them, the commutation path of the converter circuit includes current path of chips, DC bus terminal and the Cu interconnection path of the upper surface of DBC.

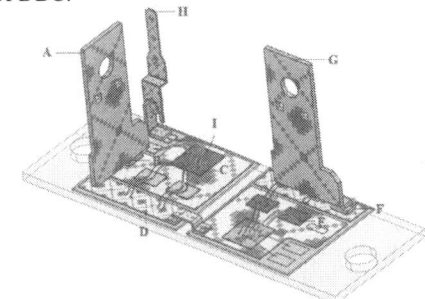

Fig3. Calculation nodes of the parasitic inductance of the power module

In order to evaluate the influencing factors of the parasitic inductance of each part, parasitic inductances were simulated by Q3D software according to the established nodes. Table 2 summarizes the results, it is pretty obvious that $L_{D-terminal}$, $L_{S-terminal}$, and L_{DBC} accounted for the major parts of the inductances of the commutation path, respectively. These parts are the preferred direction for inductance optimization of power module, However, due to the requirements of the application environment, terminal optimization is limited, so the packaging structure based on bond wire is difficult to provide more direction for improvement. Meanwhile, the interconnection between chips and bond wires on DBC is less restricted, the flexibility of optimization is greater.

TABLE I.　　MAIN PARAMETERS OF THE SIMULATION MODEL

Name	Material	Conductivity (s/m)	Dimension (mm)
Bonding wire	aluminium	3.8E7	Diameter: 0.25 Space: 1
MOSFET	SiC	On: 5.8E7 s/m Off: 0	8.8×8.8×0.14
SBD	SiC	On: 5.8E7 s/m Off: 0	6.3×4.5×0.12
DBC	Cu Al2O3 Cu	/	Cu: 28×21×0.2 28×6×0.2 Al2O3: 30×30×0.38 Cu: 30×30×0.2
Base plate	Cu	5.8E7 s/m	91×31×2

TABLE II.　　INDUCTANCE OF EACH NODE

Name		Node		Inductance（200KHz）
L_{tot}	$L_{D-terminal}$	A- B- C- D- E- F- G	A-B	11.982nH
	L_D		B-C	1.532nH
	L_S		C-D	5.261nH
	L_{DBC}		D-E	6.746nH
	L_A		E-F	3.919nH
	$L_{S-terminal}$		F-G	11.996nH
L_G		H-I		28.542nH

(The Inductance of A-B through F-G is summarized as 44.385nH for L_{tot}.)

III. OPTIMIZATION OF POWER MODEL

In the inverter system, the MOSFET/IGBT chips require multiple diode chips in reverse parallel to form a current loop, so the MOSFET chips and the diode chips should package on the same DBC substrate to form a half-bridge or full-bridge power module. As shown in Fig. 4, there are two types of interconnection: 1) Chip-Chip: SiC-MOSFET chip and diode chips are directly bonded by wires; 2) Chip-DBC: SiC-MOSFET Chip and diode Chips are respectively bonded with DBC by wires. Their differences are listed as follows:

Chip-Chip (Fig. 4 (a)): The source of MOSFET chip is directly connected with the negative electrode of diode by bonding wires, and then connected together to the DBC substrate.

Chip-DBC-1, Chip-DBC-2 (Fig. 4 (b),(c)): The source of MOSFET chip and the negative electrode of diode are bonded in the DBC substrate by bonding wires, and then connected by using the upper copper layer of DBC substrate.

Hybrid-interconnect(Fig. 4 (d)): The source of MOSFET chip is individually bonded to the DBC substrate, and the diode chips are interconnected through the bonding wires and then together bonded to the DBC to form a parallel structure with the MOSFET chip.

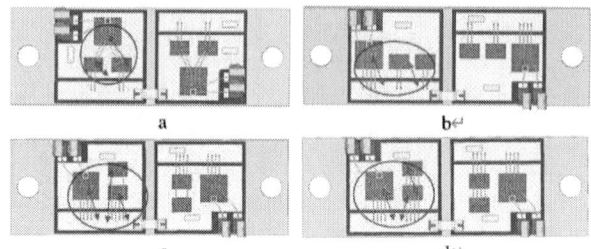

Fig4. Structure diagram of four packaging schemes (a: Chip-Chip, b: Chip-DBC-1, c: Chip-DBC-2, d: Hybrid-interconnect)

Fig. 5 shows the simulation comparison results of inductances of the four structures. It is observed that the inductance of the commutation path of the Chip-Chip circuit is 44.385nH, which is nearly 30% higher than the parasitic inductances of the other three models. At the same time, when MOSFET chip is off, the freewheel diode need to be open, so there is a problem of current uniformity exists between two parallel diodes. The larger inductances cause the higher impedances, so the current will flow excessively to one of the diodes which with the lower impedance of bonding wires, which will cause current density distribution and accelerate the diode chip aging. Therefore, it is required that the parasitic inductance of the two diodes (L_A) should be as close as possible.

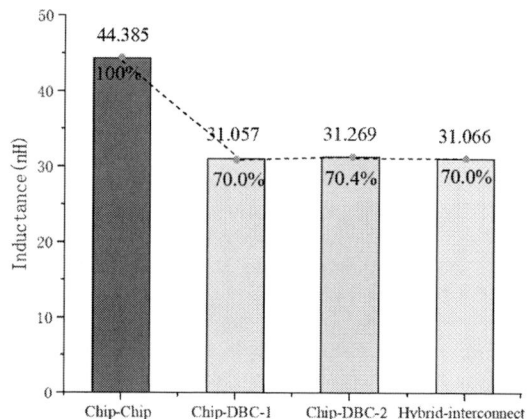

Fig5. Simulation comparison results of inductances of the four structures (AC 200kHz)

Table 3 shows the parasitic inductance of the E-F nodes of four structures, the numerical simulation results that the Chip-DBC-1 package structure can effectively control the uniformity of current density on two parallel diode chips ($\Delta L_{diode} < 0.18\%$). Fig. 6 shows the current density distribution at the contact point between the diode and the bond wires of the four structures, m1, m2, m3, m4, m5, m6, m7 and m8 are the central positions of the connecting surfaces. Compared with the definition points at the same conditions, the difference of current density at each junction point is less than 0.1% when using Chip-Chip and Chip-DBC-1 packaging structure. Integrated analysis imply that the Chip-DBC-1 structure is selected as the package design scheme of SiC-MOSFET half-bridge module under the condition that the size of substrate is not limited.

TABLE III.   PARASITIC INDUCTANCES OF THE E-F NODES OF FOUR STRUCTURES

Name	Chip-Chip	Hybrid-interconnect	Chip-DBC-1	Chip-DBC-2
Diode-1	3.345nH	2.262 nH	3.293 nH	3.642 nH
Diode-2	3.350nH	4.645 nH	3.299 nH	7.363 nH
ΔL_{diode}	0.1%	105.4%	0.1%	102.2%

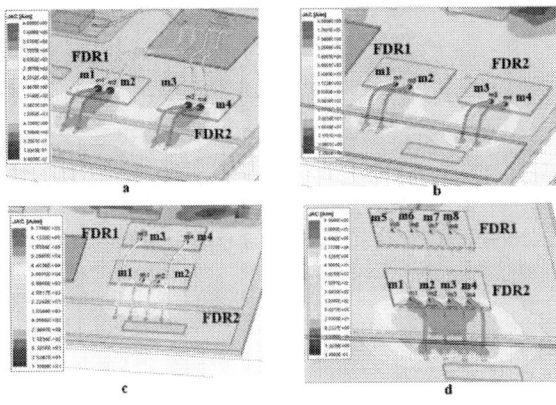

Name	m1	m2	m3	m4	m4	m6	m7	m8
Chip-Chip (A/m)	4.20E+05	3.94E+05	3.55E+05	3.79E+05				
Chip-DBC-1 (A/m)	3.70E+05	3.88E+05	3.91E+05	3.19E+05				
Chip-DBC-2 (A/m)	6.16E+05	6.36E+05	2.47E+05	3.02E+05				
Hybrid-interconnect (A/m)	3.25E+05	3.41E+05	3.21E+05	3.53E+05	3.18E+03	3.55E+03	2.91E+03	2.56E+03

Fig6. Current density distribution between the diode and the bond wires (a: Chip-Chip, b: Chip-DBC-1, c: Chip-DBC-2, d: Hybrid-interconnect)

IV.   FAILURE DETECTION APPLICATION

A. Bond wire failure simulation

Based on the Chip-DBC-1 packaging structure, we designed an analysis 3D model of partial bond wires broken to get further insight into the variation of parasitic inductances. In order to better understand the variation rule of parasitic inductance after bonding wire failure, eight parallel bonding wires are set on the SiC-MOSFET chip. Fig. 7 shows the current density distribution and the contact point between the surface of MOSFET chip and the bond wires under AC working condition. It can be seen that the current density distribution at the junction of each bonding wire is uneven, and its distribution is: M8 > M1 > M7 > M6 > M2 > M3 > M5 >M4, while the power device needs to bear the impact of temperature and current for a long time, and the failure probability is higher when the current density is concentrated. According the current density distribution, we set the breakage of the bond wires in turn and extract the parasitic inductance between the D-S terminals, respectively. Fig. 8 shows the change results of parasitic inductance after the fracture of the bond wires. As the number of the broken bond wires increases, the inductance between the D-S terminals of the power module presents an obvious trend of increasing.

When the broken wires are less than 3, the inductance increased by only 0.783nH and a growth rate of less than 6%, it is not significantly obvious that the inductance change between D-S terminals. But as the number of the broken bond wires increases, the inductance increment rate increases to more than 10%, this change can be accurately captured by using the specialized equipment. Base on this rule, we can propose a failure detection method for SiC power devices.

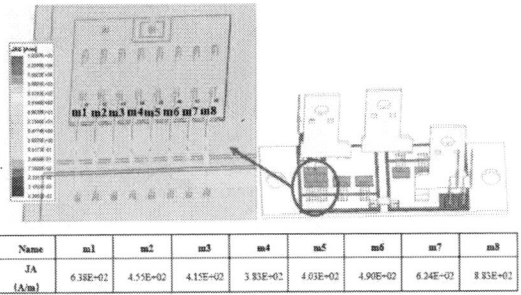

Name	m1	m2	m3	m4	m5	m6	m7	m8
JA (A/m)	6.38E+02	4.55E+02	4.15E+02	3.83E+02	4.03E+02	4.90E+02	6.24E+02	8.83E+02

Fig7. Current density distribution between the MOSFET and the bond wires

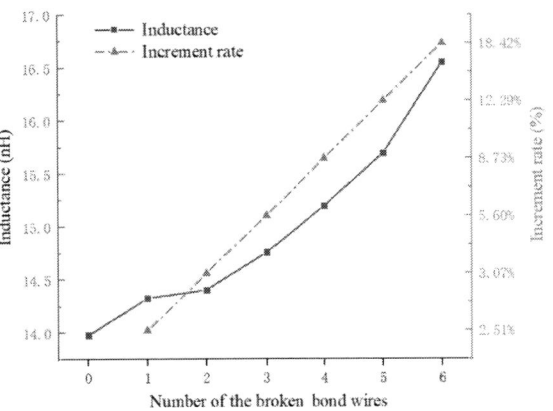

Fig8. Inductances vary with the number of broken bond wires

B. Experimental verification

To verify the feasibility of the new reliability evaluation method, a characterization technique to extract the parasitic inductances was developed based on two-port S-parameters measurement with VNA. A TO-247-3 package power device was used to complete the experiment. According to the requirement of parasitic inductance extraction in datasheet, we designed the test fixture, the specific experimental method and calculation theory are strictly referred to [8], the MOSFET small-signal equivalent circuit under zero biasing condition was shown as a two-port network in Fig. 9 with S–G as Port 1 and D–G as Port 2. First, we used the laser equipment to remove the plastic sealing layer of the sample to expose the bond wires, and then break the bond wires one by one. Meanwhile, the impedance information was calculated by using the S-parameters, which obtained from VNA under the 10MHz-300MHz test frequency. According to [8], the damage of bond wires only affects the change of Z_{11} impedance parameters. Fig. 10 shows the comparison results of sample Z_{11} parameters under different damage conditions. With the number of the broken bond wires

978-1-6654-1392-3/21 $31.00 © 2021 IEEE

increasing, the imaginary part of Z_{11} presents a same increasing trend, increasing 1.1dB and 4.22dB, respectively. Using (1) - (5) to calculate the parasitic inductance of the good sample, L_S=11.3nH, the comparison error between the numerical calculation results and the datesheet data (L_S=13nH) is less than 1.7nH. The value of the datesheet is an approximation, which is only used to limit the L_S under 13nH, so our extraction method is considered to be effective.

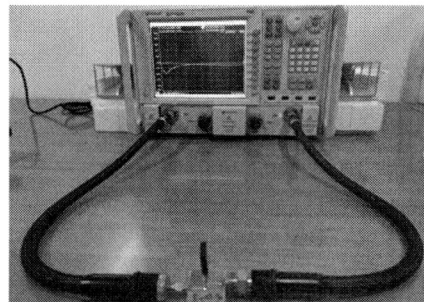

Fig9. Two-port S-parameters measurement by VNA

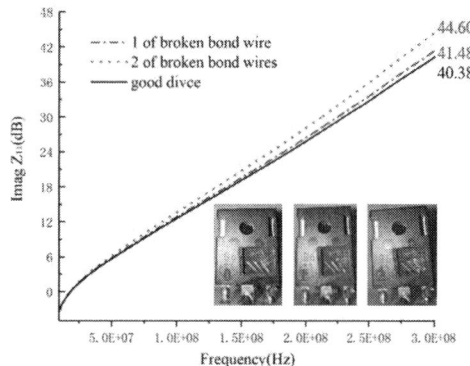

Fig10. Z_{11} parameters vary with the number of broken bond wires

$$Z_{11}=X_{LS}+X_{LG} \quad (1)$$

$$Z_{12}=X_{LG} \quad (2)$$

$$Z_{21}=X_{LG} \quad (3)$$

$$Z_{22}=X_{LD}+X_{LG} \quad (4)$$

$$L=Im(Z)/jw \quad (5)$$

Calculation results indicated that when one of bond wires is broken, the L_{D-S} becomes 14.973nH, it is a 4.69% increase compared to the good device. When two of bond wires are broken, the L_{D-S} becomes 16.546nH, the inductance increments to 15.69%. Therefore, it can be clearly observed that the L_{D-S} gradually increases with the gradual broken of the bond wires, which can be used to predict the reliability of the power device based on the inductance changed of the bond wires.

V. CONLUSION

In this paper, a detailed calculation nodes of each part of SiC half-bridge modules had been established, and the Q3D software were used to characterize the parasitic parameters of each nodes. Simulation results showed that the parasitic inductance can be optimized by changing the package structure, so Die-DIE, Die-DBC-1, Die-DBC-2 and hybrid

interconnect structures were designed. Comparison results indicated that the chip-DBC-1 design can reduce the parasitic inductance as well as effectively control the uniformity of current density on two parallel diode chips. Based on this model, the correlation mechanism between the partial bond wires broken and the variation of parasitic inductance of the D-S current loop had been developed. Then, a risk evaluation method for bond wire was proposed, which does not depend on the working condition and can avoid the influence of temperature on the test. Finally, a two-port S-parameters measurement technique was used to characterize the parasitic parameter of the TO-247-3 package power discrete device and prove the feasibility of the method.

ACKNOWLEDGMENT

This research has been supported by the National Natural Science Foundation of China (No. 61865004), the Key R & D Plan Project of Guangxi Province (grant No. GuiKe AB20159038), the Innovation Project of GUET Graduate Education, the Innovation Driven Development Project of Guangxi Province (No. AA182420), and the Guilin Science Research and Technology Development Program (No. 2020010302).

REFERENCES

[1] S. Q. Ji, S. Zheng, F. Wang, L. M. Tolbert, "Temperature-dependent characterization, modeling, and switching Speed-limitation analysis of third-generation 10-kV SiC MOSFET," IEEE T. Power Electr, vol. 33, pp. 4317-4327, July 2017.

[2] H. Lee, V. Smet, R. Tummala, "A review of SiC power module packaging technologies: Challenges, advances, and emerging issues," IEEE J. Em Sel Top P, vol. 8, pp. 239-255, November 2019.

[3] F. Z. Hou, W. B. Wang, L. Q. Cao, j. Li, M. Y. Su, T. Y. Lin, et al. "Review of packaging schemes for power modules," IEEE J. Em Sel Top P, vol. 8, pp. 223-238, March 2020.

[4] U. M. Choi, F. Blaabjerg, K. B. Lee, "Study and handling methods of power IGBT module failures in power electronic converter systems," IEEE T. Power Electr, vol. 30, pp. 2517-2533, December 2018.

[5] K. B. IF, R. Stark, M. Guacci, J. W. Kolar, U. Grossner, "Parasitic extraction procedures for SiC power podules," CIPS 2018, Stuttgart. Germany, pp. 343-348, March 2018.

[6] F. Yang, Z. Liang, Z. Wang, F. Wang, "Parasitic inductance extraction and verification for 3D planar bond all module," 3D-PEIM, Raleigh, NC, USA, pp. 1–11, June 2016.

[7] L. Yang, W. G. H. Odendaal, "Measurement-based method to characterize parasitic parameters of the integrated power electronics modules," IEEE T. Power Electr, vol. 22, pp. 54-62, January 2007.

[8] T. J. Liu, T. T. Y. Wong, Z. J. Shen, "A New Characterization Technique for Extracting Parasitic Inductances of SiC Power MOSFETs in Discrete and Module Packages Based on Two-Port S-Parameters Measurement," IEEE T. Power Electr, vol. 33, pp. 9819-9833, January 2018.

[9] K. X. Wei, M. X. Du, L. L. Xie, J. Li, "Study of bonding wire failure effects on external measurable signals of IGBT module," IEEE T. Device. Mat. Re, vol. 14, pp. 83-89, May 2014.

[10] C. l. Chen, V. Pickert, M. Al-Greer, C. J. j, C. N, "Localization and detection of bond wire faults in multi-chip IGBT power modules," IEEE T. Power Electr, vol. 35, pp. 7840-7815, January 2020.

[11] P. G. Sun, C. Gong, D. Xiong, Y. Z. Peng, B. Wang, L. W. Zhou, "Condition monitoring IGBT module bond wires fatigue using short-circuit current identification," IEEE T. Power Electr, vol. 32, pp. 3777-3786, June 2017.

[12] I. Ndip, A. Őz, H. Reichl, K. D. Lang and H. Henke, "Analytical models for calculating the inductances of bond wires in dependence on their shapes, bonding parameters, and materials". IEEE Trans. Electromagn. Compat, vol. 57, pp. 241-249, April 2015.

Simulation Analysis of Residual Stress of Sintered Nano-Silver under Multilayer Stacked Module

Zhentang Liang
School of Electronics & Information
South China University of Technology
Science and Technology on Reliability Physics and Application
of Electronic Component Laboratory
China Electronic Product Reliability and Environmental Testing
Guangzhou, China

Hongyue Wang
Science and Technology on Reliability Physics and Application
of Electronic Component Laboratory
China Electronic Product Reliability and Environmental Testing
Research Institute
Guangzhou, China

Bin Zhou
Science and Technology on Reliability Physics and Application
of Electronic Component Laboratory
China Electronic Product Reliability and Environmental Testing
Research Institute
Guangzhou, China
zhoubin722@163.com

Guoyuan Li
School of Electronics & Information
South China University of Technology
Guangzhou, China
phgyli@scut.edu.cn

Abstract—**In this paper, the residual stress of the multi-layer stacking structure is analyzed, the material selection of different layers is discussed, and the optimization scheme is given. The finite element analysis (FEA) software ANSYS is used for calculation and analysis, the ANAND constitutive model is adopted to describe the visco-plastic behavior of Au80Sn20 and sintered nano-silver, the bilinear isotropic hardening is adopted to describe the plastic deformation of materials, and the thermal-mechanical coupling is used to realize the multi-scale coupling between the soldering layer and other function layers. It was found that the maximum deformation is concentrated at the edge of each layer and the residual stress is much serious with the Au80Sn20 compared with the sintered nano-silver.**

Keywords—*residual stress, package, solder, nano-silver*

I. INTRODUCTION

The wide-bandgap power device is developing in the direction of lightweight, micro systematization and building blocks.[1] Power module consist of different components, which will provide the electrical, thermal and mechanical protections.[2] Thus, power modules are mostly form of lamination by means of adhesive bonding, reflow soldering, low temperature sintered and so on. Due to the coefficient of thermal expansion (CTE) mismatch, various components stacked under different process will lead to mechanical deformation, which can introduce the unpredictable problems of reliability.

Cross-verification is mainly carried out through theoretical calculation and finite element simulation for the reliability problems of CTE mismatch under the device assembly. Chen et al, performed the aging experiments and analyze the interconnect of package base on the LED with sintered nano-silver.[3] Yang et al., focus on the soldering parameters of the reflow soldering process in the insulated gate bipolar transistor (IGBT) module packaging technology, the finite element model is carried out the thermal-structural coupling analysis.[4] Tan et al., conducted an experimental and simulation cross analysis on the reliability of bonding of IGBT under thermal cycle, and optimized the parameters of bonding process.[5] Xu et al., verified the validity of the finite element model with theoretical analysis, and predicted the fatigue life of IGBT based on the simulation.[6] The present research only consider the influence of the bonding process in stack layer,

while the optimization of the bonding material and bonding thickness still need a further discussion.

Thermal-mechanical stress and thermal-induced deformation in multilayer stacking model are closely related to the consideration of materials and solder. In this paper, based on GaN power module, the influence of different materials on the multi-layer stacked model is discussed, and the optimal scheme is given.

II. FINITE ELEMENT ANALYSIS

A. Finite Element Analysis Model

A typical GaN power module package includes the top cover, silicon grease, connecting pins, GaN power IC, high thermal conductivity solder, substrate and case. Therefore, the simulation model in this paper focus on main structures, as shown in Fig. 1.

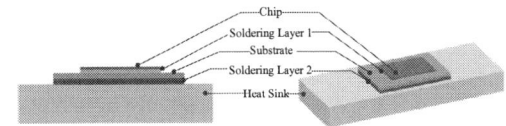

Fig. 1. Mainly structure

B. Material Property

For transition thermal analysis, the element type is Solid70 and Visco107. The property of the materials are shown in TABLE I.

TABLE I. MATERIAL PROPERTY[7]

Material	Elastic Module (GPa)	Poisson's Ratio	Coefficient of Thermal Expansion (10⁻⁶/℃⁻¹)	Yield Strength (Pa)
GaN	295	0.183	5.6	15×10⁹
Au80Sn20	70.9@25℃ 62.5@50℃ 45.5@100℃ 28.5@150℃ 11.5@200℃	0.3	16	275×10⁶@23℃ 217×10⁶@100℃ 165×10⁶@150℃
Sintered nano-silver paste	9.01 @-40℃ 7.96 @0℃ 6.28 @25℃ 4.25 @60℃ 2.64 @120℃ 1.58 @150℃	0.37	19	/
CPC	310	0.3	7.9	650×10⁶

978-1-6654-1392-3/21 $31.00 © 2021 IEEE

ANAND constitutive model is adopted to describe the visco-plastic behavior of $Au_{80}Sn_{20}$ and sintered nano-silver. The relate equations are given below[8]:

$$\sigma = cs \quad (1)$$

Where, σ is equivalent stress; s is deformation impedance; c is material parameter. At a constant strain rate, the equation is:

$$c = \frac{1}{\xi}\sinh^{-1}\left[\left(\frac{\varepsilon_p}{A}\exp\left(\frac{Q}{RT}\right)\right)^m\right] \quad (2)$$

Where, A is constant; Q is gas activation energy; R is gas constant; T is absolute temperature; ε_p is plastic strain rate; m is strain sensitivity index; ξ is stress multiplier. Equation (3) can be derive from equation(2):

$$\varepsilon_p = A\exp\left(-\frac{Q}{RT}\right)\left[\sinh\left(\xi\frac{\sigma}{s}\right)\right]^{1/m} \quad (3)$$

The static plastic flow of ANAND constitutive model can be describe as:

$$\varepsilon_p = A\exp\left(-\frac{Q}{RT}\right)\left[\sinh\left(\xi\frac{\sigma^*}{s^*}\right)\right]^{1/m} \quad (4)$$

$$s = \left(h_0|B|^\alpha\frac{B}{|B|}\right)\varepsilon_p \quad (5)$$

$$B = 1 - \frac{s}{s^*}$$

$$s^* = \hat{s}\left[\frac{\varepsilon_p}{A}\exp\left(\frac{Q}{RT}\right)\right]^n \quad (6)$$

Where, s^* is the saturation value of the internal variable at a given temperature strain rate; σ^* is the saturation equivalent force value; $\hat{s}$ is the coefficient; h_0 the hardening/softening coefficient; n is the exponential; α is the cyclic strain hardening index. Parameters of the two ANAND constitutive models involved in this paper are shown in TABLE II.

TABLE II.   THE PARAMETER OF ANAND MODEL[8][9]

Parameter	$Au_{80}Sn_{20}$	Sintered nano-Silver Paste
s_0/MPa	69.99	2.768
$(Q/R)/K$	7578.66	5706
$A/(s^{-1})$	93.07	9.81
ξ	11	11
m	0.573	0.6572
h_0/MPa	306280	15800
s/MPa	446.13	67.389
n	0.046	0.00326
α	1.402	1

The bilinear isotropic hardening is adopted to describe the plastic deformation of materials, and the thermal- mechanical coupling is used to realize the multi-scale coupling between the soldering layer and other function layers.

Due to the large difference between the solder layer size and the main structure size, convergence analysis of the solder layer mesh are carried out to balance the accuracy and the calculation time.

C. The Simulation Scheme

In this paper, the simulation scheme includes pre-processing, solution and post-processing. The detail is shown in Fig. 2.

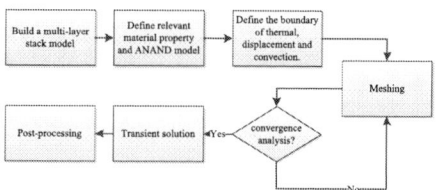

Fig. 2. The procedure of simulation

D. Boundary Conditions and Loads

Since the environment of the soldering process is complicated, the boundary of constrain and thermal convection are defined, the load of heat flux is defined. The displacement constrain is fix at the bottom of the heat sink, the ambient temperature is set to 30°C, and the thermal convection is applied to all external surfaces by setting the coefficient of thermal convection in 10 W·m^{-2}·K^{-1}.

In order to meet the soldering process, the transient analysis will conduct the following step: Firstly, the 310°C thermal load applied to substrate and soldering layer 1 ,the 150°C thermal load applied to chip. Secondly, the thermal load of the first step should be deleted to meet the cooling requirement from 310°C down to 150°C. Thirdly, the 310°C thermal load applied to heat sink and solder layer 2, the 150°C thermal load applied to substrate and soldering layer 1. Finally, the thermal load of the third step should be deleted to meet the cooling requirement from 310°C down to 30°C.

III. OPTIMIZAITON OF SOLDERING SCHEME

In this work, the sintered nano-silver paste is introduce to the soldering optimization. Sintered nano-silver paste is a kind of lead-free and low-temperature sintered of micro-nano metal particle thermal interface material (TIM). Its sintering temperature is about 180-380°C. The melting temperature of the sintered joint is similar to that of bulk silver (961 °C). The optimization scheme makes full use of the following characteristics: low-temperature sintering, high-temperature serving. Therefore, the sintered nano-silver with high melting point must be soldering in the first step during the assembly process ,which can avoid the re-melting problem. At the same time, the residual stress of the solder layer with different area, thickness and materials in the multi-layer stack module are considered, the optimization of soldering material, thickness and chip area is shown in TABLE III. and TABLE IV.

TABLE III.   THE OPTIMIZATION OF SOLDERING MATERIAL

Condition	Soldering Layer 1	Substrate1	Soldering Layer 2	Substrate 2
I	$Au_{80}Sn_{20}$	CPC	$Au_{80}Sn_{20}$	CPC
II	Sintered nano-silver	CPC	Sintered nano-silver	CPC
III	Sintered nano-silver	CPC	$Au_{80}Sn_{20}$	CPC

TABLE IV.   THE OPTIMIZATION OF SOLDERING THICKNESS

Condition		Soldering Layer 1	Thickness	Soldering Layer 2	Thickness
Large chip Area (6.1×1.73mm)	1	Sintered nano-silver	20/60/100	Sintered nano-silver	20/60/100
	2	Sintered nano-silver	20/60/100	$Au_{80}Sn_{20}$	20/60/100
Small chip Area (2×0.5mm)	3	Sintered nano-silver	20/60/100	Sintered nano-silver	20/60/100
	4	Sintered nano-silver	20/60/100	$Au_{80}Sn_{20}$	20/60/100

IV. RESULT AND DISCUSSION

A. Simulation Results

According to the description of this project above, there are corresponding numeric model which is simulated and analyzed by ANSYS. The conditions of optimization of soldering materials and thickness are analyzed. The residual stress distribution and mechanical strain are obtained when the temperature of the model has reached room temperature as shown in Fig. 3 and Fig. 4.

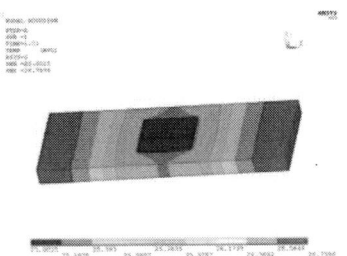

Fig. 3. The temperature distribution of the cooling model

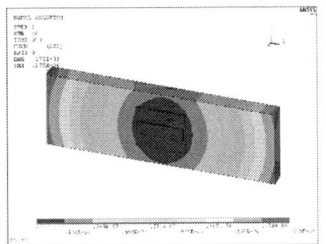

Fig. 4. The deformation distribution of the cooling model

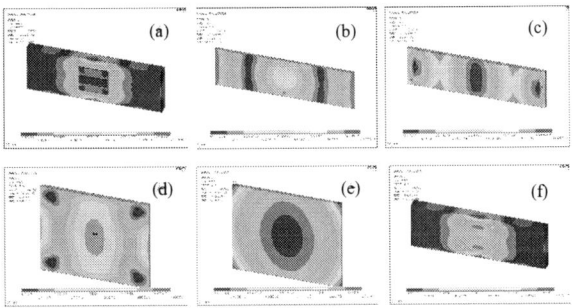

Fig. 5. The residual stress distribution of the cooling model:(a) Total model (b) GaN Chip (c) Soldering Layer 1 (d) Substrate (e) Soldering Layer 2 (f) Heat Sink

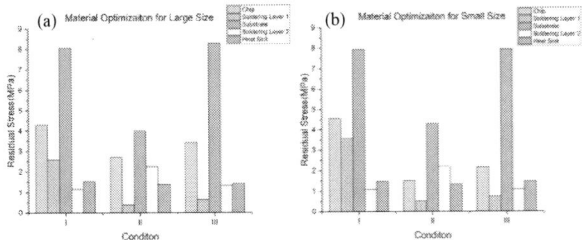

Fig. 6. The result of optimization of soldering material: (a) material optimization of large chip (b) material optimization of small chip

B. Discussion

As shown in Fig. 4, due to the slight deformation, the total model of the residual stress is in equilibrium internally. It means that the thermal mechanical stress do not exceed the yield stress in soldering procedure and the model is still in the elastic condition(Fig. 7 dark zone), it can be describe by equation (7).

$$E = \frac{\sigma}{\varepsilon} (7)$$

Where σ is stress, ε is strain. If the stress exceed σ_0, the material will enter the plastic zone which causes deformation.

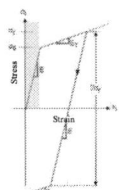

Fig. 7. Bilinear isotropic hardening diagram of the model

As shown in Fig. 5.(a), the maximum residual stress is located in the edge of the heat sink, the minimum residual stress is located in the GaN chip. The detail of each layer will be discussed later. The level of residual stress is shown in Fig. 6 (a-b), no matter whether a large area condition or a small one, the overall residual stress of condition II is the smallest, and the overall residual stress of condition I is the largest. It can be inferred from TABLE I. that the elastic modulus of $Au_{80}Sn_{20}$ is larger than that of sintered nano-paste, the residual stress level of the model connected by $Au_{80}Sn_{20}$ is higher than that of sintered nano-paste.

The optimization of soldering thickness in different area is also discussed. For further discussion, the meaning of the X-axial in Fig. 8 and Fig. 9 is shown below. For example, ag20ausn100 means that the soldering layer 1 is sintered nano-silver paste and thickness is 20um, the soldering layer 2 is $Au_{80}Sn_{20}$ and thickness is 100um.

- For the condition of both layer are sintered nano-silver:

Firstly, the soldering area dramatically decrease the residual stress with comparing the Fig. 8(a-b), the larger the size of nano-silver, the more sensitive the change of residual stress with thickness. Secondly, the residual stress of chip and layer 1 are simultaneously decrease with increasing of layer 1 thickness. It also can derive the same result with the changing of layer 2. Thirdly, the residual stress of the substrate is not only the largest but also not affected by the size of the soldering layers, so the material selection of the substrate needs to be further studied. Finally, due to large stiffness of the heat sink, it suffer little effect from vary thickness of soldering layers.

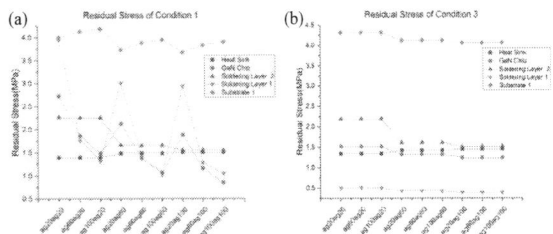

Fig. 8. The optimization of soldering thickness: (a) Residual stress of condition 1(large area); (b) Residual stress of condition 3(small area).

- For the condition of layer 1 is sintered nano-silver, layer 2 is $Au_{80}Sn_{20}$:

978-1-6654-1392-3/21 $31.00 © 2021 IEEE

Firstly, the changing of soldering area has a slight effect on residual stress with comparing the Fig. 9(a-b). Secondly, the residual stress of chip layer shows obviously change with vary thickness of layer 1 and 2. It is owing to the large elastic module of GaN, which is sensitive to stress and strain describing by equation (7). Thirdly, substrate suffer the largest residual stress comparing with all of the conditions, which has reached about 8MPa. Finally, the residual stress of soldering layer 1, soldering layer 2 and heat sink has slightly change in soldering procedure.

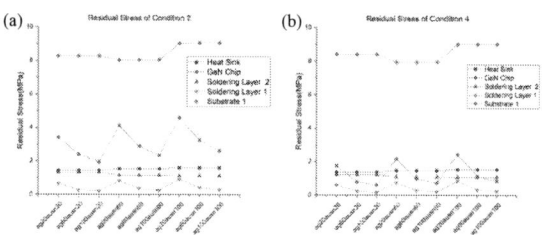

Fig. 9. The optimization of soldering thickness: (a) Residual stress of condition 2(large area); (b) Residual stress of condition 4(small area).

V. CONCLUSION

According to the simulation conditions, the results could be concluded that:

For the optimization of different soldering size, small soldering size has the lowest residual stress; For the optimization of different soldering material, sintered nano-silver paste has lower residual stress than that of the $Au_{80}Sn_{20}$ after the soldering procedure, which shows the ability of relieving and absorbing stress.[10]

For the optimization of different soldering thickness, sintered nano-silver paste is more sensitive to vary thickness and its residual stress decrease with the increasing of the thickness; The substrate is also a critical factor of the multi-layer stack model, which suffer largest residual stress in the whole model. Even though the condition of 2 & 4 are more stable than that of the condition of 1 & 3, the substrate suffer large residual stress after the soldering. Thus, it is necessary to introduce the sintered nano-silver to relax the residual stress in multi-layer soldering procedure.

ACKNOWLEDGMENT

This work was supported by CEPREI's special project under Grant: 20D16, and the national basic scientific research project under Grant: JBF202800010.

REFERENCES

[1] Z Zeng, "Packaging characterization and integration of SiC power device," Science Press, p94-95, 2020.

[2] Sheng, W. W. , and R. P. Colino . "Power Electronic Modules Design & Manufacture.",2016.

[3] J Chen, "Study on the performance and Reliability of High Power density COB packaged LED Module". Tianjin University, 2016.

[4] G DYang, "Research on package and Reliability of 650V-150A IGBT half-Bridge Power Module", Huaqiao University, 2020.

[5] Y F Tan, "Research on the reliability of bonding line based on a new cascaded DBC hybrid package power module", Huazhong University of Science and Technology, 2019.

[6] L Xu, "Study on package Reliability of IGBT Power Module", Huazhong University of Science and Technology, 2016.

[7] Y H Mei, "Study on electromigration and bonding thermal bending properties of low temperature sintered nanometer silver solder paste ", Tianjin University, 2010.

[8] D J Yu, X Chen, G Chen, et al. "Applying Anand model to low-temperature sintered nanoscale silver paste chip attachment", Materials and Design,2009,30(10).

[9] G S Zhang, "Study on mechanical properties of 80Au/20Sn solder alloy", Tianjin University, 2010.

[10] J RGroza, R J Dowding, "Nanoparticulate materials densification", Nanostructured Materials, 1996, 7(7):749-768.

Low Temperature Bonding Polycrystalline Diamond to Si by Using Au Thin-layer for High-power Semiconductor Devices

Yi Zhong
School of Electronic Science an Engineering, Xiamen University,
Xiamen 361005, China
Xiamen Sky Semiconductor Technology Co., Ltd.,
Xiamen 361005, China
zhongyi@xmu.edu.cn

Shuchao Bao
School of Electronic Science an Engineering, Xiamen University,
Xiamen 361005, China
Xiamen Sky Semiconductor Technology Co., Ltd.,
Xiamen 361005, China
1030299505@qq.com

Ke Li
School of Electronic Science an Engineering, Xiamen University,
Xiamen 361005, China
Xiamen Sky Semiconductor Technology Co., Ltd.,
Xiamen 361005, China
1173754797@qq.com

Mingchuan Zhang
Xiamen Sky Semiconductor Technology Co., Ltd.,
Xiamen 361005, China
zhangmc@sky-semi.com

Daquan Yu
School of Electronic Science an Engineering, Xiamen University,
Xiamen 361005, China
Xiamen Sky Semiconductor Technology Co., Ltd.,
Xiamen 361005, China
yudaquan@xmu.edu.cn

Abstract—**As the semiconductor devices are getting higher frequency, higher power and smaller size, the management of thermal dissipation becomes a big challenge. The application of diamond as heat dissipation substrate for high-power semiconductor devices has been placed great expectation due to its ultra-high thermal conductivity. In this study, Au thin layer was used for the low temperature bonding of polycrystalline diamond and Si. The Au-Au atomic diffusion bonding was successfully achieved. Clean processes were optimized. Scanning acoustic microscope (SAM) was used to determine the bonding porosity, which typically exceeded 10%. Atomic force microscope (AFM) tests indicated the diamond surface roughness (Ra>1nm). The poor surface flatness of diamond contributed to the degradation of bond-ability. The technological route of diamond heat dissipation substrate for high-power semiconductor devices needs more optimization.**

Keywords—Diamond, Porosity, Au-Au bonding, Surface roughness

I. INTRODUCTION

As the semiconductor devices are getting higher frequency, higher power and smaller size, the management of thermal dissipation becomes a big challenge. Thermal accumulation increases rapidly in the active region of electronic devices, forming local hotspots and leading to the significant degradation of performance [1]. Therefore, how to manage the heat dissipation becomes one of the key technical bottlenecks restricting the further development of high-power semiconductor devices. Recently, a falling price of artificial polycrystalline diamond was achieved through mass production. The application of diamond as heat dissipation substrate has been placed great expectation due to its ultra-high thermal conductivity, which could reach up to ~2000

W/m·K [2-5]. Fig.1 shows the schematic diagram of a diamond based high-power semiconductor device.

Although diamond can be directly grown on the semiconductor wafers by chemical vapor deposition (CVD), the growth of diamond needs an environment of more than 700℃ [3], which is unacceptable for semiconductor devices to withstand such high temperature. Surface Activated Bonding (SAB) technology can bond a single-crystal diamond to device at room temperature by using Ar beams for activation [4]. But it needs extremely smooth surface, which is very challenging for polycrystalline diamond. Another method is soldering diamond to semiconductor devices [5]. However, the thermal conductivity of solder layer is typically two orders of magnitude smaller than that of diamond, and the bonding requires thick filler of metal layers. That introduces huge thermal resistance and thus significantly degrades the heat dissipation performance of diamond.

In this study, novel Au thin layer was used for the low temperature bonding of polycrystalline diamond and Si. The Au-Au atomic diffusion bonding was successfully achieved. Scanning acoustic microscope (SAM) was used to determine the bonding porosity. The effects of surface roughness and

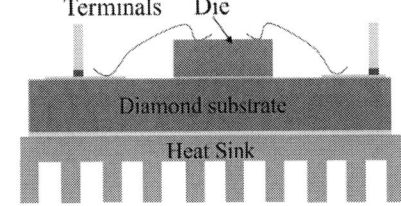

Fig. 1. Schematic diagram of a diamond based high-power semiconductor device.

This work was supported by the National Natural Science Foundation of China (Grant No. 61974121).

flatness on the porosity were discussed. The technological route of diamond heat dissipation substrate for high-power semiconductor devices needs more optimization.

II. EXPERIMENTAL PROCEDURES

The CVD deposited polycrystalline diamond films were 0.7 mm in thickness and were cut into plates with size of 10mm×10mm. The diamond plates had being ground and polished. Then they were carefully cleaned. Physical vapor deposition (PVD) method was used to deposit 5nm Ti/200nm Cu/5nm Ti/20nm Au on diamond plates. At the same time, Si plates (10mm×10mm×0.6mm) were cleaned and deposited with 5nm Ti/20nm Au. Diamond-Si pairs were alignment together and pressed by hand. They could be pre-bonded at room temperature. Then they were bonded in a vacuum chamber with 200°C and 6MPa. As control experiments, Si-Si pairs were bonded with the same parameters. Scanning acoustic microscope (SAM) was used to determine the bonding porosity. Atomic force microscope (AFM) was used to characterize the surface roughness.

III. RESULTS AND DISCUSSION

A. Cleaning diamond

The surface morphology of an as-received sample is shown in Fig.2(a). Massive black spots were observed and distributed randomly. Therefore, additional cleaning process was needed.

Firstly, the surfaces were wiped carefully using the clean cloth which dipped with acetone. Second, a standard semiconductor wetting clean process (H_2SO_4:H_2O_2=3:1, at 240°C) was used. Third, the samples were cleaned with acetone and alcohol in a ultrasonic bath, respectively. Lastly,

cleaned in a plasma (100sccm O_2+ 100sccm Ar, at 200W) chamber for 2 min.

The surface morphology of a sample after the cleaning process was shown in Fig.2(b). Most of the black spots have been removed. However, the edge and corner region of the diamond plate still distributed with micron scale black spots, which may be caused by the laser damage during diamond cutting. Noting that, polishing a diamond is very difficult to achieve a smooth/flat surface without any grooves because of its highest hardness and chemical inertness.

B. Surface roughness and flatness

Au-Au bonding based on the solid-state diffusion. Thus, atomic flat surface was needed. In this study, AFM was used to characterize the surface roughness before bonding. Fig. 3(a) shows typical AFM results of diamond surface after cleaning. In an area of 20μm×20μm, the surface roughness was approximately Ra 1.122nm. After PVD Ti/Cu/Ti/Au layer, the surface roughness was approximately Ra 3.156nm, as shown in Fig. 3(b). Noting that, these were not in-situ tests. Other regions were also tested and the results show that it was difficult to achieve surface roughness Ra<1nm for diamond.

The total thickness variation (TTV) of diamond has been measure by a silicon thickness measuring instrument for 5 points (4 corner and 1 middle regions) each. Three samples were checked, and their TTV were 8, 2 and 6.5μm, respectively. The total TTV for those three was 24μm. The TTV data could acted as cursory quantitative indicators to reflect the surface flatness.

Fig. 4(a) shows AFM results of Si surface after cleaning. In an area of 20μm×20μm, the surface roughness was approximately Ra 0.895nm. After PVD Ti/Au layer, the surface roughness was approximately Ra 0.898nm, as shown in Fig. 4(b). The typical TTV for Si plates could less than 3μm. The surface flatness was much better than that of diamond, due to the mature polishing and cleaning processes for Silicon.

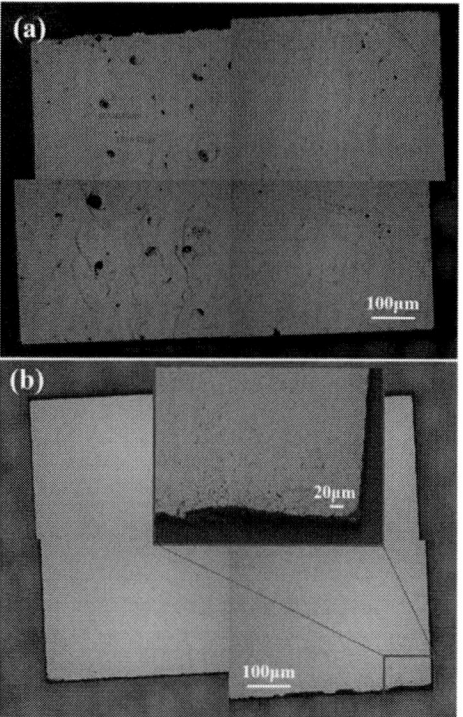

Fig. 2. The surface morphology of (a) an as-received diamond, (b) after cleaning processes.

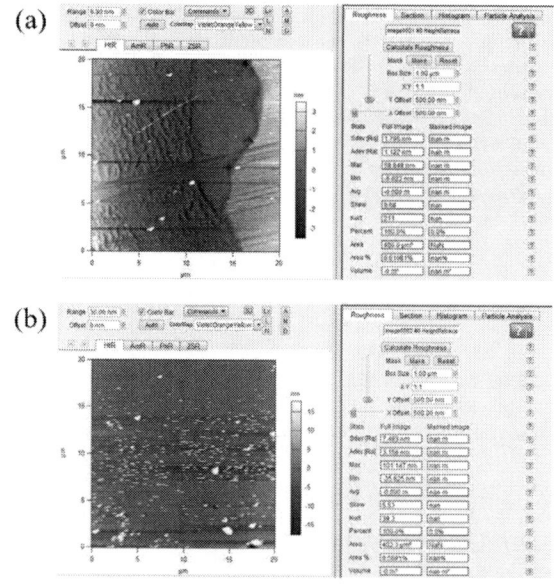

Fig. 3. AFM results of diamond surface: (a) after clearning, (b) after PVD.

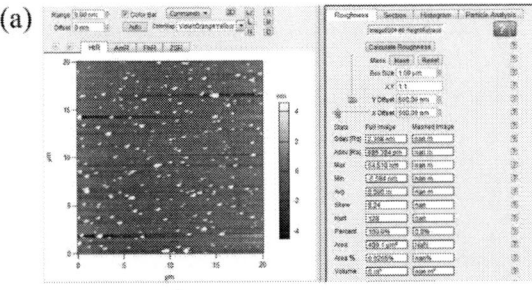

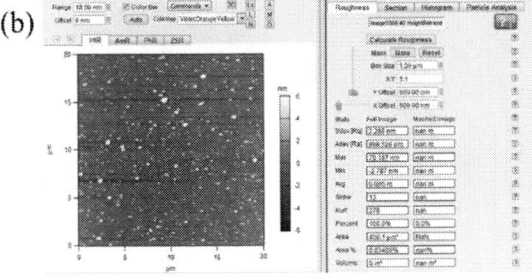

Fig. 4. AFM results of Si surface: (a) after clearning; (b) after PVD.

C. Bonding porosity

The Au-Au atomic diffusion bonding was successfully achieved. SAM was used to determine the bonding porosity, Typical SAM measurement results are shown in Fig. 5. The Si-Si samples were bonded pretty good with tiny voids distributed randomly. The porosity was measured by about 0.55%. While the porosity for Diamond-Si bonders were approximately 14.45%. Large poor-bonded area were distributed at the edge and corner regions. Si-Si bonding quality was much better than that of Diamond-Si. This was due to the poor surface roughness and flatness of diamond, especially at the edge and corner.

Unlike the soldering process which contained solid-liquid-solid phase transitions, Au-Au bonding based on the solid-state diffusion and thus can't endure much surface roughness. The effect of surface roughness on solid-state diffusion bonding can be discussed in terms of elastic deformation and energy [6,7]. The elastic energy must be smaller than the work of adhesion (W) to achieve good solid-state diffusion bonding:

$$\frac{R^2}{\lambda} < \frac{2(1-v^2)}{\pi E} W \qquad (1)$$

where R and λ are the surface roughness and wavelength of the bonding surface. E and v are Young's modulus and Poisson's ratio, respectively.

Wavelength λ is assumed to correspond to the average grain size. According to Eq. (1), surface roughness R is a critical parameter for Au-Au bonding. Better surface roughness needs lower work of adhesion, i.e., lower bonding pressure and temperature. Thus, the future work should aim to flatten the diamond surface by polishing or introducing other intermate layer.

IV. CONCLUSIONS

In this study, Au thin layer was used for the low temperature bonding of polycrystalline diamond and Si. The Au-Au atomic diffusion bonding was successfully achieved. Clean processes were optimized. SAM was used to determine the bonding porosity, which typically exceeded 10%. AFM tests show that the typical diamond surface roughness was Ra>1nm. In addition, the poor surface flatness (TTV≥2μm) of diamond contributed to the degradation of bond-ability. The effect of surface roughness on solid-state diffusion bonding was discussed. Future work needs to flatten the diamond surface and study the influence of diamond thickness and thermal conductivity on the heat accumulation effect of semiconductor devices. Develop and optimize the technological route of diamond heat sink for high-power semiconductor devices.

ACKNOWLEDGMENT

The authors would like to thank the engineering team in Xiamen Sky Semiconductor Technology Co., Ltd.

REFERENCES

[1] Y. Han, B.L. Lau, G. Tang, X. Zhang. Thermal management of hotspots using diamond heat spreader on Si microcooler for GaN devices. IEEE Transactions on Components, Packaging and Manufacturing Technology, 5(12), 1740-1746, 2015.

[2] T. Matsumae, Y. Kurashima, H. Umezawa, K. Tanaka, T. Ito, H. Watanabe, H. Takagi. Low-temperature direct bonding of β-Ga₂O₃ and diamond substrates under atmospheric conditions. Applied Physics Letters, 116(14), 141602, 2020.

[3] M. Rabarot, J. Widiez, S. Saada, J.P. Mazellier, C. Lecouvey, J.C. Roussin, J.P. Roger. Silicon-On-Diamond layer integration by wafer bonding technology. Diamond and related materials, 19(7-9), 796-805, 2010.

[4] Y. Minoura, T. Ohki, N. Okamoto, A. Yamada, K. Makiyama, J. Kotani, N. Nakamura. Surface activated bonding of SiC/diamond for thermal management of high-output power GaN HEMTs. Japanese Journal of Applied Physics, 59(SG), SGGD03, 2020.

[5] D. Francis, F. Faili, D. Babić, F. Ejeckam, A. Nurmikko, H. Maris. Formation and characterization of 4-inch GaN-on-diamond substrates. Diamond and Related Materials, 19(2-3), 229-233, 2010.

[6] H. Takagi, R. Maeda, T.R. Chung, N. Hosoda, T. Suga. Effect of surface roughness on room temperature wafer bonding by Ar beam surface activation. Japan Journal of Applied Physics, 37, 4197-4203, 1998.

[7] M. Yamamoto, T. Matsumae, Y. Kurashima, H. Takagi, T. Suga, S. Takamatsu, E. Higurashi. Effect of Au film thickness and surface roughness on room-temperature wafer bonding and wafer-scale vacuum sealing by Au-Au surface activated bonding. Micromachines, 11(5), 454, 2020.

Fig. 5. SAM images of (a) Si-Si sample, (b) Diamond-Si sample.

Research on Reliability Life Evaluation Method Based on Airborne T/R Components

1st Rui Deng[1]
Component Testing Center
China Electronic Product Reliability and Environment Testing Research Institute
Guangzhou, China
dengrui@ceprei.com

2nd Chuanjin Deng[2]
Component Testing Center
China Electronic Product Reliability and Environment Testing Research Institute
Guangzhou, China
tjudeng@126.com

3rd Ruolei Wang*
Component Testing Center
China Electronic Product Reliability and Environment Testing Research Institute
Guangzhou, China
wrl19931011@163.com

4th Weixi Gong
Component Testing Center
China Electronic Product Reliability and Environment Testing Research Institute
Guangzhou, China
375781132@qq.com

5th Zongqi Cai
Component Testing Center
China Electronic Product Reliability and Environment Testing Research Institute
Guangzhou, China
zqcai.uestc@hotmail.com

6th Kun Jiang
Component Testing Center
China Electronic Product Reliability and Environment Testing Research Institute
Guangzhou, China
jiangkun2086@126.com

Abstract

In order to ensure the using life requirements of the whole machine, it is necessary to predict the reliability of the internal chip components of the whole machine under the same environmental stress. According to the experimental results, we can try to eliminate the problem that the service life of the whole machine is reduced due to the failure of the chip component. The reliability life expectancy can increase the failure process of internal chip components and can reflect the problems of the chip more quickly. For a new chip component, the reliability index is unknown, which needs us to study and determine the stress cycle type and specific stress parameters of the environmental test, to judge its reliability. This article aims the airborne T/R component chip as an example, and studies, determines and evaluates its reliability index under Aircraft installed conditions. According to different operating conditions, the environmental stress screening test and the environmental adaptability test is carried out firstly. Then the stress cycle method of the stress reliability test is studied according to the airborne environment spectrum. The reliability test is carried out after the test stress cycle is determined which accelerates the deterioration. The experimental results are analyzed for the study of the failure mechanism. The reliability test process in this paper provides a reference for the follow-up reliability test research, which has great significance to the development of installed chip components.

Keywords—Reliability, T/R Components, Stress cycle, Stress profile, Environmental test

I. INTRODUCTION

As the "eyes" of aircraft and fighter jets, airborne phased array radar has the ability to explore, identify, and track. The key to realizing the function of phased array radar lies in its phased array antenna, and the T/R component is the core component of the phased array antenna. A T/R component mainly completes the amplification of the transmitted/received signal, and realizes the functions of phase shift and beam control required for antenna beam scanning. It has the characteristics of high integration, wide installation coverage, multiple functional models, and complex structure[1].

With the continuous updating and improvement of airborne control array radars, the core T/R components of a type of airborne phased array radar suitable for production equipment must have high reliability. Generally, the service life of an aircraft is about 30 years. In order to evaluate whether the service life of the T/R components of the airborne phased array radar can meet the same MTBF (Mean time between failures) as the whole aircraft, we need to carry out relevant life expectancy. Since it is impossible to take reliability tests similar to the overall environment, relying on GJB 899A-2009 [2]statistical principles and schemes to carry out reliability life tests to verify whether the reliability of T/R components meets the design requirements has become the most effective verification method at present. Domestic research on the reliability of T/R components focuses on component-level failure mechanism analysis and complete machine-level accelerated life test and lacks a systematic discussion of component-level life testing and research on test methods. Usually in order to realize the light weight and miniaturization of T/R components, its internal structure is more delicate and complicated, as shown in Figure 1. In the T/R component , the microwave transmitter chipset adopts highly integrated GaAs MMIC(Monolithic Microwave Integrated Circuit)chips and bare chip silicon control ASIC (Application Specific Integrated Circuit)chips[3]. The bare chip used in the multi-chip module can only be subjected to non-energized tests in the chip production stage, and cannot be energized and aged screening, temperature shock screening and mechanical screening. The early failure of the chip cannot be eliminated during the chip production test, so it needs to be tested at the component-level after assembly. In addition, other components inside the T/R components such as amplifier tubes, circulators, isolators, resistors, and capacitors, are all in the form of surface mounts to reduce volume and weight[4]. Therefore, for T/R components with high integration, high power consumption, and miniaturization, component-level reliability testing can be more accurately reflect the rationality of the internal layout structure, and It can also realize the assessment of the quality of the internal components and the level of craftsmanship.

978-1-6654-1392-3/21 $31.00 © 2021 IEEE

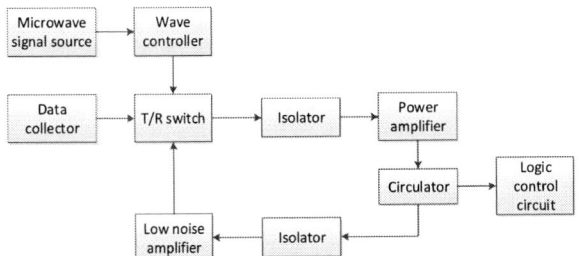

Fig. 1. Basic structure of T/R components

This article takes a certain type of the T/R component chipset as an example to determine whether it meets the aircraft installation conditions. To determine and evaluate the reliability index of T/R component chipset, this article first carry out environmental stress screening and environmental adaptability test to eliminate early defects, reduce the impact on reliability due to accidental factors such as process and quality, and test whether T/R components can cope with environmental damage in the airborne environment. Then, according to the characteristics of the airborne environment, the analysis of the use of airborne equipment and the relevant standards，this article gives the airborne T/R component reliability assessment test plan, stress cycle method, test profile and other contents and methods. Finally, the reliability test and reliability evaluation under laboratory conditions are realized, and the specific process is shown in Figure 2.

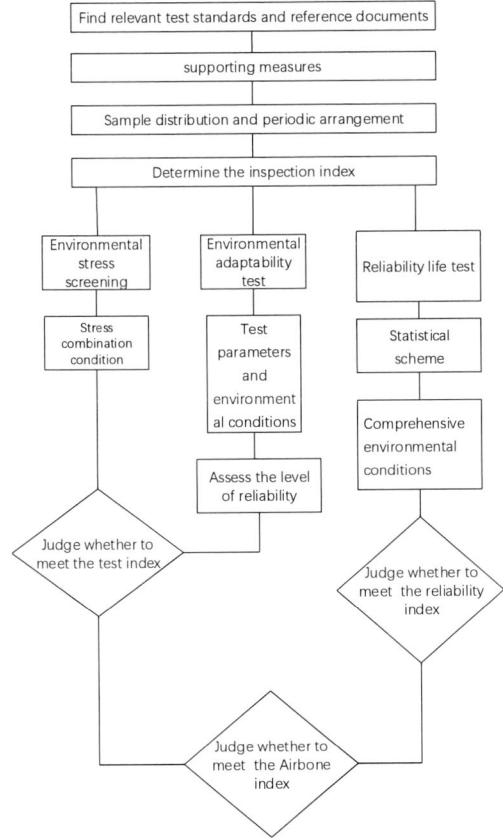

Fig. 2. Basic structure of T/R components

II. Test Principle and Method

A. Environmental stress screening

The stress screening of T/R components is based on GJB1032-1990[5] to carry out random test and temperature (high and low temperature) cycle test. Generally, the internal components of T/R components will be surface-mounted due to the miniaturization design requirements, and vibration screening test can eliminate components that do not meet the assembly process. At the same time, the chipsets in the T/R components are mostly multi-layer ceramics as the carrier and a large number of bare chips are integrated. Random vibration can also eliminate the chipsets in the T/R components with insufficient strength and substandard flatness. The transmitting channels of the T/R components have a large number of power devices, which consume a lot of power and have a large heat flux. Therefore, the assembly requirements of the power chip are very strict in the process. The welding void rate of the power chip does not meet the chip heat dissipation requirements and can be eliminated during the temperature cycle. The failure mechanism of the MMIC is mainly related to the high temperature stress[6]. The high temperature environment can test whether the internal MMIC quality meets the standard. The airborne T./R components are usually in a high-altitude and low-temperature environment. The temperature cycle can truly imitate the real temperature changes carried by the components, causing the internal chipset to change phases, so as to screen out components with substandard assembly structures. Environmental stress screening is one of the necessary links in the reliability test, and it is also a process in the product manufacturing process. Its purpose is to ensure the follow-up reliability test, and help guide reliability design improvement and product upgrade optimization.

B. Environmental adaptability test

The purpose of the environmental adaptability test is to detect whether the T/R components can work normally under various airborne environmental conditions encountered. The essence of the test is to study the destructive effects of environmental factors as external factors and the material and structure of the equipment itself as internal factors. The environmental adaptability test combines the main environmental factors that T/R components may experience during the service period, and mainly have a greater impact on the airborne equipment The test items include temperature, humidity, vibration, shock, mold, salt spray and other corrosive environment tests during the storage, transportation, and use stages of the equipment[6]. The specific test parameters are shown in Table 1.

TABLE I.   ENVIRONMENTAL ADAPTABILITY TEST CONDITIONS

test	Reference standard	condition：	Criteria：
Low temperature storage and work	GJB150.4A-2009	Storage phase： -55℃, 48h Work phase： -55℃, 2h	Meet electrical performance indicators
High temperature storage and work	GJB150.3A-2009	Storage phase: +85℃, 48h Work phase: +70℃, 2h	Meet electrical performance indicators
Temperature shock or cycle	GJB150.5A-2009	Low temperature:-55℃ High temperature:70℃	Meet electrical performan

978-1-6654-1392-3/21 $31.00 ©2021 IEEE

			Keep: 0.5h Transform: ≤5min cycles: 10	ce indicators
Mechanical shock	GJB150.18 A-2009		Shape: Half sine wave Acceleration: 15g Keep:11ms Direction & frequency: ±X×3、 ±Y×3、 ±Z ×3	Meet electrical performance indicators
Mechanical vibration	GJB150.16 A-2009		Direction: Triaxle; Vibration time;1h/ Axial。	Meet electrical performance indicators
Temperature-altitude test	GJB150.6A-2009		Table 2 Step;1a →3 →4 →10	Meet electrical performance indicators
Salt spray test	GJB150.11 A-2009		Concentration;5% ± 1%; PH: 6.5~7.2; Temperature:35 ℃; Deposition rate: (1~3) ml/80cm².h; Time Spray 24h,Dry 24h,Cycle: 2times	Shell without damage

TABLE II.   TEMPERATURE-ALTITUDE TEST CONDITIONS

experiment procedure	1a	3	4	10
Temperature (℃)	25	-40	-10	55
Altitude (m)	6000	3000	Ground	3000
Hold time (h)	1h;	20min	4h	4h
Power-on and power-off requirements	off	Depressurization after power on at low temperature	on and off 3 times	Depressurization after power on at high temperature

C. Reliability life test

Reliability life test is to restore the airborne stress level of T/R components under laboratory conditions, and calculate the reliability index of T/R components through experiments and theoretical models. In order to ensure that the test results are consistent with the actual situation, we usually assume that the failure of the reliability life test is consistent with the regular failure law, while satisfying the theory of damage accumulation. The reliability life test can accelerate the accumulation of damage and reflect the problems of the T/R components, and finally obtain Life characteristics under normal working environment. For a new T/R component, the reliability index is unknown, and we need to determine the stress cycle type and specific stress parameters of the environmental test. The overall idea of reliability life in this article is as follows

- According to the MTBF requirements of airborne T/R components (including product life distribution, life confidence interval, and confidence level), determine the airborne T/R component test statistical programs,

total test time and test profiles in accordance with GJB 899A-2009 [2].

- Find out the most typical and representative use conditions that affect the reliability of airborne T/R components, find the key stimulating environmental factors, and determine the type of stress. Then combine the corresponding airborne reliability test experience and failure analysis to determine the stress index.

- On the basis of the assumption that the degradation mechanism is consistent and the statistical distribution is ideal, the combined failure model of various sensitive environmental stresses is used to calculate the test time and stress value. The conventional stress test profile is converted into an equivalent comprehensive stress profile, and reliability tests are carried out with this information.

- Evaluate the reliability of airborne T/R components based on the final test time and test results.

From service to retirement, an aircraft needs to undergo delivery training, transition, mission execution, daily maintenance, parking, maintenance, and retirement. Depending on the take-off and landing location, the nature and difficulty of the flight mission are different, and under different mission conditions The stress profile is more complicated. In addition, airborne equipment and accessories also have tasks that are different from the aircraft's service phase, including ground testing, equipment maintenance, installation and use, etc. The life cycle of airborne equipment and accessories should be considered in the actual stress profile. In order to make the experiment more representative, the calculation and experiment design are simplified at the same time. In this test profile, we consider the dominant take-off, cruising and landing behaviors. At the same time, t The radar is only powered on during the pre-takeoff self-check and loading mission and turned on during the cruising phase. T/R components apply electrical stress during ground parking and cruising phases.

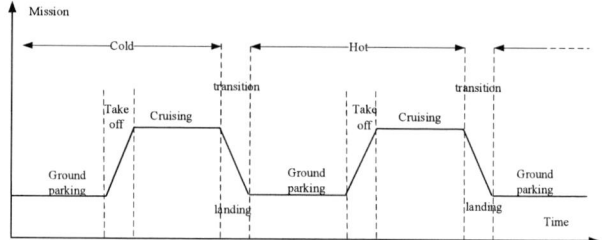

Fig. 3. Mission profile of airborne T/R components

It can be seen from Figure 3 that the test profile simulates the situation of aircraft parking, take-off and climb, air cruise execution, and landing in cold and hot days. At the same time, the landing phase is used as a transition to achieve the switching from hot weather. to cold weather. The influence of reliability test on T/R components is mainly the coupling effect and cumulative damage caused by the continuous application of the above three environmental stresses. The environmental effects of single stresses such as low temperature or high temperature are the same as those of low and high temperature tests. The environmental effects of double stress and multiple stresses are basically the same as the comprehensive environmental effects in the temperature-vibration-work test.

TABLE III.   TEMPERATURE-ALTITUDE TEST CONDITIONS

	Mission	Type of stress	index	Percentage of time
A	Ground parking	Low temperature &Electrical stress	Maximum working temperature	5%
B	Take off	Rapid temperature change & vibration	Actual temperature change rate and vibration stress during climb	5%
C	Cruising	Low temperature & vibration &electric stress	Temperature under working conditions Vibration stress during cruising	30%
D	Landing (transition)	Rapid temperature change & vibration	Actual temperature change rate and vibration stress during the descent phase	10%
E	Ground parking	High temperature & electrical stress	Maximum working temperature	5%
F	Take off	Rapid temperature change & vibration	Actual temperature change rate and vibration stress during the climbing phase	5%
G	Cruising	High temperature & vibration stress & electrical stress	Temperature under working conditions Equipment measured vibration	30%
D	Landing (transition)	Rapid temperature change & vibration	Actual temperature change rate and vibration stress during the descent phase	10%

The test profile adopts a combined test method. Single stress test and multiple stress test are applied in a certain order to form a comprehensive test profile. Its task stage can be divided into two parts: cold cycle and hot cycle, including the taking-off and landing process and the cruise phase, the radar is turned on to perform the combat mission process. The mission profile specifically includes the temperature profile, vibration profile and electrical stress profile as shown in Figure 4.

Fig. 4.   Mission profile of airborne T/R components

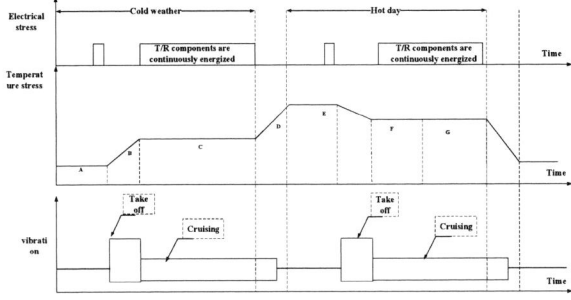

III.   CONCLUDING

Based on the requirements for the reliability index verification of airborne T/R components, this paper determines the reliability test profile based on relevant standards and actual specifications, studies the reliability test theory of airborne T/R components, and formulates the accelerated test profile. The reliability life test method is provided, which provides a reference for evaluating the reliability index of airborne T/R components.

[1] YANG Peng, "Spaceborne T/R Module Accelerated Life Testing Method," J. Aerospace Control and Application, vol. 3, pp. 38-43, May 2016 .

[2] Reliability appraisal and acceptance test. National Defense Science and Technology Committee. Electronic Information and Basic Department of the Equipment Department of the Chinese People's Liberation Army, GJB 899A, 2009.

[3] ZHOU J and GAO L, "Research on power MMIC chip accelerated life testing method," J. The Journal of New Industrialization, vol. 7, pp. 35-40, April 2014

[4] GUO Qing and LV Shen-gang, "A Design of C Band Solid T / R Module," J. JOURNAL OF MICROWAVES, vol. 29, pp. 69-73, Aug 2013

[5] WANG Xiao, "Reliability Evaluation of Spaceborne T/R Module," J. Environmental Adaptability&Reliability, vol. 37, pp. 16-19, April 2019

[6] Environmental stress screening of electronic products. National Defense Science and Technology Committee, GJB1032-90, 1990.

[7] Environmental test methods for military equipment laboratories. Electronic Information and Basic Department of the Equipment Department of the Chinese People's Liberation Army, GJB 150A, 2009

Effect of grain size on dielectric properties and reliability for ultra-thin MLCCs

Kulun Jiang[a,b]
[a]Tsinghua Shenzhen International
Graduate School,
Tsinghua University
[b]Shenzhen Institute of Advanced
Electronic Materials, Shenzhen
Institute of Advanced Technology
Chinese academy of sciences
Shenzhen, China
jkl20@mails.tsinghua.edu.cn

Lei Zhang[b*]
Shenzhen Institute of Advanced
Electronic Materials, Shenzhen
Institute of Advanced Technology
Chinese academy of sciences
Shenzhen, China
zhanglei@siat.ac.cn

Bo Li[a*]
Tsinghua Shenzhen International
Graduate School,
Tsinghua University
Shenzhen, China
boli@mail.tsinghua.edu.cn

Rong Sun[b]
Shenzhen Institute of Advanced
Electronic Materials, Shenzhen
Institute of Advanced Technology
Chinese academy of sciences
Shenzhen, China
rong.suni@siat.ac.cn

Shuhui Yu[b]
Shenzhen Institute of Advanced
Electronic Materials, Shenzhen
Institute of Advanced Technology
Chinese academy of sciences
Shenzhen, China
sh.yu@siat.ac.cn

Zhenxiao Fu[c*]
State Key Laboratory of Advanced
Materials and Electronic Components,
Guangdong Fenghua Advanced
Technology Holding Co., Ltd
Zhaoqing, China
fuzx@china-fenghua.com

Xiuhua Cao[c]
State Key Laboratory of Advanced
Materials and Electronic Components,
Guangdong Fenghua Advanced
Technology Holding Co., Ltd
Zhaoqing, China
caoxh@china-fenghua.com

Abstract—BaTiO$_3$-based ultra-thin (< 1 μm) multilayer ceramic capacitors (MLCCs) are fabricated with various grain sizes by sintering at different temperature under reduction atmosphere. Effects of the average crystal size on microstructure, tetragonality, dielectric performances, ferroelectric properties and reliability of MLCCs are studied via Raman spectrum, temperature coefficient of capacitance (TCC) curves, polarization-electric field (PE) loop and Weibull distribution of breakdown voltage (BDV). It is demonstrated that compared with coarse-grain sample, the fine-grain sample of 160 nm yields a higher TCC at the temperature range of -55~85 °C (X5S, EIA, Electronic Industry Alliance), owing to the low defect contribution. Meanwhile, the fine-grain sample exhibits a smaller remnant polarization but a little higher saturated polarization due to its low tetragonality. According to Weibull distribution of BDV results, fine-grain ceramics possess high grain-boundary density, contributing to relatively high reliability.

Keywords—ultra-thin MLCCs, grain size, tetragonality, dielectric properties, reliability

I. INTRODUCTION

As one type of the most promising electronic components, high-level performance multilayer ceramic capacitors (MLCCs) is in urgent need owing to the rapid development of modern electronics. In recent years, the ultra-thin MLCCs with < 1 μm layer thickness, high dielectric properties and reliability are in urgent demand in the semiconductor packages industry, which requires that the grain size of the nano-crystallization ceramics is about 200 nm or smaller. It is required at least 5 grains across each layer thickness to ensure the reliability [1]. Simultaneously, grain size can dramatically tailor piezoelectric constant, energy storage characteristics and the dielectric properties of MLCCs by changing internal stress, tetragonality and the phase transition temperature.

Kinoshita *et al*. [2] found that it can be achieved the largest dielectric constant for BaTiO$_3$ ceramic as the grain size is approximately 1 μm. Curecheriu *et al*. [3] concluded that when the grain size is reduced to 90 nm and below, the dielectric constant of fine-BaTiO$_3$ is reduced due to the decreased nonlinearity of ferroelectric core and the increased amount of the nonferroelectric grain boundaries. Chen *et al*.[4] studied BaTiO$_3$ ceramics ranging from 330 nm to 1.05 μm and found a clear low-frequency relaxation behavior in orthorhombic phase in fine BaTiO$_3$. Wang *et al*. [5] reported that MLCCs have excellent reliability when the thin BaTiO$_3$ dielectric layer has 5~7 grains across the thickness (grain size was approximately 100 nm). However, systematic investigations focusing the effects of grain size on the dielectric performances, ferro-electricity properties and reliability of BaTiO$_3$-based ultra-thin (< 1 μm) MLCCs have been rarely reported as so far.

978-1-6654-1392-3/21 $31.00 © 2021 IEEE

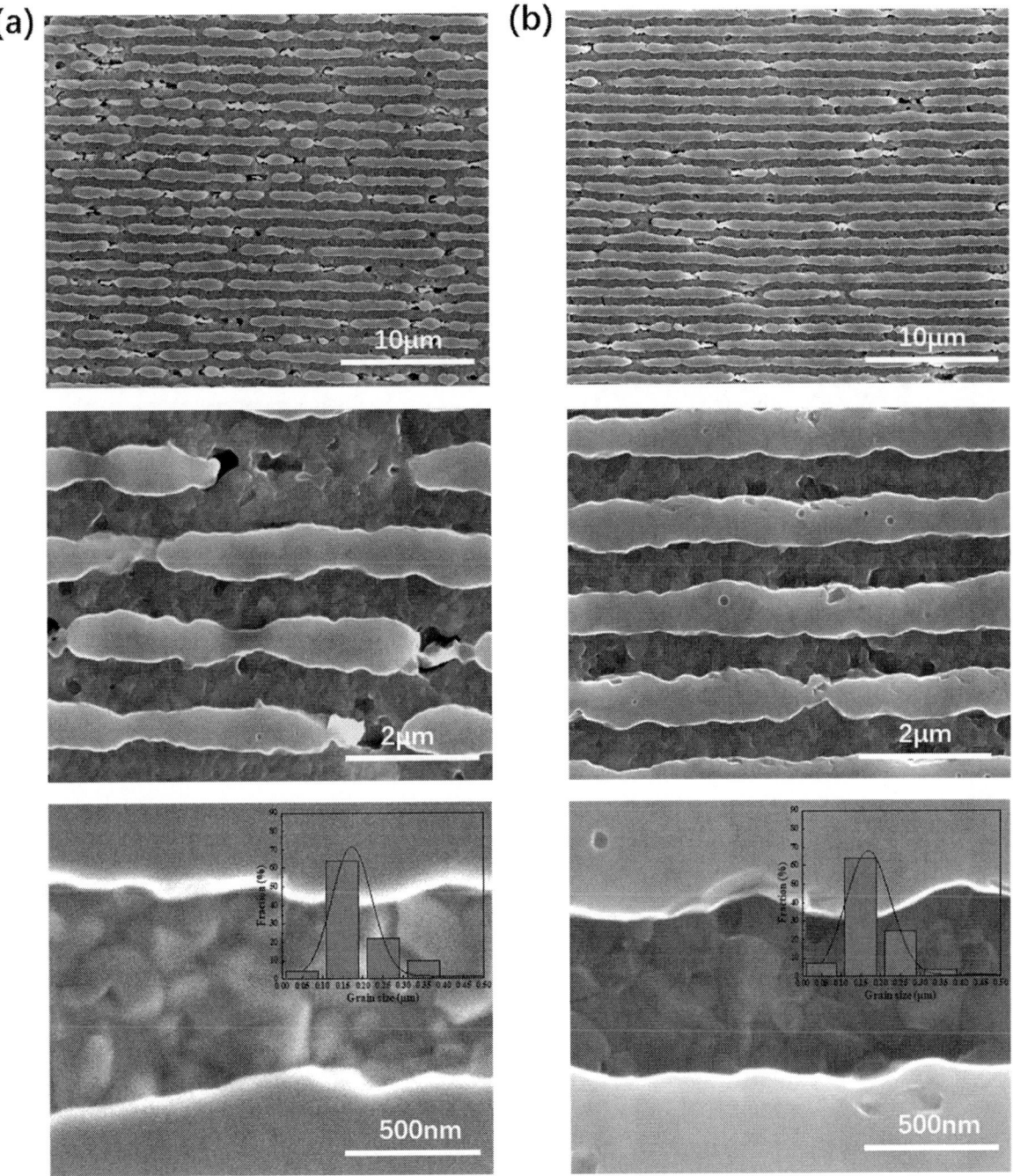

Fig. 1. Microstructure of the dielectric layers and Ni electrodes of (a) S1 and (b) S2, collected by scanning electron microcopy (SEM)

In this work, the ultra-thin (< 1 μm) MLCCs samples with different grain sizes in dielectric layers were synthesized by sintering in a reducing atmosphere. Dielectric and ferro-electric parameters, and breakdown voltage of samples were obtained and calculated from experimental results. The impacts of average size of $BaTiO_3$ grains was discussed systematically to enable high electrical properties and reliability of $BaTiO_3$-based ceramics for 0201 (0.6* 0.3*0.3 mm) MLCCs application.

978-1-6654-1392-3/21 $31.00 © 2021 IEEE

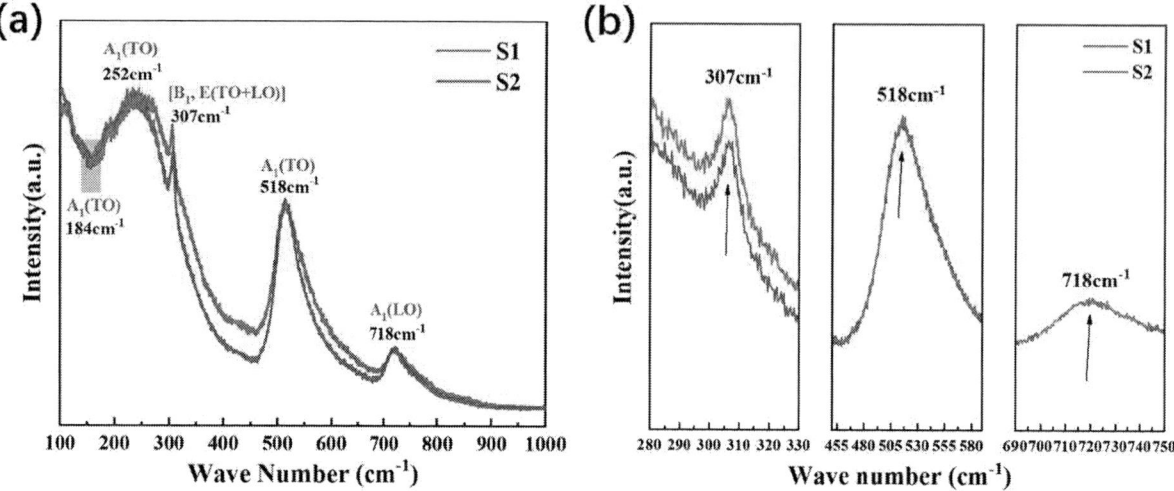

Fig. 2. Raman spectrum of S1/S2 at (a) wide range and (b) positions of 307 cm^{-1}, 518 cm^{-1}, 718 cm^{-1}.

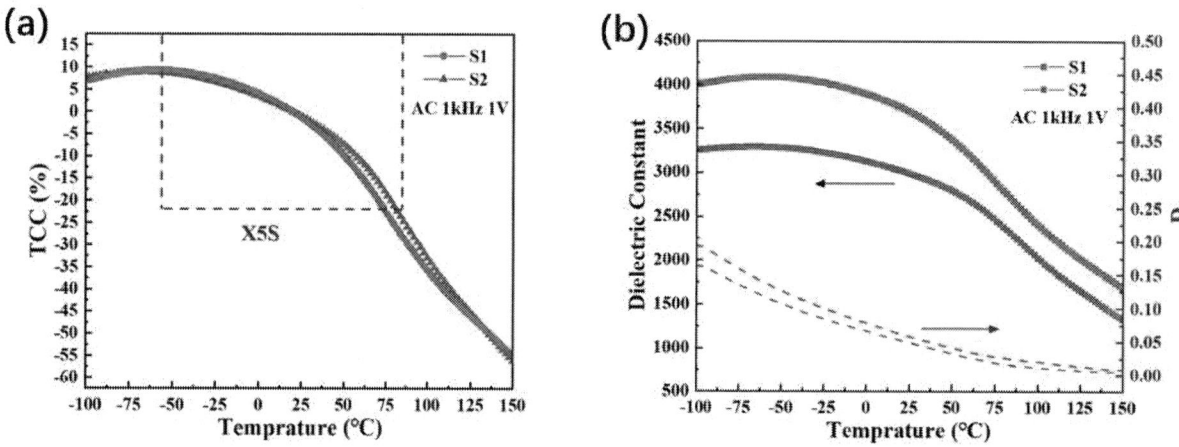

Fig. 3. (a)The temperature coefficient of capacitance (TCC) and (b) measured dielectric constant and dissipation factor of the samples at temperature range of -100~150 °C in 1 kHz

II. EXPERIMENT

A. Sample preparation

The raw materials were solid-state synthesized ~130 nm grade BaTiO$_3$ powder (Fujian Basic Electronic Materials Co., LTD, Changting, China), doping additives were reagent grade BaCO$_3$, SiO$_2$, MnO$_2$, and V$_2$O$_5$ powders in common. Then the powders were ball-milling in ethanol and mixed with yttrium-stabilized ZrO$_2$ working as ball-milling medium, followed by tape-casting into green sheets. A prototype Ni internal electrode-MLCC samples with a size of 0.6 mm*0.3 mm was prepared, in which the number of dielectric layers was 30, and the thickness after sintering was about 1.0μm. Perform standard multi-layer processes, including casting, screen printing nickel electrodes on green sheets, stacking, pressing and cutting, sintering at 1150 °C and 1120 °C, respectively, which were followed by reoxidation annealing. Therefore, two MLCC samples with different grain sizes were prepared, named S1 and S2, respectively.

B. Characterization

The average grain size of BaTiO$_3$-based MLCC samples was obtained by image analysis software. Raman spectroscopy (LabRAM HR Evolution, HORIBA Scientific, Paris, France) was performed by 633 nm excitation light to determine the tetragonality.

The dielectric constant and dissipation factor (DF) were measured by a LCR meter (4980A, Keysight, Palo Alto, CA) at 1 V ac driving filed from -100 °C to 150 °C at 1 Hz, 10 Hz, 100 Hz, 1 kHz, 10 kHz, 100 kHz. The polarization (P)-electric field (E) loops were obtained at room temperature by Radiant Precision Multi-ferroic using max field of 30 V/μm. The dc breakdown test was carried out by a withstanding voltage tester (CS9912BX, CHANGSHENG, Nanjing, China) through breaking 10 samples of each MLCC with a boost voltage rate of 1 V/s.

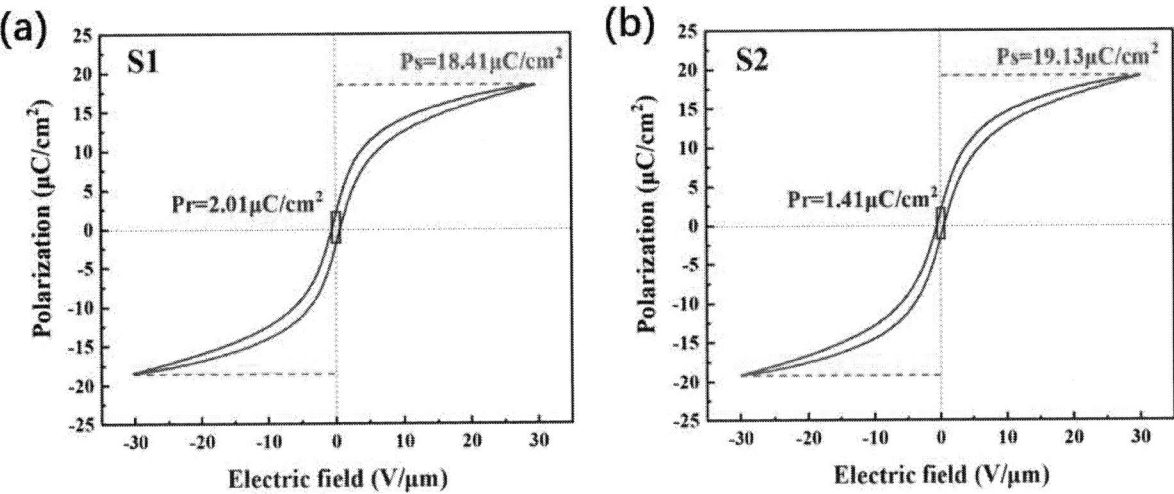

Fig. 4. Polarization-electric field loop of (a) S1 and (b) S2 at 25 °C with max field of 30V/μm

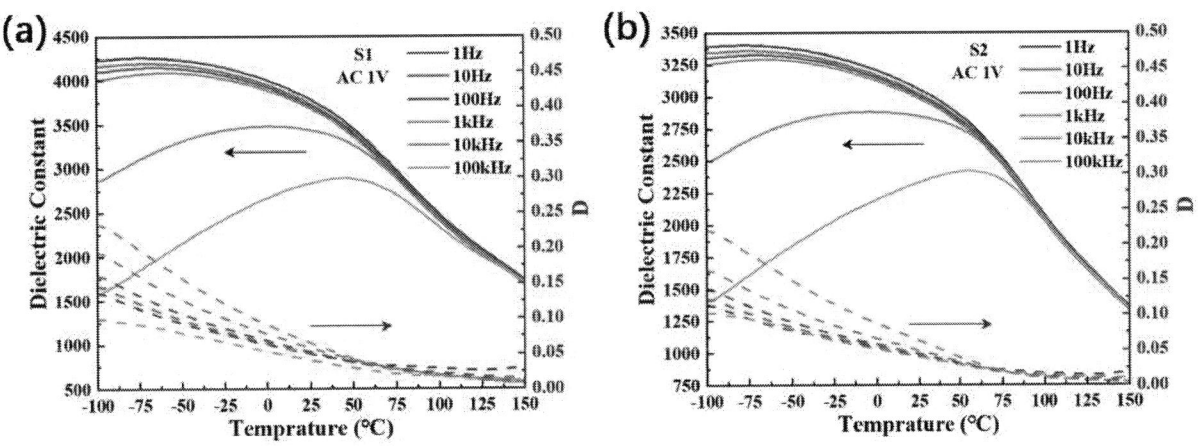

Fig. 5. Dielectric constant and dissipation dependent to a temperature range of -100~150 °C in 1 Hz, 10 Hz, 100 Hz, 1 kHz, 10 kHz, 100 kHz of (a) S1 and (b) S2.

III. RESULTS AND DISCUSSIONS

A. Microstructure and phase

The microstructure of samples is illustrated in Fig. 1. The inner electrode layer and dielectric layer are continuous with uniform thickness of about 1 μm. Furthermore, the average grain size is estimated to 200 nm and 160 nm corresponding to S1 and S2, respectively. In comparison with S1, S2 exhibits no obvious pores and defects , which implies the abnormal particle growth of S1 during the higher sintering temperature.

Taken the huge difference in the local structure and tetragonality into account, the Raman spectra of samples were studied. Raman peaks near 252 cm^{-1}, 307 cm^{-1}, 518 cm^{-1} and 718 cm^{-1} are marked at Fig. 2. It is well known that the cubic phase structure of BaTiO$_3$ has no Raman active modes whereas the tetragonal phase exhibits significant Raman activity, and 307 cm^{-1} is characteristic peak of tetragonal BaTiO$_3$ [6]. The characteristic peak of 307 cm^{-1} of tetragonal BaTiO$_3$ is presented in both S1 and S2, which means that both samples have tetragonal phase structure. It will be further confirmed by XRD patterns in the future. Especially, the intensity of S1 is larger than that of S2, suggesting the tetragonality of S2 is enhanced with the increasing of the grain size [7].

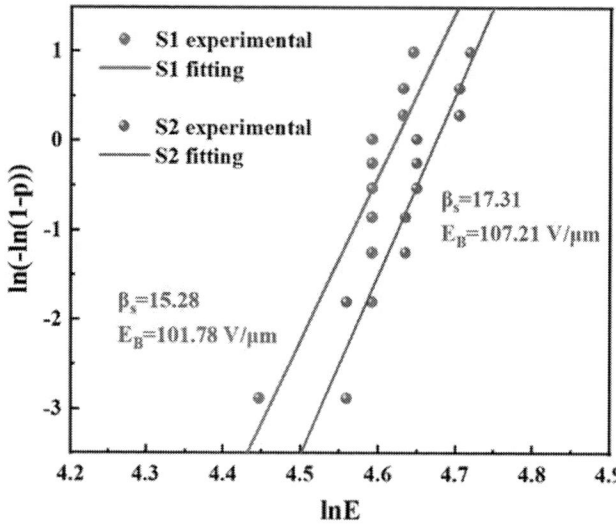

Fig. 6. Weibull distribution of breakdown voltage of S1 and S2

B. Dielectric properties of ultra-thin MLCCs

Fig. 3(a) shows that the samples exhibit the "core-shell" structure of BaTiO$_3$-based ceramics, which is necessary for a temperature-stable dielectrics [1]. Furthermore, at the ferroelectric-constant dielectric transition near 120 ° C of the core, S1 shows a higher relative permittivity peak than S2. At the same time, the peaks produced by the diffusion phase transition of shell near room temperature move S2 to higher temperatures. In other words, S1 and S2 are the same, so S2 has a higher proportion of shells than S1. Doping concentration. As a result, the dielectric constant at the whole temperature range for S1 is much higher than that of S2 (see in Fig. 3(b)). In addition, S1 possesses higher dissipation factor than S2, but at around 150 °C the loss curve of S1 exhibits a significant "upturn".

C. Ferroelectricity of ultra-thin MLCCs

Fig. 4 shows the Polarization-electric filed (PE) loop of the samples. It can be seen that S1 has higher remnant polarization (P$_r$) at 0 dc field than S2, which can be contributed to the higher tetragonality induced by larger grain size. At the same time, the lower saturated polarization (P$_s$) at the max field of S1 results from the ferroelectric domains in tetragonal phase, which were pinned by high dc field and paraelectric phase took a dominant role [8]. Especially, S1 reaches polarization saturation earlier than S2, because in the sample with larger crystal size, domain alignment with the c-axis parallel to the dc field is more likely to occur [9]. Moreover, all the samples exhibit the feature of relaxor ferroelectrics.

To further unveil the characteristics of relaxor ferroelectrics, TCC curves with different frequencies were collected and shown in Figs. 5(a, b). As the temperature is lower than the characteristic temperature of sample (~100 °C for S1 and ~85 °C for S2, respectively), the dielectric constant of specimen decreases with the increment of driving field's frequency, and almost unchanged as the temperature rising exceeds the characteristic temperature. This suggests that the samples are relaxor ferroelectrics rather than domain-switching ferroelectrics produced through grain refinement and doping [10].

D. Reliability of ultra-thin MLCCs

Fig. 6 shows the Weibull distribution of breakdown voltage (BDV) of samples. The breakdown voltage of S2 (107.21 V/μm) is greater than that of S1 (101.78 V/μm) while breakdown characteristic of specimen is similar. It is well known that grain boundaries with lower resistance than grains in the dielectric layer could work as "bridges" between Ni inner-electrodes in ultra-thin MLCCs [11]. Moreover, the defects of the dielectric ceramics layer such as intracrystalline pores and breakpoints would provide easy breakdown paths [12]. As a result, the higher breakdown voltage is observed in S2 due to the smaller grain size and less defects (see in Fig. 1). Accelerated Aging Life Tests (HALT) will be conducted in the future to investigate their reliability.

IV. CONCLUSIONS

BaTiO$_3$-based ultra-thin MLCCs with coarse-grain (approximately 200 nm) and fine-grain (approximately 160 nm) were successfully prepared by sintering under reduction atmosphere. The average crystal size dependence of the tetragonality, dielectric performances, ferro-electricity properties and reliability were studied in detail. Compared with coarse-grain sample, the fine-grain sample with the grain size of 160 nm yields a higher TCC at the temperature range of -55~85 °C, owing to its larger fraction of shell and a lower defect contribution. Moreover, the fine-grain sample exhibits a lower remnant polarization , higher saturated polarization and slower process to saturation with dc field increasing due to its low tetragonality. In addition, the Weibull distribution of BDV results reveal that the fine-grain sample possess higher grain-boundary density and less defects, contributing to relatively high reliability. Based on the above improvements, fine-grain BaTiO$_3$-based sample is promising for ultra-thin MLCCs with thickness less than 1 μm, leading to greater dielectric nonlinearity and reliability.

ACKNOWLEDGMENT

This work was supported by the National Science Foundation of China (No. 51802142), Foundation of State Key Laboratory of New Electronic Components and Materials (Grant Nos. FHR-JS-202011012, FHR-JS-202011013 and FHR-JS-202011014, and Joint Innovation Center of Advanced Electronic Components and Materials (Grant No. FHR-JS-202103001).

REFERENCES

[1] H. X. Wang, B. B. Liu, and X. H. Wang, "Effects of dielectric thickness on energy storage properties of surface modified BaTiO3 multilayer ceramic capacitors," (in English), Journal of Alloys and Compounds, Article vol. 817, p. 7, Mar 2020, Art no. 152804.

[2] K. Kinoshita and A. Yamaji, "GRAIN-SIZE EFFECTS ON DIELECTRIC PROPERTIES IN BARIUM-TITANATE CERAMICS," Journal of Applied Physics, vol. 47, no. 1, pp. 371-373, 1976 1976.

[3] L. Curecheriu, M. T. Buscaglia, V. Buscaglia, Z. Zhao, and L. Mitoseriu, "Grain size effect on the nonlinear dielectric properties of barium titanate ceramics," Applied Physics Letters, vol. 97, no. 24, Dec 13 2010, Art no. 242909.

[4] Y. Chen, H. Ye, X. Wang, Y. Li, and X. Yao, "Grain size effects on the electric and mechanical properties of submicro BaTiO$_3$ ceramics," Journal of the European Ceramic Society, vol. 40, no. 2, pp. 391-400, Feb 2020.

[5] X. H. Wang, R. Z. Chen, Z. L. Gui, and L. T. Li, "The grain size effect on dielectric properties of BaTiO$_3$ based ceramics," Materials Science and Engineering B-Solid State Materials for Advanced Technology, vol. 99, no. 1-3, pp. 199-202, May 25 2003.

[6] L. Zhang et al., "Enhanced dielectric properties of BaTiO3 based on ultrafine powders by two-step calcination," Physica B: Condensed Matter, vol. 560, pp. 155-161, 2019.

[7] C. Q. Zhu, Q. C. Zhao, Z. M. Cai, L. M. Guo, L. T. Li, and X. H. Wang, "High reliable non-reducible ultra-fine BaTiO$_3$-based ceramics fabricated via solid-state method," (in English), Journal of Alloys and Compounds, Article vol. 829, p. 9, Jul 2020, Art no. 154496.

[8] T. Hoshina, S. Hatta, H. Takeda, and T. Tsurumi, "Grain size effect on piezoelectric properties of BaTiO$_3$ ceramics," (in English), Japanese Journal of Applied Physics, Article vol. 57, no. 9, p. 5, Sep 2018, Art no. 0902bb.

[9] S. H. Yoon, M. Y. Kim, and D. Kim, "Influence of tetragonality (c/a) on dielectric nonlinearity and direct current (dc) bias characteristics of $(1-x)$BaTiO$_3$-xBi$_{0.5}$Na$_{0.5}$TiO$_3$ ceramics," (in English), Journal of Applied Physics, Article vol. 122, no. 15, p. 7, Oct 2017, Art no. 154103.

[10] L. E. Cross, "RELAXOR FERROELECTRICS," (in English), *Ferroelectrics,* Article vol. 76, no. 3-4, pp. 241-267, 1987.

[11] R. Waser and R. Hagenbeck, "Grain boundaries in dielectric and mixed conducting ceramics," Acta Materialia, vol. 48, no. 4, pp. 797-825, Feb 25 2000.

[12] T. Sada, K. Izawa, N. Fujikawa, and Y. Fujioka, "Mechanism of prebreakdown process in Ni-BaTiO$_3$ multilayer ceramic capacitors," Japanese Journal of Applied Physics, vol. 57, no. 11, Nov 2018, Art no. 11uc02.

A quick correction method for the board-level finite element analysis of QFP device under vibration

Zicheng Sa
State Key Laboratory of Advanced Welding and Joining
Harbin Institute of Technology
Harbin, P. R. China
sazc@hit.edu.cn

Shang Wang
State Key Laboratory of Advanced Welding and Joining
Harbin Institute of Technology
Harbin, P. R. China
wangshang@hit.edu.cn

Jiayun Feng
State Key Laboratory of Advanced Welding and Joining
Harbin Institute of Technology
Harbin, P. R. China
fengjy@hit.edu.cn

Guangliang Yu
Development Center of On-Board Computer and Electronics
Beijing Institute of Control Engineering
Beijing, P. R. China
ygl_222@126.com

Ning Zhang
Development Center of On-Board Computer and Electronics
Beijing Institute of Control Engineering
Beijing, P. R. China
94115170@qq.com

Yanhong Tian*
State Key Laboratory of Advanced Welding and Joining
Harbin Institute of Technology
Harbin, China
tianyh@hit.edu.cn

Abstract—**Electronic devices often work under a variety of vibration loads, especially for aerospace devices. For this reason, the reliability under vibration loading has become a critical problem of modern electronic devices. In recent years, the finite element analysis method develops rapidly. Many researchers conducted vibration analysis through simulation methods. However, due to the material and device structure difference, there is a significant error between the simulated results and the actual value. In this paper, we report a quick correction method, eliminating the errors caused by material properties and structure. We use this method to correct the board model of the quad flat package (QFP) device. The results show that the relative error was within 2% after correction. This method greatly improves the accuracy of the simulation results and provides a new route for the board-level vibration analysis of devices.**

Keywords—FEA, Vibration, QFP, Board-level

I. INTRODUCTION

Various vibration loads are received in the aerospace equipment during transportation and launch. And the vibration fatigue damage is the main failure form of the structure[1]. The structure and function of devices in aerospace are becoming more and more complex, and the diversification of the structure makes it more and more challenging to analyze the mechanical characteristics. The vibration test is usually employed to investigate the dynamic character and reliability of devices on the printed circuit board (PCB) [2]. However, it is not only extending the time of product design but also consuming a lot of samples.

Moreover, the deformation of devices during vibration is hard to be measured. In recent years, the finite element analysis (FEA) method develops rapidly [3-8], many researchers conducted vibration analysis with the assistance of simulation methods. The FEA technology is simply a process of discrete, separate solutions and combined results. Kim et al. [9] applied the finite element analyses to the solder joint reliability of plastic ball grid array packaging (PBGA) under random vibration loadings. Samavatian et al. [10] used FEA to determine the optimal system design for a longer fatigue lifetime. The stress distribution and dangerous position could be presented clearly by the FEA method.

However, the above research conducted a detailed analysis of vibration without the model correction. There are many

factors that affect the vibration state of the packages, such as mass, Young's modulus and damping coefficient. Hence, a significant error between the simulated results and the actual value will be produced with the uncorrected model. Unfortunately, there were little researches on model correction based on sinusoidal response data. The development of model correction methods based on sinusoidal vibration test data was demanded to improving the FEA in aerospace engineering practice.

II. MAIN METHOD

The damped structure is affected by mass matrices (M), stiffness matrices (K), and damping matrices (C). The dynamic equation of a multi-degree of damped freedom structure can be calculated by Eq.(1)

$$M\ddot{x}(t) + C\dot{x}(t) + Kx(t) = F(t) \qquad (1)$$

where $x(t)$ is the displacement of the node, and $F(t)$ is the stress. So this quick correction method contains three steps, mass correction, natural frequency correction, and structural damping correction.

A. Mass Correction

The mass of the model will directly determine the shape of the vibration mode and the distribution of natural frequencies. At the same time, the difference in mass will also affect the process of thermodynamic analysis. In this section, we will mainly adjust the density of the material to correct mass.

The initial density of each material is determined through literature review and experiments. And the actual weight of the device and the entire board is obtained by weighing. Then the density parameters of each material were finely tuned in the software. We can iterate to make the simulated value keep close to the actual value until the same.

B. Natural Frequency Correction

The frequency of the simple harmonic vibration can not be directly changed in the initial conditions. But it is related to the natural characteristics of the system, which is called the natural frequency. There are many influencing factors of natural frequency, such as mass, Young's modulus and model shape. The natural frequency with large Young's modulus is high in the same shape, and the frequency with large mass is low. Since the mass of the material has been correct in section

978-1-6654-1392-3/21 $31.00 © 2021 IEEE

A, the primary adjustment factor is the Young's modulus of the materials .

The process is similar to mass correction. Through testing the board, the first-order natural frequency of board can be obtained. Subsequently, we can iterate to make the first-order natural frequency obtained by the simulation keep close to the real value until to the same by changing the Young's modulus parameters of the material.

C. Structural Damping Correction

Finally, the structural damping correction which is also called the acceleration-displacement correction is performed. The structural response mainly depends on the damping characteristics under the dynamic load. Damping will reduce the resonance amplitude of the mechanical structure during vibration, thereby avoiding structural damage to a certain extant. During finite element analysis the theoretical amplitude of the model will reach infinity without damping, so it is necessary to set the correct damping coefficient. Therefore, the main correction coefficient is the damping. The classification of damping in ANSYS includes Rayleigh damping, structural damping, constant damping, modal damping and element damping. There will be some differences in the correction of different damping categories. Since there was only a simple interconnection structure in this model, structural damping was used for calculation. The specific correction process will be described in the following section.

III. FINITE ELEMENT ANALYSIS OF QFP

A. FEA Model

The full-size 3D finite element model was built as show in Fig. 1. The divided mesh shape was tetrahedral structure, and the element type was SOLID187 which could analyze the vibration structure. The solder element with the real shape at the corners was divided finest than the other parts. The materials in different part of the QFP package were marked with different color.

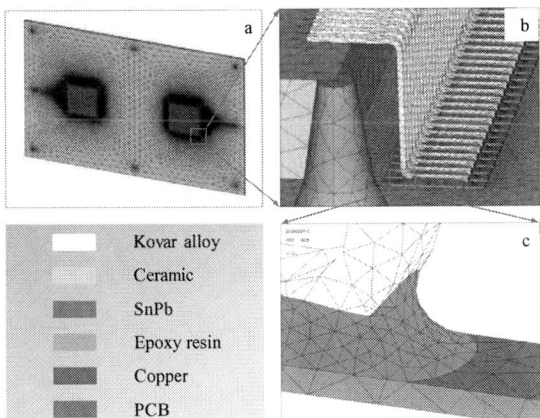

Fig. 1. Display of model and solder joint

The assumptions of the analysis were list as below. The environment temperature was set as 298K. The inner of solder was uniform without void, IMC, or crack et al. The interface of every connection was perfect.

B. Model Correction

The first step was the mass correction. The weight of the single device was 0.021 kg, the PCB board is 0.150 kg, and the whole board is 0.19245 kg. After adjusting the density parameters of various materials, the simulated value was consistent with the actual value. The result was shown in Fig. 2 (The unit in the figura was kg).

Device
SUMMATION OF ALL SELECTED VOLUMES
TOTAL VOLUME = 0.58423E-05
TOTAL MASS = 0.21000E-01
CENTER OF MASS: XC= 47.772   YC=-0.21693E-03   ZC= 0.33944E-02

PCB
SUMMATION OF ALL SELECTED VOLUMES
TOTAL VOLUME = 0.75792E-04
TOTAL MASS = 0.15000
CENTER OF MASS: XC= 55.000   YC= 0.0000   ZC=-0.10681E-02

Entire Board
SUMMATION OF ALL SELECTED VOLUMES
TOTAL VOLUME = 0.87721E-04
TOTAL MASS = 0.19245
CENTER OF MASS: XC= 30.529   YC=-0.26777E-03   ZC= 0.0000

Fig. 2. Model mass in simulation software

Next step was natural frequency correction. Because the vibration mode was mainly affected by the FR-4 board, the Young's modulus coefficient of the FR-4 board was adjusted according to the real frequency. The first-order modal frequency of the model was 220 Hz through experiment with the eight holes fixed. After adjusting the Young's modulus of the PCB, the first-order natural frequency obtained by the simulation was 220.53 Hz. The difference between the simulated result and the actual result was 0.24%, within the allowable range of error. The first four natural frequencies and mode shapes obtained by simulation are shown in Table I and Fig. 3. Since this experiment was tested by percussion method, only the first-order natural frequency could be obtained. It can also be observed from the Fig. 2 that the mode shape of the first-order natural frequency presents a tendency of fluctuations in the center of the plate. This phenomena provided a reference for the following damping correction test position.

TABLE I.   NATURAL FREQUENCY OF FIXED MODE

Order	1	2	3	4
Simulation Frequency	220.53 Hz	297.48 Hz	435.31 Hz	454.88 Hz
Actual Frequency	220 Hz	-	-	-

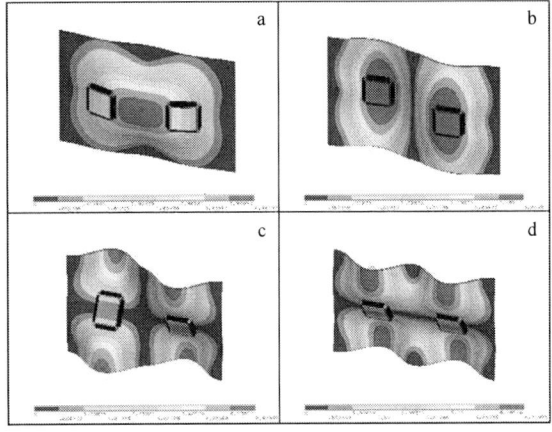

Fig. 3. Mode shapes of (a) 1st order; (b) 2nd order; (c) 3rd order; (d)4th order

After mass correction and natural frequency correction, the material parameters were determined. The final material parameters are shown in Table II.

TABLE II.   MATERIAL PROPERTIES

Materials	Properties			
	Young's Modulus (GPa)	Shear Modulus (GPa)	Poisson's ratio	Density (Kg·m^{-3})
Kovar alloy	138	-	0.32	8.20
Ceramic	270	-	0.27	3.54
SnPb	30.55	-	0.35	8.46
Epoxy resin	8	-	0.38	1.30
Copper	128.93	-	0.34	8.94
PCB	16.5 (X&Y)	7.55 (XY)	0.11 (XY)	1.98
	7.2 (Z)	3.25 (YZ&XZ)	0.39 (YZ&XZ)	

Then the damping correction was employed. As the simulation method is complete analysis, the relative damping apply method was chosen as the structural damping. According to actual experimental conditions, when a sinusoidal load with an acceleration of 0.5 g (this "g" is acceleration of gravity) was applied at the first-order vibration frequency, the acceleration at the middle position of the entire board was 47 g. So a first-order frequency (220.53 Hz) sinusoidal vibration load was applied with an acceleration of 0.5 g. The Final damping of the structure was 0.0173.

C. Verification of Method Suitability

The material parameters obtained from the calibration were verified after the above correction. To examine the accuracy of the model, a free modal analysis was performed on the board, which means no position constraints on the model. The simulation results were shown in Table III. Through experiments, it was found that the first-order natural frequency of the board in the free mode was 113.61 Hz, which was 1.74% away from the simulated. It was within the allowable error of 5%. The result proved that the quick correction method was reliable.

TABLE III.   NATURAL FREQUENCY OF FREE MODE

Order	1	2	3	4
Simulation Frequency	111.63 Hz	131.34 Hz	258.60 Hz	305.20 Hz
Actual Frequency	113.61 Hz	-	-	-

IV. CONCLUSIONS

In this paper, we introduced a quick correction method for the board-level finite element analysis. This method contains three steps, mass correction, natural frequency correction and structural damping correction. We used this method to correct the board model of the QFP device. After the correction process, the material parameters were determined.

Through the verification of the corrected material parameters, it could be found the relative error was within 2% after tuning by the model correction method. This quick correction method provided a basis for subsequent vibration simulation and random vibration reliability life prediction.

ACKNOWLEDGMENT

This work was supported by the National Key R&D Program of China (No. 2018YFB1307501), and Heilongjiang Touyan Team.

REFERENCES

[1] M. Ernst, E. Habtour, A. Dasgupta, et al, "Comparison of Electronic Component Durability Under Uniaxial and Multiaxial Random Vibrations," Journal of Electronic Packaging, vol. 1, pp. 0110091, March 2015.

[2] F. X. Che, and J. H. L. Pang, "Study on reliability of PQFP assembly with lead free solder joints under random vibration test," Microelectronics Reliability, vol.12, pp. 2769-2776, December 2015.

[3] F. X. Che, and J. H. L. Pang, "Vibration reliability test and finite element analysis for flip chip solder joints," Microelectronics Reliability, vol. 7, pp. 754-760, July 2009.

[4] A. Perkins, and S. K. Sitaraman, "Vibration-induced solder joint failure of a ceramic column grid array (CCGA) package," 54th Electronic Components & Technology Conference. Las Vegas, pp.1271-1278, February 2004.

[5] Y. Zhou, M. Al-Bassyiouni, and A. Dasgupta, "Harmonic and Random Vibration Durability of SAC305 and Sn37Pb Solder Alloys," Components and Packaging Technologies, vol. 2, pp.319-328, July 2010.

[6] F. Liu, and G. Meng, "Random vibration reliability of BGA lead-free solder joint," Microelectronics Reliability, vol. 1, pp.226-232, January 2014.

[7] H. Zhang, Y. Liu, J. Wang, et al, "Failure study of solder joints subjected to random vibration loading at different temperatures," Journal of Materials Science, vol. 4, pp. 2374-2379, April 2015.

[8] T. Lu, B. Zhou, K. Pan, et al, "Harmonic Vibration Analysis and S-N Curve hstimate of PBGA Mixed Solder Joints," ICEPT. Chengdu, pp. 778-782, August 2014.

[9] Y. K. Kim, and D. S. Hwang, "PBGA packaging reliability assessments under random vibrations for space applications," Microelectronics Reliability, vol. 1, pp. 172-179, January 2015.

[10] M. Samavatian, L. K. Ilyashenko, A. Surendar, et al, "Effects of system design on fatigue life of solder joints in BGA packages under vibration at random frequencies," Journal of Electronic Materials, vol. 11, pp. 6781-6790, November 2018.

Preparation and microwave properties of TiO₂/PTFE composites reinforced by mullite fibers

Zhangzhao Weng
China Electronic Product Reliability and Environmental Testing Research Institute
Guangzhou, China
347643043@qq.com

Jun Luo
China Electronic Product Reliability and Environmental Testing Research Institute
Guangzhou, China
kyea168@126.com

Hongfeng Lv
China Electronic Product Reliability and Environmental Testing Research Institute
Guangzhou, China
Chinatiger2012@sina.com

Shuai Zhou*
China Electronic Product Reliability and Environmental Testing Research Institute
Guangzhou, China
zs5h@163.con

Xiaoqiang Wang*
China Electronic Product Reliability and Environmental Testing Research Institute
Guangzhou, China
ps_800@126.com

Daojun Luo
China Electronic Product Reliability and Environmental Testing Research Institute
Guangzhou, China
luodj@ceprei.com

Abstract—In this paper, flexible TiO₂/PTFE composite materials with different mullite fiber content were prepared. The effects of mullite fiber addition on microwave dielectric properties, tensile properties and other physical properties of TiO₂/PTFE composites were studied. The experimental results show that the dielectric constant of TiO₂/PTFE=20/26 (wt%) composite material decreased from 8.2 to 4.82, and its dielectric loss was also significantly reduced, from 12.8×10^{-4} drops to 8.11×10^{-4}, when the content of mullite fiber increased from 0 wt% to 4.wt%. In addition, the tensile strength first rises and then decreases. Especially, when the mullite content is 2.wt%, the dielectric material obtains the strongest tensile strength (14.75Mpa). On the other hand, when the content of mullite fiber increased from 5.7wt% to 11.3wt%, the dielectric constant of TiO2/PTFE=20/33(wt%) composites slowly decreased from 6.18 to 5.63, and the dielectric loss increased from 7.62×10^{-4} increase to 9.33×10^{-4}, whereas the tensile strength decreases from 19.63 Mpa to 12.68 Mpa. Compared with the traditional introduction of glass fibers to improve the mechanical properties of microwave composites, TiO₂/PTFE composite reinforced by mullite fiber maintains relatively good microwave dielectric properties and tensile strength.

Keywords—Dielectric properties, Composites, PTFE

I. INTRODUCTION (HEADING 1)

In our previous work, TiO₂/HDPE composites have excellent microwave dielectric properties, whereas its high CTE limits the application of microelectronic substrate in a wide temperature range, which is mainly directly related to the high CTE of HDPE substrate[1]. In this work, PTFE with superior thermodynamic stability [2,3]is always selected as the base material, and rutile titanium dioxide[4] is used as the filler. In view of the poor mechanical properties of PTFE-based sintered composites, especially PTFE emulsion as starting material, the mullite fiber will select to enhance its mechanical properties the PTFE-based composite. Herein, the influence of different mullite fiber content on microwave dielectric properties and tensile strength of TiO₂/PTFE composites was studied.

II. PREPARATION OF TiO₂/PTFE MICROWAVE COMPOSITES

The ratio of TiO₂/PTFE is fixed at 20/26 mass ratio, and low content mullite fiber from 0 to 4.wt% is added. As the content of mullite fiber further increases, mixing becomes

more difficult. Therefore, it is necessary to increase the content of PTFE emulsion. The ratio of TiO₂/PTFE is fixed at 20/33 mass ratio, and a higher content (5.7~11.3wt%) of mullite fiber is added. The specific formulations of TiO2/PTFE microwave composite reinforced with different mullite fiber contents are shown in Table 1.

TABLE I.   THE COMPOSITIONS OF MULLITE FIBER REINFORCED TiO₂/PTFE COMPOSITES

TiO₂/PTFE=20/26（wt%）			
Sample	TPTFE-f0	TPTFE-f1	TPTFE-f2	TPTFE-f4
Mullite fiber(.wt%)	0	1	2	4
TiO₂/PTFE=20/33（wt%）			
Sample	TPTFE-f5.7	TPTFE-f7.5	TPTFE-f9.4	TPTFE-f11.3
Mullite fiber(.wt%)	5.7	7.5	9.4	11.3

A. Mixing

Mullite fibres with a length of 15 mm to 40 mm are accurately weighed and mixed in a PTFE ball mill tank containing a certain mass ratio of titanium dioxide powder and PTFE emulsion. Here, mechanical stirring can also be used.

B. Pressing

After ball milling and mixing at room temperature, the "dough" mixture with a certain degree of ductility is put into the mold for repeated compression. If necessary, it can also be heated to 80   °C to accelerate the volatilization of the low-melting point organic solvent in the PTFE emulsion .

C. Sintering

In order to develop a reasonable sintering regime for PTFE-based microwave composite media materials, the TG-DTA curve of the PTFE dispersion was tested. When the temperature is increased from room temperature to 100°C, the weight of the PTFE emulsion is rapidly reduced by up to 51% weight loss due to the large amount low-boiling solvent evaporation in the emulsion. When the temperature was increased from 100°C to 500°C, the weight loss of the sample was only about 4%. At around 180°C, there is an endothermic peak on the corresponding DSC curve. This may be because the low molecular weight PTFE monomer is cross-linked and cured, and the cross-coupling of small gels into larger molecules requires heat absorption. In addition, there is also a small endothermic peak near 330°C, which is the melting point of PTFE. The infrared spectra of PTFE gels and the ATR spectra at different temperatures were also compared in our previous work. The intensity of the reflection peaks of PTFE gels are weaker than those after heat treatment at 100°C, 150°C, 250°C and 350°C, which is due to the low crosslink density of the samples at low temperatures, mostly in the form of monomers.

The TiO$_2$/PTFE sintering system was finally determined as shown in Fig.1. The weight loss rate was as high as 51% within 100°C. Therefore, it was not easy to raise the temperature too fast in this temperature range, otherwise it would be easy to cause warping of the media material due to rapid solvent evaporation. In addition, when the temperature is cooled, it should be extended as long as possible to improve the crystallinity near the melting point of PTFE (about 320 °C).

$$\text{Room temperature} \xrightarrow{1°C/min} 380°C \xrightarrow{3h} 380°C \xrightarrow{1°C/min} 320°C \xrightarrow{1h} 320°C \xrightarrow{1°C/min} 200°C \xrightarrow{\text{furnace cooling}} \text{Room temperature}$$

Fig.1 Sintering process of mullite reinforced TiO$_2$/PTFE microwave composite.

III. THE PHASE, MICROMORPHOLOGY AND DENSITY ANALYSIS OF TiO$_2$/PTFE COMPOSITE

Fig.2 shows the XRD patterns of TiO$_2$/PTFE composite media materials with different contents of mullite fibres. Except for the diffraction peak near 18°, which is PTFE (PDF#54-1595), the other diffraction peaks are completely matched with the rutile TiO$_2$ (standard JCPDS card No. 21-1276). No diffraction peaks associated with mullite fibers are found here.

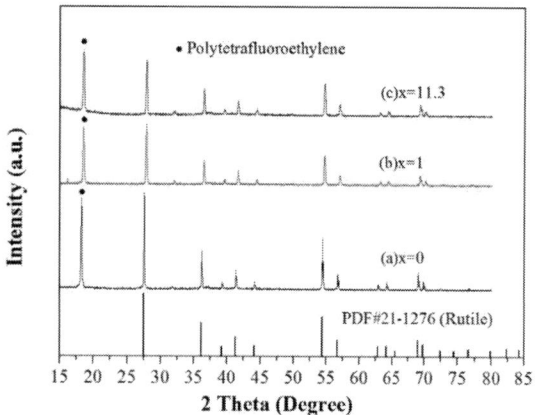

Fig.2 The XRD patterns of TiO$_2$/PTFE+x.wt% mullite fiber composite: (a) x=0, (b) x=1, (c) x=11.3.

The density curves of TiO$_2$/PTFE composite media materials reinforced with mullite fibres of different contents are shown in Fig.3. It can be seen from the figure that the density of TiO$_2$/PTFE composite media material decreases with the increase of mullite fibre content. The theoretical density are calculated according to the material compound formula (1).

$$\rho = \rho_{TiO_2} v_{TiO_2} + \rho_{PTFE} v_{PTFE} + \rho_{mullite} v_{mullite} \quad (1)$$

Where: v_{TiO2}, v_{PTFE}, and $v_{mullite}$ are the volume fractions of TiO$_2$ particles, polytetrafluoroethylene and mullite fibers, respectively. The theoretical densities of PTFE and TiO$_2$ are 2.305 g/cm^3 (PDF#51-1595) and 4.25 g/cm^3 (PDF#21-1276), respectively. The density of the mullite fiber of this formula (Al$_2$O$_3$/SiO$_2$=52/47 mt%) is unknown. As can be judged from the trend in Fig.3, the density of mullite fibre is lower than the theoretical density of the TiO$_2$/PTFE = 20/33 (wt%), and TiO$_2$/PTFE = 20/26 (wt%) composite. The theoretical densities of TiO$_2$/PTFE = 20/33 (wt%) and TiO$_2$/PTFE = 20/26 (wt%) composites can be calculated as 2.7862 g/cm^3 and 2.8776 g/cm^3 respectively according to the composite material composite equation (1). In fact, the density of the TiO$_2$/PTFE = 20/26 (wt%) composite is 2.64755 g/cm^3, which corresponds to 92% of the theoretical density, due to the rheological problems caused by PTFE having a high melt viscosity (10^{10} to 10^{12} Pa•s) [6], resulting in obtaining void-free PTFE composites is very difficult. The density of the TiO$_2$/PTFE=20/26 (wt%) composite media material decreases faster as the mullite fibre content increases, related to the fact that it is difficult to disperse the mullite fibres uniformly using the ball milling process, resulting in increased voids. In the process of ball milling and mixing, the PTFE dispersion is both a mixture and a ball-milling solvent. Solvent volatilization and polymerization reaction occur in the PTFE dispersion, which causes the mixture to become thicker and more difficult to uniformly disperse fibers. The more solvent, that is, the more PTFE content formula is more conducive to the uniform dispersion of mullite fiber.

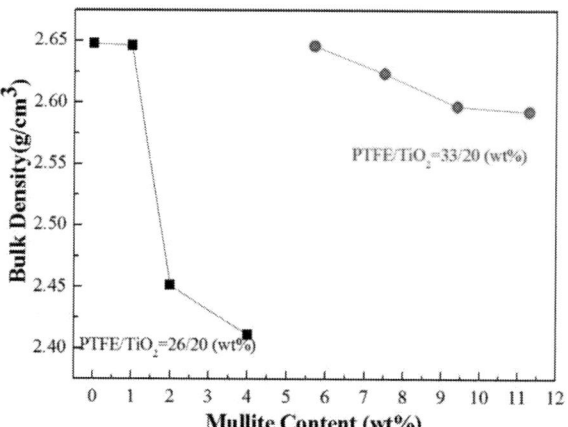

Fig.3 Density of TiO$_2$/PTFE composite with different mullite contents

The SEM image of the mullite fiber reinforced TiO$_2$/PTFE composite is shown in Fig.4. As can be seen from Fig.4, the white dispersed phase is TiO$_2$ particles with a diameter of about 2 μm~3 μm, and the darker continuous phase is PTFE, where the TiO$_2$ particles are relatively uniformly dispersed in the PTFE matrix. As the mullite fibre content increases, it is also easier to observe the mullite fibres, which can be judged to be relatively homogeneously dispersed in the PTFE matrix. Fig.5 shows the elemental distribution of the TiO$_2$/PTFE composite media material (i.e. sample TPTFE-f11.3) at a mullite fibre content of 11.3 wt%, where the mullite fibre and TiO$_2$ powder distribution can be clearly determined. Due to the low surface energy of PTFE, it is difficult for PTFE to have good adhesion with TiO$_2$ filler. As a result, no ceramic particles were observed to adhere to PTFE, indicating a weak interfacial strength between the ceramic filler and the PTFE matrix.

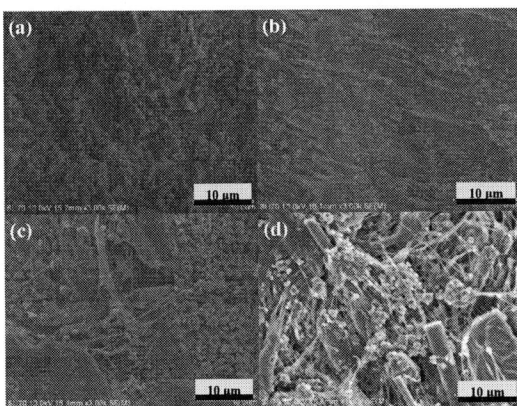

Fig.4 SEM image of TiO$_2$/PTFE+x.wt% mullite fiber composite:(a) x=0, (b) x=1, (c) x=4, (d) x=11.3.

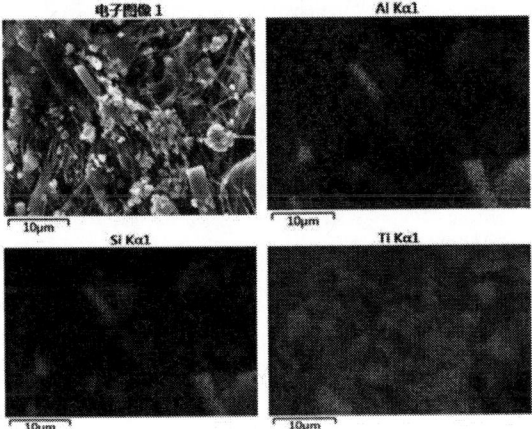

Fig.5 Elemental distribution of TiO$_2$/PTFE composites with mullite fiber content of 11.3 wt%.

IV. MICROWAVE DIELECTRIC PROPERTIES OF TiO$_2$/PTFE COMPOSITES

Fig.6 shows the plots of the dielectric constant and loss angle tangent curves at 1.8 GHz for TiO$_2$/PTFE composites with different contents of mullite fibres. The relationship between the permittivity of the TiO$_2$/PTFE composites and the mullite fibre content is consistent with the density of the mullite fibre content reinforced TiO$_2$/PTFE composites shown in Fig.3. As the mullite fibre content increased from 0 wt% to 4.wt%, the dielectric constant of the TiO$_2$/PTFE = 20/26 (wt%) composite decreased from 8.20 to 4.82, and its dielectric loss also decreased significantly, from 12.8 ×10^{-4} to 8.11×10^{-4}. With the increase of mullite fiber content from 5.7wt% to 11.3wt%, the dielectric constant of TiO$_2$/PTFE=20/33(wt%) composite dielectric material slowly decreases from 6.18 to 5.63. Although more PTFE dispersants are beneficial to the uniform dispersion of mullite fibers, the dielectric loss still slowly increases from 7.62×10^{-4} to 9.33×10^{-4}. Compared with conventional microwave composite dielectric materials with the introduction of glass fibres to achieve improved mechanical properties [5], mullite fibre reinforced TiO$_2$/PTFE composites showed lower dielectric losses, maintaining better microwave dielectric properties, probably related to the lower dielectric losses of mullite fibres compared to the traditionally used E-glass fibres, although the specific microwave dielectric properties of this type of mullite fibre are still not known. At present, there are few literature

reports concerning the dielectric properties of mullite. In conjunction with the curve in Fig.6 and the relationship between the dielectric constants of the composites, it can be judged that the dielectric constant of mullite is below 5.65. Zhang Fu-kuan et al [7] used mullite ceramics as an impedance transformation layer for wave absorbing materials to improve the absorbing properties of the materials. When the dense density of mullite ceramics increased from 91.45% to 97.00%, the real part of its complex dielectric constant increased from 3.85 to 5.87 and the imaginary part increased from 0.04 to 0.11. Numerous factors influence the loss factor of the composite, in addition to the uncertainty of the dielectric properties of the mullite fibres caused by the different mullite compositions, it is also related to the shape, size and dispersion of the fillers (including rutile TiO$_2$ and mullite fibres) and the interfacial bonding properties with the PTFE matrix. Herein, several types of commonly used models are not used to predict the dielectric constant of the composite medium, is related to the fact that the introduction of mullite fibres affects the homogeneous mixing effect of the ball milling process, in addition to the fact that all models are based on very ideal assumptions.

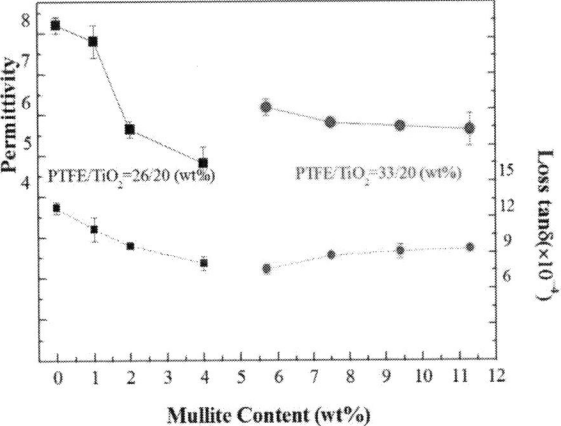

Fig. 6 Variation of dielectric constant and loss tangent at 1.8 GHz of TiO$_2$/PTFE composites with different contents of mullite fiber.

V. TENSILE STRENGTH OF TiO$_2$/PTFE COMPOSITE

Fig.7 shows the variation of the tensile strength of the TiO$_2$/PTFE composite with the mullite fibre content. With the increase of mullite fiber content from 0wt% to 4.wt%, the tensile strength of TiO$_2$/PTFE=20/33(wt%) composite dielectric material showed a tendency to increase first and then decrease. The strongest tensile strength (14.75Mpa) of the composites material is obtained when the mullite content is 2.wt%, mainly related to the effect of the fibre reinforcement. In addition to being affected by the nature of the material itself, the morphology, dispersion, content, combination with the interface of the filler, and the mechanical properties of the reinforcing material (such as fiber) will also affect the tensile strength of the composite. The tensile strength of the TiO$_2$/PTFE=20/33 (wt%) composite material tends to decrease as the mullite fibre content increases from 5.7wt% to 11.3wt%, with the tensile strength decreasing from 19.63 Mpa to 12.68 Mpa. It is mainly related to the difficulty of uniform mixing caused by excessive introduction of mullite fibers, particles and mullite fibers are prone to aggregation and uneven distribution, thereby resulting in the reduction of tensile strength. Comparing the tensile strength of TiO$_2$/PTFE composites reported in the literature [5], the tensile strength of unmodified TiO_2/PTFE was less than 11 MPa and the tensile strength of modified TiO_2/PTFE was less than 12.7 MPa when the rutile content was 40.wt%. In this work, the tensile strength of TiO_2/PTFE composites were significantly improved by using mullite fibers.

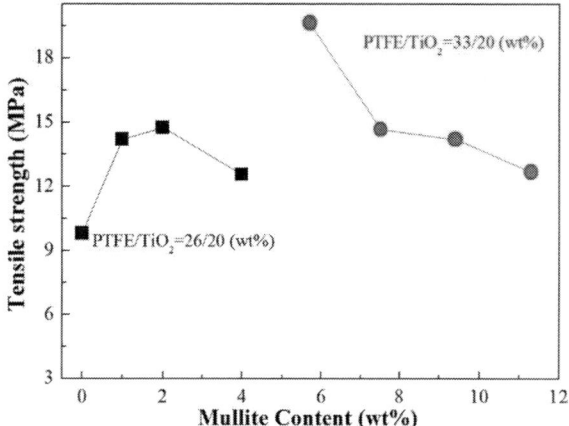

Fig.7 Variation of tensile Strength for TiO_2/PTFE Composite with Mullite Fiber Content

VI. CONCLUSION

In this work, flexible TiO_2/PTFE composite with different mullite fiber content were prepared. The effects of mullite fiber addition on microwave dielectric properties, tensile properties of TiO_2/PTFE composites was investigated. The experimental results show that the density of the TiO_2/PTFE composite decreases with the increase of the mullite fiber content. The mechanical properties of TiO_2/PTFE composite were obvious enhanced by the mullite fibers. When the content of mullite fiber increased from 0 wt% to 4.wt%, the dielectric constant of TiO_2/PTFE=20/26 (wt%) composite decreased from 8.2 to 4.82, its dielectric loss was also reduced significantly, from 12.8×10^{-4} to 8.11×10^{-4} and the tensile strength first rises and then decreases. When the mullite content is 2.wt%, the dielectric material obtains the strongest tensile strength (14.75Mpa). When the content of mullite fiber increased from 5.7wt% to 11.3wt%, the dielectric constant of TiO_2/PTFE=20/33(wt%) composite dielectric material slowly decreased from 6.18 to 5.63, the dielectric loss decreased from 7.62×10^{-4} Slowly increase to 9.33×10^{-4}, and the tensile strength decreased from 19.63 Mpa to 12.68 Mpa. Compared with the traditional introduction of glass fibers to improve the mechanical properties of microwave composites, mullite fiber reinforced TiO_2/PTFE composite dielectric materials maintain relatively good microwave dielectric properties and tensile strength.

ACKNOWLEDGMENT

The authors would like to thank the Key-Area Research and Development Program of Guangdong Province (No. 2019B010145001) and Guangzhou Science and Technology Plan Project (No.201904010457) for supporting this research.

REFERENCES

[1] Lin H, et al. Preparation and Microwave Dielectric Properties of Polyethylene/TiO2 Composites[J]. Journal of Electronic Materials, 2019, 48(10):6771-6776.

[2] Feng X, et al. Preparation and dielectric properties of bismuth-based dielectric/PTFE microwave composites[J]. Journal of the European Ceramic Society, 2006.

[3] Yuan Y, et al. TiO2 and SiO2 filled PTFE composites for microwave substrate applications [J]. Journal of Polymer Research, 2014, 21(2): 380-7.

[4] Weng Z, et al. Low temperature sintering and microwave dielectric properties of TiO2 ceramics[J]. Journal of the European Ceramic Society, 2017,37(15):4667-4672.

[5] RAJESH S, et al. Rutile filled PTFE composites for flexible microwave substrate applications [J]. Materials Science & Engineering B, 2009, 163(1): 1-7.

[6] ARIAWAN A B, et al. Properties of polytetrafluoroethylene (PTFE) paste extrudates [J]. Polymer Engineering & Science, 2002, 42(6): 1247-59.

A Segmented Plasma Etching Method for 2.5D/3D Through Silicon Vias

Yuanwei Lin
Department of Semicondutor Etching
NAURA Technology Group Co., Ltd.
Beijing, P. R. China
Email address: yuanweilin@pku.edu.cn
ORCID: 0000-0002-5293-1777

Abstract—**Advanced packaging is the growth engine of the semiconductor industry, and 2.5D/3D packaging is arising to maintain the Moore's Law. Through silicon vias (TSVs) are commonly used structures in 2.5D/3D packaging, which are fabricated by plasma etching. In this work, we demonstrated a segmented plasma etching method, and the issues in 2.5D/3D TSVs (such as, sidewall roughness and notch) were addressed by using this modified Bosch process. Thus, this method is of significance to improve the etch profiles in discrete devices, microelectromechanical systems (MEMS) and advanced packaging.**

Keywords—*Plasma etching, Bosch process, Through Silicon Vias, Segmented etching, Advanced packaging*

I. INTRODUCTION

The development of modern micro-electronics industry has faced a slowing down in 2020, especially for the tough time under COVID-19 pandemic. However, as a branch of the micro-electronics industry, the advanced packaging, which integrates/strengthens electronic devices, has been witnessing an accelerated development. According to the Moore's law, the minimum line width of the chip is shrinking. But beyond a certain size, the quantum effect will appear, which changes the physical laws that the chip currently follows. This size reduction is mainly in the planar area. For advanced packaging, if the three-dimensional space can be used, the integration of the chip will be increased, which can be regarded as a continuation of the Moore's Law. The use of three-dimensional space for stack packaging is called 3D integration, which has higher capacity, better performance, and higher yield. Through silicon via (TSV) is commonly used to perform the interconnection between layers in 3D integration. In addition to 3D, 2.5D uses an interposer to achieve the system in a package (SiP). Due to the additional interposer used, the critical dimension (CD) of the 2.5D TSV is larger than that of the 3D TSV, and the depth of the 2.5D TSV is also larger than that of the 3D TSV.

To obtain TSV structures, plasma etching is used, where the ionized gas can etch the wafer by physical bombardment and chemical reaction [1-3]. For deep silicon etching, it becomes more difficult as the etching depth increases, because a) the reactive plasma or etchant is hard to enter the deep microstructure, inhibiting the etching at the bottom of the microstructure; b) the product of the etching formed at the bottom is also difficult to be carried away from the reaction system, suppressing etching according to chemical equilibrium [4]. Thus, to obtain deep silicon microstructures with high aspect ratio and high verticality, the ramping technique [4] is used as following:

$$P = P_{initial} + (P_{final} - P_{initial})*n/n_{total} \qquad (1)$$

where n, n_{total}, P, $P_{initial}$, and P_{final} are cycle number at present, total cycle number, any parameter in Bosch process, initial parameter and final parameter, respectively.

In this study, a segmented plasma etching method has been proposed. Using this segmented Bosch process, the sidewall roughness at the top of 2.5D TSV and the notch issue at the bottom of 3D TSV are optimized. This modified Bosch process is of significance to obtain high quality silicon microstructures in discrete devices, microelectromechanical systems (MEMS) and advanced packaging.

II. PROBLEM DESCRIPTION

For 2.5D TSV, due to the relative larger CD than that of the 3D TSV, which is convenient for the etchant entrance/product exit, the sidewall is rough at the top (the scallop size will be large when etch effect is strong [7]). If the aspect ratio further increases, the undercut forms [8]. After plasma etching, the sidewall roughness introduced by the scallop structure is harmful for the following process, such as the oxide layer growth and the TiN barrier layer deposition. As shown in Fig.1, the scallop size at the TSV top is about 64.5 nm, using the traditional Bosch process without segmentation.

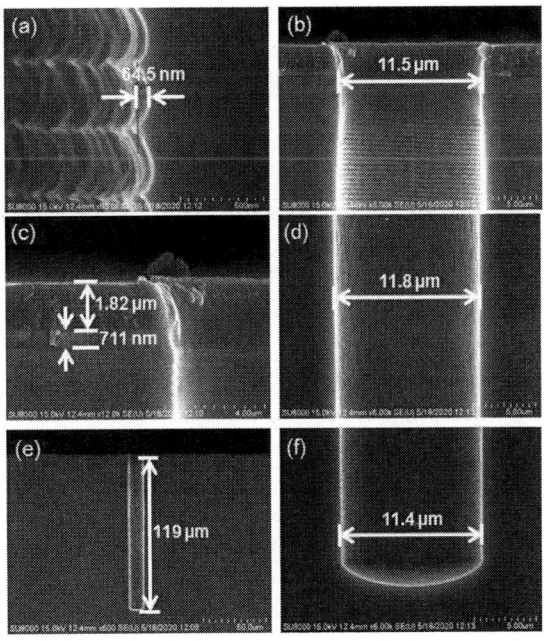

Fig. 1. (a) Cross-sectional SEM image of the wafer before etching. Scale bar: 1 μm. (b) Schamatic of the cross-sectional view of the deep silicon trench.

978-1-6654-1392-3/21 $31.00 © 2021 IEEE

For 3D TSV, due to the narrow CD and relative high aspect ratio, it is easy to form the notch at the bottom of the TSV structure if a etch stop layer is used. Duty cycle is found to be an effective method to eliminate the notch [9], but it is invalid for the microstructures with high aspect ratio and narrow CD. This is because the etchant is hard to exit the high aspect ratio microstructure, and the duty cycle, which decreases the ion energy [10], can solely address the charge accumulation issue. As shown in Fig. 2, the notch is as large as 275 nm, using the traditional Bosch process without segmentation. For large notch size, the oxide layer cannot be grown in the notch structure, increasing the leakage current of the device.

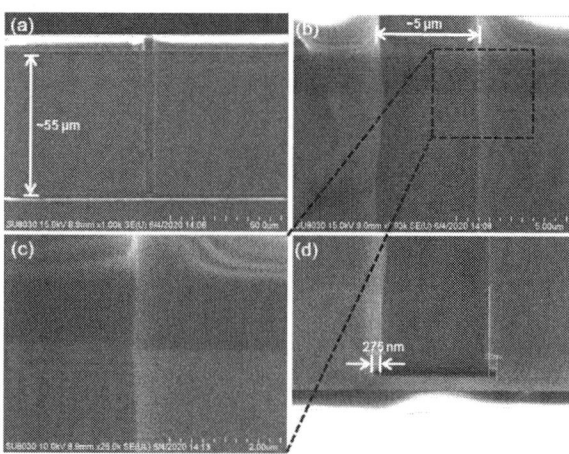

Fig. 2. Cross-sectional SEM images of the 3D TSV. (a) The entire view of the TSV; (b) The enlarged view of the TSV top; (c) The enlarged view of the one side of the TSV top showing no undercut; (d) SEM image showing the notch size at the TSV bottom.

III. EXPERIMENTAL SECTION

Commercially available 12 inch silicon wafer with (100) face upward was used in this study. SiO_2 hard mask (HM) with thickness of ~0.7 μm was grown on the silicon wafer by PECVD using a NMC EPEE 550. The HM was opened by using Ar and CF_4 plasma in a GDE chamber of a NMC MASE P230 etcher after defined by photo-lithographically, and the thickness of the remaining photoresist (PR) is ~4 μm. The deep dry silicon etching was then performed in a HSE chamber of a NMC MASE P230 etcher. The scanning electron microscope (SEM) images were taken by using Hitachi SU8000.

IV. RESULTS AND DISCUSSION

To optimize the roughness of the sidewall (reduce the scallop size), minimize the deposition step time and etch step time simultaneously is an effective way [7]. For the problem described in this work, the scallop size is large solely at the top of the TSV structure, so a segmented plasma etching can be considered. We can re-write the formula (1) as follows:

$$P_{hikl} = P_{hikl\ initial} + (P_{hikl\ final} - P_{hikl\ initial}) * n/n_{total\ h} \qquad (2)$$

where h, i, k and l mean the segmentation number of the Bosch process, the parameter type with ramping (such as bias power and step time), the step type (deposition or etch step), the order of the deposition/etch step, respectively.

The current Bosch process is a special case where $h=1$. Then, an algorithm is proposed to make the segmented etching smoothly. First, the conversion factor α_{hm} (m= initial or final) is defined to convert the step time. For instance, if the etch step time requires reduction, α_{hm} is less than 1, and vice versa. The total cycle number should be $n_{total\ h}/(\Pi\ \alpha_{hm})^{0.5}$ to make the total etch time unchanged. Second, the correction factor β_{hm} is defined to refine the total time, because additional step time is introduced when the process is switched between deposition step and etch step. For simplification, β_{hm} defaults to be 1 in this work. Third, the segmentation factor γ_h is defined to be the proportion of the segmentation h. For instance, if the etch process is divied into two segmentation evenly, then $\gamma_1=\gamma_2=0.5$. Obviously, the sum of the proportion of each segmentation equals 1 ($\Sigma\gamma_h=1$).

To make the algorithm easy to be understood, the deposition/etch step time t_{hklm} is extracted from P_{hiklm} separately. For the first segmentation, $t_{1kl\ initial} = t_{0kl\ initial} * \alpha_{1\ initial}$. Because the parameter of previous segmentation should be as the same as the initial parameter of the latter segmentation, $t_{(h+1)kl\ initial} = t_{hkl\ final}$. For the last segmentation, $t_{hkl\ final} = t_{0kl\ initial} + (t_{0kl\ final} - t_{0kl\ initial}) * \Sigma\gamma_h * \alpha_{h\ final}$. Generally, $t_{hkl} = t_{hkl\ initial} + (t_{hkl\ final} - t_{hkl\ initial}) * n/ (n_{total\ 0} * \gamma_h/\Pi_m (\alpha_{hm} * \beta_{hm})^{0.5})$, and so are the other parameters P_{hiklm}. Thus, the whole algorithm for the segmented Bosch process can be expressed in Fig. 3.

The segmentation can also be performed for multiple times, where the segmentation factors are multiplied, as shown in Fig. 4.

978-1-6654-1392-3/21 $31.00 © 2021 IEEE

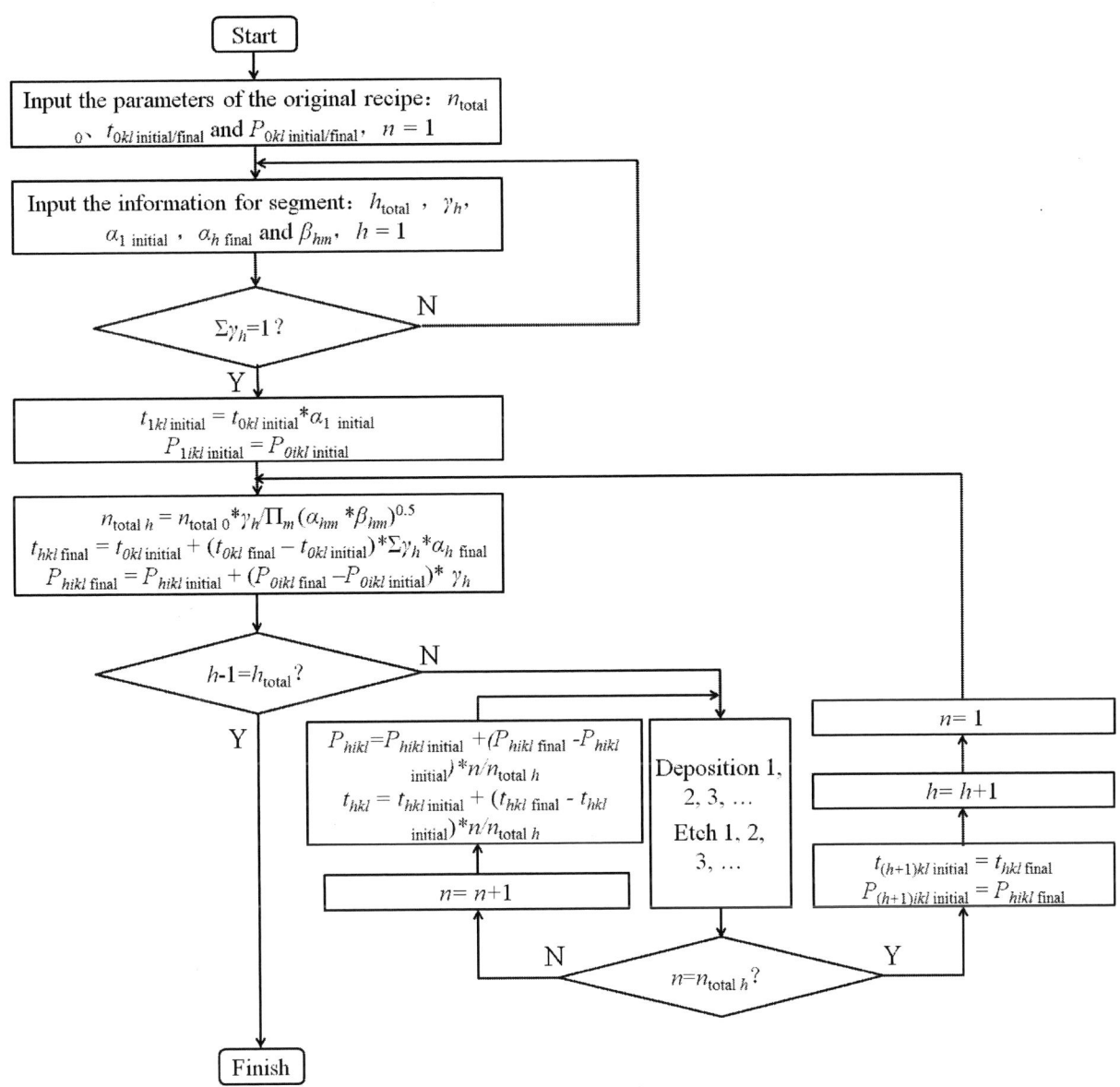

Fig. 3. Flow chart of the segmented Bosch process.

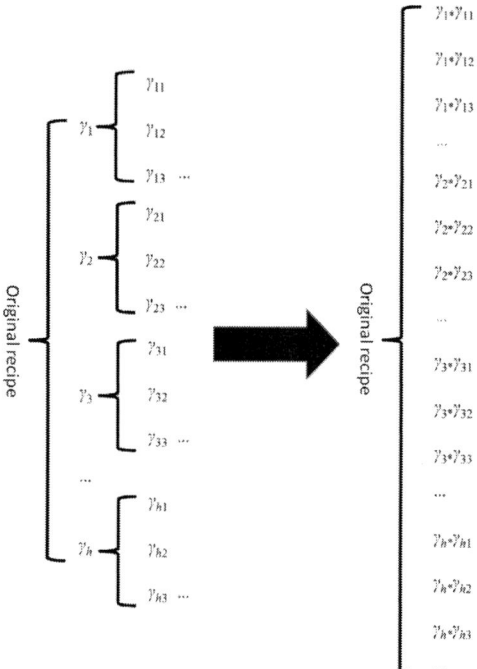

Fig. 4. The segmentation factor for multi-segment.

Using the segmented Bosch process, the sidewall roughness at the top of the TSV structure can be optimized. As shown in Fig. 5, the scallop size is reduced form 64.5 nm to 32.0 nm.

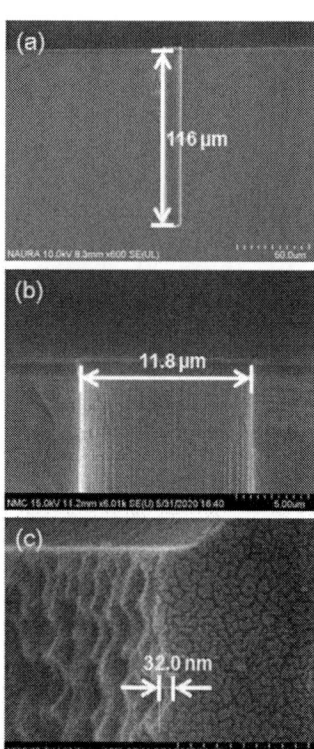

Fig. 5. Cross-sectional SEM images of the 2.5D TSV. (a) The entire view of the TSV showing the etch depth; (b) The enlarged view of the TSV top; (c) SEM image showing the scallop size at the TSV top.

The segmented Bosch process recipe used for the 2.5D TSV optimization is shown in TABLE I.

TABLE I. SEGMENTED BOSCH RECIPE FOR 2.5D TSV

Step	Recipe (total cycle number: 100)					
	Chamber pressure	Source power	C_4F_8 flow	SF_6 flow	Bias power	Time
Deposition-1	40 mTorr	2200 W	150 sccm	0	1 W	1~2 s[a]
Etching-1	50 mTorr	3000 W	0	200 sccm	75 W ~ 70 W[b]	1.2 s ~ 2.5 s[c]
Step	Recipe (total cycle number: 200)					
	Chamber pressure	Source power	C_4F_8 flow	SF_6 flow	Bias power	Time
Deposition-2	40 mTorr	2200 W	150 sccm	0	1 W	2 s
Etching-2	50 mTorr	3000 W	0	200 sccm	70 W ~ 65 W[b]	2.5 s ~ 2 s[c]

[a.] Ramping of deposition time.

[b.] Ramping of bias power.

[c.] Ramping of etching time.

The original recipe (before segmentation) of TABLE I is shown in TABLE II. Compared with the original recipe, the plasma etching process is divided into two sections, and the segmentation parameters have the values of $\alpha_{1\ initial}=0.5$, $\alpha_{1\ final}=1$, $\alpha_{2\ final}=1$, $\beta_1=\beta_2=1$, $\gamma_1=0.27$, $\gamma_2=0.73$, $h_{total}=2$. $\alpha_{1\ initial}<1$ makes the depostion and etch time at the beginning are both less than that of the original recipe, which is benificial to the reduction of the scallop size.

TABLE II. ORINGINAL BOSCH RECIPE FOR 2.5D TSV

Step	Recipe (total cycle number: 275)					
	Chamber pressure	Source power	C_4F_8 flow	SF_6 flow	Bias power	Time
Deposition	40 mTorr	2200 W	150 sccm	0	1 W	2 s
Etching	50 mTorr	3000 W	0	200 sccm	75 W ~ 65 W[a]	2.5 s ~ 2 s[b]

[a.] Ramping of bias power.

[b.] Ramping of etching time.

Similarly, using the segmented Bosch process, the notch issue at the bottom of the TSV structure can be optimized. As shown in Fig. 5, the notch size is reduced form 275 nm to 67.0 nm. This notch size is not harmful to the following oxide layer growth.

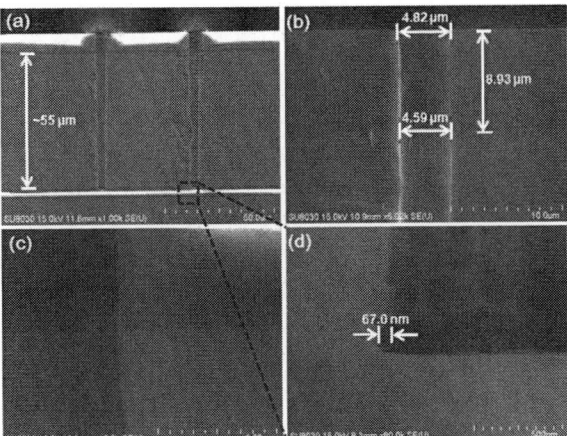

Fig. 6. Cross-sectional SEM images of the 3D TSV. (a) The entire view of the TSV; (b) The enlarged view of the TSV top; (c) The enlarged view of the one side of the TSV top showing no undercut; (d) SEM image showing the notch size at the TSV bottom.

The segmented Bosch process recipe used for the 3D TSV optimization is shown in TABLE III.

TABLE III. SEGMENTED BOSCH RECIPE FOR 3D TSV

Step	Recipe (total cycle number: 100, duty cycle: 20%)					
	Chamber pressure	Source power	C_4F_8 flow	SF_6 flow	Bias power	Time
Deposition1-1	50 mTorr	2200 W	250~ 270 sccm[a]	0	1 W	1.0~ 1.2 s[b]
Etching1-1	70 mTorr	2800 W	0	300 sccm	350 W	0.4 s ~ 0.6 s[c]
Etching1-2	70 mTorr	2800 W	0	300 sccm	180 W	0.6 s ~ 1.1 s[c]
Step	Recipe (total cycle number: 200, duty cycle: 20%)					
	Chamber pressure	Source power	C_4F_8 flow	SF_6 flow	Bias power	Time
Deposition2-1	50 mTorr	2200 W	270~ 300 sccm[a]	0	1 W	1.2~ 0.7 s[b]
Etching2-1	70 mTorr	2800 W	0	300 sccm	350 W	0.6 s ~ 0.4 s[c]
Etching2-2	70 mTorr	2800 W	0	300 sccm	180 W	1.1 s ~ 0.6 s[c]

[a.] Ramping of deposition flow.

[b.] Ramping of bias power.

[c.] Ramping of etching time.

The original recipe (before segmentation) of TABLE I is shown in TABLE IV. Compared with the original recipe, the plasma etching process is also divided into two sections, and the segmentation parameters have the values of $\alpha_{1\ initial}=1$, $\alpha_{1\ final}=1$, $\alpha_{2\ final}=0.5$, $\beta_1=\beta_2=1$, $\gamma_1=0.9$, $\gamma_2=0.1$, $h_{total}=2$. $\alpha_{2\ final}<1$ makes the deposition and etch time at the end are both less than that of the original recipe, which avoids the accumulation of the charge and the etchant at the bottom of the TSV structure. This is benificial to the reduction of the notch size

TABLE IV. ORINGINAL BOSCH RECIPE FOR 3D TSV

Step	Recipe (total cycle number: 275, duty cycle: 20%)					
	Chamber pressure	Source power	C_4F_8 flow	SF_6 flow	Bias power	Time
Deposition1-1	50 mTorr	2200 W	250~ 300 sccm[a]	0	1 W	1.0~ 1.3 s[b]
Etching1-1	70 mTorr	2800 W	0	300 sccm	350 W	0.4 s ~ 0.7 s[c]
Etching1-2	70 mTorr	2800 W	0	300 sccm	180 W	2.5 s ~ 2 s[c]

[a.] Ramping of deposition flow.

[b.] Ramping of deposition time.

[c.] Ramping of etching time.

Thus, the segmented plasma etching method is useful for both 2.5D and 3D TSV in advanced packaging. For the segmented plasma etching method used in this work, the final parameter of the previous stage of the recipe is as the same as the initial parameter of the latter stage of the recipe, which minimizes the occurrence of abnormalities in etch profile as much as possible.

V. CONCLUSION

In summary, through a segmented plasma etching method, we demonstrated the sidewall roughness at the top of 2.5D TSV and the notch issue at the bottom of 3D TSV can be optimized. This modified Bosch process is of significance to obtain high quality silicon microstructures in discrete devices, MEMS and advanced packaging.

ACKNOWLEDGMENT

The author thanks Mr. Jie Chen from 58th Institute of China Electronic Technology Group Co., Ltd. for beneficial discussions on this work.

REFERENCES

[1] F. Lärmer, and A. Schilp, Patents DE 4241045, US 5501893 and EP 625285, 1996.

[2] Y. Lin, R. Yuan, X. Zhang, Z. Chen, H. Zhang, Z. Su, S. Guo, X. Wang, and C. Wang, "Deep dry etching of silicon with scallop size uniformly larger than 300 nm," *Silicon*, vol. 11, pp. 651-658, 2019.

[3] Z. Dong, and Y. Lin, "Ultra-thin wafer technology and applications: a review," *Mat. Sci. Semicon. Proc.*, vol. 105, pp. 104681, 2020.

[4] Y. Lin, "Towards microstructures with ultrahigh aspect-ratio and verticality in deep silicon etching," *2020 China Semiconductor Technology International Conference*, DOI: 10.1109/CSTIC49141.2020.9282553, July 2020.

[5] Y. Zhao, and Y. Lin, "Estimating the etching depth limit in deep silicon etching," *2019 China Semiconductor Technology International Conference*, DOI: 10.1109/CSTIC.2019.8755766, March 2019.

[6] Y. Tang, A. Sandoughsaz, K. J. Owen, and K. Najafi, "Ultra deep reactive ion etching of high aspect-ratio and thick silicon using a ramped-parameter process," *J. Microelectromech. S.*, vol. 27, pp. 686-697, 2018.

[7] Y. Lin, R. Yuan, C. Zhou, Z. Dong, Z. Su, H. Zhang, Z. Chen, Y. Li, and C. Wang, "The application of the scallop nanostructure in deep silicon etching," *Nanotechnology.*, vol. 31, pp. 315301, 2020.

[8] I. Saraf, M. Goeckner, Brian Goodlin, Karen Kirmse, and L. Overzet, "Mask undercut in deep silicon etch," *Appl. Phys. Lett.*, vol. 98, pp. 161502, 2011.

[9] N. Hershkowitz, and M. K. Harper "Elimination of notching and ARDE by simultaneous modulation of source and wafer RF," *1999 IEEE International Conference on Plasma Science*, DOI: 10.1109/PLASMA.1999.829348, June 1999.

[10] D. C. Kwon, M. Y. Song, and J. S. Yoon, "Numerical investigation of RF pulsing effet on ion energy and angular distributions," *2013 IEEE International Conference on Plasma Science*, DOI: 10.1109/PLASMA.2013.6633364, June 2013.

Study on Thermal cycle and Insulation Characteristics of High Reliable IGBT Module

Tao Chen
Beijing Spacecraft
China Academy of Space Technology
Beijing, China
chentaoshe@163.com

Minghua Zhang
Beijing Spacecraft
China Academy of Space Technology
Beijing, China
75254847@qq.com

Binbin Zhang
Beijing Spacecraf
China Academy of Space Technology
Beijing, China
zhang0001234@163.com

Qing Chen
Beijing Spacecraft
China Academy of Space Technology
Beijing, China
2822508@sina.com

Kun Tian
Beijing Spacecraft
China Academy of Space Technology
Beijing, China
akun_126@qq.com

Yan Li
Beijing Spacecraft
China Academy of Space Technology
Beijing, China
393871627@qq.com

Abstract—With the increasing complexity of the new generation of high-quality weapon system, more and more IGBT modules are used under the condition of high voltage and large current, and the reliability of IGBT directly affects the stability of equipment system. IGBT module is composed of a variety of packaging materials, the packaging material, packaging process and packaging structure of IGBT are the key factors that affect its service life and reliability. Thermal cycle test and dielectric voltage resistance test were carried out to compare the reliability difference of the four IGBT modules which encapsulated different ceramic substrates. The results show that the thermal cycle life and insulation performance of Si3N4 ceramic substrate are better than those of other ceramic substrates withstand voltage test and appearance are passing after 500 thermal cycle. The geometry parameter and materials of DBC ceramic substrate are the key parameters which affect the reliability. Si_3N_4 ceramic substrate has the best performance, and Si_3N_4 is best used as substrate material for high reliability IGBT modules.

Keywords—IGBT, thermal cycle, insulation

I. INTRODUCTION

IGBT is a new type of power semiconductor device, which integrates the advantages of BJT and MOSFET. It has the advantages of high voltage, large current, large input impedance, small driving power, fast on-off speed and so on. It is widely used in rail transit, smart grid, aerospace, ocean engineering, weaponry and other industries. IGBT is the core component of power electronic device, and its reliability determines the lifetime of the whole device to a great extent [1,2]. With the increase of the working voltage and current of the IGBT module and the decrease of the chip size, the chip power density increases sharply. Heat dissipation and reliability are key issues that must be addressed. Ceramic substrate is the most widely used key material for IGBT module. It has excellent thermal conductivity, heat resistance, insulation and low coefficient of expansion. It is suitable for aluminum wire bonding. Ceramic copper clad substrate is composed of metal circuit layer and ceramic layer. Due to the large thermal expansion difference between ceramic and metal, the thermal stress generated during the service process will cause the substrate cracking and failure [3]. Cracks usually occur in the stress concentration or high strain area of the material. After experiencing enough cycles, the crack initiates in the stress concentration or high strain area of the material. Under the further action of the cyclic load, the crack expands until the material is completely fractured.

Therefore, it is of great significance to study the thermal cycling reliability of ceramic substrate.

II. STATUS OF CERAMIC SUBSTRATE MATERIALS

There are three main types of IGBT module DBC substrate materials, which are Al_2O_3 、 AlN and Si_3N_4 ceramic substrate [4]. Table 1 lists the properties of the three materials.

TABLE I.   PHYSICAL PROPERTIES OF COPPER-CLAD CERAMIC SUBSTRATES

Material	Al_2O_3	AlN	Si_3N_4
Thermal conductivity/$W \cdot (m \cdot K)^{-1}$	24	130-180	60
Bending strength/MPa	400	350	850
Fracture toughness/$MPa \cdot m^{-1/2}$	3.3	2.7	5.0
Thermal expansion/$10\text{-}6 \cdot K^{-1}$	7.1	4.1	2.7
Density /$g \cdot cm^{-3}$	3.86	3.28	3.2
Thickness /mm	0.38	0.6	0.32

Al_2O_3 is the most commonly used material, with good insulation, chemical stability, mechanical properties, relatively mature manufacturing process, low cost. However, the thermal expansion coefficient of Al_2O_3 is not well matched with the thermal expansion coefficient of semiconductor chip (Si base is generally $2.8 \times 10^{-6} \cdot K^{-1}$), so it is suitable for medium and low power IGBT modules.

The thermal conductivity of AlN is about 6 times that of Al_2O_3, and the thermal expansion coefficient matches the semiconductor chip well. But it is difficult to apply copper directly on its surface, and the cost is about 4 times that of Al_2O_3. AlN may be decomposed into alumina hydrate at higher temperature and higher humidity, and the bending strength and fracture toughness are relatively low, leading to easy cracking in the thermal cycle after welding, affecting the reliability of the whole power module, which is suitable for high-power IGBT module.

Thermal expansion coefficient of Si_3N_4 is the best match with the semiconductor chip. The bending strength of Si_3N_4 are more than 2 times that of Al_2O_3 and AlN, and the thermal conductivity is 2.5 times that of Al_2O_3. Thermal standing ability and shocking ability are good. and the cost is about 2.5 times that of Al_2O_3. For high-power IGBT module, Si_3N_4 is the best material at present.

978-1-6654-1392-3/21 $31.00 © 2021 IEEE

III. THERMAL CYCLE TEST

The internal structure of 650V/200A IGBT module is shown in Figure 1.

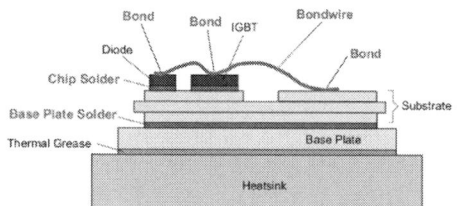

Fig. 1. Diagram of the IGBT module multi-layer structure

In order to ensure that IGBT passes the JM2 grade assessment requirements, it is necessary to evaluate the reliability of IGBT. At present, the commonly used method is thermal cycle test[5]. The whole IGBT is heated and cooled by temperature shock test chamber. According to the GJB128A "Test Method for Semiconductor Disjunct Devices" Method No. 1051 Condition G, the temperature range is -55℃～150℃, the transfer time is not less than 1min, and the holding time should not be less than 10min. IEC60749-25 "Semiconductor devices - Mechanical and climatic tests - Part 25: Thermal cycle holding time shall be greater than or equal to 15min. Therefore, the thermal cycle test holding time was increased to 30min to verify the reliability of the ceramic substrate.

A. Test purpose

The influence of different ceramic substrates on the voltage resistance of IGBT insulation when exposed alternatively to this limit temperature was researched. The failure phenomenon of IGBT under harsh use and storage conditions appeared over time. The test instrument is Temperature impact test chamber TSH-4-10-10-WC, This is shown in Figure 2.

Fig. 2. IGBT module in the thermal cycle tester

B. Test conditions

Before the test, confirm that the equipment is within the validity period of calibration to ensure the reliability of the test results. The temperature of the model test was -55 $_{-10}^{+0}$ ℃～150 $_{-0}^{+15}$ ℃, and the remaining time was 30 minutes, with a total of 1000 cycles. The transfer time between the hot zone and the cold zone should not exceed 1 minute.

After the end of the test, the insulation voltage test should be carried out within 8 hours to be effective..

C. The failure mechanism

In the process of thermal cycle test, when the module is subjected to temperature load varying with time, the mismatch of thermal expansion coefficient between copper and ceramic will lead to stress concentration at the interface. When the external temperature load is 150℃, the copper layer on the ceramic substrate will generate plastic deformation. In the process of thermal cycle, the plastic deformation of the copper layer will accumulate greatly, and stress concentration and crack initiation will generate at the edge of the copper-ceramic interface. At the same time, the ceramic substrate experienced a great difference from 1066℃ to room temperature in the manufacturing process, and there was a certain residual stress in the substrate, which would lead to the crack initiation and then propagate into the ceramic base material. In addition, ceramics are sintered by powder, there are very small cracks or cavities and other inherent defects, these inherent defects will also act as the weak point of ceramic base material and induce the crack to expand. The crack continues to propagate along the interface after a certain length, and finally leads to the complete fracture of the substrate.

D. Test process

A total of 20 IGBT modules of AlN, Si_3N_4, Al_2O_3 doped 9% Zr and Al_2O_3 ceramic substrates were used to conduct 1000 thermal cycle tests (JM2 grade). Before the test, the module shall be tested for insulation and voltage resistance. After the 100 time, the module shall be tested for insulation and voltage resistance every 50 times, up to 1000 times, is shown in Figure 3.

Fig. 3. The test module

In the 200th time, one ALN module was found to be unqualified in insulation and voltage withstand. In the 250th time, 2 modules failed in insulation voltage withstand test, and in the 300th time, 2 modules failed, so far 5 modules are not qualified in the insulation voltage withstand test. For the 500th time, 3 Al_2O_3 modules were found to be unqualified in insulation and voltage withstand. Si_3N_4, Al_2O_3 (9% Zr) ceramic substrate after 1000 cycles of temperature insulation and voltage are all qualified. The change of leakage current of ceramic substrate is shown in Fig. 4 ～Fig. 7. Figure 8 and Figure 9 respectively show the discharge conditions of AlN substrate and Al_2O_3 substrate. It is proved that the theoretical analysis of crack propagation of ceramic substrate is reasonable. The reliability of aluminum nitride is inferior to that of silicon nitride and Al$_2$O$_3$, and the reliability of Al$_2$O$_3$ is inferior to that of silicon nitride.

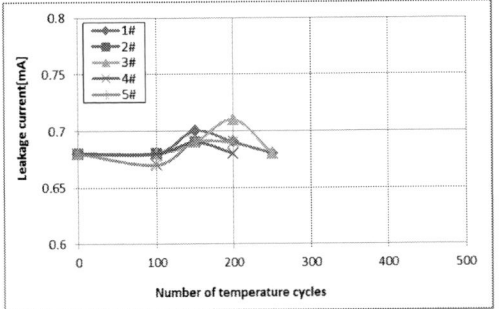

Fig. 4. Leakage current changes of AlN substrate

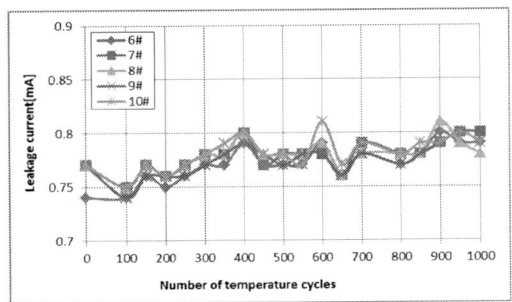

Fig. 5. Leakage current changes of Si$_3$N$_4$ substrate

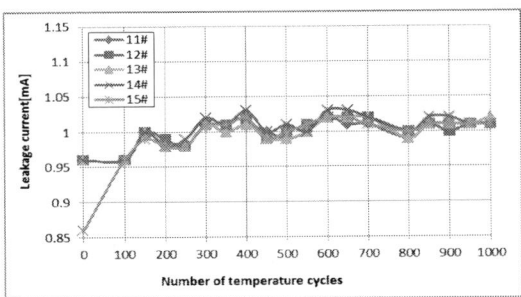

Fig. 6. Leakage current changes of Al$_2$O$_3$ doped 9% Zr substrate

Fig. 7. Leakage current changes of Al$_2$O$_3$ substrate

Fig. 8. Discharge of AlN substrate

Fig. 9. Discharge of Al2O3 substrate

IV. SIMULATION

Taking 650V/200A IGBT module as the research object, the steady temperature field of different ceramic substrates was simulated and analyzed by ANSYS finite element method, and the thermal resistance of different substrates with widely used thickness was compared to provide the best thermal conduction scheme.

The highest steady-state working temperature of the IGBT chip using Al2O3 is 125.39℃, the bottom temperature is 103℃, and the thermal resistance is 0.022℃/W. The highest temperature of the FRD chip in steady state operation is 89.953℃, the bottom temperature is 65.211℃, and the thermal resistance is 0.049℃/W.

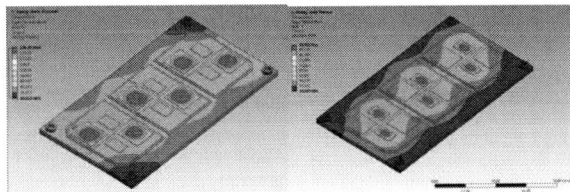

Fig. 10. The steady-state operating temperature distribution of Al2O3 IGBT chip and FRD chips on Al$_2$O$_3$ substrate

Under the same power and heat transfer conditions, the highest steady-state working temperature of the IGBT module using Si$_3$N$_4$ is 117.75℃, bottom temperature is 104.74℃, and the thermal resistance is 0.013℃/W. The highest temperature of the FRD chip in steady state operation is 82.079℃, bottom temperature is 64.651℃, and the thermal resistance is 0.036℃/W.

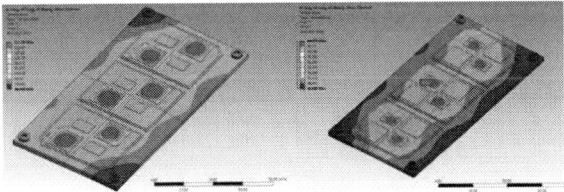

Fig. 11. The steady-state operating temperature distribution of IGBT chip and FRD chips on Si$_3$N$_4$ substrate

Under the same power and heat transfer conditions, the highest steady-state working temperature of the IGBT module using ALN is 116.76℃, the bottom temperature of 101.1℃, and the thermal resistance is 0.015℃/W. The highest temperature of the FRD chip in steady state operation is 80.934, the bottom temperature is 63.815, and the thermal resistance is 0.034℃/W.

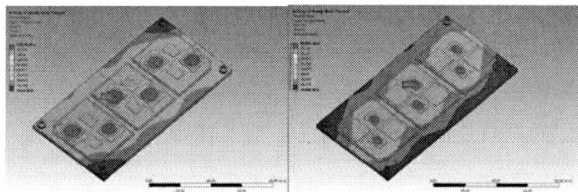

Fig. 12. The steady-state operating temperature distribution of IGBT chip and FRD chips on AlN substrate

TABLE II.   CONTRAST OF DIFFERENT SUBSTRATE MATERIALS

		AlN	Al_2O_3	Si_3N_4
IGBT	Highest temperature/℃	116.76	125.39	117.75
	Lowest temperature/℃	101.1	103	104.74
	Thermal resistance/℃/W	0.015	0.022	0.013
FRD	Highest temperature/℃/	80.934	89.953	82.079
	Lowest temperature/℃/	63.815	65.211	64.651
	Thermal resistance/℃/W	0.034	0.049	0.036

By comparing the IGBT structure and thermal resistance of different ceramic materials, it can be seen that the thermal resistance of AlN and Sin4 is similar, while the thermal conductivity of Al_2O_3 is relatively poor and the thermal resistance value is higher.

V.   CONCLUSION

In this paper, the thermal cycle test of 650V/200AIGBT module is carried out. According to the test results, the following conclusions are drawn: (1) The ceramic substrate failure occurred in the thermal cycle test, which occurred on the ceramic side near the interface at the edge of the substrate.(2) Considering the manufacturing process. AlN substrate is twice as thick as that of Si_3N_4 substrate, the thermal resistance models of different ceramic substrates were established by using ANSYS finite element method. The calculation results show that the thermal resistance of AlN substrate is nearly the same with that of silicon nitride substrate.(3) Si3N4 ceramic substrate has the best performance, and Si3N4 is used as substrate material for high reliability IGBT modules.

[1] VANESSA S, FRANCOIS F, JEAN-JACQUES H, et al. Ageing and failure modes of IGBT modules in high-temperature power cycling[J]. IEEE Transactions on Industrial Electronics, 2011,58(10): 4931–4940.

[2] QIU Zhijie, ZHANG Jin, NING Puqi, et al. Lifetime evaluation of inverter IGBT modules for electric vehicles mission-profile[C]. Electrical Machines and Systems（ICEMS）, China, Japan, 2016.

[3] McCluskey,P.Reliability of power electronics under thermal loading[C].In 7th International conference on integrated power electronics systems,2012;1-8.

[4] Andreas Volke, Michael Hornkamp.IGBT Modules Technologies, Driver and Application[M]. Second Edition. Machinery industry Press,2016.

[5] Josef Lutz,Heinrich Schlangenotto,Uwe Scheuermann,Rik De Doncker.Semiconductor Power Device, Physics, Characteristics, Reliability [M], Second Edition, Spinger, 2018.

Warpage Behavior Study and Optimization for Ultra-thin POP Memory with Multi-stacked Chips

Dongmei Xia
Technology Research and Central Lab,
Shenzhen Kaifa Technology Co., Ltd.
Shenzhen, China
Dongmeixia@kaifa.cn

Bei Wang
Technology Research and Central Lab,
Shenzhen Kaifa Technology Co., Ltd.
Shenzhen, China
Beiwang@kaifa.cn

Chengyu Liao
Advanced Packaging Research Center,
Payton Technology(Shenzhen) Co., Ltd.
Shenzhen, China
Billy_Liao@payton.com.cn

Zuyao Liu
Technology Research and Central Lab,
Shenzhen Kaifa Technology Co., Ltd.
Shenzhen, China
Zuyaoliu@kaifa.cn

Hongwen He, Ph.D.
Advanced Packaging Research Center,
Payton Technology(Shenzhen) Co., Ltd.
Shenzhen, China
Michael_He@payton.com.cn

Lu Liu
Technology Research and Central Lab,
Shenzhen Kaifa Technology Co., Ltd.
Shenzhen, China
Luliu@kaifa.cn

Abstract—**The POP advanced electronic packaging type is widely used in high-end chips because of its huge advantages comparing to the traditional packaging. Electrically, POP offers benefits by minimizing track length between different chips or packages. This brings better electrical performance of devices, since shorter routing of interconnections yields faster signal propagation and reduced noise and cross-talk. Furthermore, it has both the flexibility of the supply chain and cost control for the system integration. Most importantly, the yield rate is guaranteed due to the bottom and upper components of the POP have been packaged and tested. However, as the chip size increasing and multi dies stacking, POP products will meet great technical challenges, among them, the warpage problem is one of the most important issues. With the ultra-thin trend of POP, the warpage will be too sensitive to predict empirically. There are many factors to affect warpage, including mold compound material and substrate selection, package structure design, and assembly process control etc. In this article, the FEA simulation model and warpage behavior are analyzed for ultra-thin POP memory. To improve the model's prediction accuracy, the viscoelastic property of the mold compound material was measured and fitted, the results from the viscoelastic warpage models show well correlation with the shadow moiré testing data. Factors sensitivity analysis are also studied for warpage prediction and optimization, which can guide the selection of the mold component material, substrate and optimization of the structure design.**

Keywords—*Warpage, POP, Simulation, Shadow moiré*

I. INTRODUCTION

The new growth point of semiconductor integrated circuits (IC) has shifted from the traditional computer and communication industries to portable mobile devices such as smart phones, tablet computers and the wearable devices. It is driving component packaging towards 3D packaging technologies to meet the specific requirements of mobile devices, such as increased functional flexibility, improved electrical performance, thinner volume, lower cost and faster delivery.

As one of the most popular 3D packaging technologies, Package-on-package (POP) technology is widely used in high-end chips. A typical POP stack up is shown in Figure 1,which allows multiple ICs to be integrated into a single package containing memory devices and logic devices in top package and bottom package, leading to increased memory density and reduced mounted area. Although the POP structure size is slightly larger than traditional 3D package, the system company can have more component suppliers. Because the top and bottom package of the POP have been packaged and tested separately, the yield is guaranteed, so the POP system integration has the advantage of both supply chain flexibility and cost control. POP technology has been proven as a low-cost solution for system integration.

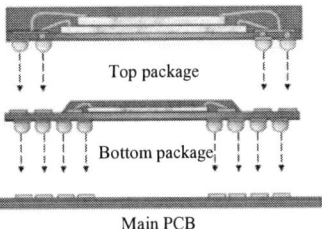

Fig. 1. Package on Package (POP)

The development tendency of the POP packaging technology can be summarized as below:

• The chip size and the ratio of chip size to package size are increasing.

• The pitch of the interconnection between the bottom and top package is reduced from 0.5mm or more to 0.4mm, and 0.3mm pitch will appear in near future.

• The thickness of each layer material is required to be thinner and thinner. The thickness of the substrate has thinned from the common 0.2mm~0.3mm to 0.13mm, or even 0.09mm.The thickness of encapsulation decreased from 0.28mm to 0.2mm and 0.13mm. As for the chip itself, the thickness has reduced to 0.05mm.

• Customized design for different applications and customer requirements make packaging difficult to adopt a traditional unified material system, but must be customized optimization.

The ultra-thin trend of POP will meet the great technical challenges, among them, warpage problem will be one of the most important issue. Excessive warpage not only causes device failure issues such as die crack and interface

978-1-6654-1392-3/21 $31.00 © 2021 IEEE

delamination, but also may cause assembly problems in the subsequent processes, such as dimension instability, non-coplanarity, etc. For the top package of POP, it must tolerate the warpage of the bottom package or remain taut to allow higher warpage in the bottom package, which poses a huge challenge for many packaging vendors. The ultra-thin POP will allow for a wide range of possible warpage size and direction, and will be too sensitive to predict empirically. Therefore, it is necessary to setup the finite element warpage model during the design phase to help predict the warpage of the final package and improve the scheme such as the thickness of each layer and the selection of materials[1,2].

In this paper, the finite element dynamic warpage models were studied for improving the prediction accuracy, and used to correlate to the shadow moiré measured data in the wide temperature range. One of the challenges for warpage modeling is the viscoelastic property of epoxy molding compound(EMC) material. Most of the existing warpage models were based on the linear elastic and could not account for the EMC viscoelastic effect. So the viscoelastic property was measured by DMA tester. Master curve and time temperature shifting function for the material was obtained by curve fitting the stress relaxation data. Then a viscoelastic finite element warpage model was developed. Another challenge for warpage modeling is the proper accounting for the residue warpage caused by the molding and post molding process. Most of the existing warpage models only did not include the chemical shrinkage effect, inevitably causing errors in the model. For above reasons, a modeling method was developed that integrated the chemical shrinkage effect into viscoelastic property to properly explain the dynamic warpage in the final package. For the ultra-thin POP, the tighter warpage control is required, the solution of warpage reduction were analyzed in design optimization and material selection.

II. VISCOELASTIC WARPAGE MODELING

A. Viscoelastic parameters for epoxy molding compound

EMC is one of the most important micro-electronic packaging material. It is one of the key material which determines the final performance of semiconductor packages. First of all, the properties of EMC should be measured by suitable equipment. EMC's stiffness behavior is similar to a viscoelastic material rather than a typical behavior of elastic-plastic material. DMA can be used to measure the material viscoelastic properties. Three-point-bending method with a sinusoidal force or displacement was applied on the EMC materials in this test. This results in a bending deformation of the sample which varies sinusoidally with time. The experiment can be scanned under a continuous frequency range (e.g., 0.01 to 100 Hz). The storage modulus and the loss modulus can be obtained by DMA test. Fig. 2 shows the storage modulus and loss modulus heating curve of cured EMC material.

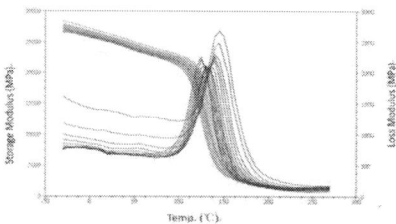

Fig. 2. Complex Modulus curves with temperature at different frequencies

Stress relaxation modulus can be calculated by below equation (1):

$$E(t) = E'(\omega) - 0.4E''(0.4\omega) + 0.014E''(10\omega) \quad (1)$$

Where, t is the time (s); ω is the angular frequency (rad/s).

Based on the Time-Temperature equivalency principle, the master curve (as shown in Fig.3) by rigid horizontal shift function can be obtained[4,5].

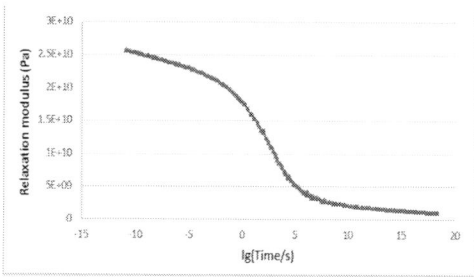

Fig. 3. Master curve of the relaxation modulus at 120 ℃

Select the Maxwell model with different relaxation time constant τ_i ranging over a decade to describe the viscoelastic constructive, and the tensile relaxation function equation as below:

$$E(t) = E_\infty + \sum_{i=1}^{n} E_i e^{(-t/\tau_i)} \quad (2)$$

where $E(t)$ is the tensile relaxation function. E_∞ is the steady state value, E_i and τ_i are the individual modulus and time constant of each Maxwell elements among the 'n' such elements. The short-term modulus $E_0 (E_{t=0})$ is given by:

$$E_0 = E_\infty + \sum_{i=1}^{n} E_i \quad (3)$$

ANSYS FEA need input the shear relaxation function and bulk relaxation function with time. Assuming that the instantaneous Poisson's ratio of EMC is a constant and the instantaneous terms (Young's modulus, shear and bulk modulus) related with each other through an elastic relationship in below Equations, the instantaneous shear and bulk modulus can be calculated [3].

$$G = \frac{E}{2(1+\mu)} \quad (4)$$

$$K = \frac{E}{3(1-2\mu)} \quad (5)$$

where μ is the Poisson's ratio.

For EMCs, the constants C_1 and C_2 can also be obtained from DMA test and to describe the temperature affecting on relaxation function

$$Lg(A(T)) = \frac{-C_1(T-T_r)}{C_2+(T-T_r)} \quad (6)$$

Where T is the experiment temperature, and T_r is the reference temperature which selected as T_g in this paper, $C_1 = 5.409, C_2 = 548.3$.

TABLE I. CURVE FITTING RESULT OF RELAXATION FUNCTION FOR EMC DMA TEST

n	α_i^G	τ_i^G	α_i^K	τ_i^K
1	1.84E-01	5.92E+02	2.20E-02	9.70E+08
2	1.43E-01	1.26E+04	1.68E-01	1.06E+01
3	4.48E-02	1.49E+07	7.14E-02	5.06E-08
4	7.14E-02	5.06E-08	8.55E-02	3.50E+05
5	1.17E-01	1.02E-01	8.27E-02	2.10E-04
6	8.27E-02	2.10E-04	4.48E-02	1.49E+07
7	2.20E-02	9.70E+08	1.84E-01	5.92E+02
8	8.55E-02	3.50E+05	1.43E-01	1.26E+04
9	1.68E-01	1.06E+01	1.17E-01	1.02E-01

B. Chemical Shrinkage Effect

For the ultra-thin POP top package, the residual warpage caused by molding and post-molding cannot be ignored in simulation analysis. But the actual curing induced warpage is a complicated process involving cure conversion and stress relaxation as the mold compound transitions from uncured stage to fully cured stage (including the in-mold curing and post-mold curing). The strip panel warpage caused by the in-mold and post-mold can be simulated and predicted by mold flow simulation, but when slicing into single package, it is difficult to predict the residue warpage on the single package according to the strip panel warpage, which makes the follow-up reprocessing method become unpredictable.

In this paper, the CureCTE and MeanCTE [6] were separately used to simulate the warpage caused by the chemical shrinkage and thermal mismatch in cooling stage. Chemical shrinkage can be equivalent to the coefficient of thermal expansion for the same volume change,

$$\varepsilon_{cure} = \Delta T * CureCTE \qquad (7)$$

CTE of EMC shows piecewise linear before and after T_g, MeanCTE is used in the cooling stage from the molding temperature to the ambient temperature.

$$MeanCTE = \frac{\alpha_1(T_g - T_{low}) + \alpha_2(T_{High} - T_g)}{T_{High} - T_{low}} \qquad (9)$$

Where

α_1 = Coefficient of thermal expansion before T_g;

α_2 = Coefficient of thermal expansion after T_g;

$T_g \doteq$ Glass transition temperature;

T_{High} = 175°C (molding temperature);

T_{Low} = 25°C (ambient temperature);

Two load steps were employed in the simulation to describe the chemical shrinkage and thermal shrinkage effects. Firstly, the mold compound encapsulation effect within the mold is addressed by giving an initial chemical shrinkage at the molding temperature using CureCTE. Then the thermal shrinkage effect is taken care of by using the MeanCTE.

C. FEA model setup

The top package of POP (POPt) with eight memory dies was used in the paper. The scheme of POPt is shown in Fig.4.

The package thickness is 0.92mm, the die thickness is 60um, and the substrate thickness is 130um.

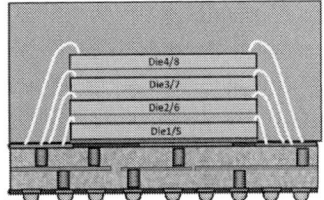

Fig. 4. Scheme of POP top package

Due to the symmetry, the quarter model (as shown in Fig.5) was developed in the simulation to reduce the calculation work. In order to improve the calculation accuracy, the full hexahedral mesh is applied. Reflow curve is as the thermal loading. Fix one point of the substrate bottom to simulate the free state in reflow process. Material properties are shown in Table II.

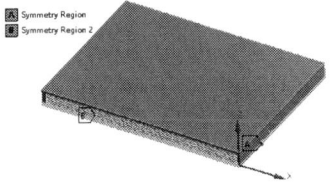

Fig. 5. Quarter symmetry model

As the substrate core gets thinner, the copper signal layers and the solder mask layers account for a more significant amount of the total substrate materials. The detail copper signal layer design layout and density have a direct impact on the package warpage and modeling substrate as a uniform composite material introduces inaccuracies in warpage prediction. The actual substrate layered-structure with detailed signal layers was modeled with the trace mapping method as shown in Fig.6.

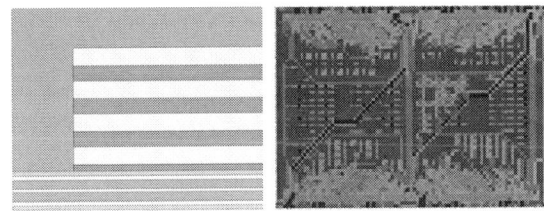

a.Layered substrate   b. Trace mapping model

Fig. 6. Layered model in the substrate

TABLE II. MATERIAL PROPERTIES

Items	Material properties			
	Tg/°C	Young's Modulus/GPa	CTE/ppm/°C	Poisson's ratio
Die	/	160	2.64	0.28
Copper	/	110	16.4	0.35
S/R mask	135	4.8	43/130	0.3
Substrate core	270	17/11	7 (XY)/25(Z)	0.24
Die attach	126	0.48	225/361	0.3
Compound	120	17/0.24	14/45	0.26

III. ANALYSIS AND VALIDATION

A. Simulation results

Select the substrate bottom surface to observation, from the simulation results (as shown in Fig.8), we can see that the package shows smiling deformation at the ambient temperature and the maximum warpage is 82 um, the package shows crying deformation at the high temperature and the maximum warpage is 106um, and separately defined as the positive warpage and negative warpage (as shown in Fig.7).

Fig. 7. Warpage direction definition

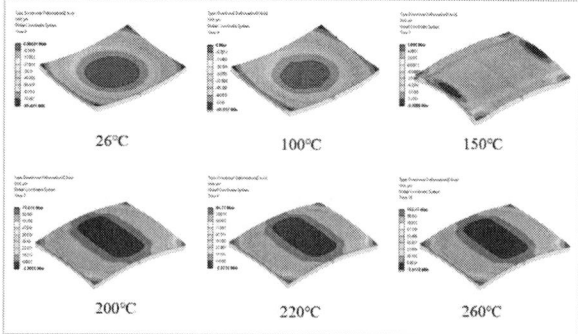

Fig. 8. Simulation deformation in reflow process

B. Verification of the model with shadow moiré test

In order to verify the accuracy of the simulation results, the POP top package sample was taken for shadow moiré test. From the comparison of simulation and shadow moiré test results(as shown in Fig.8 and Fig.9), the same warpage trend and smaller gap (as shown in Fig.10) in the dynamic temperature range, especially well consistence at the room temperature and the high temperature, the error is less than 10%, which shows that there is a positive guiding significance.

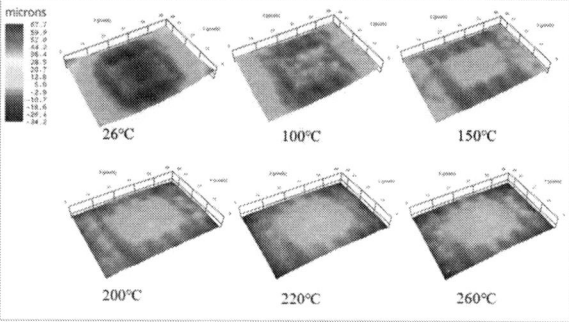

Fig. 9. Shadow moiré test results

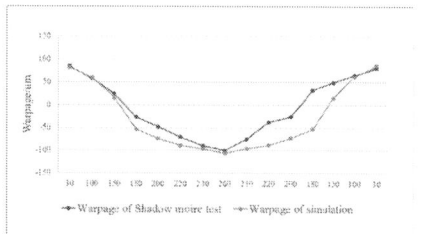

Fig. 10. Comparison between simulation and test

Compared with the viscoelastic model, the typical elastic model shows bigger warpage gap in the whole dynamic reflow process(as shown in Fig.11). For the ultra-thin POP, the viscoelastic model shows higher accuracy.

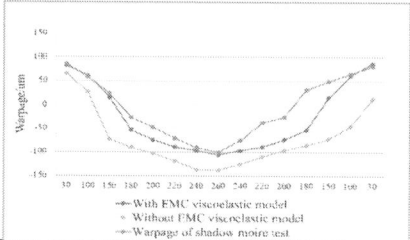

Fig. 11. Warpage effect of viscoelastic model vs. elastic model

With the die size and the ratio of die size to package size are increasing, the chemical shrinkage induced warpage can't be ignored. The residue warpage calculated by the chemical shrinkage(CureCTE) and thermal mismatch(MeanCTE) can be well consistent with the actual test results.

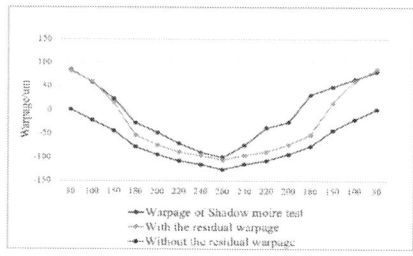

Fig. 12. Effect of the residual warpage

IV. WARPAGE CONTROL ANALYSIS

For the ultra-thin POP package, the finer solder pitch requires the tigher warpage control. For this package, the residue warpage in room temperature and the warpage at reflow high temperature need to be strictly controlled in the design optimization and material selection. And the above model is used for further warpage control analysis.

A. Warpage Control at the room temperature

Increasing the EMC mold shrinkage can reduce the residual warpage at the room temperature. The higher EMC mold shrinkage can cause the bigger shrinkage force in the molding process to resist the deformation, so can reduce the residual warpage.

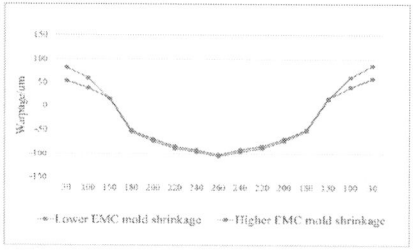

Fig. 13. EMC mold shrinkage effect

When higher residue warpage was predicted, replacing with the thinner substrate was also an effective method to reduce the warpage at the room temperature.

978-1-6654-1392-3/21 $31.00 © 2021 IEEE

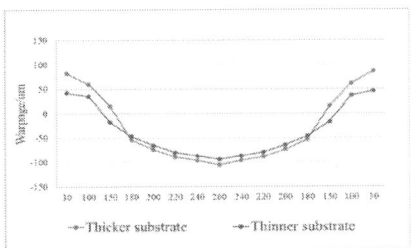

Fig. 14. Substrate thickness effect

B. Warpage Control at the high temperature

From the warpage trend in reflow process, increasing compound CTE after Tg can obvious reduce the warpage in high temperature.

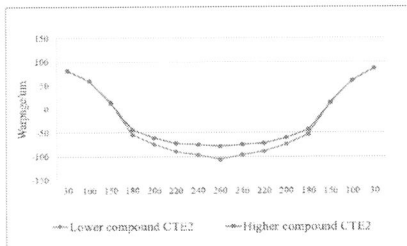

Fig. 15. Compound CTE effect

Now, to reduce the warpage, the substrate core with lower CTE is used for the ultra-thin POP, Fig.12 shows that lower core CTE can reduce the high temperature warpage.

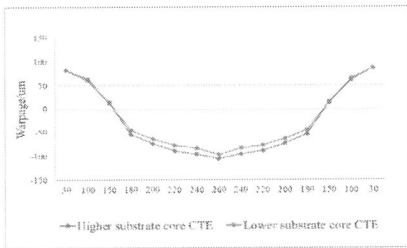

Fig. 16. Substrate core CTE effect

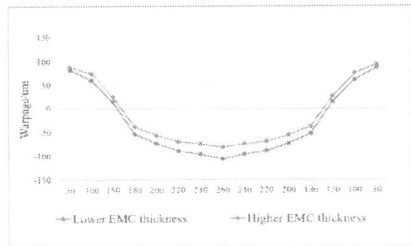

Fig. 17. EMC thickness effect

Increasing the EMC thickness can reduce the high temperature warpage, but at the same time it causes higher residue warpage at the room temperature.

CONCLUSION

With the ultra-thin POP development trend, warpage control is increasingly tighter. The package vendor meet the great challenges in design optimization and material selection. In this paper, to improve the warpage model prediction accuracy, a modeling method was developed that integrated the chemical shrinkage effect into viscoelastic model to explain the dynamic warpage in the final package.

The simulation results of the improved warpage model showed well consistence with the shadow moiré test results in the wide temperature range, especially accurately predict the room temperature warpage and the high temperature warpage, the error is less than 10%. For the new package design in the early stage, we can rely on the simulation results to select BOM effectively or do some design alteration quickly.

Further warpage control analysis give the guidance to reduce the residual warpage from increasing EMC mold shrinkage or replacing the thinner substrate, and reduce the high temperature warpage from increasing the CTE after Tg of EMC, reducing the substrate core CTE or thickness EMC thickness.

REFERENCES

[1] Wei L , Min W L . PoP/CSP warpage evaluation and viscoelastic modeling[C]// IEEE. IEEE, 2008.

[2] Tan Lin, Chen Chuan, Li Jinrui. Viscoelastic Simulation and Verification of FBGA Packaging Warpage [J]. Semiconductor Technology, 2015, 040(002):142-147.

[3] Dr. Caiying He. Thermo-mechanical Simulation and Optimization Analysis for Warpage-induced PBGA Solder Joint Failures[C]// SMTA International technical conference.

[4] Lim, M. , Chua, L. Y. , Yeo, A. , & L Ee , C. . (2006). Impact of Mold Compound Cure Shrinkage on Substrate Block Warpage Simulation. (pp.196-201).

[5] Beijer, J. , Janssen, J. , Bressers, H. , Driel, W. V. , Jansen, K. , & Yang, D. G. , et al. (2005). Warpage minimization of the HVQFN map mould.

[6] Qiu, & Wang. (2011). The effect of initial warpage of top component on POP assembly. International Conference on Electronic Packaging Technology & High Density Packaging. IEEE

978-1-6654-1392-3/21 $31.00 © 2021 IEEE

Research on Power Device Structure Based on FO Package Method

Guan qiang Song
Product Research and Development
Sky Chip Interconnection Technology
CO.,LTD
Shenzhen, China
songgq@scc.com.cn

Jia ren Huo
Product Research and Development
Sky Chip Interconnection Technology
CO.,LTD
Shenzhen, China
huojiar@scc.com.cn

Juntao Wang
Product Research and Development
Sky Chip Interconnection Technology
CO.,LTD
wuxi, China
wangjt@scc.com.cn

Debo Liu
Product Research and Development
Sky Chip Interconnection Technology
CO.,LTD
Shenzhen, China
dayper @scc.com.cn

Chenwei Zhang
Southern University of Science and
Technology
Shenzhen, China
11930218 @mail.sustech.edu.cn

Jing Jiang
Product Research and Development
Sky Chip Interconnection Technology
CO.,LTD
Shenzhen, China
king @scc.com.cn

Huaiyu Ye
Shenzhen Institute of Wide-Bandgap
Semiconductors; School of
Microelectronics, Southern University
of Science and Technology
Shenzhen, China
h.ye@tudelft.nl

Abstract—Most of the existing power chips are connected by wire bonding or copper sheet welding. With the development of high performance, small size, modularity and high power density of power devices, the packaging method starts to develop towards FO(Fan-out)-based packaging technology. This scheme can be designed with complex line layout and flexible metal shape linkage method. And it is a high efficiency package method with low parasitic capacitance, inductance, very low R_{dson} and excellent heat dissipation at high frequency.

The thermal stress problem in reflow soldering is a key research object in device development, and this paper mainly focuses on the research of DFN3030 product structure based on FO package thermal stress. Using finite element software to establish a three-dimensional model, the main factors affecting the metal shape of the chip surface, copper thickness, and plastic sealing material thickness are simulated and analyzed. The simulation is used to determine the relationship between different factors affecting thermal stress, and two different structural solutions are selected for sample verification to confirm whether the simulation structure matches the actual product result trend. Through the actual product reliability verification, it is confirmed that the thermal stress trend of the actual product matches with the theoretical results. And through the sample verification, the product with FO package DFN3030 meets the reliability requirements of MSL3, TC 500cycles@(-55~125℃), PCT 96H@(120℃, 100%RH), HTSL 1000H@150℃, THB 1000H@(85℃, 85%RH).

Keywords—power devices, thermal stress, FO packages, thermal simulation

INTRODUCTION

As the trend of high performance and miniaturization of smart terminal devices such as cell phones, wearable devices and IOT(Internet of Things) is becoming more and more obvious, customers' requirements for low cost and miniaturization of accessories are also increasing, prompting the industry to continuously improve packaging technology to meet the market demand for products. Fan-out Panel Level Package (FOPLP) has lower cost, larger size and 3D integration process compared with FOPLP. In addition, the package solution adopts thick copper and shaped metal block to meet the requirements of power packaging devices for current resistance and heat dissipation, and can be used in "component level" packages for power and analog applications with a small number of I/Os to achieve high current and high heat dissipation for power modules.

The FOPLP process and new product development of the third generation of semiconductor power devices and modules. To provide customers with a very different packaging solutions and product upgrades to meet the needs of customers in the power device packaging. To provide customers with reliable quality of low-cost, highly integrated module packaging solutions.

INTRODUCTION OF FOPLP STRUCTURE

FOPLP is a new and expandable advanced power device package. The technology is connected through shaped metal block and thick copper layer on the top of the chip, and the thick copper layer is connected to the bottom pins through deep via .Due to the presence of metal porous and thick copper layer with high thermal conductivity on the chip, it can enhance the heat dissipation capability on the top of the chip. In addition, the outer package has a high copper content and a short current flow path, which can realize a high-efficiency package of power devices with low resistance, low parasitic capacitance and inductance. Fig.1 is the structure of the package products.

Fig. 1. FOPLP product structure diagram

Table 1 shows the results of FOPLP compared with conventional wire bonding and copper solder package scheme. The resistance value of FOPLP is 0.22mΩ, which is close to that of copper soldering (0.2mΩ). Fig. 2 shows the different resistance values of the same chip package DFN3030 using FOPLP and WB scheme, where R_{dson}@ID=30A, V_{gs}=10V, FOPLP scheme is 3.17mΩ ,lower than WB scheme (R_{dson}=4.15mΩ) by 0.975mΩ. Similarly for R_{dson}@I_D=20A ,V_{gs}=4.5V, the FOPLP scheme is 1.04mΩ lower than the WB scheme.

TABLE I. COMPARISON RESULTS OF DIFFERENT PACKAGING SCHEMES RS

-	FOPLP@DFN5060	DPAK@PDFN5060	LFPAK@PDFN5060
Model			
Resistance(mΩ)	0.22	1.5	0.2

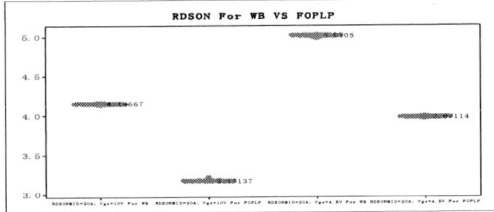

Fig. 2. The comparison of FOPLP and WB scheme R_{dson}

MODELING AND SIMULATION

Heat Transfer and Thermal Stress

Thermal indicators [1]

$$R_{th} = \sum_{i=1}^{6} \frac{h_i}{k_i A_i}$$

Where h_i is the thickness of each encapsulation layer; A_i is the equivalent heat transfer area of i-layer material; k_i is the thermal conductivity of i-layer material.

For the encapsulant, the interface stress equation is [2]

$$D_W = \frac{\sum_{i=1}^{6} E_i \alpha_i V_i}{\sum_{i=1}^{6} V_i}$$

Where α_i is the coefficient of thermal expansion of layer i material, E_i is the Young's modulus of layer i material, V_i is the volume of layer i material.

Boundary conditions

The DFN3030 device is selected as the object of study, and the PCB board is built according to the JESD51-3 [3] standard, with dimensions of 76.2 × 114.3 mm × 1.6 mm, and the line layer is modeled after thermal equivalence treatment according to the JEDEC standard, with dimensions of 50 mm × 50 mm × 0.071 mm. And the model didn't have symmetry, so the model is used 1: 1 for analysis. Fig. 3 shows the overall simulation model of DFN3030 package.

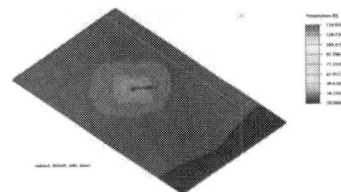

Fig. 3. DFN3030 package overall simulation model

In order to facilitate comparative analysis, the simulation uses the steady-state thermal analysis module and the steady-state structural analysis module of the finite element analysis software, which only simulates the stress of the device at 150 ℃ .The thermal diffusion and thermal stress material properties are shown in Table 2.

TABLE II. THERMAL DIFFUSION AND THERMAL STRESS MATERIAL PROPERTIES

Material Category	Density (Kg/M3）	Isotropic Thermal Conductivity （W/m.℃）	Specific Heat （J/Kg.℃）
Conductive adhesive	3.7	2.5	/
Copper	8300	401	385
FR-4	1840	0.38@X 0.38@Y 0.30@Z	/
EMC	1550	0.65	/
Silicon	2330	124	702
Tin	10900	35	139
Material Category	**CTE** （PPM/℃）	**Young's Modulus (GPa)**	**Poisson's Ratio**
Conductive adhesive	40@0℃ 150@120℃	4.41@65℃ 3.93@25℃ 2.0@150℃ 0.303@250℃	0.3
Copper	18.0	1.10×1011	0.34
FR-4	12.5@X 11.4@Y 82.0@Z	20.4@X 18.4@Y 15.0@Z	0.11@XY 0.09@YZ 0.14@XZ
EMC	18.0@0℃ 48.0@160℃	9.0@25℃ 0.0120@250℃	0.35
Silicon	2.46@20℃ 3.61@250℃	/	/
Tin	37.0	1.20×109	0.30

The material properties are derived from the supplier datasheet.

Simulation scheme

1) Fig. 4 shows the simplified structure of FOPLP package body. The thermal stress simulation mainly focuses on the chip interface stress condition, and three factors are selected for thermal stress:shaped metal block, bottom copper thickness and dielectric thickness on the chip.And make a set of simulations for shaped metal block, and make another set of simulations with the bottom copper thickness and the on-chip dielectric thickness as different influencing factors.

Fig. 4. The simplified structure of FOPLP package body

2) Shaped metal block simulation scheme

The fixed bottom copper thickness is 210μm, the dielectric thickness and the metal thickness on the chip are 40μm, and three models of the metal shape on the chip are designed for simulation. Fig.5 shows the simulation scheme of the metal shallow via matrix, where the shallow via spacing is designed to be 0.35mm and the shallow via diameter is 0.15mm. Fig.6 is a simulation scheme of metal groove, in which the groove length is 0.85mm, the width is 0.15mm, and the spacing is 0.35mm. Fig.7 is the simulation scheme of a metal copper block, in which the length of the copper sheet is 1.2mm and the width is 0.85mm.

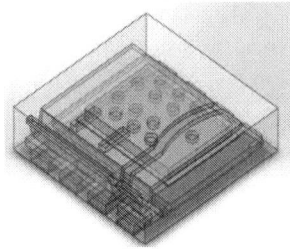

Fig. 5. The simulation scheme of the metal shallow via matrix

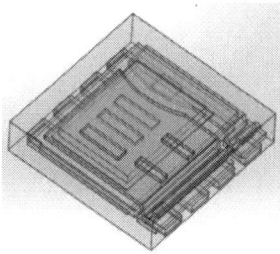

Fig. 6. The simulation scheme of metal groove

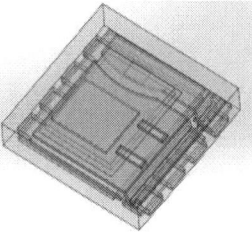

Fig. 7. The simulation scheme of a metal copper block

3) Copper of chip bottom and dielectric thickness on the chip simulation scheme .

Taking into account the cost, package thickness and heat dissipation requirements, we set its chip bottom thickness range as follows:

$$x \in [35\mu m, \ 250\mu m]$$

For the thickness of the media on the chip, the thickness setting range is limited by the process capability such as blind hole aperture and aperture depth capability:

$$y \in [40\mu m, \ 120\mu m]$$

The DOE schedule of copper at the bottom of the chip and dielectric thickness on the chip is shown in Table 3, where the hole spacing is modeled with reference to the above optimal results.

TABLE III.   COPPER THICKNESS AND DIELECTRIC THICKNESS DOE SCHEDULE FOR THE CHIP

StdOrder	RunOrder	CenterPt	Blocks	x	y
1	1	0	1	150	80
2	2	1	1	250	120
3	3	1	1	50	120
4	4	1	1	50	40
5	5	-1	1	8.58	80
6	6	-1	1	291.42	80
7	7	0	1	150	80
8	8	0	1	150	80
9	9	0	1	150	80
10	10	0	1	150	80
11	11	1	1	250	40
12	12	-1	1	150	23.43
13	13	-1	1	150	136.57

DEVICE SIMULATION RESULTS AND DISCUSSION

Shaped metal block simulation results

Table 4 shows the simulation results of the surface stress on the chip with different metal shapes. The stress mainly occurs at the contact position of the chip and the contact metal.

1）The first column in the table is the stress simulation result of the shallow via matrix scheme. The maximum stress at the four corners of the metal block is 397.4Mpa, 435.5Mpa, 417.0Mpa, 395.2Mpa, and the average maximum stress is 411.3Mpa. The stress in the shallow via at both ends is greater than the stress in the middle shallow via.

2) The second column in the table is the stress simulation result of the metal groove scheme. The stress values at the four corners of the hole and groove are the largest, and they are 522.8Mpa, 368.7Mpa, 580.8Mpa, 536.1Mpa, and the maximum stress is 502.1Mpa.

3）The value in the third column of the table is the stress simulation result of the metal block scheme. The maximum stress occurs around the copper block. The four corner stresses are 652.0Mpa, 569.7Mpa, 497.5Mpa, 718.9Mpa, and the average maximum stress is 609.5Mpa.

4）The smaller the stress, the better the structure, which is analyzed from the above simulation results. Therefore, the preferred solution is the metal shallow via matrix solution, and the worst is the metal copper block solution. The thermal stress difference between the two reaches 198.2Mpa, the difference range reaches 48.1%.

TABLE IV. THE SIMULATION RESULTS OF THE SURFACE STRESS ON THE CHIP WITH DIFFERENT METAL SHAPES

Description	Metal shallow Via matrix	Metal groove	Metal copper block
The simulation results	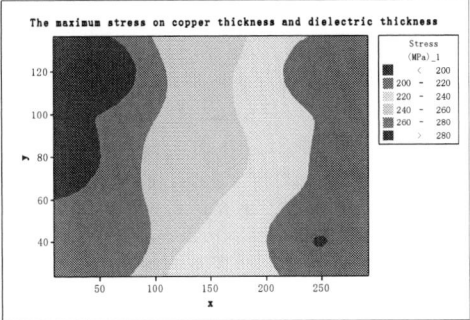		
The maximum stress	397.4Mpa 435.5Mpa 417.0Mpa 395.2Mpa	552.8Mpa 368.7Mpa 580.8Mpa 536.1Mpa	652.0Mpa 569.7Mpa 497.5Mpa 718.9Mpa

Table 5 shows the simulation results of the lower surface stress of the chip with different metal shapes. The maximum stresses of the three schemes are: 253.1Mpa~291.7Mpa, 267.2Mpa~290.6Mpa, 256.4Mpa~289.4Mpa. It is found that adjust the different shapes of the metal on the top surface of the chip has little effect on the stress on the bottom of the chip.

TABLE V. THE SIMULATION RESULTS OF THE BOTTOM SURFACE STRESS ON THE CHIP WITH DIFFERENT METAL SHAPES

Description	Metal shallow Via matrix	Metal groove	Metal copper block
The simulation results			
The maximum stress	253.1Mpa 291.7Mpa 268.3Mpa	267.2Mpa 285.9Mpa 290.6Mpa	256.4Mpa 269.2Mpa 289.4Mpa

Based on the above analysis, there is a positive correlation between the stress and the size of the metal shape. The larger the area, the greater the maximum stress. This phenomenon occurs mainly because the expansion coefficients of the two materials are quite different between the chip interface, and the interface will produce greater stress when the temperature changes. In addition, the thermal conductivity of the chip and the copper layer is higher than that of the plastic molding compound. When the temperature changes, there is a larger thermal gradient between these materials and the molding compound, which will aggravate the degree of deformation of the material and further aggravate the interfacial stress.

Simulation results of copper thickness at the bottom of the chip and its dielectric thickness on the chip

The model is established based on the above analysis results, the metal shape on the chip adopts shallow via matrix design. Table 6 shows the bottom chip surface simulation results of different copper and on the chip dielectric thickness. Fig. 8 shows the equivalence of the maximum stress about copper and dielectric thickness, and the analysis from the figure confirms that the maximum stress results show a positive relationship with copper thickness and a negative relationship with dielectric thickness. Through data analysis, the mathematical model is as follows:

$$\sigma = 243.54 - 52.41x + 8.27y + 9.02x^2 - 7.30y^2 - 3.72xy$$

TABLE VI. THE BOTTOM CHIP SURFACE SIMULATION RESULTS OF DIFFERENT COPPER THICKNESS AND ON THE CHIP DIELECTRIC THICKNESS:

StdOrder	x	y	Stress (MPa)
12	150	80	243.54
4	250	120	208.50
3	50	120	295.61
1	50	40	279.18
5	8.58	80	297.14
6	291.42	80	205.40
11	150	80	243.54
10	150	80	243.54
13	150	80	243.54
9	150	80	243.54
2	250	40	199.50
7	150	23.43	227.40
8	150	136.57	242.51

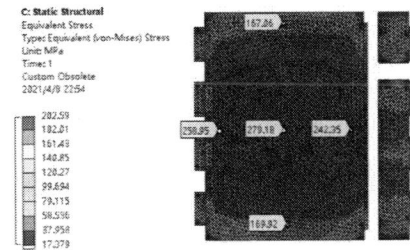

Fig. 8. The equivalence of the maximum stress on copper thickness and dielectric thickness

The maximum stress change occurs mainly at the bottom of the chip as the thickness of the copper layer and the medium changes. As the thickness of the copper layer decreases, the mean and maximum values of thermal stress are decreasing. Fig. 9 shows the simulation results for the bottom copper thickness of 50μm and dielectric thickness of 80μm, the maximum thermal stress at its center is 279.18MPa, the average thermal stress around is 205.84MPa, and the overall stress area is relatively large. Fig. 10 shows the simulation results when the bottom copper thickness is 250μm and the dielectric thickness is 80μm, the maximum thermal stress at the center is 199.53MPa, the average thermal stress around is 156.10MPa, and the maximum stress area is relatively small.

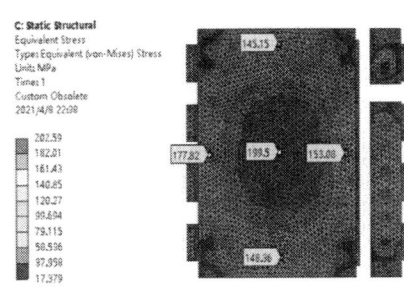

Fig. 9. Bottom stress diagram of thin copper bottom

Fig. 10. Bottom stress diagram of thick copper bottom

Through the mathematical model analyzed above, considering the feasibility of processing, in the actual product test, the metal on the chip adopts shallow via matrix design. Therefore, we choose two schemes in Table 7 for the subsequent scheme verification.

TABLE VII.   TWO EXPERIMENTAL VALIDATION SCHEMES

Solution	Hole spacing in X-direction	Hole spacing in X-direction	Copper thickness	Dielectric thickness
Min	0.29	0.42	250	80
Max	0.29	0.42	50	80

EXPERIMENTAL VALIDATION RESULTS

· Table 8 shows the thermal stress verification according to JESD22-A113E MSL3 test standard, where the product performance test is OK after lead-free reflow of min solution. The products are 100% tested with NG after lead-free reflow of max solution, which mainly has a large R_{dson} bias.

TABLE VIII.   THE TEST RESULTS ACCORDING TO JESD22-A113E MSL3 TEST STANDARD

Solutions	Number of tests	Standard	Test results	Test method
Min	77	JESD22-A113E	100%OK	1.Baking the board at 125℃ for 24h first.
Max	77		77pcs R_{dson} NG	2.MSL3 level, 30℃, 60%RH, 168H; 3.Over lead-free reflow, 260℃, 3 times.

Fig. 11 shows the post-slice diagram of the min solution product MSL3. Fig. 12 shows the chip electron microscope analysis of the blind hole edge and the position of the blind hole in the min solution. According to the analysis, no sealing layer or fracture problem is found at the interface of each layer. Therefore, min solution products meet MSL3 requirements.

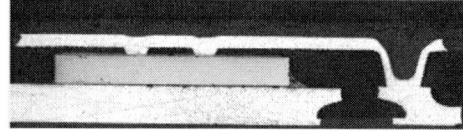

Fig. 11. illustrates the section after the MSL3 of the Min solution of the product

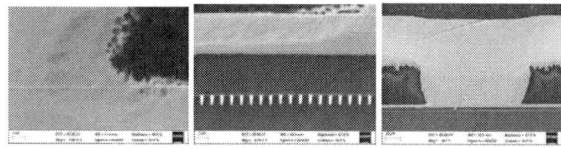

Fig. 12. The Min scheme for SEM

Fig. 13 shows the section after MSL3 of the max solution of the product, and Fig. 14 shows the electron microscope analysis of the bottom of the max solution chip. From the analysis results, the main reason for R_{dson} NG is the delamination of the bottom of the chip.

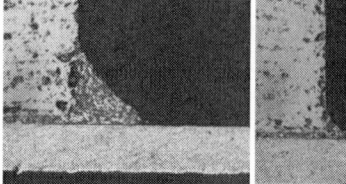

Fig. 13. Illustration of the section after the MSL3 of the max solution of the product

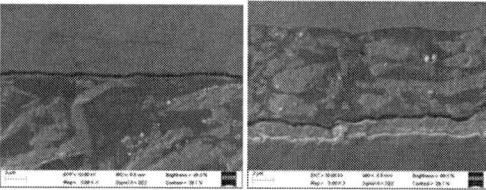

Fig. 14. illustrates the electron microscope analysis of the max solution

According to the above two test verification solutions, the simulated stress center point of the max solution product is 279.18 MPa and the surrounding area is 205.84 MPa, and the simulated stress center point of the Min solution product is 199.53 MPa and the surrounding area is 156.10 MPa. In the actual product process, the Min solution sample is OK and the Max solution sample has the NG state, and the results are basically consistent with the trend of the simulation results.

Table 9 shows the reliability verification of Min solution TC, PCT, THB, and HTSL. And the results meet the requirements of TC 500cycles@(-55~125 ℃), PCT 96H@(120℃, 100%RH), HTSL 1000H@150℃, and THB 1000H@(85℃, 85%RH).

TABLE IX.   THE RELIABILITY VALIDATION OF TC, PCT, THB, AND HTSL FOR THE MIN SOLUTION

Testing items	Number of samples	Standard	Test method/parameter	Pass/fail criteria
TC/TS	77	JESD22A104C	-55℃—125℃; 500 cycles	77/0
PCT	77	JESD22-A102	121℃/100%RH,2ATM,96H	77/0
THB	77	JESD22-A101	85℃/85%,1000H	77/0
HTSL	77	JESD22-A103	150℃,1000H	77/0

SUMMARY

- Metal shape on the top surface of the chip，the preferred solution is the metal shallow via matrix solution.

- Bottom copper and dielectric thickness factor have the greatest effect on the stress corresponding to copper thickness. The thicker the copper, the lower the stress.

- Through the verification of the actual product of the two structural solutions, the product is verified as NG in the sample reflow thermal stress when the simulated stress is large.

978-1-6654-1392-3/21 $31.00 © 2021 IEEE

- The sample reflow thermal stress verification is OK when the simulated stress is small, and the simulation results follow the same trend as the actual product results.

- The reliability of the product is verified., and it can meet the requirements of TC 500cycles@(-55 ~ 125 ℃), PCT 96H@ (120 ℃, 100% RH), HTSL 1000H@150 ℃, THB 1000H@ (85 ℃, 85% RH).

ACKNOWLEDGMENT

This work was supported by the National Key R&D Program of China (2018YFE0204600), the Key-Area Research and Development Program of GuangDong Province (2019B010131001), and the Shenzhen Key Project for Basic Research (JCYJ20200109140822796).

[1] [1]WONG E H,KOH S W,LEE K H,et al.Advanced moisture diffusion modeling &characterization for electronic packaging[C]//Proe of Electronic Components and Technology conf,Difornia,2002;1297-1303

[2] [2]Chen X,Zhao S F,Zhai L.Moisture absorption and diffusion characterization of molding compound.ASME Journal of Electronic Packaging,2005,127:460-465.

[3] [3] EIA/JESD51-3A.Integrated Circuits Thermal Test Method Environmental Conditions-Natural Convection (Still Air) [S].

A Broadband Model of Stacking TSV Channels for Nondestructive Defect Localization in 3D ICs and Microsystem

Chenbing Qu
Science and Technology on Reliability
Physics and Application of Electronic
Component Laboratory
China Electronic Product Reliability
and Environmental Testing
Research Institute
Guangzhou, China
quchenbing@126.com

Liwei Wang
Science and Technology on Reliability
Physics and Application of Electronic
Component Laboratory
China Electronic Product Reliability
and Environmental Testing
Research Institute
Guangzhou, China
wanglw@ceprei.com

Si Chen
Science and Technology on Reliability
Physics and Application of Electronic
Component Laboratory
China Electronic Product Reliability
and Environmental Testing
Research Institute
Guangzhou, China
806532073@qq.com

Chen Sun
Science and Technology on Reliability
Physics and Application of Electronic
Component Laboratory
China Electronic Product Reliability
and Environmental Testing
Research Institute
Guangzhou, China
smilesunc@163.com

Zhiwei Fu
Science and Technology on Reliability
Physics and Application of Electronic
Component Laboratory
China Electronic Product Reliability
and Environmental Testing
Research Institute
Guangzhou, China
fzw19940124@163.com

Guan-Lin Feng
（corresponding author）
China Electronic Product Reliability
and Environmental Testing
Research Institute
Guangzhou, China
1548381305@qq.com

Abstract—The defect localization is a key link and bottleneck for the failure analysis of 3D interconnects. This paper proposes a combined-cylindrical-shape defect model for the nondestructive defects testing and fault localization of TSV channels in 3D stacked dies. The TSV equivalent circuits and defective fault equivalent circuit are established to explain the fault principle. A simple scheme of GSG TSV channels is made, and their simulated results are shown with the frequency from 100MHz to 20GHz. By fitting the S-parameters of defect TSV chain, the open-circuit defect size and the defect module localization are estimated based on signal integration. The scalable defect localization method is used for multiple stacked dies scheme based on silicon substate.

Keywords—*3D IC; through silicon via; nondestructive defect localization; equivalent circuit model; signal integration*

I. INTRODUCTION (*HEADING 1*)

With the rapid development of three-dimensional integrated circuit(3D IC) products, the demand for reliability research of their interconnects and package has become increasingly prominent. Through silicon via(TSV) technology is one of the core technologies of 3D IC[1]. In their manufacturing process, many factors including nonuniform filling, incompletely chemical mechanical polishing lead to shorted effect and open defect.

Under the micro nano size, TSV process defects are caused by process level and process fluctuation[2]. Internal and external stress, copper creep and other factors lead to interconnect damages. Defects such as copper bulge, cracks, voids and solder joint voids lead to open circuit or high resistance fault of TSV 3D interconnections. It has a very prominent impact on the signal integrity of RF front-end circuit. Figure 1 shows the SEM photos of several common defects of 3D TSV interconnects. Thus, efficient and accurate noninvasive defect analysis methods are beneficial

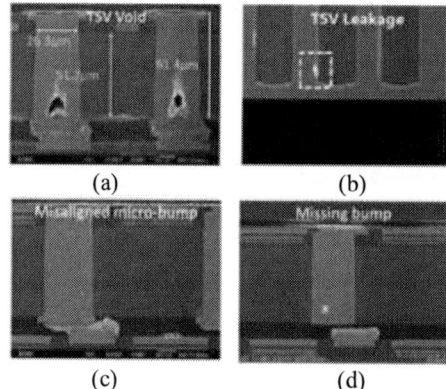

(a)   (b)

(c)   (d)

Fig. 1 SEM photos of several common defects of 3D TSV interconnects. (a)TSV void; (b)TSV leakage; (c)misaligned micro-bump; (d)missing bump[1][4].

to reduce unpredictable electrical faults. Common kinds of defects, as shown in Fig. 1, are summarized, and the hole-TSV void or misaligned micro-bump will lead to the open circuit. So, the open-circuit fault is one common fatal fault and its defect always occurs in the junction of the two connections of TSV and RDL.

To enhance the reliability of TSVs, the fault-tolerance process is classified into three major phases: detection, localization (or diagnosis), and recovery[5]. The defect diagnosis technology of high-density chain TSV interconnects mainly includes microscopic imaging, digital circuit signal acquisition and RF S-parameters testing, which are applied to pre- or post-bonding processes[6]. The TSV signal integration of electronic model and S-parameters can cover all frequency-dependent characteristics of TSV defects.

978-1-6654-1392-3/21 $31.00 © 2021 IEEE

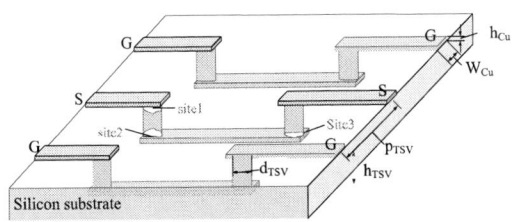

Fig. 2 Scheme of GSG TSV channels and open-circuit defect sites.

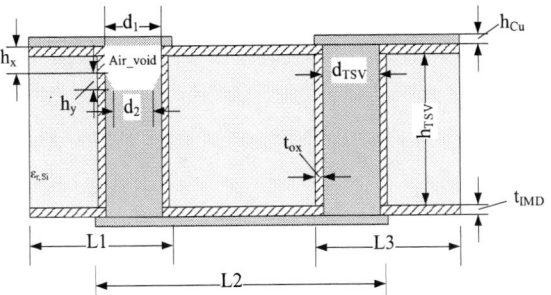

Fig. 3 Cross section scheme of transmission channel and proposed TSV void.

The currently used methods of diagnosing TSV fault with RF S-parameters are mainly limited to the estimation of open or short circuit. It has attracted many researchers to explore in-depth mechanism for practical nondestructive detect detection technology[7][6][8]. However, there are few literatures about the defect size and the micron defect's localization for long chain TSVs. The defect localization is a bottleneck for the failure analysis of 3D interconnects. It is an effective method of extracting fault electrical characteristics for 3D failure analysis.

In this paper, a combined cylindrical shape 3D TSV defect model is explored for the nondestructive fault diagnosis and localization of TSV channels in 3D stacked dies. The GSG daisy TSV channels are simulated by 3D full wave simulator. The methods of open-circuit fault diagnosis and localization are presented by the comparisons of extracted S-parameters. What's more, the equivalent circuit of TSV defect is established to explain the fault principle.

II. TSV DEFECT MODEL

Under the micro nano size, TSV process defects are caused by process level and process fluctuation. In this part, a structure of 3D TSV void model and its predicted fault characteristics are shown.

A. Open-circuit TSV fault construction

The micro-interconnects include TSVs, RDLs and micro-bumps. There are at least two stacked dies. Fig.2 shows a simple typical testing construction scheme of TSV channels. GSG microwave probes and a vector network analyzer are applied to get RF S-parameters. Fig.3 shows the cross sections and physical dimensions.

In this work, $d_{TSV} = 30$ μm and $p_{TSV} = 250$ μm are the diameter and center-to-center pitch between two adjacent of Cu TSV plugs, as shown in Fig.2. The height of the silicon substrate is $h_{TSV} = 94$ μm with dielectric constant $\varepsilon_{r,Si} = 11.9$

and conductivity $\sigma_{Si} = 10$ S/m. Ones use the silicon oxide as insulator layers of TSVs, and the oxide silicon and Cu are selected as the insulation and redistribution layers(RDLs), respectively. The thicknesses of TSV insulator(t_{ox}) and IMD (t_{IMD}) are 0.5 μm and 3 μm, respectively. $w_{Cu} = 30$ μm and $h_{Cu} = 3$ μm are the width and height of the RDLs, respectively. The RDLs connect TSVs on and down of substrate with L1=300 μm, L2=500 μm, L3=300 μm, as shown in Fig.3. Open fault sites are preset in different representative positions of the signal TSV channel.

B. 3D modle of TSV void defect

Common kinds of defects from SEM photos are show that the open-circuit fault always occurs in the junction of the two connections of TSV and RDL. Thus, we explore a 3D TSV defect model. The defect model consists of a circular truncated cone and a cylinder, as shown in Fi.3. The air void defect thickness and diameters are donated by hx, hy, d1, and d2, respectively.

As the size of the void changes, we can simulate the characteristic parameters of the actual defect. Since the open-circuit fault of TSV channels is directly related to the air void diameter, we assume $d_2 = 6$ μm, hx = hy = 5 μm for comparison.

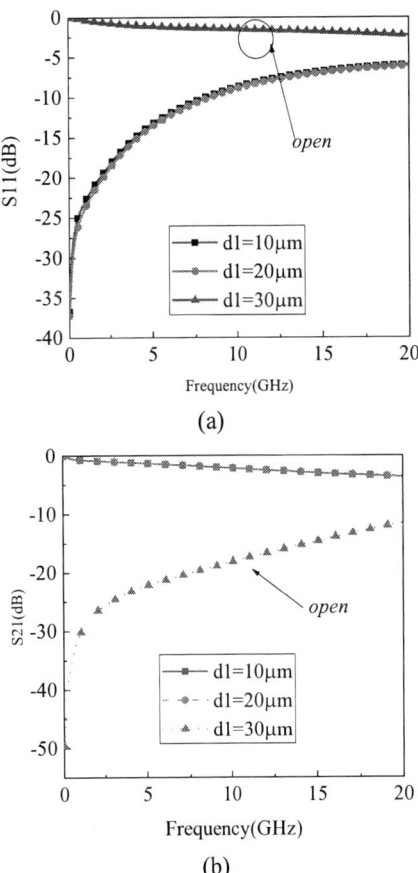

Fig. 4 Simulated S-parameters of TSV channel varying with air void defect diameter. (a) S11 and (b) S21 results.

Fig. 4 shows the simulated transmission performance with d1 from10 μm to 30 μm. It is shown that air void of the same diameter as the TSV can cause an open circuit.

978-1-6654-1392-3/21 $31.00 © 2021 IEEE   1041

C. Failure mechanism analysis

The electrical characters are analyzed through full-wave simulation and cascaded network calculation. For the TSV void defect, A scalable equivalent circuit model from the physical structure is established to explain its failure mechanism, as shown in Fig.5. The parasitic parameters can be calculated by [9][5].

Air void with the same diameter with TSV Cu plug causes the metal to break with RDL. This slit creates a parasitic capacitance C_{void}. At the same time, metal voids cause uneven distribution of interconnect materials and the change of TSV height. Then, the parasitic resistance and inductance are changed during this process. For TSV channels, each defect equivalent model is a part of the whole transmission network. So, the two-port network of daisy TSV channels is related with each circuit model. The transmission performance of the whole TSV channel varies with the air void dimension and position. The equivalent circuits based on silicon substate can be used in the multiple 3D stacked dies scheme.

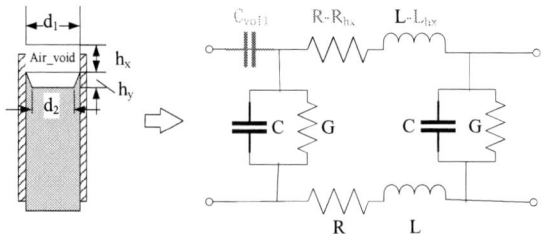

Fig.5 TSV void defect and the equal circuit model.

III. METHOD OF DEFECT MODULE DIAGNOSION

The defect type can be determined according to the characteristics of return and insertion loss from S-parameters. A simple scheme of GSG TSV channels is simulated, and their results are shown with the frequency from 100MHz to 20GHz. Combined with the finite element simulation model, the reflection characteristics of the transmission chain structure are analyzed. The defect size and defect localization are considered. By fitting the S parameters of defect TSV chain, the size of TSV open circuit defect and the localization of defect module are estimated.

S-parameters of the GSG-type fault TSVs with different design parameters are shown in Fig. 6. The changes of S21 are not remarkable, which due to the difference of a capacitance value. S11 with different height of TSV void are shown in Fig. 6(a). As the thickness of air void increasing, the S11 decreases. We can find that there will be a curve similar to the transmission performance of the faulty interconnects. Then, the dimensions of TSV air void can be inferred approximately.

Based on the simulated dimensions of TSV air void, the return loss is the main diagnosis basis of fault localization. It can be explained through circuit network. The distance from the incident port makes the reflection coefficient change. We select several typical localization coordinates. Note that the second lines accord well with the faulty symbolic line.

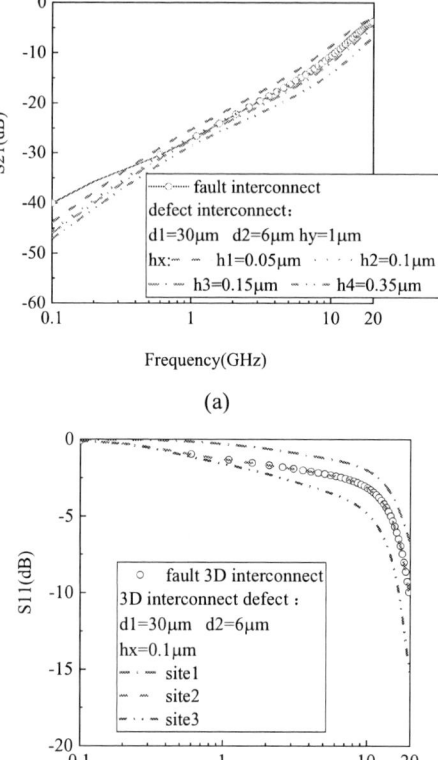

(a)

(b)

Fig.6 S-parameters of GSG-type TSV channels varying with (a) TSV defect height and (b) TSV defect locations.

IV. CONCLUSION

This paper proposed a combined cylindrical shape 3D TSV defect model for the nondestructive fault diagnosis and localization of TSV channels in 3D stacked dies. The GSG daisy TSV channels were simulated by 3D full wave simulator. What's more, the equivalent circuit of TSV defect was established to explain the fault principle. According to the tested S-parameters, the signal integration of TSV transmission chain with different open-circuit faults was analyzed, which greatly sped up the efficiency of defect localization. A method of combining circuit characteristics was proposed through finite element simulations. Considering the defect size and localization, the coordinate defect module model was established to fit the defect circuit characteristics.

ACKNOWLEDGMENT

Project supported by the Project supported by Guangdong Basic and Applied Basic Research Foundation (Grant No. 2021A1515011996), Science and Technology Program of Guangzhou, China (Grant No. 202102020520), the Key Laboratory Foundation (Grant No. 6142806200104) and the National Natural Science Foundation of China (NSFC)(Grant No. 61804032).

REFERENCES

[1] J. Cho et al., "Modeling and analysis of through-silicon via (TSV) noise coupling and suppression using a guard ring," IEEE Trans.

Compon., Packag., Manuf. Technol., vol. 1, no. 2, pp. 220–233, Feb. 2011.

[2] F. Wang, X. Liu and J. Liu, "Effect of Stirring on the Defect-Free Filling of Deep Through-Silicon Vias," IEEE Access, vol. 8, pp. 108555-108560, April, 2020.

[3] Y.-H. Lin et al., "Parametric delay test of post-bond through-silicon vias in 3-D ICs via variable output thresholding analysis," IEEE Trans. Comput.Aided Design Integr. Circuits Syst., vol. 32, no. 5, pp. 737–747, May 2013.

[4] X. Fang, Y. Yu, K. K. Xu, et al., "TSV-defect modeling, detection and diagnosis based on 3-D full wave simulation and parametric measurement," IEEE Access, vol. 6, no 1, pp. 72415-72426, Nov., 2018.

[5] K. N. Dang, A. B. Ahmed, A. B. Abdallah, et al., "TSV-OCT: A Scalable Online Multiple-TSV Defects Localization for Real-Time 3-

D-IC Systems," IEEE Trans. Very Large Scale Integr. (VLSI) Syst., vol. 28, no. 3, pp. 672–385, Mar. 2020.

[6] R. Rodríguez-Montañés, et al. "Postbond Test of Through-Silicon Vias With Resistive Open Defects," IEEE Trans. Very Large Scale Integr. (VLSI) Syst., vol. 27, no. 11, Nov. 2019.

[7] D. H. Jung, et al, "Through Silicon Via (TSV) Defect Modeling, Measurement, and Analysis," IEEE Trans. Compon., Packag., Manuf. Technol.,vol. 7, no. 1, pp. 138–152, Jan. 2017.

[8] Y. Shang, et al. "Detection and Diagnosis of TSV Composite Faults in 3D Integrated Circuits," Semiconductor Testing and Equipment, vol. 44, no. 12, pp. 976-982, Dec., 2019.

[9] Z. Mei, G. Dong, and Y. Yang, "Analysis of the Coupling Capacitance Between TSVs and Adjacent RDL Interconnections," IEEE Trans. Electromagn. Compat., vol. 61, no. 2, pp. 512-520, April, 2018.

A hybrid degradation modeling of light-emitting diode using permutation entropy and data-driven methods

1st Minzhen Wen[#]
Institute of Future Lighting, Academy for Engineering & Technology, Fudan University, Shanghai, China. #: first author

2nd Zhou Jing[#]
College of Mechanical and Electrical Engineering, Hohai University, Changzhou, China. #: first author

3rd Mesfin S. Ibrahim
College of Engineering, Kombolcha Institute of Technology-Kombolcha, Wollo University, Ethiopia

4th Jiajie Fan*
[1] *Institute of Future Lighting, Academy for Engineering & Technology, Fudan University,* Shanghai, China
[2] *EEMCS Faculty, Delft University of Technology,* Delft, The Netherlands
*Corresponding: jiajie_fan@fudan.edu.cn

5th Guoqi Zhang
[1] *EEMCS Faculty, Delft University of Technology,* Delft, The Netherlands
[2] Shenzhen Institute of Wide-Bandgap Semiconductors, China

Abstract—The LED degradation failure is highly dependent on temperature and this degradation failure is an irreversible energy dissipation process in thermodynamics. In this paper, the entropy generation is used to quantify the energy dissipation, which is regarded as one of the main performance characteristics of LED's degradation process. Considering the thermodynamic characteristics of entropy generation in the LED failure, a hybrid degradation prediction model based on the permutation entropy (PE) and data-driven methods was proposed. Firstly, a thermal aging test was designed for white LEDs in which the entropy generation rates (EGRs) of LEDs were extracted from the online collected thermoelectric performance parameters. Then, the EGRs of LEDs were treated as a time-series signal to perform phase space reconstruction and calculate PEs. Finally, both neural network model and Wiener process based data-driven methods were used to process the PEs. This hybrid model links the thermodynamic entropy of LEDs with its optical performance. The results show that: (1) Entropy generation based on thermodynamics can characterize the degradation process of LEDs; (2) The proposed hybrid degradation prediction model based on the PE and Wiener method can achieve early failure warning of LEDs before the actual failure occurs.

Keywords: Light-emitting diode, Entropy generation, Permutation entropy, Neural network, Wiener process

I. INTRODUCTION

The development of light-emitting diodes (LEDs) has attracted the attention of experts in the past few decades, and shown greater progress in the lighting industry. However, the failure mechanism and reliability modeling of LEDs from chip to system is too complicated to develop an accurate physics-based degradation model. Therefore, the data-driven methods are more widely used in LED reliability prediction. Specifically, the degradation modeling prediction refers to estimating the remaining useful life (RUL) or predicting the further reliability based on previous and current degradation data, as well as assessing the degree of deviation from the normal specified operation index [1]. Data-driven based degradation modeling methods mainly utilize amount of degradation data collected by sensors as training data to obtain degradation models of products and systems. Whereas, there are still some problems with data-driven methods used in LED degradation modeling, primarily because most data-driven approaches only take the degradation of LEDs, i.e., light output and color shift, but lack of considering its failure mechanism into modeling [2].

There are many parameters, such as temperature, humidity and electrical parameters, those can influence the degradation failure of LEDs and they can be measured online and effectively linked to RUL during the degradation process. Mostly, the degradation failure mechanism of LEDs is highly related to the temperature dependent physical or chemical reactions [3]. So, the effects of temperature on the degradation process should be first considered. In thermodynamics, temperature dependent degradation damage is an irreversible energy dissipation process. Such as, Cai *et al.* [4] pointed out that temperature was actually not a good representation of the temperature transmission state and they proposed another characterization method based on the principle of entropy increase, which treated entropy generation and EGR as new quantification models of material damage and energy dissipation. In 2015, Imanian and Modarres [6] proposed a reliability analysis and evaluation method based on the corrosion fatigue entropy generation function, which showed that a unified damage index could be defined under the concept of entropy generation. Cuadras *et al.* [7] presented a model to evaluate the thermal deterioration of LEDs considering EGR. They found that there was a strong correlation between lumen depreciation and EGR. Therefore, the concept of entropy generation can be used to quantify the energy dissipation, and is regarded as one of the key performance indicators of the degradation of LEDs.

Recently, several scholars have used data-driven methods, including machine learning algorithms and stochastic process methods, to achieve rapid and high-precision lifetime prediction of LEDs for general lighting [8]. For instance, Fan *et al.* [9] utilized neural network to predict the spectral characteristic parameters of white LEDs packages, and improved the prediction results by integrating genetic algorithms. Liu *et al.* [10] used two artificial neural network algorithms to simplify the lifetime prediction of the LEDs lighting system and achieved more accuracy. In terms of stochastic methods, Fan *et al.* [11] applied Wiener and Gamma methods to predict the lifetimes of ultraviolet LEDs with the improvement of precision and confidence intervals. Huang *et al.* [12] employed the Wiener method with attempt to characterize the degradation of mid-power white LEDs (WLEDs). In general, although there are some researches on the analysis of entropy or data-driven methods independently, a hybrid model for predicting the reliability of LEDs that incorporates thermodynamic uncertainties has not been explored. In this work, the entropy generation data is processed by the permutation entropy (PE) method to extract the physics-based failure mechanism of the LEDs. Combined

978-1-6654-1392-3/21 $31.00 © 2021 IEEE

with the data-driven methods, a hybrid degradation prediction model for LEDs based on PE of entropy generation data is proposed.

The remaining parts of the paper are arranged as follows: Section II presents the theory of hybrid degradation modeling for LEDs, and Section III describes the experimental setup, test samples and testing conditions. Section IV summarizes and analyzes the prediction results and the final section draws conclusions.

II. THEORY AND MODELING

In this section, it basically explains the entropy generation of LEDs and the specific calculation method of PE, as well as the neural network and Wiener process methods used in the degradation modeling.

A. Entropy generation and PE

The EGR can be obtained from the measurements of voltage, current and temperature. There is a certain correlation between the EGR and the degradation of optical performance [7]. This method has been evaluated in resistors, capacitors and LEDs, therefore, it is confirmed that the EGR can be used to describe the degree of degradation. Entropy generation can be stated as:

$$S = \int \mathrm{d}Q/T \qquad (1)$$

where Q is the heat dissipation of the system, and T is the junction temperature of the LED. The EGR refers to:

$$\dot{S} = \frac{\mathrm{d}S}{\mathrm{d}t} = \frac{P}{T} = \frac{UI}{T} \qquad (2)$$

where P is the thermal power or input power of the system, U is the input voltage, and I is the input current.

The PE algorithm is a signal mutation detection method proposed for the spatial characteristics of time series [13], which has received widespread attention as early as the beginning of the 21st century. Liu et al. [14] introduced it into the field of mechanical equipment fault diagnosis, and observed the influence of rolling bearing vibration signals under different working conditions, the effect of embedding dimension and time delay on machinery, along with the impact of the equipment state change and the vibration shock on the PE calculations. The results show that the PE could effectively detect the state change of the mechanical equipment as one of the most significant parameters. Afterwards, Nair U et al. [15] collected the noise signal generated during metal cutting, calculated the PE of the signal, and accurately located the starting moment of chatter during cutting. The method was seemingly ideal for online monitoring of metal cutting chatter. In short, PE might conveniently and precisely locate the moment of sudden change in the system, and has amplifying effect on small changes in signals.

In this study, the EGR data of the LED degradation process could be calculated. With the permutation entropy algorithm, a new performance parameter was extracted, which can be processed by data-driven methods and was related to the lumen maintenance lifetime of LED. The algorithm of basic principle is shown as follows:

The phase space is remade to acquire a matrix from a time series $\{x(i), i = 1,2,3,..,n\}$:

$$\begin{bmatrix} x(1)x(1+\tau)\cdots x[1+(m-1)\tau] \\ x(2)x(2+\tau)\cdots x[2+(m-1)\tau] \\ x(j)x(j+\tau)\cdots x[j+(m-1)\tau] \\ \vdots \\ x(K)x(K+\tau)\cdots x[K+(m-1)\tau] \end{bmatrix}, j = 1,2,\cdots,K \quad (3)$$

where m is embedding dimension and τ delay time; A row vector could be regarded as a reconstruction component, whose total number is K, and $K = n-(m-1)\tau$..

Next, rearrange the j-th reconstructed component in ascending order according to the numerical value:

$$x[i + (j_1 - 1)\tau] \leqslant x[i + (j_2 - 1)\tau] \leqslant \cdots \leqslant x[i + (j_m - 1)\tau] \quad (4)$$

Therefore, a row vector after reconstruction can attain a set of symbol sequences for each time series $x(i)$,:

$$S(l) = (j_1, j_2, \cdots, j_m) \qquad (5)$$

$S(l)$ is treated as a permutation. Assumed the probability of each symbol sequence is recognized as P_1, P_2, ..., or P_k, afterwards, according to the Shannon entropy form, PE expression concerning the k different symbols of the time sequence $S(l)$ can be stated as:

$$H_{PE}(m) = -\sum_{j=1}^{k} P_j \ln P_j \qquad (6)$$

The value of $H_{PE}(m)$ characterize the randomness of time series. The more disorder time series are closer to random. Meanwhile, the trends also reflect and amplify the minute changes in the time series.

B. Degradation modeling with back propagation neural network

In recent years, many scholars have applied machine learning or neural network prediction methods to accurately and rapidly predict the degradation of LEDs [16]. Yuan et al.[17] used deep learning methods to model the correlation between the thermal aging load and light output of LEDs. However, when dealing with degraded data mixing up time series, the intelligent algorithms including artificial neural networks also have some disadvantages, such as slow learning speed and easy to fall into local minima, which greatly affects the prediction accuracy[Error! Reference source not found.]. The long short-term memory neural network algorithm proposed by Jing et al. [18] effectively solved the accuracy bottleneck in the field of ultraviolet LED lifetime prediction.

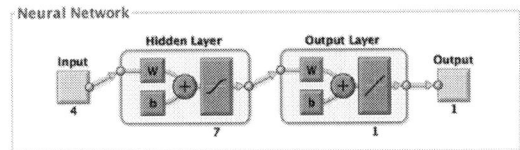

Fig. 1. Basic structure of BPNN

As shown in Fig.1, the back propagation neural network (BPNN) has strong nonlinear mapping ability. A 3-layer BPNN can approximate any nonlinear function. The output results of the model are forward propagation while errors are reversed. The BPNN, as one of the most common neural networks, has three layers with input layer, hidden layer and output layer in general mode. The main features are demonstrated: (1)The input layer inputs data and the output layer outputs results; (2)Neurons of the previous layer are connected to neurons of the next layer with the weight value, which is the meaning of back propagation [19].

In this work, we constructed a BPNN by arranging the entropy value, solder joint temperature, voltage, current, and output as the luminous flux. It can be seen that the number of input, output and hidden layer nodes is 4, 1 and 7 respectively.

C. Degradation modeling with stochastic Wiener process

The stochastic Wiener process method is commonly applied to model degradation processes. Often, the Wiener process method with drift is defined as follows [11]:

$$X(t) = \mu t + \sigma B(t) \qquad (7)$$

where μ and σ is the drift and diffusion parameter, and $B(t)$ the standard Wiener process.

One of the most significant properties of the Wiener process with drift $\{X(t), t \geq 0\}$ could be stated: the increment $\Delta X(t)$ with the time increment Δt obeys a normal distribution, which could be expressed: $X(t+\Delta t) - X(t) \sim N(\mu \Delta t, \sigma^2 \Delta t)$ for all $t, \Delta t \geq 0$.

In this study, the condition is that the deterioration of performance charactesrics of LEDs follows the Wiener process with the corresponding failure threshold ρ ($\rho > 0$). Fitting the normal distribution of the increments during the performance degradation is helpful to estimate μ and σ. The probability density function (PDF) of lifetime could be further defined [11]:

$$f_T(t) = \frac{\rho}{\sqrt{2\pi\sigma^2 t^3}} \exp\left[-\frac{(\rho - \mu t)^2}{2\sigma^2 t}\right] \qquad (8)$$

It is assumed that the lifetime distribution obeys an inverse Gaussian distribution [20]. The two equations, $v = \rho/\mu$ and $\lambda = \rho^2/\sigma^2$, could be substituted into $IG(v, \lambda)$. After substituting Eq. (8), the result is as follows:

$$f_T(t) = \sqrt{\frac{\lambda}{2\pi t^3}} \exp\left[-\frac{\lambda(t-v)}{2v^2 t}\right] \qquad (9)$$

III. EXPERIMENTAL DATA COLLECTING AND PROCESSING

In this section, it mainly introduces the test samples, experimental setup and data collection, along with preliminary calculation of PE value based on the accelerated degradation test (ADT) data of the test samples.

A. Test samples

The type of LED used in this work is a 3W high-power WLED light source, and the part number is ASMT-JN31-NTV01[Error! Reference source not found.]. The set of test samples consists of InGaN based phosphor converted white LEDs as shown in Fig. 2. A total of 9 sets of samples were measured and tested, while each set of samples is strictly carried out in accordance with the experimental requirements and guidelines to minimize human and environmental factors.

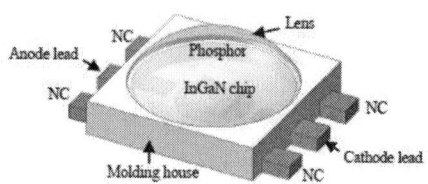

Fig. 2. LED package architecture schematic view[Error! Reference source not found.]

B. Experimental setup and data collection

According to the relevant recommended test conditions given by the selected light source (drive current $I_c = 350$ mA; forward voltage $V_F = 3.2$ volts), an ADT with high temperature condition was designed. The experimental setup is shown in Fig.3.

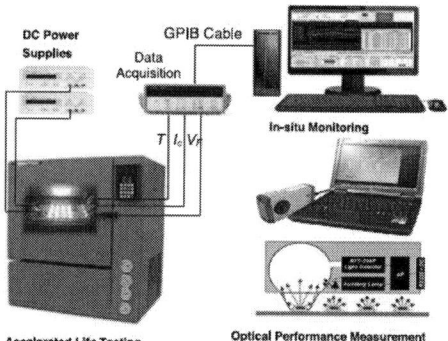

Fig. 3. Accelerated degradation test setup [23]

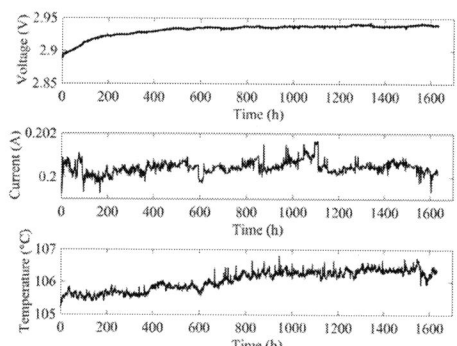

Fig. 4 . The collected thermal and electrical performance data of LEDs [23]

In this study, the test samples were stably driven by the DC electrical current ($I_c = 200$ mA) which was provided by a DC power supply (Agilent E3611A). A constant aging temperature ($T_a = 90$ °C) was kept in the chamber. The performance characteristics data of the LEDs including the lumen maintenance and color shift were taken using a Gigahertz-Optik BTS256-LED device. After the lumen maintenance data were collected, the test samples were returned to the thermal chamber to repeat the aging process every 23 hours. Electrical and thermal data (voltage, current, and solder joint temperature) were also measured, as depicted in Fig. 4.[Error! Reference source not found.] Finally, a total of nine groups of sample data were selected for the further modeling analysis.

C. Data processing with PE

PE values could be calculated via the time series of WLED's EGRs. The function phaseSpaceReconstruction(X) in MATLAB was utilized to estimate the two parameters in the Eq. (3), embedding dimension m and delay time τ, with the EGR data of the 9 LED test samples. The reconstruction parameters of each sample are shown in Table I. Then, a sliding window was set according to the parameters, and the PE values of the entropy production rate data $\dot{S}$ within the length of the window was sequentially calculated, recorded as PE($\dot{S}$) shown in Fig. 5(a). Similarly, the PE values of the increment data $\Delta\dot{S}$ of EGR were also calculated, recorded as PE($\Delta\dot{S}$) shown in Fig. 5 (b).

TABLE I. EGR AND ITS INCREMENTAL PHASE SPACE RECONSTRUCTION PARAMETERS

Samples	PE($\dot{S}$)		PE($\Delta\dot{S}$)	
	m	τ	m	τ
No.1	4	10	4	2
No.2	4	10	4	2
No.3	4	8	4	4
No.4	4	10	4	2
No.5	4	10	4	3
No.6	4	10	4	2
No.7	4	5	4	4
No.8	4	10	4	3
No.9	4	10	4	5

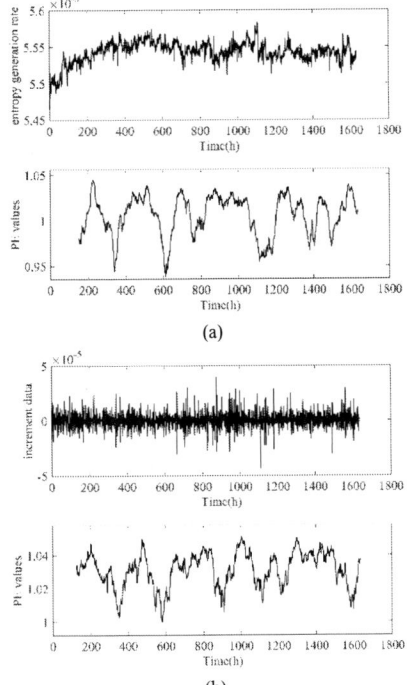

Fig. 5. (a) PE values of EGR of sample No.1, (b) PE values of the increment of EGR of sample No.1

IV. RESULTS AND DISCUSSION

A. Degradation modeling with BPNN

In this section, a BPNN was constructed by 4 input parameters, including PE($\dot{S}$) value, solder joint temperature, voltage, and current, along with one output as lumen maintenance. Among them, a total of 520 sets of data for the first 1-8 white LEDs were used as the training set, and 65 sets of data for the No.9 LED were test set. The prediction results are as follows:

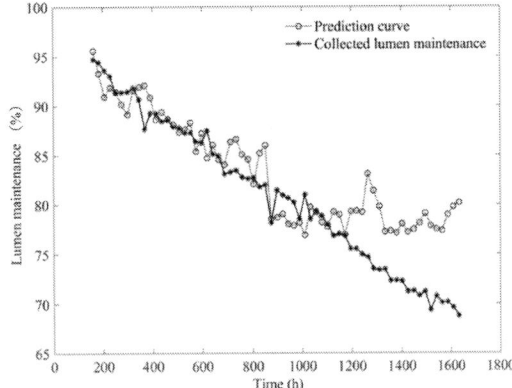

Fig. 6. BPNN prediction results from EGR

Judging from the results of this experiment, whether it is directly calculating PE value from the EGR or the increment data, no obvious powerful correlation of PE has been found, but the latter can show PE is potentially related to LEDs' life prediction. The possible reasons are as follows:

1) The data sample size is too small

In the neural network method, the larger the amount of data, the higher the accuracy. In the designed ADT, the total sample is only 1634 EGRs, which may not be enough for the actual sample size.

2) The design of relevant parameters in BPNN is unreasonable

In the BPNN algorithm, the number of hidden layers and nodes determine the complexity and precision of the neutral networks. It is generally suggested that as the number of hidden layers increases, errors could be reduced, which means improving accuracy but complicated network model. It means that extra training time of the model and the trend of "overfitting" appear. When it comes to obtain higher accuracy, the approach to increase the number of nodes is much easier to achieve than hidden layers.

3) The distribution of BPNN is not properly constructed

This paper selected solder joint temperature, voltage, current, and PE as input, while luminous flux as output to construct a BPNN. That is relatively straightforward and simple. In fact, life prediction is related to many factors, not limited to the above.

4) Multiple data processing reduces the accuracy

The entropy generation was calculated through current, voltage, and junction temperature. In this study, PE was calculated again from the entropy generation, and these data were re-used as the BPNN input. Multiple data processing increase the processing complicated and reduces the prediction accuracy.

Thus, in further research, we attempt to further explore the practical feasibility of this algorithm by adjusting the parameters, and then improve the accuracy of PE algorithm.

B. Degradation modeling with stochastic Wiener process

Taking the increment data $\Delta\dot{S}$ of the EGRs as an example, the variation of permutation entropy values was regarded as a stochastic process. Assuming that it satisfies the Wiener process, as in Eq. (7), the drift parameter μ was close to 0. The degradation increments $\Delta\text{PE}\left(\Delta\dot{S}\right)$ were tested for satisfying the normal distribution. And the parameters μ and σ were estimated by fitting the normal distribution.

In this paper, the early failure detection threshold ρ was defined as twice the difference between the upper and lower limits, as shown in the Fig. 7. Thanks to the corresponding failure threshold ρ, the estimated values ν and λ could be acquired in the inverse Gaussian distribution that the lifetime follows. The estimated Wiener process parameters of PE($\Delta\dot{S}$) data for each sample are shown in Table II.

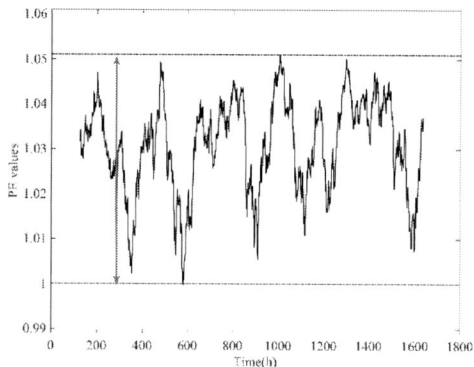

Fig. 7. PE($\Delta\dot{S}$) data and failure threshold of No. 1 LED sample

TABLE II.   THE ESTIMATED WIENER PROCESS PARAMETERS OF PE($\Delta\dot{S}$) DATA AND DETECTION LIFETMIE FOR EACH LEDS

Samples	ν	λ	Detection time (h)	Failure time (h)
No.1	23042.4	4199.7	1394.9	1518
No.2	6693.7	3258.5	1059.1	1518
No.3	276924.3	4066.2	1355.5	1518
No.4	41096.4	3725.1	1240.7	1518
No.5	6815.7	3052.0	995.7	1518
No.6	127071.4	3268.1	1089.4	1518
No.7	11075.1	4084.8	1341.7	1518
No.8	103856.4	3874.7	1291.5	1472
No.9	7492.3	3142.0	1027.7	1518

Therefore, the PDFs of the LED samples were obtained according to Eq. 9. Take the No. 1 LED sample as an example, its PDF curve was shown in the Fig. 8. The lifetime corresponding to the maximum probability density was the early failure detection time with the hybrid method with PE and stochastic Wiener process. The corresponding early failure detection time of all samples were predicted and recorded in Table II. As shown in Fig. 9, the early failure detection time was about 70-90% of the LED lumen maintenance lifetime L_{70}. It means that an early anomaly warning of LEDs can be achieved before the actual failure occurs.

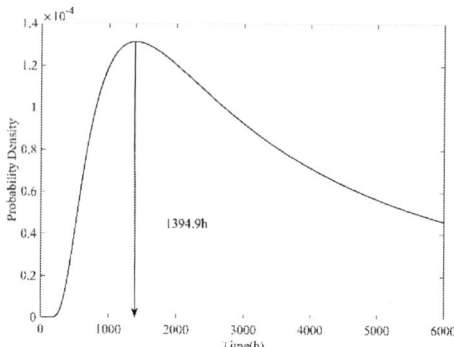

Fig.8. PDF curve of No. 1 LED sample with Wiener process modeling

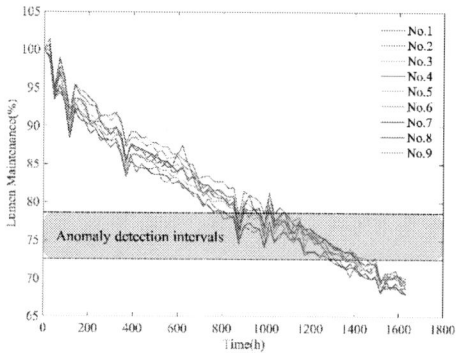

Fig.9. The corresponding lumen maintenance intervals of LEDs' under the hybrid model predicted the lifetimes

V. CONCLUSIONS

In this study, a hybrid degradation prediction approach, combining the permutation entropy (PE) with data-driven methods, is proposed to consider the temperate dependent degradation process. The result shows that: (1) Through real-time monitoring of electrical-thermal parameters, considering the thermodynamic uncertainty of the LED degradation process, the PE method with sliding window can be used to process the LED EGR signals, that can be regarded as a new LED degradation performance indicator; (2) The proposed hybrid model based on the PE and stochastic process method can achieve a well description of the LED degradation process, and an early failure detection and warning for LEDs.

ACKNOWLEDGMENT

This work was partially supported by the Key-Area Research and Development Program of GuangDong Province (2019B010131001) and the National Natural Science Foundation of China (51805147).

REFERENCES

[1] Tsai C. C., Tseng S. T., and Balakrishnan N., "Mis-specification analyses of gamma and Wiener degradation processes," Journal of Statistical Planning & Inference, vol. 141, no. 12, pp. 3725-3735, 2011.

[2] Ibrahim M.S, Fan J, Yung WKC, Prisacaru A, Driel W, Fan X, Zhang G. 2020. "Machine Learning and Digital Twin Driven Diagnostics and Prognostics of Light‐Emitting Diodes". Laser & Photonics Reviews. 14(12):2000254.

[3] Rammohan A., Ramesh Kumar C., Chandramohan V.P., "Experimental analysis on estimating junction temperature and service life of high power LED array," Microelectronics Reliability. 120:114121, 2021.

[4] Cai M , Liang Z , Tian K , et al. Junction Temperature Prediction for LED Luminaires Based on a Subsystem-Separated Thermal Modeling Method[J]. IEEE Access, 2019, PP(99):1-1.

[5] Cai M, Cui P, Qin Y, et al. "Entropy Generation Methodology for Defect Analysis of Electronic and Mechanical Components—A Review." Entropy, 2020, 22(2): 254.

[6] Imanian A, Modarres M. "A thermodynamic entropy approach to reliability assessment with applications to corrosion fatigue". Entropy, 2015, 17(10): 6995-7020.

[7] Cuadras A, Yao J, Quilez M. "Determination of LEDs degradation with entropy generation rate". Journal of Applied Physics, 2017, 122(14): 145702.

[8] Sun B., Jiang X., Yung K., Fan J., and Pecht M. G., "A Review of Prognostic Techniques for High-Power White LEDs," IEEE Transactions on Power Electronics, vol. 32, no. 8, pp. 6338-6362, 2017.

[9] Fan J., Li Y., Fryc I., Qian C., Fan X., and Zhang G., "Machine-Learning Assisted Prediction of Spectral Power Distribution for Full-Spectrum White Light-Emitting Diode," IEEE Photonics Journal, vol. 12, no. 1, pp. 1-18, 2020.

[10] Liu H.W. et al., "Lifetime prediction of a multi-chip high-power LED light source based on artificial neural networks," Results in Physics, vol. 12, pp. 361-367, 2019.

[11] Fan J, Jing Z, Cao Y, et al. "Prognostics of radiation power degradation lifetime for ultraviolet light-emitting diodes using stochastic data-driven models," Energy and AI, 2021, 4: 100066.

[12] Huang J, Golubović D S, Koh S, et al. Degradation modeling of mid-power white-light LEDs by using Wiener process. Optics express, 2015, 23(15): A966-A978.

[13] Feng F. Z., Rao G. Q., Si A. W., Sun Y. "Application and Development of Permutation Entropy Algorithm," Journal of Armored Force Engineering Institute. 26(02):34–38, 2012.

[14] Liu Y.B. Long Q., Feng Z. H., Liu W. L., "Detection Method for Nonlinear and Nonstationary Signals." Journal of Vibration and Shock.(12):131-134+176, 2007.

[15] Nair U, Krishna BM, Namboothiri VNN, Nampoori VPN. "Permutation entropy based real-time chatter detection using audio signal in turning process". Int J Adv Manuf Technol. 46(1–4):61–68, 2010.

[16] Prisacaru A., Guerrero E. O., Gromala P. J., Han B., and Zhang G. Q., "Degradation Prediction of Electronic Packages using Machine Learning," in 2019 20th International Conference on Thermal, Mechanical and Multi-Physics Simulation and Experiments in Microelectronics and Microsystems (EuroSimE), 2019, pp. 1-9: IEEE.

[17] Yuan CCA, Fan J, Fan X. "Deep machine learning of the spectral power distribution of the LED system with multiple degradation mechanisms". Journal of Mechanics 37:172-183.

[18] Jing, Z. Liu, M.S. Ibrahim, J. Fan, X. Fan and G. Zhang, "Lifetime Prediction of Ultraviolet Light-Emitting Diodes Using a Long Short-Term Memory Recurrent Neural Network," IEEE Electron Device Lett., vol. 41, no. 12, pp. 1817-1820, Dec. 2020.

[19] Lu K., Zhang W., and Sun B., "Multidimensional Data-Driven Life Prediction Method for White LEDs Based on BP-NN and Improved-Adaboost Algorithm," IEEE Access, vol. 5, pp. 21660-21668, 2017.

[20] Chhikara R. The inverse Gaussian distribution: theory: methodology, and applications. CRC Press, 1988.

[21] Ibrahim MS, Fan J, Yung WKC, Wu Z, Sun B. 2019. "Lumen Degradation Lifetime Prediction for High-Power White LEDs Based on the Gamma Process Model". IEEE Photonics J. 11(6):1–16.

[22] ASMT-Jx3x, Data Sheet: 3W Mini Power LED Light Source; Avago Technologies: San Jose, CA, USA, 2012.

[23] Fan J, Cheng Q, Fan X, et al. "In-situ monitoring and anomaly detection for LED packages using a Mahalanobis distance approach," International Conference on Reliability Systems Engineering. IEEE, 2016.

Research on adjustment of the electrical resistivity of Aluminum Nitride ceramic

Jinhu Fan[1*]

School of Chip Industry

Hubei University of Technology

Wuhan, China

20181114@hbut.edu.cn

Zirong Tang[2]

School of Mechanical Science and Technology

Huazhong University of Science and Technology

Wuhan, China

zirong@hust.edu.cn

Jie Wang[3]

School of Material Science and Technology

Huazhong University of Science and Technology

Wuhan, China

jwong@hust.edu.cn

Abstract-Rahbek electrostatic chucks (J-R ESC) with a surface layer with lower resistivity could produce a bigger clamping force under low applied voltage and has been widely studied and used. In this paper, the adjustment of resistivity of aluminum nitride ceramic has been studied. Taking Y_2O_3 and TiN as the modifier, two groups experiment, group A and B, among which group A takes 2wt. %, 6wt. % Y_2O_3 as additives while group B with 2wt. %, 6wt. % $MoSi_2$ and 2% Y_2O_3 as additives. After ball-milling and tape casting, they were kept d at 1800℃ 2∼5h to acquire a sintered sample with a diameter of 50mm and a thickness of 1mm by SPS method. The result of the resistivity measurement showed that the resistivity of the two samples in group A had a small decrease, while the samples in group B decreased drastically to the degree of $10^{10}\Omega\bullet cm$.

Keywords—aluminum nitride; electrostatic chuck; resistivity; Y_2O_3; $MoSi_2$

I. INTRODUCTION

In semiconductor manufacturing industry, silicon wafer always need to be fixed in the etching, washing, and other processes. The stationary and clamping of the wafer has a significant dependence on the quality of the wafer. Electrostatic chuck can produce homogeneous and strong enough electrostatic force to clamp the wafer, which avoids the wafer break and wafer pollution caused by mechanical clamping and vacuum clamping [1][2][3]. There are two different kinds of electrostatic chuck, Coulomb style and Johnsen-Rahbek style. Among them, the Johnsen-Rahbek electrostatic chucks (J-R ESC) has a surface layer with lower resistivity and could produce a bigger clamping force under low applied voltage than that of Coulomb style, which has been widely studied and used as the mainstream of the market development [4][5][6] .

Due to its expansion coefficient closing to that of wafer, high thermal conductivity and good corrosion resistance, aluminum nitride ceramic could be a good choice to prepare electrostatic chuck. However, according to the principle of electrostatic clamping, the resistivity of the top layer is the main determine factor the clamping force [7]. As an insulator, the adjustment of the resistivity of the aluminum nitride was a critical issue. But there have scarcely been study and papers on the adjustment of resistivity. In this paper, the principle of the resistivity adjustment has been studied.

Considering the special sintering requirement of AlN ceramic, the effect of atmosphere and additives such as rare earth oxide and alkaline oxide on the resistivity must be considered, typically like Y_2O_3, Sm_2O_3, CuO, etc. [8].

Jin-Wook Lee g et al. found that the volume resistivity of AlN decreased when added Y_2O_3, Al_2O_3 and CaO to AlN power, and sintered in graphite crucible under nitrogen atmosphere due to the generation of $Y_3Al_5O_{12}$ and $CaAl_4O_7$ [9]. Jun Yoshikawa et al. reduced the electrical resistivity to $10^{10} \sim 10^{12}\Omega\bullet cm$ through adding 1.0–2.9 wt. % Sm_2O_3 to AlN ceramic and sintered together at more than 1800 ℃, with a lower resistivity phase-Sm-β-alumina phase which could decrease the resistivity of aluminum nitride ceramic [10].

Takafumi Kusunose et al. found a drastical decrease in electrical resistivity taking Y_2O_3 powder as additives and attributed it to the generation of three dimensional (3-D) grain boundary phase [Y (O, C) $_{1-x}$] which have an excellent electrical conductivity of $7.1\cdot 10^3$ S/cm [11] [12].

Huina Ma et al. found that when the content of conductive particles such as tungsten in AlN matrix below the certain threshold value a conducting pathway formed, which will cause the decrease of electrical resistivity of the composites [13].Amir Azam Khan and Jean Claude Labbe found a sharp decrease over a series of samples with varying molybdenum concentration, whose volume fraction increased from 0.2 to 0.22 (0.75×10^9 ohm cm and 3.95×10^{-3} ohm cm, respectively) based on the percolation phenomena [14].

Fig. 1. The Aluminum Nitride electrostatic chuck after resistivity adjustement

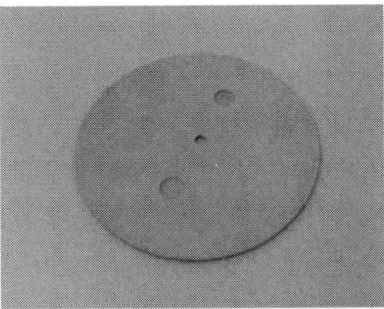

Based on the primary investigation, we consider that it would be a useful way to add a low electrical resistivity phase into AlN matrix. Then its resistivity could be controlled at the presupposed range [15]. Among so many refractory metal, molybdenum seems to one good choice because of its good thermal and electrical properties, and its thermal expansion coefficient close to that of aluminum nitride. However, due to its high melting temperature (2610 °C), we choose molybdenum disiticide ($MoSi_2$) as the resistivity modifier to replace it. $MoSi_2$ is a compound possessed double peculiarity of metal and ceramic which also has excellent thermal-electric conductivity, and a low thermal expansion coefficient like AlN. Additionally we choose Y_2O_3 as the sintering aid. It also can decrease the resistivity of AlN ceramic in some extent.

II. EASE OF USE

A. Sample preparation

An available commercially AlN powder (Aluminum KK Corp., Tokyo) was used as the starting material. The purity is high to 99.9%. The particle sizes ranges in 0.5~2μm with an average granularity of 1μm. A 99.99% pure Y_2O_3 and $MoSi_2$ powder (Sin pharm Chemical Reagent Co., Ltd) was used as additives to modify the resistivity because of the resistivity adjustment effect of $MoSi_2$ and the sintering promotion of Y_2O_3.The mass ratio of them is listed in Table 1. After precise weighting, these powder were mixed by the conventional wet ball milling method with Al_2O_3 ball and absolute ethyl alcohol for 24h in a polyethylene bottle to obtain a homogeneously mixed powder. Then the mixed powder was dried in a vacuum drying oven, and pressed into 30 mm diameter discs with a thick of 5mm under a uniaxial pressure of 50 MPa and an isotactic pressure of 200 MPa. The powder compacts were placed in a graphite die and

sintered by spark plasma sintering (SPS) method in a nitrogen atmosphere. The sintering temperature is kept at 1800 °C for 4h with a heating rate of 100 °C/min. Before measurement, the surface carbon layer should be removed away. The sintered body of AlN Chuck is shown as Fig. 1.

TABLE 1 THE COMPOSITION OF THE FOUR GROUPS

Number		Y_2O_3	$MoSi_2$	AlN
A	1	2wt%	0	98wt%
	2	6wt%	0	94%wt
B	3	2wt%	2wt%	96wt%
	4	2wt%	6wt%	92wt%

B. Characterization

The phase composition was characterized by X-ray diffract meter (XRD-7000S, Shimadzu CO., LTD) with Cu Kα radiation. The relative density was measured through drainage method based on Archimedes principle. The volume resistance of the sintered AlN samples was measured by Digital High Resistance Micro-Current tester instrument (EST121, Beijing municipal institute of labour protection). Then the volume resistivity of the samples was calculated according to the Classic formula, as in:

$$\rho = RS/L = R\pi D^2/4L$$

III RESULTS AND DISCUSSIONS

A. Relative density

We measured the quality of the AlN samples before and after dipped in the water. The results are listed in the Table 2.

TABLE 2 QUALITY MEASUREMENT RESULT

Number	Actual density(g/cm³)	Theoretical density(g/cm³)
1	3.298	3.283
2	3.340	3.318
3	3.358	3.314
4	3.86	3.378

According to the principle of drainage method, the relative density value could be calculated out. The density value of the four samples, η1, η2, η3, η4 are respectively 99.46%, 100.66%, 100.03%, 100.24%. Compared the density value of 2# with other groups, we found that the promoting sintering effect of Y_2O_3 did not increase as its content increases. According to the comparision between 1#, 2# and 3#, the results showed that $MoSi_2$ could also promote the sintering and densifying of aluminum nitride ceramic. Furthermore, $MoSi_2$ had a stronger promotion effect on the sintering degree of AlN samples than Y_2O_3. In fact, the actual volume is lower than that the theoretical value due to the particle interval and the reaction during sintering, the theoretical density value is low, while the density value is high.

B. Resistivity

Table 3 shows the results of the resistivity measurement. The results showed that both Y_2O_3 and $MoSi_2$ had an influence on the electrical resistivity and the latter had a stronger impact, just as Hiroaki SAKAI and Yuji KATSUDA's experiment [16].

TABLE 3. RESISITIVITY VALUES OF THE FOUR

GROUPS

Number	Resistive(Ω)	Calculated resistivity($\Omega\cdot cm$)
1	7×10^{11}	6×10^{12}
2	8×10^{10}	6×10^{11}
3	2×10^{10}	1×10^{11}
4	2×10^{9}	4×10^{9}

In order to further analyze the mechanism of resistivity changing of the four groups samples, of XRD analysis was conducted to detect the composition of the samples after sintering, as the results presented in *Fig.2*. We found that the phase compositions of different groups are almost the same but a new phase-$Y_3Al_5O_{12}$ appeared as a generation of the reaction between Y_2O_3 and Al_2O_3. It's considered that $Al_5Y_3O_{12}$ existed as a glass state and had a slightly lower resistivity than AlN, which could only slightly increase the electron concentration in the grain boundary at some extent. It also showed that $MoSi_2$ did not change during the sintering process and still keep existing as a good electric conductor with a conductivity of $2.15 \times 10^{-5} \Omega\cdot cm$. Thus it can produce more free electrons and drastically reduce the resistivity of AlN matrix sintering samples than $Y_3Al_5O_{12}$ to a degree below 10^{10} $\Omega\cdot cm$.

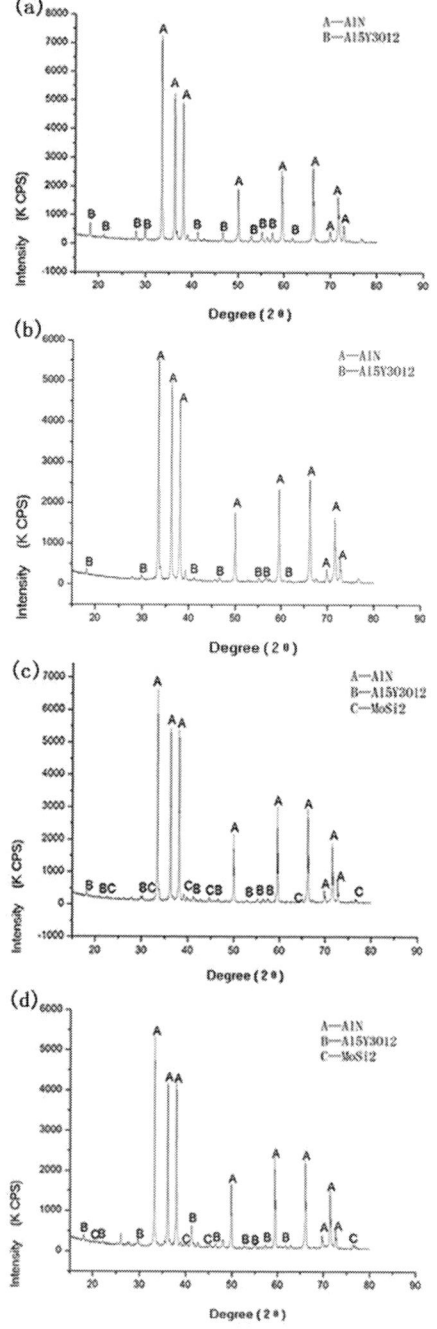

Fig.2 XRD diagrams of the four samples: (a) 1#; (b) 2#; (c) 3#; (d) 4#

C. Micro-Structure

In the four groups samples, only 4# achieved the resistivity ruquirement for J-R ESC. Then we observed its cross-section morphology through Field Emission Scanning Electron Micro Scope. As shown in *Fig.3*, the sample was densely sintered and with some light phase at the grain boundaries, which might be $MoSi_2$ and need further more precise composition analysis.

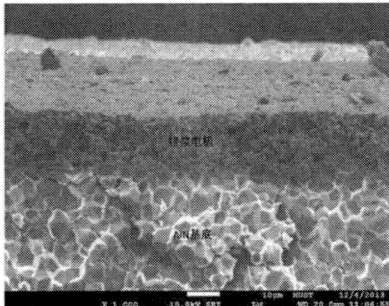

Fig.3 the micro morphology of 4#

IV. SUMMARY

It is possible to control the electrical resistivity of insulating AlN ceramics in a wide range without losing its inherent excellent property like intrinsic high thermal conductivity by adding a conductive phase such as $MoSi_2$ as a modifier. Furthermore, Y_2O_3 also could also reduce the resistivity of AlN in some extent by the reaction product through reaction with the Al_2O_3 located at the surface of AlN particle. Therefore, it can be expected that p electrical resistivity of AlN are controlled by varying the amount of added low resistivity phase or conductive phase.

REFERENCES

[1] Yoshikazu S., Toru S. The Ceramic Society of Japan. Advanced Ceramic Technologies & Products: 417-420. Springer, 2012.

[2] George A. Wardly. Electrostatic wafer chuck for electron beam micro fabrication. Review of Scientific Instruments, 2003, 44(10): 1506-1509.

[3] Asano K, Hatakeyama F, Yatsuzuka K. Fundamental study of an electrostatic chuck for silicon wafer handling. Industry Applications, IEEE Transactions on, 2002, 38(3): 840-845.

[4] Yatsuzuka K, Toukairin J, Asano K, et al. Electrostatic chuck with a thin ceramic insulation layer for wafer holding. Industry Applications Conference, 2001. Thirty-Sixth IAS Annual Meeting. Conference Record of the 2001 IEEE. IEEE, 2001, 1: 399-403.

[5] Kusunose T, Sekino T, Niihara K. Fabrication and Microstructure of Electrically Conductive AlN with High Thermal Conductivity. Key Engineering Materials, 2011, 484: 57-60.

[6] Lee, H. J., Kim, S. W., Ryu, S. S.. Sintering behavior of aluminum nitride ceramics with MgO–CaO–Al2O3–SiO2 glass additive. International Journal of Refractory Metals & Hard Materials, 2015, 53: 46–50.

[7] Tamagawa K, Yonekura H. Electrostatic chuck: U.S. Patent 8,023,248. 2011-9-20.

[8] Hwang, J. G., Oh, K. S., Chung, T. J., Kim, T. H., Paek, Y. K.. Low-Temperature Sintering Behavior of Aluminum Nitride Ceramics with Added Copper Oxide or Copper. Journal of The Korean Ceramic Society, 2019, 56(1): 104–110.

[9] Lee J.W., Lee W.J., Lee S. M. Electrical behavior of aluminum nitride ceramics sintered with yttrium oxide and titanium oxide. J. Korean Ceram. Soc., 2016, 53: 635-640

[10] Kaya, P., Suyolcu, Y. E., Aken, P. A. van, Turan, S., Gregori, G., Maier, J.. Grain Boundary Blocking Effects in Sm/Yb-doped AlN Ceramics. Journal of The European Ceramic Society, 2021, 41(9): 4870–4875.

[11] Kusunose T, Sekino T, Niihara K. Production of a grain boundary phase as conducting pathway in insulating AlN ceramics. Acta Materialia, 2007, 55(18): 6170-6175.

[12] Yang N, Tan X, Ma Z, et al. Fabrication and Characterization of $Ce_{0.8}Sm_{0.2}O_{1.9}$ Microtubular Dual - Structured Electrolyte Membranes for Application in Solid Oxide Fuel Cell Technology. Journal of the American Ceramic Society, 2009, 92(11): 2544-2550.

[13] Ma H, Yang Z, Du J. Influence of tungsten particles on the electrical properties of AlN ceramic. Journal of Materials Science: Materials in Electronics, 2012, 23(12): 2181-2185.

[14] Hwang, J. G., Oh, K. S., Chung, T. J., Kim, T. H., Paek, Y. K.. Low-Temperature Sintering Behavior of Aluminum Nitride Ceramics with Added Copper Oxide or Copper. Journal of The Korean Ceramic Society, 2019, 56(1): 104–110.

[15] Tangen I L, Yu Y, Grande T, et al. Preparation and characterization of aluminum nitride–titanium nitride composites Journal of the European Ceramic Society, 2004, 24(7): 2169-2179.

[16] Sakai H, Katsuda Y, Masuda M, et al. Effects of adding Y_2O_3 on the electrical resistivity of aluminum nitride ceramics. Journal of the Ceramic Society of Japan, 2008, 116(1352): 566-57

Solder Preforms Composed of High Cu-content Sn–xCu Alloys for Power Electronic Packaging and Characterization of the Processing Performance and Joint's Properties

Ru-Zeng Shi
Lab of Smart Materials and Electronic Packaging in School of Materials Science & Engineering, and Guangdong Provincial Engineering Technology R&D Center of Electronic Packaging Materials and Reliability
South China University of Technology
Guangzhou, 510640, China
msshiruzeng@mail.scut.edu.cn

Ming-Qiang Chen
Lab of Smart Materials and Electronic Packaging in School of Materials Science & Engineering, and Guangdong Provincial Engineering Technology R&D Center of Electronic Packaging Materials and Reliability
South China University of Technology
Guangzhou, 510640, China
msmingqiangchen@mail.scut.edu.cn

Hai-Jun Huang
Lab of Smart Materials and Electronic Packaging in School of Materials Science & Engineering, and Guangdong Provincial Engineering Technology R&D Center of Electronic Packaging Materials and Reliability
South China University of Technology
Guangzhou, 510640, China
mshjhuang@mail.scut.edu.cn

Min-Bo Zhou*
Lab of Smart Materials and Electronic Packaging in School of Materials Science & Engineering, and Guangdong Provincial Engineering Technology R&D Center of Electronic Packaging Materials and Reliability
South China University of Technology
Guangzhou, 510640, China
msmbzhou@scut.edu.cn

Xin-Ping Zhang*
Lab of Smart Materials and Electronic Packaging in School of Materials Science & Engineering, and Guangdong Provincial Engineering Technology R&D Center of Electronic Packaging Materials and Reliability
South China University of Technology
Guangzhou, 510640, China
mexzhang@scut.edu.cn

Abstract—With the rapid development of power electronics, how to make the power chips run stably at high temperature has become an important issue. This paper reports our latest work on development of Sn–xCu solder preforms with high Cu-contents (4 to 12 wt.%) for fabricating Cu/Sn–xCu-preform/Cu die-attachment joints whose interface and matrix are composed of Cu-Sn intermetallic compound (IMC) skeleton to ensure the joints capable of withstanding high temperature impact. Results show that the amount of Cu-Sn IMC skeleton in the Sn–Cu alloy can be changed by increasing Cu content. The fineness and uniformity of dispersion of the Cu-Sn IMC skeleton in Sn–xCu solder preforms with different Cu contents can be tailored by non-continuous rolling. Sandwich structure Cu/Sn–xCu-preform/Cu die-attachment joints are prepared by means of a conventional pressure reflowing process, and microstructures of solder preforms and die-attachment joints are characterized. The results show that the fraction of β-Sn decreases and the amount of Cu_6Sn_5 increases with increasing Cu content in Sn–xCu alloys. During subsequent non-continuous rolling process, the microstructure of Sn–xCu solder preforms becomes finer, meanwhile the dispersion of IMC is improved during the rolling process. Reflowed Cu/Sn–xCu-preform/Cu die-attachment joints consist of the Cu-Sn IMC skeleton filled with the β-Sn phase and exhibit high shear strength over 50 MPa.

Keywords—power electronics, high Cu-content Sn–xCu alloy, solder preform, intermetallic compound, die-attachment

I. INTRODUCTION

Power semiconductor devices are responsible for the core functions of voltage transformation, frequency conversion and DC/AC interchange in power electronic systems, and are widely used in various applications, such as new energy vehicles, industrial controllers and power converters [1, 2].

Recently, in order to meet the requirements of high power, high frequency and high integration in the power electronics, there has been an increasing trend in the replacement of conventional semiconductors such as silicon (Si) with an operating temperature threshold of only 150 °C by third-generation semiconductor materials, typically silicon carbide (SiC) and gallium nitride (GaN), which can work normally at 300 °C [3]. Therefore, the die-attachment materials with high temperature working stability have become an urgent need to match the excellent performance of high-power electronics. At present, a large number of soldering materials are still employed in power device packaging. High Pb-content solder pastes or preforms are still used in the high-power electronic packaging because thus far there are no suitable high melting point lead-free Sn-based solders as a substitute. Some high melting point Au-based solder alloys, such as Au–20Sn with melting point of 280 °C and Au-3Si at 363 °C, have been adopted in power device packaging. However, they are not widely used because of high material cost. The eutectic Zn–6Al alloy with melting point of 380 °C can be a candidate, but it is easily oxidized and corroded [4]. In addition, high melting point of the solders and thereby high temperature process lead to increased requirements of the soldering equipment, so these solders are not widely used in power device packaging. To fully utilize the excellent performance of SiC power devices, there is an urgent need to develop die-attachment materials that can be used for soldering at relatively low temperature, while the soldered joint can serve at high temperature. A recent study [5] showed that when the gap-size of the lead-free solder joints is reduced to 10 μm level, the upper and lower interfacial intermetallic compound (IMC) layers of the joint can gradually grow to contact and interact directly, resulting in formation of the full IMC joint, which is regarded as being able to withstand the impacts of high working temperature. However, making the attachment joint consisting entirely of IMC phase requires an extremely long-term heating process and minimal gap-size level, which certainly increases the difficulty and cost of the power electronic packaging process.

In a typical power electronic packaging structure, the die-attachment material is located at the interface between die and substrate, as depicted in Fig. 1, and the die-attachment joint plays the role in realizing electrical interconnection, mechanical connection and heat transfer. There are several conventional die-attachment methods, including the soldering by using high melting point solders, electrically conductive adhesive (ECA) bonding, thermocompression bonding, metal nanoparticle sintering and transient liquid phase (TLP) bonding. Notably, TLP is based on the principle of diffusion and interfacial eutectic reaction between two metals with different melting points. For the TLP bonding couple with double-sided Cu metallization (substrate) and the Sn-based alloy in the middle, Cu and Sn atoms diffuse each other and react to form Cu-Sn IMC with much higher melting point than that of the Sn-based alloy, so as to achieve the purpose of low-temperature reflowing and high-temperature service. A recent study showed the attempt to realize TLP bonding for die-attachment by using electroless tin-coated copper particles (i.e., Cu@Sn powder preform) [6], and the solder joint composed of Cu-Sn IMCs (Cu_6Sn_5 and Cu_3Sn) skeleton filled with Cu particles was obtained, which could bear the high temperature. However, both the process complexity and manufacturing cost for fabrication of such Cu@Sn core-shell powders are very high. In the present work, we report a lead-free Sn–Cu based solder preform with high Cu content (i.e., Sn–xCu) fabricated by using a facile casting and rolling process, which is employed to prepare the Cu/Sn–xCu-preform/Cu die-attachment joint, and the interface and matrix of the solder joint are composed of Cu-Sn IMC skeleton to ensure the solder joint capable of withstanding effectively high temperature impact.

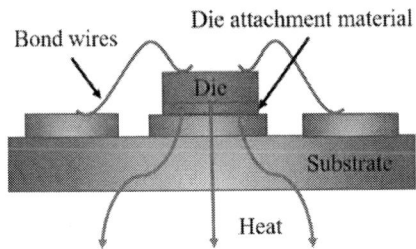

Fig. 1. Schematic of a power electronic package structure.

II. EXPERIMENTAL PROCEDURES

Sn–xCu alloys with different Cu contents of 0.7, 4, 6, 8, 10 and 12 wt.% were smelted, and microstructures of Sn–xCu alloys were analyzed. Sn–xCu solder preforms were fabricated by non-continuous rolling and Cu/Sn–xCu-preform/Cu die-attachment joints were prepared by using the optimized Sn–xCu alloy preforms. The microstructures of solder preforms and die-attachment joints were characterized; meanwhile the joints were tested to obtain the shear strength. Combined with the comprehensive analyses, a solder preform of high Cu-content Sn–xCu alloy has been developed, which shows great potential for realizing "reflow at low temperature

and service at high temperature" as required by power electronic packaging.

A. Sn–xCu Alloy Smelting

According to the Sn–xCu binary phase diagram, the Cu content in the Sn–xCu alloy was preliminary selected to be 4, 6, 8, 10, 12 wt.%, respectively, and the eutectic Sn–0.7Cu alloy was also smelted for a comparison. After removing surface oxides from the high-purity Sn blocks and Cu plates, they were cleaned ultrasonically in a bath of ethanol. The alloy was smelted at 500 °C for 30 min, and during the dwelling process the alloy melt was mechanically stirred regularly. Finally, the evenly melted alloy was poured into a rectangular steel mold and cooled down in the air to obtain the Sn–xCu alloy ingot.

B. Solder Preform Rolling

To facilitate the die-attachment soldering, Sn–xCu alloy ingots were made into solder preforms by non-continuous rolling using a two-roller hot-calender machine at temperature of 100 °C with a rolling speed of 2.0 ~ 4.0 m/min. The final thickness of the solder preform is about 0.2 ± 0.02 mm.

C. Preparation of Cu/Sn–xCu-preform/Cu Joints

Sandwich structure Cu/Sn–xCu-preform/Cu die-attachment joint was assembled by using a dummy die (simplified as Cu sheet) with dimension of 3 mm × 3 mm × 0.8 mm, Sn–xCu preform with size of 3 mm × 3 mm × 0.2 mm and Cu substrate with geometry of 10 mm × 10 mm × 1 mm, as illustrated in Fig. 2. The assembled sample was reflowed at a constant temperature of 255 °C for 15 s under a small pressure of 0.02 MPa. At reflow temperature of 255 °C, tin in the preform melted, while the shape of IMCs skeleton in the solder preform remained largely unchanged.

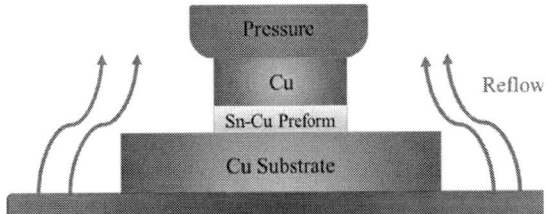

Fig. 2. Schematic of preparation of sandwich structure Cu/Sn–Cu-preform/Cu die-attachment joint.

D. Analysis of Microstructure and Composition

The as-casted Sn–xCu alloys and rolled solder preforms as well as reflowed die-attachment joints were embedded in epoxy resin by cold mounting process for metallographic analysis. After grinding, polishing and etching, the microstructures and compositions of Sn–xCu solder alloys, solder preforms and die-attachment joints were analyzed by scanning electron microscope (SEM, Phenom ProX) equipped with energy-dispersive spectrometer (EDS). In addition, to observe the morphological distribution of IMC in the solder preforms and reflowed joints more directly, deep selective etching was carried out to etch away most of the tin on the surface of the solder preforms and joints, so as to better observe the three-dimensional structure of the IMC skeleton. The deep etching solution was 3%HCl+7%HNO$_3$+90% deionized water, and the deep etching lasted for 30 min.

E. Shear Test

Shear strength of die-attachment joints was measured by a multifunctional mechanical testing machine (MFM1200, TRY Precision). For die-attachment joints by using solder preforms with different Cu contents, at least ten identical samples with the same Cu content in the solder preform were tested so as to obtain the average shear strength of the joints. A loading rate of 100 μm/s and a loading height of 0.3 mm were employed during shear tests, and the schematic diagram of the shear test is shown in Fig. 3.

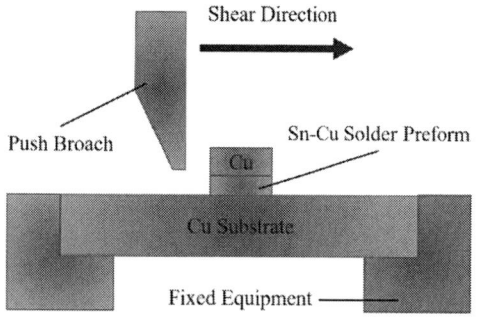

Fig. 3. Schematic diagram of the shear test.

III. RESULTS AND DISCUSSION

A. Microstructures of Sn–xCu Alloys

Fig. 4 shows the microstructures of as-casted Sn–xCu alloys. Elemental analysis was conducted at the points in Fig. 4 (e2), and the results are shown in Table I. The results indicate that the pale gray phase is the β-Sn phase, the medium gray phase is the Cu_6Sn_5 phase, and the darkest phase is the Cu_3Sn phase. It can be seen clearly from Fig. 4 that the area percentage of β-Sn phase in the Sn–xCu alloy decreases with the increase of Cu content, and meanwhile the Cu_6Sn_5 primary phase appears and gradually aggregates into a fishbone morphology. Moreover, when the Cu content is not higher than 8 wt.%, only β-Sn and Cu_6Sn_5 phases exist in Sn–xCu alloys. For high Cu contents of 10 wt.% and 12 wt.%, the Cu_3Sn phase is formed in Sn–10Cu and Sn–12Cu alloys, but it only locates in the center of the Cu_6Sn_5 phase. Particularly, in the Sn–12Cu alloy, the Cu-Sn IMC phases (Cu_6Sn_5 and Cu_3Sn) largely exist in the form of strips and occupy most of the alloy area compared with other Sn–xCu alloys, as presented in Fig. 4(f1) and (f2). In addition, when the Cu content is less than 6 wt.%, the distribution of IMC phases is too dispersed to form the skeleton structure, but when the Cu content reaches 8 wt.%, the amount of IMCs and the distribution density are significantly increased. Therefore, Sn–xCu alloys with Cu contents of 8 wt.%, 10 wt.% and 12 wt.% are preferentially used for making solder preforms by rolling.

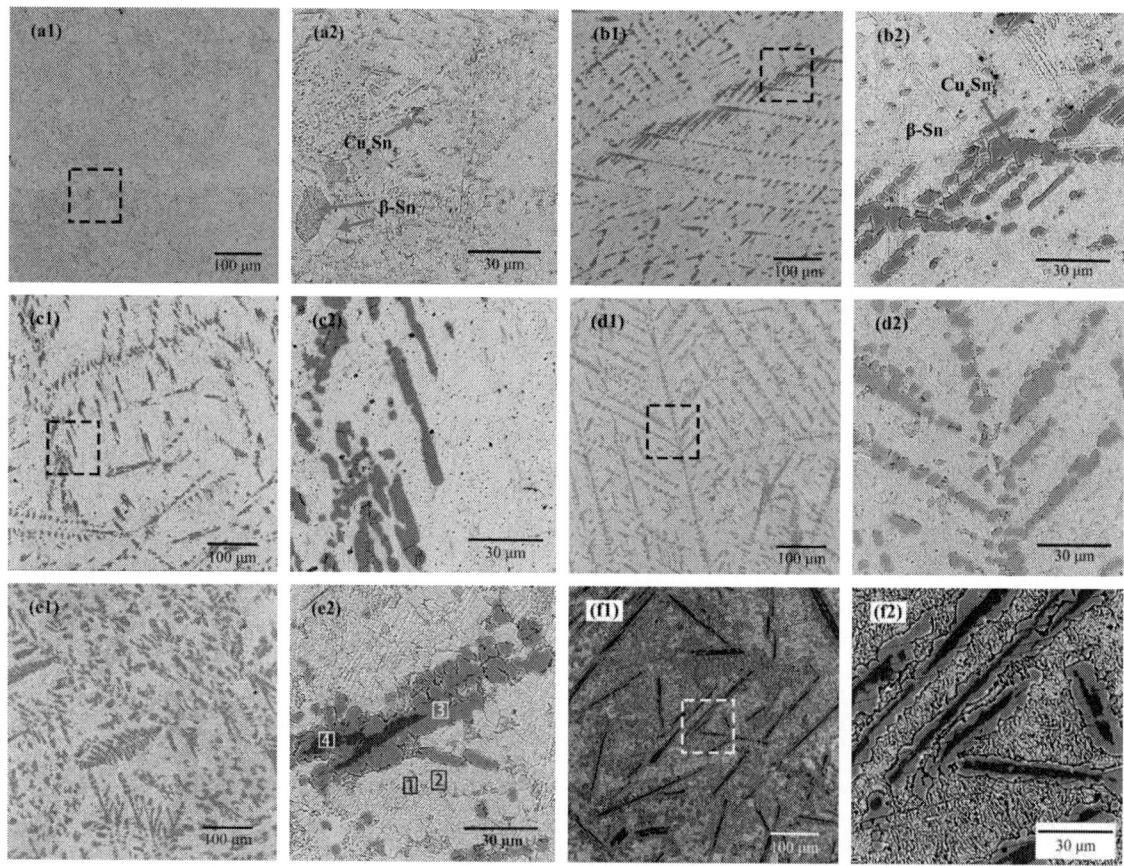

Fig. 4. Microstructures of as-casted Sn–xCu alloys with different Cu contents: (a1, a2) 0.7, (b1, b2) 4, (c1, c2) 6, (d1, d2) 8, (e1, e2) 10, (f1, f2) 12 wt.%, and (a2–f2) enlargened views of marked areas in (a1–f1).

978-1-6654-1392-3/21 $31.00 © 2021 IEEE

TABLE I ENERGY SPECTRUM RESULTS AT MARKER POINTS IN FIG. 4 (e2)

Position	Cu (at.%)	Sn (at.%)	Phase
1	0	100.0	β-Sn
2	0	100.0	β-Sn
3	47.0	53.0	Cu_6Sn_5
4	66.2	33.8	Cu_3Sn

B. Microstructures of Sn–xCu Solder Preforms

Fig. 5 presents the microstructure of Sn–xCu solder preforms with different Cu contents after metallographic etching and deep etching, respectively. EDS energy spectrum analysis was performed at each of marked points in Fig. 5 (c3), and the results are shown in Table II. Apparently, Point 1 that locates in the center of IMCs corresponds to the Cu_3Sn phase, Point 2 on the surface of IMCs denotes the Cu_6Sn_5 phase and Point 3 stands for the β-Sn phase that has not been etched away. Compared with as-casted Sn–xCu solder alloys with the same compositions, the microstructures of IMCs in the Sn–xCu solder preforms are finer and exhibit more uniform dispersion than that of as-casted ones (Fig. 4). Further, it can be seen from Fig. 5 (c1) that the phase length of IMCs in the Sn–12Cu solder preform is reduced, while most of IMCs still show strip morphology. In addition, a detailed comparison among Fig. 5 (a1), (b1) and (c1) indicates that with the increase of Cu content, the amount of Cu-Sn IMC phases in Sn–xCu solder preforms increases significantly, and IMC phases become more compact, in particular, IMCs occupy most of the matrix of solder preforms and are interconnected to form a three-dimensional skeleton structure. Also, it is interesting to see that in Fig. 5 (a3) to (c3) some large IMCs are coated with tiny rods of Cu_6Sn_5 IMC.

C. Microstructures of Cu/Sn–xCu-preform/Cu Joints

Fig. 6 shows microstructures of Cu/Sn–xCu-preform/Cu die-attachment joints with different Cu contents. Clearly, a thin scallop-shape Cu_6Sn_5 layer is formed at the interface between the Cu substrate and the solder preform, and Cu_3Sn is not found between the interfacial Cu_6Sn_5 layer and the Cu substrate. Seemingly, all the interfacial IMC layers exhibit a similar thickness about 1.4 μm, which means that increasing the Cu content in the solder does not appear to lead to increase in thickness of the interfacial IMC layer. However, the gap height of the die-attachment joints increases with the increasing Cu content, for example, the height of Cu/Sn–8Cu/Cu joint is about 93 μm, which is much smaller than 140 μm and 180 μm of Cu/Sn–10Cu/Cu and Cu/Sn–12Cu/Cu joints, respectively. This may be caused by relatively weak support of the un-melted IMC skeleton in the Sn–8Cu solder preform during the pressure soldering process. Besides, some voids are also found inside the solder joints adjacent to the interfacial IMC, as exhibited in Fig. 6 (a1) to (c1). The voids

may be formed when part of β-Sn reacts with Cu substrate while there is not sufficient supply of liquid β-Sn during the pressure soldering process. Noteworthily, by comparing Fig. 5 (a3)–(c3) with Fig. 6 (a2)–(c2), it can be seen that Cu/Sn–xCu-preform/Cu joints (x=8, 10 and 12 wt.%) show obviously higher area percentage than the corresponding solder preforms, i.e., the amount of IMC skeleton in the joints increases after the pressurized reflow soldering process. In addition, part of the interfacial IMC grains coalesced with the IMC skeleton in the solder matrix to form a fully interconnected network. Understandably, with increasing Cu content, the IMC skeleton in solder preforms is more likely to connect to the interfacial IMC grains, resulting in highly compact interconnection network in Cu/Sn–10Cu/Cu and Cu/Sn–12Cu/Cu die-attachment joints, which develops easily from the upper interface to the lower interface of the joints and leads to formation of Cu-Sn IMC skeleton dominated interconnect joints.

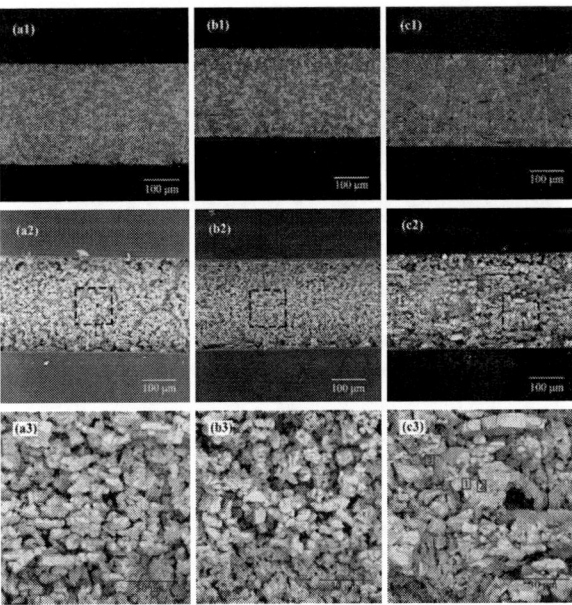

Fig. 5. Microstructures of Sn–Cu solder preforms with different Cu contents of (a1, a2) 8 wt.%, (b1, b2) 10 wt.%, and (c1, c2) 12 wt.% after metallographic etching (a1–c1) and deep etching (a2–c2), and (a3–c3) enlarged views of the marked areas in (a2–c2).

TABLE II ENERGY SPECTRUM RESULTS AT EACH OF MARKER POINTS IN FIG. 5 (C3)

Position	Cu (at.%)	Sn (at.%)	Phase
1	71.63	28.37	Cu_3Sn
2	55.25	44.75	Cu_6Sn_5
3	0	100.00	β-Sn

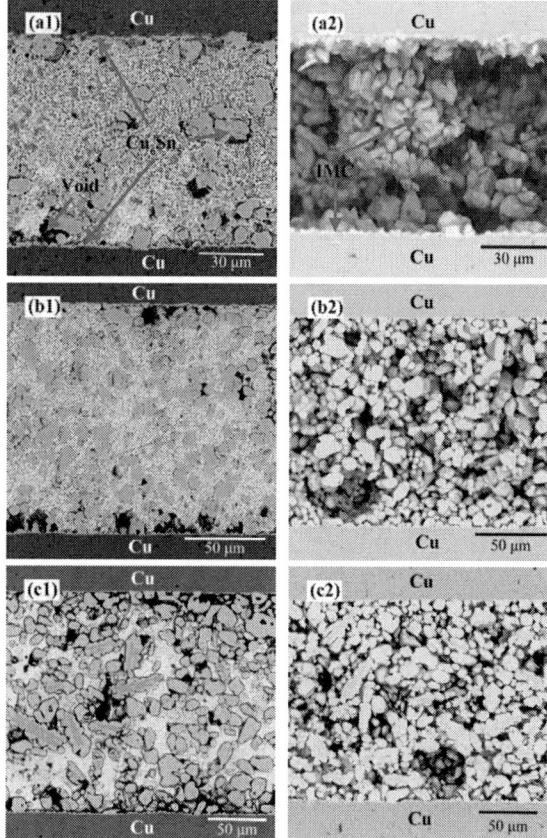

Fig. 6. Microstructures of Cu/Sn–xCu-preform/Cu die-attachment joints using solder preforms with different Cu contents of (a1, a2) 8 wt.%, (b1, b2) 10 wt.% and (c1, c2) 12 wt.% after metallographic etching (a1–c1) and deep etching (a2–c2).

D. Shear Strength of Cu/Sn–xCu-preform/Cu Joints

Fig. 7 depicts the change of shear strength of die-attachment joints using Sn–xCu preforms with different Cu contents (x=0.7, 8, 10 and 12 wt.%). Evidently, all Cu/Sn–xCu-preform/Cu joints exhibit the shear strength higher than 50 MPa and Cu/Sn–0.7Cu/Cu joint shows the highest shear strength of 62.5 MPa.

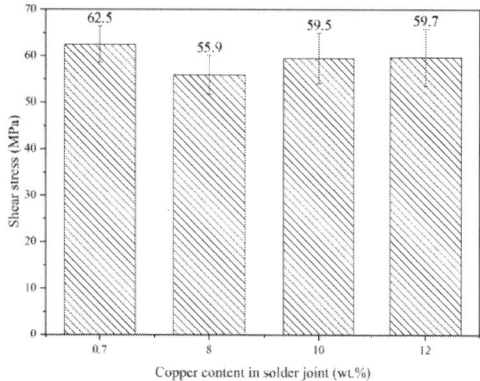

Fig. 7. Shear stress of Cu/ Sn–xCu-preform/Cu die-attachment joints with different Cu contents (0.7, 8, 10 and 12 wt.%).

When using solder preforms with Cu content between 8 wt.% and 12 wt.%, the shear strength of joints increases slightly with Cu content. It is worth noting that Cu/Sn–xCu-preform/Cu (x=8, 10 and 12 wt.%) joints have lower shear strength than that of Cu/Sn–0.7Cu/Cu joint, which is probably because the latter has fewer processing defects (such as voids) and consists of mainly β-Sn with dispersed IMCs, thereby being able to withstand high load; by contrast the former contain a large amount of IMCs skeleton and some shrinkage voids (cavities) and thus tend to fracture somewhat by brittle mode with relatively low shear strength.

IV. CONCLUSIONS

Aiming at power electronic packaging, in this work Sn–xCu solder preforms with high Cu content have been fabricated successfully. Microstructures and mechanical performance of reflowed Cu/Sn–xCu-preform/Cu die-attachment joints are characterized systematically. Main conclusions obtained in the present study are as follow:

1) The fineness and uniformity of dispersion of IMC in Sn–xCu solder preforms can be increased by rolling. Increasing Cu content in Sn–xCu solder preforms brings about an increased amount of IMC skeleton with highly compact microstructure.

2) With the sufficient supply of Cu-Sn IMC in solder preforms subjected to severe plastic deformation by non-continuous rolling, Cu/Sn–xCu-preform/Cu die-attachment joints consist of the Cu-Sn IMC skeleton filled with β-Sn phase and show excellent shear strength higher than 50 MPa.

3) Sn–xCu solder preforms with high Cu content (8 to 12 wt.%) are capable of reflow soldering at low temperature with slight pressure, and the technique has a great potential to be used for power electronics packaging.

ACKNOWLEDGMENTS

The present work is supported by the National Natural Science Foundation of China under grant Nos. 51775195 and 51405162, and Guangzhou City Science & Technology Planning Scheme under grant No. 201807010028.

REFERENCES

[1] X.R. Guo, Q. Xun, Z.X. Li, and S.X. Du, "Silicon carbide converters and MEMS devices for high-temperature power electronics: A critical review," Micromachines, vol. 10, pp. 406, June 2019.

[2] J. Broughton, V. Smet, R.R. Tummala, and Y.K. Joshi, "Review of thermal packaging technologies for automotive power electronics for traction purposes," Journal of Electronic Packaging, vol. 140, pp. 040801, December 2018.

[3] Y. Mikamura, K. Hiratsuka, T. Tsuno, H. Michikoshi, S. Tanaka, et al, "Novel designed SiC devices for high power and high efficiency systems," IEEE Transactions on Electron Devices, vol. 62, pp. 382–389, 2015.

[4] W.P. Liu, P. Bachorik, and N.C. Lee, "A composite solder alloy preform for high temperature Pb-free soldering applications," Welding Journal, vol. 91, pp. 50–58, April 2012.

[5] A.M. Gusak, K.N. Tu, and C. Chen, "Extremely rapid grain growth in scallop-type Cu₆Sn₅ during solid–liquid interdiffusion reactions in micro-bump solder joints," Scripta Materialia, vol. 179, pp. 45–48, April 2020.

[6] T.Q. Hu, H.T. Chen, M.Y. Li, and Z.Q. Zhao, "Cu@Sn core–shell structure powder preform for high-temperature applications based on transient liquid phase bonding," IEEE Transactions on Power Electronics, vol. 32, pp. 441–451, January 2017.

The mechanical and physical properties of the phases in the microstructure of Sn-Bi solder alloy

Chuyi LEI
Engineering Research Center of Electronic Information Materials and Devices, Ministry of Education, Guilin University of Electronic Technology
Guilin, China
19012201022@mails.guet.edu.cn

Xinghe LUAN
Engineering Research Center of Electronic Information Materials and Devices, Ministry of Education, Guilin University of Electronic Technology
Guilin, China
banbiandian@163.com

Zhigao LIU
School of Mechanical and Electronic Engineering, Guilin University of Electronic Technology
Guilin, China
792179425@qq.com

Hongbo QIN*
Guangxi Key Laboratory of Manufacturing System and Advanced Manufacturing Technology, Guilin University of Electronic Technology
Guilin, China
*qinhb@guet.edu.cn

Bin HOU*
China-Ukraine Welding Institute, Guangdong Academy of Sciences
Guangzhou, China
*houb@gwi.gd.cn
*corresponding author

Wangyun LI
Guangxi Key Laboratory of Manufacturing System and Advanced Manufacturing Technology, Guilin University of Electronic Technology
Guilin, China
li.wangyun@guet.edu.cn

Abstract—In this study, Sn58Bi solder alloy and solid solution samples of Sn-rich phase and Bi-rich phase were prepared, and their mechanical and physical properties were tested. The elastic moduli of Sn58Bi, Sn-rich phase and Bi-rich phase are 42.30 GPa, 61.10 GPa, and 31.90 GPa, respectively, and their hardnesses are 0.33 GPa, 0.37 GPa, and 0.28 GPa, respectively. The resistivity of Bi-rich phase is much higher than that of Sn-rich phase. As the test temperature was increased from 300 K to 400 K, the resistivity of Bi-rich phase increases slightly, while that of Sn-rich phase is almost unchanged. Different with the resistivity, the specific heat capacity of Sn-rich phase increases slightly with the increase of temperature, while the specific heat capacity of Bi-rich phase is quite stable. The thermal conductivity of Sn-rich phase is much higher than that of Bi-rich phase, indicating that the Sn-rich phase in the eutectic structure has better heat dissipation performance. Meanwhile, the coefficient of thermal expansion (CTE) of Sn-rich phase is larger than that of Bi-rich phase.

Keywords—Sn58Bi, Sn-rich phase, Bi-rich phase, mechanical property, physical property

I. INTRODUCTION

Microstructures of alloys are composed of various phases. A phase refers to a part of material with a certain crystal structure and chemical composition, and phases are separated from each other by phase interface. Solid solution is a typical substance for a phase in an alloy, which usually has a uniform composition [1]. In the field of electronic packaging, Sn-Bi solder alloy has been used in low-temperature soldering because of its lead-free, good wettability, high yield strength and low cost [2, 3]. The microstructure of eutectic Sn-Bi alloy is composed of Sn-rich phase (i.e., crystal structure is basically same as Sn, but contains Bi atoms) and Bi-rich phase (i.e., crystal structure is basically same as Bi, but contains Sn atoms or atomic groups) [4, 5]. With the growth of high-density packaging technology, the size of solder joint has been reduced to tens of micrometers, and the eutectic phases in the microstructure have shown obvious inhomogeneity. Fig. 1 exhibits the inhomogeneous phases in a Sn-Bi alloy joint, in which the gray areas are Sn-rich phases as well as the white areas are Bi-rich phases. For the different crystal structures and compositions of the two eutectic phases, it can infer that maybe there are differences between the mechanical and physical properties of two eutectic phases. Previous study has found that elastic modulus and hardness of Sn-rich phase were quite different from those of Bi-rich phase [6]. In addition, the resistivities of Sn-rich and Bi-rich phases are also different, which lead to inhomogeneous distribution of current densities and temperature gradients in different phases, which further affects the electromigration (EM) behavior in solder joints [4]. Thus far, the differences of various physical properties of the phases in the microstructure of solder alloy have rarely been considered, and the solder joint has long been treated as a homogeneous material. Considering the two phases are inhomogeneously distributed in the eutectic structure, understanding difference of their mechanical and physical properties is essential to clarify failure mechanism in the case of inhomogeneity. Because of extremely small characteristic size, the mechanical and physical properties related to two phases cannot be directly measured.

In this study, in order to obtain the mechanical and physical properties of Sn-rich phase and Bi-rich phase, single-phase solid solution samples, which have the same crystal structures and chemical compositions as the phases in the microstructure of Sn-Bi alloy, were experimentally prepared and tested.

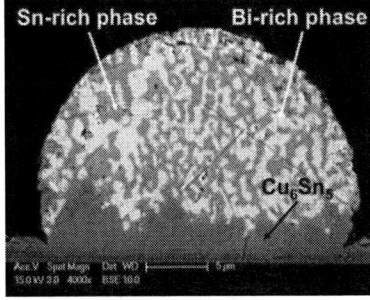

Fig. 1. Inhomogeneous distribution of eutectic phases in Cn/Sn-Bi solder joints [7].

II. EXPERIMENTAL METHOD

A. Preparation of Single-Phase Solid Solution

The solid solubility of Sn in Bi was less than 0.18 at.% Sn and the solid solubility of Bi in Sn was 1.5 at.% Bi [8, 9]. For the solid solution samples of Sn-rich phase, the masses of Sn and Bi were 300.000 g and 8.040 g in the alloy smelting, respectively. For the solid solution samples of Bi-rich phase, the masses of metallic Bi and Sn are 300.005 g and 0.307 g, respectively. Sn58Bi alloy and single-phase solid solutions were smelted in a protective atmosphere of rosin and heated in a tin furnace. For Sn58Bi alloy and solid solution of Sn-rich phase, the smelting temperature was set at 250 °C. However, due to the higher melting point of Bi, when smelting the Bi-rich phase solid solution, the temperature was 300 °C. After

the two solid solution samples were obtained, they were cut into many specimens by electrical discharge machining (WEDM) for experimental tests. Figure 2(a) displays a large and a small specimens used for the measurements of specific heat capacity and coefficient of thermal expansion (CTE), respectively. The heights of large sample and small sample were 18.0 mm and 7.6 mm, respectively, and their bottom diameters were 20.0 mm and 3.0 mm, respectively. Figures 2(b) and (c) are scanning electron microscope (SEM) images of microstructure of two solid solution samples, respectively.

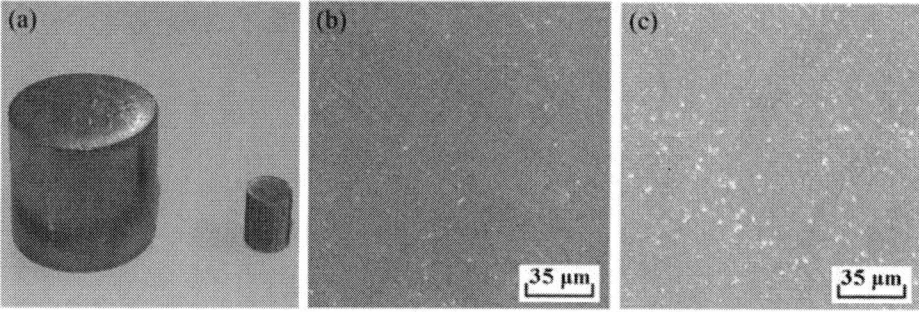

Fig. 2. The specimens and microstructures of two solid solution samples: (a) two specimens, (b) the microstructure of Sn-rich phase solid solution, and (c) the microstructure of Bi-rich phase solid solution.

B. Experiment of Mechanical Properties

The nanoindentation instrument (Nano Indenter XP, Mechanical Testing & Simulation, American) was used to measure the elastic moduli and hardnesses of Sn58Bi alloy as well as two single-phase solid solutions. First, these samples were cut into small pieces by WEDM. Then, for the convenience of flattening the surface, the small pieces of samples were inlaid into epoxy resin and then ground and polished. Figure 3(a) shows an inlaid sample. In order to prevent impurities from affecting the nanoindentation test, the inlaid Sn58Bi, Sn-rich phase and Bi-rich phase were cleaned with an ultrasonic cleaner before the test. Last, the specimens of Sn58Bi, Sn-rich phase and Bi-rich phase were tested via Berkovich indenter. The testing points of the three samples were all five, and the center distances between any two indentations were larger than 60 μm in order to avoid the stress interference. The indentation depth was 500 nm, and the loading strain rate was 0.05 s⁻¹. The indentations of Berkovich indenter in Sn-rich and Bi-rich phases are shown in Figs. 3(b) and (c).

C. Experiment of Physical Properties

The large specimen (i.e., the height and diameter were 18.0 and 20.0 mm, respectively) was applied to determine the density of sample. The large specimen was ground, polished,

and then ultrasonically cleaned. Then, high-precision electronic balance and vernier calipers were used to measure the masses and volumes of the specimens, respectively. The densities of two single-phase solid solution samples are listed in Table I.

The PPMS-9 measuring system (PPMS-9, Quantum Design, American) was adopted to measure the resistivities of two solid solution samples, and the Kelvin connection was applied under a vacuum of 5mTorr. Because the resistivities of the single-phase solid solutions were very small, the lead resistance and contact resistance could no longer be ignored. Therefore, four-wire method was used to test the resistivities of the two solid solutions, as seen from Fig.4(a). In the resistivity test, two solid solution samples were cut into small specimens of $7 \times 3 \times 2$ mm³, and the surface of the specimen was cleaned and polished to reduce measuring errors.

The STA 449 F3 Jupiter synchronous thermal analyzer (STA 449 F3, NETZSCH-Gerätebau GmbH, Germany) was used to test specific heat capacity. In specific heat capacity test, the large specimens of two solid solution samples shown in Fig. 2(a) were selected, and the heat capacity of sample was measured by differential scanning calorimetry (DSC).

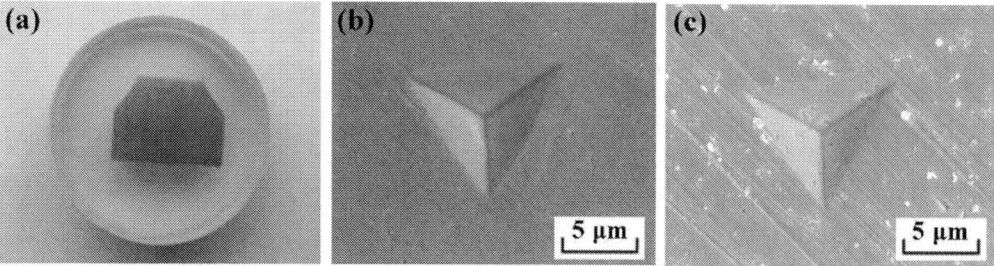

Fig. 3. Spacemen and indentation in the nanoindentation test: (a) an inlaid sample, (b) an indentation in the Sn-rich phase and (c) an indentation in the Bi-rich phase.

TABLE I. Phase Density.

Materials	Sample size and density					
	Length (mm)	Average diameter (mm)	Volume (m³)	Mass (kg)	Density (kg/m³)	Density average (kg/m³)
Sn-rich-1	18.14	19.717	5.539×10^{-6}	4.054×10^{-2}	7320.1	7319.75
Sn-rich-2	18.17	19.697	5.537×10^{-6}	4.052×10^{-2}	7319.4	
Bi-rich-1	18.10	19.937	5.651×10^{-6}	5.515×10^{-2}	9760.5	9773.60
Bi-rich-2	18.09	19.910	5.651×10^{-6}	5.512×10^{-2}	9786.7	

The hot disk thermal constant analyzer (Hot Disk, K-Analysis Trading, Sweden) was used to test the thermal conductivity, as shown in Fig. 4(b). In this study, the large sample exhibited in Fig. 2(a) was used, and the thermal probe of the analyzer was tightly clamped in the middle of the specimen. Thermomechanical analyzer (TMA Q400, TA instruments, American) was used to measure the CTEs of the two single-phase solid solutions, see Fig. 4(c), and the small sample shown in Fig. 2(a) was selected. When measuring the CTEs, the pure Al sample came with the instrument was used for calibration, and the test was started after the calibration was correct. In the experiment, there were two loading modes: 1) the initial loading temperature was 20 °C, and then temperature was increased to 150 °C; 2) the initial loading temperature was 40 °C, and then temperature was raised to 150 °C.

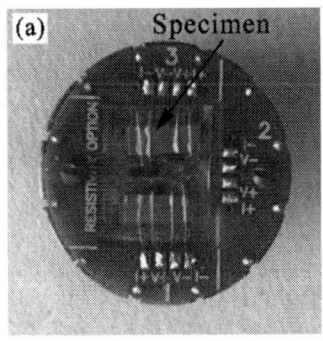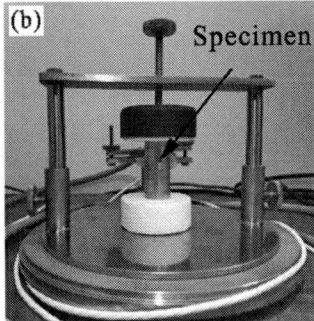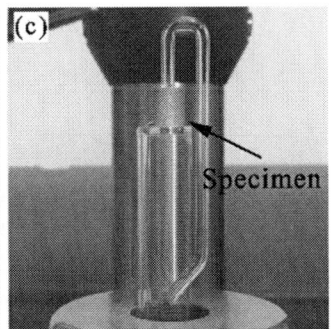

Fig. 4. Specimens in the instruments of (a) PPMS-9, (b) Hot Disk, and (c) Q400 TMA.

III. RESULTS AND DISCUSSION

A. The Mechanical Properties of Samples

Table II gives the average elastic moduli and the average hardnesses of Sn58Bi, Sn-rich and Bi-rich phases. According to the test results of five points, for Sn58Bi, the average value of elastic modulus is 42.30 GPa, and the elastic modulus ranges from 39.9 GPa to 42.8 GPa; the average value of hardness is 0.33 GPa, and the hardness ranges from 0.31 GPa to 0.33 GPa. For the Sn-rich phase, the average value of elastic modulus is 61.10 GPa, and the average value ranges from 59.60 GPa to 62.00 GPa; the average value of hardness is 0.37 GPa, and the hardness is between 0.33 GPa and 0.37 GPa. For the Bi-rich phase, the average value of elastic modulus is 31.90 GPa, and the elastic modulus is between 29.60 GPa and 31.30 GPa; the average value of hardness is 0.28 GPa, and the hardness is between 0.26 GPa and 0.30 GPa.

B. The Physical Properties of Samples

Figure 5 shows the resistivities of two solid solution samples from 300 K to 400 K. The test results indicate that the resistivity of Bi-rich phase is much higher than that of Sn-rich phase as well as increases slightly with the increase of temperature. It is worth noting that because the resistivities of the two samples are very small, the resistivities of the two solid solutions fluctuate to different degrees during the measurement process. At room temperature (i.e., 300 K), the resistivities of Sn-rich and Bi-rich phases are $11.3 \times 10^{-8} \, \Omega \cdot m$ and $66.6 \times 10^{-8} \, \Omega \cdot m$, respectively.

TABLE II. Elastic Modulus (E) And Hardness (H), GPa.

Materials	Mechanical property	
	E	H
Sn58Bi	42.3 ± 2.4	0.33 ± 0.02
Sn-rich phase	61.1 ± 1.5	0.37 ± 0.04
Bi-rich phase	31.9 ± 2.3	0.28 ± 0.02

Table III lists the specific heat capacity of the test sample by selecting five temperature points of 20 °C, 40 °C, 60 °C, 80 °C and 100 °C, and results imply that the specific heat capacity of Sn-rich phase increases from 200 J/kg·K to 240 J/kg·K with the increase of temperature, while the specific heat capacity of Bi-rich phase is a constant of 120 J/kg·K. Obviously, the specific heat capacity of Sn-rich phase is higher than that of Bi-rich phase.

TABLE III. Specific Heat Capacity, J/kg·K.

Materials	Temperature and specific heat capacity				
	20 °C	40 °C	60 °C	80 °C	100 °C
Sn-rich	200	200	210	220	240
Bi-rich	120	120	120	120	120

Table IV gives the thermal conductivities of the two solid solution samples. The analysis shows that the thermal conductivity of Sn-rich phase is much higher than that of Bi-rich phase, which indicates that Sn-rich phase in the eutectic structure has better heat dissipation performance. At 20 °C, the average values of thermal conductivities of Sn-rich phase and Bi-rich phase are 70.86 W/m·K and 7.677 W/m·K, respectively.

Table V presents the CTE test results. It can be seen that, whether initial loading temperature is 20 °C or 40 °C, the final test results are not much different for these two samples, indicating that the initial loading temperature has little effect on the test results. CTE of Sn-rich phase is larger than that of Bi-rich phase, and the average CTEs of Sn-rich phase and Bi-rich phase are $23.736 \times 10^{-6}/°C$ and $11.963 \times 10^{-6}/°C$ at 20 °C, respectively.

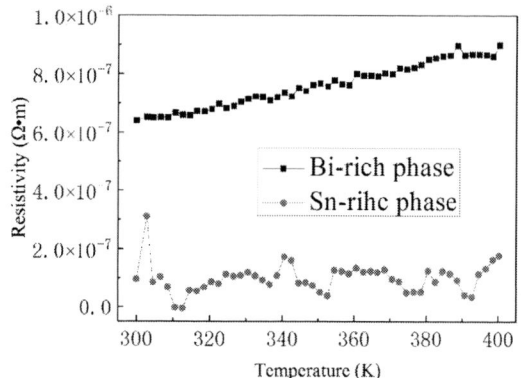

Fig. 5. Resistivity test results.

TABLE IV. Thermal Conductivity, W/m·K.

Materials	Test point and thermal conductivity								
	1	2	3	4	5	6	7	8	AVG
Sn-rich	68.12	68.24	68.24	68.52	68.80	68.71	69.63	86.58	70.86
Bi-rich	7.637	7.645	7.575	7.601	7.543	7.608	7.899	7.897	7.677

TABLE V. Coefficient Of Thermal Expansion.

Materials	20 °C-150 °C		40 °C -150 °C	
	Length (mm)	CTE (/°C)	Length (mm)	CTE (/°C)
Sn-rich-1	7.6822	23.873×10^{-6}	7.6834	22.463×10^{-6}
Sn-rich-2	7.6571	23.599×10^{-6}	7.6550	23.254×10^{-6}
Bi-rich-1	7.5911	11.369×10^{-6}	7.6050	11.780×10^{-6}
Bi-rich-2	7.5877	12.557×10^{-6}	7.5838	12.470×10^{-6}

IV. CONCLUSION

Solid solution samples of Sn-rich phase and Bi-rich phase were prepared, and their mechanical and physical properties were tested. Best on the test results, the following conclusions can be drawn:

1) The elastic moduli of Sn58Bi, Sn-rich phase and Bi-rich phase are 42.30 GPa, 61.10 GPa, and 31.90 GPa, respectively, and their hardnesses are 0.33 GPa, 0.37 GPa, and 0.28 GPa, respectively.
2) The resistivity of Bi-rich phase is much higher than that of Sn-rich phase as well as increases slightly with the increase of temperature.
3) The specific heat capacity of Sn-rich phase is higher than that of Bi-rich phase. The specific heat capacity of Sn-rich phase increases with the increase of temperature, while the specific heat capacity of Bi-rich phase is a constant.
4) The thermal conductivity of Sn-rich phase is much higher than that of Bi-rich phase, indicating that Sn-rich phase in the eutectic structure has better heat dissipation performance.
5) The CTE of Sn-rich phase is larger than that of Bi-rich phase.

ACKNOWLEDGMENT

This study was sponsored by the National Natural Science Foundation of China (NSFC) under grant. Nos. 51505095, 51805103 and 52065015; Guangxi Natural Science Foundation under grant. Nos. 2018GXNSFAA281222 and 2021 GXNSFAA075010; Science and Technology Planning Project of Guangxi Province under Grant Nos. GuiKeAD AD18281022 and 18281021, Director Fund Project of Guangxi Key Laboratory of Manufacturing System and Advanced Manufacturing Technology Nos. 19-050-44-003Z and 20-065-40-002Z, Self-Topic Fund of Engineering Research Center of Electronic Information Materials and Devices Nos. EIMD-AB202005 and EIMD-AB202007. Innovation Project of GUET Graduate Education under grant No. 2020YCXS001 and 2021YCXS006.

REFERENCES

[1] Y. Zhang, Y. Zhou, J. Lin, P. Liaw, and G. Chen, "Solid-solution phase formation rules for multi-component alloys," Advanced Engineering Materials, Vol. 10, pp. 534-538, 2008.

[2] F. Wang, H. Chen, L. Liu, Y. Huang, and Z. Zhang, "Recent progress on the development of SnBi based low-temperature Pb-free solders," Journal of Materials Science: Materials in Electronics, Vol. 30, pp. 3222-3243, 2019.

[3] L. Liu, S. Xue, and S. Liu, "Mechanical property of Sn58Bi solder paste strengthened by resin," Applied Sciences, Vol. 8, 2024, 2018.

[4] H. Qin, B. Li, W. Yue, M. Zhou, C. Ke, and X. Zhang, "Interaction effect between electromigration and microstructure evolution in Cu/Sn58Bi/Cu solder interconnect," IEEE 64th Electronic Components and Technology Conference (ECTC), 2014, pp. 2249-2254.

[5] B. Silva, G. Reinhart, H. Nguyen-Thi, A. Garcia, N. Mangelinck-Noël, and J. Spinelli, "Microstructural development and mechanical properties of a near-eutectic directionally solidified SnBi solder alloy," Materials Characterization, Vol. 107, pp. 43-53, 2015.

[6] U. Kang and Y. Kim, "The microstructure characterization of ultrasmall eutectic BiSn solder bumps on Au/Cu/Ti and Au/Ni/Ti under-bump metallization," Vol. 33, pp. 61-69, 2004.

[7] M. Roh, J. Jung, and W. Kim, "Microstructure, shear strength, and nanoindentation property of electroplated SnBi micro-bumps," Microelectronics Reliability, Vol. 54, pp. 265-271, 2014.

[8] H. Kaya, "Dependency of microstructural parameters and microindentation hardness on the temperature gradient in the In-Bi-Sn ternary alloy with a low melting point," Metals and Materials International, Vol. 14, pp. 575-582, 2008.

[9] L. Shen, P. Lu, S. Wang, and Z. Chen, "Creep behaviour of eutectic SnBi alloy and its constituent phases using nanoindentation technique," Vol. 574, pp. 98-103, 2013.

Scalable Modeling and Analysis of TSV Using Bumpless Interconnects Technology

Zewei Li
National and Local Joint Engineering Laboratory of RF Integration and Micro-Assembly Technology
Nanjing University of Posts and Telecommunications
Nanjing, China
18852733661@163.com

Zhikuang Cai
National and Local Joint Engineering Laboratory of RF Integration and Micro-Assembly Technology
Nanjing University of Posts and Telecommunications
Nanjing, China
whczk@njupt.edu.cn

Lei Pan
National and Local Joint Engineering Laboratory of RF Integration and Micro-Assembly Technology
Nanjing University of Posts and Telecommunications
Nanjing, China
p1227346849@163.com

Lu Liu
National and Local Joint Engineering Laboratory of RF Integration and Micro-Assembly Technology
Nanjing University of Posts and Telecommunications
Nanjing, China
liulu@njupt.edu.cn

Binbin Xu
Nantong Institute of Nanjing University of Posts and Telecommunications
Nanjing University of Posts and Telecommunications
Nanjing, China
xubinbin@njuptnti.com

Yufeng Guo
National and Local Joint Engineering Laboratory of RF Integration and Micro-Assembly Technology
Nanjing University of Posts and Telecommunications
Nanjing, China
yfguo@njupt.edu.cn

Abstract—Chiplet becomes a key technology to continue Moore's law, but the interconnection between dies is one of its difficulties. Bumpless interconnects technology reduces the impedance of the TSV interconnects with no bumps and Ultra-thinning of wafers enables maintain total height while supporting more layer counts. Instead of bump, small TSV passes through substrate and underfill whose thickness are comparable. Considering this structural feature, the equivalent-circuit model of bumpless TSV is proposed and the analytical expressions of proposed model are studied. Up to 100 GHz, the maximum error of S11 and S21 between proposed model and the physical model built in HFSS less than 3% and 9%, which demonstrates the accuracy of equivalent-circuit model. For the structural feature of bumpless TSV, the effect of thickness of underfill and silicon substrate that varies with TSVs height are analyzed respectively. The two conditions are studied in the frequency domain analysis and the electrical behavior differing from traditional TSV is found while scalability is verified.

Keywords—Chiplet, 3DIC, bumpless, TSV, equivalent-circuit model

I. INTRODUCTION

In recent years, there is a rapid growth of 5G, AI, and HPC market. These new application scenarios have created a requirement for high-efficiency chips. However, system performance gains have been abating owing to the slow growth rate of interconnection and packaging technologies in Post-Moore Era. A critical demand for innovation in silicon ancillary technologies is created. [1] A breakthrough is made by the newly heterogeneous integration technology called Chiplet. Chiplet has advantages and potential in aspects of chip performance, power optimization and cost reduction. It provides an efficient and low-cost implementation way for the development of chips in various fields such as CPU, FPGA and networking processor. Chiplet technology is like building blocks, some pre-produced dies that can achieve specific functions are integrated into a system through advanced package technology (such as 3D integration, etc.). As a result, interconnection and interface technology is vital importance to applicability of Chiplet. As a new IP reuse pattern, Chiplet has gradually become a research hotspot in academia and industry, which involves interconnection, package and DFT.

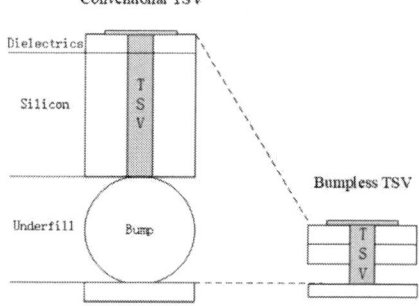

Fig. 1. Schematic diagram of bumpless interconnects technology.

The bumpless interconnects technology has been applied to low-temperature System-on-Integrated Chip (LT-SoIC) [2], which enables lower power consumption, higher bandwidth, and achieve multilayers' stacking for future applications related to Chiplet.

As shown in the Fig. 1, a 300-mm wafer can be thinned down to 5.6 μm with no degradation of retention characteristics by using ultra-thinning technique [3]. When the thicknesses of the device layer and underfill layer are 5 μm and 2 μm respectively, the aspect ratio (depth-to-diameter ratio) of a TSV is only 1-2 for diameter of TSV ranging from 5 μm to 10 μm. Instead of bump, small TSV directly connects to the pad of the next die through the underfill. The TSV density is subject to the pitch of the bumps, so the bumpless process provides a higher density and short transmission distance. With the use of small TSVs and bumpless feature, stress induced by a mismatch in the coefficients of thermal expansion between TSVs and bumps decreases [4], so it enables maintain total height while supporting more layer counts.

In aspect of performance, the bumpless process and ultra-thinning will realize vertical interconnects with RC values orders of magnitude lower. High-density TSVs increase the total bandwidth even at the same data rate per TSV. Finally, the power efficiency will increase on account of the combination of multiple channels and high density bumpless interconnects. This is helpful to reduce the thermal issues and

978-1-6654-1392-3/21 $31.00 © 2021 IEEE

TABLE I.   MATERIAL PARAMETERS AND SYMBOLS

Symbol	Parameter	Symbol	Parameter
ρ_{TSV}	Resistivity of TSV	ε_{ox}	Relative permittivity of insulator and IMD
σ_{Si}	Conductivity of silicon substrate	ε_{Si}	Relative permittivity of silicon substrate
μ_{TSV}	Relative permeability of TSV	$\varepsilon_{Underfill}$	Relative permittivity of underfill
μ_0	Permeability of vacuum	ε_0	permittivity of vacuum

power consumption for multi-package modules such as 3D-DRAMs and MPUs [5].

However, miniaturized size and complex structure will lead to strict and changeable environment of signal transmission. A scalable equivalent-circuit model of bumpless TSVs is easier to understand the effects of material properties, dimensions, and geometrical arrangement on electrical properties. This is very important to aid system design and performance estimation. Many modeling methods of traditional TSV, like analytical methods [6]-[8], numerical-, and measurement-based methods [9] have been studied. In 2011, Joohee Kim proposed the scalable high-frequency equivalent-circuit model of TSV with bump and good agreement was got [10]. In [11], the influence of TSVs geometric parameters on electrical properties is discussed.

In this paper, the equivalent-circuit model of bumpless TSV is proposed on the basis of bumpless structure and the scalable electrical model of TSVs with bump in [10]. The electrical parameters are extracted according to the theory of analytical methods. To experimentally verify the accuracy of proposed model, the S-parameter extracted from proposed model is compared with the physical model constructed in High Frequency Structure Simulation (HFSS) and the equivalent-circuit model without considering the capacitance of underfill. Furthermore, the scalability that the proposed model can capturing the electrical behavior of bumpless TSVs with different profile is validated. For the structural features of bumpless TSV, the effect of thickness of underfill and silicon substrate that varies with TSVs height are analyzed respectively. The two conditions are studied in the frequency domain analysis, in which electrical behavior of bumpless TSV are found while scalability is verified.

II. SCALABLE MODELING OF TSV USING BUMPLESS INTERCONNECS TECHNOLOGY

A. Equivalent-circuit model of TSVs

Analytical methods based on material properties, dimensions, and geometrical arrangement are used to search for a rapid and efficient extraction method of TSV using bumpless interconnects technology. The profile of bumpless TSVs shown in Fig. 2 use a pair of identical signal and ground (GS) TSVs in consider of current TSV application in 3DIC. TSV extends through the underfill in that the remove of bump and electrically isolated from the underfill by oxide liners.

Expect for the resistance and inductance of the TSVs, the capacitance and conductance of the silicon substrate and the capacitance of insulator in traditional TSV [6], the capacitance of underfill and inter-metal dielectric (IMD) are also considered in the equivalent-circuit model shown in Fig. 3. The capacitance of underfill is modeled as $C_{Underfill}$ paralleled with C_{Si} and in series with parallel $C_{insulator}$ and the capacitance of IMD is modeled as C_{IMD} paralleled with the aforementioned

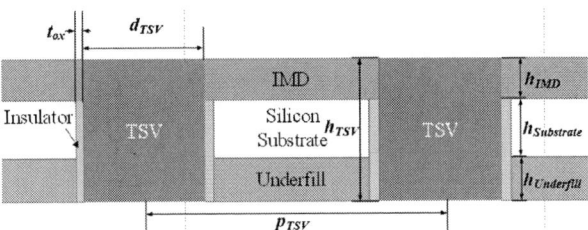

Fig. 2.   Structure of TSVs using bumpless interconnects technology.

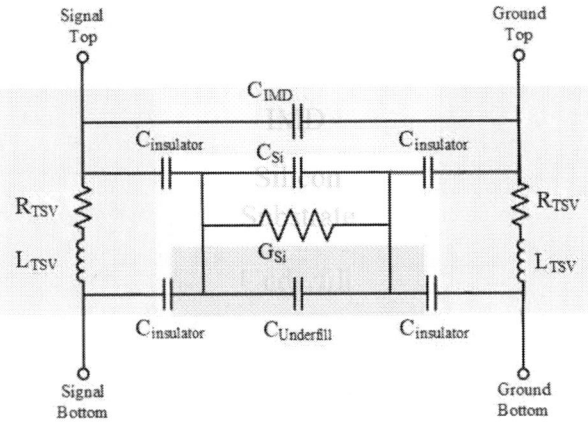

Fig. 3.   Equivalent-circuit model of TSVs.

circuit. In addition, what calls for special attention is that Si substrate is not biased in this paper, thus the MOS capacitance between TSV and silicon substrate is not taken into consideration.[12].

B. Extraction of electrical parameters

In this section, analytical expressions for extraction are developed for quick calculation. Material parameter used in following expressions are showed in Table I.

The resistance of TSV is mainly made up of DC and AC resistances. The DC resistance is derived from Ohm's law as shown in (1).

$$R_{dc,TSV} = \rho_{TSV} \times \frac{h_{TSV}}{\pi(d_{TSV}/2)^2} \tag{1}$$

High-frequency current will distribute on the edge of the conductor because of skin effects, thus increasing the AC resistance. The expression account for this increase is shown as follows:

$$R_{ac,TSV} = \frac{\rho_{TSV} h_{TSV}}{\pi d_{TSV} \delta_{TSV}} \tag{2}$$

978-1-6654-1392-3/21 $31.00 © 2021 IEEE

Where $\delta_{TSV} = \frac{1}{\sqrt{\pi \mu_0 \mu_{TSV} f \sigma_{TSV}}}$. Proximity effect does not be considered in that other TSVs are distributed near the signal TSV in actual application. As a result, the resistance is given by (3).

$$R_{TSV} = \sqrt{\left(R_{dc,TSV}\right)^2 + \left(R_{ac,TSV}\right)^2} \qquad (3)$$

The inductance of TSV can be divided into external inductance and self-inductance. The external inductance is derived from the loop inductance model of two-wire transmission line [13]. Since GS TSVs space closely enough that currents will distribute toward the facing sides, we use the exact per-unit-length inductance of a two-wire transmission line showed in (4).

$$L_{TSV,ex} = \frac{\mu_0 \mu_{TSV}}{\pi} \ln\left(\frac{p}{d_{TSV}} + \sqrt{\left(\frac{p}{d_{TSV}}\right)^2 - 1}\right) \qquad (4)$$

To capture the frequency-dependent inductance of GS TSVs, the self-inductance interrelated with resistance are taken into consideration and the total inductance of single TSV [14] in this circuit model showed in (5).

$$L_{TSV} = \frac{h_{TSV}}{2} L_{TSV,ex} + \frac{R_{ac,TSV}}{2\pi f} \qquad (5)$$

These expressions of resistance and inductance are more accuracy compared with other methods for extracting RLCG parameters of TSVs [15].

According to the Fig. 2, both of silicon substrate and underfill are isolated from the underfill by oxide liners. The single TSV conductor and the oxide liners are in a coaxial configuration, so the coaxial capacitance [16] of oxide liners is calculated using the following expression.

$$C_{insulator} = \frac{1}{2} \frac{2\pi \varepsilon_0 \varepsilon_{ox}(h_{TSV} - h_{IMD})}{\ln\frac{d_{TSV}/2 + t_{ox}}{d_{TSV}/2}} \qquad (6)$$

The capacitance of IMD layer is obtained from the model of the parallel-wires capacitance [16] as follows:

$$C_{IMD} = \frac{\pi \varepsilon_0 \varepsilon_{ox} h_{IMD}}{\ln\left(\frac{p}{d_{TSV}} + \sqrt{\left(\frac{p}{d_{TSV}}\right)^2 - 1}\right)} \qquad (7)$$

The capacitance of silicon substrate and underfill account for thickness of oxide liners showed in (8) and (9) [11].

$$C_{si} = \frac{\pi \varepsilon_0 \varepsilon_{si} h_{Substrate}}{\ln\left(\frac{p}{d_{TSV}+2t_{ox}} + \sqrt{\left(\frac{p}{d_{TSV}+2t_{ox}}\right)^2 - 1}\right)} \qquad (8)$$

$$C_{Underfill} = \frac{\pi \varepsilon_0 \varepsilon_{Underfill} h_{Underfill}}{\ln\left(\frac{p}{d_{TSV}+2t_{ox}} + \sqrt{\left(\frac{p}{d_{TSV}+2t_{ox}}\right)^2 - 1}\right)} \qquad (9)$$

The conductance of silicon substrate is calculated as follows:

$$G_{Si} = C_{Si} \frac{\sigma_{Si}}{\varepsilon_0 \varepsilon_{Si}} \qquad (10)$$

$$G_{Si} = \frac{\pi \sigma_{Si} h_{Substrate}}{\ln\left(\frac{p}{d_{TSV}+2t_{ox}} + \sqrt{\left(\frac{p}{d_{TSV}+2t_{ox}}\right)^2 - 1}\right)} \qquad (11)$$

III. VERIFICATION AND ANALYSIS OF ELECTRICAL CHARACTERISTICS

To experimentally verify the validity of the scalable equivalent-circuit model proposed in this paper, frequency domain simulation of S-parameter with the equivalent-circuit model and 3D field solver (HFSS) are compared. The electrical behavior of bumpless TSVs is analyzed using proposed model and HFSS, so that the scalability of proposed model can be demonstrated. Benzocyclobutene(BCB) is used as the material of the underfill [17]. The values of material properties are shown in TABLE II and Fig.4 shows the physical model of bumpless TSVs designed in HFSS.

TABLE II.   MATERIAL PROPERITIES OF THE DESIGNED TEST VEHICLES

Symbol	value
ρ_{TSV}	$1.72 \times 10^{-8}\ \Omega \cdot m$
σ_{Si}	$3.03\ S/m$
μ_{TSV}	1
μ_0	$4\pi \times 10^{-7}\ H/m$
ε_{ox}	3.9
ε_{Si}	11.9
$\varepsilon_{Underfill}$	2.6
ε_0	$8.854 \times 10^{-12}\ F/m$

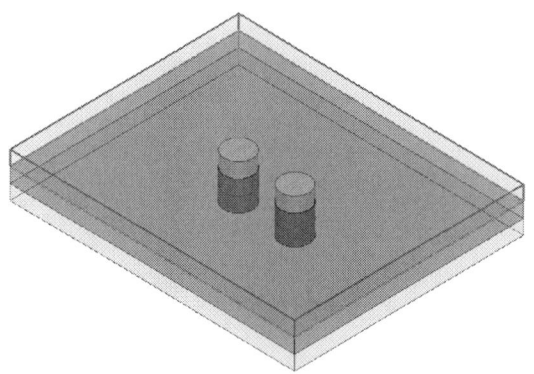

(a) Schematic view of bumpless TSVs

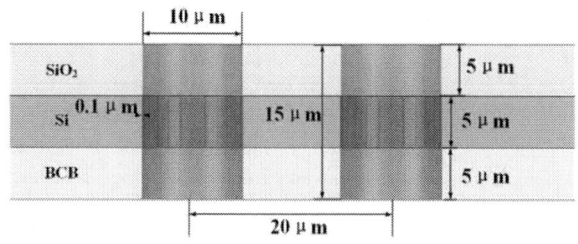

(b) Cross-sectional view and initial values of geometric parameters

Fig. 4.   Physical model constructed in HFSS.

978-1-6654-1392-3/21 $31.00 © 2021 IEEE

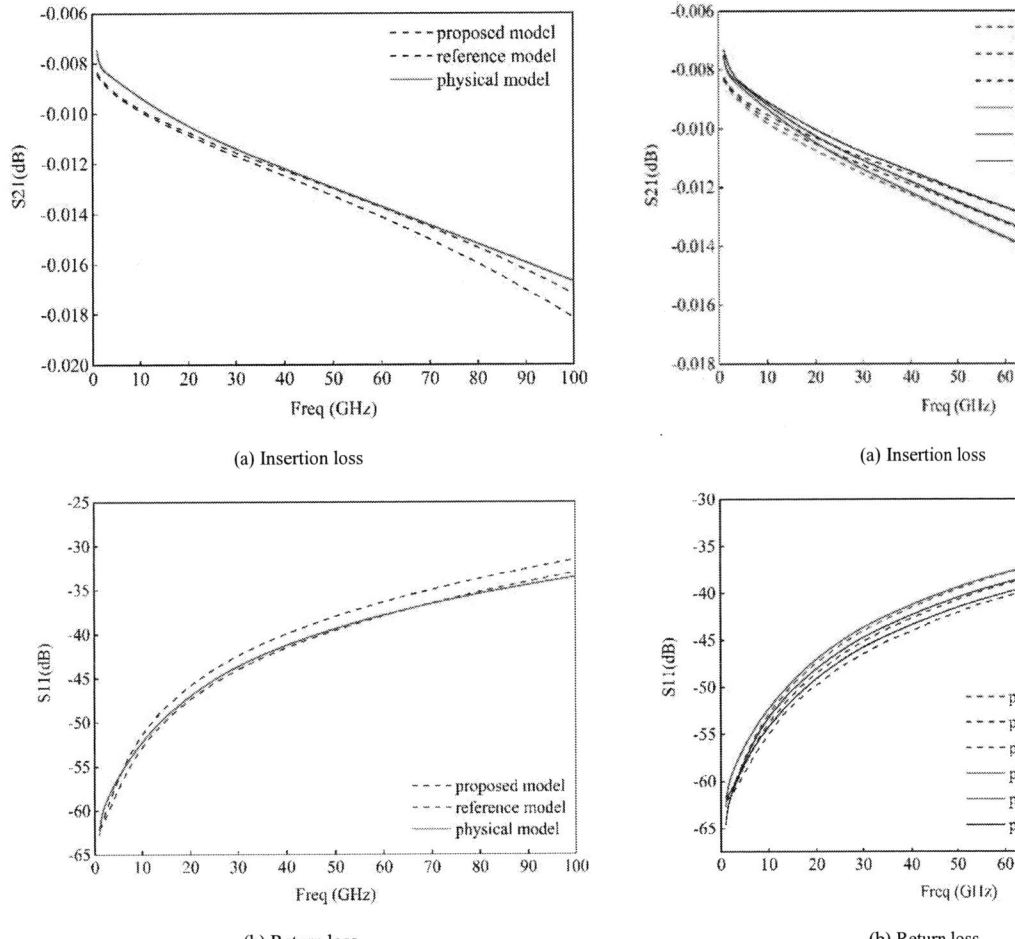

(a) Insertion loss

(b) Return loss

Fig. 5. Comparion of S-parameter in physical model, proposed model and reference model.

(a) Insertion loss

(b) Return loss

Fig. 6. Comparion of S-parameter in physical model, proposed model varying thickness of underfill with TSVs height.

We use the lumped model wherein h_{TSV} is short enough to be modeled as lumped element up to 100GHz. Hence, the S-parameter is simulated from 1 GHz to 100 GHz and excellent insertion loss is got for the low R_{TSV} and G_{Si} of bumpless TSVs. It can be seen from Fig.5 that good agreement is got between proposed model and physical model, and the maximum error for S11 and S21 is less than 3% and 9%. The proposed model without $C_{Underfill}$ is used as a reference model to study the effect of underfill. As show in Fig.5, the proposed model gets better match than reference model, which concludes that the effect of underfill needs to be considered while analyzing the electrical behavior of bumpless TSV.

To experimentally demonstrated the scalability of proposed model, S-parameter of proposed model and physical model are respectively simulated with variation of d_{TSV} and p_{TSV}, and the electrical behavior corresponds to the conclusions in [11]. However, the h_{TSV} needs further research due to the structural feature that TSV passes through the underfill instead of bump for the bumpless process. Considering that h_{TSV} is consist of h_{IMD}, $h_{Substrate}$ and $h_{Underfill}$, we respectively analyze the effect of $h_{Substrate}$ and $h_{Underfill}$ that varies with h_{TSV} with the constant value of h_{IMD}.

The equivalent-circuit model is verified in Fig. 6 when $h_{Underfill}$ varies with h_{TSV}, from 5 μm to 3 μm, and the h_{TSV} varies from 15 μm to 13 μm. There is also a good agreement

between the proposed model and physical model, and the insertion loss and return loss decrease from 1 GHz to 100 GHz. The proposed model catches the trend of change in magnitude well compared with S-parameter simulated from HFSS with aforementioned variation. From the analysis of the variation of electrical parameters, the decreased R_{TSV} and L_{TSV} are the main factor for this result while the variation of $C_{Underfill}$ is too small to obviously effects S-parameter.

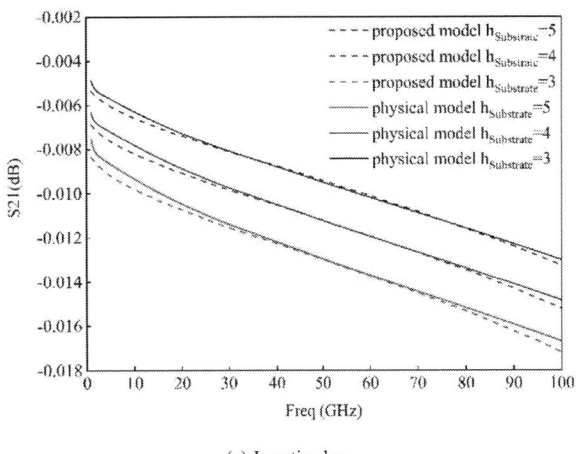

(a) Insertion loss

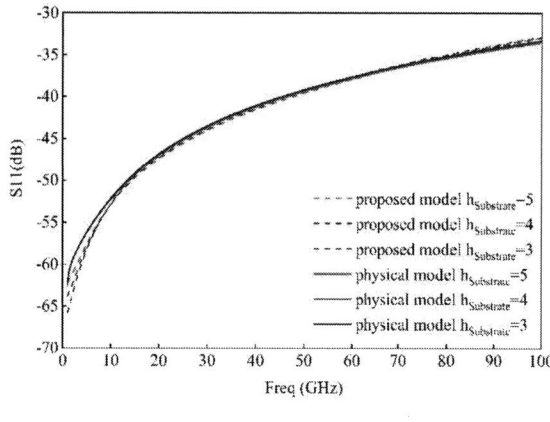

(b) Return loss

Fig. 7. Comparion of S-parameter in physical model, proposed model varying thickness of substrate with TSVs height.

The equivalent-circuit model is verified in Fig. 7 when $h_{Substrate}$ varies with h_{TSV}, from 5 μm to 3 μm, and the h_{TSV} varies from 15 μm to 13 μm. The insertion loss decreases even more over the entire frequency range due to the increasement of Gsi, but there is no obvious change in the return loss. This has not been common in researches of TSVs to date, so we analysis the results and get the conclusion as follow. After ultra-thinning of silicon substrate, the variation of $h_{Substrate}$ is bigger than the variation in traditional TSV [6]. For example, as h_{TSV} varies from 15 μm to 14 μm, the deduction of $h_{Substrate}$ is 1/5 in bumpless TSV and about 1/15 in traditional TSV. Hence, the effect of deduction of C_{Si} and L_{TSV} on return loss are comparable.

IV. CONCLUSION

A scalable equivalent-circuit model of bumpless TSV is proposed in this paper and its electrical parameters are extracted from analytical expressions. The proposed model of the bumpless TSVs is validated by frequency domain simulation up to 100 GHz with the physical model designed in HFSS. The error of S11 and S21 is less than 3% and 9% and the effect of underfill needs to be considered for greater accuracy. Furthermore, the effect of thickness of underfill and silicon substrate that varies with TSVs height are analyzed respectively while scalability is verified. As $h_{Underfill}$ varies with h_{TSV}, from 5 μm to 3 μm, and the h_{TSV} varies from 15 μm to 13 μm, the insertion loss and return loss decrease from 1 GHz to 100 GHz due to the deduction of R and L. As $h_{Substrate}$ varies with h_{TSV}, from 5 μm to 3 μm, the insertion loss has much more decrease over the entire frequency range due to the increasement of Gsi. There is no obvious change in the magnitude of S11 for the comparable effect of deduction of C_{Si} and L_{TSV} on return loss.

ACKNOWLEDGMENT

This work was supported by the General Program of National Natural Science Foundation of China, grant number 61974073; the Natural Science Foundation of the Jiangsu Higher Education Institutions of China, grant number 19KJB510007.

REFERENCES

[1] P. A. Thadesar, X. Gu, R. Alapati and M. S. Bakir, "Through-Silicon Vias: Drivers, Performance, and Innovations," in IEEE Transactions on Components, Packaging and Manufacturing Technology, vol. 6, no. 7, pp. 1007-1017, July 2016.

[2] M. F. Chen et al., "SoIC for Low-Temperature, Multi-Layer 3D Memory Integration," 2020 IEEE 70th Electronic Components and Technology Conference (ECTC), 2020, pp. 855-860.

[3] Y. S. Kim et al., "A robust wafer thinning down to 2.6-μm for bumpless interconnects and DRAM WOW applications," 2015 IEEE International Electron Devices Meeting (IEDM), Washington, DC, USA, 2015, pp. 8.3.1-8.3.4.

[4] C. Lee et al., "Temperature Cycling Reliability of WOW Bumpless Through Silicon Vias," 2019 International 3D Systems Integration Conference (3DIC), 2019, pp. 1-4.

[5] S. Sugatani, N. Chujo, K. Sakui, H. Ryoson, T. Nakamura and T. Ohba, "Vertically Replaceable Memory Block Architecture for Stacked DRAM Systems by Wafer-on-Wafer (WOW) Technology," in IEEE Transactions on Electron Devices, vol. 67, no. 11, pp. 4606-4610, Nov. 2020.

[6] R. Wang, G. Charles and P. Franzon, "Modeling and compare of through-silicon-via (TSV) in high frequency," 2011 IEEE International 3D Systems Integration Conference (3DIC), 2011 IEEE International, 2012, pp. 1-6.

[7] Y. P. R. Lamy, K. B. Jinesh, F. Roozeboom, D. J. Gravesteijn, and W. F. A. Besling, "RF characterization and analytical modelling of through silicon vias and coplanar waveguides for 3D integration," IEEE Trans. Adv. Packag., vol. 33, no. 4, pp. 1072–1079, Nov. 2010.

[8] G. Katti, M. Stucchi, K. De Meyer, and W. Dehaene, "Electrical modeling and characterization of through silicon via for three-dimensional ICs," IEEE Trans. Electron Devices, vol. 57, no. 1, pp. 256–262, Jan. 2010.

[9] Z. Xu, X. Gu, and J.-Q. Lu, "Parasitics extraction, wideband modeling and sensitivity analysis of through-strata-via (TSV) in 3D integration/packaging," in Proc. 22nd Annu. IEEE/SEMI ASMC, May 2011, pp. 1–6.

[10] J. Kim et al., "High-Frequency Scalable Electrical Model and Analysis of a Through Silicon Via (TSV)," in IEEE Transactions on Components, Packaging and Manufacturing Technology, vol. 1, no. 2, pp. 181-195, Feb. 2011.

[11] J. Hu, L. Wang, L. Jin and H. Z. JiangNan, "Electrical modeling and characterization of through silicon vias (TSV)," 2012 International Conference on Microwave and Millimeter Wave Technology (ICMMT), 2012.

[12] A. E. Engin and S. R. Narasimhan, "Modeling of Crosstalk in Through Silicon Vias," in IEEE Transactions on Electromagnetic Compatibility, vol. 55, no. 1, pp. 149-158, Feb. 2013.

[13] C. R. Paul, Inductance—Loop and Partial. New York, NY, USA: Wiley, 2010.

[14] Z. Xu and J. Lu, "Through-Strata-Via (TSV) Parasitics and Wideband Modeling for Three-Dimensional Integration/Packaging," in IEEE Electron Device Letters, vol. 32, no. 9, pp. 1278-1280, Sept. 2011.

[15] I. Ndip et al., "Analytical, Numerical-, and Measurement–Based Methods for Extracting the Electrical Parameters of Through Silicon Vias (TSVs)," in IEEE Transactions on Components, Packaging and Manufacturing Technology, vol. 4, no. 3, pp. 504-515, March 2014.

[16] D. H. Cheng, Fundamentals of Engineering Electromagnetics. MA: Addison-Wesley, 1992.

[17] N. Araki, S. Maetani, K. Young Suk, S. Kodama and T. Ohba, "Development of Resins for Bumpless Interconnects and Wafer-On-Wafer (WOW) Integration," 2019 IEEE 69th Electronic Components and Technology Conference (ECTC), 2019

978-1-6654-1392-3/21 $31.00 © 2021 IEEE

The Influence of Molding Compound Properties on System-in-Package Reliability for 5G Application

Dashun Liu
Center for Engineering Materials and Reliability
Guangzhou HKUST Fok Ying Tung Research Institute
Guangzhou, China
dsliu@ust.hk

Kai Chen
Center for Engineering Materials and Reliability
Guangzhou HKUST Fok Ying Tung Research Institute
Guangzhou, China
chenkai@ust.hk

Yijing Qin
Center for Engineering Materials and Reliability
Guangzhou HKUST Fok Ying Tung Research Institute
Guangzhou, China
yjqin@ust.hk

Yong Zhong
Center for Engineering Materials and Reliability
Guangzhou HKUST Fok Ying Tung Research Institute
Guangzhou, China
yongzhong@ust.hk

Zhaorong Wan
Center for Engineering Materials and Reliability
Guangzhou HKUST Fok Ying Tung Research Institute
Guangzhou, China
zrwan@ust.hk

Richeng Liu
Center for Engineering Materials and Reliability
Guangzhou HKUST Fok Ying Tung Research Institute
Guangzhou, China
rcliu@ust.hk

Dong Lu
Center for Engineering Materials and Reliability
Guangzhou HKUST Fok Ying Tung Research Institute
Guangzhou, China
maeld@ust.hk

Ke Xue
School of System Design and Intelligent Manufacturing
Southern University of Science and Technology
Shenzhen, China
xuek@sustech.edu.cn

Jingshen Wu
Center for Engineering Materials and Reliability
Guangzhou HKUST Fok Ying Tung Research Institute
Guangzhou, China
mejswu@ust.hk

Abstract—With the development of electronic packaging technology, devices become much smaller size and higher integration, and the reliability of devices has become one of the most concerned issues. Interfacial delamination has become one of the significant factors influencing the reliability of devices in System-in-Package (SiP) module with the rapid development of the fifth generation (5G) technology. In this work, a method for interface delamination modeling considering cure shrinkage and viscoelasticity was proposed, which could predict the interfacial delamination accurately. Combining the theoretical calculation formula and the warpage test results, the cure shrinkage rate of the Epoxy molding compound (EMC) was calculated. The parameters of EMC viscoelastic constitutive model (Maxwell model) were obtained by Dynamic mechanical analysis (DMA) test. Based on the Finite element analysis, the warpage deformation during reflow process was simulated by adjusting the CTE value of the EMC until it was consistent with the actual results from of shadow moiré test. Furthermore, using the cohesive force zone modelling (CZM), simulation for interface delamination of Cu/EMC was carried out by using ABAQUS. The influences of residual stress of cure shrinkage and viscoelastic of EMC at high temperature on the delamination failure of the device were also considered.

Keywords—*5G, Epoxy Molding Compound, Cohesive Zone Model, Delamination, Viscoelasticity*

I. INTRODUCTION

The fifth generation (5G) revolution have brought lots of innovative technologies, including Sub-6GHz (lower than 6 GHz) and millimeter wave (higher than 28 GHz) spectrum, beam forming, multi-connectivity and edge computing [1]. Furthermore, 5G connectivity application significantly promotes the development of System-in-Package (SiP) module design in Internet of Things (IoT) system to fabricate the products as compact as possible for multi-functional and

specific applications [2]. However, the reliability of SiP-based electronic devices would be threatened by the complex circuit design and large power of chips during working. Among this, the bi-material interfacial delamination is one of the significant failure modes in reliability issues, especially in the interface between the epoxy molding compound (EMC) and its adjacent material.

Generally, the interfacial delamination failure can be caused by the thermal mismatch of the adjacent materials, the cure shrinkage or the moisture absorption of EMCs [3, 4]. Accordingly, the delamination may occur in the packaging assembly stage or the device use stage, especially during the process of assembling the printed circuit board (PCB). It is because that the device must undergo a high reflow reflux temperature up to 260 °C, during which the coefficient of thermal expansion (CTE) mismatch of EMC and copper-based lead fame (Cu) usually results in the thermal stress. If some moisture is absorbed by EMCs, the diffusion of moisture and the vapor pressure in the interface would lead to a degradation of interfacial adhesion, and accelerate the interfacial delamination [5]. In addition, the residual stresses from the transfer molding processes and the post cure shrinkage of EMCs are also the important factors of delamination.

To predict the onset and growth of interfacial delamination of electronic components prior to the mass production, more attention has been paid on the Fracture Mechanics' energy criterion approach. In this criterion approach, it is assumed that the crack propagates once the strain energy release rate (SERR) G reaches its critical value G_c, where SERR denotes the driving force for the crack propagation, G_c means the critical strain energy release rate [6]. To apply this approach, the different fracture mode tests to characterize the interface strength are necessary, such as Four-Point Blend (4PB) Delamination Test [7], Double

978-1-6654-1392-3/21 $31.00 © 2021 IEEE

Cantilever Blend (DCB), Three-Point End Notched Flexure (3-ENF) Delamination Test [8], and Four-Point End Notched Flexure (4-ENF) Delamination Test [9]. Moreover, the Virtual Crack Closure Technique (VCCT) technology has been widely used for computing SERR due to its advantages over other Finite Element (FE) analysis-based methods [10]. The following empirical equation is proposed by Hutchinson and Suo, which can be used to obtain the interfacial fracture toughness under different mode angle [6]. Considering that most engineering materials are not complete brittle failure, Cohesive Zone Model (CZM) are mainly used to evaluate the cohesive forces.

$$G_c = G_{c0}[1 + tan^2(\psi(1 - \lambda))] \qquad (1)$$

where ψ is the mode angle, λ is an adjustable parameter, G_{c0} is the fracture energy when $\psi = 0$.

In this work, the interface strength characterization of Four-point Bending tests was performed on various interface samples. Based on the test results, the VCCT technology was used in the simulation to infer the fracture energy Gc and mode angle ψ of each interface. In addition, the CZM parameters of each interface was deduced, and the simulation via ABAQUS software and experimental regression verification of different System in SiP modules were carried out. This research combined the VCCT with CZM provides a pathway to accurately obtain Gc and predict the interfacial delamination risk by means of the damage threshold in complicated SiP for 5G application.

II. MODELING AND TESTING METHOD

Cohesive zone model for interface delamination

Cohesive zone models (CZM) as an important tool are widely used for describing fracture in engineering materials and structures, especially in interfacial delamination. In cohesive zone models, the fracture behavior is generally characterized by a traction–separation law as Fig. 1 with a positive slope, followed by a negative slope indicating a decreasing resistance during separation. Area beneath the traction-separation law is equivalent to critical energy release rate G_c.

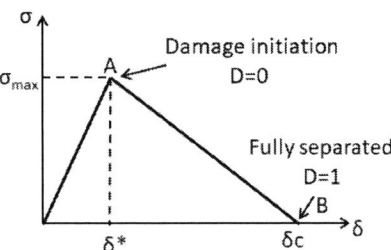

Fig. 1. Bilinear traction-separation law, where δ means the displacement and δc denotes the critical separation displacement, δ∗ is the critical interfacial separation displacement.

Damage (D) is initiated when the critical stress is reached. The evolution of D is defined as:

$$D = \begin{cases} 0 & if \ \delta \leq \delta* \\ (\frac{\delta-\delta*}{\delta})(\frac{\delta c}{\delta c-\delta*}) & if \ \delta* < \delta < \delta c \\ 1 & if \ \delta \geq \delta c \end{cases} \qquad (2)$$

Interface toughness testing

The interface fracture toughness can be obtained by different fracture mode tests, such as DCB test and ENF test for pure mode I (G_I) and pure mode II (G_{II}), respectively. In this work, we used 4PB testing method because it is independent of crack length and no crack length measurement is required to determine the fracture toughness. The test machine is shown in Fig. 2.

The critical forces, which were achieved from the four point bending (4PB) experiments, were used to obtain the critical SERRs G_c (i.e. interfacial fracture toughness) of the interface between substrates and EMC. They were determined by the FE simulations using the VCCT. Based on the critical SERRs obtained from different specimens tested at different temperatures, the temperature dependence of the interfacial fracture toughness was investigated.

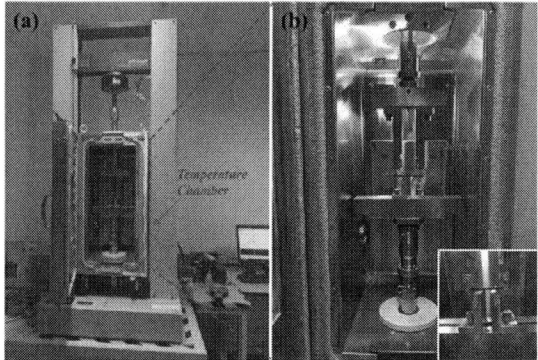

Fig. 2. Universal testing machine for 4PB Delamination Test.

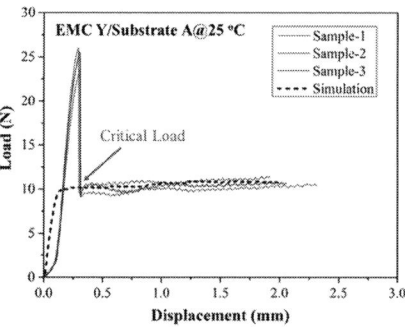

Fig. 3. A comparison of the experimental and simulation results of EMC Y/Substrate A at 25 °C.

According to the fracture energy G_c and mode angle ψ by VCCT technology, as well as the critical load of interface cracking at different temperatures obtained by the interface strength characterization test, one can determine the type I critical fracture energy release rate G_{1c} of each interface by simulation and calculation. Meanwhile, based on inverse method to adjust the parameters of the normal stiffness K_{nn} and the maximum cracking stress σ_{max}, the CZM parameters can be determined from the simulation of the 4PB load-displacement curves. Fig. 3 of EMC Y/Substrate A at 25 °C was exemplified, where three samples for each group were measured to ensure data repeatability. The simulation results are marked by the black dotted line, showing a great agreement with the experimental data. All the obtained parameters at different temperatures of the EMC/Substrates are given in Fig. 4, which were used in the simulation part. The reference data are from EMC X/Substrate A at 25 °C, where the G_{1c}, σ_{max} and δ_{1c} are respectively about 50 J/m, 38 MPa and 2.7×10^{-3} mm.

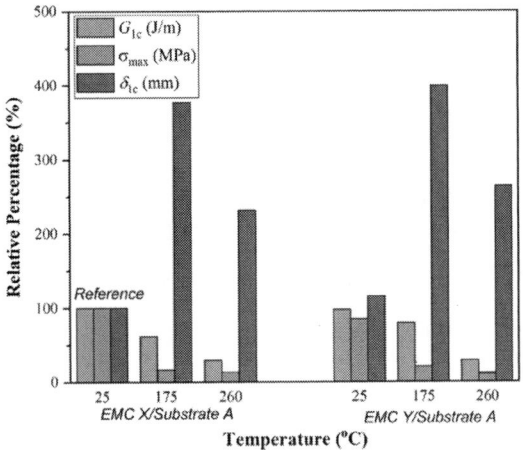

Fig. 4. G_{1c} and CZM parameters for two types of EMC/substrate interface at different temperatures.

III. SIMULATION AND VIRIFICATION

A. SAM Observations

A SiP module is designed to verify the accuracy of the predication on the delaminated interface through the CZM. The C-SAM was performed to detect the actual delaminated interface after post curing and after the first reflow at 260 °C in the TherMoiré PS200. The second reflow at 280 °C would be conducted if no interfacial delamination could be found after the first reflow. Here the reflow heating rate was set as 5 °C /min.

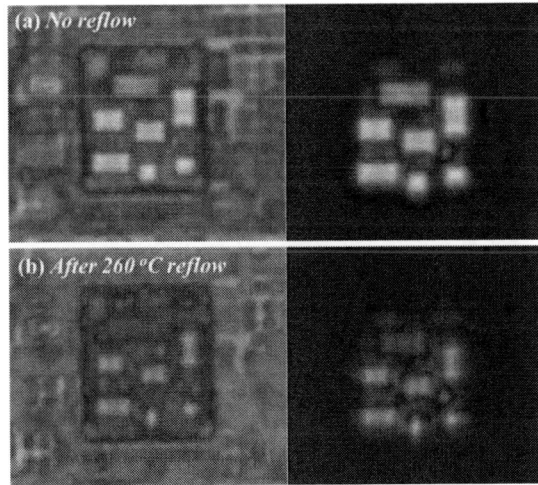

Fig. 5. Partial enlarged pictures of EMC Y/Substrate A presenting the solder flow after reflow at 260 °C.

To make a clear explanation, the partial enlarged pictures of EMC Y/Substrate A before and after reflow are displayed in Fig. 5. Compared with the morphology after molding and post curing in Fig. 5(a), the changed solder profile in Fig. 5(b) hints the solder flow. This indicates that the interface at the bottom of the device has delamination and cracking, which makes the solder flow into the cracked interface after high temperature melting during reflow at 260 °C, thus showing the change of the solder profile.

B. Optical Microscopy (OM) Observations

To make a clear determination of the interfacial delamination, the morphologies of the SiP samples were collected using the Leica OM. To be brief, the OM morphologies of EMC Y/Substrate A are presented in Fig. 6 as the representatives, in which the interfacial delamination between the EMC and soldermask on the substrate can be observed.

Fig. 6. OM picture of EMC Y/Substrate A with the interfacial delamination between the EMC and soldermask after reflow at 280 °C.

C. Comparision between Simulation and Testing

To perform the interfacial delamination simulation analysis, the linear elastic full model considering the cure shrinkage of EMC was used. The modulus of EMCs and substrate A for SiP system are shown in Fig. 7, where the 100 % reference point is approximately 3×10^4 MPa from Substrate A at 25 °C. The simulation results of the warpage deformation for EMC X/Substrate A and EMC Y/Substrate A at 260 °C and 280 °C are plotted in Fig. 8 and Fig. 9, respectively.

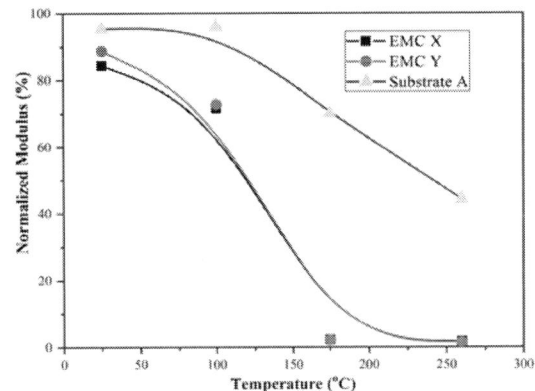

Fig. 7. The normalized modulus of EMCs and substrate A as a function of temperature.

In Fig. 8 and Fig. 9, with the "negative" warpage, the maximum deformation appears at the corner of the samples, while the minimum one appears in the center for all the samples. In addition, at the higher reflow temperature, i.e., 280 °C, a more significant warpage gradient can be found, indicating a larger deformation of the samples. The stress distributions of pure EMC X on the side of component and on the side of Substrate A are provided in Fig. 10, where the maximum stresses can be seen at the corner of the EMC. The results suggest that the initial cracks may appear at the corner,

978-1-6654-1392-3/21 $31.00 © 2021 IEEE

and then spread along the EMC plane to the entire device, which is consistent with the OM observations in Fig. 6.

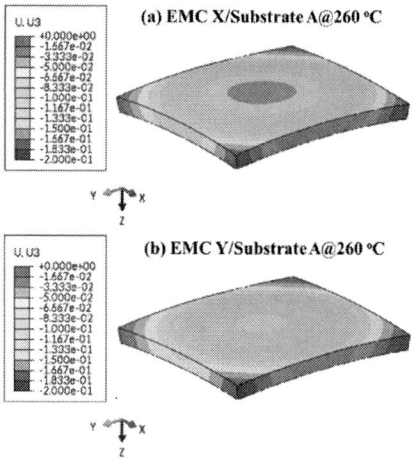

Fig. 8. Warpage simulation results of EMC X/Substrate A and EMC Y/Substrate A at 260 °C.

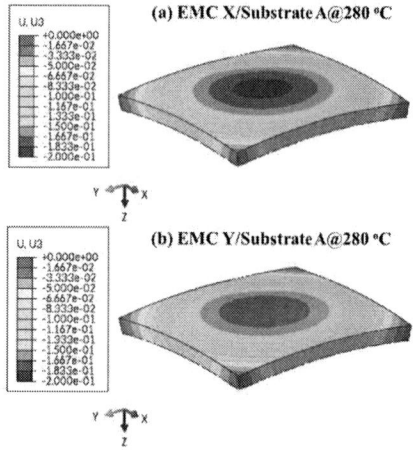

Fig. 9. Warpage simulation results of EMC X/Substrate A and EMC Y/Substrate A at 280 °C.

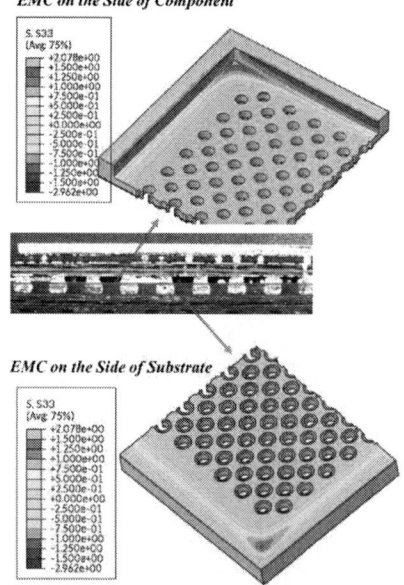

Fig. 10. Stress distribution of EMC X on the side of component and on the side of Substrate A.

To explain the interfacial delamination at different temperatures as given in Section B, the damage variation as a function of temperature in Fig. 11 are derived from the simulation results, where C1, C2 and C3 represent the Component 1, Component 2 and Component 3 located in different position in the SiP, respectively. The red dotted lines indicate the damage threshold with the damage value of 0.85, which implies that once the damage exceeds this value, the interfacial delamination in the component will be observed. The simulation results agree well with the observed morphologies.

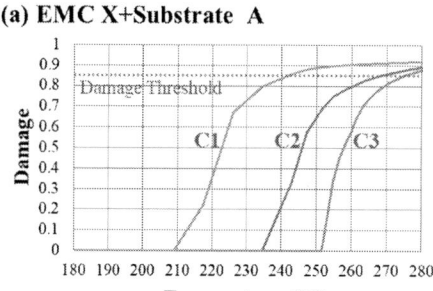

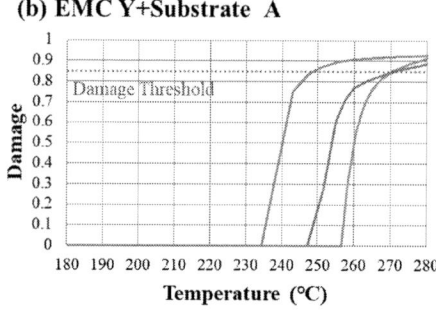

Fig. 11. Maximum damage value as a function of temperature derived from simulation.

D. EMC Properties Effect on the Warpage and Delamination

a) Viscoelastic effect

In the simulation, the viscoelastic effect of EMC materials should be considered. Dynamic Mechanical Analysis (DMA) was used to evaluate the viscoelastic properties of the EMCs. Based on the time-temperature superposition a master curve was gained and fitted by Prony coefficients for the FE software ABAQUS. A model was selected to analyze the EMC stress distribution on the side of substrate at 260 °C. Fig. 12 shows that there is no significant difference of the warpage for linear elastic model and viscoelastic model. From the stress comparison analysis in Fig. 13, the stress of EMC is larger in viscoelastic model than that in the linear elastic model with the difference about 6 %. Considering the computing time cost, it is efficient to simulate of interfacial delamination using the linear elastic model, and then to make the risk assessment by the method of safety margin.

978-1-6654-1392-3/21 $31.00 © 2021 IEEE

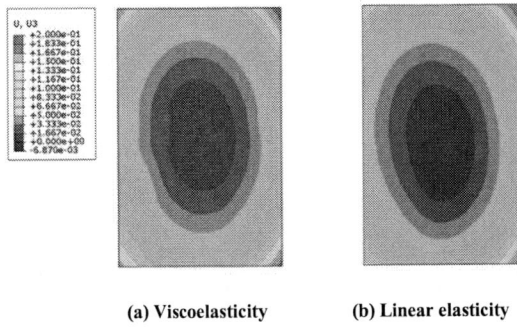

(a) Viscoelasticity   **(b) Linear elasticity**

Fig. 12. Comparison of the displacement of U3 for the viscoelastic model and linear elastic model.

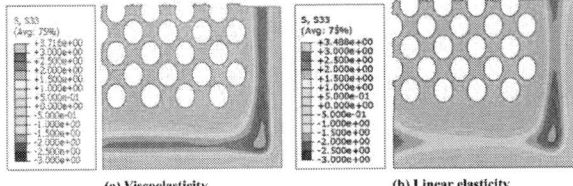

(a) Viscoelasticity   **(b) Linear elasticity**

Fig. 13. A comparison of the stress distribution on the EMC between the viscoelastic model and linear elastic model.

b) Cure shrinkage effect

Due to the chemical reactions, the EMC shrink during the transfer molding process. The decrease in volume is known as the cure shrinkage and causes a deflection of the package system. In order to investigate the effect of chemical cure shrinkage, a special bi-material Cu/EMC sample was designed and carried the warpage test by Shadow Moiré. In the simulation, by changing the CTE value of the molding compound, it can produce a shrinkage which is consistent with the cure shrinkage within a certain temperature range. First, an initial hypothetical CTE value of the molding compound was assigned to the model. Then simulate the warpage test, and correct the initial CTE value until the simulated warpage value is consistent with the actual measurement. The results of warpage from FE simulation together with the corresponding experiments are shown in Fig. 14. It shows that the simulation warpage with EMC cure shrinkage is consistent with the Shadow Moiré results. The fact that the simulation considering EMC cure shrinkage could accurately predict the warpage at the selected time and temperatures was a strong reason that the residual stresses were also calculated correctly. This is an important issue, especially when fracture mechanics is used for the prediction of interfacial delamination in SiP system.

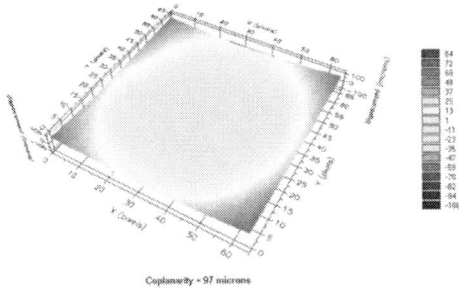

(a) Warpage of SiP package at 25 °C

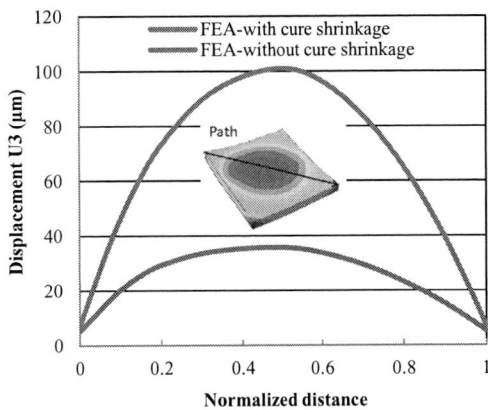

(b) Displacements of U3 along the path

Fig. 14. Warpage of SiP package by Shadow Moiré (a) and results from FE simulation at 25 °C (b).

IV. CONCLUSIONS

In this research, the 4PB delamination test was used to evaluate the critical load of interface cracking at different temperatures. Then through the combination of VCCT technology and CZM method, the fracture energy and CZM parameters were obtained. For EMC deformation simulation in the engineering applications, compared with the viscoelastic model, it is more efficient to select the linear elastic model and then make the risk assessment according to the set safety margin. The FE model should account for the cure shrinkage to accurately capture thermo-mechanical stresses in the SiP system. Based on the FE with cure shrinkage and linear elastic model, the interfacial delamination risk between the EMC and the bottom of component in SiP module can be quickly determined based on the damage threshold. This systematic simulation method provides an effective and desirable way to predict the interface delamination risk of SiP module for 5G application.

ACKNOWLEDGMENT

The authors would like to sincerely acknowledge the financial support from the Department of Science and Technology of Guangdong Province (Project No. 2020B010179002); the Guangzhou Science, Technology and Innovation Commission (Project No. 201904010279) and the 2018 Guangzhou International Postdoctoral Exchange Fellowship Program.

REFERENCES

1. Li, J., et al. *EMI Shielding Technology in 5G RF System in Package Module.* in *2020 IEEE 70th Electronic Components and Technology Conference (ECTC).* 2020.
2. Tsai, M., et al. *Innovative Packaging Solutions of 3D System in Package with Antenna Integration for IoT and 5G Application.* in *2018 IEEE 20th Electronics Packaging Technology Conference (EPTC).* 2018.
3. Zhu, W.H., et al. *Cure shrinkage characterization and its implementation into correlation of warpage between*

simulation and measurement. in *2007 International Conference on Thermal, Mechanical and Multi-Physics Simulation Experiments in Microelectronics and Micro-Systems. EuroSime 2007.* 2007.

4. Tran, H.T., et al., *Temperature, moisture and mode-mixity effects on copper leadframe/EMC interfacial fracture toughness.* International Journal of Fracture, 2014. **185**(1): p. 115-127.

5. Ferguson, T.P. and Q. Jianmin. *Moisture absorption in no-flow underfill materials and its effect on interfacial adhesion to solder mask coated FR4 printed wiring board.* in *Proceedings International Symposium on Advanced Packaging Materials Processes, Properties and Interfaces (IEEE Cat. No.01TH8562).* 2001.

6. Hutchinson, J.W. and Z. Suo, *Mixed Mode Cracking in Layered Materials*, in *Advances in Applied Mechanics*, J.W. Hutchinson and T.Y. Wu, Editors. 1991, Elsevier. p. 63-191.

7. Charalambides, P.G., et al., *A Test Specimen for Determining the Fracture Resistance of Bimaterial Interfaces.* Journal of Applied Mechanics, 1989. **56**(1): p. 77-82.

8. Shirangi, M.H., *Simulation-based Investigation of Interface Delamination in Plastic IC Packages under Temperature and Moisture Loading.* 2010.

9. Turon, A., et al., *Simulation of delamination in composites under high-cycle fatigue.* Composites Part A: Applied Science and Manufacturing, 2007. **38**(11): p. 2270-2282.

10. Krueger, R., *Virtual crack closure technique: History, approach, and applications.* Applied Mechanics Reviews, 2004. **57**(2): p. 109-143.

Thermal resistance of eutectic Ga-In-Sn/particles binary thermal interface materials

Wendong Wang
Institute of Metal Research, Chinese Academy of Sciences(CAS)
School of Materials Science and Engineering, University of Science and Technology of China(USTC)
Shenyang 110016, China
wdwang19b@imr.ac.cn

Meijuan Lv
Institute of Metal Research, Chinese Academy of Sciences(CAS)
School of Materials Science and Engineering, University of Science and Technology of China(USTC)
Shenyang 110016, China
mjlv18s@imr.ac.cn

Jingdong Guo
Institute of Metal Research, Chinese Academy of Sciences(CAS)
School of Materials Science and Engineering, University of Science and Technology of China(USTC)
Shenyang 110016, China
jdguo@imr.ac.cn

Abstract—With the continuous improvement of the integration of microelectronic chips, the heat dissipation of microelectronic chips is becoming more and more challenging. The interface thermal resistance is particularly prominent in the heat dissipation, and the use of thermal interface materials (TIMs) is generally considered to be an effective way to reduce interface thermal resistance. Low melting temperature alloy (LMTA) is an ideal thermal interface material due to its low interface thermal resistance and good thermal conductivity. However, LMTAs are always prone to failure or cause circuit failure due to its good fluidity and low viscosity in work environment.

The eutectic Ga-In-Sn low melting temperature alloy (Ga-In-Sn LMTA) has a melting point of about 11 ℃ and a thermal conductivity of about 39 W/(m • K), which make it applicable as a TIM. In this paper, thermal conductive particles, such as diamond and Tungsten, were added into the Ga-In-Sn LMTA to fabricate composite TIMs. By this method, the overflow of the LMTA may be effectively retard, and the thermal conductivity of the TIMs can also be enhanced. To improve the wetting contact property and the interface combination status between diamond fillers and liquid metal matrix, chromium transition layer was coated on the surfaces of diamond particles by magnetron sputtering method. The interfacial microstructure between diamond and the LMTA is analyzed by field emission scanning electron microscope (SEM) and x-ray energy dispersive spectroscopy (EDS). Thermal resistance of the composite TIMs is measured by a steady-state heat flow analysis using a specific layer structure sample, and a corresponding theoretical simulation model is constructed subsequently. Meanwhile, the effect of agglomeration of thermal conductive particles on the thermal flux density of LMTA/particles composite TIMs is also explored. The results show that addition of chromium-coated diamond particles and tungsten microparticles can dramatically increase the thermal conductivity of eutectic Ga-In-Sn LMTA at room temperature. The SEM images show that the diamond and tungsten particles were coated effectively by the eutectic Ga-In-Sn LMTA, which suggest good wetting and good interfacial contact. Particularly, the wettability between chromium coated diamond and eutectic Ga-In-Sn LMTA is significantly enhanced, indicating that chromium can be used as the medium layer between heat-conducting particles and the metal substrates to maintain long-term reliable service of eutectic Ga-In-Sn LMTA TIMs. Thermal flux results of finite element analysis show that the efficient heat transfer channels were established by connecting the heat-conducting fillers with eutectic Ga-In-Sn LMTA, and play a significant role in the composite TIMs.

Keywords—TIMs, Thermal resistance, Ga-In-Sn, Diamond, Packaging Materials

I. INTRODUCTION

With the rapid development of microelectronics field and continuous improvement of integration of electronic components, the power density of microelectronic devices has been dramatically improved in the past few years, which severely limited the performance of the microelectronic devices. Therefore, heat dissipation of microelectronic components becomes a significant field to keep the device operating within a safe temperature range. Thermal interface materials (TIMs) are played a significant role in filling the micro-gaps between chip surfaces and heat sinks to lower the temperature of microelectronic devices by reducing the thermal resistance caused by air in micro-gaps between contact surfaces, as shown in Fig. 1.

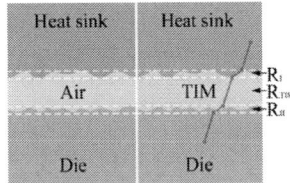

Fig.1. Schematic diagram showing that the real area of contact is less than apparent area and an ideal TIM could completely fill the gap between die and heat sink

The actual total thermal resistance at the interface of two solid surfaces between microprocessor chips and heat sinks including volume thermal resistance of the TIM and interface contacts thermal resistance (ICTR) between the TIM and the two solid contact surfaces. So, the equation of thermal resistance can be written as:

$$R = R_{\mathrm{I}} + R_{\mathrm{TIM}} + R_{\mathrm{II}} \tag{1}$$

$$R_{\mathrm{TIM}} = \frac{BLT}{\lambda_{\mathrm{TIM}}} \tag{2}$$

where R is the total thermal resistance of TIM, R_{TIM} is the volume thermal resistance of TIM, R_{I} and R_{II} are the interface contacts thermal resistance of each side of TIM, BLT is the bond line thickness of TIM, λ_{TIM} is thermal conductivity of TIM.

At present, commercially typical TIMs including greases, gels, thermal pads, carbon based TIMs and phase change materials (PCMs) [1-5]. Greases are one of the most widely used TIMs due to it' s easy operation and low viscosity. However, BLT of greases TIM is easy to change from different pressure due to its low viscosity, result in that greases can extrude edge. Thermal pads with high flexibility and good

978-1-6654-1392-3/21 $31.00 © 2021 IEEE

mechanical properties are easy to handle, meanwhile, advantage with good thermal conductivity and high viscosity makes sure that it's an alternative product to replace greases. But thermal resistance of thermal pads range from 100 K·mm2/W to 400 K·mm2/W is slightly higher than greases range from 10 K·mm2/W to 100 K·mm2/W [6-7]. Carbon based materials due to pure carbon material with extremely high thermal conductivity, on the one hand it as an ideal filler such as diamond, graphene (as filler in composite pastes), carbon nanotubes (CNTs) and graphite for TIMs, on the other hand itself with high performance such as graphene paper and 3D graphene for TIMs [8-10]. However, extremely high thermal interface contacts resistance between carbon based materials and surface of heat sinks or dies caused by the different physical mechanism of heat conduction is one of the difficult problems to be solved urgently. Phase change materials with both the solid states properties in the room temperature like thermal pads and paste states properties over melting temperature like thermal greases. The melting temperature of phase change materials limits it operating in commerce.

Low melting temperature alloy (LMTA) is liquid states in the room temperature, which is known as liquid metal or fusible alloy, containing metals such as gallium, indium, tin and bismuth, such as eutectic Ga-In-Sn and eutectic Bi-In-Sn. LMTAs has two significant properties: a low melting temperature and the ability to control volume changes as they solidify. First recoded use of LMTAs dates back to 1928 [11]. LMTA not only has superior thermal and electrical conductivity, low viscosity and superior fluidity, but also has a significant effect on temperature uniformity. In addition, roughness of surface of heat sinks and dies has little effect on the interface contacts thermal resistance between LMTA and heat sinks or dies. Therefore, LMTA and LMTA TIMs are becoming ideal thermal management materials for heat conduction in the microelectronics field [12-14].

Nowadays, heat dissipation requirements of LMTA TIMs are gradually urgent with the improvement of the properties of TIMs. Improving thermal conductivity, viscosity and reducing interface contacts thermal resistance, fluidity of LMTAs is an obstacle to the practical application and development of a new type of LMTA TIM. In numerous studies, it is easy to achieve the interface contacts thermal resistance as low as 5 K·mm2/W by using pure LMTA [15-16]. However, it is an extreme disadvantage that the high fluidity and electrical conductivity of pure LMTAs are easy to overflow the edge and damage the circuit. Adding high thermal conductivity or reinforced phase particles into pure LMTA to fabricate LMTA/particles composite TIMs is a novel method. Compared with pure LMTA, it effectively reduces the fluidity and improves the thermal conductivity of TIMs [17]. Both metal particles such as copper or tungsten and inorganic particles with high conductivity such as diamond, silicon carbide (SiC) or boron nitride (BN) can significantly improve the thermal conductivity of LMTAs [18-20].

In our work, thermal conductive particles, such as diamond and Tungsten, were added into the Ga-In-Sn LMTA to fabricate composite TIMs. Thermal conductivity and interface contacts thermal resistance of the TIMs were measured, the effect of heat dissipation structure and agglomeration of LMTA/particles on the thermal resistance of the TIMs were analyzed by experiments and finite element analysis.

II. Experimental

Gallium, indium, tin (all metals were 99.99% purity) were employed as raw materials of the eutectic Ga-In-Sn low melting temperature alloy (Ga-In-Sn LMTA, the theory of melting point of about 11 ℃ and a thermal conductivity of about 39 W/(m · K)), which was prepared to begin with gallium, after that indium was melted in the Ga matrix, and finally tin was melted in the Ga-In matrix. A bulk Ga-In-Sn LMTA sample with a mass of about 30mg was tested for melting point using differential scanning calorimetry (DSC, Q1000). The as-received chromium-coated diamond particles (CCDPs, particle size≈40μm, HFD-D type, Henan Huanghe Whirlwind Co., China) and tungsten microparticles (TMPs, average particle size ≈ 1.53μm) were employed as thermal conductive particles. Fabrication of CCDP follows a complex process that diamond particles were coated with chromium by using magnetron sputtering method with the basic vacuum degree of $5×10^{-3}$Pa at 300°C. The particles size of TMPs follows a normal distribution and average are about 1.53μm. CCDPs and TMPs were tested for qualitative analysis using X-ray diffraction (XRD, D8 Advance).

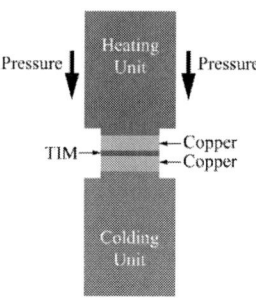

Fig. 2. Schematic of a typical ASTM D5470 standard setup

Here, a LMTA/particles (CCDPs and TMPs) mixture with high thermal conductivity was fabricated. In a porcelain mortar, Ga-In-Sn LMTA and thermal conductive particles were stirred with volume mixing ratio of 25:75，60:40 and 75:25 until well dispersed and blended. Morphology of clad layer was characterized by field emission scanning electron microscope (FE-SEM; Zeiss Supra 55 Sapphire). The thermal resistance of LMTA-particles mixtures with different volume mixing ratios was measured by thermal conductivity tester (DRL-III), which the test principle based on the steady-state thermal methods following the ASTM D5470 standard, as shown in Fig.2.

The finite element simulation of the heat flux and the temperature distribution of LMTA-particles mixtures was carried out. The finite element simulation model of LMTA-particles mixtures was built to begin to write code by using MATLAB.

III. Results And Discussion

A. DSC of eutectic Ga-In-Sn LMTAs

Eutectic Ga-In-Sn LMTA phases transition temperature is a significant material parameter. Fig. 3(a) and (b) indicates the phase diagrams of eutectic Ga-In-Sn LMTA, Fig. 3(c) shows the heating DSC curves for eutectic Ga-In-Sn LMTA from -60℃ to 70℃. The DSC curves show one peak at about 12℃, which correspond to the melting point seen from ternary phase diagrams and vertical section from A to B in Fig. 3(a). It means that the ternary eutectic reaction L=(Ga)+In3Sn+(Sn) occurs at this temperature.

B. Microstructures of LMTA/particles composite TIMs

SEM characterization of the TMPs shows that it has heterogeneous particle size, as shown in Fig. 4(a). The average particle size of the TMPs is about 1.53μm, as shown in Fig. 4(j). Figure 4 (d-f) shows that when TMPs are mixed with eutectic Ga-In-Sn LMTA, TMPs are covered by eutectic Ga-In-Sn LMTA after thorough stirring, indicating the wetting ability of tungsten with eutectic Ga -In- Sn LMTA is pretty good. However, pure TMPs are prone to agglomerate in eutectic Ga-In-Sn LMTA due to its heterogeneous particle size, and this phenomenon become obvious with the increase of the volume mixing ratio of the eutectic Ga-In-Sn LMTA, as shown in Fig. 4(f). It is mainly due to the fact that the particle size of TMPs is only 1.53μm, which cause the TMPs has high specific surface area and surface energy. As the volume mixing ratio of eutectic Ga-In-Sn LMTA increases, it is difficult to separate TMPs during stirring. The qualitative analysis of pure TMPs has clear diffraction peaks at 2θ of 40.3°, 58.3°, 73.2°and 87°, as shown in Fig. 4(k).

SEM characterization of the CCDPs shows that particle size is about 40μm, and the particle size is homogeneous, as shown in Fig. 4(b). The chromium coating on the diamond surface obtained by magnetron sputtering methods is quite flat and uniform, suggesting that this surface metallization technology is mature for diamond. When CCDPs are mixed with eutectic Ga-In-Sn LMTA, after fully stirring, CCDPs are completely covered with eutectic Ga-In-Sn LMTA, indicating that the wetting ability of chromium coating with eutectic Ga-In-Sn LMTA is fairly good, which may effectively improve the phonon mismatch between heterogeneous materials. Compared with the TMPs, the specific surface area and surface energy of CCDPs are much smaller, so CCDPs is uniformly distributed in eutectic Ga-In-Sn LMTA, as shown in Fig. 4(g-i). As shown in Fig. 4(k), the qualitative analysis of pure CCDPs has clear diffraction peaks of diamond at 2θ of 43.9°, 75.3° and 91.5°, and the diffraction peaks of chromium at 2θ of 44.4°, 64.6°, 81.7°and 98.1°, which is not obvious compared to diamond because the thickness of chromium coating on diamond surface by magnetron sputtering methods is too shin. The purpose of chromium coating on the diamond surface is on the one hand to form a significant wetting effect with eutectic Ga-In-Sn LMTA, on the other hand to enhance the bonding strength between diamond and eutectic Ga-In-Sn LMTA. It is measured that as the weight increment of diamond particles is 3‰, the thickness of chromium coating on diamond surfaces is approximately 9.73 nm.

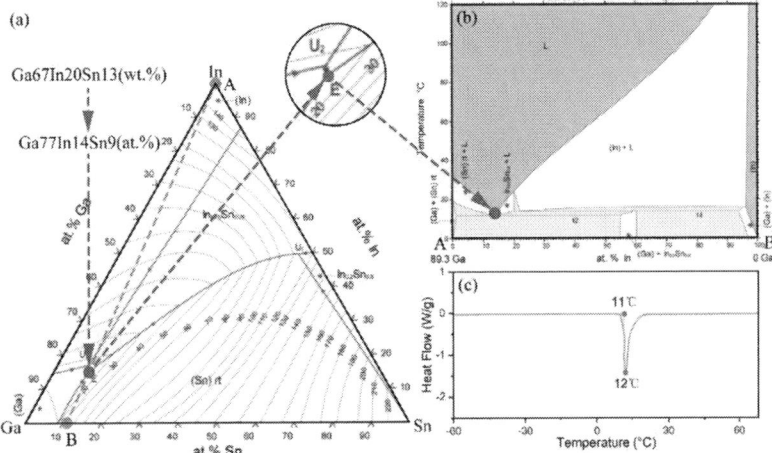

Fig. 3. Phase diagrams and Differential scanning calorimetry (DSC) of eutectic Ga-In-Sn LMTA, (a)projection of the liquidus surface[21], (b)vertical section of A-B[22], (c)heating DSC curve between 70℃ and -60

C. Thermal resistance of LMTA/particles composite TIMs

Fig. 5 shows the thermal properties of LMTA/CCDP and LMTA/TMP composite TIMs (Each error bar is derived from the standard deviation of the measured data for three samples). It is clearly that the increase of eutectic Ga-In-Sn LMTA proportion has a significant effect on the thermal conductivity of the TIMs, as shown in Fig. 5(a). It presents a bell-shaped trend curve, which means thermal conductivity of TIM does not always increase with the increase of the proportion of thermal conductive particles. The LMTA/particles composites achieved the highest thermal conductivity when the volume mixing ratio of LMTA: particles is 60:40. Compared with pure eutectic Ga-In-Sn LMTA, the thermal conductivity increased by 1.6 times (TMP) and 2.7 times (CCDP), respectively, as shown in Fig. 5(b). Meanwhile, Compared with commercial TIMs, thermal conductivity still has obvious advantages when the volume mixing ratio of LMTA: particles is 60:40, as shown in Fig. 5(c).

Thermal resistance and ICTR (including R_I and R_{II}) of LMTA/particles composite TIMs are shown in Fig. 5(c). It can be seen that with the increase of the liquid metal content, thermal resistance and ICTR of LMTA/particles composite TIMs are significantly decreased. The maximum value of thermal resistance and ICTR are observed at a 25% eutectic Ga-In-Sn LMTA volume mixing ratio. On the one hand, this rapid increase of thermal resistance can be attributed to insufficient wetting and poor coverage of eutectic Ga-In-Sn LMTA on thermal conductive particles, which caused poorly conducting particle–particle contacts. On the other hand, air trapped in between loosely bound thermal conductive particles is a likely source of lower thermal conductivity for high volume mixing ratios of thermal conductive particles. The minimum value of thermal resistance and ICTR are observed at a 60% eutectic Ga-In-Sn LMTA volume mixing ratio. This can be attributed to sufficient wetting and good coverage of eutectic Ga-In-Sn LMTA on thermal conductive particles. The excess eutectic Ga-In-Sn LMTA can form a liquid bridge and eliminate air trapped between the thermal conductive particles. At a 75% TMPs volume mixing ratio, the small increase of thermal resistance and ICTR can be observed, which means that the decrease in the content of TMP with a particle size of only 1.53μm has a significant impact on the thermal conductivity. The decrease in thermal conductivity causes a slight increase in thermal resistance.

The effect of pressure on the total thermal resistance of LMTA/CCDPs and LMTA/TMPs composite TIMs is shown in Fig. 5(d). When the LMTA: particle volume mixing ratio is 25:75, thermal resistance and thickness of LMTA/TMPs composite TIMs show a downward trend with increasing pressure. For LMTA/TMPs composite TIMs, as the pressure range increases from 100KPa to 500KPa, the thickness decreases by 5.58%, but the thermal resistance sharply decreases by 25.53%. While for LMTA/CCDPs composite TIMs, the thickness decreases by 5.8%, but the thermal resistance only decreases by 47.31%, as the pressure range increases from 100KPa to 500KPa. It can be predicted that thermal resistance of LMTA/TMPs composite TIMs will decrease with further increasing pressure. However, as

pressure continues to increase, the thermal resistance of LMTA/CCDPs composite TIMs decreases at a significantly slower rate. There are two reasons for this phenomenon. Firstly, the size of the TMPs is very small, and it is easier to slip under the action of pressure. Secondly, the influence of pressure on the total thermal resistance of the LMTA/CCDPs composite TIMs is not sensitive, this is because the limited pores in the LMTA/CCDPs composite TIMs have been compressed to the maximum extent, and the thickness change is very small, which also proves this.

D. Finite element analysis of LMTA/particles composite TIMs

In order to study the internal heat transfer mechanism of LMTA/particles composite TIMs, the LMTA: particles volume mixing ratio of 60:40 was selected as the finite element analysis object. The finite element model of LMTA/CCDPs and LMTA/TMPs composite TIMs with a LMTA: particles is shown in Fig. 6 (a) and (e). The results show that after mixing and stirring, the thermal conductive particles are randomly distributed in the eutectic Ga-In-Sn

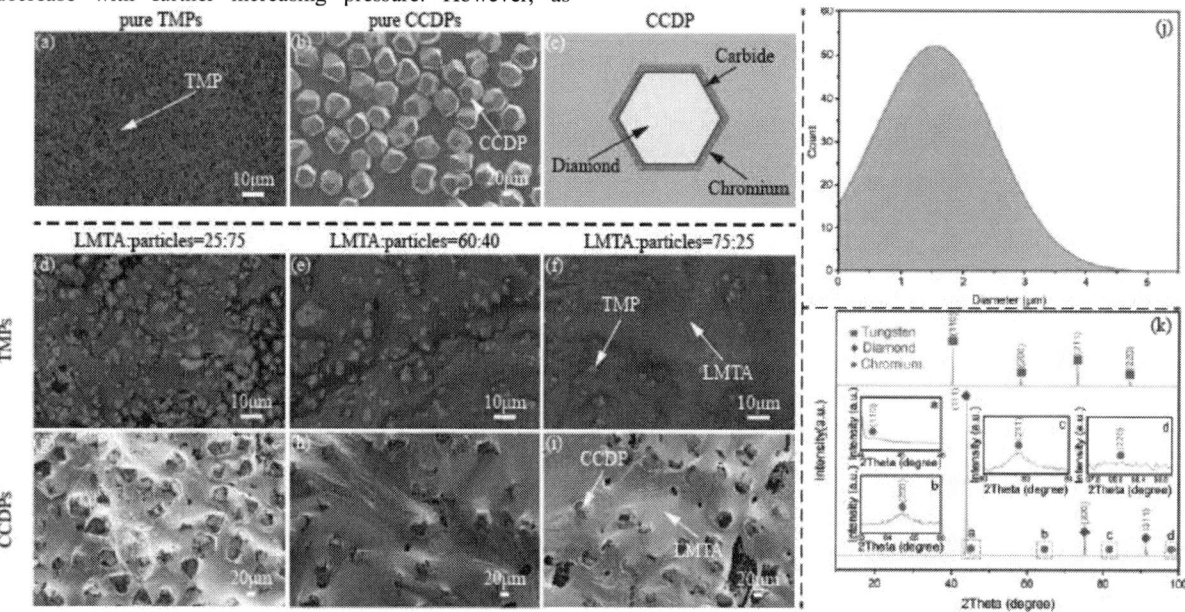

Fig. 4. Scanning electron microscopy (SEM) and X-ray diffraction (XRD) characterization of the chromium-coated diamond particles (CCDPs) and the tungsten microparticles (TMPs), (a)SEM of the pure TMPs, (b)SEM of the pure CCDPs, (c) Diagram of CCDP, (d)SEM of LMTA : TMPs = 25:75, (e)SEM of LMTA : TMPs = 60:40, (f)SEM of LMTA : TMPs = 75:25, (g)SEM of LMTA : CCDPs = 25:75, (h) SEM of LMTA : CCDPs = 60:40, (i)SEM of LMTA : CCDPs = 75:25, (j)distribution histogram of the TMPs size, (k)XRD characterization of the CCDPs and the TMPs

LMTA. Therefore, in order to make the model similar to the experimental sample, a model of the random distribution of thermal conductive particles in the eutectic Ga-In-Sn LMTA was designed and established using MATLAB software.

The temperature gradient distribution of LMTA/particles composite TIMs is shown in Fig. 6(b) and (f). The isotherms of LMTA/TMPs composite TIMs are relatively smooth. Obviously, the isotherms of LMTA/CCDPs composite TIMs have a tendency to distribute along the diamond contour, as shown in Fig. 6(f).

The thermal flux density results of LMTA/particles composite TIMs are shown in Fig. 6(c) and (g). It can be seen that the distribution of heat flux density inside the LMTA/particles composite TIMs has obvious directivity. And from the cloud chart of the heat flux density, the color of the

heat conducting particles is the same, which shows that the thermal flux is mainly concentrated in the region of the thermal conductive particles. Adjacent thermal conductive particles will be connected in series to form a chain-type heat transfer path, which is beneficial to the transfer of thermal flux. When the distance between two thermal conductive particles is very close, the thermal flux density between the two particles increases significantly, which shows that the distance between the thermally conductive particles has an impact on the heat flow transmission. Combined with the cloud chart of the temperature gradient (as shown in Fig. 6(b) and (f)), it can be seen that the chain structure formed by the thermal conductive particles is the main path of thermal flow transmission. Fig. 6(d) and (h) shows the cloud chart of thermal flux vector of LMTA/particles composite TIMs, and it is not difficult to see that the thermal flow vector is mainly

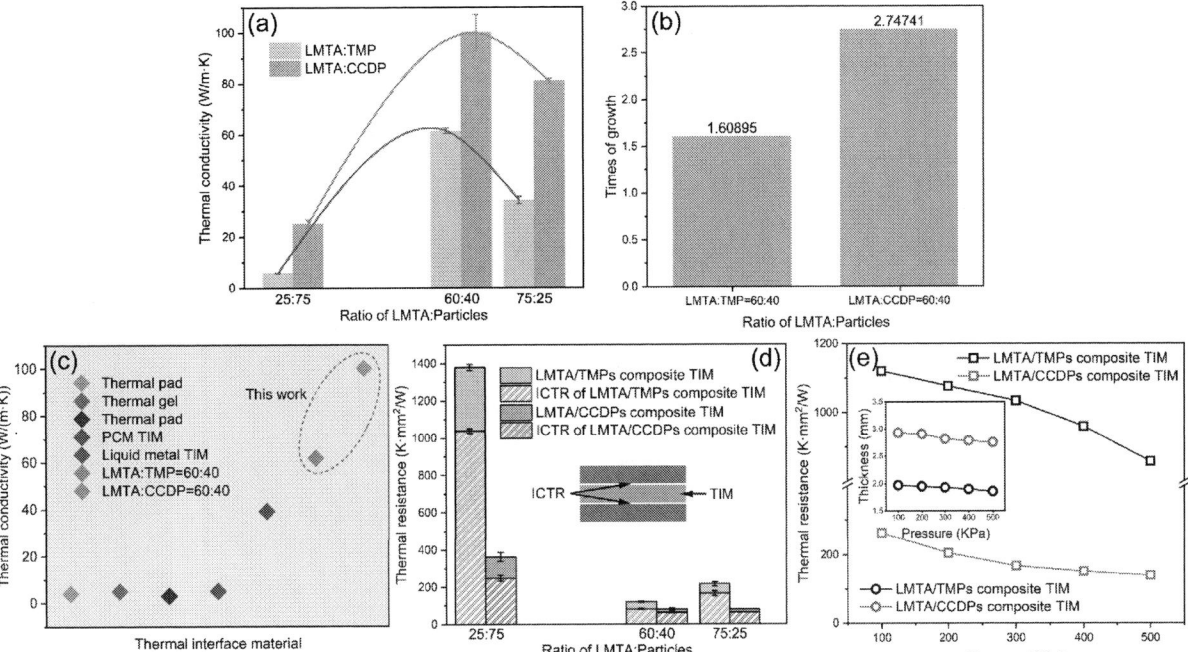

Fig. 5. Thermal properties of LMTA/CCDP and LMTA/TMP composite TIMs in the different volume mixing ratio, (a)Thermal conductivity of LMTA/CCDP and LMTA/TMP composite TIMs, (b)Times of growth of thermal conductivity, (c)Thermal conductivity of commercial TIMs and this work[2], (d)Thermal resistance of LMTA/CCDP and LMTA/TMP composite TIMs, (e)Effect of pressure on thermal resistance and thickness of LMTA/CCDP and LMTA/TMP composite TIMs when the volume mixing ratio of LMTA: particles is 25:75

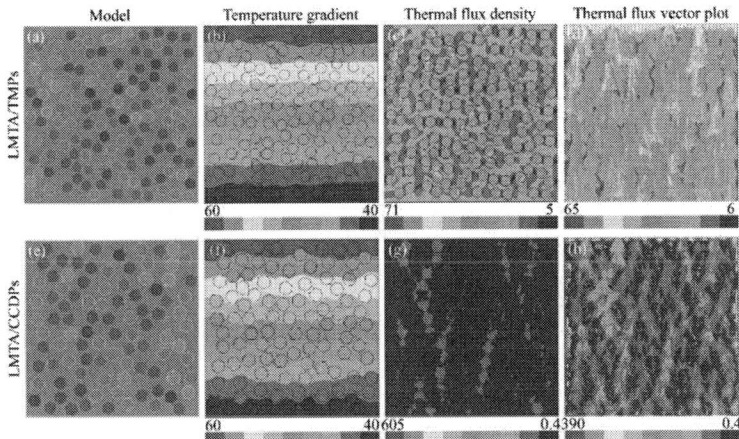

Fig. 6. Finite element analysis results of LMTA/CCDPs and LMTA/TMPs composite TIMs when the volume mixing ratio of LMTA: particles is 60:40, (a)Finite element model of LMTA/TMP composite TIMs, (b)temperature gradient of LMTA/TMPs composite TIMs, (c)Thermal flux density of LMTA/TMPs composite TIMs, (d)Thermal flux vector plot of LMTA/TMPs composite TIMs, (e) Finite element model of LMTA/CCDP composite TIMs, (f)temperature gradient of LMTA/ CCDPs composite TIMs, (g)Thermal flux density of LMTA/ CCDPs composite TIMs, (h)Thermal flux vector plot of LMTA/ CCDPs composite TIMs

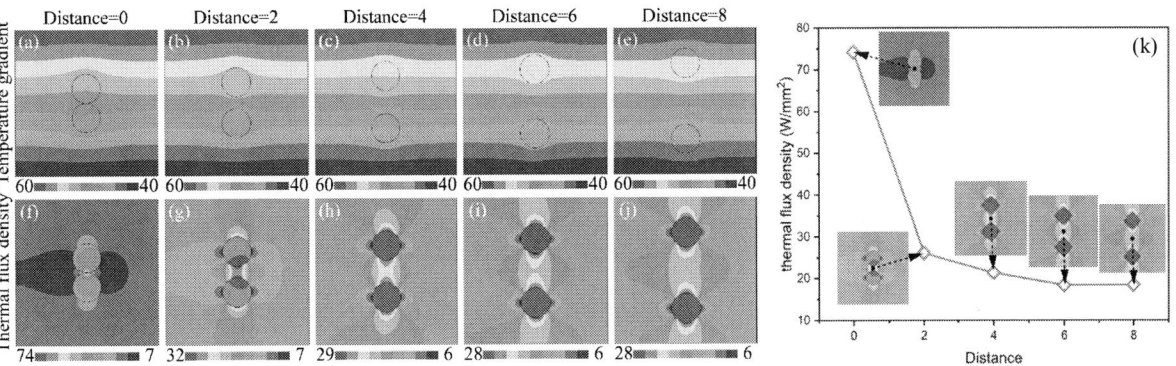

Fig. 7. Finite element analysis results of the influence of TMPs agglomerate by two tungsten particles model, (a-e) The temperature gradient of two tungsten particles model with different distance, (f-j) The thermal flux density of two tungsten particles model with different distance, (k) Effect of distance on thermal flux density of two tungsten particles

978-1-6654-1392-3/21 $31.00 © 2021 IEEE

concentrated in the region of the thermal conductive particles, which is consistent with the thermal flux density cloud diagram, proving that the chain structure formed by the thermal conductive particles is effective.

In order to explore the influence of thermal conductive particles agglomeration on the thermal flux density of LMTA/particles composite TIMs, the model of two tungsten particles with different distance was established using finite element analysis, the results were shown in Fig. 7.

The temperature gradient cloud chart of two tungsten particles model with different distance are shown in Fig. 7(a-e), and it is clearly to see that as the distance between the two tungsten particles increases, the temperature gradient between the heating unit and the colding unit changes significantly.

When the distance between two tungsten particles exceeds a certain value (distance=6), the temperature gradient between the two particles does not change obviously, as shown in Fig. 7(d) and (e).

As shown in Fig. 7(f)-(j), the thermal flux density cloud chart of two tungsten particles models with different distance shows that the effect of tungsten particles agglomeration on the thermal flux density is obvious. When two tungsten particles are in contact, the thermal flux value in the contact region is the largest, indicating that the two tungsten particles have established a chain heat transfer structure. When the distance between the two tungsten particles gradually increases, the thermal flux density in the region of the two tungsten particles gradually decreases. This is consistent with the change of the thermal flux at the center of the two tungsten particles region, as shown in Fig. 7(k). But when the distance between the two particles exceeds a certain value (distance=6),the thermal flux density at the center position hardly changes, which shows that the chain heat transfer structure has been broken.

IV. Conclusion

In summary, the specific conclusions are as follows:

(1)Chromium and tungsten could achieve good wettability with eutectic Ga-In-Sn LMTA. The size of the thermal conductive particles affects its distribution in eutectic Ga-In-Sn LMTA. Pure TMPs with heterogeneous particle size is easy to agglomerate and the homogeneous CCDPs is uniformly distributed in eutectic Ga-In-Sn LMTA.

(2)The thermal conductivity does not always increase with the increase of the proportion of thermal conductive particles, it presents a bell-shaped trend curve. When the volume mixing ratio of LMTA: particles is 60:40, The LMTA/particles composite TIMS achieved the highest thermal conductivity.

(3)The finite element analysis results show that the thermal conductive particles establish a chain structure inside the LMTA/particles composite TIMs to undertake the main heat transfer assignment, and the distance between the thermal conductive particles greatly affects the establishment of this chain structure.

Acknowledgment

We gratefully acknowledge the financial support from the National Natural Science Foundation of China (No. 51971231).

References

[1] Kafil M. Razeeb, Eric Dalton, Graham Lawerence William Cross and Anthony James Robinson. "Present and future thermal interface materials for electronic devices," International Materials Reviews, 2018, 63(1), pp. 1-21.

[2] Josef Hansson, Torbjörn M. J. Nilsson, Lilei Ye and Johan Liu. "Novel nanostructured thermal interface materials: a review," International Materials Reviews, 2018, 63(1), pp. 22-45.

[3] Jens Due and Anthony J. Robinson. "Reliability of thermal interface materials: A review," Applied Thermal Engineering, 2013, 50(1), pp. 455-463.

[4] HOU Si-yu,YAN Huan-huan, REN Fang and DI Ying-ying. "Research progress on thermal conducting polymer composites," Synthetic Materials Aging and Application, 2020,49(06),pp. 135-138+83.

[5] Junaid Khan, Syed Abdul Momin and M. Mariatti. "A review on advanced carbon-based thermal interface materials for electronic devices," Carbon, 2020, 168, pp. 65-112.

[6] enjamin Sponagle and Dominic Groulx. "Measurement of thermal interface conductance at variable clamping pressures using a steady state method," Applied Thermal Engineering, 2016, 96, pp. 671-681.

[7] D. Blazej, "Thermal interface materials," Electron. Cool. 9 (2003), pp. 14–21.

[8] Biercuk M J, Llaguno M C, Radosavljevic M, JK Hyun, AT Johnson, JE Fischer. "Carbon nanotube composites for thermal management," Applied Physics Letters, 2002, 80(15), pp. 2767-2769.

[9] Kim P, Shi L, Majumdar A and McEuen P L. "Thermal transport measurements of individual multiwalled nanotubes," Physical review letters, 2001, 87(21), pp. 215502.

[10] Yu Choongho, Shi Li, Yao Zhen, Li Deyu and Majumdar Arunava. "Thermal conductance and thermopower of an individual single-wall carbon nanotube," Nano letters, 2005, 5(9), pp. 1842-6.

[11] W Bannard. "Properties of fusible alloy," Brass World and Platers' guide, 1928.Vol. 24.

[12] Bo, Ren, Xu, Du and Dou. "Recent progress on liquid metals and their applications," Advances in Physics: X, 2018, 3(1).

[13] Shiqian Liu, Keith Sweatman, Stuart McDonald and Kazuhiro Nogita. "Ga-Based Alloys in Microelectronic Interconnects: A Review," Materials, 2018, 11(8).

[14] Qian Wang, Yang Yu and Jing Liu. Preparations, "Characteristics and Applications of the Functional Liquid Metal Materials," Advanced Engineering Materials, 2018, 20(5), pp. n/a-n/a.

[15] A Hamdan,A Mclanahan, R Richards and C Richards. "Characterization of a liquid–metal microdroplet thermal interface material," Experimental Thermal & Fluid Science, 2011, 35(7), pp. 1250-1254.

[16] Martin Y and Kessel T V. "High Performance Liquid Metal Thermal Interface for Large Volume Production"// IMAPS international symposium on microelectronics. IBM - T.J. Watson Research Center 1101 Kitchawan Rd Yorktown Heights, NY 10598; IBM - T.J. Watson Research Center 1101 Kitchawan Rd Yorktown Heights, NY 10598;, 2007.

[17] BOOTH R B, GRUBE G W, GRUBER P A. "Liquid metal matrix thermal paste" :US5198189 A[P].1993-03-30.

[18] Narayanan, P Ramesh, Sharma, Deepak, Kumar, Praveen, Tiwari, R Kumar and Rohan. "Two-Phase Metallic Thermal Interface Materials Processed Through Liquid Phase Sintering Followed by Accumulative Roll Bonding," IEEE Transactions on Components, Packaging and Manufacturing Technology, 2016.

[19] W Kong, Z Wang, M Wang, KC Manning and K Rykaczewski. "Oxide‐Mediated Formation of Chemically Stable Tungsten‐Liquid Metal Mixtures for Enhanced Thermal Interfaces," Advanced Materials, 2019, 31(44).

[20] Raj, P.M, Gangidi, P.R, N Nataraj and N Kumbhat. "Coelectrodeposited Solder Composite Films for Advanced Thermal Interface Materials," in IEEE Transactions on Components, Packaging and Manufacturing Technology, vol. 3, no. 6, pp. 989-996, June 2013.

[21] Ga-In-Sn Liquidus Projection of Ternary Phase Diagram.

[22] Ga-In-Sn Vertical Section of Ternary Phase Diagram.

Research on BEOL Failures of the Chip-Package Interaction by Shear Tests of the Bumps

Shizhao Wang
School of Power and Mechanical Engineering
Wuhan University
Wuhan, China
wangshizhao@whu.edu.cn

Lianghao Xue
School of Power and Mechanical Engineering
Wuhan University
Wuhan, China
lance_xue@whu.edu.cn

Hongjie Wang
TongFu Microelectronics Co., Ltd.
TongFu Microelectronics 288.
Nantong, China
scip1093@sina.cn

Rui Li
The Institute of Technological Sciences
Wuhan University
Wuhan, China
2019106520010@whu.edu.cn

Can Sheng
School of Mechanical Science and Engineering
Huazhong University of Science and Technology
Wuhan, China
cansheng_chongqing@163.com

Yameng Sun
The Institute of Technological Sciences
Wuhan University
Wuhan, China
2019106520010@whu.edu.cn

Sheng Liu
School of Power and Mechanical Engineering
Wuhan University
Wuhan, China
shengliu @whu.edu.cn

Abstract—**The continuous demand for high-speed, low-power consumption and multi-functionality of new generation electronic products has led to an exponential increase for high-yield and high-reliability packaging technologies, which has promoted the application of the 7nm node process technology to promote higher I/Os and smaller bump pitches. The advantages of copper pillar bump, which was widely used in high-density packaging, include fine pitch, high strength, low interconnection resistance, excellent electromigration performance, and lead-free packaging solutions. The low-k/ultra-low-k (LK/ULK) material medium used in back-end of the line (BEOL) can effectively reduce the parasitic capacitance without reducing the wiring density.**

Therefore, to improve thermomechanical and electrical performance, chips with a 42nm node and beyond usually integrate LK/ULK structure and metal line in BEOL. However, the encapsulation of the chip exceeding the 28nm node would lead to a sharp increment in the number of BEOL interconnection layers (vertical) and high-density copper traces (horizontal). The thermomechanical stress caused by the mismatch of the CTE of wafer and packaging material may cause LK/ULK delamination, bump cracks, and UBM peeling failures. Therefore, the accurate analysis and calculation of the thermomechanical stress in the film structure has always been the focus of engineering and academia.

The integrity of the bumps and microstructure of BEOL can be evaluated effectively by shear test method, which was suitable for structural inspection. In this work, the shear test simulation was applied for single bump to characterize the failures of BEOL structures, e.g., fractures of ULK/LK. Owing to the fragile characteristics of LK/ULK, BEOL was susceptible to external loads. When the copper pillar bumps above BEOL were subjected to shearing forces caused by thermal mismatch, the failure would occur in some microstructures within BEOL, especially in ULK/LK interfaces. Because the bumps near die corner suffered a
critical shear load, so the single bump of the aforementioned area was investigated by a shear simulation model. Studies had shown that the shear rate had less effect on the maximum shear stress, but the increment of the shear height brought about more fractures of LK, which showed that reducing the bump height helps to reduce the risk of BEOL damage under thermomechanical loads. Generally, the shear test of copper pillar bumps could effectively evaluate the strength and adhesion of the BEOL film interfaces. As an early evaluation under extreme conditions, this simulation was designed to check the integrity of ULK/LK stack film and the strength of the bump structure through the chip package interaction without underfill protection and finally find the optimal packaging solution.**

Keywords—*BEOL; ULK/LK; copper pillar bump; failure; shear test*

I. INTRODUCTION

Although the size of transistors that require higher performance in logic and memory devices has been scaling down, global interconnect resistance-capacitance (RC) delay on semiconductor devices increased significantly [1,2]. To reduce the RC delay, dual damascene Cu, LK, and ULK processes are being used, but reliability problems such as delamination and low-k dielectric breakdown begin to appear with a high probability, seriously affecting the yield and reliability [3–5]. Due to the strong interaction between the highly sensitive copper interconnect structure of the BEOL stack and the influence of other external stresses, the fracture and delamination of different layers such as ULK and TEOS continue to emerge. The material interface between the ILD material and the copper vias and wires often appears as initial crack points. Under shear load conditions, the cracks near the copper interface and inside the ILD material continue to grow, failing [6]. Single-material and dual-material fracture mechanics methods have laid a good foundation for evaluating crack processes and optimization, improving reliability and robust design. The extended finite element method (XFEM) is a numerical method to solve fracture mechanics problems. Its theory was firstly proposed in 1999. The idea of separation solves the problem of crack propagation while retaining all the advantages of traditional finite element methods, it does not need to prefabricate the defects such as cracks in the structure. The application shows the method in finding reasonable crack(Fig.1).

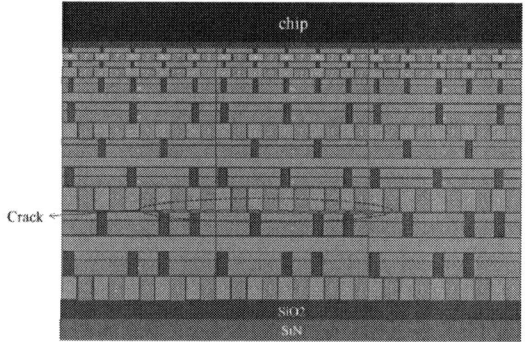

Fig. 1. Reasonable crack path near Cu bump in the BEOL-stack (cross-section)

The LK/ULK material was used in BEOL, which can effectively reduce the parasitic capacitance without reducing the wiring density. However, it significantly reduces the strength of ULK than before, causing many problems [7-8]. BEOL involves many kinds of materials, and the shear-induced delamination caused by CTE mismatch is one of the typical problems observed since ULK introduced IC chips. The integrity and microstructure of BEOL bumps can be effectively evaluated by the shear test method, which is suitable for structural inspection [9-10].

However, experimental methods seem to be difficult to find the main cause of the shear-induced delamination. One of the most difficult issue is that many process parameters in the shearing process depend on each other. In this case, the FEM is very useful [11], where the influence of each parameter can be analyzed separately. For this, we had done some finite element research on shear and found out the main reason for the shear-induced delamination in the Cu/ULK interconnection layer through stress simulation analysis. Usually characterizing the interfacial strength of Cu/ULK interconnection by bump shear test. The goal of this manuscript is to evaluate the reliability of Cu/ULK interconnection by the FEM.

In this work, a bump shear test was used to characterize the failure of BEOL structures, such as the fracture of LK/ULK. Since the bumps near the corner of chip bear the critical shear load, so the shear test was carried out on the single bumps in the above regions. In the FEM, the effects of the shear height and shear rate were studied.

II. FINITE ELEMENT ANALYSIS

When electronic components were subjected to thermo-mechanical loads, due to the CTE mismatch between different materials, failures usually occur near the corners and edges of bump. The shear stress around the corners or edges of bump is the main cause of delamination failure. Therefore, the finite element simulations and

parameters studies of shear process were carried out to evaluate the reliability of low-k structures and explore the failure mechanism.

Due to the difficulty of stress measurement, a finite element simulation was performed to evaluate the distribution of stress in the BEOL under the bump under shear load. As shown in Fig. 2, the typical principle model of bump shear. Three types of main methods shown in Fig. 2, like substrate, chip, and bump pillars. Due to the mismatch of the CTE of each material, when the structure receives a temperature load change, it will cause a stress difference between the materials, which will cause the key structure to receive the effect of shear. Firstly, when a temperature difference is generated, a shear force between die and substrate will occur (Fig.2.a), which can be simplified as the external shear force on the copper pillar (Fig.2.b)). The shearing force will cause the shearing phenomenon of the interconnection layer. Finally, the fragile LK and ULK will fail and affect reliability. The materials included in the 3D model can be concluded: Cu pillar, PEOX (-SiO$_2$), TEOS, LK/ULK, Si, TaN, TiN, Ta, and Al (Fig.2. c)). Due to the symmetrical geometry, only a half of structure was modeled for finite element analysis. At present, the cause of delamination is not fully understood. From previous experiments, it can be known that delamination usually occurs inside the LK/ULK, the copper, and the interface between ULK and the copper. The possible starting mode of delamination can be seen as the following three situations: Mode 1: Internal damage of LK/ULK; Mode 2: Crack at the ULK/Cu interface; Mode 3: Internal failure of Cu bump. Mode 1 and Mode 3 are failures within a single material. Generally, brittle materials such as SiO$_2$, SiCN, SiCOH, and TEOS are used as LK/ULK. Brittle materials have stress-based failure criteria, such as the maximum normal stress criterion (Rankine criterion and Mohr's theory). Mode 2 fails in the interface, and many scholars use the combination of interface normal stress and interface shear stress to analyze interface failure. A well-known criterion is the following quadratic criterion.

$$\left(\frac{\sigma_{11}}{S_{11}}\right)^2 + \left(\frac{\tau_{12}}{S_{12}}\right)^2 + \left(\frac{\tau_{13}}{S_{13}}\right)^2 = 1, [\langle x \rangle = |_{x\ x>0}^{0\ x\leq0}].$$

Where, σ_{11}, τ_{12}, and τ_{13} are the interface normal stress and the two directional interface shear stresses. S_{ii} is the interface strength in the ii directions. However, applying the quadratic criterion requires a lot of experimentation. This work attempts to draw meaningful conclusions through the analysis of interface normal stress and interface shear stress.

Fig. 3. shows a schematic diagram of the basic package model used for thermal and mechanical simulation analysis. The parameters affecting failure include: Nano-indenter shear height, crack location. These parameters have obvious effects on the failure of BEOL. First, the shear height is defined as the gap between the bottom of the nano-indenter and the BEOL surface, as shown in Figure 3. Determine the appropriate shear height by finite element method for the following modeling.

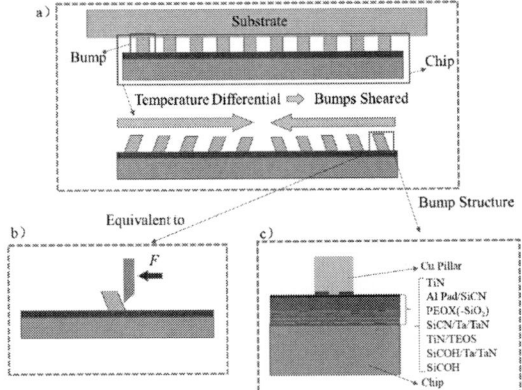

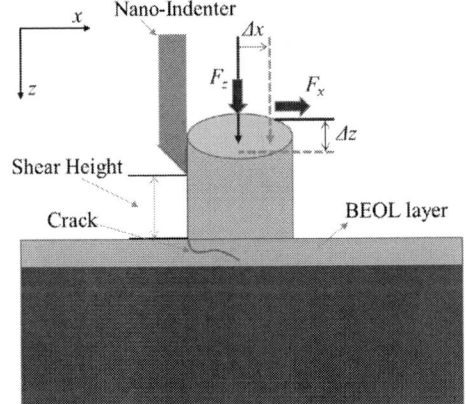

Fig. 2. Schematic of bump shear sample.

Fig. 3. Schematic diagram of the basic package model used for thermal and mechanical simulation analysis

To improve the accuracy of the finite element analysis results, meanwhile taking into account the diameter of the bumps, a three-dimensional finite element (3D) model is used, as shown in Figure 4. In the FEM model, BEOL uses the extremely dense mesh of C3D8R for modeling. The number of elements is 41384. The finite element analysis includes the plastic behavior of Cu, Al, Ta/TaN and TiN materials. Since Si, SiCN, and LK/ULK (SiO2, SiCOH, TEOS, PEOX) materials are very brittle, they are modeled as linear elastic materials.

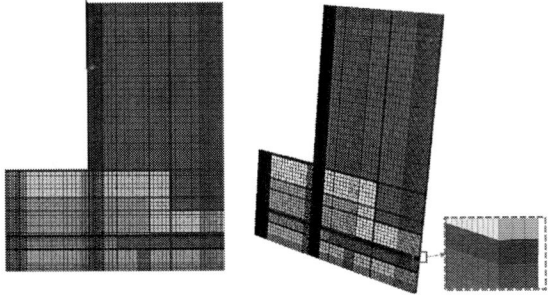

Fig. 4. The finite element model of BEOL in shear test

III. RESULTS OF SHEAR TESTS AND ANALYSIS

Fig. 5 shows the shear force-displacement curve. We can find that the shear height has a significant influence on the shear force. Studies have shown that the shear rate had less

effect on the maximum shear stress, but the increment of the shear height brought about more fractures of LK, which showed that reducing the bump height helps to reduce the risk of BEOL damage under thermomechanical loads.

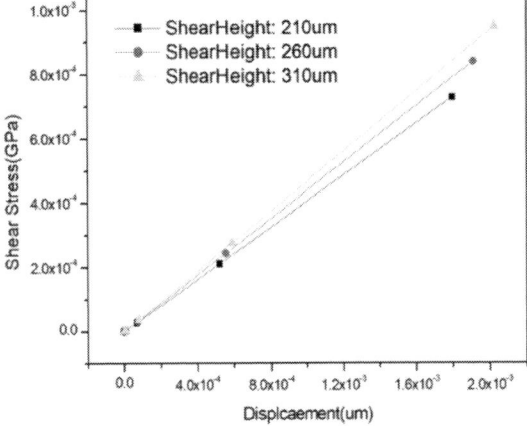

a). different shear height.

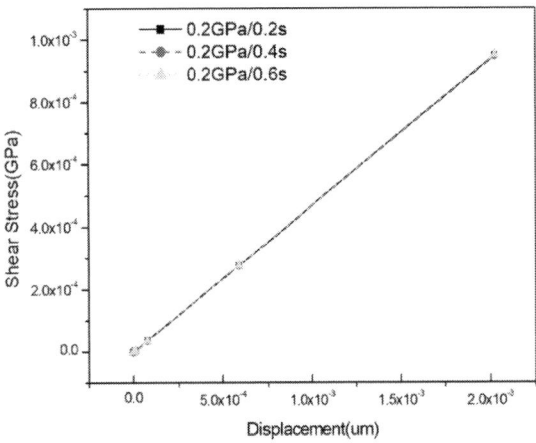

b). different shear ratio.

Fig. 5. Shear force-displacement curves.

The high shear force leads to high-stress levels in LK/ULK structures. Fig. 6 shows the critical stress and displacement distribution for BEOL structures during -45°C~125°C.

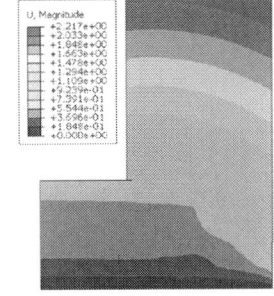
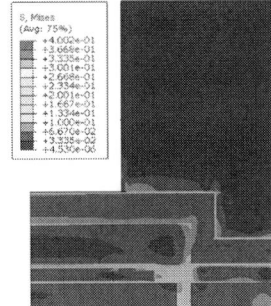

Fig. 6. Von Mises stress and displacement distribution for different material from -45°C to 125°C

Taking into account the fragile characteristics of LK/ULK dielectric materials, it can be seen that cracks are very likely to occur under lower stress. According to the first strength theory of brittle materials, we know that the material will fracture when the first principal stress reaches the critical value. Increasing the shear height while keeping the shear force constant will result in a higher maximum stress in BEOL. Fig. 7 shows the three modes of failure of the Cu/ULK structures of BEOL at the shear displacement 2.21µm.

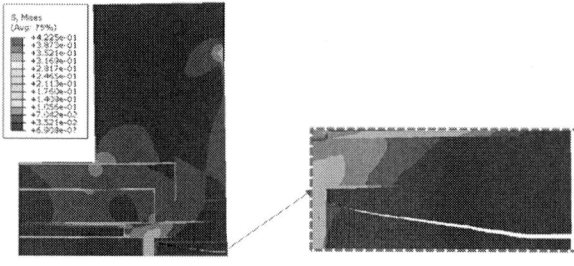

a) Mode 1: Crack in SiCOH

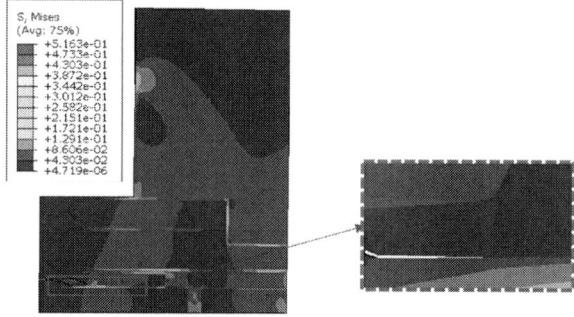

b) Mode 2: Crack in interface

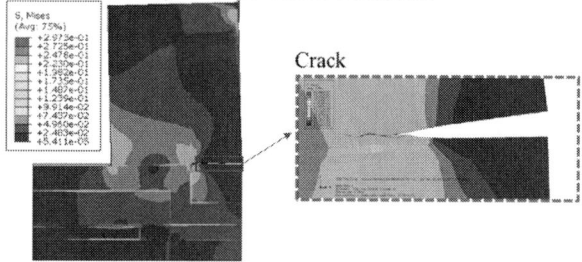

c) Mode 3: Crack in Cu.

Fig. 7 Three modes of failure of the Cu/ULK structures (unit: GPa)

IV. CONCLUSION

The 3D FEM was developed for the reliability assessment of the Cu/ULK structures. When the copper bumps above BEOL were subjected to shearing forces caused by the thermal mismatch. The failure would occur in some microstructures within BEOL, especially in ULK/LK. Usually, the delamination failure of three-mode can be observed from FEM analysis.

The result of FEM analysis shows that the stress is much larger on the stretched side, especially near the bump root. The Cu/ULK structure of BEOL is more susceptible to damage. When the shear height increases, the direction of tensile stress will rotate to the normal direction of the interface. In the shear test with a larger shear height, cracks often occur.

Studies had shown that the shear height will cause more Cu/ULK fracture, which shows that reducing the bump height helps to reduce the risk of BEOL damage under thermo-mechanical load. Generally speaking, the shear test of bumps can effectively evaluate the strength and adhesion of the BEOL interface. As an early evaluation under extreme conditions, this simulation aims to check the integrity of the ULK/LK stacked film and the strength of the bump structure through the chip package interaction without underfill protection, and finally find the best packaging solution.

ACKNOWLEDGMENT

This work was supported by the National Natural Science Foundation of China (Grant Nos. 51727901 and U1501241), the National Key R&D Program of China (No. 2017YFB1103904), and the Hubei Provincial Natural Science Foundation of China under Grant No. 2020CFA032.

REFERENCES

[1] ITRS, International Technology Roadmap for Semiconductors 2016 Edition Interconnect, 2016.

[2] C. Wu, Y. Li, M.R. Baklanov, K. Croes, Electrical Reliability Challenges of Advanced Low-K Dielectrics, ECS J. Solid State Sci. Technol. 4, N3065–N3070, 2015.

[3] H. Ceric, H. Zahedmanesh, K. Croes, Analysis of Electromigration Failure of Nanointerconnects Through A Combination of Modeling and Experimental Methods, Microelectron. Reliab. 100–101, 2019.

[4] L. Lin , J. Wang , L. Wang , F. Xiao , and W. Q. Zhang, The Stress Analysis and Parametric Studies for the Low-K Layers of a Chip in the FlipChip Process, Microelectron. Reliab., 65, pp. 198–203. 2016,

[5] K. Okada, Fundamental FEOL Reliability: Defect Generation in Gate Dielectrics to induce Dielectric Breakdown and Device Degradation (tutorial), IEEE Int. Reliab. Phys. Symp, pp. 1–577, 2017.

[6] K. Boon Yeap, M. Gall, Z. Liao, C. Sander, U. Muehle, P. Justison, O. Aubel, M. Hauschildt, A. Beyer, N. Vogel, E. Zschech, In Situ Study on Low-k Interconnect Time-Dependent-Dielectric-Breakdown Mechanisms, J. Appl. Phys. 115 124101, 2014.

[7] X.-J. Long, J.-T. Shang, and L. Zhang, Design Optimization of Pillar Bump Structure for Minimizing the Stress in Brittle Low K Dielectric Material Layer, Acta Metallurgica Sinica-English Letters, vol. 33, no. 4, pp. 583-594, 2020.

[8] M. He, C. Gaire, G. C. Wang, and T. M. Lu, Study of Metal Adhesion on Porous Low-k Dielectric Using Telephone Cord Buckling, Microelectronics Reliability, vol. 51, no. 4, pp. 847-850, 2011.

[9] C. Yang, L. Wang, K. H. Yu, J. Wang, F. Xiao, and W. Q. Zhang,Assess Low-k/Ultralow-k Materials Integrity by Shear Test on Bumps of a Chip, J. Mater. Sci., Mater. Electron., 29(19), pp. 16416–16425, 2018.

[10] S. G. Garreignot, , N. Benzima, E. Benmussa, C. Moutin, and P. O. Bouchard, Qualification of Bumping Processes: Experimental and Numerical Investigations on Mechanical Stress and Failure Modes Induced by Shear Test, Microelectron. Reliab., 55(6), pp. 980–989, 2015.

[11] K. Agathos, G. Ventura, E. Chatzi, and S. P. A. Bordas, Stable 3D XFEM/vector Level Sets for Non-planar 3D Crack Propagation and Comparison of Enrichment Schemes," International Journal for Numerical Methods in Engineering, vol. 113, no. 2, pp. 252-276, Jan 13 2018.

Evaluation of solder joints reliability of ball grid array assembly in astronavigation modules

W. L. Qin
The 24th Institute of China Electronics
Technology Group Corp.
Chongqing, P. R. China
Chanfoune@163.com

Z. P. Yan
The 24th Institute of China Electronics
Technology Group Corp.
Chongqing, P. R. China
Chanfoune@163.com

M. H. Wu
The 24th Institute of China Electronics
Technology Group Corp.
Chongqing, P. R. China
Chanfoune@163.com

Abstract—The microstructral characterization and thermomechanical reliability investigation of the Sn-3Ag-0.5Cu pump were implemented after the thermal shock of the lead-free and hybrid BGA assemblies. The lead-free process was assembled with Sn-3Ag-0.5Cu solder alloy, and the hybrid process was assembled with Sn-36Pb-2Ag solder alloy. Both of the lead-free and hybrid assemblies were subjected to a temperature shock range of 218K to 398K. The intermetallic compound between the Sn-3Ag-0.5Cu solder and the electroless nickel/immersion gold (ENIG) pads were composed of two regions, i.e., a continuous region of $(Cu,Ni)_3Sn_4$ and discontinuous particles of Ag_3Sn. In comparison, the homogenous lead-rich phase distribution was found in the bulk Sn-36Pb-2Ag solder. In order to obtain the solder joints fatigue failure mechanism of the modules, finite element analysis was introduced. It is found that thermally induced solder fatigue sharply increase in the hybrid BGA assembly, which determine the thermomechanical reliability of the modules.

Keywords—*ball grid array, reliability, solder joint, thermal shock cycling, finite element analysis*

I. INTRODUCTION

Ball grid array (BGA) package provides excellent interconnection performance, high integrated I/O density, and high signal frequency benefit from the shortest electrical interconnection path length. The solder joints play a significant role in physical supports as well as electrical interconnections, thus result in the structural integrity a significant reliability concern for the high integration density microelectronic package and modules. Solder joints are prone to fatigue due to large stress and deformation generated by mismatch of thermal expansion coefficient(CTE) between the package and the substrate. Thermal stress activated solder joint fatigue and mechanically fracture at the intermetallic compound (IMC) interface are the major fatigue failure modes in the BGA package.The fatigue failures occur due to the solder joints damages caused by thermal cycling strains. Because of the compliant property of solder, the BGA solder joints are susceptible to creep deformation and failures may occur under the excessive compressive load. Previous studies on the BGA solder joints have indicated that inelastic strain power have a significant influence on the fatigue lifetime of microelectronic package, and the formation of large area of IMC layers is the predominant cause of the solder joint reliability degeneration. However, most of the experimental investigations and mathematical simulations such as finite element analysis are view apart from only lead-free or hybrid assembly. Comparison of the lead-free and hybrid assemblies is consequential for the evaluation of the BGA solder failure.

In this paper, the microstructral characterization and thermomechanical reliability investigation of Sn-3Ag-0.5Cu pump were implemented after the thermal shock process of the lead-free and hybrid BGA assemblies. The lead-free process was assembled with Sn-3Ag-0.5Cu solder alloy, and the hybrid process was assembled with Sn-36Pb-2Ag solder alloy. Both of the lead-free and hybrid assemblies were subjected to a temperature shock range of 218K to 398K. The intermetallic compound between the Sn-3Ag-0.5Cu solder and electroless nickel/immersion gold (ENIG) pads were composed of two regions, i.e., a continuous region of Ag_3Sn and discontinuous particles of $(Cu,Ni)_6Sn_5$. In comparison, the homogenous lead-rich phase distribution was found in the bulk Sn-36Pb-2Ag solder. In order to obtain the solder joints fatigue failure mechanism of the modules, finite element analyses were introduced. It is found that thermally induced solder fatigue sharply increase in the hybrid BGA assembly, which determine the thermomechanical reliability of the modules.

II. EXPERIMENTAL AND ANALYSIS

For the purpose of this paper, a 27mm $\times$ 27mm and 900 I/O lead-free (SAC305) plastic ball grid array (PBGA) component was devoted to lead-free and hybrid BGA assemblies, respectively. The diameter of soldering ball was 0.40mm and the pitch was 0.80mm. The lead-free process was assembled with Sn-3Ag-0.5Cu (in at.%) solder alloy, and the hybrid process was assembled with Sn-36Pb-2Ag (in at.%) solder alloy. The bumping of the BGA components was prepared using an 0.2 μ m immersion Au/6 μ m electroless Ni – P. The Cu pad in the PCB substrate was deposited with immersion Au/electroless Ni (ENIG). In order to evaluate the thermomechanical reliability of solder joint, the BGA assemblies were subjected to a thermal shock test at the temperature range from 218K to 398K (30 minutes cycle time, air atmosphere, 15 minutes dwelling time).

After the thermal shock test, the 30 ball grid cross-section was performed through the first latitude cross of the BGA package. The microstructure of solder joint alloy was observed with SEM, and the IMC layer were measured by EPMA.

Fig.1 BGA assembly specimen and solder joints

978-1-6654-1392-3/21 $31.00 © 2021 IEEE

Finite element analyses were carried out to evaluate the deformation behaviors within the BGA solder joints which are stressed by the thermal cycling strains from the environment. ANSYS software, release 14.0 was used as the finite element platform. The cross-sectional plane morphology of solder joints along the latitude is shown in Fig.1. Furthermore, the 3-dimensional elements were employed to simulate the solder joint and the other components containing Cu pad and Au/electroless Ni－P layer, etc., respectively. The zero-stress temperature was set to 298 K, and the center of the package was designated as the zero-stress region.

III. RESULTS AND DISCUSSION

In order to study the microstructural variation under the temperature cycling strains, cross-sectional investigation were performed. Fig. 2 shows the SEM micrographs of latitude cross-sectional in the lead-free assembly solder joints after 100 thermal shock tests. Fig. 2(a) and 2(b) indicates the regular and magnified views the solder bumps at the edge side, respectively. As shown in Fig. 2(b), the solder bump was well formed and firmly bonded to both the package side and the substrate metallization pad side. At the package side interface between the solder alloy, a continuous region of (Cu,Ni)3Sn4 and discontinuous particles of Ag3Sn were formed. However, the reaction alloy crystal between the electroless Ni－P layer and the package solder bump interface was few (Ni,Cu)3Sn4. In the buck of solder joint, Ag3Sn particles were found, these particles were regarded as a major factors determining the physical properties and the solder reliability[9,10].

Fig. 2(c) and (d) show the regular and magnified microstructure of the first solder bump of the hybrid assembly after 100 thermal shocks. In these micrographs, At the package side interface between the solder a continuous layer of (Cu,Ni)3Sn4 and discontinuous particles of Ag3Sn were formed. However, the homogenous lead-rich phase distribution was found in the bulk Sn-36Pb-2Ag solder. Therefore, more investigations were conducted to study further microstructural variations and the crack initiation.

Fig. 2 Microstructure of lead-free assembly ((a), (b)) and hybrid assembly ((c), (d)) solder joint interface after thermal shocks

The influence of microstructure variation on solder joints reliability were concerned significantly due to the miniaturization trend of microelectronic package assembly. It was shown that the failure mode depend on the intermetallic compound (IMC) microstructure variation in the thermal stress concentration regions, and the crack growth come along the recrystallization boundaries. Futher study indicated that the tensile strength of the IMC declined significantly on account of the propagation of Ag3Sn IMC discontinuous particles in the interface during thermal cycles. However the predicted fatigue lifetime of solder joints is dependent on the thermal cycles stress in terms of IMC thickness and composition.

Above all, a representative thermally generated fracture failure of BGA solder joints was investigated. However, more research should be carried out with further thermal shocks and sufficient crack length to confirm the failure mechanism. Simultaneously, the fatigue crack was located in the component package side and propagated alone the IMC crystal interface. On account of the stress concentration in the interface of bump metallization and the solder alloy, where brittle IMC was generated and propagated during the thermal shocks. The fatigue crack propagation was the primarily cause for the early failure in the BGA package assemblies, and thus called the original interfacial fracture, which have a profound difference with the thermally generated fatigue cracks. The fatigue cracks in the solder of package side were generated at the tip of the IMC region. Whereas, the cracks did not propagate along the solder joint/bump metallization interface but inside the solder alloy bulk with a path of paralleling with the bump. According to fatigue crack growth kinetics, the cracks propagate along a strictly horizontal path following the direction of the highest strain energy region. Therefore, it could be concluded that thermally activated fatigue crack generation is a major failure mechanism of BGA bump and solder joint during the thermal shocks, and crack propagation along the highest strain interface have a profound influence on the package reliability.

The low-cycle fatigue lifetime of BGA package and solder joint during the thermal shocks was described by Coffin-Manson model:

$$N_f^m \Delta \varepsilon_p = C \tag{1}$$

While N_f is low-cycle fatigue life, $\Delta \varepsilon_p$ is inelastic deformation, m and C are constants.

According to Young's equation, the low-cycle fatigue lifetime of solder joint could be calculated by formula (2):

$$N_f = 1/2(\frac{\Delta \gamma}{2\varepsilon_f})^{1/c} \tag{2}$$

Table I. Material properties of BGA assemblies

Materials	Young's modulus (Gpa)	Poisson's ratio	CTE (ppm/K)
Solder (Sn-3Ag-0.5Cu)	(218K): 55.4; (298K): 41.7;	0.35	24.5

Solder (Sn-36Pb-2Ag)	(398K): 22.2; (218K): 55.4; (298K): 41.7; (398K): 22.2;	0.35	24.5
Copper	155	0.34	16.3
BT substrate	(X,Y): 26; Z: 11	(X,Y): 0.39; Z: 0.11	(X,Y): 15; Z: 53
Die(Silicon)	191	0.28	(218K): 1.3; (298K): 2.6; (398K): 3.1;
PCB (FR-4)	(X,Y): 20; Z: 9.8	(X,Y): 0.28; Z: 0.11	(X,Y): 18; Z: 51
Mold compound	16	0.25	15

A summary finite element analysis was carried out for the purpose of analyzing the failure mechanisms of solder joint after the thermal stress. The simulation of inelastic stain energy density and tensile stress after thermal shocks 100 thermal shocks were showed in Fig.3-4. The simulation results show that the distributions of inelastic stress and strain are most crucial and concentrated in the first solder bump. This indicates that solder bump in the edge suffer the highest energy storage for the crack propagation, for this reason the fatigue failure of the solder bumps is conclusively aroused by the accumulation of the thermomechanical stress. As shown in Fig.3, the inelastic stress and inelastic strain were catastrophically accumulated in the interface of package side, the crack immediately then propagated along the IMC interface toward the direction of the Ni－P bump, dwelling on the highest energy storage for the crack generation and propagation in the interface of substrate pad and solder bump. The experimental conclusion is highly consistent to the simulation results. Compared to hybrid assembly with Sn-36Pb-2Ag solder alloy, lead-free assembly with Sn-3Ag-0.5Cu solder alloy generates more brittle IMC layer, which cause higher strain energy in solder bump and accelerates the solder joint fatigue failure accordingly.

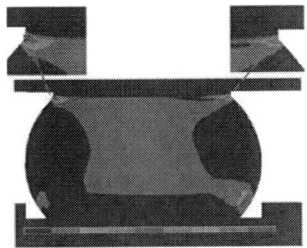

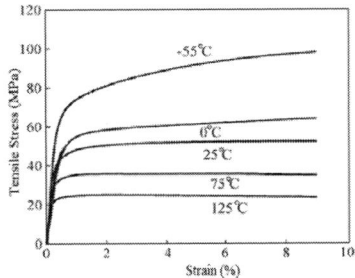

Fig 3 Inelastic stain energy density (a) after thermal shocks

Fig 4 Tensile stress (b) after thermal shocks at different temperatures

IV. CONCLUSION

In this study, the microstructure characterization of lead-free/leaded solder and IMC layer were studied in detail, and the thermal mechanical reliability of BGA package after thermal shock test was evaluated. The reaction products of bump alloy and electroless plating Ni-P processes was composed of two layers, namely continuous $(Cu,Ni)_3Sn_4$ layer and discontinuous Ag_3Sn particles. In comparison, the homogenous lead-rich phase distribution was found in the bulk Sn-36Pb-2Ag solder. A brittle interfacial failure mode occurs occasionally, but the failure assemblies are almost thermal fatigue cracks. The cracks propagate along the highest strain interface which have a profound influence on the package reliability. The finite element analysis shows that the crack is caused by the accumulation of inelastic stress and creep strain. Therefore, it should be emphasized that thermal activated solder fatigue failure is the main failure mode of BGA package assemblies, and inelastic strain accumulation mainly affects crack initiation and propagation time.

REFERENCES

[1] J.W. Kim, S.B. Jung, Material Science and Engneering: A Structure Material 371 (2004) 267-274.

[2] D.G. Kim, J.W. Kim, J.G. Lee, S.B. Jung, J. Alloys Compound 395 (2005) 80-84.

[3] F. X. Che, H. L. Pang, IEEE Transaction on Device and Materials Reliability 1 (2013) 36-49.

[4] K.C. Chang, K.N. Chiang, IEEE Transactions on Component and Package Technology 27 (2004) 373-379.

[5] J.W. Kim, S.B. Jung, Material Science and Engneering: A Structure Material 397 (2005) 185-190.

[6] K.C. Wu, S.Y. Lin, T.Y. Hung, IEEE Transaction on Device and Materials Reliability 3 (2015) 437-442.

[7] T.C. Chiu, D. Edwards, M. Ahmad, IEEE Transaction on Device and Materials Reliability 3 (2010) 324-337.

[8] L. Nie, D. Edwards, M.G.Pecht, IEEE Transaction on Device and Materials Reliability 2(2010) 276-286.

978-1-6654-1392-3/21 $31.00 © 2021 IEEE

Novel Design of SiC MOSFET Active Drive Circuit Based on Improved Auxiliary Branch Method

Li Yuhong
Tiangong University
School of Electrical and Electronic
Engineering
Tianjin Key Laboratory of intelligent
control of electrical equipment
Tiangong University
Tianjin, China
928578979@qq.com

Niu Pingjuan
Tiangong University
School of Electrical and Electronic
Engineering
Tianjin Key Laboratory of intelligent
control of electrical equipment
Tiangong University
Tianjin, China
niupingjuan@tiangong.edu.cn

Mei Yunhui
Tiangong University
School of Electrical and Electronic
Engineering
Tianjin Key Laboratory of intelligent
control of electrical equipment
Tiangong University
Tianjin, China
meiyunhui@163.com

Ning Pingfan
Tiangong University
School of Electrical and Electronic
Engineering
Tianjin Key Laboratory of intelligent
control of electrical equipment
Tiangong University
Tianjin, China
ningpingfan@tiangong.edu.cn

Zhao Di
Tiangong University
School of Electrical and Electronic
Engineering
Tianjin Key Laboratory of intelligent
control of electrical equipment
Tiangong University
Tianjin, China
zd314450936@163.com

Bai Jie
Tiangong University
School of Electrical and Electronic
Engineering
Tianjin Key Laboratory of intelligent
control of electrical equipment
Tiangong University
Tianjin, China
baijie_tgu@163.com

Abstract—**Excessive voltage and current spikes, oscillation, crosstalk and electromagnetic interference will be caused by the increase of switching frequency and switching speed in the application background of high-speed switching of SiC MOSFET. In order to solve these problems, SiC MOSFET active drive circuit is studied in this paper. Based on the dual-pulse test circuit platform with parasitic inductance, the SiC MOSFET voltage and current spike oscillation/crosstalk problems, transfer/output characteristics, turn-on/turn-off dynamic characteristics are simulated experiments, and the improved SiC MOSFET active drive circuit is innovatively designed. These simulation experiments show that the switching process of SiC MOSFET devices is accompanied by a higher rate of change of voltage and current. It is more prone to spikes and oscillations, and the influence of parasitic inductance in the circuit is not negligible. Meanwhile, the dynamic characteristics of SiC MOSFETs are compared and studied. On the basis of minimizing the sacrificing SiC MOSFET switching loss, a novel SiC MOSFET active drive circuit based on an improved auxiliary branch is proposed. Relying on the double-pulse test circuit platform, the design of the improved auxiliary branch and the calculation of related parameters have been completed, which have proved that the optimization effect of the oscillation and crosstalk of voltage and current spikes has reached 1/3. Compared with the traditional RCD absorption circuit and the typical active drive circuit, the SiC MOSFET active drive circuit based on the auxiliary branch designed in this paper can effectively suppress the voltage and current spikes and oscillation problems while reducing the complexity of the drive circuit. Meanwhile, on the basis of minimizing the sacrificial switching loss, the bridge crosstalk problem is effectively alleviated by adding a self-regulating mechanism. Finally, taking the improved synchronous Buck converter as an example, the improved design of the SiC MOSFET active drive circuit based on the auxiliary branch is verified for drive protection and anti-interference ability in actual application scenarios. Moreover, the conversion efficiency of the improved synchronous Buck converter is increased by 7.25%.**

Keywords—*SiC MOSFET, active drive circuit, auxiliary branch, spikes, oscillation suppression*

I. INTRODUCTION

As a high-frequency power electronic switching device, SiC MOSFET wants to give full play to its excellent performance in practical application [1-4]. The key is to design the driving circuit to be high temperature resistant, du/dt resistant and di/dt resistant, so that SiC MOSFET can realize high-frequency switching under high voltage and high temperature. During the turn-on process of SiC MOSFET, when commutating with freewheeling diode, the drain current is affected by the reverse current of freewheeling diode, and SiC MOSFET will have current spike and oscillation [5-6]; during the turn-off process of SiC MOSFET, due to the influence of stray parameters in the circuit loop, the drain source voltage of SiC MOSFET will have voltage spike and oscillation [7-8]. Due to the basic characteristics of SiC MOSFET, such as low threshold gate voltage, high internal gate resistance and so on [9], the crosstalk problem is a serious problem for SiC MOSFET, so there must be an appropriate solution to suppress it.In this paper, through the comparative analysis of the basic characteristics of SiC MOSFET, considering the voltage and current spikes, oscillation and crosstalk during the switching process of the device, an improved active driving circuit suppression method of auxiliary branch is proposed. Finally, the improved active driving circuit of auxiliary branch is verified to have superior suppression ability on the double pulse test platform and synchronous buck converter prototype.

II. BASIC CHARACTERISTICS OF SiC MOSFET

A. Analysis of dynamic characteristics

Based on the dual pulse driving test platform, the dual pulse turn-on and turn-off waveforms of SiC MOSFET are compared and analyzed. During the turn-on process, the drain current spike and oscillation time of SiC MOSFET are small, mainly due to the SiC MOSFET Due to the small gate source capacitance, the rise time of gate source voltage is short; due to the weak reverse recovery ability of bulk diode, the drain current spike is small, as shown in Fig. 1(a). During the turn off process, the drain source voltage oscillation amplitude of

SiC MOSFET is small, as shown in Fig. 1(b). The above comparison shows that SiC MOSFET is superior to other switching devices in switching characteristics, but there are still some problems such as large peak and oscillation.

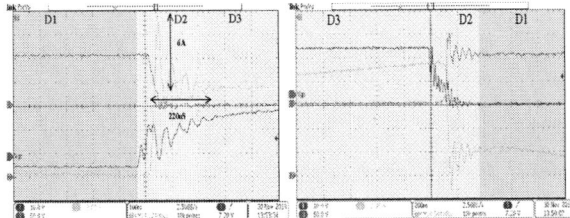

Fig. 1. (a) Turn on process of SiC MOSFET; (b) Turn off process of SiC MOSFET.

B. Analysis of the spike and oscillation in switching process

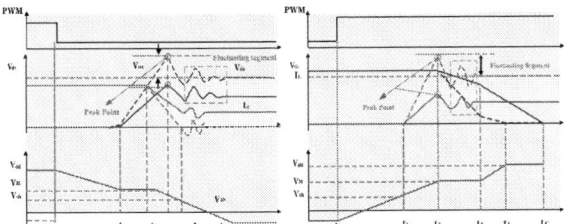

Fig. 2. (a) SiC MOSFET spikes and oscillations during turn-on, (b) SiC MOSFET spikes and oscillations during turn-off.

The stage 1 is [T0, T1]. The SiC MOSFET receives the high level, and the driving voltage charges the input capacitor C_{iss} until the gate source voltage V_{gs} of the SiC MOSFET rises to the on voltage V_{th}, while the drain source voltage V_{ds} and drain current ID remain basically unchanged.

The stage 2 is [T1, T2]. V_{gs} exceeds the turn-on voltage V_{th} and rises to Miller platform voltage vmiller. The drain source voltage V_{ds} remains unchanged and the drain current begins to rise to i_{peak}.

The stage 3 is [T2, T3]. The gate source voltage V_{gs} remains unchanged in the Miller plateau period, and the reverse recovery current of freewheeling diode decreases from the maximum value to 0, so the drain current ID decreases and the drain source voltage V_{ds} decreases.

The stage 4 is [T4, T5]. When the SiC MOSFET is turned on, the drain source voltage V_{ds} drops to 0 and remains at the on voltage, the drain current remains unchanged at the load current, and the gate source voltage V_{gs} rises to the driving voltage V_{gg}.

C. Research on crosstalk of bridge circuit

The bridge circuit is a common structure in SiC MOSFET [10-12]. The above process of switching on and off is taken as an example, as shown in Fig. 3. First, at the moment when the upper tube is turned on, the channel of the upper tube exchanges current with the VD_L of the lower tube, the V_{ds} of the upper tube decreases, the V_{ds} of the lower tube rises, and the C_{gd} of the lower tube begins to charge. The current direction is shown in Fig. 3(a), and flows through the gate resistance and the parasitic capacitance C_{gsL} respectively, so that the V_{gs} of the lower tube exceeds the turn-on voltage, making the device turn on. At the moment when the upper tube is turned off, the channel of the lower tube exchanges current with the V_{dh} of the upper tube, the V_{ds} of the upper tube rises, the V_{ds} of the lower tube falls, and the C_{gd} of the lower

tube begins to discharge. The current direction is as shown in Fig. 3(b). The V_{gs} of the lower tube exceeds the negative safety voltage and damages the lower switch tube.

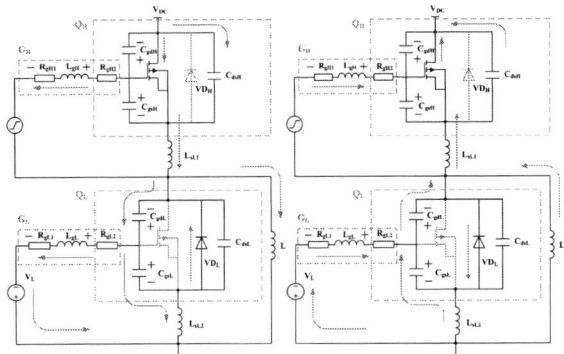

Fig. 3. Crosstalk effect of bridge circuit.

III. DESIGN OF IMPROVED ACTIVE DRIVING CIRCUIT BASED ON AUXILIARY BRANCH

A. Basic principle of improved drive

This paper proposes an improved auxiliary branch of active driving circuit. In the crosstalk stage of the turn-on process, a branch providing a parallel resistor at the gate source shares a part of the current, and the SiC MOSFET in this process The gate source voltage is clamped at both ends of the resistance voltage; in the crosstalk stage of the turn off process, a parallel voltage source is provided at the gate source, and the branch of series resistance shares part of the current, and the gate source, voltage is clamped at both ends of the voltage source series resistance voltage. In the switching process without crosstalk, it is the same as the conventional driving circuit, so as to ensure that the switching speed and switching loss are not sacrificed as much as possible.

B. Design of active drive suppression circuit

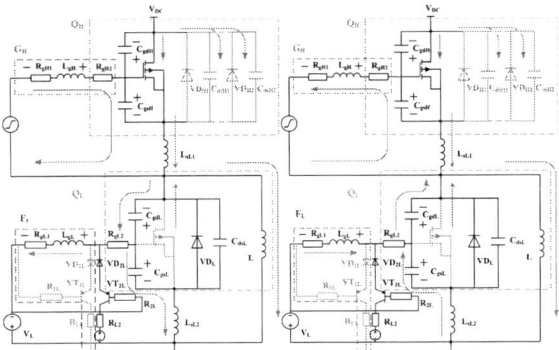

Fig. 4. (a) SiC MOSFET spikes and oscillations during turn-on, (b) SiC MOSFET spikes and oscillations during turn-off.

In order to solve the problems of grid source shunt capacitor, larger switching loss, increasing negative pressure of grid turn off, and limited voltage range, an improved design method of crosstalk suppression drive circuit is proposed under the premise of negative pressure turn off, as shown in Fig. 4. The specific working principle is as follows: when crosstalk occurs, by controlling the turn-on and turn off of triode, the voltage range is limited, In this way, the crosstalk problem can be effectively suppressed. At the same time, the influence of the shunt capacitor of the auxiliary branch on the switching performance of SiC MOSFET can be reduced by controlling the transistor. The improved driving circuit has the characteristics of small switching loss, short delay time and relatively simple control.

C. Design of auxiliary branch drive circuit

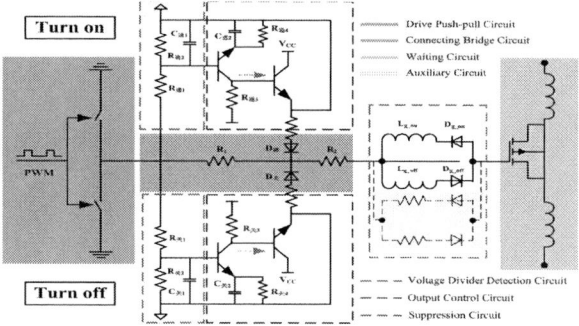

Fig. 5. Design of improved driving circuit.

Firstly, in order to suppress the current spike and oscillation in the process of open circuit, an improved SiC MOSFET active driving circuit based on auxiliary branch is composed of driving push-pull circuit, connecting bridge circuit, waiting circuit, auxiliary circuit, voltage divider detection circuit, output control circuit and suppression circuit on the premise of current injection gate, as shown in Figure 5.

The specific working principle of suppressing current spikes and oscillation problems during the turn-on process is as follows: When the drive circuit generates a +20V voltage signal, the resistance voltage divider detection circuit monitors the gate-source voltage in real time, and the voltage divider detection circuit controls it according to the voltage signal obtained by the gate-source The n1 transistor is turned on and off. When $U_{n1bc}>0.7V$, the n1 transistor is turned on, and the P_1 transistor is also turned on. At this time, the gate current is drawn; after the capacitor C_1 is discharged, when $U_{P1bc}<0.7V$, the P_1 transistor is turned off.

The working principle of suppressing voltage spike and oscillation in turn off process is: the driving circuit generates - 5V voltage signal, and the voltage divider detection circuit controls the turn-on and turn off of triode N2 according to the voltage signal obtained by grid source. When N2 is connected, the external power supply charges the capacitor C2 and the triode N3 is connected. The external power supply injects current into the gate to reduce di / dt and suppress voltage spike and oscillation; When the detected circuit detects that the gate source voltage of the sicmosfet is less than Vth, the triode N1 and N2 are closed at the same time, and N3 is closed at the same time. At this time, C1 starts to charge rapidly and C2 starts to discharge rapidly.

This improved driving method does not need to generate a separate control pulse, which reduces the complexity of circuit control. In order to reduce the impact of common source parasitic inductance on the switching characteristics, the drive circuit uses a larger capacitor to decouple it [13-14].

D. Design of scheme parameters

Firstly, in order to simplify the analysis and ignore the influence of parasitic parameters, during the turn-on and turn off phase of the upper transistor, when the active clamping auxiliary branch of the lower transistor is not working, the lower transistor will turn into working state when the upper transistor is switched on and off, and the Miller current of

charge and discharge will produce a voltage drop on the grid resistance of the lower transistor, which makes the triode positive bias and the Schottky diode positive conduction. The value of driving R_{gL} resistance should satisfy certain inequality relationship with V_{2L} and V_{gsL}[15-16], as in:

$$\Delta V_1 = \left(\frac{V_{2L} + V_{gsL}}{R_{gL} + R_{gL(in)}} \right) \times R_{gL} \geq U_{BE} + U_f \quad (1)$$

As in (1), U_{BE} is the on voltage of triode and U_f is the on voltage drop of Schottky diode.

When the auxiliary branch is working, the on resistance of transistor and diode is ignored. When the upper tube is in the opening process, the equivalent circuit parameters of the lower tube should meet the following requirements:

$$\frac{V_1(t)}{R_1} + C_{gsL}\frac{dV_{gs}(t)}{dt} = C_{gdL}\frac{d\left[at - V_{gs}(t)\right]}{dt} \quad (2)$$

As in (2), a is the switching speed, which is about 40ns.

$$\frac{V_1(t)}{R_1} \times \left[R_1 + R_{(in)} \right] = V_{gs}(t) \quad (3)$$

Combining (2) and (3), as in:

$$\frac{(C_{gsL} + C_{gdL})\left[R_1 + R_{(in)} \right]}{R_1} \times \frac{dU(t)}{dt} + \frac{U_1(t)}{R_1} = aC_{gdL} \quad (4)$$

The voltage formula at both ends of R_1 is obtained by solving (4), as in:

$$U_1(t) = aC_{gdL}R_1\left[1 - \exp\left(-\frac{V_{DC}}{a\left(C_{gsL} + C_{gdL}\right)\left(R1 + R_{(in)}\right)} \right) \right] \quad (5)$$

When the upper tube is in the shutdown process, the equivalent parameter formula of the lower tube is obtained:

$$\frac{(C_{gsL} + C_{gdL})(R_2 + R_{(in)})}{R_2} \times \frac{dU_2(t)}{dt} + \frac{U_2(t)}{R_2} = aC_{gdL} \quad (6)$$

The voltage formula at both ends of resistance R_1 is obtained by solving (6) as follows:

$$U_2(t) = \left(aC_{gdL}R_2 + 5\right)\left[1 - \exp\left(-\frac{V_{DC}}{\left(C_{gsL} + C_{gdL}\right)\left(R_2 + R_{(in)}\right)a} \right) \right] \quad (7)$$

In order to suppress crosstalk, according to $U_1(t)$ and $U_2(t)$ need to be less than 3.9V, and through (5) and (7) calculation, when R_1 and R_2 are respectively 400Ω to 1000Ω, the curve shows a smooth trend. Therefore, the resistance R_1 and R_2 should be 1000Ω.

IV. OPTIMIZATION VERIFICATION OF IMPROVED DRIVING CIRCUIT

A. Double pulse driving test platform

The double-pulse test experimental platform is shown in Fig. 6. The main circuit includes high-voltage DC power supply, low-voltage DC power supply, DSP control board, PC, drive board, supporting capacitor, magnetic loop inductance, etc. The test equipment mainly adopts MDO3024 model Tektronix oscilloscope, Agilent N2791A model isolated voltage probe, and Tektronix TCP0030 current probe. The high-voltage DC power supply provides DC voltage, and the low-voltage DC power supply provides 12V voltage for the drive board. The PC controls the DSP control board to output 5V pulse signals. The drive board is the drive circuit designed in the previous section. The SiC MOSFET adopts CREE company C2M0080120D type device, and the supporting capacitor adopts The six electrolytic capacitors are connected in series and parallel to reduce the inductance of the commutation loop. The inductance is a self-made 500uH magnetic loop inductance. The main experimental parameters are: DC bus voltage is 400V, load inductance is 500uh, switching frequency is 100kHz, SiC MOSFET adopts C2M0080120D MOS transistor, SiC diode adopts C4D20120 diode.

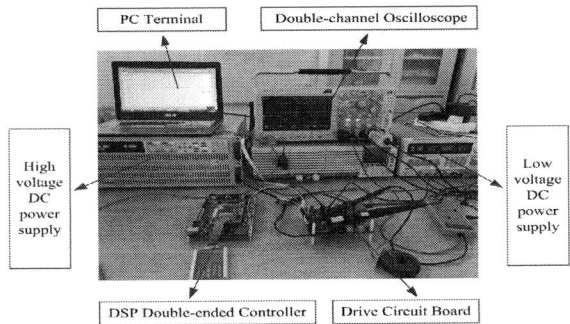

Fig. 6. Double pulse test experimental platform.

B. Verification of peak and oscillation suppression performance

In order to verify the effectiveness of the improved driver circuit to suppress the peak oscillation in the switching process of the device, combined with the experimental conditions, based on the double pulse test platform.

The traditional RCD circuit, the typical active driving circuit and the improved active driving circuit based on auxiliary branch are compared in voltage and current spikes and oscillation suppression performance. The main experimental parameters are as follows: DC bus voltage 400V, load inductance 500uh, switching frequency 100kHz. The on- off waveforms of the three experimental circuits are shown in Fig.

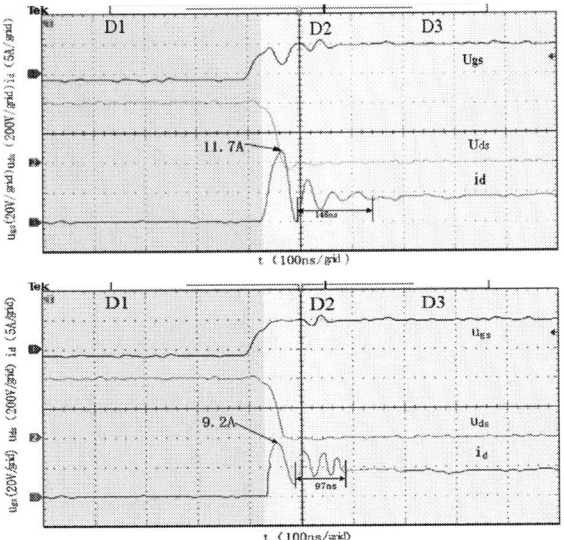

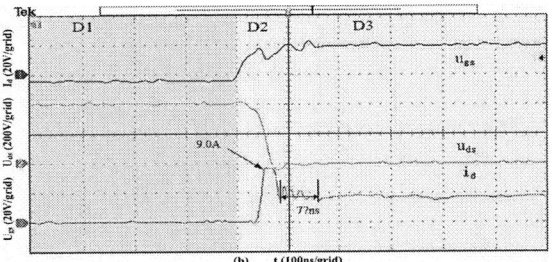

Fig. 7. (a) Turn on waveform of traditional driving circuit, (b) Turn off waveform of traditional driving circuit.

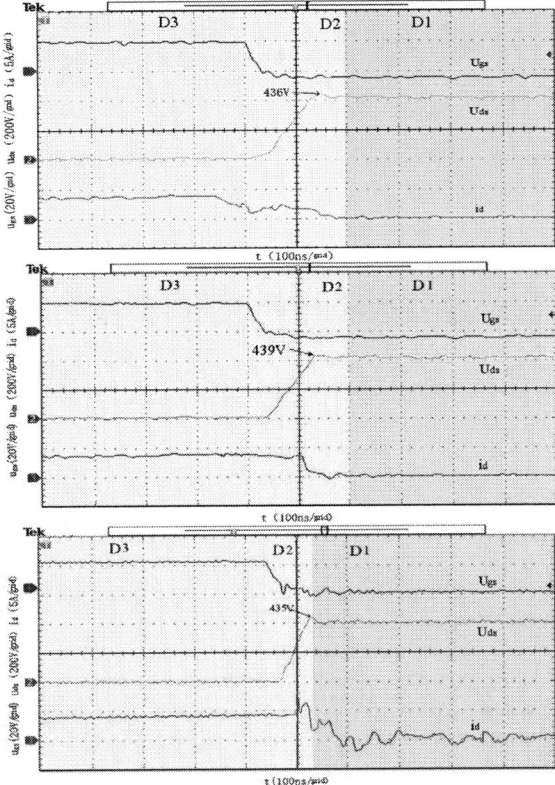

Fig. 8. (a) Turn on waveform of traditional driving circuit, (b) Turn off waveform of traditional driving circuit.

7 and Fig. 8.

According to the experimental results, compared with the typical method of using active driving circuit to suppress current and voltage spikes, the current spikes, voltage spikes, oscillation and switching loss of the improved driving circuit proposed in this paper are almost the same. Therefore, the improved driving circuit method proposed in this paper has a considerable suppression effect on current and voltage spikes and oscillation. As shown in Tab. 1, based on the above analysis, the proposed improved drive circuit can not only effectively suppress voltage and current spikes and oscillations, but also ensure that the switching loss of SiC MOSFET is not sacrificed as much as possible. This paper proposes that the improved drive circuit uses a triode to complete the control of gate current injection, and compared with the typical active drive, there is no need to use the improved drive circuit SiC MOSFET, no need to produce a separate control signal, so the system is quite simple.

978-1-6654-1392-3/21 $31.00 © 2021 IEEE

TABLE I. COMPARISON OF EXPERIMENTAL DATA UNDER DIFFERENT EXPERIMENTAL CONDITIONS

Active Driving Method	Voltage / Current Parameters				
	Current spike/A	voltage spike/V	Current /Voltage oscillation time/ns	Turn-on loss/µJ	Turn-off loss/µJ
traditional drive	12.3 A	149 ns	34.5µJ	435V	0 ns
typical drive	9.2 A	98 ns	26.2µJ	440V	0 ns
improved driver	9.0 A	75 ns	24.8µJ	438V	0 ns

Therefore, the proposed improved driver circuit combines the clamp circuit with the current injection type to suppress voltage spikes The voltage signal of the gate source is detected by the voltage divider detection circuit in the clamping circuit to control the turn-on and turn off of the triode, and then the current is injected into the gate to effectively suppress the voltage spike without changing the switching resistance and generating a separate control signal. Finally, the effectiveness of the improved driving circuit is verified by experiments.

C. Verification of crosstalk suppression performance

In order to verify the peak value and oscillation suppression performance of the improved active drive circuit, two C2M0020180D MOSFETs are used to compare the traditional drive suppression method, typical drive circuit method and the improved active drive circuit proposed in this paper. The parameters of the auxiliary branch of three different drive circuits are shown in Tab. 2.

TABLE II. RELATED PARAMETERS OF DIFFERENT DRIVE CIRCUITS

Active Driving Method	Resistance Parameters			
	R_{g1}/Ω	$R_{1L}, R_{2L}/\Omega$	C_L/nF	$R_1, R_2/\Omega$
traditional drive	10Ω	—	100Ω	—
typical drive	10Ω	0	100Ω	—
improved driver	10Ω	0	—	1000Ω

For crosstalk suppression, traditional driving circuit, typical driving circuit and improved active driving circuit based on auxiliary branch are used for comparison. According to the experimental waveforms and the data in Tab. 3, it can be known that the peak value of the forward voltage of the traditional driving circuit is -1.4V, and the peak value of the

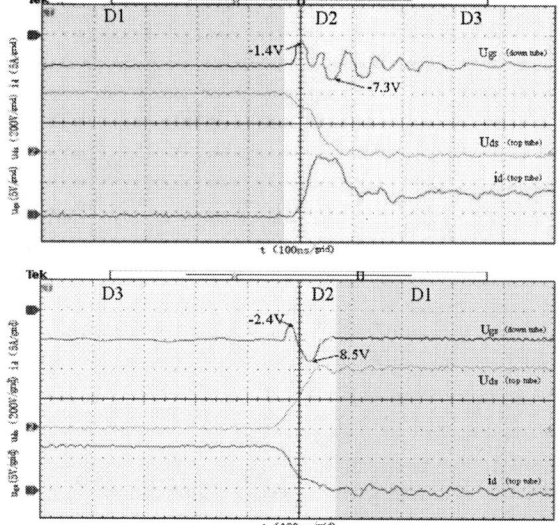

Fig. 9. (a) Turn on waveform of traditional driving circuit, (b) Turn off waveform of traditional driving circuit.

negative voltage is -8.5V, which does not exceed the threshold voltage (on voltage) and the negative safety voltage of the device. can effectively suppress the crosstalk problem and reduce the impact on the loss of the device during the turn-on and turn off process.

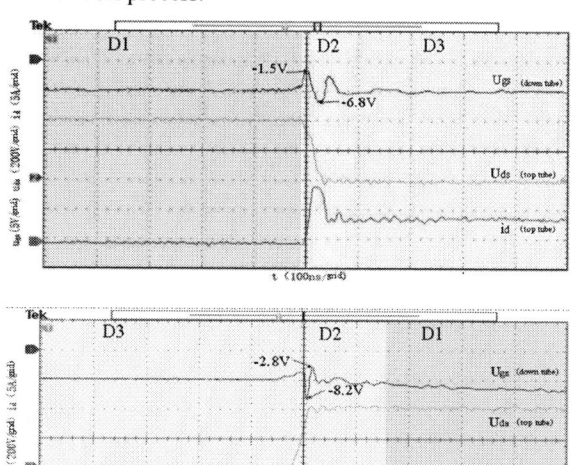

Fig. 10. (a) Turn on waveform of typical driving circuit, (b) Turn off waveform of typical driving circuit.

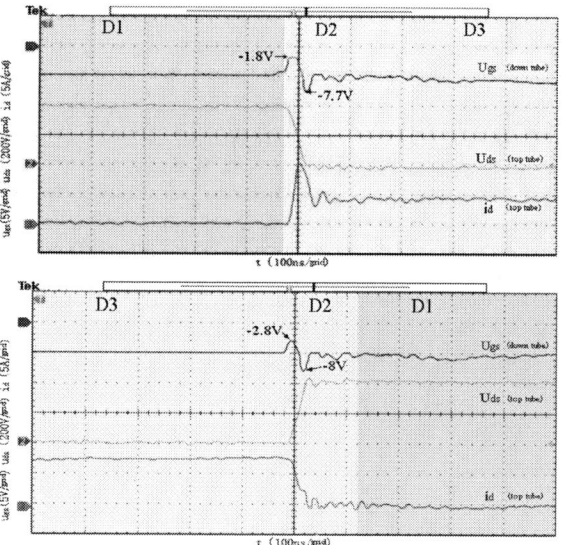

Fig. 11. (a) Turn on waveform of improved driving circuit, (b) Turn off waveform of improved driving circuit.

Therefore, the traditional driving circuit can effectively suppress the crosstalk problem. The peak voltage of typical driving circuit is -1.5V, and the negative voltage is -8.2v, which is equivalent to the traditional driving circuit, but the switching loss is significantly reduced. Compared with the above two suppression methods, the improved driving circuit.

TABLE III. EXPERIMENTAL RESULTS OF DIFFERENT DRIVING CIRCUITS

Active Driving Method	Loss / Peak Parameters			
	Turn-on loss/μJ	Turn-off loss/μJ	Positive peak/V	Negative peak/V
traditional drive	91.2μJ	165.4μJ	-1.4V	-8.5V
typical drive	37.8μJ	70.2μJ	-1.5V	-8.3V
improved driver	32.8μJ	55.8μJ	-1.9V	-8.0V

To sum up, the traditional driving circuit and typical driving circuit have good anti-interference ability. The capacitance of the auxiliary branch of the gate source has a certain impact on the switching characteristics of the device in the typical driving circuit design. Therefore, the switching loss is slightly larger than that of the improved driving circuit. However, the traditional driving circuit is over-voltage because of the on-off state

V. EXPERIMENT OF IMPROVED SYNCHRONOUS BUCK CIRCUIT

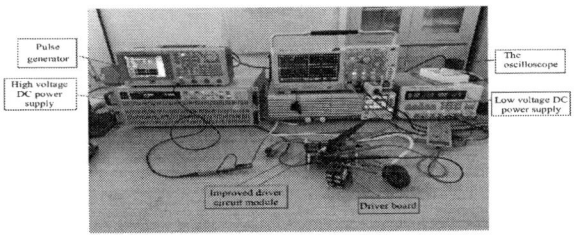

Fig. 12. (a) Turn on waveform of improved driving circuit, (b) Turn off waveform of improved driving circuit.

In order to verify the effectiveness of the improved drive circuit, a detection platform for synchronous buck converter is built, as shown in Fig. 12. The filter inductor adopts the existing magnetic ring inductor with inductance value of 500uh, and its rated current is 20A; The filter capacitor adopts six aluminum electrolytic capacitors with series connection and parallel connection of 150uf and voltage withstand of 400V; The input voltage of high voltage power supply is 399.99v, the input current is 1.71a, and the input power is 682.3w; The load resistance of DC electronic load is 10.016 Ω In constant resistance mode, the output voltage is 80.57v, the output voltage floats up 0.57v, the output current is 7.913a, the output power is 620.28w, and the efficiency is about 90.94%.

Through the synchronous buck detection platform, under the condition that the driving circuit is inconsistent and other conditions are consistent, the efficiency is 90.94% by using the traditional driving circuit board as shown in Figure 5-19, and the efficiency is 98.19% by using the improved driving board based on the auxiliary branch, and the efficiency is increased by 7.25%, Therefore, it is proved that the efficiency of the converter can be effectively improved by using the improved active driving circuit based on the auxiliary branch.

ACKNOWLEDGMENT

This study was supported by the Tianjin Science and technology plan support project (No. 20YDTPJC00740).

REFERENCES

[1] C.H. Li, Z.D. Zhang, Y.F. Liu, Y.P. Si, and Q. Lei, Smart Self-Driving Multilevel Gate Driver for Fast Switching and Crosstalk Suppression of SiC MOSFETs, 1st ed., vol. 8. IEEE Journal of Emerging and Selected Topics in Power Electronics, 2020, pp.442-453.

[2] Y. Zhou, X. Wang, L. Xian, and D. Yang, Active Gate Drive With Gate-Drain Discharge Compensation for Voltage Balancing in Series-Connected SiC MOSFETs, 5th ed., vol. 36. IEEE Transactions on Power Electronics, 2021, pp.5858–5873.

[3] F. Zhang, Y. Ren, X. Yang, W.J. Chen and L.L. Wang, A Novel Active Voltage Clamping Circuit Topology for Series-Connection of SiC-MOSFETs, 4th ed., vol. 36. IEEE Transactions on Power Electronics, 2021, pp.3655–3660.

[4] Y. Wen, Y. Yang, and Y. Gao, Active Gate Driver for Improving Current Sharing Performance of Paralleled High-Power SiC MOSFET Modules, 2nd ed., vol. 36. IEEE Transactions on Power Electronics, 2021, pp.1491–1505.

[5] V.K. Miryala, and K. Hatua, Low-cost analogue active gate driver for SiC MOSFET to enable operation in higher parasitic environment, 3rd ed., vol. 13. IET Power Electronics, 2020, pp.463–474.

[6] H.T. Tang, H.S.H. Chung, J.W.T. Fan, R.S.C. Yeung, and R.W.H. Lau, Passive Resonant Level Shifter for Suppression of Crosstalk Effect and Reduction of Body Diode Loss of SiC MOSFETs in Bridge Legs, 7th ed., vol. 35. IEEE Transactions on Power Electronics, 2020, pp.7204–7225.

[7] B. Cougo, H.H. Sathler, R. Riva, V. Dos Santos, N. Roux, and B. Sareni, Characterization of Low-Inductance SiC Module With Integrated Capacitors for Aircraft Applications Requiring Low Losses and Low EMI Issues, 7th ed., vol. 36. IEEE Transactions on Power Electronics, 2021, pp.8230–8242.

[8] J.X. Wei, S.Y. Liu, R.C. Lou, L.Z. Tang, R. Ye, L. Zhang, X.B. Zhang, W.F. Sun, and S. Bai, Investigation on the Degradation Mechanism for SiC Power MOSFETs Under Repetitive Switching Stress, 2nd ed., vol. 9. IEEE Journal of Emerging and Selected Topics in Power Electronics, 2021, pp.2180–2189.

[9] J.X. Wei, S.Y. Liu, H.B Zhao, H. Fu, X.B. Zhang S.Y. Li and W.F. Sun, Verification of Single-Pulse Avalanche Failure Mechanism for Double-Trench SiC Power MOSFETs, 2nd ed., vol. 9. IEEE Journal of Emerging and Selected Topics in Power Electronics, 2021, pp.2190–2200.

[10] H. Li, Y.F. Jiang, Z.D. Qiu, Y.T. Wang, and Y.H. Ding, A Predictive Algorithm for Crosstalk Peaks of SiC MOSFET by Considering the Nonlinearity of Gate-Drain Capacitance, 3rd ed., vol. 36. IEEE Transactions on Power Electronics, 2021, pp.2823–2834.

[11] J.O. Gonzalez, and O. Alatise, Impact of BTI-Induced Threshold Voltage Shifts in Shoot-Through Currents From Crosstalk in SiC MOSFETs, 3rd ed., vol. 36. IEEE Transactions on Power Electronics, 2020, pp.3279–3291.

[12] D.K. Yuan, Y.M. Zhang, X.H. Wang, and J.X. Gao, A Detailed Analytical Model of SiC MOSFETs for Bridge-Leg Configuration by Considering Staged Critical Parameters, vol. 9. IEEE Access, 2021, pp.24823–24847.

[13] H. Li, Y. Zhong, R.Z. Yu, R. Yao, H.Y. Long, X. Wang, and Z.J. Huang, Assist Gate Driver Circuit on Crosstalk Suppression for SiC MOSFET Bridge Configuration, 2nd ed., vol. 8. IEEE Journal of Emerging and Selected Topics in Power Electronics, 2020, pp.1611–1621.

[14] T. Gao, Z. Cheng, Q. Wang, and Y. Yang, Driver circuit to eliminate bridge leg crosstalk in SiC MOSFETs, 2nd ed., vol. 20. Journal of Power Electronics, 2020, pp.634–643.

[15] Z.Z. Dong, X.K. Wu, and K. Sheng, Suppressing Methods of Parasitic Capacitance Caused Interference in a SiC MOSFET Integrated Power Module, 2nd ed., vol. 7. IEEE Journal of Emerging and Selected Topics in Power Electronics, 2019, pp.745–752.

[16] Y. Li, M. Liang, J.G. Chen, T.Q. Zheng, and H.B. Guo, A Low Gate Turn-OFF Impedance Driver for Suppressing Crosstalk of SiC MOSFET Based on Different Discrete Packages, 1st ed., vol. 7. IEEE Journal of Emerging and Selected Topics in Power Electronics, 2019, pp.353–365

Optical Path design of a high coupling efficiency DFB laser package based on collimator

Xiaomeng Lv
Sichuan Province Engineering Research Center for Broadband Microwave Circuit High Density Integration
The 29th Research Institute of China Electronics Technology Group Corporation
Chengdu, China
342721062@qq.com

Ao Liao
Sichuan Province Engineering Research Center for Broadband Microwave Circuit High Density Integration
The 29th Research Institute of China Electronics Technology Group Corporation
Chengdu, China
zyliaoao@qq.com

Weihua Xu
Sichuan Province Engineering Research Center for Broadband Microwave Circuit High Density Integration
The 29th Research Institute of China Electronics Technology Group Corporation
Chengdu, China
1091574787@qq.com

Yangzhi Liu
The 29th Research Institute of China Electronics Technology Group Corporation
Chengdu, China
397996251@qq.com

Yilong Wu
The 29th Research Institute of China Electronics Technology Group Corporation
Chengdu, China
yilong_wu@126.com

Lan Qiao
The 29th Research Institute of China Electronics Technology Group Corporation
Chengdu, China
26390826@qq.com

Yong Zhao
The 29th Research Institute of China Electronics Technology Group Corporation
Chengdu, China
308319510@qq.com

Junli Yang
The 29th Research Institute of China Electronics Technology Group Corporation
Chengdu, China
253172513@qq.com

Abstract—**In this article, we present a high coupling efficiency package structure for DFB lasers based on the single lens-collimator coupling structure, with which the theoretical coupling efficiency can be improved to 88%. Then we build an optical coupling system based on the tolerance analysis for the optical path, and the actual test results show coupling efficiency higher than 75%, which verify the rationality and reliability of this coupling structure.**

Keywords—Fee-space optical interconnects, DFB laser package, collimator

I. INTRODUCTION

With the rapid development of optical communication devices, narrow linewidth DFB lasers have been widely used for their stable and highly reliable single-mode operation in optical system [1,2]. The DFB laser resonant cavity constitutes a Bragg grating with periodic changes in refractive index, which is equivalent to multiple miniature resonators. Multiple restrictions suppress mode jumps and enable stable single-mode output for DFB lasers over a wide temperature range [3]. Due to the complicated structure and complex production process, its divergence angle is usually large, and the difference between the horizontal and vertical directions is about 10°. In this case, how to continuously improve the coupling efficiency with optimized coupling system and package has attracted great interest. In this article, we present a highly coupling efficient package structure for DFB lasers based on the single lens-collimator coupling structure, demonstrate with simulation and experimental results how to design the collimator and tolerance analysis of the link to optimize the link design to realize high coupling efficiency.

II. OPTICAL COUPLING STRUCTURES

Free-space optical coupling for DFB lasers are usually realized by a lens combination to couple the laser light emitted by the DFB into an optical fiber. There are three main types of combined lens package optical paths: single coupling lens, double lens coupling system and single lens-collimator coupling system. The single coupling lens has simple structure and easy process realization, but the coupling efficiency is low while the working distance of the lens strictly limited. The double lens coupling system will enable extremely high the coupling efficiency with a more complicated structure. The single lens-collimator coupling system is similar to the optical path structure of a double lens but with a simpler structure and easier to realize. The divergent light emitted by the DFB is integrated into collimated light by the first collimator lens, and then passes through the collimator then directly coupled to optical fiber, as shown in Figure 1. The distance between the first lens and the collimator can be adjusted in a large range, and optical devices can be added. The collimator integrates the second lens and the output fiber, which simplifies the double lens structure and reduces the difficulty of process realization, but compared with double lens coupling system, the coupling efficiency will be lower. We will demonstrate in following how to improve the coupling efficiency with an optimized design.

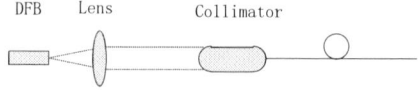

Fig. 1 Single lens-collimator coupling system

III. COLLIMATOR DESIGN

The collimator is used to converge the collimated light integrated by the first lens into an optical fiber, and it's basically composed of a C-lens and a capillary tube, as shown in Figure 2 for the collimator structure [4]. After selecting the DFB chip and the first lens, the system coupling efficiency could be optimized by tuning the structure of the collimator, such as the radius of curvature of the C-lens, the length, and the distance between the capillary and the C-lens [5].

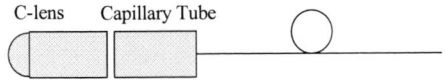

Fig. 2 Collimator structure

The optical path, composed of a DFB laser, a collimator lens and a collimator, is simulated and analyzed using simulation software as shown in Figure 3.

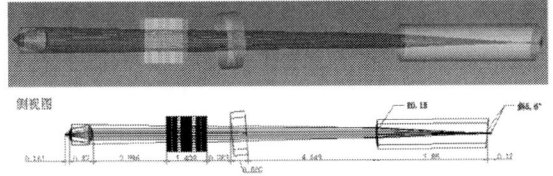

Fig.3 The optical path in the laser package

According to the selected optical path structure, the key parameters of the collimator are optimized to improve the coupling efficiency, as shown in Figure 4.

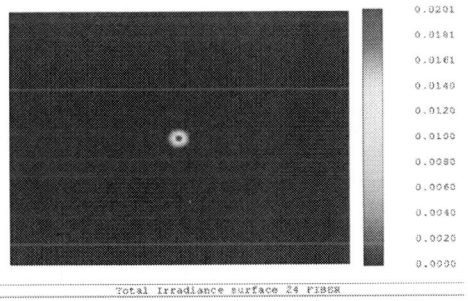

Fig. 4 Optimized coupling efficiency results

The horizontal and vertical divergence angles of DFB lasers at FWHM are 27° and 37° , respectively. Due to this difference, the spot pattern focused by the lens is an ellipse with major axis 870μm and minor axis 600μm, which leads to the spot pattern mismatch with the inherent mode of receiving fiber, and the coupling efficiency cannot be improved to higher than 90%. The theoretical coupling efficiency of the collimator is normally around 70%. After optimizing the C-lens structure, the receiving mode field of the collimator and the output mode field of the first lens can be matched much better. After the C-lens optimization, the theoretical coupling efficiency of the link can reach 88%, which is close to the coupling efficiency of the two-lens system. In order to find the best coupling point, the collimator needs to rotate -0.39° centered by the Y axis.

IV. COUPLING EFFICIENCY ANALYSIS

Based on the optimized optical path, the coupling efficiency could be improved theoretically, but the influence of the assembly process and displacement of each dimension should also be considered. We then further analyzed the coupling tolerance of the link, found the sensitive dimensions, and assemble the coupling optical path with a high-precision adjustment frame.

During the coupling process of the collimator, the mode field mismatch will easily cause the drop of coupling efficiency. Mode field mismatch has four main contributors: axial mismatch, radial mismatch, angular mismatch and mode field size mismatch [5]. The tolerances of these four mismatch phenomena are simulated and analyzed separately as shown in Fig.5-Fig.8.

Axial tolerance refers to the adjustment tolerance of the axial distance between the lens and the collimator. Fig. 5 shows the variation of the coupling efficiency with the axial displacement. It can be seen that in the range of axial distance from 2 mm to 100 mm, the change of coupling efficiency is less than 3%. Therefore, the link is not sensitive to axial adjustment, and the axial distance can be adjusted in a large range.

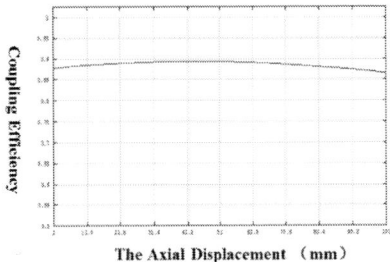

Fig. 5 Coupling efficiency versus the axial displacement

The radial tolerance refers to the displacement adjustment tolerance in the X or Y axis direction between the lens and the collimator. Figure 6 shows the coupling efficiency versus radial displacement. It can be seen from the figure that the best ideal position for coupling is at 0.02mm, and the coupling efficiency drops to 70% when the offset is 0.1 mm. Therefore, the coupling efficiency is much more sensitive to the radial displacement, thus the deviation should be precisely controlled within 0.05 mm.

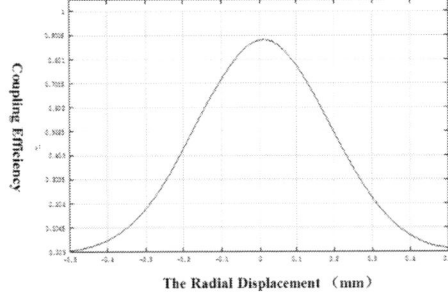

Fig. 6 Coupling efficiency versus radial displacement

Angular tolerance refers to the displacement tolerance in the rotation angle of the collimator, Figure 7 shows the variation of coupling efficiency with angular displacement. It can be seen from the figure that the best ideal position for coupling is -0.39°. When the offset is 0.3°, the coupling efficiency drops to 80%. Therefore, the link coupling efficiency is also sensitive to the angular displacement of the collimator, and the deviation should be controlled within 0.1°.

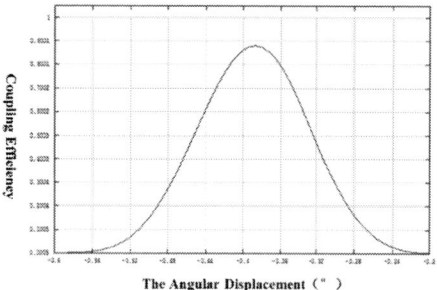

Fig. 7 Coupling efficiency versus angular displacement

For the fine tuning of the mode field mismatch tolerance, since the mode field is fixed after the collimator is designed, the only tunable factor is the mode field diameter of the DFB laser after passing through the first lens. Since the mode field size cannot be directly monitored in real time during the coupling process, we can change the axial displacement of the first lens to tuning the output optical mode field diameter. Therefore, the tolerance on the axial displacement of the first lens can reflect the tolerance on mode field mismatch. Figure 8 shows the variation of the coupling efficiency with the axial displacement of the first lens. The reference point of the displacement is the light-emitting surface of the DFB laser. It can be seen from the figure that when the lens is at 0.214 mm, the system could be in the best coupling state with a coupling efficiency of 88%. When the offset is 0.002 mm, the coupling efficiency drops to about 70%. This indicates that the system is extremely sensitive to the axial displacement of the first lens, and the offset should be controlled within 0.5 μm.

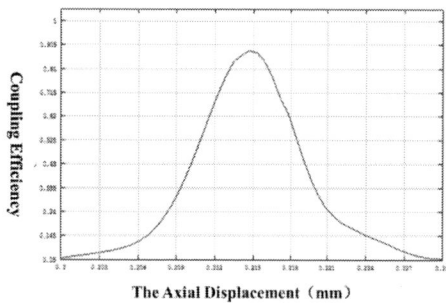

Fig. 8 Coupling efficiency with axial displacement

In summary, the minimum radial adjustment tolerance is 0.5 μm, the axial adjustment tolerance is 1 mm, and the angular adjustment tolerance is 0.1°.

V. SUMMARY

Based on the tolerance analysis, the adjustment accuracy of the axial and radial direction is 0.5 μm, and the angle adjustment accuracy is 50".

The DFB laser, lens and collimator were coupled and assembled in the coupling system. The horizontal and vertical beam divergence angles are 30° and 39°, respectively. The numerical aperture of the lens is 0.6 and spot diameter 700μm. The working distance and beam waist of the C-lens are 115mm and 280μm respectively. The test results with 6 different samples are shown in Table 1. The coupling efficiency of the system can reach about 75%, which indicates 13% (0.6 dB) difference compared with theoretical coupling loss. The main contributors for the difference are process error, surface reflection and material absorption loss, which are within a reasonable range. Therefore, by optimizing the design of the optical path, a collimator coupling system can be used to realize the high coupling efficiency packaging of the DFB laser.

Table 1 Coupling efficiency test results

No.	Pout of Laser (mW)	Pout of system (mW)	Coupling Eff.
1	22.1	16.1	73%
2	21	16.3	77%
3	23	16.9	73.50%
4	22.5	16.7	74%
5	22	16.7	76%
6	22.6	17	75%

REFERENCES

[1] F. Wei, F. Yang, Xi Zhang, D. Xu et al, Subkilohertz linewidth reduction of a DFB diode laser using self-injection locking with a fiber Bragg grationg Fabry-Perot cavity, Optics Express, 2016, pp 17406-17415.

[2] P. Wang, J. Xiong, T. Zhang, D. Chen, P. Xiang et al, Frequency tunable optoelectronic oscillator based on a directly modulated DFB semiconductor laser under optical injection, Optics Express, 2015, pp 20450-20458.

[3] D. Wang, N. Zhou, R. Zhang, X. Huang, L. Li et al, High speed and wide temperature range uncooled 1.3-μm ridge waveguide DFB lasers, Chinese Optics Letters, pp 809-811.

[4] 曾冰梅.基于 DFB 阵列可调谐激光器耦合系统的光学设计[D].华中科技大学,2017.

[5] 丁宗玲,孙进,魏蒙恩.激光光束准直器与耦合器的设计方法研究[J].大学物理实验,2017(1):12-15.

[6] 王彦晓,裴立明,陈盼.光纤准直器的耦合效率[J].科技资讯,2013(19):36-38.

978-1-6654-1392-3/21 $31.00 © 2021 IEEE

SiO₂/SiO₂ Bonding Technology Research on Wafer-level 3D Stacking

Zhang Peng
CETC 58
Wuxi，Chian
15961796508@163.com

Chengyu Yu
CETC 58
Wuxi，China
ycy901120@126.com

Kai Cen
CETC 58
Wuxi，China
565148313@qq.com

Jie Pu
CETC 58
Wuxi，China
pujie15306271639@163.com

Pengcheng Xia
CETC 58
Wuxi, China
xpc1011@163.com

Chengqian Wang*
CETC 58
Wuxi, China
chengqiankk@126.com

Abstract—With Moore's Law moving to the limit and the wave of miniaturization, intelligence and multi-functional development of electronic products is launched, expanding the packaging dimension of chips from two-dimensional to three-dimensional is recognized by the microelectronics industry as an effective way to shorten the interconnection length and an ideal solution to improve the chip's functional density. As wafer-level permanent bonding is the key process, this paper studies the SiO₂/SiO₂ bonding technology based on 12inch wafer. First, the preparation process of BSI-CIS sample is introduced. Then, the low-temperature SiO₂/SiO₂ bonding mechanism is analyzed, and the bonding process is optimized and determined. The experimental results show that activation time A, annealing temperature B, bonding time C, and bonding force D is the optimal bonding condition based on the 12inch wafer. Finally, the bonding layer is only SiO₂, which is a relatively stable inorganic medium and can provide bond strength more than 2 J/m². The bond strength has reached to the industry-recognized requirement for wafer level bonding.

Keywords—SiO₂/SiO₂ bonding, direct bonding interconnect, bond strength

I. INTRODUCTION

In the past nearly 50 years, the development of the microelectronics industry has been driven by Moore's Law, committed to continuously optimizing the size of lithography lines and reducing process nodes, in order to achieve high-performance, multi-functional integrated structures on two-dimensional chips, but this law will eventually come to an end due to the limits of technology and cost[1]. As a world-renowned strategic consulting company focusing on semiconductor micro-manufacturing market and technology research, Yole Développement pointed out a point of view in the "2.5D/3D TSV and Wafer-Level Stacking: Technology and Market" report updated in January 2019. The chip stacking technology which extends the interconnection dimension to three dimensions is recognized in the industry as the only solution that meets the performance requirements of today's mid-to-high-end applications such as artificial intelligence and data centers[2].

3D(Three-dimensional) integration technology greatly shortens the interconnection length by stacking semiconductor units in the vertical direction, thereby reducing signal delay and parasitic capacitance, reducing noise and effectively increasing the density of chip functions. The 3D integration includes 3D IC packaging, 3D IC integration and 3D Si integration , and its maturity is shown in Figure 1[3]. Considering integration density and functional diversity, 3D Si integration without micro bumps and underfill is the ultimate development trend to achieve high interconnect density.

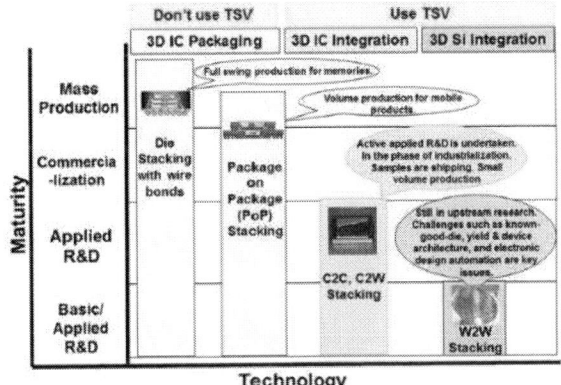

Fig. 1. 3D integration technology and maturity[3]

Three-dimensional Si integration technology is currently a hot research field in CIS(Contact Image Sensor) products. With the development of CIS, the size of pixels is gradually reduced. The traditional FSI (front-side illumination) process will reduce the light intensity of the sensor because themetal wire is located above the sensor, and cause the light to scatter and even the crosstalk effect. The BSI (back-side illumination) sensor avoids the obstruction of the sensor by the metal line by irradiating light to the side without the wiring layer. Combining the requirements of BSI-CISdevice stacking technology, this paper focuses on the low-temperature bonding technology using inorganic dielectric layer SiO₂ and metal Cu as the intermediate layer, including explores the SiO₂/SiO₂ DBI（Direct Bonding Interconnect）mechanism, and determines the optimal bonding process conditions.

II. FABRICATION TECHNOLOGY

A. Fabrication of SiO₂/Cu Bonded Sample

We use 12inch device wafer as the original material. The preparation of BSI-CIS stacking chip shown in Figure 2, which mainly includes the following processsteps:

978-1-6654-1392-3/21 $31.00 © 2021 IEEE

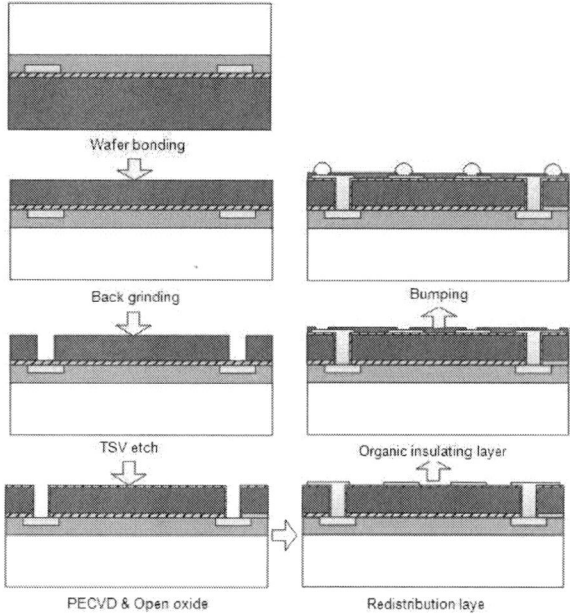

Fig. 2. Flow diagram of the BSI-CIS stacking chip preparation

Wafer-level bonding is the core technology in the above roadmap, which mainly includes three parts: film deposition and surface treatment before bonding, bonding process and post-bonding effect inspection. The specific process steps include thin film deposition, CMP(Chemical Mechanical Polishing) treatment, wet cleaning, surface activation, bonding alignment, pre-bonding, annealing and bonding strength testing.

(a) Thin film deposition: the silicon nitride film is deposited on the outermost metal by chemical vapor method, which has excellent compactness and water resistance. It can effectively prevent the diffusion of copper atoms to the outside and prevent external water vapor from entering the metal layer to protect the interconnection lines from external influences. Then deposit about 1um silicon oxide on the surface of the silicon nitride as a bonding dielectric layer;

(b) CMP treatment: the CMP process can provide ideal flatness and roughness for the bonding interface. Its working mechanism is based on mechanical polishing by adding corresponding chemical additives as required to achieve the effect of enhanced polishing and selectivepolishing;

(c) Wet clean: clean with some organic acid and deionized water to ensure the cleanliness of the bonding interface;

(d) Surface activation: N_2/O_2 plasma was selected to activate the surface of the wafer. The working principle is that the plasma impact will energize the unstable non-bridging oxygen atoms on the surface, causing them to leave the original bonded silicon atoms to form dangling bonds. In addition, plasma treatment can destroy and remove hydrocarbons on the surface, increase the number of hydroxyl groups, and achieve the purpose of activating the surface;

(e) Bonding alignment: using the characteristic of infrared light to penetrate the wafer, the EVG Smart View machine is used to align the alignment marks on the two layers of wafers.

(f) Pre-bonding: a certain external force is applied to the aligned wafers at room temperature to promote the complete contact of the wafers at the molecular level through surface activation bonds. At the same time, certain measures are taken to avoid adding dust or particles on the bonding surface.

(g) Anneal: through the annealing process at low temperature, the dangling bonds on the bonding surface are fully reorganized to generate a large number of silicon-oxygen-silicon bonds, and impurities such as water vapor molecules on the bonding surface are removed.

B. Bond strength Calculation

There are many methods for bonding strength calculation, including crack propagation method, infrared detection method, tensile force detection method, etc. The crack propagation method selected in this paper is one of the most commonly used surface energy characterization methods, also known as the double cantilever test method or the blade insertion method. The principle of this method is shown in Figure 3.

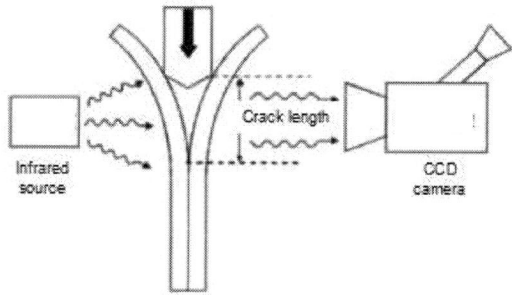

Fig. 3. Schematic diagram of crack propagation method

Accurately insert a blade of a certain thickness into the bonding interface of the two wafers at a constant speed, which will cause a crack at the interface, and the bonding strength can be calculated using the following formula:

$$\gamma = \frac{3E_1 d_1^3 E_2 d_2^3 h^2}{16(E_1 d_1^3 + E_2 d_2^3) L^4}$$

Among above formula, d_1 and d_2 are the thickness of the 12inch wafer, E_1 and E_2 are the Young's modulus of the material (along the direction of crack generation), L is the length of the crack, and h is the thickness of the blade.

C. Optimization of SiO_2/SiO_2 bonding process

The principle of SiO_2/SiO_2 bonding is: the surface of the activated wafer is highly clean, with a large number of hydroxyl (-OH) suspended on the surface, which adsorbs an additional layer of water molecules on the wafer surface through hydrogen bonds and Van der Waals forces. Water molecules play a key role in hydrophilic bonding, and the following reactions can occur through hydrogen bonding.

$$\text{Si-OH} + x\text{H2O} + \text{OH-Si} \leftrightarrow \text{Si-O-Si} + (x+1)\text{H2O}$$

The reaction process at the interface during pre-bonding is reversible. The silicon-oxygen-silicon bond can also be broken into the state before bonding, and the generated water vapor cannot be effectively discharged. An annealing process at low temperature must be used to make the bonding surface The dangling bonds are fully reorganized to generate a large number of silicon-oxygen-silicon bonds[4-5].

The effects of plasma activation time, annealing temperature, bonding force and bonding time on the bonding strength were studied. The activation time is A-60, A and A+60, the anneal temperature is B-150, B-50 and B, the bonding time is C-10, C and C+20, and the bonding force is D-5, D and D+10, respectively. Then, the SEM and EDS (Energy Dispersive Spectrometer) were used to analyze the micro-morphology of SiO_2/SiO_2 bonding layer to determine the best bonding process conditions.

III. RESULTS AND DISCUSSION

A. SiO_2/SiO_2 bonding process optimization results

As bond strength is one of the most important bond index, so this paper uses it to evaluate the bond effect. The experimental conditions and experimental results of SiO_2/SiO_2 bonding process are shown in Table 1.

TABLE I.   EXPERIMENTAL RESULTS OF CU/SN BONDING

No.	CMP	Activation time (s)	Anneal temperature (K)	Bonding time (min)	Bonding force (kN)	Bond strength (J/m^2)
1	×	A	B	C	D	NA
2	√	A-60	B	C	D	0.89
3	√	A	B	C	D	2.11
4	√	A+30	B	C	D	2.05
5	√	A	B-150	C	D	1.50
6	√	A	B-50	C	D	1.78
8	√	A	B	C-10	D	1.89
9	√	A	B	C+20	D	2.13
10	√	A	B	C	D-5	2.03
11	√	A	B	C	D+10	2.09

Experiment 1 in Table 1 shows that in wafer-level SiO_2/SiO_2 bonding, the CMP process is essential.The use of silica slurry and support rings can remove all irregularities on the surface of thin film to meet the flatness requirements of the wafer-level bonding process. The average oxide thickness on the bottom wafer surface after grinding is 975nm, and the thickness on the top wafer surface is 963nm, which meets the expected design requirements.

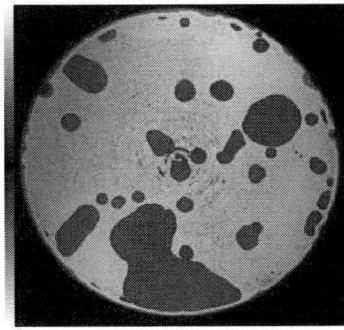

Fig. 4. CSAM image of sample #1

Experiment 2-4 in Table 1 shows that the averagesurface roughness of the unactivated SiO_2 interface is 0.50 nm. After activation by N_2/O_2 plasma, the value of Ra which measures the surface roughness decreases rapidly. When the activation time is A, the surface roughness of the bonding interface reaches the minimum , meanwhile the bonding strength reaches maximum. This is because SiO_2/SiO_2 bonding is based on the principle of hydrophilicity. It mainly relies on short-range Van der Waals force and hydrogen bonding to achieve the connection between the wafers at room temperature. In order to ensure the range of force, the surface roughness of the wafer must be minimized and the surface height consistency is best.

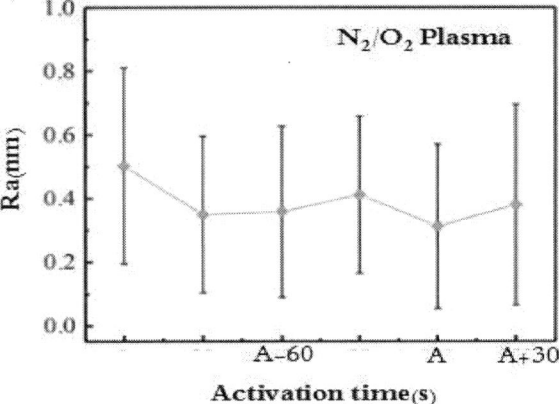

Fig. 5. SiO_2 surface Ra changes with plasma activation time

Experiment 3, 5-6 in Table 1 shows that when the temperature is lower than B K, the interface has discontinuous micro-voids and when the temperature rises to B K, the bonding interface is completely interconnected.

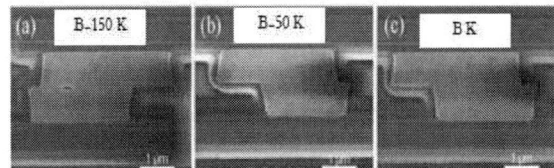

Fig. 6. SEM image of interface at different annealing temperatures

The experimental results in sample 7-11 show that the increase of bonding time has certain benefits to the bonding strength after annealing, and the change of bonding pressure has little change in the bonding result, which is related to the DBI reaction mechanism.

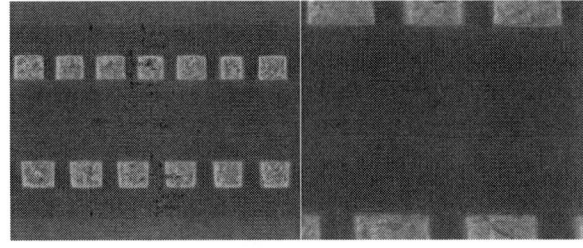

Fig. 7. SEM image of sample #3

B. Bonding layer microstructure and composition analysis

The wafer-level SiO_2/SiO_2 bonding is to make two very flat and clean wafer surfaces contact to form Van der Waals force, which can act between silicon oxide and silicon oxide or silicon and silicon. During the bonding process, as the oxygen between the two surfaces close to each other is converted into solid silicon oxide, the volume previously occupied by the oxygen is greatly reduced, thereby forming a partial vacuum between the two wafer surfaces, and the wafers are brought into close contact by the external atmospheric pressure. The bonding process at this time is reversible, and further annealing is required to convert the hydrogen bond at the contact interface into a stronger Si-O-Si covalent bond, so as to achieve the required strength[6]. The required bond strength is greater than 2 J/m^2, which is the industry-recognized requirement for wafer levelbonding.

Figure 8 shows that the material generated at the bonding interface was silica, and no other components were found, indicating that no other impurities were added during the bonding process. Analysis of Si and O atomic percentages shows that the number of Si and O atoms in the bonding interface layer is close to 1:2, indicating that it mainly exists in the form of SiO_2, which is in line with experimental expectations. At the same time, SiO_2 is a relatively stable inorganic medium, which can provide high bonding strength.

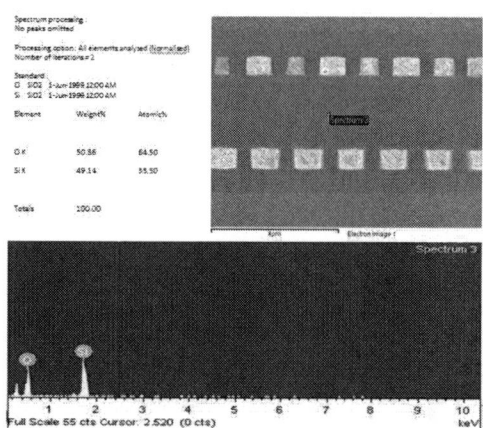

Fig. 8. SEM and EDS image of optimal bonded sample

IV. CONCLUSIONS

This paper studies the SiO_2/SiO_2 bonding technology applied to the BSI-CIS sample, which is the key process. First, the preparation process of SiO_2/SiO_2 bonded BSI-CIS sample is introduced. Then, the low-temperature SiO_2/SiO_2 bonding mechanism is analyzed, and the bonding process is optimized and determined. The experimental results show that activation time A, annealing temperature B, bonding time C, and bonding force D is the optimal bonding condition based on the 12inch wafer. Finally, the bonding interface layer is only SiO_2, which is a relatively stable inorganic medium and can provide high bonding strength. The required bonding strength is greater than 2 J/m^2.

ACKNOWLEDGMENT

The authors acknowledge the support of NCAP and CKS engineering team for wafer process and FA team for help on characterization and analysis.

REFERENCES

[1] A. Wang, C. Qi, L. Cheng, et al. "More-Than-Moore: 3D heterogeneous integration into CMOS technologies"C. IEEE International Conference on Nano/micro Engineered & Molecular Systems, 2017:205-208.

[2] M. Ibrahim, " 2.5D/3D TSV & Wafer-level stacking technology & marke updates 2019," J. Yole Développemen t, 2019.

[3] J. Lau. "Overview and outlook of three-dimensional integrated circuit packaging, three-dimensional Si integration, and three-dimensional integrated circuit integration," J. Journal of Electronic Packaging: Transactions of the ASME, 2014, 18(1): 32-36.

[4] G. Gao, L. Mirkarimi, T. Workman, et al. "Low temperature Cu interconnect with chip to wafer hybrid bonding"C. 2019 IEEE 69th Electronic Components and Technology Conference, 2019: 628-635.

[5] P. Gueguen, C. Ventosa, L. Cioccio, et al. "Physics of direct bonding : applications to 3D heterogeneous or monolithic integration," J. Microelec Engineer, 2010, pp58-

[6] J. Utsum, K. Ide, Y. Ichiyanagi. "Room temperature bonding of SiO_2 and SiO_2 by surface activated bonding method using Si ultrathin films," J. Japanese Journal of Applied Physics, 2016, 55(2):026503.1-026503.4.

High Speed Cu Plating Technology for Wafer Level Packaging

Jian Wang
ACM Research
Shanghai, P.R.China
jian.wang@acmrcsh.com

David Wang
ACM Research
Shanghai, P.R.China
david.wang@acmrcsh.com

Zhaowei Jia
ACM Research
Shanghai, P.R.China
zhaowei.jia@acmrcsh.com

Abstract—According to Moore's Law, the wafer level packaging and fan-out packaging market is expected to witness a compound annual growth rate (CAGR) of 18% during the forecasted period (2021 - 2026). And there are also some challenges for the high speed plating technology in the production of wafer level advanced packaging. Recently after a long time development, ACM announces high-speed copper (Cu) plating technology, which is now available for its ECP ap system. The tool supports Cu pillar bumping for copper(Cu), nickel (Ni) and tin-silver (SnAg) plating. The unique paddle design enables the new high-speed electroplating technology to support the copper electroplating chamber with a stronger mass transfer capacity during the electroplating process. Also, the channel of diffusion layer becomes shorter and the diffusion efficiency of cation is improved. The second anode well control the electric field distribution of cations on the wafer surface, and with the paddle stirring action, the cathode polarization phenomenon is reduced, which improved the electroplating efficiency and effectively reduced the boundary layer thickness to ensure the electroplating quality. Not only does it perform well in electroplated copper pillars, it can still maintain good uniformity and coplanarity in the electroplating of oversized bumps with higher aspect ratios and more stringent requirements for electroplating speed. The test results of ACM Ultra ECP ap series equipment meet the requirements of high speed volume production and the uniformity is also at leading edge level.

Keywords—wafer level packaging, high speed plating, pillar bump

I. INTRODUCTION

In the future, the development trend of intelligent life is irreversible, and the ever-changing electronic devices put forward more and higher requirements on chips. Most certainly, the progress of a modern society is closely related to the development of semiconductor technology. Integrated circuit(IC) manufacturing technology has exploded in the footsteps of Moore's Law for more than sixty years. Today, the nanometer-scale microscopic size is close to the atomic-scale size limit. It is no longer appropriate that continuously reducing the feature size of integrated circuits to meet the needs of industrial development by additions of excessive capital. Not only is the high cost, but the complex process is also a pain point that restricts the development of traditional chip manufacturing. In the face of market demand for miniaturization, high density, low energy consumption, and high functionality, technology nodes are shifting to back-end of line (BEOL), and advanced packaging is the most critical link [1-3].

In the post-Moore era, chip manufacturing costs continue to rise, prompting the industry to start relying on integrated circuits packaging to expand profits. Relevant statistics show that the advanced packaging market will continue to grow, with compound annual growth rate(CAGR) of 8% from 2018 to 2024, while the CAGR of the traditional packaging market is only 1.1%. In addition, the market size of advanced packaging is expected to exceed the scale of traditional packaging in 2022, and the market size of advanced packaging will reach US$44 billion by 2024. Advanced packaging technology can preferentially assemble a variety of active electronic components with different materials and different functions with optional passive devices, as well as other devices such as MEMS or optical devices. Among different advanced packaging technologies, 3D through-silicon via (TSV) and fan-out wafer-level packaging (Fan-out) will grow at a rate of 23% and 36%, respectively. The horizontal expansion and vertical stacking of chips brought about by the development of advanced packaging technology can greatly reduce the difficulty of design and manufacturing of semiconductor products, and realize the double improvement of transistor density and overall performance of the chip. Not only can it meet the needs of lightness, thinness, and multi-function, it is also considered an important way to surpass Moore's Law [2,4,6].

The advanced packaging process includes UBM (Under Bump Metal) sputtering, thick resist photolithography, electroplating, debonding and UBM etching, among which bottom-up copper deposition is the key to achieving high-performance chips. However, compared to the Dual Damascus sub-micron trench, the copper deposition in the advanced packaging process requires a larger aperture (a few micrometers to a dozen micrometers) and a deeper trench (a dozen micrometers to a few hundred micrometers). In the electrodeposition process, such extreme depth holes and other holes with high aspect ratio place very stringent requirements on electroplating equipment, and at the same time, it is necessary to ensure the copper plating rate to achieve rapid deposition and meet mass production requirements. Furthermore, with the continuous development of advanced packaging technology, higher requirements are put forward for electroplating equipment and production efficiency. One of the main challenges for high-depth 3D electroplating applications is the need to maintain high speed and bump uniformity when electroplating metal films in deep through holes or trenches with a depth of more than 200 microns. Through long-term technical exploration, ACM successfully developed the interconnection copper bumps, rewiring layers, tin-silver plating, and high-density fan-out （HDFO） packaging processes that can support copper, nickel (Ni) and tin-silver (SnAg) electroplating. The high-speed electroplating technology is mainly to enhance the mass transfer process of the electroplating solution ion substance transmission through the reciprocating stirring motion, supplemented by multi-scale control of second

978-1-6654-1392-3/21 $31.00 © 2021 IEEE

anode, the optimization of the stirring motion and the function of the flow field plate, which greatly improves the copper plating rate and the quality of the bumps [4-6]. ACM's ECP Ultra ECP ap have entered a number of well-known packaging manufacturers for mass production.

II. CORE TECHNOLOGY

Among the wafer level packaging manufacturing process steps, including sputtering barriers, PR coating, exposure, developing, wet etching, the deposition of conductive seed layers, the copper electroplating, the most difficult step to control is the electroplating process.

Various process factors such as voltage, current, bath characteristics, temperature, solution convection transport state, vertical and horizontal electric field distribution, additive adsorption behavior, etc., have a certain degree of influence on bump quality [7]. The key point is that large holes with high depth or high aspect ratio have large surface tension and extremely uneven current density distribution on the inner and outer surfaces, which makes it difficult for the material transfer and convection in the hole, which makes copper deposition difficult and costly. Moreover, the deposition rate is slow, and it is difficult to achieve complete filling, which will inevitably affect the quality and performance of 3D chip integration. In other words, if the problem of convection mass transfer and changes in electrical filed during copper electroplating process could be solved, then more than 70% of the advanced packaging problems will be basically figure out. After continuous exploration and experimentation, ACM has developed and used paddle and second anode accessories. The second anode can adjust the distribution of metal cation near the surface of wafer and improve the uniform distribution of electric field across the whole wafer. As we all known, the cathode polarization caused by ion concentration difference has a profound effect on the quality of electroplating. The movement of ion in plating solution is mainly carried out by diffusion, convection and electromigration. If the mass transfer is too weak, the driving force for metal ions to reach the near surface interface of wafer will be insufficient, so that the discharge cations in the interface liquid layer can not be supplied in time. The concentration of metal cations in the interface liquid layer is lower than that in the depth of the plating solution, resulting in concentration polarization. The application of external stirring effect on the plating solution can improve the mass transfer rate, timely supplement the metal cation consumed in the liquid layer of the cathode interface, and reduce the concentration polarization. In this way, the cathode current density can be allowed to rise within a certain range, and the electroplating rate can be improved, and the possibility of coating burning can be reduced. The special structure of the paddle and the continuously refined process conditions enable our high-speed electroplating technology to enhance the transfer of copper cations when depositing copper films, and cover all bumps on the entire wafer with the same electroplating rate, during high-speed electroplating, better uniformity is achieved inside the wafer and inside the chip.

III. RESULTS AND DISCUSSION

We mainly conducted experiments on the actual effect of the second anode and paddle. First of all, the second anode designed to be added above the main anode shields the influence of notch. Even if using the traditional physical

shielding method, current disturbance will still occur in the vicinity, resulting in abnormal height of the bumps around it. Besides, essentially, the paddle with enhanced reciprocating oscillation improves ion transmission efficiency, and we have also verified its true effect. Without paddle, when the current density increases to the range of high-speed copper electroplating, the deposited bumps will cause serious tilt problems. Due to the uneven and inadequate distribution of additives inside the photoresist bump openings, the accelerator and leveler/suppressor cannot effectively function during the bump growth process, and the deposited bump morphology will tilt seriously. The uniformity is also relatively poor. The high-speed copper electroplating technology with the addition of paddle has greatly improved the bump morphology and the uniformity within the wafer. The real result analysis is as follows:

Materials and Methods

The pre-wetting, plating and cleaning processes involved in the electroplating process are all performed by ACM ULTRA ECP ap equipment. We used the electroplating solution SC40 produced by Enthone Inc. to carry out conventional copper electroplating experiments. For macro copper pillars, Dow Chemical's IV 9000 electroplating solution was used. The process conditions of the electroplating process are the best known method of ACM plating tool, and the current density is 8-12 ASD. The SnAg plating solution is TS140 produced by Ishihara Chemical, the plating current density is from 3-5 ASD. Paddle stirring speed must reach more than 6 Hz to achieve the effect. It should be noted that the following tests to explore the paddle effect were all performed as using the second anode.

The bump profile images were measured by Keyence's white light interferometer. Automatic optical inspection machine (AOI) from Camtek EagleT-AP was used to measure the height and coplanarity (COP) of the bumps.

Second Anode Technology

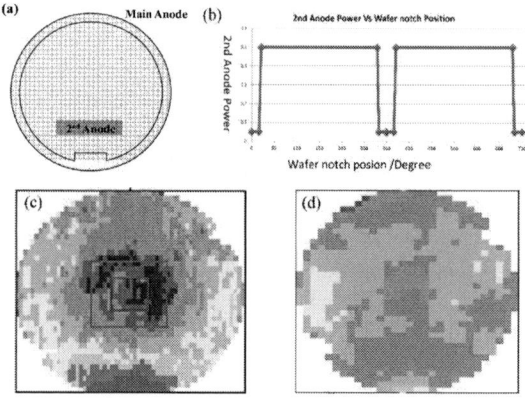

Fig. 1. Simple illustration and test results: (a) Anode location image, (b) Diagram of notch position and power, (c) Bump contour image of normal plating, (d) Bump contour image using the second anode.

As shown in Fig. 1, through the control of multi-scale factors, including current power, angle, etc., the second anode optimized edge plating profiles and non-continuous edge die pattern that there is almost no abnormal increase in bumps near the notch. Of course, the second anode has many other advantages that cannot be reflected in the test, such as simple daily maintenance, suitable for the various products with different notch area size just by change of power table, etc.

Test of bump profile

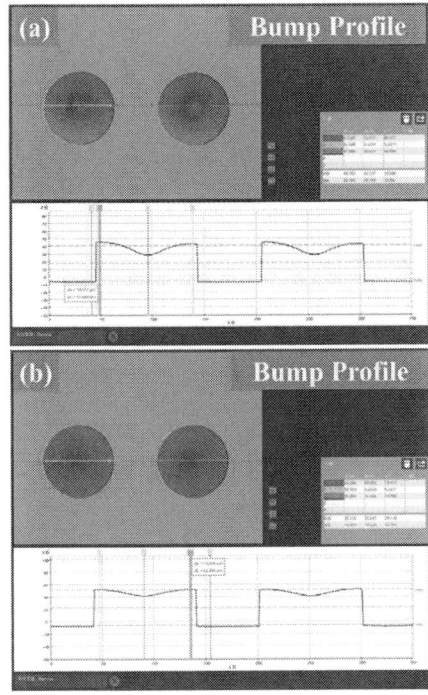

Fig. 2. Side profile of copper pillar: (a) without paddle, (b) with paddle.

TABLE I   BUMP PROFILE DATA

Point	1	2	3	4	5	6	7
with paddle (µm)	0.73	0.92	0.32	0.91	0.82	0.30	0.78
W/o paddle (µm)	3.51	5.77	3.94	5.32	6.15	3.48	4.24
Point	8	9	10	11	12	13	Avg.
with paddle (µm)	0.26	1.37	1.18	0.21	0.33	0.43	0.66
W/o paddle (µm)	5.39	4.03	1.15	5.58	3.09	5.54	4.40

Fig. 2 shows the topography of copper pillars. Without paddle oscillation, the transmission of copper ions will be restricted. Various additives such as levelers, suppressors, accelerators, etc. have a greater impact on the bumps during deposition. The edges of the openings are conducive to the growth of copper ions, resulting in more obvious central pits. After adding paddle, not only can copper ions be replenished in time at the cathode seed layer, but also various additives can be evenly distributed in the plating solution, which minimizes the influence of additive distribution on the plating layer, so it is obvious in the right picture that we can find the improvement in bump tilt and also the dishing value.

The tilt condition of the copper pillar is listed in Table I. Comparing the values, it can be found that the bump tilt value decreased significantly, and the average value reduced from 4.40 µm to 0.66 µm. Therefore, we can conclude that the paddle used by ACM high-speed copper plating equipment has greatly improved the previous terrible situation, and the copper pillar bump profile and inclination value can be in production spec.

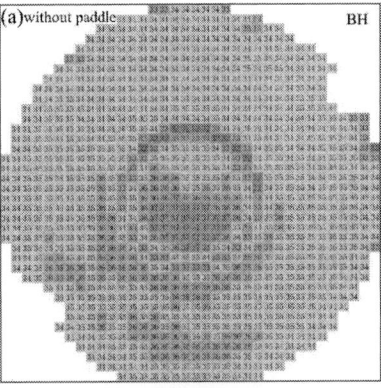

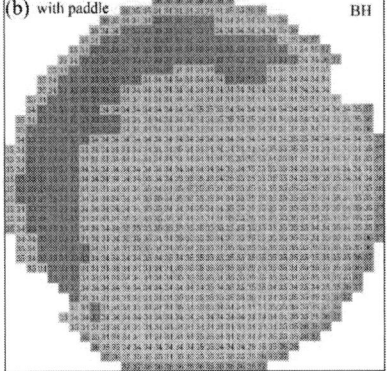

Fig. 3. Copper pillar height chart: (a) without paddle, (b) with paddle.

TABLE II   BUMP HEIGHT DATA OF CU PILLAR

BH (µm)	MAX	MIN	Avg.	Nu	Tilt
without paddle	37.33	31.65	33.76	8.23%	5.68
with paddle	34.87	32.58	33.88	3.38%	1.29

As shown in the Fig. 3 and Table II, the BH nonuniformity (Nu) value of the bumps reduced from 8.23% to 3.38%, and the Tilt value reduced from 5.68 µm to less than 1.50 µm.

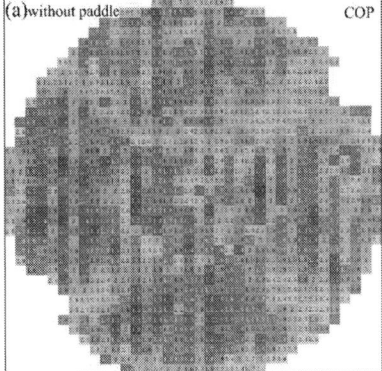

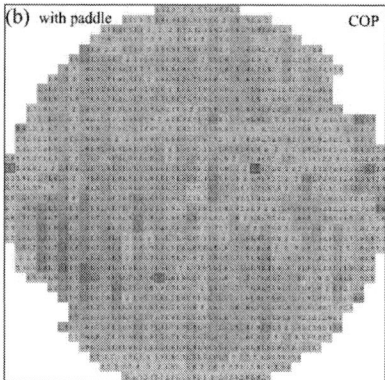

Fig. 4. Copper pillar COP chart: (a) without paddle, (b) with paddle.

TABLE III COPLANARITY DATA OF CU PILLAR

COP (μm)	MAX	MIN	Mean
with paddle	0.74	0.06	0.40
without paddle	3.40	0.9	2.48

The COP value of the bump was shown in Fig. 4 and TABLE III. After installing the paddle, the COP reduced by nearly 500%, from 2.48 to 0.40. The obvious changes in BH and COP prove that paddle improved uniformity of bumps.

Macro pillar

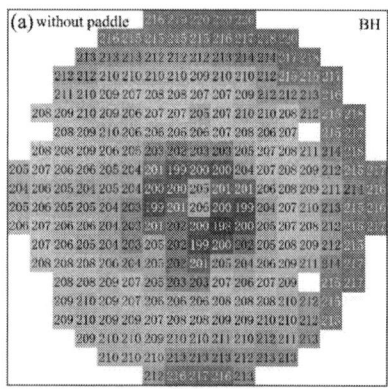

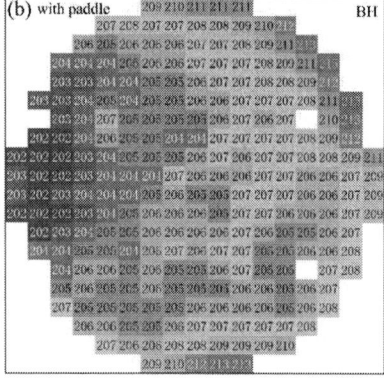

Fig. 5. Macro copper pillar height chart: (a) without paddle, (b) with paddle.

TABLE IV BUMP HEIGHT DATA OF CU PILLAR

BH (μm)	MAX	MIN	Ave	Nu	Tilt
without paddle	218.3	195.6	206.8	5.51%	11-13
with paddle	213.4	199.6	204.3	3.37%	< 2

In the above, it is analyzed that the uniquely designed paddle of ACM in the general copper pillar electroplating ensures the quality of the bumps, especially the compensation for the center height. In order to further verify the great role of this Paddle, tests were also carried out in the macro pillar dummy wafers with larger through-hole aspect ratio and higher plating rate requirements. Fig. 4 and TABLE IV show the BH value before and after installing the paddle. Nu decreased from 5.51% to less than 5%, which was 3.37%. The tilt reduced from 11-13 μm to 1-2 μm, the height distribution range of the bumps reduced, which could ensure the good contact of the chip and improve the performance.

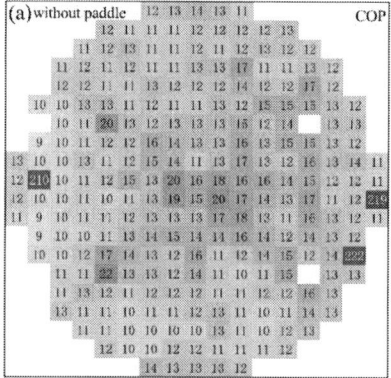

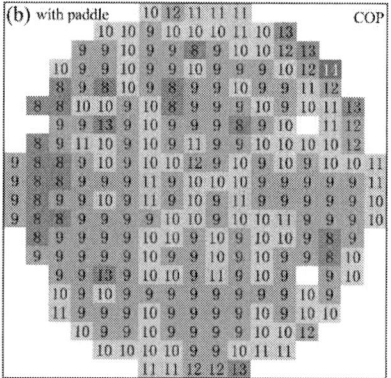

Fig. 6. Macro copper pillar COP chart: (a) without paddle, (b) with paddle.

TABLE V COPLANARITY DATA OF CU PILLAR

COP (μm)	MAX	MIN
without paddle	20.3	8
with paddle	13	7

As shown in the COP results in Fig. 6 and TABLE V, the uniformity of copper bumps was effectively controlled under the favorable action of paddles. COP reduced from 20.3-8 to 13-7 and there is no missing bump, and the effective control of levelness shows that our high-speed copper plating equipment can be applied to a higher upper limit of the depth-width ratio, and can fit a variety of packaging process flow.

Experiments on electroplating SnAg products

As we all know, copper has excellent electrical conductivity, thermal performance, reliability and high migration and other characteristics are the most important electroplating material in chip packaging industry. And some other metals or alloys are also commonly used in packaging plating, so we have also carried out plating tests on other more specific plating solutions. The results are as follows:

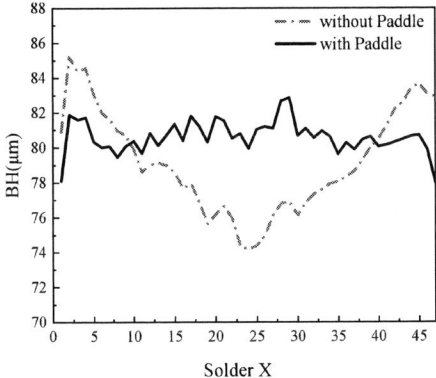

Fig. 7. Side profile of SnAg-plated Solder product cross wafer diameter

From the solder product (Fig. 7), after adding paddles, the height of the SnAg bump center was significantly improved， the bump height variation range reduced from 11μm to 5μm.

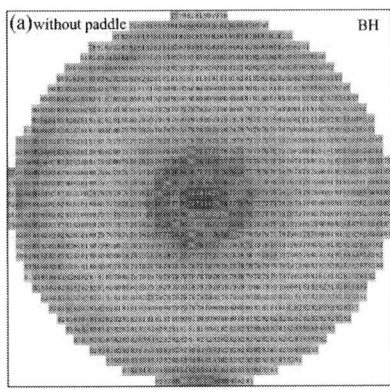

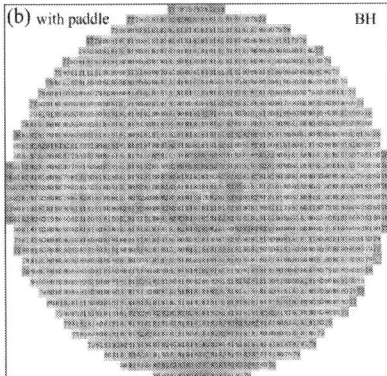

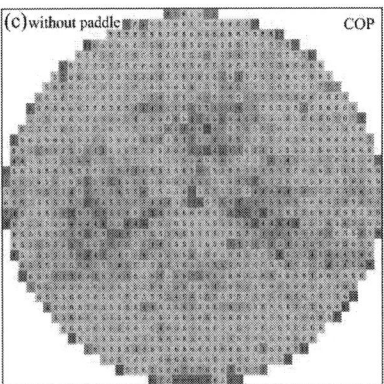

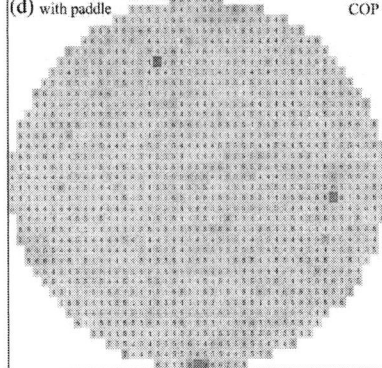

Fig. 8. SnAg pillar data chart：

BH: (a) without paddle, (b) with paddle；

COP: (c) without paddle, (d) with paddle；

TABLE VI SnAg pillar data

	Description	BH (μm)			
		max	*min*	*range*	*average*
Wafer	*without paddle*	83.2	75.7	7.5	80.6
	with paddle	89.7	73.2	16.5	80.7
	Description	COP (μm)			
		max	*min*	*range*	
	with paddle	106.6	3.4	103.2	
	without paddle	9	3.5	8.7	

As shown in Fig. 8 and TABLE VI, the addition of paddles by Snag makes the mass transfer more uniform and reduces the uniformity of solder BH to less than 5%. In addition, Solder COP around wafer edge was reduced from 10-13 to 5-6 μm, meeting the requirements.

IV. CONCLUSION

ACM's high-speed plating technology uses paddle with special design to enhance the efficiency of cation transfer during the deposition of the copper pillar deposit in the opening at almost double plating rate compared with low speed plating, resulting in better uniformity in the wafer and inside the chip during high-speed plating. At the same time, the second anode adjusts the current distribution and effectively controls the electric field in the electroplating wafer notch area. The results showed that the uniformity of the wafer treated by this technique could be controlled below

5% successfully under the same plating condition, which was obviously improved compared with the test results of other methods. Not only that, in the macro pillar and SnAg electroplating test, our electroplating technology still has good performance, to ensure the center height of the convex point and product uniformity, but also to ensure better coplanarity performance and higher productivity. There seems every reasonable to believe that our high-speed electroplating technology can meet the diversified market needs.

ACKNOWLEDGMENT

This work is sponsored by China National Major Project (Large Scale Integrated Circuit Manufacturing Technology and Complete Process). Thanks to the Shanghai Municipal Government for its support to ACM's construction of Shanghai Municipal Key Laboratory.

REFERENCES

[1] R. R. Tummala, "In Moore's Law for Packaging to Replace Moore's Law for ICS," 2019 Pan Pacific Microelectronics Symposium (Pan Pacific), USA, p. 1-6, February 2019.

[2] L. England and I. Arsovski, "Advanced packaging saves the day! — How TSV technology will enable continued scaling," 2017 IEEE International Electron Devices Meeting (IEDM), USA, p. 3.5.1-3.5.4, December 2017.

[3] S. Elisabeth, "Advanced RF Packaging Technology Trends, from WLP and 3D Integration to 5G and Mmwave Applications," 2019 International Wafer Level Packaging Conference (IWLPC), USA, p. 22-24, October 2019.

[4] K. Kholostov, A. Klyshko, D. Ciarniello, P. Nenzi, R. Pagliucci, R. Crescenzi, D. Bernardi, and M. Balucani, "High uniformity and high speed copper pillar plating technique," 2014 IEEE 64th Electronic Components and Technology Conference (ECTC), USA, p. 1571-1576, May 2014.

[5] L. Y. Kao, H. T. Hung, Y. H. Chen, and C. R. Kao, "Bonding of Copper Pillars Using Electroless Cu Plating," 2019 International Conference on Electronics Packaging (ICEP), Japan, p. 220-222, April 2019.

[6] C. Zhou, J. Tao, B. Yao, A. B. Wang and H. Zhang, "Investigations of Copper Electrodeposition for High Aspect Ratio TSVs," 16th International Conference On Mechatronics Technology, China, p. 35-38, October 2012.

[7] H. Ling, H. Cao, Y. Guo, H. Yu, M. Li, and D. Mao, "Influence of leveler concentration on copper electrodeposition for through silicon via filling," 2009 International Conference on Electronic Packaging Technology & High Density Packaging, China, p. 860-862, August 2009.

ATE board design solution of small PIN pitch DUT

Fei Pan
IC Test Department
SKY CHIP INTERCONNECTION
TECHNOLOGY CO.,LTD
Wuxi, China
panf@scc.com.cn

Qing Zhou
IC Test Department
SKY CHIP INTERCONNECTION
TECHNOLOGY CO.,LTD
Wuxi, China
zhouqing@scc.com.cn

Jie Zhou
IC Test Department
SKY CHIP INTERCONNECTION
TECHNOLOGY CO.,LTD
Wuxi, China
zhoujie@scc.com.cn

Abstract—With the development of the semiconductor industry towards Moore's law, more and more IC with higher integration but smaller size appears in the market, which also means that small PIN pitch IC will become an irresistible trend. This also brings great difficulty to the design and manufacture of ATE (Automatic Test Equipment) board. ATE board is the interface of DUT (Device Under Test) and ATE, it is a special PCB using in the process of IC test. If the PIN pitch less than 0.35mm, the via hole size always should less than 0.15mm, and ATE board thickness normally should more than 3mm. If aspect ratio more than 27:1, that will be a big challenge of manufacture. However, more and more small pitch DUT appears and the application always request high board thickness, how to solve this problem on design solution side is the key point of this paper. The example we use is a 0.3mm PIN pitch DUT, board thickness is 3.6mm, via hole diameter is 0.13mm. This parameter is difficult for many factories to manufacture. We provide a solution to make most factories can manufacture it. The solution is use SVH (Stack Via Hole), the board is divided into two parts for secondary compression, upper side thickness is 3mm, down side thickness is 0.6mm, then, upper side layers can use small hole size vias, down side layers should use more big size via (also can fanout by small size via from upper side). Based on this solution we completed the design and the design can be manufactured. Our paper will provide detail design solution, simulation result and manufacture parameters for providing a solution reference of small PIN pitch DUT ATE board design.

Keywords—*Small PIN pitch, DUT, ATE, PCB, High board thick diameter ratio*

I. INTRODUCTION

Before a semiconductor chip goes into the market, there are 4 main steps in the process of production:

1. Wafer fabrication
2. Wafer probe test
3. Packaging
4. Final test

Wafer probe test and final test are the steps that select qualified products to go to next production step. Therefore, these 2 steps are the key step to ensure the quality of product. The test is based on the ATE (Automatic Test Equipment), and we should use ATE board to load test circuit and achieve the interconnection for test objects and ATE as follows:

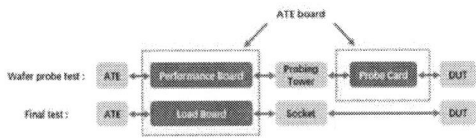

Fig. 1. Relation schema of ATE and ATE board

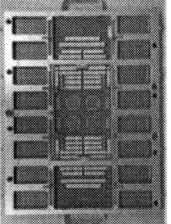

Fig. 2. Load board

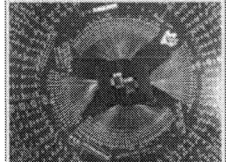

Fig. 3. Probe Card

Fig 4 shows a typical layout of DUT area in a ATE board. Pin pitch of DUT and anti-pad size are the most important parameters that we should consider in ATE board design. If the PIN pitch less than 0.35mm, the via hole size always should less than 0.15mm, and ATE board thickness normally should more than 3mm for stress requirement. Then high aspect ratio problem coming. If aspect ratio more than 25:1, that will be a big challenge of manufacture. However, more and more small pitch DUT appears, how to solve this problem on design solution side is the key point of this paper.

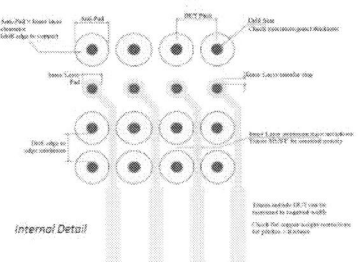

Fig. 4. Relevant parameters of DUT

II. LAYER STRUCTURE

Fig.5 shows the manufacturability of different aspect ratio in our company. When you start a small pin pitch ATE board design, the first thing you should do is researching about the manufacturability. From the chart we can see when the hole size should less than 0.15mm, the aspect ratio should less than 27:1. When aspect ratio more than 27:1, our factory only can do small scale manufacture or even cannot manufacture.

978-1-6654-1392-3/21 $31.00 © 2021 IEEE

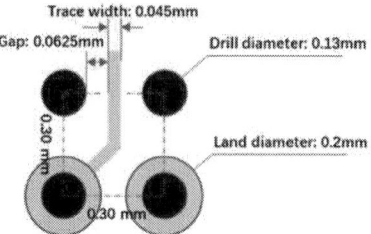

Fig. 8. Signal layer manufacturing parameters

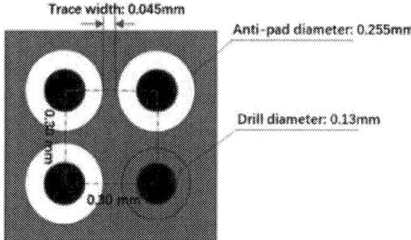

Fig. 9. Power/GND layer manufacturing parameters

Aspect ratio	Board thickness(mm)							
Drill diameter		2.0	3.0	4.0	5.0	6.0	7.0	8.0
	0.30mm	7:1	10:1	13:1	17:1	20:1	23:1	27:1
	0.25mm	8:1	12:1	16:1	20:1	24:1	28:1	32:1
	0.20mm	10:1	15:1	20:1	25:1	30:1	35:1	40:1
	0.15mm	13:1	20:1	27:1	33:1	40:1	47:1	53:1
	0.10mm	20:1	30:1	40:1	50:1	60:1	70:1	80:1

Legend: Mass, Small scale, R&D, Pre-R&D, Future

Fig. 5. Manufacturability of different aspect ratio

Board thickness always determined by the ATE system request. Basically should more than 3mm.

After you get the information about minimum hole size, manufacturability and board thickness, you can calculate aspect ratio, if it can meet factory aspect ration request, that will be fine. However, if it can not meet factory manufacturability, how to treat it? That is the key point this paper want research.

Fig.6 is the example of this situation. In this case if we use 0.13mm hole, aspect ratio will more than 27:1.

DUT PIN PITCH	0.3 MM
MINIMUM HOLE SIZE	0.13 MM
BOARD THICKNESS	3.6 MM
ASPECT RATIO	27.7 : 1

Fig. 6. Example case of high aspect ratio

For this example case, hole size and board thickness can not change, so we consider the solution of SVH(Stack Via Hole) to solve this problem. As Fig.7 shows the layer structure, from L1 to L12 is SVH layers, and thickness of these layer only 3mm, if 0.13mm hole only apply in these layers, the aspect ratio can meet manufacturability. Meanwhile, for other hole that hole size more than 0.15 mm, it can be manufactured by through hole.

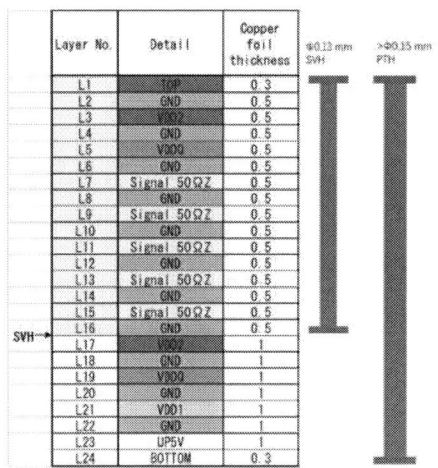

Fig. 7. Layer structure of example case

III. LAYOUT

DUT PINs can be divided into 3 type: power PIN, signal PIN, and GND PIN. For different PIN type, layout consideration will be different. Fig.8 and Fig.9 shows the manufacturing parameters of each type layers in our example case. We should confirm with factory first about whether between 2 holes can passing trace or plane. If it is OK, that will be fine , however, if will have manufacture problems, we may should consider reduce hole size or do a special fanout.

A. Power

In our example case, power PINs have high current requirement, that means 0.13mm holes are not enough. So, as Fig.10, for power PINs and GND PINs, if they are in the outer side of DUT, we fanout them by 0.2mm vias.

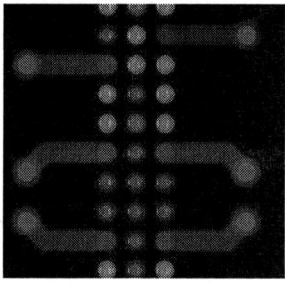

Fig. 10. Power/GND PINs fanout

About power plane layout, basically should more wider. But too wide plane may also not good due to we should ensure power plane path all will have capacitors under it. Fig.11 is the example layout. Meanwhile, if the power current will very large, that is better use more than one layer to trace it and copper thickness also can consider to increase, VDD2 and VDDQ in Fig.7 is example.

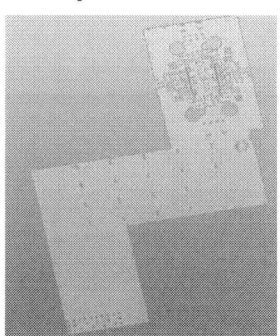

Fig. 11. Power plane layout

B. Signal

For small PIN pitch DUT, neck-down trace can not avoid, these neck-down trace will not meet impedance control request, so we should try to reduce the length of them. Meanwhile add teardrops to smooth the line width change.

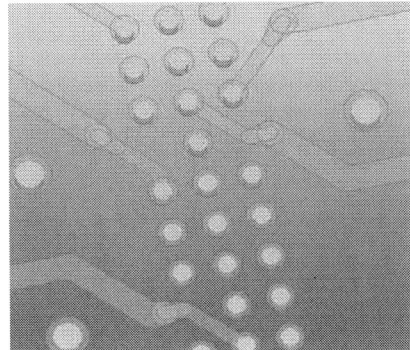

Fig. 12. Signal trace layout

C. GND

ATE board have too many GND plane layers, for the connection of GND net is easy, we just need add GND plane in each GND layers is fine.

IV. PERFORMANCE VERIFICATION

After basic layout complete, we should start consider performance verification and modify the design to let get more good performance. In this section, we should use simulation tools.

A. Power Integrity

For ATE Board PI consideration, normally divide to 2 part of AC and DC. AC part is indicating by PDN impedance, DC is indicating by DCR or IR-drop.

For ATE board, sometimes, we need keep same for all DUT, so, we always need adjust power shape according to the DCR results to avoid have big tolerance between different DUT.

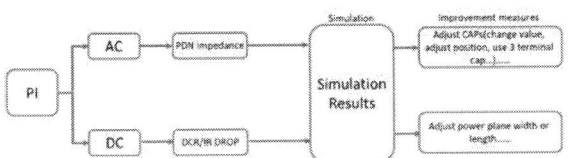

Fig. 13. Power integrity consideration

For DC part, as Fig.14, after simulation, we can get the IR-drop results, it should less than 5% at least, otherwise, we should consider how to reduce it.

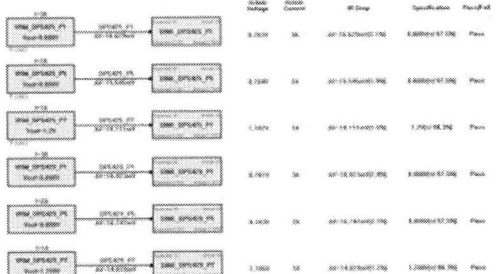

Fig. 14. IR-drop

For PDN impedance, on board level, we mainly focus on the frequency that below 100M. PDN impedance mainly related to the capacitors and power/GND plane structure.

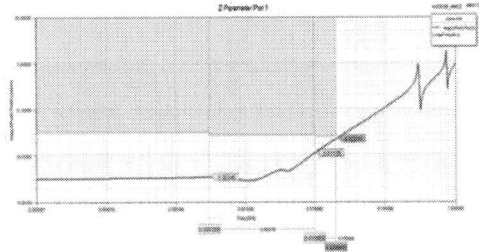

Fig. 15. PDN impedance

B. Signal Intergrity

For ATE Board SI consideration, normally we are focus on following 3 items, they are reflect the signal quality. We get the results by simulation tools, and based on the results to find out improvement measures if some place is not good.

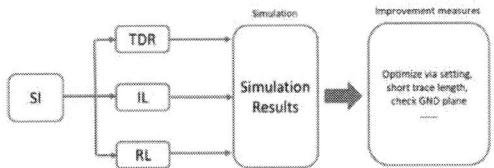

Fig. 16. PDN impedance

For DUT PIN side design always need optimize by following steps:

*1) Consider about PIN pad size:*Normally, DUT pin side impedance are always low than our expectation, reduce the parasitic capacitance of the DUT PIN can improve impedance control. As the TDR, insertion loss and return loss result of Fig.17 and Fig.18, we can see a little improving by increase the PIN pad size.

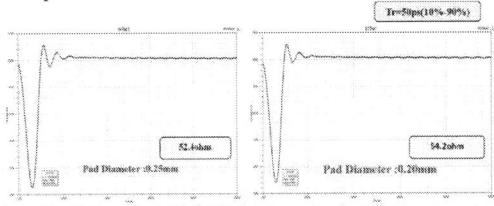

Fig. 17. TDR result after pad size change

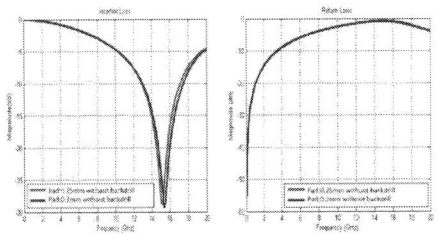

Fig. 18. Insertion loss and return loss result after pad size change

2) Consider about anti-pad size: Increase the anti pad size can increase the impedance, according this to adjust the anti-pad size. From Fig.19 and Fig.20, we can see optimization by anti-pad size change. The insertion loss improve nearly 2.5dB at 10GHZ.

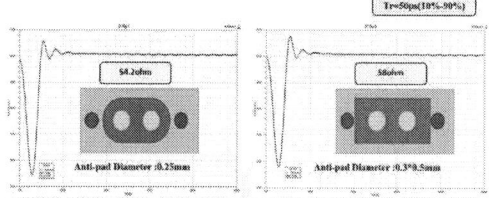

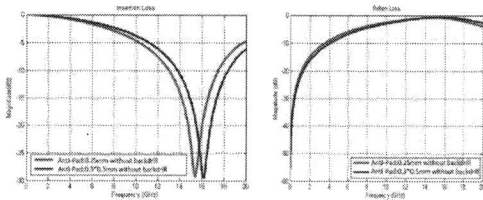

Fig. 19. TDR result after anti-pad size change

Fig. 20. Insertion loss and return loss result after anti-pad size change

3) Consider about backdrill: the backdrill will have big effort to the design, SVH also can do backdrill. As Fig.21 shows, after back drill, the impedance increased 24.5 ohm. And also contribute great effort to the insertion loss and return loss as Fig.22.

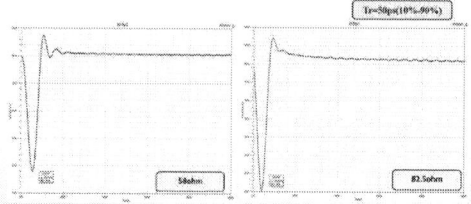

Fig. 21. TDR result after backdrill

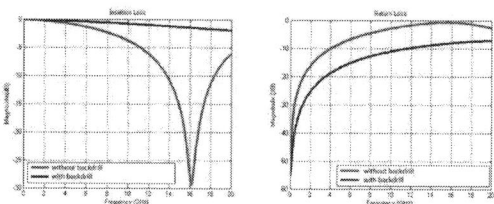

Fig. 22. Insertion loss and return loss result after backdrill

REFERENCES

[1] Heidi Barnes、Jose Moreira, "Measurement-based modeling for High speed semiconductor Test Interface Boards", Signal Integrity Workshop, November 30th 2017

[2] Eric Bogatin, "Signal Integrity: Simplified", Publishing house of electronics industry, 2004

[3] Jose Moreira, Ming Tsai, Don Faller, "PCB load board Design Challenges for Multi-Gigabit Devices in Automated Test Applications" Design con 2016

[4] Lyons R G. "Understanding Digital Signal Processing" Beijing Science Press, 2003

[5] Ken Martin. "Digital Integrated Circuit Design" Oxford University Press, 2000

A Study of Copper Oxidization Mechanism at Metal Interface

Ruolin Zhang
Packaging Engineering
Western Digital Coporation
Shanghai 200241, China
Ruolin.Zhang@wdc.com

Xu Wang
Packaging Engineering
Western Digital Coporation
Shanghai 200241, China
Xu.Wang@wdc.com

William Cao
Packaging Engineering
Western Digital Coporation
Shanghai 200241, China
William.Cao@wdc.com

Ying Tang
Packaging Engineering
Western Digital Coporation
Shanghai 200241, China
ying.tang@wdc.com

Pradeep Rai
Packaging Engineering
Western Digital Coporation
Milpitas, CA, 95035, US
Pradeep.Rai@wdc.com

Abstract—In modern semiconductor industry, copper is widely used as interconnecting material in ultra-large-scale integration (ULSI) devices due to its high thermal and electrical properties. However, the formation of copper oxide layer will cause a significant decrease in its thermal and electrical conductivity, as well as the degradation of interconnection capability. In one package, copper oxidization occurred at the interface between metal and copper. The influence factors were analyzed by conducting mechanism study on both packaged samples and glass carriers. The copper oxide was evaluated by the resistance, oxide structure, oxide thickness, Cu valence state and the atomic ratio of Cu/O. The results showed that copper oxide formation mainly depended on the condition of *Process_1*. *Process_1* in condition *A* is mainly responsible for serious oxidation, and *Process_1* in condition *B* can almost entirely suppress the oxidation. The plasma step also influences the oxide removal and reduction abilities during plasma treatment, and these abilities exist with *X* and *Y*. *Plasma(X)* impact results in oxide layer thinning and *Plasma(Y)* is able to transform the oxide to crystal Cu, which can effectively reduce the copper oxidation. The surface color observation of the glass samples was also performed to understand the formation of oxidation layer along with different processes.

Keywords—Copper oxidation; Resistance; Plasma

I. INTRODUCTION

In modern manufacturing of ultra-large-scale integration (ULSI) devices, copper has developed as the most promising interconnecting material. This mainly attributes to its high thermal and electrical performance. However, copper can be easily oxidized even under vacuum environment. Oxidization not only limits the use of copper, but also degrades copper's thermal and electrical performance.

Regarding oxide formation, thin oxide layers easily form on copper surface once contact with air. The typical types of oxides are Cu_2O (cuprous oxide), CuO (cupric oxide) and Cu_4O_3. Cu_4O_3 is usually regarded as the mixture of CuO and Cu_2O in a certain ratio. The oxidization process of bulk-copper surface has been studied in many researches. It has been proved that oxidation occurs through inside to surface Cu migration. Due to a built-in electric field at the copper-oxygen interface, the migration is accelerated at high temperature, but becomes slow at lower temperature. Ilia Platzman [1] found that a surface oxide layer around 1~8 nm is formed at ambient

conditions. Based on his research, even though the oxidation starts immediately after exposed to air, it will take several months to grow a few nanometers.

Valladares [2] analyzed the electrical resistivity variation along with oxide evolution in a $Cu/SiO_2/Si$ thin film after thermal oxidation. The resistivity got larger with increased temperature from 150 °C to 1000 °C. The phase evolution $Cu \rightarrow Cu + Cu_2O \rightarrow Cu_2O \rightarrow Cu_2O + CuO \rightarrow CuO$ occurred. In Takayuki's study [3], he also found that oxide formed at low temperature consists of amorphous structure and crystal Cu_2O.

Plasma treatment was deemed as an effective method to reduce copper oxidation. M.A. Badillo-Avila [4] found that a crystal CuO thin film can be rapidly reduced to metallic Cu under microwave Ar plasma. This effect was verified by variation of electrical resistivity which ranged from 30 μΩ·cm to 500 Ω·cm. The stable Cu can be obtained by applying longer processing time. In Kelly and Rothman's studies [5, 6], under high energy plasma, the amorphous oxide can be transformed to crystal structure oxide due to the thermal-spike and ion-beam heating effects.

Cu oxidization is a serious problem in electronic package, so we need to understand key influence factors in specific processes and conditions. In this paper, we studied the formation and reduction of copper oxide, and proposed a systemic method to quantify oxidation level for future applications.

II. PHENOMENON

A. Problem Statement

In one package, copper oxidization was found at metal/ Cu interface after completed all process flow. The oxide layer showed distinct structures in different process conditions, as displayed in Fig. 1. However, the detailed influence of each process step on this interface oxidation is still unclear. It is necessary to figure out the concrete mechanism, not only for the optimization of current process, but also for the feasibility of new process.

978-1-6654-1392-3/21 $31.00 © 2021 IEEE

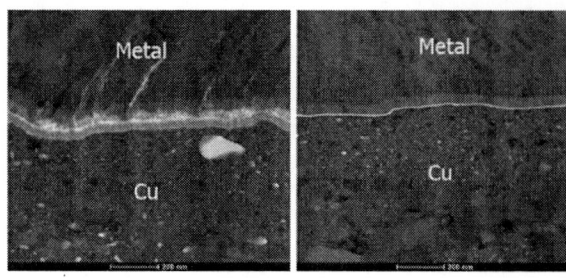

Fig. 1. The copper oxide structure (white layer) at the interface with different *Processes*

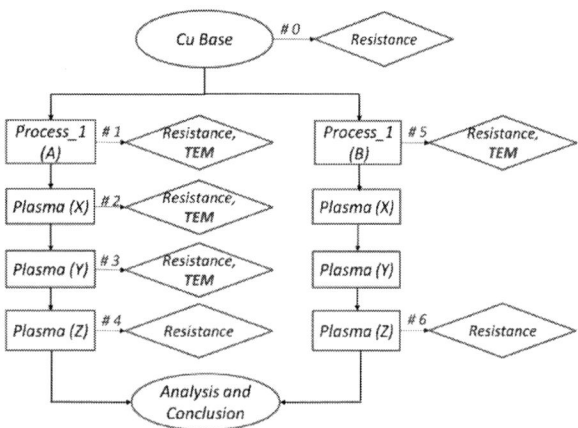

Fig. 2. The process flow of DOE with packaged samples

TABLE I   DETAILED DOE MATRIX

Run	Process Flow Steps
#0	*Cu Base Only*
#1	*Process_1(A)*
#2	*Process_1(A)+Plasma(X)*
#3	*Process_1(A)+Plasma(X)+ Plasma(Y)*
#4	*Process_1(A)+Plasma(X)+ Plasma(Y)+Plasma(Z)*
#5	*Process_1(B)*
#6	*Process_1(B)+ Plasma(X)+ Plasma(Y)+Plasma(Z)*

B. Experiment Process Flow

The experiment flow is showed in Fig. 2. The process starts with copper base material, then is divided into two routes: One route goes through *Process_1(A)-Plasma(X)-Plasma(Y)-Plasma(Z)*; the other route goes through *Process_1(B)-Plasma(X)-Plasma(Y)-Plasma(Z)*. *Process_1* *"A"* or *"B"* is precondition of samples, and *"A"* and *"B"* stand for difference precondition parameter settings. "X, Y, Z" are different plasma treatment processes to remove impurities and contaminants from surfaces. To protect further oxidation in room temperature, a metal layer was added after all plasma processes.

To study the oxidization process, 6 observation points were chosen. The 6 checking points were marked as Run#0 to #6. For each of these points, 5 pcs samples were separated from the group and sent for interface resistance measurement. Later, 1 out of 5 pcs samples was sent for TEM analysis in Run#1,2,3,5. The details are shown in TABLE I.

The resistance was measured by Keysight 34470A Digital Multimeter to evaluate oxidation degree. TEM material analysis was performed by FEI Talos F200.

III. MECHANISM INVESTIGATION

After deep dive and check, we found that *Process_1* condition, plasma (X) and plasma (Y) are the most important factors.

A. The resistance tendency with process flow

As shown in Fig. 3, the resistance values are high after *Process_1(A)*. It suggests the oxidization starts from the step *Process_1(A)*. The wide range of measured values suggests the interface condition is ununiform. The values sharply decrease post *Plasma(X)*, and further decrease after *Plasma(Y)*. On the other side, the average resistance after *Process_1(B)* is the lowest.

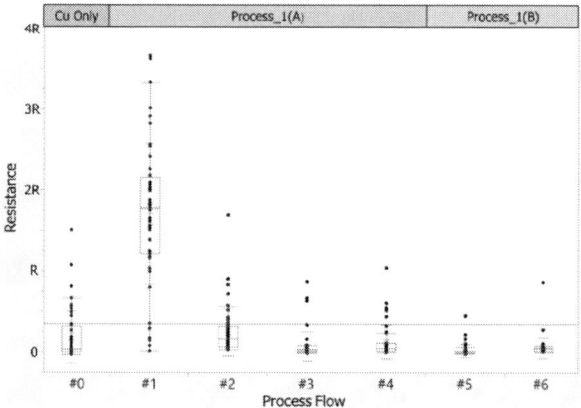

Fig. 3. The resistance evolution with the process flow

B. The oxide structure and phase characterization

The high resolution TEM microstructure is shown in Fig. 4. After Run#1 *Process_1(A)*, the formed Cu oxide is composed of two Zones. The loose feather-like Zone A is about 30~40 nm, and oxidization is inhomogeneous. The plate-like Zone B is about 30~40 nm. ZoneA and B's total thickness is over 70 nm. Both two zones of oxide are a mixture of little black crystal and most white amorphous. After Run#2 *Plasma(X)*, Zone A is removed, but plate-like Zone B still exists. A thin amorphous film Zone C is formed at metal side (3~5 nm). After Run#3 *Plasma(Y)* treatment, the amorphous film Zone C gets thicker (10~15 nm) and the total thickness is about 50~55 nm. The amounts and size of crystal dramatically increase. However, in the other route, Run#5 showed only an ultra thin oxide film post *Process_1(B)*. The oxdation layer thickness is less than 5 nm. According Ilia Platzman's studies, this ultra thin layer was formed in ambient condition [1]. Hence, no oxidization occurs in condition *B*, the amorphous film has already been formed when Cu is exposed to air prior to *Process_1(B)*.

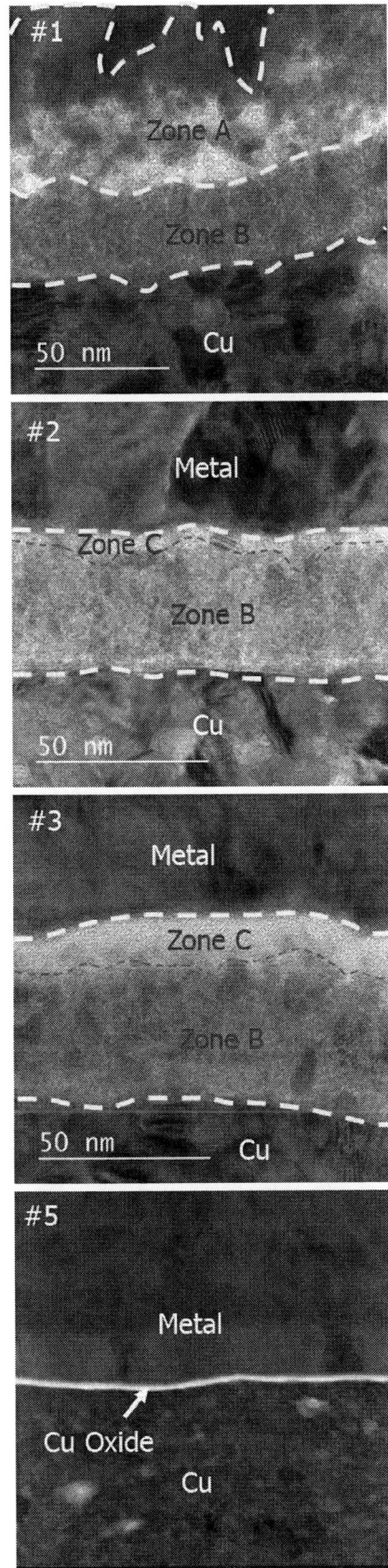

Fig. 4. The Cu oxide structure at the interface post different processes

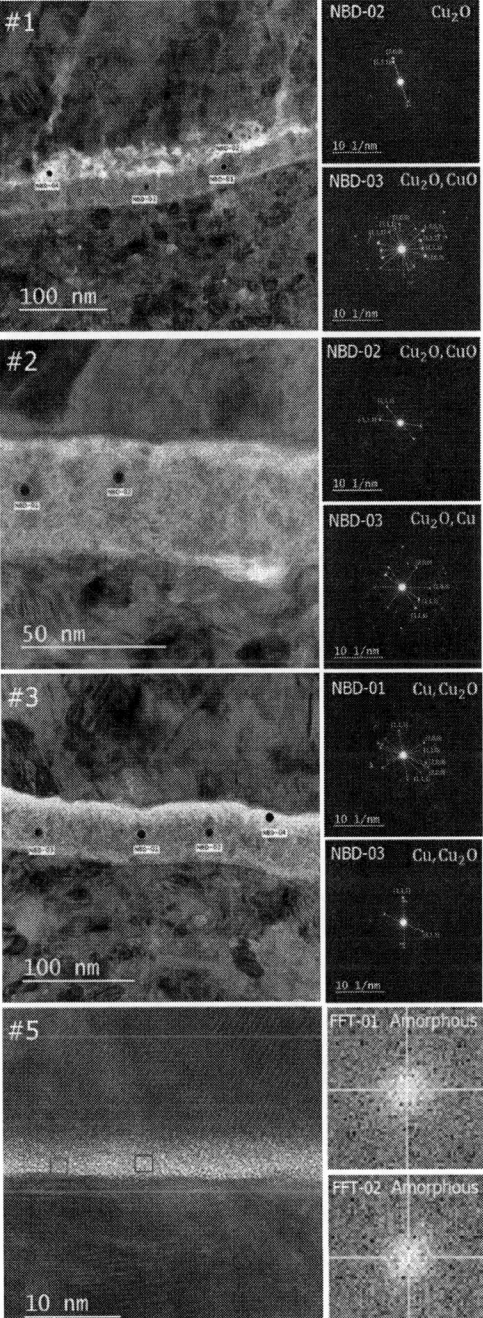

Fig. 5. The phase characterization by NBD and FFT of TEM

The detailed evolution of Zone B crystal types and amounts can be seen in Fig. 5. Based on NBD (Nano Beam Diffraction) and FFT (Fast Fourier Transformation), post Run#1 *Process_1(A)*, Zone B is a mixture of Cu_2O + CuO. After Run#2 *Plasma(X)*, Cu shows up, and the mixture becomes Cu_2O + CuO + Cu . Post Run#3 *Plasma(Y)*, abundant Cu forms with larger size and amounts and the mixture is composed of Cu_2O + Cu. Follow the process flow (Run #1~4), the interface initially oxidizes to form an oxide layer, then the oxide is gradually reduced. This process is also reflected by the resistance variation since the values depend on Cu valence. As for *Process_1(B)* (#5~6), no crystal exists in the amorphous oxide. The relationship between the phase types and the resistivity is shown in TABLE II. Less CuO and more Cu lead to lower resistance.

TABLE II   CU-O OXIDE VALUE [2]

Phase types	Cu/O Atomic Ratio	Color	Resistivity ($\Omega\,m$)
Cu	>2	Yellow	1.75×10^{-8}
Cu_2O	≈2	Red	0.55
CuO	≈1	Black	6000

C. The element distribution analysis

The element distribution can display the oxide structure more clearly, as depicted in Fig. 6. Some voids and gaps exist between the feather-like Zone A and plate-like Zone B. With the reduction effect of Cu^{2+} and Cu^{+} in oxide, the produced O escape towards the oxide edge by diffusion and volatize to outside. But some O still remain at the edge and form a thin O-rich amorphous film. The stronger the reduction effect, the thicker this film is, but it only has minor influence on the resistance due to the limited thickness.

Fig. 6. The element distribution mapping image post different processes

The Cu/O composition fluctuates in the oxide area within a certain range, and above phase judgement can be supported by their atomic ratio. Before and post *Plasma(X)*, the ratio changes from 1.37 to 1.39, showing no change. Cu/O atomic ratio is raised to 2.91 after *Plasma(Y)*, which suggests large amounts of Cu formation.

To observe and compare the interface evolution better, glass carrier samples were also prepared. The surface of glass sample was pre-sputtered with thin copper film to simulate the original state of the interface, as displayed in Fig.8. The glass carrier samples followed the same process flow as Run#0,1,2,3. The surface color evolution along with the process flow is shown in Fig. 9. It is consistent with the results of phase characterization.

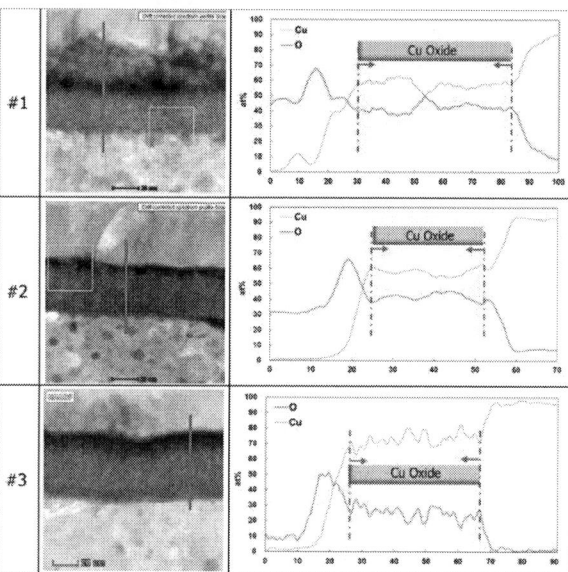

Run	Composition (at%)		
	Cu	O	Cu/O Ratio
# 1	57.64	42.11	1.37
# 2	58.11	41.69	1.39
# 3	72.92	25.02	2.91

Fig. 7. The composition of Cu/O ratio in the interface Cu oxide film

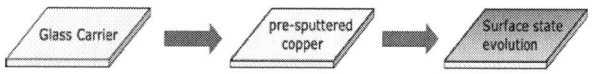

Fig. 8. The schematic of glass carrier sample

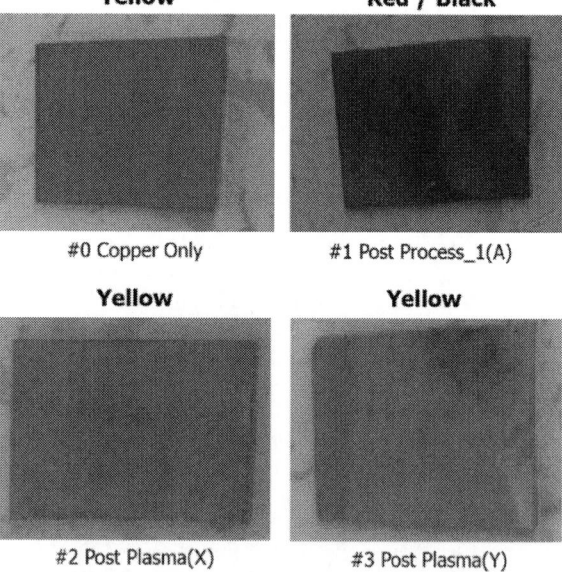

Fig. 9. The evolution of surface color along with the process flow

Process	Process_1(A)
● Amorphous ◐ Crystal CuO ◑ Crystal Cu2O ○ Crystal Cu Schematic of mechanism	Serious oxidization 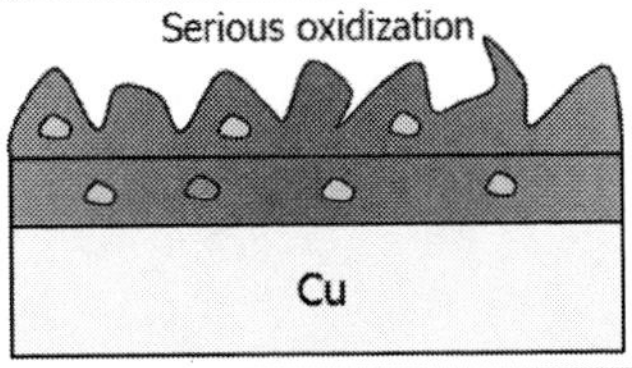 Cu
Thickness and Structure	• Zone A 30~40nm • Zone B 30~40nm
Crystal in Cu Oxide	• Small amounts crystal • Amorphous+CuO+Cu2O
Cu/O Ratio	• Cu/O: 1.37 (1~2)
Process	Process_1(A)+Plasma(X)
Schematic of mechanism	Plasma(X) etching and oxide removal 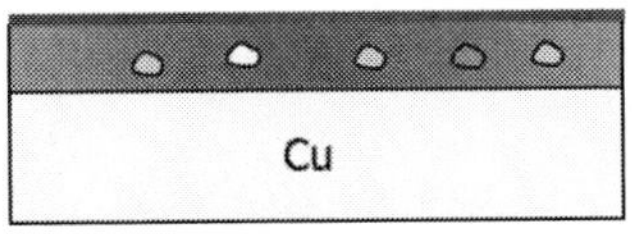 Cu
Thickness and Structure	• Zone B 30~40nm • Zone C 3~5nm
Crystal in Cu Oxide	• Small amounts crystal • Amorphous+CuO+Cu2O+Cu
Cu/O Ratio	• Cu/O: 1.39 (1~2)
Process	Process_1(A)+Plasma(X)+Plasma(Y)
Schematic of mechanism	Plasma (Y) reduces Cu to lower valence 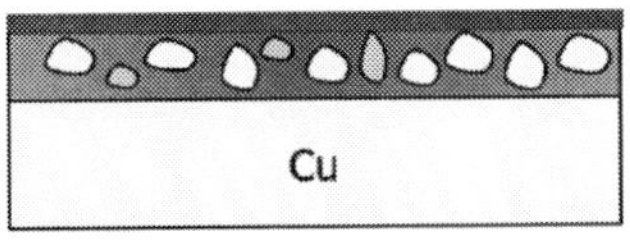 Cu
Thickness and Structure	• Zone B 30~40nm • Zone C 10~15nm
Crystal in Cu Oxide	• Large amounts crystal • Amorphous+Cu2O+Cu
Cu/O Ratio	• Cu/O: 2.91 (2~3)
Process	Process_1(B)
Schematic of mechanism	No oxidization 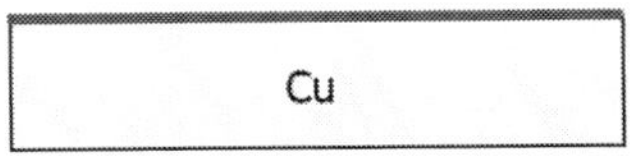 Cu
Thickness and Structure	• Amorphous 3~5 nm
Crystal in Cu Oxide	• No crystal
Cu/O Ratio	• No accurate signal in thin film

Fig. 10. The mechanism and characteristic summary of Cu oxidization

D. The mechanism of copper oxidization, removal and reduction

The key oxidization processes are summarized in Fig. 10, including the oxide thickness, oxide structure, crystal types and Cu/O atomic ratio.

- Oxide formation: *Process_1(A)* are mainly responsible for serious oxidation. The surface oxide is composed of most amorphous and little crystal. The formed crystal is a mixture of abundant Cu_2O and little CuO. And *Process_1(B)* can suppress the oxidation.

$$Cu + O_2 \xrightarrow{Oxidization} Cu^+ (amorphous)/Cu_2O \ (crystal) \qquad (1)$$

$$Cu + O_2 \xrightarrow{Oxidization} Cu^{2+} (amorphous)/CuO \ (crystal) \qquad (2)$$

- Oxide removal: *Plasma(X)* has strong etching ability in this plasma mode, so its bombardment removes the loose feather-like Zone A oxide.

- Oxide reduction: the reduction process have two steps. In step 1, *Plasma(Y)* makes Cu valence in amorphous lower, and the reduction reactions are shown in (3-5). In step 2, since *Plasma(Y)* has larger weight energy, the amorphous can be furtherly transformed to crystal under its impact. This thermal-spike and ion-beam heating effect has been reported previously [5, 6]. The formation of abundant crystal Cu, with low resistance, builds local connection from metal to Cu. Therefore, the interface resistance can finally display a good performance.

$$4Cu^{2+} + 2O^{2-} + 4e^- + 4Ar^+ \xrightarrow{Reduction} 4Cu^+ + O_2 + 4Ar \qquad (3)$$

$$2Cu^{2+} + 2O^{2-} + 4e^- + 4Ar^+ \xrightarrow{Reduction} 2Cu + O_2 + 4Ar \qquad (4)$$

$$4Cu^+ + 2O^{2-} + 4e^- + 4Ar^+ \xrightarrow{Reduction} 4Cu + O_2 + 4Ar \qquad (5)$$

IV. CONCLUSION

In this work, DOE was conducted to analyze the copper oxidization mechanism at Cu/metal interface. The primary factors of copper oxidization include *Process_1* condition and plasma step. The results from both packaged and glass samples can achieve mutual confirmation. The conclusions are summarized as below:

(1) The leading factor of oxide formation is the condition of *Process_1*. Condition *A* will result in serious oxidation, but no oxidization occurs in condition *B*.

(2) The subsequent improvement is obtained post plasma treatment. Plasma *X* and *Y* are endowed with oxide etching and reduction ability respectively.

(3) The loose part of oxide is firstly removed under the impact of *Plasma(X)*, and then abundant crystal Cu is produced post *Plasma(Y)*. The thickness thinning and crystal phase transition of the oxide are the root cause of dropped resistance.

(4) A systemic TEM characterization method for package samples was proposed. We mainly focus on the interface resistance, oxide structure /thickness, crystal types /amounts and Cu/O atomic ratio. Another method with glass sample was also put forward, and the surface color can well support the TEM results.

ACKNOWLEDGMENT

Thanks Rui Guo, Kevin Du, Xundi Zhang, Fan Yang, Gary Zheng and Claire Liu for valuable inputs during this study.

REFERENCES

[1] Platzman I, Brener R, Haick H, Rina T. Oxidation of polycrystalline copper thin films at ambient conditions, The Journal of Physical Chemistry C, 2008

[2] Valladares L D L S, Salinas D H, Dominguez A B, D. A Najarro e, S.I. Khondaker f, T. Mitrelias, et al. Crystallization and electrical resistivity of Cu_2O and CuO obtained by thermal oxidation of Cu thin films on SiO_2/Si substrates, Thin Solid Films, 2012

[3] Ohba T, Sugiyama M, Yamashita K. High resolution analysis of Cu thin oxide formed on Cu in 32-nm node Cu/Low-k application, International Conference on Solid-state and Integrated Circuit Technology, 2006

[4] Badillo M A , R. Castanedo Pérez, J. Márquez-Marín, D.E. Guzman-Caballero b, G. Torres-Delgado. Fast rate oxidation to Cu_2O at room temperature of metallic copper films produced by the argon-plasma bombardment of CuO films, Materials Chemistry and Physics, 2019

[5] Naguib H M, Kelly R. The crystallization of amorphous ZrO2 by thermal heating and by ion bombardment. Journal of Nuclear Materials, 1970

[6] Hussein M Na. Ion-bombardment-induced structural transformations in oxides, 1971

Irregular heat dissipating cover based on AlN material

1st Jianhui Liu
R&D Design Department
SKY CHIP INTERCONNECTION
TECHNOLOGYCO.,LTD
Shen Zhen, China
Liujh@scc.com.cn

2st Yuxin Guo
R&D Design Department
SKY CHIP INTERCONNECTION
TECHNOLOGYCO.,LTD
Shen Zhen, China
Guoyx@scc.com.cn

3st Xiaowei Zhao
R&D Design Department
SKY CHIP INTERCONNECTION
TECHNOLOGYCO.,LTD
Shen Zhen, China
Zhaoxiaow@scc.com.cn

Abstract—With the development of integrated circuits, SiP(system in package) technology has been widely used in recent years. SiP often integrates multiple chips. Although this kind of packaging method is highly integrated, it also brings heat dissipation problems. When the package contains both high-power FC and WB chips, how to export FC chips effectively needs to be focused on. This paper introduces a kind of irregular heat cover based on AlN material, it can be seen from the simulation this heat dissipation cover can effectively export the heat of FC chip without affecting the structure of other chips. AlN has good material properties and mature processing, which can be widely used in high-power multiple chips SiP in the future.

Keywords—SiP, high-power FC, multiple chips, heat dissipating cover

I. INTRODUCTION

In recent years, with the rise of 5G and the IOT, SiP has gained more momentum of development. SiP developed from single chip to multi-chip, from single process to mixed package process integration. SiP encapsulation architecture is evolving towards higher integration and higher performance. Along with the development of the problem, is the power dissipation problem.

Thermal as the bottleneck of SiP packaging always affected the development of SiP. Recent years the heat Thermal of SIP has been promoted in many directions depend on the development of materials and structures of package. The main ways usually used like substrate with high thermal conductivity or packaging material; heat dissipation metal matrix; heat sink or cover. The method of heat dissipation is based on different package type.

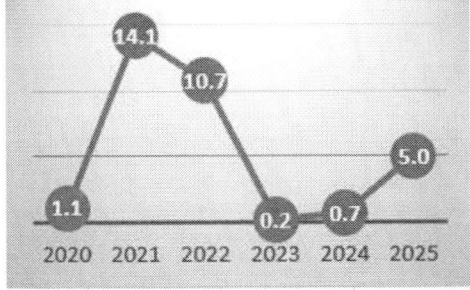

Fig. 1. Estimated growth rate of FCBGA

FCBGA packaging revenue is expected to reach 12 billion US dollars by 2025 from 10 billion US dollars in 2020，

Driven by AI (Artificial Intelligence), data center and HPC (High Performance Computing)[1]. FC chip is developing rapidly on the momentum and number of high power dissipation FC chip has arisen. The TDP (thermal design power) can reach 70W or even 100W. In this situation, Heat cover is more effective than Molding.

The material of heat covers on the market have Metal (such as copper, aluminum), alloy (such as Kovar, Mo-Cu) and other materials (AlN).Characteristics of the materials have diversity effects. Copper or aluminum heat cover is widely used in business based on low cost and high thermal conductivity. However it's CTE does not match the chip which can cause chip crack. Kovar is also commonly used as heat cover in ceramic package, with low thermal conductivity[2].Mo-Cu material performance appropriate but inapplicability of mass production due to price[3].material comparison refer to TABLE Ⅰ.

TABLE I. PROPERTIES OFDIFFERENT MATERIAL

Material	CTE	Conductivity (W/mk)	Young's Modulus	Price
Cu	17	380	110（GPa）	Low
Al	23	237	72（GPa）	Low
Kovar	5.86	17.3	138（GPa）	Low
Mo-Cu	8.0	170	--	High
AlN	4.7	150	320（GPa）	Middle
Molding	--	0.9	--	Low
Si	2.8	147	112（GPa）	

Nowadays common structure of heat cover cannot meet the needs of SiP with multiple modules. In this study, devised a kind of irregular heat cover based on AlN material. This irregular heat cover can diffuses the heat for FC chip with high TDP and compatible complex structure of multi modules SiP. With the method of sintering manufacturing AlN cover can produce complex cavity structure and the cover is more appropriate comprehensive properties and price.

II. STRUCTURAL DESIGN

To achieve better results, the back of FC needs to be connected to heat cover with adhesive. See Fig.2(a) for common structures. However, this structure cannot be used when other structure requirements are included in SiP, For example, there are stack-chips conflict with structure, See Fig.2 (b)

978-1-6654-1392-3/21 $31.00 © 2021 IEEE

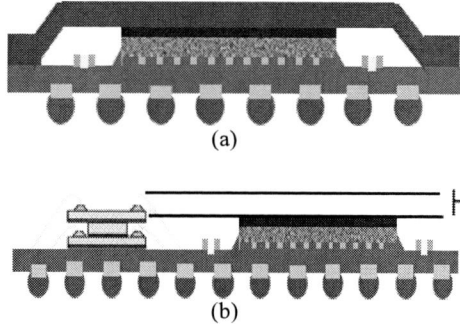

(a)

(b)

Fig. 2. Common structure and special structure requirements

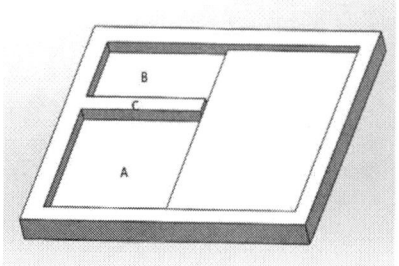

Fig. 5. Description of cavity

Fig. 1. The integration maybe include RF modules, analog modules or optical modules. In this study also have several modules as Fig.3, it needs to be structural isolation between modules. The irregular heat cover can provide a compartment and AlN support local metal plating to form a shielding, the specific structure shown as Fig.4.

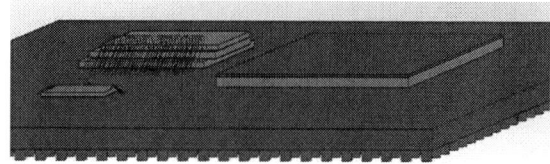

Fig. 3. Requirements for SiP system

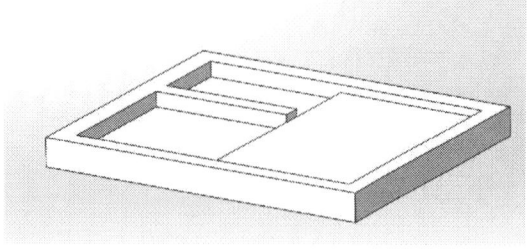

Fig. 4. The structure of Irregular heat cover

A. Structural design

FC part adopts cover sinking design to avoid other conflict areas. Integrate the production accuracy of AlN and assembly about SiP, the gap between FC and the heat cover is about 80-100um(taking into account assembly height deviation of the heat cover). Cavity's aspect ratio and area for separated Sensitive module within 50% is ideal, this area refers to the total area of cavity A and cavity B in Fig.5.

According to production experience, the cavity sintering deformation accuracy of AlN is about 1%, Therefore, the cavity with large size should be avoided in the design. Irregular structure may lead to warping of heat cover during sintering, To prevent deformation, pay attention for two caveats, One is cavity depth is generally thinner than the thickness of the fundus, the other is wall thickness as C in Fig.5 of septum thicker than 3mm. Cavity wall insulation can provide support and effectively reduce the size of the cavity.

B. Substrate Design

As a high temperature ceramic material, AlN can be used in ceramic package, also can be used in substrate package. Since CTE mismatching between substrate material and AlN, corresponding adjustment is needed in the assembly area of AlN heat cover.

a) Etch or shape with electrical property not allowed under installation area like Fig.6(a);

b) No isolated or small area of copper in the installation area to avoid material lamination

c) The copper in the installation area can be X hatch treated refer to Fig.6(b)

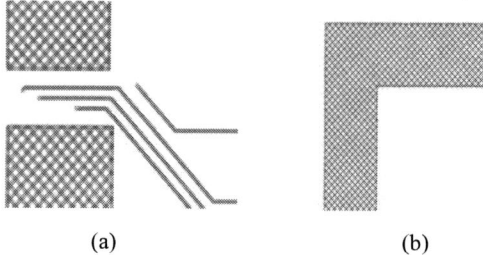

(a) (b)

Fig. 6. Conductor in the assembly area and processing schematic of shape

AlN has a lower density($3.26g/cm^3$)than Cu($8.96g/cm$)^3. Irregular heat cover based on AlN. AlN is lighter than Cu if the heat cover with same structure. But the stack-up of substrate for heat cover which used AlN should have almost the same thickness as Cu heat cover. This is done to ensure the flatness of the substrate. Substrate structure for SiP contains FC chip is generally complex. In this study, it need 8-layers substrate as TABLE II to meet behavior of electricity.

TABLE II. STACK-UP OF 8-LAYERS SUBSTRATE

Layer		Thickness(um)	Material
	Solder mask	20	
Top	Cu	15	
	PP	30	GZ41
Gnd2	Cu	15	
	PP	30	GZ41
Sig3	Cu	15	
	PP	30	GZ41
Power4	Cu	20	
	Core	800	E705G
Power5	Cu	20	
	PP	30	GZ41
Sig6	Cu	15	
	PP	30	GZ41
Gnd7	Cu	15	
	PP	30	GZ41
Bottom	Cu	15	
	Solder mask	20	

High speed signal and RF signal coexistence in the study, material type need support to run up to 8GHz rate .Core thickness up to 800um can maintain a rigid construction and flatness. This ensures the yield of FC chip welding and provides reliability for heat cover assembly.

III. THERMAL SIMULATION ANALYSIS

AlN has excellent thermal, electrical and mechanical properties. The kernel performance is in the high thermal conductivity and Si matching CTE[4]. Materials that can satisfies both at the same time are few and far between. Although Mo-Cu is satisfied, it's difficult to be processed into irregular structure as a result of hardness. For compatible structural and performance requirements, Kovar materials are usually used. Kovar has CTE close to AlN, thermal simulation analysis is carried out for the heat dissipation performance of these two materials conclusion.

The packaged thermal resistance θ Ja represents thermal resistance between the node of the chip and the outside air, and the unit is °C/W. The size of the thermal resistance θ ja is usually used to judge whether the chip heat dissipation performance is good or bad[5]. The calculation formula of the chip thermal resistance θ Ja:

$$\theta\ Ja = (Tj-Ta)/P$$

where θ ja is the thermal resistance (°C/W) from the chip node to the ambient air, Tj is the maximum temperature of the die of the chip, and Ta is the ambient air temperature ,P is the heat consumption of chip Die (W). For packages with a smaller size, in the actual heat dissipation analysis, natural convection is considered when considering the best heat dissipation method. Therefore, the chip is placed in a sea led test box of JEDEC standard[5], and the chip is dissipated by natural cooling, that is, the outside wind speed is 0, Calculate the thermal resistance, as shown in Fig.7.

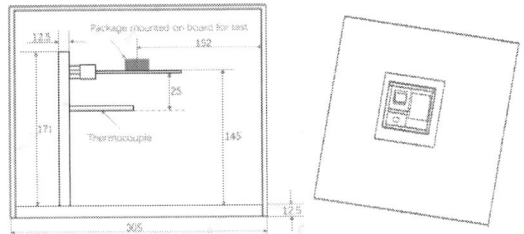

Fig. 7. Model for SiP

The simulation conditions and parameters are shown in the TABLE III。

TABLE III. CONDITIONS THERMAL SIMULATION ANALYSIS

Ambient temperature	FC dissipation	Stack-dies dissipation	Standard
25°C	12.8w	<1.2 W(total)	JA

This article uses the JEDEC standard static air environment domain simulation, ignoring the effects of heat radiation and wire bond alloy wire, only considering natural convection heat dissipation. AlN heat cover and Kovar heat cover temperature rise for reference Fig.8. and Fig.9.

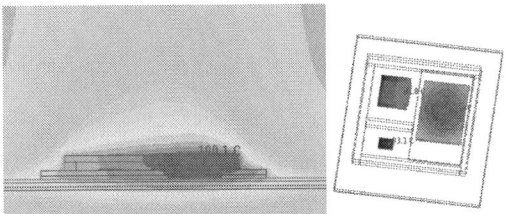

Fig. 8. Temperature rise of Kovar heat cover

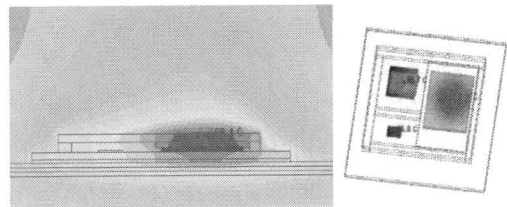

Fig. 9. Temperature rise of Kovar heat cover

The results show that the temperature rise of AlN(100℃) is less than that of Kovar(114 ℃) without any external measures. Because the power consumption of FC chip is much higher than other chips, so the highest temperature is at the location of the FC chip. The change of material has little effect on other chips because there is no direct contact between them. The thermal resistance of the two materials can be calculated as TABLE IV.

TABLE IV.THE THERMAL RESISTANCE AND TEMPERATURE

Ambient temperature	25°C	
	Tj(°C)	θ Ja(°C/W)
AlN	100.2	5.86
Kovar	114.4	6.98

The greater power dissipation of the FC chip, the more difference in temperature rise. AlN material can be more effectively the heat of the chip to the surface of the package, if other measures of thermal conductivity can be added the advantage is even more obvious. And the material parameters of Kovar have been relatively stable, and the thermal conductivity has not changed much. AlN material also has the potential to be optimized, by adjusting the proportions of raw materials. Now the thermal conductivity of AlN in domestic is 150(W/mk), it can reach to 180(W/mk) of some imported materials [6], with the development of material its thermal conductivity will be better.

IV. CONCLUSIONS

According to the particularity of the SiP package structure requirements analyze the properties of different materials, this irregular heat cover structure based on AlN was designed. This structure makes full use of the characteristics of the AlN, effectively avoids structural conflicts in the package, and has high manufacturability. This structure can be adjusted according to requirements, with a high degree of design freedom. The structure is suitable for the situations such as the high TDP for high performance FC, the variety and special structure customization modules. As SiP evolves and demands, Irregular heat cover based on AlN material will get more applications.

ACKNOWLEDGEMENT

This work was supported by the National Key R&D Program of China (2018YFE0204600), the Key-Area Research and Development Program of GuangDong Province (2019B010131001), and Shenzhen Fundamental Research Program）(JCYJ20200109140822796).

REFERENCES

[1] SSD Fans, Unprecedented growth in FCBGA packaging! Relying on AI, cars, data center chips will get more applications (Market size of FCBGA). 2021.

[2] M. X. Xia., Research progress of Mo-Cu alloys 2013.

[3] K. RAVI, A.S. MUJUMDAE. Thermal analysis of a flip chip ceramic ball grid array （CBGA） package [J].

[4] J. Y. Liu, W. C. Sun, H. Cui, Design and fabrication of AlN multilayer co-fired ceramic microwave shell Art. no. 1001-2028.2012.07.001

[5] H. P. Zhou, Y. C. Liu, Y. Wu Research and application of aluminum nitride ceramics

[6] Z. J. Yu, D. Yu, and Y. Ren. Application of Flo THERM simulation technology in extracting dual thermal resistance model of integrated circuit package [J]. Science and Technology Vision, 2016(5):104-105

[7] D. M. Yan, X. J. Gao, Research progress of high thermal conductivity aluminum nitride ceramics, Art. no. GSYT.0.2011-03-027

Evaluation and Reduction of Optical Crosstalk in Quantum Dot Color-Converted Mini/Micro-LED Displays

Yuanjie Cheng
Department of Mechanical and Aerospace Engineering
Hong Kong University of Science and Technology
Clear Water Bay, Kowloon, Hong Kong, China
ychengao@connect.ust.hk

Jeffery C. C. Lo
HKUST Foshan Research Institute for Smart Manufacturing
Hong Kong University of Science and Technology
Clear Water Bay, Kowloon, Hong Kong, China
jefflo@ust.hk

Xing Qiu
Department of Mechanical and Aerospace Engineering
HKUST Shenzhen-Hong Kong Collaborative Innovation
Research Institute, Futian, Shenzhen
Hong Kong University of Science and Technology
Clear Water Bay, Kowloon, Hong Kong, China
epfelix@ust.hk

S. W. Ricky Lee*
Department of Mechanical and Aerospace Engineering
HKUST Foshan Research Institute for Smart Manufacturing
HKUST LED-FPD Technology R&D Center at Foshan
HKUST Shenzhen-Hong Kong Collaborative Innovation
Research Institute, Futian, Shenzhen
Hong Kong University of Science and Technology
Clear Water Bay, Kowloon, Hong Kong, China
*rickylee@ust.hk

Abstract— **Quantum dot (QD) color-converted mini/micro-LED displays are considered as the most promising technology for next-generation display with the benefits of high production yield, low cost and good process reproducibility. However, it is still facing the challenge of optical crosstalk among adjacent subpixels, which degrades the display resolution and worsens the color rendering. To reduce the optical crosstalk, this paper presents an effective structure to confine the blue backlight paths by a deep reactive ion etching (DRIE)-fabricated silicon light confining grid. By optical simulations, this paper validates the effectiveness of silicon light confining grid on optical crosstalk reduction, and the influence on pattern design of silicon light confining grid is also discussed. A prototype display based on designed structure is assembled and the reduced optical crosstalk effect is confirmed by a microscope. Both simulation and experiment results demonstrate the effectiveness of the crosstalk-reduced design.**

Keywords— optical crosstalk; silicon light confining grid; full-color mini/micro-LED display

I.   INTRODUCTION

A growing demand has been witnessed for better display performance in terms of high dynamic range (HDR), high brightness and refresh rate for applications in gaming, home theaters and wearable devices. GaN-based full-color mini/micro-LED displays are considered as the most promising technology to fulfill these demands, owing to their excellent performance compared with other existing technologies, such as organic light-emitting diodes (OLEDs) and liquid crystal displays (LCDs) [1].

Mini/micro-LED displays can be realized with the combination of red, green and blue (RGB) LED chips. However, RGB chips-based mini/micro-LED displays suffer from numerous problems preventing the technology from being embraced by major vendors. One main problem is that the traditional fabrication method, the mass transfer of RGB LED chips, suffers from low transfer efficiency and transfer yield as the pixel size scales down [2]. In addition, RGB LED chips-based mini/micro-LED displays also suffer from

luminous efficacy mismatch and drive voltage mismatch problem, which complicates which complicates the drive circuit design [3].

To overcome these problems, development of alternative quantum dot color conversion (QDCC) technology is underway. In QDCC technology, each blue mini/micro-LED chip pumps a subpixel in a patterned QDCC layer, and an absorptive color filter (CF) array is registered above to absorb unconverted blue light. A QDCC layer offers enhanced optical performance in terms of color gamut, power efficacy and response time [4]. And the fabrication of monochromatic blue LED backlight arrays by forming a mesa-typed LED array on GaN epi-wafers produces a much higher yield compared with mass transfer of RGB chips.

As the technology to fabricate monochromatic blue LED backlight arrays with down-scaled pixels has become more mature, achieving patterned QDCC layer with desirable converting lights has attracted increasing attention. QDCC layers patterned by jet printing [4] or photolithography [5] with controllable monochromatic blue LED backlight arrays are the two major techniques reported in the literatures for producing full-color mini/micro-LED displays. However, the two techniques still suffer from low optical performance in terms of color gamut owing to insufficient absorption of excitation backlight, and low production efficiency and stability are also challenges for mass production.

To improve the optical performance of patterned QDCC layers and simplify the fabrication procedure, our team conducted preliminary studies and introduce the laser patterning approach into the patterning of QD film [6]. As shown in Fig.1(a), a resulting sandwich structure of the laser-patterned QDCC with a matrix RGBB was produced. As we were able to design and fabricate a sufficiently thick QDCC layer, we achieved better blue light absorption and the optical characterization results showed better color gamut, at 128% sRGB standard, which is shown in Fig.1(b). The laser patterning approach also possesses the advantages of high production efficiency, robust fabrication and process scalability.

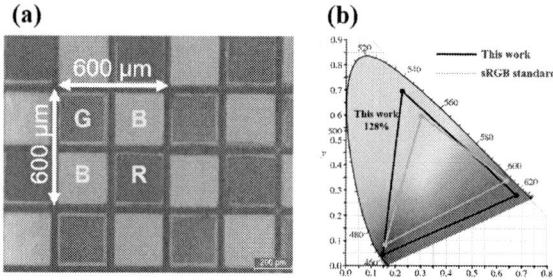

Fig.1. (a) Laser-patterned QDCC layer with pixel size 600x600 μm using RGBB matrix design. (b) Color gamut of excited QDCC layer compared with sRGB standard.

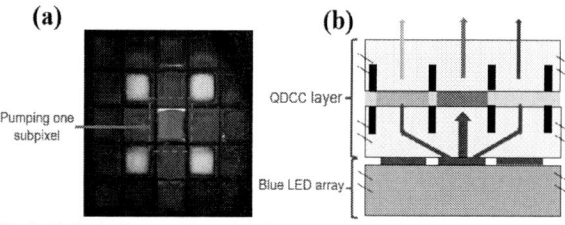

Fig.2. (a) Optical crosstalk can be observed under microscope when pumping QD subpixel in center. (b) Schematic diagram of light path results in optical crosstalk.

However, one major problem remains – the presence of optical crosstalk. Optical crosstalk is a common issue for QDCC technology, which is mainly associated with the optical structure, and it degrades the display resolution and worsens the color rendering. As shown in Fig.2(a), optical crosstalk is confirmed by pumping one QD subpixel under an optical microscope. Fig.2(b) indicates that the optical structure needs to be further improved in order to guide the light path. In this paper, we present an effective structure to confine the blue backlight paths by deep reactive ion etching (DRIE)-fabricated silicon light confining grid. Both optical simulation and visual observation are conducted in this study.

II. OPTICAL MODELLING AND SIMULATION

This part mainly focuses on optical modelling for studying and eliminating the optical crosstalk. In our preliminary study [6], a QDCC layer with pixel size 600x600 μm and subpixel size 250x250 μm was fabricated. Based on this design, we introduce a DRIE-fabricated silicon light confining grid to confine the light paths. Optimized hole pattern design of DRIE-fabricated silicon light confining grid will be introduced in this part.

A. Optical Crosstalk Modelling

To study the optical crosstalk effect of the current case, TracePro software (Lambda Research Corporation, Littleton, MA, USA), a commercial ray tracing program, was used to trace the simulated rays based on the basic laws of geometric optics [7]. As shown in Fig.3(a), a simulation model of the display with 9 subpixels was created. A sandwich-structured QDCC layer was created and placed on the top of the blue mini-LED array. In this QDCC layer, the thickness of top and bottom glass was set to 300 μm, while the depth of the trenches on each glass layer was set to 150 μm. The QD film was placed between two glass layers with the thickness of 25 μm. 9 blue LED chips was placed on a 300 μm thick sapphire

substrate with a pitch 300 μm. As shown in the Fig.3(b), the size of blue LED chips was set to 150x150 μm and the size of subpixels was set to 250x250 μm. The width of the trenches was set to 50 μm.

The emitting layer of the blue LED chip in the central subpixel was set as the top layer assuming it was a vertical structure blue LED chip. The peak wavelength of the blue LED chip was 450 nm and the angular distribution of light was set to be Lambertian. The QD film was set as a bulk scatter material with anisotropy factor 0.9 and scattering coefficient 4/cm, which aligns with the scattering media 1wt% SiO_2 with diameter 300 μm. The black matrix material in the trenches was set as perfect absorber.

To analyze the light distribution exiting from the QD film, the surface of irradiance map was set at top of the QD film, which is shown in Fig.3(c). The value of irradiance in the irradiance map was normalized to the peak irradiance.

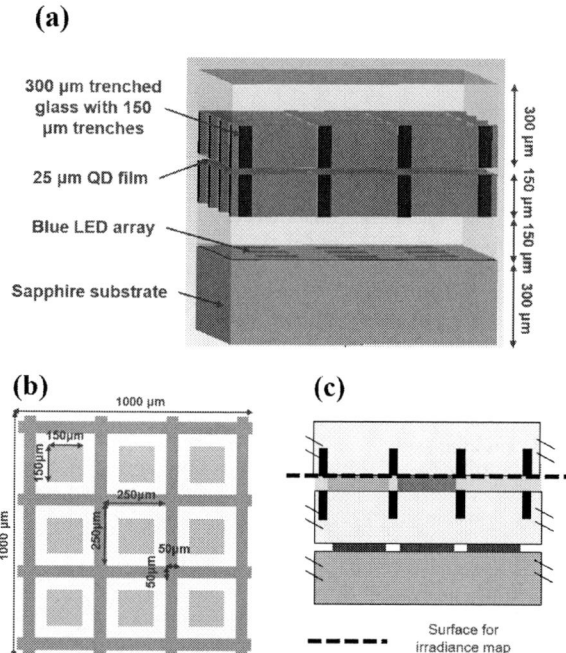

Fig.3. (a) TracePro model based on current design for optical crosstalk validation. (b) Top view of the simulation model (c) Surface for the irradiance map of the light exiting from QD film.

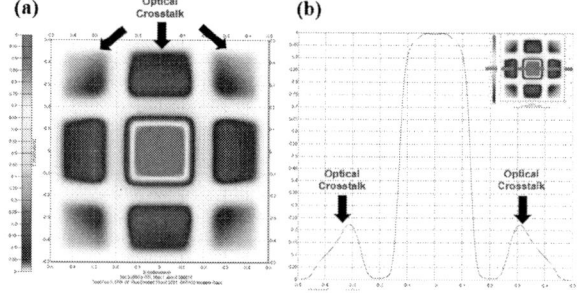

Fig.4. (a) Irradiance map of the light exiting from QD film (normalized to peak irradiance). (b) 2D-profile of the irradiance map along central axis (normalized to peak irradiance)

B. Validation of Optiacl Crosstalk Effect

The result of irradiance map is shown in Fig.4(a). From the result, the color of the area surrounding the central subpixel appears to be blue, which indicates there is severe

978-1-6654-1392-3/21 $31.00 © 2021 IEEE

optical crosstalk adjacent to the central subpixel. As shown in Fig.4(b), the 2D-profile shows that the irradiance of the optical crosstalk area reaches about 20% of the peak irradiance, which is unacceptable for display application.

In conclusion, the optical crosstalk of current design is validated by Tracepro simulation result, which indicates that it is of importance to reduce the optical crosstalk by the backlight confinement.

C. Backlight Confinement Modelling

To reduce the optical crosstalk, we proposed to use silicon light confining grid to achieve backlight confinement. As shown in Fig.5(a), a silicon light confining grid patterned by deep reactive-ion etching (DRIE) was proposed and placed between the QDCC layer and the LED array. Based on the schematic diagram, a 200 μm-thickness silicon light confining grid was added into the previous simulation model, and the new model is shown in Fig.5(b).

The pattern design of the through-holes array formed by DRIE was the key for backlight confinement. The targets for pattern design is to achieve no optical crosstalk as well as a sufficient light transmission through the holes. The hole size and the hole shape were the two key factors affecting above two targets. Therefore, we studied the square hole array with the hole dimension 50x50, 75x75, 100x100, 125x125 and 150x150 μm, and the circular hole array with hole diameter 50, 75, 100, 125 and 150 μm, which structures are shown in the Fig.5(c). To reduce the light reflection between light confining grid and sapphire substrate, down-side of the light confining grid was set as perfect absorber. The surface of irradiance map remained same.

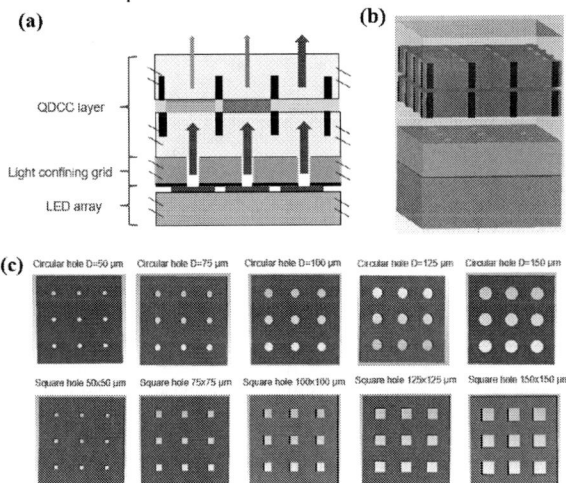

Fig.5. (a) Schematic diagram of light path confined by light confining grid. (b) Simulation model with light confining grid (c) light confining grid with different hole shapes and hole dimensions.

D. Validation of Optiacl Crosstalk Reduction

The results of optical crosstalk reduction using different light confining grid pattern design are shown in Fig.6. Compared to the case shown in Fig.4(a), it can be demonstrated that the light confining grid with through-hole array can effectively reduce the optical crosstalk. However, with the increase of hole size, the optical crosstalk effect tends to be more severe. Only circular hole array with hole diameter 50, 75 and 100 μm can achieve zero optical crosstalk (shown in red dashed box).

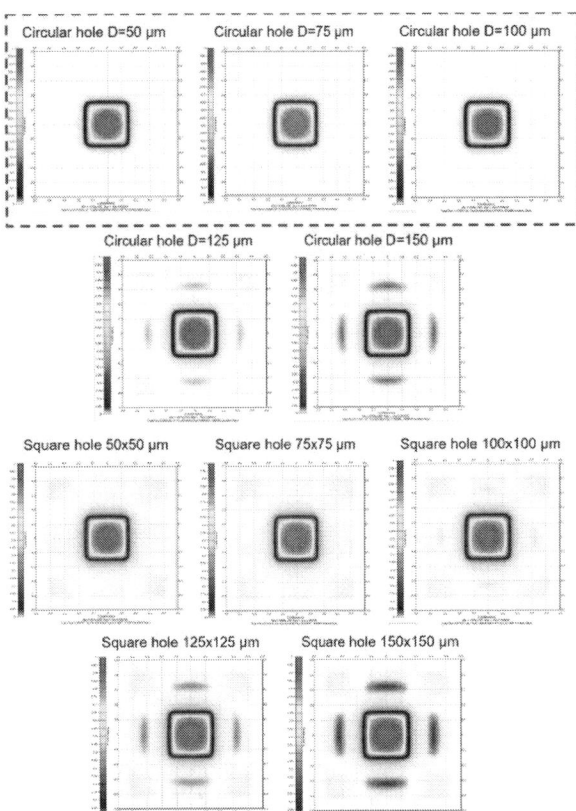

Fig.6. Irradiance map of the light exiting from QD film (normalized to peak irradiance) using different light confining grid pattern design.

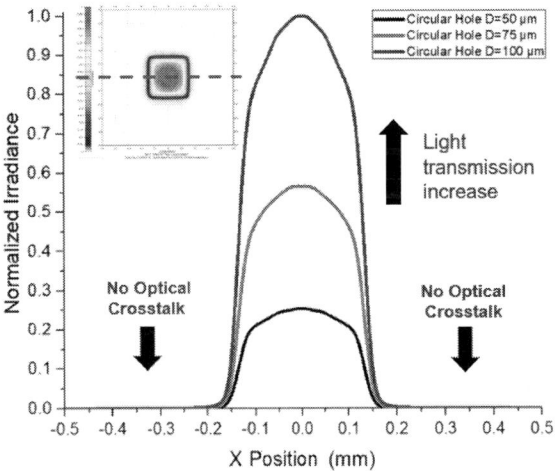

Fig.7. 2D-profile of the irradiance map along central axis (normalized to peak irradiance) using different hole array pattern. No optical crosstalk effect from the simulation result.

To figure out the most suitable pattern from above three patterns, we studied the 2D-profile of the irradiance map of above three patterns. As shown in Fig.7, it can be demonstrated that there is no optical crosstalk of above three patterns. Since another important target is to get sufficient light transmission through the holes, it can be easily figured out that the pattern of circular hole array with hole diameter 100 μm achieves largest light irradiance among three patterns.

In conclusion, by optical simulations we validate the effectiveness of silicon light confining grid on optical crosstalk reduction. The pattern of circular hole array with hole diameter 100 μm can achieve no optical crosstalk and high light transmission simultaneously.

III. OPTICAL REDUCTION BY DRIE-FABRICATED SILICON LIGHT CONFINING GRID

This part mainly focuses on fabrication of light confining grid by DRIE on silicon wafer, and studies the optical crosstalk effect under microscope observation.

A. Fabrication of Silicon Light Confining Grid

Based on the simulation result, the array of circular through-holes with hole diameter 100 μm and pitch size 300 μm was determined. As shown in Fig.8(a), DRIE patterning process was conducted on a 200 μm silicon wafer with designed pattern. After etching process, down-side of the silicon light confining grid was spin-coated by 1 μm black matrix to reduce the optical crosstalk, while the top-side of the plate was sputtered by 0.5 μm aluminum to increase the light extraction efficiency. Fig.8(b) shows the microscope images of the DRIE-fabricated light confining grid with hole diameter 100 μm and pitch size 300 μm.

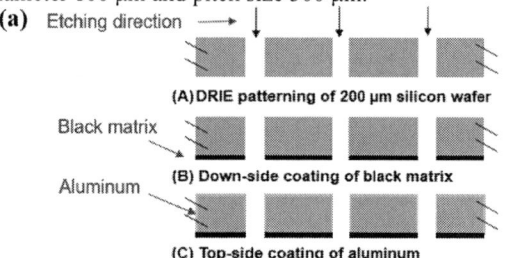

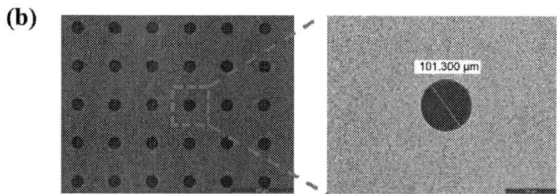

Fig.8. (a). Fabrication process flow of silicon light confining grid. (b) silicon light confining grid with circular hole array with hole diameter 100 μm and pitch size 300 μm

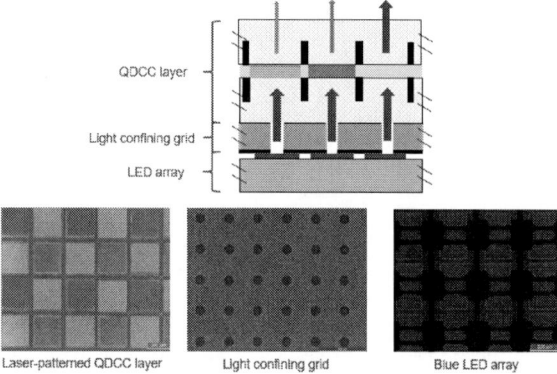

Fig.9. A display demo was assembled by laser-patterned QDCC layer, light confining grid and blue LED array.

B. Assembly of Display Demo

As shown in Fig.9, a display demo was assembled laser-patterned QDCC layer, light confining grid and blue LED array. The QDCC layer was fabricated by laser-patterning technology [6], while the blue backlight LED array, a passive-matrix addressing mini-LED array with pitch size 300 μm, was fabricated from a GaN epi-wafer using the standard process. The alignment of these three parts was achieved by a flip chip bonder (FineTech Co. Ltd), then sticked with each other by a UV-curable glue. Therefore, a display demo with pixel pitch 600x600 μm was fabricated for evaluation of optical crosstalk reduction.

C. Results on Optical Crosstalk Reduction

The results on optical crosstalk reduction by microscope observation are shown in Fig.10. From the comparison, it can be demonstrated that the DRIE-fabricated silicon light confining grid can effectively reduce the optical crosstalk. By applying the silicon light confining grid, it can be achieved a clear separation among neighboring subpixels.

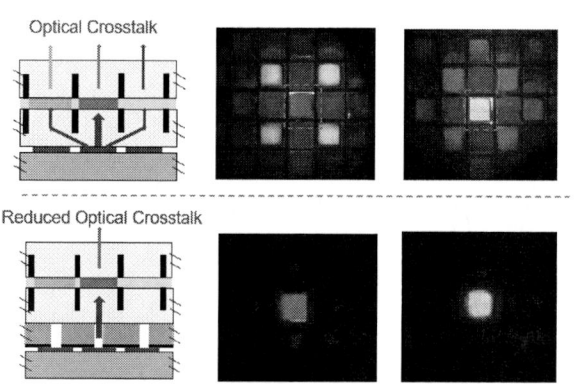

Fig.10. A comparison between device with optical crosstalk effect and device eliminates the optical crosstalk by DRIE-fabricated silicon light confining grid.

IV. CONCLUSIONS

This paper introduces an effective structure to confine the blue backlight paths by a deep reactive ion etching (DRIE)-fabricated silicon light confining grid. Circular through-holes array with hole diameter 100 μm and hole pith 300 μm is validated by both simulation and visual observation to be the best pattern for optical crosstalk reduction.

This study demonstrates that silicon light confining grid is a simple and cost-effective method to achieve low-crosstalk micro-display.

ACKNOWLEDGMENT

This research was funded by the Foshan Government through a grant to HKUST Foshan Research Institute for Smart Manufacturing.

REFERECES

[1] Wu, Tingzhu, et al. "Mini-LED and micro-LED: promising candidates for the next generation display technology." Applied Sciences 8.9 (2018): 1557.

[2] Virey, Eric H. and Nicolas Baron. "45‐1: Status and Prospects of microLED Displays," SID Symposium Digest of Technical Papers.

[3] Liu, Zhaojun, et al. "Micro-light-emitting diodes with quantum dots in display technology." Light: Science & Applications 9.1 (2020): 1-23.

[4] Han, Hau-Vei, et al. "Resonant-enhanced full-color emission of quantum-dot- based micro LED display technology." Optics express 23.25 (2015): 32504-32515.

[5] Zhang, Xu, et al. "23-5: Late-News Paper: High-Resolution Monolithic Micro-LED Full-color Micro-display." SID Symposium Digest of Technical Papers. Vol. 51. No. 1. 2020.

[6] Cheng, Yuanjie, et al. "Quantum Dot Film Patterning on a Trenched Glass Substrate for Defining Pixel Arrays of a Full-color Mini/Micro-LED Display." 2020 21st International Conference on Electronic Packaging Technology (ICEPT). IEEE, 2020.

[7] Qiu, Xing, et al. "UV LED Assisted Printing Platform for Fabrication of Micro-Scale Polymer Pillars." Journal of Microelectromechanical Systems 29.6 (2020): 1523-1530.

Study on a kind of thermal source chip for the performance analysis of micro channel heat sink: simulation and experimental validation

Ming Zhao*
Southwest China Research Institute of Electronic Equipment
Chengdu, China
zhaoming1990@163.com

Jian Zhang
Southwest China Research Institute of Electronic Equipment
Chengdu, China
mse_zhj@163.com

Qian Lu
Southwest China Research Institute of Electronic Equipment
Chengdu, China
1315734050@qq.com

Weiwei Xiang
Southwest China Research Institute of Electronic Equipment
Chengdu, China
xiangww2010@qq.com

Yangyang Li
Southwest China Research Institute of Electronic Equipment
Chengdu, China
liyang617730@163.com

Miaomiao Jiang
Southwest China Research Institute of Electronic Equipment
Chengdu, China
283415226@ qq.com

Huijie Ye
Southwest China Research Institute of Electronic Equipment
Chengdu, China
13006106279@163.com

Ting Peng
Southwest China Research Institute of Electronic Equipment
Chengdu, China
pengting2002@163.com

Abstract—Micro channel cooling technology is one kind of forced heat exchange cooling technology, and has good potential applications in high power electronic devices and microsystem integration field. However, the performance analysis is a technical issue for the application of micro channel heat sink. A thermal source chip for analyzing the performance of micro channel heat sink accurately and conveniently is presented in this paper. In order to analyze the performance of thermal source chip, the thermodynamics simulation is carried out by using finite element software. Then, the thermal source chip is fabricated, and the thermal measuring experiment is performed to verify the simulation results. The simulation results indicate that the surface temperature of thermal source chip increases with the applied voltage, and the maximum heat flux of thermal source chip appeared at the 80 V applied voltage. The experiment results show good agreement with the simulation.

Keywords—3D integration technology, Micro channel cooling technology, thermal source chip

I. INTRODUCTION

With the development of 3D integration technology toward miniaturization and high density, the heat flux of microelectronic devices has increased dramatically[1-3]. The problem of thermal failure has become one of the bottlenecks that restricts the development 3D integration technology. Micro channel cooling technology is one kind of forced heat exchange cooling technology, and has good potential applications in high power electronic devices and microsystem integration field[4,5]. By integrating micro channel heat sink into the electronic system, the efficiency of heat dissipation can be greatly improved. However, the performance of micro channel heat sink is affected by many factors, such as structural shape, distribution, cross section size, inlet and outlet position of micro channel heat sink. So, in order to analyze the performance of micro channel heat sink accurately and conveniently, it is necessary to design the

thermal analysis method that matching micro channel heat sink.

At present, the common analysis methods including thermal simulation analysis and infrared thermography technology[6-8]. For example，Feng proposed GPU-based thermal simulation methods for fast thermal analysis of 3D ICs with integrated liquid-cooled microchannels[6]. By using infrared thermography technology, Saad studied the thermal-hydraulic performance of sCO_2 for uniform single wall heat flux boundary condition in rectangular channels [8]. However, the thermal simulation analysis cannot really reflect the performance of micro channel heat sink, the further experiments should be performed. Infrared thermography technology is an effective method, and has good precision for the thermal analysis. However, the bulk of infrared thermography system is large, it is difficult to realize the integration with micro channel heat sink. For the performance study of micro channel heat sink, the micro temperature sensors such as the fluorescence sensor and the ultrasonic wave sensor have been developed[9,10]. However, these temperature sensors are difficult to integrate with thermal source, which limits the application to a certain extent. Thin film resistive metal heater not only can provide the heating, but also can measure the temperature, which realizes the monitoring and controlling of temperature simultaneously [11,12]. And it has been proven to be the most suitable choice for such localized heating applications within integrated microsystems.

Based on the thin film resistive metal heater, a kind of thermal source chip for analyzing the performance of micro channel heat sink accurately and conveniently is presented in this paper. In order to quickly analyze the performance of thermal source chip, the computer-aided analysis is performed. Finally, the thermal source chip is fabricated and the simulation results are verified by the experimental results.

978-1-6654-1392-3/21 $31.00 © 2021 IEEE

II. SIMULATION ANALYSIS

A. Geometry Model of Numerical Simulation

The geometry model of thermal source chip is established and is shown in Fig.1. It is consisted of silicon substrate, platinum metal film used to produce heat. The overall size of chip is 5×5×0.16 mm, the thickness of silicon substrate and the metal film are100 μm and 600 nm, respectively. The platinum metal film is composed of three parallel metal lines and two pads. The metal lines distribute in S shape, the line width of metal is 100μm, and the gap between two metal lines is 20 μm. The two pads are used to provide the working voltage of metal lines.

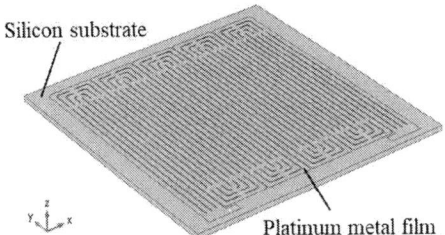

Fig. 1. The geometry model of thermal source chip.

B. Theoretical Model

The thermal source chip utilizes the principle of film resistance radiation, and it adjusts the heat flux by changing the working voltage. So the joule heating model is used to describe the working principle. For joule heating, the heat source Q_j of metal film comes from the electric current, it can be expressed as follows:

$$Q_j = J \cdot E \quad (1)$$

Where J and E is the current density and electric field strength, respectively.

According to the heat transfer in solids, the heat transfer in the metal film and silicon substrate can be described as[13]:

$$\rho C_p \frac{\partial T}{\partial t} - \nabla \cdot (k \nabla T) = Q \quad (2)$$

Where ρ is the density of material, C_p is the heat capacity, k is the thermal conductivity, T is temperature and Q is heat source.

The thermal source chip adjusts the heat flux by changing the working voltage, so in the model, one of the electrodes is set as 0V, and the other one is set as different electric potentials. In the simulation, the electric potentials are set as 20V, 40V, 60V and 80V, respectively.

In the process of thermal test, the thermal source chip is usually integrated on the heat sink. So there are two ways to cooled the thermal source chip, one is by the air on the metal film side, another is by the heat sink on the back side of chip. In the model, the boundary between metal film and air, and the boundary between silicon substrate and heat sink are defined by convective cooling. The heat flux of convective cooling is described by the equation:

$$-n \cdot (-k \nabla T) = h(T_{out} - T) \quad (3)$$

Where h is the heat transfer coefficient and T_{out} is temperature of external media. The main simulation parameters are shown in Table I.

TABLE I. SIMULATION PARAMETERS.

	Heat transfer coefficients of air	Heat transfer coefficients of fluid	Sheet resistance of platinum	Temperature of air	Temperature of fluid
value	5 W/(m²K)	20 W/(m²K)	0.3 Ω/□	23℃	23℃

The simulation is carried out by using finite element software. After calculation, the potential and temperature distribution of the chip can be obtained by the post processing.

C. Simulation Results and Discussion

The electrical performance has a direct effect on the performance of thermal source chip, so the electrical performance is investigated firstly. Fig.2(a) shows the simulation result of potential distribution on the surface of platinum metal film. The working voltage is set as 20 V. The result shows that from one end of applied voltage to the ground strap, the variation of potential gradient is continuous and uniform. Meanwhile, there are no strong current density concentration around the corner of the platinum metal film, which means that the designed thermal source chip has a good electrical performance. The simulation result of temperature distribution is shown in Fig.2(b). The result shows that the temperature of the chip is high in the middle and low on the edge. And the maximum temperature of the whole chip is only 0.8°C more than the minimum temperature. This means that the thermal source chip can provide uniform heating for the analyzing the performance of micro channel heat sink.

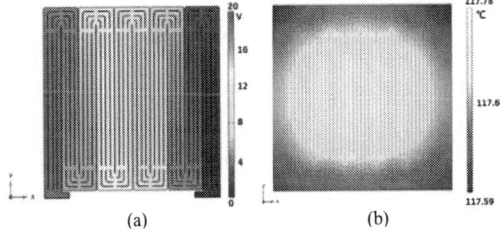

(a)   (b)

Fig. 2. Simulation results (a) potential distribution; (b) temperature distribution.

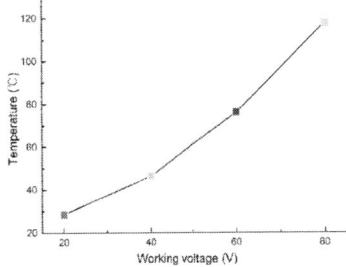

Fig. 3. Mean temperatures with different working voltages.

Besides, the thermal source chips are simulated with different working voltage. The working voltage are 40V, 60V, 80V, respectively. The results show that the temperature distributions are similar. But the temperature increases with the increase of working voltage. In order to quantitatively evaluate the performance of thermal source chips, the mean temperature of chip has been taken. Fig.3 shows the statistical results of mean temperature. The mean temperatures are 28.4℃, 46.3℃, 76.0℃ and 117.7℃ with the working voltage of 20 V, 40V, 60V, 80V, respectively. The variation tendency of mean temperature indicates that the heating power of chip increases with working voltage. And it means that the chip can realize the controlling of temperature continuously.

III. EXPERIMENTS

In order to verify the feasibility and practicality of thermal source chip, the thermal source chip was fabricated and the thermal measuring experiment is performed. To measure the temperature of the chip surface, a temperature sensor was designed and placed in the center of the chip, as shown in Fig.4. Using the four-wire resistance measurement, the temperature of the chip surface can be measured accurately.

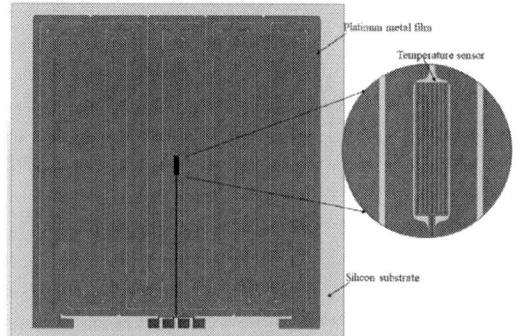

Fig.4. The designed thermal source chip.

A. Fabrication of Thermal Source Chip

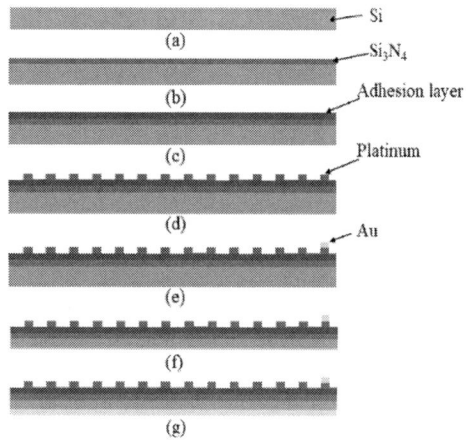

Fig.5. Fabrication steps of thermal source chip.

The fabrication steps of thermal source chip are schematically depicted in Fig.5. (a) The silicon wafer was thoroughly cleaned by using RCA standard cleaning method. After that, the silicon wafer was baked in drying oven. (b) As passivation layer, silicon nitride thin film grown on silicon wafer through PECVD technology. (c) the adhesion layer was sputtered on passivation layer by the sputtering system. (d)

Then, the platinum metal film grown on adhesion layer by CVD technology, and the parallel metal lines and six pads were fabricated by photolithography and etching technology. (e) the Au metal film was fabricated on the surface of six pads by using photolithography and thin film deposition technology. (f-g) finally, the thickness of silicon substrate was thinned from back by using etching technology and the Au metal film was fabricated on the back by film deposition technology. The fabricated thermal source chip is shown in Fig.6.

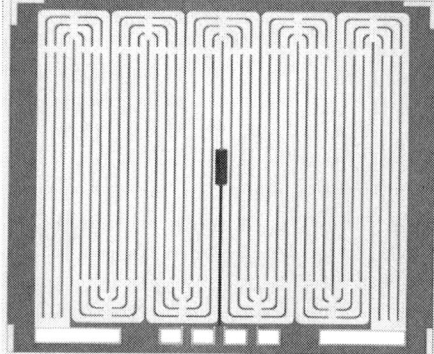

Fig.6. The fabricated thermal source chip.

B. Thermal Test

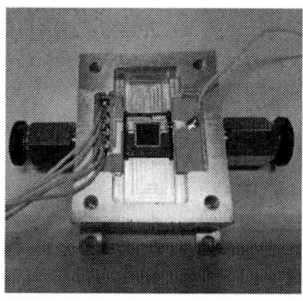

Fig.7. The testing sample.

The testing sample are shown in Fig.7. It is integrated by using micro-packaging technology and consisted of a thermal source chip, a micro channel heat sink, three thin film circuits, and a metal box. The thermal source chip provides different heat fluxes to the thermal test, and the micro channel heat sink is used to cool the thermal source chip. The thin film circuit provides the bonding pads to the electrical interconnection between thermal source chip and the outside testing circuit. Before the test experiment, the wire bonding was used to realize the electrical interconnection.

First, the TCR(temperature coefficient of resistance) of temperature sensor was measured by the testing experiments. In the experiments, the temperature sensor of the thermal source chip was connected to the ohmmeter, and put in the programmed oven. After that, different temperatures of 20℃, 30℃, 40℃, 50℃, 60℃, 70℃, 80℃, 90℃ were adjusted by the programmed oven, and the resistance values of the temperature sensor were recorded when the fluctuation of reading became less. Finally, the TCR of temperature sensor was calculated.

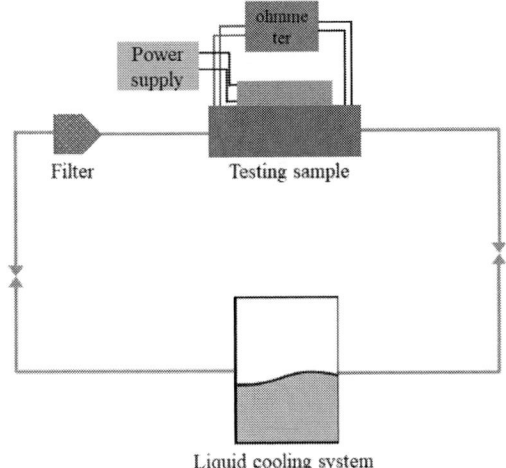

Fig.8. Schematic diagram of the equipment for thermal testing.

Fig.8 shows the schematic diagram of the equipment for thermal testing. It includes the testing sample, the liquid cooling system, the ohmmeter and the power supply. The temperature sensor and the thermal source chip were connected to the ohmmeter and power supply. And the pump of the liquid cooling system was also connected to a power supply. Before the measurement, the liquid cooling system should be opened firstly. The flow of the liquid cooling system during measurement was set 150mL/min. Then the different heat fluxes of thermal source chip were output by adjusting the working voltage, and the corresponding resistance values of the temperature sensor were recorded. The selected working voltage was 20V, 40V, 60V and 80V, respectively. Finally, the temperatures of the thermal source chip were calculated by the relation between the resistance values and the TCR.

C. Experimental Results and Discussion

TABLE II. TEST DATA OF TCR.

Temperature (°C)	Resistance (Ω)
20.4	713.34
29.7	731.32
39.6	754.02
50.2	775.58
60.2	797.31
70.0	818.26
80.3	840.05
90.1	860.86

The test data of TCR are shown in Table II. The resistance of platinum metal varies linearly with temperature, so the test data is linear fitted. And the slope of the fitted curve is the value of TCR. The fitted curve is shown in Fig.9, and the value of TCR is 2985.4ppm/°C.

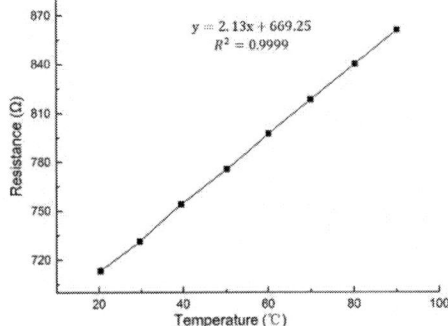

Fig.9. Fitted curve.

TABLE III. EXPERIMENTAL RESULTS OF THERMAL TESTING.

Voltage (V)	Heat flux (W/cm^2)	Temperature (°C)
0	0	20.0
20	70.4	26.1
40	267.2	45.2
60	561.6	74.3
80	924.8	112.1

The experimental results of thermal testing are shown in Table III. The experiment results show that under the applied voltage of 20 V, 40 V, 60 V, 80 V, the average heat fluxes of thermal source chip are 70.4 W/cm^2, 267.2 W/cm^2, 561.6 W/cm^2, 924.8 W/cm^2 respectively. And the center temperatures are 26.1°C, 45.2°C, 74.3°C and 112.1°C respectively. The results of the experiment show good agreement with the simulation. Compared with the experimental values, the minimum deviation of the simulated temperature is about 2.3%, and the maximum deviation is less than 8.8%. the deviation may come from the simplification of model. The thermal source chip provides a new option for analyzing the performance of micro channel heat sink accurately and conveniently.

IV. CONCLUSION

A thermal source chip for analyzing the performance of micro channel heat sink accurately and conveniently is presented in this paper. First, the computer-aided analysis is performed to analyze the performance of thermal source chip. Then, the thermal source chip is fabricated, and the experiment is performed. The simulation results indicate that the surface temperature of thermal source chip increases with the applied voltage, and the maximum heat flux of thermal source chip appeared at the 80 V applied voltage. The experiment results show good agreement with the simulation. And the maximum average heat flux is 924.8 W/cm^2. The thermal source chip realizes the monitoring and controlling of temperature simultaneously, which provides a new option for analyzing the performance of micro channel heat sink accurately and conveniently.

REFERENCES

[1] S. V. Garimella, A. S. Fleischer, J. Y. Murthy, A. Keshavarzi, R. Prasher, C. Patel, S. H. Bhavnani, R. Venkatasubramanian, R. Mahajan, Y. Joshi, B. Sammakia, B. A. Myers, L. Chorosinski, M. Baelmans, P. Sathyamurthy and P. E. Raad, "Thermal Challenges in

978-1-6654-1392-3/21 $31.00 © 2021 IEEE

Next-Generation Electronic Systems,"IEEE Trans. Compon. Packag. Technol, Vol.31, pp. 801-815, November 2008.

[2] D. Chaudhuri, D. N. Das, H. Rahaman and T. Ghosh,"Heat Mitigation in 3D ICs by Improvised TTSV Structure," 2020 International Symposium on Devices, Circuits and Systems, March 2020.

[3] Z. Wang, "Microsystems using three-dimensional integration and TSV technologies: Fundamentals and applications," Microelectron. Eng., vol. 210, pp 35-64, April 2019.

[4] R. V. Erp, R. Soleimanzadeh, L. Nela, G. Kampitsis and E. Matioli, "Co-designing electronics with microfluidics for more sustainable cooling," Nature, vol. 585, pp. 211-216, September 2020.

[5] S. Wang, Y. Yin, C. Hu and P. Rezai, "3D Integrated Circuit Cooling with Microfluidics," Micromachines, vol. 9, no. 6, pp. 287, June 2018.

[6] Z. Feng and P. Li, "Fast Thermal Analysis on GPU for 3D ICs With Integrated Microchannel Cooling," IEEE Transactions on Very Large Scale Integration (VLSI) Systems, vol. 21, no. 8, pp. 1526-1539, August 2013.

[7] B. Szymanik, T. Chady and K. Gorący, "Numerical modelling and experimental evaluation of the composites using active infrared thermography with forced cooling," Quant. InfR.Therm. J., vol. 17, no.2, pp. 107-129, April 2020.

[8] S. A. Jajja, B. M. Fronk, "Investigation of near-critical heat transfer in rectangular microchannels with single wall heating using infrared thermography," Int. J. Heat. Mass. Tran., vol 177, pp. 121470, October 2021.

[9] C. Hoera, S. Ohla, Z. Shu, E. Beckert, S. Nagl and D. Belder, "An integrated microfluidic chip enabling control and spatially resolved monitoring of temperature in micro flow reactors," Anal. Bioanal. Chem., vol.407, no. 2, pp. 387-396, November 2015.

[10] G. Yaralioglu, "Ultrasonic heating and temperature measurement in microfluidic channels," Sensor. Actuat. A-Phys., vol. 170, no. 1-2, pp. 1-7, November 2011.

[11] D. Resnik, J. Kovač, D. Vrtačnik, M. Godec, B. Pečar and M. Možek, "Microstructural and electrical properties of heat treated resistive Ti/Pt thin layers," Thin Solid Films, vol. 639, pp. 64-72, October 2017.

[12] R. M. Tiggelaar, R. G. P. Sanders, A. W. Groenland and J. G. E. Gardeniers, "Stability of thin platinum films implemented in high-temperature microdevices," Sensor. Actuat. A-Phys., vol. 152, no. 1, pp. 39-47, May 2009.

[13] H. D. Baehr, K. Stephan, Heat and Mass Transfer, second ed., Springer: Berlin, 2006, pp. 106-120.

Plane Position Measurement for μLED Based on Single Camera

Jie Bai
School of Mechanical Engineering
Tiangong University
Tianjin, People's Republic of China
baijie_tgu@163.com

Pingjuan Niu
School of Electrical and Electronics
Engineering
Tiangong University
Tianjin, People's Republic of China
niupingjuan@tiangong.edu.cn

Erdan Gu
School of Electrical and Electronics
Engineering
Tiangong University
Tianjin, People's Republic of China
erdan.gu@strath.ac.uk

Abstract—μLED, whose size is less than 100 μm, is the LED technology of miniaturization and matrixing.The prospect of μLED technology is widely used, such as micro sensors and display screen. But in actual application, it needs to be transferred to a large number of circuit substrates. This subject uses monocular vision to identify and locate μLED. The focus is on sub-pixel edge detection technology. Compared with traditional edge detection technology, sub-pixel edge detection technology can improve the accuracy by 2~10 times, and the accuracy is more higher. Hough transform, least squares and corner detection methods were compared in detail in straight line fitting. Weighted least squares were selected according to the characteristics of the μLED chip image. By image pre-processing, image segmentation, feature recognition, target positioning and other operations are performed on the circuit board and μLED chip pictures, respectively, to obtain micron-level position information. The magnification of the system can be determined by resolution board with a group of bars. The uncertainty of the results is analyzed, such as the tilt angle between camera and μLED, lens distortion and the edge of the image. The results show that the uncertainty is less than 5μm.

Keywords—μLED, sub-pixel detection, image

I. INTRODUCTION

μLED technology refers to the LED miniaturization and matrix technology, that is, a high-density micro size led array integrated on a chip, each μLED pixel size is generally less than 100 μm [1~3]. As shown in Fig. 1. But now the bottleneck technology is the transfer of micro led, it needs to be transferred to the circuit substrate.

Fig.1 The image of an array of μLED

Supported by Tianjin science and technology support major project (Grant No 18ZXCLGX00090).

A single vision is used to guide the transfer of micro led. Images can provide position information to detect and control small objects and mechanical operations [4~6].

II. THE MEASUREMENT SYSTEM

A. The measurement device

To highly precise measurement of position for μLED based on single camera, a measurement system was designed as shown in Fig. 2. To perform the proposed method, a microscope (MX20) with amplification factor 20 is used.

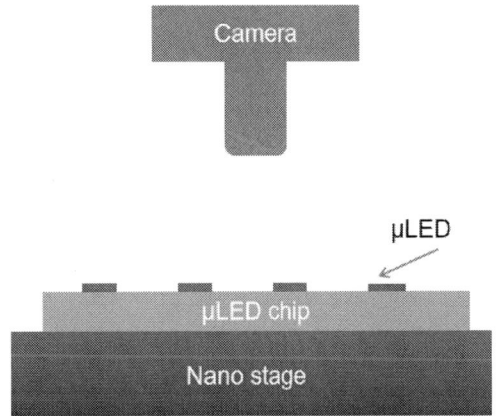

Fig.2. Schematic diagram of μLED position measurement system

B. Image processing

To get the position of μLED, the image should be processed before calibration and calculation. The image processing mainly include image enhancement, image transformation, image denoising, image filtering, image segmentation, image feature extraction, and object positioning, as shown in Fig.3.

C. The calibration of system

The magnification of the system can be determined by resolution board with a group of bars[7~8]. Canny algorithm is used to extract the image edge, and then fit the three groups of parallel lines. The distance of each group of parallel lines is d_n, the average of the three groups is d. The actual width of this group of parallel lines in the resolution board is D. The magnification of the system β, can be obtained:

$$\beta = \frac{d}{D} = 0.31 \text{ pixel/μm} \qquad (1)$$

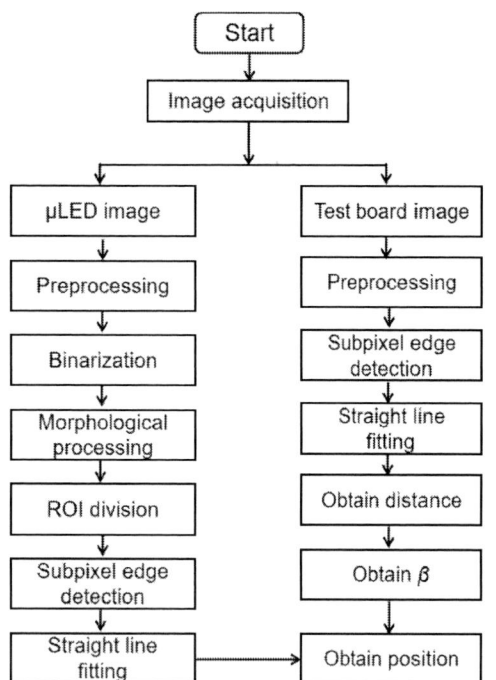

Fig.3. The flow chart of image process

III. MEASUREMENT RESULT AND UNCERTAINTY ANALYSIS

A. Measurement result

According to the image processing and calibration results. The results of image process is shown in Fig.4.

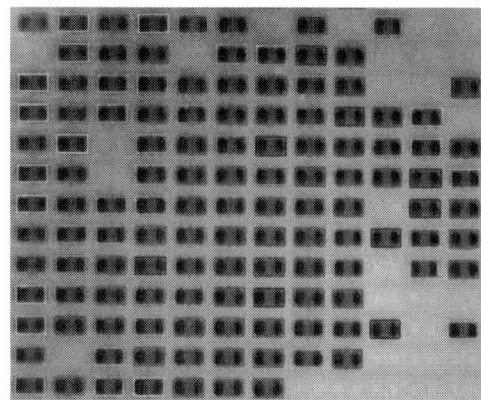

Fig.4. Image process results

Take one of measured data of μLED as an example, as shown in tab. I

TABLE I. IMAGE PROCESS RESULTS

	The Row coordinates of the center point	The column coordinates of the center point	The angle of the rectangle	The length of the rectangle	The width of the rectangle
Measured data/pixel	25.298	416.146	0.012	16.800	9.981
The actual data /μm	81.593	1342.195	0.012	54.185	32.193

B. Uncertainty sources

The tilt angle between camera and μLED can be adjusted by a rotation stage under μLED, with a resolution of 0.5°. In this system, the position of μLED and camera is fixed, it can be negligible.

Lens distortion is mainly caused by the aberration of the camera lens. It is mainly divided into pillow distortion and barrel distortion, which can be eliminated by some calibration methods. The uncertainty can be less than one pixel.

The resolution board used for calibration also will lead to an error, which is about 0.001mm.

Because the μLED chip is very thin and less than 2.5 μm in thickness, the edge of the μLED chip in the image is very thin, and the image is translucent, so the boundary will be error within 1 pixel.

The edge of the image is extracted with sub-pixel accuracy, the accuracy is very high, and the uncertainty can reach 0.1 pixels.

C. Mathematical Model

Under the condition that the relative position of the measuring object and the camera remains unchanged, the light and other conditions remain unchanged, the actual length of the μLED rectangle is

$$L = \frac{l}{\beta} \tag{2}$$

where L is the actual length of the μLED rectangle, l is the pixel number of μLED, and β is the magnification of the microscope camera.

Since the mathematical model only includes the form of product and quotient, it is expressed by relative uncertainty, and the measured relative combine uncertainty is

$$u_{c,rel}^2(L) = u_{rel}^2(l) + u_{rel}^2(D) + u_{rel}^2(d) \tag{3}$$

D. Uncertainty analysis

(a) LED fitting pixel length, including camera distortion error, shooting error due to its own nature, sub-edge detection error, shape fitting error.

The distortion error of the camera is not more than 1 pixel, and it is estimated from the normal distribution to get

$$u_1(l) = \frac{1}{2} = 0.5 \, pixel \tag{4}$$

The repeatability is about no more than 1 pixel, estimated with a normal distribution, and we get

$$u_2(l) = \frac{1}{2} = 0.5 \, pixel \tag{5}$$

The error of sub-edge detection is approximately no more than 0.1 pixels. Estimated with a normal distribution, we get

$$u_3(l) = \frac{0.1}{2} = 0.05 \, pixel \tag{6}$$

The shape fitting error is approximately no more than 0.5 pixels. Estimated with a normal distribution, we get

$$u_4(l) = \frac{0.5}{2} = 0.25 \, pixel \tag{7}$$

Combine these four uncertainty, the uncertainty of the fitted pixel length is

$$u(l) = \sqrt{u_1^2(l) + u_2^2(l) + u_3^2(l)} = 0.7517 \quad (8)$$

The relative uncertainty is

$$u_{rel}(l) = \frac{0.7517}{16.8001} = 4.474\% \quad (9)$$

(b) The actual width of the resolution board

$$u_{rel}(D) = \frac{1\mu m}{111\mu m} = 0.9009\% \quad (10)$$

(c) Measured width of resolution board

Three measurements were taken, which are

$$d_1 = 34.4119, d_2 = 34.4661, d_3 = 34.3682 \quad (11)$$

Since the number of measurements is less, the range method is used to calculate the standard deviation. The confidence interval C for three measurements is 1.69. The standard deviation is obtained.

$$\sigma = \frac{R}{C} = \frac{d_{\max} - d_{\min}}{C} = \frac{34.4661 - 34.3682}{1.69} = 0.05793 \quad (12)$$

The corresponding standard uncertainty is

$$u_1(d) = \frac{0.05793}{\sqrt{3}} = 0.03355 \quad (13)$$

The error of sub-edge detection is approximately no more than 0.1 pixels. Estimated with a normal distribution, we get

$$u_2(d) = \frac{0.1}{2} = 0.05\, pixel \quad (14)$$

Because the width of the bars in the resolution board is measured in the central part of the image, the camera distortion is very small and can be ignored. The bars in the resolution board are very clear and regular, the fitting error can also be ignored.

The combined uncertainty is

$$u(d) = \sqrt{u_1^2(d) + u_2^2(d)} = 0.06021 \quad (15)$$

The relative uncertainty is

$$u_{rel}(d) = \frac{0.06021}{34.4154} = 0.1750\% \quad (16)$$

E. Combined Uncertainty

The relative standard uncertainty is

$$u_{c,rel}(L) = \sqrt{u_{rel}^2(l) + u_{rel}^2(D) + u_{rel}^2(d)} = 4.567\%$$

The measurement result is 54.185μm，So the combined standard uncertainty is

$$u_c(L) = L \times u_{c,rel}(L) = 2.475\,\mu m$$

Take the inclusion extend k=2 directly, so the expanded uncertainty is

$$U(L) = k u_c(L) = 4.949\,\mu m$$

The length of the actual LED chip:

$$L = (54.185 \pm 4.949)\,\mu m$$

The expanded uncertainty is obtained by multiplying the standard uncertainty by the extend factor k=2.

Compared with traditional edge detection technology, sub-pixel edge detection technology can improve the accuracy by 2~10 times, and the accuracy is more higher.

IV. CONCLUSION

In this paper, the μLED chip is positioned by single camera. Future research requires miniaturization of the system so that the system can be applied in a manipulation system and the measurement process could be completed in parallel by FPGA. The proposed method will be used in the transfer printing of μLED.

ACKNOWLEDGMENT

The authors sincerely thanks to Professor Le Song of Tianjin University for his critical discussion and reading during manuscript preparation.

REFERENCES

[1] S.-I. Park, Y. Xiong, R.-H. Kim, P. Elvikis, M. Meitl, D.-H. Kim, J. Wu, J.Yoon, C.-J. Yu, Z. Liu, Y. Huang, K.-C. Hwang, P. Ferreira, X. Li, K.Choquette, J.A. Rogers. Printed assemblies of inorganic light-emitting diodes for deformable and semitransparent displays. Science, 325 (2009), pp. 977-981.

[2] M.A. Meitl, Z.T. Zhu, V. Kumar, K.J. Lee, X. Feng, Y.Y. Huang, I. A desida, R.G. Nuzzo, J.A. Rogers. Transfer printing by kinetic control of adhesion to an elastomeric stamp. Nat. Mater., 5 (2006), pp. 33-38

[3] M. Feng, M.A. Meitl, A.M. Bowen, Y.Huang, R.G.Nuzzo, J.A.Rogers. Competing fracture in kinetically controlled transfer printing. Langmuir, 23 (2007), pp. 12555-12560

[4] Trivedi M M , Mills J K. Centroid calculation of the blastomere from 3D Z-Stack image data of a 2-cell mouse embryo[J]. Biomedical Signal Processing and Control, 2020, 57:101726.

[5] Grassi, Ana Perez; Frolov, Vadim; Leon, Fernando Puente, Information fusion to detect and classify pedestrians using invariant features, 12(4): 284-292, 2011

[6] Farid M S, Mahmood A, Al-Maadeed S A. Multi-focus image fusion using Content Adaptive Blurring[J]. Information Fusion, 2018, 45:96-112.

[7] David, Folio, Antoine, et al. Two-Dimensional Robust Magnetic Resonance Navigation of a Ferromagnetic Microrobot Using Pareto Optimality[J]. IEEE Transactions on Robotics, 2017.

[8] Zheng Yelong, Song Le, Huang Jingxiong, Zhang Haoyang, Fang Fengzhou. Detection of the three-dimensional trajectory of an object based on a curved bionic compound eye. OPTICS LETTERS, 2019,44(17): 4143-4146.

978-1-6654-1392-3/21 $31.00 © 2021 IEEE

Cure Shrinkage Characterization and Warpage Simulation Optimization of Epoxy Molding Compound for 5G Application

Kai Chen
Center for Engineering Materials and Reliability
Guangzhou HKUST Fok Ying Tung Research Institute
Guangzhou, China
chenkai@ust.hk

Dashun Liu
Center for Engineering Materials and Reliability
Guangzhou HKUST Fok Ying Tung Research Institute
Guangzhou, China
dsliu@ust.hk

Yijing Qin
Center for Engineering Materials and Reliability
Guangzhou HKUST Fok Ying Tung Research Institute
Guangzhou, China
yjqin@ust.hk

Yong Zhong
Center for Engineering Materials and Reliability
Guangzhou HKUST Fok Ying Tung Research Institute
Guangzhou, China
yongzhong@ust.hk

Zhaorong Wan
Center for Engineering Materials and Reliability
Guangzhou HKUST Fok Ying Tung Research Institute
Guangzhou, China
zrwan@ust.hk

Richeng Liu
Center for Engineering Materials and Reliability
Guangzhou HKUST Fok Ying Tung Research Institute
Guangzhou, China
rcliu@ust.hk

Dong Lu
Center for Engineering Materials and Reliability
Guangzhou HKUST Fok Ying Tung Research Institute
Guangzhou, China
maeld@ust.hk

Ke Xue
School of System Design and Intelligent Manufacturing
Southern University of Science and Technology
Shenzhen, China
xuek@sustech.edu.cn

Jingshen Wu
School of System Design and Intelligent Manufacturing
Southern University of Science and Technology
Shenzhen, China
wujingshen@sustech.edu.cn

Abstract—With the development of electronic packaging technology in 5G era, System-in-Package (SiP) module is developing into smaller size and higher integration to realize the multi-functional applications, and the reliability of SiP module, especially the interfacial delamination has become one of the most concerned issues. To predict interfacial delamination precisely through modeling and simulation methodologies, accurate warpage of SiP module should be calculated accordingly since warpage deformation is one of the main reasons of interface delamination. For plastic packaged devices, like most SiP modules, coefficient of thermal expansion (CTE) mismatch between epoxy molding compound (EMC) and its adjacent materials will lead to warpage during the process of transfer molding and post-mold cure of EMC. In addition to this, another important reason of warpage is the chemical shrinkage of EMC during curing process. In this work, a method of characterizing cure shrinkage of EMC was proposed by using bi-material test method. The cure shrinkage rate of EMC was calculated combining the simulation methods and the warpage test results, which was applied in optimizing the SiP module warpage simulation during reflow process. Furthermore, the effect of viscoelasticity of EMC on SiP module warpage was also investigated in this work.

Keywords—5G; Epoxy molding compound; Cure shrinkage; Warpage simulation; Viscoelasticity

I. INTRODUCTION

With the development of 5G technology, smart electronic devices represented by mobile phones have higher requirements for chip performance and power consumption, which accelerated the development of advanced packaging technology, including System-in-Package (SiP) packaging technology. Nowadays SiP packaging technology has become one of the most important technologies in the semiconductor industry during 5G era. However, the combination of both miniaturization and function integration trends drives microelectronics technology into an unknown level of complexity, characterized by multi-material/interface, multi-damage and multi-failure mode. As a consequence, the reliability analysis of SiP to guarantee the yield and service life is crucial.

Thermo-mechanical reliability of microelectronics devices is one of the major concerns in the industry, it is found that reliability problems are often triggered by various thermal and mechanical loadings associated with manufacturing processes. SiP products will appear warpage during the packaging process, especially during the molding and post-mold cure (PMC) process. In addition to the mismatch of the thermal expansion coefficient of epoxy molding compound (EMC) and the substrate, cure shrinkage of the EMC is also an important factor to cause warpage, which is always ignored.

The volume shrinkage of EMC after molding process contains cooling shrinkage and cure shrinkage (thermochemical shrinkage). Cooling shrinkage is caused by the drop in temperature during the cooling process, while cure shrinkage is the chemical shrinkage caused by the cross-linking and the formation of the polymer networks under high curing temperature during the molding and post-mold cure process.

As shown in previous studies, the effect of cure shrinkage on initial warpage of the package is obvious, and various methods for characterizing cure shrinkage rate have been proposed [1-5]. In this paper, warpage test of bi-material specimens using Shadow Moiré combined with finite element analysis is employed to obtain the cure shrinkage rate of EMC.

In addition, the corresponding warpage simulation method is applied to SiP products during reflow process and the results correspond well to the actual situation.

II. MODELING AND TESTING METHOD

Bi-material test is chosen as the method of characterizing cure shrinkage of EMC in this paper. Two types of commercial EMC named EMC-A and EMC-B are investigated. Cropped prepreg sheets are used as the substrate in bi-material specimens instead of copper sheets due to the internal stress of copper sheets during processing which may affect the warpage of bi-material specimens and lead to inaccurate cure shrinkage calculation of EMC. Using transfer molding machine, bi-material specimens are prepared by molding the EMC on the substrate at 175°C for 200 seconds, in which the in-mold pressure is 7 MPa, and then baking them at 175°C for 4 hours for post-mold cure.

After the specimens are cooled to room temperature, warpage are measured using shadow moiré, with the TherMoiréPS200, as shown in Fig. 1.

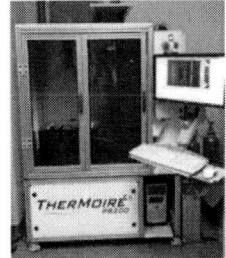

Fig. 1. Shadow Moiré warpage measurement equipment.

To improve the accuracy of the test, all bi-material specimens are sprayed with a thin layer of high-temperature paint on the surface to form a good diffuse reflection. In addition, specimens will undergo additional baking at 125°C for 4 hours after sprayed, the purpose of which is to dry the specimens and the surface paint to prevent the sample from generating water vapor at high temperature (260°C), or part of the solvent that may volatilize at high temperatures and adhere to the surface of the moiré glass, ultimately reducing the test accuracy. The test process and data processing of bi-material specimens are shown in Fig. 2.

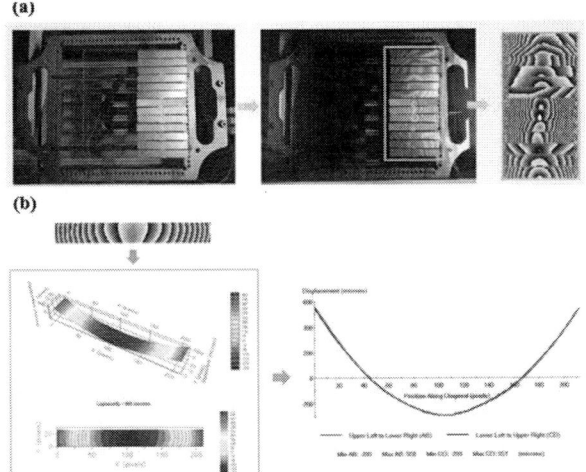

Fig. 2. Process of the bi-material warpage test. (a) Test process; (b) Data processing.

The temperature loading in shadow moiré is exhibited in Fig. 3, the temperature rises from 34°C to 260°C at a rate of 5°C/min, then drops to room temperature at a rate of about 2°C/min. In this process, the warpage of the specimen will be measured once at an interval of 20°C, Fig. 4 illustrates the typical warpage results of EMC-A/Prepreg at room temperature.

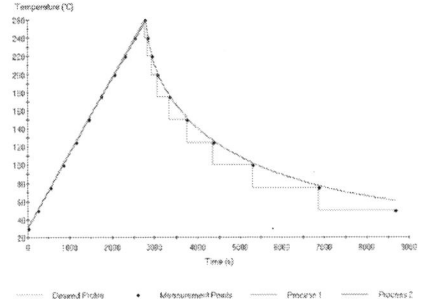

Fig. 3. Temperature loading in shadow moiré.

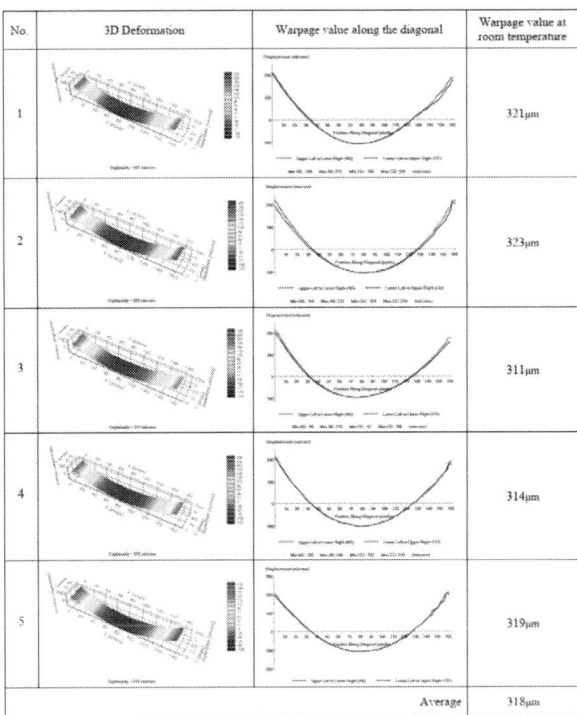

No.	3D Deformation	Warpage value along the diagonal	Warpage value at room temperature
1			321μm
2			323μm
3			311μm
4			314μm
5			319μm
		Average	318μm

Fig. 4. Warpage measurement results of EMC-A/Prepreg specimens at room temperature.

III. SIMULATION AND VIRIFICATION

EMC occurs cure shrinkage during the molding and post-curing process, which will induce shrinkage strain and correspondingly affect the initial warpage of the bi-material specimen and realistic SiP product. To accurately simulate the initial warpage after post-curing, the cure shrinkage of EMC needs to be considered and represented in a way.

In this paper, equivalent CTE of cure shrinkage is proposed to represent the degree of cure shrinkage of different EMC and easily consider and calculate the shrinkage strain in commercial simulation software, and the temperature range is defined from 175°C to 174.9°C (175°C is the curing temperature and chosen as the reference temperature). In this way, the shrinkage strain caused by cure shrinkage during molding and post-curing process is substituted by additional thermal strain caused by thermal expansion from 175°C to 174.9°C.

To obtain the value of equivalent CTE of cure shrinkage, the finite element model of bi-material specimen with the same structure size is established. Cure shrinkage and cooling process from 175°C to 25°C are considered, in which the superposition of shrinkage strain and thermal strain causes warpage, as shown in Fig. 5. By changing the value of equivalent CTE of cure shrinkage in the model and compare the warpage results between shadow moiré test at room temperature and simulation, the finally determined value of EMC-A and EMC-B is 1000 ppm/K and 7650 ppm/K respectively. In addition, cure shrinkage rate can be calculated and the value is 0.03% and 0.23% correspondingly.

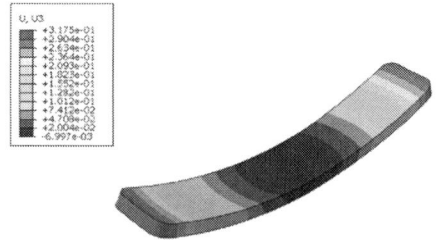

Fig. 5. Deformation contour plot of EMC-A/Prepreg specimens at room temperature.

Fig. 6 exhibits the measurement and simulation results of warpage of bi-material specimens during reflow process, in which red dotted line represents the simulation results. As can be seen from the Fig, the measurement results correspond well to the simulation results considering cure shrinkage. One thing should be noted is that there exist larger deviations between the simulation and measured results in the cooling stage compared to the heating stage (obvious in EMC-B), this is mainly due to the difference of measured CTE between heating and cooling process, while only the CTE measured during heating process is used in simulation model, and hence cause the deviations in cooling stage.

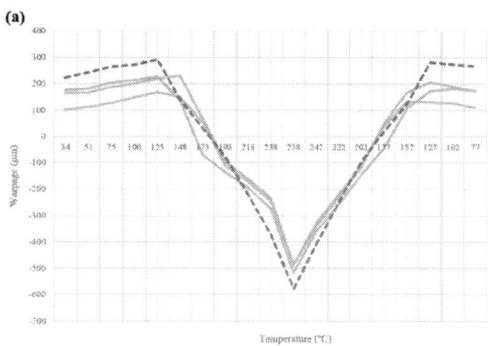

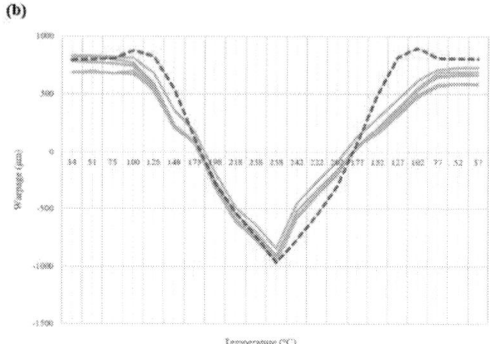

Fig. 6. Comparison of simulation and measurement results. (a) EMC-A/Prepreg; (b) EMC-B/Prepreg.

Viscoelasticity of EMC will cause stress relaxation and affect the warpage of its plastic packaging structure. To analyze the influence of the viscoelasticity of EMC-A and EMC-B on structural warpage, bi-material model, simplified SiP model and complete SiP model are established respectively to compare with the models using linear elastic constitutive, as shown in Fig. 7.

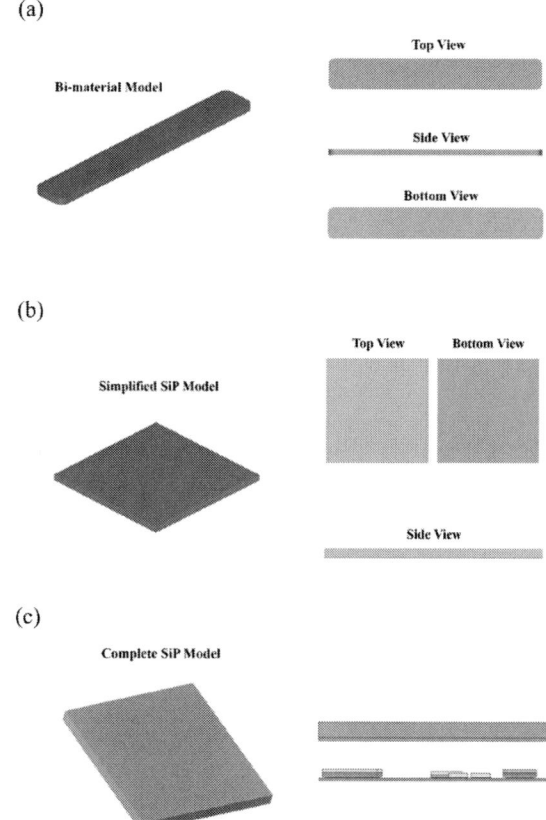

Fig. 7. Simulation models. (a) Bi-material model; (b) Simplified SiP model; (c) Complete SiP model.

Fig. 8 exhibits the comparison results of three simulation models using viscoelasticity and linear elasticity, and it shows that EMC adopts linear elastic or viscoelastic constitutive to have little effect on simulation results of warpage. In subsequent analysis, EMC can directly use linear elastic material parameters for warpage simulation.

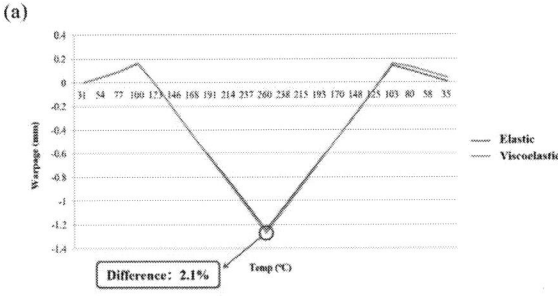

(a)

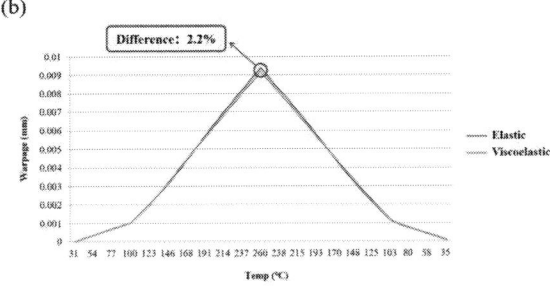

(b)

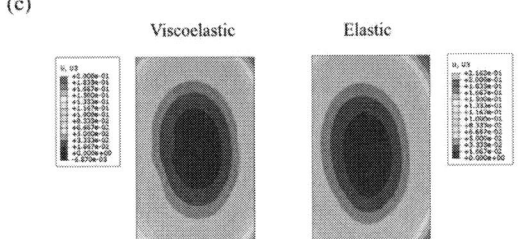

(c)

Viscoelastic   Elastic

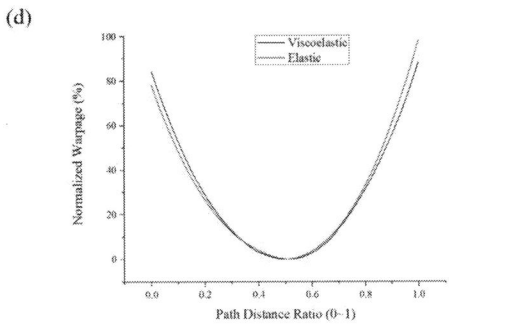

(d)

Fig. 8. Comparison of simulation results using viscoelasticity and linear elasticity. (a) Bi-material model; (b) Simplified SiP model; (c) Deformation contour plot of complete SiP model at 260℃; (d) Warpage on the diagonal of the model.

To verify the accuracy of equivalent CTE of cure shrinkage and simulation method of warpage, the determined equivalent CTE of cure shrinkage of EMC-A and EMC-B and the corresponding simulation method are applied to actual SiP products. The molded and post-cured SiP specimen is measured for the warpage during the reflow process, the moiré on the surface at the beginning of the test is shown in Fig. 9.

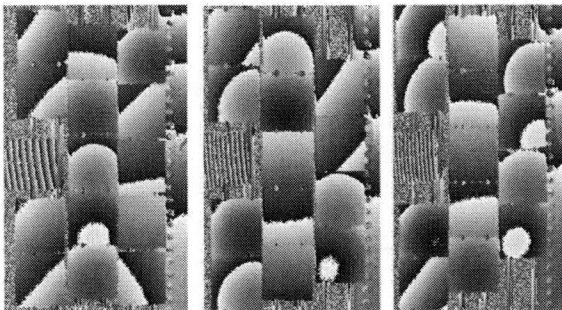

Fig. 9. Surface moiré of SiP specimen before testing.

Part of the warpage test and simulation results are shown in Fig. 10, and the accuracy of simulation is given in Table 1. It can be seen from the results that the accuracy of the simulation is considerable, which verifies the reliability of the equivalent CTE of cure shrinkage of EMC and the warpage simulation method for SiP.

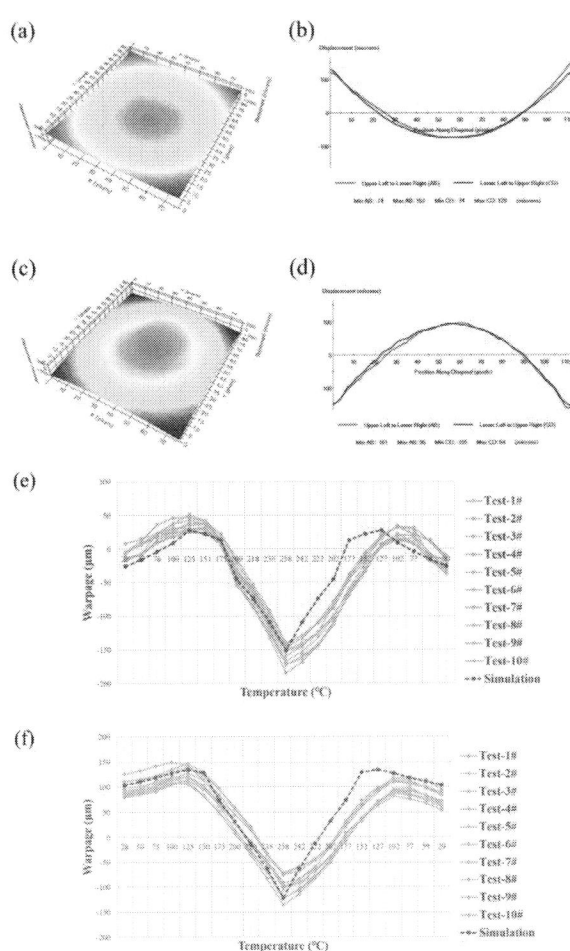

Fig. 10. Warpage test and simulation results of SiP. (a) Deformation contour plot of warpage test at 34℃; (b) Warpage results on the diagonal at 34℃; (c) Deformation contour plot of warpage test at 259℃; (d) Warpage results on the diagonal at 259℃; (e) Comparison of simulation and measurement results (SiP with EMC-A); (f) Comparison of simulation and measurement results (SiP with EMC-B).

TABLE I. ACCURACY OF THE SIMULATION

Temperature	EMC-A		EMC-B	
	123°C	*258°C*	*100°C*	*258°C*
Measured Results (*μm*)	73	-101	250	-249
Simulaion Results (*μm*)	80	-97	295	-213
Accuracy	90.4%	96.0%	82.0%	85.5%

IV. CONCLUSIONS

Equivalent CTE of cure shrinkage is proposed and used to simulate the initial warpage. Based on the bi-material warpage test, values of the equivalent CTE and the cure shrinkage rate of EMC can be quantitatively characterized, and the corresponding warpage simulation method can be obtained. In addition, the determined EMC cure shrinkage simulation parameters and warpage simulation method are applied to SiP products in the reflow process, and the simulation results show great agreement with the measured results, which verifies that this method can effectively simulate the effect of cure shrinkage on warpage.

ACKNOWLEDGMENT

The authors would like to sincerely acknowledge the financial support from the Department of Science and Technology of Guangdong Province (Project No. 2020B010179002); the Guangzhou Science, Technology and Innovation Commission (Project No. 201904010279) and the 2018 Guangzhou International Postdoctoral Exchange Fellowship Program.

REFERENCES

[1] Emst, L. J., et al. "Fully Cure-Dependent Modeling and Characterization of EMC's with Application to Package Warpage Simulation." 2006 7th International Conference on Electronic Packaging Technology. IEEE, 2006.

[2] De Vreugd, J., et al. "Prediction of cure induced warpage of micro-electronic products." Microelectronics Reliability 50.7 (2010): 910-916.

[3] Chiu, Tz-Cheng, Hong-Wei Huang, and Yi-Shao Lai. "Warpage evolution of overmolded ball grid array package during post-mold curing thermal process." Microelectronics Reliability 51.12 (2011): 2263-2273.

[4] Hu, G., Luan, J., and Chew, S. (February 13, 2009). "Characterization of Chemical Cure Shrinkage of Epoxy Molding Compound With Application to Warpage Analysis." ASME. J. Electron. Packag. March 2009; 131(1): 011010.

[5] W. H. Zhu et al., "Cure shrinkage characterization and its implementation into correlation of warpage between simulation and measurement," 2007 International Conference on Thermal, Mechanical and Multi-Physics Simulation Experiments in Microelectronics and Micro-Systems. EuroSime 2007, 2007, pp. 1-8.

Failure Mechanism Study for Flip Chip QFN Crack Issue under Temperature Cycling Test

Ke Xue
School of System Design and Intelligent Manufacturing
Southern University of Science and Technology
Shenzhen, China
xuek@sustech.edu.cn

Dong Lu
Center for Engineering Materials and Reliability
Guangzhou HKUST Fok Ying Tung Research Institute
Guangzhou, China
maeld@ust.hk

Kai Chen
Center for Engineering Materials and Reliability
Guangzhou HKUST Fok Ying Tung Research Institute
Guangzhou, China
chenkai@ust.hk

Yijing Qin
Center for Engineering Materials and Reliability
Guangzhou HKUST Fok Ying Tung Research Institute
Guangzhou, China
yjqin@ust.hk

Dayang Li
Foundry Department
Chengdu Monolithic Power Systems Co., Ltd.,
Chengdu, China
Dylan.Li@monolithicpower.com

Qi Tang
Foundry Department
Chengdu Monolithic Power Systems Co., Ltd.,
Chengdu, China
Tommy.Tang@monolithicpower.com

Abstract—In this work, finite element analysis and failure analysis are performed to analyze the crack failure in flip chip quad flat no-lead (Flip Chip QFN). Although a QFN is a traditional package, the reliability requirements of Flip Chip QFN in power applications have become strict and more and more challenging to meet. This paper elaborates the thermo-mechanical reliability challenges of Flip Chip QFN associated with its unique package construction features, and investigate the failure mechanisms of package cracking issue found in thermal cycling test. Parametric analysis is conducted to determine the optimal design in the Flip Chip QFN to combat crack failure. This investigation can help improve the product reliability by preventing crack failures in the Flip Chip QFN package.

Keywords—Flip Chip QFN, Thermal cycling test, Thermo-mechanical reliability, Die crack

I. INTRODUCTION

Advanced FC (Flip Chip) QFN package integrates fine pitch Cu pillar BOL (Bump on Lead) interconnect in a QFN (Quad Flat No-leads) body, which can enhance both electrical and thermal efficiency of the QFN package [1]. Combing with other advantages such as shorter assembly cycle time and small chip-to-package ratio closest to WCSP in terms of package footprint, flip chip QFN package is becoming prevalent and well-received in market with applications on power management [2]. But as the increasing demand for power load capacity and performance in the market, the chip size will inevitably increase. It will trigger challenging thermo-mechanical reliability issues like catastrophic crack failure [3].

In this study, a Flip Chip QFN 5x5-30 package (shown in Figure 1) with high chip-to-package ratio (>86%) was observed electrical failure during ATE test after temperature cycling and associated with passivation crack and even serious die crack. Some samples were found external mold crack after closer visual inspection. But not every mold crack caused electrical failure. The failure position was relatively fixed at nearby a certain I/O pad. Cross-section, SEM analysis and thermo-mechanical simulation were done to investigate the failure mechanism. Delamination between side wall of leadframe pad and molding compound was noticed at T0 fresh units, and simulation results confirmed a high stress concentration around the suspect I/O pad. The T0 delamination between leadframe and molding compound would propagate as the temperature cycle increase, until it fully cracked the interface between Cu pillar and molding compound. This would trigger severe thermo-mechanical stress on the BEOL layer under the Cu pillar bump which caused passivation crack and even vertical die crack. If the adhesion between pad top surface and molding compound was enhanced by leadframe roughening treatment, the initial delamination would only propagate into mold body rather than the interface between Cu pillar and molding compound, which could lower but cannot eliminate the die crack risk.

Simulation analysis also revealed that this accumulated die stress cannot be reduced by adding polyimide (PI) layer and it was proven by experiments that adding PI layer cannot resolve the crack issue. Simulation parametric study showed that with thinner die and thinner mold body could reduce the thermo-mechanical stress and mitigate the initial delamination propagation risk. Besides, since interface delamination is the key problem in the production process when cutting encapsulation QFN packages[4], sawing process optimization was also advised in order to eliminate the T0 delamination between leadframe and molding compound. Finally, the proposed solutions were verified by TCT experiments.

Fig. 1. Schametic View of Flip Chip QFN Package

978-1-6654-1392-3/21 $31.00 © 2021 IEEE

II. TESTING & EXPERIMENTATION

A. Testing Samples

A Flip Chip QFN 30LD 5mm x 5mm package was selected as the test vehicle. A large die with 4.6mm x 4.6mm x 0.2mm in size is connected to 0.2mm lead frame by total 72 copper pillar bumps in this package. The bumps are 120μm in diameter and 65μm in height, underneath of which is 10μm thickness RDL layer (refer to Figure 2).

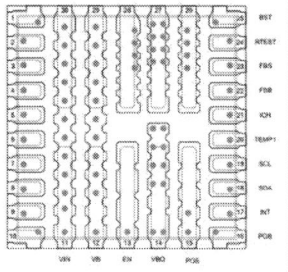 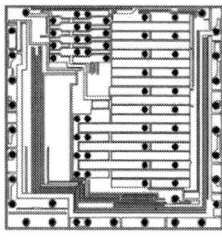

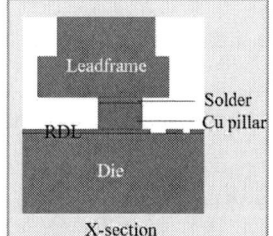

Fig. 2. Package and Die Layout of Flip Chip QFN 5x5 Package

B. Testing Evaluation Process

Reliability stress assessment was performed based on the air to air temperature cycling (TC) at -65°C to 150°C that includes pre-conditioning (PC) at moisture sensitivity level 3 (MSL3) with 3x reflow at 260°C (peak level temperature). From T0 and succeeding stress read points after PC, 240 500 750 and 1000 cycles electrical testing was examined.

C. Testing Results

Electrical failures were found after 240 cycles and the failure rates increased as temperature cycling accumulated (Table I). Physical decap indicated that die or passivation crack was the dominant failure mode, while no such failures were observed for T0 and after preconditioning samples (in Figure 3).

TABLE I. THERMAL CYCLING TESTING DATA FOR ORIGINAL DESIGN

	TC 240	TC 500	TC 750	TC 1000
Lot #1	1/85	3/81	/	18/77
Lot #2	1/85	2/81	9/78	21/69

Due to the high die-to-package ratio, silicon die would accumulate high thermal-mechanical stress during temperature cycling. One of the solutions to reduce die stress is to add a polyimide (PI) layer on the surface of the chip. It is generally believed that the PI film can play a buffer role, thereby reducing the accumulation of thermal stress caused by the CTE (coefficient of thermal expansion) mismatch

between EMC and silicon die [5]. So the original package design was modified by adding a PI layer above RDL on top surface of silicon die.

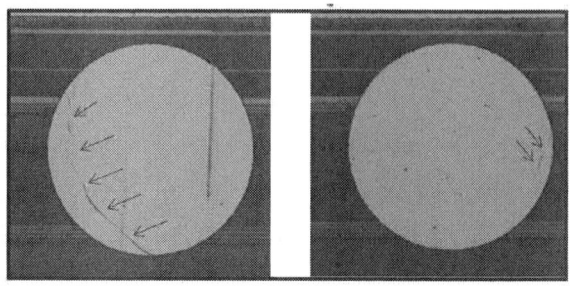

a) Passivation Crack

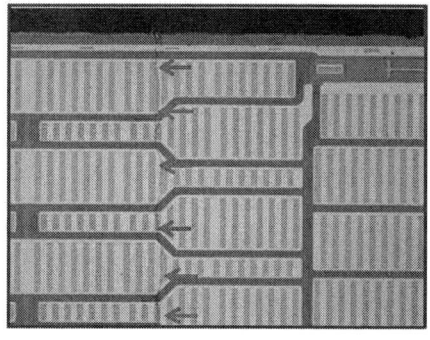

b) Die Crack

Fig. 3. Decap Verfication Results

TABLE II. THERMAL CYCLING TESTING DATA COMPARISON BETWEEIN ORIGINAL AND MODEIFIED DESIGNS

		TC 240	TC 500	TC 750	TC 1000
Original design	Lot #1	1/85	3/81	/	18/77
	Lot #2	1/85	2/81	9/78	21/69
+PI	Lot #3	4/85	2/81	5/79	/
	Lot #4	5/85	7/79	/	/

Thermal cycling test results revealed that no obvious improvement on TC failure rate by adding PI layer (in Table II), die crack still prevailed in failed samples and even mold crack of some failed samples was found by a detailing failure analysis. Failures mainly occurred near SW/VBO pins (in Figure 4).

 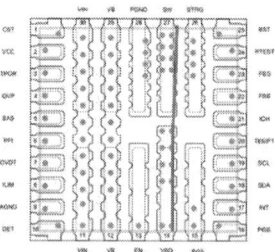

a) Die Crack

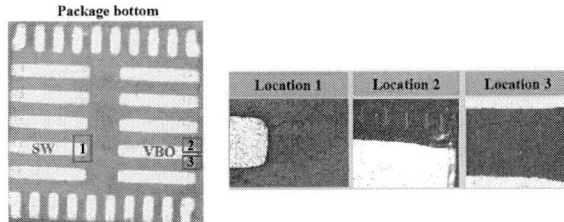

b) Package Mold Crack

Fig. 4. Failure Locations

To further understand the failure mechanisms, cross-section (or X-section) analysis for both failure samples and T0 fresh samples were performed. X-section results showed that the crack was originated from leadframe-mold interface, and propagated along the side surface of Cu pillar. The passivation layer underneath this Cu pillar was found crack as well, probably due to severe thermo-mechanical stress accumulated during TC test (in Figure 5).

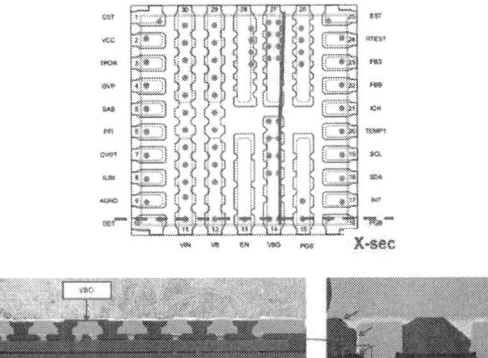

Fig. 5. X-section Photos for TC Failed Sample

Regarding T0 fresh samples, delamination between leadframe and molding compound was noticed and 11/22 units were suspected after SAM C-Scan. X-section confirmed delamination happened on sidewall of SW/VBO pins and no obvious delamination on other pins, this delamination was probably triggered by sawing singulation of molded strip into single package units (in Figure 6).

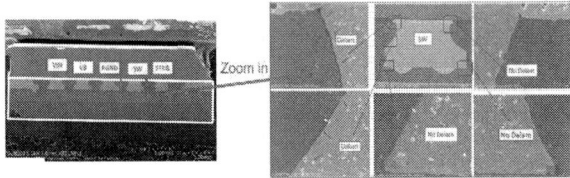

Fig. 6. X-section Photos for T0 Fresh Sample

So the whole failure picture was clear: slight delamination between EMC and sidewall of a certain pin would be triggered during package singulation process. As the number of temperature cycle accumulates, this delamination will propagate along the lead frame and EMC interface, and then crack the Cu Pillar and EMC interface. After complete delamination between Cu Pillar and EMC interface occurs,

high thermal stress will be built in the passivation layer of the die. Once exceeding the strength of the passivation, the passivation crack or chip crack will happen. On the other hand, the original delamination between lead frame and EMC interface will also penetrate deep into the EMC body, resulting in package body cracks. Next, simulation analysis was performed to verify this failure mechanism and find possible solutions.

III. MODELING AND SIMULATION

A. Modeling Description

Finite element method (FEM) was used for the package stress analysis, with the focus on die surface under the Cu pillar bumps. The package model was simulated with ANSYS WORKBENCH commercial software, where the models of PI and RDL layers were identical to their layout description (in Figure 7).

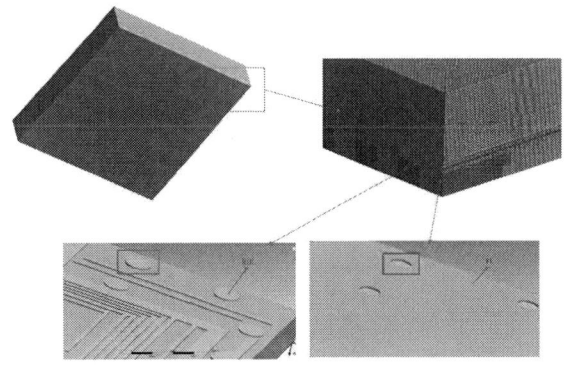

Fig. 7. Simulation Model for Flip Chip QFN Package

In case of the molding compound, temperature dependent linear elastic material properties have been used for simplicity although the material has viscoelastic behavior. Other package materials were also assumed linear elastic properties (Table III). A cooling down step was used for the analysis, meaning that thermal mismatch was the main reason causing the stresses in the package. The stress free temperature was set at molding temperature 175°C. Due to geometrical and material complexity, only the elastic range was analyzed in this study. Therefore, maximum principal stresses have been observed and fracture mechanics was not used in this study.

TABLE III.   MATERIAL PROPERTIES EMPLOYED FOR SIMULATIONS

Name	Young's modulus (GPa)		Poisson's Ratio	CTE (ppm/°C)	
Die	131		0.278	2.8	
RDL	121		0.34	16.3	
PI	3.5		0.34	35	
Solder	48		0.35	21.7	
Bump	121		0.34	16.3	
LF	121		0.33	18.2	
EMC	-65°C	30.6	0.35	-65°C	7.23
	25°C	28.7			
	80°C	26.9		103.5°C	7.23
	110°C	18.8			
	150	3.55		104.5°C	43.1
	175	2.83		260°C	43.1
	260	1.15			

978-1-6654-1392-3/21 $31.00 © 2021 IEEE

Five scenarios were simulated to analyze the influence of PI layer and pad delamination on die stress, including modeling of FC-QFN 1) without PI layer and NO delamination; 2) with PI layer and NO delamination; 3) without PI layer and having delamination on the side wall of critical pad; 4) without PI layer and having delamination on three surfaces of critical pad; and 5) without PI layer and having delamination on both three surfaces of critical pad and cylindrical surface of Cu pillar bump. Frictionless contact on the interface surface was set in this delamination modeling (Figure 8).

a) 2 Side Surfaces of Critical Pad with Delamination

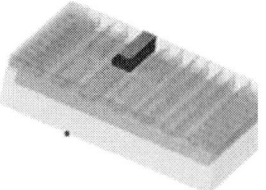

b) b) 3 Surfaces of Critical Pad with Delamination

Fig. 7. Modeling of Interfacial Delamination on Critical Pad

B. Results and Analysis

The 1^s principal stress distribution on silicon die during cooling down to –65°C for each scenarios were calculated and compared, the typical stress contour is shown in Figure 9.

From the simulation results, it was found that die maximum 1st principal stress happened in the spots under the peripheral Cu pillar bumps located along die bottom edge, and the stress even increased by adding PI layer. It corroborated the testing findings that die crack issue cannot be fixed by adding PI layer.

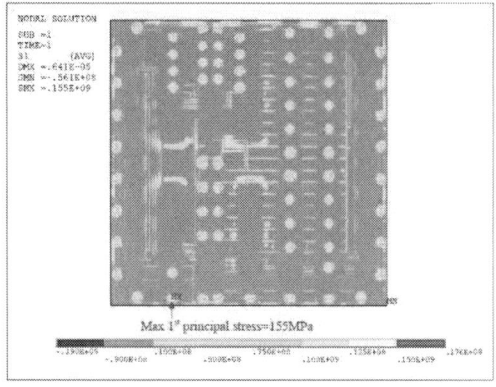

a) Without PI Layer and NO Delamination

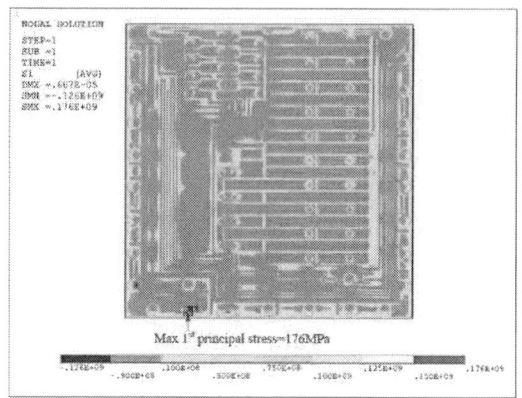

b) With PI Layer and NO Delamination

Fig. 8. Die 1st Principal Stress Distribution at -65°C

By comparing the results of all five simulation scenarios (in Table IV), the conclusion can be drawn that die should be on highest risk of fracture when crack penetrates the whole interface between leadframe pad and EMC, and as well as the interface between Cu pillar bumps and EMC during TC test. It validates the failure analysis findings in Section II.

TABLE IV.   COMPARISON OF SIMULATION RESULTS

Scenario	w/, w/o PI	Delamination	Die Max. 1st principal stress (MPa) at -65°C
#1	w/o	NO	155
#2	w	NO	176
#3	w/o	2 side surfaces of critical pad	158
#4	w/o	3 surfaces of critical pad	172
#5	w/o	3 surfaces of critical pad and cylindrical surface of Cu pillar bump	198

C. Further Design Study

Further analysis was related to how to mitigate the die crack risks during thermal cycling test. One possible solution is to implement roughen treatment for leadframe surfaces in order to enhance the resistance for delamination propagation; others are including package structure optimization to reduce thermo-mechanical stress accumulated during temperature cycling. So the influence of package thickness and die thickness was investigated as well base on simulations.

Simulation results in Figure 10 showed that the accumulated thermo-mechanical stress decreased when reducing both EMC thickness and die thickness, which means that the risks for both leadframe/EMC delamination propagation and die crack should be lower when applying a thinner package and thinner die. Besides that, thinner die usually has higher flexural strength [6], which are beneficial for preventing die crack failure as well.

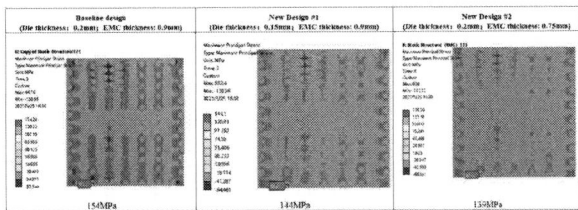

a) 1ˢᵗ Principal Stress on die for Package and Die thickness Variation

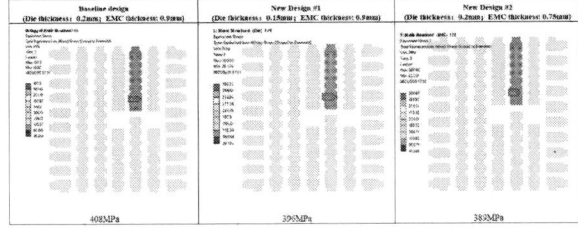

b) Leadframe von-Mise Stress for Package and Die thickness Variation

Fig. 9. Stress Comparison for Different Package and Die thickness

TABLE V. THERMAL CYCLING TESTING DATA COMPARISON BETWEEIN DIFFERENT DESIGNS

		TC 240	TC 500	TC 750	TC 1000
Original design	Lot #1	1/85	3/81	/	18/77
	Lot #2	1/85	2/81	9/78	21/69
+PI	Lot #3	4/85	2/81	5/79	/
	Lot #4	5/85	7/79	/	/
Roughen LF	Lot #5	0/100	3/98	3/90	7/82
Roughen LF+Die thickness reduction (0.2mm to 0.15mm)	Lot #6	0/100	0/96	0/88	0/84

Finally, the simulation results were verified by real samples. TCT testing results (in Table V) showed that the failure rate during thermal cycling test could be reduced by applying roughen leadframe, and all sample passing TC 1000 cycles was achieved by utilizing both thinner die and roughen leadframe.

IV. CONCLUSION

Based on all the results and data gathered from each experiments and simulations, it was found out that the delamination or crack propagation in Flip Chip QFN during thermal cycling test could finally trigger die crack issue and lead to electrical failures. This failure risk can be mitigated by applying delamination control measures (e.g., roughen leadframe instead of standard leadrame), and package structure optimization for lower thermo-mechanical stress accumulation. In this case, a decrease in the die thickness and molding compound thickness are preferred for reducing die crack risk during thermal cycling test. This failure study provides a practical guideline for prevent die crack failure when developing new Flip Chip QFN products.

REFERENCES

[1] Mccann D.R. , Ha S. H., 2002. "Package characterization and development of a flip chip QFN package: fcMLF," in in Proc. IEEE Electron. Comp. Technol. Conf., 2002, pp. 365-371.

[2] Baello J. R. , Colte J. , Quiazon R., "Resolving key manufacturing challenges in Flip Chip QFN package," 2016 International Conference on Electronics Packaging (ICEP), 2016.

[3] Wu, T. Y., Tsukada, Y., & Chen, W. T. 1996. "Materials and mechanics issues in flip-chip organic packaging," in Proc. IEEE Electron. Comp. Technol. Conf., 1996, pp. 524-534.

[4] L. Vigneswaran and Chong Chooi Mei, "Lead delamination correlation with Sn layer melting during sawing in power QFN package," 2010 12th Electronics Packaging Technology Conference, 2010, pp. 373-378.

[5] Patten, D. , and J. Phou . "Elimination of polyimide stress buffer on integrated circuits using advanced packaging materials." International Electronics Manufacturing Technology Symposium IEEE, 2002.

[6] Min, Y. K. , and J. W. Byeon . "Evaluation of Flexural Strength of Silicon Die with Thickness by 4 Point Bending Test." Journal of the Microelectronics & Packaging Society 18.1 (2011).

Study on the Antioxidation of Coated Nano-copper

1st Weijie Zhang
R&D Department
Sky Chip Interconnection Technology
co.,LTD.
Shenzhen, China
zhangwj@scc.com.cn

2nd Quan zhou
College of Optoelectronic Engineering
Chongqing University
Chongqing, China
LucyZhou@cqu.edu.cn

3rd Chenshan Gao
R&D Department
Sky Chip Interconnection Technology
co.,LTD.
Shenzhen, China
gaocs@scc.com.cn

4th Debo Liu
R&D Department
Sky Chip Interconnection Technology
co.,LTD.
Shenzhen, China
dayper@scc.com.cn

5th Jun Li
R&D Department
Shennan circuits co.,LTD.
Shenzhen, China
Lij@scc.com.cn

6rd Huaiyu Ye
Shenzhen Institute of Wide-Bandgap
Semiconductors;
School of Microelectronics, Southern
University of Science and Technology
Shenzhen, China
h.ye@tudelft.nl

Abstract— **Ensuring that there is no oxidation on the surface of the nano-copper particles is an important task to make sure the subsequent sintering performance. In order to reduce the surface oxidation of nano-copper particles, this research mainly studies the benzotriazole (BTAH) coating. Corresponding XRD, TEM and XPS test results show that after BTAH treatment, the oxidation degree of copper particles decreases. At the same time, organics related to BTAH is detected on the surface of the treated copper particles. This fully shows that the BTAH coating can effectively prevent the surface oxidation of copper nanoparticles, which lays a good foundation for further sintering experiments.**

Keywords—nano-copper, BTAH coating, reducing oxidation

I. INTRODUCTION

In recent years, due to the toxicity of lead in lead-containing solders, there is an urgent need for a lead-free solder in chip interconnection [1]. The birth of nano-silver paste brings hope to the solution of this problem [2, 3]. However, due to the electromigration of silver and relatively high price, there are still certain problems in its promotion and use. The nano-copper paste can solve the problems of nano-silver, thereby bringing new opportunities for the research and development of new sintered materials [4, 5].

In the preparation process of nano-copper paste, it is an extremely important step to protect the nano-copper particles by physical and chemical means to avoid or reduce their oxidation. This can ensure that the sintered nano-copper has sufficient strength, thereby ensuring the reliability of sintering. Studies have shown that organic coating can greatly reduce the oxidation behavior of nanoparticles. It is pointed out that adding a small amount of NiO particles during the preparation of nano-copper paste can effectively prevent copper oxidation. The deterioration of the interconnection performance of the copper paste caused by the sintering process produces Cu and Ni (Cu$_x$) solid solutions, which can not only effectively remove the copper oxide on the surface of the nano-copper, but also enhance the strength of the matri[6]. Clean the copper nanoparticles with formic acid or oxalic acid solution before preparing the copper paste, so that the surface of the copper nanoparticles can react with the acid solution to form copper carboxylate. In the subsequent sintering process, the copper carboxylate undergoes high temperature decomposition to form copper [7, 8]. It can also greatly reduce the oxidation of the surface of the copper nanoparticles.

In this study, the main effects of BTAH-treated copper powder and organic acid-treated copper powder on the sintering performance were mainly studied. Through XRD, we found that the organic acid removes the oxides on the surface of the copper powder. The copper powder after removing the oxide layer is treated with BTAH solution, and the copper powder has no oxides. It can be seen that the copper powder after BTAH treatment has a certain degree of oxidation resistanc. Through the results of TEM and TGA, we found that the copper powder treated with BTAH can effectively prevent the oxidation behavior of nano-copper powder; In the subsequent sintering experiment, we found that with the passage of time, the shear strength of the copper plate sintered with organic acid-treated copper powder decreases faster than the shear strength of the copper plate sintered with BTAH-coated nano-copper powder. Therefore, the copper powder treated with BTAH is less susceptible to environmental factors than the copper powder treated with organic acid. For long-term storage, it has more advantages.

II. EXPERIMENTAL PROCESS

Experimental procedures: first, remove the oxides on the surface of the copper powder in an organic acid solution. In anhydrous ethanol solution, let the molar mass ratio of hydrogen ion and copper be 1:1 to clean the copper powder. After ultrasonic treatment for 1 hour, wash the copper powder several times with deionized water, and clean the copper powder with anhydrous ethanol. Then, the treated copper powder is placed in a 50℃ vacuum drying oven for 2 hours. According to the mass ratio of 2:1 (M2), 10:1 (M10) and 80:1 (M80), the obtained copper powder and BTAH are dissolved in anhydrous ethanol solution, and the solution is fully stirred at 600 rpm for 6 hours.

The equipment involved in this experiment are: magnetic stirrer, vacuum drying oven, ultrasonic cleaning machine, etc. The test equipment involved this time includes: scanning electron microscope (FE-SEM, Zeiss), field emission projection electron microscope (TEM, Talos

978-1-6654-1392-3/21 $31.00 © 2021 IEEE

F200S, FEI), X-ray diffractometer (XRD, Rigaku Smartlab, Rikagu) and X-ray photoelectron spectroscopy (XPS, ESCALAB250Xi). SEM and TEM are used to characterize the surface micromorphology and the particle size of copper nanoparticles; In addition, XRD and XPS are used to analyze the oxidation and crystal structure of the particles. The X-ray range is set to 10-80 degrees, and the scanning speed is 0.2 degrees/second; the thermal behavior of nano-copper particles is studied by using a synchronous thermal analyzer (TGA/DSC, 1600LF, Mettler). The temperature rises from room temperature to 500 degrees Celsius in an atmosphere of N_2 or air at a rate of 10 K/min.

III. RESULTS AND DISCUSSION

We first analyze the physical properties of the copper powder. Fig.1 (a) reflects the microscopic morphology information of copper powder. It can be seen from the figure that the shape of the copper particles is approximately spherical and relatively regular, and the particle size distribution is uniform, but there are still a few larger particles. The particle size is approximately 100 nm. Fig.1 (b) demonstrates the SEM image of the copper particles treated with organic acid under high magnification. There is no obvious change in their shape.

Fig.1 SEM image of (a) untreated CuNPs and (b) CuNPs with organic acid treatment

Fig.2 shows the XRD pattern of untreated copper particles. It can be clearly seen from the figure that there are three main peaks at 2θ = 43.3, 50.4 and 74.8 degrees, corresponding to the (100), (111) and (110) planes of the face-centered cubic structure of copper, respectively. Besides, there is a small diffraction peak at 2θ = 36.5 degrees, which corresponds to the diffraction peak of cuprite (Cu_2O) in comparison with the standard. Therefore, the main component of untreated copper powder is copper with a certain degree of oxidation. The copper powder after organic acid treatment shows higher diffraction peaks among the three main peaks, indicating that its crystallinity is better. More importantly, there is no obvious Cu_2O diffraction peak at 2θ = 36.5 degrees. It can be seen that the organic acid will reduce the copper and cleanly remove the oxides on the surface of the powder. After treating the copper powder with the BTAH solution, the oxide layer on the surface of the copper powder is removed. It can be seen that the copper powder treated with BTAH also has a certain degree of oxidation resistance.

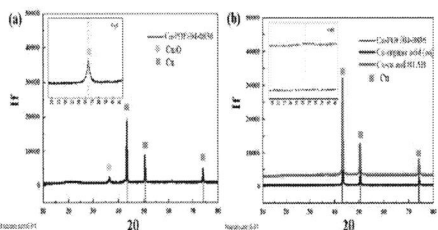

Fig.2 XRD pattern of (a) untreated CuNPs and (b) CuNPs with organic acid and BTAH treated

XPS can qualitatively and quantitatively analyze the elements and their chemical states within 10 nm of the material surface. In Fig.3 (a)-(d), we can see the XPS spectrum of the original copper powder and BTAH treatment. For the original nano-copper particles, the Cu2p3 spectrum has a characteristic peak at 933.15 eV. Since there is no significant difference in the binding energy of the copper that peaks in different valence states, the Auger analysis of the peaks is continued. Also, the kinetic energy spectrum corresponds to the CuO. Therefore, the untreated copper powder does have obvious oxidation, and the peak of divalent copper ions can be seen near the position of 943 eV. From the characteristic peak of C, its characteristic peaks are located at 284.8 eV, 286.02 eV and 288.84 eV, corresponding to the characteristic peaks of C—C, C—O and C=O bonds, respectively. It can be seen that there is a small amount of organics on the surface of the untreated copper nanoparticles. C-N can be detected on the surface of the copper powder treated with BTAH at 285.6 eV, which is mainly derived from BTAH, and thus BTAH is detected on the surface of nano-copper particles. In Fig.3 (d), after BTAH treatment, there are two clear and asymmetric peaks in the nanoparticles, which can be fitted to multiple peaks. For the characteristic peak appearing at the position of 932.51 eV, since the binding energy of pure Cu and Cu_2O is close at this position, it is difficult to determine a single XPS. There is no Cu_2O peak in the combined XRD pattern, so this peak corresponds to the characteristic peak of Cu. According to relevant literature, the characteristic peak at 935.06 eV corresponds to the divalent copper ion of Cu(II) copper salt (reference value is 934.7 eV). Therefore, after the BTAH treatment, the oxidation degree of the copper particles decreases, mainly from CuO to Cu_2O and Cu, and divalent copper ions appear on the surface. Moreover, BTAH-related organics are also detected on the surface of the treated copper particles.

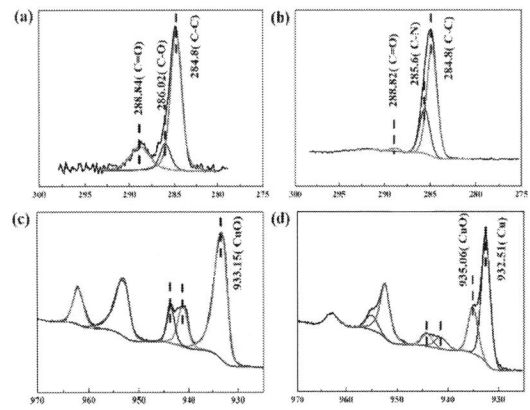

Fig.3 (a) the C 1s and (c) the Cu 2p are XPS patterns of the original CuNPs; (b) the C 1s and (d) the Cu 2p are XPS patterns of the CuNPs with BTAH treated

From Fig.4 (a)-(c), we can more intuitively see the microscopic morphology of copper powder treated with different concentrations of BTAH. As shown in Fig.4 (d), the untreated copper particles are approximately spherical and the sizes are close to 100±36 nm. This is consistent with the previous analysis. The edge of the copper particles after BTAH treatment is wrapped with a uniform transparent film. The thickness of the transparent film of copper powder with M2 concentration is about 10 nm, and the thickness of copper powder with M10 concentration is about 5 nm. When the concentration of BTAH is reduced to M80, the film thickness is about 3-5 nm, and the thickness is not uniform. It can be seen that the uniform packaging cannot be achieved when the concentration is too low. In order to explore the composition of the film, we perform an energy spectrum scan (as shown in Fig.4 (e)-(h)). From the energy spectrum, the internal composition of the particles is Cu, and the N element mainly from BTAH can be detected in the outermost part. Therefore, BTAH reacts with copper, and BTAH-related substances are detected on the surface of copper particles.

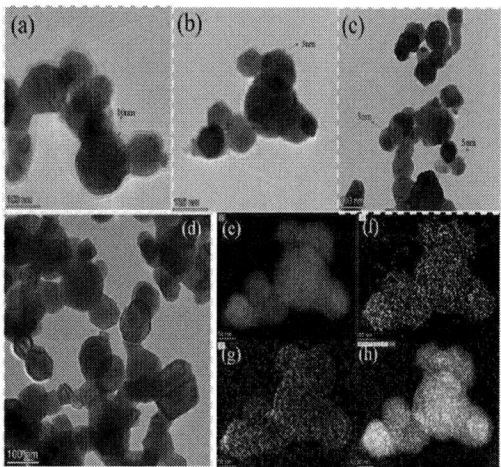

Fig.4 TEM images of different coating thicknesses: (a) M2; (b) M10; (c) M80 and (d) original CuNPs, (e) to (h) are the element mapping scanning of Cu, N, C and all

After that, we further analyze the thermal behavior of the BTAH-treated copper particles, and conduct a comparison experiment between the BTAH coating and the original particles. First, as shown in Fig.5 (a), the TG curve of pure BTAH in N_2 atmosphere has two strong endothermic peaks at 100°C and 250°C. It can be seen that BTAH starts to melt at about 100°C. At this time, the TG did not drop significantly. When the temperature reaches about 160°C, the TG of BTAH starts to decrease, and when the temperature reaches about 250°C, the TG stops decreasing. This curve means that the temperature has reached the boiling point of BTAH, and the TG of BTAH drops rapidly as the temperature rises. Therefore, there is an obvious endothermic peak between 160 and 250 degrees Celsius.

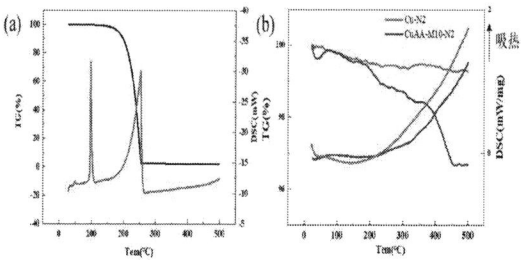

Fig. 5 TG-DSC of (a) BTAH and (b) CuNPs with BTAH treated

Then, in Fig.5 (b), the TG of the original copper powder in the N_2 atmosphere slowly decreases with the increase of temperature, which may be due to the decomposition of the surface oxide layer, so the whole process is accompanied by exothermic phenomenon. For the particles coated with BTAH, when the temperature is below 160°C, the decrease of the TG is consistent with that of the original copper powder, but at 160°C, the two curves start to separate, the TG of the coated one decreases more obviously, which may be related to the decomposition of BTAH physically adsorbed on the surface of the particles. The second gradient of the TG drop occurs when the temperature rises to about 230°C, and the drop is more gentle than the first gradient. At this point, the BTAH on the surface of the particles may have been decomposed. At about 350°C, the rate of TG decline begins to increase again, and only when the temperature is above 400°C does the rate begin to decrease, and then stabilize. The reason behind the change in the curve at this stage is not the decomposition of BTAH. It is very likely that some bonds in the compound formed by BTAH and copper are broken, which means that the copper salt gradually decomposes at this temperature. Therefore, the compound formed by BTAH and copper can protect the surface of nano-copper particles.

In order to study the main influence of BTAH-treated copper powder and organic acid-treated copper powder on the sintering performance, we made corresponding sintering experiments. In the sintering process, a copper plate is used to replace the chip for a sintering experiment, where the size of the upper copper plate is 3mm×3mm, and the size of the lower copper plate is 5mm×5mm. The organic acid-treated and the BTAH-treated copper powder are mixed uniformly with terpineol to obtain a simple paste. The solid content is 71% to 73%.

The paste storage condition is in vacuum and low temperature. The conditions used in this sintering are nitrogen atmosphere, sintering at 280 °C for 10 minutes. After sintering, we conducted push-pull experiments on the bonding force of the upper and lower copper plates. The basis for judging whether the sintering result is good or not is: if under the same push-pull test conditions, the sintered copper plate with higher shear strength has better sintering performance, and vice versa. We conducted a push-pull force test on the sintered copper plate after the sintered organic acid treated copper powder (Fig. 6), and the shear strength it can withstand is 50.26Mpa; another group of the same copper plate is treated with organic acid. Afterwards, the BTAH-coated nano-copper powder was sintered for research. In order to ensure the comparability of the two experiments before and after, we used the same experimental conditions except for the different coating states of the same nano particles. It was found that the copper plate sintered with nano copper powder coated with BTAH can withstand the maximum shear strength of 33.44Mpa. The speculative analysis of the above sintering results can lead to the following conclusions: the surface of the copper powder after BTAH treatment has a protective layer, and it is the protective layer that hinders the sintering process. Next, we put the copper powder treated with organic acid and the copper powder treated with BTAH in the same environment. Afterwards, take it out in a different time and do the same sintering experiment as above. In the experiment, we found an interesting phenomenon: with the passage of time, the shear strength of the copper plate sintered with the organic acid-treated copper powder is rapidly reduced after the push-pull force test. After 11 days, its shear strength was reduced to half of its original value, and after 20 days, it remained basically unchanged. At this time, the shear strength is only about one-fourth of the initial shear strength. After the push-pull force test of the sintered copper sheet after the BTAH treatment of the copper powder, its shear strength decreased slowly with time, and it decreased by about 9 Mpa after 18 days. After 27 days, the shear stress is still one-third of the original data. These results indicate that the copper powder treated with BTAH is less susceptible to environmental factors than the copper powder treated with organic acid. Considering long-term storage and use, the copper powder treated with BTAH is relatively more advantageous.

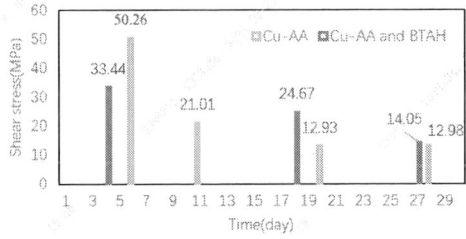

Fig. 6 Shear strength of copper-copper interconnects after organic acid treatment and BTAH treatment changes with time

IV. CONCLUSION

Through experiments, it is found that after the test of XRD, TEM and XPS test of copper nanoparticles coated with BTAH, the degree of oxidation is significantly lower than that of untreated particles. This fully shows that the BTAH coating can effectively prevent the surface oxidation of copper nanoparticles, through the sintering experiment, we found that although the organic acid treated copper powder has stronger shear strength after sintering with the copper flakes, the BTAH treated copper powder is less susceptible to environmental factors than the organic acid treated copper powder. Therefore, it is more advantageous in terms of long-term storage and use, which lays a good foundation for further sintering research.

ACKNOWLEDGMENT

This work was supported by the Key-Area Research and Development of GuangDong Province (2019B010131001) and the Shenzhen Key Project for Basic Research (JCYJ20200109140822796).

REFERENCES

[1] X. B. Wang, Void of leaded and lead-free solder paste on different PCB surface pad finish under various reflow conditions - Further investigation on void of SnInAgBi lead-free soldering paste (Proceedings of the 7th Electronics Packaging Technology Conference, Vols. 1 and 2). 2005, pp. 347-352.

[2] C. Qian et al., "Characterization and reconstruction for stochastically distributed void morphology in nano-silver sintered joints," Materials & Design, vol. 196, Nov 2020, Art. no. 109079.

[3] Y. Tan, X. Li, G. Chen, Q. Gao, G.-Q. Lu, and X. Chen, "Effects of thermal aging on long-term reliability and failure modes of nano-silver sintered lap-shear joint," International Journal of Adhesion and Adhesives, vol. 97, Mar 2020, Art. no. 102488.

[4] J. Qian et al., Effect of pressure on nano-copper sintering in interconnections of power device (Icept2019: The 2019 20th International Conference on Electronic Packaging Technology). 2019.

[5] V. Madhur, M. Srikanth, A. R. Annamalai, A. Muthuchamy, D. K. Agrawal, and C.-P. Jen, "Effect of Nano-copper on the Densification of Spark Plasma Sintered W-Cu Composites," Nanomaterials, vol. 11, no. 2, Feb 2021, Art. no. 413.

[6] T. Satoh and T. Ishizaki, "Enhanced pressure-free bonding using mixture of Cu and NiO nanoparticles," Journal of Alloys and Compounds, vol. 629, pp. 118-123, Apr 25 2015.

[7] J. Liu, H. Chen, H. Ji, and M. Li, "Highly Conductive Cu-Cu Joint Formation by Low-Temperature Sintering of Formic Acid-Treated Cu Nanoparticles," Acs Applied Materials & Interfaces, vol. 8, no. 48, pp. 33289-33298, Dec 7 2016.

[8] Y. Mou, Y. Peng, Y. Zhang, H. Cheng, and M. Chen, "Cu-Cu bonding enhancement at low temperature by using carboxylic acid surface-modified Cu nanoparticles," Materials Letters, vol. 227, pp. 179-183, Sep 15 2018.

The Impact of Packaging Materials on Thermo-mechanical Reliability of FC-LGA (Flip-Chip Land Grid Array) Package for 5G Application

Dayuan Wan
School of System Design and Intelligent Manufacturing
Southern University of Science and Technology
Shenzhen, China
wandy@mail.sustech.edu.cn

Ke Xue
School of System Design and Intelligent Manufacturing
Southern University of Science and Technology
Shenzhen, China
xuek@sustech.edu.cn

Weijing Dai
School of System Design and Intelligent Manufacturing
Southern University of Science and Technology
Shenzhen, China
daiwj@sustech.edu.cn

Yijing Qin
Center for Engineering Materials and Reliability
Guangzhou HKUST Fok Ying Tung Research Institute
Guangzhou, China
yjqin@ust.hk

Dong Lu
Center for Engineering Materials and Reliability
Guangzhou HKUST Fok Ying Tung Research Institute
Guangzhou, China
maeld@ust.hk

Zhiqi Wang
Chengdu Monolithic Power Systems Co., Ltd
Chengdu, China
Dax.Wang@monolithicpower.com

Yi Chen
Chengdu Monolithic Power Systems Co., Ltd
Chengdu, China
Ethan.Chen@monolithicpower.com

Abstract—With the advent of the 5G era, power devices and telecom equipment require small form factor and excellent thermal performance. Flip-chip Land Grid Array (FC-LGA) is rapidly becoming the package choice for those devices, because of its low RDS (ON) and high frequency operation. However, it also brings great challenges to reliability design. Large number of flip chip solder joints between die and substrate lead to a more rigid package body, which will accumulate strong thermo-mechanical stress under alternating temperature and so inevitably trigger package reliability issues. Excessive warpage and body crack will occur if without appropriate package design consideration.

In this study, the thermo-mechanical behaviors of FC-LGA were investigated by finite element simulation under different TCC conditions. Due to the complexity of the LGA substrate, the simulation with detail substrate model in use will incur expensive computation. This work proposed a simplification method considering a substrate of complicated structure as an equivalent anisotropic volume. With simulation models simplified, DoE study was performed to investigate the influence of packaging materials and geometry on the thermo-mechanical reliability of FC-LGA packages in high efficiency. A higher die-to-EMC thickness ratio will result in a more severe package convex warpage, while using a thinner die and a thicker EMC layer can prevent package body from cracking during heating or cooling processes. Besides packaging geometry, the influence of substrate layouts and EMC mechanical properties were also studied to find the optimal packaging design to mitigate the risks of excessive package warpage and body crack.

Keywords—Flip-Chip land grid array; Thermo-mechanical reliability; Finite element method; Warpage; Crack

I. INTRODUCTION

Electronic packaging technology has developed significantly in the past few decades. From through-hole technology before 1980s to surface mounted technology (SMT) in the 1980s, and further to ball grid array package (BGA) and chip scale package (CSP) in the 1990s, the efficiency and reliability of packaging have been greatly improved [1]. With the demand for miniaturization of devices and increased variety of product functionality, like mobile phones, smart wearable devices and 5G devices, advanced packaging technology, has attracted people's attention, such as System in Package (SiP) and Flip-Chip Land Grid Array (FC-LGA) [2-5].

Although advanced packaging technology enables function diversity of packaging modules, the internal structures of these heterogeneous integration systems become rather complicated, bringing great challenges to reliability design. Finite element analysis is a widely used tool to facilitate design for packaging reliability without the need of expensive prototype testing. With well-specified design variables and design objectives, the Design-of-Experiment (DoE) technique can be easily implemented with finite element analysis for packaging design. By simulating and analyzing the performance of different combination of design variables, the optimized design can be extracted. However, the increasing complexity of packaging modules continuously elevates the computation burden and therefore degrades the DoE efficiency.

To resolve this issue, we proposed and validated a simplification methodology to accelerate finite element simulation for packaging, and thus the effectiveness of DoE could be maintained. Conventionally, it is advised to include detailed geometry of a given package in finite element simulation to keep the best accuracy. However, it needs engineers and researchers to spend a large amount of time on modeling and meshing. In our cases, the packaging substrates contain three-dimensional copper traces and laminate

978-1-6654-1392-3/21 $31.00 © 2021 IEEE

structure. Thus, to avoid bearing the need to build model for this fine structure, a simplification method was demanded to homogenize the substrates as equivalent media but with anisotropy. Their Young's modulus, shear modulus and Poisson's ratio of three main axial directions were experimentally characterized, and the corresponding material properties inputs for finite element simulation were numerically calculated according to the experiments. The validity of these equivalent material properties for the packaging substrate were verified by the consistency between experimental measurement and numerical result.

With these equivalent properties available, the 3D finite element model is further simplified into 2D finite element model for strain energy release rate (SERR) calculation by using Virtual Crack Closure Technique (VCCT) which is calculated in 3D model as

$$G_{\mathrm{I}} = -1/(2\Delta A) \, [Z_{\mathrm{Li}}(w_{\mathrm{L}} - w_{\mathrm{L}*})] \qquad (1)$$

with G_{I} representing the SERR of mode I. ΔA is the area virtually closed by $\Delta A = \Delta ab$, where Δa is the length of the elements at the delamination front, and b is the width of the elements. Z_{Li} denotes the forces at the delamination front, and w_{L} and $w_{\mathrm{L}*}$ denote the displacement of the upper face node and the lower face node, respectively [6]. Then, we study the effect of geometric parameters and material properties on warpage and body cracking of the chip through finite element simulation. Eventually, we found that it's effective to save the cost of simulation using simplification method for reliability design, and the die thickness, EMC thickness and their coefficient of thermal expansion (CTE) have enormous impact on warpage and cracking property.

II. METHODOLOGY

A. Flexural modulus characterization method

To obtain the equivalent flexural modulus of the substrate used in this study, we conducted three-point bending test for these substrates by Thermomechanical Analyzers (TMA) under different temperatures. The flexural moduli of X direction and Y direction were obtained by changing the direction of fixtures and loading in the three-point bending test, as shown in Figure 1.

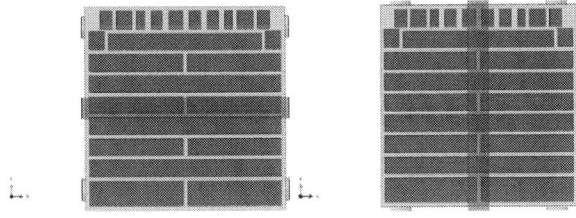

Figure 1. The diagram of three-point bending test of X direction (left) and Y direction (right).

After three-point bending test, the flexural modulus can be calculated by the formula as follows:

$$E_{\mathrm{bend}} = L^3 F/4wh^3 d \qquad (2)$$

where L is the distance between two fixtures, w is the width of the substrate, h is the thickness of the substrate, and d is the deflection of the point force applied at. According to *Hooke's law* [7], the constitutive relationship about anisotropic materials between stress and strain can be expressed as:

$$\begin{Bmatrix} \varepsilon_1 \\ \varepsilon_2 \\ \varepsilon_3 \\ \gamma_{23} \\ \gamma_{13} \\ \gamma_{12} \end{Bmatrix} = \begin{bmatrix} \frac{1}{E_1} & -\frac{v_{21}}{E_2} & -\frac{v_{31}}{E_3} & 0 & 0 & 0 \\ -\frac{v_{12}}{E_1} & \frac{1}{E_2} & -\frac{v_{32}}{E_3} & 0 & 0 & 0 \\ -\frac{v_{13}}{E_1} & -\frac{v_{23}}{E_2} & \frac{1}{E_3} & 0 & 0 & 0 \\ 0 & 0 & 0 & \frac{1}{G_{23}} & 0 & 0 \\ 0 & 0 & 0 & 0 & \frac{1}{G_{13}} & 0 \\ 0 & 0 & 0 & 0 & 0 & \frac{1}{G_{12}} \end{bmatrix} \begin{Bmatrix} \sigma_1 \\ \sigma_2 \\ \sigma_3 \\ \tau_{23} \\ \tau_{13} \\ \tau_{12} \end{Bmatrix} \qquad (3)$$

where E_1 and E_2 are the equivalent flexural modulus of X direction and Y direction of the substrate, respectively, E_3 is the flexural modulus along thickness direction, G_{12}, G_{23} and G_{13} are the shear modulus of three directions of the substrate, and the v denotes the Poisson's ratio of the substrate. We assumed that the three values of the Poisson's ratio are identical because of their weak effect on deformation. As for G_{12}, G_{23} and G_{13}, they can be calculated by the relationship as follows:

$$1/G_{12} = 2(1/E_1 + v_{12}/E_2) \qquad (4)$$

$$1/G_{13} = 2(1/E_1 + v_{13}/E_3) \qquad (5)$$

$$1/G_{23} = 2(1/E_2 + v_{23}/E_3) \qquad (6)$$

Table 1 presents the equivalent properties obtained from the three-point bending test, which will be imported into the finite element simulation.

Table 1. Equivalent substrate property.

Temp.	E1	E2	E3	G12	G13	G23
-65	51333	34100	100000	19727	23277	15961
25	45783	31517	100000	17738	20971	14824
150	35767	26867	100000	14123	16689	12748
260	20100	12767	100000	7643	9662	6224
Temp. unit: ℃; Flexural modulus: MPa; All the directions of Poisson's ratio are 0.2.						

B. Finite element simulation with simplified model

In our case, the packaging substrate contains three-dimensional copper traces and laminate structure, thus, building a finite element model for this complicated structure is time-consuming. It is necessary to apply the flexural modulus to simplify the corresponding model. To validate the feasibility of this model simplification, we established both 3D detailed finite element models, shown in Figure 2, and 3D equivalent homogeneous finite element model. These two types of models were compared to the corresponding experiments under the same condition.

Figure 2. FEM model of substrate structure.

Additionally, the elastic and elastoplastic property of copper traces was separately used in the detailed finite element simulation. As in Figure 3, comparing these three types of simulation with the experiment, we found that the result of

978-1-6654-1392-3/21 $31.00 © 2021 IEEE

simulation with equivalent material property is in well agreement with the experimental data. Therefore, it's practicable to regard a complex substrate as a homogeneous medium with properly characterized anisotropic material property in finite element simulation.

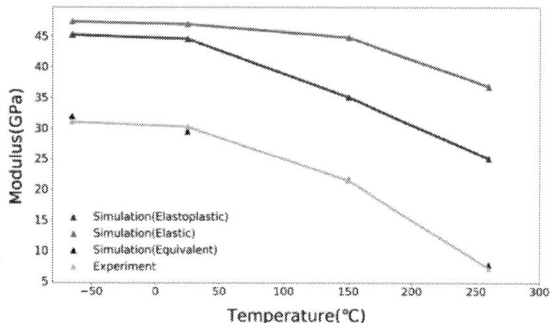

Figure 3. Flexural modulus obtained from experiment and simulation.

Because the package structure is approximately identical along one axis, using an appropriate 2D finite element model, as shown in Figure 4, can further increase the efficiency of finite element simulation.

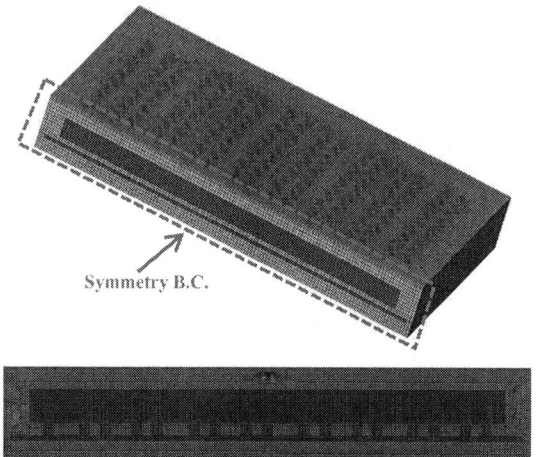

Figure 4. The 3D (top) and 2D (bottom) finite element models with equivalent substrate.

Thereby, we examined the SERR of both 2D and 3D model using VCCT to verify how much deviation is resulted from the abstraction from 3D model to 2D model. Table 2 compares the SERR result of the simulation, and we found that deviation occurred but within an acceptable range. Both the 2D model and 3D model present similar SERR variation against the temperature. In consideration the computational efficiency, the 2D model was used in this study.

Table 2. Warpage and SERR of simulation.

Temperature (°C)	Model dimension	G_1 (kJ/m²)
-55	3D	0.1273
25		0.0053
150		0.0049
-55	2D	0.1097
25		0.0460
150		0.0049

-55	Deviation	13.85%
25		13.45%
150		0.11%

III. RESULT AND DISCUSSION

A. Parametric study on warpage deformation

With the equivalent substrate properties employed, three parameters are focused in this session to investigate their influence on warpage, including die thickness, mold cap thickness and the CTE of molding materials. Due to the symmetrical structure of the package, only half of the model is implemented. For each run of simulation, only one parameter is changed from its nominal value. The typical finite element model with meshes is shown in Figure 4 (3D).

A typical finite element modeling result is shown in Figure 5, and the maximum absolute Z-direction nodal displacement is used to indicate the warpage magnitude. In general, positive values mean concave warpage, while negative ones suggest convex warpage.

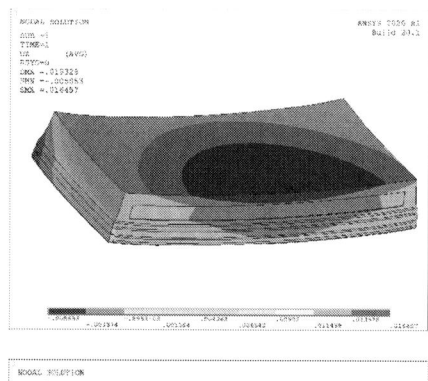

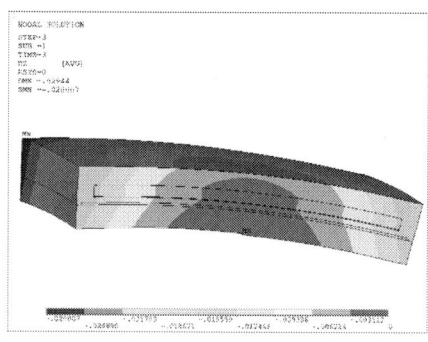

Figure 5. Typical nodal displacement contours of die thickness = 150 μm, Top – concave warpage, bottom– convex warpage.

Table 3 summarizes the warpage values extracted from a series of finite element simulation. The first row shows the temperature condition, and the first column indicates the value of the investigated parameter. The rest cells are the corresponding warpage magnitude.

Table 3. Warpage of package with different parameters.

		Temperature (°C)		
		25	150	-55
Die thickness (μm)	150	-16.6	1.4	-28
	300	-25.1	-3.2	-39.3
	400	-25.9	-4.42	-39.8

978-1-6654-1392-3/21 $31.00 © 2021 IEEE

	400	-26.9	-2.9	-42.2
Mold cap thickness (µm)	450	-23.6	-1.8	-37.7
	550	-17.4	-0.9	-28.9
	20	-19.9	-3.3	-30.7
Mold cap CTE (10^{-6})	40	5.3	0.7	8.7
	60	17.9	3.4	25.9

For the influence of die thickness, its increase will strengthen the intensity of the convex warpage. This is because the CTE of die is one order of magnitude smaller than other materials in the simulation. Thus, since the volume of die increases in this scenario, the equivalent CTE of the composite die-EMC layer is expected to decrease. This effect results in the increasing mismatch of CTE between the EMC/die layer and the substrate. Therefore, the relative shrinkage of the latter one is larger than the former one. Another phenomenon to notice is that the concave-to-convex transition occurs at 150 °C when the die thickness increase from 150 µm to 300 µm. This is mainly because the CTE of EMC is larger than the other materials at temperature higher than 150 °C. However, this effect will be diminished as the volume of die increases.

In considering the influence of EMC thickness, the overall trend of warpage variation due to the increasing mold cap thickness is opposite to the previous discussion. Since the volume of EMC increases, the equivalent CTE of EMC/die layer increases as well, reducing CTE mismatch between the EMC/die layer and the substrate layer. For the same reason, the convex-to-concave transition happens, rather than the concave-to-convex transition.

When varying the CTE of EMC, referred the last three rows in Table 3, the warpage exhibits a clear convex-to-concave transition at all temperatures. The reason is that the variation of CTE mismatch between the EMC/die layer and the substrate layer is decreased and then increased with the CTE of EMC increasing.

B. Parametric study and optimization for simulated EMC crack properties

According to the fracture mechanics, the critical SERR of a given material determines whether a crack will develop at a specific condition. The crack will continue to invade the material if the current energy release rate of the crack becomes larger than the critical one. Therefore, if the critical energy release rate is unknown, the larger magnitude of the current energy release rate indicates more likeliness to crack. In this consideration, finding the combination of the investigated parameters to reduce the energy release rate calculated by the VCCT method is the aim of this parametric study. Additionally, three thermal profiles, i.e., TCC condition (150 °C >> -65 °C), TCG condition (125 °C >> -40 °C) and reflow condition (25 °C >> 260 °C) were used in this finite element simulation. For each condition, the initial temperature was set as the reference temperature in the finite element simulation.

As the previous session (II) proves the validity of using 2D simplified model to calculate SERR, this session further implements the 2D model to conduct parametric study for investigating the properties of EMC crack with different parametric setting.

The material properties and key geometrical dimension are listed in Table 4 to Table 6. The crack length is fixed as 0.025 mm. The EMC thickness is selected from the range between 0.38 mm and 0.58 mm, and the die thickness is selected from the range between 0.15 mm and 0.35 mm. This selection ensures the crack do not penetrate into the die in the finite element simulation. Eight types of EMC materials are chosen to investigate their influence on crack properties. The energy release rate calculated by the VCCT method is used as the indicator of crack properties. In addition, as the crack only develops due to convex warpage, the energy release rate will only be extracted from the simulation exhibiting convex warpage.

Table 4. Geometry dimension.

Crack length (mm)	0.025				
EMC thickness (mm)	0.38	0.43	0.48	0.53	0.58
Die thickness (mm)	0.15	0.2	0.25	0.3	0.35

Table 5. General material properties

	Die	Copper
Young's modulus (MPa)	131000	121000
Poisson ratio	0.278	0.34
CTE (ppm)	2.8	16.3

Table 6. EMC properties.

	M1	M2	M3	M4	M5	M6	M7	M8
CTE-a1 (ppm)	11	14	25	17	15	15	13	14
CTE-a2 (ppm)	44	49	90	65	47	47	48	53
Tg (°C)	160	115	130	140	140	140	135	130
Flexural Modulus (GPa) @ 25 °C	20	18.1	11	17.5	30	30	20	19
Flexural Modulus (GPa) @ 260 °C	1.00	0.15	0.15	0.29	1.00	1.00	0.80	0.50
Poisson ratio	0.35	0.35	0.35	0.35	0.35	0.35	0.35	0.35

Regarding to the effect of geometry, a trend shared by every case is that the diagonal region of the current EMC thickness/Die thickness matrix shows more intensive SERR as demonstrated in Figure 6 and Figure 7. The (EMC, die) thickness combination of (0.38 mm, 0.25 mm), (0. 38 mm, 0.3 mm), (0.43 mm, 0.3 mm), (0.43 mm, 0.35 mm) and (0.48 mm, 0.35 mm) in general present largest SERR result in every material combination. In addition, with the thinner die, the thicker EMC can prevent cracking. It is due to the thicker EMC will alleviate CTE mismatch between EMC/Die layer and substrate layer. In contrast, using the thicker die, the thinner EMC can prevent cracking. This is because the thinner EMC layer makes the crack close to the Die, which is much less willing to deform. Thus, the development of crack is restricted by this effect. These two findings can be further used in designing the thickness ratio between EMC and Die. Interestingly, the (EMC, die) thickness combination of (0.53 mm, 0.15 mm), (0.58 mm, 0.15 mm) and (0.58 mm, 0.2 mm) present smallest SERR in every material combination. This result is well consistent with the analysis above.

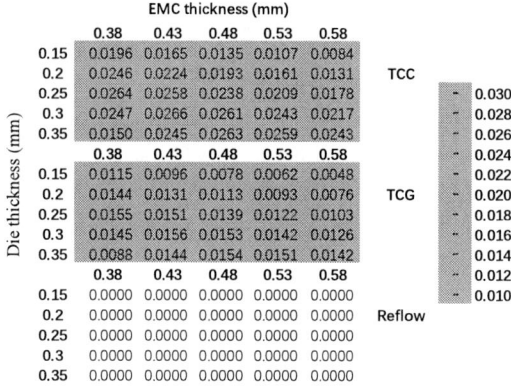

EMC thickness (mm)					
	0.38	0.43	0.48	0.53	0.58
0.15	0.0196	0.0165	0.0135	0.0107	0.0084	TCC
0.2	0.0246	0.0224	0.0193	0.0161	0.0131
0.25	0.0264	0.0258	0.0238	0.0209	0.0178
0.3	0.0247	0.0266	0.0261	0.0243	0.0217
0.35	0.0150	0.0245	0.0263	0.0259	0.0243
	0.38	0.43	0.48	0.53	0.58
0.15	0.0115	0.0096	0.0078	0.0062	0.0048	TCG
0.2	0.0144	0.0131	0.0113	0.0093	0.0076
0.25	0.0155	0.0151	0.0139	0.0122	0.0103
0.3	0.0145	0.0156	0.0153	0.0142	0.0126
0.35	0.0088	0.0144	0.0154	0.0151	0.0142
	0.38	0.43	0.48	0.53	0.58
0.15	0.0000	0.0000	0.0000	0.0000	0.0000	Reflow
0.2	0.0000	0.0000	0.0000	0.0000	0.0000
0.25	0.0000	0.0000	0.0000	0.0000	0.0000
0.3	0.0000	0.0000	0.0000	0.0000	0.0000
0.35	0.0000	0.0000	0.0000	0.0000	0.0000

Figure 6. SERR calculated case by the EMC with M1

EMC thickness (mm)					
	0.38	0.43	0.48	0.53	0.58
0.15	0.0632	0.0506	0.0000	0.0000	0.0000	TCC
0.2	0.0830	0.0731	0.0612	0.0000	0.0000
0.25	0.0926	0.0878	0.0793	0.0684	0.0000
0.3	0.0908	0.0944	0.0905	0.0831	0.0733
0.35	0.0554	0.0910	0.0950	0.0919	0.0854
	0.38	0.43	0.48	0.53	0.58
0.15	0.0276	0.0227	0.0178	0.0000	0.0000	TCG
0.2	0.0350	0.0315	0.0269	0.0221	0.0000
0.25	0.0381	0.0368	0.0338	0.0297	0.0251
0.3	0.0366	0.0386	0.0376	0.0350	0.0314
0.35	0.0219	0.0365	0.0387	0.0379	0.0357
	0.38	0.43	0.48	0.53	0.58
0.15	0.0000	0.0000	0.0000	0.0000	0.0000	Reflow
0.2	0.0000	0.0000	0.0000	0.0000	0.0000
0.25	0.0000	0.0000	0.0000	0.0000	0.0000
0.3	0.0000	0.0000	0.0000	0.0000	0.0000
0.35	0.0000	0.0000	0.0000	0.0000	0.0000

Figure 7. SERR calculated case by the EMC with M3.

In addition, the clearance between the top of dies and the top of EMC layers is also used as another factor to perform the parametric study. From the matrix, with the same die thickness, the SERR exhibits a maximum value with the increase of EMC-Die clearance in the current variation range. A clearance around 0.1 mm to 0.15 mm may give the highest SERR. While with a fixed clearance, the SERR shows a monotonical increase until saturation with thickening the dies in the current variation range. As the die thickness increases while the clearance is fixed, the volume of die improves, which aggravates the mismatch between EMC and Die. Consequently, it is necessary to balance the proportion of adjacent components, whose gap of CTE is too big.

IV. CONCLUSION

Simplified substrate and equivalent substrate property imported are effective for enhancing the efficiency of simulation and saving the cost of finite element analysis. Thus, a complete DoE study can be performed to search the optimal parameters for the required package application. In our case, the size and mechanical properties of components in a package have a significant influence on warpage and body crack of the package. A higher die to EMC clearance would lead to a more severe convex warpage, while the thinner die and thicker EMC could prevent body cracking during temperature changes. Accordingly, it's important to design a proper internal structure in the FC-LGA package for 5G application, which is critical to ensure the thermo-mechanical reliability.

ACKNOWLEDGMENT

The authors sincerely acknowledge the financial support from the Department of Science and Technology of Guangdong Province (Project No. 2020B010179002).

REFERENCES

[1] Rao, T. "Fundamentals of Microsystems Packaging." Microelectronics International 20.1(2001).

[2] Tai, K. L. System-In-Package (SIP): challenges and opportunities. IEEE Computer Society, 2000.

[3] Geng, P. "Structural Design of Land Grid Array Loading Mechanisms for Intel Central Processor Unit Stack Retention." Journal of Electronic Packaging 141.1(2019):010801.1-010801.8.

[4] Corbin, et al. "Land grid array sockets for server applications. " IBM Journal of Research & Development 46.6(2002):763-763.

[5] Hejase, J. A., et al. "A hybrid land grid array socket connector design for achieving higher signalling data rates." 2017 IEEE 26th Conference on Electrical Performance of Electronic Packaging and Systems (EPEPS) IEEE, 2017.

[6] Krueger, R. "Virtual crack closure technique: History, approach, and applications." Applied Mechanics Reviews 57.1(2004).

[7] Fung, Y. C. "International Series on Dynamics. (Book Reviews: Foundations of Solid Mechanics)." Science 152(1966).

Failure Analysis and Reliability Improvement of Crimping Assembly of Copper-Clad Aluminum Conductors for Aerospace

Rui Cao
China Aerospace Components Engineering Center,
China Academy of Space Technology,
Beijing, China
caorui0306@163.com

Meng Meng
China Aerospace Components Engineering Center,
China Academy of Space Technology,
Beijing, China
macromeng@163.com

Zhibin Wang
China Aerospace Components Engineering Center,
China Academy of Space Technology,
Beijing, China
wzb_1984@126.com

Yarong Chen
Beijing Spacecraft,
China Academy of Space Technology,
Beijing, China
yarongch@126.com

Yu Ye
Guizhou Space Appliance Company Limited,
Guiyang,China
381937717@qq.com

Xueyin Huang
China Electronics Standardization Institute,
Beijing, China
huangxy@cesi.cn

Abstract—The copper-clad aluminum (CCA) composite wire combines the excellent electrical conductivity of copper and the light weight of aluminum, and replacing the copper-core wire can effectively reduce the weight of cable network. This paper adopts the methods of assembly performance test, metallographic inspection, FIB and SEM in-situ inspection, to analyze the reasons for the failure of CCA assembly after environment test. When the sealing of the assembly is not good, the crimping position of the CCA wire communicates with outside by forming an air leakage channel, which will cause corrosion of the aluminum core and lead to the failure of the assembly. By improving the strain matching of the materials at crimping position and insuring the integrity of sealing structure, the sealing crimping can be achieved, effectively improving the reliability of CCA assembly.

Keywords—Copper clad aluminum wire; Crimping assembly; Space cable; Failure analysis; Reliability

I. INTRODUCTION

The copper-clad aluminum (CCA) composite wire combines the excellent electrical conductivity of copper and the light weight of aluminum, and replacing the copper-core wire can effectively reduce the weight of cable network. However, the potential difference between the aluminum core and the copper layer of the copper-clad aluminum wire is relatively large. If traditional crimp terminals are used for assembly, the aluminum core is prone to galvanic corrosion when exposed to the atmosphere. Aluminum and its oxides have poor weld ability. At the same time, the high temperature of welding easily causes the growth of intermetallic compounds at the copper-aluminum interface, which reduces the mechanical and electrical properties of the wire. Therefore, in high-reliability and long-life applications such as aerospace field, copper-aluminum composite wires are generally assembled by sealed crimping.

ESA and NASA have used copper-clad aluminum wires to replace copper wires to increase the payload of spacecraft. Among them, NASA civil communication satellites use copper-clad aluminum wires to achieve a weight reduction of 10%. The application of CCA wires in China started late and is mainly promoted in the civilian field. The demand for effective load of spacecraft is gradually increasing, and there is an urgent need to reduce the weight of spacecraft cables through high-reliability CCA wires.

At present, high-reliability CCA wire crimping terminals mostly use pure silver crimping sleeves as the intermediate layer of the conductive terminal and the wire crimping part, which can ensure the electrical conductivity of the assembly and realize the sealed crimping of the terminal and the wire. At present, systematic research has not been carried out on the structural evaluation and reliability improvement of the copper-clad aluminum conductor crimping assembly.

This paper adopts the methods of assembly performance test, metallographic inspection, EBSD, FIB and SEM in-situ inspection, to analyze the reasons for the failure of CCA assembly after environment test. When the sealing of the assembly is not good, the crimping position of the CCA wire communicates with outside by forming an air leakage channel, which will cause corrosion of the aluminum core and lead to the failure of the assembly. By improving the strain matching of the materials at crimping position and insuring the integrity of sealing structure, the sealing crimping can be achieved, effectively improving the reliability of CCA assembly.

II. MATERIALS AND EXPERIMENTAL

A. CCA wire

AWG20 and AWG24 CCA wires were used, in which the AWG20 core conductor is a twisted structure of 19 silver-plated CCA alloy wires, and the AWG24 core conductor is a silver-plated copper alloy wire plus 6 plated CCA wires. The wire insulation layer uses polyimide and polytetrafluoroethylene as the wrapping insulation material, and the wire structure and lateral metallographic morphology are shown in Fig.1.

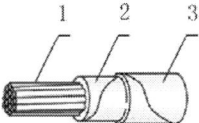

1. CCA conductor 2. Polyimide composite tape 3. PTFE tape

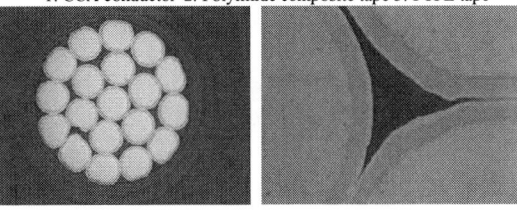

Fig. 1 Schematic diagram and metallographic morphology of CCA wire

B. Sealed crimp terminal

The test uses 2020 and 2024 sealed crimp terminals matched with AWG20 and AWG24 CCA composite wires respectively. The terminals are composed of copper alloy terminals and silver sheaths. The schematic diagram of the

978-1-6654-1392-3/21 $31.00 © 2021 IEEE

structure is shown in Fig. 2. The sealing performance requirement of the terminal crimping end is its key performance that is different from the copper wire terminal. In order to prevent external water vapor from entering the crimping terminal and causing corrosion of the aluminum core, the silver sheath adopts a thin-walled blind hole structure. When crimping and assembling, the core wire of the silver sheath and the insulating layer need to be crimped synchronously to realize the sealing of the lead off position.

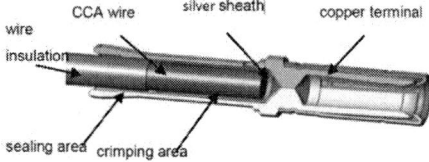

Fig. 2 Schematic diagram of the crimp terminal structure of a CCA wire

C. Experimental methods

The research uses a stereo microscope to conduct visual inspection of CCA wire assembly; MPT250 tensile machine to test the tensile strength; use a special testing device to test the tightness of the assembly; use metallographic analysis and focused ion beam (FIB) and Scanning electron microscope (SEM) analyzes the internal structure and bonding interface of the assembly.

III. RESULTS AND DISCUSSION

A. Tensile strength test

Table I shows the tensile strength test results of the 2020 terminal press-fit AWG20 wire and the 2024 terminal press-fit AWG24 wire assembly. It can be seen from the results that the tensile strength of all 2020 terminals and some 2024 terminals does not meet the design requirements of the corresponding assembly, and all the test assembly samples are broken at the core crimping position.

Table I Tensile strength test results of 2020 terminal-AWG20 wire/2024 terminal-AWG24 wire crimping assembly joint

Terminal	Tensile strength (>6.1Kgf)	Terminal	Tensile strength (>3.2Kg)
2020	4.7 (Fail)	2024	4.2 (Pass)
2020	4.3 (Fail)	2024	3.5 (Pass)
2020	5.1 (Fail)	2024	2.8(Fail)
2020	6.0 (Fail)	2024	3.5 (Pass)
2020	5.2 (Fail)	2024	3.2 (Fail)

B. Sealing test

Table II shows the sealing test results of the 2020 terminal and 2024 terminal crimp assembly. All 2020 terminal are unqualified and the amount of air leakage is large. The position of the leakage is located in the insulation lamination area and the contact observation hole. The 2024 terminal sealing test is less than 1mL, and the leaking position of the terminal is at the observation hole of the contact.

Table II Sealing test results of 2020 terminal-AWG20 wire/2024 terminal-AWG24 wire crimping assembly joint

Terminal	Sealing Test (≤1ml)	Leakage Are	Terminal	Sealing Test (≤1ml)	Leakage Area
2020	>12(Fail)	Insulation area and observation hole	2024	<1(Pass)	None
2020	>12(Fail)		2024	<1(Pass)	
2020	>12(Fail)		2024	<1(Pass)	
2020	>12(Fail)	Insulation area	2024	<1(Pass)	Observation hole
2020	8 (Fail)		2024	<1(Pass)	

C. Failure analysis

In order to further analyze the failure reasons of the low tensile strength of the copper-clad aluminum wire crimping assembly and the excessive air leakage of some samples in the sealing test, some samples of the same specification were selected for visual inspection and profile metallographic inspection.

Fig. 3 shows the appearance, overall morphology of the axial section, and radial metallographic morphology of the marked position of the 2020 terminal assembly. The observation hole of the contact avoids the core crimping deformation zone. Metallographic inspection shows that the silver sheath can cover the copper-clad aluminum wire core in the core crimping area and remains intact after crimping. The deformation of the copper alloy terminal, silver sheath and copper-clad aluminum wire core in the core crimping area is relatively sufficient. The copper terminal at the six-side clamping position in the conductor insulation lamination area is fully deformed and is tightly combined with the insulation layer.

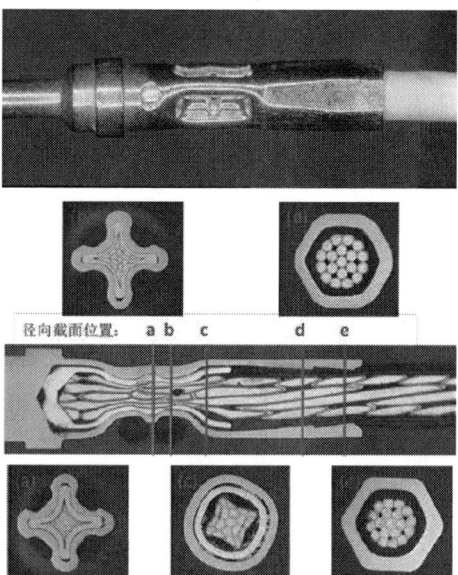

Fig.3 Appearance and metallographic morphology of 2020 terminal-AWG20 wire assembly

Fig. 4 shows the partial cross-sectional morphology of the 2020 terminal with obvious air leakage in the observation hole or the position of the insulating layer. It can be seen from Fig. 4(a) that there is a gap between the crimping interface between the silver sheath and the copper terminal, which may cause the terminal's tensile strength to be low and the observation hole will cause the core to form an air leakage channel. In the position shown in Fig. 4(b), the lead wire is too long to cause the crimping part to fail to achieve sufficient crimping of the insulation, resulting in poor sealing of the insulation lamination position.

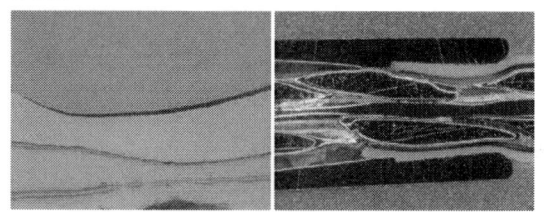

(a) Core crimping zone   (b) Insulation crimping zone

Fig.4 Morphology of 2020 terminal core crimping zone and insulation crimping zone

Fig. 5 shows the appearance, overall morphology of the axial section, and radial metallographic morphology of the marked position of the 2024 terminal assembly. The observation hole of the contact piece does not avoid the core crimping deformation zone. The metallographic examination showed that the silver sheath can cover the copper-clad aluminum core in the core crimping area, and the crimping interface is tightly bonded, but the silver sheath at the marked position b in the Fig.5 is totally broken. The copper terminal at the crimping position of the conductor insulation lamination area is fully deformed and tightly combined with the insulating layer, without causing overvoltage damage to the insulating layer.

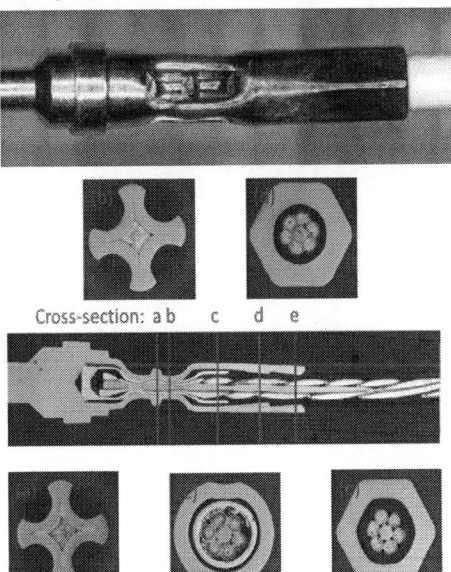

Cross-section: a b c d e

Fig.5 Morphology of 2020 terminal core crimping zone and insulation crimping zone

Fig.6 shows the axial cross-sectional morphology of the 2024 terminal observation hole and the fracture location of the silver sheath. As shown in Fig.6 (a), when the terminal observation hole is located at the core crimping position and the edge of the observation hole cuts the silver sheath seriously, it will cause partial damage to the silver sheath and affect the sealing performance of the assembly. Fig. 6(b) shows the position where the crimping deformation is the largest, and the inner wall of the copper terminal is tightly combined with the silver sheath, but the silver sleeve is partially broken at the crimping position, which affects the tightness and tensile strength of the assembly. Hidden hazards, which may cause failure phenomena such as core corrosion or pull-off in subsequent use.

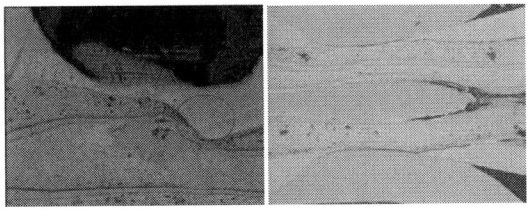

Fig.6 Sectional morphology of damaged area of 2024 terminal silver sheath

D. Reliability Improvement

The wire stripping position and assembly depth will affect the insulation lamination position after crimping. At the same time, the plastic deformation of the core crimping zone will cause the insulation layer to shift axially relative to the initial position. In view of the air leakage failure caused by insufficient insulation and lamination of some 2020 terminals, by controlling the stripping length and the core assembly depth, the consistency of the terminal crimping position can be effectively ensured. Fig. 7 is the profile morphology of the position of the terminal insulation layer that controls the wire tapping length and the core crimping assembly depth. The insulation lamination zone has a margin, uniform deformation and no damage, which can achieve tight crimping.

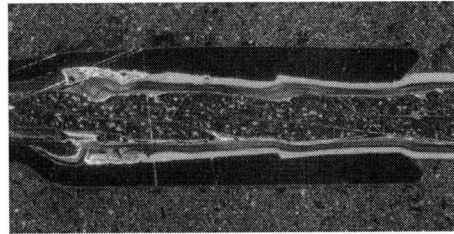

Fig.7 Cross-sectional morphology of the insulation crimping zone after controlling wire-off length

The observation hole of the terminal can check that the silver sheath is in place, and at the same time enhance the circulation of the plating solution to ensure the integrity of the plating on the inner wall of the crimping hole. But the observation hole of the AWG2024 terminal coincides with the position of the wire core crimping area. Thus there is a hidden danger that the edge of the observation hole may be damaged by squeezing the silver sheath. Moving the observation hole forward by 0.5mm while moving the crimping position back by 0.5mm can help to avoid the overlap between the observation and the core crimping area. Fig. 8 shows the appearance of the improved 2024 terminal after crimping and assembling.

Fig.8 The crimping appearance of 2024 terminal after improving the process hole position

The metallographic inspection showed that the core and the terminal of the 2020 terminal were deformed, but there was a gap between the silver sheath and the copper terminal; the core crimping area of the 2024 terminal was tightly bonded with the copper terminal, but the silver sheath was partially broken. When crimping and assembling, the copper clad aluminum core, silver sheath and copper alloy terminal in the core crimping area are all plastically deformed. When the strain matching between materials is good, plastic deformation can be achieved under appropriate crimping parameters, and tight compression can be achieved. Fig.9a shows the axial overall EBSD morphology of the crimping zone of the terminal core before the improvement. Fig.9b~9d are the local EBSD morphologies of aluminum core, silver sheath and copper terminal respectively. It can be seen from Fig.9b that the crystal grains of the aluminum wire core have undergone significant plastic deformation along the axial direction, and the crystal grains are crimped into strips, indicating that the aluminum core has been fully deformed. The grains of the silver sheath and the copper terminal are relatively small and the grains at the crimping position maintain equiaxed crystal morphology. As shown in Fig.9c, part of the silver sheath has undergone grain refinement or grain deformation. Fig.9d shows the copper terminal only undergoes grain refinement near the crimp interface. The strain state of the three metals in the crimping area is quite different before and after crimping. Thus, the strain matching of the wire core, silver sheath and copper terminal is poor. When the crimping force is small, it is difficult to coordinate deformation to form a tight bond. When the crimping force is large, the wire core or the silver sheath may be damaged due to excessive deformation.

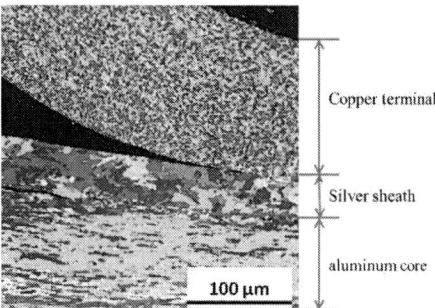

(a) overall EBSD morphology of the crimping zone

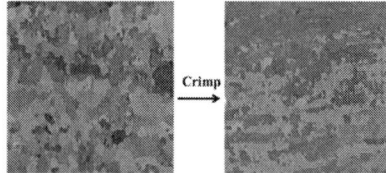

(b) EBSD morphology of aluminum core before and after crimping

(c) EBSD morphology of silver sheath before and after crimping

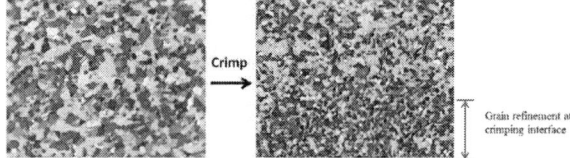

(d) EBSD morphology of copper terminal before and after crimping

Fig.9 The crimping appearance of 2024 terminal after improving the process hole position

In order to improve the crimp matching of the silver sheath, copper terminal and CCA core, the silver sheath is annealed to reduce the hardness. Partial annealing is performed on the end of the terminal to ensure the strength of the material of the terminal to the plug end, while reducing the hardness of the crimping end and improving the crimping matching. Fig.10 shows the SEM morphology of the FIB section of the improved terminal and CCA core before and after crimping. It can be seen from the figure that all positions before crimping are in a relatively large-sized equiaxed crystal state, and obvious plastic deformation has occurred after crimping. The aluminum wire core undergoes grain refinement, and the internal grains are transformed into small-sized axial strip grains. The silver sheath and the copper terminal undergo sufficient plastic deformation, and dislocation movement and pinning occur in the crystal grains.

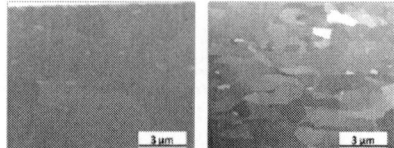

(a) FIB morphology of aluminum core before and after crimping

(b) FIB morphology of silver sheath before and after crimping

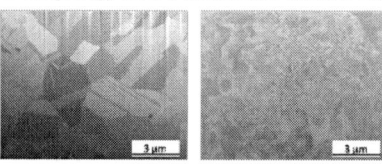

(c) FIB morphology of copper terminal before and after crimping

Fig.10 SEM morphology of FIB cut section before and after terminal crimping

Fig.11 shows the metallographic morphology of the improved terminal core crimping area. The observation holes of the two specifications of terminals are not located in the core crimping area to prevent the copper terminal from cutting the silver sheath. After crimping, the inner core of the terminal, the silver sheath and the copper terminal interface are well combined and the silver sheath remains intact. The improved terminal was tested for tensile strength and sealing and the test results were all qualified.

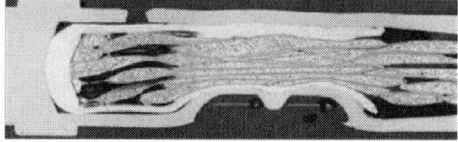

(a) Improved 2020 crimp terminal

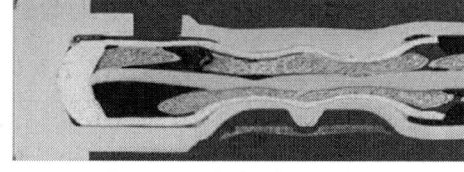

(b) Improved 2024 crimp terminal

Fig.11 Metallographic morphology of core crimping area after terminal improvement

Table III Test results of 2020 terminal-AWG20 wire/2024 terminal-AWG24 wire crimping assembly joint after terminal improvement

Terminal	Tensile strength (>6.1Kgf)	Sealing Test (≤1ml)	Terminal	Tensile strength (>3.2Kgf)	Sealing Test (≤1ml)
2020	9.2 (Pass)	<1(Pass)	2024	5.2 (Pass)	<1(Pass)
2020	9.6 (Pass)	<1(Pass)	2024	5.4(Pass)	<1(Pass)
2020	9.5 (Pass)	<1(Pass)	2024	4.9 (Pass)	<1(Pass)

IV. CONCLUSION

In this study, the structure, tensile strength and tightness of the aerospace CCA wire crimp assembly are tested, combined with the failure cause analysis and improvement methods of unqualified samples, and the following conclusions are drawn:

1) Moving the contact observation hole out of the core crimping area to avoid damage to the sealed silver sheath. Controlling the length of the stripped insulation and the penetration depth of the wire can ensure that the insulation lamination position forms a sealed crimp.

2) When the inter-metal crimping strain matching of the core crimping zone is poor, it is difficult to simultaneously produce matching plastic deformation. When the crimping force is small, it is difficult to coordinate the deformation to form a tight bond. When the crimping force is large, the wire core or the silver sheath may be damaged due to excessive deformation.

3) Annealing treatment reduces the hardness of the crimping position of the copper terminal and the silver sheath, which can improve the crimp matching with the CCA core. When crimping, both the terminal and the wire core undergo plastic deformation, thereby ensuring that the tensile strength and sealing performance of the assembly are qualified.

REFERENCES

[1] M.R. Patel, Spacecraft Power Systems. Boca Raton, Florida: CRC Press, 2005.

[2] Hug E, Bellido N. Brittleness study of intermetallic (Cu, Al) layers in copper-clad aluminium thin wires [J]. Materials Science and Engineering: A, 2011, 528(22-23): 7103-7106.

[3] Su, YaJun, Liu, et al. Interfacial Microstructure and Bonding Strength of Copper Cladding Aluminum Rods Fabricated by Horizontal Core-Filling Continuous Casting [J]. Metallurgical & Materials Transactions A, 2011, 42(13): 4088-4099.

[4] D. CHU, J. ZHANG, Cu/Al interfacial compounds and formation mechanism of copper cladding aluminum composites. Transactions of Nonferrous Metals Society of China. 2017, 27(11): 2521-2528

[5] W.B. LEE, K.S. BANG, S.B, JUNG. "Effects of intermetallic compound on the electrical and mechanical properties of friction welded Cu/Al bimetallic joints during annealing". Journal of Alloys & Compounds, 390(1/2), 2005, pp. 212-219.

[6] H.U. YUAN, Y.Q. CHEN, L. LI, H.D. HU. "Microstructure and properties of Al/Cu bimetal in liquid-solid compound casting process". Transactions of Nonferrous Metals Society of China, 26(6), 2016, pp.1555-1563.

[7] R.Kauffman, R. Makote, P. Youngerman, S. Sathish, "Research for the Electrical Wiring Interconnection System Program" , Washington, DC, U.S.Department of Transportation Federal Aviation Administration, 2007.

[8] Sarraf D., Schmidt H. Electrical Crimp Consolidation [J]. SAE International Journal of Engines, 2017, 10(4): 2046-2050

[9]Gueydan A., Hug E. Secondary creep stage behavior of copper-clad aluminum thin wires submitted to a moderate temperature level [J]. Mater Sci Eng A-Struct Mater Prop Microstruct Process, 2018, 709:134-138.

[10] Acarer M. Electrical, Corrosion, and Mechanical Properties of Aluminum-Copper Joints Produced by Explosive Welding [J]. Journal of Materials Engineering & Performance, 2012, 21(11): 2375-2379.

A 3D-MEMS Based Architecture for CS-CMOS Heterogeneous Integration

Min Huang
*National Key Laboratory of Science
and Technology on Monolithic
Integrated Circuits and Modules*
Nanjing, China
179278182@qq.com

Pengfei Liu
Nanjing Electronic Devices Institute
Nanjing, China

Hongze Zhang
*National Key Laboratory of Science
and Technology on Monolithic
Integrated Circuits and Modules*
Nanjing, China

Yan Wu
Nanjing Electronic Devices Institute
Nanjing, China

Jian Zhu
Nanjing Electronic Devices Institute
Nanjing, China

Abstract—**Integration of high performance III-V compound semiconductor circuits with CMOS provides optimum performance and compact size for RF modules. A 3D-MEMS based architecture is presented in this paper to integrate compound semiconductor devices with CMOS circuits. Devices made by different materials can be integrated in a 3-interposer stacked configuration. The 3D-MEMS based architecture is suitable for RF modules up to 40GHz. Maximum module size supported is up to 30 mm by 30 mm, providing enough design flexibilities for complex circuit functionality.**

Keywords—heterogeneous integration, MEMS, RF system, CMOS, compound semiconductor, Si interposer

I. INTRODUCTION

Complementary metal oxide semiconductor (CMOS) offers ultimate advantages in terms of maturity, complexity and integration density among semiconductor technologies, especially in digital function applications [1]. III-V compound semiconductor (CS) technology is dominating radio frequency (RF) applications due to their superior transport properties. CS devices offer better RF performance in power, efficiency, bandwidth, dynamic range and frequency, comparing to their Si based competitors [2, 3]. Therefore, integration of high-performance CS circuits with CMOS is essential for a RF module with both optimum performance and compact size. Conventional printed circuit board (PCB) or low temperature co-fired ceramic (LTCC) based assemblies are cost-effective and reliable solutions for CS and CMOS integration. However, drawbacks in pattern resolution, stacking accuracy, and interconnect size restrict these techniques in higher frequency and more compact applications [4, 5]. Integration of CMOS and CS devices on a common Si interposer is emerging as an attractive approach [6, 7]. Si interposer can serve as a high-density substrate with redistribution layer (RDL). Unlike traditional packages, the fine pitch made by Si process can easily match advanced IC technology nodes. Combined with Through Silicon Via (TSV) and wafer level stacking, Si based package has highly efficient interconnections between interposers. Si interposer enables integrated passive devices (IPD) and is compatible with MEMS process. In addition, Si has relatively large thermal conductivity, making it a good candidate for high power applications.

In this paper, we present a 3D-MEMS based architecture to integrate CS devices with CMOS circuits for unprecedented circuit performance. Three layers of Si interposers are stacked vertically using wafer level packaging technique. The 3D-MEMS based architecture is suitable for RF modules up to 40GHz. CS and CMOS devices can be interconnected with on-wafer RDL and cross-wafer TSV, resulting in reduced time delay and RF signal loss. Furthermore, CMOS enables more functions that CS doesn't have. For example, control circuits, calibration functions, and memory functions in CMOS can be integrated to improve IC utility and analog performance, leading to significantly improved design flexibility, performance, and compact module size. Maximum module size supported by this architecture is 30mm by 30mm, thus more complex circuit functionality can be integrated if needed.

II. 3D-MEMS ARCHITECTURE AND PROCESS

A. Proposed 3D -MEMS Architecture

Schematic view of the 3D-MEMS integration architecture is shown in Fig. 1 (a). It consists of three silicon interposers. Interposer is made of high-resistivity silicon (HRS) to reduce RF insertion loss. Silicon wafer resistivity is chosen to be ~10000 Ω·cm. Si Interposers are of 400 μm in thickness with TSVs of 80 μm in diameter. RDLs are patterned on both side of the interposer. Minimum pitch and pattern size of the transmission lines are 5 μm. Interposers are stacked vertically using wafer level bonding technique. Interconnecting precision is within 2μm between Si interposers. RDLs combined with TSVs build a 3D transmission network for both RF and digital signals.

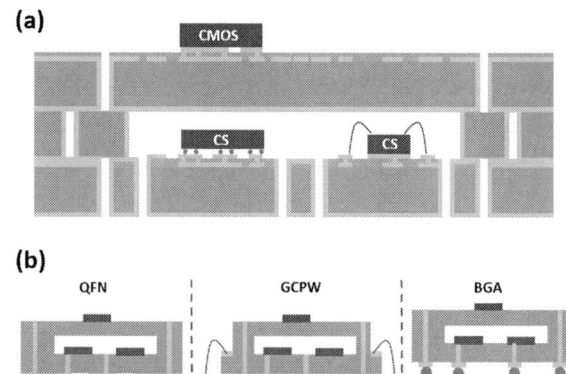

Fig. 1. (a) Cross-sectional view of 3D-MEMS based heterogeneous integration architecture, (b) Signal lead-out types supported by 3D-MEMS based integration architecture.

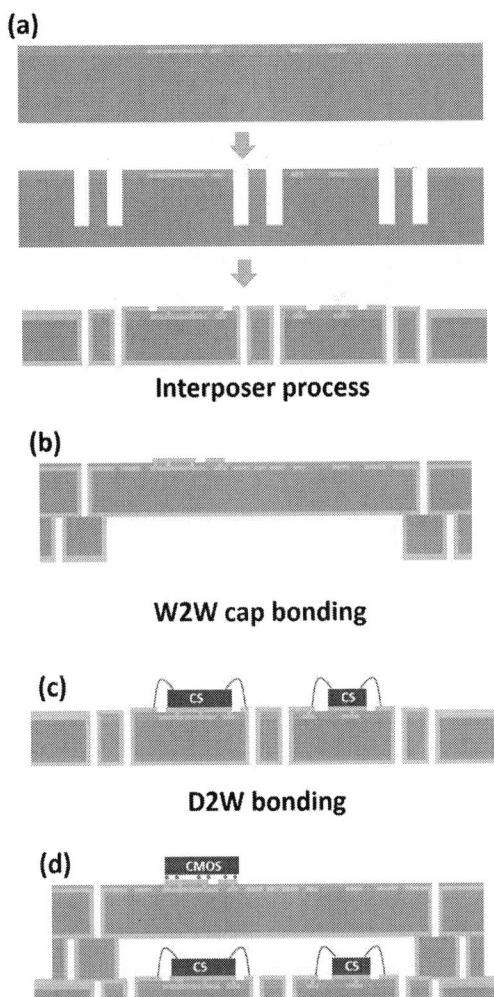

Interposer process

W2W cap bonding

D2W bonding

W2W stacking and D2W

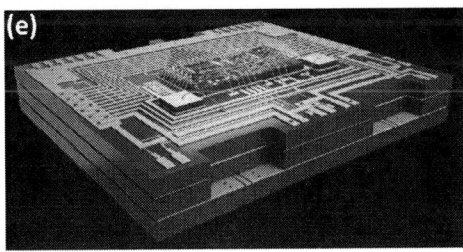

Fig. 2. 3D-MEMS Fabrication Process. (a) MEMS interposer process. (b) Wafer to wafer fusion bonding process to form a cap. (c) Die to wafer process. (d) Wafer to wafer eutectic bonding process. (e) 3D model of the TSV-MEMS configuration.

CMOS and CS devices can be heterogeneously integrated onto the Si interposer with various die to wafer (D2W) techniques: flip chip, surface mount, ultrasonic bond, bump, wire bond, die-attach epoxy, et al. Typically, CS devices are placed on the bottom interposer, while CMOS circuits are integrated on the top interposer. RF transmission losses are minimized and CS devices are shielded in the cap formed by top and middle interposers in this configuration. As shown in Fig. 1 (b), Multiple Signal lead-out types are supported by our

architecture: quad flat no-lead package (QFN), ball grid array (BGA), and grounded coplanar waveguide (GCPW), depending on different applications and design.

This 3D-MEMS based architecture supports RF modules up to 40 GHz. CS devices enable better RF performance while CMOS provide digital control functions. Combination of CS and CMOS devices in this heterogeneous integrated configuration provide enough design flexibility in a compact module volume. Size of the module is limited to 30 mm by 30 mm due to the stress mismatch between silicon and conventional PCB substrate.

B. 3D-MEMS Fabrication Process

The process flow is shown in Fig. 2. The whole process can be divided into three parts: interposer fabrication, D2W integration and wafer to wafer (W2W) stacking. Every interposer shares the same fabrication process.

Interposer fabrication flow is shown in Fig. 2 (a). We start with a standard 8-inch double side polished silicon wafer. Resistivity of the wafer is $10000\ \Omega \cdot cm$. Wafer is cleaned using RCA first. The first layer of RDL is patterned. Followed by silicon oxide layer deposition and via etch. Then TSVs are formed by photo lithography and deep reactive ion etching (DRIE). DRIE is also applied to etch cavities if necessary. The TSVs are 80 μm in diameter and 400 μm in depth. After that, the wafer is thinned to 400 μm. A silicon oxide layer is deposited on both side of the wafer. Followed by 1000 Å titanium (Ti) and 2000 Å gold severing as seed layer. After that, 1.2 μm gold layer is electroplated on both sides of the wafer, and inside the TSVs as well. Finally, the top layer of RDL is patterned by lithography and dry etch. Bottom solder mask can be added if using BGA as lead-out. Two interposers are bonded using wafer level bonding process, as shown in Fig 2 (b). Au/Au fusion bonding is applied at 300 °C for 2 hours in this case. Bonding accuracy is within 3 μm. Cavity in one interposer combined with the other interposer forms a cap, providing electromagnetic shielding for the integrated CS devices. Devices are integrated onto the interposer using D2W process. Various integration techniques can be employed . Fig. 2 (c) shows a typical method for MMIC chip integration. The chip can be epoxy attached and then wire bonded onto the substrate. Finally the cap and bottom interposer are stacked

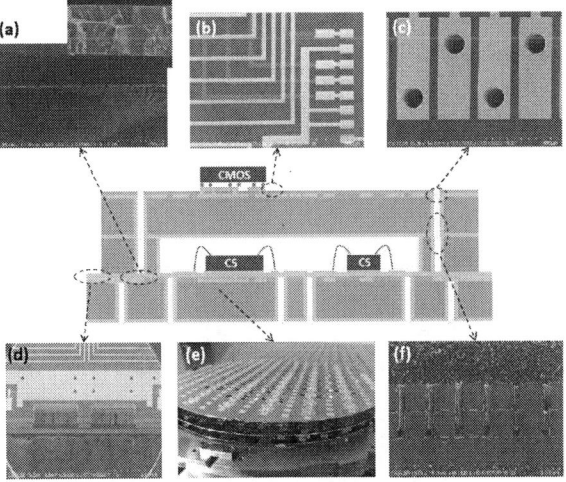

Fig. 3. (a) Cross sectional view of W2W bonding area. (b) Top view of RDL. (c) Top view of TSV and bonding area. (d) SEM image of GCPW lead-out. (e) W2W stacking. (f) Cross section view of TSV alignment.

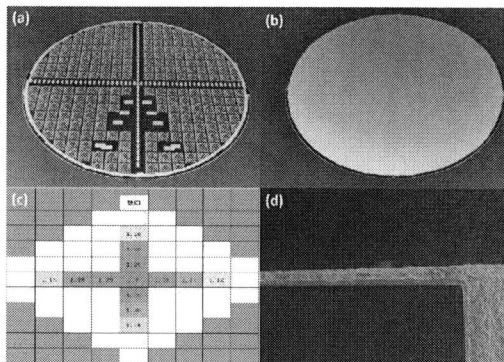

Fig. 4. (a) Front side view of a interposer wafer, (b) Back side view of a interposer wafer, (c) Within wafer uniformity of plating, (d) SEM image of the plated gold layer.

TABLE I. Au/Su ratio at bonding interface

Position	Au Atomic %	Sn Atomic %
1	94.927%	5.073%
2	90.507%	9.493%
3	91.417%	8.583%
4	85.481%	14.519%
5	87.761%	12.239%
6	90.231%	9.769%
7	97.138%	2.862%
8	96.647%	3.353%

using low temperature W2W technique, as shown in Fig. 2 (d). Other devices can be integrated on top of the cap. All the devices are connected through RDLs and TSVs. Fig. 2 (e) shows a 3D animation of a typical RF module build based on this 3D-MEMS architecture. GCPW lead-out is applied in this case.

III. 3D-MEMS Configuration in Details

A. 3D-MEMS Architecture in details

Fig. 3 show a typical RF module that supported by this 3D-MEMS architecture. Details of the configuration are presented. Fig. 3 (a) shows the cross sectional view of the W2W bonding area. Au/Sn low temperature bonding technique is used. Scanning electron microscope (SEM) image shows a well bonded interface between two interposer. Inset in Fig. 3 (a) reveals that three metal layers are fused together. No void or crack is observed on the bonding area. Fig. 3 (b) shows the double layer RDLs. More layers can be added by Cu damascene technique. Interconnections between RDLs and TSVs are shown in Fig. 3 (c). Top layer of the RDL is connected to the TSV through gold pads. Typically, control signals sent from CMOS chips go through the TSVs and then reach the CS devices. Fig. 3 (d) shows the GCPW lead-out. Top two layers of interposers are partially etched to reveal the ground-signal-ground (GSG) pads. The RF module is grounded by the TSVs on substrate, and the signal input/output are realized by wire bonding onto other substrate. Fig. 3 (e) is a schematic view of wafer level stacking. Fig. 3 (f) show the cap bonding interface. Alignment accuracy is within 2 μm due to wafer level bonding technique.

B. Key process modules

Many novel fabrication techniques are implemented in this 3D-MEMS process, including double side plating, Au/Sn eutectic bonding, wafer level testing, et. al. Double side plating is used in the process. Front/back side of the wafer and the inner walls of TSVs are plated at the same time. Greatly

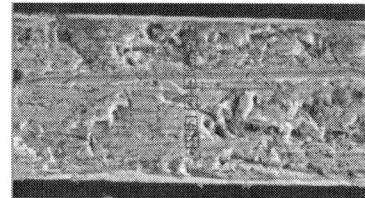

Fig. 5. Cross sectional view of Au/Sn eutectic bonding interface.

increase the efficiency for mass production. Fig. 4 (a) and 4 (b) show the front and back side of the plated interposer, respectively. Fig. 4 (c) shows the plating uniformity. 13 points are measured within a wafer. The gold thickness is between 1.1 μm and 1.3 μm. Cap to substrate stacking is realized using wafer level Au/Sn eutectic bonding. Bonding is performed at 280 °C to protect the integrated devices. SEM image of the bonding area is shown in Fig. 5. Metal layers are melted and form a eutectic system. No boundary is observed. Further measurement is done by using elemental analysis. Table I shows the Au and Sn ratio at different positions marked in Fig. 5. Sn gradually decrease from the bonding interface, forming a solid eutectic system. Several reliability tests are applied to see if our 3D-MEMS configuration meets industrial requirements. Our module passes all the following tests: high temperature storage at 125 °C for 1000 hours, -40 °C to 150 °C temperature cycle for 500 times, constant acceleration at 29400 m/s^2 and mechanical vibration for 100 times.

IV. Conclusion

We present a Si based integration technique that integrates CS (GaAs, GaN, InP, et al.) with CMOS. RF modules made using this technology can take advantage of the high frequency, high power and broadband performance of CS devices while maintaining the unique qualities of Si CMOS, such as maturity and complex digital functionalities at a low cost. Pattern resolution, stacking accuracy and thermal conductivity in our technique is better than PCB or LTCC, making it more suitable for high frequency and high power RF applications. The architecture support varieties of D2W heterogeneous integration techniques and lead-out types, leading to significantly improved design flexibility. The whole fabrication process is built on mature and straightforward MEMS process. It is highly feasible and cost effective for mass production.

References

[1] I. Abdomerovic, W. D. Palmer, P. M. Watson, R. Worley and S. Raman, "Leveraging Integration: Toward Efficient Linearized All-Silicon IC Transmitters," IEEE Microwave Magazine, vol. 15, pp. 86-96, 2014.

[2] S. Rangan, T. Rappaport and E. Erkip, "Millimeter-wave cellular wireless networks: Potentials and challenges," Proc. IEEE, vol. 102, pp. 366-385, 2014.

[3] K. B. Cooper and G. Chattopadhyay, "Submillimeter-wave radar:Solid-state system design and applications," IEEE Microwave Magazine, vol. 15, pp. 51-67, 2014.

[4] K. K. Samanta, "Ceramic high performance RF front-ends at mmW and beyond," IEEE European Microwave Conference, 2015.

[5] T. V. Heikkila and M. Lahti, "LTCC 3-D integration platform for microwave and millimeter wave modules," Proc. IEEE Int. Symp. Radio-Frequency Integration Technology, pp.95-97, 2012.

[6] Y. Zhao and S. Wang, "A novel 3D T/R Module with MEMS technology," IEEE international Conference on Integrated Circuits and Microsystems(ICICM), pp. 286-289, 2017.

[7] K. K. Samanta, "Pushing the envelope for heterogeneity," IEEE Microwave Magazine, vol. 17, pp. 28-43, 2017

Design and Optimization of Temperature Sensor Based On LGA Package Structure

1st Jianhui Liu
R&D Design Department
SKY CHIP INTERCONNECTION
TECHNOLOGY CO., LTD
Shen Zhen, China
liujh@scc.com.cn

2st Juntao Wang
R&D Design Department
SKY CHIP INTERCONNECTION
TECHNOLOGY CO., LTD
Shen Zhen, China
wangjt@scc.com.cn

3st Jun Li
Shennan circuits co.,LTD.
Shenzhen, China
Lij@scc.com.cn

Abstract—With the rapid development of wearable devices, the IC temperature sensors which used LGA Package have been widely used to meet the needs of product miniaturization. There are three heat conduction modes in LGA package: heat conduction, air convection and radiation, and thermal conductivity directly affects the temperature sensor's induction speed and measurement accuracy. This paper mainly introduces the heat conduction structure, heat transfer mechanism and thermal resistance calculation method of temperature sensor based on LGA package; the finite element analysis software used to model the structure of the IC temperature sensor module, and simulates the thermal resistance value and temperature rise. By analyzing the heat conduction between the inner structures of the packaging module, the thermal conductivity structure of the module is optimized, to improve the induction speed and measurement accuracy of the IC temperature sensor. By replacing part of the resin material with stainless steel, increasing the thermal conductivity area of the stainless steel material and sticking the temperature sensor chip on the stainless steel, the design and optimization of the thermal conductivity effect were carried out. The simulation analysis showed that the third thermal conductivity effect was the best.

Keywords—LGA Package, IC Temperature sensor, Temperature, Heat conduction, Thermal resistance

I. INTRODUCTION

When people get information from the outside world, they must rely on their own sensory organs. But in the study of natural phenomena and laws, their functions cannot meet the needs. In order to adapt to this situation, sensors came into being, and temperature sensor is one of them [1]. Temperature is a physical parameter reflecting the cold and hot state of objects, which is closely related to human living environment. In order to detect the temperature, people begin to use temperature sensor detection, whether industry, agriculture, commerce, scientific research, medicine and so on have a close relationship with temperature [2]. With the development of semiconductor technology, semiconductor thermocouple sensor, PN junction temperature sensor and integrated temperature sensor have been widely used [3].

There are four types of temperature sensors: thermocouple, thermistor, RTD and IC temperature sensor [4]. This paper mainly introduces the IC temperature sensor, that is, digital integrated temperature sensor. Its shape is very small, and it can be widely used in production practice, providing countless convenience for our life. In particular, the intelligent temperature sensor chip has the advantages of low cost, low power consumption, compatible with standard digital technology and small chip area [5].

At present, IC temperature sensor has been widely used in portable human body temperature measurement products. This type of product mainly uses a temperature sensor which has good thermal contact between the temperature measuring element and the measured object, and achieves thermal balance through heat conduction. This paper mainly introduces the thermal conduction structure of temperature sensor based on LGA package and optimizes the structure design of IC temperature sensor package device by using finite element analysis method, so that the temperature can be transferred to the temperature sensing area of chip more effectively.

II. PRODUCT STRUCTURE AND TECHNOLOGY

LGA packaging is a relatively mature chip packaging process at present. It has the characteristics of small product size and excellent thermal conductivity, which can meet the requirements of miniaturization of temperature sensor packaging in portable temperature measuring products. The packaging structure of the temperature sensor chip is shown in Figure 1 below. The temperature chip is placed in the center of the LGA substrate, and the chip pad is connected to the substrate by Wire Bonding (Gold wire), and then the chip is protected by molding compound (Resin).

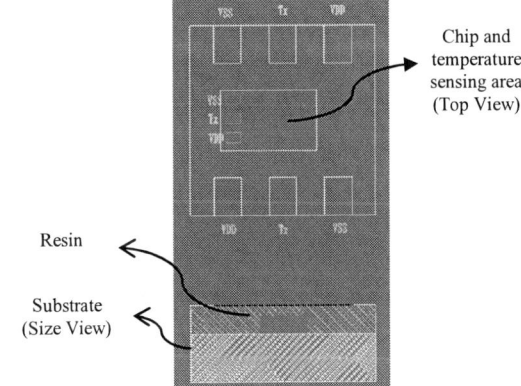

Fig. 1. Package structure.

The temperature sensing area is 0.1*0.1*0.01mm, located in the center of the chip surface. With the enhancement of product performance and package miniaturization, the accuracy and stability of the data obtained by experiments cannot meet the needs of the products. It is an important part to use the advanced finite element analysis software for thermal analysis and design of the temperature sensor chip package, improve design correctness.

978-1-6654-1392-3/21 $31.00 © 2021 IEEE

III. THERMAL DESIGN THEORY AND THERMAL RESISTANCE CALCULATION

A. Thermal Design Theory

There will be heat transfer between any two different objects and different temperature regions of the same object. The basic law of heat transfer is that heat is transferred from high temperature region to low temperature region. The basic expression of heat transfer is as follows (1):

$$Q=KA\Delta t \qquad (1)$$

Q-Heat flux，the unit is W；K-Heat transfer coefficient，the unit is W/(m^2 • °C); A-Heat exchange area，the unit is m^2; Δt-Temperature difference between high and low temperature materials，the unit is °C.

Heat transfer includes three basic modes: heat conduction, convection and heat radiation. Heat conduction and convection need heat conduction medium, and radiation transfers energy directly in the air [6]. In order to facilitate the comparison of experimental data, this analysis ignores the influence of convection and thermal radiation, and only analyzes the heat conduction between the internal structures of packaging devices.

B. Thermal Resistance Calculation

The heat conduction of microcircuits is a complex process. The most basic calculation formula for each layer of heat conducting medium is as follows (2)：

$$R=L/(k \cdot A) \qquad (2)$$

L is the thickness of the heat transfer path, the unit is m；k is the thermal conductivity of the material，the unit is W(m•K)-1；A is the cross-sectional area perpendicular to the direction of heat flow，the unit is m^2。

IV. MODELING AND SIMULATION ANALYSIS

A. Finite Element Model

The finite element model of LGA Package is shown in Figure 2, which mainly includes chip, substrate and resin. The bottom of the chip is bonded to the substrate with insulating adhesive. Combined with the application scenario, the temperature sensor chip has no heating, and the LGA packaging surface is set at constant 37 °C, the temperature is transferred from top to bottom to the chip temperature sensing area.

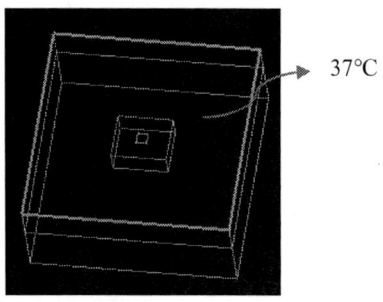

Fig. 2. Finite element model a.

The surface temperature is mainly transferred to the chip temperature sensing region by heat conduction through the resin.

The material parameters and size of each layer structure are shown in Table I, and the thermal conductivity of material is closely related to temperature and density. In this project, the influence of temperature and density on the thermal conductivity of material is ignored.

TABLE I. MATERIAL AND SIZE

Structure	Material	Conductivity/W (m • K) -1	Specific heat J/kg-K	Density kg/m3	Size/mm
Chip	Si	148	556.9	2330	0.4*0.5*0.2
Molding	Resin	0.25	1000	980	1.5*1.5*0.3
Sheet metal	Stainless steel	14.6	502	8000	3.0*3.0*0.3
SUB	HL832	0.8	795	1900	3.5*3.5*0.5

B. Simulation Analysis

A temperature monitoring point is set in the 0.1 * 0.1 mm temperature sensing area of the chip center to monitor the temperature change. According to the application requirements, set the ambient temperature at 25 °C and the constant temperature of the surface at 37°C for 5s, and analyze the transient state of the temperature sensor chip within 5S. The simulation data are shown in Table II, figure 3, figure 4 and figure 5.

TABLE II. 0~5S TEMP

Time	1s	2s	3s	4s	5s
Temp/°C	34.24	35.68	36.36	36.69	36.85

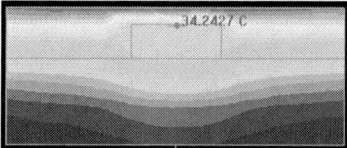

Fig. 3. 1s Temp.

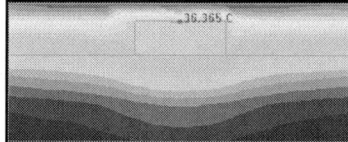

Fig. 4. 3s Temp.

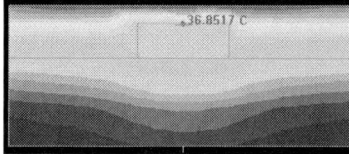

Fig. 5. 5s Temp.

According to the simulation results, the temperature sensor chip reaches 36.85 °C in 5 seconds, which is 0.15 °C different from the target temperature of 37 °C. It can be observed from the time-dependent data and temperature variation diagram that the temperature conduction on the top of the chip is slow, which is caused by the low thermal conductivity of the resin material. It is necessary to optimize the LGA package structure to improve the temperature sensing accuracy.

C. Structural Optimization and Simulation Analysis

It can be seen from (2) that the thermal resistance is related to the thermal conductivity, thickness and area of the medium. In order to improve the thermal resistance of LGA package, we can optimize the structure of the cooling channel.

Optimized scheme：

1) Some resin materials are replaced by stainless steel with high thermal conductivity to improve the thermal conductivity of LGA packaging materials, to achieve the purpose of reducing the thermal resistance R without changing the L and A, as shown in Fig. 6 Structure b.

2) Reduce the area of resin material, increase the heat conduction area of stainless steel material, and then improve the thermal conductivity of LGA packaging material, as shown in Figure 7 Structure c.

3) The temperature sensor chip is pasted upside down on the stainless steel sheet, and the temperature is directly transferred from the stainless steel to the chip, as shown in Fig. 8 Structure d.

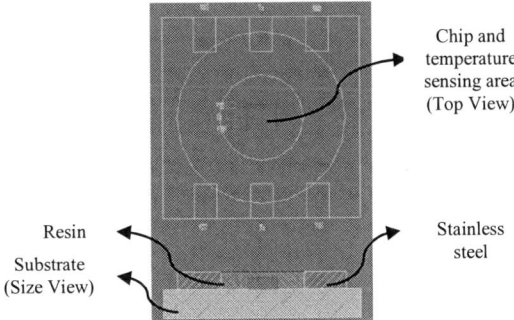

Fig. 6.   Structure b.

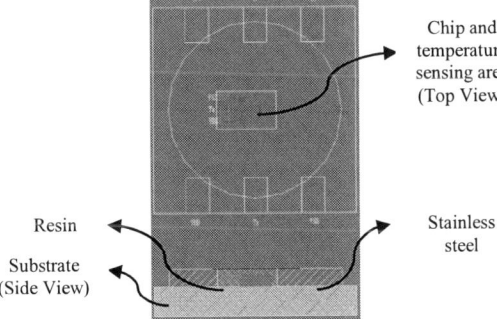

Fig. 7.   Structure c.

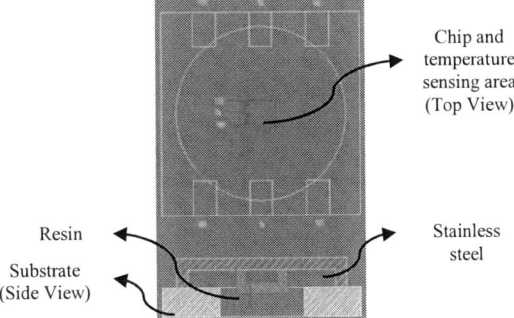

Fig. 8.   Structure d.

Compared with model a, model b and c increase the heat conduction channel of stainless steel, and model d increases the overall heat conduction capacity. According to (2), the theoretical thermal resistance values of each model are shown in Table III.

TABLE III.   THEORETICAL THERMAL RESISTANCE

Model	a	b	c	d
R/°C•W-1	44	11	4.1	1.0

The theoretical thermal resistance of b, c and d is 75%, 90.6% and 97% lower than that of a, b, c and d simulation models were established for simulation analysis. The models is shown in Figure 9 ~ 11.

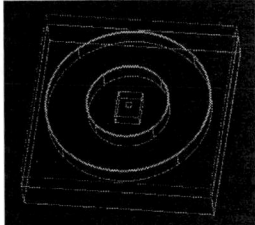

Fig. 9.   Model b.

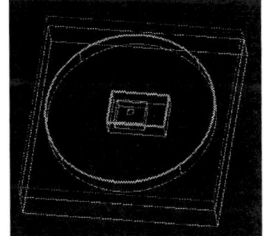

Fig. 10. Model c.

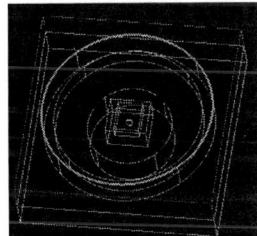

Fig. 11. Model d.

The simulation method is the same as model a. The temperature change within 5s of the temperature sensing chip is shown in Table IV, and the heat conduction in 5s is shown in Fig. 12 ~ 14.

TABLE IV.   0~5S TEMP

Time	b	c	d
1s	35.45°C	36.39°C	36.99°C
2s	36.59°C	36.87°C	37°C
3s	36.86°C	36.96°C	37°C
4s	36.95°C	36.98°C	37°C
5s	36.98°C	36.99°C	37°C

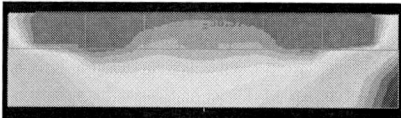

Fig. 12. 5S Temp of model b.

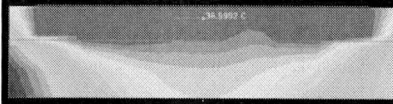

Fig. 13. 5S Temp of model c.

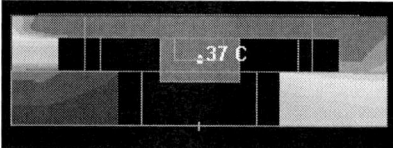

Fig. 14. 5S Temp of model d.

According to the simulation data, model b is 36.98 °C in 5S, model c is 36.98 °C in 4S and model d is 37 °C in 2S. The induction rates of b, c and d are 40%, 60% and 90% higher than that of a, respectively. The thermal conductivity of structure d is better than that of the other three structures, but the packaging complexity is higher.

V. CONCLUSIONS

This paper mainly introduces the thermal conductivity structure, heat transfer mechanism and thermal resistance calculation method of the temperature sensor based on LGA Package. The structure modeling and simulation analysis of IC temperature sensor were carried out by using the finite element software, and three optimization structure which could effectively improve the IC temperature sensor induction speed and measurement accuracy were given. The LGA package structure and simulation analysis method used in this paper can be applied to the packaging of high-precision IC temperature sensor, optimize package design and improve product performance.

ACKNOWLEDGEMENT

This work was supported by the National Key R&D Program of China (2018YFE0204600), the Key-Area Research and Development Program of GuangDong Province (2019B010131001), and Shenzhen Fundamental Research Program）(JCYJ20200109140822796).

REFERENCES

[1] Zhang ping. "Research and Design of CMOS Integrated Temperature Sensor," Xi'an Electronic Technology University, 2009, pp.5-8.

[2] Liu zhiyong. "Current Situation and Prospect Analysis of Sensor Industry Development," Heilongjiang Science and Technology Information, 2013, pp.20.

[3] Hu ling. "Research and Development of Optical Fiber Temperature Sensor," Science & Technology Association Forum, No.5, 2010, pp.90.

[4] Huang zexian. "Recent Developments in Temperature Sensors," Instrument materials, 1987.

[5] Cao xinliang, Yu ningmei, Wei qinxiao. "Integrated CMOS Temperature Sensor Design Experiment and Test," Chinese Journal of Sensors and Actuators, vol.23 No.1, 2010, pp.39-40.

[6] JOINER B, ADAMS V. "Measurement and simulation of junction to board thermal resistance and its application in thermal modeling," Proceedings of the 15th Annual IEEE Semiconductor Thermal Measurement and Management Symposium. USA, pp. 212-220, 1999.

Higher Aspect Ratio TSV Structure ECP Bottom-Up Plating Process

Yinuo Jin
Copper Process Department
ACM Research (Shanghai), Inc
Shanghai, P.R. China
yinuo.jin@acmrcsh.com

Bo Zheng
Copper Process Department
ACM Research (Shanghai), Inc
Shanghai, P.R. China
bo.zheng@acmrcsh.com

Jian Wang
CEO Office
ACM Research (Shanghai), Inc
Shanghai, P.R. China
jian.wang@acmrcsh.com

David H. Wang
B.D.O
ACM Research (Shanghai), Inc
Shanghai, P.R. China
dwang@acmrcsh.com

Qixing Yu
Copper Process Department
ACM Research (Shanghai), Inc
Shanghai, P.R. China
qixing.yu@acmrcsh.com

I. ABSTRACT

With the rapid development of the electronic industry, the demand on mini-size, low power consumption and high reliability becomes inevitable to electronic products. Based on Moore's law decreasing the feature size of the integrated circuits is approaching a bottleneck. Recently wafer-level vertical miniaturization 3D Through-Silicon-Via (TSV) package integration becomes an alternative solution to breakthrough the bottleneck of Moore's law down scaling in design, process and cost. Correspondingly, due to copper's higher conductivity, less resistant to electromigration, copper is widely used material to fill the TSV via.

Conventionally the copper metal layer deposition and planarization process contains several process steps: PVD, ECP, annealing, and CMP. Electrochemical plating (ECP) is a key process for copper via filling, which is required to achieving void-free and seam-free via filling, and with a minimized the copper overburden thickness. In wafer level mass production, the thinner overburden is beneficial to reduce wafer stress and shorten process time and reduce process cost.

The ECP process's performance is dependent on seed layer pre-treatment, chemistry, chemistry mass transfer, and plating current etc [1]. In this study, the 300mm wafer is pre-treated in a pre-wet chamber by DIW; after that the wafer will be located in a horizontal plating chuck and face down immersed in the plating chemistry with protecting voltage and current; the plating chamber will control the flow rate distribution, plating current, distance between anode and plating rotation speed. These parameters dominated via filling, overburden thickness and within wafer plating uniformity.

Keywords—Through Silicon Via (TSV), Electrochemical Plating (ECP), Copper, Via fill, Bottom-Up Plating

II. INTRODUCTION

With the semiconductor industry development, 3D package is an option for advance process integration, which becomes an alternative solution to breakthrough the bottleneck of Moore's law down scaling in design, process and cost. In the TSV process, the via filling step—which is commonly performed using copper electrochemical plating (ECP) accounts for almost 40% of the total cost. As the core and critical technologies of TSV, defect-free filling with minimization of process time and cost has attracted much attention[1].

In the 3D TSV and 2.5D Interposer process integration, the copper deposition is a critical process for higher aspect ratio via structure. The copper deposition process contains pre-wet treatment, electrochemical plating, post plating clean and anneal process. In this study, the ECP process test is on a 300mm horizontal type ECP tool "Ultra ECP 3d Tool" and use copper methane sulfonate as the plating solution. Based on the TSV plating theoretical and simulation study, to search the optimal process condition for 300mm wafer TSV by the experiment.

III. EXPERIMENT

A. Process Sequence

In the automatic ECP tool "Ultra ECP 3d Tool", the 300mm post PVD (Physical Vapor Deposition) wafer will be loaded from a loadport into the aligner to centering the wafer notch, the wafer will be transferred intro the pre-wet chamber. In the pre-wet chamber, the wafer will be handled in vacuum environment and pre-wetted by DIW (Deionized Water), to remove the air inside the TSV blind-vias and let the DIW fully fill in the TSV via. The moist wafer with DIW will be located in the plating chamber, which flipped by transfer robot and facing down into the plating chamber to complete the elctrochemical plating process. After that the post plating wafer will be transferred into the clean chamber, to remove plating solution residual on the wafer surface. In this study, the anneal process was skipped, focusing on the plating process experiential study. Figure1 shows the TSV ECP process sequence.

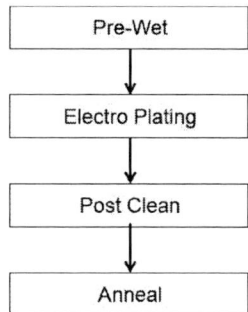

Fig.1.TSV ECP Process Sequence;

B. Electroplating Chamber Apparatus

In this study, the plating process was run in a 300mm wafer horizontal type plating chamber, which consisted of anode, ion membrane and diffusion plate. Figure2 shows the TSV electroplating chamber diagrammatic sketch. The TSV wafer was carried by the plating chuck and immersed in the plating chamber with the chuck. The copper methane sulfonate plating solution with the additive circulating and filling the plating chamber. During the plating process the wafer was horizontally rotated, and the plating current was controlled by the DC power supplier. In the experiment, the plating additive concentration, the flow rate, plating current density, and chamber circulating flow were as the main parameter to study the TSV via filling performance.

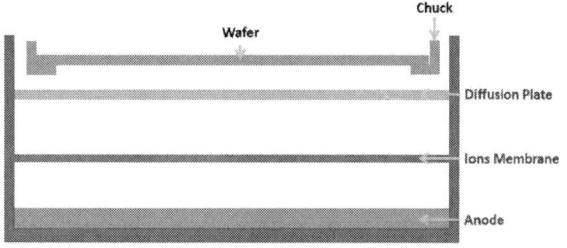

Fig. 2. Electroplating Chamber Diagrammatic Sketch;

C. Sample and Analysis Method

The 300mm TSV pattern wafer were used to study the effect of various process parameters and optimization. Various feature sizes,ranging from 5*65um，12*50um, and 7.7*60um in depth were electroplating using different chemical additive concentration/current density/flow rate, and the quality of the final deposit and fill characteristics was inspected using the FIB (focused ion beam) for sample preparation and SEM (scanning electron microscope) for via cross section. Otherwise, the X-Ray was introduced to observe wafer level process results. To improve the fill performance, additional processes, such as seed layer enhancement or pre-wet were applied to improve the quality of seed layer or wettab`ility of the features, respectively.

IV. RESULTS AND DISCUSSION

A. Pre-Treatment for Plating

Figure3 shows the cross-sectional SEM images for the TSV filling during electrodeposition without pre-wet treatment before plating process. A long and huge void defect was observed in the bottom of the via, as shown Fig.3(a). When the bottom of the via was zoomed in, there

was no electroplating in this area, due to the plating solution could not fill in.

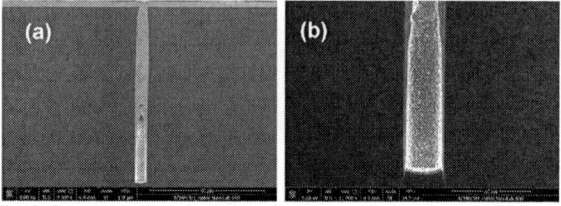

Fig.3.Cross-sectional SEM images of 5*65um TSV: (a) Without pre-wet overview; (b) Without pre-wet via bottom local zoom;

B. Plating Solution Additive Proportion.

Figure4 shows the cross-sectional SEM images for the TSV filling during electrodeposition at chemical additive proportion A/B of 3:10, 3:12 and 3:15; The copper methane sulfonate plating solution $Cu(CH_3SO_3)_2$ was made of 55 g/L Cu^{2+}, 50 mg/L Cl^-, the additive A contained the accelerator, and the additive B was the mixture of suppressor and leveler. The different proportion can be seen that a change in the additive proportion caused a change in the filling model. At the proportion A/B of 3:10, a long line defect was observed at the center of the via, as shown in Fig. 4(a). However, upon increasing the Additive A/B to 3:12 ,the voids turn to tiny. When the Additive A/B is 3:15, the TSVs were fully filled almost without any defects.

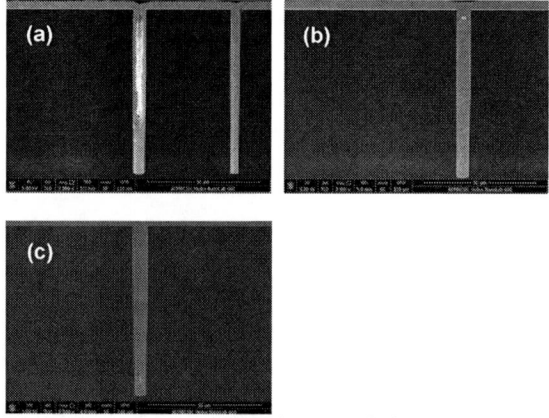

Fig.34.Cross-sectional SEM images of 5*65um TSV:

(a) A/B=3:10; (b) A/B=3/12; (c) A/B=3/15;

C. Plating current density

Figure5 shows the cross-sectional SEM images for the TSV filling during electrodeposition at current densities of 0.5ASD, 0.4ASD; It can be seen that a change in the current density caused a change in the filling model. At a higher current density (0.5ASD), a seam defect was observed at the centre of the via, as shown in Fig. 5(a). However, upon increasing the current density to 0.4ASD, the TSVs were fully filled without any defects, as shown in Fig. 5(b).

Reletive lower current density 0.4ASD induce Cu film having higher deposition rate in vertical direction, and lower deposition rate in horizontal direction, especially restrain the deposition rate in the overhang, finally reduce the probability of void formation. Due to the increased current crowding at the top of the via and more significant mass transport limitations towards the bottom of via, when plating at higher current densities [2].

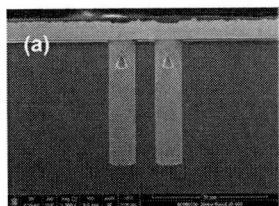 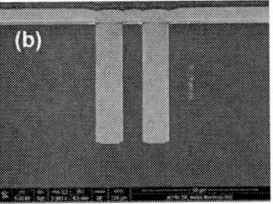

Fig.5. Cross-sectional SEM images of 12 * 50um blind-vias electrodeposited by direct current with different average current densities, (a)0.5ASD, (b) 0.4ASD;

D. Plating Flow Rate.

Figure6 shows the cross-sectional SEM images for the TSV filling during electrodeposition at flow rate of 30lpm and 35lpm; It can be seen that a change in the flow caused a change in the filling model. At the flow of 30lpm, a long line defect was observed at the center of the via, as shown in Fig. 6(a). However, upon increasing the flow to 35lpm ,the void was free.

 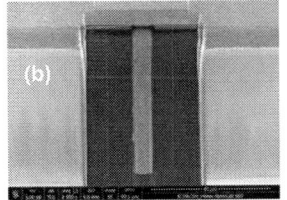

Fig.6.Cross-sectional SEM images of 7.7*60um TSV composited by different flow rate: (a) 30L/min; (b) 35L/min;

E. Dynamic TSV Filling Process with Medium Current Density.

Dynamic TSV filling process with low current density. According to previous reports, a low currentdensity results in a low deposition rate but seldom causes seam defects. To understand how a seam defect is formed, experiments with deposition times of 10, 30, 70, 110 and 150min were carried out at 0.2ASD.

In this manner, the dynamic filling process could be observed clearly, as shown in Fig.7. The thicknesses of deposited copper on the bottom are higher when time is longer, and bottom-up filling does is observed in any of the cases. More precisely, during the first 30 min of electrodeposition, the plating thickness at the bottom was more greater than that at the top. During the last 100 min of electrodeposition, the TSV was fully filled in.

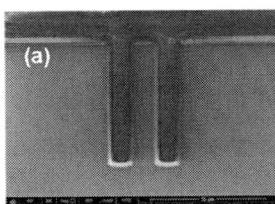 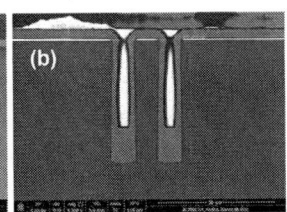

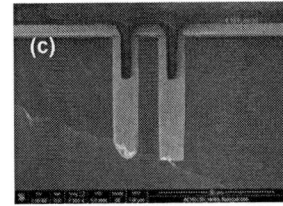 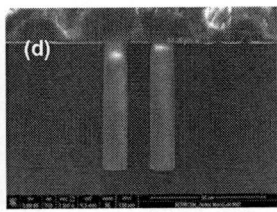

Fig.7. Cross-sectional SEM images of 12*50um blind-vias electrodeposited for the indicated periods of time with a constant average current density of 0.2ASD. The process of feature filling is captured at different time intervals:(a) 10min; (b)30min; (c) 70min; (d)110min;

F. TSV Filling Process in Wafer Level Control

Figure8 shows the X-Ray images for the TSV filling during electrodeposition in wafer level scale. A post plating 5*65um TSV structure pattern with the inspection locations was as shown in Fig. 8(a).

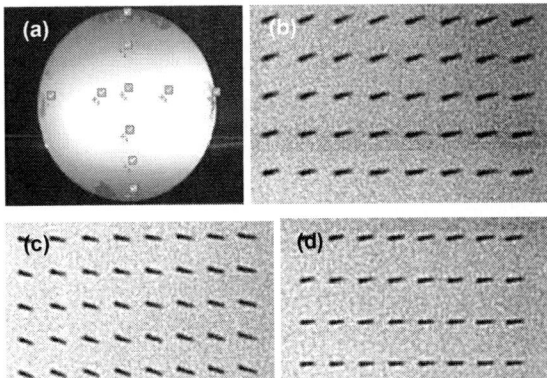

Fig.8. X-Ray Inspection images of 5*65um blind-vias electrodeposited for the indicated wafer level scale via filling; (a) inspection point location; (b) Location1wafer edge; (c) Location2 wafer middle ; (d) Location3 wafer center;

Figure9 shows the cross-sectional SEM images for the TSV filling during electrodeposition in wafer different locations. It can be seen that different locations' via filling and surface overburden thickness. The overburden thickness is listed in the Table1, the overburden thickness represented the within-wafer lever process uniformity, the data could be contrast to wafer surface copper film RS (film resistance) data.

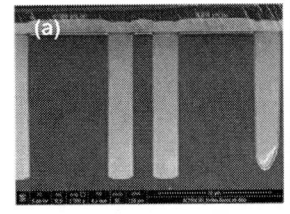 

Fig.9.Cross-sectional SEM images of 12*60um TSV via filling process in different wafer location: (a) wafer center; (b) wafer middle; (c) wafer edge;

Table 1 Overburden Thickness of Different location

Overburden Thickness of Different Location on the Wafer			
Overburden(um)	Center	Middle	Edge
	4.86	4.63	4.77

V. CONCLUSIONS

Copper electrodeposition has been proven to be a feasible solution to filling TSV blind-vias structure, applicable over a wide range of features, especially for higher aspect ratio blind-vias.

For the TSV wafer process development, pre-treatment such as pre-wet process is an essential process step to ensure the plating solution could be convect and diffuse to the bottom of blind-vias.

For the electroplating process required void and seam free deposition, the plating solution additive proportion, current density, plating solution circulating flow rate are the critical parameter for experiment results. A optimal additive proportion with lower current density in a higher flow rate will achieve an significately bottom-up plating process. The idle bottom-up plating process will have a wider process window for wafer level via filling control.

For TSV mass production, beside the via filling, the overburden thickness control is another process specification.

The wafer level uniformity is determined by electrochemical plating chamber's flow field distribution and current distribution. A thinner and uniform overburden copper will improve the throughput of ECP and the following process step CMP (Chemical Mechanical Planziation), to achieve a lower process cost.

VI. ACKNOWLEDGMENTS

The authors would like to acknowledge SMNC thin film department and CXMT BCC department for providing the blind-vias TSV test wafer, Singyang R&D lab for plating solution and additive concentration analysis, and our colleague ACM FA lab their continuous help with sample preparation and FIB/SEM analysis.

VII. REFERENCE

[1] Kondo, K., Suzuki, Y., Saito, T., Okamoto, N. & Takauchi, M. High Speed Through Silicon Via Filling by Copper Electrodeposition. Electrochemical and Solid-State Letters 13, D26 (2010).

[2] Lutz Hofmann, Ramona Ecke, Stefan E. Schulz a, Thomas Gessner. Investigations regarding Through Silicon Via filling for 3D integration by Periodic Pulse Reverse plating with and without additives. Microelectronic Engineering, Volume 88, Issue 5, May 2011, Pages 705-708

Monte Carlo based stochastic finite element model for uncertainty quantification in flip chip BGA electronic packaging

Liu Chu
School of transportation
Nantong University
Nantong, China
chuliu@ntu.edu.cn

Jiajia Shi
School of transportation
Nantong University
Nantong, China
shijj@ntu.edu.cn

Robin Braun
School of Electrical and Data
Engineering
University of Technology Sydney
Ultimo, NSW 2007, Australia
robin.braun@uts.edu.au

Abstract — **The Ball grid array (BGA) electronic packaging method is an efficient and appropriate way for high density integrated circuits. The uncertainties in the material and geometrical parameters to product reliability and safety under the real operating situation are the crucial issues deserved more attention. In this paper, the Monte Carlo based stochastic finite element model (MC-SFEM) is proposed for uncertainty quantification in flip chip BGA electronic packaging. The Monte Carlo stochastic sampling process is combined with the finite element computation for the resonant frequencies response of the flip chip BGA electronic packaging. The uncertainties in material and geometrical parameters of flip chip BGA electronic packaging are performed and propagated by advanced Monte Carlo method (Latin Hypercube sampling method). Four different BGA configurations are compared and discussed. The proposed model and the computational results in this paper provide meaningful references to the electronic package reliability prediction.**

Keywords — *Monte Carlo, Stochastic finite element model, Uncertainty quantification, Electronic packaging component*

I. INTRODUCTION

The significant characteristics of ball grid array (BGA) packages are small size and high level of integrated density [1-2], making it one of the most promising packaging technologies in automotive industry [3]. The BGA package is a possible and feasible solution for increasing density of electronics on vehicles. However, the uncertainty in material properties and manufacturing process is an essential factor with evident impacts in the electronic package quality and reliability.

It is low-efficient and cost expensive to have tremendous experimental tests for the research of uncertainty quantification. The experimental test preparation is also complicated and time cost for engineers. In addition, the uncertainties caused by the different sample supplier, preparation and manufacture methods, and also different measurement methods and equipment can lead to incomparable results. Therefore, it is very necessary to develop an efficient and feasible numerical method to perform the uncertainty quantification and propagation.

The numerical models for the mechanical analysis of flip chip BGA electronic package are reported in the published literatures. For example, Han et al [1] developed a prognostics

model to study the mechanism of the precursory resistance pattern in the BGA assemblies. Liu et al [2] predicted the fatigue life of BGA under random vibration loading by calculation of response power spectral density (PSD) of the printed circuit board (PCB) assembly. The group of Liu also used non-contact TV Laser holography technology to understand the dynamic characteristics of the BGA packages [3]. Wu et al [4] used Monte Carlo simulation for uncertainty analysis of the BGA fatigue life in the global–local perspective.

This paper is motivated to provide a reliable and efficient model for uncertainty quantification in flip chip BGA electronic packaging. The description of the BGA geometrical configuration and finite element model in electronic package, as well as the flowchart of Monte Carlo based finite element procedure are presented in Section 2. The validation of the finite element model computation results for resonant vibration is performed in Section 3. In addition, the stochastic probability results of four different BGA configurations are analyzed and compared in Section 4. The major conclusions are presented in Section 5.

II. METHODS

A. Parameter description

In order to discuss the effects of related material and geometrical parameters to the resonant frequencies of the BGA in electronic package, four different configurations are introduced in this paper as in Fig.1.

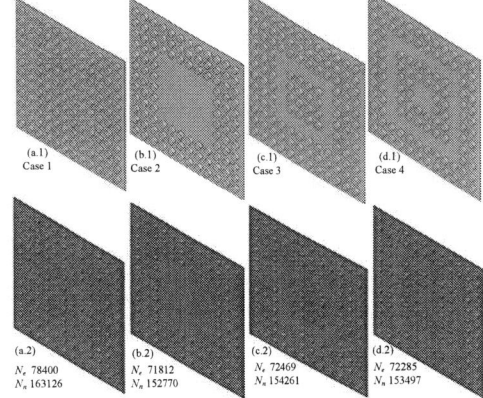

Fig. 1. The FEM for four different configurations of BGA in electronic package

TABLE I. THE NOMENCLATURE OF VARIABLES AND SYMBOLS

The project is supported by NSFC (61901235), NSF of Jiangsu (BK20200971), NSF of Jiangsu Higher Education Institutions (19KJB130001). The authors also thank the Nantong Fujitsu Microelectronics Company and the Jiangsu Key Laboratory of ASIC Design.

Symbols	Definitions	Symbols	Definitions
ρ_{PCB}	Density of PCB	f1, f2, f3, f4, f5	The first to fifth order of resonant frequency
E_{PCB}	Modulus of PCB	BGA	Ball grid array
υ_{PCB}	Poisson's ratio of PCB	LGA	Land grid array
ρ_S	Density of solder	PBGA	Plastic ball grid array
E_S	Modulus of solder	PSD	Power spectral density
υ_S	Poisson's ratio of solder	PCB	Printed circuit board
D1	The upper height of solder ball	FEM	Finite element model
D2	The lower height of solder ball	KSM	Kriging surrogate model
P1	The pitch along X direction	Ne	The number of elements
P2	The pitch along Y direction	Nn	The number of nodes
d	The diameter of the solder ball		

FEM as a sophisticated numerical method to simulate the vibration behavior and mechanical performance of the BGA is used in this work based on ANSYS parameter design language (APDL). For the results accuracy of FEM, the combination structure of BGA and PCB is meshed with sufficient elements and nodes, as presented in Fig.1.

TABLE II.   THE MATERIAL PARAMETER OF BGA PACKAGE

	ρ_{PCB} (kg/m3)	E_{PCB} (GPa)	υ_{PCB}	ρ_S (kg/m3)	E_S (GPa)	υ_S
Park [13]	2477	31.9	0.15	8490	29.4	0.36
Yu [14]	3400	25.0	0.28	7440	53.3	0.35
Liu [10]	2680	22.5	0.12	8410	42.5	0.4
Suh [15]	---	---	---	7400	51.0	0.36
Present	2800	25.0	0.18	8000	45.0	0.36
Interval	2000-4000	20-40	0.1-0.3	6000-9000	20-60	0.25-0.45

Furthermore, the nomenclature of variables and symbols is also provided in Table 1 with supplementary information in Table 2 and Table 3. The deterministic values and specific intervals for each material parameter in this paper are provided in Table 2. Both the deterministic values and certain intervals for material and geometrical parameters are settled in the same magnitude according to that in the reported literatures [4-7]. It is possible to have parallel comparison between the computational results of FEM in this paper with that in the published works. Besides, the radius (or diameter) of solder ball is introduced as the basic variable, which is related with the pitch size and ball height.

For the boundary condition of the FEM, the degree of freedom of nodes in the four side of PCB are controlled and settled as zero. Besides, the parameter original values for the deterministic finite element model of electronic package are 0.3mm, 0.4mm, 0.5mm for the radius, the upper height and lower height of the solder balls in BGA, respectively.

TABLE III.   THE GEOMETRICAL PARAMETER OF BGA PACKAGE

	Package size (mm*mm)	Ball counts	Pitch size (mm)	Ball diameter (mm)	Ball height(mm)
Yu [14]	12*12	---	0.5	0.3	---
Liu [10]	11*13	103	---	0.8	---
Wu [12]	15*15	225	1.27	0.76	0.46
Present		144	0.9	0.6	0.4
Interval	L1*L2	n1*n2	(1.1-2.5)*d	0.2-2	(0.2-0.45)*d

The interval for the radius (0.5*d) of solder ball is 0.2 mm to 1 mm.

B. Monte Carlo based finite element model

Monte Carlo based finite element model is a combination of Monte Carlo stochastic sampling procedure with finite element computation. In this work, Monte Carlo stochastic sampling procedure is performed to provide the effective samples for geometrical and material parameters in the specific intervals given in Table 2 and Table 3. The probability density distribution for the related parameters of BGA are uniform to have the multiple dimension homogeneous filled range. In order to ensure sampling efficiency and samples' quality, Latin Hypercube sampling is used in the sampling process. The samples of each related parameter of BGA are transferred into the FEM for numerical computation of resonant vibration. The flowchart of Monte Carlo based finite element model is presented in Fig.2.

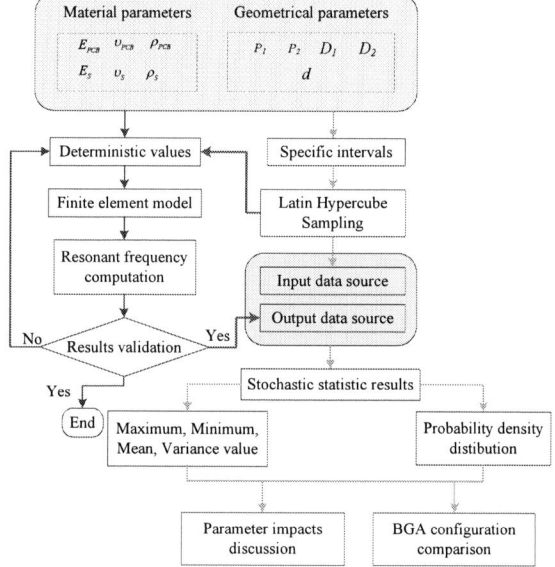

Fig. 2.   The flowchart of Monte Carlo based stochastic finite element model

As presented in Fig.2, the material parameters (Young's modulus, poison ratio and physical density of BGA solder balls and PCB) and geometrical parameters (the BGA pitch along X and Y direction, the radius, as well as the upper and lower height of solder ball) are given as deterministic values and specific intervals. The deterministic values are introduced to the deterministic FEM, also acted as the general traditional method, while the specific intervals are transferred into the Monte Carlo based stochastic finite element method with the green arrow as a guide in the flowchart. Furthermore, the samples provided by Latin Hypercube sampling process according to the parameter related intervals are also passed into the FEM of BGA electronic package for numerical computation. In some sense, the general FEM computation is one specific loop of Monte Carlo based finite element procedure. Monte Carlo based finite element model of BGA electronic package is the comprehensive expansion of general FEM in the parameter range space with the introduction of uniform probability density distribution.

III. VALIDATION

For the model validation, the deterministic FEM of BGA electronic package is computed to check the numerical results. The contour displacement results of first and second order resonant vibration in Fig.3 have a good qualitative agreement in displacement contour distribution with that in the work of Liu's group [10]. The maximum displacement is located in the center for the first resonant vibration, and symmetrically distributed in the two side of the structure for the second resonant vibration, which are well consistent with that in the reported work [10,13]. Therefore, the deterministic FEM of BGA electronic package is verified as the original reliable model for the following further work.

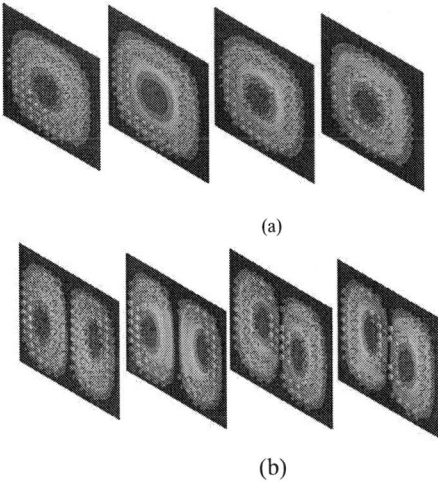

(a)

(b)

Fig. 3. The The displacement results of four BGA configurations ((a) and (b) are for the first and second resonant vibration, respectively)

IV. RESULTS AND DISCUSSION

By the Latin Hypercube sampling method, each parameter is sampled according to the certain intervals in Table 2 and Table 3. Based on the finite element computation, the resonant frequencies of four different BGA configurations are tracked and recorded for the stochastic samples. The mean value, maximum and minimum value, and standard variance of resonant frequency for different BGA configurations are presented and compared in Fig.4. The second BGA configuration has the largest mean, maximum, minimum and the standard variance from the first to the fifth order resonant vibration. More attention is required when the second BGA configuration is selected and used in the electronic package design.

In addition, for the probability density distribution results in Fig.5, the second BGA configuration has a wider probability density distribution with lower peak when compared with the other BGA configuration. It is more obvious in the low order resonant vibration than the high order. Besides, the coincidence of probability density distribution results for the third and fourth configuration are also observed in Fig.4. Furthermore, all the probability density distribution results for four BGA configurations are presented as a sharp

peak with the long tails. Therefore, the resonant frequencies of BGA in electronic package is limited in a narrow range even the geometrical and material parameter fluctuated in the corresponding intervals, but the extreme conditions for the high resonant frequencies can happen with a small possibility. The situation with the small probability are the essential factor should be taken into consideration in the process of BGA design and optimization.

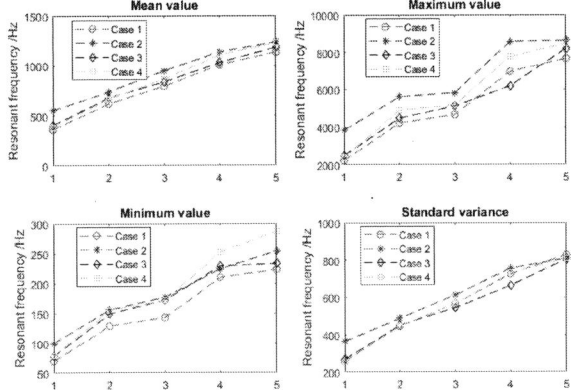

Fig. 4. The stochastic results (mean value, maximum and minimum value, standard variance) of resonant frequency

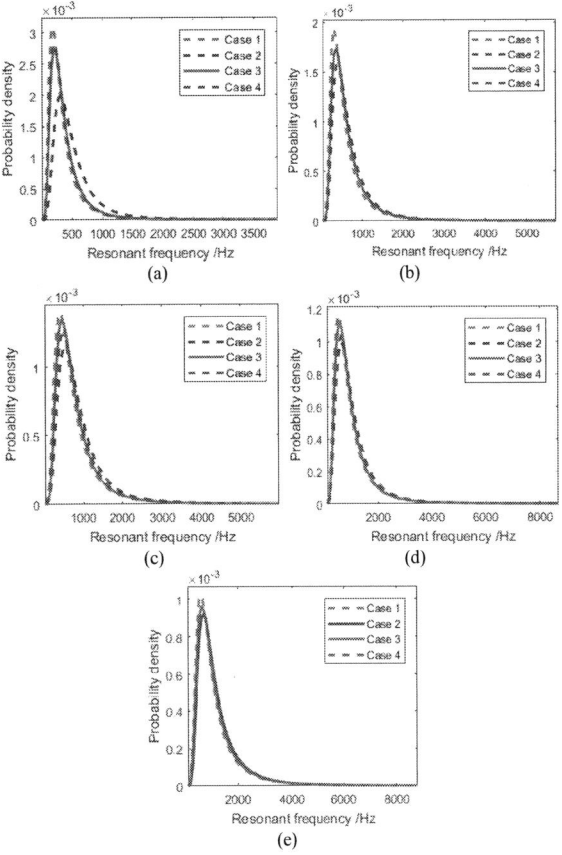

Fig. 5. The probability density distribution of resonant frequency (from (a) to (e) are for the first to the fifth order)

V. CONCLUSION

The Monte Carlo based stochastic finite element model is an effective and feasible method to perform uncertainty quantification and propagation for the mechanical analysis of BGA electronic package. According to the BGA configuration comparison and result discussion, the results of initial FEM for the BGA electronic package has good agreement with the reported literatures in the terms of displacement response for resonant vibration. Furthermore, the extremely high resonant frequencies of BGA electronic package happens with the small possibility. The proposed model and the computational results in this paper provide meaningful references to the electronic package reliability prediction.

REFERENCES

[1] Han, C., Oh, C. M., & Hong, W. S. (2012). "Prognostics model development of BGA assembly under vibration environment," IEEE Transactions on Components, Packaging and Manufacturing Technology, 2(8), pp. 1329-1334, 2012.

[2] Liu, F., Lu, Y., Wang, Z., & Zhang, "Numerical simulation and fatigue life estimation of BGA packages under random vibration loading," Microelectronics Reliability, vol. 55(12), pp. 2777-2785, 2015.

[3] Liu, F., & Meng, G. "Random vibration reliability of BGA lead-free solder joint," Microelectronics Reliability, vol. 54(1), pp. 226-232, 2014.

[4] Wu, M. L., & Barker, D. "Rapid assessment of BGA life under vibration and bending, and influence of input parameter uncertainties," Microelectronics reliability, vol. 50(1), pp. 140-148, 2010.

[5] Park, T. Y., Park, J. C., & Oh, H. U. "Evaluation of structural design methodologies for predicting mechanical reliability of solder joint of BGA and TSSOP under launch random vibration excitation," International Journal of Fatigue, vol. 114, pp. 206-216, 2018.

[6] Yu, D., Al-Yafawi, A., Nguyen, T. T., Park, S., & Chung, S. "High-cycle fatigue life prediction for Pb-free BGA under random vibration loading," Microelectronics Reliability, vol. 51(3), pp. 649-656, 2018.

[7] Suh, Daewoong, et al. "Effects of Ag content on fracture resistance of Sn–Ag–Cu lead-free solders under high-strain rate conditions," Materials Science and Engineering, vol. 460, pp. 595-603, 2007.

Effect of bump shapes on the electromigration reliability of copper pillar solder joints

Zhekun Fan
College of Mechanical and Electrical Engineering
Central South University
Changsha,China
f747991731@163.com

Zhankun Li
College of Mechanical and Electrical Engineering
Central South University
Changsha,China
lzk13589362696@163.com

Junhui Li*
College of Mechanical and Electrical Engineering
Central South University
Changsha,China
lijunhui@csu.edu.cn

Jinqing Xiao
College of Mechanical and Electrical Engineering
Central South University
Changsha,China
xiaojq_xjq@163.com

Yunpeng Liu
The 43rd Research Institute
China Electronics Technology Group Corporation
Hefei,China
liuyunpeng1990@126.com

Junfu Liu
The 43rd Research Institute
China Electronics Technology Group Corporation
Hefei,China
liujunfu0110@163.com

Taotao Chen
The 43rd Research Institute
China Electronics Technology Group Corporation
Hefei,China
15256985225@163.com

Abstract—As the distance between the copper pillars and the size of the solder continues to shrink, the solder joints are subjected to extremely high current density and thermal energy density. This paper is aiming to analyze the electromigration reliability of solder joint with different shapes. Both experiment and finite element method (FEM) analysis were performed to investigate the Mean-time-to-failure(MTTF) of solder joints. Firstly, finite element models of solder joints were established. And then, the current density, temperature, and stress distribution in the solder joints were analyzed according to the model. After that, the chip samples were made by thermal compression bonding. The electromigration experiments were carried out at 150℃, and the interface evolution of solder joints with different shapes was analyzed. Finally, the modified Black's formula was adopted to calculate the MTTF of three different solder joints. According to the calculation results, it can be obtained that cylinder-shaped solder joints have better electromigration reliability than hourglass-shaped bumps and barrel-shaped bumps.

Keywords—*Solder joints shapes; Electromigration; Lead-free solder; MTTF*

I. INTRODUCTION

As the development of electronic products toward miniaturization and functionalization, the size of the chip is continuously shrinking. The diameter of solder joints and the distance between the copper pillars continue to shrink, which will lead to the solder joints are subjected to extremely high current density and thermal energy density [1]. In addition, due to the increase of Joule heat, the ambient temperature of the chip continues to increase, which accelerates the electromigration reaction [2]. The failure caused by electromigration is becoming serious, which has become the key factor in determining the life of solder joint.

When the current density exceeds $10^4 A/cm^2$, electromigration will occur in the solder layer. Also, the higher the current density, the faster the reaction rate of the intermetallic compound (IMC). Electromigration will thicken the IMC at the anode of solder joints and cause holes and cracks in the cathode. Therefore, the mechanical reliability of the solder joints is severely weakened[3]. The reliability of solder joints under current stressing is affected by many factors, such as the solder shape, under bump metallization, size, and material[4]. The shape of the solder joint is a key factor that determines the reliability of the solder joint. Chen et al. analyzed the magnitude of the maximum current density of different BGA solder joints through FEM. And the results show that the hourglass-shaped solder joint with an appropriate waist radius has the lowest current density and the longest MTTF[5].

The above research shows that the shape of the solder joint is a key factor to determine the reliability of electromigration, and the hourglass solder joint has good reliability. In addition, copper pillars can effectively retard current crowding and improve the resistance to electromigration of the solder joint [6]. Therefore, the influence of solder joint shape on the reliability of copper pillar interconnection structure under current stressing needs further study.

II. EXPERIMENTAL AND FEA SIMULATION

A. Experimental samples and experimental conditions

The flip chips used in the experiment are shown in Fig. 1. The chip and substrate are made of silicon. And the solder is made of Sn3.5Ag. The size of the chip is 6mm×6mm×0.5mm. The diameter of the copper pillar bump is 100μm, and the height is 45μm. A solder layer with a height of 25μm was electroplated on the surface of the upper copper pillar. The pitch between adjacent copper pillars is 200μm, and copper traces connect the two adjacent copper pillars. The height of the copper traces is 5μm. The substrate size is 12mm×12mm×0.5mm. The diameter of the copper pillar is 100μm, and the height is 30μm. Samples were prepared by thermal compression bonding. From Fig. 2, by controlling the thermal compression bonding process, three solder joints with different shapes were obtained. The height and waist radius of the three solder joints are marked in Fig. 2.

978-1-6654-1392-3/21 $31.00 © 2021 IEEE

The chip was loaded with current at 150℃. The current density used in the experiment was $2 \times 10^4 A/cm^2$. The samples tested for 10h, 20h, and 40h were taken out and sealed with epoxy resin to observe the growth of IMC.

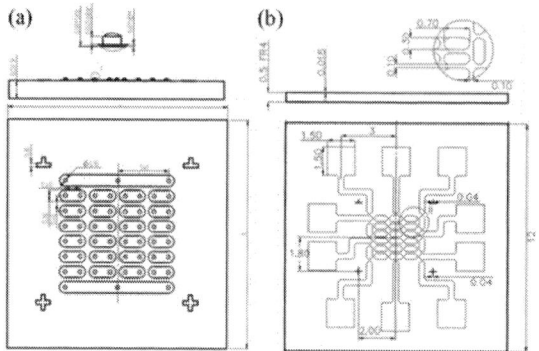

Fig. 1. Flip Chip, A: Chip; B substrate

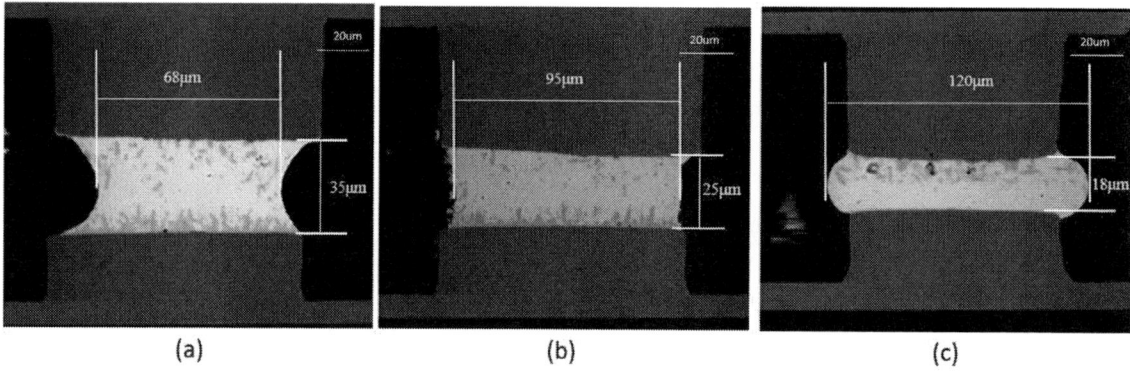

Fig. 2. SEM images of solder joints, (a) hourglass-shaped, (b) cylinder-shaped, (c) barrel-shaped

B. Simulation

As shown in Fig. 3, according to the actual shapes of the solder joints, three-dimensional models were established. Considering the nonlinear mechanical behavior characteristics of the solder joint material, the Anand viscoplastic model was used to characterize it. The material parameters of the unified viscoplastic Anand model of $Sn_{3.5}Ag$ used in this paper are shown in Table 1[7], and the specific performance parameters of each material are shown in Table 2 [8]. The rest of the material properties are linear elastic models, and all materials are considered isotropic. The convective heat transfer coefficient is 20W/m²K at 150℃.

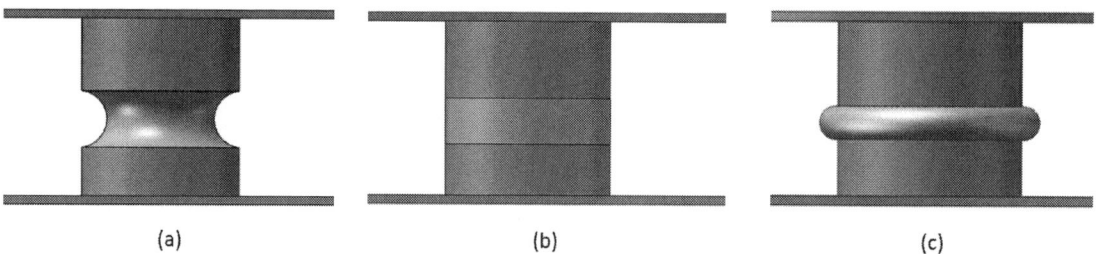

Fig. 3. Finite element model, (a) hourglass-shaped, (b) cylinder-shaped, (c) barrel-shaped

978-1-6654-1392-3/21 $31.00 © 2021 IEEE

TABLE I. $Sn_{3.5}Ag$ SOLDER ANAND MODEL PARAMETERS

model parameters	Value	model parameters	Value
Initial value of deformation resistance: s_0 (MPa)	2.3165	Hardening/softening constant: h_0 (MPa)	27782
Activation energy than Boltzmann constant: Q/R	10278.9	Saturation coefficient of deformation resistance: s (MPa)	52.4
Creep rate Coefficient: A (1/s)	177016	Sensitivity to deformation resistance: n	0.0177
Stress multiplier: ξ	7	Hardening sensitivity: a	1.6
Stress sensitivity: m	0.207		

TABLE II.   ELECTRICAL-THERMAL PERFORMANCE PARAMETERS OF MATERIALS

Material	Resistivity (Ωm)	Thermal conductivity (W/m·K)	Thermal expansion coefficient (ppm/K)	Poisson's ratio	Elastic Modulus (GPa)
Copper	1.72e-8	400	16.3	0.34	129
Sn3.5Ag	1.1e-7	64	24	0.35	Anand model
Si	-	150	2.5	0.28	163
Ni	7e-8	91	13.4	0.31	200

III. RESULTS AND DISCUSSION

A. Current density, temperature, and stress distributions of solder joint

The distribution of maximum current density, temperature, and stress of three kinds of solder joints by electrothermal-mechanical coupling FEM. The corresponding current density is 1×10^4 A/cm². It can be observed from Fig. 4 that the maximum current densities of the three different structures all occur at the junction of the copper trace and the copper pillar on the substrate. And it is not difficult to find that the three maximum current densities are basically the same.

Fig. 5 shows the current density distributions of the three solder joints. From Fig. 5 (b),5 (c), the current density at the current entrance of the copper pillar on the substrate is the largest. Because the copper pillars on the substrate are shorter, which cannot effectively retard the current crowding compared to the copper pillars on the chip. It can be seen from Fig. 5 (a) that the current density at the waist of the hourglass-shaped solder joint is the largest. Among the three kinds of solder joints, the rotation angle of motion trail of the electrons in the hourglass-shaped solder joint is the smallest, which effectively avoids the current crowding effect[9]. However, the copper pillar can effectively weaken the current crowding effect. Therefore, the excellent anti-electromigration performance of the hourglass-shaped solder joints is weakened. Besides that, because the cross-sectional area of the waist is much tinier than the area at both ends, the current

density of the waist is too large. In contrast, the current density at the waist of the barrel-shaped solder joint is much smaller than at both ends.

It can be seen from Fig. 6 that the maximum temperature of the three solder joints all appear at the top of the solder. In addition, it is not difficult to find that the temperature of the hourglass-shaped solder joints is the highest among the three solder joints. Because it has the highest resistance and the highest maximum current density, the temperature rise is higher. Furthermore, due to the different heat dissipation areas of the chip and the substrate, temperature differences appear in the solder joints of the three shapes. After calculation, the temperature gradients of the three solder joints are 20.00℃/cm, 12.00℃/cm, 11.11℃/cm, which are much lower than the threshold temperature gradient of the thermal migration effect. In addition, the temperature of the surface of the chip and the substrate was measured by a thermocouple, and the calculated result was consistent with the simulation result. Hence there is no thermal migration in the copper pillar interconnection structure.

Fig. 7 shows the stress distribution of three different solder joints under thermoelectric coupling. Because the thermal expansion coefficient between the solder material and the copper pillar does not match, thermal stress is generated in the solder joint. From Fig. 7 (b),7 (c), the maximum stresses of the cylinder-shaped and barrel-shaped solder joints all appear at the interface between the solder joints and the copper pillars. They are 5.14MPA and 6.84MPA. However, according to Fig.7(a), in the hourglass-shaped solder joint, the maximum stress appears at the waist of the solder joint, which is 3.38 MPA.

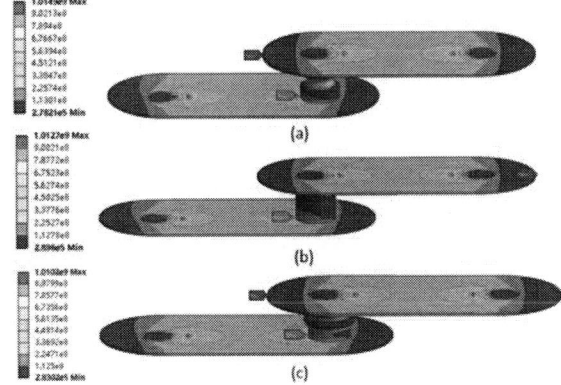

Fig. 4.   Current density distributions of copper pillar interconnect structure, (a) hourglass-shaped, (b) cylinder-shaped, (c) barrel-shaped

978-1-6654-1392-3/21 $31.00 © 2021 IEEE

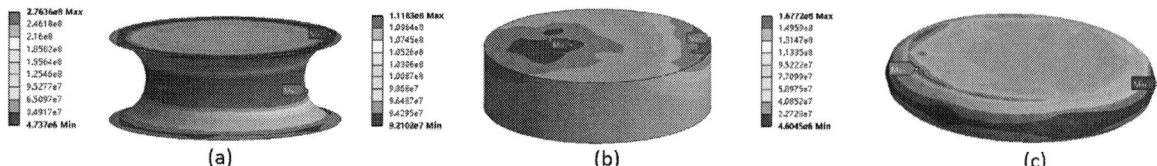

Fig. 5. Current density distributions in solder joints, (a) hourglass-shaped, (b) cylinder-shaped, (c) barrel-shaped

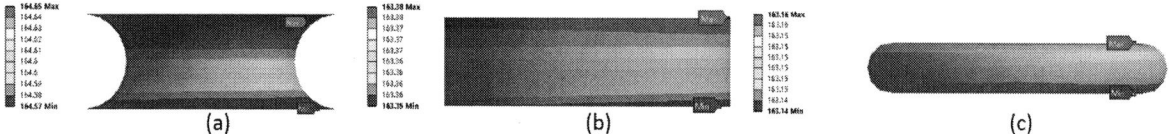

Fig. 6. Temperature distributions in solder joints, (a) hourglass-shaped, (b) cylinder-shaped, (c) barrel-shaped

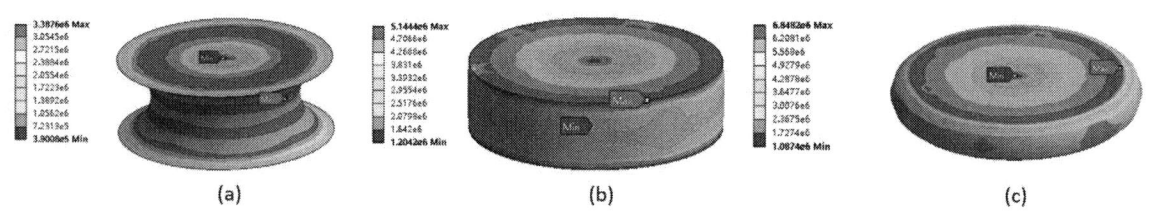

Fig. 7. The equivalent stress distributions in solder joints, (a) hourglass-shaped, (b) cylinder-shaped, (c) barrel-shaped

B. Growth and evolution of IMC in three solder joints

Fig. 8 shows the SEM images of three different solder joints after different electromigration experiment times(10h, 20h, 40h). The direction of electron flow and electromigration time is marked in the figure. It can be found that the three kinds of solder joints all have a thick Cu_6Sn_5 layer formed on the anode side when the electromigration time is 10h. In addition, in the cylinder-shaped and barrel-shaped solder joints, a thin Cu_3Sn layer formed on the anode and cathode sides, as shown in Fig. 8 (b1) and (c1).

As the electromigration time increases, the IMC layer of the three solder joints continues to thicken until it occupies the entire solder layer. It can be seen from Fig. 8 (a2), (b2), and (c2), when the electromigration time increases to 40h, the solder has been completely transformed into IMC. Due to the lack of Sn atoms, Cu_6Sn_5 reacts quickly with migrating Cu atoms to form Cu_3Sn. As a result, a thicker Cu_3Sn layer appears in both the anode and the cathode. In addition, it can be found that Kirkendall voids appear in Cu_3Sn and the interface between Cu_3Sn and copper pillar. This is because the diffusion rate of Cu atoms in Cu_3Sn is several orders of magnitude higher than that of Sn atoms in Cu_3Sn, and the diffused Sn atoms cannot make up for the vacancies left by the diffusion of Cu atoms. Then the vacancies gradually form Kirkendall voids. As shown in Fig. 8 (a2), the middle of the copper pillar on the hourglass-shaped solder joint is severely eroded. In addition, it is not difficult to find that on the right

side of the solder, part of Cu_6Sn_5 is converted to Cu_3Sn, and Kirkendall voids appear. According to Fig. 5 (a), the current density at the edge of the waist of the hourglass-shaped solder joint is large, which may accelerate the migration of Cu atoms. By comparing Fig. 8 (b2) and (c2), it can be found that when both are used as cathodes, the corrosion of the lower copper pillar is more serious than that of the upper copper pillar. Since the lower copper pillars are shorter than the upper copper pillars, the current crowding effect cannot be effectively delayed. Therefore, this will result in extremely high current density, and the copper pillar is seriously eroded.

As shown in Fig. 8 (a3), when the electromigration time is 60h, the Cu_3Sn layer is thicker, and a lot of Kirkendall voids appear on both sides of the waist of the solder. At the same time, a crack appeared in the middle of the solder, resulting in a serious failure of the entire interconnection structure. According to the previous conclusion from Fig.7 (a), the maximum stress appears in the middle of the solder layer, where the cross-sectional area is the smallest. Therefore, when the solder layer is completely converted to IMC under the action of current stressing, the waist of the interconnect structure is a weak spot and is more prone to fracture. The thickness of the IMC layer in the barrel-shaped and cylinder-shaped solder joints increases significantly, and Kirkendall voids are further increased. From Fig. 8 (c3), the Kirkendall voids are transformed into a crack, making the solder joint fracture and fail.

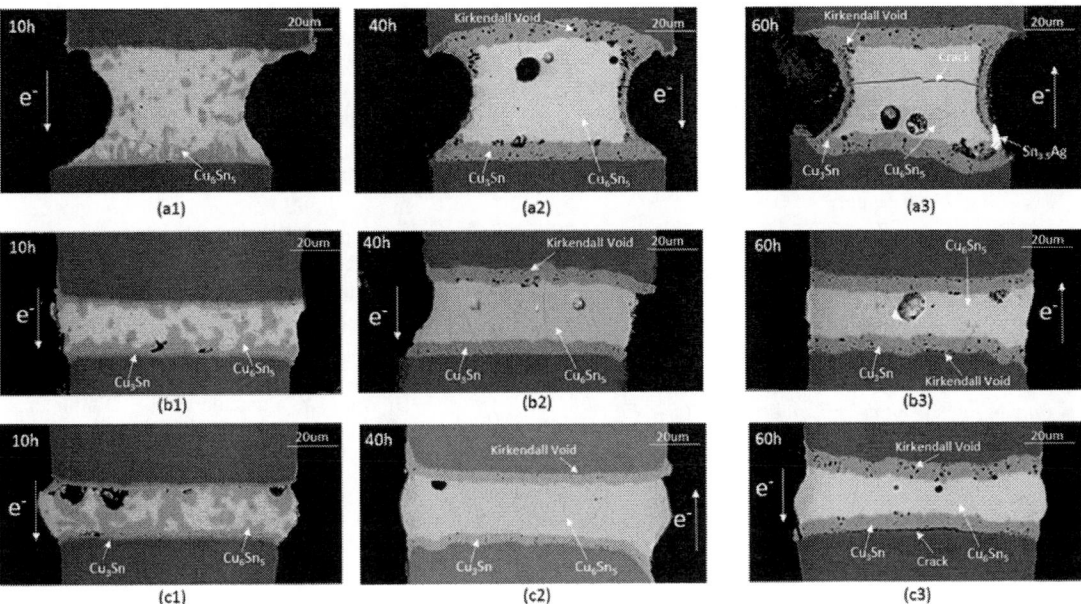

Fig. 8. SEM images of three shapes，(a) hourglass-shaped, (b) cylinder-shaped, (c) barrel-shaped

C. MTTF of solder joint

The MTTF is an essential indicator for evaluating the reliability of electronic components in the electronics industry and has been used for many years. Back's equation has been widely used since its proposed date. W. J. Choi pointed out that Black's equation could not well fit the actual life of solder joints under different current conditions. The MTTF calculated at low current is lower than the actual MTTF, and the MTTF calculated at high current is higher than the actual MTTF. For this reason, he modified Back's equation to obtain (1)[10].

$$MTTF=A\frac{1}{(cj)^n}\exp\left[\frac{Q}{k(T+\Delta T)}\right] \quad (1)$$

In (1), A is a constant, j is the current density, c is a constant introduced to modify the current density, Q is the activation energy, k is Boltzmann's constant, T is the Kelvin temperature of the solder joints, and ΔT is the temperature rise caused by Joule heating. After comparison with the experimental results, this paper adopts c=0.1 and ΔT=2. The MTTFs of the three solder joints at the two current densities are listed in Table 3.

TABLE III. SUMMARY TABLE OF RELEVANT PARAMETERS OF MTTF

Current density(A/cm²)	Bump Shape	Max. current density (A/cm²)	Max.temperature (K)	MTTF (h)
1×10⁴	Hourglass	2.763×10⁵	439.80	641.22
	Column	1.118 ×10⁵	438.53	3537.24
	Barrel	1.677×10⁵	438.31	1709.66
2×10⁴	Hourglass	5.380 ×10⁵	478.91	34.03
	Column	2.217×10⁵	474.62	203.40
	Barrel	3.213 ×10⁵	473.44	108.78

From Table 3, we can draw a conclusion that when the maximum current density of the three solder joints is doubled, the MTTF will decrease by one order of magnitude. In addition, it can be obtained from Table 3 that under the two current densities, the MTTF of the cylinder-shaped solder joint is the largest.

IV. CONCLUSION

This paper comprehensively analyzes the electromigration reliability of three different solder joint shapes. The results show that, compared with the other two solder joints, the cylinder-shaped solder joint has better electromigration reliability and longer MTTF. Here are some conclusions:

(1). The maximum current density of the hourglass-shaped solder joint is higher than the others. Because the cross-sectional area of the waist of the hourglass-shaped solder joint is small, resulting in excessive current density.

(2). The maximum temperatures of the three structures are about 164℃. The maximum stresses of the cylinder-shaped solder joint and the barrel-shaped solder joint are 5.14 MPA and 6.84 MPA. The maximum stress of the hourglass-shaped solder joint is 3.38 MPA.

(3). The thinner the copper pillar of the cathode, the heavier the copper pillars are eroded. The reason is that the thin copper pillars cannot effectively retard the current crowding effect, resulting in excessive current density and serious corrosion of the copper pillar. It can be observed that a crack appears on the waist of the hourglass-shaped solder joint, and cracks appear at the barrel-shaped solder joint after 60 hours of the electromigration experiment.

(4). After calculation, the cylinder-shaped solder joint has the largest MTTF. The cylinder-shaped solder joint is a good choice for copper pillar interconnect structures to improve electromigration resistance.

ACKNOWLEDGMENT

This work was supported by National Natural Science Foundation of China (No. 51975594), National Natural Science Foundation of China joint fund for regional innovation and development（No.U20A6004）.

REFERENCE

[1] H. Zhang, J. Li, and W. Zhu, "Electromigration in flip chip with Cu pillar having a shallow Sn-3.5Ag solder interconnect," 2018, pp. 1653-1656.

[2] C. L. Kao and Y. S. Lai, "Electrothermal coupling analysis of current crowding and Joule heating in flip-chip package assembly," in *Electronics Packaging Technology Conference, 2004. EPTC 2004. Proceedings of 6th*, 2004.

[3] B. J. Kim, G. T. Lim, J. Kim, K. Lee, and Y. C. Joo, "Intermetallic Compound Growth and Reliability of Cu Pillar Bumps Under Current Stressing," *Journal of Electronic Materials,* vol. 39, no. 10, pp. 2281-2285, 2010.

[4] W. C. Kuan, S. W. Liang, and C. Chen, "Effect of bump size on current density and temperature distributions in flip-chip solder joints," *Microelectronics Reliability,* vol. 49, no. 5, pp. 544-550, 2009.

[5] Y. Li, X. C. Zhao, Y. Liu, and H. Li, "Effect of Bump Shape on Current Density and Temperature Distributions in Solder Bump Joints under Electromigration," *Advanced Materials Research,* vol. 569, pp. 82-87, 2012.

[6] S. Lee, Y. X. Guo, and C. K. Ong, "Electromigration effect on Cu-pillar(Sn) bumps," in *Electronic Packaging Technology Conference,* 2005, pp. 135-139.

[7] J. Li, Y. Zhang, H. Zhang, Z. Chen, and W. Zhu, "The thermal cycling reliability of copper pillar solder bump in flip chip via thermal compression bonding," *Microelectronics Reliability,* vol. 104, pp. 113543-, 2020.

[8] D. Wang, Y. Yuan, and L. Le, "Failure analysis of Sn-3.5Ag solder joints for FCOB using 2-D FEA model," *IEEE,* 2010.

[9] Y. W. Chang and C. Chen, "Optimal Design of Passivation/UBM Openings for Reducing Current Crowding Effect Under Electromigration of Flip-chip Solder Joint," in *Electronics Packaging Technology Conference,* 2010.

[10] W. J. Choi, E. Yeh, and K. N. Tu, "Mean-time-to-failure study of flip chip solder joints on Cu/Ni(V)/Al thin-film under-bump-metallization," *Journal of Applied Physics,* vol. 94, no. 9, pp. 5665-5671, 2003.

Study on hermetic package of antiradiation direct-head transmitter

Chuanwei Wang*
The 38th Research Institute of CETC
Hefei, China
lemonwcw@163.com

Yukun Wu
The 38th Research Institute of CETC
Hefei, China
shiyi@163.com

Jiabo Zhang
The 38th Research Institute of CETC
Hefei, China
zhangjiabo@163.com

Kuang Pan
The 38th Research Institute of CETC
Hefei, China
pankuang@163.com

Daochang Wang
The 38th Research Institute of CETC
Hefei, China
dc_wang@189.cn

Yi Liang
The 38th Research Institute of CETC
Hefei, China
liangyi@163.com

Abstract—**As an important component in antiradiation direct-head, the uniformity and reliability of transmitter paly the key role in the electrical performance. The characteristic of missile transmitter with high integration is plenty sidewall connectors, larger size, and the 4047 aluminum alloy plate is easy to cause weld airtight failure in helium leak detection process. To keep an excellent microwave properties and high reliability, it is of great importance to insure the gas tightness of transmitter. For this reason, higher requirements are proposed for the weld penetration and sealing quality, while the temperature rise should be under the melting point of low temperature brazing weld. In this paper, the laser welding technology in different energies was employed, weld penetration and sealing quality of the components were investigated after the laser welding process, meanwhile the temperature around the RF connector was also tested. On the basis of this, the appropriate weld penetration and corresponding laser energy for large-size transmitter are obtained, so that the gas tightness performance of the components is satisfied and superior to the requirements of GJB548B.**

Keywords—transmitter, aluminum, laser welding, hermetic package

I. INTRODUCTION

Antiradiation direct-head transmitter is the core component of Missile Radar Seeker. Due to the dual polarization and multi-channel design, the layout of transmitter is very compact and the assembly density is very high. On the premise of ensuring the realization of high performance indicators, the transmitter should also meet the long-term use in the storage environment, with high uniformity and reliability. Therefore, in order to ensure that the transmitter can withstand the harsh environment of missiles, and also have good microwave properties and high reliability in space, the transmitter must be hermetically sealed.

At present, the outer frame material of the microwave components is usually 6061 aluminum alloy, while the cover plate is 4047 aluminum alloy. The airtight welding of the outer frame and the cover plate is realized by laser technology[1-3]. The combination of 6061-4047 can meet the requirements of conventional seal welding, and is widely used in the shell manufacturing of various microwave components. However, in Missile Radar Seeker, the size of

transmitter is often larger than that of microwave components. In the process of helium leak detection, large-size cover plate is prone to elastic deformation or even plastic deformation, resulting in weld airtight failure, which puts forward higher requirements for weld penetration and sealing quality. On the other hand, RF connectors distribute on the side wall of the outer frame near the laser welding seam, so there are strict requirements for the temperature rise near the connectors, and the temperature must be controlled below the melting point of the brazing seam around the connectors. That is to say, the laser welding seam should not only meet the requirements of penetration, but also not cause the brazing seam remelting and air tightness reduction due to too high temperature rise. In this paper, according to the characteristics of missile transmitter, laser welding is selected for hermetically sealing of the transmitter. The influence of laser energy on the depth of weld pool, sealing quality and temperature of RF connector after laser sealing process are mainly studied.

II. EXPERIMENTAL PROCEDURES

The shell of the transmitter consists of two parts: the outer frame and the cover plate part, as shown in Fig. 1(a). The size of the transmitter is $115 \times 85 \times 8$ mm. The welding joint of the outer frame and the cover plate adopts the embedded butt joint, and the size of the lap width at the joint and the thickness of the cover plate are shown in Fig. 1(b).

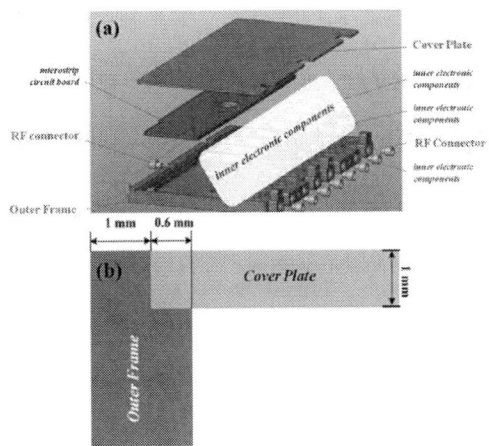

Fig. 1. The diagram of sample structure (a) and welding joint (b).

6061 aluminum alloy is selected for outer frame and 4047 aluminum alloy is selected for cover plate. The chemical compositions are illustrated in Table 1.

TABLE I. CHEMICAL COMPOSITIONS OF 6061 AND 4047 ALUMINUM (WT. %).

	Si	Fe	Cu	Mn	Zn	Mg	Al
4047	12.0	0.8	0.3	0.15	0.2	0.1	Bal.
6061	0.6	0.7	0.3	0.15	0.25	1.1	Bal.

Laser sealing equipment is mainly composed of Nd:YAG pulse laser system, glove box system, industrial computer control system and three-dimensional workbench. The laser welding arrangement used for this work is simply shown in Fig. 2. The laser processing parameters set as wavelength 1064nm, maximum peak power 10kW, pulse duration 5.0 ms, focal spot size 0.6 mm, frequency 18 Hz. High purity argon gas was supplied as a shielding gas in the glove box. During laser welding, the workbench with numerical control carried out the movement of the transmitter to adjust the sealing path. The moving speed of the stage in XY plane was set as 6.0 mm/s, while in the Z-direction the stage was stationary to maintain the defocusing amount (1 mm).

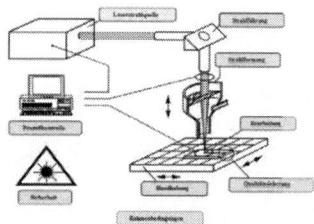

Fig. 2. Schematic diagram of the laser welding apparatus.

After laser welding, the gas tightness of the transmitter was tested by using helium leak detector. The test standard was based on GJB 548b-2005 "Test Methods And Procedures For Microelectronic Devices". At the same time, the surface and cross-section of the weld seam were also observed after laser sealing.

III. RESULTS AND ANALYSES

A. Leakage detection and deformation of 4047 cover plate

Laser welding technology used for microwave components is usually applied to the product with size less than 70×60 mm. However, for the missile transmitter with a size of $115\times85\times18$ mm, it is found that due to the large size and the low weld penetration, its cover plate is prone to bulging deformation during helium leak detection due to vacuum pumping.

The results of simulation and measurement show that the maximum bulging deformation is about 0.53mm, as shown in Fig. 3. After leak detection, the weld often fails to be airtight due to deformation. The original weld has been difficult to meet these requirements, and this puts forward higher requirements for weld penetration and quality. The simulation results show that the weld penetration should be at least 0.59mm to meet the anti deformation requirements during the leak detection process.

Fig. 3. Simulation diagram of cover plate deformation.

B. Influence of laser energy on weld penetration

Aiming at the higher penetration requirement of this kind of large-size transmitter, the welding penetration optimization test is carried out. The welding energy of conventional microwave modules is single pulse 6.0J. At present, the parameters of single pulse energy of 8.0J, 9.0J, 10.0J, 11.0J are separately used for welding test. After the test, the penetration at the starting position, middle position and end position in the welding path is measured respectively.

It can be seen from Table 2 that with the increase of laser energy, the weld penetration increases gradually. At the same time, due to the heat accumulation in the welding process, the penetration at the end position is significantly deeper than that at the start position, and the difference of penetration is proportional to the laser energy.

TABLE II. RESULTS OF PENETRATION TEST.

Specimen groups	Single Pulse /J	Weld Penetration/mm			Penetration Difference /mm
		Starting Position	*Middle Position*	*End Position*	
No.1	8.0	0.46	0.48	0.51	0.05
No.2	9.0	0.64	0.70	0.74	0.10
No.3	10.0	0.71	0.82	0.91	0.20
No.4	11.0	0.83	0.96	1.14	0.31

Accordingly, the weld surface after leak detection was observed and compared, as shown in Fig. 4. As the weld penetration is less than 0.59mm in sample No.1, it cannot effectively resist the deformation of cover plate in the leak detection process. Local cracks were then observed in the weld of sample No.1 (Fig. 4 (a)), which would also led to the air tightness failure of transmitter. With the increase of laser energy, the weld depth increases gradually and its strength increases correspondingly[4,5], which can react on the bulging deformation and avoid the generation of cracks, as shown in Fig. 4 (b) and (c). It is worth noting that with the further increase of laser energy, although there is no obvious cracks in the weld, local metal vaporization is caused by too high energy[6,7], which eventually leads to a large number of pores and other defects on the weld seam of sample No.4 (Fig. 4 (d)).

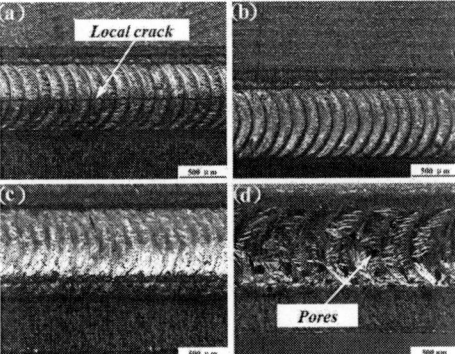

Fig. 4. The weld surface under different laser energies: (a) 8J, (b) 9J, (c) 10J, (d) 11J.

C. Influence of laser energy on sealing quality

Although the weld depth of sample No.2, No.3 and No.4 is greater than 0.59mm, which has met the requirements of anti bulging deformation, the packaging reliability is also affected by the synthetic quality of the weld. Therefore, the weld cross-section and air tightness of four groups of samples were compared to find the variation characteristics of sealing quality with laser energy.

TABLE III. RESULTS OF LEAKAGE DETECTION TEST.

Specimen groups	Single Pulse /J	Helium Leakage Rate /Pa·m3/s
No.1	8.0	5.0×10^{-5}
No.2	9.0	7.0×10^{-9}
No.3	10.0	4.0×10^{-9}
No.4	11.0	6.0×10^{-7}

As can be seen in Fig. 5 (a), when the single pulse energy is 8.0J, a crescent-shaped welding seam was observed. Some shrinkage cavities were also noticed in the weld (Fig. 5 (b)), which would degrade the sealing quality as well as the shallow penetration. Therefore, the air tightness of sample No.1 is only 5.0×10^{-5} Pa · m^3/s, as shown in Tab. 3.

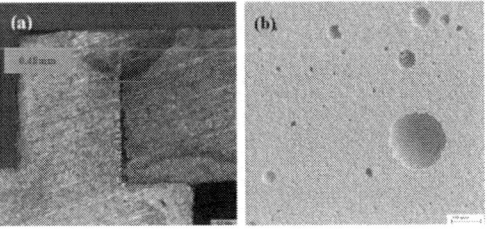

Fig. 5. The weld seam of sample No.1: (a) cross-section appearance, (b) shrinkage cavities.

When the single pulse energy increases to 9.0J, it shows the characteristics of deep penetration welding (Fig. 6 (a)). As shown in Fig. 6 (b), the weld surface at the same position presents a dense and smooth fish scale weld appearance. The micrograph of the weld shows that the microstructure is very compact and thus has good airtight quality.

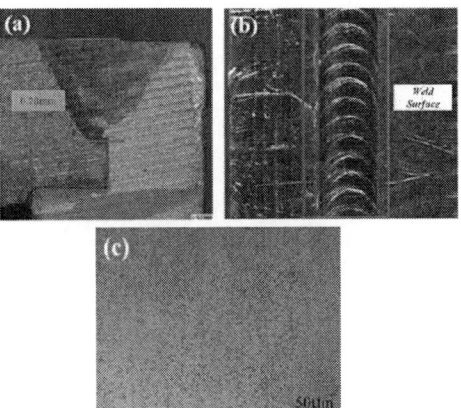

Fig. 6. The weld seam of sample No.2: (a) cross-section appearance, (b) weld surface, (c) microstructure.

With the gradual increase of laser energy, the surface temperature of weld pool increases correspondingly[8,9]. When the surface temperature of weld pool is too high, spatter appears in the welding process, which affects the quality of welding surface and even the internal quality of weld.

When the single pulse energy is 10.0J, the penetration of starting position is 0.71mm (Fig. 7 (a)), while that of end position is 0.91mm (Fig. 7 (b)). It can be seen from the figure that although the weld penetration has been improved, spatter appears at the end position. As a result, local weld is sunken, and some metal particles formed by splashing appear on the surface (Fig. 7 (c)), which will affect the later assembly.

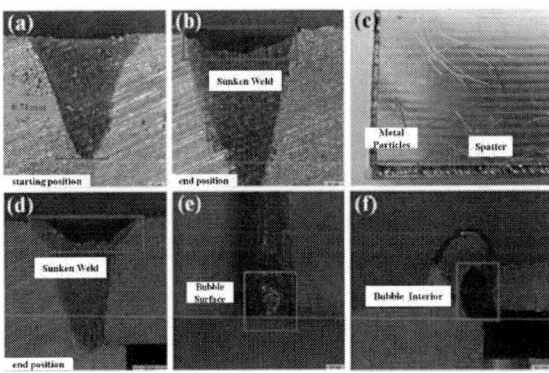

Fig. 7. The weld seam of sample No.3 and No.4: (a) cross-section appearance of No.3, (b) sunken weld of No.3, (c) spatter and metal particles, (d) sunken weld of No.4, (e) and (f) bubbles of No.4.

As might have been expected, when the energy is 11.0J, the metal spatter in the welding process is further intensified, and the weld surface is seriously depressed (Fig. 7 (d)). At the same time, bubbles appear in the weld (Fig. 7 (e) and (f)), and correspondingly the sealing quality of transmitter decreases to 6.0×10^{-7} Pa·m^3/s.

D. Sealing verification test

Based on the above experiments, it can be seen that when the single pulse energy is 9.0J, the weld penetration meets the simulation requirements, and the sealing quality is good. Consequently, this parameter is selected for further sealing verification test. At the same time, temperature thermocouples are connected to the RF connector to monitor the temperature change of the connector during welding. The

978-1-6654-1392-3/21 $31.00 © 2021 IEEE

test results show that the maximum temperature of the connector during the welding is 143 ℃ at the laser energy 9.0J. It is lower than the melting point (183 ° C) of brazing seam around the RF connector. It can be seen that the laser energy 9.0J can be selected as the safety parameter. Finally, after the helium leak detection, no bulging deformation was observed on the cover plate of the transmitter, and the air tightness was up to 7.0×10^{-9}Pa·m^3/s, meeting and even better than the requirements of GJB548B.

IV. CONCLUSIONS

The aluminum transmitter was sealed with different laser energies, and the weld penetration and sealing quality were studied and analyzed. The conclusions are as follows:

1)   With the increase of laser energy, the weld penetration increases, and there is a penetration difference between the start position and the end position, which is proportional to the laser energy.

2)   When the laser energy is too low, local cracks will appear on the weld surface after leak detection. With the increase of laser energy, the penetration and deformation resistance of the weld are improved, but when the laser energy is too high, pores appear on the weld surface.

3)   When the laser energy is 10.0 J and 11.0 J, although the weld depth is much greater than 0.59mm, spatter particles appear on the weld surface, and even bubbles appear inside the weld, which affect the sealing quality of the transmitter.

4)   By setting laser energy reasonably, the weld penetration, connector temperature and air tightness can meet the requirements of large-size transmitter. In this study, the relatively optimal laser energy is 9.0J.

V. ACKNOWLEDGEMENTS

This article was supported by the Major Science And Technology Projects In Anhui Province and National MCF Energy R&D Program (2019YFE 03100100).

REFERENCES

[1]   H. Zhou, N. Sun, H. Shan, D. Ma, X. Tong and L. Ren: 'Bio-inspired wearable characteristic surface: Wear behavior of cast iron with biomimetic units processed by laser', Appl. Surf. Sci., 2007, 253, 9513-9520.

[2]   T. Ishizaki and M. Sakamoto: 'Facile formation of biomimetic color-tuned superhydrophobic magnesium alloy with corrosion resistance', Langmuir, 2011, 27, 2375–2381.

[3]   C. Wang, H. Zhou, Z. Zhang, Y. Zhao, D. Cong, C. Meng, P. Zhang and L. Ren: 'Mechanical property of a low carbon steel with biomimetic units in different shapes', Opt. Laser Technol., 2013, 47, 114–120.

[4]   M. Wollgarten, M. Beyss, K. Urban, H. Liebertz and U. Köster: 'Direct evidence for plastic deformation of quasicrystals by means of a dislocation mechanism', Phys. Rev. Lett., 1993, 71, 549–552.

[5]   S. Ni, Y. B. Wang, X. Z. Liao, S. N. Alhajeri, H. Q. Li, Y. H. Zhao, E. J. Lavernia, S. P. Ringer, T. G. Langdon and Y. T. Zhu: 'Strain hardening and softening in a nanocrystalline Ni–Fe alloy induced by severe plastic deformation', Mater. Sci. Eng. A, 2011, 528, 3398–3403.

[6]   C. France, H. Klöcker, J. Lecoze and A. Fraczkiewicz: 'Nitrogen strengthening of a martensitic steel: relation between microstructure and mechanical behaviour', Acta Mater., 1997, 45, 2789-2799.

[7]   H. Yan, A. H. Wang, X. L. Zhang, Z. W. Huang, W. Y. Wang and J. P. Xie: 'Nd:YAG laser cladding Ni base alloy/nano-h-BN self-lubricating composite coatings', Mater. Sci. Technol., 2010, 26, 461-468.

[8]   T. Lee, C. H. Park, D. L. Lee and C. S. Lee: 'Enhancing tensile properties of   ultrafine-grained medium-carbon steel utilizing fine carbides', Mater. Sci. Eng. A, 2011, 528, 6558–6564.

[9]   B. Chen, P. D. Wu and H. Gao: 'A characteristic length for stress transfer in the nanostructure of biological composites', Compos. Sci. Technol., 2009, 69, 1160–1164.

Study on die-bonding key technology of UV LED packaging

Wei Liu
General Manager's Office
Shenzhen Huizhi Optoelectronic
Technology Co., Ltd
Shenzhen, China
purpleliu@126.com

Chunfang Zi
Department of Microelectronics and
Solid electronics
Shenzhen Graduate School of PKU
Shenzhen, China
purplelw@126.com

Qingping Lin
Department of Microelectronics and
Solid electronics
Shenzhen Graduate School of PKU
Shenzhen, China
hz1189@126.com

Abstract—Due to its potential market value, the packaging and application technology of UV LED are paid more and more attention. In this paper, the die-bonding interface and thermal resistance of UV LED were investigated by using different TIMs, and the influence of junction temperature on photochromic properties was studied. The results show that the traditional silver glue as a thermal interface material has the highest junction temperature, and its advantage is that it is cheap. Compared with SnAgCu solder, Nano Ag paste has better thermal conductivity and better thermal stability, but the price is higher. Moreover, the reflow process of Sn based solder is more mature. Compared with AuSn and SnAgCu solder, the research results show that the problem of solid-state interface of Sn based solder is poor thermal stability. Under the action of electric stress and thermal stress, the interface will have atomic migration. After high temperature aging, the interface of AuSn solder-boning interface even has serious atom migration and relatively large voids. The study indicated that with the increase of junction temperature, the radiation power and photochromic characteristics of all tested samples decrease.

Keywords—UV-LED; packaging; bonding process; interfacial; IMC

I. INTRODUCTION

Solid state lighting is another great invention after incandescent lamp and fluorescent lamp. In 2014, blue LED won the Nobel Prize in physics. So far, LED has been widely used, and the packaging technology of white LED has been fully developed. In recent years, LED packaging manufacturers in order to obtain better profit growth. Ultraviolet LED and infrared LED have attracted people's attention, and have been widely used in curing, disinfection, sterilization, security and other fields. However, these applications require higher photo-thermal characteristics and reliability of UV LED [1-2]. It is predicted that the market demand of UV LED will continue to grow in the next few years. Its biggest application field is UV curing. Its main market is nail curing lamp, UV curing and other applications. UV sterilization is also a potential application market. At present, the main technical bottlenecks of UV LED packaging are light efficiency and heat dissipation, as well as the reliability problems caused by these two problems [3-4].

At present, the luminous efficiency of UV LED is low, which can not meet the technical requirements of large-scale line light source and surface light source. Usually, it is necessary to integrate multiple chips to form a light source module, so as to obtain higher radiation power density. Like white LED, reducing the thermal resistance of the solid crystal interface and improving the heat dissipation ability of the solid crystal layer are the important technical ways to improve the radiation power density of UV LED packaging. In this study, we focus on the bonding materials and solidification process of UV LED packaging. In this paper, we compare the effects of nano silver paste, conductive silver glue and concentrated solder on the photoelectric and thermal properties of UV LED packaging devices [5].

II. EXPERIMENTS

In this study, the effects of solid-state materials and solid-state structure on the performance of near ultraviolet LED packaging devices are studied. The solid-state materials are SnAgCu solder, AuSn eutectic solder, nano silver paste and traditional solid-state silver glue, and the packaging structures are front mount structure, flip mount structure and new type packaging structure. LED chip substrate material, electrode material, heat sink material and substrate material also have great influence on the performance of the device. In order to study the consistency, the peak wavelength of the UV LED chip is 366nm, the substrate material is Si substrate, the electrode material is Cu, the heat sink material and substrate material is Al. At the same time, the vertical structure of UV LED chip with the peak wave length of 398nm is investigated and the heat sink is AlN ceramic substrate. The optical, electrical and thermal performance parameters under different substrate temperature and different injection current were compared. The junction temperature under different conditions is measured by forward voltage method. The junction temperature test system is shown in Figure 2-1. The main steps of junction temperature test are as follows: Firstly, the thermal control platform is set in the incubator, and the tested samples are placed on the heating platform. Then, different driving currents can be applied to the samples. Finally, the computer is connected with the junction temperature tester to collect data.

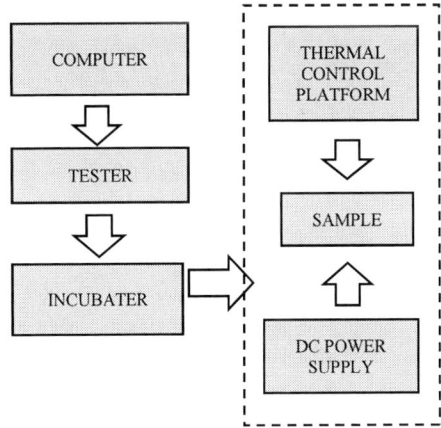

Figure 2-1 Schematic diagram of junction temperature test system

The photoelectric parameters of tested samples under different test conditions are measured by photoelectric parameter test system. The schematic diagram of the measurement system is shown in Figure 2-2. The specific test steps are as follows: Firstly, the tested sample is placed in the integrating sphere, then the temperature of the substrate is controlled by the temperature control platform, and the driving current is set by the programmable power supply. After the sample is electrified, and the substrate temperature reaches a stable state, the computer is connected with the integrating sphere to collect the photoelectric parameters such as luminous flux, radiation power and forward voltage.

The parameters including luminous flux, radiation power and forward voltage are obtained by the test system. In this paper, the aging test of the tested samples is also carried out. Different injection current is selected and aging is carried out under different substrate temperature. The samples are taken out regularly to test the photoelectric parameters such as luminous flux, radiation power, color coordinate and half wave width.

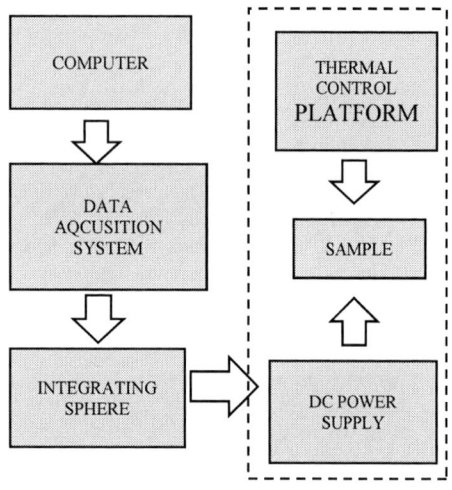

Figure 2-2 Schematic diagram of photoelectric parameter test system

III. RESULTS AND DISCUSSIONS

The luminous principle of UV LED chip is that when the forward voltage is applied to the PN junction, the minority carrier recombination generates light, so the electro-optic conversion efficiency of LED is very low, and most of the input electric energy is converted into heat energy. If the heat generated by LED chips can not be dissipated in time, it will further affect the luminous performance of LED devices and reduce the life of LED devices. Solid crystal interface is the most important cooling channel, which is the closest to the LED chip. In addition, optimizing the solid crystal material, heat sink and substrate, optimizing the packaging process and structure can effectively improve the luminous performance of UV LED devices. The thermal characteristic parameters of UV LED mainly refer to the junction temperature and thermal resistance. The main factors affecting the junction temperature are the driving current, the overall heat dissipation capacity of the packaged device and the ambient temperature. Because the junction temperature has an important influence on the photoelectric performance and lifetime of UV LED, the junction temperature of LED should not exceed 60 ℃ in theory [6-7].

In this paper, the junction temperatures of samples with different solid crystal materials are measured by voltage method. The solid crystal materials are SnAgCu solder, nano silver solder and traditional silver adhesive. The method of this paper is to fix the substrate temperature and test the junction temperature of the sample with different injection current. The specific test conditions are as follows: the substrate temperature is 25 ℃ and 70 ℃, the injection current is 150mA, 250mA, 350mA, 450mA and 550mA respectively. As shown in Figure 3-1, the junction temperature increases with the increase of driving current, and the junction temperature of nano silver is the lowest. With the increase of injection current, the junction temperature changes more gently, which is consistent with the theoretical model. Part of the electric energy injected into LED devices is converted into heat and part into light, and the electro-optic conversion efficiency decreases with the increase of junction temperature. The junction temperature of LED device increases with the increase of injection current, as shown in formula (3-1) [8-10].

$$T_j = RV_f \eta I_f + T_a \qquad (3-1)$$

Where, T_j is the junction temperature, T_a is the ambient temperature, η Is the heat conversion efficiency. According to formula (3-1), the junction temperature of LED device increases with the increase of injection current, and the higher the substrate temperature is, the higher the junction temperature will be.

The results show that the junction temperature of traditional silver adhesive is the highest, that of nano silver is the lowest, and that of SnAgCu solder is between them. The results show that the thermal resistance values are 15.8K.W⁻¹, 11.5K.W⁻¹ and 13.21K.W⁻¹, respectively.

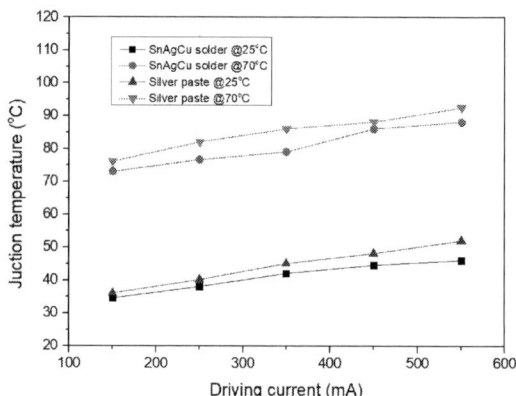

Fig. 3-1 Junction temperature under different driving current and substrate temperature

The influence of forward structure and flip structure on the junction temperature of UV LED devices is also reported in the literature. Our experimental results also verify that the flip structure has better heat dissipation performance due to the simplified packaging process and structure, short heat conduction path and the same packaging material and substrate material. The higher the thermal resistance, the higher the junction temperature [11]. The thermal resistance of UV LED is calculated as follows:

$$R = \frac{T_j - T_a}{P} \qquad (3-2)$$

978-1-6654-1392-3/21 $31.00 © 2021 IEEE   1184

$$P = V \cdot I \qquad (3\text{-}3)$$

T_j in formula (3-2) is the junction temperature, T_a is the ambient temperature. P is the dissipative power, V is the positive voltage, and I is the working current.The change trend of substrate temperature and peak wavelength is also studied under the same injection current. The principle of ultraviolet LED is the same as that of LED, and its peak wavelength is determined by energy gap just as shown in formula (3-4) [12].

$$E_g = hv = \frac{1.24}{\lambda} \qquad (3\text{-}4)$$

The results show that at the same injection current, the peak wavelengths of all samples are red shifted with the increase of substrate temperature, and the variation range is several nanometers. The relationship between LED peak wavelength and junction temperature is shown in formula (3-5).

$$\lambda(T_2) = \lambda(T_1) + \Delta TK \qquad (3\text{-}5)$$

Among them, $\lambda(T_1)\ and\ \lambda(T_2)\ is$ the peak wavelength at temperature T_1 and T_2, K is the coefficient of wavelength change with temperature. There are many reasons for wavelength redshift, among which thermal effect is one of the most important factors. Researchers believe that the final redshift effect is due to the competition of many factors. In addition, the aging test was carried out to study the initial interface morphology and the interface morphology after aging, as well as the photoelectric properties of each sample during aging. The substrate was controlled at 50 ℃, the injection current was 350mA, and the aging time lasted for 2000h. The samples were taken out every 200h for photoelectric performance test. Fig.3-2 shows the variation of luminous flux of three samples [13].

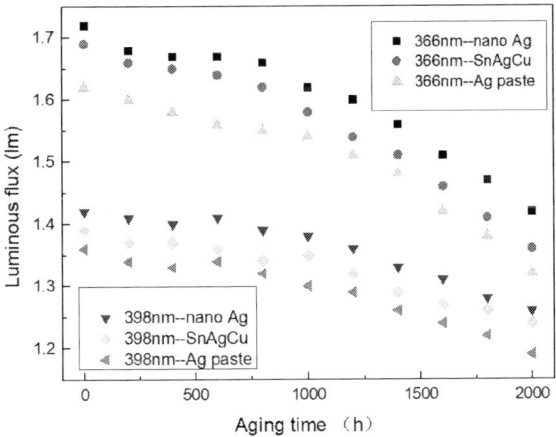

Fig.3-2 Variation of luminous flux of three samples during aging process

The research results show that the luminous flux of nano-silver decays the slowest with the increase of aging time, while the luminous flux of traditional silver glue decays rapidly with the increase of aging time. This is due to the rapid aging of silver glue itself in the high temperature aging process of traditional silver glue, which leads to the increase of interface thermal resistance, As the junction temperature of the whole device increases, the light flux decays faster. The luminous flux of SnAgCu solder changes

slowly at the beginning with the increase of aging time, and the attenuation rate increases rapidly in the second half of aging. In this study, the photochromic characteristics of radiation power, forward voltage, color coordinate, peak wavelength and half wave width of UV LED devices are also investigated in detail. The results show that almost all the photoelectric properties decrease with the increase of junction temperature, which is consistent with the results in the literature [5,6,11]. In this study, nano Ag has the best thermal conductivity, the lowest junction temperature and stable photochromic properties.

In this paper, the initial state and aging state of the solid-state interface were studied by SEM, as shown in Figure 3-3 and 3-4. The results show that the initial state of the interface is smooth and the void ratio is very low just as shown in Figure 3-3. However, with the increase of high temperature aging time, compounds and voids begin to appear in the interface, especially AuSn solder-bonded interface, thus resulting in the increase of thermal resistance and junction temperature of LED devices just as shown in Figure 3-4. In Figure 3-3 (a), further analysis confirmed that the composition of the interface compound is mainly Cu/Ni/Sn compound, and its more accurate composition and structure need to be further analyzed and confirmed by other analysis methods. Fig. 3-3 (b) shows the welding interface aged at 200 ℃ for 1h, which indicated that the compound thickness of the interface does not increase significantly, and the Ag_3Sn is not coarsened, but almost equal to the thickness of the welding layer in the length direction. The size of the cavity increases. This suggests that the diffusion and migration of atoms take place at the interface during aging reaction. In Figure 3-4, the energy spectrum analysis shows that there is a Ni coating and an Ag coating. There are two distinct layers in the solder joint interface. The bright area is mainly Au-rich phase, and the gray area is Sn-rich phase.

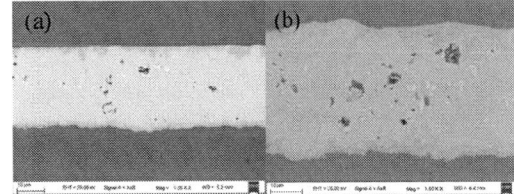

Fig.3-3 SEM image of UV-LED packaging interface by using SnAgCu solder: (a) @ 240 ℃ / 6S, (b) aged for 1h @ 200 ℃

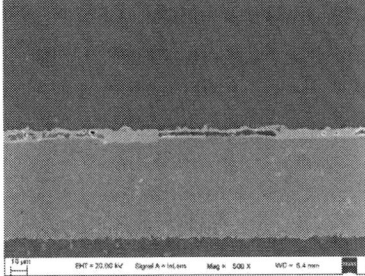

Fig. 3-4 SEM image of UV-LED packaging interface by using eutectic AuSn solder after aged for 2000h with the injection current of 350mA and the substrate temperature of 50℃

It is reported in the literature that the forward voltage, radiation power and color coordinate also change with the increase of aging time. On the whole, the device packaged with nano silver is more stable than the traditional silver glue and SnAgCu solder, and the flip chip structure is more stable than the forward structure. This is consistent with the theoretical research results of forward voltage in literature, as shown in formula (3-6).

$$V_f = \frac{nkT}{q} ln\frac{I}{A} + \frac{E_G(T)}{q} \qquad (3\text{-}6)$$

Where A is a constant, which is related to the LED chip itself, and $E_G(T)$ is the width of the band. It can be seen from the above formula that the forward voltage decreases obviously with the increase of temperature. This is due to the decrease of internal quantum efficiency of LED chip and electro-optic conversion efficiency of LED device with the increase of temperature [14-17].

IV. CONCLUSIONS

In this paper, the effects of nano silver solder, SnAgCu solder and traditional conductive silver adhesive on the photoelectric and aging properties of LED devices were studied. The results show that the nano silver solder has the lowest junction temperature, thermal resistance and the most stable photoelectric performance. The properties of SnAgCu solder are better than those of traditional silver adhesive. The initial state and the interface state after aging are studied by SEM. It is found that the increase of interface voids and compounds after aging leads to the increase of interface thermal resistance and the decrease of photoelectric properties of UV LED devices. In addition, the photoelectric performance of flip-flop structure is more stable than that of normal structure. Junction temperature and thermal resistance are important thermal parameters of UV LED devices. Only by optimizing packaging materials, packaging structure and packaging process, and controlling ambient temperature, can the thermal resistance of UV LED devices be effectively reduced and the overall stability of devices be improved.

ACKNOWLEDGMENT

In this research, solder pastes were provided by Mr. Wang. Thank the assistance of engineer Lin in sample preparation. The research was supported by the National Natural Science Foundation of China and Shenzhen Science and Technology Innovation Committee.

REFERENCES

[1] Ploch N L, et al. Investigation of the temperature dependent efficiency droop in UV LEDs. Semicond. Sci Technol. 2013, 28(12) : 125021-1-4.

[2] Schneider M., Leyer B., Herbold C., et al. Thermal improvements for high power UV LED Clusters[C], 61st Electronic Components and Technology Conference. Lake Buena Vista: IEEE, 2011.

[3] Schneider M., Leyer B., Herbold C., et al. Index matched fluidic packaging of high power UV LED clusters on aluminum substrates for improved optical output power[C], 62nd Electronic Components and Technology Conference. San Diego, CA: IEEE, 2012.

[4] Calata J N, Lei T G, Lu G Q, Sintered nanosilver paste for high-temperature power semiconductor device attachment, International Journal of Materials and Product Technology, 2009, 34(1): 95-110.

[5] Wang T, Chen X, Lu G Q, Low-temperature sintering with nano-silver paste in die-attached interconnection, Journal of Electronic Materials, 2007, 36(10):1333-1340.

[6] Lu D, Wong C P, Materials for advanced packaging, New York: Springer, 2009:143-171.

[7] Calata J N, Lei T G, Lu G Q, Sintered nano-silver paste for high-temperature power semiconductor device attachment, International Journal of Materials and Product Technology, 2009, 34(1): 95-110.

[8] Wang T, Chen X, Lu G Q, Low-temperature sintering with nano-silver paste in die-attached interconnection,Journal of Electronic Materials, 2007, 36(10):1333-1340.

[9] Bai J G, Zhang Z Z, Calata J N, Low-temperature sintered nanoscale silver as a novel semiconductor device-metallized substrate interconnect material,IEEE Transactions on Components and Packaging Technologies, 2006, 29(3): 589-593.

[10] Lei T G, Calata J N, Lu G Q, Low-temperature sintering of nanoscale silver paste for attaching large-area(>100 mm2) chips, IEEE Transactions on Components and Packaging Technologies, 2010, 33(1): 98-104.

[11] Zhang Z, Lu G Q, Pressure-assisted low-temperature sintering of silver paste as an alternative die-attach solution to solder reflow, IEEE Transactions on Electronics Packaging Manufacturing, 2002, 25(4): 279-283.

[12] Knoerr M, Kraft S, Schletz A, Reliability assessment of sintered nano-silver die attachment for power semiconductors,12th Electronics Packaging Technology Conference (EPTC), 2010:56-61.

[13] Grzanka S, Franssen G, Targowski G, et al, Role of the electron-blocking layer in the low-temperature collapse of electroluminescence in nitride light-emitting diodes, Applied Physics Letters, 2007, 90(10): 103-507.

[14] Lee S N, Cho S Y, Ryu H Y, et al, High-power Ga N based blue-violet diodes with AlGa N/Ga N multiquantum barriers, Applied Physics Letters, 2006, 88(11):1111 01-111103.

[15] Wang J, Huang X, Liu L, et al, Effect of temperature and current on LED luminous efficiency, Chinese Journal of Luminescence, 2008, 29(2): 358-362.

[16] Song G, Song J, Miao J, et al, Study on energy conversion efficiency of LED,Semiconductor Technology, 2008, 33(7): 592-595.

[17] Wen H J, Sun H, Backlight LED pulse drive method and luminous efficiency, 4th International Conference on Image Analysis and Signal Processing (IASP), 2012:1-6.

Study on Performance Optimization of Nanometer Copper Paste

Qiang Liu[1#]
[1]State Key Laboratory of Precision
Electronic Manufacturing Technology
and Equipment
Guangdong University of Technology
Guangzhou, China
892636543@qq.com

Jian Wen[2#]
[2]Jihua Laboratory
Foshan, P.R China
wenjian@jihualab.com

Yu Zhang[1*]
[1]State Key Laboratory of Precision
Electronic Manufacturing Technology
and Equipment
Guangdong University of Technology
Guangzhou, P.R China
zhangyu@gdut.edu.cn

Yu Liu[1]
[1]State Key Laboratory of Precision
Electronic Manufacturing Technology
and Equipment
Guangdong University of Technology
Guangzhou, China
105522284@qq.com

Zhongwei Huang[1]
[1]State Key Laboratory of Precision
Electronic Manufacturing Technology
and Equipment
Guangdong University of Technology
Guangzhou, China
3111238543@qq.com

Chengqiang Cui[1*]
[1]State Key Laboratory of Precision
Electronic Manufacturing Technology
and Equipment
Guangdong University of Technology
Guangzhou, China
cqcui01@qq.com

Guannan Yang[1]
[1]State Key Laboratory of Precision
Electronic Manufacturing Technology
and Equipment
Guangdong University of Technology
Guangzhou, China
ygn@gdut.edu.cn

Abstract—In this paper, in order to improve the sintering property of copper paste as interconnection material, the influence of solvent content and solvent type in copper paste on sintering strength was studied, so as to select the solvent type and solvent content with the best sintering effect. A certain amount of solvents such as ethylene glycol, diethylene glycol, glycerol and terpineol were added into the copper nanoparticles with an average particle size of 100nm-400nm. After mixing and vacuum homogenization, the nano-copper paste with solid content of 85% was obtained. Four kinds of copper pastes were sintered in 260 ℃ sintering temperature, 2MPa sintering pressure, 5% hydrogen and 95% argon mixture for 30min to obtain interconnection joints. Furthermore, the shear strength of the sintered joints was tested. The shear strength of the copper paste prepared with ethylene glycol as solvent is 50.1 MPa, while the shear strength of the copper paste prepared with the other three solvents is less than 50 MPa. Then, ethylene glycol was added into copper nanoparticles as solvent to prepare copper paste with solid content of 55%, 70% and 85%, respectively. Pure copper nanoparticles were used as control group. The above four groups of samples were sintered in 260 ℃ sintering temperature, 2MPa sintering pressure, 5% hydrogen and 95% argon mixture for 30min to obtain the interconnection joint. Finally, the shear strength of four kinds of interconnects was tested, and the shear strength of the interconnects sintered with 85% solid content copper paste was the highest, reaching 50.1MPa. In conclusion, the best sintering performance can be obtained by adding ethylene glycol as solvent into copper powder to form a paste with solid content of 85%, which can reach 50.1MPa. The surface morphology of the failure surfaces of the interconnection joints was observed by scanning electron microscopy.

Keywords: *nano-copper，solvent，shear strength，solid content，sintering*

I. INTRODUCTION

With the development of science and technology, the development of power semiconductors such as SiC and GaN that are suitable for working in high-temperature, high-power, and high-frequency environments has attracted increasing attention[1-4].Traditional electronic packaging interconnect materials mainly use solder paste or conductive glue, but they cannot meet the requirements of high-power devices[5]. Therefore, nano-metal materials with unique physical and chemical properties have become a research hotspot in packaging interconnect materials[6]. After being sintered under low temperature conditions, the nano metal can be close to the melting point, electrical conductivity and thermal conductivity of the bulk metal, which can better meet the working environment of high temperature service. Yan Jian feng et al. analyzed the sintering performance of nano-silver particles and found that nano-silver was sintered at 250°C to form a sintered layer, and the sintered layer would only melt at 960°C[7]. Wang Shuai et al. achieved a pressureless connection of silver-plated copper sheets at a sintering temperature of 200 ℃, with a shear strength of up to 60 MPa[8]. However, nano-silver has problems such as higher cost and poor resistance to electromigration[9-10]. However, copper stands out because of its low cost, easily available materials, low electrical mobility, and conductivity and thermal conductivity close to silver[11]. The researchers mixed a certain amount of organic solvent into the copper powder and configured it into a nano-copper paste with a certain solid content, which was used in the research of package interconnection of power devices. This article mainly explores the influence of copper paste's own properties on sintering interconnection. Specifically, the influence of different solvents and different solid content of copper paste on the sintering effect is studied, and the copper paste with the best sintering effect is obtained through testing and analysis.

II. EXPERIMENTAL PART

A. Experimental materials

Dilute sulfuric acid, nano copper powder and deionized water used in this experiment were prepared in the laboratory，Anhydrous ethanol, ethylene glycol, propylene glycol, glycerol, terpineol were purchased from Shanghai Macklin Biochemical Co., Ltd. All chemicals were uesd as received without further purification.

B. Experimental method

The nano copper powder prepared in the laboratory was divided into four groups, and a certain amount of ethylene glycol, propylene glycol, glycerol, and terpineol were added to prepare a nano copper paste with a solid content of 85%. Use a coating device to coat copper paste on a copper simulation substrate with a size of 10mm×10mm×0.8mm, and then cover the coated position with a copper simulation chip with a size of 4mm×4mm×0.8mm to form a sandwich structure. Then it was sintered in a hydrogen-argon mixed gas atmosphere (5% hydrogen and 95% argon) at 260°C, with an auxiliary pressure of 2 MPa, a sintering time of 30 minutes, and a heating rate of 6°C/min to form a Cu-Cu joint. Test the shear strength of the interconnection joint with an IC package soldering strength tester, and it can be concluded that the solvent with the best comprehensive performance is ethylene glycol. The solvent was mixed into the copper powder to prepare three groups of copper pastes with a solid content of 55%, 70%, and 85% respectively, and the copper powder without any solvent was used as a control group. Coating the above four groups of samples with a coating device on a copper simulation substrate with a size of 10mm×10mm×0.8mm, and then covering the coated position with a 4mm×4mm×0.8mm copper simulation chip to form a sandwich structure. Sintering under the same conditions to obtain Cu-Cu interconnection joints.

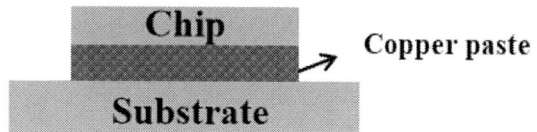

Fig. 1. Diagram of Cu-Cu interconnection structure

C. Experimental method

An IC package soldering strength tester (Nordson, DAGE-4000) was used to measure the shear strength of interconnect joints at a shear rate of 200 μm·s-1.Scanning electron microscopy (SU8220, Hitach) was used to scan the size and surface morphology of copper nanoparticles. The composition of Cu NPs was determined by X-ray diffractometry (Bruker D8 Advance).

III. RESULT AND DISCUSSION

A. Characterization of Cu Nanoparticles

Figure 2 is the SEM image and particle size distribution diagram of the copper nanoparticles prepared by the liquid phase reduction method in the laboratory. It can be seen from the figure that the particle size distribution of the copper nanoparticles is relatively concentrated, the average particle size is about 100nm-300nm, and the particle morphology is Spherical, with good dispersibility. Because the process of preparing the nano copper particles did not add organic substances such as dispersants and coating agents that are difficult to remove at 260°C. Therefore, the nano-copper particles cleaned with ethanol are very pure, which is very suitable for exploring the influence of different solvents on sintering.

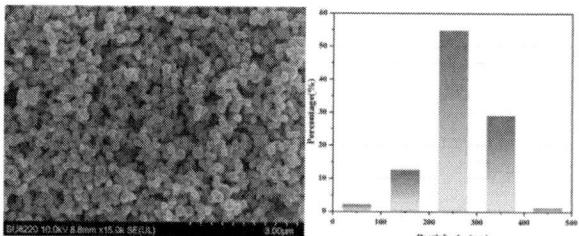

Fig. 2. Nanoparticles and their size distribution

The prepared nano-copper particles were characterized by XRD, and the phase composition and structure of the nano-copper were analyzed by XRD diffraction pattern. The obtained XRD diffraction pattern is shown in Figure 3. In the figure, there are only three sharp peaks at 2θ=43.5°, 50.6° and 74.3°, which correspond to the diffraction of {111}, {200} and {220} crystal faces of face-centered cubic Cu, respectively.In addition, there are no other impurity peaks, which indicates that the prepared copper nanoparticles are not oxidized and are suitable as interconnecting materials for sintering.

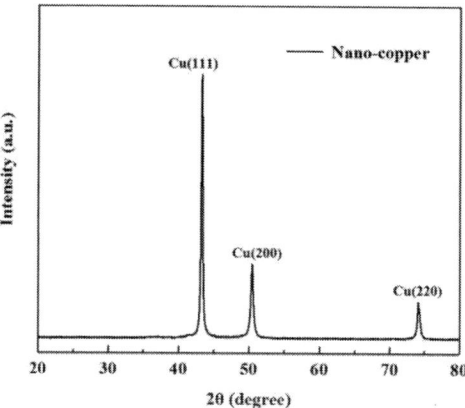

Fig. 3. XRD pattern of nanometer copper powder

B. Influence of copper paste with different solvents on sintering

Four copper pastes with different solvents are formed into a sandwich structure with the copper simulation substrate and the copper simulation chip. Sintering was carried out for 30 minutes under the conditions of a sintering temperature of 260°C, a sintering pressure of 2 MPa, and a sintering atmosphere of hydrogen and argon mixed gas. Four different interconnection joints were obtained. The shear strength of the interconnection joint was tested with an IC package welding strength tester, and the obtained shear strength is shown in Figure 4. It can be seen from the figure that the shear strength of the interconnection joint obtained with the copper paste with ethylene glycol as the solvent is the highest, which can reach 50.1 MPa. Ethylene glycol has a boiling point of 193°C, contains two hydroxyl groups, and is reductive. At a certain temperature, ethylene glycol will first be oxidized to glycolic acid and then oxalic acid, which helps sintering. The interconnection joints obtained by sintering copper paste with glycerol as the solvent have the lowest shear strength, with a shear strength of 35.6MPa. Since the boiling point of glycerol is 290°C, the volatilization rate is too slow when sintered at 260°C. This is the reason why the sintering effect is relatively poor. The shear strength of interconnected joints obtained from copper paste with propylene glycol as the solvent can reach 46.7MPa, and the boiling point of propylene glycol is about 210°C, which meets the requirements for sintering at 260°C, but propylene glycol has strong moisture retention and is easy to absorb water. Is not conducive to the preservation of copper paste. The shear strength of the interconnecting joint obtained by using terpineol as a solvent is 38.9 MPa. The boiling point of terpineol is 217°C. The reason for the low shear strength may be that it has only one hydroxyl group, and the sintering effect is relatively poor.

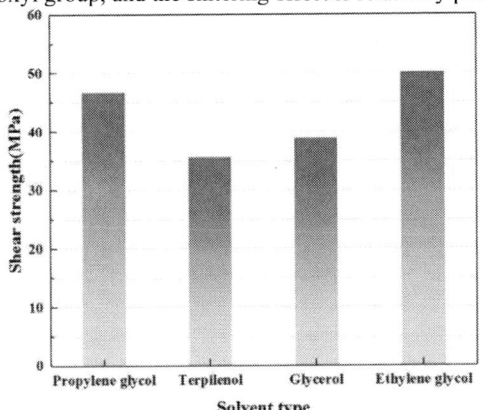

Fig. 4. Comparison of shear strength of copper paste with different solvents

Field emission scanning electron microscopy was used to observe the failure surface of the sintered layer. Figure 5 shows the SEM images of the failure surfaces of the four sintered layers. In Figure 5, a, b, c, and d are respectively the SEM images of the failure surface obtained after sintering the copper paste prepared with propylene glycol, terpineol, glycerol, and ethylene glycol as the solvent. It can be seen from the figure that the four failure surfaces have very obvious tensile traces and very dense dimple structures. From the figure d, it is obvious that the ductile fracture of the sintered layer is adhered to adjacent sintered necks. This is one of the reasons for the highest shear strength of copper paste with ethylene glycol as the solvent. The tensile traces in figures b and c are relatively small, and the sintered necks are relatively scattered, which also reflects the low shear strength. It can be seen that ethylene glycol has the best effect.

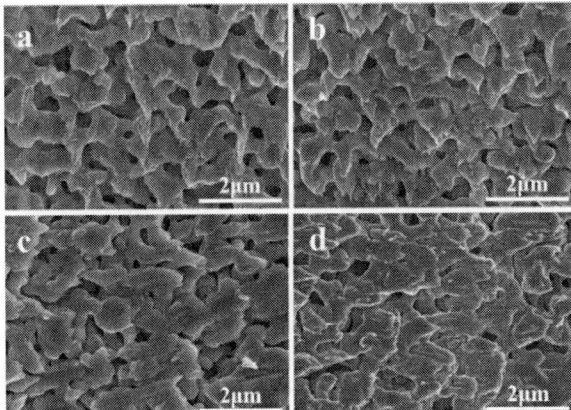

Fig. 5. SEM image of sintering fracture of copper paste prepared with four solvents at 260°C

C. Influence of copper slurry with different solid content on sintering

A certain amount of ethylene glycol was mixed into the copper powder to form a copper paste with a solid content of 55%, 70%, and 85% respectively. At the same time, the copper powder without solvent was used as a control group. These four groups of samples were formed into a sandwich structure with a copper simulation substrate and a copper simulation chip respectively. Then, under the conditions of sintering temperature of 260°C, sintering pressure of 2MPa, and sintering atmosphere of hydrogen and argon mixed gas, four different interconnection joints were obtained after sintering for 30 minutes. The shear strength of the interconnection joint was tested with an IC package soldering strength tester. The obtained shear strength is shown in Figure 6. It can be seen from the figure that the copper paste with a solid content of 85% has the highest shear strength after sintering, which can reach 50.1MPa. The sintered shear strength of copper paste with a solid content of 55% is 32.3 MPa, the shear strength of copper paste with a solid content of 70% is 35.7 MPa, and the shear strength of copper powder with a solid content of 100% is 47.6. MPa.

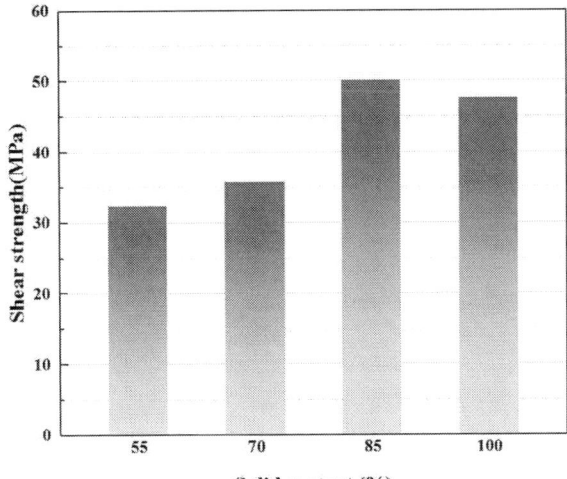

Fig. 6. Comparison of shear strength of copper paste with different solid content

The failure surface of the interconnection joint was tested by SEM, and the test result is shown in Figure 7. In Fig. 7, a, b, c, and d are respectively the SEM images of the failure

978-1-6654-1392-3/21 $31.00 © 2021 IEEE

surface of the copper paste with a solid content of 55%, 70%, 85% and 100% after sintering. It can be seen from the figure that when the solid content is 55%, the sintered layer has very many voids, very large voids, and not obvious dimples, so the shear strength is the lowest and the sintering effect is the worst. When the solid content of the copper paste is 70%, the dimple structure increases, but there are still many voids, so the shear strength is slightly increased. Because the solid content of the above two copper pastes is relatively low, the shear strength is not much different. When the solid content of the copper paste reaches 85%, the degree of mixing of copper powder with a particle size of 100nm-400nm and ethylene glycol has reached the limit, and it is difficult to increase the solid content. If the proportion of ethylene glycol continues to be reduced, the ethylene glycol and copper powder cannot be uniformly mixed. From Figure 7c, we can see that there are many obvious dimple structures, which have obvious tensile deformation and fracture, so the increase in solid content also improves the shear strength. The copper paste with a solid content of 85% has the best sintering effect among the four groups of copper pastes. D in Figure 7 is the SEM image of the failure surface after dry powder sintering. It can be seen from the figure that the SEM image of the failure surface of this interconnection joint has many dimple structures and relatively obvious tensile traces. However, the shear strength of this interconnection joint is lower than that of the interconnection joint after sintering a copper paste with a solid content of 85%. This is because the dry powder does not have better viscosity and plasticity compared with the paste, and cannot be filled well between the analog chip and the analog substrate, so the solid content is the highest, but the sintering effect is relatively poor.

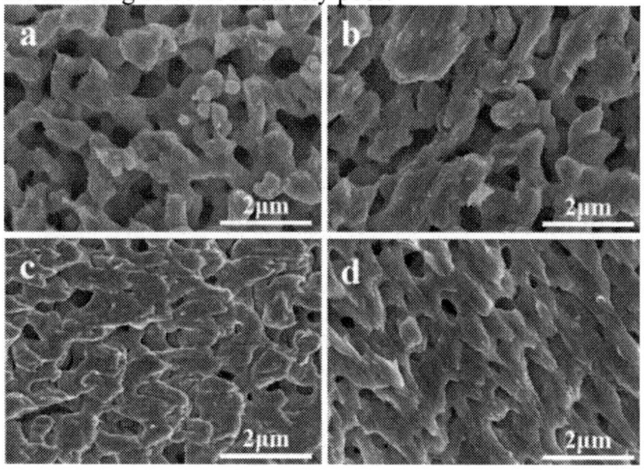

Fig.7. SEM image of sintered fracture of copper slurry prepared with four kinds of solid content copper paste at 260°C

IV. CONCLUSION

This experiment tried to use propylene glycol, terpineol, glycerol, and ethylene glycol as solvents to prepare four copper pastes and sinter them under the same conditions. The shear strength of the obtained interconnection joints was tested, and the reasons for the different shear strengths were analyzed. Among the four copper pastes, the sintering effect of the copper paste with ethylene glycol as the solvent was the best, reaching 50.1Mpa. Then, using ethylene glycol as a solvent, four copper pastes with different solid content were prepared. When the solid content was 85%, the mixing limit of copper powder with a particle size of 100nm-400nm and ethylene glycol was reached, and the sintering effect was the best.

\# The authors contributed equally to this work.

ACKNOWLEDGMENT

This work was partially supported by the Guangdong Basic and Applied Basic Research (2021A1515011642) and theNational Key R&D Program of China(2018YFE0204601).

REFERENCES

[1] Maboudian R , Carraro C , Senesky D G , et al. Advances in silicon carbide science and technology at the micro- and nanoscales[J]. Journal of Vacuum Science & Technology A Vacuum Surfaces & Films, 2013, 31(5):050805-050805-18.

[2] Kong D , Liu Y , Wang W . A Survey of Wide Bandgap Power Electronic Devices and Applications(Part Ⅱ)－－Gallium Nitride Devices[J]. The World of Inverters.

[3] Buttay C , D Planson, Allard B , et al. State of the art of high temperature power electronics[J]. Materials Science & Engineering B, 2011, 176(4):283-288.

[4] Navarro L A , Perpina X , Godignon P , et al. Thermomechanical Assessment of Die-Attach Materials for Wide Bandgap Semiconductor Devices and Harsh Environment Applications[J]. IEEE Transactions on Power Electronics, 2014, 29(5):2261-2271.

[5] Clemente R , Tolentino E N , Azman M A . Reliability considerations of sintered silver paste on clip semiconductor packages[C]// 2016 IEEE 37th International Electronics Manufacturing Technology (IEMT) & 18th Electronics Materials and Packaging (EMAP) Conference. IEEE, 2016.

[6] Liu J , Chen H , Ji H , et al. Highly Conductive Cu-Cu Joint Formation by Low-Temperature Sintering of Formic Acid-Treated Cu Nanoparticles[J]. Acs Applied Materials & Interfaces, 2016:33289.

[7] Yan J , Zou G , Hu A , et al. Preparation of PVP coated Cu NPs and the application for low-temperature bonding[J]. Journal of Materials Chemistry, 2011, 21.

[8] Shuai, Wang, Mingyu, et al. Rapid pressureless low-temperature sintering of Ag nanoparticles for high-power density electronic packaging[J]. Scripta Materialia, 2013.

[9] Mou Y , Cheng H , Peng Y , et al. Fabrication of Reliable Cu-Cu Joints by Low Temperature Bonding Isopropanol Stabilized Cu nanoparticles in Air[J]. Materials Letters, 2018, 229(OCT.15):353-356.

[10] Lee Y , Choi J R , Lee K J , et al. Large-scale synthesis of copper nanoparticles by chemically controlled reduction for applications of inkjet-printed electronics[J]. Nanotechnology, 2008, 19(41):415604.

[11] Wu C J , Cheng S L , Sheng Y J , et al. Reduction-assisted sintering of micron-sized copper powders at low temperature by ethanol vapor[J]. RSC Advances, 2015, 5(66):53275-53279.

Study on the Effect of Assembly Errors on the Electrostatic Tuning Ability in Micro Umbrella Shell Resonators

Lu Xu
Key Laboratory of MEMS of Ministry of Education
Southeast University
Nanjing, China
220181401@seu.edu.cn

Bin Luo
Key Laboratory of MEMS of Ministry of Education
Southeast University
Nanjing, China
luob@seu.edu.cn

Jintang Shang*
Key Laboratory of MEMS of Ministry of Education
Southeast University
Nanjing, China
*jshang@seu.edu.cn

Zhaoxi Su
Key Laboratory of MEMS of Ministry of Education
Southeast University
Nanjing, China
zhaoxi_su@seu.edu.cn

Shouyu Han
Key Laboratory of MEMS of Ministry of Education
Southeast University
Nanjing, China
1647664628@qq.com

Yinghui Zhang
Key Laboratory of MEMS of Ministry of Education
Southeast University
Nanjing, China
2413327649@qq.com

Abstract—This paper investigates the effects of resonator-electrode assembly errors on the electrostatic tuning ability in micro umbrella shell resonators. Theoretically analyzed the resonator-electrode capacitance effect, which refers to the stiffness adjustment of the shell by applying a voltage between the electrode and the shell. Then the electrostatic tuning models are established in the COMSOL Multiphysics, including spherical, cylindrical and bottom electrodes. Based on the electrostatic tuning models, two resonator-electrode assembly error models are established, including parallel moving error and tilting error, and the capacitance change between the resonator and the electrode caused by assembly errors is studied by finite element method. The simulation results show that the parallel moving error affects the electrostatic tuning ability of the spherical and cylindrical electrodes more than that of the bottom electrodes, while tilting error has less effects on the electrostatic tuning ability of the spherical and cylindrical electrodes compared with that of the bottom electrodes.

Keywords—Assembly error, electrostatic tuning, parallel moving error, tilting error, finite elements method

I. INTRODUCTION

As an inertial sensor for measuring rotation angle or angular velocity, hemispherical resonator gyroscope (HRG) has the advantages of high precision, good reliability and low power consumption [1,2]. Micro shell resonator gyroscope (mSRG) is not only more competitive than the traditional gyro in cost, size, weight and power (CSWaP), but also opens a brand new application market for filter and micro-positioning and navigation systems [3].

The structural symmetry of mSRG has a great influence on its performance. It is difficult to eliminate the structural asymmetry caused by fabrication mismatches. Structural asymmetry makes the resonant frequency of the primary mode different from that of the secondary mode, resulting in a frequency split Δf, thus affecting the performance of mSRG. At present, the main methods used to reduce Δf are mechanical tuning [4-8] and electrostatic tuning [9,10]. Compared with mechanical fine-tuning, electrostatic fine-tuning does not cause permanent damage to the structure of the micro shell resonator. By applying a voltage between the resonator and the electrode to adjust the shell stiffness [11], Δf can be effectively reduced. However, assembly error, which affects

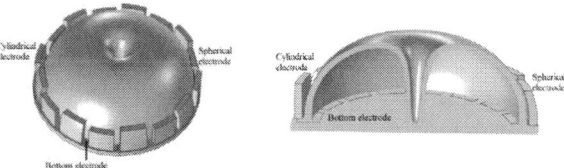

Fig. 1. Three electrode configurations for micro umbrella shell resonators.

the uniformity of the gap between the resonator and the electrode, will inevitably be introduced when the resonator is assembled with the electrode, thus affecting the electrostatic tuning ability of the micro shell resonator. This problem exerts a non-trivial influence on the gyro performance, and needs further study.

On the basis of micro umbrella shell resonators prepared by our research group [12,13], three electrode schemes, including the spherical, cylindrical and bottom electrodes, are modeled in this paper (as shown in Fig.1). Then, based on the electrostatic tuning model, two resonator-electrode assembly error models are established, including parallel moving error and tilting error.

II. THEORETICAL ANALYSIS AND SIMULATION

A. Negative stiffness adjustment

The equation of a single degree of freedom resonator can be expressed as [14]

$$m\ddot{x} + c\dot{x} + kx = F_x \qquad (1)$$

where m is the effective mass of the resonator, c and k represent the damping coefficient and stiffness coefficient respectively, x is the amplitude of the resonator, F_x is the electrostatic force. Assuming that the amplitude x is much less than the gap d between the resonator and the electrode, then the resonator and the electrode form a plane-parallel capacitor. Edge effects of adjacent electrodes are ignored. A voltage is applied to the capacitor plate to generate an electrostatic force,

$$F_x = \frac{\partial E}{\partial x} = -\frac{\varepsilon_0 \varepsilon_r S V^2}{2d^2} + \frac{\varepsilon_0 \varepsilon_r S V^2}{d^3} x \qquad (2)$$

where ε_0 is the vacuum dielectric constant, ε_r is the dielectric constant of material, S is the area of the capacitor plate, V is

978-1-6654-1392-3/21 $31.00 © 2021 IEEE

voltage applied. Substitute Equation (2) into Equation (1), the equation (1) can be expressed as

$$m\ddot{x} + c\dot{x} + \left(k - \frac{\varepsilon_0 \varepsilon_r S V^2}{d^3}\right) x = -\frac{\varepsilon_0 \varepsilon_r S V^2}{2d^2}. \qquad (3)$$

According to Equation (3), the natural frequency can be expressed by Equation (4)

$$\omega_{eff}^2 = \frac{k_{eff}}{m} = \frac{kd^3 - \varepsilon_0 \varepsilon_r S V^2}{md^3}. \qquad (4)$$

According to Equation (4), when the DC voltage V increases, the effective stiffness k_{eff} decreases and the resonant frequency ω_{eff} decreases. Therefore, the negative stiffness adjustment of the resonator can be realized by applying a voltage between the resonator and the electrode.

B. Electrostatic tuning model

In our previous research, a foaming process was developed for micro shell resonators [12]. On this basis, firstly, the fabrication process of glass shell (material: Borofloat®33) is modeled in COMSOL Multiphysics by physical field of Laminar Two-Phase Flow. The moving mesh is used to track the interface position, and the micro shell resonators with different thickness and aspect ratio are obtained. The aspect ratio of the micro shell resonator is defined as the ratio of the shell height to the shell radius. The parameters of micro shell resonator used in this paper are shown in Table I.

Table I. Resonator parameters used in simulation

thickness t(μm)	height h(mm)	radius r(mm)	aspect ratio ($AR=h/r$)
300	2.78	3.14	0.88

Then electrostatic tuning model of the electrode is established through the physical fields of Solid Mechanics and Electrostatic, including three electrode schemes (material: Au): spherical, cylindrical and bottom electrode. Finally, the electrostatic tuning ability of different electrode schemes is studied by finite element method. In this paper, the electrostatic tuning ability is defined as the normalized frequency split change per tuning voltage.

$$\text{Normalized } \Delta f \text{ change} = \frac{\Delta f_c}{f_0} \times 10^6 = \frac{\Delta f_0 - \Delta f_V}{f_0} \times 10^6 \ (ppm) \ (5)$$

Where, Δf is the frequency split of the micro shell resonator, f_0 is the resonant frequency of the micro shell resonator, Δf_c is the change value of frequency split, Δf_0 is the intrinsic frequency split of the micro shell resonator, Δf_V is tuned frequency split of the resonator, equal to $(f_y)_V - (f_x)_V$, $(f_y)_V$ and $(f_x)_V$ are the tuned resonant frequency of the secondary mode and the primary mode respectively.

C. Assembly error models

The principle of electrostatic tuning is electrostatic force is applied to the resonator by capacitor to change the effective stiffness of resonator, so that the frequency split of the resonator can be reduced. However, when the resonator is assembled with the electrode, the assembly error will inevitably be introduced, which will cause the 16 capacitors formed between the resonator and the electrode to be no longer uniform. The assembly errors studied in this paper are parallel moving error and tilting error. Parallel moving error mainly studies the capacitor inhomogeneity caused by the central pillar of the resonator deviating from the origin of the electrode substrate during the assembly process of the resonator and the electrode. Tilting error mainly studies the

capacitor inhomogeneity caused by the tilting of the resonator when the resonator is assembled with the electrode. Since the direction of parallel moving and tilting cannot be determined, in order to simulate the actual situation and simplify the modeling process, parallel moving assembly error mainly studies four kinds of parallel moving errors, namely up/down/left/right parallel moving, while tilting assembly error mainly studies two kinds of tilting errors, namely left/right tilting.

In COMSOL Multiphysics, parallel moving error model is established based on electrostatic tuning model. In the "geometry" module, the micro shell resonator is moved to form parallel moving error, and parallel moving error is shown in Fig. 2. In parallel moving error model, 16 electrodes were numbered from 1-16 in the counterclockwise direction of the XY plane during capacitance calculation, and "boundary" terminal conditions are imposed on each electrode under the physical field of Electrostatic to meet the electrical conditions required for the formation of capacitance between the resonator and the electrode. The electrode numbering is shown in Fig. 3.

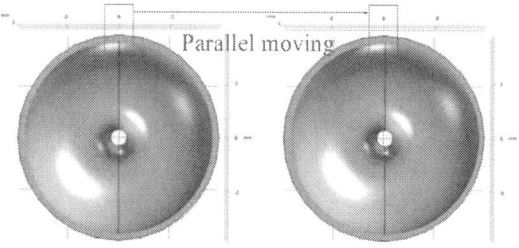

Fig. 2 Parallel moving error model

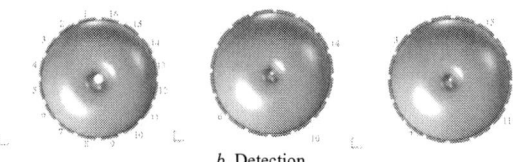

b. Detection

a. Electrode numbering compensation electrode c. Detection electrode

Fig. 3 Electrode numbering of assembly error model. (a) is the numbering of 16 electrodes, (b) and (c) are tuned electrode used in this paper.

In COMSOL Multiphysics, tilting error model is established based on electrostatic tuning model. According to the theoretical analysis, the resonator is tilted while the electrode remains unchanged in the "geometry" module to simulate tilting error. However, in order to reduce the calculation error caused by grid division in the COMSOL Multiphysics and ensure that the grid division of resonator is a regular hexahedron, this paper adopts the tilt of the electrode while resonator remains unchanged to simulate the tilting error. The tilting error model is shown in Fig. 4. The electrode numbering of tilting error model is the same as that used in

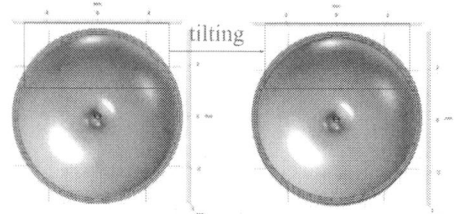

Fig. 4 Tilting error model

parallel moving error model during capacitance calculation, as shown in Fig. 3.

III. SIMULATION RESULTS AND ANALYSIS

A. Parallel moving error

When parallel moving error is simulated, seven control groups are designed, including: deviation 0μm (no deviation), deviation 1μm, deviation 2μm, deviation 3μm, deviation 4μm, deviation 5μm and deviation 6μm. In this paper, the change of capacitance and the change of electrostatic tuning ability of spherical, cylindrical and bottom electrodes are studied after parallel moving error is introduced.

In Fig. 5, the four figures show the changes of 16 capacitors between the resonator and the spherical electrode caused by the up/down/left/right parallel moving errors in turn. The ordinate is the capacitance value, and the abscissa is the electrode numbering. As can be seen from the figure, when there is no parallel moving error (deviation 0μm), 16

capacitors between the resonator and the spherical electrode are basically the same, showing a straight line. However, with the increase of parallel moving error (deviation 1/2/3/4/5/6μm), 16 capacitances are not uniform. Compared with the absence of parallel moving error, the capacitance increases and decreases, and the capacitance increase is greater than the capacitance decrease. The four figures in Fig. 6 show the effect of the up/down/left/right parallel moving error on the electrostatic tuning ability of the spherical electrode. The ordinate is the normalized frequency split change, and the abscissa is the tuning voltage. It can be seen from the figure that for a certain frequency split change, with the increase of parallel moving error, the smaller the tuning voltage required, the stronger the electrostatic tuning ability of the spherical electrode.

The four figures in Fig. 7 show the changes of 16 capacitors between the resonator and the cylindrical electrode caused by the up/down/left/right parallel moving errors in turn. The ordinate is the capacitance value, and the abscissa is the electrode numbering. When there is no parallel moving error,

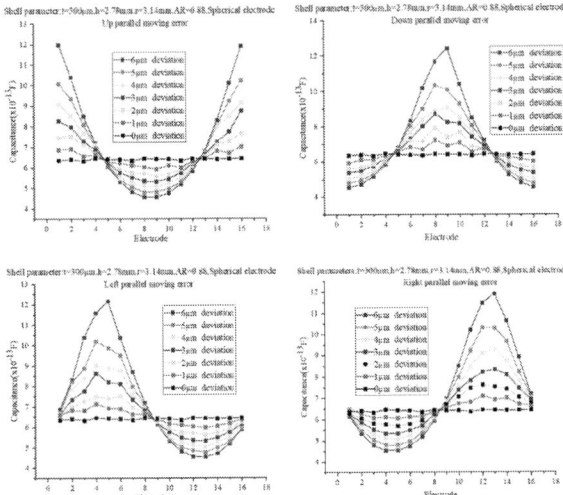

Fig. 5 Effect of parallel moving error on capacitance when the resonator is assembled with the spherical electrode. The gap between the resonator and the spherical electrode is d=12μm, the height of the spherical electrode is h=860μm, and the radian of the spherical electrode is rad=18°.

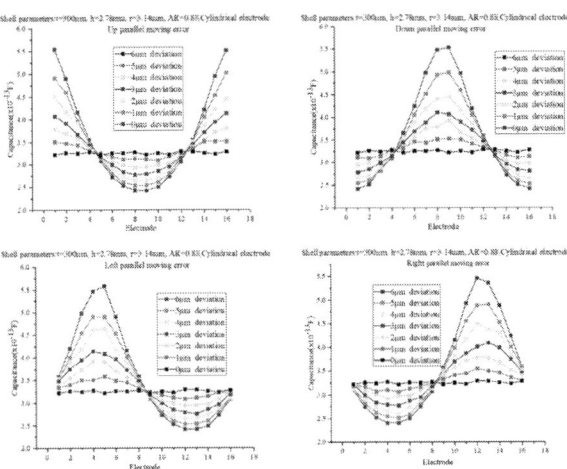

Fig.7 Effect of parallel moving error on capacitance when the resonator is assembled with the cylindrical electrode. The gap between the resonator and the cylindrical electrode is d=10μm, the height of the cylindrical electrode is h=645μm, and the radian of the cylindrical electrode is rad=15.5°.

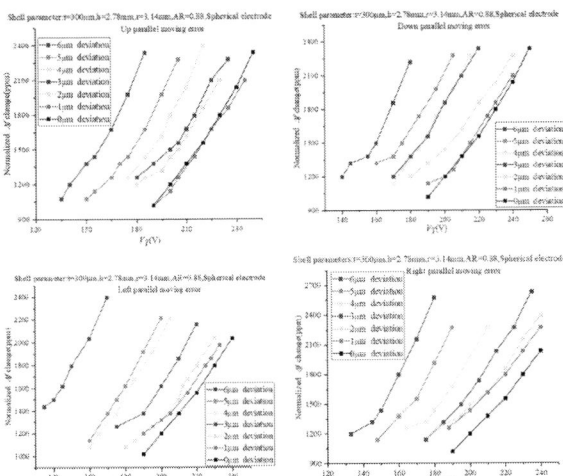

Fig. 6 Effect of parallel moving error on electrostatic tuning ability when the resonator is assembled with the spherical electrode. The gap between the resonator and the spherical electrode is d=12μm, the height of the spherical electrode is h=860μm, and the radian of the spherical electrode is rad=18°.

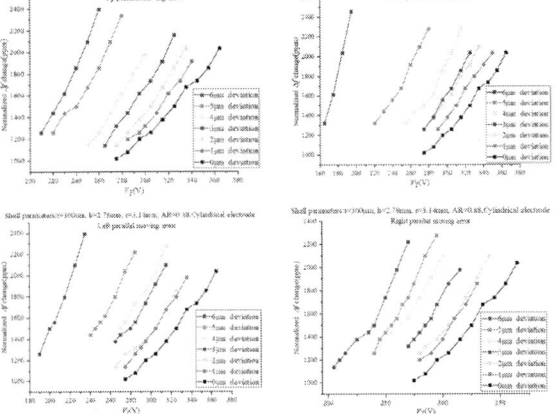

Fig.8 Effect of parallel moving error on electrostatic tuning ability when the resonator is assembled with the cylindrical electrode. The gap between the resonator and the cylindrical electrode is d=10μm, the height of the cylindrical electrode is h=645μm, and the radian of the cylindrical electrode is rad=15.5°.

16 capacitors are basically equal, which is shown as a straight line in the figure. As the deviation increases, the 16

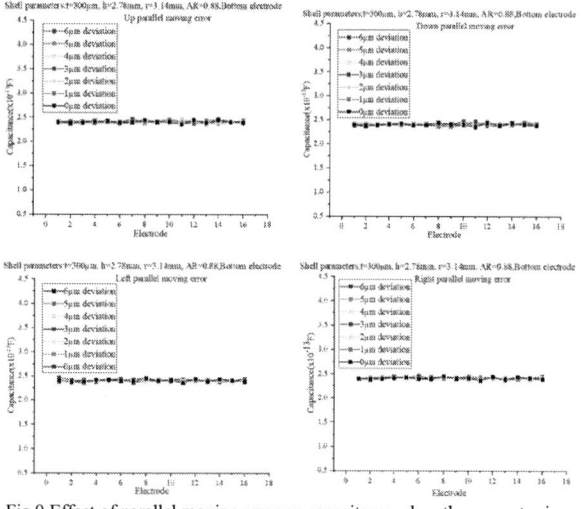

Fig.9 Effect of parallel moving error on capacitance when the resonator is assembled with the bottom electrode. The gap between the resonator and the bottom electrode is d=10μm，and the radian of the bottom electrode is rad=14.6° (the electrode area is S=8.22mm^2).

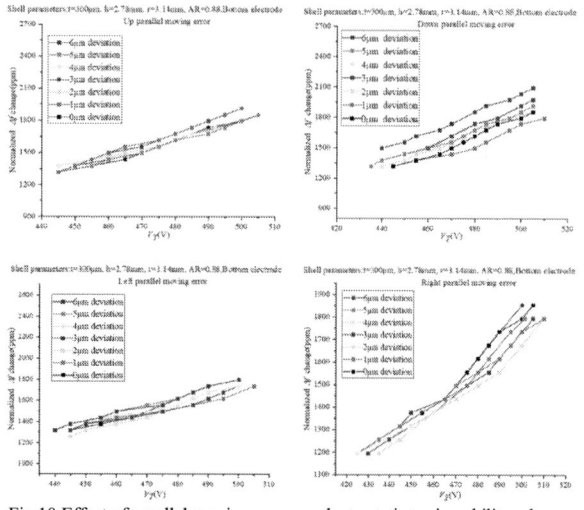

Fig.10 Effect of parallel moving error on electrostatic tuning ability when the resonator is assembled with the bottom electrode. The gap between the resonator and the bottom electrode is d=10μm，and the radian of the bottom electrode is rad=14.6° (the electrode area is S=8.22mm^2).

capacitances change correspondingly. It can be seen from the figure that the capacitance increase is greater than the capacitance decrease. The four figures in Fig. 8 show the effect of the up/down/left/right parallel moving error on the electrostatic tuning ability of the cylindrical electrode. The ordinate is the normalized frequency split change and the abscissa is the tuning voltage. As can be seen from the figure, for a certain frequency split change, with the increase of the deviation, the required electrostatic tuning voltage decreases, and the electrostatic tuning ability of increases of the cylindrical electrode.

The four figures in Fig. 9 show the changes of 16 capacitance between the resonator and the bottom electrode caused by the up/down/left/right parallel moving errors in turn. The ordinate is the capacitance value, and the abscissa is the electrode numbering. As can be seen from the figure, the

capacitance of the bottom electrode basically does not change along with parallel moving error, and is shown as a straight

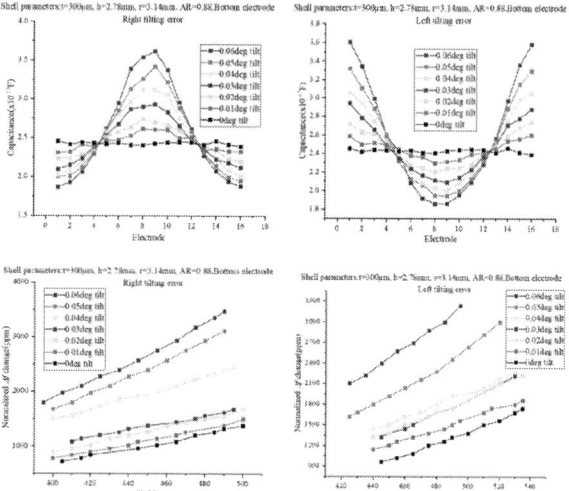

Fig.11 Effect of tilting error on capacitance and electrostatic tuning ability when the resonator is assembled with the bottom electrode. The gap between the resonator and the bottom electrode is d=10μm，and the radian of the planar electrode is rad=14.6° (the electrode area is S=8.22mm^2).

line of coincidence. The four figures in Fig. 10 show the effect of the up/down/left/right parallel moving error on the electrostatic tuning ability of the bottom electrode. The ordinate is the normalized frequency split change and the abscissa is the tuning voltage. As can be seen from the figure, for a certain frequency split change, with the increase of parallel moving error, the required tuning voltage and electrostatic tuning ability of the bottom electrode remain basically unchanged.

B. Tilting error

In the modeling and simulation of tilting error, the effect of tilting error on bottom electrode is only studied in this paper. According to simulation results of parallel moving error, the simulation results of the assembly error of spherical and cylindrical electrodes are different from the simulation results of bottom electrodes. Therefore, we can infer the effect of tilting error on electrostatic tuning ability of spherical and cylindrical electrodes from simulation results of tilting error of bottom electrodes. When tilting error model is simulated, seven control groups are designed as follows: tilt 0° (no tilt), tilt 0.01°, tilt 0.02°, tilt 0.03°, tilt 0.04°, tilt 0.05° and tilt 0.06° (Equivalent to moving 3μm).

The upper two figures in Fig. 11 show the capacitance change between the resonator and the bottom electrode caused by right/left tilting error. The ordinate is the capacitance value, and the abscissa is the electrode numbering. As can be seen from the figure, when there is no tilting error, 16 capacitors are basically the same, which is presented as a straight line. As tilting error increases, 16 capacitors are no longer uniform. Compared with the no tilting error, 16 capacitances increase and decrease, and the capacitance increase is greater than the capacitance decrease. The lower two figures show the variation of the electrostatic tuning ability of the bottom electrode caused by right/left tilting error. The ordinate is the normalized frequency split change and the abscissa is the tuning voltage. As can be seen from the figure, for a certain frequency split change, with the increase of tilting error, the smaller the required tuning voltage is, the better the electrostatic tuning ability of the bottom electrode is.

IV. CONCLUSION

In this paper, based on the electrostatic tuning model, two assembly error models between the resonator and the driving/sensing electrode are established, including parallel moving error and tilting error. The simulation results show that under parallel moving error model, the total capacitance of spherical and cylindrical electrodes increases with the increase of deviation, and the electrostatic tuning ability is enhanced. The total capacitance of the bottom electrode does not change with the deviation, and the electrostatic tuning ability does not change. Under tilting error model, the total capacitance of the spherical and cylindrical electrodes does not change with tilting error, and the electrostatic tuning ability is basically unchanged. The total capacitance of the bottom electrode increases with the increase of tilting error, and the electrostatic tuning ability is enhanced. In a word, the capacitance increase caused by assembly errors accounts for the enhancement of electrostatic tuning ability. Therefore, to achieve better electrostatic tuning ability of micro umbrella shell resonators, increasing resonator-electrode capacitance used for electrostatic tuning and decreasing intrinsic frequency split of resonator can be taken into consideration.

ACKNOWLEDGMENT

This work is supported by National Science Foundation of China under Grant 51675102. The authors would like to thank Shouyu Han for providing micro umbrella shell resonator parameters.

REFERENCES

[1] Z. Wei, G. Yi, Y. Huo, Z. Qi and Z. Xu, "The Synthesis Model of Flat-Electrode Hemispherical Resonator Gyro," Sensors, vol. 19, no. 7, p. 1690, 2019.

[2] F. Delhaye, "HRG by SAFRAN: The game-changing technology," 2018 IEEE International Symposium on Inertial Sensors and Systems (INERTIAL), 2018, pp. 1-4.

[3] S. Singh, J. Woo, G. He, J. Y. Cho and K. Najafi, "0.0062 °/√hr Angle Random Walk and 0.027 °/hr Bias Instability from a Micro-Shell Resonator Gyroscope with Surface Electrodes," 2020 IEEE 33rd International Conference on Micro Electro Mechanical Systems (MEMS), 2020, pp. 737-740.

[4] C. Fox, "Analysis and correction of imperfection in vibrating cylinders and rings," in Proc. IFToMM Ninth World Congress on the Theory on Machines and Mechanisms, 1995, pp. 1126-1130.

[5] B. Gallacher, J. Hedley, J. Burdess, A. Harris and M. McNie, "Multimodal tuning of a vibrating ring using laser ablation," Proceedings of the Institution of Mechanical Engineers, Part C: Journal of Mechanical Engineering Science, vol. 217, no. 5, pp. 557-576, 2003.

[6] D. M. Rozelle, "The hemispherical resonator gyro: From wineglass to the planets," in Proc. 19th AAS/AIAA Space Flight Mechanics Meeting, 2009, vol. 134, pp. 1157-1178.

[7] D. M. Schwartz, D. Kim, P. Stupar, J. DeNatale and R. T. M'Closkey, "Modal parameter tuning of an axisymmetric resonator via mass perturbation," Journal of Microelectromechanical Systems, vol. 24, no. 3, pp. 545-555, 2015.

[8] W. Li, et al., "Micro shell resonator with T-shape masses for improving out-of-plane electrostatic transduction efficiency," 2016 IEEE International Symposium on Inertial Sensors and Systems (INERTIAL), 2016, pp. 78-80.

[9] B. Gallacher, J. Hedley, J. Burdess, A. Harris, A. Rickard and D. King, "Electrostatic tuning of a micro-ring gyroscope," in Tech. Proc. 2004 NSTI Nanotechnology Conference and Trade Show, 2004, vol. 1, pp. 430-433.

[10] C. H. Ahn, et al., "Mode-matching of wineglass mode disk resonator gyroscope in (100) single crystal silicon," Journal of Microelectromechanical Systems, vol. 24, no. 2, pp. 343-350, 2014.

[11] A. Darvishian, et al., "Effect of Electrode Design on Frequency Tuning in Shell Resonators," 2019 IEEE International Symposium on Inertial Sensors and Systems (INERTIAL), 2019, pp. 1-4.

[12] B. Luo, J. Shang and Y. Zhang, "Hemispherical glass shell resonators fabricated using Chemical Foaming Process," 2015 IEEE 65th Electronic Components and Technology Conference (ECTC), 2015, pp. 2217-2221.

[13] B. Luo, J. Shang, Z. Su, J. Zhang and C.-P. Wong, "Height adjustment of 3-D axisymmetric microumbrella shells for tailoring wineglass frequency," IEEE Transactions on Components, Packaging and Manufacturing Technology, vol. 9, no. 3, pp. 567-574, 2019.

[14] C. Acar and A. Shkel, MEMS vibratory gyroscopes: structural approaches to improve robustness. Springer, Boston, MA, 2009.

Research on Integrated Metasurface Lens for High Gain Multibeam System in Package Application

Yuxiang Zheng
Institute of Microelectronics of the Chinese Academy of Sciences
University of Chinese Academy of Sciences
Beijing, China
zzysrzys@outlook.com

Weikang Wan
Institute of Microelectronics of the Chinese Academy of Sciences
University of Chinese Academy of Sciences
Beijing, China
wanweikang@ime.ac.cn

Qidong Wang*
Institute of Microelectronics of the Chinese Academy of Sciences
Beijing, China
wangqidong@ime.ac.cn

Liqiang Cao
Institute of Microelectronics of the Chinese Academy of Sciences
Beijing, China
caoliqiang@ime.ac.cn

Abstract—**An integrated metamaterial-based planar lens fed by linearly polarized patch driven metasurface antenna is presented and characterized for spatial beamforming and multibeam system. A double-layered Jerusalem cross unit cell is adopted for design of metasurface lens. The integrated ultrathin lens with the size of $9.4\lambda_0 \times 9.4\lambda_0 \times 0.097\lambda_0$ (λ_0 is the free-space wavelength at the frequency of 29 GHz) is achieved to operate over the range from 26.8 GHz to 31.2 GHz with a gain ripple of 3 dB and a maximum gain of 22.54 dBi. By using discrete antenna sources separately distributed at different locations, the proposed lens antenna achieves a wide beam scanning range about $\pm 33°$ with a gain ripple of 4.2 dB. The proposed ultrathin metasurface lens without airgap realizes high gain and wide scanning angle, which is promising for the 5G millimeter wave system in package application.**

Keywords—*metasurface lens, lens antenna, multibeam, beam scanning, system-in-package*

I. INTRODUCTION

High gain antennas with high directivity, multibeam and beamforming will meet the demands of signal coverage and telecommunications, which are widely used in radar systems, space explorations and modern communication systems [1]. Researches on high gain antennas such as active phased array antennas, parabolic antennas, reflect-array antennas and conventional lens antennas promote the development of 5G millimeter wave wireless communication [2-5]. The active phased array antennas generate multibeam nimbly with the complex lossy feeding networks and expensive transceiver components. Therefore they are unsuitable to integrate with other RF modules for system-level package applications [2]. Parabolic antennas have bulky volume and slow response of beam scanning on account of mechanical rotation [3]. Reflect-array antennas possess the feeding source which blocks the outward electromagnetic (EM) wave reflected from the bottom array [4]. The conventional lens antennas demanding precise manufacturing accuracy and hybrid materials are incompatible for planar manufacturing process [5]. The demand of lightweight high gain multibeam antenna with simple manufacturing process and low cost becomes urgent for 5G wireless communication system.

Metamaterial is an artificial material with sub-wavelength dimension, which has been used to design miniaturized antenna in recent years. The metamaterial changes the magnitude and phase of EM wave due to its unique electromagnetic characteristics. Arranging periodic planar metamaterial units constitutes the metasurface with desirable electromagnetic behaviors [6].

The metalens collimating the spherical wave emitted from feeding source is the substitution of traditional lens. High gain pencil beams with different deviation angles can be generated by placing the feeding elements at different locations. Exciting different feeding elements simultaneously will realize multibeam to cover wide region of space. Independently exciting single feeding source generates discrete beams with different directions [7-8]. The total thickness of metalens is set around one fifth to half wavelength in free space with airgap or multilayered pattern to achieve fully controllable planar wavefront [9]. Those metalens structures are relatively large, and the airgap also brings the potential risks with contamination, corrosion and heat dissipation.

In this paper, an ultrathin integrated metasurface lens is presented. A linearly polarized patch driven antenna [10] is set as a feeding source of the metalens. The metamaterial unit cell in the metalens serves as phase shifter to correct the phase difference among lens surface. For simple design process, a metamaterial-based Jerusalem cross unit cell is used to control the phase shift as well as transmission amplitude. Placing the unit cell with larger phase shift away from the center of the metalens will make up the phase delay caused by path difference. Therefore, the spherical EM wave front can converge to planar wave front through the metalens. The multibeam and beam scanning are implemented by the high gain pencil beam with planar wave front.

This paper is organized as the follow sequence. The design of metalens antenna is illustrated in Section II. The simulation results are presented in Section III. Section IV gives the conclusion of metalens antenna.

II. DESIGN OF METALENS ANTENNA

A. Design of Jerusalem cross unit

978-1-6654-1392-3/21 $31.00 © 2021 IEEE

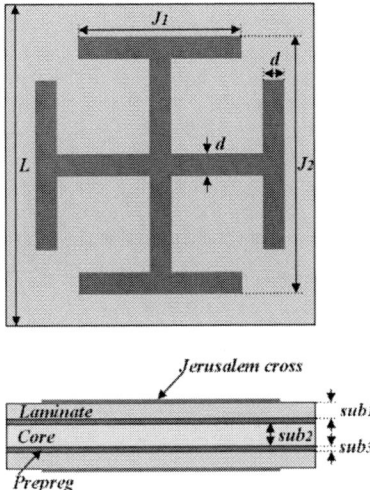

Fig. 1. Dimensions and compositions of JC unit cell.

A Jerusalem cross (JC) is used in this design. The dimensions and compositions of JC unit cell are shown in Fig.1. The JC unit cell is composed of five dielectric layers without airgap and two identical copper layers separately located at top and bottom of the substrate. The core dielectric layer is Rogers 4350B, the relative permittivity is 3.66 and the dissipation factor is 0.0037. The laminates on top and bottom layers are HL972, the relative permittivity is 3.6 and the dissipation factor is 0.005. The prepreg close to each laminate is GHPL-970 acting as buildup dielectric, the relative permittivity is 3.4 and the dissipation factor is 0.004.

The JC unit cell serves as a transceiver which transmits the EM wave efficiently with a certain phase shift. Through placing JC unit cells with different phase shifts in the metalens appropriately, the phase difference caused by EM path difference is counterbalanced. The spherical wave front is transformed into planar wave front, thus producing pencil beam with high gain. The way to change phase shift without impeding transmissivity can be achieved by tuning the length of J_1, as shown in Fig.2. The transmission coefficient is high in major variation range of J_1. Also the transmission coefficient is better than -5 dB in bilateral regions. The range of transmission phase reaches up to 230°. The wide phase range is sufficient for phase compensation of planar lens design.

The conversion from spherical wave front to planar wave front is shown in Fig.3. The phase difference can be decided by

$$\Delta\varphi = -\frac{2\pi}{\lambda_0} \times \Delta d \pm 2n\pi \qquad (1)$$

where $\Delta\varphi$ is the phase difference of EM waves between the off-center unit cell and the center unit cell in the metalens, and Δd is extra wave path corresponding to the phase difference. The unit cells with smaller transmission phases are placed near the center of the metalens, and the unit cells with larger transmission phases will be placed away from the center of the metalens. With the phase compensation of unit cell, the approximately uniform phase distribution of lens surface concentrates the spherical wave front into planar wave front. Tuning the unit cell dimensions at each location of lens

surface, the whole configuration is realized with 23×23 metamaterial-based JC unit cell array, the top view of metalens is shown in Fig.4.

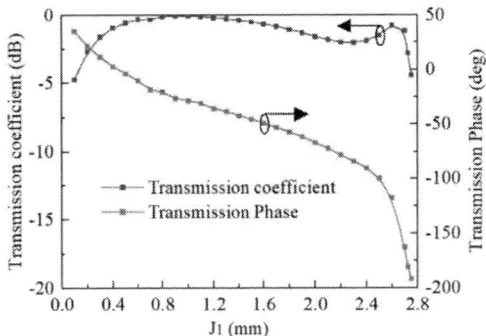

Fig. 2. Transmission coefficient and transmission phase of JC unit cell with change of J_1.

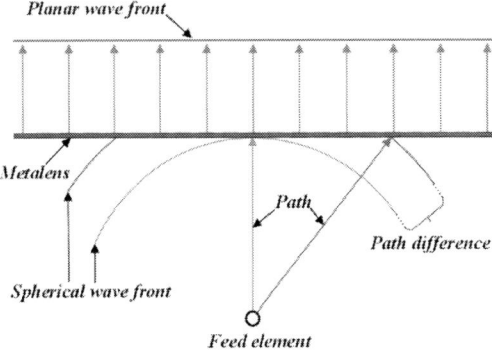

Fig. 3. Spherical wave front is converted into planar wave front.

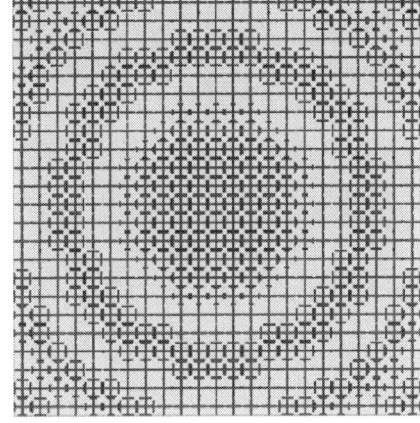

Fig. 4. Top view of metasurface lens.

B. Selection of feeding elements

The linearly polarized patch driven antenna is served as feeding element, as shown in Fig.5. The obtained average gain is 7.63 dBi from 24 GHz to 31.5 GHz, and the variation of gain is from 6.57 dBi to 8.53 dBi. The -10 dB impedance bandwidth of the antenna is 19.1% from 25.1 GHz to 30.4

GHz. The cross polarization level of patch driven antenna is less than 30 dB by the merit of aperture coupling. The efficiency of feeding antenna is above 80%. The patch driven antenna with broadside radiation pattern will be approximately regarded as the point source which radiates the spherical wave. The spherical wave front is transformed into planar wave front with the aid of metamaterial unit cells.

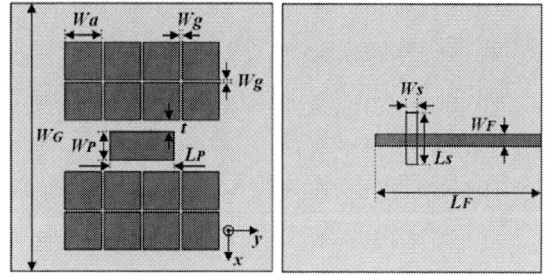

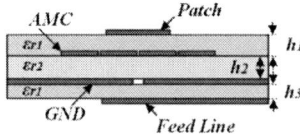

Fig. 5. Geometry of linearly polarized patch driven antenna, where $W_G = 12.84$, $W_P = 1.3$, Wa = 1.32, Wg = 0.05, $L_P = 2.5$, t = 0.4, Ws = 0.18, $W_F = 0.076$, Ls = 1.9, $L_F = 7.52$, $h_1 = 0.04$, $h_2 = 0.422$ and $h_3 = 0.04$, all dimensions are in mm.

C. Overall structure of lens antenna

The total configuration of metasurface lens antenna is shown in Fig.6.

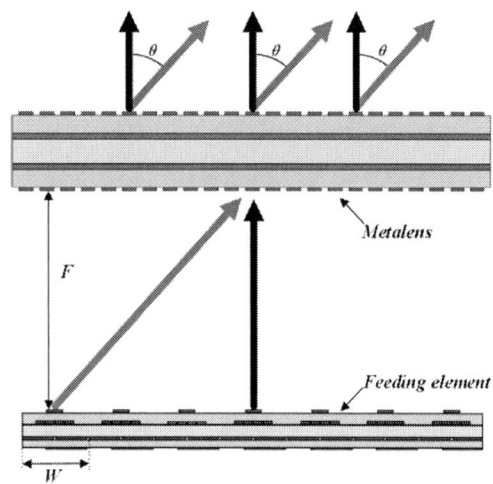

Fig.6. Total configuration of metasurface lens antenna.

The upper constructure is metalens which converges the omnidirectional spherical wave into directional planar wave. The feeding elements under the metalens generate EM waves from discrete locations. Then the EM waves can be transformed into discrete straight beams deviating from normal direction of metalens surface with angle θ.

A protype of 1 × 7 feeding source array is used. The center space between feeding sources is W. The center feeding element is placed right beneath the lens at a distance of F. The distance between metalens and feeding source is selected to maintain the identifiable beams as well as wide scanning range. The beam with normal direction of lens surface is formed by the center feeding source. The feeding elements at other locations generate certain beams deviating from the normal direction. Controlling single feeding element working independently can realize the beam scanning with desired deviation angle. Exciting those feeding elements at the same time result in multibeam to coverage wide area of space.

The metalens antenna is simulated by ANSYS HFSS with partial parameters given in Table I.

TABLE I. PARTIAL PARAMETERS OF SIMULATION

Symbol	Explanation	Parameter value
L	Period of JC unit cell	4.2 mm
J_1	Length of copper strip	2.5 mm
J_2	Length of copper box	4.1 mm
d	Width of copper strip	0.3 mm
sub_1	Thickness of laminate	0.25 mm
sub_2	Thickness of core	0.42 mm
sub_3	Thickness of prepreg	0.04 mm
F	Vertical distance from metalens to feeding source	48.3 mm
W	Offset between adjacent feeding source	11 mm

III. SIMULATED RESULTS AND DISCUSSIONS

A. $|S_{11}|$ of patch driven antenna

The reflection coefficients of individual patch driven antenna and feeding antennas with different scanning angles are shown in Fig.7. The shape of $|S_{11}|$ almost remains unchanged. There are some tiny fluctuations among the curve group. Back scattering of metalens surface has a few effects on the patch driven antenna. Thus the -10 dB impedance bandwidth shifts slightly towards higher frequency. The sources with different locations have similar -10 dB impedance bandwidth. Therefore, the patch driven antenna excites metalens effectively regardless of scanning angle.

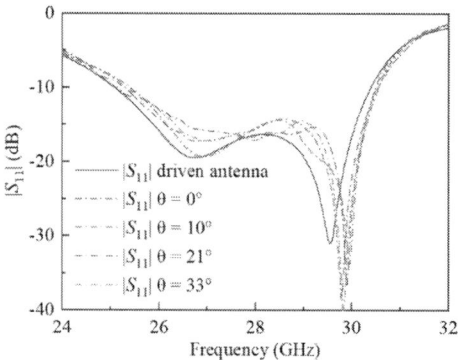

Fig.7. Scattering parameters of individual driven antenna and feeding antennas with different scanning angles.

B. Gain of lens antenna

The gains of patch driven antenna and metalens antenna are shown in Fig.8. It can be found that the metalens antenna achieves a flat gain variation around 29 GHz, and the maximum gain is obtained in 29.2 GHz with 22.54 dBi. The aperture efficiency is 16.2%. The -3 dB gain bandwidth of metasurface lens antenna is 15.2% from 26.8 GHz to 31.2 GHz. The metasurface lens brings about an improvement of 14.24 dB with regard to patch driven antenna. The high gain metalens antenna approximately equals to the combined gains of 27 patch driven antennas without considering the lossy feeding network.

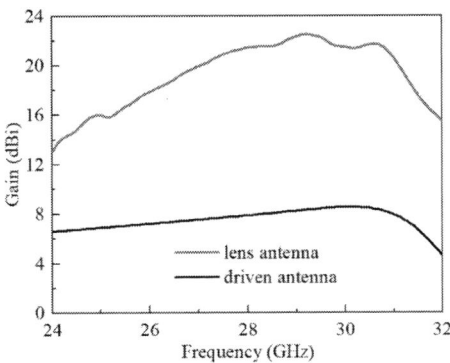

Fig.8. Gains of metalens antenna and patch driven antenna.

C. Beam performance of lens

The omnidirectional beam will be collimated by metalens and transformed into directional beam. The metalens designed by (1) is based on the center feeding source right beneath the center of lens. The displacement of feeding source relative to the center source would change wave-path difference on the surface of the metalens. The change of path difference leads to variation of phase difference. As a result, the wave front transmitted from lens surface becomes non-flat slightly. Consequently only normal beam with scanning angle θ=0° is predesigned, other beams have partially deterioration in total gains due to the nonideal planar wave fronts, as shown in Fig.9.

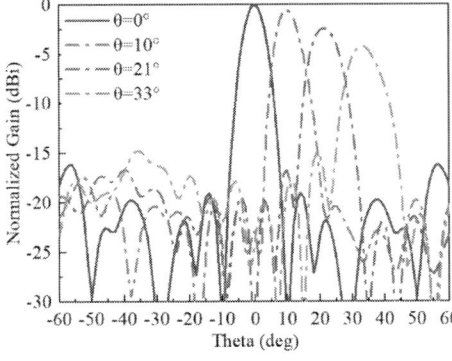

Fig.9. Gains of metalens antenna with different beam scanning angles.

By placing the feeding sources at the distance of W, 2W and 3W away from center feeding source, the generated beams by these sources point to the directions of 10°, 21° and 33°

separately deflecting from normal direction. The gain of scanning beam with θ = 33° decreases by 4.2 dB compared to normal beam with θ = 0°. Accordingly, the gain of beam with deviation angle θ = 33° is 18.34 dBi, which is high enough in wireless communication system. The side lobe level of normal beam is -19 dB. The half power beam width is 7°.

The metalens antenna is simulated and the results are listed in Table II.

TABLE II. SIMULATED RESULTS OF METALENS ANTENNA

Performance indexes	Value
Maximum gain	22.54 dBi@29.2 GHz
Gain improvement	14.24 dB
-3 dB gain bandwidth	26.8 GHz-31.2 GHz (15.2%)
Scanning range	± 33°
Scanning loss	4.2 dB
Side lobe level	-19 dB
Half power beam width	7°
Aperture efficiency	16.2%

IV. CONCLUSION

A high gain integrated metasurface lens fed by linearly polarized patch driven antenna is proposed. A prototype of 23 × 23 JC unit cell array is designed. The metamaterial-based JC unit cell serves as phase shifter to adjust the phase of EM wave transmitted through the metalens. The high gain pencil beams are created with the aid of uniform phase distribution on the lens surface. The patch driven antennas beneath the lens are used as feeding sources. By exciting the sources at different locations, discrete beams with different directions are generated. The multibeam and beam scanning are implemented by these discrete beams. The maximum gain of lens antenna is 22.54 dBi and the -3dB gain bandwidth is 15.2%. The scanning range of the lens antenna reaches up to ± 33° with a gain reduction of 4.2dB. The compact ultrathin metalens without airgap is compatible for standard PCB process. The metalens is a promising candidate in future radar system and 5G wireless communication system.

REFERENCES

[1] Hong, Wei, Zhi Hao Jiang, Chao Yu, Jianyi Zhou, Peng Chen, Zhiqiang Yu, et al., "Multibeam antenna technologies for 5G wireless communications," IEEE Transactions on Antennas and Propagation, vol. 65, no. 12, pp. 6231-6249, 2017.

[2] Yeh, Yi-Shin, and Brian A. Floyd, "Multibeam phased-arrays using dual-vector distributed beamforming: architecture overview and 28 GHz transceiver prototypes," IEEE Transactions on Circuits and Systems I: Regular Papers, vol. 67, no. 12, pp. 5496-5509, 2020.

[3] Mehrabani, Ali, and Lotfollah Shafai, "Compact dual circularly polarized primary feeds for symmetric parabolic reflector antennas," in IEEE Antennas and Wireless Propagation Letters, vol. 15, pp. 922-925, 2016.

[4] J. Thornton and K. C. Huang, "Modern lens antennas for communications engineering," John Wiley & Sons, Inc, 2013.

[5] R. Sauleau, "Lens antennas for MM and SUB-MM wave applications," International Conference on Mathematical Methods in Electromagnetic Theory, pp. 18–23, 2006.

[6] Maci, S., G. Minatti, M. Casaletti, and Marko Bosiljevac. "Metasurfing: addressing waves on impenetrable metasurfaces," IEEE Antennas and Wireless Propagation Letters, vol. 10, pp. 1499-1502, 2011.

[7] Jiang, Mei, Zhi Ning Chen, Yan Zhang, Wei Hong, and Xiaobo Xuan, "Metamaterial-based thin planar lens antenna for spatial beamforming and multibeam massive MIMO," IEEE Transactions on Antennas and Propagation, vol. 65, no. 2, pp. 464-472, 2017.

[8] Li, Shunli, Zhi Ning Chen, Teng Li, Feng Han Lin, and Xiaoxing Yin, "Characterization of metasurface lens antenna for sub-6 GHz dual-polarization full-dimension massive MIMO and multibeam systems," IEEE Transactions on Antennas and Propagation, vol. 68, no. 3, pp. 1366-1377, 2020.

[9] A. H. Abdelrahman, A. Z. Elsherbeni, and Y. Fan, "Transmitarray antenna design using cross-slot elements with no dielectric substrate," IEEE Antennas and Wireless Propagation Letters, vol. 13, pp. 177-180, 2014.

[10] Wan, Weikang, Mei Xue, Liqiang Cao, Tianchun Ye, and Qidong Wang, "Low-profile broadband patch-driven metasurface antenna," IEEE Antennas and Wireless Propagation Letters, vol. 19, no. 7, pp. 1251-1255, 2020.

Study on thermal stability of all copper interconnect structures under thermal shock

Hao Li
State Key Laboratory of Mechanical Transmission
College of Material Science and Engineering, Chongqing University,
Chongqing,400044,China，
330597587@qq.com

Jun Shen*
State Key Laboratory of Mechanical Transmission
College of Material Science and Engineering, Chongqing University,
Chongqing,400044,China，
shenjun@cqu.edu.cn

Jiacheng Xie
State Key Laboratory of Mechanical Transmission
College of Material Science and Engineering, Chongqing University,
Chongqing,400044,China，
524586242@qq.com

Abstract—Copper nanoparticles have become a new generation of chip interconnection materials with considerable application prospects due to their excellent electrical and thermal properties, good electromigration resistance, and low economic cost. Different all copper interconnects were prepared by air sintering and vacuum sintering respectively, and the thermal shock test was carried out. The mechanical and electrical properties of the interconnects after thermal shock cycle were tested. Combined with microstructure characterization and theoretical analysis, the influence of thermal shock conditions on the thermal stability of all-copper interconnects was studied. The experimental results indicated that the bonding strength of the all-copper interconnection structure decreased first and then increased slightly with the number of thermal shock cycles. After 1000 cycles, the interconnection structures prepared by air sintering and vacuum sintering still maintain good bonding strength, and the average shear strength is respectively 23.24 MPa and 25.01 MPa.

Keywords—*copper nanoparticle; all-copper interconnections structure; thermal shock; shear strength.*

I. Introduction

In recent years, with the increasing application of high-power devices in all walks of life, the traditional packaging materials and processes have been difficult to meet the higher packaging performance and reliability requirements of power devices [1]. Among them, the most critical is how to ensure the interconnection reliability of high-power chip packaging, and the development of new interconnection technology and materials has become an important break through Aiming at the performance requirements of high power devices such as high temperature service and high power density, the new interconnection technology of sintered metal nanoparticles has become a solution to replace the traditional solder welding process [2, 3]. Due to the high thermal conductivity and electric conductivity of copper nanoparticles, good resistance to electric migration and low cost, the low temperature copper-copper interconnection technology based on sintered copper nanoparticles has become the first choice in high-density packaging interconnection of high-power chips [4, 5].

As for low temperature copper-copper interconnection technology, scholars from the United States, Japan, the United Kingdom and other countries have confirmed the feasibility of sintering copper nanoparticles at low temperature on copper substrate to prepare all-copper interconnection structure through a large number of studies. In addition, Lockheed Martin [6] and IBM [7] in the United

States have applied copper-copper interconnection technology to the actual packaging and manufacturing of chips, and successfully developed related electronic product. Huazhong university of science and technology [8], Harbin industrial university [9], Chongqing university [10，11] and many other colleges and universities also studied the copper to copper interconnect technology, changes of copper and copper connection strength of joint, the microstructure evolution of sintering organization and oxidation resistance of copper nanoparticles was carried out by a large number of experiments, and combining the data are analyzed. However, there are few studies on the influence of temperature shock load on the thermal stability of all-copper interconnect structures. Therefore, the all-copper interconnect structure was prepared by low-temperature sintering process. Thermal shock experiments were designed to characterize the mechanical properties and microstructure of the interconnect structure before and after thermal shock, and to analyze the thermal stability of the all-copper interconnect structure under thermal shock conditions.

II. Experimental

A. Materials

Nano-copper paste was formed from Cu nanoparticles (mean diameter 50 nm) and propylene glycol in a certain ratio. Additional dispersant and antioxidant were added to the paste to prevent the Cu nanoparticles from agglomeration and excessive oxidation. Two Cu disks with diameters of $\phi 3$ m mand $\phi 10$ mm (both 3 mm thick) were used for the sintering process.

B. Characterization and Measurement

The shear strength of the joints was assessed byimplementing a shear tester (CMT5105) at a veloc-ity of 1 mm/min. X-ray diffraction (XRD) analysis tothe phase composition of the as-sintered structurebefore and after isothermal aging was conducted byan x-ray diffractometer (PANalytical Empyrean).Morphological features of the Cu nanoparticles, thefracture surface, and cross-sections of Cu-Cu jointswere observed by field-emission scanning electronmicroscopy (SEM; Zeiss Auriga FIB-SEM) equippedwith energy dispersive spectroscopy (EDS).

C. Sintering and Thermal Shock Process

First, a batch of all-copper interconnection structure samples were prepared by low temperature sintered nanometer brazing paste for testing. The specific sintering

978-1-6654-1392-3/21 $31.00 © 2021 IEEE

process is as follows: the heating rate is 10°C/min, the sintering temperature is 280°C, the sintering time is 15min, the sintering pressure is 6 MPa, and the sintering environment is set as vacuum environment and atmospheric environment respectively. The air-sintered sample is naturally cooled to room temperature in the air environment, and the vacuum-sintered sample is cooled to room temperature in the furnace before being taken out. Referred to JESD22-A104-B standard and combined with the actual situation of the high and low temperature thermal impact test chamber equipment, the following thermal impact test scheme was designed. The thermal shock test temperature was set as -40°C~+125°C, the high and low temperature zones were kept for 10 min, and the thermal shock environment was atmospheric. The specific experimental parameters were shown in Table 1, and the diagram of the thermal shock test curve was shown in Fig. 1. After every 0, 250, 500, 750, 1000 cycles, 5 samples of vacuum sintered samples and 5 samples of air sintered samples were taken out respectively. The samples were placed at room temperature and stored in the sample box when the temperature of the samples approached the room temperature. After the thermal shock test was completed, the shear strength and electrical conductivity of all samples were tested.

Table 1 Various experimental parameters in thermal shock tests.

Temperature range /°C	Heating rate/（°C/min）	Cooling rate / （°C/min）	Holding time /min
-40~+125	33	33	10

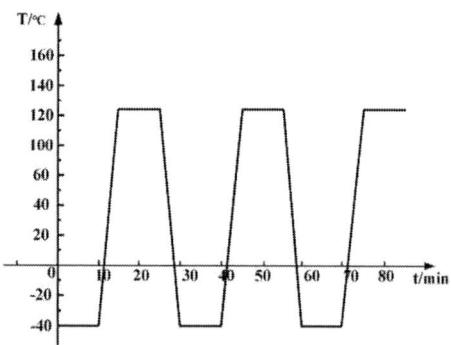

Fig. 1 Schematic diagram of thermal shock experiment curve.

III. RESULTS AND DSICUSSION

The vacuum sintered samples and air sintered samples subjected to thermal shock test were subjected to shear strength test. The specific average shear strength values of different samples varying with the number of thermal shock tests are given in Fig. 2 (a). As can be seen from the figure below, both vacuum sintered samples and air sintered samples show a trend of first decrease and then increase in connection strength as the number of thermal shocks increases. The average shear strength of air sintered samples (36.47 MPa) is significantly higher than that of vacuum sintered samples (28.87 MPa) when the samples have not been tested by thermal impact. When the samples were subjected to thermal shock conditions, the strength values decreased significantly. When the number of thermal shock reached 500 times, the average shear strength of air sintered samples and vacuum sintered samples decreased to the minimum value of 15.55 MPa and 18.42 MPa, respectively.

As the number of thermal shocks continues to increase, the strength of air sintered samples fluctuates greatly, showing a trend of first increase and then a small decline, while the vacuum sintered samples show a trend of continuous increase. When the number of thermal shocks reaches 1000 times, the average shear strength of air sintered and vacuum sintered samples is 23.24 MPa and 25.01 MPa, respectively.

At the same time, the conductivity of the remaining samples of air sintering and vacuum sintering was tested, and the average resistance value of the interconnection structure was calculated by the resistance value measured by the multimeter several times, as shown in Fig. 2 (b). It can be seen from the average resistance values that the average resistance values of air sintered and vacuum sintered samples are both lower before thermal shock test, being 2.93 Ω and 3.76 Ω, respectively. When subjected to thermal shock test, after 250 cycles, the average resistance of both air sintered samples and vacuum sintered samples increased significantly to 69.07 Ω and 42.5 Ω, respectively. As the number of cycles continues to increase, the on-resistance of the interconnect structure also increases. When the number of cycles reaches 500, the on-resistance of the interconnect structure reaches the peak value. Then, when the number of cycles reaches 750, the resistance of the interconnected structure begins to decline, and when the number of cycles reaches 1000, the resistance value of the air sintered sample drops to 63.57 Ω, and that of the vacuum sintered sample drops to 22.47 Ω.

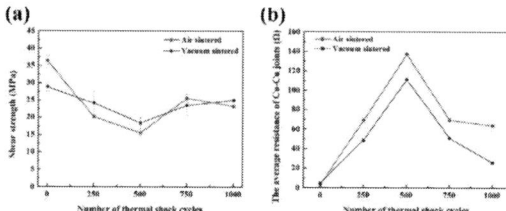

Fig. 2 (a) Shear strengths and (b) The average resistance of interconnection structures under different sintering conditions with the number of thermal shock cycles.

According to the above results of shear strength variation, the strength values of the samples under different sintering conditions have certain differences after thermal impact test. In order to analyze the influence rule of thermal shock conditions on the micromorphology of sintered structures of interconnected structures, the shear fractures of air sintering and vacuum sintering without thermal shock test were firstly observed. The microscopic morphology of the fractures and the corresponding sintered structures are shown in Fig. 3. The Fig. 3 (a, b) shows that both air sintering samples or vacuum sintering samples, after the shear test on copper plate connection kept most of the sintering layer, on the surface of copper plate edges under vacuum sintering part of sintering layer peeling, this shows that the sintering layer and is superior to copper plate of the connection strength of sinter layer and the copper plate, when shear failure occurs, Cracking occurs at the interface between the sintered layer and the connecting surface of the lower copper plate; It can also be observed that a small amount of residues of organic volatilization exist on the surface of the copper plate in Fig. 3 (a), while it can be observed that small lump-like microstructure appears in the sintered layer in Fig. 3 (b), which may be caused by hindered volatilization of organic matter in a low oxygen environment.

The microscopic sintering microstructure morphology of the fracture was further observed. According to Fig. 3 (c), when the sintering temperature was 280 , the sintering time was 15 min, and the sintering environment was in the air, continuous small peak-like slender pit structure was observed in some areas of the sintering layer, showing a dense sintering microstructure morphology. This indicates that the interconnection degree between the copper nanoparticles is high under this condition, and the sintered layer has a quasi-plastic deformation during the shear process, so that the interconnection structure bears a larger shear load, corresponding to a higher connection strength. Compared with air sintered samples, as shown in Fig. 3 (d), copper nanoparticles in the sintered layer under vacuum sintered only form a certain degree of interconnection between sintering necks, while there are micro-pores between sintering necks, resulting in a certain porosity, and the corresponding interconnection structure connection strength is lower than that of air sintered samples. This is due to the low oxygen/anaerobic conditions, the organic layer on the surface of the copper nanoparticles can not evaporate quickly and efficiently, some volatile organic matter form the micro pore morphology, after the rest of the organic layer hinders the atom diffusion between particles, particles formed between sintering neck structure, but growth is not sufficient, on the macro and the interconnection structure of the connection strength slightly lower. A large number of micropores will reduce the electrical conductivity of the interconnect structure to a certain extent and increase its on-resistance.

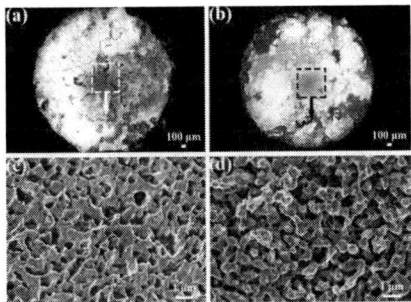

Fig. 3 The macro-morphologies (a, b) and corresponding microstructure morphologies (c, d) of the fracture surfaces of interconnection structures under air sintering and vacuum sintering respectively.

In order to further compare the interconnection degree and oxidation degree of sintered tissue, EDS surface scanning analysis was carried out on the cross sections of air sintered and vacuum sintered samples. The results are shown in Fig. 4. The microstructures with spherical particle interconnection were selected for comparative observation, and it was found that in Fig. 4(a), as more copper nanoparticles on air conditions in gathering and diffusion bonding, part of the micro pore was subsequently filled, the formation of micro interconnection structure is bigger than the size of vacuum sintering organization in Fig. 4(b).

According to EDS surface scanning results, oxygen content in air sintered tissue (29.5%) is relatively higher than that in vacuum sintered tissue (19.0%) due to the higher oxygen content in air environment. The oxygen content in air sintered tissue is composed of a certain degree of oxidation and residual oxygen after the volatilization of organic matter in micropores. However, the oxygen content in the vacuum sintered tissue is mainly due to the surface organic coating of

a large number of copper particles, which also indicates that the oxidation degree of copper nanoparticles in the air sintered tissue is higher than that in the vacuum sintered tissue.

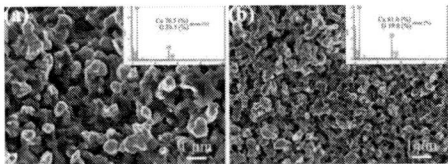

Fig. 4 SEM images and EDS results of the fracture surfaces of interconnection structures sintered in (a) air and (b) vacuum respectively

After comparing the differences between the air sintered and vacuum sintered microstructures, in order to further explore the relationship between the changes of mechanical and electrical properties of the interconnected structures after thermal shock test and the micromorphology of the sintered structures, the shear sections of the interconnected structures under different cycles were observed under scanning electron microscopy. Fig. 5 shows the micromorphology of the shear fracture of the air sintered interconnected structure under different cycles and the corresponding EDS surface scanning results. As can be seen from Fig. 5 (a), after 250 cycles of the interconnected structure, oxygen content in the sintered layer (31.3%) increased to a certain extent compared with that before impact (29.5%). Some particles oxidized and generated typical oxide micropore structure, resulting in an increase in porosity. As shown in Fig. 5 (b, c), cycles during 250 to 750, the sintering of residual oxygen consumption and escape, oxide continue to grow, the consumption of copper in the copper core of atoms to form micro holes gradually fills with surface oxide layer, formed in the form of oxide interconnection between particles of large-area interconnection microstructure. When the number of cycles reaches 1000 times, as shown in Fig. 5 (d), the oxygen content (30.2%) is relatively reduced. Most of the oxygen content on the surface at this time is the result of oxidation of sintered tissue after 1000 thermal shock cycles. As a whole, it presents a relatively dense interconnect microstructure, and the decrease of porosity is beneficial to reduce the on-resistance of the interconnect structure to a certain extent. At the same time, the formation and growth of oxides show the morphology changes of first loose and then dense, which also leads to the result that the connection strength of air sintered interconnect structure decreases at first and then slightly increases.

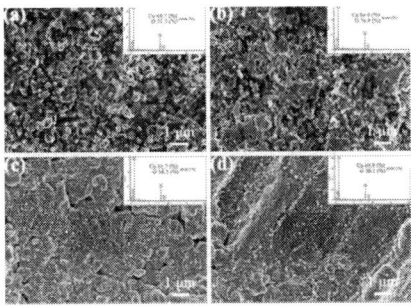

Fig. 5 The micro-morphologies of fracture surfaces of air-sintered interconnection structures and the corresponding EDS results under different cycles (a) 250 cycles, (b) 500 cycles, (c) 750 cycles, (d) 1000 cycles.

Fig. 6 shows the micromorphology of the shear fracture of the vacuum sintered interconnect structure under different cycles and the corresponding EDS surface scanning results. As shown in Fig. 6 (a), after 250 cycles of the interconnected structure, oxygen content (26.0%) increased to a certain extent compared with that before the impact (19.0%). Most of the surface of particles were still coated by organic matter, which played a certain anti-oxidation effect. There is also a small amount of residual oxygen in the micropores of the sintered tissue. The atomic diffusion in the copper core combined with a small amount of residual oxygen and oxidized to form the micropore structure of the oxide. As shown in Fig. 6 (b, c) can be observed that cycles from 250 to 750 the process of growing, with the constant impact of high and low temperature load, the slow decomposition of organic layer on the surface of the particles, particles between the copper continued to mutual diffusion, continue to work with a small amount of environmental oxygen reaction generated oxide at the same time, also fill in micro holes and micro pore, The consumption of ambient oxygen in sintered tissue also leads to the trend of oxygen content increasing first and then decreasing. It is worth noting that when the number of cycles reached 1000 times, the oxygen content (5.1%) significantly decreased, indicating that the oxidation degree of the vacuum sintered tissue was very low. Moreover, it can also be observed in Fig. 6 (d) that most of the micropores and pores disappeared, and the sintered tissue presented a large and dense interconnected microstructure. This is due to the sufficient interdiffusion of a large number of copper atoms in a low oxygen environment. In addition, the very low porosity and oxidation degree in the vacuum sintered structure also lead to a significant decrease in the on-resistance of the interconnected structure.

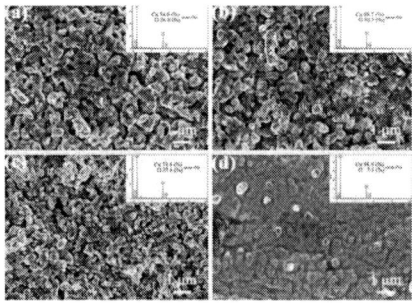

Fig. 6 The micro-morphologies of fracture surfaces of vacuum-sintered interconnection structures and the corresponding EDS results under different cycles (a) 250 cycles, (b) 500 cycles, (c) 750 cycles, (d) 1000 cycles.

Combined with the changes of connection strength, resistance value and microstructure morphology, it is found that the combined action of temperature load and oxygen can affect the mechanical and electrical properties of the all-copper interconnect structure. In this study, a two-dimensional model was built to explain the micro-interconnection behavior of copper nanoparticles after sintering and thermal shock, and to explain the relationship between the evolution of sintering microstructure and the interconnection properties of interconnection structure. The schematic diagram of the model is shown in Fig 7. As shown in the figure below, before the sintering experiment, the surface of copper nanoparticles was coated with an organic layer and had a certain dispersion. However, due to the size of the nanoparticles at the nanometer level, agglomeration was unavoidable. The spherical particles were close to each other and there were a large number of micro-pores, in which a certain amount of oxygen remained. Copper nanoparticles in the sintering process, mutual diffusion bonding between

particles, particles on the surface of the organic volatile, bare copper surface contact with residual oxygen oxidation reaction will happen, to generate a certain thickness of oxide layer, the area of the sintering group increased, the position of the micro pore was dominated by constantly and smaller size, results in the decrease of porosity. Then, in the process of thermal shock of high and low temperature, the oxide continues to grow under the joint action of temperature and ambient oxygen. Because the oxide layer is brittle, the connection performance of the interconnect structure will be weakened, and the strength of the interconnect structure will decrease with the increase of the oxide content. Moreover, the conductivity of the oxide is poor. With the increase of the oxide, the overall conductivity of the interconnect structure decreases, which is consistent with the resistance value of the interconnect structure mentioned above. However, as the further effect of impact load, temperature sintering of residual oxygen consumption and escape, oxide growth slowing, mutual diffusion between particles fully connected, micro porosity also shrinking until disappear, reduce the porosity, prompted the sintering densification of organization, and formed a copper + copper oxide mixed micro interconnection structure. The densification of the interconnect microstructure enhances the connection strength of the interconnect structure to a certain extent, and finally leads to the trend of the connection strength of the interconnect structure decreases first and then increases in the process of thermal shock. In conclusion, the degree of oxidation and porosity of the sintered structure together affect the mechanical and electrical properties of the all-copper interconnect structure.

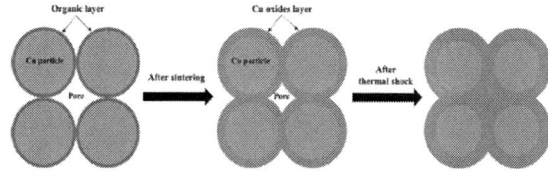

Fig. 7 Schematic diagram of micro-interconnection behavior of copper nanoparticles under thermal shock condition.

IV. CONCLUSION

In summary，different Cu interconnection structures were prepared by air sintering and vacuum sintering, and the thermal shock tests were carried out. Combined with microstructure characterization and theoretical analysis, the influence of thermal shock conditions on the thermal stability of all-copper interconnect structures was studied. With the increase of the number of thermal shock cycles, the connection strength of the all-copper interconnect structure decreases to a certain extent, and then increases slightly. The average shear strength of air sintered sample and vacuum sintered sample is 23.24 MPa and 25.01 MPa, respectively, when the number of thermal shock cycles reaches 1000.After 1000 cycles of thermal shock, the resistance value of air sintered sample (63.57 Ω) is significantly higher than that before the shock (2.93 Ω), and the resistance value of vacuum sintered sample (22.47 Ω) is slightly higher than that before the shock (3.77 Ω).The degree of oxidation and porosity of the sintered structure will affect the interconnection properties of the all-copper interconnect structure. With the increase of the number of thermal shock, the particle surface and oxygen effect will continue to generate oxide, copper continuous mutual diffusion between particles at the same time, the micro interconnection area expands unceasingly, The micropore structure of the oxide and the micropores between the particles decrease continuously, which leads to the morphology evolution of the sintering microstructure from loose to dense, and finally forms the mixed interconnection microstructure of Cu + Cu oxide.

ACKNOWLEDGMENT

This research is supported by a Fundamental Research Funds for the Central Universities of China (Grant No. 2018CDGFCL0003).

REFERENCES

[1] Marzoughi A, Burgos R, Boroyevich D. Characterization and comparison of latest generation 900-V and 1.2-kV SiC MOSFETs[C]// Energy Conversion Congress & Exposition. IEEE, 2017.

[2] Kirchhof M J, Förster H, Schmid H J, et al. Sintering kinetics and mechanism of vitreous nanoparticles[J]. Journal of Aerosol Science, 2012, 45: 26-39.

[3] Wang S, Li M, Ji H, et al. Rapid pressureless low-temperature sintering of Ag nanoparticles for high-power density electronic packaging[J]. Scripta Materialia, 2013, 69(11): 789-792.

[4] Park B K, Jeong S, Kim D, et al. Synthesis and size control of monodisperse copper nanoparticles by polyol method[J]. Journal of Colloid and Interface Science, 2007, 311(2): 417-424.

[5] Carro L D, Zürcher J, Drechsler U, et al. Low-Temperature Dip-Based All-Copper Interconnects Formed by Pressure-Assisted Sintering of Copper Nanoparticles[J]. IEEE Transactions on Components, Packaging and Manufacturing Technology, 2019, 9(8): 1613-1622.

[6] Schnabl K, Wentlent L, Mootoo K, et al. Nanocopper Based Solder-Free Electronic Assembly[J]. Journal of Electronic Materials, 2014, 43(12): 4515-4521

[7] Replacing Solder with All-Copper Interconnects [EB/OL]. [2015-5-29]. http://ibmresearchnews.blogspot.com.

[8] Mou Y, Peng Y, Zhang Y, et al. Cu-Cu bonding enhancement at low temperature by using carboxylic acid surface-modified Cu nanoparticles[J]. Materials Letters, 2018, 227: 179-183

[9] Liu Jingdong, Ji Hongjun, Wang Shuai, Li Mingyu. The Low Temperature Exothermic Sintering of Formic Acid Treated Cu Nanoparticles for Conductive Ink[J]. Journal of Materials Science: Materials in Electronics,2016,27(12) :13280-13287.

[10] Zuo Y, Shen J, Xie J, et al. Influence of Cu micro/nano-particles mixture and surface roughness on the shear strength of Cu-Cu joints[J]. Journal of Materials Processing Technology, 2018, 257: 250-256.

[11] J Xie, J Shen, J Deng, et al. Influence of Aging Atmosphere on the Thermal Stability of Low-Temperature Rapidly Sintered Cu Nanoparticle Paste Joint[J]. Journal of Electronic Materials, 2020, 49(4): 2669-2676.

Machine Learning based Prediction of Wire Bonding Profile in 3D stacked integrated microelectronic packaging

Zhengping Ou
State Key Laboratory of Precision Electronic Manufacturing Technology and Equipment
Guangdong University of Technology
Guangzhou, China
ouzhengping2021@outlook.com

Junyu Long
State Key Laboratory of Precision Electronic Manufacturing Technology and Equipment
Guangdong University of Technology
Guangzhou, China
JuneYu_Long@foxmail.com

Shuquan Ding
State Key Laboratory of Precision Electronic Manufacturing Technology and Equipment
Guangdong University of Technology
Guangzhou, China
shuquanding@163.com

Yun Chen*
State Key Laboratory of Precision Electronic Manufacturing Technology and Equipment
Guangdong University of Technology
Guangzhou, China
chenyun@gdut.edu.cn

Maoxiang Hou
State Key Laboratory of Precision Electronic Manufacturing Technology and Equipment
Guangdong University of Technology
Guangzhou, China
maoxiangh@gdut.edu.cn

Yunbo He
State Key Laboratory of Precision Electronic Manufacturing Technology and Equipment
Guangdong University of Technology
Guangzhou, China
heyunbo@gdut.edu.cn

Xin Chen
State Key Laboratory of Precision Electronic Manufacturing Technology and Equipment
Guangdong University of Technology
Guangzhou, China
Chenx@gdut.edu.cn

Jian Gao
State Key Laboratory of Precision Electronic Manufacturing Technology and Equipment
Guangdong University of Technology
Guangzhou, China
gaojian@gdut.edu.cn

Abstract—Wire bonding (WB) have been widely used in 3D stacked integrated microelectronic packaging because of its high reliability and low-cost. In 3D stacked packaging, the required size for microelectronic devices was much smaller than before, which led the shape of wire profile becoming more critical as it directly affects the total thickness of the chip package. However, there is no facile means to predict the wire profile. In this study, machine learning (ML) was proposed to predict the wire profile of wire bonding under various conditions. After collecting abundant relevant information using finite element simulation experiments, the support vector regression (SVR) was selected as the predictor while support vector machine (SVM) was used as the classifier to the train the ML model, thereby, the wire bonding profile that can meet the requirements of microelectronic industry can be obtained. This study can provide useful insights for developing wire bonding processing techniques.

Keywords—prediction, machine learning (ML) model, wire bonding, 3D stacked integrated microelectronic packaging

I. INTRODUCTION

With the increasing demand for multifunctional chips and the development of chip integration technology, developing advanced microelectronic packaging technology is urgently required[1]. Wire bonding is one of the most important interconnection method in microelectronic packaging. It had attracted much attention due to its excellent reliability, flexibility and low cost, especially suitable for the current three-dimensional stacked packaging[2-4]. Wire bonding can utilize various large span and versatile loops to maximize chip integration, improve reliability of connections, reduce overall device size, and shorten development cycles, which makes it to be widely used in the complex 3D packaging.

In 3D stacked packaging, the shape of wire profile is critical as it directly affects the total thickness of chip packaging[5, 6]. As requirements of microelectronic devices continually increase in size and integration density, the wire bonding was required to completes the chip integration in limited space and complex connections. It brings more challenges to the processing and the design of wire bonding, which is urgently requires the wire profile prediction technology[7-9]. At present, Finite Element Analysis (FEA) is used as the main prediction technology with high accuracy in industry.

However, the cost of FEA is unfavorable due to large amounts of calculations, which significantly prolongs the processing cycle of wire bonding. In recent years, Machine learning (ML) is widely used in data mining, computer vision or natural language processing[10-12]. ML algorithm can build model based on learning and training certain amount of valid data to obtain result, which makes it possible to bypass complex mechanisms to draw conclusions[13]. Long et al.[14] was used SVM algorithm through extracting feature from images to build a precision recognition and classification system for bonding joint of ultrasonic heavy aluminum wire. Chen[15] proposed a data-driven method building SVM model to find wire bonding defects quickly form integrated circuit (IC) X-ray images. Using the ML model by learning and training large amounts of data can realize the rapid prediction. Therefore, the data of wire profile obtained by FEA can be used to build a ML model for wire profile prediction rapidly.

978-1-6654-1392-3/21 $31.00 © 2021 IEEE

In this paper, a model based on support vector machine (SVM)[16] and support vector regression (SVR)[17, 18] algorithm in machine learning is proposed to predicting the wire profile of wire bonding. This model can calculate the wire profile rapidly through modifying three initial parameters. Additionally, for evaluating the quality of the wire bonding after processing, zero deflection angle as an index was designed in this study. This study provides a novel and effective method for predicting the profile of the processed wire.

II. DESIGN AND DEVELOPMENT OF ML MODEL

A. Characteristics of the model

In order to reduce the wire loop height, the trajectory of the wire bonder head was adopted from flat-looping mode (as shown in Fig. 1 (a)). Based on the prior knowledge, three parameters (the first reverse angle θ1, the second reverse angle θ2 and the highest point (TH) as shown in Fig. 1 (b)) were designed in this experiment. The purpose of selecting θ1 and θ2 was to control the shape of the wire more conveniently, and TH was selected to control the height of the highest point of the wire. After processing in flat-looping mode, the coordinates of Kink 1st, Kink 2nd and Kink 3rd as main features were extracted from more than 100 finite element simulation experiments. These features were mainly used as the training and learning objects of the algorithm. On the basis of this process, the coordinates of Kink 1st, Kink 2nd and Kink 3rd obtained from finite element simulation experiments can be used to build a machine learning model.

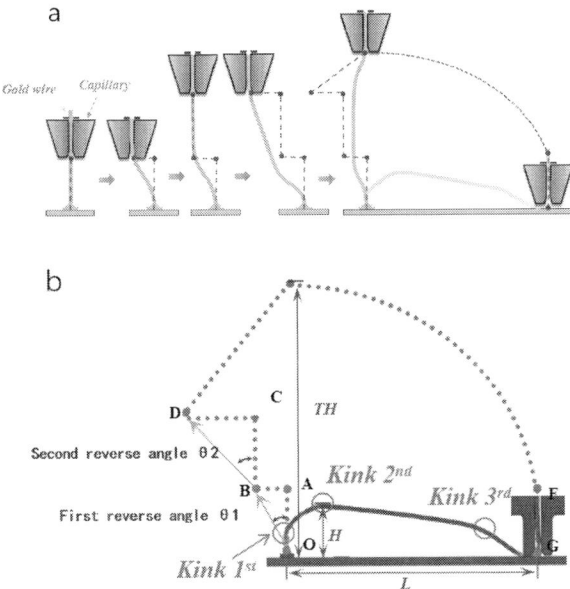

Fig. 1. a is flat-looping mode; b are important parameters in flat-looping trajectory design, which are the first reverse angle θ1, the second reverse angle θ2 and the highest point (TH).

B. The development of ML mode

The whole process of wire bonding was a complex mechanical and empirical process. It was difficult to design a suitable theoretical model according to the traditional design of experiments (DOE). Therefore, it is considered a feasible method to select machine learning algorithm to predict the wire profile. Fig 2 shows the process of specific ML modeling.

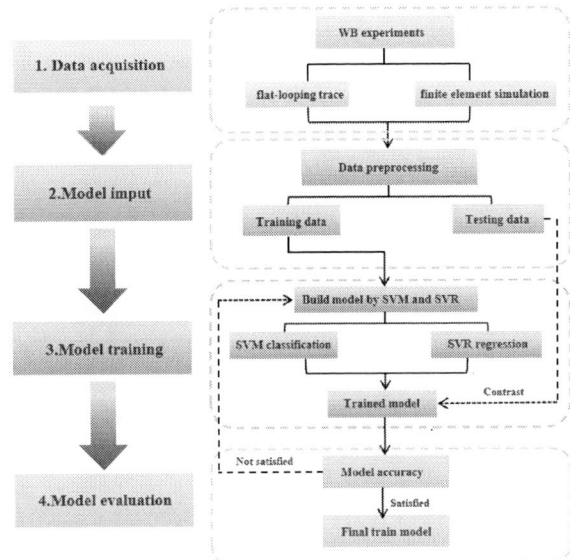

Fig. 2. Flow chart of ML model establishment.

For obtaining the data suitable for training, the orthogonal test method of three factors and three levels was adopted in this experiment, which is an efficient, fast and economical method of experiment design. For verifying the correlation between the initial three parameters and kink 1st, kink 2nd and kink 3rd, Pearson correlation coefficient method was used in this study. Fig 3 shows the relativity of the data, the closer to 1, the stronger the positive correlation; the closer to -1, the stronger the negative correlation. It can be seen that the correlation of TH was the strongest, and that of θ2 was the weakest, which verifies the original purpose of parameters design, θ1 and θ2 were to ensure the shape of the wire, and the coordinates of each point are mainly affected by TH.

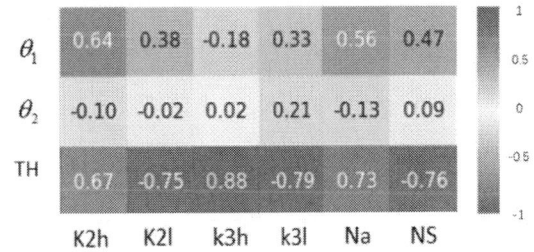

Fig. 3. the degree of correlation, which indicates the degree of correlation between θ1, θ2, TH and the ordinate and abscissa of kink 2nd (K2h, K2l), kink 3rd (K3h, K3l), the neck deflection angle (Na) and the neck strain (Ns).

For further training, the results were divided into training data and test data. The coordinates of kink 2nd and kink 3rd were predicted by using SVR algorithm, and kink 1st was classified by using SVM algorithm, and then predicted by using SVR algorithm, which is due to the different methods of calculating coordinates among kink 2nd, kink 3rd and kink 1st. The coordinates of kink 2nd and kink 3rd can be predicted by SVR algorithm directly. Because of the original design, kinks1st has an offset angle. The left deviation was called positive deviation angle, the right deviation was called negative deviation angle, and the angle without deviation angle or with smaller deviation angle was zero deviation angle (show as Fig 4). Zero deflection angle was more needed in industry, so all zero deflection angles are defined as zero in experiment, positive and negative deflection angles were non-zero. According to a large number of simulation experiments, it was found that when the span was 630 μ m, the ordinate of kink 1st changes little, about 49 μ m, so the abscissa of kink 1st $x1 = 49 \tan \theta$ (θ is the value of the neck deflection angle).

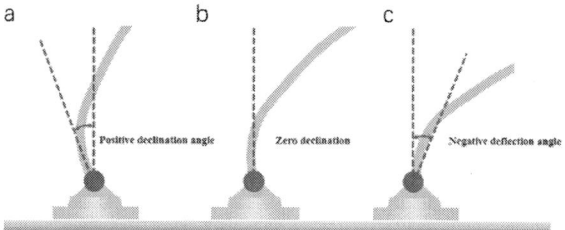

Fig. 4.   a is positive deviation angle; b is zero deflection angle; c is negative deviation angle.

III.   RESULT AND DISCUSSION

In a word, the characteristics of Kink1st (zero and non-zero values must be judged) were different from Kink 2nd and Kink 3rd. Kink 2nd and Kink 3rd were regression models, while Kink 1st was a typical classification model. According to the above characteristics, Kink 2nd and Kink 3rd were predicted by SVR. The predicted and actual results of the coordinates of kink 2nd and kink 3rd by SVR algorithm were shown in Fig 5.

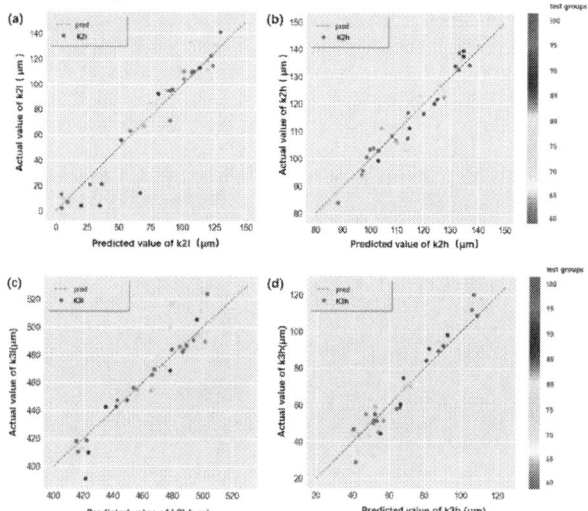

Fig. 5.   (a) and (b) were abscissa (k2l) and ordinate (K2h) of Kink 2nd; (c) and (d) were abscissa (k3l) and ordinate (K3h) of Kink 3rd.

The closer to the dotted line indicates that the higher the accuracy of prediction, it can be seen that they agree well which demonstrates the accuracy of the developed model. Then Kink 1st was classified by SVM classifier (as shown in Fig 6 (a) and (b)), yellow represents zero and red represents non-zero value and (a) is the actual value and (b) is the SVM classification value), and the accuracy of classification results reaches 90%. After classification, the model will automatically judge whether the abscissa of kink 1st (x1) is zero or non-zero, and if it is zero, x1=0, if it is non-zero, then the SVR algorithm will be used to predict (show as Fig 6 (c)). Finally, in order to predict the stress of the wire, it is necessary

to predict NS(neck strain),which is the maximal strain in the wire and the prediction results were shown in Fig6 (d).

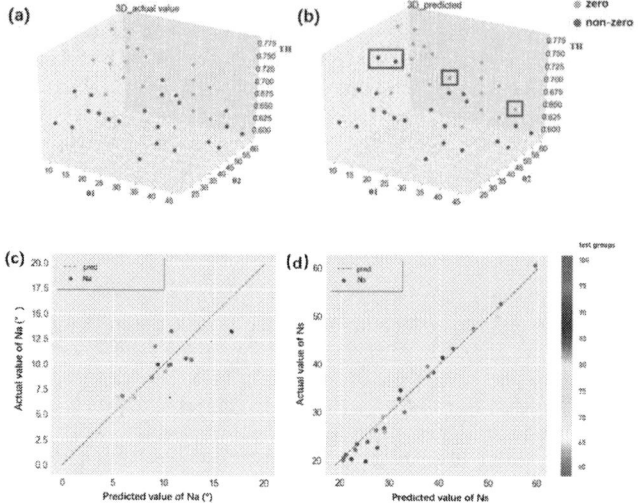

Fig. 6.   (a) is actual state of Na value, (b) is classification result chart of Na. (c) is prediction of Na after classification. (d) is prediction of Ns after classification.

The ML model of wire bonding was shown in Fig 7 (a), which only input the three parameters (the first reverse angle $\theta1$, the second reverse angle $\theta2$ and the highest point (TH)), and the model predicts the wire bonding profile very quickly. Comparing the predicted results of the model with the FEA of wire bonding, the comparison results were shown in Fig 7 (b), and it can be seen that the approximate wire profile was similar.

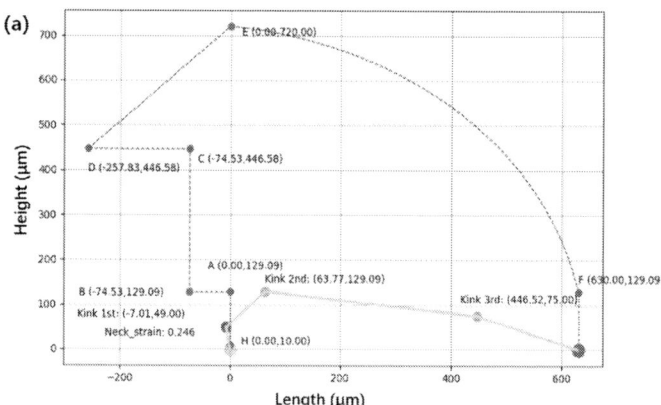

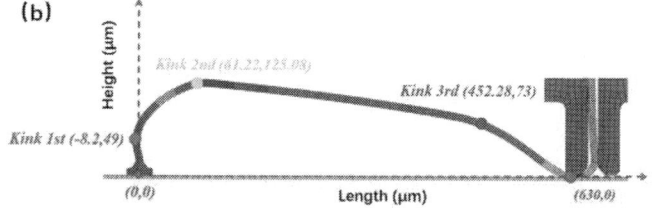

Fig. 7.   (a) is the wire bonding simulation diagram based on machine learning model. (b) is FEA of wire bonding.

IV. CONCLUSION

In this work a ML based wire bonding prediction method was designed and developed, and a machine learning model based on a large number of finite element experiments was established, The model determines the coordinates of the key nodes that affect the wire bonding profile which are Kink 2nd coordinate and Kink 3rd coordinate that can be predicted directly, and Kink 1st coordinate that can be obtained by classification and prediction according to the characteristics of wire deflection. Compared with the traditional finite element simulation, it greatly reduces the prediction time and improves the efficiency. At present, there is not enough theory to explain the formation mechanism of wire bonding profile, so it is possible to establish a model from the perspective of data. It provides a new possible way to predict more complex wire profile in the future.

ACKNOWLEDGMENT

The work presented in the paper is supported the National Natural Science Foundation of China (51975127, U20A6004), the Fund of Research and Development Program of Guangdong Province (2020A0505140008) and Key-Area Research and Development Program of Guangdong Province (2018B090906002).

REFERENCES

[1]. Dreslinski, R.G., et al., Near-Threshold Computing: Reclaiming Moore's Law Through Energy Efficient Integrated Circuits. Proceedings of the IEEE, 2010. 98(2): p. 253-266.

[2]. Lu, Z., et al., Performance Characterization of Micromachined Inductive Suspensions Based on 3D Wire-Bonded Microcoils. Micromachines, 2014. 5(4): p. 1469-1484.

[3]. Wang, T. and Y. Lu, Fast and Accurate Frequency-Dependent Behavioral Model of Bonding Wires. IEEE Transactions on Industrial Informatics, 2017. 13(5): p. 2389-2396.

[4]. Kratt, K., et al., A fully MEMS-compatible process for 3D high aspect ratio micro coils obtained with an automatic wire bonder. Journal of micromechanics and microengineering, 2010. 20(1): p. 015021.

[5]. Lee, S.H., K. Chen and J.J. Lu, Wafer-to-Wafer Alignment for Three-Dimensional Integration: A Review. Journal of Microelectromechanical Systems, 2011. 20(4): p. 885-898.

[6]. Lu, J., 3-D Hyperintegration and Packaging Technologies for Micro-Nano Systems. Proceedings of the IEEE, 2009. 97(1): p. 18-30.

[7]. Qin, I., et al., Wire Bonding Looping Solutions for Advanced High Pin Count Devices, in Electronic Components and Technology Conference. 2016, IEEE COMPUTER SOC: LOS ALAMITOS. p. 614-621.

[8]. Fischer, A.C., et al., Unconventional applications of wire bonding create opportunities for microsystem integration. JOURNAL OF MICROMECHANICS AND MICROENGINEERING, 2013. 23(0830018): p. 083001.

[9]. Hou, T., et al., An integrated system for setting the optimal parameters in IC chip-package wire bonding processes. The International Journal of Advanced Manufacturing Technology, 2006. 30(3-4): p. 247-253.

[10]. Yao, L., et al., Recommendations on the Internet of Things: Requirements, Challenges, and Directions. IEEE Internet Computing, 2019. 23(3): p. 46-54.

[11]. Wei, L., et al., Machine Learning Optimization of p‐Type Transparent Conducting Films. Chemistry of materials, 2019. 31(18): p. 7340-7350.

[12]. Domingos, P., A few useful things to know about machine learning. Communications of the ACM, 2012. 55(10): p. 78-87.

[13]. Manzhos, S. and P. Golub, Data-driven kinetic energy density fitting for orbital-free DFT: Linear vs Gaussian process regression. The Journal of Chemical Physics, 2020. 153(7): p. 074104.

[14]. Long, Z., et al., Recognition and Classification of Wire Bonding Joint via Image Feature and SVM Model. IEEE Transactions on Components, Packaging and Manufacturing Technology, 2019. 9(5): p. 998-1006.

[15]. Chen, J., Z. Zhang and F. Wu, A data-driven method for enhancing the image-based automatic inspection of IC wire bonding defects. INTERNATIONAL JOURNAL OF PRODUCTION RESEARCH, 2020.

[16]. Tian, Y., Q. Zhang and D. Liu, v-Nonparallel support vector machine for pattern classification. Neural Computing and Applications, 2014. 25(5): p. 1007-1020.

[17]. Jahed Armaghani, D., et al., Examining Hybrid and Single SVM Models with Different Kernels to Predict Rock Brittleness. Sustainability, 2020. 12(6): p. 2229.

[18]. Smola, A.J. and B. Schölkopf, A tutorial on support vector regression. Statistics and computing, 2004. 14(3): p. 199-222.

The design of a voltage conversion SIP based on electrothermal coupling method

Hao-hang SU
Beijing Institute of Space Mechanics & Electricity
Key Laboratory for Optical Remote Sensor Technology of CAST
Beijing , China
xiaosu0741@163.com

Shuai FU
Beijing Institute of Space Mechanics & Electricity
Key Laboratory for Optical Remote Sensor Technology of CAST
Beijing , China
shuaifu@163.com

Abstract—With the continuous improvement of demand and indicators, miniaturization, high performance, and high reliability are the development trends of remote sensing cameras. So the SIP technology has been instead of the traditional PCB design method in the design of remote sensing cameras more and more. In this paper, 10-channel voltage conversion SIP module has been designed, which supply power for the focal plane of the remote sensing camera detector, and the ALN integrated forming package has been choose to SIP module. In the SIP design process, the electro-thermal coupling simulation analysis method is used to predict and analyze the design temperature, optimize the layout and power/ground network design, and analyze the impact of the package on the module temperature. At last, the temperature comparison has been made between the simulated and measured. Through three times electro-thermal coupling iteration, the maximum simulated temperature of the SIP substrate was 47.6℃, and the maximum temperature of the substrate was 46.3℃ which measured by the infrared thermal imager. The relative error is 2.8%, which satisfies the design requirement. The electro-thermal coupling method can ensure the performance analysis of SIP, which no longer isolated the temperature and electrical characteristics, the result of which are more accurate and effective to guide the SIP design, and guarantees the performance of SIP products.

Keywords—SIP, Electrothermal coupling, integrated forming package, Joule heat

I. INTRODUCTION

With the development of space exploration technology, remote sensing cameras are developing in the direction of high resolution and wide coverage. New generation of remote sensing cameras not only increase the working frequency, signal transmission rate, but extremely demanding the requirements of the compact in volume, light in weight, and small in size. The functional density and performance of remote sensing cameras electronics have been greatly increasing, which has increased the scale of circuits closely integrated by dozens of times. Traditional PCB level electronic design techniques have been unable to meet the needs of the development of camera electronics system, so the miniaturization and integrated design has become an inevitable trend for remote sensing cameras, Therefore, it is necessary to use SIP technology to module the camera.

Although SIP module greatly compress the volume of cameras, but there are a lot of incidental problems by which. For example, the focal plane power supply circuit is the core part of the remote sensing camera. Its main function is to complete the multi-channel power supply to the detector. The noise of which is a key factor in whether the camera can transmit high-quality pictures. The SIP module of the focal plane power supply circuit has about 4 times higher functional density than that of the traditional PCB design, and reduced by 80% in volume, which greatly reduces the weight and size of the remote sensing camera. However, the power consumption in unit volume increase sharply, which resulting in a greatly rise in the temperature in the module, and the local heat of the SIP would cause the structural deformation. Also, a large amount of heat will destroy the temperature field of the optical system of the remote sensing camera and cause the degradation of the image quality. Therefore, it is necessary to carry out effectively thermal design and thermal management during the SIP design process.

In traditional thermal design, electronic and thermal are designed and simulated separately, and the mutual influence be ignored. For the power supply SIP module, drastically voltage drops and large amount of switching current would be caused during camera working condition, and which will cause large Joule heating in the SIP substrate. If the traditional design method be used, the Joule heat would be ignored, and the temperature change in package would not be accurately analyzed, and heat dissipated poorly, which would cause the power system noise be increased, and eventually affect the image quality of the camera, limit the system performance. The electro-thermal coupling analysis technology proposes a new type of analysis method that takes into account both electrical performance and thermal effects. The Joule heating factor is introduced into the thermal analysis to design an accurate and practical electrical system.

The electro-thermal coupling method is a hot spot of research in recent years, and which is mainly focused on PCB design. WAN Ran and SUN Wenb [1] used electro-thermal coupling model of switched-mode power supply based on Isight , which effective to improve the simulation accuracy of switching power supply system. Jia Yingjie and Xiao Fei[2] proposed an electro-thermal co-simulation method based on field-circuit coupling, constructed a circuit model based on IGBT physical model and a FEM-based thermal model in Simulink and COMSOL respectively, and realized the co-simulation under the multi-rate simulation strategy through a control file of Matlab script. And verified by switch transient test and short-circuit test on a high power IGBT module. Tianjian Lu and Jian-Ming Jin[3] used electrical-thermal co-simulation for DC IR-Drop Analysis of Large-Scale Power Delivery. The Joule heating effect is considered for an accurate prediction of voltage distribution and temperature rise in the integrated circuits. Xin Ai and An-Yu Kuo[4] presented a numerical approach to simulate the coupling between electrical and thermal solutions, and

978-1-6654-1392-3/21 $31.00 © 2021 IEEE

accurately provide current, voltage, and temperature solutions at each and every component, via, solder ball/bump, wire and pin, which precise control of electrical and thermal design performance, avoid of smoke/fire hazards of an electronics product, and lower overall costs due to shorter design cycles and more efficient designs.

In this paper, the power supply circuit of the remote sensing camera is modularized based on SIP design technology. Ten LDO chips are integrated in the module. In the SIP design process, the electro-thermal coupling analysis method is used to predict and analyze the design temperature, optimize the layout and power/ground network design, and analyze the impact of the package on the module temperature. At last, the comparison has been made between the simulation temperature and measure temperature, which showed the electro-thermal coupling analysis method is more accurate and effective to guide the SIP design, and guarantees the performance of SIP products.

II. SIP DESIGN

This module is a voltage converter with ten independent and adjustable outputs, which integrated ten LDO die and passive components, and which would supply power for the focal plane of the remote sensing camera detector. Dies are independent of each other and do not affect each other, which could be chose any channel rout according needs. Each LDO chip has enable control and power output indication pins to control the power-on sequence among ten power supplies channels. When the enable control terminal is low, the output is turn off, and when the enable control terminal is high, the output is turn on. The output voltage can be set though the adjustment resistor. Protection functions including short-circuit protection, over-temperature protection, over-current protection have been integrated in this SIP module.

Since the main function of the SIP module is voltage changes and carries a large amount of current flips, the package of which must has good feature of heat dissipation. The structure and materials of package must be considered, at same time the performance at high temperatures of substrate would not be ignored. With the good thermal conductivity, aluminum nitride (ALN) ceramic is the ideal material of integrated circuit substrates and electronic packaging. ALN integrated forming package has been used and shown in Figure 1. The integrated forming package consists of four parts: ALN multilayer substrate, ring, cover plate, and pin grid array, as shown in Figure 1.

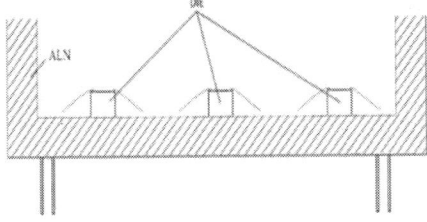

Fig. 1. SIP assembly

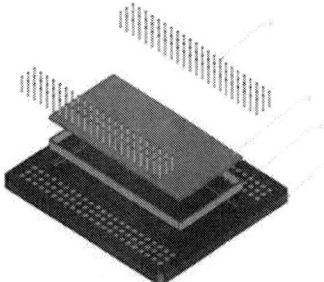

Fig. 2. The integrated package design

III. ELECTRO-THERMAL COUPLING ANALYSIS MECHANISM AND MODELING

In steady state, the voltage distribution equation is shown in formula 1

$$\nabla\left(\frac{1}{\rho(x,y,z)}\nabla\varphi(x,y,z)\right) = 0 \qquad (1)$$

Where $\rho(x,y,z)$ and $\varphi(x,y,z)$ in the formula respectively represent the resistivity and voltage distribution affected by temperatdure.

In the steady state, the heat distribution equations of solids and fluids are expressed by formula 2 and formula 3:

$$\nabla[K(x,y,z) \cdot \nabla T(x,y,z)] = -P(x,y,z) \qquad (2)$$
$$\sigma c_P \bar{v}(x,y,z) \cdot \nabla T(x,y,z) = \nabla(K_f \nabla T(x,y,z)) \qquad (3)$$

The definition of each parameter in the formula is as follows:

$K(x,y,z)$：Solid thermal conductivity；

$T(x,y,z)$：Temperature；

$P(x,y,z)$：Heat source；

σ：Fluid density；

c_P：Fluid heat capacity；

$\bar{v}(x,y,z)$：Fluid velocity；

K_f：Fluid thermal conductivity。

In formula 2, $P(x,y,z)$ is the overall heat source excitation, which includes the device heat and the Joule heat generated by the current flowing through the conductor in the PDN. The device heat can be obtained from the device technical manual. The Joule heat can be calculated by formula 4:

$$P_{Joule}(x,y,z) = \bar{J} \cdot \bar{E}(x,y,z) \qquad (4)$$

The definition of each parameter in the formula is as follows:

$\bar{J}$：Current density；

$\bar{E}(x,y,z)$：Voltage distribution。

The influence of temperature on electrical properties can be calculated by the conductivity based on temperature changes. The conductivity of the conductor at each temperature is obtained, and the new supply voltage is obtained. Equation 5 gives the new calculation method of resistivity:

$$\rho = \rho_0[1 + \alpha(T - T_0)] \qquad (5)$$

The definition of each parameter in the formula is as follows:

ρ_0: the resistivity at the T_0;

α: the factor that the resistivity is affected by the temperature;

ρ: the temperature slope compensation coefficient of the conductor.

Taking into account that the resistivity has a resistivity that changed with temperature, and the current flowing through the conductor will generate Joule heat, the coupling relationship between the electrical heating on the SIP is organized, as shown in Figure 3.

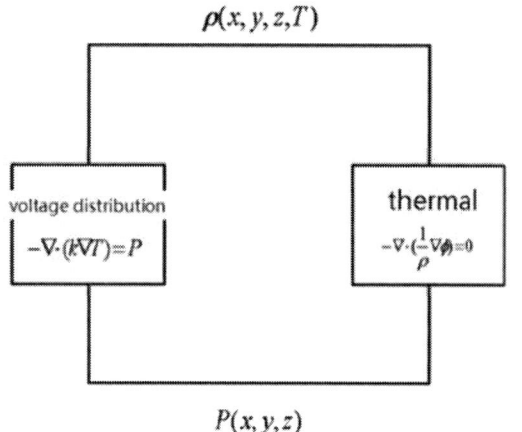

$$\rho(x, y, z, T)$$

voltage distribution
$$-\nabla \cdot (k\nabla T) = P$$

thermal
$$-\nabla \cdot (\frac{1}{\rho}\nabla\phi) = 0$$

$$P(x, y, z)$$

Fig. 3. Relationship between electrical and thermal fields

In order to obtain the voltage distribution with the influence of Joule heating and thermal, it is necessary to solve the nonlinear electro-thermal equation at the same time. The electrical analysis and thermal analysis are coupled through (1) and (5). The iterative scheme of concurrently is shown as the following steps:

Step1: Set up SIP stack, initialized material properties, excitation, and boundary conditions for electro-thermal analysis under steady-state conditions;

Step2: Perform electrical analysis at steady state, obtain distributions including voltage, current, power consumption, and calculate Joule heat;

Step3: Update the Joule heat in the electrical analysis to the steady-state thermal analysis, and perform thermal analysis including thermal conduction and thermal convection to obtain thermal distribution and electrical conductivity affected by temperature;

Step4: Update the conductivity of the conductor and determine whether the convergence condition is reached;

Step5: Specify the convergence conditions of the temperature and voltage distribution, and the analysis iteration stops until the convergence conditions are met.

IV. EXPERIMENTAL RESULTS AND ANALYSIS

For the remote sensing camera, heat dissipation is mainly conduction and radiation. The heat generated by the device or the circuit is transferred to the structural frame of camera though PCB. Therefore, the accurate analysis of the circuit temperature and the control of the heat flow direction during the design process are the main basis for thermal design.

In this SIP module, 10 LDO die on the substrate are connected by welding. The layout design and the stack of the multilayer substrate are shown in Figures 4 and 5. The substrate material is ALN, and the conductor material is Wu. The material of the ring, the cover plate and pin grid array is Kovar.

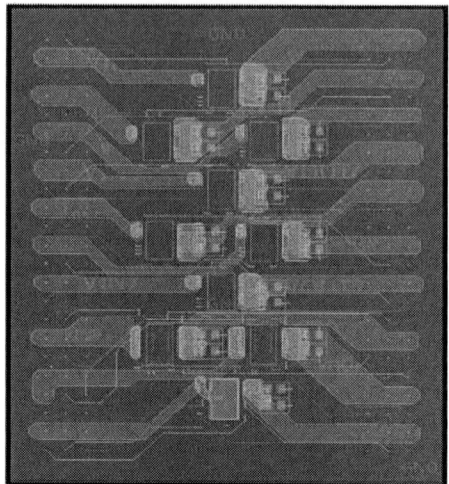

Fig. 4. SIP layout

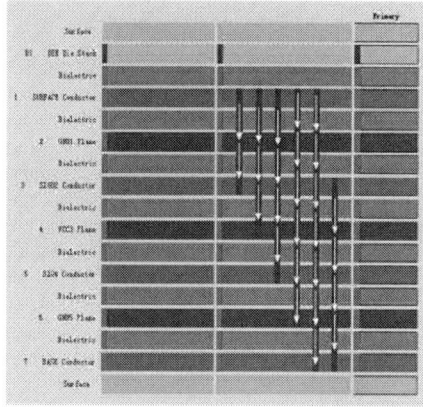

Fig. 5. The stack of the multilayer substrate

The SIP module is welding on the test PCB during application period, and heat dissipation pads and heat dissipation via are designed at the contact position between the PCB and the package cover of this SIP. The heat is conducted to the metal structure frame around the PCB through the copper layer in the PCB. In this paper, the SIP electromagnetic simulation software is ANSYS HFSS 3D, and the thermal simulation software is ICEPAK. The ambient temperature is 22°C. The heat dissipation method is only conduction and radiation. The four boundaries between the test PCB and structure frame are set as "wall", which is set to a constant temperature of 22°C. The power consumption of the LDO is 0.8W, the input voltage is 3.3V, and the output voltage is 2.5V. Each LDO is powered separately, and the maximum current of each die is 1A.

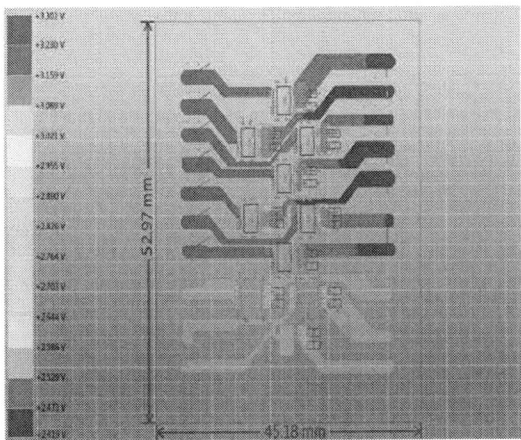

Fig. 6. IR-drop

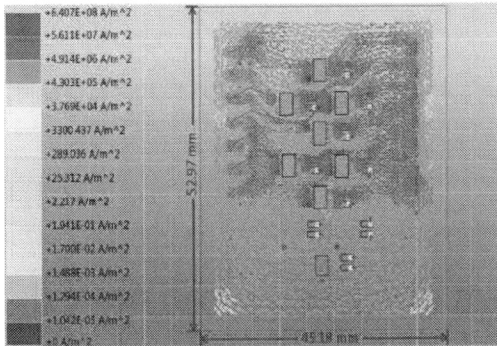

Fig. 7. the current density distribution

The temperature of the die on the substrate is calculated through the between electro-thermal iterative simulation. Figures 6 and 7 show the simulation results of IR-drop and current density. The maximum voltage drop of 3.3V power on SIP is 86.59mV, which is 2.6% of the 3.3Vsupply voltage. The maximum voltage drop of 2.5V power is 78.05mV, which is 3.1% of the 2.5V supply voltage, which is less than 5% of the total voltage, which meets the design requirements.

The SIP temperature is calculated by iterations between three times by electro-thermal method. Figure 8 shows the comparison of temperature distribution cloud diagrams between "pure thermal" analysis and electro-thermal coupled analysis. The temperature of the substrate and the device is higher than the simulation result under the "pure heat" condition. The temperature increased obviously of the ALN substrate and die because of Joule heat.

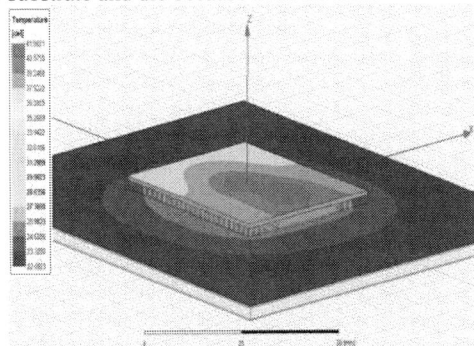

(a) pure thermal analysis

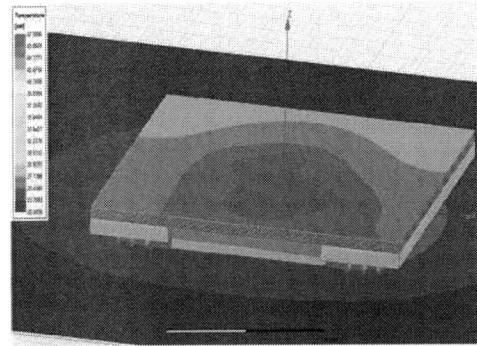

(b) the Electrical-thermal coupling analysis

Fig. 8. the comparision between two results

After the iteration reaches 3 times, the temperature change tends to be stable. Table 1 shows the temperature change of the LDO die, substrate and cover as the test points. From the calculation results in Table 1, it can be seen that Joule heating has a significant effect on the temperature of the device. Therefore, if the effect of electro-thermal coupling is not considered in the thermal analysis, the calculation results will lack accuracy.

TABLE I. THE TEMPERATURE COMPARE BETWEEN RESULTS

	die (℃)	substrate (℃)	cover (℃)
"Pure heat"	42.1	41.2	39.5
Electro-thermal	48.3	47.6	42.3
Joule heating (ΔT)	6.2	6.4	2.8

Finally, the working temperature of this SIP module is measured by the infrared thermometer, as Figure 9 shown. The measured maximum temperature was 46.3℃. The relative error of electro-thermal coupling simulation results and measured resulted was 2.8%. The relative error of "pure heat" simulation results and measured resulted was 11%. The data has further proved the accuracy of the electro-thermal coupling design simulation method, and ensures the reliability of the product.

Fig. 9. SIP module

Fig. 10. the measure results

V. CONCLUSION

Based on the study of electro-thermal coupling mechanism, the electro-thermal coupling design simulation has been processed in this paper, and establishes an analysis model for SIP module, which realizes the coupling between the electromagnetic field and the thermal field. The electro-thermal coupling method accurately calculates the "heat source" distribution in the SIP module at design, which can effectively guide the electrical design, and the SIP heat dissipation, reduce the temperature of key components, and extend the service life of the components

REFERENCES

[1] ZHOU Yue-ge1, ZHENG Hui-ming1,et al, "Coupled Electro-Thermal Simulation of Switched Mode Power Supply Based on Isight", Electric Machines and Control , 2017,VOL.21, NO.10, pp. 23-26.

[2] Jia Yingjie, Xiao Fei, et al, "Multi-Rate Electro-Thermal Simulation Method for High Power IGBT Based on Field-Circuit Coupling", Transactions of China Electrotechnical Society, 2020, VOL.35, NO.9, pp. 23-26.

[3] Tianjian Lu, Jian-Ming Jin, "Electrical Thermal Co-simulation for Large-scale Power Delivery Analysis of DC IR-drop". IEEE Transactions on Components, 2011, VOL.22, NO.1, pp. 79-84.1

[4] Xie J, Swaminathan M. , "Electrical Thermal co-simulation and Joule Heating Effects of 3D Integrated Systems". IEEE Transactions on Components Packaging & Manufacturing Technology, 2011, VOL.1 NO.2, pp. 234-246.

[5] Xin Ai, An-Yu Kuo, Mazen Baida, Yun Chase, "An Accurate and Fast 3D Numerical Approach to Power/Signal and Thermal co-simulation", IEEE 61st Electronic Components and Technology Conference (ECTC), Lake Buena Vista, United States, 2011, pp. 484-487

[6] Su Haohang, "Optimization Design of High-speed Circuit Power Network Based on Hybrid Simulation", Space Return and Remote Sensing, 2017, VOL.38, NO.(5), , pp.50-56.

[7] Zhou Liping, Song Yanping, "Thermal Design and Thermal Analysis of Spaceborne Traveling Wave Tube Power Supply", Space Electronic Technology, 2009, VOL.6, NO.3, pp. 73-77.

[8] Tan Zongxiao, Qi Ying, "Application of Thermal Design Analysis in Radar Antenna ", Space Electronic Technology, 2011, VOL.2, NO.3, pp. 55-59.

[9] Tong Yelong, Li Guoqiang, Geng Liyin, "Research Status of Spacecraft Precision Temperature Control Technology", Space Return and Remote Sensing, 2016, VOL.37, NO.2, pp. 55-59.

Reliability analysis on wafer bonding process of MEMS circulator

Dongxue Luo
Nanjing Electronic Devices Institute
Nanjing, China
have11zhang@163.com

Yan Wu
Nanjing Electronic Devices Institute
Nanjing, China
346329354@qq.com

Junfeng Sun
Nanjing Electronic Devices Institute
Nanjing, China
870915508@qq.com

Miao Yu
Nanjing Electronic Devices Institute
Nanjing, China
yumiao1230@163.com

Abstract—**MEMS circulator is a promising substitute of the traditional circulator to cascade RF modules and realize the miniaturization of RF microsystem for radar and communication system. The silicon-based MEMS circulator is assembled with silicon chip, ferrite and magnetic steel. The silicon chips for 6 mm × 6 mm MEMS circulator were prepared by wafer bonding using Au-Au and Au-Sn as bonding material and signal interconnection with bonding area of 8.55 mm². And the bonding behaviors of two different metal bonding processes were compared and analyzed in this paper. The results illustrated that the shear forces of both bonding methods were > 21kg·F (the shear strength > 24 MPa). In shear force test of process control monitor (PCM) cells with different bonding areas, the shear force of the chip was promoted with the increase of bonding area. The shear force is supposed to be > 2 kg·F (the shear strength > 20 MPa) when the bonding area is > 1.00 mm². The insertion loss of the chips using both processes were less than 0.5 dB. It has been validated that both bonding methods are feasible and reliable for MEMS circulators.**

Keywords—wafer bonding, MEMS circulator, bonding strength

I. INTRODUCTION

Circulator is a nonreciprocal multi-port passive device. With the function of the magnetic material under the combined effect of the external microwave field and DC steady magnetic field, RF signals in the circulator can be transmitted in one direction and isolated in the opposite direction in a circle. Circulators are widely used in radar, microwave communication, electronic systems and other fields[1-3]. Compared with the traditional circulator, MEMS circulator has the characteristics of small size, high precision and mass production. It is expected to replace the traditional microstrip circulator isolator to meet the higher integration of the radio frequency system. In the field of MEMS, wafer bonding technology is mainly used to realize wafer level packaging and interconnection[4-6]. In order to provide both good air tightness and electrical function, metal bonding technology is the most suitable way of bonding, and diffusion bonding and eutectic bonding have been massively applied [7-9]. In diffusion bonding, two metal layers contact tightly under the condition of heat and pressure, the atoms from two layers diffuse and rearrange to form the bonding at the interface. And in eutectic bonding, eutectic alloy forms between two metal layers above the eutectic temperature and fills the gap of two wafers.

In this paper, the Au-Au diffusion bonding and Au-Sn eutectic bonding processes were compared and analyzed. The bonding strength was evaluated using shear force test and the bonding interface was analyzed by scanning electron microscopy (SEM) and energy dispersive spectroscopy (EDS).

PCM cells with different bonding areas were measured in the shear force test, and the effect of bonding area on bonding strength was discussed. The bonding behaviors and electrical performance of the two bonding metals were analyzed, and the feasibility of the two bonding modes has been verified.

II. CIRCULATOR FABRICATION

MEMS circulator is assembled with silicon chip, ferrite and magnetic steel, and the structure schematic is shown in Fig.1.

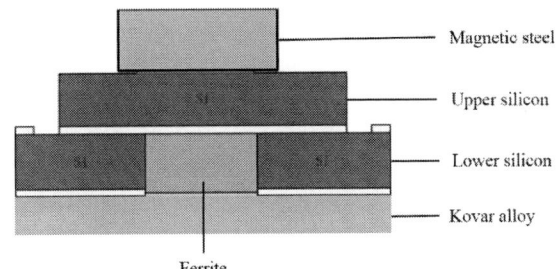

Fig. 1. Structure schematic of MEMS circulator

The circulator chips in this paper were fabricated using two 400 μm high resistance silicon wafers stacked with bonding area of 8.55 mm². The manufacturing process of MEMS circulator includes three main processes: metallization, deep reactive ion etching (DRIE) and wafer bonding. Fig.2 shows the process flow of (a) Au-Au bonding and (b) Au-Sn bonding.

A. Metallization

The 4 μm gold bonding layers were electroplated on upper and lower silicon wafers for both bonding processes, and additional 2 μm Sn was evaporated subsequently after DRIE of RF input / output (I/O) ports on the upper wafer for Au-Sn bonding process. In order to prevent the oxidation of the Sn layer, a 50 nm Au was evaporated on Sn as a passivation layer. The bonding interfaces are required to be clean before bonding process. The interfaces were pretreated by oxygen plasma to remove the contamination and particles effectively. In addition, it improves the wettability of Sn layer and reduces the formation of interface void.

978-1-6654-1392-3/21 $31.00 © 2021 IEEE

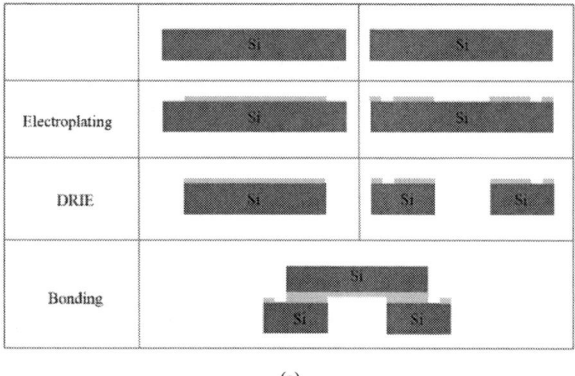

(a)

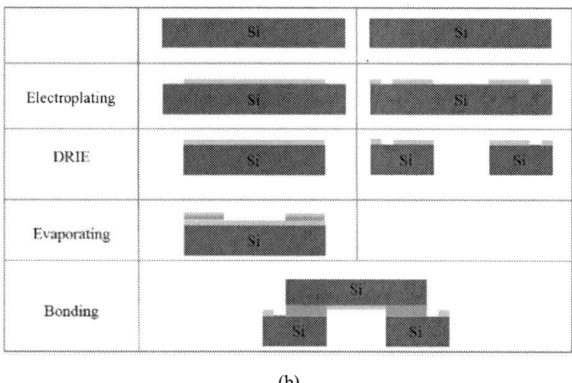

(b)

Fig. 2. Schematic of (a) Au-Au and (b) Au-Sn wafer bonding process flow

B. DRIE

The etching patterns define RF I/O ports in the upper wafer and ferrite cavity in the lower wafer in both process flows. The wafers were etched through vertically using DRIE.

C. Wafer bonding

Thermal-compression-bonding of wafers was conducted using the bonding pressure of 15 kN for 30 min at 320 °C in Au-Au bonding process and 10 kN for 5 min at 320 °C in Au-Sn bonding process.

III. EXPERIMENT AND DISCUSSION

A. Bonding strength of circulator chips

The shear force to de-bond the chip is a critical parameter to evaluate the bonding strength. The shear force was measured using a shear force tester to calculate the bonding strength. The shear force of the chips prepared by both processes was out of the range of shear force tester, and no chip was de-bonded in the test. The measurement results illustrated that the shear forces of both bonding methods were > 21 kg·F (the shear strength > 24 MPa) as it is listed in Table 1, which is much higher than the inspection standard of the shear strength of chips with the same bonding area.

TABLE I.   MEASUREMENT RESULTS OF SHEAR FORCE TEST

No.	Bonding method	Shear force (kg·F)				
		Single cell (top)	Single cell (bottom)	Single cell (center)	Single cell (left)	Single cell (right)
1	Au-Au	21	21	21	21	21
2	Au-Sn	21	21	21	21	21

Two different MEMS circulator chips both cracked in silicon substrate instead of delamination as Fig.3 when they were separated by a sharp knife pressing at the interface. The metal layers were torn off with silicon fragments in Fig.3(a) and some silicon fragments remained at the fracture interface in Fig.3(b). It indicates that bonding strengths of both bonding methods are higher than that of Si-Si bonding.

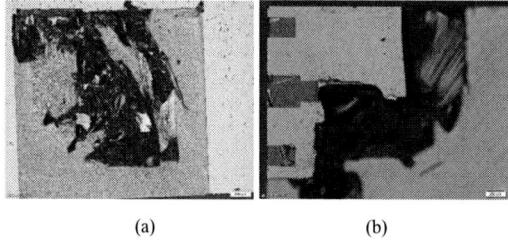

(a)   (b)

Fig. 3. The microstructure of (a) Au-Au and (b) Au-Sn bonding surface cracked by knife

B. Interface analysis of silicon chips

The quality and reliability of bonding processes largely depend on the interface microstructure for both bonding methods and the intermetallic compounds (IMC) especially for eutectic bonding. A good bonding structure forms a compact interface in most bonding area. The bonding performance can be judged qualitatively by the interfacial structure and element concentration analysis. The bonding interfaces of the chips were observed using SEM. More than 7/10 Au-Au bonding and Au-Sn bonding chips were in good condition and no delamination along the interface was observed, as the interfaces were presented in Fig.4. A very small number of cracks in Au-Au and Au-Sn bonding interfaces were shown in Fig.5.

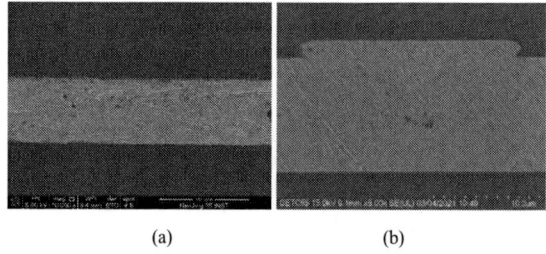

(a)   (b)

Fig. 4. The morphology of (a) Au-Au and (b) Au-Sn bonding interface in good condition

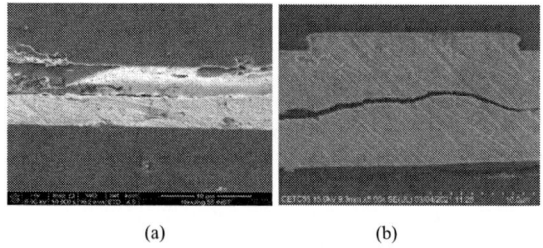

(a)   (b)

Fig. 5. The morphology of (a) Au-Au and (b) Au-Sn bonding interface with cracks

Au-Au diffusion bonding is a typical thermal-compression-bonding process. Au atoms diffuse and form new bonding at the interface in the process, and it results in excellent interface purity and air tightness. The metal does not melt in the bonding process therefore the bonding surfaces require to be in close contact without contaminations.

In the process of Au-Sn bonding, a transient liquid phase of Sn appears. Due to high diffusion rate of Au, a large amount of Au atoms diffuse into the liquid Sn when the two wafers contact. With the diffusion reaction going on and the temperature gradually rising, the whole interface forms the multilayer structure of Au-IMC-Au[10]. Taking IMC formation condition into consideration, the thickness accuracy and uniformity are important parameters of the bonding layers. And it is necessary to pretreat the surfaces to remove the oxidation and contaminations. The critical bonding process requirements are extremely high temperature and pressure uniformity to form a stable IMC. The element concentrations at the interface were measured by EDS as Fig. 6 presents. The points distributing uniformly in a line across the interface were selected for the analysis, and Fig. 7 shows the points and the atomic ratios of Sn and Au. It is demonstrated that the proportion of Sn in the bonding region presents high in the middle and low on both interfaces and Sn exists in the form of IMC instead of pure Sn.

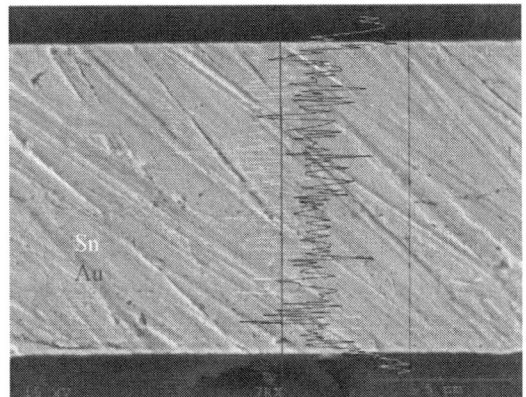

Fig. 6. The line scanning results of Au-Sn bonding interface

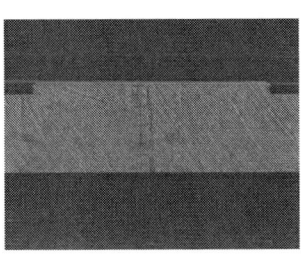

Atomic (%)	Sn	Au
1	5.48	94.52
2	3.37	96.63
3	4.17	95.83
4	18.36	81.64
5	12.61	87.39
6	15.94	84.06
7	11.88	88.12
8	6.29	93.71
9	2.80	97.20

(a)   (b)

Fig. 7. The EDS results of (a) morphology (b) the atomic ratio of Sn and Au of Au-Sn bonding interface

Equilibrium phase diagram of Au-Sn binary alloy is shown in Fig. 8[11]. When the atomic mass ratio of Sn is 29%, the eutectic reaction occurs at 278 °C. ζ (Au$_5$Sn) phase and δ (AuSn) phase are two main IMCs at the bonding temperature. When the atomic proportion of Sn across the interface is less than 16%, and IMCs are mainly in ζ (Au$_5$Sn) phase and β (Au$_{10}$Sn) phase. It means the majority of the IMCs after wafer bonding are stable below 522 °C.

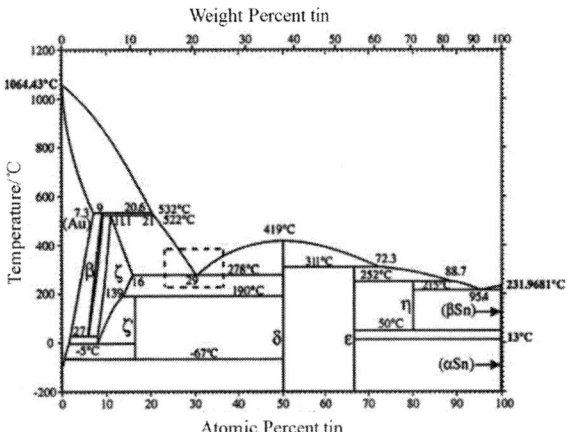

Fig. 8. The equilibrium phase diagram of Au-Sn binary alloy

C. Within-wafer shear force distribution

In order to study the effect of bonding area on bonding strength, 1 mm × 1 mm sheet chips with 0.36 mm², 0.64 mm² and 1.00 mm² bonding area for shear force test were designed in process control monitor (PCM) cells. Within-wafer shear force distributions were analyzed by 2 groups of different bonding methods. The statistical results were in normal distribution, and two groups of data were close in value as shown in Fig.9.

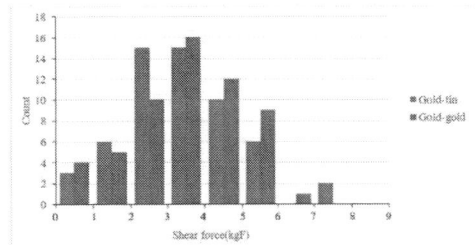

(a) bonding area of 0.36mm²

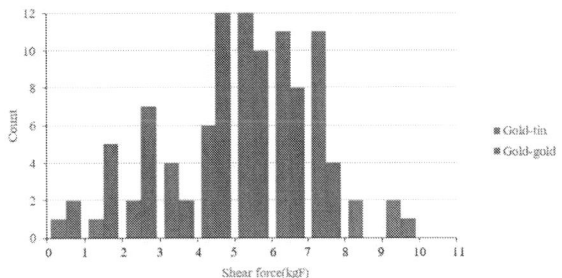

(b) bonding area of 0.64mm²

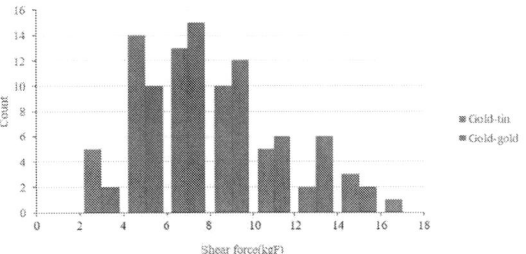

(c) bonding area of 1.00mm²

Fig. 9. The with-in wafer shear force distributions with the bonding areas of (a) 0.36 mm², (b) 0.64 mm² and (c) 1.00 mm² by Au-Sn and Au-Au bonding

TABLE II.   MEASUREMENT RESULTS OF PCM SHEAR FORCE TEST OF AU-AU BONDING

No.	Bonding area (mm²)	Average shear force (kg·F)	Average shear strength (MPa)
1	0.36	2.85	79
2	0.64	4.22	65.9
3	1.00	6.11	61.1

TABLE III.   MEASUREMENT RESULTS OF PCM SHEAR FORCE TEST OF AU-SN BONDING

No.	Bonding area (mm²)	Average shear force (kg·F)	Average shear strength (MPa)
1	0.36	3.08	85.6
2	0.64	4.46	69.7
3	1.00	5.97	59.7

The shear force test results of PCM cells with different bonding methods are shown in Table 2 and 3. As can be seen, the shear force of the chips will be promoted with the increase of bonding area. The shear force is supposed to be > 2 kg·F (the shear strength > 20 MPa) when the bonding area is > 1.00 mm². It has been validated that both bonding methods are feasible and reliable for MEMS circulators.

IV. ELECTRICAL VALIDATION

The RF electrical performance of the MEMS circulator was measured using a network analyzer in the design frequency range from 15.0 to 18.0 GHz. Comparison of the measurement results is shown in Fig.10. It can be seen from the figure that the insertion loss of both kinds of chips are less than 0.5 dB, and both curves are relatively consistent. Due to the isolation and the standing wave ratio depending more on magnetic steel and ferrite than on the bonding process , and therefore those specifications are not compared here.

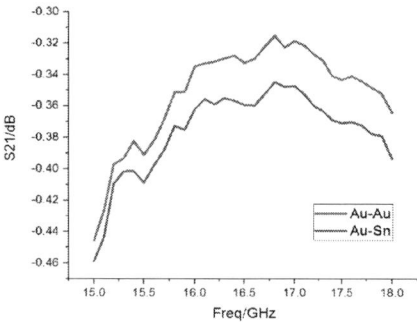

Fig. 10. The insertion loss of MEMS circulator

V. CONCLUSIONS

Massive experiments in this paper were performed to study the reliability on two metal bonding processes of MEMS circulator. The conclusions which will be useful for the design and fabrication of MEMS circulator and other multilayer wafer bonding devices are drawn basing on the experimental data as follows:

1) Both bonding processes result in good bonding strength. Comparing with Au-Au bonding, lower bonding pressure and shorter process time are required in the Au-Sn bonding.

2) The within-wafer shear force distributions using two bonding processes are both in normal distribution, and the distributions of both bonding processes with 3 different bonding areas are close.

3) The bonding area is a critical parameter that the bonding strength depends on. The shear force of the chip will be promoted with the increase of bonding area.

4) Both bonding processes are compatible for RF signal transmission along the interface of bonded wafers.

REFERENCES

[1] T. Jensen,V. Krozer,C. Kjrgaard. Realisation of microstrip junction circulator using LTCC technology[J]. Electronics Letters,2011,47(2).

[2] Microelectromechanical Systems; Researchers from Carnegie Mellon University Detail Findings in Microelectromechanical Systems (Magnet-Less Circulator Using AlN MEMS Filters and CMOS RF Switches)[J]. Journal of Technology & Science,2019.

[3] C.M. Waits,B. Morgan,M. Kastantin,R. Ghodssi. Microfabrication of 3D silicon MEMS structures using gray-scale lithography and deep reactive ion etching[J]. Sensors & Actuators: A. Physical,2004,119(1).

[4] Francis E.H. Tay. Production scheduling of a MEMS manufacturing system with a wafer bonding process[J]. Journal of Manufacturing Systems,2002,21(4).

[5] Viorel Dragoi,Gerald Mittendorfer,Franz Murauer,Erkan Cakmak,Eric Pabo. Metal Wafer Bonding for MEMS Applications[J]. MRS Proceedings,2008,1139.

[6] M. Reiche,M. Haueis,J. Dual,C. Cavalloni,R. Buser. Multiple Wafer Bonding for MEMS Applications[J]. MRS Proceedings,2001,681.

[7] Viorel Dragoi,Gerald Mittendorfer,Franz Murauer,Erkan Cakmak,Eric Pabo. Metal Wafer Bonding for MEMS Applications[J]. MRS Proceedings,2008,1139.

[8] Jian-Qiang Lu,J. Jay McMahon,Ronald J. Gutmann. 3D Integration Using Adhesive, Metal, and Metal/Adhesive as Wafer Bonding Interfaces[J]. MRS Proceedings,2008,1112.

[9] Roy Knechtel. International conference on wafer bonding for MEMS technologies and wafer level integration[J]. Springer Berlin Heidelberg,2018,24(1).

[10] X L Wei,J Q Liu,H F Liu,W J Wu,J Fan,L C Tu. Electroplating of 3D Sn-rich solder for MEMS packaging applications[J]. Journal of Micromechanics and Microengineering,2019,29(4).

[11] Massalski T.Binary alloy phase diagram, 2 nd edn //Sylvester D, Hu C. Analytical modeling and characterization of deepsubmicrometer interconnect. ASM International, Materials Park, OH, Proc IEEE, 1990, 89(5): 634-664.

Experimental Tests and Stress Analysis of SnPb Solder Joints in a Ceramic PoP Device

Kaiyu Guo
Department of Materials Science
Fudan University
Shanghai, China
gkykelly@163.com

Honglei Ran
The 13th Research Institute of CETC
Shijiazhuang, China
hongleiran@163.com

Jun Wang*
Department of Materials Science
Fudan University
Shanghai, China
jun_wang@fudan.edu.cn

Abstract—The solder joint reliability has been a research focus for three-dimensional (3-D) integrations. SnPb (63Sn37Pb) solder still has an irreplaceable place in aerospace applications, mostly in 2-D packages, due to its accumulated experiences and good performance in reliability. In this paper, the mechanical performance and reliability of SnPn solder balls in a board-level assembled ceramic PoP 3-D device were studied. Firstly, the shear test and pull test for a single SnPb solder joint near the corner or edge of PoP were carried out. The varied shear height and solder joint diameter were considered in tests and followed a failure analysis. And the appropriate pull test setting was discussed. Secondly, a finite element analysis for the board-level assembled PoP was performed to reveal the critical solder joint. With the measured modulus of print circuit board (PCB) and a plastic model for SnPb solder, the finite element model of the assembly was established and the stress field by a thermal cycle with a range from 85°C to -55°C was computed. The results showed that stresses are much higher at the edge of the solder joint array, especially near to PCB. The influence of PCB's thickness on the two-layer solder joints in the PoP assembly are not negligible.

Keywords—SnPb solder joint reliability, finite element analysis, testing, PoP

I. INRODUCTION

As an emerging technology, three-dimensional (3-D) packaging actualizes highly integrated systems by stacking functional components vertically via solder joint arrays[1]. However, the solder joint reliability along with the higher integration of 3-D packaging is critical, which affects the performance of components and even the microelectronic equipment. The reliability of solder joints is related to their structures, materials and environmental conditions. In the trend of package miniaturization, smaller solder joints are easier to deform in the same condition. Despite the widespread use of lead-free solder for packaging devices, SnPb solder still has an irreplaceable place in priority fields nowadays such as aerospace and national defense due to their excellent mechanical properties and reliability in some complex or extreme environments but mostly used in 2-D packaging. However, the 3-D packaging technologies, e.g. PoP with SnPb solder joints etc., is still not popular in the aerospace applications. Since the packaging structures are improved in a higher integration, the reliability study of solder ball is needed for a board-level PoP assembly before they are used practically.

Although there have been modeling and testing efforts on SnPb solder joints in the past two decades, most of the researches were based on 2-D packages, such as BGA devices

[2-8]. Ghaffarian evaluated the reliability of a PoP device and a system-in-package (SiP) device containing SnPb solder joints in fine pitch ball grid array (FPGA) under thermal cycles, and performed the analysis by scanning electron microscope (SEM), 3-D X-ray, cross-sectioning, etc.[9, 10]. Nevertheless, the impact of thermal loadings on SnPb joints in 3-D packaging and board-level assembly are not abundant quantitatively by simulations. Meanwhile, more tests are required to reveal mechanical properties of solder joint, especially for a real PoP package.

In this stuty, the mechanical performance and reliability of SnPn solder balls in a board-level assembled ceramic PoP 3-D device were investigated. The pull test and shear test for a single SnPb solder joint near the corner or edge of PoP were carried out firstly. The impacts of the shear height and solder joint diameter were demonstrated experimentally. Then a finite element analysis for the board-level assembled PoP was performed to reveal the positions of critical solder joint. With measured modulus of print circuit board (PCB) and an elastoplastic model for SnPb solder, the finite element model of the assembly was established and the stress fields for PCB with three thicknesses by a thermal cycle with a range from 85℃ to -55℃ was analyzed and discussed.

II. EXPERIMENT PROCEDURE

A. Shear/ Pull Testing

When the 3-D package is deformed under loading, the solder joints within the package may sustain tension or

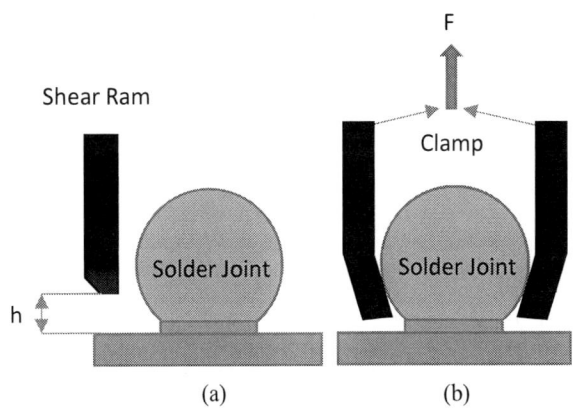

Fig. 1. (a) Solder-ball shear testing, (b) Solder-ball pull testing.

978-1-6654-1392-3/21 $31.00 © 2021 IEEE

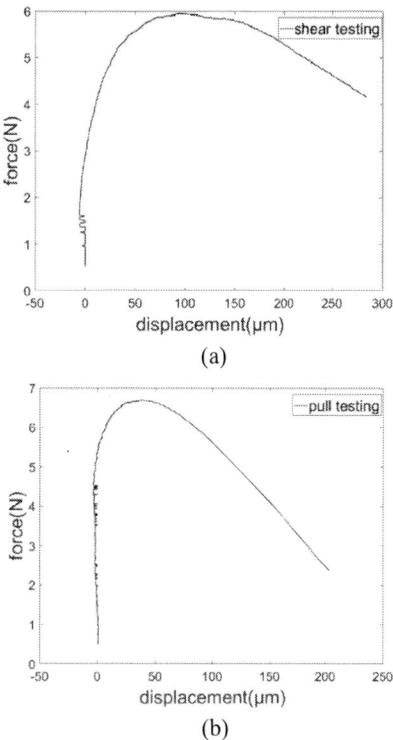

(a)

(b)

Fig.2. Typical force-displacement curves by the shear test and pull test, respectively.

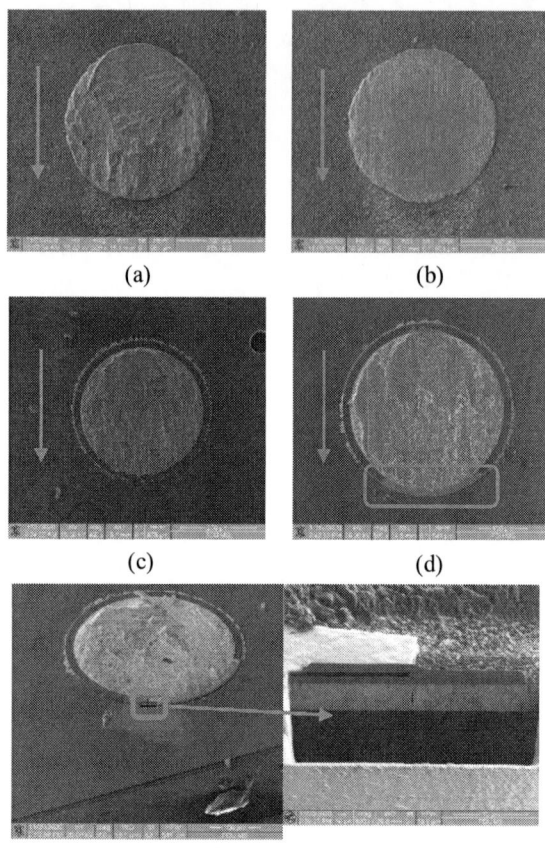

(a)   (b)

(c)   (d)

(e)

Fig. 3. SEM photos of shear failure assessment for SnPb solder joints: (a) 500μm sample, shear height greater than 5μm, (b) 500μm sample, shear height less than 5μm, (c) 400μm sample, shear height greater than 5μm, (d) 400μm sample, shear height less than 5μm, (e) 400μm sample, shear height less than 5μm.

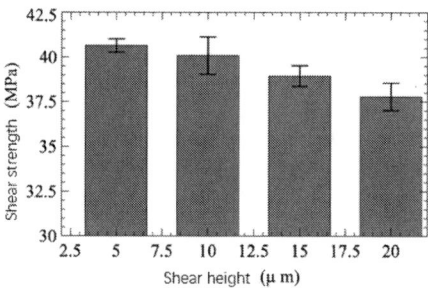

Fig. 4. Shear strength at different shear heights, 500μm SnPb solder joint samples.

shearing, especially for the solder joints near the corner or edges. Hence, the single solder joint tests, i.e. pull tests and shear tests, are helpful to understand the solder joint's strength with respect to the simple loading, which is a reference for the critical solder joint in a simulation of the entire device model.

In addition to mechanical stress, owing to the thermal mismatch between solder and other materials connected with, the device is prone to warping in the procedure of packaging while the solder ball has to sustain large strain. Actually, the above can be simplified as a shearing action towards the ball. Under many usage circumstances (such as drop and impact) where solder joint fractures have been observed, the primary loading orientation is tensile rather than shearing, analogous to the pulling process[11]. As fractures occurring at the pad/solder region in particular, mainly account for the failure of the SnPb solder joints, the shear and pull tests are crucial for assessing their quality.

The shear and pull tests for a single SnPb solder joint on ceramic substrate were conducted at room temperature in this study employing the Dage4000 push and pull testing machine developed by Dage Precision, Ltd. The shear testing, sketched in Fig. 1(a), was conducted on solder balls of two different diameters (400 μm and 500 μm), during which a shear ram with a 750 μm wide flat (standard) face pushed the solder ball to failure. The vertical offset between the substrate surface and shear tool tip ranged from 1.5 μm to 20 μm, while shear speeds varied from 100 μm/s to 700 μm/s. The impact of different shear speed and shear heights, i.e. 5 μm, 10 μm, 15 μm and 20 μm, on the strength were studied. The pull test, illustrated in Fig. 1(b), was carried out with 700 μm/s using a tweezer-like clamp. The failure modes were investigated after the shear and

pull tests through the scanning electron microscope (SEM) and the focused ion beam (FIB).

B. Test Results

The typical force-displacement curves by shear test and pull test are demonstrated in Fig. 2(a) and (b), respectively. The fluctuation at the beginning of the curves was caused by local unloading and a vibration of shear head.

Normally, brittle fractures occurring at the pad/solder region can be classified into three basic types: pad lift, interfacial failure and bulk solder failure[11]. According to

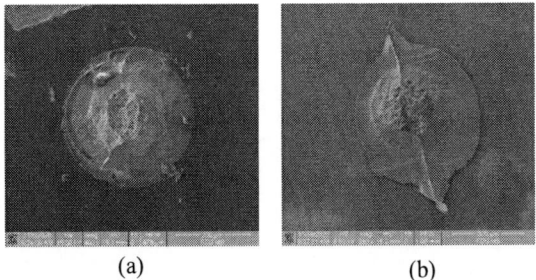

(a)   (b)

Fig. 5. Pull failure assessment for SnPb solder joints, SEM photos: (a) The transverse force of the clamp was appropriate, (b) The transverse force of the clamp was inappropriate.

SEM results (Fig. 3(a)-(c)), fractures were mainly observed at SnPb bulk solder in this study, indicating that the solder and the pad are well bonded to each other. The pad was damaged only in the 400 μm sample when the shear height was less then 5 μm, as shown in Fig. 3(d). However, the fracture of the pad didn't lead to further damage to the ceramic substrate, demonstrated by the FIB analysis (Fig. 3(e)). The direction of the blue arrow in the figures represents the shear direction.

The shear strength increased when shear heights decreased or the shear speeds increased from 100 μm/s to 700 μm/s. Referring to the given column chart for 500μm SnPb solder joint samples (Fig. 4), the shear strength increased as shear heights declining from 20 μm to 5 μm, closer to the strength of the interface between the solder ball and the solder pad. It is worth noting that larger shear heights result in the damage of solder rather than the failure of interface according to the previous SEM analysis. Hence the vertical offset between the substrate surface and the shear tool tip should be controlled within a relatively small range in order to obtain the interfacial strength.

The pull test was conducted on SnPb solder joints with the diameter of 400 μm, in which fractures were taken place at bulk solder (Fig. 5). As a matter of fact, the improper size of the clamp might lead to excessive transverse force, thus affecting the test results of the pull strength. The pull strength was obtained as follows: 83.36 MPa, 81.49 MPa and 79.76 MPa, when the transverse force was appropriate.

III. FINITE ELEMENT MODEL

A. FEM Modeling

Finite element analysis (FEA) is useful to obtain thermo-mechanical stress and strain distribution in complex package constructions. It enables researches to investigate the

reliability of novel devices efficiently, minishing time, manpower and cost for conducting tests.

An elastoplastic finite element model of a 3-D PoP packaging device with two layers of SnPb solder joints interconnection structures was established for thermal stress analysis by ABAQUS in this study. The length and width of the printed circuit board (PCB) are 70mm and 35mm, while three kinds of thickness (1.0mm, 1.6mm, 2.0mm) were taken into account respectively. The mold and its meshes are illustrated in Fig. 6. More than five hundred of SnPb joints were set up in the model to simulate a practical PoP device.

As the complexity of the structure might lead to an enormous amount of computation, the model was simplified in two aspects. First, only solder joints at the edge of the device were meshed intensively in that those in the middle area was less prone to failure according to preliminary analysis results. Second, the PCB was modeled by shell elements and the connections between solid and shell elements were set by equivalent constrains. Hence, the calculation amount of the overall model was controlled while the accuracy of the solder joints' stress was maintained.

The model underwent a thermal process in one temperature cycle. It was assumed as free-stress at room temperature, and the environment temperature rose to 85°C and then dropped to -55°C, in accordance with one of the conditions in JEDEC standards.

B. Material Properties

The PCB is mainly composed of FR4, Young's modulus of which was measured through dynamic thermomechanical analysis (DMA). As shown in Table I, Young's modulus of the PCB decreases as the temperature or the thickness increasing.

TABLE I.   YOUNG'S MODULUS (GPA) OF PCB

Thickness(mm)/ Temperature(°C)	1.0	1.6	2.0
25	25.2	19.5	16.5
35	25.1	19.2	16.4
45	24.8	18.9	16.2
55	24.5	18.5	16.0
65	23.9	18.2	15.9
75	23.7	17.9	15.8
85	23.4	17.6	15.7

Compared to the elastic model, the elastoplastic model is more accurate as the plastic behavior of the metal or alloy materials is considered. Moreover, such model requires less computation in comparison with the viscoelastic model. Consequently, the elastoplastic model has dominant advantages in stress analysis of encapsulated microstructures. For an elastoplastic model, when the stress is less than the yield stress of the material, the material will deform elastically at first. As the stress reaches the yield stress, the plastic deformation will increase until failure of the material occurs. Table II lists the rest of materials properties used in the 3-D packaging device model, including the yield stress of SnPb

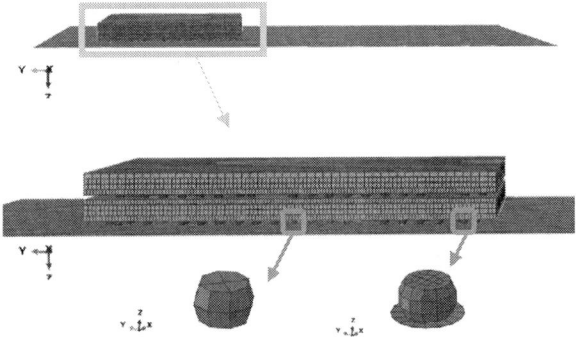

Fig. 6. Board-level assembly model and its meshes

978-1-6654-1392-3/21 $31.00 © 2021 IEEE

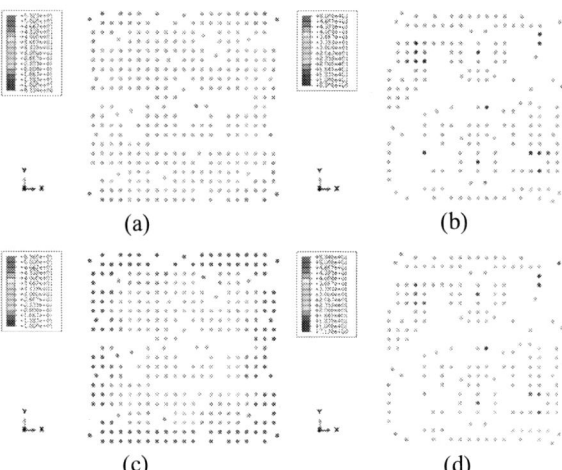

Fig. 7. Stess distribution (MPa) of SnPb solder joints: (a) Solder joints connected to the PCB at 85℃, (b) Solder joints within the device at 85℃, (c) Solder joints connected to the PCB at -55℃, (d) Solder joints within the device at -55℃.

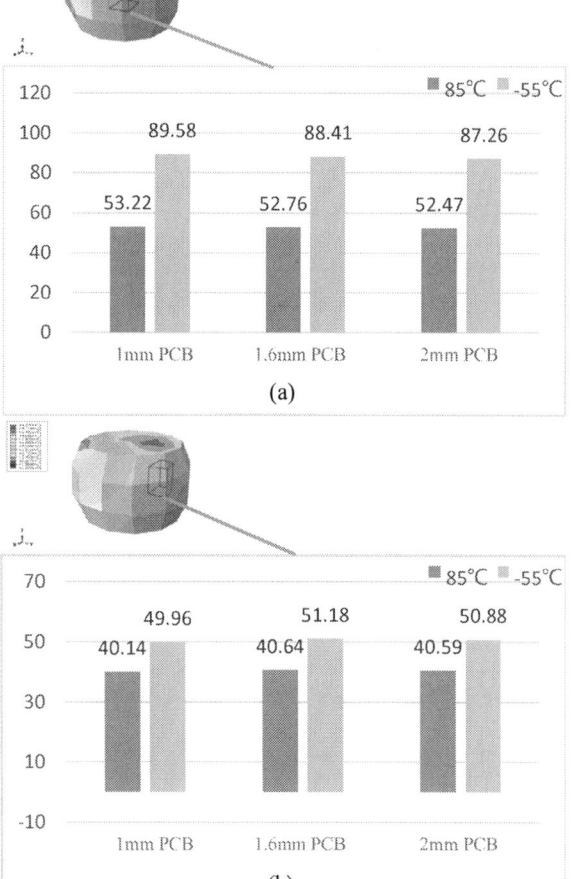

Fig. 8. Stress variation (MPa) of specific elements under different thickness of PCB: (a) Solder joint at the corner of the bottom layer, (b) Solder joint at the corner of the top layer

and Cu. The grey, red, green and blue parts in Fig. 6 correspond to ceramic, SnPb, Cu and FR4.

TABLE II.   MATERIAL PROPERTIES IN THE 3-D PACKAGING DEVICE

Materials	Young's modulus (GPa)	Yield stress (MPa)	Poisson's ratio	CTE (ppm/°C)
ceramic	299	/	0.3	7.5
SnPb	43	36.4	0.36	25.4
Cu	58	240	0.3	17.7
FR4	Table I	/	0.28	15.8

C. FEM Results

The stress distribution of SnPb solder joints in the ceramic PoP device (the thickness of PCB is 1mm) is shown in Fig. 7. The result is similar when the PCB is of other thickness (1.6mm, 2.0mm). As depicted in Fig. 7, stress of SnPb ball arrays meets the regularity of smaller distribution at the center and larger in the corners. It is demonstrated that the solder layer linked to the PCB (Fig. 7(a), (c)) undertakes more thermal stress than the other one (Fig. 7(b), (d)) at the same temperature as a result of thermal mismatch between FR4 and ceramic.

There is lower stress at 85℃ than -55℃ during the temperature cycle. For one thing, the temperature difference between the latter case and the room temperature state (assumed as free-stress) is greater while PCB has greater modulus at high temperature as well. For another, residual stress generated by the heating process has a certain influence on the cooling process, which is dominant in solder joints at the corner. The maximum stress is 53.23 MPa, 50.00 MPa, 89.65 MPa and 53.40 MPa for solder joints connected to the PCB at 85℃, solder joints within the device at 85℃, solder joints connected to the PCB at -55℃ and solder joints within the device at -55℃ respectively, all found at the corner of the layer.

In order to study the stress variation of specific solder joints in low temperature and high temperature under different thickness of PCB, the stress of a certain position (the finite element marked in red, Fig. 8) of the joints at the corner in both layers was calculated. The solder joint undertakes larger stress at -55℃ and the effect of temperature change on the solder joint in the bottom layer is obviously greater than that on the joint within the device. Referring to the given column chart (Fig. 8(a)) for the solder joint connected to the PCB, the stress value mildly falls as the thickness of PCB increasing from 1 mm to 2 mm. The impact of the PCB's thickness on the solder joint within the device is much slighter.

IV.   RESULTS AND DISCUSSION

During the shear/pull study on 400 μm and 500 μm SnPb solder joint samples, fractures mainly occurred at SnPb bulk solder, which demonstrated a tight integration between the solder and the pad. The shear strength increased when shear heights decreased or the shear speeds increased from 100 μm/s to 700 μm/s. The fracture of the pad was observed in 400 μm samples when the shear height was less than 5 μm. The interfacial strength was obtained in the circumstances where the vertical offset between the substrate surface and the shear tool tip was controlled within a relatively small range in the shear test as well as the transverse force of the clamp was appropriate in the pull test.

The FEM model of a ceramic PoP device with PCB of different thickness (1.0mm, 1.6mm, 2.0mm) was established. It is demonstrated that solder joints at the edge of the device are prone to failure as a consequence of concentrated stress. The model reveals that the SnPb solder layer linked to the PCB undertakes more thermal stress than the one within the device during the thermal impact as a result of thermal mismatch between FR4 and ceramic. The model sustains lower stress at 85°C than -55°C during the temperature cycle. In comparison with solder joints within the device, PCB's thickness and temperature variation have a more dominant influence on the ones connected to the PCB.

Interfacial strength on either side of the critical solder joints with concentrated stress at -55°C in top and bottom layers was calculated after simulation (Table III). According to the FEM results, the stress distribution on both sides of the same solder joint is similar. Stress on the side closer to the PCB (A2, B2, C2, D2) is slightly greater. Compared to the previous shear test results in which interfacial strength σ_c was measured around 40 MPa, all the FEM results are close to σ_c, suggesting the solder joints are prone to fracture under thermal impact.

TABLE III. INTERFACIAL STRENGTH (MPA) FROM FEM RESULTS

PCB's thickness(mm)/ Position		1.0	1.6	2.0
Solder joint at the corner of the bottom layer	A1	38.55	37.33	36.24
	A2	38.91	37.69	36.57
	B1	37.93	37.35	36.28
	B2	39.15	37.71	36.61
Solder joint at the corner of the top layer	C1	36.15	36.72	36.21
	C2	36.23	37.01	36.50
	D1	34.04	36.25	35.65
	D2	36.23	36.52	35.93

V. CONCLUSION

In this study, both shear/pull tests and FEA were conducted. The strength of single solder joint was assessed by varying test conditions. It can be considered that fractures occurring at bulk solder mainly account for the failure of SnPb solder joints as the 3-D package sustains tension or shearing. Meanwhile, the finite element analysis was performed to reveal the stress distribution in the two layers of solder joint array in a ceramic PoP device. Based on the investigation, SnPb solder joints at the edge of array, especially near to PCB,

is critical to the failure of the device due to concentrated stresses. Solder joints connected to the PCB are more responsive to the PCB's thickness and the temperature variation. Among the devices with the three different PCB thickness of 1.0mm, 1.6mm and 2.0mm, the 3-D assembly with 2.0mm PCB has a lower stress than that of the other two cases. It is still of significance to evaluate and optimize SnPb solder joint reliability for 3-D packaging devices.

ACKNOWLEDGEMENT

The authors gratefully acknowledge the National Nature Science Foundation of China (Grant No.61774044).

REFERENCES

[1] J. Lu, "3-D Hyperintegration and Packaging Technologies for Micro-Nano Systems," Proceedings of the IEEE, vol. 97, pp. 18-30, 2009.

[2] A. Yeo, C. Lee and J. H. L. Pang, "Flip chip solder joint fatigue life model investigation," in 4th Electronics Packaging Technology Conference, Singapore, 2002, pp. 107-114

[3] F. Song, S. W. R. Lee, K. Newman, B. Sykes, and S. Clark, "Brittle Failure Mechanism of SnAgCu and SnPb Solder Balls during High Speed Ball Shear and Cold Ball Pull Tests," in 2007 Proceedings 57th Electronic Components and Technology Conference Sparks, NV, USA, 2007.

[4] J. Jang, A. P. De Silva, J. E. Drye, S. L. Post, N. L. Owens, J. Lin, and D. R. Frear, "Failure Morphology After Drop Impact Test of Ball Grid Array (BGA) Package With Lead-Free Sn−3.8Ag−0.7Cu and Eutectic SnPb Solders," IEEE Transactions on Electronics Packaging Manufacturing, vol. 30, pp. 49-53, 2007.

[5] S. R. Vempati, N. Su, C. H. Khong, Y. Y. Lim, K. Vaidyanathan, J. H. Lau, B. P. Liew, K. Y. Au, S. Tanary, A. Fenner, R. Erich, and J. Milla, "Development of 3-D silicon die stacked package using flip chip technology with micro bump interconnects,", 2009, pp. 980-987.

[6] J. Chen, W. Lv, B. An, L. Zhou, and Y. Wu, "Influence of rapid thermal cycling on the microstructures of single SnAgCu and SnPb solder joints," in 2013 14th International Conference on Electronic Packaging Technology, Dalian, China, 2013, pp. 848-852.

[7] H. Li, T. An, X. Bie, G. Shi, and F. Qin, "Thermal fatigue reliability analysis of PBGA with Sn63Pb37 solder joints," in 2016 17th International Conference on Electronic Packaging Technology, Chinese Inst Elect, Elect Mfg & Packaging Technol Soc, Wuhan, China, 2016, pp. 1104-1107.

[8] S. Su, F. Akkara, T. Sanders, J. Zhang, J. Evans, and G. Harris, "Reexamination of Thermal Cycling Reliability of BGA Components with SNAGCU and SnPb Solder Joints on Different Board Designs," in 2020 Pan Pacific Microelectronics Symposium (Pan Pacific), 2020, pp. 1-8.

[9] R. Ghaffarian, "Reliability of Package on Package (PoP) Assembly Under Thermal Cycles," in 2019 18th IEEE Intersociety Conference on Thermal and Thermomechanical Phenomena in Electronic Systems Las Vegas, NV, USA, 2019.

[10] R. Ghaffarian, "System in Package (SiP) Assembly and Reliability," in 2019 18th IEEE Intersociety Conference on Thermal and Thermomechanical Phenomena in Electronic Systems Las Vegas, NV, USA, 2019.

[11] K. Newman, "BGA brittle fracture - alternative solder joint integrity test methods," in 55th Electronic Components and Technology Conference, Lake Buena Vista, FL, USA, 2005, pp. 1194-1201 Vol. 2.

Analysis of factors affecting the outgassing rate of MEMS vacuum packaging materials

Yong Yang
College of Mechanical and Control Engineering
Guilin University of Technology
Guilin, China
Science and Technology on Reliability Physics and Application
of Electronic Component Laboratory，CEPREI
Guangzhou, China
m13995991817@163.com

Bin Zhou
Science and Technology on Reliability Physics and Application
of Electronic Component Laboratory，CEPREI
Guangzhou, China
zhoubin722@163.com

Xuanjun Dai
College of Mechanical and Control Engineering
Guilin University of Technology
Guilin, China
daixuanjun@glut.edu.cn

Yun Huang
Science and Technology on Reliability Physics and Application
of Electronic Component Laboratory,CEPREI
Guangzhou, China
huangyun@ceprei.com

Abstract—The finite element Comsol Multiphysics software was used to analyze the thermal desorption on the surface of the material in vacuum packaging, and the destruction of the vacuum. In this study, three influencing factors, the concentration of initial adsorbed molecules on the surface, temperature and surface roughness, are selected to study the vacuum degradation in vacuum packaging. The orthogonal test analysis of three levels and three factors shows that the initial value of the surface adsorption of the packaging material, temperature, and surface roughness affect the outgassing performance of the package as follows: the value of the initial adsorption> surface roughness> temperature.

Keywords—vacuum package， outgassing， Comsol Multiphysics，thermal desorption

I. INTRODUCTION

The surface outgassing of materials is one of the common reasons for the degradation of vacuum in vacuum packaging[1-3]. The main functional units of MEMs devices or systems include sensors and drivers, and the existence of units that need to be moved is a typical feature of most MEMs devices. Moving parts would be subject to air damping in an air environment, and vacuum packaging provides a partial vacuum environment, so that moving parts are subject to as little air damping as possible. However, even if a good package can be formed in the packaging process in vacuum packaging, the pressure in the vacuum packaging will cause the vacuum to be damaged due to gas penetration and material outgassing. With the miniaturization of vacuum packaging, the slight outgassing on the surface of the material have a significant impact on the vacuum degree of the vacuum package, which will cause the parameter drift of the MEMs vacuum device[4-7]. This has become a serious problem for the reliability of vacuum packaging.

The material used in the vacuum packaging material will adsorb gas in the atmospheric environment and during the production process. When the material is placed in a vacuum environment, the material will desorb and outgas, which leads to the deterioration of the vacuum in the vacuum packaging[8]. If the vacuum degree in the vacuum package cannot be maintained, its gas damping will be severely increased. The

vacuum degree is an important technical indicator to measure its performance and service life, and the gas outflow rate of the vacuum package material is one of the main factors affecting the vacuum degree of the vacuum package. Therefore, the outgassing rate of the packaging material used in the institute is of great significance to ensure its mechanical motion performance and service life. It can provide an important basis for the design of the vacuum chamber, the optimization of the vacuuming process, and the selection of getter. so that we can alleviate the deterioration of vacuum level to a certain extent, and finally achieve long-term high-efficiency vacuum degree and mechanical properties of its moving parts.

II. THEORETICAL CONSIDERATIONS

A. Adsorption and desorption

A surface covering σ molecules per unit area, if it is in a vacuum, the adsorbed molecules will gradually desorb due to thermal movement, which is thermal desorption[9]. The rate of thermal desorption is first proportional to the size of σ, σ can be expressed by following equation:

$$-\frac{d\sigma}{dt}=k\sigma \qquad (1)$$

Where σ is the number of molecules adsorbed per unit area (per cm^2), K is proportional coefficient, and t is time.

We assume that the initial adsorption capacity is σ_0 when t=0, the solution of the above formula can be expressed as :

$$\sigma=\sigma_0\exp(-kt) \qquad (2)$$

The expression shows that σ decreases exponentially with time. We set k=1/τ, then τ has the meaning of the time required to decay to 1/e of the original value, which is equivalent to average adsorption time. According to Franck's formula, we can calculate that :

$$\frac{1}{k}=\tau=\tau_0\exp(E_d/RT) \qquad (3)$$

Where E_d is activation energy of desorption (erg/mol), τ_0 is the vibration period of the adsorbed molecule perpendicular to the surface, its value is about 10^{-13}s, R is gas constant [erg/(K·mol)], and T is absolute temperature (K).

Therefore, we can write equation as :

$$\frac{d\sigma}{dt}=-\frac{\sigma}{\tau_0}\exp(-E_d/RT) \qquad (4)$$

978-1-6654-1392-3/21 $31.00 © 2021 IEEE

The equation above can be rewritten as :

$$\frac{d\sigma}{dt}=-\sigma u_f \exp(E_d/RT) \tag{5}$$

Where u_f can be expressed as :

$$u_f=\frac{1}{\tau}[s^{-1}] \tag{6}$$

The u in the equation is also called the first-order desorption rate constant, and its value is about $10^{13}/s$.

There is no interaction between the adsorbed molecules, and it is correct when they are dissociated. Therefore, it is only suitable for the desorption of physically adsorbed or non-dissociated chemically adsorbed molecules (single-atom molecules).

Most of the chemically active gases in vacuum technology are diatomic molecular gases, such as H_2, N_2, O_2, CO, etc. When they are adsorbed on the metal surface, they will dissociate (except for CO). This is because their decomposition heat is less than the adsorbing heat of atoms. This dissociatively adsorbed gas on the surface must first collide and combine two atoms on the surface into molecules during desorption, and then leave the surface. We assume that σ_1 represents the number of adsorbed atoms per unit area, and σ_2 represents the number of desorbed molecules per unit area. Because the probability of two atoms colliding on the surface is proportional to σ_1^2. So the desorption rate can be represented as:

$$\frac{d\sigma_2}{dt}-\sigma_1^2 u_s \exp(-E_d/RT) \tag{7}$$

Where σ_1 is the number of atoms adsorbed per unit area($1\,cm^2$), σ_2 is the number of molecules desorbed per unit area ($1\,cm^2$); and u_s is the secondary desorption rate constant ($cm^2 \cdot s$), the meaning of other symbols is the same as before.

Under normal circumstances, the surface of the material under the atmosphere will contain hundreds of layers of gas molecules. The first layer is intercepted by a strong chemical adsorption force, and the subsequent layers are physically adsorbed. As the thickness of the deposited layer increases, the bonding strength reduce. Once this material is exposed to a vacuum, the desorption of the gas will happen automatically until equilibrium is reached.

B. Pressure in the case of the vacuum package

Ideal gas pressure can be expressed as

$$p=\frac{1}{3}nm_0 v_s^2 \tag{8}$$

where n is the number of molecules per unit volume, which is the molecular number density of the gas $[m^{-3}]$, m_0 is the mass of gas molecules [kg], and v_s is the root mean square velocity of gas molecules [m/s].

Root mean square velocity:

We can add up the square of the velocity of all gas molecules, then divide by the total number of molecules, and then take the square to get the root mean square velocity v_s[10]. The root mean square can be expressed as

$$v_s=\sqrt{\frac{3RT}{M}}=4.994\sqrt{\frac{T}{M}} \tag{9}$$

where v_s is root mean square velocity of gas molecules [m/s], R is molar gas constant, its value is about 8.314 J/(mol·K), T is thermodynamic temperature [K], and M is gas molar mass [kg/mol].

We can deduce from equations (8) and (9)

$$p=4.005c^2 T \cdot 10^{14} \tag{10}$$

where c is the concentration in the vacuum chamber. The formula shows that the pressure is only related to the molar concentration and temperature. The simulation result shows the concentration of molecules in the case, so we will pay attention to the concentration and temperature of gas molecules in the vacuum package case in the follow-up simulation results.

III. SIMULATION ANALYSIS

A. Simulation process

Structural design:

Adopting a specific airtight package of MEMS devices, with device wafers, cap wafers, MEMS devices and other structures. The schematic diagram of the device packaging structure is shown in the figure. The core task of the wafer-level vacuum packaging of MEMS devices is to design and implement a sealed structure to provide a sealed vacuum case of 4mm×4mm×0.5mm for the MEMS device. The internal vacuum has very high requirements. The chip is embedded with the substrate and occupies a part of the sealed case.

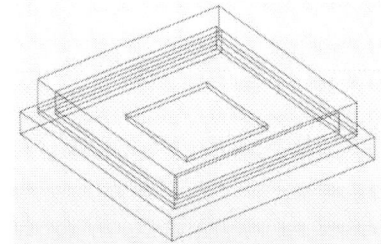

Fig. 1.   3D schematic diagram of vacuum package

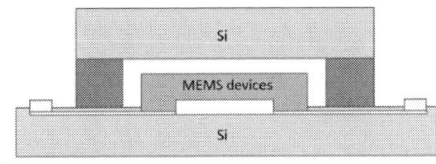

Fig. 2.   Cross section of MEMS device[11]

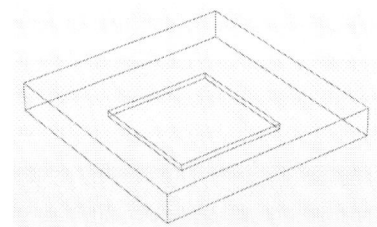

Fig. 3.   Sealed vacuum case

B. Simulation settings and process

The structures of vacuum packaging components are relatively complex, and there are large orders of magnitude difference in size. Moreover, due to the limitations of the Comsol Multiphysics software, the structure needs to be simplified. The main research in this paper is the influence of the thermal desorption of the adsorbed gas on the internal surface of the case on the pressure inside the vacuum package case, so structures such as getter, solder ring, and metallization layer are ignored. Finally, we build a model and analyze the vacuum case of the MEMS device. In order to simplify the analysis model in the transfer of dilute substances, the relevant literature was referred to and the following settings were made to build the establishment the model.

The simulation settings and the boundary conditions of the simulation analysis are as follows:

- In this paper, the vacuum case interior is modeled, and the chip composed of multi-interface materials is simplified to silicon material.

- Once the molecules are desorbed from the inner surface of the case, due to the small internal volume of the case, the internal gas molecules move to any position inside the case in a short time

- The initial pressure in the vacuum case is 0Pa.

- The cap is formed by multi-layer electroplating and deposition, and only the properties of the Au layer directly in contact with the vacuum package case are considered.

- The initial pressure in the vacuum case is 0Pa.

- All outer boundaries of the model entity are set as adiabatic boundary conditions.

TABLE I.   MATERIAL COMPARISON TABLE OF EACH PART STRUCTURE

Structure in the package	Material
Cap	Au
Substrate	Si
Wafer	Si

TABLE II.   CHARACTERISTIC PARAMETERS OF GOLD AND SILICON MATERIALS

Attributes(Unit)	material	Value
Thermal expansion coefficient (1/K)	gold	2.6e-6
	silicon	14.2e-6
Constant pressure heat capacity (J/(kg · K))	gold	129
	silicon	678
Density (kg/m³)	gold	19300
	silicon	2320
Thermal Conductivity (W/(m · K))	gold	317
	silicon	34
Young's modulus (Pa)	gold	70e9
	silicon	160e9
Poisson's ratio (1)	gold	0.44
	silicon	0.22
Conductivity(S/m)	gold	45.6e6

C. Simulation results

Based on the finite element model built above, the Diluted Matter Transport module of Comsol Multiphysics software is used to analyze the gas desorption on the inner surface of the case. The simulation results are shown in the figure below.

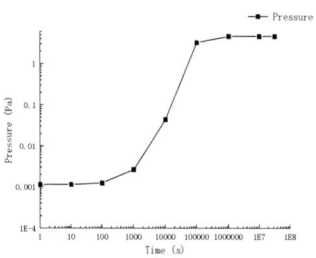

Fig. 4.   The pressure in the case gradually reaches a steady state

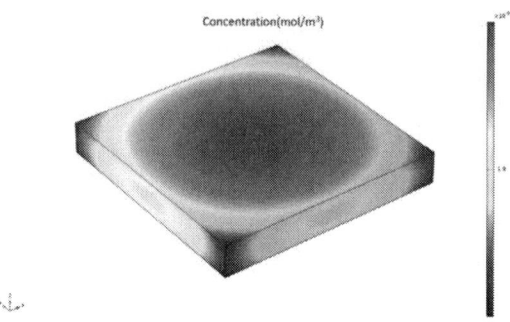

Fig. 5.   Pressure distribution in the case

Figure 4 shows that the temperature T=393.15K, the concentration=1e-3mol/m³, the surface roughness is 1, and the pressure in the case gradually reaches a steady state as time increases. From the beginning until the pressure in the case of 1e6 s increases rapidly, from 1e6 to 1e7 s, the increase is very small. From t=1e7 to t=3.16e7 s, the increase of pressure in the case is almost stagnant. The follow-up test is required to take 1e7 s as the stabilization time of the pressure in the case.

Figure 5 shows the temperature T=393.15K, the concentration=3e-3mol/m³, and the surface roughness is 1. When the time t=1E7s, the concentration of gas molecules in the case. The pressure in the middle of the case is slightly higher than that of the surroundings, but it is basically the same as the surroundings.

D. Comparative test analysis

Through the above analysis, it can be seen that the initial value and temperature of the adsorbed gas inside the case will affect the pressure of the vacuum package case. The research of L. Yan[12] and HF Dylla[13] found that the surface roughness and temperature will affect the outgassing rate of the material, but what are the specifics? The factors that affect the material outgassing degree are relatively large, and it needs to be analyzed through comparative experiments. In this paper, the initial value of surface adsorption, temperature, and surface roughness are selected for analysis. As shown in Table III, Table IV, Table V, the specific values of the three variables, the simulation analysis results are shown in the figure 6.

TABLE III.   THE INITIAL PRESSURE OF THE INNER SURFACE OF DIFFERENT CAVITIES

Initial value(mol/m2)	1e-3	2e-3	3e-3	4e-3	5e-3
Stable case pressure（Pa）	4.63	18.64	41.5	74.01	115.76

TABLE IV.   PRESSURE AT DIFFERENT TEMPERATURES

Temperature （K）	293.15	323.15	353.15	383.15	413.15
Stable case pressure （Pa）	41.5	46.23	50.52	55.4	59.73

TABLE V.   PRESSURE OF DIFFERENT SURFACE ROUGHNESS

Surface roughness	1	2	3	4	5
Stable case pressure （Pa）	4.63	18.64	41.5	73.97	115.76

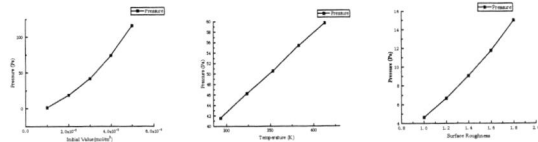

Fig. 6. The relationship between the initial adsorption value, temperature, surface roughness and pressure in the case

The analysis result is: as the initial adsorption value of the inner surface of the case increases, the temperature increases, the surface roughness increases, and the pressure of the case becomes higher when it is stable. As the initial adsorption value of the surface or the roughness of the case surface increases, the pressure in the case increases. Moreover, the pressure increases faster and faster, which becomes more significant with the increase of the initial adsorption value.

IV. ORTHOGONAL TEST

A. Orthogonal experimental design

It can be known from the previous comparative experiments that the initial value of surface adsorption, temperature, surface roughness and other parameters may affect the pressure inside the case, which in turn affects the motion performance of the moving parts in the MEMS. However, we need to study which parameters have a greater influence on the pressure in the case, and how much influence is it. Based on the orthogonal experiment, this chapter changes the parameters of the vacuum package case, uses the finite element analysis software to analyze the case in the MEMS device, and finds the influence relationship after the solution.

After the comparison through preliminary simulation analysis, it can be known from the results that the initial value, temperature, and surface roughness of the internal surface adsorption of the case will affect the internal pressure of the case. In the scheme, the initial value of the surface adsorption of the case is in the range of 0.001-0.005 mol/m², the design range of the temperature is 293.15-393.15K, and the design range of the surface roughness is 1-1.5. Take 3 factors, each of which takes 3 levels, and the design plan of the factor level is shown in Table VI below. According to the factor level table, we use the orthogonal experiment method to design 9 sets of orthogonal experiments.

In Table VI, a is the temperature, b is the initial value of adsorption inside the case, and c is the surface roughness.

TABLE VI.   ORTHOGONAL TEST FACTOR LEVEL TABLE

Factor	Code	Level 1	Level 2	Level 3
Temperature (K)	a	293.15	343.15	393.15
Initial value(mol/m²)	b	1e-3	3e-3	5e-3
surface roughness	c	1	1.25	1.5

B. Orthogonal test data range analysis

The first step is to combine the factors and levels that affect the outgassing parameters. First determine the boundary conditions of the simulation analysis, and then perform the simulation. Complete 9 tests in sequence, and then extract the pressure inside the case when it is stable, and perform range analysis as the result. The range analysis obtains the statistical test results of each level of each factor in a set of test data. The difference between the maximum and minimum mean values in the comparison results. If the difference is greater, it means that the influence of this factor is greater. On the contrary, it shows that the influence of this factor is smaller. The results of data range analysis are shown in the table below.

TABLE VII.   ORTHOGONAL TEST RESULTS

No	a	b	c	t = 1E7, the pressure in the vacuum case (Pa)
Test1	293.15	1e-3	1	4.6
Test2	293.15	3e-3	1.25	64.84
Test3	293.15	5e-3	1.5	260.45
Test4	343.15	1e-3	1.25	8.53
Test5	343.15	3e-3	1.5	110.85
Test6	343.15	5e-3	1	136.37
Test7	393.15	1e-3	1.5	14.18
Test8	393.15	3e-3	1	56.84
Test9	393.15	5e-3	1.25	246.92
K_1	329.89	27.31	197.81	*
K_2	255.75	232.53	320.29	*
K_3	317.94	643.74	385.48	*
k_1	109.96	9.10	65.94	*
k_2	85.25	77.51	106.76	*
k_3	105.98	214.58	128.49	*
Range	24.71	205.48	62.56	*

In Table VII, a is the temperature, b is the initial value of adsorption inside the case, and c is the surface roughness. Kn represents the sum of all orthogonal test results corresponding to the factor level n, and kn represents the average value of the test results corresponding to the factor level n. For example, for the factor a, the corresponding indicators at each level are K_1, K_2, K_3, k_1, k_2 and k_3 values. The calculation process is as follows:

The index sum corresponding to factor a at level 1 is:

$$K_1=4.6+64.84+260.45=329.89$$

The index sum corresponding to factor a at level 2 is:

$$K_2=8.53+110.85+136.37=255.75$$

The index sum corresponding to factor a at level 3 is:

$$K_3=14.18+56.84+246.92=317.94$$

According to the above formula, the average value of the test at the factor level can be obtained, as shown in the following formula:

$$k_1= K_1/3=329.89/3=109.96$$

$$k_2= K_2/3=255.75/3=82.25$$

$$k_3= K_3/3=317.94/3=105.98$$

According to the calculation method of a factor, the range index value corresponding to each level of the two factors b and c can be obtained. The results are shown in Table 9.

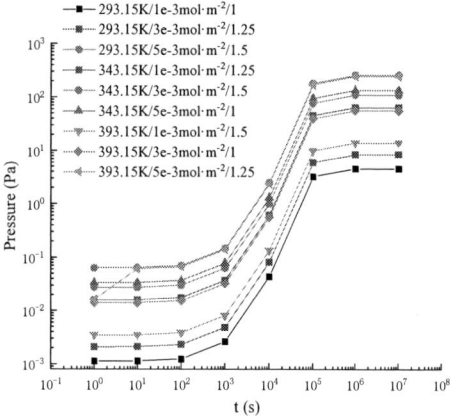

Fig. 7. The pressure in the case increases with time

The legends in Figure 7 are temperature, initial adsorption concentration and surface roughness, which is separated by "/". The analysis result shows that the variable that has the greatest influence on the pressure of the vacuum chamber is the initial adsorption value of the inner surface of the case. The influence of temperature and surface roughness is small, and the influence of temperature on the outgassing rate gradually becomes smaller with time. The degree of influence of these parameters selected according to actual conditions on the pressure of the vacuum package case is as follows: initial adsorption value of the case surface> case surface roughness> temperature.

C. Limit value of surface adsorption

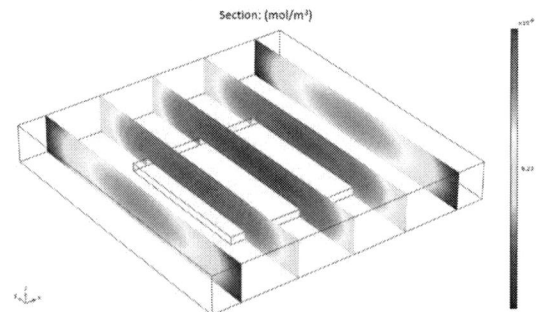

Fig. 8. A cross-sectional view of the concentration in the case

The temperature and surface roughness in Figure 8 are 293.15K and 1, respectively. The initial concentration on the set boundary is 1.47e-2 mol/m². When the concentration in the case reaches a stable level, the concentration in the case is 9.23e-9 mol/m³, equivalent to 1000Pa. The vacuum package will not reach the failure threshold within 3.16e7 seconds, which is one year. According to Tang's designing[11], the internal vacuum degree of the case is not higher than 1000Pa. What's more, we neglected the basic vacuum and other factors when designing the experiment, and the initial concentration of adsorbed molecules on the surface of the case should be much less than 1.47e-2 mol/m².

V. CONCLUSION

This paper first studies the principle of the change in the pressure of the vacuum package case, and then determines the three factors that affect the pressure increase: the initial

adsorption value of the case surface, the surface roughness of the case, and the temperature.

Secondly, through comparative experiments, under the same conditions, changing a single variable, the initial adsorption value of the case surface, the surface roughness of the case, and the influence of temperature on the pressure change in the case, according to the analysis results, with the initial adsorption of the case surface As the value increases, the surface roughness of the case becomes larger, the temperature increases, and the pressure in the case becomes higher when it is stable.

Finally, an orthogonal experimental design is used to study the relationship between the initial adsorption value of the case surface, the surface roughness of the case, the temperature and the pressure in the vacuum package case. According to the results of the simulation analysis and the range analysis of the data, the results show that the initial adsorption value of the case surface has the greatest influence on the pressure inside the case. Therefore, the degree of influence of the selected three factors on the pressure of the case of the vacuum package is as follows: the initial adsorption value of the case surface> the surface roughness of the case> temperature.

The purpose of MEMS vacuum packaging is to provide a case in a vacuum environment for MEMS devices. Since the initial value of the adsorbed gas on the case surface has the greatest impact on the vacuum degree of the case, the case surface should be fully degassed before vacuum packaging, and the entire packaging process should be completed in a vacuum environment. The initial concentration of adsorbed molecules on the surface of the case should be less than 1.47e-2 mol/m², so that the MEMS device can work reliably throughout its life cycle.

ACKNOWLEDGMENT

This research was supported by the National Basic Research Program of China under Grant No. JBF202800010.

REFERENCES

[1] R Pastore, A Delfini, M Albano, A Vricella, F Piergentili. "Outgassing effect in polymeric composites exposed to space environment thermal-vacuum conditions." Acta Astronautica, 2020, 170.

[2] C Zou, Y Song, H Wu, G Shen, W Wu, K Lu, J Wei. "Outgassing measurements for the turn insulation of CFETR poloidal field coils." Fusion Engineering & Design, 2016, 105(apr.), pp. 101-103.

[3] K Battes, C Day, V Hauer. "Outgassing rate measurements of stainless steel and polymers using the difference method." Journal of Vacuum Science & Technology A Vacuum Surfaces and Films, 2015, 33(2):021603.

[4] Schlppi, B, et al. "Influence of spacecraft outgassing on the exploration of tenuous atmospheres with in situ mass spectrometry." Journal of Geophysical Research Space Physics, 2010, 115(A12), pp. 1648-1660.

[5] JS Dyer, R C Be Nson, TE Phillip, JJ Guregian. "Outgassing analyses performed during vacuum bakeout of components painted with Chemglaze Z306/9922." Proceedings of Spie the International Society for Optical Engineering, 1992.

[6] M Ichimura. "Introduction to Vacuum Science and Technology for Absolute Beginners." Journal of the Vacuum Society of Japan, 2015, 58(8), pp. 273-281.

[7] L Chen, Y L Li, W J Sun, M Dong, YJ Cheng. "Prediction of In-Orbit Power on Time for Transformer Based on Gas Permeation Analysis of the Seal Case." MAPAN-Journal of Metrology Society of India, 2020, 35(1).

[8] R Grinham, DA Chew . "A review of outgassing and methods for its reduction." Applied Science and Convergence Technology 2017, 26.

[9] Y Z Wang, X Chen. "Vacuum Technology (2nd edition)." Beijing University of Aeronautics and Astronautics Press, 2007, pp. 129-131.

[10] D A Da. " Design Handbook (3rd edition).". National Defense Industry Press, 2004, pp. 54-55.

[11] G R Tang. "MEMS wafer-level vacuum packaging structure design and process integration." Huazhong University of Science and Technology.

[12] Y Luo, X Wu, K Wang, W Kuibo. "Comparative Study on Surface Influence to Outgassing Performance of Aluminum Alloy." Applied Surface Science , 2019, 502:144166.

[13] Dylla, F H. "Correlation of outgassing of stainless steel and aluminum with various surface treatments." Journal of Vacuum Science & Technology A Vacuum Surfaces & Films, 1993, 11(5), pp. 2623-2636.

Fatigue Life Predictions of SnPb Solder Ball in A Ceramic PoP Device

Yu Yao*
Department of Materials Science
Fudan University
Shanghai, China
2911471544@qq.com

Zirui Cui*
Department of Materials Science
Fudan University
Shanghai, China
1605052659@qq.com

Honglei Ran*
The 13th Research Institute of CET
Shijiazhuang, China
hongleiran@163.com

Jun Wang^
Department of Materials Science
Fudan University
Shanghai, China
jun_wang@fudan.edu.cn

Abstract—In this study the fatigue life of the critical SnPb solder ball of a ceramic package-on-package (PoP) device containing two layers of SnPb solder balls was predicted via a suitable fatigue model based on nonlinear finite element analysis (FEA). The viscoplastic constitutive relationship was applied for SnPb solder and the other materials have linear elastic behaviors in FEA. The evolution of the stress in the solder balls under thermal-cycle loadings (-55℃ to 125℃) was simulated by ABAQUS. Using the stress analysis results, we predicted the fatigue life of the ceramic PoP device by the empirical Darveaux-based lifetime model. The fatigue life of the critical solder ball in the top layer is over 5 folds than that of the critical solder ball in the bottom layer. The computation results illustrated that the creep energy density plays an important role in fatigue life. The analysis showed that the critical solder ball in PoP may not suffer serious fatigue problems under thermal cycling loads comparing with the 2D packaging case that removed the top component.

Keywords—SnPb solder ball, Finite Element Analysis, fatigue life, PoP.

I. INTRODUCTION

Nowadays, with the trend of miniaturization and multi-functional requirements of devices, novel 3-D high-density electronic packaging technologies, e.g. PoP (Package on Package) etc., are increasingly used in many consumer products. As the demands of lightweight and high integration, the PoPs are being introduced for aerospace applications recently. Generally, SnPb (63Sn-37Pb) solder in PoPs is employed as electronic connecting material owing to its excellent mechanical properties. The studies of SnPb solder balls in 2-D electronic devices revealed that the temperature is one of the major factors contributing to the fatigue damage of SnPb solder balls[1], [2]. The finite element analysis (FEA) for a power module illustrated that the creep strain in the solder balls is accumulated during thermal cycling[3]. The maximum stress in thermal cycle test was appeared in the interfaces between the solder balls and substrate, and the stress trapped in the interface tended to initiate small cracks [4]. Since the inelastic behavior of solder balls is significant, the viscoplastic constitutive relationship, i.e. Anand model, is adopted to optimize structural parameters and predict thermal fatigue life of solder balls[5]. The pre-cracks initiation in thermal cycling decreases both drop and vibration reliability of board-level solder balls[6].

Stress is not only dependent on material properties, but also on device structures. The structural complexity of the 3-D PoP may reduce the reliability of SnPb solder balls in the device. Furthermore, it is challenging for devices used in aerospace application to work well in extreme and complex environments. Hence, it is necessary to study the fatigue reliability of SnPb solder balls in the ceramic PoP before the device was used in aerospace applications.

In this study, the fatigue life of SnPb micro solder ball in a ceramic PoP device was predicted via a suitable fatigue model based on finite element analysis. The ceramic PoP device containing two layers of SnPb solder balls was modeled as a 3D case and simulated by ABAQUS. Fine and uniform meshes were implemented for the critical solder balls. Besides, the Anand model was applied to the SnPb solder and the materials parameters were from the papers[7], [8]. To keep a moderate computational cost, the printed circuit board (PCB) was modeled as shell elements. The stress distribution and nonlinear energies in the critical solder balls of PoP device were calculated when the device was subjected to thermal cycles (-55℃ ~125℃). On the basis of stress analysis, the fatigue life of critical solder ball was predicted. Moreover, the contribution from the energies respectively related to creep strain and plastic strain were discussed.

II. FINITE ELEMENT MODEL

A. FEM Modeling

The 3D PoP finite element model, including ceramic substrates, ceramic coverplates, copper pads, SnPb solder balls and PCB, was setup by ABAQUS. In the model, the diameter of solder ball is 0.40 millimeters and the height is 0.2365 millimeters. The sizes of other components are listed in Table I.

TABLE I.   COMPONENT PARAMETERS

components	Length (mm)	Width (mm)	Height (mm)
Ceramic coverplate	20	20	0.75
Ceramic substrate	20	20	0.25
PCB	70	35	1

---

*These authors contributed equally to this work.
^Corresponding author.

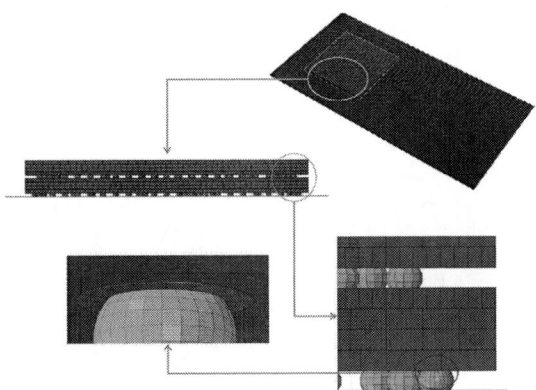

Fig. 1. The model of simulation

3D 6-node and 8-node continuum elements with hourglass controlled reduced integration were employed in the model. As the PCB was modeled as a shell, 4-node quadrilateral shell elements with large-strain formulation and reduced integration were used in meshing. The meshes of the model are demonstrated in Fig. 1. The figure illustrates the entire meshes and the local meshes for solder balls.

B. Material Properties

The isotropic and elastic behaviors are assumed in the simulation for all materials except the SnPb solder. The martial properties are shown in Table II.

TABLE II. MATERIALS PROPERTIES

Materials	Young's modulus (GPa)	CTE (ppm/℃)	Poisson's Ratio	yield stress (MPa)	plastic strain
Cu	58	17.7	0.3	240	0
				720	0.04
ceramic	299	7.5	0.3	-	-
SnPb	43	25.4	0.36	36.4	0
				253.5	0.123
PCB (FR-4)	25.20@25℃	15.8	0.28	-	-
	25.23@30℃				
	25.13@35℃				
	25.01@40℃				
	24.84@45℃				
	24.68@50℃				
	24.50@55℃				
	24.12@60℃				
	23.92@65℃				
	23.73@70℃				
	23.68@75℃				
	23.53@80℃				
	23.40@85℃				

Anand model can describe the viscoplastic behavior of SnPb solder accurately because the model has two basic characteristics[9]. First, there is no definition of yield surface in Anand model, and there is no requirement of loading and unloading criteria in the process of loading and unloading. Second, a single internal variable (deformation impedance s, with stress dimension) is used in Anand model to represent the average impedance of the isotropic reinforcement to the macroscopic plastic flow, which is related to solid solution strengthening, dislocation density and grain size effect. Hence, Anand model can describe viscoplastic behavior of SnPb solder not only briefly but also accurately.

In this study, we used Anand viscoplastic model[7] to describe the thermomechanical behavior of SnPb solder balls. The parameters of Anand model are shown in Table III.

TABLE III. PARAMETERS OF ANAND EQUATION[7]

Parameter	Units	Value
S_0	MPa	16
Q/k	K	10560
A	s^{-1}	3250000
ξ	-	5
m	-	0.62
$s^{\wedge}$	MPa	45
h_0	GPa	800
n	-	0.032
a	-	3.37

The Anand model contains nine parameters in total. The initial framework without modifications of Anand model remains as follows:

$$\sigma = cs \#(1)$$

where, s denotes the deformation resistance with the dimension of stress, σ the values of stress, and c is defined by a function as follows:

$$c = \frac{1}{\xi}\sinh^{-1}\left[\left(\frac{\dot{\varepsilon}_p}{A}\exp\left(\frac{Q}{RT}\right)\right)^m\right]\#(2)$$

The steady-state plastic flow measured by $\dot{\varepsilon}_p$ is defined as follows:

$$\dot{\mathcal{E}}_p = A\exp\left(-\frac{Q}{RT}\right)\left[\sinh\left(\xi\frac{\sigma}{s}\right)\right]^{1/m}\#(3)$$

where, A denotes a material constant in this equation, $\dot{\varepsilon}_p$ is the inelastic strain rate, m the exponent of strain sensitivity, T the absolute temperature, ξ the stress multiplier, σ the equivalent stress of the steady plastic flow, Q the activation energy, R the universal gas constant. The deformation resistance s can be written as

$$\dot{s} = h(\sigma, s, T)\dot{\varepsilon}_p, \#(4)$$

where, the function h is defined as

$$h = h_0\left|1 - \frac{s}{s^*}\right|^a sign\left(1 - \frac{s}{s^*}\right)\#(5)$$

with

$$s^* = \hat{s}\left[\frac{1}{A}\dot{\varepsilon}_p\exp\left(\frac{Q}{RT}\right)\right]^n\#(6)$$

Where, s^* denotes the saturation value of internal variables, a is the exponent related to hardening and softening, h_0 the deformation hardening–softening constant. $\hat{s}$ and n are the coefficient of deformation resistance and the rate sensitivity exponent, respectively.

C. Loading and Boundary Conditions

In this study, the stress-free condition was assumed to be at 25 ℃ and interface contacts are tied for each material. The profile of temperature in thermal cycling, from -55 ℃ to 125 ℃ with dwell time 10 minutes, is illustrated in Fig. 2. The increase and decrease rate of temperature are 20℃ per minute. The length of each cycle is 38 min and the temperature of the whole device changes evenly during the rising and the dropping process of temperature.

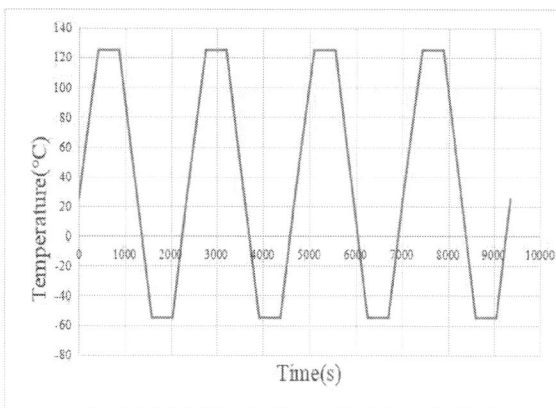

Fig. 2. Temperature cycling profile

The assembled PoP is assumed on a smooth plane without friction. Hence, three nodes on the bottom of PCB are respectively constrained in directions to eliminate the rigid translations and rotations.

D. Fatigue model

An energy-based empirical Darveaux lifetime equation, which considers the processes of crack initiation and crack growth respectively, was employed to forecast fatigue life of SnPb solder balls in this PoP device. The fatigue model estimates lifetime of the crack initiation by equation (7) and the lifetime of crack growth by equation (8), respectively.

$$N_0 = K_1 (\Delta W_{ave})^{K_2} \#(1)$$

$$\frac{da}{dN} = K_3 (\Delta W_{ave})^{K_4} \#(2)$$

$$N_f = N_0 + \frac{a}{da/dN} \#(3)$$

where, N_f is the characteristic life at 63.2% failure rate, ΔW_{ave} refers to the average inelastic strain energy density increment per cycle, N_0 is the fatigue life of crack initiation process, a is the diameter of solder ball at the interface, which is 0.325mm, $\frac{da}{dN}$ represents the speed of crack growth, $K_1 \sim K_4$ are parameters that are obtained from experiments where the crack growth rate is measured, which are listed in Table IV. Noting that the parameters $K_1 \sim K_4$ are expressed in imperial units, the unit conversion should be done before calculation.

TABLE IV.   PARAMETER K1~K4

K_1	K_2	K_3	K_4
22400	-1.52	5.86E-7	0.98

III. RESULTS

The stress distribution of the top and bottom layer solder ball array is shown in Fig. 3 and Fig. 4. According to the stress distribution of solder ball arrays, the maximum Von Mises stress in the upper solder ball layer is much smaller than that in the bottom layer of solder ball array. The stress distribution in the critical solder ball, which is at the corner of the bottom layer solder ball array, is illustrated in Fig. 5. Moreover, the maximum Mises stress in edge of interface between solder ball and ceramic substrate is higher, which is denoted in Fig. 5. The results indicate that the most vulnerable solder ball is at the corner of bottom layer, and

crack will initiate around the interface of the ceramic substrate and the critical solder balls.

The inelastic energy density, including plastic strain and creep strain energy density, was measured from the outer circle elements of the interface between critical solder ball and ceramic substrate in the bottom layer of PoP device so as to forecast the fatigue life of solder ball. The energy accumulation as a function of time in thermal cycles is shown in Fig. 6. The average inelastic energy densities for the first four cycles are listed in Table V, respectively.

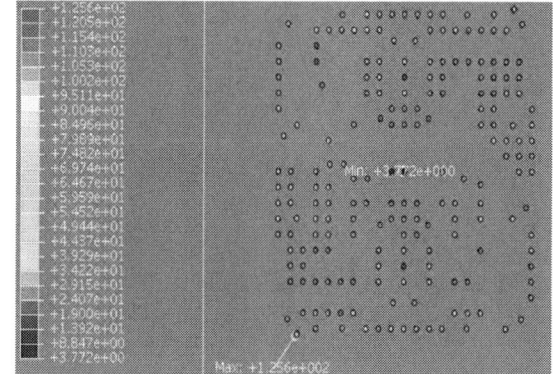

Fig. 3. Stress distributions of solder ball array in the top layer

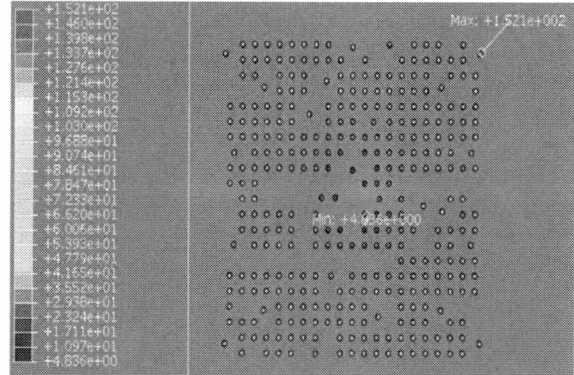

Fig. 4. Stress distributions of solder ball array in the bottom layer

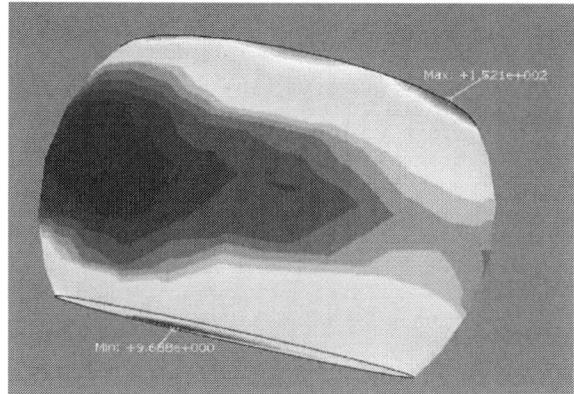

Fig. 5. Stress distributions of a solder ball at the corner of bottom

TABLE V.   MAXIMUM INELASTIC ENERGY DENSITY

W_1(MPa)	W_2(MPa)	W_3(MPa)	W_4(MPa)
3.016443	6.339475	9.429836	12.341262

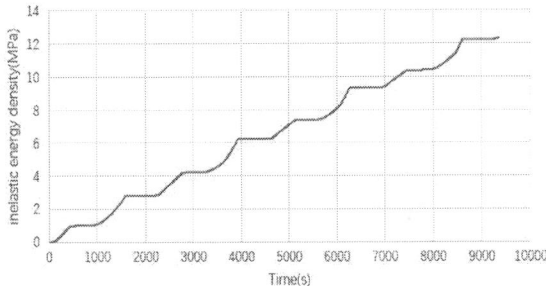

Fig. 6. Average inelastic energy density of the critical solder ball in the bottom layer as a function of time

Although inelastic strain energy density accumulates after each thermal cycle, its increment, ΔW_{ave}, converges to a stable value after a few thermal cycles. In this case, the ΔW_{ave} is ~3.09 MPa. According to equations (7)-(9), the fatigue life of crack initiation N_0 and the characteristic life N_f can be calculated as 2.10 and 57.25, respectively.

IV. DISCUSSION

To identify the contribution of different kinds of energy densities, the average creep and plastic energy density versus the time is plotted in Fig. 7. The creep energy density is higher than the plastic energy density, and the increment of two energy densities converged to a stable value after several temperature cycles. In general, the creep energy density is higher than the plastic energy density in each element that experienced thermal cycles, which is demonstrated in Fig. 8. Thereby, in this study, the creep energy accumulated under thermal cycle contributes more to the damage of the solder ball than the plastic energy does.

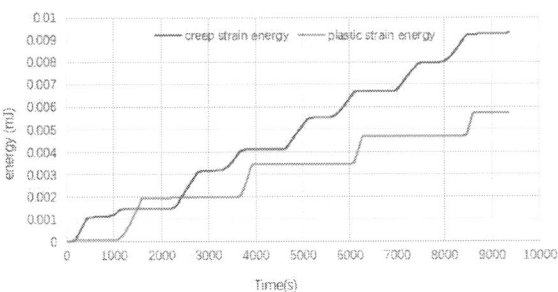

Fig. 7. Average creep and plastic energy of each element of the critical solder ball in the bottom layer as a function of time

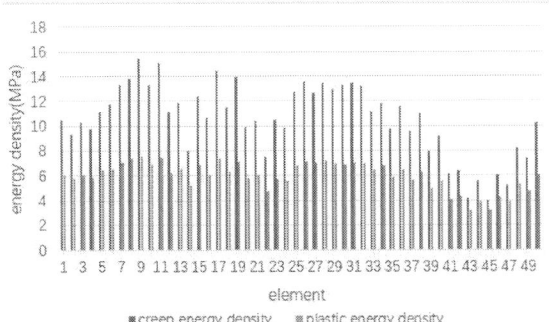

Fig. 8. Creep and plastic energy density of each element of the critical solder ball in the bottom layer

Because the stresses of the top layer solder balls are smaller than that of the bottom layer solder balls, it is intuitive to assume that the fatigue life of the critical solder ball in the top layer should be higher than that of the counterpart in the bottom layer. Using the same method of fatigue life prediction, the fatigue life of the critical solder ball in the top layer can be calculated. The fatigue life of crack initiation N_0^{top} and the characteristic fatigue life N_f^{top} of the critical solder ball are 25.49 and 305.34, respectively.

The results can be understood because the mismatch of coefficient of thermal expansion (CTE) between the PCB and the ceramic is much larger than that between the top and bottom ceramic components. This difference leads to the fact that the thermal strain, stress and inelastic energy density are much higher in the solder balls in the bottom layer, especially the one near the corner. Thus, the average increment of the maximum inelastic energy density per cycle ΔW_{ave} of the critical solder ball in the bottom layer is higher, the fatigue life is shorter compared with the critical solder ball in the top layer.

In order to compare with the 2D packaging case, a model with only the bottom layer of PoP was constructed, in which all the material properties, component sizes, boundary conditions and meshes were kept in the same as the counterpart of the ceramic PoP device. After the same thermal loading is applied, the fatigue life of the critical solder balls is predicted using the same fatigue life prediction method. The results showed that the fatigue life of crack initiation N_0^{scp} and the characteristic life N_f^{scp} for the critical solder ball are 1.6 and 47.9, respectively.

The three cases, the critical solder ball in the top layer or in the bottom layer, and the case when the stacked top component is removed, are compared. The average increment of the maximum inelastic energy densities per cycle, ΔW_{ave}, in the outer circle elements of the critical solder ball for the three cases are plotted in Fig. 9. The higher values of the density are, the shorter fatigue life will be. The fatigue life comparison for the three cases is demonstrated in Fig. 10. The fatigue life of the critical solder ball in the bottom layer of PoP is slightly better than that of the third case without the top component.

When the stacked top component was removed, the stiffness of the whole structure is decreased, but the stresses in the solder balls that connected the PCB are increased. The rising of the stress improves the inelastic energy and reduces the fatigue life of the critical solder ball. This result indicates that the stacked 3D packaging, e.g. PoP, may not suffer serious solder ball fatigue problems compared with the 2D packaging under thermal cycling loads.

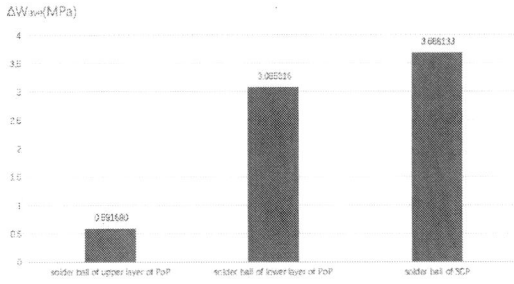

Fig. 9. The comparison of ΔW_{ave} for the three cases

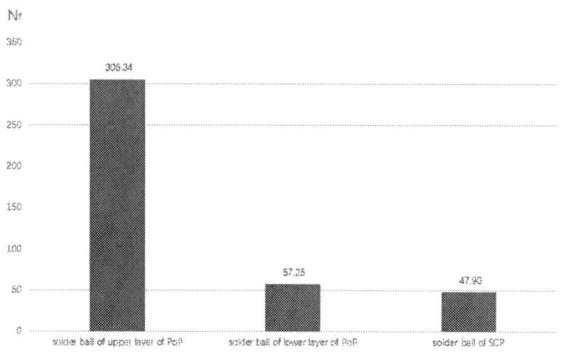

Fig. 10. The fatigue life comparison for the three cases

V. CONCLUSIONS

The finite element analysis was applied to a ceramic PoP device assembly with Sn63Pb37 solder using Anand model and the fatigue life of the critical solder ball was predicted through Darveaux-based lifetime equation.

The FEA showed that stresses in the solder balls of bottom layer in the ceramic PoP device are much higher than that in the solder balls of top layer. The critical solder balls are located near the corner of the solder ball array in each layer. A fracture is likely to take place along the interface between the PCB and the solder ball. The further analysis revealed that the creep strain energy has more impact on the fatigue life of the solder ball compared with the plastic strain energy. The predicted fatigue life of the critical solder ball in the top layer is about five folds than that in the bottom layer. It's indicated that improving reliability in bottom layer solder balls is a key point in the PoP device.

Although the stacked components in PoP make the structure more complex, the thermal fatigue life of the critical solder ball is not severely degraded when the top stacked component is added. Presumably the reason is the PoP packaging improves the structural stiffness, which may alleviate the thermal fatigue damages of solder balls.

ACKNOWLEDGEMENT

The authors gratefully acknowledge the National Nature Science Foundation of China (Grant No.61774044).

REFERENCES

[1] H. Jiaoying, C. Yang, and G. Cheng, "Fatigue failure of microelectronic packaging solder joints: a review," *Electronic Components & Materials*, vol. 39, no. 10, pp. 11-16,24, 2020.

[2] B. Vandevelde, M. Gonzalez, P. Limaye, P. Ratchev, and E. Beyne, "Thermal cycling reliability of SnAgCu and SnPb solder joints: A comparison for several IC-packages," *Microelectronics Reliability*, vol. 47, no. 2–3, pp. 259–265, Mar. 2007, doi: 10.1016/j.microrel.2006.09.034.

[3] A. Surendar, K. H. Kishore, M. Kavitha, A. Z. Ibatova, and V. Samavatian, "Effects of Thermo-Mechanical Fatigue and Low Cycle Fatigue Interaction on Performance of Solder Joints," *IEEE Transactions on Device and Materials Reliability*, vol. 18, no. 4, pp. 606–612, Dec. 2018, doi: 10.1109/TDMR.2018.2879123.

[4] C.-S. Lau, M. Z. Abdullah, and F. C. Ani, "Computational fluid dynamic and thermal analysis for BGA assembly during forced convection reflow soldering process," *Solder. Surf. Mt. Technol.*, vol. 24, no. 2, pp. 77–91, 2012, doi: 10.1108/09540911211214659.

[5] Y. Wan, H. Huang, and M. Pecht, "Thermal Fatigue Reliability Analysis and Structural Optimization Based on a Robust Method for Microelectronics FBGA Packages," *IEEE Transactions on Device and Materials Reliability*, vol. 15, no. 2, pp. 206–213, Jun. 2015, doi: 10.1109/TDMR.2015.2417888.

[6] J. Gu, J. Lin, Y. Lei, and H. Fu, "Experimental analysis of Sn-3.0Ag-0.5Cu solder joint board-level drop/vibration impact failure models after thermal/isothermal cycling," *Microelectronics Reliability*, vol. 80, pp. 29–36, 2018, doi: 10.1016/j.microrel.2017.10.014.

[7] X. Long, Z. Chen, W. Wang, Y. Fu, and Y. Wu, "Parameterized Anand constitutive model under a wide range of temperature and strain rate: experimental and theoretical studies," *J. Mater. Sci.*, vol. 55, no. 24, pp. 10811–10823, Aug. 2020, doi: 10.1007/s10853-020-04689-1.

[8] R. Darveaux, "Effect of simulation methodology on solder joint crack growth correlation and fatigue life prediction," *J. Electron. Packag.*, vol. 124, no. 3, pp. 147–154, Sep. 2002, doi: 10.1115/1.1413764.

[9] S. B. Brown, K. H. Kim, and L. Anand, "An internal variable constitutive model for hot working of metals," *International Journal of Plasticity*, vol. 5, no. 2, pp. 95–130, Jan. 1989, doi: 10.1016/0749-6419(89)90025-9.

Design and fabrication of a soft micro-actuator based on distributed magnetic composite

Langkun Wang
State Key Laboratory of High Performance Complex Manufacturing. College of Mechanical and Electrical Engineering, Central South University
Changsha 410083, China
1043840074@qq.com

Shimei Liu
Changsha Research Institute of Mining and Metallurgy Co.,Ltd.
Changsha 410083, China
358971237@qq.com

Hu He
State Key Laboratory of High Performance Complex Manufacturing. College of Mechanical and Electrical Engineering, Central South University
Changsha 410083, China
hehu.mech@csu.edu.cn

Abstract—Benefit from the development of micro/nano manufacturing and flexible electronics technology, soft micro-actuators which could be potentially applied in drug delivering and environment monitoring have developed rapidly in recent years. Due to the simplicity of fabrication process and flexibility of control strategies, magnetic-controlled soft actuators based on magnetic-responsive composites have attracted intensive interests. In literature, lots of works focus on the structure and manufacturing process of the micro-actuators in order to enhance their driving performance. However, as fully soft micro-actuators, the driving efficiency of the soft parts is comparatively low, and the stability is poor in a liquid environment. In this work, a micro-actuator combined with rigid and flexible parts was presented to improve the driving force as well as the stability. We proposed a method of using substrates with higher stiffness to manufacture the nonuniform stiffness composites. The magnetic unit is made of magnetic particles mixed with rigid substrate, and then embedded in or connected with the flexible body. Following this strategy, the rigid sections take responsible for quick magnetic response while the soft area have desired deformation performance. Firstly, the magnetic response of magnetic composite made of epoxy resin was analyzed. Then, hysteresis loop test was deployed to investigate the effect of magnetic particle concentration on magnetic properties. Furthermore, the deformation of the composite actuator is analyzed by simulation with respect to the different positions and quantity of the rigid parts. Finally, the fabricated soft micro-actuator is controlled by external magnetic field to demonstrate the potential applications. It was believed that the combination of rigid and flexible components in micro electro mechanical system is beneficial to improve the driving ability and broaden the application range of micro actuators.

Keywords—Soft robotics; Rigid parts; Structural design; Non-uniform stiffness

I. INTRODUCTION

In recent years, with the rapid development of flexible micro/nano manufacturing technology and the emergence of innovations based on materials, molding processes, driving and control methods, micro soft untethered robots have emerged, which have potential applications in many fields, such as targeted transportation, environmental detection, biotechnology, and minimally invasive medicine[1]. They are small, easily manipulated by changing the shape of their flexible bodies, and can enter narrow spaces and complete certain tasks. The driving mode of this kind of soft robot mainly depends on the intelligent materials, which can induce the driving according to the changes of the external field,

thereby completing some basic functions[2].

External driving fields, such as electric[3], magnetic[4], optical[5], thermal[6], etc.,are widely deployed to control micro soft robots. In particular, magnetic field being capable of wireless control, responsive, and low requirement for magnetic field strength that can be safely penetrated in biological environment[7] has progressed rapidly to realize micro magnetic-driving soft actuators performing specific functions in complex environments[8].The prevailing magnetic micro-actuators are mainly made by mixing of hard magnetic particles with flexible substrates and program the local magnetization to accomplish predictable deformation movements. Joyee *et al.*[9] developed a biomimetic inchworm-like robot that moves by creeping and bending of the body. Kim *et al.*[10] prepared a small soft continuum robot for active steering and navigation, which is magnetized at the nozzle and can move forward in a complex environment under the control of a small external magnetic field. Several researchers managed to separate inductive components from moving parts. Du *et al.*[11] proposed a untethered millimeter scale robot with a head as the magnetic driver and a tail as the pure flexible material. The robot can be controlled, driven and sensed by its magnetic composite cooperated with intelligent flexible material. Jeon *et al.*[12] built a micro navigation robot with a small magnet embedded in the PDMS beam, and carried out its steering control and 3D tracking. However, several problems that hindered the development and wide application still exist in this field. Firstly, the modulus and remanence of the composites increase as the increase of the mass fraction of magnetic particles for homogeneous magnetic composites. Thus, it is infeasible to improve the mechanical and magnetic properties simultaneously with uniform structure. Moreover, the mass fraction of magnetic particles in the flexible substrate is generally less 50%, thus, a higher magnetic field intensity is required in the magnetization process to make it remain a higher remanence after demagnetization[13]. Finally, for the homogenized composite magnetic system, it is difficult to preset different magnetization directions in different regions and maintain high accuracy for complex programming design.

Here, we proposed a magnetic unit based on a hard substrate, fabricated by mixing and curing magnetic particles with a hard substrate, as a driving part in a microrobot system, and connect these magnetic units with flexible films to form the distributed magnetic composite system. The magnetic units are stimulated by the external field while the flexible film between the magnetic units transforms the load of the magnetic response into deformation, enabling the controllable movement of the entire composite structure. The deformation effect of this distributed magnetic composite is qualitatively analyzed by simulation model. The test samples were made and the actuation experiments were conducted. Experiments show that the hard magnetic unit can be magnetized under relatively small magnetizing field, driven by response excitation under minimal external magnetic field intensity, and has programming ability of complex magnetization.

II. METERIALS AND METHODS

A. Selection of hard substrate

Thermoset polymer is a kind of high molecular polymer material. The molecular chains are linked together by chemical crosslinking to form a rigid three-dimensional network structure. As the most widely used thermoset polymer, epoxy resin has excellent comprehensive properties, including high strength, good heat resistance, excellent electrical performance, corrosion resistance, aging resistance, good dimensional stability, etc. Epoxy resin has low density, simple process of curing and can hold a higher filler concentration while maintaining certain rigidity and stable shape. Therefore, in this study, we chose epoxy resin as the hard substrate.

B. Preparation of magnetic composite ink

The magnetic composite ink is needed as basic material. As shown in Fig. 1(a), it was obtained by mixing the resin matrix with magnetic particles(Neodymium iron boron particles, average particle size:38 μm) and ultrasonically stirring for 10 min, then adding curing agent(weight ratio of the resin matrix and curing agent as 4:1), and at last placing it in a vacuum deaeration machine and stirring at 2000rpm for another 10 min.

C. Fabrication and magnetization of the magnetic units

As a magnetic unit that receiving external field response, its shape should be simple and its size should be small, and its driving mode should be designed by combing the deformation effect of multiple magnetic units. Therefore, we designed a rectangular thin block with the size of 5*5* 0.5mm. In order to shape the size and structure of the hard magnetic unit, a soft mold needs to be made. Fig. 1(b) shows the manufacturing process of magnetic units. First, a CAD model is designed using Solidworks, and the hard mold is fabricated by a stereo lithography apparatus (SLA) printer. After mixing with its curing agent, silicon rubber (Ecoflex 00-30) is poured into the mold. After 60 minutes of heating in the oven at 55°C, the soft mold can be peeled off from the hard one. Next, pour the magnetic composite ink into the indentation of the soft mold, scrape off the excess ink with a scraper, and keep the surface of the composite ink flat. After 60 minutes of heating in the oven at 60°C, the hard magnetic units can be removed from the soft mold. Hard magnetic units with mass ratios of 20% to 80% have been prepared by this method. They will be experimented and discussed in later chapters.

The polarity direction of magnetic particles in epoxy resin determines the magnetization direction of the magnetic unit. In the absence of an external magnetic field, neodymium iron boron particles are randomly dispersed in the epoxy resin, and their magnetic moment varies randomly due to thermal fluctuations. Therefore, as shown in Fig. 1(c), the magnetic units were magnetized by placing on the surface of a permanent magnet (Sintered NdFeB N52, surface magnetic flux density is about 550mT) in the length direction. Finally, in Fig. 1(d), a mold was customized to join magnetic unit with flexible film, the magnetic units were placed in a certain direction in the mold, then Ecoflex was added between them, excess solution was removed with scraper, and heat it in a 60°C oven for 60 minutes. After the flexible film cured, the whole actuator can be removed.

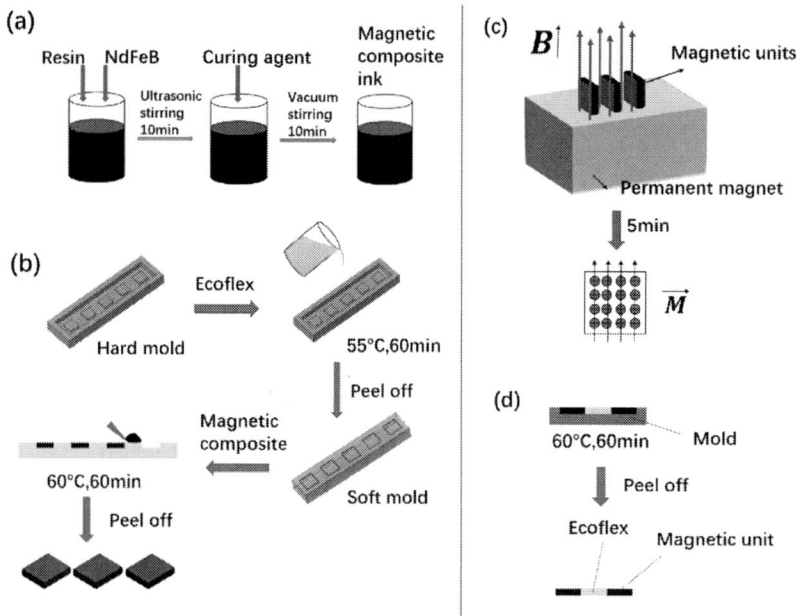

Figure 1. Design and fabrication of the magnetic units. a) Preparation of magnetic composite ink. b) Schematic diagram of the fabrication process of the magnetic units. c) Schematic diagram of the simple magnetizing method. d) Schematic diagram of the assembly method of magnetic unit and flexible film.

III. RESULT AND DISCUSSION

A. Theory of actuation

While the magnetic units are locating in the range of magnetic field, they experience force and torque given by[14]:

$$F = M \cdot \nabla B \qquad (1)$$
$$T = M \times B \qquad (2)$$

where F and T are the force and torque vectors respectively. M is the dipole moment of the magnetic units, and B is the magnetic field formed by the external permanent magnet. Since low magnetic field strength, we mainly consider the effect of magnetic moment. The magnetic unit can be approximated by a collection of magnetic dipoles which receive and transforms the excitation from the external field into force and torque to drive the magnetic unit. First, the magnetic unit needs to be magnetized in a strong magnetic field to obtain polarity. Because the Young's modulus of the flexible film is much smaller than that of the magnetic unit, the load is mainly borne by the flexible film and the deformation is produced. The distributed magnetic composite will respond quickly and make predictable actions when the external magnetic field changes. Following this principle demonstrated in Fig. 2(a)(b), predictable motions can be easily made to drive the entire structure.

B. Characterization of materials

We comprehensively considered various properties of hard magnetic units with different magnetic particle concentrations, and found that the higher the mass ratio of magnetic particles, the better the magnetic response effect of the magnetic units. However, it will reduce the fluidity of the magnetic composite ink, resulting in a rough surface of the magnetic units, affecting the quality of the product. Taking the above considerations together, we chose the magnetic composite with 75% mass fraction to prepare the magnetic unit. In the following, the magnetic unit based on 75% mass fraction was characterized.

1) Mechanical characterization

The mechanical property of the fabricated magnetic unit was tested on a universal material testing machine. The samples were molding into dog bone–shaped specimens and were uniaxially stretched at a low stretching speed of 0.5mm/min. From the stress-strain curve in Fig. 3, the tensile strength, yield strength and Young's modulus of the material are 19.35MPa, 1.24 MPa and 3921.01Mpa, respectively.

2) Magnetic characterization

The magnetization of the magnetic unit was measured by the vibrating sample magnetometer (VSM, maximum magnetic field as 3T). The magnetic hysteresis loops in Fig. 4 were obtained, indicating the magnetization and coercivity for the magnetic materials to be 60.3emu/g and 874.9mT respectively.

3) SEM observation

As shown in Fig. 5(a)(b), the surface as well as the sections morphology of the samples were observed by SEM. From the images, it can be concluded that the Neodymium magnetic particles differ in size but are well distributed in the resin.

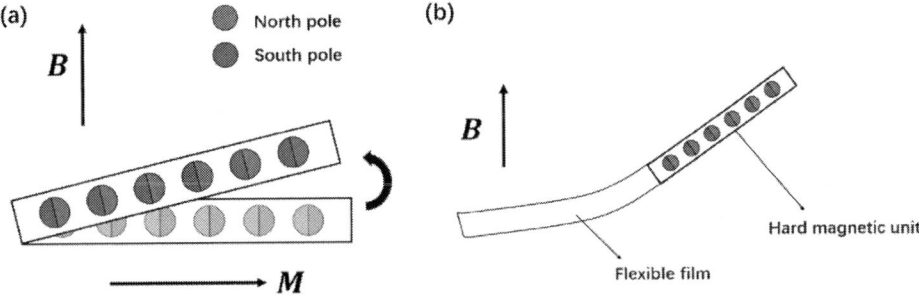

Figure 2. Schematic diagram of a) magnetic unit after magnetizing, b) soft micro-actuator driven by external magnetic field.

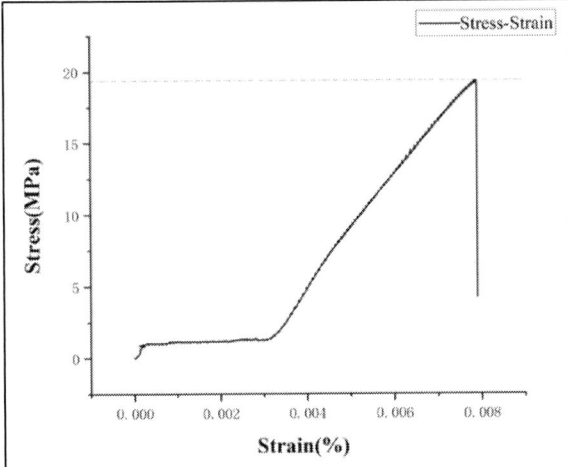

Figure 3. Stress-strain curve of the sample.

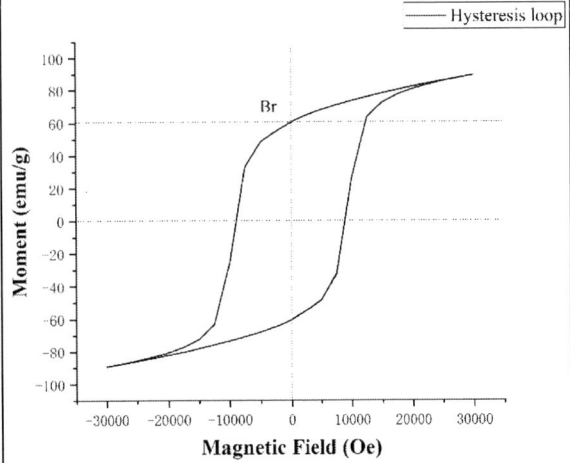

Figure 4. Hysteresis loop of magnetic units.

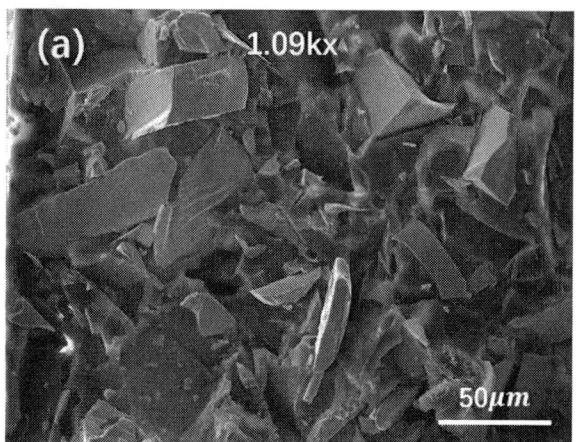
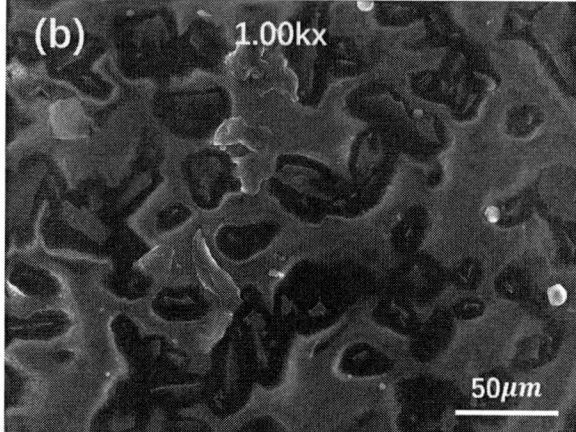

Figure 5. Distribution of magnetic particles in magnetic units: a) Fracture, b) surface morphology, observed by SEM.

C. Actuation test and demonstration

The samples were all fabricated to the same size. As shown in Fig. 6(a), a height-adjustable bracket is set, a piece of glass is placed on the bracket, a permanent magnet (PM, surface magnetic field strength as 100 mT) is placed on the desktop, and the magnetic flux density at the glass plane is adjusted by changing the height of the bracket, since the magnetic flux density on the glass surface is relatively low, we can consider it as a uniform field.

Fig. 6(b) shows a performance measure for actuation: the minimum magnetic flux density that keeps the magnetic unit in a stable upright position. The bracket height was adjusted from low to high, slow down the adjustment when the magnetic unit progressively destabilized, and the intensity of magnetic sensing at the surface of the glass plate was recorded using a gauss meter (Table I). From the experimental results, a magnetic unit with a higher content of magnetic particles has a smaller "stable magnetic field", under which the magnetic unit will move together with the fine-tuning of the direction of the external magnetic field.

Finally, in Fig. 7(a), we used the same equipment to carry out experiments on a micro-actuator containing 4 magnetic units. Fig. 7(b)(c) shows that the desired deformation results appeared when the magnetic flux density reached about 12mT.

D. Finite element simulation

To predict the magnetically actuated deformation trend of the distributed magnetic composite material, we designed and built a simulation model by using the COMSOL Multiphysics. In order to simplify the model, the magnetic units and the nonmagnetic films of soft micro-actuator were assumed to be isotropic, and the model is derived under two-dimensional conditions. The integral of Maxwell's stress tensor is used to calculate the electromagnetic force of the magnetic units, and the electromagnetic force is given to the solid mechanics module as a volume force to calculate the overall deformation. For non-magnetic thin films, a new Hooke energy function coupled with magnetic potential is used to describe the nonlinear deformation under magnetic actuation of the system. Due to the high nonlinearity of the coupling of hyper-elastic material with the magnetic field-driven load, we only make qualitative analysis on the simulation model.

First, in Fig. 8(a) we established a model of the permanent magnet in the air. The distribution of magnetic induction lines indicates that magnetic flux density reaches its maximum at the edge and corner of the permanent magnet, but the magnetic field is closest to the static field at the center of the surface. For magnetic units, the materials' shear moduli, Poisson's ratios, and remanence were set respectively, magnetic unit cells were modeled with dimensions 5 mm by 5mm by 0.5mm, and they were connected by flexible film with 0.5mm thickness. The external magnetic field intensity was defined by changing the remanence of a permanent magnet in order to observe the deformation trend of this actuator at different field intensity. We bulit models with one(Fig. 8b) and two magnetic units(Fig. 8c) to demonstrate the simulation.

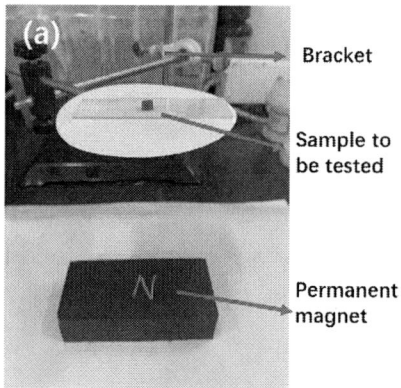

Bracket

Sample to be tested

Permanent magnet

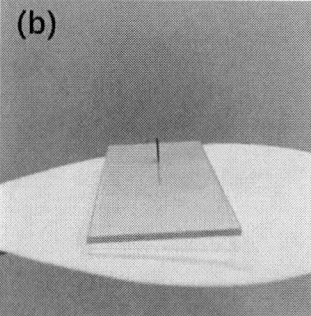

Figure 6. a) Platform for testing driving effects. b) The magnetic unit remains upright.

TABLE I.   TEST RESULT OF MINIMUM MAGNETIC FIELD

Mass fraction（%）	20	30	40	50	60	70	75	80
Minimum（mT）	15.81	6.94	5.43	4.25	3.59	3.31	3.26	1.80

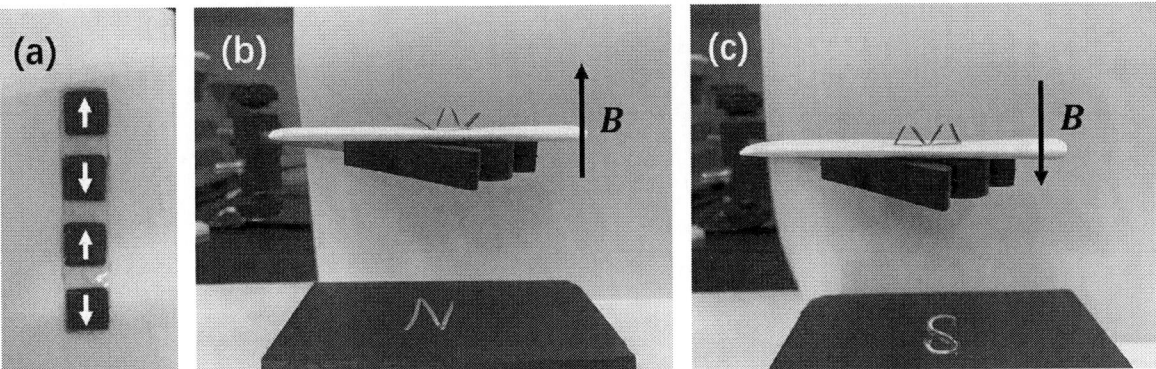

Figure 7. a) Micro-actuator with 4 magnetic units, white arrows indicate the initial magnetizing directions (Top view). b) deformation of micro-actuator under an upward, c) downward magnetic field (Side view).

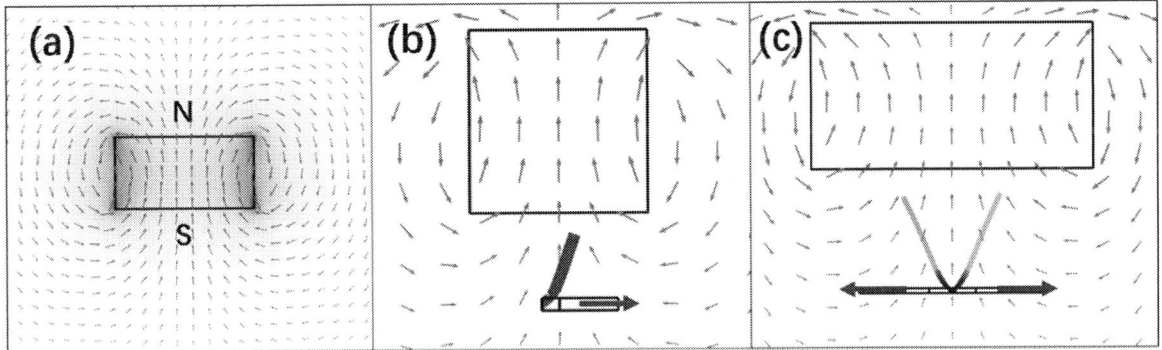

Figure 8. a) Distribution of magnetic induction lines in air for permanent magnet. Deformation of micro-actuator with b) single magnetic unit, c) two magnetic units under magnetic field generated by permanent magnet, black arrows indicate the initial magnetizing directions.

IV. CONCLUSION

In this work, we designed and manufactured a distributed structure of magnetic units connected with flexible film, which separated the magnetic components and flexible components. The magnetic components receive the external magnetic field excitation, generate instantaneous magnetic torque to promote the overall rotation, flexible components bears bending deformation, provides a variety of possible motion modes. The magnetic unit based on hard substrate greatly improves its response ability in the magnetic field by increasing the mass fraction of magnetic particles. Compared with the actuators based on integrated homogeneous magnetic soft composite, its driving performance and control stability are better. In this study, the experiment verified that the magnetic unit only needs to be magnetized for 5 minutes on the surface of a 500 mT permanent magnet to meet the basic control requirements, without the use of expensive high-power magnetization equipment. After magnetizing, the magnetic unit can be manipulated under a weak magnetic field intensity (less than 4mT).Moreover, the magnetization programming process had been simplified due to the independence of the magnetic unit, it can be magnetized separately, magnetization direction of each magnetic unit, and finally connected with the flexible film in a certain way. Through the above efforts, it is expected to promote the development of rigid-flexible composite structure and improve the comprehensive especially the driving performance of flexible micro actuator. Following works like reducing the size of the magnetic unit, improving the complexity of mechanical structure design and improving simulation quality will be carried out in the future. We believe that the experimental attempts of distributed magnetic composite and combination of rigid and flexible components in MEMS are beneficial to improve the driving ability of micro actuators and broaden the application range of micro actuators.

ACKNOWLEDGMENT

This work was supported by Natural Science Foundation of Hunan Province (2020JJ5728), Innovation-Driven Project of Central South University (2020CX05), National Natural Science Foundation of China (U20A6004), and Postgraduate Research Innovation of Central South University (2021zzts0628).

REFERENCES

[1] J. Hwang, J.-y. Kim, H. Choi, A review of magnetic actuation systems and magnetically actuated guidewire- and catheter-based microrobots for vascular interventions, Intelligent Service Robotics 13(1) (2020) 1-14.

[2] H.W. Huang, M.S. Sakar, A.J. Petruska, S. Pane, B.J. Nelson, Soft micromachines with programmable motility and morphology, Nat Commun 7 (2016) 12263.

[3] R.R. Ribeiro de Almeida, L.R. Evangelista, E.K. Lenzi, R.S. Zola, A. Jákli, Electrical transport properties and fractional dynamics of twist-bend nematic liquid crystal phase, Communications in Nonlinear Science and Numerical Simulation 70 (2019) 248-256.

[4] J.M. Silveyra, E. Ferrara, D.L. Huber, T.C. Monson, Soft magnetic materials for a sustainable and electrified world, Science 362(6413) (2018) :eaao0195.

[5] J.S. Sodhi, P.R. Cruz, I.J. Rao, Inhomogeneous deformations of Light Activated Shape Memory Polymers, International Journal of Engineering Science 89 (2015) 1-17.

[6] Q. Zhang, L. Liu, C. Pan, D. Li, G. Gai, Thermally sensitive, adhesive, injectable, multiwalled carbon nanotube covalently reinforced polymer conductors with self-healing capabilities, Journal of Materials Chemistry C 6(7) (2018) 1746-1752.

[7] S. Jeon, S. Kim, S. Ha, S. Lee, E. Kim, S.Y. Kim, S.H. Park, J.H. Jeon, S.W. Kim, C. Moon, B.J. Nelson, J.-y. Kim, S.-W. Yu, H. Choi, Magnetically actuated microrobots as a platform for stem cell transplantation, Science Robotics 4(30) (2019) :eaav4317-.

[8] X.-Z. Chen, M. Hoop, F. Mushtaq, E. Siringil, C. Hu, B.J. Nelson, S. Pané, Recent developments in magnetically driven micro- and nanorobots, Applied Materials Today 9 (2017) 37-48.

[9] E.B. Joyee, Y. Pan, A Fully Three-Dimensional Printed Inchworm-Inspired Soft Robot with Magnetic Actuation, Soft Robot 6(3) (2019) 333-345.

[10] Y. Kim, G.A. Parada, S. Liu, X. Zhao, Ferromagnetic soft continuum robots, Sci Robot 4(33) (2019) :eaax7329.

[11] X. Du, H. Cui, T. Xu, C. Huang, Y. Wang, Q. Zhao, Y. Xu, X. Wu, Reconfiguration, Camouflage, and Color-Shifting for Bioinspired Adaptive Hydrogel-Based Millirobots, Advanced Functional Materials (2020), 30(10).

[12] A.K. Hoshiar, S. Jeon, K. Kim, S. Lee, J.-y. Kim, H. Choi, Steering Algorithm for a Flexible Microrobot to Enhance Guidewire Control in a Coronary Angioplasty Application, Micromachines (2018), 9(12).

[13] L. Descamps, S. Mekkaoui, M.-C. Audry, A.-L. Deman, D. Le Roy, Optimized process for the fabrication of PDMS membranes integrating permanent micro-magnet arrays, AIP Advances 10(1) (2020) :01521.

[14] J. Park, C. Lee, J. Lee, J.I. Ha, H. Choi, H.C. Jin, Magnetically Actuated Forward-Looking Interventional Ultrasound Imaging: Feasibility Studies, IEEE Transactions on Biomedical Engineering 67(6) (2020) 1797-1805.

Electroless Copper Deposition with Pyramidal Micro-cones Morphology for Low-temperature Cu-Cu Bump Interconnections

Yiming Chen
School of Mechanical and Electrical Engineering, Central South University
ChangSha, China
cym310@csu.edu.cn

Yiqiao Wei
School of Mechanical and Electrical Engineering, Central South University
Changsha, China
1164912932@qq.com

Zhuo Chen
State Key Laboratory of High Performance Complex Manufacturing, Central South University
School of Mechanical and Electrical Engineering, Central South University
Changsha, China
zhuochen@csu.edu.cn

Wenjing Zhang
School of Aerospace Engineering and Applied Mechanics, Tongji University
Shang Hai, China
zhangwenjing@tongji.edu.cn

Fuliang Wang
State Key Laboratory of High Performance Complex Manufacturing, Central South University
School of Mechanical and Electrical Engineering, Central South University
Changsha, China
wangfuliang@csu.edu.cn

Wenhui Zhu
State Key Laboratory of High Performance Complex Manufacturing, Central South University
School of Mechanical and Electrical Engineering, Central South University
ChangSha, China
zhuwenhui@csu.edu.cn

Abstract—Interconnection between IC chip and substrate is the core function of microelectronic packaging. Direct copper-to-copper interconnection has received extensive attention owing to its superior electrical and mechanical properties. At present, the bonding temperature of Cu-Cu interconnection is generally as high as 300℃-400℃, which is unfavorable for the integrity of thermally fragile ICs. There have been methods of forming non-solder metal interconnections between chip- and substrate-side metallizations using electroless Cu or Ni deposition in recent years, which provide strong bonding strength through low-temperature metal growth process and interfacial merging. However, the possible presence of interfacial voids caused by insufficient cupric ion supply and peripheral closure under high deposition rate has limited the practical productivity of electroless interconnection to a low value. In this study, in pursuit of a fast and reliable electroless interconnection process, we propose the method of depositing Cu with a special micro-cone morphology for forming all-copper interconnections at the rate of >15μm/h. By using certain crystallization modifier, the growth of Cu deposit can be tuned to highly anisotropic such that sharp and fast-growing micro-cones can be formed and the merging of opposite Cu bumps can take place through simultaneous fusing at multiple locations, leaving sufficient space for the electroless deposition solution to flow through to allow subsequent merging. A microfluidic channel device was utilized to pump the plating solution, which enhanced the ion supply and guaranteed high growth rate of Cu. The cross-sectional morphology of the joints was examined microscopically, the electrical performance, as well as the mechanical strength of the joints, were also evaluated. It was found that high shear force joints of 1N per joint was obtained through deposition time of 1h. The merging mechanism of the micro-cone-morphology Cu layer is discussed. This method is therefore demonstrated as a potential candidate for the high-quality fine-pitch interconnection process, especially for 3D integration.

Keywords—electroless deposition, high-quality interface, low-temperature interconnection

I. INTRODUCTION

The core of packaging technology is the interconnection between the IC chip and the substrate. In addition to providing the electrical connection between the IC and the outside world and realizing the scale transition from the micro-nano-scale I/O connection points to the component pins, the interconnection also undertakes part of the mechanical support and heat dissipation function. How to realize the interconnection with fine-pitch and high-quality interface, especially at low temperature, has become a major technical problem in advanced packaging. The demand for low temperature interconnection comes from the following aspects: 1. Thermal damage of devices caused by high temperature; 2. CTE-induced misalignment; 3. Interface stress and failure due to thermal mismatch; 4. The high density of interconnection makes the molten solder is no longer applicable [1].

Interconnecting by solder is the preferred method to make high-performance mechanical and electrical flip-chip connections. However, due to the uneven distribution of current density in solder joints, the reliability of solder joints is severely challenged under the effect of thermoelectric coupling. Meanwhile, in the various bonding methods of 2.5D/3D IC interconnection, when the bump pitch is reduced to 20 microns, the molten solder bonding is no longer applicable, and it can only be achieved by solid-state bonding. Copper-to-copper interconnections between the bumps can eliminate many of electrical and mechanical problems of the solder interconnection mentioned above. Copper has excellent electrical conductivity and resistance to electromigration, higher yield stress and Young's modulus of Cu than solder, and will not form brittle intermetallic compounds. However, the bonding temperature of Cu-Cu interconnection is usually between 300°C and 400°C, and the bonding time is between several minutes to several hours, which are unfavorable for the integrity of thermally fragile ICs.

In the development of low-temperature Cu-Cu bonding technology, Suga's team proposed the method of surface activation [2], when Cu-Cu contacts each other, the atoms on both sides of the surface can automatically form bonding. This method needs to be carried out in an ultra-vacuum environment. Meanwhile, it is too expensive to large-scale production. Yang et al reached Cu-Cu bonding by inserting [3], but this process is complex and it produces great internal stress at the contact position of the interconnection. Ghosh et al. achieved Cu-Cu bonding by using a self-assembled monolayer which prevented the adverse oxidation of Cu, but the removal of SAM was troublesome and the bonding temperature remained very high [4].

Using electroless Cu deposition is another method for low-temperature Cu-Cu interconnection. Electroless Cu deposition has been proved to be a feasible method for interconnecting all-copper IC chip and substrate, which not only allows the fabrication of high aspect ratio structures, but also has a simple process. Osborn used electroless plating and annealing treatment to achieve a high-strength (>150Mpa) all-copper chip-to-substrate connection, but the high annealing temperature will cause some thermal stress inside the joins [5,6]. Koo used two facing Cu bumps and electroless plating to interconnect two bumps to achieve a seamless interconnection [7], but the requirement of the distance between bumps is high, and the operation requirements are also high. Yang used the flow of the plating solution to realize interconnection of small-pitch bumps, but it will leave a gap in the interconnection which is difficult to eliminate [8]. At the same time, the electroless Cu deposition for interconnection is a slow process, mainly because of the shortage of ions in the deposition solution and the difficulty in supplying the reactants to the confined micro-gaps which obstructs rapid deposition.

In this paper, the influencing factors of forming a special electroless micro-cones morphology are studied [9], and an efficient interconnection has been realized by using electroless Cu deposition with this micro-cones morphology. We achieved a high-speed interconnection with a rate of $>$ 15μm/h in this way. Although there are still few voids in the joints obtained, the shear force achieved by this way is much higher than that of solder and we can realize the ultra-low-temperature interconnection.

II. EXPERIMENTAL

A. Deposition of copper micro-cones

Cold rolled copper sheets with dimensions of 20 mm × 10 mm × 0.2 mm were used as substrates. Pretreatment: The copper substrates were degreased by ultrasonic cleaning with absolute ethanol for 3 min, and cleaned with $10\%H_2SO_4$ for 1min to remove the surface oxide layer, and then immediately activated 90s with 0.1g/L $PdCl_2$ solution. In between all steps, the substrates were rinsed with deionized water for 1min. The composition of the electroless deposition solution is in accordance with literature report [10.1016/j.apsusc.2012.05.096]. Saturated sodium hydroxide solution(12mol/L) was used to adjust the pH=9.5. Finally, the activated Cu substrates were electroless deposited at 55°C,65°C and 75°C for 20 min. In addition to this, the electroless deposition solution of the formaldehyde system was used for electroless deposition at 65°C, which was considered as a control group.

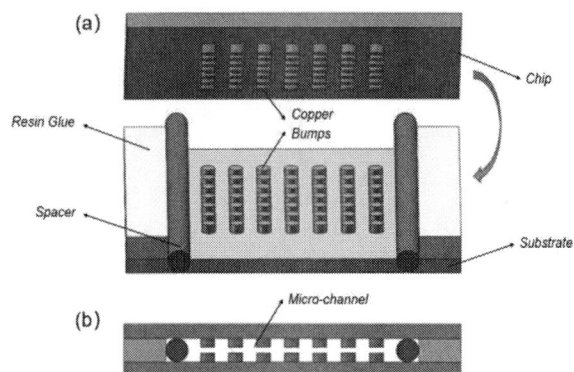

Fig. 1. (a)Alignment of test vehicle;(b) cross-section view of test vehicle and micro-channel.

B. Test vehicle

The chip and substrate used in this experiment are completely designed independently, the size of the chip is 6mm×6mm×0.5mm, and the height of the bumps of the top chip and the bottom substrate is 25μm and 10μm respectively. The diameter of the Cu bumps is 40μm and the pitch is 200μm. Firstly, the chip and substrate were cleaned by absolute ethanol and $10\%H_2SO_4$, and then the chip was flipped, aligned and fixed by CB600 bonder, so that the top and bottom bumps were aligned in the horizontal direction and a little distance was maintained in the vertical direction by using insulating wires with a certain diameter as spacers and making the spacer contact with the chip and substrate closely, then the aligned chip and substrate were fixed by resin glue to obtain the test vehicle, as shown in Fig 1(a). It can be seen that there was a micro-channel left in the middle of the test vehicle, as shown in Fig 1(b). The micro-channel could allow the electroless deposition solution to pass through during subsequent interconnecting, so that the bumps surface could fully contacted with the electroless deposition solution.

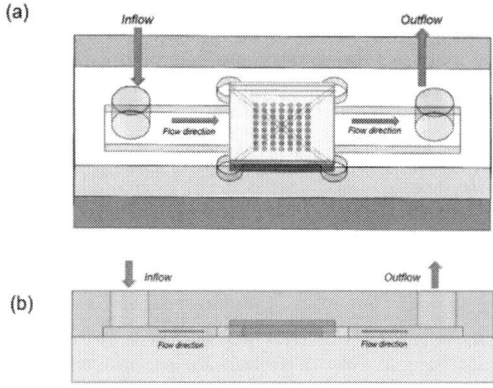

Fig. 2. (a) Microfluidic channel device and flow of electroless deposition solution;(b) cross-section view of the device.

C. Electroless copper deposition for interconnection

In order to promote the flow of deposition solution in the micro-channel of the test vehicle, we designed a microfluidic channel device, as shown in Fig 2. The device consisted of two parts: the upper part was a channel with special shape and the lower part was a flat glass plate, then the upper and lower parts were fixed together by screws. The internal size of the microfluidic channel was exactly the same as the size of the test vehicle, when the test vehicle was put into the channel, the cavity in the channel could just accommodate the test vehicle without any extra space. Before the electroless Cu deposition, a microflow pump was used to pass the pretreatment solution from one end of the microfluidic device to clean the oil and remove the oxide layer and active the surface of the bumps. Deionized water was passed to remove the residual reagents between each step. Then, the whole microfluidic device with test vehicle was placed in water at 75°C and 65°C, and the electroless Cu deposition solution was passed though the microfluidic device for 1h. During the interconnection, the vertical distance between the test vehicle was always maintained at 55μm and the flow rate of the microflow pump was 4.67 mL/min(10rpm). At the same time, the interconnection was obtained in the formaldehyde system through the microfluidic channel device at 65°C for 1h and 2h, which were used as the control groups.

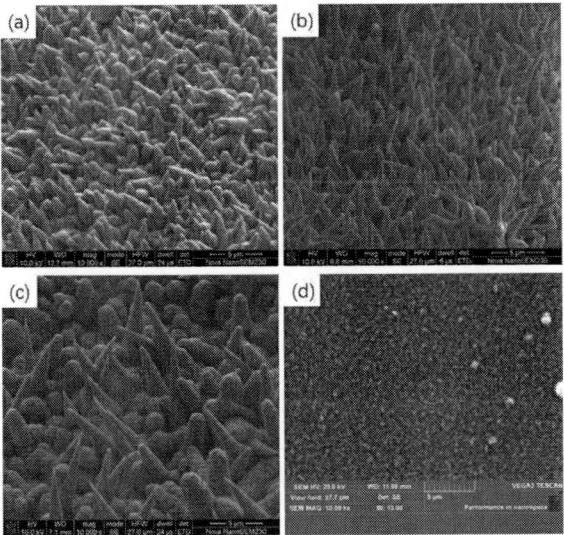

Fig. 3. Electroless copper deposition with different micro-cones morphology at different temperatures (a) 55°C; (b) 65°C; (c) 75°C, and (d) deposited copper in formaldehyde system at 65°C.

III. RESULTS AND DISCUSSION

A. Micro-cones morphology

In order to obtain the most suitable Cu micro-cones morphology as the interconnection interface, the morphology of deposited Cu micro-cones at different temperatures was studied under the same solution with PH=9.5. Fig 3 shows the scanning electron microscope images of the deposited Cu micro-cones morphology at (a)55°C, (b)65°C and (c)75°C. We can see that at 55°C, there are only a small amount of Cu micro-cones formed, the direction of the Cu micro-cones is not regular and half of the micro-cones lack sharp cones at the tip and the distribution of them is also uneven. It is because that when the temperature is too low, it is not conducive to the supply of metal ions, so the Cu deposits too slowly and the deposited layer grain size is small, which cannot form complete micro-cones. At 65°C, the deposited Cu micro-cones are not only complete in structure, but also very uniform and well oriented. They are pyramidal Cu micro-cones with a height of 3-5μm and a bottom diameter of about 1μm. At 75°C, the deposition reaction is so intense that the adjacent Cu cones are contacting each other, forming a

larger circular bulge, resulting in the growth of grain size and the loss of sharp micro-cones, so both pyramidal micro-cones and circular bulges are formed, and they are not uniform. Fig 3(d) shows the surface of Cu obtained in formaldehyde system without any micro-cones because of no PEG as the crystallization modifier. In summary, when the temperature does not reach 65°C, the grain size and the pyramidal morphology of the deposited Cu micro-cones increases with the temperature increases, and so do the uniformity and directionality. When the temperature is too high (over 65 °C), the pyramidal micro-cones will disappear and the good uniformity and directionality of the Cu micro-cones will also disappear, while some circular bulges will appear with the growth of grain size. Therefore, the optimal temperature for the pyramidal Cu micro-cones with good uniformity and directionality is 65°C, the Cu micro-cones have good uniformity and growth orientation at this temperature, which will promote the interconnection for Cu-Cu bumps.

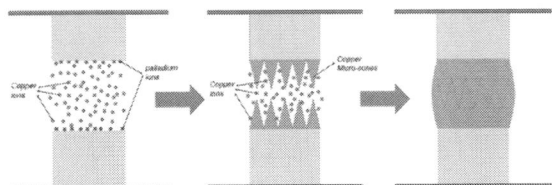

Fig. 4. An ideal model for the formation of interconnection interface by using electroless Cu deposition.

B. Electroless copper deposition interconnection

Fig 4 is an ideal model for the formation of interconnection interface by using electroless Cu deposition. Firstly, the activated Cu surface adsorbs a large amount of Pd. Pd is used as an activator to promote the electroless Cu deposition by reducing Cu ions. Then a thin layer of Cu with micro-cones will be grown on the surface of the Cu bumps, after that the micro-cones will continue to depositing and growing in the vertical direction, finally the top and bottom Cu micro-cones contact with each other and merge to form a complete interconnection interface.

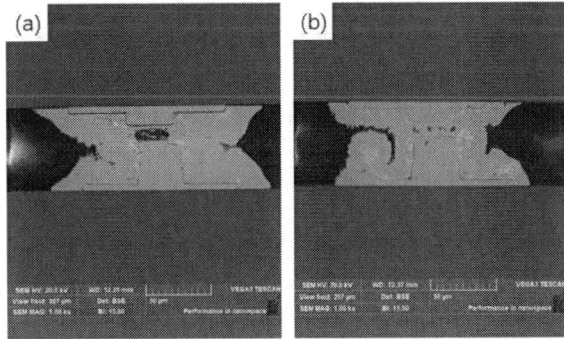

Fig. 5. Cross-section views of the interconnection interface of electroless copper deposition with micro-cones morphology for 1h at (a) 75°C and (b) 65°C by the microfluidic channel device.

Fig 5(a) shows that the pads beside bumps were interconnected completely, but a large void left between two bumps, which is undesirable for us. There are three main reasons for this phenomenon: 1.The gap between the pads is much larger than that between the bumps, so activation and the electroless deposition solution is easier to pass through between the pads, resulting in a strong reaction on the pads instead of bumps; 2.The rate of electroless deposition at the edge of the bumps is faster than that at the center, which leads to depositing occur at the periphery of the bumps firstly, causing the gap at the edge of the bumps to close, and preventing the solution entering the center of the bumps, so the center of bumps cannot be deposited. Therefore, the electroless Cu deposition occurs on the pad and the edge of the bumps firstly, then they will contact each other, which prevents depositing at the center of the bumps. Thirdly, because the temperature is high, the rate of reaction speed is fast, which increases the different rates between the center and edge of the bump, so it will leave a large void in the center of the bump. In order to reduce the void at the center of the bumps, it is necessary to reduce the different speeds between the bumps and pads, so we can reduce the temperature to get a higher quality interface.

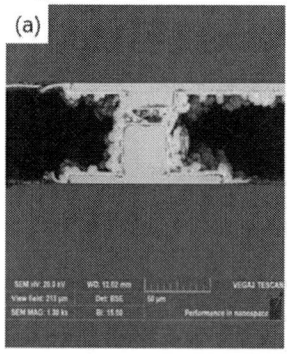 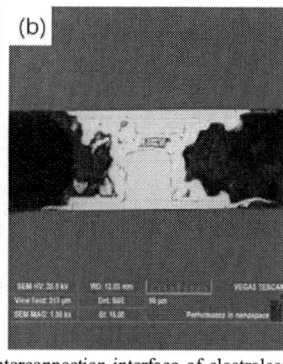

Fig. 6. Cross-section views of the interconnection interface of electroless copper deposition without micro-cones morphology for (a) 1h and (b) 2h by the microfluidic channel device at 65℃.

Therefore, we reduce the temperature to 65℃, as shown in Fig 5(b). It can be seen that there are only few of voids between the bumps, and the obvious large void has disappeared, as shown in Fig 5(b). We can find that when the temperature is reduced, the speed of deposition also decreases from the thickness of the deposition comparison of Fig 5(a) and Fig 5(b). The different speeds between the bumps and pads are also greatly reduced, which is good for interconnection. The reason of the residual small voids is mainly that during the process of electroless Cu deposition, some hydrogen is generated, and the different speeds of electroless Cu deposition between the edge and the center of bumps has not been completely eliminated, so the generated hydrogen will be trapped in the center of the bumps, then few of small voids will generate. Nevertheless, we can see that reducing the temperature can greatly improve the quality of interconnection interface. At the same time, no obvious fusion intersection line can be seen in the interface between different bumps, which indicates that the Cu with pyramidal micro-cones can promote the opposite merging, which is conducive to improve the quality of interconnection interface. However, how to remove the adverse effects completely of hydrogen produced still needs further research.

By this way, though at a low temperature, the quality of joints is high and the rate of electroless Cu deposition can reach >15μm/h. However, there is a large void that cannot be eliminated in the joints by using electroless Cu deposition without micro-cones morphology, which is much bigger than that with micro-cones morphology for 1h, as shown in Fig 6(a). The reason for this phenomenon is that when Cu with micro-cones is about to contact with each other, their merging area is larger because of micro-cones, the good uniformity

and growth orientation also play a positive role for merging, then the merging of opposite Cu bumps can take place through simultaneous fusing at multiple locations, which can promote to reduce the voids. Therefore, using micro-cones as the interconnect interface can achieve a higher quality than that without micro-cones. We can also see from Fig 5(b) and Fig 6(b) that the deposition rate of the Cu with micro-cones is about two to three times that of Cu without micro-cones, it can be seen from the thickness of deposited Cu. The reason is that the former has a larger surface area because of its micro-cones, which can attract more Cu ions for reacting. Therefore, the deposition rate and interconnection efficiency will be greatly improved by this. In summary, using deposited Cu with micro-cones can not only improve the quality of interface but also the speed of the deposition.

C. Shear test

Shear test was carried out on the interconnected chip by using electroless Cu deposition. It was found that there were two failure modes for the interconnected chip: 1. The Cu bumps themselves broke from the bottom ;2. The fracture occurs at the intersection between the electroless Cu deposition and the top of the Cu bumps in the interconnect joints. Fig 7(a-b) show the substrate- and chip-side images under the first failure mode. From which can we see that the column of the Cu bumps themselves have been cut off, and the Cu bumps on the chip-side have been left on the substrate-side because of bonding force after being cut off, as shown in Fig 7(a). Fig 7(c-d) show the substrate- and chip-side under the second failure mode. They show that when the fracture occurs, the bumps are intact, the interconnection joints was cut off on the top of bumps, leaving a trace of fracture on the top of bumps.

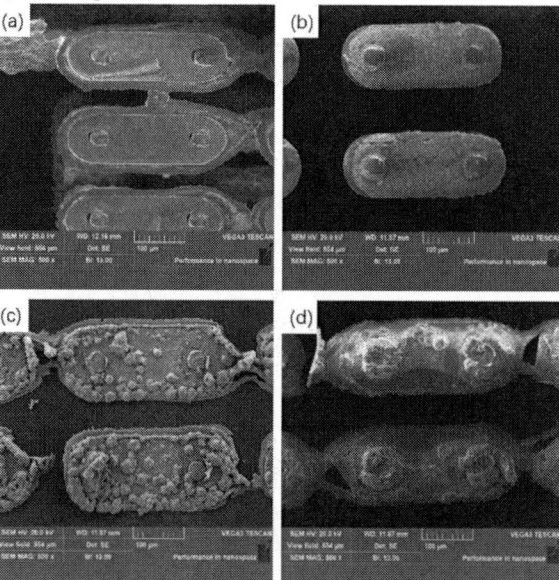

Fig. 7. (a)Chip- and (b)substrate-side in the first shear failure mode; (c)Chip- and (d)substrate-side in the first shear failure mode

In order to further understand the two failure modes, the models under the two failure modes are built in Fig. 8. It can be considered that the first failure mode is mainly due to the fact that there are no large voids in the interconnection joints during the electroless deposition process, or only few of small voids are generated. The opposing merging of the electroless

Cu deposition between the bumps is very good, and the electroless Cu is fully in contact with the top of the Cu bumps and forms great bonding force. Meanwhile, there are a lot of extra interconnections because the bumps sidewalls and pads also form interconnections and bonding force, which is much bigger than the cut off force of the bumps, so the failure occurs at the bottom of the bump. The second failure mainly because the interconnection is not sufficient, there are still many bubbles remaining in the interconnection joints, which prevents the merging between the electroless Cu and the top of the Cu bumps(It is guessed that the reason for the residual bubbles is that the flow of the deposition solution in the microchannel of the test vehicle is not completely uniform, so that part of the Cu bumps cannot be completely contacted with the electroless deposition solution, leading to the generated bubbles cannot be discharged in time) . At this time, the shear force in the joints is not as high as that of the Cu bumps themselves, so it will break at the connection between the electroless Cu and the top of the bumps under the shear test.

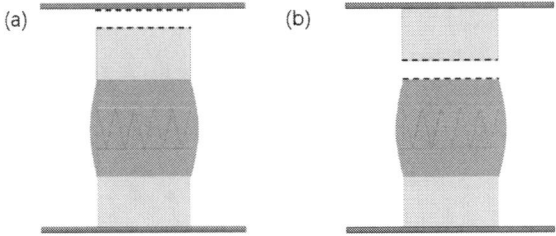

Fig. 8. Model in the (a) first and (b) second failure mode.

The failure results show that the failure shear force of the chip is as high as 227.7N under the above two mixed failure modes, which is much higher than by using solder. Therefore, we can confirm that, although the chip has two failure modes, the bonding force of Cu-Cu bumps interconnection using electroless deposition is high enough for 3D integration. And if we can further reduce the voids in the interconnection joints, this shear force will increase even further.

IV. CONCLUSION

1) The surface micro-cones morphology of electroless Cu deposition is greatly affected by the temperature of solution. When the temperature of solution is 65℃ and pH=9.5, the Cu wieh pyramidal micro-cones morphology with complete structure and good uniformity and directionality can be obtained, which is the most suitable for the interconnection interface of Cu-Cu bumps.

2) There will be a large void in interconnection layer between the Cu-Cu bumps obtained by electroless deposition at high temperature(75℃). By reduce thetemperature can indeed improve the quality of the interconnect interface and obtain a interconnect joint with only few of voids and residual bubbles. Using Cu with micro-cones morphology for interconnection can get improve speed and efficiency of interconnection, and the rate of electroless Cu deposition can reach ＞15μm/h. Because the merging of opposite Cu bumps can take place through simultaneous fusing at multiple locations with the Cu micro-cones morphology and the area of deposition is big enough, which will increase the interconnection rate greatly.

3) Shear tests show that there are two shear failure modes in the interconnected chips: 1. The Cu bumps themselves broke from the bottom ;2. The fracture occurs at the intersection between the electroless Cu deposition and the top of the Cu bumps in the interconnect joints.The different failure modes are determined by the quality of the interconnection joints. Although there are two failure modes, the bonding force of the interconnection interface by electroless deposition is high enough for 3D integration.

ACKNOWLEDGMENT

This work was supported by National Natural Science Foundation of China (No. Grant No. 51605498 and U20A6004), National Basic Research Program of China (973 Program, Grant No. 2015CB057206), and State Key Laboratory of High Performance Complex Manufacturing (No. ZZYJKT2020-08).

REFERENCES

[1] Ko, C.-T. and K.-N. Chen, "Low temperature bonding technology for 3D integration Microelectronics Reliability," 52(2):302-311,2012.

[2] Kim,T. H., et al, "Room temperature Cu-Cu direct bonding using surface activated bonding method," Journal of Vacuum Science and Technology A: Vacuum, Surfaces and Films 21(2):P449-453,2003.

[3] Yu-Tao Yang, Tzu-Chieh Chou, Ting-Yang Yu, Yu-Wei Chang.,et al. "Low-Temperature Cu-Cu Direct Bonding Using Pillar-Concave Structure in Advanced 3-D Heterogeneous Integration," Senior Member, IEEE TRANSACTIONS ON COMPONENTS:P1560-1566,2017.

[4] Tamal Ghosh, K. Krushnamurthy,Asisa Kumar Panigrahi.,et al, "Facile non thermal plasma based desorption of self assembled monolayers for achieving low temperature and low pressure Cu-Cu thermocompression bonding," RSC Adv, 2015.

[5] Tyler Osborn.,et al, "Electroless Copper Deposition with PEG Suppression for All-Copper Flip-Chip Connections," Journal of The Electrochemical Society, 156 (7)D226-D230,2009.

[6] Tyler Osborn.,et al, "All-Copper Chip-to-Substrate Interconnects Part I. Fabrication and Characterization," Journal of The Electrochemical Society, 155(4)D308-D313,2008.

[7] Hyo-Chol Koo.,et al, "Electroless Copper Bonding with Local Suppression for Void-Free Chip-to-Package Connections," Journal of The Electrochemical Society, 159 (5) D319-D322,2012.

[8] Sean Yang.,et al, "Bonding of Copper Pillars Using Electroless Ni Plating," ICEP 2016 Proceedings.

[9] Wenjing Zhang.et al, "Influence of PEG molecular weight on morphology, structure and wettability of electroless deposited Cu–Ni–P films," Applied Surface Science 258 ,8814–8818,2012.

The effect of silicon anisotropy on the thermal stress of TSV structure of 3D packaging chip under thermal cyclic loads

Jingyang LIANG
Engineering Research Center of Electronic Information Materials and Devices, Ministry of Education, Guangxi Key Laboratory of Manufacturing System and Advanced Manufacturing Technology,Guilin University of Electronic Technology,
Guilin, China
liang_jingyang@126.com

Minjie NING
Reliability Research and Analysis Center, No.5 Electronics Research Institute of the Ministry of Industry and Information Technology, Advanced IC Reliability Engineering Research Center of Guangdong
Guangzhou, China
ningmingjie@126.com

Chao DING
Engineering Research Center of Electronic Information Materials and Devices, Ministry of Education, Guangxi Key Laboratory of Manufacturing System and Advanced Manufacturing Technology,Guilin University of Electronic Technology,
Guilin, China
19012302007@mails.guet.edu.cn

Tianhan LIU
Reliability Research and Analysis Center, No.5 Electronics Research Institute of the Ministry of Industry and Information Technology, Advanced IC Reliability Engineering Research Center of Guangdong
Guangzhou, China
1801302029@mails.guet.edu.cn

Zongbei DAI*
Reliability Research and Analysis Center, No.5 Electronics Research Institute of the Ministry of Industry and Information Technology, Advanced IC Reliability Engineering Research Center of Guangdong
Guangzhou, China
dai_zongbei@126.com
* corresponding author

Hongbo QIN
Engineering Research Center of Electronic Information Materials and Devices, Ministry of Education, Guangxi Key Laboratory of Manufacturing System and Advanced Manufacturing Technology,Guilin University of Electronic Technology,
Guilin, China
qinhb@guet.edu.cn

Abstract—Due to the existence of anisotropy, the previous isotropic assumption of silicon-based materials can no longer accurately characterize the mechanical behavior of materials. In this study, finite element analysis (FEA) method was adopted to research the effect of silicon anisotropy on the thermal stress of through silicon via (TSV) structure including copper pillar arrays in stacked 3D packaging chips under thermal cyclic loads. The simulation results show that, in both isotropic and anisotropic cases, the high thermal stress is always located at the interface between two materials in the structure. Compared to the isotropic case, the magnitude of thermal stress is always larger in the anisotropic case. In both isotropic and anisotropic cases, the maximum thermal stress in the bottom TSV copper column arrays is always slightly larger than that in the upper layer arrays, and the thermal stress of the copper column at the farthest diagonal line is most concentrated.

Keywords—TSV, finite element analysis, anisotropic, thermal stress, thermal cyclic load

I. INTRODUCTION

Through silicon via (TSV) interconnection is one of the most important technologies to achieve 3D integration. It shortens the interconnection length by vertical interconnection and thus reduces signal delay and capacitance/inductance, which can achieve low power consumption and high-speed communication [1]. TSV has greatly promoted the high packaging density of integrated circuit. Attributed to the mismatch of the coefficient thermal expansion (CTE) between the filler metal and the matrix material, under thermal cyclic loads, the TSV structure will subject to thermal stress, which may causes voids and cracks in the TSV structure, leading to its failure [2]. Due to miniaturization of the TSV structure, the reliability test and failure analysis of TSV structure are difficult to perform. The existing literature shows that the analysis of TSV failure is basically carried out by numerical simulation analysis methods, including phase field method, crystal plasticity finite element method, finite element method

and so forth [3]. This is because for experimental testing, the current destructive testing methods and micro-characterization methods are difficult to implement due to the lack of efficacious and reliable TSV test device and invalidation analysis apparatus [4]. Selvanayagam et al. [5] analyzed and predicted the thermal stress failure of copper electroplating in TSV by establishing a finite element model of a two-dimensional symmetrical structure. He et al. [6] studied the thermal reliability of TSVs filled with copper, solder and copper core solder. In previous studies, the silicon-based materials in the TSV structure is usually based on the assumption of isotropy. However, in the actual packaging application, the silicon substrate shows obvious anisotropy. It is clear that maybe there are many errors in the results of analysis. Dai et al. [7] studied the anisotropy of silicon and explained its influence on TSV crack growth and other failure phenomenons by using the energy release rate (ERR). Fan et al. [8] researched the impact of the stress and strain distribution of the TSV microstructure. Obviously, the anisotropic properties play a crucial role in the result. Through the FEA method, the changing behavior of interfacial cracks was also studied. However, reports about the influence of anisotropy on the thermal stress of TSV structure including copper pillar arrays under thermal cyclic loads are quite rare. Therefore, in our work, the influence of silicon anisotropy on the mechanical behavior of the TSV structure including copper pillar arrays under thermal cyclic loads was studied by using the finite element analysis method.

II. SIMULATION METHODS

In this study, finite element analysis (FEA) software ABAQUS was employed to characterize the thermal stress of TSV structure including copper pillar arrays in a three-layer stacked 3D packaging chip under a thermal cyclic load. The three-layer stacked chip is exhibited in Fig. 1(a) [9]. In order to further improve the calculation efficiency, considering the central symmetry of the model, a 3D-1/8 finite element model with 10×10 copper-filled TSV structure arrays, was used, see Fig. 1(b). At the same time, the defects of packaging materials on the finite element analysis were ignored, and the impact of the thickness and morphology of the inter metallic compound (IMC) layer in the micro solder joint on the reliability was ignored, the material properties of the IMC layer were regarded to be the same as copper. Symmetrical constraints were imposed on the symmetry plane of the entire model, and full constraints were set to the degrees of freedom at the bottom to simulate the chip being fixed on the circuit board with fillers. The diameters of the copper pillar arrays and copper pads were set to 20 μm and 30 μm, respectively. The height of the interconnecting micro-bumps and distance of the micro-bumps were 20 μm and 100 μm. The thickness of the silicon chip and the side length of the chip were set to 36 μm and 1000 μm, respectively. Existing studies have shown that the upper and lower contact angles of the micro-bump interconnects can be treated as equal. In this study, both of them were set to 45°. For element type, the element library for standard analysis was selected, and the element employed contains 20-node quadratic hexahedral elements to reduce integration.

Fig. 1. Three-layer stacked thin TSV chip and its finite element model: (a) sequential reflowed 3-layer stacks of thinned TSV chip [9]; (b) 3D 1/8 symmetric finite element model .

According to JESD22-A104D standard, the finite element models were loaded with an accelerated temperature cycle (herein, 25 °C to 125 °C, then 125 °C to -40 °C, 5 min heating / 5 min cooling / 5 min heat preservation). The material properties required for modeling are shown in Table 1 [10], where T denotes the absolute temperature, indicating that the material properties are related to real-time temperature.

Table I. MATERIAL PARAMETERS USED IN FINITE ELEMENT ANALYSIS MODEL[10]

Material properties	Si wafer	Copper	Sn3.5Ag	Epoxy
Elastic modulus (GPa)	162.7	128.9	52.58-0.075(T+273)	15
Thermal expansion coefficient ($10^{-6}K^{-1}$)	2.8	$13.80+9.44×10^{-3}T$	$21.858+20.39×10^{-3}T$	13.8
Poisson's ratio	0.36	0.34	0.36	0.25

In this study, the Johnson-Cook constitutive model was used to represent the mechanical behavior of Sn3.5Ag micro-bumps in the plastic phase.

$$\sigma = [A + B(\varepsilon^p)^n](1+C\ln\dot{\varepsilon}^*)(1-T^{*m}) \tag{1}$$

where σ, ε^p, and $\dot{\varepsilon}$ are the von Mises flow stress, the equivalent plastic strain, and the dimensionless strain rate, respectively. $\dot{\varepsilon}^*$ can be defined as $\dot{\varepsilon}^* = \dot{\varepsilon}/\dot{\varepsilon}_0$, where $\dot{\varepsilon}_0$ means reference strain rate and its value is set to 0.001 s^{-1}. Herein A is the yield stress defined by the quasistatic compressive strain-stress data, B indicates the effects of strain hardening, C is the parameter used to describe the strain rate effect, m represents the effect of thermal softening, n has the same meaning as B. T^{*m} is regarded as the homologous

temperature , which can be calculated by $T^{*m} = (T-T_r)/(T_m-T_r)$, herein T_r and T_m are the reference temperature and melting temperature of the solder materials, respectively, as shown in Eq. (2). In this model, the reference temperature is regarded as equivalent to room temperature. The Johnson-Cook constitutive model parameters of Sn3.5Ag solder are shown in Table II [11].

$$T^{*m} \equiv \begin{cases} 0 & \text{for} \quad T < T_r \\ (T-T_r)/(T_m-T_r) & \text{for} \quad T_r \le T \le T_m \\ 1 & \text{for} \quad T > T_m \end{cases} \tag{2}$$

For the purpose of describing the crystal-orientation-dependent material properties, the crystal plane and its directions can be characterized by the Miller index, which are three-integer triples ("hkl") used to describe a certain crystal face family in a certain kind of lattice lattice. Generally, through determining the intercept point of a plane on the three axes of the rectangular coordinate system, and measuring the corresponding intercept with the unit of lattice constant. Then taking the reciprocal of the intercept, then reduce it to the simplest integer ratio, and denote this result as hkl, which is the Miller exponent of this plane. The important planes [12] for silicon are exhibited in Fig. 2.

Table II. THE JOHNSON-COOK CONSTITUTIVE MODEL PARAMETERS OF SN3.5AG SOLDER [11]

Solder	A(MPa)	B(MPa)	C	n	m	$T_m(K)$
Sn3.5Ag	29	243	0.0956	0.70	0.8	494

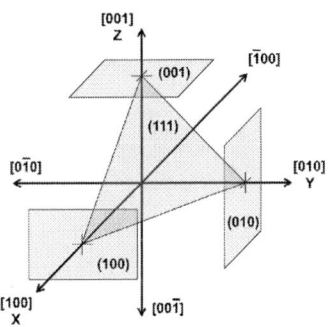

Fig. 2. The important planes for silicon [12].

Orthotropic material refers to the existence of three mutually perpendicular symmetry planes at any point through this material. For orthotropic elastomers, there are nine independent elastic constants. Orthotropic anisotropy is used to describe material properties without plenty of difficult discussion about crystal orientation and tensor rotation, thus the relationship of stress/strain could be easily gotten from the matrix. In order to provide the elasticity values which is [110], [$\bar{1}$10], and [001], the orthotropic expression of silicon is used to describe it, which is also its most common application. Here, the orthotropic stress-strain matrix of silicon was used as follows [12]:

$$\begin{bmatrix} \sigma_1 \\ \sigma_2 \\ \sigma_3 \\ \sigma_4 \\ \sigma_5 \\ \sigma_6 \end{bmatrix} = \begin{bmatrix} 194.5 & 35.7 & 64.1 & 0 & 0 & 0 \\ 35.7 & 194.5 & 64.1 & 0 & 0 & 0 \\ 64.1 & 64.1 & 165.7 & 0 & 0 & 0 \\ 0 & 0 & 0 & 79.6 & 0 & 0 \\ 0 & 0 & 0 & 0 & 79.6 & 0 \\ 0 & 0 & 0 & 0 & 0 & 50.9 \end{bmatrix} \begin{bmatrix} \varepsilon_1 \\ \varepsilon_2 \\ \varepsilon_3 \\ \varepsilon_4 \\ \varepsilon_5 \\ \varepsilon_6 \end{bmatrix}$$

III. RESULTS AND DICUSSION

A. Thermal Stress Distribution and Magnitude of 3D-1/8 Finite Element Model under Thermal Cyclic Loads

Fig. 3 show the thermal stress (herein, von Mises stress) of the finite element model that regards silicon as isotropic material and anisotropic material respectively at 125°C. Results show that thermal stress concentration areas are located at the copper/solder and copper/silicon interfaces. This is because the CTE of copper is obvious different with silicon and solder alloy, and the different CTEs will induce the thermal stress concentration. At 125°C, the maximum thermal stress in the model regarding silicon as an isotropic material (i.e., 508.9 MPa) is smaller than the model treating silicon as an anisotropic material (i.e., 532.1 MPa).

Fig. 4 (a) and (b) present the thermal stress distribution and magnitude of entire model at -40 ℃ in the cases of isotropic and anisotropic silicon materials, respectively. Clearly, the maximum thermal stress in the structure at -40 ℃ is obviously decreased compared with that at 125 ℃, and the maximum thermal stress in the anisotropic case (i.e., 170.5 MPa) is larger than that in the isotropic case (i.e., 167.4 MPa). Similarly, the thermal stress concentration area is still located at the interface between the two materials in the TSV structure.

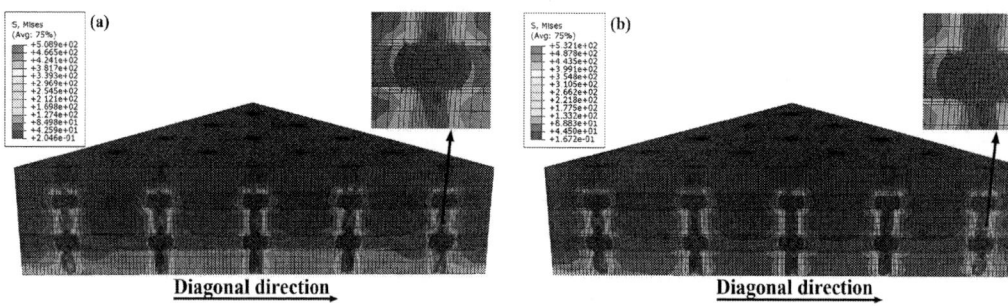

Fig. 3. Thermal stress of TSV structure at 125 °C, MPa: (a) considering silicon as an isotropic material; (b) considering silicon as an anisotropic material.

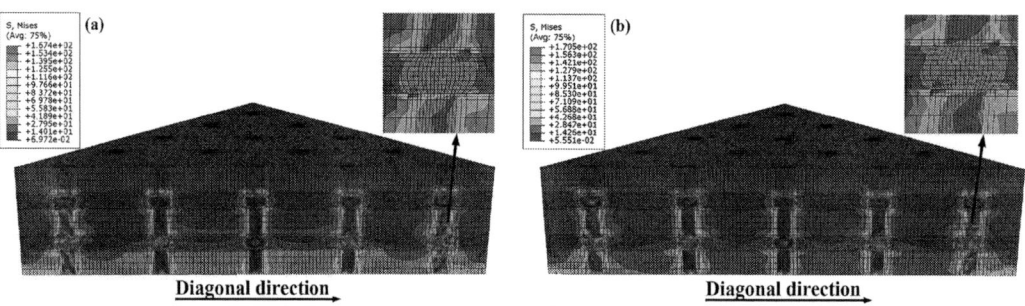

Fig. 4. Thermal stress of TSV structure at -40 °C, MPa: (a) considering silicon as an isotropic material; (b) considering silicon as an anisotropic material.

B. Distribution and Magnitude of Thermal Stress of TSV Copper Pillar Arrays under Thermal Cyclic Loads

Simulation shown that, for each copper pillar, the thermal stress concentration area is located at the interface of copper pillar/silicon and copper pillar/solder at high and low temperatures, meanwhile there are also obvious thermal stress concentrations at the corners of the copper pillars and copper pads. Previous study reported that, the expansion of the copper pillar to the side was restricted and could only move to the relatively free end, so there was obvious copper extrusion at both ends of the copper pillar. The micro-bump solder joints are connected to the copper pillars through the copper pads, thus copper extrusion phenomenon will indirectly affect the micro-bump solder joints, which are extremely easy to fail, and greatly reduce the reliability of the TSV microstructure. Fig. 5(a) and (b) show the thermal stress of the copper pillar arrays at 125°C in the cases of isotropic and anisotropic silicon materials, respectively. At 125°C, the maximum thermal stress in the isotropic case (i.e., 350.8 MPa) is smaller than the result in the anisotropic case (i.e., 371.1 MPa). In the anisotropic case, the maximum thermal stress in the bottom copper pillar arrays (i.e., 371.1 MPa) is larger than that of the upper arrays (i.e., 258.3 MPa), and the maximum thermal stress of the bottom copper column arrays is located at the end of the diagonal of the array.

Fig. 6(a) and (b) show the results of magnitude and distribution of thermal stress in the copper pillar arrays at -40 °C. Obviously, the thermal stress distribution in these two cases is similar to the result at 125 °C. At the same time, the value of maximum thermal stress of the copper column arrays is also less than that at 125 °C. Similarly, at -40 °C, the maximum thermal stress in the isotropic case (i.e., 102.4 MPa) is smaller than the result when silicon is anisotropic (i.e., 109.0 MPa). Besides, in the anisotropic case exhibited in Fig. 6(b), the maximum thermal stress in the bottom copper pillar array (i.e., 109.0 MPa) is slightly larger than that in the upper arrays (i.e., 75.12×10^8 MPa), and the maximum thermal stress in the bottom copper pillar arrays is located at the end of the diagonal of the array.

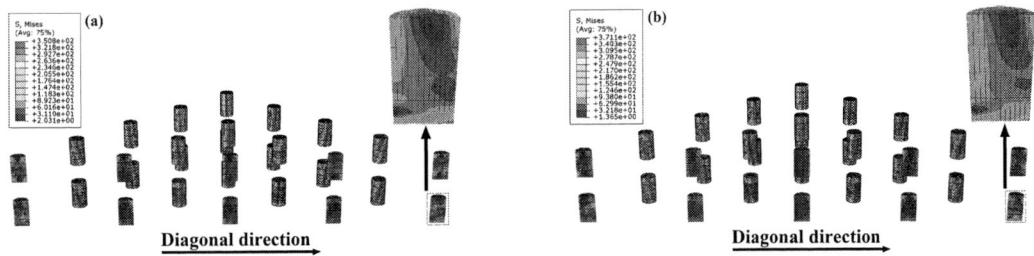

Fig. 5. Thermal stress of the copper pillar array at 125 °C: (a) considering silicon as an isotropic material; (b) considering silicon as an anisotropic material.

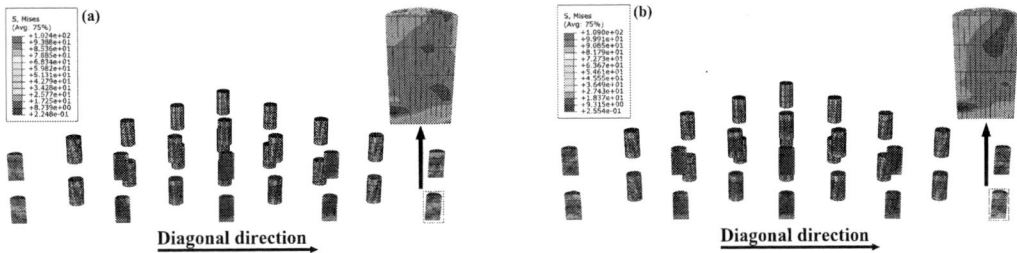

Fig. 6. Thermal stress of the copper pillar array at -40 °C: (a) considering silicon as an isotropic material; (b) considering silicon as an anisotropic material.

IV. CONCLUSIONS

In this paper, the influence of the anisotropic properties of silicon on the thermal stress of TSV microstructure under thermal cyclic loads was investigated by finite element analysis method. The results represent that during the temperature cyclic process, high thermal stress areas are always located at the corners of the copper/silicon interface, the copper/solder interface and the copper pad. Compared with the isotropic model, the maximum thermal stress generated under thermal cyclic loads at the same temperature is always larger than that in the anisotropic model, and the anisotropic characteristics of silicon have obvious impact on the reliability of TSV structure. Moreover, in the TSV copper column array, the maximum thermal stress in the bottom copper column arrays is always slightly larger than that in the upper layer array, and the thermal stress of the copper pillar at the farthest diagonal line is most concentrated.

ACKNOWLEDGMENT

This study was sponsored by Science and Technology Program of Guangzhou under grant. No. 202102020008; Guangxi Natural Science Foundation under grant. Nos. 2018GXNSFAA281222 and 2021 GXNSFAA075010; Science and Technology Program of Guangzhou under grant. No. 202102020008. Science and Technology Planning Project of Guangxi Province under Grant Nos. GuiKe AD AD18281022 and 18281021, Director Fund Project of Guangxi Key Laboratory of Manufacturing System and Advanced Manufacturing Technology Nos. 20-065-40-002Z and 19-050-44-003Z, Self-Topic Fund of Engineering Research Center of Electronic Information Materials and Devices Nos. EIMD-AB202005 and EIMD-AB202007. Innovation Project of GUET Graduate Education under grant No. 2020YCXS001，2021YCXS006 and 2021YXW06.

REFERENCES

[1] L.W. Kong, A. Rudack, P. Krueger, E. Zschech, S. Arkalgud, et al. "3D-interconnect: Visualization of extrusion and voids induced in copper-filled through-silicon vias (TSVs) at various temperatures using X-ray microscopy," Microelectronic Engineering, vol. 92, pp. 24-28, 2012

[2] J. Burns, L. McIlrath, C. Keast, C. Lewis, A. Loomis, et al. "Three-dimensional integrated circuits for low-power, high-bandwidth systems on a chip," IEEE International Solid-State Circuits Conference, vol. 1, pp. 268-269(2), 1999.

[3] Z.W. Fan, Y. Liu, X. Chen, Y. Jiang, S.F. Zhang, et al. "Research on fatigue of TSV-Cu under thermal and vibration coupled load based on numerical analysis," vol. 106, pp. 1-19, 2020.

[4] I.D. Wolf, K. Croes, E. Beyne, "Expected failures in 3-D technology and related failure analysis challenges," Packaging and Manufacturing Technology, vol. 8, pp. 711-718, 2018.

[5] C.S. Selvanayagam, J.H. Lau, X.W. Zhang, S.K.W. Seah,K. Vaidyanathan, et al. "Nonlinear thermal stress/strain analyses of copper filled TSV (through silicon via) and their flip-chip microbumps," vol. 32, pp. 720-728, 2009.

[6] R. H, H.J. Wang, J. Zhou, X.P. Guo, D.Q. Yu, et al. "Nonlinear thermo-mechanical analysis of TSV interposer filling with solder, Cu and Cu-cored solder," International Conference on Electronic Packaging Technology and High Density Packaging, Shanghai, China, pp. 224-227, 2011.

[7] Y.W. Dai, M. Zhang, F. Qin, P. Chen, T. An, "Effect of silicon anisotropy on interfacial fracture for three dimensional through-silicon-via (TSV) under thermal loading," vol. 209, pp. 274-300, 2019.

[8] Z.W. Fan, X. Chen, Y. Liu, Y. Jiang, Y.A. Zhang, "Effects of anisotropy on the reliability of TSV microstructure," vol. 114, pp. 1-6, 2020.

[9] J. Busby, D. Hawken, E. Perfecto, B. Dang, J Shah, et al. "C4NP lead free solder bumping and 3D micro bumping," IEEE/SEMI Advanced Semiconductor Manufacturing Conference, Cambridge, MA, USA, 2008.

[10] S. Park, H. Bang, H. Bang, J. You, "Thermo-mechanical analysis of TSV and solder interconnects for different Cu pillar bump types ," Microelectronic engineering, vol. 99, pp. 38-42, 2012.

[11] F. Qin, T. An, N. Chen, "Strain rate effects and rate-dependent constitutive models of lead-based and lead-free solders," Journal of Applied Mechanics, vol. 77, pp. 1-11, 2010.

[12] M.A. Hopcroft, W.D. Nix, T.W. Kenny, "What is the Young's Modulus of Silicon?", Journal of microelectromechanical systems, vol. 19, pp. 229-238, 2010.

High Yield and High Throughput Lithography Solution for Emerging High Density Fan-Out Panel Level Packaging

Junbo Jiang
Chengdu ESWIN SIC Technology Co., Ltd.
Chengdu, China
jiangjunbo@eswin.com

Kang Zhang
Chengdu ESWIN SIC Technology Co., Ltd.
Chengdu, China
zhangkang@eswin.com

Di He
Chengdu ESWIN SIC Technology Co., Ltd.
Chengdu, China
hedi@eswin.com

Chen Xiang
Chengdu ESWIN SIC Technology Co., Ltd.
Chengdu, China
xiangchen@eswin.com

Cheng-Tar Wu*
Chengdu ESWIN SIC Technology Co., Ltd.
Chengdu, China
terry.wu@eswin.com

Minghao Shen
Chengdu ESWIN SIC Technology Co., Ltd.
Chengdu, China
shenminghao@eswin.com

Abstract—The fan-out packaging (FOP) catches the general interest of microelectronics packaging, which has several advantages such as smaller formfactor, more efficient heat dissipation, better signal integrity, and higher reliability, matching with the requirements of advanced consumer electronics. It is, however, the reconstituted die shift and the carrier warpage have yet to overcome. These critical issues tighten the lithography overlay tolerance. Warpage issue can be controlled by suitable carrier selection. On the other hand, die shift can be well managed when following factors are taken into account: (1) carrier coefficient of thermal expansion (CTE) and thickness; (2) die/mold thickness and fan-out ratio; (3) shrinkage/expansion of involved materials.

Fan-out panel level packaging (FOPLP) is considered a cost-effective solution owing to higher carrier usage ratio. However, the aforementioned challenges are aggravated by the fact that the panel is not centrosymmetric and has larger die placement area with respect to the traditional wafer form carrier. Mask-less laser direct imaging (LDI) lithography technology has been adopted as a potential solution to overcome large die shift by relaxing lithography overlay constraint. But the low throughput of LDI undermines the cost benefit of PLP.

In this paper, we implement an off-line mapping system to collect and analyze the die shifts on a panel with size of 510mm × 515mm. The mapping data are subsequently introduced into a stepper to guide the PLP exposure process and achieve more than double the throughput. With proper die shift compensation, on top of the integrated mapping/exposure system, the throughput by using the full reticle exposure can be further enhanced to 20 times than that of die-by-die exposure. The integrated system can be utilized to predict lithography yield under the combination of different exposure field size and different overlay spec. By using the same system, we achieve a lithography yield of 99.9% within ±12um lithography overlay spec at the maximum field size.

Keywords—Fan-out panel level packaging, Die shift, Lithography yield

I. INTRODUCTION

With the emerging demand for higher performance and reliability over consumer electronics, especially in mobile sector, the importance of advanced packaging becomes clear as it can provide compact package size as well as lower power consumption. Driven by the advances in mobile device, there is a transition towards advanced packaging, migrating from leadframe, organic substrate-based wire-bonding or flip-chip packages, to wafer level fan-in and fan-out packaging, among which the substrate-less fan-out (FO) technology is garnering industrials' attention, due to the advantages of smaller formfactor, lower thermal resistance, better performance because of shorter interconnects, and higher reliability.

Fan-out wafer level packaging (FOWLP) was first introduced by Infineon, known as embedded Wafer Level Ball grid array (eWLB). eWLB was then licensed by leading OSATs, such as STATS ChipPAC, ASE, and NANIUM, for high-volume-manufacturing (HVM). eWLB has been the most prevalent FO packaging technology until the presence of integrated fan-out wafer-level packaging (InFO-WLP) by TSMC [1].

For cost-effective consideration, a higher area utilization packaging, fan-out panel-level packaging (FOPLP), was then proposed by J-Device. Fraunhofer and SPIL have examined the FOPLP in different dimensions [2-3]. Until recently, FOPLP has been proven successful to HVM by Samsung and PTI.

In FOWLP/PLP, various package formation has been investigated, such as chip-first with die face-down, chip-first with die face-up, and chip-last, among which chip-first approach is the most sensitive to the properties of underneath temporary glue, coefficient of thermal expansion (CTE) of temporary carrier, and the parameters of following molding process. Die placement control, including shift and rotation, is a critical issue even though a high accuracy die bonder is implemented. The occurrence of die shift is resulted from the naturally shear induced by the compression mold flow, creating significant challenges in subsequent via to pad alignment. Despite the die shift compensation has been introduced, lithography compensation is still needed to be considered to get a better yield on the exposure process of a reconstituted wafer/panel.

Currently, maskless laser direct imaging (LDI) and step-and-repeat stepper are common lithography patterning techniques that applied in FO packaging. No matter using a LDI or a stepper, pre-exposure automatic optical inspection (AOI) measurement is a must to collect the die shift data. With the AOI data, LDI is capable to reroute the full pattern, and the stepper is also capable to expose a smaller repeating pattern across the carrier. However, LDI relies on the mature algorithm of the pattern to obtain stable results, but has poor adjustments for dealing with non-systemic distortions and routing algorithm. Moreover, LDI does not have higher laser power for dealing with the fine-line pattern, especially on liquid photoresist. In contrast, stepper is more flexible in dealing with distortions, which caused by processing and patterning mismatch.

Here, we present a unique prediction methodology of via to pad overlay yield on a reconstituted 510mm × 515mm glass panel with the chip-first with die face-up approach. The die position and rotation are measured by the exposed alignment patterns. The data is then feedforwarded to a software algorithm to synthesis the photolithography yield under corresponding exposure field sizes.

II. EXPERIMENTAL

A. Test Vechicle Design

To demonstrate the overlay compensation performance, a test vehicle (TV) package with a size of 7.1 mm × 7.1 mm was designed, as shown in Fig. 1.

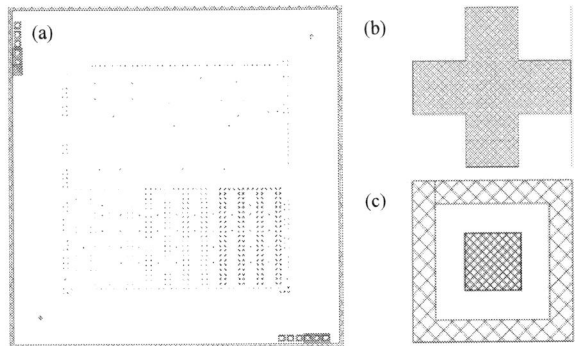

Fig. 1. Test vechicle layout of (a) die pattern, (b) alignment mark for mapping, and (c) pad (green mesh)/via (grey mesh) overlay test pattern

B. Preparation of a Reconstituted Panel

A 12-inch bumped TV wafer was thinned down to 250um and singulated into TV dice. 3,600 units of TV die were attached on an adhesive-coated 510 mm × 515 mm glass panel. A compression molding process was employed followed by a grinding process. The alignment mark and pad patterns were then exposed for mapping and pad overlay measurement, respectively.

C. Relaibilaity of Mapping System Demostrated by Stepper and AOI

A stepper and an off-line AOI were used for the mapping system in this study. Designated offsets (X, Y, and θ) of the 1st layer pad were created to mimic the post-molding/grinding die shift. Fig. 2 shows 17 groups of designated offsets of the 1st layer pad pattern applying on corresponding region of the panel. The stepper was firstly employed as the mapping tool to measure the offsets, as shown in Fig. 3. The reported measurement result corresponded well with intended offsets, indicating the stepper's capability for the following overlay metrology. On the other hand, identical results were obtained while an off-line AOI used for the mapping.

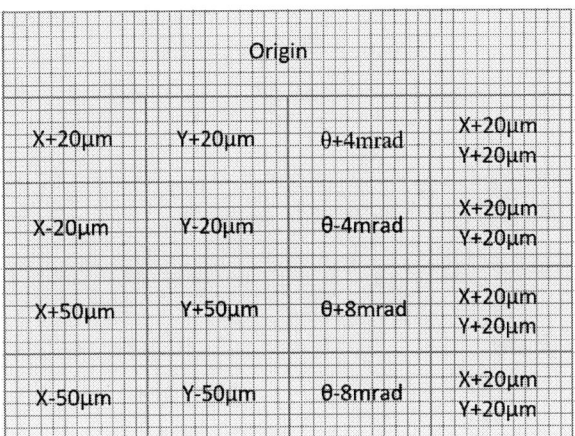

Fig. 2. Designated pad offsets with corresponding region on a panel

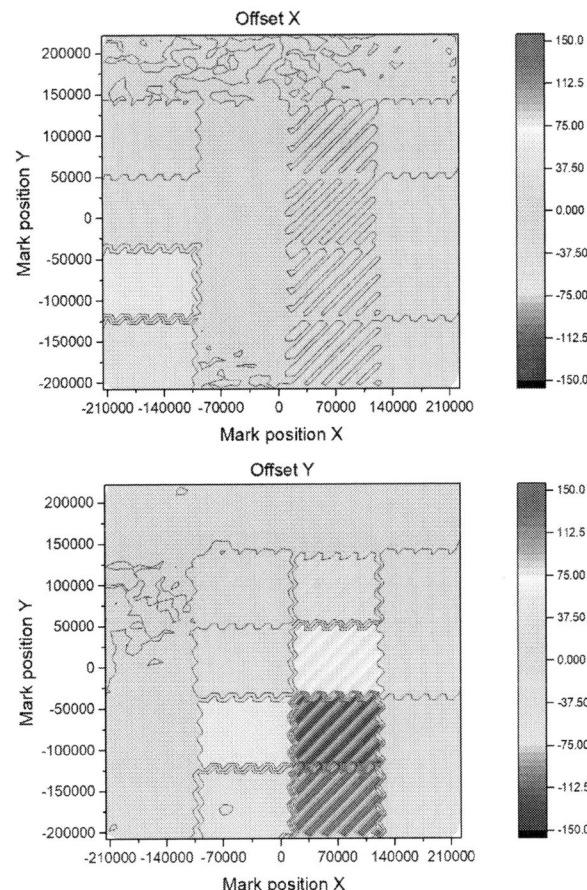

Fig. 3. Mapping performance in terms of offest variations carried out by the stepper

Die shift measurement of the feedforward compensation for the stepper of the 2nd layer via pattern was collected by using either the stepper or the AOI on the shift in respect to the 1st layer pad. A TPCD was used to measure the via to pad overlay as shown in Fig. 4. Both stepper and the AOI show good overlay residual within ±1.5μm, showing the capability of both tools used for the mapping system in this study are in the same level.

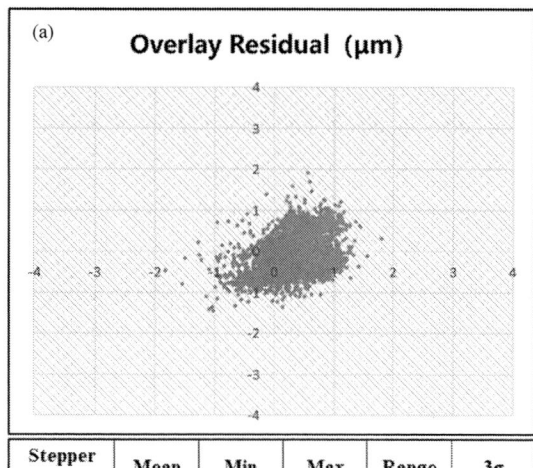

Stepper Only	Mean	Min	Max	Range	3σ
dx	0.41	-1.83	1.81	3.64	1.08
dy	-0.12	-1.46	1.90	3.36	1.01

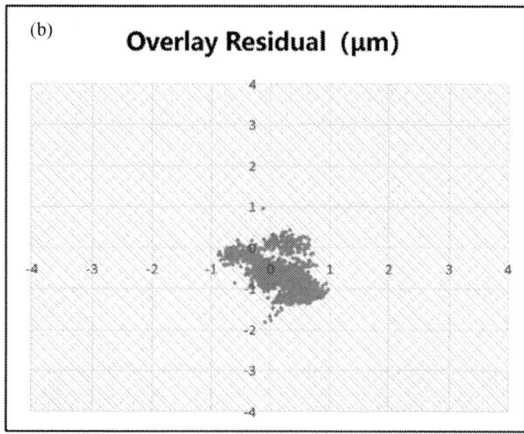

Parallel processing	Mean	Min	Max	Range	3σ
dx	0.19	-0.87	0.97	1.84	0.98
dy	-0.69	-1.82	0.96	2.79	1.12

Fig. 4. Overlay residual carried out by different mapping system (a) stepper, and (b) AOI

D. TV Die Measurement

Eliminating the basis between the stepper and the off-line AOI, the stepper was the only mapping system throughout the TV die measurement for the compensation of the 2nd layer. The heat map of the die shift and rotation are shown in Fig. 5. It is important saying that die shift without any compensation on a compression-molded reconstituted panel is hundred microns, and the rotation is far below the rotation limit of the stepper stage.

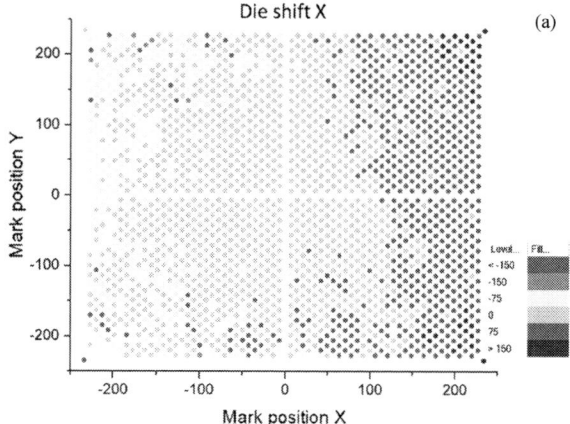

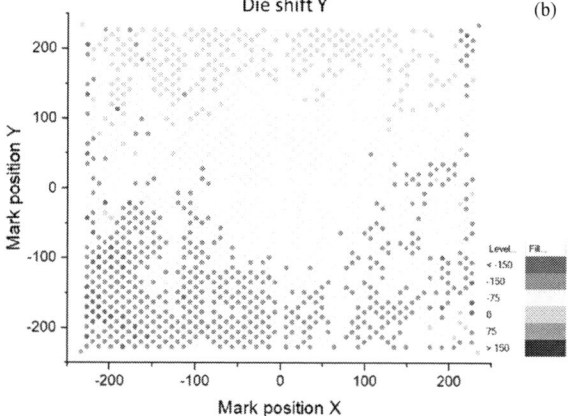

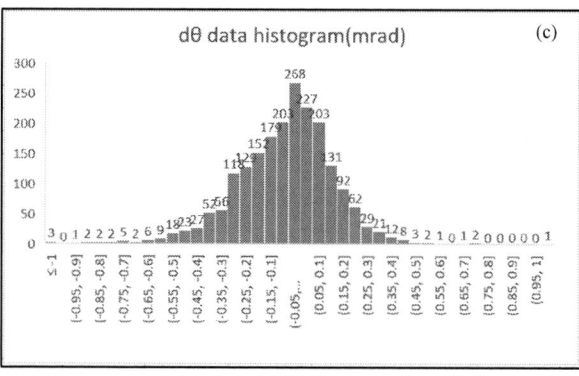

Fig. 5. Die mapping results of a reconstitued panel: (a) dx heat map of TV die shift (μm), (b) dy heat map of TV die shift (μm), (c) TV die rotation

III. RESULTS AND DISCUSSION

A. Yield Prediction

Here, by importing the TV die measurement data to a software algorithm, the lithography overlay yield can be synthesized under different exposure field sizes. Depending on the overlay requirements, different exposure field sizes can be manipulated. 1×1, 2×2, 4×4, 4×7, and 14×14 die array based on the package size of 7.1 mm × 7.1mm were examined for the software algorithm simulation.

Fig. 6 shows the overlay histograms in X and Y direction under different exposure field sizes with corresponding die array and are summarized in Fig. 7. It is noted that the 3σ of the simulated exposure area of 14×14 die array is smaller than ±15μm, which is in line with the industrial overlay specification, indicating a good process window were made for fabricating the reconstituted panel in this study.

Overlay histogram in Y is concentrated than that of X because the expansion/shrinkage in X and Y is different due to the panel dimension in X and Y direction are different (X × Y = 515mm × 510mm).

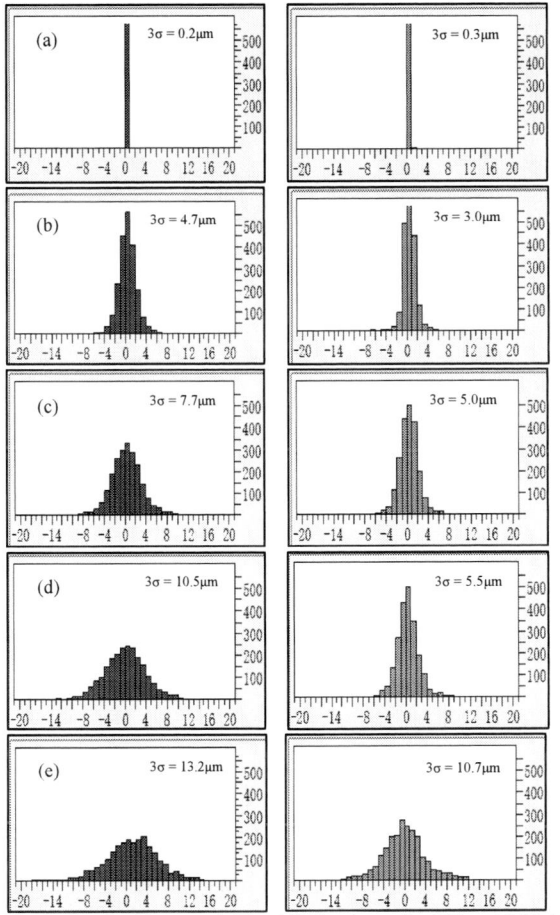

Fig. 6. Overlay histogram in X and Y direction under different field sizes: (a) 1 × 1, (b) 2 × 2, (c) 4 × 4, (d) 4 × 7, and (e) 14 × 14 die array

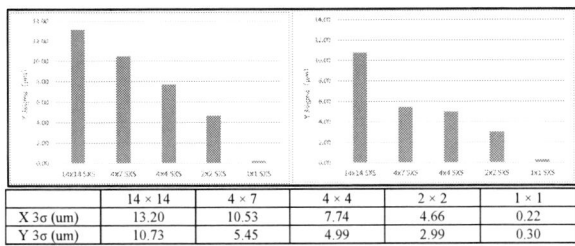

	14 × 14	4 × 7	4 × 4	2 × 2	1 × 1
X 3σ (um)	13.20	10.53	7.74	4.66	0.22
Y 3σ (um)	10.73	5.45	4.99	2.99	0.30

Fig. 7. A summary of X and Y overlay under different field sizes with site-by-site (S×S) alignment

B. Throughput Improvement

Yield prediction on different combination of exposure field sizes and overlay requirements are illustrated in Fig. 8. The larger the field size, the higher the lithography overlay yield is obtained. In addition, 1 × 1 exposing clearly shows the highest yield of 100% and has the highest overlay tolerance as

compared with others. It is, however, very time-consuming and not practical for HVM. Therefore, a hypothesis of a lower overlay yield of 99.90% was made for HVM consideration. Here, the 4 × 4 die array alignment demonstrates the best optimized stepping process throughout simulation, achieving the highest overlay tolerance of ±12μm, meanwhile maintain the lithography yield at 99.90%. Even though a larger exposure field of 14 × 14 die array was exercised, a relatively high overlay yield of 98.97% can still be obtained.

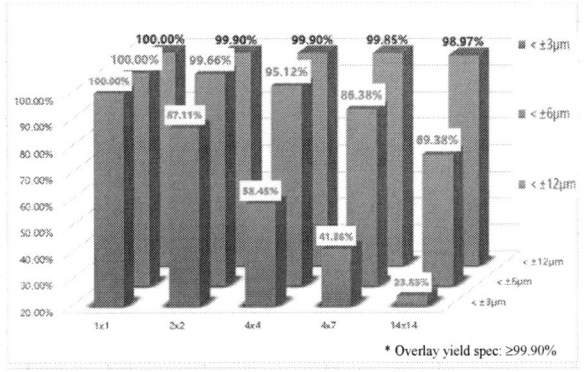

Die Array Layout	Overlay < ±3μm	Overlay < ±6μm	Overlay < ±12μm
1 x 1	100.00%	100.00%	100.00%
2 x 2	87.11%	99.66%	99.90%
4 x 4	58.45%	95.12%	99.90%
4 x 7	41.36%	86.38%	99.85%
14 x 14	23.83%	69.38%	98.97%

Fig. 8. Overlay yield corrisponding to different die array exposure fields and overlay specifications

An attempt on the throughput improvement is made by changing the serial die shift measurement and exposure process to a parallel approach, i.e. an off-line AOI for mapping was utilized to replace the metrology done by the stepper. The throughput of the die shift measurement on 1 × 1 exposing can be improved to ca. 2.6 times. Based on the parallel mapping/exposing process over die-by-die exposure, with the exposure field size further increasing to 4 × 4 and 4 × 7 die array, the throughput can be further enhanced to ca. 12 times and 20 times, respectively.

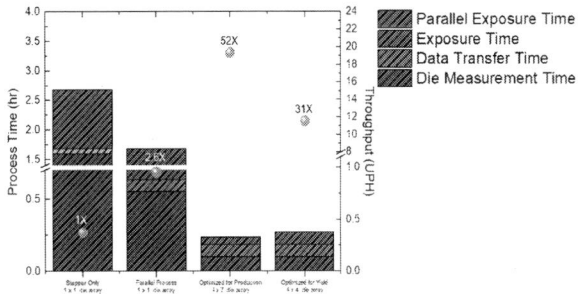

Fig. 9. Throughput comparison on different mapping and exposure stratgey

IV. CONCLUSION

In this paper, a mapping system to collect and analyze the die shift on a 510mm × 515mm reconstituted panel is implemented. We successfully demonstrate that hundred microns die shift can easily be compensated by lithography, yielding an excellent overlay at a decent exposure field size. A simulated lithography yield of 99.9% within ±12um lithography overlay specification is synthesized on a field size of 4 × 4 die array with a package size of 7.1 mm × 7.1 mm. Under the same TV panel, a relatively high overlay yield of 98.97% can still be attained when enlarging the field size to 14 × 14 die array.

It is noted that, the yield prediction function can prevent lithography rework during the fabrication process, as well as to provide a guideline for via to pad overlay design at various pad sizes of chips accordingly.

The lithography throughput can be greatly improved to 20 times than that of 1 × 1 exposing while introducing an off-line mapping system. With proper die shift compensation, on top of the integrated mapping/exposure system, the long-standing die shift issue on FOWLP/PLP can be easily overcome, realizing a more competitive production cost.

REFERENCES

[1] C. Tseng, C. Liu, C. Wu and D. Yu, "InFO (wafer level integrated fan-out) technology," 2016 IEEE 66th Electronic Components and Technology Conference (ECTC), 2016, pp. 1-6.

[2] T. Braun et al., "Challenges and opportunities for fan-out panel level packing," 2014 9th International Microsystems, Packaging, Assembly and Circuits Technology Conference (IMPACT), 2014, pp. 154-157.

[3] H.-D. Chang et al., "Development and characterization of new generation panel fan-out (P-FO) packaging technology," 2014 IEEE 64th Electronic Components and Technology Conference (ECTC), 2014, pp. 947-951.

An optimization method of ultra hign speed differential structure for BGA package *

1st Nong Jin
The 13th Research Institute of CETC
China Electronics Technology Group
Corporation
Shijiazhuang, China
646509144@qq.com

2st Zhizhuang Qiao
The 13th Research Institute of CETC
China Electronics Technology Group
Corporation
Shijiazhuang, China
qiaozz@cetc13.cn

3st Linjie Liu
The 13th Research Institute of CETC
China Electronics Technology Group
Corporation
Shijiazhuang, China
lj.liu@cetc13.cn

4st Ke Wang
The 13th Research Institute of CETC
China Electronics Technology Group
Corporation
Shijiazhuang, China
wangke@cetc13.cn

5st Yangfan Zhou
The 13th Research Institute of CETC
China Electronics Technology Group
Corporation
Shijiazhuang, China
zhouyf@cetc13.cn

6st Zan Ren
The 13th Research Institute of CETC
China Electronics Technology Group
Corporation
Shijiazhuang, China
renz@cetc13.cn

Abstract—With the advent of the 5G era, the clock frequency continues to increase, and the system's requirements for signal transmission rates are getting higher and higner. High frequency and high speed chip technology is developing vigorously, and the demand for hign frequency and high speed packaging enclosures to match it is strong. Among them, the BGA package structure is favored by people due to its advantages such as high density and miniaturization. However, the impedance control introduced by the vertical interconnect holes and BGA balls in the ultra high speed BGA package structure has brought serious problems to the signal integrity. How to solve the mismatch problem of the differential pair in the ultra high speed BGA package is the key. This article provides an optimization method for the differential structure of an ultra high speed BGA package, through vertical via impedance adjustment technology, BGA ball matching compensation technology,etc.to achieve impedance matching and reduce loss. It is the earliest to realize ultra high speed differential BGA transmission within 50GHz bandwidth, with less than 1.8dB the difference loss, greater than the return loss, and the theoretical maximum transmission rate of 100Gbps.

Keywords—BGA, hign frequency, high speed, differential structure

I. INTRODUCTION

In recent years, the amount of communication required for information interaction has exploded, and the quality requirements for signal transmission have become higher and higher. One of the hottest high-frequency and high-speed technologies[1-3]. The key to high-frequency and high-speed is not only the design of high-speed chips, packaging, as the last step in the practical application of modules, is of paramount importance for the device to achieve a good frequency response. Among them, the BGA(Ball grid array) packaging form has received key attention due to its miniaturization, large number of terminals, and easy integration. However, the BGA structure is a vertical transmission structure[4], and the introduced vertical interconnect holes and BGA solder balls will cause impedance mismatch, resulting in a series of signal integrity problems, which seriously restricts the improvement of the differential signal transmission rate.

This article mainly focuses on the optimization analysis of vertical vias and BGA ball implants, and proposes impedance adjustment methods and matching compensation techniques. Through the establishment of via equivalent models, the impact of relevant parameters on the impedance of vias is analyzed, and the structure is optimized and used Three-dimensional electromagnetic simulation software HFSS establishes three-dimensional model simulation optimization, and finally prepares samples for testing and verification. In the 0~50GHz frequency band, the differential transmission loss is ≤1.8dB, the return loss is ≥15dB, and the theoretical value of the transmission rate can reach 100Gbps.

II. TRANSMISSION STRUCTURE

Establish a three-dimensional transmission structure, switch from the surface layer differential line to the vertical interconnection hole and then connect to the test substrate through the BGA ball. The packaging material is low-temperature co-fired ceramics, the dielectric constant is 5.8, the single-layer dielectric thickness is 0.1mm, a total of 10 layers, the aperture is 0.13mm, and the test substrate is a PCB substrate. The three-dimensional structure diagram is shown in Figure 1

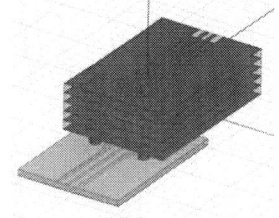

Figure 1 Schematic diagram of differential transmission structure

III. IMPEDANCE MATCHING AND COMPENSATION

A. Impedance matching of vias

Use Q3D parameter extraction software to extract via parameters, including the influence of dielectric thickness, aperture, pad, anti-pad and other parameters on via parasitic parameters, as follows:

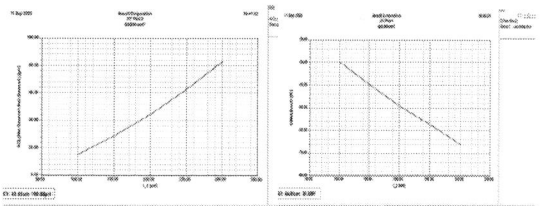

The influence of dielectric thickness on inductance and capacitance

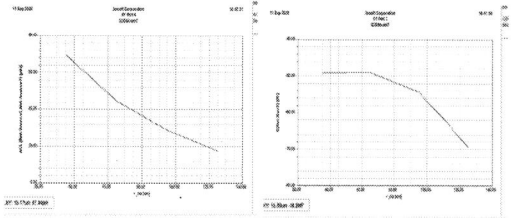

The influence of via diameter on inductance and capacitance

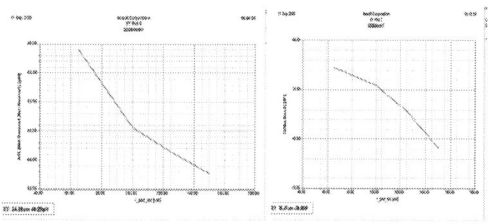

The influence of pad diameter on inductance and capacitance

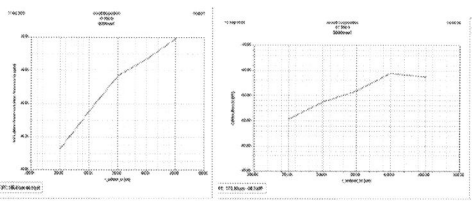

The influence of antipad diameter on inductance and capacitance

Figure 2 The influence of different factors on parasitic parameters

We know that loss is divided into three categories, conductor loss, dielectric loss and radiation loss. When the frequency is high, the conductor loss cannot be ignored. Affected by the skin effect, the higher the frequency, the more current distribution is concentrated on the surface, which is equivalent to resistance. Increase, increase loss. The conductor loss is obvious, and when the material is fixed, it can only be solved by increasing the surface area as much as possible. By establishing the equivalent model of vias and modeling and simulation in ADS, it can be found that parasitic inductance and capacitance will greatly affect the equivalent impedance of vias. There are two main ways to reduce parasitic parameters of vias: reduce Equivalent capacitance; appropriately increase the inductance when the capacitance cannot be reduced. The parameters that have a greater impact on the capacitance are the thickness of the dielectric and the size of the anti-pad, and the parameters that have a greater impact on the inductance are the aperture and length of the via. At present, for the LTCC process, the thickness and aperture of the single-layer dielectric are fixed, and it is difficult to adjust through the process. Therefore, we can increase the thickness of the dielectric equivalently by increasing the distance between the two metal stratums and increase the length of the via at the same time In this way, the parasitic capacitance is reduced and the parasitic inductance is increased to achieve the purpose of adjusting the impedance of the via.

In addition, the vertical via can be regarded as a coaxial-like structure. The distance between the surrounding ground hole and the middle signal hole and the size of the anti-pad directly affect the impedance of the via[6-10]. The related parameter adjustment method has been described in many related documents. Of course, each parameter has a different degree of impact on impedance, and often trades off one by one. Therefore, it is necessary to optimize one by one and finally determine the parameter value.

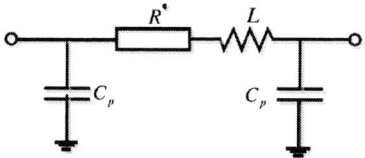

Figure 3 Equivalent circuit of vertical via

B. Matching adjustment of BGA ball

Compared with the via hole, the BGA ball has a large size, and the impedance at the ball is sharply reduced due to the material, and the impedance needs to be increased through structural adjustment. Through the foregoing analysis, we know that the parasitic capacitance at the planting ball is very large, and the impedance can be increased by increasing the inductance value. If the aperture cannot be reduced, the large-area ground layer on the back can be changed to a single pad form, as shown in the figure 4

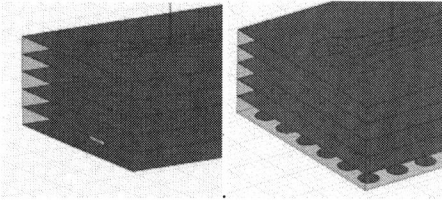

Figure 4 Schematic diagram of the large area and a single pad

In this way, the capacitance value between the ball and the bottom metal layer is reduced, and at the same time, the parasitic inductance can be adjusted by adjusting the thickness of the last layer of the medium to achieve the effect of impedance adjustment. The size of the anti-pad above the bump is also the key. Generally, increasing the size of the bottom anti-pad can effectively reduce the equivalent impedance of the solder ball. However, due to the pad pitch, the anti-pad size cannot be large enough, and the bottom is changed to a single The pad form is no longer made into an anti-pad structure, and the penultimate layer of anti-pad is used to adjust the impedance of the solder ball, so that the size is not limited, and the modulation effect is obvious. The comparison results of S-parameters after matching and optimizing the two methods are as follows:

978-1-6654-1392-3/21 $31.00 © 2021 IEEE   1256

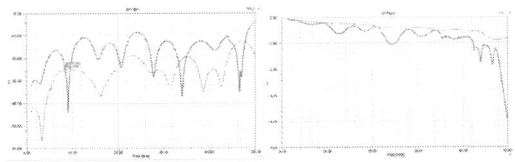

Figure 5 Comparison of s-parameters of the large area and the single pad

The samples were prepared according to the optimized structural parameters of the simulation, and the actual measurement verification was carried out. The results are in good agreement with the simulation results. The differential transmission loss is ≤1.8dB and the return loss is ≥15dB in the 0~50GHz frequency band. The transmission bandwidth is up to 50GHz. According to Nyquist theorem, the transmission rate B=2W bandwidth, so the theoretical transmission rate can reach 100Gbps.

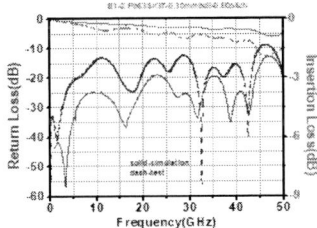

Figure 6 Comparison of actual measurement and simulation

IV. Conclusion

This paper analyzes the related parameters of vertical vias and BGA solder balls, and proposes impedance matching by changing the structure of the dielectric layer and the metal layer. The time-differential transmission bandwidth is up to 50GHz and the theoretical transmission rate is up to 100Gbps.

References

[1] G. Eason, B. Noble, and I. N. Sneddon, "On certain integrals of Lipschitz-Hankel type involving products of Bessel functions," Phil. Trans. Roy. Soc. London, vol. A247, pp. 529–551, April 1955. (references)

[2] Jacob Minz, Eric Wong. Placement and Routing for 3D System-On-Package Designs[J], IEEE TRANSACTIONS ON COMPONENTS AND PACKAGING TECHNOLOGIES, 2005(1):1-14

[3] Heyen, J. , von Kerssenbrock, T. , Chemyakov,A. . Novel LTCC-BGA modules for highly integrated millimeter-wave transceivers[J]. IEEE Trans. Microw. Theory Tech. , 2003(51) : 2589-2596

[4] T. Kangasvieri, J. Halme, et al. An ultrawideband BGA-via transition for high—speed digital and millimeter wave packaging applications[C], IEEE MTT-S Dig, Honolulu, 2007 : 1637-1640

[5] C.Schuster, W.Fichtner. Corrections to "Parasitic modes on printed circuit boards and their effects on EMC and signal integrity". IEEE Trans.Electromagnetic Compatibility, 2003, 45(4): 664

[6] Show-Gwo H, Ruey-Beei W. Full wave characterization of a through hole via using the matrix-penciled moment method[J], IEEE Transactions on Microwave Theory and Techniques, 1994, 42(8): 1540-1547

[7] Q.Gu, Y.E.Yang M.A.Tassoudji.Modeling and analysis of vias in multilayered integrated circuits. IEEE Trans.Microwave Theory Tech., 1993, 41(2): 206-214

[8] Yao-Jiang Z, Zaw Z O, Xing-Chang W, et al. Systematic Microwave Network Analysis for Multilayer Printed Circuit Boards With Vias and Decoupling Capacitors[J]. IEEE Transactions on Electromagnetic Compatibility, 2010, 52(2): 401-409

[9] Chada A R, Yaojiang Z, Gang F, et al. Impedance of an infinitely large parallel-plane pair and its applications in engineering modeling[C]. IEEE International Symposium on Electromagnetic Compatibility, Austin, 2009, 78-82

[10] HE H, WU P, LIU F, et al. Parasitic parameter extraction and modeling of via of high speed differential pair [C]// Pro- ceedings of 2015 International Conference on Electronic Packag-ing Technology. Changsha, China : IEEE, 2015 : 619-624.

Reflow Soldering Process Optimization Based on Surface Evolver Solder Joint Shape Simulation and Finite Element Analysis of PCB Assembly

Xing Jin
The 20th Research Institute of China Electronics Technology Group Corporation
Xi'an, China
jx_ustb@163.com

Wenlong Wang
The 20th Research Institute of China Electronics Technology Group Corporation
Xi'an, China
5220338@qq.com

Wenzhong Zhao
The 20th Research Institute of China Electronics Technology Group Corporation
Xi'an, China
1603550958@qq.com

Yuting Zhang
School of Microelectronics University of Xidian
Xi'an, China
390510546@qq.com

Xiaopeng Tan
The 20th Research Institute of China Electronics Technology Group Corporation
Xi'an, China
2955321037@qq.com

Shuai Chen
The 20th Research Institute of China Electronics Technology Group Corporation
Xi'an, China
chenshuai19871219@163.com

Abstract—In view of the complexity and high cost of temperature curve setting in reflow soldering process of multi variety and small batch products, and the existing process simulation methods of PCB components lack of comprehensive consideration of solder paste printing volume and IMC growth and other internal factors resulting in unstable simulation results, the Surface Evolver interactive software is used to establish the physical shape simulation model of solder joints of components, and the IMC growth dynamic model is established based on the principle of atomic diffusion in the reflow process. Then it is creatively combined with the simulation results of reflow soldering process. Through the simulation curve output simulation solder joint model, and then establish the best process parameters window according to the solder joint appearance size, IMC thickness and other parameters, which provides a new research method for PCB assembly reflow soldering process simulation optimization.

Keywords—*PCB assembly, Process simulation, Verification test, Surface mount technology (SMT), Reflow soldering*

I. INTRODUCTION

With the rapid increase in the assembly density of modern electronic products, a printed board assembly may contain tens of thousands of solder joints at most. However, the defect of a solder joint will cause the whole circuit to fail. In the process of reflow soldering, the direct factor that determines the quality of solder joint is reflow temperature curve, so the setting of process parameters of reflow soldering determines the reliability of PCB components. The traditional method of optimizing the reflow curve by multiple temperature measurement using test board and thermometer can satisfy the production of a single type of PCB components to a certain extent, but it is far from meeting the requirements for the production of multi-variety and small batch products. The existing simulation schemes for the process of PCB components are mainly a single reflow soldering process, without considering the comprehensive influence of internal factors such as solder paste type and solder paste printing quantity. The simulation results are quite different from the real results, which cannot truly reflect the reflow process.

In this paper, the Surface Evolver interactive software is used to establish the physical shape simulation model of solder joints of components, and the IMC growth dynamic model is established based on the principle of atomic diffusion in the reflow process. On this basis, the iterative optimization of experimental verification is carried out to obtain the reliability simulation solder joint model. Then it is creatively combined with the simulation results of reflow soldering process. Through the simulation curve output simulation solder joint model, and then establish the best process parameters window according to the solder joint appearance size, IMC thickness and other parameters, which provides a new research method for PCB assembly reflow soldering process simulation optimization.

II. ESTABLISHMENT OF SOLDER JOINT MODEL

A. Establishment of physical form model

Surface Evolver software is used to model the physical shape of solder joints. Based on the principle of minimum energy, gradient descent method is used to calculate the minimum state of energy and predict the final shape of solder joint [1-4]. In the process of reflow soldering, due to the direct effect of energy (hot air, infrared, etc.), the temperature change of resistance and capacitance components, QFP, SOP and other devices with exposed pins is relatively controllable, and the soldering process window is large enough. For BGA devices, because the solder joint array is distributed at the bottom of the device, the energy source of solder joint during soldering is mainly heat conduction and partial convection heat transfer, and the heat transfer of the solder joint in the middle of the device is slow, which is usually the key device type in setting the reflow soldering process curve.

978-1-6654-1392-3/21 $31.00 © 2021 IEEE

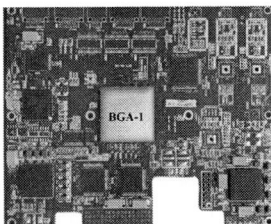

Fig. 1. Physical drawing of X-type PCB assembly

As shown in Fig.1, the main BGA device (BGA-1) in the module has a corresponding solder ball diameter of 0.6mm. In the process of creating solder joint models for the two devices based on the principle of minimum energy, solder joint models are established by comprehensively considering solder paste thickness, reflow temperature curve and other factors, and appropriate solder paste thickness and temperature curve are selected for verification. The results are shown in Fig.2

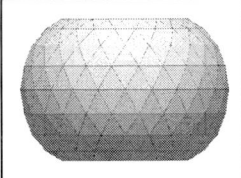 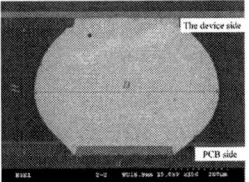

(a) Simulation results,　(b) Verification results.

Fig. 2. Comparison of simulation results and verification results of BGA solder joints under specific process parameters

There are many elements involved in the simulation of physical shape of solder joint. In order to accurately express the degree of fit between simulation results and verification results, this paper specifies three dimensions of data of a single solder joint for characterization, which are the maximum diameter D, height h and pad diameter D of solder joint, as shown in Fig.2b), The corresponding index is that under the same process parameters, the error between the simulation solder joint and the verification results is less than 10%. Through a large number of verification tests, the simulation shape of solder joint is optimized iteratively. According to the verification rules, the simulation results of solder joint physical shape under other process conditions are obtained, and finally the simulation solder joint model library meeting the index requirements is obtained

B. Establishment of IMC growth kinetics model for solder joints

The establishment of solder joint model includes not only the establishment of physical morphology model, but also the establishment of IMC growth kinetics model during reflow soldering. The thickness of IMC between pad and solder directly determines the strength of solder joint when the solder joint of BGA device has no obvious defects. The thickness of IMC is too thin, the reflow time is short or the peak temperature is low, the diffusion is insufficient, and the solder joint strength is low; if IMC is too thick, the reflow time is too long or the reflow temperature is too high, Cu_3Sn will continue to grow on the interface Cu_6Sn_5, the IMC grain is larger, the brittleness is improved, and the solder joint strength is reduced. Therefore, the combination of IMC growth kinetics model and physical morphology model can obtain the solder joint model with guiding significance.

The dynamic model of solder joint IMC growth is based on the interfacial reaction between liquid solder and pad and Nernst-Shchukarev equation [5,6,9,12,13]. The preliminary theoretical formula needs to consider too many variables, and the calculation process is complex. According to the window of reflux temperature curve and the range of solder paste thickness, the model is simplified. At the same time, a large number of experimental results are verified and iteratively modified. Finally, the IMC growth kinetic model is obtained, as shown in (1).

$$\partial(t,T) = c_h[(0.0027t + 0.8966) \times (-9.0336 \times 10^{-5}T^2 + 0.0472T - 5.0844) + b] \quad (1)$$

Where: ∂ (T, t) - IMC thickness after reflow soldering

　　　 t - out of liquidus time (over 183℃ in reflux temperature curve)

　　　 T - Peak temperature of t-reflow curve

　　　 c_h - ratio coefficient

(Note: the ratio coefficient c_h is used to reflect the influence of solder paste thickness on IMC thickness. In this paper, the thickness range of solder paste is 100~200μm. Since the thickness of IMC is approximately proportional to e^(1/h), the ratio coefficient c_h obtained from the reference case is calculated as follows(The base case is that the scale factor is 1 when the paste thickness is 150μm):

$$e^{\frac{1}{h}} : e^{\frac{1}{150}} = c_h : 1 \quad (2)$$

Where: c_h is the proportion coefficient when the thickness of solder paste is h)

Based on this, the solder joint model can be established, and the corresponding physical shape simulation results and IMC thickness results can be obtained under the condition of specific process parameters, which has a clear guiding role for the rationality judgment of process parameters.

III. SIMULATION OF REFLOW SOLDERING PROCESS

The simulation of reflow soldering process is mainly based on the different types of components on PCB components and the different heat exchange modes between reflow soldering equipment and components [7,8]. The research object of this paper is 12 temperature zone hot air reflow soldering equipment. The selected components are X-type PCB components as shown in Fig.1. The equipment modeling is completely based on its own structural parameters and heat transfer mode. The component modeling is created with reference to printed circuit board design documents, component packaging material attribute parameters and structural parameters [10,11], as shown in Fig.3 and Fig.4 respectively.

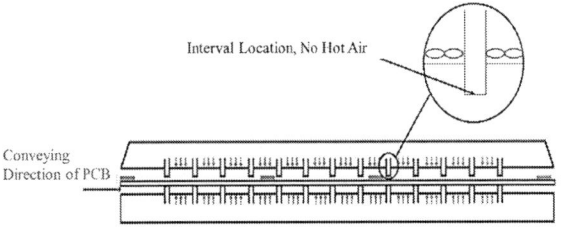

a) Schematic diagram of overall model of reflow soldering equipment

978-1-6654-1392-3/21 $31.00 © 2021 IEEE

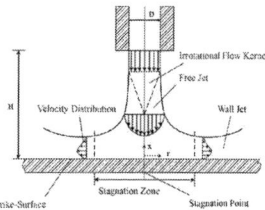

b) Schematic diagram of jet flow field of round nozzle in furnace cavity

Fig. 3. Schematic diagram of reflow soldering equipment model

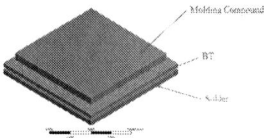

a) Schematic diagram of key device model

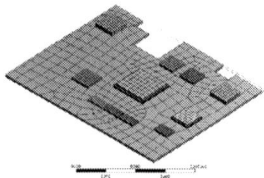

b) Schematic diagram of PCB component modeling and grid generation

Fig. 4. Schematic diagram of PCB assembly simulation model

According to the setting requirements of each temperature zone of the reflow soldering simulation model, the temperature of 12 temperature zones is input to simulate the reflow soldering process of the printed circuit board assembly under specific process conditions, and the temperature change curve of the key temperature measuring points (the position shown in the red dot in Fig.5) of the device BGA-1 in the simulation process is extracted. At the same time, according to the temperature setting used in the simulation process, the test pieces with the same process parameters are made, and the actual soldering temperature curve of the same key temperature measuring points of BGA-1 is measured, which is compared with the simulation results. Through the iterative optimization of the simulation process, the consistency deviation between the measured temperature curve and the temperature curve extracted from the simulation process is less than 5 ℃ under the same temperature measurement point and the same process conditions, which ensures the reliability of the simulation results of PCB reflow soldering process.

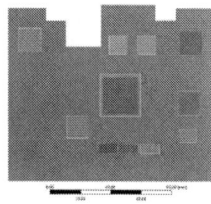

Fig. 5. Schematic diagram of key temperature measurement points in reflow soldering process simulation of components

IV. OPTIMIZATION OF REFLOW SOLDERING PROCESS

Based on the accuracy of the above three simulation models, the set value of temperature zone in Table I is used to simulate the reflow soldering process of X-type components. The relationship between the simulation process temperature data and time of key temperature measuring points of BGA-1 device on the components is extracted, and the simulation reflow temperature curves corresponding to temperature measuring points 1, 2 and 3 are formed, as shown in Fig.6.

TABLE I. TEMPERATURE SETTING OF REFLOW SOLDERING PROCESS SIMULATION

Temperature Setting	Warm Area 1	Warm Area 2	Warm Area 3	Warm Area 4	Warm Area 5
1	160	160	170	170	170
2	120	140	145	150	150
Temperature Setting	Warm Area 6	Warm Area 7	Warm Area 8	Warm Area 9	Chain Speed /cm • min⁻¹
1	180	180	245	250	72
2	190	230	265	265	72

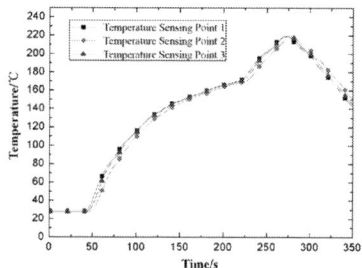

a) Temperature curve of temperature measuring point under temperature setting 1

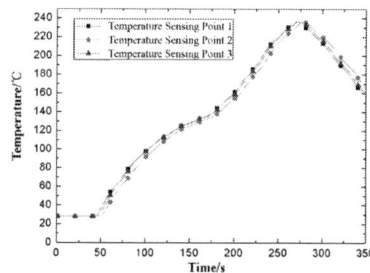

b) Temperature curve of temperature measuring point under temperature setting 2

Fig. 6. Temperature curve of temperature measuring point in process simulation of reflow soldering

The temperature curves of temperature measuring points obtained from the simulation of reflow soldering process under two different temperature settings in Fig. 6 are analyzed. The time beyond liquidus (t) and peak temperature (T) are shown in Table II.

TABLE II. TIME OF TEMPERATURE CURVE EXCEEDING LIQUIDUS AND PEAK TEMPERATURE IN REFLOW SOLDERING PROCESS SIMULATION

Temperature Sensing Point	Temperature Setting 1		Temperature Setting 2	
	Beyond the Liquidus of Time t/s	Peak Temperature T/℃	Beyond the Liquidus of Time t/s	Peak Temperature T/℃
1	81	219.12	109	236.37
2	81	218.38	114	234.66
3	81	219.58	114	236.74

According to (1), the thickness of IMC at the temperature measuring point of reflow soldering process simulation is calculated, and the results are shown in Table III.

TABLE III.   IMC THICKNESS OF TEMPERATURE MEASURING POINT IN REFLOW SOLDERING PROCESS SIMULATION

Temperature Sensing Point	The Size of IMC in Temperature Setting 1/μm	The Size of IMC in Temperature Setting 2/μm
1	1.73	1.92
2	1.72	1.93
3	1.73	1.94

Note: The thickness of solder paste is 150μm, c_h = 1; The pad coating is considered according to HASL, b = 0.7

It can be seen from the results in Table III that the IMC thickness of solder joints obtained by simulation of reflow soldering process with temperature setting 1 and 2 meets the IMC thickness requirements for forming good solder joints, that is, for X-type printed circuit board components, the two temperature settings can achieve good soldering effect.

According to the two kinds of temperature settings, the process parameters of reflow soldering equipment are adjusted, and the X-type PCB assembly test pieces are made. The DPA analysis of BGA-1 device and the actual reflow soldering temperature measurement of key temperature measuring points are carried out for the test pieces under the two groups of parameters. The comparison between the measured temperature and the simulated temperature of the key temperature measuring points of BGA-1 device is shown in Fig.7 (key points 1 and 2 are selected here for comparison, and the temperature curve of key point 3 is basically consistent with that of key point 1). It can be seen that the temperature curve of the key temperature measuring points obtained by simulation and the measured temperature curve show good fitting effect. The SEM image of BGA-1 device is shown in Fig.8, and the average thickness of interface IMC is basically consistent with the simulation results.

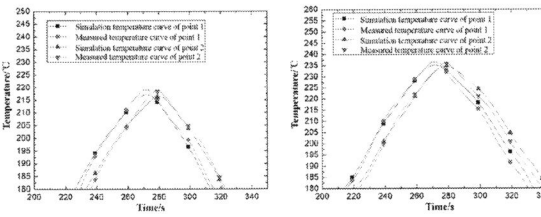

a)   Temperature setting 1   b)   Temperature setting 2

Fig. 7.   Comparison of measured and simulated temperatures of key temperature measuring points of BGA-1

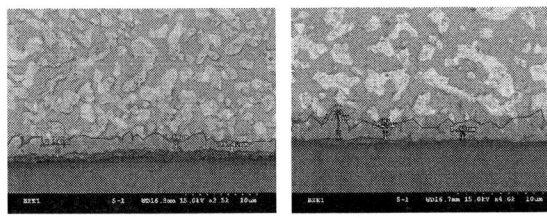

a)   Temperature setting 1   b)   Temperature setting 2

Fig. 8.   IMC scanning electron microscopy imaging of BGA-1

V. CONCLUSION

The physical shape simulation model and IMC growth dynamic model of solder joint optimized by iteration are combined with the simulation analysis results of reflow soldering process, within optimizing the reflow curve by inputting the parameters such as the appearance size and IMC thickness of the solder joint model outputted from the simulation curve, can obtain a more reasonable and accurate process window. It can greatly improve the reflow process of multi-variety and small batch products, reduce the cost of curve optimization and improve the product quality. At the same time, it provides a new idea for SMT process simulation optimization method.

REFERENCES

[1]   Liu X H, Zhou D J. Research on modeling and simulation of PCBA reflow soldering process based on the ICEPAK. Aviation Precision Manufacturing Technology, 2013.

[2]   Chen X Y, Zhou D J, Zhao-Hua W U. A study on the data processing of SMT 3-D solder joint shape based on AutoCAD[J]. Journal of Guilin University of Electronic Technology, 2004.

[3]   Zhao H B, Zhou D J. Study on the conversion method of Three-Dimensional shape prediction model of BGA solder joints[J]. Advanced Materials Research, 2013, 706-708(2):1693-1696.

[4]   Sakai, H, Yamaoka, N, Ujiie, K, Motegi, M, Shiratori, M, & Qiang, Y (2000). Shape prediction and residual stress evaluation of BGA and flip chip solder joints. Conference on Thermal & Thermomechanical Phenomena in Electronic Systems. IEEE.

[5]   Lee J H, Kim Y S. Kinetics of intermetallic formation at Sn-37Pb/Cu interface during reflow soldering [J]. Journal of Electronic Materials, 2002, 31(6):576-583.

[6]   Yoon J W, Lee C B, Jung S B. Growth kinetics of IMC formed between Sn3.5Ag0.75Cu BGA solder and electroless NiP/Cu substrate by solid-state isothermal aging [J]. Materials Science Forum, 2004, 449-452:893-896.

[7]   Fan, Q, Han, G M, Huang, B Y, & Mao, X L. (2004). Study on simulation of SMT. Electric Welding Machine.

[8]   Na L I, Yuan J, Tian X M. Temperature field simulation analysis for lead-free solder in the reflow welding of PCBA adding the tooling [J]. Electric Welding Machine, 2014.

[9]   Zhou M B, Ma X, Zhang X P. Premelting behavior and interfacial reaction of the Sn/Cu and Sn/Ag soldering systems during the reflow process [J]. Journal of Materials Science Materials in Electronics, 2012, 23(8):1543-1551.

[10]   Mao X L, Han G M, Huang B Y. Modeling and simulating of reflow soldering in SMT. Welding Technology, 2004.

[11]   Luo X, Wang Z, Deng W. Optimization model design for temperature curve of reflow furnace[J]. Journal of Physics: Conference Series, 2021, 1802(2):022020 (7pp).

[12]   Yin, Z, Sun, F, Liu, Y, & Liu, Y. (2018). Growth kinetics of IMC at the solid Cu/liquid Sn interface. Soldering and Surface Mount Technology, SSMT-02-2017-0004.

[13]   Dariavach, N, Callahan, P, Liang, J, & Fournelle, R. (2006). Intermetallic growth kinetics for Sn-Ag, Sn-Cu, and Sn-Ag-Cu lead-free solders on Cu, Ni, and Fe-42Ni substrates. Journal of Electronic Materials, 35(7), 1581-1592.

A Novel Bumping Method for Flip-Chip Interconnection

Kun Li
State Key Laboratory of Functional Materials for Informatics, Shanghai Institute of Microsystem and Information Technology (SIMIT), Chinese Academy of Sciences (CAS) CAS Center for Excellence in Superconducting Electronics (CENSE)
Shanghai, China
University of Chinese Academy of Sciences (UCAS)
Beijing, China
likun@mail.sim.ac.cn

Gaowei Xu*
State Key Laboratory of Functional Materials for Informatics, Shanghai Institute of Microsystem and Information Technology (SIMIT), Chinese Academy of Sciences (CAS) CAS Center for Excellence in Superconducting Electronics (CENSE)
Shanghai, China
University of Chinese Academy of Sciences (UCAS)
Beijing, China
*Email: xugw@mail.sim.ac.cn

Quan Zhou
State Key Laboratory of Functional Materials for Informatics, Shanghai Institute of Microsystem and Information Technology (SIMIT), Chinese Academy of Sciences (CAS) CAS Center for Excellence in Superconducting Electronics (CENSE),
Shanghai, China
zhouq@mail.sim.ac.cn

Wei Gai
State Key Laboratory of Transducer Technology, Shanghai Institute of Microsystem and Information Technology (SIMIT), Chinese Academy of Sciences (CAS)
Shanghai, China
gaiwei@mail.sim.ac.cn

Yanhong Wu
State Key Laboratory of Transducer Technology, Shanghai Institute of Microsystem and Information Technology (SIMIT), Chinese Academy of Sciences (CAS)
Shanghai, China
wuyh@mail.sim.ac.cn

Jie Ren
State Key Laboratory of Functional Materials for Informatics, SIMIT ,CAS CAS Center for Excellence in Superconducting Electronics (CENSE)
Shanghai, China
University of Chinese Academy of Sciences (UCAS)
Beijing, China
jieren@mail.sim.ac.cn

Zhen Wang
State Key Laboratory of Functional Materials for Informatics, SIMIT ,CAS CAS Center for Excellence in Superconducting Electronics (CENSE)
Shanghai, China
University of Chinese Academy of Sciences (UCAS)
Beijing, China
zwang@mail.sim.ac.cn

Abstract—In this paper, a novel bumping method for the flip-chip interconnection is proposed, which can be used in the multichip modules (MCM) of superconducting circuits. The silicon bump was prepared by the etching method, and surface was metalized by sputtering niobium metal, which was used for the interconnection between the chip and the substrate. Through electromagnetic simulation, the silicon bump can achieve the transmission performance of $S_{11} \leqslant$ -15dB and $S_{21} \geqslant$ -1dB in the range of 1-100GHz. By using superconducting materials, superconducting interconnection can be realized. And its excellent high-frequency transmission performance shows that silicon bump can be used for multi-chip interconnection of superconducting integrated circuits.

Keywords—interconnection, flip-chip, multichip module, silicon bump, superconducting integrated circuits

I. INTRODUCTION

It is well known that the superconducting integrated circuits (ICs) based on the single flux quantum (SFQ) technology have great advantages in high speed operation and low power consumption, which is an important research direction for the solution of "more than Moore's Law"[1]. The multichip module (MCM) based on superconducting ICs has

a strong requirement for electronic packaging. Interconnection technology has a significant impact on the high-frequency performance of the entire system. In order to ensure the advantages of high-speed transmission for superconducting ICs, it is needed to find a high speed and low loss interconnection technology. Wire bonding is widely used for inter-chip interconnection. But when its operating frequency reaches the gigahertz range, serious parasitic problems will occur. Among the high-frequency range, no matter how well the process parameters are controlled, stronger parasitic parameters will be generated, which limits the application of wire bonding interconnection in high-frequency signal transmission[2]. The flip-chip interconnection is used as one of the most attractive candidates compared to other schemes with low reflection and low insertion loss due to the lower parasitic involved, that provide good performance in high-frequency packaging[3]. Therefore, flip-chip bonding technology has been extensively studied due to its advantages for superconducting ICs.

Among them, bump technology is the key technology of flip-chip interconnection. The quality of bump preparation directly affects the high-frequency signal transmission performance. The existing bump materials mainly include solder bumps, metal bumps, etc., which can be made by the

This work was supported by the Strategic Priority Research Program of Chinese Academy of Sciences (No. XDA18020300).

978-1-6654-1392-3/21 $31.00 © 2021 IEEE

evaporation method, the electroplating method, the laser jet method, etc., each has its advantages and disadvantages[4] [5]. The quality of chip bump preparation directly affects the quality of flip-chip bonding and the reliability of flip-chip bonding[6]. Therefore, this paper developed a method for preparing a bump structure with simple process, good consistency and good shape retention.

II. PROCESS DESIGN

In this paper, we developed a silicon bumping method, which was obtained by etching on the silicon wafer substrate to obtain conical silicon bumps.

The specific bumping process is as follows: 1) On the silicon substrate, the wet etching or dry etching can be prepared to obtain the silicon bump main body. Then, silicon dioxide or silicon nitride is prepared for the insulating layer by thermal oxidation or deposition. 2) Depositing metal niobium. The silicon bump covered niobium and get the metal wiring layer at the same time. The superconducting transition temperature of niobium is 9.2K, so it can be used for superconducting interconnection at low temperature. 3) Due to the niobium is easily oxidized, a layer of silicon dioxide is required to be protected as a passivation layer. At the same time, open a passivation layer on the silicon bump for interconnection. 4) The thin layer of metal indium is evaporated on the silicon bump. Due to high melting point metal, high hardness, there is a certain difficulty during subsequent welding. And indium by evaporated, which has good plasticity and cold-welding performance [7]. Indium can also realize superconducting interconnection at a lower temperature, which to meet the special requirements in superconducting applications. 5) The silicon bumps were coated with a layer of extremely thin solder or conductive glue. By using the ultrasonic bonding or cold pressure bonding methods, the chips were bonded to the silicon substrate by the silicon bumps.

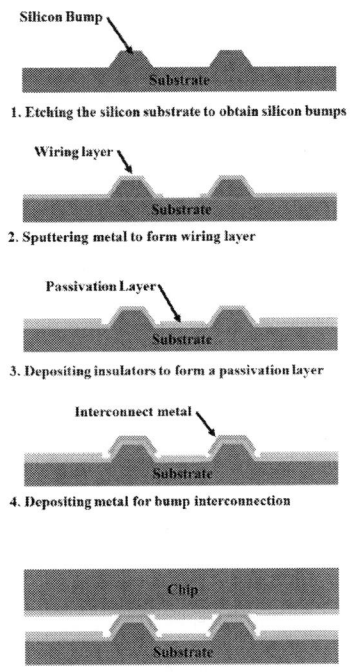

1. Etching the silicon substrate to obtain silicon bumps

2. Sputtering metal to form wiring layer

3. Depositing insulators to form a passivation layer

4. Depositing metal for bump interconnection

5. Bonding the chip and the substrate

Fig.1. Process flow the silicon bump.

The bumping process was simple and the main body of the bump was made of silicon material, which has good shape retention. It was very robust to bonding pressure compared with solder material, reduced the risk of the circuits shorted which caused by the adjacent bump collapsed.

III. SIMULATION

The designed interconnect structure of CPW-to-CPW (coplanar waveguide) was simulated using electromagnetic simulation software.

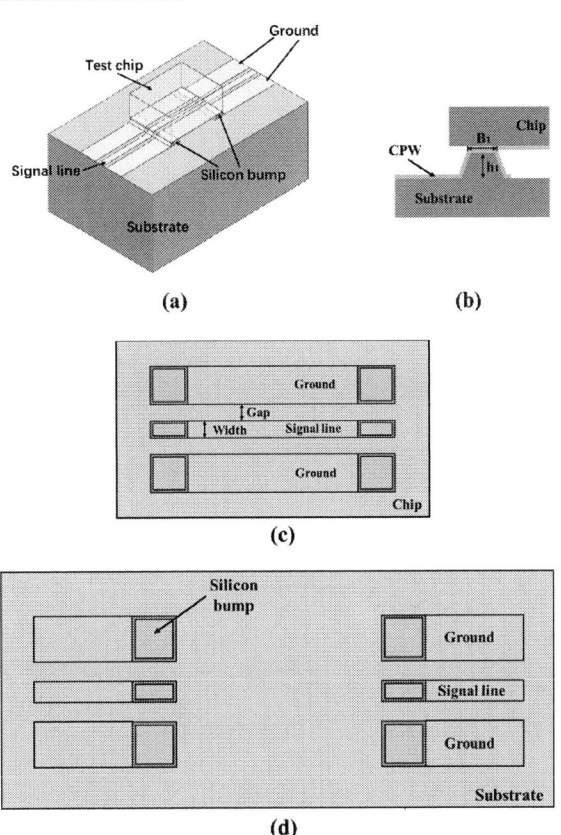

Fig.2. The interconnect structure of CPW-to-CPW: (a) Schematic diagram of model structure; (b) The CPW on the substrate is interconnected with the CPW on the chip through silicon bumps. B_1 is the interconnect length between silicon bump and chip, h_1 is the height of silicon bump; (c) The layout structure of CPW on the chip; (d)The layout structure of CPW on the substrate. Gap is the distance between the ground plane and the center signal line, and Width is the width of the center signal line.

A CPW structure with the characteristic impedance of 50Ω was designed, in which the metal layer was set as an ideal conductor to simulate the superconducting environment in the simulation design. The Gap of CPW was designed to be 55um through calculation, the Width of the center signal line was 90um, and the modeling length was 1mm. Through simulation, the designed CPW structure has the transmission performance of $S_{11} \leq$-10dB and $S_{21} \geq$-1dB in the range of 1-100GHz, which verifies its high frequency transmission performance. Therefore, based on the designed CPW structure, the CPW-to-CPW interconnection structure was also designed to interconnect the CPW on the substrate and the CPW on the chip through bumps to verify the influence of the bumps insert on the signal transmission of the flip-chip structure.

978-1-6654-1392-3/21 $31.00 © 2021 IEEE

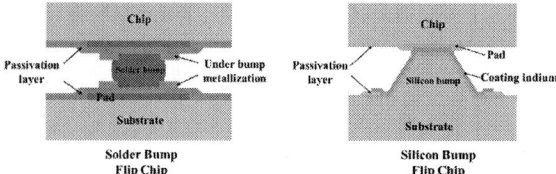

Fig.3. Schematic diagram of two different bump bonded structures.

In the following modeling and simulation, we designed two different bump interconnection structures. One is the solder bump structure, the other is the silicon bump structure. Which to simulate the influence of different bump structures on the signal transmission.

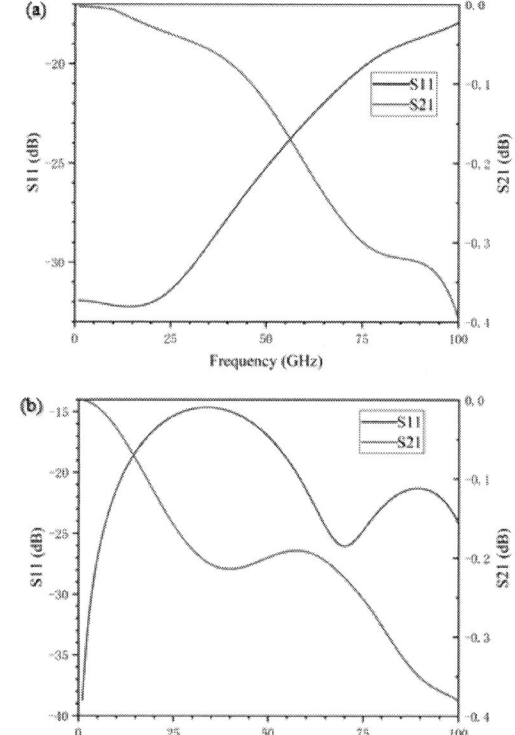

Fig.4. S-parameter: (a) solder bump; (b) silicon bump.

The results showed that the flip-chip structure has excellent transmission performance up to 100GH. Compared with the traditional solder bump flip-chip bonding process, the silicon bump can also transmit high frequency signals, and realized the transmission performance of $S_{11} \leqslant -15dB$ and $S_{21} \geqslant -1dB$ in the range of 1-100GHz. Because of its simple process, and the height of the bump was controllable and could be prepared with extremely small thickness, suitable for low loss and low delay transmission of high frequency and high bandwidth signals. The advantages of silicon bump indicated that the silicon bump can be used for multi-chip interconnection in superconducting ICs.

IV. CONCLUSION

A silicon bumping method was developed, which can be used for the multi-chip interconnection of superconducting ICs. The silicon bump is directly formed by substrate micromachining, and the bumping process is simple. The bump material is silicon, which is very robust to bonding pressure and can reduce the risk of short circuits caused by bump collapse. Simulation results showed that excellent transmission performance up to 100GHz can be obtained.

ACKNOWLEDGMENT

The authors would like to thank Ms. Minghui Niu, Ms. Huanli Liu, Ms. Hui Zhang of CAS Center for Excellence in Superconducting Electronics (CENSE), the superconducting design process test platform and Superconducting Electronics Facility (SELF) of CENSE, for their helpful technical supports.

REFERENCES

[1] Heinrich W. The flip-chip approach for millimeter wave packaging. Microwave Magazine IEEE 2005; 6(3):36-45.

[2] Hirose T. High-frequency IC packaging technologies. Indium Phosphide and Related Materials 2003.

[3] Wong Y L, Patrison L, Linton D. Flip chip interconnect analysis at millimeter wave frequencies. High Frequency Postgraduate Student Colloquium 1999.

[4] Jentzsch A, Heinrich W. Theory and measurements of flip-chip interconnects for frequencies up to 100 GHz. Microwave Theory & Techniques IEEE Transactions 2001; 49(5):871-878.

[5] Hsieh M C, Tzeng S L. Design and stress analysis for fine pitch flip chip packages with copper column interconnects. 15th International Conference on Electronic Packaging Technology. IEEE, 2014.

[6] Gao S. New technologies for lead-free flip chip assembly. Imperial College London 2005.

[7] Ikeda A, Kajiwara K, Watanabe N, et al. High-frequency signal transmission characteristics of coplanar waveguides with cone bump interconnections. Tencon IEEE Region 10 Conference. IEEE, 2011.

Design and Test of Transmission Line in SFQ Circuit

Quan Zhou
State Key Laboratory of Functional Materials for Informatics, Shanghai Institute of Microsystem and Information Technology (SIMIT), Chinese Academy of Sciences (CAS) CAS Center for Excellence in Superconducting Electronics (CENSE) School of Materials Science and Engineering, Shanghai University
Shanghai, China
zhouq@mail.sim.ac.cn

Gaowei Xu*
State Key Laboratory of Functional Materials for Informatics, Shanghai Institute of Microsystem and Information Technology (SIMIT), Chinese Academy of Sciences (CAS) CAS Center for Excellence in Superconducting Electronics (CENSE)
Shanghai, China
University of Chinese Academy of Sciences (UCAS)
Beijing, China
*Email: xugw@mail.sim.ac.cn

Kun Li
State Key Laboratory of Functional Materials for Informatics, Shanghai Institute of Microsystem and Information Technology (SIMIT), Chinese Academy of Sciences (CAS) CAS Center for Excellence in Superconducting Electronics (CENSE)
Shanghai, China
University of Chinese Academy of Sciences (UCAS)
Beijing, China
likun@mail.sim.ac.cn

Lingyun Li
State Key Laboratory of Functional Materials for Informatics, SIMIT ,CAS CAS Center for Excellence in Superconducting Electronics (CENSE)
Shanghai, China
University of Chinese Academy of Sciences (UCAS)
Beijing, China
lilingyun@mail.sim.ac.cn

Le Luo
State Key Laboratory of Functional Materials for Informatics, SIMIT ,CAS CAS Center for Excellence in Superconducting Electronics (CENSE)
Shanghai, China
University of Chinese Academy of Sciences (UCAS)
Beijing, China
leluo@mail.sim.ac.cn

Jie Ren
State Key Laboratory of Functional Materials for Informatics, SIMIT ,CAS CAS Center for Excellence in Superconducting Electronics (CENSE)
Shanghai, China
University of Chinese Academy of Sciences (UCAS)
Beijing, China
jieren@mail.sim.ac.cn

Zhen Wang
State Key Laboratory of Functional Materials for Informatics, SIMIT ,CAS CAS Center for Excellence in Superconducting Electronics (CENSE)
Shanghai, China
University of Chinese Academy of Sciences (UCAS)
Beijing, China
zwang@mail.sim.ac.cn

Xiaoming Xie
State Key Laboratory of Functional Materials for Informatics, SIMIT ,CAS CAS Center for Excellence in Superconducting Electronics (CENSE)
Shanghai, China
University of Chinese Academy of Sciences (UCAS)
Beijing, China
xmxie@mail.sim.ac.cn

Abstract—In this paper, the transmission line structure applied to the SFQ circuit was designed and tested. The 4.7Ω impedance MSL designed and prepared based on the Nb-based superconducting process can ensure the PTL circuit to have a current bias margin of 58%, and achieve 0 error rate during the transmission of the SFQ signal. The prepared 50Ω impedance MSL and CPW have been measured for low-temperature and high-frequency S-parameters. The results show that the structure has excellent transmission performance with $S_{11} \leqslant$ -15dB and $S_{21} \geqslant$ -1dB at 0-18GHz.

Keywords—*Transmission line, SFQ circuit, High frequency interconnection*

I. INTRODUCTION

SFQ (Single Flux Quantum) circuit has great application prospects in the field of supercomputing in the future due to its advantages of high speed and low power consumption. For the interconnection of SFQ circuits as an indispensable part in the research process, there are currently two main methods,

namely JTL (Josephson Transmission Line) and PTL (Passive Transmission Line). Among them, because PTL transmits SFQ signals through the transmission line structure, it has the advantages of low circuit power consumption and small propagation delay[1][2]. Therefore the transmission line structure based on the Nb-based superconducting process was designed and simulations and tests were carried out.

In this paper, based on the design and simulation of the Nb-based superconducting process, 4.7Ω impedance MSL (Microstrip line) was prepared, and an impedance-matched SFQ-Driver/Receiver was designed and fabricated at both ends of the PTL. Through the size of the current bias margin of Driver/Receiver under 0 symbol error rate, the transmission performance of PTL to SFQ signal is judged and verified. The test result of "Driver+PTL+Receiver" under 4.2K shows that: under the condition of ±58% current bias margin, Driver/Receiver maintains 0 error rate, proving that PTL has good SFQ signal transmission performance. In order to further verify the design of the SFQ circuit transmission line, 50Ω MSL and CPW (Coplanar waveguide) transmission

---

This work was supported by the Strategic Priority Research Program of Chinese Academy of Sciences (No. XDA18020300).

lines and matching test pads were designed based on the Nb superconducting process, and electromagnetic field simulation and microwave test verification were carried out. The S-parameter test was carried out on the low-temperature microwave probe station, and the results showed that the insertion loss of MSL and CPW was ≤1dB and the return loss was ≥15dB (0-18GHz) at 4.2K, indicating that MSL and CPW based on Nb process have good performance at high frequency Signal transmission and can realize chip-to-chip high-frequency interconnection.

II. DESIGN AND SIMULATION

The signal transmission interconnection in the SFQ circuit mainly relies on two basic unit circuits: Josephson Transmission Line (JTL) and Passive Transmission Line (PTL). Among them, the short-distance signal transmission within the circuit is mainly realized through JTL, while the long-distance signal transmission between the circuits requires the use of PTL [3].

PTL uses transmission lines, such as Microstrip Line (MSL), Coplanar Waveguide (CPW), to transmit SFQ signals. In order to distinguish it from the active transmission line JTL, they are collectively referred to as PTL. The PTL transmission circuit is composed of three parts, namely the transmitter (Driver), receiver (Receiver) and PTL, as shown in Fig.1.At the transmitter end, because the sum of the SFQ input pulse signal and the bias current I_B exceeds the critical current Ic of the transmitter junction, the transmitter junction generates an SFQ pulse signal, which is continuously transmitted from the PTL to the receiver. Then the sum of SFQ pulse current and the bias current of the receiver exceeds the critical current value of the receiver junction, so the receiver junction generates an SFQ pulse, and the signal continues to be transmitted to the subsequent circuit[4].

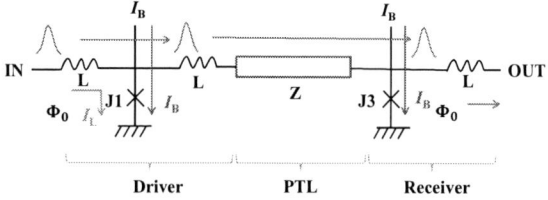

Fig.1. PTL circuit schematic.

Since PTL is a superconducting transmission line, the resistance is 0, so there is no RC delay. The model can be equivalent to an LC model, so compared to JTL, the delay per unit length of PTL is very short. And only the transmitter and receiver have Josephson junction, so the power consumption and timing error will not increase with the interconnection length, which is suitable for long-distance interconnection. And because the SFQ pulse signal itself has the characteristics of high frequency, so the superconducting PTL can propagate with picosecond width SFQ pulse during the wiring process, and there should be no obvious loss[5].

In this paper, the SIMIT-Nb03 process was used to make superconducting circuit structures[6]. The cross-sectional TEM image of the process is shown in Fig.2, and the information of each layer is shown in Table I. When designing the PTL structure for the SFQ circuit, the impedance of the transmission line needs to match Driver and Receiver. The impedance of Driver and Receiver designed

and prepared under the Nb03 process is 4.7Ω, by using these parameters MSL with characteristic impedance of 4.7Ω was designed. At the same time, in order to verify the transmission performance of the transmission line, MSL and CPW with a simulated characteristic impedance of 50Ω were designed and the low temperature and high frequency S parameters were tested.

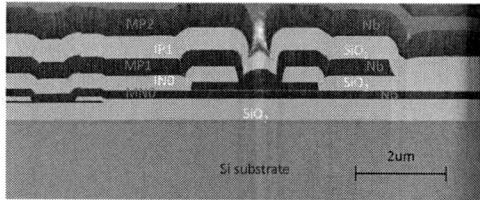

Fig.2. TEM profile of SIMIT-Nb03 process PTL circuit schematic.

Electromagnetic field simulation software HFSS was used to assist in the design of transmission line structure. The simulation parameters of the transmission line structure with a characteristic impedance of 50Ω are shown in Table II. The cross-sectional diagram of CPW and MSL is shown in Fig.3. The simulation uses perfect conductor conditions to simulate superconductors and the simulation frequency is 0-18GHz.

TABLE I.   PARAMETERS OF EACH LAYER OF SIMIT-Nb03 PROCESS

Layer	Material	thickness(nm)
PP2	Au	100±5
MP2	Nb	500±25
IP1	SiO₂	400±20
MP1	Nb	300±15
IN0	SiO₂	250±10
MN0	Nb	150±10

TABLE II.   CPW AND MSL PARAMETERS OF DIFFERENT SIGNAL LINE METAL LAYERS

Parameter	CPW	MSL
GND	MP2+MP1+MN0	MP2
Dielectric layer	Si substrate	IN0+IP1
Signal layer	MP2+MP1+MN0	MN0
Signal line width	90um(Gap:55um)	1um
Line length	1mm	1mm

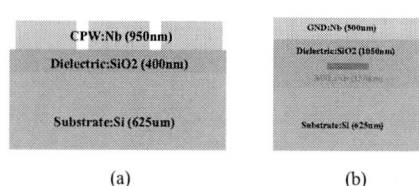

Fig.3. CPW and MSL cross-section diagram: (a) CPW; (b) MSL

978-1-6654-1392-3/21 $31.00 © 2021 IEEE

The simulation result is shown in Fig. 4. The designed CPW and MSL structure can achieve $S_{11} \leqslant$ -20dB and $S_{21} \geqslant$ -1dB at 0-18GHz, with excellent high-frequency signal transmission performance.

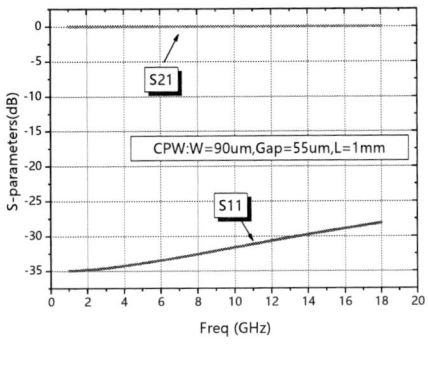

(a)

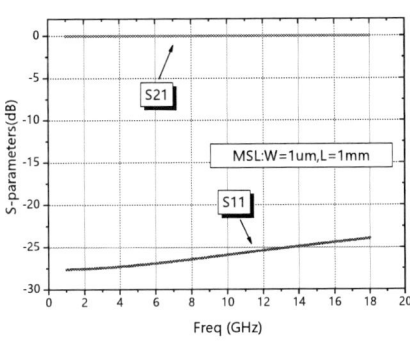

(b)

Fig.4. CPW and MSL simulation S parameters: (a) CPW; (b) MSL.

III. EXPERIMENTS AND RESULTS

A. PTL logic function test

The low-temperature superconducting digital circuit test system, as shown in Fig. 5, was used to test the logic function of the PTL structure. The PTL circuit structure(Fig. 6) was in a liquid helium environment(4.2K), appropriate current bias is given. By inputing a specific waveform at the input terminal, and detecting the output waveform at the output terminal to determine whether the logic function of the circuit is correct[7].

Fig. 5. Superconducting digital circuit test system.

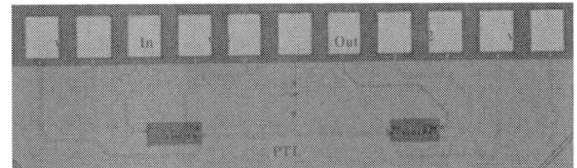

Fig. 6. PTL circuit photo.

The correct PTL logic function is shown in Fig. 7. The output signal is reversed when the rising edge of each input signal arrives. The collection of all current bias points that can make the PTL circuit work normally is called the working margin. The larger the margin, the circuit structure will be more stable. The bias current within the margin range can enable the PTL circuit to achieve 0 bit error rate signal transmission.

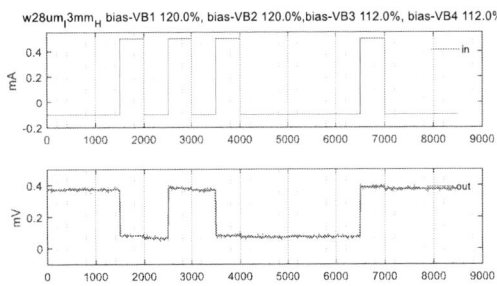

Fig.2. TEM profile of SIMIT-Nb03 process PTL circuit schematic.

The test results show that the prepared PTL structure with a margin of ±58% can enable the PTL circuit to achieve 0 bit error rate signal transmission, indicating that the designed and prepared 4.7Ω MSL has excellent transmission performance for high-frequency SFQ signals.

B. Transmission line low temperature S parameter test

In order to further verify the low-temperature and high-frequency transmission performance of the designed and prepared transmission properties, the S parameters of the CPW and MSL structures with a characteristic impedance of 50Ω were tested. The test is carried out in the system shown in Fig. 8. The low temperature S parameters of CPW and MSL at 0-18 GHz were measured by a low temperature probe station and a vector network analyzer.

Fig. 8. Low temperature and high frequency S-parameter test system.

The test results show (Fig. 10) that the prepared CPW can achieve 0-18GHz, $S_{11} \leqslant$ -20dB, $S_{21} \geqslant$ -1dB signal transmission, and the prepared MSL can achieve 0-18GHz, $S_{11} \leqslant$ -15dB, $S_{21} \geqslant$ -1dB signal transmission. Both have excellent low temperature and high frequency signal transmission performance.

Comparing the measured results and the simulation results, the main difference is that the measured reflectance is larger than the simulation results. This difference may be due to the inability of the simulation software to fully simulate the superconductor and the error of the test system.

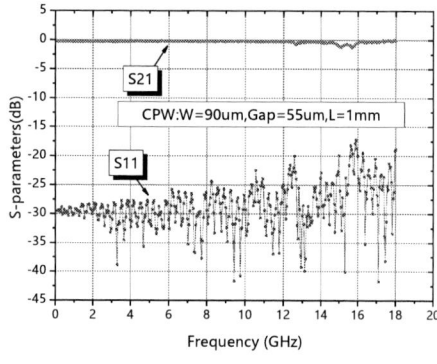

(a)

(b)

Fig. 9. Low temperature and high frequency S-parameter:(a) CPW;(b) MSL.

IV. CONCLUSION

In this paper, the transmission line structure applied to the SFQ circuit was designed and tested.

1. The CPW and MSL structures were designed based on the SIMIT-Nb03 process. The simulation results show that 50 Ω CPW and MSL can achieve 0-18GHz, $S11 \leqslant$-15dB, $S21 \geqslant$-1dB signal transmission.

2. The 4.7Ω impedance MSL was tested at a low temperature of 4.2K so that the Driver+PTL+Receiver structure can ensure 0 error rate transmission of the on-chip SFQ signal under the condition of a current bias margin of ±58%.

3.The designed 50Ω CPW can achieve transmission of $S_{11} \leqslant$-20dB and $S_{21} \geqslant$-1dB for 0-18GHz high frequency signals in a 4.2K low temperature environment, and the designed 50Ω MSL can achieve transmission of $S_{11} \leqslant$-15dB and $S_{21} \geqslant$-1dB for 0-18GHz.The experimental results shows show that CPW and MSL can be used for SFQ, CMOS and other low temperature Chip-to-chip high-frequency interconnection.

ACKNOWLEDGMENT

The authors would like to thank the superconducting design process test platform of the Center for Excellence in Superconductivity Electronics, Chinese Academy of Sciences.

REFERENCES

[1] Takagi K , Tanaka M , Iwasaki S , et al. SFQ Propagation Properties in Passive Transmission Lines Based on a 10-Nb-Layer Structure[J]. IEEE Trans.appl.supercond, 2009, 19(3):617-620.

[2] M R Rafique, I Kataeva, H Engseth. Optimization of superconducting microstrip interconnects for rapid single-flux-quantum circuits[J]. Superconductor Science & Technology, 2005.

[3] Likharev K K , Semenov V K . RSFQ logic/memory family: a new Josephson-junction technology for sub-terahertz-clock-frequency digital systems[J]. IEEE Transactions on Applied Superconductivity, 2002, 1(1):3-28.

[4] Duzer T V , Zheng L , Whiteley S R , et al. 64-kb hybrid Josephson-CMOS 4 Kelvin RAM with 400 ps access time and 12 mW read power[J]. IEEE Transactions on Applied Superconductivity, 2013, 23(3):1700504-1700504.

[5] Bunyk P , Likharev K , Zinoviev D . RSFQ technology: Physics and devices[J]. International Journal of High Speed Electronics and Systems, 2001, 11(1).

[6] Ying L , Zhang X , Niu M , et al. Development of Multi-Layer Fabrication Process for SFQ Large Scale Integrated Digital Circuits[J]. 2020.

[7] Zinoviev D Y , Polyakov Y A . Octopux: an advanced automated setup for testing superconductor circuits[J]. Applied Superconductivity IEEE Transactions on, 1997, 7(2):3240-3243

The Influence Of Soldering Voids In Power Devices

Jianming Fang
Reliability Analysis and Research Centre
China Electronic Product Reliability and Environmental Testing Research Institute (CEPREI)
Guangzhou, China
fangjm08@126.com

Min Wang
Reliability Analysis and Research Centre
China Electronic Product Reliability and Environmental Testing Research Institute (CEPREI)
Guangzhou, China

Xuanlong Chen
Reliability Analysis and Research Centre
China Electronic Product Reliability and Environmental Testing Research Institute (CEPREI)
Guangzhou, China

Abstract—**The smaller size, higher integration, higher power density and higher reliability has become the development trends of power electronics, but at the same time, the heat dissipation issue of power devices is more serious. In order to improve system reliability, the heat generated by the power devices should be quickly and efficiently transferred to the external environment. The soldering interface of the power chip has the function of conduction, support and heat dissipation. The soldering interface of the die is the key interface in the heat flow path from the die to the heat sink. If there are voids or delamination on the soldering interface, the heat transfer will be blocked and the thermal resistance of the device will increase, causing local overheating of the chip, and ultimately causing the power module to activate thermal protection or the power chip to overheat and burn out. In this paper, a failure analysis has been carried out on power electronic module whose failure was due to die soldering voids, the (3D)X-Ray and finite element simulation had been used. The purpose is to clarify the failure mechanism, summarize the influence law of soldering voids, and propose improvement measures (vacuum reflow soldering, low-temperature sintering nano-silver technology, etc.) to improve product reliability.**

Keywords—Power Electronics; Soldering Voids; Heat Dissipation; Reliability; Failure Analysis; (3D)X-Ray; Finite-Element Simulation

I. INTRODUCTION

In recent years, with the emergence of wide-bandgap semiconductor (SiC, GaN) power devices, the switching frequency and power density of power electronic modules have been greatly improved, and the smaller size, higher integration, higher power density and higher reliability has become the development trends of power electronics, but at the same time, the heat dissipation issue of power devices is more serious [1]. The energy loss of power devices includes conduction loss and switching loss. These energy losses are converted into heat energy inside the module, making the chip work in a high temperature environment.

The mean time between failures (MTBF) of an electronic system increases exponentially with the reciprocal of temperature, indicating that the lower the operating temperature of the system, the greater the MTBF, and the higher the system reliability [2]. Therefore, the heat generated by the power devices should be quickly and efficiently transferred to the external environment.

In this paper, a failure analysis had been carried out on power electronic module whose failure was due to die soldering voids, the (3D)X-Ray and finite element simulation had been used. The purpose is to clarify the failure mechanism, summarize the influence law of soldering voids, and propose improvement measures (vacuum reflow soldering, low-

temperature sintering nano-silver technology, etc.) to improve product reliability.

II. THEORY AND METHOD

In the package structure, the power chip is usually soldered on the substrate with solder paste, then the substrate dissipates heat to the outside through the heat sink, as shown in Fig.1. The soldering interface of the power chip has the function of conduction, support and heat dissipation. The soldering interface of the die is the key interface in the heat flow path from the die to the heat sink [3]. If there are voids or delamination on the soldering interface, the heat transfer will be blocked and the thermal resistance of the device will increase, causing local overheating of the chip, and ultimately causing the power module to activate thermal protection or the power chip to overheat and burn out.

The X-Ray inspection is one of the important means of failure analysis, which can penetrate silicon dies, plastic packaging materials, and ceramic materials, non-destructive inspection of the void distribution of the soldering interface. The X-Ray equipment can be divided into (2D)X-Ray and (3D)X-Ray [4]. The (2D)X-Ray can be used to check the overall frame structure of the device, but the images of each layer are superimposed; However the (3D)X-Ray could scans and reconstructs the overall three-dimensional frame structure of the device, and could be used to check the structure of any cross-section in the device, it is especially suitable for checking the void distribution of the soldering interface.

We can also use finite element simulation to more intuitively characterize the influence of factors such as the size, shape and distribution of the soldering interface voids on the heat flow distribution and temperature distribution of the system, which can further be used to guide the thermal design of the power module.

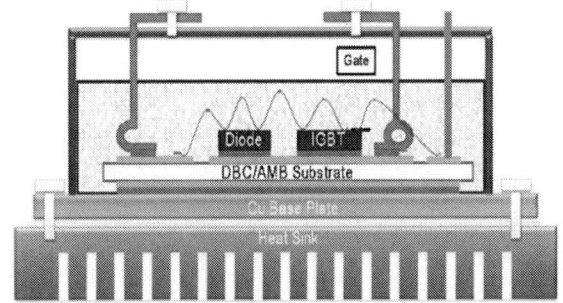

Fig.1 The typical package structure of power module

978-1-6654-1392-3/21 $31.00 © 2021 IEEE

III. CASE ANALYSIS AND DISCUSSION

A. Case Study I

When a motor drive module (see Fig.2) had been running for several hours, the output of phase A would be interrupted, but it could still work normally after re-power on.

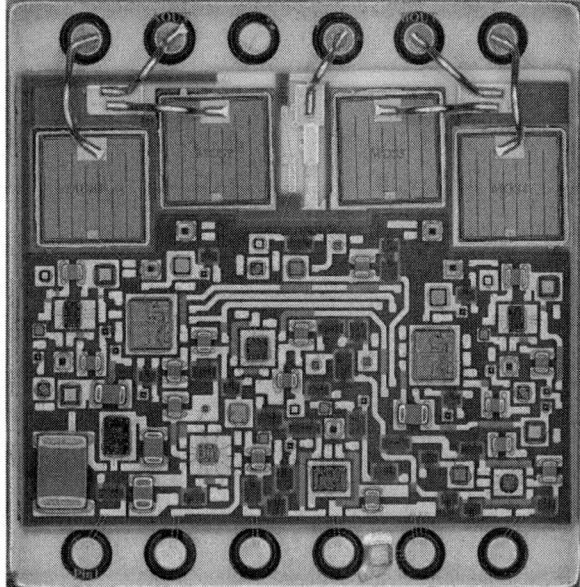

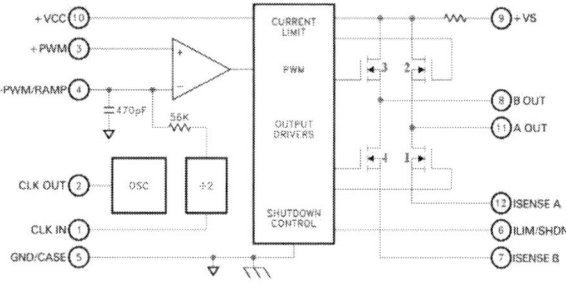

Fig.2 Module internal structure and functional schematic diagram

(3D)X-Ray inspection was performed on the soldering interface of the four MOS dies inside the module. It was found that the void ratio of the soldering interface of the MOS2 in the A phase was about 25.1%, which was obviously larger. The void ratio of the soldering interface of the other MOS were only about 10%, as shown in Fig.3.

Fig.3 The (3D)X-Ray image of MOS die soldering interface

In order to evaluate the impact of die soldering voids on heat dissipation, thermal simulation was performed using finite element analysis software (Ansys). The soldering interface of MOS2 and MOS3 were selected for 1:1 modeling and simulation, and compared with the ideal situation which the soldering void ratio was 0%. The simplified CAD model was shown in Fig.4. The power consumption of the MOS was set to 20W, and the heat transfer coefficient of the bottom surface of the heat dissipation substrate was set to 1400W/m². ℃. The typical simulation results are shown in Fig.5 and Table 1.

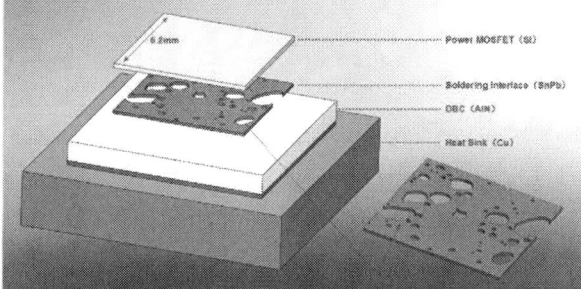

Fig.4 The simplified CAD model

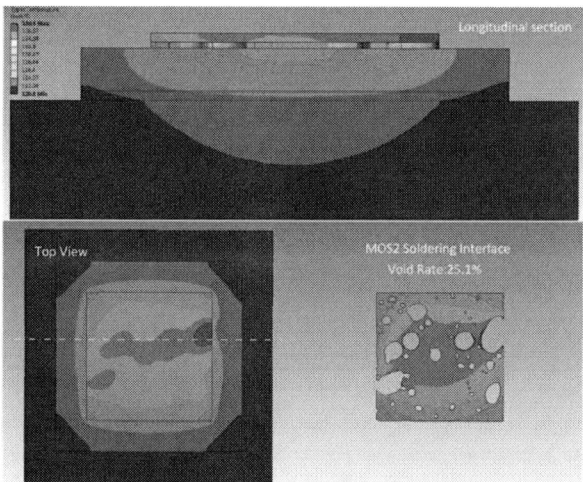

Fig.5(a) The thermal simulation results of MOS2

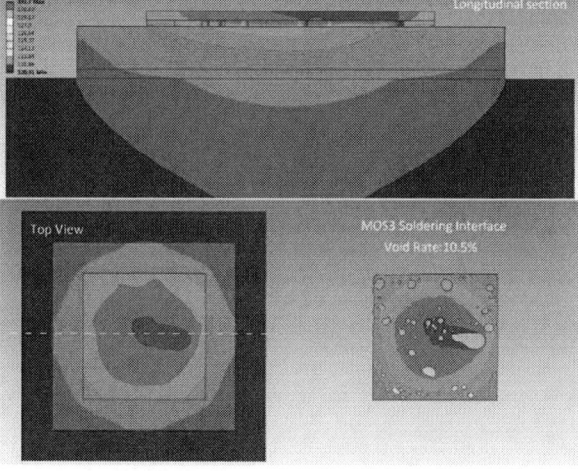

Fig.5(b) The thermal simulation results of MOS3

978-1-6654-1392-3/21 $31.00 © 2021 IEEE

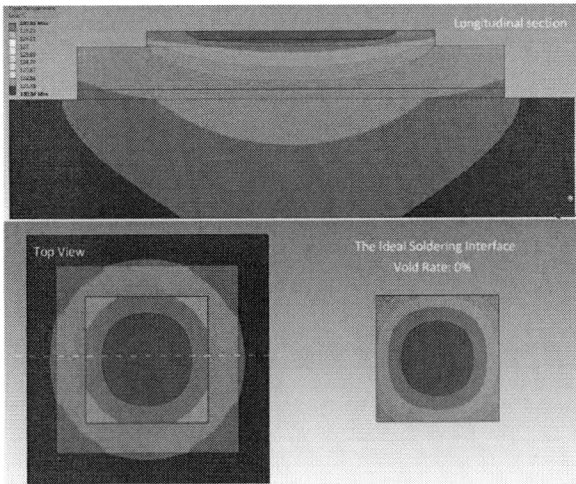

Fig.5(c) The thermal simulation results of ideal situation

Table 1 Thermal simulation data statistics

CAD Model	Soldering Void Ratio	Maximum Temperature Difference of Module	Maximum Temperature Difference of MOS
MOS2	25.1%	18.3℃	9.6℃
MOS3	10.5%	11.4℃	5.2℃
Ideal Situation	0%	10.0℃	3.9℃

The simulation results showed that the die would have local hot spots above the voids in the soldering interface, and the impact of edge voids was greater; the higher the void rate, the greater the temperature difference and thermal resistance of the die, which might easily cause local overheating.

In this case, the solder void rate of the A-phase MOS2 reached about 25.1%, which was obviously higher, would cause poor heat dissipation and local hot spots. If it was used for a long time, the junction temperature of the die would continue to rise, triggering the protection function inside the module and affecting its normal output.

B. Case Study II

A motor drive module (see Fig.6) found an abnormal output during the external test.

Fig.6 Module functional schematic diagram

(3D)X-Ray inspection was performed on the soldering interface of the four MOS inside the module, and found that there were obvious voids on the soldering interface of the MOS1, MOS3 and MOS4. Then the internal inspection found that the MOS was overheated and burned on the position of soldering voids, as shown in Fig.7.

Due to the obvious soldering voids in MOS, it was easy to cause local hot spots on the die during operation. Eventually, the die would fail due to uncontrolled overheating, and causing high current to burn out [5].

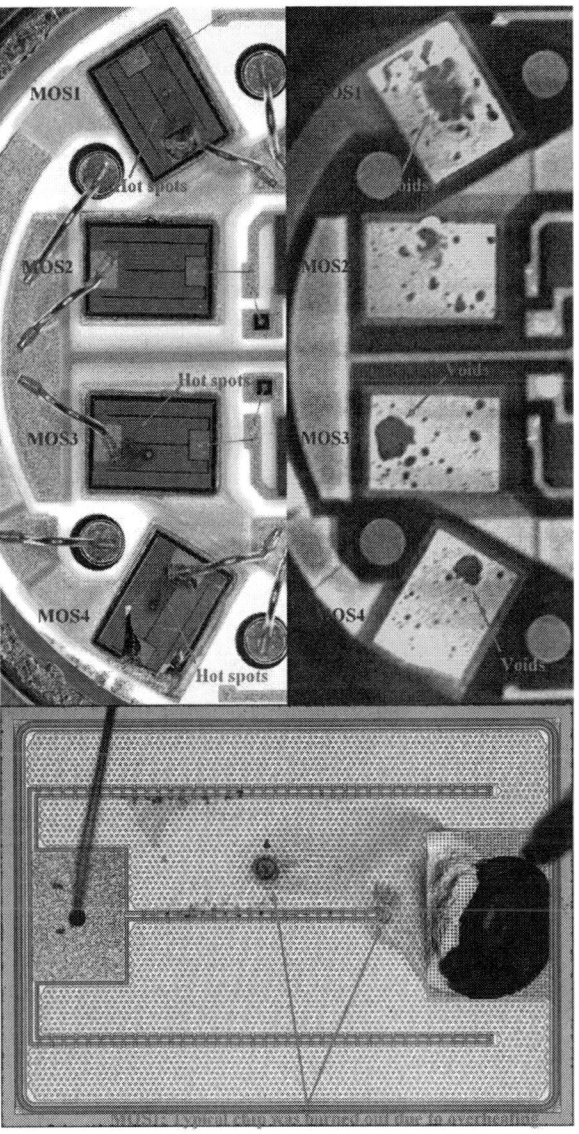

Fig.7 MOS burned by overheating and soldering voids image

C. Improvement Measures

Through the above two typical cases, we know the hazards of soldering voids in power chips. From the perspective of the soldering process, since the solder paste contains flux, a gaseous state is formed during the reflow soldering process. If the gas cannot be discharged in time, the soldering voids are formed. Therefore, in order to avoid the generation of soldering voids, there are two methods:

978-1-6654-1392-3/21 $31.00 © 2021 IEEE

1) Using the vacuum reflow oven: Due to the addition of the air extraction process during the reflow soldering process, the gas formed during the reflow soldering process can be discharged in time, avoiding the generation of soldering voids.

2) Using the low-temperature sintered nano-silver: (a) Excellent electrical and thermal conductivity, which is 3 to 5 times that of ordinary solder; (b) It can work stably at a high temperature above 200°C, suitable for high temperature, high power and high density package; (c) It is a green electronic packaging material, free of lead and other toxic metals [6].

IV. CONCLUSION

This paper analyzes the cases of power electronic module whose failure were due to soldering voids through experiment and simulation, and concludes:

1) How to use (3D)X-Ray and finite element simulation to evaluate the impact of soldering voids.

2) Above the soldering voids, the die will form local hot spots, and the edge voids have a greater impact.

3) The higher soldering void rate, the larger die temperature difference and thermal resistance will be. It is easy to cause local overheating, and eventually cause the power module to start thermal protection or the power chip directly overheats and burns out.

4) In order to reduce the influence of soldering voids, vacuum reflow soldering or low-temperature sintering nano-silver technology can be used to improve product reliability.

ACKNOWLEDGMENT

This work is financially supported by the National High Quality Development Program of China under Grant No. 2020-0093-2-1, the Key R&D Program of Guangong under Grant No. 2019B010143002

REFERENCES

[1] Chen Z. "Electrical Integration of SiC Power Devices for High-Power-Density Applications", VIRGINIA POLYTECHNIC INSTITUTE AND STATE UNIVERSITY, 2013.

[2] Richard K.Ulrich and William D.Brown. "Advanced Electronic Packaging", the Institute of Electrical and Engineers, 2006.

[3] Younes Shabany. "Heat Transfer: Thermal Management of Electronics", the China Machine Press, 2013.

[4] Sylvester Y, Johnson B, Estrada R, et al. "3D X-Ray microscopy: A nondestructive high resolution imaging technology that replaces physical cross-sectioning for 3D IC packaging", the Advanced Semiconductor Manufacturing Conference, 2013.

[5] Yunfei E, Ping L and Shaoping L. "Failure Analysis Technology of Electronic Components", the House of Elecronics Industry, 2015.

[6] Zhang Z Z, Lu G Q. "Pressure-assisted low-temperature sintering of silver paste as an alternative die-attach solution to solder reflow", Electronics Packaging Manufacturing, IEEE Transactions on, 2002, 25(4): 279-283.

Double-sided Electroplating Process for Through Glass Vias (TGVs) Filling

Ke Li
Xiamen University
Xiamen Sky Semiconductor
Xiamen, China
lik@sky-semi.com

Heng Wu
Xiamen Sky Semiconductor
Xiamen, China
wuh@sky-semi.com

Weijian Chen
Xiamen Sky Semiconductor
Xiamen, China
chenwj@sky-semi.com

Daquan Yu*
Xiamen University
Xiamen Sky Semiconductor
Xiamen, China
yudq@sky-semi.com

Abstract—**This article reported a through-glass vias(TGVs) plating technology based on glass substrates. As one of the important technologies for continuation of Moore's law, through holes play the role of electrical connection on both sides. The TGVs were formed by laser drilling on the glass substrate, sputtering titanium as a barrier layer, and copper as a seed layer, and its diameter-to-depth ratio was about 1:4. Due to the difference in the layout of the metal wiring on both sides of the wafer, there is a difference between the plating area on both sides of the wafer and the current distribution during electroplating. Therefore, the double-sided electroplating technology reported in this article has successfully achieved void-free hole filling through convection strength, additives, current density, conduction mode, and can ensure that the uniformity on both sides meets the requirements. This electroplating technology has been used in mass-produced products, and can also realize copper filling in through holes of other structures.**

Keywords — through glass via, double-sided electroplating, void-free hole filling

I. INTRODUCTION

The through-silicon-via (TSVs) is considered to be the most critical technology in 3D technology[1]. Three-dimensional interconnection is a configuration and a technology that integrates multiple chips in the third dimension (vertical direction) through wafer bonding and uses TSV as an electrical connection chip for vertical interconnection[2]. With the development of three-dimensional packaging technology, as a possible alternative to silicon-based transfer boards, through-glass via (TGV) interconnection technology has application advantages such as excellent high-frequency electrical characteristics, low cost, simple process flow, and strong mechanical stability. So it has a wide range of application prospects in the fields of radio frequency devices, micro-electromechanical system (MEMS) packaging, and optoelectronic system integration[3, 4].

The types of hole involved in the chip packaging structure are divided into blind holes and through holes. The current density inside the blind hole is low, the material exchange rate is slow, and the plating speed at the bottom of the blind hole is slow, especially in the blind hole with a high aspect ratio structure, a cavity is formed in the deep hole

during the electroplating process. Therefore, it is necessary to use complex plating additives and equipment to achieve bottom-up filling, so the blind hole plating is still one of the most complex and costly processes in 3D integration. For this reason, researchers use through holes to solve blind holes with high aspect ratios. This kind of through hole structure is to make blind holes first, and then make through holes through the blind holes through the backside thinning technology. In order to achieve high spect ratio hole filling, only the copper seed layer is sputtered on the back of the wafer, and then the copper layer at the hole is deposited at a faster rate to seal the through hole opening on the back to form a blind hole, so that the sealed copper can be used as a seed layer.

Although the through hole can be filled with a high aspect ratio according to this single-sided electroplating method, it takes a long time to fill the entire hole, and there is no copper seed layer on the sidewall which has less restriction on the thermal expansion of copper, so its reliability needs to be verified. At present, double-sided electroplating is used for TGVs. In the double-sided electroplating process, it is difficult to ensure the quality of the surface copper while filling the holes, so the Chemical mechanical polishing (CMP) process is often used to improve it. So in this paper, the double-sided electroplating process can simultaneously achieve the filling of the through-hole and ensure the quality of the surface copper. What is more, the patterns on both sides of the wafer are different, which poses a greater challenge to the copper plating process.

II. EXPERIMENT

The basic solution used in the TSV filling experiment consisted of 200 g/L $CuSO_4 \cdot 5H_2O$, 50 g/L H_2SO_4 and 50 ppm chloride ion. Additives were also added to the basic solution. The filling of the hole is detected by x-ray (nordson，DAGE XD7500VR Jade plus).

The TGV manufacturing process is shown in the following Fig.1. After the glass substrate is cleaned, the TGV is manufactured by laser drilling, and then the TGVs with the diameters of 60μm and depths of 220μm inside, is obtained through HF etching solution. The wire distribution on the surface is achieved by dry film and photolithography processes. Due to the different wire distribution on the two sides, the plating area on the two sides is different. In this paper, the ratio of the front and back surfaces to be plated reaches 3:1.

Due to process and reliability requirements, it is necessary to deposit a diffusion barrier layer and a seed layer on the surface of the insulating layer before electroplating. The diffusion barrier layer is used to prevent metal copper from diffusing into the insulating layer and affecting the performance of the insulating layer. Generally, materials such as Ti, Ta, TiN, and TaN are selected. In order to match the crystal lattice of the seed layer, copper is generally selected. For diffusion barrier layer and seed layer deposition, the current mainstream process method is PVD technology, which achieves continuous coverage of metal materials on deep holes through technologies such as ionized metal ions and secondary sputtering to ensure the effective progress of subsequent electroplating processes. The PVD process generally can only achieve a coverage effect of about 1% of the metal layer, so a certain thickness of the seed layer is required to ensure the quality of the electroplating filling process. In this paper, the thickness of the titanium layer is 2000A, and the thickness of the copper layer is 10000A.

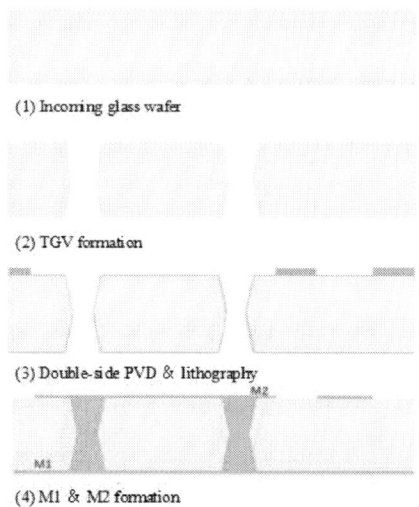

(1) Incoming glass wafer

(2) TGV formation

(3) Double-side PVD & lithography

(4) M1 & M2 formation

Fig.1. Schematic diagram of TGV process

III. RESULT AND DISCUSSION

The surface of the wafer is oxidized and dirty during the transfer process, so it needs to be cleaned. As shown in Fig. 2, there is a fault in the hole, which is caused by foreign matter in the hole, including dirt and air bubbles. Common cleaning methods include wet cleaning and dry cleaning. Wet cleaning is carried out with water and other chemicals, and then dried after cleaning. This method is characterized by a high cleaning selection ratio, but it is prone to incomplete cleaning. Dry cleaning is to use plasma for cleaning, so the energy and other parameters need to be controlled to achieve cleaning and excessive damage to the coating. After preliminary process exploration, this article adopted the method of dry cleaning.

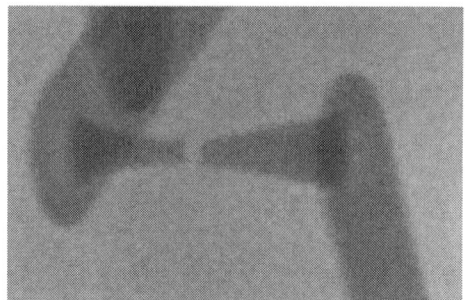

Fig.2. X-ray image of holes caused by poor cleaning and pretreatment

The stirring and circulation of the equipment will affect the convection intensity of the plating solution, thereby affecting the concentration distribution of each component of the plating solution. In order to obtain the required filling effect, it is also necessary to add a variety of organic additives to the electroplating solution, generally including inhibitors, accelerators and leveling agents. The inhibitor is a large molecule, which diffuses slowly, and mainly inhibits the deposition speed at the opening position of TSV. The accelerator is a small molecule, which diffuses faster, so it can enter the inside of the pore, and only adsorb at the position where there is no inhibitor. As a result, by combining with the inhibitor, they succeeded in obtaining the plating speed at the bottom of the hole faster than others. The leveling agent is a macromolecule with a small diffusion coefficient, which can inhibit the plating speed at a higher position and form a leveling effect. The three kinds of additives cooperate with each other to achieve the required filling effect.

At present, with the development of TSV in the direction of small aperture and high-diameter ratio, the accelerators and inhibitors play a leading role in the electrochemical reaction of copper plating inside the through holes. After the holes are completely filled, the leveling agent plays a role on the surface. Leading role[5]. In order to fill the hole completely, the inhibitor and leveler must have a strong ability to "squeeze" the accelerator to the bottom of the hole to achieve seamless filling. In double-sided electroplating, the fixture is only in contact with the edge of the wafer 3 mm, and the other places are in a suspended state. In addition, due to the different patterns on both sides, the flow rate of the plating solution is weak and the flow rates on both sides are different during double-sided electroplating. By adjusting the flow rate of the plating solution, hole-free filling can be achieved (shown as Fig.3).

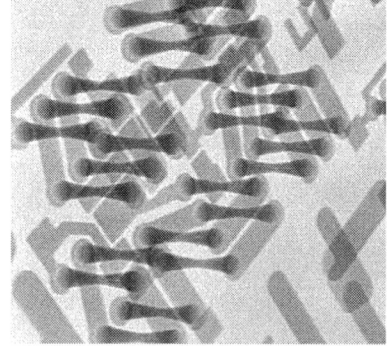

Fig.3. X-ray image of no voids in the hole after adjusting the flow rate of the plating solution

The filling method of double-sided electroplating is different from that of blind holes. Through hole copper filling and blind holes copper filling use high concentration copper and low concentration acid system. Although the same copper plating system is used, the copper plating additive formula used is different. The reason is that mass transfer and fluid mechanics cause through holes and blind holes. The difference of holes filling. The Factors governing filling of blind via and through hole in electroplating[6], and the mechanism is the adsorption, consumption, and diffusion mode of additives, while copper filling of blind holes uses bottom -up filling mode, so blind holes and through-hole copper filling require different additive formulations. When a "butterfly-shaped" copper layer is formed in the middle of the through hole, it becomes two symmetrical blind holes, and the copper filling mode becomes bottom-up filling.

In order to achieve a fast intermediate coating speed without sealing, the taper of the hole should be considered during TGV manufacturing. The appropriate taper can balance the difficulty of filling the through hole and the performance of signal transmission. At the same time, consider the PVD sputtering barrier layer and the seed layer. At the same time, we must also consider the PVD sputtering barrier layer and the seed layer. In addition, the process capability of chemical etching to make TGV must also be considered. In terms of current regulation, two aspects need to be considered. On the one hand, the thickness of the plating layer in the hole is inconsistent with the thickness of the surface copper. In order to reduce the cost, the CMP process is not used. Therefore, the thickness and uniformity of the surface copper must be strictly controlled while filling the hole. On the other hand, the pattern distribution on both sides of the wafer is different. This is a great challenge for double-sided electroplating, because the distribution and area of the plated surface are different at this time, which puts high requirements on current control. For this reason, it is necessary to adjust the current density in different stages. To this end, we have made special designs on both the power supply and the fixture. In the Fig.4 below, after adjusting the process parameters of the current, plating solution flow rate and other, the plating situation can be seen that there are Void-free hole filling.

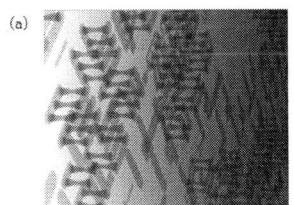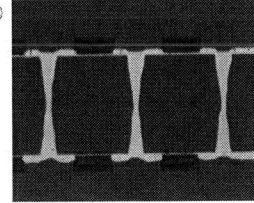

Fig.4. The diagram of void-free hole filling. (a) x-ray image, (b) metallographic photo

Since the current density at the TGV orifice is higher than that in the center of the hole, in order to achieve defect-free filling, electroplating is often carried out with a small current. However, low-current electroplating not only causes the electroplating time to be too long, which is not conducive to mass production requirements, but also causes the surface of the copper to be not bright (shown as Fig.5). In order to obtain a bright plating layer, the plating layer must have fine crystals, which requires that sufficient copper nuclei be produced as seeds for copper growth during electroplating, and the formation of copper nuclei needs to be under a

certain over-potential. Therefore, it is impossible to blindly pursue excellent filling results with low current. This process is particularly important to control opportunity for the conversion to a higher current after filling can be realized.

An important indicator of surface copper quality is uniformity, and the difference between different parts is generally required to be less than 10%. However, the distribution of through holes and metal wiring are not same on the entire wafer, especially when there are wiring on both sides. There are two differences between double-sided electroplating and single-sided electroplating. First of all, the flow rate of the plating solution in the double-sided electroplating process is slow and inconsistent, which will cause the concentration of the plating solution to be inconsistent and cause the uniformity of the plating layer to be different. On the other hand, due to the different plating area, the current distribution is uneven on both sides of the wafer. In order to solve this problem, adjusting the size of the cathode shielding ring is one of the effective ways.

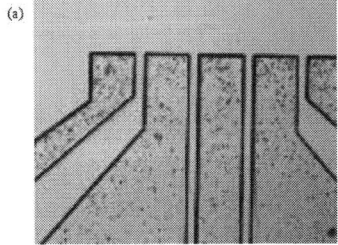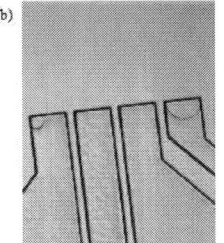

Fig.5. Metallographic microscope photos under different current densities. (a) abnormal roughness at low current density, (b) normal copper layer at higher current density.

IV. CONCLUSION

The TGV filling process is one of the most important processes for 3D packaging. In this article, we introduce the double-sided electroplating process when the patterns on both sides of the wafer are different. In order to achieve non-hole filling, strict control is required from the TGV production process, cleaning, and pretreatment before electroplating. During the electroplating process, because the two sides of the wafer are different, the current needs to be adjusted at different stages of the electroplating process, and the plating solution flow rate will also affect the filling effect. With the development of three-dimensional packaging technology, TGV has a deep hole structure with sub-micron pore size and ultra-high aspect ratio, which is a huge challenge for double-sided electroplating.

ACKNOWLEDGMENT

This research is supported by the National Natural Science Foundation of China (Grant No. 61974121).

REFERENCES

[1] H. Xiao, F. Wang, Y. Wang, H. He, and W. Zhu, "Effect of Ultrasound on Copper Filling of High Aspect Ratio Through-Silicon Via (TSV)," Journal of The Electrochemical Society, vol. 164, no. 4, pp. D126-D129, 2017.

[2] Z. Wang, "Microsystems using three-dimensional integration and TSV technologies: Fundamentals and applications," Microelectronic Engineering, vol. 210, pp. 35-64, 2019.

978-1-6654-1392-3/21 $31.00 © 2021 IEEE

[3] V. Sukumaran et al., "Design, Fabrication, and Characterization of Ultrathin 3-D Glass Interposers With Through-Package-Vias at Same Pitch as TSVs in Silicon," IEEE Transactions on Components, Packaging and Manufacturing Technology, vol. 4, no. 5, pp. 786-795, 2014.

[4] V. Sukumaran, T. Bandyopadhyay, V. Sundaram, and R. Tummala, "Low-Cost Thin Glass Interposers as a Superior Alternative to Silicon and Organic Interposers for Packaging of 3-D ICs," IEEE Transactions on Components, Packaging and Manufacturing Technology, vol. 2, no. 9, pp. 1426-1433, 2012.

[5] L. Hofmann, R. Ecke, S. E. Schulz, and T. Gessner, "Investigations regarding Through Silicon Via filling for 3D integration by Periodic Pulse Reverse plating with and without additives," Microelectronic Engineering, vol. 88, no. 5, pp. 705-708, 2011.

[6] C. Cui, M. Wu, and J. Wang, "Factors governing filling of blind via and through hole in electroplating," Circuit World, vol. 40, no. 3, pp. 92-102, 2014.

Sawing Investigation for Thin Wafer Laminated with Die Attach Film

Haiyan Liu
NXP Semiconductors
Tianjin, China
Haiyan.liu@nxp.com

Qingyu Pan
NXP Semiconductors
Tianjin, China
Qingyu.pan@nxp.com

Sean Xu
NXP Semiconductors
Tianjin, China
Sean.xu@nxp.com

Jianhong Wang
NXP Semiconductors
Tianjin, China
Jianhong.wang@nxp.com

Lu Li
NXP Semiconductors
Tianjin, China
Lu.L@nxp.com

Xueting Wu
NXP Semiconductors
Tianjin, China
Xueting.wu@nxp.com

Abstract—**The drive for smaller size and thickness has created new package assembly process challenges. Stacked die packages are increasing in interest due to the resultant smaller package and increased functionality. The requirements for a stacked package is a thinner wafer that enables more die to be stacked in a package. The die are bonded using Die Attach Film (DAF) that is laminated on the back of a wafer prior to singulation. The singulation of DAF laminated wafers with wafer thickness <3 mils is challenging, especially for devices used in automotive applications. Part of the singulation challenge is that it involves two different hardness materials (Si and the DAF polymer). Previous studies have shown that for thin die, a step-cut saw process is better than single cut but backside chipping and residual DAF whiskers must be resolved. In this study, the saw singulation process was optimized through a series of wafer saw singulation experiments. The results showed that the depth of cut for both Z1 and Z2 is a key process parameter to improve the backside chipping and resolution of DAF residual/whisker formation. Singulated die were evaluated optically and by SEM to characterize the dicing performance, including chipping and cracking. With optimized parameters, the backside chipping can meet backside chipping requirements with no DAF residual material or whisker formation in a high volume production monitor. Packaged stacked die also passed all AEC Grade 1 requirements with no chipping related defect.**

Keywords—thin die, DAF, sawing, backside chipping

I. INTRODUCTION

The drive for smaller package size and package thickness has created new package assembly process challenges. The stacked die package is gaining more attention because it is a smaller package and more functionality. The stacked package requires thinner die that enables more die to be stacked in the package.

For wafers <100um, the quality factors of die attach process using epoxy includes: die tilt, bond line thickness (BLT), coverage and fillet height, are critical requirements for stack die package. Poor dispending system might cause excessive bleed out and fillet height that could lead to device failure.

Die attach film (DAF) material is a good alternative to dispensed epoxy die attach. The DAF material can be laminated on the wafer backside together with dicing tape. The wafer can be singulation with the DAF attached. The laminated DAF die is the bonded in the package at elevated temperatures The generic structure of a DAF-dicing tape material is shown in below Fig 1. It consists dicing tape, adhesive layer and DAF layer.

Fig.1. DAF-Dicing Tape Material Structure

Sawing DAF die is different than sawing bare silicon because it includes materials of different composition and hardness. The silicon die is hard and brittle. The DAF is a polymer that is soft and has a lower glass transitional temperature. The heat generated by dicing process can soften the DAF.

The conventional silicon wafer dicing technology has been used for <40 years. The typical blade used for singulation is nickel bonded diamond grit blade with a single cut or step cut mode.

The dicing blade is composed of grit and bond. The grit is what abrades and cuts the wafer. The bond material bonds the grit to the metal blade surface. When abrasive blades cut or groove the wafer, they are actually grinding and removing material. The blade dicing is illustrated in Fig.2.

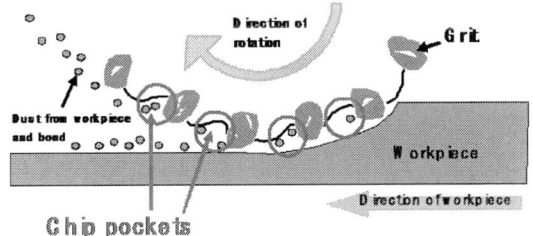

Fig.2. Blade Sawing Motion

One of the most important aspects of abrasive blades is their self-sharpening mechanism, also known as autogenous sharpening. The bond, which holds the grit, is gradually worn away during processing by pressure and friction with the workpiece. As the bond wears away, the grit falls away from the blade. New grit is exposed that then begins to cut, thereby maintaining the cutting power of the blade. The blade self-sharpening mechanism is shown in Fig.3.

978-1-6654-1392-3/21 $31.00 © 2021 IEEE

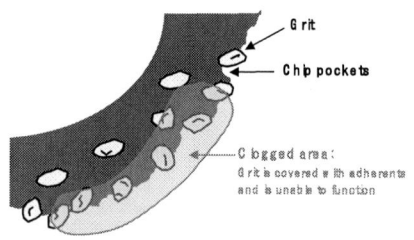

Fig.3. Blade Self-sharpening

During DAF dicing, the polymeric material will laminate on the blade surface, covering the grit and fill the chip pockets that is known as blade loading. Fig.4 is an illustration of blade loading that hinders blade self-sharpening. The efficiency to remove the polymeric materials determine the dicing quality. The issues is further challenged with thinner die when singulate wafers with DAF.

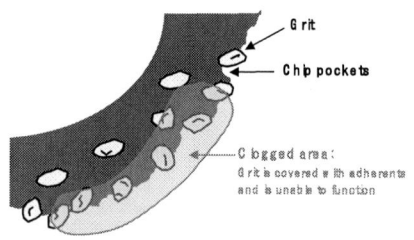

Fig.4. Blade loading

II. BACKGROUND

Experience from previous work[1-3] showed that when dicing thin Wafer Laminated with Die Attach Film, the single pass saw process encountered lateral crack lines and back-side chipping across the die surface. The crack may extend from one corner to the corner of the die[4]. These issues arise due to the different hardness of the Si and the DAF. Step-cut singulation was investigated for thin wafer with DAF to improve the sawing quality.

For this study, the wafer thickness was 75um and laminated with non-conductive DAF material. The wafer tech is 130nm Si technology, and die size is 2.15X1.45mm.

In the initial development phase, based on previous experience on thin wafer dicing are shown in Table I.

TABLE I.   SAW SINGULATION PARAMETER SETTING

Items	Saw Singulation Parameter Setting		
	Key Parameter	Z1	Z2
1	Blade Height	165um	70um
2	Spindle Speed	50k	40k
3	Feed Speed	20mm/s	

Post dicing topside chipping/ backside chipping/ sidewall crack/wafer skeleton post die pick were evaluated. Many dicing issues were encountered in the first trail run.

A. DAF Whisker

DAF whiskers were found on the wafer skeleton during visual inspection and during die bonding and wire bonding processes. The typical pictures are shown in Fig.5.

(a) Whisker on die   (b) Whisker left on tape

Fig.5. Typical pictures of DAF Whisker

The DAF whisker could cause potential quality concern and yield loss during mass production.

B. DAF Residule

Post die pick, DAF residual were found on the wafer skeleton. Si residue was also be found in some locations. DAF residue might cause die bond coverage concerns.

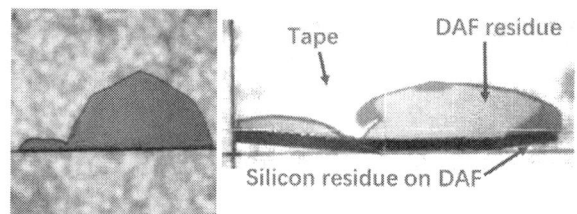

Fig.6. Typical pictures of DAF Residual

C. Lateral Cracks

After dicing, the DAF material was chemically etched to remove it from the backside of the die and the die evaluated for backside chipping using optical microscopy and scanning electron microscopy (SEM). The three dimensions of backside chipping are defined in Fig.7.

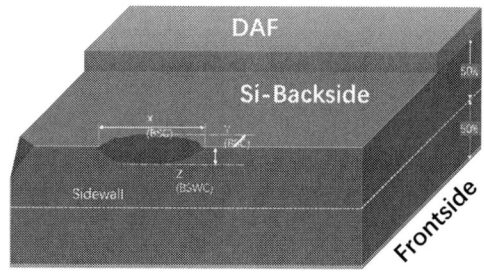

Fig.7. Explanation of backside chipping dimensions

The backside chipping dimensions can meet requirements but a lateral crack was observed on the die sidewall. A worst-case image is shown in Fig.8.

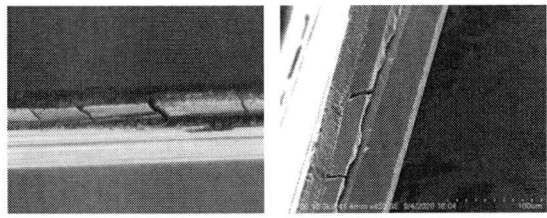

Fig.8. Lateral crack on die sidewall

III. PARAMETER OPTIMIZATION AND DISCUSSION

To reduce dicing defects noted in Section II, parameter optimization was performed. Topside chipping was not an issue and passed 100% AOI with no failures so this did not need to be addressed. The primary defects from the first pass evaluation were on the wafer backside, and was thought to be related to the Z2 singulation setting.

A. DAF Whisker

One sample with DAF whisker was selected to check the side wall. The sidewall picture is shown in Fig.9. The DAF residue was stuck on the die sidewall and could have occurred during the Z2 cuts.

Fig.9. Die sidewall with DAF whicker

The blade conditions were checked using an SEM. The blade condition is shown in Fig.10. Polymeric material is laminated on the blade surface. The diamonds and chip pockets are covered (clogged blade), which hinders blade self-sharpening. A clogged blade has difficulty to expose new diamonds to penetrate and remove cutting debris from the work piece (the wafer). This blade buildup causes the blunt blade and its debris to push against the work piece material, creating added stress and high blade loads, and can result in die chipping.

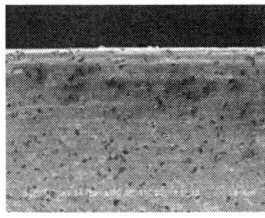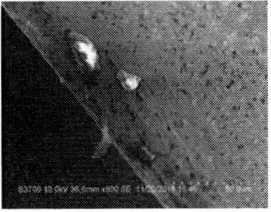

Fig.10. Blade condition under SEM

The blade can be easily be clogged because the adhesive film is viscous. In order to lower the risk, it is better to dicing less of the adhesive film. The original blade height setting of Z2 cuts into the base film layer through the entire adhesive

film. The optimized Z2 blade height cuts just into adhesive layer, not through the entire layer. A diagram comparing the Z2 blade height before and after optimization is shown Fig.11.

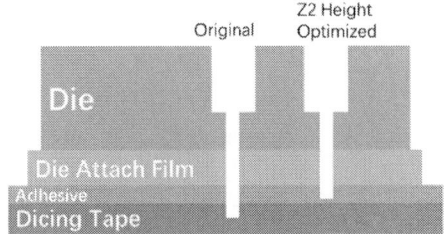

Fig.11. Die sidewall with DAF whisker

With the optimized Z2 blade height, one wafer was diced. No DAF whiskers were found in <3k units. The failure rate of DAF residual and lateral crack is also lower, but not fully solved.

B. DAF Residual and lateral crack

When checking the DAF residual on wafer skeleton, the silicon residual is also can be found that likely was caused by backside chipping/lateral cracking. When wafer chipping occurs, the DAF is laminated into the tape along the saw lane , and partially tears off during die pickup.

The position of residual DAF material and Si particles on the wafer skeleton was checked. Nine wafer skeletons were checked and all the DAF/Si residue was found on the first saw channel and this is summarized in Fig. 12.

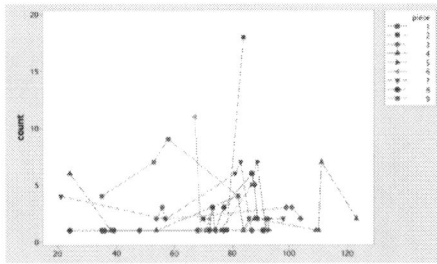

Fig.12. DAF residual and silicon residual positions

The data shows that the DAF/Si residue is randomly located. The residue is related to the Z2 blade cutting condition. When dicing DAF and adhesive materials, the Z2 blade can easily be clogged and cutting debris is not efficiently removed. When this occurs backside chipping and lateral cracks can occur. To improve the blade self-sharpening process, actions were taken to efficiently remove blade build-up.

In-line dressing using a dressing board is normally used for dicing a thin wafer laminated with DAF. The purpose of in-line dressing is to expose the grit fully and form chip pockets. Increased in-line dressing frequency could help.

The original in-line dressing frequency was 1 line for each channel with a dressing position at the very beginning of each channel. From the DAF/Si residual material location study, there is no residual material in the first 20 lines. The proposed in-line dressing setting was one dressing for every 15 saw lanes.

978-1-6654-1392-3/21 $31.00 © 2021 IEEE

The Z1 blade heigh was also adjusted to leave more silicon for the Z2 to cut. The silicon will also help to remove blade build-up and expose chip pocket. A diagram to compare the Z1 blade height before and after optimization is shown Fig.13.

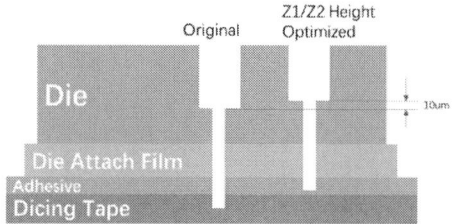

Fig.13. Die sidewall with DAF whisker

C. Confirmation run with new parameter

The new parameter are shown in Table II.

TABLE II.   OPTIMIZED SAW SINGULATION PARAMETER SETTING

Items	Saw Singulation Parameter Setting		
	Key Parameter	Original	Optimized
1	Z1 Blade Height	165um	175um
2	Z2 Blade Height	70um	110um
3	In-line Dressing	1 time	8 times

A total of 25 wafers were evaluated after singulation with the new parameters shown in Table II. No DAF whiskers were found, and no DAF/Si residue was observed after high magnification inspection.

The dicing blade, post saw, was checked using an SEM. Figure 14 shows that there is no polymeric material on the blade surface.

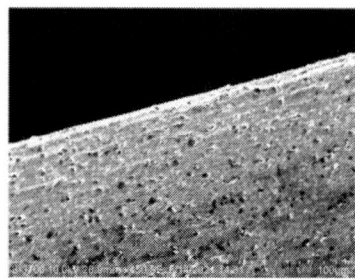

Fig.14. Blade condition under SEM

30 samples were chemically etched to remove DAF from the backside of the die. Backside chipping (BSC-Y and BSC-Z) data were collected using an optical microscope, Figure 15. All the backside chipping met requirements (<37.5um).

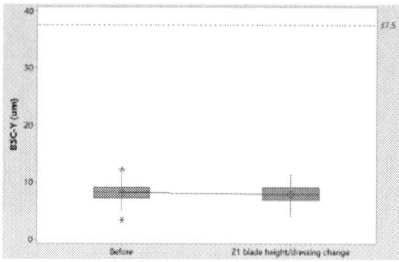

(a)   Backside chipping Y direction

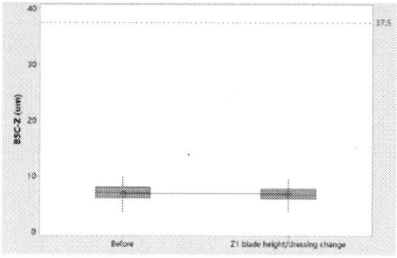

(b)   Backside chipping Z direction

Fig.15. Backside chipping comparison

The die sidewall was also checked using an SEM. A typical image is shown in Fig.16, there are no lateral cracks observed and the chipping is minor, meeting all requirements.

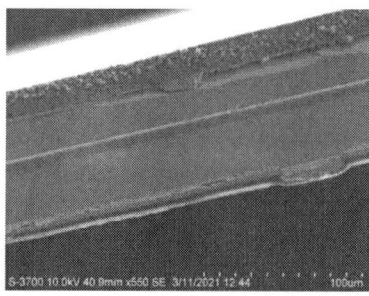

Fig.16. Die sidewall SEM check

D. Reliability Evaluation

Total 160 dies were packaged into a 32LD SOIC stack die package. Post assembly, CSAM and electrical test were performed on the assembled parts at T0, post MSL3/260degree C, post 264h HAST (110°C/85%RH), and post TC2000cycles (-55°C to 150°C). All units passed ATE test post stress with no failures observed.

IV. CONCLUSION

The depth of both the Z1 and Z2 blade during singulation of a DAF coated wafer is a key process parameter to improve the backside chipping and DAF residues.

Adjust Z2 blade height to cut into adhesive layer can improve the DAF whisker issue. Adjusting Z1 blade height to leave more silicon for Z2 can help Z2 self-sharpening as does in-line dressing.

With the best optimized parameter, the backside chipping can meet the requirements of no chips >37.5um. No DAF residue was found in high volume monitors. The die was packaged into a stack die package and it passed all AEC Grade 1 requirements without no die chipping related defects.

978-1-6654-1392-3/21 $31.00 © 2021 IEEE

ACKNOWLEDGMENT

The authors would like to thank J.Z Yao from NXP Semiconductor and Assembly Packaging Innovation (API) for support in this project. Thanks to NXP PA UG group for experience sharing and supporting. Thanks to dicing machine vendors for technical support.

REFERENCES

[1] SW Wang, MC Yo, "4 mil DAF Die Thickness Sawing Capability Study," IEMT 2006, Putrajaya, Malaysia

[2] Chetan Paydenkar, Anindya Poddar, Haryanto Chandra, "Wafer Sawing Process Charakterization for Thin Die (75 micron) Applications, IEMT 2004

[3] YH Chiew, JY Liong, FF Tan "Mechanical Dicing Challenges and Developments on 50um Saw Street with Wafer Backside Coating (WBC),".

[4] Hoh Huey Jiun, Ibrahim Ahmad, Azman Jalar, "Alternative Double Pass Dicing Method for Thin Wafer Laminated with Die Attach Film," ICSE2004 Proc 2004, Kuala Lumpur, Malaysia

Blade Dicing on Wafer Saw Study

Qiuchen Zhang
Package R&D
Nexperia(China) Ltd.
Dongguan, China
edison.zhang@nexperia.com

Hongbin Xia
Package R&D
Nexperia(China) Ltd.
Dongguan, China
hongbin.hb.xia@nexperia.com

Shwu Miin Tan
Package R&D
Nexperia(China) Ltd.
Dongguan, China
shwu.miin.tan@nexperia.com

Abstract—Wafer sawing is one of the back-end technologies of advanced packaging. Higher quality and narrower saw street can improve the density of die in one wafer, thus reduce the wafer cost. In this semiconductor industry, there are different wafer dicing methods, blade dicing, laser dicing, stealth dicing and plasma dicing. Plasma dicing gives the best chipping performance, almost zero chipping defect, but it requires high running cost. Laser dicing uses laser ablation to separate units, however, Heat Affect Zone (HAZ) on the sidewall is incurred, which will then lower the die strength. The conventional wafer dicing method, blade dicing, it induces mechanical stress that leads to chipping. The advantage of this technology is mature, low cost comparing with other methods, and it is still the most widely used method in this semiconductor industry. It is believed that with a process optimization, blade dicing is the best option among all.

In this paper, we focus on replacing the plasma dicing technology with blade dicing in order to further reduce the processing costs. Result shows: (a) For sidewall chipping, slower feed rate and lower blade height Z1, chipping will be better. (b) For backside chipping, slower spindle speed Z2 and feed rate, lower blade height of Z1, backside reject rate will be decrease. (c) The backside and sidewall chipping size are within the quality specification range through process optimization. (d) All topside chipping are inspected by Automated Optical Inspection (AOI) and meet the standards (>99.8%). (e) Fly die issue can be solved by baking the wafer before sawing.

Keywords—Blade dicing, Spindle speed, Feed rate, Blade height

I. INTRODUCTION

This paper focuses on wafer dicing, one of the back-end technologies of advanced packaging. Silicon wafer sawing is the first step in the "back end" assembly process. The process divides the wafer into individual chips for subsequent processes, like die bonding, wire bonding and testing.

Blade dicing is one of the options in sawing. A rotating blade completes slicing. A spindle rotates the blade at a high speed of 30000-60000rpm (linear speed of 83-175m/ sec). The blade is made of a ground diamond embedded in an electroplated nickel matrix adhesive (Shown in Fig.1)

During the chip segmentation, the blade crushes the base material (wafer) while removing the resulted debris. The material removal occurs along the special sawing line (trace) between the active regions of the dice (shown in Fig.2). Coolant (usually is deionized water) is channeled into the sawing seam to improve the sawing quality and extend the blade life by helping to remove debris. The width (notch) of each street is proportional to the thickness of the blade.

There are two common ways of blade dicing: 1) full cut 2) step cut. (shown in Fig.3)

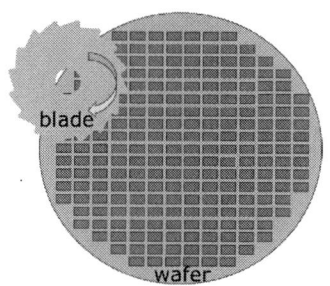

Fig. 1. Blade dicing 1

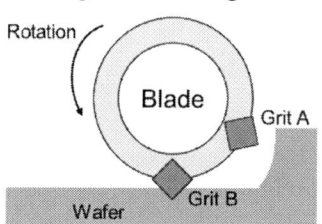

Fig. 2. Blade dicing 2

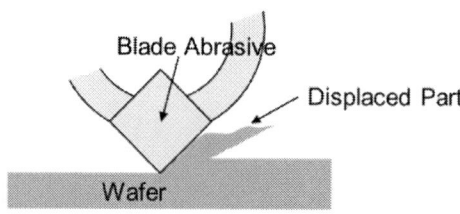

(a)Full cut

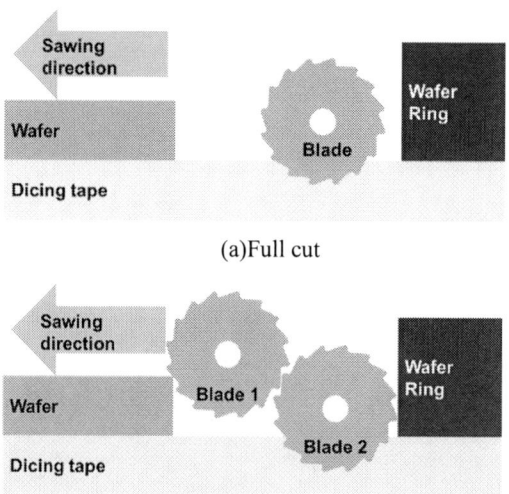

978-1-6654-1392-3/21 $31.00 © 2021 IEEE

(b) Step cut

Fig. 3.Blade dicing

Step cut is using two blades (blade Z1, blade Z2) cut at different heights, which helps to reduce the overload of the blade and reduce chipping. Step cut was used in this experiment.

After determining the sawing mode, this paper carries out a series of DOE (design of experiment), the purpose is to select the appropriate dicing tape, blade type and dicing parameters (spindle 1 speed, spindle 2 speed, feed speed, blade 1 dicing height), and finally find the best experimental scheme. The experimental results show that the chipping of grain is within the quality standard. The relationship between the four sawing parameters and the size of fracture is found out.

II. EXPERIMENT SET UP

A. Selecting a Template (Heading 2)

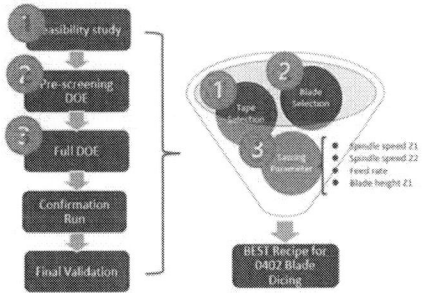

Fig. 4.Design of Experiment (DOE)

Our Design of Experiment(DOE) include 5 steps, starts with feasibility study, pre-screen DOE, full DOE, confirmation trial and lastly conduct final validation (shown in Fig. 4). In the feasibility study, we firstly select the right dicing tape from 5 different kinds of dicing paper (including UV and non-UV tape) to avoid tape burr residue, as residue accumulated on the die backside will lead to the failure of assemble subsequently. In the pre-screening DOE, we select 3 different kinds of blades (its width and density) according to the incoming wafer thickness and material. From DOE result, we select narrow and low density blade. In full DOE, further sawn 18 pieces wafers, using the blade we selected, and collect more than 2000 die samples in different area randomly, to measure the chipping size. The collected data were then analyzed by Minitab, an optimum sawing parameters and key factors from data graph (including pareto charts, main effect& interaction plot, overlaid contour plot) is selected. Further validate the optimum parameter, using selected blade type and dicing tape, two batches of verification experiments were conducted in confirmation trial and final validation. As a result, through the 5 steps, the best recipe is ready which the topside, backside and sidewall defect is within acceptable specification, meeting die strength test, as well as fly die issue is solved.

In the selection of dicing tape, we have selected four kinds of dicing tape A, B, C and D (show in Tab.1). The wafer was mounted on the tape and the was diced with the same sawing parameters. After dicing, we took 30 dies randomly on the wafer surface and measured the chipping of top side, back side and sidewall. It was found that the chipping size was within the standard range. However, we found that tape burr residue issue in the tape skeleton except for tape A, which indicated that the dicing tape was not well combined with the back of wafer, which would affect the die attach process (show in Tab.2).

Table 1. Tape information

Tape model		A	B	C	D
Tape type		Non UV	UV	UV	UV
Tape thickness/ μm		>70	>80	>90	>50
Adhesion mN/25mm	Before UV	120	6300	5000	5
	After UV	120	720	200	0.0375

Table 2. Tape burr residue

Tape A	Tape B	Tape C	Tape D

To sum up, tape A was selected as dicing tape.

Die fly issue is observed in the feasibility study, due to when the dicing blade is sawing the wafer, a huge shear stress is created, making the die spatter from the wafer surface. If die spatter collides with the wafer, it will scratch on the wafer surface and damage the circuit; If the spatter collides with the dicing blade, the blade will be damaged. First time encountered this scenario, could it be the die size is too small 0.400*0.200mm, as the same dicing tape, we did not observe this with 0.600*0.300mm chip size. This data tells us the dicing tape could not hold firm 0.400*0.200mm small chip size. Baking the wafer at >80 ℃ for not less than 15min minutes before sawing to increase the stickiness of the dicing tape is the counter action for die fly issue. (shown in Fig.4)

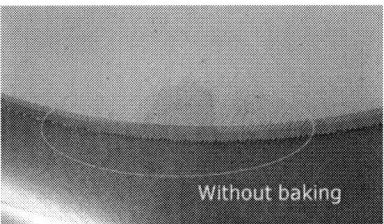

(a) Without baking

(b) Baking

In pre-screening, we choose four different blades (A, B, C, D) according to the material and thickness of the water (show in Tab.3), use tape A as the dicing tape, and adopt the step cut (Z1 + Z2) sawing method to carry out six groups of dicing experiments (shown in Tab.4)

Table 3. Blade information

Blade type	I	II	III
Blade	Z1: Blade A Z2: Blade B	Z1: Blade C Z2: Blade B	Z1: Blade A Z2: Blade D
Spindle speed/Kprm	Z1:III Z2: I		
Feed rate mm/s	II		
Blade Z1 height/um	H1(tape A thickness + 2/3 wafer thickness) H2(tape A thickness + 1/3 wafer thickness)		

Table 4. Pre-screening DOE

Leg	Step cut	Spindle speed/ Kprm	Feed rate mm/s	Blade Z1 height/μm
1	III			H2
2	II			H2
3	II	Z1:III Z2: I	II	H1
4	III			H1
5	I			H1
6	I			H2

After six groups of pre-screening DOE, we randomly took 30 die from different positions of each wafer and test the backside and sidewall chipping. The statistical charts of boxplot and main effects plot (show in Fig.5 and Fig.6) are given.

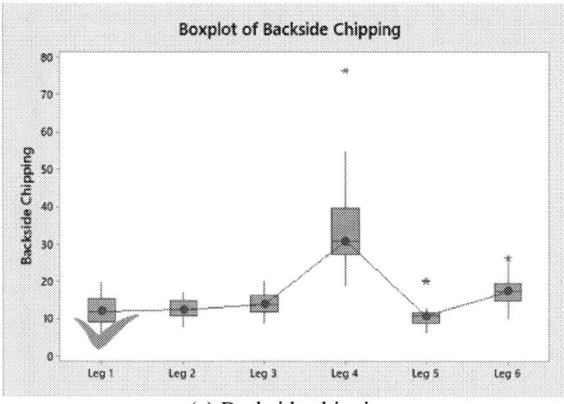

(a) Backside chipping

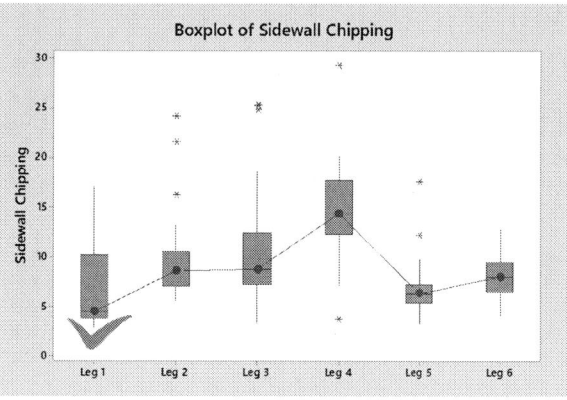

(b) Sidewall chipping

Fig.5.Boxplot

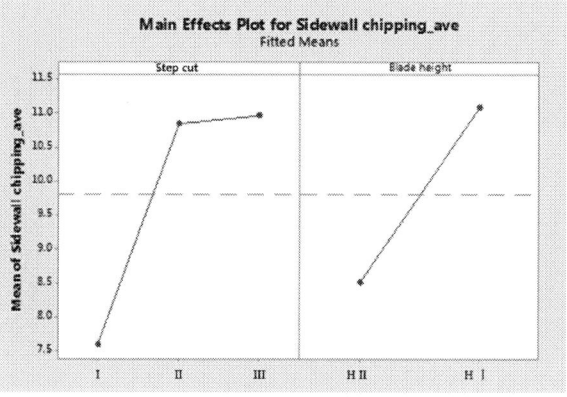

(a) Sidewall chipping

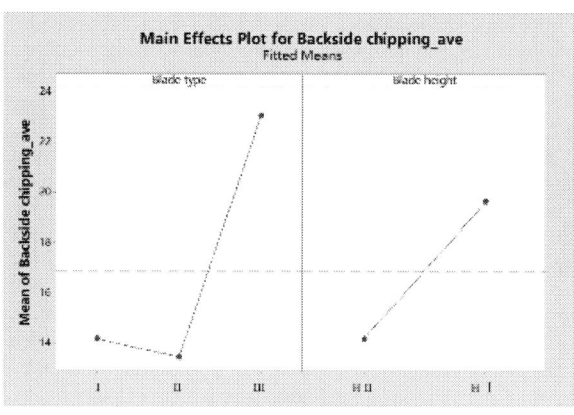

(b) Backside chipping

Fig. 6 Main effects plot

Combining Fig.5 and Fig.6, we can find that using step cut I, blade height Z1 is H2. The chipping of leg1 of is the smallest. Therefore, in the subsequent Full DOE, we determined Z1: Blade A Z2: Blade B as the dicing blade.

In DOE, we set four kinds of sawing parameters: spindle speed Z1 (I~III, I: low level, II: middle level, III: high level), spindle speed Z2 (I~III, I: low level, II: middle level, III: high level), feed rate (I~III, I: low level, II: middle level, III: high level) and blade Z1 height (H I~H III, H I: tape A thickness + 1/3 wafer thickness, H II: tape A thickness + 1/2 wafer thickness, H III: tape A thickness + 2/3 wafer thickness) according to the thickness and material of wafer and the working performance of dicing machine. The DOE table was generated by statistical software, and 18 groups of dicing experiments were carried out. 90 dies were randomly selected from each wafer after swing and measured (show in Tab. 5). Main effects plot (show in Fig.7), interaction plot (show in Fig.8) overlapping contour plot (show in Fig.9) are generated.

Table 5. Full DOE

Leg	Spindle speed Z1/Kprm	Spindle speed Z2/Kprm	Feed rate mm/s	Blade Z1 height /µm	Backside chipping Ave/µm	Backside chipping reject Qty.(>25µm)	Sidewall chipping Ave/µm	Topside chipping Reject Qty.
1	III	I	III	H III	15.7	5	12.5	17
2	I	III	III	H III	16.4	5	14.8	11
3	III	III	I	H I	20.3	19	8.3	18
4	I	III	III	H I	20.3	18	9.4	13
5	I	I	III	H I	15.8	7	12.1	6
6	III	III	I	H III	13.0	2	9.2	29
7	I	I	III	H III	13.0	0	12.7	698
8	II	II	II	H II	14.4	2	11.3	173
9	III	I	III	H I	15.2	0	10.9	28
10	I	I	I	H III	13.9	2	13.8	6
11	I	III	I	H III	15.1	5	12.5	242
12	III	I	I	H III	13.4	0	8.3	37
13	I	I	I	H I	14.3	0	11.6	83
14	I	III	I	H I	15.0	2	8.3	169
15	III	III	III	H I	14.9	3	14.5	4
16	III	III	III	H III	13.2	1	15.4	4
17	II	II	II	H II	11.9	3	13.9	19
18	III	I	I	H I	11.8	1	17.7	214

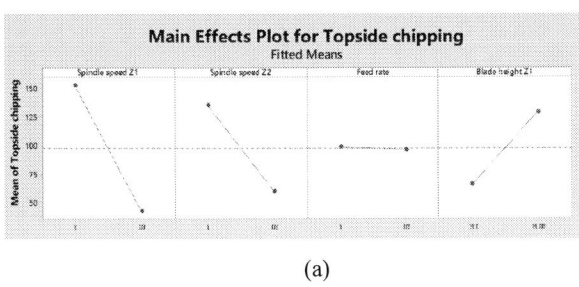

(a)

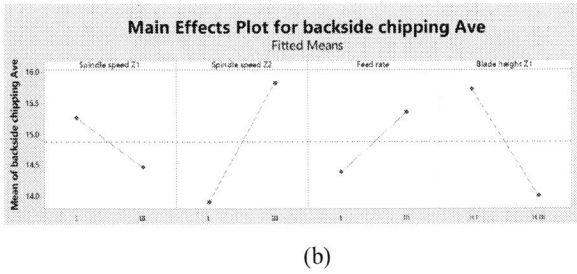

(b)

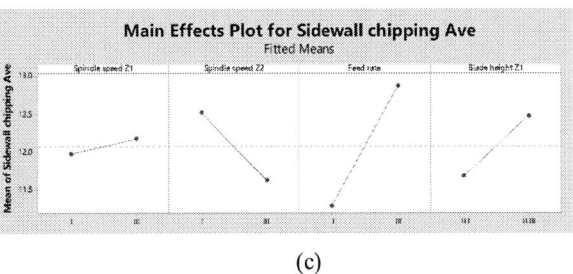

(c)

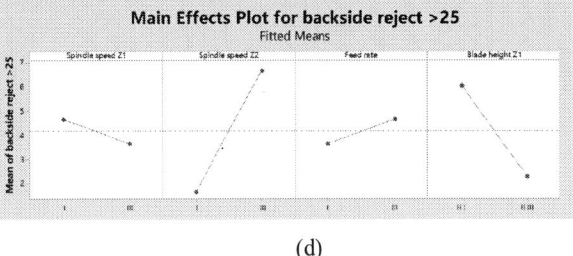

(d)

Fig. 7.Main effects Plot

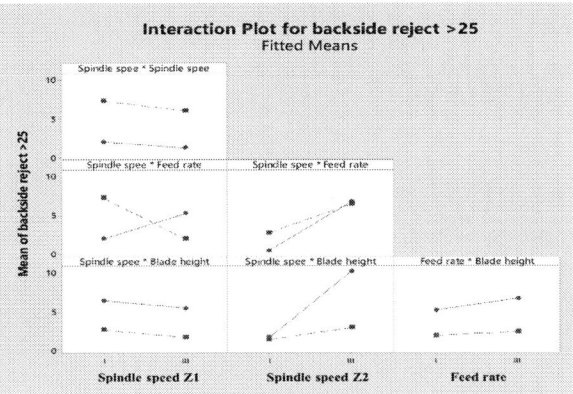

Fig. 8.Interaction plot

Combining Fig.7 and Fig.8, we can get the following conclusions

- For topside chipping, faster spindle speed Z1, topside chipping will be better.

- For backside chipping, slower spindle speed Z2 and feed rate, chipping will be better

- For backside reject quantity, slower spindle speed Z2 and feed rate, higher blade height of Z1, backside reject number will be lower

- For sidewall chipping average, slower feed rate and lower blade height Z1, chipping will be better

We can express the above conclusion in a table （shown in Tab.6）

Table 6. Main effects plot & Interaction plot analysis

Factor	Low Level	High Level	Center Level
Spindle speed Z1/Kprm			II
Spindle speed Z2/Kprm	I	III	

Feed rate mm/s	I		II
Blade Z1 height/μm	H I		H II

From Tab.6, we get four ranges of sawing parameters. Next, we use overlap contour plot (show in Fig.9) to determine the sawing process parameters.

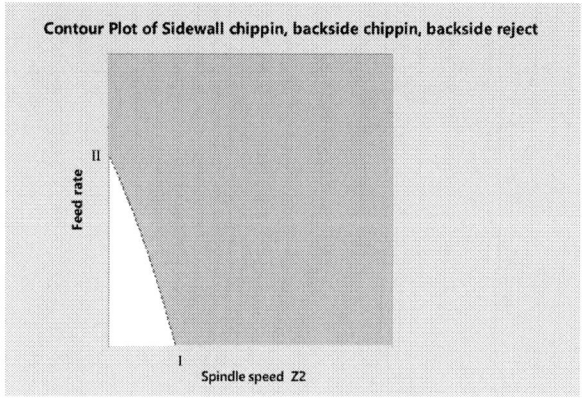

Fig. 9 Overlaid contour plot

From Fig.9, white is the effective value area, we can see that the spindle speed Z2 < I, feed rate <II So we further narrow the process window (show in Tab. 7)

Table 7. Overlaid contour plot analysis

Factor	Low Level	High Level	Center Level
Spindle speed Z1/Kprm			II
Spindle speed Z2/Kprm	I		
Feed rate mm/s			II
Blade Z1 height/μm	H I		H II

From Tab. 7, we can see that blade Z1 height has not been determined. We will continue the verification in the subsequent confirmation run.

In confirmation run, we have carried out experiments on blade Z1 height (shown in Tab.8)

Table 8. Confirmation DOE

Parameters	C1	C2
Spindle speed Z1/ Kprm	II	II
Spindle speed Z2/ Kprm	I	I
Feed rate/ mm/s	II	II
Blade Z1 height/ μ m	H II	H I

We measured the top side, back side and sidewall chipping of two diced wafers (C1, C2)(shown in Tab.9, Fig.10 and Fig.11）

Table 9. Topside scan data

	Scanned dies	Faults	Bad dies	Yield/%	Chipping
C1	22560	93	80	99.96	6
C2	225571	130	107	99.95	8

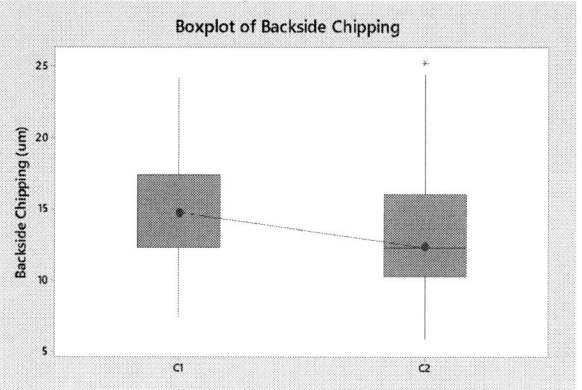

Fig. 10.Boxplot of backside chipping

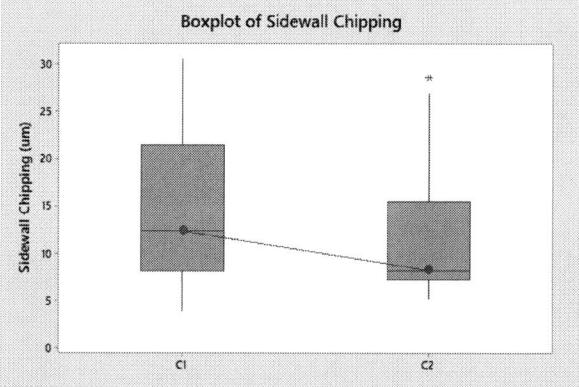

Fig. 11.Boxplot of sidewall chipping

Combined with Tab. And Fig. above, it is found that chipping is within the standard range, and there is not much difference. Observe and monitor the unqualified data in tapping machine (shown in Tab.10).

Table 10. Tapping machine test data

	C1	C2
Backside long edge chipping yield loss/%	0.13	1.70
Backside short edge chipping yield loss/%	0.09	0.23
Sidewall chipping yield loss/%	0.03	1.11
Chipping yield/%	1.71	2.26

In Tab.10, we can see that the yield loss of C1 is lower than that of C2, so blade Z1 height is H II.

In order to ensure the sufficiency of the experiment, we carried out final validation. We do the final validation .We diced one wafer (F1) with the parameters obtained from confirmation, and also detected the backside chipping and sidewall chipping after dicing (shown in Tab.11, Fig.12 and Fig.13).

Table 11. Final validation DOE

Parameters	F1
Spindle speed Z1/ Kprm	II
Spindle speed Z2/ Kprm	I
Feed rate/ mm/s	II
Blade Z1 height/ μ m	H II

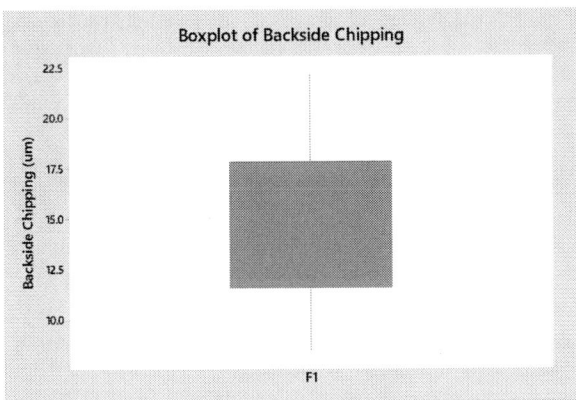

Fig. 12.Boxplot of backside chipping

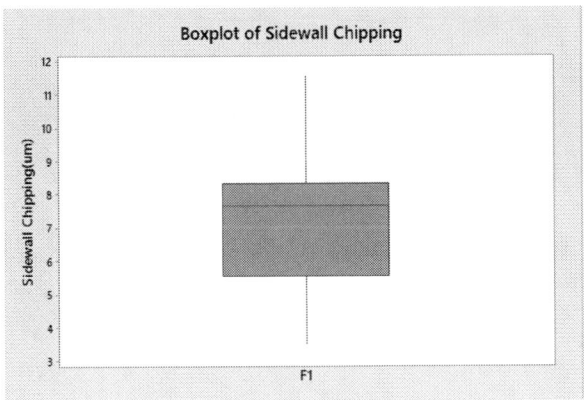

Fig. 13.Boxplot of sidewall chipping

Combined with the above chart, we can see that when blade Z1 height is set to H II, chipping are within the standard range.

By synthesizing the whole DOE process, we get the best parameters as follows (show in Tab.12)

Item	Parameters
Tape	A
Blade	Z1: A Z2: B
Spindle speed Z1/ Kprm	II
Spindle speed Z2/ Kprm	I
Feed rate/mm/s	II
Blade Z1 height/ μ m	H II

III. CONCLUSION

Through this DOE, the best recipe and right blade is decided, which the chipping criteria is within the accepted level. The chipping PPM with respect to dicing parameters like spindle speed, feed rate and blade height set up is studied in this paper.

ACKNOWLEDGEMENT

The authors acknowledge supports from ATGD wafer sawing staff on trial supports.

REFERENCES

[1] Kong X , Wang X , Wu C . Diamond sawing process of 12 inch low-K silicon wafer applied in smart card[C]// International Conference on Electronic Packaging Technology. IEEE, 2014.

[2] Wang Z J , Wang J H , Lee S , et al. 300-mm Low- Wafer Dicing Saw Development[J]. Electronics Packaging Manufacturing IEEE Transactions on, 2007, 30(4):313-319.

[3] Sanamthong W , Chuti Ma P . Chipping Size Reduction on Ultra-Thin Wafers and Narrow Saw-Streets for Wafer Sawing Process[J]. Solid State Phenomena, 2020, 305.

3D Dislocation Multi-stack Fan-out Package of Ultra-thin Dies for Heterogeneous Integration

Lijun Chen
National Center for Advanced Packaging
Wuxi, China
lijunchen@ncap-cn.com

Feng Chen
National Center for Advanced Packaging
Wuxi, China
fengchen@ncap-cn.com

Fengwei Dai
National Center for Advanced Packaging
Wuxi, China
fengweidai@ncap-cn.com

Abstract—It's believed that 3D integration technology is a solution to beyond More-than-Moore law and to bring incremental cost-power-performance value. An advanced 3D dislocation multi-stack integration structure by fan-out wafer-level packaging has been developed for 3D integration. Four ultra-thin NAND Flash chips which thickness is about 40um can be integrated by Package-on-Package (PoP) method. The package size is 18mm×12mm×0.89mm，and BGA pitch is 1mm. The 3D dislocation multi-stack integration technology can be used as an alternatives solution to wire bonding and through-silicon-via (TSV) technology for NAND flash packaging.

Keywords—heterogenous integration; 3D dislocation multi-stack; package-on-package; NAND Flash; fan-out wafer level packaging

I. INTRODUCTION

Moore's law is slowing down since it's more and more difficult and expensive to reduce the feature size. One promising way is to integrate dissimilar chips with different materials and functions to gain space, performance and reduced cost. The solution can be agreed with fan-out technology, which can realize many integration possibilities for different size dies, flip chips or stacked chips. Fan-out technology can embed multi-chips with different size on the same plane and realize better connection. However, it's difficult to realize 3D integration, no matter by chip-first (die face-up or die face-down) and RDL-first (chip-last) process. Three-dimensional package can be realized by Package-on-Package (PoP) structure, which has been used to house the application processor (AP) and memory for the mobile application [1, 2].

It's reported by TSMC in 2019, two 3D Multi-stack (MUST) packages were developed for the memory module integration [2]. LDDR5 chips were packaged by 3D MUST-in-MUST (3D-MiM) integration technology. Compared to the FC-PoP package, the 3D-MiM package has a lot of benefits: thickness decrease 50%, bandwidth increase 100%, electrical loss decrease 66%, latency decrease 80% and the PDN impendence decrease 90%. However, up to now, very little research has been carried out on the fan-out packaging of NAND Flash.

NAND Flash stacking by wire bonding has been in high volume production for the smartphones, tablets and solid state drives. The thickness of each chip is only 40um [3]. In order to increase the bandwidth, density and performance, the memory chips are stacked through micro bumps and TSV (through-silicon via) [4-6]. Both the wire bonding and TSV

stacking are assembled by chip to chip, and the packaged unit are tested one by one. While the wafer-level packaging can increase the output and save the test time.

In this paper, an advanced 3D dislocation multi-stack integration structure by fan-out wafer-level packaging (FOWLP) technology has been developed for NAND Flash heterogeneous integration. Four ultra-thin NAND Flash chips can be integrated to one package by PoP.

II. PACKAGE DESIGN AND PROCESS FLOW

Fig. 1 shows the schematic of 3D dislocation multi-stack fan-out package. The package which size is about 18mm× 12mm×0.89mm was assembled by PoP. The size of top package is 16mm×10mm×0.14mm (not including BGA height), while the size of bottom package is 18mm×12mm× 0.14mm (not including BGA height). The PoP package contains four ultra-thin chips which thickness is about 40um.

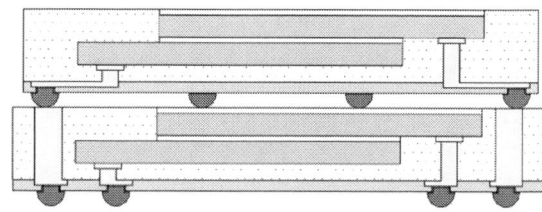

Fig. 1. schematic of 3D dislocation multi-stack fan-out package

Fig. 2 shows the process flow of PoP top package. Step 1, Cu pillar plating on the incoming wafer, grind the wafer to 40um and singulate to dies. Step 2, die attachment on the temporary glass carrier. Step 3, compression molding on the glass carrier. Step 4, grind the molded wafer to reveal the Cu pillars. Step 5, passivation and RDL formation. Step 6, BGA, debond the temporary carrier and singulate into top packages.

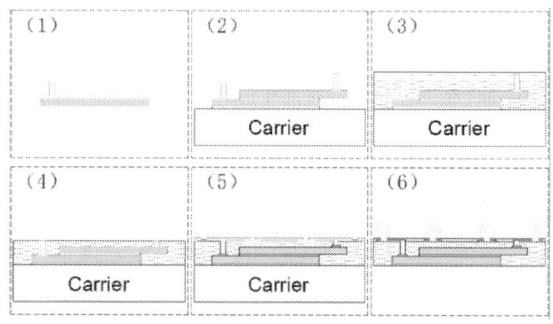

Fig. 2. process flow of PoP top package

978-1-6654-1392-3/21 $31.00 © 2021 IEEE

Fig.3 shows the process flow of PoP bottom package. Step 1, pad and passivation layer formation on the temporary carrier. Step 2, mega Cu pillar growth. Step 3, die attachment on the carrier. Step 4, compression molding on the glass carrier. Step 5, grind the molded wafer to reveal the Cu pillars and mega pillars. Step 6, passivation and RDL formation. Step 7, debond the temporary carrier and electroless plant Ni/Au on the pad. Step 8, second temporary bonding and BGA. Step 9, remove the temporary carrier and singulate into bottom packages.

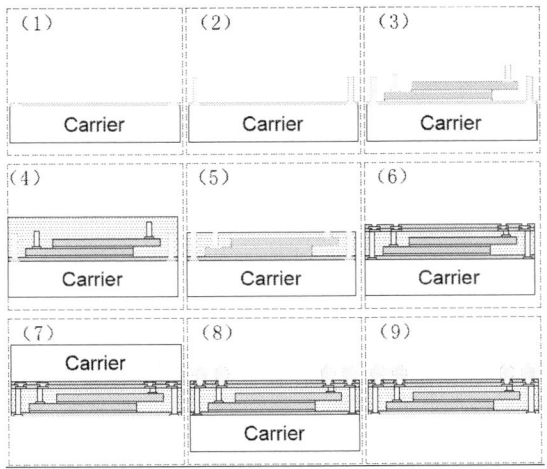

Fig. 3.   process flow of PoP bottom package

III. KEY PROCESS MODULES.

Considering there are some similar process between the top package and the bottom package, here we only present the key process modules of the PoP bottom package.

A. Ultra-thin die and temporary carrier

The incoming 12" wafer contains about 840 chips which size is about 11mm×7mm. There are 48 pads for wire bonding, and the minimum pad pitch is 85um. Cu pillars were electroplated by wafer level packaing. The AOI result shows the height distribution of all Cu pillars, and the average height is about 73um. Then the wafer was grinded to 40um and singulated into chips with Die Attach Film (DAF) backside.

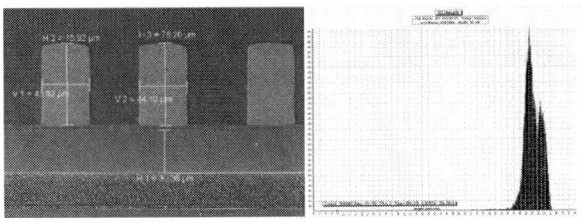

Fig. 4.   Ultra-thin chips and statis of pillar height

The temporary wafer was prepared by RDL-first fan-out technology. Firstly, the laser release layer was spin coated on the glass carrier. Secondly, the pad was electroplated on the release layer, then covered with passivation layer. Thirdly, mega pillars with diameter 200um were electroplated on the PI opening position. Fig.5 shows the cross-section of mega pillar which height is about 155um. The right graph shows the

surface of temporary carrier. The pads (under the PI layer), align mark for die attachment and mega pillars are indicated.

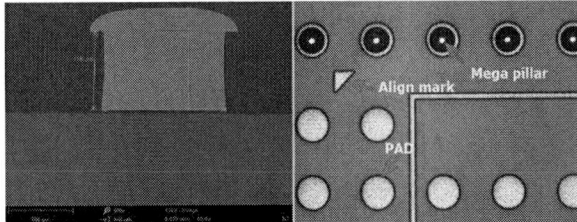

Fig. 5.   Mega pillars and prepared tempoary carrier

B. Die attchment and compression molding

The ultra-thin dies were designed for wire bonding, and all the pads were located on the same side of the die. The ultra-thin dies were picked up from the frame with DAF backside, then were stacked dislocation with the align mark on the carrier. As shown in Fig. 6, the overhang of the dislocation dies is about 400um. Fig. 7 shows the offset of X/Y direction and the rotation angle of attached chips. The mean offset of X/Y is about 0.01um, and mean rotation angle is about 0.001°. The maximum of X/Y offset is less than 5um, while the maximum rotation angle is about 0.03°.

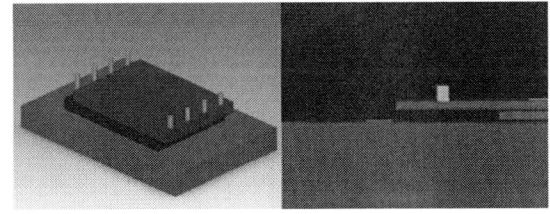

Fig. 6.   Schematics of 3D dislocation multi-stack structure.

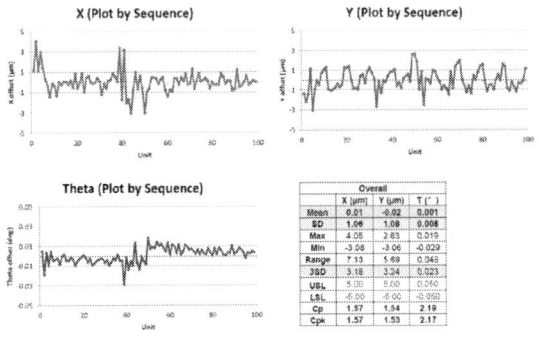

Fig. 7.   Die attachment measurement.

The attached dies were embedded in Epoxy Molding Compound (EMC) by compression molding. To obtain better flow ability for large surface area and thin film molds, the liquid molding compound was provided for wafer level compression molding. The molding diameter is 297mm, a little bit smaller than the glass carrier size. Fig. 8 shows the wafer before compression molding and the thickness of molded wafer. The average thickness of nine points is 0.243mm. The maximum thickness is 0.254mm while the minimum thickness is 0.232mm. The measured warpage of molded wafer after hard bake is less than 1mm.

978-1-6654-1392-3/21 $31.00 © 2021 IEEE

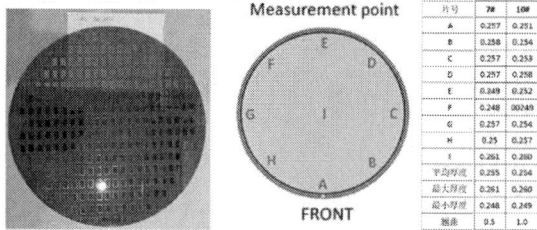

Fig. 8. Molding thickness measurement

C. Cu pillars revealing and RDL

The molded wafer should be grinded to reveal all Cu pillars. As shown in Fig. 9, the thickness of glass carrier is about 700um (A), the thickness of EMC is about 250um (B), and target thickness of remaining EMC is about 130um (C). The remaining wafer thickness should be controlled carefully. When Cu pillars of top chips and bottom chips are revealed, the upper die surface cannot be damaged.

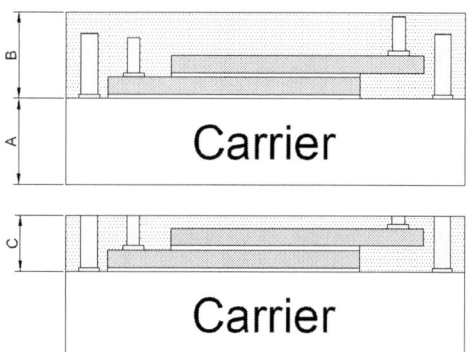

Fig. 9. Grinding EMC to expose Cu pillars

Fig.10 shows the surface of grinded wafer. The purple chips can been seen under the transparent EMC. Thickness of remaining EMC on the upper dies is about 5um. From left to right, we can see the Cu pillars revealing from bottom dies, from top dies, and mega pillars from carrier, respectively.

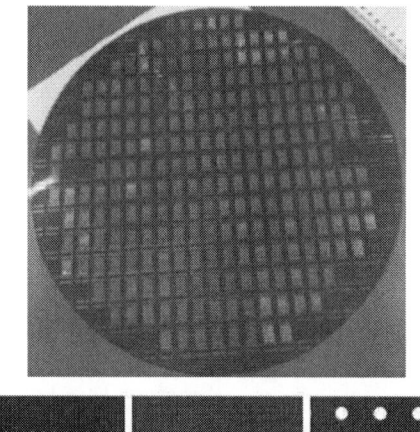

Fig. 10. Grinded EMC wafers and revealed Cu pillars

The bottom package has two layers of RDL near the BGA balls. To obtain the target impedance, the width and the length of RDL should be optimized. As shown in Fig.11, when the Cu pillars and the mega pillars were revealed, they can be connected by RDL.

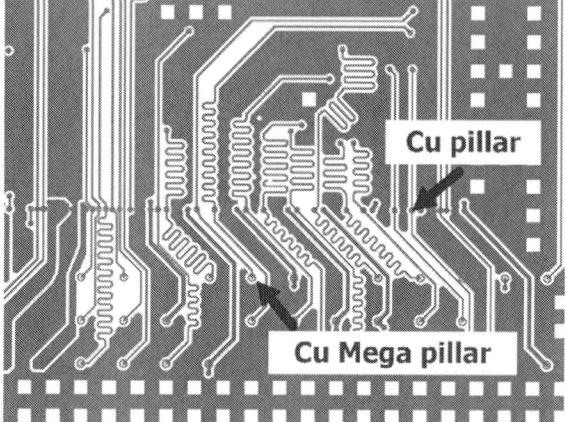

Fig. 11. Interconnection between Cu pillar and mega pillar by RDL

D. 3D PoP and vertical conncetion

Fig. 12 shows the appearance of 3D-PoP package. There are 132 BGA balls with pitch 1mm. There are four chips embedded in the package which thickness is about 886um. Fig. 13 shows the SEM image of a PoP package. It can been seen that: (A) the tall Cu pillars on the bottom dies from bottom package and top package, respectively; (B) the center of PoP package, the enlarge photos show that the thickness of embedded chips is about 40um; (C) the short Cu pillars on the top dies from bottom package and top package, respectively.

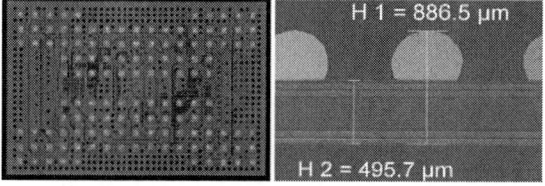

Fig. 12. Overview of package

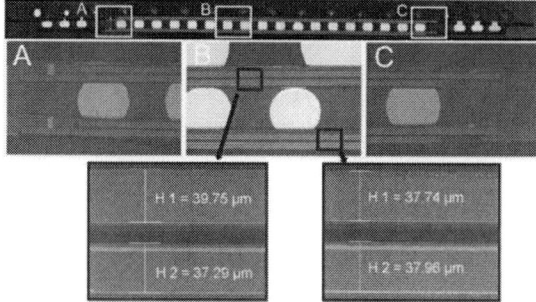

Fig. 13. Cross-section of package and ulta-thin chips inside

The PoP package contains four ultra-thin dies, and all the pads of four dies should be connect to the BGA balls of bottom package. Fig.14 shows the interface between the top package and bottom package. The signals of the top package transfer in turn from the Cu pillars, RDL and BGA balls of the top package, to mega pillars, RDL and BGA balls of the bottom package. Up to now, the 3D-PoP NAND Flash package can perform read and write.

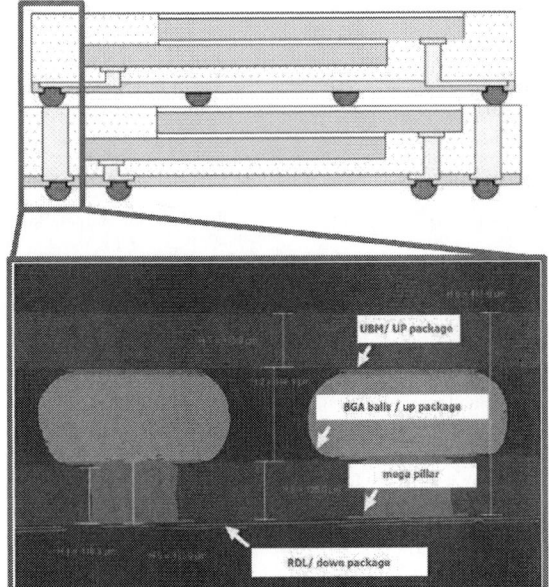

Fig. 14. Interconnection between top packge and bottom package

IV. CONCLUSION

In this study, the structure, process flow and key modules of 3D heterogeneous integration of NAND Flash by FOWLP has been reported. An advanced 3D dislocation multi-stack structure can embed four ultra-thin NAND Flash chips which thickness can be controlled about 40um. Up to now, the 3D PoP NAND Flash package can perform read and write.

ACKNOWLEDGMENT

This work was supported by the National Science and Technology Major Project with contract No. 2017ZX02301004-003 under taken by the National Center for Advanced Packing.

We would like to thank ASM pacific, Towa, Panosonic Industry, and Han' laser for their support for this work.

REFERENCES

[1] Yole, "Equipment and Materials for Fan-out Packaging 2019-Market and Technology Report," 2019.

[2] J. H. Lau, Fan-Out Wafer-Level Packaging. Springer Singapore, 2018.

[3] B. Milton et al., "Smart Wire Bond Solutions for SiP and Memory Packages," in 2019 IEEE 69th Electronic Components and Technology Conference (ECTC), 28-31 May 2019 2019, pp. 55-62.

[4] T. Onagi, C. Sun, and K. Takeuchi, "Impact of through-silicon via technology on energy consumption of 3D-integrated solid-state drive systems," in 2015 International Conference on Electronics Packaging and iMAPS All Asia Conference (ICEP-IAAC), 14-17 April 2015 2015, pp. 215-218.

[5] T. Hatanaka, K. Johguchi, and K. Takeuchi, "Experimental Investigation of Program Voltage (20 V) Generation With Boost Converter for 3-D-Stacked NAND Flash SSD," IEEE Transactions on Components, Packaging and Manufacturing Technology, vol. 5, no. 2, pp. 188-193, 2015.

[6] M. Nakanishi, Y. Adachi, C. Matsui, Y. Sugiyama, and K. Takeuchi, "Application-oriented wear-leveling optimization of 3D TSV-

integrated storage class memory-based solid state drives," in 2018 International Conference on Electronics Packaging and iMAPS All Asia Conference (ICEP-IAAC), 17-21 April 2018 2018, pp. 27-32.

The multi-dimension co-design for the package of the high-speed multi-function chip

Yuan Guan
Beijing smartchip microelectronics technology company limited
Haidian district, Beijing，China
guanyuan@sgitg.sgcc.com.cn

Yubo Wang
Beijing smartchip microelectronics technology company limited
Haidian district, Beijing，China
wangyubo@sgitg.sgcc.com.cn

Dejian Li
Beijing smartchip microelectronics technology company limited
Haidian district, Beijing，China
lidejian@sgitg.sgcc.com.cn

Shunfeng Han
Beijing smartchip microelectronics technology company limited
Haidian district, Beijing，China
hanshunfeng@sgitg.sgcc.com.cn

Bofu Li
Beijing smartchip microelectronics technology company limited
Haidian district, Beijing，China
libofu@sgitg.sgcc.com.cn

Dameng Li
Beijing smartchip microelectronics technology company limited
Haidian district, Beijing，China
lidameng@sgitg.sgcc.com.cn

Fengman Liu
Institute of Microelectronics of Chinese Academy of Sciences
Chaoyang district, Beijing, China
liufengman@ime.ac.cn

Huimin He
Institute of Microelectronics of Chinese Academy of Sciences
Chaoyang district, Beijing, China
hehuimin@ ime.ac.cn

Abstract—in this paper, a kind of high-speed multi-function chip is packaged. Different from the general chip, the chip includes multiple high-speed parts and extremely high-density pins. Based on the package of the chip, a multi-dimension co-design method is proposed, which includes chip-package-PCB co-design and electrical/thermal co-design of the package. Ultimately, the packaged chip is tested and it could normally read and write data from the registers at the 25 ℃ ambient temperature, which indicates that the multi-dimension package co-design is feasible for the chip.

Keywords—multi-function chip, multi-dimension co-design method, multi-dimension co-design method, electrical/thermal co-design

I. INTRODUCTION

The development of the chip manufacture technology conforms to the Moore's law so that the number of the transistors on a chip is increasing and the performance of the chip improves obviously. Besides, the secondary equipment is the core equipment of the intelligent power grid. With the development of the intelligent power grid, the high-performance secondary equipment is necessary. And the processing chip is the core device of the secondary equipment. Considering on the rapid development of the chip and the urgent requirement of the processing chip in the intelligent power grid, a high-speed multi-function processing bare chip is developed and it includes multiple high-speed functions such as DDR controller, USB/GPIO/ UART interface and so on. Since the packaged chip could be applied to the equipment, the chip should be packaged carefully to avoid the influence on the performance of the chip.

The traditional design of a product is as shown in Fig.1. The design of the chip, the package and the PCB is conducted independently. The design of the chip is the first, the design of the package is the second and the design of the PCB is the end. For the complicated chip, the traditional design has several shortcomings. Firstly, the traditional design leads to the insufficient design margin so that the design of the product is perhaps unsuccessful. The I/O configuration designs of the chip and the package have the impact on the designs of the package and the PCB respectively. The chip with too small spacing increases the design difficulty of the package and the PCB. Secondly, the traditional design has not taken enough emphasis on the simulation of the package. With the rapid development of the high-performance chip, the method of the traditional packaging design is not appropriate. Thirdly, the period of the traditional design is so long to extend the time-to-market of the product. Therefore, the traditional method is not suitable for the design of the high-speed multi-function chip and a multi-dimension co-design method is applied in the high-speed multi-function chip.

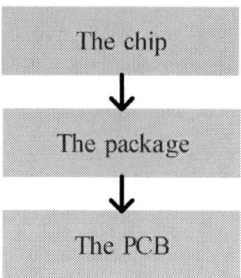

Fig.1 The traditional design of a product

The multi-dimension co-design method of the high-speed multi-function chip includes the chip-package-PCB co-design and the electrical/thermal co-design of the chip-package-PCB. For the chip-package-PCB co-design, the package connects the package and the PCB, and the designs of the chip and the PCB are considered simultaneously with the design of the package. Besides, the package is designed by multi-physics field simulation to ensure the electrical and the thermal performance. The electrical/thermal performance of the package is simulated with the design parameters of the chip and the PCB. Therefore, the chip-package-PCB co-design of the high-speed multi-function chip is presented in Section II. The thermal co-design and the test verification of the chip-package-PCB is presented in Section III. The electrical co-design, and the test verification of the performance of the chip-package-PCB is presented in Section IV. The conclusion is presented in Section V.

II. CHIP-PACKAGE-PCB CO-DESIGN

The chip-package-PCB co-design is that the package is designed with the designs of the chip and the PCB. It includes the chip-package co-design and the package-PCB co-design.

For the chip-package co-design, the I/O module, I/O distribution and the size of the chip are optimized considering the layout of the package. Besides, the way that the chip mounts on the package, the size and the stack-up of the package and so on are determined by the various parameters of the chip. For the package-PCB co-design, the property of the packager substrate, the way that the package assembles on the PCB, the I/O pin-map and so on have influence on the PCB. Besides, the design of the package is restricted by the property of the PCB, the layout and so on. The method of the chip-package-PCB co-design is illustrated in Fig.2.

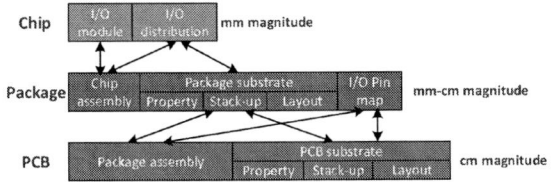

Fig.2 Chip-package-PCB co-design

The high-speed multi-function chip includes mainly three I/O modules which are DDR controller, USB and other low-speed interfaces. Among them, the parts of the DDR controller and USB have the large power consumption and the large transmission rate (1333Mbps). To reduce the temperature and the crosstalk of the package, the DDR controller are placed in the left and the lower-left corner of the chip, and the USB is placed in the upper right corner of the chip. For the other low-speed parts, the I/Os are close to the transistors of the module. Because the number of I/Os is 567, they are distributed in the 4 sides of the chip.

Since the I/Os are distributed in the 4 sides of the chip, the chip is connected to the package substrate by the bonding wires. To reduce the size of the chip, the I/Os in the chip are distributed at two rings. Besides, for the high-speed pins, a reference ground/power appears for each three high-speed signals, and the high-speed signals at the outer ring take the close pads at the inner ring as the reference signal. The design of the high-speed I/Os is to ensure the signal quality in the package. The distribution of the I/Os in the chip is shown in Fig.3. Finally, the size of the optimized chip is 5123 um*5150um.

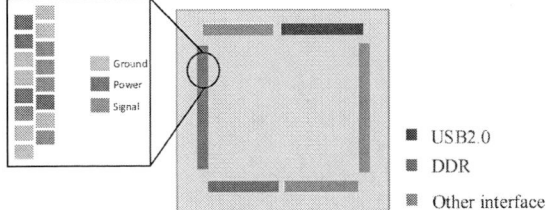

Fig.3 The distribution of the I/Os

The bonding fingers in the package substrate are also considered by the several factors. Firstly, the line width/space in the package substrate is larger than that in the chip so that the bonding fingers are not placed in one line. Secondly, the bonding wires will go against the rule of the bonding wire with two rings of the bonding fingers. Finally, the DDR module has three rings of the bonding fingers and the other module has two rings of the bonding fingers which has the size with 50um space and 60um width. The designed bonding fingers are shown in Fig.4.

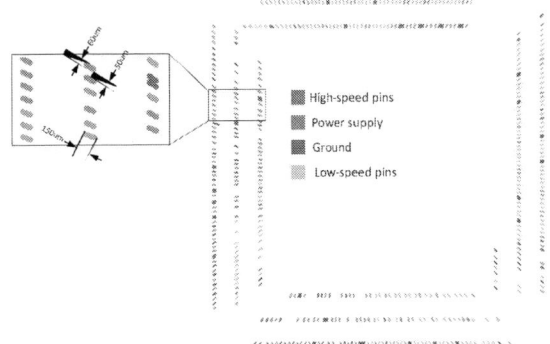

Fig.4 The designed bonding fingers

Besides the above parts, the I/O design under the package substrate and the stack-up of the package substrate are also determined considering the chip and the PCB. According to the application requirement, the package is assembled to the PCB by the BGA. The BGA is distributed as the hollow square to interconnect successfully in PCB. The power/ground signals are arranged in the center of the BGA and the power signals with the same electrical property are put together to reduce the dropping voltage and the fluctuation of the voltage. In addition, the high-speed signals are around the package and the reference ground/power signal is close to the high-speed signal to ensure the signal integrity. The BGA pin-map under the package substrate is shown in Fig.5.

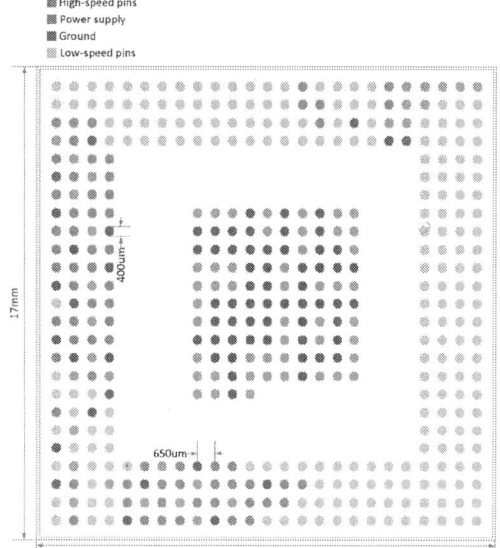

Fig.5 The BGA pin-map under the package substrate

For the stack-up of the package substrate, the high-frequency material is adopted so that the high-speed lines are located at the top and the bottom layer of the stack-up. The specific stack-up with the signal-ground-power-signal layer as shown in Fig.6 is the symmetrical structure. Moreover, the rate of the copper cladding at every layer is the symmetrical. Such design reduces the warpage of the package, which leads to the broken circuit at the BGA interconnection. Based on the stack-up, the single-end 50 Ω micro-strip with the 40um width are designed, and the differential 100 Ω micro-strip with the 35um width and the 100um space are designed. Finally, the size of the package is 17mm*17mm.

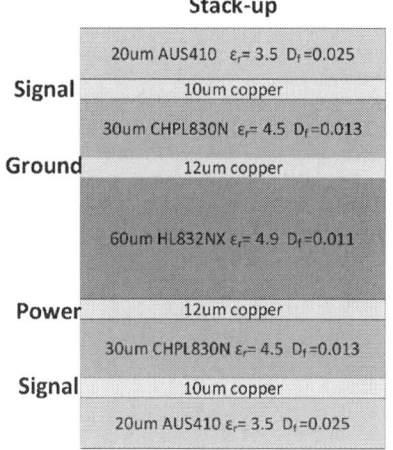

Fig.6 The stack-up of the package substrate

III. THE THERMAL DESIGN OF THE PACKAGE

To ensure the normal operation of the chip, the temperature of the package should maintain in the operating temperature of the chip. Therefore, the thermal of the package is designed and simulated. Besides, the thermal of the package is tested and verified by the thermocouple probe.

The thermal of the package is simulated. Firstly, to ensure the accuracy of the simulation, the structure that the

package mounts on the PCB is simulated. Thus, the heat can be dissipated from the top of the chip and the bottom of the PCB. Besides, under the natural convection, the thermal simulation is done with the current of the power supply at the ambient temperature 25 ℃. The power consumption of the chip is 1.5W. By simulation, the junction temperature is 53.24°C as shown in Fig.7 and it is under the operation temperature of the chip, which proves that the chip can work at the ambient temperature.

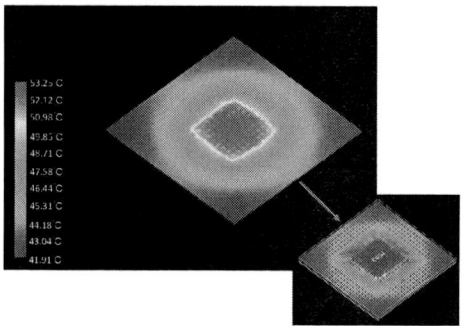

Fig.7 The thermal distribution of the simulation

To prove that the thermal simulation is accurate, the thermal of the packaged chip is tested by the thermocouple probe. The test points are marked in Fig.8. The comparison of the test and the simulation temperature is illustrated in Table1. The maximum difference between the simulation and the test is 1.81 ℃, which prove that the simulation method is creditable.

Fig.8 The thermal test points of the chip

Table 1 The comparison of the simulation and test temperature

Test point	Simulation(℃)	Test(℃)	Difference
1	44.83	43.3	1.53
2	46.22	44.5	1.72
3	45.14	43.9	1.24
4	45.96	41.7	1.24
5	52.90	51.5	1.7
6	46.41	44.6	1.81
7	45.14	43.8	1.34
8	46.18	44.6	1.58

| 9 | 45.10 | 43.7 | 1.4 |

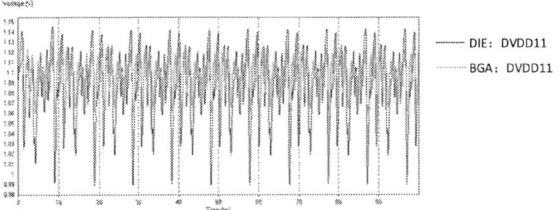

Fig.11 The power ripple of the simulation for DVDD11

IV. THE ELECTRICAL DESIGN OF THE PACKAGE

For the high-speed multi-function chip, the maximum transmission rate of the I/Os is 1333Mbps, which is the DDR ports. Other I/Os is below 500Mbps. Therefore, the DDR part is simulated from the S parameter of the package, the eye pattern and the power ripple of the system.

The chip includes 20 supply voltages and 9 grounds. To ensure the normal power supply, the DC/AC voltage and the power ripple are simulated. Firstly, the current density and DC voltage drop of the supply voltage are simulated. The current density is optimized by adjusting the distribution of the power plane and increasing the vias of the power. The via optimization at the BGA place is shown in Fig.9. By the simulation optimization, the current density is minimized under 200 A/mm², Meanwhile, all the voltage drops are simulated and verified after the optimization of the power and ground links, and the result indicates that all the voltage drops are within 2% of the supply voltage, as shown in Fig.10. Among the all voltages, the DVDD11 has the largest voltage drops, which is because of the large current. Besides, considering the AC noise of the power, the several decoupling capacitors with different capacitance are added in the main supply voltages. By simulation, the value and the number of the capacitor are optimized. In addition, the ground and power vias which connect to the plane are in pair and rather close. On the basis of the DC and AC power optimization, the power ripple with the chip-package-PCB power trace are simulated in the time domain. The ripple voltage of all the voltage is within 20% and the simulation result of the ripple voltage with the largest voltage drop is illustrated in Fig.11.

In the chip, the DDR controller has the largest transmission rate with 1333Mbps. Therefore, in the frequency domain, the high-speed transmission paths are simulated under 2GHz which is about the triple-harmonic of the 1333Mbps. Firstly, the S parameter of the high-speed transmission path with the bonding wire, buried/blind vias and the BGA is simulated. The worst S parameter with a byte is shown in Fig.12. The result indicates that the insertion loss is below -1dB within 2GHz and the largest return loss is -13dB, which is rather small. In addition, the largest far-end crosstalk is -20dB, which results from the restriction of the package size and the tight coupling of the two adjacent lines. Next, the eye pattern of the system data link with the IBIS model of the chip and the S parameter of the transmission link in the package substrate and the PCB is simulated in the time domain. By simulation, it is found that the eye width and height reach the optimal value at the 60Ω ODT which is close to the characteristic impedance of the transmission line. At the 60Ω ODT, the eye pattern of the worst byte is shown in Fig.13. The picture represents the clear eye pattern of the byte data with the 562.22ps eye width and the 1.1V eye height.

At the end, the packaged chip mounts on the PCB and the reading and writing functions of the DDR are measured at the 25°C ambient temperature. By the measurement, the read data is the same with the written data in the registers, which proves the chip could work normally and the co-design method is feasible.

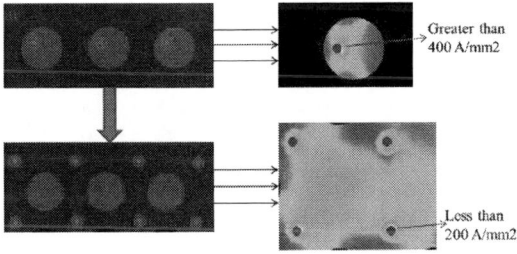

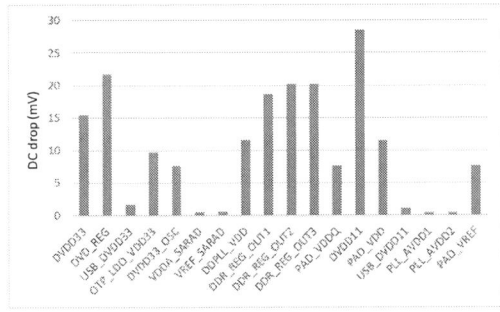

Fig.9 The via optimization at the BGA place

Fig.10 All the voltage DC drops

S Amplitude (dB)

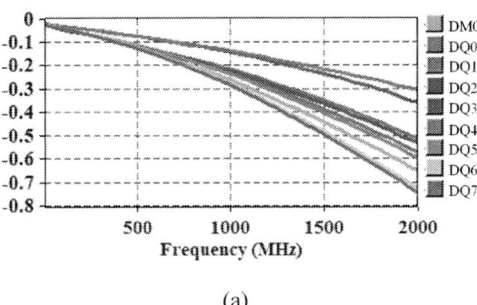

(a)

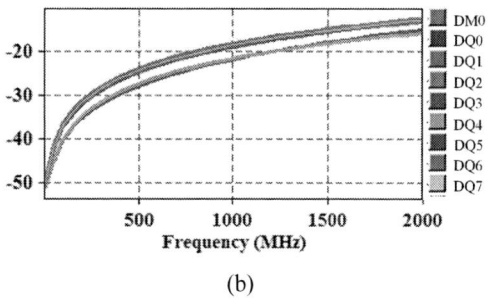

(b)

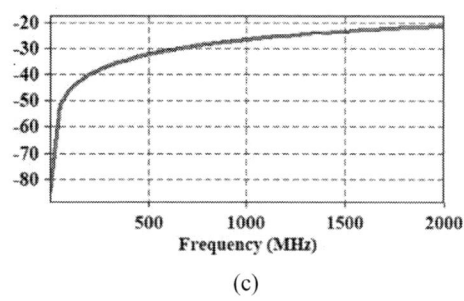

(c)

Fig.12 The worst simulation S parameter with a byte

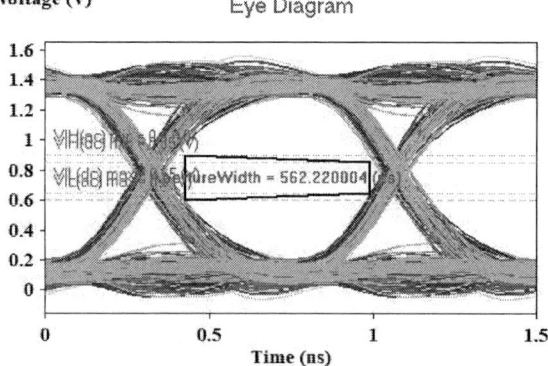

Fig.13 The simulation eye pattern of the worst byte

V. CONCLUSION

In the paper, the multi-dimension co-design method is adopted for the package of the multi-function high-speed chip. Firstly, the chip-package-PCB co-design is done to guide the design of the chip, complete the design of the package and consider the design of the PCB. The chip is connected to the package by bonding wires, and the bonding fingers are determined. The package of the chip is assembled to the PCB by the BGA. The size and pin map of the solder balls are determined. Besides, the stack-up and the routing of the package substrate are designed taking all the aspects into account. The size of the designed package is 17mm*17mm. Secondly, the thermal characteristics of the packaged chip is simulated and tested. The results indicate the simulation and the tested temperature is identical, and at the 25°C ambient temperature the junction temperature is about 53.24°C which is within the operating temperature (-40°C-105°C). Thirdly, the power and signal integrity of the packaged chip is simulated in the frequency and the time domain. By

simulation, the ripple voltage of all the voltage is within 20% and the clear eye pattern of the data at 1333Mbps is presented. Finally, the packaged chip mounts on the PCB and it could read and write the data from the registers the 25°C ambient temperature.

ACKNOWLEDGMENT

This paper is funded by the project of the State Grid Corporation of China in 2019, "The key technologies research of the special multi-core SoC architecture for relay protection equipments (5700-201941501A-0-0-00)".

REFERENCES

[1] Allan A . International Technology Roadmap for Semiconductors (ITRS)[J]. Journal of Applied Physics, 2015, 86(17):045406.

[2] Kim J H , Schmitt R , Dan O , et al. Design of low cost QFP packages for multi-gigabit memory interface[C]// Electronic Components & Technology Conference. IEEE, 2009.

[3] Lin Y , Lee W , Horng T , et al. Full Chip-Package-Board Co-Design of Broadband QFN Bonding Transition Using Backside via and Defected Ground Structure[J]. IEEE Transactions on Components, Packaging, and Manufacturing Technology, 2017, 4(9):1470-1479.

[4] Dinh T V , D Lesénéchal, B Domengès, et al. Modeling and Characterization of Bond-Wire Arrays for Distributed Chip-Package-PCB Co-Design[C]// European Microwave Conference. IEEE, 2015.

[5] A. Han, V. Zaderej and R. Fitzpatrick, "Application Specific Electronics Package – Electronics Manufacturing Technology of the Future," 2021 14th International Congress Molded Interconnect Devices (MID), Amberg, Germany, 2021, pp. 1-6, doi:10.1109/MID50463.2021.9361621.

[6] K. X. Cai and S. Y. Ji, "Cost competitive PI-SI co-design for DDR interfaces," 2015 IEEE International Symposium on Electromagnetic Compatibility (EMC), Dresden, Germany, 2015, pp. 645-649, doi: 10.1109/ISEMC.2015.7256239.

[7] V. Rossi, C. Somma, G. Graziosi and A. Sanna, "BGA Package for DDR3 Interface – 4 vs 6 Layers Design Strategy and Electrical Performance Comparison," 2020 IEEE 8th Electronics System-Integration Technology Conference (ESTC), Tønsberg, Norway, 2020, pp. 1-6, doi: 10.1109/ESTC48849.2020.9229667.

[8] Joonghyun Baek, Byungse So, Taekoo Lee, Yunhyeok Im and Seyong Oh, "Thermal characterization of high speed DDR devices in system environments [DRAM modules]," Ninteenth Annual IEEE Semiconductor Thermal Measurement and Management Symposium, 2003., San Jose, CA, USA, 2003, pp. 138-143, doi: 10.1109/STHERM.2003.1194352.

[9] Na Suo and Lina Guo. RESEARCH ON THERMAL DESIGN CONTROL AND OPTIMIZATION OF RELAY PROTECTION AND AUTOMATION EQUIPMENT[J]. Thermal Science, 2020, 24(5B) : 3119 - 3128.

Effect of plasma and staging time on the underfill voids in fine pitch flip-chip package

1st Saif Wakeel
Centre of Advanced Materials,
Department of Mechanical
Engineering, University of Malaya,
Kuala Lumpur-50603, Malaysia.
saifwakeel@gmail.com

2nd Dominic Koey Poh Meng
Assembly process innovation
NXP Semiconductors, Selangor-47300
Malaysia.
dominic.koey@nxp.com

3rd Stella Wong Wun Chin
Assembly process innovation
NXP Semiconductors, Selangor-47300
Malaysia.
stella.wong@nxp.com

4th Jos Philipsen
Product Diagnostic Center
NXP Semiconductors, Nijmegen-6534,
Netherlands.
j.h.m.philipsen@nxp.com

5th Pieter Gommers
Product Diagnostic Center
NXP Semiconductors, Nijmegen-6534,
Netherlands.
pieter.gommers@nxp.com

6th Annelies Joosten
Product Diagnostic Center
NXP Semiconductors, Nijmegen-6534,
Netherlands.
annelies.joosten-diesveld@nxp.com

Abstract— **Underfill voids have been one of the major reliability concerns in flip-chip packaging specially in a fine pitch. Therefore, monitoring of process based root causes of the void formation is significantly important. In this study, 7.5x7.5 Si die was bonded on two 17 mm *17 mm^2 substrates viz. S$_1$ and S$_2$ to prepare a ~150 μm pitch flip chip package. Effect of two process parameters such as plasma and staging time was observed on void formations. Plasma cleaning of both substrate was done for 1 minute. Before underfill dispensing, staging of units was performed for 4, 6 and 8 hours, and void formation was observed using confocal scanning acoustic microscopy (C-SAM). Differences in void formations among S$_1$ and S$_2$ are related to the change in surface of solder mask. Therefore, optical profiler was used to detect the changes in surface roughness of solder mask with plasma and staging time. Whereas, formation of different chemical bonds before and after plasma was analyzed using x-ray photoelectron spectroscopy (XPS). Also, water contact angle on S$_1$ and S$_2$ was evaluated before and after plasma cleaning. As a result of this study, percentage unit failure was increased with increasing staging time. This is attributed to changes in surface roughness, and amount of O-C=O, C-O, C-C bonds on solder mask. Contact angle on secondary non-bump area may or may not be representative of underfill wettability due to surface roughness difference with bump area on one substrate supplier but similar surface roughness between bump and non-bump area on another substrate supplier.**

Keywords— *Underfill voids, plasma time, staging time, Optical profiler, XPS, wettability.*

I. INTRODUCTION

Underfill is used to reduce the thermo mechanical stress transfer between die and substrate due to CTE mismatch. With fine pitch technology, advancement in underfill has been made. For example, soldering industries have developed underfill with high glass transition temperature, compatible with copper pillar, lower CTE, higher elastic modulus [1]-[3]. However, behavior of underfill in package depends on various design, process and materials parameters. Many process parameters such as prebaking time, plasma time, staging time, underfill dispensing have significantly affect

the properties and flowability of underfill. Improper flowability will develop the voids and delamination [4], [5]. Root causes of underfill voids and delamination are expressed in fish bone diagram.

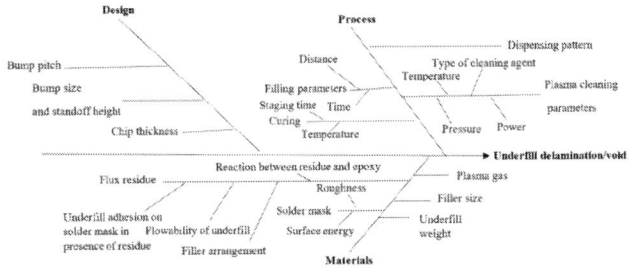

Figure 1: Fish bone diagram for root causes of underfill void and delamination [5], [6].

From Fig.1, it can be observed that process and materials related parameters are majorly affecting the underfill voids. For example, plasma cleaning is a significant factor to remove impurities from substrate surface and increase substrate surface energy.

Various works have been done on the plasma cleaning of substrate and their effect on underfill adhesion. Noh et al. [7] observed the effect of two gaseous treatments namely Ar/O$_2$ and Ar/H$_2$ on adhesion of three underfills on FR-4 and copper substrate, respectively. Contact angle of underfill on substrates was evaluated. Then, change in the substrate surface before and after plasma treatment was observed using AFM and XPS analysis. Result showed that adhesion strength of underfill is greatly affected by change in surface of substrate. Formation of hydrophilic carbonyl functional group (C=O, C-O3) after plasma treatment led to higher adhesion of underfill. Also, plasma treatment reduced surface roughness which improved the wetting of underfill on FR-4 and copper substrate. Liu et al. [8] studied the optimization of plasma parameters such as power, time and pressure and proposed the optimum combination to improve the stud pull strength and reliability of film BGA. XPS of polymer BGA substrate was done before and after plasma. Results showed that hydrophilic C=O bond increased the adhesion of film BGA package. Surface energy of C=O is not known but dissociation energy of this bond is very high which leads to surface energy improvement of BGA substrate [9]. Teo et al. [10] investigated the effect of plasma time (0, 15 and 30 second) and plasma gas (O2 and mixture) on moisture sensitivity performance of flip-chip packaging. With plasma time of 15 sec. new functional groups were found such as (C=O, O-C=O, O-C-O) and these functional groups are hydrophilic in nature which improved the wetting. Plasma treatment enhanced the MSL performance due to adhesion enhancement by hydrophilic groups which improves the surface energy. All these literature showed that plasma time has significant effect on the adhesion of underfill, die pull strength and MSL reliability of flip-chip package.

To the best of author's knowledge, effects of plasma and staging time on underfill voids formation in fine pitch flip-chip package have not been covered yet. Thus, objective of this study is to do failure analysis of underfill voids formed during plasma and staging.

II. EXPERIMENTAL PROCEDURE

A. Materials

Two 17×17 mm^2 organic substrate (S_1 and S_2) with different processing methods were used. Die bonding was performed using 7.5×7.5 mm^2 Si die with water soluble flux. Properties of underfill used in this study is given in Table 1.

Table 1: Properties of underfill

Viscosity (Pa-s)	Glass transition temperature (°C)	CTE <Tg (ppm/°C)	Storage modulus <Tg (GPa)
50	119	29	11

Die bonding was performed using reflow profile shown in Fig. 2.

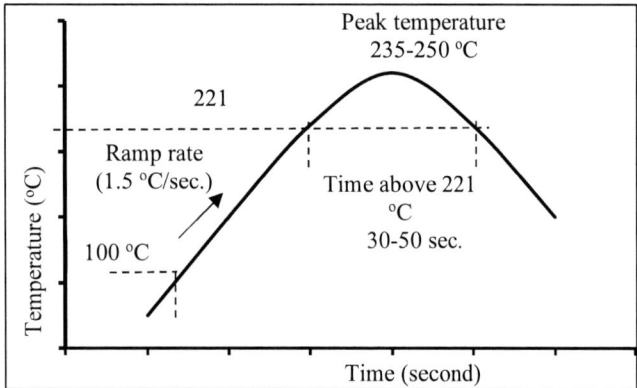

Figure 2: Reflow profile used in this study

Overview of type and number of sample for each test are given in Table 2.

Table 2: Test, sample type and number of sample used in this study

Test	Sample type	Number of sample per substrate
Plasma cleaning	Die bonded units without underfill	20
Post Plasma Staging	Die bonded and plasma-ed units with underfill	60
Contact angle	Non-bump area	8
Surface roughness	Die sheared off bump area + non-bump area	3 + 4
X-ray photoelectron spectroscopy (XPS)	Die sheared off bump area	3

B. Underfill plasma and staging

Before underfill dispensing, plasma cleaning of units was performed for 1 minute using plasma gases argon (Ar) and oxygen (O_2) in ratio of 6:1. Plasma chamber was fully loaded with dummy units in pallets besides actual units for study to mimic actual production run. Staging of the units was done at room temperature for 0, 4, 6 and 8 hours. Then, underfilling was done at 75 °C and curing was performed for 6 hours at 150 °C. Further, C-SAM was done to observe the voids.

C. Contact angle

Water contact angle of non-bump area of substrate was measured using needle drop test method. In this test, one point on the non-bump surface of substrate was selected and water droplets were dropped using advanced optima machine. This test was performed after plasma and staging.

D. Surface roughness

Physical changes on the surface of solder mask after plasma cleaning were observed by surface roughness measurement. Total 3 location on center of bump area and 4 location on non-bump area were selected and surface roughness was determined using Bruker optical profiler. Root mean square value of roughness (Sq) in given area was calculated. Surface roughness was evaluated after plasma cleaning only because controlling the exact staging time is difficult. One unit was selected to perform this test.

E. X-ray photoelectron spectroscopy (XPS)

Effect of the plasma gas bombarding on chemical state of solder mask surface was observed using XPS analysis. In this test, one die attached unit was taken into analysis. For die attach unit, die was sheared using with internal dry decapsulation tool to access the die cage area. For each unit, three points on bump area were selected for XPS.

III. RESULTS

As per industrial failure criterion, percentage amount of units failure based on C-SAM images is listed in Table 3.

Table 3: Percentage units failure matrix for S_1

Substrate S_1	Plasma cleaning time (minute)
Staging time (hours)	t1
T4	0%
T6	0%
T8	75%

Table 4: Percentage units failure matrix for S_2.

Substrate S_2	Plasma cleaning time (minute)
Staging time (hours)	t1
T4	0%
T6	20%
T8	100%

From Table 3, 0% failure was found for S_1 at 4 and 6 hrs of staging. For 8 hours of staging, number of units failure increased to 75%. In case of S_2, increase in staging time increases void %. Also, 100% void was found for 8 hours. Overall, percentage unit failure for S_1 was significantly lower than S_2 for each staging time.

Water contact angle measured on non-plasma and plasma cleaned units with staging is presented in Table 5. Contact angle measurement is taken at secondary non-bump area of substrate as primary substrate bump area is covered with die.

Table 5: Water contact angle after plasma and staging

Staging time (min.)	Plasma time (min.)	
	t1	
	Contact angle (θ)	
	S_1	S_2
T0	12.4	5.2
T4	17.4	11.9
T6	20.0	17.2
T8	30.8	18.2

From the Table 5, it can be observed that increasing the staging time increases the contact angle for S_1 and S_2.

Surface roughness of bump central area for both substrate is plotted in Fig. 2.

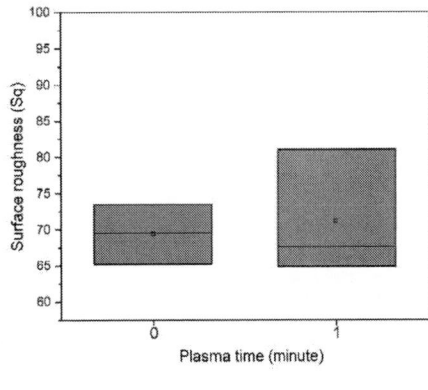

(a)

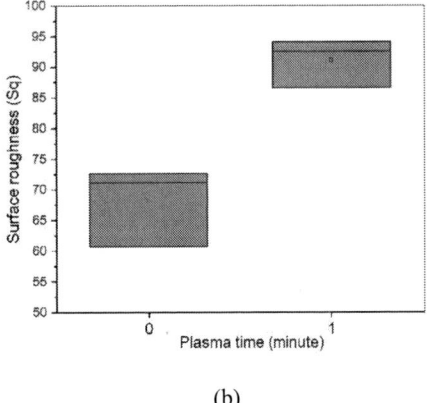

(b)

Figure 2: Root mean square surface roughness on bump area of (a) S_1 and (b) S_2

Surface roughness of S_1 and S_2 for 1 min. plasma is higher than non-plasma units. Also, average surface roughness of S_2 for 1 minute plasma was higher than S_1 as shown in Fig.2. This is unusual as both substrate suppliers are using same solder resist/ mask material.

Surface roughness of non-bump area is shown in Figure 3.

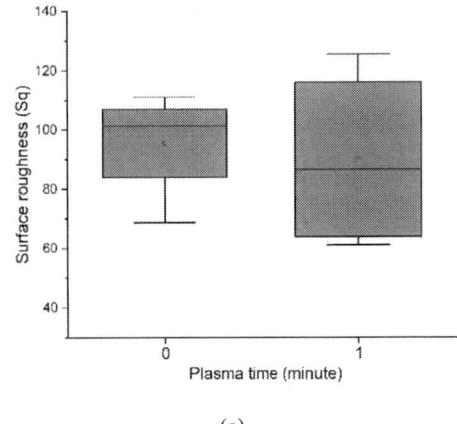

(a)

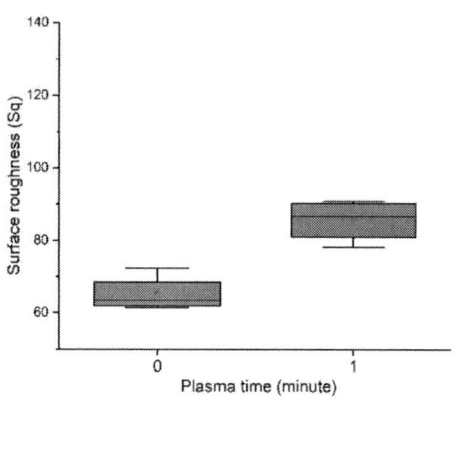

(b)

Figure 3: Root mean square surface roughness on non-bump area of (a) S_1 and (b) S_2

Non-bump substrate surface roughness on S_2 for 1 min. plasma is higher than non-plasma units. This non bump surface roughness trend is the same as S_2 bump area substrate surface roughness trend with minimal roughness difference between non bump and bump area for both non plasma and plasma cells.

However, large variation in non-bump area surface roughness of S_1 was found after plasma treatment. Additionally, the S_1 surface roughness of non-bump area non-plasma and plasma cells were significantly higher than bump area due to the higher variability of non-bump area non-plasma and plasma cells.

Besides, average and maximum surface roughness of S_1 for 1 min. plasma and non-plasma were higher than S_2 as shown in Fig. 2.

It is interesting to observe different substrate surface roughness performance on bump vs non-bump area for S_1 but not on S_2. This is unusual as both substrate suppliers are using the same solder resist/ mask material.

Carbon spectra (C1s) of bump area on S_1 and S_2 are shown in Figure 3.

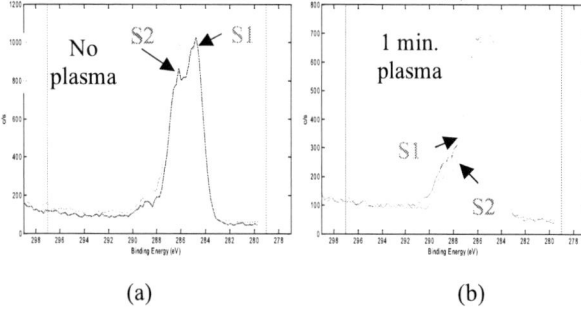

(a)   (b)

Figure 3: C1s spectra of S1 and S2 on bump area for (a) no plasma, (b) 1 min.

Peaks of C1s spectra with specific binding energy show the presence of C-C, O-C=O, C-O functional groups on the solder mask surface of S_1 and S_2. For no plasma and 1 min. plasma spectra, percentage amount of each functional group found is listed in Table 6.

Table 6: Functional groups on die cage area

Substrate	Plasma	C-C (%)	C-O (%)	O-C=O (%)
S_1	no	54	41	4
	1min	44	37	19
S_2	no	50	46	3
	1min	48	43	9

Percentage of hydrophilic carbonyl (O-C=O) functional group for 1 min. plasma is higher than non-plasma units. Also, amount of O-C=O is two times higher for S_1 as compared to S_2. This is odd as both substrate suppliers are using same solder resist/ mask material.

IV. DISCUSSION

Results showed the effect of change in solder mask surface on underfill voids with staging and plasma. Overall, S_2 has higher percentage of units' failure as compared to S_1 for each plasma and staging time matrix as presented in Table 3 and 4. This is attributed to higher hydrophilic functional group (O-C=O) and lower surface roughness on S_1 bump area surface as compared to S_2. These are explained as follows:

Bombarding of plasma gas changes the surface of solder mask in two ways. First, it changes the surface roughness. Second, it develops and alter the amount of hydrophilic functional groups such as O-C=O and C-O. Carbonyl (O-C=O) bond has higher dissociation energy which improves the surface energy of solder mask surface [7], [8]. Therefore, this will increase underfill wettability on solder mask.

Relation between surface roughness and contact angle of liquid/solid interface can be found using modified young dupre's equation [11].

$$\cos \emptyset = r \cos \propto \qquad (1)$$

Where $\emptyset$ is corrected contact angle, r is roughness factor and $\propto$ is contact angle.

This equation shows that corrected contact angle would be higher on less rough surface. However, equation 1 cannot be validated through this study even though bump and non-bump area surface roughness were measured as no contact angle measurement was done on primary bump area to compare with non-bump area contact angle due to presence of die on bump area. The equation can only be used as inference in this study

Average surface roughness of bump area for S_1 is lower than S_2; hence reduced barrier for underfill flow through capillary effect on S_1 besides higher surface energy promoting better underfill wettability on S_1 than S_2. However, equation 1 above implies that secondary non-bump area contact angle for S_1 cannot be used as indirect underfill wettability response on bump area as there is significant difference in trend and distribution of roughness between non-bump area and bump area for S_1. On the other hand, secondary non-bump area contact angle for S_2 can be used as indirect underfill wettability response on bump area because of same roughness distribution and trend between non-bump and bump area for S_2. Further investigation on mechanical properties post substrate processing comparison between the 2 substrate suppliers will be needed to explain the differences between their roughness performance as both as using the same solder resist/ mask material

V. CONCLUSIONS

This study systematically investigated the effect of plasma and staging time on physical and chemical changes on two substrates. Then, variation in underfill voids was demonstrated.

- Different substrate supplier even though using same solder resist/ mask material exhibits different solder resist surface roughness. S_1 supplier bump area has significantly lower roughness than S_2 supplier after plasma. Hence, barrier to underfill flow by capillary effect is greatly reduced for S_1 than S_2. Further investigation on mechanical properties post substrate processing comparison between the 2 substrate suppliers will be needed to explain the differences between their roughness performance as both as using the same solder resist/ mask material

- Even though plasma increases roughness, it also increases carbonyl functional group to increase surface energy which assist underfill flow by capillary effect. In this study, S_1 having higher carbonyl bonds than S_2 translated to higher surface energy. This coupled with lower roughness than S_2 resulted in better UF wettability on S_1. Also, S_1 is less susceptible to UF void occurrence with longer staging time than S_2.

- Increase in staging time increases percentage of units failure for both substrates. Therefore, lower staging time such as 4 hours is recommended as practical manufacturing control even though S_1 can be staged up to 6hours post plasma before underfill than S_2

- Assumption that contact angle measurement taken at secondary substrate non-bump area represents contact angle at primary bump area is no longer valid across substrate suppliers because this study shows different roughness performance between

substrate supplier S_1 and S_2 on bump and non-bump area. Any new substrate supplier even with the same solder resist type requires a bump and non-bump area roughness check to ascertain if contact angle at the secondary non-bump area can be used as indirect UF wettability measurement.

ACKNOWLEDGMENT

Authors greatly acknowledge the financial support and research facilities from NXP, Semiconductors.

REFERENCES

[1] Orii, Yasumitsu, Kazushige Toriyama, Sayuri Kohara, Hirokazu Noma, Keishi Okamoto, Daisuke Toyoshima, and Keisuke Uenishi. "Micro structure observation and reliability behavior of peripheral flip chip interconnections with solder-capped Cu pillar bumps." *Transactions of The Japan Institute of Electronics Packaging* 4, no. 1 (2011): 73-86.

[2] Kim, Byoung-Joon, Gi-Tae Lim, Jaedong Kim, Kiwook Lee, Young-Bae Park, Ho-Young Lee, and Young-Chang Joo. "Microstructure evolution in Cu pillar/eutectic SnPb solder system during isothermal annealing." *Metals and Materials International* 15, no. 5 (2009): 815-818.

[3] Aksöz, Sezen, Yavuz Ocak, Kâzım Keşlioğlu, and Necmettin Maraşlı. "Determination of thermo-electrical properties in Sn based alloys." *Metals and Materials International* 16, no. 3 (2010): 507-515.

[4] Ma, Hui-Cai, Jing-Dong Guo, Jian-Qiang Chen, Di Wu, Zhi-Quan Liu, Qing-Sheng Zhu, Jian Ku Shang, Li Zhang, and Hong-Yan Guo. "Reliability and failure mechanism of copper pillar joints under current stressing." *Journal of Materials Science: Materials in Electronics* 26, no. 10 (2015): 7690-7697.

[5] Ong, Xuefen, Soon Wee Ho, Yue Ying Ong, Leong Ching Wai, Kripesh Vaidyanathan, Yeow Kheng Lim, David Yeo et al. "Underfill selection methodology for fine pitch Cu/low-k FCBGA packages." *Microelectronics Reliability* 49, no. 2 (2009): 150-162.

[6] Ho, P. S., Z. P. Xiong, and K. H. Chua. "Study on factors affecting underfill flow and underfill voids in a large-die flip chip ball grid array (FCBGA) package." In *2007 9th Electronics Packaging Technology Conference*, pp. 640-645. IEEE, 2007.

[7] Noh, Bo-In, Chang-Sung Seok, Won-Chul Moon, and Seung-Boo Jung. "Effect of plasma treatment on adhesion characteristics at interfaces between underfill and substrate." *International journal of adhesion and adhesives* 27, no. 3 (2007): 200-206.

[8] Liu, Wen-Jen, Xing-Jian Guo, and Chun-Han Chuang. "The effects of plasma surface modification on the molding adhesion properties of Film-BGA package." *Surface and Coatings Technology* 196, no. 1-3 (2005): 192-197.

[9] Greenwood, O. D., R. D. Boyd, J. Hopkins, and J. P. S. Badyal. "Atmospheric silent discharge versus low-pressure plasma treatment of polyethylene, polypropylene, polyisobutylene, and polystyrene." *Journal of adhesion science and technology* 9, no. 3 (1995): 311-326.

[10] Teo, Mary, Ka Yau Lee, Alex Chew, Simon Lim, Charles Lee, and Masaru Nonomura. "Plasma surface modification and impact on MSL performance for flip chip packaging." In *2007 9th Electronics Packaging Technology Conference*, pp. 657-663. IEEE, 2007.

[11] Veselý, P., D. Bušek, O. Krammer, and K. Dušek. "Analysis of no-clean flux spatter during the soldering process." *Journal of materials processing technology* 275 (2020): 116289.

The Study of Far-UVC 222-nm Excilamp and Germicidal-UVC 254-nm Low-pressure Hg Lamp: Optical Characteristics and Service Life

Fanny Zhao
GK Technology Co., Ltd.
Foshan, China
zhaohs@fsldctr.org

Guoshuai Dong
GK Technology Co., Ltd.
Foshan, China
donggs@fsldctr.org

Hao Wu
HKUST LED-FPD Technology R&D Center
Foshan, China
wuh@fsldctr.org

Guoming Yang
HKUST LED-FPD Technology R&D Center
Foshan, China
yanggm@fsldctr.org

Brian Shieh
GK Technology Co., Ltd.
Foshan, China
shiehyr@fsldctr.org

S. W. Ricky Lee
HKUST LED-FPD Technology R&D Center
Foshan, China
HKUST Foshan Research Institute for Smart Manufacturing
Hong Kong University of Science and Technology
Hong Kong, China
leesw@fsldctr.org

Ronghua Deng
Foshan Zhiye Photoelectric Technology Co., Ltd.
Foshan, China
Shenzhen LightS Technology Co., Ltd.
Shenzhen, China
51361593@qq.com

Abstract—**Optical characteristics and service life of the Germicidal-UVC (GUVC) low-pressure Hg lamp and the Far-UVC KrCl excimer lamp (FUVC excilamp) were investigated. The results showed that UV irradiance at the center of the FUVC excilamp and the GUVC lamp was 0.22 mW/cm² and 9.27 mW/cm² at a working distance of 5 cm, respectively. To achieve the same virus inactivation effect, the harmful power in range of 230~280 nm should be more than 20 times that of the FUVC excilamp without considering the UVA (320~400 nm) and UVB (280~320 nm). Furthermore, the service life was assessed under room temperature 30 ± 5 ℃. It was found that the average service life at L50 (with the irradiance decay of 50%) of the FUVC excilamp and GUVC lamp was about 2,600 and 4,700 hours, respectively.**

Keywords—Far-UVC, KrCl excilamp, Germicidal-UVC, Disinfection, Antiviral, Service life

I. INTRODUCTION

Although Germicidal-UVC (GUVC) light (with peak wavelength of 254-nm) is effective in disinfecting viruses, there exists a concern on human body safety. Far-UVC (FUVC) light (with peak wavelength of 222-nm) addresses this issue as its short wavelength does not penetrate the outer cell layer of skin or eye tissue [1]. There is evidence that 222-nm FUVC light, may cause less damage to the skin, eyes and DNA than the 254-nm GUVC, although long-term safety data is not available [2] and [6]. Kouji Narita, et al. concluded that 222-nm FUVC is able to inactivate a wide spectrum of microbial pathogens. Although the germicidal effect to the fungal hyphae and spores is low, 222-nm FUVC exhibited stronger germicidal effect to the bacterial endospores than conventional 254 nm GUVC [3]. The study of Luke Horton, et al. revealed that the virus inactivation doses extrapolated from the relative sensitivities of viruses. The doses are normalized against that at 254 nm and represent the respective efficacy to achieve 1/e (63%) reduction in the viral load [4]. According to H. Kitagawa, et al. 99.7% of the SARS-CoV-2 was inactivated after just a 30-second exposure to 222-nm irradiation at 0.1 mW/cm² [5].

In the present study, the optical characteristics and service life of both the 254-nm GUVC lamp and the 222-nm FUVC excilamp were investigated.

II. OPTICAL CHARACTERISTICS

A. Appearance and Schematic Diagram

The picture of the FUVC excilamp used in our study is shown as in Fig. 1a. From Fig. 1b, the cross-sectional schematic view shows that the FUVC excilamp is made of a sealed quartz tube filled with a mixture of krypton and chlorine molecular to produce the KrCl excimer and radiating peak wavelength at 222 nm. Its external metallic mesh surrounding the quartz tube is the ground electrode and an internal cylinder as the electrode is connected to the high-voltage power supply providing AC voltage 2kV with frequency of 28 kHz.

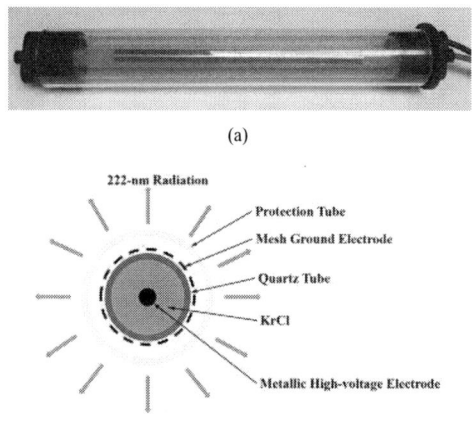

(a)

(b)

Fig. 1 222-nm KrCl FUVC Excilamp (a) Actual Lamp Sample,
(b) Schematic Diagram of Cross-section

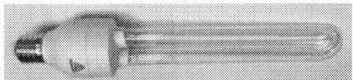

Fig. 2 254-nm GUVC Lamp

Also, the 254-nm GUVC lamp driven by AC220V is shown as in Fig. 2 for comparisons.

B. Measurement of Optical Characteristics

Fig. 3 UV Irradiance Meter   Fig. 4 Compact Array Spectrometer

The irradiance of both the FUVC excilamp and GUVC lamp was measured using a UV irradiance meter (Model: OHSP-350UVS) with a measuring range of 200~400 nm and a spectral interval of 1 nm, as shown in Fig. 3. Furthermore, spectrum was obtained by a high-resolution compact array spectrometer (Model: CAS 140 CT) with a range of 200~800 nm and an interval of 0.6 nm, as shown in Fig. 4.

The results showed that UV irradiance at the center of the FUVC excilamp and the GUVC lamp was 0.22 mW/cm^2 and 9.27 mW/cm^2 at a working distance of 5 cm, respectively.

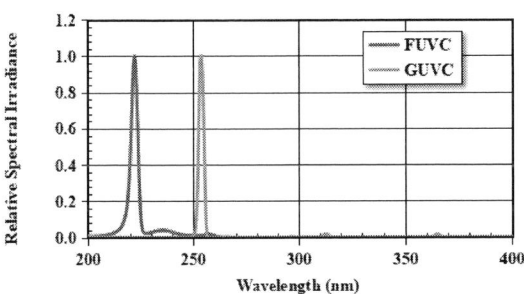

Fig. 5a Relative Spectral Irradiance Distribution of the KrCl FUVC Excilamp and the GUVC Lamp in Linear Scale.

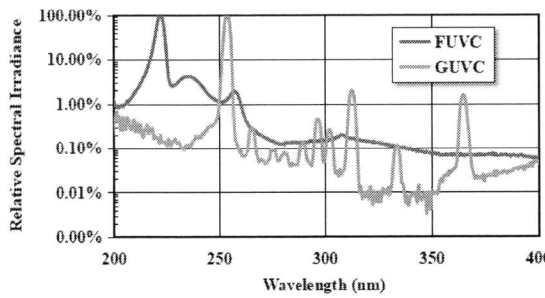

Fig. 5b Relative Spectral Irradiance Distribution of the FUVC Excilamp and the GUVC Lamp in Logarithmic Scale.

Based on the measurement of spectrum, the FUVC excilamp shows a Full Width Half Maximum (FWHM) of 3.6 nm at the 222 nm waveband as well as additional smaller

emission peaks at 235 nm and 257 nm, as shown in Fig. 5a and b.

As we suggested, in the range of UVC (200~280 nm), the radiation waveband (230~280 nm) is considered to be much harmful to human. The power ratios of the harmful waveband (230~280 nm) to that in the total UVC range (200~280 nm) were shown as in Table Ⅰ. Luke Horton, et al. showed that the radiation dose of 222-nm is about one fifth of the dose of 254-nm to achieve the equivalent virus inactivation effect [4]. That infers the GUVC lamp, based on the same virus inactivation effect, the harmful power in range of 230~280 nm should be more than 20 times that of the FUVC excilamp without considering the UVA and UVB.

TABLE I.   POWER RATIO OF THE HARMFUL WAVEBAND TO THE UVC WAVEBAND

Wavelength / Lamp	Suggested Harmful Waveband in UVC （230-280 nm）	UVC (200-280 nm)
GUVC	97.51%	100.00%
FUVC	22.85%	100.00%

III. SERVICE LIFE

A. Experimental Setup and Measurement

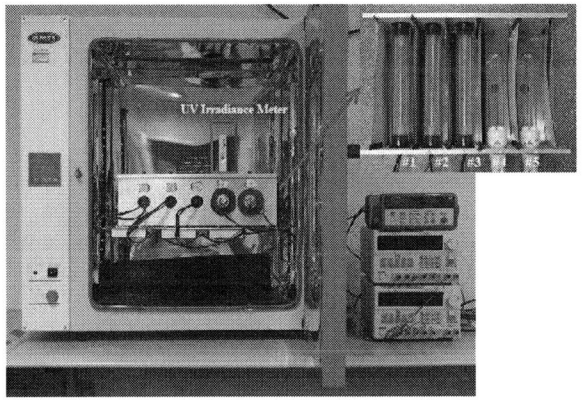

Fig. 6 Experimental Setup

The experimental setup for service life test was shown as in Fig. 6. In our experiment, three FUVC excilamps (#1, 2 and 3) and two GUVC lamps (#4 and 5) shown with the close-up view as in Fig. 6 were aging under rated operating condition at room temperature 30±5 ℃. And the attenuation of UVC irradiance was used for benchmarking. During the test program, the value of irradiance was evaluated at 96/168/264/336/432/504 hours.

Fig. 7 shows the maintenance of irradiance E_x/E_0 versus operating time for all five lamps. Here, E_x represents the irradiance after aging for hours of x and E_0 represents the irradiance before aging. For the two GUVC lamps, the E_x/E_0 witnessed a slightly fluctuation of within 5% during the test program, reaching the average irradiance decay of 4.3% at 500h. For the three FUVC excilamps, while #2 showed a minor increase in E_x/E_0, approximate 5% at the end of the life test, #1 and 3 saw the same steady decline, about 15% for each. Based on the assessment method on life time of IES TM-21-11, the irradiance evaluated above can help to predict that the average L50 service life of the studied FUVC excilamps and the GUVC lamps should be 2,600 and 4,700 hours, respectively, as shown in Fig. 8.

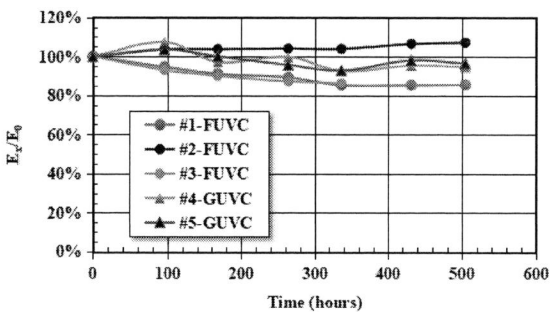

Fig. 7 Maintenance of Irradiance versus Operating Time

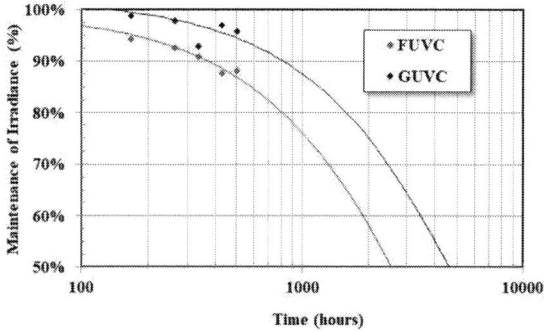

Fig. 8 Service Life Curve Based on IES TM-21-11

Current of the three FUVC excilamps remained steady with a slightly fluctuation of within 1% as they were operated through a constant voltage of 12 V, while the electric power input of the two GUVC lamps decreased significantly with an approximate 5.4% of decay, which is closed to the average irradiance decay of 4.3%. This infers that the decay of irradiance of the tested FUVC excilamps was mainly related to the lamp itself. Instead, that of the tested GUVC lamps was basically caused by the power input.

IV. Conclusion

Irradiance of the studied FUVC excilamps was significantly lower than those of the GUVC lamps. However, based on the same virus inactivation, the harmful power (230~280 nm) of the GUVC lamp should be more than 20 times that of the FUVC excilamp without considering the UVA and UVB. Moreover, although the decay of irradiance of the tested FUVC excilamps was mainly related to the lamp itself, while that of the tested GUVC lamps was basically caused by the power input (decay of 5.4%), the average L50 service life of the FUVC excilamps and GUVC lamps can be expected to be about 2,600 and 4,700 hours, respectively.

References

[1] J.-Y. Zhang and I. W. Boyd, et al. Lifetime investigation of excimer UV sources. Applied Surface Science, 14-Dec-2000; DOI: 10.1016/S0169-4332(00)00628-0.

[2] UV Lights and Lamps: Ultraviolet-C Radiation, Disinfection, and Coronavirus, https://www.fda.gov/medical-devices/coronavirus-covid-19-and-medical-devices/uv-lights-and-lamps-ultraviolet-c-radiation-disinfection-and-coronavirus.

[3] Kouji, Narita; Asano, Krisana; Naito, Keisuke. 222-nm UVC inactivates a wide spectrum of microbial pathogens. Journal of Hospital Infection, 23 March 2020; DOI:10.1016/j.jhin.2020.03.030.

[4] Luke Horton, Angeli Eloise Torres, et al. Spectrum of virucidal activity from ultraviolet to infrared radiation. The Royal Society of Chemistry and Owner Societies 2020, August 12, 2020.

[5] H. Kitagawa, T. Nomura, T. Nazmul, K. Omori, N. Shigemoto, T. Sakaguchi, and H. Ohge. Effectiveness of 222-nm ultraviolet light on disinfecting SARS-CoV-2 surface contamination. American Journal of Infection Control, 04-Sep-2020; [Online]. DOI: https://doi.org/10.1016/j.ajic.2020.08.022. [Accessed: 07-Oct-2020].

[6] Martin Hessling, Robin Haag, Nicole Sieber, and Petra Vatter. The impact of far-UVC radiation (200–230 nm) on pathogens, cells, skin, and eyes – a collection and analysis of a hundred years of data, GMS Hygiene and Infection Control 2021, Vol. 16, ISSN 2196-5226.

Effectiveness Validation on Equivalent Model for Wafer-Level Warpage Prediction

Guoli Sun
Faculty of Materials and Manufacturing
Beijing University of Technology
Beijing, China
GL-SUN@emails.bjut.edu.cn

Jiahui Wei
Faculty of Materials and Manufacturing
Beijing University of Technology
Beijing, China
13269632099@163.com

Fei Qin
Faculty of Materials and Manufacturing
Beijing University of Technology
Beijing, China
qfei@bjut.edu.cn

Yanwei Dai
Faculty of Materials and Manufacturing
Beijing University of Technology
Beijing, China
ywdai@bjut.edu.cn

Kui Li
Xi'an Microelectronics Technology Institute
Xi'an, China
likui285@sina.com

Baoxia Li
Xi'an Microelectronics Technology Institute
Xi'an, China
libaoxia163@163.com

Abstract—Due to high density requirement in electronics packaging, heterogeneous integration technology is used widely. Heterogeneous integration is always based on through silicon via (TSV) interposer integrating. Silicon interposer is currently the mainstream and most mature interposer technology, which has been studied widely and presents extensive applications. In this paper, the structure and process of wafer-level TSV interposer is proposed. This article presents an equivalent method to predict the warpage of wafer-level silicon interposer. A process-dependent simulation methodology is performed, which integrates element birth and death technique as well as restart technique. The validation of actual model and warpage of wafer during fabrication process are investigated finally.

Keywords—equivalent model, process, numerical simulation, warpage

I. INTRODUCTION

With the rise of the smart terminals, and industrial intelligence, the demands for high-density integration, multi-function, and low power consumption of electronic devices have become an important topic [1]. Due to the limitation of physical sides, three-dimensional packaging integration provides a new technical route to improve and enhance the electronic devices in recent years, and the three-dimensional stack packaging based on TSV silicon interposer is a recognized feasible method [2].

The electrical connection between the chips is achieved through RDL and micro bumps interconnections. The chip and substrate on the back of the interposer connects through TSV inside the interposer [3]. However, if the warpage is too large, it is impossible to operate wafer-level TSV silicon interposers during fabrication processes such as steppers, physical vapor deposition (PVD), electrochemical deposition and etching[4]. Recently, a process-dependent simulation methodology is introduced which found that gravity loading has an significant effect on wafer warpage [5].

Wafer warpage is a serious problem during the process. Large mismatch of coefficient of thermal expansion (CTE) between the polyimide (PI) and silicon wafer continuously results in high warpage and thermal stress from process temperature to room temperature. Besides, the warpage changes with the process temperature, especially during RDL process[6]. Predicting the evolution of the wafer warpage during the whole packaging process would be beneficial to the reliability evaluation of the TSV interposer.

This study validates the effectiveness of the equivalent method to predict the evolution of the warpage for wafer-level TSV interposer is established based on Finite Element Analysis (FEA). Due to symmetry, only a quadrant FE model is chosen to characterize the feature of the whole wafer for the consideration of conserving computer resources and time. "Element Birth and Death" and "Restart" technology is used to simulate the fabrication process, using ANSYS commercial software to simulate the wafer-level interposer fabrication process.

II. STRUCTURE AND PROCESS OF WAFER-LEVEL TSV SILICON INTERPOSER

A. Package Information

Fig.1 shows the layout of TSV silicon interposer on 8-inch wafer. The thickness of wafer is 750μm. The package size of single interposer is 32660 μm×25160 μm×200 μm, and there are 1988 TSV with a diameter =20 μm. The cross section of the interposer is schematically shown in Fig.2. It can be seen that there are there RDLs and the thickness of RDL1 and RDL2 is 5 μm and of RDL3 is 10 μm. The polyimide(PI) of DL1, DL2, and DL3 are 15μm each and of DL4 is 15 μm. The key structural dimensions are shown in Table 1.

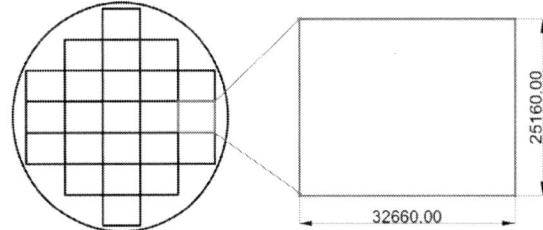

Fig. 1. The layout of TSV interposer

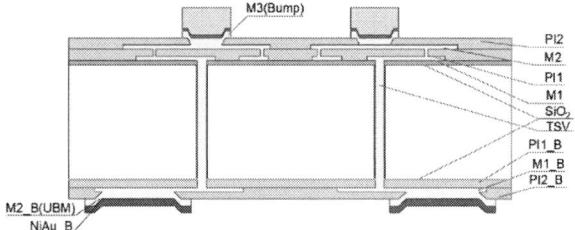

Fig. 2. Cross sectional view of TSV interposer

TABLE I.   MAIN PARAMETER OF PACKAGING STRUCTURE

Interposer structure	Dimensions
Wafer thickness	750 μm
Bonding glue	90 μm
Bump height	32 μm
Via depth/diameter	200/20 μm
RDL linewidth/spacing	10μm/10 μm

B. Silicon Interposer Process Flow

The process flow of wafer-level TSV silicon interposer is demonstrated in Fig.3. The fabrication process mainly involves the following key process stages. At first, TSV holes with a depth of 200 μm are etched on 8-inch wafers. Then 800 nm Cu seed layer and 500nm Ti barrier layer are deposited through physical vapor deposition (PVD) process. However, the thickness of Ti/Cu layer is too thin to be considered. A Cu layer with the thickness of about 5 μm is formed on the surface of the wafer and the inside of the TSV hole by electroplating. The annealing temperature of copper is 350°C. Then spinning and lithography method is employed to produce a layer of PI1 with a thickness of 10 μm, baking at a temperature of 210°C for 1 hour. The PI is a liquid-like material.

The PVD process is chosen to deposit Ti/Cu layer on the PI1. After PI2 is cured, 150 nm Ti layer and 300 nm Cu layer is deposited on the surface of RDL2. Then the bump are created by electroplating at the interposer top side. After the completion of bumps, the temporary bonding process is carried out between wafer and glass carrier wafer at 220°C. DISCO's backgrinding machine is utilized to grinding the wafer from the back side until TSV copper is exposed.

The first back side PI(BPI1) layer with the thickness of 5 μm is spin-coated and patterned on back side of the wafer and the solder mask layer is formed after baking and curing. Then the BRDL1 is created by photolithography and electroplating on the PI at room temperature. BPI2 is fabricated with the same thickness and process as BPI1. Next under bump metallization (UBM) pattern is manufactured vias using electroplating. The debonding of the glass carrier is by laser scanning from the glass carrier side, and finally, wafer dicing.

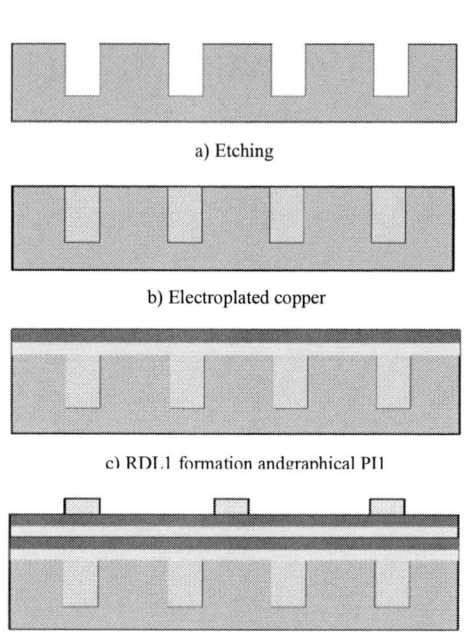

a) Etching

b) Electroplated copper

c) RDL1 formation andgraphical PI1

d) RDL2 formationand graphical PI2

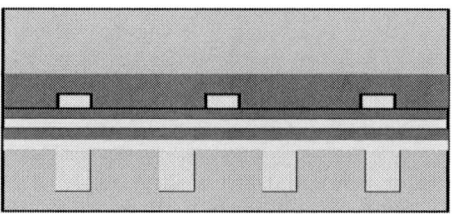

e) Bonding to another wafer

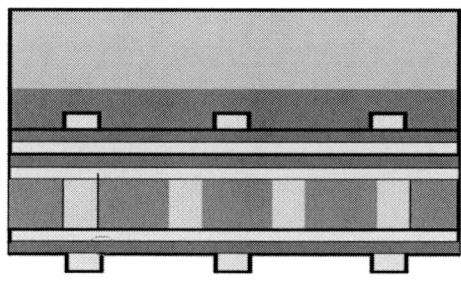

f) Back graphical PI1 and RDL1

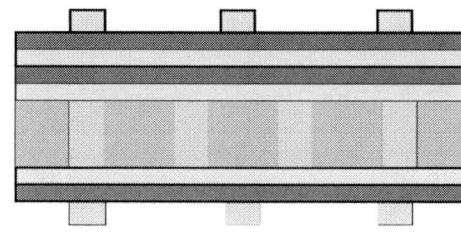

g) Debonding

Fig.3. Process flow of the proposed TSV silicon interposer

III. FINITE-ELEMENT MODEL

A. Actual And Equivalent Finite Element Model

Fig.4 shows the detailed finite element model of a part of TSV interposer which is composed of 50 cell models and PI layer with a thickness of 15 μm. The dimension of a single cell

978-1-6654-1392-3/21 $31.00 © 2021 IEEE   1306

model is 0.3 mm×0.3 mm×0.2 mm. The TSV is located at the geometric center of each cell model. There are 50 TSVs in the detailed finite element model. In order to reduce the calculation scale, an effective finite element model is established based on the same deformation response, as shown in Fig.5. The dimensions of both models are 3×1.5×0.215 mm. According to the composite material mechanics analysis method[7], the effective material parameter calculation formula is:

$$E_z = E_f c_f + E_m c_m, \quad E_x = \frac{E_m E_f}{E_f c_f + E_m c_m} \quad (1)$$

$$v_{zx} = \frac{K_f v_f c_f + K_m v_m c_m}{K_f c_f + K_m c_m} \quad (2)$$

$$v_{xy} = v_f c_f + v_m c_m \left[\frac{1 + v_m - v_{zx}(E_m / E_m)}{1 - v_m^2 + v_m v_{zx}[E_m / E_z]} \right] \quad (3)$$

$$G_{zx} = \frac{G_f G_m}{G_f(1 - c_f) + G_m c_f}, \quad G_{xy} = G_{yx} = \frac{E_x}{2(1 + v_{xy})} \quad (4)$$

$$\alpha_z = \frac{\alpha_f E_f c_f + \alpha_m E_m c_m}{E_f c_f + E_m c_m} \quad (5)$$

$$\alpha_x = (1 + v_f)\alpha_f c_f + (1 + v_m)\alpha_m c_m - \alpha_z(v_f c_f + v_m c_m) \quad (6)$$

where E_m and c_m represent the elastic modulus and volume fraction of silicon, E_f and c_f are the elastic modulus and volume fraction of copper, v_f and v_m are the Poisson's ratio of copper and silicon, respectively. G_m and G_f are the shear modulus of copper and silicon, α_f and α_m are the thermal expansion coefficient of copper and silicon, respectively.

The material properties of Cu and PI as listed in Table II for detailed TSV model. Table III shows the effective material properties of the equivalent model. The curing temperature of PI with the thickness of 15 μm is 210°C as the reference temperature. Loading conditions is the cooling process down to the room temperature. The symmetry boundary condition is defined on the two symmetric planes. Besides, constraints all degrees of freedom of nodes located at the intersection of two symmetry planes[8]. Element type solid 185 is used, and the two finite-element models are meshed with 501,960 elements and 72,000 elements respectively.

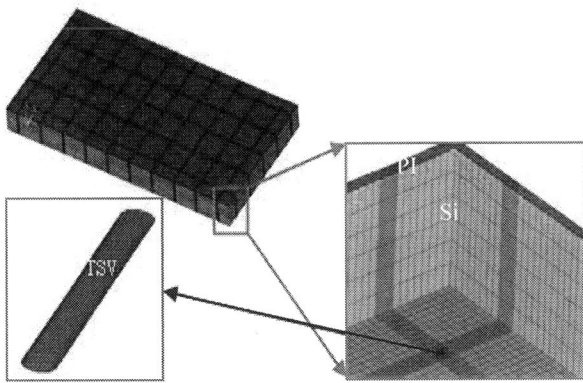

Fig. 4. Actual finite element model

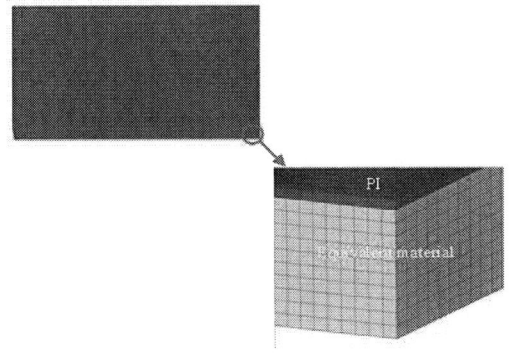

Fig. 5. Equivalent finite element model

TABLE II. MATERIAL PROPERTY

Material	Yong's Modulus(MPa)	Poisson's Ratio	CTE (ppm/°C)
Cu	155470	0.326	16.7
Silicon	131000	0.35	2.3
PI	2500	0.3	54
Glass	75000	0.25	3
SiO₂	70000	0.38	0.5
Glue	Above Tg:1200 Below Tg:400	0.38	Below Tg:54 Above Tg:160

TABLE III. EQUIVALENT MATERIAL PROPERTY

Direction	Equivalent Yong's Modulus (MPa)	Equivalent Poisson's Ratio	Equivalent CTE (ppm/°C)	Equivalent Shear modulus (MPa)
Z direction	131085.400	0.349	2.359	48586.138
X、Y direction	131071.998	0.349	1.349	48581.170

B. Process-Dependent Finite Element Modeling

Due to the symmetry, only a quadrant of the wafer-level interposer assembly is considered in the FE modeling, and the symmetry boundary condition is applied to the corresponding planes. The displacement and rotation of the center are zero, as displayed in Fig.6. The three finite element model of the molded wafer assembly is shown in Fig.7, consisting of 191,240 elements and 201,324 nodes. Temperature load changes with loading steps is described in Fig. 8. And the maximal temperature is 220°C which is the curing temperature of PI, the lowest is 25°C. What's more, Table II illustrates the six material parameters involved in simulation, namely copper, silicon, PI, glass, glue and silicon dioxide. Based on the element death-birth technique and restart technique, wafer-level TSV silicon interposer warpage is predicted by simulation.

978-1-6654-1392-3/21 $31.00 © 2021 IEEE

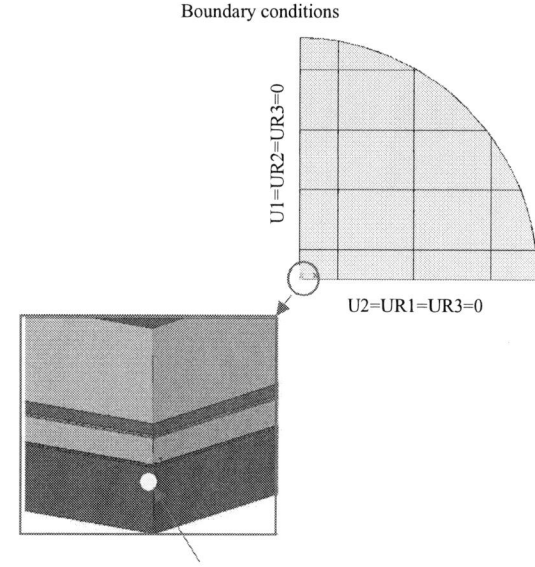

Fig.6. Boundary conditions of quarter model

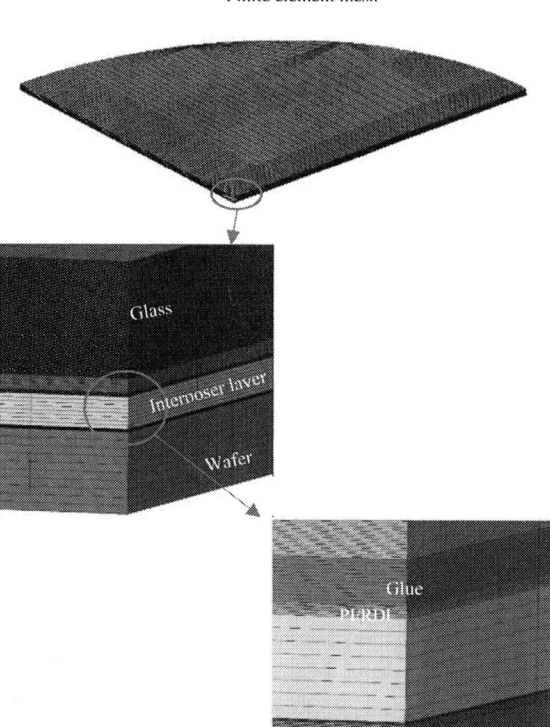

Fig.7. Finite-element meshes of quarter model

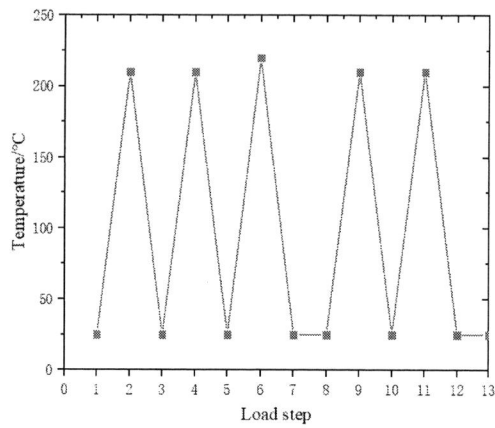

Fig.8. Temperature load changes with loading steps

IV. RESULTS AND DISCUSSION

A. Validation of equivalent model

The simulation results of the actual model and the effective model are shown in Fig.9 and Fig.10. It can be seen that the maximum displacement in the z direction of the actual model and equivalent mode are 2.296 μm and 2.289 μm respectively. Obviously, The maximum warpage between the two models is only 0.007μm, demonstrating the validation of this equivalent method. Fig.10 shows the comparison of warpage of two models. Based on this equivalent method, it provides a basis for simplifying the package structure of the wafer-level TSV interposer.

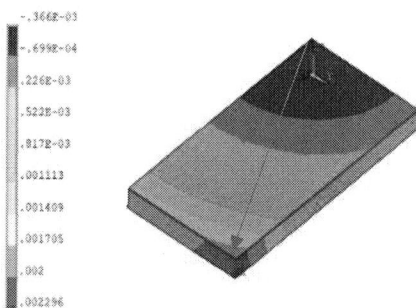

Fig.9. Displacement in z direction cloud of the actual FE models

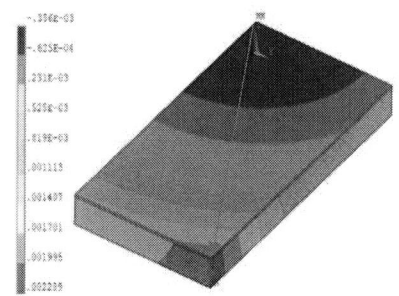

Fig.10. Displacement in z direction cloud of the equivalent FE models

978-1-6654-1392-3/21 $31.00 © 2021 IEEE

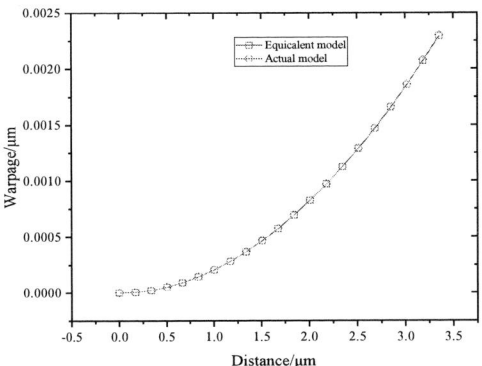

Fig.11. The warpage results of the equivalent model and the real model

B. Effect of copper volume fraction on warpage

The influence of the volume fraction of copper on the warpage of the interposer is also investigated in this study. If one changes the diameter of TSV from 20 μm to 30 μm, thereby adjusting the volume fraction of copper. The volume fraction of copper changed from 0.35% to 0.79%. The boundary conditions, reference temperature and element type of finite element model are the same as Fig. 4. Fig. 12 depicts the simulated deformation at 25℃. The maximum warpage is 2.293 μm less than 2.296μm.

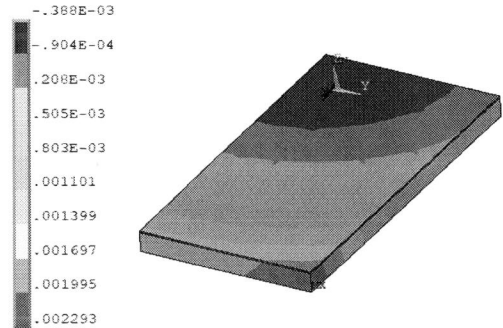

Fig.12. Displacement in Z direction of simulation result at 25℃

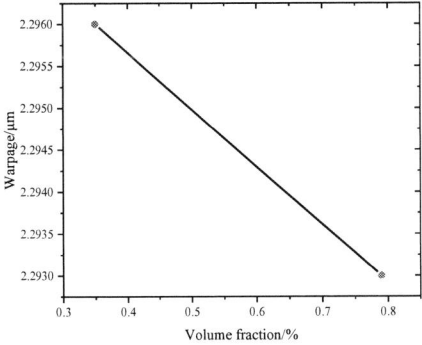

Fig.13. Warpage of different copper volume fraction

C. Warpage evolution during fabrication process

The calculated warpage of the wafer-level silicon interposer during the fabrication process is showed in Fig. 14. The definition of the warpage orientation of the wafer is given in Fig.15, where the convex arc (smiling surface) is positive and vice versa. The warpage of wafer significantly develops with the process steps. What's more, the direction of warpage

will change throughout the whole process. It is clear that the wafer warpage value reaches the maximum at the 6th load step, and the maximum is -4.85 mm. The simulated warpage contour distribution at step 13 is illustrated in Fig.16.

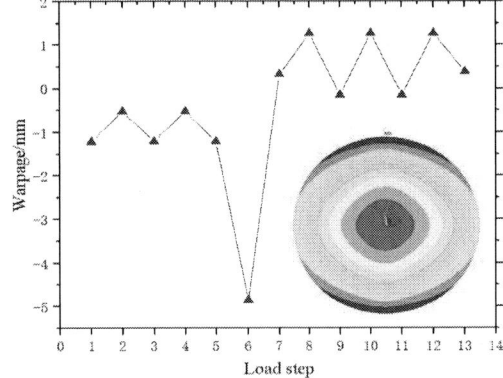

Fig.14. The warpage evolution over the loading steps

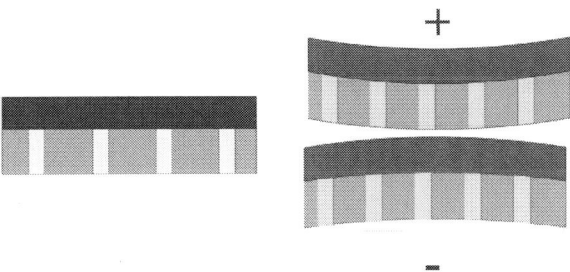

Fig.15. Schematic of warpage orientation

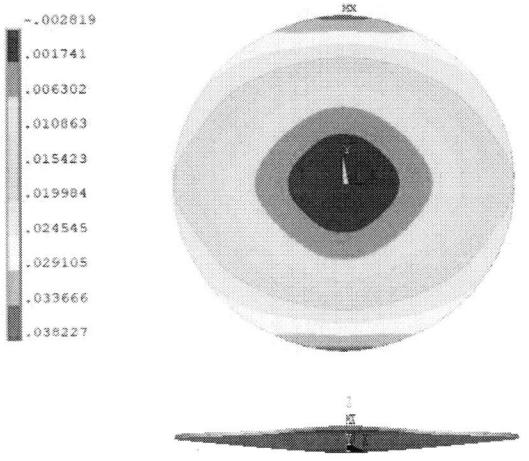

Fig.16. The simulated warpage contour distribution at step 13

V. CONCLUSION

In this paper, an actual model and effective model are established and compared. After validating the accuracy of the effective model, the actual FE model can be replaced by an equivalent FE model. Beside, an decreasing copper volume fraction would lead to a reduced warpage of the interposer. Based on the same deformation response, a quarter FE model is carried out to simulate wafer warpage during the fabrication process based on element death-birth technique. In addition, maximum process warpage occurs at bonding process (e.g. step 6) rather than at the end of the fabrication process(e.g. step 13).

ACKNOWLEDGMENT

The authors would like to thank Xi'an Microelectronics Technology Institute for providing process flow.

REFERENCES

[1] H. C. Cheng, Z. D. Wu, and Y. C. Liu, "Viscoelastic Warpage Modeling of Fan-Out Wafer-Level Packaging During Wafer-Level Mold Cure Process." IEEE Transactions on Components, Packaging, and Manufacturing Technology, 2020, pp.99.

[2] D. Malta, E. Vick, S. Goodwin, C, Gregory, M. Lueck, A. Huffman, D. Temple, "Fabrication of TSV-based silicon interposers." 3d Systems Integration Conference IEEE, 2010.

[3] G Katti, SW Ho, LH Yu, S Zhang, R Dutta, R Weerasekera, KF Chang, SR Vempati, S Bhattacharya, "Fabrication and Assembly of Cu-RDL-Based 2.5-D Low-Cost Through Silicon Interposer (LC‑TSI)." IEEE Design & Test, 32(4), 2015, pp.23-31.

[4] JH Lau, L Ming, D Tian, N Fan，Qingxiang Yong. "Warpage and Thermal Characterization of Fan-Out Wafer-Level Packaging." IEEE Transactions on Components, Packaging, and Manufacturing Technology, 2017, pp.99.

[5] HC Cheng, YC Liu. "Warpage Characterization of Molded Wafer for Fan-Out Wafer-Level Packaging." Journal of Electronic Packaging, 2020, pp.42-51.

[6] M Detalle，B Vandevelde，P Nolmans，JD Messemaeker，E Beyne. "Minimizing interposer warpage by process control and design optimization." 2014 IEEE 64th Electronic Components and Technology Conference (ECTC) IEEE, 2014.

[7] RM Christensen, Mechanics of Composite Materials, Dover Publications, Mineola, NY, 2015.

[8] M. Yang, F. Qin, D. Yu. "Numerical simulation on wafer warpage during molding process of WLCSP." 19th International Conference on Electronic Packaging Technology (ICEPT) 2018.

Multi-chips High-density Interconnection Design on InFO Platform

Jiang Qiang[2]*
Department of Packaging and Testing
ZTE Corporation
Shenzhen,China
jiang.qiang1@sanechips.com.cn

Zhou Guodan[2]*
Department of Packaging and Testing
ZTE Corporation
Shenzhen,China
zhou.guodan@sanechips.com.cn

Wang Zongwei[2]*
Department of Packaging and Testing ZTE
Corporation
Shenzhen,China
wang.zongwei3@sanechips.com.cn

Pang Jian[2]
Department of Packaging and Testing
ZTE Corporation
Shenzhen,China
pang.jian@sanechips.com.cn

Sun Tuobei[2]
Department of Packaging and Testing
ZTE Corporation
Shenzhen,China
sun.tuobei@sanechips.com.cn

Keqing Ouyang[1]
State Key Laboratory of Mobile Network
and Mobile Multimedia Technology
Shenzhen,China
ouyangkeqing@sanechips.com.cn

Abstract—*With the increase in the difficulty of process breakthroughs, advanced packaging has become a breakthrough for Moore's law to continue. 2.5D advanced packaging technologies can package together processors, memory and other IPs under different processes, which not only avoids the increase in die size, but also With the decline of the yield rate, IP can be selected according to the function, which reduces the development cost and speeds up the product launch. Compared with the silicon interposer board, the InFO package has a greatly reduced cost, and compared to the 2D package, it can achieve multi-chip packging while having good electrical performance. The package design is that one main ASIC chip is connected with two auxiliary ASIC chips, and the two auxiliary ASIC chips are connected. The auxiliary ASIC chip is used to enhance the function of the main ASIC chip. The high-density interconnection adopts the signal layer 2um/2um and the power layer 5um/5um 5-layer RDL wiring method, the frequency is 2.4Ghz, and the data rate is 4.8Gbps. The wiring method adopts the ground isolation reference wiring to ensure high-speed signal integrity. At the same time comparing a variety of wiring structure, the electrical performance under this wiring method is the most superior.*

Keywords—*2.5Dpackage;fine pitch;electrical performance*

I. INTRODUCTION

With the advent of the post-Moore era, the role of 2.5D/3D packaging technology has gradually increased, and it has become the commanding heights of major design house, fabs, and assembly house. The reason why advanced packaging occupies such an important position , First, because the output ratio brought by the improvement of the process is getting smaller and smaller, the second is that the larger the chip, the lower the yield, and the higher the price cost. The third is the sealing of multiple chiplets, which can effectively reduce the project development cycle and reduce development costs[1]. Currently 2.5D packages manufactured by various packaging factories can be divided into two technologies: TSV Technology (INTERPOSER) and TSV-less Technology. The most representative ones are Intel's EMIB, TSMC's integrated fan-out (InFO_oS), ASE's FOCoS, and Substrate-level multi-die packaging technology FCMCM[2][3]. In addition to manufacturing, the DTD block docking protocol is also very important. There are serial and parallel interfaces. Intel has successively proposed the AIB interface protocol and the MDIO protocol, and TSMC also proposed the LIPINCON2 protocol*. XLINX's Open HBI3, SAMSUNG's SFD 2D, etc. Their are used for die to die

high-bandwidth density interconnection protocols[4][5].It is no longer computing power that restricts the integrated computing speed of current AI systems. Computing power can often be increased by stacking computing cores, but the bandwidth density with data transmission has become a bottleneck, making computing speed difficult to break through. The InFO_oS used in this article is based on a 7nmASIC core chip interconnected with two 5nm auxiliary chips for data transmission. The die-to-die data transmission rate reaches 4Gbps, with 2um/2um line width and 5-layer RDL routing.

II. INFO PACKAGE PHYSICAL DESIGN

A. InFO Layout design

This article uses InFO_oS technology as the assembly for multi-chip packaging. A 7nm ASIC core chip is interconnected with two 5nm auxiliary chips for data transmission. The multi-chip package diagram is shown in Figure 1. In addition to the DTD block above chip2, there is one block on the left and right side respectively, which can realize the docking between two chip2s without designing a new die, resulting in the improvement of the reusability of the die.DTD block that are not used on both sides are directly processed for power supply suspension, and DTD block does not work;

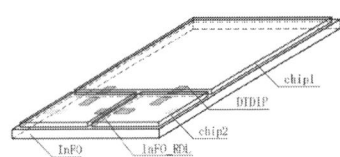

Fig. 1. Multi-chip package diagram

B. Bump distribution of DTD

DTD block uses a separate power supply method for analog power supply and digital power supply. It uses 0.75V high-level swing and 0.3V low-level swing power supply respectively to reduce DTD power consumption while ensuring power integrity,the ground is merged into one type on InFO,independent of the core ground, while there are 6 rows of signal bumps. While ensuring signal integrity, the bandwidth density must be ensured. The signal transmission rate is 4.8Gbps and the bandwidth density is 1Tbps/mm.

978-1-6654-1392-3/21 $31.00 © 2021 IEEE

These signals together form a channel and use multiple same numbers. The channel of the data signal can form a column. The distribution of a single channel bump is shown in Figure 2.

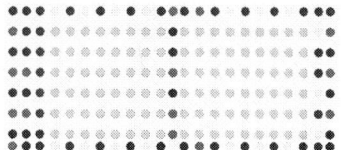

Fig. 2. bump distribution

C. RDL routing scheme on InFO platform

The InFO package wiring scheme uses adjacent routing as ground-signal-ground isolated routing. Since the bandwidth of the effect of density, can not use the full ground plane for signal isolation, it will result in reducing the total number of I/O signal lines 50%. However, in order to ensure signal integrity, RDL is selected as the upper and lower reference of the signal-ground-signal incomplete reference layer. To ensure that all signals can be completely shielded, a row of ground lines is added around the signal hole for shielding , Use dielectric material to fill between metals. The RDL line width and line spacing are 2um/2um, and the signal line length of each column is the same. The last layer of ground plane, the line width is 20um, which can play a ground plane reference function for the upper layer signal while ensuring the current density. The signal RDL wiring structure is shown in Figure 3.

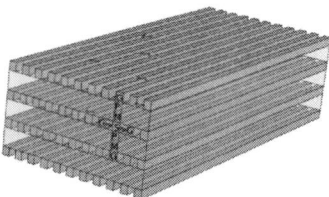

Fig. 3. signal wiring structure

In order to improve the power consumption of the block and consider the integrity of the power supply, DTD block adopts the method of processing the high and low-level power supplies separately on the die. Therefore, in order to meet the power supply scheme of DTD, according to the characteristics of the power bump being scattered around the signal bump, we made a low-level swing power mesh on the periphery of the RDL4 layer, and arranged a high-level swing power mesh on the RDL5 layer, and the power network maintained closed state can effectively balance the voltage. The power network distribution is shown in the figure 4.

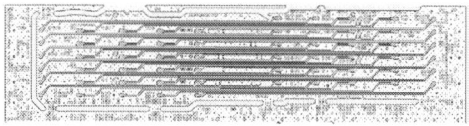

(a)

(b)

Fig. 4. power distribution network (a)RDL4(b)RDL5

III. SIGNAL AND POWER SIMULATION

A. Signal integrity of DTD block interconnect

The S parameter is the Scatter parameter. The S parameter describes the way the interconnection affects the incident signal. The S parameter describes the characteristics of the transmission channel in the frequency domain. It is a parameter that SI engineers must use when analyzing the signal integrity of the serial link. Therefore, it is very important to obtain accurate S-parameters[6][7]. Through S-parameters, we can see almost all the characteristics of the transmission channel, such as reflection, crosstalk, loss and other signal integrity issues. At the same time, we can further predict the output waveform, such as eye diagrams[8].

For the interconnection of the DTD block on the InFO, the input and output port impedance is 50 ohm, which is a linear passive element, and the frequency of the scattered wave will be exactly the same as the incident wave. The only parameters that can change the scattered wave are amplitude and phase. Each S parameter is the ratio of the output sine wave to the input sine wave:

$$S = \frac{\text{out sine wave}}{\text{inner sine wave}} \tag{1}$$

In the design made in this article, the line width and line spacing are both 2um. While fully considering the shielding protection of the signal line, it is necessary to ensure that the signal bandwidth density exceeds 1Tbps/mm.

When the wiring type and DTD block are determined, the position distance of the dual die on the InFO will directly affect the length of the interconnection line, and the signal transmission quality will deteriorate. This article compares the distance between chip1 and chip2 on the signal integrity Impact, Figure 5~8 I is the signal line length of 1880um, II is the signal line length of 1250um IL, RL, FEXT, NEXT simulation results

The commercial software HFSS performs port docking S parameter simulation on the DTD block interconnect structure. The simulation parameters are shown in Table I.

TABLE I.   SIMULATION PARAMETER

layer \ Parameter	Dk	Df	roughness/um	Process deviation
Dielectric layer	3.3	0.02	0.1	±0.1
metal			0.1	±0.1

978-1-6654-1392-3/21 $31.00 © 2021 IEEE

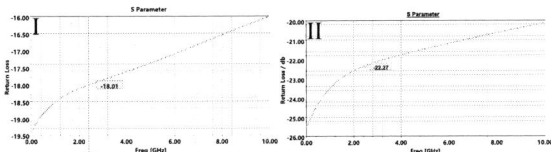

Fig. 5. I.RL:-18.01@2.4GHz,II. -22.37dB@2.4GHz

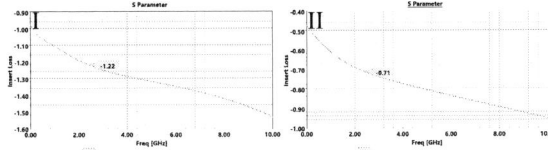

Fig. 6. I.IL -1.22@2.4GHz II.-0.71dB@2.4GHz

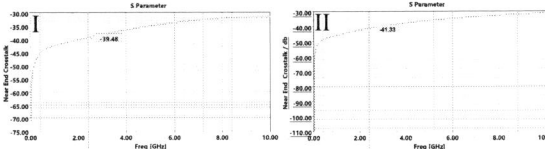

Fig. 7. I.NEXT -39.48@2.4GHz II.-41.33dB@2.4GHz

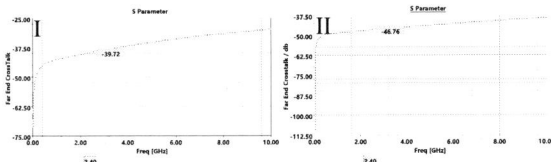

Fig. 8. I.FEXT-39.72dB@2.4GHz II.-46.76db@2.4GHz

The simulation results of the signal line interconnection channels on the InFO shown in Figures 5 to 8 are the worst S-parameter port results. When the signal line length is 1880um and the frequency is 2.4Ghz, RL is -18.01db, IL is -1.22db, NEXT And FEXT are-39.48db, -39.72db, when the signal line length is 1250um, and the frequency is 2.4Ghz, RL is -22.39db, IL is -0.71db, NEXT and FEXT are-respectively 41.33db, -46.76db . It can be clearly seen that the shorter the signal line length, the better the signal transmission quality, especially the insert loss from 1880um to 1250um, a decrease of 41.8%.According to the rule of thumb,the length of the signal line a signal line 1880um whether or 1250um, but in the case of satisfying the condition, we change the distance between the chip, the signal transmission loss can be minimized in the package。

B. Current distribution of DTD block

When designing a DTD block power supply, there are two ideas to choose from. One is to combine the power supply on the InFO, but the core power noise will be added to the DTD block power supply; the other is to separate the DTD block power supply from the core power supply. It will increase the complexity of power supply design[9]. Since DTDIP has two power supplies of high and low-level swing power, when the power is merged, it will cause power integrate risks. In addition, in order to reduce the impact of the introduction of external power on DTD block, we choose to separate the DTD block power supply from the core power supply, and the high and low voltage swing are separately

powered up, due to limited spaces available of DTD block, the high voltage swing power supply network design is carried out on the RDL5 layer, and the low-level swing power supply is designed in the peripheral area of the RDL4 layer.

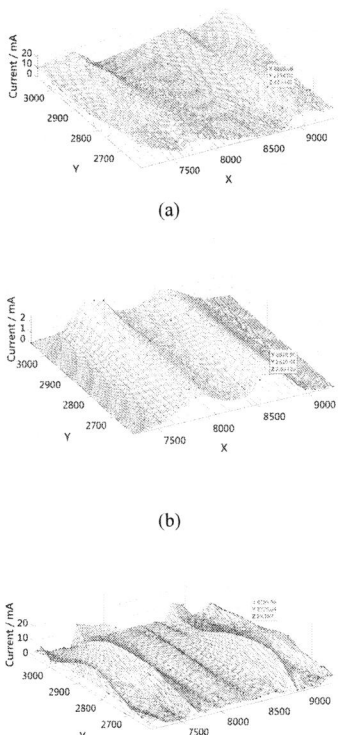

(a)

(b)

(c)

Fig. 9. current distribution: (a)high-level swing current distribution max current 22.418mA (b)low-level swing current distribution max current 2.634mA (c)ground current distribution max current 29.787mA

Use commercial software to simulate the static current of the power supply bump under the conditions of 105 ℃ working environment and all I/O flips. From Figure 6, it can be seen that the maximum current of the high-level swing is 22.418mA, and the maximum current of the low-level swing is 2.634 mA, evenly distributed on DTD block, the maximum ground current is -29.787mA. The size of the ground current is adapted to the power distribution. The maximum current that InFO RDL can provide at 105 ℃ is 74mA, and the power supply network can well meet DTD block power supply.

IV. CONCLUSION

The InFO package uses a 2/2μm ultra-fine pitch 5-layer RDL routing method. After fully considering the signal integrity during the wiring, the two dies are successfully interconnected, and the ground, isolation, and upper and lower layer reference wiring of the 5 signal lines can be achieved within a signal bump pitch. Meanwhile, At the same time, the power network distribution design of separate high and low-level power supply for DTD block on InFO platform is realized.

Through the simulation analysis of IL, RL, FEXT, NEXT, the routing method of ground, isolation, and upper and lower reference shielding structures can meet the signal integrity requirements. The shorter the signal line length, the better the signal transmission quality can be seen through S parameters .Through the current analysis of the power distribution network of high and low-level swing power, The distribution network structure on InFO platform can meet the power supply requirements of DTD block.

ACKNOWLEDGMENT

The author would like to acknowledge the support of the department of packaging and testing of ZTE corporation. Institute of microelectronics offer great assistance on signal and power simulation.

REFERENCES

[1] C. Wang and D. Yu, "Signal and Power Integrity Analysis on Integrated Fan-Out PoP (InFO_PoP) Technology for Next Generation Mobile Applications," 2016 IEEE 66th Electronic Components and Technology Conference (ECTC), Las Vegas, NV, USA, 2016, pp. 380-385, doi: 10.1109/ECTC.2016.130.

[2] H. Onozeki et al., "Study of Fine Pitch RDL First FO-PLP/WLP," 2018 International Wafer Level Packaging Conference (IWLPC), San Jose, CA, USA, 2018, pp. 1-7, doi: 10.23919/IWLPC.2018.8573282.

[3] K. T. Chang et al., "Ultra High Density IO Fan-Out Design Optimization with Signal Integrity and Power Integrity," 2019 IEEE 69th Electronic Components and Technology Conference (ECTC), Las Vegas, NV, USA, 2019, pp. 41-46, doi: 10.1109/ECTC.2019.00014.

[4] C. Liu, J. Botimer and Z. Zhang, "A 256Gb/s/mm-shoreline AIB-Compatible 16nm FinFET CMOS Chiplet for 2.5D Integration with Stratix 10 FPGA on EMIB and Tiling on Silicon Interposer," 2021 IEEE Custom Integrated Circuits Conference (CICC), 2021, pp. 1-2, doi: 10.1109/CICC51472.2021.9431555.

[5] H. Lee et al., "Design and signal integrity analysis of high bandwidth memory (HBM) interposer in 2.5D terabyte/s bandwidth graphics module," 2015 IEEE 24th Electrical Performance of Electronic Packaging and Systems (EPEPS), San Jose, CA, USA, 2015, pp. 145-148, doi: 10.1109/EPEPS.2015.7347149.

[6] Bogatin E . Signal and Power Integrity - Simplified[J]. Pearson Schweiz Ag, 2010.

[7] C. Gonzalez, H. Liu, M. Noh, E. Karl, T. Toifl and S. Hsu, "F5: Enabling New System Architectures with 2.5D, 3D, and Chiplets," 2021 IEEE International Solid- State Circuits Conference (ISSCC), San Francisco, CA, USA, 2021, pp. 529-532, doi: 10.1109/ISSCC42613.2021.9365834.

[8] L. Kangrong and M. D. Rotaru, "Frequency Domain Methodology for Evaluating Signal Integrity Performance of Logic to Logic and HBM Interconnect Models for Chiplet Packaging," 2020 IEEE 22nd Electronics Packaging Technology Conference (EPTC), Singapore, Singapore, 2020, pp. 147-151, doi: 10.1109/EPTC50525.2020.9315036.

[9] P. Chang, C. Hsieh, C. Chang, C. Chuang and C. Chiang, "Signal and Power Integrity Analysis of InFO Interconnect for Networking Application," 2018 IEEE 68th Electronic Components and Technology Conference (ECTC), 2018, pp. 1720-1725, doi: 10.1109/ECTC.2018.00258.

First-principles study on the mechanical properties of Cu₃Si compound

Jian WANG
Engineering Research Center of Electronic Information Materials and Devices, Ministry of Education, Guangxi Key Laboratory of Manufacturing System and Advanced Manufacturing Technology,Guilin University of Electronic Technology,
Guilin, China

Xiaowei XU
Reliability Research and Analysis Center, No.5 Electronics Research Institute of the Ministry of Industry and Information Technology, Advanced IC Reliability Engineering Research Center of Guangdong Province
Guangzhou, China

Chao DING
Engineering Research Center of Electronic Information Materials and Devices, Ministry of Education, Guangxi Key Laboratory of Manufacturing System and Advanced Manufacturing Technology,Guilin University of Electronic Technology,
Guilin, China

Tianhan LIU
Reliability Research and Analysis Center, No.5 Electronics Research Institute of the Ministry of Industry and Information Technology, Advanced IC Reliability Engineering Research Center of Guangdong Province
Guangzhou, China

Zongbei DAI*
Reliability Research and Analysis Center, No.5 Electronics Research Institute of the Ministry of Industry and Information Technology, Advanced IC Reliability Engineering Research Center of Guangdong Province
Guangzhou, China
*dai_zongbei@126.com
* corresponding author

Hongbo QIN*
Engineering Research Center of Electronic Information Materials and Devices, Ministry of Education, Guangxi Key Laboratory of Manufacturing System and Advanced Manufacturing Technology, Guilin University of Electronic Technology
Guilin, China

Abstract—In this paper, the mechanical properties of Cu₃Si were researched by the first-principles method. And the method based on density functional theory (DFT). The exchange correlation energy function adopts the generalized gradient approximation (GGA). Approximately calculate the bulk modulus, Young's modulus, shear modulus and Poisson's ratio of Cu3Si crystal by Voigt-Reuss. The B/G value of Cu₃Si is 3.66, and it can be considered that Cu₃Si is a ductile material. The anisotropy index A^U is 6.169, indicating that the anisotropy of Cu₃Si is extremely obvious. The anisotropy indexes A_B and A_G are 0.004 and 0.381, respectively, revealing the weak anisotropy for bulk modulus and obvious anisotropy for shear modulus. The shear anisotropy indexes A_1, A_2 and A_3 are 0.605, 0.605, 1.005, respectively, showing that the Cu₃Si crystal exhibits obvious shear anisotropic characteristics at {100} and {010} planes, and there is almost no anisotropy at {001} plane. Moreover, The orientation of the bulk modulus, Young's modulus and shear modulus, as well as their maximum and minimum values were identified.

Keywords—Cu₃Si, first principles, mechanical properties, anisotropy

I. INTRODUCTION

Through silicon via (TSV) is playing an important role in the core technology of three-dimensional packaging. In TSV structure, a barrier layer containing SiO_2 and TaN is formed on the internal surface of Si hole. Usually, TSV are filled with Cu. Actually, in integrated circuits, TSV failures may cause serious reliability problems, especially chip reliability problems[1]. Because Si and Cu have quite different coefficients of thermal expansion (CTE), the failure of barrier layer may occur during thermal cycling and high temperature ageing. As a result, Cu atoms will penetrate the barrier layer, and then chemically react with Si atoms to form intermetallic compound Cu₃Si [2], see Fig 1(a). Lee et al.[3] reserched the mechanical properties of Cu/Si films by using a nanoindentation instrument and found that Cu₃Si compound formed near the indenter tip. Thus far, the reliability of TSV

structure in 3D packaging has been widely been investigated. Dai et al. [4] researched the influence of the anisotropy of Si on TSV under thermal load, and indicated that the anisotropy of silicon is one of the most important in the ERR of the three-dimensional interface crack front. Fan et al. [5] studied the micron-scale TSV structure and found that the silicon crystal orientation and copper crystal grain orientation have a major influence on the stress and strain distribution of the structure and the extention of interface cracks. It should be noted that, Sonawane et al found that numerous micro-cracks present to nucleate and developed near Cu₃Si in Si wafer during thermal aging, which can induce TSV failure [2]. Clearly, the investigation of Cu₃Si is helpful to understand the failure mechanism of TSV. However, thus far, the mechanical properties of Cu₃Si compound have been little researched, so does the study of their anisotropy.

In this paper, the mechanical properties, and anisotropic characteristics were calculated by the first-principles method. This research can help understand the mechanical properties of Cu₃Si, which is benefit to the reliability analysis of TSV structure.

II. METHODS AND COMPUTATIONAL DETAILS

Cu₃Si crystal model, as shown in Fig. 1(b), was established in Materials Studio (MS) software which using Cambridge Serial Total Energy Package (CASTEP) code [6]. The Cu₃Si crystal has a trigonal structure and the space group is P-3M1(164). For its lattice, $a=b=4.052$ Å, $c=7.321$ Å, $\alpha=\beta=90°$, and $\gamma=120°$. The atom locations of Si atoms are (0.3333, 0.6667, 0.8280) while Cu atoms are (0, 0, 0), (0, 0, 0.333), (0.3333, 0.6667, 0.1592) and (0.3333, 0.6667, 0.4981). The value of kinetic energy convergence cutoff plane wave basis was set to 408.2 eV. In this study, the k-point of first irreducible zone 9×9×7 was adopted for structure optimization. The value of the energy convergence tolerance and the self-consistent field (SCF) convergence threshold were set at $1.0×10^{-5}$ eV/atom and $1.0×10^{-6}$ eV/atom, respectively. The max. force and the max. stress were set at 0.03 eV/Å and 0.05 Gpa, respectively.

III. RESULTS AND DISCUSSION

A. Lattice constants

The calculated lattice constants of Cu_3Si compared with experimental data [7] are listed in TABLE I. The differences between the calculated lattices constant and the experimental values are 1.10 % and 3.06 %, and the volume error is 5.34%, which are in the typical volume errors for GGA-based DFT calculations. Thus, the optimized lattice structure was used for further calculations.

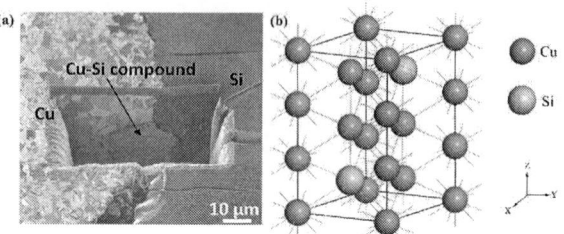

Fig. 1. Cu_3Si compound: (a) SEM image, (b) crystal structure.

TABLE I. THE LATTICE CONSTANTS OF Cu_3Si COMPOUND AND EXPERIMENTAL DATA

Material	Lattice Constants						Note
	Crystal system	Space group	k-point mesh	a=b(Å)	c(Å)	V	
Cu_3Si	Trigonal	P-3M1(164)	9×9×7	4.097	7.545	109.66	This work
	Trigonal	P-3M1(164)		4.052	7.321	104.10	Exp[7]
	/	/	/	1.10	3.06	5.34	error/%

B. Elastic Properties

Elastic constants is one of the important role in the microscopic properties and macroscopic mechanical behavior of materials. Hooke's law $\sigma_{ij} = C_{ijkl}\varepsilon_{ij}$ can be used to determine the stress-strain relationship, where C_{ijkl} represents the elastic constant, also known as the stiffness matrix. According to the Cu_3Si single crystal lattice symmetry, the stress-strain matrix can be written as follows:

$$
\begin{pmatrix} \sigma_1 \\ \sigma_2 \\ \sigma_3 \\ \tau_1 \\ \tau_2 \\ \tau_3 \end{pmatrix} = \begin{pmatrix} C_{11} & C_{12} & C_{13} & C_{14} & 0 & 0 \\ & C_{22} & C_{23} & C_{24} & 0 & 0 \\ & & C_{33} & 0 & 0 & 0 \\ & & & C_{44} & 0 & 0 \\ & Sym. & & & C_{55} & C_{56} \\ & & & & & C_{66} \end{pmatrix} \begin{pmatrix} \varepsilon_1 \\ \varepsilon_2 \\ \varepsilon_3 \\ \gamma_1 \\ \gamma_2 \\ \gamma_3 \end{pmatrix}
$$

where σ_i, τ_i, ε_i, and γ_i are normal stress and shear stress corresponding normal stress strain and shear strain, respectively. In this study, the calculated elastic constants of Cu_3Si are listed in TABLE II. Since the particularity of symmetry and trigonal system, only six values given here (C_{11}, C_{12}, C_{13}, C_{14}, C_{33}, C_{44}), where $C_{11}=C_{22}=222.92$ GPa, $C_{12}=92.46$ GPa, $C_{23}=C_{13}=93.85$ GPa, $C_{24}=-C_{14}=34.77$ GPa, $C_{33}=184.59$ GPa, $C_{44}=C_{55}=33.27$ GPa, $C_{56}=C_{14}=-34.77$ GPa, $C_{66}=(C_{11}-C_{12})/2=65.23$ GPa.

The entire elastic constant calculated by first-principles should meet the following mechanical stability conditions:

$$C_{11} > |C_{12}|, \quad (C_{11}+C_{12})C_{33} > 2C_{13}^2, \quad (C_{11}-C_{12})C_{44} > 2C_{14}^2 \quad (1)$$

There is a certain relationship between the elastic flexibility matrix S_{ij} and the elastic stiffness matrix $[C_{ij}]$, and the expression can be written as $[S_{ij}] = [C_{ij}]^{-1}$. The elastic constants C_{ij} and S_{ij} can be used to calculate the bulk modulus B, shear modulus G, Young's modulus E, and Poisson's ratio v of Cu_3Si. In this study, the B, E, and G of Cu_3Si are approximated by using the Voigt-Reuss method. The v of

Cu_3Si determined by B and G. These properties were calculated by the following equations:

$$9B_V = (C_{11}+C_{22}+C_{33}) + 2(C_{12}+C_{23}+C_{31}) \quad (2)$$

$$15G_V = (C_{11}+C_{22}+C_{33}) - (C_{12}+C_{23}+C_{31})$$
$$+3(C_{44}+C_{55}+C_{66}) \quad (3)$$

$$1/B_R = (S_{11}+S_{22}+S_{33}) + 2(S_{12}+S_{23}+S_{31}) \quad (4)$$

$$15/G_R = 4(S_{11}+S_{22}+S_{33}) - 4(S_{12}+S_{23}+S_{31})$$
$$+3(S_{44}+S_{55}+S_{66}) \quad (5)$$

$$B = \frac{1}{2}(B_V + B_R) \quad (6)$$

$$G = \frac{1}{2}(G_V + G_R) \quad (7)$$

$$\upsilon = \frac{3B-2G}{6B+2G} \quad (8)$$

$$E = \frac{9GB}{G+3B} \quad (9)$$

where B_V and B_R represent the upper and lower limits of B, G_V and G_R represent the upper and lower limits of G, respectively. The values of B_V and B_R are 132.3 GPa and 131.32 GPa, and the values of G_V and G_R are 49.70 GPa and 22.27 GPa. The bulk and shear moduli calculated in this research were taken the arithmetic average of the two extreme values. In TABLE III, the calculated B, E, and G are listed, and their values are 131.81 GPa, 35.99 GPa and 98.95 GPa, respectively.

Among them, the ratio of B/G is an important indicator for judging the the ductility of the material [8]. The critical value of B/G is 1.75. If the ratio (i.e., B/G) is large, indicating that the material of Cu_3Si compound has good ductility, if not, the material is not ductile and exhibits brittleness. From TABLE III, The calculated value of Cu_3Si is 3.77, and the ratio of B/G

is obviously much greater than 1.75, so it can conjecture that Cu₃Si has good ductility.

TABLE II.   THE VALUE OF ELASTIC CONSTANTS C_{ij}, GPa

Compound	C_{11}	C_{12}	C_{13}	C_{14}	C_{33}	C_{44}
GGA	222.92	92.46	93.85	−34.77	184.59	33.27

TABLE III.   CALCULATED MATERIALS PROPERTIES OF CU3SI

Phase	Materials properties								
	B_V, GPa	B_R, GPa	G_V, GPa	G_R, GPa	B, GPa	G, GPa	E, GPa	v	B/G
GGA	132.30	131.32	49.70	22.27	131.81	35.99	98.95	0.37	3.66

C. Elastic Anisotropy

The anisotropy of elastic properties is a very important index for predicting the fracture toughness of materials, and it has a great relationship with the difficulty of micro-crack formation in polycrystalline materials. Usually, the anisotropy index (A^U, A_B and A_G) is used to describe the anisotropic of the crystals [9]. The index A^U presents the universal anisotropic and the compression and shear anisotropy indexes are represented by A_B and A_G, respectively. The equations as follows:

$$A^U = 5\frac{G_V}{G_R} + \frac{B_V}{B_R} - 6 \tag{10}$$

$$A_B = \frac{B_V - B_R}{B_V + B_R} \tag{11}$$

$$A_G = \frac{G_V - G_R}{G_V + G_R} \tag{12}$$

In the above indicators, a value of 0 means isotropy. As the value increases, the anisotropy of the material rises. The calculated value of A^U is 6.169, which shows that Cu₃Si has obvious anisotropy. The value of A_B 0.003 indicates that the compression anisotropy of Cu₃Si is very limited, and the value of A_G 0.381 signifies that the shear anisotropy of Cu₃Si is obvious.

The shear anisotropy factor is used to measure the degree of anisotropy of the bonding between atoms in different crystal planes. The three indices (A_1, A_2 and A_3) represent the shear anisotropy factors of {100}, {010} and {001} planes, respectively. The shear anisotropic elements (A_1, A_2, A_3) are as follows[9]:

$$A_1 = \frac{4C_{44}}{C_{22} + C_{33} - 2C_{13}} \tag{13}$$

$$A_2 = \frac{4C_{55}}{C_{33} + C_{11} - 2C_{23}} \tag{14}$$

$$A_3 = \frac{4C_{66}}{C_{11} + C_{22} - 2C_{12}} \tag{15}$$

TABLE IV.   THE CALCULATED ANISOTROPIC INDEX OF CU₃SI

Anisotropic index	A^U	A_B	A_G	A_1	A_2	A_3	A_{Ba}	A_{Bc}
Value	6.169	0.004	0.381	0.605	0.605	1.005	1.000	1.407

If the value of the shear anisotropy factor is 1, it means that the crystal exhibits isotropic properties. The degree of anisotropy depends on the deviation of A1, A2 and A3 from 1[9]. In this work, the value of A_3 is 1.005, and A_1 and A_2 are equal to 0.605. This shows that the Cu₃Si crystal exhibits obvious shear anisotropy on the {100} and {010} planes, and there is almost no anisotropy at {001} plane. In addition, the anisotropy indexes A_{Ba} and A_{Bc} of the bulk modulus along the a-axis and the c-axis relative to the b-axis are equations (17) and (18), respectively.

$$A_{Ba} = \alpha \tag{16}$$

$$A_{Bc} = \beta \tag{17}$$

where

$$\alpha = \frac{(C_{11} - C_{12})(C_{33} - C_{13}) - (C_{23} - C_{13})(C_{11} - C_{13})}{(C_{33} - C_{13})(C_{22} - C_{12}) - (C_{13} - C_{23})(C_{12} - C_{23})} \tag{18}$$

$$\beta = \frac{(C_{22} - C_{12})(C_{11} - C_{13}) - (C_{11} - C_{12})(C_{23} - C_{12})}{(C_{22} - C_{12})(C_{33} - C_{13}) - (C_{12} - C_{23})(C_{13} - C_{23})} \tag{19}$$

The results listed in TABLE IV show that A_{Ba}=1, A_{Bc}=1.47, indicating there is no isotropy for the bulk modulus along the a-axis while isotropy exists for the bulk modulus along the c-axis. In order to further study the anisotropy characteristics of Cu₃Si crystals, the elastic anisotropy of the crystals was characterized by three-dimensional surface. The relationship between the bulk modulus, Young's modulus and shear modulus and the direction of the trigonal crystal system can be expressed as equations (20) to (22) [9]:

$$\frac{1}{B} = \left(S_{11} + S_{12} + S_{13}\right)l_1^2 + \left(S_{12} + S_{22} + S_{23}\right)l_2^2 + \left(S_{13} + S_{23} + S_{33}\right)l_3^2 \quad (20)$$

$$\frac{1}{E} = l_1^4 S_{11} + l_2^4 S_{22} + l_3^4 S_{33} + 2l_1^2 l_2^2 S_{12} + 2l_1^2 l_3^2 S_{13} + 2l_2^2 l_3^2 S_{23} + l_2^2 l_3^2 S_{44} + l_1^3 l_3^2 S_{55} + l_1^2 l_2^2 S_{66} \quad (21)$$

$$\frac{1}{G}(\boldsymbol{n},\boldsymbol{m}) = 4\left[2S_{12} - \left(S_{11} + S_{22} - S_{66}\right)\right]n_1 m_1 n_2 m_2 + 4\left[2S_{23} - \left(S_{22} + S_{33} - S_{44}\right)\right]n_2 m_2 n_3 m_3$$
$$+4\left[2S_{31} - \left(S_{33} + S_{11} - S_{55}\right)\right]n_3 m_3 n_1 m_1 + 4\left(n_1 m_2 + n_2 m_1\right)\left[\left(S_{16} - S_{36}\right)n_1 m_1 + \left(S_{26} - S_{36}\right)n_2 m_2\right]$$
$$+4\left(n_2 m_3 + n_3 m_2\right)\left[\left(S_{24} - S_{14}\right)n_2 m_2 + \left(S_{34} - S_{14}\right)n_3 m_3\right] + 4\left(n_3 m_1 + n_1 m_3\right)\left[\left(S_{35} - S_{25}\right)n_2 m_3 + \left(S_{15} - S_{25}\right)n_1 m_1\right]$$
$$+S_{44}\left(n_2 m_3 - n_3 m_2\right)^2 + S_{55}\left(n_3 m_1 - n_1 m_3\right)^2 + S_{66}\left(n_1 m_2 - n_2 m_1\right)^2 + 2S_{45}\left(n_2 n_3 + n_3 m_2\right)\left(n_3 m_1 + n_1 m_3\right)$$
$$+2S_{56}\left(n_3 m_1 + n_1 m_3\right)\left(n_1 m_2 + n_2 m_1\right) + 2S_{64}\left(n_1 m_2 + n_2 m_1\right)\left(n_2 m_3 + n_3 m_2\right) \quad (22)$$

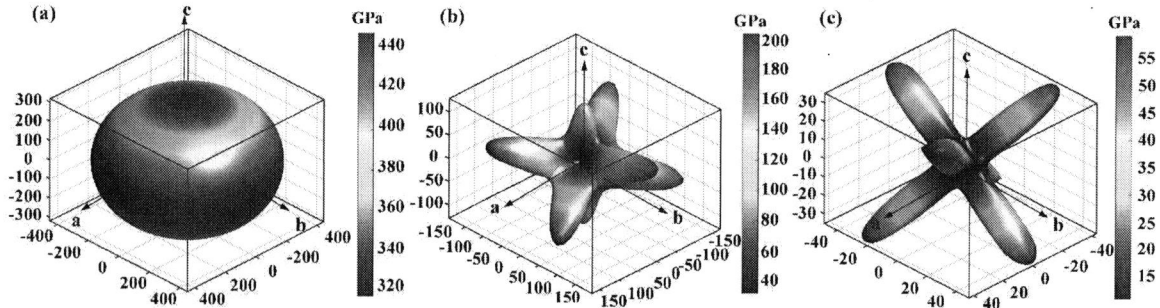

Fig. 2 Directional dependence of bulk modulus (a), Young's modulus (b) and shear modulus (c)

where l_1, l_2 and l_3 represent the three different cosines derections(a,b and c) of the Cu₃Si lattice. Then, using matlab software to generate three-dimensional modulus maps to characterize its anisotropy in three-dimensional space, as shown in Fig. 2.

As mentioned in the above text, for Cu₃Si, the bulk modulus B exhibited weak anisotropy, and Figure 2(a) shows that the spatial three-dimensional (3D) surface of bulk modulus of Cu₃Si is an ellipsoid. The maximum value is 447.39 GPa and located in the directions at (001) plane, while the minimum value is 317.98 GPa and situated in the orientation of [001] and [00$\bar{1}$]. Figure 2(b) exhibits the directional dependence of Young's modulus, in which the largest value is 205.31 GPa and located in three crystal directions [32$\bar{1}$], [$\bar{3}$$\bar{2}$1], [3$\bar{2}$1], [$\bar{3}2\bar{1}$], [031] and [0$\bar{3}$$\bar{1}$], while the smallest value is 32.10 GPa and is situated in the orientations [323], [$\bar{3}$2$\bar{3}$], [054], [0$\bar{5}$$\bar{4}$], [$\bar{5}$24], and [5$\bar{2}$$\bar{4}$]. Fig. 3(c) displays the directional dependence of shear modulus, where the maximum shear modulus is 59.12 GPa and is in directions [$\bar{4}$23], [4$\bar{2}$$\bar{3}$], [0$\bar{3}$2], [03$\bar{2}$], [423] and [$\bar{4}$$\bar{2}$$\bar{3}$], and the minimum shear modulus is 10.98 GPa. It can be seen that the anisotropy in the Young's modulus is more obvious than the shear modulus.

IV. CONCLUSIONS

Using first-principles and density functional calculation methods, the mechanical properties and elastic anisotropy of trigonal Cu₃Si single crystals have been extensively studied. Compared with the experimental data, the calculated lattice constants has a very small error and a good agreement, which shows the reliability of the theoretical model. In this work, the lattice constant of Cu₃Si is calculated, and the bulk modulus, Young's modulus and shear modulus, as well as their anisotropy index are calculated by the stiffness matrix. The orientation of bulk modulus, Young's modulus and shear modulus and their maximum and minimum values are determined. The following conclusions can be drawn:

1) As described in the above text, the bulk modulus, Young's modulus, shear modulus and Poisson's ratio of Cu₃Si crystal are 131.81 GPa, 98.95 GPa, 35.99 GPa and 0.37, respectively.

2) Through the ratio of B and G to reflect the ductility of the material, the value of B/G Cu₃Si is 3.66, indicating that Cu₃Si is a ductile material.

3) The calculation results show that the bulk modulus anisotropy of Cu₃Si is week, and obvious anisotropy in Young's modulus and shear modulus, especially, in Young's modulus.

4) The maximum values of bulk modulus, Young's modulus and shear modulus are 447.39 GPa, 205.31 GPa and 59.12 GPa, respectively, and the minimum values are 317.98 GPa, 32.10 GPa and 10.98 GPa, respectively. The directions of both maximum and minimum value are identified.

ACKNOWLEDGMENT

This study was sponsored by Science and Technology Program of Guangzhou under grant. No. 202102020008; Guangxi Natural Science Foundation under grant. Nos. 2018GXNSFAA281222 and 2021 GXNSFAA075010; Science and Technology Program of Guangzhou under grant. No. 202102020008. Science and Technology Planning Project of Guangxi Province under Grant Nos. GuiKeAD AD18281022 and 18281021, Director Fund Project of Guangxi Key Laboratory of Manufacturing

System and Advanced Manufacturing Technology Nos. 20-065-40-002Z and 19-050-44-003Z, Self-Topic Fund of Engineering Research Center of Electronic Information Materials and Devices Nos. EIMD-AB202005 and EIMD-AB202001. Innovation Project of GUET Graduate Education under grant No. 2020YCXS001 and 2021YCXS006.

REFERENCES

[1] T. Frank, S. Moreau, C. Chappaze, et al., "Reliability of TSV interconnects: Electromigration, thermal cycling, and impact on above metal level dielectric," Microelectron. Reliab., vol. 53, no. 1, pp. 17-29, 2013.

[2] D. Sonawane and P. Kumar, "New insights into fracture of Si in Cu-filled through silicon via during and after thermal annealing," Eng Fract. Mech., vol. 238, p. 107281, 2020.

[3] W. S. Lee, T. H. Chen, C. Lin, and Y. L. Chuang, "Nano-mechanical behaviour and mmicrostructural evolution of Cu/Si thin films at different annealing temperatures," Int. J. Eng. Sci., vol. 2, no. 3, pp. 207-215, 2012.

[4] Y. Z. Dai, Min, Qin, Fei, Chen, Pei, An, Tong, "Effect of silicon anisotropy on interfacial fracture for three dimensional through-silicon-via (TSV) under thermal loading," Eng. Fract. Mech., vol. 209, pp. 274-300, 2019.

[5] Z. Fan, X. Chen, Y. Liu, Y. Jiang, and Y. A. Zhang, "Effects of anisotropy on the reliability of TSV microstructure," Microelectron. Reliab., vol. 114, p. 113745, 2020.

[6] F. Segbers, "First principles methods using CASTEP," Z. Krist-cryst. Mater., vol. 220, no. 5-6, pp. 567-570, 2005.

[7] N. Mattern, R. Seyrich, L. Wilde, C. Baehtz, M. Knapp, and J. Acker, "Phase formation of rapidly quenched Cu–Si alloys," J. Alloy. Compd., vol. 429, no. 1-2, pp. 211-215, 2007

[8] F. S. Pugh, "XCII. Relations between the elastic moduli and the plastic properties of polycrystalline pure metals," Philos. Mag., vol. 45, no. 367, pp. 823-843, 2009.

[9] S. I. Ranganathan and M. Ostoja-Starzewski, "Universal elastic anisotropy index," Phys. Rev. Lett., vol. 101, no. 5, p. 055504, 2008.

978-1-6654-1392-3/21 $31.00 © 2021 IEEE

Study on the Effect of Silica Submicron Particle Size and Content on Fracture Toughness of Filled Epoxy Resin

Qin Zhou
School of Materials Science and Engineering
South China University of Technology
Guangzhou,China
zq2469552742@163.com

Xuecheng Yu
Shenzhen Institute of Advanced Electronic Materials
Shenzhen Institute of Advanced Technology, Chinese Academy of Scineces
Shenzhen, China
xc.yu@siat.ac.cn

Leicong Zhang
Shenzhen Institute of Advanced Electronic Materials
Shenzhen Institute of Advanced Technology, Chinese Academy of Scineces
Shenzhen, China
lc.zhang@siat.ac.cn

Pengli Zhu*
Shenzhen Institute of Advanced Electronic Materials
Shenzhen Institute of Advanced Technology, Chinese Academy of Scineces
Shenzhen, China
pl.zhu@siat.ac.cn

Rong Sun
Shenzhen Institute of Advanced Electronic Materials
Shenzhen Institute of Advanced Technology, Chinese Academy of Scineces
Shenzhen, China
Rong.sun@siat.ac.cn

Abstract—**Epoxy resins with excellent physical, mechanical properties and electrical insulation are used in a wide variety of fields. In order to overcome their inherent brittleness, silica particles, one of the most intriguing inorganic particles, have been extensively used as modification fillers of epoxy composites, and various toughening mechanisms have been brought out. On the base of former studies and theories, different sizes and contents of silica particles on submicron scale, ranging from 100 nm to 1 μm and 0 to 60 wt.% respectively, were used in this study to toughen epoxy resins, and the effect of particle size and filler content on the mechanical properties, especially fracture toughness (K_{IC}) of silica-epoxy composites were investigated. The size and morphology of silica particles were observed by scanning election microscopy (SEM). Fracture toughness (K_{IC}) of silica-epoxy composite was measured by a single-edge notch three-point-bending (SEN3PB) test. The fracture surfaces were observed by SEM to investigate and determine the toughening mechanisms of silica-epoxy composites. The effect of filler size and content on fracture toughness has been further investigated and understood.**

Keywords—*Silica submicron particles, Epoxy resins, Fracture toughness*

I. INTRODUCTION

Epoxy resins are widely used in various fields where high performance is required, such as coatings, composite materials, bonding agents and electronic encapsulation materials, because of their excellent physical, mechanical properties and electrical insulation. However, most epoxies are inherently brittle because of their cross-linked structures. Therefore, fillers are introduced to overcome their brittleness and increase fracture toughness. As a relatively new, low-budget modification method of resins while remaining their glass transition temperature, filling inorganic particles to strengthen and toughen epoxy resins has drawn much researchers' attention. [1] Compared to microparticles, particles on submicron and even nano scales have higher specific surface area, which can promote the transfer of external stress from matrix to fillers and allow particles to improve the strength of the epoxy matrix to a higher level.

Among the inorganic fillers, silica submicron particles with its superior thermal and electrical resistance performance were obtained in reinforcing and toughening epoxy resins in order to meet the needs of the electronics packaging industry. Due to the small size of the fillers, their high surface energy, and the interface layer effect, physical and mechanical capacities of silica-epoxy composites can both be uplifted. To explain these enhanced properties of epoxy resins filled with silica submicron particles, various mechanisms and theories have been suggested. These mechanisms include crack pinning/bowing, plastic void growth model, crack deflection, microcracking and crack bridging, regarding fracture toughness. In many conducted surveys, researchers found that toughening process of epoxy resins can be rather complex, often the result of synergy of multiple mechanisms[2]. The effect of particle size on fracture toughness has been studied as well, however, discrepancies still exist. The role that filler size played in toughening epoxy matrix is still under discussion and not well understood[1]. Therefore, acquiring an in-depth understanding in the effect of filler size and different particle size distribution on material properties is of necessity, which can provide guidance and allow us to design composite materials in electronic applications.

In this study, the effect of particle size and filler content on the mechanical properties, especially fracture toughness, of silica-epoxy composites are studied by using silica particles with different particle sizes on submicron scale, from 100 nm to 1 μm, and filler contents ranging from 0 to 60 wt.%, respectively. The fracture surfaces were observed by scanning election microscopy (SEM) to determine the toughening mechanisms of silica-epoxy composites. The effect of filler size and content on fracture toughness has been further investigated.

978-1-6654-1392-3/21 $31.00 © 2021 IEEE

II. Experiment

A. Materials.

Monodispersed silica particles with different particle sizes were synthesis by Stöber method. Bisphenol F was used as epoxy matrix. Diethyl methyl benzene diamine （DETDA）was used as the curing agent.

B. Preparation of silica-epoxy composites.

Epoxy composites were prepared by using silica particles with different particle sizes on submicron scale as fillers respectively. Particle sizes of silica were 100 nm, 300 nm, 500 nm, 800 nm, 1 μm. Curing agent was added into the silica-epoxy mixture of an appropriate amount. TABLE I shows the composition of each set of epoxy composites. After that, the mixture was put into a vacuum speed mixer at 1500 rpm for 2 minutes and 2000 rpm for another 2 minutes under vacuum. Repeat this procedure several times until there were no trapped bubbles. Finally, the resin was poured into a mold and cured at 160 °C for 2 hours, before gradually cooled to room temperature. The cured composite was released from the mold with the help of coated mold releasing agent, and then sanded until the surface is smooth and capable for further tests.

TABLE I.   COMPOSITIONS

Content (wt.%)	Silica submicron particles (g)	Neat epoxy (g)	Curing agent (g)	Weight (g)
0	0	7.14	2.86	10
30	3	5	2	10
40	4	4.29	1.71	10
50	5	3.57	1.43	10
60	6	2.86	1.14	10

C. Characterization.

The particle sizes of silica submicron particles was determined through dynamic light scattering (DLS) with a Zetasizer Nano ZS.

Fracture toughness (K_{IC}) was measured using single-edge notch three-point-bending (SEN3PB) type specimen. In accordance with the ASTM D5045 standard, a pre-crack was made by trapping a fresh razor blade between adjoining plates. At least two single-edge-notched specimens with dimension of 36 mm × 8 mm × 4.0 mm were tested at a rate of 1 mm/min, and the span was set as 32 mm. K_{IC} of all composites was determined using the relationship in the following equation [3].

$$K_{IC} = Y \frac{6P_f S}{4tw^2} \sqrt{a}$$

where Y is the shape factor, P_f is the load at the break, S is the length of the span, w is the width of the sample, t is the thickness of the sample, and a is the crack length.

The particle size of silica particles dispersed in epoxy matrix and the fracture surfaces of all composites were examined by scanning election microscopy (FEI, Nova 450). In order to facilitate electrical conduction on the surface, all samples were coated with a thin layer of sputtered gold.

III. Results and Discussion

A. Morphology of silica submicron particles with different sizes.

The SEM micrographs of different sized submicron-silicas are shown in Fig. 1(a-e). Silica submicron particles are all spherical with smooth surface. It can be seen from the SEM morphology that particles in each figure present uniform distribution in size, from 100 nm to 1 μm. Each set of submicron-silicas shows well dispersion. No signs of aggregation are observed.

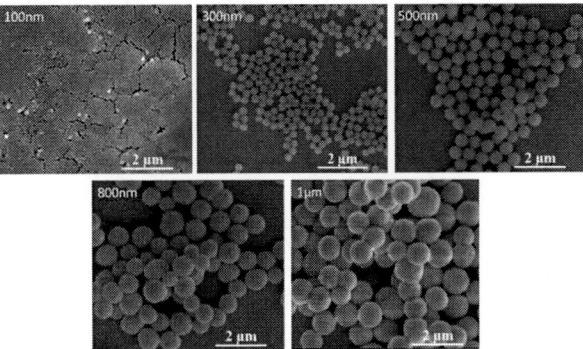

Fig. 1. SEM micrographs of silica submicron particles with different particle sizes. a) 100 nm. b) 300 nm. c) 500 nm. d) 800 nm. e) 1 μm.

B. Effect of silica submicron particles with different particle sizes and contents on toughening behavior.

TABLE II shows the values of fracture toughness , of composites which fillers were silicas with a submicron particle size of 100 nm, 300 nm, 500 nm, 800 nm, 1μm, respectively. In order to illustrate this visually, a diagram drawn from the experimental data is shown in Fig. 2. increases from initial value of 0.69 $Mpa \cdot m^{1/2}$ for neat epoxy resin to 2.49 $Mpa \cdot m^{1/2}$ at 50 wt.% filler content for particle size of 800 nm. In terms of overall trends, the fracture toughness increases with greater amounts of fillers. However, when it comes to 800 nm to 1μm in size of submicron-silicas at the content of 50 wt.% to 60 wt.%, a noticeable inflection point appears. For fillers which size of 500 nm, the effect of toughening is also hindered by the high filler content. It is likely that an upper threshold on the filler volume exists. Before the limit is reached, the toughening effect increases steadily when more particles are added. Large volumes of fillers tend to form severe agglomerations, which influences the dispersion in epoxy matrix to a limited extent, reducing the interface between fillers and epoxy, and in turn impede and even decrease the toughening behavior.

TABLE II. K_{IC} VALUES

Silica size / Content	100 nm	300 nm	500 nm	800 nm	1 μm
0 (Neat epoxy resin)			0.689424		
30 wt.%	1.1990	1.1686	1.3805	1.1865	1.4994
40 wt.%	1.4137	1.5101	1.6887	1.6017	1.5985
50 wt.%	1.8143	1.7852	1.7022	2.4943	2.2505
60 wt.%	2.1651	2.1732	1.7859	2.3329	1.8148

With the same increase filler content, fracture toughness grows almost monotonically for smaller particle sizes of 100 nm and 300 nm. The growth rate of K_{IC} values is higher for the larger particle diameter of 500 nm at the beginning, and decreases when the content of filler is over 40 wt.%. This variation is more signifitant for 800 nm and 1 µm particle sizes, as their slopes decrease to negative values at 50 wt.% filler content, therefore shows the turning point. Within experimental error, larger particle size leads to a more severe obstruction at high filler content to the toughening effect of silica-epoxy composites, for the slope after the inflection point of 1 µm has a greater absolute value than 800 nm , which shows a faster decline rate in the K_{IC} values.

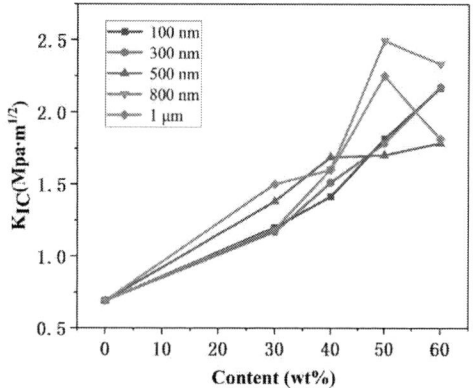

Fig. 2. Comparison between K_{IC} of silica-epoxy composites with different filler sizes and contents.

To better understand this eccentric "turning point" and construction phenomenon, a more in-depth investigation of the toughening mechnism of epoxy filled submicon-silicas was carried out.

C. Toughening mechanisms for epoxy filled silica submicron particles.

Fig. 3(a-b). show the same part on fracture surfaces of epoxy composites containing 60 wt.% of silica submicron particles with a size of 300 nm and 1 µm. Voids can be observed in each composite, and the diameter of the cavities is approximately equal to that of the particles, which indicates that particles fall off by external force from where they used to stay on the epoxy matrix. These result can be a support for plastic void growth model proposed by Johnsen et al[4]. The smaller the particle size, the fewer voids are presented, and this might explain the decrement of values from 1 µm to 300 nm. However, the images show that only a few amount of particles debonded from the matrix.

Silica nanoparticles are uniformly dispersed in the epoxy matrix at a low weight content, and a few voids are observed. Cracks spread over the matrix bifurcate when they reach the nanoparticles embedded in the matrix, supporting the crack pining effect brought up by Lange to explain the toughening mechanism[5]. According to this concept, filler particles serve as bridges or pins in the epoxy matrix and restrict the further expansion or extension of the micro-cracks.

These SEM images show that the toughening mechanism of silica submicron particles on epoxy resins is the result of both crack pining and plastic void growth. At lower fillers content, crack pining effect is dominant, and the plastic void growth model plays a crucial part when it comes to higher content.

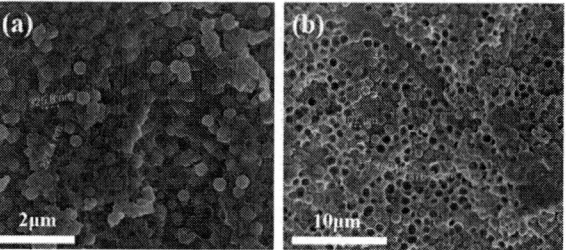

Fig. 3. SEM micrographs for fracture surface of silica-epoxy composites. a) the middle part of fracture surface of composite containing 60 wt.% 300 nm submicron-silica. b) the middle part of fracture surface of composite containing 60 wt.% 1 µm submicron-silica.

IV. CONCLUSION

The effect of silica-epoxy composites with different particle sizes from 100 nm to 1 µm, filler contents from 30 wt.% to 60 wt.% (and neat epoxy as comparison), on fracture toughness (K_{IC}) was investigated. The mechanism of toughening behavior was also studied.

For the filler content, before reaching a certain volume of filler content, the toughening effect increases steadily when more particles are added. More fillers than the critical value tend to impede and even decrease the toughening behavior. For size effect, larger particle size leads to a more severe obstruction at high filler content to the toughening effect of silica-epoxy composites

The toughening mechanism of silica submicron particles on epoxy resins was observed and determined using SEM. Both crack pining and plastic void growth work together to toughen the silica-epoxy composites. At lower fillers content, crack pining effect is dominant, and the plastic void growth model plays a crucial part when it comes to higher content.

ACKNOWLEDGMENT

The authors acknowledge the support of the National Key R & D Project from Minister of Science and Technology of China (2020YFB0311800), Shenzhen basic research plan (JCYJ20190807154409372).

REFERENCES

[1] Dittanet, P., and Pearson, R.A.: 'Effect of silica nanoparticle size on toughening mechanisms of filled epoxy', Polymer, 2012, 53, (9), pp. 1890-1905

[2] Adachi, T., Osaki, M., Araki, W., and Kwon, S.-C.: 'Fracture toughness of nano- and micro-spherical silica-particle-filled epoxy composites', Acta Materialia, 2008, 56, (9), pp. 2101-2109.

[3] 'Deformation and Fracture Mechanics of Engineering Materials', Materials Evaluation, 2020, 78, (10), pp. 1108-1108.

[4] Johnsen, B.B., Kinloch, A.J., Mohammed, R.D., Taylor, A.C., and Sprenger, S.: 'Toughening mechanisms of nanoparticle-modified epoxy polymers', Polymer, 2007, 48, (2), pp. 530-541.

[5] Lange, F.F., and Radford, K.C.: 'FRACTURE ENERGY OF AN EPOXY COMPOSITE SYSTEM', Journal of Materials Science, 1971, 6, (9), pp. 1197-1203.

Effect of Temperature Cycling, High Temperature Storage and Steady-State Operation Life Test on Reliability of GaN HEMTs

Mao Mao
Components and Materials Research Department,
China electronic product reliability and environmental testing research institute
Guangzhou,China
maomao@ceprei.com

Sha Tang
Electronic Component Inspection and Testing Center,
China electronic product reliability and environmental testing research institute
Guangzhou,China
tangsha@ceprei.com

Zhizhe Wang*
Components and Materials Research Department,
China electronic product reliability and environmental testing research institute
Guangzhou,China
zhizhewang@yeah.net

Rui Deng
Electronic Component Inspection and Testing Center,
China electronic product reliability and environmental testing research institute
Guangzhou,China
dengrui@ceprei.com

Chuanjin Deng
Electronic Component Inspection and Testing Center,
China electronic product reliability and environmental testing research institute
Guangzhou,China
tjudeng@126.com

Si Chen*
Science and Technology on Reliability Physics and Application of Electronic Component Laboratory,
China electronic product reliability and environmental testing research institute
Guangzhou,China
chensiceprei@yeah.net

Yan Ren
Components and Materials Research Department,
China electronic product reliability and environmental testing research institute
Guangzhou,China
184465031@qq.com

Abstract—With the advantages of high frequency and high power performance capabilities, AlGaN/GaN high-electron-mobility transistors (HEMTS) have attracted more and more attention. However, the reliability under stress conditions limits the application of GaN HEMTs in microwave power markets. In this paper, the effect of temperature and voltage on GaN HEMTs reliability were studied. A temperature cycling test with the temperature range from -55℃ to 175℃, a high temperature storage test of 175℃ and a steady-state operation life test with the channel temperature of 175℃ have been conducted on GaN HEMTs. The variation of key parameters such as drain current (I_{DS}), threshold voltage ($V_{GS(th)}$), gate-to-source leakage current (I_{GSS}) have been analyzed before and after the test.

Keywords—GaN HEMTs, reliability, temperature cycling test, high temperature storage test, steady-state operation life test.

I. INTRODUCTION

As third generation semiconductor material, GaN exhibits excellent performance, such as wide band gap, high breakdown field strength, excellent thermal resistance. More and more attention is paid to GaN-based high electron mobility transistors (HEMTs) due to the advantage of high temperature, high frequency, high power and high breakdown voltage. However, the reliability problems of GaN HEMTs have not yet been solved, which limits the large-scale application of GaN HEMTs in microwave power markets[1,2]. As high-power microwave devices, GaN HEMTs are often suffered to high temperature environment, however, the effective heat dissipation area of the device is gradually reduced, which leads to higher and higher device temperature. According to the Arrhenius model, the failure rate of devices will increase by an order of magnitude for the increase of every 10℃, which named "10℃ rule". Therefore, more and more attention are paid to the high-temperature reliability of GaN HEMTs[3-7]. However, the majority of previous research has focused on the effect of high temperature reverse bias (HTRB) stress or high temperature operation (HTO) stress on the key parameters of AlGaN/GaN HEMTs. The research on the reliability of GaN HEMTs under alternate extreme temperature and high storage temperature has been relatively few. The effect of temperature and voltage on GaN HEMTs reliability have not yet been comprehensively studied.

In this paper, temperature cycling, high temperature storage and steady-state operation life test of GaN HEMTs were carried out respectively. The samples were divided into three groups, with 11 samples in each group. The effect of different tests on key characteristics of GaN HEMTs were studied. The result may be helpful for the improvement of the reliability for GaN HEMTs.

II. TEMPERATURE CYCLING TEST

A. Test Condition

In order to evaluate the reliability of the device under alternate extreme temperature, temperature cycling test was

This research was supported by the Guangzhou Science and Technology Plan Project under Grant No. 201904010333 and Grant No. 201904010457, Key-Area Research and Development Program of Guangdong Province under Grant No. 2020B010173001 and No. 2018B010142001, National Natural Science Foundation of China (NSFC) under Grant No. 61804032, the Academician Fund under Grant No. ZHD201806.

carried out for GaN HEMTs. 11 samples numbered 28-38 were selected for 225 temperature cycles. According to the condition c of the method 1051 in GJB128A-1997, the minimum temperature of -55°C and the maximum temperature of 175°C were selected, with the holding time of 10min and the transfer time of less than 1min. The drain current (I_{DS}), threshold voltage ($V_{GS(th)}$), gate-to-source leakage current (I_{GSS}) of GaN HEMTs after 0 temperature cycles, 25 temperature cycles and 225 temperature cycles were measured respectively.

B. Results Analysis

I_{DS} was measured under the measure conditions of V_{DS} = 6V and V_{GS} = -1V. The change of I_{DS} for different numbers of temperature cycles was shown in figure 1. I_{DS} of all the devices has risen with the increase of the number of temperature cycles.

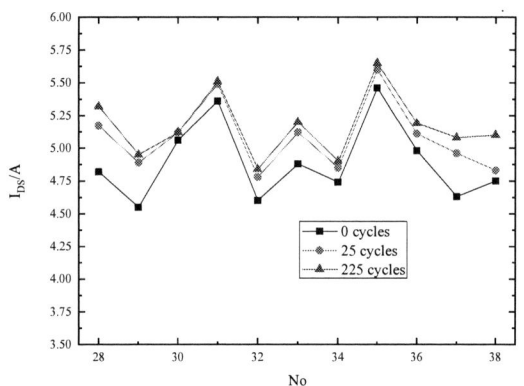

Fig. 1. I_{DS} of different samples after 0, 25 and 225 temperature cycles.

$V_{GS(th)}$ was measured under the conditions of V_{DS} =6V and I_D =6mA. As shown in Fig.2, the threshold voltage of GaN HEMTs has drifted negatively with the increase of the number of the temperature cycles. I_{GSS} was measured under the conditions of V_{DS} = 0V and V_{GS} =-6V. It was observed in Fig.3 that the change of I_{GSS} of GaN HEMTs was relatively small, indicating that the schottky characteristic of GaN HEMTs kept relatively stable during the temperature cycling test.

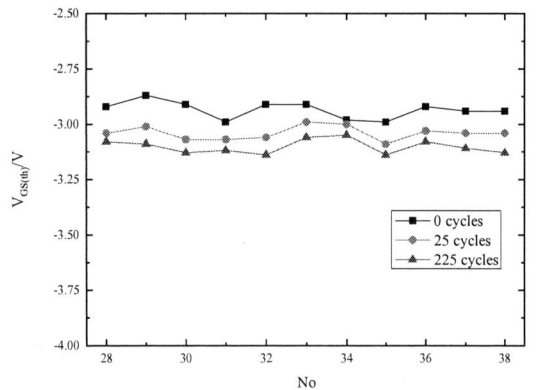

Fig. 2. $V_{GS(th)}$ of different samples after 0, 25 and 225 temperature cycles

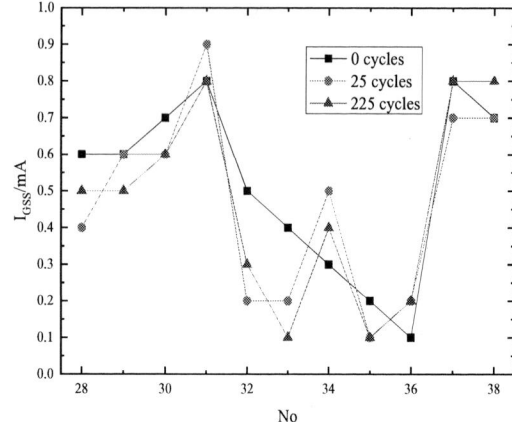

Fig. 3. I_{GSS} of different samples after 0, 25 and 225 temperature cycles.

III. STEADY-STATE OPERATION LIFE TEST

A. Test Condition

According to the condition c of the method 1027 in GJB128A-1997, the steady-state operation life test of 500h for GaN HEMTs was carried out with the channel temperature of 175°C. 11 samples numbered 50-60 were selected and the key parameters of devices after the test time of 0h, 340h and 500h were measured respectively.

B. Results Analysis

$V_{GS(th)}$ was also measured under the conditions of V_{DS}=6 V，I_D=6 mA. As shown in Fig.4, $V_{GS(th)}$ of all the devices has changed slightly, which indicated that the effect of the steady-state operation life test on $V_{GS(th)}$ could be neglected.

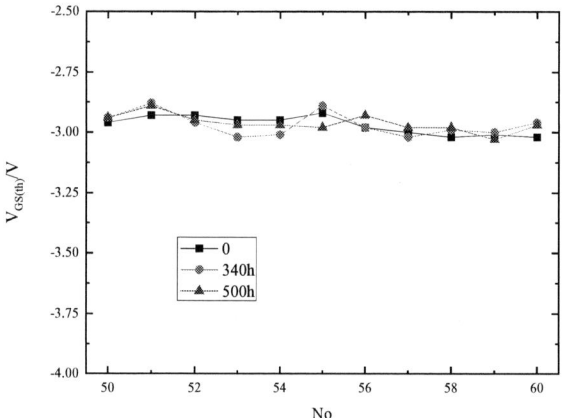

Fig. 4. $V_{GS(th)}$ of different samples after 0h, 340h, and 500h steady-state operation life test.

I_{DS} was measured under the conditions of V_{DS}=6 V, V_{GS}=-1V. It could be indicated that the drain current of all the GaN HEMTs have risen after the steady-state operation life test, which may be caused by the rising concentration of 2-dimensional electron gas in the channel.

Po at three frequency points of f=3.7 GHz, 4.0 GHz and 4.2 GHz was measured respectively under the conditions of V_{DS}=28 V, I_{DSQ}=60mA~160mA, Pin=31.1dBm. As shown in Fig.6, Po of GaN HEMTs at different frequency has decreased with the increase of the steady-state operation life test time.

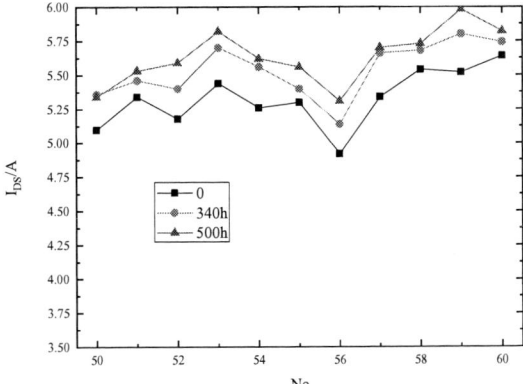

Fig. 5. I_{DS} of different samples after 0h, 340h, and 500h steady-state operation life test.

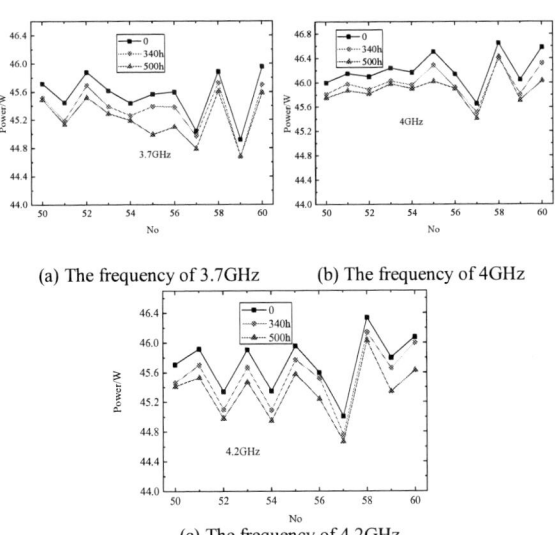

(a) The frequency of 3.7GHz   (b) The frequency of 4GHz

(c) The frequency of 4.2GHz

Fig. 6. P_o at different frequency points after 0h, 340h, and 500h steady-state operation life test.

IV. HIGH TEMPERATURE STORAGE TEST

A. Test Condition

According to the method 1032 of GJB 128A-1997, the high temperature storage test was carried out. 11 samples numbered 1-11 was stored for 340h at ambient condition of 175℃. The key parameters such as $V_{GS(th)}$, I_{DS}, and I_{GSS} of GaN HEMTs before and after the test were measured.

B. Results Analysis

It could be observed in Fig.7 that the drain current of all the GaN HEMTs have risen after the high temperature storage test. It may be caused by the rising concentration of 2-dimensional electron gas in the channel. As shown in Fig.8, the threshold voltage of GaN HEMTs has decreased after the high temperature storage test.

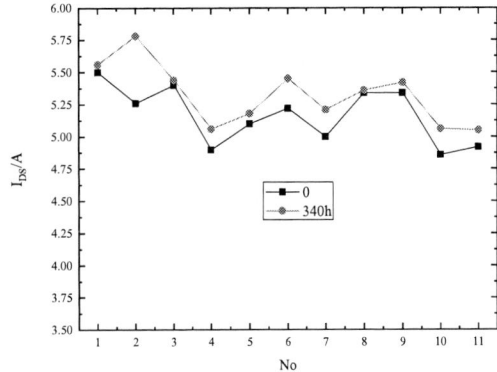

Fig. 7. I_{DS} of GaN HEMTs before and after 340h high temperature storage test.

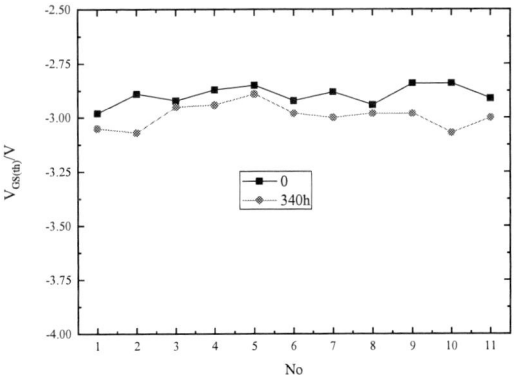

Fig. 8. $V_{GS(th)}$ of GaN HEMTs before and after 340h high temperature storage test.

It was shown in Fig.9 that the trend of the gate-to-source leakage current for GaN HEMTs during the high temperature storage test may not be obvious.

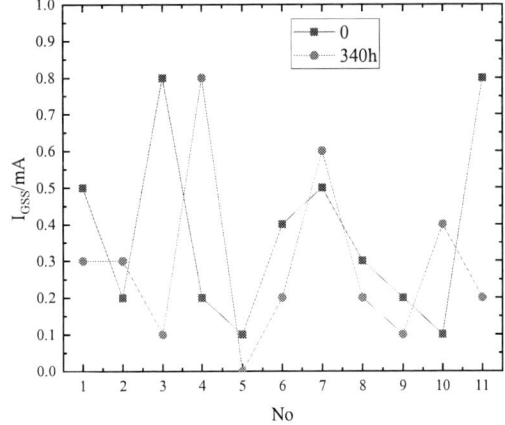

Fig. 9. I_{GSS} of GaN HEMTs before and after 340h high temperature storage test.

V. CONCLUSIONS

In this paper, three kinds of environmental adaptability experiments were carried out on GaN HEMTs, including a temperature cycling test with the temperature range from -55℃ to 175℃, a high temperature storage test of 175℃ and a steady-state operation life test with the channel temperature of

978-1-6654-1392-3/21 $31.00 © 2021 IEEE

175℃. The influence of environmental adaptability test on key characteristics of GaN HEMTs was analyzed. The research results may provide an introduction for the evaluation and improvement for the reliability of GaN HEMTs. According to the research, the conclusions were as follows:

1) The drain current of all the GaN HEMTs have risen after the temperature cycling, high temperature storage and steady-state operation life test, which may be caused by the rising concentration of 2-dimensional electron gas in the channel.

2) The threshold voltage of GaN HEMTs has decreased after the temperature cycling, and high temperature storage test when the threshold voltage of GaN HEMTs has kept constant nearly after the steady-state operation life test.

3) There was no obvious trend of the gate-to-source leakage current variation for GaN HEMTs before and after the high temperature storage or temperature cycling test.

ACKNOWLEDGMENT

This research was supported by the Guangzhou Science and Technology Plan Project under Grant No. 201904010333 and Grant No. 201904010457, Key-Area Research and Development Program of Guangdong Province under Grant No. 2020B010173001 and No. 2018B010142001, National Natural Science Foundation of China (NSFC) under Grant No. 61804032, the Academician Fund under Grant No. ZHD201806.

REFERENCES

[1] B. Padmanabhan, D. Vasileska and S. M. Goodnick, "Reliability of GaN HEMTs: Current degradation in GaN/AlGaN/AlN/GaN HEMT,"2012 15th International Workshop on Computational Electronics, 2012, pp. 1-4.

[2] M. Ťapajna, O. Hilt, E. Bahat-Treidel, J. Würfl and J. Kuzmík, "Gate Reliability Investigation in Normally-Off p-Type-GaN Cap/AlGaN/GaN HEMTs Under Forward Bias Stress," IEEE Electr. Device L., vol. 37, no. 4, pp. 385-388, April 2016.

[3] C. Zeng, Y. Wang, X. Liao, R. Li, Y. Chen, P. Lai, et al. "Reliability assessment of Algangan Hemts for high voltage applications based on high temperature reverse bias test," 2014 10th International Conference on Reliability, Maintainability and Safety (ICRMS), 2014, pp. 298-301.

[4] Y. Wang, X. Hong, C. Zeng, P. Lai and Y. Huang, "Reliability evaluation and failure analysis of AlGaN/GaN high electron mobility transistor by photo emission microscope," Proceedings of the 20th IEEE International Symposium on the Physical and Failure Analysis of Integrated Circuits (IPFA), 2013, pp. 259-262.

[5] R.Li, Y. Wang, C. Zeng, X. Liao, P. Lai and Y. Huang, "Impact of high temperature reverse bias (HTRB) stress on the degradation of AlGaN/GaN HEMTs," 2014 IEEE International Conference on Electron Devices and Solid-State Circuits, 2014, pp. 1-2.

[6] T. J. Anderson, M. J. Tadjer, J. K. Hite, J. D. Greenlee, A. D. Koehler, K. D. Hobart, et al., "Effect of Reduced Extended Defect Density in MOCVD Grown AlGaN/GaN HEMTs on Native GaN Substrates," IEEE Electr. Device L., vol. 37, no. 1, pp. 28-30, Jan. 2016.

[7] C. Zeng, Y. Wang, X. Liao, R. Li, Y. Chen, P. Lai, et al., "On the degradation kinetics and mechanism of AlGaN/GaN HEMTs under high temperature operation (HTO) stress,"2014 IEEE International Conference on Electron Devices and Solid-State Circuits, 2014, pp. 1-2.

Effect of IMC morphology on the current density and temperature gradient of line-type Cu/Sn/Cu solder joint

Jiaqi TANG
Engineering Research Center of Electronic Information Materials and Devices, Ministry of Education, Guangxi Key Laboratory of Manufacturing System and Advanced Manufacturing Technology, Guilin University of Electronic Technology,
Guilin, China

Tianhan LIU
Reliability Research and Analysis Center, No.5 Electronics Research Institute of the Ministry of Industry and Information Technology, Advanced IC Reliability Engineering Research Center of Guangdong Province
Guangzhou, China

Chao DING
Engineering Research Center of Electronic Information Materials and Devices, Ministry of Education, Guangxi Key Laboratory of Manufacturing System and Advanced Manufacturing Technology, Guilin University of Electronic Technology,
Guilin, China

Jian WANG
Engineering Research Center of Electronic Information Materials and Devices, Ministry of Education, Guangxi Key Laboratory of Manufacturing System and Advanced Manufacturing Technology, Guilin University of Electronic Technology,
Guilin, China

Hongbo QIN*
Engineering Research Center of Electronic Information Materials and Devices, Ministry of Education, Guangxi Key Laboratory of Manufacturing System and Advanced Manufacturing Technology, Guilin University of Electronic Technology
Guilin, China
qinhb@guet.edu.cn
* corresponding author

Wangyun LI
Engineering Research Center of Electronic Information Materials and Devices, Ministry of Education, Guangxi Key Laboratory of Manufacturing System and Advanced Manufacturing Technology, Guilin University of Electronic Technology
Guilin, China

Abstract—With the development of miniaturization and multifunction of electronic package, the reliability problem of electromigration (EM) and thermomigration (TM) in interconnect solder joints has been prevailing. The purpose in the present study is to probe into the impact of intermetallic compound (IMC) on morphology electromigration and thermomigration, the finite element (FE) model was used to analyze electromigration and thermomigration behavior under different IMC layer morphology and loading conditions in this study. The results indicate the current density has little correlation with the morphology of the IMC layer, while the IMC thickness has a slight influence on the temperature gradient. In addition, the values of the maximum and the minimum current density appear on the sunken zone and projection parts of IMC layer, respectively. With regard to temperature gradient, the maximum temperature gradient and minimum temperature gradient are located at the solder and the IMC layer, respectively. the temperature difference can lead to large temperature gradients.

Keywords—IMC, microscale solder, electromigration, thermomigration, finite element model

I. INTRODUCTION

Because of the microminiaturization of electronic equipment and device, the magnitude of solder joints is gradually reduced[1]. It is worth mentioning that due to the reduction of solder joints magnitude, the mechanical, electrical and thermal loads per unit volume are greater than before. The failure of one solder joint usually determines life of the whole device. Thus, from here we see that the failure of solder joint has become one of the critical problems in the field of electronic packaging[2]. Meanwhile, the intermetallic compound (IMC) formed by the interfaces reaction has more obvious influence on electromigration and thermomigration behavior of solder joints[3]. The

miniaturization of solder joint leads to the raise of IMC proportion in solder joint and causes a sharp increase in current density in the solder joints, which can result in severe electromigration phenomena, such as whiskers, micro-voids and phase separation[4]. For example, Wang et al.[5] investigated the effects of different current stressing on electromigration on solder/copper substrate were studied. The results show that with the raise of current density, the more defects can be observed. At present, many experiences focused on the relation between electromigration effects and IMC growth.

Due to the limitation of experimental methods and devices, many investigators ignored the effect of solder matrix structure on electromigration and thermal migration. In our work, the thermo-electrical two-dimensional (2D) FE models were employed to analyze electromigration and thermomigration behavior under different IMC layer morphology and loading conditions. In the symmetrical Sn based solder joints. especially, the maximum position of different IMC morphologies will also be analyzed.

II. SIMULATION METHODS

The reliability of line-type solder joints with different IMC morphology was characterized by Finite element analysis (FEA). Due to the small proportion and slow formation during the growth process of IMC [6], Cu_3Sn was ignored, IMC mainly refers to Cu_6Sn_5 in this study. Image recognition method was used to extract the model of IMC layers obtained from classical diffusion theory [7], the red part is Sn base solder, the green part is Cu_6Sn_5, and the blue part is copper plate, besides, the light blue part is Cu_3Sn, which can be ignored in this paper, as shown in Fig.1 (a) and (b). At the same time, the FE models of the symmetrical Sn based solder joint with IMC layer structure were established.

see Fig.1 (c) and (d), where the green and grey parts are Cu_6Sn_5 and Sn, respectively. The both ends of the solder joint are Cu. In the 2D FE model, the length and width of the solder matrix were 40 and 30 μm, respectively. The thickness of the Cu_6Sn_5 layer was set as 1, 2 and 5 μm, individually. And the mesh number of the FE model was 19,200 (160 × 120). The different IMC morphology FE models were analyzed under three loading conditions: (1) The temperature on the top and bottom of the copper block is set as 20°C, and the current density on both sides of the copper is 1.5×10^8 A/m². (2) the different temperature at the top and below sides was 0.02806 °C, and the current density wasn't applied; (3) The temperature loading condition was the same as condition (2) and the current density condition was in line with condition (1). In addition, the material properties used in the finite element model were displayed in Table I. The simulation was implemented by ANSYS 18.0, and adopted element type PLANE67.

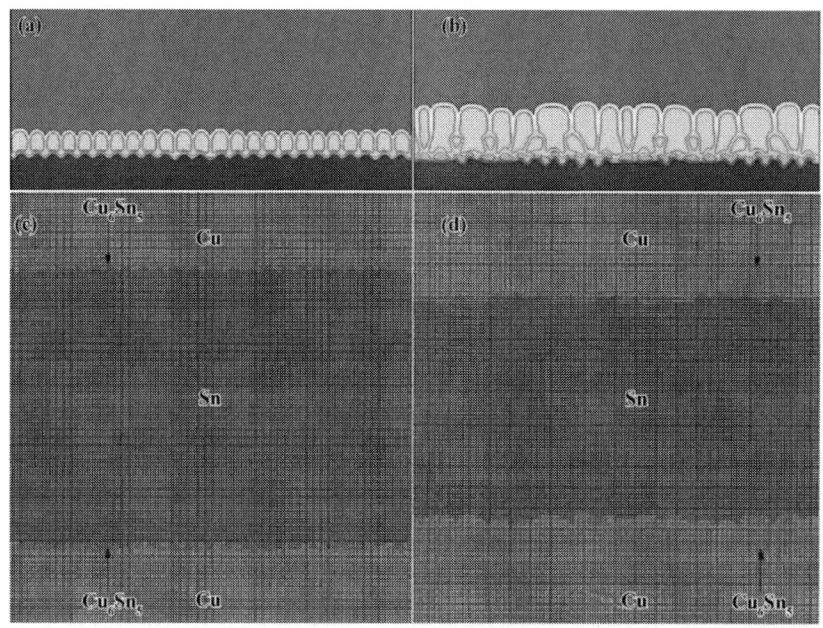

Fig.1 Microstructural evolution of IMC grains and FE models of solder joint employed in this simulation: (a) the little scallop-shaped IMC grains; (b) the big scallop-shaped IMC grains; (c) the FE model of (a); (d) the FE model of (b)

TABLE I. MATERIAL PROPERTIES IN FE MODELS

Materials properties	Materials and phases		
	Cu	Sn	Cu₆Sn₅
Specific heat capacity [J/(kg·K)]	385.2	222.1	286.0
Density (kg/m³)	8.92	7.38	8.28
CTE (×10⁻⁶/K)	16.6	22.4	16.3
Poisson's ratio	0.34	0.36	031
Young's modulus (GPa)	117.0	37.4	85.6
Thermal conductivity [W/(m·K)]	398	57.3	34.1
Electrical resistivity (×10⁻⁸ Ω·m)	1.7	11.5	17.5

III. RESULTS AND DISCUSSION

A. Equations under current stressing and temperature loading

Under current stressing loading condition, the diffusion flux of atom in solder joint in EM (J_{EM}) is given by Eq. (1) [8]:

$$J_{EM} = C \frac{D}{\kappa T} Z^* \rho e j \qquad (1)$$

where C is the number of atoms per unit volume, D is defined as the diffusivity, κ is the Boltzmann constant, T is the absolute temperature, Z^* is the effective nuclear charge number, ρ is the electrical resistivity, e is the elementary charge, and j is the current density. Clearly, EM phenomenon is primarily affected by current density under the given condition. Furthermore, the atomic diffusion flux in TM, J_{EM}, can be calculated using Eq. (2) [9]:

$$J_{TM} = C \frac{D}{\kappa T} \frac{Q^*}{T} (-\frac{\partial T}{\partial x}) \qquad (2)$$

where Q^* is defined as the heat transfer, and $\partial T/\partial x$ is the temperature gradient. Obviously, J_{TM} is mainly determined by $\partial T/\partial x$.

B. The magnitude and distribution of current density and temperature gradient under different conditions

If condering the IMC layer as a flat layer (the thickness of IMC is 1 μm) under simulation experiment (1), the current density in the solder joint is same as the applied current density, which is still 1.5×10^8 A/m².and the current density of the flat layer is low compared to that the scallop-shaped IMC morphology solder joints. Meanwhile, in FE model containing the scallop-shaped IMC layers under loading condition (1), when the thickness of IMC layer increases to 2 μm, the current density is from 1.22×10^8 A/m² to 1.94×10^8 A/m², and in the model with thickness of 5 μm, the current density varies from 1.24×10^8 A/m² to 2.08×10^8 A/m², as shown in Fig. 3 (a) and (c). The minimum current density of two IMC morphology is almost the same. In addition, Fig. 3 (a) and (c) show the locations of the maximum current density and the minimum current density both appear on the IMC layer. The minimum current density

978-1-6654-1392-3/21 $31.00 © 2021 IEEE

is located projection parts of the IMC layer, while the maximum current density is in the sunken zone of IMC layer. In addition, no matter what kind of IMC morphology, the current density at the IMC layer is close to 1.5×10^8 A/m^2. The current density is evenly distributed throughout the solder joint. Clearly, the current density difference can be ignored, the IMC morphology has little influence on the current density of solder joint.

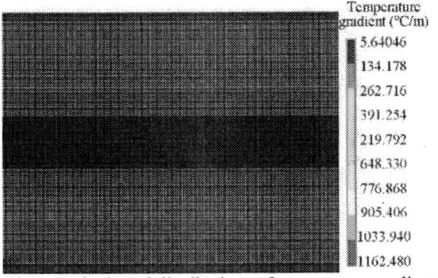

Fig.2 The magnitude and distributions of temperature gradients in solder joint with flat IMC layer under loading condition (1).

In the FE model containing flat layer, the temperature gradient is from 5.64046 °C/m to 1162.48 °C/m, see Fig. 2. Fig. 3 (b) and (d) show that the distributions of temperature gradients in FE model containing the scallop-shaped IMC layers. The temperature gradient is among 5.407 °C/m with 1199.03 °C/m, when the thickness of IMC layer is 2μm. Meanwhile, as the thickness is 5 μm, the temperature gradient is from 4.94801 °C/m to 1318.41 °C/m. The

maximum and minimum temperature gradient are located at the IMC layer and solder joint, respectively, see Figs. 3 (b) and (d). An interesting discovery is that the closer the middle of the solder, the lower the temperature gradient, the closer the interface of IMC layer/Cu, the higher the temperature gradient. As can be seen from Figs. (b) and (d), The temperature gradient difference of the IMC layer of both scallop layers is in a quantitative level.

In loading condition (2), the temperature gradient varies from 697.576 °C/m to 1186.14 °C/m in the containing flat IMC layer FE model, as illustrated in Fig. 4 (a). When the IMC morphology is scallop-shaped, the maximum temperature gradient are 1284.19 °C/m and 1230.09°C/m, the minimum temperature gradient are 594.939 °C/m and 556.092 °C/m, see Fig. 4 (b) and (c). The minimum temperature gradients under loading condition (2) are larger than those under loading condition (1). Moreover,similar to loading condition (1), The distribution of maximum and minimum temperature gradients are Sn-based solder and interface of IMC layer/Cu, respectively. It is worth mentioning that the temperature gradient difference gradually increases with the increase of the thickness of IMC layer. Obviously, the location of maximum and minimum temperature gradient is independent of the morphology of IMC layer., Besides, the temperature difference can lead to large temperature gradients.

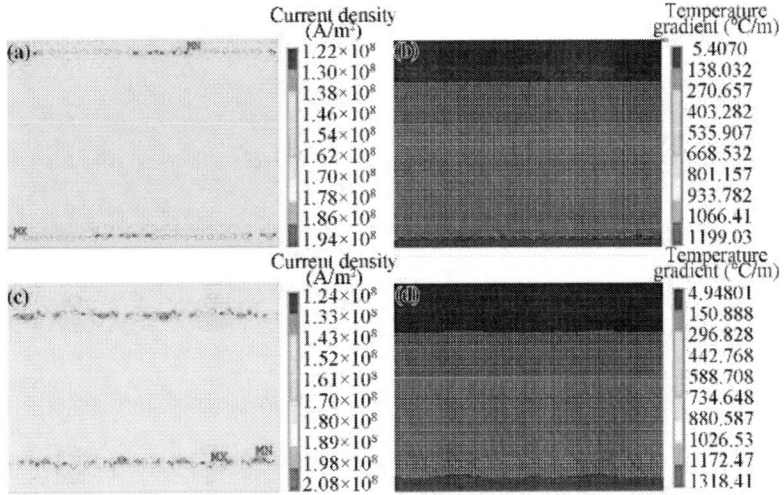

Fig.3 The magnitude and distributions of current density and temperature gradient in solder joint with scallop-shaped IMC layers under loading condition (1): (a) the little scallop-shaped IMC grains, current density; (b) the little scallop-shaped IMC grains, temperature gradient; (c) after the big scallop-shaped IMC grains, current density; (d) the big scallop-shaped IMC grains, temperature gradient.

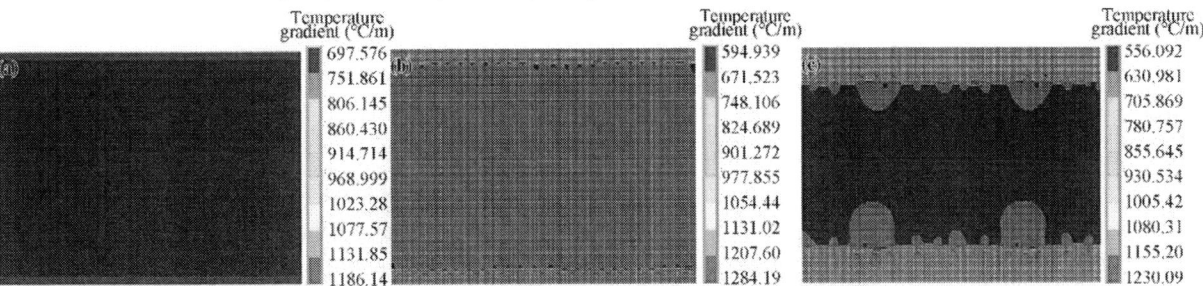

Figs.4 The distributions of temperature gradient under loading condition (2): (a) the flat IMC grains, temperature gradient; (b) the little scallop-shaped IMC grains, temperature gradient; (c) the big scallop-shaped IMC grains, temperature gradient;

In loading condition (3), the magnitude and distribution of the current density in the solder joint with scallop-shaped IMC layers is in keeping with that in the condition (1), see Figs.6(a) and (c). In the FE model considered flat layer, the temperature gradient varies from 9.22822 °C/m to 2348.42 °C/m, as illustrated in Fig. 5. Meanwhile, the temperature gradient is between 12.0764 °C/m and 2438.31 °C/m as the thickness is 2 μm. In the case of a thickness of 5 μm, the temperature gradient is from 6.83187 °C/m to 2384.46 °C/m, as displayed in Fig. 6 (b) and (d). Apparently, the maximum temperature gradients for both flat and scallop-shaped IMC layers are larger than those in the condition (1). In terms of the flat IMC layer, the maximum temperature gradient can be approximated as the sum of the maximum temperature gradients under the former two loading conditions. In addition, the location of maximum temperature gradient under the condition (3) are similar to that under the condition (1) and condition (2), which both are near the IMC layer/Copper interface. Clearly, In the case of electrothermal coupling field, especially, the loading temperatures at both ends of the copper plate are different, the temperature gradient variation is large, when IMC morphology changes.

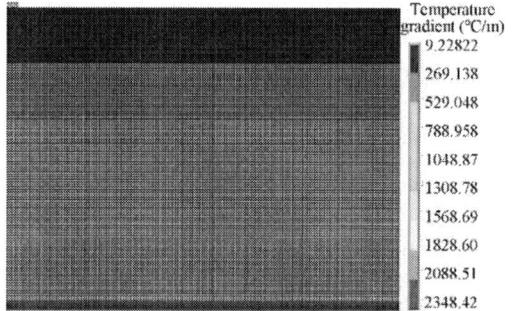

Fig.5 The distributions of temperature gradients in solder joint with flat IMC layer under loading condition (3)

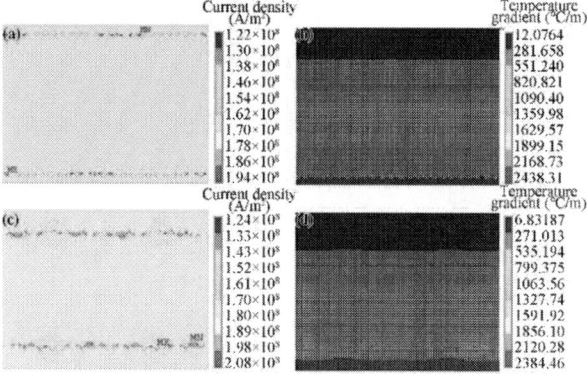

Fig.6 The distributions of current density and temperature gradient in solder joint with scallop-shaped IMC layers under loading condition (3): (a) the little scallop-shaped IMC grains, current density; (b) the little scallop-shaped IMC grains, temperature gradient; (c) the big scallop-shaped IMC grains, current density; (d) the big scallop-shaped IMC grains, temperature gradient.

IV. CONCLUSIONS

In this paper, FE models of the symmetrical Cu/Sn/Cu solder including the microstructure of IMC layer were built, and the electromigration and thermomigration behavior

under different IMC thickness were discussed. The results indicate that the morphology of IMC layers have little effect on the current density, while they weakly affect the temperature gradient. In addition, the maximum values of current density and temperature gradient are located at the IMC layer and solder joint near the solder/Cu interface, respectively. In three simulated samples, the closer to the middle of Sn base solder, the smaller the temperature gradient. Electrothermal coupling under loading conditions, the temperature gradient of the Sn based solder is almost the sum of the temperature gradient under the first two loading conditions. In addition, the temperature difference can lead to large temperature gradients. In future studies, the influence of IMC layer on electromigration behavior can be ignored.

Acknowledgment

This study was sponsored by the National Natural Science Foundation of China (NSFC) under grant. Nos. 51505095, 51805103 and No. 52065015; Guangxi Natural Science Foundation under grant. Nos. 2018GXNSFAA281222 and 2021 GXNSFAA075010; Science and Technology Program of Guangzhou under grant. No. 202102020008. Science and Technology Planning Project of Guangxi Province under Grant Nos. GuiKeAD AD18281022 and 18281021, Director Fund Project of Guangxi Key Laboratory of Manufacturing System and Advanced Manufacturing Technology Nos. 20-065-40-002Z and 19-050-44-003Z, Self-Topic Fund of Engineering Research Center of Electronic Information Materials and Devices Nos. EIMD-AB202005 and EIMD-AB202001. Innovation Project of GUET Graduate Education under grant No. 2020YCXS001 and 2021YCXS006.

REFERENCES

[1] A.M. Maniatty, J. Ni, Y. Liu, H. Zhang, Effect of Microstructure on Electromigration-Induced Stress, Journal of Applied Mechanics 83(1) (2016) 13-13.
[2] B. Knowlton, J.J. Clement, C.V. Thompson, Simulation of the effects of grain structure and grain growth on electromigration and the reliability of interconnects, Journal of Applied Physics 81(9) (1997) 6073-6080.
[3] Electromigration-induced growth mode transition of anodic Cu6Sn5 grains in Cu|SnAg3.0Cu0.5|Cu lap-type interconnects, Journal of Alloys & Compounds 703 (2017) 1-9.
[4] S.W. Chen, C.M. Chen, W.C. Liu, Electric current effects upon the Sn/Cu and Sn/Ni interfacial reactions, Journal of Electronic Materials 27(11) (1998) 1193-1199.
[5] Z.Y. Wang, N.M. Dang, P.H. Wang, T.Y.F. Chen, M.T. Lin, Study on Electromigration Effects and IMC Formation on Cu-Sn Films Due to Current Stress and Temperature, Applied Sciences-Basel 10(24) (2020).
[6] J. Son, D.Y. Yu, M.S. Kim, Y.H. Ko, J. Bang, Nucleation and Morphology of Cu6Sn5 Intermetallic at the Interface between Molten Sn-0.7Cu-0.2Cr Solder and Cu Substrate, Metals - Open Access Metallurgy Journal 11(2) (2021) 210.
[7] M.S. Park, S.L. Gibbons, R. Arróyave, Phase-field simulations of intermetallic compound evolution in Cu/Sn solder joints under electromigration, Acta Materialia 61(19) (2013) 7142-7154.

[8] K. Zeng, K.N. Tu, Six cases of reliability study of Pb-free solder joints in electronic packaging technology, Materials Science & Engineering R 38(2) (2002) 55-105.

[9] C. Chen, H.Y. Hsiao, Y.W. Chang, F. Ouyang, K.N. Tu, Thermomigration in solder joints, Materials Science & Engineering R 73(9-10) (2012) 85-100.

Design and Analysis of MOSFET Based on Fan-out Panel-Level Package Technology

Zhi Liang
SHENZHEN SIPTORY
TECHNOLOGY CO., LTD
SHENZHEN, China
zhiliang0628@163.com

Dongdong Shao*
SHENZHEN SIPTORY
TECHNOLOGY CO., LTD
SHENZHEN, China
derek@siptory.com

Kunpeng Ding*
SHENZHEN SIPTORY
TECHNOLOGY CO., LTD
SHENZHEN, China
kemp@siptory.com

Chuang Tian
SHENZHEN SIPTORY
TECHNOLOGY CO., LTD
SHENZHEN, China
t.chuang@siptory.com

Abstract—Nowadays, fan-out package has received extensive attention from enterprises and researchers because it possesses lower cost, smaller size, higher packaging efficiency, and better electrical and thermal performance. As one of the fan-out packages, panel-level fan-out package has advantages in single package cost, board utilization, and packaging efficiency. In this paper, based on the panel-level fan-out package process, a power MOSFET device is designed, which is composed of pads, solder, chip, circuits and molding compound. The drain of the chip is fixed on the copper pad by solder, and the gate and source are interconnected with the pads through laser drilling and electroplating processes. Finally, realize the electrical connection of the chip. In addition, the thermal resistance of the MOSFET device is analyzed by the commercial software ICEPAK. The results show that the thermal resistance of MOSFET based on fan-out panel-level package process is 43.44℃/W, which is 11.4% lower than the traditional wire-bonding process. The MOSFET based on fan-out panel-level package process has good heat dissipation performance.

Keywords—fan-out panel-level package, MOSFET, ICEPAK, thermal resistance, heat dissipation

I. INTRODUCTION

Semiconductor packaging is the process of leading out and redistributing chip terminals. It provides electrical pathways, signal pathways, heat dissipation pathways, as well as mechanical support and environmental protection for the chip. As Moore's Law is gradually approaching the physical limit, how to solve the increasingly difficult problem of reducing the size of the semiconductor process has become a current research hotspot[1]. Fan-out package is considered to be one of the important technical means to continue and surpass Moore's law due to its lower cost, smaller size, higher packaging efficiency, good electrical and thermal properties[2,3].

So far, many fan-out package technologies have been developed[4-8]. For example, Infineon's[4] embedded Wafer-Level Ball Grid Array (eWLB), Freescale's[5] redistribution chip package technology (RCP), TSMC's[6] integrated fan-out package technology (InFO), Huatian Technology's [7]silicon-based fan-out wafer package (eSiFO), ASE's[8] Fan-out Chip on Substrate (FOCos), and so on.

With the rise of mobile phones, the Internet of Things and wearable devices and the strong demand for chip miniaturization, integration, and intelligence, The further development of fan-out package technology has resulted in large-size panel-level package, which is called fan-out panel-level package (FOPLP)[9]. FOPLP has lower cost, higher carrier utilization, higher packaging efficiency and better electrical and thermal performance. At present, many domestic and foreign companies have begun to develop FOPLP, such as Fraunhofer IZM, Xinxing Electronics and Huajin Semiconductor[10].

Power semiconductor devices are the foundation of power electronic technology and the core devices of power electronic conversion devices[11]. As a type of power device, metal-oxide semiconductor field effect transistors（MOSFET）are widely used in analog circuits and digital circuits. At present, according to the bonding method of the chip, MOSFET package forms include wire bonded package, wire bounded and cu clip combination package and cu clip package, as show in figure.1.

In this paper, based on FOPLP, a power MOSFET device is designed, which is composed of pads, solder, chip, circuits and molding compound. Using ICEPAK software to analyze the thermal resistance of the MOSFET device, the results show that compared with the traditional wire bonding process, the MOSFET device based on FOPLP has lower thermal resistance and better heat dissipation performance.

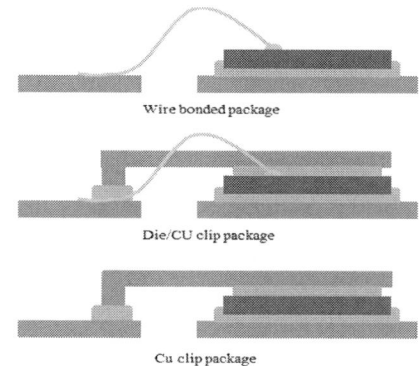

Fig. 1. Common encapsulation form of MOSFET devices

II. STRUCTURE

The structure of MOSFET is shown in figure 2. which is composed of pins, die attach, chip, circuit and molding compound. The drain at the bottom of the chip is soldered to the copper pad, and the gate and source are interconnected with the pad by laser vias and electroplating, thereby realizing electrical and signal transmission. Compared with the traditional wire bonding process, this method can improve the thermal conductivity of the chip, reduce the thermal resistance, improve production efficiency and reduce costs.

In addition, by using the electroplating process, multi-chip system integration can be realized, as shown in the figure 3. They are single-chip package, dual-chip package and six-chip package respectively.

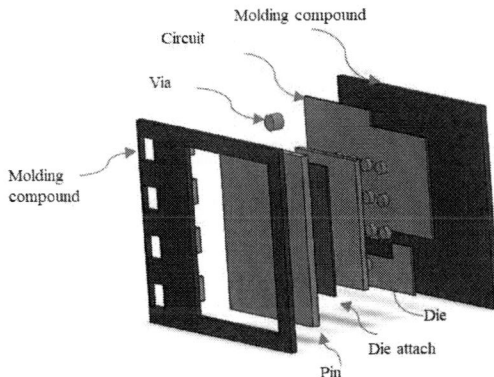

Fig. 2. Structure of MOSFET based on FOPLP process

(a) single-chip package

(b) dual-chip package

(c) six-chip package

Fig. 3. Structure of multi-chip integration MOSFET based on FOPLP process

III. THERMAL RESISTANCE ANALYSIS SETING

Thermal resistance is the ratio of the temperature difference between the two ends of the object and the power of the heat source when heat is transferred over the object. Due to its own characteristics, power MOSFET devices generate large amounts of heat during operation, and the increase in temperature will increase the failure rate and shorten the life of the device. As an important parameter to characterize the heat dissipation capacity of a device, thermal resistance is of great significance to the study of device performance. The calculation equation of thermal resistance is as follows:

$$R_{ja} = (T_j - T_a)/P \qquad (1)$$

In which Rja is the thermal resistance, Tj is the junction temperature of the chip, Ta is the ambient air temperature, and P is the thermal waste power of the chip.

In this paper, using the ICEPAK software to simulate and analyze the thermal resistance of MOSFET devices with a size of 5mm*6mm*0.7mm. The MOSFET device is placed in a sealed test case of JEDEC standard, and calculate the thermal resistance of the device under natural cooling conditions, as shown in the figure 4. The internal model of the case is composed of MOSFET devices, a board and a support. The size of the case is 305mm*305mm*305mm, its six sides are set as walls, and the heat transfer coefficient is 5W/(m2*k). The board is a four-layer board with a size of 114.3mm*76.2mm*1.6 mm. The thickness of the copper foil of the top and bottom layers is 0.07mm, the thickness of the middle two layers of copper foil is 0.035mm. The thermal conductivity of the tangential direction and the normal direction of the circuit board are 25.763W/(m*k) and 0.383W/(m*k) respectively, and the thermal conductivity of the support is 0.2W/(m*k). The material parameters of MOSFET device are shown in the table I, and the thermal power of the chip is 1W. The simulation model is shown in figure 5, which (a) is the MOSFET device based on FOPLP process and (b) is the MOSFET device based on wire bonded process.

In the basic parameter setting, the radiation model is the discrete ordinates radiation model (DO model). The air flow pattern is the Turbulent model. The calculation equation is the zero equation. A gravitational acceleration of -9.8m/s2 is applied in the z direction. the ambient temperature is 25°C. the number of iteration steps is set to 300, and the flow residual and the energy residual are 0.001 and 1e-7, respectively.

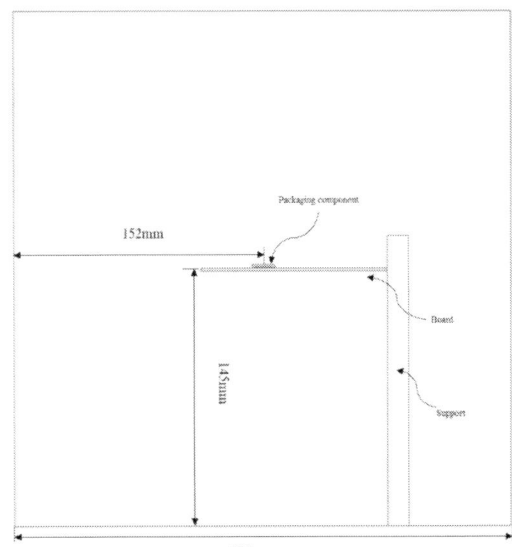

Fig. 4. Sealed test case of JEDEC standard

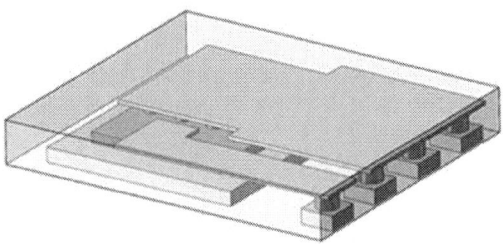

(a) MOSFET based on FOPLP process

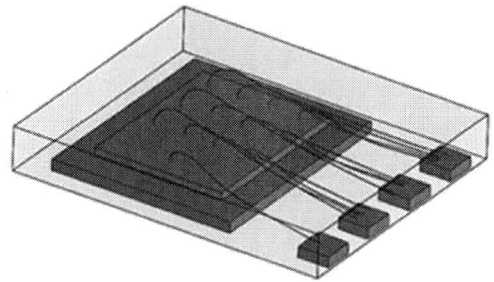

(b) MOSFET based on Wire bonded process

Fig. 5. The simulation model of MOSFET devices

Table I . MATERTAL PARAMETERS OF MOSFET DEVICE

components	Materials	Thermal conductivity (W/(m*k))
Molding compound	epoxy	0.8
Pin	Cu-pure	387.6
Die attach	Pb90Sn10	25
Die	Si	180
Circuit	Cu-pure	387.6
Via	Cu-pure	387.6

IV. RESULT AND DISCUSSION

The calculation result is shown in figure 6, (a) is the temperature distribution of MOSFET based on FOPLP process and (b) is the temperature distribution of MOSFET based on wire bonding process. The heat generated by the chip is transferred to the air through the PCB and the molding compound. The junction temperature of the device based on wire bonded process is 74.01°C, while the junction temperature of the device based on FOPLP process is 68.44°C. According to equation (1), the thermal resistance of the MOSFET device of the two packaging processes is 49.01°C/W and 43.44°C/W. The thermal resistance of MOSFET in FOPLP process is 11.4% lower than that in wire bonding process. Therefore, the MOSFET based on the FOPLP process has better heat dissipation performance.

In addition, compared with the wire bonding process, the temperature distribution of the MOSFET based on the FOPLP process is more uniform. As can be seen from the temperature distribution, the vias and the circuit layer provide a good heat transfer channel, and the heat generated by the chip can be transferred to the air well.

(a)Temperature distribution of MOSFET based on FOPLP process

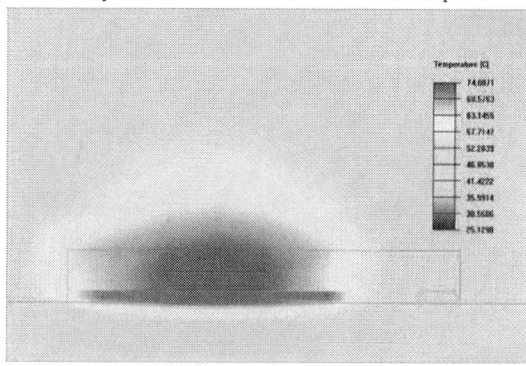

(b)Temperature distribution of MOSFET based on wire bonded process

Fig. 6. Temperature distribution of MOSFET devices

V. CONCLUSION

In this paper, a MOSFET device based on panel-level fan-out packaging technology is designed. The MOSFET device composed of pads, solder, chip, circuits and molding compound. The drain of the chip is fixed on the copper pad by solder, and the gate and source are interconnected with the pads through laser drilling and electroplating processes to realize the transmission of chip electricity and signals. The thermal simulation of MOSFET device was carried out by using ICEPAK software. The results show that compared with the traditional wire bonding process, the MOSFET based on FOPLP technology has lower thermal resistance and better heat dissipation performance.

REFERENCES

[1] Jiang D. Semiconductor efficient expansion beyond Moore's law. Applications of IC, 2015, 000(006):28-29.

[2] BRUNNBAUER M, FURGUT E, BEER G, et al. An embedded device technology based on a molded reconfigured wafer. 56th Electronic Components and Technology Conference 2006. 2006, 5.

[3] JI Yong, WANG Chengqian, LI Yang. Development, Challenges and Opportunities of Fan-out Packaging. Electronics & Packaging, 2020,20(8):080101.

[4] M. Brunnbauer, E. Fürgut, G. Beer, and T. Meyer, "Embedded wafer level ball grid array (eWLB)," in Proc. 8th IEEE Electron. Packag. Technol. Conf. (EPTC), Singapore, Dec. 2006, pp. 1–5.

[5] B. Keser et al., "Advanced packaging: The redistributed chip package," IEEE Trans. Adv. Packag., vol. 31, no. 1, pp. 39–43, Feb. 2008.

[6] C.-F. Tseng, C.-S. Liu, C.-H. Wu, and D. Yu, "InFO (wafer level integrated fan-out) technology," in Proc. 66th IEEE Electron Compon. Technol. Conf. (ECTC), Las Vegas, NV, USA, May 2016, pp. 1–6.

[7] MA S, WANG C, ZHENG F, et al. Development of Wafer Level Process for the Fabrication of Advanced Capacitive Fingerprint Sensors Using Embedded Silicon Fan-Out (eSiFO (R)) Technology. 2019 IEEE 69th Electronic Components and Technology Conference (ECTC) . 2019:28-34.

[8] Lin, Y. T. , et al. "Wafer Warpage Experiments and Simulation for Fan-Out Chip on Substrate." Electronic Components & Technology Conference IEEE, 2016.

[9] Sadeghinia M, Jansen K M B, Ernst L J, et al. Characterization of the viscoelastic properties of an epoxy molding compound during cure . Microelectronics Reliability, 2012, 52(8) : 1711－1718

[10] Takahashi H, Noma H, Suzuki N, et al. Large Panel Level Fan Out Package Built up Study with Film Type Encapsulation Material. Electronic Components and Technology Conference. 2016:134-139.

[11] Wang Bo. Review of power semiconductor device reliability for power converters . CPSS Transactions on Power Electronics and Applications, 2017, 2 (2) : 101-117

Adsorption behavior of HCOOH on the crystal surfaces of Cu(111) and (100)

Liheng JIANG
Guangxi Key Laboratory of Manufacturing System and Advanced Manufacturing Technology, Engineering Research Center of Electronic Information Materials and Devices, Ministry of Education, Guilin University of Electronic Technology,
Guilin, China

Tianhan LIU
Reliability Research and Analysis Center, No.5 Electronics Research Institute of the Ministry of Industry and Information Technology, Advanced IC Reliability Engineering Research Center of Guangdong Province
Guangzhou, China

Siliang HE
Guangxi Key Laboratory of Manufacturing System and Advanced Manufacturing Technology, Engineering Research Center of Electronic Information Materials and Devices, Ministry of Education, Guilin University of Electronic Technology,
Guilin, China

Hongbo QIN*
Guangxi Key Laboratory of Manufacturing System and Advanced Manufacturing Technology, Engineering Research Center of Electronic Information Materials and Devices, Ministry of Education, Guilin University of Electronic Technology,
Guilin, China
qinhb@guet.edu.cn

Wangyun LI
Guangxi Key Laboratory of Manufacturing System and Advanced Manufacturing Technology, Engineering Research Center of Electronic Information Materials and Devices, Ministry of Education, Guilin University of Electronic Technology,
Guilin, China

Daoguo YANG
Guangxi Key Laboratory of Manufacturing System and Advanced Manufacturing Technology, Engineering Research Center of Electronic Information Materials and Devices, Ministry of Education, Guilin University of Electronic Technology,
Guilin, China

Abstract—In order to investigate the adsorption mechanism of formic acid (HCOOH, FA) on the surface of Cu substrate during soldering, the adsorption energies of formic acid gas at the Top, Bridge and Fcc sites of Cu(111) and (100) planes and the surface energies of Cu(111) and (100) were identified by employing first-principles method relied on density functional theory (DFT). These findings demonstrate that the adsorption energy of HCOOH is the highest at the Bridge site of (100) plane, and the adsorption type is physical adsorption, which indicates that the HCOOH molecule is more easy to adsorb at the Bridge site of (100) plane. In addition, the surface energy of (100) plane is higher than that of (111) plane, revealing that (100) plane is more stable than that of (111) plane.

Keywords—*first-principles calculation, formic acid, Cu(111), Cu(100), adsorption behavior*

I. INTRODUCTION

For a long time, Sn-Pb eutectic solder has been widely employed in electronic packaging because of its good wettability, high mechanical performance, suitable melting point and low cost. However, due to Pb is a toxic matter, it not only seriously pollutes the environment but also does great harm to organisms. Countries around the world have promulgated relevant laws and regulations to restrict the use of Pb [1]. For example, the EU implemented the Restriction of Hazardous Substances (RoHS) in 2006 to restrict the employ of Pb in electrical and electronic products. At the same time, lead-free alloy solder has been implemented since July 2006 in China. In the last few years, Sn-Ag-Cu (SAC) solder has become the most promising lead-free solder, and SAC alloys have been demonstrate well done in reflow soldering, wave soldering and hand soldering [1]. However, the wettability problem between the unleaded solder and the Cu substrate is still one of the problems that need to be solved during the soldering process.

The flux not only promotes the solder melting and wetting on the surface of the metal to be soldered, enhance the wettability, reduce the surface tension of the molten solder, but also remove the oxides on the surface of the metal substrate and solder. The traditional rosin fluxes induce flux residues after welding, that are corrosive and will affect the reliability of the solder joints, so they must be removed [2]. The traditional chlorofluorocarbon (CFC-113) cleaning agent has a destructive effect on the ozone layer [3] and has been banned by the Montreal Convention. In addition, when the solder paste of the no-clean flux is soldered, the flux residues will be spattered on in-circuit testing pads [4], which will reduce the optical efficiency of the optoelectronic product. As the electronic packaging industry continues to face challenges in manufacturing faster, smaller and smarter devices, certain applications need to provide good solder joints in condition of fluxing soldering [1]. As a commonly used reagent for reducing metal oxides, formic acid (HCOOH) vapor has attracted great attention of researchers. Fosca Conti et al. focused on the compounds formed during SAC welding under formic acid vapor [1]. Another method of a residue-free soldering process by employing flux-free solder in HCOOH vapor atmosphere has been proposed by Hanss et al.[5]. Formic acid gas has also been used to reduce the solder oxide layer in the flux-free chip bonding process [6]. Moreover, the effect of HCOOH reflux solder on the protrusion of eutectic solder joints by experiments was discussed by Lin et al [7]. It has been reported that flux-free solder has wide application prospects [8]. However, thus far, the adsorption behavior of HCOOH on the crystal surfaces of Cu(111) and (100) is still unclear.

The purpose of this work is to describe the adsorption behavior of formic acid gas on the surface of Cu substrate, in this paper, first-principles method based on density functional theory (DFT) was employed to comparatively the adsorption behavior of HCOOH on the surface of Cu(111) and (100).

II. SIMULATION METHODS

In our context, first-principles method were performed by using the geometry optimization function of the Dmol3 module in Materials Studio software. The traditional generalized gradient approximation (GGA) usually underestimates the adsorption energy [9], thus the Perdew-

978-1-6654-1392-3/21 $31.00 © 2021 IEEE

Burke-Ernzerhof (PBE) function was selected as the exchange correlation function to compute the exchange interaction. Thanks to the interaction of van der Waals, the grimme custom method for DFT-D correction was chosen [10]. For geometry optimizations, the convergence errors of energy, force, and maximum displacement were 1×10^{-5} Ha, 0.004 Ha/Å, and 0.005 Å, respectively, and the maximum number of iteration steps was set to 1000. To ensure that the simulation accuracy is within the acceptable range, the double numerical plus d-functions (DND) basis set was performed to expand the electronic wave function. In addition, for the geometric optimization and electronic structure calculation on the crystal plane of Cu(111) and (100) [9], Monkhorst-Pack form of $4 \times 4 \times 1$ K point grid was carried out Brillouin zone integration. Considering the concentration effect, 3×3 crystal structure models of Cu(111) and Cu(100) were established. In order to avoid periodic interaction, a vacuum layer of 15 Å was made in the model. When performing geometric optimization, the bottom three layers of atoms were fixed to provide a substrate environment, and the surface atoms were in a relaxed state. The structures of Cu(111) and Cu(100) planes are shown in Figure 1 (a) and (b).

To ensure the reliability of the molecular system of HCOOH [11], the structural parameters of HCOOH molecule before and after geometric optimization were compared. The optimized bond lengths of H-O, C-H, C-O, and C=O were 0.992, 1.320, 1.095 and 1.219 Å, respectively, which were basically consistent with the previous experimental results, showing that the selected model is reliable. Fig. 1 (d) exhibits the HCOOH molecular model after optimized. Taking into account that different adsorption configurations have different adsorption effects [12], three adsorption positions of FA on the surface of Cu(111) and Cu(100) were studied, which are the top site, Bridge site and Fcc site, respectively, which are plotted in Fig. 2. The FA gas molecules were placed horizontally directly above three position, 3 Å away from the Cu surface atoms.

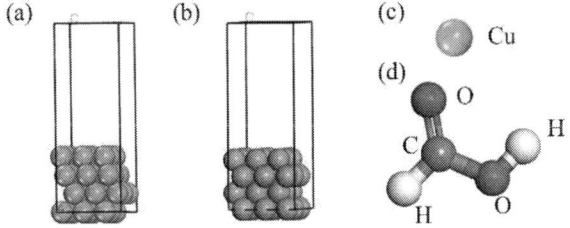

Fig. 1. Structures of models: (a) Cu(111), (b) Cu(100), (c) copper atom, and (d) HCOOH.

To analyze the adsorption performance of the HCOOH on different planes of Cu substrate, the adsorption energy (E_{ads}), equilibrium distance (d), and Hirschfeld charge transfer (Q_c) of FA on Cu(111) and (100) were calculated. With the value of E_{ads} increases, the HCOOH molecules is easier to adsorb on the substrate. Q_c affects the electrical properties of materials. When Q_c is a negative value, it reveals that the charge is transferred from the copper substrate to the formic acid molecule, and the d indicates the closest distance between the formic HCOOH and the copper substrate. It can be used to determine the type of adsorption. The adsorption energy was defined as follows:

$$E_{ads} = E_{(molecule+Cu)} - E_{Cu} - E_{molecule} \qquad (1)$$

in which $E_{(molecule+Cu)}$, E_{Cu}, and $E_{molecule}$ denote the total energy of FA molecules on the Cu, Cu(111) or Cu(100) surface, and HCOOH molecule, respectively.

In order to characterize the surface stability and wettability of HCOOH on the surface of Cu(111) and (100), the surface energy which is a measure of surface stability and wettability of the Cu(111) and (100) was also calculated by first-principles calculations. The higher the value, the better the stability. The calculation can be expressed as:

$$E_{surface} = \frac{1}{2A}(E_{slab} - NE_b) \qquad (2)$$

where A is defined as the surface area. E_{slab} is presented as the total energy and N is represent the number of copper atoms, respectively, and E_b represent the bulk energy per copper atom.

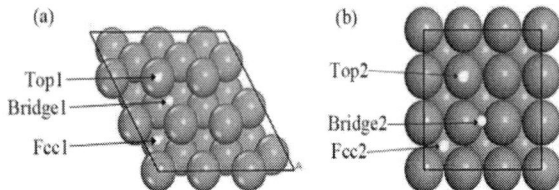

Fig. 2. Three adsorption sites on the surface of (a) Cu(111) and (b) Cu(100).

III. RESULTS AND DISSCUSION

The final optimization structure of HCOOH molecules on Cu (111) and (100) surface is plotted in Fig. 3. TABLE I lists the adsorption parameters of HCOOH molecules adsorbed on Cu (111) and (100) planes, for example, adsorption energy E_{ads}, equilibrium distance d and Hirschfeld charge transfer Q_c. Comparing the calculation results of FA adsorbed on Cu(111) and (100) surfaces respectively, it can be found that the d of HCOOH gas molecules on Cu(111) and (100) surfaces are all greater than 2 Å, and after optimization, the atoms closest to the surface of Cu are the double-bonded oxygen atoms in the formic acid molecule. The Q_c of HCOOH molecules on the Cu surface is all less than 0.1 e. In addition, the E_a of Cu(100) surface is greater than that of Cu(111), which indicates that HCOOH molecules are more easily adsorbed on Cu(100) surface. The position of Cu(100) surface with the largest E_{ads} is at the Bridge position, that is, when the formic acid gas molecule is at the Bridge position, the adsorption configuration is the most stable. The E_{ads}, Q_c, and d at Bridge site are -0.371 eV, -0.040 e, and 2.177 Å, respectively. Furthermore, the length of the equilibrium distance d is greater than the covalent radius of Cu and O atoms ($l_{Cu=O} = 1.75$ Å) [13], which imply that the HCOOH molecules are physically adsorbed on the Cu(100) surface through van der Waals force. It can be seen that the HCOOH molecule is adsorbed on the Cu(100) surface at a high adsorption rate without the formation of chemical bonds. The adsorption is reversible and can be quantitatively desorbed under certain conditions.

The surface energy are displayed in TABLE II, where the surface energy of (111) and (100) surface are 1.8077 J/m² and 1.9779 J/m², respectively. Obviously, the surface energy of Cu(100) is a bit higher than (100) plane, so formic acid gas molecule is more easy to adsorb on the Cu(100) surface.

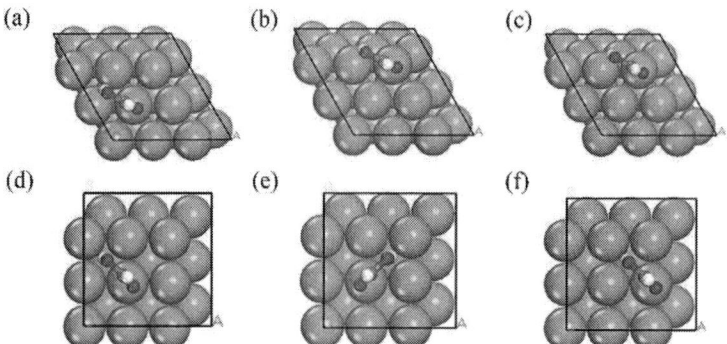

Fig. 3. The adsorption site of optimized configurations of the HCOOH:(a) Cu(111) Top site, (b) Cu(111) Bridge site, (c) Cu(100) Fcc site, (d) Cu(100) Top site, (e) Cu(100) Bridge site and (f) Cu(100) Fcc site.

TABLE I. $E_{ADS}, Q_C,$ AND D OF THE GAS ADSORPTION SYSTEMS OF HCOOH ADSORPTION ON THE CU(111) AND CU(100)

substrate	Adsorption site	Eads(eV)	Qc(e)	d(Å)
Cu(111)	Top	-0.286	-0.032	2.215(O=Cu)
	Bridge	-0.285	-0.033	2.218(O=Cu)
	Fcc	-0.287	-0.031	2.211(O=Cu)
Cu(100)	Top	-0.369	-0.039	2.179(O=Cu)
	Bridge	-0.371	-0.040	2.177(O=Cu)
	Fcc	-0.370	-0.042	2.173(O=Cu)

TABLE II. SURFACE ENERGIES OF CU(111) AND (100).

surface	A (10^{-20} m^2)	E_{slab} (eV)	E_b (eV)	$E_{surface}$ (J/m^2)
Cu(111)	5.3543	−10335.3243	−1476.6475	1.8077
Cu(100)	6.1826	−10335.0060	−1476.6475	1.9779

To further investigate the adsorption mechanism of FA on (111) Fcc site and (100) Bridge site, the total density of states (DOS) of the adsorption system of HCOOH on Cu(111) and (100) were calculated. Fig. 4 indicates that the influence of HCOOH on electronic levels of absorption system is mainly concentrated in the range of -7.5 eV to -0.5 eV in the valence band, not at the position of Fermi level. This verify that there is no chemical absorption, and HCOOH is adsorbed on the Cu substrate with weak adsorption force.

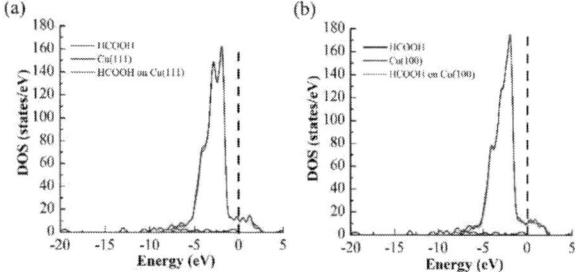

Fig. 4. Total density of states (DOS) for the Cu substrate(red curve), HCOOH-Cu system (blue curve), and HCOOH molecules (black curve). (a) Cu(111) DOS, (b) Cu(100) DOS. The dash lines means the Fermi level.

IV. CONCLUSIONS

Herein, First-principles method was utilized to calculate the adsorption performance and electronic properties of

HCOOH on the Cu substrate. It is possible to conclude that HCOOH is more easy to adsorb on Cu(100) surface by means of physical adsorption through van der Waals force, its adsorption rate is high, and the desorption effect is better than Cu(111). The surface energy of (111) is a bit lower than (100).

ACKNOWLEDGMENT

This paper was sponsored by the project of National Natural Science Foundation of China (NSFC) under grant. Nos. 51505095, 51805103 and No. 52065015; Guangxi Natural Science Foundation under grant. Nos. 2018GXNSFAA281222 and 2021 GXNSFAA075010; Science and Technology Planning Project of Guangxi Province under Grant Nos. GuiKeAD AD18281022 and 18281021, Director Fund Project of Guangxi Key Laboratory of Manufacturing System and Advanced Manufacturing Technology Nos. 19-050-44-003Z and 20-065-40-002Z, Self-Topic Fund of Engineering Research Center of Electronic Information Materials and Devices Nos. EIMD-AB202005 and EIMD-AB202007. Innovation Project of GUET Graduate Education under grant No. 2020YCXS001 and 2021YCXS006.

REFERENCES

[1] F. Conti, A. Hanss, and O. Mokhtari, "Formation of tin-based crystals from a SnAgCu alloy under formic acid vapor," New.J. Chem. vol. 42, pp. 19232-19236, 2018.

[2] D. Bušek, K. Dušek, D. Růžička, and M. Plaček, "Flux effect on void quantity and size in soldered joints," Microelectron. Reliab. vol. 60, pp. 135-140, May 2016.

[3] Q. B. Lu, L. Sanche, "Effects of cosmic rays on atmospheric chlorofluorocarbon dissociation and ozone depletion," Phys. Rev. Lett. vol. 87, pp. 078501-078505, August 2001.

[4] K. Dušek, and D. Bušek, "Problem with no-clean flux spattering on in-circuit testing pads diagnosed by EDS analysis," Microelectron. Reliab. vol. 56, pp. 162-169, January 2016.

[5] M. Monta, K. Okiyama, and T. Sakai, "Formation of solder cap on Cu pillar bump using formic acid reduction," 2012 IEEE Electronics Packaging Technology Conference (EPTC), Singapore, pp. 602-607. December 2012.

[6] P. S. Lim. and M. Z. Ding, "Development of fluxless flip chip reflow process for high density flip chip interconnect," 2015 IEEE Electronics Packaging Technology Conference (EPTC), Singapore, pp. 1-6. December 2015.

[7] Y. S. Lin, C. H. Shih, and W. Chang. "Fluxless reflow of eutectic solder bump using formic acid." Proceedings of the 2009 12th International Symposium on Integrated Circuits, Singapore, pp. 514-517. 2009.

[8] L. Wei, Y. C. Lee, "Study of fluxless soldering using formic acid vapor." IEEE T. Adv Packaging, vol. 22, pp. 592-601, 1999.

[9] R.S. Meng, M. Cai, J.K. Jiang, and Q.H. Liang, "First principles investigation of small molecules adsorption on antimonene," IEEE Electr. Device L., vol. 38, pp. 134-137, 2017.

[10] S. Grimme, "Semiempirical GGA - type density functional constructed with a long - range dispersion correction," J.Comput.Chem., vol 27, pp 1787-1799, September 2006.

[11] Z. Jiang, P. Qin, and T. Fang, "Decomposition mechanism of formic acid on Cu (111) surface: a theoretical study," Appl Surf Sci 396 857-864.2017.

[12] T. H. Liu, H. B. Qin, and D. G Yang, "First principles study of gas molecules adsorption on monolayered β-SnSe," Coatings, vol. 9, pp.390-399, June 2019.

[13] P. Pyykko, M. Atsumi, "Molecular single-bond covalent radii for elements 1-11," Chem. vol. 15, pp. 186-97, 2009.

Inner microcrack induced by intermetallic compound in right-angle Au/Sn3.0Ag0.5Cu/Au solder joints

Wu Yue*
School of Materials Engineering,
Lanzhou Institute of Technology
Lanzhou 730050, China
*Email: yuewu@lzit.edu.cn

Zhenyu Zhang
School of Materials Engineering,
Lanzhou Institute of Technology
Lanzhou 730050, China
Email: 853966410@qq.com

Bunv Liang
School of Materials Engineering,
Lanzhou Institute of Technology
Lanzhou 730050, China
Email: 1250842311@qq.com

Jing Li
School of Materials Engineering,
Lanzhou Institute of Technology
Lanzhou 730050, China
Email: 19721881@qq.com

Cheng Xue
School of Materials Engineering,
Lanzhou Institute of Technology
Lanzhou 730050, China
Email: 43122539@qq.com

Abstract—With the promptly development of electronic products toward miniaturization and multi-functionalization, the volume of solder joints become smaller and smaller, and which result in a serial of reliability and processing issues, for example, the higher ratio intermetallic compound (IMC), the lower mechanical properties, the soldering process and the higher rejection rate, etc.. Previous literature on the shocking resistance of right-angle Au/Sn3.0Ag0.5Cu(SAC305)/Au solder joints has shown that with the decreasing of the volume, the ratio of the brittle phase – IMC in the matrix of solder joints obviously increases, which further result in the poorer mechanical shocking resistance of right-angle solder joints fabricated by the solder balls with more smaller diameter. It is imperative to further investigate that the mechanical properties and the reliability fabricated by the more smaller solder balls for the electronic packing.

In this study, the inner microstructure of right-angle Au/SAC305/Au solder joints fabricated by the smaller solder ball with the diameter of 50 µm were investigated by employing scanning electron microscopy (SEM) and energy-dispersive X-ray spectroscopy (EDX), and the deterioration of the mechanical properties of solder joints was analyzed on the basis of the inner microstructure. Owing that the solder ball become smaller, there is an obviously sagging curve on the surface of solder joints. The SEM results of cross section of solder joint show that the inner microcracks easily formed at the top corner of the right-angle solder joints, and there are different IMCs at both sides of the microcracks at there. The analysis manifest that the disregistry between the different IMCs and the difficult feeding are the key factors of the inner hidden microcrack.

Keywords—*inner microcrack, solder joint, intermetallic compound (IMC), disregistry*

I. INTRODUCTION

With the promptly development of electronic products toward miniaturization, multi-functionalization and portability, the dimension of the electronic components and devices has been decreasing dramatically. As an essential key, the solder joints become smaller and smaller, and the packaging technology become more and more difficult [1]. Thus, it is urgent to research the higher packaging density, the higher reliability and the novel packing technology for the

next generation electronic products. However, this trend result in a serial of the packaging process issue and the higher risk of solder joints, for example, the poor wetting, the higher ratio intermetallic compound (IMC), the misalignment between two bonding pads, the bridging defect, the lower mechanical reliability, the weaker anti-electromigration and the higher rejection rate [2-4], etc. Owing to the unique advantages of localized heating and untouched soldering, the laser jet solder ball bonding (LJSBB) is widely used to packaging some three dimensional (3D) solder joints for the , such as the right angle solder joint [5]. Previous literature has reported that when the diameter of solder balls decreased from 70 µm to 60 µm, the shocking resistance of solder joints obviously become worse, and the analysis manifest that the higher ratio of the brittle phase – IMC in the solder matrix is the essential factor, which can lead to the lower mechanical properties and the weaker reliability [2]. However, the miniaturization trend is inevitable, and it can be predicted that there are many challenges in the soldering process and the reliability of the smaller solder joints. Thus, it is imperative to further investigate that the properties and the reliability fabricated by the more smaller solder balls for the future electronic packing.

In this study, the section microstructure of right angle Au/SAC305/Au solder joints fabricated by SAC305 solder balls with the diameter of 50 µm were investigated by employing scanning electron microscopy (SEM) and energy-dispersive X-ray spectroscopy (EDX), and the deterioration of the mechanical properties of solder joints was analyzed on the basis of the inner defects and microstructures.

II. EXPERMENTAL PROCEDURE

The right angle Au/SAC305/Au solder joints was fabricated by LJSBB in this study, figure 1 is the corresponding schematic of LJSBB process. It is noted that the diameter of solder ball is 50 µm. In addition, the width of both bonding pads is cut to 45 µm and to match the smaller solder ball. The energy of the laser beam irradiated from CO_2 laser generator (Xell-2000) is preset to 4.3 mJ and the irradiating time is set to few milliseconds during LJSBB process. In order to avoid the higher fraction of Au-Sn IMC, the thickness of Au film on the bonding pads are decreased to 0.8 µm. Other detail can refer to our previous study [2].

National Natural Science Foundation of China (grant No. 51565024), Gansu Provincial key talent projects in 2020 and 'Kaiwu' Innovation Team Support Project of Lanzhou Institute of Technology (grant No. 2018KW-05).

978-1-6654-1392-3/21 $31.00 © 2021 IEEE

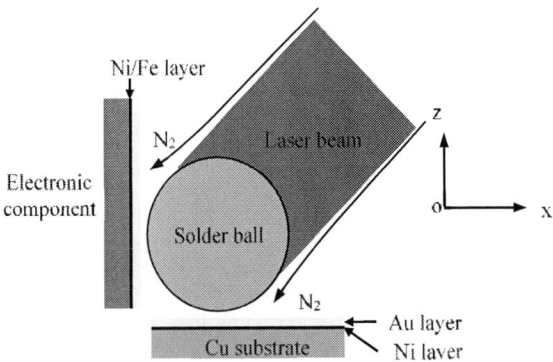

Fig. 1. Schematic of pad structure of right angle solder joint and the irradiation direction of LJSBB

After LJSBB, the outline of Au/SAC305/Au solder joints was checked by optical microscopy (OM) and scanning electron microscopy (SEM). Then, the inner microstructure and defects of solder joints were investigated by employing SEM, at the same time, the elements of different microstructures were characterized by energy-dispersive X-ray spectroscopy (EDX). On the basis of above results, the deterioration of the mechanical properties of solder joints was analyzed.

III. RESULTS AND DISCUSSION

A. Inner microstructure and cracks

Fig. 2(a) presents the overview of as-reflowed right angle Au/SAC305/Au solder joints. Obviously, there are two sagging curve nearby both interfaces of each solder joint, as marked by the arrow. Nevertheless, the sagging curve at the bottom interface is serious than the top one for every solder joint. Compared with the overview of solder joints fabricated

by 70 μm solder ball, seen in Fig.2(b) [2], the bottom sagging curve seem like a step. Consequently, it is suspicious that there are the inner microcracks in the solder matrix of solder joints, which is necessary to further investigate and confirm the inner defects of solder joints from viewpoint of the reliability.

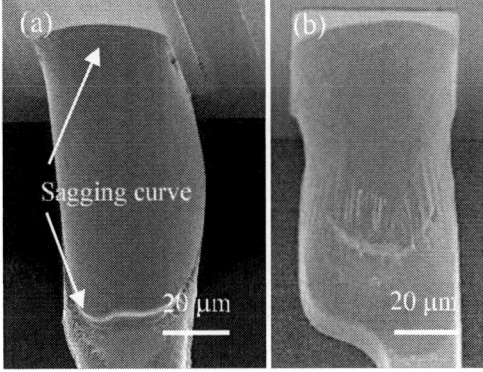

Fig. 2. Overview of as-reflowed right angle Au/SAC305/Au solder joints fabricated by the different solder balls: (a) 50 μm and (b) 70μm.

In order to confirm the inner microstructures of as-reflowed right angle Au/SAC305/Au solder joint in Fig. 2(a), the cross sectional samples of were prepared and the corresponding BSE images are presented in Fig. 3. Clearly, an obvious microcracks occurred at the upper corner of solder joints, while there is not any microcracks at the bottom corner and it can be judged that the suspicious crack in Fig. 2(a) is a step. The different contrasts in Figs. 3(b)-(d) manifest that there are different kinds IMCs nearby both interfaces of solder joint. More important, although the contrasts at the both sides of the microcrack seem the same, but the orientation of IMCs are clearly different, which implies that the inner microcrack is a typical intergranular fracture and occurred in the solidification course of solder.

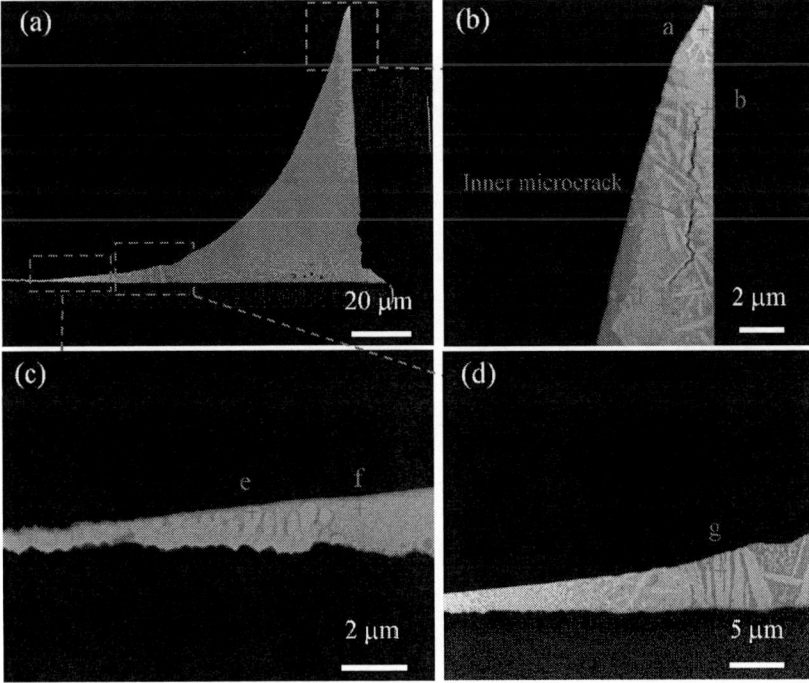

Fig. 3. (a) Cross-sectional microstructure of right angle solder joints fabricated by solder balls with diameter 50 μm and (b)-(d) higher magnification images of the corresponding red dash frame in (a).

978-1-6654-1392-3/21 $31.00 © 2021 IEEE

On the basis of EDX results, table I listed out the elements of some typical points a-g marked by red "+" in Fig.3 (b) – (d). According to the Au-Sn atomic ratio, the phases at the points a-g can be orderly identified to Au_2Sn, $AuSn$, $AuSn_2$, $AuSn_2$, Au_2Sn, Au_2Sn, and $AuSn_7$ [6]. However, there are not any trace of the microcrack at the bottom corner of right angle solder joints in Fig. 3(c) and (d).

TABLE I.   ELEMNTS ANALYSIS FOR POINTS A-G IN FIGS. 3(B)-(D)

Point	Atomic Ratio		
	Au	Sn	Ag
a	69.58	30.42	-
b	50.34	49.66	-
c	29.28	68.37	2.35
d	24.41	72.94	2.95
e	70.32	29.68	-
f	69.39	30.61	-
g	12.05	82.74	5.21

In order to further confirm whether or not there are the inner microcrack nearby the bottom corner of solder joint, another cross-sectional sample of solder joint with serious sagging curve was prepared, and the SEM images are presented in Fig.4. Although there is not any microcrack at the bottom corner, it is obvious that there is a severe shrinkage at there, which is corresponding to the bottom sagging curve. In addition, the orientation of IMCs nearby the shrinkage are perpendicular distributed each other, as marked by the red line in Fig. 4 (c). The regular distribution of IMCs is formed in the solidification course of the molten solder alloy. Even the shrinkage is very serious, there is not any microcrack in the cross-sectional microstructure.

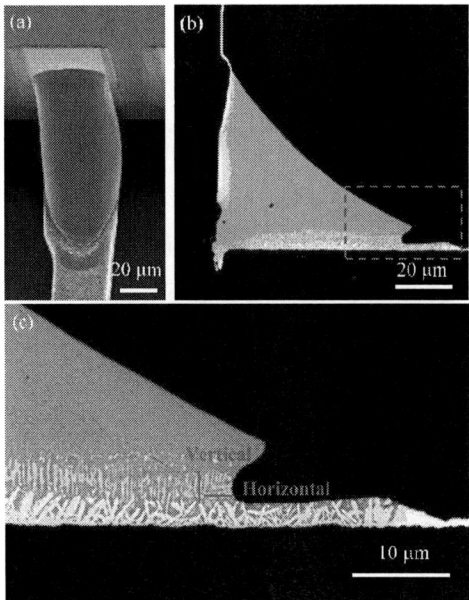

Fig. 4. (a) Outlines of another as-reflowed Au/SAC305/Au solder joint fabricated by 50 μm solder ball, (b) cross-sectional microstructure of sample in (a) and (c) higher magnification image of red dash frame in (b).

B. Analysis

During the solidification course, the liquid solder alloy not only was cooled, but metallurgically reacted with Au layer on the bonding pad. Owing to the gravity effect, there is enough liquid solder alloy to feeding the shrinkage at the bottom corner of solder joint. Therefore, even though the sagging curve is severe at here, there is not any microcrack. However, the liquid solder alloy at the top corner can promptly transfer toward the bottom of solder joint, that is, there is not enough liquid alloy to feeding the atomic shortage during the courses of the solidification and metallurgical reaction. At the same time, the disregistry among the same IMC phases with the different orientations or the different IMCs formed in the solidification course can bring to the higher stress gradient. Under the couple effect of the disregistry and the insufficiently liquid solder ally, the microcrack easily occurred at top corner of solder joints.

IV. CONCLUSIONS

(1) Due to the gravity, the liquid solder alloy at the top part easily transfer to the bottom of right angle solder joints and further lead to the unbalance of the feeding.

(2) The smaller solder joint easily lead to the prompt solidification and the simultaneous growth of different IMCs along the different orientations.

(3) Under the couple effect of the disregistry between the different orientation IMCs and the difficult feeding, the inner microcracks easily occurred at the top corner of right angle solder joint during the prompt solidification, and the reliability of joints is severe damaged.

REFERENCES

[1] Y.C.Chan,; D. Yang. "Failure mechanisms of solder interconnects under current stressing in advanced electronic packages". Prog. Mater. Sci. 2010, 55, 428–475.

[2] W. Yue, M.B. Zhou, and X.P. Zhang, "Influence of solder ball volume on mechanical shock reliability of right-angle solder interconnects", Proceedings of 19th International Conference on Electronic Packaging Technology, Shanghai, China, pp.1472–1475, August 2018.

[3] W. Yue, J.X. Zhang, C.C. Gong, M.B. Zhou and X.P. Zhang, "Identification of essential factors causing solder bridging of right-angle solder interconnects in laser jet solder ball bonding proces", Proceedings of 20th International Conference on Electronic Packaging Technology, Hongkong, China, August 2019.

[4] W. Yue, M.B. Zhou, and X.P. Zhang, "Effect of the Au bonding pad contamination on the wettability of Au/Sn−3.0Ag−0.5Cu/Au solder joints in flux-free laser jet solder ball bonding proces", Proceedings of 17th International Conference on Electronic Packaging Technology, Wuhan, China, pp.1201−1205, August 2016.

[5] W. Liu, C.Q. Wang, M.Y. Li, and L. Ling, "Effect of laser input energy on wetting areas of solder and formation of intermetallic compounds at Sn−3.5Ag−0.75Cu/Au right-angled joint interface", Proceedings of International Conference on Asian Green Electronics, Shanghai, China, pp. 197−201, March 2005.

[6] H.J. Ji, Y.Y. Ma, M.Y. Li, and C.Q. Wang, "Effect of the silver content of SnAgCu solder on the interfacial reaction and on the reliability of angle joints fabricated by laser-jet soldering", J. Electron. Mater, vol. 44, pp.733-743, 2015.

A compact chip filter for 5G communication front-end based on Glass IPD technology

Zhitao Zhang
College of Electronic and Optical Engineering & College of Microelectronics
Nanjing University of Posts and Telecommunications
Nanjing, 210023, China
1219023518@njupt.edu.cn

Ling Gu
College of Electronic and Optical Engineering & College of Microelectronics
Nanjing University of Posts and Telecommunications
Nanjing, 210023, China
1219023426@njupt.edu.cn

Shanwen Hu
College of Electronic and Optical Engineering & College of Microelectronics
Nanjing University of Posts and Telecommunications
Nanjing, 210023, China
shanwenh@njupt.edu.cn

Yuehua Zhang
College of Electronic and Optical Engineering & College of Microelectronics
Nanjing University of Posts and Telecommunications
Nanjing, 210023, China

Xinlei Zhang
College of Electronic and Optical Engineering & College of Microelectronics
Nanjing University of Posts and Telecommunications
Nanjing, 210023, China

Abstract—**A glass-based IPD band-pass chip filter for 5G N77 application is presented in this paper. This filter is composed of five shunt LC resonators. The low loss passband performance is realized by two LC resonators connected in shunt near the input and output terminals. The high rejection property for the stopband is realized by three LC resonators connected in series on the main branch of this filter. This filter is successfully manufactured with a commercial glass IPD process. Its chip size is 1.6×0.8mm². The measurement results show that, the working frequency band of this filter is 3.2-4.0GHz, the maximum passband loss is only 2.0dB, and the stopband rejection is higher than 30dB at the designed zero-point frequencies.**

Keywords—chip filter; 5G filter; band-pass filter; Glass IPD

I. INTRODUCTION

5G mobile communication is now growing in rapid demand due to its capacity of high-speed, high data-rate, and low latency. In the meantime, WiFi, Bluetooth, GPS and other communication systems are also fully utilized in mobile devices to meet the requirement of short-range high speed data transfer, positioning or other communication services. There are more than 50 bands from 600MHz to 6GHz in the mobile device as shown in Fig.1 [1]. To select the desired signal for these multiple communication channels, the band-pass filter is becoming the key component to select the passband signal and reject the sideband interference, which size should be very compact due to the limited circuit board space on a mobile device.

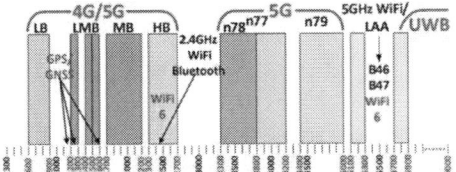

Fig. 1. Sub-6G wireless communication spectrum

A band-pass IPD filter implemented on high resistivity silicon substrate is reported in work [2] for 5G N77 band communication. The maximum loss for the passband is no more than 1.8dB. The minimum rejection for the stopband between 0GHz and 2.7GHz is 36dB, and the minimum rejection for the stopband between 5.2GHz and 5.5GHz is 33dB. However, the chip size is large as 2×1.25 mm². A 5G N78 band-pass filter is reported in work [3] which is also based on silicon IPD process. The insertion loss is less than 1.8dB covering the frequency range from 3.3GHz to 3.8GHz, and the sideband rejection is higher than 30dB at 2.7GHz and 4.9GHz. The chip layout size is 1.6×0.8mm². The Silicon-based IPD filter shows the properties of low cost and compact size due to the capacity of silicon semiconductor process. However, the rejection for the stopband is not very high to meet the requirement of 5G communication devices. The reason is that the quality factor of the inductor on silicon substrate is still limited. To achieve high Q passive device, the highly isolated glass substate is then adopted to manufacture RF filters. Three glass-based IPD band-pass chip filters are reported in work [4-6] for 2.0GHz, 2.4GHz and 4.2GHz respectively. In this work, the Glass IPD technology is adopted to design and implement a 5G N77 band-pass filter which covers 3.3-4.2GHz frequency band. This filter is manufactured on glass substrate with 1.6mm×0.8mm chip size. The targeted passband width is up to 900MHz with no more than 2.0dB loss, and the stopband rejection is high as 42dB at 1.8GHz and 50dB at 5.2GHz.

II. GLASS IPD TECHNOLOGY

The proposed glass IPD filter is designed and implemented on a 230um glass substrate with thick backside metal. A simplified cross-sectional view of the glass IPD process used in this work is shown in Fig.2.

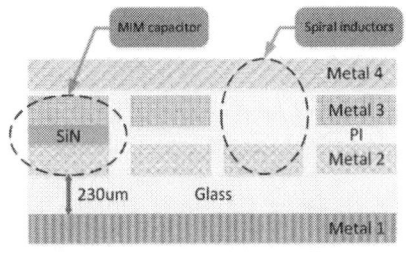

Fig. 2. Structure of Glass IPD

978-1-6654-1392-3/21 $31.00 © 2021 IEEE

A highly isolated glass substrate is adopted to realize high Q passive devices. The thickness of the glass substrate is 230um. A backside metal (Metal 1) is used to further improve RF performance of the devices, which is under the bottom of the glass substrate. Passive capacitors and inductors can be manufactured using dielectric layers and metal layers on this glass IPD substrate. As shown in Fig.2, the capacitor is realized with the top metal (Metal 3), the bottom metal (Metal 2), and the low loss dielectric layer (SiN). The spiral inductor can be realized with the bottom metal layer (Metal 2) or the top metal layer (Metal 4). Metal 2 and Metal 4 can be connected together by drilling through a via hole in-between these two metal layers. A thick metal is then available to be used as the wire of the spiral inductor.

III. FILTER DESIGN

The topology view of the filter circuit is illustrated in Fig.3. The proposed band-pass filter is composed of five shunt LC resonators. Resonators 1&2&3 are used to generate three zeros for the filter to improve the stopband rejection. Resonators 4&5 are used to generate two poles for the filter to improve the passband behavior.

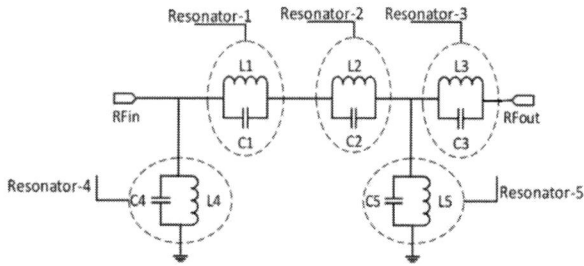

Fig. 3. Circuit topology of the IPD chip filter

The zero points of this IPD chip filter are dependent on the inductors and capacitors of the zero resonators as shown in equation (1). The good stopband rejection can be realized by tuning the value of the passive elements in the zero resonators to change the zero-point frequencies. L1 and C1 generate the first zero frequency to improve the sideband rejection for the frequency range higher than the passband. The shunt oscillation frequency is 5.8GHz, which is the frequency band for 5.8GHz WiFi applications. L2 and C2 generate the second zero frequency to improve the sideband rejection for the frequency range lower than the passband. The shunt oscillation frequency is 2.4GHz, which is the frequency band for 5.4GHz WiFi applications. L3 and C3 generate the third zero frequency to improve the sideband rejection for the frequency range lower than 2GHz. The shunt oscillation frequency is 1.8GHz, which is the frequency band for 3G and 4G applications. Finally, the resonate frequencies are set as 1.8GHz and 2.4GHz and 5.8GHz to achieve good rejection performance for this 5G n77 band-pass filter to suppress interference signals from 3G, 4G-LTE and WiFi channels on the mobile device.

$$ f_{Z1} = \frac{1}{2nf\sqrt{L_1C_1}} \quad f_{Z2} = \frac{1}{2nf\sqrt{L_2C_2}} \quad f_{Z3} = \frac{1}{2nf\sqrt{L_3C_3}} \quad (1) $$

The pole points of this IPD chip filter are dependent on the inductors and capacitors of the pole resonators as shown in equation (2). The good passband performance can be realized by tuning the value of the passive elements in the pole resonators to change the pole-point frequencies. L4 and C4 generate the first pole frequency to generate a pole at 3.6GHz

for the passband. L5 and C5 generate the second pole frequency to generate a pole at 4.0GHz for the passband. Therefore, the 5G N77 passband (3.3-4.2GHz) can be realized by carefully tuning these two poles of this design.

$$ f_{p1} = \frac{1}{2nf\sqrt{L_4C_4}} \quad f_{p2} = \frac{1}{2nf\sqrt{L_5C_5}} \quad (2) $$

The 5G N77 glass IPD filter is finally designed and simulated based on above architecture and design methodology. The S parameters simulation result is shown in Fig. 4. The 1-dB frequency band covers from 3.3GHz to 4.2GHz. S21 is less than 2.0dB during the whole passband. The stopband rejection is very low as -43dB@1.8GHz, -32dB@2.4GHz, and -36dB@5.8GHz. The impedance matching is very good with a S11 lower than -20dB during the passband. Consequently, a band-pass filter for 5G N77 application is well designed and realized base on glass IPD technology with good passband and stopband performance.

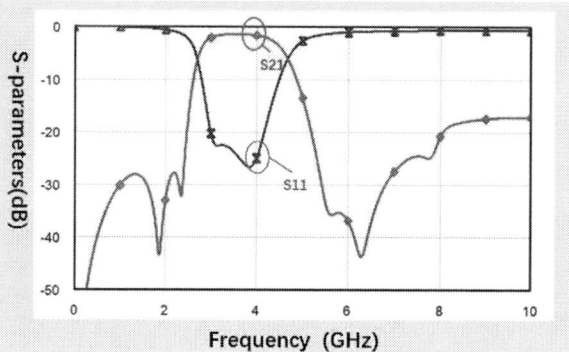

Fig. 4. Simulation result of the proposed filter

IV. IMPLEMENTATION AND MEASUREMENT

The 5G N77 band-pass filter is manufactured on a commercial IPD process on a 230um glass substrate with thick backside metal. The backside meatal is Metal 1 which thickness is 7μm. The bottom metal plate for passive capacitor is Metal 2 which thickness is also 7μm. The top metal plate for passive capacitor is Metal 3 which thickness is 5μm. The top metal for passive inductor is Metal 4 which thickness is also 5μm. This 5G N77 band-pass filter is designed based on the architecture discussed in above sections, and it has been successfully manufactured on this process as illustrated in Fig.5. The total area size of this 5G N77 band-pass filter including scribe line is only 1.6×0.8mm². There are four bonding pads on the fabricated chip filter. The pad at the left bottom edge of the chip is used as input pad to feed the input signal into the filter. And the output pad is located at the right bottom edge of the chip to send out the signal through the filter. There are two pads located at the left and right top edge of the chip to connect ground through bonding wires.

Fig. 5. Photo of the 5G N77 band-pass filter

This 5G N77 band-pass filter is measured on a printed circuit board. The proposed chip is connected to the circuit board through bonding wires. RF SMA connectors are used to connect the circuit board and the network analyzer. S parameters of this 5G N77 filter are measured as illustrated in Fig. 6. The passband frequency range of this filter covers from 3.2GHz to 4.0GHz. The total range is 800MHz, which is narrower than the 900MHz design target (3.3-4.2GHz). The first reason is that the whole passband shift about 100MHz to low frequency band due to the mismatching between the Electromagnetic (EM) simulation results and the practical testing results. The second reason is that S21 decreases immensely at high frequency because of that quality factor of the inductors and capacitors on the glass substrate gradually become smaller with the increasing frequency. The maximum value of S21 during the passband is about 2.0dB, and it is only 1.6dB at the center frequency of the passband. S11 of this 5G N77 band-pass filter is lower than -10dB from 3.2GHz to 4.0GHz, and the minimum value is only -23dB at 3.6GHz. Consequently, the input reflection of this filter is very low. Three transmission zeros are located at 1.8GHz, 2.3GHz and 5.2GHz respectively. The zero point at 2.4GHz is shifted to 2.3GHz, and the zero point at 5.8GHz is shifted to 5.2GHz. The reason is also caused by that quality factor of the inductors and capacitors on the glass substrate gradually become smaller with the increasing frequency. The realized stopband rejection is -42dB at 1.8GHz, -33dB at 2.3GHz and -51dB at 5.2GHz.

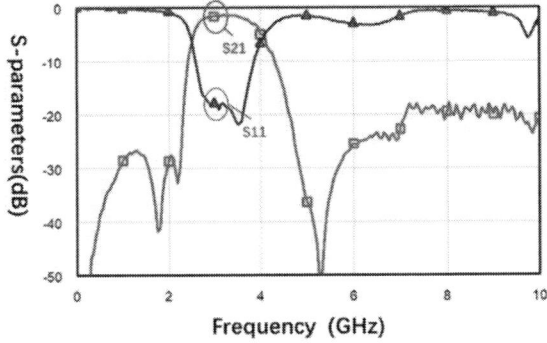

Fig. 6. Measurement results of the proposed filter

CONCLUSION

A glass IPD filter is presented for 5G N77 mobile communication in this work. Five shunt LC resonators is adopted to build the main circuit topology of this filter. Two LC resonators are used as pole resonators to ensure the passband performance, and the other three LC resonators are used as zero resonators to improve the stopband rejection properties. The architecture of this 5G N77 band-pass filter is carefully designed and analyzed. This filter is then successfully manufactured on a glass substrate. The total area size of this 5G N77 band-pass filter including scribe line is only $1.6 \times 0.8 \text{mm}^2$. The passband frequency range of this filter covers from 3.2GHz to 4.0GHz. The maximum value of S21 during the passband is about 2.0dB, and it is only 1.6dB at the center frequency of the passband. S11 of this 5G N77 band-pass filter is lower than -10dB from 3.2GHz to 4.0GHz, and the minimum value is only -23dB at 3.6GHz. The realized stopband rejection is -42dB at 1.8GHz, -33dB at 2.3GHz and -51dB at 5.2GHz.

REFERENCES

[1] Balteanu F., Modi H., Choi Y., et al. 5G RF Front End Module Architectures for Mobile Applications, 49th European Microwave Conference (EuMC). 2019.

[2] X. Li, Song J., Wang D , et al. A Highly Selective and Compact 5G n77 Band Pass Filter Based on HRS IPD Technology, IEEE MTT-S International Microwave Workshop Series on Advanced Materials and Processes for RF and THz Applications (IMWS-AMP). IEEE, 2020.

[3] K. R. Shin and K. Eilert, Compact low cost 5G NR n78 band pass filter with silicon IPD technology, IEEE 19th Wireless and Microwave Technology Conference (WAMICON), 2018.

[4] C.-H. Huang, C.-H. Chen, et al., Compact bandpass filter using novel transformer-based coupled resonators on integrated passive device glass substrate, Microw. and Opt. Tech. Lett. 54 (1) (2012) 3-7.

[5] C.-H. Chen, C.-S. Shih, et al., Very Miniature Dual-Band and Dual-Mode Bandpass Filter Designs on an Integrated Passive Device Chip, Prog. in Electromag. Reser. 119 (2011) 461-476.

[6] Y.-C. Tseng, T.-G. Ma, On-Chip GIPD Bandpass Filter Using Synthesized Stepped Impedance Resonators, IEEE Microw. Wireless Compon. Lett. 24 (3) (2014) 140-142.

Encapsulation Techniques of Perovskite Solar Cells

Qi Wu
School of Electronic Science and Engineering,
Xiamen University,
Xiamen, China
wuqii@stu.xmu.edu.cn

Wenfeng Li
School of Electronic Science and Engineering,
Xiamen University,
Xiamen, China
36120201150364@stu.xmu.edu.cn

Mengyu Chen*
School of Electronic Science and Engineering,
Xiamen University,
Xiamen, China
mychen@xmu.edu.cn

Cheng Li*
School of Electronic Science and Engineering,
Xiamen University
Xiamen, China
Future Display Institute of Xiamen
Xiamen , China
chengli@xmu.edu.cn
ORCID: 0000-0003-0282-7899

Abstract—Due to the rapid development of solar cells based on organic metal halide perovskite as light-absorber, the power conversion efficiency (PCE) of perovskite solar cells (PSCs) has increased from 3.8% to 25.5%, which exceeds that of silicon ones. In the meantime, the stability of the device has also made breakthrough progress. However, the degradation of perovskite is the primary reason which largely limits large-scale application. PSCs quickly chemically react with the water/oxygen in the air and decompose when exposed to ambient condition. It is urgently necessary to develop appropriate encapsulation techniques to isolate perovskite from external stimuli including moisture, oxygen and UV light to effectively extend lifetime. In this paper, the influences and physical mechanisms of the stability of perovskite are summarized. Then, the main encapsulation methods and technical approaches to improve and enhance the stability of perovskite are also reviewed, including traditional rigid-substrate-based encapsulation, thin film encapsulation, and advanced encapsulation strategies. Finally, it overviews the challenges and further development strategies faced by PSCs encapsulation techniques in practical applications.

Keywords—Perovskite solar cells; Encapsulation; Stability

I. INTRODUCTION

With the brilliant optoelectronic properties of perovskite materials and the efforts from researchers, the National Renewable Energy Laboratory (NREL) announced that the latest certification efficiency of single-junction perovskite solar cells (PSCs) have reached 25.5%, far exceeding the other third-generation solar cells[1]. Meanwhile, the cost of PSCs has made a breakthrough, reducing to half of the cost of silicon-based solar cells[2]. The preparation process of PSCs is simple, low-cost, and can be prepared in solution. Therefore, PSCs have demonstrated a quite promising industrialization prospect as a strong competitor of existing commercial silicon based solar cells[3]. PSCs are still facing many difficulties to be solved urgently, such as poisonous, humidity sensitivity and difficulty of large area preparation, especially poor stability[4]. In addition, the grain boundary and surface defects of perovskite film and the interface defects between hole transporting layers, severely restrict the performance and stability of PSCs[5]. In the case, the stability of the perovskite material has been improved by many relevant works. The stability and photovoltaic performance of PSCs have been improved and enhanced through passivating the perovskite

film grain boundaries and structure defects by adding metal ions, organic molecules or hydrophobic group materials[6].

Nevertheless, there are several factors that influence the stability of PSCs. For example, moisture and oxygen permeating from ambient atmosphere can react chemically with perovskite materials, which cause irreversible degradation of the perovskite. PSCs not only need to be resistant to high temperature, but also need to endure alternating heat and cold. In addition, high temperature accelerate the degradation process of PSCs. As a result, temperature matters a lot in the stability of PSCs. Surprisingly, ultraviolet light in sunlight also make a difference. Based on the current research, on the one hand, UV light causes more defects in electron transport layer. On the other hand, it also leads to ion migration in the perovskite light absorption layer which degrades the performance of the device[7].

Therefore, reliable encapsulation technology is the effective way to maintain and even enhance device stability. In this paper, the mechanisms of degradation of PSCs are summarized. Then, the main encapsulation methods and technical approaches to improve and enhance the stability of perovskite are also reviewed, including traditional rigid-substrate-based encapsulation, thin film encapsulation, and advanced encapsulation strategies. Finally, by comparing various encapsulation strategies on PSCs, not only the challenges faced by PSCs encapsulation techniques in practical applications are overviewed but also new ideas and development directions for encapsulation strategies of PSCs are proposed.

II. DEGRADATION MECHANISM

As the PCE and cost of PSCs have met the requirements of commercialization, now the stability is the key to realize large-scale commercialization. The degradation mechanism of PSCs is mainly divided into two respects: internal degradation mechanism and external degradation mechanism.

In addition to the instability of perovskite materials, the different functional layers such as electron transport layer (ETL), hole transport layer (HTL) and the interface between each other also affects the stability. Currently, there have been lots of research works to solve problems in the internal degradation mechanism. For instance, Hitoshi Nishino et al. used Sb_2S_3 nanocrystals modified $TiO_2/MAPbI_3$ interface to significantly slow down the photodecomposition of $MAPbI_3$,

978-1-6654-1392-3/21 $31.00 © 2021 IEEE

thereby improving the light stability of the device[8]. Chen Wei et al. employed p-type $CuGAO_2$ nanosheet to replace the traditional hole transport material 2,2',7,7'-Tetrakis[N,N-di(4-methoxyphenyl)amino]-9,9'-spirobifluorene(Spiro-OMeTAD) and prepared a planar PSC with $TiO_2/CH_3NH_3PBI_{3-x}Cl_x/CuGAO_2$ structure. The high conductivity of $CuGAO_2$ makes the PCE as high as 18.51%. After 30 days of storage in the indoor environment, the PCE only decayed by 10%[9].

However, the external degradation mechanism of PSCs also have an effect. Perovskite materials are easily to denature or degrade in the condition of external environment. The main external environment affection factors of stability of PSCs are as follows: a) water, b) oxygen, c) temperature. Most of the effects caused by external degradation mechanism can be addressed by high quality and compact encapsulation[10].

a) Water. Water vapor is the main factor in the irreversible degradation of perovskite. Walsh et al. proposed a degradation mechanism of perovskite[11]. In the condition of H_2O, the perovskite ($CH_3NH_3PbI_3$) would degrade to produce HI, CH_3NH_2 and PbI_2. The HI and CH_3NH_2 would continue to dissolve in the extra water, accelerate the degradation reaction, lead to its irreversibility, and then cause the degradation and death of solar cells device performance.

b) Oxygen. Under light illumination, the perovskite produces photo-generated carriers. Since perovskite in the excited state is very sensitive to oxygen, perovskite is prone to cause photo-oxidation reaction under light[12]. The specific photo-oxidation mechanism is as follows (as shown in Fig.1.(a))[13]: (1) Diffusion and incorporation of oxygen into the lattice, (2) creation of electrons and holes by photoexcitation of $CH_3NH_3PbI_3$, (3) formation of superoxide from O_2, and (4) PbI_2, H_2O, I_2 and CH_3NH_2 by reaction and degradation

c)Temperature. Organometal halide perovskite materials generally have lower crystal formation energy and lower chemical bond energy than conventional photovoltaic semiconductor materials, and are susceptible to thermal interference and change. Divitini et al. observed the in-situ thermal degradation of perovskite (as shown in Fig.1. (b)). Even at low temperatures, iodide ions can migrate into the hole transport layer[14].

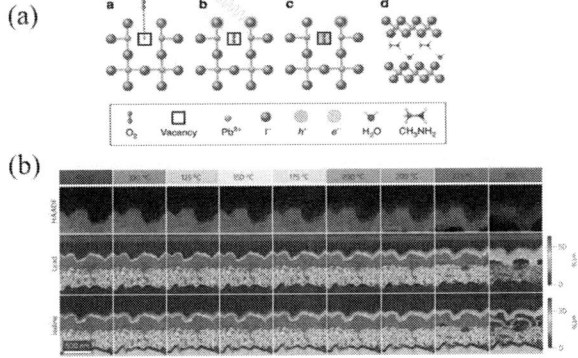

Fig. 1. a)Schematic representation of the reaction steps of O_2 with $CH_3NH_3PbI_3$. Reprinted with permission from [13]. b)HAADF images and EDX elemental maps of iodine and lead from sample after heating at different temperatures. Reprinted with permission from [14].

III. ENCAPSULATION TECHNIQUES OF PSCs

As the poor stability of perovskite has attracted widespread attention, scientists have paid effort to improve the

stability of PSCs and proposed various effective strategies, which can be roughly divided into the following three ways: (a) Crystal engineering, that is, through doping to improve its structural stability; (b) Surface engineering, that is, through surface modification to improve stability; (c) Device encapsulation, that is, employing suitable encapsulation process to extend the working life of the device.

At present, the stability of PSCs is indeed improving, but it is still far from reaching the standards for commercial applications. Therefore, it is necessary to extend the life of the device by encapsulation. The encapsulation techniques of PSCs can be mainly divided into two ways: traditional rigid-substrate-based encapsulation which imitates the silicon-based photovoltaic devices, and the other one is the thin-film encapsulation (TFE).

A. Traditional rigid-substrate-based encapsulation

Due to the chemical instability of organic materials, most organic optoelectronic devices are highly susceptible to corrosion by water and oxygen in the air, resulting in a rapid decline in device performance. The temperature and high relative humidity (RH) environment further accelerate the degradation effects. In order to ensure the long-term working life of the device, it needs to be encapsulated to isolate water and oxygen from the outside. In addition, the use of high-transmittance materials for encapsulation can maximize work efficiency for optoelectronic devices with photoelectric conversion as a working mode. The traditional encapsulation method is to employ a rigid substrate (such as a transparent glass cover) to encapsulate. The traditional rigid-substrate-based encapsulation is divided into full-encapsulated structure (as shown in Fig.2. (a)) and half-encapsulated structure (as shown in Fig.3. (a)).

The full-encapsulated structure and the half-encapsulated structure have their own advantages and disadvantages. The full-encapsulated structure equipped with stronger bonding ability, could withstand greater impact force and be operated in the air. However, it costs more materials, resulting in poor heat dissipation performance. Although the half-encapsulated structure owns lower material cost and better heat dissipation performance, compared with the former, its bonding ability and impact force are poorer, and the encapsulation process cost is higher.

1) Full-encapsulated structure

The full-encapsulated structure with encapsulation adhesive and backplane refers to the glass, hot melt adhesive film, solar cell, and hot melt adhesive film backplane placed in sequence on the laminator including heating the laminator to make the solar cell, glass and the backplane firmly bond together. Fig.2. (a) shows an typical encapsulation process of full-encapsulated structure.

Fu et al. proposed a method of encapsulating PSCs by hot melt adhesive film with glass cover and completed a stability test for more than 2000 h under outdoor conditions. Compared the performance of various encapsulation structures, encapsulation materials (polyurethane, PU; polyolefin elastomer, POE; ethylene vinyl acetate, EVA) and encapsulation processes, they employed the full-encapsulated structure and PU film to encapsulate printable PSCs made by screen-printing techniques, as shown in Fig.2. (b)[15]. The PCE of encapsulated large area (100 cm2) printable PSCs only decayed by 2.48% after 2136 h of outdoor work. They believed that between the changes of PSC interfaces and degradation of perovskite layer both caused by high temperature, the former is the mainly reason for the decrease of PSC efficiency.

The structure used by McGehee et al.[16] is still a full-encapsulated structure, but the difference from the above is that they deposited PSC directly on the glass substrate and then employed transparent indium tin oxide (ITO) electrode as an isolation layer to separate PSCs from metal. ITO performs as a barrier to ensure that there are 2 mm lateral space between PSCs and electrode (as shown in Fig.2. (c)), which would prevent organics from escaping into perovskite layer and reacting with Ag electrode. The factors caused PSCs degradation including reaction with metals, as well as moisture[17]. Consequently, the problems of minimization of moisture permeation and elimination of vacancy formed by cell movement are solved by direct deposition of electrical feedthroughs on one side of glass cover. Their encapsulated devices not only maintained the original efficiency after 200 thermal cycling tests, but also the PCE decayed less than 3% after 1000 h damp heat test.

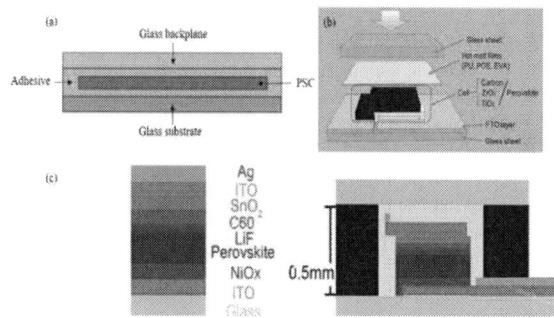

Fig. 2. a) Cross section of the full-encapsulated structure. b)Schematic representation of the encapsulation of printable PSCs. Reprinted with permission from [15]. c) Left : PSC stack in this study, Right: Cross section of encapsulation structure. Reprinted with permission from [16].

2) Half- encapsulated structure

The half-encapsulated structure with encapsulation adhesive and backplane is typically employed by organic light emitting diode (OLED). The half-encapsulated structure refers to the substrate and the backplane are bonded together with encapsulation adhesive on the blank area around the device prepared on the substrate. There is desiccant attached to backplane. Fig.3.(a) shows a typical encapsulation process of half-encapsulated structure.

The ternary mixed perovskite $Cs_{0.05}(MA_{0.17}FA_{0.83})_{0.95}Pb(I_{0.83}Br_{0.17})_3$ is optimized to $Rb_{0.05}Cs_{0.05}FA_{0.75}MA_{0.15}Pb(I_{0.95}Br_{0.05})_3$ perovskite through doping with Rb by Tai et al. , which significantly improved the thermal stability of the perovskite materials. They placed glass cover with water-absorbing sealant (HD-S051414W-40, Dynaic) on the device in condition of dry air (<2%RH). And then the substrate and glass cover were bonded by epoxy resin for 15 min with UV treatment (as shown in Fig.3.(b)). In spite of 1000 h of storage at 85°C/85% RH, the device can still maintain 92% of its initial efficiency [18].

Qi Dong et al. deposited a 50 nm of SiO_2 layer by using electron beam deposition, which can protect electrode effectively and a thin layer of graphene oxide in sequence on the active of the PSC, and then covered a glass with a piece of desiccant to protect the PSC and finished encapsulation process (as shown in Fig.3.(c))[19]. In addition, they

compared three different epoxies to be the sealing materials including UV-curable epoxy, AB epoxy glue (obtained from Super Glue Corp), and thermal curing epoxy in encapsulation strategies. It turns out that the UV-curable epoxy lead to the best performance, and after exposed to light soaking at 858°C for 48 h , the PCE of PSCs can still maintain about 80% of initial efficiency.

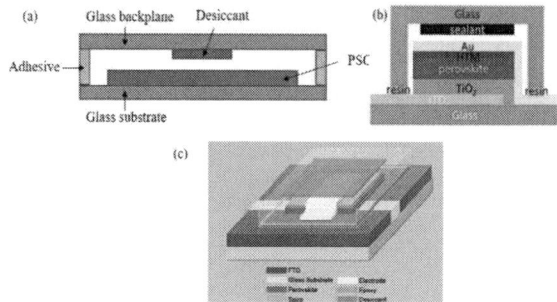

Fig. 3. a) Cross section of the half-encapsulated structure. b) Cross section of encapsulated cell. Reprinted with permission from [18]. c) Schematic representation of the encapsulated cell. Reprinted with permission from [19].

B. Thin film encapsulation

In traditional encapsulation strategies, the glass substrate and the encapsulated glass cover account for more than 90% of the overall thickness[20]. In addition, traditional encapsulation strategies request higher material and process cost. In recent years, with the popularization of ultra-thin flexible devices, the method of glass-to-glass encapsulation is no longer applicable sufficiently. In addition to the organic composition, there is also a high demand for the water vapor transmittance rate (WVTR) of the encapsulation method for flexible PSCs device. More importantly, flexible applications require the encapsulation materials with corresponding bending radius[21]. Due to insufficient flexibility of the glass cover, it is quite inappropriate for flexible devices applications. Furthermore, the encapsulation adhesive used in traditional encapsulation will become one of the paths for water vapor to penetrate from edge sides, which limits the WVTR of the whole encapsulation system.

Therefore, it is necessary to explore new encapsulation methods, that is, thin film encapsulation(TFE) [20]. Compared with traditional encapsulation, TFE directly deposits encapsulation materials on the device to avoid the use of encapsulation adhesive. Therefore, TFE has gradually become the first primary choice for flexible PSC devices to block water and oxygen erosion, and it has also improved the stability of PSCs effectively. The process of TFE has experienced the upgradation from the early sputtering[22] to plasma chemical vapor deposition (PCVD)[23], and the film density and encapsulation performance have been greatly improved. In further research and development, atomic layer deposition (ALD) has been proved to replace PCVD as a reliable method for growing high quality inorganic encapsulation films in the laboratory[24,25]. However, the main bottleneck problem at present is the low growth rate.

1) Spin-coating

Spin coating is the simplest laboratory method for rapidly producing thin films. The precursor is added to the surface of substrate, and then the solution spread in a thin film driven by centrifugal force, forming a uniform coating layer on the substrate. A novel and simple passivation process overcoming poor stability in ambient atmosphere by using a hydrophobic polymer layer for PSCs is reported by Hwang et al. They successfully enhanced the water resistance of PSCs by depositing a Teflon hydrophobic polymer to cover PSCs with spin-coating process (as shown in Fig.4 (a)[26]. Furthermore, X-ray diffraction, light absorption spectrum, and quartz crystal microbalance are employed to detect changes in the perovskite layer, so as to get more details about the degradation of PSCs. The PCE of PSCs only decayed by 5% after storage in the ambient air for 30 days.

Bella et al. coated a self-cleaning multifunctional fluorinated photosensitive polymer on the front surface of perovskite solar cells (as shown in Fig.4.(b))[27], which can convert UV light into visible light to avoid the damage from UV light and enhance the anti UV ability of devices. In the meantime, a strong hydrophobic photosensitive polymer coated on the back of the device enhanced the water resistance ability of the device and significantly improved the stability of the device. The efficiency of the device decreased less than 5% after being stored outdoors for 3 months.

2) Plasma chemical vapor deposition

The plasma chemical vapor deposition (PCVD) is a method to deposit a thin film on a substrate by chemically reacting gaseous substances containing the atoms of the film by means of plasma. The PCVD makes it possible to deposit denser, pinhole-free films of higher quality than traditional chemical vapor deposition (CVD) at lower temperature, which avoid the degradation of PSCs caused by exposure of oxygen, moisture, UV, and chemicals. As a consequence, PCVD is an ideal choice for encapsulating solutions of flexible PSCs. Submicron-thick organosilicate barrier films are deposited on the surface of PSCs with a scalable spray plasma process by Rolston et al. After over 3000 h of storage in dry heat (85℃, 25%RH), the PCE of encapsulated PSCs just decayed by 8%(as shown in Fig.4.c)[28]. In the meanwhile, there are not visible cracking or delamination of flexible PSCs after 10 000 bending cycles where the bending radius is 1 cm, proving that stability of PSCs are enhanced significantly by PVCD. Moreover, the spray plasma vapor deposition process can produce an uniform layer with a width of 1 m, which is promising for industrial mass production.

Although the inorganic encapsulation films processed by low-temperature have made remarkable progress, these films as encapsulation barriers still have inherent defects, which make quite bad influence on the performance of PSCs. Moisture and oxygen can easily diffuse through these defects into PSCs. Multi-layer coating barrier structure can maintain stable moisture and oxygen permeability, so that the high-performance of encapsulation barriers can be obtained. In other words, alternating use of organic and inorganic multilayer films is a promising method to decrease and prevent intrinsic pinholes.

3) Atom layer deposition

Similar to the PCVD process, atomic layer deposition (ALD) is a method of introducing different reaction precursor by alternately pulsed gas phase into reaction chamber, chemically absorbed and reacted on the deposition substrate to form a dense thin film in the form of a monoatomic film. The thickness and uniformity controliability of films prepared by ALD process make it advantageous for encapsulation[29].

Ruud E. I. Schropp et al. found that after coating ultra-thin Al_2O_3 on the surface of the perovskite layer, the PSCs were effectively protected and showed excellent device performance[30]. The PCE of PSCs reached 18%, while the hysteresis loss was also significantly reduced. After exposure to the devices containing 10 cycles of ALD AL_2O_3 for 70 days in humid environment, the PCE measurement showed a significant delay in humidity-induced degradation.

Considering that most optoelectronic devices cannot withstand high-temperature processing and the high thermal sensitivity of PSCs, the temperature of ALD process must be further reduced when preparing encapsulation films so as to avoid the damage caused by high temperature to devices. The organic and inorganic films composed of poly (1,3,5-trimethyl-1,3,5-trivinyl cyclotrisiloxane) (pV3D3) /Al_2O_3 layer are alternatively deposited by LEE et al. on the PTAA-based PSCs by CVD and ALD process under the low temperature at 40℃ and 60℃, respectively (as shown in Fig.4.(d)). Eventually, the WVTR of devices reached 10-4gm-2d-1 in the condition of 38 ℃ and 90% RH. Additionally, despite 300 h of stability test, PSCs still maintain 97% of their initial PCE[31].

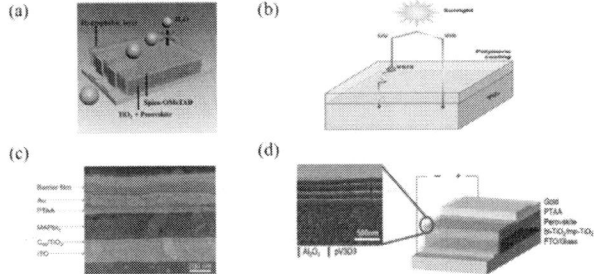

Fig. 4. (a) A PSC with hydrophobic polymer deposited on top by a spin-coating method. Reprinted with permission from [26]. (b) Scheme of the UV-coating operating principle. Reprinted with permission from [27]. (c) A cross-sectional SEM image of the PSCs with submicron-thick organosilicate barrier films. Reprinted with permission from [28]. (d) Left: The cross-sectional SEM image of the TFE, Right: Cross section of the encapsulated PSC. Reprinted with permission from [31].

ALD can form a dense layer to prevent diffusion of oxygen and moisture, coupled with the chemical attack of the decomposed components of the perovskite layer on the metal electrode. Nevertheless, the high equipment cost and time-consuming deposition limit the commercial application of ALD.

C. Advanced encapsulation

Emami et al. reported an advanced laser-assisted glass-frit encapsulation technique to encapsulate PSCs without HTL (as shown in Fig.5.(a))[32]. The HTL-free PSCs encapsulated with glass frit have completed 70 thermal cycles (-40℃-85℃) and damp heat (85℃, 85%RH) tests for 50 h, resulting in 2.3% increase and 0.9% decrease in PCE, respectively. When exposed to humid environment (80±5% RH), the PCE of the sealed PSCs maintained constant for 500 h. The big limitation of laser-assisted glass frit encapsulation is the high process temperature (100℃), which makes extremely bad influence on the PSCs with poor thermal stability. The material cost of glass frit extremely hinders its commercial application in encapsulation, more attention needs to be paid to lowering down the encapsulation temperature in the future.

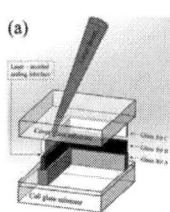

Fig. 5. a) Schematic view of the triple layer glass frit laser-assisted sealing configuration. Reprinted with permission from [32].

The semitransparent PSCs sputtered transparent conductive oxide (TCO) as top electrode are proved to be stable in long-term operation. Cheng et al. proposed a method of depositing a ITO solid electrode on PSCs using room temperature sputtering technique[33], which can prevent the "metal-perovskite" reaction[34] and improve the stability of the device effectively. As the traditional vacuum sputtering of ITO needs a higher deposition process temperature (over 300°C), the perovskite film could be severely damaged at this temperature. Therefore, they developed a room temperature sputtering ITO technology suitable for PSCs to replace metal electrodes. The sunlight from front and back could be absorbed by double-sided PSCs with translucent-ITO-based electrode. Therefore, PSCs still exhibit excellent stability after storage under ambient light in the glovebox filled with N2 for more than 2000 h.

IV. DISCUSSION

Although the intrinsic factors for degradation of PSCs are unavoidable, the encapsulation still contribute in the stability of PSCs. By using appropriate encapsulation techniques, the decline of PCE can be slowed down in a certain extent. The above reports indicate that stabilizing the chemical property of perovskite, modifying interfaces and developing robust electrodes, along with reliable encapsulation strategies are effective ways to enhance the stability of PSCs, which drive PSCs fast toward commercialization.

Rigid glass and encapsulation adhesive are used as the barrier to oxygen and moisture in traditional encapsulation strategies. Traditional encapsulation owns the advantages of low cost, reliable stability, and technical maturity, etc. However, in recent years, with the popularization of ultra-thin flexible devices, the method of glass-to-glass encapsulation is no longer applicable sufficiently. The encapsulation of flexible PSC has attracted more attention from the society.

The encapsulation processing temperature of TFE need to be accurately controlled within the processing temperature range of organic and polymer substrate materials. It is hard to control the permeation of localized pin-holes which is unavoidable in the barrier film. So new materials and process that meet the demands of encapsulation are required to isolate oxygen and moisture for the stability of PSCs. The encapsulation with multilayer polymer films can isolate water vapor and oxygen by blocking the pin-holes in films. However, the cost of depositing multilayer thin film encapsulation structures under vacuum is too expensive. Solution process deposition can be a hopeful alternative technique for low-cost thin multilayer polymer films encapsulation. In addition, the quality of encapsulation layers also matters a lot to ensure the stability of PSCs.

Additionally, we are also looking for advanced encapsulation employed for PSCs. For example, combination of encapsulation with light confinement structures maybe a promising solution to reduce optical loss or enhance light-harvesting ability in order to improve stability. Employing new encapsulation materials, trying advanced encapsulation structure are the mainly ways to propose advanced encapsulation solution for PSCs.

Not only materials cost and processing cost need to be considered in the encapsulation of commercial application, but also processing time, which also seriously restraints large-scale application. In the future, attention should be paid to find more appropriate encapsulation materials and structures on the basis of existing research achievements. Additionally, encapsulation materials are required to be with high transmittance, low Young's modulus, low WVTR, and harmless byproduct. With the progress made in the past few years, the optimistic outlook of PSCs is expected. There is no doubt that effective encapsulation will accelerate the commercialization process of PSCs.

V. CONCLUSION AND OUTLOOK

In conclusion, PSCs have attracted extensive attention as a low-cost and efficient solar cells, and become the focus of commercialization gradually. Recently, with the rapid increase in PCE of PSCs, the stability issues turn into the main obstacle of large-scale commercialization. In this paper, the influences and physical mechanisms of the stability of perovskite are summarized. Then, the main encapsulation methods and technical approaches to improve and enhance the stability of perovskite are also reviewed, including traditional rigid-substrate-based encapsulation, thin film encapsulation, and advanced encapsulation strategies. Finally, it overviews the challenges and further development strategies faced by PSCs encapsulation techniques in practical applications.

ACKNOWLEDGMENT

This work is financially supported by National Natural Science Foundation of China (61974126, 51902273, 62005230) and Fundamental Research Funds for the Central Universities, No.20720200086.

REFERENCES

[1] National Renewable Energy Laboratory. Best Research-Cell Efficiency Chart (National Renewable Energy Laboratory, 2021). https://www.nrel.gov/pv/assets/images/efficiency-chart.png.

[2] Z. N. Song et al., "A technoeconomic analysis of perovskite solar module manufacturing with low-cost materials and techniques," Energy Environ. Sci., vol. 10, pp. 1297-1305, June 2017.

[3] K. Domanski, E. A. Alharbi, A. Hagfeldt, M. Gratzel, and W. Tress, "Systematic investigation of the impact of operation conditions on the degradation behaviour of perovskite solar cells," Nat. Energy, vol. 3, pp. 61-67, January 2018.

[4] M. A. Green, A. Ho-Baillie, and H. J. Snaith, "The emergence of perovskite solar cells," Nat. Photonics, vol. 8, pp. 506-514, July 2014.

[5] J. Huang, Y. Yuan, Y. Shao, and Y. Yan, "Understanding the physical properties of hybrid perovskites for photovoltaic applications," Nat. Rev. Mater., vol. 2, pp. 58-62, 2017

[6] Z. J. Shi et al., "Lead-Free Organic-Inorganic Hybrid Perovskites for Photovoltaic Applications: Recent Advances and Perspectives," Adv. Mater., vol. 29, p. 28, 2017.

[7] N. Li, X. Niu, Q. Chen and H. Zhou, "Towards commercialization: The operational stability of perovskite solar cells," Chem. Soc. Rev., vol.49, pp. 8235-8286, November 2020.

[8] S. Ito, S. Tanaka, K. Manabe, and H. Nishino, "Effects of Surface Blocking Layer of Sb2S3 on Nanocrystalline TiO2 for CH3NH3PbI3 Perovskite Solar Cells," J. Phys. Chem. C, vol. 118, pp. 16995-17000, July 2014.

[9] H. Zhang, H. Wang, W. Chen, and A. K. Jen, "CuGaO2 : A Promising Inorganic Hole-Transporting Material for Highly Efficient and Stable Perovskite Solar Cells," Adv. Mater., vol. 29(8), p. 8, 2017.

[10] Y. H. Cheng, Q. D. Yang, and L. M. Ding, "Encapsulation for perovskite solar cells," Sci. Bull., vol. 66, pp. 100-102, January 2021.

[11] Gao F, Zhao Y, Zhang X, You J et al. Recent Progresses on Defect Passivation toward Efficient Perovskite Solar Cells[J]. Adv.Energy Mater., 2019, 10(13):1902650.

[12] H. Kautsky, "Quenching of luminescence by oxygen," Trans. Faraday Soc., vol. 35, pp. 216-219, 1939.

[13] N. Aristidou et al., "Fast oxygen diffusion and iodide defects mediate oxygen-induced degradation of perovskite solar cells," Nat. Commun., vol. 8, p. 10, 2017.

[14] G. Divitini, S. Cacovich, F. Matteocci, L. Cina, A. Di Carlo, and C. Ducati, "In situ observation of heat-induced degradation of perovskite solar cells," Nat. Energy, Article vol. 1, p. 6, 2016, Art. no. 15012.

[15] Z. Y. Fu et al., "Encapsulation of Printable Mesoscopic Perovskite Solar Cells Enables High Temperature and Long-Term Outdoor Stability," Adv. Funct. Mater., vol. 29, p. 7, 2019.

[16] R. Cheacharoen et al., "Encapsulating perovskite solar cells to withstand damp heat and thermal cycling," Sustain. Energy Fuels, vol. 2, pp. 2398-2406, November 2018.

[17] A. Uddin, M. B. Upama, H. M. Yi, and L. P. Duan, "Encapsulation of Organic and Perovskite Solar Cells: A Review," Coatings, vol. 9, p. 17, 2019.

[18] T. Matsui et al., "Compositional Engineering for Thermally Stable, Highly Efficient Perovskite Solar Cells Exceeding 20% Power Conversion Efficiency with 85 degrees C/85% 1000 h Stability," Adv. Mater., vol. 31, p. 6, 2019.

[19] Q. Dong et al., "Encapsulation of Perovskite Solar Cells for High Humidity Conditions," Chemsuschem, vol. 9, pp. 2597-2603, September 2016.

[20] S. Lee, J. H. Han, S. H. Lee, G. H. Baek, and J. S. Park, "Review of Organic/Inorganic Thin Film Encapsulation by Atomic Layer Deposition for a Flexible OLED Display," Jom, vol. 71, pp. 197-211, January 2019.

[21] Birey et al., "Ion‐beam‐sputtered AlOxNy encapsulating films," J. Vac. Sci. Technol., vol. 16, pp. 2086-2089, December 1979.

[22] R. Z. Sang et al., "Thin Film Encapsulation for OLED Display using Silicon Nitride and Silicon Oxide Composite Film," in: IEEE International Conference on Electronic Packaging Technology & High Density Packaging, pp. 1175-1178, 2011

[23] S. Lee, J. H. Han, S. H. Lee, G. H. Baek, and J. S. Park, "Review of Organic/Inorganic Thin Film Encapsulation by Atomic Layer Deposition for a Flexible OLED Display," Jom, vol. 71, pp. 197-211, January 2019.

[24] J. S. Park, H. Chae, H. K. Chung, and S. I. Lee, "Thin film encapsulation for flexible AM-OLED: a review," Semicond. Sci. Technol., vol. 26, p. 8, 2011.

[25] I. Hwang, I. Jeong, J. Lee, M. J. Ko, and K. Yong, "Enhancing Stability of Perovskite Solar Cells to Moisture by the Facile Hydrophobic Passivation," ACS Appl. Mater. Interfaces, vol. 7, pp. 17330-17336, August 2015.

[26] F. Bella et al., "Improving efficiency and stability of perovskite solar cells with photocurable fluoropolymers," Science, vol. 354, pp. 203-206, October 2016.

[27] N. Rolston et al., "Improved stability and efficiency of perovskite solar cells with submicron flexible barrier films deposited in air," J. Mater. Chem. A, vol. 5, pp. 22975-22983, November 2017.

[28] R. L. Puurunen, "Growth per cycle in atomic layer deposition: A theoretical model," Chem. Vap. Deposition, vol. 9, pp. 249-257, October 2003.

[29] D. Koushik et al., "High-efficiency humidity-stable planar perovskite solar cells based on atomic layer architecture," Energy Environ. Sci., vol. 10, pp. 91-100, January 2017.

[30] Y. I. Lee et al., "A Low-Temperature Thin-Film Encapsulation for Enhanced Stability of a Highly Efficient Perovskite Solar Cell," Adv. Energy Mater., vol. 8, p. 8, March 2018.

[31] S. Emami, J. Martins, D. Ivanou, and A. Mendes, "Advanced hermetic encapsulation of perovskite solar cells: the route to commercialization," J. Mater. Chem. A, vol. 8, pp. 2654-2662, February 2020.

[32] Y. H. Cheng et al., "High-power bifacial perovskite solar cells with shelf life of over 2000 h," Sci. Bull., vol. 65, pp. 607-610, April 2020.

[33] C. C. Boyd, R. Cheacharoen, T. Leijtens, and M. D. McGehee, "Understanding Degradation Mechanisms and Improving Stability of Perovskite Photovoltaics," Chem. Rev., vol. 119, pp. 3418-3451, March 2019.

[34] J. L. Li et al., "Encapsulation of perovskite solar cells for enhanced stability: Structures, materials and characterization," J. Power Sources, vol. 485, p. 15, 2021.

Effect of Ni-CNTs on wetting properties, microstructure, and creep resistance of Sn58Bi-0.1Er composite solder

1st Qi li
Guangdong Provincial Key Laboratory
of Advanced Welding Technology,
China-Ukraine Institute of Welding,
Guangdong Academy of Sciences
Guang dong, Guangzhou, 510650,
China
liq@gwi.gd.cn

2st FengMei Liu*
Guangdong Provincial Key Laboratory
of Advanced Welding Technology,
China-Ukraine Institute of Welding,
Guangdong Academy of Sciences
Guang dong, Guangzhou, 510650,
China
Corresponding author:liufm@gwi.gd.cn

3st Yaoyong Yi
Guangdong Provincial Key Laboratory
of Advanced Welding Technology,
China-Ukraine Institute of Welding,
Guangdong Academy of Sciences
Guang dong, Guangzhou, 510650,
China
yiyy@gwi.gd.cn

4st Xueying Zhang
Guangdong Provincial Key Laboratory
of Advanced Welding Technology,
China-Ukraine Institute of Welding,
Guangdong Academy of Sciences
Guang dong, Guangzhou, 510650,
China
zhangxy@gwi.gd.cn

5st Haitao Gao
Guangdong Provincial Key Laboratory
of Advanced Welding Technology,
China-Ukraine Institute of Welding,
Guangdong Academy of Sciences
Guang dong, Guangzhou, 510650,
China
gaoht@gwi.gd.cn

Abstract—Sn-58Bi-0.1Er solder alloys with different Ni-CNTs contents were prepared by vacuum melting, the influence of Ni-CNTs content on the wetting properties of Sn-58Bi-0.1Er solder was studied. The interface morphology of IMC at Sn-58Bi-0.1Er/Cu joint and the creep resistance of Sn-58Bi-0.1Er/Cu joints with different contents of Ni-CNTs were analyzed. The results showed that when added 0.01~0.05 wt% Ni-CNTs, it enhanced the wettability of composite solder alloy on Cu plate, and the intermetallic compound of Sn58Bi/Cu interface changed from sawtooth Cu6Sn5 to thin layer (Cu,Ni)6Sn5. With the increase amount of Ni-CNTs enhanced particles, it can effectively reduce the thickness of IMC layer at the Sn58Bi-0.1Er/Cu interface. By adding Ni-CNTs particles, the creep fracture life of Sn58Bi-0.1Er/Cu joints was greatly improved. The creep fracture life of the joint was the longest when 0.03 wt%Ni-CNTsadded, which was 25650s. The addition of Ni-CNTs reinforcement particles effectively improved the mechanical properties of the Sn58Bi joint.

Keywords—Sn58Bi-0.1Er, Composite solder alloy, Creep resistance, Intermetallic compound, Ni-CNTs

I. INTRODUCTION

Sn-Pb solder has been widely used in flip chip in integrated circuit manufacturing industry. It plays the role of chip connection, signal transmission, current bearing, mechanical support and has been widely recognized[1,2]. Due to environmental and human health problems, the status of traditional Sn-Pb solder has been challenged in recent years[3]. Because of the toxicity of Pb, it can harm the central nervous system and cause mental decline, which will bring harm to people's health and environment. Therefore, the development and stability of a new type of lead-free solder was the focus of current research in the field of solder. Since the ultimate goal was to replace Sn-Pb solder, the new lead-free solder was still dominated by Sn based solder, adding one or more elements on the basis of Sn to meet the needs of electronic components by adjusting the proportion of components[4]. The most common lead-free solder in low temperature welding was Sn-Bi solder, which was due to the low melting

point and high mechanical properties[5].In addition, as a nontoxic element, Bi is rich in resources and can meet the input of raw materials for manufacturing solder. The ductility of Sn58Bi solder is poor,because the brittleness of Bi will reduce the properties of Sn-Bi solder. During the aging process, Bi is easy to crystallize and form irregular shapes, which resulting weakened the properties of solder joints and decrease the reliability of solder joints[6]. These problems limit the wide use of Sn-Bi solder in the packaging industry. In order to solve the above problems, we need to introduce an enhancement phase to optimize the performance of Sn58Bi solder.

Wu and Li et al.[7] added rare earth elements Er in the Sn58Bi, founded that adding an appropriate amount of Er can enhance the wettability of the solder to the Cu substrate, increase the shear strength of the solder, and increase the creep rupture life of the solder. Existing studies have shown that adding Ni and Ni-CNTs reinforcing particles can improve the microstructure and mechanical properties of Sn-58Bi solder alloy. With the addition of Ni elements and Ni-CNTs particles, it can greatly reduced the Cu_6Sn_5 phase produced in the welding process, it suppresses the formation and growth of Cu_3Sn phase during high temperature aging process, and improves the service life of joints under high temperature conditions[8,9].

In the early stage, the influence of Er content on the structure, wettability and tensile properties of Sn58Bi solder was discussed, and the influence of nickel carbon nanotube content on the microstructure and mechanical properties of Sn58Bi-0.1Er solder was also discussed[10,11]. Therefore, based on the former results, in this paper, the addition of Ni-CNTs enhanced particles on the wettability properties of Sn58Bi-0.1Er composite solder, the influence of Ni-CNTs on the microstructure and shear properties of the IMC of Sn58Bi0.1Er/Cu joints were discussed.

978-1-6654-1392-3/21 $31.00 © 2021 IEEE

II. EXPERIMENTAL MATERIALS

Sn、Bi pure metal, Sn10Er master alloy and Ni-CNTs particles with purity of 99.9 wt% were used as raw materials. the raw materials were placed in an ultrasonic cleaning machine and cleaned with acetone and alcohol for 10 min, respectively to remove organic matter and oil stains on the surface of the raw materials. Each solder alloy melts 1000 g in proportion to its design composition. The experimental materials Sn10Er, Sn, Bi and Ni-CNTs particles were accurately weighed on the BSA224S-CW electronic balance with an accuracy of 0.1mg. After weighing, put the solder into the crucible of the VIF-10 vacuum melting furnace, evacuated and rushed into the argon gas for protection to prevent the splashing of droplets during the melting process. The solder alloy was observed to melt into a liquid state through the vacuum melting furnace After the metal, it is kept at low power for 30 minutes to ensure the uniformity of the alloy composition. After the metal, it is kept at low power for 30 minutes to ensure the uniformity of the alloy composition.After the smelting is completed, the molten metal is poured into a stainless steel mold and cooled and solidified in the air. The cast solder is cylindrical with a diameter of 20 mm. Fig.1(a) shows the SEM image of NiCNTs. Fig.1(b) shows an enlarged view of Ni-CNTS. The content of Ni-CNTs in the experiment is between 0.01 and 0.1wt%, and the solder alloy composition design is shown in Table 1.

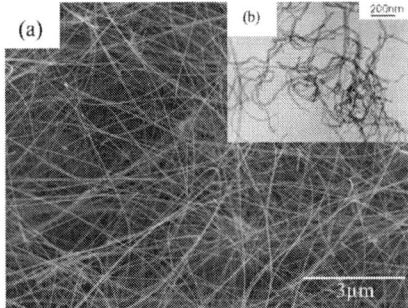

Fig. 1. SEM images of (a) multi-layer GNSs, (b) magnified layered GNSs.

TABLE I. SERIES OF Sn58Bi0.1Er-x(Ni-CNTs) SOLDER ALLOYS (WT.%)

Samples	Solder alloy				
	1	2	3	4	5
Ni-CNTs(%)	0	0.01	0.03	0.05	0.1
Initial alloy	Sn58Bi0.1Er				

The Cu substrate used in the spreading rate experiment was cut into 30 mm×30 mm×1 mm. Before the experiment,the anaerobic copper sheet pretreatment was carried out according to Appendix 4 of the JIS-Z-3284[12]. The flux (25 Wt% Grade 2 Colophony +75wt% C12H26O3) was obtained as a soldering flux. The calculated volume and density of the solder alloy were placed in the center of the processed copper substrate. Using a micro syringe, 0.02ml flux was dropped onto the solder alloy to cover the surface of the solder alloy with a layer of flux. Put it into ZB2015 HL precision lead-free reflow soldering furnace for wetting and spreading, and prepare 5 pieces of each alloy composition. The heating temperature curve is shown in Fig.2. After heating and cooling, take out the welded Cu substrate, and clean the flux residue on the surface of Cu substrate with circuit board cleaning agent. The spreading height of solder alloy was measured by vernier caliper, and the wettability of solder alloy was characterized by spreading rate S_R (%)[13].

$$S_R=(D-H)/D\times100\% \qquad (1)$$

S_R is spreading rate, %; H is the height of paved brazing material, mm; D is wet paving as spherical paving diameter of brazing material, mm, $D=1.24V^{1/3}$; V is the mass / brazing density in test, mm^3.The thickness of the measured H was replaced with formula (1) to obtain the final spreading rate of the brazing alloy.

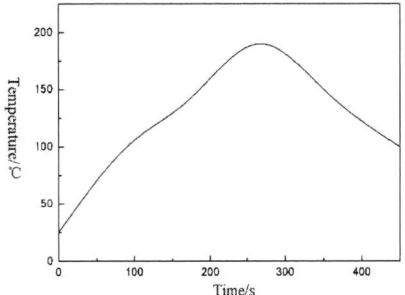

Fig. 2. Reflow soldering curve of composite solder

The substrate size used in the shear properties test is shown in the Fig. 3. Remove the oxide scale on the surface of the copper substrate to be welded, polish it with a series of sandpaper, keep it flat, and clean it with tap water and alcohol in turn. The cutting size of steel wire is Φ 5mm × For 0.45mm disc welding alloy, clean the oil stain with acetone, and polish the surface with No. 800 and No. 1000 sandpaper. In the welding area, use a syringe to drop 0.02ml of flux onto the top of the solder alloy to cover the surface of the solder alloy. The reflow soldering process curve of the sheared sample and the remaining wet and spreading sample were consistent. The creep properties of welded joints were studied by GWT2105 high temperature creep strength tester. Each component of solder alloy was tested under 80 ℃ temperature and 4 MPa stress, and the test results were the average of the test data of 5 samples in each group.

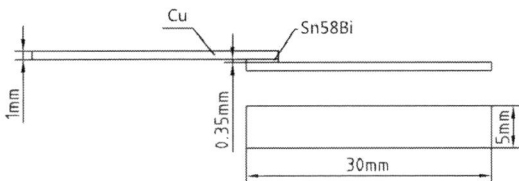

Fig. 3. Dimensions of creep specimen

The microstructure, phase structure and composition of each solders were examined by Quantm 250 scanning electron microscope (SEM) and energy spectroscopy (EDS).

III. RESULTS AND DISCUSSIONS

A. Effect of Ni-CNTs enhanced particles on wetting properties of Sn-58Bi-0.1Er solder

The experimental results of expansion rate of composite solder are shown in the Fig. 4. When the content of Ni-CNT particles is 0.01wt%, the diffusion rate of solder is 79.26%, which is higher than that of sn58bi-0.1er solder by 73.81%. When the content of Ni-CNTs reinforced particles is 0.03wt%, the diffusion rate of the solder decreases slightly to 78.65%, but it is still higher than that of the solder without Ni-CNTs reinforced particles. When the content of Ni-CNT particles is 0.05%, the diffusivity of the solder reaches 80.84%, which is 7.03% higher than that of Sn58Bi-0.1Er solder. With the addition of Ni-CNTs reinforced particles, it reduces the surface tension of the liquid solder on the Cu substrate, improves the fluidity of the liquid solder, and thus improves the wetting performance of the solder. However, when adding Ni-CNTs reinforced particles When the content is 0.1 wt%, the spreading rate of the brazing filler metal is 80.21%. It means that the spreading rate of the brazing filler metal begins to decrease when the content of particles continues to be added exceeds 0.05%. This is because when too much Ni-CNTs reinforcing particles are added, due to the existence of van der Waals forces and nano-scale effects between molecules, the Ni-CNTs reinforcing particles will agglomerate and hinder the fluidity of the liquid solder. In addition, at a certain welding temperature, the larger the melting range of the solder, the worse the fluidity of the solder. When too many Ni-CNTs reinforcing particles are added, the melting range of the solder increases. The worse the fluidity. Therefore, the two factors together lead to a decrease in the wetting performance of the solder.

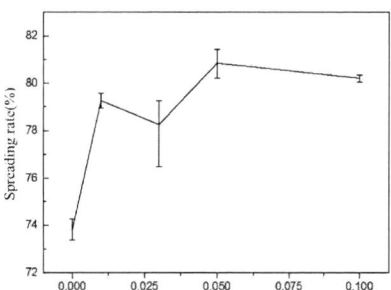

Fig. 4. Effect of Ni-CNTs contents on spreading rate of Sn-58Bi-0.1Er solder alloys

B. Sn58Bi0.1Er-x(Ni-CNTs)/Cu interface microstructure morphology

The intermetallic compounds at the joint interface were analyzed by EDS, and the composition was determined according to the composition ratio of various elements. Fig.5 shows the EDS spectra of the compounds at the interface of the welded joint when 0.03% Ni-CNTs are added. The EDS analysis results of IMC intermetallic compounds at the joint interface are shown in Table 2. It can be seen from the table that the atomic mass ratio of copper to tin is 55.02:41.51, and there are some Ni atoms in the interface structure. According to the research results of Zhou et al[14], the microstructure and energy spectrum analysis show that the IMC phase at the interface of Sn-58Bi /Cu joint is (Cu, Ni)$_6$Sn$_5$ when Ni-CNTs reinforced particles are added. The results are similar to the experimental results. Therefore, it can be judged that the intermetallic phase at the interface of Sn58Bi-0.1Er/Cu joint is (Cu, Ni)$_6$Sn$_5$.

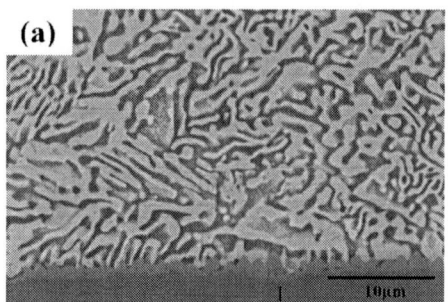

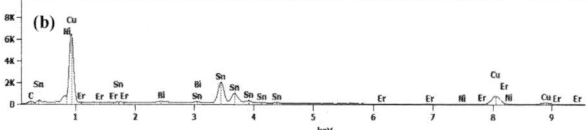

Fig. 5. SEM image of interfacial compounds of Sn58Bi0.1Er-0.03(Ni-CNTs)/Cu solder joints and EDS spectrum

TABLE II. EDS ANALYSIS RESULTS OF AREA 1

Area	Elements	Atoms%	Wt%
1	Ni	0.17	0.21
	Cu	55.11	64.76
	Sn	43.07	34.44
	Er	0.16	0.07
	Bi	1.49	0.53

The microstructure and morphology of Sn58Bi-0.1Er-x (Ni-CNT)/Cu interface after reflow soldering are shown in Fig. 6. It can be seen that when Ni-CNTs are added, the intermetallic compound formed by the reaction of composite solder and Cu substrate is no longer scallop-shaped Cu$_6$Sn$_5$, but a thin layer (Cu ,Ni)$_6$Sn$_5$. The thickness of (Cu,Ni)$_6$Sn$_5$ intermetallic compound layer decreases with the increase of Ni-CNT particle content. When the content of Ni-CNTs is 0.01wt% and 0.03wt%, the diffusion activation energy of (Cu,Ni)$_6$Sn$_5$ intermetallic compound is 34.6 kJ/mol less than the diffusion activation energy of Cu$_6$Sn$_5$ intermetallic compound is 58.6 kJ/mol, resulting in the easier preferential formation of (Cu,Ni)$_6$Sn$_5$ intermetallic compound layers[15]. Also, it can be observed that with the addition of Ni-CNTs, the thickness of the (Cu,Ni)$_6$Sn$_5$ interface phase decreases first and then increases, and the thickness was the smallest at the addition amount of 0.03wt%. There were two reasons for this optimization effect: First, it is caused by the uniform heterogeneous nucleation of Ni$_3$Sn$_4$ in the composite solder. This is because when Ni$_3$Sn$_4$ is present in the solder, it can inhibit the function of Sn58Bi solder mesh structure coarsenin; another reason is that carbon nanotubes can inhibit the grain growth in the flux[16]. However, as shown in Fig. 6(d), when the Ni-CNTs content is increased to 0.1wt%, the grain spacing of the Sn58Bi0.1Er-0.1CNTs solder increases, indicating that the microstructure of the composite solder becomes coarser. This phenomenon is caused by the increase of Ni$_3$Sn$_4$ in solder and the aggregation effect of Ni CNTs. When the content of Ni-CNTs is low, Ni$_3$Sn$_4$ can be evenly distributed in tin phase and dual phase to inhibit the growth of grains. However, when the content of Ni-CNTs is high, the nucleation of Ni$_3$Sn$_4$ will produce caking effect, and the aggregated Ni$_3$Sn$_4$ can not effectively inhibit the grain growth[11]. Therefore, the microstructure of composite solder joints are relatively rough.

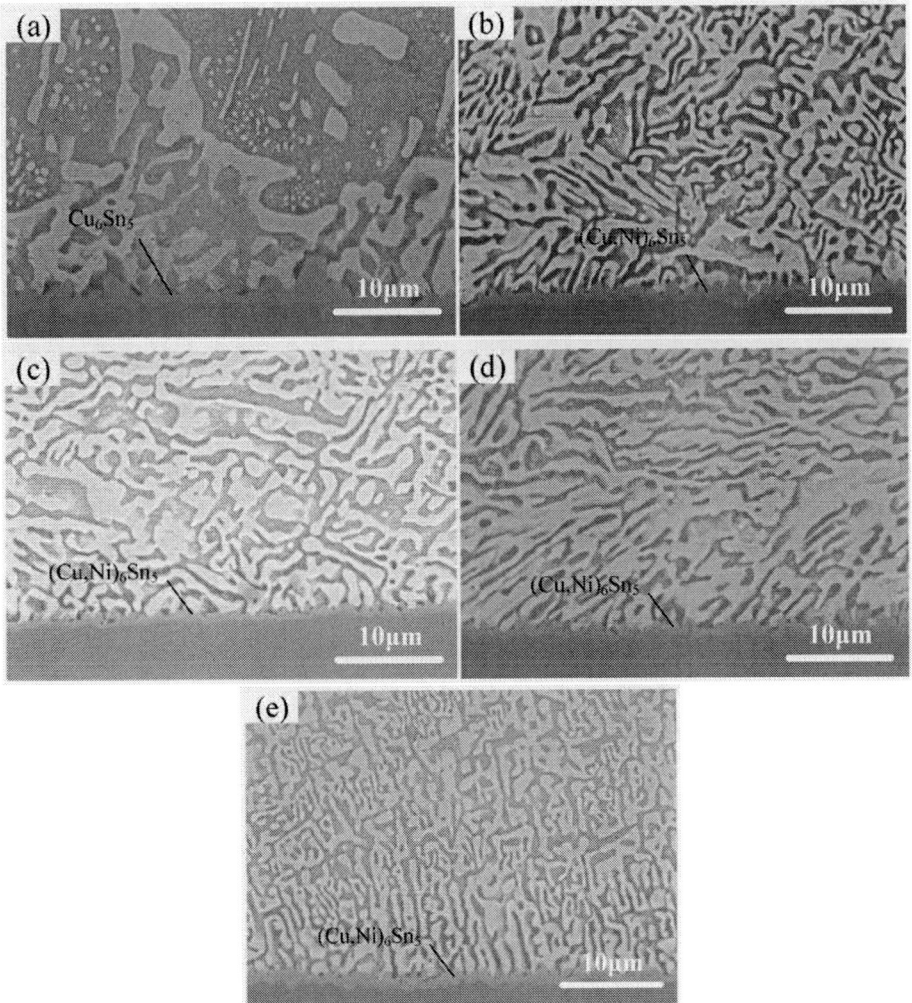

Fig. 6. Interfacial microstructures of Sn-58Bi-0.1Er-x(Ni-CNTs)/Cu solder joints

C. Effect of Ni-CNTs reinforced particles on creep resistance of Sn-58Bi-0.1Er solder joints

Sn-58Bi-0.1Er-x(Ni-CNTs)/Cu welded joint creep experimental results were shown in Fig.7. By adding Ni-CNTs particles to the solder, creep fracture life of Sn-58Bi-0.1Er-x(Ni-CNTs)/Cu welded joints was significantly improved. And when added 0.03 wt% Ni-CNT, Sn-58Bi-0.1Er-0.01(Ni-CNTs)/Cu welded joint creep fracture life increased from 10620s to 23130s, compared with Sn-58Bi-0.1Er/Cu welded joint. And when added 0.03 wt% Ni-CNTs, creep fracture life was 25650s,which was the best of these solders. When the addition amount of Ni-CNTs reinforcement particles is 0.05% and 0.1%, the creep rupture life of the weld decreases slightly, which is 25344s and 21708s, respectively. Combined with the morphology of the interface structure of the composite solder joint, it can be seen that by adding an appropriate amount of Ni-CNT reinforcing particles to the composite solder, an appropriate amount of Ni3Sn4 phase can be evenly distributed in the matrix, thereby suppressing the coarseness of the network structure in the solder. Ni-CNTS can refine the structure of the weld body, reduce the size of the IMC boundary strain

and interfacial tension, and and hindering the movement of grain slip and dislocations. The addition of Ni-CNTS reinforcement particles reduces the cracks during the creep fatigue process, prevents the propagation of micro-cracks during the creep process, and improves the creep resistance of the welded joint.

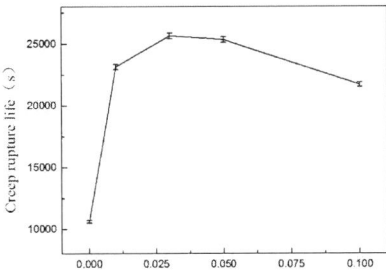

Fig. 7. Effect of Ni-CNTs contents on creep rupture life of Sn-58Bi-0.1Er-x(Ni-CNTs)/Cu solder joints

IV. CONCLUSIONS

This study systematically studied the properties of Sn58Bi-0.1Er composite solder under different amounts of Ni-CNTs. The conclusions that can be drawn are as follows:

1. Adding an appropriate amount of NI-CNTs particles can reduce the surface tension of the liquid solder, thereby improving the wetting and spreading performance of the composite solder alloy. Excessive addition of Ni-CNTs particles will cause the aggregation of Ni-CNTs, hinder the fluidity of the liquid solder, and reduce the wettability of the solder.

2. With the addition of Ni carbon nanotube particles, the intermetallic compound at the joint interface changes from serratedCu_6Sn_5 to a thin layer $(Cu, Ni)_6Sn_5$. By adding an appropriate amount of Ni carbon nanotube particles, the thickness of the intermetallic compound layer can be effectively reduced.

3. The addition of Ni-CNTs particles greatly improves the creep rupture life of composite solder joints. When 0.03wt% Ni-CNTs is added, the creep rupture life of the joint is the longest, which is 25650s. With the increase of Ni-CNTs particles, the creep rupture life of the joint shows a downward trend at this time.

ACKNOWLEDGMENT

The authors gratefully acknowledge the financial support by the GDAS· Project of Strategic Pilot Science and Technology (2020GDASYL-20200301001), Guangzhou Science and Technology Project (Foreign Science and Technology Cooperation) (201807010028.201807010029).

REFERENCES

[1] Goosey M, "Environment-friendly electronics: lead free technology," Circuit World, vol. 4, pp. 10-20, August 2002.

[2] Glazer J, "Metallurgy of low temperature Pb-free solders for electronic assembly," International Materials Reviews, vol. 2, pp. 65-93, Nov 2013

[3] Minhua Lu,Da-Yuan Shih,Paul Lauro , et al, "Effect of Sn grain orientation on electromigration degradation mechanism in high Sn-based Pb-free solder," Applied Physics Letters,vol. 21, pp. 211909-211909-3, May 2008.

[4] Zhang X , Matsuura H , Tsukihashi F , et al, "Wettability of Sn-Zn, Sn-Ag-Cu and Sn-Bi-Cu alloys on copper substrates," MATERIALS TRANSACTIONS, vol. 5, pp. 926-931, March 2012.

[5] Jo J L, Nagao S, Hamasaki K, et al, "Mitigation of Sn Whisker Growth by Small Bi Additions," Journal of Electronic Materials, vol. 43, pp. 1-8, September 2013.

[6] Chen X, Xue F, Zhou J, et al, "Effect of In on microstructure, thermodynamic characteristic and mechanical properties of Sn‐Bi based lead-free solder," Journal of Alloys and Compounds, vol.633, pp. 377-383, June 2015.

[7] Li, Q., Xiong, M., Liu, F., et al, "Effect of Er content on the interfacial microstructure, shear properties and creep properties of Sn58Bi joints," In IOP Conference Series: Earth and Environmental Science,Vol.714, pp. 032017, March 2021.

[8] Gain A K, Zhang L, "Growth mechanism of intermetallic compound and mechanical properties of nickel (Ni) nanoparticle doped low melting temperature tin–bismuth (Sn–Bi) solder," Journal of Materials Science Materials in Electronics, vol. 27, pp. 781-794, September 2015.

[9] Yang L , Wei Z , Liang Y , et al, "Improved microstructure and mechanical properties for Sn58Bi solder alloy by addition of Ni-coated carbon nanotubes," Materials Science & Engineering A, vol. 642, pp. 7-15, Aug 2015.

[10] Li, Q., Wu, H., Xiong, M , et al, " Effect of Er on microstructure and mechanical properties of Sn58Bi based lead-free solder," In 2020 International Conference on Artificial Intelligence and Electromechanical Automation (AIEA), IEEE, pp. 648-653, June 2020 .

[11] Li, Q., Gao, H., Xiong, M. , et al, "Effect of Ni-CNTs on microstructure, thermodynamic characteristic and mechanical properties of Sn58Bi-0.1 Er based lead-free solder," In 2020 21st International Conference on Electronic Packaging Technology (ICEPT), IEEE, pp. 1-6, August 2020 .

[12] Takemoto T. Introduction of JIS related to lead-free solder and soldering [C]// Electronic Packaging Technology, 2005 6th International Conference on. 2005.

[13] Wuhua Chen, Liu Fengmei, Li Qi, et al, "Er effect on microstructure and properties of Sn-58Bi solder alloys," Hot working process, vol. 49,pp. 25-28, March 2020.

[14] Zhou Shiyuan, Yang Li, Zhang Yaocheng, et al, "Study on microstructure and mechanical properties of solder joints reinforced Sn-58Bi Ni-coated carbon nanotubes composite solder joints," Hot working process, vol.47 ,pp. 81-83,87, March 2018.

[15] Yang Lizhuang, "Microstructure and Mechanical Properties of Nickel-plated Carbon Nanotubes Reinforced Sn58Bi Lead-Free Solder," Tianjin University ,2014.

[16] Yang L , Wei Z , Li X , et al, "Effect of Ni and Ni-coated Carbon Nanotubes on the interfacial reaction and growth behavior of Sn58Bi/Cu intermetallic compound layers," Journal of Materials Science Materials in Electronics, vol.27,pp. 1-7, July 2016.

A Process Improvement in Silver-indium Transient Liquid Phase Bonding Method for the High-Power Electronics and Photonics Packaging

Donglin Zhang
School of Materials
Science&Engineering
Beijing Institute of Technolo
Beijing, China
xzsxgxhh@163.com

Xiuchen Zhao
School of Materials
Science&Engineering
Beijing Institute of Technolo
Beijing, China
zhaoxiuchen@bit.edu.cn

Yingxia Liu
School of Materials
Science&Engineering
Beijing Institute of Technolog
Beijing, China
yingxia.liu@bit.edu.cn

Ying Liu
School of Materials
Science&Engineering
Beijing Institute of Technolog
Beijing, China
yingliu@bit.edu.cn

Yongjun Huo
School of Materials
Science&Engineering
Beijing Institute of Technolog
Beijing, China
huoyongjun@bit.edu.cn

Abstract—It has been demonstrated that the silver-indium (Ag-In) transient liquid phase (TLP) method [1] can be used in the heterogeneous integration between the chemical vapor deposition (CVD) diamond and Cu, whereby forming an advanced composite heat spreading submount for the future high-power electronics and photonics. However, some issues regarding to the O_β species oxygen group trapped within grain boundaries of silver layer and Ag tarnishing during electroplating process were found in the previous research work[2]. Above issues resulted in an adversary effect on the bonding strength of Ag-In TLP joint.

In this current paper, to further explore the Ag-In TLP bonding method in the utility of high-power electronics packaging, the authors have deposited multilayer metal film structures on diamond and copper substrates, with a combined process of the electroplating method and E-beam evaporation deposition. In order to eliminate the isolated voids on the bonding interface caused by undersupply of the indium molten phase during the TLP process, a vacuum annealing treatment was introduced to the E-beam deposited and electroplated silver layer, prior to the bonding process. We have changed the recipe of indium electro-plating solution, to a sulfur-element free one. Therefore, the previous reported oxide trapping and Ag tarnishing issues during the indium plating have been successfully resolved. In conclusion, this research work will pave the path for the future utility of Ag-In TLP method in the advanced high-power electronics and photonics packaging[3].

Keywords—High-power electronics; Transient liquid phase bonding; Silver-indium; Indium electroplating.

I. INTRODUCTION

The emergence of wide-band gap (WBG) semiconductors such as silicon carbide (SiC) has broken through the physical limits of the traditional semiconductor material silicon, making the development of power electronics devices has made more significant progress, a new generation of power electronics devices have higher operating temperatures and energy efficiency. At the same time, electronic devices are gradually to the direction of miniaturization and higher power density, which puts forward higher requirements for high-power electronic devices and optoelectronic devices packaging technology. High-power electronic devices require more heat-resistant, more reliable bonding methods, and make bonding joints can be timely device work heat out, which presents a serious challenge to the substrate material and chip connection technology. To better perform the function of high-power electronic devices, it is important to develop more promising bonding methods for high-power electronic and optoelectronic devices, in addition to using materials with higher thermal conductivity such as chemical vapor deposition (CVD) diamond as a heat dissipation substrate.

Obviously, in the field of high-power electronics, traditional chip bonding techniques such as tin-lead solder and lead-free solder are no longer applicable, and their low melting temperatures (below 250°C) cannot withstand the high temperatures generated during the operation of high-power electronics, which operate at temperatures even higher than 350°C [4]. Moreover, since copper is commonly used as a substrate for heat dissipation and electrical connections in electronics manufacturing, the large difference in thermal expansion coefficients between copper and chemical vapor deposited (CVD) diamond can easily cause thermal mismatch. Therefore, having a joint with good electrical connection and high thermal conductivity is the key technology to make this configuration possible. Conventional techniques use gold-tin eutectic solder as the thermal interface material, but the high raw materials cost for gold largely limits its use [5].

In recent years, nanoscale and micron-sized particles of silver paste and various silver-tin compositions of solder paste are widely used in electronic packaging, these types of solder paste can form interconnect joints at relatively low temperatures, solder paste usually consists of nanoscale and micron-sized silver particles with the addition of certain components of flux to help bonding, silver solder paste due to its lower production cost and more flexible and convenient

978-1-6654-1392-3/21 $31.00 © 2021 IEEE

use has been massively Applications, silver nanoparticle pastes have a much wider operating temperature range compared to conventional solders, as silver nanoparticles can be sintered into a sintered metal with a bulk metal melting temperature [6], but their drawbacks are also evident: the exhaust of fluxes and additives of organic composition during the sintering process can lead to the formation of pores in the soldered joint, posing a serious challenge to the conductivity and thermal conductivity of the joint. More seriously, the electrochemical migration phenomenon of pure silver poses a serious challenge to the reliability of silver nanoparticle silver pastes, as shown in the water drop test (WDT) in Fig. 1, where a large number of silver whiskers grew at the cathode of the sintered silver particle silver paste under bias voltage.

Fig. 1. Silver whisker growth at the sintered silver paste in the water drop test (WDT), the substrate is AlN

The transient liquid phase (TLP) technology, which uses silver and some low melting point metals such as tin and indium to form an interlayer of high melting point metals and low melting point metals, utilizes the interlayer with a lower melting temperature as the joining metal, and the liquid interlayer metal wets the high melting point metal and reacts to form a new phase with a high melting point, is a promising bonding technology for high power devices in recent years. Transient liquid phase has the characteristics of low temperature connection and high temperature service, and its bonding temperature can be as low as about 180°C, and form a solid joint without voids or tiny pores, and can withstand higher operating temperatures. Ag-In system transient liquid phase bonding has been proved to be a reliable low temperature bonding system. It is capable of forming bonded joints with high strength, thermal stability and fatigue resistance [7]. The Ag-In systems can perform bonding processes in the temperature range of 180°C, while the final formed joints can remain thermally stable at much higher temperatures (> 300°C). This is due to the low melting point of indium (156.6°C), which enables transient liquid-phase interconnections with silver at temperatures below 180°C. From the phase diagram of Ag-In binary alloy shown in Fig. 2, there is a eutectic reaction between the two in the low temperature phase. intermetallic compounds formed between Ag-In are Ag_2In as well as $AgIn_2$, making the bonded joint formed rich in silver, and the eutectic reaction between silver and indium can obtain good soldering performance at low temperature and make the remelting temperature higher than 700°C. By designing a multilayer metal structure, Ag-In low-temperature transient liquid phase bonding can achieve excellent performance in optoelectronic devices such as high-power LEDs and high-power semiconductor lasers.

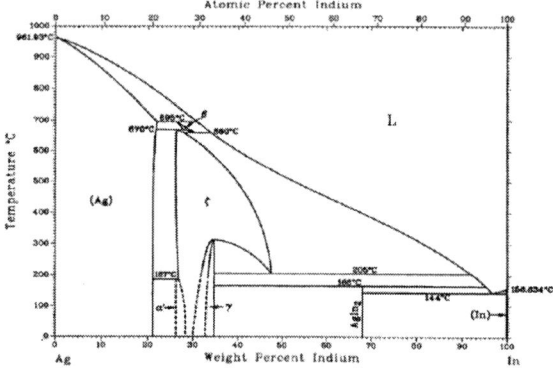

Fig. 2. Silver-indium binary phase diagram

Using the architecture shown in Fig. 3, a layer of chromium and gold is first pre-deposited on the chemical vapor deposited diamond surface by electron beam evaporation (E-beam), which is intended to enhance the bonding of silver and diamond. A layer of silver is then vaporized as a seed layer for the silver plating. The treatment of the copper substrate is simpler, and different thicknesses of silver and indium are directly plated on the surface of the copper substrate, the thickness of which can be adjusted according to the silver-indium phase diagram in order to explore the bond strength and the composition and structure of the joint in the subsequent bonding process with different ratios of silver-indium multilayer structures.

Fig. 3. The silver-indium multilayer structure

II. MATERIALS AND METHODS

A. Silver Electroplating

A silver plating solution with $K_4[Ag_2(CN)_6]$ as the active ingredient was used to electroplate the silver layer, and the drugs required for the preparation of the plating solution are shown in Table 1. For example, 30 g of $K_4[Fe(CN)_6]$ and 60 g of K_2CO_3 were dissolved in distilled water, boiled and mixed, 40 g of AgCl was added slowly with constant stirring, boiled for 2 hours, and then Fe^{2+} were completely converted into $Fe(OH)_3$ and precipitated, filtered to obtain a light yellow transparent liquid, and finally 150 g of KSCN was added and dissolved. was added and dissolved, and finally the solution was fixed to 1L.

TABLE I. DRUG TYPE AND CONTENT (AG)

AgCl	K_2CO_3	$K_4[Fe(CN)_6]$	KSCN
40 g/L	80 g/L	80 g/L	150 g/L

The chemical reactions that occur during the preparation of the plating solution are:

$$2AgCl+K_4[Fe(CN)_6]=K_4[Ag_2(CN)_6]+FeCl_2 \quad (1)$$

$$FeCl_2+H_2O+K_2CO_3=Fe(OH)_2+2KCl+CO_2 \quad (2)$$

$$2Fe(OH)_2+1/2O_2+H_2O=2Fe(OH)_3 \quad (3)$$

The scanning electron microscope image shown in Fig. 4 is the silver plating layer obtained by using the above plating solution, the cathode current density is 0.2A/dm², it can be seen that the plating layer of silver plating is of good quality, and the surface roughness of the plating layer is about 1 um. This plating solution is a cyanide-containing plating solution, which is relatively simple and easy to obtain a dense and bright plating layer, but the cyanide-containing plating solution has a greater impact on the environment, and nowadays it is mostly used instead of cyanide-containing plating solution, such as silver fluoroborate plating solution. However, cyanide-containing electroplating solution has a greater impact on the environment, and Nowadays, more cyanide-free plating solutions such as silver fluoroborate and silver sulphamate plating solutions are used instead.

Fig. 4. The Scanning electron microscope image of silver electroplating layer, the substrate is copper.

B. Indium Electroplating

Compared with silver, there are fewer plating solution systems for indium plating, mainly indium sulfate, indium fluoroborate and indium sulfamate. In order to exclude the influence of sulfur elements in the electroplating solution on silver plating, indium fluoroborate electroplating solution is selected for indium plating. The composition of indium fluoroborate electroplating solution and the main parameters of plating are shown in Table 2, and its main component is indium fluoroborate ($In(BF_4)_3$), and the pH value of the electroplating solution is adjusted by ammonium fluoroborate (NH_4BF_4) and boric acid (H_3BO_3). 1.0, and use indium metal as the anode. Plating with indium fluoroborate can get fine-grained indium plating, the composition of the plating solution changes less, and 100% anode current efficiency can be achieved when pure indium is used as the anode. The main disadvantage of this method is the low cathode current efficiency, which is 40%~75%.

TABLE II. DRUG TYPE AND CONTENT (IN)

In(BF$_4$)3	NH$_4$BF$_4$	H$_3$BO$_3$
236 g/L	40~50 g/L	22~30 g/L

III. ANNELING OF SILVER

A. Ag-In Transient Liquid Phase

In Ag-In transient liquid-phase bonding, the reaction between silver and indium is very rapid, and even at room temperature, a 5um thick indium layer reacts completely with silver to form an intermetallics compound (IMC) within one day [8]. The formation of intermetallic compounds Ag2In and AgIn2 between silver and indium is a very fast process. Indium diffuses into silver mainly by grain boundaries, while silver diffuses into indium through interstitial positions. In the Ag-In transient liquid phase system, maintaining the amount of liquid phase and indium plays a very important role in the quality of the formed joint. Insufficient supply of liquid phase during the soldering process is a serious problem that can lead to the formation of interfacial voids. To slow down the rate of indium diffusion through grain boundaries to silver during the soldering process, increasing the grain size of the silver layer by annealing and thus slowing down the consumption of indium is a simple and feasible method.

B. Vacuum Annealing

The annealing process of silver can be divided into air annealing and vacuum annealing. The purpose of the annealing process before bonding is to increase the grain size of silver, thus slowing down the rate of indium diffusion to the silver grain boundaries and preventing the bonded joints from appearing cavities due to insufficient liquid phase, which affects the quality of the joints. The annealing temperature higher than 350°C will cause bubbles on the surface of the silver layer, therefore, the annealing process generally does not exceed 350°C. When annealing in air, the resulting reorganization of the porous silver microstructure and surface state will affect the subsequent bonding process; while vacuum annealing will not have this problem and will only cause the growth of silver grains without any porosity, so we use the process of vacuum annealing at 350°C for 0.5 hours with the vacuum level controlled at 0.1 Torr.

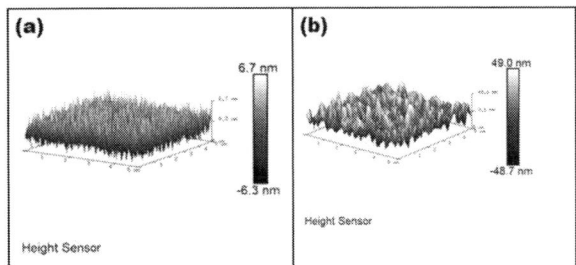

Fig. 5. The AFM image of silver plating: (a) Un-annealed silver plating and (b) Annealed silver plating

As an example, silver thin films prepared by electron beam evaporation (E-beam) were plated with 50 nm thick Cr and 200 nm thick Ag on the silicon wafer surface sequentially using the E-beam evaporation method. Fig. 5 shows the atomic force microscopy (AFM) images of the silver plated layer without annealing process and after vacuum annealing at 300°C for 0.5 hours. The surface roughness of the unannealed silver plating is about 12 nm and the surface grain size is about 55 nm, while the surface roughness of the silver plating increases to 100 nm and the surface grain size is about 90 nm after vacuum annealing at 300°C for 0.5 hours. It can be seen that the annealing process before bonding can make the silver grains grow and slow down the diffusion of indium through the silver grain boundaries during transient liquid phase bonding.

IV. CONCLUSION

The vacuum annealing process effectively slows down the rate of indium diffusion through grain boundaries to silver, avoiding the problem of insufficient liquid phase during transient liquid phase bonding. However, vacuum annealing also makes the silver layer more susceptible to sulfide, which reduces the shear strength of the bonded joint. We know that sulfides have particularly poor mechanical properties and can be a factor in crack expansion. The introduction of sulfide is mainly due to the silver and indium plating solutions containing sulfur. To address this problem, we have improved the plating process by using a new sulfur-free indium plating solution, indium fluoroborate, which reduces the effect of sulfide to some extent. The vacuum annealing process was introduced to avoid the formation of silver porous structures on the one hand, and to prevent oxygen atoms from passing through the silver grain boundaries to form oxides and thus reduce the shear strength of the joints on the other. In the future work, besides continuing to investigate the various factors affecting the quality of bonded joints, the quality of the obtained plating will be further examined using test methods such as GIXRD, and the reaction kinetics at the Ag-In interface will be investigated by DSC to establish the microscopic mechanism of bonding of Ag-In transient liquid phase system, so that it can provide theoretical guidance for heat treatment and bonding temperature and time.

ACKNOWLEDGMENT

The authors thank Analysis & Testing Center at Beijing Institute of Technology for the electron beam evaporation (E-beam) equipment, and the SEM/AFM works done in Experimental Center of Advanced Materials at Beijing Institute of Technology.

REFERENCES

[1] R. W. Chuang and C. C. Lee, "Silver-indium joints produced at low temperature for high temperature devices," in IEEE Transactions on Components and Packaging Technologies, vol. 25, No. 3, pp. 453–458, 2002

[2] R. Sheikhi, Y. Huo, C. H. Tsai, C. R. Kao, F. G. Shi, and C. C. Lee, "Prior-to-bond Annealing Effects on the Diamond-to-copper Heterogeneous Integration using Silver-indium Multilayer Structure", Journal of Materials Science: Materials in Electronics, vol. 31, pp. 8059-71, 2020.

[3] R. Sheikhi, Y. Huo, F. G. Shi, and C. C. Lee, "Low Temperature VECSEL-to-Diamond Heterogeneous Integration with Ag-In Spinodal Nanostructured Layer", Scripta Materialia, vol. 194, No. 113628, pp. 1-4, 2021.

[4] H. S. Chin, K. Y. Cheong, and A. B. Ismail, "A review on die attach materials for SiC-based high-temperature power devices," Metallurgical and Materials Transactions B, vol. 41, No.4, pp.824-832, 2010.

[5] C.C. Lee, C.Y. Wang, G.S. Matijasevic, A new bonding technology using gold and tin multilayer composite structures. IEEE Trans. Compon. Hybrids Manuf. Technol. 14(2), 407–412 (1991).

[6] M. Maruyama, R. Matsubayashi, H. Iwakuro, S. Isoda, T. Komatsu, Appl. Phys. A 93 (2008) 467.

[7] Y.Y. Wu, C.C. Lee, The strength of high-temperature Ag-In joints produced between copper by fluxless low-temperature processes. J. Electron. Packag. Trans. ASME 136(1), 011006 (2014).

[8] Y.Y. Wu, W.P. Lin, C.C. Lee, A study of chemical reactions of silver and indium at 180 °C. J. Mater. Sci. Mater. Electron. 23(12), 2235–2244 (2012).

Research on SiP Signal Integrity Based on Ansys SIwave in Wearable Medical Systems

Zheng Yang
Beijing Institute of Control Engineering
China Academy of Space Technology
Beijing, China
yangzheng.zz@pku.edu.cn

Yingke Gao
Beijing Institute of Control Engineering
China Academy of Space Technology
Beijing, China
gaoyingke@126.com

Shenglong Li
Beijing Institute of Control Engineering
China Academy of Space Technology
Beijing, China
497300406@qq.com

Chuanchuan Sun
Beijing Institute of Control Engineering
China Academy of Space Technology
Beijing, China
18600207816@163.com

Yunfu Zhao
Beijing Institute of Control Engineering
China Academy of Space Technology
Beijing, China
21782459@qq.com

Abstract—One of the main challenge for wearable medical systems is how to integrate more functional ICs in a very small space. As a kind of new packaging and system integration technology, SiP (System in Package) simplifies the system design and complies with the requirements of device miniaturization with multi-component integration. However, high layout and routing density in SiP substrate tend to cause SI (Signal Integrity) and PI (Power Integrity) issues. This paper introduces the main reasons and improvement measures for signal and power integrity issues in SiP designs. Based on ANSYS SIwave simulation platform, the simulation process of major parameters for signal and power integrity and related optimization methods are discussed in the design process of SiP.

Keywords—SiP, Signal integrity, Power integrity, Ansys SIwave, Wearable medical systems

I. INTRODUCTION

With rapid growing of wearable electronics, medical semiconductors are developing towards higher integration, miniaturization and higher energy efficiency. With SiP technology, it is possible to integrate multiple integrated circuits into a single packaged system component, reducing area on printed circuit board. Thereby Realize more functions on one single lighter, thinner and shorter terminal product. The more the products miniaturization, the better the consumers' experience, and wearable products really will be smart devices for daily life.

As one of the newest electronic packaging and system integration technology, SiP has the following advantages:

(1) SiP integrates multiple ICs and passive components in one package. The implementation could be a stack of dies or a stack of packages. The interconnection could be multi-chip interconnection, flip-chip, etc. SiP slims the amount of external electrical connection ports to a large extent, reduces package size and improves system integration effectively;

(2) SiP encapsulates ICs of different processes and materials into one system, achieves system-level heterogeneous integration efficiently and reliably;

(3) SiP is a kind of single package with system functions, which solves the issue that SoC cannot integrate analog, radio frequency and digital functions. It is a kind of semiconductor technology that realizes system functions in packaging field.

SiP devices also has drawbacks. SiP substrate has high layout and routing density as well as high output switching speed, tend to cause SI (Signal Integrity) issues such as delay, reflection, crosstalk, and PI (Power Integrity) issues such as power noise and ground bounce. All these lead to system instability, even loss of function. Therefore, the key for a successful design is comprehensive consideration of SI PI factors and effective control measures in both system design and layout design.

As mentioned above, for a successful wearable medical SiP project, SI simulation, PI simulation, thermal analysis, thermal-mechanical coupling simulation, etc. should be carried out during the design phase, by which electrical issues, temperature concentration and stress problems may show up in advance. Based on this, electrical reliability verification processes and methods for SiP design are introduced in this paper. First of all, the reasons that cause SI and PI issues are discussed. Then based on ANSYS SIwave platform, the simulation process, key parameters analysis and optimization method of SI and PI issues in SiP design are introduced.

II. SIGNAL AND POWER INTEGRITY ANALYSIS

Signal integrity refers to signal quality on transmission lines, in other words, the signal is able to respond with correct timing and voltage amplitude in the circuit. If the signals reach the receiver with required timing, duration, and voltage amplitude, that is, the circuit has good signal integrity. Conversely, if the signals do not respond normally, signal integrity issues arise. With continuous acceleration of clock rate and fast iteration of semiconductor process, the signal edge rate in digital circuit system has increased to picosecond or even higher, which results in SI issues such as reflection, crosstalk, and synchronous switching noise in transmission lines getting worse.

In PI analysis, quantitative analysis will be carried out from three aspects: power supply voltage drop, power current density, and PDN (Power Distribution Network) impedance. Power voltage drop analysis is to detect whether there is an excessive voltage drop in power plane, and the criterion is less than 5% of the power voltage amplitude. The purpose of power density analysis is to obtain the current density distribution. Excessive current density may cause a large temperature rise and lead to package warping.

978-1-6654-1392-3/21 $31.00 © 2021 IEEE

In PDN system design, in order to suppress power noise, it's necessary to analysis and optimize PDN impedance to make it lower than the target impedance within a certain frequency [1-3]. In PDN filtering frequency band allocation, the filter band of VRM (Voltage Regulation Module) is less than 100KHz, on-chip capacitor is greater than 100MHz, and board and package level is between 100KHz and 100MHz. Therefore, in this SiP, the frequency range for target impedance optimization is 100KHz~100MHz.

A. Reflection

When signals travel along transmission lines, they will feel the transient impedance at every step of their transmission path. If the transient impedance changes, part of the signal will reflect opposite to the original direction. The magnitude of reflection is determined by how much the impedance mismatch. The greater the impedance mismatch, the greater the reflection. The schematic diagram of impedance matching is shown in Fig. 1. Z_t represents load impedance, and Z_0 is transmission line impedance. The magnitude of signal reflection can be expressed by reflection coefficient. In Fig. 2, the reflection coefficient can be acquired by (1). Γ is for the reflection coefficient. When $Z_0=Z_t$, load impedance equals to transmission line impedance, the reflection coefficient is 0. That is, no reflection occurs; when $Z_0 \neq Z_t$, reflection arises. Reflection is the immediate cause of overshoot, undershoot and ringing current, and it's the most common SI issue in high-speed digital circuits [4].

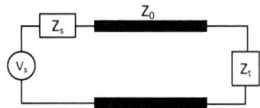

Fig. 1.   Reflection.

$$\Gamma = \frac{Z_t - Z_0}{Z_t + Z_0} \tag{1}$$

It can be seen from Fig. 1 and (1) that reflection is caused by the impedance mismatch between source, load and transmission line impedance. The main theoretical basis for suppressing reflection is to make load or source reflection coefficient zero by proper impedance matching. Generally, two ways are utilized for impedance matching, parallel termination and serial termination. Parallel termination is to add pull-up or pull-down resistors R_p close to the load termination. In this way, parallel termination matches the load equivalent impedance with the characteristic impedance of the transmission line. So, parallel termination is also called load termination matching. Serial termination is realized by inserting resistor R_s close to the source end. Serial termination is to match the source impedance with the transmission line characteristic impedance, and it is also known as source termination matching [4].

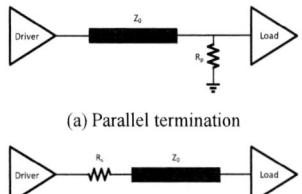

(a) Parallel termination

(b) Serial termination

Fig. 2.   Impedance matching schematic.

B. Crosstalk

Crosstalk is energy coupling from one transmission line to another. Alternating electric field produces magnetic field. And crosstalk arises when the magnetic fields generated by different transmission lines interact, which can be described in Fig. 3. In Fig. 3, there are two adjacent parallel transmission lines with the attack line on the right and the victim line on the left. When current passes through the attack line, a certain strength magnetic field appears in space. As the current changes, an alternating magnetic field will be generated near the attack line and the magnetic field lines cut the victim line. In this way, induced electromotive force generated, and finally formed induced current [4].

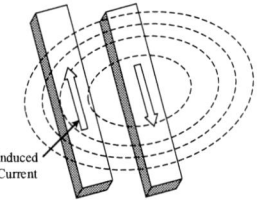

Fig. 3.   Generation of crosstalk.

In multi-transmission SiP systems, coupling between a large number of transmission lines (including capacitive coupling and inductive coupling) results in two harmful effects. Firstly, crosstalk changes characteristic impedance and transmission speed on lines, which brings adverse effects on system timing; secondly, crosstalk generates noise on other transmission lines, further reducing signal quality and noise margin.

In view of the causes of crosstalk, several methods to reduce crosstalk are summarized:

(1) Increase the spacing between signal paths and reduce the parallel coupling length;

(2) Termination matching is able to reduce or eliminate reflections, thereby reducing crosstalk;

(3) Take planes as return paths instead of lines;

(4) Protective wiring, that is inserting isolation lines between attack lines and victim lines.

C. IR Drop

The function of power supply system is to provide a stable voltage within a normal range for each device. When DC current flowing from power system to ICs through the interconnection network, there is a voltage drop because of the interconnection resistance, which is called "IR drop". Therefore, the actual voltage amplitude in IC power pin must be lower than the nominal value VRM. An excessively high IR drop is called "rail collapse", which may cause IC power pin voltage be lower than the minimum voltage required by the device, leading to circuit failure [5].

D. PDN Impedance

In actual case, the operating current in ICs fluctuates within a certain spectral width. And the interconnection impedance of PDN is not simple resistive. There are also inductive and capacitive impedance that vary with frequency. Therefore, the power impedance seen from the

978-1-6654-1392-3/21 $31.00 © 2021 IEEE   1362

IC power pin is frequency dependent. Keeping the PDN impedance below the target impedance value is the guideline in PDN design [5, 6]. In target impedance design method, PDN impedance should be controlled at a low value range within the bandwidth to ensure that the power noise stay in a reasonable range. The target impedance can be calculated by (2),

$$Z_{target} = \frac{V_{ripple}}{I} \qquad (2)$$

V_{ripple} represents power ripple, the amplitude usually 5% of the nominal supply voltage. I is the worst-case transient operating current. In practical calculation, it is usually 50% of the maximum operating current.

III. SIMULATION RESULTS

A. Reflection analysis

In this SiP, the longest and most complex path A9 is selected. To analyze the reflection of A9 path, SIwave S parameter simulation tool is utilized. The reflection result and insertion loss result are shown in Fig. 4. According to the device datasheet, the reflection should be lower than -15dB and the insertion loss higher than -0.5dB/mm. The length of A9 path is 36.16mm, so the insertion loss should be higher than -18.08dB. As shown in Fig. 3, the reflection is -31.5dB and the insertion loss is -0.22dB, which are all compliance with design specifications.

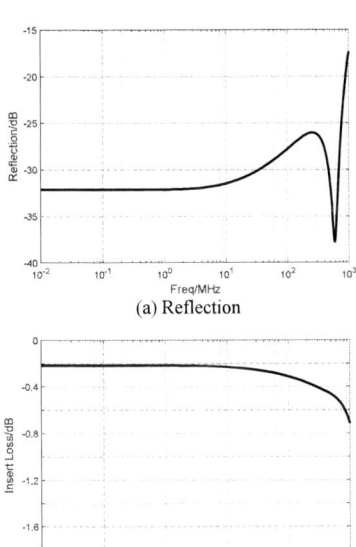

(a) Reflection

(b) Insert loss

Fig. 4. A9 reflection simulation results.

B. Crosstalk analysis

In this SiP layout, three paths with the shortest distance and the longest coupling length A8, A9, and A10 are selected for crosstalk analysis, as shown in Fig. 5. A8, A10 are attack lines and A9 is victim line. Based on SIwave S parameter simulation tool, the near-end crosstalk and far-end crosstalk simulation results shown in Fig. 6 are obtained. According to the specifications in datasheet, both near-end crosstalk and far-end crosstalk should be lower than -30dB. As shown in Fig. 6, the near-end crosstalk of A9 is -

37.62dB, and the far-end crosstalk is -51.51dB. The simulation results are compliance with design requirements.

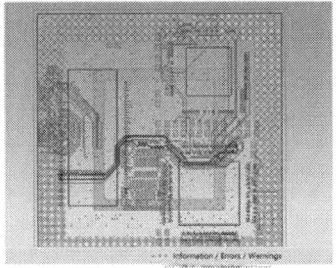

Fig. 5. A10/A9/A8 signal selection.

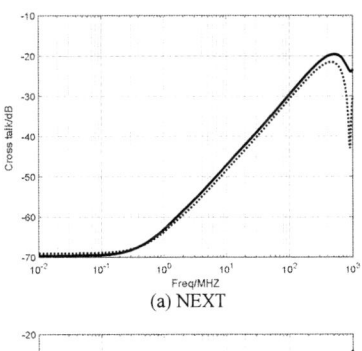

(a) NEXT

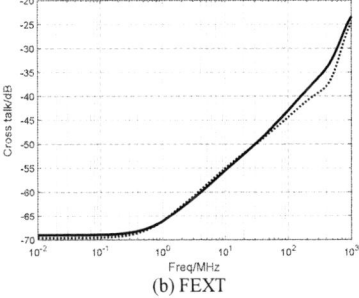

(b) FEXT

Fig. 6. A10/A9/A8 crosstalk simulation results.

C. Voltage, current density and temperature analysis

SIwave DC analysis tool creates accurate resistance models from the complex physical structure and calculates the IR drop from VRM to IC power pins. Allocate voltage and current sources for each IC according to their actual working conditions, and do DC simulation in SIwave. The voltage and current distributions are obtained in Fig. 7 and Fig. 8. Table I shows the power specification of the ICs from the datasheets. The top requirement for 1.2V power supply is a fluctuation of 4%, and for 3.3V is 3%.

TABLE I. VOLTAGE REQUIREMENTS.

ICs	1.2V	3.3V
U10	4%	3%
U1	NONE	9%
U6	NONE	9%
U9	10%	10%

978-1-6654-1392-3/21 $31.00 © 2021 IEEE

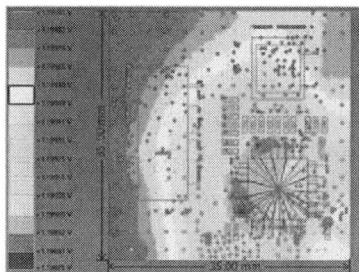

(a) 1.2V IR Drop

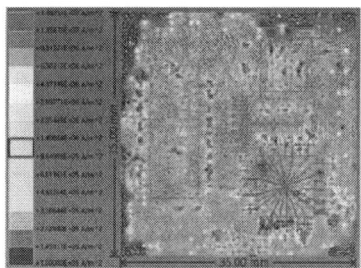

(b) 3.3V IR Drop

Fig. 7. IR Drop results.

Fig. 7 (a) shows IR Drop distribution for 1.2V power plane. The blue part is the minimum voltage, and the red part is the maximum. The minimum voltage of 1.2V power plane is 1.19875V (>1.2V * 96%). Fig. 7(b) is for 3.3V power plane. The minimum voltage of the 3.3V power plane is 3.2982V (>3.3V * 97%). The results indicate that 1.2V and 3.3V DC voltage distributions comply with the standard shown in Table I.

(a) 1.2V current density

(b) 3.3V current density

Fig. 8. Current density results.

As mentioned above, excessive current density may cause rises in temperature, hence package warping. Refers to IPC222-1A, the generic standard on printed board design, the reference value for current density is 1.01E+8A/m². As shown in Fig. 8, the maximum current density is 1.9826 E+7A/m² and 2.0584 E+7A/m² for 1.2V and 3.3V power plane respectively, both of which comply with the standard.

As mainstream thermal simulation software, Icepak is oriented for heat dissipation in electronic devices, from chip level to system level. Icepak Solver can be invoked in SIwave directly. According to the operating voltage and current listed in Table II, power dissipation is allocated to each IC. And temperature distribution is obtained in Fig. 9. It is clear that the highest temperature appears at U10. The highest temperature is 90.4°C and 78.7°C for 1.2V and 3.3V power supply, within a safe range.

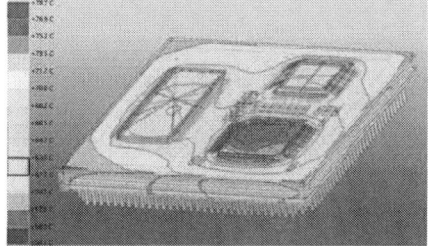

(a) 3.3V temperature distribution

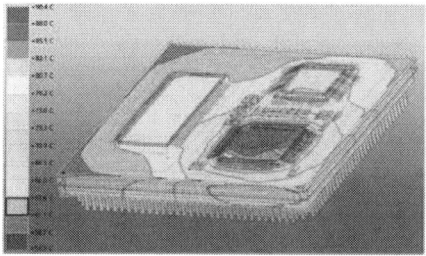

(b) 1.2V temperature distribution

Fig. 9. Temperature distribution in Icepak.

D. PDN impedance analysis

TABLE II.   VOLTAGE, CURRENT AND TARGET IMPEDANCE.

ICs	1.2V	3.3V	Z(1.2V)	Z(3.3V)
U10	1200 mA	300 mA	0.1 Ω	1.1 Ω
U1	0 mA	30 mA	None	11 Ω
U6	0 mA	25 mA	None	13.2 Ω
U9	20 mA	200 mA	6 Ω	1.65 Ω
U5	0 mA	150 mA	None	2.2 Ω

According to (2), target impedance for each IC can be obtained from their operating voltage and current, as illustrated in Table 2. Based on SIwave S parameter analysis tool, AC simulations are carried out for 1.2V and 3.3V power supplies within 100MHz in this SiP. The impedance distribution results are shown in Fig. 10. Comparing the 3.3V impedance analysis result shown in Fig. 10(a) with the target impedance value listed in Table II, it is clear that the PDN impedance of each IC is lower than the target impedance within 100MHz, which meets the design requirements. However, there is a peak impedance of 0.175Ω at 50MHz in U10 PDN illustrated in Fig. 10(b), which exceeds the target impedance value of 0.1Ω. Therefore, further optimization should be done to eliminate the hidden power integrity risk exists in U10 1.2V PDN.

978-1-6654-1392-3/21 $31.00 © 2021 IEEE

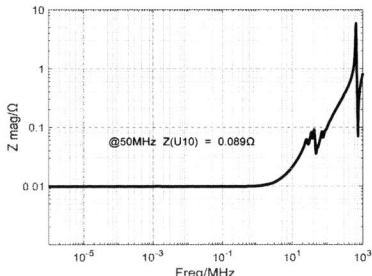

Fig. 11. U10 1.2V PDN impedance optimizing options.

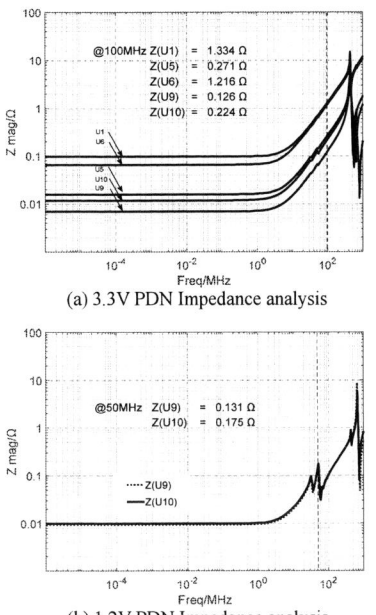

(a) 3.3V PDN Impedance analysis

(b) 1.2V PDN Impedance analysis

Fig. 10. Impedance analysis results.

The PI Advisor tool in SIwave is for decoupling network optimizing. It provides a variety of optimization solutions for designers to choose. As shown in Fig. 11, firstly, select the PDN network and capacitors to be optimized in PI Advisor; then set the target impedance mask according to the requirements in Table 2; and finally allocate capacitor models for the capacitors to be optimized and set simulation parameters. After several iterations of optimization, a variety of optimization schemes show up in PI advisor. In Fig. 11, scheme 9 best meets the target impedance requirement within the operating frequency range out of 10 optimized options, which can be used as the final scheme of this optimization process. Substituting scheme 9 into the circuit and do impedance analysis of 1.2V PDN again. Fig. 12 indicates the PDN impedance curve. The peak impedance is 0.089Ω within 100MHz, which shows that U10 PDN is compliance with the target impedance requirements after optimization.

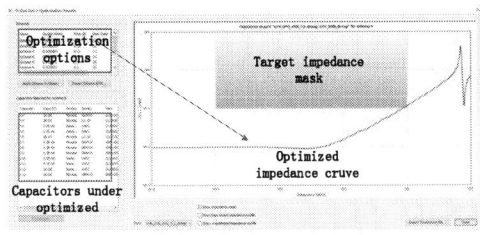

Fig. 12. U10 1.2V PDN optimized impedance result.

IV. CONCLUSION

As a kind of low-cost system integration technology, SiP integrates various functions on wearable devices. SiP enhances the portability, wirelessness and immediacy of wearable devices without additional volume. However, due to high layout and routing density and high output switching speed, signal and power integrity problems are more likely to arise in SiP. A high-performance, high-density SiP is illustrated in this paper. And the reasons that cause signal power integrity problems and control methods are discussed. Based on ANSYS SIwave, critical signal paths and power planes are verified and optimized. By effective simulations, design risks can be discovered in advance, and signal and power integrity problems will be eliminated in design stage, thereby avoiding wastage of time and cost, speeding up project progress.

REFERENCES

[1] M. Chandana, J. Mervin and David Selvakumar. Power integrity analysis for high performance design. 2015 International Conference on Control, Electronics, Renewable Energy and Communications (ICCEREC). 2015:48-53.

[2] Wang Li-xin, Zhang Yu-xia and Zhang Gang. Power Integrity Analysis for High-Speed PCB. 2010 First International Conference on Pervasive Computing, Signal Processing and Applications. 2010:414-418.

[3] J. Drewniak. Power integrity concepts for high-speed design on multi-layer PCBs. 2017 IEEE International Symposium on Electromagnetic Compatibility & Signal/Power Integrity (EMCSI).2017:1-41.

[4] Eric Bogatin. Signal and Power Integrity - Simplified, 3rd Edition. Pearson, 2018.

[5] Yang Zheng, Kan Hongwei, Liu Tiejun. Power Integrity Analysis of FPGA Heterogeneous Accelerate Board. Journal of Beijing University of Posts and Telecommunications, 2018, 41(4): 76-80.

[6] Poornima A. Khened and Sujata D. Badiger. Power integrity analysis for solid state drive PCB. 2016 International Conference on Emerging Technological Trends (ICETT). 2016:1-4.

978-1-6654-1392-3/21 $31.00 © 2021 IEEE

Lightweight Silver Nanowire Aerogel for Electromagnetic Interference Shielding

Fei Peng
Sauvage Laboratory for Smart Materials, School of Materials Science and Engineering
Harbin Institute of Technology (Shenzhen)
Shenzhen, China
hitpengfei@163.com

Yi Fang
Sauvage Laboratory for Smart Materials, School of Materials Science and Engineering
Harbin Institute of Technology (Shenzhen)
Shenzhen, China
fangyi@stu.hit.edu.cn

Zhehao Han
Sauvage Laboratory for Smart Materials, School of Materials Science and Engineering
Harbin Institute of Technology (Shenzhen)
Shenzhen, China
18344474766@163.com

Wenbo Zhu*
Sauvage Laboratory for Smart Materials, School of Materials Science and Engineering
Harbin Institute of Technology (Shenzhen)
Shenzhen, China
zhuwenbo@hit.edu.cn

Mingyu Li*
Sauvage Laboratory for Smart Materials, School of Materials Science and Engineering
Harbin Institute of Technology (Shenzhen)
Shenzhen, China
myli@hit.edu.cn

Abstract- The fast development of highly integrated circuit packaging electronics generates electromagnetic pollution, causing harmful effects on the performance of the facility and the health of human. Thus, there is an increasing demand for electromagnetic interference (EMI) shielding materials. Metals are traditional EMI shielding materials with extremely high EMI shielding effectiveness (SE) because of their high electrical conductivity. However, the disadvantages of metals, including high density, poor corrosion resistance and poor workability, restrict their use. At present, carbon materials (such as graphene, CNF, and CNTs) are widely used to prepare EMI materials because of their high flexibility and conductivity. However, the complex synthesis process of carbon-based fillers and poor dispersion limit their application. Recently, silver nanowires are considered excellent fillers for the preparation of EMI materials due to their excellent conductivity, dispersibility and flexibility. Herein, we prepared silver nanowire aerogels by the method of freeze casting and freeze drying. The aerogel has an ultralow density below 5 mg/cm^3 and the porosity is as high as 99.3%. The conductivity of aerogels is further enhanced with the treatment of NaCl solution. The EMI SE of the aerogel approaches 75.2 dB at a thickness of 4 mm due to the interconnected three-dimensional network structure and excellent conductivity of 28368 S/m.

Keywords—silver nanowire aerogel; sintering; electromagnetic interference shielding;

I. INTRODUCTION

The rapid development of science and technology has promoted the use frequency of electronics equipment, which has significantly improved the quality of people's life, but it has also brought severe electromagnetic interference (EMI) problems. Excessive electromagnetic interference will affect the operation of electronic facility and may bring about the leakage of important information, which poses a threat to national security. In addition, the electromagnetic wave radiation may cause potential threats to the human body. Therefore, it is extremely essential to develop electromagnetic shielding materials with high performance. Minimizing the

interference of electromagnetic waves is the primary purpose of electromagnetic shielding. Therefore, the electromagnetic interference shielding materials need to have high conductivity so that most of the electromagnetic waves can be absorbed instead of transmitting.

Commonly used electromagnetic shielding materials include metals, carbon materials, magnetic materials and polymer materials. Metals display superb EMI shielding performance because of their high conductivity, whereas their shortcomings of high density, low flexibility and poor corrosiveness, etc[1]. Carbon nanostructures (graphene, CNTs, CNF, and carbon black) are considered excellent candidates for electromagnetic shielding materials due to their excellent conductivity and flexibility, corrosion resistance and low density, but the complex synthesis process and difficulty in dispersion restrict their application[2]. Recently, the silver nanowire is considered a potential filler to fabricate EMI shielding materials on account of its high conductivity, good flexibility, outstanding machinability, ease to synthesis and dispersion[3]. However, the improvement of conductivity of EMI shielding material is limited due to the connect resistance between the silver nanowires, which limiting the enhancement of EMI shielding performance. Furthermore, the EMI shielding materials using silver nanowires in the current literature are mostly two-dimensional materials, or composites. The connections between silver nanowires are not tight, so the electromagnetic waves can not be consumed by multiple reflections inside the EMI shielding materials. Aerogels are unique materials with some interesting characteristics, such as low density, high strength-to-weight ratio and high surface area. Recent years, aerogels have attracted considerable attention. Metal aerogel is an emerging aerogel material that combines the advantages of aerogels and metals. Metal aerogels have the characteristics of high conductivity and high porosity, which make them have apparent advantages in electromagnetic shielding.

Recently, impressive progress has been made in the development of EMI shielding materials with low density and high performance using porous structures. For instance, CNT sponge was prepared by the method of CVD. Because of the low density of CNT and porous structure, the density of CNT sponge is as low as 20 mg/cm^3. Due to the high conductivity and ultralow density, the EMI SE and specific SE (SSE) of the

---

The authors acknowledge the support from the Science and Technology Innovation Committee of Shenzhen (JCYJ20180306172006392), the grant from State Key Laboratory of Advanced Welding and Joining (AWJ-21M0), and the support from National Nature Science Foundation of China (No.52075125).

978-1-6654-1392-3/21 $31.00 © 2021 IEEE

CNT sponge approached 20 dB and 1100 dB·cm^3/g, respectively, and showing an absorption mechanism. Reduced graphene oxide/lignin-derived carbon composite aerogels were fabricated by freeze-drying method[4]. The aerogels have uniform micro pores and cell walls, forming synergistic effects and leading to high EMI SE of 21.3 dB to 49.2 dB at the density of 2 to 8 mg/cm^3. In addition to carbon aerogels, metal foams were also used to prepare EMI shielding materials. Cu-Ni foam integrated with CNTs was fabricated the electrophoretic deposition method[5]. With the increase of density and thickness, the EMI SE of Cu-Ni composite increased. Specifically, a sample with a thickness of 1.5 mm and a PPI of 110 showed a maximum and average EMI SE of 54.6 dB and 47.5 dB, respectively. The current research on porous shielding materials mainly focuses on carbon aerogels, while there is little research on metal aerogel shielding materials.

Herein, we prepared silver nanowire aerogels by freeze-casting and freeze-drying method. Subsequently, the silver nanowire aerogels were sintered at room temperature using electrolyte solution. The aerogels we prepared have a density as low as 5 mg/cm^3 and porosity as high as 99.3%. EMI SE of the aerogel approaches 73.5dB at a thickness of 4mm due to interconnected three-dimensional network structure and excellent conductivity of 28368 S/m.

II. EXPERIMENTAL PART

A. Materials

Poly-(vinylpyrrolidone) (PVP) (Mw = 55000) and iron (III) chloride (FeCl$_3$) were supplied by Sigma-Aldrich. Silver nitrate (AgNO$_3$) was purchased from Sinopharm Chemical Reagent Co., Ltd. NaCl and ethylene glycol (EG) were purchased from Aladdin Biochemical Technology Co., Ltd.

B. Synthesis of AgNW Gels and Aerogels

The silver nanowires were prepared by hydrothermal reaction. Firstly, 0.2g PVP was slowly added to 30 ml EG and dissolved by magnetic stirring. When the PVP is completely dissolved, 0.5g AgNO$_3$ was added to the PVP solution and dissolved by magnetic stirring. Afterward, 2 ml FeCl$_3$ (1mM) EG solution was added into the above solution and stirred for a certain minutes. Then, the above solution was transferred to a reaction kettle, subsequently be placed in a blast drying oven for heating reaction with 140 °C for 5 hours. After the reaction was completed, we obtained silver nanowires. The silver nanowires were centrifuged and cleaned with deionized water at 3000 rpm/min for 5 min several times to remove PVP on the surface of nanowires. After that, use a vortexer to redisperse the AgNWs in deionized water, the suspension was subsequently frozen in liquid nitrogen. AgNW aerogels were then fabricated by freeze-drying the mixture at -80°C and 0.05 mTorr. The prepared AgNW aerogels were put into NaCl solution (0.3M) for 1 hour to sinter the nanowires. Finally, the sintered aerogels were obtained by the same freeze-drying method.

C. Characterization and testing

The morphologies of the obtained AgNW aerogels were observed by SEM with an acceleration voltage of 5 kV (MDTC-EQ-M18-01, SU8010). The electrical conductivity of the AgNW aerogels was measured by a four-probe tester equipped with a digital source meter (2450 series produced by the American Keithley company). The correlated electromagnetic shielding performance was investigated using a Vector network analyzer (ZNA43 produced by ROHDE&SCHWARZ).

III. RESULTS AND DISCUSSION

Fig. 1 shows the as-prepared silver nanowire suspension, AgNW hydrogel and aerogel. Firstly, a homogenous solution including PVP, EG, AgNO$_3$, and FeCl$_3$ was added into a reactor and then heated in an oven. After the reaction has been completed, the color of the solution changed to silver-gray as shown in Fig. 1a, from which we judged the formation of silver nanowires. The AgNW suspension was freeze-dried and the product was placed in an electrolyte solution for processing, as shown in Fig. 1b. It can be seen that the hydrogel is suspended above the solution, while the morphology and structure of hydrogel will not change after a long time. The aerogel is off-white and does not shrink after freeze-drying, as shown in Fig. 1b. It can be placed on the petals without bending them, indicating that the density of aerogel is ultralow. It should be noted that the shape of aerogel is determined by the freeze casting container, which can be changed as needed. For instance, we can easily fabricate different shapes of AgNW hydrogels (such as cube, cylinder, and sphere) by using various shapes of containers. The impurities and remaining solvents were removed using plenty of DI water. Put the wet gels in the NaCl solution (1M) as shown in Fig. 1b, and the AgNWs were sintered at room temperature. Furthermore, the structure of wet gel in the solution has not changed after a few days, indicating the structure of the gels is stable. Finally, the aerogels were obtained by freeze-drying the wet gels and the morphology of aerogel is shown in Fig. 1c. The lightweight aerogel can be placed on the petals and the petals will not be bent, indicating the ultralow density of the AgNW aerogels. Compared with AgNW aerogels reported in other literature[6], the morphology of AgNW aerogel prepared in this paper is smoother. Furthermore, the size of the aerogel can be as large as a few hundred millimeters or as small as a few millimeters as needed.

Fig. 1. Morphology of the as-prepared (a) silver nanowire suspension; (b) AgNW hydrogel; (c) AgNW aerogels

The internal microstructure of the silver nanowire aerogel is shown in Fig. 2. As depicted in Fig. 2, AgNW aerogel has an cross-linked porous network. The shape of pores is irregular, while the size is within the range from a few microns to tens of microns. The aerogel consists of ultra-thin nanowire network without obvious bundles. After the aerogel is treated in an electrolyte solution, the PVP on the surface of silver nanowires is thoroughly removed, while the silver nanowires are sintered at places where they overlap. As shown in Fig. 2b, there are chemical and physical cross-linking in aerogels. As displayed by the black circles, nanowires contact on the surface of another, forming physical cross-linked points. As displayed by the red circle, fused spherical structures are formed at the place where two or more silver nanowires overlap each other, implying the formation of chemical cross-linked points. Due to the ultra-high porosity of aerogels, the density is lower than 5 mg/cm^3, about 2000 times lighter than bulk silver (10.49 g/cm^3). The contact resistance between silver nanowires has been reduced because of the physical and chemical cross-linking, so that the conductivity of aerogels is greatly improved, and the structural strength is further enhanced. Since electrical conductivity is a key factor affecting the electromagnetic shielding performance of materials, the improvement of conductivity is of great significance for improving the EMI shielding performance of materials.

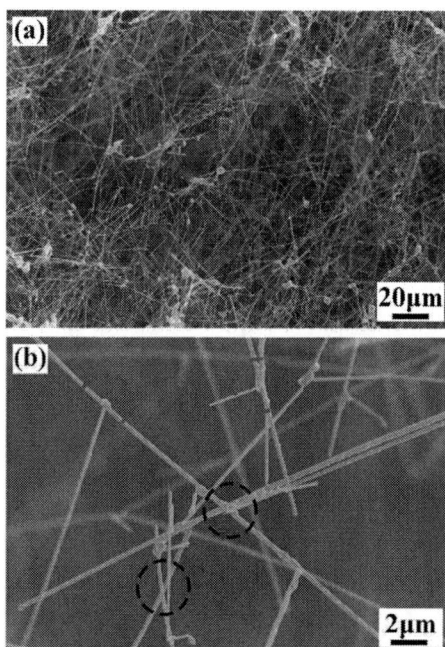

Fig. 2. SEM morphology of AgNW network in the AgNW aerogel

The cross-linking points of the aerogels make the structure stable and robust and improve the conductivity. The AgNW aerogels have excellent electrical conductivity as shown in Fig. 3. There is a good linear relationship between electrical conductivity and density. The electrical conductivity increases from 4023 to 28368 S/m with the increase of density of aerogel from 4.25 to 23.86 mg/cm^3. Compared to other metal aerogels at similar densities, such as AgNW aerogels (1667 S/m)[6] and CuNW aerogels (116 S/m)[8], the AgNW aerogels prepared in this work have much-improved conductivities. The significant enhancement of electrical conductivity can be attributed to the interconnected network and junction welding, which greatly reduce the contact resistance between silver nanowires and provide an abundant conductive network.

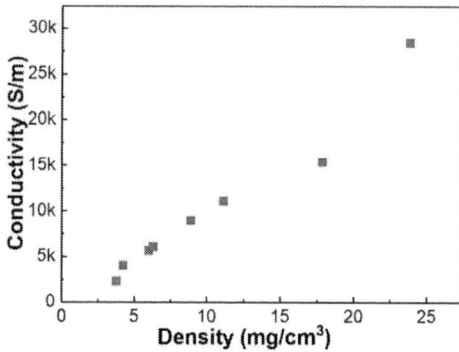

Fig. 3. The relationship of conductivity and density of AgNW aerogel

Electromagnetic interference (EMI) shielding effectiveness (SE) is usually used to quantitatively describe the shielding ability and effect of shielding materials. The total shielding effectiveness of shielding materials is expressed by SE$_T$, including absorption (SE$_A$), reflection (SE$_R$) and multi-reflection (SE$_{MR}$). For high-performance shielding materials, SE$_{MR}$ can be ignored when the value of SE$_A$ is greater than 15 dB, because multi-reflections will be absorbed eventually. Therefore, the total EMI SE can be given by the following formula:

$$SE_T\ (dB) = SE_R + SE_A \qquad (1)$$

The reflection loss (SE$_R$) can be evaluated by Fresnel's equations as follows:

$$SE_R\ (dB) = 168 - 10log\frac{f\mu}{\sigma} \qquad (2)$$

Absorption loss (SE$_A$) for non-magnetic and conducting shielding can be expressed as follows:

$$SE_A\ (dB) = 131.4d\sqrt{f\mu\sigma} \qquad (3)$$

where σ, d, and μ represent the conductivity, thickness, and the magnetic permeability of the shielding materials, respectively. f represents the frequency of the incident electromagnetic waves.

Silver nanowire aerogels have high porosity and large specific surface aera, providing many reflective interfaces. Therefore, when the incident electromagnetic waves penetrate into the AgNW aerogels, multi-reflections will occur in the porous architectures. Most multi-reflected electromagnetic waves can be eventually absorbed by shielding materials, resulting in an increase in the total EMI SE.

EMI shielding efficiency of AgNW aerogels with different thicknesses is shown in Fig. 4. Fig. 4a shows SE$_T$ of the aerogels from 8.2 to 12.4 GHz. SE hardly changes with frequency in test frequency band, indicating the uniform microstructure and electrical conductivity of the aerogels. When the thickness increases from 2 mm to 4 mm, the SE$_T$ increases from 42.1 dB to 75.2 dB. In addition, the specific shielding efficiency (SSE) of the AgNW aerogel reaches 14061.9 dB·cm^3/g, which is at a relatively high level in the reported literature. For AgNW aerogels, sintering improves the conductivity of the AgNW skeleton, which enhances EM waves reflection and conducting loss.

Fig. 4b shows the changes of SE$_R$, SE$_A$ and SE$_R$ of the AgNW aerogel from the thickness of 2 mm to 4 mm. SE$_R$ hardly changes with the change of thickness, all around 20 dB. According to equation (2), SE$_R$ is related to f, μ and σ, which only dependents on the nature of shielding materials and the frequency of the incident wave, and has nothing to do with thickness. Therefore, for aerogels of the same density, μ and σ are constant and SE_R will not change. On the contrary, SE_A changes dramatically with the change of thickness, as shown in Fig. 4b. The SE_A increases from 21.8 dB to 54.5 dB, while the thickness of the AgNW aerogel increases from 2 mm to 4 mm. The increase of SE_A almost entirely provides the increase of SE_T. On the one hand, for AgNW aerogels, the multi-reflections of electromagnetic waves inside the material increase with the increase of aerogel thickness. The shielding material absorbs the electromagnetic wave after multi-reflections. On the other hand, according to equation (3), SE_A is proportional to shielding materials' thickness. Therefore, SE_A changes linearly with the thickness.

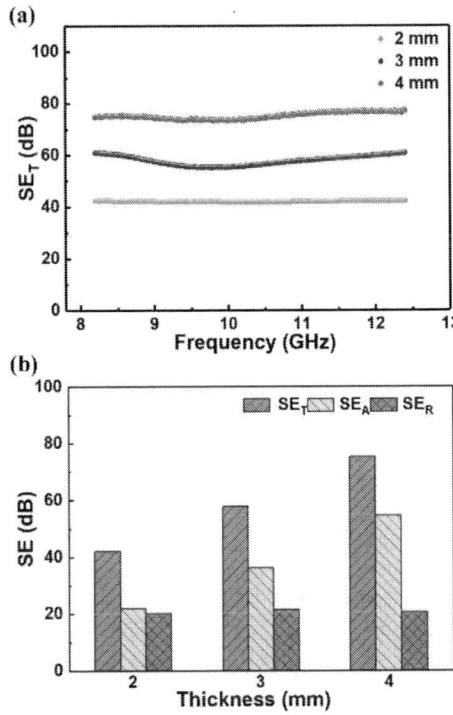

Fig. 4. EMI SE of the AgNW aerogel with different thickness: (a) SE_T at 8.2-12.4 GHz; (b) SE_T, SE_A and SE_R of the AgNW aerogel

Evaluation of SE_T for our AgNW aerogels depends on the aerogel thickness. The values of SE_T for the 4 mm thick AgNW aerogel specimen (Fig. 4) correspond to 75 dB in the whole 8.2-12.4 GHz range, where absorption dominants the attenuation mechanism. Since the AgNW aerogel we prepared possesses ultralow densities lower than 5 mg/cm^3, comparisons with other lightweight and porous shielding materials are listed in Table. 1, including EMI SE and specific SE (SSE). The EMI SE and SSE for our AgNW aerogel approach to 75.2 dB and 14061.9 dB·cm^3/g, respectively, which are higher than those reported porous shielding materials as shown in Table. 1.

Table. 1. EMI shielding performance of various porous materials reported

Materials	EMI SE (dB)	Thickness (mm)	SSE (dB·cm^3/g)	Ref.
CNT sponge	22	2.38	1100	[4]
Graphene foam	25.5	0.3	420	[4]
Graphene aerogel	22.3	2	4956	[4]
CuNi foam	15-25	1.5	63-104	[5]
AgNW@C sponge	37.9	1	9921	[9]
AgNW aerogel (present work)	75.2	4	14061.9	/

IV. CONCLUSION

AgNW aerogels were fabricated by freeze casting, freeze drying, and subsequent electrolyte solution sintering. AgNW aerogel has an outstanding conductivity of 28368 S/m with an ultralow density of 28.36 mg/cm^3. Benefit from the high conductivity and porosity of AgNW aerogels, the total EMI SE reaches 75.2 dB at a low AgNW aerogel density of 5.15 mg/cm^3 and a thickness of 4 mm. The SSE approaches to 14061.9 dB·cm^3/g. AgNW aerogel also shows high flexibility. This work broadens the avenue to prepare high-performance EMI shielding materials for outcoming flexible electronics.

V. ACKNOWLEDGMENT

The authors acknowledge the support from the Sauvage Laboratory for Smart Materials of Harbin Institute of Technology (Shenzhen).

REFERENCES

[1] D.W. Jiang, V. Murugadoss, Y. Wang, et al. "Electromagnetic Interference Shielding Polymers and Nanocomposites-A Review". Polym. Rev.. 2019, vol. 59, pp. 280-337.

[2] L. Wang, X.T. Shi, J.L. Zhang, Y.L. Zhang, J.W. Gu. "Lightweight and robust rGO/sugarcane derived hybrid carbon foams with outstanding EMI shielding performance-ScienceDirect". J. Mater. Sci. Technol., 2020, vol. 52, pp. 119-126.

[3] X.Z. Zhu, J. Xu, F. Qin, Z.Y. Yan, A.Q. Guo and C.X. Kan. "Highly efficient and stable transparent electromagnetic interference shielding films based on silver nanowires" [J]. Nanoscale, 2020, vol.12.

[4] Z.H. Zeng, C.X. Wang, Y.F. Zhang, P.Y. Wang, S.I.S. Shahabadi, Y.M. Pei. "Ultralight and highly elastic graphene/lignin-derived carbon nanocomposite aerogels with ultrahigh electromagnetic interference shielding performance". ACS Appl. Mater. Interfaces, 2018, vol. 9, pp. 8205-8213.

[5] K.J. Ji, H.H. Zhao, J. Zhang, J. Chen, Z.D. Dai. "Fabrication and electromagnetic interference shielding performance of open-cell foam of a Cu–Ni alloy integrated with CNTs". Appl. Surf. Sci., 2014, vol. 311, pp. 351-356.

[6] F. Qian, P.C. Lan, M. Freyman, W. Chen, T. Kou, T.Y. Olson. "Ultralight Conductive Silver Nanowire Aerogels". Nano Lett., 2017, 7b02790.

[7] P.L. Yan, E. Brown, Q. Su, J. Li, J. Wang, C.X. Xu. "3D Printing Hierarchical Silver Nanowire Aerogel with Highly Compressive Resilience and Tensile Elongation through Tunable Poisson's Ratio". Small, 2017, vol. 13.

[8] S. M. Jung, D.J. Preston, H.Y. Jung, Z.T. Deng, E.N. Wang, J. Kong. "Porous Cu Nanowire Aerosponges from One-Step Assembly and their Applications in Heat Dissipation". Adv. Mater., 2016.

[9] Y.J. Wan, P.L. Zhu, S..Yu, R.Sun, C.P.Wong, W.H.Liao. "Anticorrosive,Ultralight, and Flexible Carbon-Wrapped Metallic Nanowire Hybrid Sponges for HighlyEfficient Electromagnetic Interference Shielding". Small, 2018, vol. 14, pp. 1800534.

A Facile Bonding Material to Enable Interconnection among complex Surfaces through AgNWs Aerogel

Zhehao Han
Sauvage Laboratory for Smart Materials, School of Materials Science and Engineering
Harbin Institute of Technology (Shenzhen),
Shenzhen, China
18344474766@163.com

Mingyu Li*
Sauvage Laboratory for Smart Materials, School of Materials Science and Engineering
Harbin Institute of Technology (Shenzhen),
Shenzhen, China
myli@hit.edu.cn

Wenbo Zhu*
Sauvage Laboratory for Smart Materials, School of Materials Science and Engineering
Harbin Institute of Technology (Shenzhen),
Shenzhen, China
zhuwenbo@hit.edu.cn

Fei, Peng
Sauvage Laboratory for Smart Materials, School of Materials Science and Engineering
Harbin Institute of Technology (Shenzhen),
Shenzhen, China
hitpengfei@163.com

Yi Fang
Sauvage Laboratory for Smart Materials, School of Materials Science and Engineering
Harbin Institute of Technology (Shenzhen),
Shenzhen, China
1149354663@qq.com

Abstract—Metal aerogels are a new material class with unique properties, including lightweight, high surface area, high reactivity, and high electrical conductivity, making them highly attractive in various applications, such as sensing, thermal insulation, energy generation/storage, and catalytic conversion. A novel flexible preformed film has been produced from the AgNW aerogel to realize the flexible integration among complex structures in this work. As the AgNW forming bulk 3D networks in the aerogel has excellent flexibility, stability and high reactivity and conductivity, the proposed preformed film can enable the interconnection between metals like Cu or Ag through self-sintering without additional flux or solders. According to the experimental data, reliable and flux-free bonds can be achieved between components with U-shaped, V-shaped, round-shaped or step-shaped structures through the preformed AgNW aerogel film, with a high shear strength over 30 MPa, a low electrical resistivity less than 4.5 μΩ·cm and a large thermal conductivity attaining 110 W/m*K, if proper temperature and stress were applied. Meanwhile, these flexible preformed films can also provide ultra-high electromagnetic shielding effect over 75 dB in the frequency range of 8-12 GHz, and be further combined with PDMS to serve as a flexible circuit with a high electrical conductivity over 1.4*10^6 S/m. In summary, this work provides novel approaches to realize the adaptive interconnection between complex structures and provide new material classes for electromagnetic shielding and flexible circuit, which may enable the extended usage of metal aerogel materials in flexible electronics and packaging.

Keywords—AgNW aerogel, preformed film, complex structures, adaptive interconnection.

I. INTRODUCTION

The next-generation automotive and aerospace industry products have to integrate electronic components or functional sensors on devices to form multi-functional high-performance systems [1-5]. Many research groups have used island-bridges structures, pleated structures or flexible board carriers to

The authors acknowledge the support from the Science and Technology Innovation Committee of Shenzhen (JCYJ20180306172006392), the grant from State Key Laboratory of Advanced Welding and Joining (AWJ-21M0), and the support from National Nature Science Foundation of China (No.52075125).

achieve large-scale deformation capabilities for high-performance silicon-based devices and produce intelligent systems on natural skin, optical lenses, robotic arms and other complex curved surfaces [6-10]. In these systems, the functional devices and the complex matrix are closely connected, which effectively suppresses sensor noise and spurious signals to realize the high-quality distributed sensing and high-precision signal detection or transmission. However, the structural features in different systems are generally different. It is difficult to accurately align or mount functional devices or signal circuits on the devices, which causes severe structural mismatches, heavy residual stress and signal noise in the system. Hence, how to realize the flexible packaging of functional components on complex structures has become a key issue restricting the manufacturing and development of a new generation of high-performance electronic systems [10-12].

The 3D integration on non-planar structures is mainly completed by adhesive bonding and eutectic soldering or nano-sintering assisted by flexible printing techniques [13-16]. However, the mechanical properties, stability, and electrical and thermal conductivity of adhesives are too poor to provide reliable interconnection under long-term service or high-temperature conditions. Hence, the application of adhesive in the 3D integration of the high-performance system is limited [17]. On the other hand, the solder pastes or nano-inks utilized for printing always have strong fluidity and contain volatile components, such as fluxes and surfactants. While it is difficult to control the shape and evenness of printed materials on the surface of complex structures, the volatile gases released from the pastes or nano-inks can easily accumulate in the interconnection, especially in the central bonding region or the U/V-shaped corner, causing large-scale voids or cracks in the joint and many reliability issues like tombstoning, stress concentration or desoldering. These problems severely restrict the manufacture and application of high-performance complex structure electronic systems. Therefore, exploring flexible connection materials and integration processes that

978-1-6654-1392-3/21 $31.00 © 2021 IEEE

can spontaneously adapt and match various complex structures has become a critical issue that needs to be resolved in the manufacturing and development of high-performance systems.

The metal aerogels are a new material class with a unique combination of properties, including lightweight, high flexibility, adiabatic ability and high electrical conductivity, making them highly attractive in various applications, such as energy generation and storage, sensing, catalytic conversion, thermal insulation, and so forth [18-21]. If flexible aerogel of high reactivity can be prepared, it is possible to realize the adaptive deformation among complex structures and enable the metallurgical interconnection through self-sintering. Hence, the authors have developed a new preformed film based on the AgNW (Ag nanowire) aerogel. A series of bonding experiments and tests have been implemented to evaluate its structural adaptability and weldability. According to the experimental results, the preformed film can produce reliable interconnections with high shear strength, low electrical resistance and high thermal conductivity between complex structures. Moreover, it is also capable of providing ultra-high electromagnetic shielding effectiveness or serving as a flexible circuit. For these unique advantages, this proposed preformed AgNW aerogel films (PAFs) should have great potential in the manufacturing and application of electronics packaging.

II. EXPERIMENTS

The AgNW aerogels were produced through a modified polyol method. For a typical synthesis, the precursor (AgNO$_3$, 99.8 %) and the developed catalyst were mixed in a flask containing surfactants, oxidants, and ethylene glycol (C$_4$H$_{10}$O$_3$, 99.9 %). After stirring for 2 hours, all the reactants were dissolved to form a light-yellow transparent solution. This solution was then put in a 160 °C oven to initiate the redox reaction, and a white jelly-like AgNW alcogel containing cross-linked AgNW network could be obtained in 4 hours of heating. Note that free Ag nanowires or nanoparticles might be formed in this process, which contributes negatively to physical performance as they do not transmit electrons yet add significant weight, meanwhile the residual reactants and EG also reduced the conductivity and solderability of the AgNW network. Hence, a necessary purification procedure is implemented following the polyol reaction to remove the residual reactants and enable the selective precipitation of free NWs and NPs. After 1 hour of purification, alcogel of almost 100 % AgNWs was obtained, which was then supercritical dried by CO$_2$ to produce the as-expected AgNW aerogel. These AgNW aerogel were diced into blocks of different size, typically 10×10×10 mm and 20×10×5 mm. Finally, PAFs of 500 μm in thickness were obtained after cold-pressing.

To evaluate the joining performance of PAFs, bonding pairs of copper plates with round-shaped, V-shaped, U-shaped and stepped-shaped interfaces were prepared by wire cutting, as shown in Fig. 1. These copper plates were electroless-plated by silver and ultrasonically cleaned in ethyl alcohol for 5 min before bonding. Then, the PAFs were inserted into the bonding pairs, forming a typical sandwiched structure, and the thermo-compression bonding was carried out in the air through an AJM XL-TC 200 hot press at the temperature of 200-300 °C. Meanwhile, the as-made PAFs were also mixed with the PMDS monomer through vacuum filtration, as such

an AgNW-PDMS composite structure was produced to realize the electromagnetic shielding and flexible circuit.

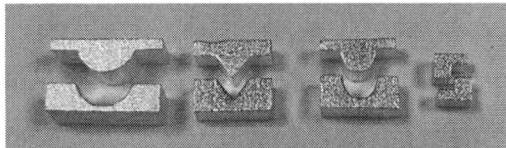

Figure 1. The Cu bonding pairs electroless-plated with Silver.

The morphologies of the obtained AgNW aerogels are observed by SEM with an acceleration voltage of 20 kV (MDTC-EQ-M18-01, SU8010). Cross-sectional samples of the PAFs interconnections were prepared and polished by an Ar2-plasma polisher (CP, SM-09010, JEOL) to prevent any distortion or particle embedding in the microstructure. The cross-sectional morphologies were characterized by SEM, and the shear strength of the interconnections was measured using a MFM1200 shear tester with a shear speed of 200 μm/s and a 50 μm shear height from the surface. Moreover, the electrical and thermal conductivity of the interconnections and AgNW-PDMS were tested using a four-probe tester and laser thermal conductivity meter, and the correlated electromagnetic shielding performance was investigated using a Vector network analyzer from ROHDE&SCHWARZ.

III. RESULTS AND DISCUSSION

The morphology of the as-prepared AgNW alcogel, aerogel and PAFs are shown in Figure 2. As illustrated in Fig. 2a and Fig. 2b, white jelly-like alcogel and the ultra-lightweight aerogel were evidently formed after polyol reaction and purification. Then, through laser-dicing and hot-pressing, aerogel blocks and PAFs of different size and shapes could be easily produced, as demonstrated in Fig. 2c. According to the microstructure of aerogel in Figure 3, a continuous 3D network of highly cross-linked AgNWs was generated in the aerogel. Different from the metal aerogels prepared by freezing-drying method in Xu's [22] study, little organic phases could be detected in the purified aerogel, making it a pure metal material with high reactivity. Meanwhile, the AgNWs in one-step produced aerogels have more uniform distribution, and hence resulting in better flexibility and stability. For these advantages, the PAFs could be utilized to fill the gaps in the non-planar bonding pairs and

Figure 2. Morphology of the as-prepared (a) AgNW alcogel, (b) AgNW aerogel and (c) the aerogel blocks and PAFs.

realize a reliable interconnection.

The SEM morphologies of the Cu bonding pairs with different shapes of interface are shown in Figure 4. Dense soldering structures without cracks and delamination were formed at bonding interfaces. For these complex structures, solder paste and traditional solder materials cannot be evenly filled to form a good solder joint. From the magnified topography, massive nano-voids and sizable sintered bulks were observed, being consistent with the morphology of

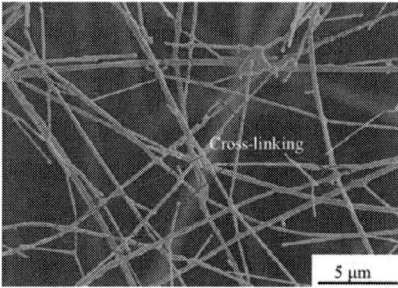

Figure 3. SEM morphology of AgNW network in the AgNW aerogel.

sintered Ag nanoparticles, which might be caused by the gaps in initial PAFs and the accumulation of vacancies in sintering. As the AgNW aerogel has excellent electrical, thermal and mechanical properties, the obtained Cu interconnections have better performance in these respects when compared to traditional interconnections formed by SAC alloys. From the shear tests, the strength of these interconnections can attain 33 MPa and 19.8 MPa at room condition and 350 ℃, respectively, when bonded at 250 ℃ for 20 min under 20 MPa, which could even satisfy the requirements of power electronics. In addition, the AgNW aerogel showed an obvious exothermic peak at 200 ℃ in DSC, which is familiar with the nano-silver paste. However, compared with nano-silver solder paste, the AgNW aerogel has more stable and better shear strength (38.03 MPa) in large-size bonding solder joints(10×10mm). Meanwhile, the electrical resistivity of the bonded region was measured to be less than 4.5 $\mu\Omega\cdot$cm (close to the resistivity of bulk silver, 1.65 $\mu\Omega\cdot$cm), and the thermal conductivity can attain 110 W/(m*K), much higher than any traditional solder alloys (better than the SAC305, 58W/(m*K)) or thermal interface materials. Therefore, the PAFs proposed

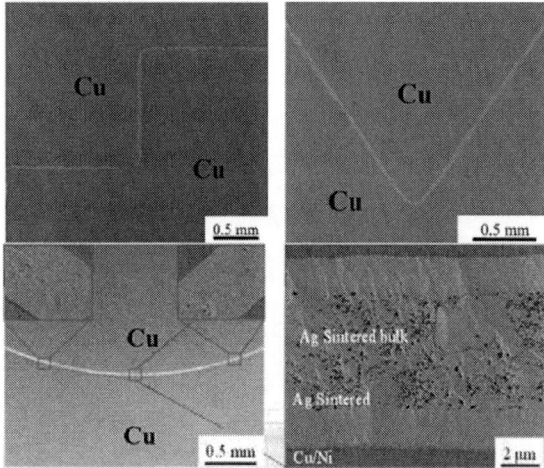

Figure 4. SEM image of APs curved joint section microstructure.

in this work can well meet the requirements of packaging on complex structures.

Figure 5 illustrates the morphology of the flexible circuit

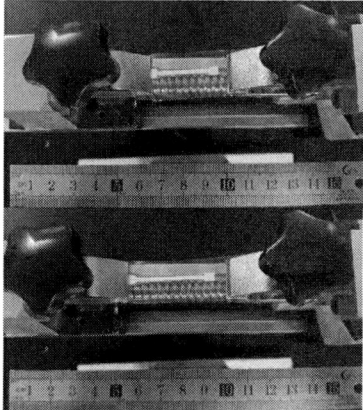

Figure 5. Morphology of the flexible circuit prepared by combining the AgNW aerogel preformed film with PDMS, which shows great flexibility and stretchability.

produced by AgNW-PDMS composite material. According to the results from repeated tensile tests, this structure was stretchable, friction-resistant, highly conductive (exhibiting a minor resistance of 21.3 mΩ, in other words, high electrical conductivity of 1.4*10^6 S/m), and remained stable even after 10000 times stretching. Hence, this AgNW-PDMS structure was also capable of serving as flexible circuits for complex electronics.

On the other hand, the electromagnetic shielding effect of the as-prepared PAFs is also investigated. As shown in Fig. 6, the shielding effectiveness of the 500 μm-thick PAF could attain over 75 dB in the frequency range of 8-12 GHz, much higher than traditional shielding film fabricated by AgNWs. This improved SE might be attributed to the cross-linked AgNW networks in PAFs, which enhanced the transmission of electrons and further improve the absorption capability of electromagnetic signals. Noted that this PAF can be easily combined with flexible substrate materials to improve its stability and environmental tolerance, such as PI, PDMS. Hence, it is also a promising material for the electromagnetic shielding of electronics.

IV. CONCLUSION

In summary, a flexible preformed film has been developed from AgNW aerogel to enable the adaptive interconnection of complex structures. Reliable bonds with dense microstructure have been successfully obtained between round-shaped, V-

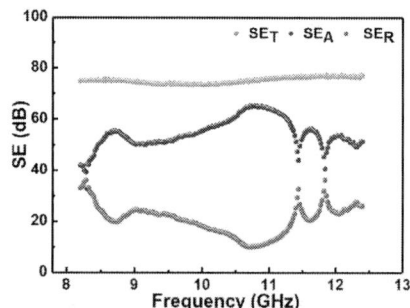

Figure 6. The electromagnetic shielding performance of the PAFs in the frequency range of 8-12 GHz.

shaped, U-shaped and step-shaped components. The shear strength of the bonded region can reach 33 MPa, and the correlated electrical and thermal conductivity attain 4.5 $\mu\Omega\cdot$cm and 110 W/m*K. Meanwhile, this preformed film can also be combined with the PDMS to provide a high electrical conductivity of over $1.4*10^6$ S/m and ultra-high electromagnetic shielding effectiveness of more than 75 dB. Given the excellent stability and environmental tolerance of the AgNW-PDMS composite material, this preformed film has shown great potential in the application of flexible circuits and signal shielding. Therefore, the proposed AgNW aerogel preformed film provides novel approaches to realize the adaptive interconnection between complex structures and provide new material classes for electromagnetic shielding and flexible circuit, which may enable the extended usage of metal aerogel materials in flexible electronics and packaging.

Acknowledgment

The authors acknowledge the support from the Sauvage Laboratory for Smart Materials of Harbin Institute of Technology (shenzhen) and the Advanced Joining Materials Co., Ltd. (shenzhen).

REFERENCES

[1] Y. Zhang, L. Zhang, K. Cui, et al. "Flexible Electronics Based on Micro/ Nanostructured Paper", Advanced Materials, 2018, 30(51): 1801588.

[2] R. D. Bringans, J. Veres. "Challenges and opportunities in flexible electronics", 2016 IEEE International Electron Devices Meeting (IEDM), San Francisco, USA, 2016, pp. 6.4.1-6.4.2.

[3] C. Wang, C. Wang, Z. Huang, et al. "Materials and Structures toward Soft Electronics", Advanced Materials, 2018, 30(50): 1801368.

[4] J. A. Rogers, X. Chen, X. Feng. "Flexible Hybrid Electronics", Advanced Materials, 2020, 32(15): 1905590.

[5] T. Huang, L. Shao, T. Lei, et al. "Robust design and design automation for flexible hybrid electronics", 2017 IEEE International Symposium on Circuits and Systems (ISCAS), Baltimore, USA, 2017, pp. 1-4.

[6] A. S. Dahiya, D. Shakthivel, Y. Kumaresan, et al. "High-performance printed electronics based on inorganic semiconducting nano to chip scale structures", Nano Converg, 2020; 7(1): 33.

[7] J. A. Rogers, T. Someya, Y. Huang. "Materials and mechanics for stretchable electronics", Science, 2010, 327(5973): 1603-1607.

[8] H. Yuk, T. Zhang, S. Lin, et al. "Tough bonding of hydrogels to diverse non-porous surfaces", Nature Materials, 2016, 15: 190-196.

[9] S. Cho, D. H. Kang, H. Lee, et al. "Highly Stretchable Sound-in-Display Electronics Based on Strain-Insensitive Metallic Nanonetworks", Adv. Sci., 2020, 8(1): 2001647.

[10] H. Wu, G. Yang, K. Zhu, et al. "Materials, Devices, and Systems of On-Skin Electrodes for Electrophysiological Monitoring and Human-Machine Interfaces", Adv. Sci., 2020, 8(2): 2001938.

[11] Y. Zhang, A. Pierre, S. Doris, et al. "Methods for fabrication of flexible hybrid electronics", Hybrid Memory Devices and Printed Circuits 2017, San Diego, USA, 2017, pp.103650C.1-103650C.6.

[12] V. V. Soman, Y. Khan, M. Zabran, et al. "Reliability Challenges in Fabrication of Flexible Hybrid Electronics for Human Performance Monitors: A System-Level Study", IEEE Transactions on Components, Packaging and Manufacturing Technology, 2019, 9(9): 1872-1887.

[13] Z. Li, T. Le, Z. Wu, et al. "Rational Design of a Printable, Highly Conductive Silicone-based Electrically Conductive Adhesive for Stretchable Radio- Frequency Antennas", Advanced Functional Materials, 2015, 25(3): 464-470.

[14] K. L. Suk, H. Y. Son, C. K. Chung, et al. "Flexible Chip-on-Flex (COF) and embedded Chip-in-Flex (CIF) packages by applying wafer level package (WLP) technology using anisotropic conductive films (ACF)", Microelectronics Reliability, 2012, 52: 225-234.

[15] M. Kim, Y. Ko, J. H. Bang, et al. "The Chip Bonding Technology on Flexible Substrate by Using Micro Lead-free Solder Bump", Journal of the Microelectronics and Packaging Society, 2012, 19(3): 15-20.

[16] H. H. Huang, J. G. Duh. "Fatigue Characterization for Flexible Circuit with Polyimide on Adhesive-less Copper", Journal of Electronic Materials, 2015, 44(10): 3934-3941.

[17] K. M. Harr, S. C. Kim, Y. M. Kim, et al. "Development of chip-on-flex bonding using Sn-based bumps and non-conductive adhesive", Microelectronics & Reliability, 2015, 55(8): 1241-1247.

[18] R.R. Mallepally, M.A. Marin, V. Surampudi, et al. "Silk fibroin aerogels: potential scaffolds for tissue engineering applications", Biomedical Materials, 2015, 10,.

[19] O. Karatum, S.A. Steiner, J.S. Griffin, et al. "Flexible, Mechanically Durable Aerogel Composites for Oil Capture and Recovery", ACS Applied Materials & Interfaces, 2015, 8: 215-240.

[20] X. Xu, R. Wang, P. Nie, et al. "Copper Nanowire-Based Aerogel with Tunable Pore Structure and Its Application as Flexible Pressure Sensor," ACS Applied Materials & Interfaces, 2017, 9: 14273-80.

[21] A.C. Pierre, G.M. Pajonk, "Chemistry of aerogels and their applications, " Chem Rev, 2002, 102: 4243-4265.

[22] X. Xu, R. Wang, P. Nie, et al. "Copper Nanowire-Based Aerogel with Tunable Pore Structure and Its Application as Flexible Pressure Sensor ", ACS Applied Materials & Interfaces, vol.9,2017, pp.14273-14280

978-1-6654-1392-3/21 $31.00 © 2021 IEEE

High-sensitivity flexible sensor based on silver nanowire aerogel

Yi Fang
Sauvage Laboratory for Smart Materials, School of Materials Science and Engineering
Harbin Institute of Technology (Shenzhen)
Shenzhen, China
fangyi@stu.hit.edu.cn

Zhehao Han
Sauvage Laboratory for Smart Materials, School of Materials Science and Engineering
Harbin Institute of Technology (Shenzhen)
Shenzhen, China
18344474766@163.com

Fei Peng
Sauvage Laboratory for Smart Materials School of Materials Science and Engineering
Harbin Institute of Technology (Shenzhen)
Shenzhen, China
hitpengfei@163.com

Mingyu Li*
Sauvage Laboratory for Smart Materials, School of Materials Science and Engineering
Harbin Institute of Technology (Shenzhen)
Shenzhen, China
myli@hit.edu.cn

Wenbo Zhu*
Sauvage Laboratory for Smart Materials, School of Materials Science and Engineering
Harbin Institute of Technology (Shenzhen)
Shenzhen, China
zhuwenbo@hit.edu.cn

Simulating various human perceptions through flexible electronic sensors is the key to the development of artificial intelligence and biomimetic robots. In this paper, a novel composite material of Silver nanowire aerogel(AgNWA) and conventional flexible polymer, such as PDMS, PI or other flexible polymers, is proposed to fabricate flexible sensors. The AgNWA used in this paper is a three-dimensional porous material self-assembled from silver nanowire, which combines the ultra-high conductivity of silver and the ultra-low density of aerogel materials as well as the flexibility. Hence, the conductivity of AgNWA would change immediately even a minor pressure applied. Based on this property, we fabricated flexible AgNWA/PDMS elaster and applied it to the field of flexible sensing. The AgNWA/PDMS elaster has shown excellent sensitivity to contact pressure, which can detect the stress in the range of 6kPa-1MPa. In addition, the resistance change rate of this composite material has a linear relationship with the stress received, so this composite material can sense the magnitude of the contact pressure. Furthermore, this composite material has good stability, and the minimum strain fatigue failure number is not less than 5000 times, maintaining surface integrity and structural flexibility. Therefore, the AgNWA/PDMS elaster has excellent sensitivity and stability, which provides a new way to realize flexible sensing.

Keywords—silver nanowire aerogel; sensitivity; flexible sensors

I. Introduction

With the rapid development of artificial intelligence and biomimetic robots, flexible sensors which can sense changes in the external environment are very important[1-3]. Piezoelectric effect, piezoresistive effect, capacitance effect and triboelectric effect are common sensing mechanisms of flexible sensors[4]. Among them, piezoresistive effect sensors have been widely applied because of their simple structure, convenient measurement and high sensitivity[5-8]. Put simply, the principle of the piezoresistive effect is that the resistance of the conductive materials changes when external pressure is applied, by which the external signals can be achieved. The

conductive materials of the piezoresistive effect sensor is commonly graphene[9], carbon nanotubes[10], metal nanowire network[11]et al., the base materials are usually polydimethylsiloxane(PDMS)[12], polyimide (PI) [13] or other flexible polymers.

Among the metal nanowire network materials, silver nanowires have outstanding electrical conductivity and mechanical flexibility, therefore they have potential applications in the fields of flexible electrodes and flexible sensors. However, since the silver nanowires are not linked with each other, the resistance between the silver nanowires is generally large and it is difficult to obtain good applications. To solve the problem of high resistance between silver nanowires, many solutions have been reported. For example, the freeze-drying solution provided by Fang [14] is to generate a single-directional temperature gradient to rearrange the silver nanowires before sintering. The conductivity of the obtained AgNWA can reach 51000S/m. However, this preparation method is complicated and the AgNWA has a low degree of cross-linking. Tokuno [15] directly used mechanical pressure to reduce the distance between the silver nanowires or overlap them so as to reduce the overall resistance. First, the silver nanowire solution was spin-coated on the plastic substrate. Then, it was pressed by mechanical pressure with 25MPa for 5 seconds. The film has a square resistance of 8.6Ω/sq. This method was limited in application scenarios and required a suitable flexible substrate, moreover, silver nanowires could only be deposited on the upper surface. Garnett [16] used light-induced plasma nano-welding technology to weld silver nanowires into a silver nanowire network. After welding, the resistance of the 5cm^2 silver nanowire network dropped from 1000kΩ to 1kΩ. Although the method proposed by Garnett welded the silver nanowires with each other, but the high resistance of 1kΩ still limits the application of this method.

Here, we propose a flexible sensor based on AgNWA and flexible PDMS substrate. Compared with the AgNW-sensor, AgNWA/PDMS elaster has higher conductivity and stability. PDMS, as a kind of rubber can provide good stability and flexibility. The Young's modulus of PDMS is generally 0.6-3.6MPa, which can be changed according to the ratio of PDMS base and curing agent as well as the curing temperature. AgNWA is self-assembled from silver nanowires. Silver

The authors acknowledge the support from the Science and Technology Innovation Committee of Shenzhen (JCYJ20180306172006392), the grant from State Key Laboratory of Advanced Welding and Joining (AWJ-21M0), and the support from National Nature Science Foundation of China (No.52075125).

nanowires in dendritic will no longer have the problem of junction resistance. According to the report of the polyol method for preparing silver nanowires, firstly, Ag nanoparticles are heterogeneously nucleated on AgCl seed crystals, and then a certain Ag nanoparticle on the seed crystals is unidirectionally grown into silver nanowires[17]. However, We prepared silver nanowire hydrogels by adding active factors that can regulate the three-dimensional growth of silver nanowires during the preparation of silver nanowires by the polyol method. The AgNWA was obtained by further supercritical drying method

II. EXPERIMENTAL PART

A. Preparation of AgNWA

Polyvinylpyrrolidone (PVP) was added to the stirring ethylene glycol solution. After dissolving for a while , the mixed solution was clear. The stirring speed of the solution was turned down. Silver nitrate was added to the mixed solution in a dark environment in stirring. Then the ferric chloride solution and the active factor were dissolved in solution. Finally, the reaction solution was subjected to liquid phase reduction to obtain the silver nanowire hydrogel. The solution replacement method was used with ethanol to replace the ethylene glycol in the hydrogel, meanwhile, the unreacted PVP was removed at the same time. Finally, the silver nanowire hydrogel was dried by supercritical drying to get the AgNWA.

B. Manufacturing of AgNWA/PDMS elaster

PDMS base (Sylgard-184 from Dow Corning) and curing agent (Sylgard-184 from Dow Corning) with mixing ratio of 10:1 were mixed thoroughly and degassed under vacuum condition. The mixed solution after vacuum degassing was mixed with the AgNWA and stood for 1 hour, in order to allow PDMS to fully penetrate into the three-dimensional network inside the AgNWA. The AgNWA/PDMS elaster was cured for 2 hours in an environment of 60°C.

C. Characterization and testing

We photographed the appearance and characterized the growth process of AgNWA as well as the cross-sectional view of the AgNWA/PDMS elaster through a scanning electron microscope (HITACHI SU8010). In addition, for testing the electrical response performance of the AgNWA/PDMS elaster, we also built a mechanical test platform based on the push-pull pressure meter (model EMS303 produced by the American MARK-10 company) and a resistance measuring instrument (2450 series produced by the American Keithley company). The push-pull pressure meter can apply a pressure of 0.001N-10N, the resistance measuring instrument adopts the constant voltage measurement mode, the voltage is set to 10mV, and the minimum time resolution is 100ms.

III. RESULTS AND DISCUSSION

Ethylene glycol is the solvent and reducing agent in the reaction process. PVP is a good coating agent which can selectively coat the surface of silver nanowires and regulate the linear growth of silver nanowires. As shown in Fig. 1a, AgCl seed crystals are produced during the reaction process. But unlike the polyol method[17], the silver nanowires grow in many directions of the AgCl seed crystals instead of unidirectional growth. During the growth process of silver nanowires, the dendritic growth of silver nanowires is the reason why the three-dimensional structure of silver nanowires can be generated. As shown in Fig. 1b, the final product is a silver nanowire hydrogel. The shape of the silver nanowire hydrogel can be controlled according to the container of the hydrothermal reaction. Ethylene glycol is the solvent in the silver nanowire hydrogel. Ethylene glycol would be replaced by solution replacement method using ethanol. Finally, silver nanowire hydrogel will be dried with supercritical drying to make AgNWA as shown in Fig. 1c. The AgNWA is off-white, and the shape is controlled by the cylindrical reaction container. AgNWA can be cut in any shape by laser.

Fig. 1. (a) The microscopic process of AgNWA growth. (b) Silver nanowire hydrogel. (c) The appearance of AgNWA.

The AgNWA prepared by this method is composed of dendritic silver nanowires. Benefit from the silver nanowires of dendritic structure, rather than simple overlap, freeze-drying or plasmonic welding, the AgNWA has better performance. The density of the silver nanowire aero the conductivity can exceed 10^6 S/m. Because of the gel can be controlled according to the concentration of the reactant. Moreover, The higher the reactant concentration, the greater the density of the aerogel. The aerogel has higher density means that the number of silver nanowire inside the aerogel is greater and the conductivity of aerogel is higher. As shown in Fig. 2, the conductivity and density of the AgNWA have a good linear relationship. With the control of the concentration and the forming process, the density range of AgNWA can reach 3.79-1000mg/cm³, and advantages of the three-dimensional network structure and high conductivity of AgNWAs, we tried to compound the AgNWA with PDMS to prepare an AgNWA/PDMS elaster.

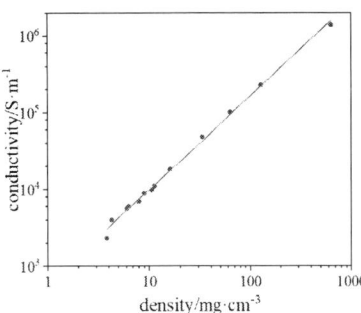

Fig. 2. The relationship between conductivity and density of AgNWA

Fig. 3a shows the appearance of the unprocessed AgNWA/PDMS elaster. After compounding PDMS, the overall strength of the AgNWA/PDMS elaster can withstand multiple bending. Fig. 3b is the result of loading the sensor 5000 times with a cyclic loading machine under the condition of 50% strain. It can still maintain surface integrity and structural flexibility after 5000 bending cycles. The cross-sectional view of the AgNWA/PDMS elaster is shown in Fig. 3c. On the one hand, the three-dimensional hole structure inside the AgNWA is filled with PDMS, which is also the reason why the AgNWA/PDMS elaster is flexible and elastic. PDMS plays a good role in structural support. On the other hand, the distribution of silver nanowires and PDMS was uniform and no obvious reunion components were observed, means that PDMS didn't damage the structure of the AgNWA. As shown in Fig. 3d, at a higher magnification, we can confirm that the internal structure of the AgNWA is well maintained and the internal silver nanowires provide excellent conductivity for the AgNWA/PDMS elaster.

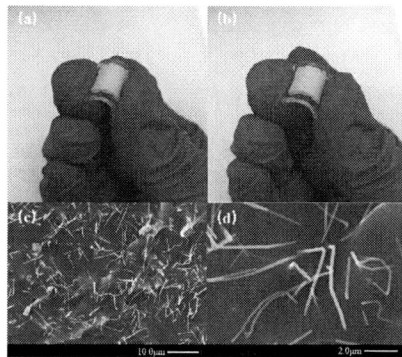

Fig. 3. (a)Unprocessed flexible sensor. (b) Flexible sensor after 50% strain 5000 times. (c) Cross-sectional view of the flexible sensor under low magnification SEM. (d) Cross-sectional view of the flexible sensor under high magnification SEM.

The AgNWA/PDMS elaster was fixed on the Printed Circuit Board(PCB) coated with silver paste as shown in Fig. 4a. The silver paste was used to lead the electrical signal through the PCB. Then, the mechanical and electrical properties of the AgNWA/PDMS elaster were explored with the aid of a mechanical testing platform. It can be seen from Fig. 4b that the AgNWA/PDMS elaster has a good linear response to stress and the sensitivity of the sensor is 0.226kPa^{-1}. Therefore, the stress applied to the sensor can be estimated by testing the current through the sensor, and the sensor's sensing performance is well realized. Furthermore, we can estimate the stress experienced by the sensor by testing the current flowing in the sensor.

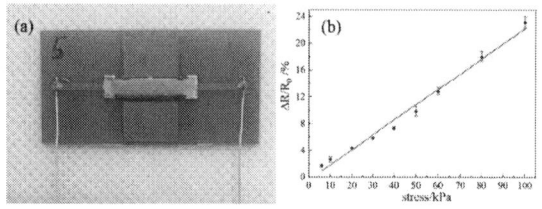

Fig. 4. (a)Flexible sensor with PCB. (b)The relationship between the resistance change rate and the stress of the sensor

Stresses of 6 kPa and 1 MPa were applied to the AgNWA/PDMS elaster, 1 MPa is the maximum stress that can be applied by the mechanical test platform. It can be seen from Fig. 5a that the resistance of the AgNWA/PDMS elaster has a mutation after the 6kPa stress was applied. The resistance of the AgNWA/PDMS elaster has increased from 739mΩ to 752mΩ and the resistance change rate is 1.8%. After removing the pressure, the resistance of the sensor returned to the initial resistance. In addition to the exploration of the small detection pressure, we also tried the testing of the maximum bearing stress of the sensor. The stress of 1MPa was applied to the sensor to observe the change of the sensor resistance. As shown in Fig. 5b, the resistance of the sensor

increased from 580mΩ to 925mΩ, the resistance change rate exceeds 60%, and it could still return to the initial resistance after the stress was released, showing that the sensor also has an advantage in detecting large stress. The change of resistance during the stress cycle application is basically the same, indicating the AgNWA/PDMS elaster based on AgNWA has outstanding reliability and repeatability in detecting pressure.

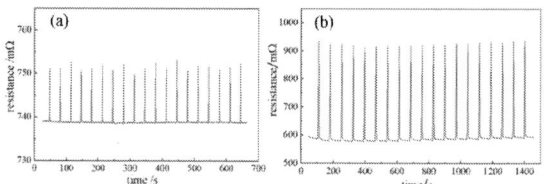

Fig. 5. Real-time resistance-time curve of load/unload sensor with (a)6kPa and (b)1MPa stress

The performance of the AgNWA/PDMS elaster at high temperature was also explored, and the electrical performance of the AgNWA/PDMS elaster at 20 °C, 60°C and 80°C under different stress is tested. As shown in Fig. 6, the sensor was at 20°C, 60°C. The sensitivities of and 80°C were 0.226 kPa^{-1}, 0.193kPa^{-1} and 0.157kPa^{-1}, which meant that as the temperature increased, the sensitivity of the AgNWA/PDMS elaster decreased.

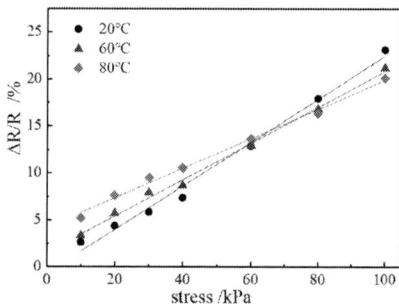

Fig. 6. The relationship between the relative resistance change rate and stress of flexible sensors at different temperatures

IV. CONCLUSION

In summary, the flexible sensor based on AgNWA shows high sensitivity and stability. The flexible sensor can be prepared by infiltrating PDMS into the AgNWA and curing the mixture. The dendritic structure of the AgNWA provides high conductivity and PDMS can improve the flexibility as well as stability of the sensor. Even if it is stretched 5000 times, the surface of the sensor is still not damaged. In addition, this sensor can detect the stress in the range of 6kPa-1MPa. Besides, the sensor has a very good linearity in the test range of 6kPa~100kPa. Therefore, the flexible sensor has excellent sensitivity and stability, which provides a new way to realize flexible sensing.

REFERENCES

[1] L. E. Osborn, D. Andrei, J. L. Betthauser, C. L. Hunt, H. H. Nguyen and R. R. Kaliki, "Prosthesis with neuromorphic multilayered e-dermis perceives touch and pain," Science Robotics, 2018, pp. eaat3818.

[2] M. Ha, S. Lee and H. Ko, "Wearable and flexible sensors for user-interactive health-monitoring devices," Journal of Materials Chemistry B, 2018, 10.1039.

[3] H. B. Wang, Massimo, Totaro, Lucia and Beccai, "Toward perceptive soft robots: progress and challenges," Advanced Science, 2018, 5, 1800541.

[4] J. C. Yang, J. Mun, S. Y. Kwon, S. Park, and S. Park, "Electronic skin: recent progress and future prospects for skin‐attachable devices for health monitoring, robotics, and prosthetics," Advanced Materials, 2019, 31(48).

[5] G. Liu, J. Qiu, B. Yang and H. D. Shieh, "Selective deposition of silver nanowires and its application for wearable pressure sensor," 2016 IEEE SENSORS, 2016, pp. 1-3.

[6] S. D. Zhang, L. Hu, S. Y. Yang, X. Z. Shi, D. B. Zhang and C. X. Shan, "Ultrasensitive and highly compressible piezoresistive sensor based on polyurethane sponge coated with cracked cellulose nanofibril/silver nanowire layer," ACS Applied Materials & Interfaces, 2019, pp. 10922-10932.

[7] Z. L. Ma, J. W. Wei, J. Z. Ma, S. Liang, H. Jiang and D. A. Dong, "Lightweight, compressible and electrically conductive polyurethane sponges coated with synergistic multiwalled carbon nanotubes and graphene for piezoresistive sensors," Nanoscale, 2018, pp. 7116-7126.

[8] Y. Y. Gao, Y. Cheng, H. C. Huang, T. Yang, T. Guo and D. Xiong, "Microchannel-confined mxene based flexible piezoresistive multifunctional micro-force sensor," Advanced Functional Materials, 2020, 30, 1909603.

[9] M. S. Lee, K. Lee, S. Y. Kim, H. Lee, J. Park, and K. H. Choi, " High-performance, transparent, and stretchable electrodes using graphene-metal nanowire hybrid structures," Nano Letters, 2013, pp. 2814-2821.

[10] Y. Y. Zhang, J. S. Chris, J. Y. Zhai, G.F. Zou, H. M. Luo and J. Xiong, "Polymer-embedded carbon nanotube ribbons for stretchable conductors," Advanced Materials, 2010, pp. 3027-3031.

[11] S. Ye, A. R. Rathmell, Z. Chen, I. E. Stewart and B. J. Wiley, "Metal nanowire networks: the next generation of transparent conductors," Advanced Materials, 2014, pp. 6670-6687.

[12] T. D. Le, J. An and Y. Kim, "Femtosecond laser direct writing of graphene oxide film on polydimethylsiloxane (PDMS) for flexible and stretchable electronics," 2017 Conference on Lasers and Electro-Optics Pacific Rim (CLEO-PR), 2017, pp. 1-4.

[13] N. S. Korivi and L. Jiang, "Metal Patterning on Polymers for Flexible Microsystems and Large-area Electronics," 2007 Thirty-Ninth Southeastern Symposium on System Theory, 2007, pp. 181-185.

[14] Q. Fang, C. L. Pui, C. F. Megan, C. Wen, T. Y. Kou and Y. O. Tammy, "Ultralight Conductive Silver Nanowire Aerogels," Nano Letters, 2017, pp. 7171-7176.

[15] T. Tokuno, M. Nogi, M. Karakawa, J. T. Jiu, T. T. Nge, and Y. Aso, "Fabrication of silver nanowire transparent electrodes at room temperature," Nano Research, 2011, pp. 1215-1222.

[16] E. C. Garnett, W. S. Cai, J. J. Cha, F Mahmood, S. T. Connor and M. G. Christoforo, "Self-limited plasmonic welding of silver nanowirejunctions," Nature Materials, 2012, pp. 241–249.

[17] W. M. Schuette and W. E. Buhro, "Silver chloride as a heterogeneous nucleant for the growth of silver nanowires," Acs Nano, 2013, pp. 3844-3853.

978-1-6654-1392-3/21 $31.00 © 2021 IEEE

Soldering of Graphene Assembled Films with Ultrasonic Assistance and Its Utilization Potentiality in Electronic Devices

Huaqiang Fu
School of Materials Science and Engineering
Wuhan University of Technology
Wuhan, China
fuhuaqiang@whut.edu.cn

Yong Xiao*
School of Materials Science and Engineering
Wuhan University of Technology
Wuhan, China
yongxiao@whut.edu.cn

Daping He
School of Science
Wuhan University of Technology
Wuhan, China
hedaping@whut.edu.cn

Abstract—In this study, inspired by ultrasonic-assisted soldering technique, we proposed the soldering method for GAF with ultrasonic assistance. Sn solder spot were firstly prepared on GAF terminates ultrasonically, and then remelted to realize soldering of GAF. As a result, GAF was successfully soldered with ultrasonic assistance. The optimal ultrasonic vibration time should be 10 seconds for the preparation of Sn solder spot. Under the action of ultrasonic waves, Sn solder was driven to not only wet and spread on GAF surface, but also squeeze into GAF interlayers, which allowed the Sn solder spots to form a mechanically interlocked tight bonding with GAF. Furthermore, a compact interface of GAF/amorphous carbon (a-C) layer/Sn was formed. Significantly, GAF soldered joints exhibited high tensile resistance and excellent electrical property closed to original GAF. All the findings can promote the application and development of GAF in the field of electronic devices.

Keywords—graphene assembled films, ultrasonic-assisted soldering, electrical property

I. INTRODUCTION

Graphene assembled film (GAF) has captured significant attention in the fields of electronic devices, especially as an alternative to replace traditional metal materials such as Cu and Al, due to its good flexibility, low density and excellent electrical conductivity[1]. For further application of GAF-based electronic devices, high-performance connection between GAF with other work pieces must be guaranteed.

Materials connection in electronic devices is generally realized by adhesive bonding or soldering. However, the existing connected methodology cannot satisfy the high-performance connection requirements for GAF in electronic devices. Traditional adhesive agent exhibits high cost, low electrical conductivity, and weak resistance to aging and oxidation, which cannot meet practical connected requirements[2]. Soldering may serve as a promising approach for GAF connection with high performance, due to it can achieve atomic bonding at soldered interface[3]. But the solders cannot wet the surface of GAF during regular soldering process[4], which hinders the further application of soldering for GAF.

In recent years soldering technology with ultrasonic assistance has drawn great concern in connection fields of materials with wetting obstacles[5]. The cavitation collapse induced by ultrasonic waves can generate shock waves and micro-jets near the interface of molten solder/solid substrate, which could generate locally extreme high temperature and pressure, and then break the oxide film of molten solder to promote the wettability of substrate. Wetting is the basis of successful soldering. Hence, soldering with ultrasonic assistance may act as a promising method for GAF to obtain high-performance connection.

Herein, in this study, inspired by ultrasonic-assisted soldering technique, we proposed the soldering method for GAF with ultrasonic assistance. Sn solder spots were firstly prepared on GAF terminates ultrasonically, and then remelted to realize soldering of GAF. Microstructures of GAF soldered joints and interfacial bonding mechanism were investigated. Physical properties including tensile breaking force and electrical conductivity were measured. The purpose of this research is to develop a new soldering method and fulfil high-performance connection for GAF.

II. EXPERIMENTAL

A. Materials

Fabricated process of GAF consists of the following three steps: Firstly, GO was homogeneously mixed into deionized water to obtained GO gel (20 mg/ml). Secondly, GO gel was scraped onto polyethylene terephthalate substrate to get GO film. Finally, the GO film was hold at 2850 ℃ for 2 hours in a vacuum furnace (CX-GFL-R20/30). Sn solder was purchased from Shenzhen Jushi Electronics Co., Ltd. GO and silver conductive adhesive (SCA) were purchased from Wuxi Chengyi Education Technology Co., Ltd.

B. Soldering process

Soldering of GAF contains two steps, preparation of Sn solder spots on GAF terminates and remelting of Sn solder spots. Before soldering, GAF was firstly cut into the size of 15 mm×5 mm by laser (LPKF ProtoLaser S). As shown in Fig. 1, preparation of Sn solder spots on GAF terminates were realized by vertical dipping of GAF terminates into melted Sn solder with ultrasonic vibration. The Sn solder was hold at 300 °C, ultrasonic frequency was 20 kHz, and ultrasonic vibration time was 5 seconds, 10 seconds and 20 seconds

978-1-6654-1392-3/21 $31.00 © 2021 IEEE

respectively. After dipping, Sn solder spots were prepared on GAF terminates. Finally, GAF soldered joints were fabricated by remelting Sn solder joints on GAF terminates.

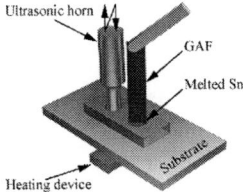

Fig. 1. Preparation process of Sn solder spots on GAF terminates.

C. Characterizations and test methods

Structures and graphitization degree of GAF were measured by X-ray diffraction (XRD) and Raman spectra. XRD of GAF were performed in a D8 advanced diffractometer, and Raman spectrum of GAF was conducted using an excitation wavelength of 457.9 nm provided by a Spectra-Physics Model 2025 argon ion laser. Morphologies and microstructures of GAF and Sn solder spots were characterized in Zeiss Ultra Plus Field emission scanning electron microscope (SEM). Interfacial characterization of GAF/Sn solder spots was taken on a transmission electron microscope (TEM, Talos F200S). The tensile breaking forces of GAF soldered joints were measured in an electromechanical universal testing machine (E44.104) at a crosshead speed of 0.1 mm/min (three times for each sample). And the electrical properties of GAF soldered joint were qualitative analyzed by comparing the voltage-current curves of GAF circuits with different kinds of joints.

III. RESULTS AND DISCUSSION

A. Characterizations of GAF

Fig. 2a is the photo of GAF. The GAF exhibited metallic luster and excellent flexibility. Based on four probe method, the electrical conductivity of GAF is approximately 1.23×10^6 S/m. Fig. 2b is the XRD pattern of GAF. An intense and sharp (0 0 2) peak at around 26.5 ° demonstrates that the GAF was reduced sufficiently, which exhibits high-oriented stacking structure of graphene multilayers, and the interlayer spacing was about 0.34 nm[6]. Fig. 2c displays the Raman spectra of GAF, in which the weak intensity of D band implies few defects inside GAF, and the intense G band further demonstrates the high-oriented stacking structure of GAF[7]. Fig. 2d is the SEM image of GAF surface. The GAF surface is not smooth, showing multiple wrinkles. These wrinkles may hinder the wetting and spreading of liquid solder on GAF surface, and even cause connected defects. Fig. 2e is the SEM images of GAF cross-section. The thickness of GAF is about 25 μm. After weighing the weight of GAF, the density of GAF was calculated as 1.8 g/cm³. Fig. 2f is a partial enlargement of the blue box in Fig. 2d, showing a highly oriented stacking structure, which is consistent with the XRD result.

B. Microstructure of Sn solder spot on GAF terminate

Wetting is the forming premise of soldered joints. However, the wetting angle of molten Sn solder on GAF surface was reported to be larger than 120 °[4], implying that generally the Sn solder cannot wet the surface of GAF. With ultrasonic assistance, Sn solder spots was successfully prepared on GAF terminate, as depicted in fig. 3. The Sn solder spot was connected with GAF firmly with a small size of 5 mm×2 mm.

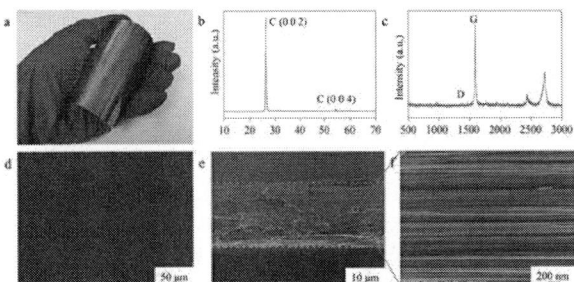

Fig. 2 (a) Photo of GAF, (b) XRD pattern of GAF, (c) Raman spectrum of GAF, (d) Scanning electronic microscope (SEM) image of GAF surface, (e, f) SEM images of GAF cross-section.

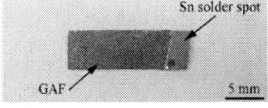

Fig. 3. Image of GAF with Sn solder spot.

Fig. 4a-c is the SEM images of GAF cross-section with Sn solder spots prepared with ultrasonic vibration time of 5 seconds, 10 seconds and 20 seconds, respectively. As the ultrasonic vibration time was 5 seconds, the Sn solder was initially penetrated into GAF interlayers. Due to oriented stacking graphene multilayer structure of GAF, the GAF terminate formed a ramified structure. But in the part far away from the ramified terminate, the interface between GAF and Sn solder spot was separated during the grinding and polishing process, which may be attributed to the poor bonding caused by the short ultrasonic vibration time. As the ultrasonic vibration time increased to 10 seconds, the penetrated distance of Sn solder into GAF interlayers elongated, and the ramified angle became larger. Moreover, after grinding and polishing, no cracks and defects appeared at interface, manifesting good combination between GAF and Sn solder spot was formed. As the ultrasonic vibration time continued increasing to 20 seconds, the ramified structure of GAF was damaged, the formed graphene nanoflakes scattered into Sn solder spot. The damaging of GAF ramified structure shortened the effective length of GAF, which may affect the performance of some electronic devices such as radio frequency devices. Based on the above analysis, it can make a conclusion that the optimal ultrasonic vibration time is 10 seconds.

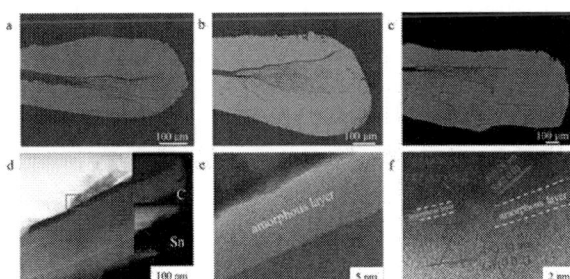

Fig. 4. (a, b, c) SEM images of GAF cross-section with Sn solder spots prepared with ultrasonic vibration time of 5 seconds, 10 seconds and 20 seconds, respectively, (d) Transmission electronic microscope (TEM) image of GAF/Sn interface, and its elemental distribution, (e, f) High-resolution TEM image of GAF/Sn interface.

In addition, the volume of Sn solder spot gradually increases as the ultrasonic vibration time extended from 5 seconds to 20 seconds, as shown in fig. 4a, 3b and 3c. This phenomenon may be resulted from destruction of ramified structure of GAF with the extension of ultrasonic vibration time. The graphene nanoflakes formed by ultrasonic destruction of GAF were scattered homogeneously into Sn solder spot under the stirring action of ultrasonic wave, which improved the viscosity of the solder[8]. The viscosity increase of the liquid solder near the GAF may be conducive to the forming of the solder spot.

Interestingly, the coefficient of thermal expansion (CTE) of a-C layer was exactly between the CTE of GAF and Sn solder[9], which may be helpful to alleviate the stress problem caused by interfacial CTE mismatch. Furthermore, the ramified structure inside Sn solder spot formed a mechanically interlocked structure, which may be beneficial to improve the tensile resistance of GAF soldered joints.

In order to investigate the interfacial bonding mechanism, a TEM sample presented the GAF/Sn interface was fabricated as shown in fig. 4d. The interface for TEM characterizing was selected between ramified GAF and Sn solder spot prepared with the ultrasonic vibration time of 10 seconds. Fig. 4e is the High-resolution TEM image of GAF/Sn interface. The phase with a lattice spacing of about 0.34 nm is (0 0 2) lattice plane of GAF. Significantly, an obvious Light gray transition layer was formed at GAF/Sn interface. According to the elemental distribution in fig. 4d measured by energy dispersive X-ray spectrometer, the transition layer is mainly consisted with C. The C transition layer exhibits no regular lattice, thus we speculate that the transition layer is amorphous carbon (a-C) layer. The bonding mode of a-C transition at interface of metal/carbon materials has been verified and reported previously[10]–[12].

The forming of a-C layer may be concerned with ultrasonic waves. During ultrasonic vibration, the cavitation effect induced by ultrasonic waves can generate shock waves and micro-jets at the interface between GAF and liquid molten Sn solder. The striking of shock waves and micro-jets towards GAF can cause Locally instantaneous high temperature (above 1600 °C) and high pressure (about 5 GPa)[5], which can destruct the surface structure of GAF, as shown in fig. 4f, a graphene nanocrystal was exfoliated from GAF surface, which may be attributed to the ultrasonic waves.

C. Tensile test and electrical property measurement

GAF soldered joints were obtained by remelting Sn solder spots on GAF terminates. As shown in fig. 5a, fracture of GAF outside the soldered joints results in failure after tensile test. Fig. 5b illustrates a typical cross-section image of GAF soldered joints after tensile test. The combination between GAF and Sn solder keeps integrity after tensile, demonstrating excellent connected stability of GAF soldered joints. As depicted in fig. 5c, the average tensile breaking force of GAF is 14.91 N, while that of GAF soldered joints was 12.46 N, which further verifies the tightly bonding between GAF and Sn solder.

Electrical properties of GAF soldered joint were qualitative analyzed by comparing the voltage-current curves of GAF circuits with different joints as shown in fig. 5d. Circuit A-Sn is a continuous GAF circuit with no GAF soldered joints, circuit B-Sn is a GAF circuit with 30 GAF soldered joints, and circuit B-SCA is a GAF circuit with 30 GAF joints bonded by SCA (which is widely used in packaging of electronic devices). Each circuit was soldered with a red LED to magnify the effect of joints resistance. Fig. 5e shows current-voltage curves of different circuits. Notably, the curves of A-Sn and B-Sn are extremely close, which

suggests that the electrical properties of GAF soldered joints is similar to GAF. Compared with circuit B-SCA, the curve of B-Sn is above, implying that under the same voltage, the current inside circuit B-Sn is larger than that inside B-SCA, which reveals that the resistance of circuit B-Sn is smaller than that of circuit B-SCA. Therefore, the GAF soldered joints exhibits excellent electrical property.

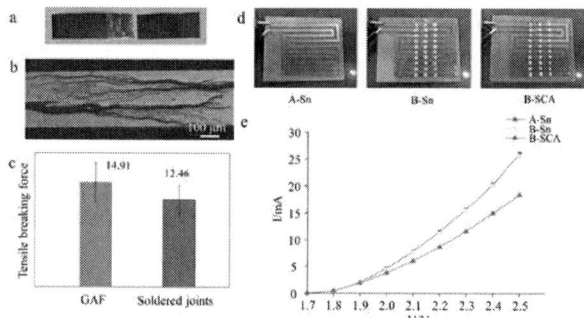

Fig. 5. (a) Failed GAF soldered joints after tensile test, (b) Typical cross-section image of GAF soldered joints after tensile test, (c) Tensile breaking force of GAF and GAF soldered joints, (d) GAF circuits designed for electrical conductivity measurement of GAF soldered joints, (e) Current-voltage curves of different circuits.

IV. CONCLUSION

In conclusion, soldering of GAF was realized by remelting Sn solder spots prepared ultrasonically on GAF terminates. Microstructures of Sn solder spots and interfacial bonding mechanism were investigated. Moreover, physical properties containing tensile breaking force and electrical property were measured. The main conclusions of this study are shown as follows:

1) Soldering of GAF was successfully realized with ultrasonic assistance.

2) Under the action of ultrasonic waves, a mechanically interlocked tight bonding structure and a compact interface between GAF and Sn solder were formed.

3) with ultrasonic assistance, GAF was connected with Sn solder with a transition layer of a-C.

4) The GAF soldered joints exhibited significant structural stability and excellent electrical conductivity, which can contribute to the further application and development of GAF in electronic devices field.

REFERENCES

[1] A. Scidà et al., "Application of graphene-based flexible antennas in consumer electronic devices," Mater. Today, vol. 21, pp. 223–230, 2018.

[2] G. M. Odegard and A. Bandyopadhyay, "Physical aging of epoxy polymers and their composites," J. Polym. Sci. Part B Polym. Phys., vol. 49, pp. 1695–1716, 2011.

[3] K. M. Razeeb, E. Dalton, G. L. W. Cross, and A. J. Robinson, "Present and future thermal interface materials for electronic devices," Int. Mater. Rev., vol. 63, pp. 1–21, 2018.

[4] S. A. Sánchez, J. Narciso, E. Louis, F. Rodríguez-Reinoso, E. Saiz, and A. Tomsia, "Wetting and capillarity in the Sn/graphite system," Mater. Sci. Eng. A, vol. 495, pp. 187–191, 2008.

[5] D. G. Shchukin, E. Skorb, V. Belova, and H. Möhwald, "Ultrasonic cavitation at solid surfaces," Adv. Mater., vol. 23, pp. 1922–1934, 2011.

[6] J. Khan, S. A. Momin, and M. Mariatti, "A review on advanced carbon-based thermal interface materials for electronic devices," Carbon, vol. 168, pp. 65–112, 2020.

[7] S. Reich and C. Thomsen, "Raman spectroscopy of graphite," Philos. Trans. R. Soc. London. Ser. A Math. Phys. Eng. Sci., vol. 362, pp. 2271–2288, 2004.

[8] A. Sharma, H. R. Sohn, and J. P. Jung, "Effect of graphene nanoplatelets on wetting, microstructure, and tensile characteristics of Sn-3.0Ag-0.5Cu (SAC) alloy," Metall. Mater. Trans. A Phys. Metall. Mater. Sci., vol. 47, pp. 494–503, 2016.

[9] D. Yoon, Y. W. Son, and H. Cheong, "Negative thermal expansion coefficient of graphene measured by raman spectroscopy," Nano Lett., vol. 11, pp. 3227–3231, 2011.

[10] C. Zhou et al., "Fabrication, interface characterization and modeling of oriented graphite flakes/Si/Al composites for thermal management applications," Mater. Des., vol. 63, pp. 719–728, 2014.

[11] W. Li, Y. Liu, and G. Wu, "Preparation of graphite flakes/Al with preferred orientation and high thermal conductivity by squeeze casting," Carbon, vol. 95, pp. 545–551, 2015.

[12] Y. Huang, Q. Ouyang, Q. Guo, X. Guo, G. Zhang, and D. Zhang, "Graphite film/aluminum laminate composites with ultrahigh thermal conductivity for thermal management applications," Mater. Des., vol. 90, pp. 508–515, 2016.

Theoretical calculation and Simulation of BGA package stress under temperature cycling load

Zhang Yueping
Science & Technology on Reliability &
Environmental Engineering Laboratory
Beijing Institute of Structure &
Environment Engineering
Beijing,China
2409577138@qq.com

Hou Chuantao
Science & Technology on Reliability &
Environmental Engineering Laboratory
Beijing Institute of Structure &
Environment Engineering
Beijing,China
houcht@qq.com

Tong Jun
General Department of test technology
Beijing Institute of Structure &
Environment Engineering
Beijing,China
tongjunwh@163.com

Wang Long
Science & Technology on Reliability &
Environmental Engineering Laboratory
Beijing Institute of Structure &
Environment Engineering
Beijing,China
lwang@spacechina.com

Abstract—Under temperature cycling load, solder joints are the most vulnerable locations for faults in electronic equipment. Damage and failure of solder joints are important factors causing faults in electronic equipment. In order to solve the problem of solder joint damage and failure caused by cycling thermal stress in instruments and equipment under temperature load, the BGA packaging circuit board of ceramic chip is taken as the research object, and the maximum shear stress of solder joint is calculated theoretically and simulated based on engineering algorithm and finite element simulation respectively. Based on engineering algorithm and finite element simulation, the maximum shear stress of solder joints is calculated and simulated respectively. An engineering algorithm for maximum shear stress at solder joints of BGA encapsulated circuit boards is presented based on the geometric relationship of each part of the model and the theory of material mechanics. To explore the effects of different constitutive models on the simulation results of weld creep characteristics, Anand viscoplastic constitutive model and Wiese two-power constitutive model are introduced into the simulation software to simulate. The results show that the two models have similar results, which indicates that the two constitutive models are both suitable for describing the creep properties of weld joints. The method presented in this paper has good engineering application value and can be used to analyze the stress level of BGA encapsulated circuit board under temperature cycling load.

Keywords—*Anand model; Wiese model; temperature cycling load; finite element; BGA package*

I. INTRODUCTION

Electronic instruments and equipment undergo temperature changes during manufacture, storage, use, transportation and maintenance. According to public reports, about 55% of the failures in environment-related electronic equipment are caused by temperature and temperature cycles. Temperature changes can result in mismatched stresses between printed circuit boards (PCBs), solder joints and chips due to different thermal expansion coefficients. During the high and low temperature stages of temperature, concave and convex warps also occur in the printed circuit board, during

which the shear stress is reversed. Temperature cycling also generates cyclic stress in the solder joints, resulting in low-cycle fatigue of the solder joints. At the same time, the low melting point (183 ℃) of the solder joints leads to creep deformation at room temperature. The damage accumulation caused by creep deformation of the solder joints under thermal cycling leads to crack initiation, crack propagation and ultimately failure and destruction of the entire solder joints at the stress concentration position.

At present, little research has been done on the damage and failure of solder joints of BGA encapsulated circuit board under temperature-dependent load. Further study on the damage mechanism and failure mode of solder joints is needed. Under temperature-dependent load, the shear stress of solder joints is one of the causes of cracking of solder joints. Its size is closely related to the life of solder joints. Some life models, such as Coffin-Mason model, use shear stress as the life calculation parameter, so it is necessary to calculate the maximum shear stress of solder joints. Based on the geometric relationship of each part of the model and the theory of material mechanics, an engineering algorithm for calculating the maximum shear stress of solder joints under temperature-dependent load is deduced in this paper. At the same time, the Anand viscoplastic constitutive model and the Wiese double power constitutive model are introduced into Abaqus software and simulated respectively. The stress-strain distribution and change trend of each solder point of the printed circuit board under temperature-dependent load can be well analyzed, which provides a basis for analyzing the damage and failure modes of the printed circuit board.

II. THEORETICAL CALCULATION OF THERMAL STRESS IN BGA PACKAGING

In this paper, the horizontal force and shear stress of solder joints of BGA encapsulated circuit board under temperature-dependent load ranging from -55 ℃ to + 125 ℃ are analyzed, and the results are compared with the results of finite element simulation to verify the accuracy of the calculation results. The device consists of three parts: ceramic packaging chip, lead-tin solder joint and PCB board.Its structure diagram is shown in Fig.1.

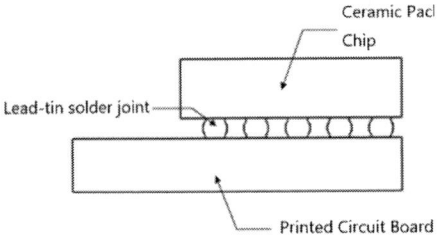

Fig. 1. Device Model 1/2 Planar Diagram

The distance between solder joints is 0.75mm. The geometric dimensions of engineering algorithm theoretical calculation and finite element simulation are shown in the Table 1:

TABLE I.   DEVICE GEOMETRY PARAMETERS

Structure	Length/ mm	Width/ mm	Height/ mm	Diameter /mm
Ceramic package chip	8	8	1.05	-
printed circuit board	12	12	0.95	-
Lead tin solder joint	-	-	0.3	0.45

The solder joints distribution are 10×10, with a total of 100 solder joints. The number and location distribution are shown in the figure.2. The dotted line in the figure is the symmetry axis of solder joints in X direction, the horizontal number is A-E, and the vertical number is 1-10. The distance between each solder joint and the symmetrical center line is shown in the Table 2.

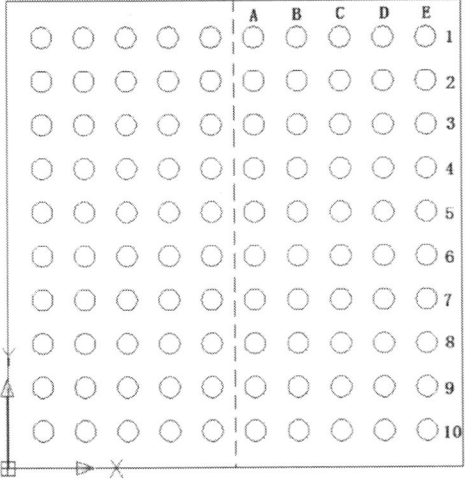

Fig. 2. Solder joint distribution and numbering of device model(Circles are solder joints)

TABLE II.   DISTANCE BETWEEN EACH SOLDER JOINT AND THE SYMMETRICAL CENTER LINE

Solder joint number	Distance from center line (mm)
A1-A10	0.375
B1-B10	1.125
C1-C10	1.875
D1-D10	2.625
E1-E10	3.375

The solder joint distribution of BGA circuit board can be regarded as a combination of several circuit boards with solder joints only on the outermost edge and no solder joints inside. It can be obtained by mathematical induction. The actual analysis model can also be simplified as a solder joint on one side of the symmetrical line, which is located at the center of gravity of the solder joint. Now all the solder joints on one side of the symmetry axis are equivalent to a whole solder joint to calculate the overall horizontal force on the solder joints on one side of the circuit board during the temperature cycling process. The simplified schematic diagram is as shown in the figure.3. All the solder joints are simplified as two whole solder joints distributed along both sides of the symmetry axis.

Fig. 3. Simplified diagram of equivalent distribution of solder joints

The distribution of the solder joints of the BGA printed circuit board can be seen as a combination of several printed circuit boards with solder joints only on the outermost edge and no solder joints inside. It can be obtained by mathematical induction. The actual analysis model can also be simplified to one solder point on one side of the symmetric line, and the position is at the center of gravity of the solder joints.All solder joints on one side of the symmetric axis are now equivalent to a single global solder joint. By calculating the global horizontal force exerted on the solder joints on one side of the circuit board during the temperature cycle, all solder joints can be simplified to two global solder joints distributed along both sides of the symmetric axis.

The distance between the overall solder point and the symmetric axis is calculated based on the center of gravity of all solder points on this side.The expression of the distance between the center of gravity of the whole solder point and the midline is:

$$X_z = \frac{0.375 \times 10 + 1.125 \times 10 + 1.875 \times 10 + 2.625 \times 10 + 3.375 \times 10}{50} = 1.875mm \quad (1)$$

According to the compatibility of deformation, the sum of chip displacement and solder displacement is equal to PCB plate displacement:

$$X_P = X_X + X_H \quad (2)$$

The force analysis shows that the horizontal force exerted by the PCB board and the chip is equal to the horizontal force exerted by the whole solder joint:

$$P_P = P_X = P_H \quad (3)$$

The thermal expansion displacement of the PCB plate along the X-axis is considered. It is the thermal expansion displacement of the chip along the X-axis. It is the thermal expansion displacement of the whole solder joints along the X-axis.

The expression for the thermal expansion of PCB plates along the X-axis is

$$X_P = \alpha_P L_P \Delta t - \frac{P_P L_P}{A_P E_P} \quad (4)$$

Where: α_P is the thermal expansion coefficient of PCB plates;

L_P is the effective length of PCB plates;

Δt is the change of temperature;

P_P is the horizontal force exerted by PCB plates;

A_P is the cross-sectional area of PCB plates;

E_P is the elastic modulus of PCB plates.

The heat expansion expression of the chip along the X-axis is

$$X_X = \alpha_X L_X \Delta t + \frac{P_X L_X}{A_X E_X} \quad (5)$$

Where: α_X is the thermal expansion coefficient of the chip;

L_X is the effective length of the chip;

Δt is the temperature change;

P_X is the horizontal force on the chip;

A_X is the cross-sectional area of the chip;

E_X is the elastic modulus of the chip.

Next, the displacement expression of the whole solder joint needs to be derived. The force diagram of the whole solder joint is as follows.

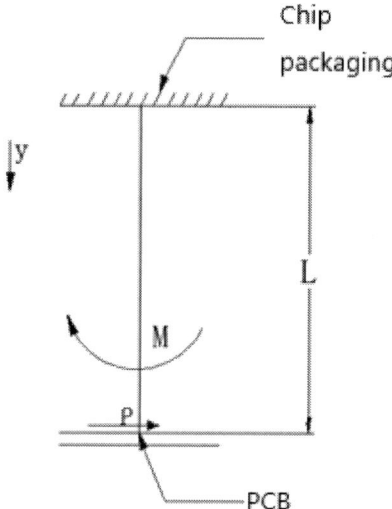

Fig. 4. Schematic Diagram of the Overall Force Analysis of Solder Joints

Set the length of the solder point to L, the chip above the solder point, the PCB plate below, and the upper part of the solder point to be fixed. Due to the different thermal expansion coefficients of solder joints, chips and PCB plates, the upper part of the solder joints is offset horizontally from the lower part in the temperature-dependent load. The upper part of the solder joints is the origin, and the lower part of the solder joints is subjected to horizontal force P and bending moment M. Based on the force analysis:

$$M_y = M - Py \quad (6)$$

$$\frac{\partial M_y}{\partial M} = 1; \frac{\partial M_y}{\partial P} = -y \quad (7)$$

Assume that the vertical rotation angle of the solder joint is 0:

$$\Delta \delta = \frac{1}{EI} \int_0^L M_y \frac{\partial M_y}{\partial M} dy = 0 \quad (8)$$

The equation is solved:

$$M = \frac{PL}{2} \quad (9)$$

$$\Delta x = \frac{1}{EI} \int_0^L M_y \frac{\partial M_y}{\partial P} dy = \frac{PL^3}{12EI} \quad (10)$$

From material mechanics, the displacement in X direction at the bottom of the solder joint is:

$$X_H = \frac{P_H h^3}{12 E_H I_H} \qquad (11)$$

By combining all the previous equations, the overall horizontal force of solder joint P_H is 151N.

The average shear stress in solder joint is expressed as

$$S_s = \frac{P_H}{A} \qquad (12)$$

Where, P_H is the overall horizontal force of solder joint 151N; A is the shear area of 50 solder joints, the value is $\pi \times 0.2^2 \times 50 = 6.28$ mm². By substituting the equation, the average shear stress $S_s = 24.04 MPa$ in the solder joint can be obtained.

III. FINITE ELEMENT SIMULATION WITH TWO VISCOPLASTIC CONSTITUTIVE MODELS

The eutectic 60Sn40Pb solder has lower melting point, obvious creep and stress relaxation at room temperature, obvious viscous effect and strong temperature-loading rate dependence. Therefore, a unified viscoplastic constitutive model should be adopted to describe the deformation behavior of SnPb solder. Anand model is the most widely used one in the unified viscoplastic constitutive model. Anand model is simple in form and has few parameters, which is widely used in electronic life prediction. Wiese model is also a constitutive model describing the creep rate of solder joints. Stoechl S et al. used Wiese model to analyze the reliability and fatigue life of solder joints using ANASYS software. The simulation results are in agreement with the experimental results.

This paper simulates the circuit board under temperature cyclic load based on Abaqus finite element platform The circuit board quarter model and constraints are shown in Fig.5. The uniform viscoplastic Anand constitutive model and Wiese double power constitutive model are used to describe the creep characteristics of solder joints at edges and corners. The maximum values and distribution positions of Mises stress, shear stress, displacement and equivalent creep in the circuit board are calculated and compared. Anand's Constitutive Model uses self-measured parameters, and Wiese's Constitutive Model parameter data comes from the values measured by Wiese et al.

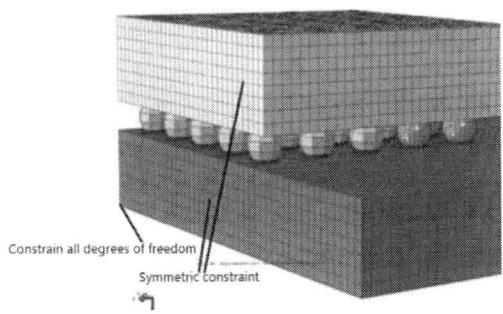

Fig. 5. Circuit Board 1/4 Model and Constraints

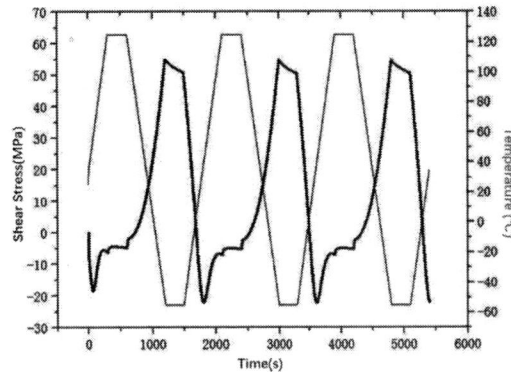

Fig. 6. Load Curve of Maximum Shear Stress with Temperature(Anand Constitutive Model)

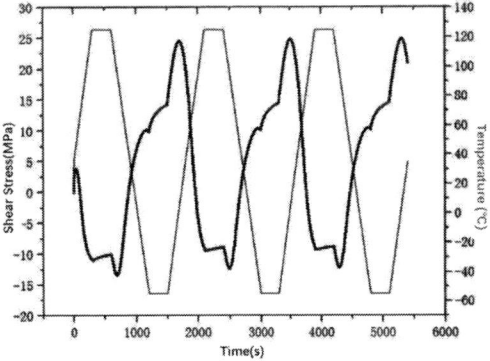

Fig. 7. Load Curve of Maximum Shear Stress with Temperature(Wiese Constitutive Model)

Fig.6 and Fig.7 show the maximum shear stress curves corresponding to the simulation results of two constitutive models. It can be seen that the trend of change is similar. During the cooling stage, the shear stress changes from negative minimum to maximum positive. In the warming phase, the opposite is true. The Anand model simulates stress relaxation at the highest heat preservation stage, but the Wiese model simulates that the shear stress will continue to drop to the minimum negative value during this period. The minimum negative shear stress in the Anand model occurs near the end of heating and decreases gradually with the increase of cycles. The minimum negative shear stress in the Wises model occurs at the end of the high-temperature insulation phase and increases with the increase of cycles at the beginning of cooling. Anand model shear stress range is -22MPa to 56MPa；The Wiese model shear stress ranges from -13MPa to 25MPa.

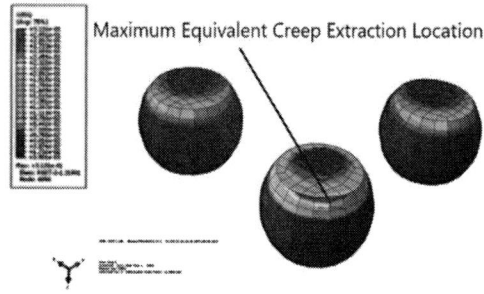

Fig. 8.   Location of Maximum Equivalent Creep Stress

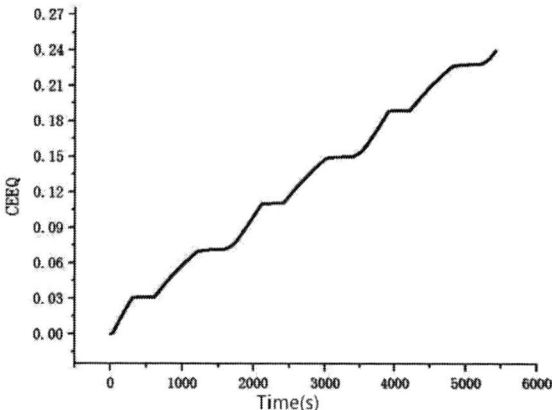

Fig. 9.   Maximum Equivalent Creep Varies with Temperature-Load(Anand Constitutive Model)

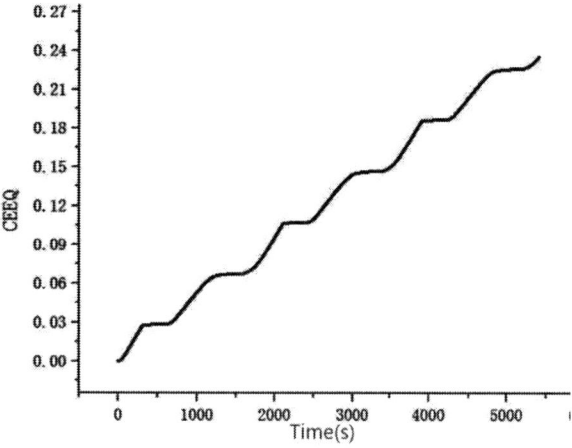

Fig. 10. Maximum Equivalent Creep Varies with Temperature-Load(Wiese Constitutive Model)

Fig.8 shows the location of the maximum equivalent creep stress.Fig.9 and Fig.10 show the maximum equivalent creep of two constitutive models with temperature-dependent load. It can be seen that the variation range and trend of the two simulation results are the same, and the range of equivalent creep is between 0 and 0.24, and it accumulates with the increase of the number of cycles. It is generally believed that as the number of cycles increases, creep strain and fatigue damage of solder joints increase cumulatively, which is well simulated by the simulation results.The average equivalent creep per cycle obtained by the Anand model is 0.08, and that by the Wiese model is 0.078.

IV.   CONCLUSION

With the development of electronic packaging technology, the distribution of solder joints used in packaging is more and more compact. In the actual use of electronic products, there are more and more failures caused by temperature changes, which attracts more and more attention.Analytical and finite element methods are often used to evaluate and analyze the failure of electronic packaging. These two methods can draw conclusions quickly and easily, and avoid huge test costs and time.

Shear stress under temperature cyclic load is the main failure mode of solder joints.The engineering algorithm proposed in this paper can calculate the maximum shear stress of solder joints quickly and easily under the temperature-dependent load of BGA packaging, which provides a basis for determining the shear stress level and life prediction of solder joints.Two constitutive models are used to simulate the circuit board, and results of the two models are approximately the same, which indicates that the two constitutive models are also suitable for describing the creep properties of solder joints.The finite element simulation method presented in this paper has good engineering application value, which can be used to analyze the stress level of BGA encapsulated circuit board under temperature cyclic load, and can also provide basis for life prediction of circuit board under temperature cyclic load.

REFERENCES

[1]   Steinberg D S. Vibration Analysis for Electronic Equipment[M]. Wiley, 1988.

[2]   Brown S B, Kim K H, Anand L. An internal variable constitutive model for hot working of metals[J]. International Journal of Plasticity, 1989, 5(2):95-130.

[3]   GZ Wang, ZN Cheng. The viscoelastic Anand constitutive equation of SnPb alloy [j]. Journal of applied mechanics, 2000, 17 (3): 133-133

[4]   L Zhang, X Chen, Nose h, et al. Anand model for predicting the stress-strain behavior of 63Sn37Pb solder [J]. Mechanical strength, 2004, 026 (004): 447-450

[5]   Stoeckl S , Yeo A , Lee C , et al. Impact of fatigue modeling on 2/sup nd/ level joint reliability of BGA packages with SnAgCu solder balls[C]// Electronic Packaging Technology Conference. IEEE, 2005.

[6]   Wiese, S. , et al. "Creep and crack propagation in flip chip SnPb37 solder joints." 1999 Proceedings. 49th Electronic Components and Technology Conference (Cat. No.99CH36299) IEEE, 2002.

[7]   David S. Steinberg, Steinberg, Chang Yong, et al. Thermal cycling and vibration fault prevention of electronic equipment [M]. Aviation Industry Press, 2012

Loss Modelling and Analysis of a High-Efficiency Wireless Power Transfer System for Automated Guided Vehicle Applications

Jincheng Yu
Integrated Circuits and Systems
Hong Kong Applied Science and
Technology Research Institute
Hong Kong SAR, China
jinchengyu@astri.org

Wai Leong Ng
Integrated Circuits and Systems
Hong Kong Applied Science and
Technology Research Institute
Hong Kong SAR, China
edwardng@astri.org

Minglu Xia
Integrated Circuits and Systems
Hong Kong Applied Science and
Technology Research Institute
Hong Kong SAR, China
mingluxia@astri.org

Ziyang Gao* (Corresponding Author)
Integrated Circuits and Systems
Hong Kong Applied Science and
Technology Research Institute
Hong Kong SAR, China
zygao@astri.org

Abstract—In this paper, the high-accuracy loss modelling and analysis of a kW-level wireless power transfer (WPT) system for automated guided vehicles (AGVs) robotic applications are presented. The proposed analysis process is universally fit for WPT system design. First, a simplified coil AC loss model is built, with consideration of skin and proximity effects. Compared with previous models, the proposed model maintains the analytical accuracy and realizes 62.5% computing time decrease in average. Also, the shielding effects and electronic losses are analyzed respectively, through both analytical deduction and finite element analysis (FEA). Influences of the parameter variation on the efficiency performances are specifically illustrated, including the input voltage and load condition. Thus, the deduced system efficiency is obtained and compared with the measured counterpart. Within the whole operating range, the system efficiency remains over 85%, with the deduction error lower than 2.3%. Specially, under rated condition, the deduction error is only 0.3%. All the results prove the high precision of the presented loss modelling and analysis for the proposed system.

Keywords—*loss modelling, wireless power transfer (WPT), high efficiency, iron loss, eddy-current loss*

I. Introduction

Due to the merits of convenience and reliability, wireless power transfer (WPT) has gained considerable recognition in robotic applications, such as automated guided vehicle (AGV), robot vacuum cleaner, etc [1], [2]. Without auxiliary of electrical cables, the maintenance of high efficiency in a wide operating range turns to be the core consideration in the loose coupling system [3], which accordingly highlights the importance of precise system loss analysis. In this paper, the loss modelling and analysis of a high-efficiency WPT system for AGV applications are proposed.

Various investigations have been carried out on the WPT system losses, mainly including the coil [3]-[7], shielding [8]-[11] and electronic losses [12], [13]. First, besides the DC coil copper loss, the mechanism of the coil eddy-current loss has been extensively explored, which is mainly caused by the skin and proximity effect [3]. To avoid the skin effect, Litz wire is introduced, and several coil models have been built accordingly, based on different transmission medium, coil configuration and relative position, etc [4]-[7]. G. Wei *at el.*

have built a coil AC resistance model in [4], where the high precision is proved. While the model is relatively complicated, which leads to heavy computing burden. In this paper, a simplified coil resistance model has been built accordingly, with simplified magnetic field strength deduction and Litz wire modelling. In this way, an 80% computing time reduction is achieved.

Also, shields are introduced to avoid the electromagnetic field (EMF) emission, usually containing the ferrite core and aluminium plate [8]. In [8], a lumped model for the shield is developed for shield loss optimization within the standard emission limit. J. Li *et al.* has investigated the mechanism of the ferrite and aluminium shielding and obtained the better shield structure [9]. Moreover, the effect of the relative location between the different shields is taken into consideration. It has been illustrated that the shielding effect has been enhanced by introducing an air-gap between the ferrite plate and metal strips, which leads to the transmission efficiency improvement [10].

Furthermore, the electronic loss modelling evolves with the upgrading of the semiconductor devices and electronic topologies [12], [13]. An electronic loss modelling has been done in [12], aiming at the WPT system with Gallium Nitride (GaN) devices. Synchronous rectifier topology is also taken into consideration under heavy-load operation [13]. Generally, the major system parameter changing will lead to the value variation of all the loss components, subsequently further influencing the loss distribution. Hence, a precise system efficiency analysis should contain the variation tendency of all the loss components, as a foundation of system efficiency optimization. J. Shi *et al.* has analysed all the loss components in a 100 W inductive coupling power transfer system considering underwater environments, without ferrite shielding [14]. Development of an MW-level WPT system for high-speed train is proposed in [15], with system efficiency deduction and analysis. While the loss distribution analysis investigations for kW-level WPT system are limited. In this paper, the proposed kW-level WPT system configuration is presented in Section II. Then, the coil, shielding and electronic losses are all combined into the loss analysis of the WPT system in Section III, with both analytical deduction and finite element analysis (FEA). Then, in Section IV, comparisons are

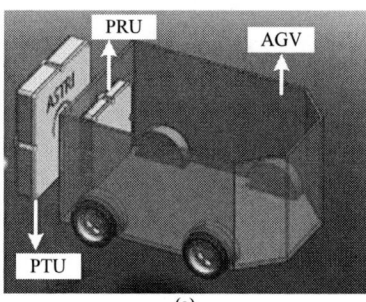

(a)

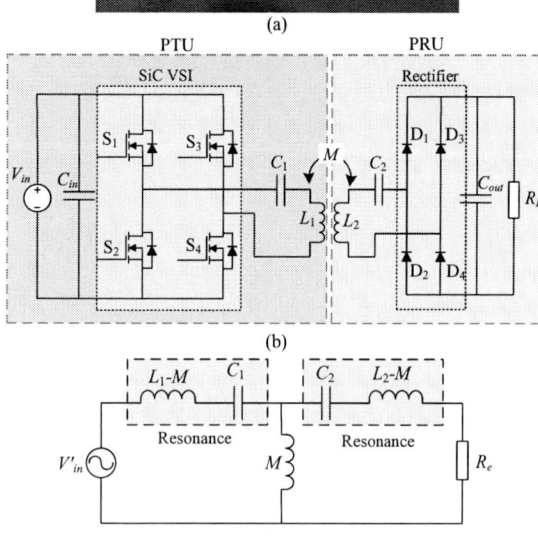

(b)

(c)

Fig. 1. Configuration of proposed kW-level WPT system for AGV application. (a) System configuration. (b) Circuit topology. (c) Equivalent circuit topology.

TABLE I.   Major Parameters of Proposed WPT System.

Parameter	Symbol	Unit	Value
Coil turn number	N	-	32
Inner coil radius	r_{in}	mm	38.0
Outer coil radius	r_{out}	mm	150.0
Filament radius of Litz wire	$r_{filaLitz}$	mm	0.05
Radius of Litz wire	r_{Litz}	mm	1.0
Filament turn number of Litz wire	$n_{filaLitz}$	-	250
Rated operating frequency	f	kHz	85
PTU, PRU coil self-inductance	L_1, L_2	μH	324
PTU, PRU resonant capacitor	C_1, C_2	nF	16.5
PTU filter capacitor	C_{in}	μF	5.0
PRU filter capacitor	C_{out}	μF	480.0

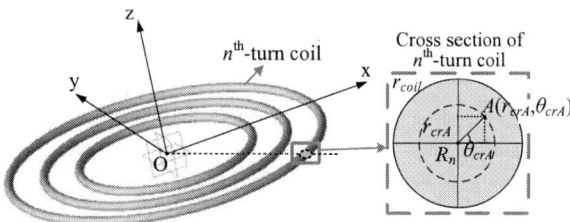

Fig. 2. Equivalent configuration of flat spiral coil.

carried out between measured and deduced results to verify the validation of the proposed loss modelling and analysis.

II. System Configuration

In this section, the proposed kW-level WPT system for AGV applications is briefly introduced. Firstly, the integral system configuration is demonstrated in Fig. 1(a). The power transmitting units (PTUs) are installed in the pre-set AGV wireless charging system, while the power receiving units (PRUs) are mounted on the back of the AGVs. With wireless charging, the aging of metallic contact between connector and AGVs can be avoided, which effectively decreases the possibility of electric shocks [16].

Fig. 1(b) demonstrates the system circuit topology, with the adoption of series-series (SS) compensation. The silicon carbon (SiC) voltage-source inverter (VSI) is selected in PTU for higher switching capability [17]. V_{in} is the DC input voltage, L_1 and L_2 denote the self-inductances of the transmitter and receiver coil, M is mutual-inductance, governed by

$$M = k\sqrt{L_1 L_2} \qquad (1)$$

where k is the coil coupling factor. C_1 and C_2 are the corresponding resonant capacitors, C_{in} and C_{out} denote the filter capacitors, R_L refers to the loading resistance. Moreover, the equivalent circuit topology is depicted in Fig. 1(c), with elimination of VSI and rectifier. The input voltage V'_{in} is directly set as sinusoidal. The equivalent loading resistance R_e is deduced as

$$R_e = 8R_L / \pi^2 . \qquad (2)$$

It can be seen that the leakage-inductance compensation is adopted to realize the zero-voltage switch (ZVS), the higher ratio between active and reactive power, as well as the steady output voltage with load variation [18], [19]. Thus, the resonant frequency f_r is governed by

$$f_r = \frac{1}{2\pi\sqrt{(L_1 - M)C_1}} = \frac{1}{2\pi\sqrt{(L_2 - M)C_2}} . \qquad (3)$$

Considering the target applications, the major system parameters are listed in Table I.

III. System Loss Modelling and Analysis

The system loss modelling and analysis are presented in this section. In general, the WPT system loss $P_{sysloss}$ is governed by

$$P_{sysloss} = P_{coil} + P_{fe} + P_{Al} + P_{electro} \qquad (4)$$

where P_{coil}, P_{fe}, P_{Al}, $P_{electro}$ denote the coil joule loss, ferrite iron loss, aluminium eddy current loss and electronic component loss. A simplified coil resistance model is proposed first. Then the principles of shielding and electronic losses are analysed respectively, with variation of input voltage and load condition.

A. A Simplified Coil Resistance Model

It has been well known that the skin and proximity effect give rise to the coil eddy-current loss [3]. To avoid the skin effect, the filament radius of the Litz wire $r_{filaLitz}$ is usually set much smaller than the skin depth d_{skin}, governed by

$$r_{filaLitz} < d_{skin} = \frac{1}{\sqrt{\pi f \mu_{coil} \sigma_{coil}}} \qquad (5)$$

where μ_{coil} and σ_{coil} denote the copper permeability and conductivity. Considering the rated frequency, $r_{filaLitz}$ is set as 0.05 mm. Hence, the AC coil resistance modelling can be simplified into the calculation of proximity effect resistance R_{prox}.

The coil electromagnetic field distribution is the core of the coil AC resistance model. To be more specific, the magnetic field strength of an arbitrary point on the coil should be clarified first. The proposed flat spiral coil is simplified into concentric annulus with the same turn number, which is depicted in Fig. 2. The origin of XYZ coordinate axis is set at the centre of the coil concentric annulus. The centre circle of each annulus locates on the XY plane. The centre circle radius of the nth-turn coil is annotated as R_n. A is an arbitrary point belongs to the nth-turn coil, and (r_{crA}, θ_{crA}) indicates its polar coordinate on the coil cross section. Under cylindrical coordinate, the field strength at point A by one current-carried coil concentric annulus with radius R_t is deduced as [4]

$$H_{At}^2\left(r_A, z_A\right) = H_{Azt}^2\left(r_A, z_A\right) + H_{Art}^2\left(r_A, z_A\right) \tag{5}$$

where $H_{Azt}(r_A, z_A)$ and $H_{Art}(r_A, z_A)$ are the axial and radius field strength components of $H_{At}(r_A, z_A)$. r_A, z_A are governed by

$$r_A = R_n + r_{crA}\cos\theta_{crA} \tag{6}$$

$$z_A = r_{crA}\sin\theta_{crA} \tag{7}$$

Thus, H_{Azt} and H_{Art} are deduced as [20]

$$H_{Azt}\left(r_{crA}, \theta_{crA}\right) = \frac{I_c R_t}{2}\int_0^\infty xe^{-x|r_{crA}\sin\theta_{crA}|}J_0\left[x(r_{crA}\cos\theta_{crA} + R_n)\right]J_1\left(xR_t\right)dx \tag{8}$$

$$H_{Art}\left(r_{crA}, \theta_{crA}\right) = \frac{I_c R_t}{2}\int_0^\infty xe^{-x|r_{crA}\sin\theta_{crA}|}J_1\left[x(r_{crA}\cos\theta_{crA} + R_n)\right]J_1\left(xR_t\right)dx \tag{9}$$

where I_c is the coil current, $J_0(x)$ and $J_1(x)$ denote the Bessel functions. Specially, H_{An}, namely the field strength by the same coil annulus point A locates in, can be deduced by Ampere's law:

$$H_{An} = \frac{I_c r_{cr}}{2\pi r_{coil}^2} \tag{10}$$

$$H_{Azn} = H_{An}\cos\theta_{crA}, H_{Arn} = H_{An}\sin\theta_{crA} \tag{11}$$

Hence, the total field strength of point A H_A is obtained as

$$H_A^2 = H_{Az}^2 + H_{Ar}^2 = \sum_{t=1}^N H_{Azt}^2 + \sum_{t=1}^N H_{Art}^2 \tag{12}$$

H_{Ar} and H_{Az} are the total radial and axial field strength components. For the proposed system, the coil turn number $N=32$. Thus, the average field strength of the nth-turn coil cross section can be deduced as

$$H_{ave-n}^2 = \frac{\iint\limits_{S_{cr-n}} H_A^2\left(r_{crA}, \theta_{crA}\right)r_{crA}dr_{crA}d\theta_{crA}}{S_{cr-n}} \tag{13}$$

where S_{cr-n} denotes the area of the nth-turn coil cross section. Thus, the AC resistance of the nth-turn coil R_{prox-n} is governed by [4]

$$R_{prox-n} \approx \frac{2\mu^2\sigma\pi^4 n_{filaLitz}r_{filaLitz}^4 R_n H_{ave-n}^2 f^2}{I_c^2} \tag{14}$$

where $n_{filaLitz}$ is the filament turn number of Litz wire. Thus, the coil joule loss P_{coil} is expressed as

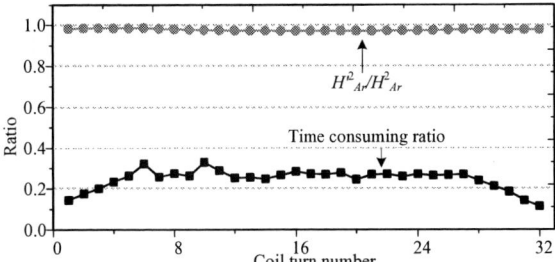

Fig. 3. Performance comparison between H_{Ar} and H'_{Ar}.

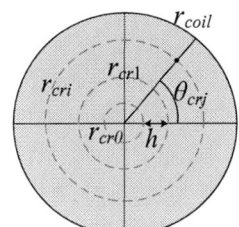

Fig. 4. Equivalent configuration of flat spiral coil.

$$P_{coil} = I_c^2\left(R_{prox} + R_{DC}\right) \tag{15}$$

where R_{DC} is the coil DC copper loss.

To release the heavy computing burden, a series of simplifications are adopted.

Firstly, based on (8) to (11), the calculation process of H_A can be simplified. In theory, the influence of all the coil turns on point A should be taken into account. However, since the larger distance between the coil and point A leads to the lower field strength influence, only considering the coil turns with major influences can effectively decrease the computing burden. Thus, the total radial field strength is simplified as

$$H'_{Ar} = \begin{cases} \sum\limits_{t=1}^{n+k_1} H_{Art}, 1 \le n \le k_1 \\ \sum\limits_{n-k_1}^{n+k_1} H_{Art}, k_1 < n \le N - k_1 \\ \sum\limits_{n-k_1}^{N} H_{Art}, N - k_1 < n \le N \end{cases} \tag{16}$$

Considering both the accuracy and computing time, $k_1=5$ is selected. Fig. 3 demonstrates the comparison between H_{Ar} and H'_{Ar}, with $r_{crA}=5r_{coil}/12$ and $\theta_{crA}=\pi/4$. With coil turn varying from 1 to 32, H'_{Ar} accommodates no apparent difference compared with H_{Ar}, while the time consumption realizes 75% reduction. Furthermore, for H_{Az}, it has been proved that all the coil turns should be included to guarantee the algorithm precision.

Secondly, integral discretization can be adopted to obtain the average field strength on the coil cross section. On basis of [4], the four-notes Newton-Cotes integration is adopted to decrease the computing burden in (12)

$$H_{ave-n}^2 \approx \frac{3h}{8}\frac{\sum\limits_{j=1}^{6} f_{0,j} + 3f_{1,j} + 3f_{2,j} + f_{3,j}}{6\pi r_{coil}^2}2\pi, f_{i,j} = H_n^2\left(r_{cri}, \theta_{crj}\right)r_{cri} \tag{17}$$

978-1-6654-1392-3/21 $31.00 © 2021 IEEE   1389

The coil cross section in Fig. 2 is redrawn in Fig. 4 accordingly. Similar as point A, a series of points (r_{cri}, θ_{crj}) have been selected for integral discretization. Thereinto, the radius and angle division are set as

$$r_{cri}=r_{cr0}+ih=r_{coil}/8+7ir_{coil}/24,$$
$$\theta_{crj}=\begin{cases} j\pi/4, j=1,2,3 \\ (j+1)\pi/4, j=4,5,6 \end{cases} \quad (18)$$

Also, based on the symmetry and linearity of the field strength, further simplification is adopted. For symmetrical feature, taking the instance of point A, the field strength satisfies:

$$\begin{cases} H_{Art}\left(\theta_{crA}=\theta_{crj}\right)=H_{Art}\left(\theta_{crA}=2\pi-\theta_{crj}\right), t \neq n, n \in 1,2,..,N \\ H_{Art}\left(\theta_{crA}=\theta_{crj}\right)=-H_{Art}\left(\theta_{crA}=2\pi-\theta_{crj}\right), t=n, n \in 1,2,..,N \\ H_{Azt}\left(\theta_{crA}=\theta_{crj}\right)=H_{Azt}\left(\theta_{crA}=2\pi-\theta_{crj}\right), t=1,2,..,N \end{cases} \quad (19)$$

Hence, the range of θ_{crj} is simplified within $[0, \pi]$. Moreover, the linearity feature leads to the following conclusions:

$$\begin{cases} H_{Art}\left(\theta_{crA}=\theta_{crj}\right) \approx H_{Art}\left(\theta_{crA}=\pi/2-\theta_{crj}\right), t \neq n, n \in 1,2,..,N \\ H_{Azt}\left(\theta_{crA}=\theta_{crj}\right) \approx H_{Azt}\left(\theta_{crA}=2\pi\right)-H_{Azt}\left(\theta_{crA}=2\pi-\theta_{crj}\right), t=1,2,..,N \end{cases} \quad (20)$$

In this way, the performance comparison between the original H_{ave-n} and the simplified H'_{ave-n} is shown in Fig. 5. It can be seen that before and after simplification, the field strength performance remains basically unchanged. The maximum field strength error is below 4%. Therefore, with maintenance of the precision, the combination of the two simplification steps realizes up to 62.5% computing time saving in average, which effectively increase the computing efficiency of the proposed algorithm.

B. Shielding and Electronic Losses

In this part, the principle of shielding and electronic losses is analysed. Both the ferrite and aluminium plate shields have been considered. The ferrite core loss P_{fe} is governed by [21]

$$P_{fe}=C_m f^\alpha B_{\max}^\beta \quad (20)$$

where C_m, α, β denote the constants depending on the ferrite material property. $B_{\max}$ is the peak value of the magnetic flux density. Also, the eddy-current loss generated by the Aluminium plate P_{Al} is expressed as [22]

$$P_{Al}=\frac{\pi^2 B_{\max}^2 d_{Al}^2 f^2}{6\rho_{Al}D_{Al}} \quad (21)$$

where d_{Al}, ρ_{Al} and D_{Al} are the thickness, material resistivity and density of the Aluminium sheet.

Furthermore, the electronic loss mainly contains the SiC MOSFET and the diode losses. Losses by one MOSFET and one diode, P_{MOS} and P_{diode}, are expressed as [23]

$$P_{MOS}=R_{ds-on}I_{MOS-rms}^2+2V_{in}I_{MOS}f\left(t_{rise}+t_{fall}\right) \quad (22)$$

$$P_{diode}=V_{th}I_{diode-ave}+R_{diode}I_{diode-rms}^2 \quad (23)$$

where R_{ds-on} denotes the on-state resistance of MOSFET. I_{MOS} is the MOSFET current, $I_{MOS-rms}$ denotes its RMS value. t_{rise} and t_{fall} are the rise and fall time of MOSFET. Full-bridge single-phase VSI and rectifier are adopted for the proposed system.

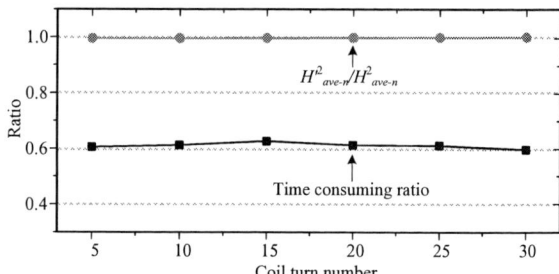

Fig. 5. Performance comparison between H_{ave-n} and H'_{ave-n}.

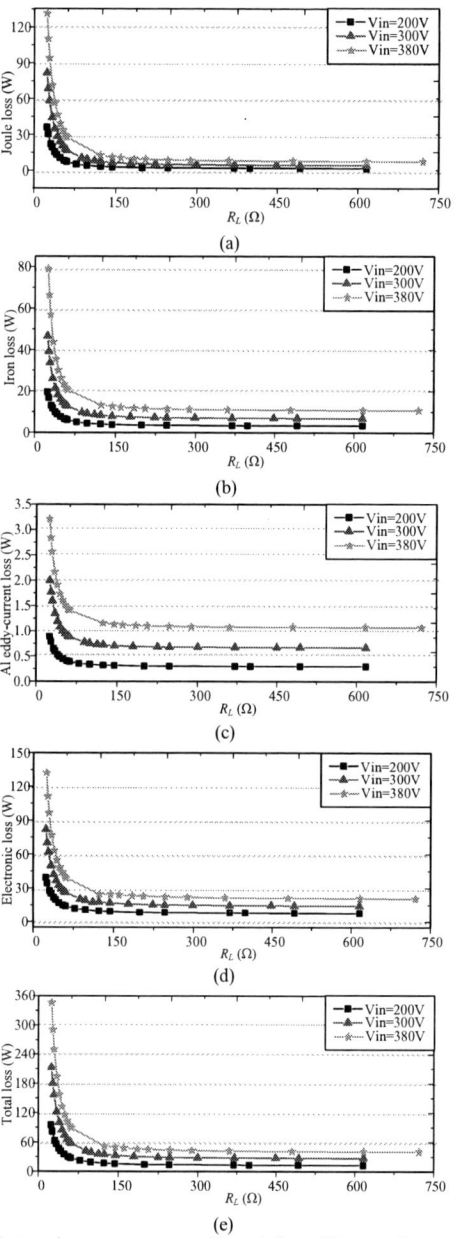

Fig. 6. System loss components with variation of input voltage and load condition. (a) Joule loss. (b) Iron loss. (c) Aluminium eddy-current loss. (d) Electronic loss. (e) Total loss.

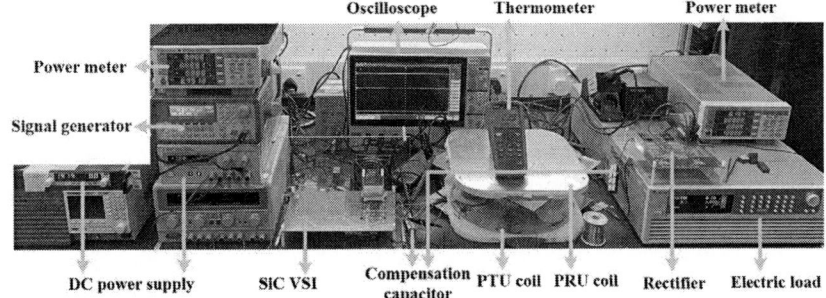

Fig. 7. Experiment setup of proposed WPT system.

C. System Loss Analysis

Based on the previous loss modelling, the system loss component analysis has been carried out in this part, including the joule loss, iron loss, Aluminium eddy-current loss, electronic loss and total loss. Fig. 6 depicts the loss component performances with variation of input voltage and load. Some conclusions are listed below.

- With the same input voltage, all the loss components increase dramatically with load increment under heavy-load range ($0<R_L<150$ Ω). While in the light-load range ($R_L>150$ Ω), the loss components remain basically unchanged.

- Based on the above statement, it can be concluded that the light-load efficiency will continuously decrease with the load reduction. While with load increment (R_L decreases), the heavy-load efficiency first rises then falls, where an optimal efficiency point exists.

- Under the same load condition, the increase of input voltage leads to all the loss component increment.

- With rated frequency, the proportion of the loss components ranking from high to low are listed as: electronic loss, joule loss, iron loss and Aluminium eddy-current loss. Hence, the optimization of power electronics and coil configuration can improve the system efficiency more effectively.

IV. SYSTEM EFFICIENCY ANALYSIS

Based on the deduced and experimental results, the system efficiency comparison is carried out. The experimental setup of the proposed WPT system is shown in Fig. 7, including the DC power supplies, power meters, a signal generator, single-phase full-bridge SiC VSI, thermometers, compensation and filter capacitors, the proposed PTU/PRU coils, full-bridge rectifier, a 2-kW electric load and an oscilloscope. Based on (4), the system efficiency η_{sys} is obtained as

$$\eta_{sys} = \frac{P_{out}}{P_{out} + P_{sysloss}} \times 100\% \qquad (24)$$

where P_{out} is the output power of the system.

Under rated switching frequency 85 kHz, the loading resistance R_L varies from 20 to 750 Ω, and the input voltage changes from 200 to 400 V, with the specific instance of 200, 250, 300 and 380 V. Fig. 8(a) depicts the deduced and measured efficiency comparison under rated input voltage 380 V. With R_L varying from the rated load 144 Ω to light load 722 Ω, the deduction error remains below 0.36%. Point a and b refer to the deduced and measured efficiency under rated condition, whose specific loss distribution and efficiency

comparison are listed in Table II. Furthermore, the system efficiency performance under the whole operating range is demonstrated in Fig. 8(b). Point c and d demonstrate the maximum error between measured and deduced result. Under the whole operating range, the measured efficiency maintains over 85%, and the deduced error is below 2.3%, which have proven the high precision of the proposed loss modelling and analysis.

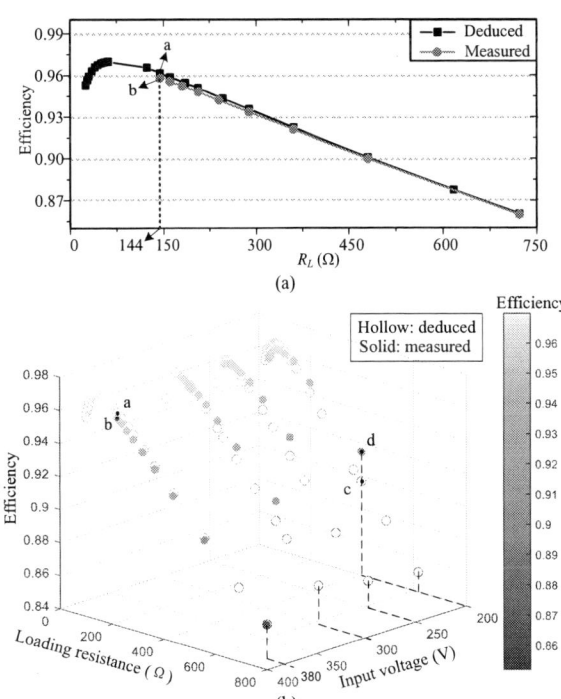

Fig. 8. System efficiency comparison between deduced and measured results. (a) Rated input voltage V_{in}=380V. (b) Whole operating range.

TABLE II.   DEDUCED LOSS DISTRIBUTION UNDER RATED CONDITION.

Losses		PTU	PRU	Total
Joule loss (W)		9.3	1.9	11.2
Iron loss (W)		10.0	2.5	12.5
Electronic loss (W)		18.9	5.7	24.6
Aluminium eddy-current loss (W)		0.8	0.3	1.1
Output power (W)		-	-	1230.0
Efficiency	measured	-	-	96.13%
	deduced	-	-	95.79%
Deduction error				0.36%

V. CONCLUSION

In this paper, the loss modelling and analysis of the proposed high-efficiency WPT system for AGV applications have been presented. Conclusions are summarized below.

Firstly, a simplified coil AC loss model has been built. The skin, proximity effects and Litz wire structure have been taken into account. Based on the modelling simplification of the field strength components H_{Ar}, H_{Az} and H_{ave-n}, the 62.5% computing time saving has been realized, with the field strength analytical error remains below 4%.

Secondly, the shielding and electronic losses have been analysed specifically, including the iron loss, Al eddy-current loss, PTU and PRU electronic losses. Both analytical deduction and FEA simulation have been presented. Based on all the loss analysis, influences of parameter variation on the AGV WPT systems have been specifically summarized, serving as effective references for other system designs.

Thirdly, the system efficiency analysis has been carried out. Measured and deduced efficiency performances have been compared with input voltage and load variation. In the whole operating range, the system efficiency maintains over 85%, while the deduction error remains below 2.3%. Especially, the deduction error is limited to 0.3% under rated condition. As a result, the high precision of the proposed loss modelling and analysis has been well validated. Also, the proposed analysis procedure is universally applicable to the WPT system design.

REFERENCES

[1] R. Wu, W. Li, H. Luo, J. K. O. Sin and C. P. Yue, "Design and Characterization of Wireless Power Links for Brain–Machine Interface Applications," *IEEE Trans. Power Electron.*, vol. 29, no. 10, pp. 5462-5471, Oct. 2014.

[2] J. Baek, C. Ahn, B. Kim, S. Choi and S. Kwak, "High frequency wireless power transfer system for robot vacuum cleaner," 2014 IEEE International Conference on Consumer Electronics (ICCE), 2014, pp. 308-310.

[3] C. Utschick, C. Merz and C. Som, "AC Loss Behavior of Wireless Power Transfer Coils," 2019 IEEE Wireless Power Transfer Conference (WPTC), 2019, pp. 120-125.

[4] G. Wei, X. Jin, C. Wang, J. Feng, C. Zhu and M. I. Matveevich, "An Automatic Coil Design Method With Modified AC Resistance Evaluation for Achieving Maximum Coil–Coil Efficiency in WPT Systems," *IEEE Trans. Power Electron.*, vol. 35, no. 6, pp. 6114-6126, Jun. 2020.

[5] Z. Yan et al., "Frequency Optimization of a Loosely Coupled Underwater Wireless Power Transfer System Considering Eddy Current Loss," *IEEE Trans. Ind. Electron.*, vol. 66, no. 5, pp. 3468-3476, May 2019.

[6] S. Mehri, A. C. Ammari, J. B. H. Slama and M. Sawan, "Design Optimization of Multiple-Layer PSCs With Minimal Losses for Efficient and Robust Inductive Wireless Power Transfer," *IEEE Access*, vol. 6, pp. 31924-31934, 2018.

[7] X. Zhang, C. Quan and Z. Li, "Mutual Inductance Calculation of Circular Coils for an Arbitrary Position with Electromagnetic Shielding

in Wireless Power Transfer Systems," *IEEE Trans. Transp. Electrification*, doi: 10.1109/TTE.2021.3054762.

[8] M. Mohammad, E. T. Wodajo, S. Choi and M. E. Elbuluk, "Modeling and Design of Passive Shield to Limit EMF Emission and to Minimize Shield Loss in Unipolar Wireless Charging System for EV," *IEEE Trans. Power Electron.*, vol. 34, no. 12, pp. 12235-12245, Dec. 2019.

[9] J. Li, F. Yin, L. Wang and L. Wang, "Research on the transmission efficiency of different shielding structures of wireless power transfer system for electric vehicles," *CSEE J. Power Energy Syst.*, doi: 10.17775/CSEEJPES.2019.00500.

[10] H. H. Park, J. H. Kwon, S. I. Kwak and S. Ahn, "Effect of Air-Gap Between a Ferrite Plate and Metal Strips on Magnetic Shielding," *IEEE Trans. Magn.*, vol. 51, no. 11, pp. 1-4, Nov. 2015.

[11] T. H. Kim, S. Yoon, J. G. Yook, G. H. Yun and W. Y. Lee, "Evaluation of power transfer efficiency with ferrite sheets in WPT system," 2017 IEEE Wireless Power Transfer Conference (WPTC), 2017, pp. 1-4.

[12] M. Wu, M. Jiang, L. Wang and G. Ning, "Loss Analysis in wireless power transfer system based on GaN," 2018 1st Workshop on Wide Bandgap Power Devices and Applications in Asia (WiPDA Asia), 2018, pp. 55-59.

[13] K. Krestovnikov, E. Cherskikh and N. Pavliuk, "Concept of a synchronous rectifier for wireless power transfer system," IEEE EUROCON 2019 -18th International Conference on Smart Technologies, 2019, pp. 1-5.

[14] J. Shi, D. Li, and C. Yang. "Design and analysis of an underwater inductive coupling power transfer system for autonomous underwater vehicle docking applications." *Journal of Zhejiang University SCIENCE C* vol. 15, no. 1, pp. 51-62, 2014.

[15] J. H. Kim et al., "Development of 1-MW Inductive Power Transfer System for a High-Speed Train," *IEEE Trans. Ind. Electron.*, vol. 62, no. 10, pp. 6242-6250, Oct. 2015.

[16] S. Huang, T. Lee, W. Li and R. Chen, "Modular On-Road AGV Wireless Charging Systems Via Interoperable Power Adjustment," *IEEE Trans. Ind. Electron.*, vol. 66, no. 8, pp. 5918-5928, Aug. 2019.

[17] L. Zhang et al., "Evaluation of Different Si/SiC Hybrid Three-Level Active NPC Inverters for High Power Density," *IEEE Trans. Power Electron.*, vol. 35, no. 8, pp. 8224-8236, Aug. 2020.

[18] Y. Wang, Y. Yao, X. Liu, D. Xu and L. Cai, "An LC/S Compensation Topology and Coil Design Technique for Wireless Power Transfer," *IEEE Trans. Power Electron.*, vol. 33, no. 3, pp. 2007-2025, Mar. 2018.

[19] W. Zhang and C. C. Mi, "Compensation Topologies of High-Power Wireless Power Transfer Systems," *IEEE Trans. Veh. Technol.*, vol. 65, no. 6, pp. 4768-4778, Jun. 2016.

[20] J. Acero, R. Alonso, J. M. Burdo, L. A. Barragan, and D. puyal, "Frequency-dependent resistance in Litz-wire planar windings for domestic induction heating appliances," *IEEE Trans. Power Electron.*, vol. 21, no. 4, pp. 856–866, Jul. 2006.

[21] M. Mohammad and S. Choi, "Optimization of ferrite core to reduce the core loss in double-D pad of wireless charging system for electric vehicles," 2018 IEEE Applied Power Electronics Conference and Exposition (APEC), 2018, pp. 1350-1356, doi: 10.1109/APEC.2018.8341192.

[22] B. Olukotun, J. S. Partridge and R. W. G. Bucknall, "Loss Performance Evaluation of Ferrite-Cored Wireless Power System with Conductive and Magnetic Shields," 2019 IEEE PES Innovative Smart Grid Technologies Europe (ISGT-Europe), 2019, pp. 1-5, doi: 10.1109/ISGTEurope.2019.8905437.

[23] A. Ramezani and M. Narimani, "A Wireless Power Transfer System with Reduced Output Voltage Sensitivity for EV Applications," 2018 IEEE PELS Workshop on Emerging Technologies: Wireless Power Transfer (Wow), 2018, pp. 1-5, doi: 10.1109/WoW.2018.8450895.

Three Dimensional Wafer-level Vacuum Packaging of MEMS Resonant Accelerometer

Ziji Wang[1], Chaoyang Xing[2], Jin Zhang[1], Zhaoxi Su[1], Wenqi Li[1], Bin Luo[1] and Jintang Shang[1*]

1 Key Lab of MEMS of Ministry of Education, Southeast University, Nanjing, 210096, CHINA

2 Department of MicroSystem Integration, Beijing Institute of Aerospace Control Devices,Beijing, 100039,CHINA

*email: jshang@seu.edu.cn

Abstract—*In this work, a novel approach for wafer-level vacuum packaging of MEMS resonant accelerometer is presented and experimentally demonstrated. 3D composite glass-silicon wafer based on glass reflow process is designed and fabricated as carrier wafer. The through glass via (TGV) which is integrated in the composite carrier wafer enables hermetically vertical electrical interconnection for the vacuum packaged MEMS resonant accelerometer. The three dimensional glass lid wafer contains several micro foaming glass caps. The fabricated micro glass caps realize a 2.5mm vertical space for packaged device. The experimental results indicate that utilizing 3D glass cap and composite carrier wafer for vacuum packaging potentially offer a low-cost and effective solution for high performance MEMS resonant accelerometers packaging.*

Keywords—*Wafer-level vacuum packaging; MEMS resonant accelerometer; 3D glass cap; Composite glass-silicon interposer*

I. INTRODUCTION

The ambient environment fluctuations will drift the resonant frequency and damage the long-term bias stability of MEMS inertial devices such like mechanical resonators, gyroscopes and accelerometers [1]. Vacuum packaging is a promising approach to further improve the resonant accelerometer performance. Over the past few years, ceramic or metal packages have been widely used in commercial MEMS device vacuum packaging. In addition, wafer-level vacuum packaging (WLVP) has been considered as an attractive solution for high performance MEMS devices packaging since its cost and size advantages over current commercial vacuum packaging approaches [2-4].

Wafer level vacuum packaging is commonly realized through a carrier wafer and a lid wafer. The carrier wafer is utilized to load the packaged devices/device wafer. In addition, vertical electrical interconnection needs to be integrated in the carrier wafer to provide electrical connections inside and outside the package. Two of the most well-known approaches of vertical through-wafer-interconnections are through silicon via (TSV) [5-6] and through glass via (TGV) [7]. The lid wafer contains several micro cavities which provides vacuum space for packaged device chips, the micro cavities can be obtained through thick wafer with deep etching [8-9]. Increasing the cap vertical space will effectively promote the versatility of WLVP.

In this work, a novel three dimensional wafer-level vacuum packaging (3D-WLVP) approach aiming for MEMS resonant accelerometer package is presented. The proposed method utilizes a lid wafer which contains several three dimensional micro thermal foaming glass caps and a silicon-glass composite carrier wafer based on glass reflow process [10][11]. The composite interposer also provides lateral/vertical interconnection by redistribution layer (RDL) and through glass via (TGV). Then the fabrication process of both 3D all-glass lid wafer and composite interposer are introduced. Finally, the proposed fabrication process is experimentally verified.

II. DESIGN AND CONSIDERATION

The cross-section schematic diagram of the proposed 3D-WLVP approach for the MEMS resonant accelerometer is shown in Fig. 1.

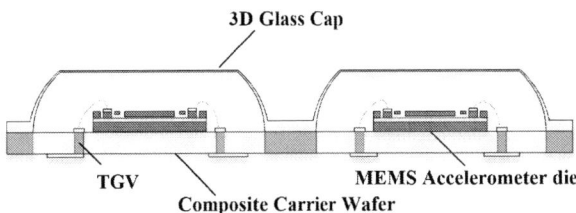

Fig. 1. Schematic of the cross section of wafer-level vacuum packaging for MEMS resonant accelerometer

The whole packaging mainly consists of three parts: MEMS resonant accelerometer with chip area of 7440μm *5100μm; silicon-glass composite carrier wafer which is used to mount MEMS accelerometer dies and provide lateral and vertical electrical interconnections; three dimensional all glass lid wafer which is used to provide steric vacuum space for the packaged device die.

A. Three dimensional micro glass cap

A novel 3D all-glass wafer is designed as the wafer-level packaging cap in this work. Borosilicate glass is chosen as the lid wafer material since its excellent chemical stability, low thermal conductivity and high electrical resistivity [10]. Meanwhile, its moderate softening point also brings convenience to wafer level 3D thermal forming [12]. Apart from the chemical foaming process, wafer-level mold is utilized to improve the anisotropy of cap geometry.

Both of the blowing pressure during thermal foaming and the shape of mold wafer will affect the three dimensional geometry of micro glass cap. The calculated cross-section

978-1-6654-1392-3/21 $31.00 © 2021 IEEE

geometry of softened glass during abovementioned thermal forming process is illustrated in Fig. 2. One of the mold shaping methods is introduced in the calculation.

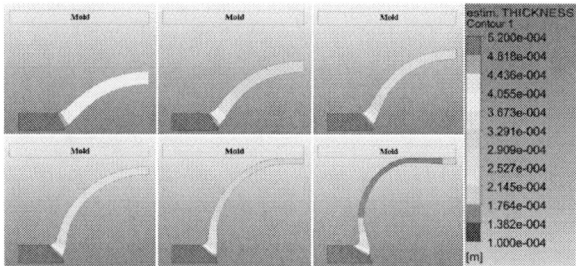

Fig. 2. Calculated two-dimensional cross-section of thermal forming process under the influence of both foaming pressure and mold shaping

The borosilicate glass is set with an initial thickness of 500µm . With the increase of foaming pressure, the top of the softened glass will rise until contacting with the mold. Hence, thickness of the glass cap varies under the effects of foaming pressure and mold height.

B. Composite glass-silicon carrier wafer

The design consideration of the carrier wafer utilized in this work is similar with the composite silicon-glass interposer [11]. Composite carrier wafer based on glass reflow process provides a low cost but effective solution for obtaining through glass via in glass substrate. Several glass panels with size of 13mm*7.5mm are embedded in the silicon wafer. Ten cylindrical highly conductive silicon interconnects with 500µm diameter are integrated into each panel to realize vertical electrical interconnection. These design parameters aim at providing high alignment tolerance during the subsequent assembly process. The packaging size can be further miniaturized through shrinking the silicon interconnects pitch and glass substrate area. Two Cr/Au redistribution layers (RDL) are deposited on both sides of the carrier wafer to obtain lateral electrical interconnection. Through ohmic contact with through glass via, the composite carrier wafer is able to provide 3D interconnection for the packaged device.

III. FABRICATION AND RESULTS

The fabrication process for three dimensional all glass lid wafer is shown in Fig. 3, which mainly consists of five steps, namely: 1) Photolithography and wet etching to form ~200µm cavities on a 4-inch silicon wafer; 2) introducing foaming agents into the cavities; 3) sealing silicon cavities with a 4-inch borosilicate glass wafer through anodic bonding; 4) forming three dimensional glass cap through glass foaming process; 5) removing silicon substrate by wet etching. With the above mentioned fabrication process, a 3D glass cap with 2.5mm vertical space is obtained through a <200µm etching process, which simplifies the process and cost.

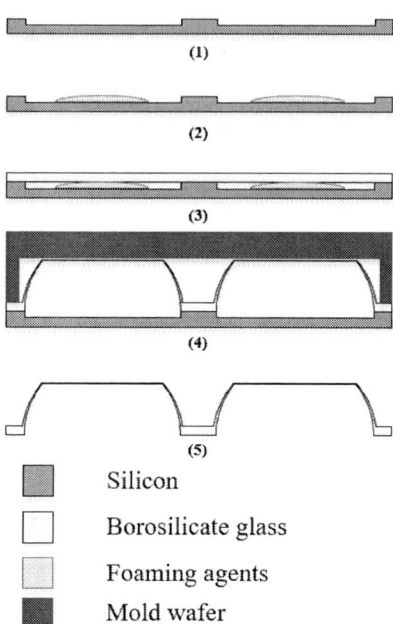

Silicon

Borosilicate glass

Foaming agents

Mold wafer

Fig. 3. Fabrication process of three dimensional all glass lid wafer with five steps: 1) photolithography and wet etching; 2) introducing foaming agents; 3) sealing silicon cavities by anodic bonding; 4) 3D glass cap thermal foaming with mold shaping; 5) silicon removing

The fabricated lid wafer with 3D micro glass caps before silicon substrate removing is shown in Fig. 4. Micro glass caps are foamed by inner pressure produced by foaming agents.

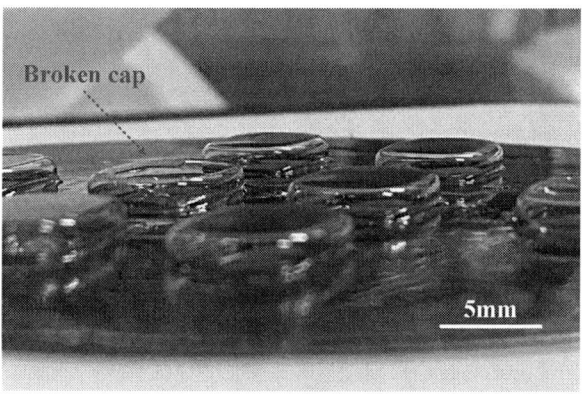

Fig. 4. Fabricated three dimensional all-glass lid wafer, one of the micro glass cap is broken since excessive undue pressure difference inside and outside cavity.

Excessive foaming agents will produce huge pressure difference inside and outside the cavity, which may lead cap break during the forming process. As illustrated from Fig. 4, one of the glass cap is broken in the cooling step. Therefore, further investigations are required for the breaking mechanism of thermal foaming glass caps.

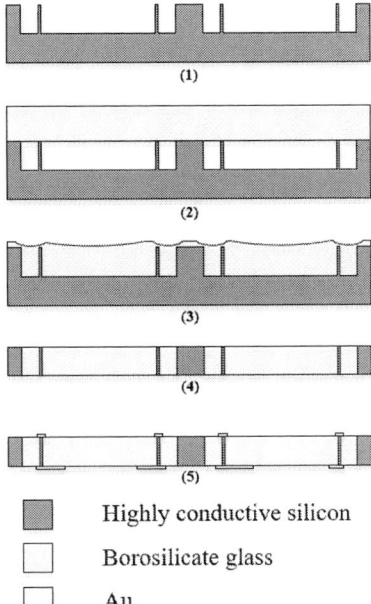

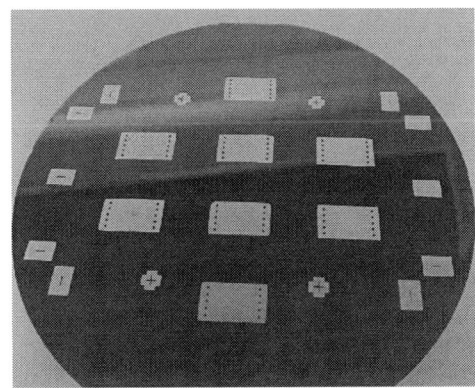

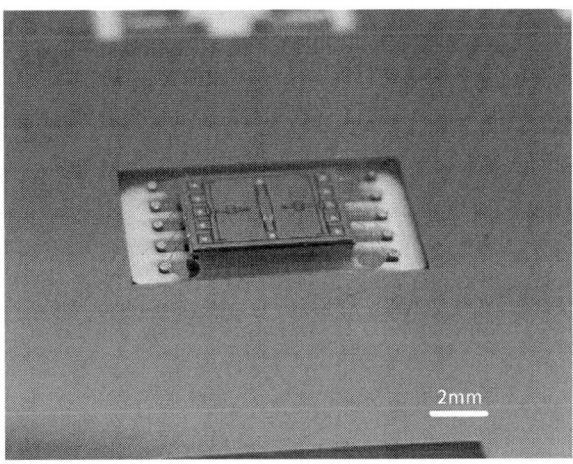

Highly conductive silicon

Borosilicate glass

Au

Fig. 5. Fabrication process of composite glass-silicon carrier wafer which mainly includes:1) deep reactive etching; 2) anodic bonding; 3) glass reflow; 4) grinding and chemical mechnical polising; 5) Cr/Au redistribution layer sputtering and wet etching.

As illustrated in Fig. 5, the fabrication process of glass-silicon composite carrier wafer starts with a 1mm thick highly conductive silicon wafer. The wafer is spin coated with positive photoresist and then dry etched by deep reactive ion etching (DRIE) to form cavities and micro silicon pillars with a depth of 500μm. The silicon cavities are then vacuum sealed by a 500μm borosilicate glass through anodic bonding. After this, the bonded wafer is delivered into the furnace to perform the reflow process under 880 °C.

Fig. 6. Fabricated composite glass-silicon 4-inch carrier wafer before redistribution layer deposition.

Increasing the reflow duration time will enhance the filling degree of softened glass into the silicon cavity, which guarantee the hermetic level of vacuum cavity formed between carrier wafer and lid wafer. Followed by cooling and annealing process, both sides of the reflowed wafer are thinning grinded and polished to remove residual glass/silicon

parts. The fabricated composite glass-silicon carrier wafer without redistribution layer is shown in Fig. 6. It realizes a seedless vertical interconnection through the exposed highly conductive micro silicon pillars. After grinding and polishing, the thickness of carrier wafer reduces from 1500μm (1000μm highly conductive silicon and 500μm borosilicate glass) to 475μm. Then, 50nm Cr layer and 100nm Au layer are deposited onto the composite wafer through sputtering. Finally, the pattern of redistribution layer is defined by one step photolithography and wet etching.

The vacuum packaging device, which is a MEMS resonator accelerometer, is shown in Fig. 6. The accelerometer is based on a silicon-glass-silicon triple stacked structure [13], the whole die area is 7440μm×5100μm with a thickness of ~600μm. The accelerometer chip is mounted onto the glass panel of the composite carrier wafer, one of the mounted accelerometer chip is shown in Fig. 7. The electrical interconnection between electrode on the chip and RDL is realized by wire bonding. With the through glass via, driving/sensing signals can hermetically transmit between both sides of the carrier wafer.

Fig. 7. MEMS resonant acelerometer mounted onto the glass panel of composite silicon-glass carrier wafer

It can be seen that bonding wires consume extra vertical space, which requires an increase of the packaging cap height. Therefore, 3D micro glass cap with adjustable vertical height is a promising solution for wafer-level low cost but high reliability MEMS device packaging which employs wire bonding process. Similar to our previous work [14], residual silicon area of the carrier wafer can be anodic bonded with the planar area of lid wafer to realize hermetic and vacuum packaging. In addition to anodic bonding，various wafer-level bonding approaches including metal-metal bonding, eutectic alloy bonding can be employed for the final vacuum bonding step for different application scenarios.

IV. CONCLUSION

A novel wafer level vacuum packaging approach for high performance MEMS resonant accelerometer is presented. Comparing with existing lid wafer approaches whose height is mainly limited by wafer thickness, three dimensional all-glass lid wafer prepared by chemical forming process is capable of providing packaging caps with enhanced vertical height range. In addition, by adjusting the cap shape through wafer molding method, the space efficiency of glass cap can be further improved. Combing with the TGV integrated composite carrier wafer, our three dimensional wafer level vacuum packaging approach is capable for low cost but high performance package of 3D MEMS devices such as micro shell resonator gyroscope (mSRG) [15], 3D MEMS alkali vapor cell integrated chip-scale atomic devices (CSAD) [16], and even three dimensional on-chip microsystems.

ACKNOWLEDGMENTS

The authors would like to appreciate the help with wire bonding provided from assistant professor Dr. Wei Li.

REFERENCES

[1] B. Lee, S. Seok, and K. Chun, "A study on wafer level vacuum packaging for MEMS devices," *Journal of Micromechanics and Microengineering.*, vol. 13, no. 5, pp. 663–669, 2003.

[2] J. Chae, J. M. Giachino and K. Najafi, "Fabrication and Characterization of a Wafer-Level MEMS Vacuum Package with Vertical Feedthroughs," *Journal of Microelectromechanical Systems*, vol. 17, no. 1, pp. 193-200.

[3] Jin J-Y, Yoo S-H, Yoo B-W and Kim Y-K A Wafer-level vacuum package using glass-reflowed silicon through-wafer interconnection for nano/micro devices *Journal of Nanoscience and Nanotechnology*, no. 17 pp.52–62

[4] M. A. Urquia et al., "High vacuum wafer level packaging for uncooled infrared sensor," *2020 Symposium on Design, Test, Integration & Packaging of MEMS and MOEMS (DTIP)*, 2020, pp. 1-5.

[5] J. Zhao, Q. Y uan, X. Kan, J. Y ang, and F. Y ang, "A low feed-through 3D vacuum packaging technique with silicon vias for RF MEMS resonators," *Journal of Micromechanics and Microengineering*, vol. 27, p. 014003, Nov. 2016.

[6] M. Sunohara, T. Tokunaga, T. Kurihara, and M. Higashi, "Silicon interposer with TSVs (through silicon vias) and fine multilayer wiring," *Proc. 58th Electron. Compon. Technol. Conf.*, Lake Buena Vista, FL, USA, 2008, pp. 847–852.

[7] M. Töpper et al., "3-D thin film interposer based on TGV (Through Glass Vias): An alternative to Si-interposer," in *Proc. 60th Electron. Compon. Technol. Conf.*, Las V egas, NV , USA, Jun. 2010, pp. 66–73.

[8] D. Xu, E. Jing, B. Xiong and Y. Wang, "Wafer-Level Vacuum Packaging of Micromachined Thermoelectric IR Sensors," in *IEEE Transactions on Advanced Packaging*, vol. 33, no. 4, pp. 904-911.

[9] J. Delrue, R. Ostholt and N. Ambrosius, "Glass Wafer Level Packaging Enabled by Laser Induced Deep Etching of Closed Cavities," *2019 22nd European Microelectronics and Packaging Conference & Exhibition (EMPC)*, 2019, pp. 1-5.

[10] P. Merz, H. J. Quenzer, H. Bernt, B. Wanger and M. Zoberbier, "A novel micromachining technology for structuring borosilicate glass substrates," *TRANSDUCERS '03. 12th International Conference on Solid-State Sensors, Actuators and Microsystems. Digest of Technical Papers (Cat. No.03TH8664)*, 2003, vol.1, pp. 258-261

[11] B. Luo, M. Ma, M. Zhang, J. Shang and C. Wong, "Composite Glass-Silicon Substrates Embedded With Microcomponents for MEMS System Integration," in *IEEE Transactions on Components, Packaging and Manufacturing Technology*, vol. 9, no. 2, pp. 201-208.

[12] J. Shang et al., "Preparation of wafer-level glass cavities by a low-cost chemical foaming process (CFP)", *Lab Chip*, vol. 11, no. 8, pp. 1532-1540.

[13] N. Li, C. Xing, P. Sun and Z. Zhu, "Silicon-Glass-Silicon Triple Stacked Structure for Fabrication of MEMS Resonator Accelerometer," *2019 20th International Conference on Electronic Packaging Technology(ICEPT)*, 2019, pp. 1-4.

[14] Z.Su, C. Xing et al., "Microsystem Wafer-level 3D Packaging Based on Composite Glass-Silicon Substrate," *Navigation and Control (in Chinese)*, vol. 18, no. 2, pp. 61-68.

[15] B. Luo, J. Shang, Z. Su, J. Zhang and C. Wong, "Height Adjustment of 3-D Axisymmetric Microumbrella Shells for Tailoring Wineglass Frequency," in *IEEE Transactions on Components, Packaging and Manufacturing Technology*, vol. 9, no. 3, pp. 567-574.

[16] Y. Ji, Q. Gan, L. Wu and J. Shang, "Geometry influence of the micro alkali vapor cell on the sensitivity of the chip-scale atomic magnetometers," *2017 IEEE 30th International Conference on Micro Electro Mechanical Systems (MEMS)*, 2017, pp. 342-345.

Glass Reflow and Thermo-Mechanical Stress Simulation for Through Glass Via in Glass-Silicon Composite Interposer

Wenqi Li[1], Chaoyang Xing[2], Jianfeng Zhang[1], Ziji Wang[1], Zhaoxi Su[1], Bin Luo[1] and Jintang Shang[1*]

1 Key Lab of MEMS of Ministry of Education, Southeast University, Nanjing, 210096, CHINA
2 Department of MicroSystem Integration, Beijing Institute of Aerospace Control Devices, Beijing, 100039, CHINA
*email: jshang@seu.edu.cn

Abstract—This paper investigates glass reflow process and thermo-mechanical stress of Through Glass Via (TGV) in glass-silicon composite interposer by Finite Element Analysis (FEA). The TGV structure in this work is composed of highly conductive silicon as vertical feedthrough and borosilicate glass as insulation. The relationship between reflow depth and reflow time is investigated in glass reflow simulation. The simulation results show that the reflow depth reaches its maximum at about 90s, and it takes about 360s for the glass reflow to fill the mold completely, where the mold is a $1.6*0.8*0.3mm^3$ cavity with a cylinder of 0.2mm in diameter and 0.3mm in height at its center. The effects of TGV structural parameters on Von Mises (VM) stress is analyzed in thermal-mechanical stress simulation. The maximum VM stress of glass-silicon composite TGV in the temperature load range of $-40\sim85°C$ is around 8 MPa, where the aspect ratio varies from 1 to 7, the via diameter ranges from 50 to 125 μm, and the pitch changes from 50 to 400 μm. The maximum VM stress can be reduced by 14% with structural parameter optimization.

Keywords—Glass Reflow, Thermo-Mechanical Stress, TGV, Finite Element Analysis (FEA)

I. INTRODUCTION

With the rapid development of microelectronics technology and the diversification of applications, integrated circuit packaging is gradually evolving from 2D to 3D [1], especially in emerging fields such as automotive electronics and smartphones, etc. The TGV interposer [2][3], as an essential component in 3D integrated packaging, has the advantages of miniaturization, high integration and multi-functionality, which is currently attracting the interest of many researchers.

TGV interposer consists of glass layer and electrical interconnection vias (EIV) where the EIV can be composed of conductive materials such as copper, nickel, highly doped silicon [4][5][6][7]etc. Highly doped silicon TGV interposer is fabricated by glass reflow process [6][8], which was first brought up by [9]. Glass-silicon composite TGV interposers are widely used in micro-electromechanical systems (MEMS) packaging [6]and optoelectronic integration [7]due to their Coefficient of Thermal Expansion (CTE) matching and high reliability [10]. There have been many experimental studies on the filling issue in glass reflow process [6][11][12][13], but these results are based on numerous experiments or previous experiences, which are not only time-consuming and labor-intensive but also non comprehensive. There are also some studies focusing on the effect of variation in glass-copper TGV interposer structural parameters on thermo-mechanical stress under thermal load through finite element

simulation [14][15], but the related studies in the field of silicon-filled TGV are scarce.

In this paper, the glass reflow model of glass-silicon composite TGV interposer and the thermo-mechanical stress model under thermal load were established by finite element analysis, respectively. The transient process of the glass reflow is analyzed, and the effect of the reflow time on the reflow depth of the glass-silicon composite TGV interposer is investigated. In addition, the relationship between the structural parameters and the distribution and magnitude of the thermo-mechanical stress of the TGV interposer under thermal load is also obtained. The above research demonstrates an intuitive process simulation of the thermal reflow process of TGV, and can also provide theoretical guidance for the structural parameter optimization design of glass-silicon composite TGV interposer.

II. MODELING

A. Glass reflow model design of glass-silicon composite TGV interposer

A non-scaled schematic diagram of the glass reflow fabrication process [6][9]for glass-silicon composite TGV interposer is shown in Fig. 1.

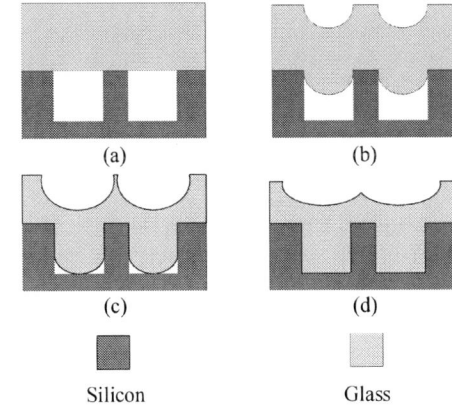

Fig. 1. Glass reflow process of glass-silicon composite TGV interposer

In order to further investigate the glass reflow process and improve its reliability, finite element simulation of the process is performed in this section. The whole structure of the model is shown in Fig. 2 (a), where the bottom mold is a polygonal structure with a rectangular cavity and a solid cylinder, and the top layer is a glass wafer. In order to improve the efficiency of the simulation, this work simulates the 1/4 structure of the model according to the symmetry of the structure. The schematic diagram of the structure after

978-1-6654-1392-3/21 $31.00 © 2021 IEEE

meshing is presented in Fig. 2 (b). Table I and Table II show the material and structural parameters involved in the glass reflow simulation, respectively [16].

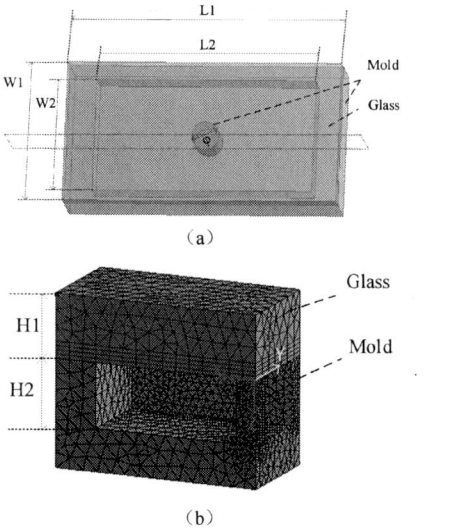

(a)

(b)

Fig. 2. Schematic diagram of the glass reflow model (a) whole structure (b) 1/4 structure after meshing

TABLE I.   MATERIAL PARAMETERS OF GLASS

Material	Density (g/cm³)	Characteristic viscosity (Pa·s)(821 ℃)
Glass	2.23	$10^{6.6}$

TABLE II.   GEOMETRIC PARAMETERS OF THE STRUCTURE

Structure	L1	L2	W1	W2	H1	H2	φ
Value (mm)	2	1.6	1	0.8	0.5	0.3	0.2

B. Thermo-mechanical stress model design for glass-silicon composite TGV interposer

Thermo-mechanical stress in the glass-silicon composite TGV interposer is investigated in this part. The planar structure of the glass-silicon composite TGV interposer is schematically shown in Fig. 3 (a), which consists of Schott Borofloat®33(BF33) glass and highly doped silicon. For the simplification of calculations based on the symmetry of the structure, finite element simulation is performed for the 1/2 model of a single glass-silicon composite TGV interposer, as shown in Fig. 3 (b). The material properties involved in the simulation are shown in Table III [16][17][18].

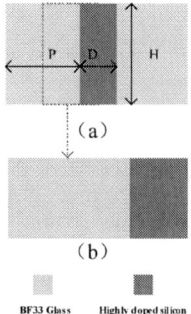

(a)

(b)

BF33 Glass   Highly doped silicon

Fig. 3. Schematic diagram of the thermo-mechanical stress model (a) whole structure (b) 1/2 structure

TABLE III.   MATERIAL PROPERTIES IN THE SIMULATION

Materials	Young modulus (GPa)	Poisson's ratio	Thermal expansion coefficient (ppm/°C)	Density (kg/m³)
Highly doped silicon	130	0.28	2.6	2330
BF33 glass	64	0.2	3.25	2230

III.   RESULT AND DISCUSION

A. Glass reflow process simulation results for glass-silicon composite TGV interposer

The following assumptions are applied to the simulation of the glass reflow process for glass-silicon composite TGV interposer:

1) The mold is invariant and constant temperature.
2) The thermal expansion pressure is 1 atm.
3) Reflow temperature is kept at 821°C.

Fig. 4 presents the deformation of the glass layer during the glass reflow process versus time, with the viewpoint of the cross section represented by the red dashed line in Fig. 2 (a). The deformation of the glass layer becomes larger as the reflow time increases, and eventually the bottom surface of the glass fits the inner surface of the mold to obtain the ideal reflow pattern. The reflow depth is defined as the depth of the glass layer reflowed to the mold, and the maximum surface deformation is the Z coordinate difference between the surface of the glass layer at the maximum deformation and the initial position. Fig. 5 shows the reflow depth and the maximum surface deformation versus reflow time. With the increase of reflow time, the reflow depth and the maximum surface deformation gradually increase, where the reflow depth reaches the maximum at about 90s, and then remains constant, while the maximum surface deformation continues to increase, and finally tends to stabilize after 360s, indicating that the reflow glass layer has completely filled the mold layer at this time and the reflow process is finished. The above simulation results are consistent with the results in the experiment [12], demonstrating the accuracy of the established model.

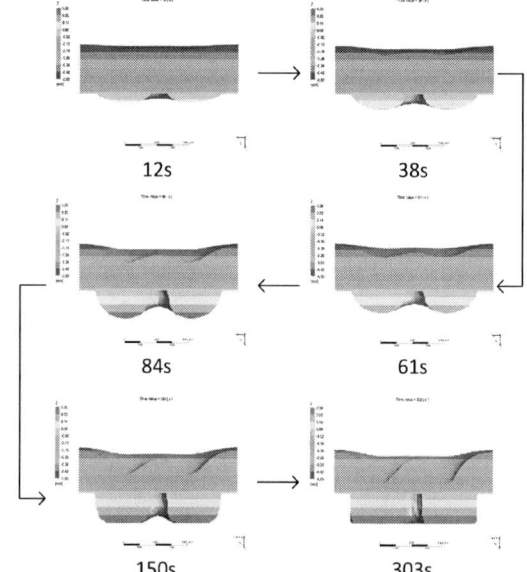

12s   38s

84s   61s

150s   303s

Fig. 4. Evolution of glass layer deformation with reflow time

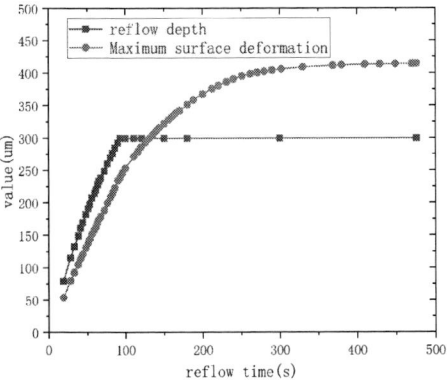

Fig. 5. Reflow depth and maximum surface deformation versus reflow time

B. Simulation results of thermo-mechanical stress of TGV interposer

The following assumptions are applied to this simulation model:

1) The initial temperature of the TGV interposer is 85°C.

2) The thermal stress of the TGV interposer at the initial temperature is zero.

3) The TGV interposer plunges from 85°C to -40°C.

The meshing of the 1/2 TGA model and the results after finite element simulation are shown in Fig. 6.

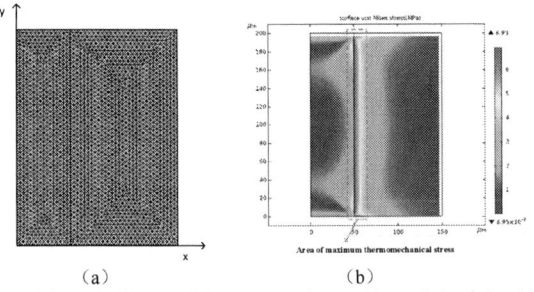

(a)   (b)

Fig. 6. Schematic diagram of the structure after meshing and simulation (a) structure after meshing (b) VM stress cloud (deformation magnified 50 times)

The maximum VM stress of the TGV under thermal load is distributed at the glass-silicon boundary, where the glass shrinks more than the highly doped silicon due to its larger CTE coefficient, creating a tensile force on the interface and forcing the TGV to bend in the direction of the glass.

The effect of structural parameters on the maximum VM stress in the TGV is also obtained, and the combination of structural parameters is shown in Table IV.

TABLE IV.   COMBINATION OF TGV STRUCTURAL PARAMETERS IN THE SIMULATION

Combination 1		Combination 2	
H/D	3	P (μm)	200
D (μm)	50/75/100/125	D (μm)	50/75/100/125
P (μm)	50/100/200/400	H/D	1/3/5/7

The simulation results of the maximum VM stress for the two combined structures are shown in Fig. 7. The maximum VM stress value of glass-silicon composite TGV is around 8 MPa, where the maximum is 8.69 MPa and the minimum is 7.50 MPa, with difference of 1.19 MPa and 14% reduction. According to reference [14], the maximum VM stress variation for glass-copper TGV ranges from 115 MPa to 240

Mpa. In comparison, the maximum VM stress for glass-silicon TGV is much smaller, due to the more matching CTE coefficients of silicon and glass.

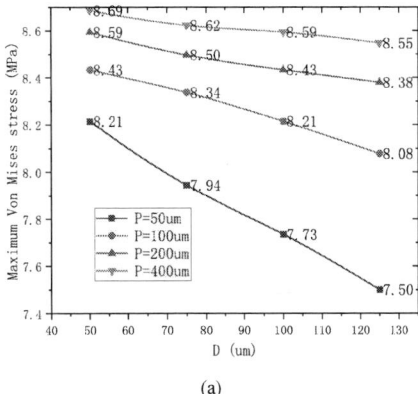

(a)

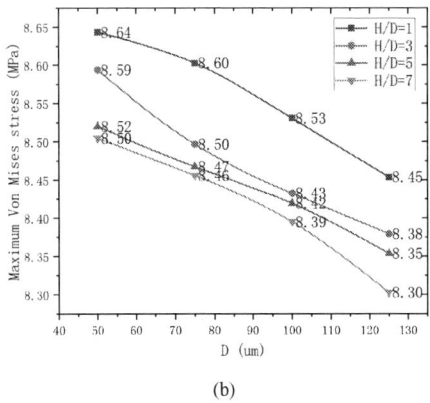

(b)

Fig. 7. Effect of structural parameter modification on maximum VM stress (a) maximum VM stress versus TGV diameter for H/D=3 (b) maximum VM stress versus TGV diameter at P=200um

Through the two graphs in Fig. 7 (a) and (b), it can be obtained that:

1) The maximum VM stress can be reduced by increasing the diameter of TGV with a fixed aspect ratio and pitch.

2) The maximum VM stress could be reduced by decreasing the TGV pitch while TGV diameter and aspect ratio are fixed.

3) The maximum VM stress can be lowered by increasing the H/D while the pitch and diameter are fixed.

IV. CONCLUSION

This paper has successfully simulated glass reflow process. Results show that the viscous glass reaches the maximum reflow depth at 90s and fills the mold completely at about 360s, which is consistent with the previous thermal reflow experiments [12], indicating that the glass reflow simulation is valid.

The thermal mechanical stress of silicon-filled TGV has also been analyzed in this paper. The maximum VM stress is around 8 MPa in the range of thermal load (85°C to -40°C), where the aspect ratio varies from 1 to 7, the via diameter ranges from 50 to 125 μm, and the pitch changes from 50 to 400μm. The maximum VM stress can be reduced from 8.69 MPa to 7.50 MPa by optimizing the structural parameters, such as increasing the via diameter, decreasing the pitch and increasing the aspect ratio.

This will provide valuable guidance for glass reflow process and structural design of glass-silicon composite TGV interposer to improve the process efficiency and structural reliability.

REFERENCES

[1] R. Tummala et al., "Glass Panel Packaging, as the Most Leading-Edge Packaging: Technologies and Applications," 2020 Pan Pacific Microelectronics Symposium (Pan Pacific), 2020, pp. 1-5, doi: 10.23919/PanPacific48324.2020.9059521.

[2] M. Töpper et al., "3-D Thin film interposer based on TGV (Through Glass Vias): An alternative to Si-interposer," 2010 Proceedings 60th Electronic Components and Technology Conference (ECTC), 2010, pp. 66-73, doi: 10.1109/ECTC.2010.5490887.

[3] V. Sukumaran et al., "Through-package-via formation and metallization of glass interposers," 2010 Proceedings 60th Electronic Components and Technology Conference (ECTC), 2010, pp. 557-563, doi: 10.1109/ECTC.2010.5490913.

[4] Bor Kai Wang, Yi-An Chen, A. Shorey and G. Piech, "Thin glass substrates development and integration for through glass vias (TGV) with copper (Cu) interconnects," 2012 7th International Microsystems, Packaging, Assembly and Circuits Technology Conference (IMPACT), 2012, pp. 247-250, doi: 10.1109/IMPACT.2012.6420306.

[5] M. J. Laakso et al., "Through-Glass Vias for Glass Interposers and MEMS Packaging Applications Fabricated Using Magnetic Assembly of Microscale Metal Wires," in IEEE Access, vol. 6, pp. 44306-44317, 2018, doi: 10.1109/ACCESS.2018.2861886.

[6] B. Luo, M. Ma, M. Zhang, J. Shang and C. Wong, "Composite Glass-Silicon Substrates Embedded With Microcomponents for MEMS System Integration," in IEEE Transactions on Components, Packaging and Manufacturing Technology, vol. 9, no. 2, pp. 201-208, Feb. 2019, doi: 10.1109/TCPMT.2018.2889368.

[7] Z. Wang and J. Shang, "3D composite photonic interposer integrated with low-cost silicon nitride optical interconnects," 2020 21st International Conference on Electronic Packaging Technology (ICEPT), 2020, pp. 1-3, doi: 10.1109/ICEPT50128.2020.9202997.

[8] Fraunhofer ISIT, Itzehoe (DE) 2005 Electrical feedthroughs using wafer-level glass-flow technology Achievements and Results Annual Report 2005

[9] P. Merz, H. J. Quenzer, H. Bernt, B. Wanger and M. Zoberbier, "A novel micromachining technology for structuring borosilicate glass substrates," TRANSDUCERS '03. 12th International Conference on Solid-State Sensors, Actuators and Microsystems. Digest of Technical Papers (Cat. No.03TH8664), 2003, pp. 258-261 vol.1, doi: 10.1109/SENSOR.2003.1215302.

[10] Ming-ai Zhang. Study on Optimized Fabrication of Glass Interposer for 3D Integration [D]. Southeast University, 2017.

[11] J. Liu, Q. Huang, J. Shang, J. Song and J. Tang, "Micromachining of Pyrex7740 glass and their applications to wafer-level hermetic packaging of MEMS devices," 2010 IEEE 23rd International Conference on Micro Electro Mechanical Systems (MEMS), 2010, pp. 496-499, doi: 10.1109/MEMSYS.2010.5442456.

[12] J. Liu, J. Shang, J. Tang and Q. Huang, "Micromachining of Pyrex 7740 Glass by Silicon Molding and Vacuum Anodic Bonding," in Journal of Microelectromechanical Systems, vol. 20, no. 4, pp. 909-915, Aug. 2011, doi: 10.1109/JMEMS.2011.2160043.

[13] A. Amnache, J. Neumann and L. G. Fréchette, "Capabilities and Limits to Form High Aspect-Ratio Microstructures by Molding of Borosilicate Glass," in Journal of Microelectromechanical Systems, vol. 28, no. 3, pp. 432-440, June 2019, doi: 10.1109/JMEMS.2019.2902066.

[14] A. Benali, M. Bouya, M. Faqir, A. El Amrani, M. Ghogho and A. Benabdellah, "Through glass via thermomechanical analysis: Geometrical parameters effect on thermal stress," 2013 8th IEEE Design and Test Symposium, 2013, pp. 1-5, doi: 10.1109/IDT.2013.6727093.

[15] Benali, A., Faqir, M., Bouya, M., Benabdellah, A., & Ghogho, M. (2016). Analytical and finite element modeling of through glass via thermal stress. Microelectronic Engineering, 151, 12-18.

[16] Schott. Thermal and Mechanical properties of Borof11oat® 33[EB/OL] https://www.schott.com/zh-cn/products/borofloat/technical-details

[17] M. A. Hopcroft, W. D. Nix and T. W. Kenny, "What is the Young's Modulus of Silicon?," in Journal of Microelectromechanical Systems, vol. 19, no. 2, pp. 229-238, April 2010, doi: 10.1109/JMEMS.2009.2039697.

[18] K. G. Lyon, G. L. Salinger, C. A. Swenson, and G. K. White, "Linear thermal expansion measurements on silicon from 6 to 340 K," J. Appl. Phys., vol. 48, no. 3, pp. 865–868, Mar. 1977.

Research on the application of feedforward + high-order iterative learning of permanent magnet linear motor in wire bonding machine

Haomiao Wu
State Key Laboratory of Precision Electronic Manufacturing Technology and Equipment
School of Electromechanical Engineering, Guangdong University of Technology
Guangzhou, P.R.China
1607165669@qq.com

Yunbo He
State Key Laboratory of Precision Electronic Manufacturing Technology and Equipment
School of Electromechanical Engineering, Guangdong University of Technology
Guangzhou, P.R.China
heyunbo@gdut.edu.cn

Xiaohui Lin
State Key Laboratory of Precision Electronic Manufacturing Technology and Equipment
School of Electromechanical Engineering, Guangdong University of Technology
Guangzhou, P.R.China
linxiaohui331@163.com

Haolin Li
State Key Laboratory of Precision Electronic Manufacturing Technology and Equipment
School of Electromechanical Engineering, Guangdong University of Technology
Guangzhou, P.R.China
hollvineli@163.com

Abstract—Permanent magnet linear motors are widely used on XY motion platforms in semiconductor packaging equipment. This module has high requirements for motion accuracy. In order to solve the requirements for higher speed and accuracy of the XY motion platform in semi-conductive packaging equipment, a composite control algorithm of feedforward + high-order iterative learning is proposed, and a speed plan based on S-curve is designed for the working characteristics of the linear motor in the wire bonding machine. In this paper, the mathematical model of linear motor is established, and the convergence conditions of the high-order iterative learning control algorithm are given. Through the establishment of MATLAB/Simulink simulation model, the system trajectory tracking and error under the control algorithm of first-order iterative learning, high-order iterative learning, feedforward + high-order iterative learning are compared and analyzed. The simulation results show that high-order iterative learning has faster convergence speed than low-order iterative learning; the control algorithm of feedforward + high-order iterative learning can effectively solve the problem of slow convergence speed in later iterations, and the number of iterations is less than that of a single iterative learning algorithm. And the accuracy is higher. It meets the requirements of semiconductor packaging equipment for the fast and high-precision performance of linear motors, which is of great significance for improving the production efficiency of semiconductor products and improving product quality.

Keywords—wire bonding machine; linear motor; high-order iterative learning; feedforward control

I. INTRODUCTION

Wire bonding is a key link in semiconductor packaging, and wire bonding technology is directly related to wire bonding quality [1]. Due to its simple structure, suitable for high-speed linear motion, easy adjustment and control, and high acceleration [2], linear motors are widely used in high-speed, high-precision semiconductor packaging equipment. It is also one of the core components that make up the XY

motion platform of the wire bonding machine. The control algorithm of this module has a direct impact on the wire bonding performance of the wire bonding machine. The traditional PID control algorithm is not easy to realize the high-speed, high-acceleration, and high-precision motion control of the linear motor. Therefore, the search for efficient and reasonable control algorithms to achieve the high-precision motion control of the linear motor is useful for the realization of higher-precision production of semiconductor packaging equipment.

The linear motor in the wire bonding machine is often in a cyclical repetitive action, and the application scenario of iterative learning is often a system that requires repetitive operations in the industry. The output of the system is realized by iterative learning of errors. Complete tracking of the desired trajectory. Reference [3] proposed a dual iterative learning control strategy for the problems of iterative learning control with many iterations and slow convergence speed. This control strategy can effectively reduce the tracking error and the number of iterations. Reference [4] proposes an iterative learning controller based on feedback control, which adjusts the tracking error of the system by changing the feedback gain. In this paper, in order to achieve better trajectory tracking effect and reduce the error in the process of linear motor operation, the error of the system when using low-order and high-order iterative learning control is compared and analyzed. At the same time, the number of iterations is reduced through the combination of feedforward controller and high-order iterative learning controller, The purpose of improving the tracking accuracy of the system and solving the slow convergence speed of the iterative learning control algorithm in the later stage is to improve the tracking accuracy of the system. In order to realize the high precision motion of linear motor, a new reference method is proposed, which provides a theoretical basis for the subsequent XY motion experiment platform.

978-1-6654-1392-3/21 $31.00 © 2021 IEEE

II. MATHEMATICAL EQUATION

A. Mathematical model of linear motor

Generally, a linear motor is regarded as a rotating electric machine that is formed by expanding in a plane in the radial direction. At the same time, in order to simplify the analysis and obtain the mathematical model of the linear motor, the following assumptions are made:

(1) Excluding core saturation; (2) Excluding eddy current and hysteresis loss; (3) There is no damping effect on the permanent magnet and secondary; (4) The back EMF is sinusoidal.

Then the linear motor voltage equation in the d-q coordinate system is:

$$\begin{cases} u_d = Ri_d + L_d \frac{di_d}{dt} - \omega_e L_q i_q \\ u_q = Ri_q + L_q \frac{di_q}{dt} + \omega_e(L_d i_d + \Psi_f) \end{cases} \quad (1)$$

Where, u_d、u_q respectively represent the voltage component of the d-q axis; i_d、i_q respectively represent the current component of the d-q axis; R is the armature winding resistance; ω_e is the electrical angle; L_d、L_q respectively represent the inductance component of the d-q axis; Ψ_f represent the permanent magnet flux linkage.

The thrust equation of the linear motor is:

$$F_e = \frac{3\pi}{2\tau}\left[(L_d - L_q)i_d + \Psi_f\right]i_q \quad (2)$$

Where, F_e is electromagnetic thrust; τ is the pole pitch of the permanent magnet.

Analyze the force of the linear motor, and obtain the mechanical motion equation of the permanent magnet linear motor from Newton's second law:

$$M\ddot{x} = F_e - F_{load} - D\dot{x} \quad (3)$$

Where, M is the mass of the motor mover; F_{load} is Load resistance; D is the system friction coefficient.

When the control mode of the permanent magnet linear motor is the vector control of , the formula (2) can be obtained:

$$F_e = \frac{3\pi}{2\tau}\Psi_f i_q \quad (4)$$

$$u_q = Ri_q + L_q\frac{di_q}{dt} + k_e v \quad (5)$$

Where, electromagnetic thrust coefficient is $k_f = \frac{3\pi}{2\tau}\Psi_f$, back EMF coefficient is $k_e = \frac{\pi}{\tau}\Psi_f$。

B. Longitudinal static end effect

The special structure of the linear motor brings more application fields to it, but also produces corresponding problems. Among them, the greater impact is the greater thrust fluctuation at the end. As a kind of thrust fluctuation, the static end effect is caused by the discontinuity of the iron core and the coil winding placed in the slot at both ends of the motor, which makes the mutual inductance between the phases unequal, which is caused by the pulsating magnetic field and the reverse magnetic field[5]. This effect has the greatest impact on the motor and will directly lead to a reduction in the efficiency of the motor. Therefore, this article mainly considers the longitudinal static end effect of the motor. The effect can be described as a periodic function of displacement:

$$F_{rip} = F_M\cos(\frac{2\pi l}{\tau} + \varphi_0) \quad (6)$$

Where, F_M is the end effect thrust wave amplitude; l is the mover displacement; τ is the pole pitch; φ_0 is the initial phase electrical angle。

C. Friction

Friction is the resistance generated by objects in relative motion or when they have a tendency to relative motion. The relative motion between the mover and the stator in a linear motor must have friction. The friction force contained in the permanent magnet linear motor servo drive system mainly includes two types of static friction force and viscous friction force. The mathematical model of static friction is:

$$F_{fric} = f_c sgn(v) \quad (7)$$

The mathematical model of viscous friction force is:

$$F_b = B_v v \quad (8)$$

Where, f_c is the Coulomb friction coefficient; B_v is the coefficient of viscous friction; v is the running speed。

III. CONTROLLER DESIGN

A. Design of iterative learning controller

The wire bonding machine is a high-precision manufacturing equipment with high speed and high acceleration. The motion performance of its linear motor will directly affect the overall wire bonding performance of the equipment and have a greater impact on the quality of the wire bonding. It is required to have high-precision motion performance in actual production.

Among the many control methods, iterative learning is a control method that corrects the next input by continuously learning the previous or previous error information. In theory, it can fully track the desired trajectory. This method is suitable for systems that require a large amount of repetitive work in the industry. The actual working state of the linear motor in the actual wire bonding machine is also the continuous cyclic wire bonding process action, so it is reasonable to apply the iterative learning control strategy to the linear motor control system.

The mathematical description of the high-order open-loop PD iterative learning controller designed in this paper is as follows:

$$u_{k+1}(t) = P_1 u_k(t) + P_2 u_{k-1}(t) + K_{P_1} e_k(t)$$

$$+ K_{P_2} e_{k-1}(t) + K_{D_1}\frac{de_k(t)}{dt} + K_{D_2}\frac{de_{k-1}(t)}{dt} \quad (9)$$

Where, $e_k(t) = y_d(t) - y_k(t)$

And, $y_d(t)$、$y_k(t)$、$u_{k+1}(t)$、$u_k(t)$、$u_{k-1}(t)$、$e_k(t)$、$e_{k-1}(t)$ respectively represent the expected trajectory, actual trajectory, current control input, previous control input, pre-previous input, previous error, pre-previous error; K_{P_1}、K_{P_2} is the proportional learning matrix parameter; K_{D_1}、K_{D_2} is the differential learning matrix parameter.

We can get the control block diagram of this controller as shown in Figure 1. Where y_d is the desired trajectory input.

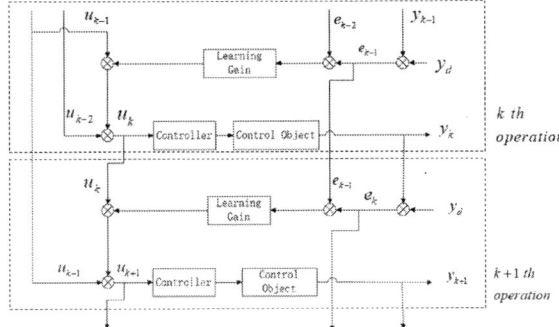

Figure 1 Block diagram of open-loop second-order iterative learning control

B. Equations Feedforward controller

The feedforward controller is one of the open-loop control methods. It mainly makes corresponding supplementary measures to input the commands that the controlled system is about to run to eliminate the occurrence of deviations, and play a role in preventing problems. It can improve the response speed of the system without affecting the stability of the system. After the speed feedforward and acceleration feedforward are introduced into the high-order iterative learning control system, under the compensation of the feedforward controller, the initial state of the system will be closer to the system planning after the iteration starts. This will cause the initial state of the iterative controller to change, make iterative convergence come earlier, and reduce the number of iterations required, so it can solve the problem of slow convergence speed in the later stage of iterative learning control. This article mainly uses velocity feedforward and acceleration feedforward to compensate the control system, thereby reducing the tracking error and speeding up the iteration speed.

By solving the inverse process of the controlled system, the transfer function of the controlled system can be obtained as:

$$P(s) = \frac{K_f}{s(Ms+D)} \tag{10}$$

Then the transfer function of the speed and acceleration feedforward controller can be obtained:

$$G_f(s) = \frac{M}{K_f}s^2 + \frac{D}{K_f}s = K_{aff}s^2 + K_{vff}s \tag{11}$$

The compound servo control structure of feedforward + iterative learning can be obtained, as shown in Figure 2. The linear motor mathematical model is in the dashed box.

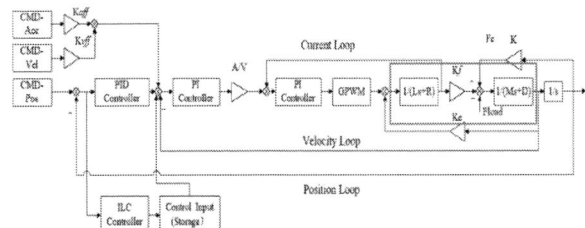

Figure 2 System control block diagram

C. Speed planning

Through the analysis of the speed planning algorithm of the linear motor commonly used in the wire bonding machine, and in order to better simulate the daily reciprocating working state of the linear motor, a speed planning curve based on the S-curve is designed, as shown in Figure 3. And use this plan as the input of the control system. Set the initial and final speed and acceleration of the motor to be zero, the maximum speed is 1m/s, and the maximum acceleration is 20g. The speed formula for each time period is shown in equation (12).

$$v(t) = \begin{cases} J\tau_2^2/2, 0 \le t < t_1 \\ JT_1(T_1 + 2\tau_2), t_1 \le t < t_2 \\ J(T_1T_2 + T_1^2/2 + T_1\tau_3 - \tau_3^2/2), t_2 \le t < t_3 \\ JT_1(T_1 + T_2), t_3 \le t < t_4 \\ J(T_1^2 + T_1T_2 - \tau_5^2/2), t_4 \le t < t_5 \\ JT_1(T_2 + T_1/2 - \tau_6), t_5 \le t < t_6 \\ J(T_1T_2 + T_1^2/2 - T_1\tau_7 + \tau_2/2), t_6 \le t < t_7 \end{cases} \tag{12}$$

Where, $J = \frac{da}{dt}$ is jerk, $T_k = t_k - t_{k-1}(k = 1, \cdots, 7)$ is the duration of each different acceleration time period, and $\tau_k = t - t_{k-1}(k = 1, \cdots, 7)$.

The S-shaped speed planning curve mainly has 7 parts, mainly including acceleration process (T1), uniform acceleration process (T2), deceleration process (T3), uniform speed process (T4), acceleration and deceleration process (T5), uniform deceleration Process (T6), deceleration process (T7). In combination with the actual wire bonding machine, the linear motor is a high-acceleration device, so the uniform acceleration and uniform deceleration processes are removed when designing the speed planning.

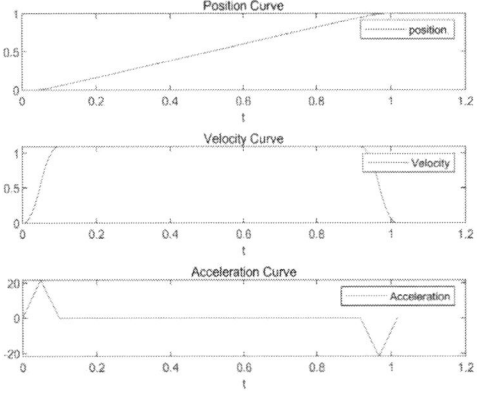

Figure 3 S-type speed planning curve

978-1-6654-1392-3/21 $31.00 © 2021 IEEE

IV. SYSTEM DESCRIPTION

Consider the following linear system:

$$\begin{cases} \dot{x}_k(t) = Ax_k(t) + Bu_k(t), x(0) = \xi^0 \\ \qquad\quad y_k(t) = Cx_k(t) \end{cases} \tag{13}$$

Where, k is the number of iterations of the system; $t \in [t, T]$ is the time constant; $x_k(t)$ is the state variable of the system; $y_k(t)$ is the output of the system; $u_k(t)$ is the input of the system; A、B、C is a constant coefficient matrix。

Analysis of convergence conditions:

When the initial condition $x(0) = \xi^0$ of the linear time-invariant system (13) is given, the convergence conditions are given for the high-order PD-type iterative learning control rate (9) proposed in this paper[6]:

$$P_1 + P_2 = I \tag{14}$$

$$\left\| P_1 - BCK_{D_1} \right\|_\infty + \left\| P_2 - BCK_{D_2} \right\|_\infty < 1 \tag{15}$$

When the system satisfies the convergence conditions shown in equations (14) and (15). For a given expected input trajectory $y_d(t)$, when $0 \le t \le T$, iterative formula (9) can guarantee that for every $t \in [t, T]$, when $k \to \infty$, there is $y_k(t) \to y_d(t)$.

V. SIMULATION ANALYSIS

In order to verify the effectiveness of the method proposed in this paper, the servo control system is simulated and analyzed by building a MATLAB/Simulink model. The tracking speed and accuracy of low-order and high-order iterative learning are compared and analyzed, and the tracking speed and accuracy of the high-order iterative learning control system after adding the feedforward controller.

The system parameters of the controlled motor in this article are shown in Table 1.

Table 1 Linear motor system parameters

Structural parameters	Value (unit)	Structural parameters	Value (unit)
Armature resistance (R)	2.5 Ω	Magnetic flux (Ψ_f)	0.107 Wb
Armature inductance (L)	8.2 mH	Motor mover mass (M)	8.2 kg
Coefficient of Viscous Friction (D)	4 $N \cdot S/m$		

By adjusting the parameters in the learning rate of each order, the trajectory tracking effects of 80 iterations of the first-order control system, 21 iterations of the second-order control system, and 17 iterations of the third-order iterative learning control system are obtained, as shown in Figure 4. And the optimal tracking error corresponding to each order is shown in Figure 5. It can be seen that the tracking effect of the third-order iterative learning control system is better than that

of the second-order iterative learning control system and the first-order iterative learning control system. The error accuracy of the first-order system needs 80 iterations to reach 1.52×10^{-5}m, while the tracking accuracy of the second-order system and the third-order system after 21 and 17 iterations can reach 7.1×10^{-6}m and 5.74×10^{-6}m, respectively. It can be proved that in terms of trajectory tracking accuracy, as the order of iterative learning increases, the better the trajectory tracking ability of the system, the higher the accuracy. However, it can also be found from Figure 5 that as the order of iterative learning increases, although the steady-state error accuracy of the system is improved, but the stability of the dynamic process is reduced. The dynamic error of the system will increase with the increase of the iterative learning order, which will adversely affect the work of the equipment in actual engineering applications.

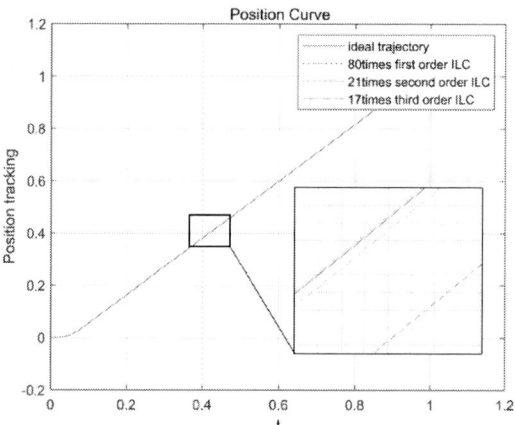

Figure 4 Trajectory tracking of each order iterative control system

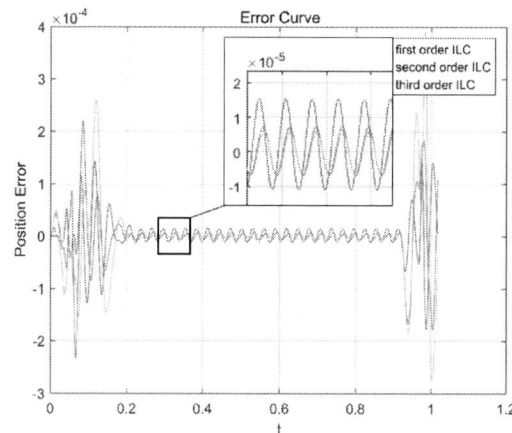

Figure 5 Trajectory tracking error of each order iterative control system

At the same time, as shown in Figure 6 and Figure 7, the trajectory tracking and error data graphs of the first, second, and third-order iterative control systems after 9 iterations are not difficult to find that, the higher of the iterative order, the faster the convergence speed. However, it can also be found that the tracking error of the third-order system at the ninth iteration is -8.65×10^{-4}m. And by observing the subsequent 9-17 iterations of the third-order system, it is found that at the 17th iteration, the tracking accuracy of the system will reach a good 5.74×10^{-6}m, but the tracking error of the system for these few times all appears around 0 oscillation. This also shows from the side that the tracking accuracy of the third-order iterative learning control system is better, but it is easy to cause system instability.

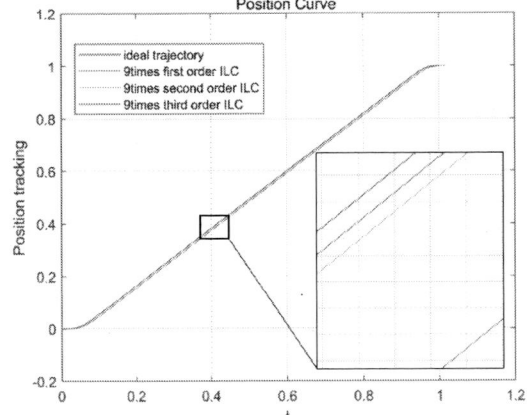

Figure 6 Trajectory tracking of each order iterative control system after 9 iterations

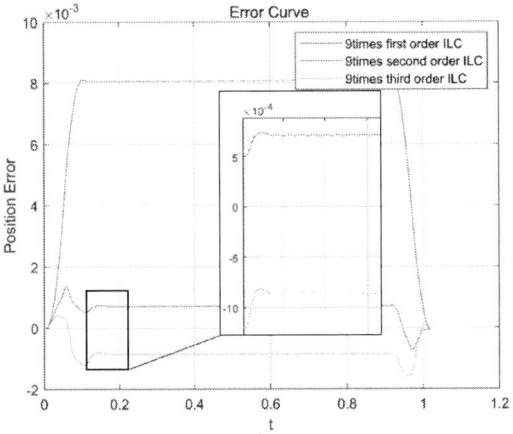

Figure 7 Trajectory tracking error of each order of iterative control system after 9 iterations

In order to ensure the stability of the system, without increasing the order of iterative learning, the purpose of accelerating the speed of iterative convergence and reducing the number of iterations is achieved. In this paper, a feedforward controller is added to the control system, and a linear motor control system with feedforward + high-order iterative learning is obtained. The simulation results are shown in Figure 8 and Figure 9.

Through Figure 8, it can be found that the trajectory tracking effect of the system is better after the feedforward controller is introduced. As shown in Figure 9, the error accuracy of the feedforward + high-order iterative learning control system after 14 iterations reached 5.4×10^{-6}m, which is better than the 7.1×10^{-6}m error of the second-order iterative learning control system after 21 iterations. And from Figure 9, it can also be found that the error of the system in the dynamic process is better than that of the second-order iterative learning control system, indicating that the control method has a certain effect on improving the stability of the system.

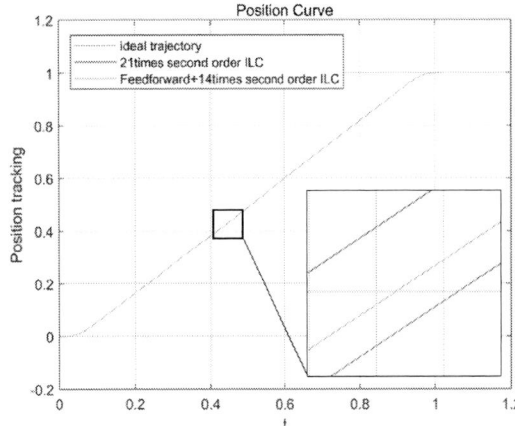

Figure 8 The trajectory of the second-order iterative learning control system after the introduction of feedforward

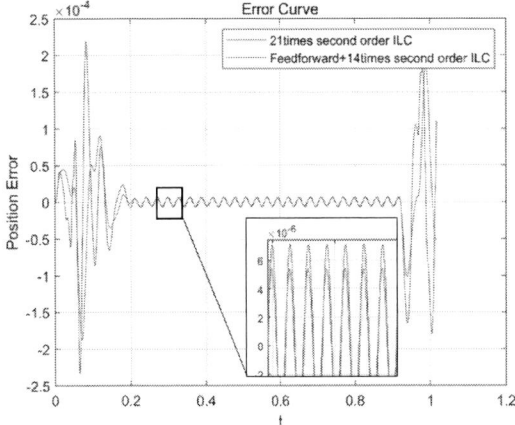

Figure 9 The trajectory error of the second-order iterative learning control system after the introduction of feedforward

By comparing the trajectory tracking data and error data of the second-order iterative learning control system and the feedforward + second-order iterative learning control system in Fig. 10 and Fig. 11. It can be found that after the feedforward controller is introduced, it only goes through three iterations. The error data has been greatly improved, which shows that the feedforward controller can effectively solve the problem of slower convergence speed inherent in the iterative learning control method, which can make the system convergence faster and the error accuracy higher.

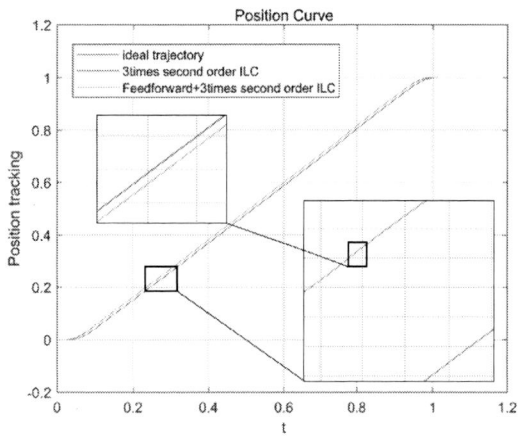

Figure 10 Second-order iterative control and feedforward + second-order iterative control trajectory tracking after 3 iterations

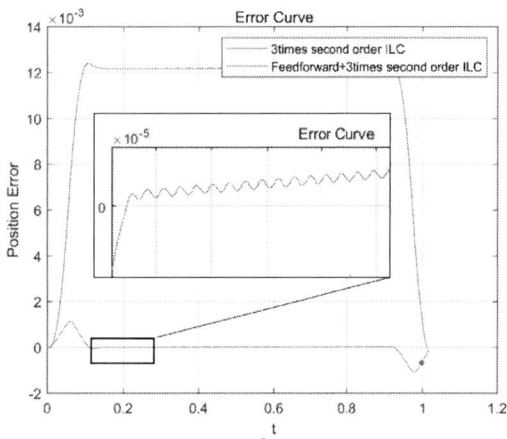

Figure 11 Second-order iterative control after 3 iterations and feedforward + second-order iterative control trajectory tracking error

VI. CONCLUSION

The feedforward + high-order iterative learning control algorithm proposed in this paper solves the problem of slow convergence speed of the iterative learning control algorithm in the later stage, reduces the number of iterations, and also verifies that the high-order iterative learning algorithm has faster convergence. But the stability will also decrease as the order of the iterative controller increases. Finally, the simulation result of the system has an error accuracy of 10^{-6}m, which provides a reference for realizing high-precision motion control of the linear motor of semiconductor packaging equipment.

ACKNOWLEDGMENT

The work presented in the paper is supported by National Natural Science Foundation of China (No.61973093, F030815), Foshan Industry-University-Research Project (No.1920001000236).

REFERENCES

[1] He Yunbo-20132013'National Seminar on Semiconductor Device Technology and Industry Development and the 6th China Micro-Nanoelectronic Technology Exchange and Academic Seminar

[2] Cheng Shixiang, Xia Lian, Han Jiang. Research on linear motor control based on ICSO fuzzy PID control[J]. Journal of Hefei University of Technology (Natural Science Edition), 2020, 43(10): 1307-1312.

[3] Chen Jiajun, He Yunbo, Li Weitian, Liao Junhong. Research on dual iterative learning controller based on XY motion platform[J]. Modular Machine Tool and Automatic Manufacturing Technology, 2020(11): 94-98.

[4] Yu Zhongwei, Chen Huitang, Wang Yuejuan. Design of iterative learning controller based on feedback control[J]. Control Theory and Application, 2001(05):785-791.

[5] Chen Zhanqin. Terminal second-order sliding mode control of permanent magnet linear synchronous motor[D]. Shenyang University of Technology, 2015.

[6] Bien Z, Huh K M. Higher-order iterative learning control algorithm[C]//IEE Proceedings D (Control Theory and Applications). IET Digital Library, 1989, 136(3): 105-112.

Machine Learning Enabled Optimization of Pick-up Process for Thin Die

Peilun Yao
Department of Mechanical and
Aerospace Engineering
Hong Kong University of Science and
Technology
Hong Kong, China
pyaoaa@connect.ust.hk

Haibin Chen
Department of Mechanical and
Aerospace Engineering
Hong Kong University of Science and
Technology
Hong Kong, China
chenhb@ust.hk

Jinglei Yang
Department of Mechanical and
Aerospace Engineering
Hong Kong University of Science and
Technology
Hong Kong, China
maeyang@ust.hk

Jingshen Wu
Department of Mechanical and
Aerospace Engineering
Hong Kong University of Science and
Technology
Hong Kong, China
mejswu@ust.hk

Abstract—Driven by the demand of artificial intelligent, high-performance computing, electric vehicle and smart city, semiconductor industry is building high performance and high integrated systems. To increase the energy efficiency of these complicated systems, particularly in power modules, reducing the power consumption is needed. Thinning the die is one of the solutions, which can provide a lot of advantages including faster heat dissipation, low junction temperature, low electrical resistance, etc. However, due to its sensitivity of stress, processing of thin die is a challenging, including die pick-up process. In this paper, we proposed a methodology using machine learning and finite element method (FEM) to optimize the pin designs with the objective to minimize the stress in the die during pick-up process. A dynamic numerical model was first built up to simulate the stress distribution in the die during pick-up process, and a Gaussian process regression was applied to generate the relationships between pin parameters and die stress. Meanwhile, a Bayesian optimization was used to optimize the parameters. The optimized pin designs using the above methodology was further validated by FEM. Compared with traditional DOE method, the proposed methodology using machine learning and FEM exhibits much higher effectiveness in parameter optimization with less testing cost and efforts.

Keywords—Machine learning, Bayesian optimization, Thin wafer, Die pick-up process

I. INTRODUCTION

Driven by an increasing demand of artificial intelligent, high-performance computing, mobile device, electric vehicle and smart city, semiconductor industry is building systems with higher performance and higher integration. To increase the energy efficiency and reduce the power consumption of these complicated systems, using thin die is one of the solutions. Besides low electrical resistance, thin die provides other advantages, such as fast heat dissipation and low junction temperature. However, due to its sensitivity of stress, processing of thin die is a challenging, including die pick-up process. During die pick-up, the die is pushed by injection pins away from sticky tape; when the die thickness is thin, a much higher stress is generated, which can lead to the occurrence of die cracking. Stress distribution with different die thicknesses during die pick-up is shown in Fig. 1, which demonstrates that when the die thickness decreases, the stress and the risk of cracking during processing increases exponentially. To reduce the risk of die crack, optimization of the die pick-up parameters was traditionally made by finite elemental method

assisted DOE methodology[1-4]. However, such approaches are appropriate only when the correlation among the parameters is minimal[5].

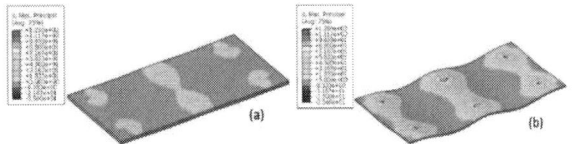

Fig.1. Effect of die thickness on stress distribution during die pick-up. (a) 100 um thick die; (b) 50 um thick die

Machine learning, driven by analytics models and algorithms, has recently been successfully applied in different fields[6-8], including parametric optimization. These techniques have enabled machine to learn the potential relationship between different parameters from the training sequences by accumulating data through algorithms.

In this paper, we proposed a methodology with a combination of machine learning and FEM for optimization of pin designs for die pick-up. A FEM model was first generated to construct the dataset of pin parameters and die stress. Afterwards, a machine learning algorithm using Gaussian process surrogate model was applied to train the relationships between pin parameters and die stress. Bayesian optimization was used to optimize the parameters to achieve the lowest die stress under a given condition, including tape adhesion strength, die dimension, etc. The optimized pin designs obtained using the above methodology was finally validated by FEM and experiments. The results from this study clearly show that compared to traditional DOE method, the machine learning assisted methodology exhibit much higher efficiently and effectiveness in parameter optimization with less testing cost and efforts. This methodology is very promising to be extended to other applications for advanced microelectronic packaging.

II. PROBLEM DESCRIPTION

A. Die pick-up process analysis

Typical die pick-up process is shown in Fig.2. By running the linear motor drive system, the pin will move and push the die up and away from the tape [2,3]. For small die usually one pin is enough, while for die with a large scale,

978-1-6654-1392-3/21 $31.00 © 2021 IEEE

more pins are needed to ensure that the die can be separated from the tape uniformly and smoothly.

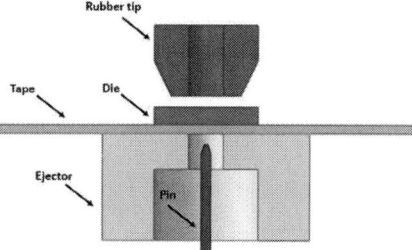

Fig.2. Schematic of die pick-up process.

A Finite element model can be built up to simulate the stress distribution in the die, as shown in Fig. 3. The six pins are fixed and a vacuum pressure with a value of -0.1MPa is applied on the bottom side of the die. Table 1 shows the geometry parameters of FE model. The objective of this numerical simulation is to establish the relationship between stress distribution and pin position, and generate a set of data for subsequent machine learning algorithm analysis.

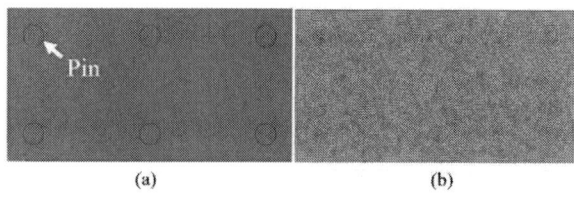

(a)   (b)

Fig.3 Simulation model for die pick-up process. (a) Geometric model, (b) FEM model

Table 1 Geometric parameters of FE model

Model	Unit	Pin radius	Die length	Die width	Die thickness
	mm	0.2	5.5	2.8	0.05

B. Parameter matrix

To optimize the die stress during pick-up process, an amount of control parameters need to be considered, which are listed in Table 2. Fig. 4 indicates the nomenclature of six pins on die use in this study.

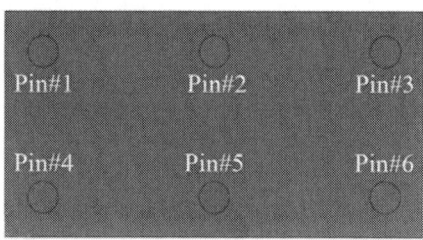

Fig.4 Pin number

Table 2 Control parameters for optimization

Parameter	Unit	Min	Max
Pin location X-coordinate	mm	0.2	5.3
Pin location Y-coordinate	mm	0.2	2.6

III. PROPOSED APPROACH FOR PREDICTION

A. Gaussian Process Theorem

Gaussian process regression (GPR) is an machine learning regression algorithm developed in recent years. As a promising technique, it has a strict statistical learning theory and has attracted widespread attention. Due to the advantages in uncertainty predictions, GPR has a good adaptability to complex problems such as high dimensionality, small samples, and nonlinearity. Meanwhile GPR has strong generalization ability[9]. Different from neural network using multiple hidden layers to handle nonlinear, complex classification and regression problems, GPR uses kernel trick which is similar to SVR or SVM, makes GPR easier to be understood and implemented in practice.

As a concept of probability statistics, Gaussian process (GP) is a kind of Stochastic process. It calculates the posterior degradation estimated by constraining the prior distribution. Similar as Gaussian distribution, GP is completely determined by its mean function and covariance function. According to the definition of GP, assuming $f(x_i)$ obey multivariate Gaussian distribution, GP can be defined as:

$$f(x) \sim N\big(m(x), k(x, x')\big) \qquad (1)$$

where $m(x)$ and $k(x, x')$ represent mean function and covariance function, respectively[10] and x, $x' \in X$ are random variables.

Since the collected data often contains noise, the Gaussian process regression model becomes:

$$y = f(x_i) + \varepsilon \qquad (2)$$

where $\varepsilon \sim N(0, \sigma_n^2)$.

According to Bayesian principle, For the observed data $D = \{x_i, y_i\}_{i=n+1}^{n+n}$, GP can get the corresponding posterior knowledge, and a joint Gaussian prior distribution of training output vector y and testing output vector y^* can be established as follows:

$$\begin{bmatrix} y \\ y^* \end{bmatrix} \sim N\left(0, \begin{bmatrix} K(X,X) + \sigma_n^2 I & K(X,x^*) \\ K(x^*,X) & K(x^*,x^*) \end{bmatrix}\right) \quad (3)$$

$$K(X,X) = \begin{bmatrix} k(x_1,x_1) & k(x_1,x_2) & \cdots & k(x_1,x_n) \\ k(x_2,x_1) & k(x_2,x_2) & \cdots & k(x_2,x_n) \\ \vdots & \vdots & \ddots & \vdots \\ k(x_n,x_1) & k(x_n,x_2) & \cdots & k(x_n,x_n) \end{bmatrix} \quad (4)$$

$$K(X,x^*) = \big[k(x^*,x_1), k(x^*,x_2)...k(x_*,x_n)\big] \quad (5)$$

$$K(x^*,x^*) = k(x^*,x^*) \qquad (6)$$

where $K(X,X)$ is symmetric positive definite covariance matrix for training data X, $K(X,x^*)$ is covariance matrix for training data X and testing data x^* and $K(x^*, x^*)$ is covariance matrix for testing data x^*.

Thus, according to conditional distribution form of multivariate Gaussian distribution, we can obtain the posterior probability formula for y^* as follows:

$$y^* \big| X, y, X^* \sim N(m(x^*), \text{cov}(y^*)) \qquad (7)$$

$$m(x^*) = K(x^*, X)(K(X, X) + \sigma_n^2 I)^{-1} y \qquad (8)$$

$$\text{cov}(y^*) = K(x^*, x^*) - K(x^*, X)(K(X, X) + \sigma_n^2 I)^{-1} K(X, x^*) \qquad (9)$$

Since the covariance matrix in GPR need to meet Mercer conditions, Equation (8) can be written as follows:

$$m(x^*) = \sum_i^n \alpha_i k(x^i, x^*) \qquad (10)$$

where $k(x^i, x^*)$ is called kernel function.

In GPR, kernel function maps the data of the non-linear relationships to a high-dimensional feature space for linear regression, and constructs the hypothesis function by defining the relationship among the data points. Therefore, the selection of the covariance kernel function is very important for Gaussian process modeling.

Commonly used kernel functions include squared exponential (SE) kernel function, rational quadratic (RQ) kernel function and Matern class kernel function[11]. The specific definition is given as following:

$$k_{SE}(x_i, x_j) = \sigma_f^2 \exp\left(-\frac{(x_i, x_j)^2}{2l^2}\right) \qquad (11)$$

$$k_{RQ}(x_i, x_j) = \sigma_f^2 \left(1 + \frac{(x_i, x_j)^2}{2\alpha l^2}\right)^{-\alpha} \qquad (12)$$

$$k_M(x_i, x_j) = \sigma_f^2 \left[\left(1 + \frac{\sqrt{3}}{l}(x_i, x_j)\right)\exp\left(-\frac{\sqrt{3}}{l}(x_i, x_j)\right)\right] \qquad (13)$$

where α, l, σ_f^2 are hyper parameters and determined by maximizing marginal likelihood method[12], as shown below:

$$L = \log p(y|\theta) = -\frac{1}{2} y^T (K + \sigma_n^2 I)^{-1} y \\ -\frac{1}{2}\log |K + \sigma_n^2 I| - \frac{n}{2}\log 2\pi \qquad (14)$$

Under some conditions, a single kernel function may not be able to accurately express the relationship among variables. In view of the shortcomings of a single kernel function, the composite kernel function may have better performances in certain fields. With the assumption of independent and identical distribution, the composite kernel function can be described as:

$$k(x_i, x_j) = \sum_{m=1}^n k_m(x_i, x_j) \qquad (15)$$

In this study, three single kernel functions and three composite kernel functions were used and evaluated.

B. GPR prediction result

1000 datasets obtained by the numerical simulation with the methodology mentioned above are divided into two sets, i.e. training set and testing set. Gaussian process regression prediction model was established through 800 training set data, and the remaining 200 data are used to verify the accuracy of the prediction model. Cross-validation method was used to ensure the results be stable.

Model performance was evaluated by Mean Absolute Percentage Error (MPAE), Root Mean Squared Error (RMSE) and Mean Absolute Error (MAE), which are defined as following:

$$MAE = \frac{1}{n}\sum_{i=1}^n \left| y_i - \hat{y}_i \right| \qquad (16)$$

$$MAPE = \frac{1}{n}\sum_{i=1}^n \frac{\left\| y_i - \hat{y} \right\|}{\left\| y_i \right\|} \qquad (17)$$

$$RMSE = \sqrt{\frac{1}{n}\sum_{i=1}^n \left\| y_i - \hat{y}_i \right\|_2^2} \qquad (18)$$

where n represents the number of samples, y_i represents actual value and $\hat{y}_i$ represents predicted value.

Six kernel functions, i.e. RBF, Matern, RQ, RBF+Matern, Matern+RQ, and RBF+RQ, were evaluated. 10 data from testing set were extracted for comparison in prediction performance. As shown in Fig.5, most of the kernel functions can match well with the actual data, and exhibit a similar trend. The RMSE, MAE and MAPE were also compared, as listed in Table 3. Though the calculation time for composite kernel function models is higher than single kernel function models, composite kernel function models do have improvements in the accuracy of prediction, as reflected by a lower RMSE (10%), MAPE (7%) and MAE (10%). Using RBF+M composite kernel function gives the best accuracy of prediction. Fig.6 shows the Gaussian process regression results using RBF + M composite kernel function for training set and testing set.

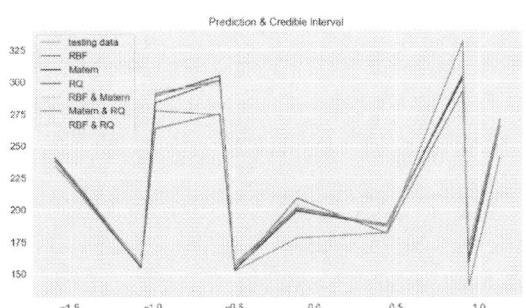

Fig.5 Predictions with different covariance functions

Table 3. Comparison of different kernels

Kernel	RMSE	MAPE	MAE	Run Time
SE	21.69	8.43%	18.38	2.02s
RQ	20.83	8.26%	18.14	4.01s
Matern(M)	20.90	8.44%	18.10	3.49s
SE+RQ	18.70	7.95%	16.02	8.82s
SE+M	18.29	7.87%	15.44	7.61s
RQ+M	18.53	8.01%	16.14	8.96s

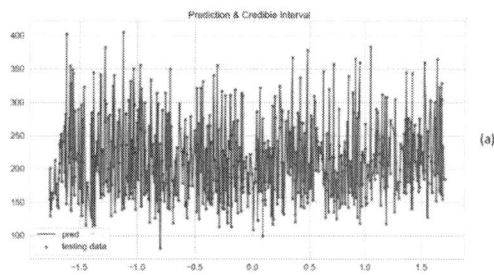

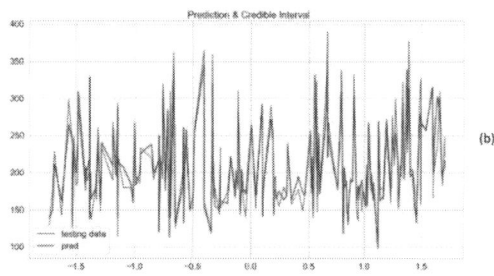

Fig.6 Gaussian process regression results.(a) Fitting result for training set. (b) predicted result for testing data.

IV. PROPOSED APPROACH FOR OPTIMIZATION

A. Bayesian Optimization(BO) with Gaussian Process(GP)

Bayesian optimization (BO) is a sequential design strategy developed from Bayes' theorem and is suitable for black-box functions. In Bayesian statistics, the uncertainty is modeled from prior probability distribution and use the known information to estimate the sampling point next, and through iterating this process to understand the potential objective function. Therefore, BO is effective when the target function is unknown and has random noise[13].

As shown in Equation (3) and (7), GP provides a posterior distribution of the black-box function, and BO, as a global optimization method, can choose the next point with high expectations to search the minimize or maximum target.

The flow chart of GPBO is illustrated in Fig.7. BO uses acquisition function to choose posterior[14], and three acquisition functions, i.e. Expected improvement (EI), Probability of improvement (POI) and Upper confidence bound (UCB), are widely used in GPBO. In this section, we choose EI as acquisition function for optimization and it is described by the following equation:

$$EI(x) = \begin{cases} (\mu(x) - f(x^+) - \xi)\Phi(Z) + \sigma(x)\phi(Z) & \sigma(x) > 0 \\ 0 & \sigma(x) = 0 \end{cases} \quad (19)$$

where ϕ is Probability density function (PDF) and Φ is Cumulative Distribution Function (CDF). ξ is a parameter to balance the exploration (seek the point with high variance) and exploitation (seek the point with low mean).

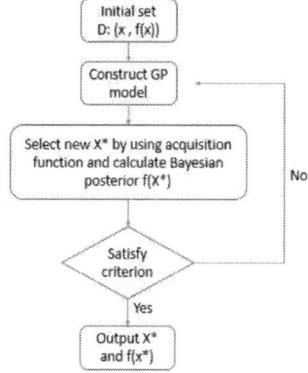

Fig.7 Flow of Bayesian optimization

B. Optimization result

The optimization of the parameters in the overall design space is shown in Fig.8. The results demonstrate that after around 25 iterations, the variation of optimized value for the most pin locations becomes stable, and after 100 iterations, most of the optimized values tends to be a fixed value. With the fast convergence of Bayesian optimization algorithm, we can obtain the target value with minimal iteration steps, as shown in Fig. 9.

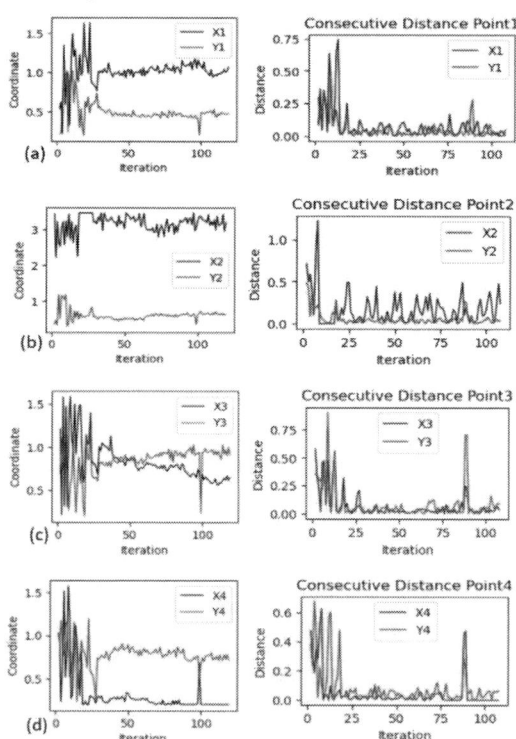

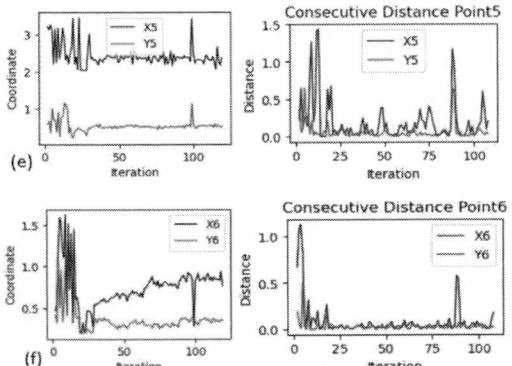

(e)

(f)

Fig.8 Convergence of optimized parameters. Coordinate of (a) Pin1, (b) Pin2, (c) Pin3, (d) Pin4, (e) Pin5, (f) Pin6.

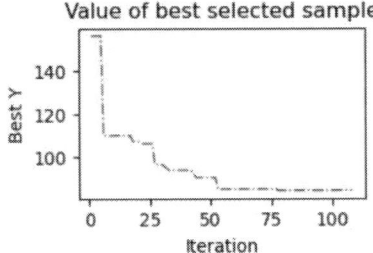

Fig.9 Convergence of target parameters

Fig.10 shows the comparison of stress distribution on die during die pick-up process between the original pin design model and the model after 100 iterations Bayesian optimization. According to simulation results, the optimized model shows a lower stress distribution. The maximum stress is reduced from 99.9 MPa in the original design to 81.3 MPa in the optimized model. There is around 20% improvement in stress reduce. It should be mentioned that the original design was already a good design based on the traditional DOE optimized methodology. Clearly, Bayesian optimization shows good optimization efficiency, including fast calculation

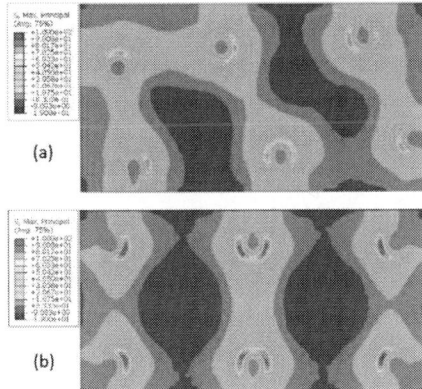

(a)

(b)

Fig.10 Stress distribution on die during pick-up. (a) optimized model, (b) original model.

speed, fast convergence speed, and good scalability. The methodology proposed in this study using machine learning enabled optimization is demonstrated to be very suitable for optimization of pick-up process.

V. CONCLUSION

In this paper, we proposed a methodology with a combination of machine learning and FEM for optimization of pin designs for die pick-up. Results showed that the Gaussian process regression have a good fitting performance and Bayesian optimization is scalable for IC packaging design or manufacture process. Since the simulation is usually time consuming and DOE analysis also requires a large number of simulation case, machine learning optimization method can greatly reduce the computation cost without losing accuracy, showing a great potential opportunity for IC design ad manufacturing.

REFERENCES

[1] Rodriguez R S, Gomez F R I. Pick and place process optimization for thin semiconductor packages[J]. Journal of Engineering Research and Reports, 2019: 1-9.

[2] Lin Y J, Hwang S J. Static analysis of the die picking process[J]. IEEE transactions on electronics packaging manufacturing, 2005, 28(2): 142-149.

[3] Shen H, Ye L, Tang L, et al. Study on thin die pick-up process based on Taguchi method[C]//2015 16th International Conference on Electronic Packaging Technology (ICEPT). IEEE, 2015: 1344-1347.

[4] Qian R, Liu Y. Thin and large die assembly pick up process optimization by dynamic modeling[C]//2016 17th International Conference on Electronic Packaging Technology (ICEPT). IEEE, 2016: 147-152.

[5] G.Taguchi and S.Konishi, Orthogonal Arrays and Linear Graphs. Dearborn, MI: American Supplier Inst.Press,1987.

[6] Hong S, Zhou Z, Lu C, et al. Bearing remaining life prediction using Gaussian process regression with composite kernel functions[J]. Journal of Vibroengineering, 2015, 17(2): 695-704.

[7] Huang F, Liu Y, Chen B. Prediction model of network traffic based on combined kernel function Gaussian regression[J]. Computer Engineering and Applications, 2015, 51(19): 93.

[8] Zhong-Da T, Shu-Jiang L, Yan-Hong W, et al. Network traffic prediction based on ARIMA with Gaussian process regression compensation[J]. Journal of Beijing University of Posts and Telecommunications, 2017, 40(6): 65.

[9] He Zhikun, Liu Guangbin, Zhao Xijing, et al. Overview of Gaussian process regression[J]. Control and Decision, 2013, 28(8): 1121–1129.

[10] Rasmussen C E. Gaussian processes in machine learning[C]//Summer school on machine learning. Springer, Berlin, Heidelberg, 2003: 63-71.

[11] Liu K, Liu B, Xu C. Intelligent analysis model of slope nonlinear displacement time series based on genetic-Gaussian process regression algorithm of combined kernel function[J]. Chinese Journal of Rock Mechanics and Engineering, 2009, 10: 2128-2134.

[12] Rasmussen C E, Nickisch H. Gaussian processes for machine learning (GPML) toolbox[J]. The Journal of Machine Learning Research, 2010, 11: 3011-3015.

[13] Frazier P I. A tutorial on Bayesian optimization[J]. arXiv preprint arXiv:1807.02811, 2018.

[14] Wilson J T, Hutter F, Deisenroth M P. Maximizing acquisition functions for Bayesian optimization[J]. arXiv preprint arXiv:1805.10196, 2018

Design and Analysis of a Fast-Speed Flip-Chip Bonding System with Force Control

Zhongyuan Zhu
Mechanical Engineering
Guangdong University of Technology
Guangzhou, China
2112001025@mail2.gdut.edu.cn

Hui Tang
Mechanical Engineering
Guangdong University of Technology
Guangzhou, China
huitang@gdut.edu.cn

Jiedong Li
Mechanical Engineering
Guangdong University of Technology
Guangzhou, China
3114000346@mail2.gdut.edu.cn

Sifeng He
Mechanical Engineering
Guangdong University of Technology
Guangzhou, China
3114000057@mail2.gdut.edu.cn

Abstract—Force sensing and control functions are very important for flip chip bonding systems for the reason of purchasing high-quality chip interconnection. But it is regrettable that, using the existing commercial technology and equipment, the bonding interconnection procedure is hard to be performed perfectly, specifically in the aspects of efficiency, accuracy and quality. There are many processes involved in the bonding interconnection procedure, so as to say, we need to take as far as possible every step and aspect into consideration. This paper proposed a high-speed flip-chip bonding system with force control, which can make the chips after alignment be bonded on substrate well as soon as possible. In the first place, benefited from the work of our predecessors, the active soft-landing (ASL) interconnection is realized by using the advantages of the monolithic force integrated flexible bonding device. However, the adopted flexure bonder is not enough to satisfy the high-efficiency flip-chip bonding requirement. Because the lightly-damping second-order spring system characteristic caused by the force detective part of the adopted flexure bonder. Secondly, based on the flip-chip bonding process, a flexure-based switch is designed to change the unidirectional stiffness of the bonder with the property of force detection for the purpose of reducing the vibration caused by the linear motor braking in the process. Thirdly, a novel closed-loop control strategy which can accommodate the requirements of position and force, namely hybrid position/force closed-loop (HPFC) control, with integrator composed of inertial filter (ICIF) is proposed to realize the high-efficiency force control under high-dynamic working conditions. By adopting these methods above, the bonding system can go through each step of chip bonding after alignment fleetest, and provide high-quality chip interconnection at the same time.

Index Terms—*vibration attenuation, ASL bonding interconnection, compliant mechanism, force sensing, position and force control*

This work was supported in part by the Natural Science Foundation of China under Grant 51975132, Grant U20A6004, in part by the Guangdong Programs for Science and Technology under Grant 2019A1515011896, Grant pdjh2020a0170, and by the National Key R&D Project under Grant 2020YF-B1712703. *(Corresponding author: Hui Tang)*

I. Introduction

In this fast-growing era, based on state-of-the-art technology and looking forward to the future, the world of chips is running from the micrometer-scale to the nanometer-scale. Abandoning the traditional reflow welding process, flip-chip direct packaging is becoming the mainstream, for the reason that it can satisfy the requirements brought from the changes of packaging objects, which are being more tiny, thin, and with complex structures. As is mentioned above, by the year 2025, die to substrate FC bump pitch will reduce from 200 μm to < 40 μm, and stacked die bump pitch will reduce from 95 μm to 10 μm [1]. Under such a background, it put forward the pressing demand on the accuracy of bonding pressure ($\le \pm 5$ N), and accuracy of positioning ($\le \pm 500$ nm) in the process of flip-chip bonding interconnection [2].Totally, seeing Fig. 1, there are two steps in the flip-chip bonding interconnection process. The first step is high-speed feeding, and the second step, consisting of low-speed searching and force bonding, is low-speed bonding. Under the influence of high-velocity and high-acceleration movement generated by the electro-mechanical actuator [3], substrate and die may bump into each other even damage may happen. Hence, by what means can we realize great bonding interconnection guaranteeing accuracy, efficiency, and reliability simultaneously become a major challenge to develop advanced flip-chip bonding system.

In order to satisfy this demand, some pioneering research has been carried out, such as the development of a bonding machine and position/force hybrid control, to improve the bonding performance [4]. Whereas, subjected to the high cost, large volume and other reasons, it is hard for the conventional bonding system to realize both position closed-loop control and force closed-loop control in one control frame, largely because their actuators are usually just the linear motor. It is a pity that, instead of utilizing the commercial force transducer, estimating and regulating the force through saturating the

978-1-6654-1392-3/21 $31.00 © 2021 IEEE

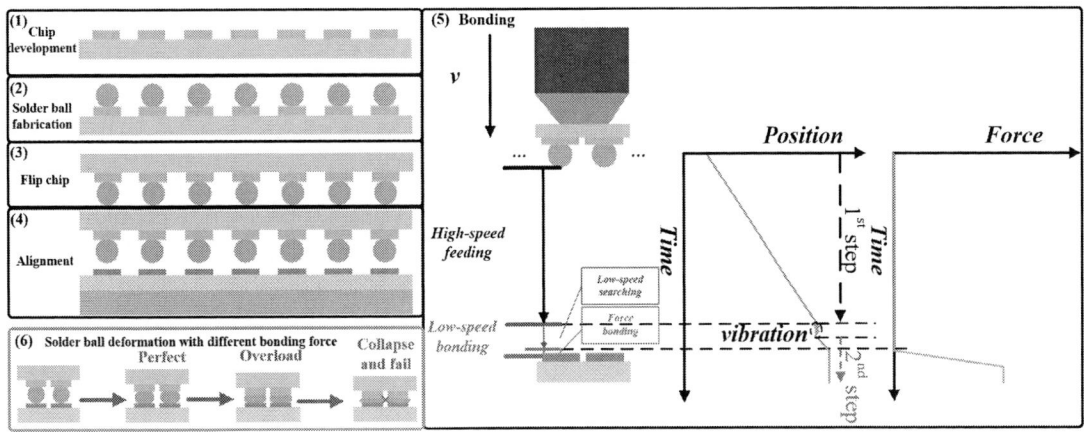

Fig. 1. Flip chip packaging process (1) - (5) and description of solder balls failure under different levels of pressure (6).

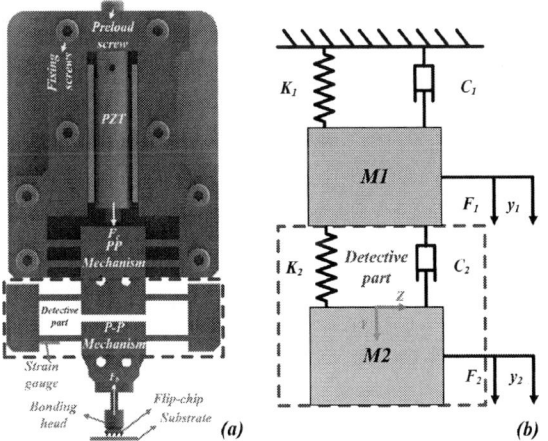

Fig. 2. (a) The schematic diagram of the previous flexure based bonder with the property of force detection; (b) Dynamic model of flexure bonder.

electric current loop is not accurate enough [5]. It cannot find a good trade off between avoiding chips' braking and high-efficiency interconnection [6].

In summary, traditional bonding systems driven by only one linear motor are difficult to perform flip-chip bonding effectively.In recent years, the flexible mechanism driven by piezoelectric ceramics (PZT) has been widely adopted in the area of precision manufacture for the reason of its features including but not limited to quick response and high thrust [7], [8]. Utilizing the stable mapping between strain of a flexure bonder and external force, a common bonding system can be empowered the potential ability of accurate force detecting [9]–[11].

In the previous work, as shown in Fig.2, Sifeng He has designed and fabricated a monolithic force sensing integrated flexure bonder [12], which gives bonding system the ability of force detection conveniently. To acquire better force feedback signal with higher Signal to Noise Ratio (SNR), the force detection part of the proposed flexure bonder tends to be designed as lower stiffness, so that more strain can be detected under the same force. However, the force detective part of the proposed flexure bonder acts like a spring (see Fig2.(b)). After

the first step (high-speed feeding), to prevent the crash between chips and substrate, the system cannot enter the second step until the vibration caused by linear motor braking dissipated. Lower the stiffness, larger the amplitude and longer the settling time of the vibration is, which is against the requirements of high-efficiency chips bonding. What's more, the inherent feature about small damping of the bonding system results from the flexure-based bonder will significantly affect the control performance, especially in the range of high frequency [13]. To address the problems mentioned above, a flexure-based switch and a HPFC control strategy with ICIF are thus proposed respectively.

In this paper, a novel flip-chip bonding system with force control is proposed. By adopting the flexure bonder and switch, the modified flexure bonder and HPFC control strategy contained in this system will guarantee the faster speed and the higher efficiency of the flip-chip bonding progress after alignment compared with the traditional flip-chip bonding system. Firstly, a flexure switch is designed and be assembled on to resolve the contradiction concerning the stiffness of the force detective part. Secondly, the whole control frame including position and force closed-loop control is constructed, which enables the proposed system to achieve its full potential of force control, is developed. Finally, a suite of confirmatory tests and performance tests are exhibited in the form of experimental data and finite element analysis.

II. DESIGN AND MODELING OF A MODIFIED FLEXURE BONDER

As is mentioned above, as a whole, there are two steps within the ASL bonding procedure, namely high-speed feeding and low-speed bonding The latter step consists of low-speed searching and force bonding two parts [12]. The bonding procedure will be repeated lots of time in every product, hence, each step should be executed reliably as rapidly as possible for yield improvement. More time the stage run at a high speed, less time the first step occupies. That is to say, we hope the flexure bonder can stop at the designated position braking at high acceleration. However, because of the force detective part of the proposed flexure bonder acts like a spring, the chips

978-1-6654-1392-3/21 $31.00 © 2021 IEEE

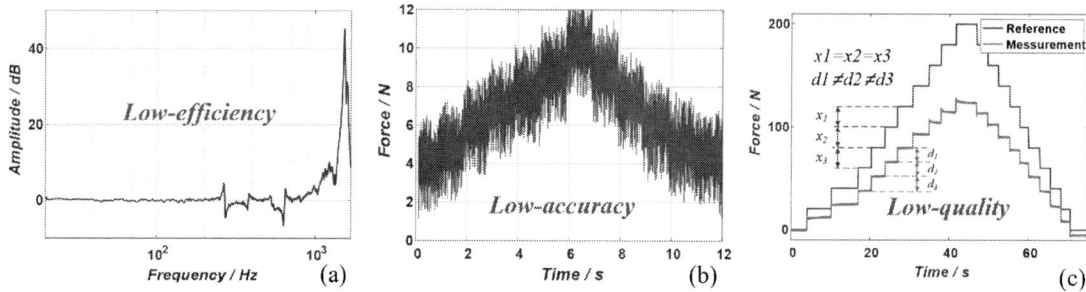

Fig. 4. Difficulties in the control of flexure-based bonder. a) Bode diagram of the flip-chip bonding system; b) low signal to noise ratio; c) Nonlinearity between output force and control signal.

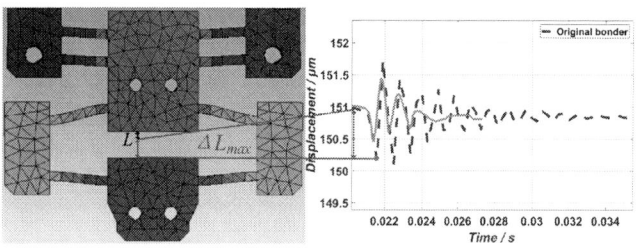

Fig. 3. Schematics of the vibration occurred after the braking.

attached to the bottom of the end of the flexure bonder will appear a vibration (see Fig.3), which may cause the crash between chips and substrate.

Furthermore, it takes surplus time before enter the second step, which drives down the production efficiency. It seems that we can increase the stiffness of the force detective part to alleviate the phenomenon, however, as see Fig.4(b), the amplitude of noise is too large.

The Signal to Noise Ratio (SNR) and resolution of the force feedback will decrease with the increasing of the stiffness, because smaller voltage change will represent the same change of force. On the other hand, as see Fig.3, during the vibrating of the end of the flexure bonder, at the moment of the red point, all the kinetic energy is converted to elastic potential energy expressed as:

$$E_k = \frac{1}{2} K \Delta L_{\max}^2 = \int F_k \, dL \qquad (1)$$

which will be dissipated by the internal damping. Where F_k is the elastic force largely produced by K_2 is expressed as follow according to the Hooke's law:

$$F_k = K \cdot \Delta L. \qquad (2)$$

Unfortunately, the compliantly mechanism is a lightly damping structure, that is why vibration energy dissipation takes such a long time.

Based on this, we can spontaneously think out the way to alleviate the vibration. Thus, a flexure switch should be designed to increase the unidirectional stiffness when m_2 is keep moving after linear motor braking. So, that the vertical deformation($\Delta L_{\max}$) will be smaller which means less energy is waiting for dissipating by the poorly internal damping.

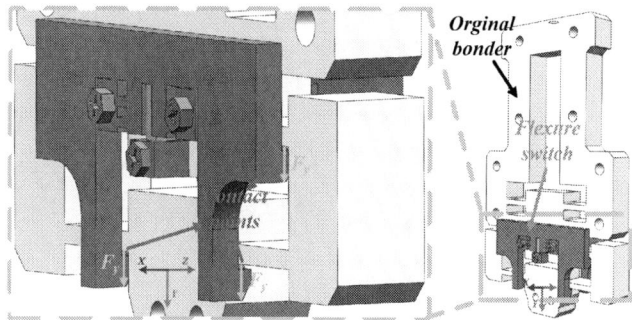

Fig. 5. assembly drawing and partial enlarged view of the modified fonder.

A. Unidirectional stiffness changing method

Based on the definition of the stiffness, which expression is shown as follow, it means that larger the stiffness, more pressure is needed under the same deformation.

$$K = \frac{\Delta F}{\Delta L}. \qquad (3)$$

So as to say, as long as an external pressure is added to the m_2, the equivalent stiffness will be larger. In this work, the equivalent stiffness only needs to be larger, when m_2 keep moving along negative Y axis, in order to make $\Delta L_{\max}$ smaller, while the stiffness should be stay the same in order not to impact the force detective action. The equivalent stiffness can be demonstrated as:

$$K_{equivalent} = \begin{cases} K_2 & , m_2 \ move \ along \ -Y \\ \frac{K_2 \cdot \Delta L + F_e}{\Delta L} & , m_2 \ move \ along \ +Y \end{cases} \qquad (4)$$

where F_e is the pressure produced by the switch.

B. Design and Modeling of the flexure switch

The switch, the blue part shown in Fig.5, connected through a P-P compliant structure to two pads, which are bolted connected to the flexure bonder, actuated by piezoelectric stack (green part in Fig.5), is restricted to one degree of freedom (vertical translation). To make piezoelectric stack can push the switch move along Y axis, another pad is bolted connected to give it a support. Thus, when piezoelectric stack elongates, the switch can provide a single positive direction (along Y axis) pressure to ensure a smaller ΔL_{max}, so that the time taken by dissipating the energy and the amplitude of vibration will be shorter. Meanwhile, the decrement of

978-1-6654-1392-3/21 $31.00 © 2021 IEEE

the amplitude of the vibration will further reduce the risk of the crush between chips and substrate. Once the vibration is dissipated completely, the piezoelectric stamp shortens and the switch recovers. Based on the designed mechanism structure, as shown in Fig.5, one switch has four contact points with flexure bonder, can only provide a unilateral restriction, so as to say, when the switch recovers to initial state, it will neither impact the force feedback action nor the closed-loop control.

In this work, owing to the practicability and stability, the cantilever beam type flexure structure is slathered in the switch as a guide mechanism. However, whatever force is applied to the unilateral cantilever beam type flexure structure, it will with high probability appear 3 degrees of freedom movements simultaneously, which will lead to unwanted off-axis deviations, resulting in an unbalanced pressure of m_2 (see Fig. 5). Therefore, a symmetric and parallel plate spring mechanism (P-P mechanism) was proposed as the basic guiding mechanism to avoid the generation of unbalanced pressure. On the basis of the flexibility theory, the kinematic relation between the force and displacement of the flexure mechanism is:

$$D = C \times F \tag{5}$$

where D is the displacement vector, C is the compliance matrix, and F is the force vector. Specific expression is

$$\begin{bmatrix} \Delta x \\ \Delta y \\ \Delta \theta \end{bmatrix} = \begin{bmatrix} C_{F_x,\Delta x} & 0 & 0 \\ 0 & C_{F_y,\Delta y} & C_{M_z,\Delta y} \\ 0 & C_{F_y,\Delta \theta} & C_{M_z,\Delta \theta} \end{bmatrix} \times \begin{bmatrix} F_x \\ F_y \\ M_z \end{bmatrix} \tag{6}$$

As shown in Fig.5, the F_y is the reactive force produced by K_2. Due to the symmetrical structure of parallel mechanism, the Δx and $\Delta \theta$ can be ignored. So, the expression can be simplified as:

$$\begin{bmatrix} \Delta x_1 \\ \Delta y_1 \\ \Delta \theta_1 \end{bmatrix} = \begin{bmatrix} 0 \\ C_{F_y,\Delta y} \cdot F_{1y} \\ 0 \end{bmatrix} \tag{7}$$

$$C_{F_y,\Delta y} = (l/2)^3/3EI \cdot 2 = l^3/12EI \tag{8}$$

where, in this equation, l is the length of the cantilever beam, E is the Young's modulus and the moment of inertia I is expressed as:

$$I = a \cdot \omega^3/12 \tag{9}$$

The displacement of the parallel mechanism can be obtained as:

$$\Delta y = \Delta y_1 = Fl^3/24EI \tag{10}$$

and the compliance of the parallel flexure is derived as:

$$C_p = \Delta y/F = l^3/24EI. \tag{11}$$

The compliance of P-P mechanism can be derived by the parallel rule is:

$$C = C_p \times 2. \tag{12}$$

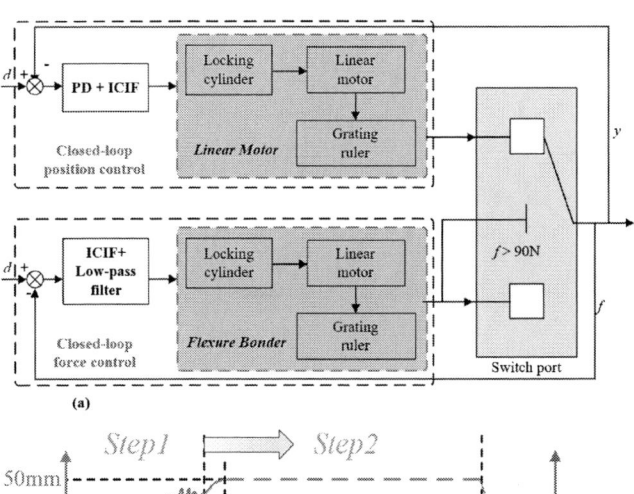

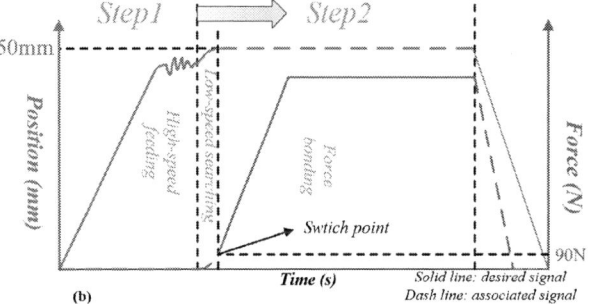

Fig. 6. The whole control frame. (a) HPFC control strategy; (b) Schematic of time-dominant actual signal.

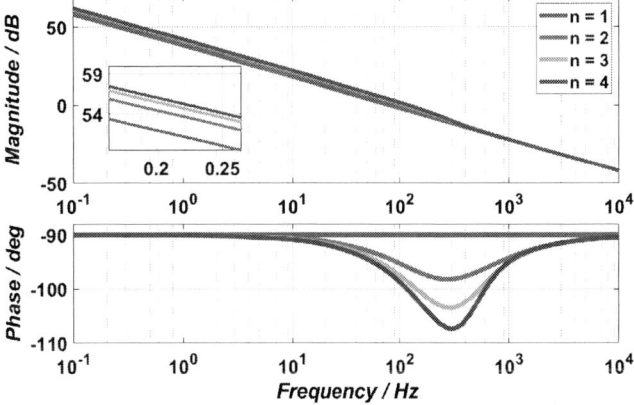

Fig. 7. Bode diagram of ICIF with different integral values n.

III. HYBRID POSITION/FORCE CLOSED-LOOP CONTROL STRATEGY

Based on the nyquist stability analysis, we can easily understand that the control gain must be small enough to avoid oscillation appearing, a novel ICIF is designed, which transfer function can be expressed as follows:

$$ICIF(s) = \frac{\frac{1}{n} \sum_{i=1}^{n} \left(\frac{1}{Ts/n+1} \right)^i}{1 - \frac{1}{n} \sum_{i=1}^{n} \left(\frac{1}{Ts/n+1} \right)^i} \tag{13}$$

where n is the order of ICIF set to be positive integral value. As shown in Fig.7, we can obtain that, along with the growth of the n, before 300 Hz, the magnitude is becoming larger

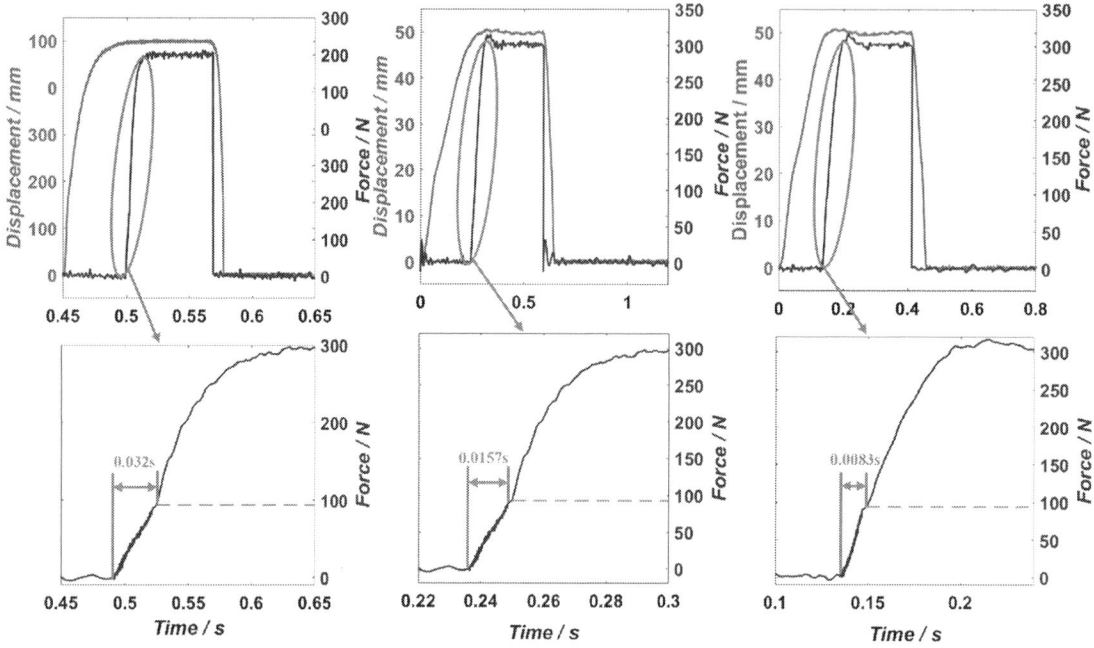

Fig. 10. HPFC control tests under three kinds of loading force at different speeds.

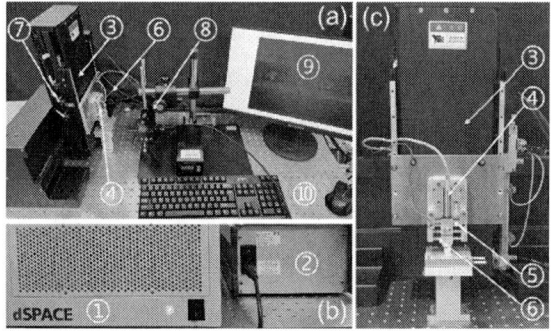

Fig. 8. Prototype system components. 1) signal I/O module, 2) PZT power driver, 3) Z -direction linear motor platform, 4) PZT, 5) flexure-based bonder head, 6) resistance strain gauge, 7) grating scale, 8) visual monitor, 9) upper computer, 10) vibration isolation platform.

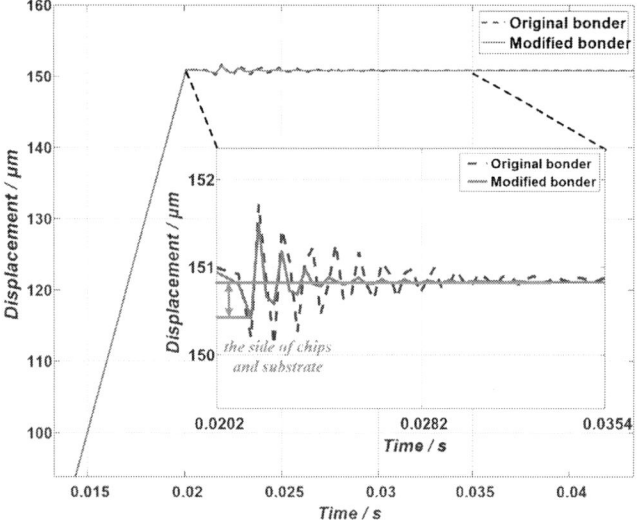

Fig. 9. The performance validation of the flexure bonder with switch through FEA results.

at the same frequency. On the other hand, compared to the traditional integrator controller, it hasn't decreased too much phase margin at the 0dB crossing line. It means a higher gain is applied to the actual control process.

IV. TESTING AND ANALYSIS

In this section, a prototype of the designed bonding system was established. The confirmatory test and Finite Elements Analysis test are carried out.

A. Prototype Fabrication and Experimental Studies

The prototype components are listed in Fig. 8, the flexure-baed bonder head is mounted on and moved through milimeter displacement by the Z-direction linear motor. The milimeter displacement in the first step is measured by the grating scale, and in the second step, the force is measured by the strain gauge and its amplifier circuit. The signal I/O module includes DS2004 high-speed A/D board and DS2102 high-speed D/A board. The visual monitor is utilized to check the bonding consequence of whether the solder balls are correctly pressed. The test items include closed-loop bonding performance validation and modified flexure bonder vibration attenuation finite element analysis.

B. Performance validation test of the modified flexure bonder

In the common flip-chip bonding system, the linear motor carrying the bonder can arrived the designated position with acceleration 2-10G(see Fig.9). To fully validate the stability of the system, the FEA analysis is implemented under the acceleration of 10G. As shown in the Fig.9, the peak-to-peak amplitude of the vibration when there is no switch on the flexure bonder achieves 1.53 μm, and settling time is 14.2379 ms, while the peak-to-peak amplitude of the vibration down to 1.08 μm, and settling time was cut to 3.488 ms, when the flexure switch act on the flexure bonder. It is worth mentioning that the side excursion on the side of substrates down to 0.41 μm, compared with the situation without switch, there is a 44% reduction in the unilateral excursion. And the reduction of settling time achieves 75%.

C. HPFC control test

In order to verify whether the proposed system including flexure-based bonder head can provide accurate force and displacement in the bonding process in actual production. Under different speed of loading force, the actual displacement and force are measured in real time and are applied to the HPFC control. And seeing Fig. 10, under different situations, the contact force can achieve 300 N quickly and stably, which directly verifies that it is capable of actual bonding process requirements.

V. CONCLUSION

A fast-speed flip-chip bonding system with force control is proposed in this paper, which successfully releases the potential abilities of the proposed bonding system. Among the system, two important parts that guarantee the bonding system operate as fast as possible are illustrated and some experiments and finite element analysis are implemented to prove the high-quality and high-efficiency that the system can achieve. Firstly, to speed up the process and prevent the crush between chips and substrate, a switch is adopted to alleviate the vibration after the braking of linear motor. Secondly, based on the ICIF, the HPFC control strategy is designed to cater for the requirements of high-efficiency, high-accuracy, and high-quality interconnection by flexure bonding system with lighting damping characteristic. A series of experiments and analysis are carried out to prove: 1) the modified flexure bonder has the same force-detection performance with the original type; 2) using the switch, the unilateral excursion and settling time of vibration declines 44% and 75% respectively; 3) under the proposed control frame (HPFC control strategy), the bonding force can achieve a given target and maintain stable even though the loading speeds are different. Thanks to the two important parts of the proposed system, the bonding system can run fast in the process and keep a precision force and displacement control, which means high efficiency.

REFERENCES

[1] K. Hollstein and K. Weide-Zaage, "Advances in Packaging for Emerging Technologies," in *2020 Pan Pacific Microelectronics Symposium (Pan Pacific)*, 2020.

[2] W. Huang, W. Wu, J. Wu, G. Dong, and C. F. Oo, "Effect of Bonded Ball Shape on Gold Wire Bonding Quality Based on ANSYS/LS-DYNA Simulation," *China Semicond. Technol. Int. Conf. 2020*, Jun. 2020.

[3] H. M. Lee, S. Y. Lee, T. J. Choi, S. I. Park, E. K. Ko, and K. W. Paik, "Effect of the Curing Properties and Viscosities of Nonconductive Films on the Solder Joint Morphology and Reliability of Chip-On-Board Packages Using Cu-Pillar/Sn-Ag Bumps," *IEEE Trans. Components, Packag. Manuf. Technol.*, vol. 10, no. 5, pp. 924-928, May 2020.

[4] Y. Zhang and Q. Xu, "Adaptive force/position control with unknown parameters estimation for a piezo-driven micromanipulator," *IEEE Reg. 10 Annu. Int. Conf. Proceedings/TENCON*, Jan. 2016.

[5] J. H. Kim, J. H. Cho, and S. K. Lee, "A Design of Hybrid Contact Detection Algorithm for Wire Bonder Machine," *IECON Proc. (Industrial Electron. Conf.*, vol. 3, pp. 2963-2967, 2003.

[6] T. Nanthavittayaporn, K. Ugsornrat, D. Klaitabtim, and P. Buathong, "Optimization Parameters of Installation Automatic Die Bonding Machine for Integrated Circuit Packaging," *15th International Conference on Electrical Engineering/Electronics, Computer, Telecommunications and Information Technology (ECTI-CON)*, pp. 266-269, Jan. 2019.

[7] J. Pinskier, B. Shirinzadeh, L. Clark, Y. Qin, and S. Fatikow, "Design, development and analysis of a haptic-enabled modular flexure-based manipulator," *Mechatronics*, vol. 40, pp. 156-166, Dec. 2016.

[8] H. C. Liaw and B. Shirinzadeh, "Neural network motion tracking control of piezo-actuated flexure-based mechanisms for micro-/nanomanipulation," *IEEE/ASME Trans. Mechatronics*, vol. 14, no. 5, pp. 517-527, 2009.

[9] S. Tayouri, S. F. Alem, I. Izadi, J. Ghaisari, and F. Sheikholeslam, "Chattering-free PD Sliding Mode Control of Piezoelectric Actuators with Hysteresis," *ICEE 2019-27th Iran. Conf. Electr. Eng.*, pp. 1064-1069, Apr. 2019.

[10] Z. Sun, X. Li, and X. Li, "The finite element analysis of elastic-plastic contact of single asperity with different materials," *7th Int. Conf. Cond. Monit. Mach. Non-Stationary Oper.*, pp. 108-114, Jun. 2021.

[11] P. Liu, Z. Lin, P. Zhu, and M. A. Soto, "Finite Element Analysis of Microbending Losses for Plain-weave GFRP Packaging Optical Fiber Sensor," *7th Int. Conf. Cond. Monit. Mach. Non-Stationary Oper.*, pp. 205-208, Jun. 2021.

[12] Y. Liu, Y. Zhang, and Q. Xu, "Design and Control of a Novel Compliant Constant-Force Gripper Based on Buckled Fixed-Guided Beams," *IEEE/ASME Trans. Mechatronics*, vol. 22, no. 1, pp. 476-486, Feb. 2017.

[13] Q. Xu, "Design and Development of a Novel Compliant Gripper with Integrated Position and Grasping/Interaction Force Sensing," *IEEE Trans. Autom. Sci. Eng.*, vol. 14, no. 3, pp. 1415-1428, Jul. 2017.

Deep Learning Product Classification Framework based on the Motivation of Target Customers

Fei Sun
School of Art & Design
Guangdong University of Technology
Guangzhou, China
1111917005@mail2.gdut.edu.cn

Ding-Bang Luh*
School of Art & Design
Guangdong University of Technology
Guangzhou, China
luhdingbang@mail.gdut.edu.cn

Qidong Wang
Institute of Microelectronics
Chinese Academy of Sciences
Beijing, China
wangqidong@ime.ac.cn

Yulin Zhao
School of Art & Design
Guangdong University of Technology
Guangzhou, China
512080653@qq.com

Yue Sun
School of Art & Design
Guangdong University of Technology
Guangzhou, China
328162766@qq.com

Abstract—Due to the dynamic and competitive market environment, it is widely recognized that the development of new products and processes has become the critical point of attention for many companies. The first step in the development process of a product is to define the nature and function of the product, which is to classify the new product. In recent years, Artificial intelligence has been applied in a broad range of fields including product classification. However, the traditional product classification only uses the product features as the input training data. Thus, these methods failed to build the connection between the product and its target customers. Therefore, such a product classification model has low accuracy and can provide little help to the design manager. In this paper, a fast and effective product classification framework is proposed, which is based on a novel attempt that embeds the innovation idea of human in machine learning technologies. It is composed of a costumer modeling method based on the deep learning technologies, and a customer information deduction method based on the Motivation Design Model (MDM), which calculates multi-level features of costumers from the limited information. This paper is the first time to add customer motivation analysis to traditional machine learning method, which has a strong application prospect.

Keywords—new product development, motivation design model, product classification, deep learning

I. INTRODUCTION

Due to the dynamic and competitive market environment, it is widely recognized that the development of new products and processes has become the critical point of attention for many companies. New item advancement (NPD) is an unpredictable field that shows system, the board, innovative work, creation, promoting, and dynamic, and requires connecting science and innovation with the commercial center. The traditional NPD process only focuses on the product and market, which is developed by designers on the basis of the products of past dynasties. This way is labor-consuming, inefficient, and has become a bottleneck or constraint for these enterprises to improve their productivities and regulate production [17,18,19]. Moreover, because the design process is dominated by the designer, there is no costumer participation, which will lead to mismatches between the new product and the needs of customers. Specifically, for companies that lack experience in past generations of products, it is struggle for them to find efficient and effective processes and management activities for NPD [20].

Along these lines, the need to take part in item advancement, notwithstanding, has gotten progressively significant as organizations face the danger of huge decay in the event that they can't stay up with changes in their enterprises. The first step in the product development process is to define the nature and function of the product, which is to classify the new product. The concept of "product classification" is to divide products into a structured category according to their specific characteristics. Many researches and organizations have put forward the method and theory of product classification. In general, other than of a normalized item characterization framework, there are additionally numerous casual techniques for item order formulated by different industry associations.

In recent years, due to the continuous upgrading of hardware and software, Artificial intelligence, or AI, has been applied in a wide range of fields to perform specific tasks. Computer based intelligence alludes the capacity of a machine to function insight of a product like human. The different assignments that are referenced in Machine learning includes the sensing, self-arranging, acknowledgment, robots controlling, estimating, forecasting, prediction and product classification. However, the traditional product classification only uses the product features as the input training data. Thus, these methods failed to build the connection between the product and its target customers. Therefore, such a product classification model has low accuracy and can provide little help to the design manager.

In machine learning, computers learn how to perform tasks from sample data. We realize that if we give a machine more experience in defining tasks, its performance will improve [1]. Previous studies have shown that there is a close relationship between the characteristics of target costumers and the characteristics of products they have purchased historically. Therefore, if costumer information and product features could be combined for model training, the accuracy of product classification will be improved. Moreover, the calculated multi-dimensional data could be provided to product managers and help them make decisions. However, due to privacy concerns, the information of target customers is hard to collect.

In this paper, a fast and effective product classification framework is proposed, which is based on a novel attempt that embeds the innovation idea of human in machine learning technologies. It is composed of a target costumer modeling method based on the deep learning technologies,

978-1-6654-1392-3/21 $31.00 © 2021 IEEE

and a customer information deduction method based on the Motivation Design Model (MDM), which calculates multi-level features of costumers from the limited information [2,3,4,19]. This paper is the first time to add customer motivation analysis to traditional machine learning method, which has a strong application prospect.

II. RELATED WORK

A. Deep learning and CNNs

For many years, it takes a lot of efforts to design a feature extractor to implement a machine learning system that converts the original data (i.e. the pixel value in the image) into detectable or classifiable patterns in a learning subsystem with feature vectors[13]. Deep learning appeared in 2006along with deep belief network (DBN), as a part of machine learning system. For supervised or unsupervised feature extraction and transformation, the system uses multi-layer nonlinear information processing as well as pattern analysis and classification.

Two new unsupervised depth models are developed after DBNs: (1) a method for learning sparse and over completed features, which applies linear encoder and decoder, and then uses sparse nonlinearity to convert code vector into quasi binary sparse code vector, and (2) an automatic encoder variant with greedy hierarchical training.

With the progress of technology inhardware and software and the evolvement of newly developed optimization algorithms [14], deep learning has achieved a new milestone, resulting in three types of deep learning approaches:

- Unsupervised deep networks (generative learning): which do not need label classes, while finding patterns between pixels by capturing high-order correlation of data.
- Supervised deep networks (discriminative deep networks): use tag information to classify the input data in theseof. This approach represents the most common form of machine learning, whether deep or non-deep [13], with more efficiency in training and testing, as well as superior flexibility in construction, and in all are more suitable for end-to-end learning in complex systems.
- Hybrid deep networks (semi-supervised methods): these methods utilize generative and discriminative model components, that is, they work regardless of tagged data (for example, generative countermeasure networks or GANs [15]). Semi supervised learning is useful in hyperspectral image classification, which can solve the limitation of training samples [16].

B. Factors for New Product diffusion

According to Everett M. Rogers' diffusion theory, that is, the diffusion process of introducing new concepts or new products, if a curve is used to describe the changes of the number and types of users with time, it will produce a curve very similar to the normal distribution curve. In addition, the percentage of different categories of adopters among all adopters is always constant. Rogers added that the rapid spread of a new product to the population depends on five factors.

- *Relative Advantage,* or the perceived desirability or benefit of the new product relative to the benefit provided by the existing product. The main cause why a good product spreads rapidly is that the first users of the product advertise the product through the customer experience.

- *Compatibility,* or an agreement between the new product ideas and the consumer's concept of cherished values. The compatibility also relates to consumer's experience, lifestyle, religion and knowledge of items.

- *Complexity,* or the access of understanding how a product works. The complexity is dependent with consumer's perception, knowledge of operation, and background.

- *Divisibility,* before consumers accept a new product, the common sense is to access the trial products. If this product offers a accessible try-out, its acceptability would likely to be raised.

- *Communicability,* which indicates whether the product is popular, also it means that the new concept must be conveyed to the target consumers with easy access, which is directly correlated to whether there are alternative communication channels.

The aforementioned arc the five well-recognized factors dominates the diffusion of a new product into the customers. Thereafter known as "primary factors". People in addition, like Onkvisk and Sham: Kun-Mo Kuo. and others proposed complimentary factors affecting the speed of diffusion, also known as "secondary factors".

III. FRAMEWORK

A. Network Structure

As aforementioned, the product classification framework proposed in this paper is a novel attempt that embeds the innovation idea of human in machine learning technologies. Unlike most product classification processes, we train the classification model from the perspective of the

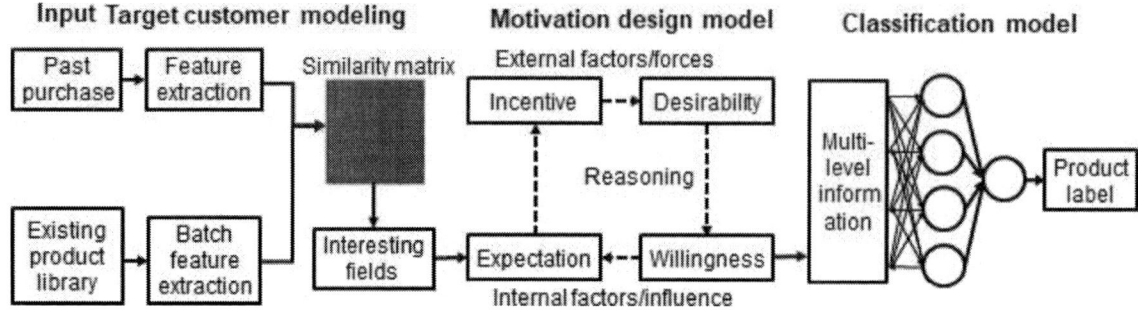

Fig. 1. An overview of the proposed framework. The main network consisting of a costumer modeling method based on the deep learning technologies, and a customer information deduction method based on the MDM, which calculates multi-level features of costumers from the limited information.

costumer. The proposed framework is composed of a target costumer modeling module and a customer information deduction module, which can be seen from Fig. 1. The target costumer modeling module utilizes the large-scale image retrieval technology on customer past purchases, which explores the relationship between customer purchase history information and product information, personalized modeling of the target customers. Then, the MDM is applied on the costumers' information [9,10,11,12]. The complaint is derived from the interest information. By doing this way, we are able to get the multi-level features of costumers from the limited information. Finally, the calculated multi-level features information is used to retrain the product classification model based on the neural networks. The details are given in the next chapter.

B. Target Costumer Modeling

In order to improve the performance of the classification model in the last step, we need to gather bigger datasets, learn more remarkable models, and utilize better strategies to prevent overfitting. However, due to privacy concerns, we have only a limited prior knowledge of target customers, which is insufficient for training. Therefore, the aim of the target costumer modeling module is to expand the scope and concept of customer experience perception according to the limited past purchase information. Then realize the expansion of target customer information.

To find out around a huge number of items from countless images, a model with large learning capacity is required. Convolutional Neural Organizations (CNNs) establish such described class of models. Their ability is manipulatable via changing their depth and breadth, and they also make solid and generally right assumptions about the nature of products. In this manner, contrasted with standard feedforward neural networks with similarly-sized layers, CNNs is superior in fewer inter-connections and less degree of freedom while thus they are simpler to prepare andbest predictable performance is likely to be degrades in a small scale.

Inspiring from the advancement of CNNs, we utilize the large-scale image retrieval technology to expand the information of target customers. We contribute to a simple yet effective supervised learning framework for rapid image retrieval, which takes the past purchases information of target customer as input, and output is all similar products in the company's existing product library. Specifically, this module extracts features from each design picture of the existing product database and store them in the database. For

the product that used to purchase by the target customer, the same feature vector is extracted. Then, the distance between the customer purchased product's vector and the vectors of the product database is calculated to find the similar pairs, and the corresponding products are the search result.

C. Motivation Design Model

In the last step, we utilize the large-scale image retrieval technology to expand the interest field of target customers. Then, the MDM is applied on the costumers' information to continue to dig customer characteristics. The complaint is derived from the interest information. By doing this way, we are able to get more comprehensive multi-level features of costumers from the limited information.

When customers classify products, the decision vary with different factors including the individual, time, place, event, thing, or situation. It is the need that guides human behavior. Some needs are extravert, which are concerned with physiological levels; some arc introvert, which are concerned with psychological ones. Additionally, these needs can be categorized as individual needs, family needs, and social needs. In spite of the fact that there are a wide range of human necessities, after considerable research only a handful of basic needs have arisen. Plentiful proof gathered by social anthropologists shows that despite the fact that there are differences between cultures in the methods used to satisfy these requirements, the major necessities of all individuals are very similar, whereas psychological needs change according to the environmental quality.

MDM is proposed based on the concept of human needs [5,6,7,8]. In this step, the MDM is applied on the costumers' information to continue to dig customer characteristics. We start with the target customer interests, then extend the information of target customers from four aspects including expectation, incentive, desirability, and willingness. By doing this, the multi-level customer information is achieved.

There are 11 main steps in MDM, including (a) persona building: subject behavior analysis; (b) subject motivation analysis: initial motivation (IM); (c) subject complaint journey analysis: complaint set (ΣCM); (d) mirroring for wishful journey: solution set; (e) selection of desirable solution (DS) as activity design guideline; (f) major stakeholder analysis: key participants (KP); (g) KP behavioral analyses: ΣCMs of each KP; (h) mirroring for solution set of each KP: work demand for each KP; (i) mission statement and activity design (AD): Planned Scenario (PS); (j) AD matching with PS for all KPs; (k) subject motivation formation in PS: designed motivation (DM).

Specifically, the proposed network utilizes scores to quantify customers' interests and complaints. Follow the definition that if the products are interested, a high score will be assigned, otherwise is a low score. The data structure is {Extracted products features, Scores}.

IV. CONCLUSION

In this paper, a fast and effective product classification framework is proposed, which is based on a novel attempt that embeds the innovation idea of human in machine learning technologies. It is composed of a costumer modeling method based on the deep learning technologies, and a customer information deduction method based on the Motivation Design Model (MDM), which calculates multi-level features of costumers from the limited information. This paper is the first time to add customer motivation analysis to traditional machine learning method, which has a strong application prospect.

ACKNOWLEDGMENT

Our work was supported in part by Design Science and Art Research Center, Guangdong University of Technology.

REFERENCES

[1] Krizhevsky A, Sutskever I, Hinton G E. Imagenet classification with deep convolutional neural networks[J]. Advances in neural information processing systems, 2012, 25: 1097-1105.

[2] Ding-Bang Luh , Yao-Tsung Ko & Chia-Hsiang Ma (2011) A structural matrix- based modelling for designing product variety, Journal of Engineering Design, 22:1, 1-29.

[3] Tan S K, Luh D B, Kung S F. A taxonomy of creative tourists in creative tourism[J]. Tourism Management, 2014, 42: 248-259.

[4] Lin C C, Luh D B. A vision-oriented approach for innovative product design[J]. Advanced engineering informatics, 2009, 23(2): 191-200.

[5] Luh D B, Lu C C. From cognitive style to creativity achievement: The mediating role of passion[J]. Psychology of Aesthetics, Creativity, and the Arts, 2012, 6(3): 282.

[6] Luh D B, Ko Y T, Ma C H. A structural matrix-based modelling for designing product variety[J]. Journal of Engineering Design, 2011, 22(1): 1-29.

[7] Luh D B, Li E C, Kao Y J. The development of a companionship scale for artificial pets[J]. Interacting with Computers, 2015, 27(2): 189-201.

[8] Ma C H, Ko Y T, Luh D B. A structure-based workflow planning method for new product development management[J]. International Journal of Management Science and Engineering Management, 2009, 4(2): 83-103.

[9] Luh D, Yang T. Museum recommendation system based on lifestyles[C]//2008 9th International Conference on Computer-Aided Industrial Design and Conceptual Design. IEEE, 2008: 884-889.

[10] Chang C L, Luh D B. User as designer: A design model of user creativity platforms[J]. Journal of Integrated Design and Process Science, 2012, 16(4): 19-30.

[11] Luh D B, Chang C L. Incorporating users' creativity in new product development via a user successive design strategy[J]. International Journal of Computer Applications in Technology, 2008, 32(4): 312-321.

[12] Luh D B, Ma C H, Hsieh M H, et al. Using the systematic empathic design method for customer-centered products development[J]. Journal of Integrated Design and Process Science, 2012, 16(4): 31-54.

[13] LeCun Y, Bengio Y, Hinton G. Deep learning[J]. nature, 2015, 521(7553): 436-444.

[14] LeCun Y, Bottou L, Bengio Y, et al. Gradient-based learning applied to document recognition[J]. Proceedings of the IEEE, 1998, 86(11): 2278-2324.

[15] Goodfellow J. Pouget-Abadie, M[J]. Mirza, B. Xu, D. Warde-Farley, S. Ozair, A. Courville, Y. Bengio, Generative Adversarial Networks, 2014.

[16] Ma X, Wang H, Wang J. Semisupervised classification for hyperspectral image based on multi-decision labeling and deep feature learning[J]. ISPRS Journal of Photogrammetry and Remote Sensing, 2016, 120: 99-107.

[17] Huang C Y, Luh D B, Ma C H, et al. The project management tool for integrating knowledge context and multidisciplinary responsibility[J]. African Journal of Business Management, 2012, 6(43): 10687-10696.

[18] Bo X Y, Luh D B. Study on the Design of Asymmetric Breast Correction Product[C]//International Conference on Applied Human Factors and Ergonomics. Springer, Cham, 2020: 175-180.

[19] Sun Y, Luh D B. Research on the Interactive System of Long-Distance Relationship for Elderly Users[C]//International Conference on Applied Human Factors and Ergonomics. Springer, Cham, 2020: 357-360.

[20] Chang C L, Luh D B. Design process and knowledge searching model based on user creativity[C]//International Conference of Design, User Experience, and Usability. Springer, Berlin, Heidelberg, 2013: 469-478.

Characterizatoin of Longitudinal Thermal Conductivity of Graphene Film

Jiajia Chen
SMIT Center, School of Mechatronics
Engineering and Automation
Shanghai University
Shanghai, China
guyuechuqi@shu.edu.cn

Xinjian Gong
SMIT Center, School of Mechatronics
Engineering and Automation
Shanghai University
Shanghai, China
gongxinjian@shu.edu.cn

Sihua Guo
SMIT Center, School of Mechatronics
Engineering and Automation
Shanghai University
Shanghai, China
shguo@shu.edu.cn

Yong Zhang
SMIT Center, School of Mechatronics
Engineering and Automation
Shanghai University
Shanghai, China
yongz@shu.edu.cn

Jin Chen
SHT Smart High Tech AB
Kemivägen 6, 412 58
Gothenburg, Sweden
jin.chen@sht-tek.com

Yuanyuan Wang
SHT Smart High Tech AB
Kemivägen 6, 412 58
Gothenburg, Sweden
yuanyuan.wang@sht-tek.com

Johan Liu
Department of Microtechnology
and Nanoscience
Chalmers University of Technology
Gothenburg, Sweden
jliu@chalmers.se

Corresponding author:
jliu@chalmers.se

Abstract—The chase of high performance by chip manufacturers has greatly increased the power consumption of integrated circuits, which brings great challenges to the heat dissipation of electronics systems. It has also slowed down following up of the Moore's Law, and it is expected to hit the wall soon [1]. Graphene film with high in-plane thermal conductivity is one of the key materials to make it possible for electronics industry to continue to follow the Moore's Law. However, there are few studies focusing on the longitudinal thermal conductivity of graphene films. The purpose of this study is to investigate the longitudinal thermal conductivity of graphene films according to ASTM D5470 [2]. The results show that the longitudinal thermal conductivity of the pressed graphene film is greater than that of the unpressurized graphene film. The longitudinal thermal conductivity is 10.6 W/m·K for the unpressurized graphene film and 20.6 W/m·K for the pressed graphene film.

Keywords—longitudinal thermal conductivity, graphene film, ASTM D5470

I. Introduction

One of the trends in the development of the microelectronics industry is the gradual miniaturization of devices, which means that the integration of transistors on-chip is getting higher and higher. If the heat can not be exported in time, too high temperature will reduce the stability of the chip, especially the excessive temperature difference between electronic module and external environment will produce thermal stress, which will directly affect the reliability of the electronic chip. Therefore, it is necessary to rely on excellent heat dissipation materials to quickly release the heat generated by the device.

Graphene is widely used in the field of chip heat dissipation due to its excellent thermal conductivity. As widely reported, graphene has a very high in-plane thermal conductivity. Based on the relationship between the temperature dependence of the G peak of single-layer graphene and the laser excitation frequency of Raman scattering, Balandin et al. calculated that the thermal conductivity of single-layer graphene in suspension state is up to 5300 W/m·K, which is much higher than that of graphite, carbon nanotubes and other carbon materials[3].

However, graphene film is a typical anisotropic film, and the in-plane thermal conductivity is much better than that in the longitudinal direction. The thermal conductivity along the longitudinal direction restricts the application of graphene film in heat dissipation materials to a certain degree. However, to the best of our knowledge, there are few studies focusing on the longitudinal thermal conductivity of graphene films. Therefore, the longitudinal thermal conductivity of the pressure and unpressed graphene films was measured according to the ASTM D5470 standard.

II. Experimental sections

In this experiment, the thermal conductivity test instrument of Xiangtan Xiangyi Instrument Co., Ltd., model: DRL-V, combined with ASTM D5470 standard, was used to measure the longitudinal thermal conductivity of two kinds of graphene film. One is pressed graphene film, the other is not pressed graphene film. These two kinds of graphene films are supplied by SHT Smart High-Tech AB, Sweden.

A. Experimental principle

ASTM D5470 is an experimental method for measuring thermal transmission properties of thermally conductive electrical insulation materials, which adopts a steady-state heat flow method. The schematic of ASTM D5470 tester is shown in Fig. 1.

Under the linear assumption, two temperature sensors installed on the meter bar measure the temperature gradient, $\Delta T/L$, in the meter bar. The heat flow is calculated according to formula 1.

$$Q = K \cdot A \cdot \frac{\Delta T}{L} \tag{1}$$

where K is the thermal conductivity of the meter bar, and A is the cross-sectional area of the meter bar. The temperature gradient is extrapolated to the interface surface of the meter bar to obtain the surface temperature T_C and T_H of the graphene film[4].

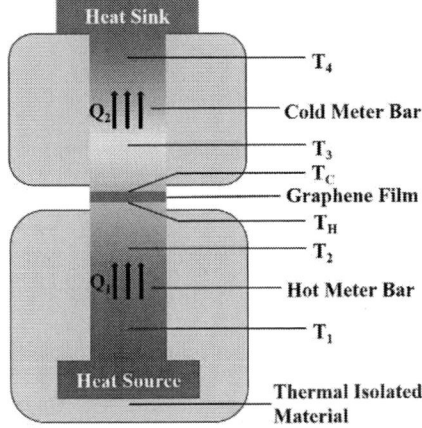

Fig. 1. Schematic of ASTM D5470 tester

Thermal impedance, R, is calculated by the heat flux, Q/A, across the sample and the temperature difference, ΔT, between the upper surface and the lower surface of the sample under the equilibrium state[2]. As shown in formula 2, the thermal impedance is equal to the temperature difference divided by the heat flux.

$$R = \frac{\Delta T}{Q/A} \tag{2}$$

The total thermal interface impedance, R, includes the thermal impedance of the graphene film itself and the contact thermal impedance at the interface between the graphene film and the two substrates, R1、R2, as shown in formula 3 [5].

$$R = R_1 + \frac{d}{\lambda} + R_2 \tag{3}$$

where λ is the thermal conductivity of the graphene film and d is the thickness of the graphene film. According to formula 3, the relationship between sample thickness and total thermal impedance is linear. Therefore, the total thermal impedance of three groups of different thicknesses is measured, and the linear fitting is carried out by using the least square method with the help of Matlab. The linear slope obtained is the reciprocal of the thermal conductivity λ.

B. Sample handling and instrument setting requirements

According to 5.4.2 of ASTM D5470, in order to effectively remove the air from the contact surface between the sample and the equipment and reduce the contact thermal impedance, the sample surface is coated with silicone grease. The thermally conductivity of the thermal conductive silicone grease used in this experiment is 8 W/m·K.

It is necessary to apply higher pressure to the sample, so as to further reduce the influence of contact thermal impedance on the experiment. Because the DRL-III thermal conductivity test instrument can not accurately control the pressure, so each experiment can only control the pressure at about 290 kPa.

According to 8.1.1.2 of ASTM D5470, the average temperature of the sample should be kept at 50 ℃. In addition, because DRL-III thermal conductivity test instrument needs to control the temperature difference between cold and hot electrodes to be 40 ℃. The hot electrode temperature is set to be 85 ℃, and the cold electrode temperature is set to be 15 ℃.

III. RESULTS AND DISCUSSIONS

A. Calibration

The accuracy of equipment measurement can be verified by measuring the thermal conductivity of standard material. Brass H62 standard material was processed into discs with diameters of a 30 mm and thickness of 0.785, 2.981 and 4.931 mm respectively. The measuring data is shown in Table I and Fig. 2.

Table I. Thermal impedance and thickness of Brass H62

Thickness (cm)	Thermal Impedance (K·cm²/W)
0.0785	0.34
0.2981	0.55
0.4931	0.69

The slope of the line fitted by the least square method in Fig. 2 is the reciprocal of the thermal conductivity of Brass H62. Intercept is the contact thermal impedance at the interface between the brass and the two meter bars. The thermal conductivity of H62 brass calculated by the least square method is 117.6 W/m·K. The published value was 116.7 W/m·K, and the variance was 0.77%.

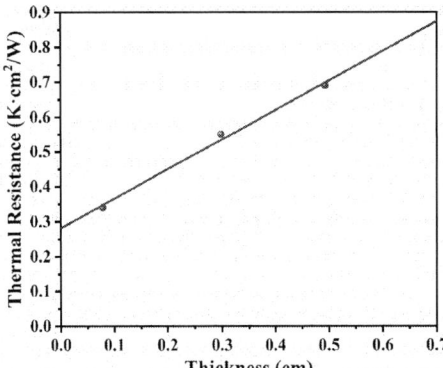

Fig. 2. Fitting diagram of thermal impedance and thickness for Brass H62

B. Thermal conductivity of unpressurized graphene films

Samples were divided into two groups, each group of samples from three different thicknesses of the film cut a 30 mm diameter circular film. The same method is used to measure the thermal conductivity of the unpressurized graphene film. The measuring data is shown in Table II and Fig. 3.

Table II. Thermal impedance and thickness of unpressurized graphene film

Group	Thickness (mm)	Thermal Impedance (K·mm^2/W)
1	0.175	53
	0.286	67
	0.410	75
2	0.175	48
	0.276	62
	0.402	70

The slopes of R and d curve in Fig. 3 are 93.2 and 95.4, corresponding to the thermal conductivity of graphene λ_1 = 10.7 W/m·K and λ_2 = 10.5 W/m·K respectively. The average of the two values, 10.6 W/m·K, is taken as the longitudinal thermal conductivity of the unpressurized graphene films.

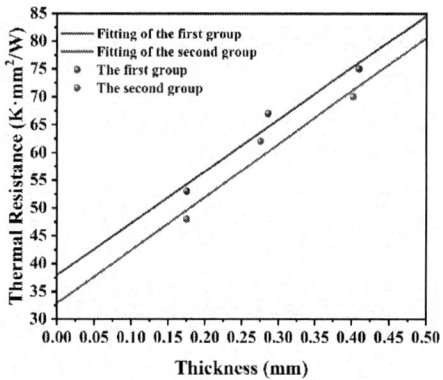

Fig. 3. Fitting diagram of thermal impedance and thickness for unpressurized graphene film

Fig. 3. also shows that the intercepts are 37.95 K·mm^2/W and 32.90 K·mm^2/W respectively, which are the total contact impedance at the interface between the graphene film and the two meter bars in two groups of experiments.

C. Thermal conductivity of pressed graphene films

Samples were also divided into two groups, each group of samples from three different thicknesses of the film cut a 30 mm diameter circular film. After measuring by the same method, the measuring data is shown in Table III and Fig. 4.

Table III. Thermal impedance and thickness of pressed graphene film

Group	Thickness (mm)	Thermal Impedance (K·mm^2/W)
1	0.098	48
	0.108	45
	0.166	50
2	0.089	45
	0.120	40
	0.155	48

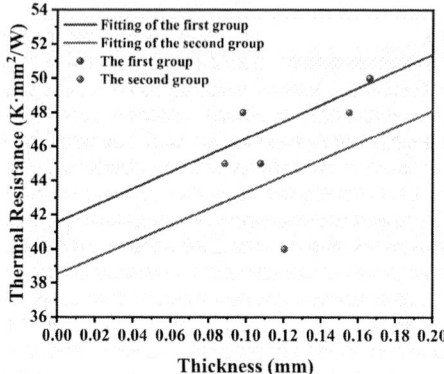

Fig. 4. Fitting diagram of thermal impedance and thickness for pressed graphene film

The slopes of R and d curve in Fig. 4 are 49.2 and 47.9, corresponding to the thermal conductivity of graphene λ_1 = 20.3 W/m·K and λ_2 = 20.9 W/m·K respectively. The longitudinal thermal conductivity of the film is 20.6 W/m·K. The contact impedance is 41.56 K·mm^2/W and 38.51 K·mm^2/W.

Theoretically, under the condition of constant thermal conductivity, the thicker the graphene film is, the higher the thermal impedance is. However, the thermal impedance of the two thin films decreases with the increase of thickness. Therefore, in addition to the above two groups of experiments, repeated experiments have been carried out many times. The result is still that the thermal impedance of the two thin films decreases with the increase of thickness. Therefore, the data is believed to be reliable.

The results show that the longitudinal thermal conductivity of the pressed graphene film is higher than that of the unpressurized graphene film. Because the unpressurized graphene film is looser than the unpressurized graphene film, the space between graphene layers is enlarged, and the internal thermal impedance is increased[6]. Therefore, the thermal conductivity of the pressed graphene film in the thickness direction is better than that of the unpressurized graphene film.

IV. CONCLUSION

According to ASTM D5470, the thermal conductivity along the thickness direction of graphene film was measured at 50 °C. The results show that the longitudinal thermal conductivity of the unpressurized graphene film is 10.6 W/m·K, and that of the pressed graphene film is 20.6 W/m·K. The result is that the pressed graphene film is more compact than the unpressurized graphene film, which reduces the internal thermal impedance and makes the thermal conductivity of the pressed graphene film better than the unpressurized graphene film in the thickness direction.

ACKNOWLEDGMENT

J. Chen, X. Gong, S. Guo, Y. Zhang, and J. Liu acknowledge the financial support from the National Natural Science Foundation of China (No:51872182). J. Chen, Y. Wang and J. Liu also acknowelge The Swedish Board for Innovation (Vinnova) Siografen program. Finally, J. Liu acknowledges the financial support from the Swedish National Science Foundation with the contract No: 621-2007-4660 as well as from the Production Area of Advance at Chalmers University of Technology, Sweden.

REFERENCES

[1] H. Ye. Research on the preparation of copper/carbon nanomaterials and their application in electronic packaging[D]. University of Chinese Academy of Sciences, 2020.

[2] Z. Liu. Practical test methods for thermal interface materials based on Steady-state theory[J]. Shanghai Measurement and Testing, 2019, 46(06):25-30.

[3] J. Wang. Research Based on Graphene preparation and its devices[D]. Beijing Jiaotong University, 2014.

[4] R.N. Jarrett, C.K. Merritt, J. P. Ross, Hisert J. Comparison of Test Methods for High Performance Thermal Interface Materials[C]. Proc of 23th IEEE SEMI-THERM Symposium. 2007:83-86.

[5] K. Hanson. ASTM D 5470 TIM material testing[C]. Semiconductor Thermal Measurement and Management Symposium, 2006 IEEE Twenty-Second Annual IEEE. IEEE, 2006.

[6] G. Shi. Study on Fabrication of In-situ CNT/Graphene Composite Film and Its Thermal Transport Properties[D]. National University of Defense Technology,2018.

Random Vibration Response Analysis of Sip Power Supply Module

1st Tao Lin
R&D Design Department
AVIC INTERNATIONAL HOLDING CORPORATION
Bei Jing, China
lint@avic-intl.cn

2st WeiYin Wang
R&D Design Department
SKY CHIP INTERCONNECTION TECHNOLOGY CO.,LTD
Shen Zhen, China
wangwg@scc.com.cn

3st Jun Li
Shennan circuits co.,LTD.
Shen Zhen, China
Lij@scc.com.cn

Abstract—**System-in-Package (SiP) aims to integrate a functional subsystem, one or more semiconductor chips and passive components into a package, so as to realize a basically complete function. The power module based on system-in-package (SiP) has many structural and performance advantages, such as miniaturization, high heat dissipation, low impedance and so on, which fits the development trend of electronic packaging towards miniaturization and high performance. In this paper, the random vibration simulation analysis of a power module is carried out by using finite element simulation software, and the influence of random vibration load onto the stress on the micro bumps on the adapter board in the power module is studied, and the potential risk points are determined before the product manufacture, thus improving the product reliability, reducing the development cost and shortening the development cycle.**

Keywords—system-in-package, random vibration, solder joint stress, power supply module

I. PREFACE

Electronic products are not only affected by power load and ambient temperature, but also by random vibration during service. These two factors have always been considered as the main factors affecting the service life of electronic products [1-2]. Electronic products in automotive electronics, aerospace and other fields can not be separated from a stable power supply system, while electronic devices such as automotive electronics and aerospace often work in vibration and impact load environment, so the power module is often affected by vibration and impact, resulting in alternating stress and strain in the power package module, which eventually leads to fatigue failure of internal devices in the power package.

In this paper, the finite element analysis software is used to analyze the influence of random vibration load on the solder ball stress on the adapter board in the package, and the distribution law of solder ball stress on the adapter board is analyzed, so as to provide design reference for improving the service life of the power module.

II. FINITE ELEMENT MODELING OF POWER MODULE

A. Power Module Packaging Structure Design

The power supply module is a four-channel DC/DC step-down μ module (micro-module) regulator, each output is 4A, and its interior includes four switch controllers, four large inductors and discrete resistive capacitors. The packaged devices are assembled together, and the overall package size is 15x9x5mm in length, width and height. Fig. 1 is a schematic assembly cross-section. The upper plate contains two QFN packaged switch controllers and some resistive capacitors. The main board contains four large inductors, two other switch controllers of the same type and the remaining resistance vessels. The signals of the upper board and the main board are communicated with each other through an intermediate adapter board containing solder balls. Figure 2 is a schematic diagram of the assembled three-dimensional structure.

Fig.1 Schematic diagram of power module assembly section

Fig. 2 Schematic diagram of three-dimensional structure of power module assembly

B. Establishment of Geometric Model of Power Module

The model mainly includes a 50x50x1.6mm PCB board, and the power module is placed centrally on the PCB, with the size of 15 x 9 x 5 mm; The upper plate size is 4.54x14.52x0.8mm；The QFN package device size is 4x4x0.8mm； Inductor size is 3.2x3.5x3mm；The size of the adapter plate is 1.6x3.6x0.8mm, and 3x8=24 solder balls with the size of 0.4mm are planted on both sides of the adapter plate. The thickness of plastic package is 4mm, and Figure 3 is a schematic diagram of geometric model of each part.

Fig.3(a) Schematic diagram of power module motherboard assembly

Fig. 3(b) Schematic diagram of upper plate assembly of power module

Fig. 3(c) Schematic diagram of power module adapter board structure

Fig. 3 Geometric model of each part of power module

When carrying out random vibration simulation analysis, it is necessary to define the material properties of each part of the material. The material properties of each part of the model are shown in Table 1[3-5].

TABLE I. THE MATERIAL PROPERTIES OF EACH PART OF THE MODEL

Materials	Elastic modulus E/Gpa	Poisson's ratio v	Density /kg/m3
Substrate	24.13	0.3	1200
resistance-capacitance	340	0.22	3600
PCB	22	0.2	2680
Plastic compound	17	0.2	1890
Inductance	220	0.295	6850
SAC305	43.7	0.36	7300

III. MODAL ANALYSIS

Modal analysis is the basis of vibration analysis. Through modal analysis, the natural frequencies and vibration modes of the whole PCB under given constraints can be obtained. By analyzing the vibration modes of each order, the most vulnerable parts of the power module under resonance frequency can be obtained. The boundary conditions of modal analysis impose fixed constraints on the mounting holes at four corners of PCB according to the actual mounting mode of PCB. For general structural vibration, the low-order vibration mode of the model is often greater than the influence of high-order vibration mode on the structure, so the dynamic characteristics of the structure are mainly determined by the low-order vibration mode [6]. The 6th-order modes are extracted as shown in Table 2.

TABLE II. THE 6TH-ORDER MODES

Modal	1	2	3	4	5	6
Natural frequency/Hz	686.25	1377.4	1382.7	1792.8	3440.5	3485.3

IV. RANDOM VIBRATION RESPONSE ANALYSIS OF POWER MODULE

A. Loading Random Vibration Load and Applying Boundary Conditions

On the basis of the above modal analysis, the finite element software is used to load random vibration load on the power module. According to the requirements of vibration test standard of related equipment environmental test methods [7], the PSD analysis function of the software is used to input the acceleration power spectrum shown in the following figure 4. The lowest frequency of random vibration is 20Hz, the highest frequency is 2000Hz, and the gravity acceleration is 9.8m/s². Excitation loads are applied to the fixed edges of the four diagonal corners of the PCB.

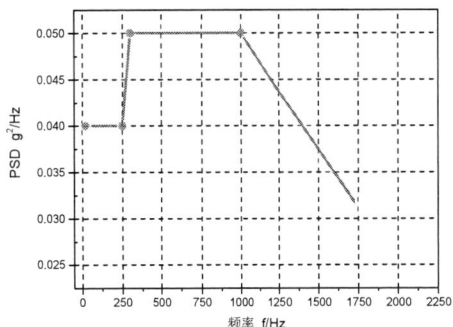

Fig.4 Acceleration power spectrum

B. Influence of Random Vibration on Stress of Solder Joint

Figure5 shows the whole stress distribution nephogram of the power module 1σ under random vibration load. It can be seen from the figure that the high stress area of the whole module is mainly distributed in the main substrate area of the power module, and the stress of the two switch controllers on the upper board is obviously less than that of the two switch controllers on the main board. This is mainly because the main board is combined with the PCB board, and the random vibration load is most easily transmitted to the main substrate of the power supply module through the PCB board. Therefore, the stress distribution of the whole power supply module has a tendency that the stress on the upper device is small and the stress on the lower device is large.

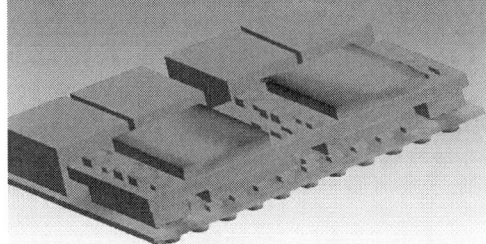

Fig.5 1σ overall stress distribution nephogram of power module

It can be seen from figure 6 that the maximum stress point of the power module appears at the corner of BGA solder joint on the main substrate, and the 11σ stress nephogram of BGA solder joint under the whole main substrate is extracted and analyzed. It can be seen that the stress of the solder joint on the main substrate is mainly concentrated on the solder ball on the outer ring, and the stress value of the solder ball in the inner middle area is small, and its maximum 11σ stress is 34.7MPa. From the nephogram distribution of the maximum stress solder joint, the solder joint failure generally occurs in the area with the maximum stress, so it can be determined that this solder joint is the key solder joint.

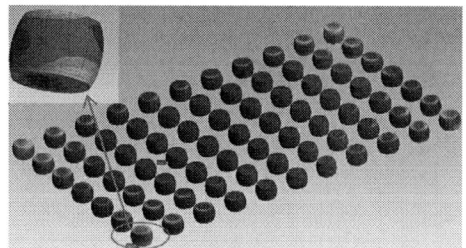

Fig.6 11σ stress nephogram of BGA solder joint of main substrate

C. Stress Analysis of Solder Joint of Adapter Plate

Due to the small size of the adapter plate, the diameters of the solder joints on the upper and lower sides are relatively small, so it is also one of the positions where the whole power module is relatively prone to failure when subjected to vibration load. From the overall nephogram distribution, the stress range of solder joints on the adapter plate is symmetrically distributed, and the stress on the inner solder joints is larger than that on the outer side, and the stress on the upper solder joints on the adapter plate is smaller than that on the lower side, which is consistent with the trend analyzed above. It can be seen from figure 7 that the maximum stress of welding spot is 15.8MPa.

Fig.7 11σ stress nephogram of BGA solder joint of adapter board

V. ASSEMBLABILITY OF POWER MODULE

Figure 8 shows the schematic diagram of the assembly process of the power module, and Figure 9 is a physical picture of the packaged finished product. In the actual production process, the power module also has good manufacturability and excellent electrical performance.

Fig.8 Schematic diagram of power module assembly process

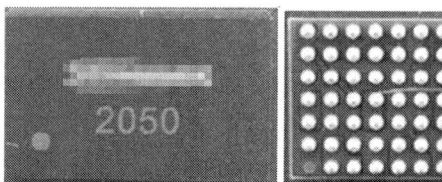

Fig.9 Package finished product of power module

CONCLUSION

In this paper, the solder joint stress of power supply module under random vibration load is simulated and analyzed. The results show that the maximum stress of solder joint appears at the corner of BGA solder joint on the main substrate, and the maximum stress is 34.7MPa ; The maximum stress of the solder joint in the weak area of the adapter plate is 15.8MPa, and the stress of the power module is within a reasonable range under a given random load.

ACKNOWLEDGEMENT

This work was supported by the National Key R&D Program of China (2018YFE0204600), the Key-Area Research and Development Program of GuangDong Province (2019B010131001), and Shenzhen Fundamental Research Program）(JCYJ20200109140822796).

REFERENCES

[1] Zhao zhi bin, Wu zhaohua, FEM Analysis of Effect of LGA Solder Joint Shape on Lifetime of Solder Joints[J] Electronics Process Technology, 2013(06): 320-322.

[2] J. Clerk Maxwell, Vibration Fatigue Life Prediction Model for Flip Chip Solder Joint[J],JOURNAL OF SHANGHAI JIAOTONG UNIVERSITY, 2001,35(12):1855-1857.

[3] Amagai M,Watanabe M,Omiya M , et al. Mechanical characterization of Sn-Ag-based lead-free solders[J]. Microelectronics Reliability, 2002, 42(6):951-966.

[4] Motalab M,Cai Z,Suhling J C,et al. Determination of Anand constants for SAC solders using stress-strain or creep data[C]// IEEE Intersociety Conference on Thermal & Thermomechanical Phenomena in Electronic Systems. IEEE, 2012.

[5] Chaillot A,Munier C , Maire O , et al. Thermomechanical simulation of a SiP (System-in-a-Package) LGA (Land Grid Array): Impact of the internal IC (Integrated Circuits) on the 2nd level solder joint reliability[C]// 2010 11th International Thermal, Mechanical & Multi-Physics Simulation, and Experiments in Microelectronics and Microsystems (EuroSimE). IEEE, 2010.

[6] Yan haidong, Research on solder joint reliability of flexible substrate laminated CSP device[D], Guilin University Of Electronic Technology.

[7] MIL-STDNAVMATP-9492[S].

The influence mechanism of process defects of plane welding on signal transmission in microwave components

Ruining Li
Key Laboratory of Electronic Equipment Structure Design,Ministry of Education
Xidian University
Xi'an, China
lirning@163.com

Jun Tian*
Key Laboratory of Electronic Equipment Structure Design,Ministry of Education
Xidian University
Xi'an, China
congsiwang@163.com

Song Xue
Key Laboratory of Electronic Equipment Structure Design,Ministry of Education
Xidian University
Xi'an, China
sxue@xidian.edu.cn

Congsi Wang*
Key Laboratory of Electronic Equipment Structure Design,Ministry of Education
Xidian University
Xi'an, China
congsiwang@163.com

Yan Wang
School of Information and Control Engineering
Xi'an University of Architecture and Technology
Xi'an, China
wangyan5169@163.com

Cheng Zhou
Xi'an Institute of Space Radio Technology
Xi'an, China
zhoucheng11n@sina.com

Jing Liu
Xi'an Institute of Space Radio Technology
Xi'an, China
liuj-may@hotmail.com

Zhihai Wang
CETC No.38 Research Institute
Hefei, China
ericwang@ustc.edu.cn

Kunpeng Yu
CETC No.38 Research Institute
Hefei, China
yukunp@hotmail.com

Haitao Shi
CETC No.38 Research Institute
Hefei, China
htshi2018@163.com

Daoheng Sun
School of Aerospace Engineering
Xiamen University
Amoy, China
sundh@xmu.edu.cn

Abstract—The grounding technology of plane welding is widely used in microwave components, and the grounding defects have a direct influence on the transmission performance of the circuit. In this paper, the three-dimensional electromagnetic simulation software HFSS is used for modeling and simulation, and the shape and performance correlation analysis of plane welding cavity is carried out by adapting orthogonal experiment. The influence mechanism of plane welding process defects on microwave signal transmission performance is explored. The influence mechanism of cavity caused by plane welding on the microwave RF performance is given qualitatively and quantitatively. The results shows that the plane welding cavity at the feed port and under the microstrip line should be avoided as far as possible. In the transmission performance of microwave signal, the change of cavity along Y direction has a greater impact on the insertion loss, and the change of cavity along X direction has a greater impact on the voltage standing wave ratio. The transmission performance of microwave signal will improve with the increase of penetration rate.

Keywords—Active phased array antenna; Microwave components; Microstrip line; Plane welding cavity; Penetration rate

I. INTRODUCTION

With the development of electronic products towards miniaturization and multi-function, the demand for interconnection between heterogeneous materials becomes more urgent, especially in micro system products[1]. As one of the typical connection methods in the microwave module commonly used in electronic information system, plane welding is mainly used to realize the grounding at the center of the substrate chip to ensure the transmission of electrical signals. In the reflow soldering process, the outgassing substance in the solder that can not be discharged, the soldering flux that is remained, and the oxidation of the interface will lead to the appearance of plane welding cavity, which will create impedance. Therefore, the heat dissipation of the microwave module will be affected, and the reliability of the device grounding will be reduced, which will further interfere with the heat conduction and conductivity of the device[2]. For high-performance microwave components, the existence of planar solder cavity has become an important factor affecting the long-term stability and high efficiency of the device.

In this paper, the typical plane welding in the interconnection process is taken as the research object. The electromagnetic simulation model of planar welding cavity is established by using HFSS simulation software. Different parameters are considered such as cavity size, cavity location, cavity number and penetration rate, orthogonal experiment is introduced to analyze the shape-performance correlation of each parameter, and the influence mechanism of plane welding process on microwave module is explored.

II. MODELING OF PANAR WELDING

A. Structure form and material property

Taking the microwave printed circuit board as an example, it consists of five parts: microstrip line, substrate, clad copper, plane welding layer and grounding shell[3], as shown in Fig. 1, Fig. 2 and Fig. 3.

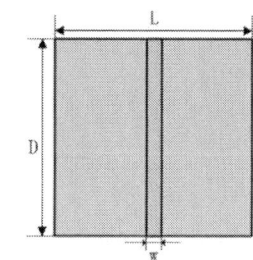

Fig. 1.   Top view of plane welding

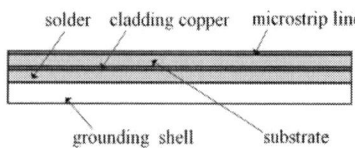

Fig. 2.   Front view of plane welding

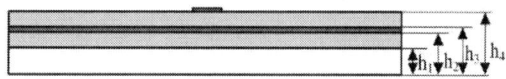

Fig. 3.   Left view of plane welding

The structural size and material properties of the selected components are shown in Table I.

TABLE I.   STRUCTURE SIZE AND MATERIAL PROPERTIES

Modules	Geometric size (Length×Width×Height) /mm³	Material
Microstrip line	0.6×20×0.003	Au
Substrates	20×20×0.635	ArlonCLTE-XT™
Clap copper	20×20×0.01	Cu
Plane welding layer	20×20×0.1	$Sn_{63}Pb_{37}$
Grounding shell	20×20×0.5	Invar alloys

In different operating frequency bands, the microwave module will be affected by the plane welding cavity[4]. In this paper, a low frequency S-band, a middle frequency X-band and a high frequency Ka band are selected to analyze the influence mechanism of planer welding cavity under different frequencies.

B. Electromagnetic modeling of planar welding process

The electromagnetic simulation software HFSS is used to do the analysis. Firstly, the solution type is set to mode driven, and the reference field is created. Then the planer welding structure model is created according to the dimension parameters and material properties designed before. Also, the vacuum cavity and wave port are established. Finally, the wave ports at both ends of the microstrip line are set as radiation boundary conditions and the excitation mode is set to wave port excitation. The electromagnetic model is shown in Figure 4.

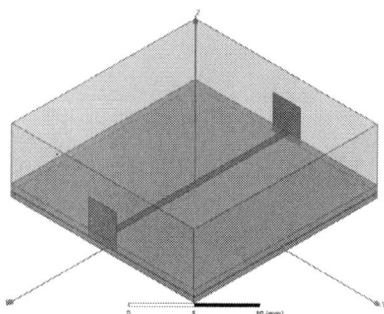

Fig. 4.   Electromagnetic model of plane welding

III. PHYSICAL VALIDATION OF THE ELECTROMAGNETIC MODEL

In the process of plane welding, it is often difficult to accurately control the cavity characteristics of plane welding, resulting in large errors. Therefore, in the test of the physical sample, the penetration rate is selected as the variable, which is easier to control. By changing the area of the plane welding layer, the physical samples with different penetration rates are made for testing and verification. By comparing the experimental data with the data from HFSS software simulation, the error is analyzed and the model is modified.

As one of the important technical indexes in plane welding process, the penetration rate directly reflects the efficiency of signal transmission. In this paper, the method of changing the area of plane welding layer is used to meet the data requirements of different penetration rates, as shown in Fig. 5.

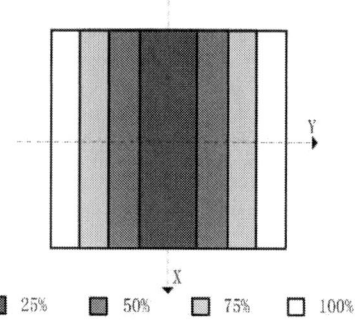

Fig. 5.   Schematic diagram of welding layer area under different penetration rates

The effects of penetration rate on VSWR and insertion loss of low frequency S-band and high frequency Ka band are calculated under four conditions of 25%, 50%, 75% and 100%. Electromagnetic modeling is carried out under different penetration conditions, and the simulation data is imported into Matlab for data post-processing. The results are shown in Fig. 6 and Fig. 7.

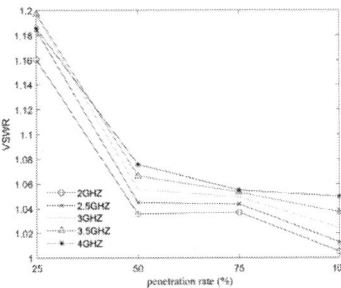

(a) Voltage standing wave ratio

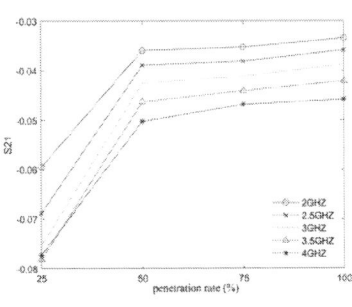

(b) Insert loss

Fig. 6. Transmission performance of S-band microwave circuits under different penetration rate

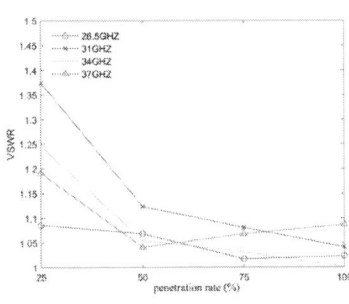

(a) Voltage standing wave ratio

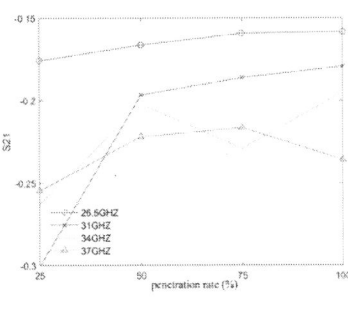

(b) Insert loss

Fig. 7. Transmission performance of Ka-band microwave circuits under different penetration rate

It can be seen from Fig. 6 and Fig. 7 that the VSWR and insertion loss will increase when the penetration rate decreases in S-band and Ka-band. So the increasing penetration rate will improve the transmission of microwave circuits. Compared with the insertion loss, VSWR is more sensitive to the penetration rate, and the influence of penetration rate on insertion loss can be ignored. Therefore, in the planar welding process, it is an important means to improve the transmission performance of microwave circuit by adopting appropriate and reasonable planar welding method to keep the penetration rate as high as possible.

TABLE II. ERROR TABLE OF SIMULATED VALUE AND MEASURED VALUE

VSWR	Penetration rate 40%	Penetration rate 60%	Penetration rate 80%	Penetration rate 100%
S-band	13%	9%	8%	2%
Ka-band	18%	16%	11%	7%
Insertion loss	Penetration rate 40%	Penetration rate 60%	Penetration rate 80%	Penetration rate 100%
S-band	19%	16%	8%	8%
Ka-band	16%	14%	13%	6%

It can be seen from table II that although there is some error between the simulated values and the measured values, they are within a reasonable range of 20%, so they have a strong consistency. In the process of actual experiment, the form of solder cannot be controlled as accurate as that in simulation analysis, so it is reasonable that there is a certain error in the actual penetration rate.

IV. INFLUENCE MECHANISM OF PLANE WELDING CAVITY ON SIGNAL TRANSMISSION

In the process of planar welding of microwave module, the distribution and shape of planar welding cavity are random, and the influence on signal transmission performance is different when they appear in different parts of the module. In this paper, the common cylindrical cavity is selected as the analysis object. The influence of the size, position and number of the cavity on the transmission performance of the plane welding in S-band is analyzed by HFSS. However, due to the problems of single, time-consuming, difficult to evaluate the correlation strength, difficult to determine the key parameters, and difficult to measure the correlation interval, the orthogonal experimental design is introduced for data analysis and processing.

A. Process defect analysis and modeling

The 1/4 model, located in the lower left corner, is selected for analysis, as shown in Fig. 8.

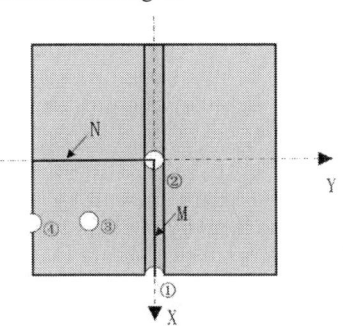

Fig. 8. Schematic diagram of the location of the cavity

Four cavity position which is being concerned in engineering are marked. Position ① is at the feed port of the microstrip substrate. Position ② is in the center below the microstrip line. Position ③ is inside the planer welding layer. Position ④ is at the edge of the welding layer. The dotted line M is the vertical line connecting the front and rear feed ports in the plane welding layer, and the dotted line N is the horizontal line passing through the center origin inside the welding layer. The electromagnetic simulation model is established according to the position of the cavities, as shown in Fig. 9.

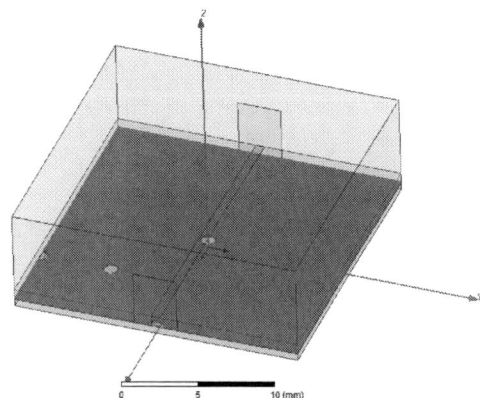

Fig. 9. Simulation model of plane welding cavity

In order to study the influence of cavity size, location and number on the transmission performance, five parameters affecting the signal transmission performance of planar welding cavity are selected for simulation analysis. The five parameters are cavity diameter, cavity position in X direction, cavity position in Y direction, cavity numbers in X direction and cavity numbers in Y direction.

TABLE III.   CONTROL TABLE OF CAVITY PARAMETERS

Order Number	Parameters	Variable name	Initial value	Range of variation
1	Cavity diameter	D	0	[0.5,2.5]
2	Cavity position in X direction	M	0	[0,10]
3	Cavity position in Y direction	N	0	[-10,0]
4	Cavity numbers in X direction	NX	1	[0,5]
5	Cavity numbers in Y direction	NY	1	[0,5]

B. Experimental design and extremum difference analysis

In this paper, orthogonal experimental design and range analysis are used to study the influence mechanism of cavity parameters on signal transmission performance. Three parameters of cavity diameter D, cavity X direction position and cavity Y direction position are selected to carry out orthogonal experiment. According to the principle of orthogonal experiment design, orthogonal table is used to carry out orthogonal experiment with three-factors and five-levels, as shown in table IV.

TABLE IV.   FACTORS AND LEVELS TABLE

Factors \ Levels	A D/mm	B M/mm	C N/mm
1	0.5	0	0
2	1	2.5	-2.5
3	1.5	5	-5
4	2	7.5	-7.5
5	2.5	10	-10

According to the factor-level table in table IV, the test data of 25 different parameter level combinations are generated, and according to these 25 kinds of cavity parameters, the simulation analysis is carried out in the 3D electromagnetic model of plane welding with cavity defects.

The results of orthogonal test were analyzed by extremum difference analysis, and the trend of extremum difference analysis results can be obtained, as shown in Fig. 10.

It can be known from Fig. 10 that when the cavity diameter d = 0.5mm, the insertion loss of the microstrip transmission line is the smallest, and the transmission performance is the best. The signal transmission performance becomes worse with the increase of the cavity diameter. When the cavity position changes along the X-axis, the insertion loss is the smallest when it is located in the middle of the microstrip transmission line. On the whole, as the cavity approaches the excitation port, the insertion loss increases and the transmission performance becomes worse. When the cavity position changes along the Y-axis, the insertion loss is the largest when the cavity is in position ②, and the influence of the cavity on the signal transmission gradually decrease with the cavity away from the microstrip line.

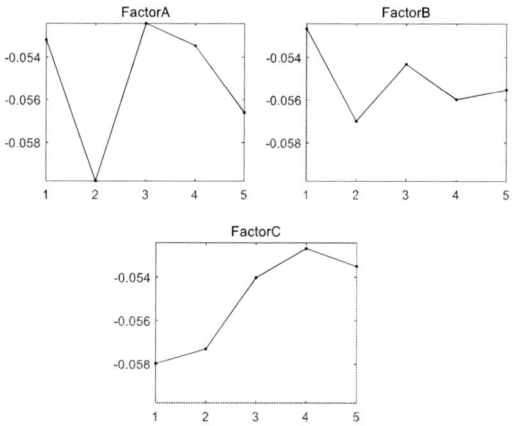

Fig. 10. Insertion loss trend chart in S-band

It can be known from Fig. 11 that when the cavity diameter d = 0.5mm, the VSWR of the microstrip transmission line is the smallest, and the transmission performance is the best. The signal transmission performance becomes worse with the increase of the cavity diameter. When the cavity position changes along the X direction, the insertion loss is the smallest when it is located in the middle of the microstrip transmission line. On the whole, as the cavity approaches the excitation port, the VSWR increases and the transmission performance becomes worse. When the cavity position changes along the Y direction, the VSWR is the largest when the cavity is in position ②, and the influence of the cavity on the signal transmission gradually decrease with the cavity away from the microstrip line.

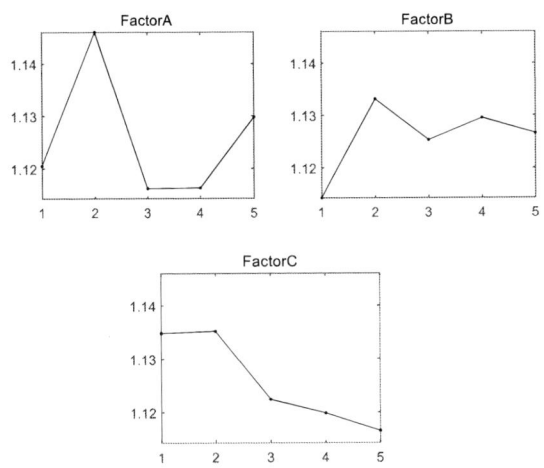

Fig. 11. Voltage standing wave ratio trend chart

As a method of analyzing experimental data, extremum difference analysis can be used to determine whether the influence of factors on experimental results is significant. Therefore, extremum difference analysis is used to determine the influence of cavity diameter D, cavity position in X-axis and cavity position in Y-axis on microwave signal transmission performance, as shown in table V.

TABLE V.   EXTREMUM DIFFERENCE ANALYSIS TABLE OF THE INFLUENCE OF CAVITY PARAMETERS ON ELECTRICAL PROPERTIES

Cavity Parameters		D/mm	M/mm	N/mm
extreme difference values	Insertion loss	0.0074	0.0043	0.0053
	VSWR	0.0298	0.0188	0.0187

It can be obtained from table V that the order of the influence of cavity parameters on insertion loss is D > N > M, which means that the cavity size is the most important parameter, and the influence of cavity position along Y direction on insertion loss is greater than that along X direction. The order of the influence of cavity parameters on VSWR is D > M > N, the cavity size has the greatest influence on the transmission performance, but the influence of cavity position along Y direction on VSWR is less than that along X direction.

Multiple binomial regression method is used to fit the effect of cavity position on the signal transmission performance of planar welding

$$y = \beta_0 + \beta_1 x_1 + \cdots + \beta_m x_m + \sum_{j=1}^{m} \beta_{jj} x_j^2 \qquad (1)$$

The prediction formula of insertion loss and VSWR of planar welding affected by cavity position are as follows:

$$\begin{cases} S_{21} = -0.0574 + 0.0012D - 7.8914\times10^{-4}M - 0.0011N - 4.2857 \\ \qquad \times10^{-4}D^2 + 6.0114\times10^{-5}M^2 - 5.6457\times10^{-5}N^2 \\ VSWR = 1.1342 - 0.0073D + 0.0044M + 0.0024N + 0.0017D^2 \\ \qquad - 3.5954\times10^{-4}M^2 + 3.3143\times10^{-5}N^2 \end{cases} \qquad (2)$$

C. The influence of the cavity numbers

In order to study the influence of the cavity numbers on the transmission performance, starting from position ②, add one cavity each time along the horizontal and vertical directions. A cylindrical cavity model with a diameter of 1 mm and a height of 0.1 mm is established for electromagnetic simulation analysis.

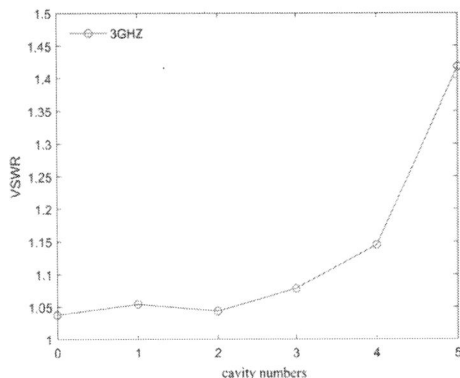

Fig. 12. The change of Voltage standing wave ratio with the increase of cavity number in Y direction

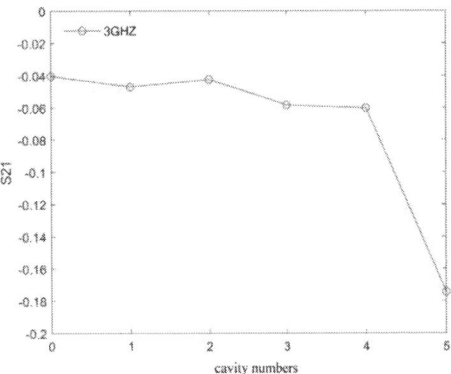

Fig. 13. The change of insertion loss with the increase of cavity number in Y direction

It can be seen from Fig. 12 and Fig. 13 that in the S-band, when the cavity number increases along the Y direction, the horizontal direction, the VSWR and insertion loss values will increase, and the influence on microwave signal transmission will increase. In order to predict the influence of the number of cavities on insertion loss and VSWR when it changes in horizontal direction, the simulation data are fitted through curve fitting:

$$\begin{aligned} Y_{S_{21}} &= -0.0053x^2 + 0.0230x - 0.0629 \\ Y_{VSWR} &= 0.0170x^2 - 0.0678x + 1.0986 \end{aligned} \qquad (3)$$

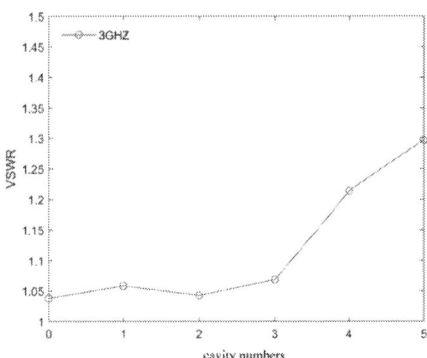

Fig. 14. The change of Voltage standing wave ratio with the increase of cavity number in X direction

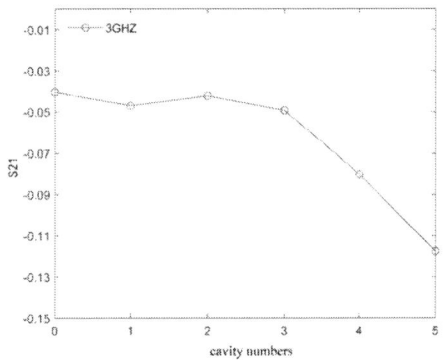

Fig. 15. The change of insertion loss with the increase of cavity number in X direction

It can be seen from Fig. 12 and Fig. 13 that in the S-band, when the cavity number increases along the X direction, the vertical direction, the VSWR and insertion loss values will increase. Similarly, the simulation data are fitted through curve fitting:

$$Y_{S_{21}} = -0.0101x^2 + 0.0499x - 0.0920$$

$$Y_{VSWR} = 0.0284x^2 - 0.1357x + 1.1741$$

(4)

By observing the change of VSWR and insertion loss when the number of cavities changes along Y direction and X direction, it can be known that VSWR fluctuates from 1.05 to1.45, and insertion loss fluctuates from -0.18dB to -0.04dB with the cavity numbers changing along Y direction. And VSWR fluctuates from 1.05 to 1.3, and insertion loss fluctuates from -0.12dB to -0.04dB with the cavity numbers changing along X direction. The fluctuation range along the Y direction is larger than that along the X direction. It is concluded that the change of the number of cavities in the horizontal direction has more influence on the transmission of microwave signal than that along the vertical direction.

V. CONCLUSION

Through the modeling simulation and experimental analysis of plane welding with process defects, the influence mechanism of the defects on signal transmission performance is studied. The following conclusions can be drawn:

1) The transmission performance of microwave signal will improve with the increase of penetration rate.

2) The transmission performance of microwave signal will degrade with the increase of the diameter of the cavity.

3) The cavity has a greater influence on the transmission performance when it is closer to the feeding port. When the cavity position changes along the direction of the microstrip line, the voltage standing wave ratio is more sensitive, and the insertion loss is more sensitive when the hole position changes along the horizontal direction

4) The amount of the cavity in signal transmission direction has a greater impact on the transmission performance compared to amount of horizontal direction.

The active devices used in practical engineering are composed of multiple devices cascaded together. The influence will be amplified hierarchically when the effect of planar welding is not ideal, resulting in self excitation phenomenon. And the influence of grounding defect on signal transmission will be more significant when the module works in higher frequency band. Therefore, in the process of microwave module plane welding connection, the cavity should be avoided under the feed port and the microstrip line. Also cavity area should be reduced as much as possible, and the penetration rate should be improved to ensure the microwave signal transmission performance, making the components grounded so that the entire system has better reliability and consistency.

ACKNOWLEDGMENT

This work was supported by National Natural Science Foundation of China under No. 51975447 and U1737211, Natural Science Foundation of Shaanxi Province under No. 2018JZ5001, Youth Innovation Team of Shaanxi Universities under No. 201926.

REFERENCES

[1] J. Li Y, Srinath P.K.M, and Goyal D, "A review of failure analysis methods for advanced 3D microelectronic packages", Journal of Electronic Materials, vol. 45. 2016, pp. 116–124.

[2] J. Hassan A, and Savaria Y, "Electronics and packaging intended for emerging harsh environment applications: A review", IEEE transactions on very large scale integration (VLSI) systems, vol. 26. 2018, pp. 2085–2098.

[3] J. Tian Jun, "Electromechanical coupling parameter identification for flexible conductor wire interconnection considering interaction effect in microwave circuits", Electronics.,vol. 10. 2021, pp. 464–464.

[4] C. Chen H, Liang J, and Gao R, "Laser soldering of Sn-based solders with different melting points", 2020 21st International Conference on Electronic Packaging Technology (ICEPT), 2020, pp. 1-5.

[5] C. Han F, Liu S, and Liu S, "Defect detection: Defect classification and localization for additive manufacturing using deep learning method", 2020 21st International Conference on Electronic Packaging Technology (ICEPT), 2020, pp. 1-4.

[6] J. Wang Z, Gao J, and Flowers G.T, "The impact of connection failure of bonding wire on signal transmission in radio frequency circuits", IEEE Transactions on Components, Packaging and Manufacturing Technology, vol. 10. 2020, pp. 1729–1737.

[7] J. Lu T, and Jin J.M, "Coupled Electrical-thermal-mechanical simulation for the reliability analysis of large-scale 3-D interconnects", IEEE Transactions on Components, Packaging and Manufacturing Technology, vol. 7. 2017, pp. 229–237.

Epoxy Flux Prevent Hot Tear at VIPPO Solder Joints

Elaina Zito
Indium Corporation
Clinton, NY, USA
ezito@indium.com

Dave Bedner
Indium Corporation
Clinton, NY, USA
dbedner@indium.com

Ning-Cheng Lee
Dr. Ning-Cheng Lee, Consultant
New Hartford, NY, USA
Nclee715@gmail.com

Abstract--VIPPO design enabled a better signal quality and speed, but it also suffered side-effect caused by this design. At VIPPO location with or without a solid-Cu-hole-fill, the coefficient of thermal expansion (CTE) typically around 17ppm/°C which is quite lower than that of non-VIPPO pads nearby, which is around 45ppm/°C. For double-sided PCB, at ramp-up stage of the second reflow, the difference in CTE of neighboring pads would result in excessive tension on the joint on VIPPO pads, and consequently resulted in Hot-Tear (HT) before reaching melting point of solder joints. Epoxy Flux (EF) was developed to eliminate HT problem. EF dip process did not work due to insufficient EF volume to bond BGA strongly to PCB to nullify the ΔCTE factor. EF dispense process wet well, exhibited low voiding, and allowed sufficient EF volume to fill the gap between BGA and PCB, thus enabled a strong bonding force to nullify the ΔCTE factor, and consequently eliminated the HT defect. The EF developed had the epoxy stress balanced around the solder joints and avoided joint-drifting and solder-extrusion problem. EF dispense process was recommended to be used without solder paste when the BGA was to be assembled. It cured fully in the reflow process, and was non-tacky on touch, and also met IPC SIR requirement. Commercial underfills tested did not work on preventing HT from happening, due to the joint-drifting and solder-extrusion problem. Edge bond or corner bond was also expected to be unable to prevent HT defect. Compared with other polymeric reinforcement materials, EF dispense process minimized the process steps hence was a lower cost valid solution than other reinforcement approaches.
Keywords
Epoxy flux, hot tear, VIPPO, BGA, soldering

I. INTRODUCTION

For telecommunication industry, improving the signal quality and speed is a never-ending quest. Recently a board design incorporating via-in-pad-plated-over (VIPPO) for BGA has been adopted to achieve the goal. This VIPPO design shortened the signal-path-length of conventional dog-bone structure, resulted in a reduced capacitance and inductance between surface signal trace line and board inner-layer, as shown in Figure 1 [1].

Back drill of VIPPO structure to remove the "un-used" portion of the via eliminated the signal reflections thus further assured to achieve a cleaner signal. With elimination of dog-bone structure, the routing between pads becomes easier and the circuitry density can be increased accordingly.

While VIPPO design enabled a better signal quality and speed, it also suffered side-effect caused by this design. At VIPPO location with or without a solid-Cu-hole-fill, the coefficient of thermal expansion (CTE) typically around 17ppm/°C which is quite lower

than that of the non-VIPPO pads nearby, which is around 45ppm/°C.

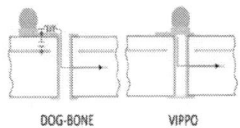

Figure 1 VIPPO design exhibits reduced capacitance and inductance when compared with conventional dog-bone design.

For double-sided PCB, at ramp-up stage of the second reflow, the difference in CTE of neighboring pads would result in excessive tension on the joint on VIPPO pads, and consequently resulted in hot-tear before reaching melting point of solder joints, as shown in Figure 2 [2].

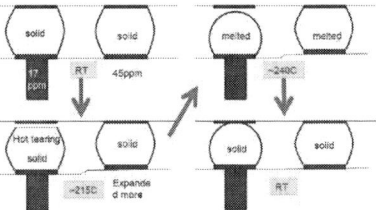

Figure 2 Mechanism of forming dome-shaped joint at VIPPO site

This hot-tear symptom may or may not be discernible after reflow by X-ray, sono-scan, or electrical-continuity testing, thus becomes a major liability problem. Several approaches have been adopted to address this hot-tear problem, including underfilling or edge-bonding after first reflow, with very limited success or a quite higher manufacturing cost due to additional process steps.

In this study, a more effective solution "Epoxy Flux" has been developed to prevent this hot tear issue. Upon SMT assembly of the first side PCB, solder paste was first printed onto PCB pads except for the BGA pads. Epoxy flux gel was then dispensed onto the BGA footprint area, followed by BGA placement and other SMT component placement, and reflow. After the second reflow, the BGA joint formed was examined for Hot-Tear, and was compared with BGA with bumps dipped in epoxy flux prior to BGA placement. BGA assembled with regular solder paste process was also used as a reference. The results on Hot-Tear will be analyzed and discussed.

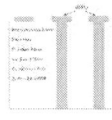

978-1-6654-1392-3/21 $31.00 © 2021 IEEE

Figure 3 Schematic of VIPPO design in this work

II. EXPERIMENTAL

1. PCB Design

The VIPPO pad design in this work is shown in Figure 3. Three types of BGA were incorporated in this study, corresponding board design were shown in Figure 4, where the pads with VIPPO structure were colored in yellow.

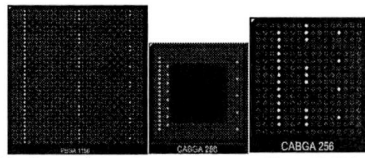

Figure 4 PCB footprints for corresponding BGAs with yellow pads being VIPPO design

The test vehicle was shown in Figure 5.

Figure 5 Test vehicle layout used for VIPPO hot tear study

2. BGA Assembly Process

(1) Conventional SMT Process

PCB printed with solder paste, followed by BGA placement and reflow. This process was used as control for HT study.

(2) BGA Assembly Using Epoxy Flux Dispensing

Figure 6 shows the schematic process of BGA assembly using dispensed epoxy flux.

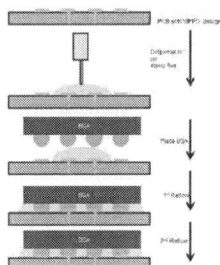

Figure 6 BA assembly using epoxy flux dispensing

(3) BGA Assembly Using Epoxy Flux Dipping

Figure 7 showed schematic process of BGA assembly using epoxy flux dipping process.

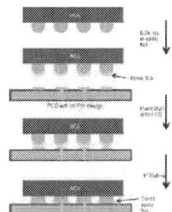

Figure 7 BGA assembly using epoxy flux dipping

3. Epoxy Fluxes Used

Two primary epoxy fluxes A and B were evaluated for their performance on preventing hot tear at VIPPO sites. The material properties were shown in Table 1.

Besides A and B, a couple of other epoxy fluxes with similar chemistry were also used for certain specific testing.

TABLE 1 Epoxy fluxes tested

Epoxy Flux	EF A	EF B
Typical Viscosity @25°C (cps)	17,500	25,500
Tg (°C)	5	45
Pot Life	7 days	16 hours
SIR	Pass	Pass

4. Wetting of Epoxy Flux

The wetting performance was assessed by printing epoxy flux on the OSP or ENIG coupon through a 6.35 mm diameter hole x 0.25mm thickness stencil, followed by placing a SAC305 ball onto the epoxy flux, then sent through reflow profile under air atmosphere. The wetting behavior was determined by the spread extent of solder ball.

5. Epoxy Flux Dispense

The epoxy flux dispense routing and dispensed pattern were shown in Figure 8. The amount of epoxy flux deposited was to show an peripheral epoxy fillet formation after BGA placement.

Figure 8 Epoxy flux dispense routing (left), dispensed pattern (center), and epoxy flux fillet appearance after BGA placement (right)

6. Reflow Profile

The reflow profile is a typical profile for BGAs, and is shown in Figure 9, followed by BGA placement.

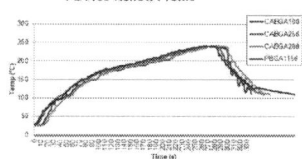

Figure 9 Reflow profile used for BGA assembly

7. Epoxy Flux Curing

Curing of epoxy flux under large BGA packages could be compromised due to temperature gradient factor and hampered outgassing. Therefore, the curing of epoxy flux under a simulated large package was checked for curing extent.

As shown in Figure 10, one gram of Epoxy Flux A coated on the laminate side of an 100mm x 100mm test coupon, another test coupon placed on top to mimic large BGA package and then reflowed 2 times. The coupons were then pried apart to check whether the epoxy flux was cured to non-sticky state upon touch.

Figure 10 Test coupons used to simulate curing of epoxy flux under large BGA packages

8. Solder Joint Drifting / Solder Extrusion / Voiding

For BGA assembled with epoxy flux, the solder joints were surrounded by epoxy, therefore may suffer solder joint drifting during reflow. Accordingly, after reflow, the BGA joints were examined with X-Ray for possible solder joint drifting or solder extrusion. Also checked was the voiding performance of joint

9. Hot Tear Assessment

The hot tear propensity was determined by running cross-sectioning across the VIPPO sites. The solder joints were examined under SEM.

10. Drop Test

The drop test was conducted on test vehicle shown in Figure 11, where the 10x10 BGA was pre-bumped with SAC305 balls. The BGA was then assembled onto substrate with three different processes prior to reflow, (1) BGA dipped into regular flux, then placed onto bare substrate, (2) BGA dipped in epoxy flux, then placed onto bare substrate, (3) epoxy flux dispensed onto substrate, followed by BGA placement.

Figure 11 Test vehicle with fixtures for drop test

The reflowed BGA test vehicle was then placed onto drop test machine and sent through drop testing until BGA fell off, as shown in Figure 12 [3,4].

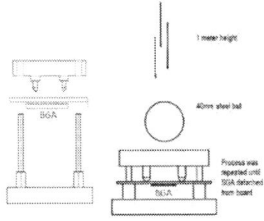

Figure 12 Drop test setup

III. RESULTS

1. Wetting of Epoxy Flux

The wetting of EF (Epoxy Flux) A was shown in Figure 13.

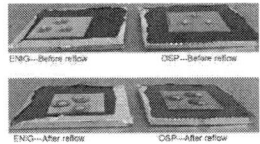

Figure 13 Wetting behavior of EF A on ENIG and OSP

The SAC305 sphere spread well on both ENIG and OSP surface finishes. Besides EF A, EF B, C, and D also spread similarly on the coupons.

2. Epoxy Flux Curing

All epoxy fluxes tested displayed a non-tacky body when cured between two 10cm x 10cm coupons after the coupons were pried apart.

The curing behavior was also examined with Differential Scanning Caloremetry (DSC), and was exemplified by Figure 14 for EF B. The curing reaction occurred at 170°C, and the Tg after curing was 45°C. Accordingly, EF B reflowed twice up to 240°C was fully cured, as evidenced by the lack of exotherm at the second heating cycle in Figure 14, and was consistent with non-tacky body observation.

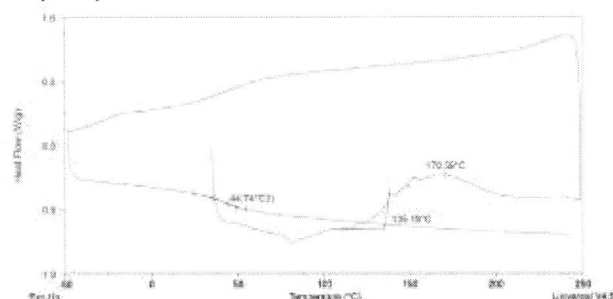

Figure 14 DSC thermograph of EF B, by scanning from room temperature up to 250°C, then cool down to -50°C, and scan up again to 250°C.

3. Solder Joint-Drifting /Solder-Extrusion/Voiding

All BGAs assembled were examined with X-ray after both first reflow and second reflow. Neither joint-drifting nor solder-extrusion was observed, and the results were demonstrated with PBGA1156, as shown in Figure 15. The amount of epoxy flux applied was around 0.25 gm/BGA for PBGA1156.

Figure 15 PBGA1156 assembled with epoxy flux

However, if the polymeric materials around the solder joint were not properly formulated, whether being underfill, edge bond, molding compound, or epoxy flux, joint-drifting or solder-extrusion may happen, as highlighted by red circles in Figure 16.

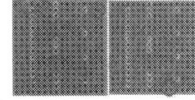

Figure 16 CABGA joint-drifting (left) and solder-extrusion (right)

Since joint-drifting or solder-extrusion was expected to be affected by viscosity of epoxy flux, and the viscosity may increase with time when being used on production line, the viscosity with increasing time was examined, as shown in Table 2 for EF B.

The joint-drifting and solder-extrusion of EF B was checked for PBGA1156 when the epoxy flux was left at RT for 0, 4, 8, and 24 hours before dispensed for BGA assembly, with results shown in Table 3. In all incidences, no defect could be discerned.

978-1-6654-1392-3/21 $31.00 © 2021 IEEE

TABLE 2 Viscosity of EF B with time at Room Temperature

Temp (°C)	RT
Time (hrs)	Viscosity (cPs)
0	26,180
2	27,220
5	28,210
6	28,500
8	28,960
24	35,300
% increase in 8hrs	10.62
% increase in 24hrs	34.84

TABLE 3 Prospect of EF B on forming displaced joint or solder-extrusion with time left at RT when assembling PBGA1156

Time (hrs)	Sample	Joint-Drifting		Solder Extrusion
		After 1st reflow	After 2nd reflow	After 2nd reflow
0	A1	0	0	0
	A2	0	0	0
	A3	0	0	0
4	B1	0	0	0
	B2	0	0	0
	B3	0	0	0
8	C1	0	0	0
	C2	0	0	0
	C3	0	0	0
24	D1	0	0	0
	D2	0	0	0
	D3	0	0	0

While examining the joint-drifting with X-ray, the voiding behavior was also checked. Figure 17 showed the X-ray image of CABGA joints assembled with EF A. No void can be discerned. This good non-voiding behavior is typical of epoxy flux family due to low outgassing of epoxy chemistry [5,6].

Figure 17 X-ray image of CABGA assembled with EF A

4. Representative VIPPO Joint Shape Related to Epoxy Flux Process

(1) Typical Solder Paste Assembly
BGA assembled with typical SMT solder paste attachment. Hot-Tear was common, as shown in Figure 18.

Figure 18 CABGA256 VIPPO joint assembled with solder paste

(2) Assembled with EF Dip Process
BGA assembled with EF dip process was also prone to have hot-tear defect, as shown in Figure 19.

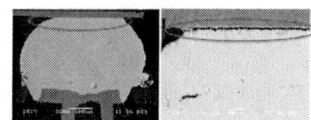

Figure 19 CABGA288 VIPPO joint assembled with EF dip process

(3) Assembled with EF Dispense Process
BGA assembled with EF dispense process did not exhibit hot-tear defect, as shown in Figure 20.

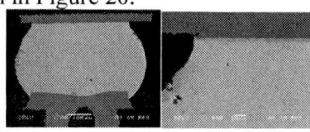

Figure 20 CABGA256 assembled with EF dispense process

5. Effect of Epoxy Flux Dip Condition on Hot-Tear (HT)
EF B (871-65-1) was tested for three dip depth, as shown in Table 4. Due to limited sample size, the effect of dip depth on HT (hot-tear) rate was not clear, except that EF dip suppressed HT rate when compared with solder paste assembly process. Epoxy Flux dip reduced HT (hot tear) failure rate when compared with solder paste assembly.

6. Effect of Epoxy Flux Dispense on HT
The effect of EF dispense on HT at VIPPO sites was investigated by assembling CABGA256 using EF B, and the results were shown in Table 5.

TABLE 4 EF B used for CABGA256 assembly using various dip depth

Dip Depth (% of bump height)	Failures /Total VIPPO joints	% Failure of VIPPO Joints	Note
78	0/9	11.11	
	1/9		
	2/9		
85	5/9	33.33	
	3/9		
	1/9		
100	N/A	N/A	Too tacky to pick up after dipping
Assembled via solder paste	9/9	100	No EF used
	9/9		
	9/9		

TABLE 5 HT failure rate of CABGA256 when assembled with EF B dispense process

Sample	HT Failures at VIPPO Site
1	0/9
2	0/9
3	0/9
4	0/9
5	0/9
6	0/9
7	0/9
8	0/9
9	0/9

10	0/9

Similar tests were also conducted on PBGA1156. No HT failure could be discerned at VIPPO sites for the 4 components tested.

7. Location of Epoxy Flux After Cure

Figure 21 Epoxy flux formed small fillets at both upper and lower corner of solder joints for dip process

The location of epoxy flux after cure was dictated by the method of deposition. Figure 21 shows that for dip process, the epoxy flux formed a small fillet around both upper and lower joint corners. On the other hand, the epoxy flux filled the space between the solder joints for dispense process, as shown in Figure 22.

Figure 22 Epoxy flux filled the space between joints for dispense process

The zero HT defect rate associated with epoxy flux dispense process as shown in Table 5 was attributed to the epoxy bonding between BGA and PCB, which reduced the impact of ΔCTE between VIPPO and non-VIPPO, thus prevented Hot-Tear from happening. The mechanism was shown in Figure 23.

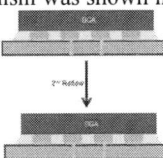

Figure 23 Epoxy flux dispense process prevent HT

On the other hand, epoxy flux dip process still suffered high HT defect rate, as shown in Table 4. This was attributed to that insufficient epoxy bonding to BGA body to reduce the impact of ΔCTE between VIPPO and non-VIPPO, thus was unable to prevent Hot-Tear from happening, as shown in Figure 24.

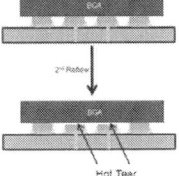

Figure 24 Epoxy Flux dip process unable to prevent HT from happening

8. Effect of Epoxy Flux Deposition on Drop Test

Use of epoxy flux for BGA assembly also affected the drop test performance of BGA. In Figure 25, the BGA assembled with EF B dispense process exhibited the highest drop number, approximately twice of that obtained from using regular flux assembly or EF B dip process.

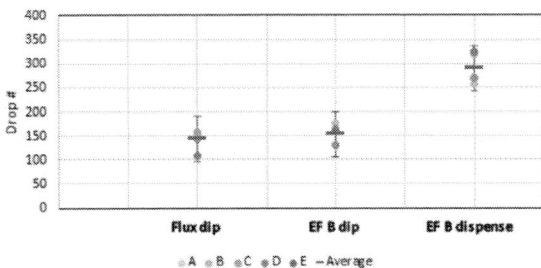

Figure 25 Effect of epoxy flux deposition method on drop test performance

9. Use of Underfill on Preventing HT at VIPPO Site

Two commercial underfills, one with filler and one without filler, were used to reinforce CABGA288 after first reflow using regular solder paste, as shown in Table 6. The BGA was then underfilled, followed by curing/second reflow. The second reflow also served as curing process for underfills. The joint-drifting and HT were examined after the second reflow.

TABLE 6 Effect of underfills on HT and joint-drifting for CABGA288 after second reflow

Reinforcement	# BGA	# VIPPO Joints	% HT Failures
Unfilled Underfill	3	81	All had Joint-Drifting
Filled Underfill	3	81	13.6

When unfilled underfill was used, all three BGAs showed joint-drifting phenomena, as shown in Figure 26. When filled underfill was used, 13.6% VIPPO joints showed HT defect.

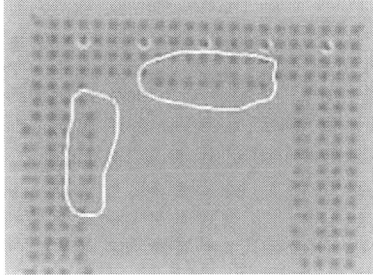

Figure 26 Joint-drifting observed at second reflow when CABGA288 was filled with unfilled underfill after first reflow

10. Peripheral Bleed Out from Epoxy Flux Dispense

On the extent of bleed out, as shown in Figure 27, it was controlled by the volume and pattern of epoxy flux dispensed. For CABGA256, the bleed out width of three packages was measured to be 1.29, 1.11, and 1.13mm. This narrow bleed out indicated that the dispense process would not interfere with neighboring components.

Figure 27 Peripheral bleed out from Epoxy Flux Dispense process

IV. DISCUSSION

1. Epoxy Flux Dip Process Limitation

To prevent HT failure at VIPPO joints, sufficient epoxy adhesion between BGA and PCB is required in order to nullify the tension on the VIPPO joint caused by the ΔCTE between VIPPO (17ppm/°C) and non-VIPPO (45ppm/°C) structures. However, dip process up to 85% ball height only resulted in small epoxy fillets at corner of solder joints, as shown in Figure 21. Obviously the weak adhesion brought by those small epoxy fillets was not enough to nullify the ΔCTE factor. If a deeper dip process is employed, such as 100% ball height dip, the tack force of epoxy flux turned out to be too high for the BGA to be picked up by the vacuum nozzle.

On the other hand, epoxy flux dispense allows sufficient epoxy to fill the space between BGA and PCB, as shown in Figure 22, thus effectively suppress the ΔCTE factor and the resultant HT failure.

2. Epoxy Flux vs Underfills

The epoxy fluxes tested in this study exhibited no joint-drifting or solder-extrusion. This was resulted from epoxy formulation optimization which prevented uneven stress distribution within epoxy layer, and consequently avoided joint-drifting or solder-xtrusion problems.

The commercial underfills without filler suffered joint-drifting problem. This was attributable to the chemistry not optimized to provide zero stress at reflow temperature, thus resulted in joint-drifting. For filled underfill, HT defect rate was still measurable. Presumably this was caused by penetration of filled underfill into the molten solder joints at second reflow, thus resulted in open joints. These opens could be avoided by curing at temperature below the solder melting temperature, then followed by second reflow. However, this would result in additional heating process step, and the associated increase in cost.

Presumably the underfills could be optimized in formulation to avoid joint-drifting or HT opens. But, the solder paste flux residue from the first reflow process still pose a concern on reliability, such as TCT reliability. Since epoxy flux enabled BGA assembled without use of solder paste, interference of regular flux residue to adhesion of epoxy flux to BGA and PCB would not exist.

3. Corner Bond or Edge Bond to Prevent HT?

Both corner bond and edge bond only bond peripheral area of the BGA body, thus failed to provide strong adhesion between BGA body and PCB, therefore neither one is considered adequate for preventing HT from happening at VIPPO joints.

4. EF A versus EF B

Both EF A and EF B considered viable solution for eliminating HT at VIPPO design. EF A exhibited a better storage stability, while EF B promised a higher mechanical strength due to its higher Tg. For both EFs, assembly of BGA was recommended to be conducted without printing solder paste, since the compatibility between any given solder paste and epoxy flux was not clarified yet.

V. CONCLUSION

VIPPO design enabled a better signal quality and speed, but it also suffered side-effect caused by this design. At VIPPO location with or without a solid-Cu-hole-fill, the coefficient of thermal expansion (CTE) typically around 17ppm/°C which is quite lower than that of non-VIPPO pads nearby, which is around 45ppm/°C. For double-sided PCB, at ramp-up stage of the second reflow, the difference in CTE of neighboring pads would result in excessive tension on the joint on VIPPO pads, and consequently resulted in Hot-Tear (HT) before reaching melting point of solder joints.

Epoxy Flux (EF) was developed to eliminate HT problem. EF dip process did not work due to insufficient EF volume to bond BGA strongly to PCB to nullify the ΔCTE factor. EF dispense process wet well, exhibited low voiding, and allowed sufficient EF volume to fill the gap between BGA and PCB, thus enabled a strong bonding force to nullify the ΔCTE factor, and consequently eliminated the HT defect. The EF developed had the epoxy stress balanced around the solder joints and avoided joint-drifting and solder-extrusion problem. EF dispense process was recommended to be used without solder paste when the BGA was to be assembled. It cured fully in the reflow process, and was non-tacky on touch, and also met IPC SIR requirement. Commercial underfills tested did not work on preventing HT from happening, due to the joint-drifting and solder-extrusion problem. Edge bond or corner bond was also expected to be unable to prevent HT defect. Compared with other polymeric reinforcement materials, EF dispense process minimized the process steps hence was a lower cost valid solution than other reinforcement approaches.

VI. ACKNOWLEDGEMENT

The authors would like to express deep appreciation to Christine LaBarbera for her strong support in SEM analysis work and to Lee Kresge and Jonathan Minter for their support in use of laboratory equipment.

VII. REFERENCE

1. S.Y. Teng, P. Peretta and P. Ton (Cisco), V. Kome-ong and W. Kamanee (Celestica), "VIA-IN-PAD PLATED OVER (VIPPO) DESIGN CONSIDERATIONS FOR MITIGATION OF A UNIQUE SOLDER SEPARATION FAILURE MODE", SMTAI 2016, APT7-P1, September 25-29, 2016, Rosemont, ILNG

2. Ning-Cheng Lee, short course "Hot Tear At VIPPO and Other Designs for 5G", CEIA K3 Symposium online, May 16, 2020.

3. Francis Mutuku, Jie Geng, Hongwen Zhang, and Ning-Cheng Lee, "Low Temperature Solder Alloy with High Reliability Performance", APEX, San Diego, California, February 24 – March 1, 2018

4. Hongwen Zhang, Samuel Lytwynec, Huaguang Wang, Jie Geng, Francis Mutuku, & Ning-Cheng Lee, "A Novel Mixed Powder Low Temperature Solder with Superior Drop-Shock Resistance", APEX March 8-12, 2021

5. Wusheng Yin, Geoffrey Beckwith, Hong-Sik Hwang, Lee Kresge, and Ning-Cheng Lee; "Epoxy Flux – An Answer For Low Cost No-Clean Flip Chip Assembly", Nepcon West/Fiberoptic Automation Expo, San Jose, CA, December 3-6, 2002

6. Wusheng Yin and Ning-Cheng Lee; "A Novel Epoxy Flux For Lead-Free Soldering", International Brazing 2003.

Process Optimization for CCGA Surface Mount Assembly Based on Physics of Failure

Hui Xiao *
China Electronic Product
Reliability and Environmental
Testing Research Institute
Guangzhou, China
xiaohui_ceprei@163.com

Weiming Li
China Electronic Product
Reliability and Environmental
Testing Research Institute
Guangzhou, China
liwm@ceprei.com

Yabing Zou
China Electronic Product
Reliability and Environmental
Testing Research Institute
Guangzhou, China
zouybing@163.com

Xiaotong Guo
China Electronic Product
Reliabilit and Environmental
Testing Researc Institute
Guangzhou, China
guoxiaotong0713@163.com

Jiahao Liu
China Electronic Product
Reliabilit and Environmental
Testing Researc Institute
Guangzhou, China
ljh071408@163.com

Abstract—There is still limited second-level interconnect reliability for large-size ceramic column grid array (CCGA) devices. In this paper, the process optimization technology based on physics of failure (PoF) was proposed and applied in the CCGA surface mount assembly, considering the structure characteristics of the package and coupled process environment. This methodology is problem-oriented, focusing on the internal physical and chemical process related to the failure of components and materials. The failure mechanism study is conducted for typical failure modes, by which the sensitive process parameters about failure can be obtained. Then the corresponding process optimization tests are conducted, ensuring the board-level products meet the electronic interconnect requirements in service. The results showed that the process optimization methodology based on PoF was a useful and efficient tool for solving the interconnect defects of CCGA surface mount assembly.

Keywords—CCGA surface mount assembly, process optimization, physics of failure

I. INTRODUCTION

Ceramic column grid array (CCGA) is a kind of high-density surface mount package, which can be connected to the printed circuit board (PCB) by solder balls as the I/O circuit ends. Because the ceramic substrate can offer the advantages of finer routing density, multiple power and ground planes, improving signal integrity, moisture resistance and thermal conductivity, CCGA packaging has been gradually used in aviation equipment and other high reliability field in recent years [1].

Fig. 1 shows the two typical structures of CCGA solder columns. The most commonly used solder column is Pb90Sn10 solder column, as shown in Fig. 1a. Another one is Pb80Sn20 solder column with a copper spiral on the surface, as shown in Fig. 1b, which could ease the column deformation during thermal cycling [2]. The solder columns are usually attached to ceramic substrate with 63Sn37Pb solder. Compared with the ceramic quad flat package (CQFP) or ceramic ball grid array (CBGA) package, the CCGA package with higher standoff is believed to have a much higher reliability under thermal cycling.

However, the CCGA package exhibits poor solder joint reliability when compared to organic packaging technologies, there is still limited second-level interconnect reliability for large-size CCGAs [3]. Solder joints' interconnect failure would often induce the failure of whole system or even severe problem, because such package is usually used as the core component, such as CPU, FPGA, DDR, and so on. As surface mount technology is the key process of the interconnection formation between the CCGA component and the printed circuit board (PCB), it is necessary and important to study such process optimization technology, and more attention should be paid to the structure characteristics of the package and coupled process environment [4].

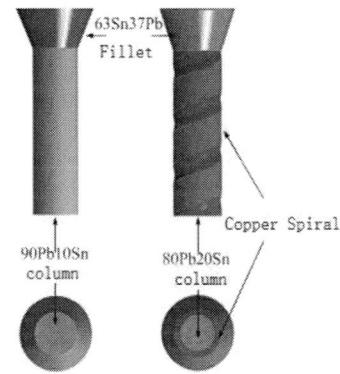

Fig. 1 Typical structure of solder columns [2]

As to the process optimization for electronic assembly, the traditional method, such as taguchi method, is based on a large number of process tests and systematically mathematical statistics analysis [5, 6]. However, for the kind of high-reliability and core component product, the traditional reliability technology is not suitable because of high test cost and long test time [7]. In this paper, the process optimization technology based on physics of failure (PoF) was proposed and applied in the surface mount assembly, considering the structure characteristics of the package and coupled process environment. This methodology is problem-oriented, focusing on the internal physical and chemical process related to the failure of components and materials. The failure mechanism study is conducted for typical modes, by which the sensitive process parameters about failure can be obtained. Then the corresponding process optimization tests are conducted, ensuring the board-level products meet the electronic interconnect requirements in service.

978-1-6654-1392-3/21 $31.00 © 2021 IEEE

II. PROCESS OPTIMIZATION METHODOLOGY BASED ON PoF

The PoF-based process optimization methodology is as shown in Fig. 1. It was conducted by theoretical research combining with experimental verification. Firstly, the mainly problem was identified by product information investigation, including process site survey and historical failure investigation. Secondly, Failure analysis was conducted to identify the failure mechanism of the corresponding typical failure modes. Then process optimization was conducted according to the specific modes and mechanism. Finally, optimization result can be verified by reliability evaluation tests.

In general, this methodology mainly contains five steps, including product information investigation, failure modes and failure mechanism research, the root cause determination, corresponding process optimization, and result verification.

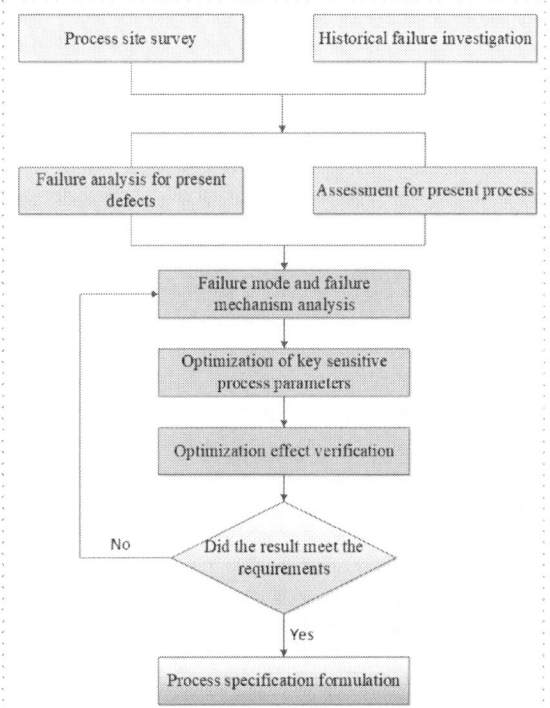

Fig. 2. The technical framework of the PoF-based process optimization methodology

III. APPLICATION OF THE PoF-BASED PROCESS OPTIMIZATION FOR THE CCGA SURFACE MOUNT ASSEMBLY

A. Samples and Background

The CCGA601 package was used in the present study, the package size of which was 40mm * 40mm. The solder column was 90Pb10Sn, and the dimension was φ0.51mm and 2.2mm high. The pad diameter of the substrate was 0.75mm and the pitch was 1.27mm. The paste used in the board-level soldering was Loctite CR32Sn63AGS89.5. The solder columns were attached to the PCB substrate with the solder paste by reflow soldering.

B. Failure Analysis Investigation

Fig. 3 to Fig. 5 shows the failure phenomenon of the CCGA board-level connection before the process optimization study. There were mainly three types of defect

modes: plenty of voids in the column solder joints, a large number of solder columns deviated from the corresponding pads' perimeter, interconnect failure during vibration reliability test.

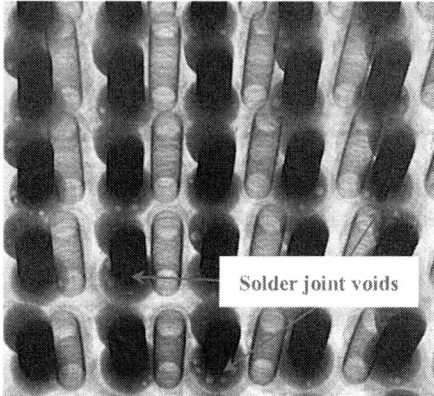

Fig. 3 Voids in the column solder joints

(a) 3D X-ray inspection

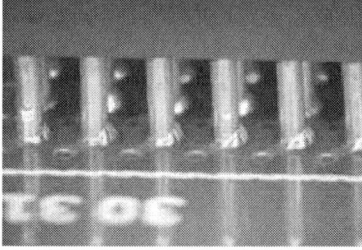

(b) Optical microscope observation

Fig. 4 Solder columns deviated from the pads under the views of Optical microscope and 3D X-ray inspection

Fig. 5 Cracking in the columns during vibration reliability test

C. Root cause analysis

Failure mechanism study was conducted to address above three types of failure modes, as shown in Fig. 3 to Fig. 5.

978-1-6654-1392-3/21 $31.00 © 2021 IEEE

(1) Solder joint voids

Solder joint voids were related to flux volatilization during reflow soldering of CCGA surface mount assembly [8]. The rate of voids could be controlled by reflow temperature profile optimization, such as pre-heat temperature, soldering temperature and soldering time.

(2) Solder columns deviation

Compared with CBGA package, self-centering of the solder column array was poorer than the solder ball array, during the reflow soldering stage of the surface mount assembly. In order to ease the solder columns deviation, there were two types of solution. One solution was the solder columns coplanarity control for the solder columns of the CCGA package, which was related to the incoming inspection. Another one was process optimization for solder paste printing and the CCGA package patch.

(3) Poor variation resistance

Variation resistance was mainly related to CCGA package's structure, especially related to the structure frangibility of the ceramic package with the long solder column array. The solution could be made to promote the deformation resistance for the structure. Corner reinforcing with glue for the CCGA package may be an effective solution.

D. Corresponding process optimization

According to the root cause analysis results, process optimization for the CCGA surface mount assembly was conducted.

A daisy-chain circuit is designed for the CCGA601 package assembly, which could be used to conduct process optimization test as well as optimization effect verification. Fig. 6 depicts schematically the daisy-chain circuit design.

There were vias and traces buried within the package for the electrical connection of the neighboring solder joints. Fig. 6b depicts the pathway of the current flow across the entire assembly.

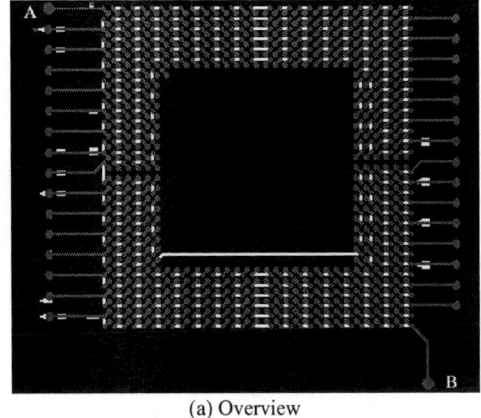

(a) Overview

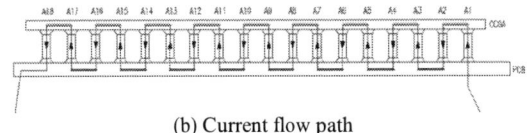

(b) Current flow path

Fig. 6 Schematic Diagram of Daisy-chain Circuit Design of the connection between the CCGA601 package and the substrate

(1) Reflow profile optimization

Fig. 7 shows reflow profile optimization for the CCGA surface mount assembly. Compared with CCGA1, the CCGA2 soldering temperature was higher and soldering time was longer, which could help the flux volatilization bubbles escape from the solder during the reflow soldering. Reflow profile optimization could be used to reduce void rate of the CCGA solder joints.

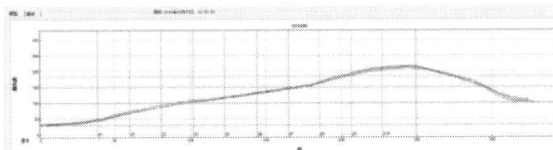

(a) Before process optimization (marked as CCGA1)

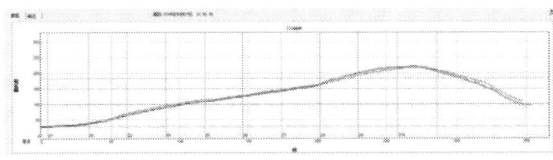

(a) After process optimization (marked as CCGA2)

Fig. 7 Reflow profile optimization for the CCGA surface mount assembly

(2) Process optimization for solder paste printing and the CCGA package patch

Process optimization for solder paste printing was aimed to control the solder paste volume in a suitable range, which would affect the centering of the solder columns during soldering stage. It contained two aspects, steel mesh opening size and solder paste printing parameters. The related parameter of the CCGA package patch was patch pressure, which would work with the weight of the CCGA package and affect the solder columns' centering.

(3) Corner reinforcing with glue for the CCGA package

Zymet2605 glue was used for corner reinforcing of the CCGA package. The dispensing process optimization was conducted according to IPC-7095C standard [8].

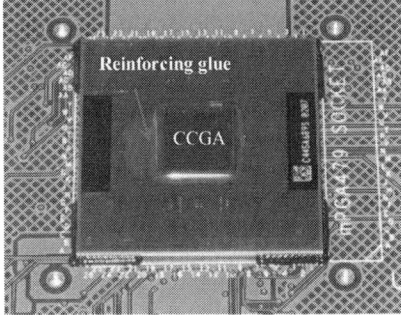

Fig. 8 Corner reinforcing with glue for the CCGA package

E. Optimization effect verification

Process optimization was conducted to solve the soldering voids, columns' deviation and board-level interconnect reliability. A serial of experimental analysis technologies and reliability qualification tests were used to verify the optimization effect, such as X-ray inspection, microsectioning, scanning electron microscope (SEM) analysis, temperature cycling test and variation test.

The comparison results of solder joints' morphology before and after process optimization were as shown in Fig. 4. Fig. 5 shows the microstructure and elements' distribution in the CCGA solder joint. It can be seen that there was not any obvious abnormality for the soldered interfacial zone microstructure.

(a) Before process optimization   (b) After process optimization
Fig. 9 Comparison results of solder joints' morphology before and after process optimization

 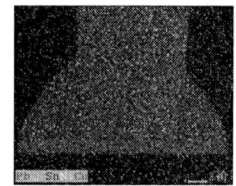

(a) Microstructure   (b) Elements' distribution
Fig. 10 Microstructure and elements' distribution in the CCGA solder joint after process optimization

The solder columns were well aligned and the voids's rate of the solder joints was much less than 25%. It was suggested that the CCGA solder joint quality met the requirements of IPC-610G and ECSS-Q-ST-70-38C [9, 10]. Moreover, there was not any solder joint failure occurred during thermal cycling and variation reliability qualification tests. This suggested that PoF was a useful and efficient tool for solving the interconnect defects of CCGA surface mount assembly.

IV. CONCLUSIONS

Process optimization was conducted to solve the soldering voids, columns' deviation and board-level interconnect reliability. These defects were mainly related to soldering process, solder paste printing process, and the die attach process. Based on PoF analysis, the key sensitive parameters could be determined, which were die attach pressure, steel mesh opening size, soldering temperature and soldering time. Corresponding process optimization and Corner reinforcing with glue for the CCGA package were conducted, the results showed that the solder joint quality met the requirements of IPC-610G and ECSS-Q-ST-70-38C. And there was not any vibration failure occurred in the solder joints. This suggested that PoF was a useful and efficient tool for solving the interconnect defects of CCGA surface mount assembly.

ACKNOWLEDGMENT

This work is financially supported by by the National Key R&D Program of China under Grant No. 2020YFB1710300, the Science and Technology Program of Guangzhou, China under Grant No. 202002030357, CEPREI Innovation and Development Fund No. 20Z32, which were acknowledged.

REFERENCES

[1] Park T. Y. et al., International Journal of Aerospace Engineering, 2018 (2018): 1687.

[2] Tong L. et al., Proceedings of the 19th ICEPT (2018): 1382-1386.

[3] Ding Y. et al., Microelectronics Reliability, 55 (2015): 2396-2402.

[4] Ghaffarian R. et al., Microelectronics Reliability, 46 (2006) 2006-2024.

[5] Hendricks C. et al., Reliability Characterization of Electrical & Electronic Systems. 2015: 27-42.

[6] Hendricks C. et al., Reliability Characterization of Electrical & Electronic Systems. 2015: 27-42.

[7] Lu T. et al., Proceedings of the 19th ICEPT (2018): 1140-1144.

[8] IPC-7095C, Design and Assembly Process Implementation for BGAs, Association Connecting Electronics Industries, 2013.

[9] IPC-610G, Acceptability of Electronic Assemblies, Association Connecting Electronics Industries, 2018.

[10] ECSS-Q-ST-70-38C, High-reliability soldering for surface-mount and mixed technology, ECSS for space product assurance, 2008.

210 °C reflow technology study in 3D Packaging

Lingyao Sun
School of Materials
Science&Engineering
Beijing Institute of Technology
Beijing, China
sunlingyao_513@163.com

Yaru Dong
School of Materials
Science&Engineering
Beijing Institute of Technology
Beijing, China
dongyaru52@163.com

Zhuangzhuang Hou
School of Materials
Science&Engineering
Beijing Institute of Technology
Beijing, China
hzzpaper@163.com

Xiuchen Zhao
School of Materials
Science&Engineering
Beijing Institute of Technology
Beijing, China
zhaoxiuchen@bit.edu.cn

Yongjun Huo
School of Materials
Science&Engineering
Beijing Institute of Technology
Beijing, China
huoyongjun@bit.edu.cn

Ying Liu
School of Materials
Science&Engineering
Beijing Institute of Technology
Beijing, China
yingliu@bit.edu.cn

Yingxia Liu
School of Materials
Science&Engineering
Beijing Institute of Technology
Beijing, China
yingxia.liu@bit.edu.cn

Abstract—Low temperature reflow is promising to relieve the warpage issue caused by the mismatch of thermal expansion coefficients (CTE) of PCB and BGA components during assembly. While Sn58Bi with a melting point of 138 °C has a low reflow temperature, its brittleness can lead to solder joint reliability problems. SAC305 solder alloy has excellent soldering yield and reliability performance, however, the reflow temperature is difficult to be reduced further by element addition or modification.

In this paper, the mixed solder of SAC305 and Sn58Bi was prepared by solid-liquid low temperature soldering. low temperature reflow is promising to relieve the warpage issue caused by the mismatch of thermal expansion coefficients (CTE) of PCB and BGA components during assembly. We found that after reflowing at 210 °C for 5 min, the solder ball and solder paste were completely mixed. This temperature is much lower than that of SAC305, which can effectively reduce the thermal warpage and thermal budget. In addition, according to the shear test data of solder joints under each reflowing condition, the shear strength increases obviously with the increase of mixing degree. The shear strength of samples refluxed at 210 °C for 5 min (63.68 MPa) is about 10 MPa higher than that refluxed for 1 min (52.42 MPa). The results show the mitigation of brittleness in Bi contained solder joints. Our research results provide a promising solution for low-temperature soldering technology in 3D package.

Keywords—3D package, Low temperature soldering, Diffusion kinetics, Sn58Bi

I. INTRODUCTION

With the development of Internet science and technology, human beings have entered the intelligent era. Chip technology, as a major technological power to promote the development of intelligence, its level of development is extremely important. As the miniaturization of Moore's Law is coming to an end, the integration of chips is required to be higher and higher to achieve More-than-Moore [1]. 3D packaging breaks through the traditional concept of planar packaging, and on the basis of 2D packaging, multiple chips, components, packages are stacked and interconnected, forming 3D packaging, which greatly improves the assembly density. Three-dimensional integrated circuit (3D IC) has the advantages of high performance, small size and high heterogeneous integration, which has become an effective way to face the challenges brought by Moore's law failure [2].

In the development of 3D IC, as the packaging structures becomes quite complicated, we are facing a more serious warpage issue compared with 2D packaging. Previously, eutectic Sn-Pb solder with a melting point of 183°C and reflow temperature of 210~220°C was widely adopted in electronic packaging technology [3]. But Pb is poisonous to environment, and was forbidden to be applied in consumer electronic [4]. Sn-Ag-Cu alloy is currently the most commonly used lead-free solder for SMT manufacturing. Among them, the eutectic alloy SAC305 (Sn96.5Ag3.0Cu0.5) is represented, which has the characteristics of good wettability, excellent fatigue resistance and excellent solder joint reliability [5,6]. However, due to the high required reflow temperature and the mismatch of the CTE of the components in the system, SAC305 solder will have warpage issue, which may further lead to bridging and open-circuit solder joint defects. The warpage issue is becoming predominant in 3D IC. To relieve the serious wraping issue in 3D IC, low reflowing technology needs to be investigated [7, 8]. Sn-58Bi alloy (w.t.%) has a low melting point (139 °C) and good mechanical properties, and is currently the most widely used low temperature solder in the field of electronic packaging. But the brittleness of Bi needs to be overcome [9]. The soldering by SAC305 and Sn58Bi has been studied, however, the real application scenario, where the alloys are in the form of ball and paste, has not been studied [10].

In our work, we studied SAC305 solder ball and Sn58Bi solder paste reflow technology. This technology can effectively reduce the reflow temperature below 220°C, while reduce the brittleness of solder joints. On this basis, SEM (regulus8230) was used to observe the microstructure of the solder joints. Finally, the strength of the joint was verified by a shear test.

978-1-6654-1392-3/21 $31.00 © 2021 IEEE

II. EXPERIMENTAL PROCEDURE

The Sn58Bi solder paste was printed on the polished and cleaned copper pad. The SAC305 solder balls (need to mention the size) were placed on the Sn58Bi solder paste and reflowed at 210 °C for 1 min, 5 min, and 10 min. After the soldering was completed, the samples were cooled to room temperature and cleaned with absolute ethanol. The samples were then fixed in epoxy resin and polished with SiC sand paper and diamond polishing spray to obtain a smooth sample cross-section.

The cross-section and microstructure of the polished sample were observed with a scanning electron microscope (SEM, Regulus 8230), and the diffusion and mixing of the two kinds of solder during reflow were analyzed.

Finally, the mechanical properties of the solder joints were evaluated. The samples reflowed at 210 °C for 1 min, 5 min and 10 min were sheared with PTR-1100 shear testing machine. The shear height was 0.04 mm and the shear strain rate was 0.5 mm/s. The samples under each reflowing condition were subjected to 10 times of shear test, and the obtained data were averaged to obtain the shear strength of the corresponding samples, which were recorded and analyzed.

III. RESULTS AND DISCUSSION

A. Microstructure and wettability of the mixed solder

In the reflow soldering test, we used the temperature higher than the melting point of Sn58Bi (139 °C) and lower than SAC305 (217 °C). In Fig. 1 below, (a) ~ (c) are the overall SEM images of bumps obtained after reflowing 1 min, 5 min, 10 min at 210 °C, respectively. It can be seen from Fig. 1 that there is still a very obvious interface between the two solders when reflowing at 210 °C for 1 min. After reflowing 5 min at 210 °C, the interface between the two kinds of solder disappeared. With the reflow time extended to 10 min, the result was similar to that at 5 min. This initially shows that the mixing degree of the two kinds of solder increases with the reflowing time at the same reflowing temperature, and when reflow at 210 °C for 5 min or more, the two kinds of solder are completely mixed to form a new solder with uniform composition. Moreover, with the extension of the reflowing time, the shape of the bump has changed significantly, as shown in (b) and (c) in Fig. 1, the spreading area at the bottom of solder is larger than (a). With the dissolution of solid SAC305 solder balls into liquid Sn58Bi solder paste, the solder balls collapse obviously, and the total height of bumps decreases. This is because the higher the reflowing temperature is, the faster the diffusion and dissolution rate between two kinds of solder is. When the reflowing temperature is extended to 5min at 210 °C, the two kinds of solder are completely mixed, the composition distribution is uniform, and the bump shape becomes scallop. When the reflowing time is extended to 10min, the bump shape and the mixing degree of the two kinds of solder are basically consistent with 5min.

Fig. 2 shows the microstructure of Sn58Bi, in which the light phase is Bi and the dark phase is Sn. The two phases are evenly distributed, which is the main reason for the good tensile properties and shear strength of Sn58Bi. However, because Bi is a brittle phase with high content, and Bi is easy to accumulate at the interface during aging treatment of Sn58Bi solder joint, the mechanical properties of solder alloy will deteriorate due to its brittleness.

The microstructure of the mixed solder of Sn58Bi solder paste and SAC305 solder ball obtained by reflowing at 210 °C for 5 min is shown in Fig. 3. The figure shows the top of the whole solder bump. There are three phases of Sn, Bi and Ag3Sn in the solder, and they are evenly distributed. This further shows that Bi has diffused to the top of the solder after reflowing at 210 °C for 5 min, indicating the solder is completely mixed, and the composition distribution is uniform. More importantly, the content of Bi in the mixed solder is obviously less than that of Sn58Bi. We predicted that the mechanical properties of the mixed solder will be better than that of Sn58Bi, which were demonstrated and explained in part B.

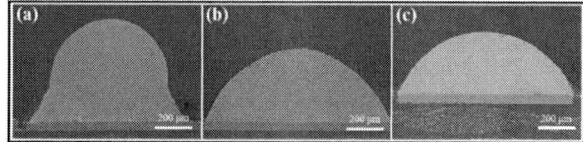

Fig. 1. SEM images of samples that were refluxed at 210 °C for (a) 1 min (b) 5 min (c) 10 min

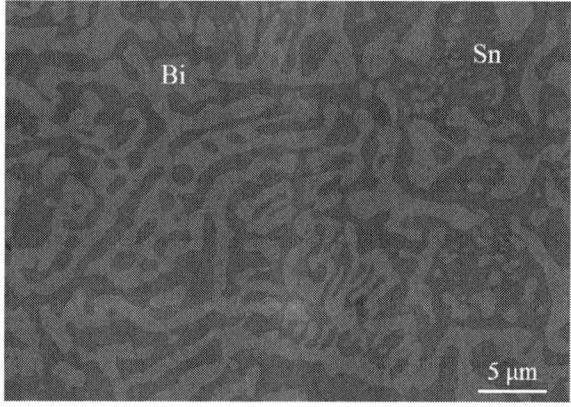

Fig. 2. Microstructure of Sn58Bi

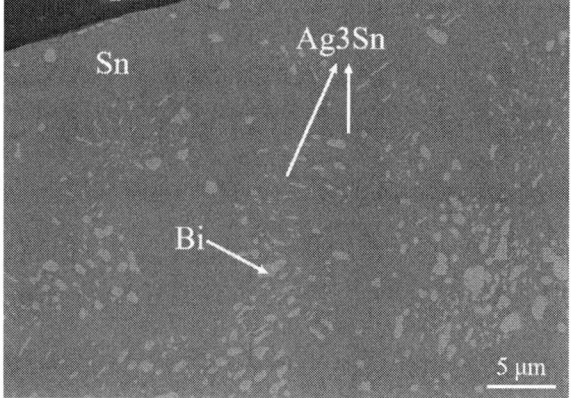

Fig. 3. Microstructure of the mixed solder

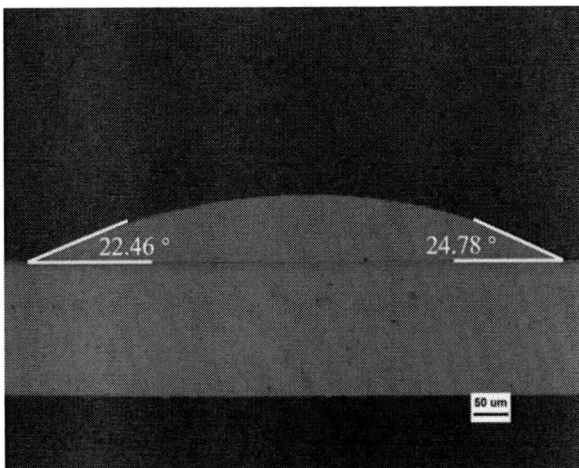

Fig. 4. Wetting angle of Sn58Bi after reflowing

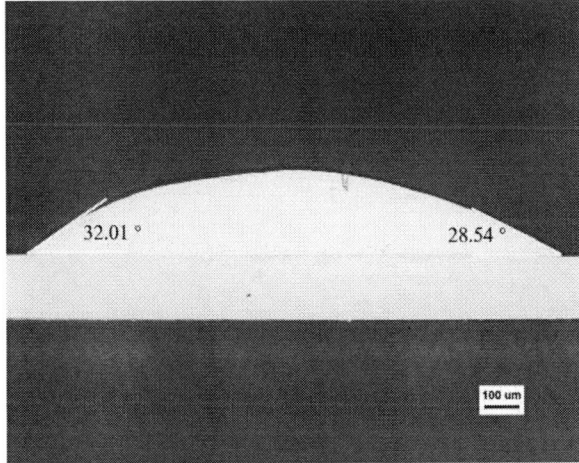

Fig. 5. Wetting angle of mixed solder after reflowing at 210 °C for 5 min

Fig. 4 shows the wettability of Sn58Bi after reflowing, and the wetting angle of the mixed solder is about 23.62 °, Fig.5 shows the wettability of the solder obtained after reflowing Sn58Bi and SAC305 at 210 °C for 5 min, and the wetting angle is about 30.27 °, a small gap from the wetting angle of the single Sn58Bi. Therefore, we can draw a conclusion that the mixed solder also has good wettability.

In a word, Sn58Bi and SAC305 can be reflowed at 210 °C for 5 min to obtain a completely mixed solder with uniform composition. The relative content of Bi can be reduced. Also, the mixed solder also has good wettability.

B. Mechanical properties

For the test of mechanical properties, we mainly did the shear test of solder joints, and studied the shear strength of the solder joints. Fig. 6 is the schematic diagram of shear test. The two kinds of solder were reflowed at 210 °C for 1 min, 5 min and 10 min respectively and repeated soldering 10 bumps under each condition. 10 groups of data were obtained under PTR-1100 shear testing machine. The corresponding shear strength of the solder joints were obtained by calculating the average value, as shown in Fig. 7. It shows that under all test conditions, the highest shear strength of the bump is 63.68 MPa after reflowing at 210 °C for 5min. The data from 210 °C

show that the shear strength of the samples reflowed for 5 min is higher than that of the samples reflowed for 1 min and 10 min, which indicates that not the longer the reflowing time is, the higher the shear strength is, and 5 min can be regarded as the best reflowing time. Moreover, the shear strength of solder joints reflowed at 210 °C is almost 2 times that of Sn58Bi (35.2 MPa) [11]. This is because the content of Bi atoms in the mixed solder is greatly reduced, which can mitigates the brittle effect caused by the Bi atoms. In this way, the mechanical property of the low reflow temperature joints can be improved, which is promising to be applied in mobile electronic devices.

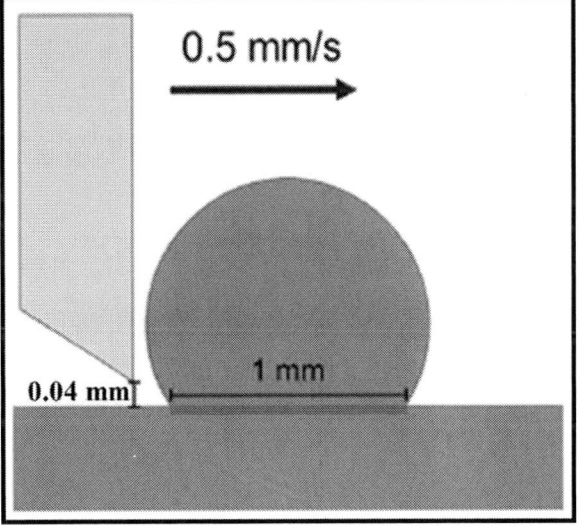

Fig. 6. Schematic diagram of shear test

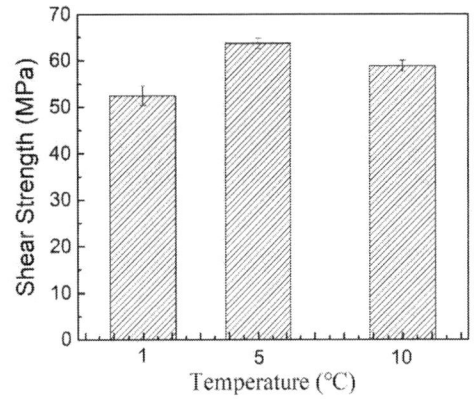

Fig. 7. The shear strength of the samples reflowed at 210 °C for 1 min, 5 min and 10 min, respectively

IV. CONCLUSION

We achieved low-temperature reflow soldering using Sn58Bi paste and SAC305 soldering balls, effectively reduced the reflow temperature below 220 °C, and solved the problem of dynamic thermal warping. Not only the excellent soldering performance of SAC305 and the low temperature weldability of Sn58Bi were used, but also the mixture of the two kinds of solder was formed in this process, which reduced the brittleness of solder and obtained the bumps with better performance.

The tests showed that the two kinds of solder were fully mixed after reflowing at 210 °C for 5 minutes, and the composition distribution is uniform. And the mixed solder not only has good wettability, but also has higher shear strength. However, it is not that the longer the reflow time is, the higher the shear strength of the solder joint is. The test results showed that 5min is the optimal reflowing time. The shear strength of the fully mixed solder formed at 210 ℃ for 5 min is the highest, which is 63.68 MPa.

Compared with SAC305 or Sn58Bi single-component solder, the proposed low-temperature reflow technology can not only reduce the thermal warping and thermal balance by reducing the reflow temperature, but also greatly improve the shear strength of the solder joints by forming the mixed solder. Our results provide a promising solution for low-temperature soldering in 3D packaging.

REFERENCES

[1] K.N. Tu, Y. Liu, and M. Li, "Effect of Joule heating and current crowding on electromigration in mobile technology", Applied Physics Reviews, 2017 4(1).

[2] K.N. Tu, "Reliability challenges in 3D IC packaging technology", Microelectronics Reliability, 2011, 51(3), pp. 517-523.

[3] Y. Liu and K.N. Tu, "Low melting point solders based on Sn, Bi, and In elements", Materials Today Advances, 2020, 8, pp. 100115.

[4] K. Suganuma, "Effects of intermetallic compounds on properties of Sn–Ag–Cu lead-free soldered joints", Journal of Alloys and Compounds, 2003, 352(1-2), pp. 226-236.

[5] H.R. Kotadia, P.D. Howes, and S.H. Mannan, "A review: On the development of low melting temperature Pb-free solders", Microelectronics Reliability, 2014, 54(6-7), pp. 1253-1273.

[6] L. Zhang, W. Long, and F. Wang, "Microstructures, interface reaction, and properties of Sn–Ag–Cu and Sn–Ag–Cu–0.5 CuZnAl solders on Fe substrate", Journal of Materials Science: Materials in Electronics, 2020, 31(9), pp. 6645-6653.

[7] S.Y. Yang, Y.D. Jeon, S.B. Lee, and K.W. Paik, "Solder reflow process induced residual warpage measurement and its influence on reliability of flip-chip electronic packages", Microelectronics Reliability, 2006, 46(2-4), pp. 512-522.

[8] G. Ren and M.N. Collins, "Improved reliability and mechanical performance of Ag microalloyed Sn58Bi solder alloys", Metals, 2019, 9(4).

[9] Scott Mokler, P.E, etc, "The application of Bi-baed solders for low temperature reflow to reduce cost while improving SMT yields in client computing systems", Proceedings of SMTA International.Rosemont, IL, USA, pp. 318-325, Sep. 25 - 29, 2016.

[10] Shen Y A, Zhou S, Li J, et al. "Sn-3.0 Ag-0.5 Cu/Sn-58Bi composite solder joint assembled using a low-temperature reflow process for PoP technology", Materials & Design, 2019, 183: 108144.

[11] H.F. Zou, Q.K. Zhang, Z.F. Zhang, "Interfacial microstructure and mechanical properties of SnBi/Cu joints by alloying Cu substrate", Materials Science and Engineering a-Structural Materials Properties Microstructure and Processing 532 (2012) 167-177.

Research on the Humidity Resistance Reliability of Different Packaging Structures

Hui Xuan
Tongfu Microelectronics Co. Ltd.
Nantong, *China*
xuan.hui@tfme.com

Zheng Yu
Tongfu Microelectronics Co. Ltd.
Nantong, *China*
yu.zheng@tfme.com

Hua Wu
Tongfu Microelectronics Co. Ltd.
Nantong, China
sam.wu@tfme.com

Wanchun Ding
Tongfu Microelectronics Co. Ltd.
Nantong, China
ding.wc@tfme.com

Guohua Gao*
Tongfu Microelectronics Co. Ltd.
Nantong, China
g.guoh@tfme.com

Abstract—Packaging process is an indispensable part in the process of electronic components manufacturing, and its packaging quality directly affects the nominal power, reliability and other functions of the product in the subsequent application process. Through the research on the humidity resistance reliability of different packaging structures, C-Mount packaging structure, TO packaging structure and butterfly packaging structure, three common packaging structures are introduced and analyzed first, and then studied the humidity resistance reliability of different packaging structures by analyzing the high and low temperature and temperature cycle reliability of different packaging structures and the thermal characteristics of different packaging structures in different humidity environments research. The results show that the butterfly package structure has the best humidity resistance reliability followed by TO package structure and C-Mount package structure was the worst.

Keywords—packaging structure, humidity resistance, reliability, electronic components

I.   INTRODUCTION

In recent years, the application and packaging process of semiconductor lasers have been improved and optimized, and by virtue of its advantages of high brightness, high power and long service life, they are gradually applied in various fields [1]. Therefore, the current requirements for packaging structures are gradually increasing, especially the power and heat dissipation of the packaging structure. If the package structure is not reasonable, the problem of non-uniformity will occur in the heat dissipation process, which will have a great negative impact on the life of the package structure. Meanwhile, this problem has become the primary problem to get rid of the nominal power bottleneck of the package structure, and has been attached great importance to by relevant researchers in this field [2]. At the same time, if there is a problem in the material and structure selection of the packaging structure, the expansion coefficient of the packaging chip will not match in the use process, which will further affect the wavelength, spectrum, threshold and other characteristics of the equipment, resulting in the inability of the generated people to get fast dispersion. In such a long period of application, the functions of the encapsulated chip will gradually decrease, and power attenuation will occur, and even affect the working life of the chip, or even cause the chip to be burned down [3]. Therefore, different package structure, size and other parameters have different degrees of influence on the moisture and heat resistance performance. At present, the research on this problem is the main research topic to realize the development of packaging technology towards the advantageous direction of high nominal power

and high brightness in the future. Based on this, this paper carries out the research on the humidity and heat resistance reliability of different packaging structures of electronic components.

II.   ANALYSIS OF DIFFERENT PACKAGING STRUCTURE

In order to realize the reliability analysis of moisture and heat resistance of different package structures, three different package structures of electronic components were taken as an example for comprehensive analysis. As for electronic components, their electrical conductivity in practical application is usually between conductor and insulator, and the semiconductor materials are usually silicon, gallium, etc. Therefore, any device with PN structure can be called electronic components. Currently, common electronic components include diodes, triodes, single crystal tubes and chips, etc. [4]. For electronic packaging process, at the same time is based on discrete device packaging process, but there are large differences between them, the main characteristics of discrete device packaging technology is that it can effectively prevent the encapsulated chip being affected by the physical factors, such as to prevent other mechanical equipment, temperature and humidity, and dust in the air and other harmful effects on the circuit, causing the serious electrical performance degradation problems [5]. The main purpose of semiconductor packaging technology is to more accurately realize the connection of the circuit, and to achieve the transmission of electrical signals. In that the power electronic components in use process will be in the form of thermal energy conversion, if the electronic components are hard to dissipate heat or uneven cooling exists, it may cause stored heat accumulating too much on electronic components and resulted in local problems appearing and increasing, and further causing catastrophic optical cavity surface suffering serious injuries and the normal use of the electronic components if more severe [6].

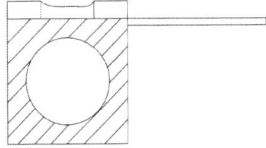

Fig. 1.   Schematic diagram of the package structure of C-mount electronic components

Currently, according to the characteristics, shapes and different functional needs of electronic components, three packaging structures have been proposed for electronic components, namely C-mount packaging structure, TO packaging structure and butterfly packaging structure. The

978-1-6654-1392-3/21 $31.00 © 2021 IEEE

following will give a detailed description of the various packaging structures proposed above.

A. C-Mount Packaging Structures

C-Mount Packaging of electronic components is a relatively mature packaging form in the current industrial production process with relatively mature technology and process. Copper metal elements with good thermal conductivity and high thermal conductivity are usually selected as the heat sink material for the packaging structure of this form [7]. At the same time, in order to facilitate the fixation of the powdery structure of electronic components, a circular pore structure is usually set in the heat sink center. Fig. 1 is a schematic diagram of the package structure of C-mount electronic components.

In the packaging structure of C-mount electronic components, the P-face of the chip is downward, the SubMount is used as the transition heat sink, welded to the upper part of the copper metal element heat sink, and then the gold wire is used to bond it and ensure the connection of the circuit [8]. At present, the electrode strip of the common packaging structure of C-mount electronic components is the negative electrode, and the heat sink of copper metal element is the positive electrode. This structure is more suitable for the chip packaging of single-tube electronic components with relatively low power at present.

B. TO Packaging Structures

TO electronic component packaging is the most common packaging form for transistors and other similar low-power components. It is composed of pin, metal shell, base, photodetector and other structures, which contains 2~4 pins with different functions, and 3 pins at the bottom of the structure [9]. The base of the TO electronic component packaging structure consists of two kinds of steel material and alloy material. Fig. 2 is the schematic diagram of the packaging structure of TO electronic components.

During the operation of electronic components, the emitted laser is usually injected into the photoelectric detection device in the package structure of the TO electronic components. The dynamic adjustment of the current value is

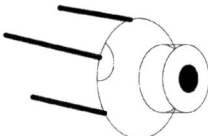

Fig. 2. Schematic diagram of the packaging structure of TO electronic components

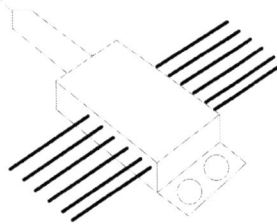

Fig. 3. Schematic diagram of package structure of butterfly electronic components

realized by using the feedback of illumination intensity, so as TO ensure the stable operation of electronic components.

C. Butterfly Package Structure

The package structure of butterfly electronic components contains more than four pins, and the overall structure looks like a butterfly, so it is called the butterfly package structure [10]. Butterfly electronic component packaging is mainly aimed at the relatively large power component packaging form, its overall volume is larger than the above two kinds of electronic component packaging structure, and in its internal usually installed refrigeration devices, so it has a good heat dissipation performance. Fig. 3 is a schematic diagram of the package structure of butterfly electronic components.

When the butterfly-shaped electronic components are packaged, the internal chip of the electronic components will not be burned due to high temperature in the working process, and it has very good stability and high reliability.

Since the size of the core component of electronic components usually only exists in a few millimeters or even tens of microns, higher requirements will be required for the dust density, temperature and humidity in the surrounding environment during the process of powder processing [11]. In the process of packaging process, the temperature and humidity are usually controlled within a certain range, the temperature is generally within the range of 16.5°C~25.5°C, the humidity is generally controlled within the range of 28.3°C~76.2°C. At the same time, because electronic components belong to weak components, and all work on the principle of PN junction. Therefore, the effect of static electricity on electronic components is often fatal. Therefore, in this paper, the anti-static technical indicators of commonly used items in packaging structures should be clarified before the subsequent research on the reliability of moisture and heat resistance of different powder structures.

TABLE I. ANTI-STATIC TECHNOLOGY INDEX TABLE OF COMMON ARTICLES IN PACKAGE STRUCTURE

The project name	Surface resistance value	Frictional voltage	Resistance to earth
Anti-static work clothes	104Ω~108Ω	<280V	——
Anti-static shoes	104Ω~108Ω	<120 V	——
Anti - static packaging table	104Ω~108Ω	<120 V	104Ω~108Ω
Anti - static packaging structure	104Ω~108Ω	<120 V	——
Packaging personnel	104Ω~108Ω	——	104Ω~108Ω

At the same time, because of the complexity of the packaging process has certain complexity, therefore, after fulfilling the packaging, there are many causes of realizing the internal electronic components and if there is something wrong with some links, it may impact on subsequent each link, and further influence the humidity resistance reliability. As a result, the following will focus on specific technological process, selection of materials of different powder structure and accuracy control of its resistance to heat and humidity to conduct a comprehensive analysis of reliability.

III. DAMP AND HEAT RESISTANCE RELIABILITY ANALYSIS OF DIFFERENT PACKAGE STRUCTURES

A. High and Low Temperature and Temperature Cycle Reliability Analysis of Different Package Structures

In view of the above three packages with different structures, the stability of the electronic components encapsulated under the long-term working state, such as high and low temperature and temperature cycle, is compared and analyzed. The main reason for setting the above two environmental conditions is to ensure that the bearing capacity of the ambient temperature of the electronic components is explored after the completion of packaging [12]. Through this article on the analysis of the temperature, choose high comprehensive performance, stability, strong reliability, equipment use strong laser as the main equipment of the study, in the process of temperature cycling on the reliability analysis, the layout of different temperature test box (high temperature and low temperature), respectively, under the condition of two kinds of temperature, perform the inspection of different packaging structure, assuming that laser precision of the above said is plus or minus 0.01 degrees Celsius, and equipment for automatic temperature protection function, namely when the outside temperature control range, the laser will automatically trigger the alarm device, and immediately cut off power supply.

Through the high and low temperature test chamber, high and low temperature tests were carried out on the above three different package structures. The high temperature was set at 56°C ± 1.5°C and the low temperature was set at - 28.5°C ± 2.5°C. When the corresponding standards of high temperature and low temperature are reached, the temperature rise and fall are automatically controlled, and the time is controlled above 15h, and the products with three different package structures are taken out. Through the microscope to observe the different packaging structure of the product is qualified, and the results are recorded. For storage in high and low temperature environment, the specific operation steps are as follows: firstly, click the MODE button in the standard state, and when the first part of PROD begins to flash, select the setting group to be modified by adding and subtraction buttons. When the value displayed by PROD is 1, you can set the parameters in the first paragraph and click the MODE button. Repeat the above operation for 20~30 times to complete the setting of two different environmental conditions, high and low temperature.

In a storage box with a temperature range from - 28.5° C to 56°C, the temperature was alternately cycled for 100 times. After the cleaning was completed, it was taken out and placed at room temperature for 2.5h. The bearing capacity of the three different package structure products on the temperature cycle was tested and the matching of the expansion coefficient was recorded.

Through the above operations, the high and low temperature and temperature cycle tests of different package structures were completed, and the test results were obtained as follows: butterfly electronic components package structure has the best reliability under high and low temperature and temperature cycle environment, followed by TO electronic components package structure, and finally C-mount electronic components package structure.

B. Analysis of Thermal Characteristics of Different Package Structures in Different Humidity Environment

After the above analysis of the high and low temperature and temperature cycle reliability of different package structures, it is also necessary to analyze the thermal characteristics of different package structures in different humidity environments, so as to obtain the final comprehensive humidity and heat resistance reliability of different package structures. In the normal operation of electronic components, part of their energy will be converted in the form of heat energy, which will improve the temperature of the whole product through conduction. In different humidity environment, the amount of thermal energy conversion of electronic components with different package structures is greatly different. Therefore, in order to further explore the thermal characteristics in different humidity environment. In this paper, the thermal analysis method is introduced to measure the relationship between thermal characteristics and humidity of products with different package structures under the program control of humidity. The steady-state thermal balance of package structure product is: input energy - output energy =0. Then, the differential formula of steady-state heat transfer is further deduced as follows:

$$\frac{\partial}{\partial x}(K\frac{\partial T}{\partial x}) + \frac{\partial}{\partial y}(K\frac{\partial T}{\partial y}) + q = 0 \qquad (1)$$

In Formula (1), x is the heat transfer distance in the horizontal direction; y is the heat transfer distance in the vertical direction; T is the heat transfer time of products with different package structures; K is the heat transfer coefficient; q denotes ambient humidity. According to the above formula (1), the corresponding finite element equilibrium equation can be obtained as follows:

$$\{K\} \cdot \{T\} = \{Q\} \qquad (2)$$

During the operation, humidity can be applied to a specific area, and uniform humidity can be applied to all nodes and does not constitute a humidity constraint condition. Assuming that the photoelectric conversion efficiency of the electronic components themselves is 50%, the room temperature is 26°C, and the nominal power is 6.5W, under different humidity control parameters, Workbench19.6 is used for steady-state thermal analysis of the electronic components with different package structure materials. Through the steady state thermal analysis of the structure got from Workbench19.6, when the temperature increased, the different packaging structure showed the trend of thermal characteristics to reduce, but the butterfly electronic packaging structure under the condition of the thermal characteristics of decreased significantly more than the other two packaging structure is not obvious and the most obvious decrease trend exists in C-Mount electronic components packaging structure of the steady state thermal analysis results. Therefore, based on the above analysis in this paper, it can be concluded that the humidity and heat resistance reliability of the package structure of butterfly electronic components is obviously better than that of the other two package structures, and it can be applied more widely. At the same time, compared with the other two package structures, the humidity and heat resistance reliability of O electronic components package structure is obviously better than that of C-mount electronic components package structure. Therefore, in practical application, according to the packaging needs of different electronic components, the above three different packaging structures can be comprehensively selected.

IV. CONCLUSION

Combined with the above study of this topic, the research of laser belongs to the important evaluation index of the photoelectric packaging structure, but due to different packaging structure of the influence of different reliability index for damp and hot is different. In order to further explore this kind of relations, in this paper, the humidity resistance reliability of different packaging structure reliability was studied, in order to ensure the accuracy of the research results, in the study, we puts forward the comprehensive characteristic testing way. After finishing researching, we analyze different packaging structures under different humidity, temperature, the comprehensive stability. Because in this paper, the research has more than 2000.0 h, it can be considered that the research results of humidity and heat resistance reliability in this paper are relatively real. Synthesizing the research results of this paper, it can also provide guidance for the production and mass construction of packaging structure equipment in the later stage, so as to ensure the sustainable development of machinery and equipment production units in China.

REFERENCES

[1] H. Guo, Y. Y. Zhou, and Q. Li, "Disinfection efficacy and cost analysis of disinfecting wipes combined with 75% ethanol gauze on heat and humidity-sensible reusable medical instruments," Chinese Nursing Research, vol. 33, no. 24, pp. 4321–4324, 2019.

[2] L. T. Li, B. Jing, J. X. Hu, Y. L. Zhang, and Z. G. Li, "Degradation law analysis of QFP package interconnection structure based on PCMD health index," Journal of Electronic Measurement and Instrumentation, vol. 33, no. 12, pp. 100–108, 2019.

[3] D. L. Zhang, Z. W. Liu, J. G. Zou, and Z. Y. Sun, "Preparation and Properties of Epoxy Pouring Sealant Modified by Humidity and Heat Resistant Flexible Polyurethane," Insulating Materials, vol. 53, no. 09, pp. 19–23, 2020.

[4] T. Huang, Q. H. Liao, W. Y. Wu, and C. Luo, "Analysis of delamination failure behavior of QFN packaging conductive adhesive with laminated chip structure," Electronic Components and Materials, vol. 39, no. 09, pp. 97–104, 2020.

[5] X. L. Lin, X. W. Zhang, and R. Gao, "A Circuit Edit Method for ICs of Flip-Chip /Multilayer Interconnected Structure," Journal of South China University of Technology(Natural Science Edition), vol.48, no. 12, pp. 63–71, 2020.

[6] W. J. Wang, S. L. Hu, Y. T. Zheng, Z. G. Zhao and H. J. Zhou, "A Parallel Preconditioned Finite Element Method and Its Applications in System – in – Package Electromagnetic Simulations," Acta Electronica Sinica, vol.49, no. 01, pp. 58–63, 2021.

[7] C. Li, X. W. Yang, Y. L. Zhao, X. Cheng, and L. Tian, "SOI pressure sensor chip suitable for leadless package," China Measurement & Test, vol. 46, no. 12, pp. 54–59, 2020.

[8] Z. Y. Ping, W. Zhao, Q. H. Zheng, and W. W. Zhou, "Spectra and Packaging Performance of BaLaLiTeO6: Eu3+ Red-emitting Phosphors for White LEDs," Chinese Journal of Luminescence, vol. 40, no. 04, pp. 432–439, 2019.

[9] H. Y. Wang, Z. B. Zhao, P. Y. Fu, J. Y. Li, and P. Zhang, "Insulating Properties of PEEK Frame in Pressed IGBT Device," Insulating Materials, vol. 52, no. 06, pp. 60–66, 2019.

[10] J. X. Hu, B. Jing, Y. F. Huang, Z. J. Sheng, Y. J. Chen, and Y. L. Zhang, "Electrical Characteristics Modeling and Degradation Analysis of QFP Package Interconnect Structure," Acta Electronica Sinica, vol. 47, no. 02, pp. 366–373, 2019.

[11] Y. B. Du, Y. H. Zhang, R. G. Zhang, and S. Lin, "The Design of COB Packaging Structure with New Phosphor Coating," China Illuminating Engineering Journal, vol. 30, no. 01, pp. 43–46, 2019.

[12] J. Yang, B. Wang, and Y. Liu, "Mechanical Reliability Analysis of the Packaging Architecture with TSV Interposer," Electronics & Packaging, vol. 19, no. 10, pp. 4–7+12, 2019.

Electroplating nanotwinned copper for ultra-fine pitch redistribution layer (RDL) of advanced packaging technology

Yu-Bo Zhang
School of Materials Science and Engineering,Harbin University of Science and Technology, Harbin150080, China
Shenzhen Institute of Advanced Electronic Materials, Shenzhen Institute of Advanced Technology, CAS
Shenzhen 518055, China

Li-Yin Gao*
Shenzhen Institute of Advanced Electronic Materials, Shenzhen Institute of Advanced Technology, Chinese Academy of Sciences
Shenzhen 518055, China
ly.gao@siat.ac.cn

Xiao Li
Shenzhen Institute of Advanced Electronic Materials, Shenzhen Institute of Advanced Technology, Chinese Academy of Sciences
Shenzhen 518055, China

Zhe Li
Shenzhen Institute of Advanced Electronic Materials, Shenzhen Institute of Advanced Technology, Chinese Academy of Sciences
Shenzhen 518055, China

Xu-Liang Ma*
School of Materials Science and Engineering,
Harbin University of Science and Technology,
Harbin 150080, China
maxuliang@hrbust.edu.cn

Zhi-Quan Liu
Shenzhen Institute of Advanced Electronic Materials, Shenzhen Institute of Advanced Technology, Chinese Academy of Sciences
Shenzhen 518055, China

Rong Sun
Shenzhen Institute of Advanced Electronic Materials, Shenzhen Institute of Advanced Technology, Chinese Academy of Sciences
Shenzhen 518055, China

Abstract—Copper is the most common interconnected materials in the field of microelectronic packaging, and it is widely used in various advanced package technologies, such as redistribution layer (RDL), copper pillar (CuP) and under bump metallization (UBM) etc. However, the rapid development of packaging technology, decreasing size and increasing Joule heat make a great challenge on the mechanical properties and thermal stability of copper interconnected materials. It was widely reported that the nanotwinned copper had very high strength and excellent conductivity, which was quite suitable for the next generation interconnected materials. However, different from electroplating thin films, the achievement of uniformity and flatness for electroplating ultra-fine pitch nanotwinned RDLs would be difficult since the line width was as low as 1.5μm. Through the study of growth mechanism and additives effect, we had invented an acid nanotwinned copper electrolyte as reported previously. So in our study, the evaluation of modified nanotwinned copper electrolyte was conducted in order to clarify its performance of on the electroplating ultra-fine pitch RDLs. Also, three kinds of commercial RDL electrolytes were served as comparisons. In detail, different electrolytes were evaluated from the aspect of microstructure, hardness, thermal stability, uniformity and flatness. Firstly, the morphology of electroplating copper film was characterized by focus ion beam (FIB), and then the hardness was measured. The thermal stability was revealed by the hardness evolution before and after heat treatment test. Finally, the 1.5μm, 4μm and 15μm RDLs were electroplated on a patterned wafer, and their uniformity and flatness were calculated. Both uniformity and flatness showed a decreasing tendency basically when the line width kept decreasing. The uniformity and flatness were 6.73~13.89% and 2.32~4.17% respectively for nanotwinned copper electrolyte, which was at the upper level compared to the commercial RDL electrolytes.

Keywords—electroplating; nanotwinned copper; redistributed layer (RDL); thermal stability; uniformity

INTRODUCTION

Copper is the most common interconnected materials in the field of microelectronic package. And it was widely used in the advanced package technologies, such as redistribution layer(RDL), copper pillar(CuP) and under bump metallization (UBM) etc[1].

During the DC electroplating process, the acid/copper ratio and the use of additives within the electroplating solution has a greater impact on the performance of electroplating films. Usually, a small amount additive plays a key role on smoothing the roughness of surface, eliminating defects and improving the filling capacity of complicated figures [2, 3]. Cheng et al. [4] studied the electrochemical behavior of three commonly used copper plating additives, and screened out an additive system with excellent leveling performance, further applied it in the production process of signal transmission lines. Tao et al.[5]studied the effect of copper sulfate and sulfuric acid on the filling effect of blind holes with certain diameter and depth in a certain high density interconnect (HDI) printed circuit board. What's more, the potential difference of E-t curves under different rotation speed was proposed to characterize the hole filling capacity [6,7].Wang et al. [8] revealed the action mechanism of additives including MPS, IML and EPE4500 separately and mutually within the copper electroplating solutions during the blind hole filling process.

Above all, the acidic copper electrolyte and its electroplating layers have been applied in the industry for decades[9]. However, the rapid development of packaging technology, decreasing size of interconnected materials and increasing Joule heat make a great challenge on the mechanical properties of copper interconnected materials [10]. Ordinary microstructure copper was prone to recrystallize under high Joule heat, which leads to a decline of mechanical properties or even fracture at the grain boundaries. Nanotwinned copper was reported to have a high strength of 1068MPa, which was about ten folds compared to the coarse grain copper. And its conductivity was about $96.9 \pm 1.1\%$ of the international annealed copper standard (IACS)[11]. However, previous studies of nanotwinned copper were mostly focus on electroplating thin films [12]. While for fine-pitch RDLs, nanotwinned copper electrolytes reported in literatures lacked key additives, including accelerator, suppressor and leveler, which can serve an important role on electroplating effect. Through the study of growth mechanism and the effect of additives within nanotwinned electrolyte, we had invented an acid nanotwinned copper electrolyte as reported previously [13-14].

In our study, in order to evaluate the performance of our modified nanotwinned copper electrolyte and other commercial electrolytes, the microstructure, hardness and thermal stability were investigated. Fine pitch RDLs with different line width (1.5μm~15μm) were electrodeposited, which uniformity and flatness were calculated in order to reveal the electroplating effects quantitatively.

EXPERIMENT

A. Elecroplating details

A customized polytetrafluoroethylene (PTFE) electroplating cell was used to electroplate the copper film and RDL in this study. And a DC power supply (Tektronix, PWS-4323) was used to offer constant current during the electroplating process. The stirring speed was set at 300 rpm, the temperature was set at $25\pm2°C$, and the current density was $20\sim30mA/cm^2$. The anode was a soluble phosphorus copper plate. The cathode was a silicon wafer seeded with 100nm Ti and 400nm Cu for microstructural observation and hardness measurement [15-16]. For the analysis of uniformity and flatness, the cathodic silicon wafer was patterned with fine pitch RDLs, which line width ranges from 1.5μm to 15μm. Before electroplating, 10% dilute sulfuric acid was used to remove the oxide film of the seed layer on the wafer.

Four copper electroplating solutions were used in this study, which were all consisted of $CuSO_4\cdot5H_2O$, H_2SO_4, NaCl and additives (including accelerator, leveler and suppressor etc). The main compositions of the four different electrolytes were listed in Table1. Notably, as state above, the NT electrolyte was a novel electrolyte modified for the RDL appliaiton based on our previous work. While the other three commercial electrolytes were foreign products, which were widely used in large packaging factories. The commercial electrolyte was denoted as CE1, CE2 and CE3 later for simplicity, while our modified nanotwinned copper electrolyte was denoted as NT[17-18].

Table 1 The compositions of different electrolytes

Contents	CE1	CE2	CE3	NT
$CuSO_4\cdot5H_2O(g/L)$	140~160	190~200	230~250	100~120
$H_2SO_4(g/L)$	120~140	80~100	40~60	20~40
NaCl(g/L)	0.05~0.1	0.05~0.1	0.05~0.1	0.05~0.1
Accelerator (mL/L)	4~8	4~8	1~5	1~5
Leveler(mL/L)	8~12	10~15	15~20	1~5
Suppressor(mL/L)	0	10~15	10~15	1~5

B. Microstructure and hardness characterization

A focused ion beam system (FIB, FEI Scios, UK) was used to observe the cross sectional copper electroplating layers. A voltage of 30 kV, a current of 50 pA and a tilt angle of 52° were used during the observation. The preparation process of cross sectional samples were listed below: The films were firstly embedded in epoxy resin and polished with 100#, 250#, 500# , 1000# and 2000# sandpapers subsequently. Then the films were polished with SiC paste before the FIB observation.

For Vickers hardness test, a thin film about 100μm thick was electroplated on the seeded silicon wafer. And a hardness tester (Mitutoyo, HM-200A) was used [19]. More than 5 points for each kind of sample were measured in order to calculate the average hardness.

C. Calculation of uniformity and flatness

The 1.5 μm, 4μm and 15μm wide RDLs were electroplated. The morphology was observed by 3D laser confocal microscopy (CLSM, KEYENCE, VK-X1100).

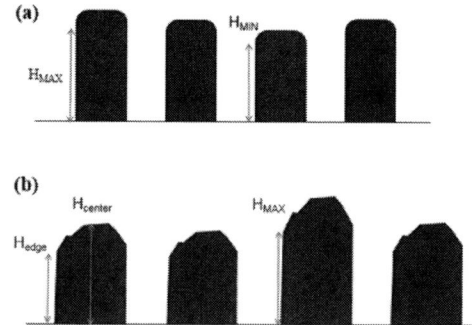

Fig.1 Schematic drawing of uniformity and flatness (a)uniformity; (b)flatness

The uniformity and flatness are calculated by the below equation in order to reveal the filling performance.

$$U = \frac{1}{2}\left[\frac{H_{MAX} - H_{MIN}}{H_{AVG}}\right] \times 100\% \qquad (1)$$

where U stands for the uniformity, H_{MAX} and H_{MIN} are the maximum and minimum heights of each line width, and H_{AVG} is the average height.

$$L = \frac{H_{center} - H_{edge}}{H_{MAX}} \times 100\% \qquad (2)$$

Where L is flatness, H_{edge} and H_{center} are the edge height and center height of a single line. The schematic drawing of uniformity and flatness is given in Fig1.

Fig.2 FIB images of cross-sectional copper films electroplating using different baths, (a) CE1; (b) CE2and (c) CE3 (d) NT.

RESULTS AND DISCUSSION

A. The microstructure of electroplating copper films

The cross sectional image of four electroplated films were observed using FIB. The copper films electrodeposited by CE1 and CE2 plating solutions were relatively uniform. A few nanotwinned boundaries can be detected within the grains. Grain size of CE2 samples seemed a little larger than CE1 samples. Notably, CE1, CE2 and CE3 samples were all consisted of equiaxed grains In contrast, the grain size of the films obtained by the CE3 plating solution was in a bimodal distribution. Diameter of the larger grains (highlighted as A) was about several microns, which contains high density of nanotwinned boundaries. Diameter of smaller grains (highlighted as B) was below 1μm. So the microstructure of CE3 samples looks nonuniform as shown in Fig.2c. The microstructure of NT samples was shown in Fig.2d. Different from the equiaxed grains of commercial electrolytes, the NT samples was consisted of columnar grains, in which high density horizontal nanotwinned boudaries were observed.

B. The hardness and thermal stability of electroplating layers

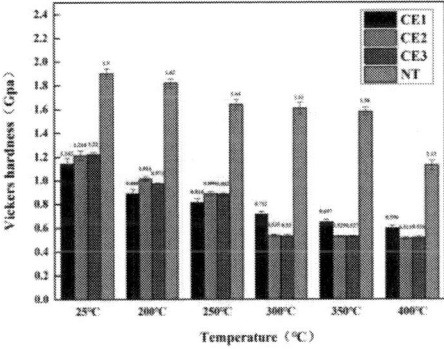

Fig. 3. The Vickers hardness of different copper electroplating layers before and after heat treatments

Figure 3 shows the Vickers hardness of 100μm thick copper films electroplated upon Cu/Ti seeded wafer substrate. The hardness of CE1 samples (1.142GPa) was a little lower than that of CE2 (1.214GPa) and CE3 (1.22GPa) in the initiate state. Notably, the hardness of NT samples was far higher than the other three samples, which was estimated at 1.9GPa. The heating temperature was set as 200°C, 250°C, 300°C, 350°C and 400°C respectively for an hour.

The hardness of copper films kept decreasing when the heat treatment temperature continuously increased from room temperature to 400°C. When the films were heated for an hour at 200°C, the hardness of CE1 samples decreased about 22.2%, and a hardness of 0.888GPa was obtained. The CE2 samples after 200° C heat treatment was 1.014GPa(16.4% decreased), and CE3 sample was 0.973GPa(20.2% decreased). In contrast, the NT samples only dropped 4.2% compared to the initiate state.

Statistically, the hardness of CE1 samples dropped 47.8% after 400° C heat treatment compared to the initial state, while the CE2 copper film dropped 57.9%, and the CE3 copper film dropped 57.7% dramatically. For CE2 and CE3 samples, the hardness firstly decreased, then remained relatively stable when the temperature reached 300℃. However, for NT sample, the hardness decreased slowly below 350°C, which demonstrated a higher thermal stability among the four samples. Since the upper temperature limit for RDL fabrication process is about 350° C, so the nanotwinned copper possessed an excellent thermal stability for the RDL application.

C. Electroplating fine pitch RDLs

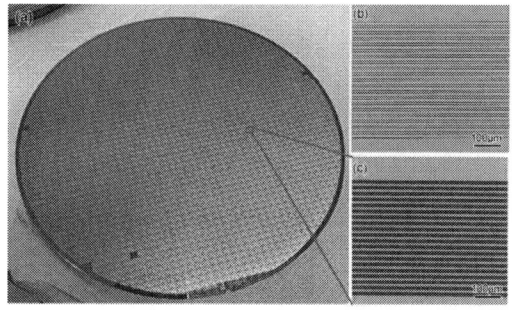

Fig.4 (a)an 8 inch wafer with fine pitch RDL patterns; (b)the fine pitch RDLbefore electroplating, (c)after electroplating.

Fig. 4a showed an 8 inch wafer with fine pitch RDL patterns and Fig.4b was the RDL's morphology of a patterned wafer before electroplating. The photoresist was fabricated upon Ti/Cu seed layer to form the RDL patterns. After 6min electroplating, the 15μm wide RDL was electroplated on the wafer without any defects as shown in the Figure 4c.

In our study, three dimensions RDL, i.e. 15μm, 4μm and 1.5μm RDLs were electroplated. Figure 5 showed the morphology of different RDL lines under 3D laser confocal microscope (CLSM). In detail, Figure 5a-d were the morphologies of 15μm wide RDLs prepared by CE1, CE2, CE3 and NT solutions, respectively. Figure 5a'-d' and a''-d'' were the situations of 4μm and 1.5 μm samples, respectively. It can be clearly seen that the surface of the RDL electrodeposited using CE1 and CE2 plating solutions were relatively smooth, while the surface of NT and CE3 samples had some bumps, which was relatively rough. It was deduced that grain size of NT and CE3 was relatively larger, and the growth steps of nanotwins led to the rough surface. Also, the difference of electroplating solutions, including copper acid ratios, content of additives, had a great impact on microstructure, which further influenced the electroplating effect of fine-pitch RDLs.

In order to exanimate different electrolytes, the height of more than nineteen RDLs for each line width were counted. Also, the maximum, minimum and average values were calculated for the uniformity. The outline morphologies of nineteen RDLs for each line width was observed by 3D laser confocal microscopy (CLSM), and the height difference at the center and edge were measured in order to calculate the flatness.

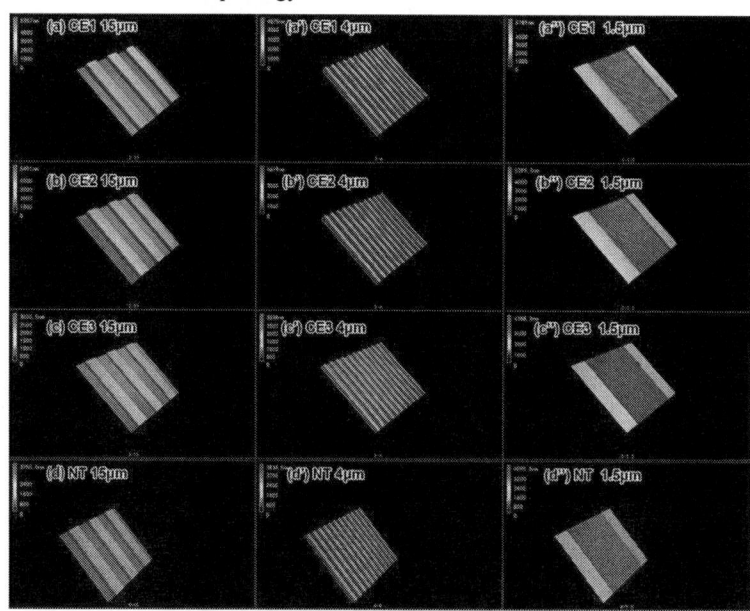

Fig.5 The laser confocal images of three RDL line widths using different plating solutions after electroplating 6 minutes : (a) CE1 15μm; (b) CE2 15μm; (c) CE3 15μm; (d) NT 15μm; (a') CE1 4μm; (b') CE2 4μm; (c') CE3 4μm; (d') NT 4μm; (a'') CE1 1.5μm; (b'') CE2 1.5μm; (c'') CE3 1.5μm; (d'') NT 1.5μm .

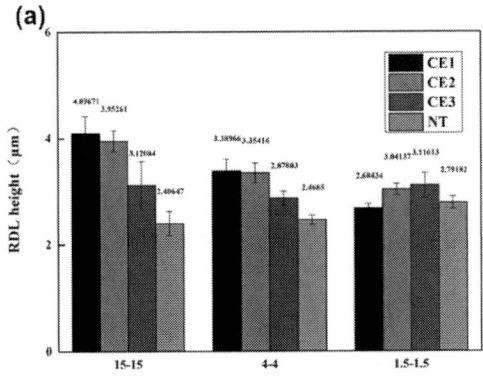

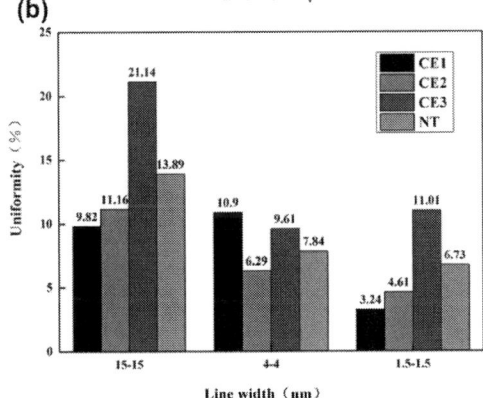

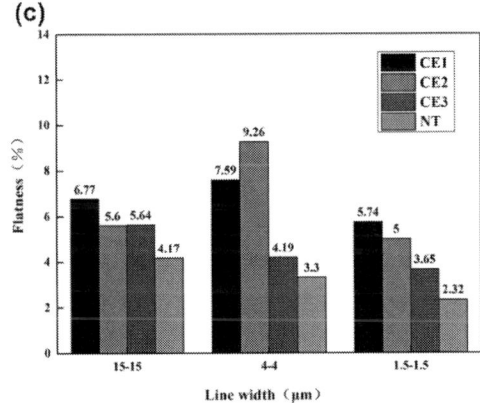

Fig.6 (a) the statistical results for the height of RDLs with different line widths; the calculated (b) uniformity and (c) flatness.

Figure 6a showed the statistic results of the height within different RDL lines. And Figure 6b and 6c showed the calculated result of uniformity and flatness. The uniformity of RDL refers to the height differences between different lines. And the flatness refers to the height differences within a single line. The uniformity of CE1, CE2, CE3 and NT samples were 3.24~9.82%, 4.61~11.26%, 9.61~21.14% and 6.73~13.89%. So the subsequence of uniformity was CE1＞CE2＞NT＞CE3. It can be seen that the RDL lines obtained from the CE1 and CE2 baths had a good uniformity since the height difference was relatively smaller. As shown in Fig.6c, the

flatness were 5.74~7.59%, 5~9.26%, 3.65~5.64% and 2.32~4.17% for CE1, CE2, CE3 and NT samples, respectively. The subsequence of flatness was NT＞CE3＞CE1＞CE2. Since both flatness and uniformity are important indexes to evaluate the uniformity, so it can be safely concluded that the NT samples with a good uniformity of 6.73~13.89% and an excellent flatness of 2.32~4.17% possessed a superior filling capability for fine pitch RDLs.

CONCLUSION

In our study, a novel nanotwinned copper electrolyte were evaluated from the aspect of microstructure, hardness, thermal stability, uniformity and flatness. For microstructural observation, different from the samples using commercial copper electrolytes, the NT samples were consisted of columnar grains with high density nanotwinned boundaries rather than equiaxed grains. What's more, it was demonstrated from the heat treatment test, the NT samples is about twice higher than the commercial samples in the initiate state, and kept an excellent thermal stability below 350°C annealing. Finally, in order to evaluate the electroplating effect, the uniformity and flatness of NT samples were 6.73~13.89% and 2.32~4.17%, which were at upper level compared to the other three commercial products. Above all, our work demonstrates the nanotwinned copper is a promising candidate for the RDL applications.

ACKNOWLEDGMENT

This work was financially supported by Guangdong Basic Research Fund (Grant No. 2019A1515110771 & 2019A1515110469) and SIAT Innovation Program for Excellent Young Researchers.

REFERENCES

[1] 金磊, 杨家强, 杨防祖, 等. 芯片铜互连研究及进展 [J]. 电化学, 2020, 26(4): 521-530.

Jin L, Yang JQ, Yang FZ, et al. Research progresses of copper interconnection in chips [J]. Journal of Electrochemistry, 2020, 26(4): 521-530.

[2] M.A. Pasquale, L.M. Gassa, A.J. Arvia. Copper electrodeposition from an acidic plating bath containing accelerating and inhibiting organic additives[J]. Electrochimica Acta, 2008, 53(20).

[3] Zhang, Q.; Yu, X.; Hua, Y.; Xue, W., The effect of quaternary ammonium-based ionic liquids on copper electrodeposition from acidic sulfate electrolyte. Journal of Applied Electrochemistry 2014, 45 (1), 79-86.

[4] 陈昀钊. 印制电路镀铜添加剂调控铜沉积平整化的研究[D]. 电子科技大学, 2019.

Chen JZ. Research on Copper Plating Additives for Printed Circuits Regulating Copper Deposition Leveling[D]. University of Electronic Science and Technology, 2019.

[5] Tao, P.; Chen, Y.; Cai, W.; Meng, Z. Effect of Copper Sulfate and Sulfuric Acid on Blind Hole Filling of HDI Circuit Boards by Electroplating. Materials 2021, 14, 85.

[6]   Elgrishi, N.; Rountree, K. J.; McCarthy, B. D.; Rountree, E. S.; Eisenhart, T. T.; Dempsey, J. L., A Practical Beginner's Guide to Cyclic Voltammetry. Journal of Chemical Education 2017, 95 (2), 197-206.

[7]   Zheng, L.; He, W.; Zhu, K.; Wang, C.; Wang, S.; Hong, Y.; Chen, Y.; Zhou, G.; Miao, H.; Zhou, J., Investigation of poly (1-vinyl imidazole co 1, 4-butanediol diglycidyl ether) as a leveler for copper electroplating of through-hole. Electrochimica Acta 2018, 283, 560-567.

[8]   王义. 电镀铜填盲孔添加剂性能及机理研究[D]. 江西理工大学, 2018.

      Wang Y.Study on the Performance and Mechanism of Additives for Filling Blind Holes in Copper Electroplating[D].Jiangxi University of Science and Technology,2018.

[9]   Brown, D. A.; Morgan, S.; Peldzinski, V.; Brüning, R., Crystal growth patterns in DC and pulsed plated galvanic copper films on (1 1 1), (1 0 0) and (1 1 0) copper surfaces. Journal of Crystal Growth 2017, 478, 220-228.

[10] Dixit, P.; Salonen, J.; Pohjonen, H.; Monnoyer, P., The application of dry photoresists in fabricating cost-effective tapered through-silicon vias and redistribution lines in a single step. Journal of Micromechanics and Microengineering 2011, 21 (2).

[11] 卢柯. 纳米孪晶纯铜的强度和导电性研究 [J]. 中国科学院院刊, 2004(5): 352-354.

      Lu K. A study on ultrahigh strength and high electrical conductivity in copper [J]. Bulletin of the Chinese Academy of Sciences, 2004(5): 352-354.

[12] Li SJ, Zhu QS, Zheng BD, et al. Nano-scale twinned Cu with ultrahigh strength prepared by direct current electrodeposition [J]. Materials Science & Engineering A, 2019, 758: 1-6.

[13] Sun FL, Liu ZQ, Li CF, et al. Bottom-up electrodeposition of large-scale nanotwinned copper within 3D through silicon via [J]. Materials, 2018, 11(2): 319.

[14] Huang J, Gao LY, Liu ZQ. The electrochemical behavior of leveler JGB during electroplating of nanotwinned copper. 2020 21st International Conference on Electronic Packaging Technology(ICEPT). 2020.

[15] Sun FL, Liu ZQ, Li CF, et al. Bottom-Up Electrodeposition of Large-Scale Nanotwinned Copper within 3D Through Silicon Via.[J]. Materials (Basel, Switzerland),2018,11(2).

[16] Sun FL, Gao LY，Liu ZQ, et al. Electrodeposition and growth mechanism of preferentially orientated nanotwinned Cu on silicon wafer substrate[J]. Journal of Materials Science & Technology, 2018, 34(10):1885-1890.

[17] Sun FL, Gao LY，Liu ZQ. Electrodeposition of Nanotwinned Copper Film as under Bump Metallization. 2017 18th International Conference on Electronic Packaging Technology. 2017.

[18] Li ZG, Gao LY, Liu ZQ. The effect of transition layer on the strength of nanotwinned copper film by DC electrodeposition. 2020 21st International Conference on Electronic Packaging Technology(ICEPT). 2020.

[19] 黄静, 李忠国, 高丽茵, 等. 亚甲基蓝对直流电镀纳米孪晶铜组织及力学性能的影响[J]. 集成技术, 2021, 10(1): 55-62.

      Huang J, Li ZG, Gao LY, et al. Effect of methylene blue on the microstructure and mechanical properties of nanotwinned copper during DC electroplating [J]. Journal of Integration Technology, 2021, 10(1): 55-62.

Mixed Cu Nanoparticles and Cu Microparticles with Promising Low-temperature and Low-pressure Sintering Properties and Inoxidizability for Microelectronic Packaging Applications

Haiqi Lai[1#]
State Key Laboratory of Precision Electronic Manufacturing Technology and Equipment
[1]Guangdong University of Technology
Guangzhou, China
1099436451@qq.com

Jian Wen[2#]
[2]Jihua Laboratory，
Foshan, China
wenjian@jihualab.com

Guannan Yang[1*]
State Key Laboratory of Precision Electronic Manufacturing Technology and Equipment
[1]Guangdong University of Technology
Guangzhou, China
ygn@gdut.edu.cn

Yu Zhang[1]
State Key Laboratory of Precision Electronic Manufacturing Technology and Equipment
[1]Guangdong University of Technology
Guangzhou, China
zhangyu@gdut.edu.cn

Chengqiang Cui[1*]
State Key Laboratory of Precision Electronic Manufacturing Technology and Equipment
[1]Guangdong University of Technology
Guangzhou, China
cqcui@gdut.edu.cn

Abstract— Nano-copper has become an important interconnection material for third-generation semiconductor devices due to its excellent electrical and thermal properties. However, nano-copper also has the characteristics of easy oxidation, which limits its application in micron packaging. In order to solve the above problems, this paper proposes to use micro-nano multi-scale composite copper powder instead of nano-copper powder. Adding micro-copper to nano-copper can improve the oxidation resistance of copper powder, while retaining a small amount of small-size nano-copper so that the copper powder remains unchanged. Possess good sintering performance. In this paper, a multi-size composite micro-nano copper powder of 200nm~1.8μm is used to prepare a micro-nano copper paste to interconnect the chip and the substrate. The hot press sintering and pressureless sintering are performed at 220°C and 260°C respectively. The results show that the micro-nano copper powder can obtain better sintering performance under low pressure, low temperature or no pressure conditions. The shear strength of 42.7 MPa can be obtained under sintering conditions of 220°C and 2 MPa, and the shear strength of 27.1 MPa can be obtained under pressureless sintering conditions of 260°C. A strong connection between the chip and the substrate is achieved. Through the XRD diffraction analysis of the pressureless sintered sample, it was found that the sintered body of micro-nano copper particles did not appear to be oxidized, and had better oxidation resistance. Under the conditions of low temperature, low pressure or no pressure, the multi-size composite micro-nano copper powder can obtain higher interconnection strength, and at the same time has stronger oxidation resistance and lower cost than nano-copper powder. This method provides a new idea for improving the oxidation resistance of nano-copper.

Keywords—Micro-nano copper powder, package interconnection, multi-size composite, low temperature and low pressure

I. INTRODUCTION

In the field of microelectronic packaging, the third-generation semiconductor devices have been facing the characteristics of high power and high heat dissipation.

Therefore, the packaging interconnection materials used for third-generation semiconductor devices should have high comprehensive characteristics. The thermal, electrical and mechanical properties of the chip connection interface and the packaging process determine the working reliability of the microelectronic system. The chip connection interface material of the third-generation semiconductor device should have the following characteristics[1-3]: a high melting point (>400°C) to improve the reliability of the connection interface under extreme high temperature environments; high thermal conductivity and low resistivity to match high thermal conductivity and low thermal conductivity On-resistance third-generation semiconductor devices; relatively low process temperature (<350°C) to reduce thermal stress in the packaging process of power electronic devices and ensure process consistency.

Copper is an excellent metal material. Its thermal conductivity is second only to silver, but its starting cost is much lower than that of metallic silver. Copper paste has been used in the microelectronic packaging and manufacturing industries[4]. Due to the excellent conductivity and heat dissipation of nano-copper. Nano-copper used for packaging interconnection has the characteristics of easy oxidation, which leads to severe sintering conditions of nano-copper paste, low strength of interconnection joints and insufficient reliability. Researchers at home and abroad mostly use bimetallic core-shell structure and organic coating of nano-copper to improve its oxidation resistance[5-6]. However, these methods are relatively cumbersome, so this article proposes to use nano-copper and micro-copper to prepare multi-size composite micro-nano copper powder with stronger oxidation resistance for microelectronic packaging interconnection. Micron copper has a higher melting point and sintering temperature, and its surface activity is relatively low. It is not easy to oxidize in the air, has good electrical and thermal conductivity, and has relatively stable chemical properties[7-8]. At present, the research on pressureless sintering of multi-size micro-nano copper particles has not

978-1-6654-1392-3/21 $31.00 © 2021 IEEE

been carried out. Therefore, this paper has carried out the research of pressureless sintering and low-temperature and low-pressure sintering of multi-size micro-nano copper paste composed of micron copper and nano-copper More energy-saving and environmentally friendly new packaging interconnection materials meet the "low-temperature interconnection, high-temperature service" working requirements of the third-generation semiconductor devices, and at the same time reduce the preparation cost of packaging materials.

II. EXPERIMENTAL PART

A. Experimental materials

The materials used in the experiment included 200 nm~1.8 μm copper powder, ethylene glycol, and absolute ethanol. The above reagents were purchased from Aladdin Reagent (Shanghai) Co., Ltd., and the dilute sulfuric acid was configured in the laboratory. The instruments used include electronic spiral micrometer, four-point probe, centrifuge, hot pressing sintering furnace, thrust testing machine, integrated ultrasonic crusher, ultrasonic cleaning machine, and rapid annealing furnace.

B. Experimental method

A commercial micro-nano composite copper powder of 200 nm to 1.8 μm was used as the experimental material, the copper powder was poured into a centrifuge tube, and anhydrous ethanol was added dropwise to prepare a multi-micro nano-copper solution. Use an integrated ultrasonic breaker to ultrasonically oscillate the solution. Centrifugation was performed after ultrasonic vibration for 4 min. The powder after centrifugation was pickled, and ultrasonic vibration centrifugation was performed at the same time. Finally, ethylene glycol was added, and a multi-size composite micro-nano copper paste was obtained after centrifugal separation.

An applicator was used to coat the copper paste on the small copper plate, cover the small copper plate on the large copper plate to form a sandwich structure, and the large and small copper plates were used to simulate the flip-chip packaging structure of the chip. Hot-pressing sintering furnace (VHP-5T-4) and rapid annealing furnace (Unitemp RTP-100) were used for hot-pressing sintering and pressureless sintering respectively.

A shear tester (Dage SERIES-4000) was used to test the thrust of the sample. A field emission scanning electron microscope (Hitach SU8220) was used to observe the morphology of the fracture surface of the sintered sample. An X-ray diffractometer (Bruker D8-ADVANCE) was used to conduct phase analysis on the sintered micro-nano copper sintered body to detect whether the powder was oxidized.

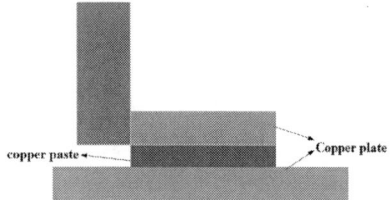

Fig. 1. Schematic diagram of thrust test of sintered specimen

III. RESULT AND DISCUSSION

A. Shear strength test

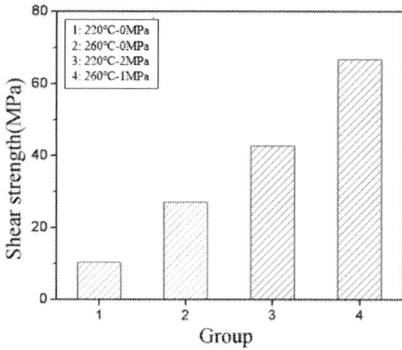

Fig. 2. Shear strength of micro-nano copper powder under different sintering conditions

The sintering temperature was set to 220°C and 260°C, and hot-press sintering and pressureless sintering were performed to compare the difference in connection performance between unpressurized and pressurized. It can be seen from the shear strength graph that the shear strength of the micro-nano copper powder for pressureless sintering at 260°C was reduced to 27.1 MPa, and the shear strength for pressureless sintering at 220°C was reduced to 10.4 MPa. Under pressureless conditions, as the sintering temperature decreases, the shear strength of the micro-nano copper powder also decreases. When the sintering pressure was 2 MPa and the sintering temperature was 220°C, the shear strength reached 42.7 MPa. Under pressurized conditions, the shear strength of pressureless sintering was increased by a higher value. Therefore, the micro-nano copper paste can obtain better interconnection strength at a lower sintering temperature under the condition of hot press sintering. The shear strength of the micro-nano copper powder reaches 66.6 MPa at a sintering temperature of 260°C. and a pressure of 1 MPa, which indicates that the copper paste can also obtain a strong sintering effect under a lower pressure. Therefore, it can be seen from the above shear strength data that the micro-nano copper powder can obtain better sintering effect at lower temperature (220°C) and lower pressure (1 MPa), and obtain stronger shear strength. A shear strength of 27.1 MPa can be obtained even under no pressure at 260°C, so the micro-nano copper paste also has better sintering performance under no pressure.

B. SEM characterization

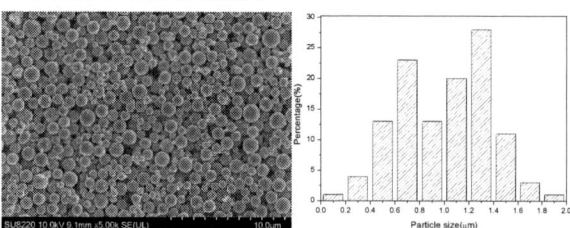

Fig. 3. Micro-nano copper particles and their size distribution map

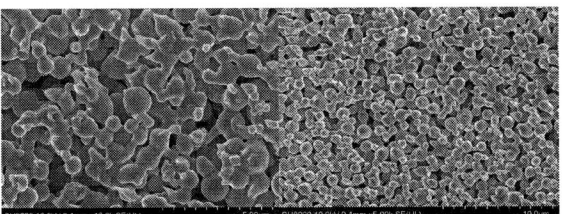

Fig. 4. SEM image of fracture of micro-nano copper after pressureless sintering at 260°C-0 MPa

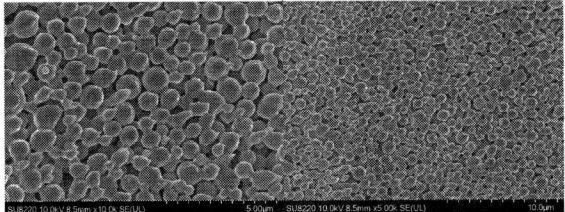

Fig. 5. SEM image of fracture surface of micro-nano copper powder after sintering at 220°C-2 MPa

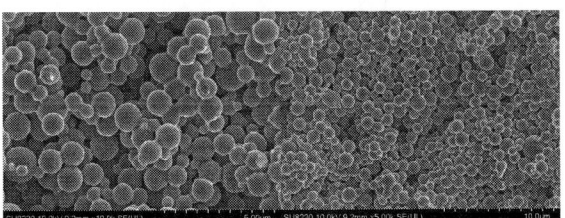

Fig. 6. SEM image of fracture surface of micro-nano copper powder after sintering at 220°C-0 MPa

Fig.3 is the SEM morphology observation of the micro-nano copper powder and the measurement of the particle size distribution range. From the particle size distribution diagram, it can be seen that the copper powder has a wide size distribution, including particles from 200 nm to 1.8 μm, and more than 60% of the particles are larger than 1μm. There are fewer pores between particles, and this dense structure enables the powder to obtain good sintering properties. Fig.4 is the fracture diagram of the micron composite copper powder after sintering at 260°C. Under the condition of pressureless sintering at 260°C, the small-sized copper particles basically disappeared, and the large-sized copper particles are clearly visible. A small amount of fracture toughness appears at this temperature, and a more obvious sintered neck is formed. But there are still more holes in the sintered body, which makes its final shear strength only 27.1 MPa. Due to the low sintering temperature of nano-copper, sufficient sintering has been achieved at 260°C and filled into the gaps of the micron copper particles through surface diffusion and volume diffusion.

Fig.5 is the fracture diagram of the micron composite copper powder after pressure sintering at 220°C. Under the sintering conditions of 220°C and 2 MPa, the pores between the particles were further reduced after pressure, the fracture after sintering is closer to the block. The sintered body showed more tearing morphology, and the flat indentation area is larger. So the contact area between the sintered body and the copper plate is larger, and higher shear strength is obtained, reaching a shear strength of 42.7 MPa. The sintered body under pressure obtains higher density and lower porosity.

Fig. 6 is the fracture diagram of sintering at 220°C without pressure. Under this condition, the interconnection performance of micro nano copper powder is poor, and the shear strength is only 10.4 MPa. Because of the unpressurized sintering process, the micro copper particles keep their initial morphology, and obvious sintering neck was formed after sintering, but there were also large cavities and no obvious tearing toughness on the fracture surface, which makes the shear strength lower under this condition.

During the sintering process, the small-sized nano-copper particles quickly melt and diffuse at low temperatures, and bond between the micron copper particles. As the sintering temperature increases, more sintering necks are formed between the micron particles, and the addition of nanoparticles helps reduce the driving force for sintering of the micron copper particles. Therefore, it is easier to obtain better interconnection performance between the micro-nano particles. Under of the above-mentioned low temperature, low pressure and no pressure conditions, the micro-nano copper powder can achieve better interconnection performance.

C. Electrical performance test

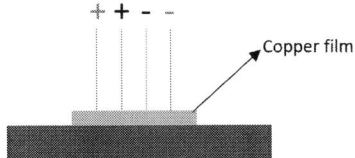

Fig. 7. Schematic diagram of resistivity measurement with four-point probe

In this paper, the electrical resistivity of 5 samples selected and sintered under pressureless conditions at 260°C-0 MPa was tested, and the average electrical resistance of the copper film was $1.80 \times 10^{-7} \Omega \cdot m$, which is about 10 times the resistivity of pure copper ($1.75 \times 10^{-8} \Omega \cdot m$). The copper film has good electrical conductivity. From the SEM of Fig. 4, it can be seen that a relatively obvious sintered neck is formed between the micron copper, which is conducive to the sintered body to obtain higher electrical conductivity.

D. XRD diffraction analysis

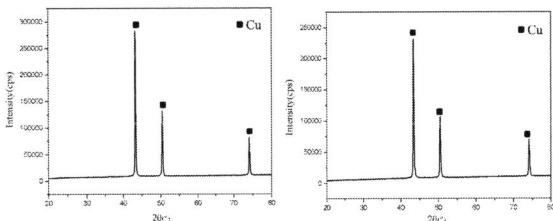

Fig. 8. XRD patterns of micro-nano copper powder in the air for 1 day and 7 days

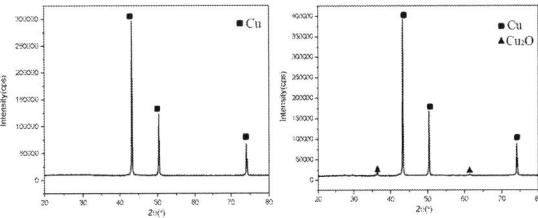

Fig. 9. XRD patterns of 100 nm Cu powder placed in the air for 1 day and 7 days

Because more than 60% of the micro-nano copper powder has a particle size of 1 μm or more, the surface energy of the micron copper particles is relatively low, and the surface atomic particles are relatively stable, so the large-size micron copper particles are not prone to oxidation in the air. The acid-washed micron copper powder and nano copper powder were placed in the air for 1 day and 7 days, and then the XRD diffraction patterns of the powder were observed. It can be seen from the above spectrum that the XRD patterns of the micron copper powder paste does not show the diffraction peaks of CuO and Cu_2O after being placed for 1 day and 7 days, which indicates that the micro-nano copper powder will not be oxidized after being placed in the air for 7 days and can remain relatively stable chemical nature. The nano copper powder with an average particle size of 100 nm did not appear to be oxidized after being placed in the air for 1 day, but after being placed in the air for 7 days, an obvious oxidation peak appeared. $2\theta=36.502°$, $61.518°$ corresponds to the simple cubic lattice of Cu_2O (111), (220) Planes. Therefore, nano-copper is easier to oxidize in the air than micro-copper. Multi-size micro-nano copper powder has stronger oxidation resistance than single-size nano-copper.

IV. CONCLUSION

Experiments showed that the multi-size copper powder of 200 nm~1.8 μm selected in this paper has better low-temperature and low-pressure sintering performance, higher density and stronger oxidation resistance.

When ethylene glycol was used as a solvent, the micro-nano copper powder can obtain a shear strength of 42.7 MPa at 220°C and a sintering pressure of 2 MPa, and a strong bond between the copper sheets was achieved. The shear strength of 27.1 MPa can be obtained under the sintering condition of 260°C without pressure. Through the electrical performance test of the sintered sample, it was found that the resistivity of the multi-size mixed micro-nano copper sintered body was 1.801×10^{-7} $\Omega \cdot m$, and good electrical conductivity was obtained.

Therefore, the multi-size copper powder prepared in this paper can meet the working requirements of microelectronic package interconnection, and can obtain better sintering performance and working stability than nano copper powder.

\# The authors contributed equally to this work.

ACKNOWLEDGMENT

This work was partially supported by the National Natural Science Foundation of China (61874155) and the Guangdong Basic and Applied Basic Research (2021A1515011642).

REFERENCES

[1] L. Jiang, T.G. Lei, D.T. Khai, "Evaluation of Thermal Cycling Reliability of Sintered Nanosilver Versus Soldered Joints by Curvature Measurement," IEEE Transactions on Components, Packaging, and Manufacturing Technology, 2014, vol. 4.5, pp.751-761.

[2] M. Knoerr, A. Schletz, "Power semiconductor joining through sintering of silver nanoparticles: Evaluation of influence of parameters time, temperature and pressure on density, strength and reliability," Integrated Power Electronics Systems (CIPS), 2010 6th International Conference on IEEE, 2010.

[3] V. R. Manikam, K. Y. Cheong, "Die Attach Materials for High Temperature Applications: A Review," IEEE Transactions on Components, Packaging and Manufacturing Technology, 2011, vol. 1.4, pp.457-478.

[4] J. J. Li, C. L. Chen, T. L. Shi, "Surface effect induced Cu-Cu bonding by Cu nanosolder paste," Materials Letters, 2016, vol. 184, pp.193-196.

[5] Y. Lee, J. Choi, K. J. Lee, "Large-scale synthesis of copper nanoparticles by chemically controlled reduction for applications of inkjet-printed electronics," Nanotechnology, 2008, vol. 19.41, p.415604.

[6] D. G. Guan, C. M. Sun，J. H. Lin, "Characterizations of Plating Structures and Electromagnetic Shielding Properties of the Micron-Sized Flaky Silver-Coated Copper Powder," Advanced Materials Research, 2011, vol. 217-218, pp.321-325.

[7] C. J. Wu, S. L. Cheng, Y. J. Sheng, and H.K. Tsao, "Reduction-assisted sintering of micron-sized copper powders at low temperature by ethanol vapor," RSC Advances, 2015, vol. 5.66, pp.53275-53279.

[8] X. Liu, H. Nishikawa, "Low-pressure Cu-Cu bonding using in-situ surface-modified microscale Cu particles for power device packaging," Scripta materialia, 2016, vol. 120, pp.80-84.

Effect of daisy chain structure on electromigration reliability of microbumps: A simulation study

1st Chenkan Yan*
Detection Department
China Electronics Technology Group
Corporation No.58 Research Institute
Wuxi, China
*yanchenkan@alumni.tongji.edu.cn

2nd Dun Wang
Detection Department
China Electronics Technology Group
Corporation No.58 Research Institute
Wuxi, China

3rd Kaihong Zhang
Detection Department
China Electronics Technology Group
Corporation No.58 Research Institute
Wuxi, China

4th Huibin Zhang
Detection Department
China Electronics Technology Group
Corporation No.58 Research Institute
Wuxi, China

5th Yongjian Yu
Detection Department
China Electronics Technology Group
Corporation No.58 Research Institute
Wuxi, China

6th Weikun Xie
Detection Department
China Electronics Technology Group
Corporation No.58 Research Institute
Wuxi, China

7th Kai Zhu
Detection Department
China Electronics Technology Group
Corporation No.58 Research Institute
Wuxi, China

8th Yongkang Wan
Detection Department
China Electronics Technology Group
Corporation No.58 Research Institute
Wuxi, China

Abstract—A simulation method was utilized to study the electromigration reliability of two daisy chain structures which were subjected to electric current of 0.05 A at environment temperature of 22 °C. The current density and temperature distribution of the two daisy chain structures were compared and the electromigration reliability of two types of daisy chain structures were analyzed. The results show that the current crowding effect of typical bump in parallel daisy chain structure was more prominent than that of typical bump in series daisy chain structure. The current density followed eight-order polynomial distribution. The hot spots of both structures coincided with the current crowding regions and the maximum temperature in typical bump in parallel daisy chain structure was obviously higher than that in typical bump in series daisy chain structure. The electromigration reliability of parallel chain structure was lower than that of series chain structure due to higher temperature and higher current density in hot spots of parallel chain structure. These findings will offer fundamental insights into electromigration and reliability theory.

Keywords—*electromigration, reliability, daisy chain, simulation*

I. INTRODUCTION

With rapid development of microelectronic science and technology, the novel types of IC devices with features of artificial intelligence, miniaturization and multi-functionality have been popular [1, 2]. Thus, high input/output (I/O) interconnect density and smaller solder bumps have been widely utilized in IC devices to meet the new trend [3, 4]. However, the decreased dimension of solder bumps, along with the higher design power dissipation, increase the current density of solder bumps significantly, which may threaten the reliability of IC packaging when the current density exceeds the critical value of 10^4 A/cm^2 [2, 5]. The critical failure mechanism related to higher current density is electromigration (EM), which is regarded as a serious issue in IC packaging [6].

Electromigration is a kind of mass transportation of irons or atoms related to the moving electrons in response to electron wind force and electric field intensity, leading to increase of resistance and voids in the bumps [7]. The ultimate failure mode is open circuit in the bumps. Moreover, current crowding effect induced by the geometry structure of the bumps may aggravate the electromigration phenomenon [8-10]. In addition, electromigration is an interactive process that may involve a series of other failure mechanisms such as stress-migration and thermal-migration [11].

Many scholars have focused on the electromigration reliability of bumps. Chiang et al. [12] investigated the relation between current crowding and electromigration of SAC bumps. The results showed that local Joule heating effect induced by current crowding influenced the electromigration reliability of the microbumps significantly. The maximum current density, measured by simulation method, was located at the entry of the bumps where the void formed. Gu et al.[11] studied the synergetic effect of electromigration and thermo-migration of a classical daisy chain structure in Sn-Bi ball grid array bumps, the simulation results showed that thermo-migration facilitated electromigration when thermo-migration and electromigration were in the same direction, while thermo-migration impended electromigration when thermo-migration and electromigration were in the reverse direction. In addition, the bumps with no current passing by were detected migration of Bi atoms. Ismathullakhan et al. [13] proposed a novel method to improve the electromigration reliability of Sn-Bi solder joint. The Ag nano particles were added to Sn-Bi alloys and no obvious intermetallic compound formed in the bumps. The reason is that, newly formed Ag$_3$Sn particles blocked the diffusion of Sn and Bi atoms during experiment process. Jiang et al. [14] investigated the electromigration reliability of Cu/SAC/Cu bumps at current density of 1.4×10^4 A/cm^2. It was observed that the electron wind force mainly contributed to the atomic migration rather than atomic gradient.

978-1-6654-1392-3/21 $31.00 © 2021 IEEE

Although many researches have been carried out on electromigration reliability of bumps, these researches mainly concentrated on single bump or simply daisy chain structure (i.e. series daisy chain structure) [11-14]. However, the parallel daisy chain structure is also practically used structure in ICs. To the best of our knowledge, no systematic researches have been made on the electromigration reliability of the two types of daisy chain. This may be because the microbumps are completely encapsulated by the IC packaging. Therefore, it is somewhat difficult to reveal the current density distribution inside the bumps. In the present work, the electric-thermal finite element models of series and parallel daisy chain structures were developed by utilizing three-dimensional simulation software Ansys. The focus of our work was to study the current density and temperature distribution of the daisy chain structures and reveal the electromigration reliability of two types of daisy chain structures. The findings will offer fundamental insights into electromigration and reliability theory.

II. MODELLING AND SIMULATION

The finite element method was utilized in our work to investigate the current density and temperature distribution of daisy chain structure. The three-dimensional model of the flip chip ball grid array (FCBGA) packaging device is shown in Fig. 1. In order to show the internal structure of the device, the BGA sold joints were set to be transparent. The model mainly consisted of die, die attach, traces, substrate, bumps, under bump metallization (UBM) and epoxy molding compound (EMC), and the geometry dimensions of the structure are listed in Table 1. The bumps and the traces constituted the daisy chain structure. In order to study the electromigration reliability of microbumps, electric stress was chosen to impose on the device. The simulation was performed using universal finite element software ANSYS 19.0 and the meshing method were sweep method and hex dominant method. To allow for the Joule heating effect of the device under electric stress, the electrical-thermal coupling model was introduced into our simulation process. The material parameters of the structure are listed in Table 2.

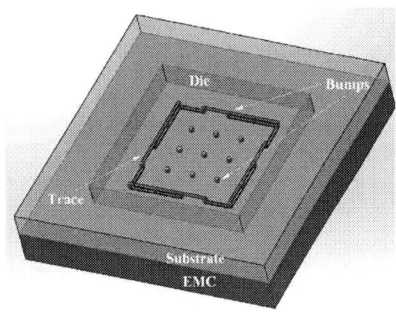

Fig. 1. Schematic illustration of the 3D device model

TABLE I. THE THREE DIMENSIONS OF THE STRUCTURE

Name	Length (μm)	Width (μm)	Thickness (μm)
EMC	4000	4000	600
Substrate	4000	4000	400
Copper Trace	500	66	20
Aluminium Trace	500	66	2
Die	3000	3000	300
Die attach	3000	3000	100
Bumps	100	100	75
UBM	66	66	3

TABLE II. THE MATERIALS PARAMETERS OF THE STRUCTURE

Name	Thermal conductivity (W/(m·°C))	Electrical resistivity ($10^{-8}\,\Omega\cdot$m)
EMC	0.7	-
Substrate	0.7	-
Copper Trace	393	1.58
Aluminium Trace	240	2.6
Die	147	-
Die attach	0.55	-
SAC Bumps	57	13
UBM	91	7

The daisy chain structures mainly contain two types, i.e. the simple daisy chain (series daisy chain) and parallel daisy chain. Fig. 2 is the schematic illustration of the two type of daisy structures. As shown in Fig. 2(a)-(b), the parallel daisy chain was the one where many bumps could provide electrons for the specified bump. The detailed series daisy chain structure of the device is depicted in Fig. 2(c)-(d). The series daisy chain was the structure that the electrons flow into the first bump in one direction, while they flowed into the second bump in the opposite direction. The two types of daisy chain structures were simulated and the electromigration reliability of the two structures were analysed in our study. In terms of the symmetry of current distribution, eight typical bumps were selected for study, which was shown in Fig. 2 (bumps A1-A4 and B1-B4). In parallel daisy chain structure, electrons flowed from the side contacting with aluminium trace to the side contacting with copper trace of bumps A4, A3 and A2, then flowed from the side contacting copper trace to the side contacting with aluminium trace of bumps A1 during simulation process. In series daisy chain structure, electrons flowed from the side contacting with aluminium trace to the side contacting with copper trace of bumps B4, flowed from the side contacting with copper trace to the side contacting with aluminium trace of B3, then flowed from the side contacting with aluminium trace to the side contacting with copper trace of B2, and flowed from the side contacting with

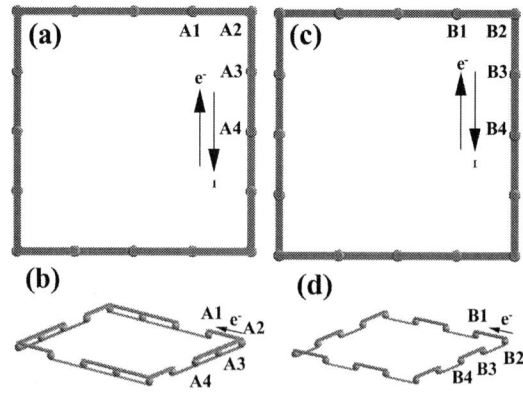

Fig. 2. Schematic illustration of the daisy chain structures: (a)-(b) parallel daisy chain structure; (c)-(d) series daisy chain structure

copper trace to the side contacting with aluminium trace of B1 during simulation process.

The detailed boundary conditions imposed on the model are: (1) the electric current of 0.05 A imposed on the daisy chain structure, which equals to the average electric current density of 1.6×10^7 A/m^2 impose on the bumps A1 and B1; (2) the convection coefficient of 10 (W/m$^2 \cdot$°C) imposed on the model; and (3) the environment temperature of 22°C.

III. RESULTS AND DISCUSSION

A. Current density distribution

Fig. 3 shows current density distribution in typical bumps in Fig. 2. It can be seen that the current density distribution in eight specified bumps was obviously non-uniform. The maximum current density was ~30 times of the average current density in both daisy chain structures. This was typical current crowding effect. Liang et al. [10, 15] designated the crowding ratio to be the index for judging the current crowding effect. The ratio could be obtained by dividing average current density into maximum current density inside the bump [15]. The crowding ratio for bump A1 was 35.3, while the crowding ratio for bump B1 was 26.3. The parallel daisy chain was the one where bumps A4, A3 and A2 could provide electrons for the specified bump A1. Consequently, the maximum electron numbers in the bump A1 was obviously more than that in other bumps and the current crowding effect was obvious. The series daisy chain was the structure that the electrons flowed into the bumps A1, A2, A3 and A4 subsequently. Thus, the current flowed through every bump in the series daisy chain was identical and the current crowding effect was not so prominent as that in parallel daisy chain.

The maximum current density in parallel daisy chain structure (1.37×10^8 A/m^2) was located at the exit spot of the bump A1 where the bump contacting with aluminium trace, while the maximum current density in series daisy chain structure (1.01×10^8 A/m^2) was located at the exit spot of bump B1 where the bumps contacting with aluminium trace. The bumps A1 and B1 had relatively large volume comparing with that of the aluminium trace or copper trace, thus, the current density inside the bumps was diluted. However, the current density at the side of the bumps contacting with the trace was higher due to the smaller cross-sectional area of the trace. The current crowded so as to flow out of the bumps. In addition, it is shown that the maximum current density was located at the side contacting with the aluminium trace rather than the side contacting with the copper trace. The weakened current crowded effect of the bumps at the copper side comparing with that at the aluminium side was due to the current dilution effect of the thicker copper trace.

It is clear that although the imposed current density was 1.6×10^7 A/m^2, the extreme current density value exceeded the threshold value of 10^8 A/m^2 for the initialization of electromigration. The current crowding effect could reduce the reliability of the bumps significantly due to the enhanced wind force in the zone. The maximum current density of the parallel chain structure was three orders of magnitude higher than the minimum current density (4.25×10^5 A/m^2). In contrast, the maximum current density of the series chain structure was two orders of magnitude higher than the minimum current density (3.65×10^6 A/m^2).

In order to study the current density distribution inside the bumps, different distribution curves were depicted, as shown in Fig. 4. Bumps A1, A4, B1 and B4 were selected for analysis. The coordinate systems and origins were displayed in Fig. 3, where the origin was located at the maximum current density spot. "A1-OY" in Fig. 3 means the Y direction inside bump A1. The current density decreased with increase of distance in curves A1-OZ, A4-OX, B1-OZ and B4-OX because the current inside the bumps spread out and drifted down. The bathtub-like curves (curves A1-OY, A4-OY, B1-OY and B4-OY) meant two extreme values existed. The extreme values on the right of the curves indicated the weakened current crowd effect of the bumps at the copper side. To quantitatively describe the non-linear distribution of current density, the functional relationship between current density and distance was established by eight-order polynomial fitting. The function relationship could be derived as follows:

$$j = 1.3E8 - 3.6E13d + 4.9E18d^2 - 3.5E23d^3 + 1.4E28d^4 - 3.3E32d^5 + 4.6E36d^6 - 3.5E40d^7 + 1.1E44d^8 \text{ (A1-OY) (1)}$$

$$j = 1.3E8 - 2.6E13d + 3.8E18d^2 - 3.1E23d^3 + 1.5E28d^4 - 4.3E32d^5 + 7.1E36d^6 - 6.4E40d^7 + 2.4E44d^8 \text{ (A1-OZ) (2)}$$

$$j = 1.2E8 - 2.1E13d + 2.9E18d^2 - 2.4E23d^3 + 1.2E28d^4 - 3.4E32d^5 + 5.7E36d^6 - 5.1E40d^7 + 1.9E44d^8 \text{ (A4-OX) (3)}$$

$$j = 1.2E8 - 3.1E13d + 3.9E18d^2 - 2.7E23d^3 + 1.1E28d^4 - 2.5E32d^5 + 3.4E36d^6 - 2.5E40d^7 + 7.6E43d^8 \text{ (A4-OY) (4)}$$

$$j = 8.2E7 - 4.8E12d - 2.1E17d^2 + 3.8E22d^3 - 1.9E27d^4 + 4.6E31d^5 - 5.7E35d^6 + 3.1E39d^7 - 3.4E42d^8 \text{ (B4-OX) (5)}$$

$$j = 8.6E7 - 1.3E13d + 1.1E18d^2 - 5.3E22d^3 + 1.6E27d^4 - 3.3E31d^5 + 4.2E35d^6 - 3.1E39d^7 + 9.9E42d^8 \text{ (B4-OY) (6)}$$

$$j = 9.9E7 - 2.1E13d + 2.4E18d^2 - 1.5E23d^3 + 5.6E27d^4 - 1.2E32d^5 + 1.6E36d^6 - 1.2E40d^7 + 3.6E43d^8 \text{ (B1-OY) (7)}$$

$$j = 9.8E7 - 9.4E12d + 4.0E17d^2 - 2.1E21d^3 - 4.9E26d^4 + 2.1E31d^5 - 3.9E35d^6 + 3.5E39d^7 - 1.3E43d^8 \text{ (B1-OZ) (8)}$$

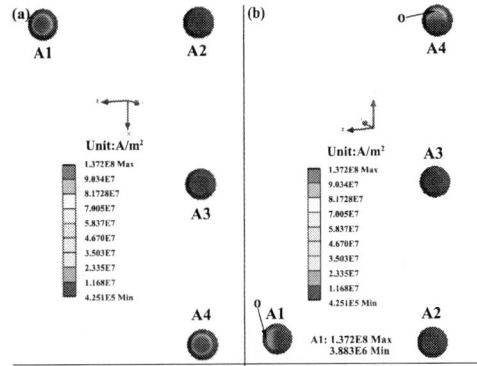

Fig. 3. Current density distribution in typical bumps in Fig. 2: (a) top view of bumps A1-A4; (b) bottom view of bumps A1-A4; (c) top view of bumps B1-B4; (d) bottom view of bumps B1-B4

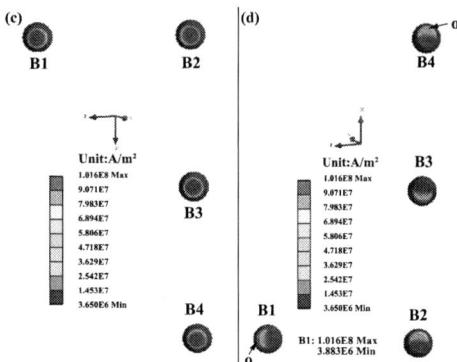

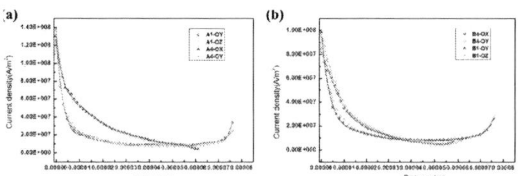

Fig. 3 *(to be continued)*

Fig.4 Current density distribution in typical bumps in Fig. 2: (a) top view of bumps A1-A4; (b) bottom view of bumps A1-A4; (c) top view of bumps B1-B4; (d) bottom view of bumps B1-B4

B. Temperature distribution

Temperature distribution in typical bumps in Fig. 2 is depicted in Fig. 5. As for parallel daisy chain structure, a hot spot existed in the exit spot of the bump A1 where the bump contacting with aluminium trace, which coincided with the current crowding region. The maximum temperature was identified as 71.94 °C, which was 4.31 °C higher than the minimum temperature. Despite the little fluctuation of temperature between hot spots and other regions, hot spots are detrimental to the reliability of bumps because the higher temperature promotes the diffusion of atoms exponentially [16]. For series parallel daisy chain structure, four hot spots in bump B1, B2, B3 and B4 coincided with the current crowding region. The maximum temperature was 56.04 °C, which was 9.57 °C higher than the minimum temperature. Our finding that the current crowding region coincided with hot spot was also verified in some series daisy chains [11, 17]. This coincidence was mainly attributed to two aspects. Firstly, the occurrence of the hot spots was attributed to the Joule heat effect induced by current crowding effect. Joule heating power of the daisy chain structure is expressed as:

$$P = I^2 R = j^2 \rho V \tag{9}$$

where P is Joule heat power, I is the current in the circuit, R is the resistance of the circuit, j is the current density in the circuit, ρ is the electrical resistivity of the material of the structure, and V is the volume of the structure. Since the joule heat power was proportional to the square of the current density, the higher current density in the current crowing region should enhance the Joule heating effect of the region notably. Secondly, the hot spots in the bumps were in contact with aluminium trace. Compared with copper trace, aluminium trace was the major Joule heating source for its

smaller cross-section area (one magnitude less than that of the copper trace) and higher electrical resistivity when they were subjected to the same current intensity. According to formula (9), it is inferred that the Joule heat power of aluminium was higher than that of copper, thus the temperature of aluminium trace was higher than that of copper trace. The newly generated Joule heat will be conducted through the packaging and absorbed [18]. To be specific, more quantity of heat conducted from the aluminium trace to the current crowding region, resulting in the coincidence of the hot spot and the current crowding region. Related study [15] shows that the dimension of aluminium trace have significant effect on joule heating and the current crowding effect could be relieved by increasing the dimensions of aluminium trace. In addition, it is worth mentioning that the maximum temperature in bump A1 was obviously higher than that in bump B1 despite the fact that both bump A1 and B1 were bumps with current crowding region. To be specific, in parallel daisy chain structure, bump A4, A3 and A2 provided electrons for the bump A1 while only bump B2 provided electrons for bump B1 in series daisy chain structure. Therefore, the current crowding ratio for bump A1 and B1 was different, as detailed in section "current density distribution".

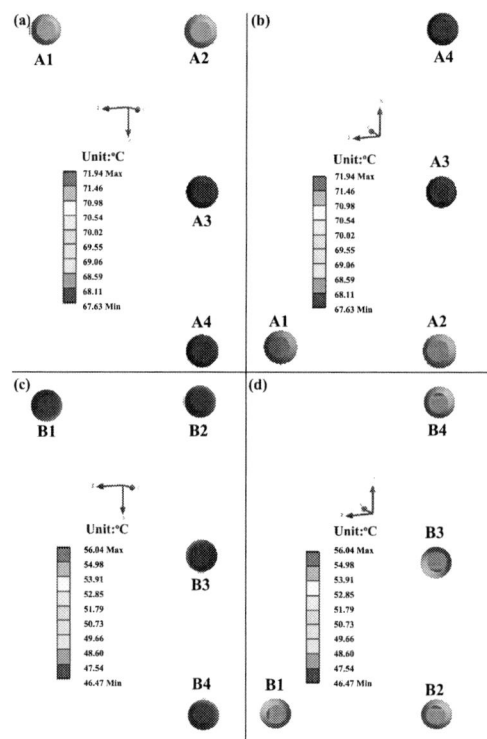

Fig.5 Temperature distribution in typical bumps in Fig. 2: (a) top view of bumps A1-A4; (b) bottom view of bumps A1-A4; (c) top view of bumps B1-B4; (d) bottom view of bumps B1-B4

C. Electromigration reliability analysis

Electromigration is the atom diffusion process induced by higher current density from cathode to anode in wires or solder joints [19-21]. Black equation, the famous equation in judging the electromigration reliability of solder joints, describes the effect of current density and temperature on mean-time-to-failure (MTTF)[18, 22-25]. The Black equation is listed as below.

978-1-6654-1392-3/21 $31.00 © 2021 IEEE

$$MTTF = Aj^{-n}\exp(E_a/kT) \qquad (10)$$

where A is the constant cross-sectional area coefficient, j is the average electric current density, n is the order of current density, E_k is the activation energy, k is constant, and T is average solder joint temperature in Kelvin scale. However, the current density and temperature distribution was non-uniform as discussed above. A modified Black equation was adopted in some scholars' study [15, 17]. The modified Black equation is listed as below.

$$MTTF = A(cj)^{-n}\exp[E_a/k(T+\Delta T)] \qquad (11)$$

where c is current crowding coefficient and ΔT is temperature increase due to Joule heating. In our study, current crowding effect and Joule heating effect was taken into consideration from the perspective of simulation, therefore, T was defined as the specific solder joint temperature in Kelvin scale obtained by simulation directly and j was defined as the specific current density in formula (10) obtained by simulation directly. As discussed above, the current crowding regions coincided with hot spots in the typical bumps, which were subjected to higher temperature and higher current density. According to Black equation (formula (10)), the hot spots in the typical bumps were also weak spots with shorter lifetime, therefore, the MTTFs of hot spots were also the MTTFs of the bumps. The order of current density n in formula (10) is equal to 1.8 for the eutectic solder joints [15]. The relative MTTF were defined as the quotient of two MTTFs as below.

$$\text{Relative MTTF} = \frac{MTTF_1}{MTTF_2} = \left(\frac{j_2}{j_1}\right)^{1.8}\exp\left(\frac{1}{T_1} - \frac{1}{T_2}\right) \qquad (12)$$

The temperature, current density of hot spots in typical bumps are listed in Table 3. In addition, the Relative MTTF of hot spot in bump B1 was defined as 1, thus that of hot spots in other bumps were calculated, as shown in Table 3.

TABLE III. RELATIVE MTTF OF HOTS SPOTS IN TYPICAL BUMPS.

Bump	Temperature(K)	Current density(A/m²)	Relative MTTF
A1	344.94	1.372E8	0.628
A4	340.63	1.224E8	0.836
B1	329.04	1.016E8	1.000
B4	329.04	1.016E8	1.000

The results show that the MTTF of bump A1 was 0.628 times of that of bump B4 and the MTTF of bump A4 was 0.836 times of bump B4. In parallel chain structure, the bump where electrons flowed into from other bumps concurrently was subjected to higher temperature and higher current density. In contrast, different bumps in series chain structure were subject to the relatively uniform temperature and current density. Therefore, the MTTF of bumps in series chain structure was larger than that in parallel chain structure. It can be concluded that the electromigration reliability of parallel chain structure was obviously lower than that of series chain structure due to higher temperature and higher current density in hot spots of parallel chain structure.

IV. CONCLUSIONS

In the present work, a simulation study was carried out on evaluate effect of daisy chain structure on electromigration reliability of microbumps. Current density distribution, temperature distribution and electromigration reliability of two daisy chain structure were discussed. Main conclusions are:

(1) The current crowding effect of typical bump in parallel daisy chain structure was more prominent than that of typical bump in series daisy chain structure. The maximum current density in both structures were located at the exit spot of the bump where the bump contacting with aluminum trace. Although the imposed current density was 1.6×10^7 A/m², the extreme current density value exceeded the threshold value of 10^8 A/m² for the initialization of electromigration for both structures. The current density followed eight-order polynomial distribution.

(2) The hot spots of both structures coincided with the current crowding regions and the maximum temperature in typical bump in parallel daisy chain structure was obviously higher than that in typical bump in series daisy chain structure.

(3) The electromigration reliability of parallel chain structure was obviously lower than that of series chain structure due to higher temperature and higher current density in hot spots of parallel chain structure.

REFERENCES

[1] S. B. Liang, C. B. Ke, W. J. Ma, M. B. Zhou, and X. P. Zhang, "Numerical simulations of migration and coalescence behavior of microvoids driven by diffusion and electric field in solder interconnects," Microelectronics Reliability, vol. 71, pp. 71-81, Apr 2017.

[2] P. Zhang, S. Xue, and J. Wang, "New challenges of miniaturization of electronic devices: Electromigration and thermomigration in lead-free solder joints," Materials & Design, vol. 192, Aug 2020, Art. no. 108726.

[3] K. N. Tu and Y. Liu, "Recent advances on kinetic analysis of solder joint reactions in 3D IC packaging technology," Materials Science & Engineering R-Reports, vol. 136, pp. 1-12, Apr 2019.

[4] J. Wang, S. Xue, P. Zhang, P. Zhai, and Y. Tao, "The reliability of lead-free solder joint subjected to special environment: a review," Journal of Materials Science-Materials in Electronics, vol. 30, no. 10, pp. 9065-9086, May 2019.

[5] M. T. Ahmed, M. Motalab, and J. C. Suhling, "Impact of Mechanical Property Degradation and Intermetallic Compound Formation on Electromigration-Oriented Failure of a Flip-Chip Solder Joint," Journal of Electronic Materials, vol. 50, no. 1, pp. 233-248, Jan 2021.

[6] A. Ishii and A. Yamanaka, "Multi-phase-field modelling of electromigration-induced void migration in interconnect lines having bamboo structures," Computational Materials Science, vol. 184, Nov 2020, Art. no. 109848.

[7] A. Kunwar, Y. A. Coutinho, J. Hektor, H. Ma, and N. Moelans, "Integration of machine learning with phase field method to model the electromigration induced Cu6Sn5 IMC growth at anode side Cu/Sn interface," Journal of Materials Science & Technology, vol. 59, pp. 203-219, Dec 15 2020.

[8] Q. Wang, X. Tao, L. Yang, and Y. Gu, "Current crowding in two-dimensional black-phosphorus field-effect transistors," Applied Physics Letters, vol. 108, no. 10, Mar 7 2016, Art. no. 103109.

[9] F. Yang, H. Chen, X. Tian, Y. Bai, and Y. Zhu, "Investigation on Current Crowding Effect in IGBTs," Ieee Transactions on Electron Devices, vol. 65, no. 2, pp. 636-640, Feb 2018.

[10] H. Lu, C. Yu, P. Li, and J. Chen, "Current crowding and its effects on electromigration and interfacial reaction in lead-free solder joints," Journal of Electronic Packaging, vol. 130, no. 3, Sep 2008, Art. no. 031008.

[11] X. Gu, K. C. Yung, and Y. C. Chan, "Thermomigration and electromigration in Sn58Bi ball grid array solder joints," Journal

978-1-6654-1392-3/21 $31.00 © 2021 IEEE

of Materials Science-Materials in Electronics, vol. 21, no. 10, pp. 1090-1098, Oct 2010.

[12] K. N. Chiang, C. C. Lee, C. C. Lee, and K. M. Chen, "Current crowding-induced electromigration in SnAg3.0Cu0.5 microbumps," Applied Physics Letters, vol. 88, no. 7, Feb 13 2006, Art. no. 072102.

[13] S. Ismathullakhan, H. Lau, and Y.-c. Chan, "Enhanced electromigration reliability via Ag nanoparticles modified eutectic Sn-58Bi solder joint," Microsystem Technologies-Micro-and Nanosystems-Information Storage and Processing Systems, vol. 19, no. 7, pp. 1069-1080, Jul 2013.

[14] Y. Jiang, H. Li, G. Chen, Y. Mei, and M. Wang, "Electromigration behavior of Cu/Sn3.0Ag0.5Cu/Cu ball grid array solder joints," Journal of Materials Science-Materials in Electronics, vol. 30, no. 6, pp. 6224-6233, Mar 2019.

[15] S. W. Liang, Y. W. Chang, and C. Chen, "Effect of Al-trace dimension on Joule heating and current crowding in flip-chip solder joints under accelerated electromigration," Applied Physics Letters, vol. 88, no. 17, Apr 24 2006, Art. no. 172108.

[16] K. Son et al., "Effect of Electromigration-Induced Joule Heating on the Reliability of Sn-Ag Microbump with Different UBM Structures," Journal of Electronic Materials, vol. 49, no. 12, pp. 7228-7237, Dec 2020.

[17] K. Son et al., "Effects of NCF and UBM Materials on Electromigration Reliabilities of Sn-Ag microbumps for advanced 3D packaging," in 2019 Ieee 69th Electronic Components and Technology Conference(Electronic Components and Technology Conference, 2019, pp. 2246-2251.

[18] K. N. Tu, Y. Liu, and M. Li, "Effect of Joule heating and current crowding on electromigration in mobile technology," Applied Physics Reviews, vol. 4, no. 1, Mar 2017, Art. no. 011101.

[19] H. Jeong, C.-J. Lee, J.-H. Kim, J.-y. Son, and S.-B. Jung, "Electromigration Behavior of Cu Core Solder Joints Under High Current Density," Electronic Materials Letters, vol. 16, no. 6, pp. 513-519, Nov 2020.

[20] S. F. Liu et al., "Electromigration behavior of Cu/Sn-58Bi-1Ag/Cu solder joints by ultrasonic soldering process," (in English), Journal of Materials Science-Materials in Electronics, Review vol. 31, no. 15, pp. 11997-12003, Aug 2020.

[21] H. R. Wang, R. Kou, T. Harrington, and K. S. Vecchio, "Electromigration effect in Fe-Al diffusion couples with field-assisted sintering," (in English), Acta Materialia, Article vol. 186, pp. 631-643, Mar 2020.

[22] K. N. Tu and A. N. Gusak, "Mean-Time-To-Failure Equations for Electromigration, Thermomigration, and Stress Migration," Ieee Transactions on Components Packaging and Manufacturing Technology, vol. 10, no. 9, pp. 1427-1431, Sept 2020.

[23] N. Wang, K. H. Bevan, and N. Provatas, "Phase-Field-Crystal Model for Electromigration in Metal Interconnects," Physical Review Letters, vol. 117, no. 15, Oct 7 2016, Art. no. 155901.

[24] H.-C. Ma et al., "Reliability and failure mechanism of copper pillar joints under current stressing," Journal of Materials Science-Materials in Electronics, vol. 26, no. 10, pp. 7690-7697, Oct 2015.

[25] C. Basaran, S. Li, D. C. Hopkins, and D. Veychard, "Electromigration time to failure of SnAgCuNi solder joints," Journal of Applied Physics, vol. 106, no. 1, Jul 1 2009, Art. no. 013707.

Novel Low-dielectric Fluorinated Carbon Fiber/Polyimide Materials with High Elongation

1st Tao Wang
Shenzhen Institute of Advanced Electronic Materials, Shenzhen Institute of Advanced Technology, Chinese Academy of Sciences
Shenzhen, China
Shenzhen College of Advanced Technology, University of Chinese Academy of Sciences
Shenzhen, China
tao.wang3@siat.ac.cn

2nd Jinhui Li*
Shenzhen Institute of Advanced Electronic Materials, Shenzhen Institute of Advanced Technology, Chinese Academy of Sciences
Shenzhen, China
jh.li@siat.ac.cn

3rd Fangfang Niu*
College of Physics and Optoelectronic Engineering
Shenzhen University
Shenzhen, China
ffn@szu.edu.cn

4th Liang Shan
Shenzhen Institute of Advanced Electronic Materials, Shenzhen Institute of Advanced Technology, Chinese Academy of Sciences
Shenzhen, China
liang.shan@siat.ac.cn

5th Guoping Zhang*
Shenzhen Institute of Advanced Electronic Materials, Shenzhen Institute of Advanced Technology, Chinese Academy of Sciences
Shenzhen, China
gp.zhang@siat.ac.cn

6th Rong Sun
Shenzhen Institute of Advanced Electronic Materials, Shenzhen Institute of Advanced Technology, Chinese Academy of Sciences
Shenzhen, China
rong.sun@siat.ac.cn

7th Ching-Ping Wong
School of Materials Science and Engineering
Georgia Institute of Technology
Atlanta, United States
cp.wong@mse.gatech.edu

Abstract—With the development of wafer-level packaging (WLP), polyimide materials with low-dielectric and high-elongation properties are urgently needed in redistribution layers (RDLs). Incorporating fillers into polyimide materials to enhance the properties of polyimide composites is one of the most compatible and lowest cost solutions. However, the degradation of composite materials performance caused by uneven filler dispersion is one of the key issues. Herein, novel low-dielectric fluorinated carbon fiber/polyimide (FCF/PI) materials were prepared by high-power ultrasonic dispersion technology to alleviate the problem of filler dispersion. The dielectric, mechanical and heat resistance properties of FCF/PI with different filler additions were characterized. The dielectric properties of 1.0 wt% FCF/PI composite film are 3.02 (Dk) and 0.0198 (Dl) at 5 GHz. And the elongation at break of 1.0 wt% FCF/PI film can reach more than 50%. In addition, the $T_{5\%}$ and T_g of all FCF/PI composites were above 550 ºC and 400 ºC, respectively. Therefore, the novel low-dielectric FCF/PI materials with high elongation have prospects in the field of advanced packaging.

Keywords—polyimide, low-dielectric, high elongation, advanced packaging

I. INTRODUCTION

Polyimide (PI) materials are widely used in the fields of microelectronics and electronics because of the excellent heat resistance, mechanical, and dielectric properties [1]. However, with the development of 5G high-frequency and high-speed communication, the semiconductor industry's demand for low-dielectric polyimide materials is becoming more and more urgent [2]. In addition, high elongation is also required by wafer-level packaging (WLP) to ensure the reliability of microelectronic devices [3]. Therefore, the research and development of polyimide materials which have low-dielectric and high-elongation performance is of great significance for WLP.

Generally, there are two major directions for the research and development of polyimides with mechanical and dielectric performance enhancements: molecular design of the monomers and micron/nanometer filler for the composite materials [4]. The performance of polyimide materials can be effectively improved by designing monomers with specific structures. For example, the introduction of fluorine into the monomer could significantly reduce the dielectric properties of polyimide. Professor Zhang's group has developed a monomer containing fluorine and an asymmetric structure, and the synthesized polyimide has a dielectric constant as low as 2.44 (@10kHz) [5]. However, the design and synthesis of new monomers are complex and show an unfavorable effect on other properties. Simultaneously, incorporating fillers into polyimide to enhance the properties of polyimide composites is one of the most compatible and lowest cost solutions. For example, Wang et. al reported the surface-modified graphene oxide/polyimide materials to obtain composite materials with low dielectric (Dk=2.0 @1 GHz)and enhanced mechanical properties [6]. However, many fillers, such as fluorine-containing fillers, are difficult to disperse due to their low surface energy, leading to degradation of composite material performance. This is also one of the biggest problems with fillers. One common method is the chemical modification of filler so as to improve its dispersibility and functionality, resulting in a great increase in cost and process. Therefore, it is challenging to research and develop low-cost PI composite materials which have high-elongation and low-dielectric performance in a more effective way.

978-1-6654-1392-3/21 $31.00 © 2021 IEEE

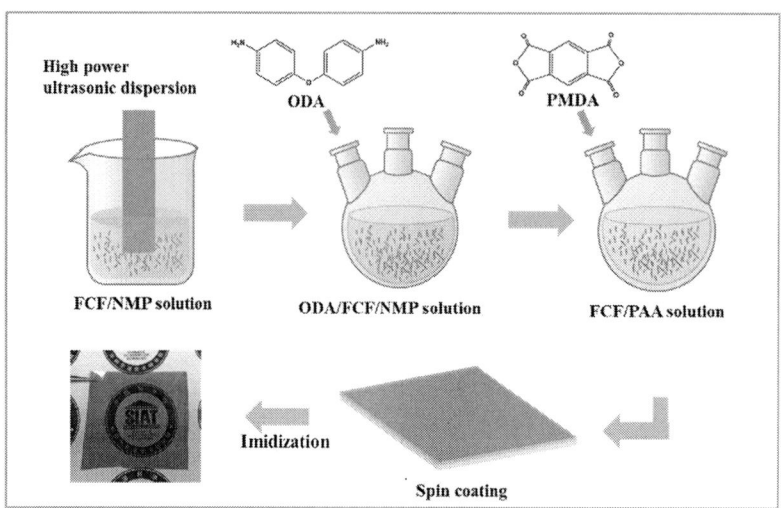

Fig. 1. Preparation of FCF/PI nanocomposite films.

In this work, fluorinated carbon fiber (FCF) filler effectively reduces the dielectric constant and increase the elongation of the polyimide composite materials due to its high fluorine content and length-diameter ratio. Especially, high power ultrasonic dispersion technology was applied to improve the dispersion uniformity of FCF in PI. The results show that FCF filler with better dispersion uniformity exhibits effectively reduction of the dielectric constant (Dk=3.02, Dl=0.0198 @5GHz) and increment of the elongation (> 50%). Therefore, the FCF/PI composite materials have prospects in the field of WLP.

II. EXPERIMENTAL

A. Materials

4,4'-oxydianiline (ODA, >98.0%(GC)(T)) and pyromellitic dianhydride (PMDA, >99.0%(T)(HPLC)) and were supplied by Tokyo Chemical Industry Co., Ltd. and heat in oven to remove moisture. 1-Methyl-2-pyrrolidinone (NMP, 99.5%, extra dry, with molecular sieves, water≤50 ppm) was supplied by Energy Chemical Co., Ltd.

B. Synthesis procedure of FCF/PI materials

The synthesis procedure of FCF/PI is shown in Fig. 1. FCF filler was dispersed in NMP solution by 3000 W high power ultrasonic at first. Then, 4,4'-oxydianiline and pyromellitic dianhydride was added into the FCF/NMP solution for *in situ* polymerization of FCF/PAA. And the FCF/PAA solution was spin-coated on glass. Finally, the film was cured at 80, 200, 300 and 350 °C for 1 h in nitrogen. The samples were named pure PI, 0.5wt% FCF/PI, 1.0 wt% FCF/PI, and 2.0 wt% FCF/PI according to the weight percent of FCF.

C. Characterization

The cross-section of samples were characterized using an SEM Thermo Scientific Apreo 2S, which was operated at an accelerating voltage of 5 kV. Thermogravimetry analysis (TGA) was tested between room temperature and 800°C in nitrogen by a TA SDTQ600 machine. The Keysight E5071C vector network analyzer was used to measure the dielectric properties of samples. The mechanical properties and Tanδ were measured by TA DMA850 machine. The mechanical properties were operated at a speed of 2 N/min and the Tanδ were measured between 40-450 °C. The coefficient of thermal expansion (CTE) was characterized by NETZSCH TMA 402 and calculated between 50-250 °C. Contact angles of samples were measured by Dataphysics-OCA20.

III. RESULTS AND DISCUSSION

A. The cross-section of FCF/PI materials

Fig. 2 shows the cross-section of FCF/PI composite materials which was employed to assess the compatibility and dispersion of FCF filler in PI. It can be seen from Fig. 2 that all the samples show a relative smooth surface which proves that the FCF fillers are dispersed uniformly. And, as the amount of FCF filler increases, few defects appeared in the cross-section of the sample (Fig. 2d). In summary, all the FCF/PI films exhibit great dispersion of FCF fillers, which would benefit for their mechanical and dielectric performance.

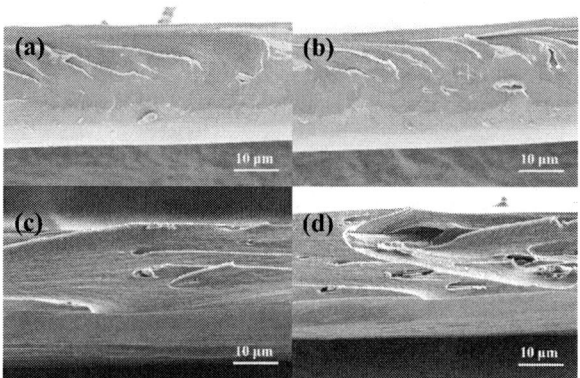

Fig. 2. The cross-section of (a) pure PI, (b) 0.5 wt%, (c) 1.0 wt% and (d) 2.0 wt% FCF/PI composite films.

B. Dielectric properties of FCF/PI materials

Fig.3 shows the dielectric constant and loss of FCF/PI nanocomposite films at 5 GHz. As the amount of FCF filler increases, the dielectric constant of samples first decreases and then increases, and the dielectric loss is almost the same at first and then significantly increases. The dielectric properties of 1.0 wt% FCF/PI nanocomposite film are as low as 3.02 (Dk) and 0.0198 (Dl) at 5GHz. And, when the FCF filler content is less than 1%, it is quite uniformly dispersed in the polyimide, which effectively reduces the dielectric constant and maintains a low dielectric loss. This is due to high power ultrasonic dispersion technology, which increases the upper

978-1-6654-1392-3/21 $31.00 © 2021 IEEE

limit of the filler content in the uniformly dispersed composite materials. However, with the further addition of FCF, the dielectric properties of 2.0 wt% FCF/PI composite film increase significantly, which is resulted from the interfacial polarization enhancement caused by the uneven dispersion of the FCF filler at large content. In this work, the high power ultrasonic dispersion technology successfully realized the dispersion uniformity of the FCF in the polyimide and reduced the dielectric constant successfully.

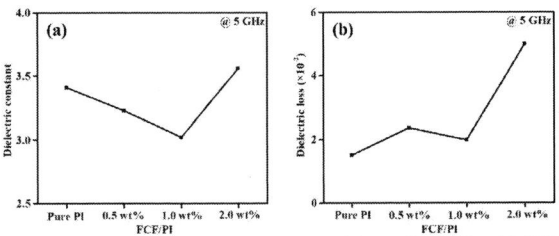

Fig. 3. (a) Dielectric constant and (b) dielectric loss of pure PI and FCF/PI.

C. Mechanical properties of FCF/PI materials

At the same time, the stress-strain curves of the FCF/PI composites were tested by dynamic thermomechanical analysis (DMA) and the corresponding data were shown in Fig. 4. The pure PI film's elongation, Young's modulus and tensile strength are 27.18%, of 2.26 GPa, and 116.71 MPa, respectively. And with the addition of FCF fillers, the mechanical properties are enhanced significantly especially of the elongation. As shown in Fig. 4b, the elongation of 1.0 wt% FCF/PI film can reach as much as 50%, which is much larger than the pure PI and it can meet the reliability requirements of WLP. However, a higher addition of FCF will cause a significant decrease in elongation and tensile strength. This is due to the uneven dispersion of FCF filler in a large amount.

D. Thermal properties of FCF/PI materials

Then, Fig. 5 (a-c) shows the thermal properties of FCF/PI nanocomposite films. The addition of FCF shows a limited impact on the thermal performance of the nanocomposite film, and the $T_{5\%}$ and T_g of all PI composites were above 550 °C and 400 °C, respectively. At the same time, the introduction of FCF filler also exhibited a limited effect on the CTE of the composite samples. And all of the CTE is between 33.18 and 37.51 ppm/K. As a result, because of the uniformity of FCF filler dispersion, all FCF/PI composite materials still maintain the excellent heat resistance of polyimide to ensure its application in WLP.

E. Hydrophobic properties of FCF/PI materials

As we all know, polyimide has a poor hydrophobic property due to the presence of amide in the molecular structure. Since the dielectric constant of water is greater than 78, the dielectric properties of the polyimide after water absorption become poor. Therefore, reducing the water absorption rate of polyimide is extremely important for its application in microelectronics. Herein, the contact angles of pure PI and FCF/PI were tested to assess the hydrophobic properties. In Fig. 6(d), with the increasement of FCF filler, the angles of samples from 81.5° increased to 89.2°, making it more advantageous than pure PI in the field of microelectronics. This is due to the hydrophobicity of the fluorine element in the FCF filler.

IV. CONCLUSSION

In summary, the novel FCF/PI composite materials with low-dielectric and high-elongation properties were developed to meet the evolving WLP. In this report, a low-cost high-power ultrasonic dispersion technology was adopted to improve the dispersion of FCF filler in polyimide. The result indicated that FCF filler with excellent dispersion effectively reduced the dielectric constant (Dk=3.02, Dl=0.0198 @5GHz) and increases the elongation (> 50%) of FCF/PI materials. In addition, the FCF/PI materials also proved excellent thermal properties. This work provides a low-cost and easy-to-industrial method to develop polyimide composite materials with performance enhancements that are expected to be used in wafer-level packaging.

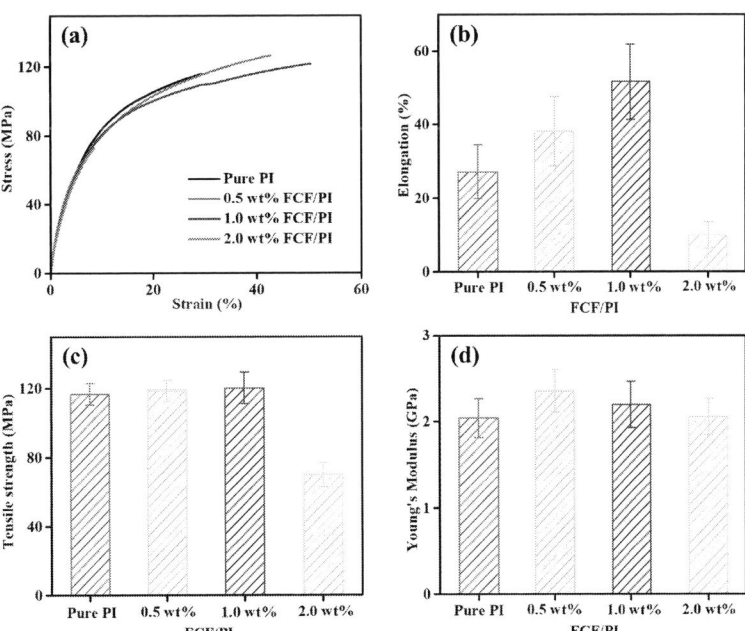

Fig. 4. (a) The stress-strain curves, (c) elongation, (b) tensile strength and (d) Young's modulus of FCF/PI nanocomposite films.

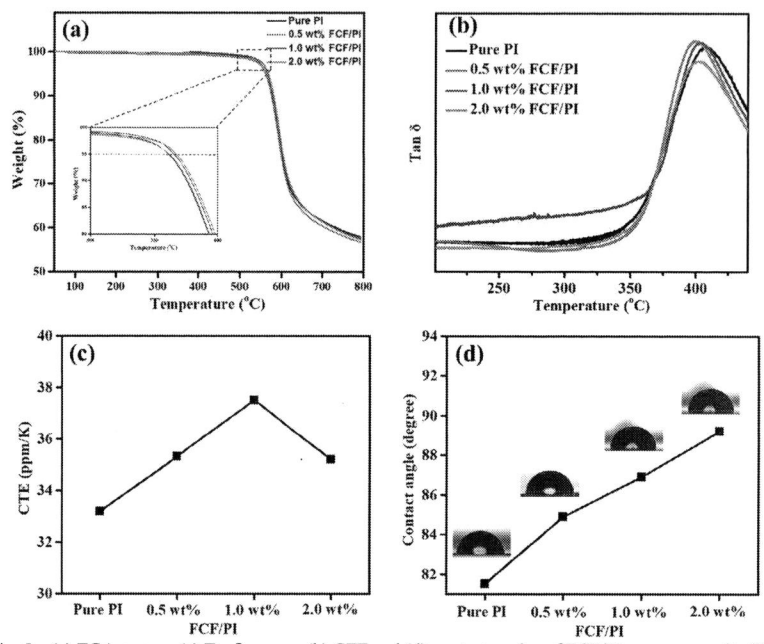

Fig. 5. (a) TGA curves, (c) Tanδ curves, (b) CTE and (d) contact angles of FCF/PI nanocomposite films.

ACKNOWLEDGMENT

We acknowledge the financial support of National Natural Science Foundation of China (61904191), Youth Innovation Promotion Association of Chinese Academy of Sciences (2017410), Key R&D Project of Guangdong Province (2020B010180001) and National Key R&D Project from Minister of Science and Technology of China (2017ZX02519).

REFERENCES

[1] F. Zhang, J. Li, F. Liu, G. Zhang, R. Sun, C. P. Wong. Intrinsic low dielectric constant and low dielectric loss polyimides: the effect of molecular structure, 2019 20th International Conference on Electronic Packaging Technology (ICEPT), IEEE, 2019: 245306.

[2] C. Huang, J. Li, G. Zhang, R. Sun, C. P. Wong. Development of low temperature curing polyimides with quinoline, 2019 20th International Conference on Electronic Packaging Technology (ICEPT). IEEE, 2019: 245169.

[3] Y. Shoji, Y. Masuda, K. Hashimoto, K. Isobe, Y. Koyama, R. Okuda. Development of novel low-temperature curable positive-tone photosensitive dielectric materials with high elongation, 2016 IEEE 66th Electronic Components and Technology Conference (ECTC). IEEE, 2016: 1707-1712.

[4] H. Kim, A. A. Abdala, C. W. Macosko. Graphene/Polymer Nanocomposites. Macromolecules, 2010, 43, 16, pp6515-6530.

[5] R. Bei, C. Qiao, Y. Zhang, Z. Chi, S. Liu, X. Chen et al. Intrinsic low dielectric constant polyimides: relationship between molecular structure and dielectric properties. Journal of Materials Chemistry C, 2017, 5, pp.12807–12815.

[6] J. Y. Wang, S. Y. Yang, Y. L. Huang, H. W. Tien, W. K. Chin, C. M. Ma et al. Preparation and properties of graphene oxide/polyimide composite films with low dielectric constant and ultrahigh strength via in situ polymerization. Journal Materials Chemstry, 2011, 21, pp.13569-13575.

Influence of Porosity on the Mechanical Properties of Hybrid Silver Sintered Joint

Zunyu GUAN
Department of Mechanical and Aerospace Engineering
Hong Kong University of Science and Technology
Hong Kong, China
zguanac@connect.ust.hk

Fred Fuliang LE
Packaging R&D
Nexperia HK Ltd
Hong Kong, China
fred.le@nexperia.com

Jingshen WU
Department of Mechanical and Aerospace Engineering
Hong Kong University of Science and Technology
Hong Kong, China
mejswu@ust.hk

Jinglei YANG
Department of Mechanical and Aerospace Engineering
Hong Kong University of Science and Technology
Hong Kong, China
maeyang@ust.hk

Rinse van der Meulen
Packaging R&D
Nexperia Nijmegen
Nijmegen, Netherlands
rinse.van.der.meulen@nexperia.com

Haibin CHEN
Department of Mechanical and Aerospace Engineering
Hong Kong University of Science and Technology
Hong Kong, China
chenhb@ust.hk

Abstract—**In this study, a type of hybrid Ag sintering paste is selected and characterized. Samples were made under different sintering conditions with rising temperature and duration. Microstructures for hybrid sintered Ag joint were characterized by SEM morphologies. The Ag particles were sintered and necked together during sintering, with the thermal resistant polymers filling in the pores around Ag particles. However, excessive high temperature will lead to oversintering and the residual polymers will decompose which result in an inhomogeneous microstructure. Meanwhile, delamination was found at the plated Ag/Cu interfaces which lead to poor shear strength at 380°C. Shear test and failure modes under different sintering conditions were then studied and analyzed. The shear strength for sintered Ag joint after one-hour sintering was lower than that after two-hours sintering, and adhesive failures were found in part of the fracture surfaces.**

Keywords— hybrid Ag sintering, sintered Ag joint, porous microstructure.

I. INTRODUCTION

Ag sintering is one of the emerging options that can be used for the die attaching material in power electronic packaging since it provides better thermal and electrical conductivity than traditional high-lead solders [1-6]. Pure Ag sintering usually uses a pressure-assisted die attaching process to achieve a porous Ag sintered microstructure, but the pores formed are empty and there is no filler (such as a polymer) inside [7,8]. Previous studies have shown that high porosity in the pure Ag sintering will cause the degradation of the die-attach joint, and make it prone to initiate cracks around the pores in reliability testing [9-13]. In recent years, the semiconductor industry has introduced hybrid Ag sintering materials to address the concerns. Compared to pure Ag sintering, a thermal resistant polymer is added into the hybrid Ag sinter paste, so after the sintering process, the formed pores are filled with polymer. However, the effect of the sintering conditions on the distribution of the polymer-filled pores and the mechanical properties of the hybrid silver sintered joints has never been studied. In this study, and with the above objective, a hybrid nano-silver sinter paste was selected. Its morphology and thermogravimetric curve were

first characterized, which provided the guidelines for the selection of processing conditions and sintering profiles. Test samples were then built and sintered using different sintering profiles. The microstructures after sintering were characterized and the strength was evaluated by shear test. The failure modes were further analyzed. The relationships among materials, processing, microstructure, and mechanical properties were discussed. The influence of sintering conditions on the porous microstructure, joint strength and failure mechanism of the hybrid Ag sintering paste were revealed.

II. EXPERIMENTAL METHODOLOGY

A. Characterization of Hybrid Ag Sintering Paste

A type of hybrid Ag sintering paste was selected and characterized. The Ag particles were characterized by scanning electron microscopy (SEM) after washing away the organic substances. As shown in Fig.1, it contains spherical Ag particles with different diameters ranging from tens of nanometer to hundreds of nanometers. The average diameter was around 100 nm.

Fig. 1. Morphology of Ag particles in the hybrid Ag sintering paste

Thermogravimetric analysis (TGA) was applied to determine the contents of the hybrid Ag sintering paste and their thermal stability, thus providing guidance on the sintering profiles. TGA tests were carried out from 25 °C to 600 °C at a heating rate of 10 °C/min under a purge of N_2 gas. As shown in Fig. 2., the solvent and organic coating in this paste completely evaporated at a tempeature below

978-1-6654-1392-3/21 $31.00 © 2021 IEEE

200°C, and the thermal resistant polymers started to decompose at around 380 °C. The loadings of Ag filler and the thermal resistance polymer were ~86 wt% and 4 wt%, respectively.

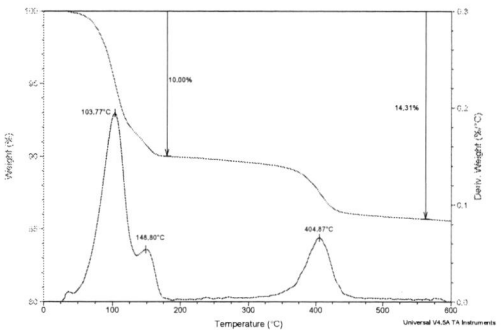

Fig. 2. TGA data for the hybrid Ag sintering paste

B. Preparation of Sintered Ag Joint

Test samples were assembled using 200 μm thick Ag plated Cu leadframe as substrate, 200 μm thick silicon die with Ag finish at backside, and hybrid Ag sintering paste as die attach material, as illustrated in Fig. 3. The detailed assembly process was: 1) printing a layer of paste on the Cu substrate; 2) placing a 2.5 mm × 2.5 mm die on the paste; 3) sintering in an oven under different temperature profiles. Fig. 4 illustrates the sintering profiles used in this study. The sintering profiles include a pre-heating stage at 60 °C for half an hour and a high temperature sintering stage. The temperatures and durations of the high temperature sintering stage are tabulated in Table 1. After sintering, the bonding line with hybrid Ag sintering paste was around 25μm.

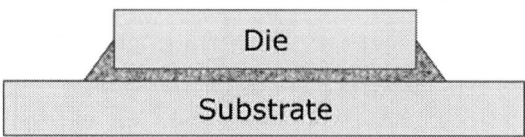

Fig. 3. Schematic of bonding structure

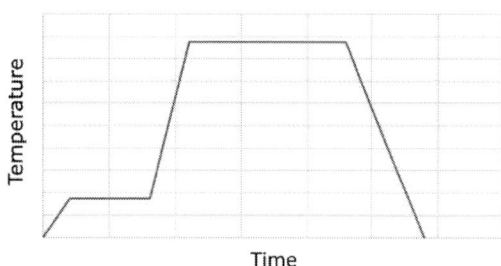

Fig. 4. General sintering profile

TABLE I. Sintering Conditions

Sample No.	Sintering Temperature(℃)	Sintering Time(Hour)
#1	200℃	1 hour
#2	250℃	1 hour
#3	300℃	1 hour
#4	200℃	2 hours
#5	250℃	2 hours
#6	300℃	2 hours
#7	380℃	2 hours

C. Microstructure Analysis and Mechanical Test

The microstructure after sintering was characterized by cross-section and SEM. Energy-dispersive X-ray spectroscopy (EDS) was employed for compositional analysis. The mechanical property was evaluated by shear test using Dage 4000 Plus. The shear rate was 500 μm/s and the shear height was 50μm. The surface after shear test was further analyzed under SEM.

III. RESULTS AND DISCUSSION

A. Microstructure of Sintered Ag Joint

The microstructures of sintered Ag joints corresponding to different sintering conditions were characterized by SEM and investigated in terms of the parameters below:

1) Structure Uniformity

The uniformity of the sintered structure is very important for the reliability of the joints. Fig. 5 shows different structure features of the sintered Ag joints, corresponding to different sintering temperatures and durations. The joints after sintering below 250°C form homogeneous microstructures, as shown in Fig. 5(a). Thermal resistant polymers are filled in the pores between necked Ag particles. As the sintering temperature rises to 300°C, as shown in Fig. 5(b), the central areas have denser sintered Ag than the edge areas. When the temperature rises to 380°C, the residual polymers further decompose. Almost complete dense Ag necking is formed at the central areas and large pores are formed at the edge areas. The phenomena can be explained in term of the compositions of the Ag sintering paste. According to the TGA data, the evaporation temperature of the solvent in the paste is around 100°C, while the onset and peak decomposition temperature of thermal resistance polymer is around 300°C and 400°C, respectively. When the sintering temperature is above 300°C, the thermal resistance polymers start to decompose, and the generated gas will spread out from the central of the joint to edge. As a result, the microstructure in the joint becomes non-uniform.

2) Microstructure Evolution

The evolution of microstructures with increasing sintering temperature and duration can be observed in Fig. 6. When the sintering temperature is 200°C and the duration is 1 hour, the formed Ag necking is loose, and the Ag particles are not completed sintered, as indicated in Fig.6(a). When the sintering temperature rises to 250°C, the Ag particles become well sintered, reflected in the growth of the particle and formed necking feature (Fig.6b). When the temperature further rises to 300°C, the Ag particles show excessive sintering with large pores in between the particles, as shown in Fig. 6(c). The phenomenon is much worse when the temperature rises to 380°C. These results indicate that the microstructure of the Ag joints can be controlled by careful choice of suitable sintering temperature and duration, which enable us to investigate the relationships between materials, processing, microstructure, and properties.

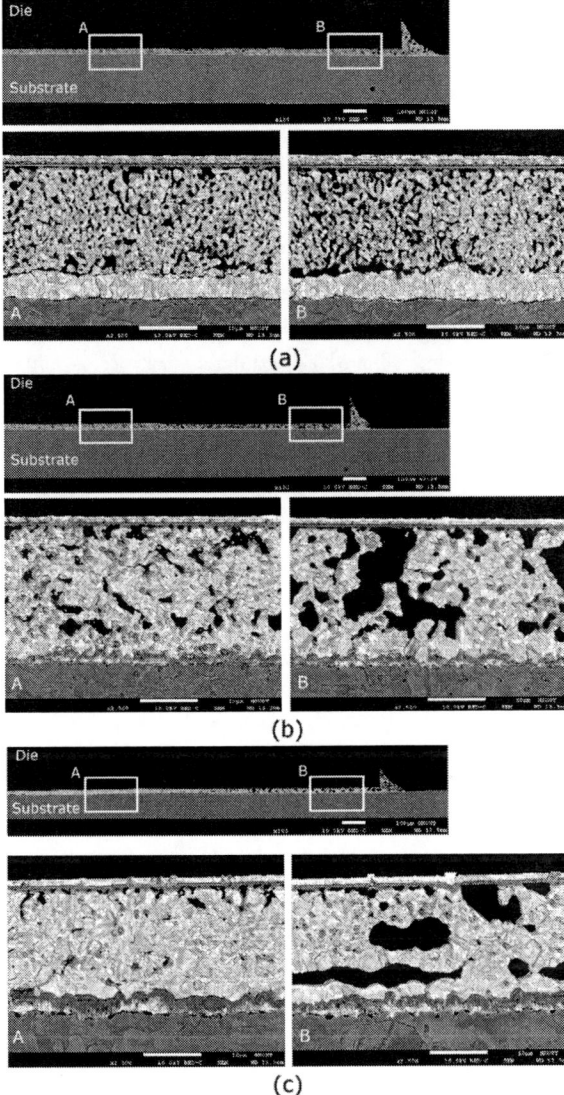

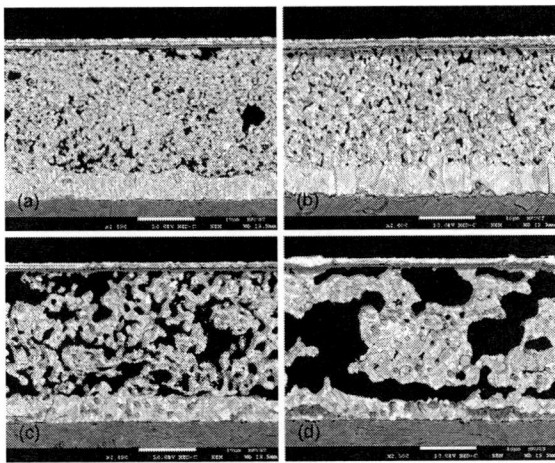

Fig. 5. Microstructures of central areas(A) and edge areas(B) for sintered Ag joints: (a) uniform porous structure after two hours sintering at 200°C, (b) denser Ag necking at central areas than edge areas after two hours sintering at 300°C and (c) completely dense Ag necking at central areas and large pores at edges after two hours sintering at 380°C.

Fig. 6. Microstructure evolution with increasing sintering temperature and time: (a)200°C/1h, (b)250°C/1h, (c)300°C/1h and (d)380°C/2h

3) Interfacical Bonding

The interface quality is of vital importance for sintered Ag joint strength. Fig. 7(a) shows typical interface morphology between the sintered Ag and die backside and Ag plated Cu substrate. The Ag particles are well sintered to these two surfaces, which indicates that the interfacial bonding is well achieved in all the sample sintered under the sintering conditions applied in this study. It is worth noting that when sintering at 380°C, there is a gap in between the plated Ag and the Cu substrate, as shown in Fig. 7(b). The inter-diffusion of Ag-Cu interface is unbalanced, and the diffusion of Cu to Ag is much faster than that of Ag to Cu [15]. As the sintering temperature and duration increase, this Kirkendall effect becomes significant. As shown in the figure, a gap is formed at the interface between Ag and Cu, which reduces the strength of the sintered joint.

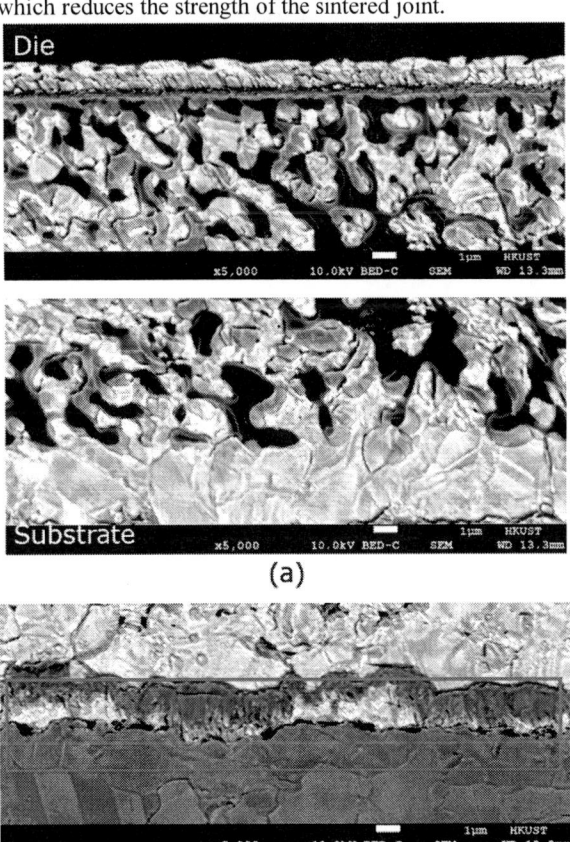

Fig. 7. (a) Typical interface morphology between the sintered Ag and die backside/substrate, and (b) interfacial separation between plated Ag and copper after two hours sintering at 380°C.

B. Shear Strength of Sintered Ag Joint

The strength of sintered Ag joints was evaluated by die shear test. Boxplots for shear force of samples sintered under different sintering conditions are shown in Fig. 8. The shear force of the samples sintered for 2 hours is higher than that of the samples sintered for 1 hour, when the sintering temperature is the same. The difference is more obvious at high temperature sintering. The shear force of the sample sintered at 300°C for 1 hour is only 25kgf, while it is 35kgf when sintering for 2 hours at the same temperature. This difference in shear force or shear strength agrees well with the microstructure evolution analysis. One-hour sintering is not enough to form well sintered joints. Moreover, as expected, the shear force of the sample sintered at 380°C is extremely low, which is a result of the weakened Cu/Ag interface as described above and will be further evidenced by failure analysis in the next part.

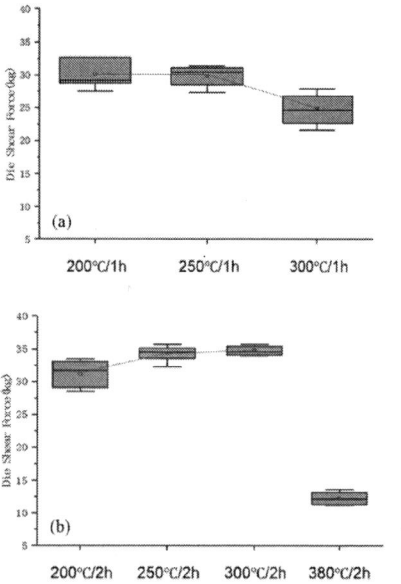

Fig. 8. Boxplot of shear force for samples under (a) one-hour sintering and (b) two-hours sintering.

C. Failure Mode Analysis

To further understand the effect of microstructure on the mechanical properties, the surface of the sample after die shear was analyzed under SEM. As illustrated in Fig. 9, three types of failure modes, i.e. cohesive failure, cohesive and adhesive mixed failure, and separation between Ag layer and Cu substrate, were observed.

1) Cohesive failure

Samples sintered at 200°C, 250°C, or 300°C for 2 hours exhibit cohesive failure. The failure takes place along the sintered Ag layer. Extensive plastic deformation and shear zone are observed in the initial shearing areas. It is believed that when the Ag jointed is well sintered, plastic deformation occurs before intergranular fracture of sintered Ag.

2) Cohesive+adhesive failure

Samples sintered at 200°C, 250°C, or 300°C for 1 hour exhibit a different failure mode, i.e. cohesive and adhesive mixed failure mode. Adhesive failure was found at the interface between sintered Ag and plated Ag layer. Under this failure mode, the plastic deformation of sintered Ag is limited.

3) Seperation between plated Ag and Cu substrate

Samples sintered at 380°C exhibit a very special failure mode, i.e. separation at Cu/Ag interface. As shown in Fig. 9(c), after die shearing, the plated Ag layer is teared off and the Cu substrate is completely exposed. This failure mode is correlated to the observation of the gap between plated Ag and Cu substrate, which is induced by high sintering temperature.

IV. CONCLUSIONS

In this study, hybrid sintered Ag joints under different sintering conditions were characterized and the strength was evaluated by shear test. The relationships among materials, processing, microstructure, and mechanical properties were discussed. The failure modes were further analyzed. The influences of sintering conditions on the porous microstructure, joint strength and failure mechanism of the hybrid Ag sintering paste were revealed. The following conclusions can be made:

(1) The microstructure of the Ag joints can be controlled by choice of suitable sintering temperature and duration. As the temperature rise and lasts longer, the Ag particles will undergo improved necking and sintering, with the thermal resistant polymers filling in the pores around the Ag particles. When the sintering temperature rises to 300°C, the thermal resistant polymers will start to decompose and generate gases which will result in an inhomogeneous microstructure with a compact center and a loose periphery.

(2) The shear force results agree well with the

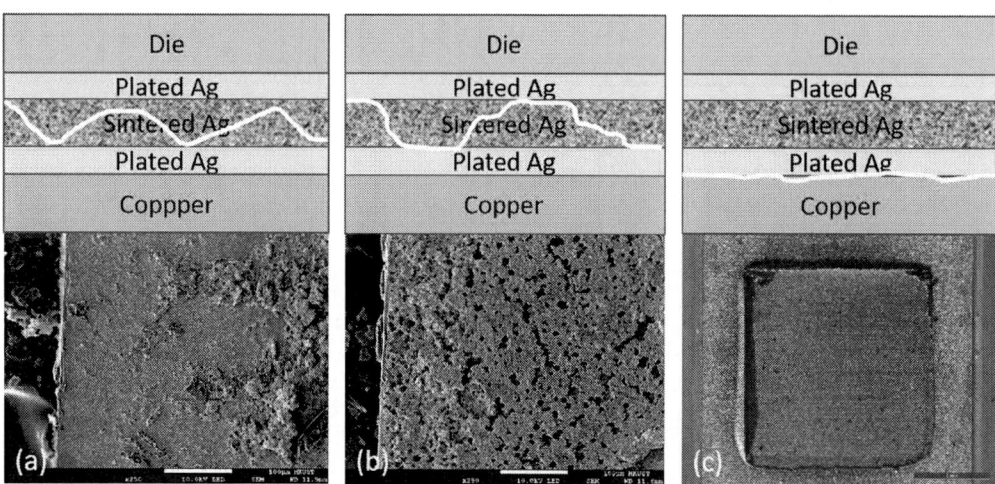

Fig. 9. Failure modes for samples after shearing: (a) Cohesive failure; (b) cohesive + adhesive failure; (c) separation between plated Ag and Cu

microstructure evolution analysis. The relationships between porous microstructure, the joint strength and corresponding failure modes are revealed. Modeling and simulation work in the future can be established based on the experimental findings in this work.

(3) The interfacial bonding is well achieved in all the samples sintered under the sintering conditions applied in this study. However, a gap is formed between the Cu substrate and plated Ag at excessive high temperatures, which weaken the sintering joint due to the non-equivalent diffusion between Cu and Ag. Two hours sintering at 250°C-300°C is ideal through this study.

V. REFERENCES

[1] Zhang H, Gao Y, Jiu J, Suganuma K. In situ bridging effect of Ag 2 O on pressureless and low-temperature sintering of micron-scale silver paste. J Alloy Comp 2017; 696:123–9.

[2] Blank T, Bruns M, Kuebel C, Leyrer B, Meisser M, Weber M, et al. Low temperature silver sinter processes on ENIG surfaces. In: CIPS 2016; 9th International conference on integrated power electronics systems. Proceedings of: VDE; 2016. p. 1–6.

[3] Wang T, Chen G, Wang Y, Chen X, Lu G-q. Uniaxial ratcheting and fatigue behaviors of low-temperature sintered nano-scale silver paste at room and high temperatures. Mater Sci Eng A 2010;527:6714–22.

[4] Yang C-X, Li X, Lu G-Q, Mei Y-H. Enhanced pressureless bonding by Tin Doped Silver Paste at low sintering temperature. Mater Sci Eng A 2016;660:71–6.

[5] Heilmann J, Nikitin I, Zschenderlein U, May D, Pressel K, Wunderle B. Advances and challenges of experimental reliability investigations for lifetime modelling of sintered silver based interconnections. Thermal, Mechanical and Multi-Physics Simulation and Experiments in Microelectronics and Microsystems (EuroSimE). In: 2016 17th International conference on. IEEE; 2016. p. 1–13.

[6] Chen S, Liu K, Luo Y, Jia D, Gao H, Hu G, et al. In situ preparation and sintering of silver nanoparticles for low-cost and highly reliable conductive adhesive. Int J Adhesion Adhes 2013;45:138–43.

[7] Chen G, Cao Y, Mei Y, Han D, Lu G-Q, Chen X. Pressure-assisted low-temperature sintering of nanosilver paste for 5*5 mm2 chip attachment. IEEE Trans Compon Packag Manuf Technol 2012;2:1759–67.

[8] Katsis DC, Zheng Y. Development of an extreme temperature range silicon carbide power module for aerospace applications. In: Power electronics specialists conference. 2008 PESC 2008 IEEE: IEEE; 2008. p. 290–4.

[9] X. Milhet, P. Gadaud, V. Caccuri, D. Bertheau, D. Mellier, M. Gerland, Influence of the porous microstructure on the elastic properties of sintered Ag paste as replacement material for die attachment. J. Electron. Mater. 44(10), 3948–3956 (2015).

[10] V. Caccuri, X. Milhet, P. Gadaud, D. Bertheau, M. Gerland, Mechanical properties of sintered Ag as a new material for die bonding: influence of the density. J. Electron. Mater. 43, 4510–4514 (2014).

[11] J. Carr, X. Milhet, P. Gadaud, S.A.E. Boyer, G.E. Thompson, P.D. Lee, Quantitative characterization of porosity and determination of elastic modulus for sintered micro-silver joints. J. Mater. Process. Technol. 225, 19–23 (2015).

[12] P. Gadaud, V. Caccuri, D. Bertheau, J. Carr, X. Milhet, Ageing sintered silver: Relationship between tensile behavior, mechanical properties and the nanoporous structure evolution. Mater. Sci. Eng. A 669, 379–386 (2016).

[13] S.A. Paknejad, G. Dumas, G. West, G. Lewis, S.H. Mannan, Microstructure evolution during 300 °C storage of sintered Ag nanoparticles on Ag and Au substrates. J. Alloys Compd. 617, 994–1001 (2014).

[14] Herboth T, Guenther M, Fix A, Wilde J. Failure mechanisms of sintered silver interconnections for power electronic applications. In: 2013 IEEE 63rd electronic components and technology conference. IEEE; 2013. p. 1621–7.

[15] Le, F.F., van der Meulen, R., Leong, Y.K., Balakrishnan, M. and Guan, Z., 2020. Evaluation of Emerging High Melting Point Lead-free Solder and Hybrid Sinter Paste as Attaching Material for Clip Bond Package. In International Symposium on Microelectronics (Vol. 2020, No. 1, pp. 000235-000241). International Microelectronics Assembly and Packaging Society.

Integration design of mm-Wave Radar Antennas based on FOWLP

Lihong Liu
Design & Development Dept.
JCET Group Co. Ltd.
Jiangyin, China
lisa.liu@jcetglobal.com

Chen Cheng
Design & Development Dept.
JCET Group Co. Ltd.
Jiangyin,China
chen.cheng@jcetglobal.com

Jiongjiong Gu
Design & Development Dept.
JCET Group Co. Ltd.
Jiangyin, China
patrick.gu@jcetglobal.com

Tianhua Shen
Design & Development Dept.
JCET Group Co. Ltd.
Jiangyin, China
sky.shen@jcetglobal.com

Quanbing Li
ICBU.
JCET Group Co. Ltd.
Jiangyin, China
quanbing.li@jcetglobal.com

Abstract— With the development of mm-Wave automotive radar, terahertz imaging industry, etc. The Antenna in Package （AIP） technology has attracted attention. However, as AIP antennas are mm band, small in size, there are great challenges from design to fabrication and test. The main work of this paper is to design mm-Wave antenna arrays for an Application Specific Integrated Circuit (ASIC) chip based on Fan-out Wafer Level Package (FOWLP) process. And the designed antenna arrays are easy processed, high gain, broadband and meet the requirements of the system, by selecting the appropriate feeding mode and optimizing the distance, length and width between the antenna patches.

Keywords—Antenna in Package, FOWLP, ASIC, mm-Wave antenna array

I. INTRODUCTION

With the development of technology, mm-wave radar is appreciated by the market ,due to its advantages, such us high resolution, strong anti-jamming performance, good detection performance and small size. It is not only widely used in the automated vehicle driving, but also widely used in home automation, security monitoring, and other scenarios. The application of mm-wave radar depends on the development of semiconductor technology. At present, there are three main technologies to achieve AIP, respectively LTCC (Low Temperature Co-Fired ceramic), HDI (High-density Interconnection) and FOWLP[1]. However, according to the current industrialization progress, we believe that FOWLP will become the mainstream process of AIP antenna. FOWLP is a package solution, which is based on BGA (Ball Grid Array) technology, regards wafer as the processing object ,and conducts packaging tests on wafer. Finally, it is cut into individual devices, which can be directly attached to the substrate or PCB board. Because WLP does not need interposer, under fill and lead frame, and eliminates the process of die bonding and wire bonding, material and labor costs can be great reduction. In addition, most WLP uses RDL (Redistribution Layer) and Bump technology as the way of I/O distribution. And FOWLP has higher integration, better heat dissipation and lower transmission loss characteristics.

Therefore, the main work of this paper is designing a 77GHz millimeter-wave radar array based on the FOWLP process. MIMO （ Multiple Input Multiple Output ） technology is used to improve the angle resolution of the radar.

II. ANTENNA ARRAY DESIGN

A. Specifications of mm-Wave Radar Antenna

In this paper, the mm-Wave radar modules consist of antenna arrays and ASIC chip. The ASIC chip manufactured by CMOS process, which can integrate RF transceiver circuit and digital processor into one chip. In this way, the radar can realize control and signal processing in local, then access to the central control platform through the standard interface, complete data fusion with other sensors, reduce system complexity. So the outside of radar chip only need few peripheral circuits, just power supply, Flash, etc. And this mm-radar module is used in home automation, which integrates two transmitting channels and four receiving channels. The peripheral circuit schematic of ASIC is as shown in Fig.1.

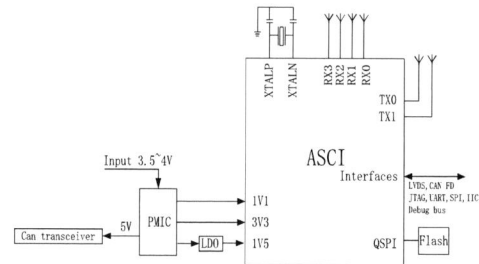

Fig. 1. The Peripheral Circuit Schematic of ASIC

To realize the transmitting function of the system, the specifications of every transmitting channel radar antenna array are as Table1,

TABLE I.   THE SPECIFICATIONS OF RADAR ANTENNA ARRAY

Specifications	Value
Work Frequency	77GHz
Bandwidth	2GHz
Gain	9dBi
3dB Beam Width on XOZ Plane	60°
Side Lobe Suppression on XOZ Plane	-15dB
Front-to-back Ratios	-20dB

978-1-6654-1392-3/21 $31.00 © 2021 IEEE

B. Antenna Array Design

1) Confirm the stack layer of package

According to the antenna theory, increasing the dielectric constant can increase the bandwidth of the antenna and reduce the size of the antenna, but it will affect the antenna gain. Increasing the thickness of the substrate can improve the antenna bandwidth and radiation efficiency, which will increase weight and volume. At the same time, the selection of substrate also needs to consider the actual process and the applicable frequency range[2]. By balancing the above factors, the mm-wave IC package structure in this paper is as shown in Fig.2. The chip is redistribution through the 3-layer WLP process, attached on the 3-layers substrate, and the total package thickness is 0. 5mm. Where, the thickness of WLP redistribution is about 40um, and the dielectric constant of passivation layer is 3.3 and loss tangent is 0.011.The FOWLP structure is thin, that is helpful to reduce the RF channel loss. The thickness of the substrate is 0.3mm. The outermost substrate layer is antenna radiators, the middle layer is ground plane, the innermost layer is signal lines. The epoxy between the antenna radiators and ground plane is 0.25mm, its dielectric constant is 3.58, and loss tangent is 0.002.

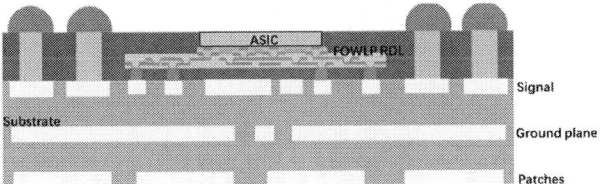

Fig. 2.   Stack-up of FOWLP Package

2) Design of Antenna Unit

After confirming the stack layers and material, the size of the patch unit should be estimated initially. Most of the mm-wave antenna arrays in AIP package are composed of microsrtip patch antennas, as shown in Fig. 3.

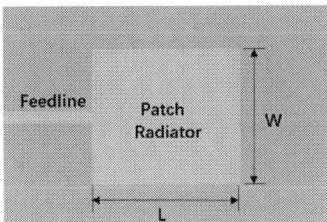

Fig. 3.   Microstrip Patch Antenna

Width estimation: The width of the rectangular patch affects the antenna pattern, radiation resistance and input impedance, the formula is as follows:

$$W = \frac{c}{2f}\left(\frac{\varepsilon_r+1}{2}\right)^{-\frac{1}{2}} \quad (4)$$

Length estimation: Due to the edge effect of the microstrip patch, the length L of the patch unit is always a little less than half of the wavelength of the microstrip medium, the formula is as follows:

$$L = \frac{c}{2f\sqrt{\varepsilon_e}} - 2\Delta L \quad (5)$$

where, the ΔL is the length of equivalent radiation slot,

$$\Delta L = 0.412\,h\frac{(\varepsilon_e+0.3)(\frac{W}{h}+0.264)}{(\varepsilon_e-0.258)(\frac{W}{h}+0.8)} \quad (6)$$

where , the ε_e is effective dielectric constant,

$$\varepsilon_e = \frac{\varepsilon_r+1}{2} + \frac{\varepsilon_r-1}{2}\left(1+\frac{12h}{W}\right)^{-\frac{1}{2}} \quad (7)$$

In(4)(5)(6)(7), dielectric constant ε_r=3.58, frequency $f = 77GHz$, after preliminary calculation, $W = 1.29mm$，$L = 0.88mm$. Then simulate for this size of patch, obtain the gain about 6.8dBi at 77GHz ,as shown in Fig.4 and Fig.5.

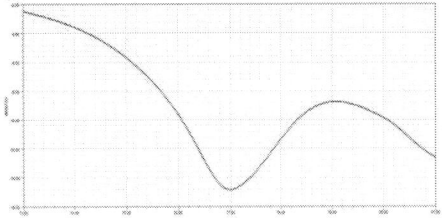

Fig. 4.   The S11 result

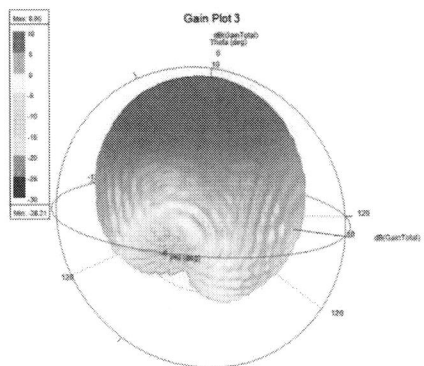

Fig. 5.   3D Radiation Pattern

However width W is related to the impedance match of the feeder, and parameters such as gain and bandwidth are mainly considered in the antenna array, so the patch size should be modulated according to the actual needs in later section.

3) Array Structure

Preliminary simulation shows that the gain of a single patch is 6.8dBi. If it needed to achieve the goal, the transmitting channel would need four patches.

In mm-wave frequency band, the loss of feedlines becomes manifest so it is particularly important to design a reasonable feeder network. In order to arrange feedlines simply, the feedline networks of array are composed of 3 dividers. And the impedance match of the network is mainly achieved by 1/4 wavelength microstrip lines.

The performances of the patch antenna arrays can achieve the set specifications by optimizing with simulation tools. The bandwidth is 2.1GHz, the gain is 9.2dBi ,the 3dB beam width and side lobe suppression on XOZ Plane is respective 57° and -15dB, the front-to-back ratio is less than -20dB.The performances are as shown in Fig.6 and Fig.7.

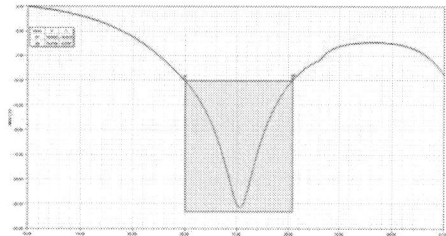

Fig. 6. S11 and Bandwidth

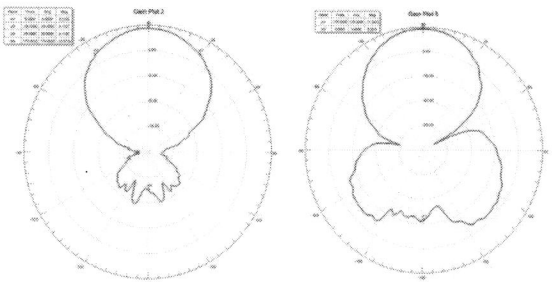

Fig. 7. Radiation Pattern: (a)XOZ Plane (b)YOZ Plane

The layout of the transmitting radar array is shown in Fig.8.

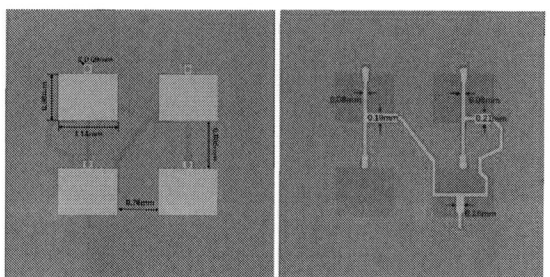

Fig. 8. The Antenna Array: (a) Top View (b)Bottom View

III. INTEGRATION OF ANTENNA ARRAYS

The above contents describe the design process of the transmitting antenna, the same method can be used to obtain the receiving antennas. Then according to the theory of MIMO radar, the distance between two transmitting antennas is $2\lambda_0$, and the distance between two receiving antennas is $\lambda_0/2$, the λ_0 is the wavelength in free space[3].Virtual synthesis of two transmitting channels, The gain and beam width of two transmitting channels are respectively increased to 12dBi，26° on XOZ Plane, which is shown in Fig.9. To keep the amplitude and phase balance, the feedlines are designed with identical length. The total plane dimension of package is 15mm*15mm, the layout of antenna array structure is as shown in Fig.10.

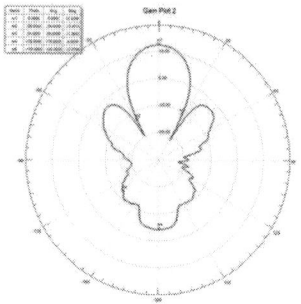

Fig. 9. Radiation Pattern on XOZ Plane for Two Virtual Synthesised Arrays

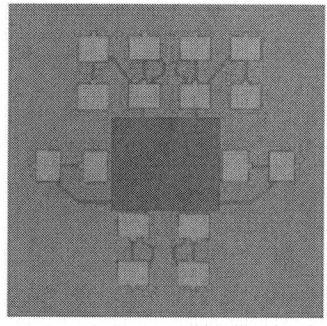

Fig. 10. The Layout of antenna arrays structure

IV. CONCLUSIONS

This paper introduced the integration design of mm-Wave radar, and designed one transmitting patch array based on FOWLP technology, which can achieve 9.2 dBi gain and 2.1 GHz bandwidth respectively. Two and four patch arrays are respectively synthesized as transmitting and receiving antennas with MIMO technology, and that is helpful to improve gain and beam width. The total dimension of package is 15mm*15mm*0.5mm, which is meeting the requirements of miniaturization and high performance.

REFERENCES

[1] "5G AIP technology FOWLP ” available On https://www.sohu.com/a/414457418_472928.

[2] I.J. Bahl and P. Bhartia, "Microstrip Antennas", Artech House,1980, pp.49-51.

[3] Haitao XinNoble, "Research of STW(Saw Tooth Wave)FMCW automotive anti-collision radar signal processing system",Xidian University,2014

[4] Zihao Chen, Xi Zhu, "Integration of mm-wave Antennas on Fan-Out Wafer Level Packaging (FOWLP) for Automotive Radar Applications", 2019 IEEE International Conference on Integrated Circuits, Technologies and Applications, pp.168-169.

Experimental and Computational Study of Shielding Effectiveness of Metal Grids

Zhiqiang LIN
Shenzhen Institute of Advanced Electronic Materials, Shenzhen Institute of Advanced Technology, Chinese Academy of Sciences
Shenzhen 518055, China
zq.lin@siat.ac.cn

Xuebin LIU
Shenzhen Institute of Advanced Electronic Materials, Shenzhen Institute of Advanced Technology, Chinese Academy of Sciences
Shenzhen 518055, China
xb.liu@siat.ac.cn

Xinrong SHI
Guangdong Institute of Metrology,
Guangzhou, 510405, China
sxrfreeman@126.com

Yougen HU*
Shenzhen Institute of Advanced Electronic Materials, Shenzhen Institute of Advanced Technology, Chinese Academy of Sciences
Shenzhen 518055, China
yg.hu@siat.ac.cn

Rong SUN
Shenzhen Institute of Advanced Electronic Materials, Shenzhen Institute of Advanced Technology, Chinese Academy of Sciences
Shenzhen 518055, China
rong.sun@siat.ac.cn

Abstract—**3D printing method is employed to fabricate silver-based metal grid with flexibility, well-ventilation, high transparency, and ultrahigh Electromagnetic Interference (EMI) shielding performance. A significant influence of the conductivity and pore area of the metal grid on its EMI shielding effectiveness is revealed, providing an intriguing metal grid design strategy. A finite element model is also used to study the EMI shielding mechanisms of the metal grid with different conductivities. The experimental and theoretical study predicts that the conductivity, pore area are the main parameters that affect the shielding effectiveness of a sample which is confirmed by the experimental and simulation results. Due to the reasonable pattern designing, the EMI shielding effectiveness of the as-prepared metal grid reaches up to 35 dB with a pore area of about 40%. Based on the results of the simulation, the metal grids show an ultrahigh average EMI SE of over 55 dB in the frequency range 0.5 to 1 GHz. The simulation and calculation study also demonstrates the relationship between the pore area and EMI SE, which will provide a guideline to design the EMI shield metal grids.**

Keywords—Electromagnetic Interference (EMI) shielding, Simulation, Metal Grids, COMSOL, Finite Element Analysis (FEA)

I. INTRODUCTION

With the rise of high-speed fifth-generation (5G) communications and highly integrated smart electronic technologies, electromagnetic interference (EMI) is becoming more and more serious, and electromagnetic interference will seriously affect the normal communication and operation of electronic and electrical equipment [1, 2]. In addition, electromagnetic (EM) wave pollution [3] is threatening human health and the surrounding environment.

Metal foils, metallic films [4], or metallic foam [5] have been employed as the conventional commercial EMI shielding materials, due to the high electrical conductive and remarkable EMI shielding performance. For electronic packaging, EMI shielding materials are required to have both excellent air-permeable and high shielding performance [6]. Such as, pore arrays always be designed and involved in the case of personal computers, due to the heat dissipation purpose. As a result, designing an air-permeable and transparent metal-based grid for high-performance EMI shielding is crucial yet challenging[7, 8].

The EMI shielding performance of metallic enclosures with apertures and have been investigated systematically[9]. The vent openings and the fan exhaust hole are the major cause of low EMI shielding effectiveness (SE) of metallic enclosures, due to the openings of enclosures will allow the penetration of the EM wave[10]. Those reports above have demonstrated that the designing of openings have a tightly relationship with the EMI SE of enclosures, and a lot of simulation results also prove this point. However, the mechanisms may be related to the shape or size of the opening, but still need further study to prove the conclusions[11].

Transparent EMI shielding films have attracted a lot of attention due to the strong demand of markets[12]. Metal meshes with high transparencey for EMI shielding have been reported with the screen printing method [13, 14]. The as-prepared metal patterns show a high EMI SE of 45 dB with high optical transmittance of 42%. And the impact of the characteristic length scale of line width and thickness on EMI SE of metal mesh has also been demonstrated. As the report, a larger line width will lead to higher EMI SE[15-17]. However, the mechanisms of line width remains to be determined.

Here, a facile 3D printing method is performed to print a series of silver/TPU mesh with a various pore sizes of 500, 900, and 1200 μm, and with a line width of about 300 μm. The relationship between pore size and have been investigated. The as-fabricated silver/TPU patterns show an EMI SE of 35 dB with a low pore area of about 40 %. Furthermore, a creative periodic unit finite element analysis (FEA) simulation model have been built by COMSOL for calculated the EMI SE of the special metal mesh with periodic structure. The simulation has guided the research and understanding of the shielding mechanism of metal mesh and help us to optimize the design of EMI shielding metal mesh. Finally, the impact of conductive metal mesh on its EMI SE is also studied by simulation method.

II. EXPERIMENTAL SECTION

A. Materials

TPU pellets (Elastollan ®1185A10) were supplied by BASF Co., Ltd. N, N-dimethylformamide (DMF, AR) were obtained from Aladdin CO., Ltd. Silver micro-flakes were purchased from CNMNX Ningxia New Materials CO., Ltd, with s average size of about 5-10 μm. The 3D printing patterns were prepared by 3D printing machine (Neotech PJ15X) with a nozzle printing head (with a diameter of 200 μm).

B. Preparation of Silver/TPU pastes (Printing inks)

The TPU particles and DMF solution were mixed according to a mass ratio of 2:3. and the TPU-DMF solution (with a TPU mass fraction of 40%) was obtained after TPU was completely dissolved under magnetic stirring at 200 rpm. The TPU-DMF solution and the flaky silver powder were mixed according to the solid phase (TPU:Ag) mass ratio of 15:85, and mixed uniformly in a high-speed mixing mixer with a rotation speed of 2350 rpm and a time of 5 min to obtain Ag/TPU conductive silver paste.

C. Characterization

The optical images of metal mesh were observed by an optical microscope (Leica DM 2700 M). The EMI SE values of metal mesh were collected at room temperature by a vector network analyzer (VNA, keysight, E5071C). The simulation results were obtained by COMSOL Multiphysics® Software (COMSOL INC.).

III. RESULTS AND DISCUSSION

As shown by Fig 1a, Silver/TPU patterns were fabricated by the 3D printing method. The insert (left of Fig 1a) is the optical image of the 3D printing machine. Due to both the the 3D printing machine console client software and COMSOL can import and read the AutoCAD DWG file, we first design the styles of patterns in AutoCAD and export the AutoCAD DWG files. And then, the DWG files were respectively imported into the 3D printing machine console client software and COMSOL (Fig 1b). Consequently, the as-printed patterns will correspond to the COMSOL simulation model perfectly.

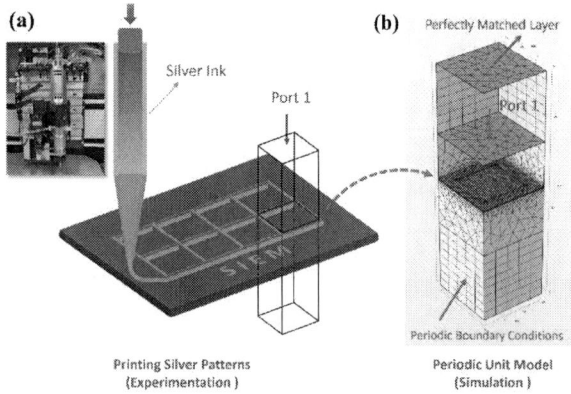

Fig 1. (a) Schematic diagram of fabrication of Silver/TPU patterns and (b) The periodic unit simulation model by COMSOL.

Due to Maxwell equations can be expressed in partial differential equations, so Finite Element Method (FEM) is one of the effective methods to solve the EM wave issue. Here,

FEM implemented in COMSOL can help to analyze the EMI issues of different materials. Moreover, the periodic structure of silver/TPU patterns allows us to simplify the simulation model as a periodic unit model by only calculating one unit structure of the patterns (Fig 1b). The boundary of the model is periodic boundary conditions rather than perfectly conductivity boundary.

The as-printed silver/TPU mesh with line periodic squared structure. As shown by Fig 2a-c the resultant silver/TPU mesh was successfully prepared with different space widths of 500, 900, and 1000 μm, and with a fixed-line width of about 300 μm. As shown by Fig 2d and 2e, the cross-section profile of the silver/TPU printed lines is measured with a height of about 180 μm and line width of about 300 μm.

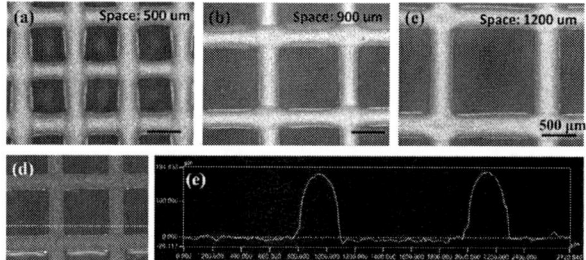

Fig 2. (a-c) as-printed silver/TPU patterns with space widths of 500, 900, and 1200 μm, respectively, the scale bar presents a length of 500 μm. (d and e) The cross-section profile of the silver/TPU printed lines.

EMI shielding performance of silver/TPU patterns with different space widths of 500, 900, and 1000 μm was evaluated by VNA (Fig 3). The silver/TPU patterns show the highest average EMI SE value of about 35 dB, which is higher than the commercial requirement (20 dB) of shielding materials. In comparison, the silver/TPU mesh with higher space of about 900 and 1200 μm exhibit lower average EMI SE values of about 23 dB and 17.5 dB, respectively. Fig 3 demonstrates that a higher space width of silver/TPU patterns will result in lower EMI shielding performance. And the simulation results, which show perfectly matching with the experimental results, also prove this conclusion. Moreover, we further study the EMI shielding performance of metal grids in the 0.5 to 1 GHz frequency band. The metal grids demonstrate a higher average EMI SE of over 55 dB in the frequency range 0.5 to 1 GHz, which means metal grids has better performance in the low frequency.

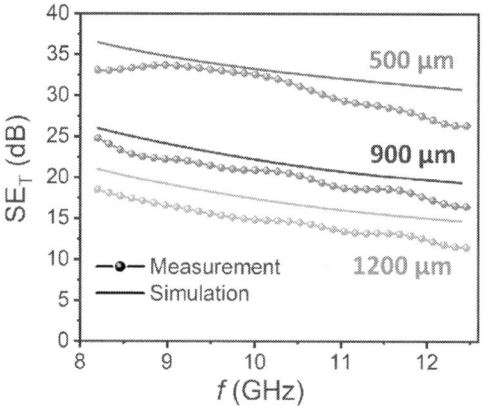

Fig 3. Comparison of EMI SE of as-printed patterns and corresponding simulation results with space widths of 500, 900, and 1200 μm, respectively.

For further investigation into the relationship between the pore area and EMI SE, we have calculated the pore area of silver/TPU patterns with different space width (500, 900, and 1000 µm, respectively). based on the one periodic unit model, The pore area was calculated according to equation (1) as follow (Fig 4a):

$$Pore\ area = \frac{l^2}{(d+l)^2} \quad (1)$$

Which, l denotes the space width of square-type hole, d presents the width of the printed line, which is about 300 µm in this paper.

As shown above, the EMI SE correlates negatively with the space width, when fixing the width of a line. However, the relationship between EMI SE and pore area has been exhibited in Fig 4b with a nearly positive linear relationship.

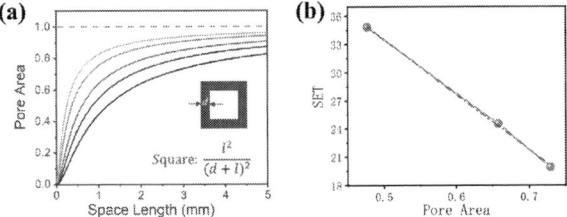

Fig 4. (a) The relationship between pore area and space width. (b) The relationship between SE$_T$ and Pore area.

The EMI SE values above are belonging to far-field shielding performance. However, the major EMI of highly integrated electronics is involved in the near-field region[18]. Therefore, the near-field shielding performance is one of the importance evaluation indexes of shielding materials. As shown by Fig 5a, the near-field testing set-up includes emission source (die), shielding materials, and detecting probe, which is used to detect the near-field magnetic field at a frequency range of 1-9 GHz. By comparison of the average near-field EMI SE mapping of signals under different shielding conditions (Fig 5b-dc), the silver/TPU mesh with a space width of 500 µm demonstrates the best near-field EMI shielding performance. It's interesting that the silver/TPU mesh shows a better near-field performance than copper foil, which possesses higher conductivity and far-field EMI SE.

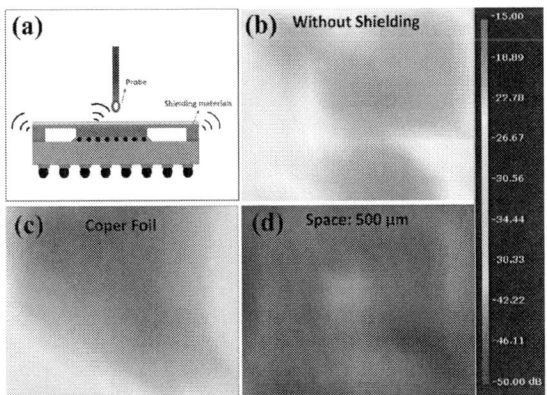

Fig 5. (a) Schematic of Near-field test set-up. (b-d) The mapping of average near-field EMI SE of blank, copper foil and silver/TPU pattern (with space width of 500 µm).

The impact of conductivity on the far-field EMI SE of both silver/TPU meshes and silver plate has been studied by simulation method. As Fig 6a shows, the lower conductivity of metal mesh will lead to a more uniform distribution of the electromagnetic induced currents in materials. It is noted that Fig 6a presents the electromagnetic induced currents of silver/TPU patterns with different conductivity under the incident EM wave frequency of 12.4 GHz. The middle square is the opening of the pattern, which is insulating material to have induced currents. A higher conductivity (10^6 S/m) of silver/TPU mesh will have high induced currents, and mainly distributed in the edge of the patterns.

The relationship between skin depth[19] and far-field EMI SE has been studied. And the calculation equation (2) of skin depth of conductivity is listed below:

$$\delta = \sqrt{\frac{2\rho}{(2\pi f)(\mu_0 \mu_r)}} \approx 503\sqrt{\frac{\rho}{\mu_r f}} \quad (2)$$

Which, the ρ denotes the resistivity of materials, and f is the frequency of incident EM wave, μ_r presents the relative permeability materials. Here, the frequency and the relative permeability of materials are fixed for further discussed the impact of conductivity on EMI SE.

As shown by Fig 6b and 6c, the EMI SE exhibits a negative relationship with the skin depth of material. By comparison, the silver plates (with a pore area of 0%) were also calculated. The EMI SE value of silver patterns is always lower than silver plates, due to the opening structure. Interestingly enough, with the increase of the conductive (decrease of skin depth), the EMI SE of silver/TPU patterns increasing slower than patterns and have leveled off. It indicates that the opening/hole structure of shielding materials is the major factor to allow the EM wave through and lead to low EMI SE.

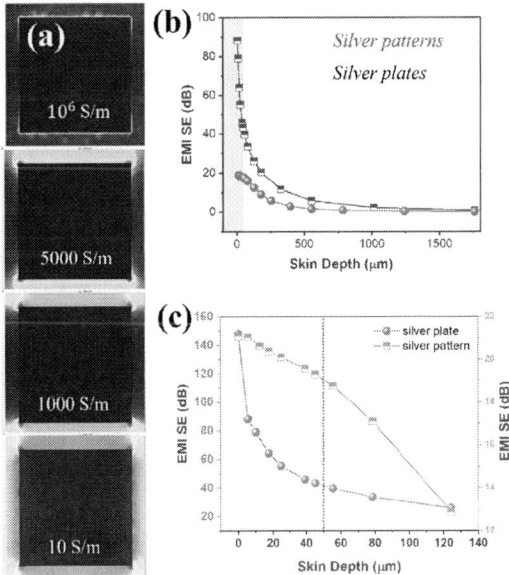

Fig 6. (a) Mapping of electromagnetic induced currents of silver/TPU patterns with different conductive. (b) The relationship between EMI SE and skin depth in a wide range of about 0 to 1750 µm. (c) The zoom-in relationship between EMI SE and skin depth in a range of about 0 to 120 µm.

IV. CONCLUSION

A silver/TPU patterns with different pore area have been successfully fabricated by the 3D printing method. The as-printed patterns show a high far-field EMI shielding effectiveness of about 32 dB with a high pore area of 40%.

978-1-6654-1392-3/21 $31.00 © 2021 IEEE

Moreover, the near-field shielding performance of silver/TPU patterns has been studied. the silver/TPU patterns show a higher near-field shielding performance than commercial copper foil. And the FEA simulation was performed to study silver plates. This study will guide the design of openings and holes of shielding materials.

ACKNOWLEDGMENT

The author are grateful for the financial support from the National Natural Science Foundation of China (62074154), China Postdoctoral Science Foundation (Grant No. 2020M682983), Guangdong Basic and Applied Basic Research Fund (2020A1515110962, 2020A1515110154), Shenzhen Basic Research Plan (JCYJ20180507182530279), and the Youth Innovation Promotion Association of the Chinese Academy of Sciences (2017411), SIAT Innovation Program for Excellent Young Researchers (E1G035).

REFERENCES

[1] Pai Y-H, Hsu M-F, Chung S-F, et al. Communication-Synthesis of Highly-Branched Silver Nanocrystals for EMI Shielding Applications [J]. Journal of the Electrochemical Society, 2021, 168(1).

[2] Chakradhary V K, Juneja S, Akhtar M J. Correlation between EMI shielding and reflection loss mechanism for carbon nanofiber/epoxy nanocomposite [J]. Materials Today Communications, 2020, 25.

[3] Xu H, Yin X, Li X, et al. Lightweight Ti2CT x MXene/Poly(vinyl alcohol) Composite Foams for Electromagnetic Wave Shielding with Absorption-Dominated Feature [J]. ACS Appl Mater Interfaces, 2019, 11(10): 10198-10207.

[4] Li P, Zhang Y, Zheng Z. Polymer-Assisted Metal Deposition (PAMD) for Flexible and Wearable Electronics: Principle, Materials, Printing, and Devices [J]. Adv Mater, 2019, 31(37): e1902987.

[5] Wan Y J, Zhu P L, Yu S H, et al. Anticorrosive, Ultralight, and Flexible Carbon-Wrapped Metallic Nanowire Hybrid Sponges for Highly Efficient Electromagnetic Interference Shielding [J]. Small, 2018, 14(27): e1800534.

[6] Lu T, Gu H, Hu Y, et al. Three Dimensional Copper Foam-Filled Elastic Conductive Composites with Simultaneously Enhanced Mechanical, Electrical, Thermal and Electromagnetic Interference (EMI) Shielding Properties [M]. 2019 Ieee 69th Electronic Components and Technology Conference. 2019: 1916-1920.

[7] Wang G, Ong S J H, Zhao Y, et al. Integrated multifunctional macrostructures for electromagnetic wave absorption and shielding [J]. Journal of Materials Chemistry A, 2020, 8(46): 24368-24387.

[8] Lian Q, Xu W, Li Y, et al. Deposited structure design of epoxy composites with excellent electromagnetic interference shielding performance and balanced mechanical properties [J]. Journal of Materials Chemistry C, 2020, 8(47): 16930-16939.

[9] Kashani H, Giroux M, Johnson I, et al. Unprecedented Electromagnetic Interference Shielding from Three-Dimensional Bi-continuous Nanoporous Graphene [J]. Matter, 2019, 1(4): 1077-1087.

[10] Kim D G, Choi J H, Choi D K, et al. Highly Bendable and Durable Transparent Electromagnetic Interference Shielding Film Prepared by Wet Sintering of Silver Nanowires [J]. ACS Appl Mater Interfaces, 2018, 10(35): 29730-29740.

[11] Li H, Yuan D, Li P, et al. High conductive and mechanical robust carbon nanotubes/waterborne polyurethane composite films for efficient electromagnetic interference shielding [J]. Composites Part A: Applied Science and Manufacturing, 2019, 121: 411-417.

[12] Vishwanath S K, Kim D-G, Kim J. Electromagnetic interference shielding effectiveness of invisible metal-mesh prepared by electrohydrodynamic jet printing [J]. Japanese Journal of Applied Physics, 2014, 53(5S3).

[13] Shi S, Peng Z, Jing J, et al. Preparation of Highly Efficient Electromagnetic Interference Shielding Polylactic Acid/Graphene Nanocomposites for Fused Deposition Modeling Three-Dimensional Printing [J]. Industrial & Engineering Chemistry Research, 2020, 59(35): 15565-15575.

the shielding mechanism of the silver/TPU patterns, including the impact of conductivity on EMI SE of both the silver/TPU patterns and

[14] Liu X, Lei Z, Rong Sun, et al. Stretchable and Printable Conductive PolymerComposites for Electromagnetic Interference (EMI) Shielding Meshes [J]. 2020 21th International Conference on Electronic Packaging Technology (Icept), 2020.

[15] Shui W, Li J, Wang H, et al. Ti(3)C(2)T(x)MXene Sponge Composite as Broadband Terahertz Absorber [J]. Advanced Optical Materials, 2020, 8(21).

[16] Han D, Mei H, Xiao S, et al. Electromagnetic shielding properties of carbon-rich chemical vapor infiltration-prone silicon carbide matrix composites [J]. Journal of the American Ceramic Society, 2018, 101(5): 1991-1998.

[17] Czeresko P J, III, Arman A S, Vogler T R, et al. EBG design and analysis for wideband isolation improvement between aircraft blade monopoles [J]. International Journal of Rf and Microwave Computer-Aided Engineering, 2020, 30(2).

[18] Gao L, Wei X-C, Huang Y-T, et al. Notice of Retraction: Analysis of Near-Field Shielding Effectiveness for the SiP Module [J]. IEEE Transactions on Electromagnetic Compatibility, 2018, 60(1): 288-291.

[19] Subramanian J, Kumar S V, Venkatachalam G, et al. An Investigation of EMI Shielding Effectiveness of Organic Polyurethane Composite Reinforced with MWCNT-CuO-Bamboo Charcoal Nanoparticles [J]. Journal of Electronic Materials, 2021, 50(3): 1282-1291.

Study on the mechanical properties of ultra-low dielectric film by tensile test

Lei Wang
Department of Microelectronics
Jiangsu Vocational College of
Information Technology
Wuxi, China
wanglei@jsit.edu.cn

Fei Xiao
Department of Materials Science
Fudan University
Shanghai, China

Jun Wang
Department of Materials Science
Fudan University
Shanghai, China

Abstract—The ultra-low dielectric constant (ULK) materials is implemented in the copper/low-k interconnection structures in industry in order to minimize the RC delay and crosstalk noise. The mechanical property of ULK dielectrics is weak compared to traditional dielectrics. So it poses a significant risk to the device reliability, not only in the fabrication process, but also in packaging processes and reliability tests. When using numerical simulations to assess the structural integrity associated with low-k integration, accurate mechanical property of ULK materials is also an important parameter that needed to be measured. In this paper, we present a method of measuring the mechanical properties of ULK materials. This method is suitable for a variety of materials which can be deposited or grown on a silicon substrate. The fabricating processes mainly include lithography, temporary bonding, grinding, Si etching and debonding. In addition, we designed two kinds of test structures. One is with pre-crack, and the other is without pre-crack. Each structure has two kinds of geometry sizes. The tensile test was done by using the in-situ tensiometer apparatus with uniaxial stretching. The stress-strain diagrams of the specimens show an elastic modulus E=1.2 GPa and tensile strength is about 13.5 Mpa for the structure without pre-crack, which is lower than that of bulk low-k material. The samples with two kinds of geometry sizes have similar modulus values which can be converted by the measured results from the diagram. While other specimens show a critical energy release rate G_c=0.25 J/m^2 for the structure with pre-crack, which is also lower than that of bulk material. The results illustrate that the ability to resist fracture of low-k thin films is weak. It provides a possible method to measure the mechanical properties of ULK thin films.

Keywords—Mechanical property; tensile test; low-k film; pre-crack; energy release rate

I. INTRODUCTION

With the development of integrated circuit, ultra-low dielectric constant (ULK) materials are implemented in the copper/low-k interconnection structures in order to minimize the RC delay and crosstalk noise[1]. The copper/low-k interconnection is a multilayered structure, in which exits various materials. Owing to these materials have different thermal and mechanical properties, it will induce significant mismatch during the fabrication, packaging and reliability tests. For example, the coefficient of thermal expansion (CTE) of silicon material is 2.6 ppm/°C. However, the CTE of the aluminum and copper is about 23.6 ppm/°C and 16.7 ppm/°C, respectively[2]. Due to low-k materials have different types and compositions, the value of their CTEs has a wide range, which is from 10 to 100 ppm/°C. Similarly, the mechanical property of low-k materials, such as elastic modulus, is lower than that of metals and silicon. In addition, low-k materials have lower adhesion strength to the barrier layers. So the

interconnection structures with various materials are more susceptible to the applied thermal mechanical stress during fabrication or in operation, which will induce crack or interfacial delamination, and finally affect the reliability of devices. The exact thermal and mechanical properties of the low-k materials is an important element in analyzing the structural integrity and device reliability, especially for the simulation model. In general, the low-k material used in inter layer dielectric (ILD) layers is a thin film, the thickness of which is submicron meter or below. So its thermal and mechanical properties are different from that of the bulk materials, which are a huge challenge for measurement of material properties for low-k thin films.

In the mechanical property measurement of low-k thin films, the Poisson's ratio, Young's modulus and fracture toughness are always needed to be obtained. At present, there are many experimental methods to measure the mechanical and thermal properties of thin films. For example, the micro-tensile test and Brillouin scattering can be used to measure the Young's modulus and Poisson's ratio according to the recorded strain–stress curves and displacement[3-5]. The CTE of thin films can be obtained by using thermo-mechanical analysis (TMA) technology or the x-ray reflectivity method [6,7].However, these methods need to meet certain conditions during the measurement. For example, the x-ray reflectivity method can be applied for on-wafer measurement, in which the thickness of films is about submicron meter or below. But other methods mentioned above cannot be applied. In addition, the samples should be free-standing films and have a minimum thickness of about several micrometers during the TMA and micro-tensile methods[8]. Due to the restrictions, it is difficult to prepare for a free-standing form of low-k materials, especially for the ultra-low-k materials[9,10]. Because the ultra-low-k materials generally porous and brittle. Even though the free-standing films can be formed, there is also a difference in thermomechanical properties between the free-standing films and on-wafer thin films[11,12]. So a simple and feasible approach for measuring mechanical properties is needed.

In this paper, we designed and fabricated two kinds of test structures and each structure has two kinds of geometry sizes. About the two kinds of test structures, one is with pre-crack, which can be used to measure the fracture toughness of the ULK film. The other is without pre-crack, which can be utilized to characterize the Young's modulus and Poisson's ratio of the ULK film. The stress vs. strain curve of the ULK film with large and small sizes is also measured to compare the effect of in-plane sizes. Then we utilized the in-situ tensiometer apparatus with uniaxial stretching to do the uniaxial tensile tests and obtained the stress-strain and tensile force-displacement diagrams, respectively. Then we can

National Nature Science Foundation of China(No.61774044)
Foundation of Jiangsu Key Laboratory of ASIC Design(No.2020KL0P011)
Research Foundation for Advanced Talents of JSIT(No.10072020028(006))

978-1-6654-1392-3/21 $31.00 © 2021 IEEE

calculate the Young's modulus, tensile strength and energy release rate(ERR) of the low-k thin film according to the diagrams obtained above. The results demonstrate that the method is feasible to measure mechanical properties of low-k thin films.

II. EXPERIMENTS

A. Design and fabrication of the test structures

Due to the ULK material used in the foundry industry, which is named "Black Diamond (BD)",is hard to obtain. In this paper, the type of used low-k material is CYCLOTENE™ 4000 serials products, which is supported by Dow Electronic Materials. The main component of the material is benzocyclobutene(BCB), which is a commonly used photosensitive polymer. The dielectric constant of the material is about 2.5, which is similar to that of the BD material. And the BCB material belongs to the negative tone. We can use spin-on method to form thin films. This material has the advantages of high degree of planarization, low curing temperature(200~300°C), negligible shrinkage during cure, and resistant to most processing chemicals.

We designed two kinds of structures to measure the mechanical properties of the ULK film. Assuming cracks exist in one structure named pre-crack type and the other is without pre-crack type. The two types of structures with detailed sizes are shown in Fig. 1. The sizes of the free-standing film without pre-crack are 20 mm*2 mm*13μm in length, width, and thickness, respectively. Moreover, the sizes of the other smaller freestanding film are 10 mm*1 mm*14 μm in length, width, and thickness, respectively, which is designed to verify the correct parameters that characterize the mechanical properties of the ULK film. Due to the material property is independent of its in-plane size, we designed the structure size mentioned above. In addition, the larger sample with pre-crack has cracks in both sides and a given crack length L=0.5 mm in one side is shown in Fig. 1(b).Similarity, a given crack length L=0.25 mm is in the smaller sample.

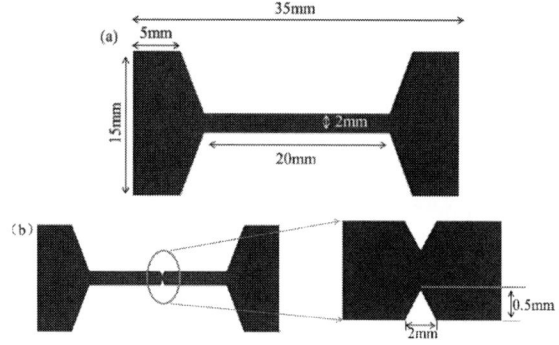

Fig. 1. Test structures with sizes, (a) without pre-crack; (b) with pre-crack

After designing the structures, we need to fabricate them. The fabricating processes mainly include lithography, temporary bonding, grinding, Si etching and debonding. The detailed process flow of the test structures is given in Fig. 2.Firstly, Ti and Cu film are deposited on the cleaning silicon substrate by using plasma vapor deposition(PVD), respectively. Then the low-k thin film is deposited by spin-coating on the Cu surface. The low-k film structures are formed by exposure and lithography. After that, the temporary bonding technique was implemented in order to grind the backside bulk silicon wafer. For the purpose of guaranteeing the holding strength for the subsequent process and also

reducing the etching time of the remained bulk Si, the residual Si thickness should keep a proper range. The final residual Si thickness is about 300 μm. Then we used wet etch method to etch the Cu and Ti film under the free-standing low-k films. Finally, the free-standing low-k films were released from the carrier by using the solution for debonding process. After clean and dry, the ultimate test structures are done, which are shown in Fig. 3. When fabrication is done, we can get samples with different structures and geometry sizes at a time because they are designed in one photomask.

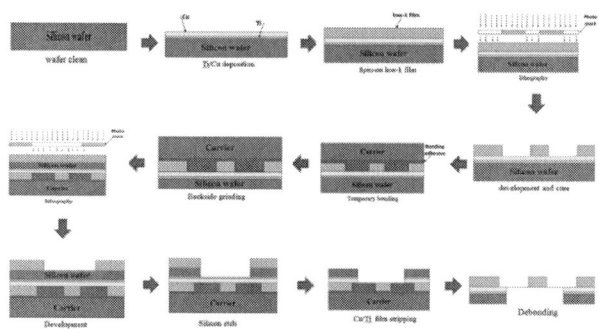

Fig. 2. The process flow for the test structures

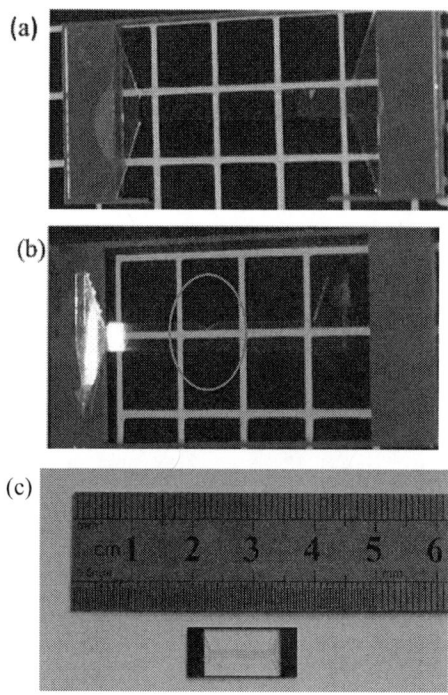

Fig. 3. The final test structures, (a) large sample without pre-crack;(b) large sample with pre-crack;(c) the small sample

B. Property characterization of low-k film

Tensile test is frequently used to measure the mechanical property. In this paper, we used the in-situ tensiometer apparatus with uniaxial stretching module to do the test and then calculated the modulus of the low-k film. Firstly, we secured the frame in the tensiometer fixture and removed the frame on both sides of the samples. Tensile tests were carried out at room temperature, and the displacement-tensile force curves were kept recording until the specimen fractured. The schematic of the measuring method is shown in Fig. 4.

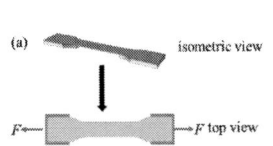

Fig. 4. The measuring method, (a) The drawing of the measuring method;(b) test sample is clamped in the fixture

C. Calculation of strain energy release rate

The strain energy release rate, which also can be simply called energy release rate, is described as the energy dissipated during fracture per unit of newly created fracture surface area[13,]. For the purpose of calculation, ERR can be calculated as (1)[15,16],

$$G=-(\partial(U-V))/\partial A \tag{1}$$

where U is the potential energy available for crack growth, V is the work associated with any external forces, and A is the crack area (for 2D problems, A is the crack length).

The criterion of ERR failure is that a crack will elongate when value of the critical ERR G_c is lower than or equal to the accessible ERR G, ($G_c \leq G$). The quantity G_c is the fracture toughness, which is also known as critical ERR. Due to it is not related to the imposed loads and the geometry of the sample, it belongs to the inherent property of the material[17].

Due to the thickness of the test structures are much smaller than the sizes in horizontal plane, this case belongs to the plane stress condition. For plane stress problems with cracks, which will elongate in a straight path, the stress intensity factor (K_I) of mode I is relevant to the ERR by (2)[18],

$$G=K_I{}^2/E \tag{2}$$

where E is the Young's modulus. For a plate with finite width and there exist cracks in both sides, the K_I can be calculated by the following (3)[19,20],

$$K_I=Y\cdot P\cdot a^{1/2}/(t\cdot w) \tag{3}$$

where Y is the function about geometry and location of the crack ($Y=f(2a/w)$), whose value can be obtained by Fig. 5. In addition, P is the tensile force, a is the crack length in both sides, t is the thickness of the film, and w is the width of the freestanding film.

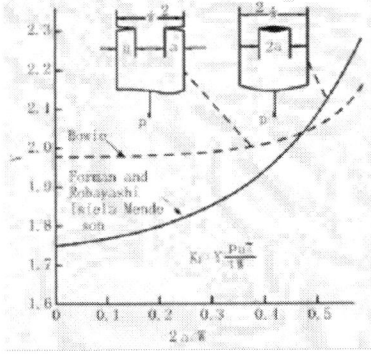

Fig. 5. Y value vs. 2a/w curve for stress intensity factor

III. RESULTS AND DISCUSSION

The stress and strain curve of the ULK film without pre-crack is presented in Fig. 6. We did the tensile test for 4 times and finally get the 4 curves of stress vs. strain. The measured results from the diagram can be converted to modulus because the modulus is defined as the ratio of stress to strain. Through calculating the ratio of stress to strain, the average modulus of the ULK film is about 1.2 GPa. The value is lower than that obtained from the supplier, which is 2.8 GPa according to the datasheet. In addition, due to the maximum tensile stress is the tensile stress of the ULK material, the tensile strength of the low-k film is about 13.5 MPa, which is also lower than that of bulk material.

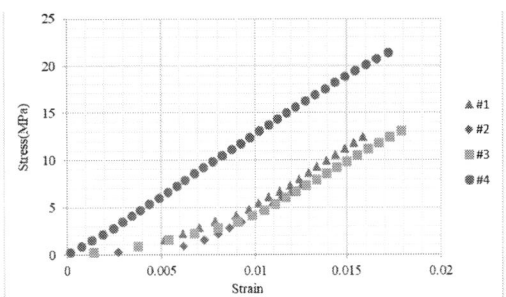

Fig. 6.The stress vs. strain curve of the ULK film

The stress and strain curve of the ULK film without pre-crack and with large and small sizes is shown in Fig. 7. It is obvious that the ratio of the curve of the large sample is agree with that of the small sample, showing that the modulus of the samples with two kinds of geometry sizes converted by the measured results from the diagram is identical because material property is independent of its in-plane size.

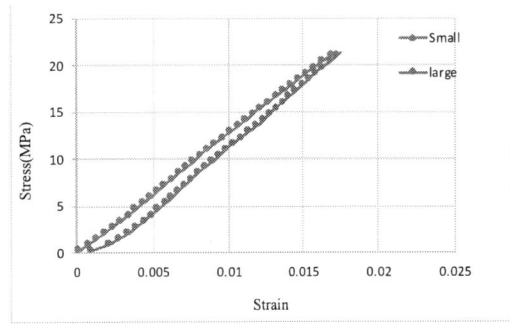

Fig. 7. The stress vs. strain curves of the ULK film with large and small sizes

The calculation method of ERR is noted above. In the equation(3), a is 0.5 mm, t is 13 μm, and w is 2 mm, respectively. So the value of the function $2a/w$ is 0.5, and the Y value is about 2.1 according to the Fig. 5. The average maximum P is 0.47 N according to the Fig. 8.Then we can calculate the K_I and then introduce the modulus value obtained by Fig. 6. Finally, the critical ERR can be deduced by the equation (2), and the average G_c is about 0.25 J/m², which is lower than that of bulk material. The results show that the ability to resist fracture of low-k thin films is weak.

978-1-6654-1392-3/21 $31.00 © 2021 IEEE

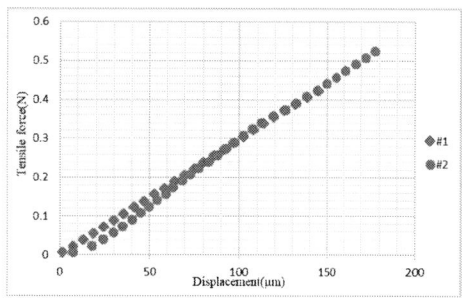

Fig. 8.The tensile force vs. displacement curve for structure with cracks

IV. CONCLUSION

In this paper, the ULK films with two kinds of test structures and different geometry sizes were made from BCB material and tested by the in-situ tensiometer apparatus with uniaxial stretching to do the uniaxial tensile tests. About the two kinds of test structures, one is with a pre-crack, which was used to measure the ERR of the ULK film. The other is without a pre-crack, which was used to measure the Young's modulus and tensile strength of the ULK film. Using the methodology, we finally obtained the elastic modulus of the low-k thin film according to the stress-strain diagrams, which is about 1.2 Gpa. The tensile strength is about 13.5 Mpa, which can be also derived from the diagram. In addition, the modulus of the large sample is agree with that of the small sample, which shows that the material property is independent of its in-plane size. Then the critical ERR $G_c=0.25$ J/m^2 was deducted using the structure with a pre-crack. The results provide that it is a feasible method to measure the mechanical properties of ULK thin films. This method is appropriate for a variety of materials which can be deposited or grown on a silicon substrate.

ACKNOWLEDGMENT

The authors are grateful for support from Dow Electronic Materials for providing the ultra-low-k material.

REFERENCES

[1] W. W. Lee and P. S. Ho, "Low-dielectric-constant materials for ULSI interlayer-dielectric applications," MRS Bulletin, vol. 22 (10), pp. 19-24, 1997.

[2] D. E. Gray, American Institute of Physics Handbook, 3rd ed.. McGraw Hill, New York, 1979.

[3] Y. L. Shen, and R. W. Johnson, "Misalignment induced shear deformation in 3D chip stacking: a parametric numerical assessment," Microelectronic Reliability, vol.53, pp.79-89, 2013.

[4] V. Gupta, J. H. Zhao, D. Edwards, C. D. Mortensen, C. Heideman, and D. C. Johnson, "Ultra low-k dielectric mechanical property characterization," Intersociety Conference on Thermal and Thermomechanical Phenomena in Electronic Systems, pp.714-719,2008.

[5] Y. S. Kang, "Microstructure and strengthening mechanisms in aluminum thin films on polyimide film," Ph.D. dissertation, University of Texas at Austin, 1996.

[6] Q. Zeng, Y. Guan, F. Su, J. Chen, and Y. Jin, "Influence of viscoelastic underfillon thermal mechanical reliability of a 3-D-TSV stack by simulation," IEEE Trans. Device Mater. Rel., vol.17, pp. 340-348, 2017.

[7] H. Nomura, M. Eguch, and Asano, "Stress in polyimide films having a rodlike molecular skeleton formed on a silicon substrate," Journal Applied Physics, Vol. 70, pp. 7085-7088, 1991.

[8] G. Carlotti, G. Socino, and L. Doucet, "Elastic properties of spin-on glass thin films," Applied Physics Letters, Vol. 66, pp. 2682-2684, 1995.

[9] G. Carlotti, L. Douce, and M. Dupeux, "Comparative study of the elastic properties of silicate glass films grown by plasma enhanced chemical vapor deposition," J. Vac. Sci. Technol. B,vol. 14, pp.3460-3464, 1996.

[10] W.-L. Wu and H.-C. Liou, "Study of ultra-thin hydrogen silsesquioxane films using X-ray reflectivity," Thin Solid Films, vol.312, pp. 73-77, 1998.

[11] P. Wang, J. Guo and S. L. Wunder, "Surface stress of polydimethylsiloxane networks," J. Polymer Science, Part B: Polymer Physics, Vol. 35, pp. 2391-2396, 1997.

[12] M. Tada, N. Inoue, Y. Hayashi, "Performance modeling of low-k/Cu interconnects for 32-nm-node and beyond," IEEE Transactions on Electron Devices, vol.56(9), pp.1852-1861, 2009.

[13] A. A. Griffith, "The phenomena of rupture and flow in solids," Philosophical Transactions, Series A, Vol. 221, pp. 163-198, 1920.

[14] C. E. Inglis, "Stresses in plates due to the presence of cracks and sharp corners," Transactions of the Institute of Naval Architects, Vol. 55, pp. 219-241, 1913.

[15] R. S. Rivlin, and A. G. Thomas, "Rupture of rubber. I. Characteristic energy for tearing," Journal of Polymer Science, vol.10(3), pp.291-318,1953.

[16] J. Ding, X. Huang, G. Zhu , S. Chen, G.C. Wang, "Mechanical Performance Evaluation of Concrete Beams Strengthened with Carbon Fiber Materials," Advances in Materials Science and Engineering,vol.2013,pp.577-585,2013.

[17] A. N. Gent, W.V. Mars, E.James, B. Erman, and M. Roland,The Science and Technology of Rubber, 4th ed., Academic Press, Boston, 2013, pp. 473–516.

[18] G.R. Irwin, "Analysis of stresses and strains near the end of a crack traversing a plate," Journal of Applied Mechanics, Vol. 24, pp. 361-364, 1957.

[19] G.R. Irwin, Fracture Dynamics. Fracturing of Metals, American Society for Metals, Cleveland, OH, 1948, pp. 147-166.

[20] E. Orowan, Fracture and Strength of Solids, Reports on Progress in Physics, Vol. XII, 1948, p. 185.

Simulation Research on Electromigration of BGA Devices

Wenchao Tian
School of Electro-Mechanical Engineering
Xidian University
Xi' an, China
wctian@xidian.edu.cn

Yiming Zhang
School of Electro-Mechanical Engineering
Xidian University
Xi' an, China
allenzhangedu@163.com

Yong Chen
Guangdong Chippacking Technology Co., Ltd
Dongguan, China
dada9004@163.com

Si Chen
China Electronic Product Reliability and Environmental Testing Research Institute
Guangzhou, China
chensiceprei@yeah.net

Hao Cui
School of Electro-Mechanical Engineering
Xidian University
Xi' an, China
hcui@stu.xidian.edu.cn

Abstract—**This paper aims to investigate the atomic migration voids formation and time-to-failure mechanism of Ball Grid Array (BGA) device. In order to analyze the parameter changes of BGA device, the actual model and corresponding equivalent model are established by using ANSYS Workbench software. In this study, the effect of various migration forces on atomic migration is determined, while the migration divergence calculated by migration force predicts the failure time of solder joints. Conclusions are made that the solder joints at the corners failed faster than those in the center, and migration caused by current crowding is the major reason for atomic migration, stress migration accelerates the process and thermal migration has little effect.**

Keywords—atomic migration, lifetime, reliability

I. INTRODUCTION

The analysis of electromigration is a hotspot, which cannot be ignored. With the decrease of size of chip and the increase of high-performance package density, electromigration becomes one of the major failure mechanisms of chip packaging. In order to improve chip interconnectivity and the reliability of packaging, an in-depth study of the mechanism is very necessary. F.Y. Ouyang proposes the phenomenon of current crowding occurs at the entrance of the solder joint, where the current density and temperature are the highest, it is called hot spot [1]. B.Y. Wu demonstrated electromigration also relates thermal stress [2]. However, very few electromigration studies report the relation between the failure time of electromigration and the physical parameter of solder tin material are not considered.

In this study, the various migration divergence of atoms is extracted based on the atomic flux divergence (AFD) method. Additionally, several simulation models including the actual model and corresponding equivalent model are established to simulate the process of void formulation and calculate the failure time of atomic migration. A calculate scheme to evaluate damage evolution and failure time of bumps is raised.

II. THEORY OF ELECTROMIGRATION

Electromigration refers to the migration of atoms in the solder joint with the direction of electron flow, resulting in voids in some parts of solder joint [3]. The migration of atoms is relating electromigration force, thermal migration force and stress migration force, all of the three provides impetus for atomic migration, in the meaning time, migration divergence calculated by migration forces can predict the failure time of solder joints.

A. Electromigration Divergence

Electromigration has both atomic flux and electron flow. Under the action of them, the current concentration or crowding abrupt current density increasing. Hence, the migration divergence caused by the increasing current density has more direct and profound impact to bump reliability. Electromigration divergence is given by the following equation:

$$J_{em} = \frac{DC}{KT} Z^* \mathrm{e} \rho j \tag{1}$$

$$div\left(J_{em}\right) = \left(\frac{E_a}{KT^2} - \frac{1}{T} + \alpha \frac{\rho_0}{\rho}\right) J_{em} \cdot \nabla T \tag{2}$$

Where D is the diffusion coefficient of the main diffusion atom, C is the atomic concentration, K is Boltzmann's constant, T is temperature, Z^* is the effective charge for the atomic migration, e is Charge, ρ is Resistivity, j is current density, E_a is activation energy.

B. Thermomigration divergence

Joule heating is a kind of waste heat in electrical conduction, while a large amount of thermal entropy is wasted, it could induce tremendous increase in the temperature of bump, which causes the generation of temperature gradient. Under the effect of temperature gradient, thermomigration divergence is given by the following equation:

$$J_{sm} = -\frac{DC}{KT} Q^* \frac{\nabla T}{T} \tag{3}$$

$$div(J_{tm}) = -\left(\frac{E_a}{KT^2} - \frac{2}{T}\right) J_{tm} \cdot \nabla T - \frac{CD}{KT^2} Q^* \nabla^2 T \tag{4}$$

Where Q^* is the heat transfer during thermal migration. With more heat transfer, thermomigration divergence increases by large temperature gradient.

C. Stress migration divergence

As mentioned previously, the production of Joule heating induces the raise of temperature. Nevertheless, bumps and copper pads have different coefficient of thermal expansion, the stress gradient is developed. As a result, the stress

978-1-6654-1392-3/21 $31.00 © 2021 IEEE

migration divergence occurs. Stress migration divergence is given by the following equation:

$$J_{sm} = -\frac{DC}{KT}\Omega\nabla\sigma \qquad (5)$$

$$div(J_{sm}) = \left(\frac{E_A}{KT^2} - \frac{1}{T}\right)J_{sm}\cdot\nabla T - \frac{\Omega D}{KT}\nabla^2\sigma_H \qquad (6)$$

Where Ω is atomic volume, σ is the average value of stress in three directions of x, y, z.

From the literatures and calculation formula [5], it is confirmed that electromigration divergence, thermomigration divergence and stress-migration divergence are all positively correlated with the temperature gradient. Thus, the ANSYS workbench is used to extract the current density, temperature gradient, and stress gradient to calculate the migration divergence. Furthermore, three type of migration divergence are added to obtain the total migration divergence.

III. ELECTROMIGRATION SIMULATION

A. Model Introduction

Fig. 1 exhibits the structure of BGA device, which consists of chip, solder joints, plastic package molding compound and PCB. Fig. 2 shows the daisy chain pattern of solder joints. Fig. 3 is the three-dimensional model of solder joints.

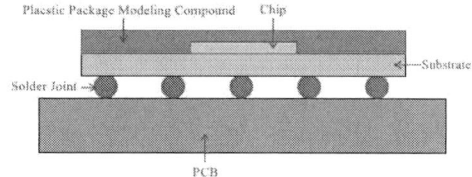

Fig. 1. Cross section of BGA package

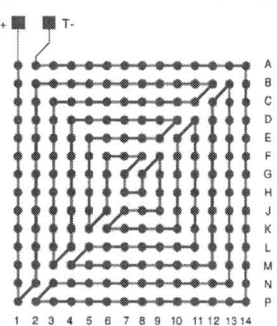

Fig. 2. Daisy Chain pattern

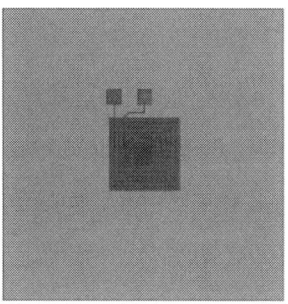

Fig. 3. The vertical view of BGA device

B. Material parameters

As shown in Table I, viscioplastic parameters for Anand model of $Sn_{63}Pb_{37}$ is used to describe the material properties of the solder joints. Table II and Table III exhibit the material properties of the remaining components, respectively.

TABLE I. VISCIOPLASTIC PARAMETERS FOR ANAND MODEL OF $Sn_{63}Pb_{37}$

s_0 (MPa)	$\frac{Q}{R}$ (K)	A (s^{-1})	ξ	m	h_0 (MPa)	$\hat{S}$ (MPa)	n	α
3.8	5509.9	392.39	5.08	0.24	4.053e4	96.07	0.012	1.6125

TABLE II. MATERIAL PARAMETERS OF SOLDER JOINTS

Item	Material	Thermal conductivity (W/m·k)	Coefficient of thermal expansion (10^{-6}/k)	Poisson's Ratio
Substrate	BT Bismaleimide Triazine	0.2	12(xy) 32(z)	0.21
Die	Silicon	98.4	2.6	0.24
Solder Joint	$Sn_{63}Pb_{67}$	53	24	0.35
BGA package	Epoxy Molding Compound	0.5	8	0.3
Pad	copper	380	17.1	0.34
Printed Circuit Board	-	1.7	16(xz) 84(y)	0.39(xy/yz) 0.11(xz)
UBM	Ni	91	13.4	0.31

TABLE III. PARAMETERS OF $Sn_{63}Pb_{37}$ DURING ELECTROMIGRATION

Electromigration parameter	Value	Symbol
Diffusion coefficient D (m^2/s)	0.016	D_0
Activation energy E (eV)	0.8	E_a
Effective charge number Z (C)	-33	Z^*
Molar thermal flow Q (eV)	0.0094	Q^*
Unit charge e (C)	1.602×10^{-19}	e
Atomic volume V (m^3/atom)	2.48×10^{-29}	V
Boltzmann constant K (eV/K)	8.7×10^{-5}	K_B

C. Loads and boundary conditions

The steady-state analysis method is used to conduct thermal-electrical-mechanical coupling analysis of BGA devices. The temperature constraint is imposed to 125°C and thermal convection constraint is imposed to natural convection heat transfer coefficient $(10 \, w/(m^2 \cdot °C))$. The positive pole is selected as current application terminal (Current=1 A), the negative pole is voltage application terminal (Voltage=0 V). Considering the symmetry of the model and the actual force, the center point of PCB is set as the reference origin to constrain displacement in all directions.

IV. RESULTS AND DISCUSSION

A. Thermo-electric-mechanical coupling analysis

As mentioned previously, the atomic migration is mainly related with current density, temperature and stress. Thus, we extract the highest temperature bump, the highest current density bump and the highest stress bump to calculate migration divergence, comparing with other two bump, the highest current density bump has larger divergence. Hence, the highest current density bump (A1) is used to further simulation.

Fig. 4 is the simulation results of the solder joint with the maximum migration divergence (A1). From the fig. 4a, it can be seen that the maximum current density of solder joint exceeded $15000 \, A/cm^2$. The maximum position of the electromigration divergence. From the simulation result, current density plays a key role during the migration. Meanwhile, other migration forces are also exerting during the electromigration process, it is not merely relying on electric field analysis to accurately forecast the location of voids. Fig. 4b shows the thermomigration divergence distribution. Due to the Joule heating effect, excessive current density aggravates the generation of Joule heat. Then, the local hot spot causes large temperature difference in solder joints, which causes the generation of temperature gradient and promotes the generation of voids. Fig. 4c illustrates the stress migration divergence distribution. By reason of the different thermal expansion coefficients of solder joints and copper pads, the relative displacement leads to the thermal stress, then the stress gradient gives rise to the stress migration.

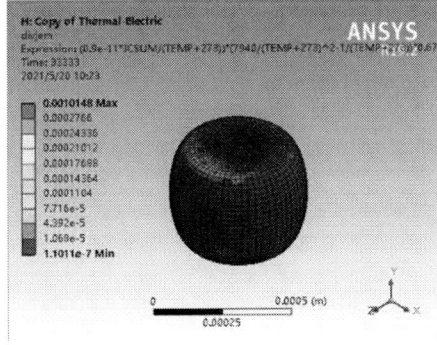

(a) Electromigration divergence distribution

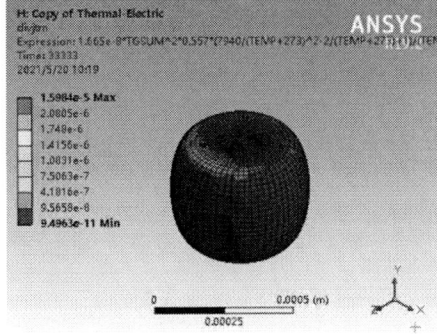

(b) Thermomigration divergence distribution

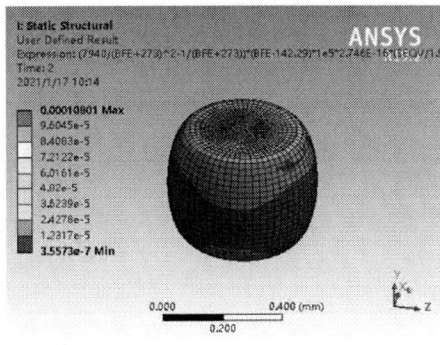

(c) Stress migration divergence distribution

Fig. 4. Simulation results of solder joint

On the basis of simulation result, the electromigration caused by current crowding is the key reason for atomic migration. Besides, stress-migration divergence is evaluated to be the similar order of magnitude with electromigration divergence, which accelerates the process, thermal migration has little effect in this simulation condition. As a comparison, fig. 5 shows the migration divergence of the solder joints at the corners (A1) is larger than the migration divergence of the solder joints at the center (G8). Table IV shows the related parameters of the corner solder joint and the center solder joint during electromigration. It is observed that electromigration divergence and stress-migration divergence of the bump at the corners are larger than the bump at the center, i.e. the effect of current density and stress is more evident for the bump at the corner. However, the bump with highest temperature (G8) has lower thermomigration divergence.

TABLE IV. PARAMETERS OF SOLDER JOINTS DURING ELECTROMIGRATION

Solder joint number	Electromigration divergence (atoms/m³·s)	Thermomigration divergence (atoms/m³·s)	Stress migration divergence (atoms/m³·s)	Total migration divergence (atoms/m³·s)
A1	1.02E-03	1.59E-05	3.47E-04	1.37E-03
G8	4.24E-04	0.54E-06	1.02E-04	5.26E-04

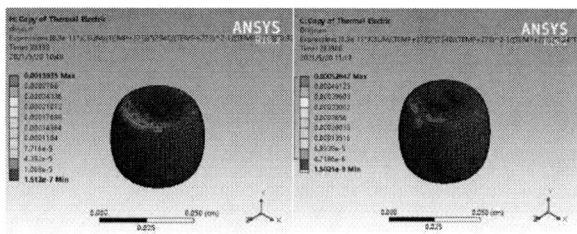

Fig. 5. Total migration divergence of A1 and G8

B. Simulation of void formation process

The expansion of the voids divides into three stages. The initial stage is the potential failure area generates initial voids, the advancement stage is the expansion of the voids, the terminal stage is the continuous expansion of voids until disconnection [6]. In the meaning time, owing to the effect of current crowding and temperature variation, the speed of atomic migration is not stable. In the initial stage, atoms move slowly, after a period of time, the migration speed increased sharply, and the failure time was greatly shortened. In this scenario, current crowding is the prime reason of atomic migration. It is primarily due to more electro-flow affects the migrate and redistribute of atoms, which makes the voids grow easily. Meanwhile, with the accumulation of voids, electric resistance is raised and the void widen further, eventually leading to the failure. Fig. 6 shows the simulation results and X-ray observation results of solid joint. It implies that the damage firstly occurs in the corner of bump near under-bump-metallization (UBM).

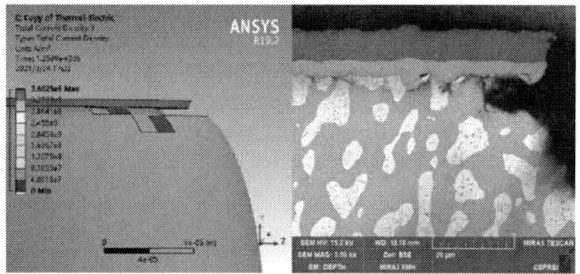

Fig. 6. Simulation results and X-ray observation results after 400h

C. Analysis of electromigration life

According to time-to-failure $\propto \frac{1}{DivJ}$, it can be known that the largest divergence element is the easiest to migrate [7]. By this criterion, using the technology of Element Birth and Death in the ANSYS workbench to set the failure elements and simulate the process of electromigration. When the ratio of the voids interface area to the voids-free interface area reaches 50%, it can be judged that the solder joint has failed [8]. The voids-free interface has 720 elements, i.e. the time consumed is the lifetime of solder joint when 360 elements migrated. The time-to-failure is given by the following equation:

$$TTF = \sum_{i=1}^{360} \frac{\ln(10)}{div(J)}$$

The failure process is divided into twelve sub-steps, each sub-step kills 30 elements, which is calculated by the divergence. The failure time of each element is multiplied by the number of elements in one sub-step, and finally the element failure time of 12 sub-steps is superimposed, which is the total life of solder joints. Consequently, the failure time of BGA device can thus be estimated. During the process of calculation, the parameters of each sub-step are shown in Table V. From the result, the total failure time of dangerous solder joints is 1019.8h.

TABLE V. PARAMETERS OF SOLDER JOINTS IN EACH STEP

Step	Divergence (atoms/m³·s)	Temperature (K)	Temperature gradient (∇ K/cm)	Current Density (A/cm²)	Failure time (h)
1	1.39E-03	417.2	32.82	15148	92.9
2	1.46E-03	419.8	43.24	14861	184.2
3	1.54E-03	421.1	54.28	13541	272.6
4	1.56E-03	422.0	64.29	15259	358.7
5	1.67E-03	422.5	73.06	15062	443.8
6	1.81E-03	422.9	81.42	15964	524.3
7	1.88E-03	423.2	90.28	15185	611.9
8	2.04E-03	423.5	100.17	16288	701.2
9	2.16E-03	423.8	110.91	24450	790.8
10	2.81E-03	423.9	122.23	22357	868.5
11	3.47E-03	424.0	135.44	21996	943.6
12	4.86E-03	424.0	153.53	37716	1019.8

Fig. 7 shows the change trends of current density and temperature during migration. In this chart, the green lines represent the maximum current density and maximum temperature of the solder joint with the greatest divergence during the electromigration process, the blue lines represent the average value and the red lines represent the minimum limits. These results suggest that eletromigration droves current density and temperature increase and indirectly influence on temperature gradient and the speed of atomic migration. Fig. 8 is the changes in current density at each stage of the atomic migration process. It is observed that electromigration failure mostly occurs at the location of current density concentrated.

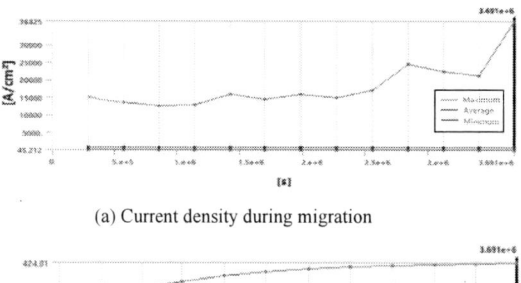

(a) Current density during migration

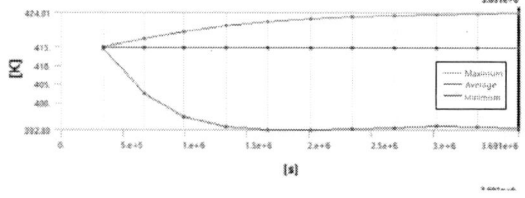

(b) Temperature during migration

Fig. 7. The trends of parameters during the migration process

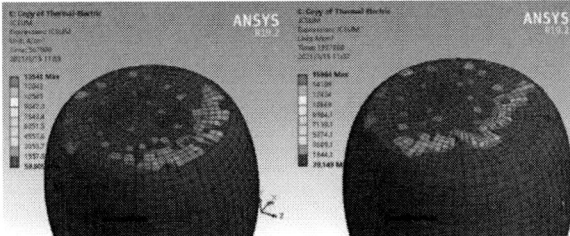

(a) Beginning of atomic migration   (b) Void generation stage

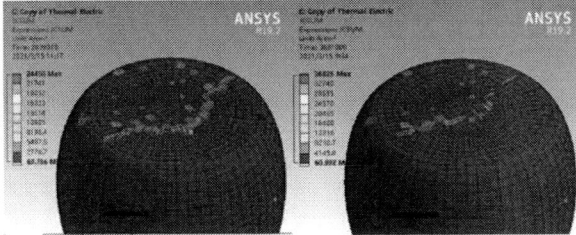

(c) Void extension stage   (d) Solder joint failure stage

Fig. 8. The changes of current density in the atomic migration process

V. CONCLUSION

This paper proposes a method to calculate the failure time of the solder joints of BGA devices under the combined action of electric field, temperature field and stress field, and predict the location of the void. Studies the electromigration in solder joint of BGA, including the formation of electromigration void and failure time, the location and failure time of atomic migration by ANSYS workbench is basically consistent with the theoretical calculation. Then gets the following conclusions in the simulation process:

- Current crowding often occurs at the corner of the UBM and solder joints' interface when electron-flow enters, voids is associated by temperature rising at the current crowding area, that deteriorates the atomic migration procedure.

- When the current density is approaching 15000A/cm² level, atomic migration caused by current crowding is the major reason for the atomic migration, stress migration accelerates the process and thermal migration has little effect.

- The total migration divergence of solder joints at the corners is greater than those in the center.

REFERENCES

[1] K. N. Tu, Yishao Lai, Ouyang Fanyi. Effect of electromigration induced joule heating and strain on microstructural recrystallization in eutectic SnPb flip chip solder joints[J]. Materials Chemistry and Physics,2012,136(1):210-218.

[2] B.Y. Wu, Y.C. Chan, H.W. Zhong. Joule Heating Enhanced Phase Coarsening in Sn37Pb and Sn3.5Ag0.5Cu Solder Joints during Current Stressing[J]. Journal of Electronic Materials, 2008, 37 (4): 469-476.

[3] Jing Yang and Yiming Jiang and Gang Chen and Ling Yang. Simulation analysis of the void location during the electromigration based on atomic concentration[J]. Journal of Mechanical Strength, 2020, 42 (3): 591-596.

[4] Yuhan Song, Xuexia Yang, Xiaochao Cui. Effect of intermetallic compound thickness on solder joint reliability under thermal cycling loading[J]. Electronic Components & Materials, 2017, 36 (7): 85-88, 92.

[5] Jong Ho Park and Byung Tae Ahn. Electromigration model for the prediction of lifetime based on the failure unit statistics in aluminum metallization[J]. Journal of Applied Physics, 2003, 93(2): 883-892.

[6] J.D. Wu et al. A study in flip-chip UBM/bump reliability with effects of SnPb solder composition[J]. Microelectronics Reliability, 2005, 46(1): 41-52.

[7] David Dalleau and Kirsten Weide-Zaage and Yves Danto. Simulation of time depending void formation in copper, aluminum and tungsten plugged via structures[J]. Microelectronics Reliability, 2003, 43(9): 1821-1826.

[8] D. J. Yao, Chin Chen, Chung-Kwuang Chou. Electromigration in Pb-free SnAg (3.8) Cu (0.7) solder stripes[J]. Journal of Applied Physics, 2005, 98 (3). 033523. 1-033523. 6.

Measurement Process Optimization in Using Lock-in Thermography for Fault Localization of CoWos Packages

Shuanshe Chao[2*]
Department of Packaging and Testing
ZTE Corporation
Shenzhen, P.R.China
10205916@zte.com.cn

Xinyi Lin[2*]
Department of Packaging and Testing
ZTE Corporation
Shenzhen, P.R.China
lin.xinyi@zte.com.cn

Dan Yang[2]
Department of Packaging and Testing
ZTE Corporation
Shenzhen, P.R.China
yang.dan97@zte.com.cn

Na Mei[2]
Department of Packaging and Testing
ZTE Corporation
Shenzhen,P.R.China
mei.na31@sanechips.com.cn

Tuo bei Sun[2]
Department of Packaging and Testing
ZTE Corporation
Shenzhen, P.R.China
sun.tuobei@sanechips.com.cn

Keqing Ouyang[1]
State Key Laboratory of Mobile
Network and Mobile Multimedia
Technology,ZTE Corporation
Shenzhen, P.R.China
ouyangkeqin@zte.com.cn

Abstract—This paper mainly investigation on defect localization application of CoWos (Chip on Wafer on Substrate) packaging through lock-in thermography. When failure samples are rare, it is necessary to use non-destructive Thermal EMMI to detect hot spots, and determine the defect Z-depth of CoWos packaging by lock-in thermography and PFA (Physical failure analysis) methods. We found that at different lock-in frequencies, the amplitudes of the hot spots were basically 10mK, but during the second exposure of 3 minutes at different frequencies, the phase differed greatly, and not even every lock-in frequency could detect the effective hot spots. It is considerable to select the appropriate lock-in frequency to get the optimal hot spot for the fault.In this paper, the concept of standard deviation is introduced into the optimal choice of hot spots. The frequency with the minimum standard deviation of phase is selected as optimal condition for failure localization. The lock-in thermography can not only quickly determine the Z-depth of fault refer to good pin or phase of each structure of CoWos chip, but also select optimal hot spot XY coordinate through the standard deviation of phase. An optimal lock-in frequency of certain hot spot on die or substrate layer are identified through a series of experiments. It is noticeable that optimal hot spot through the minimum standard deviation of phase can impove the accuracy of fault location and the success rate of subsequent physical analysis.

Keywords—*CoWos,ThermalEMMI, Fault Localization, Lock-in Thermography，Phase, Standard Deviation*

I. INTRODUCTION

In the era of integrated circuits for the fifth generation mobile communication and artificial intelligence, this vast market has lead chips with increasing functionality at a lower cost. With the help of advanced packaging technology, various chips made by different processes can be flexibly integrated together to become a high-performance large chip without paying too much cost. Advanced packaging can improve the efficiency of chip computing. Under the traditional packaging, a lot of power consumption required by chip computing is wasted. With the help of 2.5D and 3D advanced packaging technology, memory, GPU and I/O are integrated on a single substrate to shorten the distance between them and the processor and improve the transmission bandwidth, which can not only save energy consumption and cost, but also improve the computing efficiency.2.5D packaging is the progress of traditional 2D IC packaging technology, which can achieve more precise circuit and space utilization[1]. In 2.5D package,the die stacks are placed side by side on top of the interposer with silicon via TSV, as shown in Fig1(b).The silicon interposer can provide interconnection between different chips. At present, the well-known 2.5D packaging technology is nothing more than Cowos, as shown in Fig. 1. The main concept of the 2.5D package is to place the processor, memory or other chips on the silicon interposer.Cowos enables multiple chips to be packaged together and interconnected through Si interposer, achieving the effect of small package size, low power consumption and few pins.Lock-in Thermography is an invaluable tool for the advanced packaging, which can be used to quickly find the hot spots and phase, and the minimum standard deviation of phase proposed in this paper can be used to optimize the hot spot and improve the accuracy of physical failure analysis.

II. FAULT LOCALITION OF DIE THROUGH LOCK IN THERMOGRAPHY

The main failure of CoWos is mainly concentrated on the die and substrate. For the failure caused by Integrated circuit design, it is very important for design modification to investigate the root cause of failure through lock-in thermography[2][3].

A. CDM test fail

ESD test is very important for large-scale and advanced package. CDM250V test of certain CoWos package chip, VOL and IOL failure occurred during DC parameter test. Obvious leakage was found in curve trace after repeated CDM250V test on the failed GPIO. There is only one failure sample, in order not to damage the chip, the Thermal Emmi is generally used for hot spot detection. It is found that stray hot spots appear in the area corresponding to the failure pin, as shown in Fig.2.

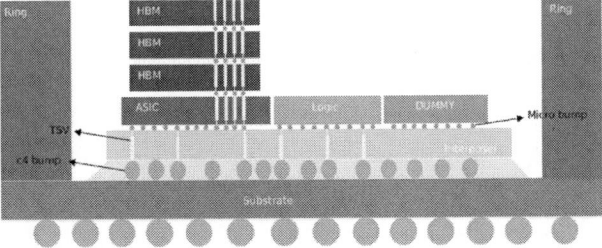

(a) Structure diagram

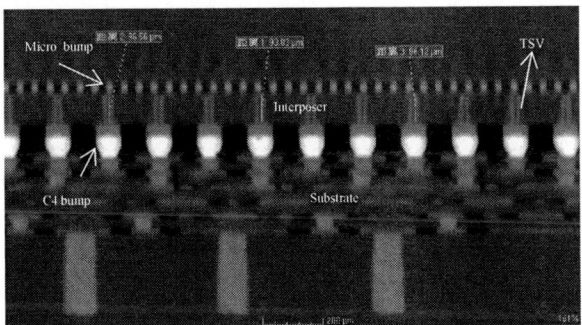

(b) 3D-xray

Fig. 1. The structure of CoWos Packaging

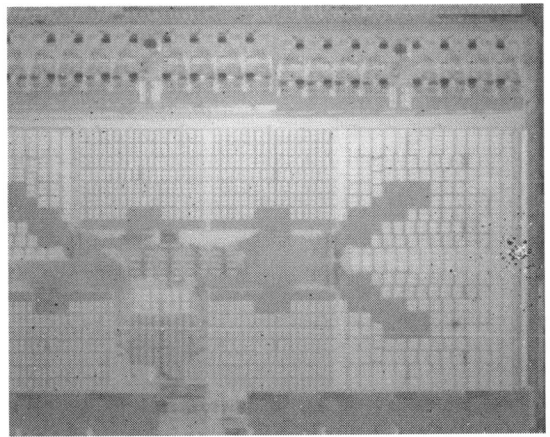

Fig. 2. Typical hotspot of failure sample by Thermal EMMI

B. select optimal hot spot through the standard deviation of phase

In this paper, we propose to introduce the standard deviation of phase into the selection of the most suitable lock-in frequency. By comparing the hot spot phase during the second exposure of 3 minutes at different frequencies, it is found that the standard deviation of the phase is the smallest at 5Hz ,as shown in table II, and the corresponding hot spot is more concentrated as shown table I . As shown in Fig.3, we can clearly find that choosing the appropriate lock-in frequency can more accurately locate the hot spots. It is noticeable that optimal hot spot through the minimum standard deviation of phase can impove the accuracy of fault location.

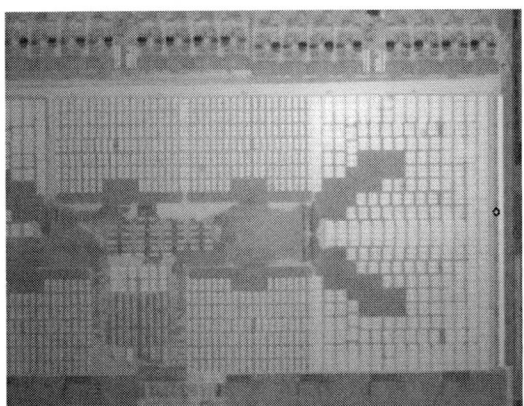

Fig. 3. Hot spot of failure sample at 5Hz with the minimum standard deviation of phase

TABLE I. PHASE AND ITS STANDARD DEVIATION OF DIFFERENT LOCK-IN FREQUENCIES

Frequency(Hz)	Phase(deg)	Standard Deviation（deg）
1	7.049	0.5854
5	27.103	0.5756
10	56.037	0.6153
16.667	92.252	0.8449
25	135.820	0.6428

C. Post-driver NMOS Failure

The hot spot is the circuit cell corresponding to GPIO，coordinates of hot spot on the layout was determined , and then delayer the metals to the contact hole.The contact layer was inspected with SEM voltage contrast[4]，as shown in Fig.4 ,red box marks are the area with abnormal voltage contrast.

D. Root cause of failure

It is found that the abnormal position of the layout is driver NMOS, which is in the same guard ring and share the drain and source with GGNMOS refer to ESD protection circuit,as shown in Fig.5.Therefore, the gate oxide layer on driver NMOS is broken down during ESD discharge, which increases the resistance value in the circuit and causes the failure of VOL and IOL.

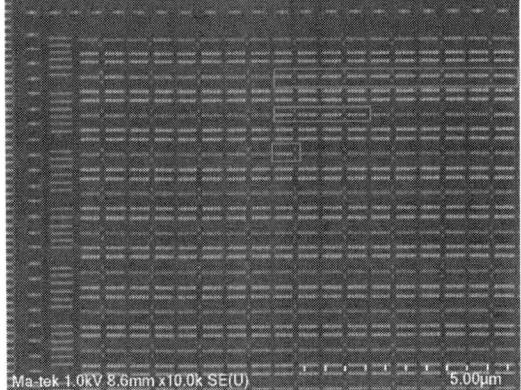

Fig. 4. SEM image on voltage contract

TABLE II.   LOCK-IN THERMOGRAPHY RESULT AT SAME SECOND EXPOSURE TIME

Frequency (Hz)	Amplitude	Phase
1		
5		
10		
16.7		
25		

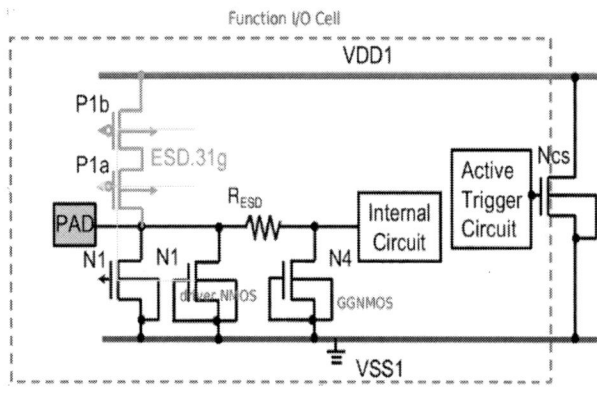

Fig. 5.   ESD protection circuit

III. FAULT LOCALITION OF SUBSTRATE THROUGH LOCK IN THERMOGRAPHY

The phase value of hot spots on the substrate is different from that on the die at the same lock in frequency,

so the fault localition of substrate can also be confirmed through lock-in thermography[3].

A. High Acceleration Stress Test

After replacing the substrate manufacturer, high acceleration stress test is carried out for CoWos chip. ATE test found that TMS pin was short , and no abnormality is found in the non-destructive analysis.Hardware faults such as test board have been eliminated, and the power on conditions and environmental stress are not abnormal. It is confirmed that the fault is the device itself.

B. Fault Localition of Substrate through Lock-in Thermography

The hot spot is not on the wafer refer to the phase,as shown in table II and III ,and the standard deviation is the smallest of hot spot is at 1 Hz,as shown in Table IV . Confirm that the XY coordinate is basically consistent with the ball of TMS failure pin, and check with 3D-xray to find that the substrate has the morphology of suspected metal migration,as shown in Fig.6.

TABLE III.   LOCK-IN THERMOGRAPHY RESULT AT SAME SECOND EXPOSURE TIME

Frequency(Hz)	phase image
1	
3.13	
5	

C. Copper Migration

The suspected location is confirmed by X-section, and delamination is found between soldermask and Prepreg,as shown in Fig.7(a) ,and EDX mapping showed that the migration metal is copper,as shown in Fig.7(b).The substrate manufacturer find that it is caused by the   control of the cleaning and spark test condition is changed from 200V to 250V.

TABLE IV. PHASE AND ITS STANDARD DEVIATION OF DIFFERENT LOCK-IN FREQUENCIES

Frequency(Hz)	Phase(deg)	Standard Deviation (deg)
1	130.87	1.3657
3.125	195.93	2.1651
5	207.64	3.0930

(a) Z-view

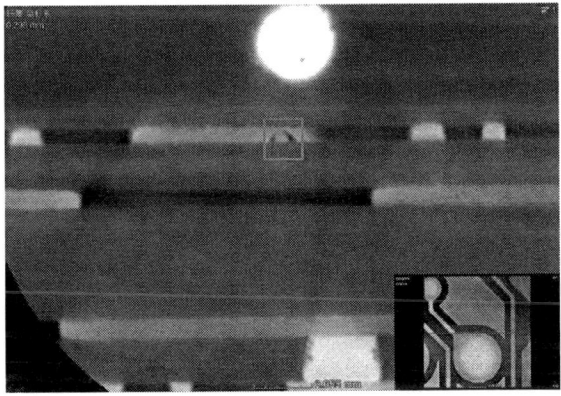

(b)X-view

Fig. 6.   Supected metal migration

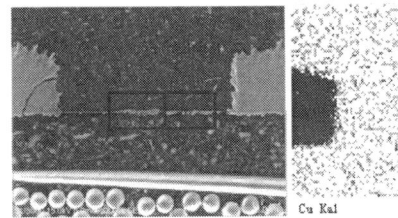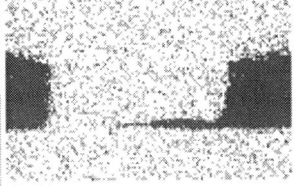

(a) metal migration   (b) EDX mapping

Fig. 7.   Copper migration

IV. SUMMARY

Cowos is an advanced packaging technology for wafer extension, that has been widely applied in communications, high performance computing and artificial intelligence products. The semiconductor chips are connected to silicon through the Chip-on-Wafer(CoW) packaging process, and

then integrate the CoWos in the substrate. In order to reduce the quality risk and improve yield, thermal EMMI is very important for failure location of expensive CoWoS chips. In this paper, through the lock-in thermography, we can measure the phase of the hot spots and infer the Z-depth of the failure location, so as to determine the substrate, and die problems. On the other hand, by selecting the suitable lock-in frequency with small phase standard deviation, the XY coordinate of the hot spots are optimized, and the positioning accuracy can be improved by combining with Z-depth to increase the success rate of subsequent physical failure analysis.In the future, we will continue to collect the phase of interposer and bump that vary in frequency to help confirm the Z-depth of the fault and find the optimal frequency of these compositions of CoWos.

REFERENCES

[1] Shin-Puu Jeng, "CoWoS™ technologies," Proceedings of Technical Program - 2014 International Symposium on VLSI Technology, Systems and Application (VLSI-TSA), 2014, pp. 1-1.

[2] K. Huang and Y. Lin, "The Application of Thermal Sensor to Locate IC Defects in Failure Analysis," 2019 IEEE 26th International Symposium on Physical and Failure Analysis of Integrated Circuits (IPFA), 2019, pp. 1-4.

[3] W. Qiu, B. Zee, H. Deslandes, B. Lai and D. Tien, "Defect Z-depth Determination in Flip-chip using lock-in thermography," 2017 IEEE 24th International Symposium on the Physical and Failure Analysis of Integrated Circuits (IPFA), Chengdu, China, 2017, pp.1-4.

[4] S.s. Chao, D. Yang, N. Mei, T. b. Sun and L. kai Yuan, "Investigation on ESD failures of RF IC," 2020 21st International Conference on Electronic Packaging Technology (ICEPT), 2020, pp. 1-3.

Microstructure and mechanical characteristics of a novel Cu/Cu₃Sn joint with Cu/Sn preform

Hongyan Xu
Beijing Engineering Laboratory of Electrical Drive System & Power Electrical Device Packaging Technology, Micro-nano fabrication technology department
Institute of Electrical Engineering, Chinese Academy of Sciences Beijing
Beijing, China
hyxu@mail.iee.ac.cn

Wei Zhang
Beijing Engineering Laboratory of Electrical Drive System & Power Electrical Device Packaging Technology, Micro-nano fabrication technology department
Institute of Electrical Engineering, Chinese Academy of Sciences Beijing
Beijing, China
weizhang@mail.iee.ac.cn

Xuan Liu
Beijing Engineering Laboratory of Electrical Drive System & Power Electrical Device Packaging Technology, Micro-nano fabrication technology department
Institute of Electrical Engineering, Chinese Academy of Sciences Beijing
Beijing, China
597878626@qq.com

Ju Xu[1,2]*
[1]*Beijing Engineering Laboratory of Electrical Drive System & Power Electrical Device Packaging Technology, Micro-nano fabrication technology department*
Institute of Electrical Engineering, Chinese Academy of Sciences Beijing
[2]*University of Chinese Academy of Sciences, Beijing, China*
xuju@mail.iee.ac.cn

Abstract—In this study, microstructure and mechanical characteristics of Cu/Cu₃Sn composite joint based on Cu/Sn core-shell micoroparticles was systematically investigated. Cu/Sn system joint structure was well-designed, in which the Cu/Sn ratio was theoretically maintained with the Cu/Sn mass ratio greater than 3:1 and Sn coating layer thickness was less than 0.206 times of Cu particle radius to obtain Cu₃Sn matrix with a dispersion of ductile Cu in the joint. Growing dynamics of Cu₃Sn in Cu/Cu₃Sn structure was deeply analyzed, indicating the lower activity energy of Cu₃Sn than that of traditional solder alloys, which make rapid formation of the joint. When the average grain diameter of Cu particle was 25μm, the microstructure evolution and shear strength of Cu/Cu₃Sn bulk with Sn coating thickness<1μm, 1~2μm and 2~2.5μm were experimentally studied, respectively, indicating that the Cu/Sn system with 1~2μm Sn coating layer has the dense joint and the largest shear strength. Zigzag fracture mechanism of Cu/Cu₃Sn joint was originated the different volume shrinkage ratio of planar and 3D network structure.

Keywords—Coating layer thickness, Microstructure, Shear strength, Theoretical calculation, Activation energy, Fracture mechanism

I. INTRODUCTION

With the trend of high power, miniaturization and high integration for electronic devices, and the demand of harsh environments (such as high temperature, large voltage, moisture and corrosion) in new automotive, aerospace, deep oil and gas drilling and energy industries, the development of high-temperature semiconductor devices is becoming increasingly important in electronics industry. Therefore, the huge thermal effect is very crucial for the long-term reliability of packages. The interconnection between the chip and the ceramic substrate provides a path for both electrical and thermal conduction in the power module, but the conventional solder joints cannot meet the requirements for high-

temperature applications. In recent years, many researches focus on the exploration of new packaging techniques and reliable Pb-free high temperature solders [1-4], where low-temperature transient liquid phase (TLP) bonding has been confirmed as a potential candidate.

Utilized the TLP bonding, a full intermetallic compounds (IMCs) joint can be obtained at low temperature through the diffusion reaction between high-melting substrate (e.g. Cu, Ag) and low-melting interlayer (e.g. Sn, In)[3, 5, 6]. Since the IMCs have much higher remelting temperature than the initial solder, the TLP-bonded joint generally possesses an excellent heat-resistant property. However, the complete consumption of low-melting metals is a time-consuming process requiring up to tens of minutes, which can lead to increased thermal stress and affect the reliability. Moreover, a joint consisting of fully intermetallic compounds (IMCs) is considered as less ductile than a bulk metallic joint obtained through conventional solder alloy soldering and emerging Ag nanoparticles sintering. To address these problems, Liu et al[7, 8] proposed a novel thermal stable Cu/Cu₃Sn composite joint based on Sn-coated Cu paste, which focused on the microstructure evolution and the high temperature stability of the bonding joint. However, the influence of Sn coating thickness on strength and toughness of joint bulk was not illustrated, and insufficient vaporization of the terpineol in paste and organic impurities in the electroless plating of Sn coated Cu particles make the void and crack formed that are likely determined to the strength of the joint. Besides, approximately 10MPa bonding pressure was used to make the particles contact with each other after paste vaporization, leading to the susceptible ceramic chip. Elakkiya et al[9] studied the solder thickness effect on the number of voids and their propagations in different solder layer thickness, which validated that optimized solder thickness can enhance the reliability of a power semiconductor. Bosco et al[10] studied the critical thickness of interlayer Sn in Cu/Sn/Cu system for averting pore formation during transient liquid phase, which was determined by the heights of the largest intermetallic grains.

National Natural Science Foundation of China

In our previous studies, the Cu/Sn preform –TLPS joint has been demonstrated as a potential candidate for high-temperature application. The Cu/Sn particles had been densified before reflowing through compressed forming under 10~20MPa avoiding high pressure utilized and voids/pores formed during binding without organic volatilization, which is beneficial to joint and chip reliability[11-13]. Nevertheless, the Sn coated layer thickness effect on the reliability of Cu/Cu₃Sn three dimensional network structure based on Cu/Sn preform have not been reported so far, the pore formation and crack evolution of the Cu/Cu₃Sn joint based on Sn coating thickness variation also were not discussed. The present work focuses on optimizing the densification and mechanical characteristics of the joint through electroplated Sn coating thickness designed, dynamic study of Cu3Sn and corresponding fracture mechanism analysis of joint.

II. EXPERIMENTAL PROCEDURES

A. *Preparation of Cu/Sn mocroparticles and preforms*

The Cu particles with different size of <15μm, 15~25μm and 30~50μm were mixed by the proportion of 1:2:3 and pretreated by a 5% hydrochloric acid solution to remove the organic impurities and oxide layer. Then, the mixed copper particles were added to the Sn electroplated solution which contained Tin Methane Sulfonate $(CH_3SO_3)_2Sn$ (10ml L^{-1}), Methane sulfonic acid CH_4O_3S (100ml·L^{-1}), resorcinol $C_6H_6O_2$ (1.0g·L^{-1}), emulsifier (OP-10) $C_8H_{17}C_6H_4O(CH_2CH_2O)_{10}H$ (1.0l L^{-1}). As shown in Fig.1. a), the electroplating bath was driven by direct current source (0.25A), which consists of Sn anode plate, agitating vane and graphite cathode plate. The mixed Cu particles was uniformly platted on cathode plate and was intermittently stirring by agitating vane every 15min at ambient temperature. The process of electroplating Sn lasted the least 0.45h to ensure that the Cu particles can be fully plated. Subsequently, the prepared Cu/Sn particles were separated and cleaned by distilled water and ethanol for five times, respectively. After cleaning, Cu/Sn particles was dried at 60℃ under vacuum atmosphere. Finally, the Cu/Sn particles were compressed under 10MPa by the mold, which was illustrated in Fig.1. b), the preform was formed with the dimension of φ9mm×150μm.

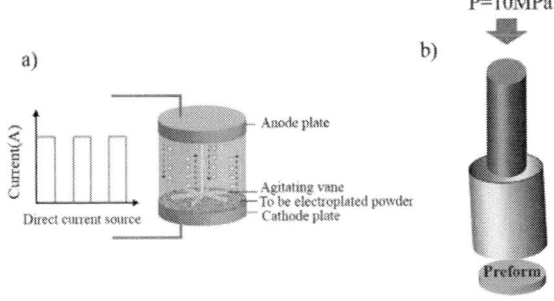

Fig.1. Schematic diagram of electroplated Cu/Sn composite particles a) and preform compressing forming b)

B. *Preparation of bonding bulk and Shear strength test sample*

The Cu substrates with dimensions of 12mm×12mm×1mm were pre-electro-plated with a layer of Sn of 2~3μm as the dummy dies and substrates for interconnection, respectively. After preform was placed on the Sn coated substrate, the prepared sample were pre-heated at 150℃ for 30s to remove the organic flux, then were heated to 180℃ for 45s under formic acid atmosphere to remove oxide layer, and finally were heated to 280℃ at 2℃/s and bonded at 280℃ for 20min in the nitrogen atmosphere under the pressure of 0.5MPa. The preforms with dimension of φ9mm×1.5mm were compressed under 10MPa for 5min, then, reflowed on the graphite substrate under the bonding process to form the bonding bulk. Shear test samples was fabricated, which size is shown in Figure 2, upper ceramic plate cladding copper (DBC) and lower DBC was connected by solder, Cu/Sn preform were used as the soldering materials. A scanning electron microscope (SEM, Sigma300) with attached electron dispersive X-ray detector (EDX) was adopted to characterize the microstructures of Cu@Sn particles and resulting bondline.

Fig.2. Shear sample of front view a) and 45° oblique view b)

III. RESULT AND DISCUSSION

A. *Theoretical calculating of Sn coating layer thickness and volume shrinkage of different joint structure*

It is highly desirable to discuss the reaction mechanism at the interface between Cu core and Sn shell and the effect of the Sn coated thickness on mechanical property of the corresponding Cu/Cu₃Sn network joint. The strength and ductility of the Cu/Cu₃Sn joint was decided by Cu₃Sn IMCs thickness resulted from Sn coated thickness of Cu/Sn core-shell particles. Bulky or blocked IMCs is brittle and easy to initiate crack during service, while thin and even distributed IMCs possess enough strength [6, 14], however, too thin IMCs layer can not interconnect efficiently. The result of previous literature references indicates that the full IMCs joint possess favorite mechanical strength, reliability and lifetime when the IMCs thickness is about 1~10μm in the planar Cu/Sn/Cu bonding structure[6, 15]. According to the Cu/Sn phase diagram (figure 3), the phase composition of the joint is only Cu₃Sn when the mass ratio of Cu/Sn is equal to 62:38, excess Cu existed can be beneficial to the joint ductility due to the relatively low elastic modulus of Cu particle. In order to get fantastic strength and ductility of Cu/Cu3Sn joint, the appropriate thickness of Sn coated layer and the ratio of Cu/Sn in the Cu/Sn particle are necessary and they can be deduced by equation (1) ~ (2). For the planar Cu/Sn/Cu structure, the thickness and phase variation of Cu-Sn before and after reflow is shown in Fig.4.

$$\frac{y \cdot \rho_{Cu}}{x \cdot \rho_{Sn}} = \frac{62\%}{38\%} \tag{1}$$

$$y \cdot \rho_{Cu} + x \cdot \rho_{Sn} = z \cdot \rho_{Cu_3Sn} \tag{2}$$

From equation of (1) and (2), ：the relationship of y and x, z and x can be obtained, respectively.

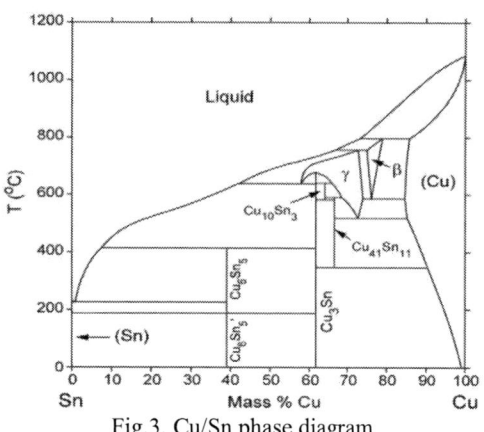

Fig.3. Cu/Sn phase diagram

$$y = 1.33x \tag{3}$$

$$z = 2.16x \tag{4}$$

Wherein, x represents initial Sn coated layer thickness of Cu/Sn particle, y is the thickness of consumed Cu layer when Sn was fully consumed to transformed Cu3Sn, and z is the thickness of resulting Cu3Sn layer. Sn layer (x) reacted with partial Cu (y) to transform Cu3Sn (z), z < x + y, about 7% volume shrinkage ratio was occurred during reflowing process. When this result was applied to Cu/Sn preform instead of single Cu/Sn particle, assuming that the Cu/Sn particle densely accumulated in the preform and Sn coating layer was fully consumed to form about 10μm Cu3Sn, it has been verified to be reliable for Cu/Sn/Cu three-layer structure [6]. According equation of (4), the initial Sn layer thickness is 4.6μm, which is evenly distributed to next to Cu particle, thus the Sn coated layer thickness of single Cu particle in Cu/Sn preform is about 2.3μm. As shown in Fig.5, red triangle in Fig.5(a) is the of Cu/Cu3Sn joint and Fig.5(b) is the magnified diagram, while Fig.5(c) is corresponding Cu/Sn preform minimum structure unit before reflowing.

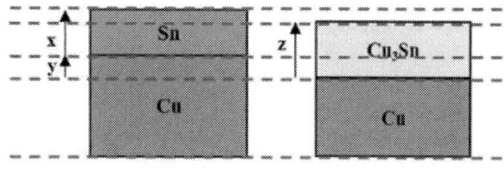

a) Cu/Sn particle   b) As-reflowed Cu/Cu₃Sn

Fig.4. Schematic diagram of thickness and phase variation of Cu/Sn before and after reflow

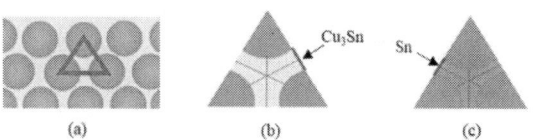

Fig.5. Schematic diagram of microstructure of Cu/Sn preform and Cu/Cu3Sn bonding bulk, (a) joint bulk after bonding; (b) triangle district after bonding; (c) triangle district before bonding

In order to verify the theory when it was applied to the Cu/Cu₃Sn three-dimension network structure, relationship of the diameter of Cu particle and Sn coating layer thickness

was deduced according to the law of conservation of mass when Cu/Sn was fully transformed to Cu3Sn.

$$\frac{4\pi r^3}{3} \cdot \rho_{Cu} = \frac{4\pi R'^3}{3} \cdot \rho_{Cu_3Sn} \cdot 0.62 \tag{5}$$

$$\left(\frac{4\pi R^3}{3} - \frac{4\pi r^3}{3} \right) \cdot \rho_{Sn} = \frac{4\pi R'^3}{3} \cdot \rho_{Cu_3Sn} \cdot 0.38 \tag{6}$$

r, R and R' is the radius of Cu, Cu/Sn and Cu3Sn, respectively. From equation of (5) and (6), R/r and R'/r was obtained, and Sn coated layer thickness x can also be got from equation (7).

$$R = \sqrt[3]{\frac{19\rho_{Cu} + 31\rho_{Sn}}{31\rho_{Sn}}} \cdot r = 1.206r \tag{7}$$

$$R' = \sqrt[3]{\frac{50\rho_{Cu}}{31\rho_{Cu_3Sn}}} \cdot r = 1.175r \tag{8}$$

$$x = R - r = 0.206r \tag{9}$$

Mean diameter of Cu particle used in this article is 25μm, the radius of Cu/Sn (R) and Cu3Sn (R') are 15.075μm and 14.688μm, respectively, according to the equation (7) and (8). The Sn coating layer thickness is about 2.6μm (equation (9)), which is nearly to the result from the Cu/Sn/Cu plating structure (2.3μm), while, $R' < R$, and the volume shrinking ratio is about 2.6%, which is much less than that of Cu/Sn/Cu planer structure (7%). Moreover, when $x < 0.206r$, the volume shrinkage ratio is still less, an excess of Cu particle is remained between Cu3Sn after Sn was fully transformed to Cu3Sn in the 3D joint. Due to excellent ductility and thermal conductivity of Cu particle, Mechanical strength, ductility and conductivity of Cu/Cu3Sn joint can be increased. When $0.26r < x < 0.417r$, only the Cu6Sn and Cu3Sn phase was existed in the joint, and the volume shrinking ratio is about 5.0%; when $x > 0.417r$, there was Sn and Cu6Sn5 phase existed in the joint, which make the lower reliability due to the unconsumed Sn. So, three-dimension network structure Cu/Cu3Sn joint possesses smaller volume shrinkage ratio, higher ductility and thermal conductivity than that of pure IMCs.

B. Interfacial reaction of Cu/Sn system and growing dynamics of Cu3Sn in 3D joint

The strength of three dimensional network structure of Cu/Cu3Sn bulk is attributed to densification and rapid formation of joint. In order to ascertain the optimal IMC thickness according to the reaction rate and reaction time and identify the reaction control factor of Cu3Sn, it is essential to investigate the reaction kinetic of Cu3Sn. The Cu3Sn thickness varies with the bonding temperature and isothermal alloying time, while Cu/Sn preform with different Sn coating layer thickness was reflowed at 250℃, 280℃ and 310℃ for 8min, 10min, 15min and 20min, respectively.

$$y = k \cdot t^n \tag{10}$$

$$k = k_0 exp(-Q/RT) \tag{11}$$

Wherein, y is the thickness of Cu3Sn, t is the reacting holding time, k is the reaction coefficient, k_0 is pre-exponential factor, n is the reaction index, T is bonding isothermal alloying temperature, R is gas constant.

The growing mechanism of Cu3Sn in Cu/Cu3Sn joint was deduced according to Cu3Sn layer thickness varied with bonding temperature and holding time, nearly linear relationship of Cu3Sn thickness and holding time was got for 250℃, 280℃ and 310℃, respectively, as shown in Fig.6. a), the error value was all less than 5% of the raw data. Reaction ratio of Cu_3Sn can be presented by equation (10), taking the logarithmic of both sides of this equation and plotting $\ln y \sim \ln t$, see Fig.6. b), the slope of linear equation is the growing index n, and the intercept is the value of $\ln k$. When the IMCs growing was controlled by the volume diffusion, interfacial reaction and grain boundary diffusion, the value of n equal to 1/2, 1 and 1/3, respectively. The value of growing index n of Cu_3Sn at 250℃, 280℃ and 310℃, respectively, were almost equal to 0.7, which included both volume diffusion and interfacial reaction during Cu_3Sn growing process. Taking the logarithmic of both sides of the Arrhenius equation (11) and plotting $\ln k \sim 1/RT$, growing activation energy Q of Cu_3Sn in the 3D network joint was about 36kJ/mol, which can be got from the slope of fitting straight line, seen in Fig.6. c). However, it was referenced about 64~85kJ/mol[5, 16, 17], 61~92 kJ/mol[5, 16] and 68~104 kJ/mol[18, 19] for Cu_3Sn in the planar Cu/Sn/Cu system, Sn3.5Ag/Cu system and Sn0.7Cu/Cu system, respectively. Growing activation energy of Cu_3Sn in the Cu/Cu_3Sn 3D network joint is the least of all the soldering material systems, which is beneficial to the rapid formation of Cu_3Sn and dense Cu/Cu_3Sn joint fabrication.

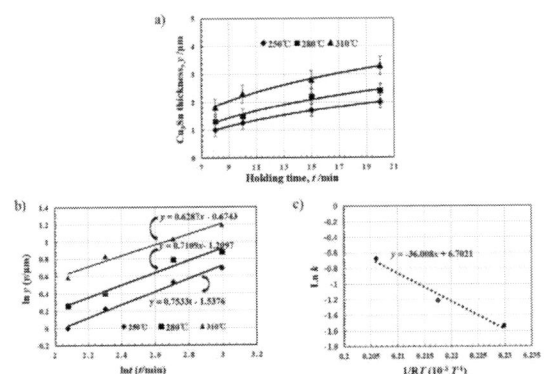

Fig.6. Reaction dynamic of Cu_3Sn, thickness of Cu_3Sn varied with holding time a), Growing index n b) and growing activity energy Q c) of Cu_3Sn in the Cu/Cu_3Sn joint

C. Microstructure evolution of Cu/Cu₃Sn bonding bulk

The key to the densification of the bonding is the size grading of Cu microparticles and Sn coating thickness of the surface of Cu microparticles. The main reasons of void formation was a lack of sufficient molten Sn on the surface of Cu microparticles for Cu/Sn microparticles with Sn coating layer <1 um, kirkendall voids and volume shrinkage was the key to void formation for Cu@Sn with Sn coating layer >2 um. The theoretic thickness of Sn coated layer is much less than 5.15um, 2.58um and 1.55um for Cu particle size of 25~50um, 15~25um and <15um, respectively, according to the equation (9), and the Sn coated layer thickness was designed to be less than 2.58 um assuming the Cu particle average size is about 25um, the resulting bonding joint was consisted of Cu_3Sn with a disperse of Cu particle.

Electroplated Sn coated Cu microparticles phase transformating before and after reflowing process is shown in Figure 7 a), the Sn coated layer was fully transformed Cu_3Sn after reflowing at 250℃ for 20 minutes,

and the Cu/Sn particle was completely transformed to Cu/Cu_3Sn. The Cu/Sn preform pressed by Cu/Sn particle under 10MPa was located between upper and lower Sn coated Cu substrates, the as-reflowed density Cu/Cu_3Sn three dimensional network structure joint was formed under the pressure of 0.5MPa, wherein the bulk is the Cu/Cu_3Sn network structure and the interface is the pure Cu_3Sn phase transforming from plated Sn/Cu substrate. the schematic diagram of Cu/Sn preform phase transformation before and after reflowing process was shown in Figure 7 b).

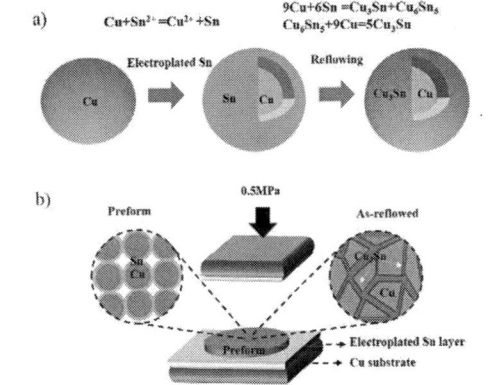

Fig.7. Schematic diagram of phase transformation of Cu/Sn particle a) and Cu/Sn preforms b) before and after reflow

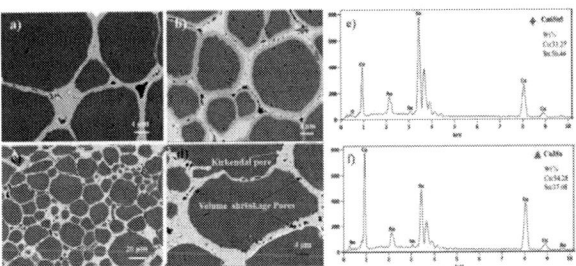

Fig.8. Microstructure and composition of Cu/Cu_3Sn bonding bulk with variating Sn coated layer thickness of <1μm a), 2~2.5 μm b) and 1~2 μm c) and d), e) and f) were phases of red marks of b)

The influence of variating Sn layer thickness of Cu/Sn particle on the microstructure and void evolution of Cu/Cu_3Sn joint was experimentally studied. When the Sn coating layer thickness is less than 1μm, the Cu_3Sn skeleton structure is too thin to interconnect Cu particle, as shown the red circle in Fig.8 a), and the small size of Cu/Sn particle (less than 10μm) was completely transformed to Cu_3Sn between the large size of Cu particles. If Cu_3Sn is much blocky together, the joint can become much brittle. Consequently, uniformly dispersed of different size Cu particles is necessary for joint reliability. When the Sn coated layer thickness is about 2~2.5μm for average diameter of Cu particle with 25μm, the volume shrinkage pores are large and widespread between Cu_3Sn layers, as shown in Fig.8 b), some Cu_6Sn_5 phase still existes between Cu_3Sn at constant reflowing time, the composition of corresponding red marks in figure 8 b) were shown in figure 8 e) and f). When the Sn coating layer thickness is 1~2μm, the dense microstructure of Cu/Cu_3Sn joint is obtained as shown in Fig.8 c). It can be seen that different size of Cu particles are uniformly dispersed in the Cu₃Sn, and little volume shrinkage pores and kirkendall pores exist between Cu particles and Cu₃Sn as shown in Fig.8 d), which are not large and linked together to form the vulnerable point of crack.

D. Fracture morphology and crack origination analysis

18 as-prepared joint samples and 18 joint samples after thermal cycling (-70~200℃) were used for shear testing at room temperature, these samples were based on Cu/Sn microparticles with different Sn coated layer thickness. In order to make the molten Sn spread on the whole surface of substrates through capillary attraction during reflowing, electroplated Sn layer was prepared on the surface of substrates, the 3D network joint was formed in consist of Cu/Cu₃Sn bulk and thin Cu₃Sn interfacial layer under the pressure of 0.5MPa. Shear strength results in this work was illustrated in Figure 9, and compared with recently reported results, which was listed in Table 1. The average shear strength of Cu/Cu3Sn joint were determined to be 140MPa, 170MPa and 220MPa with Sn layer thickness <1 um, 1~2 um and 2~2.5 um, before thermal cycling, respectively, and the shear strength dramatically decreased after thermal cycling, especially after 400 thermal cycles they were only up to 40MPa, 70MPa and100MPa, respectively, which were still higher than that of Ag@Sn preform, other Cu@Sn preform

Fig.9. Shear strength of Cu/Cu₃Sn joint based on Cu/Sn particles with different Sn coating layer thickness at different temperature cycles (-70~200℃)

Fracture morphology of Cu/Cu₃Sn joint after 400 thermal cycling indicated that the crack originated between Cu₃Sn shells of Cu cores when the initial Sn coated layer thickness <1 um, as shown in Fig.10(a), due to the thin soldering layer was hard to make the densified joint and it was so vulnerable to interconnect Cu particles, especially the trident junction of Cu particles, its shear strength is also the smallest of all the joint with different Sn coated thickness, and the error value was about ±10MPa. When the Sn coated layer thickness was increased to 2~2.5um, the kirkendall voids increased in size and linked together during reactions of Cu-Sn, which was caused by the dominant Cu element diffusive during reflowing processing, and the crack was initiated between Cu-Cu₃Sn and Cu₃Sn-Cu₃Sn, as shown in Fig.10 (b), and the later was caused by the volume shrinkage of in-situ formed Cu₃Sn, which was due to too thick Sn layer transformed to the corresponding thick Cu₆Sn₅ layer, and then the metastable Cu₆Sn₅ transformed to stable Cu₃Sn, accompanying with Cu diffusion and large volume shrinkage. Shear strength of Cu/Cu₃Sn with Sn layer thickness 1~2um was the largest (error value ±8MPa) at different thermal cycling, and its fractography was shown in Fig.10 (c), the crack originated from the interface of preform and Sn coating Cu substrate, and propagated across

Table 1 Shear strength before and after thermal cycling and comparisons with high temperature-resistant joints

Material	Bonding tempeture/℃	Bonding time/min	Thermal cycling No.	Sn layer thickness/um	Shear strength/MPa	Refs.
Cu/Sn preform	250	20	400	<1	30	This paper
	250	20	400	1~2	78	
	250	20	400	2~2.5	50	
Cu/Sn preform	250	20	0	0.5~3	29.4	[1]
Cu/Sn paste	200	20	0	~1	25.8	[7]
Ag/Sn preform	250	5	0		26.3	[20]
High Pb alloy	350	3	0		25	[21, 22]
Sintered nano-Ag	180	5	0		30~40	[23]

and Cu@Sn paste, high Pb alloy and sintered nano-Ag.

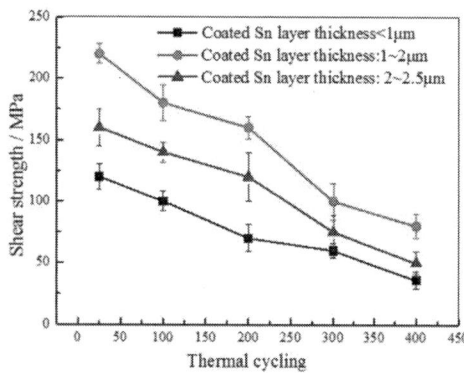

the preform to the other Cu substrate until the final failure, the zigzag cracking path will consume a large amount of energy before the final fracture, improving the ability of the joint to absorb stress. When the fracture zone was magnified, seen in Fig.10 (d), it was found that the cracks in the bulk of samples propagated mainly along Cu₃Sn IMCs. The number of voids at the interface was more than that in the interior of the bondline, thin planar Cu₃Sn was formed in the interface of the preform and substrate, the interfacial volume shrinkage ratio was about 7% according to Fig.4, which was larger than that of bulk(<2.6%), which explained the zigzag fracture mode from interface across preform to the other interface, and only small voids with sizes of 0.5~1 um are observed in the bulk because of Cu diffusion and volume shrinkage during reactions between Cu-Sn, but the voids were small and not linked together, this behavior explained a higher shear strength of Cu/Sn preform with 1~2 um Sn coating layer.

978-1-6654-1392-3/21 $31.00 © 2021 IEEE

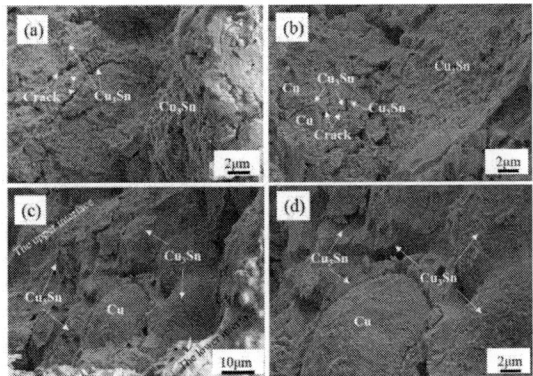

Fig.10 Fractography after 400 thermal cycling (-70~200℃) based on Cu/Sn microparticles with different Sn coating thickness, (a) Sn coating <1μm; (b) Sn coating: 2~2.5 μm; (c) Sn coating:1~2 μm, (d) High- magnification image of (c)

IV. CONCLUSIONS

1) Electroplated Sn coated layer thickness was theoretical calculated based on Cu/Sn microparticle and Cu/Cu$_3$Sn three-dimensional network structure joint. Sn coated layer theoretical thickness is <0.206 times of radius of Cu particle, while the Sn coated layer is <2.5μm when the mean diameter of Cu particle is 25μm.

2) Reaction order n of Cu$_3$Sn in Cu/Cu$_3$Sn joint is between 0.5 and 1, which is controlled by both interfacial reaction and volume diffusion. Reaction activity energy of Cu$_3$Sn is 36kJ/mol, which is far less than that of Sn3.5Ag, Sn0.7Cu and other alloy solders, indicating the superior structural characteristics.

3) The density Cu/Cu$_3$Sn joint was made based on Cu/Sn microparticles with Sn coating layer 1~2μm, Cu/Cu$_3$Sn interface is not interconnected efficiently when coated Sn layer thickness is <1μm, and pore ratio is too large and link together, which make the joint not reliable when coated Sn layer thickness is >2μm.

4) Shear strength of Cu/Cu$_3$Sn joint based on Cu/Sn particles with 1~2μm Sn coated layer is the largest before and after thermal cycling, and the fracture morphology analysis results indicated that the crack originated interface, and propagated across the preform to the other Cu substrate until the final failure, the zigzag cracking path will consume a large amount of energy before the final fracture, improving the ability of the joint to absorb stress.

ACKNOWLEDGMENT

This work was supported by National Natural Science Foundation of China under Grant No. 51777203 and National key research & develop plan "New energy vehicles special project" (20 16YFBO 100600).

REFERENCES

[1] H. Chen, T. Hu, M. Li, Z. Zhao, Cu@Sn Core–Shell Structure Powder Preform for High-Temperature Applications Based on Transient Liquid Phase Bonding, IEEE Transactions on Power Electronics, 32 (2017) 441-451

[2] H.S. Chin, K.Y. Cheong, A.B. Ismail, A Review on Die Attach Materials for SiC-Based High-Temperature Power Devices, Metallurgical and Materials Transactions B, 41 (2010) 824-832.

[3] B.-S. Lee, S.-K. Hyun, J.-W. Yoon, Cu–Sn and Ni–Sn transient liquid phase bonding for die-attach technology applications in high-temperature power electronics packaging, Journal of Materials Science: Materials in Electronics, 28 (2017) 7827-7833.

[4] H. Zhang, J. Minter, N.-C. Lee, A Brief Review on High-Temperature, Pb-Free Die-Attach Materials, Journal of Electronic Materials, 48 (2018) 201-210.

[5] J.F. Li, P.A. Agyakwa, C.M. Johnson, Interfacial reaction in Cu/Sn/Cu system during the transient liquid phase soldering process, Acta Materialia, 59 (2011) 1198-1211.

[6] K. Chu, Y. Sohn, C. Moon, A comparative study of Cn/Sn/Cu and Ni/Sn/Ni solder joints for low temperature stable transient liquid phase bonding, Scripta Materialia, 109 (2015) 113-117.

[7] X.D. Liu, S.L. He, H. Nishikawa, Thermally stable Cu3Sn/Cu composite joint for high-temperature power device, Scripta Materialia, 110 (2016) 101-104.

[8] X. Liu, S. He, H. Nishikawa, Low temperature solid-state bonding using Sn-coated Cu particles for high temperature die attach, Journal of Alloys and Compounds, 695 (2017) 2165-2172.

[9] E. R, G. Kavithaa, V. Samavatian, K. Alhaifi, A. Kokabi, H. Moayedi, Reliability Enhancement of a Power Semiconductor With Optimized Solder Layer Thickness, IEEE Transactions on Power Electronics, 35 (2020) 6397-6404.

[10] N.S. Bosco, F.W. Zok, Critical interlayer thickness for transient liquid phase bonding in the Cu-Sn system, Acta Materialia 52 (2004) 2965-2972.

[11] F. Yu, H. Chen, C. Hang, M. Li, Fabrication of high-temperature-resistant bondline based on multilayer core–shell hybrid microspheres for power devices, Journal of Materials Science: Materials in Electronics, 30 (2019) 3595-3603.

[12] H. Shao, A. Wu, Y. Bao, Y. Zhao, G. Zou, L. Liu, Novel transient liquid phase bonding through capillary action for high-temperature power devices packaging, Materials Science and Engineering: A, 724 (2018) 231-238.

[13] F. Yu, H. Liu, C. Hang, H. Chen, M. Li, Rapid Formation of Full Intermetallic Bondlines for Die Attachment in High-Temperature Power Devices Based on Micro-sized Sn-Coated Ag Particles, Jom, 71 (2019) 3049-3056.

[14] H. Shao, A. Wu, Y. Bao, Y. Zhao, G. Zou, Interfacial reaction and mechanical properties for Cu/Sn/Ag system low temperature transient liquid phase bonding, Journal of Materials Science: Materials in Electronics, 27 (2016) 4839-4848.

[15] R. Ghosh, A. Kanjilal, P. Kumar, Effect of type of thermo-mechanical excursion on growth of interfacial intermetallic compounds in Cu/Sn-Ag-Cu solder joints, Microelectronics Reliability, 74 (2017) 44-51.

[16] A. Paul, C. Ghosh, W.J. Boettinger, Diffusion Parameters and Growth Mechanism of Phases in the Cu-Sn System, Metall Mater Trans A, 42A (2011) 952-963.

[17] Q. Q. Li, Y. C. Chan, growth kinetics of the Cu3Sn phase and void formation of sub-micrometre solder layers in Sn-Cu binary and Cu-Sn-Cu sandwich structure, Journal of Alloys and Compounds 576 (2013) 47-53.

[18] D.R. Flanders, E.G. Jacobs, R.F. Pinizzotto, Activation energies of intermetallic growth of Sn-Ag eutectic solder on copper substrates, Journal of Electronic Materials, 26 (1997) 883-887.

[19] T.Y. Lee, W.J. Choi, K.N. Tu, J.W. Jang, S.M. Kuo, J.K. Lin, D.R. Frear, K. Zeng, J.K. Kivilahti, Morphology, kinetics, and thermodynamics of solid-state aging of eutectic SnPb and Pb-free solders (Sn-3.5Ag, Sn-3.8Ag-0.7Cu and Sn-0.7Cu) on Cu, J Mater Res, 17 (2002) 291-301.

[20] F.W. Yu, B. Wang, Q. Guo, X. Ma, M.Y. Li, H.T. Chen, Ag@Sn Core-Shell Powder Preform with a High Re-Melting Temperature for High-Temperature Power Devices Packaging, Advanced Engineering Materials, 20 (2018).

[21] K. Chiong, H.W. Zhang, S.P. Lim, High Lead Solder Failure and Microstructure Analysis in Die Attach Power Discrete Packages, El Packag Tech Conf, (2016) 545-550.

[22] F. Dugal, M. Ciappa, Study of thermal cycling and temperature aging on PbSnAg die attach solder joints for high power modules, Ieee Int Symp Phys, (2015) 8-8.

[23] F. Yang, W.B. Zhu, W.Z. Wu, H.J. Ji, C.J. Hang, M.Y. Li, Microstructural evolution and degradation mechanism of SiC-Cu chip attachment using sintered nano-Ag paste during high-temperature ageing, Journal of Alloys and Compounds, 846 (2020).

Thickness measurement of multi-layer thin film alloy by X-ray fluorescence spectrometer

Sung-Hua Zhong
Siliconware Precision
Taichung, Taiwan, China
sunghuazhong@spil.com.tw

Liang-Pin Chen
Siliconware Precision
Taichung, Taiwan, China
MichaelChen@spil.com.tw

Abstract—*The Nickel-Vanadium (Ni-V) metal alloy, Silver (Ag) and Titanium (Ti) thin films are sputtered on the silicon die surface to increase the adhesion between Thermal Interface Material (TIM) and die during package assembly process [1]. However, the silver film will react and be consumed completely by TIM if the film thickness is too thin. The delamination or void defect will be induced between TIM and metal thin film interface at the same time. Hence, the metal thin film thickness must be monitored during process. The X-ray fluorescence spectrometer (XRF) is a widely used technology to measure metal thin film thickness and to analyze element composition. In general, the element composition ratio of alloy film must be acquired before thickness measurement. In this article, the XRF utilizes lower element composition ratio vanadium to analysis entire Ni-V alloy thickness and use general thin-film analysis method to establish the calibration curve and measure the metal thickness.*

Keywords—*X-ray fluorescence (XRF), thermal interface material, film thickness*

I. Introduction

XRF has been widely used to analysis thin film thickness, density and element composition [2]. In addition, XRF is non-contact, fast and in-line measurement machine. Traditionally, metal thin film thickness is use cross-section SEM to determine exact value, although it can get accurate information, however it will damage the product and cost a lot of time cause it impossible apply in manufacturing process control. XRF is main metrology instrument in the semiconductor industry to control manufacturing process of package assembly [3]. There are some advantages of XRF, firstly it is can measure multi metal layers in one time with relatively fast scan time. Secondly, it is non-contact method to measure thickness and composition of variety of both single and multi-layer thin film. In this paper demonstrate accurate, non-destructive and in-line method for multi thin film thickness measurement on silicon and heat sink (nickel coating on copper) which used in assembly package.

XRF use a primary x-ray beam with the high energy x-ray photons inject to the specimen surface. When specimen irradiated by high energy photons, some electron will excited to the higher energy level, and it will create empty hole in electronic orbital, but these excited are unstable. Then outer orbitals electrons will fill those empty hole for stable state. Since the outer orbital electron have a higher energy, the electron transitions accompanied by energy emission in the form of secondary X-ray photons, which is call fluorescence

Fig.1 [4]. Every atoms have their unique fluorescence spectrum, and the XRF detector detect the fluorescence and strength to determine the metal thin film thickness.

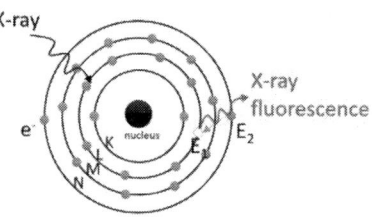

Fig.1 XRF theory

II. Experiment

There are three layer metal specimen sputtered on substrate. The bottom layer is Titanium (Ti) with 50nm thickness, the middle layer is Nickel/Vanadium with 300nm thickness, the top layer is silver with 500nm as show in Fig.2. The sputtered surface area is about 15 cm2 and sputtered on heat sink with nickel coating. The XRF utilize polycapillary optics to collect a large amount x-ray beam result in intensities greater than collimator system and the measurement time is shorter and x-ray beam size is smaller [5]. The XRF measurement parameter is 50KeV and acquisition is 60 seconds. Eventually, use the focus ion beam (FIB) to verify the XRF results accuracy.

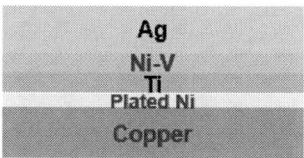

Fig.2 The cross-section structure of three sputtered metal layers on copper heat sink

III. RESULT AND DISSCUSSION

This paper is concentrate on nickel/vanadium layer thickness measurement due to the vanadium composition ratio is less than 10%. XRF utilize the higher composition element

978-1-6654-1392-3/21 $31.00 © 2021 IEEE

ratio to calculate the film layer thickness due to the fluorescence strength is higher and stable in generally. However, the heat sink will planting nickel about 3um to prevent surface oxidation in atmospheric environment. Therefore, it cannot utilize element nickel stand for nickel/vanadium layer thickness due to the X-ray fluorescence spectrum of nickel will contain nickel/vanadium by sputtered and nickel plated on heat sink.

XRF utilize fluorescence spectrum and strength to measure metal layer thickness. First of all, XRF measurement the nickel/vanadium alloy element concentration and the average concentration ratio of nickel is 92.41%, vanadium is 7.54% in table I. The vanadium concentration ratio is much lower than nickel obviously. This is very important to measure alloy concentration, because the different ratio will cause the wrong metal film thickness. Secondly, XRF measure the vanadium fluorescence spectrum of Nickel/vanadium, and it utilize ratio 7.54% to calculate entire Nickel/vanadium layer thickness.

Table I. Nickle/Vanadium concentration ratio in each specimen by XRF

Sample	#1	#2	#3	#4	#5	Average
Nickel ratio (%)	92.36	92.38	92.41	92.43	92.49	92.41
Vanadium ratio (%)	7.64	7.62	7.59	7.57	7.51	7.54

Thickness measure result were performed at the same location by FIB and XRF in order to verify truth and accuracy of XRF result. Table II. shows the FIB x-section result and layer thickness measured by XRF. The thickness range is 270.8nm~334.5nm by XRF, and the FIB range is 279.6~316.4nm. The average gap ratio between XRF and FIB is 3.3%.

Table II. Detail measurement result of nickel/vanadium layer thickness

Sputter material	Nickel/Vanadium					
Sample	#1	#2	#3	#4	#5	Average
XRF value (nm)	297.4	270.8	334.5	312.0	298.3	302.6
FESEM value (nm)	293.0	287.9	316.4	288.2	279.6	293.02
Gap (nm)	4.4	-17.1	18.1	23.8	18.7	9.58
Gap ratio (%)	1.5	-6.0	5.7	8.3	6.7	3.3

The silver and titanium layer is pure element in the sputter process, so it does not need to measure composition ratio firstly. Table III. shows the FIB x-section result and layer thickness measured of silver by XRF. The silver thickness range is 482.1nm~531.6nm by XRF, and the FIB range is 489.2~608.8nm. The average gap ratio between XRF and FIB is 1.8%.

Table III. Detail measurement result of silver thickness

Sputter material	Silver					
Sample	#1	#2	#3	#4	#5	Average
XRF value (nm)	531.6	482.1	588.0	509.9	506.8	523.6
FESEM value (nm)	506.6	489.2	608.8	530.5	531.5	533.3
Gap(nm)	25.0	-7.1	-20.8	-20.6	-24.7	-9.6
Gap ratio (%)	4.9	-1.4	-3.4	-3.9	-4.6	-1.8

Table IV. shows the FIB x-section result and layer thickness measured of titanium by XRF. The titanium thickness range is 30.1nm~40.3nm by XRF, and the FIB range is 489.2~608.8nm. The average gap ratio between XRF and FIB is 2.8%.

Table IV. Detail measurement result of titanium thickness

Sputter material	Titanium					
Sample	#1	#2	#3	#4	#5	Average
XRF value (nm)	30.1	31.6	40.3	31.6	30.7	32.8
FESEM value (nm)	31.3	31.9	39.6	34.6	31.7	33.8
Gap (nm)	-1.2	-0.3	0.6	-3.0	-1.0	-0.9
Gap ratio (%)	-3.7	-1.1	1.6	-8.7	-3.0	-2.8

VI. CONCLUSION

The XRF utilize fluorescence spectrum to obtain metal thickness and composition ratio. This paper study lower composition ratio metal thickness measurement method which use the composition ratio stand for entire layer thickness measurement. XRF is capable for multi-layer thin film alloy thickness measurement, due to the average accuracy is less than 3.3%. This method is suitable for measuring metal thin film thickness in assembly process.

REFERENCES

[1] Liu, J., Olorunyomi, M., Lu, X., Wang, W., Aronsson, T., & Shangguan, D. (2006). New Nano-Thermal Interface Material for Heat Removal in Electronics Packaging. 2006 1st Electronic Systemintegration Technology Conference. doi:10.1109/estc.2006.279970

[2] Field, A. H. (1997). Process monitoring of ultrathin oxides using surface charge analysis. In-Line Characterization Techniques for Performance and Yield Enhancement in Microelectronic Manufacturing. doi:10.1117/12.284682

[3] Wyon, C., Delille, D., Gonchond, J., Heider, F., Kwakman, L., Marthon, S., . . . Tokar, A. (2004). In-line monitoring of advanced microelectronic processes using combined X-ray techniques. Thin Solid Films, 450(1), 84-89. doi:10.1016/j.tsf.2003.10.159

[4] Vrielink, J., Tiggelaar, R., Gardeniers, J., & Lefferts, L. (2012). Applicability of X-ray fluorescence spectroscopy as method to determine thickness and composition of stacks of metal thin films: A comparison with imaging and profilometry. Thin Solid Films, 520(6), 1740-1744. doi:10.1016/j.tsf.2011.08.049

[5] Kreiner, M., Ohgaki, M., & Shinohara, K. (2019). The Latest in XRF Coatings Analysis Equipment for Micro-Scale Semiconductor Packaging. 2019 International Wafer Level Packaging Conference (IWLPC). doi:10.23919/iwlpc.2019.8914056

Nano silver particles prepared by spark ablation as packaging interconnection material

Jin Tong[1]
State Key Laboratory of Precision Electronic Manufacturing Technology and Equipment
[1]*Guangdong University of Technology*
Guangzhou, China
2111712327@qq.com

Yu Zhang[1]*
State Key Laboratory of Precision Electronic Manufacturing Technology and Equipment
[1]*Guangdong University of Technology*
Guangzhou, China
zhangyu@gdut.edu.cn

Peilin Liang[1]
State Key Laboratory of Precision Electronic Manufacturing Technology and Equipment
[1]*Guangdong University of Technology*
Guangzhou, China
724531474@qq.com

Zhongwei Huang[1]
State Key Laboratory of Precision Electronic Manufacturing Technology and Equipment
[1]*Guangdong University of Technology*
Guangzhou, China
3111238543@qq.com

Chengqiang Cui[1]*
State Key Laboratory of Precision Electronic Manufacturing Technology and Equipment
[1]*Guangdong University of Technology*
Guangzhou, China
cqcui@qq.com

Keju Zhong[2]*
State Key Laboratory of Advanced Materials and Electronic Components
[2]*Guangfdong Fenghua Adv.Technol.Holding Co.,Ltd*
Zhao Qing,China
zkj1987920@163.com

Abstract—**With the increasing requirements of the electronic industry for the performance of packaging and interconnection materials, it is the driving force for the use of metal nanoparticles as chip interconnection materials in the field of microelectronics. Due to the small size effect, metal nanomaterials have the characteristics of "low temperature sintering, high temperature service", and are considered as a good substitute for tin based solder. High purity and small size metal nanoparticles can be prepared by spark ablation without any chemical reaction. In addition, aerosol lines for gas production are continuous and have the potential to expand production. This method can be used for the preparation of semiconductor, metal, oxide, multicomponent mixed nano materials and alloys. In this paper, the preparation of silver nanoparticles by spark discharge ablation of silver target in inert gas was studied systematically. By adjusting the spark discharge parameters and gas flow rate, dispersed and non agglomerated particles or aggregates with an average size of about 20 nm were prepared. The size distribution, morphology, structure, conductivity and thickness of silver layer were characterized by scanning electron microscope (SEM), four point probe resistance test and profilometer. The experimental results show that the silver nanoparticles produced by high purity gas are very clean and can be sintered at low temperature and low pressure, which can meet the requirements of packaging interconnection performance.**

Keywords—Interconnection material, Spark ablation, Silver nanoparticles, Low temperature and low pressure sintering

I. INTRODUCTION

In recent years, the third generation semiconductor materials and technologies represented by wide band gap compounds, such as GaN and SiC, have been regarded as strategic technologies with great influence and attracted worldwide attention. The new third generation semiconductor power devices need to work at high temperature. The traditional tin based solder, conductive adhesive and other interconnection materials are gradually eliminated due to their poor high temperature resistance, which is the main driving force of metal nanoparticles as new interconnection materials. Due to the size effect, good conductivity and thermal conductivity[1-2], metal nanomaterials can be used as interconnection bonding layer to achieve "low temperature sintering, high temperature service", and meet the requirements of the third generation semiconductor for bonding materials.

At present, the synthesis of nanomaterials is mainly focused on chemical method, which provides great potential for adjusting the size, morphology and composition of nanoparticles. However, toxic and polluting chemicals such as precursors, reagents and solvents are usually required[3], which seriously affects the purity and stability of the prepared nanoparticles.

Physical method is widely used in industrial production because of its simple operation, low cost, high purity and easy control of particle size[4-5]. However, the requirements for experimental equipment are very high, and it needs to be completed under strict experimental conditions. The spark ablation method overcomes all kinds of problems existing in the current physical methods[6-7].It can realize the controllable preparation of nanoparticles without high pressure and high temperature, and it can be applied to any conductive materials including semiconductors. In recent years,it has been favored by researchers.However,few scholars at home and abroad have applied the nanomaterials prepared by this method to the field of electronic packaging.

Based on this, the effects of voltage and current coupling parameters on the distribution of silver nanoparticles were systematically studied. The size, morphology, structure, conductivity and thickness of silver layer were characterized by scanning electron microscope (SEM), four point probe resistance test and profilometer.

II. EXPERIMENTAL PART

A. Experimental equipment and materials

The Nanoparticle generator (VSPARTIAL B.V. / Holland, Voltage: 0-1.36kV, Current: 0-15.6mA), gas (high purity N_2: ≥ 99.99%, Ar: ≥ 99.999%, Guangzhou Dacheng Gas Co., Ltd.), mass flow controller (MFC, Beijing adim Control Equipment Co., Ltd), self-assembled stamping deposition device (vacuum chamber Nozzle (size: r=310um), vacuum meter, vacuum pump (Leroy Somer Electro Technology (Fuzhou) Co., Ltd.), silicon wafer (Shunsheng Electronic Technology), electrode (6H-Ag,3S-Ag electrode, purity: 99.99%, Fuzhou Kunpeng photoelectric technology), copper plate (10*10, 3*3mm).

978-1-6654-1392-3/21 $31.00 © 2021 IEEE

B. Preparation of silver nanoparticles

As shown in Figure 1 is the flow chart of nano silver preparation and stamping deposition.The spark ablation equipment (VSP-G1, hereinafter referred to as G1) purchased from the Netherlands is used to prepare nano silver particles and the self-assembled stamping device is used to deposit nano silver particles. Before the preparation of silver nanoparticles, nitrogen or argon should be introduced in advance to exhaust the air in the system. Meanwhile, the electrode in G1 should be ablated for a period of time. Preheating the equipment can avoid the oxidation of the electrode and silver nanoparticles at high temperature. According to the relevant literature parameters, when the voltage value is constant, with the increase of current value, the spark will change to arc, and the generation of arc will seriously affect the efficiency of electrode ablation. In order to ensure stable spark and maximize the yield of silver nanoparticles, the current should be increased to close to the limit value of arc generation.

In this paper, the voltage value is mainly between 1.1kV-1.3kV.Under a certain voltage,the control of the corresponding parameters on the size and morphology of silver nanoparticles is explored by changing the current value. The prepared silver nanoparticles are pressed and deposited on the substrate (10*10mm silicon or copper plate) by self-assembly device, and the deposition time is 20 min or 1 h.

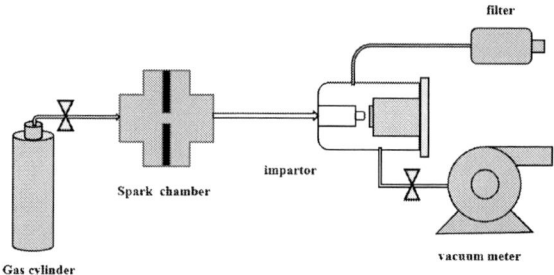

Fig. 1. Flow chart of spark ablation preparation and deposition

C. Characterization and testing

Field emission scanning electron microscopy (SU8220, Hitachi Japan) was used to characterize the structure, size and micro morphology of the prepared silver nanoparticles and sintered samples. The thickness of nano silver layer was measured by profilometer (ET-150, Kosaka Japan). Four point probe was used to measure the conductivity of the unsintered and pressureless sintered nano silver layers. The shear force of sintered samples was measured by IC package welding strength tester.

III. RESULT AND DISCUSSION

A. Shear Effect of voltage and current coupling parameters on initial size distribution of silver nanoparticles

As shown in Figure 2, under the same conditions of other parameters (gas flow rate: 1L/min,electrode: 6H-Ag,N_2, nozzle model:125,deposition time:20min), changing the spark ablation parameters (voltage, current) of G1, the distribution diagram of the initial particle size statistics of the prepared nano silver is obtained. It can be seen from the figure that under the same voltage, with the increase of current, the initial size of silver nanoparticles increases at first and then decreases. Through the real-time monitoring of the voltage and current

values on the G1 interface, it can be found that there is a best matching current value under a certain voltage, and the fluctuation range of voltage and current is the smallest, and the influence on the size distribution of silver nanoparticles is also the smallest. When the voltage is 1.1kV, the best matching current is 8 mA. At this time, the average size and distribution range of the initial particles are the best. However, the yield of nano silver is very low, which is not conducive to the subsequent sintering experiments. When the voltage is 1.3 kV and the best matching current is 12mA, the average size of silver nanoparticles is about 18 nm. However, due to the voltage close to the critical value of the device, the voltage and current will be temporarily unstable from time to time. Therefore, when the voltage is 1.2 kV and the optimal current is 10mA, the average size of silver nanoparticles is about 15nm, and the size distribution range is relatively narrow, which is more conducive to the controllable preparation of the initial particle size.

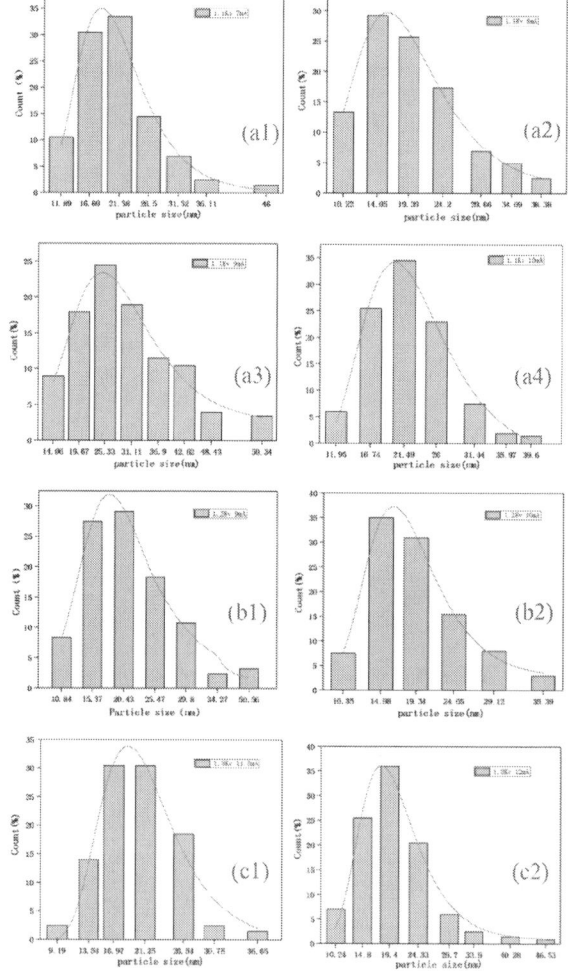

Fig. 2. Distribution of initial particle size of silver nanoparticles under different voltage and current (1.1 kV:a1~a4;1.2 kV:b1~b2;1.3 kV:c1~c2)

B. Conductivity of nano silver layer

By exploring the influence of voltage and current coupling on the initial particle size of silver nanoparticles, it can be found that the best spark ablation coupling parameters are 1.2 kV and 10 mA. In order to avoid the external gas entering the vacuum deposition chamber and reduce the loss of silver nanoparticles, the gas flow into the preparation system is set at 1.2 L/min (more than the maximum flow of nozzle 0.99 L/min). In addition, due to the influence of the type and purity of the gas on the ablation efficiency of the electrode, 99.999% argon gas was selected to prepare nano silver films in the conductivity test, and the stamping deposition time was 20 min. The resistivity of nano silver layer before and after sintering was measured by four point probe (preparation condition:125 1.2kV 10mA 1.2L/min Ar 20min).The thickness of the silver layer measured by the profilometer is shown in Fig. 3 (a), and the thickness of the film is about 5um after 20 min. The results show that the resistivity of in-situ deposited nano silver layer is 2.92×10^{-6} Ω·m. It is 177 times of that of fast silver (resistivity of bulk silver: 1.65×10^{-8} Ω·m). This indicates that the prepared silver nanoparticles have been sintered at room temperature. However, due to the existence of pores in the silver layer deposited by stamping and the micro silver produced by spark discharge instability, these factors seriously affect the conductivity of the silver layer, resulting in the high resistivity of the silver layer before sintering.

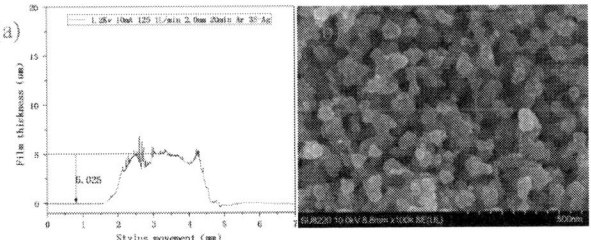

Fig. 3. (a) Measurement of film thickness with Profilometer test, (b) Electron microscope of nano silver deposited in situ by spark ablation distribution map

The samples were annealed at 260 ℃ and 300 ℃ for 30 min. After testing, the resistivity of silver layer is 5.74×10^{-7} Ω·m、1.73×10^{-7} Ω·m. The results show that the resistivity of the silver layer is 35 and 11 times higher than that of the bulk silver, respectively. The resistivity of the silver layer decreases obviously after sintering. Compared with Fig. 3 (b) and Fig. 4, it can be found that the sintering neck has been formed by the diffusion of silver nanoparticles under hot pressing.

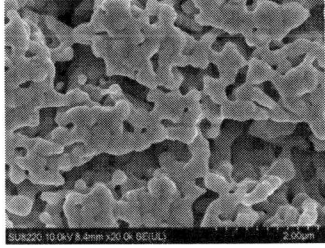

Fig. 4. electron microscope of sintering without pressure at 260℃ for 30min (preparation parameters: 1.2kV 10mA 1.2L/min 1mm 1h Ar 3S-Ag)

C. Sinterability of nano silver layer

The silver nanoparticles prepared by spark ablation (preparation parameters: 125 1.2 kV 10mA,1.2L/min 1mm 1h Ar 3S-Ag) were pressed and deposited on the large copper plate (10mm*10mm) for 60min, and the small copper plate (3*3mm) was covered on the large copper plate. Nano silver paste with solid content of 80% was prepared from commercial nano silver and nano silver prepared by liquid phase reduction method (hereinafter referred to as self-made

silver). The interconnection layer was formed by screen printing method, and the small copper plate was covered on the upper layer to form a sandwich structure. Sintering at 260 ℃ and 2 MPa for 30min, the sintering atmosphere is 5% H_2 + Ar. It can be seen from Figure 5 that the shear strength of commercial silver and self-made silver is 12.6 MPa and 22.2 MPa respectively, while the shear strength of nano silver prepared by spark ablation method is 32.74 MPa under the same hot pressing conditions, which is obviously better than that of commercial silver and self-made silver. It can be seen from the cross-section electron microscope after sintering in Fig. 6 that after sintering of commercial nano silver and self-made nano silver, due to the auxiliary effect of hot pressing, only the diffusion and fusion of nano silver particles exist in the local area, and there are many pores. The nano silver prepared by spark ablation has the characteristics of small size and high purity. The sintering neck is formed by the diffusion between the particles under the assistance of hot pressing, which fills the pores existing before sintering. Therefore, the shear property of nano silver is better than that of commercial and self-made nano silver.

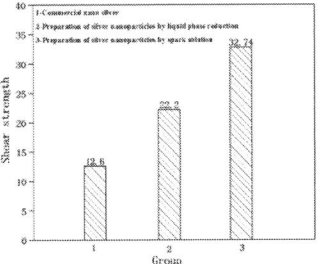

Fig.5. Shear strength of nano silver prepared by different methods after sintering at 260℃ and 2 MPa

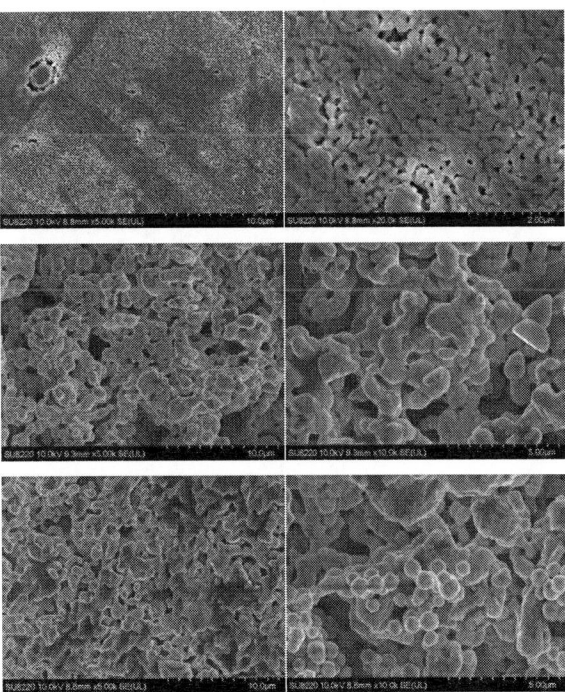

Fig. 6. Cross section electron microscope of nano silver prepared by different methods after sintering at 260 ℃ and 2 MPa(a1-a2: nano silver prepared by spark ablation, sintered by dry powder; b1-b2: commercial nano silver, solid content is 80%, c1-c2: nano silver prepared by liquid phase

reduction method, solid content is 80%, 1 and 2 are morphologies of different magnification.)

IV. CONCLUSION

In this paper, the effects of voltage and current coupling parameters on the initial size distribution of silver nanoparticles in spark ablation are systematically studied, which can realize the controllable preparation of silver nanoparticles. The conductivity of the unfired nano silver layer is $2.92 \times 10\text{-}6 \ \Omega \cdot m$. It is 177 times that of bulk silver. Because of the small size and high purity of silver nanoparticles prepared by spark ablation, they have been sintered at room temperature. After annealing at different sintering temperatures (260 ℃ and 300 ℃), the resistivity decreases greatly, which is 35 and 11 times of that of bulk silver, respectively. At the same time, under the same hot pressing conditions (260 ℃, 2MPa, 30min), the shear strength of nano silver prepared by spark ablation, commercial nano silver and self-made nano silver are 12.6 MPa, 22.2 MPa and 32.74 MPa, respectively. The sintering strength of nano silver prepared by spark ablation is much higher than that of commercial nano silver and self-made nano silver. It can be seen from the micro morphology that only the diffusion and fusion of silver nanoparticles exist in the local area, and there are many pores after the commercial silver nanoparticles and self-made silver nanoparticles are sintered. The nano silver particles prepared by spark ablation diffuse to form sintering neck under the assistance of hot pressing, which fills the pores existing before sintering. Therefore, the new nano silver interconnection material prepared by spark ablation method can replace the traditional interconnection solder to realize low-temperature and low-voltage transient interconnection and meet the requirements of package interconnection.

ACKNOWLEDGMENT

This work was partially supported by the National Key R&D Program of China (2018YFE0204601) and the Guangdong Basic and Applied Basic Research (2021A1515011642).

REFERENCES

[1] SUGANUMA K, JIU J. Advanced bonding technology based on nano- and nicro-metal pastes[M]//Materials for Advanced Packaging,2017 : 589-626.

[2] Knoerr, KAMYSHNY A, MAGDASSI S. Conductive nanomaterials for printed electronics[J]. Small,2014,10(17) : 3515-3535.

[3] PAMELA M-B. Review of SERS Substrates for Chemical Sensing [J]. Nanomaterials, 2017, 7(6).

[4] Chazelas, C., Coudert, J. F., Jarrige, J., & Fauchais, P. (2006). Synthesis of ultra fine particles by plasma transferred arc: Influence of anode material on particle properties. Journal of the European Ceramic Society, 26, 3499–3507.

[5] Förster, H., Wolfrum, C., & Peukert, W. (2012). Experimental study of metal nanoparticle synthesis by an arc evaporation/condensation process. Journal of Nanoparticle Research, 14(7), 926.

[6] S. Schwyn, E. Garwin, A. Schmidt-Ott, Aerosol generation by spark discharge, J.Aerosol Sci. 19 (1988) 639–642.

[7] Pfeiffer, T. V., Feng, J., & Schmidt-Ott, A. (2014). New developments in spark production of nanoparticles. Advanced Powder Technology,25 (2014) 56–70.

Silver film prepared by spark ablation for conductive pattern

Peilin Liang[1]
State Key Laboratory of Precision Electronic Manufacturing Technology and Equipment
[1]*Guangdong University of Technology*
Guangzhou, China
724531474@qq.com

Yu Zhang[1]*
State Key Laboratory of Precision Electronic Manufacturing Technology and Equipment
[1]*Guangdong University of Technology*
Guangzhou, China
zhangyu@gdut.edu.cn

Jin Tong[1]
State Key Laboratory of Precision Electronic Manufacturing Technology and Equipment
[1]*Guangdong University of Technology*
Guangzhou, China
2111712327@qq.com

Zhongwei Huang[1]
State Key Laboratory of Precision Electronic Manufacturing Technology and Equipment
[1]*Guangdong University of Technology*
Guangzhou, China
3111238543@qq.com

Chengqiang Cui[1]*
State Key Laboratory of Precision Electronic Manufacturing Technology and Equipment
[1]*Guangdong University of Technology*
Guangzhou, China
cqcui01@qq.com

Keju Zhong[2]*
State Key Laboratory of Advanced Materials and Electronic Components
[2]*Guangfdong Fenghua Adv.Technol.Holding Co.,Ltd*
Zhao Qing, China
zkj1987920@163.com

Abstract—**In this work, spark ablation method is used to generate silver nanoparticles, which are deposited on a silicon substrate to prepare a silver conductive film. The element, purity, film thickness, microstructure and electrical conductivity of silver conductive film were measured by X-ray diffractometer, energy dispersive spectrometer, step profiler, four-point probe tester and scanning electron microscopy. The conductive film is heated by thermal sintering to promote the agglomeration of silver nanoparticles to reduce the resistivity of the film. The influences of sintering temperature on the conductivity of the films were investigated. The results showed that when the sintering temperature is 300℃ and the sintering time is 30min, the silver conductive film deposited for 20min with a thickness of 4 μm and the electrical conductivity of 4.31×10^6 S m^{-1} can be attained.**

Keywords—spark ablation; silver conductive film; electrical conductivity

I. INTRODUCTION

A conductive film is an essential component of various electronic devices all the time. With the rapid development of electronic products, conductive film in this field have widely used in flexible electrodes[1], radiofrequency identification (RFID) tags[2],photodiodes[3]. Traditionally, photolithography is usually applied to the fabrication of conductive thin films in electronic devices. However, this method involves many difficult steps such as etching, electroplating and development. Moreover, this process is time consuming and expensive.

Printing process is a method of printing conductive materials or conductive solutions onto substrates by means of printing tools to prepare conductive films[4]. The process could be able to carry out utilizing various printing techniques including screen printing, inkjet printing, spray printing and other methods. In recent years, a number of scholars have studied the preparation of conductive films by printing method due to its advantages of low cost and low waste. However, conductive inks as conductive materials that are widely used in the printing process that are based on suspension solutions of metal nanoparticles[5] or organometallic complexes[6]. Generally, metal nanoparticles synthesized by liquid phase reduction are more complex, and kind of toxic substances are generated during these steps. In addition, as a protective agent or dispersant in the nanoparticle solution, organic compounds will reduce the contact density and electrical conductivity.

However, few studies have been done on preparing conductive films directly with no requirement for solution-based metal nanoparticles. In this work, spark ablation was used to generate pure silver nanoparticles in argon atmosphere. Through the action of inertial impact, the silver nanoparticles are deposited directly on the silicon substrate to prepare the silver conductive film. The element, microstructure, film thickness and electrical conductivity of the silver conductive film are investigated and studied. Compared with the printing method, the spark ablation method has the advantages of simplified process, simple operation and pure material production, which is capable of preparing efficiently conductive films expected to be used in optoelectronic parts, solar cells and other electronic devices.

II. EXPERIMENTAL PART

A. Nanoparticle synthesis and film preparation

Nano-silver thin films were prepared from nanoparticles produced by spark ablation: a high-purity evaporation condensation technique. In brief, two silver electrodes in a nanoparticle generator automatically adjust their spacing by setting parameters such as voltage, current and type of gas. When the breakdown voltage between the two electrodes is greater than the capacitance voltage, the gas in the electrode gap will be ionized into spark plasma to impact the two electrodes, thus forming a discharge channel. The temperature around spark plasma of up to 20,000 K causes vapor cloud to form on the surface of the electrode(see Fig.1). These vapor cloud, along with the flow of carrier gas, rapidly cool to room temperature to evolve atomic clusters, primary particles, aggregates and agglomerates [10](see Fig.2). By adjusting the velocity of inert gas, the impact pressure and the height from nozzle to substrate, the silver nanoparticles are directly deposited onto the silicon substrate for 20min by focused inertial impaction. The films are sintered in the furnace at different temperatures for 30 min to obtain silver conductive films with different electrical conductivity.

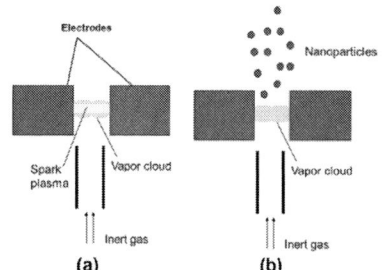

Fig.1. Nanoparticles formed by spark ablation[19].

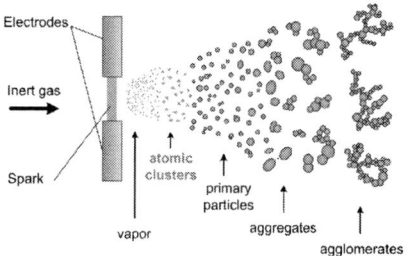

Fig.2. States of particle formation by spark discharge[19].

B. Silver conductive film and nanoparticle analysis

The crystal structure of silver conductive thin film prepared by spark ablation was analyzed by X-ray diffractometer(XRD; Bruker AXS D8 Advance). The morphology and microstructure of sintered samples at different temperature were characterized by scanning electron microscopy (SEM; Hitachi SU 228) operated at 10kV. To obtain the silver film thickness, we used step profiler (kosaka,ET-15)to test it at 50 μm/s. The chemical elements of silver film were analyzed by energy spectrum analyzer(EDS, Bruck Quantax200). A four-point probe tester (Fluck, Fluck 289) was used to measure the resistivity of the sample and the average film ` thickness was taken after 5 tests.

III. RESULT AND DISCUSSION

A. XRD characterization and analysis

Here, the crystalline structures of the silver conductive film prepared by spark ablation was characterized by X-ray diffraction(XRD), as shown in Fig.3. In the XRD pattern, significant reflections can be detected at 38.1°, 44.4°, 64.7° and 77.5°, which represent the (111), (200), (220), (311) planes of silver crystal, respectively. The other three diffractions according to the pattern represent planes of silicon crystal, respectively. The silicon element characterized by XRD is consistent with the substrate material used to deposit silver film. Therefore, we can conclude that the silver film prepared by spark ablation is composed of pure silver without other metal impurities.

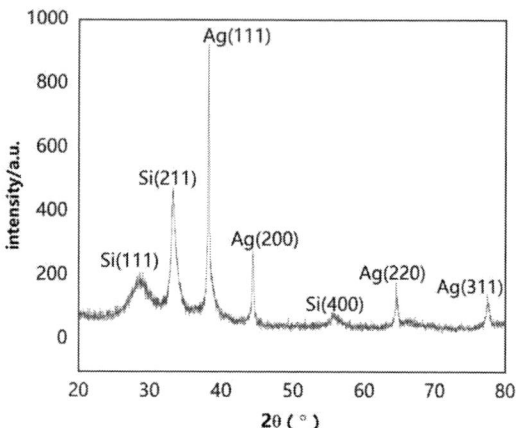

Fig.3. XRD pattern of silver conductive film.

B. Step profiler measurement and EDS analysis

The film thickness measured by step profiler is shown in Fig.4. 4μm thickness film was prepared through depositing for 20min in argon atmosphere and its roughness below 1μm. From Fig.5 the full EDS spectrum pattern showed the element of silver film and silicon substrate. The mass and atomic fraction of C, O, Si and Ag from the EDS spectra are 5.95%, 0.42%, 89.12%, 4.51%, respectively. None of other substances are founded expect the materials of silver film and silicon substrate. Therefore, the silver film can be conform as silver and silicon material exists in form of simply substance without any impurity. The result is consistent with the characterization of XRD.

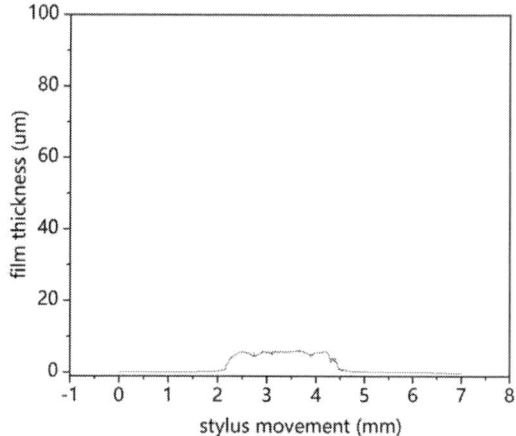

Fig.4. Film thickness measured by step profiler.

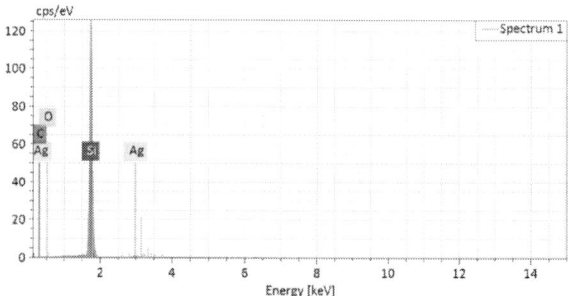

Fig.5. EDS spectrum image of sliver film.

C. Effect of sintering temperature of film conductivity

TABLE I. ELECTRICAL CONDUCTIVITY OF DIFFERENT SINTERED SAMPLES

Samples	Electrical conductivity of different sintered samples		
	Annealing	*Thickness*	*conductivity*
A	30min@100℃	4um	3.37×10^5 S m^{-1}
B	30min@150℃	4um	7.38×10^5 S m^{-1}
C	30min@200℃	4um	1.01×10^6 S m^{-1}
D	30min@250℃	4um	1.59×10^6 S m^{-1}
E	30min@300℃	4um	4.31×10^6 S m^{-1}

The influence of sintering temperature on the conductivity of silver film prepared by spark ablation is shown in Table 1. The experiment results show that the electrical conductivity of silver film will increase as the sintering temperature increases when sintering time retain a constant value. The conductivity of silver film increases from 3.37×10^5 S m^{-1} at 100℃ to 4.31×10^6 S m^{-1} at 300℃ after annealing for 30min. The highly conductive silver film can be attained by setting the sintering parameters at 300℃ for 30min and its resistivity was only 14.5 times that of bulk silver.

D. SEM characterization and analysis

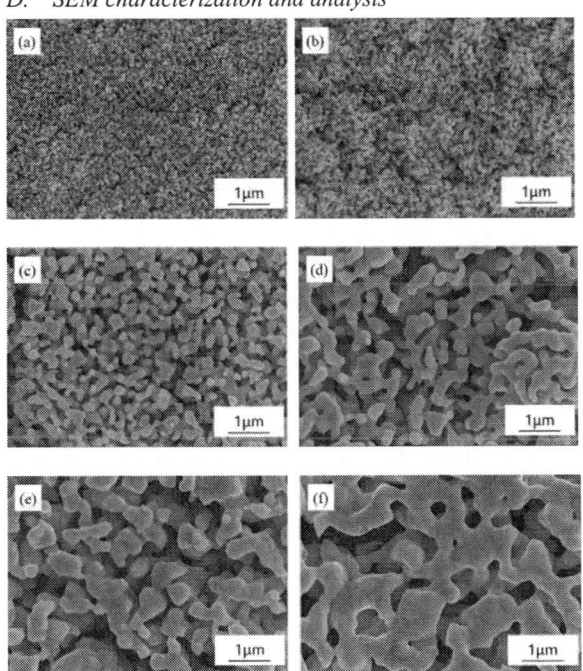

Fig.6. SEM characterization of silver films with different sintering process. **a** no sintering, **b** 30min at 100℃, **c** 30min at 150℃, **d** 30min at 200℃, **e** 30min at 250℃, **f** 30min at 300℃

The morphology of the conductive silver film sintered at different temperature for 30 minutes is shown in Fig. 6. The diameter of silver nanoparticles prepared by spark ablation is less than 100nm. The microstructure of the silver film shows that the adjacent particles was agglomerated and the sintering neck was formed when the sintering temperature raised from 100℃ to 150℃. The sintering neck between particles can be grown continuously and a interlinked network structure can be formed when the sintering temperature raised from 200 ℃ to 300℃. This is because the higher sintering temperature and 30min sintering time allow the flow of the structure and the coalescence between particles. These interconnected network structures establish a large number of continuous channels for electron transport, thus improving the conductivity of the silver film.

IV. CONCLUSION

In summary, the silver film was prepared by spark ablation, which generates silver nanoparticles to deposit them onto substrate in Ar atmosphere. A high-purity silver film was attained and no other impurities were found through XRD and EDS analysis. The electrical conductivity of silver film was enhanced by optimizing sintering parameters. As the silver film deposited for 20min is 4μm, the electrical conductivity of silver film sintered at 300℃ for 30min increase from 3.37×10^5 S m^{-1} to 4.31×10^6 S m^{-1}, which is only 14.5 times lower than that of bulk silver. Therefore, the silver film prepared by spark ablation expected to be used in conductive pattern.

ACKNOWLEDGMENT

This work was partially supported by the National Key R&D Program of China (2018YFE0204601) and the Guangdong Basic and Applied Basic Research (2021A1515011642).

REFERENCES

[1] C.F. Guo, T. Sun, Q. Liu, Z. Suo, Z. Ren, Highly stretchable and transparent nano-mesh electrodes made by grain boundary lithography, Nat. Commun. 5 (2014) 1.

[2] K.E. Belsey, A.V.S. Parry, C.V. Rumens, M.A. Ziai, S.G. Yeates, J.C. Batchelor,S.J. Holder, Switchable disposable passive RFID vapour sensors from inkjet printedelectronic components integrated with PDMS as a stimulus responsive material, J.Mater. Chem. C 5 (2017) 3167–3175.

[3] A. Falco, L. Cinà, G. Scarpa, P. Lugli, and A. Abdellah, "Fully-sprayed and flexibleorganic photodiodes with transparent carbon nanotube electrodes," ACS Appl. Mater.Interfaces, vol. 6, pp. 10593–10601, 2014.

[4] A. Kamyshny, S. Magdassi, Conductive nanomaterials for printed electronics, Small10 (17) (2014) 3515–3535.

[5] X.-F. Tang, Z.-G. Yang, W.-J. Wang, A simple way of preparing high-concentration and high-purity nano copper colloid for conductive ink in inkjet printing technology, Colloids Surf. A Physicochem. Eng. Asp. 360 (1–3) (2010) 99–104.

[6] Y. Farraj, M. Grouchko, S. Magdassi, Self-reduction of a copper complex mod ink for inkjet printing conductive patterns on plastics, Chem. Commun. 51 (2015) 1587–1590.

[7] T.V. Pfeiffer, J. Feng, A. Schmidt-Ott, New developments in spark production of nanoparticles, Adv. Powder Technol. 25 (2014) 56–70.

Simulation and optimum design of Al-Si alloy by double-sided laser welding

Liangchen Tan
Nanjing Research Institute of Electronics Technology
Nanjing, China
tlc_cetc@163.com

Shanshan Li
Nanjing Research Institute of Electronics Technology
Nanjing, China
935403687@qq.com

Jianfeng Zhong
Nanjing Research Institute of Electronics Technology
Nanjing, China

Abstract—**The finite element method was used to simulate the process of two-sided laser sealing of Al - Si alloy, and the optimization design method of the structure was proposed. The experimental and simulation results show that there is a convex influence law on the flatness of the sealed shell by double-sided laser welding, and the flatness of the shell bottom will be improved after two times of welding. The thickness of the bottom cover plate and the depth of the stress relief groove are the core factors affecting the stress distribution of the laser sealing welding.**

Keywords—*Double-sided laser sealing welding, Aluminum-silicon alloy, Thermal-mechanical coupling, Optimization design*

I. INTRODUCTION

With the growing trend of miniaturization and lightweight of equipment, higher requirements are demanded for active phased array radars. As the core part of active phased array radar, the highly reliable package of T/R modules is essential for radar. At present, most of the high power T/R modules adopt Al-Si alloy for packaging because of its low density, light weight, good thermal conductivity and low thermal expansion coefficient. However, due to the brittleness of Al-Si alloy, the large and rapid temperature variation in the laser sealing welding process is easy to cause the crack or the plastic deformation, the laser sealing welding structure should be designed reasonably[1-2]. The research shows that the welding wall thickness should be less than 1.5mm, and the narrow and long welding structure is conducive to the welding reliability[3-4]. However, the current study focused on Al-Si shell single seal welding process and weld appearance[5-8], lack of the casing deformation caused by residual stress after welding. With the development of packaging technology, More and more complex cavity sealing structures appeared[9]. Unreasonable design would result in low product rate or short fatigue failure period. Therefore, it is necessary to study the structure of multiple laser sealing welding.

In this paper, a thermal-structural multi-field coupling model was established to calculate the stress and strain distribution of the high integrated double-sided laser sealing welding structure. Next, the main structural factors affecting the stress distribution of two-sided laser welding were analyzed. Finally, response surface methodology (RSM) was applied to optimize the packaging structure and verified by experiments.

II. FINITE ELEMENT MODELING

A. Geomtric model

The model includes shell and cover plate, shell material is 50% Al/Si and geometric size is 50.6*48*8.5 (mm*mm*mm), upper cover plate material is 27% Al/Si, geometric size is 48.6*46*0.7 (mm*mm*mm)，lower cover plate material is 27% Al/Si, The geometrical size is 19*18*0.5mm, as shown in Fig. 1 below.

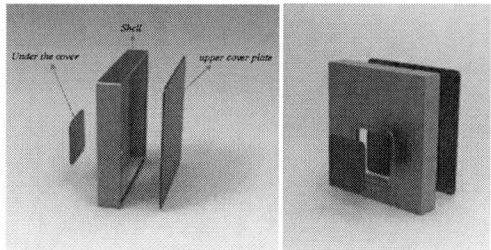

Fig. 1. Axonometric drawing of sealing welding structure

B. Assumptions and boundary conditions

The essence of laser welding process is the cooling process after the base metal reaches high temperature rapidly. In this paper, Gaussian heat source was used to simulate the laser energy. Assumed that the parts were placed on the horizontal welding platform, the shell cover plate was set in a constant temperature environment during the welding process. Finite element modeling adapt roller constraint to simulate the case of small displacement of the bottom surface of the shell, and heat flux and thermal conductivity coefficient were set to simulate the heat conduction of the shell to the environment.

C. Material properties

The material parameters of shell and cover plate are shown in the following table.

TABLE I. ELASTIC AND THERMAL MATERIAL PROPERTIES

	50% Al/Si	27% Al/Si
Thermal conductivity (W/(m •K))	130	150
Density (kg/m^3)	2500	2600
Coefficient of thermal expansion (ppm/K)	13.1	17
Poisson's ratio	0.25	0.28
Young's modulus (Gpa)	121	90
Heat capacity at constant pressure (J/kg•K)	800	850

978-1-6654-1392-3/21 $31.00 © 2021 IEEE

D. Results

The simulation results are shown in Figure 2 below. After the welding of the upper and lower cover plates, the maximum deformation of the shell was 0.18mm, and the overall appearance was convex. The maximum stress appeared at the rounded corner of the lower cover plate of the shell, and the stress was 285MPa, which exceeded the yield strength of Al/Si materials.

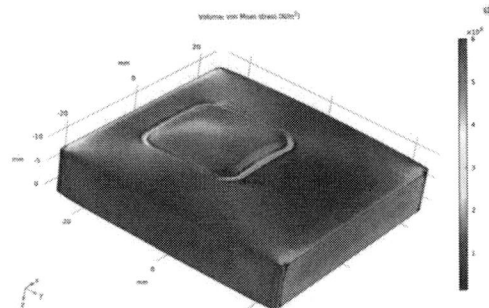

Fig. 2. Thermal stress distribution, maximum stress was 285Mpa

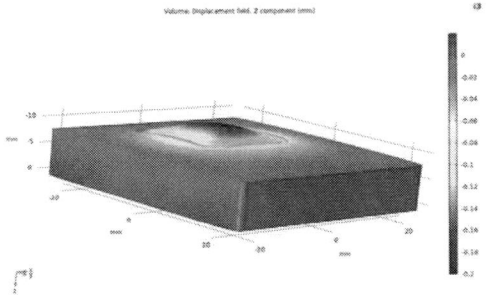

Fig. 3. Strain distribution on bottom of shell, maximum deformation was 0.18mm

III. EXPERIMENTS AND DISCUSSION

After the shell and plate machining, stress-free annealing was carried out to minimize the influence of residual stress on the welding process. Nd:YAG pulse tunable laser was used to seal welding in the glove box filled with high purity nitrogen. The weld morphology was observed under a 30x microscope, and cracks appeared at the rounded corners of the lower cover plate, as shown in the figure below., the deformation of the bottom surface of the shell was measured under the CMM(coordinate measuring machine), which was compared with the simulation results. As shown in the figure below, the maximum experimental deformation was 0.12mm, and the deformation trend was consistent with the simulation results, showing a convex wave morphology.

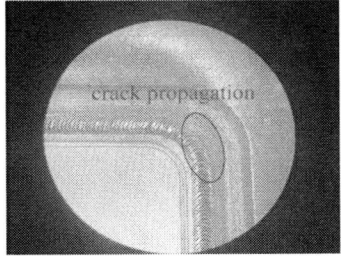

Fig. 4. Crack at fillet corner of shell's bottom

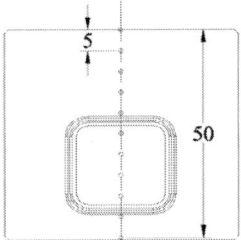

Fig. 5. Measuring point position by CMM

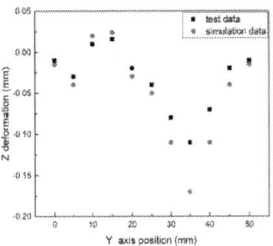

Fig. 6. Comparison of experimental deformation and simulated

IV. RESPONSE SURFACE METHODOLOGY

Due to the brittleness of Al-Si metal, it is easy to appear the penetration crack under the rapid change of temperature in the laser welding process.(Fig. 4) In this paper, the response surface method was used to optimize the design of the double-side laser fusion welding structure. Response surface method (RSM) is a method in modern quality engineering. It adopts the sequential idea. Firstly, appropriate parameters are designed, and then the RSM model is established by regression fitting according to the test results, which can approximately reflect the functional relationship between the target variables and the design variables.

$$y = \beta_0 + \sum_{i=1}^{k} \beta_i x_i + \sum_{i=1}^{k} \beta_{ii} x_i^2 + \sum\sum_{i<j} \beta_{ij} x_i x_j + \varepsilon \qquad (1)$$

In the equation, β_0 denotes the linear effect, β_{ij} denotes the linear interaction, ε denotes the error term, and the variation part that cannot be included by y. It is assumed that they are independent from each other in different tests, and obey the mean value of 0 and the variance is normally distributed σ^2

A. Design variable

According to the simulation calculation, the parameters shown in the figure below were selected as the design variables.

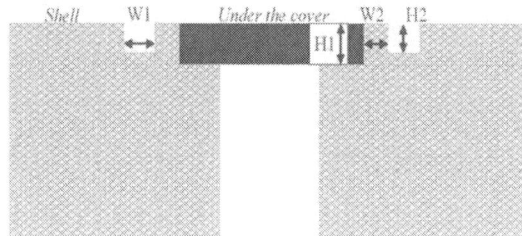

Fig. 7. The distribution diagram of design variables

B. Target variable

The flatness of the shell was selected as the target variable.

Based on the design principle of Box-Behnken response surface method, a response surface analysis experiment with four factors and three levels was designed.

TABLE II. ORTHOGONAL EXPERIMENTAL TABLE OF SHELL PARAMETERS

	W2(mm)	H1(mm)	H2(mm)	W1(mm)	F(mm)
1	0.8	0.6	0.4	1.5	0.172
2	0.8	0.4	0.6	1.5	0.115
3	0.8	0.5	0.5	1.5	0.15
4	0.8	0.5	0.6	2	0.154
5	0.8	0.5	0.5	1.5	0.152
6	0.8	0.4	0.5	2	0.1
7	0.8	0.5	0.5	1.5	0.153
8	0.7	0.5	0.4	1.5	0.14
9	0.8	0.6	0.6	1.5	0.169
10	0.7	0.5	0.5	2	0.145
11	0.8	0.5	0.4	2	0.156
12	0.7	0.5	0.5	1	0.143
13	0.9	0.5	0.5	2	0.157
14	0.8	0.5	0.5	1.5	0.154
15	0.8	0.5	0.6	1	0.153
16	0.8	0.4	0.5	1	0.114
17	0.7	0.4	0.5	1.5	0.12
18	0.7	0.5	0.6	1.5	0.147
19	0.8	0.6	0.5	2	0.17
20	0.9	0.4	0.5	1.5	0.123
21	0.7	0.6	0.5	1.5	0.17
22	0.9	0.5	0.6	1.5	0.158
23	0.8	0.6	0.5	1	0.171
24	0.8	0.4	0.4	1.5	0.11
25	0.8	0.5	0.5	1.5	0.154
26	0.9	0.5	0.4	1.5	0.161
27	0.9	0.5	0.5	1	0.154
28	0.8	0.5	0.4	1	0.154
29	0.9	0.6	0.5	1.5	0.171

C. Optimized results

The least square method is used to obtain the coefficients of each polynomial in Equation (1), and the multiple regression formula (2) for the maximum deformation of the bottom surface of the shell and the design variables is established.

$$F = -0.38 + 0.13W2 + 1.34H1 + 0.27H2 - 0.01W1 - 0.05W2H1 - 0.25W2H2 - 0.2W2W1 - 0.2H1H2 + 0.03W2^2 - H1^2 + 0.03H1^2$$

(2)

The fitting model established by formula (3) was tested for adaptability, and R^2 test was selected as the fitness test standard.

$$R^2 = 1 - \frac{\sum_{i=1}^{N}(y_{real}(i) - y(i))^2}{\sum_{i=1}^{N}(y(i) - \bar{y})^2}$$

(3)

In the equation, $y_{real}(i)$ and $y(i)$ are respectively the finite element calculation values and response surface calculation values of each point in the design space; $\bar{y}$ is the mean value of the truth values of each point in the design space; here is the average value of the finite element calculation; N is the number of test points in the design space; here is 29.

By calculating equation (3), the fitness of 87.73% is obtained, which proves that the fitting model is suitable and effective. The response surface established by Equation (2) was used to find the highest point of the surface and its corresponding parameter values, and the optimized parameters were obtained as shown in the table below

TABLE III. PARAMETER OPTIMIZATION RESULTS OF RESPONSE SURFACE METHOD

Parameters	W2(mm)	H1(mm)	H2(mm)	W1(mm)
	0.76	0.4	0.67	1.79

D. Experimental verification

Considering the possibility of machining, the optimized results of response surface method were rounded reasonably, and the welding experiment was carried out under the condition that other structural parameters and welding process conditions remained unchanged.

The maximum deformation of the bottom surface of the shell tested by CMM was 0.094mm, which was 21.7% better than before optimization. No welding cracks were observed under a 30x microscope

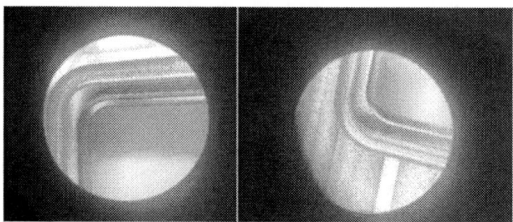

Fig. 8. Optimize welding test results

V. CONCLUSIONS

a) The melting and curing process of heat-affected zone materials in welding process can be effectively simulated by using the life and death element method, and the simulation accuracy is good. The error between the maximum

978-1-6654-1392-3/21 $31.00 © 2021 IEEE

deformation and the experimental value of the shell with typical structure size of 48*48.6*8.5mm is less than 25%.

b) In the case of laser sealing welding of upper and lower cavities with different surface areas, welding process will result in reversed deflection due to the mismatch of thermal expansion coefficient, and the improved effect is not obvious by adjusting the packaging sequence.

c) When sealing welding is carried out on the reverse side, the thickness of the lower cover plate has the greatest influence on the welding stress distribution, and the thickness of the lower cover plate should be reduced under the condition of ensuring the welding strength.

REFERENCES

[1] Dae-Won Cho, Won-Ik Cho, Suck-Joo Na, "Modeling and simulation of arc: Laser and hybrid welding process," Journal of Manufacturing Processes, vol. 16, pp. 26-55, 2014.

[2] L.J. Zhang, J.X. Zhang, A. Gumenyuk, " Numerical simulation of full penetration laser welding of thick steel plate with high power high brightness laser," Journal of Materials Processing Technology, vol. 214, pp. 1710-1720, 2014.

[3] FANG Kun, WANG Chuanwei, LIANG Ning, "Stress analysis and optimum design of T/R box by seal welding," Transactions of the China Welding Institution , vol. 37(5), pp. 19-22, 2016.

[4] Masoud Mohammadpour, Nima Yazdian, Guang Yang, "Effect of dual laser beam on dissimilar welding-brazing of aluminum to galvanized steel," Optics and Laser Technology, vol. 98, pp. 214–228, 2018

[5] Cui Xiaoli, Wu Yuying, Liu Xiangfu, " Effects of grain refinement and boron treatment on electrical conductivity and mechanical properties of AA1070 aluminum," Materials & Design ,vol. 86, pp. 395-397, 2016

[6] Sun Peng, Zhang Ting, Zhang Yanan, "Laser welding technology on high-Silicon aluminum alloy and its welded joint," Journal of Nanjing University of Aeronautics & Astronautics, vol.51, pp.44-48,2019.

[7] ZHU Xuewei, WANG Richu, PENG Chaoqun, "Microstructure and thermal expansion behavior of hypereutectic Al-Si alloy," Journal of Central South University (Science and Technology), vol.47(5), pp.1500-1505, 2011.

[8] W. Piekarska, M. Kubiak, "Three-dimensional model for numerical analysis of thermal phenomena in laser–arc hybrid welding process," International Journal of Heat and Mass Transfer, vol.54, pp.4966–4974, 2011.

[9] Gu Ye-qing, Yao Ye, Wang Chao, "Integration Design of Active Phased Array Antenna," Electro-Mechanical Engineering, vol.32(6), pp. 29–32, 2016.

Significant Effect of Temperature and Solders on The Growth Behavior of Cu_6Sn_5 on (110) Cu Single Crystal

1ˢ Chong Dong
School of Materials Sceience
Dalian University of Technolog
DaLian, China
dongchongwyy@ gmail.com

2ⁿᵈ Min Shang
School of Materials Sceience
Dalian University of Technology
DaLian, China
shangxiaomin2019@ mail.dlut.edu.cn

3ʳᵈ Ying Guo
School of Materials Sceience
Dalian University of Technolog)
DaLian, China
guoyingdut@ 163.com

4ᵗʰ Xiangxu Chen
School of Materials Sceience
Dalian University of Technolog
DaLian, China
xiangxuchen@ mail.dlut.edu.cn

5ᵗʰ Jun Zhang
Dalian AVIC Gangyan Superalloy Co.,
Ltd,
Dalian, China
13998553983@ 163.com

6ᵗʰ Changlong Dong
Dalian AVIC Gangyan Superalloy Co.,
Ltd,
Dalian, China
dcl1983@ 163.com

7ᵗʰ Haoran Ma*
School of Microelectronics
Dalian University of Technology
DaLian, China
*mhr@ dlut.edu.cn

8ᵗʰ Haitao Ma**
School of Materials Sceience
Dalian University of Technology
DaLian, China
**htma@ dlut.edu.cn

Abstract—The package interconnect in integrated electronic components is realized by Interfacial intermetallic compounds (IMC). Meanwhile, the downsizing of μ-bumps to tens of microns in advanced 3D packaging will result in the overgrowth of IMC and the number of grains contained in UBM decreased sharply, which bring new challenge for the reliability of solder joints. For this condition, numerous researches about the growth of IMC generated at the solder/Cu single crystal interface have been reported. For this paper, the growth of IMC at the interface between different solder, i.e., Sn, Sn-2Ag, Sn-3Ag, and (110) Cu single crystal at different temperature, i.e., 230°C, 250°C, 300°C, were investigated. Scanning electron microscope (SEM) and electron backscattered diffraction (EBSD) was used to characterize the morphology and orientation of Cu_6Sn_5, respectively. Results shows that the Cu_6Sn_5 formed on (110) Cu exhibit worm type at lower temperature and were tends to transform into facet type at higher temperature. Furthermore, the reflow temperature make a difference in the formation of Cu_6Sn_5 orientation. Besides, the coarsening rate of Cu_6Sn_5 was promoted following the increase of the Ag. The results have a significant meaning in controlling the orientation and improving the reliability of micro solder joints.

Keywords—Cu_6Sn_5, orientation, coarsening, single crystal copper

I. INTRODUCTION

IMC generated at the solder/Cu interface play an important role in the reliability of electronic devices in terms of electrical, thermal and mechanical properties [1]. The growth of IMC on Polycrystalline Cu has been investigated extensively in the past [2-4]. Meanwhile, the microstructure evolution of Cu_6Sn_5 is affected by solders and reaction temperature. The difference amount of Ni addition in solder can affect the type of reaction products [5]. The diffusion of Cu atoms can be influenced by the addition of Ag [6]. The orientation and morphology of Cu_6Sn_5 formed on the Sn/ Polycrystalline Cu interface is sensitive to reaction temperature [7].

The continuous development of micro solder joints results in the extensive research on the interface reaction on Cu single crystal. The Cu_6Sn_5 formed on the (001)Cu and (111)Cu surface elongated two perpendicular directions and elongated three directions at angles of 60 degrees from each other, respectively, and the strong orientation texture will gradually disappear as the reaction time increased [8]. the morphology of Cu_6Sn_5 grains generated on (110) Cu was typically scallop-type, but there are still some preferential orientations and the evolution of orientation in aging process is highly related to the orientation of Cu_6Sn_5 at nucleation stage [9]. The above results indicated that the reaction temperature and solders is a key step that controls the growth and orientation of Cu_6Sn_5.

For this paper, the nucleation of Cu_6Sn_5 at the interface between different solder, i.e., Sn, Sn-2Ag, Sn-3Ag, and (110) Cu single crystal at different temperature, i.e., 230°C, 250°C, 300°C, were investigated, respectively. The results have a significant meaning in controlling the orientation and improving the reliability of micro solder joints.

II. EXPERIMENTAL METHODS

Pure Sn and Ag were used to prepare for Sn, Sn-2Ag and Sn-3Ag solder ball in a diameter of 100μm. (110) Cu single crystal in dimensions of 7000μm×5000μm×200μm was used as substrate. The solders/(110)Cu bumps reflowed at 230°C, 250°C and 300°C for 1s, respectively. Then, the samples were cooled in water. After that, 10% HNO_3 (wt%) was employed to remove excess solder to expose the top morphology of IMC. The sample also need to polished by argon ion emission technologies to satisfy the flatness requirements of orientation detection. The orientation and morphology of Cu_6Sn_5 was characterized by EBSD and scanning electron microscope (SEM). The schematic diagram of reflowing is presented in Fig. 1.

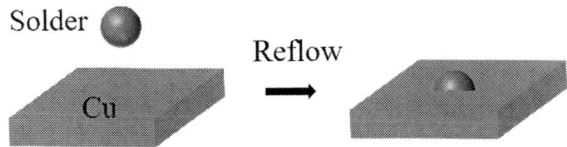

Fig.1 Schematic diagrams of reflow

III. RESULTS AND DISCUSSION

Fig. 2 the top morphology of Cu_6Sn_5 generated at the interface between Sn-based solders and (110) Cu single crystal at different temperature for 1s. obviously, Cu_6Sn_5 grains display worm-type and distribute homogenously on the whole interface at lower reflow temperature (230°C), as shown in fig.2 ($a_1 - c_1$). The shape of Cu_6Sn_5 grains in all the joints transformed from warm-type to facet-type as the reflow temperature increased, as displayed in fig.2 ($a_3 - c_3$). This phenomenon can be explained by Jackson's parameter (α). When the value of Jackson's parameter is smaller than 2, the crystal has a rough surface, but when it is larger than 2, the crystal has a facet surface. it can be expressed by this equation:

$$\alpha = \frac{\Delta H}{RT} \xi$$

Where ΔH is the change of entropy when IMC form on the interface between the liquid solder and the substrate, T is the reflow temperature, R is the gas constant and ξ is the fraction of total number of nearest neighbors in a plane parallel to the interface under consideration. The value of (α) is highly related to the reaction temperature, which increases with increasing temperature [10]. Therefore, the morphology of Cu_6Sn_5 tends to transform from worm-type to facet-type as the temperature increased.

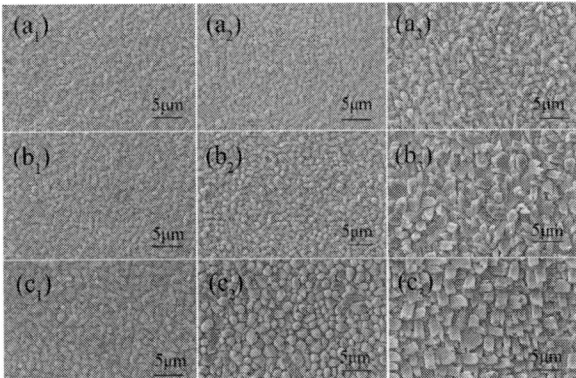

Fig. 2 the top morphology of Cu_6Sn_5 reflowed at (110) Cu single crystal at different temperature for 1s. a, b and c represent Sn, Sn-2Ag and Sn-3Ag, respectively. The subscript numbers of 1, 2 and 3 represent 230°C, 250°C and 300°C, respectively.

Obviously, the size of Cu_6Sn_5 grains increases as the reaction temperature increased. The growth of Cu_6Sn_5 is closely related to the diffusion of Cu. The higher nucleus growth rate and the higher diffusion rate of Cu atoms at the high reaction temperature leads to the increase of Cu_6Sn_5 size. Besides, the size of Cu_6Sn_5 increase as the Ag amount increased, as shown in Fig. 3. The existence of Ag_3Sn clusters in Sn-2Ag and Sn-3Ag solders inhibits the annexation and grain boundary migration between grains, and can maintain more grain boundaries resulting in the higher diffusion rate of Cu atoms.

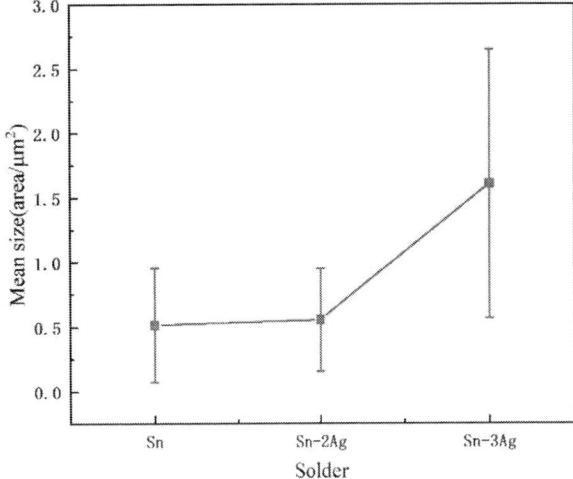

Fig. 3 the mean size of Cu_6Sn_5 formed on the solder /(110) Cu interface at 250°C for 1s

To elucidate the influence of solder and temperature on the orientation of Cu_6Sn_5 formed on the Sn-based solder/(110)Cu interface, EBSD was employed to characterize the evolution of orientation, as shown in fig. 4 and fig. 5. From the results, it can be found that the addition of Ag has little effect on the Cu_6Sn_5 orientation at low temperature (230°C), while the orientation of Cu_6Sn_5 formed on the Sn/(110)Cu interface more concentrated than that of Cu_6Sn_5 formed on the Sn-xAg/(110)Cu interface at 250°C. As displayed in fig. 4 and fig. 5, The reaction temperature have an significance influence in the growth of Cu_6Sn_5. The distribution of orientations of Cu_6Sn_5 becomes more concentrated in various joints as temperature increased. Besides, the Cu_6Sn_5 orientation tends to be distributed at the edge of the polar figure (PF) indicating that the (0001) crystal plane of Cu_6Sn_5 tends to be parallel to the reaction interface and the ND of Cu_6Sn_5 is tend to be $[10\bar{1}0]$. The change of Cu_6Sn_5 orientation may significantly affect the reliability of solder joints owing to its anisotropy.

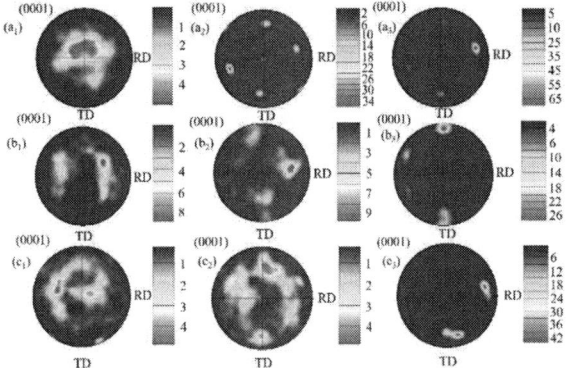

Fig. 4 the PF of Cu_6Sn_5 grains reflowed at different temperature for 1s. a, b and c represent Sn, Sn-2Ag and Sn-3Ag, respectively. The subscript numbers of 1, 2 and 3 represent 230°C, 250°C and 300°C, respectively.

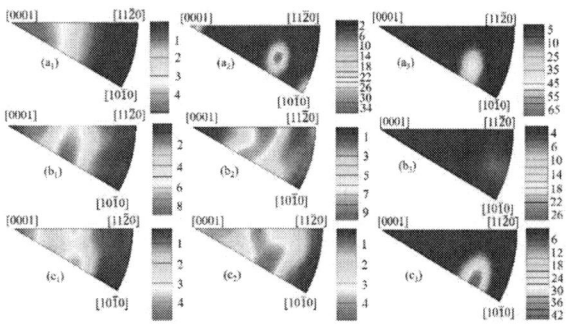

Fig. 5 the IPF of Cu_6Sn_5 grains reflowed at different temperature for 1s. a, b and c represent Sn, Sn-2Ag and Sn-3Ag, respectively. The subscript numbers of 1, 2 and 3 represent 230°C, 250°C and 300°C, respectively.

IV. CONCLUSION

This study experimentally investigated the effect of solder and reaction temperature on the orientation of Cu_6Sn_5 formed on (110) Cu. Results show that the Cu_6Sn_5 formed on (110) Cu exhibit worm type at lower temperature and were tend to transform into facet at higher temperature. Furthermore, the reflow temperature make a difference in the formation of Cu_6Sn_5 orientation. Besides, the size of Cu_6Sn_5 was promoted following the increase of the Ag and the reaction temperature. These results provide a reference for understanding IMC growth on (110)Cu single crystal surface, optimizing manufacturing process and evaluating reliability.

ACKNOWLEDGMENT

This work was supported by the National Natural Science Foundation of China (Grant No. 51871040).

REFERENCES

[1] W.K. Choi, S.Y. Jang, J.H. Kim, K.W. Paik, H.M. Lee, J. Mater. Res. Vol.17, pp. 597–599, 2002.

[2] Shi Chen, Ning Zhao, Yuanyuan Qiao, Yunpeng Wang, Haitao Ma, C.M.L. Wu, Proc. - Electron. Components Technol. Conf. 2019-May pp, 1629–1634. 2019

[3] Q.K. Zhang, W.M. Long, Z.F. Zhang, Journal of Alloys Compound. Vol. 646, pp,405–411. 2015

[4] Fengfeng Guo, A. Kunwar, Chengrong Jiang, Yunpeng Wang, Ning Zhao, Mingliang Huang, J. Mater. Sci. Mater. Electron. pp.135–140. 2021.

[5] Z.L. Li, Yanhong Tian, H.J Dong, X.J. Guo, X.G. Song, H.Y. Zhao, J.C. Feng, Scr. Mater. Vol.156, pp.1–5. pp. 2018.

[6] M. Yang, H.J Ji, Shuai Wang, Yong-Ho Ko, Chang-Woo Lee, M.Y Li，J. Alloys Compound. Vol. 679, pp,18 -25, 2016.

[7] Mingyu Li, Ming Yang, Jongmyung Kim. Materials Letter, Vol. 66, pp.135-137, 2012.

[8] J.O. Suh, K.N. Tu, N. Tamura. Applied Physics Letter, Vol. 91(5), p. 051907. 2007.

[9] Zou H F, Yang H J, Zhang Z F. Journal of Applied Physics, Vol.106(11), p. 95. 2009.

[10] Kunwar A, An L, Liu J. J. Mater. Sci. Technol., Vol. 50, pp. 115-127,2020.

A stretchable high-sensitive capacitive sensor for aerodynamic pressure measurement

Xiaofeng Yang
The 5th Electronic Research Institute of the Ministry of Industry and Information Technology
Guangzhou, China
yxf004@hotmail.com

Jian Chen
The 5th Electronic Research Institute of the Ministry of Industry and Information Technology
Guangzhou, China
18360713622@163.com

Si Chen
The 5th Electronic Research Institute of the Ministry of Industry and Information Technology
Guangzhou, China
chensiceprei@yeah.net

Bin Zhou
The 5th Electronic Research Institute of the Ministry of Industry and Information Technology
Guangzhou, China
zhoubin722@163.com

Yijun Shi
The 5th Electronic Research Institute of the Ministry of Industry and Information Technology
Guangzhou, China
syj20094870@sina.com

ZhiWei Fu*
The 5th Electronic Research Institute of the Ministry of Industry and Information Technology
Guangzhou, China
fzw19940124@163.com

Abstract—Stretchable pressure sensors have been widely studied over past years, but there are still exist many challenges to make flexible pressure sensors with great measure range and high sensitivity. In this paper, a high flexibility superb-capacitive pressure sensor has been proposed for aerodynamic pressure measurement. The innovative pressure sensor contains a PVA/KOH dielectric layer, and sandwiched by two transparent flexible electrodes. The capacitance of the proposed sensor increase with the aerodynamic pressure. The aerodynamic can be measured by the capacitance of the sensor. The proposed sensor provided a higher sensitivity (31.02/kPa) in comparison to the other pressure sensors. In addition, the fabrication processed of the sensor network is also developed.

Keywords—Flexible sensor, Electronic double layer, Aerodynamic pressure, Capacitive sensor

I. INTRODUCTION

Recently, due to the fast development of the hypersonic aircraft, aerodynamic pressure measurement has got many attentions for its broad application in aircraft design [1-3]. According to the different sensing mechanism, the method of aerodynamic pressure measurement can be classified into three categories: small pressure hole method, pressure sensitive paint technology (PSP) and computational fluid dynamic (CFD) [3, 4]. Through these method have been developed many years, there are still have many defects, both in the theoretical calculation and experimental research. The experimental method of pressure measurement hole is very complex. In addition, the accuracy of aerodynamic pressure measurement would be affect by the pressure measurement hole. PSP technology can be used to measure the global aerodynamic pressure measurement of aircraft. Nevertheless, the PSP technology can only tested in wind tunnel experiments. CFD simulation method also has many disadvantages, such as the numerical simulation method may result in distortion, generating a pseudo-physical conclusion. Therefore, it is of great important to develop a new method for real-time aerodynamic pressure measurement, which can further help us to study the aerodynamic shape possessing great properties.

Currently, flexible pressure sensors have been developed fast in the whole world [5-7]. Due to the characteristic of flexible, the flexible sensors can be attached on the fuselage of aircraft to measure the size and direction of surface pressure. Based on the sensing mechanism, the flexible pressure sensors can be classified into three kinds, including piezoresistive, piezoelectric and capacitive. The capacitive pressure sensor is one of the most popular due to its long-time drift stability, low power consumption, fast response and simple device construction as compared to piezoelectric and piezoresistive sensors [8, 9].

Suo et. Al. has developed an flexible ionic skin sensor , which can measure the strains changes from 1% to 500% with a small drift over more than one thousand cycles [10]. Meanwhile, Pan et al. proposed an flexible ionic pressure sensor using ionic liquid as dielectric layer, which has a relativity high sensitivity [11]. However, this high sensitivity can only be maintained within in a small pressure range. To develop a flexible capacitive pressure sensor with high sensitivity for aerodynamic pressure measurement is significant.

In this paper, we have developed a flexible capacitive pressure sensor for the aerodynamic pressure measurement. The capacitive sensor consists of a porous dielectric layer, and sandwiched by two flexible conductive (ITO-PET) films. The capacitive sensor's performance of pressure measurement along with sensing film, temperature and humidity were also been investigated.

II. EXPERIMENTAL

A. Materials and apparas

Polyvinyl alcohol (PVA) 1788 was supplied by Shandong Usolf Chemical Technology Co. Ltd. KOH was purchased from Jinan Qingtian Chemical Technology Co. Ltd. The indium tin oxide (ITO) coated Polyethylene terephthalate (PET) was supplied by Luoyang Tengchang Xukun Biological Technology Co., Ltd.

Impedance analyzer (E4900A, Agilent) was used to measure the sensor's capacitance change under different pressure. The microstructure of porous membrane was investigated by Scanning electron microscopes (SU70, Japan). A micro fatigue elongation machine (MMT-100NV-10, Japan) was used to supplied a accurate pressure applied on the capacitive sensor .

B. Sensor design and implementation

As Shown in Figure 1(a), the porous ionic dielectric layer has been fabricated by solution casting method. The PVA particles was dissolved in deionized water. Moreover, the mixed solution has been stirring 120 minutes at 85 ℃. And then, KOH particles has been prepared and added into the mixed solution (PVA:KOH=1:2, wt%). Finally, the mixed solution of PVA/KOH was casted at a plastic model. Figure 1(b), (c) show the microstructure of the porous PVA/KOH ionic membrane. It is obvious that the PVA/KOH film is very flexible and porous which can utilize for many flexible device application.

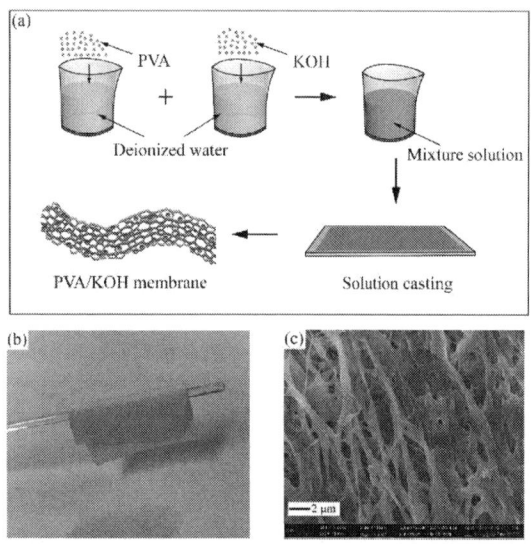

Figure 1. The fabrication process and the microstructure of the PVA/KOH ionic membrane.

In this paper, the ionic PVA/KOH film was used as the dielectric layer of the flexible capacitive pressure sensor. For a single pressure sensor, it contains a ionic PVA/KOH dielectric layer with 50 μm, and two ITO-PET electrodes, on the top and bottom of PVA/KOH membrane. The sensing mechanism of the of the developed sensor is based on the electronic double capacitors between the PVA/KOH dielectric layer and ITO-PET electrodes, as shown in Figure 2. The electronic double capacitors is formed at the surface of the dielectric layer by electrons on the electrodes and the ions from the ionic membrane. When the external pressure applied on the top ITO-PET electrode of the sensor, the ionic dielectric layer is compressed and the contact area between ITO-PET electrodes and ionic dielectric membrane increases. As a result, the capacitance of the sensor will increase rapidly due to more charges move to the interface of the porous two electrodes and PVA/KOH dielectric layer.

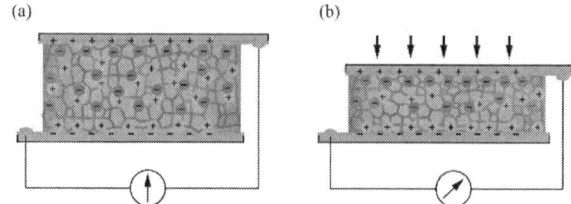

Figure 2. The schematic diagram of the sensor.

III. RESULTS AND DISCUSSIONS

A. Relatipon between capacitance and driving frequence

The capacitive pressure sensor has been attached on an aluminum plate to measure the surface pressure. When different pressure press on the surface of sensor, the WK6500B impedance analyzer was used to measure the capacitance of the sensor in real-time. Figure 3 shows the capacitance of the sensor changes with the driving frequency at 0 Pa and 5 kPa. It is indicate that, the electrical double capacitors were formed on the interface between electrodes and PVAS/KOH ionic dielectric layer.

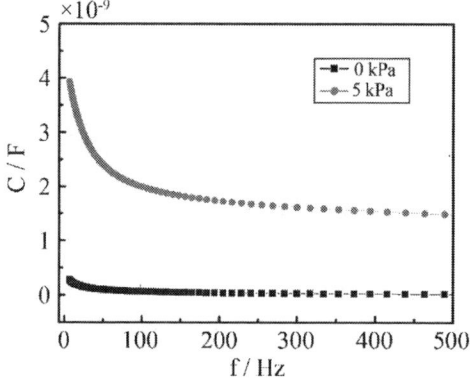

Figure 3. The capacitance changes of the sensor with driving frequency at 0 kPa and 5 kPa.

B. Relation between pressure and capacitance

As shown in Figure 4, the capacitance changes with different pressure press on the surface of sensor. It is obvious that the capacitance change of the sensor depends on the external pressure. When different pressure pressure on the surface of the sensor, the PVA/KOH membrane was compressed, the capacitance increases immediately. The sensitivity of the sensor is 31.02/kPa at 0 to 3.0 kPa, while drop to 15.06/kPa from 3.0 to 28.5 kPa.

978-1-6654-1392-3/21 $31.00 © 2021 IEEE

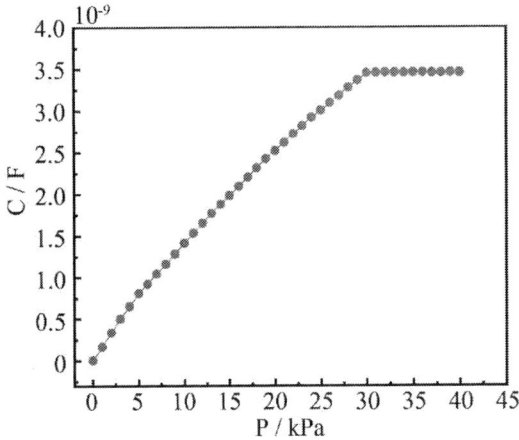

Figure 4. The relationship between the capacitance of the sensor and different pressure.

Furthermore, a pressure of 5 kPa at frequency 50 Hz was also applied on the sensor. As shown in Figure 5, the sensing performance of the sensor was stable. The sensor's relaxation timer and response time were 43 ms and 52 ms.

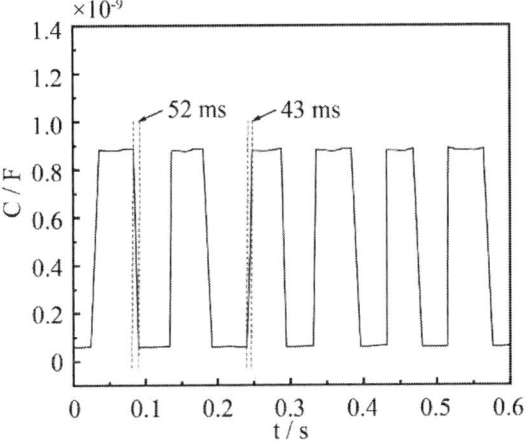

Figure 5. The relationship between the capacitance of the sensor and 5 kPa at frequency 50 Hz.

C. The influence of temperature and humidity

Because the flexible electronics' service environment is very complex, the flexible electronic must work well at different vibration, humidity and temperature. The influence of temperature and humidity was investigated. As shown in Figure 6, the capacitance of the sensor was measured at the temperature range from 25 °C to 45 °C. It is clear that, the flexible sensor used to measure the pressure at the temperature range of 25 °C to 30 °C.

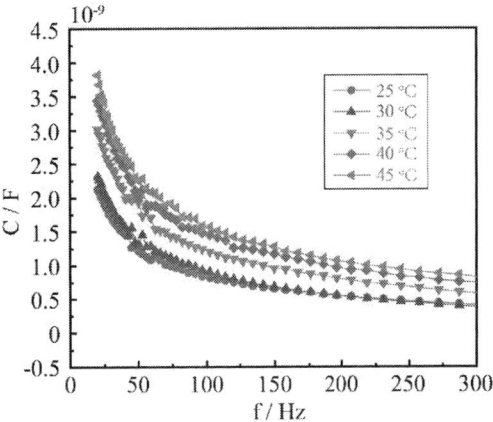

Figure 6. Relationship between Capacitance of the sensor and frequency at 25 °C to 45 °C .

In addition, the influence of humidity also has been investigated. As shown in Figure 7, the humidity do not affect the sensor's capacitance.

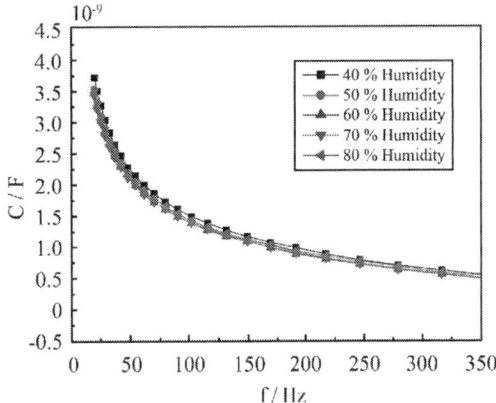

Figure 7. Relationship between Capacitance of the sensor and frequency at different humidity.

IV. APPLICATION OF SENSOR NETWORK

To study the ability of the flexible pressure sensor for sensing the distribution and magnitude of the aerodynamic pressure, we have made a 3×3 sensor network to measure the surface aerodynamic pressure. The 3×3 sensor network has been attached on an aluminum plate, as shown in Figure 8(a). We have do three experiments to verify the ability of the flexible sensor network: 1) single load was applied on the center of the 3×3 sensor network; 2) two load was applied on the diagonal of the 3×3 sensor network; 3) a dynamic load was applied on the center of the sensor network. As shown in Figure 8(b), (c), the capacitance of the sensor near the load location increased dramatically. As shown in Figure 8(d), the flexible pressure sensor can measure the dynamic pressure well. The capacitance of the sensor increased/decreased with the external pressure load/unload on the senor. According to the experiment results, the flexible sensor network can be used for measuring the location and size of the aerodynamic pressure.

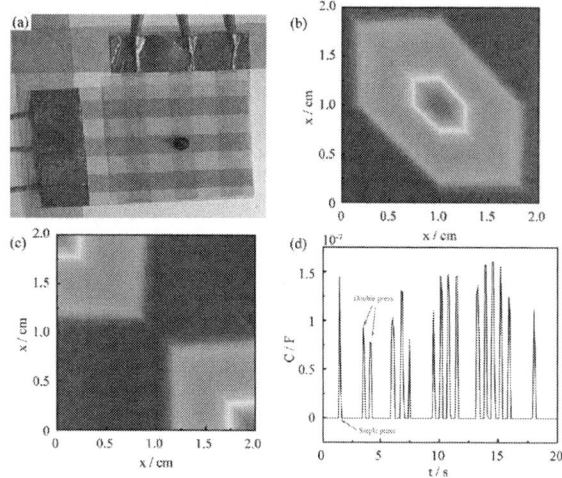

Figure 8. The 3 ×3 sensor network for pressure distribution measurement.

V. CONCLUSIONS

A stretchable capacitive pressure sensor with high sensitivity, simple structure and low cost has been proposed. The innovative sensor shows a good sensing performance. The main conclusions are described as follows:

1) The proposed flexible sensor with a high sensitivity (31.02/kPa) and fast response (52 milliseconds).

2) The proposed flexible pressure sensor can work at different humidity.

3) The sensor network can be used to measure the distribution and seize of the external pressure.

ACKNOWLEDGMENT

The authors are grateful for the support by the National Natural Science Foundation of China (No. 11772279) and CEPREI Innovation and Development Fund (No. 20Z32).

REFERENCES

[1] W. Shyy, L. Bernal, D. Yeo and E. Atkins, Aerodynamic Sensing for a Fixed Wing UAS Operating at High Angles of Attack, AIAA Atmospheric Flight Mechanics Conference 2012.

[2] N. M. Guerreiro and J. E. Hubbard, Flight Testing of a Model Aircraft with Trailing-Edge Flaps Optimized for Lift Distribution Control, AIAA Atmospheric Flight Mechanics Conference and Exhibit 2008.

[3] C. Cox, A. Gopalarathnam, C.E. Hall, Flight Test of Stable Automated Cruise Flap for an Adaptive Wing Aircraft, J. Aircraft., 47(4), pp 1178-88, 2010.

[4] J. Quindlen, J. Langelaan, Flush Air Data Sensing for Soaring-Capable UAVs, Aiaa Aerospace Sciences Meeting 2013.

[5] X. F. Yang, Y. S. Wang, and X. L. Qing, A Flexible Capacitive Pressure Sensor Based on Ionic Liquid, Sensors, vol. 18, no. 7, pp. 2395, Jul 23, 2018.

[6] L. K. Wu, W. F. Yuan, N. Hu, Z. C. Wang, C. L. Chen, J. H. Qiu, J. Ying and Y. Li, Improved piezoelectricity of PVDF-HFP/carbon black composite films, Journal of Physics D: Applied Physics, vol. 47, no. 13, pp. 135302, 2014.

[7] X. F. Yang, S. Chen, Y. J. Shi, Z. W. Fu and B. Zhou. A flexible highly sensitive capacitive pressure sensor[J]. Sensors and Actuators A: Physical, vol. 324, pp. 112629, 2021.

[8] X. F. Yang, Y. S. Wang, and X. L. Qing, A flexible capacitive sensor based on the electrospun PVDF nanofiber membrane with carbon nanotubes[J]. Sensors and Actuators A: Physical, vol. 299, pp. 111579, 2019.

[9] X. Yang, Y. Wang, H. Sun and X. L. Qing, A Flexible Ionic Liquid-Polyurethane Sponge Capacitive Pressure Sensor, Sensors and Actuators A: Physical, vol. 285, pp. 67-72, 2019

[10] J.Y. Sun, C. Keplinger, G.M. Whitesides, Z. Suo, Ionic skin, Advanced Material, vol. 26, pp. 7608-7614, 2014

[11] R. Y. Li, Y. Si, Z. J. Zhu, Y. J. Guo, Y. J. Zhang, N. Pan, G. Sun and T. R. Pan, Supercapacitive Iontronic Nanofabric Sensing, Advanced Materials, vol 29, pp. 1700253, 2017.

3D Wafer Level Packaging for SAW Filter Using Thin Glass Capping with Through Glass Vias

Zuohuan Chen
Department of Microelectronics and Integrated Circuit
Xiamen University
Xiamen, China
chenzh@sky-semi.com

Jin Zhao
Institute of Electronics Packaging Technology and Reliability
Faculty of Materials and Manufacturing
Beijing University of Technology
Beijing, China
zhaoj@sky-semi.com

Feng Jiang
Product Development Department
Xiamen Sky Semiconductor Technology
Co.Ltd.
Xiamen, China
jiangf@sky-semi.com

Heng Wu
Product and Development Department
Xiamen Sky Semiconductor Technology
Co.Ltd.
Xiamen, China
wuh@sky-semi.com

Mingchuan Zhang
Research and Development Department
Xiamen Sky Semiconductor Technology
Co.Ltd.
Xiamen, China
Zhangmc@sky-semi.com

Daquan Yu*
Department of Microelectronics and Integrated Circuit
Xiamen University
Xiamen Sky Semiconductor Technology
Co.Ltd.
Xiamen, China
yudaquan@xmu.edu.cn

Abstract—The advanced 5G wireless mobile phones have pushed surface acoustic wave (SAW) filters quickly developed in the direction toward miniaturization, high performance and low cost. Three-dimensional wafer-level packaging (3D-WLP) is a promising solution for SAW filter. It needs to be developed and replace conventional package technologies.

In this paper, wafer level SAW filter package is designed using glass capping technology. Thin glass with through glass vias (TGVs) is formed through laser-induced chemical etching. Following, the prepared glass capping dies are bonded on the corresponding SAW filter substrate to provide a safe environment for the device function area from corrosions. As compared with thin-film packaging that will cause the material outgassing at high temperatures, resulting in the contamination of the IDT, glass capping technology can avoid those problems. Finite element simulation model is built to calculate stress value to make the final package pass the reliability test. Our experiments indicated that the improved packages pass the reliability test, such as pre-conditional level 3, temperature cycling test (TCT).

Keywords—SAW filter; Wafer level package; Glass capping technology; Chip to wafer, BGA

I. INTRODUCTION

Surface acoustic wave (SAW) filter is a semiconductor device used to filter out the desired frequency. There are advantages to having the SAW filters, low insertion and design flexibility [1][2]. Based on SAW filter's features, these chips require cavity structure, comparing with IC package, SAW filter package more like MEMS package [3][4].

Conventional SAW filter uses a high temperature confirmed multilayer ceramics (HTCC) package with cavity structure and metal sealing to achieve high reliability, which limits device miniaturization and low cost. In 2000 or so, chip level package (CSP) technology has been applied in SAW filter packaging process. The chips are connected to an HTCC substrate[5], which is a typical packaging way that only used in the low frequency device.

As the cavity size of the SAW filter increases, hermetic packaging is a challenging subject. One of the key requirements is the robustness of the encapsulation. And the packaging process must be as cheap as possible. Therefore, a wise choice must be made between the available materials and technologies. Recently, there are two main packaging approaches for SAW filters: wafer bonding [1] and thin-film lamination packaging [2][5], as shown in Table 1. Wafer bonding technology, thin glass capping die to wafer bonding, seems to be more attractive due to its higher robustness which can greatly reduce the manufacturing cost. In the film lamination packaging, as shown in Fig. 1, besides the hermetic need, the outgassing behavior of film materials after the sealing greatly affects the chip's performance. It may cause pollution or damage to MEMS or interdigital transducers (IDT) [6].

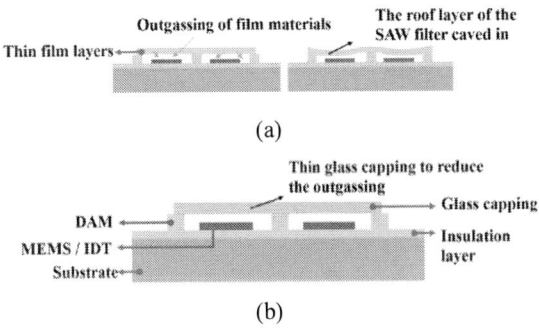

Fig. 1. (a) Thin film lamination package; (b) 3D WLP SAW filter package by thin glass capping technology.

The paper proposes a new packaging process flow for SAW filter and aims to optimize the package structure to pass the corresponding performance test. Toward this aim, a series of investigations on the reliability of this three-dimensional wafer-level packaging (3D WLP) are presented. In section II, a SAW filter package is designed and studied by finite element molding (FEM). Meanwhile, this model is built to calculate the maximum stress point in each step of manufacturing. In Section III, based on the designed packaging structure in section II, a novel process flow is presented. Section IV shows all the samples pass the reliability experiment, no samples failed. At last, some remarks are concluded in Section V.

978-1-6654-1392-3/21 $31.00 © 2021 IEEE

TABLE I.   3D-WLP FOR TYPICAL SAW FILTER USING DIFFERENT PROCESS FLOW

Typical	LiTaO3 capping technology	Thin film lamination technology	HTCC PKG without cavity structure
Process flows	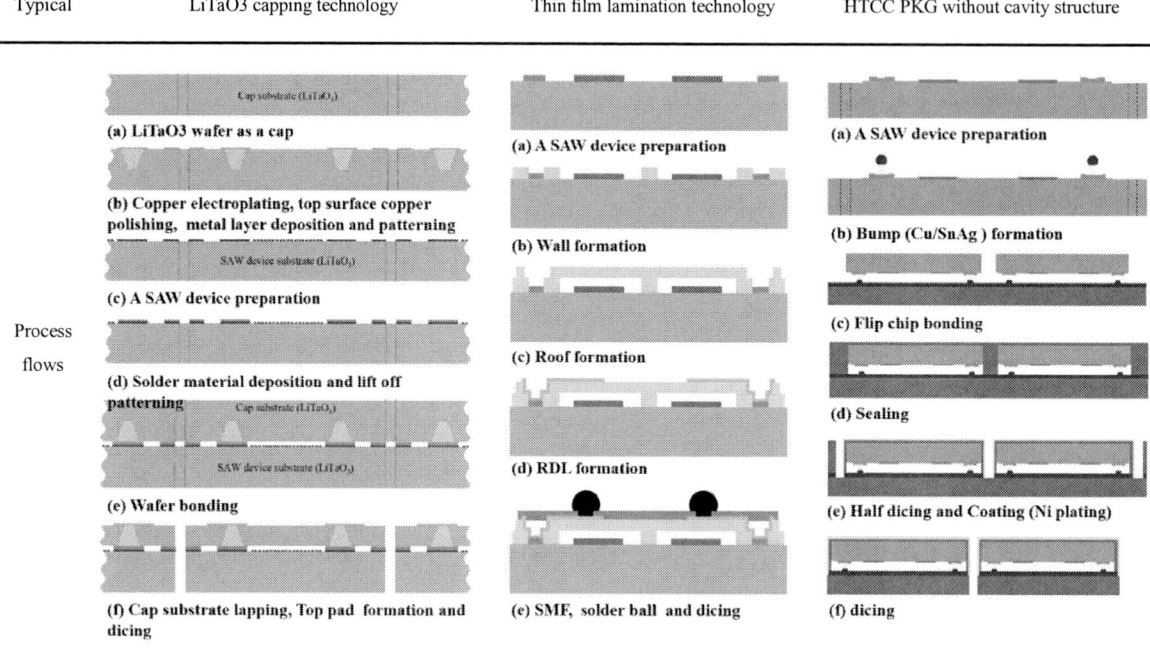		

II. SAW FILTER PACKAGE DESIGN

A. 3D thin glass cap WLP structure

Fig. 2 shows cross-section image of 3D-WLP for SAW filter. The glass cap chip with through glass vias (TGVs) is bonded to LT substrate to form a cavity structure. The signal can be transmitted through TGVs to realize the interconnection of chip's pad and solder ball.

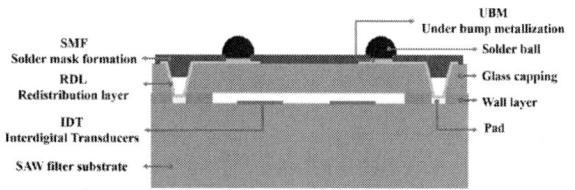

Fig. 2.   The cross-section of SAW filter wafer level packaging.

B. Vias sidewall Cu electroplating optimization

Since the performance of the package may be influenced by design structure and material parameters, it needs to analyze the effect of these factors, such as the thickness of vias sidewall Cu plating and material properties both wall and roof layer.

To achieve high reliability, FEM was built to calculate the maximum principal stress in each step of manufacturing process flows, mainly focused on the joint between the vias of wall and roof layer at the standard temperature cycling test (TCT) (-40℃~125℃). In this model, as shown in Fig. 3, the package structure model is built and simplified into four parts including the redistribution layer (RDL), polyimide 2, polyimide 1, and device (top to bottom).

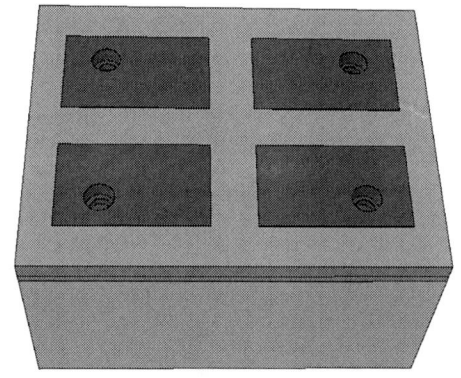

Fig. 3.   Finite element modeling.

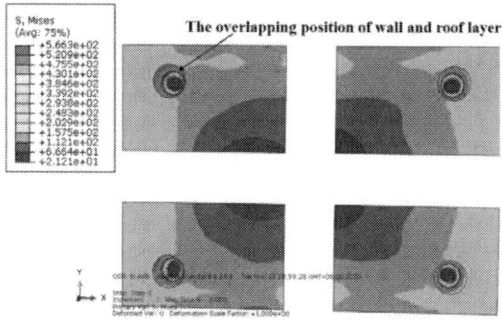

Fig. 4.   Stress distribution of the thickness of the Cu plating is 2 μm on TCT.

Fig. 4 shows the max miser stress appearing the via sidewall in the TCT (125 ℃ to -40 ℃). A crack will be produced at this position in actual fabrication process. When the Cu thickness of the via sidewall is setting as 2 μm, this will result in the crack occurs at the joint between the via of wall and roof layer. In a follow-up designing structure, the actual Cu thickness of via sidewall is set to 10 μm.

III. PACKAGING PROCESS

A. Fabrication process

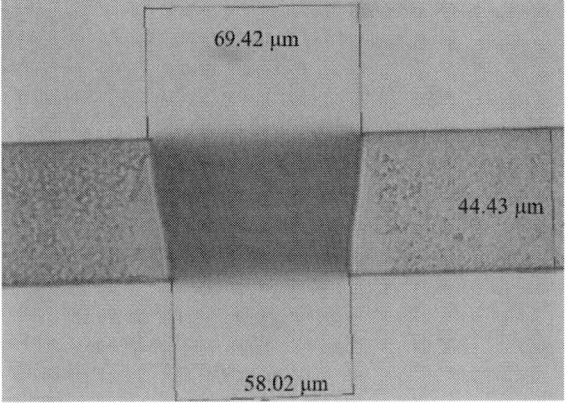

(1) Glass capping formation

(2) Wall formation

(3) Chip to wafer

(4) RDL formation

(5) SMF

(6) UBM, Solder ball and dicing

Fig. 5. Simple schematic of process implementation

WLP offers the fastest path to high reliability, low cost and small form factor. Thin glass capping technology is used to create a cavity structure for SAW filter cavity packages and creating microstructure, which is an emerging, cost-effective WLP process.

Fig. 5 shows the process flow of the 3D WLP for SAW filter. In this process flow, key processes include glass capping die formation, chip to wafer, RDL formation, ball grid array (BGA) and dicing.

1. Glass capping formation

Laser-induced chemical reaction etching includes two steps. It can produce high precision vias without using masking and lithography. First of all, the fast laser tool moves on the glass substrate, and the laser processing parameters are controlled to meet different glass appearances. Secondly, due to the significant difference in corrosion rate between the modified and unmodified regions. Laser-modified glass has a

higher etching rate. According to it, we can fabricate different TGVs. In this paper, the TGVs with a certain depth and diameter are fabricated by the laser-induced chemical etching process. Fig .6 shows the cross-section view of TGV with trapezoid geometry. The opening size and depth of the TGVs are 70/60 μm (upper and bottom opening) and 45 μm, respectively.

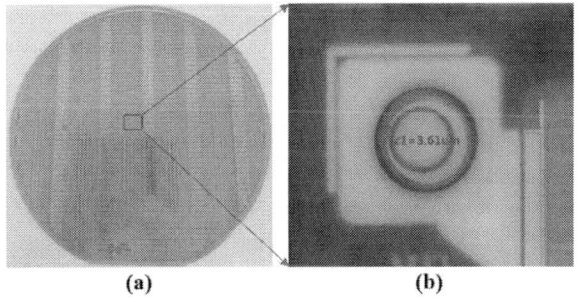

Fig. 6. The TGV profile with the specific size after laser induced chemical etching.

2. Cavity formation

First, an organic thin film is used to form the vias on the LT substrate based on the lithography technology. Then the known good glass capping die with the specific size is packed and placed to the SAW filter substrate with high attachment accuracy (shift <7 μm). Fig. 7 shows the whole wafer after the hip to wafer (C2W) attachment.

Fig. 7. (a) Top view of the whole wafer after C2W attachment. (b) The shift value of wall and roof layer

3. Metal layer patterning and BGA formation

Ti barrier (300 nm) and Cu seed (500 nm) layer are deposited on the glass cap and vias after the C2W process. Next, RDL is fabricated by standard process photoresist lithography, electroplating, photoresist strip, and etching. Then solder mask formation is performed. Subsequently, electroless plating of Ni/Au is achieved with a thickness of about 3/ 0.04 μm, respectively. Finally, in Fig. 8, a 3D WLP SAW filter package is fabricated after BGA and dicing.

978-1-6654-1392-3/21 $31.00 © 2021 IEEE

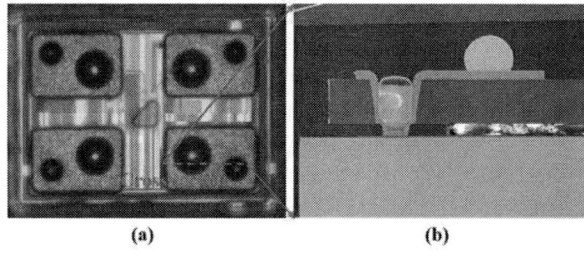

(a)   (b)

Fig. 8. The SEM view of 3D WLP SAW filter. (a) the SAW filter package. (b) the SEM view of TGV profile after BGA formation

IV. RELIABILITY EVALUATION

The performance of the package is evaluate based on the response of center frequency. Hermetic is a key factor. It can influence the inherent characteristics of the IDTs. In order to study the reliability of the package based on glass to LT wafer bonding technology, In Table 2, the samples are first baked at 125℃/24H, and then soaked 30℃/60%/192h. At last, the samples are reflowed at 260℃ (+5/-0) 3 times. TCT (1000cycles, -40℃ -125℃) is conducted for reliability evaluation. There are no significant frequency response change occurred after the TCT.

TABLE II.   ITEMS AND RESULTS OF RELIABILITY

Items	Conditions			Results
Pre-Con Level 3	Bake	Bake	125°C/24H	pass
	Soak	Soak	30°C/60%/192h	pass
	Reflow	Reflow	260°C(+5/-0) 3 times	pass
TCT	1000cycles, -40°C-125°C			pass

V. CONCLUSIONS

In summary, we present a 3D WLP SAW filter fabrication process using thin glass capping technology with TGVs. The reliability rest items are as follows pre-conditional level 3 and TCT are implemented. Main conclusions are:

(1) Thin glass capping dies are fabricated by laser-induced chemical etching and dicing. Following, the prepared glass capping dies are bonded on the corresponding LT device wafer and through TGVs to achieve interconnection between chip's pad and external circuit, which is important for a stable SAW filter signal transmission.

(2) The glass capping technology can full today's requirements for the SAW filter package in size and cost reduction.

(3) For SAW filter WLP using thin-film lamination, organic material will cause the material outgassing at high temperature, which can result in the contamination of the IDTs. The glass capping bonding technology can avoid that problem.

ACKNOWLEDGMENT

This research is supported by the Science and Technology Major Project of Xiamen City (3502Z20201004).

REFERENCES

[1] H. Campanella, Y. Qian, C. Romero, J. J. Giner and R. Kumar, "Integrated Filter and RF Front-end Module on RF SOI," 2019 Electron Devices Technology and Manufacturing Conference (EDTM), Singapore, Singapore, 2019, pp. 395-397, doi: 10.1109/EDTM.2019.8731172.

[2] Z. Chen, M. Zhang, K. Zhu, F. Jiang and D. Yu, "Development of 3D Wafer Level Package for SAW Filters Using Thin Film Lamination," 2020 21st International Conference on Electronic Packaging Technology (ICEPT), Guangzhou, China, 2020, pp. 1-5, doi: 10.1109/ICEPT50128.2020.9202920.

[3] Z. Wu, J. Malinowski and D. Questad, "Enhanced WLCSP Reliability for RF Applications," 2020 4th IEEE Electron Devices Technology & Manufacturing Conference (EDTM), Penang, Malaysia, 2020, pp. 1-4, doi: 10.1109/EDTM47692.2020.9118036.

[4] T. H. Kim et al., "Design and Characterization of Wafer Level SAW Filter Package Using LT-LT Wafer Structure," 2006 International Conference on Electronic Materials and Packaging, 2006, pp. 1-5, doi: 10.1109/EMAP.2006.4430597.

[5] Li Q, Goosen J F L, Van Beek J T M, et al. Outgassing of materials used for thin film vacuum packages[C]//2009 International Conference on Electronic Packaging Technology & High Density Packaging. IEEE, 2009: 802-806.

[6] R. Santos, J. -P. Delrue, N. Ambrosius, R. Ostholt and S. Schmidt, "Processing Glass Substrate for Advanced Packaging using Laser Induced Deep Etching," 2020 IEEE 70th Electronic Components and Technology Conference (ECTC), 2020, pp. 1922-1927, doi: 10.1109/ECTC32862.2020.00300.

978-1-6654-1392-3/21 $31.00 © 2021 IEEE

Warpage Characteristic of Glass Interposer with Different CTE's and Thickness

Jin Zhao
Institute of Electronics Packaging Technology and Reliability
Faculty of Materials and Manufacturing
Beijing University of Technology
Beijing, China
zhaoj@sky-semi.com

Fei Qin
Institute of Electronics Packaging Technology and Reliability
Faculty of Materials and Manufacturing
Beijing University of Technology
Beijing, China
qfei@bjut.edu.cn

Daquan Yu
School of Electronic Science and Engineering
Xiamen University
Xiamen Sky Semiconductor
Xiamen, China
yudaquan@xmu.edu.cn

Zuohuan Chen
School of Electronic Science and Engineering
Xiamen University
Xiamen Sky Semiconductor
Xiamen, China
chenzh@sky-semi.com

Shuai Zhao
Institute of Electronics Packaging Technology and Reliability
Faculty of Materials and Manufacturing
Beijing University of Technology
Beijing, China
zhaoshuai@emails.bjut.edu.cn

Abstract—In recent years, glass interposer has emerged as a superior alternative to organic and silicon-based interposers for advanced 2.5D or 3D IC, due to its attractive advantages such as excellent electrical isolation, better radio frequency performance, better feasibility with the CTE and could be fabricated with large size. However, the warpage for wafer level 2.5D or 3D packaging using glass interposer is a serious issue. In this paper, A new type of 3D wafer level packaging for MEMS based on glass interposer is presented and the warpage during the molding process is investigated based on finite element simulation. The effect of different material parameters and thickness of glass on warpage characteristics is investigated. The results show that when the CTE increases from 3 ppm/℃ to 10 ppm/℃ for glass interposer, the wafer warpage variation from "smile" to "cry". Our experiments indicated that the optimized glass interposer successfully passed the reliability test of pre-condition level 3. It is a conclusion that the process flow of the glass interposer and FEM simulation process offers a significant reference for 3D product applications. Moreover, in the next step, we will discuss how to modify the material parameters and thickness of the EMC to improve the performance of the glass interposer.

Keywords—Glass interposer; 2.5D; Finite element modeling; Warpage

I. INTRODUCTION

System-level 3D integration has become the main trend of Micro-Electro-Mechanical System (MEMS) packaging, which meets the urgent demand for electronic products with high-density integration and multifunction. Moreover, the development of electronic products is towards miniaturization and function integration. Interposer can realize the three-dimension (3D) integration in the future development. At present, the silicon interposer is widely used in the industry, and has achieved compelling results in the field of 2.5D or 3D integration. However, its manufacturing cost is high and has poor electrical performance like inset loss, and low productivity [1][2]. Compared with the glass interposer, it has superior electrical insulation performance, adjustable the coefficient of thermal expansion (CTE), and manufacturability with a large panel size [3][4]. Of course,

glass via formation capabilities has been developed significantly in recent years.

Recently, scholars at home and abroad mainly focus on four issues in glass interposer [5-7]: (1) ultra-fine pitch via fabrication, which is actually more difficult than TSVs due to the unfavorable laser drill and wet etch process. (2) multilayer redistribution Layer (RDL) with find pitch for wiring L/S= 5/5 μm or even 2/2 μm which becomes much more complicated than conventional wiring L/S =10/10 μm. (3) via filling process, which becomes difficult to fill the RDL trenches through damascene plating. (4) reliability and warpage. It is caused by poor mechanical strength of glass.

To investigate the warpage evolution of the fabrication processes of glass interposer, in this paper, a series of research works are presented. In section II, a new 3D glass interposer structure is proposed. Meanwhile, a 1/4 finite element modeling (FEM) is established as a warpage prediction model. Based on the optimized glass interposer structure by FEM, in section III, a novel process flow is proposed. At last, some remarks are concluded in section IV.

II. GLASS INTERPOSER PACKAGE DESIGN

A. Proposed 3D glass interposer structure

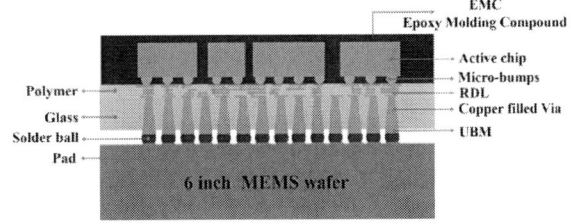

Fig. 1. The cross section image of glass interposer modules

In Fig. 1, the cross-section image of glass interposer modules with active chips, and 6inch MEMS wafer is designed. Some active dies with copper-SnAg micro-bump will be jointed to the glass interposer by copper-SnAg-copper bonding. The size of the package is 98.5 mm × 98.5 mm × 150 μm. The diameter of TGV is 50 μm with 150 μm depth.

978-1-6654-1392-3/21 $31.00 © 2021 IEEE

After that, the ASIC chips will be assembled on the designed glass wafer by solder ball.

In the work, the package is simplified into three components: the glass interposer, known good die (KGD) and epoxy molding compound (EMC). Meanwhile, the mechanical and thermophysical parameters of the materials are shown in Table 1.

TABLE Ⅰ. GEOMETRIC AND MATERIAL PARAMETERS

Material	Elasticity Modulus (GPa)	Poisson's Ration	CTE(ppm/°C)	Tg(°C)
Silicon 1	131	0.278	2.8	-
Glass 1	64	0.2	3.25	525
Glass 2	72.9	0.208	7.2	557
Glass 3	74	0.238	8.7	621
Glass 4	71	0.22	9.4	542

Fig. 2 shows a 1/4 FEM and displacement boundary condition. The internal packaging structure of the model is shown in Fig.2. Both symmetry planes are loaded by symmetric boundary conditions (U1 = UR2 = UR3 = 0; U2 = UR1 = UR3 = 0). The initial origin is fully constrained. During the molding process, the temperature increased from 25℃ to 150℃ during 45 minutes, and then keep the temperature constant for 1 hour, the temperature reduced from 150℃ to 70℃ during 1 hour. This leads to serious problems of warpage and residual stress. The process is named post-molding compound (PMC). During the FEM, solid 185 elements are employed, and the finite element model is divided into 245700 nodes and 223512 elements.

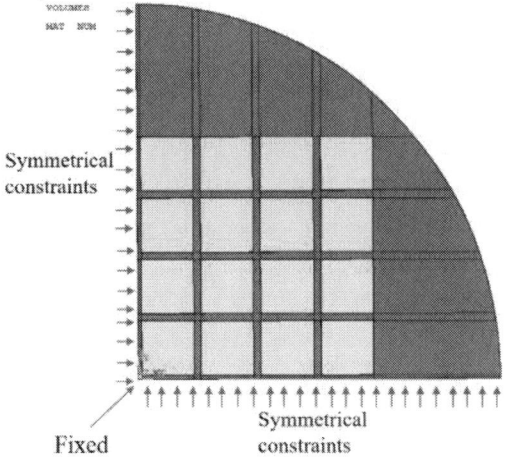

Fig. 2. FEM and displacement boundary conditions.

B. Effect of Material parameter

Three factors have an impact on warpage and stress: types and thickness of glass and types of EMC. In the simulation process, the gap between EMC and the chip is set as a fixed value, where is 75 μm. The single factor on the warpage and residual stress was studied. In FEM, each factor takes four or five levels to analyze.

According to the actual selection of substrate, the warpage variation with the different materials is studied based on the FEM. From Fig. 3, we can know the warpage orientation of the wafer varied from "smile" to "cry" with the increase of the CTE.

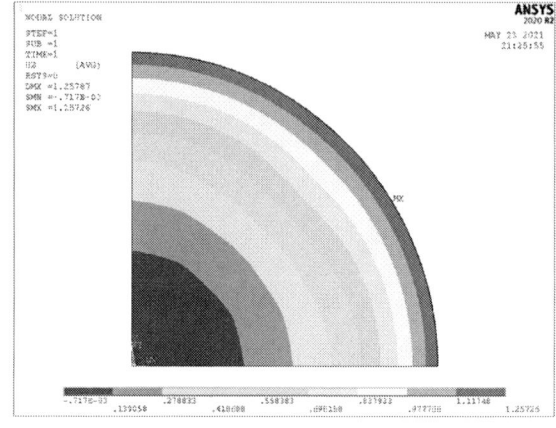

(a) warps in the smiling face direction (material: silicon, CTE 2.7)

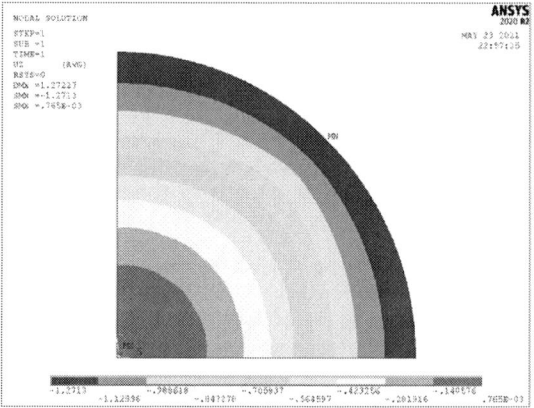

(b) warps in the crying face direction (material: glass, CTE 7)

Fig. 3. Effect of silicon and glass CTE on wafer warpage variation during molding

C. Effect of EMC material parameters and glass thickness

The influence of different glass thickness is also worth considering for the reshaped wafer. Glass with CTE 7 is selected as the substrate. The warpage value decreased from 1.27 mm to 1.16 mm with the thickness of the glass drops from 0.4 mm to 0.25 mm. From it, reducing the thickness of substrate has little effect on the improvement of warpage. During the actual packaging processes, thin glass is at risk of splintering. Based on this question, changing EMC materials is also a focus of research. In this work, it is found that when the CTE of EMC increases from 7 ppm/℃ to 15 ppm/℃, the chip warpage increases about three times.

The main cause of warpage is the CTE mismatch between the substrate and EMC. In the actual packaging, some factors should be considered, such as permittivity, thermal resistance and reliability. Under situation that process flow allow, we should make ensure that the CTE of the substrate and EMC are consistent. Summary, in Fig. 4, when the CTE increases from 3 ppm/℃ to 10 ppm/℃ for glass interposer, the wafer warpage variation from "smile" to "cry".

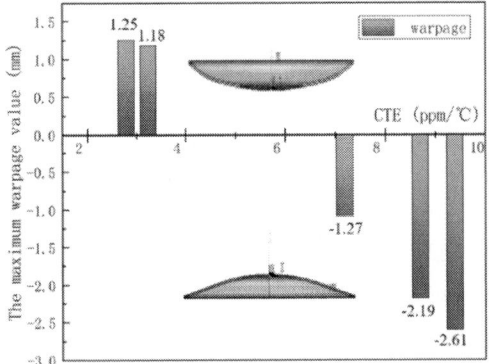

Fig. 4. Effect of CTE for the reshaped wafer warpage after molding

III. PACKAGING PROCESS

A. Manufacture process

(1) Multilayer RDL formation

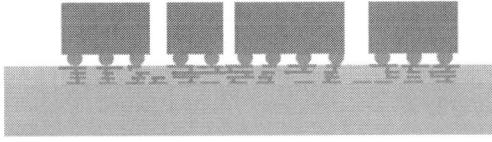

(2) Chip bonding on interposer

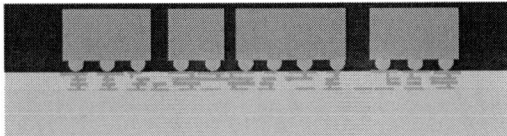

(3) Molding

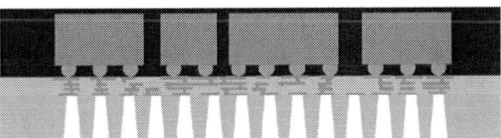

(4) TGVs formation based on laser drilling technology

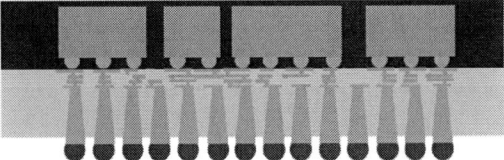

(5) Copper filled vias & UBM &solder ball

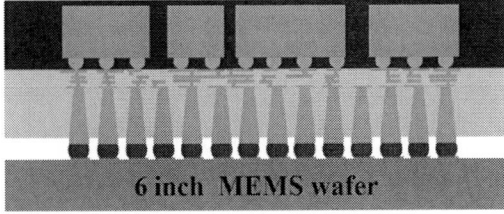

6 inch MEMS wafer

(6) MEMS wafer to Glass interposer

Fig. 5. Schematic fabrication process for glass interposer modules

In Fig. 5, the process flow for glass interposer modules is presented. Key processes include multilayer RDLs formation, chip bonding on the interposer, molding, TGVs fabrication, copper-filled vias and MEMS wafer to glass interposer modules.

Firstly, the carrier wafer is selected using the 220 μm glass, and the Ti mark is sputtered on the backside of the glass wafer to identify the wafer position for subsequent handing active chip bonding. Secondly, the under bump metallization (UBM), 4 layers RDL trenches and passivation are patterned on the glass wafer, as shown in Fig. 5(2). Thirdly, Fig. 5(3, 4) shows the active chips are attached to the glass wafer with multilayer wiring using flip chip technology, and wafer-level molding compound is laminated on chips to prevent die shift and die protrusion. Fourthly, the vias are formed on glass wafer by laser drill technology. Fifthly, Ti barrier (200 nm) and copper seed layer (1 μm) are sputtered. The TGV copper electroplating is performed through bottom to up after 5% H_2SO_4 and vacuum pretreatment; are shown in Fig. 5(5). Finally, Fig. 5(6) shows the glass interposer modules assembly process executes from the flip-chip bond on MEMS wafer by solder joint.

B. Key process results

1. Multilayer RDLs formation

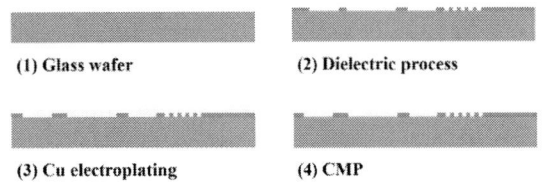

Fig. 6. Manufacturing process flows of sigle layer Copper damascene RDL.

There is the most key characteristics, such as copper traces, in high performance modules for the 2.5D interposer. Fig. 6 shows the manufacturing process flows of the single layer copper damascene RDL. Firstly, a photolithography process is implemented to pattern the micro-vias and RDL trenches. Secondly, Ti/ copper barrier and seed layer is sputtered on the top of the whole wafer. Thirdly, electroplating copper is performed to fill micro-vias and RDLs. Lastly, copper chemical mechanical polishing(CMP) process is performed to remove the excessive copper and barrier layer. After copper CMP planarization, the wafers surface flatness can be ensured to proceed the second or multilayer RDL trenches. There are four layers of RDL stacked to be achieved in this paper. In this process, the main challenge is to build lithography capability. Copper CMP is another challenge. All the copper layer on the surface of the wafer should be removed by CMP process, otherwise it will cause a short circuit

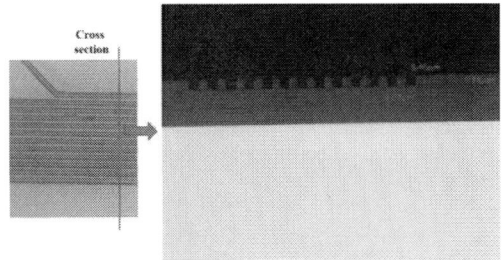

Fig. 7. Cross section image of RDL with find pitch for wiring L/S= 5/5 μm

978-1-6654-1392-3/21 $31.00 © 2021 IEEE

2. Die attach and molding

Then, the known good die is installed on the air-bearing table of the mounting machine. When taking the chip, the ejector pin will jack up the chip from the backside of the chip to make it easy to get rid of the tape and at the same time, the suction nozzle picks up the chip from the top. The carrier is transferred to the patch platform (Die Bond Table) of the mounter and the platform is heated to 120℃ which preventing the carrier from absorbing moisture and allowing the chip to pre-cure after mounting). After attach, put into an oxygen-free oven and baked at 150° for 4 hours to volatilize the solvent in the glue. The glue is completely solidified and the chip is firmly attached to the substrate. The oven is filled with nitrogen to prevent oxidation. The chip is protected with an epoxy molding compound.

Fig.8 shows the top view of 64ea ASIC chips with a size of 11mm × 11mm that has been bonded to glass wafer using a flip chip bonding with the high-accuracy die attachment data. (shift ＜5 μm). Fig. 9 shows the wafer characteristics after molding. The results of our experiments indicate that the maximum warpage value is less than 1mm (Glass 1: CTE is 3.25 ppm/°C) after molding, which agrees with simulation results well.

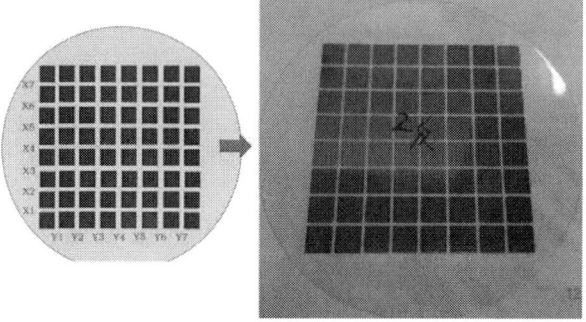

Fig. 8. Top view of the wafer after chip to wafer (C2W) attachment

Fig. 9. Warpage characteristics after molding.

The liquid compound is filled in the middle and around of the chips to form a reshaped wafer. Meanwhile, The reshaped wafer is putted into the oven for baked. Next, the formation of via is the use of the laser drill technology on the glass substrate. Meanwhile, the TGVs are filled copper through bottom-up electroplating after barrier layer Ti (200 nm)and seed layer copper (1 μm) are deposited. CMP is adopted to remove excessive copper on the top of the glass substrate. At last, the glass interposer module is assembled to 6 inch MEMS wafer by solder joint.

IV. CONCLUSION

In this work, warpage and reliability characteristic of silicon and glass interposer with different CTE's and thickness is researched from the experiment and model simulation. Main conclusions are:

(1) The 1/4 wafer level warpage models are built to simulate the actual fabrication process for glass interposer. Our simulations indicated that the type of glass has the greatest influence on warpage because of different CTE. In contrast, the thickness of the glass has less effect on the wafer warpage value.

(2) A glass interposer with a size of 98.5 mm × 98.5 mm × 150 μm is designed and fabricated by wafer level packaging technology, in this paper, which can control lower warpage to lay the foundation for 3D IC with excellent electrical isolation better radio frequency performance.

ACKNOWLEDGMENT

This research is supported by the National Natural Science Foundation of China (Grant No. 61974121). The authors would like to thank the members of the engineering teams at the Sky Semiconductor.

REFERENCES

[1] F. Qin, S. Zhao, Y. Dai, M. Yang, M. Xiang and D. Yu, "Study of Warpage Evolution and Control for Six-Side Molded WLCSP in Different Packaging Processes," in IEEE Transactions on Components, Packaging and Manufacturing Technology, vol. 10, no. 4, pp. 730-738, April 2020, doi: 10.1109/TCPMT.2020.2975571.

[2] Demir K, Sukumaran V, Sato Y, et al. Reliability of fine-pitch through-vias in glass interposers and packages for high-bandwidth computing and communications[J]. Journal of Materials Science: Materials in Electronics, 2018, 29(15): 12669-12680.

[3] Kuramochi S, Kudo H, Akazawa M, et al. Glass interposer for advanced packaging solution[C]//2016 6th Electronic System-Integration Technology Conference (ESTC). IEEE, 2016: 1-6.

[4] Huang T, Chou B, Sundaram V, et al. Novel copper metallization schemes on ultra-thin, bare glass interposers with through-vias[C]//2015 IEEE 65th Electronic Components and Technology Conference (ECTC). IEEE, 2015: 1208-1212.

[5] Wu M L, Lan J S. Simulation and experimental study of the warpage of fan-out wafer-level packaging: The effect of the manufacturing process and optimal design[J]. IEEE Transactions on Components, Packaging and Manufacturing Technology, 2018, 9(7): 1396-1405.

[6] Wei T, Wang Q, Cai J, et al. Performance and reliability study of TGV interposer in 3D integration[C]//2014 IEEE 16th Electronics Packaging Technology Conference (EPTC). IEEE, 2014: 601-605.

[7] Chen Z, Zhang M, Zhu K, et al. Development of 3D Wafer Level Package for SAW Filters Using Thin Film Lamination[C]//2020 21st International Conference on Electronic Packaging Technology (ICEPT). IEEE, 2020: 1-5.

AUTHOR INDEX

Ai, Binjie 443, 695
Ai, Daifeng 763
An, Tong 530, 555, 714
Bai, Jie 1131
Bai, Lijuan 493
Bai, Xue 1, 763, 832
Ban, Yu 986
Bao, Shuchao 999
Baojun, Qiu 148
Bedner, Dave 1435
Bi, Yuhao 705
Bo, Sun 646
Boowei, Tan 963
Braun, Robin 1169
Cai, Miao 420, 443, 854, 990
Cai, Zhikuang 1064
Cai, Zongqi 244, 1002
Cao, Chunyan 928
Cao, Guangqiang 248
Cao, Jiakai 479, 626
Cao, Liqiang 493, 1196
Cao, Rui 1153
Cao, Sicheng 497, 722
Cao, Ting 398, 671
Cao, William 1111
Cao, Wuxing 709
Cao, Xiuhua 1006
Cao, Zhijun 292
Cen, Kai 1097
Chang, Chao 823
Chang, Haixin 475
Chang, Xiaotong 809, 823, 836
Chang, Xufeng 827
Chao, Shuanshe 1494
Chen, Chender 226
Chen, Feng 1288
Chen, Gui 311, 331, 340
Chen, Haibin 1407, 1473
Chen, Hao 21
Chen, Hongtao 229, 381
Chen, Jiajia 1422
Chen, Jian 1521
Chen, Jing 121
Chen, Jin 1422
Chen, Kai 1069, 1134, 1139
Chen, Kun 755, 928
Chen, Leida 489
Chen, Liang-Pin 1504

Chen, Lijun 1288
Chen, Lin 732
Chen, Longfei 29
Chen, Mengyu 1346
Chen, Ming-Qiang 880, 1054
Chen, Mingxiang 34, 265
Chen, Pei 470
Chen, Phoebe 185
Chen, Qing 1025
Chen, Shiqi 776, 859
Chen, Shuai 67, 107, 1258
Chen, Si 48, 381, 767, 1040, 1323, 1489, 1521
Chen, Taotao 233, 393, 1173
Chen, Tao 958, 978, 1025
Chen, Tiezhu 961
Chen, Weijian 1273
Chen, Wen 864
Chen, Xiangxu 169, 1518
Chen, Xindong 420
Chen, Xinpeng 159
Chen, Xin 940, 1206
Chen, Xuanlong 1269
Chen, Yan-Ting 613
Chen, Yarong 1153
Chen, Yiming 1241
Chen, Yi 1148
Chen, Yong 1489
Chen, Yun 940, 1206
Chen, Yuqian 958, 978
Chen, Zhifeng 932
Chen, Zhiwen 58, 535
Chen, Zhuo 88, 585, 1241
Chen, Zuohuan 1525, 1529
Cheng, Chen 1478
Cheng, Hao 71
Cheng, Nan 326
Cheng, Xingwang 827
Cheng, Yuanjie 1121
Chenjun, Zhu 950
Chi, Panwang 679
Chin, Stella Wong Wun 1297
Chow-Khong, Tan 963
Chu, Baojin 239
Chu, Liu 1169
Chuantao, Hou 1382
Chunyue, Huang 257
Chunyue 371
Cu, Chengqiang 1459

Cui, Chengqiang 75, 1187, 1507, 1511
Cui, Hao 44, 1489
Cui, Zirui 1230
Dai, Fengwei 1288
Dai, Weijing 726, 1148
Dai, Xuanjun 1224
Dai, Yanwei 568, 1305
Dai, Zongbei 1246, 1315
Daojun, Luo 148
Deng, Chuanjin 972, 1002, 1323
Deng, Fei 1
Deng, Jinrong 928
Deng, Ronghua 1302
Deng, Rui 972, 1002, 1323
Deng, Yun-Kai 847
Deng, Zihao 634, 655
Di, Zhao 1088
Ding, Chao 1246, 1315, 1327
Ding, Kunpeng 1332
Ding, Lei 7, 121
Ding, Shuquan 1206
Ding, Wanchun 1449
Ding, Xinrui 634
Ding, Yuhan 967
Dong, Changlong 1518
Dong, Chong 169, 1518
Dong, Dong 950
Dong, Guoshuai 1302
Dong, Yaru 736, 1445
Dong, Yi 525
Du, Guoping 539
Du, Jianyu 517
Du, Mingyong 462, 479, 626
Du, Xiaomeng 462, 479, 626
Duan, Xiaolong 428
Fan, Chenrui 928
Fan, Haibo 699
Fan, Hongjin 99, 219
Fan, Jiajie 1044
Fan, Jingyu 180
Fan, Jinhu 1050
Fan, Tao 958, 978
Fan, Wenbin 320
Fan, Yuqing 103
Fan, Zhekun 1173
Fang, Chao 936
Fang, Jianming 1269
Fang, Mingang 88
Fang, Qu 376
Fang, Yi 1366, 1370, 1374
Fei, Jiu-Bin 870
Feng, Changqi 990

Feng, Chenzefang 12
Feng, Colin 185
Feng, Guan-Lin 1040
Feng, Jiayun 1012
Feng, Junbo 25
Feng, Xuegui 44
Fu, Dongzhi 506
Fu, Huaqiang 1378
Fu, Shuai 1210
Fu, Xianzhu 483
Fu, Xing 381
Fu, Zhenxiao 1006
Fu, Zhiwei 48, 381, 572, 603, 1040, 1521
Gai, Wei 1262
Gan, Guisheng 776, 859
Gao, Chenshan 1144
Gao, Guohua 1449
Gao, Haitao 1352
Gao, Jian 115, 305, 771, 940, 1206
Gao, Li-Ying 525
Gao, Li-Yin 1453
Gao, Liming 679
Gao, Ling 416, 424
Gao, Meng 177, 864
Gao, Qiu 7
Gao, Rui 440
Gao, Shiyi 292
Gao, Xu 517
Gao, Yingke 1361
Gao, Yingying 549, 617
Gao, Ziyang 1387
Ge, Bangtong 25
Geng, Fei 55
Gommers, Pieter 1297
Gong, Jinfeng 371
Gong, Weixi 1002
Gong, Xinjian 1422
Gong, Yanpeng 530, 555, 714
Gu, Erdan 1131
Gu, Jiabao 295
Gu, Jionajiong 1478
Gu, Ling 1343
Gu, Zhenyu 189
Guan, Yuan 1292
Guan, Zunyu 1473
Guangjie, Liu 912
Guo, Chunbing 564
Guo, Huaixin 572
Guo, Jingdong 1075
Guo, Kaiyu 1219
Guo, Sihua 1422
Guo, Xiaotong 229, 576, 642, 1441

Guo, Ying 809, 1518
Guo, Yufeng 1064
Guo, Yuhua 501
Guo, Yuxin 1117
Guodan, Zhou 1311
Han, Meng 512, 763
Han, Shouyu 1191
Han, Shunfeng 493, 1292
Han, Yang 689
Han, Yinhui 885
Han, Zhehao 1366, 1370, 1374
Hang, Yuan 185
Hao, Lichao 902
He, Aaron 185
He, Bin 763
He, Daping 1378
He, Di 1250
He, Guanghui 576, 642
He, Hengjian 854
He, Hongwen 1029
He, Hong 512
He, Huimin 493, 1292
He, Hu 233, 393, 1235
He, Jinming 17
He, Laisheng 25
He, Sifeng 1412
He, Siliang 705, 1336
He, Yunbo 144, 320, 356, 365, 1206, 1401
He, Zhiyuan 902
Hou, Bin 1059
Hou, Fan 466
Hou, Maoxiang 940, 1206
Hou, Shuhan 136
Hou, Xuewei 44
Hou, Zhuangzhuang 736, 1445
Hu, Deming 755, 928
Hu, Huaying 454
Hu, Jin 103
Hu, Qinghua 832
Hu, Shanwen 1343
Hu, Shaowei 944, 947
Hu, Wei-Lin 847, 870
Hu, Xianqin 638
Hu, Yougen 17, 483, 891, 1481
Hu, Yuehua 434
Hua, Hu Qing 898
Hua, Xueming 967
Huang, Chun-Yue 346
Huang, Chunyue 126, 131
Huang, Dayong 689
Huang, Feifei 936
Huang, Guochi 521

Huang, Hai-Jun 907, 1054
Huang, Haojie 94
Huang, Jiabin 326, 351, 360
Huang, Jianhong 791
Huang, Jiaqiang 823
Huang, Mingliang 936
Huang, Mingqi 560, 759
Huang, Min 466, 1158
Huang, Peng 932
Huang, Shijun 659
Huang, Tao 809, 823, 836
Huang, Tian 859
Huang, Wei 63
Huang, Xiangmiao 689
Huang, Xingjia 663
Huang, Xueyin 1153
Huang, Xu 572
Huang, Yixiu 663
Huang, Yiyong 193, 791
Huang, Yuhua 12
Huang, Yun 381, 1224
Huang, Zhiheng 607
Huang, Zhongwei 75, 1187, 1507, 1511
Huddar, Vinod Arjun 85
Hui, Li 950
Huo, Jia Ren 1034
Huo, Ruixia 159
Huo, Yinachao 458
Huo, Yongiun 736
Huo, Yongjun 1357, 1445
Ibrahim, Mesfin S. 1044
Ji, Liangzheng 796
Ji, Weiwei 248
Jia, Fei 684
Jia, Jinhao 591
Jia, Zhaowei 1101
Jian, Maoliang 454
Jian, Pang 1311
Jiang, Chao 613
Jiang, Feng 1525
Jiang, Jing 1034
Jiang, Junbo 1250
Jiang, Kulun 1006
Jiang, Kun 972, 1002
Jiang, Liheng 1336
Jiang, Miaomiao 1126
Jiang, Ruoyu 287
Jiang, Weiting 224
Jiang, Wenyu 80
Jiang, Zhaoqi 776, 859
Jiaxin, Liu 918
Jie, Bai 1088

Jin, Nong	1255	Li, Jiayi	634
Jin, Xing	1258	Li, Jiedong	1412
Jin, Yinuo	1165	Li, Jiexin	634, 655
Jin, Yufeng	543	Li, Jing	219, 1340
Jinfeng, Gong	257	Li, Jinhui	283, 301, 560, 746, 954, 1469
Jing, Chen	898	Li, Jinming	12
Jing, Zhou	1044	Li, Jinyang	420, 722
Jo, Eunsol	852	Li, Jin	885
Joosten, Annelies	1297	Li, Junhong	388
Jun, Tong	1382	Li, Junhui	1173
Jung, Cheong-Ha	852	Li, Junjie	408, 412
Kaixue, Ma	148	Li, Junlong	875
Kang, Jiajie	517	Li, Junwei	512, 539
Kang, Qiushi	621	Li, Jun	1144, 1161, 1426
Kao, Nicholas	598	Li, Kai	767
Ke, Chang-Bo	870	Li, Ke	403, 999, 1273
Kim, Gu-Sung	852	Li, Kui	1305
Kuang, Xianjun	316	Li, Kun	1262, 1265
Lai, Aaron	185	Li, Liang	763, 864
Lai, Canxiong	440	Li, Lingyun	1265
Lai, David	598	Li, Lu	1277
Lai, Haiqi	1459	Li, Maolin	346
Lai, Yuanting	336	Li, Menglin	29
Lan, Xin	38	Li, Mingyu	944, 947, 1366, 1370, 1374
Le, Fred Fuliang	1473	Li, Ming	679
Lee, Ning-Cheng	1435	Li, Nannan	543
Lee, S. W. Ricky	1121, 1302	Li, Qizhuo	607
Lei, Chuyi	1059	Li, Qi	1352
Lei, Wenyu	475	Li, Quanbing	1478
Lei, Zuomin	17, 483	Li, Ruining	1429
Lezhi, Ye	912	Li, Rui	1081
Li, Baoxia	450, 1305	Li, Shanshan	1514
Li, Bin	902, 972	Li, Shenglong	1361
Li, Bofu	493, 1292	Li, Shuang	265
Li, Bo	1006	Li, Shuwang	440
Li, Cai-Fu	659	Li, Tao	103
Li, Ce	827	Li, Tina	185
Li, Changqing	746, 954	Li, Wangyun	497, 695, 705, 722, 843, 1059, 1327, 1336
Li, Chaofan	239	Li, Weihao	517
Li, Cheng-Bo	907	Li, Weili	316
Li, Chenglong	261, 274, 278, 287	Li, Weiming	1441
Li, Cheng	1346	Li, Wei	791
Li, Dameng	493, 1292	Li, Wenfeng	1346
Li, Dayang	208, 1139	Li, Wenqi	1393, 1397
Li, Dejian	493, 1292	Li, Xiaodong	479
Li, Gang	261, 278, 630, 750, 781, 813, 923	Li, Xiao	525, 1453
Li, Ge	621	Li, Xing	244
Li, Guoyuan	767, 995	Li, Xujun	581
Li, Haolin	1401	Li, Yadong	470
Li, Hao	1201	Li, Yangyang	1126
Li, Huacong	75	Li, Yanning	568
Li, Jiasheng	634, 655	Li, Yan	1025

Li, Yesu	679
Li, Yingying	746, 954
Li, Yong	642
Li, Yulong	718
Li, Yun-Wei	880
Li, Zesheng	144, 356, 365
Li, Zewei	1064
Li, Zhankun	1173
Li, Zhe	525, 1453
Li, Zhipeng	581
Li, Zijian	940
Li, Zongtao	634, 655
Liang, Bunv	1340
Liang, Chen	530, 555, 714
Liang, Jiayong	634, 655
Liang, Jingyang	1246
Liang, Peilin	75, 1507, 1511
Liang, Ting	763
Liang, Yihang	403
Liang, Yi	1179
Liang, Zeng Xiao	898
Liang, Zhentang	995
Liang, Zhi	1332
Liao, Ao	1094
Liao, Cc	226
Liao, Chengyu	1029
Liao, Huilong	144, 365
Liao, Shuaidong	126, 131, 371
Liao, Wenyuan	440
Lin, Chen	740
Lin, Haoliang	781, 813, 923
Lin, Junshu	177
Lin, Qingping	1183
Lin, Shaopan	521
Lin, Shengru	679
Lin, Tao	1426
Lin, Tingyu	136, 521
Lin, Xiaohui	1401
Lin, Xinyi	1494
Lin, Yuanwei	1019
Lin, Zhiqiang	483, 1481
Liu, Cong	859
Liu, Dashun	726, 1069, 1134
Liu, Debo	1034, 1144
Liu, Dongcheng	38
Liu, Feixiang	722
Liu, Fengman	493, 1292
Liu, Fengmei	1352
Liu, Gai	895
Liu, Guanghui	800
Liu, Haiyan	1277
Liu, Hao	229

Liu, Hefeng	21
Liu, Huan	55
Liu, Huicong	805
Liu, Hui	674
Liu, Jiahao	229, 642, 1441
Liu, Jianhui	1117, 1161
Liu, Jiaxin	34
Liu, Jie	376, 986
Liu, Jinglong	265
Liu, Jing	1429
Liu, Jinshan	283, 301
Liu, Johan	180, 1422
Liu, Junfu	233, 393, 1173
Liu, Kai	7, 121
Liu, Lihong	1478
Liu, Lin-Jie	416, 424
Liu, Linjie	385, 895, 1255
Liu, Li	408, 535
Liu, Lu	1029, 1064
Liu, Mingjie	99
Liu, Min	381
Liu, Pan	786, 817
Liu, Pengfei	1158
Liu, Peng	219
Liu, Qiang	75, 283, 581, 1187
Liu, Richeng	1069, 1134
Liu, Sheng	58, 153, 535, 1081
Liu, Shimei	233, 1235
Liu, Shnjin	144
Liu, Shoufu	126, 131, 371
Liu, Shujin	356, 365
Liu, Shu	360
Liu, Tianhan	1246, 1315, 1327, 1336
Liu, Wansheng	159
Liu, Weidong	376, 398, 671
Liu, Wei	1183
Liu, Wen	283, 301
Liu, Xiaoyan	489
Liu, Xiaoying	51
Liu, Xin	197
Liu, Xuan	1498
Liu, Xuebin	1481
Liu, Xun	408, 412
Liu, Yachao	305
Liu, Yangzhi	1094
Liu, Yingxia	736, 1357, 1445
Liu, Ying	736, 1357, 1445
Liu, Yiping	679
Liu, Yongchao	638, 651
Liu, Yuan	475
Liu, Yunpeng	233, 393, 1173
Liu, Yu	1187

Liu, Zhi-Quan	525, 1453
Liu, Zhidan	67, 107
Liu, Zhigao	1059
Liu, Zilian	295
Liu, Ziyu	732, 740
Liu, Zuyao	1029
Lo, Jeffery C. C.	1121
Long, Junyu	1206
Long, Wang	1382
Long, Xu	638, 651, 809, 823, 836
Lou, Liang	805
Lu, Dong	1069, 1134, 1139, 1148
Lu, Guangsheng	443, 695, 843
Lu, Guoguang	440
Lu, Jibao	261, 274, 278, 287, 326, 351, 360, 638, 651, 667, 718
Lu, Jicun	679
Lu, Lu	278
Lu, Pei	180
Lu, Qian	1126
Lu, Xiangjun	594, 603
Lu, Xiaoxin	326, 351, 360
Luan, Huakai	466
Luan, Xinghe	674, 1059
Luh, Ding-Bang	1418
Luo, Bin	1191, 1393, 1397
Luo, Daojun	244, 1015
Luo, Dongxue	1215
Luo, Jun	244, 1015
Luo, Le	1265
Luo, Ruidong	659
Luo, Suibin	239
Luo, Xiaoting	607
Luo, Xiao	603
Luo, Yan	7, 121
Luo, Yuheng	771
Lv, Hongfeng	1015
Lv, Meijuan	1075
Lv, Mingtao	233
Lv, Xiaomeng	1094
Ma, Haitao	169, 1518
Ma, Haoran	169, 1518
Ma, Jusha	967
Ma, Peng	776
Ma, Qiangquiang	388
Ma, Shenglin	543
Ma, Shuying	506
Ma, Xiaojian	376, 398, 671
Ma, Xiao	907
Ma, Xu-Liang	1453
Ma, Yong	298
Ma, Yupa	94
Ma, Zhaolong	827
Mao, Mao	1323
Mao, Xingchao	229
Mao, Zhiyuan	316
Maolin, Li	257
Mei, Na	1494
Meng, Dominic Koey Poh	1297
Meng, Meng	1153
Meng, Yuezhong	607
Miao, Min	428, 885
Min, Zhixian	58
Mingxiang, Chen	918
Mo, Fuyao	316
Mou, Yun	34
Nan, Xujing	489
Neng, Liqiang	982
Ng, Wai Leong	1387
Ni, Liangyi	475
Ni, Yiqing	576
Ning, Minjie	1246
Nishikawa, Hiroshi	705
Niu, Fangfang	301, 1469
Niu, Leyi	684
Niu, Pingjuan	1131
Niu, Xulei	521
Oian, Qihao	144
Ou, Changping	663
Ou, Zhengping	1206
Ouyang, Keqing	111, 336, 982, 1311, 1494
Ouyang, Xing	630
Pan, Fei	1107
Pan, Huiming	58, 535
Pan, Kai-Lin	63
Pan, Kuang	1179
Pan, Lei	1064
Pan, Liu	898
Pan, Qingyu	1277
Pan, Yan	630
Pang, Jian	111, 336, 982
Pang, Yunsong	1, 213, 763
Park, Jung-Rae	852
Peng, Bo	295
Peng, Chao	638
Peng, Cheng	393
Peng, Fei	1366, 1370, 1374
Peng, Liang	630
Peng, Tao	261, 630, 923
Peng, Ting	1126
Peng, Xiaohui	630
Peng, Yang	265
Peng, Zhang	1097
Philipsen, Jos	1297

Pingfan, Ning...1088
Pingjuan, Niu...1088
Pingsheng, Zhang 950
Pu, Jie...1097
Qi, Zhangguo...898
Qian, Bin...967
Qian, Jiyu...94
Qian, Qihao..356, 365
Qiang, Jiang..1311
Qiao, Lan...1094
Qiao, Zhizhuang 385, 895, 1255
Qin, Fei.............470, 530, 555, 568, 714, 1305, 1529
Qin, Haotong 958, 978
Qin, Hongbo 1059, 1246, 1315, 1327, 1336
Qin, W. L..1085
Qin, Yijing 1069, 1134, 1139, 1148
Qin, Yikang..420, 722
Qing, Wang..918
Qingqing, Sun...740
Qiu, Guofu..320
Qiu, Xing..1121
Qiu, Yihua..655
Qu, Chenbing.......................................603, 1040
Qu, Fang...398, 671
Rai, Pradeep...1111
Ran, Honglei..1219, 1230
Ran, Teng..958
Ren, Jie..1262, 1265
Ren, Kuili...193
Ren, Linlin...........1, 173, 213, 253, 274, 287, 388, 539, 763
Ren, Shan..607
Ren, Xiaolei..51
Ren, Yan...902, 1323
Ren, Yulong...55
Ren, Zan...1255
Rong, Sun...........................213, 326, 351, 360
Ruan, Jianjun.......................................479, 626
Sa, Zicheng..1012
Shan, Liang..270, 1469
Shang, Jintang...................1191, 1393, 1397
Shang, Min...169, 1518
Shang, Panju..875
Shao, Dongdong 1332
Shao, Guangping..982
Shen, Chen...967
Shen, Jun..1201
Shen, Minghao..1250
Shen, Tianhua..1478
Shen, Ziyi...836
Sheng, Can...1081
Shi, Dachuang..940
Shi, Dianyang...564

Shi, Gaoming..316
Shi, Haitao..1429
Shi, Hongbin.....................................809, 823, 836
Shi, Jiajia...1169
Shi, Ru-Zeng...1054
Shi, Weiguang...932
Shi, Xinrong...1481
Shi, Yidian...393
Shi, Yijun..............................48, 381, 594, 1521
Shi, Yuning...699
Shieh, Brian...1302
Shoufu, Liu...257
Shu, Xiayun...928
Shuai, Zhou...148
Si, Weikang..549, 617
Si, Yu..189
Song, Guan Qiang...1034
Song, Lei...854
Song, Tingting..982
Su, Dezhi..71, 80
Su, Hao-Hang...1210
Su, Ping..466
Su, Tianxiong...823
Su, Yangquan..193
Su, Yunpeng...388
Su, Yutai...809, 823, 836
Su, Zhaoxi...1191, 1393, 1397
Suga, Tadatomo..875
Sun, Chao..466, 535
Sun, Chen..1040
Sun, Chuanchuan..1361
Sun, Daoheng...1429
Sun, Deliang...................560, 581, 746, 759
Sun, Fei...1418
Sun, Guoli...1305
Sun, Jianjun..336
Sun, Junfeng...1215
Sun, Liang..428
Sun, Lingyao.......................................736, 1445
Sun, Li...870
Sun, Ning..376, 398, 671
Sun, Peng..55, 543
Sun, Qingqing...732
Sun, Rong.............. 1, 17, 173, 239, 253, 261, 270, 274,
......278, 283, 287, 301, 388, 408, 412, 458, 462, 483, 512, 525,
539, 560, 581, 626, 630, 638, 651, 667, 718, 750, 763, 832,
891, 923, 1006, 1320, 1453, 1469, 1481
Sun, Tuo Bei...1494
Sun, Tuobei..111, 336
Sun, Xiaofeng...140
Sun, Xiaoyao..479
Sun, Yameng..1081

Sun, Yue	1418
Sun, Zhaoning	576
Sun, Zhefei	839
Tan, Daniel Q.	292
Tan, Liangchen	1514
Tan, Louise	226
Tan, Shwu Miin	1282
Tan, Xiaopeng	1258
Tang, Chu	88, 585
Tang, Hui	1412
Tang, Jiaqi	1327
Tang, Jiuyang	786
Tang, Linjiang	140
Tang, Qinglin	38
Tang, Qi	1139
Tang, Sha	244, 1323
Tang, Ying	1111
Tang, Yufeng	776
Tang, Zirong	1050
Tian, Chuang	1332
Tian, Dingkun	891
Tian, Jun	1429
Tian, Kun	1025
Tian, Ruyu	447
Tian, Wenchao	44, 1489
Tian, Xiaodi	684
Tian, Yanhong	447, 621, 1012
Tian, Yanzhong	555
Ting, Cao	376
Tong, Jin	75, 1507, 1511
Tong, Jun	21
Tong, Zhihao	805
Tu, Bingyi	809, 836
Tu, Qixuan	180
Tu, Wendian	539
Tuobei, Sun	982, 1311
Van Der Meulen, Rinse	1473
Wakeel, Saif	1297
Wan, Chengan	140
Wan, Dayuan	1148
Wan, Tao	202
Wan, Weikang	1196
Wan, Yang	674, 689
Wan, Yongkang	298, 1463
Wan, Zhaorong	1069, 1134
Wang, Bei	1029
Wang, Bingguang	827
Wang, Bin	674, 781, 813, 923
Wang, Cen	71, 80
Wang, Chengqian	1097
Wang, Chenxi	621
Wang, Chuanwei	1179

Wang, Chunlei	447
Wang, Congsi	1429
Wang, Daochang	1179
Wang, David H.	1165
Wang, David	1101
Wang, Dawei	591
Wang, Dazheng	549, 617
Wang, Dun	1463
Wang, Fangcheng	99, 219
Wang, Fuliang	88, 1241
Wang, Fuxin	80
Wang, Haozhe	287, 638, 651, 667, 718
Wang, Hong-Guang	870
Wang, Hongjie	1081
Wang, Hongkun	71
Wang, Hongyue	48, 594, 995
Wang, Huanhuan	854
Wang, Huihui	180
Wang, Jianhong	1277
Wang, Jian	1101, 1165, 1315, 1327
Wang, Jiao	506
Wang, Jie	1050
Wang, Junhao	732
Wang, Juntao	1034, 1161
Wang, Jun	1219, 1230, 1485
Wang, Ke	385, 895, 1255
Wang, Kuangyu	270
Wang, Langkun	1235
Wang, Lei	1485
Wang, Lichun	7, 121
Wang, Liuxin	763
Wang, Liwei	1040
Wang, Mei	189, 902
Wang, Ming-Sheng	839
Wang, Min	967, 1269
Wang, Nanxin	543
Wang, Ning	462, 479, 626
Wang, Pengchang	454
Wang, Pengfei	594, 603
Wang, Ping'An	189
Wang, Qiangwen	501
Wang, Qidong	1196, 1418
Wang, Qing	34
Wang, Rui	501
Wang, Ruolei	972, 1002
Wang, Shang	1012
Wang, Shinan	298
Wang, Shizhao	1081
Wang, Tao	283, 301, 1469
Wang, Tinglei	466
Wang, Weiyin	1426
Wang, Wei	55, 517

Wang, Wendong 1075
Wang, Wenlong 67, 1258
Wang, Xiaoqiang 244, 972, 1015
Wang, Xinjie 229
Wang, Xiyou 497, 695, 843
Wang, Xu 1111
Wang, Yan 311, 331, 1429
Wang, Yinghui 875
Wang, Yong 891
Wang, Yu Po 598
Wang, Yuanyuan 591, 1422
Wang, Yubo 493, 1292
Wang, Yuming 163, 202
Wang, Yunpeng 51
Wang, Yunxia 261, 274, 278
Wang, Zetian 517
Wang, Zhenyu 173, 213, 253
Wang, Zhen 1262, 1265
Wang, Zhibin 1153
Wang, Zhichao 967
Wang, Zhihai 1429
Wang, Zhiqin 58
Wang, Zhiqi 208, 1148
Wang, Zhizhe 1323
Wang, Zhuo 346, 371
Wang, Ziji 1393, 1397
Wang, Zixu 521
Wei, Jiahui 1305
Wei, Jianghao 163, 202
Wei, Jianhong 483
Wei, Jing 613
Wei, Qiang 376, 398, 671
Wei, Song 990
Wei, Tao 94
Wei, Wei 346
Wei, Xiangli 695, 843
Wei, Yiqiao 1241
Wei, Yuanyang 115
Wen, Jian 75, 1187, 1459
Wen, Minzhen 1044
Wen, Xiaokun 475
Wen, Zhibin 173
Wenchao, Wang 740
Weng, Zhangzhao 1015
Wong, Ching-Ping 261, 274, 278, 287, 326, 351, 360,
.... 638, 651, 667, 718, 1469
Wu, Cheng-Tar 1250
Wu, Daowei 159, 450
Wu, Fengshun 674, 689
Wu, Haomiao 1401
Wu, Hao 1302
Wu, Heng 1273, 1525

Wu, Houya 750, 781, 813, 923
Wu, Hua 1449
Wu, Jingshen 1069, 1134, 1407, 1473
Wu, Lv 613
Wu, M. H. 1085
Wu, Majiaqi 454
Wu, Qi 1346
Wu, Shaocheng 169
Wu, Wenyu 726
Wu, Xin 292
Wu, Xudong 292
Wu, Xueting 1277
Wu, Yanchen 854
Wu, Yanhong 1262
Wu, Yanpei 809, 823, 836
Wu, Yan 1158, 1215
Wu, Yilong 1094
Wu, Yukun 1179
Wu, Zhaohui 902
Wu, Zhen 659
Wu, Zhipeng 805
Xia, Dongmei 1029
Xia, Hongbin 1282
Xia, Jianwen 759
Xia, Minglu 1387
Xia, Pengcheng 1097
Xia, Wei 153
Xia, Xinnian 253
Xiang, Chen 1250
Xiang, Weiwei 1126
Xiao, Fei 1485
Xiao, Hui 229, 642, 1441
Xiao, Jinbo 791
Xiao, Jinqing 1173
Xiao, Ming 208
Xiao, Xiaoyu 311, 331, 340
Xiao, Yong 1378
Xiao, Zeping 29
Xiaoqiang, Wang 148
Xie, An 928
Xie, Jiacheng 1201
Xie, Weikun 197, 248, 298, 1463
Xie, Xiaoming 1265
Xing, Chaoyang 543, 1393, 1397
Xiong, Chunshui 153
Xu, Binbin 1064
Xu, Gaowei 1262, 1265
Xu, Guanzhe 177
Xu, Guoliang 535
Xu, Hongyan 800, 1498
Xu, Huanxiang 295
Xu, Jian-Bin 1

Xu, Jianbin 326, 351, 360, 832
Xu, Jile 594, 603
Xu, Ju 800, 1498
Xu, Kexin 381
Xu, Liang 458
Xu, Lu 1191
Xu, Qianzhu 776, 859
Xu, Sean 1277
Xu, Sha 564
Xu, Shen 326, 360
Xu, Weihua 1094
Xu, Xiangtao 776, 859
Xu, Xiaowei 1315
Xu, Yadong 483, 891
Xu, Yang 875
Xu, Yonglun 1, 213, 763
Xuan, Hui 1449
Xuanjie, Song 912
Xue, Cheng 1340
Xue, Dongpeng 224
Xue, Kai 403
Xue, Ke 726, 1069, 1134, 1139, 1148
Xue, Lianghao 1081
Xue, Shirui 497, 843
Xue, Song 1429
Xue, Xiangdong 163, 202
Yan, Chenkan 1463
Yan, Hui 607
Yan, Qiucheng 450
Yan, Shuxia 932
Yan, Tingnan 591
Yan, Yamei 311, 331, 340
Yan, Yan 936
Yan, Z. P. 1085
Yang, Bing-Xian 847, 870
Yang, Cheng 99, 219
Yang, Dali 726
Yang, Dan 1494
Yang, Daoguo 420, 443, 497, 695, 705, 722, 843,
 .... 854, 990, 1336
Yang, Donghua 958, 978
Yang, Guang 674
Yang, Guannan 75, 1187, 1459
Yang, Guoming 1302
Yang, Hao 336
Yang, He 549, 617
Yang, Huan 376, 398, 671
Yang, Huihui 80
Yang, Jianxin 163
Yang, Jiao 709
Yang, Jinbao 630, 923
Yang, Jinglei 1407, 1473

Yang, Junli 1094
Yang, Kai 689
Yang, Liang 755, 928
Yang, Lianqiao 454
Yang, Liu 709
Yang, Li 475
Yang, Nana 932
Yang, Qian 864
Yang, Shaohua 440
Yang, Wanchun 944, 947
Yang, Xiaofeng 48, 572, 767, 1521
Yang, Yajing 630
Yang, Yiren 990
Yang, Yong 1224
Yang, Yuanyuan 750, 781, 813, 923
Yang, Yuchi 517
Yang, Zhen-Tao 416, 424
Yang, Zheng 1361
Yao, Peilun 1407
Yao, Yimin 1
Yao, Yue 458
Yao, Yu 1230
Yao, Zhijun 229
Yaoyang, Shen 646
Ye, Huaiyu 1034, 1144
Ye, Huijie 1126
Ye, Le 177, 864
Ye, Wenbo 173, 213, 253
Ye, Yu 1153
Ye, Zhenwen 759
Yen, Freedman 598
Yi, Yaoyong 1352
Yi, Yuxi 560
Yilin, Wu 918
Yilong, Wu 950
Yin, Zhihao 800
Yu, Chengyu 1097
Yu, Daquan 193, 283, 403, 791, 999, 1273, 1525, 1529
Yu, Dianru 800
Yu, Fei 416, 424
Yu, Guangliang 1012
Yu, Jincheng 1387
Yu, Kunpeng 1429
Yu, Miao 1215
Yu, Qixing 1165
Yu, Rongying 177
Yu, Shuhui 239, 1006
Yu, Tian 403
Yu, Xuecheng 1320
Yu, Ye Huai 898
Yu, Yongjian 197, 298, 1463
Yu, Zheng 1449

Yuan, Yulei	408, 412	Zhang, Leicong	626, 1320
Yue, Wu	1340	Zhang, Lei	1006
Yue, Yu-Qing	613	Zhang, Lejun	71, 80
Yueping, Zhang	1382	Zhang, Li	111
Yuhong, Li	1088	Zhang, Luhui	17
Yumin, Zhao	912	Zhang, Mingchuan	193, 999, 1525
Yun, Minghui	990	Zhang, Minghua	140, 1025
Yun, Zhanfei	497	Zhang, Ning	450, 1012
Yunhui, Mei	1088	Zhang, Pei	709
Zehai, Wen	950	Zhang, Pengzhen	475
Zeng, Baoshan	12	Zhang, Ping	512
Zeng, Xiangliang	173, 213, 253	Zhang, Qiuchen	1282
Zeng, Xiaoliang	1, 173, 213, 253, 388, 539, 763, 832	Zhang, Qi	489
Zeng, Yanping	38	Zhang, Ruolin	1111
Zeng, Zejun	817	Zhang, Shuo	292
Zhai, Xiang	958	Zhang, Songsong	805
Zhan, Bihong	153	Zhang, Weijie	48, 1144
Zhan, Boyu	305	Zhang, Wei	1498
Zhang, Baotan	891	Zhang, Wenfeng	475
Zhang, Binbin	1025	Zhang, Wenjing	1241
Zhang, Chenwei	1034	Zhang, Xianshun	29
Zhang, Chenxu	512	Zhang, Xiaowei	674
Zhang, Chi	454	Zhang, Xin-Ping	847, 870, 880, 907, 1054
Zhang, Chongming	535	Zhang, Xinlei	1343
Zhang, Chunhong	978	Zhang, Xueying	1352
Zhang, Cong	224	Zhang, Yakun	530
Zhang, Donglin	1357	Zhang, Yan	180
Zhang, Elley	224	Zhang, Yao	709
Zhang, Fei	67, 107	Zhang, Ye	963
Zhang, Guoping	270, 283, 560, 581, 746, 759, 954, 1469	Zhang, Yiming	1489
Zhang, Guopinz	301	Zhang, Yinghui	1191
Zhang, Guoqi	420, 443, 786, 990, 1044	Zhang, Yi	689
Zhang, Hao	434	Zhang, Yong	1422
Zhang, Hongze	1158	Zhang, Yu-Bo	1453
Zhang, Houdun	791	Zhang, Yuehua	1343
Zhang, Huaiquan	126, 131, 371	Zhang, Yuexing	512
Zhang, Huibin	1463	Zhang, Yuting	1258
Zhang, Jiabo	1179	Zhang, Yu	75, 1187, 1459, 1507, 1511
Zhang, Jianfeng	1397	Zhang, Zebo	177
Zhang, Jiangtao	111	Zhang, Zhenyu	1340
Zhang, Jianping	479	Zhang, Zhihao	839
Zhang, Jian	63, 1126	Zhang, Zhitao	1343
Zhang, Jindi	771	Zhang, Zhou	689
Zhang, Jing	786, 796, 817	Zhang, Zhuanzhuan	428
Zhang, Jin	1393	Zhang, Zhuo	99, 219
Zhang, Jun	709, 1518	Zhangzhao, Weng	148
Zhang, Kaihong	197, 298, 1463	Zhao, Chen	912
Zhang, Kailin	443, 990	Zhao, Dinglei	58
Zhang, Kai	55	Zhao, Fanny	1302
Zhang, Kang	1250	Zhao, Guangyao	99
Zhang, Kun	642	Zhao, Guolin	750
Zhang, Lanyu	115, 305, 771	Zhao, Haoran	517

Zhao, Heng 25
Zhao, Jin 1525, 1529
Zhao, Junxiang 71
Zhao, Ming 1126
Zhao, Ning 51
Zhao, Shuai 568, 1529
Zhao, Tao 458, 462, 479, 626, 891
Zhao, Wenzhong 1258
Zhao, Xiaowei 1117
Zhao, Xiuchen 736, 1357, 1445
Zhao, Yong 1094
Zhao, Yue 121
Zhao, Yulin 1418
Zhao, Yunfu 1361
Zhao, Zhenbo 576
Zhao, Zhiping 107
Zheng, Bingjie 572
Zheng, Bo 1165
Zheng, Deyin 517
Zheng, Dongfei 29
Zheng, Libing 549, 617
Zheng, Lihua 38
Zheng, Qi 493
Zheng, Wei 947
Zheng, Yi 287, 667, 718
Zheng, Yuxiang 1196
Zhiyuan, Zhu 740
Zhiyue, Wang 912
Zhong, Ao 283, 746
Zhong, Caiden 224
Zhong, Cheng 261, 274, 278, 667
Zhong, Jianfeng 1514
Zhong, Keju 1507, 1511
Zhong, Sung-Hua 1504
Zhong, Yi 999
Zhong, Yongbin 115
Zhong, Yong 1069, 1134
Zhonz, Cheng 287
Zhou, Bin 48, 572, 767, 995, 1224, 1521
Zhou, Cheng 1429
Zhou, Chunming 434
Zhou, Jie 1107
Zhou, Jinzhu 189
Zhou, Liang 961
Zhou, Longzao 674, 689
Zhou, Min-Bo 870, 880, 907, 1054
Zhou, Peng 434
Zhou, Qing 1107
Zhou, Qin 1320
Zhou, Quan 1144, 1262, 1265
Zhou, Rui 714
Zhou, Shicheng 621

Zhou, Shuai 1015
Zhou, Yangfan 385, 895, 1255
Zhou, Yi 7
Zhou, Zhiwei 771
Zhou, Zhou 699
Zhu, Deliang 17
Zhu, Fulong 12
Zhu, Gang 295
Zhu, Jian 1158
Zhu, Kai 197, 1463
Zhu, Pengli 261, 278, 412, 458, 462, 479, 626,
.... 630, 750, 781, 813, 923, 1320
Zhu, Shengcong 663
Zhu, Wenbo 1366, 1370, 1374
Zhu, Wenhui 88, 311, 331, 340, 393, 585, 718, 750, 1241
Zhu, Zhiyuan 732
Zhu, Zhongyuan 1412
Zi, Chunfang 1183
Zito, Elaina 1435
Zongwei, Wang 1311
Zou, Longjiang 51
Zou, Xinrui 674
Zou, Yabing 1441
Zuo, Xinlang 642

2021 22nd International Conference on Electronic Packaging Technology (ICEPT 2021)

Xiamen, China
14 – 17 September 2021

Pages 1-766

IEEE Catalog Number: CFP21553-POD
ISBN: 978-1-6654-1392-3

**Copyright © 2021 by the Institute of Electrical and Electronics Engineers, Inc.
All Rights Reserved**

Copyright and Reprint Permissions: Abstracting is permitted with credit to the source. Libraries are permitted to photocopy beyond the limit of U.S. copyright law for private use of patrons those articles in this volume that carry a code at the bottom of the first page, provided the per-copy fee indicated in the code is paid through Copyright Clearance Center, 222 Rosewood Drive, Danvers, MA 01923.

For other copying, reprint or republication permission, write to IEEE Copyrights Manager, IEEE Service Center, 445 Hoes Lane, Piscataway, NJ 08854. All rights reserved.

*** This is a print representation of what appears in the IEEE Digital Library. Some format issues inherent in the e-media version may also appear in this print version.

IEEE Catalog Number: CFP21553-POD
ISBN (Print-On-Demand):   978-1-6654-1392-3
ISBN (Online):   978-1-6654-1391-6

Additional Copies of This Publication Are Available From:

Curran Associates, Inc
57 Morehouse Lane
Red Hook, NY 12571 USA
Phone: (845) 758-0400
Fax: (845) 758-2633
E-mail: curran@proceedings.com
Web: www.proceedings.com

TABLE OF CONTENTS

EVALUATION OF AGING PERFORMANCE OF THERMAL GEL SUBJECTED TO LASER FLASH TESTS.. 1

    Yimin Yao, Yonglun Xu, Xue Bai, Yunsong Pang, Linlin Ren, Xiaoliang Zeng, Fei Deng, Jian-Bin Xu, Rong Sun

RESEARCH ON CONFORMAL PHASED ARRAY T/ R MODULE BASED ON LCP SUBSTRATE.. 7

    Yan Luo, Kai Liu, Qiu Gao, Lei Ding, Yi Zhou, Lichun Wang

MECHANICAL RESPONSE OF BNNS-REINFORCED ALUMINUM COMPOSITES UNDER UNIAXIAL COMPRESSION... 12

    Jinming Li, Yuhua Huang, Baoshan Zeng, Chenzefang Feng, Fulong Zhu

FABRICATION OF SOFT CONDUCTIVE MICROSPHERES AND THEIR APPLICATION IN ELECTROMAGNETIC INTERFERENCE SHIELDING SHEETS 17

    Luhui Zhang, Zuomin Lei, Jinming He, Deliang Zhu, Yougen Hu, Rong Sun

FATIGUE LIFE EVALUATION AND TEST METHOD FOR REPRESENTATIVE PRINTED CIRCUIT BOARD 21

    Jun Tong, Hao Chen, Hefeng Liu

METHOD FOR PREPARING SILICON PHOTONIC CHIP EDGE PACKAGING STRUCTURE BASED ON INCLINED DEEP ETCHING PROCESS 25

    Heng Zhao, Laisheng He, Junbo Feng, Bangtong Ge

LIFE PREDICTION OF GOLD-ALUMINUM BONDING SYSTEM BASED ON FAILURE PHYSICS UNDER MULTI-STRESS COUPLING 29

    Menglin Li, Longfei Chen, Xianshun Zhang, Dongfei Zheng, Zeping Xiao

LOW-TEMPERATURE BONDING OF HIGH-POWER DEVICE USING CU-AG COMPOSITE NANOPARTICLE PASTE 34

    Jiaxin Liu, Qing Wang, Yun Mou, Mingxiang Chen

RESEARCH ON THE UNIFORM TEMPERATURE OF HEAT DISSIPATION FOR THE REVERSE OBLIQUE MICROCHANNEL... 38

    Qinglin Tang, Dongcheng Liu, Yanping Zeng, Xin Lan, Lihua Zheng

A METHOD OF RESEARCH FOR THE RELIABILITY OF SOLDER JOINT SHAPE 44

    Wenchao Tian, Xuewei Hou, Hao Cui, Xuegui Feng

INVESTIGATION OF THE RDL RELIABILITY BASED ON RF CHARACTERIZATION 48

    Hongyue Wang, Weijie Zhang, Yijun Shi, Si Chen, Zhiwei Fu, Xiaofeng Yang, Bin Zhou

EFFECT OF Ni_3Sn_4 NANOPARTICLES ON GRAIN REFINEMENT IN SAC305 FREESTANDING SOLDER BALLS AND SAC305/CU BGA JOINTS 51

    Xiaolei Ren, Xiaoying Liu, Longjiang Zou, Yunpeng Wang, Ning Zhao

RESEARCH ON EFFECT OF ANNEALING ON COPPER DEPOSITED BY ELECTROPLATING IN HIGH DENSITY TSV 55

    Wei Wang, Fei Geng, Peng Sun, Yulong Ren, Huan Liu, Kai Zhang

DESIGN AND OPTIMIZATION OF A PNEUMATIC DOD SOLDER BALL 3D PRINTING SYSTEM 58
Zhixian Min, Huiming Pan, Dinglei Zhao, Sheng Liu, Zhiqin Wang, Zhiwen Chen

STRAIN RATE AND TEMPERATURE EFFECTS ON TENSILE PROPERTIES OF MONOCRYSTALINE CU6SN5 BY MOLECULE DYNAMIC SIMULATION 63
Jian Zhang, Wei Huang, Kai-Lin Pan

EFFECT OF ELECTROMIGRATION ON INTERFACIAL REACTION IN NI/SN63PB37/CU BGA SOLDER JOINTS 67
Fei Zhang, Shuai Chen, Zhidan Liu, Wenlong Wang

OPTIMIZATION OF SOLDER HEIGHT FOR STENCIL PRINTING PROCESS PERFORMANCE ON LENGTH-WIDTH RATIO 71
Dezhi Su, Cen Wang, Hongkun Wang, Junxiang Zhao, Hao Cheng, Lejun Zhang

HIGH STRENGTH AND DENSITY CU-CU JOINTS FORMATION BY LOW TEMPERATURE AND PRESSURE SINTERING OF DIFFERENT MASS RATIO OF CU MICRON-NANOPARTICLES PASTE 75
Zhongwei Huang, Jian Wen, Yu Zhang, Qiang Liu, Huacong Li, Jin Tong, Peilin Liang, Guannan Yang, Chengqiang Cui

RELIABILITY ASSESSMENT IN WELDING PROCESS OF SIP WITH DUAL-CHAMBER BY FINITE ELEMENT ANALYSIS 80
Dezhi Su, Fuxin Wang, Lejun Zhang, Cen Wang, Huihui Yang, Wenyu Jiang

ON-DIE CLOCK TREE LOW PSIJ THROUGH PDN OPTIMIZATION 85
Vinod Arjun Huddar

THE SHAPE CONTROL PROCESS OF A CU/SNAG SOLDER JOINT WITH A NI INSERTION USING THERMO-COMPRESSION BONDING 88
Mingang Fang, Zhuo Chen, Fuliang Wang, Chu Tang, Wenhui Zhu

DESIGN AND FABRICATION OF MULTI-LAYER SILICONE MICROCHANNEL COOLER FOR HIGH-POWER CHIP ARRAY 94
Tao Wei, Haojie Huang, Yupa Ma, Jiyu Qian

LASER RAPID SYNTHESIS OF ULTRA-SMALL NI NANOPARTICLES EMBEDDED GRAPHENE FOR HIGH-PERFORMANCE SUPERCAPACITORS 99
Fangcheng Wang, Zhuo Zhang, Guangyao Zhao, Mingjie Liu, Hongjin Fan, Cheng Yang

SIGNAL INTEGRITY DESIGN AND ANALYSIS OF HIGH BANDWIDTH MEMORY ON SILICON INTERPOSER* 103
Jin Hu, Tao Li, Yuqing Fan

STUDY ON THE TRANSPORT PERFORMANCE OF MICROSTRIP CIRCUIT BOARD WITH VOIDS IN SOLDER LAYER 107
Zhidan Liu, Zhiping Zhao, Fei Zhang, Shuai Chen

112G HIGH SPEED INTERFACE PACKAGE DESIGN AND SIMULATION 111
Jiangtao Zhang, Li Zhang, Jian Pang, Tuobei Sun, Keqing Ouyang

MICRO-VISION IMAGE STITCHING SYSTEM FOR LARGE-SCALE AND FINE-FEATURED CIRCUIT SUBSTRATES 115
Yuanyang Wei, Jian Gao, Yongbin Zhong, Lanyu Zhang

STUDY IN MULTILAYER WIRING TECHNOLOGY ON HIGH-HEAT-CONDUCTION SUBSTRATES 121

Lei Ding, Jing Chen, Kai Liu, Yan Luo, Yue Zhao, Lichun Wang

METHOD OF PREDICTING THE MAXIMUM STRESS OF BGA SOLDER JOINTS BASED ON BP NEURAL NETWORK 126

Huaiquan Zhang, Chunyue Huang, Shuaidong Liao, Shoufu Liu

THERMAL STRESS STUDY OF 3D IC BASED ON TSV AND VERIFICATION OF THERMAL DISSIPATION OF STI 131

Shuaidong Liao, Chunyue Huang, Huaiquan Zhang, Shoufu Liu

A HUMIDITY-SENSITIVE CAPACITOR BASED ON FAN-OUT PANEL LEVEL PACKAGE TECHNOLOGY 136

Shuhan Hou, Tingyu Lin

MICROSTRUCTURES PROPERTIES OF BARIUM-STRONTIUM TITANATE (BST) CERAMICS DOPED WITH B-LI GLASSES FOR LTCC TECHNOLOGY APPLICATIONS 140

Linjiang Tang, Xiaofeng Sun, Minghua Zhang, Chengan Wan

SPECTRUM ANALYSIS AND APPLICATION OF XY PLATFORM SERVO SYSTEM OF THE HIGH-PRECISION PACKAGING EQUIPMENT 144

Shnjin Liu, Yunbo He, Zesheng Li, Qihao Oian, Huilong Liao

DESTRUCTIVE PHYSICAL ANALYSIS METHODS OF FLIP CHIP PACKAGING DEVICES FOR HIGH RELIABILITY 148

Zhou Shuai, Weng Zhangzhao, Qiu Baojun, Luo Daojun, Wang Xiaoqiang, Ma Kaixue

SIMULATION ANALYSIS OF COUPLING COIL OF 13.56MHZ MAGNETIC COUPLING RESONANT WIRELESS ENERGY TRANSMISSION SYSTEM 153

Bihong Zhan, Wei Xia, Chunshui Xiong, Sheng Liu

STRESS ANALYSIS OF CU/SN BUMP EUTECTIC BONDING INTERFACE 159

Xinpeng Chen, Ruixia Huo, Daowei Wu, Wansheng Liu

CHALLENGEABLE MECHANICAL ISSUES IN MICROELECTRONIC PACKAGES FOR DEVELOPMENTS 163

Xiangdong Xue, Jianghao Wei, Yuming Wang, Jianxin Yang

THE INTERFACIAL REACTION OF CU/RNULTILAYER SN -CU -SN / CU JOINT IN SOLDERING 169

Min Shang, Chong Dong, Xiangxu Chen, Shaocheng Wu, Haoran Ma, Haitao Ma

STUDY ON THE INFLUENCE OF DIFFERENT FILLER FRACTIONS ON THE PROPERTIES OF THERMAL INTERFACE MATERIALS 173

Wenbo Ye, Zhenyu Wang, Xiangliang Zeng, Linlin Ren, Rong Sun, Zhibin Wen, Xiaoliang Zeng

PACKAGING OF A MEMS SENSOR IN AN ACTIVE INTERVENTIONAL BLOOD PRESSURE MONITORING CATHETER 177

Guanzhe Xu, Junshu Lin, Rongying Yu, Zebo Zhang, Meng Gao, Le Ye

HEAT TRANSFER ANALYSIS OF PHASE CHANGE MATERIALS WITH METAL FOAMS 180

Yan Zhang, Huihui Wang, Pei Lu, Jingyu Fan, Qixuan Tu, Johan Liu

DELAMINATION REDUCTION BY MATERIAL AND PROCESS OPTIMIZATION 185

Tina Li, Aaron He, Phoebe Chen, Aaron Lai, Yuan Hang, Colin Feng

ELECTROMECHANICAL CO-DESIGN AND EXPERIMENTAL TESTING OF PACKAGE
LAYER IN STRUCTURALLY EMBEDDED PHASED ARRAY ANTENNA.. 189
Jinzhu Zhou, Zhenyu Gu, Yu Si, Mei Wang, Ping'An Wang

THE FORMATION OF CN-SN IMC INTERCONNECTION BY SOLID-LIQUID
INTERDIFFUSION BONDING FOR 3D GLASS WAFER STACKING 193
Yangquan Su, Kuili Ren, Yiyong Huang, Mingchuan Zhang, Daquan Yu

RESEARCH ON THE RELIABILITY OF BOARD LEVEL INTERCONNECT SOLDER JOINTS
UNDER THERMAL-MECHANICAL COUPLING 197
Xin Liu, Yongjian Yu, Weikun Xie, Kaihong Zhang, Kai Zhu

APPROACH TOWARDS ACCURATE MODELING OF THERMAL RESISTANCE IN
THERMAL MANAGEMENT OF PCB 202
Jianghao Wei, Tao Wan, Xiangdong Xue, Yuming Wang

STUDY ON WARPAGE AND PEELING MITIGATION OF WAFER LEVEL DURING METAL
PLATING PROCESS 208
Dayang Li, Ming Xiao, Zhiqi Wang

THE STUDY OF EFFECTS TO THE THERMO-MECHANICAL PERFORMANCE OF THE
FIRST LEVEL THERMAL INTERFACE MATERIALS 213
Zhenyu Wang, Linlin Ren, Xiangliang Zeng, Wenbo Ye, Yonglun Xu, Xiaoliang Zeng, Yunsong
Pang, Sun Rong

FACILE PREPARATION OF COBALT HYDROXIDE BASED SUPERCAPACITOR WITH
HIGH VOLUMETRIC ENERGY DENSITY AT HIGH VOLUMETRIC POWER DENSITY 219
Peng Liu, Fangcheng Wang, Zhuo Zhang, Jing Li, Hongjin Fan, Cheng Yang

DIE CHIPPING FDC DEVELOPMENT AT WAFER SAW PROCESS 224
Dongpeng Xue, Caiden Zhong, Elley Zhang, Weiting Jiang, Cong Zhang

HYBRID-EMBEDDED SIP PACKAGE DESIGN 226
Louise Tan, Chender Chen, Cc Liao

A NOVEL CU@SN@AG CORE-SHELL PARTICLES FOR DIE ATTACHMENT IN POWER
DEVICE PACKAGING 229
Jiahao Liu, Hui Xiao, Xiaotong Guo, Xinjie Wang, Zhijun Yao, Xingchao Mao, Hao Liu,
Hongtao Chen

EFFECT OF DIMENSION OF BOARD AND MICRO-BUMPS ON INTERCONNECTION
STRESS UNDER DROP TEST 233
Mingtao Lv, Shimei Liu, Taotao Chen, Yunpeng Liu, Junfu Liu, Hu He

THERMOMECHANICAL AND ELECTRICAL PROPERTIES OF THE $SIO_2/ZRW_2O_8/EPOXY$
COMPOSITE 239
Chaofan Li, Suibin Luo, Shuhui Yu, Baojin Chu, Rong Sun

HIGH EFFICIENCY TESTING SYSTEM FOR 5G POWER AMPLIFIER 244
Zongqi Cai, Sha Tang, Jun Luo, Xing Li, Xiaoqiang Wang, Daojun Luo

RESEARCH ON HIGH-SPEED SERDES INTERFACE TESTING TECHNOLOGY 248
Weikun Xie, Guangqiang Cao, Weiwei Ji

HIGH-PERFORMANCE THERMAL GREASE WITH THE ADDITION OF SILVER PARTICLES 253
 Xiangliang Zeng, Zhenyu Wang, Wenbo Ye, Linlin Ren, Xiaoliang Zeng, Xinnian Xia, Rong Sun

SOP WELDING JOINT BENDING STRESS FINITE ELEMENT ANALYSIS AND OPTIMIZATION 257
 Gong Jinfeng, Huang Chunyue, Li Maolin, Liu Shoufu

COMPARATIVE ANALYSIS OF TEMPERATURE-INDUCED MICRO-SCALE DEFORMATION OF PACKAGE BY EXPERIMENT AND FINITE ELEMENT ANALYSIS 261
 Cheng Zhong, Chenglong Li, Tao Peng, Yunxia Wang, Gang Li, Pengli Zhu, Jibao Lu, Rong Sun, Ching-Ping Wong

ACTIVE HEAT DISSIPATION BY CHIP ON THERMOELECTRIC COOLER FOR HIGH-POWER LED 265
 Shuang Li, Jinglong Liu, Yang Peng, Mingxiang Chen

TUNING THE CURING TEMPERATURE OF POLYIMIDE PRECURSOR: PLOY AMIDE ESTER 270
 Kuangyu Wang, Liang Shan, Guoping Zhang, Rong Sun

VISCOELASTIC CHARACTERIZATION AND SIMULATION OF THERMAL INTERFACE MATERIALS 274
 Cheng Zhong, Chenglong Li, Yunxia Wang, Jibao Lu, Linlin Ren, Rong Sun, Ching-Ping Wong

CHARACTERIZATION AND VERIFICATION OF VISCOELASTIC CONSTITUTIVE PARAMETERS OF UNDERFILL MATERIAL 278
 Cheng Zhong, Chenglong Li, Lu Lu, Yunxia Wang, Gang Li, Pengli Zhu, Jibao Lu, Rong Sun, Ching-Ping Wong

SYNTHESIS AND PROPERTIES STUDY OF A THERMOPLASTIC POLYIMIDE WITH HIGH GLASS TRANSITION TEMPERATURE FOR WAFER LEVEL PACKAGE 283
 Wen Liu, Jinhui Li, Jinshan Liu, Tao Wang, Ao Zhong, Guoping Zhang, Qiang Liu, Rong Sun, Daquan Yu

THE EFFECT OF THERMAL-INDUCED WARPAGE AND DEGENERATION OF THERMAL INTERFACE MATERIALS ON THE THERMAL PERFORMANCE OF A FLIP-CHIP PACKAGE 287
 Ruoyu Jiang, Cheng Zhonz, Haozhe Wang, Chenglong Li, Yi Zheng, Linlin Ren, Jibao Lu, Rong Sun, Ching-Ping Wong

ENHANCED DISCHARGED ENERGY DENSITY IN POLYETHERIMIDE COMPOSITES BY BORON NITRIDE/ALUMINUM NITRIDE HYBRID FILLERS 292
 Xudong Wu, Shiyi Gao, Xin Wu, Shuo Zhang, Zhijun Cao, Daniel Q. Tan

THE STUDY ON THERMAL AGING MECHANISM OF SILICONE MATERIALS FOR LED ENCAPSULATION 295
 Jiabao Gu, Huanxiang Xu, Bo Peng, Zilian Liu, Gang Zhu

DEFECT LOCALIZATION AND OPTIMIZATION OF PIND FOR LARGE SIZE CQFP DEVICES 298
 Shinan Wang, Yong Ma, Kaihong Zhang, Yongjian Yu, Yongkang Wan, Weikun Xie

EXPLORATION OF THE SYNTHESIS METHOD OF QUATERNARY COPOLYMERIZED THERMOPLASTIC POLYIMIDE 301

Jinshan Liu, Jinhui Li, Fangfang Niu, Tao Wang, Wen Liu, Guopinz Zhang, Rong Sun

IMPACT FORCE CONTROL OF HIGH-SPEED WIRE BONDING MACHINE BASED ON FUZZY ACTIVE DISTURBANCE REJECTION CONTROLLER 305

Yachao Liu, Jian Gao, Boyu Zhan, Lanyu Zhang

FAILURE ANALYSIS OF ANISOTROPIC CONDUCTIVE ADHESIVE PACKAGES IN NARROW-PITCH FLIP CHIP PACKAGING 311

Gui Chen, Yan Wang, Xiaoyu Xiao, Yamei Yan, Wenhui Zhu

FAILURE MECHANISM OF NICKEL-CHROMIUM THIN FILM CHIP RESISTORS 316

Zhiyuan Mao, Gaoming Shi, Weili Li, Xianjun Kuang, Fuyao Mo

FUZZY TUNING ALGORITHM FOR FEEDFORWARD PARAMETER BASED ON IC PACKAGE FOR MASS TRANSFER OF MICRO-LED EQUIPMENT XY MOTION PLATFORM 320

Wenbin Fan, Yunbo He, Guofu Qiu

NUMERICAL ANALYSIS OF THE MICROSCOPIC FACTORS INFLUENCING THE THERMAL CONDUCTIVITY OF AL_2O_3/AlN POLYMER COMPOSITES 326

Nan Cheng, Xiaoxin Lu, Jiabin Huang, Jibao Lu, Shen Xu, Sun Rong, Jianbin Xu, Ching-Ping Wong

THE INFLUENCE ANALYSIS OF GEOMETRY ON VOID IN MOLDED UNDERFILL FOR FLIP CHIP 331

Yamei Yan, Gui Chen, Xiaoyu Xiao, Yan Wang, Wenhui Zhu

A COST-SAVING THERMAL TEST CHIP DESIGN IN A TEST VEHICLE OF LARGE BGA 336

Jianjun Sun, Yuanting Lai, Hao Yang, Jian Pang, Tuobei Sun, Keqing Ouyang

THERMODYNAMIC SIMULATION AND ANALYSIS OF METAL BUMPS IN FLIP-CHIP MICRO-LED PACKAGING 340

Xiaoyu Xiao, Yamei Yan, Gui Chen, Wenhui Zhu

STRESS-STRAIN STUDY OF QFN SOLDER JOINTS WITH DIFFERENT STRUCTURAL PARAMETERS UNDER RANDOM VIBRATION LOADING 346

Maolin Li, Chun-Yue Huang, Zhuo Wang, Wei Wei

COMPARISON BETWEEN TWO NUMERICAL METHODS FOR THE COMPUTATION OF THERMAL CONDUCTIVITIES OF PARTICULATE COMPOSITES: FEM AND GEODICT 351

Xiaoxin Lu, Jiabin Huang, Jianbin Xu, Jibao Lu, Sun Rong, Ching-Ping Wong

RESEARCH ON POINT-TO-POINT MOTION CONTROL OF PACKAGING EQUIPMENT 356

Zesheng Li, Yunbo He, Shujin Liu, Qihao Qian

NUMERICAL ANALYSIS ON THE EFFECT OF MICROSTRUCTURES ON THE THERMAL AND MECHANICAL PROPERTIES OF CARBON FIBER/AL_2O_3 THERMAL PAD 360

Shu Liu, Xiaoxin Lu, Jiabin Huang, Jibao Lu, Shen Xu, Sun Rong, Jianbin Xu, Ching-Ping Wong

QUALITY INSPECTION OF OPTICAL LENS IN IC PACKAGING EQUIPMENT BASED ON MTF 365

Qihao Qian, Yunbo He, Zesheng Li, Shujin Liu, Huilong Liao

STRESS ANALYSIS AND PARAMETER OPTIMIZATION OF FINE-PITCH BGA SOLDER JOINTS UNDER CANTILEVER PLATE TORSION CONDITIONS 371

Zhuo Wang, Chunyue, Jinfeng Gong, Huaiquan Zhang, Shuaidong Liao, Shoufu Liu

THE INFLUENCE AND OPTIMIZATION OF DESIGN PARAMETERS ON INTEGRATED CIRCUITS PACKAGE WARPAGE 376

Qiang Wei, Cao Ting, Weidong Liu, Ning Sun, Qu Fang, Jie Liu, Huan Yang, Xiaojian Ma

THE IN-SITU OBSERVATION OF MICROSTRUCTURE, GRAIN ORIENTATION EVOLUTION AND ITS EFFECT ON CRACK PROPAGATION PATH IN SAC305 UNDER EXTREME TEMPERATURE CHANGES 381

Kexin Xu, Xing Fu, Min Liu, Zhiwei Fu, Si Chen, Yijun Shi, Yun Huang, Hongtao Chen

A VERTICAL TRANSMISSION LEADLESS SURFACE-MOUNTABLE CERAMIC PACKAGE WITH HIGH CORE PROPORTION 385

Zhizhuang Qiao, Linjie Liu, Yangfan Zhou, Ke Wang

RELIABILITY AND THERMAL DEGRADATION OF FIRST-LEVEL THERMAL INTERFACE MATERIALS 388

Yunpeng Su, Junhong Li, Qiangquiang Ma, Linlin Ren, Xiaoliang Zeng, Rong Sun

COUPLING DAMAGE ACCUMULATION OF DIE-ATTACH SOLDER LAYER WITH DISTRIBUTED VOID DEFECTS FOR POWER ELECTRONICS 393

Yidian Shi, Cheng Peng, Wenhui Zhu, Taotao Chen, Yunpeng Liu, Junfu Liu, Hu He

RESEARCH ON WIRE SWEEP OF INTEGRATED CIRCUIT PACKAGING BASED ON THREE-DIMENSIONAL FLOW SIMULATION 398

Fang Qu, Ting Cao, Huan Yang, Ning Sun, Qiang Wei, Xiaojian Ma, Weidong Liu

CHARACTERISTICS OF 10–110GHZ TRANSMISSION LINES ON FUSED SILICA SUBSTRATE FOR MILLIMETER-WAVE MODULES 403

Tian Yu, Kai Xue, Ke Li, Yihang Liang, Daquan Yu

LOW TEMPERATURE BONDING BY SINTERING OF AG NANOPARTICLE PASTE WITH THE ASSISTANCE OF MOD 408

Xun Liu, Yulei Yuan, Junjie Li, Li Liu, Rong Sun

CU-CU JOINT FORMATION BY SINTERING OF SELF-REDUCIBLE CU NANOPARTICLE PASTE ASSISTED BY MOD UNDER AIR CONDITION 412

Yulei Yuan, Xun Liu, Junjie Li, Pengli Zhu, Rong Sun

RESEARCH ON THE BOARD LEVEL RELIABILITY OF CQFJ CERAMIC PACKAGE 416

Zhen-Tao Yang, Fei Yu, Lin-Jie Liu, Ling Gao

THERMAL AND OPTICAL MODELING ON INTELLIGENT LED HEADLIGHTS 420

Yikang Qin, Miao Cai, Xindong Chen, Jinyang Li, Daoguo Yang, Guoqi Zhang

RESEARCH ON THE DESIGN AND PROCESSING TECHNOLOGY OF CQFJ CERAMIC PACKAGE 424

Fei Yu, Zhen-Tao Yang, Lin-Jie Liu, Ling Gao

RESEARCH ON DOUBLE-LAYER NETWORKS-ON-CHIP FOR INTER-CHIPLET DATA SWITCHING ON ACTIVE INTERPOSERS 428

Xiaolong Duan, Min Miao, Zhuanzhuan Zhang, Liang Sun

INVESTIGATION OF THE INFLUENCES OF THERMAL STRESSES AND JOULE HEATING WITHIN A PIEZORESISTIVE MEMS PRESSURE SENSOR USING THE FINITE ELEMENT MODELING 434

Chunming Zhou, Peng Zhou, Yuehua Hu, Hao Zhang

ACCELERATED AGING AND LIFETIME EVALUATION OF POLYURETHANE PACKAGING MATERIAL FOR OPTICAL FIBER HYDROPHONE 440

Wenyuan Liao, Canxiong Lai, Rui Gao, Shaohua Yang, Guoguang Lu, Shuwang Li

CHARACTERIZING THE DIE ATTACH LAYER DELAMINATION EFFECT ON THE HEAT TRANSFERRING PERFORMANCE IN LED PACKAGE WITH ENTROPY GENERATION ANALYSIS 443

Binjie Ai, Miao Cai, Daoguo Yang, Guangsheng Lu, Kailin Zhang, Guoqi Zhang

SHEAR PROPERTIES AND FRACTURE BEHAVIORS OF CU/SN-37PB/CU SOLDER INTERCONNECTIONS AT CRYOGENIC TEMPERATURES 447

Ruyu Tian, Chunlei Wang, Yanhong Tian

RESEARCH ON 3D INTERPOSER/CHIP STACKING TECHNOLOGY AND RELIABILITY 450

Ning Zhang, Baoxia Li, Qiucheng Yan, Daowei Wu

HIGHLY CONDUCTIVE SILVER NANOWIRE TRANSPARENT ELECTRODES HYBRIDIZED WITH LAMINATED MULTI-LAYER MXENE 454

Pengchang Wang, Maoliang Jian, Chi Zhang, Majiaqi Wu, Huaying Hu, Lianqiao Yang

SYNTHESIS OF AIR-SINTERABLE COPPER NANOPARTICLES FOR DIE-ATTACHMENT 458

Yue Yao, Liang Xu, Pengli Zhu, Tao Zhao, Rong Sun, Yinachao Huo

THE PARTICLE INTERACTION ANALYSIS FOR NANOPARTICLES IN UNDERFILL FOR FLIP-CHIP PACKAGING- 462

Mingyong Du, Ning Wang, Xiaomeng Du, Tao Zhao, Pengli Zhu, Rong Sun

A 3D TSV-MEMS BASED HETEROGENEOUS INTEGRATION TECHNOLOGY FOR RF APPLICATION 466

Min Huang, Tinglei Wang, Fan Hou, Ping Su, Chao Sun, Huakai Luan

THE EFFECT OF ANNEALING TIME ON THE MECHANICAL PROPERTIES OF TSV-CU 470

Yadong Li, Pei Chen, Fei Qin

STUDY ON CURRENT CARRYING CAPACITY OF A NOVEL INTERCONNECT MATERIAL ZRTE$_3$ 475

Xiaokun Wen, Liangyi Ni, Wenyu Lei, Li Yang, Yuan Liu, Pengzhen Zhang, Haixin Chang, Wenfeng Zhang

INTERACTION OF SILANE COUPLING AGENTS WITH NANO-SILICA PROBED BY NANO-IR 479

Pengli Zhu, Jianjun Ruan, Mingyong Du, Ning Wang, Xiaomeng Du, Tao Zhao, Xiaodong Li, Jiakai Cao, Jianping Zhang, Xiaoyao Sun

LIGHTWEIGHT AND COMPRESSIBLE EXPANDABLE POLYMER MICROSPHERES/SILVER FLAKES COMPOSITES FOR HIGH-EFFICIENCY ELECTROMAGNETIC INTERFERENCE SHIELDING 483

Jianhong Wei, Yadong Xu, Zhiqiang Lin, Zuomin Lei, Xianzhu Fu, Yougen Hu, Rong Sun

STUDY OF EFFICIENT AUTOMATIC DETECTION OF COPLANARITY AND POSITION OF CCGA DEVICES 489

Qi Zhang, Xiaoyan Liu, Leida Chen, Xujing Nan

MODELING AND SIMULATION OF INTERCONNECTION STRUCTURE COMPENSATION DESIGN IN HIGH-SPEED MODULES... 493
Qi Zheng, Huimin He, Lijuan Bai, Yubo Wang, Dejian Li, Shunfeng Han, Bofu Li, Dameng Li, Fengman Liu, Liqiang Cao

STUDY ON GOLD WIRE SWEEP IN CANTILEVER CHIP-STACKED PACKAGE DURING MOLDING PROCESS.. 497
Sicheng Cao, Daoguo Yang, Wangyun Li, Xiyou Wang, Shirui Xue, Zhanfei Yun

X-SHAPED THROUGH GLASS VIA AND ITS TRANSMISSION PERFORMANCE IN KA BAND... 501
Qiangwen Wang, Yuhua Guo, Rui Wang

PROCESS DEVELOPMENT AND FAILURE ANALYSIS OF SUPER-SIZE EMBEDDED SILICON FAN-OUT PACKAGE 506
Dongzhi Fu, Shuying Ma, Jiao Wang

SURFACE MODIFICATION OF GRAPHITE AND ITS EFFECT ON THERMAL AND MECHANICAL PROPERTIES OF GRAPHITE-BASED THERMAL INTERFACE MATERIALS 512
Yuexing Zhang, Hong He, Junwei Li, Chenxu Zhang, Rong Sun, Meng Han, Ping Zhang

DESIGN, FABRICATION, AND TEST OF AN EMBEDDED SI-GLASS MICROCHANNEL HEAT SINK FOR HIGH-POWER RF APPLICATION 517
Jianyu Du, Weihao Li, Xu Gao, Deyin Zheng, Yuchi Yang, Zetian Wang, Haoran Zhao, Jiajie Kang, Wei Wang

A HIGH-Q INDUCTOR BASED ON FAN-OUT PANEL LEVEL PACKAGE TECHNOLOGY 521
Xulei Niu, Zixu Wang, Shaopan Lin, Guochi Huang, Tingyu Lin

EFFECTS OF CETYLTRIMETHYLAMMONIUM BROMIDE (CTAB) ON ELECTROPLATING TWIN-STRUCTURED COPPER INTERCONNECTS................................. 525
Yi Dong, Zhe Li, Li-Ying Gao, Xiao Li, Zhi-Quan Liu, Rong Sun

ANALYSIS FOR THERMAL CONTACT RESISTANCE OF PRESS-PACK IGBTS 530
Yakun Zhang, Tong An, Fei Qin, Yanpeng Gong, Chen Liang

ANALYSIS ON THE THERMAL STRESS OF AL-SI THIN FILM USING DIC METHOD 535
Huiming Pan, Guoliang Xu, Zhiwen Chen, Chongming Zhang, Chao Sun, Sheng Liu, Li Liu

FLEXIBLE THERMAL INTERFACE MATERIALS THROUGH CONTROLLING THE RATIO OF SILICONE OIL FUNCTIONAL GROUPS 539
Wendian Tu, Linlin Ren, Guoping Du, Rong Sun, Xiaoliang Zeng, Junwei Li

SIMULATION AND OPTIMIZATION OF 3D HETEROGENEOUS INTEGRATION OF INERTIAL MICRO-SYSTEM... 543
Nanxin Wang, Shenglin Ma, Yufeng Jin, Chaoyang Xing, Nannan Li, Peng Sun

STUDY ON IMAGE ALIGNMENT TECHNOLOGY BASED ON CCD THERMAL REFLECTION METHOD.. 549
He Yang, Weikang Si, Dazheng Wang, Yingying Gao, Libing Zheng

A THERMAL NETWORK MODEL FOR THERMAL ANALYSIS IN AUTOMOTIVE IGBT MODULES... 555
Yanzhong Tian, Tong An, Fei Qin, Yanpeng Gong, Chen Liang

NOVEL WATER-SOLUBLE PROTECTIVE ADHESIVE FOR WAFER'S LASER DICING 560
Deliang Sun, Jinhui Li, Yuxi Yi, Guoping Zhang, Rong Sun, Mingqi Huang

A TGV-BASED ANTENNA IN PACKAGE FOR 5G MM-WAVE APPLICATION..................................... 564
Sha Xu, Dianyang Shi, Chunbing Guo

LOADING RATE ON MODE II FRACTURE TOUGHNESS OF SINTERED SILVER............................. 568
Yanning Li, Yanwei Dai, Fei Qin, Shuai Zhao

COMPARATIVE RESEARCH OF INFRARED THERMOGRAPHY AND ELECTRICAL
MEASUREMENT METHOD FOR THE THERMAL CHARACTERISTICS TEST OF GAN
HEMT DEVICES.. 572
Zhiwei Fu, Bingjie Zheng, Xu Huang, Bin Zhou, Xiaofeng Yang, Huaixin Guo

RESEARCH PROGRESS OF EXTREME LOW TEMPERATURE RELIABILITY OF TYPICAL
ELECTRONIC INTERCONNECTION STRUCTURES 576
Zhaoning Sun, Xiaotong Guo, Zhenbo Zhao, Yiqing Ni, Guanghui He

PROPERTIES OF ROOM TEMPERATURE BONDED AND UV CURED TEMPORARY
BONDING ADHESIVE FOR ULTRA-THIN WAFER'S HANDLING 581
Xujun Li, Qiang Liu, Deliang Sun, Zhipeng Li, Guoping Zhang, Rong Sun

EFFECT OF THERMOMIGRATION ON EVOLUTION OF INTERFACIAL INTERMETALLIC
COMPOUNDS IN CU/NI/SN-3.5AG MICROSOLDER JOINTS FOR 3D INTERCONNECTION 585
Chu Tang, Wenhui Zhu, Zhuo Chen

IMPROVEMENT OF AU ELECTRODE BY GLASS OPTIMIZATION FOR LTCC
APPLICATION.. 591
Tingnan Yan, Dawei Wang, Yuanyuan Wang, Jinhao Jia

STUDY ON THE HEAT DISSIPATION PERFORMANCE OF SYMMETRICAL BROKEN-
LINE MICROCHANNEL RADIATOR 594
Pengfei Wang, Hongyue Wang, Yijun Shi, Xiangjun Lu, Jile Xu

FCCSP(MUF) MOLD-FLOW VOID RISK PREDICTION WITH DIFFERENT SUBSTRATE
SURFACE AND BUMP HEIGHT DESIGN.. 598
Freedman Yen, Nicholas Kao, David Lai, Yu Po Wang

RESEARCH ON ELECTROMIGRATION BEHAVIOR OF CU PILLAR BUMPS UNDER
PULSE CURRENT STRESS 603
Jile Xu, Xiangjun Lu, Zhiwei Fu, Chenbing Qu, Xiao Luo, Pengfei Wang

SIMULATION ON TSV PROTRUSION FROM ATOMIC TO MICRON SCALES 607
Xiaoting Luo, Zhiheng Huang, Yuezhong Meng, Shan Ren, Hui Yan, Qizhuo Li

BGA CHIP TORSION FINITE META ANALYSIS AT HIGH TEMPERATURE....................... 613
Yu-Qing Yue, Chao Jiang, Yan-Ting Chen, Jing Wei, Lv Wu

THERMAL ANALYSIS OF HIGH-POWER LIGHT-EMITTING DIODE USING
THERMOREFLECTANCE THERMOGRAPHY 617
Dazheng Wang, Libing Zheng, Weikang Si, He Yang, Yingying Gao

LOW-TEMPERATURE CU/SIO$_2$ HYBRID BONDING USING A NOVEL TWO-STEP
COOPERATIVE SURFACE ACTIVATION 621
Qiushi Kang, Chenxi Wang, Ge Li, Shicheng Zhou, Yanhong Tian

KEY FACTOR ANALYSIS OF NANO SILICA ON THE DISPERSION IN UNDERFILL 626
Xiaomeng Du, Ning Wang, Mingyong Du, Leicong Zhang, Tao Zhao, Pengli Zhu, Rong Sun,
Jiakai Cao, Jianjun Ruan

EFFECT OF FILLER, TOUGHENING AGENT AND COUPLING AGENT ON THE CURING SHRINKAGE OF EPOXY-BASED UNDERFILLS 630

Xiaohui Peng, Jinbao Yang, Tao Peng, Pengli Zhu, Xing Ouyang, Yan Pan, Gang Li, Rong Sun, Yajing Yang, Liang Peng

SOLID-LIQUID MIXING-STATE ORGANIC LENSES FOR DEEP-ULTRAVIOLET LIGHT-EMITTING DIODES TO ENHANCE THE LIGHT-EXTRACTION EFFICIENCY 634

Zihao Deng, Jiexin Li, Jiayong Liang, Jiayi Li, Jiasheng Li, Xinrui Ding, Zongtao Li

NUMERICAL SIMULATION ANALYSIS OF FLEXIBLE PRINTED CIRCUITS UNDER BENDING CONDITIONS 638

Yongchao Liu, Haozhe Wang, Xianqin Hu, Chao Peng, Xu Long, Jibao Lu, Rong Sun, Ching-Ping Wong

TENSILE DEFORMATION MECHANISM OF SN-37PB SOLDER ALLOY AT CRYOGENIC TEMPERATURES 642

Xiaotong Guo, Kun Zhang, Jiahao Liu, Yong Li, Xinlang Zuo, Hui Xiao, Guanghui He

THE RELIABILITY ASSESSMENT OF PULSE-DRIVEN LIGHT EMITTING DIODES 646

Shen Yaoyang, Sun Bo

SEQUENTIAL ANALYSIS OF DROP IMPACT AND THERMAL CYCLING OF ELECTRONIC PACKAGING STRUCTURES 651

Yongchao Liu, Xu Long, Haozhe Wang, Jibao Lu, Rong Sun, Ching-Ping Wong

ENHANCED OPTICAL PERFORMANCE AND THERMAL STABILITY OF QUANTUM DOT CONVERTERS FOR LASER SOURCE 655

Jiayong Liang, Jiexin Li, Zihao Deng, Yihua Qiu, Zongtao Li, Jiasheng Li

MICRON-SCALE SILVER FLAKE PASTE SINTERING WITHOUT PRESSURE FOR POWER ELECTRONIC DIE ATTACHMENT 659

Shijun Huang, Ruidong Luo, Zhen Wu, Cai-Fu Li

WARPAGE MEASUREMENT OF SUBSTRATES AND PRINTED CIRCUIT BOARDS WITH SHADOW MOIRÉ 663

Xingjia Huang, Changping Ou, Shengcong Zhu, Yixiu Huang

RELIABILITY ANALYSIS OF THERMAL INTERFACE MATERIALS (TIMS) IN LARGE SIZE FCBGA PACKAGE 667

Yi Zheng, Cheng Zhong, Haozhe Wang, Jibao Lu, Rong Sun, Ching-Ping Wong

STUDY ON WARPAGE AFTER POST SOLIDIFYING OF ULTRATHIN FINGERPRINT PACKAGE PRODUCTS 671

Ning Sun, Xiaojian Ma, Fang Qu, Qiang Wei, Huan Yang, Weidong Liu, Ting Cao

INFLUENCE OF IMC MORPHOLOGY ON FATIGUE STRESS, STRAIN AND LIFE OF SOLDER LAYER BETWEEN SIC CHIP AND DBC SUBSTRATE IN IGBT UNDER THERMAL CYCLING 674

Guang Yang, Fengshun Wu, Longzao Zhou, Xinghe Luan, Xinrui Zou, Hui Liu, Yang Wan, Xiaowei Zhang, Bin Wang

RESEARCH ON RELIABILITY OF SOLDER LAYER IN IGBT MODULE PACKAGING 679

Panwang Chi, Shengru Lin, Yesu Li, Yiping Liu, Jicun Lu, Ming Li, Liming Gao

THE STUDY ON ELECTROMIGRATION OF SOLDER JOINTS UNDER THERMAL CYCLING LOAD 684

Leyi Niu, Xiaodi Tian, Fei Jia

EVALUATION OF FATIGUE CRACK GROWTH IN SOLDER LAYER OF IGBT MODULE
UNDER POWER CYCLE BY USING J-INTEGRAL METHOD 689
Kai Yang, Longzao Zhou, Fengshun Wu, Yi Zhang, Yang Han, Zhou Zhang, Yang Wan,
Xiangmiao Huang, Dayong Huang

STUDY ON LEADFRAME OVERFLOW PREVENTION OF SOLDERING PASTE USING
FLUID-STRUCTURE COUPLING ANALYSIS 695
Guangsheng Lu, Daoguo Yang, Wangyun Li, Xiyou Wang, Xiangli Wei, Binjie Ai

SIMULATION STUDY ON THERMOMECHANICAL RELIABILITY IN EMBEDDED DIE
PACKAGE FABRICATION PROCESS 699
Zhou Zhou, Haibo Fan, Yuning Shi

WETTABILITY IMPROVEMENT OF SOLDER IN FLUXLESS SOLDERING UNDER
FORMIC ACID ATMOSPHERE 705
Yuhao Bi, Siliang He, Wangyun Li, Daoguo Yang, Hiroshi Nishikawa

THE EFFECT OF FLUX ON SI-AL WIRE BONDING RELIABILITY 709
Yao Zhang, Wuxing Cao, Jun Zhang, Liu Yang, Pei Zhang, Jiao Yang

FINITE ELEMENT ANALYSIS OF THERMAL CONTACT RESISTANCE IN PRESS-PACK
IGBT MODULE 714
Tong An, Rui Zhou, Fei Qin, Yanpeng Gong, Chen Liang

ORTHOGONAL EXPERIMENT FOR ANALYZING THE IMPACT OF THERMAL STRESS
ON THE RELIABILITY OF AN EMC PACKAGE.. 718
Yulong Li, Yi Zheng, Haozhe Wang, Jibao Lu, Rong Sun, Wenhui Zhu, Ching-Ping Wong

OPTICAL PERFORMANCE ANALYSIS OF UVC-LED PACKAGE STRUCTURE BASED ON
RAY-TRACING SIMULATION 722
Jinyang Li, Wangyun Li, Daoguo Yang, Yikang Qin, Sicheng Cao, Feixiang Liu

DETERMINATION OF PARAMETERS IN MIXED-MODE COHESIVE ZONE MODELS FOR
MODIFIED BUTTON SHEAR TESTS BY PARTICLE SWARM OPTIMIZATION 726
Wenyu Wu, Ke Xue, Weijing Dai, Dashun Liu, Dali Yang

DESIGN AND SIMULATION OF 3D ANTENNA BASED ON CONICAL VIA STRUCTURE 732
Ziyu Liu, Junhao Wang, Zhiyuan Zhu, Lin Chen, Qingqing Sun

COMPARATIVE STUDY ON THE EFFECTS OF FE AND NI ADDITIONS ON THE
ELECTROMIGRATION PROPERTIES OF SN58BI SOLDER JOINTS 736
Zhuangzhuang Hou, Yaru Dong, Lingyao Sun, Ying Liu, Yongiun Huo, Yingxia Liu, Xiuchen
Zhao

LOW TEMPERATURE AND SHORT TIME AU/SN SOLID-LIQUID DIFFUSION BONDING
FOR 3D INTEGRATION 740
Ziyu Liu, Wang Wenchao, Zhu Zhiyuan, Chen Lin, Sun Qingqing

COPPER ADHESION PROMOTERS FOR POLYIMIDE: HETEROCYCLIC COMPOUNDS
ADDITIVES CONTAINING AMINO AND HYDROXYL GROUPS 746
Yingying Li, Guoping Zhang, Jinhui Li, Changqing Li, Ao Zhong, Deliang Sun

UNDERFILL FILLER SETTLING EFFECT ON THE ADHESIVE FORCE OF FLIP CHIP
PACKAGES 750
Guolin Zhao, Houya Wu, Yuanyuan Yang, Gang Li, Pengli Zhu, Rong Sun, Wenhui Zhu

THE INFLUENCE OF DIFFERENT PHOSPHOR COATING METHODS ON THE
TEMPERATURE OF LED 755
    Kun Chen, Deming Hu, Liang Yang

AN INFRARED LASER TEMPORARY BONDING MATERIAL USED FOR DEVICE WAFER
THINNING AND COMPLETION OF BACKSIDE PROCESSING TECHNOLOGY 759
    Zhenwen Ye, Deliang Sun, Mingqi Huang, Guoping Zhang, Jianwen Xia

THE STUDYS OF ADHESION AND CONTACT THERMAL RESISTANCE OF TIM1 763
    Yunsong Pang, Meng Han, Ting Liang, Xue Bai, Liang Li, Yonglun Xu, Bin He, Daifeng Ai,
    Liuxin Wang, Linlin Ren, Xiaoliang Zeng, Rong Sun

DESIGN AND VERIFICATION OF TDDB TEST STRUCTURES FOR TSV 767
    Kai Li, Si Chen, Xiaofeng Yang, Guoyuan Li, Bin Zhou

REVIEW ON ERROR COMPENSATION AND COOPERATIVE CONTROL OF MULTI-AXIS
MOTION PLATFORM 771
    Zhiwei Zhou, Jian Gao, Lanyu Zhang, Jindi Zhang, Yuheng Luo

THE EFFECT OF DUAL ULTRASONIC-ASSISTED SOLDERING PROCESS ON THE
PROPERTIES OF CU/40%ZN+60%SAC0307 POWDER/AL JOINT 776
    Zhaoqi Jiang, Guisheng Gan, Qianzhu Xu, Shiqi Chen, Peng Ma, Yufeng Tang, Xiangtao Xu

EFFECTS OF SURFACE OXIDATION TREATMENTS ON THE INTERFACIAL ADHESION
BETWEEN COPPER AND UNDERFILL 781
    Bin Wang, Haoliang Lin, Houya Wu, Yuanyuan Yang, Gang Li, Pengli Zhu

FEASIBILITY ANALYSIS OF CRACK INITIATION IDENTIFICATION OF SINTERED
SILVER FOR A FAST LIFETIME PREDICTION 786
    Jiuyang Tang, Jing Zhang, Guoqi Zhang, Pan Liu

INVESTIGATION ON THE WARPAGE OF FAN-OUT WAFER-LEVEL PACKAGING USING
CURING REACTION KINETICS OF COMPOSITES 791
    Wei Li, Jianhong Huang, Jinbo Xiao, Yiyong Huang, Houdun Zhang, Daquan Yu

INORGANIC STABILIZER INTRODUCED FLUX SYSTEM WITH HIGH TACKINESS: AN
EFFICIENT AND NOVEL MATERIAL SOLUTION FOR THE MICRO LED MASS
TRANSFER 796
    Liangzheng Ji, Jing Zhang

STUDIES ON THE EFFECTS OF SOLDERING LAYER STRUCTURES ON TEC MODULE
PERFORMANCE AND THERMAL STRESS 800
    Dianru Yu, Zhihao Yin, Guanghui Liu, Hongyan Xu, Ju Xu

A FLEXIBLE ULTRASONIC SENSOR BASED ON PIEZOELECTRIC MICROMACHINED
ULTRASONIC TRANSDUCERS (PMUTS) 805
    Zhihao Tong, Zhipeng Wu, Songsong Zhang, Huicong Liu, Liang Lou

EFFECT OF TEMPERATURE ON THE FATIGUE DAMAGE OF SAC305 SOLDER 809
    Xu Long, Ying Guo, Xiaotong Chang, Yutai Su, Hongbin Shi, Tao Huang, Bingyi Tu, Yanpei
    Wu

ENHANCEMENT OF SILANE COUPLING AGENTS ON THE UNDERFILL ADHESION
UNDERGOING HYDROTHERMAL AGING 813
    Haoliang Lin, Yuanyuan Yang, Bin Wang, Gang Li, Houya Wu, Pengli Zhu

REVIEW OF COPPER-SILVER CORE-SHELL SINTERING PASTES: TECHNOLOGY AND FUTURE TRENDS 817

Zejun Zeng, Jing Zhang, Pan Liu

MECHANICAL BEHAVIOR OF LEAD-FREE SOLDER AT HIGH TEMPERATURES AND HIGH STRAIN RATES 823

Xu Long, Tianxiong Su, Chao Chang, Jiaqiang Huang, Xiaotong Chang, Yutai Su, Hongbin Shi, Tao Huang, Yanpei Wu

EMPLOYING SINGLE-CRYSTAL COBALT SUBSTRATES TO CONTROL βSN GRAIN ORIENTATIONS IN SOLDER INTERCONNECTIONS 827

Ce Li, Xufeng Chang, Bingguang Wang, Zhaolong Ma, Xingwang Cheng

TOPOLOGY OPTIMIZATION DESIGN OF META-MATERIAL HEAT SPREADER 832

Xue Bai, Qinghua Hu, Xiaoliang Zeng, Rong Sun, Jianbin Xu

EFFECT OF SURFACE STRESS ON INDENTATION RESPONSE OF ELASTIC-PLASTIC MATERIALS 836

Xu Long, Ziyi Shen, Xiaotong Chang, Yutai Su, Hongbin Shi, Tao Huang, Yanpei Wu, Bingyi Tu

SURFACE PASSIVATION OF CU NANOFIBER FILMS FABRICATED BY ELECTROSPINNING FOR TRANSPARENT CONDUCTIVE FILMS 839

Zhefei Sun, Ming-Sheng Wang, Zhihao Zhang

RESEARCH ON THERMAL CHARACTERISTICS OF HIGH POWER 3D MICROCHANNEL MULTICHIP PACKAGE 843

Xiangli Wei, Daoguo Yang, Wangyun Li, Xiyou Wang, Guangsheng Lu, Shirui Xue

A COMPREHENSIVE SIMULATION STUDY OF WARPAGE OF FAN-OUT PANEL-LEVEL PACKAGE USING ELEMENT BIRTH AND DEATH TECHNIQUE 847

Yun-Kai Deng, Bing-Xian Yang, Wei-Lin Hu, Xin-Ping Zhang

STUDY OF PACKAGE RELIABILITY ACCORDING TO THE EPOXY MOLDING COMPOUND 852

Eunsol Jo, Jung-Rae Park, Cheong-Ha Jung, Gu-Sung Kim

SIMULATION ANALYSIS OF THE INFLUENCE OF TEST CONDITIONS ON THE BONDING STRENGTH OF PCB PADS 854

Huanhuan Wang, Miao Cai, Daoguo Yang, Hengjian He, Yanchen Wu, Lei Song

EFFECT OF SAC0307 CONTENT ON PROPERTIES OF CU/ZN POWDER/AL JOINT BY ULTRASONIC EXCITATION AT LOW TEMPERATURE 859

Qianzhu Xu, Guisheng Gan, Zhaoqi Jiang, Shiqi Chen, Tian Huang, Cong Liu, Xiangtao Xu

A BUBBLE-FREE ELECTROOSMOTIC PUMP WITH POLYANILINE-WRAPPED PLATINUM-COATED TITANIUM MESH ELECTRODES 864

Qian Yang, Liang Li, Wen Chen, Meng Gao, Le Ye

FINITE ELEMENT SIMULATION STUDY OF INTERFACIAL CRACK PROPAGATION IN THE UNDERFILLED FC-BGA PACKAGE 870

Hong-Guang Wang, Min-Bo Zhou, Jiu-Bin Fei, Li Sun, Bing-Xian Yang, Wei-Lin Hu, Chang-Bo Ke, Xin-Ping Zhang

ENHANCEMENT OF COPPER NANOPARTICLE PASTE BY PRESSURE-LESS SINTERING ON DIFFERENT SUBSTRATES IN PT-CATALYZED FORMIC ACID ATMOSPHERE 875

Junlong Li, Yang Xu, Panju Shang, Yinghui Wang, Tadatomo Suga

INFLUENCES OF THE SOLDER SIZE ON GROWTH OF INTERFACIAL CU_6SN_5 AND MECHANICAL PERFORMANCE OF SN-3.0AG-0.5CU/(111)CU JOINTS SUBJECTED TO MULTIPLE REFLOW SOLDERING 880
Ming-Qiang Chen, Min-Bo Zhou, Yun-Wei Li, Xin-Ping Zhang

SIMULATION OF ELECTROMAGNETIC WAVE PROPAGATION IN 3D INTEGRATED MODULE BASED ON 3D ADI-FDTD ALGORITHM.. 885
Yinhui Han, Min Miao, Jin Li

A COMPARATIVE STUDY ON THE INFLUENCES OF VARIOUS NICKEL POWDERS ON THE EMI SHIELDING PERFORMANCE OF CONDUCTIVE POLYMER COMPOSITES 891
Yong Wang, Dingkun Tian, Yadong Xu, Baotan Zhang, Tao Zhao, Yougen Hu, Rong Sun

A X BAND CERAMIC PACKAGE WITH KILOWATT-LEVEL HIGH-POWER AND LOW LOSS 895
Yangfan Zhou, Linjie Liu, Zhizhuang Qiao, Ke Wang, Gai Liu

ANISOTROPIC BN NANOSHEET/POLYMER COMPOSITE BULK MATERIAL: A STUDY ON MECHANICAL PROPERTY 898
Chen Jing, Zeng Xiao Liang, Zhangguo Qi, Hu Qing Hua, Ye Huai Yu, Liu Pan

SOFT DEGRADATION AND RECOVERY UNDER ESD STRESS OF E-MODE GAN HEMTS WITH P-GAN GATE.. 902
Mei Wang, Zhiyuan He, Yan Ren, Lichao Hao, Zhaohui Wu, Bin Li

FABRICATION OF FLEXIBLE PRINTED CIRCUITS ON POLYIMIDE SUBSTRATE BY USING AG NANOPARTICLE INK THROUGH 3D DIRECT-WRITING AND RELIABILITY OF THE PRINTED CIRCUITS 907
Cheng-Bo Li, Xiao Ma, Hai-Jun Huang, Min-Bo Zhou, Xin-Ping Zhang

STUDY ON THE ELECTROMAGNETIC AND THERMAL CHARACTERISTICS OF AEROSTATIC SPINDLE FOR WAFER GRINDING.. 912
Chen Zhao, Ye Lezhi, Song Xuanjie, Wang Zhiyue, Liu Guangjie, Zhao Yumin

COPPER FILLING OF HIGH ASPECT RATIO THROUGH CERAMIC HOLES: EFFECT OF CONVECTION ON ELECTROCHEMICAL BEHAVIOR OF ADDITIVES 918
Wang Qing, Liu Jiaxin, Wu Yilin, Chen Mingxiang

THE EFFECT OF TOUGHENING AGENTS ON CAPILLARY UNDERFILL IN THE FLIP CHIP PACKAGE 923
Yuanyuan Yang, Houya Wu, Tao Peng, Jinbao Yang, Bin Wang, Haoliang Lin, Gang Li, Pengli Zhu, Rong Sun

PREPARATION AND PROPERTIES OF LOW MELTING POINT SN-P-F-O-MATRIX PHOSPHOR-IN-GLASS FOR WHITE LED 928
Deming Hu, Liang Yang, An Xie, Chunyan Cao, Xiayun Shu, Chenrui Fan, Jinrong Deng, Kun Chen

ARTIFICIAL NEURAL NETWORKS MODELING TECHNOLOGY FOR SUBSTRATE INTEGRATED SUSPENDED LINE... 932
Shuxia Yan, Nana Yang, Zhifeng Chen, Peng Huang, Weiguang Shi

NON-CYANIDE ELECTROPLATING OF GOLD-TIN EUTECTIC ALLOY FOR FLIP-CHIP PACKAGING OF LED.. 936
Mingliang Huang, Chao Fang, Feifei Huang, Yan Yan

FABRICATION OF SIC NANO-PORE ARRAYS STRUCTURE BY METAL-ASSISTED PHOTOCHEMICAL ETCHING 940
Zijian Li, Dachuang Shi, Yun Chen, Maoxiang Hou, Jian Gao, Xin Chen

SIMULATION AND OPTIMIZATION OF INKJET-PRINTED OUTLINES TO IMPROVE PATTERN FIDELITY 944
Shaowei Hu, Wanchun Yang, Mingyu Li

SYNTHESIS AND CHARACTERIZATION OF AG-37AT.%CU SOLID SOLUTION NANOPARTICLES FOR HIGH RELIABILITY PACKAGING 947
Wanchun Yang, Shaowei Hu, Wei Zheng, Mingyu Li

EXPERIMENTAL ANALYSIS FOR BUMP SHEAR METHOD OF BSOB WIRE BONDING PROCESS.. 950
Zhu Chenjun, Zhang Pingsheng, Wen Zehai, Li Hui, Wu Yilong, Dong Dong

LOW TEMPERATURE CURING COPOLYIMIDE WITH MONOMER CONTAINING PYRAZINE MOIETY 954
Changqing Li, Guoping Zhang, Jinhui Li, Yingying Li

MICROSTRUCTURE EVOLUTION AND INTERFACIAL REACTION OF CO-P/SAC305/CO-P SOLDER JOINTS AT HIGH CURRENT DENSITY 958
Tao Fan, Donghua Yang, Haotong Qin, Yuqian Chen, Tao Chen, Xiang Zhai, Teng Ran

RESEARCH ON FAILURE MECHANISMS OF LOCALIZED IGBT DRIVE BOARD........................... 961
Tiezhu Chen, Liang Zhou

SUBSTRATE SOLDER RESIST CRACK ROOT CAUSE INVESTIGATION THROUGH FINITE ELEMENT ANALYSIS 963
Ye Zhang, Tan Boowei, Tan Chow-Khong

MICROSTRUCTURE AND MECHANICAL PROPERTIES OF JOINTS BETWEEN GAAS SOLAR CELL ELECTRODE AND AG INTERCONNECTOR UNDER TEMPERATURE THERMAL CYCLE.. 967
Yuhan Ding, Zhichao Wang, Xueming Hua, Chen Shen, Min Wang, Jusha Ma, Bin Qian

DESIGN METHOD OF TRIAXIAL VIBRATION FIXTURE FOR COMPLEX INTEGRATED CIRCUITS.. 972
Xiaoqiang Wang, Bin Li, Chuanjin Deng, Rui Deng, Ruolei Wang, Kun Jiang

GROWTH BEHAVIOR OF INTERFACIAL INTERMETALLIC COMPOUNDS OF CO-20%P/SOLDER JOINT UNDER TEMPERATURE GRADIENT 978
Haotong Qin, Donghua Yang, Tao Fan, Tao Chen, Chunhong Zhang, Yuqian Chen

AN ACCURATE SIMULATION METHOD OF PACKAGE WARPAGE EXPERIMENTAL RESULTS BASED ON FEM.. 982
Liqiang Neng, Tingting Song, Guangping Shao, Jian Pang, Sun Tuobei, Keqing Ouyang

SIMULATION AND EXPERIMENTAL ANALYSIS OF A COST-EFFECTIVE MINIATURIZED TRANSCEIVER FOR X-BAND APPLICATION 986
Yu Ban, Jie Liu

EXTRACTION, OPTIMIZATION AND FAILURE DETECTION APPLICATION OF PARASITIC INDUCTANCE FOR HIGH-FREQUENCY SIC POWER DEVICES 990
Minghui Yun, Kailin Zhang, Miao Cai, Yiren Yang, Changqi Feng, Song Wei, Daoguo Yang, Guoqi Zhang

SIMULATION ANALYSIS OF RESIDUAL STRESS OF SINTERED NANO-SILVER UNDER MULTILAYER STACKED MODULE 995
  Zhentang Liang, Hongyue Wang, Bin Zhou, Guoyuan Li

LOW TEMPERATURE BONDING POLYCRYSTALLINE DIAMOND TO SI BY USING AU THIN-LAYER FOR HIGH-POWER SEMICONDUCTOR DEVICES 999
  Yi Zhong, Shuchao Bao, Ke Li, Mingchuan Zhang, Daquan Yu

RESEARCH ON RELIABILITY LIFE EVALUATION METHOD BASED ON AIRBORNE T/R COMPONENTS 1002
  Rui Deng, Chuanjin Deng, Ruolei Wang, Weixi Gong, Zongqi Cai, Kun Jiang

EFFECT OF GRAIN SIZE ON DIELECTRIC PROPERTIES AND RELIABILITY FOR ULTRA-THIN MLCCS 1006
  Kulun Jiang, Rong Sun, Xiuhua Cao, Lei Zhang, Shuhui Yu, Bo Li, Zhenxiao Fu

A QUICK CORRECTION METHOD FOR THE BOARD-LEVEL FINITE ELEMENT ANALYSIS OF QFP DEVICE UNDER VIBRATION 1012
  Zicheng Sa, Shang Wang, Jiayun Feng, Guangliang Yu, Ning Zhang, Yanhong Tian

PREPARATION AND MICROWAVE PROPERTIES OF TIO₂/PTFE COMPOSITES REINFORCED BY MULLITE FIBERS 1015
  Zhangzhao Weng, Jun Luo, Hongfeng Lv, Shuai Zhou, Xiaoqiang Wang, Daojun Luo

A SEGMENTED PLASMA ETCHING METHOD FOR 2.5D/3D THROUGH SILICON VIAS 1019
  Yuanwei Lin

STUDY ON THERMAL CYCLE AND INSULATION CHARACTERISTICS OF HIGH RELIABLE IGBT MODULE 1025
  Tao Chen, Minghua Zhang, Binbin Zhang, Qing Chen, Kun Tian, Yan Li

WARPAGE BEHAVIOR STUDY AND OPTIMIZATION FOR ULTRA-THIN POP MEMORY WITH MULTI-STACKED CHIPS 1029
  Dongmei Xia, Bei Wang, Chengyu Liao, Zuyao Liu, Hongwen He, Lu Liu

RESEARCH ON POWER DEVICE STRUCTURE BASED ON FO PACKAGE METHOD 1034
  Guan Qiang Song, Jia Ren Huo, Juntao Wang, Debo Liu, Chenwei Zhang, Jing Jiang, Huaiyu Ye

A BROADBAND MODEL OF STACKING TSV CHANNELS FOR NONDESTRUCTIVE DEFECT LOCALIZATION IN 3D ICS AND MICROSYSTEM 1040
  Chenbing Qu, Liwei Wang, Si Chen, Chen Sun, Zhiwei Fu, Guan-Lin Feng

A HYBRID DEGRADATION MODELING OF LIGHT-EMITTING DIODE USING PERMUTATION ENTROPY AND DATA-DRIVEN METHODS 1044
  Minzhen Wen, Zhou Jing, Mesfin S. Ibrahim, Jiajie Fan, Guoqi Zhang

RESEARCH ON ADJUSTMENT OF THE ELECTRICAL RESISTIVITY OF ALUMINUM NITRIDE CERAMIC 1050
  Jinhu Fan, Zirong Tang, Jie Wang

SOLDER PREFORMS COMPOSED OF HIGH CU-CONTENT SN-XCU ALLOYS FOR POWER ELECTRONIC PACKAGING AND CHARACTERIZATION OF THE PROCESSING PERFORMANCE AND JOINT'S PROPERTIES 1054
  Ru-Zeng Shi, Ming-Qiang Chen, Hai-Jun Huang, Min-Bo Zhou, Xin-Ping Zhang

THE MECHANICAL AND PHYSICAL PROPERTIES OF THE PHASES IN THE MICROSTRUCTURE OF SN-BI SOLDER ALLOY 1059
Chuyi Lei, Xinghe Luan, Zhigao Liu, Hongbo Qin, Bin Hou, Wangyun Li

SCALABLE MODELING AND ANALYSIS OF TSV USING BUMPLESS INTERCONNECTS TECHNOLOGY.................... 1064
Zewei Li, Zhikuang Cai, Lei Pan, Lu Liu, Binbin Xu, Yufeng Guo

THE INFLUENCE OF MOLDING COMPOUND PROPERTIES ON SYSTEM-IN-PACKAGE RELIABILITY FOR 5G APPLICATION.................... 1069
Dashun Liu, Kai Chen, Yijing Qin, Yong Zhong, Zhaorong Wan, Richeng Liu, Dong Lu, Ke Xue, Jingshen Wu

THERMAL RESISTANCE OF EUTECTIC GA-IN-SN/PARTICLES BINARY THERMAL INTERFACE MATERIALS 1075
Wendong Wang, Meijuan Lv, Jingdong Guo

RESEARCH ON BEOL FAILURES OF THE CHIP-PACKAGE INTERACTION BY SHEAR TESTS OF THE BUMPS.................... 1081
Shizhao Wang, Lianghao Xue, Hongjie Wang, Rui Li, Can Sheng, Yameng Sun, Sheng Liu

EVALUATION OF SOLDER JOINTS RELIABILITY OF BALL GRID ARRAY ASSEMBLY IN ASTRONAVIGATION MODULES 1085
W. L. Qin, Z. P. Yan, M. H. Wu

NOVEL DESIGN OF SIC MOSFET ACTIVE DRIVE CIRCUIT BASED ON IMPROVED AUXILIARY BRANCH METHOD 1088
Li Yuhong, Niu Pingjuan, Mei Yunhui, Ning Pingfan, Zhao Di, Bai Jie

OPTICAL PATH DESIGN OF A HIGH COUPLING EFFICIENCY DFB LASER PACKAGE BASED ON COLLIMATOR 1094
Xiaomeng Lv, Ao Liao, Weihua Xu, Yangzhi Liu, Yilong Wu, Lan Qiao, Yong Zhao, Junli Yang

SIO$_2$/SIO$_2$ BONDING TECHNOLOGY RESEARCH ON WAFER-LEVEL 3D STACKING.................... 1097
Zhang Peng, Chengyu Yu, Kai Cen, Jie Pu, Pengcheng Xia, Chengqian Wang

HIGH SPEED CU PLATING TECHNOLOGY FOR WAFER LEVEL PACKAGING.................... 1101
Jian Wang, David Wang, Zhaowei Jia

ATE BOARD DESIGN SOLUTION OF SMALL PIN PITCH DUT 1107
Fei Pan, Qing Zhou, Jie Zhou

A STUDY OF COPPER OXIDIZATION MECHANISM AT METAL INTERFACE.................... 1111
Ruolin Zhang, Ying Tang, Xu Wang, William Cao, Pradeep Rai

IRREGULAR HEAT DISSIPATING COVER BASED ON A1N MATERIAL.................... 1117
Jianhui Liu, Yuxin Guo, Xiaowei Zhao

EVALUATION AND REDUCTION OF OPTICAL CROSSTALK IN QUANTUM DOT COLOR-CONVERTED MINI/MICRO-LED DISPLAYS.................... 1121
Yuanjie Cheng, Jeffery C. C. Lo, Xing Qiu, S. W. Ricky Lee

STUDY ON A KIND OF THERMAL SOURCE CHIP FOR THE PERFORMANCE ANALYSIS OF MICRO CHANNEL HEAT SINK: SIMULATION AND EXPERIMENTAL VALIDATION 1126
Ming Zhao, Jian Zhang, Qian Lu, Weiwei Xiang, Yangyang Li, Miaomiao Jiang, Huijie Ye, Ting Peng

PLANE POSITION MEASUREMENT FOR μLED BASED ON SINGLE CAMERA 1131
Jie Bai, Pingjuan Niu, Erdan Gu

CURE SHRINKAGE CHARACTERIZATION AND WARPAGE SIMULATION
OPTIMIZATION OF EPOXY MOLDING COMPOUND FOR 5G APPLICATION 1134
Kai Chen, Dashun Liu, Yijing Qin, Yong Zhong, Zhaorong Wan, Richeng Liu, Dong Lu, Ke Xue, Jingshen Wu

FAILURE MECHANISM STUDY FOR FLIP CHIP QFN CRACK ISSUE UNDER
TEMPERATURE CYCLING TEST 1139
Ke Xue, Dong Lu, Kai Chen, Yijing Qin, Dayang Li, Qi Tang

STUDY ON THE ANTIOXIDATION OF COATED NANA-COPPER 1144
Weijie Zhang, Quan Zhou, Chenshan Gao, Debo Liu, Jun Li, Huaiyu Ye

THE IMPACT OF PACKAGING MATERIALS ON THERMOMECHANICAL RELIABILITY
OF FC-LGA (FLIP-CHIP LAND GRID ARRAY) PACKAGE FOR 5G APPLICATION 1148
Dayuan Wan, Ke Xue, Weijing Dai, Yijing Qin, Dong Lu, Zhiqi Wang, Yi Chen

FAILURE ANALYSIS AND RELIABILITY IMPROVEMENT OF CRIMPING ASSEMBLY OF
COPPER-CLAD ALUMINUM CONDUCTORS FOR AEROSPACE 1153
Rui Cao, Meng Meng, Zhibin Wang, Yarong Chen, Yu Ye, Xueyin Huang

A 3D-MEMS BASED ARCHITECTURE FOR CS-CMOS HETEROGENEOUS INTEGRATION 1158
Min Huang, Pengfei Liu, Hongze Zhang, Yan Wu, Jian Zhu

DESIGN AND OPTIMIZATION OF TEMPERATURE SENSOR BASED ON LGA PACKAGE
STRUCTURE 1161
Jianhui Liu, Juntao Wang, Jun Li

HIGHER ASPECT RATIO TSV STRUCTURE ECP BOTTOM-UP PLATING PROCESS 1165
Yinuo Jin, Bo Zheng, Jian Wang, David H. Wang, Qixing Yu

MONTE CARLO BASED STOCHASTIC FINITE ELEMENT MODEL FOR UNCERTAINTY
QUANTIFICATION IN FLIP CHIP BGA ELECTRONIC PACKAGING 1169
Liu Chu, Jiajia Shi, Robin Braun

EFFECT OF BUMP SHAPES ON THE ELECTROMIGRATION RELIABILITY OF COPPER
PILLAR SOLDER JOINTS 1173
Zhekun Fan, Zhankun Li, Junhui Li, Jinqing Xiao, Yunpeng Liu, Junfu Liu, Taotao Chen

STUDY ON HERMETIC PACKAGE OF ANTIRADIATION DIRECT-HEAD TRANSMITTER 1179
Chuanwei Wang, Yukun Wu, Jiabo Zhang, Kuang Pan, Daochang Wang, Yi Liang

STUDY ON DIE-BONDING KEY TECHNOLOGY OF UV LED PACKAGING 1183
Wei Liu, Chunfang Zi, Qingping Lin

STUDY ON PERFORMANCE OPTIMIZATION OF NANOMETER COPPER PASTE 1187
Qiang Liu, Jian Wen, Yu Zhang, Yu Liu, Guannan Yang, Zhongwei Huang, Chengqiang Cui

STUDY ON THE EFFECT OF ASSEMBLY ERRORS ON THE ELECTROSTATIC TUNING
ABILITY IN MICRO UMBRELLA SHELL RESONATORS 1191
Lu Xu, Bin Luo, Jintang Shang, Zhaoxi Su, Shouyu Han, Yinghui Zhang

RESEARCH ON INTEGRATED METASURFACE LENS FOR HIGH GAIN MULTIBEAM
SYSTEM IN PACKAGE APPLICATION 1196
Yuxiang Zheng, Weikang Wan, Qidong Wang, Liqiang Cao

STUDY ON THERMAL STABILITY OF ALL COPPER INTERCONNECT STRUCTURES UNDER THERMAL SHOCK.. 1201
  Hao Li, Jun Shen, Jiacheng Xie

MACHINE LEARNING BASED PREDICTION OF WIRE BONDING PROFILE IN 3D STACKED INTEGRATED MICROELECTRONIC PACKAGING 1206
  Zhengping Ou, Junyu Long, Shuquan Ding, Yun Chen, Maoxiang Hou, Yunbo He, Xin Chen, Jian Gao

THE DESIGN OF A VOLTAGE CONVERSION SIP BASED ON ELECTROTHERMAL COUPLING METHOD.. 1210
  Hao-Hang Su, Shuai Fu

RELIABILITY ANALYSIS ON WAFER BONDING PROCESS OF MEMS CIRCULATOR 1215
  Dongxue Luo, Yan Wu, Junfeng Sun, Miao Yu

EXPERIMENTAL TESTS AND STRESS ANALYSIS OF SNPB SOLDER JOINTS IN A CERAMIC POP DEVICE 1219
  Kaiyu Guo, Honglei Ran, Jun Wang

ANALYSIS OF FACTORS AFFECTING THE OUTGASSING RATE OF MEMS VACUUM PACKAGING MATERIALS.. 1224
  Yong Yang, Bin Zhou, Xuanjun Dai, Yun Huang

FATIGUE LIFE PREDICTIONS OF SNPB SOLDER BALL IN A CERAMIC POP DEVICE 1230
  Yu Yao, Zirui Cui, Honglei Ran, Jun Wang

DESIGN AND FABRICATION OF A SOFT MICRO-ACTUATOR BASED ON DISTRIBUTED MAGNETIC COMPOSITE 1235
  Langkun Wang, Shimei Liu, Hu He

ELECTROLESS COPPER DEPOSITION WITH PYRAMIDAL MICRO-CONES MORPHOLOGY FOR LOW-TEMPERATURE CU-CU BUMP INTERCONNECTIONS........................ 1241
  Yiming Chen, Yiqiao Wei, Zhuo Chen, Wenjing Zhang, Fuliang Wang, Wenhui Zhu

THE EFFECT OF SILICON ANISOTROPY ON THE THERMAL STRESS OF TSV STRUCTURE OF 3D PACKAGING CHIP UNDER THERMAL CYCLIC LOADS................................ 1246
  Jingyang Liang, Minjie Ning, Chao Ding, Tianhan Liu, Zongbei Dai, Hongbo Qin

HIGH YIELD AND HIGH THROUGHPUT LITHOGRAPHY SOLUTION FOR EMERGING HIGH DENSITY FAN-OUT PANEL LEVEL PACKAGING.. 1250
  Junbo Jiang, Kang Zhang, Di He, Chen Xiang, Cheng-Tar Wu, Minghao Shen

AN OPTIMIZATION METHOD OF ULTRA HIGN SPEED DIFFERENTIAL STRUCTURE FOR BGA PACKAGE 1255
  Nong Jin, Zhizhuang Qiao, Linjie Liu, Ke Wang, Yangfan Zhou, Zan Ren

REFLOW SOLDERING PROCESS OPTIMIZATION BASED ON SURFACE EVOLVER SOLDER JOINT SHAPE SIMULATION AND FINITE ELEMENT ANALYSIS OF PCB ASSEMBLY... 1258
  Xing Jin, Wenlong Wang, Wenzhong Zhao, Yuting Zhang, Xiaopeng Tan, Shuai Chen

A NOVEL BUMPING METHOD FOR FLIP-CHIP INTERCONNECTION.. 1262
  Kun Li, Gaowei Xu, Quan Zhou, Wei Gai, Yanhong Wu, Jie Ren, Zhen Wang

DESIGN AND TEST OF TRANSMISSION LINE IN SFQ CIRCUIT 1265
  Quan Zhou, Lingyun Li, Zhen Wang, Gaowei Xu, Le Luo, Xiaoming Xie, Kun Li, Jie Ren

THE INFLUENCE OF SOLDERING VOIDS IN POWER DEVICES............................ 1269

Jianming Fang, Min Wang, Xuanlong Chen

DOUBLE-SIDED ELECTROPLATING PROCESS FOR THROUGH GLASS VIAS (TGVS)
FILLING 1273

Ke Li, Heng Wu, Weijian Chen, Daquan Yu

SAWING INVESTIGATION FOR THIN WAFER LAMINATED WITH DIE ATTACH FILM 1277

Haiyan Liu, Qingyu Pan, Sean Xu, Jianhong Wang, Lu Li, Xueting Wu

BLADE DICING ON WAFER SAW STUDY 1282

Qiuchen Zhang, Hongbin Xia, Shwu Miin Tan

3D DISLOCATION MULTI-STACK FAN-OUT PACKAGE OF ULTRA-THIN DIES FOR
HETEROGENEOUS INTEGRATION.. 1288

Lijun Chen, Feng Chen, Fengwei Dai

THE MULTI-DIMENSION CO-DESIGN FOR THE PACKAGE OF THE HIGH-SPEED MULTI-
FUNCTION CHIP.. 1292

*Yuan Guan, Yubo Wang, Dejian Li, Shunfeng Han, Bofu Li, Dameng Li, Fengman Liu,
Huimin He*

EFFECT OF PLASMA AND STAGING TIME ON THE UNDERFILL VOIDS IN FINE PITCH
FLIP-CHIP PACKAGE.. 1297

*Saif Wakeel, Dominic Koey Poh Meng, Stella Wong Wun Chin, Jos Philipsen, Pieter
Gommers, Annelies Joosten*

THE STUDY OF FAR-UVC 222-NM EXCILAMP AND GERMICIDAL-UVC 254-NM LOW-
PRESSURE HG LAMP: OPTICAL CHARACTERISTICS AND SERVICE LIFE 1302

*Fanny Zhao, Guoshuai Dong, Hao Wu, Guoming Yang, Brian Shieh, S. W. Ricky Lee,
Ronghua Deng*

EFFECTIVENESS VALIDATION ON EQUIVALENT MODEL FOR WAFER-LEVEL
WARPAGE PREDICTION.. 1305

Guoli Sun, Jiahui Wei, Fei Qin, Yanwei Dai, Kui Li, Baoxia Li

MULTI-CHIPS HIGH-DENSITY INTERCONNECTION DESIGN ON INFO PLATFORM 1311

Jiang Qiang, Zhou Guodan, Wang Zongwei, Pang Jian, Sun Tuobei, Keqing Ouyang

FIRST-PRINCIPLES STUDY ON THE MECHANICAL PROPERTIES OF CU_3SI COMPOUND........... 1315

Jian Wang, Xiaowei Xu, Chao Ding, Tianhan Liu, Zongbei Dai, Hongbo Qin

STUDY ON THE EFFECT OF SILICA SUBMICRON PARTICLE SIZE AND CONTENT ON
FRACTURE TOUGHNESS OF FILLED EPOXY RESIN 1320

Qin Zhou, Xuecheng Yu, Leicong Zhang, Pengli Zhu, Rong Sun

EFFECT OF TEMPERATURE CYCLING, HIGH TEMPERATURE STORAGE AND STEADY-
STATE OPERATION LIFE TEST ON RELIABILITY OF GAN HEMTS.............................. 1323

Mao Mao, Sha Tang, Zhizhe Wang, Rui Deng, Chuanjin Deng, Si Chen, Yan Ren

EFFECT OF IMC MORPHOLOGY ON THE CURRENT DENSITY AND TEMPERATURE
GRADIENT OF LINE-TYPE CU/SN/CU SOLDER JOINT 1327

Jiaqi Tang, Tianhan Liu, Chao Ding, Jian Wang, Hongbo Qin, Wangyun Li

DESIGN AND ANALYSIS OF MOSFET BASED ON FAN-OUT PANEL-LEVEL PACKAGE
TECHNOLOGY.. 1332

Zhi Liang, Dongdong Shao, Kunpeng Ding, Chuang Tian

ADSORPTION BEHAVIOR OF HCOOH ON THE CRYSTAL SURFACES OF CU(111) AND (100) 1336

Liheng Jiang, Tianhan Liu, Siliang He, Hongbo Qin, Wangyun Li, Daoguo Yang

INNER MICROCRACK INDUCED BY INTERMETALLIC COMPOUND IN RIGHT-ANGLE AU/SN3.0AG0.5CU/AU SOLDER JOINTS 1340

Wu Yue, Zhenyu Zhang, Bunv Liang, Jing Li, Cheng Xue

A COMPACT CHIP FILTER FOR 5G COMMUNICATION FRONTEND BASED ON GLASS IPD TECHNOLOGY 1343

Zhitao Zhang, Ling Gu, Shanwen Hu, Yuehua Zhang, Xinlei Zhang

ENCAPSULATION TECHNIQUES OF PEROVSKITE SOLAR CELLS.. 1346

Qi Wu, Wenfeng Li, Mengyu Chen, Cheng Li

EFFECT OF NI-CNTS ON WETTING PROPERTIES, MICROSTRUCTURE, AND CREEP RESISTANCE OF SN58BI-0.1ER COMPOSITE SOLDER 1352

Qi Li, Fengmei Liu, Yaoyong Yi, Xueying Zhang, Haitao Gao

A PROCESS IMPROVEMENT IN SILVER-INDIUM TRANSIENT LIQUID PHASE BONDING METHOD FOR THE HIGH-POWER ELECTRONICS AND PHOTONICS PACKAGING 1357

Donglin Zhang, Xiuchen Zhao, Yingxia Liu, Ying Liu, Yongjun Huo

RESEARCH ON SIP SIGNAL INTEGRITY BASED ON ANSYS SIWAVE IN WEARABLE MEDICAL SYSTEMS.. 1361

Zheng Yang, Yingke Gao, Shenglong Li, Chuanchuan Sun, Yunfu Zhao

LIGHTWEIGHT SILVER NANOWIRE AEROGEL FOR ELECTROMAGNETIC INTERFERENCE SHIELDING 1366

Fei Peng, Yi Fang, Zhehao Han, Wenbo Zhu, Mingyu Li

A FACILE BONDING MATERIAL TO ENABLE INTERCONNECTION AMONG COMPLEX SURFACES THROUGH AGNWS AEROGEL 1370

Zhehao Han, Mingyu Li, Wenbo Zhu, Fei Peng, Yi Fang

HIGH-SENSITIVITY FLEXIBLE SENSOR BASED ON SILVER NANOWIRE AEROGEL.................... 1374

Yi Fang, Mingyu Li, Zhehao Han, Wenbo Zhu, Fei Peng

SOLDERING OF GRAPHENE ASSEMBLED FILMS WITH ULTRASONIC ASSISTANCE AND ITS UTILIZATION POTENTIALITY IN ELECTRONIC DEVICES 1378

Huaqiang Fu, Yong Xiao, Daping He

THEORETICAL CALCULATION AND SIMULATION OF BGA PACKAGE STRESS UNDER TEMPERATURE CYCLING LOAD 1382

Zhang Yueping, Hou Chuantao, Tong Jun, Wang Long

LOSS MODELLING AND ANALYSIS OF A HIGH-EFFICIENCY WIRELESS POWER TRANSFER SYSTEM FOR AUTOMATED GUIDED VEHICLE APPLICATIONS................................ 1387

Jincheng Yu, Wai Leong Ng, Minglu Xia, Ziyang Gao

THREE DIMENSIONAL WAFER-LEVEL VACUUM PACKAGING OF MEMS RESONANT ACCELEROMETER 1393

Ziji Wang, Chaoyang Xing, Jin Zhang, Zhaoxi Su, Wenqi Li, Bin Luo, Jintang Shang

GLASS REFLOW AND THERMO-MECHANICAL STRESS SIMULATION FOR THROUGH GLASS VIA IN GLASS-SILICON COMPOSITE INTERPOSER... 1397

Wenqi Li, Chaoyang Xing, Jianfeng Zhang, Ziji Wang, Zhaoxi Su, Bin Luo, Jintang Shang

RESEARCH ON THE APPLICATION OF FEEDFORWARD + HIGH-ORDER ITERATIVE
LEARNING OF PERMANENT MAGNET LINEAR MOTOR IN WIRE BONDING MACHINE 1401
Haomiao Wu, Yunbo He, Xiaohui Lin, Haolin Li

MACHINE LEARNING ENABLED OPTIMIZATION OF PICK-UP PROCESS FOR THIN DIE............ 1407
Peilun Yao, Haibin Chen, Jinglei Yang, Jingshen Wu

DESIGN AND ANALYSIS OF A FAST-SPEED FLIP-CHIP BONDING SYSTEM WITH
FORCE CONTROL 1412
Zhongyuan Zhu, Hui Tang, Jiedong Li, Sifeng He

DEEP LEARNING PRODUCT CLASSIFICATION FRAMEWORK BASED ON THE
MOTIVATION OF TARGET CUSTOMERS 1418
Fei Sun, Ding-Bang Luh, Qidong Wang, Yulin Zhao, Yue Sun

CHARACTERIZATION OF LONGITUDINAL THERMAL CONDUCTIVITY OF GRAPHENE
FILM 1422
Jiajia Chen, Xinjian Gong, Sihua Guo, Yong Zhang, Jin Chen, Yuanyuan Wang, Johan Liu

RANDOM VIBRATION RESPONSE ANALYSIS OF SIP POWER SUPPLY MODULE........................ 1426
Tao Lin, Weiyin Wang, Jun Li

THE INFLUENCE MECHANISM OF PROCESS DEFECTS OF PLANE WELDING ON
SIGNAL TRANSMISSION IN MICROWAVE COMPONENTS 1429
*Ruining Li, Jun Tian, Song Xue, Congsi Wang, Yan Wang, Cheng Zhou, Jing Liu, Zhihai
Wang, Kunpeng Yu, Haitao Shi, Daoheng Sun*

EPOXY FLUX PREVENT HOT TEAR AT VIPPO SOLDER JOINTS 1435
Elaina Zito, Dave Bedner, Ning-Cheng Lee

PROCESS OPTIMIZATION FOR CCGA SURFACE MOUNT ASSEMBLY BASED ON
PHYSICS OF FAILURE.. 1441
Hui Xiao, Weiming Li, Yabing Zou, Xiaotong Guo, Jiahao Liu

210ºC REFLOW TECHNOLOGY STUDY IN 3D PACKAGING... 1445
*Lingyao Sun, Yaru Dong, Zhuangzhuang Hou, Xiuchen Zhao, Yongjun Huo, Ying Liu, Yingxia
Liu*

RESEARCH ON THE HUMIDITY RESISTANCE RELIABILITY OF DIFFERENT
PACKAGING STRUCTURES 1449
Hui Xuan, Zheng Yu, Hua Wu, Wanchun Ding, Guohua Gao

ELECTROPLATING NANOTWINNED COPPER FOR ULTRAFINE PITCH
REDISTRIBUTION LAYER (RDL) OF ADVANCED PACKAGING TECHNOLOGY 1453
Yu-Bo Zhang, Li-Yin Gao, Xiao Li, Zhe Li, Xu-Liang Ma, Zhi-Quan Liu, Rong Sun

MIXED CU NANOPARTICLES AND CU MICROPARTICLES WITH PROMISING LOW-
TEMPERATURE AND LOW-PRESSURE SINTERING PROPERTIES AND
INOXIDIZABILITY FOR MICROELECTRONIC PACKAGING APPLICATIONS 1459
Haiqi Lai, Jian Wen, Guannan Yang, Yu Zhang, Chengqiang Cu

EFFECT OF DAISY CHAIN STRUCTURE ON ELECTROMIGRATION RELIABILITY OF
MICROBUMPS: A SIMULATION STUDY 1463
*Chenkan Yan, Dun Wang, Kaihong Zhang, Huibin Zhang, Yongjian Yu, Weikun Xie, Kai Zhu,
Yongkang Wan*

NOVEL LOW-DIELECTRIC FLUORINATED CARBON FIBER/POLYIMIDE MATERIALS WITH HIGH ELONGATION 1469

Tao Wang, Jinhui Li, Fangfang Niu, Liang Shan, Guoping Zhang, Rong Sun, Ching-Ping Wong

INFLUENCE OF POROSITY ON THE MECHANICAL PROPERTIES OF HYBRID SILVER SINTERED JOINT 1473

Zunyu Guan, Fred Fuliang Le, Jingshen Wu, Jinglei Yang, Rinse Van Der Meulen, Haibin Chen

INTEGRATION DESIGN OF MM-WAVE RADAR ANTENNAS BASED ON FOWLP 1478

Lihong Liu, Chen Cheng, Jionajiong Gu, Tianhua Shen, Quanbing Li

EXPERIMENTAL AND COMPUTATIONAL STUDY OF SHIELDING EFFECTIVENESS OF METAL GRIDS 1481

Zhiqiang Lin, Xuebin Liu, Xinrong Shi, Yougen Hu, Rong Sun

STUDY ON THE MECHANICAL PROPERTIES OF ULTRA-LOW DIELECTRIC FILM BY TENSILE TEST 1485

Lei Wang, Fei Xiao, Jun Wang

SIMULATION RESEARCH ON ELECTROMIGRATION OF BGA DEVICES 1489

Wenchao Tian, Yiming Zhang, Yong Chen, Si Chen, Hao Cui

MEASUREMENT PROCESS OPTIMIZATION IN USING LOCK-IN THERMOGRAPHY FOR FAULT LOCALIZATION OF COWOS PACKAGES 1494

Shuanshe Chao, Na Mei, Xinyi Lin, Tuo Bei Sun, Dan Yang, Keqing Ouyang

MICROSTRUCTURE AND MECHANICAL CHARACTERISTICS OF A NOVEL CU/CU_3SN JOINT WITH CU/SN PREFORM 1498

Hongyan Xu, Wei Zhang, Xuan Liu, Ju Xu

THICKNESS MEASUREMENT OF MULTI-LAYER THIN FILM ALLOY BY X-RAY FLUORESCENCE SPECTROMETER 1504

Sung-Hua Zhong, Liang-Pin Chen

NANO SILVER PARTICLES PREPARED BY SPARK ABLATION AS PACKAGING INTERCONNECTION MATERIAL 1507

Jin Tong, Yu Zhang, Peilin Liang, Zhongwei Huang, Chengqiang Cui, Keju Zhong

SILVER FILM PREPARED BY SPARK ABLATION FOR CONDUCTIVE PATTERN 1511

Peilin Liang, Yu Zhang, Jin Tong, Zhongwei Huang, Chengqiang Cui, Keju Zhong

SIMULATION AND OPTIMUM DESIGN OF AL-SI ALLOY BY DOUBLE-SIDED LASER WELDING 1514

Liangchen Tan, Shanshan Li, Jianfeng Zhong

SIGNIFICANT EFFECT OF TEMPERATURE AND SOLDERS ON THE GROWTH BEHAVIOR OF CU_6SN_5 ON (110) CU SINGLE CRYSTAL 1518

Chong Dong, Min Shang, Ying Guo, Xiangxu Chen, Jun Zhang, Changlong Dong, Haoran Ma, Haitao Ma

A STRETCHABLE HIGH-SENSITIVE CAPACITIVE SENSOR FOR AERODYNAMIC PRESSURE MEASUREMENT 1521

Xiaofeng Yang, Jian Chen, Si Chen, Bin Zhou, Yijun Shi, Zhiwei Fu

3D WAFER LEVEL PACKAGING FOR SAW FILTER USING THIN GLASS CAPPING WITH THROUGH GLASS VIAS.. 1525
Zuohuan Chen, Jin Zhao, Feng Jiang, Heng Wu, Mingchuan Zhang, Daquan Yu

WARPAGE CHARACTERISTIC OF GLASS INTERPOSER WITH DIFFERENT CTE'S AND THICKNESS 1529
Jin Zhao, Fei Qin, Daquan Yu, Zuohuan Chen, Shuai Zhao

Author Index

Evaluation of Aging Performance of Thermal Gel Subjected to Laser Flash Tests

Yimin Yao
[1] *Shenzhen Institute of Advanced Technology, Chinese Academy of Sciences, Shenzhen, China*
[2] *Department of Electronics Engineering, The Chinese University of Hong Kong, Hong Kong, China*
ym.yao@siat.ac.cn
0000-0002-6567-1396

Yonglun Xu
Shenzhen Institute of Advanced Electronic Materials
Shenzhen, China
yl.xu3@siat.ac.cn

Xue Bai
Shenzhen Institute of Advanced Electronic Materials
Shenzhen, China
xue.bai@siat.ac.cn

Yunsong Pang
Shenzhen Institute of Advanced Electronic Materials
Shenzhen, China
ys.pang@siat.ac.cn

Linlin Ren*
Shenzhen Institute of Advanced Electronic Materials
Shenzhen, China
ll.ren@siat.ac.cn

Xiaoliang Zeng
Shenzhen Institute of Advanced Technology, Chinese Academy of Sciences
Shenzhen, China
xl.zeng@siat.ac.cn

Fei Deng
Shenzhen CONE Tech. Co., Ltd, 51&52 Building, Software Town of Shenzhen Universiade, Shenzhen, 518172, China
dengfei@chinacone.com

Jian-Bin Xu*
Department of Electronics Engineering, The Chinese University of Hong Kong
Hong Kong, China
jbxu@ee.cuhk.edu.hk

Rong Sun*
[1] *Shenzhen Institute of Advanced Technology, Chinese Academy of Sciences*
[2] *Shenzhen Institute of Advanced Electronic Materials*
Shenzhen, China
rong.sun@siat.ac.cn

Abstract—As an important class of thermal management material, thermal gel is widely used to fille the gaps between heat generating devices and the heat sink. The performance of thermal gel varies when withstanding high environment temperature. Real-time monitoring the intrinsic thermal conductivity (k) of thermal gels, and the variation of interfacial thermal resistance (R_i) at the thermal gel/substrate interface during the aging process is urgently-needed for exploring the anti-aging mechanism and further improve the overall transfer performance. In this study, laser flash method was applied as to characterize the thermal performance. We experimentally demonstrate that there exists strong dependence of k-thickness in single-layered test and thus introducing great data error. Double-layered test could largely weaken the k-thickness dependence of thin thermal gel layer and provide the detailed R_i value for conference simultaneously. The results illustrate that the double layers-based laser flash method could be a crucial tool to real-time characterize k and R_i of thermal gel-metal system during the aging process.

Keywords—thermal gel, thermal conductivity, thermal resistance, laser flash method

I. Introduction (*Heading 1*)

Thermal gel, an important class of thermal interface materials (TIMs), is an important component in thermal management solutions that provide a path of low thermal resistance between heat generating devices and the heat spreader/sink [1-6]. Inorganic/organic hybrid thermal gels are of increasing interest since they offer the prospect of combining the mechanical toughness and flexibility of organic component with the thermal conductivity (k) of the inorganic component [7, 8]. However, the hardness of hybrid thermal gels would increase after long-time working under high temperature, thus resulting in undesirable larger interfacial thermal resistance (R_i). The intrinsic k of thermal gels simultaneously varies along with the changing hardness due to the different dispersion state of filler. Improving the anti-aging performance of thermal gels is essential to ensure the efficient heat transfer. Therefore, real-time monitoring the intrinsic k of thermal gels, and the variation of R_i at the thermal gel/substrate interface during the aging process is urgently-needed for exploring the anti-aging mechanism and further improve the overall transfer performance.

The ASTM D5470 standard is commonly applied to test the thermal performance of thermal gels [9, 10]. Longwin

978-1-6654-1392-3/21 $31.00 © 2021 IEEE

TIM LW-9389 from Taiwan is the representative equipment. The sample was put into the process of steady heat transfer as k value was calculated on the basic of Fourier formula. This method is particularly suitable for isotropic composites with relatively low k. However, this standard has a main drawback to monitor the real-time parameters of thermal gels when the external temperature varies. Meanwhile, the test temperature cannot be accurately fixed, thus the k values at different temperatures cannot be obtained. On the other hand, the test sample normally bears external pressure to maintain desired contact with the upper and the below surface during the characterization. The inaccuracy of thickness measurement will bring great error to the result, especially for thin films with the thickness less than 100 μm.

Laser flash test as a representative transient method has been widely used to characterize highly thermally conductive carbon-based or metal materials [11-13]. A beam of laser pulse irradiates the lower surface of the sample to allow the surface temperature rise instantaneously after absorbing the energy. The lower surface then acts as the hot end to transfer the energy to the cold end in a one-dimensional heat conduction mode. The thermal diffusivity (α) is calculated by fitting the half heating time (t_{50}) in equation 1

$$\alpha = \frac{0.1388 \times d^2}{t_{50}} \qquad (1)$$

where d is the thickness of the sample. This equation works under some ideal conditions, e.g. the width of the laser pulse is nearly infinitesimal or can be neglected relative to t_{50} of the sample. t_{50} should be controlled in a reasonable range by firstly selecting appropriate d. Highly thermally conductive sample with $\alpha > 50$ mm²/s should possess larger d, e.g. the recommended d for metal and carbon-based materials falls in the range of 2 ~4 mm. On the other hand, polymer composites with lower α should be casted into thinner sample for test. However, there are some restrictions in practice. It is extraordinary time-consuming to fabricate graphene films with millimeter d. Meanwhile, some polymer films with small d can't be self-supported on the carrier and could be easily soften or melted after absorbing the energy of the laser pulse. Therefore, an improved method should be proposed.

In this work, we introduce double layers-based laser flash method as a thermal measurement mean, and also a strategy to control the sample temperature and imitate the actual aging environment under continuous temperature cycle. LFA 467 HyperFlash equipment (NETZSCH, Germany) was employed on the basic of ASTM E1461 standard. One of the commercial thermal gels that we have developed, named SIAT-1, has been chosen as the experimental sample. The gel was hot-pressed on silicon wafer to form double-layered structure. In order to evaluate the thermal performance of SIAT-1, the samples with different thicknesses are put in air atmosphere and subjected to a temperature cycling test (TCT). We found that the intrinsic k (mainly from the thermal diffusion coefficient α) of thermal gel obtained by single-layer mode varies greatly with the film thickness. On the other hand, the dependence of d-k was largely weakened by applying double layers-based model. As d of thermal gel is much lower than that of the silicon layer, the data error is greatly reduced due to the fact that the uniform metal bulk is able to cover most of the error produced by inhomogeneous thermal gel. After eight cycles, SIAT-1 exhibited relatively stable α values in the range of $1.89\pm1.3147e{-002}$ mm²/s at room temperature. Interfacial thermal resistance (R_i) values for SIAT-1/silicon interface fall in the range of 1.1328e-007 to 4.0218e-007 Km²/W. In addition, the sample is able to be heated by laser, and being maintain under a fixed temperature for certain time, providing k and R_i variation during the aging process. The results illustrate that the double layers-based laser flash method is a crucial tool to real-time characterize k and R_i of TIMs-metal system. The results illustrate that the double layers-based laser flash method is a crucial tool to real-time characterize k and R_i of TIMs-metal system.

II. EXPERIMENTAL

A. Preparation of double-layered structure

The thermal gel SIAT-1 was firstly dispensed on the square silicon wafer (Figure 1). Release membrane was then put on the surface of thermal gel, followed by hot-pressing at 15 psi and 140 °C for 1h. d of thermal gel could be varied from 30 to 150 μm. After thermal curing, the spillover part of thermal gel outside the silicon wafer was removed. The release membrane was carefully peeled off to form thermal gel-silicon wafer double-layered structure.

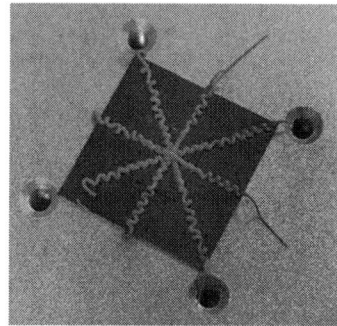

Figure 1. Pattern of dispensing thermal gel on silicon wafer.

B. Double layers-based laser flash characterzation

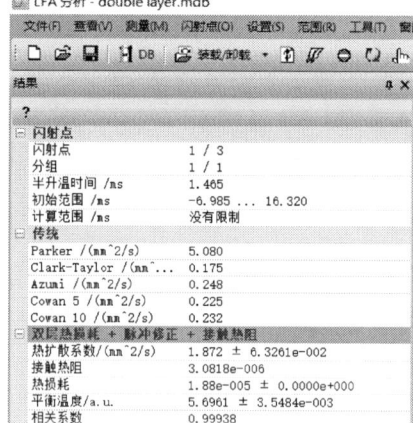

Figure 2. Typical interface of simulation results in double-layered test.

α of the square silicon wafer was first obtained by single-layer model of laser flash method. The specific heat capacity (C_p) of thermal gel SIAT-1 or square silicon was characterized by Q2000 differential scanning calorimeter (TA, United States) via sapphire method. Coefficients of thermal expansion (CTE) of the samples at different temperatures were obtained by TMA 402 F1/F3 Hyperion (NETZSCH, Germany). Density (ρ) was characterized by density meter (METTLER TOLEDO, Switzerland).

978-1-6654-1392-3/21 $31.00 © 2021 IEEE

Double layers-model was chosen for the test. α, d, ρ, C_p and CTE of silicon wafer were pre-input in the system. d, ρ, C_p and CTE of thermal gel were also filled in simultaneously. Both sides of the double-layered structure were sprayed with carbon powder. The side of silicon wafer is down in order to maintain the absorption of laser. The typical interface of experimental data is shown in Figure 2. The results were processed *via* 'Double-layered heat loss & Pulse correction+ Contact thermal resistance' model. After the test, the analysis software will give detailed α and R_i values. For Temperature cycling test (TCT), the sample is heated to 200 °C with heating rate then cooled to room temperature. The corresponding α along with interfacial thermal resistance (R_i) values are set to be recorded every 25 °C. k values of the samples were then calculated by

$$k = \alpha \rho C_p \tag{2}$$

For reference, k of thermal gel SIAT-1 was also measured by LW-9389 TIM Thermal Resistance and Conductivity Measurement Apparatus (Long Win Science and Technology, Taiwan) on the basic of steady-state method at room temperature. k was calculated by

$$k = -\frac{QL}{A\Delta T} \tag{3}$$

Herein, Q is the heat flux, L is the thickness of the sample, A is the area of the sample, and ΔT is the temperature difference between temperature sensors of the hot meter bar.

III. RESULTS AND DISCUSSION

A. *Dependence of d-k in single-layered test*

α is one of the basic properties of materials as it should not be changed with the varying d. However, we found that there exists obvious dependence of d-k for multiple materials in single-layered model. Thermal gel SIAT-1 with different d values were characterized, as shown in Figure 3a. All tests are effective in consideration of the time-signal curves. The dependence of d-α existing in SIAT-1 is obvious. SIAT-1 with d value of 0.58 mm exhibits α of 2.262 mm²/s. As d further increases to 2.45 mm, α value sharply reduces to 0.933 mm²/s, indicating a 58.75% attenuation when compared to the initial value. It is speculated that the variety mainly comes from the thickness calibration when calculating α in single-layered model. In-plane test is carried on under ideal situation where the through-plane diffusion time can be neglected. However, the diffusion time will certainly vary with different d. Meanwhile, there might happen that smaller d below the critical value results in ultrafast absorption time of laser which is less than the detection limit. C_p values of SIAT-1 with different d were also obtained (Figure 3b). As d increases from 0.58 to 2.45 mm, C_p shows only a 6.92% increasement, demonstrating the test error of k is mostly caused by the dependence of d-α. Temperature cycling test (TCT) was conducted to further explore the dependence of d-α. SIAT-1 samples with different d were heated to 200 °C then cooled to room temperature. It is noted that sample with d of 0.58 mm exhibits much rough variation in absolute value. Specially, α values of SIAT-1 (0.58 mm) at 50, 75 and 100 °C demonstrate huge errors during TCT. Herein, vibration coefficient of α (Δ_α) is calculated by

$$\Delta_\alpha = \left| \frac{\alpha_{th} - \alpha_{tc}}{\alpha_{th}} \right| \times 100\% \tag{4}$$

Where α_{th} is the corresponding α under a fixed temperature during the heating process, α_{tc} is the counter α related to the cooling process. The detailed Δ_α values extracted from Figure 3c for SIAT-1 with d of 0.58 and 2.45 mm are summarized in Table 1.

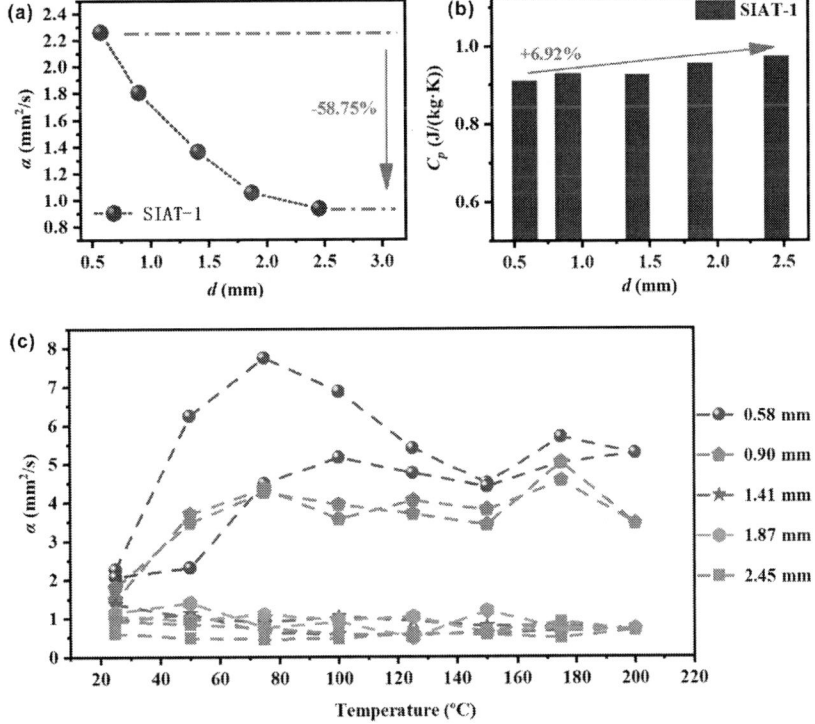

Figure 3. Thermal characterization of SIAT-1 by single-layered model. (a) Variation of d-α; (b) Variation of d-C_p; (c) TCT results of SIAT-1 with different d values. The sample is heated to 200 °C then cooled to room temperature.

Table 1. Detailed Δ_α values for SIAT-1 in single-layered test.

Temperature (°C)	d=0.58 mm	d=2.45 mm
25	8.62%	35.69%
50	62.96%	43.01%
75	42.11%	38.78%
100	24.86%	22.22%
125	11.99%	12.15%
150	2.14%	13.76%
175	11.67%	45.92%
Average	23.48%	30.22%

Table 1 shows the relatively large data deviation in single-layered test. The average Δ_α value as d is 0.58 and 2.45 mm reaches 23.48% and 30.22%, respectively. Therefore, the dependence of d-k in single-layered test need to be weaken to get more accurate results.

B. Double-layered test

Most of the industrial application scenarios requires thermal gel layer with d below 100 μm. Single-layered model is not appropriate to be applied in this situation. Therefore, an addition metal layer is chosen to attached to the surface of thermal gel to form double-layered structure (Figure 4). The metal layer is designed to act as ideal laser absorber and buffer layer to reduce the data error. Moreover, the introduction of metal layer is imitating the actual interface of thermal gel/hot sink simultaneously. Thus, k along with R_i could be real-time monitored, providing experimental support for anti-aging research. In this study, SIAT-1 was dispensed on silicon wafer followed by hot-pressing. d of silicon wafer is 505 μm.

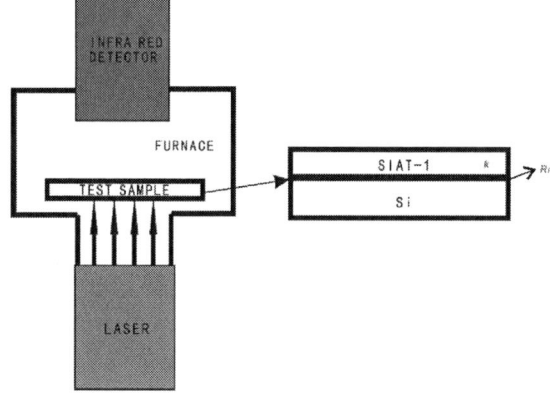

Figure 4. Illustration of double-layered test.

The dependences of α-d and R_i-d were examined as shown in Figure 5. As d increases from 37 to 103 μm, the α value stays relatively steady, exhibiting much weaker dependence of α-d in double-layered test when compared to single-layered model. Significantly, the maximum α value in single-layered test (2.262 mm^2/s) is larger than that of double-layered test (1.932 mm^2/s), indicating the existence of great data error in single-layered test. Unlike metal, polymer composites materials are inhomogeneous, thus direct irradiation of laser on different areas would results in distinct laser absorptions. It is speculated that the silicon wafer acts as the buffer layer for SIAT-1 which can weaken the impacts from laser scattering and penetration. Moreover, silicon possesses much larger d than that of SIAT-1, thus d variation of SIAT-1 brought less impact for the double-layered structure and reduced the data error. Figure 5b shows the R_i values at the SIAT-1/silicon interfaces. The numerical value basically maintains at the same order of magnitude ($\sim 10^{-7}$), indicating the good stability of double-layered test. The values are in accordance with previous reports. To further illustrate the data stability, TCT was conducted as shown in Figure 5c. SIAT-1 samples with three different d values all exhibit obvious downward trend with the increasing temperature. This is due to the characteristic of alumina. The detailed Δ_α values for SIAT-1 in double-layered test are summarized in Table 2. It is noted that the average Δ_α value is reduced by an order of magnitude when compared to the value in single-layered test. Specifically, SIAT-1 with d of 57 μm shows a Δ_α value of only 1.71% falling in the normal system error range. For data reference, we also characterize SIAT-1 sample via the ASTM D5470 standard. For steady state method, the k value reaches 4.05 W/mK. For our double-layered test, the k value calculated by multiplying α, ρ and C_p reaches 4.23 W/mK. It is usual that data difference exists between steady state and laser flash methods. The above results indicate that double-layered test could obtain precise and steady data.

As an extension of this work, sandwich structure was also explored. This kind of structure bring more procedures to calculate α and R_i values by firstly fitting parameters for one controllable interface. Related contents will be discussed in our following papers.

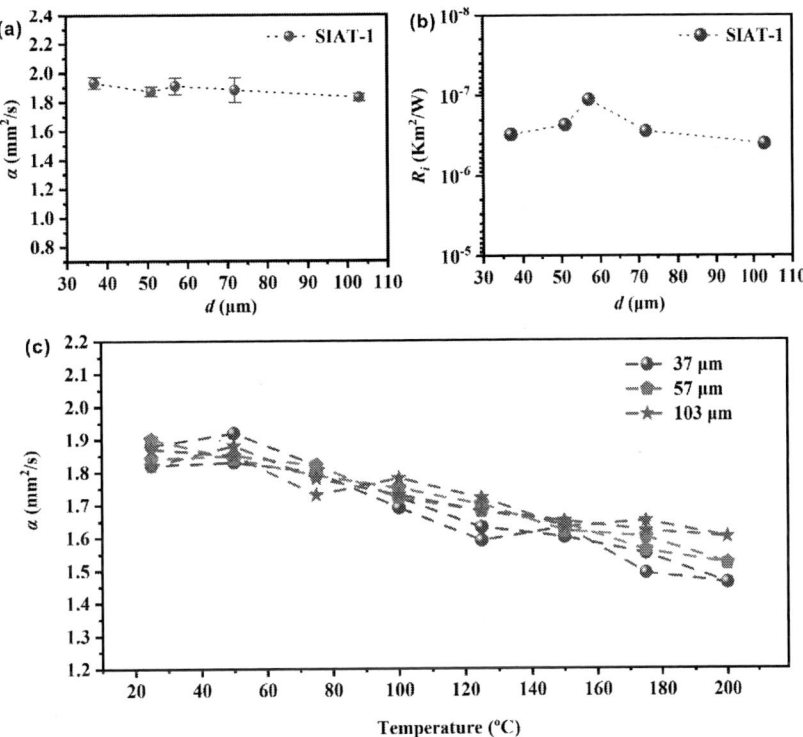

Figure 5. Thermal characterization of SIAT-1 by double-layered model. (a) Variation of d-α; (b) Variation of d-R_i; (c) TCT results of SIAT-1 with different d values. The sample is heated to 200 ℃ then cooled to room temperature.

Table 2. Detailed Δ_α values for SIAT-1 in double-layered test.

Temperature (℃)	d=37 μm	d=57 μm	d=103 μm
25	3.20%	3.16%	2.67%
50	4.69%	0.54%	1.62%
75	1.01%	1.68%	2.89%
100	1.74%	1.71%	2.81%
125	2.45%	1.12%	2.33%
150	1.89%	1.23%	1.23%
175	3.87%	2.50%	1.82%
Average	2.69%	1.71%	2.20%

To illustrate the application of double-layered test, SIAT-1 with d of 57 μm was put into TCT for eight circles. Herein, vibration coefficient of R_i (Δ_R) is also introduced by

$$\Delta_R = \left| \frac{R_{ih} - R_{ic}}{R_{ih}} \right| \times 100\% \qquad (5)$$

where R_{ih} is the corresponding R_i under a fixed temperature during the heating process, R_{ic} is the counter α related to the cooling process. Δ_α and Δ_R values of SIAT-1 during TCT were recorded in Figure 6. Δ_α value varies in the range of 1%~2.5%, indicating SIAT-1 shows basically constant α during the TCT process. The result demonstrates that SIAT-1 possesses good anti-aging performance. On the contrary, Δ_R value varies with larger amplitude. Considering the magnitude of Δ_R value (10^{-7}), the value exhibiting larger changing amplitude is inevitable. We assume R_i remains steady as it's stable in the order of magnitude. Further exploration is needed to obtain stable and precise R_i values.

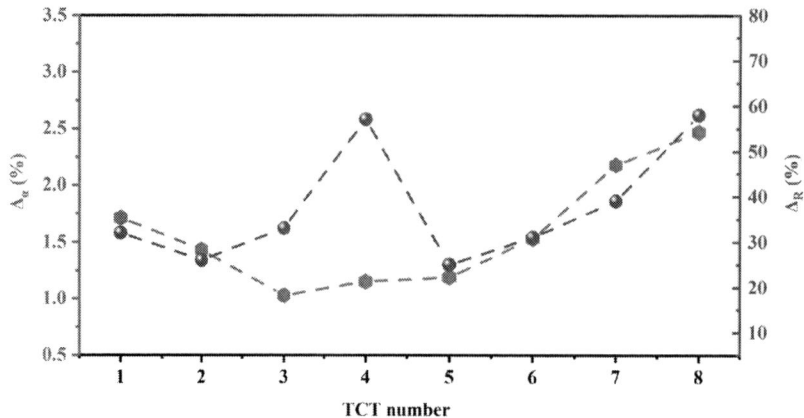

Figure 6. Δ_α and Δ_R values of SIAT-1 during TCT.

IV. CONCLUSION

In this study, we have explored the data stability and accuracy of single-layered and double-layered laser flash tests. For some thin isotropic composite films with relatively low thermal conductivity, the single-layered test shows strong dependence of α-d, resulting in large k error. Moreover, thin layer of thermal gel with d below 100 μm cannot be characterized by single-layered fixture as it's not self-supported. Thus, double-layered test is developed. To imitate the real application scene, a layer of silicon wafer was attached to the surface of thermal gel. It shows that the dependence of α-d is largely weaken, leading to stable and accurate data. It is noted that the average Δ_α value is reduced by an order of magnitude when compared to the value in single-layered test. α and R_i values can be real-time monitored, providing experimental data for exploring the anti-aging mechanism and further improve the overall transfer performance.

ACKNOWLEDGMENT

The authors would like to acknowledge the financial support from National Science Foundation of China (No. 62004210), Hong Kong Scholars Program funded by China Postdoctoral Science Foundation (No. XJ2019032), Key-Area Research and Development Program of Guangdong Province (No. 2020B010190004), Guangdong Basic and Applied Basic Research Foundation (No. 2019A1515110845).

REFERENCES

[1] Y. M. Yao, J. J. Sun, X. L. Zeng, R. Sun, J. B. Xu, and C. P. Wong: 'Construction of 3D Skeleton for Polymer Composites Achieving a High Thermal Conductivity', Small, 2018, 14, (13), 1704044.

[2] Y. M. Yao, X. D. Zhu, X. L. Zeng, R. Sun, J. B. Xu, and C. P. Wong: 'Vertically Aligned and Interconnected SiC Nanowire Networks Leading to Significantly Enhanced Thermal Conductivity of Polymer Composites', ACS Appl. Mater. Interfaces, 2018, 10, (11), pp. 9669-9678.

[3] Y. M. Yao, Z. Q. Ye, F. Y. Huang, X. L. Zeng, T. Zhang, T. Y. Shang, M. Han, W. L. Zhang, L. L. Ren, R. Sun, J. B. Xu, and C. P. Wong: 'Achieving Significant Thermal Conductivity Enhancement via an Ice-Templated and Sintered BN-SiC Skeleton', ACS Appl. Mater. Interfaces, 2020, 12, (2), pp. 2892-2902

[4] Y. Yao, X. Zeng, K. Guo, R. Sun, and J.-B. Xu: 'The Effect of Interfacial State On the Thermal Conductivity of Functionalized Al_2O_3 Filled Glass Fibers Reinforced Polymer Composites', Compos. Part A, 2015, 69, pp. 49-55.

[5] Y. F. Zhang, D. Han, Y. H. Zhao, and S. L. Bai: 'High-Performance Thermal Interface Materials Consisting of Vertically Aligned Graphene Film and Polymer', Carbon, 2016, 109, pp. 552-557.

[6] H. Hong, J. U. Kim, and T.-I. Kim: 'Effective Assembly of Nano-Ceramic Materials for High and Anisotropic Thermal Conductivity in a Polymer Composite', Polymers, 2017, 9, (9), 9090413.

[7] K. Chano, G. M. Poliskie, and J. Fregoso: 'Rheology of Thermal Interface Materials Composed of Silicone Gels', Ieee Transactions on Components Packaging and Manufacturing Technology, 2017, 7, (2), pp. 217-220.

[8] B. Pena, M. Maldonado, A. J. Bonham, B. A. Aguado, A. Dominguez-Alfaro, M. Laughter, T. J. Rowland, J. Bardill, N. L. Farnsworth, N. A. Ramon, M. R. G. Taylor, K. S. Anseth, M. Prato, R. Shandas, T. A. McKinsey, D. Park, and L. Mestroni: 'Gold Nanoparticle-Functionalized Reverse Thermal Gel for Tissue Engineering Applications', ACS Appl. Mater. Interfaces, 2019, 11, (20), pp. 18671-18680.

[9] C. K. Roy, S. Bhavnani, M. C. Hamilton, R. W. Johnson, J. L. Nguyen, R. W. Knight, and D. K. Harris: 'Investigation Into the Application of Low Melting Temperature Alloys as Wet Thermal Interface Materials', Int. J. Heat Mass Transfer, 2015, 85, pp. 996-1002.

[10] C. R. Yang, C. D. Chen, C. Cheng, W. H. Shi, P. H. Chen, and T. P. Teng: 'Thermal Conductivity Enhancement of AlN/PDMS Composites Using Atmospheric Plasma Modification Techniques', Int. J. Therm. Sci., 2020, 155, 106431.

[11] W. N. dos Santos, P. Mummery, and A. Wallwork: 'Thermal Diffusivity of Polymers by the Laser Flash Technique', Polym. Test., 2005, 24, (5), pp. 628-634.

[12] Y. H. Zhao, Z. K. Wu, and S. L. Bai: 'Thermal Resistance Measurement of 3D Graphene Foam/Polymer Composite by Laser Flash aAnalysis', Int. J. Heat Mass Transfer, 2016, 101, pp. 470-475.

[13] G. Ferrarini, A. Bortolin, G. Cadelano, L. Finesso, and P. Bison: 'Multiple Shots Averaging in Laser Flash Measurement', Appl. Opt., 2020, 59, (17), pp. 72-79.

Research on Conformal Phased Array T/R Module Based on LCP Substrate

Yan Luo
Micro-electronics business department
Shanghai Aerospace Electronic and
Communicaion Equipment Reaserch
Institute
Shanghai, China
luoyan1120@qq.com

Kai Liu
Micro-electronics business department
Shanghai Aerospace Electronic and
Communicaion Equipment Reaserch
Institute
Shanghai, China
k2006l@126.com

Qiu Gao
Micro-electronics business department
Shanghai Aerospace Electronic and
Communicaion Equipment Reaserch
Institute
Shanghai, China
gaoqiu1993@126.com

Lei Ding
Micro-electronics business department
Shanghai Aerospace Electronic and
Communicaion Equipment Reaserch
Institute
Shanghai, China
youyouzi.teng@163.com

Yi Zhou
Micro-electronics business department
Shanghai Aerospace Electronic and
Communicaion Equipment Reaserch
Institute
Shanghai, China
498097792@qq.com

Lichun Wang
Micro-electronics business department
Shanghai Aerospace Electronic and
Communicaion Equipment Reaserch
Institute
Shanghai, China
wanglichun0482@163.com

Abstract—In the conformal phased array radar, the flexible surface T/R module based on LCP substrate was studied to solve the problems such as the mismatching between components and the antenna plane and the system instability caused by the increase of plug and unplug of cable connection. The LCP laminating quality control and the chip embedding was investigated. The assembly stress on LCP multi-layer substrate was analyzed with ANSYS. The results show that optimized lamination process with auxiliary material makes the substrate flat without cavity inside. The air tightness of LCP substrate embedded with MMIC is 7.4×10^{-7} Pa · m³/s. The simulation results show that the stress increases rapidly between temperature of -55 and 80 degree centigrade when the LCP substrate's bending angle is greater than 10 degrees. The conformal T/R module remains well performance with LCP substrate bending 10 degrees. The LCP substrate has a good potential in conformal phased array.

Keywords—flexible T/R module, LCP substrate, laminating process, chip embedding, stress simulation

I. INTRODUCTION

Compared with high-fired HTCC and low-fired LTCC substrates, organic material substrates have the advantages of wide dielectric constant selection range (2~10), low-cost PCB technology, and complex multilayer structures. It is one of the widely used microwave circuit substrates. For decades, PTFE based materials have been the main choice for high-performance microwave organic substrates. However, in the last ten years, LCP substrate has received continuous attention. It exhibits excellent dielectric properties comparable to PTFE materials in the microwave to millimeter wave (up to 110GHz) frequency bands [1-3]. The characteristics of excellent thermal stability, high modulus, adjustable thermal expansion coefficient and low moisture absorption rate have shown broad application prospects in radio frequency systems [4-5].

Various application researches such as microwave /millimeter wave filters, VCOs, T/R modules, RF MEMS, SIP, and SOP developed on LCP substrates have been reported in succession [6-9]. With the continuous improvement of the working frequency of radio frequency communication,

the size of passive components such as antennas has been greatly reduced [10]. It is necessary to achieve conformal modules and antennas to meet the development of miniaturization and multi-functionality.

II. EXPERIMENTAL SETUP

A. Active Sub-array Package Structure Design

A curved T/R module is designed based on the LCP substrate. As shown in Fig. 1, there are 5 layers of LCP substrates and 4 connecting layers. The thickness of the LCP substrate is 100 μm, the thickness of the connection layer is 25 μm, and the thickness of the copper clad layer of the substrate is 12 μm. The bottom layer is the ground layer, followed by the MMIC buried layer, which is hermetically sealed by LCP. The electrical interconnection and grounding between the layers are realized based on the LCP through-hole plating technology. The MMIC is mounted on the MoCu carrier to achieve structural support and heat dissipation. The multi-layer structure and circuit are designed through HFSS simulation.

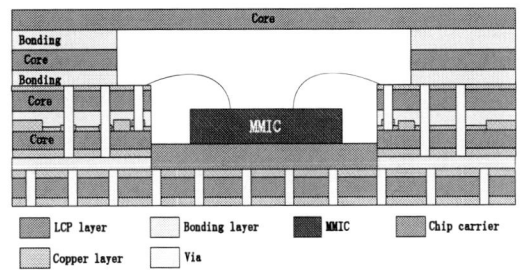

Fig. 1. The structure diagram of LCP T/R module.

B. The Packaging Process of LCP T/R module

According to the T/R module structure, the packaging process is shown in Fig. 2. First, each layer of LCP performs photolithography according to the circuit diagram. The wiring plates with Ni/Pd/Au film to improve the bonding strength. Then, the connection layer and the LCP substrate are pre-cured under pressure at temperature of 130°C for 20

978-1-6654-1392-3/21 $31.00 © 2021 IEEE

seconds. In order to realize the MMIC embedding, the embedding groove is processed by laser. The bottom LCP substrate realizes the grounding and heat dissipation by distributing a large number of ground holes. The buried holes, blind holes, and through holes are realized by bonding first and then hole processing through multiple lamination. Finally, the planar T/R module is prepared by LCP substrate airtight cover. Use bending tooling for the bending of the substrate. The substrate is bent and annealed to remove the internal stress. Then it is mounted on the conformal platform, and the connectors are soldered to obtain the conformal active sub-array.

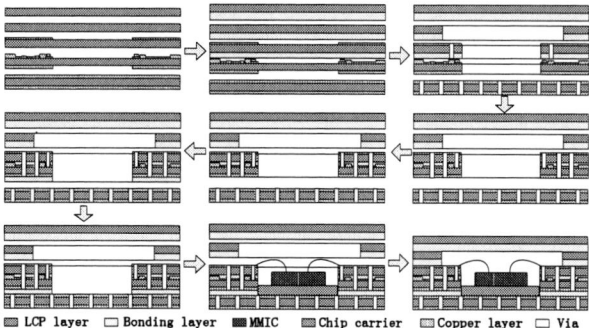

☐ LCP layer   ☐ Bonding layer   ☐ MMIC   ☐ Chip carrier   ☐ Copper layer   ☐ Via

Fig. 2. The packaging process.

C. Finite Element Model Establishment

When the substrate is bent, the film will generate stress. And in severe cases, it may cause the substrate to delamination or the film to crack and fail. The substrate model is established by ANSYS software, and the simulation material parameters of the LCP substrate are shown in Table 1.

TABLE I.   STRESS SIMULATION PARAMETERS OF LCP SUBSTRATE

LCP	Value	Unit
Density	1.4	g/cm^3
Young's Modulus	4.5	Gpa
Poisson's Ratio	0.3	
Coefficient of thermal expansion X direction	18	ppm/℃
Coefficient of thermal expansion y direction	18	ppm/℃
Coefficient of thermal expansion Z direction	209	ppm/℃
Compressive Ultimate Strength	150	Mpa

The simulation setting temperature is the thermal shock condition from -55℃ to 80℃ while the material starting temperature is 22℃. To investigate the influence of curvature on LCP substrate in environmental applications and the influence trend, choose five values (5°, 10°, 20°, 43°, 65°) of the curvature setting with the substrate length 34 mm. The simulation substrates are shown in Fig. 3. In order to peel off the interference of the bottom plate's own stress on the accuracy of the simulation data, the LCP constraint is set to be fixed.

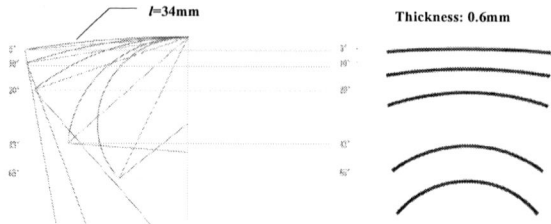

Fig. 3. Schematic diagram of substrate bending angle.

III. RESULTS AND DISCUSSION

A. Multilayer LCP Substrate Laminate Technology

Due to the flexible nature of the LCP substrate and the fusion bonding method of the connecting layer, the LCP is prone to quality problems such as deformation, warpage, voids, and liquid leakage during the bonding process, which is not conducive to maintaining the accuracy of the subsequent process, and even affects the final total shape size.

Due to the limitations of process and equipment, the temperature of the bonding machine cannot be controlled steadily and continuously near the melting point temperature. In order to accurately control the quality of LCP bonding and reduce thermal deformation, a gradient heating mode to control the quality of lamination is adopted. As shown in Fig. 4, the recommended temperature is T1 150°C for pre-bonding with holding time t1 20min, T2 180°C for bonding with holding time t2 60min, and then cooling down. The LCP bonding result obtained according to this temperature profile is shown in Fig. 5(a). It shows that the LCP substrate can be peeled off. Due to the temperature transfer delay and heat loss from the platform, the actual temperature is lower than the device display. If the temperature is too high, the conductor drifts as shown in Fig. 5(b).

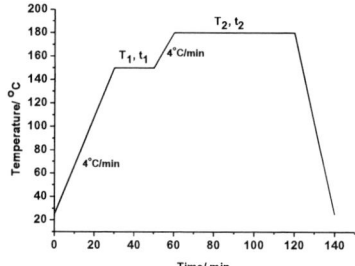

Fig. 4. Gradient heating curve.

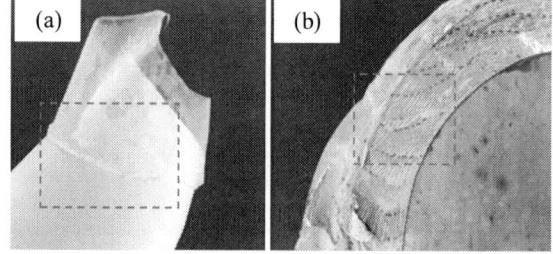

Fig. 5. LCP lamination defect, (a) Bonding layer pealed off, (b) Copper conductor outflow.

Set up different heating temperature curves to study the quality of LCP bonding. Since the pre-bonding temperature T1 does not require high accuracy, only the bonding temperature T2 is studied. The bonding temperature and results are shown in Table 2 while the bonding pressure is set to 1000 kg.

TABLE II.   BONDING TEST TEMPERATURE PARAMETER TABLE

Number	T2 /℃	t2 /min	Results
1	180	60	Not bonded
2	200	60	Not bonded
3	220	60	Partially bonded
4	240	60	Bonded
5	260	60	LCP material flows out at the edge
6	280	60	Copper conductor drifts out

Due to the lack of covering ability of Kapton tape, the laminated substrate has tiny wrinkles. The ultrasonic scanning micrograph result is shown in Fig. 6. It shows cavities between the substrates at the wrinkles which affecting the alignment of the circuit patterns on the substrate. Therefore, it is necessary to control the uniformity and flatness of the lamination pressure.

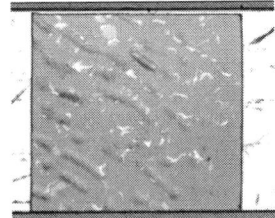

Fig. 6. Ultrasonic scanning microscope picture of laminated substrate.

Based on traditional heating and pressing, a lamination auxiliary material is added between the LCP and the tooling. As shown in Fig. 7, the lamination auxiliary material includes a high temperature pad, a release film, a filling film and an aluminum foil. The high temperature pad has the characteristics of uniform heat conduction, balanced pressure, and high temperature resistance. It has the function of equalizing pressure and thermal buffer in the lamination, and effectively controls the problems of white corners, white edges, and wrinkled bubbles. The filling film has the characteristics of uniform thickness, good filling and covering effect, and strong board surface expansion and contraction control ability. Its dimensional stability is high, XY deformation is small, and the Z direction can meet the needs of filling and covering type, and the filling height is greater than 18μm. The release film has good peeling performance, high temperature resistance, and dimensional stability. It provides a super smooth surface when LCP is laminated. Aluminum foil also has a certain degree of covering properties and plays a role in reducing deformation in LCP lamination. After the auxiliary material is coated, the substrate is relatively flat, and the laminated substrate is photographed with an ultrasonic scanning microscope. The result is shown in Fig. 8, which shows that this method can improve the flatness of the substrate.

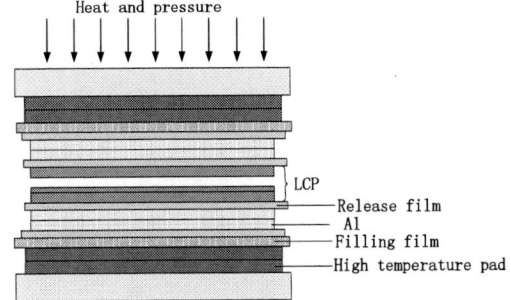

Fig. 7. Schematic diagram of LCP laminated auxiliary material structure.

Fig. 8. Ultrasonic scanning microscope picture of laminated substrate with laminated auxiliary material.

Through the pressure and uniformity control, the multi-layer LCP substrate of the integrated divider is laminated to obtain a flat LCP substrate as shown in Fig. 9.

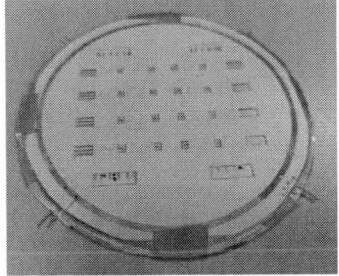

Fig. 9. Multilayer LCP substrate.

B. MMIC Embedding in LCP Substrate

During the working process of the high-frequency chip, a large amount of heat must be released. The LCP as the material of the system body is a high molecular polymer, which has poor thermal conductivity and cannot export heat in time which making the chip very easy to occur burning and falling off under long-term and full-load working conditions. A copper structure under the chip intends to increase the heat dissipation. Since the high-frequency multi-function chip is mostly GaAs material whose CTE is $5.8 \times 10^{-6} \cdot K^{-1}$, while the CTE of copper is $17.6 \times 10^{-6} \cdot K^{-1}$, problems such as thermal stress imbalance may easily occur during system operation. In response to such problems, a metal Mo layer is added in the middle of the copper carrier. Using the CTE of Mo $5.1 \times 10^{-6} \cdot K^{-1}$ to adjust the overall CTE of the carrier and match the CTE of the GaAs chip to form Cu-Mo-Cu integrated heat dissipation structure.

The initial plan is to assemble the Cu-Mo-Cu carrier on the copper layer of the lowermost LCP substrate. The thickness of the LCP copper layer is 12μm, and then copper is electroplated to thicken it. In the process of transfer and cleaning of the test substrate, the copper layer is too thin to be easily damaged. So a layer of LCP substrate is added as the ground below. Compared with the initial solution, the system has a total of 9 layers, as shown in Fig. 10.

The Cu-Mo-Cu pad is used to improve the heat dissipation capacity of the chip, and at the same time reduce the impact of bending stress on the chip when the substrate is bent. The LCP substrate improves the grounding performance through the grid ground design on the one hand, and on the other hand ensures the bonding strength between the LCP layers, as shown in Fig. 11. The bottom LCP substrate improves the heat dissipation capacity through the ground hole.

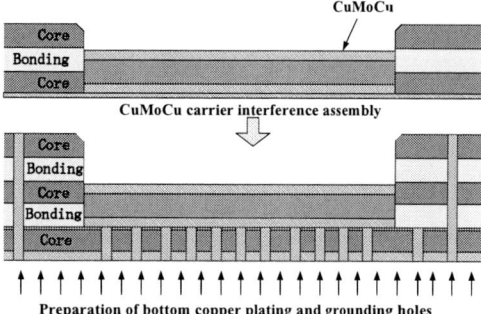

Fig. 10. Interference CuMoCu carrier.

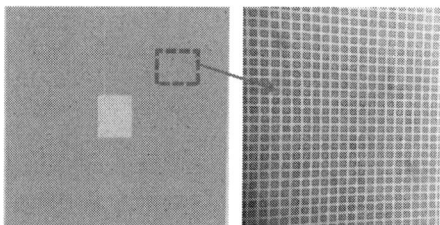

Fig. 11. The grid ground.

Due to the flexible characteristics of the LCP substrate, problems such as warping and deformation occur during vacuum adsorption, which will cause the offset of the substrate positioning coordinate system, which will seriously affect the COB placement accuracy. In order to achieve high-precision and low-drift chip placement, the local adsorption pressure is evened through high-density micro-vias. The chip placement is shown in Fig. 12.

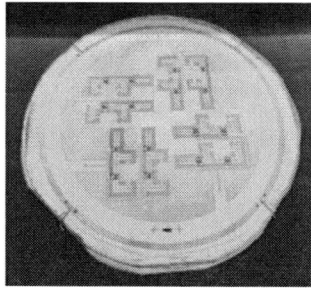

Fig. 12. Chip mount substrate.

C. Stress Analysis of Curved Assembly Substrate

When the substrate is bent, the film will generate stress, which may cause the substrate to delamination or the film to crack and fail. According to the film stress Stoney formula,

$$\sigma_f = \left(\frac{E_s}{1 - V_s} \right) \frac{t_s^2}{6 R t_f} \qquad (1)$$

where σ_f is the film stress, E_s is the elastic modulus, V_s is the Poisson's ratio of the substrate, t_s is the thickness of the substrate, t_f is the thickness of the film, and R is the radius of curvature of the substrate. It can be seen that the greater the number of layers of the substrate and the larger the radius of curvature, the greater the film stress. The bending stress of the substrate needs to be simulated.

The stress of substrate in thermal shock cycle is analyzed to verify whether the substrate is reliable in the application environment. The simulation process is shown in Fig. 13, and the stress distribution results are obtained as shown in Table 3. The five sets of stresses are all less than the ultimate stress of 150Mpa, and the minimum safety factor of the LCP substrate at each angle is 5.3 which is larger than 1, meet the requirements of mechanical reliability.

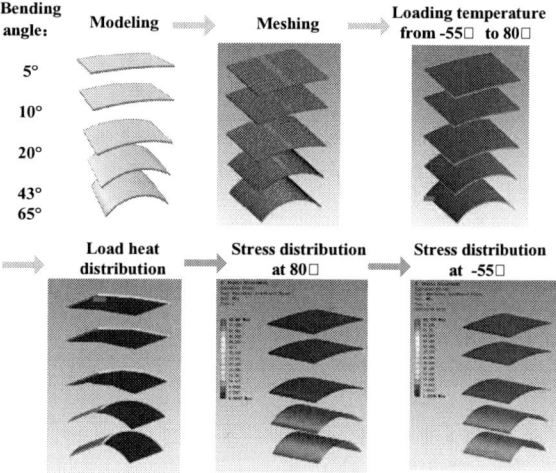

Fig. 13. Simulation flowchart.

TABLE III.   STRESS DATA OF EACH BENDING DEGREE

Angle	Maximum stress value at -55℃	Maximum stress value at 80℃	Unit
5°	11.033	8.055	MPa
10°	11.432	8.083	MPa
20°	18.758	12.95	MPa
43°	50.987	39.70	MPa
65°	82.337	62.02	MPa

Fig. 14 shows the stress trend. It can be seen that when the angle is less than 10°, the stress is at a low level and much smaller than the ultimate stress; but when the angle is greater than 10°, the stress deteriorates sharply with the increase of the angle.

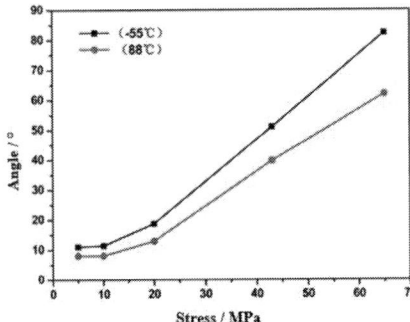

Fig. 14. Stress trend.

A multi-point coating method is designed to assemble the curved LCP substrate on the shell. Assemble the 4-channel LCP curved sub-array according to the tooling whose bending angle is 10°. The sub-array sample is shown in Fig. 15. The samples have completed typical simulation environment verification, including temperature shock, sinusoidal vibration and random vibration tests, and the electrical performance test results meet the index requirements.

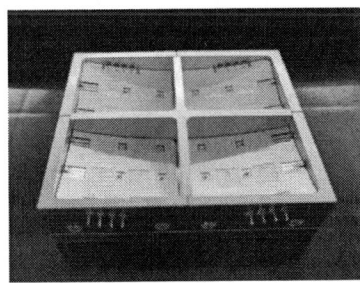

Fig. 15. 16-channel curved LCP sub-array.

IV. CONCLUSION

1) By optimizing the lamination temperature and pressure, the substrate with good bonding force can be obtained at 220℃~230℃ temperature and 1 MPa pressure. Aluminum foil and other laminated auxiliary material are used to eliminate the wrinkles of the substrate, and the substrate is flat without cavity inside.

2) The chip embedded slot is cut by laser and the MoCu pad is put inside to let heat out and release the stress on the chip when the substrate is bent.

3) Under the temperature cycle of -55℃~80℃, when the bending angle of LCP substrate less than 10 degrees, the stress is low and far less than the ultimate stress. But the stress at angle larger than 10 degrees increases sharply with the increasing bending angle.

4) The 16-channel conformal T/R module remains well performance with LCP substrate bending 10 degrees in environmental application verification. The LCP substrate has a good potential in conformal phased array..

REFERENCES

[1] Thompson D, Manos M, John P. "ackaging of MMICs in multilayer LCP substrates," IEEE Microw Wirel Co Lett 2006, 16: 410-412.

[2] Zhang Y, Martin R, Shi S, et al. "95-GHz front-end receiving multichip module on multilayer LCP substrate for passive millimeter-wave imaging," IEEE Transactions on Components, Packaging and Manufacturing Technology 2018, 8(12): 2180-2189.

[3] Manos M, Joy L, John P, et al. "3-D-integrated RF and millimeter-wave functions and modules using liquid crystal polymer (LCP) system-on-package technology," IEEE Transactions on Advanced Packaging 2004, 27(2): 332-340.

[4] Li S, Chi T, Park J, et al. "A fully packaged D-band MIMO transmitter using high-density flip-chip interconnects on LCP substrate," IEEE MTT-S International Microwave Symposium (IMS) 2016, San Francisco, CA, USA.

[5] Imran K, Geetha D, Sudhindra K, et al. "A novel LCP substrate dual band antenna loaded with T-shaped resonator and circular slot," International Conference on Advances in Electronics, Computer and Communications (ICAECC) 2018, Bangalore, India.

[6] Yang L, Xia L, Qiu Y. "Broadband transition structures using LCP packaging technology," IEEE International Conference on Communication Problem-Solving (ICCP) 2015, Guilin, China.

[7] Li S, Yi M, Spyridon P, et al. "Investigation of surface roughness effects for D-band SIW transmission lines on LCP substrate." IEEE Radio and Wireless Symposium (RWS) 2017, Phoenix, AZ, USA.

[8] Kutay C, Nursel A. "LCP substrate based crescent shaped microstrip patch array antenna design for 5G applications," 3rd International Symposium on Multidisciplinary Studies and Innovative Technologies (ISMSIT) 2019, Ankara, Turkey.

[9] Khaled A, Hong J. "LCP ultra-wideband self-packaged microstrip to stripline transition with multilayer balun demonstration," 18th Mediterranean Microwave Symposium (MMS) 2018, Istanbul, Turkey.

[10] Muller J, Welker T, Schmitt K. "LTCC-like multilayer LCP-technology for flexible RF-circuits," 7th Electronic System-Integration Technology Conference (ESTC) 2018, Dresden, Germany.

Mechanical Response of BNNS-reinforced Aluminum Composites under Uniaxial Compression

Jinming Li
School of Mechanical Science and Engineering
Huazhong University of Science and Technology
Wuhan, China
lijinming516@163.com

Yuhua Huang
School of Mechanical Science and Engineering
Huazhong University of Science and Technology
Wuhan, China
huangyuhua@hust.edu.cn

Baoshan Zeng
School of Mechanical Science and Engineering
Huazhong University of Science and Technology
Wuhan, China
zengbaoshan0714@163.com

Chenzefang Feng
School of Mechanical Science and Engineering
Huazhong University of Science and Technology
Wuhan, China
fairy945@hust.edu.cn

Fulong Zhu*
School of Mechanical Science and Engineering
Huazhong University of Science and Technology
Wuhan, China
zhufulong@hust.edu.cn

Abstract—Boron nitride nanosheet (BNNS) has been widely used as the reinforcing material in metal matrix composites because of excellent thermal, electrical and mechanical properties. Chirality directions of boron nitride nanosheet include armchair and zigzag. In this paper, the influence of temperature on the mechanical reponses of aluminum (Al) and BNNS/Al composite under uniaxial compression loads haa been investigated by molecular dynamics (MD) simulations. MD simulation results show that there is a negative relationship between temperature and ultimate stress of the above material, and ultimate stress decreases as the temperatue increases. The relationship between critical strain and temperature is the same response. In addition, BNNS plays a significant role in improving ultimate stress and Young's modulus of the nanocomposite, despite its small volume fraction. However, compression loads along different chiral directions can also result in some diversities of mechanical response. For different temperature conditions, the ultimate stress of zigzag BNNS/Al composite is bigger than others in general, and armchair BNNS/Al composite shows the better temperature stability.

Keywords—boron nitride nanosheet, aluminum, composite, compression, molecular dynamics

I. INTRODUCTION

Among the conventional metal materials, aluminum (Al) is one of the most commonly used materials due to its low density, excellent thermal conductivity, and suitable formability [1]. However, the application of Al is limited due to its mechanical properties [2]. In recent years, metal matrix composites (MMCs) have been further studied because of their good mechanical performance, high electrical conductivity and outstanding thermal ability [3]. Compared with pure Al, aluminum matrix composites (AMCs) present many unique properties [4]. Aerospace industries, automobile applications, and electronic packaging are among the important cases of their implement [5]. When the size of MMCs get down to the nanoscale state, many properties such as mechanical performance, thermal conductivity and electrical ability appear obvious differences with macroscopic state. This phenomenon has been confirmed by many studies. So it is very critical to further understanding the mechanical properties of AMCs at the nanoscale, such as elastic modulus, critical strain and ultimate stress.

Many experiments have been succeeded in investigating difference mechanical performances of AMCs with different reinforcements. Boron nitride nanosheet (BNNS) and nanotube (BNNT) can be considered as ideal reinforcements, and have been used to reinforce Al matrix. Many researchers have been demonstrated the bend strength of the AMCs with adding only 1.0 wt% BNNSs was enhanced significantly compared with the alumina [6]. Although some experiments have confirmed that both BNNTs and BNNSs are feasible reinforcements to metal matrix, there are many factors which limit the application of these two nanomaterials. Currently ,the synthesis of the nanometer materials exists some problems. Now, Some properties of BNNS/Al composite in some experiments were proposed, but more theoretical studies have been reported by molecular dynamics (MD) methods. For example, the impacts of multi-walled BNNTs on AMCs performance under uniaxial compression loadings were analyzed in the viewpoint of the microscopic deformation mechanism. These results indicate that the elastic modulus improved obviously by adding long nanotubes [7]. Some tensile properties of Al composites with adding BNNS were investigated, using a series of MD simulations. BNNSs were superior to BNNTs in the terms of improving elastic modulus and plastic deformation of Al matrix nanocomposites [8].

These available experimental and theoretical research works indicate the outstanding roles of BNNS in strengthening Al matrix materials. However, it deserves to be more studies on the mechanical reponses of BNNS/Al nanocomposite under compression loading conditions. Especially the temperature effect is still a lack of investigations for Al and BNNS/Al composite under compression load. This study investigates influence of the temperature on mechanical characteristics of BNNS/Al composite under compressive loading by MD simulations. The curves of stress and strain energy changing with strain at specified temperature were obtained and the influence of temperature was analyzed.

II. COMPUTATIONAL METHODS

As mentioned earlier, MD accurately calculates the mechanical response of the material at the nanoscale, based on Newtonian classical mechanics. In the MD simulation, it is very essential to describe the interactions between substances

978-1-6654-1392-3/21 $31.00 © 2021 IEEE

such as the atom/molecular interaction, van der Waals force, and Coulomb force. Choosing suitable potential functions is a key point for a precise MD simulation. In the MD simulations, various interatomic potentials were adopted to describe the interaction forces between different atoms in the microstructure. To be specific, the interatomic interactions between aluminum atoms was described by the embedded-atom method (EAM) potential, and the parameter values were provided by Mendelev et al [9]. The Tersoff potential was adopted to express the interaction between B-B, N-N, B-N atom pairs in the BNNS nanostructures [10]. Moreover, the Lennard-Jones (LJ) potential was used to express the weak interactions between BNNS and Al matrix, which were considered as the van der Waals force. The LJ parameters were got from Vijayaraghavan et al [11].

As shown in Fig. 1, the models of BNNS/Al composite and monocrystalline Al were divided into three parts in the direction of compression (X direction). The fix layer and the loading layer were at the left and right end of the models. The fixed layer remained stationary, and the loading layer was imposed displacement to compress the models. The orientation of Al along the compression was set as <100>. The chirality of BNNS includes zigzag and armchair. The models with zigzag and armchair BNNS were constructed in the simulations. The size of models was 3.85 nm × 3.85 nm × 11.55 nm. The corresponding Al matrix geometry was created to investigate the role of BNNS in improving mechanical properties.

In this work, the compression responses of BNNS/.Al composite and nanocrystalline Al were analyzed by MD simulations. First, the initial configurations underwent relaxation process for 50 picosecond (ps) under the certain temperature and ensemble, resulting in stable structures and the energy minimization. The temperatures were set from 100 K to 400 K by Nose-hoover method to investigate the effect of temperature on BNNS/Al composite and monocrystalline Al. The non-periodic conditions were used in the X-axis, Y-axis and Z-axis. The strain rate was controlled at 10^8/s. Subsequently, the compression load was applied to the loading layer under the corresponding temperature until the certain strain was achieved. Balancing the computational costs and accuracy, the time-step was 0.0005 ps. Both the relaxation and compression processes were under NVT ensemble. Considering the variation curves of stress and strain energy with strain, the contrast between Al and BNNS/Al composite was studied. And the visualization of different samples during compression process was supported by the Open Visualization Tool (Ovito).

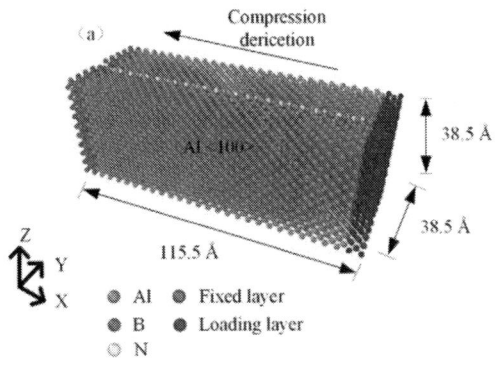

(a) BNNS/Al composite

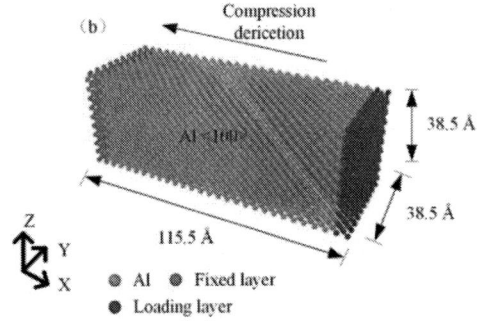

(b) Monocrystalline Al

Fig. 1. MD models of BNNS/Al composite and monocrystalline Al

III. RESULTS AND DISCUSSION

The temperature was set to 100 K, 200 K, 293 K, 400 K, respectively, to explore the influence of temperature on compressive performance. Fig. 2 shows stress versus strain curves of Al under different temperatures. By observing the stress versus strain curves of Al, it can be considered that the relationship between stress and stran is linear in the elastic stage at different temperatures, that is, the stress increases as the strain increases. When the strain increases to a critical strain, the stress drops sharply due to the plastic deformation. The ultimate stress of Al is about 3.9 GPa, and the critical strain is approximately 0.052 at 293 K. By fitting the initial linear phase of the curve with the least square method, the elastic modulus of Al is 78.3 Gpa. The fitted value is basically consistent with the existing experimental values and simulation results of 74.8-76.1 GPa [12,13].

According to the chirality, BNNS can be divided into armchair BNNS (ABNNS) and zigzag BNNS (ZBNNS). Fig. 3 shows that these stress versus strain curves of ABNNS/Al nanocomposite coincides with ZBNNS/Al composite at different temperatures. Thus, in the work, it only chooses 293 K as an example to explain the relationship between the strain and stress. It can be seen from the stress versus strain curves of Fig. 3 that the strain and stress shows linear relations at the beginning of the compression. Compared with Al, the slope of curve of ABNNS/Al composite and ZBNNS/Al composite is higher in the linear phase, which indicates an increase in elastic modulus of two BNNS/Al composites. After this, Al matrix produces 45° slip resulting in a decrease in stress. As the plastic deformation of Al matrix increases, the stress fluctuates constantly. During this process, BNNS mainly bears compression load and is shrinking gradually. When the strain reaches approximately 0.61, the stress drops rapidly again. This is because BNNS appears wrinkle under the shear force. Then the materials fails and the stress is very low.

The relationship between the strain energy of Al and strain is shown in Fig. 4. These Strain energy versus strain curves present the same trend at different temperatures. In these curves of 100 K, 200 K, 293 K, 400 K, the strain energy increases quadratically with the strain in the initial stage undergoing uniaxial compression. When the strain energy increases to the critical value, the sudden drop in strain energy occurs. This indicates that all models of monocrystalline Al occur plastic deformation, releasing the strain energy. In addition, from these curves of 100 K, 200 K, 293 K, and 400 K, it can be seen that the lower the setting temperature , the more obvious the change in strain energy of monocrystalline Al.

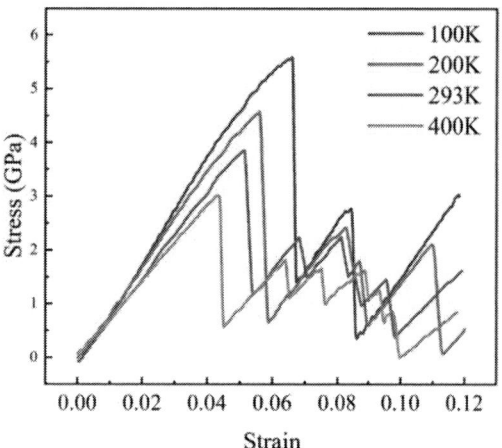

Fig. 2. Stress versus strain curves of Al at temperatures of 100 K, 200 K, 293 K, 400 K

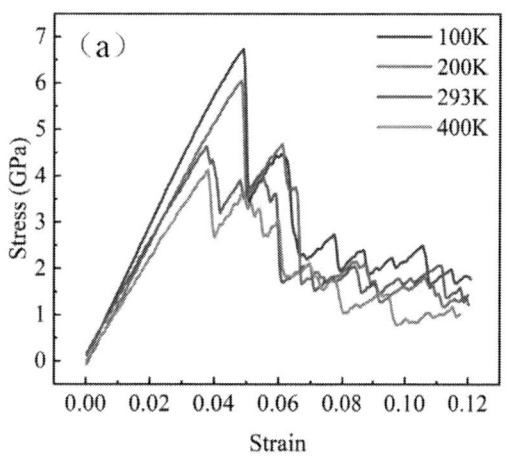

(a) ABNNS/Al composite

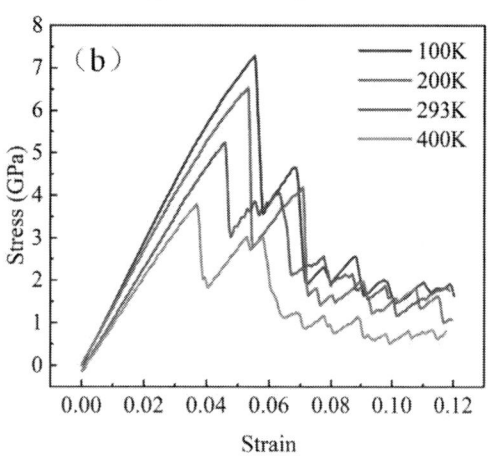

(b) ZBNNS/Al composite

Fig. 3. Stress versus strain curves of BNNS/Al composite at temperatures of 100 K, 200 K, 293 K, 400 K

Compared with the temperature effect of strain energy of monocrystalline Al, Fig. 5 shows the strain energy versus strain curves for BNNS/Al composite under specified temperature conditions. In these curves of four setting temperature - 100 K, 200 K, 293 K, 400 K, four strain energy curves of BNNS/Al composite show the first rapid drop at the

critical strain, and it is very clear that Al matrix could undergo yield deformation. Then the strain energy of BNNS/Al composite models continues to increase obviously because BNNS is still in the elastic region. When BNNS curls up severely, the strain energy decreases rapidly for the second time. And Fig. 6 shows that when BNNS/Al composite is compressed to a certain strain at the temperature of 100 K, BNNS curls up seriouly. From these stress versus strain curves in Fig. 3 and strain energy versus strain curves in Fig. 5, it is obviously that the stress and strain energy vary simultaneously.

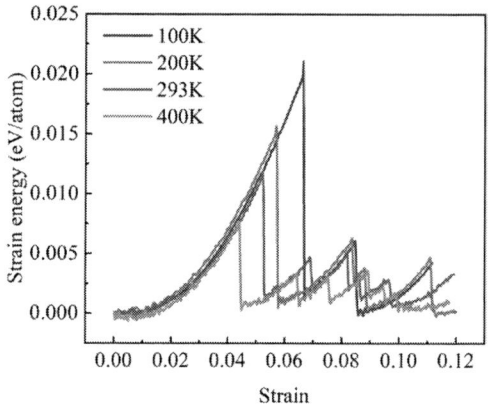

Fig. 4. Strain energy versus strain curves of Al at at temperatures of 100 K, 200 K, 293 K, 400 K

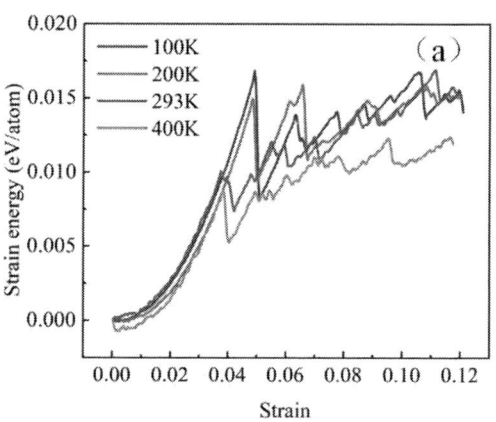

(a) ABNNS/Al composite

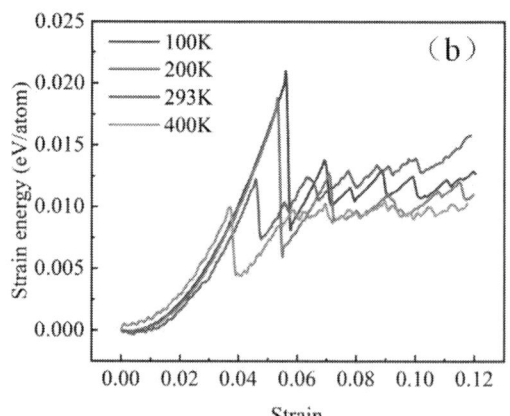

(b) ZBNNS/Al composite

Fig. 5. Strain energy versus strain curves of BNNS/Al composite at temperatures of 100 K, 200 K, 293 K, 400 K

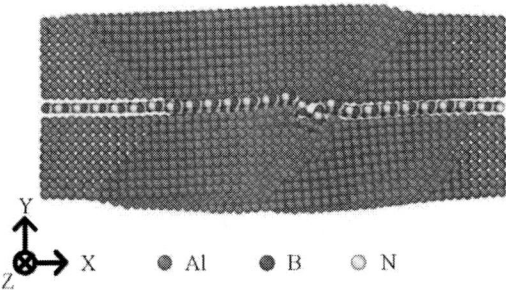

Fig. 6. Snapshot of BNNS/Al composite at the temperature of 100 K when BNNS curls up

The effect of temperature on the critical strain of three different materials - ABNNS/Al composite, ZBNNS/Al composite and monocrystalline Al is also examined. As shown in Fig. 7, the critical strain of three kinds of materials decreases as the temperature increases. However, the critical strain of ABNNS/Al nanocomposite is less sensitive to the change of temperature than the other two materials-ZBNNS/Al composite and monocrystalline Al. Even the critical strain of ABNNS/Al composite at 400 K is slightly higher than the critical strain at 293 K, whereas ZBNNS/Al composite and monocrystalline Al have a significant decrease in critical strain at high temperature.

Fig. 8 compares the trend of the ultimate stress of monocrystalline Al and BNNS/Al composite at different temperatures, indicating the higher temperature, the worse bearing capacity of three kinds of materials - ABNNS/Al composite, ZBNNS/Al composite and monocrystalline Al. At the same time, these Stress versus strain curves in Fig 2. and Fig. 3 also demonstrate that Young's modulus of three kinds of materials decreases with increasing temperature. However, In the ultimate stress versus temperature curves of Fig 7, the downward tendency of ultimate stress is slightly different, that the stress of ZBNNS/Al composite decreases faster with the temperature increasing than the other two materials-ABNNS/Al composite and monocrystalline Al. In addition, Fig. 5 also shows that small addition of BNNS can enhance the mechanical performance of material under uniaxial compression. And the load capacity of ABNNS/Al composite is better than ZBNS/Al composite at 400 K temperature, whereas the situation is opposite at low and room temperature.

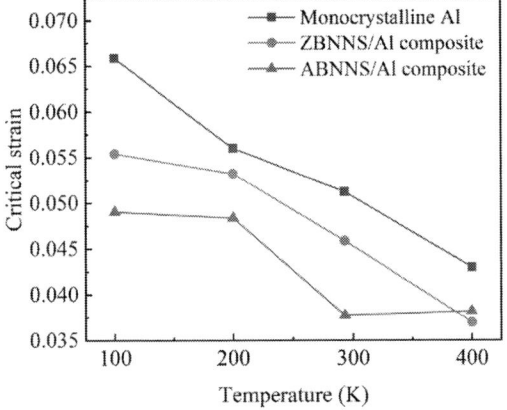

Fig. 7. Critical strain versus temperature curves of ABNNS/Al composite, ZBNNS/Al composite and monocrystalline Al

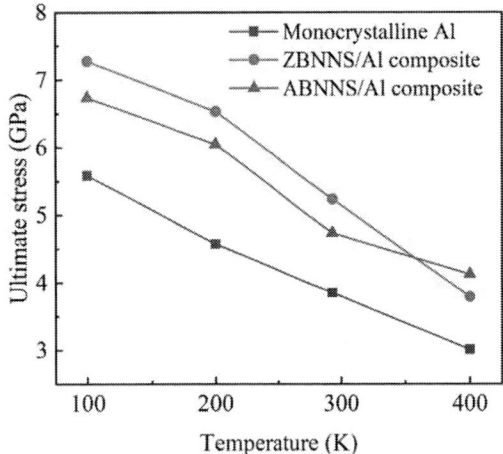

Fig. 8. Ultimate stress versus temperature curves of ABNNS/Al composite, ZBNNS/Al composite and monocrystalline Al

IV. CONLUSION

In a word, the influence of setting temperature on the compression response of monocrystalline Al and BNNS/Al composite under uniaxial compression is achieved comprehensively with the molecular dynamics methods. These stress versus strain and strain energy versus strain curves indicate that the setting temperature can significantly change the carrying capacity of ABNNS/Al composite, ZBNNS/Al composite and monocrystalline Al. Under the compressing load, Al and BNNS/Al composite show the temperature dependance, which is lower setting temperature and the better compression performance. For Al and BNNS/Al composite, the stress and strain energy vary simultaneously during the compression process, and the ultimate stress and critical strain obviously improve at the lower setting temperature of 100 K. In addition, the effect of BNNS on the composite is also analyzed. BNNS has the obvious effect on the ultimate stress and Young's modulus of BNNS/Al composite, despite BNNS is small volume fraction in this composite. In the elastic region of BNNS/Al composite, both Al matrix and BNNS bear compression load together, after reaching the critical strain, Al matrix occurs plastic deformation, whereas BNNS continues to bear compression load. Moreover, compression along different chiral directions of BNNS can also result in some diversities of mechanical response. For different setting temperature, the ultimate stress of zigzag BNNS/Al composite is bigger than others in general. What's more, armchair BNNS/Al composite has a better temperature stability.

ACKNOWLEDGMENT

This work is supported by the National Natural Science Foundation of China (Number: 52075208 and U20A6004).

REFERENCES

[1] J. Wang, Z. Li, G. Fan, H. Pan, Z. Chen, and D. Zhang, "Reinforcement with graphene nanosheets in aluminum matrix composites," Scripta Materialia, vol. 66, pp. 594-597, 2012.

[2] Z. L. Chao, L. C. Zhang, L. T. Jiang, J. Qiao, Z. G. Xu, H. T. Chi, and G. H. Wu, "Design, microstructure and high temperature properties of in-situ Al3Ti and nano-Al2O3 reinforced 2024Al matrix composites from Al-TiO2 system," Journal of Alloys and Compounds, vol. 775, pp. 290-297, 2019.

[3] P. Poddar, V. C. Srivastava, P. K. De, and K. L. Sahoo, "Processing and mechanical properties of SiC reinforced cast magnesium matrix

composites by stir casting process," Materials Science and Engineering: A, vol. 460-461, pp. 357-364, 2007.

[4] R. Q. Han, H. Y. Song, J. Y. Wang, and Y. L. Li, "Strengthening mechanism of Al matrix composites reinforced by nickel-coated graphene: Insights from molecular dynamics simulation," Physica B: Condensed Matter, vol. 601, p. 412620, 2021.

[5] S. T. Mavhungu, E. T. Akinlabi, M. A. Onitiri, and F. M. Varachia, "Aluminum Matrix Composites for Industrial Use: Advances and

[6] W. Wang, G. Sun, Y. Chen, X. Sun, and J. Bi, "Preparation and mechanical properties of boron nitride nanosheets/alumina composites," Ceramics International, vol. 44, pp. 21993-21997, 2018.

[7] Z. Cong and S. Lee, "Study of mechanical behavior of BNNT-reinforced aluminum composites using molecular dynamics simulations," Composite Structures, vol. 194, pp. 80-86, 2018.

[8] P. Sedigh, A. Zare and A. Montazeri, "Evolution in aluminum applications by numerically-designed high strength boron-nitride/Al nanocomposites," Computational Materials Science, vol. 171, p. 109227, 2020.

[9] M. I. Mendelev, M. J. Kramer, C. A. Becker, and M. Asta, "Analysis of semi-empirical interatomic potentials appropriate for simulation of crystalline and liquid Al and Cu," Philosophical Magazine, vol. 88, pp. 1723-1750, 2008.

[10] J. Tersoff, "New empirical approach for the structure and energy of covalent systems," Physical Review B, vol. 37, pp. 6991-7000, 1988.

[11] V. Vijayaraghavan and L. Zhang, "Tensile Properties of Boron Nitride-Carbon Nanosheet-Reinforced Aluminum Nanocomposites Using Molecular Dynamics Simulation," JOM, vol. 72, pp. 2305-2311, 2020.

[12] M. A. Haque and M. T. A Saif, "Mechanical behavior of 30−50 nm thick aluminum films under uniaxial tension," Scripta Materialia, vol. 47, pp. 863-867, 2002.

[13] S. Jiang, H. Zhang, Y. Zheng, and Z. Chen, "Atomistic study of the mechanical response of copper nanowires under torsion," Journal of Physics D: Applied Physics, vol. 42, p. 135408, 2009.

This work is supported by the National Natural Science Foundation of China (Number: 52075208 and U20A6004

Fabrication of Soft Conductive Microspheres and Their Application in Electromagnetic Interference Shielding Sheets

Luhui Zhang
Shenzhen Institute of Advanced Electronic Materials, Shenzhen Institute of Advanced Technology, Chinese Academy of Sciences College of Materials Science and Engineering, Shenzhen University
Shenzhen 518055, China
lh.zhang@siat.ac.cn

Zuomin Lei
Shenzhen Institute of Advanced Electronic Materials, Shenzhen Institute of Advanced Technology, Chinese Academy of Sciences
Shenzhen 518055, China
zm.lei@siat.ac.cn

Jinming He
Shenzhen Institute of Advanced Electronic Materials, Shenzhen Institute of Advanced Technology, Chinese Academy of Sciences
Shenzhen 518055, China
jm.he@siat.ac.cn

Deliang Zhu
College of Materials Science and Engineering, Shenzhen University
Shenzhen 518055, China
dlzhu@szu.edu.cn

Yougen Hu *
Shenzhen Institute of Advanced Electronic Materials, Shenzhen Institute of Advanced Technology, Chinese Academy of Sciences
Shenzhen 518055, China
yg.hu@siat.ac.cn

Rong Sun
Shenzhen Institute of Advanced Electronic Materials, Shenzhen Institute of Advanced Technology, Chinese Academy of Sciences
Shenzhen 518055, China
rong.sun@siat.ac.cn

Abstract—Conductive fillers are one of the key factors to determine the conductive and shielding properties of conductive polymer composites. Herein, an emulsion polymerization method was used to prepare soft polydimethylsiloxane (PDMS) microspheres. And the soft conductive fillers of Ag coated PDMS microspheres (PDMS@Ag) were fabricated by electroless plating technique. The EMI shielding sheets were further fabricated by mixing the soft PDMS@Ag fillers with liquid silicone precursor and curing agent, followed by rolling and heat curing process. The compression ratio of the EMI sheets with 66 wt% of PDMS@Ag is 51.2% at 40 psi. Moreover, the EMI shielding sheets present a high EMI shielding effectiveness (SE) value of ~120 dB with 66 wt% of PDMS@Ag fillers. Our work proved that the soft conductive fillers present great advantages in compression and EMI shielding performances of the conductive polymer composites, which has promising potential in electronic packaging applications.

Keywords—electromagnetic interference (EMI) shielding; conductive polymer composites; soft PDMS@Ag microspheres; electroless plating; compressibility

1. Introduction

Nowadays, with the rapid development of electronic equipment and electronic communication technologies, the quality of human life has been gradually improved. Meanwhile a large amount of electromagnetic radiation has diffused from these devices and presented a significant impact on human health[1]. Therefore, developing high performance electromagnetic interference (EMI) shielding materials has become an urgent need to meet the increasing requirements for electromagnetic radiation protection[2].

At present, conductive polymer composites (CPCs)[3] have become an important development direction in the field of EMI shielding materials[4] due to their excellent processability, adjustable shielding performance and excellent moisture sealing property. Among which the conductive fillers are one of the key factors to determine the electrical and shielding performance of CPCs[5]. The traditional conductive fillers such as pure metal of Ag[6], Cu, Ni and metal coated particles (Ag coated hollow glass[7], Ni coated graphite sheets[8], etc) exhibit high conductivity and excellent dispersion in the polymer matrix, however, these rigid fillers usually lead to the low compressibility of the as-prepared CPCs that easily causes the leakage of electromagnetic wave in some application scenarios. Using soft particles as conductive fillers is one of the solutions to improve the compressibility of the CPCs. For example, coating conductive silver on the surface of polymer core with appropriate hardness[9,10]. At present, the coating methods on the surface of particles are mainly electroless plating and electroplating[11,12]. Hao et al[13]. deposited on the surface of SiO2/Ag through the spontaneous polymerization of dopamine. The results show that the SiO2/Ag/PDA microspheres bond closely with the silicone rubber and distribute evenly. Chen et al[14]. fabricated PDMS microspheres using ultrasound, and then the electroless plated silver filled with elastomer to prepare a sensor with good performance.

In this work, an emulsion polymerization method was utilized to prepare soft PDMS microspheres. And the soft conductive fillers of Ag coated PDMS (PDMS@Ag) microspheres were fabricated by electroless plating technique. Compared with the conventional rigid conductive fillers, the as-prepared PDMS@Ag microspheres exhibited

978-1-6654-1392-3/21 $31.00 © 2021 IEEE

obviously high compressibility in low pressure. The EMI shielding sheets were further fabricated by mixing the soft PDMS@Ag fillers with liquid silicone precursor and curing agent, followed by rolling and heat curing process. The compression ratio of the EMI shielding sheets with 66 wt% of PDMS@Ag microspheres was about 51.2% at 40 psi. Furthermore, the EMI shielding sheets presented high EMI shielding effectiveness values of ~100 and 120 dB with 63 wt% and 66 wt% of PDMS@Ag microsphere loading, respectively. Our work proves that the soft conductive fillers provide great advantages in compression and EMI shielding performances of the conductive polymer composites, which has promising potential in electronic packaging applications.

2. MATERIALS AND EXPERIMENTS

2.1 Materials

Liquid PDMS silicone elastomer base and curing agent (Sylgard 184) were purchased from Dow Corning company. Bicomponent silica gel (CX 3561A/B) was commercially purchased from Guangzhou Trancytech Co., Ltd. Tin chloride dehydrate (AR，98%), silver nitrate (AR, 99.8%) and potassium sodium tartrate tetrahydrate (AR，99%) were acquired from Aladdin Co., Ltd. Ammonia solution was achieved from Shanghai Lingfeng Chemical Reagent Co., Ltd. Hydrochloric acid was obtained from Dongjiang Reagent Co., Ltd. Polyvinyl alcohol (PVA, 1788) was purchased from Shanghai Macklin Biochemical Co., Ltd.

2.2 Preparation of PDMS microspheres

PVA aqueous dispersion with mass ratio of 2.0 wt% was formulated by dispersed PVA powder into deionized water with magnetic stirring of 500 rpm for 12 h at room temperature. The liquid PMDS base and curing agent with various mass ratios (5:1, 10:1, 20:1, 25:1) were mixed in a speed mixer at 1500 rpm for 1.5 min, follow by degassing under vacuum for 5 min to remove the trapped air bubbles. The mixed PDMS was poured into the PVA aqueous dispersion at a weight ratio of 1:25. This mixture was manually mixed by with glass rod, followed by pouring into the membrane emulsifier (FM0210/500M) for emulsification into small droplets. The emulsified dispersion was then heated at 95 °C for 2 h to solidify the liquid PDMS. The PDMS microspheres were finally obtained after centrifugalization and washing by water for three times.

2.3 Preparation of PDMS@Ag microspheres

2.0 g of the as-prepared PDMS microspheres were dispersed into 400 mL of deionized water, and 400 mL of stannous chloride solution (10 mg/mL) were added into the PDMS microsphere dispersion for surface sensitization with magnetic stirring at 600 rpm for 1 h. After filtration and washing, 300 mL of silver ammonia solution (14 mg/mL) and 300 mL of potassium sodium tartrate solution (20 mg/mL) were mixed, and then added into the sensitized PDMS microsphere dispersion under stirring at 600 rpm for 1h. The PDMS@Ag microspheres were obtained by after centrifugation three times and drying in a vacuum oven at 80 °C for 6 h.

2.4 Preparation of EMI shielding sheets

The two components (A/B) of silica gel were firstly mixed at a weight ratio of 1:1, followed by addition of the PDMS@Ag microspheres and mixing in a speed mixer at 1500 rpm for 2 min. Then the mixture paste was cast into a grooved mold with the size of $30 \times 30 \times 1$ mm^3 and scraped smooth to prepare the silica gel/PDMS@Ag EMI shielding sheet followed by heat curing in a convection oven at 130 °C for 2 h.

2.5 Characterization

The micromorphology and composition of PDMS and PDMS@Ag microspheres and EMI shielding sheets were characterized using a scanning electron microscopy (FEI Nova Nano SEM 450). The compression performance of single microsphere was tested by a Micro/Nano Hardness Tester (HM2000). The mechanical-electrical properties of the samples were tested by a precision pressure tester connected with a resistance instrument (34401A). A vector network analyzer (VNA, Keysight, E5071C) was utilized measuring the EMI SE of the as-prepared sheets in the frequency of 8.2-12.4 GHz.

3. RESULT AND DISCUSS

The preparation process of PDMS@Ag microspheres is shown in Fig. 1. Firstly, the PDMS base and curing agent in proportion are mixed and added into the PVA dispersion solution. Then, the mixed solution is emulsified three times by the membrane emulsifier, and heat cured in a water bath to obtain PDMS microspheres. Finally, the PDMS@Ag microspheres are obtained by the electroless plating of PDMS microspheres.

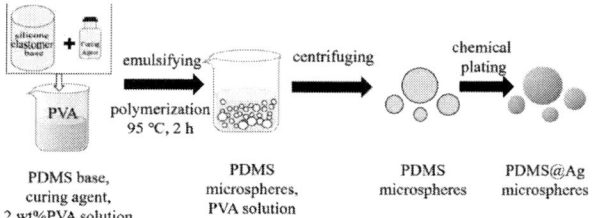

Fig. 1. Preparation process of PDMS@Ag microspheres

The SEM images of PDMS microspheres, PDMS@Ag microspheres and EMI shielding sheets are shown in Fig. 2. It can be observed from Fig. 2a that the PDMS microspheres prepared by emulsion polymerization method present good spherical and dispersibility, with an average particle size of ~15 μm. Fig. 2b shows the morphology of single microsphere. The as-prepared PDMS microspheres exhibit a good pellet-forming property and the smooth surfaces are also observed. The PDMS@Ag microspheres are presented in Fig. 2c and the enlarged SEM image of single microsphere is also shown in Fig. 2d. It can be clearly seen that a dense Ag layer was successfully coated onto the PDMS microspheres without obvious silver aggregation. On the one hand, the surfaces of PDMS microspheres are sensitized by the sensitization treatment, which is favorable for the deposition of Ag atom on the surfaces of microspheres. On the other hand, the concentration of silver ammonia solution is well controlled, leading to no more free silver observed between the PDMS@Ag microspheres. Besides, the silver shell is well bonded with the microspheres without falling off, and the dispersibility is better than that of the PDMS microspheres. Fig. 2e-f depict the cross-sectional SEM images of the EMI shielding sheets with PDMS@Ag mass fraction of 60 and 66 wt%, respectively. Obviously, the PDMS@Ag microspheres were uniformly dispersed in silica gel matrix without agglomeration, which is beneficial to obtain a stable compression performance for the as-prepared sheets. Additionally, it is found that the concentration of PDMS@Ag microspheres shows no obvious change in the SEM images when the mass fraction reaches from 60 wt% to 66 wt%, indicating that the filling amount of microspheres has reached to their upper limition.

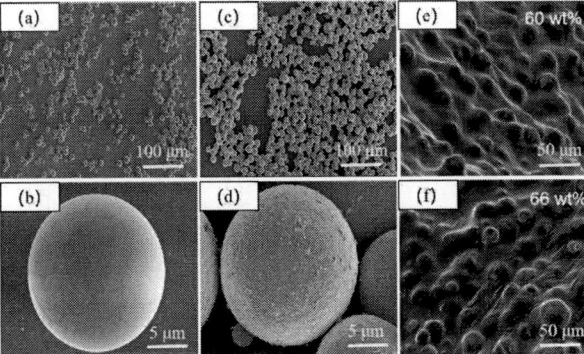

Fig. 2. (a-b) SEM images of PDMS microspheres; (c-d) SEM images of PDMS@Ag microspheres; (e-f) Cross-sectional SEM images of EMI shielding sheets filled with 60 and 66 wt% of PDMS@Ag, respectively.

The main elements and distribution of PDMS and PDMS@Ag microspheres can be seen from the SEM/EDS and the corresponding map of elements. Fig. 3a shows that the main element of PDMS microspheres is Si element, and the surface element of PDMS@Ag microspheres shown in Fig. 3b is Ag element. Analysis from the figure also proves that the surfaces of the PDMS@Ag microspheres are quite dense with silver coating. At the same time, it can be seen from the EDS of the as-prepared sheets (Fig. c-d) that the PDMS@Ag microspheres are well dispersed in the polymer matrix, and the silver shell does not fall off after suffering of high speed stirring.

Fig. 3. (a) EDS mapping of PDMS microspheres; (b) EDS mapping of PDMS@Ag microspheres; (c, d) EDS mapping of EMI shielding sheets.

Fig. 4a shows the single microsphere compression test of PDMS, PDMS@Ag and commercial available Ag coated glass microspheres, respectively. As the hardness of Ag coated glass microspheres is high enough, the external pressure is near linearly correlation to its strain before the pressure reaches to 370 mN. The hollow-glass microsphere breaks directly and the external pressure keeps stable when the displacement is beyond 3.2 μm. For PDMS microsphere, owing to its feature of elastic, the as-prepared polymer microsphere will present a good compression performance.

The PDMS@Ag microsphere (10:1) show a lower compression ratio than pure PDMS microsphere due to the covering of the incompressible silver layer. Fig. 4b shows the compression comparison of the EMI shielding sheets made by silica gel and the PDMS@Ag microspheres with different hardness at the same mass fraction of 66 wt%. The hardness of PDMS microspheres can be adjusted by the mass ratio of PDMS base to curing agent and its hardness also puts a decisive effect on the hardness of PDMS@Ag microspheres. As can be seen from the figure, the compression ratio increases with the decrease of the hardness of PDMS@Ag microspheres. The sheets filled with lowest hardness of PDMS@Ag microspheres (mass ratio of PDMS base to curing agent, 25:1) exhibit the highest compression ratio, with a compression ratio of 76% at 500 psi. In all, the compression property of the EMI shielding sheets can be adjusted by the selection of the PDMS@Ag microspheres according to their hardness.

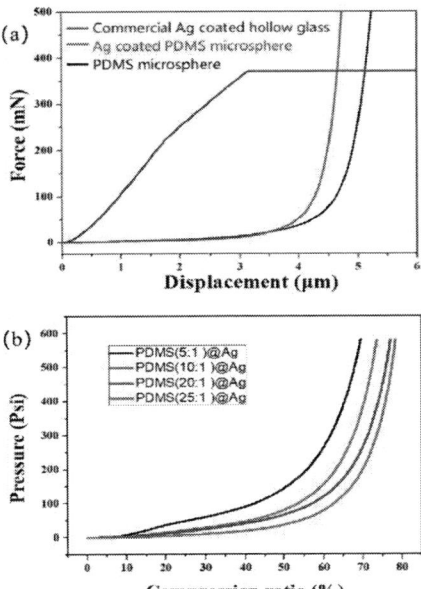

Fig. 4. (a) Force-displacement relationship of single microsphere of PDMS, PDMS@Ag and commercial available Ag coated hollow glass; (b) The compression ratio of the sheets made by silica gel and different hardness of PDMS@Ag microspheres at the same mass fraction of 66 wt%.

The PDMS@Ag microspheres are electrically conductive and the as-prepared sheets will exhibit electromagnetic shielding ability once forming of conductive network in the composites. Fig. 5 shows the EMI shielding performances of the as-prepared sheets filled with different mass fractions of PDMS@Ag (10:1) microspheres. The EMI shielding effectiveness (EMI SE) of the as-prepared sheets increases the increase of mass fraction of PDMS@Ag microspheres, with obtainment excellent EMI SE of about 95, 100 and 120 dB when increasing mass fraction of 60, 63 and 66 wt%, respectively, showing super-high EMI shielding performance for the as-prepared sheets.

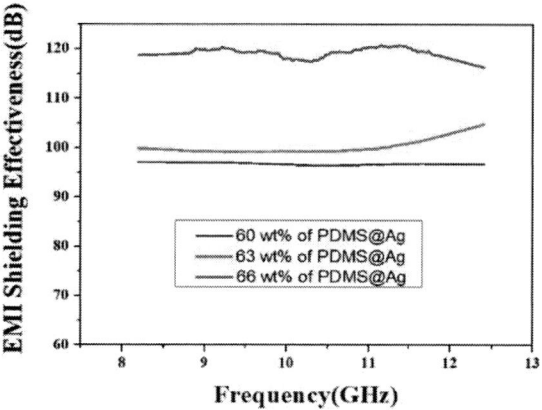

Fig. 5 EMI shielding effectiveness of the as-prepared sheets filled with different contents of PDMS@Ag microspheres.

4. CONCLUSION

In summary, a novel method for preparing soft conductive powder is proposed. The PDMS base and curing agent were mixed and added into the PVA dispersion solution, and then emulsified by membrane emulsifier. Then, the PDMS microspheres were obtained by water bath heat curing. The PDMS@Ag microspheres were obtained by surface sensitization and electroless plating. Owing to the elasticity property of internal polymer microspheres, the PDMS@Ag microspheres exhibit excellent compression performance. The shielding sheets made of PDMS@Ag microsphere and silica gel present an excellent EMI SE of 120 dB at a mass fraction of 66 wt%, and the compression ratio reach to 51.2% at 40 psi.

ACKNOWLEDGMENT

The authors are grateful for the financial support from the National Natural Science Foundation of China (62074154), China Postdoctoral Science Foundation (2020M672887), and Shenzhen Basic Research Plan (JCYJ20170818162548196).

REFERENCES

[1] M. Khodadadi Yazdi,et al., Preparation and EMI shielding performance of epoxy/non-metallic conductive fillers nano-composites [J]. Progress in Organic Coatings (2020), 145: 105674.

[2] Farbod Sharif,et al., Segregated Hybrid Poly(methyl methacrylate)/Graphene/Magnetite Nanocomposites for Electromagnetic Interference Shielding [J]. ACS Appl. Mater. Interfaces (2017), 9: 14171-14179.

[3] Lin Lin,et al., Towards Tunable Sensitivity of Electrical Property to Strain for Conductive Polymer Composites Based on Thermoplastic Elastomer [J]. ACS Appl. Mater. Interfaces (2013), 5(12): 5815-5824.

[4] Zhang, yang, et al., A Novel Polyaniline-Coated Bagasse Fiber Composite with Core–Shell Heterostructure Provides Effective Electromagnetic Shielding Performance [J]. ACS Appl. Mater. Interfaces (2017), 9(1): 809-818.

[5] Luo Junchen, et al., Mechanically Durable, Highly Conductive, and Anticorrosive Composite Fabrics with Excellent Self-Cleaning Performance for High-Efficiency Electromagnetic Interference Shielding [J]. ACS Appl. Mater. Interfaces (2019), 11(11): 10883-10894.

[6] Asheesh Kumar, et al., Enhanced carbon dioxide hydrate formation kinetics in a fixed bed reactor filled with metallic packing [J]. Chemical Engineering Science (2015), 122, 78-85.

[7] Li Dianyan, et al., Synthesis of Uniform-Size Hollow Silica Microspheres through Interfacial Polymerization in Monodisperse Water-in-Oil Droplets [J]. ACS Appl. Mater. Interfaces (2010), 2(10): 2711-2714.

[8] Yuan Haoran, et al., Nickel Nanoparticle Encapsulated in Few-Layer Nitrogen-Doped Graphene Supported by Nitrogen-Doped Graphite Sheets as a High-Performance Electromagnetic Wave Absorbing Material [J]. ACS Appl. Mater. Interfaces (2018), 10(1): 1399-1407.

[9] Tang, Yao, et al., Optical Micro/Nanofiber-Enabled Compact Tactile Sensor for Hardness Discrimination [J]. ACS Appl. Mater. Interfaces (2021), 13: 4560-4566.

[10] Pal R, et al., MnO2-Magnetic Core-Shell Structured Polyaniline Dependent Enhanced EMI Shielding Effectiveness: A Study of VRH Conduction [J]. Chemistry Select (2019), 4(31): 9194-9210.

[11] Hu, Yougen, et al., A low-cost, printable, and stretchable strain sensor based on highly conductive elastic composites with tunable sensitivity for human motion monitoring [J]. Nano Research (2018), 11(4): 1938-1955.

[12] Zhang, Liying., et al., Mussel-inspired polydopamine coated hollow carbon microspheres, a novel versatile filler for fabrication of high performance syntactic foams [J]. ACS Appl Mater Interfaces (2014), 6(21): 18644-52.

[13] Hao Mingzheng, et al., Surface Modification of As-Prepared Silver-Coated Silica Microspheres through Mussel-Inspired Functionalization and Its Application Properties in Silicone Rubber [J]. Industrial & Engineering Chemistry Reserch (2018), 57: 7486-7494.

[14] Pan, Chengfeng, et al., Silver-Coated Poly(dimethylsiloxane) Beads for Soft, Stretchable, and Thermally Stable Conductive Elastomer Composites [J]. ACS Appl. Mater. Interfaces (2019), 11: 42561-42570.

Fatigue Life Evaluation and Test Method for Representative Printed Circuit Board

Jun Tong
General Department of test technology
Beijing Institute of Structure and Environment Engineering
Beijing, China
tongjunwh@163.com

Hao Chen
General Department of test technology
Beijing Institute of Structure and Environment Engineering
Beijing, China
hct@spacechina.com

Hefeng Liu
General Department of test technology
Beijing Weiwan Hengrui Technology Co., Ltd
Beijing, China
thbwh@126.com

Abstract—**Different from the strength design method, the environmental design method is traditionally used for instruments and equipment, which relies more on the idea of statistical probability, and the failure mode and failure mechanism are not clear. With the development of computational mechanics and measurement methods, the modeling technology and measurement technology for the complex system of instruments and equipment have made great progress. The strength based design method has been applied in the development of instruments and equipment, such as the life evaluation method based on failure physics. In this paper, with the help of analytical method and finite element simulation numerical method, the stress analysis and life evaluation of typical instrument structure are carried out. Based on the equilibrium equation, geometric equation, constitutive equation and deformation compatibility equation of mechanics, the thermal stress on the solder joint is obtained, and the structural life is evaluated according to the S-N curve. The stress analysis of various packaging forms of shell under temperature cycling load is carried out by using the finite element method. There are some problems in the optical measurement method, such as the micro deformation image is difficult to collect, the micro speckle with high contrast and high adhesion is difficult to make, and the reflection of hemispherical solder joint is difficult to suppress. The micro DIC device suitable for micro thermal deformation measurement and the manufacturing technology of anti reflective micro speckle are developed. By making high contrast anti reflective micro speckle in the test area, the micro DIC test system combined with micro lens is used to measure the micro scale deformation near the solder joint of PCB, and the correctness and accuracy of the calculation model are verified. two kinds of commonly used fatigue life models are listed, and the fatigue life of typical packaged circuit board under thermal environment is evaluated. The above analysis method is verified by the means of test and inspection. Finally, master the fatigue life assessment and test verification technology of typical instrument structure under thermal environment. Realize the ability to carry out life assessment and test verification of typical instruments and equipment in thermal environment.**

Keywords—PCB;Thermal fatigue;Micro-DIC;FEM

I. INTRODUCTION

The actual combat and informatization of weapon system are closely related to the environmental adaptability and reliability of equipment. While，The traditional environmental design method is used in the instrument and equipment. Different from the strength design method, it relies more on the idea of statistical probability and does not consider the failure mode and failure mechanism. With the development of computational mechanics, measurement means and methods, the modeling technology and measurement technology of complex systems such as instruments and equipment have made great progress. The

strength based design method has been applied in the development of foreign instruments and equipment. For example, a lot of research work has been carried out abroad on the life assessment method of instruments and equipment based on failure physics. In the 1990s, the reliability Manual of Rome laboratory in the United States gave various failure models of circuit board fixed joints [1], and began to study the finite element modeling method of instruments and equipment. On this basis, the finite element modeling of PCB was summarized in 1992, and five schemes were put forward. In the early 21st century, the response of electronic components under various load environments was analyzed [2]. In the aspect of experiment, the optical method of nondestructive testing was applied to the study of micro deformation of electronic packaging board in the 1980s, and which mainly analyzed the reliability of thermodynamics[3]. In the early 1990s, shadow moire method was used to measure the thermal deformation of electronic packages (BGA, PCB, TBGA, etc)[4].

This paper firstly introduces the stress analysis method of PCB in thermal environment, and then evaluates the solder joint life of common packaging shell by using engineering algorithm and finite element analysis method. Finally, the temperature cycle test of cqfp68 package shell is designed to verify the feasibility of the calculation method. It has formed a relatively complete set of life assessment and test verification technology for board level instruments and equipment in thermal environment, and initially has the ability of failure physical mode prediction, stress life assessment and test verification for typical board level instruments and equipment in thermal environment.

II. STRESS ANALYSIS TECHNOLOGY UNDER THERMAL LOAD ENVIRONMENT

A. Engineering Algorithm

The stress level of the packaging structure is mainly obtained according to the deformation compatibility equation and balance equation. The typical deformation diagram of the packaging structure [5] is shown in Figure 1.

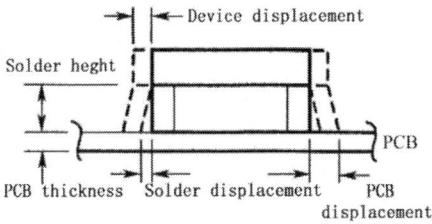

Fig.1 Deformation Coordination of Typical Package Structure

The displacement of the device is caused by the superposition of two parts. The first part is caused by the thermal expansion of the device itself, and the second part is caused by the force of the solder joint connected with the device. When the coefficient of thermal expansion of PCB is greater than that of device, tensile deformation occurs, otherwise compressive deformation occurs. According to the balance equation, the force of solder joint can be obtained.

$$P_s = (\alpha_p L_G \Delta t - \alpha_c L_G \Delta t) / (\frac{L_G}{A_p E_p} - \frac{L_G}{A_c E_c} - \frac{h_s}{A_s G_s}) \tag{1}$$

Where α_p. LG, AP and EP are the thermal expansion coefficient, length, cross-sectional area and elastic modulus of PCB respectively; PS is the shear force of solder joint; HS is the height of solder joint; As is the area of solder joint; GS is the shear modulus of solder joint; αC is the thermal expansion coefficient of the device; EC is the elastic modulus of the device; AC is the cross-sectional area of the device; ΔT is the temperature change.

After the force is calculated, the stress of the solder joint is further calculated, and then the life is evaluated according to the S-N curve.

B. Finite Element Analysis Method

Taking a typical cqfp68 package as an example, the calculation model is shown in Figure 2. The structure is symmetrical, and two symmetrical boundary conditions are imposed in the finite element model. The applied temperature cycle load is - 60 ~ 150 ℃, the rate of temperature rise and drop is 10 ℃ / min, the high temperature holding time is 30 minutes, and the low temperature holding time is 30 minutes.

Fig.2 Schematic Diagram of Finite Element Calculation Model

The creep distribution of Sn Pb solder is shown in Figure 3. It can be seen from Figure 3 that due to the different linear expansion capacity of each component, the incompatible deformation occurs, resulting in the large inelastic strain (mainly creep strain) of tin lead solder under temperature cycling load, and the maximum inelastic strain occurs at the corner solder joint. With the increase of temperature cycles, the structural stress level of Sn Pb solder will gradually increase, and the creep accumulation phenomenon will appear.

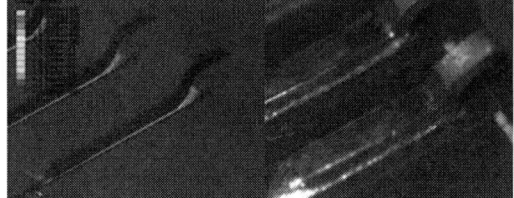

Fig.3 Creep strain nephogram

III. FULL FIELD MEASUREMENT TECHNOLOGY OF SOLDER JOINT STRAIN UNDER THERMAL LOAD

A. Full Field Measurement Technology of Solder Joint Thermal Strain Based on Micro DIC

Digital image correlation (DIC) is a non-contact, non-destructive, full field displacement and strain measurement method. Based on the principle of image correlation, the surface speckle field of the specimen is used as the deformation carrier to track the position change of the same speckle before and after deformation to obtain the displacement information of the point. The full field displacement of the specimen surface is obtained by calculating all the speckle fields. As shown in Figure 4, DIC strain measurement can be derived from the displacement measurement value.

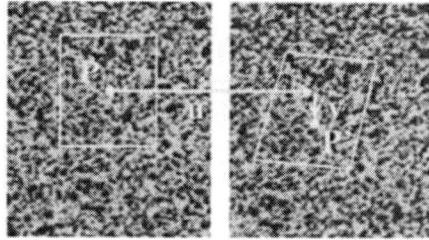

Fig.4 DIC Displacement Measurement

Micro scale deformation measurement based on digital image correlation, and the experiment was divided into two parts.

a) Fabrication of micro nano speckles: firstly, the surface of the test area is coated with black pigment (which can be wiped off after the experiment) for anti reflective treatment; At the same time, the micro nano high temperature resistant cobalt oxide and zirconia particles are uniformly adsorbed on the test area to complete the micro nano speckle fabrication and anti reflective treatment.

b) Experimental device: considering the small area to be measured, CCD camera and micro lens are used to collect the speckle images of the specimen before and after heating. The sample was heated by a small heating furnace refitted from a high precision micro melting point instrument. The experimental setup is shown in Figure 5.

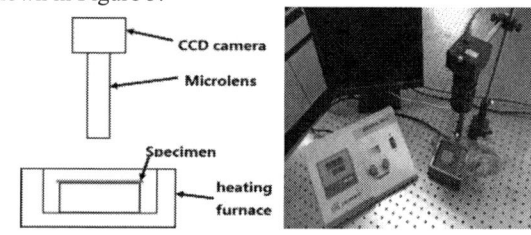

Fig.5 Thermal Strain Full Field Test Device

B. Full Field Measurement Technology of Solder Joint Strain

In order to verify the accuracy of the finite element model, micro nano speckle technology was used to measure the thermal strain of the solder joint. When the room temperature is lower than 20 ℃, the surface image of the specimen is collected as the reference image; When the temperature is raised to 150 ℃, the surface speckle pattern of the specimen is collected every 10 ℃ as the deformation image. The strain field measured at different temperatures of the solder joint is shown in Fig. 6. When the temperature is 80 ℃, the comparison of strain nephogram in X direction of solder joint is shown in Fig. 7. The calculated and experimental tensile and compressive strain curves of solder joints at different temperatures are shown in Figure 8. From the comparison, it can be found that the laws of calculation and test are basically consistent, which verifies the correctness and accuracy of the calculation model.

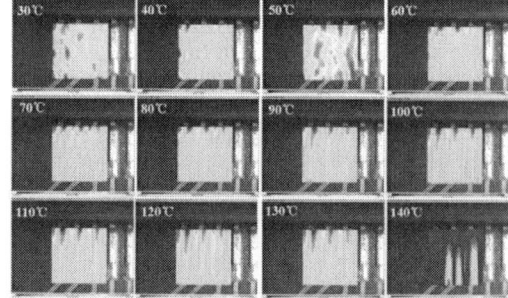

Fig.7 Nephogram of strain field at solder joint measured under different temperature fields.

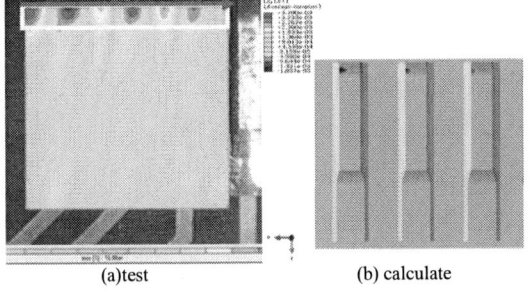

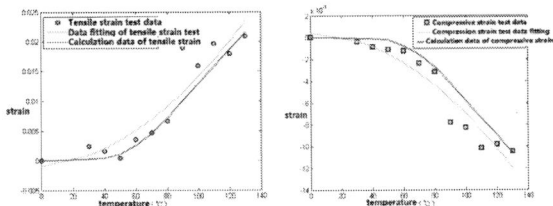

(a)test   (b) calculate

Fig.8 Comparison of X-direction Strain Nephogram of Solder Joint at 80℃

Fig.9 X-direction Strain Contrast Curve Near Solder Joint

IV. SOLDER JOINT LIFE EVALUATION MODEL

At present, the main fatigue models of solder joints [6-10] can be divided into stress-based fatigue model, plastic strain based fatigue model, creep strain based fatigue model, energy based fatigue model, damage based fatigue model and fracture mechanics based fatigue model. The above models reflect the fatigue law of solder joints in a certain range. According to the failure mode and deformation characteristics of solder joints in this paper, two commonly used life assessment models are selected:

a) According to the creep strain life model:

$$N_f = (c\varepsilon_{acc})^{-1} \qquad (2)$$

Where N_f is fatigue life; ε_{acc} is the equivalent creep strain accumulated in one cycle;c is the parameters of Solder joint material. And for sn63pb37 solder, it can be 0.0513.

b) According to the shear strain life model:

$$\frac{\Delta\gamma}{2} = \frac{\tau'_f}{G}\left(2N_f\right)^{b_0} + \gamma'_f\left(2N_f\right)^{c_0} \qquad (3)$$

Where $\frac{\Delta\gamma}{2}$ is the total shear strain amplitude; τ'_f is the shear fatigue strength coefficient; γ'_f is the shear fatigue ductility coefficient; b_0 is the shear fatigue strength index; c_0 is the shear fatigue ductility index; G is the shear modulus; N_f is the cycle of cycle failure, which is fatigue life. The related parameters of sn63pb37 solder are given [11]: $\frac{\tau'_f}{G}$=0.0024983, γ'_f=1.872, b_0=-0.04745, c_0=-0.6979.

V. HERMAL FATIGUE TEST AND SEM OBSERVATION OF SOLDER JOINT DAMAGE

In order to verify the correctness of the theoretical analysis, simulation calculation and constitutive model, a typical circuit board structure of cqfp is designed and thermal fatigue test is carried out. After the circuit board was heated in the incubator for a period of time, it was taken out of the incubator and the cracks were observed under the electron microscope. Figure 9 shows the test results of cqfp68 packaged circuit board under the electron microscope after 40 cycles in the temperature environment of - 60 ～ 150 ℃. It can be seen from Figure 9 that some parts have begun to wrinkle, the solder joints at the corners will crack, and the solder joints

in the middle are in good condition. Figure 10A shows the result after 50 cycles. Cracks have appeared on some pins, and the middle position is still intact. Therefore, it can be judged that under the temperature cycling load of - 60 ～ 150 ℃, the life of cqfp68 ceramic shell is about 50 times. Figure 10B shows the test results of cqfp68 packaged circuit board under electron microscope after 200 cycles at - 45 ～ 75 ℃. As can be seen from figure 10B, cracks have appeared on some pins, and most of the pins at the corners have wrinkled.

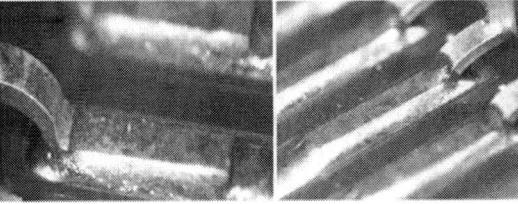

Fig.10 The Results Detected by Electron Microscope after 40 Cycles

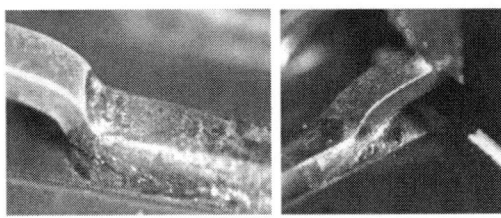

a) 50 times (- 60 ～ 150 ℃)   b) 220 times (- 45 ～ 75 ℃)

Fig.11   Results of Electron Microscopic Examination of CQFP68 after Temperature Cycling

VI. STATISTICAL ANALYSIS OF THERMAL FATIGUE LIFE EVALUATION AND TEST VERIFICATION RESULTS OF CIRCUIT BOARD

The statistical results of cqfp package life under different environmental loads are shown in Table 1. It can be seen from the table that the life of typical board level equipment predicted by theoretical analysis method and finite element simulation analysis method is in good agreement with the test results. Because the test and calculation is a typical package structure, and there is only one component on a PCB board, the structure is relatively simple, so the calculation accuracy is relatively high. At the same time, the load environment simulated by test and calculation is the condition of bottom cycle fatigue, and the life dispersion is relatively small. The application of these methods in practical engineering can be further verified in the follow-up work.

TAB.1   THERMAL FATIGUE LIFE ASSESSMENT OF TYPICAL CIRCUIT BOARDS AND STATISTICS OF TEST RESULTS

Loading conditions	Life assessment based on creep strain	Life assessment based on shear strain	Engineering algorithm	Test
–45~75℃	195	255	197	About 220
–60~150℃	49	55	48	40~50

VII. Conclusion

In this paper, with the help of analytical method and finite element simulation numerical method, the stress analysis and life evaluation of typical instrument structure are carried out. In this paper, the equilibrium equation, geometric equation, constitutive equation and deformation compatibility equation of mechanics are used to carry out the theoretical derivation, and the thermal stress on the solder joint is obtained, and the structural life is evaluated according to the S-N curve. The finite element method is used to analyze the stress of the tube shell with various packaging forms under temperature cycling load. In order to solve the problems of optical measurement, such as the difficulty of micro deformation image acquisition, the difficulty of making high contrast and high adhesion micro speckles, and the difficulty of suppressing the reflection of hemispherical solder joints, a micro DIC device and anti reflective micro speckles technology for micro thermal deformation measurement are developed in this paper. By making high contrast anti reflective micro speckles in the area to be tested, the micro DIC test system combined with micro lens is used to measure the micro scale deformation in the area near the PCB solder joint, which verifies the correctness and accuracy of the calculation model. Finally, two commonly used fatigue life models are listed, and the fatigue life of a typical packaged circuit board under thermal environment is evaluated. The above analysis method is verified by means of test and inspection. Finally, the technical approach of fatigue life assessment and test verification of typical instrument structure under thermal environment is mastered. It has the ability to carry out life assessment and test verification of typical instruments and equipment in thermal environment.

REFERENCES

[1] Chen W T, Nelson C W. Thermal stresses in bonded Joints[J]. IBM Journal of Research and Development, 1979, 23(2): 178-188.

[2] Yamada S A. Mechanism for board warpage by thermal expansion ofsurface mounted connector[J]. IEEE Transactions on Components, Hybrids and Manufacturing Technology, 1986, 9(4): 508-512.

[3] Yamada S A. A bonded joint analysis for surface mount components[J]. ASMe Journal of Electronic Packaging, 1992, 114(1): 27-31.

[4] Sundararajan R, Mc Cluskey P, Azarm S. Semi analytic model for thermal fatigue failure of die attach in power electronic building blocks[C]. Albuquerque: Fourth International High Temperature Electronics Conference, 1998.

[5] Hou Chuantao, Tong Jun, Rong Kelin. Fatigue life analysis of LCCC package[J]. Structure and Environment Engineering, 2014, 41(3): 51-57.

[6] Tong Jun, Hou Chuantao, Jia Liang. Cause analysis of bulge in thrust chamber of an engine[J], Structure and Environment Engineering, 2013, 40(1): 52-55.

[7] Wang Guozhong, Chen Zhaonian. Viscoplastic Anand constitutive equation of SnPb solder alloy[J]. Chinese Journal of Applied Mechanics, 2000, 17(3): 133-139.

[8] Chu Weihua, Li Shuchen. Fatigue failure and crack propagation analysis of solder joints in vibration environment[J]. Structure and Environment Engineering, 2012, 39(4): 56-63.

[9] Tong Jun, Hou Chuantao, Rong Kelin. Life evaluation of PCB solder joints under temperature cycling load[J]. Structure and Environment Engineering, 2014, 41(4): 48-52.

[10] Chen Chuanrao. Fracture and fatigue[M].Wuhan: Huazhong University of science and Technology Press, 2012.

[11] Guo Hongqiang. Torsional low cycle fatigue properties of 63Sn-37Pb and sn-3ag-0.5cu alloy solders[J]. Mechanical engineering materials, 2014, 38(8): 65-69.

Method for preparing silicon photonic chip edge packaging structure based on inclined deep etching process

1st Heng Zhao, 2nd Laisheng He, 3rd Junbo Feng, 4th Bangtong Ge
Chongqing United Microelectronics Centere
Chongqing, China
heng.zhao@cumec.cn heng; laisheng.he@cumec.cn; junbo.feng@cumec.cn; bangtong.ge@cumec.cn

Abstract—Packaging technology is the core technology for silicon photonic chip engineering applications. The edge structure of silicon photonic chip has the advantages of large bandwidth, low loss, small package volume, and polarization insensitivity, which makes people increasingly hope to make breakthroughs in edge packaging. This paper proposes a method for fabricating silicon photonic chip edge packaging structure based on bevel deep etching process, which can effectively reduce packaging cost. The prior art realizes precise control of the mold spot conversion structure on the edge of the silicon photonic chip through a vertical etching process. The innovation of this paper is that, one is to use the slope formed by plasma etching to match the packaging surface of the optical fiber array to reduce the coupling reflection between the edge of the silicon photonic chip and the optical fiber packaging. The second is to prepare deep silicon etching grooves by a combination of bevel etching process and bosch deep silicon etching process, which improves the redundancy of wafer scribing, and eliminates it by controlling the scribing position in the deep silicon etching groove the step bumps have an adverse effect on the optical fiber packaging, which ultimately reduces the process steps of grinding and polishing the step bumps on the chip end surface in the subsequent packaging process of the edge inverted tapered packaging surface and the inclined array fiber head, so as to completely solve the problem. At this stage, it is necessary to carry out high-precision grinding and polishing for the coupling and packaging of the inverted tapered structure of the silicon photonic chip and the optical fiber array. After testing, the loss of the edge coupler at the wavelength of 1550nm is within the range of 1.5-2.0dB. In addition, the packaging structure can be prepared in the wafer tape-out stage, has the advantages of large-scale production, effectively reducing the cost of silicon photonic chip edge packaging, and has a wide range of engineering application prospects.

Keywords—*silicon photonic chip, edge coupler, inclined deep etching, edge packaging, wafer, grinding, polishing.*

I. INTRODUCTION

In recent years, silicon photonics has made considerable progress. This is because the manufacturing process of silicon photonic integrated optical devices is fully compatible with microelectronics processes, and the carrier light waves transmitted in silicon photonic chips are extremely high-frequency and capable of signal Transmission of electromagnetic waves that provide extremely large bandwidth [1-4]. However, since the mode spot diameter of the mode field diameter in the edge facet waveguide of the silicon photonic chip is about 0.3 μm [5], the inverted cone structure of the edge facet of the waveguide is generally used to achieve the enlargement of the spot to match the mode field in the single-mode fiber [6-9], the mode field amplification structure requires precise (micron level) control of the distance from the tip of the inverted cone to the edge facet of the chip. This index can be achieved by conventional etching processes [10], but after etching, steps will be formed after dicing, which will block the The structure is coupled to the outside, especially the coupling to the fiber surface of the array. The edge facet grinding and polishing process is usually used to eliminate the edge facet steps [11]. The current grinding and polishing process has low accuracy, which is the biggest bottleneck that limits the edge facet packaging of silicon photonic chips [12,13]. Therefore, how to effectively solve the problem of the edge facet packaging of the array optical fiber and the silicon-based optoelectronic chip has become an urgent technical problem to be solved by those skilled in the art and a focus of research.

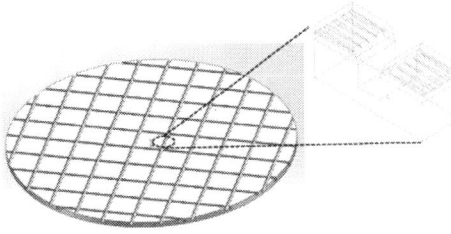

Figure 1 Schematic diagram of wafer etching

978-1-6654-1392-3/21 $31.00 © 2021 IEEE

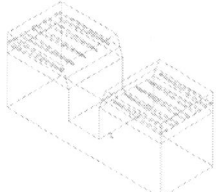 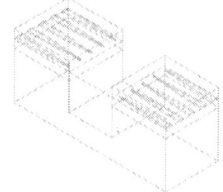

(a)Edge facet free grinding and polishing package structure   (b)Existing edge facet deep eroded structure

Figure 2 Structure comparison between this scheme and the existing scheme

II. METHOD

The main idea of the method is to improve the redundancy of scribing through a deep etching process, and to prepare a structure that can be directly packaged with external devices in combination with an inclined surface etching process. The main advantage is that the packaging surface does not require grinding and polishing process steps and can reduce reflection, which can be realized at the wafer level to greatly reduce packaging costs. Firstly, the slope is prepared by plasma etching, and the silicon substrate is used as an etching stop layer of silicon oxide, so that the etching depth and angle of the slope can be accurately controlled to form an ideal packaging surface. Then the silicon substrate is subjected to multiple Bosch etchings [14,15] to form trenches with wavy vertical sidewalls. By controlling the number of etchings, a deep silicon trench with a specific depth can be prepared. The specific process steps are shown in Figure 3. Show. Finally, use a dicing machine to perform high-precision scribing in the trench, so that the edge of the deep silicon etching step formed after scribing is at the downward extension of the inclined etching surface and the stop surface of the deep silicon etching trench. Within the line, it is possible to use the array fiber with the packaging head close to the inverted tapered packaging surface of the silicon optical chip without any need for end-face grinding and polishing of the deep-etched steps to achieve high-efficiency coupling packaging.

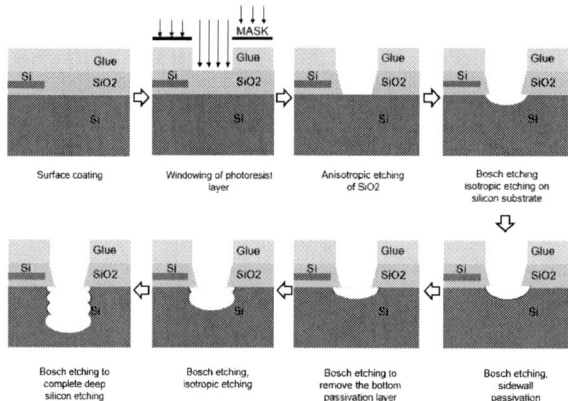

Figure 3 Schematic diagram of etching process steps

III. RESULTS AND DISCUSSIONS

The silicon photonic chip edge facet packaging structure prepared by this method, as shown in FIG. 4, improves the dicing redundancy, and no grinding and polishing process steps are required in the edge facet packaging process. When dicing the wafer with a dicing machine, the dicing position needs to be controlled within the distance L2 from the vertical sidewall surface. The required L2 parameter can be calculated by controlling the vertical height h2 of the vertical sidewall surface of the deep silicon etching and the inclined angle θ of the inclined surface.

$$L2 = h2 * tg\theta$$

After the dicing is completed, the step protrusion length of the chip end surface is less than L2, so that the chip end surface and the array fiber packaging surface can be completely attached, and high-efficiency coupling and packaging can be performed.

Four channel array fiber and four edge couplers on silicon photonics chip were used for testing. The loss of edge coupler was measured by means of averaging multiple measurements. After testing, the loss of the edge coupler at the wavelength of 1550nm is within the range of 1.5-2.0dB. Such as Figure 6.

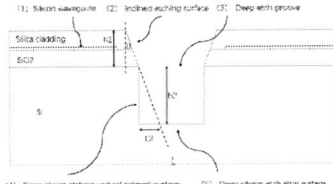

Figure 4 Schematic diagram of non-grinding and polishing package structure

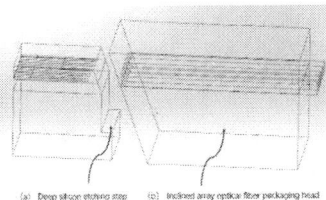

Figure 5 Schematic diagram of tilted etching surface and optical fiber array packaging head packaging

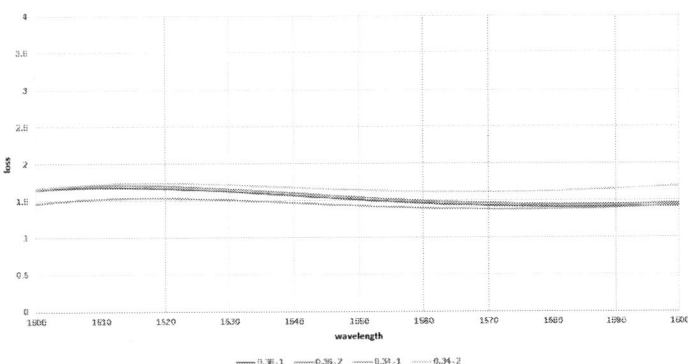

Figure 6 The loss of edge coupler

IV. COCLUSIONS

Compared with the prior art, the method in this paper uses a combination of a bevel etching process and a deep silicon etching process to improve the dicing redundancy. In the subsequent packaging process of the edge facet inverted cone packaging surface and the bevel array fiber head In this method, the chip edge facet grinding and polishing process steps are reduced, so as to completely solve the problem that the existing silicon optical chip edge facet inverted cone structure must be packaged with high precision grinding and polishing. This method realizes the waveguide edge facet inversion at the wafer tape-out stage. The preparation of the cone-free packaging structure has the outstanding advantages of large-scale production, low coupling reflection, high coupling efficiency and easy packaging, and has a wide range of engineering application prospects.

ACKNOWLEDGMENT

This work is mainly funded by the independent investment project of United Microelectronics Center Co., Ltd. (CUM-2019-ZS-RD-008-GG).

REFERENCES

[1] Peter J. Winzer, Scaling optical fiber networks: challenges and solutions, Opt. Photon. News (2015) 28–35.

[2] C. Gunn, CMOS photonics for high-speed interconnects, IEEE Micro 26 (2) (2006) 58–66.

[3] Erica R. H. Fuchs, E. J. Bruce, and Randolph E. Kirchain, "Process-Based Cost Modeling of Photonics Manufacture: The Cost Competitiveness of Monolithic Integration of a 1550-nm DFB Laser and a Electro absorptive Modulator on an InP Platform," JOURNAL OF LIGHTWAVE THCHNOLOGY, VOL. 24, NO.8, AUGUST(2006).

[4] B. Jalali and S. Fathpour, "Silicon photonics," J. Lightwave Technol.24, 4600–4615 (2006).

[5] D.R. Rowland, Y. Chen, A.W. Snyder, Tapered mismatched couplers, J. Lightw. Technol. 9 (5) (1991) 567–570.

[6] K. Kasaya, O. Mitomi, M. Naganuma, Y. Kondo, and Y. Noguchi, "A simple laterally tapered waveguide for low-loss coupling to singlemode fibers," IEEE Photon. Technol. Lett. 5, 345–347 (1993).

[7] V. R. Almeida, R. R. Panepucci, and M. Lipson, "Nanotaper for compact mode conversion," Opt. Lett. 28, 1302–1304 (2003).

[8] L. Vivien, S. Laval, E. Cassan, X. L. Roux, and D. Pascal, "2-D taper for low-loss coupling between polarization-insensitive microwaveguides and single-mode optical fibers," J. Lightwave Technol. 21, 2429–2433 (2003).

[9] A. Sure, T. Dillon, J. Murakowski, C. Lin, D. Pustai, and D. Prather, "Fabrication and characterization of three-dimensional silicon tapers," Opt. Express 11, 3555–3561 (2003).

[10] J. Cardenas, et.al., "High Coupling Efficiency Etched Facet Tapers in Silicon Waveguides", IEEE Photon. Tech. Lett. 26,2380(2014).

[11] Riccardo Marchetti, Cosimo Lacave, Lee Carroll, Kamil Gradkowski, Paolo Minzioni, "Coupling strategies for silicon photonics integrated chips," PHOTONICS Research, Vol. 7, No.2, February (2019).

[12] G. Roelkens, P. Dumon, W. Bogaerts, D. van Thourhout, and R. Baets, "Efficient silicon-on-insulator fiber coupler fabricated using 248-nm-deep UV lithography," IEEE Photon. Technol. Lett. 17, 2613–2615 (2005).

[13] T. Barwicz, et al., Assembly of mechanically compliant interfaces between optical fibers and nanophotonic chips, in: Proc. IEEE Electron. Compon. Technol. Conf., Lake Buena Vista, FL, USA, May 27–30, 2014, pp. 179–185.

[14] Chienliu Chang, Yeong-Feng Wang, Yoshiaki Kanamori, Ji-Jheng Shih, Masayoshi Esashi,, "Etching submicrometer trenches by using the Bosch process and its application to the fabrication of antireflection structures," Journal of Micromechanics & Microengineering (2005).

[15] F. Laerme, A. Schilp, K. Funk, M. Offenberg, "Bosch deep silicon etching: improving uniformity and etch rate for advanced MEMS applications," Micro Electro Mechanical Systems(1999).

Life prediction of gold-aluminum bonding system based on failure physics under multi-stress coupling

1st Menglin Li
Xi'an Institute of Microelectronics
Technology
Xi'an, China
zjdamon@sina.com

2nd Longfei Chen
Xi'an Institute of Microelectronics
Technology
Xi'an, China
18509263012@163.com

3rd Xianshun Zhang
Xi'an Institute of Microelectronics
Technology
Xi'an, China
18392068234@163.com

4th Dongfei Zheng
Xi'an Institute of Microelectronics
Technology
Xi'an, China
duffzheng@163.com

5th Zeping Xiao
Xi'an Institute of Microelectronics
Technology
Xi'an, China
18200289097@163.com

Abstract—**Gold-aluminum bonding is the main method of electrical interconnection in the electronic packaging industry, and its reliability is widely concerned. At present, most researches on the reliability of gold-aluminum bonding systems at home and abroad stay in qualitative research under single stress conditions, and lack the basis for quantitative reliability prediction under multiple stresses. Based on the requirements for high reliability and long life of Military electronic packaging, this article focuses on the failure mechanism of performance degradation of the gold-aluminum bonding system under stresses such as temperature and current, and conducts accelerated life tests under multi-stress coupling, The gold-aluminum bonding daisy chain test sample which is special prepared can accurately characterize the degradation of the gold-aluminum bonding system, and Finally, a life prediction model for the gold-aluminum bonding system based on the TNT-Weibull degradation distribution is formed, which is used to quantitatively predict the service life of the system under different temperatures and different electrical stresses, and provide a basis for the reliability prediction of the electronic system.**

Keywords—*Gold-aluminum bonding, Multi-stress coupling, Daisy chain samples, Life prediction model*

I. INTRODUCTION

SiP (Sysyem in Package) integrates a variety of functional chips, including processors, memory and other functions in a package, so as to achieve a basically complete function. As shown in Figure 1, the gold wire ball thermosonic bonding Technology occupies an important role in SiP packaging due to its wide application range and simple process realization.

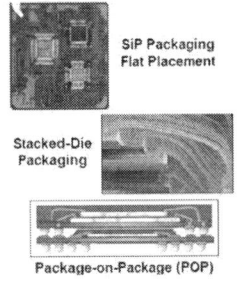

FIGURE 1. SIP STRUCTURE AND PROCESS

The thermo-ultrasonic bonding of gold wire balls has the advantages of fast speed and easy arc forming.

978-1-6654-1392-3/21 $31.00 © 2021 IEEE

However, because the surface of the chip is mostly an aluminum layer, the gold-aluminum bonding interface will inevitably form intermetallic compounds and Kirkendall holes under a certain stress. Intermetallic compounds will not only increase the actual contact resistance of the bonding interface, but also their brittle properties will also cause the tension value of the bonding point to decrease, which has an important impact on the drift of the electrical performance parameters of the device and the circuit life and reliability. Scholars have done a lot of research on the intermetallic compounds of the gold-aluminum bonding system, but they mostly stay in the qualitative research under single stress conditions, and lack more accurate reliability prediction basis under multiple stresses. In this paper, airtight encapsulated SiP electronic components in the aerospace and aviation fields are used as the object. The ambient temperature and interface loading current are used as the accelerated stress of the gold-aluminum bonding interface, and the gold-aluminum based on the TNT-Weibull degradation distribution is constructed through experimental data. The life prediction model of the bonding system is used to quantitatively predict the service life of the system under different temperatures and different electrical stresses, and provide a basis for the reliability prediction of the electronic system.

II LIFE EVALUATION METHOD BASED ON FAILURE PHYSICS

A. Gold-aluminum bond failure physics

Physics of Failure (POF) is a product reliability engineering method that describes product failure behavior and reliability through "physical" theoretical and methodological research. It mainly focuses on the study of product failure mechanisms and establishes quantitative failure physics The model describes the occurrence process of this failure mechanism, and the failure physics reflects the intrinsic nature of product failure.

The gold-aluminum bonding system produces 5 kinds of intermetallic compounds under the action of temperature and current: AuAl, AuAl2 Au2Al, Au5Al2, Au4Al. The lattice constants and thermal expansion coefficients of these intermetallic compounds are different, and they are all brittle. , Easy to break and other characteristics, more importantly, due to the change in volume, the contact resistance of the bonding interface changes greatly, and in severe cases, it causes open circuit failure. For SiP products, the change in contact resistance of the gold-aluminum bonding interface affects the product's electrical resistance. The parameters have an important influence, so this paper chooses the gold-aluminum bonding interface contact resistance as the characterizing parameter of the failure physics.

B. Accelerated life test

Accelerated life test refers to the use of high-level stress to accelerate product degradation without changing the failure mechanism of the product, so as to shorten the failure time of the product, and use the product life characteristics under high-level stress to infer the product under normal horizontal stress. Product life characteristics. The accelerated life test ensures that the degradation process of the tested product obeys the random process of the same family distribution under various accelerated stress conditions, that is, when the test stress conditions change, the product degradation only changes the process parameters, and the random process does not change.

C. The physical model of gold-aluminum bond failure

The failure of the gold-aluminum bonding system is mainly characterized by changes in contact resistance and reduced bonding strength, while the formation of intermetallic compounds is mainly affected by temperature and current. The current mature degradation prediction models for temperature and current mainly include Arrhenio Arrhenius model and inverse power law model.

a）Arrhenius model

The Arrhenius model is a life prediction model based on the reaction rate. It can well describe the relationship between the rate of a chemical reaction and the temperature. By determining the activation energy of the material, the life of the reaction at a specific temperature can be predicted. The specific model is as formula (1):

$$\ln \xi = \frac{E}{K_T} + b \qquad (1)$$

In formula (1): ξ represents product life; E is activation energy; K is Boltzmann's constant, which is $8.617 \times 10-5eV$ / ℃; T is thermodynamic temperature; b is a constant.

b) Inverse power law model

The inverse power law model is an empirical acceleration model, mainly for voltage, current and other stress types. The specific model is as shown in formula (2):

$$\xi = Bp^{-D} \qquad (2)$$

In formula (2): ξ represents the product life; B is a normal number; p is the electrical stress, common voltage, current, etc.; D is a normal number related to activation energy.

The above two models are degradation models under a single stress. This article intends to establish a life prediction model for the gold-aluminum bonding system under dual stresses of temperature and current. The TNT (Thermal-Non Thermal) temperature non-thermal stress model is to A coupling model superimposed on Uth and the inverse power law model can well describe the degradation of electronic products under both temperature and current stress. This article is based on the TNT model to establish a gold-aluminum bonding system under current and temperature stress. The life prediction model of, the specific model is shown in formula (3):

$$L(U,V) = \frac{C}{U^n e^{-\frac{B}{V}}} \qquad (3)$$

In formula (3), U represents non-thermal stress, V represents temperature stress, B and n represent constants related to activation energy, and C represents dual stress coupling constant.

The above constants are related to various factors such as materials, process parameters, and interface conditions, and cannot be calculated and solved theoretically. This article intends to use design experiments and test data to reverse the above constants to obtain the life prediction of the gold-aluminum bonding system Accurate model parameters.

III TEST PROCESS

A. Sample preparation

Since the contact resistance of the gold-aluminum bonding interface is small, the resistance of a single interface cannot be accurately measured, and a special daisy chain test sample needs to be prepared. In this paper,

a special daisy chain chip is designed. This chip adopts the typical chip manufacturing process in the SiP module. The built-in heating resistor simulates the heating of the chip. As shown in Figure 2, the chip size is 3180μm×3000μm. It is designed with 12 daisy chain PAD structures, heating resistors and temperature measuring diodes. The spacing between the two PADs used for bonding is 1mm.

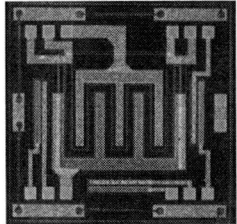

FIGURE 2. DAISY CHAIN CHIP

In order to facilitate the "daisy chain" resistance test, select a suitable housing and substrate to assemble the daisy chain chip into a sample circuit. The layout of the daisy chain chip in the package is shown in Figure 4, with 5 chips in each column, a total of 12 When the chip is distributed, the chip spacing must be ensured that the bonding wires are the same length when the daisy chain is connected.

Use gold wire with a diameter of 25μm for chip bonding. According to the test results, a total of 20 chips in four rows are connected in series as a daisy chain test sample. A total of 100 gold wires are bonded on each test sample, and there are 198 gold aluminum With heterogeneous bonding interface, 3 daisy chain test samples can be assembled in each tube case. Both ends of the daisy chain link are directly bonded to the outer pins of the case for daisy chain resistance test. The actual sample is shown in Figure 3.

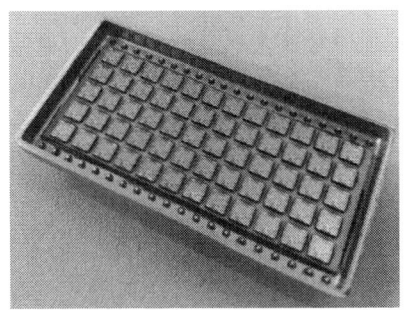

FIGURE 3. GOLD-ALUMINUM BONDING TEST SAMPLE

B. Stress selection

a) Temperature stress selection

According to the basic theory of the Arrhenius model, we choose three sets of constant temperatures as the accelerated stress of the gold-aluminum bond life evaluation test, that is, L=3. According to the bonding reliability test, it is difficult to produce Au-Al bonding failure in a short time under the temperature stress of 125℃, and the failure time is at least 2000h or even 4000h. According to the literature and actual conditions, considering the test efficiency, Choose 150 ℃ as the lowest accelerating stress, namely S1=150℃. At the same time, combined with the actual situation of the sample itself, we choose 195° C as the highest accelerating stress, that is, Sl=195° C. therefore

$$\Delta = (\frac{1}{S_1} - \frac{1}{S_l})/(l-1) = (\frac{1}{150} - \frac{1}{195})/(3-1) = 0.000786 \tag{4}$$

$$\frac{1}{S_2} = \frac{1}{S_1} - \Delta = \frac{1}{150} - 0.000786 = 0.00588 \tag{5}$$

That is, S2=172°C, combined with the actual situation, choose 175°C as the accelerating stress, and take S2=175°C.

In summary, three sets of temperature stresses are selected as the temperature accelerated stresses for the gold-aluminum bonding life evaluation test. The three sets of stresses are: S1 = 150°C, S2 = 175°C, and Sl = 195°C.

b) Current stress selection

In this paper, 25μm wire diameter gold wire is used for bonding. According to the calculation formula of gold wire rated current, the rated current of 25μm gold wire is 300mA, and 300mA is selected as a set of current stress; according to the actual overcurrent situation of SiP device gold wire, we Choose 120mA as the second group of current stress.

In summary, two sets of current stresses are selected as the current acceleration stresses for the gold-aluminum bond life evaluation test. The two sets of stresses are respectively: S1 = 120 mA, S2 = 300 mA.

C. Sample grouping

According to the temperature stress and current stress, the test samples are divided into 12 groups, of which 25℃,

0mA is the reference group, no acceleration stress is applied, 9 samples in each group, and the Agilent 4263B LCR tester is used to chrysanthemum link after bonding Contact resistance is tested. The LCR meter uses a four-terminal resistance measurement method with a measurement accuracy of 1μΩ. The results are shown in Table 1. It can be seen that the average value of the resistance of each group of 9 samples is basically the same.

TABLE I.   STRESS GROUPING OF TEST SAMPLES

Temperature stress (℃) / Current stress (mA)	25°C	150°C	175°C	195°C
0mA	11.867Ω	11.877Ω	11.869Ω	11.875Ω
120mA	11.857Ω	11.863Ω	11.860Ω	11.884Ω
300mA	11.864Ω	11.857Ω	11.874Ω	11.883Ω

D. Failure criterion

According to the IPC-9701A "Surface Mount Solder Parts Performance Test Method and Appraisal Requirements" issued by the International Electronics Industry Connection Association, the failure of the solder joint test sample is defined as: (1) The resistance value of the daisy chain resistance of the solder joint of the test sample Increased by more than 20% in the test; (2) The resistance value of the daisy chain resistance of the test sample solder joint becomes infinite in the test (that is, the test sample daisy chain is interrupted).

E. Test point selection

Set the accelerated stress test time to 4032 hours. If the number of failed samples fails during the test, the group of tests ends. It is known from relevant literature that the life characterization parameters degrade faster in the early stage of the test, and become slower in the middle and late stages of the test. Therefore, when selecting test points, the test points should be dense in the early stage and evacuated in the later stage, combined with GJB548B method 1005.1 to stabilize. Regarding the selection of intermediate test points in the state life test, we stipulate that in the first (0~1008) hours, test once every 168 hours, and in (1009~2016) hours, test once every 336 hours. After 2017 hours, Test every 504 hours, therefore, the test points are set to:1680+72h、3360+72 h、5040+72 h、

6720^{+72} h、8400^{+72} h、10080^{+72} h、13440^{+120} h、16800^{+120} h、20160^{+120} h、25200^{+168} h、30240^{+168} h、35280^{+168} h、40320^{+168} h，A total of 13 test points.

IV. TEST RESULTS AND DISCUSSION

Carry out the accelerated life test of gold-aluminum bonding according to the above settings and regularly sample the degradation data. The test is carried out for 4033.5h. At the end of the test, all the test samples under 195℃ and 300mA current stress will fail, and under other current stresses at 195℃ and 175℃ Some samples failed under 300mA and 120mA current stress at ℃, and none of the samples failed under other temperature and current stress, and the growth trend of contact resistance was in line with the linear increase.

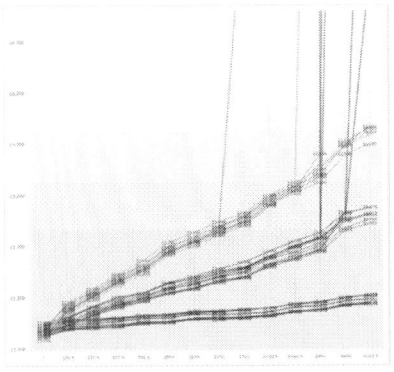

FIGURE 4. LINEAR GROWTH TREND OF CONTACT RESISTANCE

Use the RISA-PofEra reliability analysis software to analyze the test data, select the accelerated non-destructive degradation project, set the current and voltage stress, import the test data, and use 120% of the initial measured average value (14.236Ω) as the critical degradation value for analysis The test data shows that its life distribution conforms to the Weibull distribution. The TNT-Weibull life prediction model is selected to analyze the failure data, and the constants B=6759.35, C=0.0205, and n=0.4539 in the TNT life distribution model are obtained.

This model is used to predict the life of the gold-aluminum bonding system under normal operating conditions of SiP products. The ambient temperature is set to 70℃ and the current is 100mA. According to the calculation of the above model, the gold-aluminum

bonding system can be obtained under 90% confidence. The average life span is 93.21 years, and the B10% life span is 28.61 years.

V CONCLUSION

In this paper, a dedicated daisy chain chip is used as the carrier, and the change of the contact resistance of the gold-aluminum bonding system under different environmental temperatures and different current stresses over time is tested, statistics and analyzed through the accelerated life test method, and the results are based on TNT -The key parameters of the Weibull life prediction model can more accurately predict the life changes of the gold-aluminum bonding system under different stress levels, and provide a basis for the life prediction and quantitative analysis of the gold-aluminum bonding system in the SiP module.

a)The contact resistance increases linearly with time during the degradation process of the gold-aluminum bonding system.Under the same type of stress, the greater the applied stress intensity, the faster the degradation rate, the faster the rate, and the shorter the life span;

b) Under temperature and electrical stress, the lifetime distribution of the gold-aluminum bonding system conforms to the Weibull distribution.

REFERENCES

[1] Su Duhuang, He Xiaoqi. Research trends in the degradation and control of gold-aluminum bonding in hybrid integrated circuits. Electronic Components and Materials, December 2008, Volume 27, Issue 12;

[2] Yang Jiansheng. Discussion on the formation of wire bond voids during high temperature aging. Electronic Quality, Issue 9, 2012;

[3] Chang Xingping. Experimental design of gold-aluminum bonding reliability in hybrid integrated circuits. Journal of Xiangfan University, August 2011, Vol. 32 No. 8;

[4] Liu Jian, Yan Qinyun, En Yunfei, Huang Yun, Yang Dan. Research on the life evaluation method of gold-aluminum bonding. Reliability physics and failure analysis technology, June 2007, Vol. 25 No. 3.

Low-temperature Bonding of High-Power Device Using Cu-Ag Composite Nanoparticle Paste

Jiaxin Liu
School of Mechanical Science and Engineering
Huazhong University of Science & Technology
Wuhan, China
stu_liujx@163.com

Qing Wang
School of Mechanical Science and Engineering
Huazhong University of Science & Technology
Wuhan, China
wangq_hust@163.com

Yun Mou*
School of Mechanical Science and Engineering
Huazhong University of Science & Technology
Wuhan, China
mouluck@163.com

Mingxiang Chen*
School of Mechanical Science and Engineering
Huazhong University of Science & Technology
Wuhan, China
chimish@hust.edu.cn

Abstract—In this work, a Cu-Ag composite nanoparticle paste was proposed for high-power packaging. The nanoparticle (15 nm) has good oxidation resistance and sintering performance. The effect of bonding temperature on the mechanical performance of Cu-Ag composite paste were researched. To explore the practicality of application as a die-bonding material, LED packaging with Cu-Ag nanoparticle paste was prepared and compared with traditional solders. The junction temperature and differential structure functions of LEDs were tested. LEDs with different die-bonding materials were aging for 100 h, and light output power is measured to compare the reliability.

Keywords— Cu-Ag nanoparticles, LED packaging, low-temperature bonding, high thermal conductivity

I. INTRODUCTION

The third-generation semiconductors exhibit superior advantages, such as wide band-gap, high transmission current, and quick conversion efficiency, which are widely used in high-power devices [1-3]. These devices need higher operating temperature than traditional Si-based devices, which is a challenge of die-bonding materials [4]. In order to reduce the chip damage and large residual stress in bondline caused by high processing temperature, interconnection materials should be bonded under low temperature and have a higher re-melting temperature during operation [5-7]. Transient-liquid-phase (TLP) bonding method satisfies the requirement. However, the process takes too much reflowing time to transform into full intermetallic interconnection layer [8]. In addition, due to low ductility of intermetallic compounds (IMCs), the interconnection layer exhibits poor thermodynamic performance [9].

Aiming at this issue, metal nanoparticles (NPs) has been developed as die-bonding materials for high-power devices. Currently, due to inherent high thermal and electrical conductivity, Cu / Ag metal nanoparticles have attracted wide attention. In addition, small size effect of nanoparticles leads to a low bonding temperature and high re-melting point, which could meet the requirements of high-power applications. However, Ag nanoparticles are so expensive with high porosity that are difficult to be used on a large scale. Meanwhile, the high electro-migration of Ag NPs would influence the electrical performance and device life in long-term using. Cu nanoparticles have poor oxidation resistance, which are difficult to store and increase the production cost [10].

Here, we propose a new Cu-Ag composite nanoparticle paste (CNP) for high-power device packaging. The nanoparticles were synthesized in air by modified polyalcohol phase reduction method. The microstructure morphology and elemental composition of Cu-Ag CNP were observed. The effect of bonding temperature on sintered joints shear strength was carried out. Further, junction temperature change and the differential structure functions of high-power LED packaging by Cu-Ag CNP were measured. The accelerated aging tests of LED was investigated and compared with traditional bonding solders.

II. EXPERIMENTS

A. Materials

Silver acetate ($AgCH_2COOH$), ethylene glycol (EG, 99%), copper acetate ($Cu(CH_3COO)_2 \cdot H_2O$), 1-amino-2-propanol (IPA), and sodium borohydride ($NaHB_4$) were purchased from Sinopharm Chemical Reagent Co., Ltd. The bonding substrate is pure Cu flake with dimension of $3\times3\times2$ mm^3 and $5\times5\times2$ mm^3.

B. Fabrication of Cu-Ag composite nanoparticle

The preparation process of Cu-Ag composite nanoparticle is displayed in Fig. 1 (in our previous work). First, 50 mL ethylene glycol and 25 mL 1-amino-2-propanol were poured into a conical flask as solution A. Then, 1 g copper acetate and 2.52 g silver acetate were dissolved in the solution A with magnetic stirring. 2.5 g sodium borohydride was added in 25 mL ethylene glycol as solution B. Subsequently, solution A and B were mixed together under 5°C for 30 min with magnetic stirring. After full reaction, precipitated Cu-Ag CPs were obtained by high-speed centrifuge. The bonding process was processed under Ar atmosphere. The time set as 30 min. And the temperature was from 225 to 300°C.

In order to study the performance of composite paste as a die-bonding material of high-power LED, Cu-Ag CNP was mounted on a thermoelectric separation Cu substrate sink and covered with a high-power LED chip with AlN ceramic substrate.

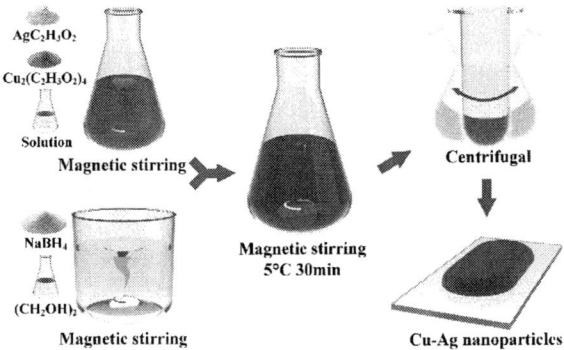

Fig. 1. Preparation flow chart of the Cu-Ag composite nanoparticles.

III. RESULTS AND DISCUSSION

The TEM image of Cu-Ag composite nanoparticles are displayed in Fig. 2(a). The particles are uniform and well dispersed with a diameter of 15 nm. Fig. 2(b) shows the high magnification image of image (a). The organic layer coating around the Cu-Ag CPs can be clearly observed, which protecting these particles from oxidation.

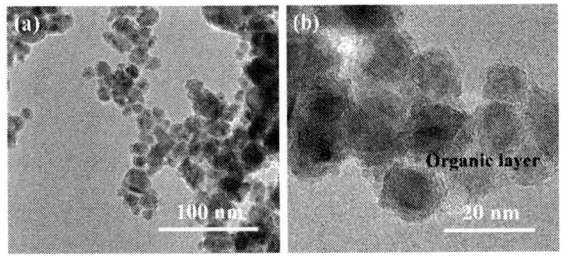

Fig. 2. (a) TEM morphology of Cu-Ag CPs and (b) high magnification image of (a).

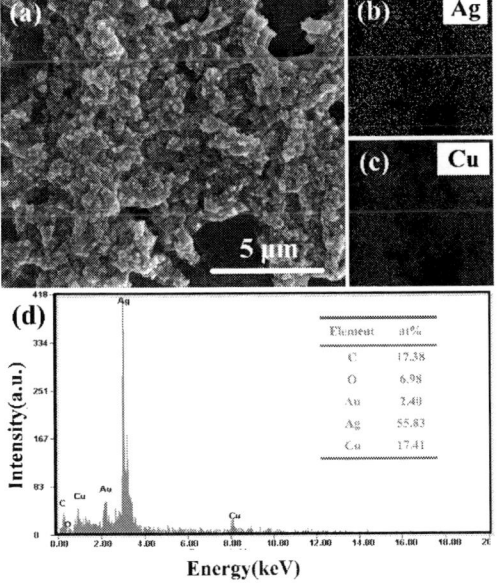

Fig. 3. (a) SEM microstructure of Cu-Ag composite nanoparticles, (b-c) elemental mapping images, and (d) EDS spectrum.

Fig. 3(a) displays the SEM microstructure of composite NPs and (b-c) are elemental mapping images of Ag and Cu. Two elements are evenly dispersed, which is conducive to

improve the antioxidant properties. The elemental spectrum of Cu-Ag CPs displays the atomic percent of Ag and Cu are 55.83 and 17.41, respectively, which complies with the proportion as mention above.

The elements composition of Cu-Ag nanoparticle on the surface was analyzed through XRD to determine the oxidation, as presented in Fig. 4. The diffraction peaks of nanoparticles represent Cu crystal and Ag crystal. No diffraction peak of Cu_2O phase appears in the pattern, which indicates that the Cu-Ag NPs have a nice oxidation-resistance.

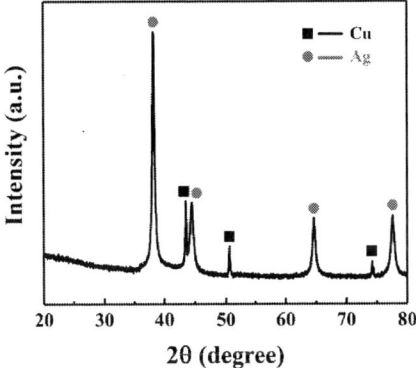

Fig. 4. XRD results of the Cu-Ag composite nanoparticles.

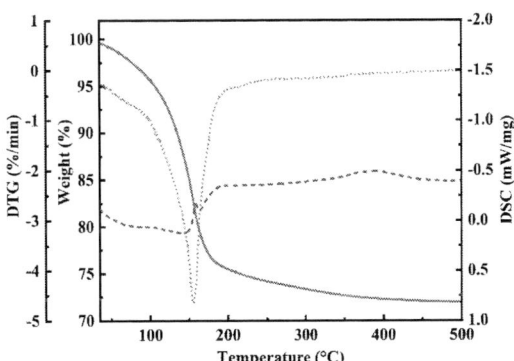

Fig. 5. TG-DSC analysis for Cu-Ag CNP.

Fig. 5 presents the thermogravimetric-differential scanning calorimetric (TG-DSC) curve of Cu-Ag CNP. The red line manifests the particles content of the paste is about 72%, and the blue line shows a heat flow change at 160°C, which represents the volatilization of organics. In order to remove the organics, a selected bonding temperature above 200°C is suitable.

Fig. 6. SEM images of Cu-Ag composite nanoparticle paste films sintering for 20 min at (a) 250°C and (b) 300°C.

To evaluate the sintering performance of Cu-Ag nanoparticles, composition paste was painted on a piece of glass and sintered at 250 and 300°C for 20 min. Fig 6 gives the microscopic morphology of sintered films. When the sintering temperature is 250°C, the grains grow significantly and sintered together. With the increasing sintering temperature, the particles grow continually and sintered into clusters.

To study the mechanical performance of the Cu-Ag CNP, shear strength of the interconnection joints with Cu-Ag CNP was tested, as shown in Fig. 7. The shear strength enhances with the increasing temperature, which is 9.23, 14.88, 19.42, and 23.74 MPa at 225, 250, 275, and 300°C, respectively.

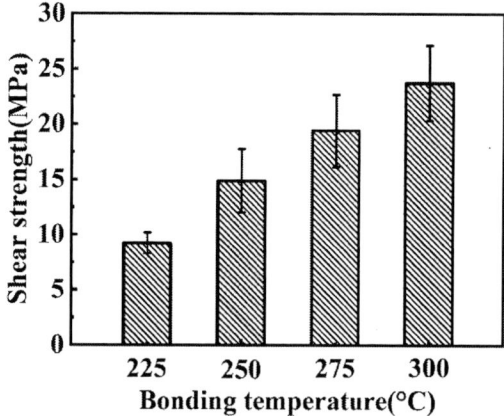

Fig. 7. Mechanical strength of the interconnection joints by Cu-Ag CNP at 225, 250, 275, and 300°C.

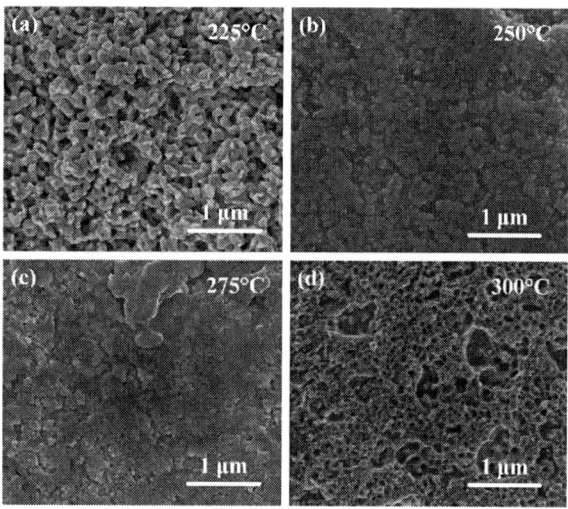

Fig. 8. Fracture surfaces of interconnection joints under 225, 250, 275, and 300°C.

Fig. 8 presents the fracture surfaces microstructure morphologies of interconnection joints under 225, 250, 275, and 300°C. As the bonding temperature enhances, the degree of coalescence and sintering phenomena increases. After sintering at 300°C, the coarsening grains interconnects and form a network structure, which can enhance the bonding ability.

Fig. 9 displays the cross-sectional morphology of LED packaged with Cu-Ag CNP and sintered Cu-Ag interconnection layer. The microstructure of the sintered interconnection layers is compact with none pores, which is helpful to heat transfer.

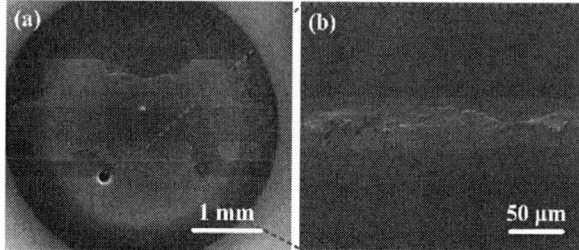

Fig. 9. Cross-sectional SEM images of (a) LED packaging with Cu-Ag CNP and (b) sintered Cu-Ag interconnection layer.

Fig. 10 presents the junction temperature change of LEDs packaging with Cu-Ag composite nanoparticle paste, SAC305, and SnCu7. The junction temperature change of these sample at a driving current of 350 mA are 5.0, 8.1, and 10.5°C, respectively. Cu-Ag nanoparticle sample has the lowest junction temperature, which indicates a best heat dissipation.

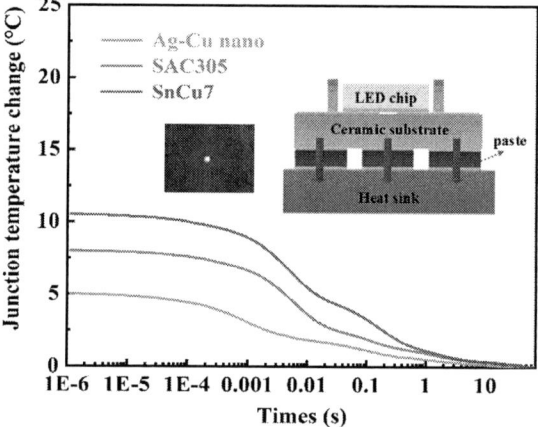

Fig. 10. Junction temperature change of LEDs packaging by Cu-Ag composite nanoparticle paste, SAC305, and SnCu7.

Fig. 11 shows the differential structure functions of high-power LEDs packaged with Cu-Ag CNP, SAC305, and SnCu7. The total thermal resistance of these LEDs are 8.21, 9.64, and 12.51 K/W, respectively. The thermal resistance of the sintered Cu-Ag composite nanoparticle paste layer is 0.58 K/W, which is 21.6% lower than the sintered SAC layer of 0.74 K/W and 36.3% lower than the sintered SnCu7 layer of 0.91 K/W. The results demonstrate that Cu-Ag composite nanoparticles have higher thermal conductivity than traditional solders.

High-power LEDs packaged by Cu-Ag CNP was aging for 100 h (@85°C, 85%RH, 700 mA) to investigate the influence of die-bonding material on light output power. With the aging current of 700 mA, the temperature of LED is 104°C. Fig. 12 presents light output power of high-power LED with different die-bonding materials before and after aging. The light output power increases with input current. Before aging, light output power of the Cu-Ag nanoparticle sample and SAC305 sample is similar, and is 1487 and 1484 mW under 1200 mA current. After aging, the light output power change to 1457 and 1382 mW, respectively, which is 30 and 102 mW lower than before. The result indicates that the Cu-Ag CMP enhances the reliability of high-power LEDs in aging process.

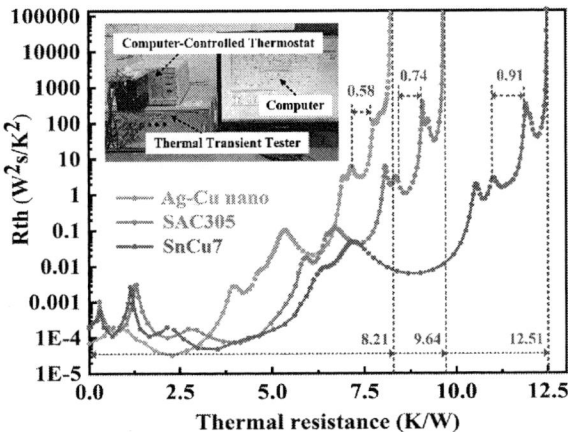

Fig. 11. Differential structure functions of LEDs packaged with Cu-Ag composite nanoparticle paste, SAC305, and SnCu7.

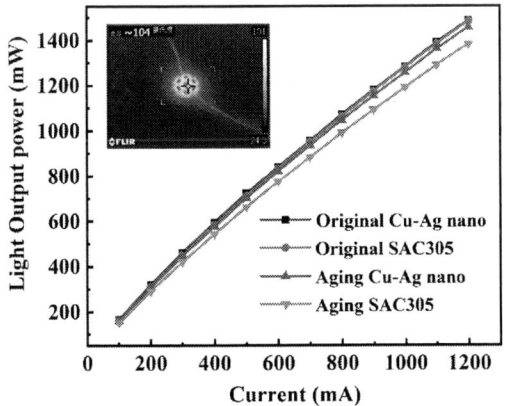

Fig. 12. Light output power of high-power LEDs with different die-bonding materials before and after aging.

IV. CONCLUSIONS

To summary, a low temperature Cu-Ag composite nanoparticle paste (CNP) was fabricated and applied to high-power LED. The Cu-Ag CPs have an excellent dispersion and

oxidation-resistance with a diameter of 15 nm. The sintering performance increase with the increasing sintering temperature, and a robust Cu-Cu joint of 23.74 MPa can obtain at 300°C. The thermal resistance of LED packaging with Cu-Ag composite nanoparticle paste is 8.21 K/W, which is lower than SAC305 sample of 9.64 K/W and SnCu7 sample of 12.51 K/W. After aging for 100 h at 700 mA current, the LED with Cu-Ag nanoparticle paste demonstrates better reliability.

REFERENCES

[1] Y. Peng, Y. Mou, J. Liu, M. Chen, "Fabrication of high-strength Cu–Cu joint by low-temperature sintering micron–nano Cu composite paste," J. Mater. Sci.-Mater. El., vol. 31, pp. 8456-8463, Jun 2020.

[2] Z. Yin, F. Sun, M. Guo, "The fast formation of Cu-Sn intermetallic compound in Cu/Sn/Cu system by induction heating process," Mater. Lett., vol. 215, pp. 207-210, Mar 2018.

[3] Y. Peng, Q. Sun, J. Liu, Y. Mou, "Fabrication of stacked color converter for high-power WLEDs with ultra-high color rendering," J. Alloy. Compd., vol. 850, pp. 156811, Jan 2021.

[4] T. Iwasaki, Y. Hoshino, K. Tsuzuki, H. Kato, T. Makino, M. Ogura, D. Takeuchi, H. Okushi, S. Yamasaki, M. Hatano, "High-Temperature Operation of Diamond Junction Field-Effect Transistors with Lateral p-n Junctions," IEEE Electron. Devices Lett., vol. 34, pp. 1175-1177, Sep 2013.

[5] F. Yu, C. Hang, M. Zhao, H. Chen, "An interconnection method based on Sn-coated Ni core-shell powder preforms for high-temperature applications," J. Alloy. Compd., vol. 776, pp. 791-797, Mar 2019.

[6] X. Liu, S. He, H. Nishikawa, "Low temperature solid-state bonding using Sn-coated Cu particles for high temperature die attach," J. Alloy. Compd., vol. 695, pp. 2165-2172, Feb 2017.

[7] Y. Mou, J. Liu, H. Cheng, Y. Peng, M. Chen, "Facile Preparation of Self-Reducible Cu Nanoparticle Paste for Low Temperature Cu-Cu Bonding," JOM, vol. 71, pp. 3076-3083, Sep 2019.

[8] H. Zhao, J. Liu, Z. Li, Y. Zhao, H. Niu, X. Song, H. Dong, "Non-interfacial growth of Cu3Sn in Cu/Sn/Cu joints during ultrasonic-assisted transient liquid phase soldering process," Mater. Lett., vol. 186, pp. 283-288, Jan 2017.

[9] J. Li, Q. Liang, T. Shi, G. Liao, Z. Tang, "Design of Cu nanoaggregates composed of ultra-small Cu nanoparticles for Cu-Cu thermocompression bonding," J. Alloy. Compd., vol. 772, pp. 793-800, Jan 2019.

[10] C. Lee, N.R. Kim, J. Koo, Y.J. Lee, and H.M. Lee, "Cu-Ag core–shell nanoparticles with enhanced oxidation stability for printed electronics," Nanotechnology., vol. 26, pp. 455601, Nov 2015.

978-1-6654-1392-3/21 $31.00 © 2021 IEEE

Research on the uniform temperature of heat dissipation for the reverse oblique microchannel

Qinglin Tang
Microsystems Center
China Electronics Technology Group
Corporation No.58 Research Institute
Wuxi, China
qltang95@163.com

Dongcheng Liu
Microsystems Center
China Electronics Technology Group
Corporation No.58 Research Institute
Wuxi, China
ldcxsp@163.com

Yanping Zeng
Microsystems Center
China Electronics Technology Group
Corporation No.58 Research Institute
Wuxi, China
zyanping123@163.com

Xin Lan
School of Energy and Power Engineering
Shandong University
jinan, China
lanxin@sdu.edu.cn

Lihua Zheng
Microsystems Center
China Electronics Technology Group
Corporation No.58 Research Institute
Wuxi, China
420380473@qq.com

Abstract—**As the power density of power devices/modules increases, power devices/modules are more demanding for heat dissipation.Liquid-cooled microchannels have strong heat transfer capabilities and will become an important way for high-power electronic devices/modules to dissipate heat. However, uneven heat dissipation of microchannels will cause excessive temperature gradients and local high temperature hot spots, and long-term operation will lead to power module solder layers fatigue, aging and degradation of its own heat dissipation capacity, which will affect the reliability of power devices/modules for long-term operation under complex working conditions.Therefore, it is of great significance to strengthen the heat dissipation capacity of the microchannel and improve the temperature uniformity of the liquid-cooled microchannel heat dissipation. In this paper, based on the microchannel enhanced heat transfer technology, We have designed a new type of liquid-cooled the ROMC(reverse oblique microchannel). Firstly, the heat dissipation characteristics of the ROMC and the RSMC (rectangular straight microchannel) are compared through experiments, and it is found that the heat transfer capacity and heat dissipation uniformity of the ROMC are significantly better than the RSMC. Next analyze the flow, heat transfer characteristics and wall temperature uniformity of the ROMC by means of numerical simulation. The results showed that: (1)When Re number, the number of oblique channel and the oblique spacing are constant, reducing the oblique angle and increasing the oblique distance can enhance heat transfer and improve the heat dissipation capacity and wall temperature uniformity of the ROMC. (2)When the number of oblique channel, oblique angle, oblique distance and oblique spacing are constant, increasing Re number can significantly improve the heat dissipation capacity and wall temperature uniformity of the ROMC. Finally, this paper uses the MOGA (multi-objective genetic algorithm) to optimize the wall temperature, temperature difference of the ROMC. Compared with the model before optimization, the maximum bottom wall temperature of the ROMC is 323.33K, the reduction rate is 0.75%, and the wall temperature difference is 10.50 K, the reduction rate is 17.52%.**

Keywords—microchannel, temperature uniformity, enhanced heat transfer

I. INTRODUCTION

With the increasing heat flux density of high-power chips, traditional heat dissipation methods cannot meet their heat dissipation requirements. At present, the indirect heat exchange technology represented by liquid-cooled microchannels has gradually attracted people's attention. The shape, structural parameters, cooling medium (nanofluid), and microchannel manufacturing technology of microchannels have been extensively studied to improve the heat transfer capacity of microchannels[1-2]. Although the liquid-cooled microchannel has strong heat exchange capability, it has the problem of uneven heat dissipation which will cause large temperature gradients and local hot spots, and the thermal reliability of chips and power modules will be seriously affected. In order to improve the temperature uniformity of heat dissipation of microchannels, a large number of studies mainly improve the temperature uniformity through structural design.

Vafai et al. [3] proposed a double-layer microchannel heat dissipation structure. The temperature gradient of the structure was significantly smaller than that of the single-layer structure through numerical simulation. Wei et al. [4] studied double-layer silicon-based structures at different flow rates through experiments and numerical simulations, and found that the arrangement of the counter-current flow direction makes the heat dissipation more uniform. Chamoli designed the top-layer truncated double-layer structure. The results show that the top-layer truncated structure can significantly improve the average temperature of the double-layer microchannels[5].

Carlos et al.[6] proposed a variable density Pin-fin structure, and found that increasing the rib density along the flow direction can increase the heat dissipation area at the exit side of the microchannel heat sink, thereby improving the heat dissipation uniformity, and its wall temperature change is less than 2°C/mm. On this basis, a staggered variable-density Pin-fin structure is proposed. Compared with a parallel arrangement, the staggered structure can reduce the overall thermal resistance of the system and further improve the temperature uniformity of microchannel heat dissipation, but the pressure drop loss increases .

Alharbi et al. [7] simulated the thermal characteristics of the branch structure under the conditions of heat flux density of 100W/cm² and volume flow rate of 10ml/s. The maximum temperature is not much different from the parallel straight channel, the wall temperature and the pressure drop are reduced by 75% and 10%, respectively. Liang et al. [8] studied the effects of volume flow and heat flux density on

978-1-6654-1392-3/21 $31.00 © 2021 IEEE

the branch structure by means of experiments and simulations. The results show that the temperature difference decreases with the increase of the inlet flow rate. Tan et al. [9] designed a multi-order branch structure and found that the third-order branch structure has more uniform heat dissipation than the ordinary branch structure, and the standard deviation of the wall temperature of the third-order branch structure increases with the heat flux density to a lesser extent.

Manoj et al. [10] studied the temperature distribution of microchannels under different hydraulic diameters through experiments, and found that reducing the hydraulic diameter will significantly improve the temperature uniformity of heat dissipation, because the smaller hydraulic diameter will make the coolant distribution more uniform. Based on the layout of flow channels and flow directions, Vasilev et al. [11] analyzed 10 types of microchannel structures.It is found that the uniform distribution of coolant can significantly reduce the wall temperature gradient and improve the wall temperature uniformity when the pumping power is constant.

Lee et al.[12] analyzed the designed unidirectional oblique structure through experiments and simulations, and found that the oblique structure will continuously develop the flow boundary layer and the thermal boundary layer, thereby enhancing the heat dissipation of the microchannel, and the oblique microstructure channels have better temperature uniformity than rectangular straight channels. Liu et al. [13] designed a cross-oblique truncated circular microchannel. Compared with the ordinary circular microchannel, this structure not only has a small total thermal resistance, but also has more uniform heat dissipation. Ahmed et al [14] designed a serpentine microchannel with a V-shaped secondary channel. Compared with the traditional serpentine channel, this structure can enhance the heat transfer of the serpentine channel, reduce the pressure drop at the inlet and outlet, and thus improve the temperature uniformity.

The uniform distribution of the coolant can significantly improve the uniformity of heat dissipation of the microchannel. Based on the unidirectional oblique truncated microchannel, the ROMC (reverse oblique microchannel) is designed in this paper, and the influence of the oblique parameters on the heat dissipation temperature uniformity of the microchannel is analyzed through experiments and numerical simulations.

II. EXPERIMENTAL AND MICROCHANNEL DESIGN

A. Experimental platform design

The microchannel heat dissipation experiment test platform in this paper is shown in Fig.1. The coolant in the liquid storage tank is transferred to the microchannel heat dissipation system by a peristaltic pump, enters and exits vertically, and finally flows into the liquid collection tank. Thermocouples were used to collect the average temperature of the inlet(T_{in}), the average temperature of the outlet (T_{out})of the microchannel and the temperature measured in the temperature measuring hole($T_{y, i}$). The DC power supply provides power to the high-temperature ceramic chip and the data acquisition module, in which the high-temperature ceramic chip is adhered to the bottom of the microchannel through thermally conductive silica gel.

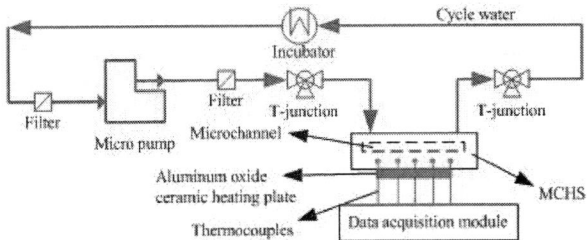

Fig.1 Schematic diagram of the microchannel heat dissipation experiment

B. Microchannel design

In this paper, two structures of the RSMC(rectangular straight microchannels) and the ROMC(reverse oblique microchannels) are manufactured. The specific structural parameters are limited by the processing technology: channel width W_c = 0.8mm, main rib width W_f = 0.8mm, channel length L_c = 20mm, channel height H_c = 1.2mm, substrate thickness H_b = 1mm, reverse oblique angle θ=30°, reverse oblique distance d=0.4mm, intercept pitch p decreases along the flow direction. The microchannel is processed by a combined process of metal etching and machining. The processed ROMC is shown in the Fig. 2a. A microchannel with a hydraulic diameter (D = 96um) is arranged in the center of the copper base plate. In order to make the cooling medium enter and exit the microchannel stably, an inlet and outlet section with a length of 8 mm is designed on the inlet and outlet side of the microchannel. On the side of the copper base plate, there is a deep hole with a diameter of 1.1mm, and the thermocouple is probed into the deep hole of the side wall to measure the temperature at the center of the channel.The final assembly of the MCHS (microchannel heat sink) is shown in Fig. 2b.

Fig.2a Actual picture of the ROMC   Fig. 2b Final assembled MCHS

III. NUMERICAL MODEL METHOD

A. Numerical model and boundary conditions

Liquid-cooled microchannel heat dissipation belongs to convective heat transfer, and a numerical analysis model for microchannel heat dissipation is established based on the law of conservation of mass, law of conservation of momentum and law of conservation of energy. The continuity differential equation can be shown in Eq.(1):

$$\frac{\partial(\rho u_x)}{\partial x} + \frac{\partial(\rho u_y)}{\partial y} + \frac{\partial(\rho u_z)}{\partial z} = 0 \tag{1}$$

where u_x, u_y and u_z stand for the velocity components in the x,y,z-direction, respectively. ρ stands for the density of the coolant. The momentum differential equation is shown in Eq.(2).Where μ denotes dynamic, p denotes pressure.

$$u_x \frac{\partial u_x}{\partial x} + u_y \frac{\partial u_x}{\partial y} + u_z \frac{\partial u_x}{\partial z} = \frac{\mu}{\rho}\left(\frac{\partial^2 u_x}{\partial x^2} + \frac{\partial^2 u_x}{\partial y^2} + \frac{\partial^2 u_x}{\partial z^2}\right) - \frac{1}{\rho}\frac{\partial p}{\partial x} \tag{2a}$$

$$u_x \frac{\partial u_y}{\partial x} + u_y \frac{\partial u_y}{\partial y} + u_z \frac{\partial u_y}{\partial z} = \frac{\mu}{\rho}\left(\frac{\partial^2 u_y}{\partial x^2} + \frac{\partial^2 u_y}{\partial y^2} + \frac{\partial^2 u_y}{\partial z^2}\right) - \frac{1}{\rho}\frac{\partial p}{\partial y} \quad (2b)$$

$$u_x \frac{\partial u_z}{\partial x} + u_y \frac{\partial u_z}{\partial y} + u_z \frac{\partial u_z}{\partial z} = \frac{\mu}{\rho}\left(\frac{\partial^2 u_z}{\partial x^2} + \frac{\partial^2 u_z}{\partial y^2} + \frac{\partial^2 u_z}{\partial z^2}\right) - \frac{1}{\rho}\frac{\partial p}{\partial z} \quad (2c)$$

For no external force to do work and no internal heat source, the fluid domain energy equation can be simplified to Eq.(3) where c_p means specific heat capacity.

$$\rho c_p \frac{\partial T}{\partial t} = k_l\left(\frac{\partial^2 T}{\partial x^2} + \frac{\partial^2 T}{\partial y^2} + \frac{\partial^2 T}{\partial z^2}\right) \quad (3)$$

Under the condition of studying steady-state heat conduction and no internal heat source, the solid domain energy equation can be shown in Eq.(4)

$$k_s\left(\frac{\partial^2 T}{\partial x^2} + \frac{\partial^2 T}{\partial y^2} + \frac{\partial^2 T}{\partial z^2}\right) = 0 \quad (4)$$

k_l and k_s is the thermal conductivity of the fluid domain and the solid domain, respectively.

The solid domain material is set to copper, and its physical property parameters are constant. The material in the fluid domain is deionized water. From the experimental data, the maximum temperature difference between the inlet and the outlet of the fluid is greater than 10K. Therefore, the variable physical properties with the temperature of the fluid need to be considered. The inlet is set as the speed inlet, T_{in} is the inlet temperature of the cooling medium, and V_{in} is the average inlet speed. The boundary conditions of the model are outer surface insulation, velocity inlet, pressure outlet and constant heat flux density. This paper studies the laminar flow calculation model, and the solver is set as a pressure-based solver. The SIMPLEC method is used to couple pressure and velocity.

B. Data processing of the experiment and simulation

In this paper, the fluid domain and solid domain are divided into tetrahedral unstructured grids, and a boundary layer is added near the wall surface. In the process of meshing the 3D numerical geometric model, the larger the number of meshes, the closer the calculation result is to the true solution. However, when the number of grids reaches a certain number, continuing to increase the number of grids has little effect on improving the calculation accuracy, but the calculation time increases substantially. Therefore, in order to improve the calculation efficiency and ensure the calculation accuracy, it is necessary to determine the appropriate number of grids, that is grid independence verification. This paper verifies the ROMC($\theta = 15°$, d=0.02mm, p=5mm) different grid numbers. The error of $T_{w,max}$ and $T_{w,d}$ is 0.10% and 2.46%, respectively, as shown in Tables 1. Therefore, the grid density of Case2 is selected for calculation in this paper.

Table. 1 Grid independence verification

Case	Num(10^4)	$T_{w,max}$(K)	Err(%)	$T_{w,d}$(K)	Err(%)
1	65.91	325.17	0.44	13.72	8.90
2	229.01	326.30	0.10	14.69	2.46
3	429.55	326.62	—	15.06	—

The temperature on the bottom wall of the MECH $T_{w,i}$ can be calculated according to Eq.(5):

$$T_{w,i} = T_{y,i} - \frac{y \cdot q}{A_h \cdot k_s} \quad (5)$$

where $T_{y,i}$ is the temperature measured in the MCHS measuring hole, A_h is the area of the heat source surface, and q is the amount of heat dissipation.

$T_{w,max}$ is the maximum bottom wall temperature, as show in Eq.(6):

$$T_{w,max} = \text{Max}\{T_{w,i}\} \quad (6)$$

$T_{w,ave}$ is the average of bottom wall temperature, as show in Eq.(7):

$$T_{w,ave} = \frac{\sum_{i=1}^{n} T_{w,i}}{n} \quad (7)$$

The standard deviation of the bottom wall temperature σ is shown in Eq.(8):

$$\sigma = \left(\sum_{i=1}^{n}\left(T_{w,i} - T_{w,ave}\right)^2 / (n-1)\right)^{1/2} \quad (8)$$

The total of thermal resistance R_t takes the followling form:

$$R_t = \frac{(T_{w,max} - T_{in})}{q} \quad (9)$$

where T_{in} is the temperature at inlet of the MCHS, q is the heat taken away by the cooling medium in MCHS, which can be calculated by Eq.(10):

$$q = UI - q_{loss} = \rho c q_v \left(T_{out} - T_{in}\right) \quad (10)$$

where U and I respectively represent the voltage and current applied by the current source; q_{loss} is the heat lost to the external environment without being taken away by the coolant; q_v is the volume flow of the coolant; T_{out} is the temperature at the outlet of the MCHS.

$T_{w,d}$ denotes the temperature difference of bottom wall, it can be calculated by Eq.(11):

$$T_{w,d} = T_{w,max} - T_{w,min} \quad (11)$$

where $T_{w,min}$ is the minimum temperature of the bottom wall, as show in Eq.(12):

$$T_{w,min} = \text{Min}\{T_{w,i}\} \quad (12)$$

Re denotes the Reynolds number, it is a dimensionless number used to characterize fluid flow. it is given by Eq.(13):

$$Re = \frac{\rho v D}{\mu} \quad (13)$$

where v and D denote flow velocity and hydraulic diameter.

IV. RESULTS AND ANALYSIS

A. Analysis of experimental results

The $T_{w,max}$ and σ of the RSMC and the ROMC have the same trend with Re, and both decrease with the increase of Re, which shows that the increase of Re can improve the uniformity of heat dissipation of the channel in Fig. 3. In addition, the heat transfer capacity and heat dissipation uniformity of the ROMC are significantly better than the RSMC. The R_t at low Re is shown in Fig. 4. The trend of the R_t of each structure changes with Re is the same, and decreases with the increase of Re, and the R_t of the ROMC is lower than that of the RSMC.

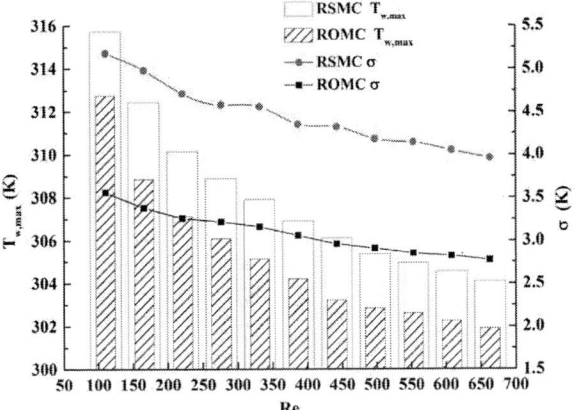

Fig.3 The $T_{w,max}$ and σ of MCHS under different Re.

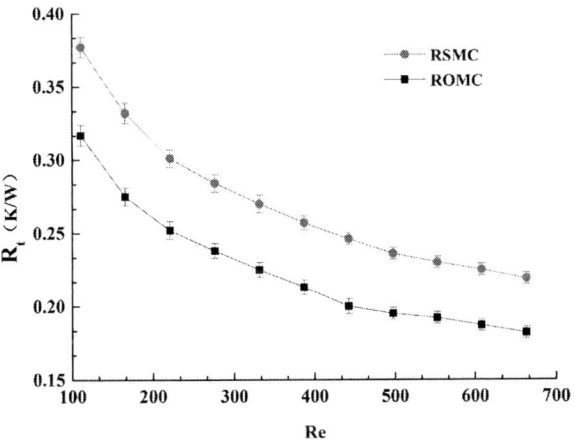

Fig. 4 Thermal resistance under different Re

B. Analysis of simulation results

The results of the $T_{w,ave}$ at different Re are obtained through experiments and simulations. It can be seen from Fig. 5 that there is little difference between the experimental and simulated results. With the increase of Re, the change trend of the $T_{w,ave}$ is consistent, and both decrease with the increase of Re.

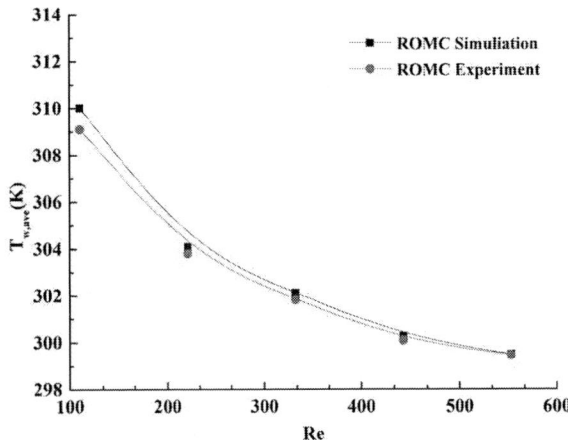

Fig. 5 The $T_{w,ave}$ under different Re

Under certain operating conditions (Re=500, d=0.2mm, p=5mm, H_c/W_c=2, W_c/W_f=1), oblique angle is different, the local velocity field at the same oblique position on the z=1.4mm cutting plane are shown in Fig.6. It can be seen from Fig.6 that the velocity at the center line of the main channel at each angle is the largest, and the flow velocity in each oblique truncated channel is lower than that of the main channel. The smaller oblique angle, the greater the flow velocity in the oblique truncated channel. Conversely, the velocity in the oblique truncated channel with a larger oblique angle is too low, and there is a low-speed vortex zone.

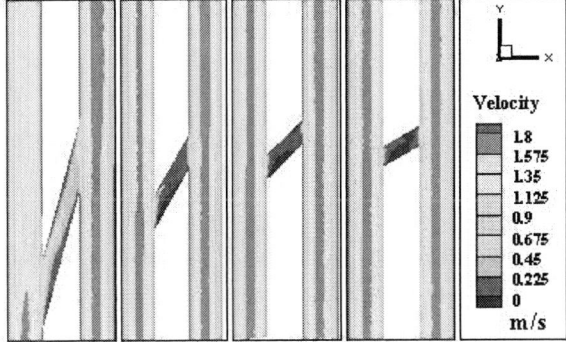

Fig. 6 Velocity distribution of the local ROMC at different oblique angles

Under certain operating conditions (Re=500 、 θ=15° 、 p=5mm 、 H_c/W_c=2 、 W_c/W_f=1), the oblique distance is different, the local velocity field at the same oblique position on the z=1.4mm cutting plane are shown in Fig.7. As can be seen from Fig.7, the larger oblique distance, the more violent the fluid fluctuations in the main channel. The velocity at the center line of the main channel decreases as oblique distance increases. When oblique distance increases, the flow velocity in the oblique truncated channel increases, and the flow velocity fluctuates greatly at the inlet and outlet. A local low-speed zone appears at the wall surface and the outlet of the oblique truncated channel near the upstream of the main channel, and the area of the low-speed zone increases with the increase of oblique distance.

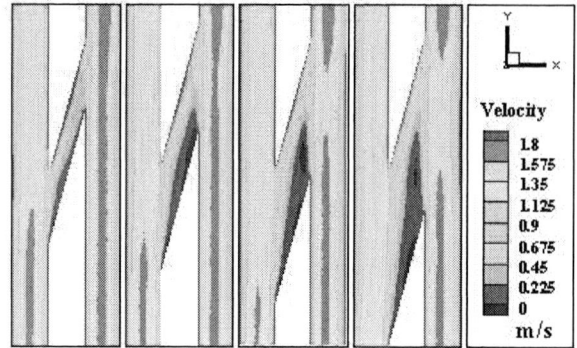

Fig.7 Velocity distribution of the local ROMC at different oblique distance

The change trend of the temperature difference with the oblique angle at each Re is the same, as shown in Fig.8. When Re is constant, the temperature difference decreases with the decrease of the oblique angle. The trend of the temperature difference with Re at each oblique angle is the same. When the oblique angle is constant, the temperature difference will decrease significantly when the oblique angle is increased. This shows that increasing Re can make the channel heat dissipation more uniform. The magnitude of the decrease in temperature difference decreases as Re increases.

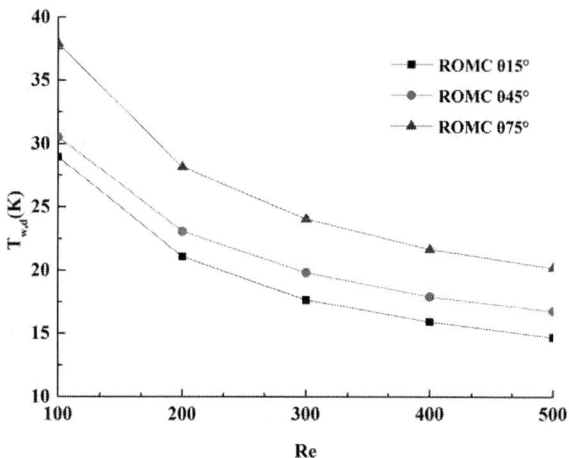

Fig.8 The changing of $T_{w,d}$, at different oblique angle

The temperature difference under each Re has the same trend with the oblique distance. When Re is constant, the larger oblique distance, the lower the temperature difference. This shows that increasing the oblique distance can make the microchannel heat dissipation more uniform. The change trend of the temperature difference of each oblique distance with Re is the same. When the oblique distance is constant, increasing Re will significantly decrease the temperature difference, as shown in Fig.9. This indicates that increasing Re can improve the uniformity of heat dissipation in the microchannel and the decrease in temperature difference decreases with increasing Re.

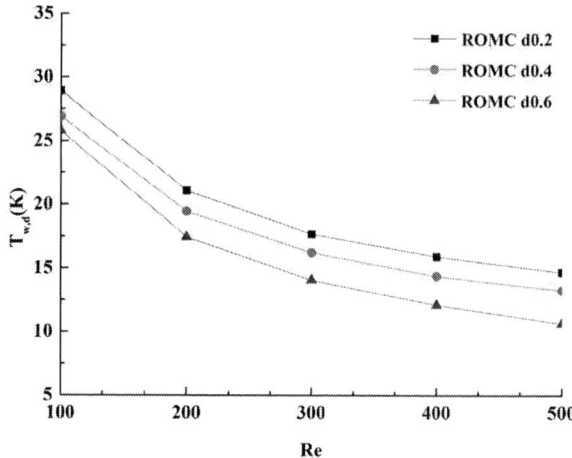

Fig.9 The changing of $T_{w,d}$, at different oblique distance

In order to design a high-efficiency microchannel radiator with good temperature uniformity, The $T_{w,max}$ and the $T_{w,d}$ are taken as the objective function of the optimization design, that is, after the optimization, The $T_{w,max}$ and the $T_{w,d}$ shall be minimized. When the number of oblique channel is fixed, the original model before optimization is a structure with the same characteristics of each oblique secondary channel. This optimization takes the oblique angle and oblique distance as design variables, where oblique angle variables are θ1, θ2, θ3, θ4 and θ5, oblique distance variables are d1, d2, d3, d4 and d 5, a total of 10 design variables.

The sensitivity analysis results of each design variable are shown in Fig.10. It can be seen from the figure that the θ5 and d5 have a greater influence on the objective function, and the influence of each design variable on the objective function can be arranged.

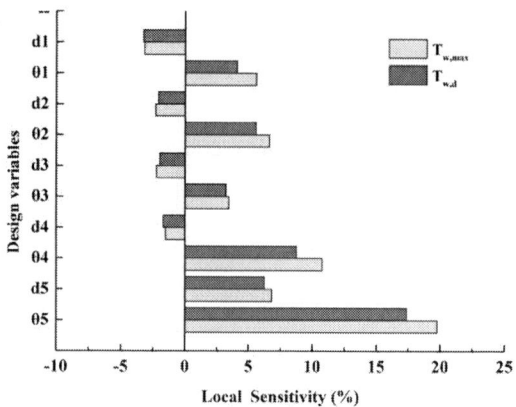

Fig.10 Local sensitivity diagram of each design variable

Taking the $T_{w,max}$ and the $T_{w,d}$ as the objective function, the MOGA(multi-objective genetic algorithm) is selected for optimization. Set the number of initial sample points to 10000, and after rounding and recalculating the optimal candidate points, the maximum wall temperature and temperature difference are 323.33K and 10.5K, respectively. Compared with the model before optimization, the optimized rounding results show that the reduction rate of the temperature difference after optimization is 17.52%, and the reduction rate of the maximum wall temperature is 0.75%.

REFERENCES

[1] N.A.C. Sidik, M.N.A.W. Muhamad, W.M.A.A. Japar, Z.A. Rasid, An overview of passive techniques for heat transfer augmentation in microchannel heat sink, INT COMMUN HEAT MASS 88(2017) 74-83.

[2] N.H. Naqiuddin, L.H. Saw, M.C. Yew, F. Yusof, T.C. Ng, M.K. Yew, Overview of micro-channel design for high heat flux application, Renewable and Sustainable Energy Reviews 82(2018) 901-914.

[3] K. Vafai, L. Zhu, Analysis of two-layered micro-channel heat sink concept in electronic cooling, INT J HEAT MASS TRAN 42(12)(1999) 2287-2297.

[4] Xiaojin Wei, yogendra Joshi, Experimental and Numerical Study of a Stacked Microchannel Heat Sink for Liquid Cooling of Microelectronic Devices, Transaction of the ASME,129(2007)1432-1444.

[5] S. Chamoli, R. Lu, H. Chen, Y. Cheng, P. Yu, Numerical optimization of design parameters for a modified double-layer microchannel heat sink, INT J HEAT MASS TRAN 138(2019) 373-389.

[6] C.A. Rubio-Jimenez, S.G. Kandlikar, A. Hernandez-Guerrero, Numerical Analysis of Novel Micro Pin Fin Heat Sink With Variable Fin Density, IEEE Transactions on Components, Packaging and Manufacturing Technology 2(5)(2012) 825-833.

[7] Ali Y. Alharbi, Deborah V.Pence, Thermal Characteristics of Microscale Fractal-Like Branching Channels,Transaction of the ASME126(2004)744-752.

[8] L. Liang, J. Hou, X. Fang, Y. Han, J. Song, L. Wang, Z. Deng, G. Xu, H. Wu, Flow characteristics and heat transfer performance in a Y-Fractal mini/microchannel heat sink, Case Studies in Thermal Engineering 15(2019) 100522.

[9] H. Tan, K. Zong, P. Du, Temperature uniformity in convective leaf vein-shaped fluid microchannels for phased array antenna cooling, INT J THERM SCI 150(2020) 106224.

[10] V. Manoj Siva, A. Pattamatta, S.K. Das, Effect of flow maldistribution on the thermal performance of parallel microchannel cooling systems, INT J HEAT MASS TRAN 73(2014) 424-428.

[11] M.P. Vasilev, R.S. Abiev, R. Kumar, Effect of microchannel heat sink configuration on the thermal performance and pumping power, INT J HEAT MASS TRAN 141(2019) 845-854.

[12] Y.J. Lee, P.K. Singh, P.S. Lee, Fluid flow and heat transfer investigations on enhanced microchannel heat sink using oblique fins with parametric study, INT J HEAT MASS TRAN 81(2015) 325-336.

[13] H. Liu, D. Qi, X. Shao, W. Wang, An experimental and numerical investigation of heat transfer enhancement in annular microchannel heat sinks, INT J THERM SCI 142(2019) 106-120.

[14] A.F. Al-Neama, N. Kapur, J. Summers, H.M. Thompson, An experimental and numerical investigation of the use of liquid flow in serpentine microchannels for microelectronics cooling, APPL THERM ENG 116(2017) 709-723.

A method of research for the reliability of solder joint shape

Wenchao Tian
School of Electro-Mechanical Engineering
Xidian University
Xi'an, China
wctian@xidian.edu.cn

Xuewei Hou
School of Electro-Mechanical Engineering
Xidian University
Xi'an, China
xwhou@stu.xidian.edu.cn

Hao Cui
School of Electro-Mechanical Engineering
Xidian University
Xi'an, China
hcui@stu.xidian.edu.cn

Xuegui Feng
Guangdong ChippackingTechnology Co.Ltd
Guangdong, China
fengxuegui@chippacking.com

Abstract—In this paper a method of research for the reliability of solder joint shapes is proposed. By using Surface Evolver software, finite element analysis, minimum energy principle and gradient descent method to calculate the minimum state of energy, the final shape of the solder joint is predicted. By studying the shape of SOP solder joints and using Surface Evolver software, the error between model parameters and slice parameters is finally less than ±10%, which reduces the error between device models and slices. The reliability and life of solder joints are more accurately predicted.

Keywords—solder joint shape, finite element analysis, minimum energy principle

I. INTRODUCTION

With the rapid development of the microelectronics industry and the continuous miniaturization of chip packaging, it not only promotes the development of printed circuit boards in the direction of high density and multilayer, but also puts forward stricter requirements on the solder ability of printed circuit boards and other processes [1].In electronic assembly, the quality of solder joints is the most important factor for the welding quality of the entire product, and the wetting angle of solder joints, soldering temperature, and soldering time are the three key factors that affect the quality of solder joints.[2]. The shape of the solder joint is the appearance structure shape of the solder joint after forming. Generally refers to the geometric size that the molten solder at the joint of the component solder foot and the printed circuit board (PCB) pad can be wetted and spread along the metal surface, as well as the metal Fillet shape of the solder in contact with the surface[3]. Research shows that: the reliability of solder joints has a corresponding relationship with the actual shape and stress distribution state of the solder joints. Research on the shape and mechanical behavior of the solder joints can reveal the relationship between them, and achieve the prediction and control the solder joint shape, the purpose of optimal design is achieved[4][5].

In this study, the solder joint shape is simulated by Surface Evolver software based on the principle of minimum energy. By changing wetting angle of solder joint, soldering temperature, soldering time, the solder joint shape has been predicted to improve the reliability of solder joint shape.

II. SOLDER JOINT SHAPES MODELING

A. Device Slicing Diagram

At the present stage, the micro section method is usually used to observe the characteristics of the solder joint morphology, but it will directly damage the sample, which is obviously not suitable when the sample amount is small. Therefore, there is an urgent need to adopt a method that can intuitively and comprehensively observe the formation and analysis of the solder joint shape. Figure 1 shows the slicing diagram of SOP device.

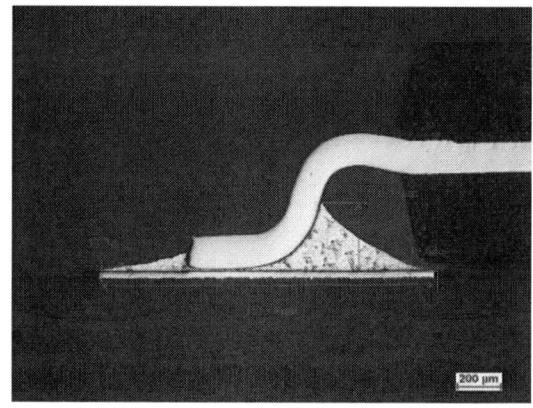

Fig.1. The slicing diagram of SOP

B. Energy Governing Equation

The system energy consists of two parts: surface potential energy and gravitational potential energy, the surface potential energy E_S can be expressed as:

978-1-6654-1392-3/21 $31.00 © 2021 IEEE

$$E_S = \iint_{A_0} \sigma dA + \sum_{i=1}^{n} \iint_{A_i} \sigma_{SLi} dA \qquad (1)$$

Where σ is the free surface tension, A_0 is the total area of the free liquid surface, and σ_i is the solid-liquid interfacial tension; n is the number of solid-liquid interfaces; A_i is the solid-liquid interface area labeled.

The gravitational potential energy of the system E_G can be expressed as:

$$E_G = \iint_V \rho gz dV \qquad (2)$$

The total energy of the system E can be expressed as:

$$E = E_S + E_G = \iint_{A_0} \sigma dA + \sum_{i=1}^{n} \iint_{A_i} \sigma_i dA + \iint_V \rho gz dV \qquad (3)$$

The amount of solder of the solder joint is usually given in advance as a constant V_0, which can be expressed as:

$$V_0 = \sum_{i=1}^{n} \iiint_{V_i} 1 dV \qquad (4)$$

According to the Lagrange conditional extreme value method, the energy functional of the system can be shown:

$$I = \iint_{A_0} \sigma dA + \sum_{i=1}^{n} \iint_{A_i} \sigma_i dA + \iint_V \rho gz dV - \lambda \left(\iiint 1 dV - V_0 \right) \qquad (5)$$

Where is the Lagrange Multiplier. When the integrand of formula 5 satisfies the Euler-Lagrange equation, the stagnation point of the functional can be obtained, which is the minimum energy state of the system. The equilibrium state of the solder joint can be solved by the numerical method of the vriational technique.

C. The Modeling Method

When modeling the solder joints, the Surface Evolver software based on the principle of minimum energy and finite element numerical analysis is used [6]. The software uses basic geometric elements, points, edges, faces, and bodies to define the initial liquid surface. The results of the shape prediction can be analyzed in any direction; the relevant geometric parameters can be extracted. The solid model and meshed mesh required by the finite element analysis software ANSYS can be generated for the mechanical analysis of the solder joints. It is used for mechanical analysis of solder joints, which realizes the integration of solder joint morphology model and reliability analysis model, which is conducive to practical engineering applications[7].

Determining the initial geometric conditions, energy form, constraint conditions, etc. are reflected in the input data file written by the Surface Evolve language. According to the initial geometric conditions, the Surface Evolve software model is used for initial cell meshing [8].Figure 2 shows the model diagram of SOP device.

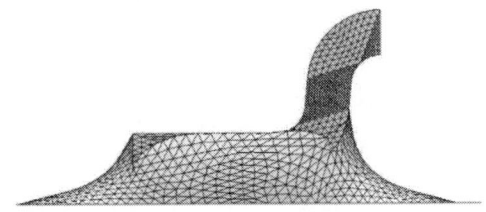

Fig.2. The model diagram of SOP

III. MODEL COEERCTION AND ERROR ANALYSIS

A. Device Slicing Parameters

Taking the SOP20 device as an example the maximum temperature during reflow soldering process is 245℃, 210℃ and 225℃, the thickness of solder paste is 0.15mm,0.15mm and 0.20mm the sample number can be marked as SOP20-0.15-245, SOP20-0.15-210and SOP20-0.20-225.The slicing diagram of the device is shown in Figure 3, Figure 4 and Figure 5.

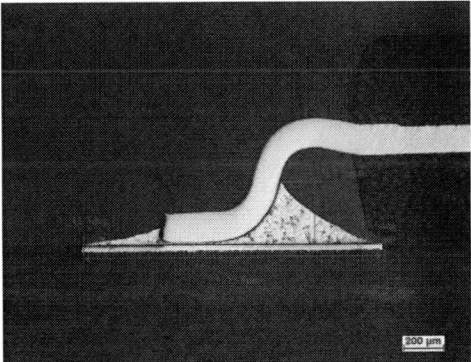

Fig.3. SOP20-0.15-245 slicing diagram

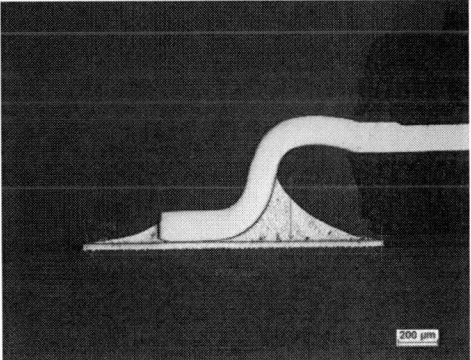

Fig.4. SOP20-0.15-210 slicing diagram

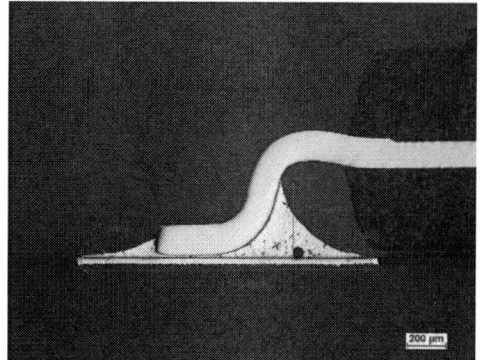

Fig.5. SOP20-0.20-225 slicing diagram

Table I shows the slicing parameter. The four parameters of the solder joint shapes of the device are measured, namely the wetting range, the climbing height, the angle of the heel and the toe.

TABLE I.　　THE SLICING PARAMETER

	20-0.15-245	20-0.15-210	20-0.20-225
Wetting range	1553.8μm	1529.4μm	1548.6μm
Climbing height	310.59μm	318.56μm	367.2μm
Heel angle	13.6°	16.7°	11.3°
Toe angle	13.3°	17.1°	11.3°

B. Model Parameters

The device model is now corrected by Surface Evolver. After the correction is completed, the measured device model diagram is shown in Figure 6, Figure 7 and Figure 8.

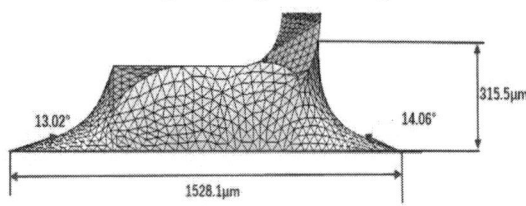

Fig. 6. SOP20-0.15-245 model diagram

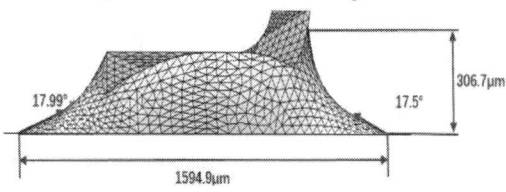

Fig.7. SOP20-0.15-210 model diagram

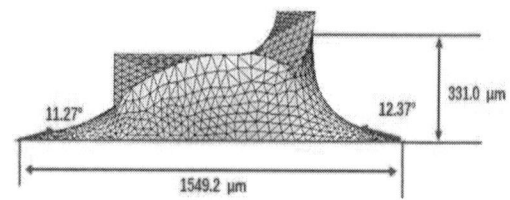

Fig.8. SOP20-0.20-225 model diagram

Table II shows the model parameter after modification. The four parameters of the solder joint shapes of the device were measured, namely the wetting range, the climbing height, the angle of the heel and the toe.

TABLE II.　　THE MODEL PARAMETER

	20-0.15-245	20-0.15-210	20-0.20-225
Wetting range	1528.1μm	1594.9μm	1549.2μm
Climbing height	315.5μm	306.7μm	331.0μm
Heel angle	14.06°	17.5°	12.37°
Toe angle	13.02°	17.99°	11.27°

C. Error Analysis

Table III shows the SOP20-0.15-245 error between the slicing parameter and the model parameter after modification. Table IV shows the SOP20-0.15-210 error between the slicing parameter and the model parameter after modification. Table V shows the SOP20-0.20-225 error between the slicing parameter and the model parameter after modification. Maximum error is less than ±10%.

TABLE III.　　SOP20-0.15-245 ERROR ANALYSIS

	Slicing parameters	Model parameters	Error
Wetting range	1553.8μm	1528.1μm	25.7μm(1.7%)
Climbing height	310.59μm	315.5μm	4.91μm(1.6%)
Heel angle	13.6°	14.06°	0.46°(3.4%)
Toe angle	13.3°	13.02°	0.28°(2.1%)

TABLE IV.　　SOP20-0.15-210 ERROR ANALYSIS

	Slicing parameters	Model parameters	Error
Wetting range	1529.4μm	1594.9μm	65.5μm(4.3%)
Climbing height	318.56μm	306.7μm	11.89μm(0.8%)
Heel angle	16.7°	17.5°	0.8°(5%)
Toe angle	17.1°	17.99°	0.89°(5.2%)

TABLE V. SOP20-0.20-225 ERROR ANALYSIS

	Slicing parameters	Model parameters	Error
Wetting range	1548.6μm	1549.2μm	0.6μm(0.04%)
Climbing height	367.2μm	331.0μm	36.2μm(9.9%)
Heel angle	11.3°	12.37°	1.07°(9.5%)
Toe angle	11.3°	11.27°	0.03°(0.27%)

IV. CONCLUSION

In this paper, a method of research for the reliability of solder joint shapes is proposed. Solder joint shape is related to the type of device, the temperature, the thickness of the solder paste, and the wetting angle. Through Surface Evolver to correct the solder joint shape parameters of the device, the error between the slice and the model parameters can be less than ±10%, the welding reliability and lifetime of the points are more accurately predicted.

REFERENCES

[1] Dong Liling, Jia Yan. Solderability test and evaluation of printed circuit boards[J]. Printed Circuit Information, 2010(11): 44-47+50.

[2] Wang Dongliang,Wang Yihan,Wei Sai.Analysis of the influence of solder joint wetting angle on soldering quality in electronic assembly[J].Electronic World,2020(21):107-109.

[3] Lu Jia.Research on the reliability of BGA solder joints based on finite element analysis[J].Microelectronics and Computer,2010,27(03):113-118.

[4] Wu Zhaohua, Zhou Dejian. SMT solder joint shape and stress analysis computer aided design. Computer aided design and manufacturing, 1996; (8): 155- 157

[5] Zhou Dejian et al. Discussion on the research method of MCM solder joint morphology[J]. Journal of Guilin Institute of Electronic Technology, 1997; 17(1): 77- 82

[6] Kenneth A. Brakke. The Surface Evolver and the Stability of Liquid Surfaces[J]. Philosophical Transactions of the Royal Society A: Mathematical, Physical and Engineering Sciences, 1996, 354(1715) : 2143-2157.

[7] Yang Jie,Zhang Keke, Zhou Xudong,Cheng Guanghui,Man Hua.Research status of micro-joint solder joint reliability[J].Electronic Components and Materials,2005(09):58-61.

[8] Liao Yongbo, Zhou Dejian, Huang Chunyue, Wu Zhaohua. Three-dimensional modeling and prediction of QFN solder joints based on orthogonal design[J]. Journal of Guilin Institute of Technology, 2007(02):274-277.

978-1-6654-1392-3/21 $31.00 © 2021 IEEE

Investigation of the RDL reliability based on RF characterization

Hongyue Wang
Science and Technology on Reliability Physics and Application of Electronic Component Laboratory
CEPREI
Guangzhou, China
wanghongyue@pku.edu.cn

Weijie Zhang
Wuxi Zhongwei High-tech electronics Co.ltd
Wuxi, China
xingxiaiyuan@163.com

Yijun Shi, Si Chen, Zhiwei Fu, Xiaofeng Yang, and Bin Zhou
Science and Technology on Reliability Physics and Application of Electronic Component Laboratory
CEPREI
Guangzhou, China

Abstract—In this letter, the RF reliability of the redistribute layer (RDL) under different environment stresses was investigated and the degradation mechanism of the RDL was analyzed. The inter-connect performance was characterized by the S parameters from 10MHz to 40GHz. The test vehicles of the RDL with coplanar waveguide (CPW) and microstrip (MS) were fabricated on Si and epoxy resin substrates. It was found that the insertion loss of CPW on Si substrate degraded seriously after reliability tests, especially at the low frequency range of 10MHz-30GHz. For other samples, the humidity stress has more impacts on the RDL reliability compared with thermal stress. Warpage was found for the RDL on the epoxy substrate, while the warpage of the RDL on the Si substrate was under detection limits. The 3D-Xray and SEM of the cross section of the samples were characterized. No declamation was found in the RDL after reliability tests. The surface of the Cu contact pad were oxidized and an obvious declamation was found in the contact region, which results in the insertion loss changing.

Keywords—Redistribution layer, RF, Reliability

I. INTRODUCTION

Because of the advantages of short wiring path and high interconnect density, the redistribute layer (RDL) based Wafer level package (WLP) can achieve high bandwidth, high I/O and high power efficiency interconnection, which is widely used in high-performance computing (HPM), 5G, radar and other fields[1-3]. The reliability of RDL under different stresses has become the focus of current research. The RDL is widely used in the Fan-out and Fan-in wafer level package. For the Fan-out wafer level packages, the RDL is usually fabricated on the epoxy resin substrates. While for the Fan-in wafer level package, the Si serve as substrate. The huge difference of the coefficient of thermal expansion (CTE) between different materials (Cu, Si, polyimide, and epoxy resin) lead to cracks, voids and delamination under thermal cycle stress[4,5]. With the interconnect line width/space scaling, the high density of the interconnect current degrades seriously on the RDL due to the electromigration effect[6].

Degradation of the RDL is usually characterized by its dc resistance (R_{dc}). However, for the high frequency applications, the transmission characteristics of RF signal are more important for the signal integrity and power efficiency[7]. Meanwhile, the R_{dc} can only detect the defects in the metal. The RF-characterization approach is demonstrated to be more sensitive to the damage in the metal lines, and it can detect

damages in the surrounding polyimides, as well as their interfaces[8]. Besides, thanks to the excellent scaling capacity, the RDL can operate at a broad frequency range, which is essential for the high frequency and high speed applications[9].

In this letter, the reliability of the RDL under different environment stresses was investigated by the RF characterization approach up to 40GHz. The RDLs were fabricated on Si and epoxy resin substrates with CPW and MS structures. The degradation mechanism of the RDL under thermal and humidity stress were analyzed.

II. VEHICLES AND EXPERIMENTS

A. Vehicles

To evaluate the high frequency reliability, a test vehicle of RDL with coplanar waveguide (CPW) and microstrip (MS) in a 65×37 mm^2 package scale was fabricated. The CPW vehicles include two layers of polyimide and one layer of Cu metal (2P1M), while the MS vehicles include three layers of polyimide and two layers of metal (3P2M). The RDLs were prepared on silicon substrate and epoxy resin substrate, both are commonly used in WLP and Fan-out WLP. The schematic view of the employed vehicles is shown in Fig. 1. Four ports transmission line (TL) with ground-signal-ground (GSG) configuration was designed for RF measurement.

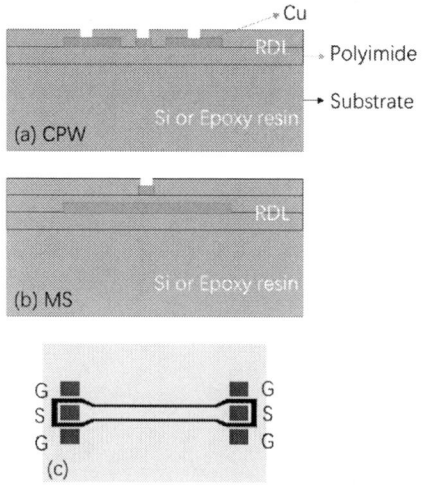

Fig. 1. *The test vehicle of RDL with (a) CPW and (b) MS. (c) The TL with GSG configuration.*

This work was supported by the Guangdong Province Basic and Applied Basic Research Fund Project under Grant 2021A1515012007, the NSFC under Grant No. 62004046 and the Basic scientific research projects in Guangzhou under Grant No. 202102020317.

B. Reliability experiments

The reliability of RDL was evaluated with three reliability testing items, which are thermal cycling (TC, -65℃ to 150℃, 500 cycles), thermal shock (TS, -65℃ to 150℃), and highly accelerated temperature and humidity stress test (HAST, 130℃, 85% humidity, 96h) according to corresponding JEDEC standards. After each test, the RF signals from 10MHz to 40GHz were used to evaluate the impacts of reliability tests on the transmission characteristics of the RDL.

III. RESULTS AND DISCUSSIONS

A. Insertion loss

The insertion loss (S21) of CPW and MS on Si and epoxy resin substrate after TC, TS and HAST are shown in Fig. 2. It was found that the insertion loss of CPW on Si substrate degraded seriously after reliability tests, especially at the low frequency range of 10MHz-30GHz, as shown in Fig.2 (a). This may result from the substrate effect considering the high CTE mismatch between silicon, metal, and polyimide. The TC and TS reliability tests impact negligibly on the MS, because the ground metallization in the MS shields the line from the substrate effects[10]. Note that the S21 of the CPW on epoxy resin substrate was improved at low frequency, which may result from the change of the transmission line impedance leading to impedance match.

The S21 of four kinds of samples degrade apparently after HAST, which indicates that humidity impacts significantly on the RDL reliability[7]. Under the HAST tests, the PI absorbs water in the air and expands, changing the electrical properties of PI, and then have a great influence on the RF signals transmission.

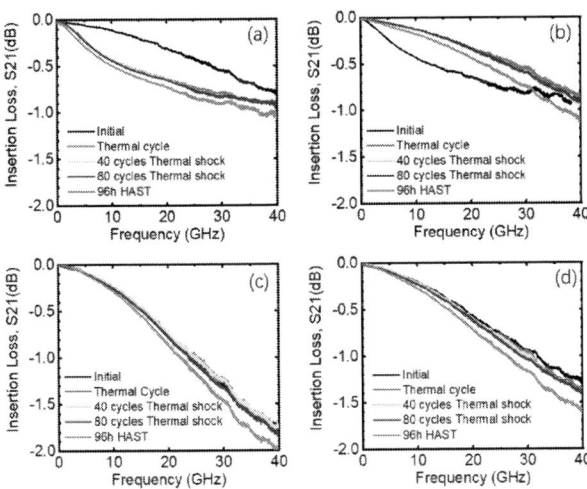

Fig. 2. The insertion loss of (a) CPW on Si, (b) CPW on epoxy resin, (c) MS on Si, (d) MS on epoxy resin before and after TC, TS and HAST reliability tests.

B. Electric field distribution

The CPWs have better transmission characteristics at higher frequency compared with the MS (the S21 of CPWs are around -1dB at 40GHz, while for the MSs are around -1.5dB to -2dB), which results from the different electromagnetic field distribution. To investigate the electromagnetic characteristics of the transmission lines, finite element simulation was conduct. The electric field distribution of CPW and MS are shown in the Fig. 3. For the CPW samples,

the electric field distributes mainly on the polyimide interface between the signal and ground. For the MS samples, the electric field distributes on the whole polyimide layer, which leads to higher electromagnetic loss.

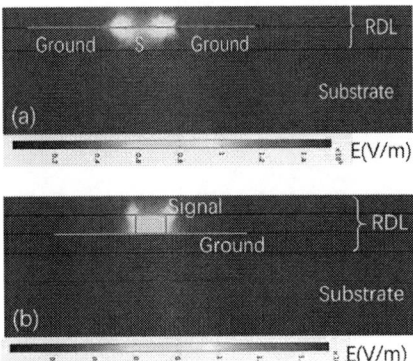

Fig. 3. The electric field distribution of (a) CPW and (b) MS.

C. Warpage

The warpage characteristics after TC and TS tests are shown in the Table I. No warpage was found for the samples with Si substrates, while the samples with epoxy resin substrates exhibit bow and twist of 0.7% - 1.1%. The warpage may result from the PI shrinking during the temperature cycling and the relaxation of residual stress, which was induced to the RDL layers during the fabrication processes, such as curing and metal deposition [11]. Low Young's modulus of epoxy resin results in higher warpage ratio compared with the Si substrate samples.

TABLE I.   THE WARPAGE CHARACTERISTICS OF THE SAMPLES AFTER RELIABILITY TESTS

Reliability tests	structure	substrate	Warpage type	Warpage ratio %
After TC	CPW (2P1M)	epoxy resin	twist	0.4
		Si	/	0
	MS (3P2M)	epoxy resin	twist	1.1
			bow	0.9
		Si	/	0
After TS	CPW (2P1M)	epoxy resin	bow	0.5
		Si	/	0
	MS (3P2M)	epoxy resin	twist	0.7
			bow	0.8
		Si	/	0

D. X-ray and SEM

The X-ray characterization was conducted on the samples. No declamation and voids was found in the RDL after reliability tests. The cross-section view of the samples characterized by SEM was shown in the Fig. 4. It was found that the RDL lines protected by PI layer exhibit no degradation, while the surface of the Cu contact pad were oxidized and an obvious declamation was found in the contact region, which may result from the increasing of stress at Cu surface.

978-1-6654-1392-3/21 $31.00 © 2021 IEEE

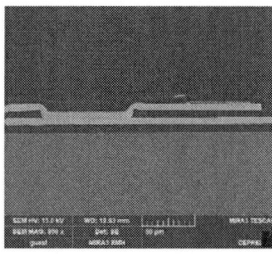

Fig. 4. *The SEM image of the RDL samples.*

IV. CONCLUSIONS

This work investigated the RDL reliability under thermal cycling, thermal shock and HAST reliability tests by the RF characterization approach. The CPW and MS test vehicles were fabricated on the Si and epoxy resin substrates. The insertion loss of the CPW fabricated on the Si substrate shows seriously degradation, while the MS exhibits little S21 degradation after TC and TS tests. All samples degrade apparently after HAST test. It was also found that the samples with epoxy resin substrates shows warpage after reliability tests, which results from PI shrinking and the residual stress relaxation. From the X-ray image, no declamation and voids was found in the RDL.

ACKNOWLEDGMENT

The author would like to thank managers and colleagues at CEPREI and Wuxi Zhongwei High-tech electronics Co. ltd for supporting this work.

REFERENCES

[1] T. G. Lim et al., "FOWLP Design for Digital and RF Circuits," in 2019 IEEE 69th Electronic Components and Technology Conference (ECTC), 2019, pp. 917-923.

[2] C. Xia, H. Wang, G. Wang, and X. Ming, "Advanced Packaging of 3D Fan-out RF Microsystem for 5G IoT Communication," in 2020 21st International Conference on Electronic Packaging Technology (ICEPT), 2020, pp. 1-4.

[3] K. Lee, "High-density fan-out technology for advanced SiP and 3D heterogeneous integration," in 2018 IEEE International Reliability Physics Symposium (IRPS), 2018, pp. 4D.1-1-4D.1-4.

[4] P. Lianto et al., "Fine-Pitch RDL Integration for Fan-Out Wafer-Level Packaging," in 2020 IEEE 70th Electronic Components and Technology Conference (ECTC), 2020, pp. 1126-1131.

[5] J. Xi et al., "Reliability of RDL structured wafer level packages," in 2013 14th International Conference on Electronic Packaging Technology, 2013, pp. 1029-1032.

[6] A. V. Vairagar et al., "In situ observation of electromigration-induced void migration in dual-damascene Cu interconnect structures," Applied Physics Letters, vol. 85, no. 13, pp. 2502-2504, 2004.

[7] P. Nimbalkar et al., "Fabrication and reliability demonstration of 5μm redistribution layer using low-stress dielectric dry film," in 2020 IEEE 70th Electronic Components and Technology Conference (ECTC), 2020, pp. 62-67.

[8] D. Kwon, M. H. Azarian, and M. G. Pecht, "Detection of solder joint degradation using RF impedance analysis," in 2008 58th Electronic Components and Technology Conference, 2008, pp. 606-610.

[9] C. Okoro et al., "Accelerated Stress Test Assessment of Through-Silicon Via Using RF Signals," IEEE Transactions on Electron Devices, vol. 60, no. 6, pp. 2015-2021, 2013.

[10] M. Wojnowski et al., "High Frequency Characterization of Thin-Film Redistribution Layers for Embedded Wafer Level BGA," in 2007 9th Electronics Packaging Technology Conference, 2007, pp. 308-314.

[11] G. Cheng et al., "Deep Understanding the role of Cu in RDL to Warpage by Exploring the Warpage Evolution with Microstructural Changes," in the 68th ECTC, 2018, pp. 2416-2421.

Effect of Ni₃Sn₄ Nanoparticles on Grain Refinement in SAC305 Freestanding Solder Balls and SAC305/Cu BGA Joints

Xiaolei Ren
School of Materials Science and Engineering
Dalian University of Technology
Dalian, China
melody0393@163.com

Xiaoying Liu
School of Materials Science and Engineering
Dalian University of Technology
Dalian, China
xiaoyliu@dlut.edu.cn

Longjiang Zou
School of Materials Science and Engineering
Dalian University of Technology
Dalian, China
zoulong@dlut.edu.cn

Yunpeng Wang
School of Materials Science and Engineering
Dalian University of Technology
Dalian, China
yunpengw@dlut.edu.cn

Ning Zhao*
School of Materials Science and Engineering
Dalian University of Technology
Dalian, China
zhaoning@dlut.edu.cn

Abstract—Sn-Ag-Cu solder joints have been reported to typically solidify from a single nucleation site, resulting in thesolder joints tend to contain a few number of twinned β-Sn grains and sometimes only one β-Sn grain. Owing to the anisotropy of tetragonal β-Sn, the solder joints will exhibit significant anisotropy in physical and mechanical properties, which will inevitably cause reliability issues. Therefore, it is necessary to develop effective method to refine the β-Sn grains in Sn-Ag-Cu solder joints. In this study, we explored the potential sites for heterogeneous nucleation in refining β-Sn grains by adding 0.5 wt.% Ni₃Sn₄ nanoparticles into 700 μm SAC305 freestanding solder balls and SAC305/Cu BGA joints. It was found that the morphology ofβ-Sn grains in the SAC305 freestanding solder balls and SAC305/Cu BGA joints was beach ball and single grain, respectively. After adding Ni₃Sn₄ nanoparticles into the SAC305 freestanding solder balls and SAC305/Cu BGA joints, the morphology of the β-Sn was multiple grains and singe grain, respectively. The reason for these phenomena were that the content of Cu was 0.8 wt.% in the SAC305/Cu BGA joints with Ni₃Sn₄ addition which was higher than 0.5 wt.% in the SAC305 freestanding solder balls with Ni₃Sn₄ addition. As a result, the Ni₃Sn₄ nanoparticles could be dissolved completely in the BGA joints and had no effect on the solidification of β-Sn; while partly dissolved in the SAC305 freestanding solder balls and the remaining Ni₃Sn₄ nanoparticles could be served as nucleation sites for β-Sn.

Keywords—Ball grid array; SAC305; Sn grain; Refinement; Nanoparticle

I. INTRODUCTION

In recent years, Sn-Pb solders have been banned owing to their toxic which was harmful to human health. Since then, several solder alloy systems were developed to replace the leaded solder, i.e, Sn-Bi , Sn-In and Sn-Ag-Cu (SAC)solders. Among these lead-free solders, Sn-Ag-Cu solders, especially Sn-3.0Ag-0.5Cu (SAC305, wt.%), have been widely used based on its favorable overall characteristics, i.e, high cost-efficiency, good wettability and excellent physical and mechanical properties.

As previously shown, Sn-based solder balls and solder joints typically solidified with large undercooling and contained a few or a single β-Sn grain[1]. The lattice of β-Sn is

body-centered tetragonal and its physical and mechanical performances are quite different along the different axis in β-Sn cell, i.e. a-axis and c-axis. When a Sn-based solder ball or solder joint contain only a few β-Sn grains, the ball/joint will exhibit highly anisotropic behavior in mechanical properties, electromigration properties and other properties[2]. This anisotropic behavior will seriously affect the service life of the solder joints. In order to solve this problem, scholars have proposed two solutions: controlling β-Sn orientation and refining β-Sn grains[1, 3]. The solution of controlling β-Sn orientation was generated a single grain joint and control the orientation of β-Sn to improve the reliability of the solder joints in specific environments. However, the solder joints produced by this method remain anisotropic, and can only be used in some certain situations but this problem does not exist in multiple grains solder joints. Therefore, the method of refining β-Sn grains has been regarded as a better way to improve the performance and service life of solder joints.

Gourlay et. al investigated the undercooling of Sn in different solder joints and found that nucleation undercooling of pure Sn soldering on Ni substrate was much lower than on Cu, Ag and Au substrates[4]. Parks et. al studied the the undercooling of SAC305-xNi solder and found that when added Ni into the SAC305 solder the undercooling of β-Sn could decreased significantly[5]. Ma et. al investigated the orientation relationship between β-Sn and Ni₃Sn₄ intermetallic compound (IMC) by remelting the Sn powder and solidified on Ni₃Sn₄ IMC Sheet[6] and found that β-Sn and Ni₃Sn₄ had the following orientation relationship: $(100)_{Ni_3Sn_4} \| (2\bar{4}1)_{Sn}$ and $[0\bar{1}0]_{Ni_3Sn_4} \| [10\bar{2}]_{Sn}$. The above phenomena and results indicate that Ni₃Sn₄ has an obvious effect on the nucleation of Sn. However, few studies have investigated the impact of Ni₃Sn₄ IMC on grain refinement of β-Sn. Therefore, SAC305 freestanding solder balls and SAC/Cu BGA joints with 0.5 wt.% Ni₃Sn₄ addition.

II. EXPERIMENTAL

Sn-3.0Ag-0.5Cu(SAC305) solder paste and home-made Ni₃Sn₄ nanoparticles were used as raw materials. The SAC305 solder paste was commercial solder paste and the Ni₃Sn₄ nanoparticles was produced via a chemical reduction method. Fig. 1(a) shows phase type test results of the chemically

---

* Corresponding author.
This work was supported by the National Natural Science Foundation of China (grant number 52075072) and the Fundamental Research Funds for the Central Universities (grant number DUT20JC46).

produced Ni_3Sn_4 nanoparticles which were identified to be highly purity and free of impurities. The SEM image of the Ni_3Sn_4 nanoparticles was shown in Fig. 1(b) and it can be found that the Ni_3Sn_4 nanoparticles had uniform size of about 20 μm in diameter.

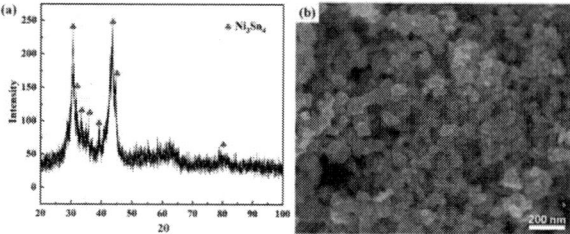

Fig. 1 (a) XRD pattern and (b) SEM image of the Ni_3Sn_4 nanoparticles

The SAC305 solder paste and the chemically produced Ni_3Sn_4 nanoparticles were mixed together by mechanical stirring to form SAC305-0.5 wt.% Ni_3Sn_4 nanocomposite solder paste. After that the SAC305pastes with and without Ni_3Sn_4 addition were printed on the Al substrate and Cu pads with a screen-printing method and reflowed by using a differential scanning calorimetry (DSC, Mettler-Toledo 822). The solder pastes were firstly heated up at a rate of 20 °C/min to 250 °C, held for 30 s and then cooled down at a rate of 90 °C/min which is usually adopted in a typical reflow process. Finally, the SAC305 freestanding solder balls (obtained on Al substrates) and SAC305/Cu BGA joints with and without Ni_3Sn_4 addition were obtained. The solder ball diameter was about 700 μm.

All the samples were wet grinded to the maximum cross section with 200 grit paper and then wet grinded by 5000 grit paper followed by polishing with colloidal silica. Hereafter, in order to investigate the change of β-Sn grains clearly, vibration polishing was applied to provide much flatter and stress-free surfaces of the solder ball and BGA joints. Then the element distribution of the freestanding solder balls and BGA joints were analyzed by electron probe micro analyzer (EPMA, JEOL JXA8500F) and the β-Sn grain orientation imaging microscopy (OIM) images were obtained by using a scanning electron microscope (SEM, SUPARR 55) equipped with an electron backscatter diffraction (EBSD, X-Max50).

III. RESULTS AND DISCUSSION

A. Morphology and grain feature of β-Sn in SAC305 freestanding solder balls and SAC305/Cu BGA joints

Fig. 2 shows the typical morphology and grain feature of β-Sn in SAC305 freestanding solder ball. The solder ball showed a uniform morphology in Fig. 2 (a). As identified by Fig. 2 (b)-(d), three β-Sn grains existed in this solder ball and were all related by ~60° rotation around a common twinning axis, i.e, <100> axis. As previously shown, Sn preferred to solidify with 57.2° by {101} or 62.8° by {301} nucleation model in SAC solder joints. It can be conclude that the three β-Sn grains in the solder ball were related by twin orientation. As a result, the SAC305 freestanding solder ball should solidify by only one nucleation and the three β-Sn grains had a beach ball morphology. However, this was a bit different from the standard beach ball morphology. Since the node positions of beach balls were formed randomly, especially in freestanding solder balls, a non-standard beach ball morphology could be observed if the solder ball was not ground to its node position.

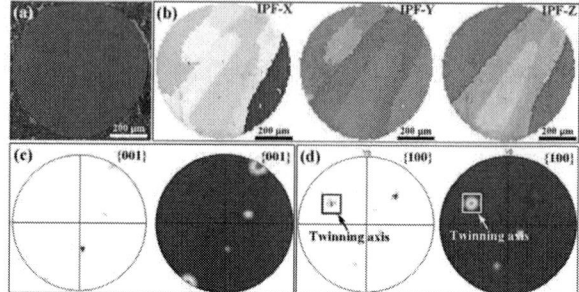

Fig. 2 (a) SEM image, (b) IPF maps and (c-d) {001} and {100} pole figures (PFs) of Sn in a SAC305 freestanding solder ball

Fig. 3 shows the typical morphology and grain feature of β-Sn in a SAC305/Cu BGA joint. From Fig. 3 (a), the solder ball were well connected to the Cu pad. As shown in Fig. 3 (b), there was a single grain in the BGA joint, which was proved by the {001} pole figures in Fig. 3 (c). Based on the theory that single grain can only be accompanied with one nucleation site, it can be concluded that the SAC305/Cu solder joint solidified by only one nucleation event. When the SAC305 solder balls were reflowed on the Cu pads, the undercooling for β-Sn solidification decreased, resulting in the nucleation model shifting from beach ball to single grain. Therefore, the β-Sn in the SAC305/Cu solder joint showed single grain feature which was different from the beach ball morphology in the case of freestanding solder ball.

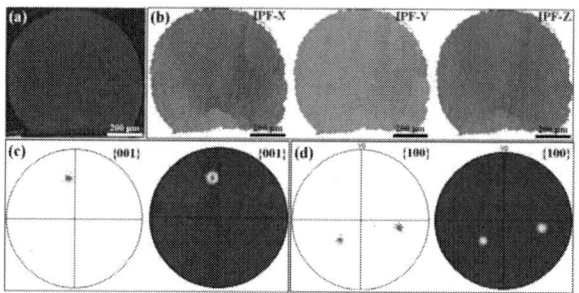

Fig. 3 (a) SEM image, (b) IPF maps and (c-d) {100} and {001} PFs of Sn in a SAC305/Cu BGA joint

B. Morphology and grain feature of β-Sn in SAC305 freestanding solder balls and SAC305/Cu BGA joints with Ni_3Sn_4 addition

Fig. 4 shows the typical morphology and grain feature of β-Sn in a SAC305 freestanding solder ball with Ni_3Sn_4 addition. Instead of beach ball, the morphology of β-Sn was

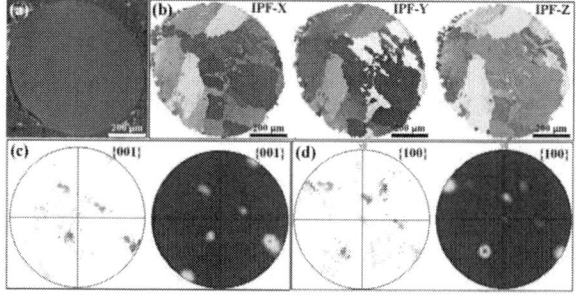

Fig. 4 (a) SEM image, (b) IPF maps and (c-d) {100} and {001} PFs of β-Sn in a SAC305 freestanding solder ball with Ni_3Sn_4 addition

multiple grains and there was no obvious orientation relationship between the grains. It can be concluded that the Ni_3Sn_4 nanoparticles can significantly increase the nucleation site in freestanding SAC305-0.5 wt% Ni_3Sn_4 solder ball during solidification.

Fig. 5 shows the typical morphology and grain feature of β-Sn in a SAC305/Cu BGA joint with Ni_3Sn_4 addition. As Fig. 5(b)-(d) shows, only a single grain exists in the BGA joint which was similar to the SAC305/Cu BGA joint. It can be concluded that SAC305/Cu BGA joint with Ni_3Sn_4 addition was solidified with one nucleation site and the effect of Ni_3Sn_4 nanoparticles on β-Sn grain was disappeared when the solder paste were reflowed on Cu substrate.

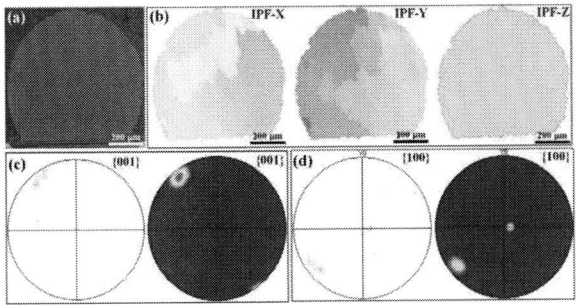

Fig. 5 (a) SEM image, (b) IPF maps and (c-d) {100} and {001} PFs of β-Sn in a SAC305/Cu BGA joint with Ni_3Sn_4 addition

C. Distribution of elements in SAC305 freestanding solder balls and BGA SAC305/Cu joints with Ni_3Sn_4 addition

Fig. 6 shows the element mappings of a SAC305 freestanding solder ball with Ni_3Sn_4 addition. From Fig. 6(b), three kinds of phases (i.e, dark, grey and bright phase) can be found in the freestanding solder ball. The table inset in Fig. 6(b) shows the quantitative analysis of the dark phase and the solder matrix. With the quantitative results, the dark phase could be identified as $(Cu,Ni)_6Sn_5$. Although no quantitative analysis was performed in the grey and bright regions, but combining Fig. 6(b)and Fig. 6(e-f), it can be found that the grey and bright phases were enriched by Sn and Ag element respectively and there was no other element. Therefore, it can be concluded that grey and bright phases were β-Sn and Ag_3Sn, respectively. In addition, it can be found that Ni was uniformly distributed in $(Cu,Ni)_6Sn_5$ in Fig. 6(f). As a result, it can be concluded that Ni_3Sn_4 nanoparticles could be dissolved into Ni and Sn atoms during the soldering process.

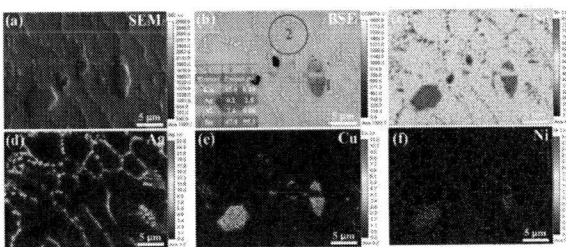

Fig. 6 Element mappings of a SAC305 freestanding solder ball with Ni_3Sn_4 addition

Fig. 7 shows the element mappings of a SAC305/Cu BGA joint with Ni_3Sn_4 addition. As shown in Fig. 7(b), three different colored phases were observed in the BGA joint, i.e, dark, grey and bright. The table inset in Fig. 7(b) shows the quantitative analysis of the interfacial compound and the solder matrix. With the quantitative analysis in the table, the interfacial compound could be identified as $(Cu,Ni)_6Sn_5$ and the solder matrix conclude 0.8 wt.% Cu which was higher than 0.48 wt.% in the freestanding solder balls due to the

dissolution Cu pads into the liquid solder. Combining Fig. 7(b) and Fig. 7(c)-(f), it can be concluded that the grey and bright phases were β-Sn and Ag_3Sn, respectively. As the Ni element mappings shows, the distribution of Ni in the interfacial $(Cu,Ni)_6Sn_5$ compound was quite uniform. As a result, it can also be concluded that the Ni_3Sn_4 nanoparticles in the SAC305/Cu BGA joint were also dissolved during the soldering process.

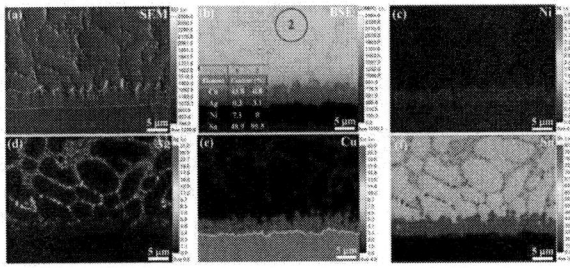

Fig. 7 Element mappings of a SAC305/Cu BGA joint with Ni_3Sn_4 addition

D. Discussion and analysis

Vuorinen et.al investigated the formation of IMCs between liquid Sn and various $CuNi_x$ metallic and found that the non-eutectic Ni_3Sn_4 could precipitated in the solder when the content of Ni was higher than 0.12 wt.% in Sn-Cu-Ni alloys[7]. According to the above analyses in this study, although the Ni_3Sn_4 nanoparticles were dissolved in the SAC305 freestanding solder ball (the content of Ni in 0.5 wt.% Ni_3Sn_4 was 0.135 wt.%), there could still exist some Ni_3Sn_4 nanoparticles during solidification, which could be confirmed by Fig. 8. With the temperature decreasing from 250 °C to room temperature, the Ni atoms (from the dissolution of the Ni_3Sn_4 nanoparticles) would combine with Sn and Cu to form Ni_3Sn_4 and $(Cu,Ni)_6Sn_5$ IMCs, and then precipitated at the same time as the solidification of β-Sn. As highly combination in certain orientations of Ni_3Sn_4 and β-Sn, the undissolved Ni_3Sn_4 particles would be the nucleation site for β-Sn. As a result, many nucleation sites would exist in the SAC305 freestanding solder ball with Ni_3Sn_4 addition, which resulted in the multiple β-Sn grain microstructure without the orientation relationship being presented in the freestanding solder balls without Ni_3Sn_4 addition.

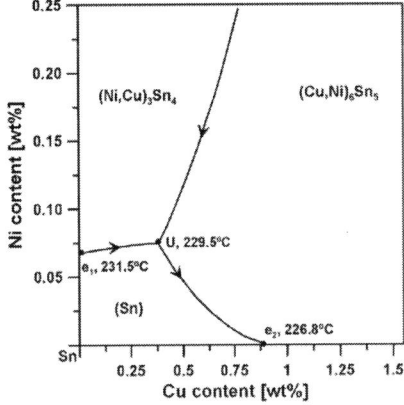

Fig. 8 Phase diagram derivation of tin-rich fraction in Sn-Cu-Ni system [7]

For the SAC305/Cu BGA joints with Ni_3Sn_4 addition, the content of Cu increased from 0.48 wt.% to 0.8 wt.%. Combining the phase diagram in Fig. 8 with the content of Ni (the Ni in 0.5 wt.% Ni_3Sn_4) and the content of Cu in the BGA joints, it can be concluded that the added Ni_3Sn_4 nanoparticles would dissolve when the temperature higher than the melting point of SAC305 solder. With the temperature reaching at the peak temperature (250 °C), the Ni_3Sn_4 nanoparticles would dissolved completely. When the temperature decreased from the peak temperature to room temperature, $(Cu,Ni)_6Sn_5$ would first precipitate and then the eutectic microstructure of β-Sn, $(Ni,Cu)_3Sn_4$ and $(Cu,Ni)_6Sn_5$. Therefore, the phase of $(Cu,Ni)_6Sn_5$ could be found at the interface between the SAC305 and Cu BGA joints with Ni_3Sn_4 addition. In addition, since the Ni_3Sn_4 nanoparticles had dissolved completely in the SAC305/Cu BGA joints, the effect of Ni_3Sn_4 nanoparticles on refining the solder matrix would disappear in the BGA joints. As a result, the morphology of the β-Sn grains in the SAC305/Cu BGA joints with Ni_3Sn_4 addition was similar to the SAC305/Cu BGA joints without Ni_3Sn_4 addition.

IV. CONCLUSION

In order to explore the impact of Ni_3Sn_4 nanoparticles on the refinement of β-Sn grains, the microstructure and grain feature of SAC305 freestanding solder balls and SAC305/Cu BGA joints with and without Ni_3Sn_4 nanoparticle addition were revealed. The following conclusions were drawn:

(1) In the SAC305 freestanding solder balls and SAC305/Cu BGA joints, the typical morphology of the β-Sn was beach ball and single grain, respectively.

(2) In the SAC305 freestanding solder balls with Ni_3Sn_4 addition, though most of the Ni_3Sn_4 nanoparticles would be dissolved, the residual Ni_3Sn_4 nanoparticles played a significant effect on the refinement of β-Sn by serving as nucleation sites.

(3) In the SAC305/Cu BGA joints with Ni_3Sn_4 addition, all of the Ni_3Sn_4 nanoparticles would be dissolved and the morphology of β-Sn in the BGA joints was similar to the SAC305/Cu solder joint without Ni_3Sn_4 addition.

(4) Due to the dissolution of Ni_3Sn_4 nanoparticles, $(Cu,Ni)_6Sn_5$ could exist in the SAC305 freestanding solder balls and SAC305/Cu BGA joints with Ni_3Sn_4 addition.

REFERENCES

[1] H. Shang, Z. L. Ma, S. A. Belyakov, and C. M. Gourlay, "Grain refinement of electronic solders: The potential of combining solute with nucleant particles," J Alloy Compd, vol. 715, pp. 471-485, 2017.

[2] Y. Qiao, N. Zhao and H. Ma, "Heredity of preferred orientation of β-Sn grains in Cu/SAC305/Cu micro solder joints," Journal of Alloys and Compounds, vol. 868, p. 159146, 2021.

[3] Z. L. Ma, S. A. Belyakov, K. Sweatman, T. Nishimura, T. Nishimura, and C. M. Gourlay, "Harnessing heterogeneous nucleation to control tin orientations in electronic interconnections," Nature Communications.

[4] C. M. Gourlay, S. A. Belyakov, Z. L. Ma, and J. W. Xian, "Nucleation and Growth of Tin in Pb-Free Solder Joints," JOM, vol. 67, pp. 2383-2393, 2015.

[5] G. Parks, A. Faucett, C. Fox, J. Smith, and E. Cotts, "The nucleation of Sn in undercooled melts: The effect of metal impurities," JOM, vol. 66, pp. 2311-2319, 2014.

[6] Z. L. Ma, J. W. Xian, S. A. Belyakov, and C. M. Gourlay, "Nucleation and twinning in tin droplet solidification on single crystal intermetallic compounds," Acta Mater, p. S1359645418301551, 2018.

[7] V. Vuorinen, H. Yu, T. Laurila, and J. K. Kivilahti, "Formation of Intermetallic Compounds Between Liquid Sn and Various CuNi x Metallizations," Journal of Electronic Materials, vol. 37, pp. 792-805, 2008..

Research on Effect of Annealing on Copper Deposited by Electroplating in High Density TSV

WeiWang
Technology Introduction Department
NATIONAL CENTER FOR
ADVANCED PACKAGING (NCAP
CHINA)
Wuxi, China
weiwang@ncap-cn.com

FeiGeng
Technology Introduction Department
NATIONAL CENTER FOR
ADVANCED PACKAGING (NCAP
CHINA)
Wuxi, China
feigeng@ncap-cn.com

PengSun
Technology Introduction Department
NATIONAL CENTER FOR
ADVANCED PACKAGING (NCAP
CHINA)
Wuxi, China
pengsun@ncap-cn.com

YulongRen
Technology Introduction Department
NATIONAL CENTER FOR
ADVANCED PACKAGING (NCAP
CHINA)
Wuxi, China
yulongren@ncap-cn.com

HuanLiu
Technology Introduction Department
NATIONAL CENTER FOR
ADVANCED PACKAGING (NCAP
CHINA)
Wuxi, China
huanliu@ncap-cn.com

KaiZhang
Technology Introduction Department
NATIONAL CENTER FOR
ADVANCED PACKAGING (NCAP
CHINA)
Wuxi, China
kaizhang@ncap-cn.com

Abstract—**Through silicon via (TSV) has proved to be a vital interconnection component in three dimensional circuit intergration and advanced packaging, for it has incomparable superiorities in device miniaturization, high density and multi-function. However, in TSV structure, the stress of thermal mismatch between silicon and electroplating deposition(ECD) copper and the accumulated residual stress in process make thermal-mechanical reliability become a main factor affecting the development of TSV technology, which may perhaps result in serious reliability problems once an appropriate destressing treatment is not taken. In order to improve the reliability of TSV, annealing is used to facilitate the recrystallization of ECD copper and reduce structure stress. In this article, a large size TSV in which the diameter is 30μm and the depth is 200μm with a high density of 0.59% was fabricated and then filled with ECD copper. After a pre-CMP process, annealing experiments at different conditions were designed to get a less residual stress structure. TSV pumping was measured by Step Profiler. Scanning Electronic Microscopy (SEM) and Electron Backscattered Diffraction (EBSD) were used to characterize ECD copper after annealing. Effect of annealing on ECD copper was studied and appropriate annealing parameters were acquired according to the measured results above.**

Keywords—*TSV; Annealing; Copper Electroplating; Pumping; Recrystallization*

I. INTRODUCTION

In advanced packaging technologies, three dimensional through-silicon via (3D TSV) occurred because of the ever increasing demands for high properties, reduced cost and diminutive size ingredients[1-2]. A representative TSV structure consists of a rather filmy silicon wafer full of through round cavities, in which the inside wall is deposited with dielectric layer silicon oxide by chemical vapor deposition (CVD) and diffusion barrier by physical vapor deposition (PVD) such as Ta, TiN and Ti[3-6]. Finally, the via is filled with ECD Cu or W as conductor. Plating Cu is widely used as filler in TSV due to its excellent electrical property, good process performance and low cost. However, mechanical stress generated around TSV on account of the enormous mismatching in the aspect of coefficient of thermal expansion (CTE) of diverse components used in TSV construction[7]. Once there is a temperature excursion from the stress free temperature, Cu pumping and a hoop stress may arise and then drive the microcracks to initiate, extend and finally form a fatal crack in the silicon interposer[8-9]. An effective solution to this issue is annealing, during which ECD Cu via may undergo a material transformation of grain growth and recrystallization and release inside thermal stress caused in the forthcoming fabrication process and other thermal budget in advance. It is reported that the repeated heat circulations behind the primary one are discovered to be composed of analogical loops in terms of profile, dimension and position, which indicates the deed of ECD Cu grows fairly steady after the original cycle that underwent the most exhaustive thermo circulation criteria[10]. There are many factors that affect the effects of annealing on TSV Cu, like the seed layer, the TSV distribution, the dielectric layer, ECD chemicals, ECD arguments and so forth. In this paper, parameters of TSV annealing are focused on in order to get rid of the potential Cu protrusion and the structure stress it might result in.

II. EXPERIMENT

There several steps of TSV structure fabrication: (a) a certain amount of 300mm diameter silicon wafers are chosen to be laser marked and then cleaned. (b) a photoresist with an appropriate thickness is spun on the surface of wafers and deep reactive ion etch (DRIE) is adopted for the sake of etching the via for its high etching rate. The via has a high density of 0.59%, a diameter of 30μm and a depth of 200μm. (c) the photoresist is stripped and TSV cleaning is taken to remove residue after etch. (d) a 2.5μm thick dielectric layer is deposited by CVD using tetraethyl orthosilicate (TEOS) as silicon source. The step coverage in the bottom of TSV should be more than 12.5%. (e) a 0.5μm Ti layer and a 3μm Cu layer are deposited subsequently onto SiO_2 by PVD. (f) ECD Cu is used as the conductor to fill TSV. And the detection mode is X-ray microscope to make sure that there is no void in TSV. (g) a pre-CMP is used to get consistent surface prepared for annealing design of experiments. (h) the wafers are annealed in the same furnance at 270°C for 1h, 270°C for 5h, 270°C for 10h, 270°C for 15h and 350°C for 1h, respectively. N_2 is used as protective gas. The ramp rate is 5°C/min and the furnance cools down naturally in the end. (i) Cu pumping is measured by step profiler and SEM and EBSD are used to observe the morphology and Cu grain size and distribution.

III. RESULTS AND DISCUSSION

The TSV Cu before annealing is shown in (a) of Fig. 1. At this moment, the protrusion of Cu is controlled at a level of 25nm when the Cu grain size is mainly up to plating chemicals, plating parameters and what is deposited as seed layer. The left part of Fig. 2 gives a visual description of Cu grain statement. After an annealing of 350°C for 1h, the top of TSV archs up and forms a protrusion of 700nm height. Meanwhile, the previously tiny crystalline grains grow up as the number of grains decreases.

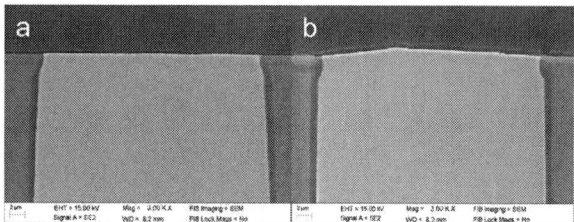

Fig. 1. SEM picture of TSV ECD Cu with different post-treatments

a. no annealing; b. annealed at 350°C for 1h

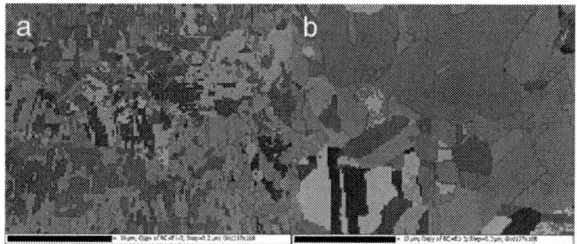

Fig. 2. EBSD picture of TSV ECD Cu with different post-treatments

a. no annealing; b. annealed at 350°C for 1h

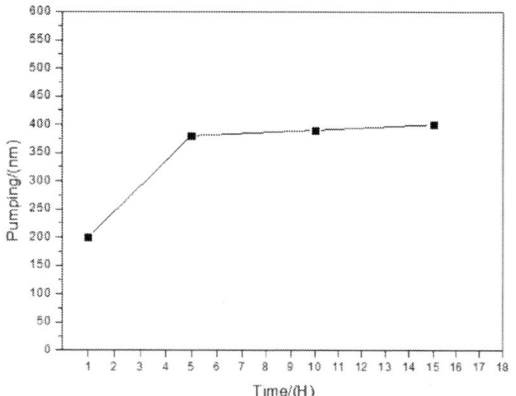

Fig. 3. Pumping of TSV ECD Cu annealed at 270°C for different duration

However, post-annealing cracks of high probability emerge when wafers are detected, which do great harm to reliability. Without changing the TSV filling material, a reduced annealing temperature is capable of shrinking temperature excursion from stress free temperature, by which the mismatch between ECD Cu and materials around it is less effective on TSV structure stress. In the meantime, in the light of subsequent thermal budget of process and packaging, 270°C is an appropriate and more gentle temperature for annealing.

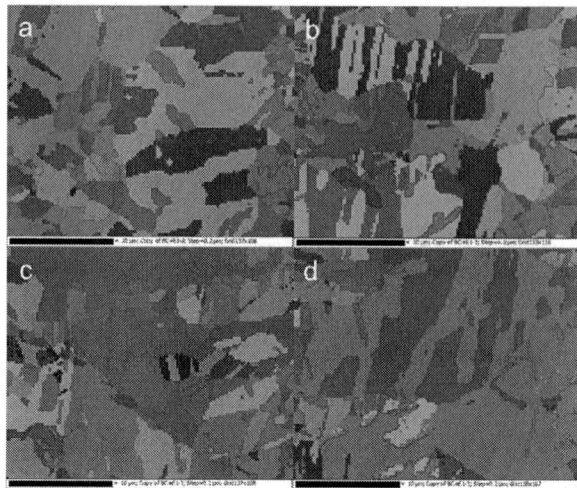

Fig. 4. EBSD picture of TSV ECD Cu with different post-treatments

a. annealed at 270°C for 1h; b. annealed at 270°C for 5h

c. annealed at 270°C for 10h; d. annealed at 270°C for 15h

In order to make ECD Cu undergo a relatively thorough recrystallization process, an augment in annealing time becomes inevitable. Annealing experiments for 1h, 5h, 10h and 15h are arranged respectively. Fig. 4 shows that Cu has a recrystallization and growth in size annealed at 270°C. The pumping height changes from 25nm to 200nm when Cu is annealed at 270°C for 1h which is shown in Fig. 3. As the annealing time comes up to 5h, the pumping height rises to 380nm. Much more duration of 5h or 10h does not show a noticeable change in Cu pumping height. When annealed at 270°C, none of wafers is found to have even a crack compared with those annealed at 350°C. In consideration of subsequent thermal budget, 270°C with 5h is sufficient to get a TSV ECD Cu recrystallization result and a thorough pumping height without crack and with less structural stress which is good for reliability in the event of a thoughtful consideration of total thermal budget.

IV. CONCLUSION

In this paper, effect of annealing on TSV ECD Cu is reported and experiments of different annealing conditions are designed to acquire a comparatively appropriate annealing parameter when it comes to a high density TSV with a large size of 30μm in diameter and 200μm in depth so as to obtain a crack free via which is beneficial to reliability. On the basis of results from SEM, EBSD and step profiler, a lower temperature with more duration is sufficiently effective to be taken as TSV annealing parameter.

ACKNOWLEDGMENT

Great thanks for the guidance and support from leaders and the help from colleagues. This paper is supported by the National Innovation Center for Integrated Circuit Technology and Packaging Testing.

REFERENCES

[1] Thomas G, Lutz H, Wei-Shan W, Mario B, Tobias S, Maik W, Stefan S. 3D integration technologies for MEMS[A]. IEEE Beijing Section. 2016 13th IEEE International Conference on Solid-State and

Integrated Circuit Technology(ICSICT) Proceedings[C]. IEEE Beijing Section: IEEE BEIJING SECTION, 2016: 4.

[2] Zheyao W. Microsystems using three-dimensional integration and TSV technologies: Fundamentals and applications[J]. Microelectronic Engineering, 2019, 210.

[3] Arunasalam, P; Ackler, HD; Sammakia, BG, "Microfabrication of ultrahigh density wafer-level thin film compliant interconnects for through-silicon-via based chip stacks", Journal of Vacuum Science & Technology B24(4):1780-1784, 2006.

[4] Jang DM, Ryu C, Lee KY, et al, "Development and evaluation of 3D SiP with vertically interconnected through silicon vias (TSV)", 57th Electronic Components and Technology Conference, Orlando, FL, pp.847-852, 2007.

[5] Ranganathan N, Ebin L, Linn L, et al, "Integration of high aspect ratio tapered silicon via for through-silicon interconnection", 58th Electronic Components and Technology Conference, Orlando, FL, pp.859-865, 2008.

[6] Rui Dong W, Cong Chun Z, Gui Fu D, Yang G. Barrier and Seed Layers Deposition in TSV Using Magnetron Sputtering[J]. Applied Mechanics and Materials, 2014, 3082.

[7] Yeonsung K, Ah-Young P, Chin-Li K, Michael S, Bryan Black, Seungbae P. Prediction of deformation during manufacturing processes of silicon interposer package with TSVs [J]. Microelectronics Reliability, 2016, 65.

[8] Chen YH, Lo WC and Kuo TY, "Thermal effect characterization of laser-ablated silicon through interconnect", Proc. Electronic systems and Technology Conference, 2006.

[9] Takahashi K and Sekiguchi M, "Through silicon via and 3D wafer/chip stacking technology", IEEE Symposium on VLSI Circuits Digest of Technical Papers, pp. 89-92, 2006.

[10] Xiangmeng J, Zhongcai N, Hu H, Wenqi Z, Ui-hyoung L. "Copper pumping of through silicon vias in reliability test", 16th International Conference on Electronic Packaging Technology, 2015.

Design and optimization of a pneumatic DOD solder ball 3D printing system

Zhixian Min
No. 38 Research Institute, Key Laboratory of Aperture Array and Space Application
China Electronics Technology Group Corporation
HeFei, China
zhixian_min@163.com

Sheng Liu
The Institute of Technological Sciences
Wuhan University
Wuhan, China
victor_liu63@vip.126.com

Huiming Pan
The Institute of Technological Sciences
Wuhan University
Wuhan, China
2018302080023@whu.edu.cn

Zhiqin Wang
No. 38 Research Institute, Key Laboratory of Aperture Array and Space Application
China Electronics Technology Group Corporation
HeFei, China
wgzhiq10@163.com

Dinglei Zhao
No. 38 Research Institute, Key Laboratory of Aperture Array and Space Application
China Electronics Technology Group Corporation
HeFei, China
510064209@qq.com

Zhiwen Chen
The Institute of Technological Sciences
Wuhan University
Wuhan, China
zwchen_lu@163.com

Abstract—This work is to set up a solder 3D printing system for solder ball printing. Effect of major factors that may affect the performance of the system has also been investigated. Therefore, performance of the system was optimized. A pneumatic DOD (drop-on-demand) solder 3D printing system was set up for solder ball printing. The effects of major factors were studied by control variable method, such as peak pressure, supply pressure and signal pulse width. Formation of liquid solder droplets during printing was simulated by 2D VOF (volume of fluid) two-phase flow model. The built DOD solder ball 3D printing system can prepare solder balls with good sphericity and uniformness. The formation of solder balls is split into four stages, ejection, necking of molten solder, fracture of liquid thread and formation of droplet. It also shows the peak pressure in ejection chamber increases alongside the rise of supply pressure and signal pulse width. The system provides a more cost-efficient solution for solder-based interconnections。

Keywords—3D printing; pneumatic drop-on-demand; solder balls; finite element simulation

I. INTRODUCTION

3D printing is developing fast due to its high flexibility and customization in producing components with complex shapes and structures. Recently, the gradual transition from rapid prototyping to rapid manufacturing in 3D printing has caused new challenges for mechanical engineers and materials scientists[1]. 3D printing is also a promising technique in electronic packaging for its potential capability of producing solder bumps. Unlike traditional lithography and electroplating methods, 3D printing is a contactless direct deposition method. Solder can be deposited directly on the substrate, creating solder bumps in the absence of masks/stencils and expensive equipment [2-6]. It has advantages of low cost, low waste, and simple procedure[3].

The basic principle of 3D printing is to create uniform droplets by applying periodic disturbance to the molten material [2]. And it falls into two categories, drop-on-demand (DOD) 3D printing and continuous-ink-jet (CIJ) 3D printing [2, 4-7]. The CIJ technology was developed based on Plateau-Rayleigh instability theory. It can achieve high efficiency and good uniformity of droplet deposition, but improving positioning accuracy of droplets is challenging. DOD technology is superior to CIJ technology in terms of positioning accuracy and cost effectiveness.

In DOD technology, various actuating methods have been developed [8-14]. Piezoelectric and Pneumatic are widely used by printer manufacturers. The inverse piezoelectric effect is utilized in the piezoelectric droplet ejection. The fluid droplet is pushed out by the pressure pulse formed by the piezoelectric effect5-18]. Pneumatic ejection technique relies on an actuator which can form air pressure of a certain value followed by pressure changing in chamber to eject droplets out of a nozzle[19-21]. Compared to pneumatic droplet ejection, piezoelectric droplet jetting technology can achieve high deposition frequency and efficiency. However, working temperature of the system is limited by the physic constraints of piezoelectric devices, so it is hard to be employed for deposition of materials with high melting point [2, 5, 7]. Otherwise, an extra cooling system is required to keep the temperature of piezoelectric module in a reasonable range, which leads to a more complex system. In contrast, there is no thermal sensitive element in pneumatic micro jetting system, which makes it more suitable for ejection of high temperature molten solder.

A 3D printing system was developed in this paper. It was pneumatic and drop-on-demand which was for solder bump deposition. Some parameters such as signal pulse width and supply pressure were investigated. Definitely, the effects on the formation of solder droplets were investigated. The formation process of solder droplets during printing was also analyzed with high speed camera and simulation.

II. EXPERIMENT PROCEDURES

Fig.1 illustrates the built pneumatic DOD solder 3D printing system in this work. It consisted of a motion platform, a droplet generator, a substrate, a motion control subsystem, a temperature control subsystem, a pneumatic control subsystem, and an in-situ imaging subsystem.

In droplet generator, molten solder is stored in storage chamber, which can then be driven into ejection chamber under air pressure. The molten solder is then ejected out of the nozzle by pulses of air pressure. Pulses of air pressure for driving molten solder out of the nozzle is controlled by pneumatic control sub-system (including compressed N2, solenoid valve, regulator and control board). When pulse signal from control board (Arduino in this work) is sent to the valve, pulses of air pressure can be generated in ejection chamber. So that molten solder can be driven out of the nozzle to form a solder droplet.

Positioning of solder droplets relies on the X-Y motion stage which moves substrate in horizontal directions. Movement along Z direction can be achieved by motion stage behind the droplet generator. Resolution of the stages is about 1.4 μm. Both the motion stages and pneumatic sub-system were linked to Arduino development board and controlled by connected computer.

A substrate is used to receive ejected molten solder droplets. Temperatures of the substrate, storage chamber and ejection chamber are kept stable by temperature control sub-system to ensure proper shape of deposited bumps. The temperature control sub-system includes heating devices, thermal couples and temperature controllers.

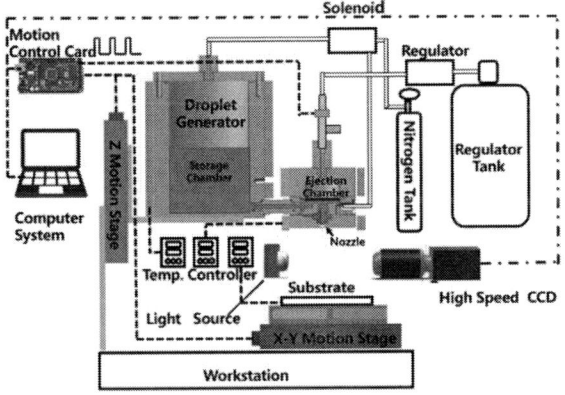

Fig.1 Schematic diagram of pneumatic DOD solder 3D printing system

The in-situ imaging sub-system consists of a camera, microscope and a illuminant. The camera is a high speed CCD camera who identification number is ORCA Flash 4.0 V3. It has max frame rate of 100 frame per second at resolution of 2048×2048. The illuminant has a accessory, fiber light guide. The camera was triggered at the same time with the illuminant to capture a chain of images of droplets.

Sn99.3Cu0.7 solder was used in solder printing experiments. Control variate method was employed when conducting the experiments. The experiments investigated the effects of some major parameters on the formation of droplets. The parameters investigated mainly included the supply pressure, the signal pulse width.

III. SIMULATION DETAILS

FLUENT was used to simulate the formation of solder droplets under air pressure pulses. Because two types of fluids (nitrogen and molten solder) are involved in droplet formation and ejection, VOF (volume of fluid, for interactions between incompatible fluids) two-phase flow model is used to study solder droplet jetting. A two dimensional axisymmetric finite element simulation model was built which consists of two parts A1 and A2(Fig. 2a). A1 is an internal flow field while A2 is an external flow field. The nozzle radius, long-width ratio of the nozzle and the contraction angle were 200μm, 3.5 and 50° respectively. The height of the internal field A1 was 25mm. Fig. 2a illustrates the boundary conditions. Axis of the

nozzle was set as the symmetry axis which was collinear with the left side of the gaseous zone. The pressure inlet was the upper boundary of the internal flow field while the pressure outlet was the right and bottom side of the gaseous zone. The side wall of the internal flow field was viewed as a no slip boundary.

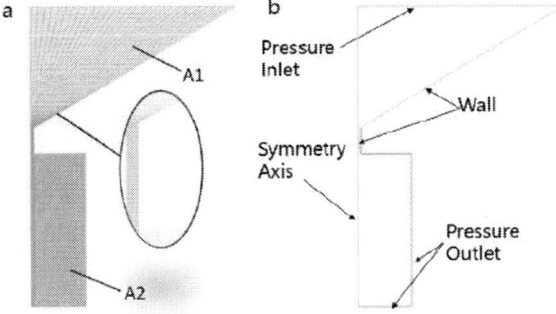

Fig. 2 Schematic diagram of the finite element model (a) 2D axisymmetric model (b) boundary conditions

Major material properties of molten Sn99.3Cu0.7 solder are summarized in Table 1. Equation (1) give the formula to compute the listed properties of liquid solder at demanding temperature (T_d). Here, T_b is the base temperature and X_b is the corresponding properties on the base temperature[22]. Base temperatures (T_b) of solder alloy and nitrogen are the melting temperature of solder (227°C) and room temperature (25°C) respectively.

$$X = X_b + (T_d - T_b)(\mathrm{d}X/\mathrm{d}T) \qquad (1)$$

where X and X_b are properties computed respectively at demanding and base temperature. The properties can be either the density, kinematic viscosity or the surface tension of liquid solder. $\mathrm{d}X/\mathrm{d}T$ is the corresponding gradient of the property with respect to current temperature. It describes the variation of material properties with different temperatures.

TABLE 1 PROPERTIES OF FLUIDS AT BASE TEMPERATURE IN MODELLING

Material	Density (kg/m³)	Viscosity (Pa·s)	Surface tension coefficient (N/m)
Sn99.3Cu0.7(227°C)	7400.5	0.0013	0.48
Nitrogen(25°C)	1.138	1.663e-5	

To evaluate the inlet conditions in simulation, variations of driven pressure in ejection chamber had been measured with piezoelectric pressure sensor in real time during solder 3D printing[6, 23]. The measured pressure data was fitted using a Fourier series with eight harmonics for inlet condition definition in simulation.

$$y(t) = a_0 + \sum_{n=1}^{\infty} a_n \sin(nwt) + \sum_{n=1}^{\infty} b_n \cos(nwt) \qquad (2)$$

where a_n and b_n are coefficients of Fourier series. t is the time variable. Fig. 3 illustrates the inlet pressure in ejection chamber formed with the supply pressure of 45kPa and signal pulse period of 0.5s. Three different stages can be identified in the curve, ascending, decreasing and stabilization within a period of 0.5s. The peak pressure reached 8.53kPa at 54ms.

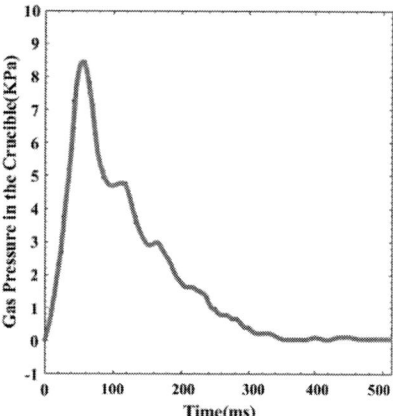

Fig. 3 Pressure-time curve in ejection chamber during solder 3D printing

To simplify the model, several assumptions were also proposed in building the model:

(1) Temperature of solder droplets remains relatively stable after detaching from the nozzle, and evolution in major material properties were negligible during jetting, including density, viscosity and surface tension of molten solder;

(2) Surfaces of nozzle that are in contact with molten solder were assumes to be smooth, such as bottom and inner surfaces of nozzle. Wetting angle of these surfaces was also constant;

(3) Influence of possible turbulence inside the storage chamber and ejection chamber on the ejection process is neglected. The formation of solder droplets could be viewed as an isothermal incompressible flow.

IV. RESULTS AND DISCUSSION

A. Effect of supply pressure on pressure in ejection chamber and droplet formation s

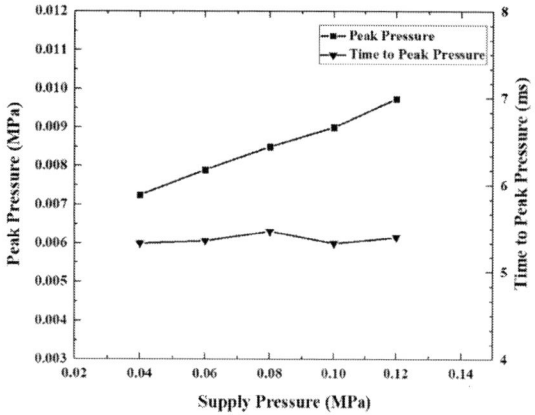

Fig. 4 Effect of supply pressure on the peak pressure and time to peak pressure

To optimize 3D printing parameters, it is necessary to analyze the correlation between supply pressure and the consequent pressure in ejection chamber and nozzle. Therefore, series of experiments were implemented with air supply pressures within the range of 0.04MPa~0.12MPa and an interval of 0.02MPa while pulse width was set to 0.5s. As indicated in Fig. 4, the peak pressure in the ejection chamber

was augmented along with the supply pressure almost linearly when the pulse width was constant. Meanwhile, the time required to reach peak pressure under different supply pressures was basically stable.

Supply pressure can also pose a significant effect on solder droplet ejection process. A chain of experiments was conducted with supply pressures different (0.02MPa, 0.04MPa, 0.08MPa and 0.10MPa) and constant pulse width. Fig. 5 shows that the limiting length and velocity of jet flow increased with higher supply pressure when the pulse width is constant. And a reasonable supply pressure range should be 0.02MPa-0.08MPa in this work. When the supply pressure was set lower than 0.02MPa, a solder droplet cannot be formed. But when it was higher than 0.08MPa, an attached droplet was formed.

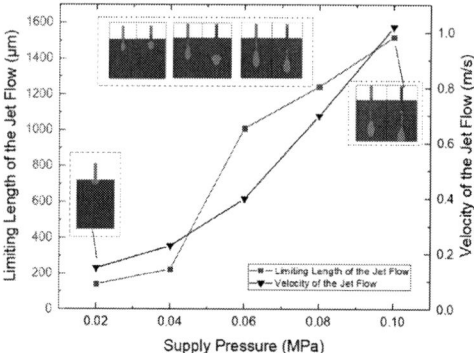

Fig. 5 Effect of supply pressure on the formation of drops

B. Effect of signal pulse width on supply pressure

To quantify how the signal pulse width affect pressure vibration, a chain of experiments was conducted with the supply pressure of 0.05MPa constantly. Five different pulse widths, 0.2ms, 0.4ms, 0.6ms, 0.8ms and 1ms. Fig.6, were used. Fig.6 shows that the peak pressure in the storage chamber increases with pulse width. It can also be found that the larger the pulse width is, the longer it takes to reach the peak pressure. But according to Fig.5, there is no notable correlation between time to peak pressure and supply pressures.

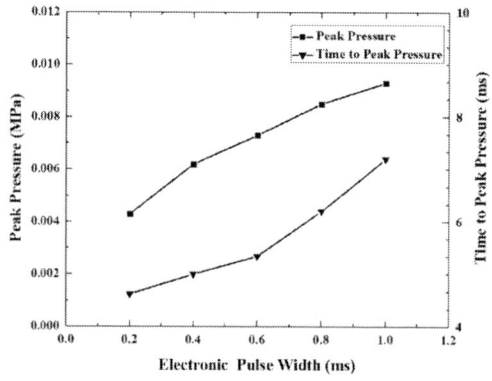

Fig. 6 Effect of pulse width on peak pressure and time of reaching peak pressure

C. Mechanism solder 3D printing

Some solder bumps are shown in Fig. 7. The bumps were all ejected by the DOD pneumatic 3D printing system. They showed good sphericity and uniform diameters (1.4 $\pm$ 0.4mm).

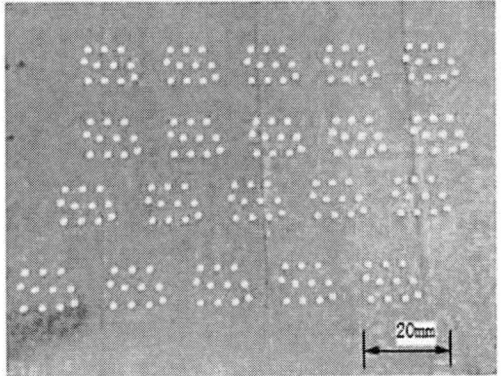

Fig. 7 Solder bumps deposited on substrate

High speed camera and simulation were employed to further investigate the solder droplet formation process during 3D printing. Some major parameters were monitored in simulation, such as the distribution of gas and liquid, pressure field change and velocity field change near the nozzle of the droplet ejection process. As shown in Fig. 8, the process of droplet formation is generally divided into four stages, which were ejection, necking of solder liquid, fracture of liquid thread and formation of droplet. The whole processes normally took about 3.4ms in our work. These agreed well with our in-situ observations with high speed camera.

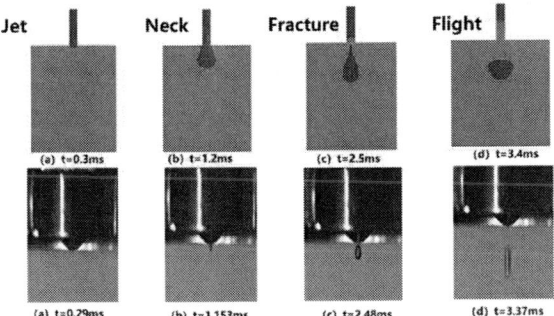

Fig. 8 Formation of solder droplets by simulation and in-situ observation with high speed camera.

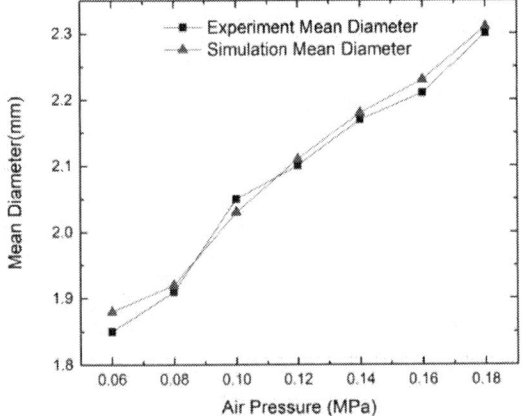

Fig. 9 Comparison between mean diameters from simulation and experiments

Fig.9 shows the comparison of average diameter of metal droplets from experiments and simulation at different air pressure. it can be found that average diameter of solder droplets generally increases with supply pressure. Furthermore, these two curves agree with each other reasonably well.

V. CONCLUSIONS

A pneumatic DOD solder 3D printing system was designed and set up in this work. Mechanisms and major factors that may affect the performance of 3D printing were also studied. The process of droplet formation is split into four periods, ejection, necking of solder liquid, fracture of liquid thread and formation of droplet. Simulation based on two-dimensional axisymmetric model showed the gas pressure vibration during the ejection. Experiments indicated the supply pressure and pulse width affects the droplet generation. It suggested the deposition performance can be improved by tuning the supply pressure and pulse width.

In future, design of nozzle and 3D printing parameters will be optimized further to reduce sizes of printed solder balls for applications that requires finer pitches. Positioning sub-system and pneumatic sub-system will also be further tuned to achieve better positioning accuracy and higher efficiency.

ACKNOWLEDGMENT

This work was funded by National Natural Science Foundation of China, grant number 61904127 and 62004144, Fundamental Research Funds for the Central Universities (young researchers), grant number 2042019kf0013, Fundamental Research Funds for the Central Universities, grant number 2042019kf1002, 202401002 and 203134004, Natural Science Foundation of Hubei Province, grant number 2018CFB212, Hubei Provincial Natural Science Foundation of China, grant number 2020CFA032, the Hubei Provincial Major Program of Technological Innovation, grant number: 2017AAA121.

REFERENCES

[1] S. C. Ligon, R. Liska, J. Stampfl, M. Gurr, and R. Mülhaupt, Chem. Rev. **117**, 10212-10290 (2017)

[2] X.-S. Jiang, L.-H. Qi, J. Luo, H. Huang, and J.-M. Zhou, Int. J. Adv. Manuf. Technol. **49** 535-541 (2010)

[3] T. Lee, T. G. Kang, J. Yang, J. Jo, K. Kim, B. Choi, et al., IEEE T. Electron. Pack. **31**, 202-210 (2008)

[4] N. Link and R. Semiat, Chem. Eng. Process. **48**, 68-83 (2009)

[5] J. Luo, L. H. Qi, S. Y. Zhong, J. M. Zhou, and H. J. Li, J. Mater. Process. Tech. **212**, 2066-2073 (2012)

[6] J. Luo, L. H. Qi, J. M. Zhou, X. H. Hou, and H. J. Li, J. Mater. Process. Tech. **212**, 718-726 (2012)

[7] D. Xie, H. H. Zhang, X. Y. Shu, J. F. Xiao, and S. Cao, Sci. China Technol. Sc. **53**, 1605-1611 (2010)

[8] J. Choi, Y.-J. Kim, S. Lee, S. U. Son, H. S. Ko, V. D. Nguyen, et al., Appl. Phys. Lett. **93**, 193508 (2008)

[9] H. Dong, W. W. Carr, and J. F. Morris, Rev. Sci. Instrum. **77**, 085101 (2006)

[10] O. Oktavianty, T. Kyotani, S. Haruyama, and K. Kaminishi, Addit. Manuf. **25**, 522-531 (2019)

[11] M.-M. Liu, X. Lian, Z.-Z. Guo, H. Liu, Y. Lei, Y. Chen, et al., Analyst **144**, 4013-4023 (2019)

[12] C. Xu, Z. Zhang, J. Fu, and Y. Huang, Langmuir, Langmuir, **33**, 5037-5045 (2017)

[13] S. Ma, F. Ribeiro, K. Powell, J. Lutian, C. Møller, T. Large, et al., Acs Appl. Mater. Inter. **7**, 21628-21633 (2015)

[14] E. Cheng, H. Yu, A. Ahmadi, and K. C. Cheung, Biofabrication, **8**, 015008 (2016)

[15] Y.-S. Chen, Y.-L. Huang, C.-H. Kuo, and S.-H. Chang, Int. J. Mech. Sci. **49**, 733-740 (2007)

[16] N. Reis, C. Ainsley, and B. Derby, J. Appl. Phys. **97**, 094903 (2005)

[17] A. Pan, E. Hanson, and M. Lee, J. Imaging Sci. Techn. **54**, 10503-1-10503-8 (2010)

[18] J. Brünahl and A. M. Grishin, Sensor. Actuat. A-Phys. **101**, 371-382 (2002)

[19] A. Amirzadeh Goghari and S. Chandra, Exp. Fluids, **44**, 105-114 (2008)

[20] X. H. Zeng, L. H. Qi, H. Huang, X. S. Jiang, and Y. Xiao, Key Eng. Mater. **419-420**, 405-408 (2010)

[21] S. X. Cheng, T. Li, and S. Chandra, J. Mater. Process. Tech. **159**, 295-302 (2005)

[22] A. A. Tseng, M. H. Lee , and B. Zhao, J. Eng. Mater.-T Asme, **123**, 74-84 (1999)

[23] S.-y. Zhong, L.-h. Qi, J. Luo, H.-s. Zuo, X.-h. Hou, and H.-j. Li, J. Mater. Process. Tech. **214**, 3089-3097 (2014)

Strain rate and temperature effects on tensile properties of monocrystaline Cu6Sn5 by molecule dynamic simulation

Jian Zhang
[1]School of Mechanical and Electrical Engineering, Guilin University of Electronic Technology
[2]Engineering Research Center of Electronic Information Materials and Devices，Ministry of Education, Guilin University of Electronic Technology
Guilin, China
544582139 @qq.com

Wei Huang*
[1]School of Mechanical and Electrical Engineering, Guilin University of Electronic Technology
[2]Engineering Research Center of Electronic Information Materials and Devices，Ministry of Education, Guilin University of Electronic Technology
Guilin, China
huang0773 @guet.edu.cn

Kai-Lin Pan
[1]School of Mechanical and Electrical Engineering, Guilin University of Electronic Technology
[2]Engineering Research Center of Electronic Information Materials and Devices，Ministry of Education, Guilin University of Electronic Technology
Guilin, China
pankl@guet.edu.cn

Abstract—In the soldering of electronic products with Cu as the substrate, intermetallic compounds (IMCs) which main component is Cu6Sn5 is usually generated. Solder joint cracking in electronic products usually occurs in the IMC layer or between IMC and solder, so the mechanical properties of IMC will affect the mechanical properties of the whole solder joint. In this paper, a 5×5×5 supercell was built, and the molecular dynamics (MD) simulations were performed using a Large-scale Atomic/Molecular Massively Parallel Simulator (LAMMPS) with the MEAM potential function. The tensile properties of single crystal Cu6Sn5 have been studied. Tensile direction in this paper is along X axis. The stress-strain curves at different strain rates and different temperatures are analyzed. The influencing factors considered in this paper are temperature and strain rate. According to the results of simulation analysis, it is found that the stress-strain curves at higher strain rate is always above the curve at lower strain rate before the IMC cracked. For the same strain rate, the higher temperature, the lower the yield strength of Cu6Sn5.The elongation before the stress - strain curve declining with larger strain rate is bigger than that with smaller one. UTS of the Cu6Sn5 increase as increment of the strain rate. Elongation corresponding to UTS also appears the same changing trend, a larger stains rate corresponds to lager elongation of the UTS.

Keywords—IMC, Cu6Sn5, Molecular dynamic simulation, Mechanical properties

I. INTRODUCTION

There are many ways to realize the physical and electrical connection of electronic components, of which reflow soldering technology is the most widely used. During the soldering process, the intermetallic compounds (IMC) are formed. The mechanical properties of the IMC layer is a prerequisite for good solder joint interconnection. The formation of IMC has an important influence on the micro and macro mechanical behavior of solder joints. It is an important method to study the failure of microelectronic solder joints by studying the damage form of the IMC layer. In the research on the reliability of electronic products, it is found that the failure of solder joints mostly occurs at the interface of the IMC layer or the solder and the IMC layer. IMC is the weakest part of the solder joint structure. Therefore, The mechanical properties of IMC will largely determine the length of the solder joint life. The composition of IMC contains Cu6Sn5 and Cu3Sn. But the growth rate of Cu6Sn5 is faster, so the main component in IMC is Cu6Sn5 [1]. In addition, the performance of Cu6Sn5 is weak, and the

damage of IMC often occurs in the Cu6Sn5 layer. The performance of Cu6Sn5 is an important feature for evaluating the crack growth and fracture resistance of solder joints [2].

Due to its many outstanding advantages, molecular dynamics methods have a wide range of applications in studying the structural evolution of matter on the atomic scale. For the calculation of macro properties, it has high accuracy and effectiveness. According to reports, researchers in China have studied the elastic modulus and the strain rate of Cu3Sn through molecular dynamics. They compared the simulation results with the results measured using nanoindentation technology, and the data are in good agreement. Davoodi et al [3]. used molecular dynamics methods to study the mechanical properties of Ag3Sn.

Therefore, this paper uses the method of molecular dynamics simulation. Specifically, the large-scale atomic/molecular massively parallel simulator (LAMMPS) was used to study the mechanical properties of single crystal Cu6Sn5 at different strain rates and different temperatures based on the MEAM potential. We discuss and analyze the obtained stress-strain curve.

II. THEORY AND SIMULATION PROCESS

A. Cu6Sn5 Crystal Structure

As we all know, Cu6Sn5 has two different structures: single crystal and polycrystalline. Generally speaking, the microstructure of polycrystalline Cu6Sn5 is scallop-shaped and the single crystal Cu6Sn5 has a ridge shape. In this study, we constructed the crystal structure of single crystal Cu6Sn5 by LAMMPS. The unit cell contains 44 atoms, with 24 Cu atoms and 20 Sn atoms[4]. As shown in Figure 1, the blue ball represents the Cu atom, and the brown ball represents the Sn atom.

In order to study its tensile properties, in this paper, as shown in Figure 2, a 5×5×5 supercell is established. We studied the effect of different strain rates and different temperatures on its tensile properties.

978-1-6654-1392-3/21 $31.00 © 2021 IEEE

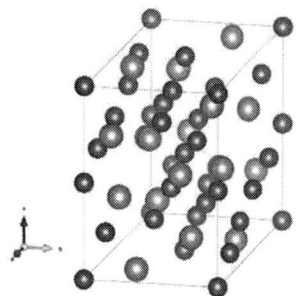

Fig.1. η' phase Cu6Sn5.

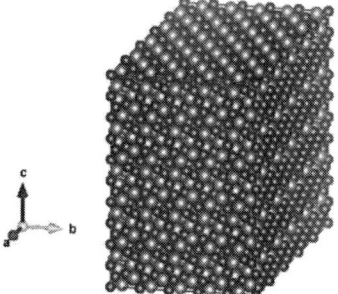

Fig.2. A 5×5×5 supercell.

B. MEAM Potential

The MEAM potential is obtained after optimizing the EAM [5]. This potential function is currently one of the most accurate theories describing the interaction between metal atoms. Compared with EAM, the effect of angular forces are added to the modified MEAM potential. Therefore, MEAM is the most commonly used potential function in LAMMPS. The total energy of the system is accumulated by the direct contributions of all atoms. The MEAM can be used to simulate the mechanical properties of alloy systems composed of different metal atoms. Its mathematical equation can be expressed as [6].

$$E_{tot} = \sum_i \left(F(\bar{\rho}_i) + \frac{1}{2} \sum_{j \neq r} \phi(r_{ij}) \right) \quad (1)$$

$$F(\bar{\rho}_i) = AE_c \frac{\bar{\rho}_i}{\rho_{i0}} \ln\left(\frac{\bar{\rho}_i}{\rho_{i0}} \right) \quad (2)$$

where in equation (1), F is the embedding energy, which is a function of the atomic electron density $\bar{\rho}_i$, r_{ij} the distance between atoms i and j, ϕ_{ij} the interaction potential between atoms i and j. In equation (2), A is an adjustable parameter; Ec the cohesive energy, ρ_{i0} the density scaling parameter. The atomic electron density $\bar{\rho}_i$ is composed of $\rho_i^{(0)}$, $\rho_i^{(1)}$, $\rho_i^{(2)}$ and $\rho_i^{(3)}$, where

$$\rho_i^{(0)} = \sum_{j \neq i} \rho_j^{a(0)}(r_{ij}) \quad (3)$$

$$\left(\rho_i^{(1)}\right)^2 = \sum_\alpha \left[\sum_{j \neq i} \rho_j^{a(1)}(r_{ij}) \chi_{ij}^\alpha \right]^2 \quad (4)$$

$$\left(\rho_i^{(2)}\right)^2 = \sum_{\alpha,\beta} \left[\sum_{j \neq i} \rho_j^{a(2)}(r_{ij}) \chi_{ij}^\alpha \chi_{ij}^\beta \right]^2 - \frac{1}{3}\left[\sum_{j \neq i} \rho_j^{a(2)}(r_{ij}) \right]^2 \quad (5)$$

$$\left(\rho_i^{(3)}\right)^2 = \sum_{\alpha,\beta,\gamma} \left[\sum_{j \neq i} \rho_j^{a(3)}(r_{ij}) \chi_{ij}^\alpha \chi_{ij}^\beta \chi_{ij}^\gamma \right]^2 \quad (6)$$

What we listed in Table I are the specific MEAM potential parameters of the Cu atom, Sn atom, and Cu6Sn5 atom used in this study.

TABLE I. THE MEAM POTENTIAL PARAMETERS USED FOR Cu6Sn5.

Element	Ec(eV)	A	$r0$(Å)	α	$\beta^{(0)}$	$\beta^{(1)}$	$\beta^{(2)}$	$\beta^{(3)}$	$t^{(1)}$	$t^{(2)}$	$t^{(3)}$	ρ_0
Cu	3.4	1.07	2.657	5.11	3.634	2.2	6.0	2.2	3.14	2.49	2.95	1.0
Sn	3.84	1.0	3.176	6.20	6.20	6.0	6.0	6.0	12.5	8.0	-0.383	1.0
Cu6Sn5	4.03		2.907	5.38								

C. NVE/NVT/NPT Ensembles

The supercell needs to relax in NVE, NVT and NPT ensembles, ensuring the total energy , temperature and pressure of the system to reach the set target. The relaxation results are shown in Figure 3. Total energy of the system is stale, temperature is about 300K, and pressure is appropriate to 0 bar. Because the single crystal Cu6Sn5 has isotropic physical properties, that is, the results obtained from different tensile directions are consistent, so for convenience, the X axis is selected as the tensile direction in this study. Strain rates in this paper are 0.001, 0.01, 0.1 and 1 ps^{-1}, and they can also be marked as 0.1%, 1%, 10% and 100% ps^{-1}. The temperature is 200K, 300K and 400K respectively.

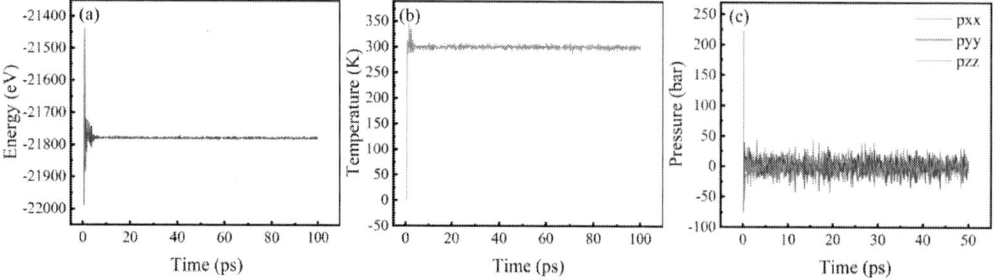

Fig.3. Relaxation results of the system.

III. RESULTS AND DISCUSSIONS

The stress-strain curves under different tensile rates were plotted together, as shown in figure 4. It is found that the stress-strain curves at higher strain rate is always above the curve at lower strain rate before the IMC cracked. Before the applied strain reaches the necking point, the slopes of the curves corresponding to different strain rates are not much different, that is, the elastic stiffness of the material has a small correlation with the strain rate. When the material reaches the necking point, the strain is increased. Then the stress will converge to a constant below the necking point. When the strain rate is 0.1%, the magnitude of the decrease in the stress value is relatively gentle. Other strain rates are less than 0.1%, and the stress value drops relatively quickly.

We can clearly find from Figure 4(a)-(c) that when the stress value converges to a constant, there will be up and down fluctuations. The reason for the above phenomenon is that due to the force loading speed is very fast, and part of the energy is stored. As the single crystal Cu6Sn5 material undergoes necking, the previously stored energy will be released, which will eventually cause stress fluctuations. The smaller the strain rate, the more significant the fluctuation of the stress value. The elongation before the stress-strain curve declining with larger strain rate is bigger than that with smaller one. When the temperature is 200K, the maximum stress values corresponding to different strain rates of 0.1%, 1%, 10% and 100% are 9.81 GPa, 10.30 GPa, 11.62 GPa and 13.99 GPa respectively.

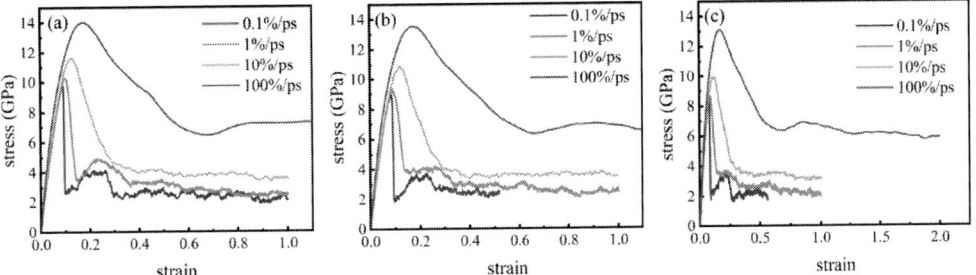

Fig.4. Stress-strain curves for (a)200K, (b)300K and (c)400K at different strain rates.

In this paper, the tensile properties of Cu6Sn5 at three different temperatures are simulated, which are 200K, 300K, and 400K. The recorded data is plotted as a stress-strain curve, as shown in the figure 5. It is obvious that with the temperature increase, the maximum stress value decreases. There are two reasons to explain the above phenomenon. On the one hand, at a higher temperature, single crystal Cu6Sn5 is more prone to plastic deformation, which leads to an increase in the elongation of the material and a decrease in the yield strength. On the other hand, a higher temperature

will increase the internal energy of the Cu6Sn5 atom, which will make the material more prone to dislocations and reduce the corresponding stress value. However, different temperatures have little effect on the downward trend of the stress value. It can be seen from Figure 5(a)-(b) that when the strain rate is low, the downward trends of the stress-strain curves corresponding to different temperatures are almost coincident. When the tensile rate is $0.001 ps^{-1}$, the ultimate strengths corresponding to the temperatures of 200K, 300K and 400K are 9.81 GPa, 9.06 GPa and 8.25 GPa, respectively.

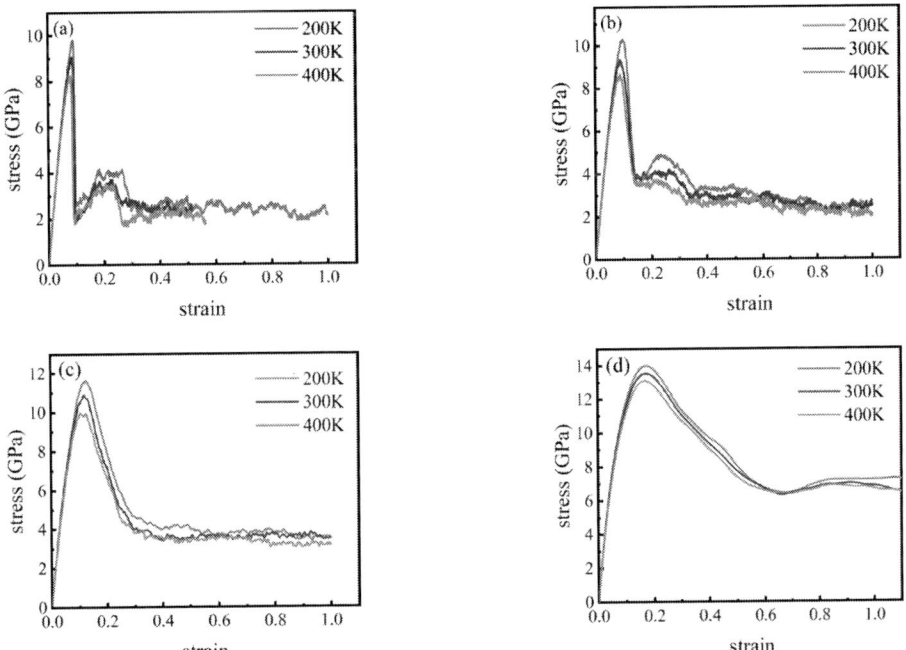

Fig.5. Stress-strain curves for (a)0.001, (b)0.01, (c)0.1 and (d)1 ps^{-1} at different temperature.

The peak point in the stress-strain curve was taken as the ultimate tensile strength (UTS), and the elongations at the UTS were also recorded. The UTSs and corresponding elongations were shown in Figure 6 and 7 respectively. As shown in Figure 7, when the temperature is 200K, the strain rates of 0.001, 0.01, 0.1 and 1 ps^{-1}correspond to elongations of 0.089, 0.098, 0.124 and 0.169, respectively. It is clear to find that UTS of the Cu6Sn5 increase as increment of the strain rate. As the strain rate is increased, the elastic stiffness will increase. Elongation corresponding to UTS also appears the same changing trend, a larger stains rate corresponds to lager elongation of the Cu6Sn5.

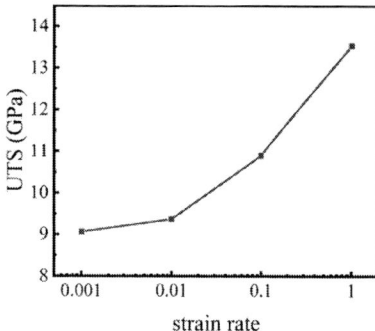

Fig.6. UTSs at a temperature of 200K and different strain rates.

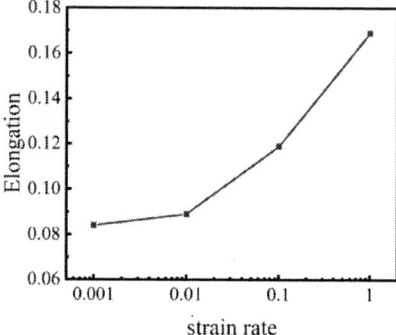

Fig.7. Elongations of the UTS at different strain rates.

IV. CONCLUSIONS

In this study, the MEAM-based lammps method was used to extensively explore the tensile properties of single crystal Cu6Sn5 at different temperatures and strain rates. By establishing a 5×5×5 super cell and tensile along the X axis. We record the stress and strain values during the whole process, and finally draw the stress-strain curve. It is clear that tensile properties of monocrystalline Cu6Sn5 can be influenced by temperatures and strain rates.

1) The larger the strain rate, the higher the corresponding stress-strain curve. Therefore, the UTS at a higher strain rate is larger than that at the lower strain rate, and the elongation at the corresponding UTS is also longer. In addition, when the strain rate is large, the stress value of single crystal Cu6Sn5 tends to drop relatively smoothly after necking.

2) The tensile properties of single crystal Cu6Sn5 will also be affected by temperature. When the temperature is higher, the UTS will decrease. But its elastic stiffness is hardly affected by temperature.

3) Under the same strain rate, load the model with different temperatures. When the temperature is higher, the single crystal Cu6Sn5 is more prone to elastic deformation and dislocation, thereby reducing its yield strength.

ACKNOWLEDGMENTS

This research was completed with the support of the School of Mechanical and Electrical Engineering, Guilin University of Electronic Technology. The author is very grateful to the above-mentioned institutions for their support of venue and computing resources.

REFERENCES

[1] M. Ramani, and N.A. Jasli, "Silver effect on the intermetallic growth in the Sn-8Zn-3Bi lead-free solder."Materials Today: Proceedings, Vol. 5, pp. 17553-17560, 2018.

[2] L.N. Van, C.S. Chung, and H.K. Kim, "Comparison of the fracture toughness of Cu6Sn5 intermetallic compound as measured by nanoindentation and other methods."Materials Letters, Vol.162, pp. 185-190, 2016.

[3] J. Davoodi, M.T. Fallahi, and H. Rafii-Tabar, "Nano-scale modelling of the mechanical properties of Pb-free solder alloys."Journal of Computational and Theoretical Nanoscience, Vol. 5, pp. 359-365, 2008.

[4] W. Huang, K.L. Pan, J. Zhang, and Y.B.Gong. "Effect of In-Doping on Mechanical Properties of Cu6Sn5-Based Intermetallic Compounds: A First-Principles Study." Journal of Electronic Materials, Vol.50, pp. 4164-4171, 2021

[5] S.M. Foiles, "Application of the embedded-atom method to liquid transition metals." Physical Review B, Vol. 32, pp. 3409, 1985.

[6] M.I. Baskes, "Modified embedded-atom potentials for cubic materials and impurities." Physical review B, Vol. 46, pp. 2727, 1992.

Effect of Electromigration on Interfacial Reaction In Ni/Sn63Pb37/Cu BGA Solder Joints

Fei Zhang
Xi'an Research Institute of NavigationTechnology
Xi'an, China
zhf121@126.com

Shuai Chen
Xi'an Research Institute of NavigationTechnology
Xi'an, China
chenshuai19871219@163.com

Zhidan Liu
Xi'an Research Institute of NavigationTechnology
Xi'an, China
798330181@qq.com

Wenlong Wang
Xi'an Research Institute of NavigationTechnology
Xi'an, China
5220338@qq.com

Abstract—In this paper, to investigate the effect of electromigration (EM) on interfacial reactions in BGA solder joints, the Ni/Sn63Pb37/Cu BGA daisy-chain flip chips were used and the experiment was conducted under a current density of $1.0 \times 10^3 A/cm^2$ at 125 ℃. It has been found that the Ni_3Sn_4 intermetallic compounds (IMCs) formed at the solder/Ni interface and the Cu_6Sn_5 IMCs formed at the solder/Cu interface in the as-reflowed state. With increasing EM time, the IMCs formed at chip interface transformed into $(Cu,Ni)_6Sn_5$ type, while the IMCs at the PCB side didn't change. During EM, at the anode side the IMCs grew and the growth behavior was linear with $t^{1/2}$. At the same side, the IMC thickness at the anode side was larger than that at the cathode side. However, the growth behavior of IMCs at the cathode side was complicated. Due to the initial IMCs in the as-reflowed were thin, under the current density, first the IMCs grew. While after the thickness of IMCs increased to a critical value, it decreased thereafter.

Keywords—Electromigration; Ni/Sn63Pb37/Cu; intermetallic compound

I. INTRODUCTION

With the rapid development of miniaturization and integration in electronic products, the package size of devices is constantly decreasing and the number of interconnection joints on per unit area continues to increase. As a result, the size of single solder joint decreases, leading to the sharp increase of the current density through a single joint when the current through it remains unchanged, and the reliability issues caused by the phenomenon, i.e. electromigration (EM), are more serious [1-3]. Electromigration refers to the directional migration of atoms in metal conductors under the action of high current density, which is caused by the interaction of electron impact force and static electric field force.

With the size of solder joint decreasing, the reliability problem caused by electromigration has been studied by more and more scholars. At the same time, with the consideration of environmental health, lead-free solders have been promoted and applied all over the world. Therefore, a large number of studies on electromigration focused on lead-free solder joints. A large number of studies found that in lead-free solder joints, EM promotes an increase in the thickness of intermetallic compounds (IMCs) at the anode interface, and accordingly, IMCs will be depleted at the cathode interface, which results in the formation of cavities and cracks at the cathode interface and finally leads to failure of the solder joints [2,3]. However, in the Sn-Zn solder joints, a completely different phenomenon occurs, that is, under the

current density with the time increasing, the thickness of the IMCs at the anode side decreases, and the thickness of IMCs at the cathode side increases. Zn element has a relatively good electromigration resistance. Due to the continuous advancement of lead-free process, the current research on EM is mainly focused on lead-free solder joints, while in high reliability fields, Sn63Pb37 solder joints are still the main soldering material, but the research on EM of Sn63Pb37 BGA solder joints was relatively few, which needs in-depth analysis.

In this paper, the effect of electromigration on interfacial reactions of Ni/Sn63Pb37/Cu BGA solder joints was investigated and the different growth behaviors of IMCs at the anode and cathode side were also discussed.

II. EXPERIMENTAL

BGA daisy-chain flip chips were used in the test, and the chip size was 15mm×15mm×1.4mm. The specific structural size is shown in Fig. 1 below, in which the solder ball composition is Sn63Pb37 and the UBM layer is Ni.

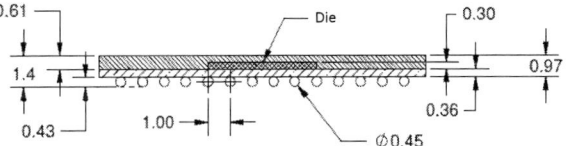

Fig. 1 Structure of Daisy chain chip

The printed circuit board (PCB) was designed according to the daisy-chain chip. The diameter of the PCB pad was 0.36mm, and the solder pad was made by Cu foil with Sn63Pb37 hot air level process. The connection relationship between the PCB and the chip after reflow with a peak temperature of 220 ℃ was shown in Fig.2.

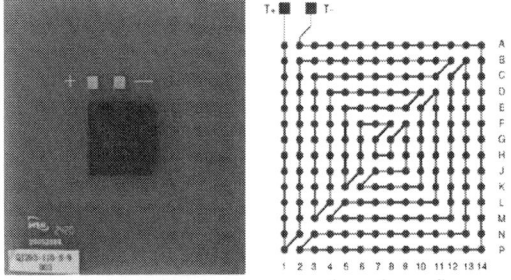

Fig. 2 BGA solder joints connection after reflow

In this paper, the EM samples were put into the heat-conducting silicone oil with 125 ℃ temperature, and the mechanical stirring method was adopted to reduce the influence of Joule heat and prevent the rapid oxidation of the solder joints under high temperature. The electromigration experiments were conducted with a current of 1 A. And the experimental period was set as 100 h, 200 h, and 400h, respectively. Through calculation, the average current density in a single BGA solder joint was $1\times10^3 A/cm^2$.

The scanning electron microscopy (SEM) was used to analyze the as-soldered and tested samples, in which the energy-dispersive X-ray (EDX) spectrometer was used to detect the compositions of the IMCs. The average thickness of the interfacial IMCs was obtained by measuring the area of the interfacial IMC with an image processing software and then dividing it by the length of the interface IMCs.

III. RESULTS

Fig. 3 shows the cross-section microstructures of an as-soldered Ni/Sn63Pb37/Cu BGA solder joint. It was observed that the different types of IMCs formed at different interfaces. The IMCs those close to the chip side were identified as Ni_3Sn_4 with the thickness of 1.03μm as shown in Fig.3 b), and those close to the PCB side was Cu_6Sn_5 with the thickness of 1.92μm as shown in Fig.3 c).

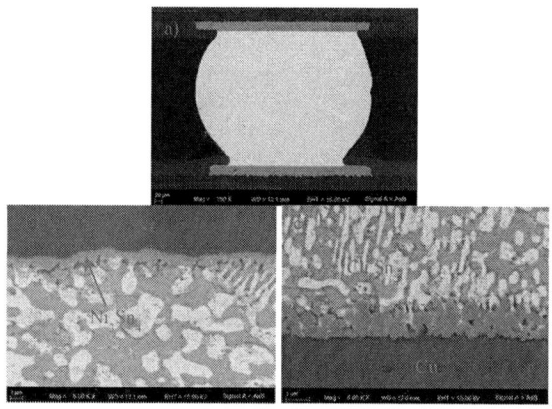

Fig.3 Interfacial microstructures of the as-soldered Ni/Sn63Pb37/Cu BGA solder joint: a) the whole solder joint; b) chip side; c) PCB side.

Fig. 4 shows the microstructures of Ni/Sn63Pb37/Cu BGA solder joints at the anode side during the EM process, where a)-c) are the microstructures at the anode side at 100h, 200h and 400h when the chips were used as the anode, while d) -f) are the microstructures at the anode side at 100h, 200h and 400h when the PCBs were used as the anode. As can be seen from Fig.4a) -c), the type of IMCs on the chip side changed from Ni_3Sn_4 to $(Cu,Ni)_6Sn_5$ after EM 100h, and the thickness of IMCs increased with the increase of EM time. The thickness of IMCs increased from 1μm after reflowed to 2.3μm after EM 100h, and that of IMCs increased to 2.45μm after EM 200h.The thickness of IMCs increased to 2.84μm after EM 400h. As can be seen from Fig. 4 d) -f), the type of IMCs on the PCB side did not change and remained Cu_6Sn_5 during the EM process, and no Ni atom was observed on the PCB side. This is because the diffusion coefficient of Ni atom in the solder is much lower than that of Cu atom. Cu atoms can be diffused from the PCB side to the chip side, while Ni atoms cannot be diffused from the chip side to the PCB side. It can be inferred that there is no dissolution of Ni UBM at the cathode interface when Ni was used as cathode.

During EM, when Ni atoms at the cathode side dissolved in solder, dissolved Ni atoms could be moved towards the anode side under the action of electron wind, and reacted with Sn atoms, which led to the IMCs form in the solder or spread to the anode interface. But in this experiment it was not identified a lot of Ni atoms whether in the solder or at the anode interface. At the same time, it was observed that Cu_3Sn formed at the Cu/Cu_6Sn_5 interface after EM 100h, and the thickness of Cu_3Sn and IMC on PCB side increased with the increase of EM time. The thickness of IMCs at the PCB side was from 1.92μm after reflowed to 4.41μm after EM 100h and 5.51μm after EM 200h and 6.43μm after EM 400h.

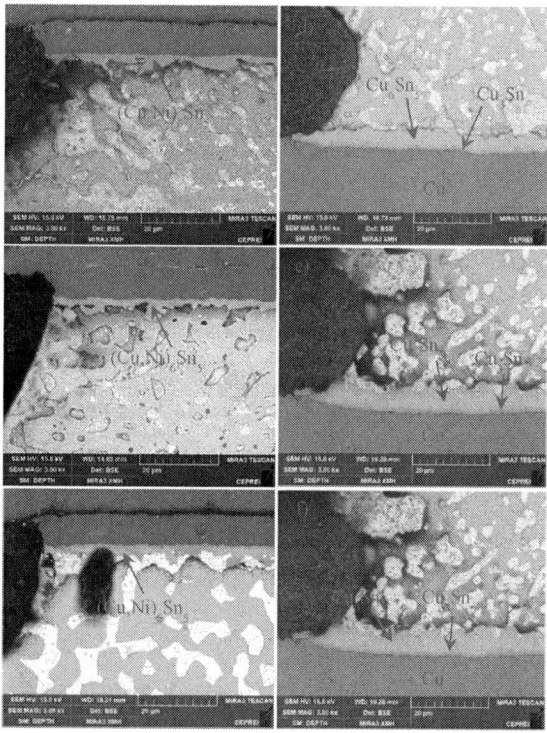

Fig. 4 Interfacial microstructures of solder joints at anode side after EM: a)~c) IMC at anode side of chip after EM 100h, 200h and 400h, respectively; d)~f) IMC at PCB anode side after EM 100h, 200h and 400h, respectively.

Fig. 5 shows the microstructures of the Ni/Sn63Pb37/Cu BGA solder joints at the cathode side during the EM process, where a)-c) are the microstructures at the cathode side at 100h, 200h and 400h when the chips were used as cathode, and d) -f) are the microstructures at the cathode side at 100h, 200h and 400h when the PCBs were used as cathode. As can be seen from Fig. 5a) -c), when the chip sides were used as the cathode, the type of IMCs also changed from Ni_3Sn_4 to $(Cu,Ni)_6Sn_5$ after EM 100h, the thickness of IMCs increased from 1μm after reflowed to 2.1μm after EM 100h, and the thickness of IMCs increased to 2.14μm after EM 200h.After 400h of electromigration, the thickness of IMCs at the interface was 2.05μm. It can be revealed that the thickness of IMCs at the interface increased significantly during EM 100h, and the thickness of IMCs changed little between EM 100h and EM 200h, then the thickness of IMC decreased slowly. Compared with the change of IMC type at the interface when the chips were used as anode, it can be seen that the IMCs at the chip side changed from Ni_3Sn_4 to $(Cu,Ni)_6Sn_5$ in the process of

978-1-6654-1392-3/21 $31.00 © 2021 IEEE

electromigration, no matter whether the chip was used as cathode or anode. This is because Cu atoms have higher solubility and diffusion coefficient in the solder compared with Ni atoms. In the reflowed process, a large number of Cu atoms in the PCB pads dissolved into the solder, resulting in a relatively rich Cu atom content in the solder. In the subsequent EM process, due to the effect of higher temperature and chemical potential gradient, the transformation of interface IMC type occured. Other scholars [4] also found that in the Ni/ solder /Cu joint structure, $(Cu,Ni)_6Sn_5$ formed at the Ni side interface after reflowed, while the Ni content was less near the Cu side. As can be seen from Fig. 5 d) -f) after reflowed, when the PCBs were used as the cathode, the type of IMCs remained unchanged in the process of electromigration, and the thickness of IMCs increased from 1.92μm after reflowed to 3.94μm after EM 100h, and increased to 4.93μm after EM 200h. After 400h of electromigration, the thickness of IMCs at the interface was 4.42μm, which was consistent with the change of IMC at the interface when the chip was used as cathode.

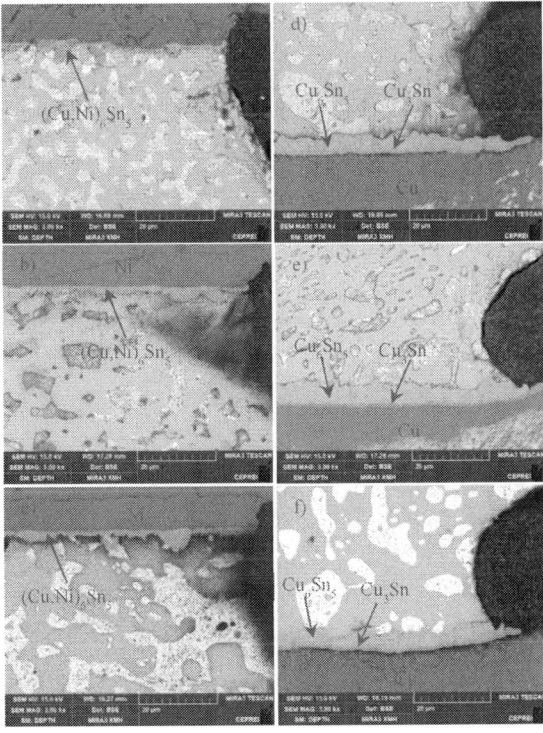

Fig. 5 Interfacial microstructures of solder joints at the cathode side after electromigration: a)~c) IMC at cathode side of chip after EM 100h, 200h and 400h, respectively; d)~f) IMC at the cathode side of PCB after EM 100h, 200h and 400h, respectively.

During EM, the growth kinetics of interfacial IMCs in the solder joins was shown in Fig. 6. It can be observed that the thickness of IMCs at the anode side, regardless of those at the PCB side or the chip side, has a linear relationship with time $t^{1/2}$, indicating that the growth of interface IMCs is controlled by the diffusion process. At the same time, it can be observed that the interfacial IMCs at the cathode side were thinner than those at the anode side, whether those were at the chip side or at the PCB side. This was because that under the action of electronic wind force, atoms were promoted to migrate to the anode interface, thus accelerating

the growth of IMCs at the anode side while inhibiting the growth of IMCs at the cathode side. At the cathode side, the thickness of the interface IMC increased at first and then decreased with the increase of the EM time. This was because the thickness of the interface IMCs at the initial stage was thin and the diffusion flux caused by the chemical potential gradient was larger, and the interfacial IMCs thickened with the increase of EM time. As the thickness of the interfacial IMCs increased, the interfacial IMCs acted as a barrier layer to hinder the diffusion of atoms from the pad to the solder, and the growth of IMCs was inhibited. And the atoms at the IMC/ solder interface diffused to the solder under the action of electronic wind force, resulting in the thinning of cathode interfacial IMCs. Under the combined actions of the two, the IMCs grew at the beginning of EM and with increasing EM time, when the thickness of the IMCs reached a critical value, the IMCs at the cathode would be depleted and then they decreased thereafter.

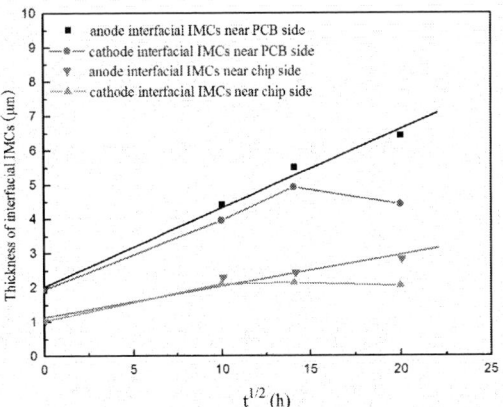

Fig. 6 Thickness of interfacial IMCs in the solder joins after electromigration

IV. CONCLUSIONS

(1) After reflow soldering, different types of IMCs formed at each side of Ni/Sn63Pb37/Cu BGA solder joint, while Ni_3Sn_4 IMCs formed at the solder/Ni interface and Cu_6Sn_5 IMCs formed at the solder/Cu interface;

(2) With increasing EM time, the IMCs formed at chip interface transformed into $(Cu,Ni)_6Sn_5$ type, while the IMCs at the PCB side didn't change. For the same interface, the interfacial IMCs at the cathode side were thinner than those at the anode side;

(3) During EM, the growth behavior of IMCs at the anode side was linear with $t^{1/2}$, and the interfacial reaction was diffuse-controlled.

(4) The IMCs at the cathode grew first and then decreased. Due to the initial IMCs in the as-reflowed were thin, with increasing EM time, when the thickness of the IMCs at the cathode reached a critical value, the IMCs would be depleted and then they decreased thereafter.

REFERENCES

[1] Mingliang Huang, Fei Zhang, Fan Yang, Ning Zhao, " Effect of electromigration on the tensile strength of Cu/Sn-9Zn/Cu solder interconnects", Proceedings of 2014 International Conference on Electronic Packaging Technology (ICEPT), Chengdu, Aug. 12-Aug. 15, pp. 1190-1193.

[2] Tu K N, " Recent advances on electromigration in very large scale integration of interconnects " , Journal of Applied Physics, vol. 94, no. 9, pp. 5451-5473, 2003.

[3] WANG C H, KUO C Y, CHEN H H, et al., " Effect of current density and temperature on Sn/Ni interfacial reactions under current stressing ", Intermetallics, vol. 19, no. 1, pp. 70-75, 2011.

[4] CHEN H T, WANG C Q, YAN C, et al., "Cross-interaction of interfacial reactions in Ni(Au/Ni/Cu)-SnAg-Cu solder joints during reflow soldering and thermal aging," Journal of Applied Physics, vol. 36, no. 1, pp. 26-32, 2007.

Optimization of solder height for stencil printing process performance on length-width ratio

Dezhi Su
Shandong Institute of Space Electronic Technology
Yantai, China
sudezhihefish@126.com

Cen Wang
Shandong Institute of Space Electronic Technology
Yantai, China
wangcen@126.com

Hongkun Wang
Shandong Institute of Space Electronic Technology
Yantai, China
wanghongkun@163.com

Junxiang Zhao
Shandong Institute of Space Electronic Technology
Yantai, China
zhaojunxiang@126.com

Hao Cheng
Shandong Institute of Space Electronic Technology
Yantai, China
chenghao0102@126.com

Lejun Zhang
Shandong Institute of Space Electronic Technology
Yantai, China
zhanglejun7583@163.com

Abstract—In this paper, various length-width ratio of printing pattern was designed on a ceramic substrate. The samples were prepared with SAC305 solder by stencil printing. The solder height was tested by coating thickness tester. It was found that squeegee load, squeegee speed, and separation speed were the main factors in the stencil printing process. The solder height was the highest when the squeegee load, squeegee speed, and separation speed were 40 N, 35 mm/s and 1.5 mm/s, respectively. According to the datasheet, a formula of quadratic equation was established. The deviation between experimental result and simulated data was below ±5%. Additionally, various length-width ratios also affected the solder height of stencil printing. When the length-width ratio exceeded 1.24, the solder height would decrease dramatically.

Keywords—*Length-width ratio, Ceramic substrate, Stencil printing*

I. INTRODUCTION

Microelectronic technology was one of the most famous technologies in global terms [1-3]. As the requests for the microelectronic technology from the common life altered, microelectronic devices with high reliability including new material based microelectronic products emerged in endlessly. The production of electronic products included circuit design, wafer fabrication, electronic packaging and testing. In the field of electronic packaging, die bonding and wire bonding were the key process.

Stencil printing was one of the steps in the die bonding process. Generally, the substrate for stencil printing was printed circuit boards (PCBs). Solder paste was firstly patterned at the surface of PCB, and then electronic components was directly mounted onto PCB. At last, all the components were firmly mounted onto PCB with welding process. In these three steps, stencil printing was the significant step, because it finally determined the filled volume, height and area of the solder paste. Rusdi et al founded that squeegee load, squeegee speed, and separation speed had important effects on filled volume, height and area of solder paste with various components [4]. However, this work took PCB as substrate and did not assess the relationship of length-width ratio and printing quality. Nourma et al [5] adopted data mining approaches through SVR with different kernel and RT. The nonlinear relationships of the stencil printing volume TE were established and the optimal printing

parameters were found. This method based on model and formula, but lacked of practical experience.

In this Paper, the substrate was took as ceramic material. Stencil of variable length-width ratio was firstly designed. Based on an length-width ratio, the effect of squeegee load, squeegee speed and separation speed on solder height was studied. Besides, models of one-factor, two-factor and three factor were also established to deep understand the process of stencil printing. At last, the relation of length-width ratio and solder height was built with quadratic trinomial.

II. EXPERIMENT

The printing graphics was designed as Fig. 1a. There were 6 kinds of shapes marked with 6 colors. The red, yellow, blue, green, purple and orange district respectively stood for the size of 0.8 mm×1.9 mm, 0.95 mm×1.9 mm, 1 mm×1.9 mm, 1.3 mm×1.9 mm, 1.9 mm×1.9 mm and 2.2 mm×1.9 mm. All samples had the same length of 1.9 mm. The length-width ratios (the length of 1.9 mm was ahead) were accordingly 2.375, 2, 1.9, 1.46, 1 and 0.86. Fig. 1b was a quarter (white dotted line in Fig. 1a) of printing sample in practice. Table 1 showed the dimensions of stencil used in the experiment.

SAC 305 (SnAg3.0Cu0.5) paste with type IV (20-38 μm) was carried out in the experiment. The length, width and thickness of the stencil were 730 mm, 730 mm and 0.2 mm, respectively. The substrate was Al_2O_3 (24 mm× 20 mm) with mass fraction of 90%. The solder paste printing system (A8, SHENZHEN ZHENGSHI AUTOMATION EQUIPMENT CO.,LTO) was used in the experiment. Measuring microscope (107J, COSSIM) was used as measuring positions were left, center and right of printing area without sintering. Then the mean value of the three values was the solder height. The variable parameters in the experiment were squeegee load, squeegee speed and separation speed. Both of experiments and tests were conducted at room temperature (25 ± 2 °C) and conventional humidity (60%-70% RH).

III. RESULTS AND DISCUSSIONS

In this section, analysis object of A, B, C and D was printing graphic with symbol of 2 which was shown in table I. Part E analyzed the relation between solder height and length-width ratio.

978-1-6654-1392-3/21 $31.00 © 2021 IEEE

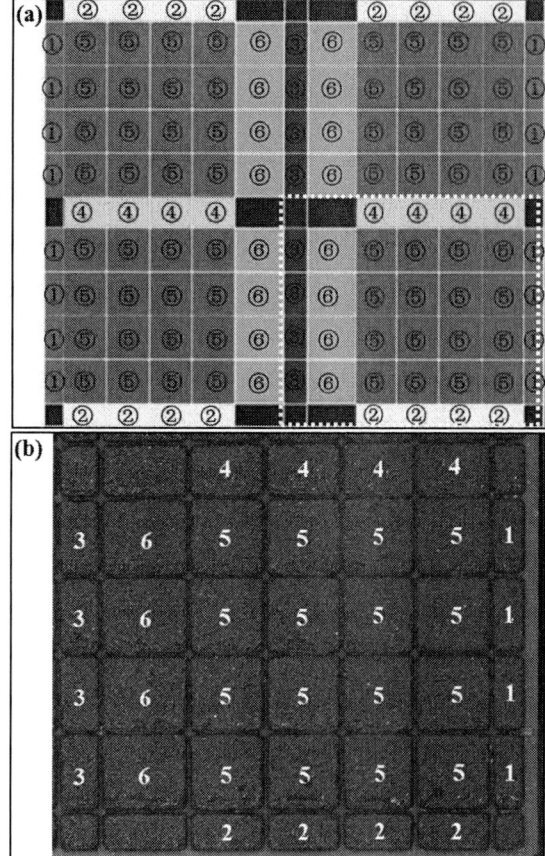

Fig. 1. (a) The schematic of printing pattern (1-red; 2-yellow; 3-blue; 4-green; 5-purple; 6-orange); (b) Optic image of stencil printing for white dotted line of Fig. 1a

TABLE I. DEMENSIONS OF STENCIL USED IN THE EXPERIMENT

Content	1	2	3	4	5	6
Color	red	yellow	blue	green	purple	Orange
Length(mm)	1.9	1.9	1.9	1.9	1.9	1.9
Width (mm)	0.8	0.95	1	1.3	1.9	2.2
L-W Ratio	2.375	2	1.9	1.46	1	0.86

A. Squeegee Load

According to Fig. 2a, in the initial stage the solder height of SAC305 increased as squeegee load enhanced. Separation speed (0.5 mm/s) and squeegee speed (35 mm/s) were fairly constant values in the experiment. When the squeegee load was 40 N, the solder height achieved the highest value of 190.47 μm. However, the solder height decreased during further improvement of squeegee pressure. The upward and downward rate of change in height were 0.56 μm/N and 2.76 μm/N, respectively. The rate of change in the descending stage was 4.93 times than that in the ascending stage. Thus, squeegee speed made a remarkable effect in the descending stage. The principles of squeegee load and solder height was shown in Fig. 2b. Too little load could not compact the solder paste due to adhesive function of solder paste, but too much load would create scooping effect [6]. Both of the two cases would lead to low printing height.

B. Squeegee Speed

According to Fig. 3a, in the initial stage the solder height of SAC305 increased as squeegee speed enhanced. In addition,

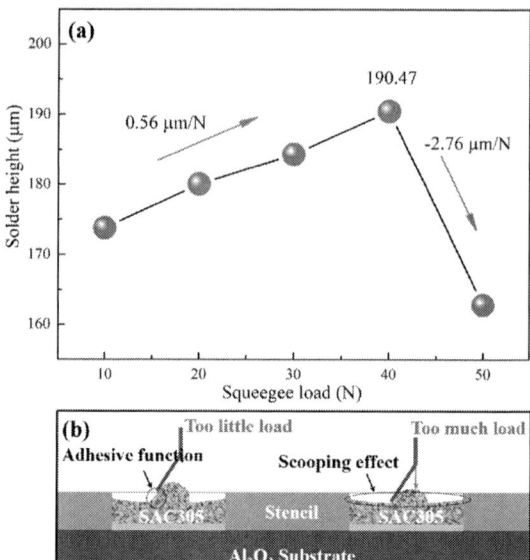

Fig. 2. (a) The practical relation and (b) principles between squeegee load and solder height

separation speed (0.5 mm/s) and squeegee load (40 N) were fairly constant values in the experiment. When the squeegee speed was 35 mm/s, the solder height achieved the highest value of 190.47 μm. However, the solder height decreased during further improvement of squeegee speed. The upward and downward rate of change in height were 0.77 μm/(mm/s) and 0.23 μm/(mm/s), respectively. The rate of change in the ascending stage was 3.35 times than that in the descending stage. Thus, squeegee speed had a noticeable effect in the ascending stage. The principles of squeegee speed and solder height was shown in Fig. 3b. Slow squeegee speed caused too much squeegee load in the printing process, and fast squeegee speed resulted in fluffy structure of SAC305 paste after stencil printing [7]. Both of the two situations would generate low solder height after stencil printing.

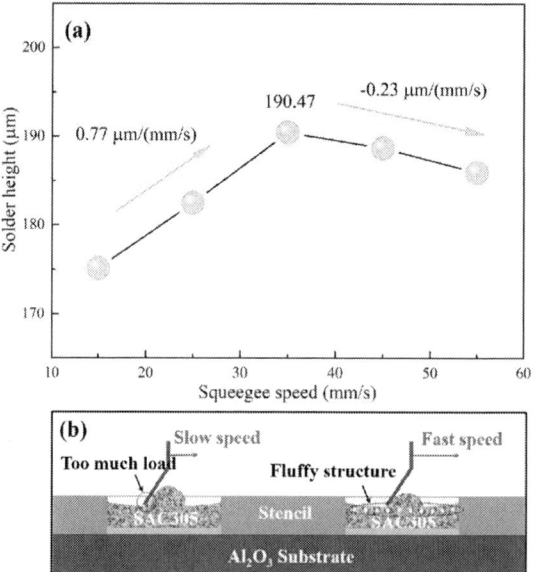

Fig. 3. (a) The practical relation and (b) principles between squeegee speed and solder height

C. Separation Speed

In this section, squeegee speed (35 mm/s) and squeegee load (40 N) were fairly constant values. Separation speed was set as variable. As shown in Fig. 4a, the curves of solder height exhibited high in the middle and low at the two sides. In the initial stage the solder height of SAC305 increased as separation speed enhanced. Separation speed with 1.5 mm/s in the above experiment resulted in the highest solder height (202.51 μm). It exceeded the thickness of stencil, because deformation of solder height was bigger after stencil printing. However, the solder height decreased during further improvement of separation speed. The upward and downward rate of change in height were 12.04 μm/(mm/s) and 7.62 μm/(mm/s), respectively. The rate of change in the ascending stage was 1.58 times than that in the descending stage. Therefore, separation speed had a distinct effect in the ascending stage. The principles of separation speed and solder height was shown in Fig. 4b. Slow separation speed caused permeation in the printing process, and fast separation speed resulted in adhesive function [8]. Both of the two phenomena would contribute to low solder height after stencil printing.

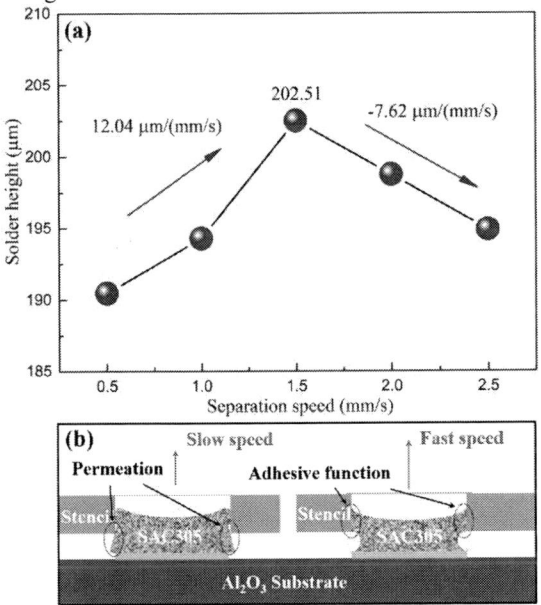

Fig. 4. (a) The practical relation and (b) principles between separation speed and solder height

D. Multi-factor

As mentioned above, squeegee load (40 N), squeegee speed (35 mm/s) and separation speed (1.5 mm/s) together resulted in the highest solder height. The optimal value was usually determined by changing single variable. In engineering practice, the solder height was depended on multi factors. Fig. 5a, 5b and 5c fixed one of the three factors, respectively. Squeegee load, squeegee speed and separation speed was 40 N, 35 mm/s and 1.5 mm/s. The relation of solder height and the other two factors was simulated and presented with single surface. From the shape of the surfaces, the value of solder height was large in the middle and small around. Separation speed and squeegee load contributed to the greatest change of solder height as shown in Fig. 5a. On the contrary, the role of squeegee speed and separation speed was

minimal. Besides, the relationship of separation speed, squeegee speed and squeegee load was established with quadratic trinomial, as in:

$$H=35.24 \times A-10.28 \times A^2+1.15 \times B-0.01 \times B^2+3.27 \times C-0.05 \times C^2 +102.32 \qquad (1)$$

In addition, A, B and C represented separation speed; squeegee speed and squeegee load, respectively. H represented the solder height. Compared experimental data with calculated result, the deviation of solder height was less than 5%.

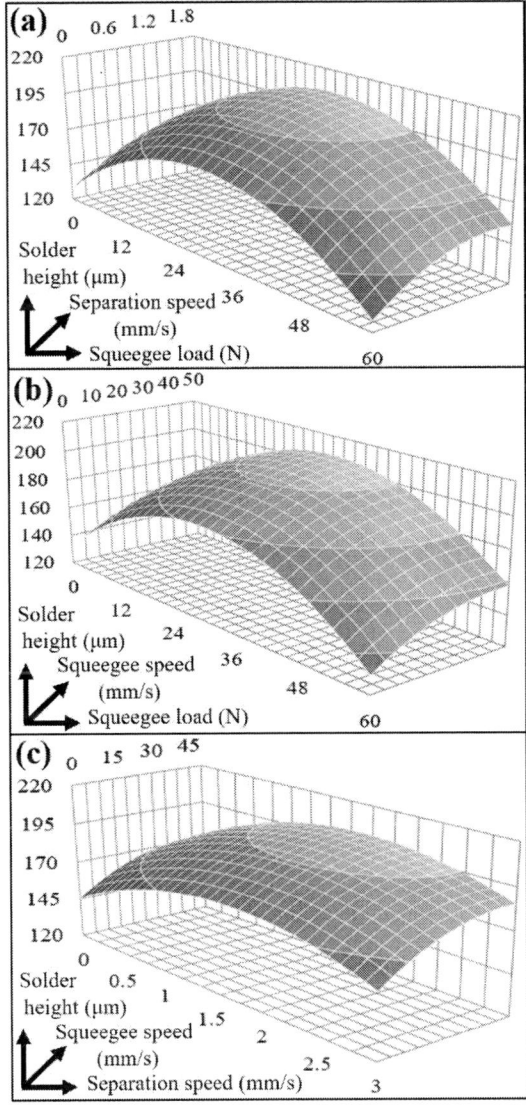

Fig. 5. Relation between solder height with two-factor: (a) separation speed and squeegee load; (b) squeegee speed and squeegee load; (c) Squeegee speed and separation speed

E. Length-width Ratio

The printing process, which was mentioned above, was analyzed with length-width of 1.9. The highest solder height was achieved by proper squeegee load, squeegee speed and separation speed. In this section, these parameters were employed on studying the relation between solder height with various length-width ratio. The curve, shown in Fig. 6, was simulated by experiment data. The solder height increased with improvement of length-width ratio. The peak of solder height was 215 μm, and the length-width was 1.24 at this time. When the length-width was above 1.24, the solder height gradually decreased. At the stage of rising, the solder height increased significantly. At the stage of declining stage, the solder height decreased slowly firstly and then quickly. Besides, the relationship of length-width ratio and solder height was established with quadratic trinomial, as in:

$$H=-129.24D^4+888.13D^3-2242.5D^2+2447.9D-760.09 \quad (2)$$

In addition, D represented length-width ratio, and H represented the solder height.

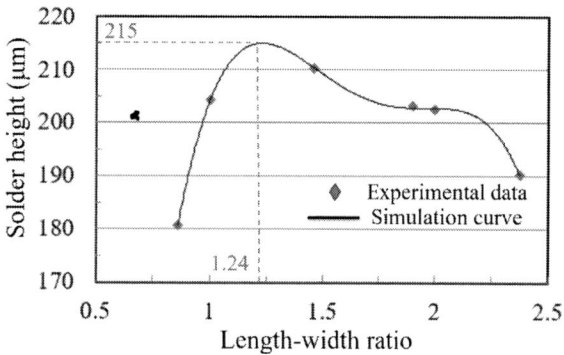

Fig. 6. Curves between solder height and length-width ratio

IV. CONCLUSION

SAC305 was graphically made on Al$_2$O$_3$ substrates with stencil printing. Based on experimental results, the relationship between the solder height and printing parameters was gradually studied. Under single factor of squeegee load, squeegee speed and separation speed, the solder height showed a trend of increasing firstly and then decreasing. The solder height, due to squeegee load, increased rapidly in the descending stage. On the contrary, the solder height which was influenced by squeegee speed and separation decreased rapidly in the ascending stage. Under the influence of two factor, the change of solder height was remarkably with separation speed and squeegee load. Besides, the polynomial of multiple factors was also fitted. This results would be instructive to the stencil printing technology in the future.

ACKNOWLEDGMENT

The authors gratefully acknowledge the support of Beijing Institute of Technology. We also wish acknowledge Chongqing University for providing simulation for this work.

REFERENCES

[1] Dezhi Su, Dan Zhao, Lejun Zhang, Huihui Yang, Cen Wang, Wenyu Jiang, "Reliability assessment of flip chip interconnect electronic packaging under thermal shocks", 21st International Conference on Electronic Packaging Technology, vol. 9202965, 2020.

[2] Dezhi Su, Peijie Guan, Dan Zhao, Qinghua Luan, Cen Wang, Yucheng Niu, "Welding reactions of lead free solder alloy for aluminum packaging electronic devices", 21st International Conference on Electronic Packaging Technology, vol. 9202503, 2020.

[3] Dezhi Su, Peijie Guan, Dan Zhao, Yucheng Niu, Quanwen Wang, Shangzhi Wang, Huihui Yang, Cen Wang, "Development of flip chip solders joint defects under temperature cycling testing", 21th International Conference on Electronic Packaging Technology, vol. 9081119, 2019.

[4] M. S. Rusdi, M. Z. Abdullah, S. Chellvarajoo, M. S. Abdul Aziz, M. K. Abdullah, P. Rethinasamy, Sivakumar Veerasamy, Damian G. Santhanasamy, "Stencil printing process performance on various aperture size and optimization for lead-free solder paste", Int J Adv Manuf Techno, vol. 102, pp. 3369-3379, 2019.

[5] Nourma Khader, Sang Won Yoon, "Stencil printing process optimization to control solder paste volume transfer efficiency", IEEE Transactions on Components, Packaging and Manufacturing Technology, vol. 8, pp. 1686-1694, 2018.

[6] Tsung Nan Tsai, "A knowledge-based system for stencil printing process planning and control", Journal of the Chinese Institute of Industrial Engineers, vol. 24, pp. 513-521, 2007.

[7] Joachim Kloeser, Katrin Heinricht, Erik Jung, Liane Lauter, Andreas Ostmann, Rolf Aschenbrenner, Herbert Reichl, "Low cost bumping by stencil printing: process qualification for 200 μm pitch", Microelectronics Reliability, vol. 40, pp. 497-505, 2000.

[8] Robert W. Kay, Gerard Cummins, Thomas Krebs, Richard Lathrop, Eitan Abraham, Marc Desmulliez, "Statistical analysis of stencil technology for wafer-level bumping", Soldering & Surface Mount Technology, vol. 26, pp. 71-78, 2014.

High strength and density Cu-Cu joints formation by low temperature and pressure sintering of different mass ratio of Cu micron-nanoparticles paste

Zhongwei Huang[1#]
[1]State Key Laboratory of Precision
Electronic Manufacturing Technology
and Equipment
Guangdong University of Technology
Guangzhou, P.R China
3111238543@qq.com

Jian Wen[2#]
[2]Jihua Laboratory
Foshan, P.R China
wenjian@jihualab.com

Yu Zhang[1*]
[1]State Key Laboratory of Precision
Electronic Manufacturing Technology
and Equipment
Guangdong University of Technology
Guangzhou, P.R China
zhangyu@gdut.edu.cn

Qiang Liu[1]
[1]State Key Laboratory of Precision
Electronic Manufacturing Technology
and Equipment
Guangdong University of Technology
Guangzhou, P.R China
892636543@qq.com

Huacong Li[1]
[1]State Key Laboratory of Precision
Electronic Manufacturing Technology
and Equipment
Guangdong University of Technology
Guangzhou, P.R China
1079534868@qq.com

Jin Tong[1]
[1]State Key Laboratory of Precision
Electronic Manufacturing Technology
and Equipment
Guangdong University of Technology
Guangzhou, P.R China
2111712327@qq.com

Peilin Liang[1]
[1]State Key Laboratory of Precision
Electronic Manufacturing Technology
and Equipment
Guangdong University of Technology
Guangzhou, P.R China
724531474@qq.com

Guannan Yang[1]
[1]State Key Laboratory of Precision
Electronic Manufacturing Technology
and Equipment
Guangdong University of Technology
Guangzhou, P.R China
ygn@gdut.edu.cn

Chengqiang Cui[1*]
[1]State Key Laboratory of Precision
Electronic Manufacturing Technology
and Equipment
Guangdong University of Technology
Guangzhou, P.R China
cqcui01@qq.com

Abstract—**High strength Cu-Cu interconnections of direct bonded copper substrate are achieved by low-temperature and low-pressure sintering of Cu micron-nanoparticles mixed in different mass ratio. The copper micronparticles with a particle size of 1μm and copper nanoparticles with a particle size of 200 nm were mixed in a certain mass ratio (for example, the mass ratio is 0/10, 1/9, 3/7, 5/5, 7/3, 9/1, 10/0). A certain amount of copper micronparticles was added to the Cu nanoparticles solution. After mixing and vacuum homogenization, a copper micron-nanoparticles paste is obtained. When sintering at the temperature of 260 °C , a sintering pressure of 2 MPa and a reduced atmosphere of 5 % hydrogen and 95% argon for 30 minutes, the interconnection joints were obtained. The shear test showed the interconnection strength of the 7 different mass ratio copper pastes. Among them, the sintering strength of the copper paste with a mass ratio of 3/7 is 43.66 MPa. In a conclusion, the microstructure analysis demonstrated that the Cu-Cu joints were formed between the copper paste and the pad. Analyzing the shear section, the fracture layer of the copper paste was in a tensile fracture state. The copper paste with a mass ratio of 3/7 had the best sintering performance realizing the dense sintering of the copper paste at low temperature and low pressure, and the shear strength can reach 43.66 MPa.**

Keywords—Mass ratio ; Cu micron-nanoparticles paste; shear strength; sintering

I. INTRODUCTION

Electronic technology is developing rapidly. The goal of electronic component packaging and interconnection is to achieve stable electrical interconnection, thermal connection and mechanical connection with sufficient strength. Stable packaging and interconnection ensure that electronic components work better [1]. The development of silicon carbide and gallium nitride high-temperature devices has enabled power electronic modules to be used at high temperatures. As traditional interconnect materials, solder paste and conductive glue can no longer meet the requirements of high-power devices. The above problems put forward higher requirements for packaging and interconnection (1) Ensure reliable mechanical connection. (2) Ensure high electrical conductivity to achieve high-speed transmission between the substrate and the chip. (3) Ensure high thermal conductivity to realize the heat dissipation of the chip and avoid the chip from being burnt due to excessive temperature [2]. (4) Low-temperature interconnection to avoid large thermal stress caused by different thermal expansions of substrates, chips and interconnecting materials due to excessive temperature changes. Nano-metal materials with special physical and chemical properties have become a research hotspot in packaging interconnect materials. Among them, copper is a good heat dissipation material and has good electrical conductivity. Its nano-particle material also has a small size effect. The smaller the particle diameter , the lower the sintering temperature [3]. Micron-nano copper metal paste has become a research hotspot due to its advantages of low cost, easy access to materials, and low electrical mobility. The existing preparation methods of micron-nano copper paste are usually liquid-phase reduction method, electrolysis method, mechanical grinding method and explosion method [4]. Among them, the most widely used is the liquid-phase reduction method to prepare micron-nano copper. The preparation method of liquid phase micron-nano copper paste generally includes the following steps (1) mixing the precursor copper source, reducing agent, dispersant or protective agent in the liquid phase to obtain the liquid phase of micron-nano copper particles. (2) Separate liquid phase and micron-nano copper particles. (3) Prepare micron-nano copper paste. However, the micron-nano copper has the disadvantage of being easily oxidized. In the process of preparing the micron-nano copper paste by liquid-phase reduction, Step (2) includes multiple cleaning and separation of the micron-nano copper particles. This step will cause the nano-copper to be exposed to the air multiple times, which intensifies the oxidation of the micron-nano copper. Although nano-copper particles can already diffuse at lower temperatures, forming Cu-Cu metal bonds [5]. However, this temperature is still too high compared to the traditional solder paste interconnection temperature. During the interconnection process, due to the different thermal expansion coefficients of the substrate, the chip and the interconnecting material, too high interconnection temperature will cause the three to expand to different degrees and form greater stress after cooling. During the working process of the chip, multiple thermal cycles will cause fatigue fracture and cause interconnection failure.

In the traditional electronic packaging interconnection process, it is usually necessary to combine the interconnection material with the substrate and the chip under a higher temperature and a certain pressure. In order to take advantage of the nano-copper particles while avoiding its shortcomings, this paper proposes a method for preparing different mass ratio(micron/nano) of copper micron-nanoparticles paste, with the copper acetate as precursor, ascorbic acid as reducing agent. The prepared copper micron-nanoparticles paste has a certain degree of antioxidant properties at room temperature. The weight loss of copper paste heated in argon was detected by thermogravimetric experiment. The copper micron-nanoparticles paste was made to samples sintering at a sandwich structure of copper substrate/copper paste/copper substrate under the condition of 260℃, pressure of 2 MPa, and 5% hydrogen + 95% argon atmosphere to obtain an interconnect structure. The shear strength of samples was detected by shear experiment. The fracture microstructure of the cross-section of the sintered samples after the shear test was charactered by field emission scanning electron microscope (FE-SEM).

II. EXPERIMENTAL

A. Materials

Cupric Acetate Monohydrate (Cu(CH$_3$COO)$_2$·H$_2$O), Ascorbic Acid (C$_6$H$_8$O$_6$) and 1μm copper particles were obtained from Shanghai Aladdin Biochemical Technology Corporation. Absolute ethyl alcohol were the product of Sinopharm Chemical Reagent Corporation.

B. Preparation of copper micron-nanoparticles and paste

In a synthetic procedure, Copper acetate is used as precursor and ascorbic acid as reducing agent. The first step is the synthesis of nano-copper particles: First, 0.2M of ascorbic acid and 0.05M of Cu(CH$_3$COO)$_2$·H$_2$O were each dissolved in two 50 ml of Ethylene glycol (EG) in the flask which was heated to 61.5℃ in water bath with mechanical stirring. Next, the solution of ascorbic acid (0.2M) was added into the solution of Cu(CH$_3$COO)$_2$·H$_2$O (0.05M) while mechanical stirring. After 15min of reaction, the color of mixed solution gradually changed from yellow to dark brown finally and the copper nanoparticles solution was obtained. The second step is the preparation of the copper micron-nanoparticles mixed solution: First a certain quality(different amount of micron-copper particles corresponded to different mass ratio) of micron-copper particles were obtained, then washed by 10% dilute sulfuric acid (H$_2$SO$_4$) for the first time, deionized water for the second time, and finally dispersed by Ultrasound treatment in Ethylene glycol (EG) and the copper micronparticles solution was obtained. Next, the solution of copper micronnparticles was added to the solution of copper nanoparticles and the solution of copper micron-nanoparticles was obtained. Third, the mixed solution was treated by ultrasound treatment for 5 mins and the mixed copper micron-nanoparticles solution was obtained. The third step is the preparation of micron-nano copper paste: When the solution temperature was closed to room temperature, the mixed particles were separated by high-speed centrifugation and the preliminary micron-nano copper paste is obtained. Then dried in a vacuum oven for 30 min, and finally a copper micron-nanoparticles paste was obtained.

C. Sintering of copper micron-nanoparticles paste

The copper micron-nanoparticles paste was made to samples sintering at a sandwich structure of copper substrate/copper paste/copper substrate. Then the copper micron-nanoparticles paste was uniformly coated on a copper block of 10 mm×10 mm with a steel screen printing plate of 4 mm×4 meshes, and a small copper block of 4 mm×4 mm was placed on the copper nanoparticles paste. The sample was sintered at a temperature of 260℃, a pressure of 2 MPa, atmosphere of 5% hydrogen + 95% argon, and holding 30 min at the sintering temperature to obtain the sintered Cu-Cu bonded sample.

D. Charaterization

The prepared copper micron-nanoparticles and the sintered Cu- Cu bonded samples were characterized. The microstructure of the prepared copper micron-nanoparticles was observed by field emission scanning electron microscope (FE-SEM). The properties and composition of copper micron-nanoparticles paste were obtained by X-ray diffractometer (XRD), the weight loss under argon atmosphere of copper micron-nanoparticles paste was measured by high temperature synchronous thermal analyzer, and the shear force of sintered sample was measured by IC package welding strength tester.

III. RESULTS AND DISCUSSION

A. Morphology of Cu nanoparticles

Figure 1 (a, b, c, d, e, f, g corresponded to the mass ratio(micron/nano) of 0/10, 9/1, 3/7, 5/5, 7/3, 9/1, 10/0)shows the scanning electron microscopy images of different mass ratio of copper micron-nanoparticles. Figure 1(h) shows the properties and composition of copper micron-nanoparticles paste of the mass ratio of 3/7 by X-ray diffractometer (XRD). It can be seen in Figure 1(c) that the micronparticles and the nanoparticles are evenly mixed, micronparticles are evenly distributed in the nanoparticles. Figure 1(g) shows that the morphology of copper micronparticles is spherical-like. The size distribution of the particles is between 800nm and 1100nm. Figure 2 shows the XRD patterns of the copper micron-nanoparticles paste placed in air for 3 days. The three diffraction angles 43.3°, 50.4°, 74.1°in Figure 2 show that the copper micron-nanoparticles in the paste is pure copper and is not oxidized.

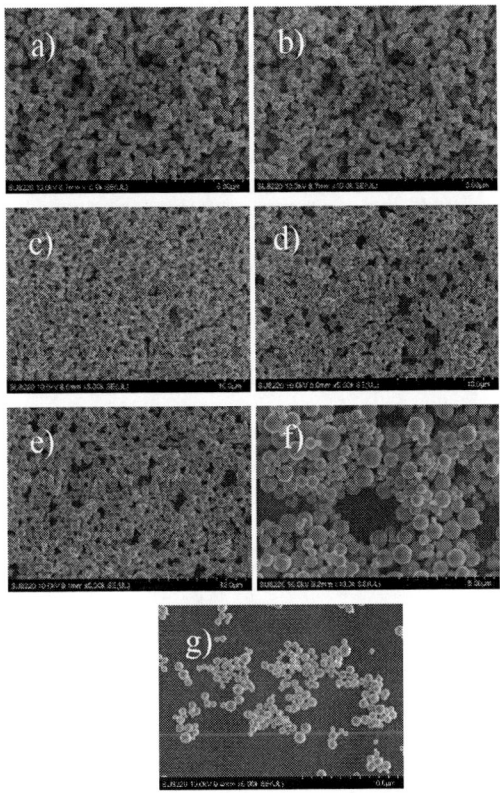

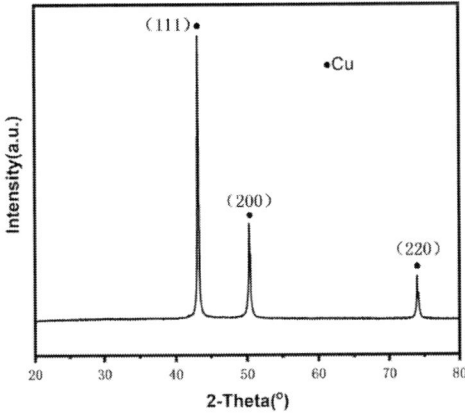

Fig.1 The figure shows the SEM images of different mass ratio(micron/nano) of copper micron-nanoparticles (a, b, c, d, e, f, g corresponded to the mass ratio of 0/10, 9/1, 3/7, 5/5, 7/3, 9/1, 10/0)

Fig.2 The XRD patterns of copper micron-nanoparticles paste in air for 3 days.

B. Bondability of Cu micron-nanoparticles paste

The prepared copper micron-nanoparticles paste was prepared for sintering of Cu-Cu joints sample. The sintering conditions were set as sintering pressure 2MPa, heating rate 5℃/min, maximum temperature 260℃, heat preservation 30 min, sintering atmosphere was air and 5% H_2+ 95%Ar. The sintered samples were tested by shear force to analyze the bonding properties after sintering of copper micron-nanoparticles paste of different mass ratio. It can be seen from figure 3 that the shear strength after sintering of the mass ratio of 3/7 copper micron-nanoparticles is the highest reached to 43.66Mpa.

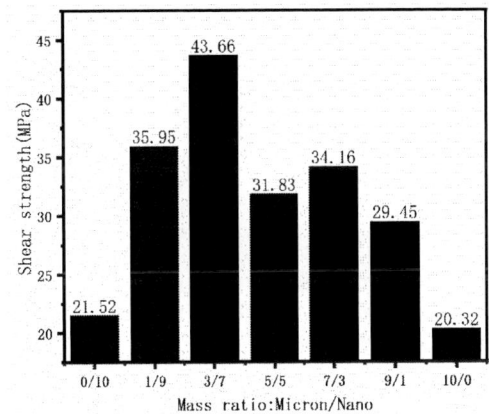

Fig.3 The shear strength of Cu-Cu joints using copper micron-nanoparticles paste.

C. Thermogravimetric analysis of Cu micron-nanoparticles paste

Figure 3 shows the TG curves of copper micron-nanoparticles paste of the mass ratio of 3/7 in N_2 atmosphere. The copper micron-nanoparticles paste of the mass ratio of 3/7 is composed of micron-nanoparticles, residual Ascorbic Acid ($C_6H_8O_6$) and ethylene glycol. From the observation of the weight loss curve in the Figure 4 , it can be seen that from 25°C to 150°C, the mass loss of copper paste reached to 17%, and from 150°C to 315°C, the mass loss of copper paste is 1%, with a very little mass loss. From 315°C to 600°C, there is almost no quality loss in copper paste. It can be seen from the endothermic curve in Figure 4 that there is an obvious endothermic peak at about 130° C. Analyzing the weight loss curve in the Figure 4, the mass loss of copper paste at 25℃-150℃ is the volatilization of ethylene glycol and the part of Ascorbic Acid ($C_6H_8O_6$) taken away by the evaporation of ethylene glycol. The mass loss at 150℃-315℃ is 1%, and this part of the mass loss is the decomposition of residual Ascorbic Acid ($C_6H_8O_6$) in the paste. At 315℃-600℃, there is no obvious quality loss of copper paste, and only copper is left in the composition of copper paste. Analyzing the Heat Flow Curve, there is an obvious heat absorption peak at 130°C, which represents the volatilization of ethylene glycol, which is consistent with the weight loss curve of copper paste.

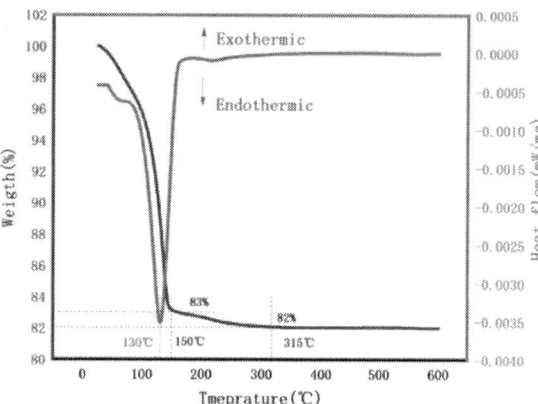

Fig.4 The TG curves of copper micron-nanoparticles paste of the mass ratio of 3/7 in N_2 atmosphere .

D. Microstructure of sintering layer

Figure 6 (a, b, c, d, e, f, g corresponded to the mass ratio(micron/nano) of 0/10, 9/1, 3/7, 5/5, 7/3, 9/1, 10/0)shows the microstructure of the cross-sectional view of the sintered samples prepared with different mass ratios of copper paste after the shear test. Figure 6(a) shows the microstructure at the fracture interface, it can be seen that the sintering necks formed between the nanoparticles, the diameter of the sintering necks is closed to the nanoparticles', indicating that strong atomic diffusion occurs between the nanoparticles, due to the sintering of nanoparticles, the nanoparticles grown bigger and meanwhile the pore of the sintering layer grown as well. It can be seen in Figure 6, as the mass ratio change the density of the sintering change as well. In the sintering process of the copper paste, the large-size particles serve as the interconnected skeleton, providing high shear strength, the small-sized particles act as a binder to connect the large-sized particles to form a complete and dense interconnection. Meanwhile the nano particles released energy when sintering, providing energy for the sintering of micronparticles. It can be seen in Figure 5(c) that the copper paste form a dense structure due to close arrangement of micron and nanoparticles.

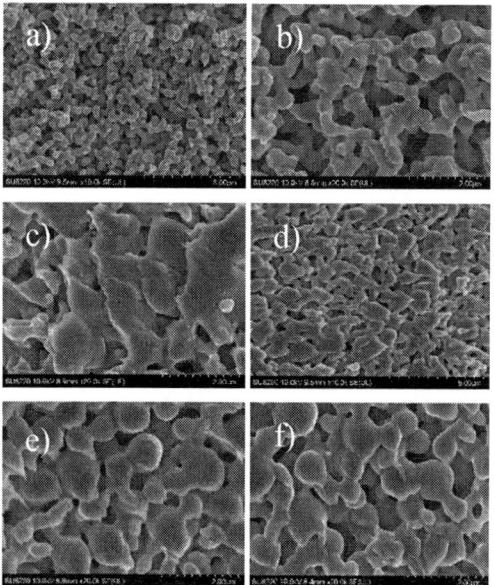

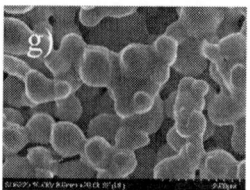

Fig.5 The figure (a, b, c, d, e, f, g corresponded to the mass ratio(micron/nano) of 0/10, 9/1, 3/7, 5/5, 7/3, 9/1, 10/0) shows the microstructure of the cross-section of the sintered samples after the shear test.

Figure 6 (a1, a2, a3, a4, a5, a6, a7,) shows the microstructure of the sintered layer of the cross-section of the copper paste with different mass ratios after the shear test. Figure 6 (b1, b2, b3, b4, b5, b6, b7) shows the corresponding pictures processed by ImageJ for statistics of the porosity of the sintered layer. In the Figure of porosity analysis, the black area represents the copper particles that have been sintered and connected, and the white area represents the pores of the sintered layer. Figure 6 (c) shows the porosity of samples prepared by the copper paste of different mass ratios. It can be seen in the Figure 6 (c) that the lowest porosity of the samples is sample prepared by the copper paste of the mass ratio of 3/7 reached to 8.45%.

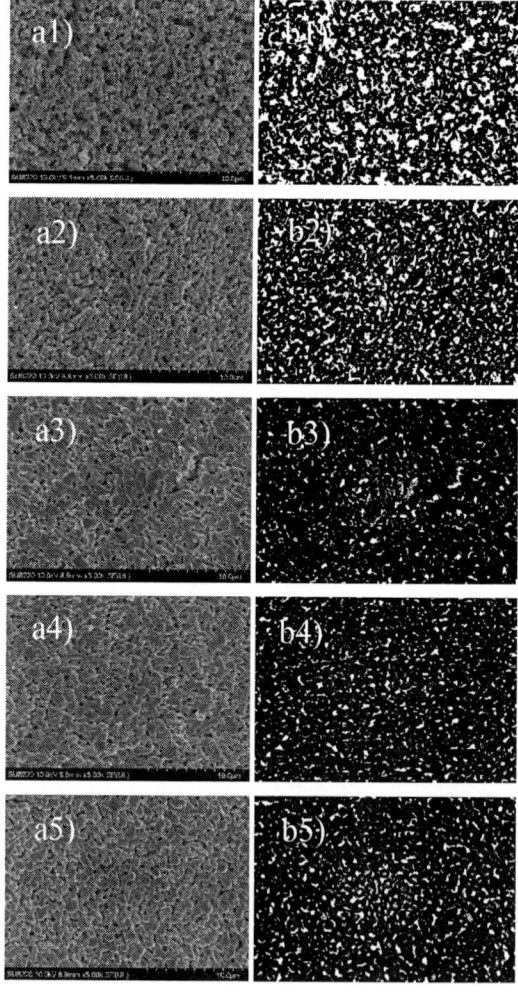

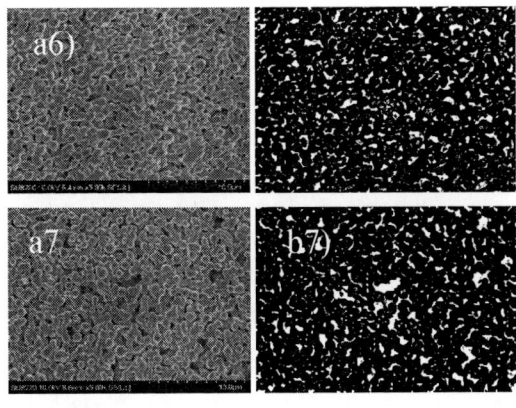

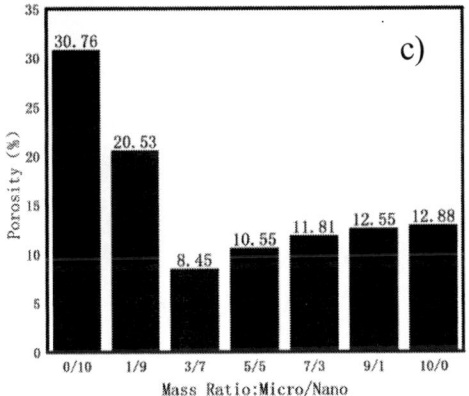

of the copper substrate/copper paste/copper substrate sandwich structure was studied. When the mass ratio is 3/7, the shear strength can be as high as 43.66 MPa. The sintering mechanism of the copper micro-nanoparticles paste of different mass ratios is proposed. Under the sintering condition of heat and pressure, the sintering neck is formed by the pre-melting of small-sized copper nanoparticles around the copper micron particles. The micro and nano particles and the substrate form a conductive network. The sintering of the nano particles released energy along with the heat and pressure causing the sintering neck to continue to grow and form a larger size sintering neck to realize the sintering of the entire copper paste and the connection with the copper substrate.

\# The authors contributed equally to this work.

ACKNOWLEDGMENT

This work was partially supported by the Guangdong Basic and Applied Basic Research (2021A1515011642) and the National Key R&D Program of China(2018YFE0204601).

REFERENCES

[1] Liu J , Chen H , Ji H , et al. Highly Conductive Cu-Cu Joint Formation by Low-Temperature Sintering of Formic Acid-Treated Cu Nanoparticles[J]. Acs Applied Materials & Interfaces, 2016:33289.

[2] Y Zuo, J Shen, J C Xie, et al. Influence of Cu micro/nano-particles mixture and surface roughness on the shear strength of Cu-Cu joints[J]. Journal of Materials Processing Technology. 2018, 257: 250-256.

[3] Mou Y , Peng Y , Zhang Y , et al. Cu-Cu Bonding Enhancement at Low Temperature by Using Carboxylic Acid Surface-Modified Cu Nanoparticles[J]. Materials Letters, 2018, 227:179-183.

[4] Jung K H , Min K D , Lee C J , et al. Pressureless die attach by transient liquid phase sintering of Cu nanoparticles and Sn-58Bi particles assisted by polyvinylpyrrolidone dispersant[J]. Journal of Alloys & Compounds, 2019, 781:657-663.

[5] Yu Zhang, Pengli Zhu, Gang Li, Zhen Cui, and et al. "PVP-Mediated Galvanic Replacement Synthesis of Smart Elliptic Cu-Ag Nanoflakes for Electrically Conductive Pastes," ACS Appl. Mater. Interfaces 2019;

11, 8382−8390.

Fig.6 The figure shows (a1, a2, a3, a4, a5, a6, a7 corresponded to the mass ratio(micron/nano) of 0/10, 9/1, 3/7, 5/5, 7/3, 9/1, 10/0) the microstructure of the cross-section of the sintered samples, (b1, b2, b3, b4, b5, b6, b7 corresponded to the mass ratio(micron/nano) of 0/10, 9/1, 3/7, 5/5, 7/3, 9/1, 10/0) corresponding pictures processed by ImageJ for statistics of the porosity of the sintered layer. (c) the porosity of samples prepared by the copper paste of different mass ratios.

In terms of the sintering energy and arrangement of micron and nanoparticles, copper micron-nanoparticles paste of the mass ratio of 3/7 has the best performance that the sintering sample reached the shear strength of 43.66Mpa and the porosity of 8.45%

IV. CONCLUSIONS

A copper micro-nanoparticles paste was prepared, taking the advantages of copper micronparticles and copper nanoparticles. The size distribution of copper nanoparticles is 200 nm±50 nm. The effect of different mass ratios of micro/nano particles on the shear strength and microstructure

Reliability assessment in welding process of SiP with dual-chamber by finite element analysis

Dezhi Su
Shandong Institute of Space Electronic Technology
Yantai, China
sudezhihefish@126.com

Fuxin Wang
Shandong Institute of Space Electronic Technology
Yantai, China
wangfuxin0328@126.com

Lejun Zhang
Shandong Institute of Space Electronic Technology
Yantai, China
zhanglejun7583@163.com

Cen Wang
Shandong Institute of Space Electronic Technology
Yantai, China
wangcen@126.com

Huihui Yang
Shandong Institute of Space Electronic Technology
Yantai, China
Yanghuihui@126.com

Wenyu Jiang
Shandong Institute of Space Electronic Technology
Yantai, China
jiangwenyu3728@163.com

Abstract—In this paper, a state-of-the art SiP with dual-chamber was established. The new structure contained bumps, chips, ceramic substrate, pins and cover plates. The deformation, principal stress and principal strain of all parts in welding process were analyzed by computer simulation and prediction techniques. The largest deformation, which was 95.97 μm, occurred in the ceramic substrate. The maximum value of principal stress was 653.59 MPa. It happened in the bumps of flip chip and exceeded the yield strength of SnAg. The maximum value of principal strain was 0.89% and it also occurred in the bumps of flip chip. The weakness of SiP in welding process was founded and it was very meaningful for future research.

Keywords—SiP; deformation; principal stress; principal strain

I. INTRODUCTION

In recent years, electrical devices vastly emerged with life style changing and cognition improving. The characteristics involved miniaturization, integration, multi-functionalization and high reliability [1-3]. Correspondingly, the technology of electrical packaging went through four stages- wire bonding, tape automated bonding, flip chip and through-silicon via [4]. The dimension decreased from millimeter to micrometer. In the post- moore law period, electrical system was divided into two directions- SoC (system on chip) and SiP (system in packaging).

SoC could appropriately realize the miniaturization of a circuit. Compared with SoC, research and development of SiP spent less time and cost [5]. SiP not only reduced the space of electrical system, but also realized the total cost of reduction. It divided the complicated circuit into modularized units with more flexibility. SiP was commonly used in the field of cell phone, smart watch, car, medical equipment and so on. The applications were very mature and commercialized. SiP made an electrical system into a packaging body. Generally, the system contained many kinds of chips, such as CPU (Central Processing Unit), DSP (Digital Signal Processing), DRAM (Dynamic Random Access Memory), FLASH and so on. With the development of electrical packaging, 3D stacking gradually became the mainstream. They mostly adopted plastic packaging form. However, requirements of excellent airtightness and stringent environment were the crucial factors that restricted the improvement of plastic packaging [6]. It was essential to design a structure of SiP for changing distinct disadvantages.

In this paper, a novel structure of SiP with ceramic packaging was firstly set up. The analyzed structures were divided into five parts- bumps, chips, cover plates, pins and substrate. The deformation, principal stress and principal strain of every part were discussed. The weak spots were pointed out, and suggestions were provided for high-risk area.

II. EXPERIMENT

At beginning, the model of SiP with dual-chamber was firstly built as shown in Fig. 1. Fig. 1a, 1b and 1c were the top, bottom and sectional view of SiP, respectively. The model not only considered accuracy and efficiency of calculation, but also optimized meshing and density distribution. The dimension of SiP was 70mm × 70mm ×15 mm. As shown in Fig. 1a, there were 3 parts including 11 chips, cover plate and heat sink on the top of ceramic substrate. The 11 chips could be divided into 4 kinds of wire bonding chip. The bottom of ceramic substrate was shown in Fig. 1b. It consisted of wire bonding chip, flip chip (FC), cover plate and many pins. The special structure could be discovered in Fig. 1c. The chip of the bottom surface was installed directly on the back of heat sink. Both of them realized with through hole in the middle of ceramic substrate. During the simulation process, the total number of calculated steps was 10^4. The maximal step-length was 600 s, and the minimal step-length was 1 μs. The material properties of SiP, analyzed in this paper, were shown in table 1.

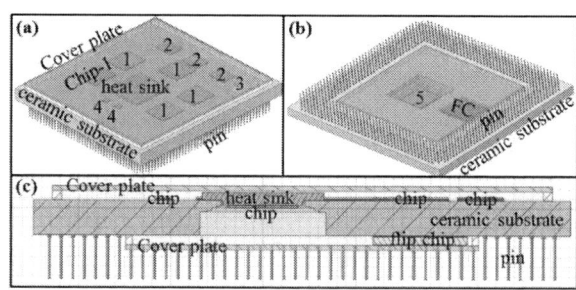

Fig. 1. (a) Top view, (b) bottom view, and (c) sectional view of SiP model

TABLE I. PARAMETERS OF MATERIAL PROPERTIES

Parts	CTE ($10^{-6} \cdot °C^{-1}$)	Temperature (°C)	Density (kg·m^{-3})
Chip/FC	2.8	25	2.3e3
Heat sink	7.4	25	2.9e3

Cover plate	7.8	25-500	8.1e3
Ceramic substrate	7.5	25	3.6e3
Pin	6.2	25-500	8.3e3
Bumps	25	25	8.4e3
	Elastic modulus (MPa)	Poisson's ratio	Flexural Strength (MPa)
Chip/FC	1.6e5	0.23	180
Heat sink	1.6e5	0.18	370
Cover plate	1.5e5	0.30	370
Ceramic substrate	3.1e5	0.27	305
Pin	1.4e5	0.31	330
Bumps	5.1e4	0.40	50

Constrained boundary was set in the middle of ceramic substrate on the bottom. The heat generated by welding was loaded into all nodes. Under the temperature load, this model in turn had 5 processes including preheating, heat preservation, heating up, welding and cooling. The detail of time and temperature was shown in Fig. 2. Besides, deformation, principal stress and principal strain were the significant parts in SiP-analyzing.

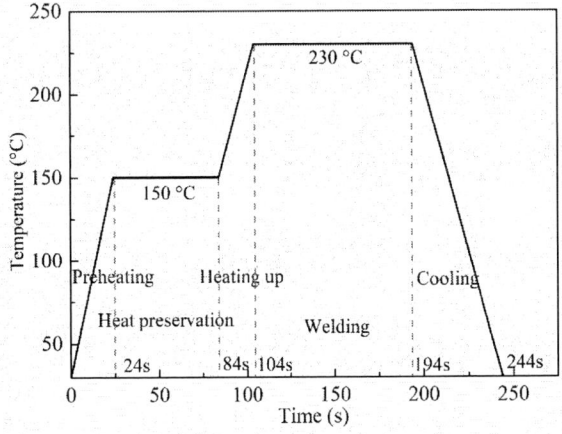

Fig. 2. The detail of time and temperature for temperature load

III. RESULTS AND DISSCUSSION

Fig. 3a, 3b and 3c elaborated the deformation, principal stress and principal strain of bumps, respectively. The trend of variation was consistent with Fig.2. From the illustrations, all of the maximum values occurred at the corner of array and in the state of welding. The maximal deformation was 24.59 μm which was 27.3% of bump's height. The maximum value of principal stress was 653.59 MPa. It exceeded the flexure strength (50 MPa) and might lead to reducing the reliability of bumps. Hence, the process of underfill was indispensable in the subsequent engineering practice [7-9]. The peak of principal strain was 0.89%.

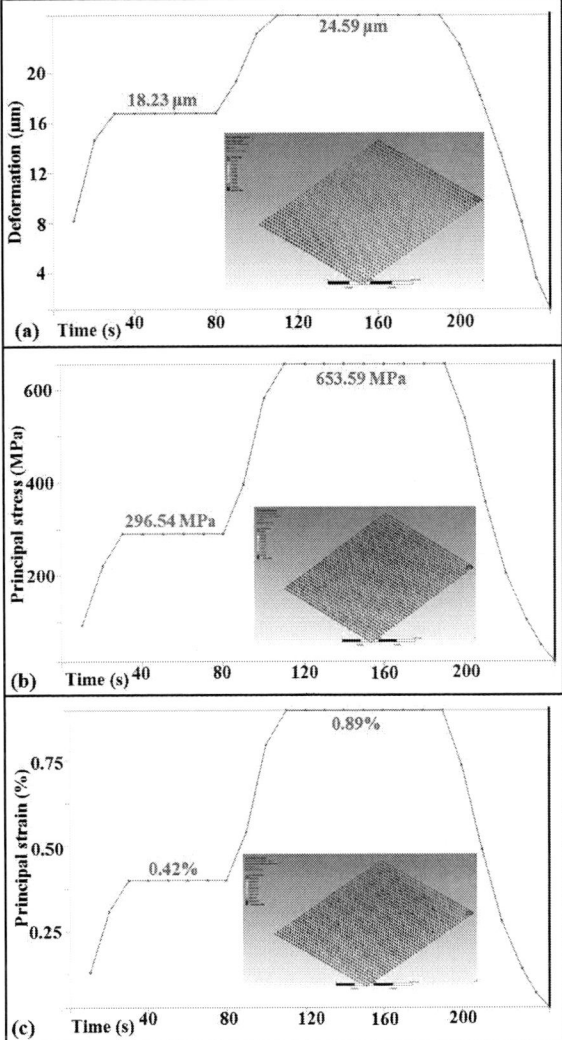

Fig. 3. Curves of bump with maximal value: (a) deformation, (b) principal stress and (c) principal strain. Illustrations were the variation of entirety at the welding process

Fig. 4a, 4b and 4c elaborated the deformation, principal stress and principal strain of chips, respectively. The trend of variation was consistent with Fig.2. Fig. 4 showed the maximal value was in the state of welding. From the illustrations, the deformation of maximum occurred with chip of No. 4, but principal stress and principal strain of maximum occurred at chip of No. 5. As shown in Fig. 4a, the maximal deformation was 59.29 μm which was 11.9% of chip's thickness. Just like Fig. 4b, the principal stress reached a peak of 147.9 MPa which did not exceed the flexure strength (180 MPa). The design of chips was reliable. As Fig. 4c showed, the principal strain was at its highest of 0.075%.

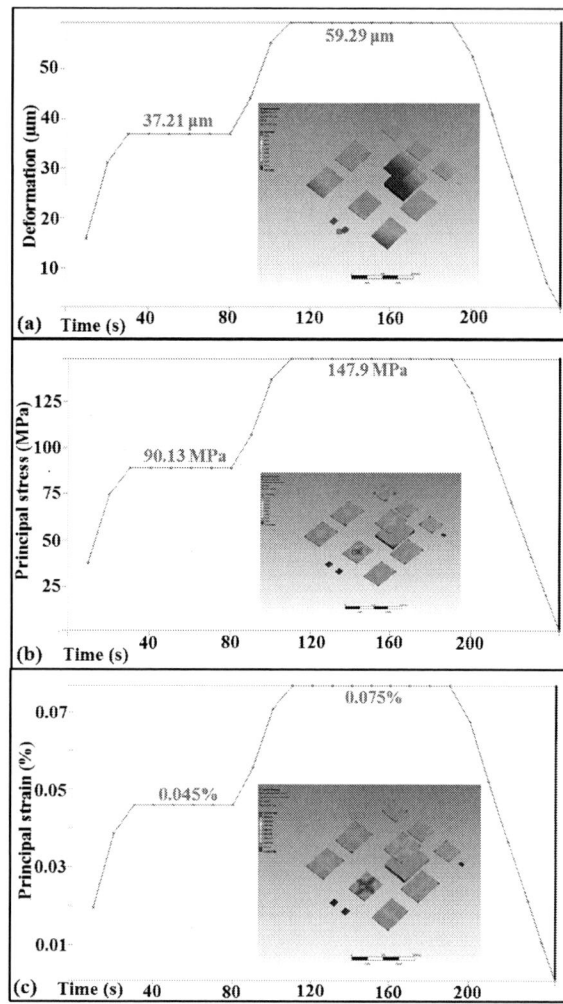

Fig. 4. Curves of chip with maximal value: (a) deformation, (b) principal stress and (c) principal strain. Illustrations were the variation of entirety at the welding process

Fig. 5a, 5b and 5c elaborated the deformation, principal stress and principal strain of cover plates, respectively. The trend of variation was consistent with Fig.2. All of the maximal values happened in the state of welding. As shown in Fig. 5a, the maximal deformation was 89.30 μm which was 6.38% of cover plate's thickness. From the illustrations, the deformation of maximum occurred at the edge of cover plate. Fig. 5b illustrated principal stress reached a peak of 107.37 MPa and it was immensely below flexure strength (370 MPa). The position was the middle of the cover plate on the top. Hence, the structure of substrate was reasonably. The principal strain was at its highest value of 0.07%, as shown in Fig. 5c.

Fig. 6a, 6b and 6c elaborated the deformation, principal stress and principal strain of pins, respectively. The trend of variation was consistent with Fig.2. All of the maximal values happened in the state of welding. Just like Fig. 6a, the deformation reached its highest level of 76.71 μm which was merely 1.39% of pin's height. From the illustrations, the deformation of maximum occurred at the corner of pins. Fig. 6b demonstrated principal stress was at its highest of 17.83 MPa and it was immensely below flexure strength (330 MPa). The position was near the corner of pins. Hence, the structure

of substrate was safety. The principal strain was at its highest value of 0.01%, as shown in Fig. 6c.

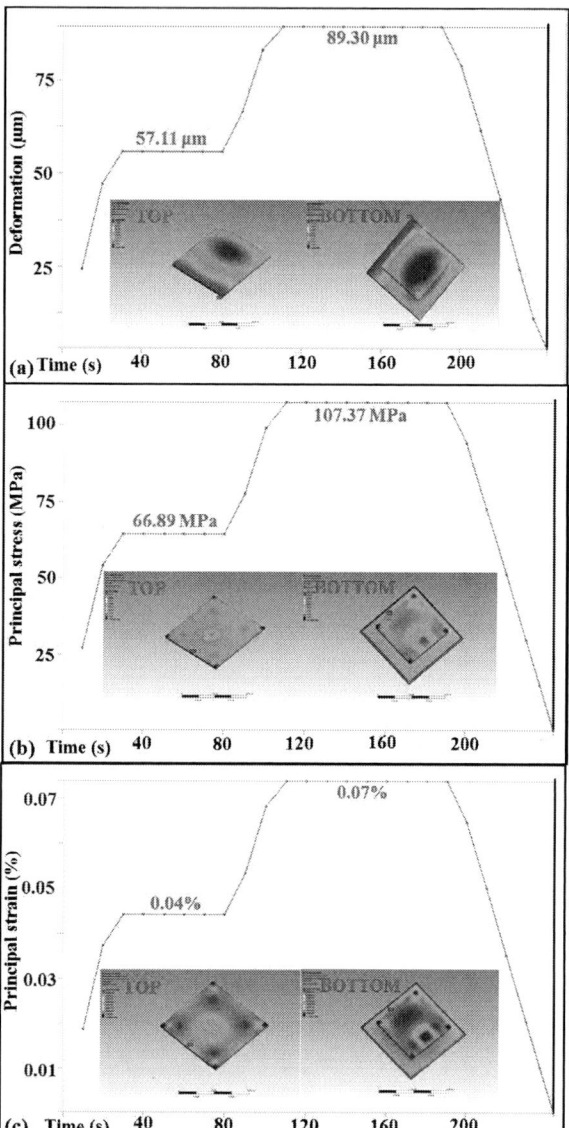

Fig. 5. Curves of cover plate with maximal value: (a) deformation, (b) principal stress and (c) principal strain. Illustrations were the variation of entirety at the welding process

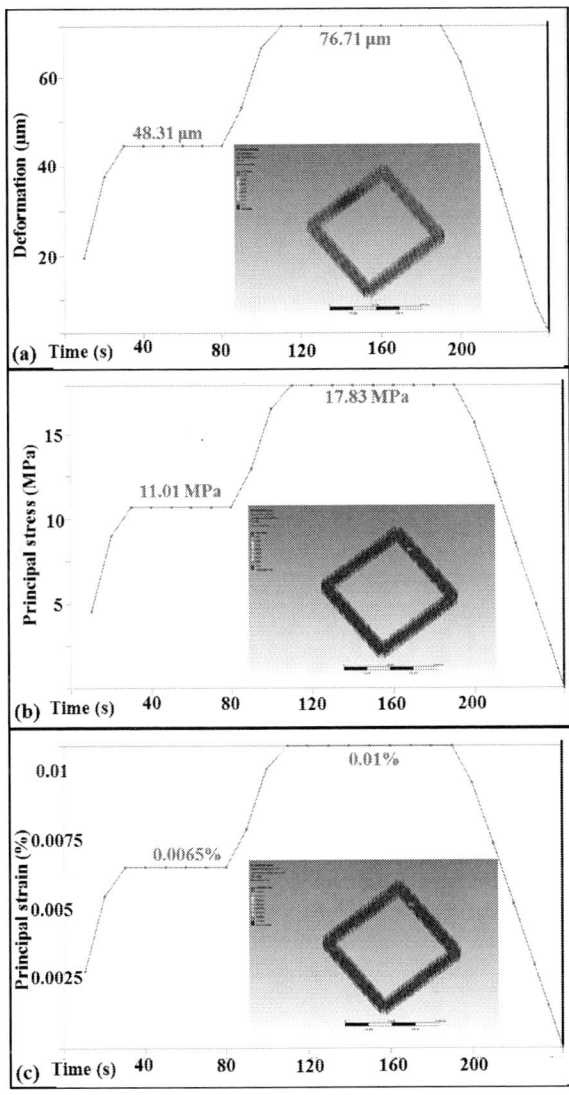

Fig. 6. Curves of pin with maximal value: (a) deformation, (b) principal stress and (c) principal strain. Illustrations were the variation of entirety at the welding process

Fig. 7a, 7b and 7c elaborated the deformation, principal stress and principal strain of substrate, respectively. The trend of variation was consistent with Fig.2. All of the maximal values happened in the state of welding. As shown in Fig. 7a, the maximal deformation was 95.97 μm which was 2.71% of substrate's thickness. From the illustrations, the deformation of maximum occurred at the edge near through-hole. The Fig. 7b shown demonstrated the highest of principal stress was 25.9 MPa and it was remotely less than flexure strength (305 MPa). The position was in the middle of the substrate. Therefore, the structure of substrate satisfied the design requirement. The principal strain reached its highest level of 0.01%, as shown in Fig. 7c.

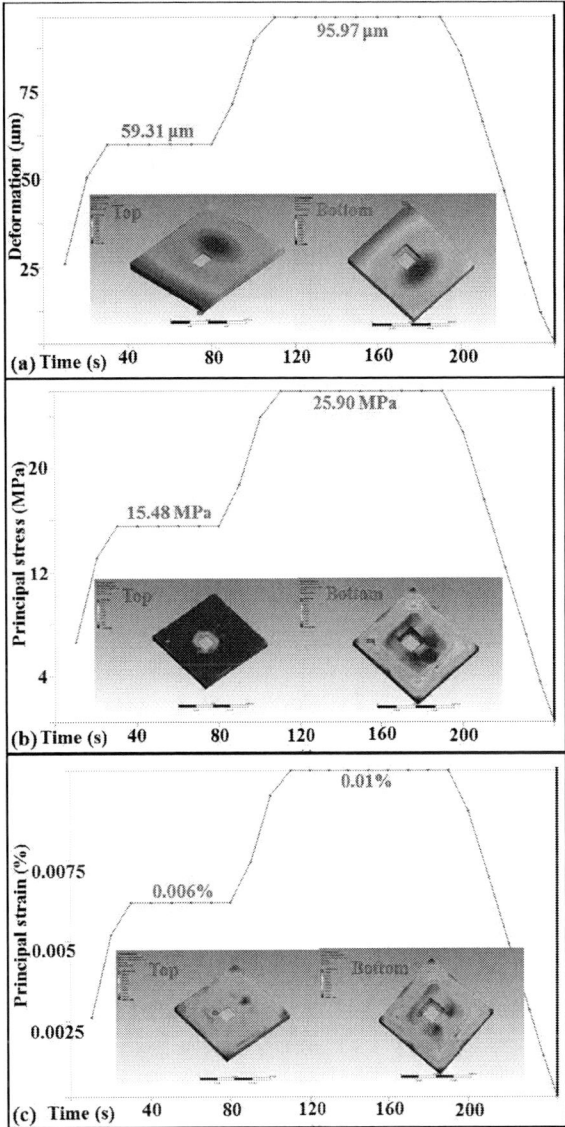

Fig. 7. Curves of substrate with maximal value: (a) deformation, (b) principal stress and (c) principal strain. Illustrations were the variation of entirety at the welding process

IV. CONCLUSION

In this paper, a state-of-the art SiP with dual-chamber was established and analyzed. The main structure included bumps, chips, substrate, cover plates and pins. The bumps had high-risk areas and should be further improved. The structure of through-hole not only appeared effective to future reduce the residual stress of ceramic substrate, but also provided another path of heat dissipation compared with conventional structure. The weakness of transformation, principal stress and principal strain was founded and it was helpful for physical design.

ACKNOWLEDGMENT

The authors gratefully acknowledge the support of Southeast University for providing simulation for this work.

REFERENCES

[1] Dezhi Su, Dan Zhao, Lejun Zhang, Huihui Yang, Cen Wang, Wenyu Jiang, "Reliability assessment of flip chip interconnect electronic

978-1-6654-1392-3/21 $31.00 © 2021 IEEE

packaging under thermal shocks", 21st International Conference on Electronic Packaging Technology, vol. 9202965, 2020.

[2] Dezhi Su, Peijie Guan, Dan Zhao, Qinghua Luan, Cen Wang, Yucheng Niu, "Welding reactions of lead free solder alloy for aluminum packaging electronic devices", 21st International Conference on Electronic Packaging Technology, vol. 9202503, 2020.

[3] Dezhi Su,Changcheng Wang, Dan Zhao, Yucheng Niu, Quanwen Wang, Peijie Guan, "Study of capillary tip states on the reliability of Al wire bonding in microelectronic package", 19th International Conference on Electronic Packaging Technology, vol. 8480546, 2018.

[4] S H B S Badri, M H A Aziz, N R Ong, Z Sauli, J B Alcain, V Retnasamy, "Ceramic ball grid array package stress analysis", 3rd Electronic and Green Materials International Conference, pp. 020291, 2017.

[5] De-Shin Liu, Chin-Yu Ni, Ching-Yang Chen, "Integrated design method for flip chip CSP with electrical, thermal and thermo-mechanical qualifications", Finite Elements in Analysis and Design, vol. 39, pp. 661-667, 2003.

[6] Kailin Pan, Peng Huang, Yu Guo, Tao Lu, Bin Zhou, "A novel packaging structure for high power LED based on chip on heat-sink method", International Conference on Information Technology and Applications, pp. 436-440, 2013.

[7] Subramanian N R, Koo Kok Kiat, Tye Ching Yun, "Non-linear deflection analysis of pin-on-package testing using FEA", 15th Electronics Packaging Technology Conference, pp. 410-414, 2013.

[8] Zhenyu Zhao, Guisheng Zou, Hongqiang Zhang, Hui Ren, Lei Liu, Y. Norman Zhou, "The mechanism of pore segregation in the sintered nano Ag for high temperature power electronics applications", Materials Letters, vol. 228, pp. 168-171, 2018.

[9] Shang-Te Tsai, Chi-Yu Lin, Sung-Mao Wu, Chung-Yao Chang, Cheng-Fu Yang, "Analyses and statistics of the electrical fail for flip chip packaging by using ANSYS simulation software and really underfill materials", Microsyst Technol, 2017.

On-Die Clock Tree Low PSIJ Through PDN Optimization

Vinod Arjun Huddar
Rambus Inc
Bangalore, India
vhuddar@rambus.com

Abstract— **On-Die Power Distribution Network (PDN) optimization for achieving low Power Supply Induced Jitter (PSIJ) on clock tree is put forth. A chip having a distribution of differently sized decoupling capacitors coupled to a distributed power supply conductor grid, where the sizing of the decoupling capacitors is selected based on the spatial locality of the noise contributed by individual clock buffers of a clock distribution network, to achieve controlled low PSIJ.**

Keywords—power integrity, jitter, clock tree, PSIJ, PDN

I. INTRODUCTION

Low PSIJ on clock trees is important for today's high-speed IO interfaces. Operating at Gbps data rates on parallel bus interfaces mandates very high-quality supply rails, which requires optimization of power supply network on silicon (Fig. 1). Power supply noise is a dominant source of clock jitter in high speed designs. As the data rate increases, smaller is the total jitter budget, thus reducing the supply noise induced jitter is goal of power integrity in high speed IO interface systems which calls for reduced ripple on clock buffer chain.

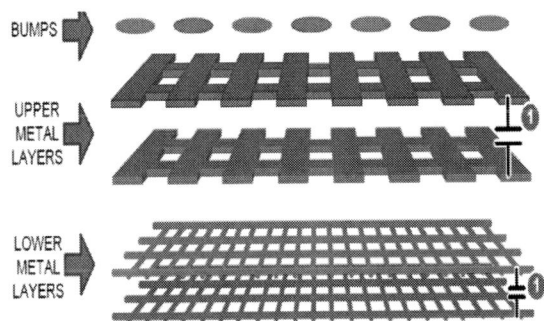

Fig. 1. Power supply network in silicon

Switching activity of on-die clock buffers results in high-frequency currents passing through PDN, thus generating voltage ripple. This ripple can be controlled using on-die capacitors (ODC). One of the major challenges in PSIJ of clock tree is, each buffer in clock tree has its own PSIJ which is not same and jitter of second buffer depends on first buffer which is driving it.

In clock buffer chain of cascaded buffers, second buffer is having two parameters which effects its jitter output i.e., its own supply jitter as well as first buffer supply jitter which is reflected in first buffer output. Both the buffers are powered from same supply, but they have different PDN network due to spatial location of buffers. This dependency on PSIJ of one buffer due to supply noise on other keeps cascading as the buffer chain increases.

II. LOW PSIJ APPROACH

Previous approaches to reduce PSIJ on clock tree were increasing ODC which resulted in overdesigning of on-die PDN and increased die area due to large number of ODC's. Few of the designs would end up with decoupling capacitors on package power grid resulting in package area increase and cost. PSIJ being random for each buffer is not controlled in previous typical approaches. Since PSIJ is random for each buffer, total PSIJ is also random. This will result in whole clock tree being analyzed for total PSIJ using circuit simulator running for days without controlling the randomness of PSIJ.

Power delivery network is mesh-like structure which can be modelled as distributed RLC network (Fig. 2). Every buffer represented as current sink in Fig. 2, sees different resistance and capacitance of PDN due to spatial location. PSIJ of each buffer can be made the same by controlling the PDN resistance and capacitance. PSIJ is decided by RC constant at the supply rail of buffer and transient current requirement. Repetitive buffers have same transient current requirements.

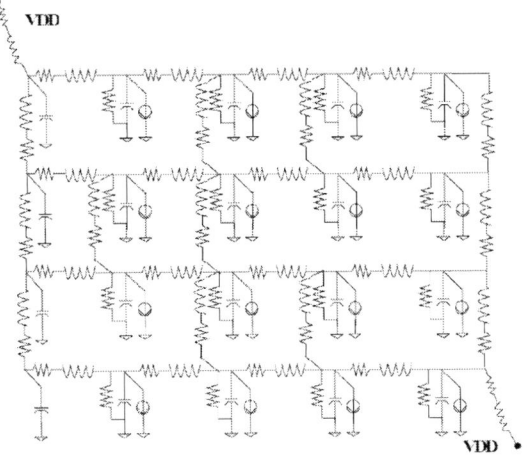

Fig. 2. Power supply mesh network distributed RLC model

RC constant of PDN for each buffer can be computed using commercial tools. Resistance of PDN is dependent on spatial location of buffer. ODC value can be modified so that ripple peak-to-peak on the supply pins of each buffer remains the same, thus making PSIJ variation across buffers to be very minimal. Sizing of the decoupling capacitors varies according to the distance of the respective clock buffer relative to the nearest voltage regulator, to reduce ripple variance between the buffers. This approach (Fig. 3) of different decoupling capacitor value for each buffer is different from typical design flow of sprinkling same decoupling capacitor value for all the buffers as part of layout design.

978-1-6654-1392-3/21 $31.00 © 2021 IEEE

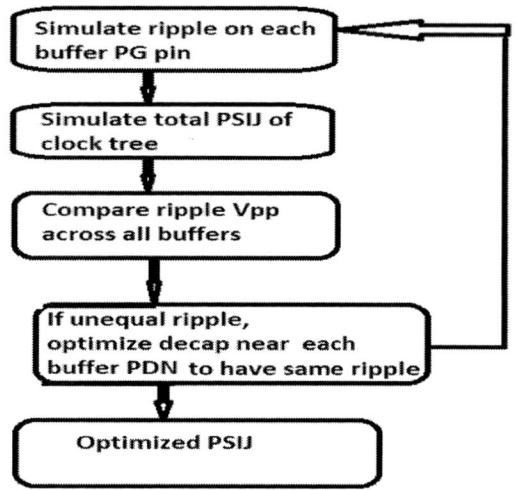

Fig. 3. High level flow chart for low PSIJ

ODC can be implemented as an array of moscap or MOM or MIM or MFC for each buffer or section of group of buffers. Array of caps can be selectable using logic implementation to plug-in only desired cap dynamically to keep the ripple on each buffer or section of group of buffers same. Each buffer PSIJ is now well controlled, thus overall PSIJ of clock tree is deterministic partially and can be kept very low by modifying capacitance. This enables very high-speed clock trees to be designed with well controlled jitter resulting in low PSIJ.

III. RESULTS

Fig. 3 flow chart is the basis for algorithm to identify the right decoupling capacitance value. To showcase results in simplified manner, two scenarios are considered for PDN optimization resulting in reduced jitter. First a two-buffer case with its own PDN is simulated with and without PDN optimization. Second a three-buffer case with its own PDN is simulated with and without PDN optimization. More the number of buffers, bigger is the simulation setup and more complex it is to arrive at right capacitance value for each of the buffer.

Fig. 4 shows simulation setup, ripple waveforms and jitter numbers for two clock buffer chain. Here 2 buffers are having 10pF decoupling capacitor each, connected to its supply. Jitter RMS is 5.38ps.

Fig. 5 shows simulation setup, ripple waveforms and jitter numbers for clock buffer chain with optimized PDN. Decoupling capacitors are now changed to 10pF and 40pF respectively for 2 buffers in the chain. Jitter RMS is reduced to 4.935ps which is 8% reduction with just two buffers in chain.

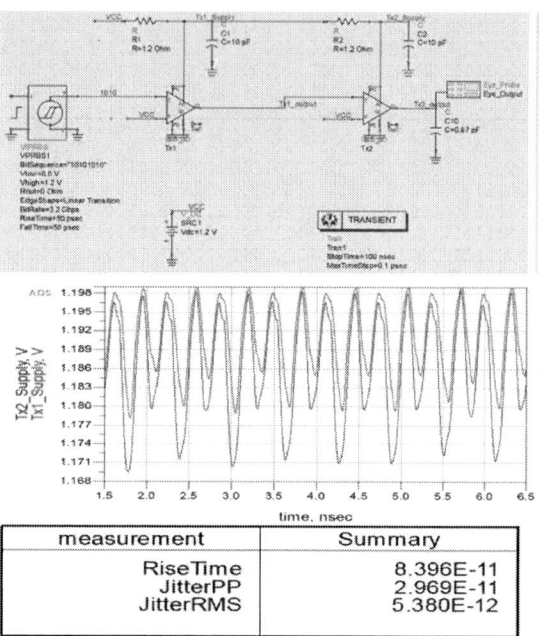

measurement	Summary
RiseTime	8.396E-11
JitterPP	2.969E-11
JitterRMS	5.380E-12

Fig. 4. Simulation setup and results for 2 clock buffer setup with 10pF capacitor on each of buffer supply rail

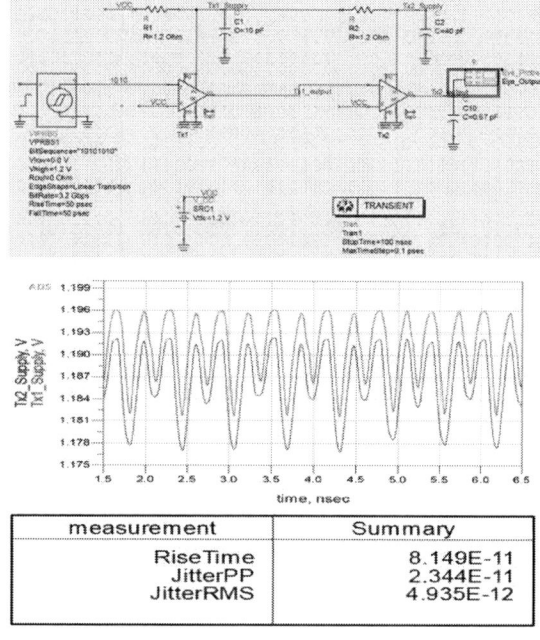

measurement	Summary
RiseTime	8.149E-11
JitterPP	2.344E-11
JitterRMS	4.935E-12

Fig. 5. Simulation setup and results for 2 clock buffer setup with 10pF and 40pF capacitor on buffer supply rail respectively

Fig. 6 shows simulation setup, ripple waveforms and jitter numbers for three clock buffer chain. Here 3 buffers are having 10pF decoupling capacitor each, connected to its supply. Jitter RMS is 5.091ps.

Fig. 7 shows simulation setup, ripple waveforms and jitter numbers for clock buffer chain with modified PDN. Decoupling capacitors are now changed to 10pF, 32pF and 22pF respectively for each of the three buffers in the chain. Jitter RMS is reduced to 4.876ps which is 4% reduction with just 3 buffers in chain.

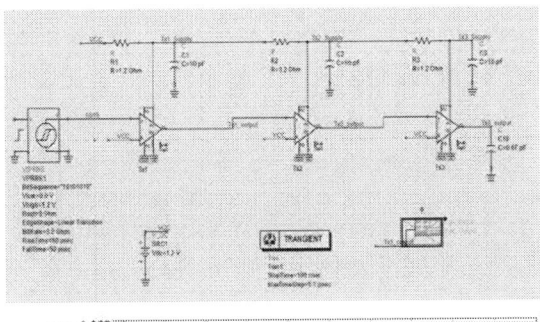

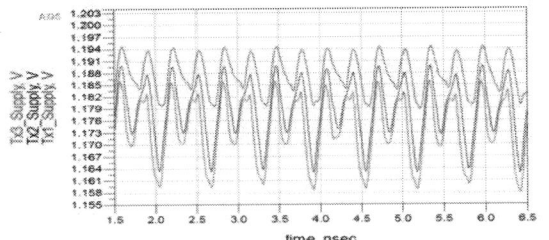

measurement	Summary
RiseTime	8.391E-11
JitterPP	3.125E-11
JitterRMS	5.091E-12

Fig. 6. Simulation setup and results for 3 clock buffer setup with 10pF capacitor on each of buffer supply rail

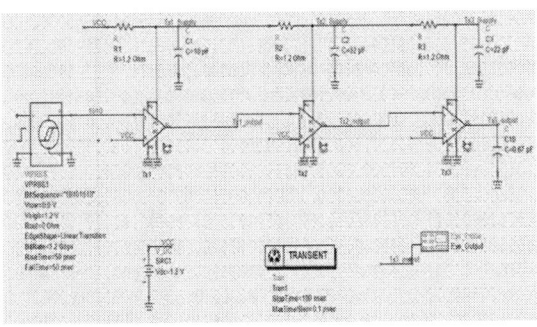

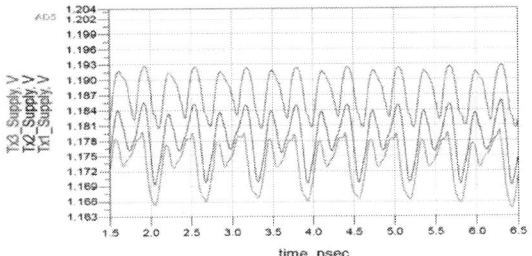

measurement	Summary
RiseTime	8.230E-11
JitterPP	2.656E-11
JitterRMS	4.876E-12

Fig. 7. Simulation setup and results for 3 clock buffer setup with 10pF,32pF and 22pF capacitor on buffer supply rail respectively

IV. SUMMARY

On-Die Power Distribution Network (PDN) optimization for achieving low Power Supply Induced Jitter (PSIJ) on clock tree was put forth. A chip having a distribution of differently sized decoupling capacitors coupled to a distributed power supply conductor grid, where the sizing of the decoupling capacitors is selected based on the spatial locality of the noise contributed by individual clock buffers of a clock distribution network, to achieve controlled low PSIJ.

Future work on further optimization of PDN can be done through resistance control which can be implemented using mosfets as resistors. Array of mosfets can be added to layout and made selectable using logic implementation to only plug-in desired resistance.

ACKNOWLEDGMENT

The authors would like to thank Dongwoo Hong, Kapil Vyas and Jonghyun Cho for their review and detailed discussions.

REFERENCES

[1] A. Strak, H. Tenhunen, "Analysis of timing jitter in inverters induced by power-supply noise," *Int. Conf. Design & Test of Integrated Systems in Nanoscale Tech.*, 2006

[2] R. Schmitt, H. Lan, C. Madden, and C. Yuan, "Investigating the Impact of Supply Noise on the Jitter in Gigabit I/O Interfaces," IEEE Conf. on Electrical Performance of Electronic Packaging, pp. 189-192, Oct. 2007.

[3] H. Lan, R. Schmitt, and C. Yuan, "Prediction and measurement of supply noise induced jitter in high-speed I/O interfaces," *Proc. of DesignCon*, 2009.

[4] R. Schmitt, H. Lan, and L. Yang, "On-Chip Power Supply Noise and Reliability Analysis for Multi-Gigabit I/O Interfaces," *Proc. of DesignCon*, 2010.

[5] Madhavan Swaminathan, and A. Ege Engin, "Power Integrity Modeling and Design for Semiconductors and Systems," Prtentice Hall, 2010.

[6] H. Lan et al., "Power supply noise induced jitter in a 6.4Gbps/Link memory interface system," presented at the IEC DesignCon, Santa Clara, CA, February 2012.

[7] Y. Shim et al., "System-level clock jitter modeling for DDR systems", in Proc. of 63rd IEEE Electronic Components and Technology Conference, Las Vegas, NV, May 2013.

[8] D. Klokotov, J. Shi, Y. Wang, "Distributed Modeling and Characterization of On-Chip/System Level PDN and Jitter Impact," *Proc. of DesignCon*, 2014.

978-1-6654-1392-3/21 $31.00 © 2021 IEEE

The Shape Control Process of a Cu/SnAg Solder Joint with a Ni insertion Using Thermo-Compression Bonding

Mingang Fang
School of Mechanical and Electrical Engineering, Central South University
ChangSha, China
mingangfang@csu.edu.cn

Zhuo Chen
State Key Laboratory of High Performance Complex Manufacturing, Central South University
School of Mechanical and Electrical Engineering, Central South University
Changsha, China
zhuochen@csu.edu.cn

Fuliang Wang
State Key Laboratory of High Performance Complex Manufacturing, Central South University
School of Mechanical and Electrical Engineering, Central South University
Changsha, China
wangfuliang@mail.csu.edu.cn

Chu Tang
School of Mechanical and Electrical Engineering, Central South University
ChangSha, China
tangchu@csu.edu.cn

Wenhui Zhu
State Key Laboratory of High Performance Complex Manufacturing, Central South University
School of Mechanical and Electrical Engineering, Central South University
Changsha, China
zhuwenhui@csu.edu.cn

Abstract—With the trend of miniaturization of electronic products, the integration and performance requirements of microelectronic chip are continuously increasing. Under this background, three-dimensional integrated circuit (3D IC) technology has received more and more attention. Fine-pitch microbump solder joint, as the vital part of 3D IC, has received extensive attention and researches. However, there is still insufficient studies on the shape control technique of Cu-pillar-based solder joint and the correlation between joint shape and the microstructure of solder layer. Hence, in this paper, three types of solder joints with different shapes were fabricated using Thermo-compression bonding (TCB) combined with displacement control in the Z-axis. Two main bonding structures, Cu/Sn and Sn/Sn, were applied in the experiment. The TCB process condition and parameters for fabricating the general type solder joints were studied first, and the adverse effects of temperature distribution inhomogeneity of TCB process on bonding quality and subsequent shape control of solder joint were discussed. Finally, through the improvement of bonding recipe, including the adjustment of bonding temperature and movement order of bonding head, a good bonding quality for general type solder joint was obtained, and the accurate control of stretch and compression type solder joints was also achieved.

Keywords—3D IC, TCB, solder joint, shape control

I. INTRODUCTION

With the miniaturization of electronic devices, the microelectronic packaging technology is developing in the direction of high density, high functionality, and high integration, the demand for 3D packaging, especially the 3D integration scheme of vertical through-die interconnections, keeps increasing[1]. In 3D integration, microbump interconnection has gradually become the mainstream technology, in which a solder layer on top of a Cu pillar is the essential part to realize the electrical interconnection of electronic chips. Traditionally, mass reflow has been the primary bonding process for solder joint in flip chip structure. However, with the reduction of package size, the mass reflow process has encountered some problems that are

difficult to overcome. On the one hand, the parallelity of the chip against the other side chip/substrate is hard to be guaranteed due to the decrease of the chip thickness, and the warpage caused by temperature variation will be more serious in the reflow process; on the other hand, as solder joint size decreases, the effect of height deviation will be magnified. As a result, the die and substrate may not be able to align under limited surface tension of the molten solder, resulting in tilt, non-contact opens (NCO) and in some cases solder ball bridging (SBB)[2].

To address the critical challenges of mass reflow, Thermo-Compression Bonding (TCB) process was developed. With special bonding process and tool, the deformation and warping problems of chip are restrained[3, 4]. However, there are still some drawbacks in TCB technology. On the one hand, low throughput is a major limitation for TCB because of the complexity of the process; on the other hand, different from the traditional mass reflow, the environment temperature of TCB process is not steady and uniform, the bonding chip keeps exchanging heat with the ambient air, as the heat source is the bonding head, there is thermal inhomogeneity in both the horizontal and vertical directions, which is easy to cause the problems of non-wetting or solder overflow during the process[5, 6].

Besides, although bonding technology keeps developing, the shape of solder joint has not changed much. Most solder joints are tire-shaped with slightly bulging edges, and the shape of solder joint cannot be controlled freely by bonding process. Huang[7] studies the influence of solder joint shape on the reliability of solder joints, they prepared four types of solder joints with different shapes, conventional type, compression type, cylindrical type and tensile type, using TCB process plus z-axis displacement, and found that the tensile type solder joint had the lowest thermal cycle life, while the compression type solder joint had the highest electromigration life. The results show that the shape of solder joint can truly affect the solder joint service behavior.

Therefore, in this paper, we mainly explored the technical scheme of the TCB technology to realize the solder joint shape control. Two main interconnection structures of TCB were used to explore the wetting behavior and thermal distribution uniformity problems in TCB process, and the solutions for these two problems were also discussed based on the results. Then, we introduced z-axis displacement control in the bonding process to realize the shape control of another two types of solder joint, the stretch and compression types. The reasons for the failure of shape control were analyzed and the method to realize the shape control was summarized.

II. MATERIALS AND EXPERIMENTS

A. Die and Substrate Configurations

In order to explore the influence of wettability and temperature inhomogeneity on the bonding quality and shape control of solder joint, two interconnection structures of TCB process were chosen for our experiments. The first structure is Cu/Sn bonding, which requires the interfacial reaction of solder and copper pillar to form a reactive bonding, as shown in Fig1(a). The second structure is Sn/Sn bonding, which only demands the fusion of solder caps, as shown in Fig1(b). There were two types of top chips and three types of substrate chips in our experiment, the dimension of top chip and substrate chip are 6mm × 6mm × 0.75mm and 12mm × 12mm× 0.75mm respectively. There are 54 bumps with a diameter of 100μm and a pitch of 400μm in both two chips. In the top chip, the structure of the bump is composed of copper pillar, nickel plating layer and solder cap which is made of Sn–2.5% wt. Ag via reflow process. In the substrate chip used in Cu/Sn bonding structure, the bump is just a copper pillar, and in another two substrate chips used in Sn/Sn bonding structure, the bump structure is similar to the top chip. The heights of different part in bumps for different chips are shown in Table I.

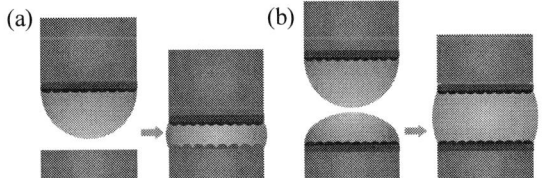

Fig.1. Bonding structures. (a) Cu/Sn bonding; (b) Sn/Sn bonding.

TABLE I.   DIMENSIONS OF DIE AND SUBSTRATE

Chip number	Composition height(μm)		
	Copper pillar	Ni barrier	Solder
Top 1	20	2	30
Top 2	20	2	15
Substrate 1	15	None	None
Substrate 2	15	2	20
Substrate 3	15	2	15

B. Bonding Process

The schematic diagram of TCB process is shown in Fig2, the substrate chip, which is usually dipped with flux, is fixed on the pedestal under vacuum adsorption, and the pedestal is heated to pre-activate the flux. Top chip is picked up and moved to the top of substrate chip by bonding head via vacuum absorption, after aligning with the substrate chip through optical recognition, the bonding head goes down and stops when the contact force between top chip and substrate chip reaches the setting value. Then the chips will be heated up quickly. When the solder melts and the top chip falls down a certain height, usually several micrometers, to ensure all solder joints are in contact, the bonding head will keep its position to prevent solder overflow. Then the temperature will be held for a while to ensure the interfacial reaction or fusion of each solder joints are finished, after that the chips will be cooled down to room temperature quickly, at the same time the vacuum will break off and the chips will be released by the bonding head, then the TCB process is completed.

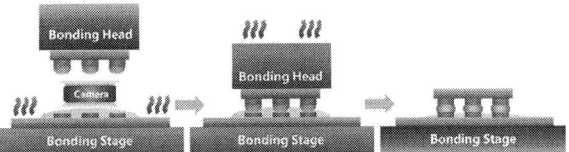

Fig.2. The schematic diagram of TCB process.

For Cu/Sn bonding, top1 and substrate1 were used. Because the wettability of Sn on Cu surface is very important to the bonding quality, and an appropriate flux is necessary for the removal of oxides and the promotion of interfacial reaction[1], the selection and usage of flux in TCB process is also very critical. Therefore, we selected three different types of flux, as shown in Fig3. All three types of flux can remove the oxides on copper surface and promote the reaction between Cu and Sn. The differences are, flux A is gelatinous at room temperature but will melt and collapse under preheating condition; flux B is solid at room temperature but will melt to liquid and adhere to the metal surface under preheating condition; flux C is gelatinous both at room temperature and preheating condition and will only collapse a little under the preheating condition.

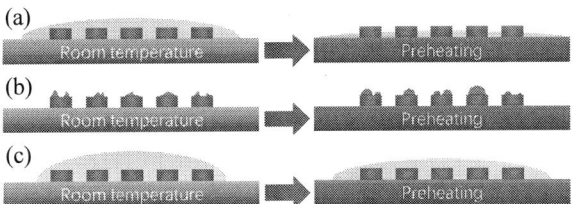

Fig.3. The transformation of three different types of flux before and after preheating. (a) Flux A; (b) Flux B; (c) Flux C.

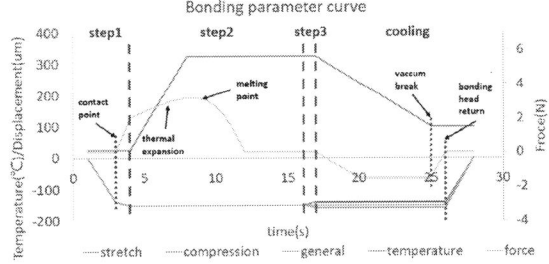

Fig.4. The bonding parameter curve of gap controllable TCB.

Before bonding, flux was applied to the substrate chip, then the substrate chip was preheated on the stage up to 150°C to activate the flux and soften the material[2]. The top chip was picked up by the ceramic tool of bonding head, after alignment using a camera, a gap controllable TCB process was applied to avoid solder overflow, the bonding parameter curves are shown in Fig4, in the first step, the top chip was pressed on the substrate for a contact force of 2N, in the second step, the bonding head was switched to the gap control mode to hold its position before heat up the chips, to diminish the thermal inhomogeneity, the heating rate was controlled at about 80°C/s[5], the holding time for peak temperature was about 10s as an experience value to ensure all solder caps melted and interfacial reaction could happen, in the third step, for general type solder joint, there was no more z-axis displacement control before cooling, for stretch type solder joint, there was a 10μm upward displacement of bonding head and 10μm downward for compression type solder joint before cooling to the room temperature.

For Sn/Sn bonding, top1 and substrate2 were used. Because there is no consideration for wettability, we only used flux C which can best keep the solder caps of substrate chip from oxidation during preheating. The bonding process of Sn/Sn bonding structure was similar to Cu/Sn bonding structure, but due to the larger overall volume of solder, the upward and downward displacement for two different types of solder joints was adjusted to 15μm.

C. Equipment and Observation Method

After TCB process, the sample was cleaned to remove the residual flux, and then cured using epoxy resin for later grinding, polishing and observation. The equipment of the TCB machine used in our experiment is CB-600 from Athlete. The SEM used for observations is Tescan Vega3 SBH.

III. RESULTS AND DISCUSSION

A. Flux-assisted wetting behavior

Firstly, we compared the effect of three different types of flux on the bonding quality of solder joint under the same bonding temperature (325°C). The results are shown in Fig5. There is no obvious interfacial reaction between Cu/Sn interface of solder joint A which used flux A, and almost all chips using flux A failed to form the interconnection between top chips and substrate chips, which is because when flux A was preheated, although the oxide layer of copper would be removed in the first time, it would collapse and expose the copper pillar to the air, then the copper surface would be oxidized again, which is the reason for interfacial reaction failed to happen. In contrast, a full interfacial reaction came into being in solder joint B which used flux B, but there were many voids in the IMC layer, which is very harmful to the mechanical reliability of solder joint, this may be caused by the scattered distribution of flux droplets. Only joint C had a good interfacial reaction without any voids. The result shows, for bonding structures with copper pillar, although the metal oxides can be removed well at the start of preheating, if the copper pillar top surface cannot be well protected during the later process, oxidation will happen again, only if the TCB machine is able to control the oxygen content of the bonding environment[1], or the substrate chip must be coated with flux, and the flux must be applied to a certain thickness which can cover all the bumps during bonding to protect the metal surface from oxidation. Besides, the flux should be even on the surface and left no gaps, otherwise it is easy to produce voids in the interface, which will jeopardize the thermal-mechanical reliability of the chip. Moreover, flux with a certain thickness may also play a role of heat transfer

medium, which can diminish the thermal inhomogeneity during the heating stage, which is another reason that the flux should not collapse during heating.

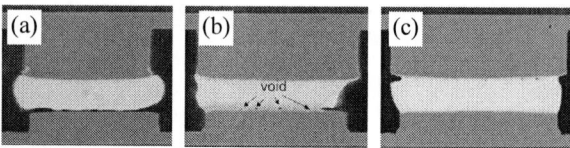

Fig.5. Solder joints using different types of flux (a) Solder joint A using flux A; (b) Solder joint B using flux B; (c) Solder joint C using flux C.

B. Thermal inhomogeneity in TCB

Further, we compared the bonding quality of solder joints under different bonding temperatures. To demonstrate the influence of thermal inhomogeneity on bonding quality, we picked a center solder joint and a corner solder joint for each bonding temperature, the results are shown in Fig6. For the bonding temperature of 350°C, both solder joints contained a thick layer of IMC, which means a good interfacial reaction of all solder joints. In contrast, for the bonding temperature of 300°C, there was no obvious IMC either in the center or corner solder joints. As for the bonding temperature of 325°C, it was obvious that the bonding quality was different in solder joints of different locations, the interfacial reaction was much more prominent in the center than corner.

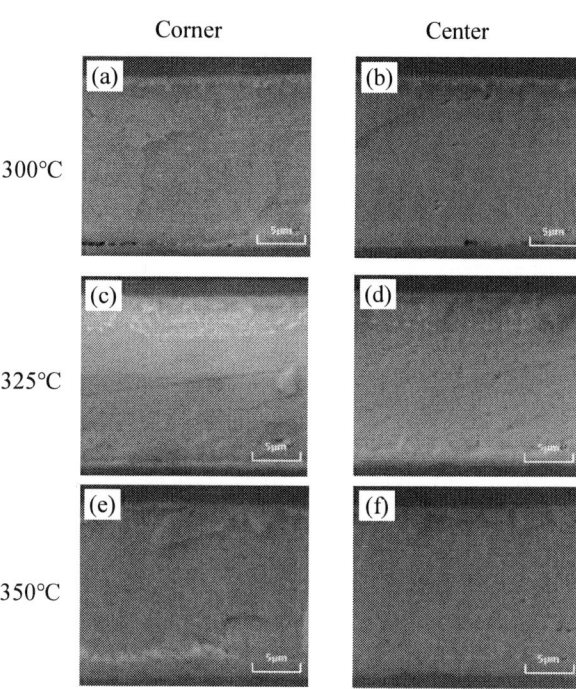

Fig.6. The solder layer in different position under different bonding temperature, where the bottom interface is the IMC layer. (a) Corner solder joint (300°C); (b) Center solder joint (300°C); (c) Corner solder joint (325°C); (d) Center solder joint (325°C); (e) Corner solder joint (350°C); (f) Center solder joint (350°C).

The IMC thicknesses of six different solder joints are shown in Fig7. The reason resulting in this phenomenon is the thermal inhomogeneity, that is, in the horizontal direction, the bonding head is the heat source, so the chip center obtains the highest temperature, and heat dissipation is relatively less here. However, the chip edge is far from the heat source and its heat dissipation is more drastic, which makes its temperature is much lower than the center[5, 6].

When the bonding temperature is too low that even the center temperature is not high enough for a valid interfacial reaction, the thermal inhomogeneity cannot be reflected from the thickness of IMC. When the bonding temperature is high enough for an obvious interfacial reaction, the thermal inhomogeneity can be demonstrated. However, when increasing the bonding temperature, the difference of the thickness of the IMC layer from different position solder joints decreased. Therefore, increasing the bonding temperature is an effective way to diminish the effect of thermal inhomogeneity, but that also means higher power consumption and lower efficiency in industrial manufacture, more than that, the circuit in die suffering high temperature is easier to fail, and too much IMC in solder joints is usually harmful to chip's reliability.

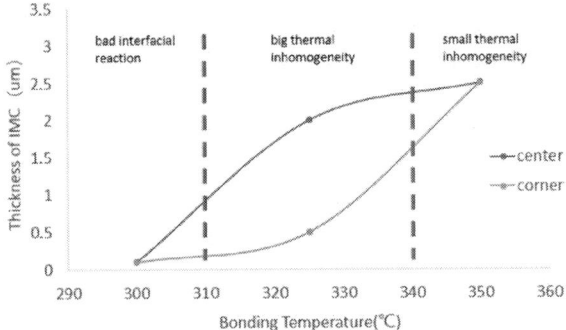

Fig.7. The thickness of IMC of solder joints with different bonding temperature and position.

In order to realize chip interconnection at a low bonding temperature and avoid the problem of interfacial reaction, we adopted the Sn/Sn bonding structure using top2 and substrate3, whose whole solder height is equal to top1. A general type solder joint fabricated under the bonding temperature of 300°C is shown in Fig8, in which the solder was fused together well. That means, with a relatively low height of solder, Sn/Sn bonding is more efficient than Cu/Sn bonding.

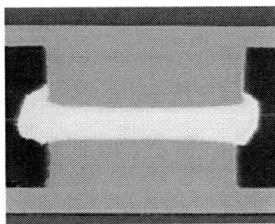

Fig.8. Solder joint of Sn/Sn bonding structure using top2 and substrate3 under the bonding temperature of 300°C.

The results of general type solder joints of Sn/Sn bonding under different bonding temperatures are shown in Fig9. It is obvious that the bottom solder did not melt in solder joint A, whose bonding temperature is 300°C, and there was a crack in the middle which means two solder caps did not fuse together. Though there was no crack in solder joint B whose bonding temperature is 325°C, it can be found from the unsmooth outline of the solder joint that the bottom solder did not melt completely. The third solder joint outline was more smooth because of a higher bonding temperature of 350°C, but the later experiment showed the bottom solder cap still did not melt completely. As for the reason, on the one hand, due to the small contact area of solder caps, the heat transfer efficiency from top to bottom is reduced to a certain extent, on the other hand, there is a temperature gradient, i.e., thermal inhomogeneity in the vertical direction[6], both two conditions result in a lower temperature of the bottom solder cap, which finally lead to an incomplete fusion. Therefore, Sn/Sn bonding structure is not suitable for high solder cap.

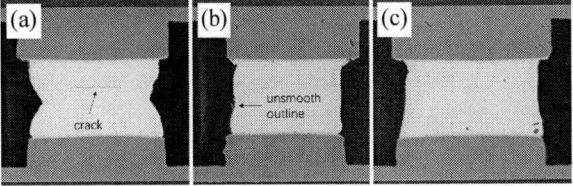

Fig.9. General type solder joints of Sn/Sn bonding structure under different bonding temperature. (a) Solder joint A(300°C); (b) Solder joint B (325°C); (c) Solder joint C (350°C).

C. Shape control

For stretch type solder joint, there were three main results under different bonding temperatures, as shown in Fig10. Solder joint A mainly came from the bonding temperature of 300°C, in which the solder cap was pulled away from the bottom copper surface because of the poor wetting behavior. Solder joint B tended to strip from the upper surface, and this type solder joint mainly came from the bonding temperature of 325°C, only solder joint C showed an ideal stretch shape, and this type solder joint mainly came from the bonding temperature of 350°C.

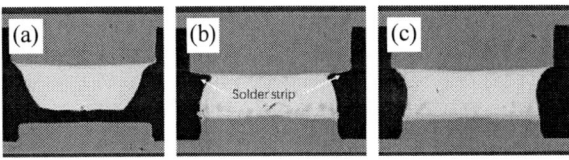

Fig.10. Three different types of stretch solder joints. (a) Solder joint A; (b) Solder joint B; (c) Solder joint C.

We further calculated the distribution of three types of stretch solder joints in different chips under different bonding temperatures, and the results are shown in Fig11. In the figure, the blue, yellow and red dots correspond to the solder joint A, B and C respectively. It is obvious that the ideal stretch type solder joints were concentrated in the area with higher temperature, which indicates that higher temperature is a necessary condition for the realization of stretch type solder joints, and the distribution of solder joints under the bonding temperature of 350°C was consistent with the temperature distribution.

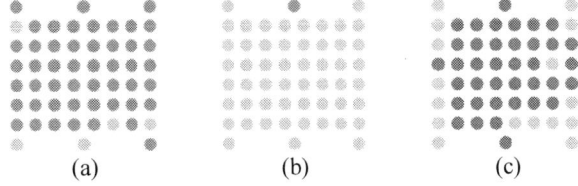

Fig.11. The distribution of three different types of stretch solder joints under different bonding temperatures. (a) 300°C; (b) 325°C; (c) 350°C.

The explanation given by us is, the shape of the solder joint is mainly affected by the comprehensive influence of surface tension, interfacial adhesion, internal stress and thermal migration. For stretch type solder joint, as shown in Fig12, the stress condition of the top half part solder can be described by the following equation.

978-1-6654-1392-3/21 $31.00 © 2021 IEEE

$$F_{IA}=F_{ST}+F_{SS}+F_{TM} \qquad (1)$$

Where F_{IA} is the interfacial adhesion of Sn/Ni interface, F_{ST} is the surface tension of solder, F_{SS} is the stretch stress in solder and F_{TM} is the kinetic force from thermal migration. When

$$F_{IA}< F_{ST}+F_{SS}+F_{TM}$$

the solder will strip from the upper surface to decrease the F_{SS} and then rebalance the equation, resulting in the solder joint in Fig10(b). Because the mechanism of the interfacial adhesion is not yet clear, in order to obtain the ideal stretched solder joint, the F_{ST}, F_{SS} and F_{TM} should be reduced. Reducing the stretch height can reduce the F_{SS} to a certain extent, and increasing the temperature can decrease the F_{ST}, and reduce the thermal gradient, which also means a small F_{TM}, at the same time, that is why the ideal stretch shape is easier to form under higher bonding temperature. As for the bottom half part solder, because of the thermal gradient mentioned above, the temperature of the top surface is always higher than the bottom, the F_{TM} always points from top to bottom, that is why the phenomenon of solder strip only happens in the top interface. However, if no interfacial reaction occurs on the Cu/Sn interface, the adhesion may be much smaller than Ni/Sn, ultimately leading to the solder joint in Fig10(a).

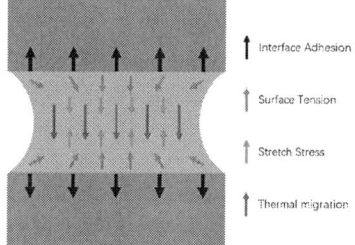

Fig.12. The stress condition of stretch type solder joint.

It is worth mentioning that the gravity of solder is not taken into account here, because we found the gravity is much less than the internal stress in the later experiment. Fig13 shows a bonding force response curve obtained by the machine sensor before and after a 25μm stretch order under gap control mode. It can be seen from the diagram that after stretching, the bonding force decreased to a negative value which means the stretch stress was generated. The difference between the pre-and post-average force is about 10mN, and the average stretch stress of each solder joint is about 200uN, which is much bigger than the gravity of a single solder joint (about 0.01uN). Therefore, the effect of gravity to equation(1) is negligible.

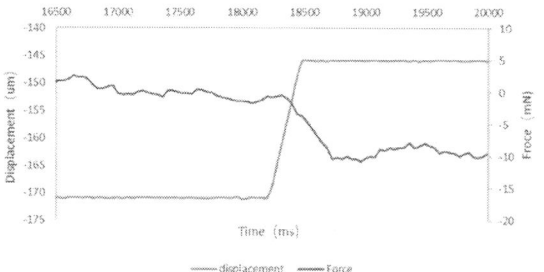

Fig.13. Bonding force response curve.

Fig.14. Three different types of compression solder joints. (a) Solder joint A; (b) Solder joint B; (c) Solder joint C.

Similar to the stretch type solder joint, there were also three main types of compression solder joints under different bonding temperatures, as shown in Fig14. Solder overflow occurred in solder joint A and B, which mainly came from the bonding temperature of 325℃ and 350℃. Interestingly, the solder overflowed in both two directions. The reason for this phenomenon is similar to that of the stretch type solder joint, but not related to the interfacial adhesion. As shown in Fig15, the stress condition of solder boundary can be described by the following equation.

$$F_{ST}=F_{CS} \qquad (2)$$

Where F_{ST} is the surface tension of solder, and F_{CS} is the compression stress in solder. When

$$F_{ST} < F_{CS}$$

the solder will overflow. As mentioned before, the influence of gravity is negligible, so the direction of solder overflow is independent of the direction of gravity. To prevent solder from overflowing, increasing the F_{ST} and decreasing the F_{CS} is the most effective way. With the same compression height, the F_{ST} can only be increased by lowering the bonding temperature, like solder joint C in Fig14(c), whose bonding temperature is 300℃.

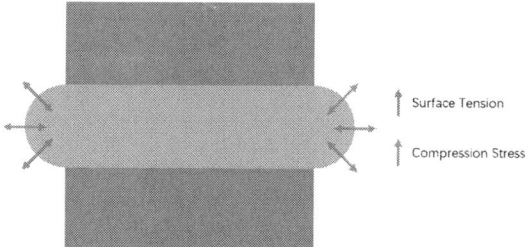

Fig.15. The stress condition of compression type solder joint.

For stretch type solder joint of Sn/Sn bonding, similar to the Cu/Sn bonding structure, the ideal stretch type solder joint did not come into being under the bonding temperature of 300℃ and 325℃. But even for the bonding temperature of 350 ℃, the result was not ideal either, as shown in Fig16(a), it is obvious that the bottom solder cap is only partially melted as mentioned above, which means a 15μm/30μm stretch rate for top solder cap, so the F_{SS} was much bigger, resulting in the strip of solder. In order to figure this problem, we induced a step-by-step displacement control bonding process, that is, after heating at 350 ℃ for 5S, the top solder cap was melted and then the solder joint was compressed for 10μm and held on for another 5S to promote the melt of bottom solder cap, then the solder joint was stretched for 25μm. Fig16(b) shows the final result, in which the whole solder joint was completely melted, due to the stretch rate was 15μm/50μm, approximately equal to 10um/30um of Cu/Sn bonding, the F_{SS} was also approximate, so the solder did not strip from the top surface, an ideal stretch solder joint of Sn/Sn bonding came into being.

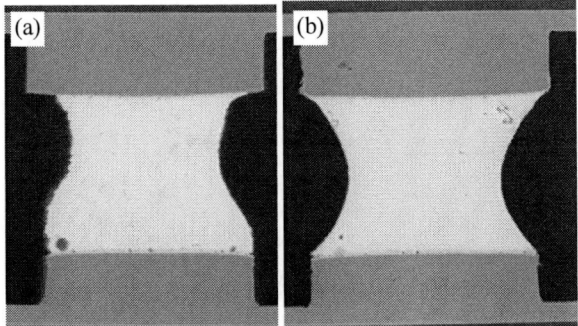

Fig.16. Stretch type solder joint of Sn/Sn bonding structure. (a) general bonding process; (b) Step-by-step bonding process.

According to the experiment results mentioned above, the compression type solder joint without solder overflow can only come into being under the bonding temperature of 300°C, but under this condition, the bottom solder cap was not able to melt completely, so we introduced the step-by-step displacement control process again, and the holding time was prolonged to raise the temperature of bottom solder as high as possible, that is, after heating at 300°C for 10S, the top solder cap was melted and then the solder joint was compressed for 10μm and held on for another 20s, then the solder joint was compressed for another 5μm. However, the bottom solder cap still did not melt, as shown in Fig17(a). Therefore, when the solder cap is too high and the bonding temperature is only a little higher than the melting point, even if the step-by-step displacement control process and prolongation of bonding time cannot melt the solder completely, which makes the compression rate directly depends on the temperature that can completely melt the whole solder joint. Finally, as shown in Fig17(b), the compression solder joint, was fabricated under the bonding time of 320 °C with a 15μm compression.

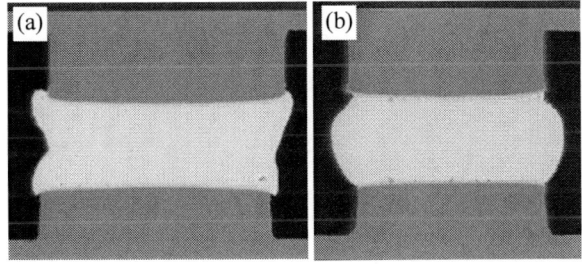

Fig.17. Compression type solder joint of Sn/Sn bonding structure using step-by-step bonding process under different bonding temperatures. (a) 300°C; (b) 320°C.

IV. CONCLUSION

This paper firstly explored the technological condition and parameters of TCB process for the controlling of solder joint shape using two main bonding structures, Cu/Sn bonding and Sn/Sn bonding. It was found that gelatinous flux, which will not collapse much under preheating condition, is more appropriate for bonding structure with a copper pillar. The thermal inhomogeneity of TCB has an important influence on the bonding quality of solder joints. In the horizontal direction, the bonding quality of edge solder joints will be worse due to the lower temperature. In the vertical direction, the excessive gap will affect the interfacial reaction in Cu/Sn bonding structure and the melt of the bottom solder cap in Sn/Sn bonding structure. But these detrimental effects can be mitigated by increasing the bonding temperature and time, and when the height is relatively low, Sn/Sn bonding structure is more effective than Cu/Sn. Then, in the process of fabricating stretch and compression type solder joints, it was found that the surface tension and thermal inhomogeneity are the critical factors for the shape control of solder joints. For stretch type solder joints, a higher temperature is needed to decrease the surface tensile and the thermal inhomogeneity to prevent solder stripping from the top interface. For compression solder joints, the temperature should be reduced to increase the surface tension to prevent solder overflowing. However, for Sn/Sn bonding, when the height of solder cap is too high, resulting in the temperature in the bump of substrate chip is too low to melt the bottom solder cap, the step-by-step displacement control can be adopted for the stretch solder joint, but for the compression solder joint, only the lowest temperature which can completely melt the solder can be selected for the fabrication process.

ACKNOWLEDGMENT

This work was supported by National Natural Science Foundation of China (No. Grant No. 51605498 and U20A6004), National Basic Research Program of China (973 Program, Grant No. 2015CB057206), and State Key Laboratory of High Performance Complex Manufacturing (No. ZZYJKT2020-08).

REFERENCES

[1] S. Lee, "Fundamentals of Thermal Compression Bonding Technology and Process Materials for 2.5/3D Packages," in 3D Microelectronic Packaging (Springer Series in Advanced Microelectronics, 2017, pp. 157-203.

[2] A. Eitan, K.-Y. Hung, and Ieee, "Thermo-compression Bonding for Fine-pitch Copper-pillar Flip-chip Interconnect - Tool Features as Enablers of Unique Technology," in 2015 Ieee 65th Electronic Components and Technology Conference (Electronic Components and Technology Conference, 2015, pp. 460-464.

[3] S. Hwang, S.-K. Choi, and Asme, OPTIMAL DESIGN OF THERMO-COMPRESSION BONDING FOR ADVANCED PACKAGING SYSTEM UNDER UNCERTAINTY (Proceedings of the Asme International Design Engineering Technical Conferences and Computers and Information in Engineering Conference, 2019, Vol 1). 2020.

[4] M. Tsai et al., Challenge and Warpage Optimization of Thermal Compression Bonding Technology on Coreless Substrates (2017 Ieee 19th Electronics Packaging Technology Conference). 2017.

[5] S. B. Jemaa, P. Gagnon, and J. Sylvestre, "In Situ Measurement Method for Temperature Profile Optimization During Thermocompression Bonding Process," IEEE Transactions on Components, Packaging and Manufacturing Technology, vol. 10, no. 11, pp. 1929-1937, 2020.

[6] J.-S. Jung et al., "A Study of 3D Packaging Interconnection Performance Affected by Thermal Diffusivity and Pressure Transmission," presented at the 2019 IEEE 69th Electronic Components and Technology Conference (ECTC), 2019.

[7] Y.-W. Huang et al., "Effect of Joint Shape Controlled by Thermocompression Bonding on the Reliability Performance of 60 mu m-pitch Solder Micro Bump Interconnections," in 2014 Ieee 64th Electronic Components and Technology Conference (Electronic Components and Technology Conference, 2014, pp. 1908-1914.

978-1-6654-1392-3/21 $31.00 © 2021 IEEE

Design and Fabrication of Multi-Layer Silicone Microchannel Cooler for High-Power Chip Array

Tao Wei
Nanjing Research Institute of Electronics Technology
Nanjing, China
toix_1015@qq.com

Haojie Huang*
Nanjing Research Institute of Electronics Technology
Nanjing, China
huang_haojie@foxmail.com

Yupa Ma*
Nanjing Research Institute of Electronics Technology
Nanjing, China
2461616815@qq.com

Jiyu Qian
Nanjing Research Institute of Electronics Technology
Nanjing, China
qianjiyu_work@163.com

Abstract—With the development of high-power and highly-integrated microwave devices with chip array, there have been plenty of methods for high heat flux dissipation. However, the existing cooling architectures mainly focused on removal of single hot spot. For microwave devices with high-power chip array, there is a co-design of heat transfer enhancement and flow homogenization. In this study, we proposed a multi-layer silicone microchannel cooler for 4×4 chip array cooling with heat flux of 500 W/cm². Using a finite-element simulation, the flow distribution optimization was conducted and an H-type bifurcation structure was obtained. The optimized cooler has three layers: the liquid supply layer, the liquid return layer, and the microchannel layer. The size of whole microchannel cooler 44 mm×44 mm×1.5 mm. The simulated results showed that a uniform velocity and pressure distribution was achieved in 4×4 microchannels with an average velocity of 0.8 m/s and total pressure drop of 0.42 bar under flow rate of 36 L/h. The maximum chip temperature rise above the inlet temperature is 39 °C with the temperature deviation less than 1.0 °C. We also fabricated a prototype of the optimized multi-layer silicone microchannel cooler using mature silicone etching and Si-Si bonding process. A thermal test system was built to evaluate the thermal performance of the cooler. The tested results showed that the maximum temperature rise was less than 40 °C with the average temperature deviation less than ±5 °C. In summary, the proposed compact multi-layer silicone microchannel cooler achieved efficient and uniform cooling of high heat-flux chip array.

Keywords—*silicone microchannel, multiple chip cooling, high heat flux*

I. INTRODUCTION

With the development of highly-integrated and multi-functioned microwave devices, the power density has increased instantly [1], [2]. Moreover, in order to gain higher output power, an array of chips is usually integrated. Due to the power dissipation of the microwave devices, the heat flux can reach 500 W/cm² at the chip level [3], resulting in multiple localized hotspots corresponding to the chip array. The peak temperature experienced in these hotspots may generally lead to a reduction of the operating efficiency, reliability of the devices [4]. These considerations thus become a limiting factor in the operation of the microwave devices, in which case there is an urgent need of an efficient cooling solution for multiple chip array with high heat flux.

In the last decades, there have been plenty of methods for high heat flux dissipation using liquid cooling with advanced architectures, including microjet with local high heat transfer coefficient by impingement cooling [5]-[8], and microchannel with high heat transfer area [9]-[13]. Walsh et al. form Massachusetts Institute of Technology embedded ring-shaped microjets in the silicone substrate to dissipate a heating power of 2 W and an area of 0.4 cm² with a temperature rise of 46 °C [14]. Ditri et al. from Lockheed Martin fabricated a palladium microjet manifold to dissipate a heat flux of 1000 W/cm² with a temperature rise of [15], [16]. Jung et al. form Stanford University incorporated an embedded silicone microchannel with a channel width of 75 μm and a three-dimensional silicone manifold to remove a heat flux of 250 W/cm² with a maximum temperature rise of 90 °C and a pressure drop less than 3 kPa at a flow rate of 0.1 L/min [17]. Altman et al. from Raytheon combined microjets etched into silicone and microchannels etched into diamond to dissipate a heat flux of 1230 W/cm² [18]. Despite the great effort in efficient cooling strategies for high-power electronics, most of the previous work is focused on high heat flux removal for single die with single hotspot. For high-power chip array, there is a co-design of heat transfer enhancement and flow homogenization.

In this work, aimed at 4×4 chip array cooling with a high heat flux of 500 W/cm² for single chip, we proposed a multi-layer silicone microchannel cooler consisting of the liquid supply layer, the liquid return layer, and the microchannel layer. The heat dissipation of the chip array was removed effectively through the distributed 4×4 microchannel array. The flow and heat transfer performances of the microchannel cooler were investigated and optimized using a finite element simulation. Then, we fabricated a prototype of the optimized multi-layer silicone microchannel cooler using mature silicone etching and Si-Si bonding process. Finally, a thermal test system was built to evaluate the thermal performance of the cooler.

II. DESIGN AND SIMULATION

Fig. 1 illustrates the schematic of the multilayer silicone microchannel cooler for cooling high-power 4x4 chip array with a heat flux of 500 W/cm² for single chip. The 16 chips are soldered to a microchannel substrate separately, leading to 16 distributed hotspots. In this case, multiple distributed microchannels with high heat transfer ability and uniform flow distribution with relatively low pressure drop are needed. To achieve this goal, the multi-layer silicone microchannel cooler is designed to consist of three layers including the liquid supply layer, the liquid return layer, and the microchannel layer. The liquid supply layer and liquid return layer play the role of uniform flow distribution and the microchannel layer plays the role of efficient cooling. The microchannel layer has 4×4 distributed parallel microchannels underneath the 4×4 chip array with microchannel width of 50 μm, height of 250μm and fin width of 50 μm. We propose four flow distribution structures, namely (a) one inlet and one outlet, b) two inlets and one outlet, (c) four inlets and one outlet, and (d) four inlets and two outlets, as shown in Fig. 2. The difference between them lies in the number of inlet and outlet. The effective flow path reduces with the rising of inlet or outlet number.

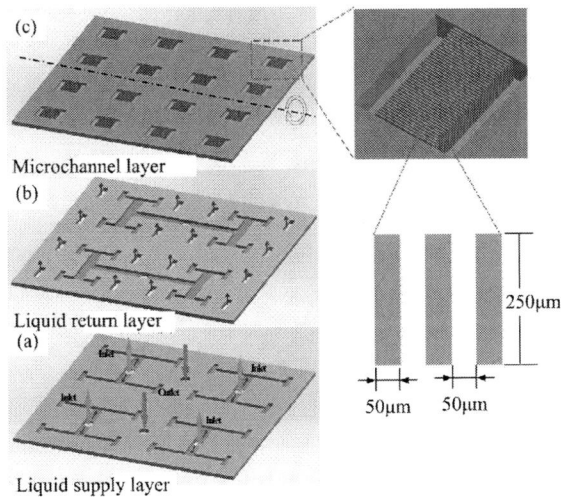

Fig. 1. The schematic of the multi-layer microchannel cooler consisting of (a) liquid supply layer, (b) liquid return layer, and (c) microchannel layer.

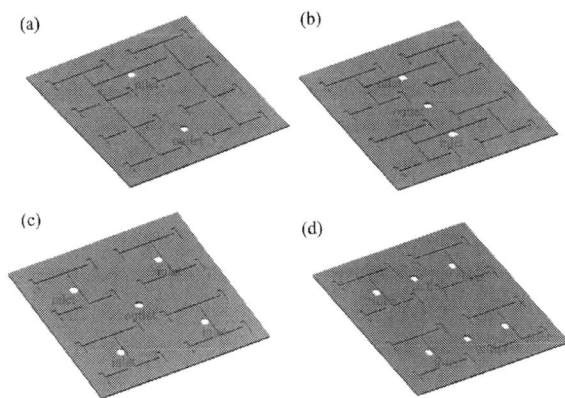

Fig. 2. The schematic of four different folw distribution structures: (a) first-one inlet and one outlet, (b) second-two inlets and one outlst, (c) third-four inlets and one outlet, and (d) fourth-four inlets and two outlets.

The flow and heat transfer performances of these four structures are investigated and optimized using a finite element simulation conducted by commercial software FloEFD. The single chip has size of 3 mm×3 mm and heating power of 45 W, corresponding to a heat flux of 500W/cm². The total heating power is 720 W. The material and the thermal-physical properties of the components of the cooler are summarized in Table l. The chips are assumed to

be uniform volumetric heat source with high thermal conductivity. The liquid supply layer, the liquid return layer, and the microchannel layer are all made of silicone with a thermal conductivity of 150 W/(m·K). The working fluid is deionized (DI) water. The inlet temperature is 40 °C and the flow rate is 36 L/h. The outlet pressure is l.0 bar. The thermal interface resistance between the chip and the microchannel is 0.01 K·cm²/W.

TABLE I.   THE MATERIAL AND THERMAL-PHYSICAL PROPERTIES OF COMPONENTS OF THE COOLER

Component	Material	Thermal conductivity (W/m·K)	Density (kg/m3)	Capacity (J/kg·K)
Chip	/	5000	/	/
Liquid supply layer	Silicone	150	2330	705
Liquid return layer				
Microchannel layer				
Working fluid	DI water	0.63	992	4179.6

III. FABRICATION AND EXPERIMENTAL TEST

In this section, a prototype of the optimized multi-layer silicone microchannel cooler was fabricated. Then, the thermal performance of the cooler was measured by establishing an experimental system.

A. Microchannel Cooler Prototype Fabrication

The fabricating processes of the microchannel cooler consisted of multi-layer silicone layer fabrication and simulated chip array fabrication. Fig. 3 shows the fabricating process of the multi-layer silicone layer. At first, three silicone wafers with thickness of 0.5 mm were prepared. The liquid supply layer has two inlets and four outlets in the back side and bifurcation structure in the front side. The inlets and outlets in the back side were fabricated by the processes of photolithography, oxide etching, and deep silicone etching. The structure in the front side was fabricated using the same processes. The thickness of the inlet and outlet was 0.5 mm and the thickness of the bifurcation structure is 0.35 mm. The liquid return layer has liquid supply structure in the back side and liquid return structure in the front side. The fabricating processes were same as that of the liquid supply layer. The thickness of the liquid supply structure was 0.5 mm and the thickness of the liquid return structure was 0.35 mm. The microchannel layer has distributed microchannels in the back side, which were fabricated by deep silicone etching. The thickness of the microchannel was 0.25 mm. The fabricated three silicone layers were then bonded using anode alignment Si-Si bonding.

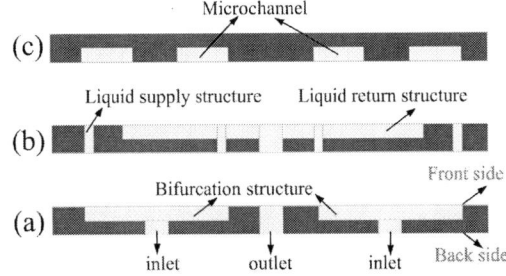

Fig. 3. The schematic of the multi-layer silicone layer fabrication: (a) the liquid supply layer, (b) the liquid return layer, and (c) the microchannel layer.

Fig. 4 shows the fabricating process of the simulated thermal chip array. First, an array of silicone thermal test chips (TTCs) with area of about 1 mm×1 mm were soldered onto the back side of the microchannel layer using nano-silver solder. Second, a printed circuit board PCB) was bonding to the back side of the microchannel layer using epoxy glue. The PCB had 4×4 holes corresponding to the 4×4 TTCs. Third, the TTCs were connected to the PCB by gold wire bonding. Fig. 5 shows the as-fabricated cooler and the enlarged single simulated thermal test chip wire-bonded to PCB

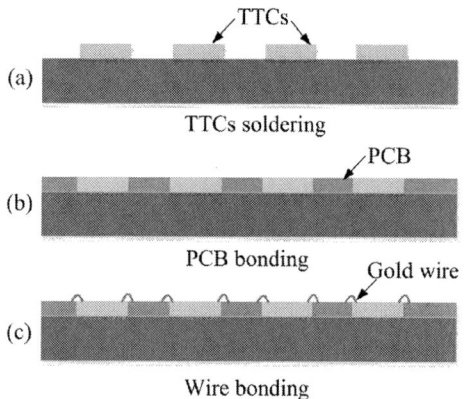

Fig. 4. The fabricating processes of the simulated thermal test chip array.

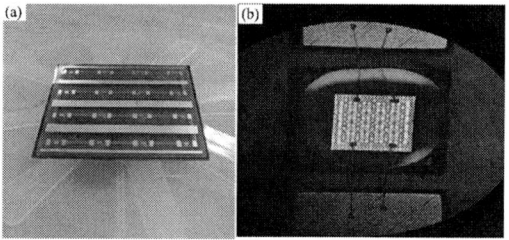

Fig. 5. The photograpg of (a) the as-fabricated microchannel cooler and (b) the enlarged single simulated thermal test chip wire-bonded to PCB.

B. Experimental Test

Fig. 4 illustrates the schematic and photograph of the experimental system to evaluate the thermal performance of the fabricated cooler. The simulated TTC array was connected to two DC power supplies. For each TTC, the voltage and current were respectively 6.4 V and 0.8 A, corresponding to a heating power of 5.1 W and an average heat flux of 500 W/cm². It should be noted that both the heating power and footprint of each chip were much lower than the simulated case due to the process and material restrictions. Despite that, the heat flux of the experiment was same as that of simulation. The experimental results can still reflect the effect of the proposed cooler. In addition, a liquid circulation system consisting of a liquid phase pump, a liquid storage tank, and a liquid collecting tank was established to provide deionized (DI) water into the cooler to dissipate the generated heat from the chip array constantly. The inlet temperature was about 20 °C and the flow rate was 36 L/h. The surface temperature distribution of the chip array were captured using an infrared (IR) imager. Both the maximum temperature and average temperature of each chip could be measured.

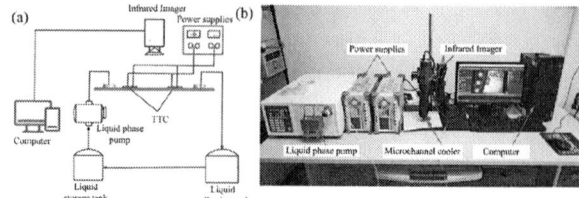

Fig. 6. The (a) schematic and (b) photopragh of the experiemntal system.

IV. RESULTS AND DISSCUSSIONS

In this part, the simulated results of four kinds of coolers were firstly discussed. Then, the experimental results of the optimized cooler (i.e. the fourth one) was presented.

A. Simulation Results

Fig. 7 shows the pressure distribution of four different flow distribution structures at a constant flow rate of 36 L/h and inlet temperature of 40 °C. The first structure with one inlet and one outlet exhibits a very high pressure drop of 4.5 bar due to the partial series and parallel connection of the 4×4 microchannels. The main pressure drop occurs at the fluid diversion and confluence point. The flow rate distribution between the distributed microchannels presents a large deviation, corresponding to a very high maximum temperature rise above the inlet temperature of 87 °C and temperature deviation of 34 °C for the chip array, as shown in Fig. 8(a).

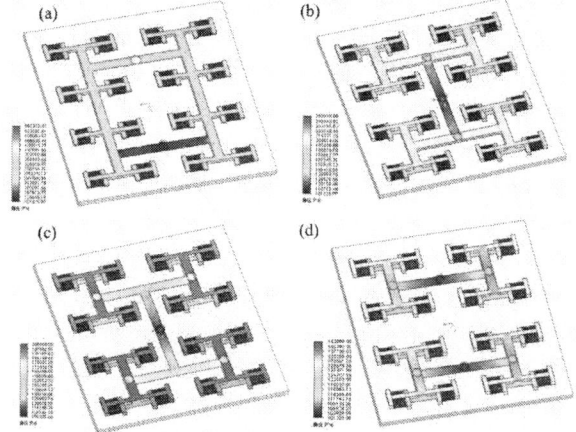

Fig. 7. The pressure distribution of four different folw distribution structures: (a) first-one inlet and one outlet, (b) second-two inlets and one outlst, (c) third-four inlets and one outlet, and (d) fourth-four inlets and two outlets..

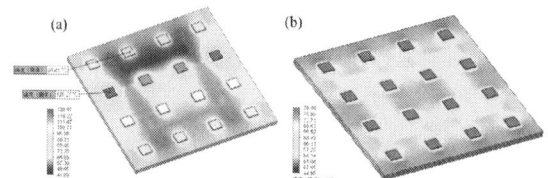

Fig. 8. The temperature distribution of (a) the first structure and (b) the fourth (optimized) structure.

The following three structures gradually reduce the proportion of series connection by adding inlet and outlet numbers. It can be seen from Fig. 7 that the more the inlet or outlet number is, the total pressure drop is. This is attributed to the increased inlet or outlet number leads to reduced effective flow path. The pressure drop of the second, third, and fourth structure is 1.52 bar, 1.03 bar, and 0.42 bar, respectively. The fourth structure with completely parallel connection characteristic exhibits the best flow and heat transfer performance. The liquid supply layer has four inlets, two outlets and 2×2 separate H-type bifurcation structures. The liquid return layer has sixteen liquid supply channels and sixteen liquid return channels. As shown in Fig 8(b), the maximum temperature rise above the inlet temperature is 39 °C, which is 48 °C lower than that of the first structure. Moreover, the temperature difference between the chip array is within 1 °C, demonstrating a very high temperature uniformity. Fig. 9 illustrates the velocity distribution of the fourth cooler. There is a uniform velocity distribution between the microchannel array with an average velocity of 0.8 m/s in the microchannel. In summary, the optimized microchannel cooler (the fourth one) can achieve simultaneous low temperature rise, low temperature deviation and low pressure drop to dissipate heat flux of 500 W/cm² for high-power chip array.

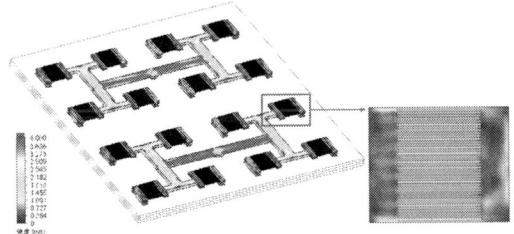

Fig. 9. The velocity of the fourth (optimized) structure..

B. Experimental Results

Fig. 5 shows the measured temperature distribution of the 4×4 thermal test chip array. It can be seen that there are 16 obvious hotspot in the filed of vision. Moreover, the temperature distribution of 16 chips was relatively uniform. Table 2 shows the measured maximum and average temperature of 4×4 chip array. The maximum temperature of the chip array was 57.7 °C at a flow rate of 36 L/h, which was 37.7 °C higher than the inlet temperature. The maximum deviation of the average temperature among these chips was 8.8 °C, which was less than ±5 °C. In summary, the proposed microchannel cooler could achieve efficient high heat-flux chip array cooing.

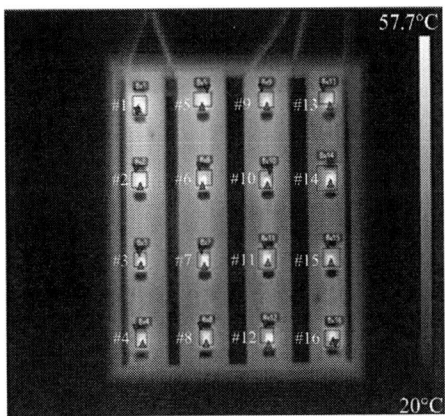

Fig. 10. The measured temperature distribution of the 4×4 thermal test chip array.

TABLE II. THE MEASURED MAXIMUM AND AVERAGE TEMPERATURE OF 4×4 THERMAL TEST CHIP ARRAY

No.	Maximum temperature (°C)	Average temperature (°C)	No.	Maximum temperature (°C)	Average temperature (°C)
#1	55.6	46.4	#9	53.3	43.5
#2	57.7	47.4	#10	53.4	42.3
#3	52.7	43.9	#11	54.5	40.1
#4	50.5	42.5	#12	55.8	41.1
#5	56.6	46.1	#13	50.0	41.6
#6	53.8	43.6	#14	54.5	38.6
#7	52.7	43.1	#15	50.5	39.2
#8	50.5	41.0	#16	49.1	40.5

V. CONCLUSIONS

In this study, we proposed a multi-layer silicone microchannel cooler for 4×4 chip array cooling with heat flux of 500 W/cm². The co-design of the heat transfer enhancement and flow homogenization was conducted using a finite-element simulation. The optimized microchannel cooler consisted of three layers, i.e., the liquid supply layer with 2×2 separate H-type bifurcation structures, the liquid return layer with sixteen liquid supply and return channels, and the microchannel layer with 4×4 distributed parallel microchannels underneath the 4×4 chip array. The results showed a uniform velocity and pressure distribution was achieved in 4×4 microchannels. And the maximum temperature rise of the chip array above the inlet temperature was less than 40°C and the average temperature deviation was less than ±5 °C. In summary, the proposed compact multi-layer silicone microchannel cooler achieved efficient and uniform cooling of high heat-flux chip array, contributing to the advancement of future high power and highly-integrated microwave devices and other power electronics.

ACKNOWLEDGMENT

None.

REFERENCES

[1] A. Horsley, P. Appel, J. Wolters. J. Achard, A. Tallaire, P. Maletinsky, and P. Treutlein, "Microwave Device Characterization using A Widefield Diamond Microscope," Phys. Rev. Appl, vol. 10, pp. 044039, 2018.

[2] N. K. Subramani, J. Couvidat, A. Al Hajjar, J. C. Nallatamby, R. Sommet, and R. Quéré, "Identification of GaN Buffer Traps in Microwave Power AlGaN/GaN HEMTs through Low Frequency S-Parameters Measurements and TCAD-based Physical Device Simulations," IEEE J. Electron Devi., vol. 5(3), pp. 175-181, 2017.

[3] N. H. Naqiuddin, L. H. Saw, M. C. Yew, F. Yusof, T. C. Ng, and M. K. Yew, "Overview of Micro-Channel Design for High Heat Flux Application," Renew. Sust. Energ. Rev., vol. 82, pp. 901-914, 2018.

[4] X. B. Luo, R. Hu, S. Liu, and K. Wang, "Heat and Fluid Flow in High-Power LED Packaging and Applications," Prog. Energ. Combust., vol. 56, pp. 1-32, 2016.

[5] S. Liu, J. Yang, Z. M. Gan, and X. B. Luo, "Structural Optimization of A Microjet based Cooling System for High Power LEDs," Int. J. Therm. Sci., vol. 47(8), pp. 1086-1095, 2008.

[6] T. Muszynski and D. Mikielewicz, "Structural Optimization of Microjet Array Cooling System," Appl. Therm. Eng., vol. 123, pp. 103-110, 2017.

[7] T. Muszynski, and R. Andrzejczyk, "Heat Transfer Characteristics of Hybrid Microjet–Microchannel Cooling Module," Appl. Therm. Eng., vol. 93, pp. 1360-1366, 2016.

[8] T. W. Wei, H. Oprins, L. Fang, V. Cherman, I. De Wolf, E. Beyne and M. Baelmans, "Nozzle Scaling Effects for the Thermohydraulic Performance of Microjet Impingement Cooling with Distributed Returns," Appl. Therm. Eng., vol. 180, pp. 115767, 2020.

[9] N. Gilmore, V. Timchenko, and C. Menictas, "Microchannel Cooling of Concentrator Photovoltaics: A Review," Renew. Sust. Energ. Rev., vol. 90, pp. 1041-1059, 2018.

[10] R. Van Erp, G. Kampitsis, and E. Matioli, "Efficient Microchannel Cooling of Multiple Power Devices with Compact Flow Distribution for High Power-Density Converters," IEEE T. Power Electr., vol. 35(7), pp. 7235-7245, 2019.

[11] A. A. Japar, N. A. C. Sidik, and S. Mat, "A Comprehensive Study on Heat Transfer Enhancement on Microchannel Heat Sink with Secondary Channel," Int. Commun. Heat Mass, vol. 99, pp. 62-81, 2018.

[12] R. Van Erp, R. Soleimanzadeh, L. Nela, G. Kampitsis, and E. Matioli, "Co-designing Electronics with Microfluidics for More Sustainable Cooling," Nature, vol. 585(7824), pp. 211-216, 2020.

[13] K. W. Jung, C. R. Kharangate, H. Lee, J. Palko, F. Zhou, M. Asheghi, and K. E. Goodson, "Microchannel Cooling Strategies for High Heat Flux (1 kw/cm2) Power Electronic Applications," in Proc. 16th IEEE Intersoc. Conf. Therm. Thermomech. Phenomena Electron. Syst. (ITherm), pp. 98-104, 2017.

[14] S. M. Walsh, B. A. Malouin, Jr., E. A. Browne, K. R. Bagnall, E. N. Wang, and J. P. Smith, "Embedded Microjets for Thermal Management of High Power-Density Electronic Devices," IEEE T. Comp. Pack. Man., vol. 9(2), pp. 269-278, 2019.

[15] J. Ditri, J. Hahn, R. Cadotte, M. McNulty, and D. Luppa, "Embedded Cooling of High Heat Flux Electronics Utilizing Distributed Microfluidic Impingement Jets," in Proc. ASME Int. Tech. Conf. Exhib. Packag. Integr. Electron. Photon. Microsyst. (INTERPACK), San Francisco, CA, USA, pp. 1–10, 2015.

[16] J. Ditri, R. Cadotte, D. Fetterolf, and M. McNulty, "Impact of Microfluidic Cooling on High Power Amplifier RF Performance," in Proc. 15th IEEE Intersoc. Conf. Therm. Thermomech. Phenomena Electron. Syst. (ITherm), Las Vegas, NV, USA, pp. 1501–1504, 2016.

[17] K. W. Jung, C. R. Kharangate, H. Lee, J. Palko, F. Zhou, M. Asheghi, E. M. Dede, K. E. Goodson, "Embedded Cooling With 3D Manifold for Vehicle Power Electronics Application: Single-Phase Thermal-Fluid Performance," Int. J. Heat Mass Tran., vol. 130, pp. 1108–1119, 2019.

[18] D. H. Altman, A. Gupta, and M. Tyhach, "Development of a Diamond Microfluidics-based Intra-Chip Cooling Technology for GaN," in Proc. ASME Int. Tech. Conf. Exhib. Packag. Integr. Electron. Photon. Microsyst. (INTERPACK), San Francisco, CA, USA, pp. 1–7, 2015.

Laser rapid synthesis of ultra-small Ni nanoparticles embedded graphene for high-performance supercapacitors

1st Fangcheng Wang
Institute of Materials Research,
Tsinghua Shenzhen International
Graduate School
Tsinghua University
Shenzhen, China
wfangcheng@sz.tsinghua.edu.cn

2nd Zhuo Zhang
Institute of Materials Research,
Tsinghua Shenzhen International
Graduate School
Tsinghua University
Shenzhen, China
zhuo-zha19@mails.tsinghua.edu.cn

3rd Guangyao Zhao
Institute of Materials Research,
Tsinghua Shenzhen International
Graduate School
Tsinghua University
Shenzhen, China
zgy19@mails.tsinghua.edu.cn

4th Mingjie Liu
Institute of Materials Research,
Tsinghua Shenzhen International
Graduate School
Tsinghua University
Shenzhen, China
liumj20@mails.tsinghua.edu.cn

5th Hongjin Fan
School of Physical and Mathematical
Sciences
Nanyang Technological University
Singapore
fanhj@ntu.edu.sg

6th Cheng Yang
Institute of Materials Research,
Tsinghua Shenzhen International
Graduate School
Tsinghua University
Shenzhen, China
yang.cheng@sz.tsinghua.edu.cn

Abstract—**Conductive carriers embedded with ultra-small metal nanoparticles have great potential in energy conversion and storage, but it is still a challenge to develop large-scale, cost-effective and rapid fabrication methods. Herein, we demonstrated that the precursors of graphene oxide and nickel salt can be easily transformed into ultra-small Ni nanoparticles (5~15 nm) anchored on reduced graphene oxide (RGO@Ni) via a simple laser instantaneous heating method. Without the need for a binder, the prepared RGO@Ni-based electrode material can be directly patterned into a micro energy storage device. As a potential application of the prepared RGO@Ni with this unique structure, we evaluated its electrochemical performance. Considering the synergistic effect of the surface oxide layer of ultra-small Ni nanoparticles and RGO, the RGO@Ni based electrodes exhibit high areal specific capacitance exceeding 51.5 mF cm^{-2}. This is much greater than that of pure RGO electrode materials. This work gives a fast and effective strategy for the rapid preparation of RGO@Ni composite materials and the elevation of capacitor performance.**

Keywords—ultra-small nanoparticles, laser reduction graphene oxide, one-step synthesis, supercapacitors

I. INTRODUCTION

The vigorous development of multi-functional electronic equipment has greatly stimulated the rapid development of high-performance energy storage systems in the fields of aerospace, smart grid, and high-end communications [1]. Supercapacitors (SCs) have attracted wide-ranging attention owing to their ultra-high power density, excellent cycle stability, environmental friendliness and wide operating temperature range. Especially in various applications such as starters that require large charge and discharge currents, supercapacitors have shown great prospects [2]. For example, three-dimensional porous carbon-based materials are promising candidate materials that meet all these requirements, and can store charges in the range of positive and negative potentials, thereby providing high voltage for SCs [3]. However, they usually show a lower capacitance because the charge is primary storage via the electric double layer. A wise approach is to introduce suitable pseudocapacitor materials to adapt to different situations, to maximize the advantages of each component, and to use the synergistic effect to further enhance the volume specific capacitance of the electric double layer capacitor (EDLC). Transition metal-based materials represent a typical pseudocapacitance material, and nickel oxides and hydroxides have become research hotspots because of their higher theoretical storage capacity, security and low cost. However, in practical applications, these pseudocapacitance electrode materials have low surface reactivity and poor electronic conductivity, and their comprehensive performance cannot meet the needs of rapidly developing electronic devices. It is worth mentioning that the Ni/hydroxide species spontaneously formed on the exposed surface of nickel particles in an air environment will also undergo redox reactions and exhibit electrochemical capacitance activity. Therefore, in the electrochemical reaction process, the metal nickel-based electrode will not only produce additional pseudocapacitance, but also the high conductivity which can further enhance the rapid transmission of electrons. This makes the nickel based nanomaterials a possible candidate for electrode materials [3,4]. The meso-scale pore structure provides an entrance for electrolyte ion diffusion, a framework for rapid charge/ion transport, and alleviates volume expansion during charge and discharge [4].

Nanoparticles (NPs), generally smaller than 10 nm, have a smaller size, larger specific surface area and faster ion/electron diffusion ability, thereby giving electrode materials higher utilization and better rate performance. To our knowledge, the preparation methods of metal-based nanoparticles embedded in carbon nanoparticles usually include multi-step chemical and physical methods, and involve some toxic reagents and special high-temperature equipment [3,4], which is usually time- and energy-

978-1-6654-1392-3/21 $31.00 © 2021 IEEE

consuming. Compared with other methods, laser processing can flexibly adjust processing parameters (such as multi-component precursors, vacuum environment, inert atmosphere and specific atmosphere) and laser parameters (such as wavelength, pulse width, pulse energy and repetition frequency), and achieve rapid synthesis of metal nanomaterials through instant heating and rapid cooling [6]. For example, a simple laser scribing can quickly convert the metal-organic framework (MOFs) into ultra-small nanoparticles (3-200 nm) embedded in conductive carbon supports [5]. However, there are still some obstacles in the practical application of carbon-based materials loaded with nickel nanoparticles prepared by this method for high-performance supercapacitors. Herein, based on the instantaneous heating effect of laser, nickel nanoparticles embedded in reduced graphene (RGO@Ni) was quickly synthesized and used for high-performance supercapacitors. Since the highly conductive three-dimensional porous structure provides active Ni centers, it can not only accelerate the transmission of electrons/ions, but also accelerate the redox reaction by enhancing the reactivity. Therefore, RGO@Ni based supercapacitors have high energy density, excellent rate performance and cycle stability, and provide a broad prospect for the preparation of carbon-based material loaded pseudocapacitor composite electrodes nanomaterials for supercapacitors with good comprehensive performance.

II. EXPERIMENTAL

A. Laser preparation of RGO@Ni electrode material

0.1 g of $NiCO_3$ (98%, NI ≥ 45%, Shanghai Aladdin Biochemical Technology Co., Ltd.) and 0.1 g of graphene oxide (GO, multilayer, Suzhou TANFENG graphene technology) were added into a reagent bottle with a 20 ml range, then 10 ml deionized water was added, and mixed for 24 hours under magnetic stirring to form a uniform precursor solution. The $GO/NiCO_3$ mixture solution was knife-coated on a stainless steel sheet, and the film was dried in a vacuum oven (60° C, 2 hours) to form a solid $GO/NiCO_3$ thin film. Next, the $GO/NiCO_3$ solid film rapidly transformed into a RGO@Ni composite film with a hierarchical porous structure under the irradiation of a focused laser beam that is positively defocused (the defocus depth is 20 mm).

B. Materials characterization and electrochemical measurements

The scanning electron microscope (HITACH S4800, Japan, working voltage is 5 kV) and transmission electron microscope (FEI Tecnai G2 F30, MDTC-EQ-M17-01, working voltage is 300 kV) have been used to study the shape and structure of RGO@Ni. XRD patterns were obtained using a Rigaku diffractometer (Bruker DS RINT2000/PC, Germany) (working at 40 kV and 120 mA in diffraction angles ranging from 10° to 80° degree). The elemental composition is analyzed by X-ray photoelectron spectroscopy (XPS, PHI 5000 Versaprobe II, Ulvac-Phi). Electrochemical performance were tested on an electrochemical workstation (VMP3, Bio-Logic, France). Electrolyte system used in all test systems is Na_2SO_4 solution (0.5 mol·L^{-1}).

III. RESULTS AND DISCUSSION

Fig. 1 is a schematic diagram of laser-induced RGO@Ni processing. The processing system using a fiber nanosecond laser equipped with a laser scanning galvanometer (CTI EC1000) directly induces the RGO@Ni film on the $GO/NiCO_3$ precursor film in an air environment. The laser scanning galvanometer system consists of X-axis, Y-axis galvanometer, field lens (JENar 170-1030 F-theta), etc. Its working principle is to control the reflection angle through a computer, and the incident laser beam can be scanned along the X and Y directions respectively, so as to control the laser beam in a two-dimensional plane. The internal deflection is finally focused on the surface of the precursor through the field lens and the direct writing of the preset pattern is completed. The scanning speed in the focal plane can reach 5000 mm·s^{-1}, and the effective processing range is 15 cm×15 cm.

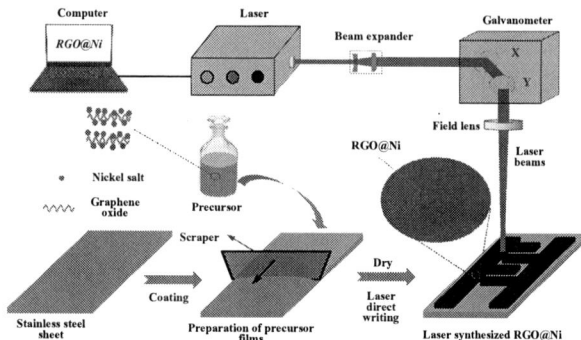

Fig. 1. Schematic diagram of the laser rapid preparation of RGO@Ni composite materials.

Fig. 2 shows the prepared RGO@Ni under different magnifications. SEM images show that the porous carbon skeletons of different pore sizes are composed of macropores and mesopores. Research results have shown that the macropores and multiple charge transfer channels of the carbon skeleton are beneficial to the migration and diffusion of ions [6]. More specifically, macropores (>50 nm) can act as ion buffer regions, accelerate ion diffusion, provide excellent mass/heat/electron transfer characteristics and a stable network framework. Mesopores (2 ~ 50 nm) can reduce ions transmission distance and can withstand volume changes of active materials. Therefore, the hierarchical pore structure with macropores and mesopores can synergize the advantages of pores of different sizes. It can be clearly observed that the surface of the RGO reduced by the laser is evenly embedded with ultra-fine Ni nanoparticles. In order to accurately measure the size of Ni nanoparticles, they were further enlarged and observed. Fig. 2 (c, d) show that the size of the nanoparticles is uniform, about 5-15 nm, indicating that Ni nanoparticles can effectively increase the specific surface area, thereby increasing the number of active sites for the catalytic reaction.

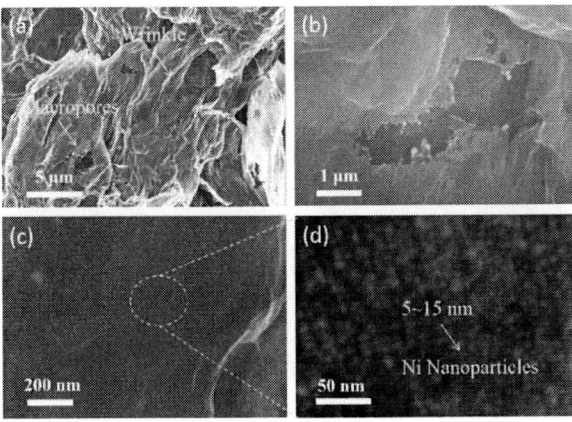

Fig. 2. (a-d) SEM images of laser synthesized RGO@Ni under different magnifications.

In Fig. 3(a), the Raman spectrum of the RGO@Ni sample shows two characteristic peaks at 1356 cm^{-1} (D bands) and 1594 cm^{-1} (G bands) [7]. Usually we use the ratio of ID/IG (ID is the area of the D band to the baseline, and so is IG) to indicate the degree of disorder of RGO. After calculation, the ID/IG ratio of RGO@Ni is 2.04, indicating that the surface of RGO has rich defect density, which is mainly due to the photo-induced plasma etching and the embedded Ni nanoparticles. These defects produced by laser-induced plasma provide abundant nucleation sites for Ni nanoparticles, while the pore defects produced by rapid release of by-product gas provide a "high-speed channel" for ion transport in electrolyte. The surface defects caused by laser-induced plasma induce the interaction between Ni nanoparticles and RGO, thereby promoting the anchoring of Ni nanoparticles [8]. The defects of RGO will be produced in the process of laser reduction of GO, so as to obtain better dispersion of Ni nanoparticles, and to prevent them from agglomerating into larger particles due to the confinement effect of the RGO template. Due to its low content, no obvious Ni peak was observed in the Raman spectrum. Therefore, X-ray diffraction (XRD) patterns is employed to strongly confirm the existence of Ni. In order to reveal the detailed information of the crystal structure of the prepared RGO@Ni sample (Fig. 2(b)). The XRD patterns at 2θ= 26.228 and 44.365 respectively coinciding with the (002) and (011) planes of the graphitized carbon of the standard PDF card (No. 96-101-1061), indicating that laser irradiation can convert GO to RGO, which is also consistent with the results of Raman spectroscopy. The diffraction peaks at 2θ=43.672°, 50.871°, 74.803° are indexed to the plane of cubic Ni (No. 96-901-3032), corresponding to the (111), (002) and (022) planes, respectively. In addition, three additional peaks at 37.095°, 43.098° and 62.590° were found in the RGO@Ni, which are correspond to the (111), (200) and (220) crystal planes of cubic NiO, which can be attributed to the Ni nanoparticles that are exposed to the air to form an oxide layer. XPS was used to analyze the surface chemical state of RGO@Ni. The C 1s XPS spectrum of RGO@Ni in Fig. 3(c) were deconvoluted into four peaks: C-C,C-H (284.8 eV), C-O (285.5 eV), C=O (286.8 eV), and O-C=O (289 eV) [8]. The Ni 2p spectrum of RGO@Ni is shown in Fig. 3(d), and the subpeak caused by the oxide layer can be detected. The high-resolution Ni2p spectrum of RGO@Ni display two main peaks at 854.02 eV and 872.03 eV, assigning to the Ni2p$_{3/2}$ and Ni2p$_{1/2}$ spectra. The two smaller satellite peaks with binding energies of 861.19 eV and 879.33 eV indicate the presence of NiO [9]. Taking into account the depth of measurement in XPS

analysis, it is believed that the thickness of the oxide layer of Ni nanoparticles is considered to be less than a few nanometers. This is because when synthesizing Ni nanoparticles below the submicron level, the surface oxide layer is inevitably formed due to the unstable energy of the surface nickel atoms.

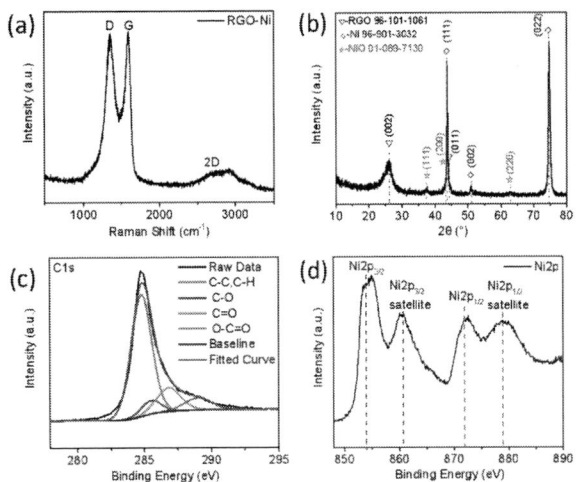

Fig. 3. (a) Raman spectra of RGO@Ni. (b) Typical XRD patterns of RGO@Ni. (c) C 1s spectra for RGO@Ni. (d) High-resolution XPS analysis of Ni 2p spectra for RGO@Ni.

In view of the many advantages of RGO@Ni, we conducted cyclic voltammetry (CV), constant current discharge (GCD) and electrochemical impedance spectroscopy (EIS) tests to analyze its electrochemical performance. Fig. 4(a) shows the CV curves of RGO@Ni at various scanning speeds in 1.0 M KOH aqueous electrolyte, showing the quasi-rectangular profile of the samples. Fig. 4(b) shows the GCD curves with current density in the range of 1.25 to 12.5 mA cm^{-2}. Fig. 4(c) further shows the correspondence between area specific capacitance and scanning speed. The corresponding plots of areal energy density and areal power density is further shown in Fig. 4(d). From the above figures, we can find that the discharge curves of SCs are always nearly linear under a wide range of various current densities. It is calculated that the areal capacitance at scanning speed of 5 mV s^{-1} is calculated as ~46.41 mF cm^{-2}. When the scanning speed is increased to 50 mV s^{-1} (~35.94 mF cm^{-2}), approximately 77% of the capacitance is still retained, demonstrating excellent rate performance. At the scanning speeds below 500 mV s^{-1}, the specific capacitance of RGO@Ni-based supercapacitors is about 5 times of that of RGO. A reasonable explanation is that ultra-small Ni nanoparticles can undergo redox reactions in the KOH aqueous electrolyte solution, and can greatly increase the capacity of supercapacitors. In addition, the typical EIS Nyquist plots of the RGO@Ni based SCs are shown in Fig. 4(e). We can estimate that the internal resistance is about 1.39 Ω, which means that the electron transmission speed is fast. This is also consistent with high-rate performance. As shown in Fig. 4 (f), our RGO@Ni based supercapacitors exhibit good electrochemical stability (capacitance remains 91.9%) at 100 mV s^{-1} scanning rate even after 10000 cycles.

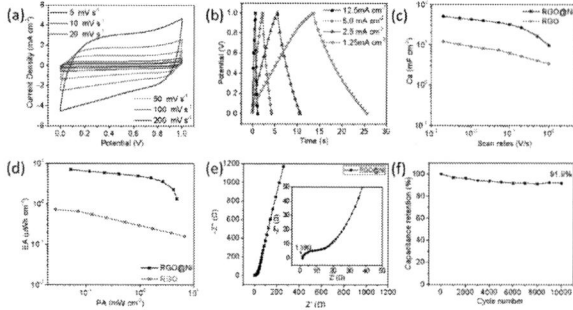

Fig. 4. (a) CV curves of the RGO@Ni-based supercapacitors at different scanning speeds in a two-electrode system. (b) Galvanostatic charge and discharge curves of RGO@Ni at varying current densities of 1.25, 2.5, 5.0, and 12.5 mA cm^{-2}. (c) The evolution trend of the areal specific capacitance of RGO@Ni and RGO at different scanning speeds. (d) Plot of area energy density versus power density of RGO@Ni and RGO. (e) Nyquist plot of impedance; inset: the enlarged view at high frequencies. (f) Cycling stability of RGO@Ni.

IV. CONCLUSION

In summary, we report a simple laser instantaneous heating to easily convert GO/NiCO$_3$ precursors into RGO@Ni composite films with ultra-small nickel nanoparticles (5~15 nm), and used to manufacture ultra-thin films with high comprehensive performance supercapacitors. Taking into account the synergistic effect of the surface oxide layer of ultra-small Ni nanoparticles and the porous structure of RGO, RGO@Ni based electrodes exhibit excellent areal specific capacitance exceeding 51.5 mF cm^{-2} and good cycle performance, which is much higher than traditional graphene-based supercapacitors. The reasons for the exciting performance of RGO@Ni can be divided into the following two factors. 1) The oxide layer on the surface of the ultra-small Ni nanoparticles can increase a part of the pseudocapacitance. 2) Three-dimensional porous structure and a large number of defects are produced in the process of laser reduction of GO, thereby providing more effective contact surface areas and improves the active sites for ion attachment between RGO@Ni and the electrolyte. We believe that the laser direct writing methods to prepare highly conductive carriers with ultra-small metal particles will become a promising material for energy storage devices in wearable and flexible electronic products.

ACKNOWLEDGMENT

The authors acknowledge the financial support from the National Natural Science Foundation of China (Project No. 52005289, 52061160482), the China Postdoctoral Science Foundation (Project No. 2020M670309), Guangdong Province Science and Technology Department (Project No. 2020A0505100014), Shenzhen Government (Project No. JSGG20191129110201725) and Tsinghua Shenzhen International Graduate School Overseas Collaboration Project, AME Individual Research Grant (Grant number: A1983c0026), Agency for Science, Technology, and Research (A*STAR) for financial supports.

REFERENCES

[1] Kyeremateng, Nana Amponsah, Thierry Brousse, and David Pech. "Microsupercapacitors as miniaturized energy-storage components for on-chip electronics." Nat. nanotech., vol. 12, no.1, pp. 7-15. November 2016.

[2] L. Christophe, J. L. Bideau and T. Brousse. "Challenges and prospects of 3D micro-supercapacitors for powering the internet of things." Energy Environ. Sci., vol. 12, no.1, pp. 96-115. Oct 2018.

[3] Y. Q. Jiang, C. Zhou and J. P.Liu. "A non-polarity flexible asymmetric supercapacitor with nickel nanoparticle@carbon nanotube three-dimensional network electrodes." Energy Storage Mater., vol. 11, pp. 75-82. Sep 2017.

[4] J. Li, Y. H. Wang, J. Tang, Y. Wang, T. Y. Wang, L. J. Zhang and G. F. Zheng. Direct growth of mesoporous carbon-coated Ni nanoparticles on carbon fibers for flexible supercapacitors. J. Mater. Chem. A, vol. 3, no.6, pp. 2876-2882. Dec 2015.

[5] D. S. Zhang, B. Gökce and S. Barcikowski. "Laser synthesis and processing of colloids: fundamentals and applications." Chem. Rev., vol. 117, no.5, pp. 3990-4103. March 2017.

[6] Z. W. Peng, R. Q. Ye, J. A. Mann, D. Zakhidov, Y. L. Li, P. R. Smalley, J. Lin and J. M. Tour. "Flexible boron-doped laser-induced graphene microsupercapacitors. " ACS nano, vol. 9, no.6, pp. 5868-5875. June 2015.

[7] X. Wang, S. X. Zhao, L. B. Dong, Q. L. Lu, J. Zhu, C. W. Nan. "One-step synthesis of surface-enriched nickel cobalt sulfide nanoparticles on graphene for high-performance supercapacitors." Energy Storage Mater., vol. 6, pp. 180-187. Nov 2016.

[8] J. Wang, Q. Zhao, H. S. Hou, Y. F. Wu, W. Z. Yu, X. B. Ji and L. D. Shao. "Nickel nanoparticles supported on nitrogen-doped honeycomb-like carbon frameworks for effective methanol oxidation." RSC adv., vol. 7, no.23, pp. 14152-14158. Feb 2017.

[9] J. Joy, A. Sekar, S. Vijayaraghavan, P. Kumar T., V. K Pillai and S. Alwarappan. "Nickel-Incorporated, Nitrogen-Doped Graphene Nanoribbons as Efficient Electrocatalysts for Oxygen Evolution Reaction." J. Electrochem. Soc., vol. 165, no.3, pp. 141-146. Feb, 2018.

Signal Integrity Design and Analysis of High Bandwidth Memory on Silicon Interposer*

Jin Hu
Jiangnan Institute of Computing and Technology
Wuxi, China
puffbar@163.com

Tao Li
Jiangnan Institute of Computing and Technology
Wuxi, China
leetel@163.com

Yuqing Fan
Jiangnan Institute of Computing and Technology
Wuxi, China
120705947@qq.com

Abstract—In this paper, the signal integrity design and analysis of HBM channel on silicon interposer is presented. Firstly, the concept of silicon interposer integrated with HBM and GPU for high bandwidth system is introduced. Secondly, the detailed stack-up of silicon interposer based on copper damascene process is proposed. And the physical dimension and the material property of the proposed silicon interposer are also provided. Thirdly, a novel signal routing pattern in silicon interposer design is presented to alleviate the problem in terms of conductor loss and crosstalk so that the signal integrity of HBM interface can be improved. Finally, the electrical performance of the HBM interface is analyzed by simulation in the frequency and time domain. The simulation results show that the electrical performance of HBM channel can meet the target specification and the proposed routing pattern design is validated.

Keywords—high bandwidth memory, silicon interposer, signal integrity, eye diagram

I. INTRODUCTION

With the rapid growth of the computing performance of high-end chip such as graphics processing unit (GPU) and artificial intelligence (AI), there has been an urgent demand for high bandwidth memory interface because memory access is becoming the bottleneck in developing high performance computing system. Over the past years, two kinds of technology strategies have been emerged to meet the ever increasing demand of implementing high bandwidth memory interface. The one is hybrid memory cube (HMC) specification [1]. It can dramatically improve the memory access bandwidth by increasing the data rate of the single link, which belongs to the serial link transmission. The other is high bandwidth memory (HBM) specification [2]. It significantly improves the memory access bandwidth by increasing the data bit width, which belongs to the parallel link transmission. As the data rate of serial links over 10Gbps, the serial link appears highly lossy. Furthermore, the power consumption of the serdes including pre-emphasis, equalization and clock circuits is increasing. So far, HMC has not been widely applied in the industry. On the contrary, HBM has been recognized by the market and achieved mass production. AMD and NVIDIA have successively launched a serial of high performance graphics processing acceleration products equipped with HBM DRAMs.

Thanks to the technology progress of advanced package and memory device, multiple high bandwidth memory DRAMs and SOC can be integrated in the silicon interposer through 2.5D package to build TeraByte memory bandwidth module, which is illustrated in Fig. 1. The HBM DRAM consists of the DRAM dies at the top of the stack and the logic die at the bottom of the stack, which are physically connected by through silicon vias (TSVs). The HBM DRAM and SOC are laterally integrated to the silicon interposer via solder ubumps typical 25 microns in diameter, which is much smaller than standard flip chip package C4 bumps. The silicon interposer is fabricated with mature process, such as 65nm, to reduce the manufacture cost. Also, the silicon interposer comprises redistribution layers (RDLs) to facilitate the high density signal routing from SOC to HBM DRAM. Then, the silicon interposer is connected to the package substrate using standard C4 bumps. Finally, the package substrate is soldered on the PCB by BGA balls to interconnect the interface signals of the SOC.

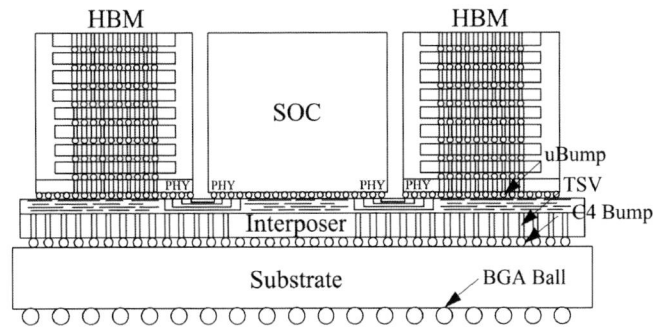

Fig. 1. The conceptual figure of 2.5D silicon interposer with HBM DRAMs and SOC.

However, when the data rate is beyond 2Gbps, HBM channel on the silicon interposer channel will suffer from crosstalk and inter symbol interference (ISI), which will result in signal integrity degradation. From this point of view, there is a need to optimize HBM channel design on silicon interposer to minimize signal distortion [3-6]. In this paper, the signal integrity design and analysis of HBM channel on silicon interposer is presented. The paper composes as follows. Section II proposes the detailed stack-up of silicon interposer based on copper damascene process. Section III introduces a novel signal routing pattern in silicon interposer design to improve the signal integrity of HBM interface. And the advantage of the proposed routing pattern is explained. The holistic simulation model contained with the driver IO buffer, the HBM channel and the receiver load is described in Section IV to implement the simulation and signal integrity analysis of the designed HBM silicon interposer. Specifically, the electrical performance of the HBM interface is analyzed by simulation in the frequency and time domain. Finally, the conclusions are shown in Section V.

978-1-6654-1392-3/21 $31.00 © 2021 IEEE

II. INTERPOSER STACK-UP

The silicon interposer has been usually fabricated by copper damascene process. The detailed stack-up of silicon interposer is illustrated in Fig. 2. And the physical dimension and the material property of the proposed silicon interposer are also provided.

The silicon interposer is composed of one-side 4 redistribution layers and silicon substrate. Metal4 and metal 2 are applied to HBM signal routing layers. Metal3 is designed for meshed ground layer, which is employed as reference return current paths for HBM signal channels. And metal1, the meshed power layer, is designed for power distribution such as IO interface power supply VDDQ and core power supply VDD.

The material property and physical dimension of the silicon interposer are usually defined based on copper damascene process. The passivation layer and the dielectric are normally fabricated by Si_3N_4 and SiO_2, which the relative permittivities are approximately 7.5 and 3.9 respectively. And the relative permittivity and the conductivity of the silicon substrate where the TSVs pass through are 11.9 and 10S/m. As for the physical dimension, the thickness of the metal and via layers approximates 1um. However, the thickness of the silicon substrate is about 100um.

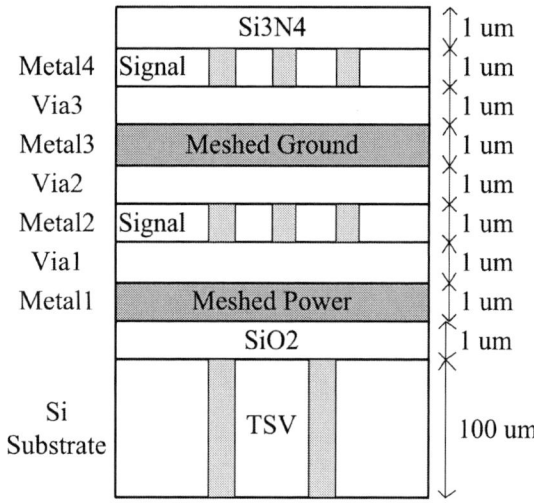

Fig. 2. The stack-up of the silicon interposer.

III. ROUTING PATTERN DESIGN

In this section, a novel signal routing pattern in silicon interposer design is proposed to improve the signal integrity of HBM interface. The presented routing pattern includes three aspects as follows. Firstly, according to the HBM DRAM ubump pitch and the routing region, the HBM DWORD and AWORD signals routing width is appropriated designed to alleviate the problem in terms of conductor loss and ensure the signal amplitude in the receiver of the HBM channel. Secondly, each HBM signal is routed symmetrically with gnd line shielded so that the horizontal crosstalk between neighbor signals within the same layer can be reduced effectively, which will optimize the eye diagram

performance of jitter, width and height. Thirdly, in order to suppress the vertical crosstalk between adjacent signals within the different layers, the arrangement of signal and gnd lines is designed so that the staggered signal and gnd routing pattern can be achieved. Therefore, not only the high density HBM signal interconnection on silicon interposer is implemented, but also the signal integrity of the HBM interface can be improved by means of the mentioned routing pattern design.

A. Appropriate Routing Trace Width

To start the HBM silicon interposer design, the signal trace routing dimension should be determined firstly. As for as the silicon interposer, the signal line width and spacing are usually defined based on copper damascene process. Normally, the accepted range of the line width and spacing is 0.4um to 5um [5]. The ubump array of the HBM DRAM stack employs a staggered pattern defined by JEDEC standard as depicted in Fig. 3. The staggered ubump is located halfway between row and column ubump array, hence its location is determined by half of the horizontal ubump pitch and vertical ubump pitch, where the horizontal and vertical ubump pitch are 96um and 55um respectively.

The typical staggered HBM ubump array illustrated in Fig. 3 consists of 6 rows with a pitch of 27.5um and 24 columns with a pitch of 48um. The first and sixth row are ground and power ubumps. And the middle four rows belong to signal ubumps, which is totally 48 signals. These 48 signals should be escaped from ubump array via the vertical routing region, which the height is 137.5um. Because two metal layers are designed for HBM signal routing, the maximum 24 signals should be routed in each metal layer. So the single trace width plus spacing is about 5.7um. Here, the HBM trace width is appropriated designed to be 2.5um considering the conductor loss and the amplitude in the receiver.

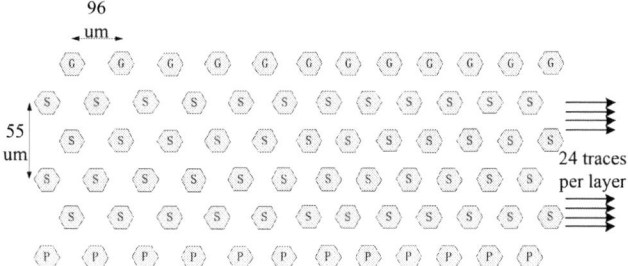

Fig. 3. The typical staggered HBM ubump array.

B. Horizontal Crosstalk Reduction

In the respect of signal integrity of HBM interface on the silicon interposer channel, crosstalk is a significant consideration, which will result in voltage and timing issues of the victim signals. So the crosstalk issue must be taken into consideration of signal integrity design of the silicon interposer. The silicon interposer consists of signal routing layers and meshed power ground layers. Trade-off should be studied between routing density and crosstalk performance.

The crosstalk includes the crosstalk from the horizontal direction between adjacent signals on the same layer and the crosstalk from the vertical direction between signals on the different layers. In the horizontal direction, each HBM

DWORD and AWORD signal is routed on wide wires, with a narrow ground shield line in parallel between the signals which is illustrated in Fig. 4. The gnd line width and the spacing between signal and gnd line are designed symmetrically so the single trace width plus spacing 5.7um can be maintained. Through the well designed of gnd line shielding, the horizontal crosstalk between neighbor signals within the same layer can be reduced effectively which will improve the receiver waveform and eye diagram performance in terms of jitter, width and height.

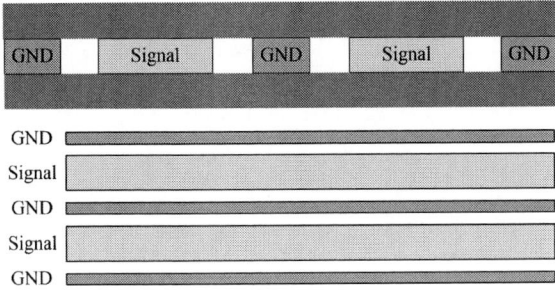

Fig. 4. Signal routing pattern with gnd line shielded.

C. Vertical Crosstalk Reduction

Furthermore, in order to suppress the vertical crosstalk between adjacent signals within the different layers, the arrangement of signal and gnd lines is proposed which results the staggered signal and gnd routing pattern can be implemented. The signals for HBM A,B,C and D channels are routed on metal4. The meshed ground layer metal3 routes the wide gnd lines parallel to the signal and shield lines for a complete gnd mesh. Then, the signal for HBM E,F,G and H channels are routed on metal2. Especially, the signal line on metal2 are underneath the metal4 shield lines with the metal4 signals over the metal2 shield lines, which is illustrated in Fig. 5. The staggered arrangement of routing pattern can minimize the vertical crosstalk between any of the signals. Therefore, not only the high density 1024 HBM signals routed between HBM DRAM and SOC die on silicon interposer is achieved, but also the signal integrity of the HBM interface can be improved by means of the mentioned routing pattern design.

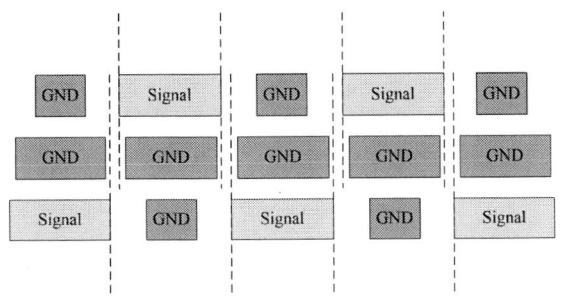

Fig. 5. The cross-section view of silicon interposer routing pattern.

IV. SIGNAL INTEGRITY ANALYSIS

As described in Section III, the high density HBM signal routing pattern on silicon interposer is proposed, in order to evaluate the performance of the presented HBM channel, the simulation model is set up for simulation and signal integrity analysis of the designed HBM silicon interposer.

Specifically, the signal integrity simulation and electrical performance analysis include frequency domain and time domain respectively. The frequency domain simulation is executed to examine the effect of crosstalk and insertion loss on signal integrity distortion. In order to assess the results of the frequency domain simulations, the time domain simulation is further conducted to evaluate the HBM signal eye diagram width and height.

A. Frequency-domain Simulation Analysis

For signal integrity analysis, 4 HBM signal lines where 2 lines routed on metal4 and 2 lines routed on metal2 are studied. All the frequency domain simulations are swept from 10MHz to 20GHz with three dimensional electromagnetic solver.

The frequency domain crosstalk simulation result is plotted in Fig. 6. Because totally 4 HBM signal lines are analyzed, totally 6 crosstalk curves which represent the effect on each other are illustrated. It can be shown that the crosstalk magnitude of all the curves is below -30dB from 10MHz to 20GHz. Hence, the crosstalk between neighbor signals within the same layer and the different layer can be reduced effectively by ground shielding and staggered signal and ground routing arrangement, which results in little effect on the signal integrity of the designed HBM silicon interposer channel.

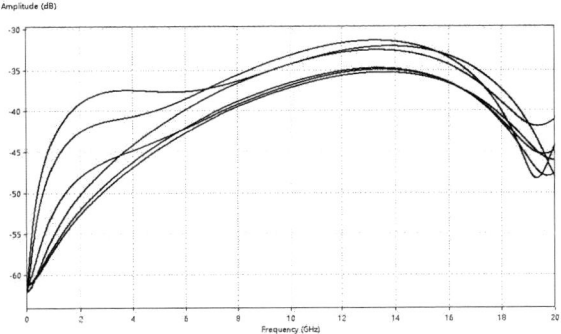

Fig. 6. The crosstalk of the silicon interposer routing.

The insertion loss of the signal lines, which will affect the signal amplitude in the receiver of the HBM channel is also simulated. The frequency domain insertion loss simulation result is plotted in Fig. 7. It can be shown that the insertion losses of all the signal lines on metal4 and metal2 are similar about -2.5dB at 1.6GHz response to the Nyquist frequency of the HBM2 specification. However, as the frequency increases, the overall insertion loss increases significantly because of the high silicon dielectric loss.

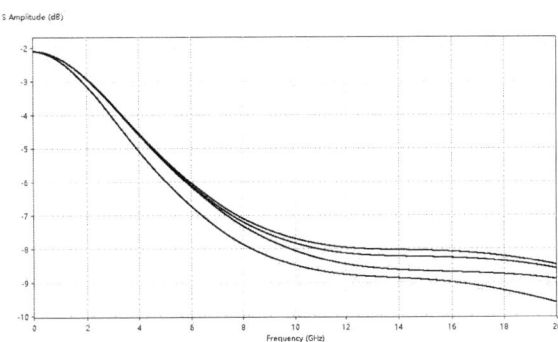

Fig. 7. The insertion loss of the silicon interposer routing.

978-1-6654-1392-3/21 $31.00 © 2021 IEEE

B. Time-domain Simulation Analysis

Based on the transmitter, the receiver and the S parameter channel model extracted from the frequency domain analysis, the transient time domain simulation can be conducted to evaluate the eye diagram performance. Similar to the frequency domain simulation, 4 HBM DWORD signal eye diagrams are studied. The pseudo random bit sequence is utilized as the input signal, the data rate and the amplitude of the input signal are 3.2Gbps and 1.2V respectively.

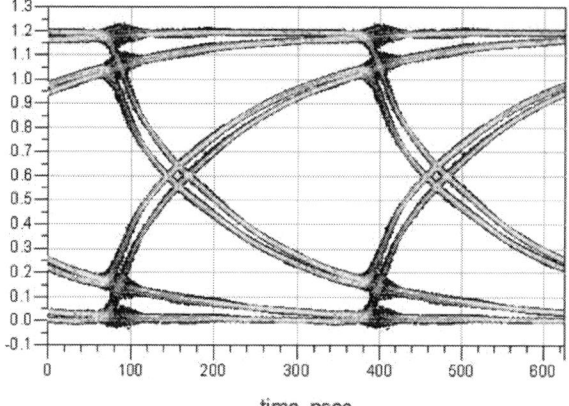

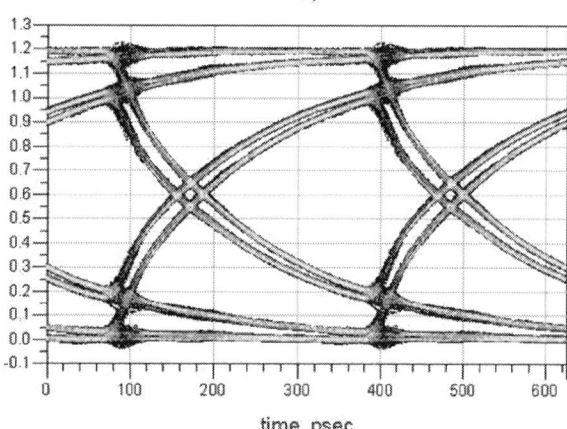

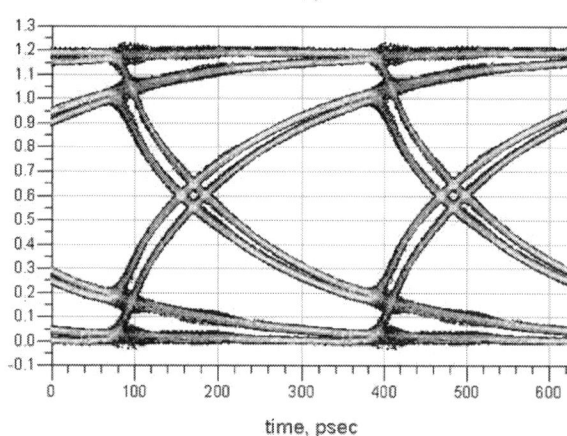

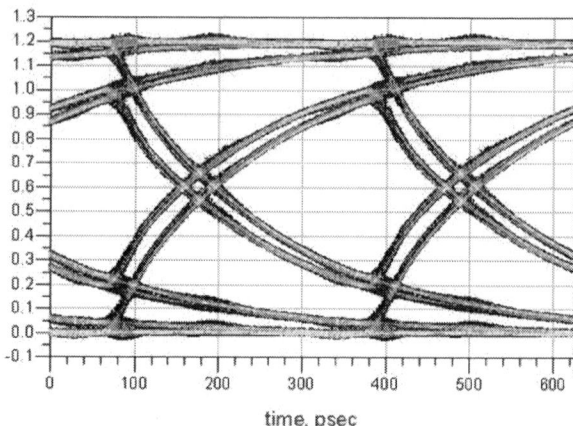

Fig. 8. The eye diagrams of the 4 HBM DWORD signals.

4 HBM DWORD signal eye diagrams in the time domain are shown in Fig. 8. As shown in the simulation results, the eye height voltages of the 4 HBM DWORD signals are 0.505V, 0.554V, 0.550V, and 0.506V respectively. And the eye widths are 278.1ps, 265.6ps, 270.3ps, and 254.7ps. As a result, the eye diagrams of all the signals are widely open and the signal integrity of the designed silicon interposer can be promised.

V. CONCLUSION

In this paper, the signal integrity design and analysis of HBM channel on silicon interposer is presented. The detailed stack-up of silicon interposer and the physical dimension as well as the material property of the proposed silicon interposer are introduced. To achieve the high density HBM channel routing, a novel signal routing pattern on silicon interposer is proposed. According to the HBM ubump pitch and routing region, the appropriate trace width is designed considering the conductor loss and the amplitude in the receiver. And the crosstalk issue can be minimized effectively by the staggered arrangement of routing pattern and ground shielding. The frequency domain and time domain signal integrity simulation validate that the electrical performance including the crosstalk, insertion loss and the eye diagram meets the target HBM specification, which illustrate the signal integrity of silicon interposer can be promised.

REFERENCES

[1] HMC Consortium, Hybrid memory cube specification 2.0, 2014.

[2] JESD235B, High bandwidth memory DRAM, 2018.

[3] K Cho, Y Kim, H Lee, et al, "Signal integrity design and analysis of differential high-speed serial links in silicon interposer with through-silicon via," IEEE Transaction on Component, Packaging and Manufacturing Technology 2019,9(1):107-121.

[4] Y Jeon, H Kim, J Kim, et al, "Design of an on-silicon-interposer passive equalizer for next generation HBM with data rate up to 8Gb/s," IEEE Transactions on Circuits and Systems-I 2018,65(7):2293-2303.

[5] K Cho, J Kim, H Lee, "Signal and power integrity design of 2.5D HBM on Si interposer," 2016 Pan Pacific Microelectronics Symposium.

[6] K Cho, Y Kim, H Lee, et al, "Signal and power integrity analysis of heterogeneous integration using EMIB technology for HBM", 2017 IEEE Electrical Design of Advanced Packaging and System Symposium.

Study on the Transport Performance of Microstrip Circuit Board with Voids in Solder Layer

Zhidan Liu
Xi'an Research Institute of Navigation Technology
Xi'an, China
798330181@qq.com

Zhiping Zhao
Xi'an Research Institute of Navigation Technology
Xi'an, China
zhaozhiping2019@163.com

Fei Zhang
Xi'an Research Institute of Navigation Technology
Xi'an, China
zhf121@126.com

Shuai Chen
Xi'an Research Institute of Navigation Technology
Xi'an, China
chenshuai19871219@163.com

Abstract—In T/R components of solid-state phased array radar, most of the microstrip printed circuit boards are grounded in large area. The quality of grounded welding directly affects the microwave performance of T/R components and the reliability and stability of components. In this paper, the influence of solder layer voids on transmission performance is analyzed by using high frequency structure simulation software to model the micro-strip circuit board and voids. Brazing voids are found to reduce the transmission performance of microwave signals and deteriorate the telecommunication performance. This effect becomes more and more significant with the increasing frequency of devices. The effect of voids near the feeding port and the edge of brazing layer on the transmission performance of microwave signal is greater than that of voids inside brazing layer. With the same brazing permeability, the void location is similar, and the scattered small voids have greater influence on the transmission performance of microwave circuit than the large voids with the same area. The effect of voids in solder layer on the transmission performance of microwave signals is that the voltage standing wave ratio is more sensitive and the effect on S_{21} is not obvious.

Keywords—Microstrip circuit board, Solder layer, Void, Transport performance

I. INTRODUCTION

T/R component is the core of phased array radar. A series of solid-state microwave integrated circuits, such as power module, power module, transmission branch module, form a planar circuit through the microstrip transmission line. The whole microstrip circuit board is fixed on the metal carrier as the ground plate to form a complete circuit[1,2]. This circuit eliminates many connectors and is widely used because of its small size, light weight and high reliability. Traditional grounding by screw connection results in crosstalk between functional blocks and increase of S_{21} due to gap between connections, as well as additional capacitance and oscillation. The commonly used method of screw compression can no longer obtain satisfactory microwave performance, which greatly increases the workload and difficulty of debugging and reduces the reliability of components. In the late 1980s, foreign countries began to try to replace the method of screw tightening with welding. By changing point contact to surface contact, the crosstalk was reduced to 0.1dB and the loss was greatly reduced.

Microwave grounding is different from DC, good grounding can suppress interference, and defective grounding can cause microwave performance degradation. The transmission loss of the microwave circuit increases, the peak transmission coefficient decreases, and the magnitude of the increase in transmission loss increases with the increase of grounding defect area. Frequency drift occurs in the S parameter of the microwave circuit, and the drift increases with the increase of the grounding defect area. When the grounding defect is located under the microwave element, transmission line and microwave passive network, it has the greatest impact. The size of grounding defects directly below the microwave elements, transmission lines and passive microwave network determines the magnitude of the variation of transmission coefficient and reflection coefficient of the microwave network[3,4].

After brazing, the micro-strip circuit board has good grounding and heat transfer characteristics, which reduces the transmission loss of the micro-strip line. However, it is found that during brazing process, it is easy to residual voids in the solder layer, resulting in uneven welding, due to the high requirement for technical personnel, bubbles due to the heat of the solder and incomplete volatilization of the solder. The brazing penetration rate is relatively low, which is one of the factors restricting the long-term reliability of microwave devices. The location and shape of the voids in the welded layer produced during packaging are random. According to the relevant literature, the voids are mostly circular or elliptical in shape. In this paper, the high-frequency structure simulation software is used to model the micro-strip circuit board and voids, and the effect of voids in the solder layer on the transmission performance is analyzed.

II. MODEL DESCRIPTION

It is difficult to quantitatively analyze the effect of the brazing voids on the transmission performance through experiments because of the random distribution of the voids in the micro-strip circuit board. In order to understand the influence of the size and location of the voids on the transmission performance of microwave signals, modeling and simulation are used. The effect of voids on the transmission performance of basic micro-strip circuit boards after large area brazing was studied. The structure consists of four parts: microstrip line, baseboard, brazing layer and grounding housing. The characteristic impedance of the microstrip plate is 50Ω. The geometric dimensions of the model are shown in Figure 1. The dimensions and materials of the components are shown in Table 1 below.

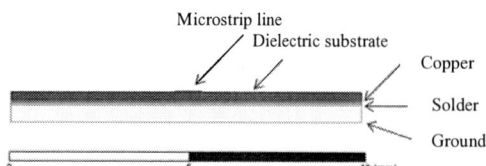

Fig. 1. The geometric dimensions of the model

TABLE 1. THE DIMENSIONS AND MATERIALS OF THE COMPONENTS

Assembly	Dimensions (Length × Width× Height) /mm	Material
Microstrip line	10×0.76×0.018	Cu
Dielectric substrate	10×10×0.254	Rogers 5880
Copper layer	10×10×0.018	Cu
Solder layer	10×10×0.08	Sn63Pb37
Ground Layer	10×10×0.5	Al

The transmission model of the micro-strip circuit board is established in the three-dimensional electromagnetic modeling software. As shown in Fig. 2, the voids of different positions and sizes are created in the brazing layer, and the electromagnetic simulation calculation is carried out to study the transmission performance. The parameters studied include VSWR and S_{21}. This paper mainly calculates the microwave circuit parameters of each model in S band (2-4GHz) and X band (8-12GHz).

Fig. 2. Transport model of microstrip circuit board

III. RESULTS AND DISCUSSION

3.1 Effect of void location on signal transmission performance

To investigate the effect of the change of void location on microwave signal transmission performance, four different locations are selected in the brazing layer. Position 1-4 above are built in different locations with a radius of 1 mm and a height of 0.08mm. Fig. 3 is a diagram of the location of a single void studied in this paper. Position 1 is the void at the microstrip line port, Position 2 is the center of the brazing layer, and directly below the microstrip line. Position 3 is the void inside the brazing layer and position 4 is the void at the side of the brazing layer.

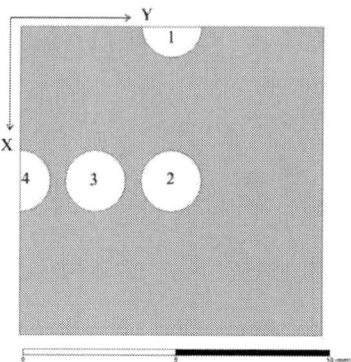

Fig. 3. Void position of solder layer.

The results of the electromagnetic modeling calculation for each case are shown in the following figure. It can be seen from the figure that the appearance of voids has an impact on the transmission performance of microwave circuits. In S band, the order of the effect of void location on the transmission performance of microwave signals is 1243 from large to small, and the forward transmission coefficient and voltage standing wave ratio increase with frequency. The VSWR is closest to the S21 when the internal void (3) in the brazing layer is fully penetrated. In X-band, the influence of cavity position on microwave signal transmission performance is 1423 in order of magnitude to smallest, and the forward transmission coefficient and voltage standing wave ratio in S-band increase compared with that in S-band.The voids have the greatest influence on the transmission performance of microwave circuit at the feeding port of brazing layer. After the voids are found, some results show that the voltage standing wave ratio and forward transmission will decrease with the increase of frequency, and the impedance matching will be better. The reason may be that the voids change the grounding capacitance of the whole circuit and make the original resonance frequency change.

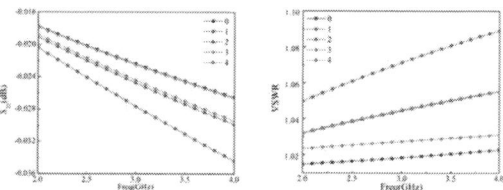

Fig. 4. S_{21} and VSWR of voids in different positions in S-band

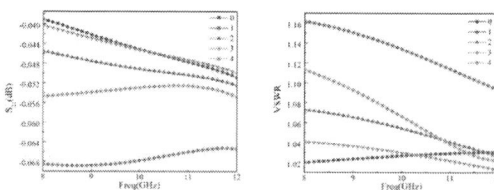

Fig. 5. S_{21} and VSWR of voids in different positions in X-band

Brazing voids have a greater impact on the transmission performance of microwave circuit at the feeding port, more severe degradation of high frequency and low frequency performance, and more sensitive to voltage standing wave, but relatively less influence on forward transmission coefficient. The presence of voids changes the grounding capacitance of the whole circuit and changes the original resonance frequency.

978-1-6654-1392-3/21 $31.00 © 2021 IEEE

3.2 Effect of void location change on signal transmission performance

In order to further explore the effect of cavity location on the transmission performance of microwave circuits, electromagnetic simulation is also performed on the transmission performance parameters of microwave circuits when the cavity changes along multiple paths X is the direction of the microstrip line and Y is perpendicular to the direction of the microstrip line. The distance from the center of the void to the edge of the strip plate is analyzed. First, the influence of void 1 moving along the X direction with a distance of 1 mm each time on the forward transmission coefficient and VSWR of the strip line is analyzed. During the simulation process, it is found that when the void is 1 mm away from the edge of the microstrip line, the cylindrical void is tangent to the upper edge of the welding layer. As shown in the figure 6, the calculation cannot be made, so no statistics are made at 1mm and 9mm.

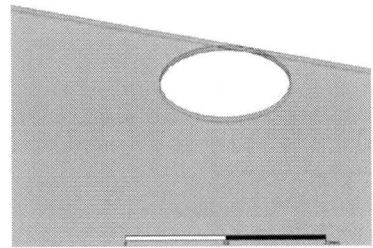

Fig. 6. Cylindrical voids tangent to the upper edge of the welded layer

The results of the electromagnetic simulation are shown in the following figure. In the above figure, the horizontal coordinate represents the location of the void in the brazing layer, coordinate 0 indicates that the void is at the feed port, and coordinate 10 indicates that the void is at the output port of the microstrip line. By comparing the above figure, it is found that when the void changes below the microstrip line, the S_{21} varies from -0.023dB to -0.028dB in the S-band and the voltage standing wave ratio varies from 1.03 to 1.07.In X-band, the VSWR varies from 1 to 1.14 and the S21 varies from -0.045dB to -0.070dB.

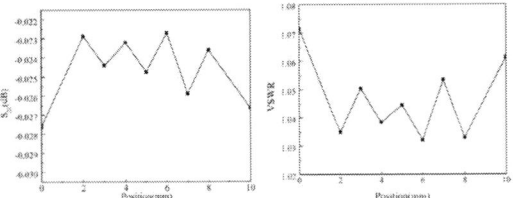

Fig. 6. S_{21} and VSWR of voids changes in X direction in S-band

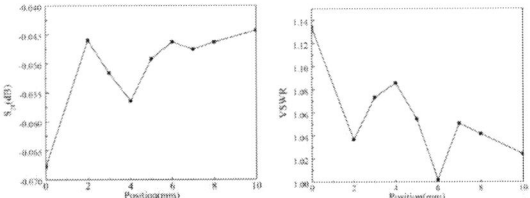

Fig. 7. S_{21} and VSWR of voids changes in X direction in X-band

It can be seen that the transmission coefficient and VSWR at S and X band input ports are significantly larger when the void changes below the microstrip line, and the void

at the output port also has a greater impact on microwave transmission performance at S band, while the higher the frequency of the void at the port, the greater the impact on Microwave transmission performance at the other locations.

From the figure below, we can see that the VSWR varies from 1.02 to 1.07 and the S_{21} varies from -0.022dB to -0.027dB in the low-frequency S-band.Within X-band, the voids at the edge of brazing layer have little influence on the transmission performance of microwave devices. The VSWR varies from 1.02 to 1.08, and the S_{21} varies from -0.045dB to -0.053dB.It can be seen that when the voids change along the Y direction (the vertical direction of the printed circuit board), the transmission coefficient and VSWR at the edge of the brazing layer are significantly larger than those at the other locations, and the other locations have little influence on the microwave transmission performance.

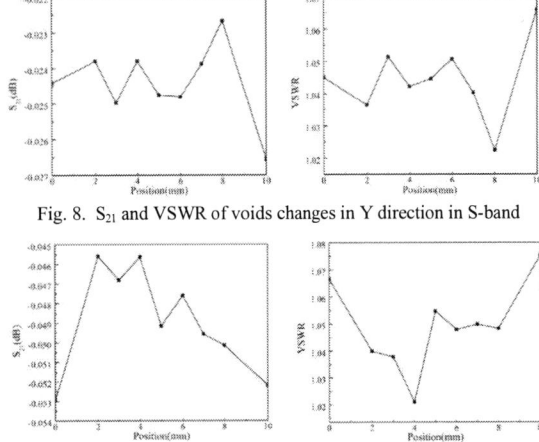

Fig. 8. S_{21} and VSWR of voids changes in Y direction in S-band

Fig. 9. S_{21} and VSWR of voids changes in Y direction in X-band

It can be seen that when the voids change along the horizontal direction below the microstrip line, the voltage standing wave is more sensitive for the passive device such as the microstrip plate, but the effect on the forward transmission coefficient is not obvious, only the voids have a greater influence on the feeding port and brazing layer edge. Generally speaking, the change of the position of a single cavity is not obvious to the transmission performance of the microwave circuit.

3.3 Effect of void distribution on signal transmission performance

In order to take the size and number of voids into account, the influence of void distribution on the transmission performance of micro-strip circuit board was studied.Based on the above research results, location 3, which has little influence on the transmission performance of the microstrip plate, is selected to create a 1 mm radius void, and four 0.5 mm voids (as shown in Fig. 10) centered on the center of location 3. The influence of the number of voids at the same location on the microwave transmission performance is studied under the same void rate..

978-1-6654-1392-3/21 $31.00 © 2021 IEEE

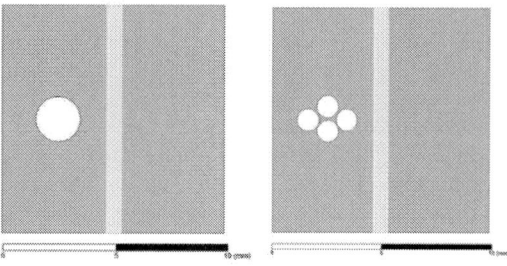

Fig. 10. Centralized and dispersed voids

The results of the electromagnetic simulation are shown in the following figure. The voltage standing wave ratio and S_{21} of four small voids are larger than those of large voids of the same area, and the gap of voltage standing wave ratio is larger. In X-band, large voids with the same area are found to have different resonance frequencies. It can be seen that with the same brazing penetration, the location of the voids is similar, and the scattered small voids have a greater impact on the transmission performance of the microwave circuit than the large voids with the same area.

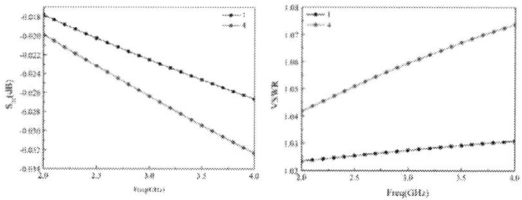

Fig. 11. S_{21} and VSWR in S-band

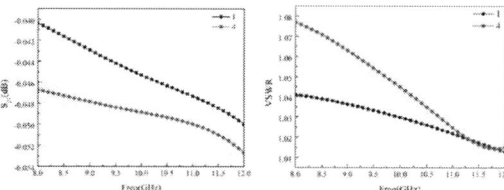

Fig. 12. S_{21} and VSWR in X-band

IV. CONCLUSIONS

From the above simulation analysis, the effect of the void near the feed port and the edge of brazing layer on the transmission performance of microwave signal is greater than that of the void inside the brazing layer. At the same brazing penetration rate, the location of the voids is similar, and the scattered small voids have a greater influence on the transmission performance of the microwave circuit than the large voids with the same area. The results show that the VSWR is more sensitive to the characteristics of brazing cavity, and the influence on the S_{21} is not obvious.

For passive devices such as microstrip plates, generally speaking, the effect of voids in solder layer on the transmission performance of their circuits is relatively small, but in practical projects, when many devices are cascaded together to form an active circuit, the grounding effect of brazing will be enlarged. When brazing is not ideal, the reliability and consistency of circuit substrates will be poor, and the grounding circuit will be increased. As the voltage loss increases, it is easy to cause the self-excitation of the active circuit. Therefore, in the process of manufacturing and packaging microwave devices, it is necessary to ensure the good grounding of the micro-strip circuit board, reduce the formation of brazing defects, improve the brazing permeability, and ensure the high consistency and reliability of the brazing of the micro-strip circuit board.

REFERENCES

[1] E. Brookner, "Developments and breakthroughs in radars and phased-arrays," in IEEE Radar Conference, Philadelphia, Pennsylvania, May ,2016, pp. 978-984.

[2] J. A. Martinez, A. Belenguer and H. Esteban González, "Highly Reliable and Repeatable Soldering Technique for Assembling Empty Substrate Integrated Waveguide Devices," in IEEE Transactions on Components, Packaging and Manufacturing Technology, vol. 9, no. 11, pp. 2276-2281, Nov. 2019.

[3] D. Xidong, L. Qi and Y. Zhenguo, "Research on Microstrip Transmission Performance in Microwave Circuit Board," 2019 International Conference on Microwave and Millimeter Wave Technology (ICMMT), 2019, pp. 1-3.

[4] J. Hsu, T. Su, X. Ye and C. Lin, "Microstrip signal integrity enhancement by using low-loss solder mask," 2017 12th International Microsystems, Packaging, Assembly and Circuits Technology Conference (IMPACT), 2017, pp. 122-125,

112G High Speed Interface Package Design and Simulation

Jiangtao Zhang
ZTE Sanechips Corporation.
Department of Packaging
&Testing
Xi'an, China
zhang.jiangtao1@sanechips.com.cn

Li Zhang
ZTE Sanechips Corporation.
Department of Packaging
&Testing
Shenzhen, China
zhang.li336@sanechips.com.cn

Jian Pang
ZTE Sanechips Corporation.
Department of Packaging
&Testing
Shenzhen, China
pang.jian@sanechips.com.cn

Tuobei Sun
ZTE Sanechips Corporation.
Department of Packaging
&Testing
Shenzhen, China
sun.tuobei@sanechips.com.cn

Keqing Ouyang
State Key Laboratory of Mobile
Network and Mobile Multimedia
Technology
Chengdu, China
ouyang.keqing@sanechips.com.cn

Abstract—**With the increasing speeding of high-speed serdes signal , IEEE and OIF Standards Institute have developed and released detailed protocol specifications, It can be seen from the protocol specification that 112G serdes has much higher fundamental frequency about 28GHz, and higher challenges for the system and package design. For example: differential IL, XTALK, differential RL, common mode noise etc, so the design and simulation of 112G high-speed serdes package are facing more and more serious challenges. In this paper, the electrical performance of common mode noise and crosstalk in the package design of 112G high-speed serdes are compared and analyzed, which will seriously affect communication system performance. Through simulation, optimization and analysis, the impact of the performance index caused by the common mode noise and crosstalk of package power supply in different ball-map layouts is quantitatively analyzed to find the optimal design. Ensure the best package routing performance of the 112G serdes design, and ensure that the performance of the package chip meets the design requirement and tap-out.**

Keywords—112G SERDES, ball-map, COMMON NOISE, XTALK

I. INTRODUCTION

With the rapid development of 5G communication, cloud computing, HD-cloud video and other high-speed communication device requirements, the signal transmission rate is also higher and higher. From 10Gbps serdes to 25Gbps serdes, 56Gbps serdes has been applied in large scale recently. Various advanced algorithms and functions support the smooth development of these products. At present, the related application and research

work of 112Gbps serdes has been developed and verified. The major IP manufacturers are scrambling for market share, and the chip R&D team is also stepping up the pace of R&D. For the research of 112G serdes, Sanechips has done some in-depth research of the development process. This paper mainly focuses on the in-depth comparative analysis of the impact of different package layout and ball-map strategies on the performance.

II. 112G SERDES PACKAGE BALL-MAP STUDY

As we all known that there are two different common arrangements of ball-map in serdes package: horizontal arrangement and vertical arrangement, as shown in Fig. 1 below. The influence of horizontal and vertical ball-map layout on package is mainly reflected in the following two aspects: SI performance and chip package size.

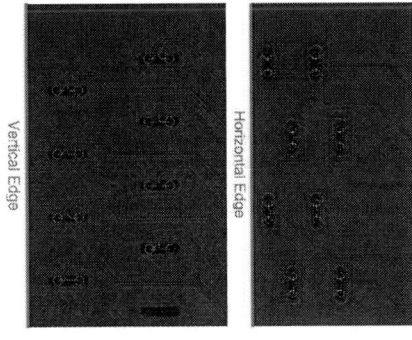

Fig. 1. *112G serdes V&H ball-map Arrangement*

For the influence of SI performance, this paper mainly makes further research and comparative analysis from the aspects of differential insertion loss(IL), return loss(RL), crosstalk and common mode noise(CMN).

978-1-6654-1392-3/21 $31.00 © 2021 IEEE

For chip package size, it has a great impact on the competitiveness of the whole package products，The chip size market always expects smaller size and stronger performance.

III. 112G SERDES BALL-MAP SIMULATION STUDY

Firstly, we compared and analyzed the performance of differential return loss and TDR effects of H-ball-map(horizontal ball-map) and V-ball-map(vertical ball-map) arrangement. It is found that the same package design has some differences in the differential RL and TDR performance. According to the analysis of the design and simulation result the mainly difference is due to H-ball-map are close to the package edge, and the impedance at the ball will increase due to the influence of the distributed capacitance of the groundless ball (due to the influence of the GND via loop of the core layer at the edge of H-ball-map). The influence of V-ball-map arrangement and H-ball-map arrangement on RL can be eliminated if the PTH hole distance are optimized,as shown in the Fig. 2 and Fig. 3 below.

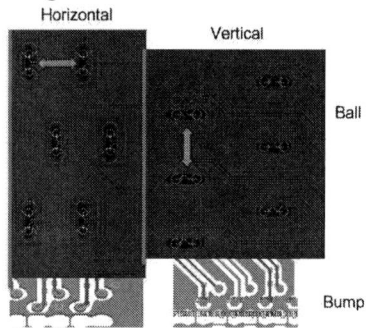

Fig. 2. *112G serdes V&H ball-map Differential Pair1*

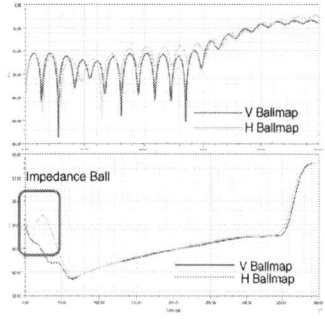

Fig. 3. *112G serdes Differential Return Loss and TDR*

According to the theoretical evaluation, the differential IL of V&H ball-map layout at package edge will cause differential IL deterioration due to edge effect, but the height of ball is only 0.45mm, and the IL affected will occur between 40GHz and 50GHz frequency band. For the frequency band before 28GHz, through simulation and comparison, it can be seen that H&V-ball-map arrangements have a very small

impact on IL, as shown in the Fig. 4 and Fig. 5 below.

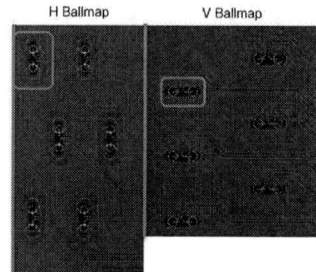

Fig. 4. *112G serdes V&H ball-map Differential Pair2*

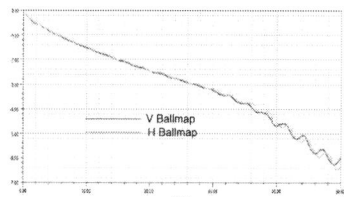

Fig. 5. *112G serdes Differential Insertion Loss*

Next, the XTALK of V-ball-map and H-ball-map arrangements will be simulated and compared. The worst case of V-ball-map arrangement is the most marginal two pairs of differential signals, whose electromagnetic field distribution is concentrated on the edge. Theoretically, it will lead to large inductive coupling, resulting in the deterioration of crosstalk performance. Through the comparison of simulation results, the XTALK of V-ball-map is higher than that of H-ball-map about 5dB@30GHz, as shown in the Fig. 6 and Fig. 7 below.

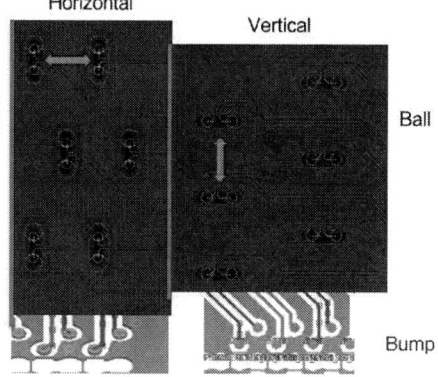

Fig. 6. *112G serdes V&H ball-map Differential Pair3*

978-1-6654-1392-3/21 $31.00 © 2021 IEEE

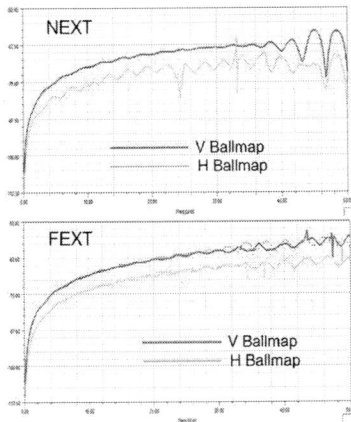

Fig. 7. 112G serdes Adjacent Differential XTALK

Next, Comparison and analysis the arrangement of V-ball-map and H-ball-map in the case of oblique pairing ball-map. Similar to the above situation, because the V-ball-map on the bump side is slightly worse than the H-ball-map, it will affect the crosstalk on the ball side, so the XTALK result of V-ball-map is very different from that of H-ball-map, about 20dB@30GHz difference. If only considering the influence of ball side, the difference will not be so big, as shown in the Fig. 8 and Fig. 9 below.

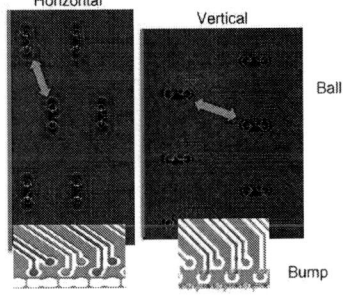

Fig. 8. *112G serdes V&H ball-map Differential Pair4*

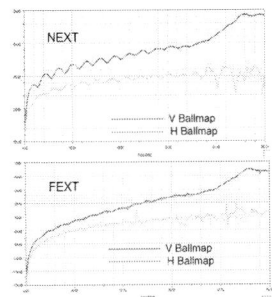

Fig. 9. 112G serdes oblique Differential XTALK

As shown in the figure simulation result below, in order to eliminate the influence of package routing on crosstalk, the routing need very short as much as possible. For the Fig. 8 ball-map layouts, through simulation and comparison, it is found that the crosstalk influence of the outer ring and inner ring is very

slight. Therefore, when the crosstalk of ball-map layout is relatively small. The influence of package routing on XTALK is relatively serious, as shown in the Fig. 10 and Fig. 11 below.

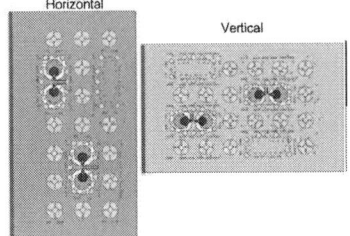

Fig. 10. *112G serdes V&H ball-map Differential Pair5*

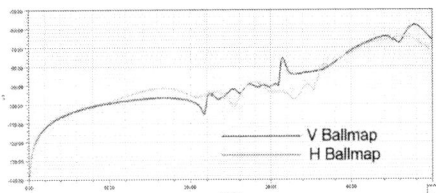

Fig. 11. *112G serdes V&H ball-map XTALK Simulation Result*

The following figure shows the comparison of XTALK power-sum results between V-ball-map and H-ball-map. The results of H-ball-map are better than V-ball-map. According to the above analysis, H-ball-map is also better than V-ball-map in terms of end-to-end crosstalk, NEXT XTALK power-sum is better 17dB@28GHz , FEXT XTALK power-sum 12dB@28GHz, as shown in the Fig. 12, Fig. 13, Fig. 14 and Fig. 15 below.

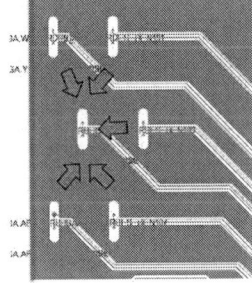

Fig. 12. *112G serdes V-ball-map XTALK Design*

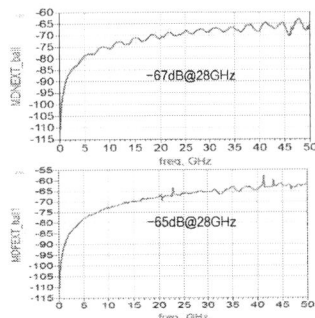

Fig. 13. *112G serdes V-ball-map XTALK Powersum Result*

978-1-6654-1392-3/21 $31.00 © 2021 IEEE

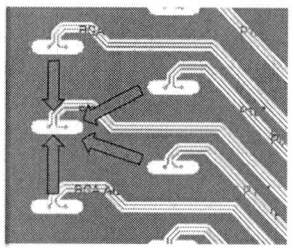

Fig. 14. *112G serdes H-ball-map XTALK Design*

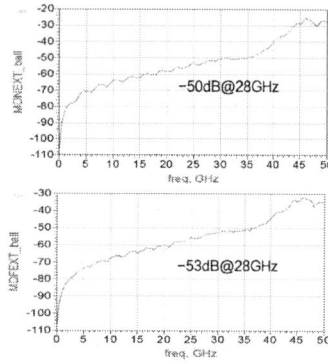

Fig. 15. *112G serdes V-ball-map XTALK Powersum Result*

For 112G serdes, IEEE and OIF actually have specifically requirements for their CMN. CMN is actually determined by common mode RL and differential mode to common mode parameters. Therefore, in the case of H-ball-map layout, although the two balls of H-ball-map layout are close to the edge, the differential mode to common mode component does not increase, as shown in the Fig. 16 below

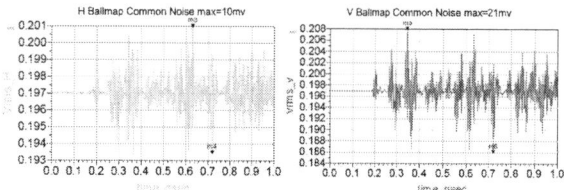

Fig. 16. *112G serdes V&H ball-map Common Noise Simulation Result*

Through the arrangement of ball-map with the same serdes differential signal number, the comparison in the below figure shows that in the same size range, the arrangement space occupied by V-ball-map is obviously smaller than that of H-ball-map. Therefore, through the above simulation and comparative analysis, although the performance of H-ball-map is significantly better than that of V-ball-map, if the XTALK SPEC meets the requirements, V-ball-map is recommended to reduce the package size and enhance the competitiveness of chip size, as shown in the Fig. 17 below.

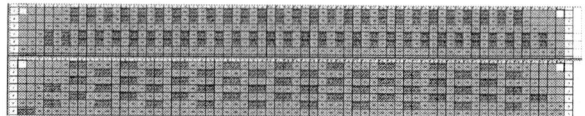

Fig. 17. 112G serdes V-ball-map Result

IV. CONCLUSIONS

According to the simulation and comparison analysis, the performance of H-ball-map layout is better than that of V-ball-map layout. First of all, there is little difference in IL and RL between H&V-ball-map layout. Secondly, the performance of H-ball-map arrangement layout on XTALK is obviously better than that of V-ball-map layout XTALK. Further analysis shows that when the crosstalk between differential pairs is relatively small, the effect of package substrate routing on crosstalk is relatively large. Finally, the CMN (10mV) of H-ball-map is less than that of V-ball-map layout (21mV). Although both ball-map layout modes can meet the SPEC requirements. However, compared with V-ball-map layout, H-ball-map layout takes up more package space, which leads to larger package size and product competitiveness reduction. For 112G serdes, after optimization and adjustment, V-ball-map layout can meet the system's requirements for package ball-map crosstalk, so H-ball-map layout is not recommended, resulting in over design and waste. In view of the advantages and disadvantages of the above V&H-ball-map layout, it is necessary to select the appropriate ball-map layout according to the actual situation of the project system requirement.

REFERENCES

[1] A. Cevrero et al., "A 100Gb/s 1.1pJ/b PAM-4 RX with Dual-Mode 1-Tap PAM-4 / 3-Tap NRZ Speculative DFE in 14nm CMOS FinFET", 2019.

[2] R. Awad et al., "A 1.7-pJ/b 112Gbps XSR Transceiver for Intrapackage Communication in 7nm FinFET technology", 2021.

[3] "Draft Standard for Ethernet Amendment 3: Media Access Control Parameters for 50 Gb/s and Physical Layers and Management Parameters for 50 Gb/s 100 Gb/s and 200 Gb/s Operation", September 2018.

[4] Amanda Dong et al., "Improved Engineering Analysis in FEC System Gain for 56G PAM4 Applications", DesignCon, 2018.

[5] Yoel Krupnik et al., "112 Gb/s PAM4 ADC Based serdes Receiver for Long Reach Channel in 10nm Process", Intel 2019 Symposium on VLSI Circuit Digest of Technical Papers.

[6] IEEE802.3cd Standard for Ethernet Amendment: Media Access Control Parameters Physical Layers and Management Parameters for 200 Gb/s and 400 Gb/s Operation, December 2018.

[7] Hong Ahn et al., "56Gbps PAM4 serdes Link Parameter Optimization for Improved Post-FEC BER", IEEE 28th Conference on EPEPS, April 2019.

[8] Y. Huang and B. Chen, "An 8b Injection-Locked Phase Rotator with Dynamic Multiphase Injection for 28/56/112Gb/s serdes Application", 2019

978-1-6654-1392-3/21 $31.00 © 2021 IEEE

Micro-vision Image Stitching System for Large-scale and Fine-featured Circuit Substrates

1st Yuanyang Wei
State Key Laboratory of Precision Electronic Manufacturing Equipment and Technology, Guangdong University of Technology
School of Electromechanical Engineering, Guangdong University of Technology
Guangzhou, China
1395650626@qq.com

2nd Jian Gao*
State Key Laboratory of Precision Electronic Manufacturing Equipment and Technology, Guangdong University of Technology
School of Electromechanical Engineering, Guangdong University of Technology, Guangzhou, China
Corresponding author :
jian_gao2004@163.com

3rd Yongbin Zhong
State Key Laboratory of Precision Electronic Manufacturing Equipment and Technology, Guangdong University of Technology
School of Electromechanical Engineering, Guangdong University of Technology
Guangzhou, China
yongb_z@163.com

4th Lanyu Zhang
State Key Laboratory of Precision Electronic Manufacturing Equipment and Technology, Guangdong University of Technology
School of Electromechanical Engineering, Guangdong University of Technology
Guangzhou, China
lyuzhang@qq.com

Abstract—**With the improvement of the manufacturing process and packaging technology of the precision electronic manufacturing industry, various precision circuits with large scale and fine features are widely used in high-end precision electronic products and instruments. At present, large-scale circuit substrates can reach very fine pitch and line width in only several micron-meter level. So the quality inspection equipment should have the inspection capabilities of large-scale and high-precision detection. Since a single view of image with high-resolution only have a small field of view, image stitching by registration and fusion is an effective method to inspect the whole large scale substrate. Hence, this paper proposes a micro vision image stitching method with high precision through coarse and fine registration. The real-time position information of the platform is used to calculate the approximate pixel offset between images to achieve coarse registration. Then, two convolutional neural regression networks connected by a spatial transformation network are used to achieve fine registration. A high-precision nano-positioning platform is used to establish image registration dataset, which is used for the training of convolutional neural regression networks. Experimental results show that under the condition of resolution of 0.5um/pixel and stitching overlap of 20%, our method can achieve less than 0.2 pixel registration error and less than 25ms registration time for a single image.**

Keywords—image stitching, image registration, microscopic vision inspection, CNN regression

I. INTRODUCTION

With the improvement of the manufacturing process and packaging technology of the precision electronics manufacturing industry, various precision circuits with large scale and fine features are widely used in high-end precision electronic products and instruments. At present, the manufacturing accuracy of large-scale precision circuit substrates has reached the micron level in line width and line spacing, which requires quality inspection equipment to have large-scale and high-precision inspection capabilities.

Image stitching is to stitch multiple images with a certain overlap into a seamless panorama. Image stitching technology is mainly divided into image preprocessing, image registration and image fusion. The quality of image registration directly affects the stitching quality. Image registration methods are mainly divided into region-based image registration methods and feature-based image registration methods.

In the 1970s and 1980s, image stitching technology was proposed in the military field. In 1975, Kuglin[1] proposed the phase correlation method. Through frequency domain analysis, the amount of translation between images with higher overlap can be solved. In 1999, Lowe[2] pioneered SIFT (Scale-Invariant Feature Transform), a multi-scale feature extraction algorithm, which was improved in 2004[3]. In 2006, Bay[4] proposed the SURF (Speeded-Up Robust Features), which improved the SIFT algorithm and greatly reduced the time-consuming algorithm. In 2011, Rublee[5] proposed ORB, which generates binary descriptors. ORB operation speed is several times faster than SURF, but it does not have scale invariance. With the development and application of deep learning in the image field, image stitching methods based on convolutional neural networks have been continuously proposed. In 2015, Simo-Serra[6] trained a twin network to distinguish the similarity between image patches, which generates 128-dimensional descriptor as an alternative to the SIFT descriptor. In 2016, Miao[7] used a series of CNN regressors to estimate the transformation parameters to achieve real-time 2D/3D image registration. In 2018, J.M Sloan[8] used jump connections to build a feature extraction network, and used CNN regressor to regress the registration parameters to achieve medical image registration. In the same year, Daniel[9] proposed a self-supervised neural network for interest point detection and descriptor generation, which can calculate the pixel-level interest point position and related descriptors through a forward propagation.

The task of image registration is to solve the parameters of the spatial transformation model between two images. Assuming that the image $f_1(x, y)$ and the image $f_2(x, y)$ have a projection transformation relationship, the transformation relationship can be expressed by a homogeneous equation as

$$\begin{bmatrix} x_2 \\ y_2 \\ 1 \end{bmatrix} = \begin{bmatrix} m_0 & m_1 & m_2 \\ m_3 & m_4 & m_5 \\ m_6 & m_7 & 1 \end{bmatrix} \begin{bmatrix} x_1 \\ y_1 \\ 1 \end{bmatrix} = M \begin{bmatrix} x_1 \\ y_1 \\ 1 \end{bmatrix} \quad (1)$$

where M is the matrix parameter of the spatial transformation model between images. The task of registration is to solve the parameters of the optimal transformation matrix M to minimize the registration error.

This paper proposes an image registration method that combines coarse registration and fine registration. The real-time position information of the platform is used to calculate the approximate pixel offset between images to achieve coarse registration. On the basis of coarse registration, two convolutional neural regression networks connected by a spatial transformation network[10] are used to achieve fine registration of images. A micro-vision image stitching system was built to verify the proposed algorithm.

The rest of the paper is as follows: Section II introduces the composition of the microscopic vision image stitching system and the image stitching method. In section III, experiments are carried out to verify the effectiveness of the image registration method that combines coarse and fine registration. Finally, a brief conclusion is made in section IV.

II. IMAGE STITCHING METHOD

A. Introduction of microscopic vision inspection system for fine feature circuit boards

The object to be tested in this paper is a high-precision flexible circuit substrate, as shown in Fig 1. Its thickness is about 30um, the minimum circuit width is 1mil (25.4um), and the minimum circuit spacing is also 1mil. Because of its small features, a microscope lens must be used for single imaging with a small field of view. The images are collected multiple times through the motion platform and then stitched together to form a complete circuit image.

Fig.1 . High-precision flexible circuit substrate.

In order to achieve a wide range of image acquisition and stitching functions, a microscopic vision inspection system was built. The schematic diagram of the image stitching system is shown in Fig 2. The computer is respectively connected to the platform controller and the image acquisition unit. After the computer sends a movement instruction, the platform moves to a stable position and sends a signal to trigger the camera to acquire image. After the

image acquisition is over, the computer sends the next movement instruction, and so on. The stitching algorithm is executed at the same time during the movement of the platform. Since the execution time of the stitching algorithm is less than the moving time of the platform, the real-time stitching of images can be realized.

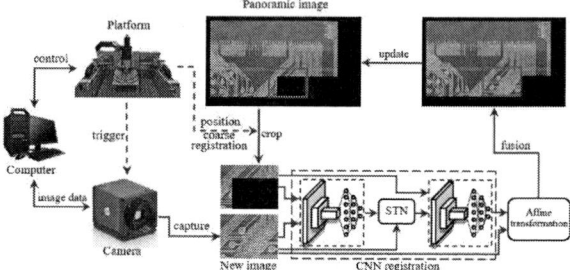

Fig.2 . The schematic diagram of the image stitching system.

The image registration algorithm is divided into two steps: coarse image registration and fine image registration. The camera and lens are calibrated by Zhang's calibration method[11]. Since the camera coordinate system has a certain deflection angle relative to the platform movement axis, the conversion relationship between the image coordinate system and the world coordinate can be calculated through calibration. The relationship between the actual displacement and the pixel displacement is established, so that the position information fed back by the grating encoder can be used for coarse image registration. The coarse registration information is used to crop the overlapping part of the newly acquired image from the panoramic image as the image input of the fine registration. The fine image registration is realized by using two convolutional neural regression networks connected by a spatial transformation network. The first convolutional neural regression network performs regression prediction on large pixel offsets. The second convolutional neural regression network makes further predictions on the basis of the first regression prediction, reaching sub-pixel level prediction errors. In order to improve the quality of the panoramic image, the brightness of the image with uneven brightness distribution is adjusted. Finally, a linear fusion strategy is used to fuse the images.

B. The method of coarse image registration

The lens is fixedly installed on the y-axis beam of the platform. It moves along with the platform to scan the circuit substrate to collect images. Since the platform still has small adjustments after rapid positioning, the position information fed back by the grating encoder fluctuates within a certain range during image acquisition. However, this position information can be used to calculate the approximate relative pixel shift between adjacent images. As shown in Fig 3, the rectangular frame is the area where the camera actually captures the image.

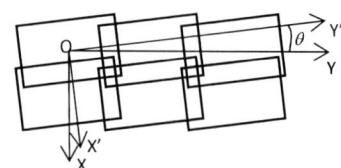

Fig.3 . The relationship between the image and the platform coordinate system.

The coordinate system $O - XY$ is the platform coordinate system, and the coordinate system $O - X'Y'$ is the image coordinate system. The angle formed between the two coordinate systems is θ.

The relationship between the image coordinate system and the platform coordinate system can be expressed as

$$\begin{bmatrix} x' \\ y' \end{bmatrix} = \frac{1}{R} * \begin{bmatrix} cos\theta & sin\theta \\ -sin\theta & cos\theta \end{bmatrix} \begin{bmatrix} x \\ y \end{bmatrix} \qquad (2)$$

where R is the pixel resolution, and its unit is um/pixel, which can be obtained by visual calibration. The parameter θ can be solved by the offset (dx, dy) between adjacent registered images, as shown in the formula

$$\theta = Atan(dy/dx) \qquad (3)$$

After the parameter θ is determined, the position information fed back by the grating encoder can be used to calculate the pixel offset between the images, so as to achieve fast image coarse registration.

C. Method of fine registration using neural network

Since the collected images are basically on the same plane, the spatial transformation model of image registration can be simplified to a translation transformation model on the plane. Therefore, it is only required to solve the relative pixel offsets(dx and dy) of the two images to complete the registration between the two images.

The image cropped in the coarse registration has a large area without information, and large image will increase the network parameters greatly, which makes the network difficult to train. Therefore, large image block pairs are cropped in the image region correspondingly as the input of the convolutional neural regression network, as shown in Fig 4. A pair of large images will generate several image block pairs, which are respectively used as the input of the network for pixel offset prediction. These offsets are averaged to get the final pixel offset.

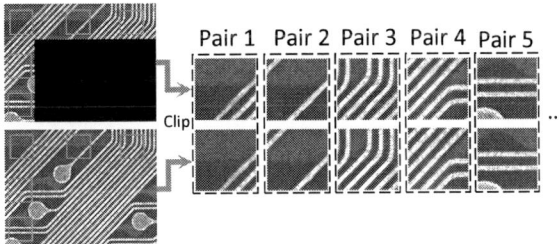

Fig.4. Schematic diagram of getting image block pairs.

Since a single convolutional neural regression network is not effective in predicting the registration parameters, two convolutional neural regression networks connected by a spatial transformation network are used to gradually predict the registration parameters, as shown in Fig 5. Each convolutional neural regression network consists of 4 convolutional layers and 3 fully connected layers. The network connects image pairs to obtain a two-channel map.

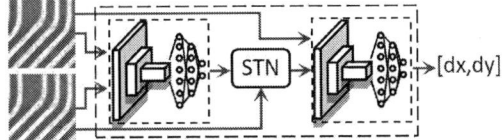

Fig. 5. Schematic diagram of the network structure.

Then, the convolution operation extracts the features, and the fully connected layers realize the regression prediction of the registration parameters.

The layer parameters of the first convolutional neural regression network are shown in Table I . The layer parameters of the second convolutional neural regression network are similar to the first one, changing the size of Conv0's Kernel to 3×3.

TABLE I. LAYER PARAMETERS OF CNN REGRESSOR

Name	Type	Kernel	Stride	Activation
Concat0	concatenate	-	-	-
Conv0	convolution	5×5	2×2	relu
Conv1	convolution	3×3	2×2	relu
Conv2	convolution	3×3	2×2	relu
Conv3	convolution	3×3	1×1	relu
Flaten0	flatten	-	-	-
FC0	fully-connected	1024	-	relu
FC1	fully-connected	128	-	relu
FC2	fully-connected	2	-	linear

The training of the network uses a data set of image pairs with real offset labels. The network training loss function is defined as

$$loss = \frac{1}{B}\sum_{i=1}^{B}\left(\left(x_i' - x_i\right)^2 + \left(y_i' - y_i\right)^2\right) \qquad (4)$$

where x_i' is the x-axis pixel offset of the network prediction, y_i' is the y-axis pixel offset of the network prediction, x_i is the real x-axis pixel offset, y_i is the real y-axis pixel offset, and B is the number of batch samples.

III. EXPERIMENTS AND RESULTS

A. Data set collection

In order to obtain the real registration information of the image pair for training the neural network, a high-precision single-axis nano-motion platform[12] is used to collect the data set, as shown in Figure 6. The nano-motion platform can achieve 50mm stroke and 43nm repeat positioning accuracy.

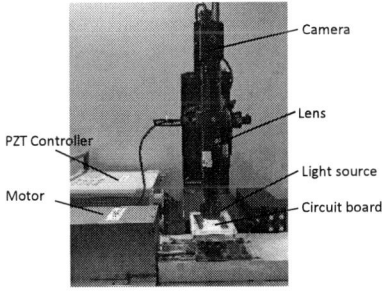

Fig. 6. Nano-motion platform for data set collection.

The overlap between the images is set to 50%. After the platform moves in place, the computer acquires the image and records the position of the grating encoder at the same time. The image data set is collected under different brightness of the light source and different magnification of the lens. A total of 2000 pairs of images with real position deviation information are obtained. According to the training data set accounted for 80%, the training data set has a total of 1600 pairs of registered images, and the test data set has 400 pairs of registered images.

B. The experiment of image registration

Image registration is an important step in image stitching. This paper proposes an image registration method that combines coarse registration and fine registration.

978-1-6654-1392-3/21 $31.00 © 2021 IEEE

According to the coarse registration method proposed in Section B of Chapter II, the parameter θ needs to be determined. On the image acquisition platform, 30 adjacent images are randomly collected for image registration, and then these registration results are used to solve the parameter θ. As shown in Fig 7, this is the calculation results of the parameter θ. Here, the mean value of the calculation results of 30 sets of experiments is taken as the value of the parameter θ. In this experiment, θ is 0.008251rad.

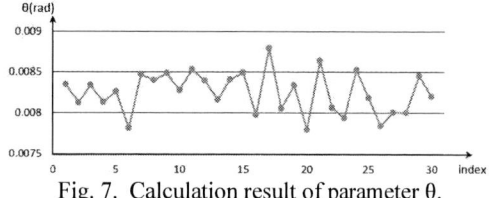

Fig. 7. Calculation result of parameter θ.

In the real-time image stitching experiment, the platform movement adopts an S-shaped trajectory to move. Each movement of the platform will only move one axis. As shown in Fig 8, this is the trajectory of the platform with a single movement distance of 0.5mm. It takes 200ms for the platform to reach the vicinity of the target position, and then make slow adjustment. The adjustment range is about 5um. In order to reduce the waiting time, the image acquisition is triggered during the slow adjustment of the platform. Therefore, when the pixel resolution is 0.5um/pixel, the coarse image registration accuracy is within 10 pixels.

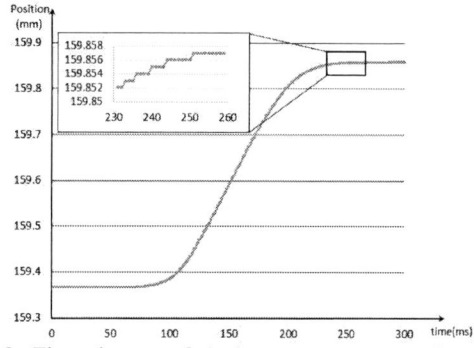

Fig. 8. The trajectory of platform movement at a distance of 0.5mm.

The training of the fine registration network uses image pairs with real position offset. Each pair of original images can generate dozens of image block pairs. In the network training stage, the second convolutional neural regression network is trained firstly. On the original registered image pair, pixel offsets are randomly generated with the size of U(-2,2) on the x- and y- axes to generate image block pairs for network training. After the second regression network is trained, we fix the parameters of the second convolutional neural regression network, and then train the first convolutional neural regression network. In the same way, pixel offsets are randomly generated on the x- and y- axes with the size of U(-10,10) to generate image block pairs for the first convolutional neural regression network training.

This paper uses the TensorFlow2(a deep learning framework) to build the network. The main configuration of the computer is as follows: CPU: Intel(R) Core(TM) i3-4160 CPU @ 3.60GHz, GPU: NVIDIA GeForce GTX 1650. The image block size of the network input is 101×101. The training of the network uses SGD with a starting learning rate of 0.0001, momentum of 0.9 and weight decay of 0.0001. The learning rate is divided by 2 if the value of the loss function does not decrease for more than 5 epochs. Two convolutional neural regression networks complete training in no more than 200 epochs. After the network training converges, only one forward propagation is needed to predict the pixel offset between a pair of image block.

In order to test the effect of network prediction, 250 pairs of image blocks were randomly generated on the test data set and input into the network for prediction. As shown in Fig 9, this is the prediction result of the first convolutional neural regression network. Fig.9(a) is a scatter plot of the network's prediction deviation on the x-axis, and Fig.9(b) is a scatter plot of the network's prediction deviation on the y-axis. The prediction result of the second convolutional neural regression network is shown in Fig 10.

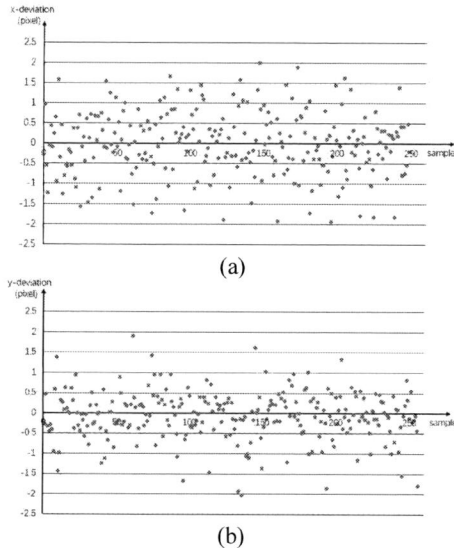

Fig. 9. The prediction deviation of the first convolutional neural regression network on x-axis(a) and y-axis(b).

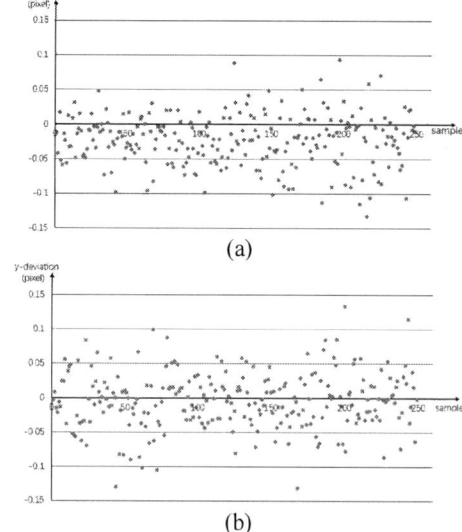

Fig. 10. The prediction deviation of the second convolutional neural regression network on x-axis(a) and y-axis(b).

According to the image block acquisition method in Section C of Chapter II, 5 pairs of image blocks are obtained and input into the network for prediction. The registration time between two images is the sum of the five prediction times. The Image registration time is shown in Fig 11.

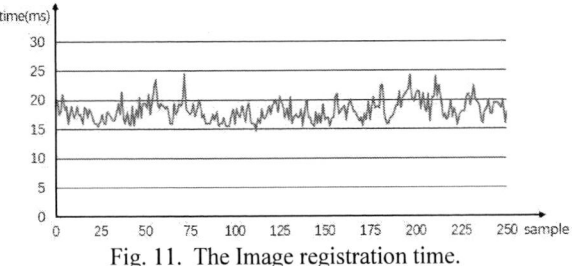

Fig. 11. The Image registration time.

C. Large-scale image stitching experiment

In order to obtain a panoramic image with uniform brightness distribution, the brightness of the image with uneven brightness is adjusted slightly before the image fusion. By collecting white paper images of different light source brightness, the light source brightness distribution law is obtained. Fig 12 shows the white paper image with different brightness and its binarized image.

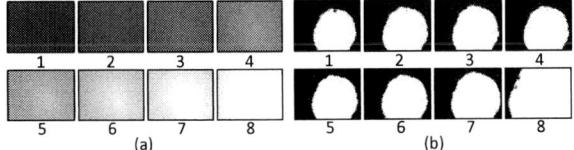

Fig. 12. The white paper image with different brightness(a) and its binarized image(b).

The brightness of the light source used in the experiment is mainly between level 3~6, and the brightness distribution of these light sources is relatively similar. The image of brightness level 4 is used as the original reference image for brightness adjustment to generate a basic template for adjusting the brightness of the image. The basic template image can be expressed as

$$t(x, y) = \lambda t'(x, y) = \lambda \left(maxValue - f(x, y) \right) \quad (5)$$

where $f(x, y)$ is the original reference image, maxValue is the maximum value of $f(x, y)$, $t'(x, y)$ is the reverse template image, $t(x, y)$ is the basic template for adjusting the brightness of the image, λ is the linear adjustment coefficient. Here, $\lambda = 0.9$.

Fig. 13 shows the results of the partial stitched image without brightness adjustment and with brightness adjustment.

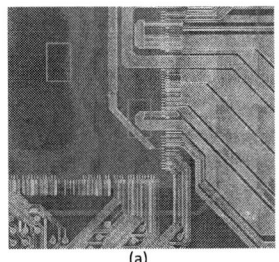 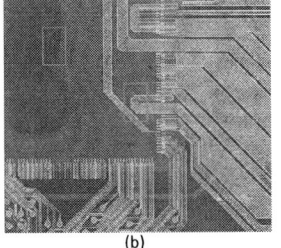

Fig. 13. Stitched image without brightness adjustment(a) and with brightness adjustment(b).

In the process of image stitching, a linear fusion strategy is used to fuse the images. When the pixel resolution is 0.5um/pix, the camera resolution is 1920×1448, and the stitching overlap is 20%, the stitching algorithm takes less than 30ms, and the total number of stitched images is more than 1000. The stitched image is shown in the Fig 14.

Fig. 14. Stitched image of the fine-featured circuit substrate.

IV. CONCLUSION

1) This paper proposes a high precision micro vision image registration method through coarse registration and fine registration. Coarse registration uses real-time position information of the platform to achieve image registration with deviations within 10 pixels. Fine registration uses two convolutional neural regression networks connected through a spatial transformation network to gradually perform regression prediction to achieve a registration error within 0.2 pixels.

2) The image stitching experiment was carried out on a large-scale and fine-featured circuit substrate. The experimental result showed that when the pixel resolution is set to 0.5um/pix, and the stitching overlap is 20%, the registration error is less than 0.2 pixel and the stitching time costs less than 30ms.

ACKNOWLEDGMENT

This work was supported in part by the National Natural Science Foundation of China under Grant 52075106, and in part by the Guangdong Provincial R&D Key Projects under Grant 2018B090906002, and 2016KZDXM049.

REFERENCES

[1] Kuglin C,Hines D, "The Phase Correlation Image Alignment Method", IEEE International Conference on Cybernetics and Society, New York, 1892.

[2] D.G. Lowe, "Object Recognition from Local Scale-invariant Features", IEEE International Conference on Computer Vision, Kerkyra, Greece, 1999.

[3] D.G. Lowe, "Distinctive Image Features from Scale-invariant Key-points", International Journal of Computer Vision, vol. 60, no. 2, pp. 91-110, 2004.

[4] Herbert Bay, Andreas Ess, Tinne Tuytelaars, et al., "Speeded-Up Robust Features (SURF) ", Computer Vision and Image Understanding, vol. 110, no. 3, pp. 346-359, 2008.

[5] Ethan Rublee, Vincent Rabaud, Kurt Konolige,et al., "ORB: An efficient alternative to SIFT or SURF", IEEE International Conference on Computer Vision, Barcelona, Spain, pp. 2564-2571, 2011.

[6] Edgar Simo-Serra, Eduard Trulls, Luis Ferraz, et al., "Discriminative Learning of Deep Convolutional Feature Point Descriptors", IEEE International Conference on Computer Vision, Santiago, Chile, pp. 118-126, 2015.

[7] Shun Miao, Z.Jane Wang, Yefeng Zheng, et al., "Real-time 2d/3d registration via CNN regression", IEEE International Symposium on Biomedical Imaging, pp. 636-648, 2016.

[8] J.M Sloan, K.A Goatman, J.P Siebert, "Learning Rigid Image Registration - Utilizing Convolutional Neural Networks for Medical

Image Registration", In Proceedings of the 11th International Joint Conference on Biomedical Engineering Systems and Technologies , pp. 89-99, 2018.

[9] Daniel DeTone, Tomasz Malisiewicz, Andrew Rabinovich, "SuperPoint: Self-Supervised Interest Point Detection and Description", IEEE/CVF Conference on Computer Vision and Pattern Recognition Workshops, pp. 337-349, 2018.

[10] Max Jaderberg, Karen Simonyan, Andrew Zisserman, et al., "Spatial Transformer Networks", Advances in Neural Information Processing Systems, pp. 2017-2025, 2015.

[11] Zhengyou Zhang, "A flexible new technique for camera calibration", IEEE transactions on pattern analysis and machine intelligence, vol. 22, no. 11, pp. 1330-1334, 2000.

[12] Lanyu Zhang, Jian Gao, Xin Chen, "A Rapid Dynamic Positioning Method for Settling Time Reduction Through a Macro–Micro Composite Stage With High Positioning Accuracy", IEEE Transactions on Industrial Electronics, vol. 65, no. 6, pp. 4849-4860, 2018.

Study in multilayer wiring technology on high-heat-conduction substrates

Lei Ding
Micro-electronics business department
Shanghai Aerospace Electronic and
Communication Equipment Research
Institute
Shanghai, China
youyouzi.teng@163.com

Jing Chen
Micro-electronics business department
Shanghai Aerospace Electronic and
Communication Equipment Research
Institute
Shanghai, China
viola9876543210@163.com

Kai Liu
Micro-electronics business department
Shanghai Aerospace Electronic and
Communication Equipment Research
Institute
Shanghai, China
k2006i@126.com

Yan Luo
Micro-electronics business department
Shanghai Aerospace Electronic and
Communication Equipment Research
Institute
Shanghai, China
luoyan1120@qq.com

Yue Zhao
Micro-electronics business department
Shanghai Aerospace Electronic and
Communication Equipment Research
Institute
Shanghai, China
565367765@qq.com

Lichun Wang
Micro-electronics business department
Shanghai Aerospace Electronic and
Communication Equipment Research
Institute
Shanghai, China
wanglichun0482@163.com

Abstract—In order to meet the requirements of high heat dissipation, high density and low coefficient of expansion for micro system miniaturization integration, a method for preparing BCB / Cu multilayer wiring based on AlSi high thermal conductivity substrate is proposed. The surface treatment of AlSi substrate, the preparation of interface buffer layer of AlSi substrate film, and the through-hole interconnection of AlSi substrate dielectric film were studied. The AlSi substrate multilayer wiring substrate samples were prepared, including three layers of BCB dielectric layer and four layers of thin film conductor wiring. After 100 temperature cycles from - 55 ℃ to 150 ℃, the qualified rate of interconnection conduction and insulation test is 100%.

Keywords—Microsystems; metal substrate; AlSi; BCB; Anodic oxidation; thin film multilayer interconnection

I. INTRODUCTION

With the rapid development of modern electronic technology, electronic system is gradually developing towards miniaturization, high integration and micro system. In military, aerospace, 5g communication, computer and other high-end applications, the chip power consumption is higher and higher, which requires packaging substrate to provide better heat dissipation capacity, more intensive wiring and lower loss.

The thin film multilayer wiring substrate using metal substrate as high thermal conductivity substrate, can not only easily achieve ultra fine wire width and narrow spacing, but also has high heat dissipation performance. At the same time, metal substrate can play the role of floating ground wire in large area to reduce the inductance, capacitance and crosstalk between signals, thus having become a research hotspot in the field of hybrid integrated circuits. The commonly used metal based materials include Al, Cu, Mo Cu, W-Cu, AlSiC, AlSi, etc. Among them, AlSi material has the advantages of adjustable coefficient of thermal expansion, light weight, high thermal conductivity, machinability and low manufacturing

cost [1-2], thus having been widely used in automobile, aerospace, power electronic packaging and other fields. In recent years, the industry has carried out extensive research work on the preparation, microstructure and properties of AlSi materials [3-5]. BCB (Benzocyclobutene) as a thin film dielectric material has the advantages of low high frequency dielectric loss, low dielectric constant, good flatness, low water absorption and good adhesion [6-7]. Above these situation, this paper proposed an innovative fabrication method of BCB / Cu thin film multilayer wiring based on AlSi high thermal conductivity substrate, to achieve high-efficiency heat dissipation and high-density wiring at the same time. The process is stable and reliable, and provides an effective solution for miniaturization of micro system integration.

II. EXPERIMENTAL SETUP

Figure 1 shows the structure of BCB/Cu multilayer wiring substrate with high thermal conductivity AlSi substrate. The substrate material of the experimental substrate is the high silicon aluminum alloy(Al-70Si), and the dielectric layer material is Dow Chemical cyclone 4026-46 series photosensitive BCB. The buffer layer between AlSi and the thin film interface is $0.15\mu m/1.5\mu m$ Ta_2O_5/Al_2O_3 dielectric film, and thin film interconnection layer is $0.15\mu m/1.5\mu m$ Ta/Al ground layer,$0.15\mu m/5\mu m/0.5\mu m/2\mu m$ TiW/Cu/Ni /Au inner conductor layer and $0.15\mu m/2\mu m$ TiW / Au outer conductor layer. The substrate uses the ordered porous structure of Ta_2O_5 / Al_2O_3 dielectric film to release and buffer the film stress in the multilayer wiring of thin film, so as to meet the high reliability of the interconnection of high thermal conductivity AlSi substrate multilayer wiring substrate.

978-1-6654-1392-3/21 $31.00 © 2021 IEEE

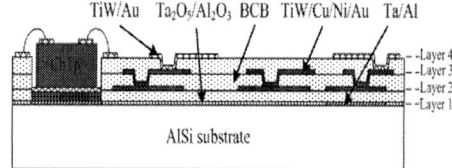

Fig. 1. Structure of BCB/Cu multilayer wiring substrate on high thermal conductivity AlSi substrate

The fabrication process of BCB/Cu multilayer wiring technology on high thermal conductivity AlSi substrate is shown in Figure 2. The specific preparation methods include: (1)Firstly, the surface of AlSi substrate was treated by thinning and polishing process to reduce the surface roughness and interface defects of AlSi substrate;(2) Thin film sputtered Ta / Al was used for selective anodization, and BCB was used for micro penetration and solidification to form a low stress AlSi film interface buffer layer;(3)TiW/Cu/Ni /Au inner conductor layer was prepared by magnetron sputtering, photolithography and electroplating;(4)Using the photosensitive properties of BCB, through-hole lithography and curing process of BCB are carried out to produce BCB insulating dielectric film with through-hole(multilayer wiring can repeat (3) and (4) operations); (5) Finally, the TiW/Au outer conductor layer was prepared by magnetron sputtering, photolithography and electroplating.

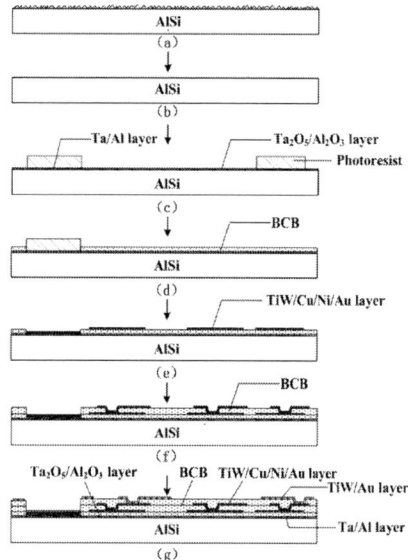

Fig. 2. Fabrication flow chart of BCB / Cu multilayer wiring technology on high thermal conductivity AlSi substrate

III. RESULTS AND DISCUSSION

A. Surface treatment of AlSi substrate

In order to obtain a good thin film wiring interface, it is necessary to grind and polish the AlSi substrate to reduce its surface roughness. In the experiment, Logitech's PM5 thinning and polishing machine was used to thin and polish the AlSi substrate 3μm、 9 μm、 20μm.In the polishing part, SF1 slurry was selected. When the thinning and polishing pressure is 6.75KPa, the rotation speed is 50r/min and the thinning time is 60min, the polishing time is 30min, 60min, 90min and 120min respectively. In the experiment, five points (center, left, right, top and bottom) on the wafer were selected, and the roughness of the polished substrate was tested by WYKO NT-2000 surface profiler. The results are shown in Figure 3.

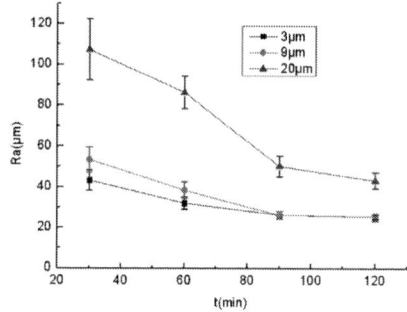

Fig. 3. Effect of different particle size and grinding time on the roughness of polished substrate

It can be seen from Figure 3 that the surface roughness of AlSi substrate decreases with the increase of polishing time. After 90 min, the surface roughness of the 3μm and 9μm particle size abrasive decreased to less than 30μm, and the surface quality was better. This is because when the abrasive particle size is larger, the number of abrasive particles in the grinding area is less at the same time, and because of the uniformity of the abrasive, there must be some small particles, and some small particles are not forced in the grinding area because of the existence of large and small particles, then the greater the partial pressure of each particle under the same total pressure, The surface roughness Ra increases with the increase of surface damage area; The smaller the abrasive particle size is, the more abrasive particles are in the grinding area at the same time, and the smaller the particle size is, the better the uniformity is. Under the same total pressure, the smaller the partial pressure of each particle is, and the better the surface quality is obtained. 3μm particle sizes and 90 min polishing time are suitable.

Figure 4 shows the surface morphology of polished AlSi substrate. It can be seen from Figure 4 that after polishing, the surface of AlSi substrate is smooth, the surface of AlSi substrate is gray white, and the average surface roughness drops to 0.038μm。

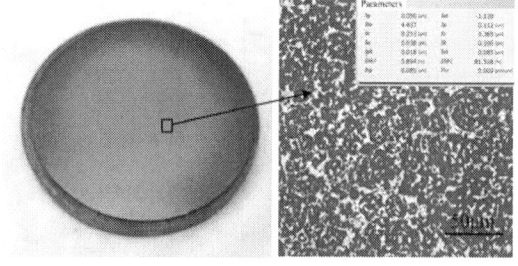

Fig.4. Surface morphology of AlSi substrate after polishing

B. Preparation of film interface buffer layer

As a metal conductive substrate with high thermal conductivity, it is necessary to fabricate a thin film interface buffer layer on the surface of AlSi before multilayer wiring. When the photosensitive BCB dielectric is used as the interface insulating layer of thin film on AlSi substrate, the thermal mismatch between the thermal expansion coefficient of BCB dielectric film and AlSi substrate (BCB:42 ppm/K, AlSi:7 ppm/K) easily leads to the stress accumulation of multilayer wiring film. In view of this, adding an anodized insulating transition layer between the BCB dielectric film and the metal AlSi substrate for stress buffering can effectively avoid the stress phenomena such as cracks in the BCB dielectric film. In this paper, the Ta/Al composite film was selected for selective anodization, and the non oxidized part was used to prepare large area metal ground. The thermal expansion coefficient of Ta/Al and AlSi substrate was matched (Ta: 6.5ppm/K, Al: 23.6 ppm/K) can reduce the film stress and improve the film adhesion; The oxidized part forms and ordered porous oxide film, and the stress is buffered and released by micro penetration and solidification of the coated BCB dielectric film.

The anodizing equipment was used in the experiment. The pretreated AlSi substrate was used as anode and oxalic acid solution was used as anodizing electrolyte. The anodizing voltage was 50V. Phosphoric acid solution was used as the electrolyte for anodic oxidation of TA, and the oxidation voltage was 180V. Fig. 5 is a micrograph of the anodized AlSi substrate. It can be seen that the color of the surface of the Al film is darker after anodization. The surface of the film is silver white and evenly distributed under the microscope. At this time, the average surface roughness is 0.034μm. There is no obvious change in the roughness after anodization.

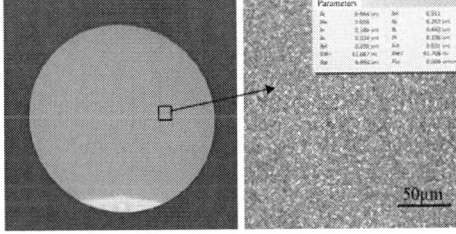

Fig.5.　Micrograph of anodized AlSi substrate

In order to further verify the effect of porous anodic oxidation on BCB stress, the AlSi substrate samples were divided into two groups. In the first group, the AlSi substrate was directly coated with BCB for three times and then cured. In the second group, the AlSi substrate was anodized and then coated with BCB for three times. Flx-2320 thin film stress tester was used to compare the deformation of AlSi substrate before and after anodization.

Figure 6 shows the deformation of AlSi substrate before and after anodization. It can be seen from Figure 6 that the surface deformation of AlSi substrate without porous oxide film is 78.33μm after BCB coating and curing. The surface deformation of porous anodized AlSi substrate is reduced to 61.33μm after BCB coating and curing, which indicates that the porous anodized film can effectively improve the

deformation of AlSi substrate after BCB coating, and the porous anodized film can buffer and release the stress of BCB dielectric film.

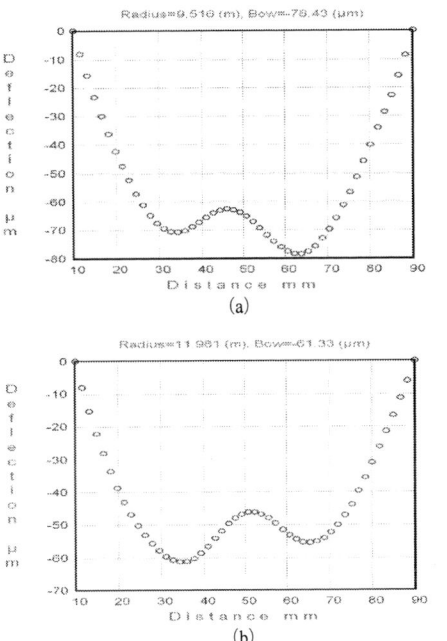

Fig. 6.　Deformation curve of BCB cured substrate,(a) no porous oxide film was prepared,(b) porous anodic oxidation was carried out

C. Through hole interconnection of dielectric film on AlSi substrate

The key of multi-layer wiring on AlSi substrate is through-hole interconnection of dielectric film. The performance of through-hole interconnection directly affects the high-speed signal transmission. Dielectric via interconnection mainly includes via fabrication and thin film interconnection layer fabrication.

First, spin coating 10μm BCB, and pre drying, exposure, pre drying and other processes, and then spray development method for through-hole development. The development temperature is 40 ℃, the development time is 3 min, and the spray pressure is 0 MPa, 10 MPa, 20 MPa and 30 MPa respectively. Ols4100 confocal laser microscope was used to observe the through-hole micrographs under different development parameters. Fig. 7 is a picture of laser confocal microscope (CLSM) after BCB through hole development. In Fig. 7 (a) and Fig. 7 (b), there are a large number of color interference fringes in the through hole of BCB. In Fig. 7 (c) and Fig. 7 (d), the contour edge of the through hole is smooth, and there is no residual BCB film in the central area of the through hole. The reason is that when the spray pressure is 0 MPa and 10 MPa, the mass transfer of developer is difficult, resulting in uneven development in the through hole; At the same time, due to the small spray pressure, the dissolved bottom film is left in the developer, resulting in repeated contamination of the through hole, resulting in BCB glue residue. When the spraying pressure reaches 20MPa or above, the BCB film begins to dissolve, and then the development difference in the through-hole can be reduced by spraying under high impact pressure, and the residual film can be removed by flushing with fresh developer [8], but excessive spraying pressure further accelerates the over development of BCB through-hole. Based on the above situation, the spraying pressure of 20MPa is the most suitable.

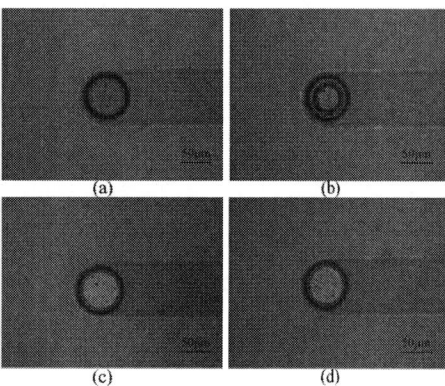

Fig. 7.   Micrograph of BCB via after development , (a) 0MPa, (b) 10MPa, (c) 20MPa, (b) 30MPa

Furthermore, after the fabrication of via, we need to fabricate the thin film interconnection layer on the BCB dielectric layer. BCB is a kind of polymer dielectric material. The deposition of metal film on the surface of BCB is accompanied by the generation and accumulation of large residual stress. When the film stress exceeds the adhesion strength of the film or the tensile strength of the BCB film, the metalized film will fracture and produce cracks. Defects such as interface cracks will appear in the film. In order to avoid this defect, the thin film was sputtered in sections at 150 ℃ for 3 min / 5 min, and the surface of TiW/Cu and TiW/Au films was free of cracks, which was conducive to the release of film stress. Fig. 8 is a micrograph of the metalized film after segmented sputtering. It can be seen from figure 8 that the film after segmented sputtering has no cracks and the film quality is good.

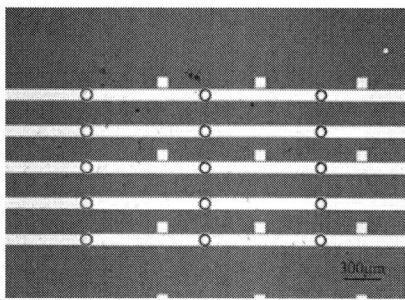

Fig. 8 .   Micrograph of metalized film after segmented sputtering

D. Physical and profile analysis of high density multilayer substrate

According to the above preparation process, the AlSi substrate multilayer wiring substrate is fabricated, including three layers of BCB dielectric layer and four layers of thin film conductor wiring, as shown in Figure 9. Table 1 shows the comparison of performance parameters of high thermal conductivity multilayer wiring substrate with PCB and LTCC. The results show that AlSi high thermal conductivity multilayer wiring substrate has the advantages of matching thermal expansion coefficient with chip, high thermal conductivity and low dielectric constant.

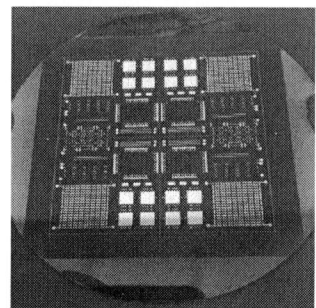

Fig. 9 .   BCB/ Cu multilayer wiring technology on high thermal conductivity AlSi substrate

TABLE I.   COMPARISON OF SUBSTRATE PERFORMANCE PARAMETERS

Parameters	LTCC	Hot-pressed AlN	High thermal conductivity Al-70Si substrate
CTE（25℃）	4-8ppm	4ppm	7ppm
Thermal conductivity	3W/mK	170-200W/mK	115-125W/mK
Firing/curing temperature	850℃	1950℃	210℃
Used metal	Au,Ag,Cu	Mo,W	Au,Cu,TiW
Dielectric Const.	6-8	8.5	2.67
50Ω line loss at 10GHz	0.01-0.02dB/mm	0.03-0.04dB/mm	0.003-0.005dB/mm

According to GJB548B-2005 standard, the temperature cycling test of AlSi substrate multilayer wiring substrate is carried out. The test conditions are shown in Table 2. After the temperature cycling test, the pass rate of the interconnection conduction and insulation test of the substrate is 100%(Test conditions: 903 test points of substrate, test voltage 250V, on resistance ≤ 10mΩ, insulation resistance ≥ 10mΩ).

TABLE II.   TEMPERATURE CYCLING TEST CONDITION

Parameters	Value
Minimum temperature	-55℃
Maximum temperature	150℃
Heating-up time	30min
High temperature residence time	10min
Cool-down time	120min
Low temperature residence time	10min
Cycle times	100

After the end of the test, take out the sample and observe the cross section. Fig. 10 shows the cross section scanning electron microscope (SEM) of the multilayer wiring interface. There is no film separation phenomenon in the wiring layer.

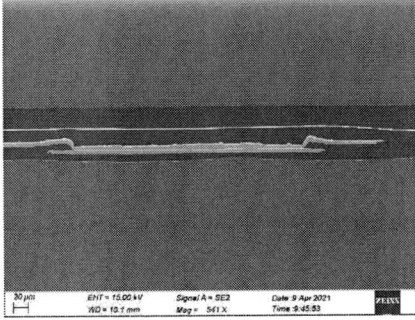

Fig. 10. Scanning electron microscopy (SEM) of the high thermal conductivity AlSi substrate

IV. CONCLUSIONS

In this paper, a method of preparing BCB / Cu multilayer wiring substrate based on AlSi high thermal conductivity substrate is proposed. Seven layers of AlSi multilayer wiring substrate are prepared. Through 100 times of temperature cycling test at - 55 ℃ ~ 150 ℃, the qualified rate of the substrate interconnection conduction and insulation test is 100%. The results show that:

(1) The average surface roughness of AlSi substrate can be reduced to 0.038μm when the slurry size is 3μm and the polishing time is 90 min;

(2) A Ta/Al anodized insulating transition layer is added between the BCB dielectric film and the AlSi substrate to buffer the stress. The surface relief of the cured BCB on the AlSi substrate is 61.33μm, and the deformation is small;

(3) A large number of color interference fringes in BCB through-hole can be effectively removed by 20MPa spray development, and no residual film is found in the through-hole after development; The residual stress accumulation of BCB dielectric film can be effectively alleviated by segmented sputtering, so that there is no crack in the metalized BCB dielectric film.

Through this method, the thermal resistance of the chip can be reduced by using AlSi high thermal conductivity substrate, and the high-efficiency heat dissipation and high-density wiring of the chip can be realized. At the same time, the stress accumulation of BCB multilayer wiring can be significantly reduced by using the anodic oxidation porous dielectric film interface. The process is stable and reliable, which provides an effective solution for the miniaturization of micro system integration.

REFERENCES

[1] Zhang Xiaohui, Wang Qiang. Research state of metal-matrix composites for electronic packaging[J]. Micronanoelectronic technology,2018，55(1):18-44.

[2] Xiang W，Castellazzi A，Zanchetta P . Observer based temperature control for reduced thermal cycling in power electronic cooling[J]. Applied Thermal Engineering, 2014, 64(1–2):10-18.

[3] Xyj A，Cfl A，Z Pg A B，et al. The characterization of Fe-rich phases in a high-pressure die cast hypoeutectic aluminum-silicon alloy[J]. Journal of Materials Science & Technology, 2020, 51:54-62.

[4] Jia Y D，Ma P，Prashanth K G，et al. Microstructure and thermal expansion behavior of Al-50Si synthesized by selective laser melting[J]. Journal of Alloys & Compounds, 2017, 699:548-553.

[5] Cui X，Cui H，Wu Y，et al. The improvement of electrical conductivity of hypoeutectic Al-Si alloys achieved by composite melt treatment[J]. Journal of Alloys & Compounds, 2019, 788:1322-1328.

[6] Shimoto, T.，Matsui, K., Utsumi, K. Cu/photosensitive-BCB thin-film multilayer technology for high-performance multichip modules[J]. IEEE Transactions on Components, Packaging, and Manufacturing Technology,1995, Vol.18(1):18-22.

[7] Fei G，Ding X，Le L . Study on a 3D packaging structure with benzocyclobutene as a dielectric layer for radio frequency application[C]// International Conference on Electronic Packaging Technology & High Density Packaging. IEEE, 2009.

[8] Lei Ding, Jing Chen,Yan Luo, et al.Fabrication technology of BCB through hole for high density multilayer wiring substrate [J].Materials science and technology，2018,Vol.26(5):66-73.

Method of predicting the maximum stress of BGA solder joints based on BP neural network

1st Huaiquan Zhang
School of Electronic Mechanical Engineering,
Guilin University of Electronic Technology
Guilin, China
1599005976@qq.com

2nd Chunyue Huang(Corresponding Author)
School of Electronic Mechanical Engineering,
Guilin University of Electronic Technology
Guilin, China
hcymail@163.com

3rd Shuaidong Liao
School of Electronic Mechanical Engineering,
Guilin University of Electronic Technology
Guilin, China
1729269081@qq.com

4st Shoufu Liu
School of Electronic Mechanical Engineering,
Guilin University of Electronic Technology
Guilin, China
1044577910@qq.com

Abstract— In the process of studying the interconnection reliability of electronic products, due to the small size of the solder joints, the solder joint stress value of the chip cannot be directly measured by conventional methods, so this paper proposes a method based on BP neural network to predict the maximum bending stress of BGA solder joints. First, the orthogonal analysis method is used to establish an orthogonal combination of three factors and two levels of solder ball diameter, solder joint height, and solder joint spacing. Then the ANSYS simulation software was used to establish a model of the orthogonal analysis results, and the stress at the center position of the chip in the X and Y directions and the maximum stress of the solder joint under the condition of applied load were measured. Finally, the BP neural network built by TensorFlow was trained and tested with the data obtained from ANSYS simulation, and the rationality of the method was verified according to the test results.

Keywords—Interconnection reliability; Solder joint stress value; BP neural network; ANSYS; TensorFlow

I. INTRODUCTION

The rapid development of electronic information industry, electronic products are gradually developing in the direction of functional diversification and structural precision [1]. In the process of design and manufacturing of electronic products, more and more attention is paid to the reliability analysis of microelectronic assemblies [2]. The solder joints formed by the soldering process are a key part of the connection between various components and the printed board, enabling the interconnection of the various integrated circuits on the board. Therefore, the mechanical and electrical properties of solder joints are key factors affecting the reliability of electronic products [3]. Electronic products in use by bending, pulling, twisting, random vibration and other loads will make the printed board deformation. Due to the different modulus of elasticity between the printed board and microelectronics, it will lead to different stress and strain between the printed board and microelectronics and makes the solder joints appear "stress concentration". The solder joints are deformed for releasing stress. This leads to the failure of the solder joint and further

causes the failure of the entire electronic product [4]. Solder joint failure is the main cause of electronic product failure. Analysis of the mechanical properties of solder joints is of great importance to improve the reliability of electronic products.

With the development of computer technology, artificial intelligence has ushered in the rapid development of machine learning technology after the two ups and downs of inference period and knowledge period [5]. Deep learning is widely used in image processing, speech recognition, data mining [6]. Deep learning can automatically dig out the relationship between data without the need to understand the mechanism of interconnection between data [7]. Secondly, due to the small size of the solder joints, it is difficult to test the stress of the solder joints in the actual operation process, but the stress value of the chip surface can be measured with strain gauges. Based on the advantages of machine learning and the characteristics of chip stress testing, this paper proposes a method based on BP (Back Propagation) neural network to predict the maximum bending stress of BGA (Ball Grid Array Package) solder joints.

II. ACQUISITION AND ANALYSIS OF TRAINING AND TEST DATA

A. Determination of NASYS simulation model

This paper analyzes the relationship between the stress value at the center of the chip and the maximum stress of the solder joints under the conditions of different solder ball diameters, solder joint heights, and solder joint spacing. The BP neural network is used to establish the transfer function between the transverse stress σ_x and the longitudinal stress σ_y at the center of the chip and the maximum solder joint stress σ_{max}, and then the maximum stress of the solder joint is predicted by experimenting the stress at the center of the chip.

Among the three factors of solder ball diameter, solder joint height, and solder joint spacing, the values of solder ball diameter are 0.28 mm and 0.30 mm, the values of solder joint height are 0.16 mm and 0.22 mm, and the value of solder joint spacing are 0.45 mm and 0.55 mm respectively.

This research is supported by the Science Foundation of Guangxi Zhuang Autonomous Region Government (No.2019JJA160101) and Science and Technology Major Project of Guangxi Zhuang Autonomous Region Government (No.AA19046004) and Innovation Project of GUET Graduate Education（2021YCXS009）

978-1-6654-1392-3/21 $31.00 © 2021 IEEE

Since the three factors have two values, the relevant theory of permutation and combination shows that the analysis model has a total of 8 situations. In order to reduce the cumbersome degree of simulation and analysis without affecting the representativeness of the model, this paper uses the orthogonal analysis method to optimize the combination of the two values of the above three factors. This paper uses the orthogonal analysis method of three factors and two levels to optimize and analyze the existing 8 combinations to obtain 4 optimized combinations. Among them, the solder ball diameter of the first group of chips is 0.28 mm, the solder joint height is 0.22 mm, the solder joint pitch is 0.45 mm, and the pad diameter is 0.22 mm. The solder ball diameter of the second group of chips is 0.28 mm, the solder joint height is 0.22 mm, the solder joint pitch is 0.55 mm, and the pad diameter is 0.22 mm. The solder ball diameter of the third group of chips is 0.30 mm, the solder joint height is 0.16 mm, the solder joint pitch is 0.55 mm, and the pad diameter is 0.22 mm. The solder ball diameter of the fourth group of chips is 0.30 mm, the solder joint height is 0.16 mm, the solder joint pitch is 0.45 mm, and the pad diameter is 0.22 mm.

B. Sample data acquisition

In this paper, the training sample data and test sample data of the BP neural network are obtained through simulation of ANSYS simulation software. According to the orthogonal analysis results, the three-point bending stress and strain finite element analysis model of 4 sets of ball grid array solder joints is established in the ANSYS simulation software, as shown in Figure 1. All ball grid array solder joints adopt a 4-level array, that is, the array is composed of 4 rows of solder joints and the number of solder joints in each row is 4. It can be seen from the figure that the chip to be tested is located at the center of the printed circuit board (hereinafter referred to as the PCB).

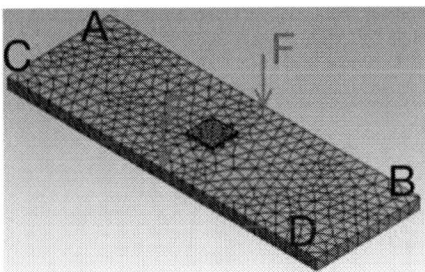

Fig. 1. Three-point bending stress and strain finite element analysis of BGA solder joints

In the ANSYS simulation analysis software, the degree of freedom of movement of the AC side of the model in the X, Y, and Z directions and the degree of freedom of rotation around the X and Y directions are restricted to zero, so that the printed board can only rotate around the side AC. Limit the freedom of movement of the BD side of the model in the Y and Z directions and the rotational freedom of the X and Y directions to 0, so that the printed board can rotate around the side BD and move along the X direction. Therefore, the force of the printed board is similar to that of a statically determinate simply supported beam structure. Then, the printed board is bent by applying multiple sets of different loads at the midpoints of the AB and CD sides of the four groups of models. Since the chip is welded to the printed board by solder joints, when the printed board is bent and

deformed, the chip will be bent and deformed, and the solder joints will be deformed to generate stress. The transverse stress and the longitudinal stress at the center of the chip and the maximum stress of the solder joints under different load conditions are measured as the training data and test data of the neural network. Due to the limited sample data in this article, the test sample data and training sample data in this article are the same set of data.

C. Correlation analysis of data

The essence of using BP neural network to predict the maximum stress of solder joints based on the transverse stress and longitudinal stress at the center of the chip is to establish the transfer function of the transverse stress and longitudinal stress at the center of the chip and the maximum stress of the solder joints. Therefore, it is possible to use BP neural network to predict the maximum stress of the solder joint only when there is a certain correlation between the transverse stress and the longitudinal stress at the center of the chip and the maximum stress of the solder joint.

This article uses SPSS to analyze the correlation between the transverse stress and the longitudinal stress at the center of the chip and the solder joint stress before using the BP neural network to analyze the data. The analysis results show that there is a correlation between the transverse stress and the longitudinal stress at the center of the chip and the maximum stress of the solder joint in the four models, so the BP neural network can be used to predict the maximum stress of the solder joint.

III. THE CONSTRUCTION OF NEURAL NETWORK AND ITS TRAINING AND TESTING

A. BP neural network construction

Using TensorFlow2.0 to build a BP neural network model is shown in Figure 2. It can be seen from the figure that the neural network is a three-layer BP neural network. The neural network in the figure consists of an input layer, a hidden layer and an output layer. The input layer is composed of two neurons, the hidden layer contains two layers of neural networks, the hidden layer one is composed of four neurons, the hidden layer two is composed of two neurons, and the output layer only contains one neuron.

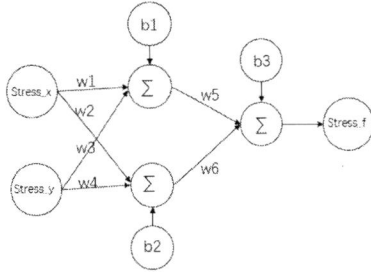

Fig. 2. Neural network model diagram

The input layer of the neural network is used as the input interface of the original data to input the lateral stress σ_x and the longitudinal stress σ_y at the center position of the chip into the neural network. The calculation model of the first layer of neural network is as follows:

$$Y_1 = \begin{bmatrix} w_1 & w_3 \\ w_2 & w_4 \end{bmatrix} * \begin{bmatrix} \sigma_x \\ \sigma_y \end{bmatrix} + \begin{bmatrix} b_1 \\ b_2 \end{bmatrix} \quad (1)$$

The output result of the transverse stress σ_x and the longitudinal stress σ_y at the center of the chip after being processed by the first layer of neural network is composed of two parameters. The calculation model of the second layer of neural network is as follows:

$$Y_2 = \begin{bmatrix} w_5 & w_6 \end{bmatrix} \begin{bmatrix} Y_{11} \\ Y_{12} \end{bmatrix} + b_3 \quad (2)$$

The output result of the first layer of neural network processed by the second layer of neural network is composed of a parameter. After the result is processed by the output layer, the maximum stress of the solder joint predicted based on the transverse stress σ_x and the longitudinal stress σ_y at the center of the chip can be obtained.

B. Neural network training

The process of training a neural network is to bring the trained sample data into the calculation model of the neural network, and then use the gradient descent method to gradually update the parameters w1, w2, w3, w4, b1, b2, and b3 in the calculation model of the neural network to obtain the best transfer function of the transverse stress, the longitudinal stress and the maximum stress of the solder joint at the center of the chip. The calculation model of the neural network is as follows:

$$stress\_out = \begin{bmatrix} w_5 & w_6 \end{bmatrix} \left\{ \begin{bmatrix} w_1 & w_3 \\ w_2 & w_4 \end{bmatrix} * \begin{bmatrix} \sigma_x \\ \sigma_y \end{bmatrix} + \begin{bmatrix} b_1 \\ b_1 \end{bmatrix} \right\} + b_3 \quad (3)$$

The mean square error of the model is as follows:

$$dev = \sqrt{\sum_{i=0}^{n}(stress\_out - stress\_max)^2} \quad (4)$$

At the beginning of training the BP network, the initial values of w1, w2, w3, and w4 are assigned with random numbers, and the initial values of b1, b2, and b3 are assigned zero. Calculate the partial derivative function expression of dev in formula (4) with respect to the parameters w1, w2, w3, w4, b1, b2, and b3 through differential operation. Bring the values of w1, w2, w3, w4, b1, b2, and b3 and the training sample data into the partial derivative function expression to calculate the partial derivative of dev relative to the current coefficient. Then update the values of parameters w1, w2, w3, w4, b1, b2, and b3 through the following expressions (5) and (6).

$$w_i' = w_i - lr * \left(\frac{\partial}{\partial w_i} dev \right) \quad (5)$$

$$b_i' = b_i - lr * \left(\frac{\partial}{\partial b_i} dev \right) \quad (6)$$

In the expression, w_i' is the value of the updated w_i, b_i' is the value of the updated b_i, and lr is a positive number that is the learning rate of the neural network. It can be seen from the above formula that when the partial derivative of the mean square error with respect to the coefficient is less than zero, the mean square error function is monotonously decreasing in the current dimension, and the coefficient will increase in the direction that reduces the mean square error. When the partial derivative of the mean square error with respect to the coefficient is greater than zero, the mean square error function increases monotonously in the current dimension, and the updated coefficient will decrease in the direction that reduces the mean square error. So when the

partial derivative of the mean square error with respect to the coefficient is greater than zero and less than zero, the coefficients will be updated in the direction that reduces the mean square error.

C. BP neural network test

Figure 3 shows the prediction results of the first group of models. Figure 3(a) is a comparison diagram of sample data and predicted data. The blue dots represent the sample data distribution diagram, and the green dots represent the predicted data distribution diagram. Figure 3(b) is a graph of the change in the prediction error rate. The green line represents the prediction error rate of each sample, and the red line represents the average value of the prediction error. It can be seen from Figure 3(a) that the prediction result curve of the first 27 groups basically fits the sample distribution curve. In the last 23 groups of data, there are individual positions that have a poor fit, but the prediction results are generally consistent with the change trend of the sample data. It can be seen from Figure 3(b) that the overall prediction result of the neural network is good, the average error rate between the prediction result and the sample data is slightly less than 7.5%, and the maximum error rate is about 17.5%.

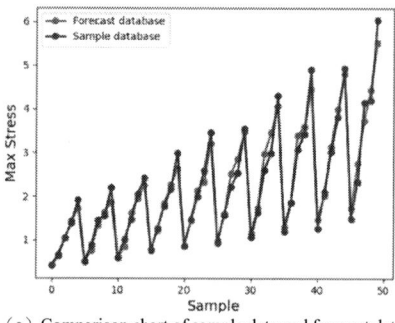

（a）Comparison chart of sample data and forecast data

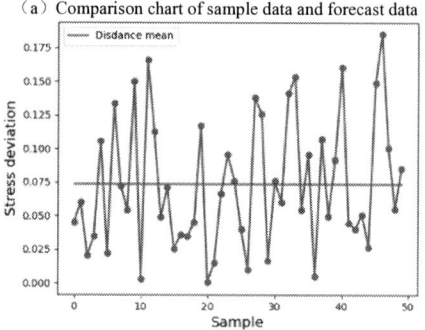

（b）Change chart of forecast error rate

Fig. 3. The first set of model training results

Figure 4 shows the prediction results of the second group of models. Figure 4(a) is a comparison diagram of sample data and predicted data. The blue dots represent the sample data distribution diagram, and the green dots represent the predicted data distribution diagram. Figure 4(b) is a graph of the change in the prediction error rate. The green line represents the prediction error rate of each sample, and the red line represents the average value of the prediction error. It can be seen from Figure 4(a) that, except for the 5 sets of data 24, 29, 34, 39, and 44, the prediction results of the other sets of data have a higher degree of fit with the sample data, and the trend of the prediction results and sample data Generally the same. It can be seen from

Figure 4(b) that the overall prediction result of the neural network is good. The average error rate between the prediction result and the sample data is about 7.5%, and the maximum error rate is slightly greater than 17.5%.

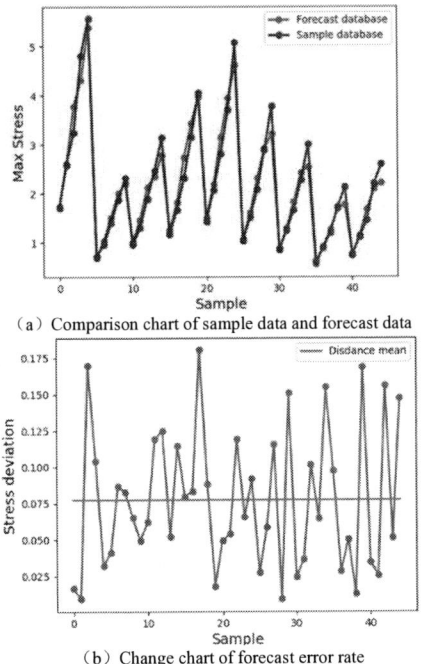

（a）Comparison chart of sample data and forecast data

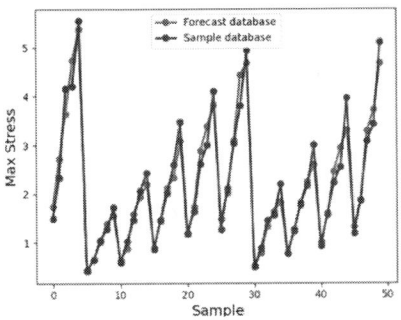

（b）Change chart of forecast error rate

Fig. 4.　The second set of model training results

Figure 5 shows the prediction results of the third group of models. Figure 5(a) is a comparison diagram of sample data and predicted data. The blue dots represent the sample data distribution diagram, and the green dots represent the predicted data distribution diagram. Figure 5(b) is a graph of the change in the prediction error rate. The green line represents the prediction error rate of each sample, and the red line represents the average value of the prediction error. It can be seen from Figure 5(a) that this set of data has a poorer fit than the first two sets of data, but the predicted results are generally consistent with the changing trends of the sample data. It can be seen from Figure 5(b) that the average error rate between the prediction results and the sample data is slightly greater than 7.5%, and the maximum error rate is slightly greater than 17.5%.

（a）Comparison chart of sample data and forecast data

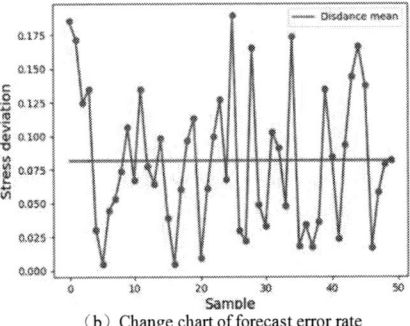

（b）Change chart of forecast error rate

Fig. 5.　The third set of model training results

Figure 6 shows the prediction results of the fourth group of models. Figure 6(a) is a comparison diagram of sample data and predicted data. The blue dots represent the sample data distribution diagram, and the green dots represent the predicted data distribution diagram. Figure 6(b) is a graph of the change in the prediction error rate. The green line represents the prediction error rate of each sample, and the red line represents the average value of the prediction error. It can be seen from Figure 6(a) that the fit of this group of data is the highest among the four sets of data. Except for the poorer fitting of points 5 and 44, the fitting of the remaining points is good. It can be seen from Figure 6(b) that the overall prediction result of the neural network is good, the average error rate between the prediction result and the sample data is slightly greater than 7.5%, and the maximum error rate is less than 17.5%.

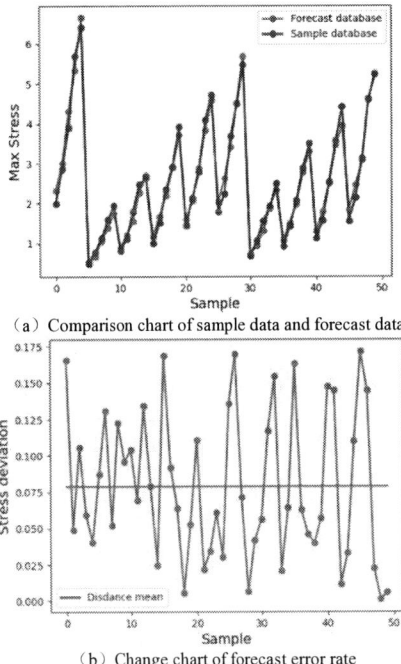

（a）Comparison chart of sample data and forecast data

（b）Change chart of forecast error rate

Fig. 6.　The third set of model training result

From the test results of the above four sets of data, it can be seen that the prediction results of the neural network are generally consistent with the change trend of the original data of the sample. The average error rate of the four sets of data is about 7.5%, the maximum error rate is about 17.5%, all within 20%, indicating that the method of using BP neural network to predict the maximum stress of solder joints has certain desirability.

IV. CONCLUSION

1）The correlation analysis between the stress values at the center of the chip surface in the X and Y directions and the maximum stress values at the solder joints shows that there is a correlation between the stress values at the center of the chip surface in the X and Y directions and the maximum stress values at the solder joints.

2）The maximum error rate of the maximum stress prediction model based on BP neural network for all four groups of models is around 17.5% and the average error rate is around 7.5%.

3）The maximum stress prediction model based on BP neural network can predict the maximum stress value σmax of the solder joint based on the first stress value $\sigma1$ and the second stress value $\sigma2$ at the center position of the chip.

REFERENCES

[1] Ying Liang, Chunyue Huang, Rui Yin, Wei Huang, Tianming Li, Hongwang Zhao, "Three-point bending stress-strain analysis of microscale ball grid array solder joints for microelectronic packaging," J. Mechanical Strength. vol. 38(04), pp. 744–748, 1955.

[2] Shi R., "Failure assessment of devices in the reliability design of electronic devices," J. Electronic Product Reliability and Environmental Testing. vol. 27(S1), pp. 155-158, 2009.

[3] Qin Hongbo, "Study of mechanical properties of lead-free microinterconnect solder joints and the size effect of fatigue and electromigration behavior," D. South China University of Technology., 2014.

[4] QI Fangjuan, "WANG Jianqiang,ZHENG Tianqun. Study on mechanical bending reliability of dispensing spherical array package assemblies," J. China Integrated Circuit. pp. 63-67, March 2007.

[5] Zhang X, Wang MH., "The development trend of artificial intelligence in China and its promotion strategy," J. Reform. pp. 31-44, September 2019.

[6] Ma S.L., Unyzhqig, Li S.P., "A review of big data and deep learning," J. Journal of Intelligent Systems. vol. 11(06) ,pp. 728-742, November 2016.

[7] Ji Shou-Ling, Li Jin-Feng, Du Tian-Yu, Li Bo, "A review of machine learning model interpretability methods, applications and security research," J. Computer Research and Development. vol. 56(10), pp. 2071-2096, 2019.

Thermal Stress Study of 3D IC Based on TSV and Verification of Thermal Dissipation of STI

1st Shuaidong Liao
School of Electronic Mechanical Engineering
Guilin University of Electronic Technology
Guilin, China
1729269081@qq.com

2nd Chunyue Huang(Corresponding Auther)
School of Electronic Mechanical Engineering
Guilin University of Electronic Technology
Guilin, China
hcymail@163.com

3rd Huaiquan Zhang
School of Electronic Mechanical Engineering
Guilin University of Electronic Technology
Guilin, China
1599005976 @qq.com

4nd Shoufu Liu
School of Electronic Mechanical Engineering
Guilin University of Electronic Technology
Guilin, China
1044577910@qq.com

Abstract—With the continuous progress of chip integration, the three-dimensional integration technology based on silicon through-hole (TSV) has emerged and become one of the key technologies to achieve high-density system integration. However, its process size and interconnect technology directly lead to severe thermal reliability problems, so it becomes urgent to study the thermal characteristics of TSV arrays in 3D integrated circuits.

In this paper, ANSYS Workbench is used to analyze the model for thermal stress. The effects of the diameter, height, spacing and filling material of the TSV on the overall thermal stress of the model were investigated using orthogonal table experiments. The orthogonal experimental analysis table was made based on the horizontal factor table of TSV, and the corresponding thermal stress values were obtained.

The data results were analyzed using extreme difference analysis to determine the optimal combination of parameters. And this set of parameters is used as the basis for further analysis. In recent years some experts and scholars proposed the Shallow Trench Isolation Technology (STI), which is an effective method to reduce thermal stress. Using the previously selected set of data, simulation analysis is performed in Workbench to compare and verify the results.

The analysis of the experimental results shows that the largest factor affecting the thermal stress of the TSV-based 3D IC is the filling material of the TSV, followed by the pitch of the TSV, the diameter of the TSV, and the smallest is the height of the TSV. Comparing the results of the experimental group shows that STI has a greater improvement on the heat deformation of the model and a significant reduction in its maximum stress, indicating that STI has a more significant improvement on the thermal stress of the TSV-based 3D integrated circuit.

Keywords—TSV; 3D IC; STI; Thermal stress; Orthogonal experimental design

---

This research is supported by the Science Foundation of Guangxi Zhuang Autonomous Region Government (No.2019JJA160101), Science and Technology Major Project of Guangxi Zhuang Autonomous Region Government（No.AA19046004）and Innovation Project of GUET Graduate Education（2021YCXS009）

I. INTRODUCTION

With the continuous progress of process manufacturing level and packaging technology, integrated circuits continue to develop in the direction of high integration, multi-function, high performance, low power consumption and low cost in accordance with Moore's Law, driving the information industry forward rapidly [1]. Reducing the size of devices and interconnects and improving circuit structures are the main means of technological innovation for traditional integrated circuits. In order to solve a series of problems in the development of traditional integrated circuits, a new interconnect design and packaging technology, which is based on silicon through-hole (TSV) three-dimensional integrated circuits came into being [2]. TSV-based 3D integrated circuits provide a completely new form of topological arrangement of the modules on the chip. The basic principle is to extend the integration from the plane to the longitudinal direction, with connections on each layer through traditional on-chip interconnects and between layers through silicon vias [3].

Three-dimensional integrated circuit in the use of the process, the biggest difference with the traditional integrated circuit is its great integration, but the excessive integration also brings greater power consumption, while the heat generated by several layers of the chip is also a large number of concentrated difficult to dissipate, so the thermal problem has become the core problem of three-dimensional integrated circuit.

Three-dimensional integrated circuit in the use of the process, the biggest difference with the traditional integrated circuit is its great integration, but the excessive integration also brings greater power consumption, while the heat generated by several layers of the chip is also a large number of concentrated difficult to dissipate, so the thermal problem has become the core problem of three-dimensional integrated circuit. The TSV technology is one of the current solutions, and many experts and scholars have conducted corresponding research [4]. For example, a finite element simulation analysis of the thermal reliability problem of TSV was carried out by Zhiqing Yang et al [5]. The results show that the use of metal-based compliant materials as metal filler materials for TSV of 3D IC can optimize the thermal stress of TSV. Fengjuan Wang

[6] et al conducted a study for the highest layer chip of 3D IC. By introducing the silicon via area scaling factor, a 3D IC temperature resolution model considering TSV is proposed, and the optimal range of the 8-layer 3D IC silicon via area scaling factor is given.

II. MODELING AND SIMULATION

A. Modeling

In this paper, we take TSV as the research object and establish the finite element analysis model as shown in Fig. 1. Multiple sets of analytical models were set up according to different TSV diameters, heights, and spacings, as well as different silicon through-hole filling materials.

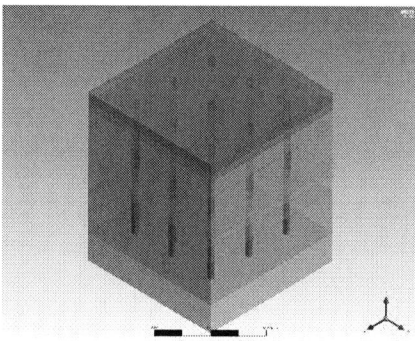

Fig. 1.   Modeling

The experimental model consists of a multilayer top and bottom structure, including the bottom heat sink layer, the chip layer, and the top package layer, where the chip layer includes the silicon substrate, the device layer, and the bonding layer in order from the bottom to the top. Compared with two-dimensional integrated circuits, three-dimensional integration is in the vertical direction of the interconnection of multiple chips, which makes the global interconnection length is reduced, but also to improve the delay and power consumption of two-dimensional integrated circuits, but at the same time is the core issue of thermal problems is also more prominent than two-dimensional integrated circuits.

The current research on the thermal characteristics of TSV of 3D integrated circuits is mainly focused on single-layer chip research or just for the study of individual TSV of multilayer chips, so the model uses two chip layers to simulate in order to explore the causes of the above problems, and their influencing factors. Since this is a simulation, the dimensions here are taken as representative data. The heat sink layer is 30mm, the silicon substrate is 50mm, the device layer is 20mm, the bonding layer is 10mm, and the packaging layer is 10mm.

After setting up the above parts, the next part is the core TSV part. This experiment uses a 3×3 matrix distribution, with the upper and lower surfaces of the TSV connected to the lower surface of the package layer and the upper surface of the heat sink layer, respectively, through the middle two chip layers. Due to the need for subsequent experiments, multiple sets of values are taken for the dimensional parameters of the TSV. TSV diameters were taken as 5μm, 10μm and 15μm, heights as 140μm, 160μm and 180μm, and pitches as 48μm, 50μm and 52μm, while its filling materials were selected as Copper, Aluminum and Tungsten for the experiments.

B. Simulation Experiment

1) Material Properties:

As an adhesive material for chip bonding used in the process of semiconductor crystals, it is necessary to have not only adequate adhesion and good cutting properties, but also to be able to achieve good peeling and to have good installation reliability. As the most widely used adhesive material for bonding, Phenolic Epoxy Resin has the above required properties to meet the requirements.

In this paper, ANSYS Workbench is used as the experimental platform. Import the built model into Workbench and set the material properties. In this model, Copper (Cu) is used for the heat sink layer, Silicon (Si) is used for the silicon substrate, Silicon Dioxide (SiO₂) is used for the device layer, Phenolic Epoxy Resin is used for the bonding and encapsulation layers, while copper, Aluminum (Al) and Tungsten (W) are used as alternative materials for the TSV. The material properties are shown in the TABLE I.

TABLE I.   MATERIAL PROPERTIES

Materials	Density (kg/m3)	Thermal conductivity (W/(m*K))	Atmospheric Pressure Heat Capacity (J/(kg*K))	Poisson's Ratio	Young's Modulus (GPa)	Coefficient of Thermal Expansion (1/K)
SiO2	2220	1.5	745	0.17	73	0.0000004
Si	2330	148	712	0.28	131	0.0000028
Cu	8933	400	385	0.34	110	0.000018
Al	2689	237.5	951	0.33	71.7	0.0000232
W	19300	174	132	0.28	344.7	0.0000045
Epoxy	1200	0.2	550	0.38	1	0.0000624

2) Simulation Parameters Setting:

The heat generated by the 3D IC mainly originates from the device layer, so the upper surface of the device layer is considered as the surface heat source in this paper to simulate the real heat generation phenomenon. In order to reduce the amount of calculation, this experiment will assume that all materials are fully bonded, the material filling is ideal filling, and the initial temperature in the simulation is set to 293.15 K.

Theoretical analysis shows that the TSV structure in the 3D IC is mainly through convective heat transfer between the surface and the surrounding air, and heat conduction between the lower surface and the substrate, while the substrate temperature is assumed to be basically constant here. Therefore, the convective heat transfer coefficient between the upper surface and the side of the model and the air is set to 5 W/(m2-°C). In this paper, a heat source with a power of 2×10-5W is used as an excitation, which is applied at the upper surface of several device layers for 1 second to simulate the temperature field of TSV structures in 3D integrated circuits.

3) Simulation Results:

The results of the finite element simulation, including temperature, total deformation and equivalent stress, are shown in Fig. 2, Fig. 3and Fig. 4. It can be seen from the results that the highest temperatures are distributed in the upper part of the model, but not on the uppermost surface. As shown in Fig. 5, the maximum stress occurs at the contact between the bottom of the TSV and the substrate of the heat sink layer. It can be seen that the three-dimensional packaging structure of the TSV has a large impact on the overall heat conduction and thermal stress distribution, so a comparative analysis of the size of the TSV can be considered.

Fig. 2.   Temperature Result

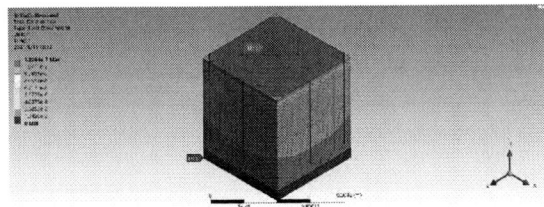

Fig. 3.   Total Deformation Result

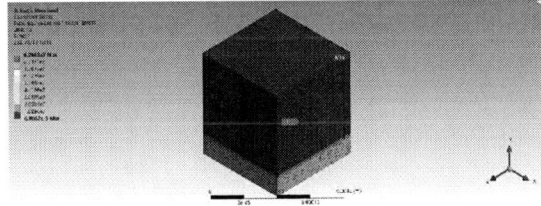

Fig. 4.   Equivalent Stress Result

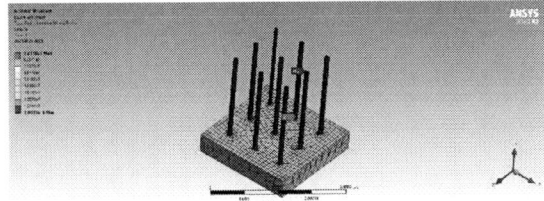

Fig. 5.   Maximum Stress

4) Orthogonal Experiments:

Orthogonal design is based on the mathematical theories and methods of probability theory and statistics to perform effective statistical analysis of data by scientifically specifying

experimental protocols. The essence is to reduce the influence of random errors on the experiment and improve the credibility of the conclusions by rationalizing the implementation scheme. The cause that affects the test result is called the influence factor, and the specific state in which it is located is called the Level. In common applications, it is often necessary to consider three or more test factors and different levels at the same time, thus boiling a sharp increase in the number of tests, making it difficult to fully implement the experiment.

Orthogonal experimental designs are often used to arrange multi-factor tests to find optimal combinations due to their high efficiency. The experimental design relies closely on the orthogonal table, which is a table based on the idea of balanced distribution, and it is this that makes the number of experiments significantly reduced.

In this paper, we use $L_9(3^4)$ orthogonal experimental table for analysis, which means the number of trials of the orthogonal experiment is 9, the number of factors that can be arranged is 4, and the number of levels of each factor is 3. The selected factors are the diameter, height, spacing and filling material of TSV, and the specific factor level table is shown in TABLE II.

TABLE II. FACTOR LEVEL TABLE

Factors	1	2	3
Diameter (μm)	5	10	15
Height(μm)	140	160	180
Spacing (μm)	48	50	52
Filling Material	Cu	Al	W

Nine sets of experiments were designed by orthogonalization, and the specific orthogonal experimental protocol table is shown in Tab. 3. A total of 9 sets of finite element simulation experiments were conducted based on this, and the deformation of the model and the stress magnitude and its distribution under this set of test parameters were derived.

TABLE III.   ORTHOGONAL EXPERIMENTAL PROTOCOL TABLE

Groups	Diameter (μm)	Height (μm)	Spacing (μm)	Filling Material
1	5	140	48	Cu
2	5	160	50	Al
3	5	180	52	W
4	10	140	50	W
5	10	160	52	Cu
6	10	180	48	Al
7	15	140	52	Al
8	15	160	48	W
9	15	180	50	Cu

III. DATA ANALYSIS

A. Analysis of Extreme Difference

Analysis of Extreme Difference determines the optimal combination of parameters by calculating the extreme difference values in each column of the orthogonal table, and can also be used to determine the order of influence of the factors on the test index. The extreme deviation value is the

978-1-6654-1392-3/21 $31.00 © 2021 IEEE

difference between the maximum value of the mean and the minimum value of a factor at each level, which indicates the fluctuation of the factor under different parameter conditions, which means the larger the value of the extreme difference, the greater its effect on the thermal stress of the TSV matrix, and conversely, the smaller the value of the extreme difference, the smaller the effect.

From the analysis results, it can be seen that the optimal combination of TSV diameter of 5 μm, height of 180 μm,

The result data of each group of previous experiments were recorded, and the data results were analyzed using the analysis of extreme differences to determine the optimal combination of parameters. The results of the extreme difference analysis are shown in TABLE IV.

spacing of 52 μm and the filling material Tungsten. And this set of parameters is used as the basis for further analysis.

TABLE IV. RESULTS OF THE ANALYSIS OF EXTREME DIFFERENCES OF ORTHOGONAL EXPERIMENTS

Groups	Diameter (μm)	Height (μm)	Spacing (μm)	Filling Material	Maximum Stress (Pa)
1	5	140	48	Cu	9.2668E+07
2	5	160	50	Al	7.2067E+07
3	5	180	52	W	6.7220E+07
4	10	140	50	W	7.7412E+07
5	10	160	52	Cu	1.3685E+08
6	10	180	48	Al	6.7363E+07
7	15	140	52	Al	7.7258E+07
8	15	160	48	W	7.1127E+07
9	15	180	50	Cu	1.2981E+08
Analysis of Extreme Differences					
1j	7.7318E+07	8.2446E+07	7.7053E+07	1.1978E+08	
2j	9.3875E+07	9.3348E+07	9.3096E+07	7.2229E+07	
3j	9.2732E+07	8.8131E+07	9.3776E+07	7.1920E+07	
Extreme Differences	1.6557E+07	1.0902E+07	1.6723E+07	4.7856E+07	

According to the extreme difference data in the above table, it can be seen that the TSV filling material has the greatest effect on the thermal stress, followed by the TSV spacing, then followed by the TSV diameter, and the smallest by the TSV height. The average value of the maximum thermal stress corresponding to each level of different factors is found, and the average main effect diagram of each factor at different levels is made. The results are shown in Fig. 6.

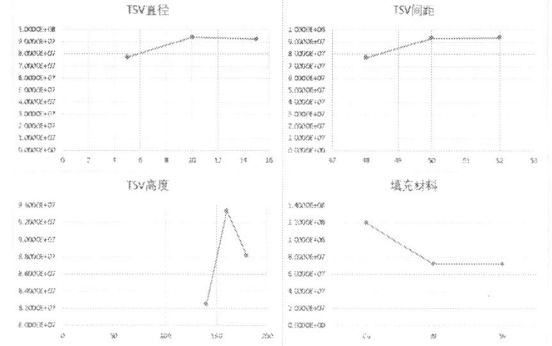

Fig. 6. Mean main effect plots of each factor at different levels

The graph shows that the maximum equivalent force tends to increase and then decrease as the TSV diameter and height continue to increase. As the TSV spacing increases, the equivalent effect force shows an increasing trend. And among the three experimental filling materials, the equivalent force of copper has a large gap with the remaining two, indicating

that its thermal conductivity has a large gap with the remaining two.

B. Further Experiment

TSV technology is the key for 3D integrated circuits, but its thermal stress affects device performance and reduces carrier mobility, leading to serious reliability problems. To reduce the influence of carrier mobility by thermal stress, the active region must be placed in the KOZ. Although such a solution avoids the effect of thermal stress, it results in a waste of area.

In recent years some experts and scholars proposed the STI, which is also an effective method to reduce thermal stress. STI is a shallow trench etched around the TSV. STI is between the TSV and the active area, and it mainly serves as a stress relief. The experimental model containing the STI structure is shown in Fig. 7.

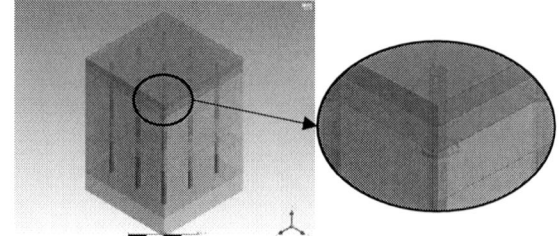

Fig. 7. Model with STI

Using the previously selected set of data, simulation analysis is performed in Workbench to compare and verify the results. The experimental results are shown in TABLE V. From the experimental results, it is clear that STI has a very significant effect on the thermal stress relief of TSV in 3D IC.

TABLE V. COMPARISON EXPERIMENTAL GROUPS WITH AND WITHOUT STI STRUCTURE

Comparison Group	Diameter (μm)	Height (μm)	Spacing (μm)	Filling Material	Maximum Deformation (μm)	Maximum Stress (Pa)
Without STI	5	140	48	W	0.991640	7.8385E+07
With STI					0.847870	6.5354E+07

IV. CONCLUTION

1) The results of the orthogonal table experimental analysis: Derived from the extreme difference analysis, show that the largest factor affecting the thermal stress of the TSV-based 3D IC is the filling material of the TSV, followed by the spacing of the TSV, again the diameter of the TSV, and finally the height of the TSV. Comparing the average maximum thermal stress values for each level of each factor, the optimal combination of parameters can be derived as TSV diameter of 5 μm, height of 180 μm, spacing of 52 μm and the filling material Tungsten.

2) Further experimental results: Using the above set of models, a comparison of the finite element analysis of the two single factors, with and without STI, shows that the deformation caused by heat in the model with STI has a greater improvement, and its maximum stress is significantly reduced. It shows that STI has a more obvious effect on improving the thermal stress of TSV-based 3D IC.

REFERENCES

[1] Wensheng Zhao, Gaofeng Wang, and Wenyan Yin, "New interconnect technologies for ICs in the post-Moore era," China Science Publishing & Media Ltd. Beijing, 2017.

[2] Ying Zhou, "Thermomechanical reliability study of three-dimensional electronic packaging based on silicon through-hole," Huazhong University of Science and Technology, 2016.

[3] Xinyue Luo, "Thermal characterization of three-dimensional integrated circuit TSV matrix," Xidian University, 2019.

[4] Mengying Fan, "Analysis of the new structure and thermal coupling characteristics of TSV in three-dimensional packages," Jiangsu Normal University, 2018.

[5] Zhiqing Yang, Zhongliang Pan, "Simulation analysis of silicon through-hole thermal stress based on metal matrix composites," J. Electronic Components & Materials, vol. 39(05), pp. 97-102, 2020.

[6] Fengjuan Wang, Zhangming Zhu, Yintang Yang, Ning Wang, "3D IC highest layer temperature model considering silicon vias," J. Chinese Journal Of Computational Physics, vol. 29(04), pp.580-584, 2012.

[7] Zhiqing Yang, Zhongliang Pan, "Thermal characterization of TSV in three-dimensional integrated circuits," J. Semiconductor Optoelectronics, vol. 40(06), pp.820-825, 2019.

[8] Shuo Wang, Kui Ma, Fashun Yang, "TSV Reliability Overview," J. Application of Electronic Technique, vol. 47(02), pp.1-6, 2021.

A Humidity-Sensitive Capacitor Based on Fan-Out Panel Level Package Technology

Shuhan Hou
School of Microelectronics
Southern University of Science and technology
Shenzhen, China
2090289071@qq.cn

Tingyu Lin
Guangdong Xinhua Microelectronics Co. , Ltd.
2/F Block A Buddha High-tech Think Tank Center Nanhai High-tech ZoneFoshan City, Guangdong Province)
Foshan, China
tingyulin@ncap-cn.com

Abstract—A highly humidity-sensitive comb capacitor using RDL (Redistribution Layer) and moisture-sensitive polyimide (MS-PI) material is constructed based on low cost Fan-Out Panel Level Package technology. The comb capacitor is design and simulated using HFSS. The length, width and spacing of the comb capacitor are 2.5mm, 0.5mm, and 0.25mm respectively. The dielectric constant of the MS-PI material using in this design ranges from 2.96 to 3.6 as the relative humidity (RH) varies from 0% RH to 100%RH. As a result the proposed comb HS capacitor achieves capacitance ranging from 4.13pF to 4.18pF(measured at 1MHz frequency) with 0.15% non-linearity. The presented HS capacitor in this paper creates a new economical method to manufacture humidity-sensitive capacitor which is the sensing component in humidity sensor.

Keywords—humidity-sensitive comb capacitor; moisture-sensitive polyimide; Fan-Out Panel Level Package; humidity sensor

I. INTRODUCTION

Nowadays, with the rapid development of the Internet of things (IoT), many sorts of things have entered the Internet era. Therefore, a higher level of underlying data acquisition is needed to adapt to the changes of the times. Humidity sensor, as an important device for humidity data acquisition, plays an increasingly important role in the network of IoT. Humidity sensors are widely used in environmental monitoring, biopharmaceutical, food processing, national defense and military industry, and other aspects of people's life. Therefore, higher requirements are put forward for its production process. Nowadays, the production process of humidity sensor is diversified. However, the existing problem is that most humidity sensors have complex production processes, complex materials, large volume and high cost. It is these reasons that limit the development of humidity sensors. [1-2].

According to the properties of different materials, humidity sensors are divided into resistive humidity sensor and capacitive humidity sensor, whose basic principle is to

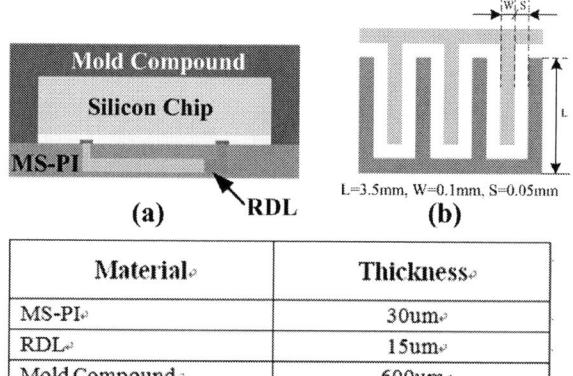

Material	Thickness
MS-PI	30um
RDL	15um
Mold Compound	600um

(c)

Fig. 1. (a) illustration of Fan-Out Panel Level Package solution of humidity sensor integrated with silicon chip and humidity-sensitive capacitor, (b) the structure of the comb capacitor, (c) the material and thickness of the Fan-Out Panel Level Package technology used in this design.

convert humidity into resistance or capacitance. Among them, electrolyte and ceramic humidity sensors as resistive type humidity sensors have huge volume, complex production materials and high price, which stop themselves from being widely used. With the progress of semiconductor technology, humidity sensor chip made by MEMS technology is derived based on CMOS technology, which can greatly reduce the size of the device. The capacitive humidity sensor made of polyimide and the MEMS technology has better performance than other resistive type. However, due to the different process of the humidity sensing part and the measurement circuit, which are mostly consist of discrete components, it is necessary to solve the interface problem between the measurement circuit and the humidity sensing unit in the MEMS process. Moreover, MEMS technology is a non-standard CMOS process, which needs additional unusual process steps and increases the cost [3-6].

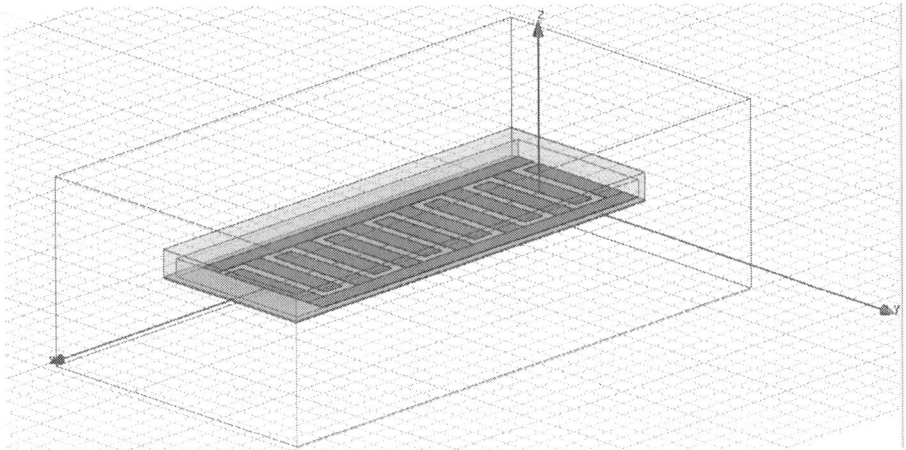

Fig. 2. the simulation model of the humidity-sensitive comb capacitor in HFSS.

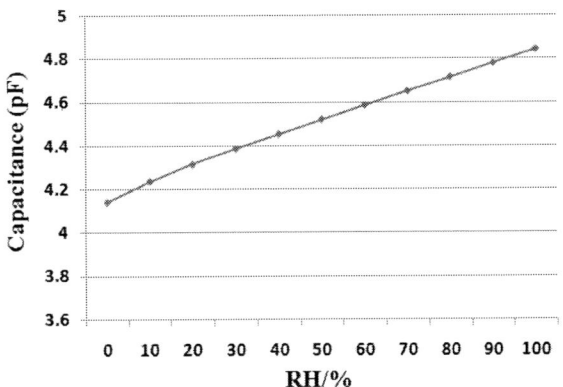

Fig. 3. variation of humidity-sensitive capacitance versus RH%.

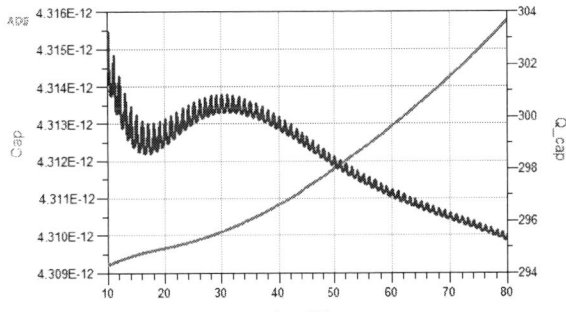

Fig. 4. simulated capacitance and quality factor of the proposed humidity-sensitive capacitor.

In order to further increasing the integration level and reducing the cost of the entire humidity sensor, the best solution is to integrate the humidity sensing component and the measurement circuit on the same chip without requiring additional process. Due to the availability of re-distribution metal layer (RDL) in modern packaging technology, it provides the possibility of producing humidity sensing capacitor during the packaging process. Fan-Out Panel Level Package (FOPLP) is one of the most prospective packaging technologies which can provide multiple RDL metal layers [7-9]. In the FOPLP package, the packaging size increases from wafer scale to panel format, which greatly brings down the cost of the package [10-14].

In this paper, a humidity sensitive capacitor using FOPLP technology is present. The capacitor is designed for humidity sensor application. The FOPLP solution for the humidity sensitive capacitor is introduced in the following section. The structure of the humidity sensitive capacitor is shown in Section III. Simulation results of the capacitor and conclusions of this research are present in Section IV and V respectively.

II. THE SOLUTION OF FAN-OUT PANEL LEVEL PACKAGE FOR THE HUMIDITY SENSITIVE CAPACITOR

FOPLP technology has the ability to supply a low-cost wafer level package solution with at least one redistribution layer (RDL) for interconnection, as shown in Fig. 1 (a). The RDL is used for humidity capacitor generation with patterned structure, as shown in Fig. 1(b). The RDL layer is embedded into a moisture sensing polyimide material (MS-PI) which is sensitive to the environmental humidity. The silicon core side of the included measuring circuit needs to be facing down and embedded in the mold compound and the entire package including the silicon chip constitutes a capacitive humidity sensor.

As shown in Fig. 1 (c), the dielectric in the FOPLP technology is made of a moisture sensing polyimide, and the thickness of it is 30um. Five faces of the MS-PI box are exposed to the air to increase the contact area between the PI material and the air, so as to increase the sensitivity of the humidity sensor. The thickness of the copper RDL is 15um and could be well controlled to be accurate enough in massive production. The humidity sensing capacitor is made of the RDL layer only, and constructed into comb type, as shown in Fig. 1(b). With the change of ambient humidity, the dielectric constant of MS-PI material will also change, which will lead to the change of dielectric constant of the material between the two comb electrodes of the capacitor directly. Thus, the capacitance of the comb capacitor will be change as the environmental humidity changes. The finger number, metal width, and the spacing between two fingers could be well designed to optimize the quality factor and the capacitance of the humidity sensitive capacitor.

978-1-6654-1392-3/21 $31.00 © 2021 IEEE   137

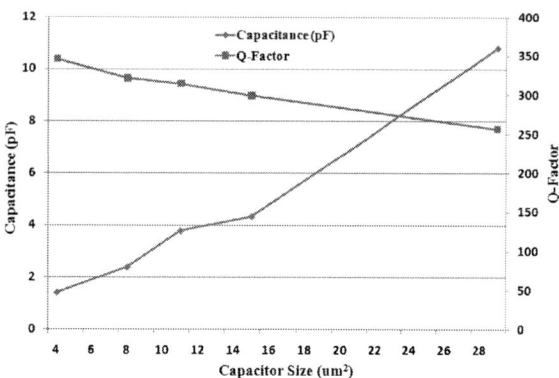

Fig. 5. simulated capacitance and quality-factor of the humidity-sensitive capacitor versus capacitor size.

III. THE PROPOSED HUMIDITY-SENSITIVE CAPACITOR STRUCTURE BASED ON FOPLP TECHNOLOGY

The capacitor can be constructed in many shapes. Among them, the comb structure is the most popular one due to its simple structure and higher contacting area. The proposed humidity-sensitive comb capacitor has seven fingers on one electrode and six fingers on the other, as shown in Fig. 2. The finger length, finger spacing, and metal width are 3.5mm, 0.05mm and 0.1mm respectively. The thickness of the RDL metal is 15um and could be modified to meet the application requirement. The spacing of the comb capacitor should be designed to meet the requirement of the response time and the capability of the manufacture. The wider the spacing the fast the response time will be.

Also, the size of the humidity-sensitive comb capacitor is constrained by the size of the signal processing chip in order to minimize the size of the entire humidity sensor. The small package size will lead to lower cost of the humidity sensor. However, the capacitance value of the humidity capacitor should be followed the requirement of the signal processing circuits in real application.

In addition, the size of the humidity sensitive comb capacitor is limited by the size of the signal processing chip to minimize the size of the entire humidity sensor. The smaller the package size, the lower the cost of the humidity sensor. However, in real application, the capacitance of humidity capacitor should meet the requirements of signal processing circuit.

IV. SIMULATION RESULTS

The humidity sensitive capacitor mentioned above is designed on the basis of fan out board level packaging technology. The material and thickness are shown in Fig. 1, and the capacitor configuration is shown in Fig. 2. The capacitor is designed into comb structure and using RDL metal layer. The dielectric constant of the moisture-sensitive polyimide ranges from 2.93 to 3.69 as the relative humidity varies from 0 to 100%. The humidity sensitive capacitor is simulated and analyzed by ANSYS HFSS, and the target working frequency is set to be greater than 10MHz. The capacitance value is designed to be in the scale of several pico-Farad to meet the requirement of normal application. Thus, the proposed capacitor is designed to have seven fingers on one electrode and six fingers on the other, and occupies around 15.2mm² area. The simulated capacitance value changes from 4.13pF to 4.18pF when the relative

humidity varies from 0%~100% RH, which is shown in Fig. 3. The humidity-sensitive capacitor achieves around 0.15% non-linearity across the entire RH range at room temperature.

The simulated quality factor of the proposed humidity-sensitive capacitor in normal situation is higher than 290, which is shown in Fig. 4. A group of humidity-sensitive capacitors with different size are designed to check the relationship of the capacitor size and its capacitance value or quality factor. As shown in Fig. 5, the capacitance of the humidity-sensitive capacitor increases roughly linear to the physical size, while the quality factor decreases as the capacitor size increases. This is reasonable, because the contacting area of the two electrodes of the humidity-sensitive capacitor is proportional to its physical size.

V. CONCLUSIONS

The designed humidity-sensitive capacitor achieves 4.13pF to 4.18pF capacitance from 0%~100% RH variation. The quality factor the proposed capacitor is greater than 250, which is usually much higher than on-chip MIM capacitor with the same capacitance value. The humidity-sensitive capacitor shows 1.7% variation in capacitance with 1% RH change. The non-linearity of the humidity-sensitive capacitor is around 0.15% across the entire RH range at room temperature. The proposed humidity-sensitive capacitor based on the FOPLP technology demonstrates good linearity versus the physical size of the capacitor. More importantly, the proposed solution used to construct the humidity-sensitive comb capacitor is economic and practical. The humidity-sensitive capacitor could be manufactured during the IC packaging stage without any additional process, which avoids using the complicated and expensive MEMS technology. Thus, it carves out a new way to develop low-cost high-performance humidity sensor, which has vast application prospect in the IoT area.

VI. ACKNOWLEDGMENT

The authors would like to thank Guangdong Fozhixin microelectronics technology research Co., Ltd. for its great contributions to the discussion on issues related to Fan-Out Panel Level packaging technology and everyone else who helped in different aspects when completing this article.

REFERENCES

[1] Wang Yang, Ke Daming, Chen Junning, Hu Jiang, Modeling of a CMOS capacitive relative humidity sensor. 2009 First International Workshop on Education Technology and Computer Science; 2009; 209-212.

[2] Jo Young Chang, Kim Kun Nyun, Nam Tae Yang. Low power capacitive humidity sensor readout IC with on-chip temerature sensor and full digital output for USN applications. IEEE SENSORS 2009 Conference. 2009; 1354-1357.

[3] Dai Ching-Liang, Lu De-Hao. Fabrication of a Micro humidity sensor with Polypyrrole using the CMOS process. Proceedings of the 2010 5th IEEE International Conference on Nano/Micro Engineered and Molecular Systems. Jan. 2010; 110-113.

[4] Chung P. J., Liang K.-C., et al., Development of a CMOS-MEMS RF-aerogel-based capacitive humidity sensor. IEEE SENSORS 2014 Conference. 2014; 1-4.

[5] Park Sujin, Lee Geon-Hwi, et al., A Capacitance-to-Digital Converter with Differential Bondwire Accelerometer, On-chip Air Pressure and Humidity Sensor in 0.18 μm CMOS. 25th Asia and South Pacific Design Automation Conference (ASP-DAC). 2020, 3-4.

[6] Burak Okcan, Tayfun Akin, A low-power robust humidity sensor in a standard CMOS process, IEEE Trans. on Electron devices, vol. 54, no. 11, Nov. 2007,3071-3078.

[7] G. Huang, Y. Dai, N. Liu, Z. Duan, Y. Wu and T. Lin, "A low cost 60GHz antenna in Fan-Out Panel Level Package for millimeter-wave radar application," 2020 21st International Conference on Electronic Packaging Technology (ICEPT), 2020, pp. 1-4.

[8] Yi Chong, Duo Wenbin, Microstrip series fed antenna array for millimeter wave automotive radar applications. 2012 IEEE MTT-S International Microwave Workshop Series on Millimeter Wave Wireless Technology and Applications;1-3.

[9] Wei Wang, Wang Xuetian, A 77GHz series fed weighted antenna arrays with suppressed side-lobes in E- and H-Planes. Process in Electromagnetic Research Letter; 72: 23-28.

[10] Ho Cheng-Yu, Hsieh Sheng-Chi, et. al., A low-cost antenna-in-package solution for 77GHz automotive radar applications. 2019 International Conference on Electronics Packaging (ICEP), 110-114.

[11] Braun Tanja, Becker Karl-Friedrich, et. al., Fan-Out Wafer and Panel Level Packaging as Packaging platform for heterogeneous integration. Micromachines 2019, 10, 342: 1-9.

[12] M. Brunnbauer, E. Furgut, G. Beer, and T. Meyer, "Embedded Wafer Level Ball Grid Array (eWLB)," in Electronics Packaging Technology Conference, 2006. EPTC '06. 8th, 2006, pp. 1 –5.

[13] K. Pressel, G. Beer, T. Meyer, M. Wojnowski, M. Fink, G. Ofner, and B. Roemer, "Embedded wafer level ball grid array (eWLB) technology for system integration," in 2010 Int. Symp. on Components, Packaging, and Manufacturing Technology, Tokyo, Japan, Aug. 2010, pp. 1–4.

[14] H. C. Lu, Y. H. Wang, J. L. Leou, Harrison Chan, Scott Chen, "Chip Last Fan-Out Packaging for Millimeter Wave Application," in Proc. 66th Electron. Compon. Technol. Conf., May 2016, pp. 1303 –1308.

Microstructures Properties of barium-strontium titanate (BST) ceramics Doped with B-Li Glasses for LTCC Technology Applications

Linjiang Tang
Beijing Spacecrafts
China Academy of Space Technology
Beijing, China
tanglinjiang2008@163.com

Xiaofeng Sun
Beijing Spacecrafts
China Academy of Space Technology
Beijing, China
motubu4166026@163.com

Minghua Zhang
Beijing Spacecrafts
China Academy of Space Technology
Beijing, China
zhengjia85678@163.com

Chengan Wan
Beijing Spacecrafts
China Academy of Space Technology
Beijing, China
shan86115114@163.com

Abstract—A new sintering aid system of boron lithium glass for LTCC layer is introduced. Due to the existence of liquid phase, the sintering temperature of barium strontium titanate decreases from 1350 °C to 950 °C. The effects of B-Li glass in LTCC devices were studied. The samples which sintering temperature at 950°C shows optimize the dielectric performance, including low dielectric constant is 368 and low dielectric loss is 0.007. Dielectric tunability of simple LTCC multilayer ceramic capacitor devices are manufactured by tape casting process. The measured capacitance increases with the increase of layers, and the tunability is stable at 12% (300 volts).

Keywords—LTCC, B-Li glass, Tape casting, dielectric properties, tunability

I. INTRODUCTION

Barium strontium titanate is an important ferroelectric material, which is widely used in many fields. As we all know, the BST has prodigious dielectric constant and tunability. Barium strontium titanate (BST) materials are become focus in the microwave field because of their tunable dielectric properties. In recent years, the study of microwave tunable materials for LTCC(Low Temperature Co-fired Ceramic) device application has become material science and microelectronics research hotspot. However, the new generation of microwave tunable devices must been designed with an ideal sintering temperature (<960 °C), excellent dielectric properties and high cost performance materials at microwave frequencies. The main challenge in designing tunable device materials is to require high dielectric tunability ($\varepsilon/\varepsilon 0 > 10\%$), low dielectric loss and medium dielectric constant at microwave frequency. Therefore, the research on barium strontium titanate-based ceramics mainly focuses on optimization of dielectric properties and reasonable reduction of sintering temperature. Published investigations attempts to optimization the dielectric constant and dielectric loss by the use of non-ferroelectric phases (MgO[1], Mg_2TiO_4[2]) and low sintering temperature lithium salts (Li_2CO_3, LiF)[3-5] and B_2O_3[6-8]. The phase diagram of Li_2O-B_2O_3 binary system [9] also shows that B-Li glass can react to form low melting point phase. B_2O_3-Li_2CO_3 co-doping is considered as a valuable sintering aid for low temperature densification. [10]

The addition of massive boron dioxide(B_2O_3) as a sintering agent causes more problems to the band casting process applied by LTCC. The preparation of slurry is difficult due to crosslinking of boron dioxide and polymer adhesive. [11], a new sintering aid or a sintering aid system of BST-based LTCC composite is required.

In this study, BST50-$SrMoO_4$-B-Li composites were synthesized by adding $SrMoO_4$ and B-Li aids to reduce dielectric constant and sintering temperature. Adding boron lithium glass into BST50-$SrMoO_4$ composite ceramics can produce dense and uniform grains at low sintering temperature of 950°C in LTCC technology and tunable applications.

II. EXPERIMENTAL PROCESSES

Barium strontium titanate ($Ba_{0.5}Sr_{0.5}TiO_3$) powder was synthesized at 1100°C and 700°C respectively with high purity barium titanate (99.9%), strontium titanate (99.9%), strontium carbonate (99.9%) and molybdenum oxide (99.99%) as starting materials. Boron-lithium glass is composed of high purity H_3BO_3 (99.5%) and Li_2CO_3 (98%). H_3BO_3 and Li_2CO_3 powders were rolled and ground about 24h. The melt was homogenized at 1100°C for 2 hours. , and then quickly take it out of the furnace and add water for quenching. BST50($Ba_{0.5}Sr_{0.5}TiO_3$) and clinker boron lithium glass with zirconium balls crushed 24h,.

TABLE I.   THE COMPOSITIONS OF SINTERING AID FOR LTCC

mol	1#glass	2#glass	3#glass	4# (H_3BO_3/Li_2CO_3)
H_3BO_3	2	2	2	2
Li_2CO_3	1.5	3	3.2	3.2

TABLE II.   THE DIELECTRIC PROPERTIES OF BARIUM-STRONTIUM TITANATE (BST)-BASED CERAMICS

Samples	Sintered/°C	Relative density	Q	ε' (20° C)	tunability(20°C) 60kV/cm
BST50	1350	97%	424	2035	23.5%
BST50	950	70%	/	/	/

978-1-6654-1392-3/21 $31.00 © 2021 IEEE

Samples	Sintered d/°C	Relative density	Q	ε' (20°C)	tunability(20°C) 60kV/cm
BST50-5%-1#glass-	950	85%	57	973	24 %
BST50-5%-2#glass	950	90%	151	1543	25%
BST50-5%-3#glass	950	96%	164	1550	29%
BST50-5%-4# (H_3BO_3/Li_2CO_3)	950	95%	97	2325	27%
BST50-SrMoO₄	1350	97%	291	898	25.6%
BST50-SrMoO₄-3#glass	950	96%	230	355	13%

Identify the phase structure by X-ray diffraction (XRD) (D8, Brooke). The microstructure is described with SEM (JSEMP-800,JEOL, Tokyo, Japan).Polish the sample to 1 & 0.15 thickness and measure the dielectric properties by splashing the gold electrodes on both sides of the sample. In the temperature range from -170°C to 140°C, The relationship between dielectric constant and temperature and the loss at 10kHz were measured by E4980A LCR measuring instrument (Agilent, Palo Alto, California). The variation of dielectric constant with electric field was measured by Keithley 2410 high voltage power supply (Cleveland, Ohio) and th2816alcr analyzer. The dielectric properties at microwave frequency were obtained by using Harke-Coleman dielectric resonator and Agilent hp8753e method. The Q value is calculated by the resonance frequency and geometry of the sample

III. RESULTS AND DISCUSSION

Scanning electron microscope micrographs of ($Ba_{0.5}Sr_{0.5}TiO_3$) BST50 ceramics and B-Li sintering aid BST50 ceramics are shown in Fig.1. The relative densities of samples which sintered at 1350°C and 950°C are listed in Table II. BST50 ceramics sintered at 1350°C without any additives (Figure 1(a)) have relatively dense microstructure. BST50 (Fig.1(e)) containing B-Li-glass can only be sintered at 950°C. It is evident that adjunction B-Li-glass increases the density of the barium strontium titanate-based ceramics in Table II. The relative density of BST50 sample with 5% lithium-boron glass at 950℃ is 96%.The results show that lithium boron glass and H_3BO_3/Li_2CO_3 can play a good role as sintering additives in $Ba_{0.5}Sr_{0.5}TiO_3$ materials, which can effectively reduce the sintering temperature of materials.

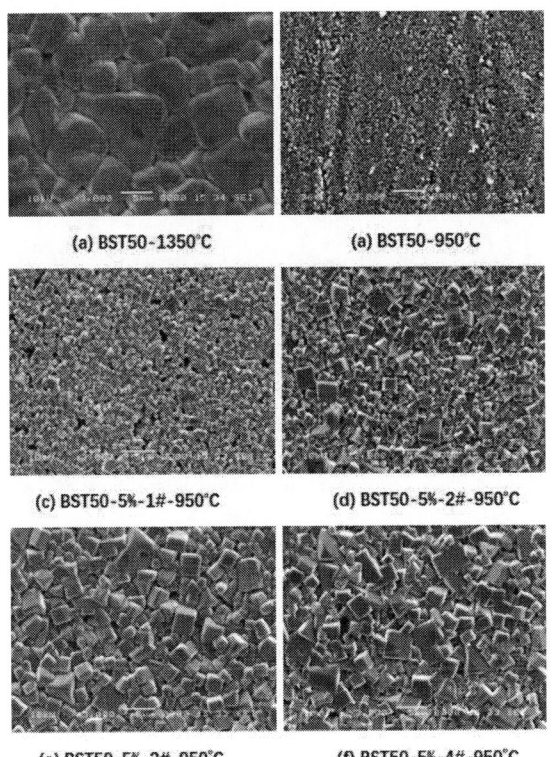

Fig. 1. SEM photos of BST50($Ba_{0.5}Sr_{0.5}TiO_3$) ceramic samples sintered at 950°C with 5% content of different sintering additives

The fig.2 shows the X-ray diffraction pattern of a ceramic sample of B-Li doped BST50($Ba_{0.5}Sr_{0.5}TiO_3$) sintered at 950 °C. 2. The X-ray diffraction pattern of B-Li doped BST50 ceramic samples is similar to pure BST50. Observe the perovskite structure and no significant secondary phase is seen in all samples. This shows that B-Li and Bst50($Ba_{0.5}Sr_{0.5}TiO_3$) can coexist kindly in the material system with no significant chemical reaction.

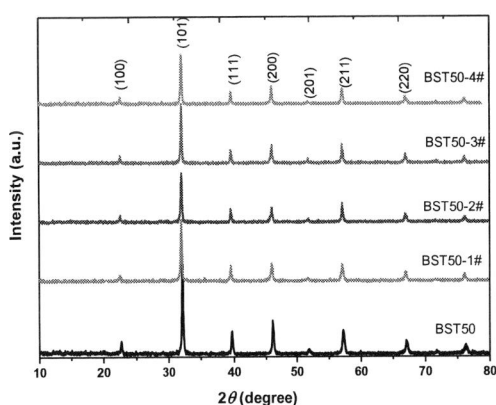

Fig. 2.XRD patterns of ceramic samples sintered at 950°C with 5% content of different sintering additives

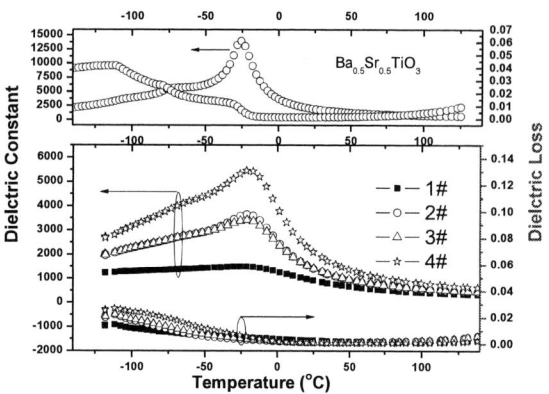

Fig. 3. Dielectric constant and dielectric loss of the samples sintered at 950°C at 10 kHz

As shown in Fig.3, It presents a dielectric temperature spectrum of a 10kHz sintering sample of a BST50($Ba_{0.5}Sr_{0.5}TiO_3$) ceramic sample mixed with different sintering additives, The low-frequency dielectric properties obtained from the temperature spectrum curve are shown in Table II. By comparing the pure BST50 with the sample, Doped with sintering additives, Discovered, The Curie peak of the sample, Doped with four sintering additives, It has been expanded, And the peak of the dielectric constant is greatly reduced, Explain that the addition of the sintering additive can reduce the sintering temperature of the BST, Also reduce the dielectric constant, And to improve the temperature stability to a certain extent, This is mainly because the doping of sintering additives causes a smaller particle size, The boundaries larger, That is, the non-electric phase changes greatly. Meanwhile, The sintering auxiliary device itself is a dielectric material with a low dielectric constant, This results in the Curie peak dispersion and a reduced dielectric constant.

However, the Q values in Table II are obviously different, which indicates that different sintering aids will affect the dielectric properties of BST50 ceramics to some extent. Therefore, it is necessary to select sintering additives reasonably on the basis of weighing sintering temperature and dielectric performance parameters. The q value of barium strontium titanate ceramics doped with lithium boron glass is obviously lower than that of pure barium strontium titanate samples. This is mainly due to the different loss mechanisms at different frequencies, and the existence of glass phase leads to the degradation of microwave performance of BST ceramics.

To sum up, by adding B-Li glass, the sintering temperature of barium strontium titanate ceramics can be effectively reduced, and dense ceramics can be obtained at 950°C. Compared with the traditional BST ceramics sintered at high temperature, the dielectric properties of BST ceramics are also optimized: lower dielectric constant, higher dielectric tunability and acceptable dielectric loss.

B-Li glass, as a sintering aid, can significantly reduce the sintering temperature of BST composite ceramics by liquid phase sintering, and obtain dense microstructure. This sintering additive is still effective for the properties of BST50-$SrMO_4$ composites, and can reduce the sintering temperature in Table II.

The microstructure of BST50-$SrMoO_4$ (fig. 4(a)) sintered at 1350°C without any additives is relatively dense. BST50-$SrMoO_4$ (fig. 4(b)) containing boron lithium glass can only be sintered at 950°C.

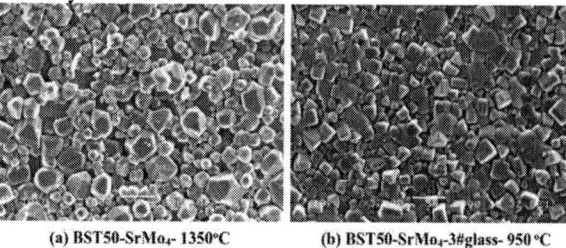

(a) BST50-SrMo₄- 1350°C   (b) BST50-SrMo₄-3#glass- 950°C

Fig. 4. The microstructure of Bst50-$SrMoO_4$ sintered at 1350°C and BSt50-$SrMoO_4$ boron-containing lithium glass sintered at 950°C.

The optimized BST50-SrMoO4-B-Li glass formula was used to prepare LTCC device tape by casting method. The conductive paste is printed on the green tape (as shown in Fig.5). In this study, 5, 10 and 15 layers of raw material tape were printed and pressed at 65°C and 20MPa for 30 minutes. The laminated capacitor sample was sintered at 550°C to burn off the adhesive and sintered at 950°C for 5 hours to get a dense structure.

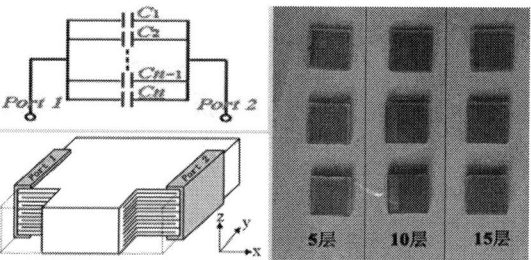

Fig. 5. Schematic diagram of low temperature co-fired (LTCC) device and photos of the prepared material

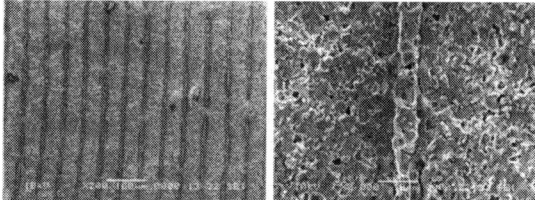

Fig. 6. SEM image of low temperature co-fired (LTCC) device section.

A cross-sectional microstructure of sintered BST50-$SrMoO_4$-B-Li multilayer ceramic capacitor is shown in fig. 6. The thicknesses of BST50-$SrMoO_4$-B-Li layer and internal electrode are about 50μm and 5μm, respectively (fig. 6). The interface between BST50-$SrMoO_4$-B-Li and the internal electrode is clear without delamination, which indicates that BST50-$SrMoO_4$-B-Li still maintains a clear interface with the internal electrode.

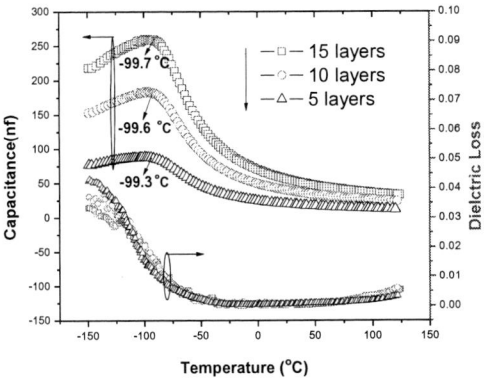

Fig. 7. Capacitance curve of low temperature co-fired (LTCC) device with temperature

Fig. 7 display the relationship between dielectric constant and loss measured from -150℃ to 130℃ at 10 kHz..BST50-SrMoO₄-B-Li LTCC multilayer capacitor measured at 10 kHz. The Curie temperatures of the 5th, 10th and 15th layers are 99.3℃, 99.6℃ and 99.7℃, respectively. The results show that, as shown in Table 2, the phase transition temperature (Tc) of LTCC is almost the same as that of BST50-SrMoO₄-B-Li ceramics. With the increase of the number of layers, the capacitance at 20℃ is 19.8nF, 40.1nF and 58.3nF,respectively. Capacitor capacity and layer number correspond to each other. As the number of layers increases, the capacitance at 20℃ is 19.8nF, 40.1nF and 58.3nF, respectively. The capacitor capacity corresponds to the number of layers.

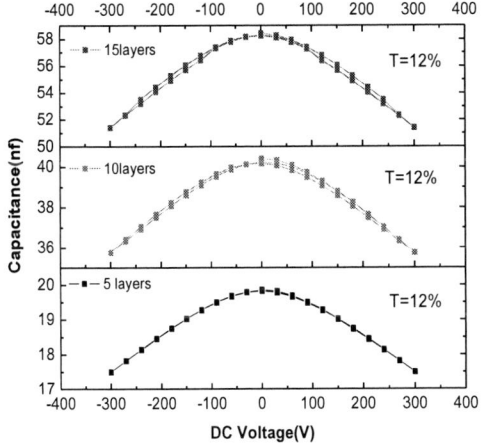

Fig. 8. The relationship between capacitance and electric field of LTCC capacitor measured at 10kHz and 300V

The relationship between capacitance and electric field of LTCC capacitor measured at 10kHz and 20 ℃ under DC bias of 300v is shown in fig. 8. The thickness of BST50-SrMoO₄-

B-Li layer is about 50μm, so the direct current electric field is about 60 kV/cm, which is applied to LTCC capacitor. The results show that all LTCC capacitors have 12% stable tunability at 300 volts in the Fig.8. The tunability does not change with the change of the number of layers of the capacitor, which indicates that the LTCC of B-Li sintering auxiliary has a promising application prospect.

IV. CONCLUSION

1)   The sintering temperature of BST can be effectively reduced from 1350 ℃ to 950 ℃ due to the existence of liquid phase in lithium boron glass.

2)   The dielectric tunability of simple LTCC multilayer ceramic capacitor devices is manufactured in the whole casting process. The measured capacitance increases with the increase of layers, and the tunability is stable at 12%(300 volts).

ACKNOWLEDGMENT

The research was supported by the Fund of GaN semiconductor power switching devices for Aerospace.

REFERENCES

[1]   U.C. Chung, C. Elissalde, M. Maglione, C. Estournes, M. Pate, J.P. Ganne, Low-losses, highly tunable Ba0.6Sr0.4TiO3/MgO composite, Applied Physics Letters, 92 (2008) 042902-042903.

[2]   X. Chou, J. Zhai, X. Yao, Dielectric tunable properties of low dielectric constant Ba0.5Sr0.5TiO3--Mg2TiO4 microwave composite ceramics, Applied Physics Letters, 91 (2007) 122908-122903.

[3]   F. Roulland, R. Terras, G. Allainmat, M. Pollet, S. Marinel, Lowering of BaB'1/3B''2/3O3 complex perovskite sintering temperature by lithium salt additions, Journal of the European Ceramic Society, 24 (2004) 1019-1023.

[4]   W. Sea-Fue, C.K.Y. Thomas, H. Wayne, P.C. Jinn, Liquid-phase sintering and chemical inhomogeneity in the BaTiO3–BaCO3–LiF system, Journal of Materials Research, 15 (2000) 1145-1148.

[5]   D.A. Tolino, J.B. Blum, Effect of Ba: Ti Ratio on Densification of LiF-Fluxed BaTiO3, Journal of the American Ceramic Society, 68 (1985) C-292-C-294.

[6]   S.M. Rhim, S. Hong, H. Bak, O.K. Kim, Effects of B2O3 Addition on the Dielectric and Ferroelectric Properties of Ba0.7Sr0.3TiO3 Ceramics, Journal of the American Ceramic Society, 83 (2000) 1145-1148.

[7]   J.Q. Qi, W.P. Chen, Y. Wang, H.L.W. Chan, L.T. Li, Dielectric properties of barium titanate ceramics doped by B[sub 2]O[sub 3] vapor, Journal of Applied Physics, 96 (2004) 6937-6939.

[8]   X.M. Shi, Q. Wang, C.X. Li, X.J. Niu, F.P. Wang, K.Q. Lu, Densities of Li2O–B2O3 melts, Journal of Crystal Growth, 290 (2006) 637-641.

[9]   J. Heli, H.U. Tao, U. Antti, L. Seppo, Ferroelectric LTCC for Multilayer Devices, Journal of the Ceramic Society of Japan, 112 (2004) S1552-S1556.

[10] T. Hu, H. Jantunen, A. Deleniv, S. Leppävuori, S. Gevorgian, Electric-Field-Controlled   Permittivity   Ferroelectric   Composition   for Microwave LTCC Modules, Journal of the American Ceramic Society, 87 (2004) 578-583.

[11] T. Hu, T.J. Price, D.M. Iddles, A. Uusimäki, H. Jantunen, The effect of Mn on the microstructure and properties of BaSrTiO3 with B2O3–Li2CO3, Journal of the European Ceramic Society, 25 (2005) 2531-2535

Spectrum analysis and application of XY platform servo system of the high-precision packaging equipment

Shujin Liu
School of Electromechanical Engineering
Guangdong University of Technology
Guangzhou, China
2543075618@qq.com

Yunbo He
School of Electromechanical Engineering
Guangdong University of Technology
Guangzhou, China
heyunbo@gdut.edu.cn

Zesheng Li
School of Electromechanical Engineering
Guangdong University of Technology
Guangzhou, China
862696000@qq.com

Qihao Qian
School of Electromechanical Engineering
Guangdong University of Technology
Guangzhou, China
1311584566@qq.com

Huilong Liao
School of Electromechanical Engineering
Guangdong University of Technology
Guangzhou, China
1471972099@qq.com

Abstract—The servo system of XY platform of high-precision packaging equipment is characterized by high requirement for motion performance, complex coupling and different load of X motor and Y motor, which makes it difficult to guarantee the mechanical installation quality of XY platform and improve the motion performance of servo system of XY platform of high-precision packaging equipment. The frequency domain characteristics directly reflect the response and inherent characteristics of the system. Obtaining the frequency spectrum of the XY platform servo system plays an important role in guaranteeing and improving the motion performance of the XY platform. In this paper, the sine frequency sweep method and PRBS signal are used as the excitation signal to obtain the frequency spectrum of the servo system of the XY platform. By inputting a sinusoidal excitation signal or PRBS excitation signal to the system, and processing the input signal and output signal of the system in the time domain, the frequency domain information of the system can be obtained. The two kinds of excitation signals have their advantages and disadvantages. Analyzing the system's frequency spectrum can be used to verify the reliability and consistency of the XY platform and improve the platform's motion performance.

Keywords—The high-precision packaging equipment, XY platform, Spectrum analysis

I. INTRODUCTION

Semiconductor packaging equipment is equipped with high speed and high precision, in which the performance of the XY platform plays a key role, so vibration testing of the XY platform is required. The vibration test is to confirm the reliability of the product and to screen out the defective products before leaving the factory in advance, and to perform frequency spectrum analysis on the products to improve the processing technology and improve the overall quality of the products. Commonly used vibration test methods are divided into sinusoidal vibration and random vibration. In the work of vibration testing, a large number of scholars have conducted research. The sine sweep signal synthesis algorithm proposed by Yang Zhidong et al. is based on the solution of the phase function differential equation. This method is more robust and accurate than the previous "phase accumulation" algorithm. Experiments have proved that this signal processing method is very suitable for the time-varying spectrum analysis of the sine frequency sweep signal, and solves the problem of the transient characteristics of the frequency sweep signal [1]. Dai

Feng used a fixed frequency scanning method to obtain the natural frequency of the bonding wire, and studied the contact problem caused by the deformation of the bonding wire in random vibration [2]. Li Hong collects vibration data through vibration sensors, and then carries out spectrum analysis on the data to obtain the data of the motor's key order. Finally, the data of the motor's key order is compared and analyzed with the data of fault order in the fault library, and the motor with similar data in the fault library is judged to be unqualified, so as to improve the quality of the product [3].

II. BLOCK DIAGRAM OF XY PLATFORM SERVO SYSTEM

XY platform of the high-precision packaging equipment is composed of two permanent magnet synchronous linear motors. The servo drive system of permanent magnet synchronous linear motor is composed of high-performance motion control card, driver, linear motor, linear grating, worktable, etc. Its control objects include drives, linear motors, and linear gratings. The high-performance motion control card has powerful data acquisition function, which can collect the planning data and the actual output data of the system in real time during the process of motor movement. This is achieved through the "Watch" function of the high-performance motion control card. By setting specific "Watch events" and "Watch variables", Data such as servo output (DAC) of motion control card, planning position, planning velocity, actual output position and actual output velocity can be collected. When the excitation signal is DAC planning data and the response signal is velocity feedback data, the open loop spectrum of the system can be obtained. The two kinds of spectral excitation signals discussed in this paper are sinusoidal excitation signal and PRBS excitation signal respectively. Both the PRBS signal and sinusoidal signal are generated by DSP in high-performance motion control card.

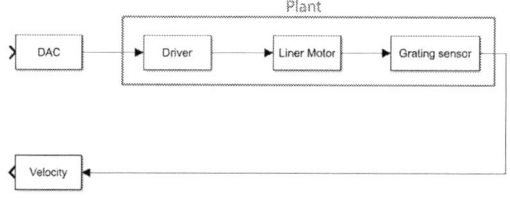

Fig. 1. The principle block diagram of frequency spectrum acquisition of permanent magnet synchronous linear motor system

III. THE TEST PRINCIPLE OF SINUSOIDAL EXCITATION SIGNAL TO ACQUIRE SPECTRUM

The sinusoidal vibration test is divided into fixed frequency test and sweep frequency test. The fixed frequency test is mainly used for vibration test with a determined resonance frequency, which is very beneficial to the exposure of product faults and defects. The sweep frequency test is mainly used for vibration test of products whose resonance frequency cannot be determined, or vibration test for product resonant search [4-5].

A. Sine sweep frequency mathematical expression

Sine sweep function expression:

$$y(t) = A(t)\sin(\varphi(t)) = A(t)\sin[2\pi f t + \varphi_0] \quad (1)$$

In the formula: $A(t)$ is the amplitude, $\varphi(t)$ is the phase of the sinusoidal signal, f is the frequency (Hz) of the sinusoidal signal, and φ_0 is the initial phase.

Sinusoidal sweep can be divided into linear sweep and logarithmic sweep according to the change of frequency. Logarithmic sweep frequency expression:

$$f(t) = 2^{rt} \cdot f_0 \quad (2)$$

$$\varphi(t) = 2\pi \int f(t)dt + \varphi_0 = 2\pi \int (2^{rt} f_0)dt + \varphi_0$$

$$= 2\pi f_0 \frac{2^{rt}}{r \cdot \ln 2} + \varphi_0 \quad (3)$$

In the formula: f_0 is the starting frequency; r is the sweep rate, which means r octave per minute (oct/min). The logarithmic sweep frequency expression is:

$$y(t) = A(t)\sin\left(2\pi f_0 \frac{2^{rt}}{r \cdot \ln 2} + \varphi_0\right) \quad (4)$$

B. The principle of sine frequency sweep test

Because the control system is a discrete system, the sinusoidal sweep signal takes the single frequency sinusoidal signal as the excitation signal of the system successively, and the frequency of the sinusoidal excitation signal increases successively. The amplitude of the sinusoidal excitation signal should be adjusted timely according to the response of the system to ensure that the sinusoidal excitation can excise the system modal. This is a sinusoidal point-by-point scanning mode, a steady-state sinusoidal frequency sweep, which requires the system to enter a stable state at every frequency point, fully exposing the system characteristics, and obtaining a very high-precision frequent-phase function. However, it is time-consuming to wait for the end of the transient response of the system after each frequency change [6]. The sinusoidal excitation signal and the response signal of the system are shown in the figure below:

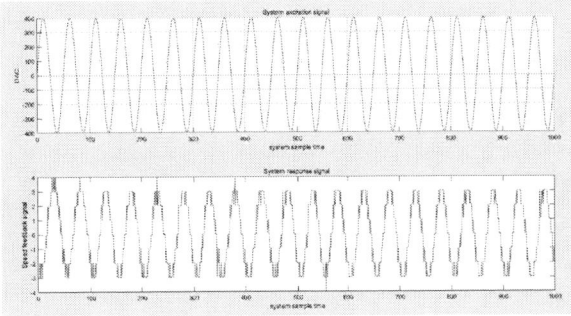

Fig. 2. Excitation signal and response signal of single sinusoidal sweep frequency

The system inputs single-frequency sinusoidal excitation signal to obtain the response signal of the system. The collected data is time-domain data, which needs to be changed into frequency-domain data. In this paper, correlation analysis method is used to transform time-domain data into frequency-domain data, because correlation analysis has good noise suppression effect, and step-by-point frequency sweep can improve identification accuracy by using correlation analysis method [7]. The principle of correlation analysis is as follows:

Assume that the unit sinusoidal input signal with the frequency of f_0, the length of the acquisition signal is N, and the interval is Δt. The calculation formula of the system response is as follows:

$$y_i(t) = A_{f_0}\sin\left(2\pi f_0 \cdot i\Delta t + \varphi_{f_0}\right) \quad (5)$$

In the formula: A_{f_0} is the amplitude of the signal to be identified, and φ_{f_0} represents the phase to be identified. Calculate the correlation functions between $y_i(t)$ and unit signals $\sin(2\pi f_0 \cdot i\Delta t)$ and $\cos(2\pi f_0 \cdot i\Delta t)$ respectively:

$$R = \frac{2}{N}\sum_{k=1}^{N} y_i \sin(2\pi f_0 \cdot i\Delta t) \quad (6)$$

$$I = \frac{2}{N}\sum_{k=1}^{N} y_i \cos(2\pi f_0 \cdot i\Delta t) \quad (7)$$

Thus, the amplitude and phase of the steady-state response of the system under the unit sinusoidal excitation signal with a frequency of f_0 are obtained, the amplitude and phase characteristics of the system are as follows:

$$A_{f_0} = \sqrt{R^2 + I^2} \quad (7)$$

$$\varphi_{f_0} = \arctan\frac{I}{R} \quad (8)$$

C. Get the frequency spectrum of a certain frequency band

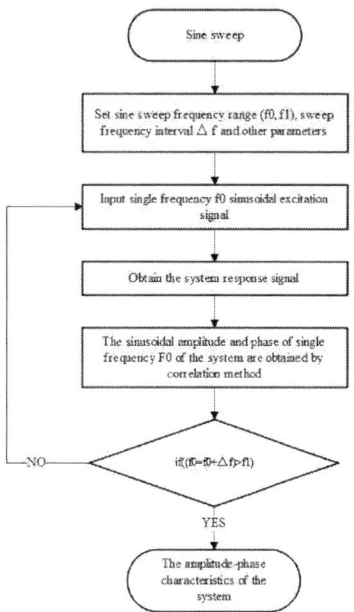

Fig. 3. Flow chart of the system spectrum obtained by sine sweep

By using the sinusoidal point-by-point scanning method and setting the appropriate sweep gain for each frequency point, the high precision spectrum of the system can be obtained. The following figure is the open-loop spectrum diagram of the sinusoidal sweep frequency of X motor at zero position:

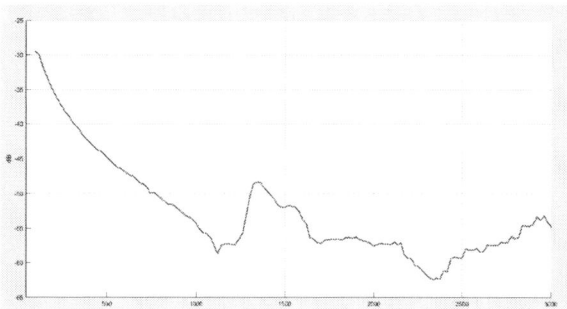

Fig. 4. System spectrum obtained by sine sweep

IV. THE TEST PRINCIPLE OF PRBS EXCITATION SIGNAL TO ACQUIRE SPECTRUM

White noise is an infinite-order continuous excitation signal, which can be used as an ideal excitation signal for any system. The energy density of all frequencies of ideal white noise is the same, and its bandwidth can be regarded as infinite, so its energy is infinite. The white noise signal is a theoretical ideal signal, which cannot be realized in actual engineering [8]. Therefore, in actual engineering applications, the approximate white noise signal is often used as the excitation signal of the system. The pseudo-random binary sequence is an ideal signal that approximates white noise [9]. The definition of pseudo-random binary sequence (PRBS) is as follows:

Suppose an N-bit sequence of $a_1, a_2, ..., a_{N-1}$, where $a_j \in \{0,1\}$, j=0,1,...,N-1, the sequence contains m=Σa_j 1 , And

N-m 0, its autocorrelation function $C(v) = \sum_{j=0}^{N-1} a_j\, a_{j+v}$ has only two values:

$$C(v) = \begin{cases} m, & if\, v \equiv 0(mod\, N) \\ mc, & otherwise \end{cases} \quad (9)$$

Where, $c = \frac{m-1}{N-1}$, c is called the duty cycle, then the sequence is called a pseudo-random binary sequence.

In this article, the PRBS signal generation function is designed as a polynomial $f(x) = x^{17} + x^{14} + 1$, and the period of the PRBS signal is $p = 2^n - 1 = 2^{17} - 1 = 131071$. The 0 in the PRBS signal is often represented by +1, and 1 is represented by -1, which multiplied by a gain value. The resulting sequence is used as the excitation signal of the system. The figure below is a section of PRBS signal data with a gain value of 400.

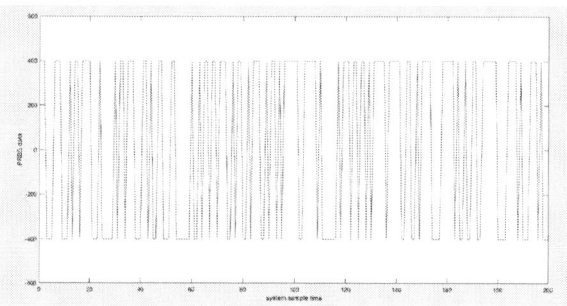

Fig. 5. The PRBS signal data

The PRBS signal is used as the excitation signal of the system to obtain the response signal of the system, and the response information in the time domain is converted into frequency domain using FFT.

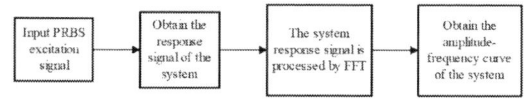

Fig. 6. PRBS as the system excitation signal spectrum processing flow chart

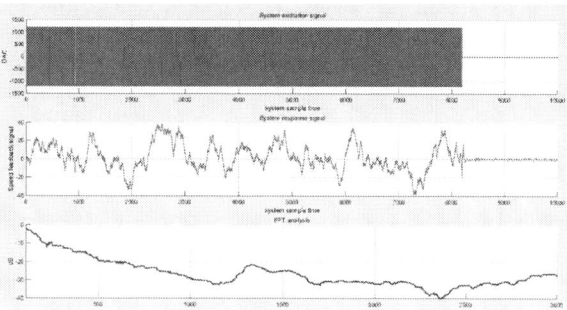

Fig. 7. PRBS excitation signal, response signal and system spectrum

The PRBS signal is a one-time input and stimulates all modes of the system. In this way, although the identification experiment process is efficient, the amount of collected data is small, and the amount of calculation in the processing process is small, the PRBS signal will also stimulate the high frequency band of the system and affect the accuracy of the test results. [10]. Generally speaking, the greater the gain of the PRBS excitation signal, the more accurate the obtained system spectrum. However, due to the travel limit of the motor and the parameter protection limit of the driver, the excitation signal of the PRBS cannot be set with too large gain. The frequency spectrum of sinusoidal sweep is used to adjust the gain value of PRBS excitation signal to ensure the correctness of the spectrum obtained by PRBS excitation signal. According to the amplitude relationship between PRBS excitation signal and response signal, the ordinate position of the spectrum was corrected. The corrected PRBS spectrum basically coincided with the spectrum of sinusoidal sweep.

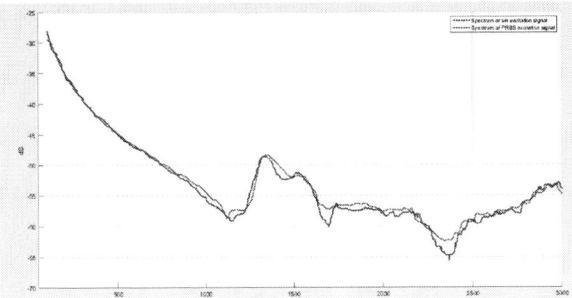

Fig. 8. Comparison of the PRBS spectrum and the sinusoidal spectrum of the system

V. CONCLUSION

It can be known from the above analysis that the system spectrum obtained by the two excitation signals is basically the same. Sine frequency sweep needs to excite and process data one by one frequency point in turn. It takes a long time to obtain spectrum, but high-precision system spectrum can be obtained; PRBS can excite all modes of the system with one input, and quickly obtain system spectrum, but it is not enough accurate. The system spectrum obtained by the two methods has the following applications::

1) The open-loop spectrum of the system can be used to view the mechanical performance of the system, find and eliminate mechanical installation problems, and ensure product consistency.

2) By analyzing the speed loop frequency spectrum of the system, using a notch filter in the control loop can reduce the

excitation of the servo system at the resonance frequency point, suppress the amplitude at the resonance frequency point, and achieve the purpose of suppressing vibration [11].

ACKNOWLEDGMENT

This paper is supported by National Natural Science Foundation of China (No. 61973093).

REFERENCES

[1] Yang Zhidong, Cong Dacheng, Han Junwei, Li Hongren,"Signal synthesis and analysis for swept-sine vibration control," in Journal of Vibration Engineering,2008(03):309-313.

[2] Dai Feng,"Contact threshold of random vibration bonding wire deformation," in Journal of Vibration and Shock,2021,40(09):228-231.

[3] Li Hong,"Application and Research of Vibration Analysis Technology in Fault Diagnosis of Permanent Magnet Synchronous Motor," in Auto Time,2021(10):170-172.

[4] Chen Zhangwei, Yu Huijun,"Present situation and development of vibration control technology,"in Journal of Vibration and Shock,2009,28(03):73-77+86+200.

[5] Wang Shucheng, Chen Zhangwei."Research on the randomization of random test in vibration control system,"in Journal of Mechanical Engineering,2005(05):230-233.

[6] Liu Rui, Zhang Chi, Shen Lin-yong, Zhao Fei, Dong Liang, Li Rong,"Linear Motor Servo System Precision Automatically Model Methed Research,"in Small & Special Electrical Machines,2016,44(04):9-12+20.

[7] Chen Lianhua, Zhao Na, Wang Yibo, Shi Guoxiang,"Fast frequency response testing method with one step,"in Electronic Measurement Technology,2011,34(02):77-79+101.

[8] Lei Luolan,"Research on Identification Technology of Tracking Control System,"Graduate School of Chinese Academy of Sciences (Institute of Optoelectronic Technology),2015.

[9] Li Ying,"Excitation parameter identification based on the M sequence of pseudo random signal,"North China Electric Power University (Beijing),2006.

[10] Tan Zhihong,"Research on excitation signal and frequency domain identification of the servo control system,"Harbin Institute of Technology,2010.

[11] He Yunbo,He Zuoxiong,"Vibration suppression of servo direct drive shaft in wafer level flip-chip equipment,"Machine Tool & Hydraulics,2020,48(14):1-5

Destructive Physical Analysis Methods of Flip Chip Packaging Devices for High Reliability

Zhou Shuai
School of Microelectronics, Tianjin University
Tianjin, China
zs5h@163.com

Weng Zhangzhao*
The Testing Center of Electronic Components
China Electronic Product Reliability and Environmental Testing Research Institute
Guangzhou, China
347632043@qq.com

Qiu Baojun
The Testing Center of Electronic Components
China Electronic Product Reliability and Environmental Testing Research Institute
Guangzhou, China
qiubaojun@ceprei.com

Luo Daojun
The Testing Center of Electronic Components
China Electronic Product Reliability and Environmental Testing Research Institute
Guangzhou, China
luodj@ceprei.com

Wang Xiaoqiang*
The Testing Center of Electronic Components
China Electronic Product Reliability and Environmental Testing Research Institute
Guangzhou, China
ps_800@126.com

Ma Kaixue
School of Microelectronics, Tianjin University
Tianjin, China
makaixue@tju.edu.cn

Abstract—How to effectively and accurately evaluate the reliability of the design, structure, material and process of flip-chip packaging devices in mass production has always been a hot topic in the industry. Currently, the destructive physical analysis (DPA) on high-reliability conventional packaging devices worldwide is carried out mainly based on the Methods of Destructive Physical Analysis for Military Electronic Components (GJB4027) or the Destructive Physical Analysis for Electronic, Electromagnetic, and Electromechanical Parts (MIL-STD-1580). However, the structure and process of flip-chip packaging devices are significantly different from those of traditional packaging devices, so traditional evaluation items are unable to effectively evaluate the reliability of flip-chip packaging devices. In this paper, typical ceramic and plastic encapsulated flip-chip packaging devices are chosen for study. The structural feature and the reliability weaknesses of flip-chip packaging devices are identified. Moreover, highly comprehensive DPA methods for flip-chip packaged devices with high applicability and efficiency are proposed by combining the typical failure modes of flip-chip packaging devices, which were verified by the examples. Herein, the reliability of key structural units such as flip-chip covers, bumps, underfill, substrates and solder balls can be effectively assessed by the new DPA method, providing a basis and assistance for revisions of advanced packaging device standards and destructive physical analysis.

Keywords—Flip-chip packaging, DPA, Structural characteristics, Failure mode

I. INTRODUCTION

Flip-chip packaging technology is an advanced form of surface mounting that directly bonds semiconductor bare chips on a printed circuit board or chip-carrier substrate. The active surface of the integrated circuit chip assembled by flip chip faces the substrate, and the assembly process is also different from the point-to-point wire bonding process. In addition, the solder bumps on the flip chip are used to connect the substrate, and the interspace between the chip and the substrate are filled with underfill material to improve packaging reliability. Compared with traditional wire bonding and tape automated bonding, flip chip has many advantages[1] such as high frequency, high interconnection density, good noise control, and the lowest device mounting height, resulting in replacing the traditional packaged integrated circuits gradually. However, these features also bring new reliability assessment challenges.

Hence, how to effectively and accurately evaluate the batch-to-batch variation of flip-chip packaged devices' quality and reliability has always been a hot topic. At present, the destructive physical analysis (DPA) of high-reliability conventional packaged devices in China and abroad is based primarily on GJB4027 and MIL-STD-1580, respectively. The test items included in current standards cannot effectively and comprehensively evaluate the reliability of flip-chip packaging devices. Current standards have many drawbacks, especially when it comes to the evaluation of some typical failure modes and mechanisms. Besides, foreign and domestic studies on flip-chip packaging devices mainly focus on failure mechanism and environmental stress adaptability, while there are few systematic destructive physical analysis (DPA) researches that combine current standards, failure mechanism, test items and evaluation procedures. As a consequence, there is no unified standard or evaluation method available for the DPA of flip-chip packaging devices, making it impossible to effectively, accurately and comprehensively grasp the reliability of flip-chip packaging devices in mass production, which brings great challenges to the application of flip-chip packaging devices to industrial fields that require high reliability. In this study, starting from the structural features of flip-chip packaging devices, based on the typical failure mechanism of flip-chip packaging and the DPA ideas of conventionally packaged integrated circuits, we designed non-destructive physical analysis (NDPA) and DPA items, figured out the relationship between the typical failure mechanism and the work item, and developed a DPA method suitable for flip-chip packaging devices to effectively evaluate the reliability of flip-chip packaging devices.

978-1-6654-1392-3/21 $31.00 © 2021 IEEE

II. Analysis of Structural Features of Flip-Chip Packaging Device

The flip-chip packaging device usually consists of the chip, bumps, under bump metallization (UBM) layer, solder balls (leading-out end), substrate and underfill, with active surface of the chip with bump electrodes on the UBM layer facing down and vertically interconnect with the wires arranged on the substrate, as shown in Fig.1. Bumps, as important media for signal transmission, are usually made of metal materials such as SnPb and lead-free soldering flux; the UBM layer is usually a metallization layer composed of an adhesive layer, a diffusion barrier layer and a wetting layer; underfilling refers to directly filling epoxy resin between the chip and the substrate to improve the reliability of the interconnection system; the substrate is usually made of ceramic or epoxy glass laminate which is arranged with metal wires to interconnect the solder joints with the internal bumps. Fig.1 shows the typical structure of the flip-chip packaging device. As shown in Fig.1, the substrate, bumps, underfill and solder balls are the core structural units of the flip chip.

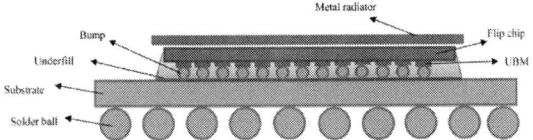

Fig.1 Typical structure of the flip-chip packaging device

III. Analysis of Typical Failure Mechanisms of Flip-Chip Packaging Device

By analyzing the structural characteristics of flip-chip packaging devices, it can be found that flip-chip packaging devices mainly fail in aspects like the bumps, underfill, chips, solder balls and packaging[2-4].

A. Bump Failure

The bump failure of the flip-chip packaging device is mainly divided into mechanical stress failure and thermal fatigue failure. Since the bumps on the chip are directly soldered to the substrate pads, when the soldering strength is unable to withstand the mechanical stress, the UBM layer will bulge, the substrate will crack or the core separation failure will occur. When it comes to the thermal fatigue failure, due to the mismatch of the thermal expansion coefficient between the chip and the substrate, under the action of the temperature cycle stress, the bumps are subjected to periodic shear stress and strain, which might lead to plastic deformation of the solder joints and continuous accumulation of cracks, thereby eventually resulting in solder joint failure.

B. Underfill Delamination Failure

The connected regions between the chip and underfill, between underfill and solder mask as well as between underfill, chip and solder joint of the flip-chip packaging device are very likely to suffer from delamination due to the thermal mismatch of the materials on both sides of the interface (there is a relatively large difference between the thermal expansion coefficients), which then triggers solder joint failure.

C. Chip Failure

Defects in the active circuit area and passivation layer on the chip constitute one of the primary reasons for parameter drift or malfunction of the device. Moreover, if the bonding strength between the chip and the substrate does not live up to the standard, the flip-chip packaging device will fail under high-intensity mechanical stress.

In addition, corrosion failure is mainly caused by external substances. Contaminated ions in the environment, e.g. H_2O and Cl^-, penetrate into the flip-chip packaging device through the underfill and react with the bumps to cause corrosion. If the chip's passivation layer is damaged or becomes incomplete, the metallization layer on the chip will also corrode, which will lead to the complete loss of the device's functions in serious cases.

D. Solder Ball Failure

The solder ball (leading-out end), as an important medium for the interconnection and transmission of signals between the flip-chip packaging device and the PCB, has a failure mechanism similar to bump failure. The solder ball failure can be mainly divided into mechanical stress failure and thermal fatigue failure. Besides, the appearance features of the solder ball, e.g. gloss, structure and integrity, also serve as variables of its failure mechanism, for the surface topography reveals the internal microstructure, and the physical properties of the solder joints on the microstructure correspond to their mechanical properties. Furthermore, the coplanarity of the solder ball is also one of the important factors affecting electronic assembly. If there is an excessively long distance from the vertex of a solder ball to the actual soldering surface, it will lead to solder skip and virtual connection on the flip-chip packaging device.

E. Encapsulation Failure

The mechanism of encapsulation failure and excessive air content of the hermetic flip-chip packaging device is basically the same as the failure mechanism of traditional DIP and SOP packaging devices. The leakage of the case or the excessively high content of water vapor, oxygen, hydrogen and carbon dioxide inside the package will cause corrosion, parameter drift and electromigration inside the device. Under the action of external factors, these hazards or potential defects will be intensified, which ultimately leads to malfunction.

IV. A DPA Method of Flip-Chip Packaging Device

According to the work items in the Methods of Destructive Physical Analysis for Military Electronic Components (GJB4027) or the Destructive Physical Analysis for Electronic, Electromagnetic, and Electromechanical Parts (MIL-STD-1580)[5], based on the structural features of the flip-chip packaging device, we figured out the corresponding relationship between work items and typical failure mechanisms, and developed the DPA method suitable for the flip-chip packaging device from the NDPA to DPA, thereby effectively evaluating the reliability of the flip-chip packaging device.

A. Design of NDPA Item

The NDPA of the flip-chip packaging device should include external visual inspection, coplanarity test conducted on the solder balls, 3D-Xray observation, scanning acoustic microscopy (SAM) and sealing test (when applicable).

978-1-6654-1392-3/21 $31.00 © 2021 IEEE

(1) External visual inspection: Not only should the external visual inspection be carried out per the criteria stated in the Test Methods and Procedures for Microelectronic Device (GJB548, 2009), attention should also be paid to the appearance and morphology features of the solder balls. A stereo microscope and a metallurgical microscope were applied to check whether the solder balls were discolored, collapsed and bridged at a magnification of 1.5 to 50 times. Fig.2 shows the typical appearance defects of the solder ball.

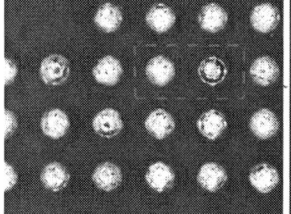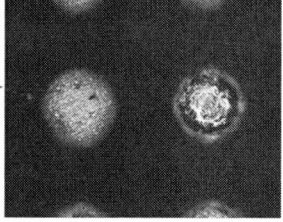

Fig.2 Typical appearance defects of the solder balls

(2) Coplanarity test conducted on the solder balls: According to GJB7677, Test Methods for Ball Grid Array (BGA), the three-point method (the triangular reference plane formed by the solder balls should include the center of gravity of the device) was used to measure the distance from the vertex of each solder ball to the reference plane, so as to check whether the solder balls lived up to the requirements of the manual or detailed specifications. Fig.3 shows some typical images of the coplanarity test conducted on the solder balls.

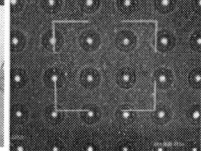

Fig.3 Typical images of the coplanarity test conducted on the solder balls

(3) 3D-Xray observation: Since there are densely distributed bumps and complicatedly arranged wires in the flip-chip packaging device, 3D-Xray observation should be conducted to check whether there are voids and cracks inside the device, and whether the bumps are missing or bridged. 3D-Xray can observe and analyze the device from any angle, which overcomes the negative effects of overlapped images on the observation and analysis. Fig.4 shows the typical images of 3D-Xray observation.

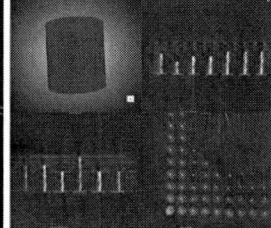

Fig.4 Typical images of 3D-Xray observation

(4) SAM: When performing SAM on the flip-chip packaging device, we should first remove the cooling fin and then check in accordance with J-STD-020 the delamination change rate of the contact surface between the internal solder mask and the resin substrate, the delamination change rate of different layers in the multi-layer substrate, whether there are delamination cracks between the underfill and the chip or between the underfill and the solder mask of the base, and whether delamination is found between the bumps and the substrate, between the bumps and the chip. Fig.5 shows the typical images of SAM.

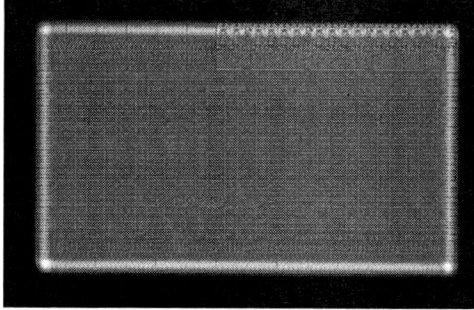

Fig. 5 Typical images of SAM

(5) Sealing and internal air content tests (when applicable): The sealing test and internal air content test should also be carried out on the flip-chip packaging device according to GJB548-1014/1018 when applicable to assess the defects of packaging.

B. Design of DPA Item

The DPA of the flip-chip packaging device includes the following main work items: bonding strength, shear/pull-off strength of solder ball, pull-off strength of the chip, metallographic phase cross-section analysis, internal visual inspection, SEM and integrity check on the glass passivation layer (when applicable).

(1) Bonding strength: Based on the experimental conditions in the GJB548 (2011), we assessed the mechanical properties of the bumps by applying an external force in the direction perpendicular to a boundary of the chip or carrier and parallel to the main substrate, to cause the shear force that would trigger the bonding failure. When the device failed, we recorded the magnitude of the failure force (shear force $\geqslant 0.05N \times$ the number of bonding) and the type of failure (failure of the bonding material or the bonding region of the substrate; cracked chip or substrate; bulge in the metallization layer).

(2) Shear/pull-off strength of solder ball: The shear/pull-off strength of the solder ball is the most important indicator to represent the joint strength of the solder ball. The pull-off and shear resistance of the solder ball was tested and evaluated according to GJB7677.

(3) Pull-off strength of the chip: The pull-off strength of the chip is the most intuitive indicator to characterize the bonding strength between the chip and the substrate. In this study, the pull-off resistance of the chip was tested according to the JESD22-109. If the force value is less than X=760 (N/cm2) × average area of bumps (cm 2) × the number of bumps, the flip-chip packaging device will fail. The failure types, e.g. separation of the interface between the bonding area and the chip, the metallization layer pulled up from the substrate, and the substrate fracture, were recorded at the same time.

(4) Metallographic phase cross-section analysis: Firstly, the flip-chip packaging device was inlaid, ground and polished. Secondly, micro-morphological features on the cross-section of the device, e.g. solder joints, bumps and

IMC, were observed with an optical microscope or an SEM, so as to further evaluate whether the device had defects in terms of process, structure and material. Fig.6 shows the typical images of metallographic phase cross-section analysis.

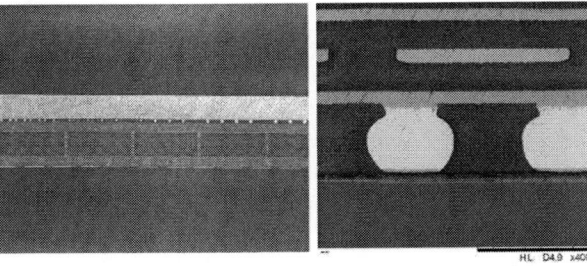

Fig.6 Typical images of metallographic phase cross-section analysis

(5) Internal visual inspection: First of all, the cooling fin or metal shell of the flip-chip packaging device was mechanically removed with the help of a heat gun. Based on the thermal characteristics of the underfill in the flip chip's interconnect structure, the underfill was removed by means of high-temperature baking, and a corresponding chemical reagent was used to clean the surface of the chip to expose the chip[6]. After that, we checked whether the chip had defects in terms of internal material, structure and process according to the GJB548 (2010). Fig.7 shows typical images of internal visual inspection.

Fig.7 Typical images of internal visual inspection

(6) SEM: Based on the GJB548 (2018), we checked defects, e.g. voids, cracks and absence, of the interconnected metallization and passivation layers on the surface of the chip. Fig.8 shows the typical SEM images.

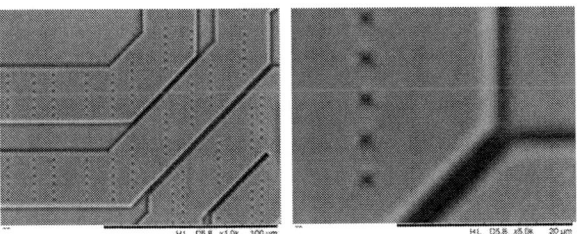

Fig. 8 Typical SEM images

(7) Integrity check on the glass passivation layer (when applicable): The material of the flip-chip packaging device absorbs moisture, so the integrity of the glass passivation layer coated on the metallization layer is of great significance. According to the method stated in GJB548 (2021), we evaluated whether there were defects in the process and material of the glass passivation layer on the aluminum metallization layer.

C. Corresponding Relationship between Work Items and Typical Failure Mechanisms

As long as the work items designed in Sections A and B are able to detect the typical failure mechanisms of the

flip-chip packaging device listed in chapter III, then the DPA method proposed in this study are effective and universal.

TABLE 1 FAILURE MECHANISMS AND THEIR CORRESPONDING WORK ITEMS

No.	Failure mechanisms	Work items
1	Bump failure	Bonding strength
		3D-Xray observation
		Metallographic phase cross-section analysis
		SEM
2	Solder ball failure	External visual inspection
		Shear/pull-off strength of solder ball
		3D-Xray observation
		Metallographic phase cross-section analysis
		SEM
		Coplanarity of the solder ball
3	Underfill Delamination failure	SAM
4	Chip failure	Internal visual inspection
		Metallographic phase cross-section analysis
		Flip chip pull-off test
		SEM
		Integrity check on the glass passivation layer (when applicable)
5	Encapsulation failure (when applicable)	Sealing and internal air content tests (when applicable)

As shown in Table 1, the works items designed in Sections A and B did not cover the evaluation of thermal fatigue failure of the bumps and solder balls, and the evaluation on thermal fatigue failure will usually be conducted in the temperature cycle or thermal shock tests, so they were not included in the DPA work items. To sum up, the work items and procedures involved in the DPA of the flip-chip packaging device are shown in Fig. 9.

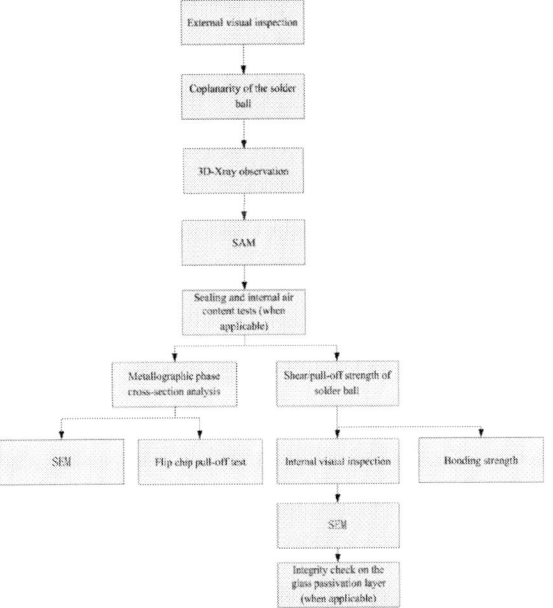

Fig.9 DPA of the flip-chip packaging device

D. Results and Analysis

The evaluation test items designed based on Sections A-B can basically cover the requirements concerning the evaluation on the reliability of flip-chip packaging devices, and have been applied to the DPA of abundant flip-chip packaging devices in our unit. However, due to the special structure of flip-chip packaging devices, we should be careful enough in conducting specific test items to avoid misjudgment. For example, to perform scanning acoustic microscopy (SAM) on flip-chip packaging devices, we usually need to remove the heat dissipation plate, but mechanical removal is very likely to produce stress concentration or excessive strength, which will lead to separation or cracking of underfill and misjudgment of delamination failure. Moreover, since the chip is mounted upside down on the substrate, before visual inspection of the chip interior, the underfill should be removed to expose the active surface of the chip. Nonetheless, removing the underfill by high-temperature aging or acid liquor cooking may lead to crack or metallization corrosion of the chip's glass passivation layer.

V.   Conclusion

The flip-chip packaging device has currently been extensively applied to industrial fields that require high reliability. At the very beginning, the application of the device was banned, but now, it is commonly used. Although the United States has officially included the flip-chip packaging device into the assessment with MIL-PRF-38535K level-Y requirements, the DPA method for flip-chip packaging devices has not yet been released. The research results achieved in this study contribute to the development of DPA methods for the flip-chip packaging device and advanced packaging devices of the same type,

and lay the foundation for the formulation of relevant standards.

Acknowledgment

The authors would like to thank the Key-Area Research and Development Program of Guangdong Province (No. 2019B010145001) and Guangzhou Science and Technology Plan Project (No.201904010457) for supporting this research.

References

[1]   Liu Peisheng, Yang Longlong, Lu Ying, et al. Development of flip chip package technology[J]. Electronic Components and Materials,2014,02:1-5+15.

[2]   Li Han，Zheng Hongyu, Zhang Xiaojun. Failure Localizations on the Open Circuit for a High Density Ceramic Flip-Chip Package After Reliability Test[J]. Semiconductor Technology, 2017, 42(09): 711-716.http://kns.cnki.net/kcms/detail/51.1746.TN.20200423.0856.003.html.

[3]   Ren Chun-ling, Gao Na-yan, Ding Rong-zheng. Study of Failure Mode and Failure Mechanisms on Flip-Chip Ceramic Packaging[J] .Semiconductor Technology, 2010,10(08):5-9.

[4]   Lin Xiaoling, Xiao Qingzhong, En YunFei, et al. Failure mechanism of FC-PBGA devices under external stress[J]. Acta Physica Sinica,2012,12:578-584.

[5]   Zhang Danqun，Zhang Sujuan. Research on DPA Experimental Process for Plastic Flip Chip[J].Semiconductor Technology,2015,40(12):950-953.

[6]   Zhou Shuai，Zheng Dayong，Wang Bin. Unsealing methods of flip chip assembly integrated circuit[J]. Journal of Terahertz Science and Electronic Information Technology,2017,15(02):328-332.

Simulation Analysis of Coupling Coil of 13.56MHz Magnetic Coupling Resonant Wireless Energy Transmission System

1st Bihong Zhan
China Ship Development and Design Center
Wuhan, China
zbhwan_2008@163.com

2nd Wei Xia
China Ship Development and Design Center
Wuhan, China
xiawei_cn@163.com

3rd Chunshui Xiong
China Ship Development and Design Center
Wuhan, China
xiongchunshui1@163.com

4th Sheng Liu*(Corresponding author)
[1]*The Institute of Technological Sciences,Wuahan University*
[2]*School of Power and Mechanical Engineering,Wuahan University*
Wuhan, China
victor_liu63@126.com

Abstract—To find a new growth point for the improvement of the coupling coil transmission efficiency in the 13.56 MHz magnetic coupling resonant wireless energy transmission (WET) system, and to provide a new solution for the improvement of the transmission efficiency of the entire system. This paper takes the coupling coil (including the primary transmitting coil and the secondary receiving coil) in the 13.56 MHz magnetic coupling resonance WET system as the basic research object. First, the Ansoft HFSS simulation software is used to digitally model the coupling coil, and the inductance value of the coupling coil and the magnetic induction intensity (denote by *B*) around a single coil are obtained through simulation calculations. Furthermore, through simulation, the distribution of magnetic induction intensity and magnetic induction intensity vector under different transmission distances between the coupling coil are obtained. Secondly, the coupling coil model established by Solidworks software was imported into Ansoft Maxwell simulation software in a certain format, and the magnetic field distribution at different vertical distances between the coupling coil and the offset distance of the central axis were obtained through simulation calculations. In this paper, the problem of the transmission efficiency of the coupling coil is transformed into the problem of the coupling coefficient, and the coupling coefficient and mutual inductance between the coupling coil at different transmission distances are obtained by Ansoft Maxwell simulation calculation, and the effects of the vertical distance between the coupling coil and the offset distance of the central axis on the coupling coefficient and mutual inductance of the coupling coil are analyzed. Finally, this article uses Ansoft HFSS to simulate and analyze the influence of the composite film on the inductance and magnetic field distribution of the coupling coil. This will provide a theoretical basis for the integration of the composite film in the 13.56 MHz magnetic coupling resonance WET system.

Keywords—*Magnetic coupling, coupling coefficient, coupling coil, 13.56 MHz, WET*

I. INTRODUCTION

With the rapid development of technology, in the past few decades, many methods have been proposed for charging flexible wearable electronics, such as: near-field inductive coupling [1], near-field capacitive coupling [2], ultrasound [3], field electromagnetic coupling [4] and far-field electromagnetic coupling [5], etc. Among them, the magnetic coupling resonance WET is the most cutting-edge research topic, and the technology application prospect is very broad, especially the MHz magnetic coupling resonance WET system is more competitive due to its small size. At present, some research results have been made in the research of MHz systems [6]-[10], module and system-level design and optimization of the system WET conducted extensive research, especially the coupling coil [11] and other improvements.

This paper aims to simulate the coupling coil of the 13.56 MHz flexible wearable electronic magnetic coupling resonant WET system, and analyze the factors affecting the transmission performance of the coupling coil. The main research contents include: coupling coil modeling and simulation calculation, and the magnetic field distribution under different working conditions between the coupled coils, the influence of the vertical distance between the coupling coil and the offset distance of the central axis on the coupling coefficient and mutual inductance, and the influence of the composite film on the coupling coil.

II. COUPLING COIL INDUCTANCE SIMULATION CALCULATION

Fig. 1 is a three-dimensional schematic diagram of the coupled coil. The primary transmitting coil is similar to the secondary receiving coil, and its dimensions are shown in TABLE I.

Fig. 2 shows the inductance value of the coupled coil calculated by HFSS simulation. It can be seen from the Fig. 2 that the inductance values of the coupling coil at the resonance frequency of 13.56 MHz are 8.4 μH and 2.4 μH, respectively. The HFSS software uses the finite element method, and the calculation results are accurate and reliable. It is an industry-recognized industry standard for the design and analysis of three-dimensional electromagnetic fields. Therefore, the coil inductance value obtained by HFSS software simulation calculation is accurate and reliable.

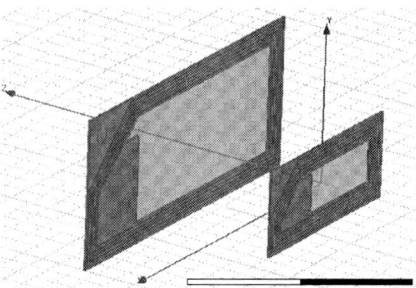

Fig. 1 Coupling coil model diagram

TABLE I. COUPLING COIL SIZE

	Primary transmitting coil (mm)	Secondary transmitting coil (mm)
Length of outer rectangle	31	18
Width of outer rectangle	14	8
Line width	0.25	0.25
Line spacing	0.35	0.35
Inner rectangle length	21	9.5
Inner rectangle width	10	4

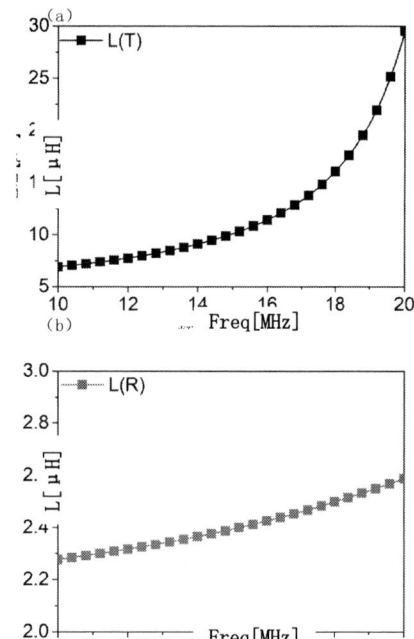

Fig. 2 (a) The inductance value of the primary transmitting coil and (b) the inductance value of the secondary receiving coil calculated using HFSS simulation

III. COUPLING COIL MODELING AND SIMULATION

Use Maxwell software to simulate the coupling coil. Select Modeler-Import in the Maxwell interface environment, import the coupled coil model shown in Fig. 1 into Maxwell, assign materials, and perform modeling and simulation. The B distribution diagram of a single coil in a plane 6 mm above the coil is shown in Fig. 3. It can be seen from the Fig. 3 that the B distribution in the same plane is uniform and regular, and presents a distribution trend of the largest in the middle and spreading to the surroundings in sequence. Therefore, the use of multi-turn double-layer coils can effectively avoid the reduction of coil quality factors due to skin effect and eddy current loss.

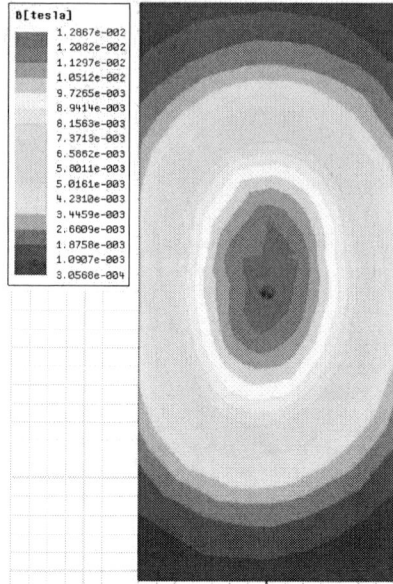

Fig. 3 Schematic diagram of the B distribution of a single coil in a plane 6 mm away from the coil

In order to analyze the coupling coil, the coupling coil are simulated and analyzed. When the primary transmitting coil (upper coil) is given a certain excitation (the excitation is 20 A in the figure), the B distribution diagram of the coupling coil is shown in Fig. 4 when the coupling coil has different transmission distance D. In the Fig. 4, when the primary transmitting coil is given a certain excitation, the B on the receiving coil gradually weakens with the increase of the transmission distance, while the B on the primary coil does not change significantly. In order to better illustrate that the B of the secondary receiving coil decreases with the increase of the transmission distance. Fig. 5 shows the magnetic induction vector distribution diagram of the B distribution between the coupling coil under different transmission distances. It can be clearly known from Fig. 5 that as the transmission distance increases, the B induced by the receiving coil gradually decreases, and the range covered by the magnetic lines of force gradually decreases. It can be proved that the B of the receiving coil decreases as the transmission distance increases.

978-1-6654-1392-3/21 $31.00 © 2021 IEEE

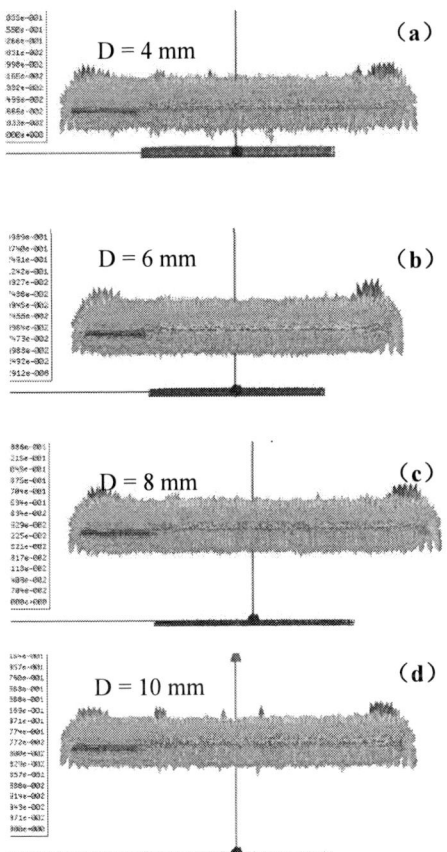

Fig. 4 Schematic diagram of the **B** distribution of the coupling coil under different transmission distances

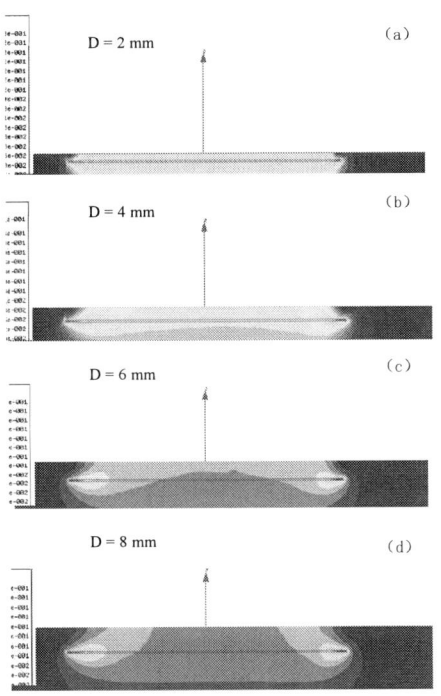

Fig. 5 The vector distribution of **B** between coupling coil under different transmission distances

IV. SIMULATION ANALYSIS OF COUPLING COIL MUTUAL INDUCTANCE AND COUPLING COEFFICIENT

For the resonant magnetic coupling system in which the coupling coil are all selected planar spiral coils, according to the analysis in the paper [12], when the shape and structure of the coupling coil, input current, voltage, and frequency have been determined, the mutual inductance between the coupling coil and the coupling coil, the vertical distance, horizontal offset distance, and angular offset are closely related to each other. In this paper, Maxwell software is used to simulate and calculate the mutual inductance and coupling coefficient under different working conditions.

In the application of wireless energy supply system, Maxwell 3D can not only simulate the direction and distribution of the magnetic field lines in the spherical surrounding space in the electromagnetic coupling system, but also calculate the mutual inductance between the coupling coil and the autonomy of the transmitting coil and the receiving coil. At the same time, it can draw energy or **B** distribution effect diagram, which is more convenient for the intuitive design of electromagnetic coupling system and the further optimization of parameters.

In the working cycle, assuming that the excitation current of the coil is I_{AC}, the mutual inductance value M_{12} of the transmitting coil to the receiving coil can be expressed as:

$$M_{12} = \frac{\psi_{12}}{I_{AC}} \qquad (1)$$

Where ψ_{12} is the mutual inductance full magnetic flux, which can be expressed as:

$$\psi_{12} = \sum_1^N \iint B_i \, dS_i \qquad (2)$$

It can be seen that when the excitation current is constant, the mutual inductance of the coupling system is determined by the **B** surrounded by the coil and the area S surrounded by the coil. The theoretical analysis of the expression of **B** is very complicated, so in engineering, for complex coupling systems, accurate finite element simulation software is usually used to accurately simulate and analyze them.

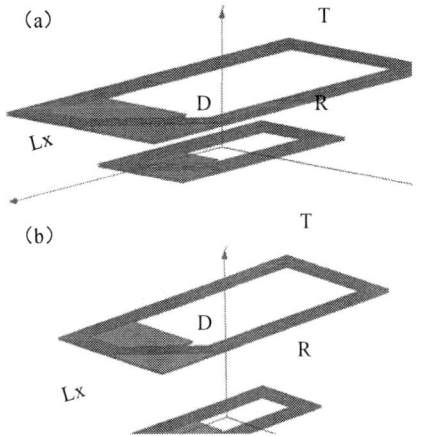

Fig. 6 Coupling coil models under different transmission distances

Fig. 6 shows the coupling coil model under different transmission distances, where D is the vertical distance between the transmitting coil T and the receiving coil R, and L_x is the offset distance of the central axis between the two coils, here refers to the coupling coil The offset distance of the center axis along the X axis. Use Maxwell simulation to analyze the relationship between the coupling coefficient and mutual inductance between the transmitting coil and the receiving coil under different transmission distances, and further analyze the magnetic field distribution between the coils at different transmission distances.

Fig. 7 is a schematic diagram showing the variation of the coupling coefficient κ and mutual inductance L_M as the vertical distance (transmission distance) D changes with the change of the vertical distance D obtained after the simulation data is processed by Origin Pro 9.0. After sorting, as shown in TABLE II. It can be learned from Fig. 7 that the coupling coefficient κ of the coupling coil and the mutual inductance L_M decrease with the increase of the transmission distance. It can be inferred from this that: the closer the distance between the coupling coil, the higher the coupling efficiency, that is, the secondary receiving coil will receive more magnetic field energy.

TABLE II. D, κ AND L_M

D(mm)	2	4	6	8	10	12	14	16
κ	0.86	0.56	0.36	0.24	0.16	0.11	0.07	0.06
L_M (μH)	3.89	2.52	1.64	1.08	0.72	0.48	0.33	0.27

Fig. 8 shows the variation of the κ and the L_M with the change of the center axis offset distance L_x. After sorting, as shown in TABLE III. It can be known from Fig. 8 that as the center axis offset distance L_x increases, the coupling coefficient and mutual inductance between the coils gradually become smaller, that is, κ and L_M are inversely proportional to L_x. Therefore, it can be inferred that when the offset distance $L_x = 0$, that is, when the central axis of the transmitting coil T and the receiving coil R are aligned, the coupling coefficient and mutual inductance L_M are the largest. At this time, the secondary receiving coil R can receive more from the primary transmitting coil T. More magnetic field energy, at this time, the magnetic coupling resonance WET efficiency is the highest.

TABLE III. L_x, κ AND L_M

L_x (mm)	2	4	6	8	10	12	14	16
κ	0.87	0.85	0.83	0.77	0.66	0.55	0.45	0.32
L_M (μH)	3.82	3.80	3.75	3.41	2.92	2.48	1.98	1.42

To further explain the relationship theoretically, the magnetic field distribution between the coupling coil under different transmission distances and offset distances is simulated, as shown in Fig. 9. It can be seen from Fig. 9 that the energy of the alternating magnetic field generated by the excitation source is mainly concentrated on the coil and the area between the coupling coil, while the area outside the two coils is less distributed. By comparing Fig. 9 (a) (b) (c), it can be learned that when D and L_x are smaller, the energy coupled between the two coils is more, which can be seen from the energy color mark on the left of the Fig. 9.

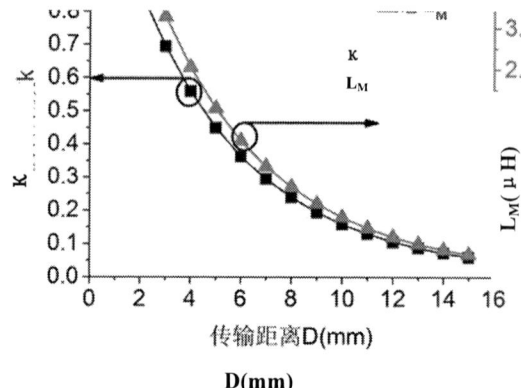

D(mm)

Fig. 7 The relationship between the coupling coefficient, mutual inductance and the vertical distance between the coils

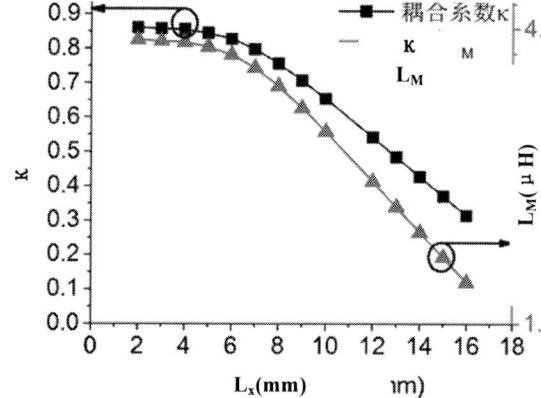

Fig. 8 The relationship curve between the coupling coefficient, mutual inductance and the offset distance of the coil center axis

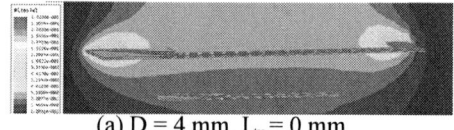

(a) D = 4 mm, L_x = 0 mm

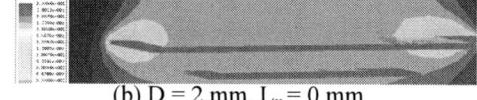

(b) D = 2 mm, L_x = 0 mm

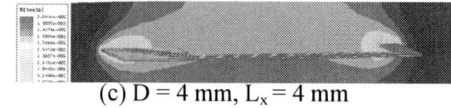

(c) D = 4 mm, L_x = 4 mm

Fig. 9 The magnetic field distribution diagram between the coupling coil under different transmission distances and offset distances

Through the above simulation analysis, the following conclusions can be drawn: (1) The coupling coefficient κ and mutual inductance L_M are related to the offset position of the central axis of the coil. When the central axis is aligned, that is, when there is no offset, κ and L_M are the largest; (2)The transmission distance is closely related to κ and L_M, and is inversely proportional; (3) The alternating magnetic field energy of the coil is mainly concentrated near the coil. It can also be known from the above simulation that the coupling coefficient between the coupled coils has nothing to do with the excitation and is not affected by the magnitude of the excitation.

V. THE INFLUENCE OF THE COMPOSITE FILM ON THE COUPLING COIL

This section mainly analyzes the influence of the composite film on the coupling coil. Based on the above design, the three-dimensional schematic diagram of the coupling coil is shown in Fig. 10(a). The total width, line width, total length, and line-to-line distance of the rectangular coupling coil are 8030 μm, 251.7 μm, 17961 μm, and 153.0 μm, respectively, and the thickness of PET is 0.05 mm, as shown in Fig. 10(b). Studying the effect of the composite film pattern on the transmission performance of the coupled coil is very important. Fig. 11 shows a coil integrating different composite films. Fig. 11(a) shows the coil without integrated composite membrane, marked as CM1. Fig. 11(b) represents that the composite film is integrated in the plane area surrounded by the upper rectangular coil and the lower rectangular coil connected by the rectangular opening on the PTE film, marked as CM2. Fig. 11(c) represents the composite membrane integrated into the entire coil, marked as CM3. Fig. 11 (d) represents that the composite film is integrated in the entire coil except the upper rectangular coil and the lower rectangular coil in the plane area enclosed by the coil, marked as CM4.

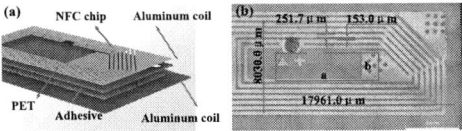

Fig. 10 (a) three-dimensional schematic diagram of the coupling coil; (b) coupling coil size.

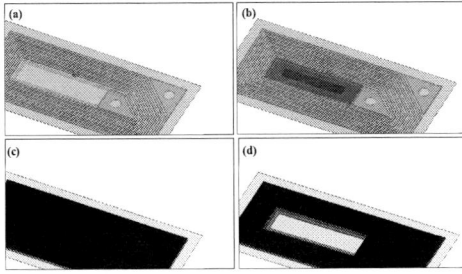

Fig. 11 Coil model diagram integrating different composite membranes

In order to evaluate the effect of the composite film integrated in different positions of the coil on the performance of the coupled coil, HFSS was used to simulate and analyze the effect of composite film on coil electrical performance. Fig. 12 shows the simulation of L and Q of the coil with frequency in the frequency range of 10 MHz to 16 MHz. From the simulation results in Fig. 12 (a), it can be learned that the inductance L value gradually increases in the frequency range, and the L value of the integrated composite membrane coil is greater than the L value of the coil without integrated composite membrane. It can be seen that the integrated composite membrane pair The inductance value L of the coil has an enhancement effect and in the case of CM3, the

inductance value L has the largest enhancement, an increase of 75.6%. When the composite film is integrated into the coil, the magnetic permeability of the entire coil increases, which leads to an increase in B. From the formula $L = \int B \cdot dS / I$,we can see that the coil inductance value L increases with the increase of B, which is conducive to the miniaturization of the coil. It can be learned from Fig. 12(b) that the quality factor Q of the coil gradually decreases with increasing frequency, revealing that the change of the quality factor Q has a relationship with the position of the composite film integrated on the coil. In the case of CM2, the Q value decreases, while in the case of CM3 and CM4, the Q value increases. The reason is that the quality factor Q is determined by the formula $Q = \omega L / R$. At low frequencies, the internal resistance R changes very little, and the quality factor is largely determined by the inductance value L. The increase of the inductance L will lead to the increase of the quality factor Q. The dielectric loss increases as the frequency increases, leading to a certain degree of increase in the internal resistance R. For CM1, before the frequency of 16 MHz, the increase of the internal resistance R is still smaller than the increase of the inductance L, resulting in the quality factor Q of CM1 is still greater than the quality factor Q value of the integrated coil without composite film. For CM2 and CM3, when the frequency exceeds a predetermined value, the increase in the internal resistance R is much larger than the increase in the inductance L, which leads to a decrease in the Q value of the quality factor.

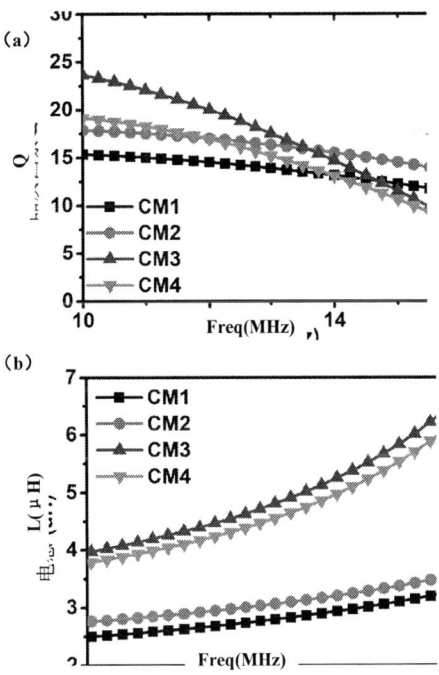

Fig. 12 (a) The inductance of different composite membrane coils with frequency changes; (b) The quality factor of the coils with different composite membrane modes changes with frequency.

Fig. 13 shows the distribution of the magnetic field above the coil under the four conditions of CM1 ~ CM4 described in the above analysis, obtained by the simulation software. Through the comparison of the simulation results, it can be known that only in the CM2 case, the B above the coil has increased(Increased by 27.3%), while in other cases, it has been reduced. The reason is: in the case of CM3 and CM4, the entire magnetic circuit exists on the composite film, which impairs the ability of the coil to radiate the magnetic field. Although the composite film enhances the magnetic flux of the coil, it also prevents the magnetic field lines from radiating to the outside space. When the composite film does not cover the center of the coil track coil, magnetic force lines generated by the coil can not form a closed loop in the magnetic film, the magnetic lines of force radiating outward thus be blocked. While increasing the magnetic flux density of the composite film of the coil, resulting in increased radiation intensity of the magnetic field. Be seen, the integration of different positions on the coil composite film has a significant effect on the radiation field strength of the coil.

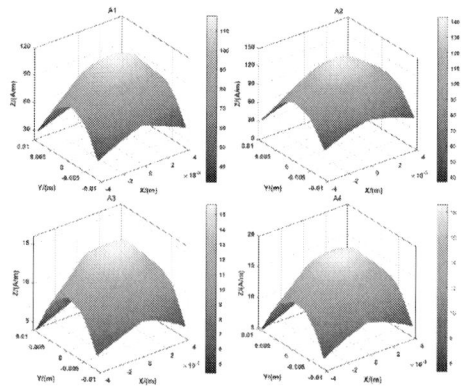

Fig. 13 Comparison of magnetic field distribution above the integrated composite film coil at different positions of the coil

The above series of simulations show that integrating the composite film into the coils can affect the magnetic field distribution, reduce magnetic leakage, make the magnetic lines of force more concentrated, reduce environmental impact, and improve the transmission efficiency between the coupled coils. This shows that in the magnetic coupling resonant WET system, by integrating the composite film on the coupling coil, especially when it is integrated in the middle plane area of the rectangular coil, the transmission efficiency of the coupling system will be significantly improved, and the efficiency of the entire system will be improved. It will provide a new solution for further miniaturization and flexibility of the coupling coil. .

VI. CONCLUSION

This paper takes the coupling coil in the 13.56 MHz magnetic coupling resonance wireless energy transmission system as the research object, and uses simulation software to model it. Through simulation calculation, the influence of factors such as the vertical distance between the coils, the offset distance of the central axis, and the composite film on the transmission performance between the coupled coils is analyzed. The simulation results will play a theoretical guiding role in the design of the coupling coil of the 13.56 MHz magnetic coupling resonance WET system.

ACKNOWLEDGMENT

R. B. G. thanks Prof. S. Liu of Wuhan University, China for help.

REFERENCES

[1] U. M. OW, M. Ghovanloo, "Modeling and optimization of printed spiral coils in air, saline, and muscle tissue environments," IEEE transactions on biomedical circuits and systems, vol. 3, pp. 339-347, 2009.

[2] A M. SODAGAR, P. Amiri, " Capacitive coupling for power and data telemetry to implantable biomedical microsystems," Neural Engineering, 2009. NER'09. 4th International IEEE/EMBS Conference on. IEEE, pp. 411-414, April 2009.

[3] S. OZERI, D. SHMILOVITZ, "Ultrasonic transcutaneous energy transfer for powering implanted devices," Ultrasonics, vol. 50, pp. 556-566, 2010.

[4] A. S. POON, S. O'Driscoll, T. H. Meng, "Optimal frequency for wireless power transmission into dispersive tissue," IEEE Transactions on Antennas and Propagation, vol. 58, pp. 1739-1750, 2010.

[5] C. LIU, Y. X. GUO, H. Sun, S. Xiao, "Design and safety considerations of an implantable rectenna for far-field wireless power transfer," IEEE Transactions on antennas and Propagation, vol. 62, pp.5798-5806, 2014.

[6] A. KARALIS, J. D. JOANNOPOULOS, M. SOLJAČIĆ, "Efficient wireless non-radiative mid-range energy transfer," Annals of physics, vol. 323, pp. 34-48, 2008.

[7] S. L. HO, J. WANG, W. N. FU, M. Sun, "A comparative study between novel witricity and traditional inductive magnetic coupling in wireless charging," IEEE Transactions on Magnetics, vol. 47, pp. 1522-1525, 2011.

[8] T. C. BEH, M. KATO, T. IMURA, S. Oh, Y. Hori, "Automated impedance matching system for robust wireless power transfer via magnetic resonance coupling," IEEE Transactions on Industrial Electronics, vol. 60, pp. 3689-3698, 2013.

[9] F. ZHANG, S. A. HACKWORTH, W. FU, C. Li, Z. Mao, M. Sun, "Relay effect of wireless power transfer using strongly coupled magnetic resonances," IEEE Transactions on Magnetics, vol. 47, pp. 1478-1481, 2011.

[10] B. L. CANNON, J. F. HOBURG, D. D. STANCIL, S. C. Goldstein, "Magnetic resonant coupling as a potential means for wireless power transfer to multiple small receivers," IEEE transactions on power electronics, vol. 24, pp. 1819-1825, 2009.

[11] W. ZHONG, C. ZHANG, X. LIU, and S. R. HUI, "A methodology for making a three-coil wireless power transfer system more energy efficient than a two-coil counterpart for extended transfer distance," IEEE Trans. Power Electron., vol. 30, pp. 933–942, Feb 2015.

[12] J. T. CONWAY, "Inductance calculations for noncoaxial coils using bessel functions," IEEE Transactions on Magnetics, 2007, vol. 43, pp. 1023-1034, 2007.

Stress Analysis of Cu/Sn Bump Eutectic Bonding Interface

Xinpeng Chen
Xi'an Microelectronic Technology Institute
Xi'an, China
chenxinpengy@qq.com

Ruixia Huo
Xi'an Microelectronic Technology Institute
Xi'an, China
h123r456x789hrx@163.com

Daowei Wu
Xi'an Microelectronic Technology Institute
Xi'an, China
wudaowei1220@163.com

Wansheng Liu
Xi'an Microelectronic Technology Institute
Xi'an, China
15162697602@163.com

Abstract—Cu/Sn wafer-level bonding is an interesting solution for wafer-to-wafer stacking technologies, due to its compatibility with 3D interconnections as well as vacuum sealing applications. The Cu/Sn eutectic bonding process is favored by industry and researchers alike for its excellent process compatibility and low cost, and has been proven for high density interconnect packaging of MEMS devices. Because of the presence of heterogeneous materials (TSV-Cu, Si substrates, Cu bumps, and Sn bumps) with large differences in coefficients of thermal expansion, the Cu/Sn interface exists varying degrees of stress and strain during the bonding process which can have an impact on the reliability of the bonding results. By optimizing the layout of the Cu/Sn bumps and the parameters of the bonding process, the stress and strain distribution during the bonding process can be reduced and the reliability of the Cu/Sn bonding interface can be improved. Based on ANSYS finite element steady-state thermodynamic analysis the stress and strain distribution at the bonding interface can be effectively obtained for different Cu/Sn bump layout, reducing the strain on the bump surface during the bonding process and improving the process reliability by optimizing the design rules.

Keywords—Cu/Sn eutectic bonding, finite element analysis, stress distribution

I. INTRODUCTION

Moore's law provided an evolution template for silicon-based chip feature size of the microelectronics industry for the past 50 years. The performance and redundancy of microelectronics system is limited by the monolithic fabrication technology. Multi-chip integrated in Z direction purveyed an alternative method in achieving high density interconnects. Integration in Z direction was obtained through stacking. direct chip to chip, it also makes the heterogeneous chip integration possible. The vertical integration was discussed both in 2.5D and 3D stacking [1,2]. 3D WLP is an emerging technology characterized by vertically stacked chips to achieve high performance, low cost and versatile packaging. 3D wafer-level package(WLP) based on TSV is a more than Moore technique to effectively continue Moore's law. The 3D WLP based through silicon via(TSV) is a system integration of flat wafers stacked to achieve reliable transmission interconnects in the Z-direction,

featuring vertically stacked chips for high performance, low cost and versatile packaging. The Core to the development of 3D integration is the achieving the effectively and reliable three-dimensional interconnects between chips to reduce interconnection path lengths, package size and power dissipation, while 3D integration makes homogeneous and heterogeneous stacking possible. The 3D package structure is shown in Fig. 1.

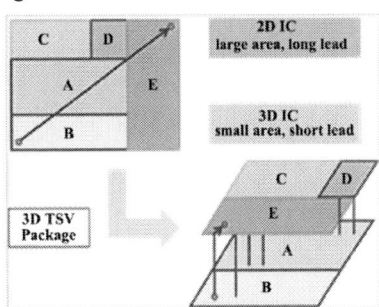

Fig.1 Schematic of 2D integration vs. 3D integration

In 3D WLP process flow, wafer-level bonding is a vital process to complete the Z-directional stacking and vertical integration of chips. Wafer bonding technology can be used for the preparation of microelectronic components and the packaging of micro and nano electronic systems. Common bonding methods used for 3D interconnections include: adhesive bonding, direct bonding, anode bonding, diffusion bonding and eutectic bonding. Cu/Sn bump eutectic bonding technology has been successfully applied to the packaging and interconnection of MEMS devices and microsystems devices due to its low cost, excellent high temperature stability, high bond strength and excellent gas tightness [3-5]

II. EXPERIMENTAL DESIGN

A. Cu/Sn eutectic bonding theroy

In the Cu/Sn bonding process used in this study, the Cu/Sn bumps are quickly pre-aligned after surface pre-treatment, allowing the Cu/Sn bumps surface contact to avoid surface oxide regeneration. The initial stage of the bonding process is usually carried out in a mixed nitrogen-hydrogen atmosphere to further remove the surface oxides. The Cu/Sn co-crystal phase diagram [5,9] shows that both Cu_3Sn and Cu_6Sn_5 can be formed at room temperature, with Cu_3Sn preferentially formed in excess of Cu and Cu6Sn5 in preference to Sn. The Cu_6Sn_5 grains gradually grow in size as the temperature increases, affecting the bonding process. If the Cu content is not sufficient, Cu_6Sn_5 will not be fully converted to Cu_3Sn and Cu_6Sn_5 will form voids in contact with the Sn surface, which will affect bond strength, device sealing and overall reliability. The bonding process temperature is typically between 232°C and 300°C, above the melting point of Sn, and the eutectic interlayer begins to form and solidify gradually as the Sn melts [4]. During the bonding process, suitable process parameters and Cu/Sn thickness ratios enable the conversion of Cu_6Sn_5 to Cu_3Sn via Equation (2), with Cu_3Sn having better strength and high temperature stability than Cu_6Sn_5. Only when the Cu/Sn atom ratio is greater than 1.3, the Cu/Sn eutectic complex could be completely transformed into Cu_3Sn [6].

$$6Cu + 5Sn \rightarrow Cu_6Sn_5 \qquad (1)$$

$$9Cu + Cu_6Sn_5 \rightarrow 5Cu_3Sn \qquad (2)$$

In the course of this study 240°C was chosen as the bonding process application temperature, which is higher than the Sn melting point (231.9°C). The process temperature allows the contact surface layer of the eutectic complex to melt and form a liquid phase alloy, further accelerating the diffusion and reaction process until the saturation composition is reached [7]. Optimization of the bonding parameter control is necessary to prevent the melt from being pressed beyond the bonding interface. Although an optimized Cu/Sn bonding process allows the bonding boundary to be fully formed with Cu_3Sn, which helps to improve the bond strength. However, the different design dimensions of Cu/Sn during the bonding process can lead to a situation where the local strain is too large, resulting in the eutectic melt being pressed outside the bonding interface to form seeping Sn. The appearance of seeping Sn can affect the electrical properties of the bonding interface and should be avoided as far as possible during the bonding process..

B. FEM anlysis

During the bonding process, the size of the bump has a significant impact on the stability of the bonding process. Simulations are carried out to analyses the stress and strain distribution at the interface in the Cu/Sn bump eutectic bonding process. Finite element analysis is the most common method of mathematical modelling for solving engineering problems. Steady-state thermodynamic simulations using ANSYS software can provide an effective understanding of the stress distribution at the Cu/Sn bonding interface.

The model dimensions of the single chip were simplified according to the chip structure, and the simplified analytical calculation model is shown in Fig.2. The detail model characteristic dimension data are shown in the TABLE I.

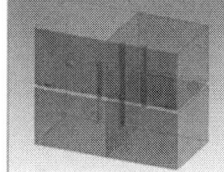

Fig. 2 Simplified model of the package for ANSYS analysis

TABLE I CONTROL GROUP DESIGNED FOR FEA
ANALYSIS EXPERIMENTS

Item	Specification
Si substrate	400μm *800μm *300μm
Bump diameter	40μm
Cu bump height	10μm
Sn bump height	5μm
TSV size	AR=5, Via diameter=30μm

In the Cu/Sn eutectic bonding process, both thermal and force loads are applied. 240°C thermal load is applied to the Si substrate according to the bonding machine structure, and the entire bonding process is in a high vacuum (vacuum <1.0E-5 Pa) condition. A uniform load of 2E-3 N is applied to the top Si substrate of the model and a full constraint is applied to the bottom Si substrate. the Cu/Sn bonding interface was friction contact, and the coefficient of friction was assumed to be 0.4. The thermal and mechanical loads are applied as shown in Fig. 3.

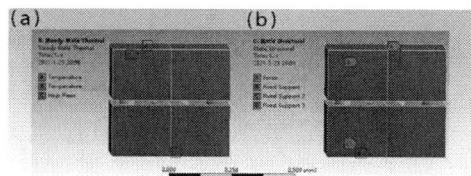

Fig.3 Chip model load distribution and boundary condition settings in ANSYS analysis

III. RESULTS AND ANALYSIS

The model was meshed with a mesh accuracy of 1 μm and a static thermal analysis was performed on the model with 240°C thermal load applied. A cloud plot of the deformation distribution of the Si substrate during the bonding process was calculated under an experimental conditions and is shown in Fig.4(a). The maximum deformation can be found to be at the center of the chip with a maximum deformation of 2.5012E-7mm. The longitudinal cross-sectional deformation distribution is shown in Fig.. 4(b). It can be found that the longitudinal Si deformation gradually decreases with the increase of the top Si thickness. The main deformation is mainly distributed on the top Si substrate.

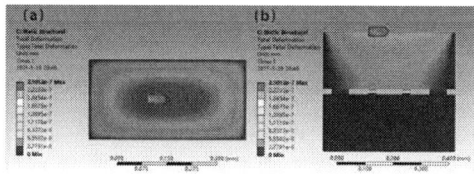

Fig.4 Cloud map of deformation distribution of Si substrate, (a) surface, (b) interior

The stress distribution in the Cu bump cross-section during eutectic bonding is shown in Fig. 5(a) below, where it can be observed that the stress distribution in the Cu bump cross-section is mainly concentrated in the direction of the two bumps adjacent to each other near the center of the model, as indicated by the red box in Fig.5(a). The stresses near the edge of the TSV Cu are significantly higher than those in the rest of the Si substrate. This is due to the fact that the coefficient of thermal expansion of Cu is 18.5 ppm/°C, whereas that of the Si substrate is only 3.61

ppm/°C. The simplified model only considers the deformation in the FEA process and the actual wafer warpage was neglected. When a thermal load of 240°C is applied, the transverse thermal expansion length value of the Cu column is higher than the Si thermal expansion length value, and the Cu column sidewall exerts a certain amount of thermal stress on the Si sidewall. The internal stress distribution in the upper Wafer is shown in Fig.5(b), which can be seen that the internal stresses in the Si substrate are mainly distributed near the TSV Cu column, which is consistent with the analysis [9].

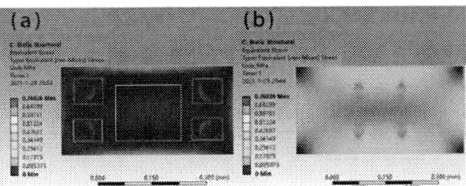

Fig. 5 Stress distribution clouds at the Cu/Sn interface, (a) interface, (b) interior

In order to decrease the maximum deformation, equivalent stress and strain from design ruler. Serval contrast experiment was set to research the stress value change in bonding process, as shown in TABLE II. The trends in the maximum deformation, maximum stress and strain values for the A, B, C and D control experimental groups were calculated from ANSYS and are shown in Fig.6. Comparing calculate results, it can be found that the maximum deformation of the Si substrate and the maximum value of von-Mise stress at the Cu/Sn interface appear decrease as the diameter of the Cu/Sn bump increases.

TABLE II CONTROL GROUP OF BUMP DIAMETER DESIGNED FOR FEA ANALYSIS EXPERIMENTS

Control group	Cu/Sn bump diameter(μm)	Cu bump height(μm)
A	40	10
B	50	10
C	60	10
D	70	10

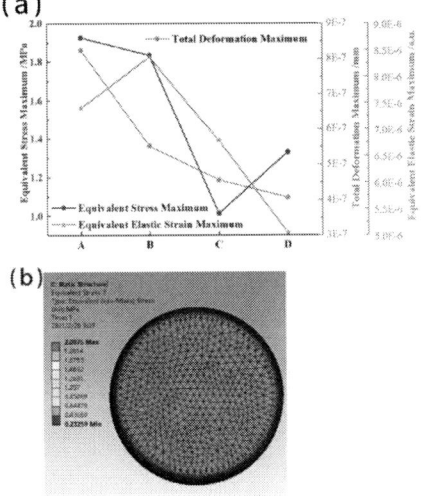

Fig.6 (a)variation of deformation, stress and strain in different Cu/Sn diameter,(b) the stress distribution pattern in the bonding interface

The Cu bump height were set to 10μm,15μm and 20 μm, and the thermal stress coupling analysis was performed according to the control design in TABLE III to calculate the von-Mise stress and strain distribution around the Si substrate, Cu/Sn bump interface and TSV under steady-state conditions, respectively. From Fig.7, it can be seen that the Cu/Sn bump interface von-Mise stress and strains reveal a sharp increase with augmenting Cu bump height. The results demonstrate the increase in Cu bump height is not conducive to reducing the bonding interface stress and strain value.

TABLE III CONTROL GROUP OF BUMP HEIGHT DESIGNED FOR FEM ANALYSIS EXPERIMENTS

Control group	Cu/Sn bump diameter(μm)	Cu bump height(μm)
A	60	10
B	60	15
C	60	20

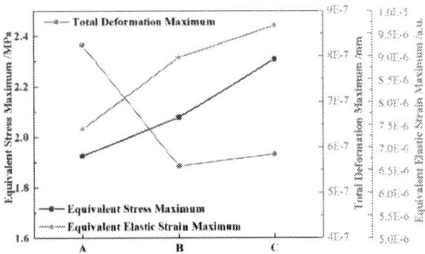

Fig.7 Variation of deformation, stress and strain in different Cu/Sn height

The Cu/Sn bump eutectic bonding process was realized according to the process flow shown in Fig.8. The diameter of the Cu/Sn bump was 60μm, the height of the Sn bump was 5μm, and the height of the Sn bump was 10μm，which ensure the complete reaction of the bonding interface to synthesize stable intermetallic Cu_3Sn.The SEM cross-section of the bump interface after bonding is shown is Fig. 9. It can be seen there is a slight offset in bonding process, which is due to the lack of accuracy. It also can be seen that there is more Sn overflow on both side of Cu_3Sn intermediate layer, which is caused by the stress concentration at the edge of the bump during the bonding process, matching the simulation analysis results and confirming the accuracy of the simplified model.

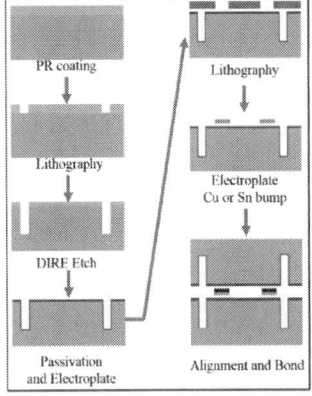

Fig. 8 Process flow of the Cu/Sn bump eutectic bond manufacturing

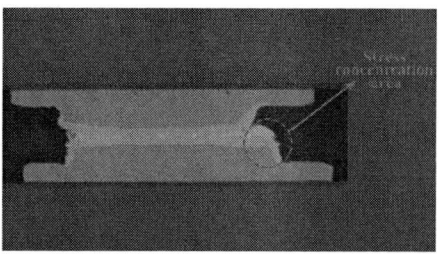

Fig. 9 SEM cross-section pattern of the Cu/Sn bump bonding interface

IV. . CONCLUSION

In this work, the ANSYS Steady State Thermodynamics and static structural module was used to analyze the stress distribution at the Cu/Sn bumps for diameters of 40μm, 50μm, 60μm and 70μm respectively, and to obtain the stress and strain value at the Cu/Sn bumps for height of 5μm, 10μm and 15μm respectively for a Cu/Sn bump diameter of 50μm. The minimum von-Mises stress and strain was obtained when the diameter/height was 60μm/5μm. The von-Mises stress distribution at the Cu/Sn bump cross-section is mainly concentrated at the two bump edges. The von-Mises stress of TSV Cu being significantly higher than those at the rest of the Si substrate. The wafer bonding experiment results of the Cu/Sn bumps eutectic bonding interface are consistent with the simulation results and verify the accuracy of the simulation results.

The experimental analysis of the ANSYS finite element simulation demonstrates that the stress distribution at the Cu/Sn bump interface is affect by both the bump height and the bump diameter. Si substrate deformation is related to the stressed contact surface and the TSV location. The maximum von-Mise stress value at the bonding interface during the bonding process can be reduced by increasing the contact area of the bonding interface and optimizing the Cu/Sn bump height. By optimizing the position distribution of the TSV structure can prevent damage from local stress concentration in the chip during the bonding process and effectively improve the bonding yield.

REFERENCES

[1] Meinshausen L. Modelling the SAC microstructure evolution under thermal, thermo-mechanical and electronical constraints[J]. Bordeaux, 2014.

[2] C. Flötgen, Corn K, Pawlak M, et al. Cu-Sn Transient Liquid Phase Wafer Bonding: Process Parameters Influence on Bonded Interface Quality[C].ECS Transactions. 2013.

[3] Lueck M R. High-Density Large-Area-Array Interconnects Formed by Low-Temperature Cu/Sn–Cu Bonding for Three-Dimensional Integrated Circuits[J]. Electron Devices IEEE Transactions on, 2012, 59(7): p.1941-1947.

[4] Aasmundtveit K E, Luu T T, Nguyen H V,Larsson A and Tollefsen T A. Intermetallic Bonding for High-Temperature Microelectronics and Microsystems: Solid-Liquid Interdiffusion Bonding[M]// Intermetallic Compounds - Formation and Applications. 2018.

[5] Sanabria C, Field M, Lee P J, et al. Controlling Cu–Sn mixing so as to enable higher critical current densities in RRP Nb$_3$Sn wires[J]. Superconductor Science and Technology, 2018, 31(6):064001 (13pp).

[6] Luu T T, Duan A, Knut E. Aasmundtveit. Optimized Cu-Sn Wafer-Level Bonding Using Intermetallic Phase Characterization[J]. Journal of Electronic Materials, 2013, 42(12).

[7] Sylvain Lemettre, Seonho Seok, Nathalie Isac, Johan Moulin and Alain Bosseboeuf. Low temperature solid–liquid interdiffusion wafer and die bonding based on PVD thin Sn/Cu films[J]. Microsystem Technologies, 2017.

[8] S. Fürtauer a, Di L, D Cupid and H. Flandorfer. The Cu–Sn phase diagram, Part I: New experimental results[J]. Intermetallics, 2013, 34:142-147.

Challengeable Mechanical Issues in Microelectronic Packages for Developments

Xiangdong Xue
School of Mechano-Electronic Engineering, Xidian University
Xi'an, China (xiangdong_xue@sina.com)

Jianghao Wei
School of Mechano-Electronic Engineering, Xidian University
Xi'an, China (2424251661@qq.com)

Yuming Wang
School of Mechanical Engineering, Tsinghua University
Beijing, China (wangyuming@tsinghua.edu.cn)

Jianxin Yang
School of Mechanical Engineering, Tsinghua University
Beijing, China (yangjianxin@tsinghua.edu.cn)

Abstract—With the slowdown of the progressive pace of miniaturization, the development of microelectronic systems shows diverse features. This leads to the emergence of new types of packages and challenges, demanding systematic approaches to sort out characteristic issues, increase insightful visions and map out progressive schemes. Focusing on mechanical characterization, this paper identifies mechanical challenges emerged in the current and future progresses of microelectronic packages. Mechanical challenges are firstly identified from the three categories of new issues, thermal interfaces and new structural patterns to highlight the key points. This is then followed by a discussion of mechanical characterization on the aspects of the domination of surface effects, the coupling of material-structural integration and the advent of multiphysical systems to describe the main schemes. Finally, mechanical issues expected in the ongoing future are analyzed through three aspects of stacked 3D structures, in-situ packages and nanomechanics to give a perspective view. Miniaturization is highlighted as the fundamental reason bringing significant changes and issues to mechanics. The interactive couplings on interfaces and between nanomaterial phases are emphasized as the phenomenal issues for mechanical investigations. By highlighting the characteristics and challenges with mechanics, this paper roughly profiles a framework of mechanical issues of microsystems.

Keywords—mechanics challenges, advent developments, perspective issues, mechanical characterization of packages

I. INTRODUCTION

System miniaturization, interdisciplinary integration and intelligent embedment are the three driving forces for the development of microsystems in the current era of information revolution, promoting phenomenal developments of portable devices, smart systems and bioengineering progresses. Among them, miniaturization provides a platform for advent developments, whereas interdisciplinary integration and artificial intelligence make use of the platform for smart developments.

The current pattern of the systems can be traced back to some half a century ago, e.g. the announcement of Moore's law [1,2] by Gordon Moore. Since then, significant changes have occurred on the patterns of microelectronic packages. For a conceptual summary, the historical developments of packages can be categorized into three domains as follows.

1) Structural Patterns: The 2D architecture, i.e. wafer based chips and PCB platform of multilayer, had never been changed, until the emergence of 3D packages [3] by PoP, PiP, TSV, SiP etc. and chip systems of 3D-IC and heterogeneous integration of chiplets in recent years. The emergence of 3D packages and new chip systems has a revolutionary impact in term of breaking out 2D architecture. The appearance of CSP, SOP and SoC prompts new flexible approaches from the conventional pattern where all modules sit on the PCB and thus conceptually opens the packaging way of individual parts for distributed systems.

2) Interconnect Methods: Interconnects may be categorized into three groups as follows, a) through hole interconnects by pins, e.g. DIP and PGA, which may be conceptually developed from traditional concepts of sockets for a solid connection; b) surface mounts on edge lines, e.g. QFP and QFN, which use lines to replace pins for increased I/O density on edges; c) surface mounts by array connections, e.g. BGA, which can be considered as a combination of PGA's array and QFP's surface mounting and offer high-dense interconnects. As possessing high thermal resistance due to the discontinuity of materials on interfaces on the one hand and having to undertake the functionalities of electronics and mechanics on the other hand, interconnects represent a critical link for mechanical behaviors.

3) Cooling Technology: Thermal energy dissipation concerns the two aspects of enduring ambient temperature and transferring internal heat. The historical development of cooling technology showed a stepped progress from external matters to internal measures. Consequently, reducing ambient temperature such as room temperature is the first choice in earlier stages. It was followed by enhancing heat transfer through constructing thermal transmission aisles and then dissipation methods by making use of thermal convections with heat sinks. After improving ambient temperature and enhancing heat transfer met difficulties in satisfying increased requirement on dissipating thermal energy, specific measures were then developed, e.g. phase change materials and embedded microchannel cooling techniques [4].

Under the push of Moore's law, a great deal of work of mechanical aspects has focused on how to satisfy the new requirements of electronic aspect following the huge achievements with miniaturization. As a result, package developments have been pushed by the tasks for ensuring reliability of microelectronics, lacking a consideration from the point of view of mechanics itself to form an integrative scheme. With the decrease of system size in miniaturization progress, there appearing more spaces for the state of the art approaches. Focusing on the need to establish a mechanical scheme, this paper reports a mechanical investigation of microelectronic packages. The challenges faced by mechanics are firstly identified, before an analysis of the mechanical characterization of packages, which are then followed by a perspective prediction of some development trends and tendencies.

978-1-6654-1392-3/21 $31.00 © 2021 IEEE

II. MECHANICAL CHALLENGES

A. Challenge 1: New Issues with New Patterns

The new issues emerged in the technique aspect are related to manufacturing techniques, multidisciplinary approaches and new structural patterns. On the aspect of manufacturing techniques, as the fabrication techniques of chips and packages differ from traditional mechanical approaches, a great number of traditional experiences and methodologies cannot be directly applied. This causes some uncertainties on reliability performances. On the aspect of multidisciplinary systems, system functionality, component fabrication and mechanical reliability conventionally belong to different disciplines. This leads to mechanical side which locates at the end process often having to accept whatever being delivered from previous processes by different disciplines, with limited spaces for systematic improvements. As a result, mechanical services have to satisfy various requirements coming from different fields, trying to solve a number of interdisciplinary issues in the end stage. On the aspect of structural patterns, as emerging structures are unfamiliar topics with experiences, confidence is lacked in dealing with. Under this situation, a common practice is using existing technologies or experiences to treat the new things through the expansion of knowledge and skills. This tends to cause some mistakes or inefficiencies before the rules and principles are truly understood.

B. Challenge 2: Thermal Interfaces

Continuum assumptions encounter interfacial challenges when system dimensions are shrunk to microscope. Three challengeable consequences are generated as follows. i) Mechanical behaviors appear a break on interfaces due to the discontinuity. Here mechanically continuum conditions refer to the two aspects of geometry and materials. On a thermal interface, both geometry and materials are different, which generate a number of problems on the cooperation of microstructures of materials and surfaces. It becomes very difficult to let different things show harmonious behaviors like a natural material. ii) Miniaturization leads to the domination of surface effects over volume effects, as the surface area to volume ratio is proportional to dimension sizes. This leads thermal interfaces to become a crucial topic in microelectronic packages. iii) Poor quality of surface fabrication in microscopes leads to low quality of interfacial fitting, resulting in high thermal resistance on thermal interfaces and the leakage of fluids in assembled microfluidic systems. A mitigating method is to use soft materials or films to fill interfacial gaps, which however increases the number of interfacial materials to three. A challenging issue on this topic is how to deal with the encamped air bubbles which increase thermal resistance but cannot be shrunk by pressure due to incompressible feature of air. The situation of three or four materials gathers in the poor fabricated interfaces aggregates the discontinuity state of interfaces, deteriorating the condition of heat transfer.

C. Challenge 3: Mechanics of Emerging Package Systems

A number of packaging patterns are emerging, such as 3D packages for stacked structures, chip system packages for mobile applications, embedded microchannel systems for cooling, fan-out packages in wafer level and chiplet assembly by heterogeneous integration. Among various advances of new packages, 3D packages and distributed packages are closely relevant to mechanics, representing a new type of platform for advancing developments. The reason for the emergence of 3D packages is due to the incapability of 2D. With the advent of microelectronic systems, the burden for applying new functionalities on conventional 2D plenary PCB systems is close to the limit. On the one hand, 2D plenary pattern is difficult to handle the increase of new system parts without scarifying performances, leading to the appearance 3D packages. On the other hand, the integrated all-in-one packages cannot be used in distributed locations for satisfying the in-situ monitoring and controlling functions for intelligent systems such as IoT systems, leading to the expectation for distributed packaging systems with specific functionalities. The above 3D packages for high efficient stacked systems and distribution packages for a dismantlement of integrated systems offer diverse progresses for microelectronic systems but also bring new types of challenges to mechanics. In this case, mechanics is required to interpret the principles, safeguard the behavior of stacked 3D layers and to investigate the environmental conditions for progressing distributed packages.

III. MECHANICAL CHARACTERIZATION OF MICROELECTRONIC PACKAGES

At present, the advances on microsystems are significant due to the change of driving forces from the one-horse race of Moore's law to a couple of paralleled key technologies. Relevant mechanical issues with immerged phenomena are hereby characterized into three types as follows.

A. Characteristic 1: Dominative Role of Surface Effects

The influence of miniaturization on mechanics is symbolized by surface-area-to-volume ratio. As surface-area-to-volume ratio is proportional to the decrease rate of dimensions, surface effect relatively increases, for example, by one thousand and one million times, respectively, when dimension sizes decrease from millimeters to microns and nanometers. This is evidenced that, for miniaturized systems in the dimension of micron or nanometers, surface effects become dominative over volumetric effects. Unlike volumetric effects governed by gravitational field and showing concentrative behaviors, surface effects are sourced by a number of interfacial factors which, to get things worse, affect with each other. Therefore interactions become a new symbolic feature of mechanics for miniaturized systems.

From a mechanical point of view, the difference between volumetric effect and surface effect comes from the force types and the integrity status of solids, discussed as follows.

- In respect of force types, mechanically volumetric effect is formed by external fields in which inertia is the major behavior. As both body forces and inertia effect due to external fields are relevant to masses, volumetric effect shows convergent behaviors and can be simply expressed by concentrated parameters at the gravitational center of the volume. As a result, the particle-based expression of Newton's laws is suitable to volumetric and inertial effects. By contrast, generated by interactive forces between surfaces, surface effect mainly comes from responsive results rather than a dominative resource. As a result, distributed expressions are required for describing surface or interface effects.

- In respect of integrity of solids, when a solid is subjected to external forces, the solid shows integrative consequences. Constrained by materials and the structure, a solid can behave as an entire object to show integrative external performances and, at the same time, distribute the forces into its body for internal interactions. The response for external performances refers to dynamic behaviors which can be simplified to a particle or a rigid body in many cases. The response for internal performances refers to stress-strain behaviors which requires to be dealt with when stresses reach damageable levels. By contrast, surface effects by interfacial interactions do not have internal constraint ability like volumes and have to take into account relative motions between surfaces. Relevant descriptions are more complex than volumetric effects.

In a summary, unlike the convergent characteristics of volumetric effect, surface effect shows diverse characteristics, leading to an increase of complexity when surface effect becomes dominative. As mechanically miniaturized systems and macro systems are respectively governed by surface effects and volume effects, different emphases or rules are required for accurate descriptions of mechanical behaviors. For microsystems, mechanical behaviors mainly show interactive features rather than integrative behaviors like in macro systems.

B. Characteristic 2: Structure Integrity with Materials

With the dominative role of surface effect endorsed by miniaturization, interfacial interactions become a phenomenal state of mechanics in microsystems. After characteristic 1 discussed above on interfacial interactions in microsystems, characteristic 2 focuses on the consequence of interfacial interactions on system behaviors. The increase of interfacial interactions in microscope brings in new features of mechanics, requiring micro-macro integration between materials and structures.

Traditionally materials and structures are considered in different disciplines. Materials provide the basic elements required for constructing a system, whereas structures use materials to make the state of the art process for building a system. In this pattern, the purpose of materials is to provide high quality of properties and thus the work of materials concentrates on a scrutinizing search of natural materials for high performances [5]. The task of a structure is how to effectively use materials to achieve the state of the art effectiveness. A key objective is to enhance the integrity of the structure for which continuous framework is pursed and a defect or weakness at local region is often considered as a potential damage for overcoming. This situation has been changed in microsystems. Due to the increased dominative role of surfaces and interfaces, weaknesses at local regions become a common phenomenon. It becomes very difficult to remove all the weaknesses generated by interfaces or other factors, creating a challengeable issue to traditional mechanics. Consequently the emphasis should be shifted to how to reduce the influence of local weaknesses and how to make use of interactions to achieve cooperative effectiveness. The relevant work requires approaches of material-structural integrity to achieve a perfection of systematic performances.

Some concepts of structural integrity with materials for earlier stage practices can be found from composite materials, smart materials and nanostructured materials. Composite materials open a way of making use of structural techniques to enhance material properties. By combining different types of materials together, i.e. metals, ceramics and polymers, to construct manmade materials for integrative properties, the state of the art of structures is embedded into composite materials. Smart materials refer to those materials behaving under different physics or showing different statuses in varying conditions. By illustrating different physical properties or flexible states corresponding to defined conditions, smart materials offer multidisciplinary effectiveness with natural ways which often show enhanced characteristic behaviors better than structural approach of individual materials. Nanostructured materials inspire a number of imaginations with superior physical properties and are actually providing a platform for universal potential applications. The marvelous features of nanostructured materials come from two aspects. One is their outstanding properties which demonstrate how structural arts have been naturally embedded into nanomaterials by nature, demonstrating a natural course for material-structure integrity. The other is that the structure of a nanomaterial is in the size of nanometers and thus shows material behaviors like natural materials without the kind of defects existed in macro dimensions. It is worth noticing that nanostructured materials possess very high geometric aspect ratio and hence require material interactions, e.g. nanocomposite materials, for applications. Composites, smart materials and nanomaterials have shined a light on the potentials of material-structural integrity. This is a promising research direction, requiring mechanical support for comprehensive developments.

With miniaturization, interfacial interactions become a key issue for microsystems, including packages, such as solder joints and thermal interfaces. Under the framework of material-structural integrity, the approaches on interfacial problems are not to eliminate the problems but alleviating the problems through a smart way of interactions.

C. Characteristic 3: Multiphysics and Manufacturing

Microsystems refer to a kind of systems shaped in miniature dimension and performing electronics and/or other disciplinary functionality, microelectronic systems hereby. Microsystems are built on mechanical foundation by means of manufacturing processes or fabricating techniques. For a reliable performance, the packages need to deal with the thermal power generated by the systems and run well in harsh environments subjected to temperatures, vibrations and impacts loads. Consequently the quality of a microsystem depends on two stages of situations. In its birth stage, a microsystem requires the harmonization of multiphysics between the functional performance with electronics and/or other disciplines involved and the body behavior with mechanics and fabrication techniques. In its service stage, microsystems are required to survive and keep reliable in the harsh conditions of temperature changes, vibration oscillations and impact scenarios. From a mechanical point of view, two types of coupling or interactive issues of multiphysics need to be taken into account. One is the disciplinary coupling of mechanics with electronics and also fabrications. The other is the loading coupling of thermal loads with vibration and impact loads. Here discusses the disciplinary coupling between mechanics and fabrications and the interactive behavior of mechanics among thermal, vibration and impact loading conditions.

Heat transfer is essentially a transformation between thermal energy and mechanical work, no matter the thermal energy generated internally or applied externally. Thermal energy is converted to kinetic energy of free electrons for transportation in conductor materials, e.g. metals, and transformed to potential energy of a series of atomic lattice deflections for wave propagation in nonmetallic materials, e.g. ceramics. As both ways of energy transportations are in microscopes, heat flux is little affected by sizes but transfer is heavily affected by interfaces and other discontinuities. Thermal loads often produce more severe consequences or damages than vibration and impact on packages of microsystems due to their miniaturized feature. The key for heat transfer is the quality condition of traveling paths or aisles. However, unlike in macro scope, the transport aisle in miniaturized systems is heavily affected by interfacial effect which increases with a decrease of system sizes. On this situation, the construction of heat transfer aisles is significant important for efficient dissipation of thermal energy. This desires independent designs from mechanical point of view, if the electronic system cannot provide an integrative aisle for heat transfer. Hence, whether a packaging design could base on mechanical principles or can only rely on the existing conditions offered by electronic systems becomes a crucial philosophical issue for packaging developments. Unlike in earlier stages where plenty of options existing due to low thermal flux, in the current stage constructing thermal aisles encounter serious resistance after conventional options have been extensively made use of.

Unlike heat transfer as an internal issue concerning material details, vibration is an external issue of the structure and concerns the entire state of the structure. When external forces or accelerations are applied on a structure, the responsive feature of the structure depends on the correlation between internal responses as a whole and the external loads, i.e. resonances when correlated on natural frequencies. The difference between vibration and thermal loads is thus related to the difference between structure-based matter and material-based matter. As a structure-based matter, the state of vibrations of a system is closely related to the size and mass of the system. This means that vibration damages tend to decrease with a decrease of system dimensions, which leads vibration to become not as crucial as thermal damage in microsystems. Impact is a phenomenon of the combination of external and internal actions. On the one hand, an impact case often originates from a local region contacted, resulting in stress wave propagation from the impact point. On the other hand, an impact load is always applied on a structure or a system from external environments. This tends to produce an entire response of the system as soon as the impact is developed. As possessing both internal and external features, impact damages need to be considered from both material and structure angles. For loading coupling, when a thermal damage achieves the critical point of the structure and when vibration or impact damage causes stress-strain concentration at a local region, a coupling is formed. Therefore, although thermal and vibration/impact do not often couple in physics due to belonging to internal loads and external loads, respectively, these three loads can however deteriorate the strength basis of other damages. For example, a plastic region formed in vibrations could increase propagation rate of thermal fatigue progression; a harden region due to impact loads may change the structural architecture of microsystems and decrease the ability of the region; a thermal damage generated in heat shocks may become the origination of fatigue damage of vibrations and impacts.

Manufacturing techniques and methods have been largely changed in miniaturized systems. One reason is that traditional machining processes lack techniques which are directly suitable to miniaturized systems. The other reason is that current miniaturized systems are usually multidisciplinary combinations where conventional machining methods designed for pure materials meet difficulties. These two reasons lead to the poor manufacturing quality of micro systems relative to their counterparts in macro scale. The effect of low-quality of manufacturing techniques is largely exaggerated with the increased surface effect roles due to miniaturization. Corresponding to the addictive ways of manufacturing in miniaturized systems, fabrication is more suitable for describing the manufacturing processes. From a mechanical point of view, the low-quality reason caused by micro fabrication comes from the discontinuity of interfaces. Due to low-quality of surfaces and the lack of processing techniques, two surfaces cannot fit well on an interface, causing both material and geometrical discontinuity. As aforementioned in section II, supplementing a soft thermal material on the interface could improve geometrical continuity of the interface but this increases the discontinuity of materials. In addition, some air-bubbles tend to encamp at the valleys of surface roughness, bringing in the fourth thermal isolated material on the interface. Containing four different materials and by nature being a geometric breaking gap, interfaces represent a challengeable problem in thermal dissipation and require a special attention from mechanics.

IV. PROGRESSIVE PERSPECTIVE OF ONGOING DEVELOPMENT

Among the three driving forces for mechanical advent of microelectronic packages, interconnect was the first driving force in history for pushing the development of packages. It was not until interconnects facing big challenges, structure then became the major driving force for promoting packages developing further. Cooling techniques have been historically an associated factor embedded in the advent of interconnects and structures, but become another driving force when interconnect met its limit. This led to the creation of parallel cooling methods such as embedded microchannel systems to alleviate the situation. It is worth noticing that a solution to a challengeable problem generally needs to experience three stages, i.e. facing the problem, diagnosing the problem and solving the problem. A problem wide influences needs to be faced if various alternative routes become not so effective. The problem of packages could be diagnosed from different angles, in which mechanical viewpoint is a basic and crucial one. In respect of driving forces, both structural patterns and cooling techniques have contained two folds of the subjects of challenges or opportunities for ongoing developments, categorized into three progressive tendencies as follows.

A. Tendency 1: Stacked System of 3D Packages

Multilayered lamination has been a dominative pattern in microelectronic packages. This reflects the requirement of the structural pattern of chips. For thin slice shaped chips, i.e. a 2D architecture, multilayer is a natural way for building system. As a result, multilayered thin-plates become the standard pattern of microelectronic systems. However, this pattern has some contradictions to system requirements. As fundamentally all systems are 3D, the 2D chip slices have to be converted to 3D in some dimension, either in component level or system level. From a systematic point of view, 2D systems possess much lower efficient than 3D, leading to low efficiency in system level when microelectronic systems become complex. From a mechanical point of view, this creates a dilemma situation. For the sake of thin-sliced 2D chips, the multilayered lamination structure has worked well, leading to historical continuous enhancements of this pattern.

978-1-6654-1392-3/21 $31.00 © 2021 IEEE

For the sake of system efficiency, the 2D structural pattern is encountering challenges to satisfy the increased requirement of microelectronic systems. This contradiction was not so serious in earlier development stages. When microsystems are well developed, there is a need to coordinate the problem. As a result, 3D systems emerged to solve this contradictive matter in microsystems.

When 3D systems are desired for dealing with the challenges faced by 2D systems, designers find some difficulties. After historically concentrated developments of 2D systems for several decades, all the existing matured technologies and techniques are coped with the requirements of 2D systems. This leads to two routes in developments. One is continuously developing new technologies and techniques to make 2D systems satisfying increased new requirements. The other is to develop 3D systems based on current existence technologies and techniques to gain improvements with not-so-much changes of the current. Driven by the latter, 3D approaches on system level, e.g. PiP, PoP, TSV, are developed. One of the reasons for current 3D packages concentrated on system levels may come from the convenience by directly using current technologies and techniques based on PCB pattern. It should be noticed however that, based on bottom-up efficiencies, the transformation to 3D from module level or component level tend to offer more effective than from system level. By then, 3D approaches through integrative processes on system, module and component levels could result in true 3D developments with significant improvements.

The current 3D packages based on stacking 2D packages bring out new issues for mechanics, discussed as follows. i) Shearing damages: With an increase of system layers in the current practices, shear deformations and relevant forces could achieve a serious state, if the stacked layers are increased to multiple levels; ii) The change of status of bending deformations: Unlike 2D systems with independent single levels, bending deformations on the entire 3D systems are difficult to appear due to interactive constraints among the layers stacked, leading to an increase of the bending stiffness of 3D systems; iii) Thermal dissipation: The traditional two-way heat transfer for 2D systems encounters difficulties in the stacked multilayer 3D systems.

The new development of 3D systems is partly triggered by the emergences of multidisciplinary approaches. It is generally an obvious challenge for adding some new disciplinary parts into well-developed systems, for which existing patterns have to be redesigned. In this case, stacking a new layer is an attractive idea. This looks like a PCB being divided into two or a few stacked parts, if the new function is supplemented in system level. However, the requirement for supplementing new modules or combining different modules is also a desire by chip level. For the purpose to flexibly enroll different chips onto a device, heterogeneous integration approaches like chiplets are developed. An interesting thing is the return of the interconnect issue with chiplets which was once the focus before structural approach, as aforementioned in the point 2 of Introduction.

B. Tendency 2: In-situ Systems, Distributive Packages and Parallel Cooling Techniques

Though respectively belonging to the domains of package structure and cooling technique, both distributive packages and parallel cooling techniques illustrate the same feature of in-situ processing: taking a part of functions of integrative systems and dispatched onto critical regions for a convenience of control and performance of specific techniques. Distributive packages and parallel cooling techniques thus refer to a new approach for distributing smartness and intelligence for system controlling.

Distributive packages can be used for performing in-situ sensing-testing and treatment of smart devices on specific functions. This can certainly shorten the transport distance of messages and increase the integration and smartness of the device as a whole. Taking an engine for example, unlike conventional systems to firstly send monitored information to central control unit and then to make treatments based on the feedback commands from the unit, the in-situ package possesses relevant simple smart functions for monitoring-collecting information and making treatments on site. This kind of distributive system packages expects to be the must-do technologies for serving the ongoing requirements of IoT and intelligent biomedical systems. The difference of the future distributive packages from the current conventional packages is somewhat like the difference of human beings from the current robots. Human bodies possess distributed intelligent systems in organs and limbs, apart from the powerful brain. This constitutes a two-level sensing-controlling system and makes human beings responses quickly and efficiently. By contrast, as a type of one-level controlling systems, current robots possess a central processor for overall responses. This one-level control system lacks some flexibility and smartness compared to two-level control system. Distributive systems provide the essential basis for two level control systems.

Parallel systems offer an approach of building multiple channel routes for an increase of transport speeds or a combination of different technologies or techniques. Due to the increase of channels, compared with a series-linked system, a parallel system possesses relatively high reliability. In principles, a system with a series of links has not possessed a redundancy, leading to any irregularity or accident having a potential for seriously retarding even breaking down the whole route. In addition, when a heat spot of a power module locates at the center of a package, the thermal energy has to pass all the way of a series of conventional links before arriving on boundaries, making thermal pollution along all the way of distance. By contrast, a parallel system can distribute the loads onto two or more routes based on the fluent and carrying methods, achieving a reliable and efficient system. Embedded microchannel cooling systems are a parallel approach. As the thermal energy generated at hot spots can be rapidly delivered through fluidic route, the efficiency is largely improved and the pollution of thermal energy on the way is removed. Mechanically this refers to a combination of heat conduction through solid paths and heat convection through the liquid path, changing the single cooling pattern of conduction or convection to the double cooling routes of conduction plus convection.

The emerged distributed systems and parallel systems indicate the advent of a new pattern in microelectronic packages. Apart from pursuing perfect integrated systems like the data center or cloud services, distributive and parallel systems make packages based on suitability and specificity. They can effectively be embedded into the physical systems served for. By this new pattern, packages become two types, conventional standalone systems for complex integrative functions and relative simple distributive packages for specific functions. As such, the future packages could be a professional system or/and a part of any smart systems.

C. Tendency 3: Nanomechanics and Discrete Description

Nanostructured materials appeared [6,7] at a crucial time when miniaturization progressed to micro- and nano-scopes and after natural materials had been extensively searched and intensively explored. Nanometerials have hence substantially filled the gap left by the exhaustion in searching natural materials and provided a strong support on nanotechnology and applications. Among various nanomaterials, carbon nanostructured materials, e.g. graphene and carbon nanotubes, have attracted great attentions on researchers and developers with their outstanding properties and extraordinary potentials in the areas of mechanics, electronics, thermal, chemistry, optics, etc [8]. Mechanically when the size reaches to a few nanometers a characteristic problem is how to describe the discrete state and atomic behaviors. It can be therefore summarized that miniaturization has brought two types of challengeable topics to mechanics. One is discrete description, progressed from continuous description, and the other is surface effects domination, replacing volumetric effects. Both topics concern the disruptive consequences following miniaturization but in different dimensional scales.

The development of nanostructured materials can be classified into three stages as follows, i) Scientific stage: focusing on the outstanding behaviors of nanostructured materials and the scientific background; ii) Technological stage: working on the approaches for effectively making use of the properties explored in scientific stage, in which nanocomposites with nanostructures as the filler phase are a major direction; iii) Engineering stage: expanding the achievements and realizations obtained in scientific and technological stages to develop engineering systems. The work in the first a couple of years since graphene was separated in laboratory on 2004 [7] concentrated on science stage. Many research focuses were then shifted to technology stage to work on nanostructured composites in the last several years. The stage of engineering is expected to arrive when the key issues in technology and science stages are overcome.

As nanostructures are related to lattice structures and atomic behaviors, mechanical description for discrete state subjected to atomic forces is required. As nanostructured composites concern the interactions of nanostructures with matrix materials in the scope of nanometers, surface effects and interface consequences are characteristic physical phenomena required for interpretations. It is worth noticing that, in the current information era, science and technology are often in an interactive status with mutual dependences and supports. Similarly, the correlation between mechanics and other disciplines is also required in an interactive pattern for interdisciplinary developments.

V. CONCLUDING REMARKS

Mechanics of microelectronic packages is investigated from the two angles of the current situations and the future expectations. Problematic issues and relevant challenges are focused for a hope to inspire further considerations and rethinks. Based on the three challenges, the three characteristics and the three perspective issues discussed with mechanics, the following points are summarily remarked.

Miniaturization is the fundamental cause for various challenges and issues emerged. On the one hand, following Moore's law, miniaturization has pushed microsystems developed over five decades at an astonishing speed, creating a golden age for microelectronic systems. On the other hand, the pace dominated by one horse has left a lot of issues required for reviewing and clarifying. A significant matter left is to enhance mechanical schemes for paralleling development along with its counterpart electronics. On this matter, surface effects and interfacial interactions become a significant factor due to the increased domination of surface effects in microsystems. With the expected development beyond Moore's law, an increase of diverse directions is undergoing, bringing in an opportunity for mechanics to set up a basis for more balanced paces in developments.

Structures and materials are the eternal theme of mechanical systems. Structures refer to a demonstration of the beauty of natural rhyme obtained through the art of design, in which mechanical equilibrium and heat fluently flowing are the two signs. After extensive efforts and achievements on 2D plenary constructions, 3D packages are emerging and desiring new patterns of the state of the art from mechanics. Materials refer to a never-ending search for better properties from nature. When the search seems close to a dead end, nanomaterials and smart materials step in to give materials a promotion. The potentials with nanostructured materials are still not so clear, which is often interpreted as a sign that the outcomes are too glorious to be easily understood. For microsystems, material-structural integration provides a way for clarifying the track of thoughts.

Parallel pattern is a concept with potential impacts on thermal management. Parallel pattern hereby refers to a fast route supplemented alongside the standard one, so that heavy heat transfer does not need to stop at every section. The new development item on this aspect is the embedded microchannel cooling systems. Apart from the fast transportation function, this also achieves a multiphysics approach for the parallel actions between heat conduction and heat convection. On this topic, nanostructured materials have the potential to provide another parallel transport of thermal energy by creating a fast transport route in solid materials alongside conventional ways.

REFERENCES

[1] G. E. Moore. "Cramming more components onto integrated circuits." Electronics Magazine, vol 38(8), April 19, 1965.

[2] G. E. Moore. "Progress in digital integrated electronics." International Electron Devices Meeting, IEEE, 1975, pp.11-13.

[3] A. Lancaster, M. Keswani. "Integrated circuit packaging review with an emphasis on 3D packaging." Integration, vol 60, pp 204-212, 2018.

[4] S. M. SohelMurshed, C. A. Nieto de Castro. "A critical review of traditional and emerging techniques and fluids for electronics cooling." Renewable and Sustainable Energy Review, vol 78, pp 821-833, 2017.

[5] A.L. Moore, L. Shi. "Emerging challenges and materials for thermal management of electronics." Materials Today, vol 17(4); pp.163-174, 2014.

[6] S. Iijima. "Helical microtubules of graphitic carbon." Nature vol 354, pp.56-58, 1991.

[7] K.S. Novoselov, A.K. Geim, S.V. Morozov, D. Jiang, Y. Zhang, S.V. Dubonos, I.V. Grigorieva, A.A. Firsov. "Electric field effect in atomically thin carbon films." Science, vol 306, pp.666–9, 2004.

[8] A.A. Balandin. "Thermal properties of graphene and nanostructured carbon materials.". Natural Materials, vol 10, pp.569-581, 2011.

The interfacial reaction of Cu/multilayer Sn -Cu -Sn / Cu joint in soldering

1st Min Shang
Shools of Matreials Science
Dalian University of Technology
DaLian,China
shangxiaomin2019@mail.dlut.edu.cn

2nd Chong Dong
Shools of Matreials Science
Dalian University of Technology
DaLian,China
dongchongwyy@gmail.com

3rd XiangXu Chen
Shools of Materials Science
Dalian University of Technology
DaLian,China
xiangxuchen@mail.dlut.edu.cn

4th ShaoCheng,Wu
Shools of Materials Science
DaLian University of Techonlog
DaLian,China
706334339@mail.dlut.edu.cn

5th HaoRan,Ma*
Shools of Materials Science
DaLian University of Technology
DaLian,China
mhr@dlut.edu.cn

6th HaiTao,Ma**
Schools of Materials Science
DaLian University of Technology
DaLian,China
htma@dlut.edu.cn

Abstract—In this paper, multilayer interfacial linear solder joints were prepared by dip soldering. The effects of soldering temperature, soldering time, soldering distance and soldering layer number on IMC growth of Cu / multilayer Sn-Cu-Sn / Cu solder joints were studied. The linear solder joint of multilayer interface was obtained by dip soldering at 250 °C for 30s. And then reflowing at 250℃, 300℃, 320℃ for 30s, 60s, 120s, 5min, 10min, 30min. EPMA results show that there are Cu_3Sn and Cu_6Sn_5 phases at the interface of each layer between solder and Cu substrates. At the same time, both reflow temperature and reflow time have effect on the interfacial compound, and the size of the interfacial compound increases with the increase of reflow temperature; With the increase of reflow time, the size of interfacial compound also increases; While decreases with the increase of dip soldering distance, the IMC of the sample with larger distance is thicker than small distance.

Keywords—Dip Soldering; Reflow; Interfacial Reaction

I. INTRODUCTION

In the rapidly changing information age, aerospace, navigation, communication and other technologies have been developed rapidly. The development of these fields is closely related to electronic communication technology. The development of electronic communication technology is inseparable from the support of electronic packaging technology. Electronic packaging technology is to realize the function of various electronic components through interconnection technology. At present, soldering technology is mainly used in the interconnection of electronic components, and Sn / Cu solder joint has become the first choice of contemporary electronic packaging due to its good mechanical, thermal and electrical properties. The essence of Sn / Cu soldering interconnection is that molten Sn solder reacts with substrate Cu to form intermetallic compound IMC, which mainly includes Cu_6Sn_5 and Cu_3Sn phases [1-2]. In the development of packaging technology, Sn-Pb solder is widely used in electronic packaging due to its good physical, chemical and other mechanical properties, which ensure the soldering quality and reliability [3]. However, since 1991, many countries have issued laws and regulations to restrict and prohibit the application of Pb and its compounds in electronic products [4-5].

Therefore, many scholars found that Sn-Ag is one of the most comprehensive lead-free solders. Because the Sn-Ag system is relatively simple and the reaction products form metallurgical bonding. In recent decades, people have a deep understanding of the basic process of interfacial reaction of soldering, the formation and growth of Cu_6Sn_5 and Cu_3Sn [6-9]. There are two soldering methods involved in this experiment. One is dip soldering[10], which is used to prepare multilayer Sn - Cu - Sn structure samples. The other is reflow soldering [11](the prepared multilayer Sn - Cu - Sn structure sample is heated to higher than the melting temperature of solder (250 °C, 300 °C, 320 °C) by constant temperature aging heating platform). Both dip soldering and reflow are two common methods in soldering[12-13], which are easy to achieve for us.

In this paper, the samples of multi-layer Cu-Sn3Ag structure linear solder joints were obtained by dip soldering and then reflow with different temperature and time, finally the microstructure evolution of IMC layer in Sn / Cu system was further studied by every parameters of experimental design.

II. EXPERIMENTAL PROCEDURES

In the experiment, An3Ag solder was melted in a tin furnace at a temperature higher than its melting point, and then the Cu / multilayer Sn Cu Sn / Cu samples were soaked at 250 °C for 30 s. We use dip soldering twice. A steel wire with a diameter of 100 μ m is tied to the middle Cu sheet, and ceramic sheets with a thickness of 650 μ m are spaced on both sides, as shown in Fig. 1 (a), for the first immersion soldering. After that, the thick ceramic pieces were removed and the second immersion welding was carried out to obtain the immersion soldering sample with 100 micron spacing of multi-layer Sn / Cu structure. The purpose of dip welding twice is to get the sample filled with uniform solder. After water cooling and drying, another dip soldering was carried out at the same temperature and time. The purpose of this process is to obtain a sample uniformly filled with Sn-3.0Ag solder. During sample preparation, steel wire (100μm) and ceramic plate (650μm) were used to control the spacing. After the first dip soldering the thicker ceramic pieces are removed and then continue the second dip soldering. The structure of the sample is shown in Fig.2. The soldered specimen was cut, polished, and then reflow with a 100μm ceramic chip. Finally, the joints were polished with 1.5mm and 0.5mm diamond polishing paste in turn, and then etched with the prepared etching solution to show the microstructure.

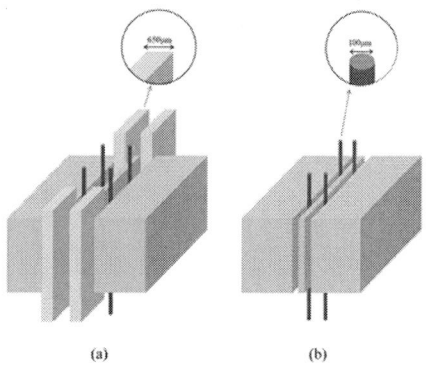

Fig.1. Dip soldering process of samples

a:the first dip soldering; b:the second dip soldering

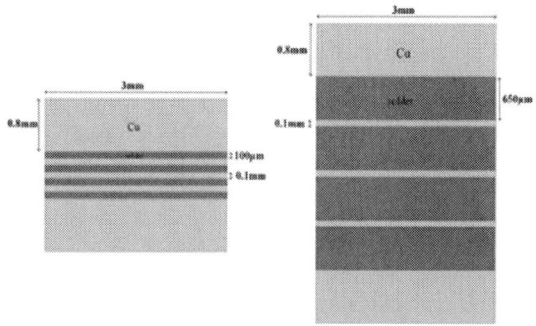

Fig.2 dipping sample and its cross section size structure after cutting.

III. RESULTS AND DISCUSSION

A. *Effect of reflow process on IMC at the interface of dip soldering samples*

Through electron probe microanalysis (EPMA), in the selected reflow process, Cu_6Sn_5 and Cu_3Sn intermetallic compound layers (IMC) were observed between each layer of Sn3Ag-Cu, as shown in Fig. 3. The size and thickness of the metal compound phase at the interface are determined by the reflow temperature and time. It is found that Kirkendall hole is not found at the interfacial area of our sample.

The results show that the reflowing temperature and time can promote the growth of intermetallic compounds, as shown in Fig. 3 (C). In both cases, the increase of the intermetallic compound layer is owing to the diffusion growth of the intermetallic compounds in the experiment.

Cu diffusion drives the growth process during the soldering reaction. Cu atoms in Cu substrate diffuse into the substrate through the end of scalloped Cu_6Sn_5, and then into liquid Sn. The end of scallop is the grain boundary separating individual scallops. Because the liquid diffuses rapidly, a large amount of copper diffuses to the top of the fan-shaped or dendritic crystal, where a quickly growing plane appears. The increase of soldering temperature increases the solubility of Cu in Sn. With the increase of temperature, more Cu atoms diffuse into liquid Sn and react with Cu_6Sn_5 to form Cu_3Sn. Therefore, the growth rate of Cu_3Sn phase increases with the increase of temperature. The solubility of Cu in Sn increases

with the increase of soldering temperature, and the thickness of IMC layer increases.

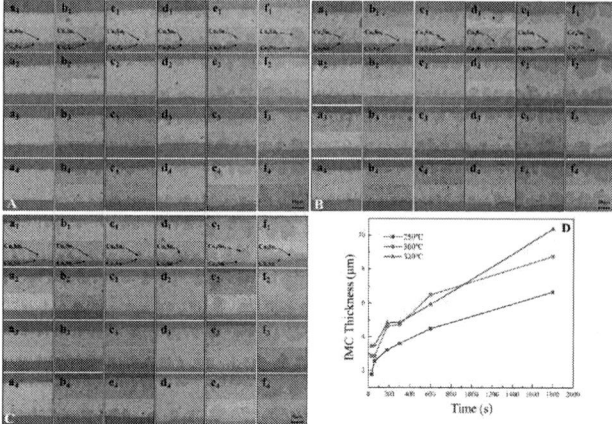

Fig.3. The interfacial morphology of four-layer Sn / Cu with 100μ m after reflowing. A:250°C; B:300; C:320 °C

(a_1-a_4: reflowing for 30s, the cross section of 8 interfaces of 4-layers in the sample is the same as the others; b_1-b_4: reflowing for 1 min;c_1-c_4: reflowing for 3 min; d_1-d_4: reflow for 5min; e_1-e_4: reflowing for 10 min; f_1-f_4: reflowing for 30min)

B. *Influence of different distance of dip soldering samples on intrefacial IMC during reflowing*

Electronic products have the trend of convenience and miniaturization, and the volume of solder joint will decrease with the diameter to the third power [14]. This will lead to the increase of IMC grain growth and thickness, which will increase the volume percentage of the whole solder joint. There is no doubt that this will lead to a strong solder joint reliability problem. Size effect has become a new basic scientific problem in interface reaction of lead-free interconnect manufacturing. For Cu matrix, the size effect means that when the size of solder joint decreases, the IMC grain size becomes coarser and the thickness increases, while the consumption of Cu matrix decreases [15-16]. Found in our research, the IMC growth thickness increases with the decrease of the dip distance (Fig.4.). With the decrease of solder joint spacing, the diffusion distance of atoms to the opposite side is greatly reduced. It leads to more significant interaction, which seriously changes the morphology and size of the compounds on both sides of the interface, and refines the IMC grains on both sides of the interface. Finally, the phenomenon of small spacing and large grain size appears. The coarsening of IMC grain size in the interface will reduce the area occupied by the grain boundary and the Cu flux from the Cu matrix to the interface. Previous studies have shown that the growth rate of IMC is positively correlated with copper concentration [16]. For the samples with smaller distance, the copper concentration in solder increases faster than that of the samples with larger distance. Because the smaller distance is affected by the higher copper concentration for a long time, the growth rate of IMC is faster

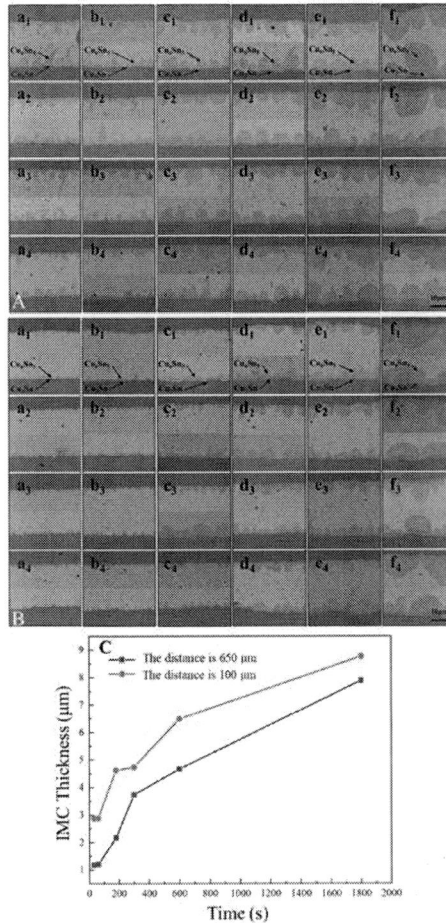

Fig.4. Interfacial morphology of samples with 100 μ m and 650 μ m distance reflowed at 300 °C. A:100μ m; B:650μ m; C: growth rate.

C. Effect on different number Sn/Cu layers on intrefacial IMC during reflowing Units

In our study, it is found that the size of IMC is negatively correlated with the number of layers of the sample. At the same temperature and time, the size of IMC generated with four layers is larger than that of eight layers (Fig.5.). Due to the large number of layers and the large size of the sample with eight layers Sn/Cu structure, there is a certain temperature difference between the bottom and top during the reflow process, which leads to this result.

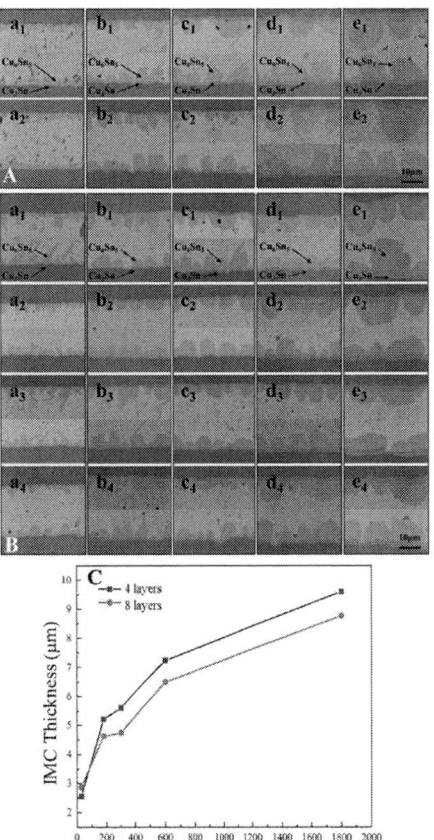

Fig.5. The interfacial morphology of dip soldering samples with different layers at 300°C.A:4 layers; B:8 layers.

ACKNOWLEDGMENT

This work was supported by the National Science Foundation of China (Grant Nos.51871040).

REFERENCES

[1] Oshaghi S . 2008.

[2] Hu X , Ke Z . Journal of Materials Science Materials in Electronics, 2014, 25(2):936-945.

[3] Ezaki H , Nambu T , Ninomiya R , et al.[J] Journal of Materials Science Materials in Electronics, 2002, 13(5):269-272.

[4] Suh J . [D]. 2006.

[5] Zhang Z , Cao H , Li M . [C]// International Conference on Electronic Packaging Technology. IEEE, 2013.

[6] Yu D Q , Duan L L , Zhao J , et al. [J]. Material Science and Technology, 2005, 13(5):532-536.

[7] Ke C B , Zhou M B , Zhang X P . [C]// International Conference on Electronic Packaging Technology. IEEE, 2014.

[8] Ma H R , Dong C , Priyanka P , et al. [J]. Materials Characterization, 2020:110449.

[9] Gagliano R A , Fine M E . [J]. Journal of Electronic Materials, 2003, 32(12):1441-1447.

[10] Hu X , Ke Z . [J]. Journal of Materials Science Materials in Electronics, 2014, 25(2):936-945.

[11] Liu C M, CE Ho, Chen W T , et al. [J]. Journal of Electronic Materials, 2001, 30(9):1152-1156.

[12] Carroll W P , Kondrat D S . [J]. US, 1939.

[13] Lee N J , Oh S S , Kim H S , et al. [J]. 한국초전도저온공학회논문지, 2010, 12(2).

[14] Abtew M . [J]. Dissertations & Theses - Gradworks, 2010.

[15] Lai Z , Wang J . [J]. Journal of Jiangsu University of ence and Technology(Natural ence Edition), 2011.

[16] Liu P S , Liu Y H , Yang L L , et al. [C]// 17th–18th Annual Conference and 6th–7th International Conference of the Chinese Society of Micro-Nano Technology. 2018.

Study on the Influence of Different Filler Fractions on the Properties of Thermal Interface Materials

1st Wenbo Ye[1,2]
1 Shenzhen Institute of Advanced Electronic Materials
Shenzhen Institutes of Advanced Technology, Chinese Academy of Sciences
Shenzhen, China
2 Nano Science and Technology Institute
University of Science and Technology of China
Suzhou, China
wb.ye@siat.ac.cn

2nd Zhenyu Wang[1]
1 Shenzhen Institute of Advanced Electronic Materials
Shenzhen Institutes of Advanced Technology, Chinese Academy of Sciences
Shenzhen, China
zy.wang4@siat.ac.cn

3rd Xiangliang Zeng[3]
3 College of Chemistry and Chemical Engineering
Hunan University
Changsha, China
xl.zeng1@siat.ac.cn

4th Linlin Ren[1*]
1 Shenzhen Institute of Advanced Electronic Materials
Shenzhen Institutes of Advanced Technology, Chinese Academy of Sciences
Shenzhen, China
ll.ren@siat.ac.cn

5th Rong Sun[1]
1 Shenzhen Institute of Advanced Electronic Materials
Shenzhen Institutes of Advanced Technology, Chinese Academy of Sciences
Shenzhen, China
rong.sun@siat.ac.cn

6th Zhibin Wen[1*]
1 Shenzhen Institute of Advanced Electronic Materials
Shenzhen Institutes of Advanced Technology, Chinese Academy of Sciences
Shenzhen, China
zb.wen@siat.ac.cn

7th Xiaoliang Zeng[1*]
1 Shenzhen Institute of Advanced Electronic Materials
Shenzhen Institutes of Advanced Technology, Chinese Academy of Sciences
Shenzhen, China
xl.zeng@siat.ac.cn

Abstract—**Thermal interface material (TIM) has attracted numerous attentions to provide excellent performance and reliability of chip. To achieve outstanding thermal conductivity, high loading thermally conductive fillers have to be added in TIM, which weak the material's processability and mechanical properties. In this work, we fabricated a novel aluminum/zinc oxide/silicone composite. We studied the viscosity, mechanical and thermal properties of TIM with different filler content, and tried to discuss the relationship between them. When the content is 70%, the tensile properties of the material is up to the maximum value 420%. This work reveals the relationship between thermally conductive fillers and the macro-mechanical properties, and provides theoretical guidance for the preparation and application of high-performance TIM.**

Keywords—thermal interface material, filler content, macro mechanical properties

I. INTRODUCTION

At the age of 5G, high-degree integration, high-frequency and high-performance development in modern electronic devices attracts numerous attentions. There is no perfect contact between the heat source and the heatsink. The gap that exists on the solid surface makes it difficult to effectively dissipate heat due to the large thermal resistance during heat transfer. Abnormal high temperature causes the weakness of the equipment performance and efficiency during electronic devices working process. Essential thermal management can provide excellent performance and high reliability of chip. The polymer thermal interface material composed of polymer and thermally conductive filler, is one of the irreplaceable components in high-power chip packaging to reduce contact thermal resistance, directly affecting performance, and prolonging chip service life[1].

The typical thermal conducting fillers are boron nitride，alumina, graphene, boron arsenide and so on[2]. Numerous researches focus on the high thermal conductivity of metal particles. Aluminum powder has the advantages of higher thermal conductivity and lower price, making it become an excellent candidate. Zhang et al. prepared graphite/aluminum composite which in-plane conductivity can reach 940 W / (m · K), by hot pressing under vacuum. The completely different properties of in-plane and out-of-plane thermal conductivity make this kind of graphite film thermal interface material broadens its application range [3]. Yang et al. prepared a kind of thermal interface material using polyurethane sponge to interconnect aluminum sheet. When the aluminum sheet content is 14%, the material can obtain 1.6 W / (m · K) thermal conductivity, and has high insulation performance[4]. To further strengthen the heat dissipation of high-power electronic equipment, zinc oxide (ZnO) was selected as the thermally conductive filler in the thermal management materials. Rajan found that compared with pure propylene glycol fluid, zinc oxide propylene glycol fluid has better heat transfer performance, and the heat transfer rate increases by 4.24%[5].

High thermal conductivity fillers are often added to low thermal conductivity organic polymers to improve composites thermal conductivity. To achieve high thermal conductivity, high loading fillers results in inevitably weakening of the material's processability, mechanical properties and the increasing of the cost. Wang et al. used the filling model to select different sizes of aluminum powder to fill the organic polymers. When the filler reaches up to 50%, the thermal conductivity is $1.381 \, W/(m \cdot K)$ [6]. Mainly researchers bend their efforts to improving the conductivity of TIMs, devoting to lower the contact thermal resistance, while ignoring the discussion on the effect of mechanical properties. In this work, a TIM was prepared with divinyl terminated silicone oil, and chain extender/crosslinker dihydrosiloxane terminated silicone oil, aluminum treated with silane coupling agent and ZnO as thermally conductive fillers. The internal dispersion, rheological machinability, tensile properties and thermal properties of aluminum/zinc oxide/silicon TIMs with different filler contents were systematically studied. Owing to the reinforcing effect of the filler itself and the pores introduced by the increases of viscosity, the tensile property of the material reaches the peak value of 420% when the filling amount of thermal conductive filler is 70%. The relationship between the thermal conductive fillers and the macro mechanical properties of materials is revealed, which provides theoretical guidance for the preparation and application of high-performance TIMs.

II. EXPERIMENTAL

A. Preparation of Silicone/aluminum thermal interface composite materials

The aluminium/zinc oxide/sillicone thermal interface materials are prepared with different amount of thermal conductive fillers. We use a precision electronic balance to weigh the Thermal conductive fillers with different mass fraction and silicone oil in turn. Then we add them to a specific mixing bottle, and mix with a high-speed planetary machine under vacuum for two minutes.

III. RESULTS

A. Morphology

Fig. 1. shows the cross section of Al / ZnO / silicone rubber quenched in liquid nitrogen, exhibiting the image of the dispersion of distinct mass fractions of thermally conductive fillers in silicone. Under low filler loading, the thermally conductive particles are divided and wrapped by silicone rubber, and the probability of contact with each other is low. When high thermally conductive fillers are added, the thermally conductive particles are contacted with each other to form the heat conduction path. The heat flow does not pass through the rubber layer with high thermal resistance, but through the filler with low thermal resistance.

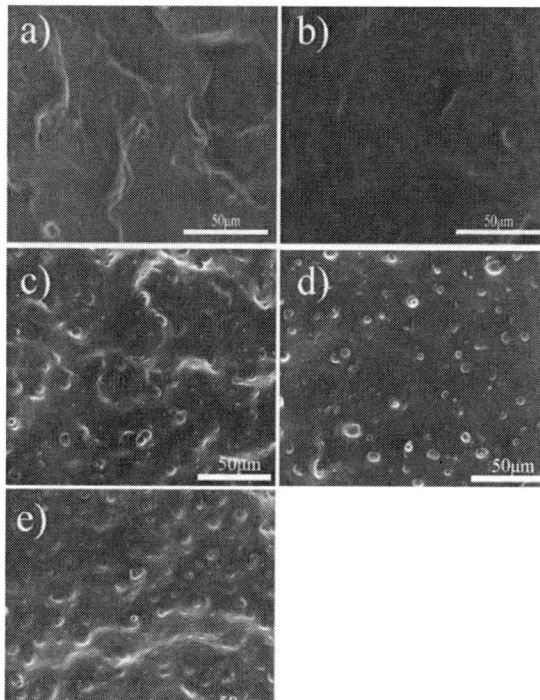

Fig. 1. SEM image of aluminium/zinc oxide/sillicone TIMs (a:30wt.%; b:50wt.%; c:70wt.%; d:80wt.%; e:90wt.%).

B. Viscosity

Fluidity is essential for the application of thermal interface materials. We studied the viscosity changes under different filler mass fractions. Fig. 2. exhibits that the shear viscosity considerably increases with the increase of filling amount. The reason is that the filler powder is uniformly dispersed and the spherical fillers hardly cause the entanglement of silicone oil molecules in the case of low filling amount. So the presence of the filler does not affect the flow ability of silicone oil. When the filler content increases, the effect of fillers enhanced, which results in the increase of viscosity.

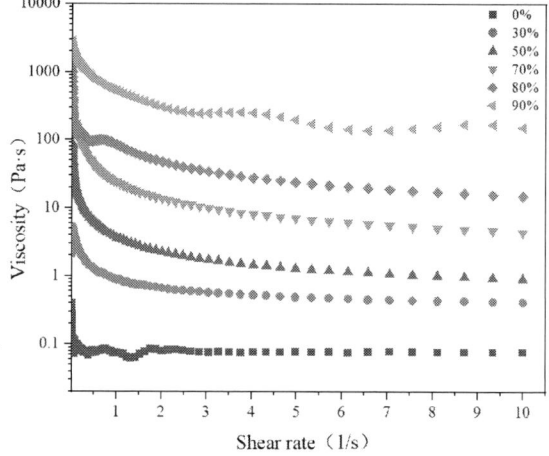

Fig. 2. Shear viscosity of thermal interface materials with different filler contents.

C. Tensile property

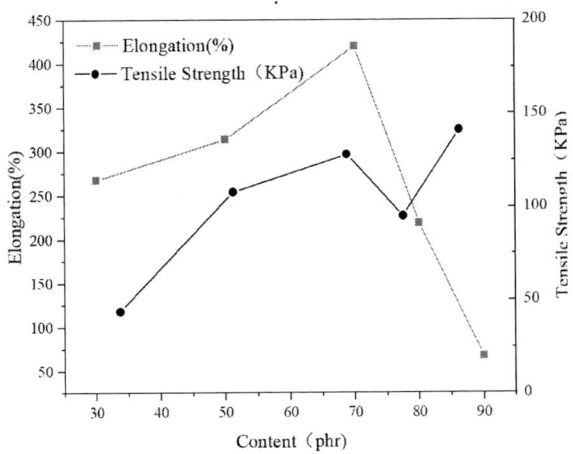

Fig. 3. Variation trend of mechanical properties with different addition content.

Figure 3 shows the change in elongation and tensile modulus of aluminum/zinc oxide/silicone with different filler contents. Because of reinforcing effect of the filler itself and the pores introduced by the increase of viscosity. The elongation at break of the composite increases initially and reaches the maximum value 420% at 70wt.%, then decreasing with the addition of filler content. With the further increase of filler content, it is difficult for silicone rubber matrix to form enough interfacial interaction on the filler surface, and even some agglomerated areas lead to greater fracture possibility. A good heat conduction path can be formed with the increase of the filler content, but the tensile performance decreased to unacceptable level.

D. Thermal conductivity

With the increase of filler content, the fillers in aluminium/zinc oxide/silicone TIMs will gradually associate together to form a thermal conduction path. The high content of filler inevitably weakens the processability and mechanical properties of the material, which has been reflected by the above rheological and mechanical properties. Fig. 4. displays heat conductivity of the aluminum/zinc oxide/silicone TIMs as the weight fraction of filling ratio. As the filler content increases, the thermal conductivity of the aluminum/zinc oxide/silicone composite material increases significantly. The maximum value is 6.26 W/(m·K) when the filler loading is 90 wt.%.

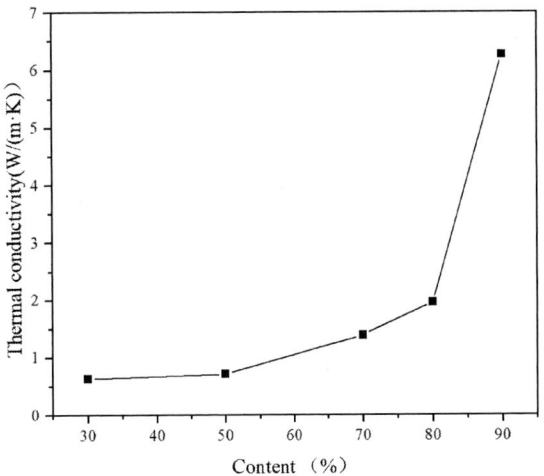

Fig. 4. Thermal conductivity of aluminium/zinc oxide/sillicone TIMs with different addition content.

E. Thermal stability

Thermal stability is the performance of composite materials against thermal degradation and aging. Fig. 5. shows the thermal weight loss curve of aluminum/zinc oxide/silicone rubber composite materials. In order to study the effect of fillers of silicone rubber thermal stability, thermogravimetric analysis tests were performed on samples which are filled with different fractions of thermally conductive fillers. It can be found that the residual mass of the sample multiplies with the increase of filler content at five percent and maximum weight loss temperature. The heat conductivity of the composites multiplies with the increase of the amount of thermal conductive filler, which can show more excellent heat transfer performance. The material is transferred inside, and the sample with filler is heated more evenly.

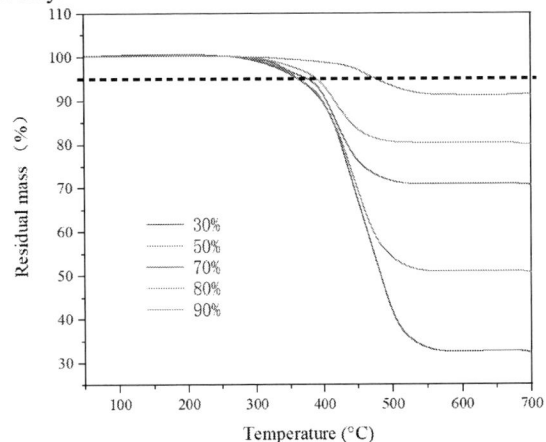

Fig. 5. TGA curves of aluminium/zinc oxide/sillicone TIMs

RESULT AND DISCUSSION

The internal dispersion, rheological machinability, tensile properties and thermal properties of aluminum/zinc oxide/silicon TIMs with different filler contents were systematically studied. The scanning electron microscope micrographs intuitively show that under higher thermal conductivity filler, a continuous and continual heat conduction path is formed in the aluminium/zinc oxide/silicone TIMs. Owing to the reinforcing effect of the filler itself and the pores introduced after the viscosity increases, when the filling amount of the thermally conductive filler is 70%, the tensile properties of the material is up to the maximum value 420%. With the increase of the filler content, a good heat conduction path can be formed, but the tensile performance decreased to unacceptable level. It reveals the balance between tensile properties and thermal conductivity of the aluminum/zinc oxide/silicone TIMs. In a word, we discussed the influence of the content of thermal conductive fillers on the viscosity, mechanical properties and thermal conductivity of thermal interface materials. The relationship between the thermal conductive fillers and the macro mechanical properties of materials is revealed, which provides theoretical guidance for the preparation and application of high-performance TIMs.

ACKNOWLEDGMENT

This work was supported by the Guangdong Basic and Applied Basic Research Fund (No. 2019A1515110845), National Natural Science Foundation of China (Grant No. 52073300), Youth Innovation Promotion Association of the Chinese Academy of Sciences (2019354), Shenzhen Science and Technology Research Fund (No. JCYJ20200109114401708 and JCYJ20180507182530279).

REFERENCES

[1] G. Xu, B. Wang, S. Song, Z. Ren, J. Li, X. Xu, W. Zhou, H. Li, H. Yang, L. Zhang and Y. Li, High-performance and robust dual-function electrochromic device for dynamic thermal regulation and electromagnetic interference shielding, Chemical Engineering Journal, vol. 422, Octmber 2021.

[2] Y. Cui, Z. H. Qin, H. Wu, M. Li and Y. J. Hu, Flexible thermal interface based on self-assembled boron arsenide for high-performance thermal management, Nature Communications, vol.12, February 2021.

[3] Y. Huang, Y. S. Su, S. Li, Q. B. Ouyang, G. D. Zhang, L. T. Zhang and D. Zhang, Fabrication of graphite film/aluminum composites by vacuum hot pressing: Process optimization and thermal conductivity, Composites Part B-Engineering, vol.107, pp.43-50, December 2016.

[4] B. J. Wei, X. Chen and S. Q. Yang, Construction of a 3D aluminum flake framework with a sponge template to prepare thermally conductive polymer composites, Journal of Materials Chemistry A, vol. 9, pp.10979-10991. May 2021.

[5] K. S. Suganthi and K. S. Rajan, Improved transient heat transfer performance of ZnO-propylene glycol nanofluids for energy management, Energy Conversion and Management, vol.96, pp.115-123, May 2015.

[6] L. Mao, J. B. Han, D. Zhao, N. Song, L. Y. Shi and J. H. Wane, Particle Packing Theory Guided Thermal Conductive Polymer Preparation and Related Properties, Acs Applied Materials & Interfaces, vol.10, pp.33556-33563,Octmber 2018.

Packaging of a MEMS sensor in an active interventional blood pressure monitoring catheter

Guanzhe Xu#
Internet of things research center
Advanced institute of information
technology, Peking University
Hangzhou, China
gzxu@aiit.org.cn

Junshu Lin#
Internet of things research center
Advanced institute of information
technology, Peking University
Hangzhou, China
jslin@aiit.org.cn

Rongying Yu
Internet of things research center
Advanced institute of information
technology, Peking University
Hangzhou, China
ryyu@aiit.org.cn

Zebo Zhang
Internet of things research center
Advanced institute of information
technology, Peking University
Hangzhou, China
zbzhang@aiit.org.cn

Meng Gao*
Internet of things research center
Advanced institute of information
technology, Peking University
Hangzhou, China
mgao@aiit.org.cn

Le Ye*
Institute of Microelectronics, Peking
University, Beijing, China
Internet of things research center
Advanced institute of information
technology, Peking University
Hangzhou, China
yele@pku.edu.cn

Abstract : In the field of interventional pressure monitoring catheter, common types of catheters mainly include invasive arterial pressure catheter, central venous pressure catheter, pulmonary artery floating catheter, etc. A common way to measure intravascular pressure with such catheters is to direct arterial or venous blood to a pressure sensor outside the body through a hole in the distal end of the catheter and a drainage canal inside the catheter. The pressure measurement method is indirect, and the accuracy of pressure measurement is easily interfered by the pipeline. In this study, an interventional blood pressure monitoring catheter with MEMS pressure sensor is prepared. The MEMS pressure sensor was placed on metal base at the head of the catheter, which can be used in direct blood pressure measuring. The sensor adopts a two-step packaging method, and the flexible transfer plate is used to realize the conversion of gold wire to copper wire, so as to reduce the material cost of production while ensuring the reliability of packaging.

Keywords : MEMS sensor packaging; Interventional catheter; Flexible printed circuit

I. INTRODUCTION

Interventional catheterization is an important measure of human hemodynamics. According to the different types of catheterizations, pulmonary artery pressure, central venous pressure, intracardiac pressure and coronary artery pressure can be monitored [1,2]. At present, there are two widely used manometric methods in clinical practice. One is intravascular perfusion of normal saline and percutaneous puncture into the blood vessel, and the blood pressure is transmitted by the normal saline to the external pressure sensor for manometric measurement. Another one is to direct the blood through the catheter out of the body to the external pressure sensor for measurement. These methods are indirect manometric methods, and the results of manometric methods are easily affected by factors such as catheter location and patient position. In vitro indirect manometric measurement

requires a higher sensor calibration system, and it is easy to cause side effects such as thrombosis and arrhythmia. In addition, the blood extraction cavity needs to be coated with heparin coating as anticoagulant [3]. If heparin coating peels off, there will be a risk of blood coagulation.

In recent years, with the development of micro/nano technology, small size micro sensors can be prepared by MEMS technology [4-6], and such small size micro sensors can be directly integrated into the measuring catheter head for hemodynamics measurement through micro/nano packaging technology, which has become the development trend of active interventional pressure monitoring catheter [7]. Direct pressure measurement by the MEMS sensor on the catheter can quickly determine whether the hemodynamic information is abnormal, and there is no need to calibrate the pressure sensor again, which reduces the occurrence of adverse complications.

In addition to the sensor signal acquisition, data storage, processing, display, sensor power supply needs the assistance of external machine. Therefore, reliable electrical extraction is an important problem to be solved for the direct integration of small size miniature sensors in the catheter head. There are two technical difficulties in sensor electrical extraction. On the one hand, the electrodes of MEMS microsensor are mostly made of gold or platinum and other precious metals, and the size of the electrode extraction area is at the micron level. The electric lead wire from the catheter head to the external machine is often 1-2 meters long, so it is costly to use precious inert metal as the lead wire. On the other hand, if copper, aluminum, and other common metals are used as the lead wire, the direct connection with gold or platinum metal is easy to produce matching problems. In this study, the method of converting copper wire from gold wire to copper wire on the connecting plate is used to solve the above problems.

II. PACKAGING OF MEMS PRESSURE SENSOR

The active pressure monitoring catheter involved in this study consists of a sensor encapsulated on the head base and a

---

This project is supported by National Key Research and Development Project of China (2019YFB2204900). #Xu and Lin contributed to this paper equally. *Corresponding author.

catheter body. Specifically, the sensor, the head base, the flexible printed circuit, and the wire together constitute the package structure. SMI SM98A was chosen as pressure sensor. It has a pressure range of 0-10 bar and a size of 1.2 mm × 1.3 mm, which are suitable for applications in hemodynamic manometry. Fig. 1 shows the structure of the MEMS sensor packaging. Position 1/2/3/4/5/6 represents the copper wires, the flexible printed circuit, the MEMS pressure sensor, the nylon catheter, the gold wires and the metal base, respectively. Fig.2 shows the structure diagram of the MEMS pressure sensor, position 1 represents the pressure sensing film, position 2/3 represents the mounting frame. Fig. 3 shows the structure of the metal base. The metal base has a groove in the middle position 2, which is used as a liquid channel for pressure monitoring. The position 1 and 3 of the metal base support the sensor and the position 4 support the flexible printed circuit. The position 5/6 is used to immobilized metal base to the nylon catheter. The metal base is made of 304 steel for the convenience of X-ray developing.

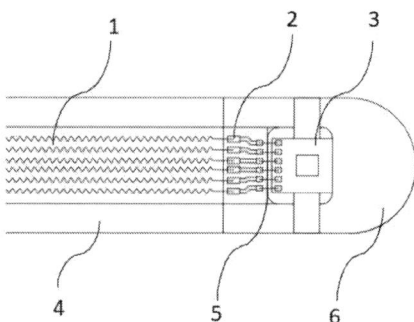

Fig. 1. Structure diagram of the MEMS sensor packaging

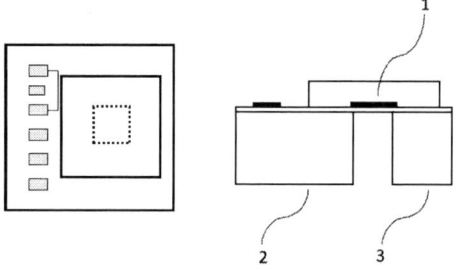

Fig. 2. Structure diagram of the MEMS pressure sensor

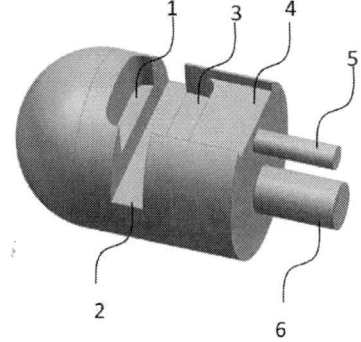

Fig. 3. Structure diagram of the metal base

Packaging of pressure monitoring catheter is a combination of pressure sensor, flexible printed circuit, metal base, wire, nylon catheter and other components. Packaging of the sensor is the vital process that affect the accuracy and precision of the pressure measurement process. In general, the first step of sensor packaging process is to glue the sensor and flexible printed circuit to the metal base (304 stainless steel). Then, we dip 0.1 microliter of Henkel 4011 adhesive onto the position 1/3/4 of the metal base with a micro dispensing system, and later moved the sensor and the flexible printed circuit onto the metal base under a microscope. At last, we wait for 3 min, the adhesive will be cured completely at room temperature.

Fig. 4 shows the photograph of metal base after the adhesion of MEMS sensor and flexible printed circuit. Both MEMS sensor and flexible printed circuit are much smaller than the metal base, so that the packaged base cannot increase the out-diameter of the nylon catheter, which can also keep the high sensitivity and stability of the sensor through the interventional catheter.

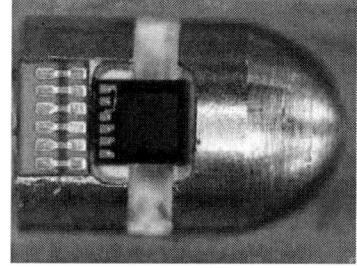

Fig. 4. Adhesion of MEMS pressure sensor and adapter plate to the metal base

After the MEMS pressure sensor and the flexible printed circuit were glued together, the lead wire was then welded. The first step was to remove the surface insulation layer at both ends of the enameled wire and connect it to the solder joint of the flexible printed circuit away from the sensor through tin soldering. The second step was to bond the gold wires between the MEMS sensor and the remaining solder spots of the flexible printed circuit through an ultrasonic gold welding machine. After the welding was completed, we use a special probe and multimeter to measure the reliability of the electrical connection between the enameled wire end and the MEMS sensor pad. The third step was to apply a few drops of UV glue on the welding site of the flexible printed circuit for light curing, which can protect the welding site. In this welding method, based on retaining the gold wire, the flexible printed circuit was used to convert the gold wire to the ordinary enameled wire, which reduced the length of the gold wire and effectively controlled the production cost of the pulmonary artery floating catheter. Fig. 5 shows the wire bonding photograph of metal base with sensor and flexible printed circuit after welding.

Fig. 5. Wire bonding of the MEMS pressure sensor

The metal base after welding wire is connected to the main body of the catheter by adhesive, and the wire is fixed in the groove of the nylon catheter. Tab. 1 shows the result of peeling experiment of the metal base and nylon catheter. It can be seen from the table, the average peeling force between metal base and nylon catheter is 42.29 N, which is higher than the minimum value (15 N) as required in interventional medical instruments.

Tab. 1. Peeling force of metal base and nylon catheter

Test No.	1	2	3	4	5	Average
Peeling Force (N)	38.64	48.68	45.66	34.44	44.05	42.29

Then we put the packaged nylon catheter with the pressure sensor into a cylindrical water container to test its voltage output for sensing calibration. Fig. 6 (a) and (b) show the photograph of the water container with the pressure catheter inside, and the data of sensor voltage output in the water container, respectively. The top water level in the container was marked as zero of the water depth of the sensor inside the container. Normally, as the position of the sensor in the container decreases, the water static pressure linearly increases. As shown in Fig. 6 (b), the sensor voltage output increases gradually and linearly with the water depth of the sensor inside the water container. The sensor in the pressure catheter can be calibrated through this data from Fig. 6 (b).

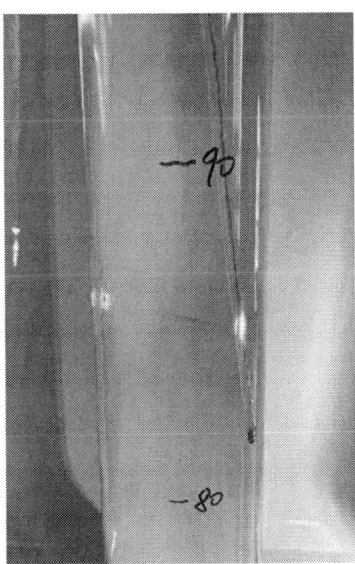

（a）

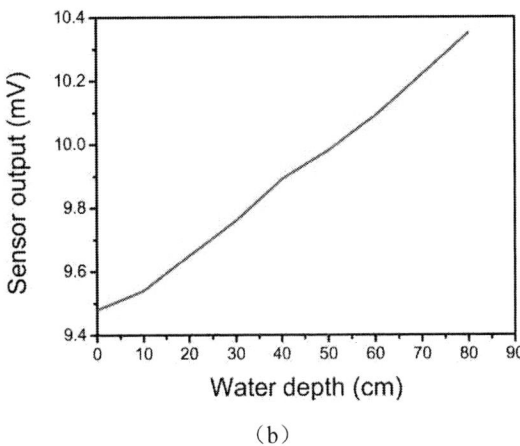

（b）

Fig. 6 (a) Photograph of the water container with the pressure catheter inside (b) Experimental results of sensor voltage output vs. water depth

III. CONCLUSIONS

In this study, we demonstrate the encapsulation process of a MEMS pressure sensor. The process includes the bonding of the sensor and the flexible printed circuit the welding of the sensor, and the assembly of the metal base. Among them, welding of sensor is the core process. In this process, the flexible printed circuit is used to convert gold wire to enameled wire, which ensures the reliability of welding wire and reduces the production cost. In general, the microscale encapsulation process of MEMS sensors involved in this study is expected to be widely used in the field of active interventional pressure catheter.

REFERENCES

[1] Thakur Y, Holdsworth D, Drangova M. Characterization of catheter dynamics during percutaneous transluminal catheter procedures. IEEE Transactions on Biomedical Engineering 2009, 56: 2410-2143

[2] Schwartz P, Scheffer C, Fourie P R and Coetzee A R. An impedance-guided intra arterial catheter. 35th Annual International Conference of the IEEE EMBS, 2013, Japan Osaka

[3] Kutsogiannis D J, Gibney N, Stollery D. Regional citrate versus systemic heparin anticoagulation forcontinuous renal replacement in critically ill patients. Kidney International, 2005,67:2361-2367

[4] Löffler M, Nierla M, Kadur M, Hoffmann M, Sutor A and Lerch R, Optimizing the dimensions of an inverse-magnetostrictive MEMS pressure sensor by means of an iterative finite-element scheme, IEEE Transactions on Magnetics, 2016, 52: 1-4

[5] Zhang Y, Zhang Z, Pang B, Yuan L and Ren T, Tiny MEMS-based pressure sensors in the measurement of intracranial pressure, Tsinghua Science and Technology, 2014,19: 161-167

[6] Ganev B, Nikolov D and Marinov M B, Performance evaluation of MEMS pressure sensors, 2020 XI National Conference with International Participation (ELECTRONICA), 2020:1-4

[7] Murtha L, McLeod D, Spratt N, Epidural intracranial pressure measurement in rats using a fiber-optic pressure transducer, Journal of Visualized Experiments, 2012,62: e3689

Heat transfer analysis of phase change materials with metal foams

Yan Zhang*
School of Mechatronics Engineering and Automation
Shanghai University
Shanghai, China
* yzhang@shu.edu.cn

Huihui Wang
School of Mechatronics Engineering and Automation
Shanghai University
Shanghai, China
wanghuihui00@163.com

Pei Lu
School of Mechatronics Engineering and Automation
Shanghai University
Shanghai, China
159054118@shu.edu.cn

Jingyu Fan
School of Mechanics and Engineering
Shanghai University
Shanghai, China
Email: jyfan@shu.edu.cn

Qixuan Tu
Sino-European School of Technology of Shanghai
Shanghai University
Shanghai, China
18621008725@163.com

Johan Liu
Department of Microtechnology and Nanoscience
Chalmers University of Technology
Gothenburg, Sweden
jliu@chalmers.se

Abstract— With the development of electronic products towards high-density integration, high performance and multifunction, the working frequencies and power consumption rate of electronic components and devices increase substantially. The resulting temperature rise has a great impact on the operation and lifetime of electronic products. Transient temperature control and efficient heat dissipation are essential to the stability and reliability of the electronic components and products. Paraffin wax, as one of the most commonly used phase change materials, has been widely applied in many products requiring transient temperature control due to its melting temperature lying in the range of electronics operation conditions. However, the applicable scopes of phase change materials were limited due to their shortcomings of low thermal conductivity and heat dissipation. In the present paper, both metal forms and carbon nanomaterials are used as thermal enhancers to increase the conduction of paraffin wax, and the heat transfer characters of the composites are investigated by numerical method. The simulation results show that the introduction of Cu or Ni foam as heat conductive enhancers can significantly increase the effective thermal conductivity of paraffin wax composite. The thermal conductivity of the composite with Ni foams is 3.684 times higher than that of the paraffin wax, and the increase is 12.485 times when Cu foam is used instead of Ni foam. Furthermore, the heat transfer of the composites can be strengthened by adding carbon nanomaterials into the paraffin wax so as to increase the thermal conductivity of the matrix. The simulation results show that the impact of dispersed carbon nanomaterials on thermal enhancement of the composites is less significant than that of metal foams.

Keywords—effective thermal conductivity, metal foam, paraffin wax, composite material, PCM

I. INTRODUCTION

Phase change material (PCM) can absorb the heat during electronic component operation and keep the product temperature within a steady range during the phase transition process. Paraffin wax is a typical PCM with advantages such as non-toxic, non-corrosiveness, chemical stability, low cost and so on, and has been widely applied in the transient temperature control of many devices and products to restrain

The work is supported by NSFC (11672171, 51872182).

temperature variation. Cao used the finite element method to establish a heat dissipation model of battery temperature control by the phase change material, and verified the simulation results by experiment measurement. In comparison to the cooling effects of natural convection and forced air flow in the heat dissipation, the introduction of phase change by a paraffin wax made the maximum battery temperature drop by 1.3°C [1]. San chose n-heneicosane as the phase change material and mixed with bentonite to prepare a composite phase change materials for architectural temperature regulation. The composite material exhibited a good thermal durability and chemical stability in the temperature cycling test. Indoor temperature regulation test indicated that the composite PCM could be used as wall panel materials and performed well in reducing the indoor temperature [2].

The disadvantage of low thermal conductivity of paraffin wax limits its heat dissipation applications in high-power electronics. Various approaches have been investigated to promote the heat transfer performance of phase change materials. Metal or carbon materials with higher thermal conductivity were used as fillers to improve the heat transfer [3]. Carbon nanomaterials of excellent features had been used as conductive enhancers to improve the mechanicaland thermal properties of composite materials. Choi prepared thermoplastic polyurethane (TPU) nanocomposite with functionalized graphene sheets (FGS). The modulus of TPU was enhanced by FGS, while the tensile strength was reduced as the graphene content increased [4]. Chan developed graphene coated nickel foams and applied them in paraffin wax. By coating graphene on the metal foam surface, the thermal conductivity increased from 3.69 W/m·K to 19.85 W/m·K, which was about 23 times higher in magnitude [5]. Chang added graphene nanosheets into paraffin and tested the heating time of graphene nanosheets/paraffin composites. The thermal conductivity of the composite materials continued to increase as the content of graphene nanosheet increased, and the dropping temperature of the material also increased[6]. Sanusi added carbon nanofibers into a paraffin matrix, and the temperature field of the material under power loadings became more uniform and the solidification time was shortened. Carbon nanofibers with different aspect ratios could enhance the thermal diffusion of composites [7]. Liu mixed paraffin wax with graphene and graphite flakes and analyzed the microstructure of the composite material. The paraffin wax was in close contact graphene and graphite flakes, and graphene and graphite flakes significantly improved the thermal conduction[8]. Warzoha studied the paraffin wax with graphite fibers, where the phase change composites both in solid and liquid were analyzed, and the experiments showed that the thermal resistance at the interface of graphite fibers and paraffin wax was twice than that between graphite fibers. The thermal resistance value became one order lower when the paraffin wax state changed from solid to liquid during the phase change procedure. Increasing the graphite fiber content obviously increased the thermal conductivity of the solid composite material, while the composite material in the liquid phase could be apparently improved only when the carbon nanofiber content exceeds the percolation threshold [9]. Rao developed graphite/paraffin composites, and the latent heat of the materials was reduced after graphite was added but the thermal conductivity was increased. Therefore, the filler proportion should be chosen according to the specific application requirements [10]. Xiao used nickel and copper foams in paraffin wax, and the melting temperature of the composites increased while the solidification temperature decreased. The thermal conductivity was increased almost 4 times when the nickel foam was used in the paraffin wax, and the use of copper foam could increase the thermal conductivity up to 15 times [11].

This paper studies the enhancement effect of metal and carbon materials on the heat conduction of paraffin waxes. By means of finite element simulations, the thermal transfer performance of paraffin waxes with metal foams (nickel foam, copper foam) was obtained, and the heat transfer characters was analyzed. Furthermore, carbon nanoplatelets were introduced into the paraffin wax matrix, and the impact of carbon nano-fillers on the heat conduction reinforcement effect of the composite material was discussed.

II. MODEL ESTABLISHMENT

A. Sample preparation

Before the numerical model establishment, experimental preparation of the composite samples and measurement of the material properties have been carried out to provide reference information for the numerical simulation.

In the experiment, a nickel foam with porosity of 94% to 98% was selected to prepare the metal foam-enhanced /paraffin wax composite. Fig. 1 is the SEM image of the nickel foam used in the sample preparation. According to the microscopic observation, the diameter of the nickel skeleton is about 48μm.

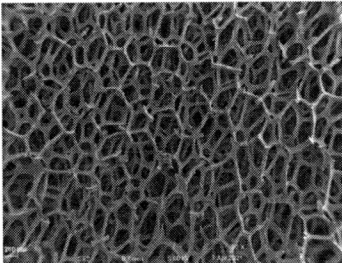

Fig. 1. SEM image of metal foam (nickel)

In the process of the sample preparation, firstly the paraffin wax was heated up until the solid melted completely into liquid. Then the nickel foam was added into the paraffin wax and the liquid paraffin wax flowed into the nickel foam to fill the pores. The composite samples was cooled down until the paraffin wax solidified.

During the preparation of composite samples with carbon nanomaterials in the matrix, carbon nanoplatelets (GNP) of 3~5 layered graphene sheets were used. The nanoplatelets were added into the paraffin wax when the paraffin wax was heated up and completely melted. The liquid composite was stirred and the nanoplatelets uniformly distributed in the paraffin wax, so that a carbon nanoplatelets/paraffin wax composite matrix was obtained. Then the composite matrix was filled into the nickel foam by the similar process as mentioned before to obtain an composite material of nickel foam+ carbon nanoplatelets/paraffin wax.

The microstructure of the metal foam maintained during the sample preparation and the interconnected nickel skeleton formed a three-dimensional conductive framework structure in the paraffin wax matrix. Various foams can be applied to achieve the conduction network in the paraffin wax composite.

B. Geometrical model

A FEM model of the representative nickel structure is established according to the geometric size of the nickel foam used in the experiment, and then the composite model is obtained. Fig.2(a) shows the nickel structure in the model, and Fig.2(b) is the composite model with paraffin wax around the nickel structure.

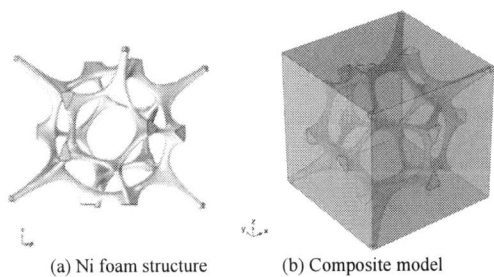

(a) Ni foam structure   (b) Composite model

Fig. 2. FEM model

The volume fraction of the nickel foam structure in the paraffin wax matrix in the model is set so as to be close to the experimental sample, so are the shape and size of the foam structure.

C. Material parameters

In the experiment, the volume of the paraffin wax doesn't change much at the addition of carbon nanoplatelets at 15wt% content, and the corresponding density values of the paraffin wax with and without nanoplatelets are 944.8 kg/m^3 and 881.8 kg/m^3, respectively. The density of paraffin wax composite increases 7.03% and the small change in density is favorable to practical application.

Thermal conductivity measurements show that the paraffin wax is about 0.4406 W/m·K and the carbon nanoplatelet/paraffin wax is about 0.6225 W/m·K. The material parameters in the numerical simulation are chosen with reference to the experimental measurements, and Table 1 is the material parameters used in the simulation.

978-1-6654-1392-3/21 $31.00 © 2021 IEEE

TABLE I. MATERIAL PARAMETERS

Material	k W/(m·K)	ρ kg/m³
Paraffin wax	0.4406	881.8
Ni	90.9	8902
Cu	401	8960
Paraffin wax+GNP	0.6225	943.8

It can be inferred from theoretical research that the foam porosity has a significantly effect on the conductive features of the phase change composite, while the pore size has little effect [11]. In this paper, the geometric dimensions of the foams keep unchanged and metal materials such as nickel and copper are considered as foams materials to compare their influences on the thermal conductive properties of the composite. In the simulation in the paper, it was assumed that the material properties of the paraffin wax and the foam did not change with temperature when the heat flow was applied on the composite models.

III. RESULTS AND DISCUSSION

During the numerical simulation, a heat flow is generated where the top surface of the model is set to 50°C and the bottom surface is 20°C, and the four sides are set to be adiathermic. After the temperature field reaches a steady status, the overall temperature field is obtained, as shown in Fig.3. And Fig.4 are the corresponding temperature distributions of metal foam structure in the composite.

For the paraffin wax with no thermal enhancer involved (neither metal foam nor carbon nanoplatelets), as in Fig.3(a), the temperature field is uniform along the horizontal direction and shows a linear change with vertical position from high temperature at the top to low temperature at the bottom. In the case of nickel foam/paraffin wax, as in Fig.3(b), temperature values in the metal form area are lower compared with the temperature at the corresponding location in the paraffin wax without metal foam, and the region of low temperature in the nickel foam/paraffin wax is larger than that in the paraffin wax. The reason is that the high thermal conductivity of the nickel foam results in an enhanced heat transfer in the paraffin wax matrix.

When the metal foam in the composite changes from nickel to copper that has an even higher thermal conductivity, the temperature of the composite material further reduced. The foams show great impact on the heat transfer. However, the comparison of temperature fields of either nickel foam/paraffin wax or copper foam/paraffin wax with and without carbon nanoplatelets in the matrix, as shown in Fig.3(b)-(c), and Fig.3(d)-(e), doesn't exhibit an obvious difference. Namely the improvement of carbon nanoplatelets in the heat transfer of the composite doesn't have such an distinct influence as the metal foams do.

Fig.5 shows the temperature distributions in the center cross-sections of paraffin wax with and without copper foam, where the metal foam outline is indicated in black lines. The horizontal temperature of the pure paraffin is uniform, as illustrated by the even temperature contours in Fig.5(a). When the metal foam is involved, the heat transfer in the foam area is locally accelerated, as shown in Fig.5(b), and the temperature contour tilted. The temperature profile in this case changes accordingly.

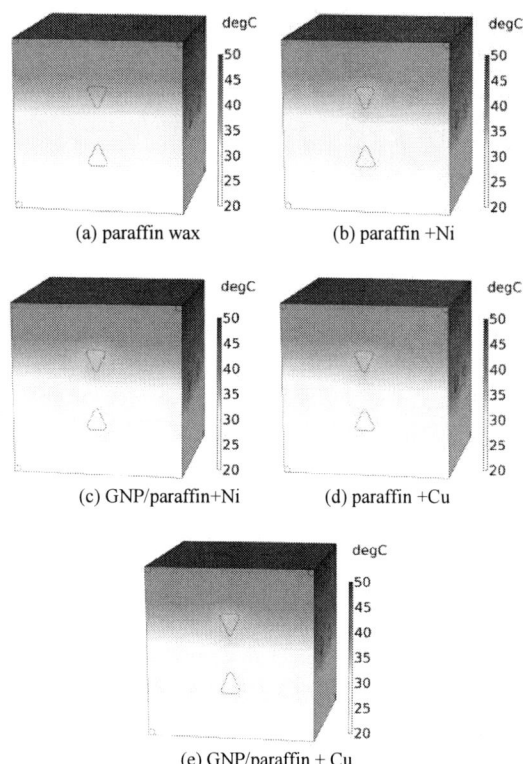

Fig. 3. Temperature distribution of composite materials

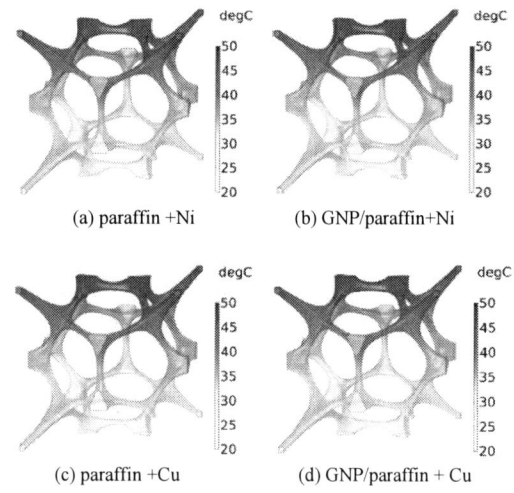

Fig. 4. Temperature distribution of metal foams

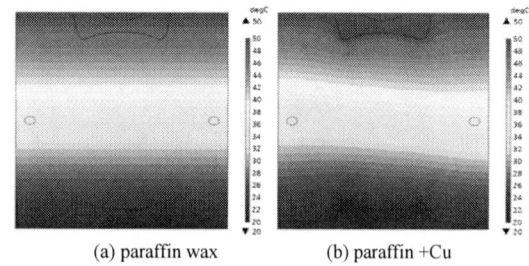

Fig. 5. Temperature distribution in central plane

The effective thermal conductivity of the composite material can be calculated by the formula as below:

$$k = \frac{Q\delta}{A(t_1 - t_2)} \qquad (1)$$

where Q is the total energy flow through the cross section, δ is the length in the heat transfer direction, A is the cross-sectional area, and t_1 and t_2 are the temperatures at the cross sections, respectively.

Fig.6 shows a comparison of the thermal conductivities of different materials. The significant difference indicates that the addition of nickel foam effectively enhances the thermal conduction in the paraffin wax. The effective thermal conductivity of the nickel foam/paraffin wax is calculated to be 1.6233 W/m·K, which is 3.684 times of the paraffin wax. According to the experimental measurement of a nickel foam/paraffin wax sample, in which the porosity of nickel foam is 95%, the thermal conductivity is 1.855 W/m·K [3]. That result agrees well with the effective thermal conductivity obtained by the present simulation and verifies the FEM modeling.

Then the pure paraffin waxes in the nickel form/paraffin wax composites are replaced by the composite paraffin wax containing carbon nanoplatelets, and the effective thermal conductivity of the nickel foam+carbon nanoplatelet /paraffin wax is 1.8227 W/m·K., which is 4.137 times of pure paraffin wax. The overall effective thermal conduction of the composite can be further improved when replacing the pure paraffin wax by the composite matrix with carbon nanoplatelets, where the increase is about 12.28%.

The introduction of metal foams can greatly improve the heat transfer of the composite material. When copper is used instead of nickel, the thermal conductivity of the copper foam/paraffin composite is 5.5007 W/m·K, which is about 12.485 times of paraffin wax. The thermal conductivity value of the composite material becomes 5.7027 W/m·K when carbon nanoplatelets are added into the matrix, which is 12.943 times of paraffin and 3.54% increase than that of copper foam/paraffin wax composite.

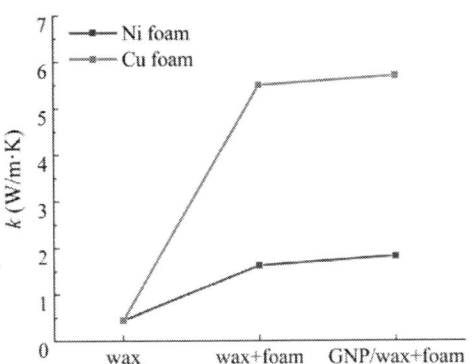

Fig. 6. Thermal conductivity of different composite materials

The effective increase in the thermal conduction by the metal foams can be attributed to that the metal foams not only have high thermal conductivity values but also provide continuous conductive pathways. The carbon nanoplatelets, on the other hand, are dispersedly distributed in the matrix. Although the addition of carbon nanoplatelets can increase

the thermal conductivity of the matrix by about 1.413 times, the magnitude is still much lower in comparison to that of the metal foams. Therefore, the metal foam plays a major role in heat transfer of the composite materials.

IV. CONCLUSION

Paraffin wax has been widely used in the transient temperature control of electronic devices and products. However, the poor thermal conductivity limits their applications in the heat dissipation of high-power electronics. Various kinds of thermal conductive enhancers have been utilized to develop paraffin wax composite materials to obtain an improved effective thermal conductivity so as to increase the heat dissipation performance.

In this paper, nickel foam, copper foam and carbon nanoplatelets are taken into consideration as the thermal enhancers of the paraffin wax to increase the thermal conduction of the composite materials. The influences of different materials are studied by means of finite element simulation. The metal foams can dramatically increase the effective thermal conductivity of composite materials. The effective thermal conductivity of the nickel foam/paraffin is 1.6233 W/m·K, which is about 3.684 times of the paraffin wax. When the matrix is replaced by the carbon nanoplatelets/paraffin wax composite material, the effective thermal conductivity can be further improved by about 12.28%, which is 4.137 times of pure paraffin wax. If the nickel form in the composite is changed to the copper foam of a higher thermal conductivity, the heat transfer can be further promoted. The copper foam/paraffin composite is 5.5007 W/m·K, which is 12.485 times of paraffin wax. And the effective thermal conductivity of the composite materials increases to 5.7027 W/m·K, which is an advance of 3.54% if carbon nanomaterials are added into the paraffin wax matrix,. The addition of carbon nanomaterials can improve the thermal conductivity of metal foam/ paraffin wax composites, but the increase effect is less significant compared with metal foams.

REFERENCES

[1] J. Cao, D. Gao, J. Liu, J. Wei and Q. Lu, "Thermal modeling of passive thermal management system with phase change material for LiFePO4 battery", 2012 IEEE Vehicle Power and Propulsion Conference, Seoul, Korea, Oct. 9-12, 2012, pp. 436-440.

[2] A. San, "Thermal energy storage properties and laboratory-scale thermoregulation performance of bentonite/paraffin composite phase change material for energy-efficient buildings", Journal of Materials in Civil Engineering, vol. 29, No.6, 2017, pp. 04017001

[3] W. Q. Xu, X. G. Yuan and Z. Li, "Study on effective thermal conductivity of metal foam matrix composite phase change materials", Gongneng Cailiao/Journal of Functional Materials, vol. 40, No.8, 2009, pp. 1329-1332+1337.

[4] J. T. Choi, D. H. Kim, K. S. Ryu, H. Lee, H. M. Jeong, C. M. Shin, J. H. Kim and B. K. Kim,"Functionalized graphene sheet/polyurethane nanocomposites: Effect of particle size on physical properties", Macromolecular Research, vol. 30, No.1, 2011,pp. 809-814.

[5] K. C. Chan, C. Y. Tso, A. Hussain and C. Chao, "A theoretical model for the effective thermal conductivity of graphene coated metal foams", Applied Thermal Engineering, vol. 161, 2019, pp. 114112.

[6] T. C. Chang, S. Lee, Y. K. Fuh, Y. C. Peng and Z. Y. Lin, "PCM based heat sinks of paraffin/nanoplatelet graphite composite for thermal management of IGBT", Applied Thermal Engineering, vol. 112, 2017, pp. 1129-1136.

[7] O. Sanusi, R. Warzoha and A. S. Fleischer, "Energy storage and solidification of paraffin phase change material embedded with graphite nanofibers", International Journal of Heat and Mass Transfer, vol. 52, No.19-20, 2011, pp. 4429-4436.

[8] X. Liu and Z. Rao, "Experimental study on the thermal performance of graphene and exfoliated graphite sheet for thermal energy storage phase change material", Thermochimica Acta, vol. 647, 2017, pp. 15-21.

[9] R. J. Warzoha, R. M. Weigand and A. S. Fleischer, "Temperature-dependent thermal properties of a paraffin phase change material embedded with herringbone style graphite nanofibers", Applied Energy, vol. 137, 2015, pp. 716-725.

[10] Z. H. Rao and G. Q. Zhang, "Thermal properties of paraffin wax-based composites containing graphite", Energy Sources Part a-Recovery Utilization and Environmental Effects, vol. 33, 2011, pp. 587-593.

[11] X. Xiao, P. Zhang and M. Li, "Preparation and thermal characterization of paraffin/metal foam composite phase change material", Applied Energy, vol. 112, 2013, pp. 1357-1366.

Delamination Reduction by Material and Process Optimization

Tina Li *
Engineering Department
Nexperia Semiconductor
Dongguan, China
Tina.c.li@nexperia.com*

Aaron He
Engineering Department
Nexperia Semiconductor
Dongguan, China
Aaron.he@nexperia.com

Phoebe Chen
Engineering Department
Nexperia Semiconductor
Dongguan, China
phoebe.chen@nexperia.com

Aaron Lai
Engineering Department
Nexperia Semiconductor
Dongguan, China
Aaron.lai@nexperia.com

Yuan hang
Engineering Department
Nexperia Semiconductor
Dongguan, China
yh.yuan@nexperia.com

Colin Feng
Engineering Department
Nexperia Semiconductor
Dongguan, China
colin.mh.feng@nexperia.com

Abstract— **With bonding wire converted from Au to Cu, corroded Cu wedge defects observed with some packages. Epoxy Mold Compound (EMC) to lead frame delamination provide penetration path for corrosive chemicals to penetrate inside and react with Cu wedge. To improve delamination performance without changing exist production and Bill of Materials, study was carried out with molding process, Lead Frame strip's structure, EMC properties. Delamination performances were collected via C-SCAN.**

In depth study on delamination mapping vs mold flow influence been carried out through detail short shot analysis. Mold parameter influence on delamination was investigated [1]. EMC flowability vs lead frame surface wetting behavior was explored. Study result shows different LF strip location design, mold process optimization can influence delamination [2]. In this paper, wedge pad surface cleanness was checked via mechanical decapsulation on package, EMC viscosity curve were analyzed to explain mold parameter settings. These all helped well explained mechanism on support delamination improvement.

This paper gives insight on material&process optimization approach to achieve minimum package delamination. Involves discussion on mold process factors, EMC material property that influence package delamination.

Keywords—Delamination; Mold Process Optimization; EMC Material

I. INTRODUCTION

Several Cu wedge corrosion defects observed during convert bonding wire from Au wire to Cu wire. Two main root causes were identified, one was package delamination between EMC and LF, the other one was corrosive chemicals penetrated inside wedge pad through delamination gap(see Fig.1). Corrosive chemical finally makes Cu wedge corroded.

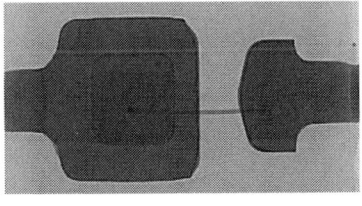

(a)

(b)

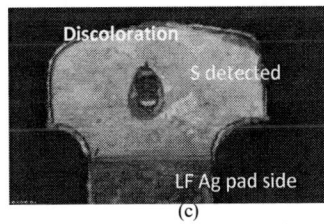

(c)

Fig. 1. Analysis defect device (a)X-ray photo (b) Csam with wedge delamination, (c) LF Ag wedge discoloration

Preventive actions were implemented, mainly from inline optimization, corrosive chemical separation, cleaning, and mechanical strength reduction. With implemented preventive actions, defect cases reduced and removed.

This study is focus on delamination improvement to further reduce risk of corroded Cu wedge defects.

II. EXPERIMENT SET UP

The study is focus on one SOT package, which have observed Cu wedge defects. With transfer molding process and EMC material.

978-1-6654-1392-3/21 $31.00 © 2021 IEEE

In-depth analysis from several defect cases were carried out, mapping defect cases with lead frame structure and mold process. Collected short shot on mold flow distribution, analyzed EMC viscosity curve and curing speed behavior, digged out dedicated material&process factors that lead to defects.

After root cause investigation, delamination reduction method with LF structure change and mold parameter optimization were validated.

Delamination performance were collected after plating by CSAM (SONOSCAN GEN5 machine). Short shot was taken on the mass production transfer molding machine, with actual mold parameters. Mechanical chemical ingression checking was made on 0hr devices after overall assembly, at least 3pcs/row/sample were collected, then photo taken under microscope. EMC flowability property were collected through Torque viscosity meter by supplier. EMC curing speed were collected by DSC test by 3rd party.

III. RESULT AND DISCUSSION

A. Defect Mapping

LF strip have 4 rows, from A1 to A4, all die pad top and wedge pad bottom sequence (Fig. 2). Defect cases were mapped by LF tracks (Fig. 3), most rejects occur on A2 and A4, no defect came from A1. Big row to row variation observed.

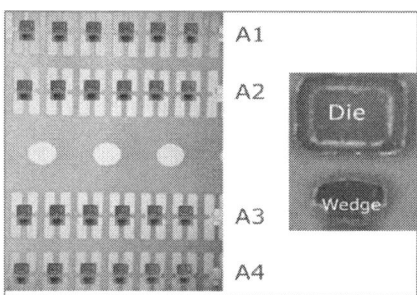

Fig. 2. Lead frame allocation

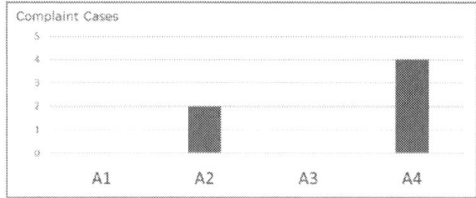

Fig. 3. Wedge reject mapping by track

Typical delamination as shown in Fig. 4. A1 and A3 shows higher DIA% on die pad, A2 and A4 much lower DIA%, but most wedge pad delamination is on A2/A4. Wedge pad delamination dominant on A2/A4(Fig.5). These are same location as defect cases (see Fig.3).

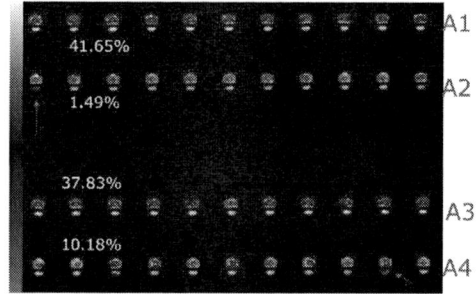

Fig. 4. CSAM mapping by track

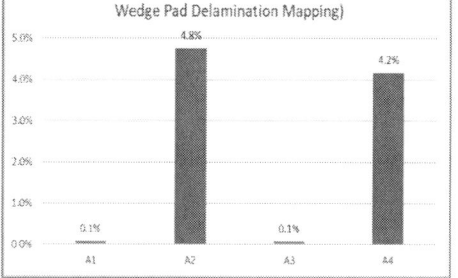

Fig. 5. Wedge pad delamination percentage mapping

Short shot was collected with half mold injection (see Fig.6). Mold filling sequence of A1/A3 was from wedge to die pad, die pad was end of mold flow, while A2/A4 was from die to wedge pad, wedge pad was end of mold flow. End of flow area matches with high DIA% area and reject location.

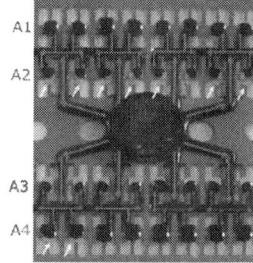

Fig. 6. Mold Flow with Short shot

To get further insights on end of flow effect. Viscosity curve were collected with 2 EMCs (see Fig.7). Under high temperature, EMC becomes melted with lower viscosity, suitable of filling into mold cavities, with time goes on, EMC polymerization continues to form 3D network and viscosity increase. Subramanian N.R. etc. all1 reported on mold flow influence on delamination performance in 2018[1]. Mold engineers always try to avoid too high viscosity at end of flow area, because high viscosity lead to poor wetting on LF surface (see Fig. 8 a). Which will lead to poor adhesion between EMC to LF, thus poor delamination. EMC type E(blue curve) curing speed is faster than EMC type G, faster viscosity increase at end of flow (see Fig.7) and confirms with curing time collected through DSC testing(see Table 1). Type E have shorter curing time than type G, 5.8min will finish full curing under DSC testing condition.

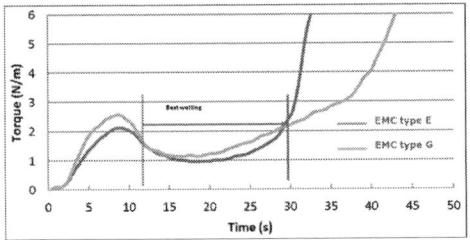

Fig. 7. Viscosity Curve EMC Type E Vs Type G

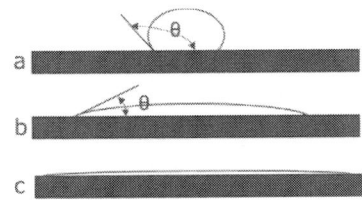

Fig. 8. Different wetting skeleton
(a) θ>90°, no wetting, (b) θ<90°, wetting, (c) θ=0° , 100% wetting

TABLE I.   CURING SPEED COMPARISION

DSC Heating rate	Reaction duration (min)	
	EMC type E	EMC type G
5℃/min	5.8	7.2

a. Data from HKUST characterization report

B. Remove Dedicated Reject Location

After understand correlation between mold end of flow at row A2/A4 and high delamination occurrence, one trial arranged with swapped die and wedge pad on A2/A4(see Fig. 9). To change wedge pad from end of mold flow to start of flow. Another method is to increase mold injection transfer speed. Fill in wedge area with lower viscosity.

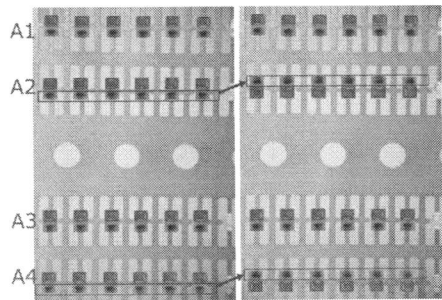

Fig. 9. Swap A2/A4 die and wedge.

• Die and Wedge Swap on Row A2 and A4

Engineering LF sample were made with swapped die and wedge on A2 and A4 rows (Fig.9).

Swapped die/wedge LF sample shows no delamination on wedge pad (see Table II, Fig.10), while control LF sample shows delamination on A2/A4 wedge pad. Other assembly performance is comparable. Chemical ingression checking shows discoloration on control samples (see Fig.11), conforms to delamination results.

TABLE II.   WEDGE DELAMINATION PERCENTAGE

Samples	Die/Wedge swap LF	No swap LF
A1	0%	0%
A2	0%	21%
A3	0%	0%
A4	0%	15%

b. 48 devices each, A2 have 7 devices' wedge shows delamination, that is 7/48=21%

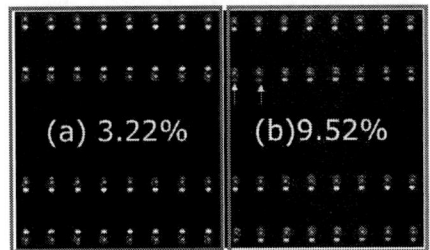

Fig. 10, CSAM photo for (a)die/wedge swap, (b)Control sample

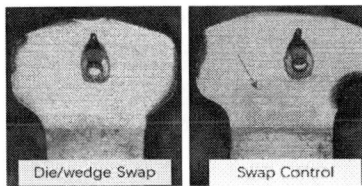

Fig. 11, Chemical ingression comparison

• Mold transfer Speed Increase

Short shot with higher transfer speed shows better cavity filling (Table III). With consideration of machine capability, only slightly increase on transfer speed from 100mm/S + 6mm/S to 130mm/S + 6.5mm/S.

Apple to apple comparison result shows reduced DIA% from 18.2% to 15.04% (see Fig. 12), while most important is that wedge pad delamination reduced from 8 per 44 devices to 2 per 44 devices.

TABLE III.   SHORT SHOT WITH TRANSFER SPEED

1st Speed	2nd Speed	Row 3 filling @ -5	Row 3 filling @ -4	Row 3 filling @ -3
150mm/S	10mm/S			
	8mm/S			
80mm/S	8mm/S			
80mm/S	6mm/S			
	4mm/S			

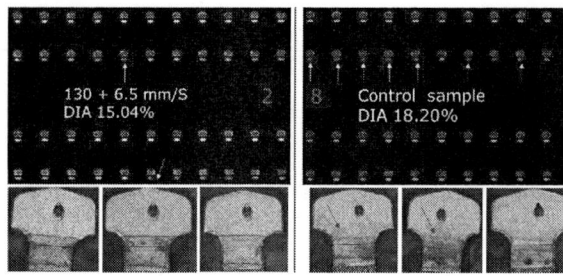

Fig. 12, CSAM photo and chemical ingression between higher speed and control sample

IV. CONCLUSION

Based on above results, with thorough reject case mapping against lead frame allocation and mold flow short shot analysis, EMC material property analysis, end of mold flow effect is the main cause of localized delamination. Which can be avoided by proper lead frame structure design and mold process set up with different EMC types.

A. Wedge pad delamination correlates well with reject cases

- Rejects could be reduced by A2/A4 die and wedge swap on LF strip.

- New package LF design and mold flow design, EMC selection should consider end of flow effect

B. Mold transfer speed optimization help reduced wedge pad delamination

- Mold process should adapt to different EMC type, pay attention to new EMC introduction.

ACKNOWLEDGMENT

Authors would like to thank Nexperia Dongguan site management support on this study, QA, FA and manufacturing team on related supports. EMC supplier on cooperation, support on improvement activities and mechanism discussion.

REFERENCES

[1] Subramanian N.R., Reynoso Dexter, Fabricante Joel, Yazid Mohamad, Evaluating delamination in molded packages with Process Simulation. 20th Electronics Packaging Technology Conference. P811-814. 2018

[2] Jang-Kyo KIM, Mohamed LEBBAI, etc. Interface adhesion between copper LF and EMC: Effect of Surface Finish, Oxidation and Dimples. Electronic Components and Technology Conference; 0-7803-5908-9. 2000

Electromechanical Co-design and Experimental Testing of Package Layer in Structurally Embedded Phased Array Antenna

Jinzhu Zhou
Key Laboratory of Electronic Equipment Structure Design, Ministry of Education, Xidian University
Xi'an, China
xidian_jzzhou@126.com

Zhenyu Gu
Key Laboratory of Electronic Equipment Structure Design, Ministry of Education, Xidian University
Xi'an, China
1600509285@qq.com

Yu Si
Key Laboratory of Electronic Equipment Structure Design, Ministry of Education, Xidian University
Xi'an, China
953954112@qq.com

Mei Wang
China Electronics Technology Group
38th Research Institute
Hefei, China
15357946086@163.com

Ping'an Wang
China Electronics Technology Group
38th Research Institute
Hefei, China
9764777@qq.com

Abstract—Structurally embedded phased array antenna (SEPAA) provides a new paradigm where the structural surface of a mobile vehicle acts as an antenna. This letter presents an optimal design method for a new SEPAA which consists of a package layer, a RF layer and a power-signal process layer. Electromechanical co-design model of the package layer was formulated to obtain an optimal mechanical and electromagnetic performance of the SEPAA. Moreover, in order to efficiently solve the model, this paper also proposes a parallel Bayesian optimization (PBO), which can utilize the parallel computing of EM simulations to alleviate the computational cost. The comparison results show that the PBO achieves better optimization results with significantly less iteration numbers, and the measured results of the fabricated Ka-band prototype demonstrate that the designed SEPAA has good mechanical load-bearing and radiation performance.

Keywords—Phased array antenna, conformal load-bearing antenna, Bayesian optimization, Package layer

I. INTRODUCTION

Recently, the growing demand for mobile data has promoted in-depth research on 5G communications, which can provide new applications such as uncompressed video streaming, mobile distributed computing and fast transfer of large files in high-speed mobile environments. However, in practice, the available space for moving vehicles is limited, so it is hoped that phased array antennas can be embedded into the structural surfaces of airplanes, high-speed trains, automobiles or ships [1].

SEPAA provides a new idea for the future development of mobile vehicle radar and wireless communications when the structural surface of a mobile vehicle becomes an antenna. SEPAA is a highly integrated phased array antenna. In order to make the highly integrated phased array antenna have a certain bearing capacity. In recent years, the antenna with bearing function has also become the focus of research. Researchers put forward many concepts such as structural integrated antenna, conformal bearing antenna and composite antenna array. However, almost all the work has been devoted to the study of the structural embedding form of passive antennas. In practical application, phased array antenna is usually needed because of its strong beam

This work was supported by National Natural Science Foundation of China (No.52175247, No.51775405), Defense Basic Research Program (No.61404130405). (Corresponding author: Jinzhu Zhou).

scanning ability.

This letter presents a new SEPAA structure with mechanical load-bearing and beam scanning capability. The SEPAA consists of a package layer, a RF layer and a power-signal process layer. The design of the package layer and RF layer involves accurate simulations of the mechanical and electromagnetic (EM) performances. Therefore, high fidelity numerical models are usually required to enhance the performance reliability of the fabricated SEPAA. However, the high-fidelity models usually cause high computational cost.

To address the expensive design problems, Bayesian optimization (BO) has been proposed in recent years [2], However, the application of BO for antenna design optimization is rare. Moreover, current BO select only one updated sample and then update the agent model with a single EM simulation each optimization cycle. When the parallel computing can be used, the solution is time-wasting.

This letter focuses on the electromechanical co-design problem of the package layer using the proposed PBO. The contributions of this work include: (1) A novel SEPAA structure is proposed. (2) An electromechanical co-design model is formulated to determine the thickness of the package layer in the SEPAA. (3) A modified BO called PBO is proposed to speed up the solution efficiency of the electromechanical co-design.

II. ELECTROMECHANICAL CO-DESIGN FORMULATION

Fig.1 presents the structural configuration of the proposed SEPAA, which consists of a package layer, an RF layer and a power-signal process layer. The package layer, which consists of the protective superstrates and encapsulate shell, can achieve impact and environmental resistances. The RF layer consists of a 52-element antenna array, tile Tx modules, microchannel heat sinks and a stripline feeding network, as shown in Fig.1 (b). How to integrate the components into a thin RF layer has been reported in our previous work [3]. The power-signal process layer, which generates the beam forming signal and power, comprises a direct current power and beam control circuit. The superstrates in the package consists of the lower foam, upper face sheet, upper foam and lower face sheet, as shown in Fig. 2. The superstrates and the RF layer are bonded to form the SEPAA by using the epoxy adhesive. The load-bearing framework and encapsulate shell are fabricated by using aluminum.

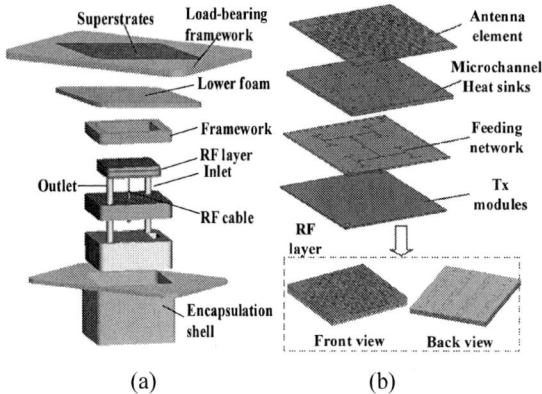

(a)   (b)

Fig. 1. Schematic diagram of structurally embedded phased array antenna. (a) Overall structural configuration. (b) Exploded view of the RF layer.

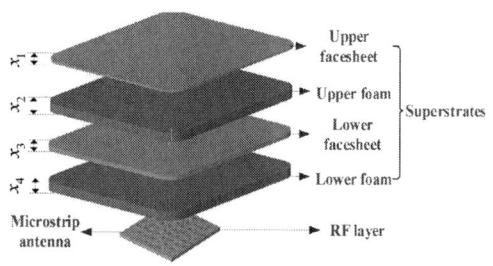

Fig. 2. Structural configuration of the RF layer and the superstrates.

The superstrates shown in Fig. 2 not only contributes to the mechanical performance, but also influences the radiation performance. Therefore, this letter proposes an electromechanical co-design method to determine the superstrates. Define the variable $x = [x_1, x_2, x_3, x_4]^T$ as the thickness of the face sheet and the foam in the superstrates, as shown in Fig. 2. Based on the optimal design of the RF layer from our previous work [3], an electromechanical co-design model is proposed to determine the thickness:

$$\text{Find: } x = [x_1, x_2, x_3, x_4]^T$$
$$\text{Min: } -G(x) \quad (1)$$
$$\text{s.t.} \begin{cases} \delta_{max}(x) - 0.1 \leq 0 \\ \sigma_{max}(x) - 10 \leq 0 \\ x_l \leq x \leq x_u \end{cases}$$

where $G(x)$ denotes the gain of the antenna array with the superstrates. $x_l = [0.1, 1, 0.1, 1]^T$ mm and $x_u = [2, 10, 2, 10]^T$ mm are the lower bound and the upper bound of the definition. The maximum displacement $\delta_{max}(x)$ and stress $\sigma_{max}(x)$ denote the mechanical stiffness and strength.

In this work, the maximum displacement and stress are determined using a mechanical model subjected to a maximum load-bearing pressure 83300 Pa. The gain of the SEPAA is calculated using an EM simulator (i.e. HFSS). Because of the EM model is complex, its computation cost is high. This letter presents a parallel computing-based BO to address the problem.

III. PREPARE SOLUTION ALGORITHM

Bayesian optimization is a typical auxiliary method for agent model. [2]. For the problem (1), a Gaussian process (GP) proxy model is established by making full use of the known data set $\{(x_i, y_i), x_i \in R^d, y_i \in R\}_{i=1}^{n-1}$ to approximate the function $y(x)$. According to, after a new design point x_n is obtained by solving an acquisition function, the response of GP proxy model at the new point obeys normal distribution, and its mean and variance functions are expressed as:

$$\mu(x_n) = k^T [K + \sigma^2 I] y$$
$$s^2(x_n) = k(x_n, x_n) - k[K + \sigma^2 I]^{-1} k^{-1} \quad (2)$$

where $k(x_i, x_j)$ is expressed as:

$$k(x_i, x_j) = \exp\left(-\frac{1}{2\theta}\|x_i - x_j\|^2\right) \quad (3)$$

where θ is the length-scale parameter.

As discussed in [4], the surrogate model $\hat{y}(x_n)$ is constantly updated according to new point x_n during each iteration. In this work, the expected improvement (EI) is chosen as the acquisition function. Utilizing the EI, the updated point x_n in each optimization cycle is expressed as:

$$x_n = \arg\max_{x \in \Gamma} E(x) \quad (4)$$

$$E(x) = (y_{min} - \hat{y}(x))\Phi\left(\frac{y_{min} - \hat{y}(x)}{s(x)}\right) + s(x)\phi\left(\frac{y_{min} - \hat{y}(x)}{s(x)}\right) \quad (5)$$

where $\Phi(\cdot)$ and $\phi(\cdot)$ denote the CDF and PDF of the normal distribution, respectively [2]. y_{min} is the current best value.

As shown in (4), under the unknown response of the first update point EM, the function is unable to select the second update point, thus hindering the parallel calculation of BO. This paper presents a simple multi-point fetch function to generate update points for parallel computation:

$$x_n^q = \arg\max_{x \in \Gamma}\left\{E(x) \cdot \prod_{i=1}^{q-1}[1 - k(x, x_n^i)]\right\} \quad (6)$$

Where x_n^q is the q-th sample chosen at the n-th iteration.

Substituting into , one can obtain a modified acquisition function. It should be noted that the acquisition function at $q = 1$ degenerates to the standard function .

As the process goes on, the q-th updating point at n iteration is identified by using , after $q-1$ points were determined. In this work, the acquisition functions are solved using a Nelder-Meads simplex algorithm in Matlab. The modified BO called parallel PBO is summarized as follows.

Algorithm: Pseudocode of the proposed PBO algorithm

Input: Maximum iteration number N, Updating number q.

1: Produce initial dataset $\{X, Y\}$;

2: **for** $j = 1, 2, \cdots, N$ **do**

3:    Construct the surrogate model $\hat{y}(x)$ using current data $\{X, Y\}$;

4:    **for** $i = 1$ to q **do**

5:        Select the update point x_n^j using ;

6:    **end for**

7: Evaluating $\left\{ y\left(x_n^1\right),\cdots,y\left(x_n^q\right)\right\}$ in parallel using EM

    simulations;

8: $X \leftarrow X \cup \left\{ x_n^1,\cdots,x_n^q\right\}$, $Y \leftarrow Y \cup \left\{ y\left(x_n^1\right),\cdots,y\left(x_n^q\right)\right\}$;

9: $y_{\min} \leftarrow \min(Y)$; $x_{\min} \leftarrow x \in X : y(x)=y_{\min}$;

10: **end for**

Output: minimum $y_{\min}$ and $x_{\min}$.

IV.   DESIGN RESULTS

This section applies the PBO to solve . Moreover, the PBO is compared with two algorithms in the field of EM design: SADEA in [5] and BO-EI in [4]. The initial design is $x = [2,10,2,10]$ mm. The same 15 initial data is applied to build the initial surrogate models of three algorithms. Fig. 3 shows the comparison of convergence trends of the objective function. Table I shows the comparisons of solution efficiency. It is observed that the PBO achieves bigger gain with less iteration number, and that the CPU time cost of the PBO reduces about 60%, compared with other algorithms.

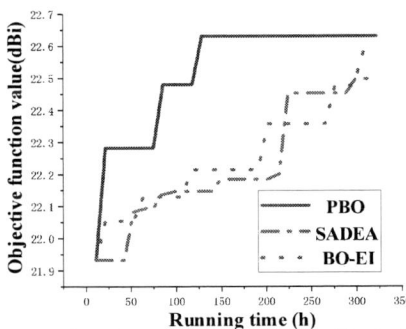

Fig. 3. Comparison of convergence trends of the objective function.

TABLE I
COMPARISONS OF SOLUTION EFFICIENCY

Algorithm	No. of iteration	No. of EM. simulation	CPU time (h)
PBO	12	24	128.6
BO-EI	30	30	318.4
SADEA	30	30	320.3

The optimal thickness of the superstrates determined by the PBO algorithm is $x = [0.25,1.02,0.25,4.67]$ mm. The radiation patterns of the antenna array with the initial and optimal superstrates were simulated, respectively. Fig. 4 (a) presents the comparisons of the simulated radiation patterns. Fig. 4 (a) shows that the gains of the antenna array with the initial and optimal design of the superstrates are 19.83 dBi and 22.63 dBi, respectively, and that the gain of the antenna array without the superstrates is 22.37 dBi. From the comparisons, it is found that the superstrates with the optimal design will not result in the reduction of the SEPAA gain. On the contrary, because of the protective superstrates, the mechanical load-bearing performance of the SEPAA can be enhanced, as shown in Fig. 5. From Fig. 5, it is also observed that show both the maximum displacement and stress are smaller than the allowable displacement 0.1 mm and stress 10 MPa. It follows that the co-design can simultaneously meet mechanical and electrical performance,

and that the simulated optimal peak gain (22.63 dBi) which is higher than the predefined specification (20 dBi).

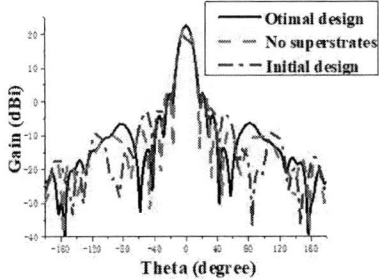

Fig. 4. Comparisons of different radiation patterns at xoz-plane at 30 GHz.

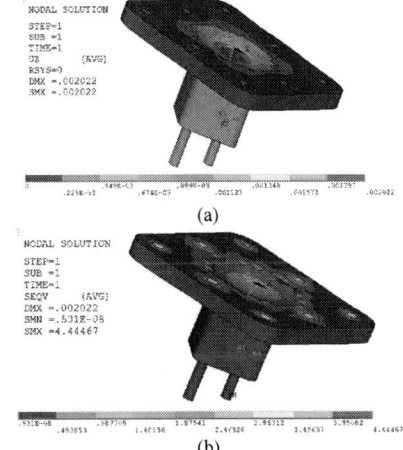

Fig.5. Mechanical performance of the optimal SEPAA subjected to a pressure 83300Pa. (a) Maximum displacement 0.002mm. (b) Maximum stress 4.44MPa.

V. EXPERIMENTAL TESTING

Utilizing the optimal design above, a Ka-band SEPAA prototype was fabricated, as shown in Fig. 5. The measured results are provided in the following.

(a)

Fig. 5. Fabricated SEPAA prototype.

A. Measured results of radiation performance

Fig. 6 contrasts the measured and simulated gain and AR at different frequency. It is observed that the measured results are in good agreement with the simulation results.

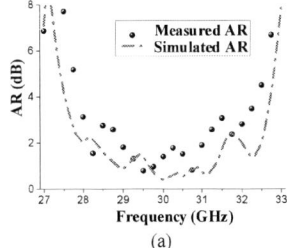

(a)

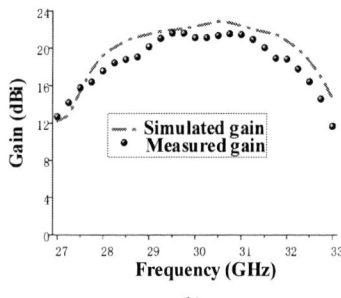

(b)

Fig. 6. Comparisons of measured and simulated gain and AR at different frequency. (a) AR. (b) Gain.

Fig. 7 contrasts the measured and simulated scanning patterns at 30 GHz. The measured scan range is −45°~ 45°, and the measured beam direction is slightly different from the expected direction (−45°~ +45°). The difference may be due to manufacturing errors. The gain fluctuation in the scanning range is less than 3.5dBi. Gain fluctuation is due to the element pattern flattening due to mutual coupling.

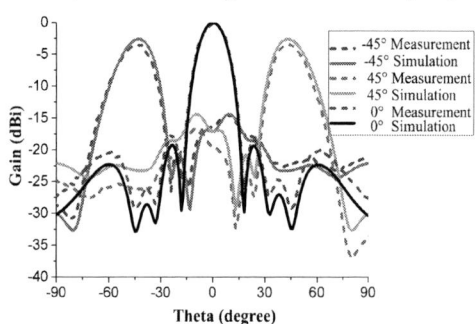

Fig. 7. Comparisons of measured and simulated scanning patterns at 30 GHz.

B. Measured results of load-bearing performance

Fig. 8 shows the impact experiments to evaluate the mechanical load-bearing function. The applied impact energies were 12.3 J and 15.2 J. From the results, it is observed that the maximum contact force is approximately 3 KN when the impact energy is 12.3 J. Under the 12.3 J impact conditions, this prototype exhibits excellent structural integrity and impact resistance. However, when the prototype was subjected to a 15.2 J of impact energy, a crack appears on the lower face sheet of the superstrates. The results imply that the package layer can effectively protect the RF layer, if the impact energy is not greater than the critical value.

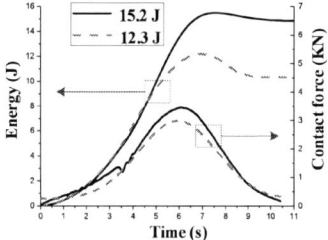

Fig. 8. Experimental results of impact test

VI. CONLUSION

This paper presents an electromechanical co-design model for designing the package layer of a new SEPAA, and the model is efficiently solved using the proposed PBO. A full-size prototype was fabricated, and its peak gain is 21.6

dBi at 30 GHz. The gain fluctuations within the scanning range of −45° to 45° are smaller than 3.5 dB, and the side-lobe level is better than -17.6 dB. The impact experiments show that the package layer can effectively protect the RF layer, if the impact energy is not greater than 15.2 J. The experimental results demonstrate that the designed SEPAA has good mechanical load-bearing and radiation performance.

REFERENCES

[1] J. S. Herd and M. D. Conway, "The evolution to modern phased array architectures," *Proceedings of the IEEE*, vol. 104, no. 3, pp. 519-529, Mar 2016.

[2] B. Shahriari, K. Swersky, Z. Y. Wang, R. P. Adams, and N. de Freitas, "Taking the Human Out of the Loop: A Review of Bayesian Optimization," *Proceedings of the IEEE*, vol. 104, no. 1, pp. 148-175, Jan 2016.

[3] J. Zhou, L. Yin, L. Kang, M. Wang, and J. Huang, "Joint Design and Experimental Tests of Highly Integrated Phased-Array Antenna with Microchannel Heat Sinks," *IEEE Antennas and Wireless Propagation Letters*, vol. 18, no. 11, pp. 2370 - 2374, Nov 2019.

[4] P. Chen, B. M. Merrick, and T. J. Brazil, "Bayesian Optimization for Broadband High-Efficiency Power Amplifier Designs," *IEEE Transactions on Microwave Theory and Techniques*, vol. 63, no. 12, pp. 4263-4272, Dec 2015.

[5] B. Liu, H. Aliakbarian, Z. K. Ma, G. A. E. Vandenbosch, G. Gielen, and P. Excell, "An Efficient Method for Antenna Design Optimization Based on Evolutionary Computation and Machine Learning Techniques," *IEEE Transactions on Antennas and Propagation*, vol. 62, pp. 7-18, Jan 2014.

The Formation of Cn-Sn IMC Interconnection by Solid–Liquid Interdiffusion Bonding for 3D Glass Wafer Stacking

Yangquan Su
Xiamen University
School of Electronic Science and Engineering
Xiamen, China
23120191150240@stu.xmu.edu.cn

Kuili Ren
Xiamen Sky Semiconductor
Xiamen, China
renkl@sky-semi.com

Yiyong Huang
Xiamen Sky Semiconductor
Xiamen, China
huangyy@sky-semi.com

Mingchuan Zhang
Xiamen Sky Semiconductor
Xiamen, China
zhangmc@sky-semi.com

Daquan Yu*
Xiamen University
School of Electronic Science and Engineering
Xiamen Sky Semiconductor
Xiamen, China
yudaquan@xmu.edu.cn

Abstract—**Wafer bonding technology promotes 3D system integration and packaing. In this paper, the Cu/Sn low-temperature bonding for 3D glass wafer stacking is studied. The effects of temperature, pressure and time on the bonding process are demonstrated by experiments and simulations. The formation of Cu$_3$Sn is incorporated into Cu/Sn bonding interface. For 5G devices, such as the applications of millimeter wave, terahertz and internet of things (IoT), advanced packaging using glass substrate with excellent electrical properties is promising. Based on the laser inducing and wet etching, through glass vias (TGV) were fabricated , and the vias wre filled with Cu. Then, the Cu/Sn RDL is electroplated to form a stacked structure for multi-layer bonding. To optimize the process, Cu/Sn bonding surface and TGV cross section are analyzed by scanning electron microscope (SEM). And the ratio of Cu/Sn alloy are measured with an energy spectrometer. Based on X-ray photography, there are no bonding defects. The seal ring formed by bonding can effectively protect the electrical signals transmitted in the TGV. There is no leakage in the seal ring, the amount of Sn overflow during the bonding process is limited, and it does not affect the TGV and metal trace near the sealing ring.**

Keywords—3D Packaging; multilayer stacked; Cu/Sn bonding; Through Glass Vias(TGVs)

I. INTRODUCTION

With the development of the internet, high-performance computing, artificial intelligence, automotive electronics, and 5G, the develop direction of microelectronic packaging technology is high integration and high performance [1-5]. Due to its small size, high-density integration, low RC delay, and other advantages, 3D-TSV technology has become a research hotspot in the current packaging industry. However, 3D-TSV technology has some weak points, such as high electrical signal loss and complex TSV interposer process.

TGV technology has many advantages, such as low substrate loss, low substrate material cost, excellent high-frequency electrical performance, and good light transmittance [19]. So it has been becoming one of the most important development direction of three-dimensional integration technology. Especially in the field of 5G (5G Sub-6GHz, 5G Internet of Things, 5G millimeter wave)

high-speed and high-frequency application requirements, millimeter-wave antennas, RF front-ends, and other large-bandwidth, low-loss high-frequency transmission devices, etc.,3D-TGV The technology has broad application prospects. The key processes of 3D-TGV technology include through-hole preparation, through-hole filling, RDL preparation, and multilayer bonding.

Bonding is one of the key technologies of 3D-TSV/TGV packaging. Currently, the most popular bonding technology including metal bonding, Si-Si bonding, and Dielectric layer bonding. Metal bonding is the most important bonding method because of its good electrical and thermodynamic properties. Among them, the most widely-used is Cu-Cu thermal bonding technology, but the bonding temperature is usually at 350℃ to 400℃. Generally, the performance and reliability of electronic device can be affected by high temperature [6]. Cu-Sn bonding belongs to low-temperature bonding, with good mechanical, thermal, and electrical properties. Especially, its metal compound Cu$_6$Sn$_5$ and Cu$_3$Sn both has high melting point. The melting point of Cu$_6$Sn$_5$ is 415°C and Cu$_3$Sn is 650°C, and both of them can meet the needs of high-temperature environments. Due to its advantages Cu-Sn bonding has potential application prospects [7]. In this article, speculative knowledge about Cu-Sn bonding is researched ply, and the stacking of glass wafer with TGV is realized by using Cu-Sn bonding technology.

II. EXPERIMENTAL PROCESS

A. Process flow

The process flow of the first layer glass wafer is shown in Fig.1.a mainly including: laser induce, HF etching, seed layer deposition, electroplating filling, CMP, RDL preparation, bump, and seed layer etching. The process flow of the second layer glass wafer is shown in Fig.1.b mainly including: laser induce, HF etching, seed layer deposition, electroplating filling, CMP, RDL preparation, temporary bonding, glass thinning, bump, seed layer corrosion. AF32, a type of 6-inch glass, was used since its performance like suitable CTE and low power signal loss. This research is based on wafer bonding of two thicknesses—300μm and

$100 \mu m$. Both layers of glass wafer have TGV. Temporary bonding is added to apparent reducing the warpage cause some process(temperature higher than 100°C) such as baking and bonding. In high temperature environment, the glass wafer regular sink into warp due to the different CTE between glass wafer and other metal layer. The stress generated by plating process can also easily cause the thin glass to warp. To realize temporary bonding, 6-inch, AF32 glass wafer with thickness of $400 \mu m$ is selected.

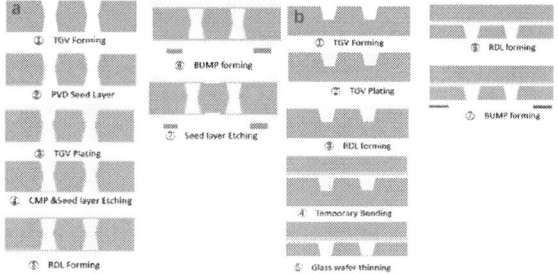

Fig. 1. process flow. (a) first layer of glass; (b) second layer of glass.

B. TGV via Laser induced and wet etching

In this paper, The glass substrates irradiated a picosecond laser with frequency of 300kHz emerge a permanent modification of the glass which triggers an anisotropic etching. Then, the 10% HF solution in water was used to etch the modified glass at room temperature. The etching time is extremely significant needed to make sure that the wanted TGVs were completely etched through. Laser-induced HF etching is used to form through vias and blind vias. Fig.2.a shows the top surface morphology of glass wafer with through vias. After etching the diameter of through-vias is $60\pm5 \mu m$, and the wafer thickness is $230\pm5 \mu m$. The sidewall morphology of the vias with a slope of $80\pm3°$ showed in Fig.2.b. Fig.2.c shows the top surface morphology of glass wafer with blind vias. After etching the diameter of blind-vias is $60\pm5 \mu m$, and the wafer thickness is $240\pm5 \mu m$. The sidewall morphology of the blind vias, as Fig.2.d shows, attain $130\pm10 \mu m$ depth and the diameter is $25\pm10 \mu m$ at $100 \mu m$ depth. Plating process is used to completely fill TGV, and then X-Ray is used to check the plating quality. X-Ray inspection of through vias is showed in Fig.2. The TGV filled was in accordance with the standard, no holes and no disconnection.

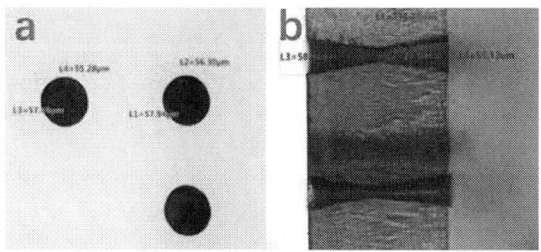

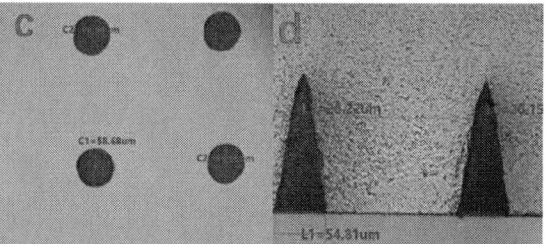

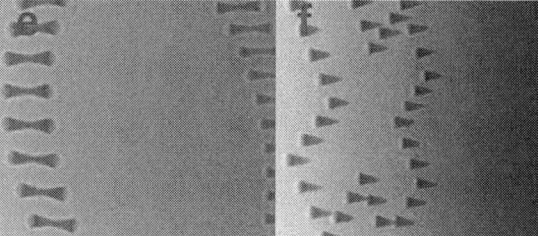

Fig. 2. The view of TGV and blind TGV. (a) and (b) show about TGV;(c) and (d) show about blind TGV;(e) and (f) show about X-RAY.

C. Formation of RDL and bump

The RDL design is showed in Fig.3, stress-relief holes are added to avoid metal delamination in the large metal area which is caused by excessive stress after plating. For Cu3Sn become the principal IMC, the ratio of Cu and Sn must above 5/3 in the system [21]. In this paper, the thickness of the Cu RDL layer is $6 \mu m$. and the height of bump is $2 \mu m/1 \mu m$(Cu/Sn). The pattern of two glass wafer is mirrored. The seed layer is formed by PVD, and the RDL layer by electroplating. The bump is developed by two steps--photolithography and electroplating again. The thickness of Cu layer and Sn layer can be to ensure that the final metal compound formed is Cu_3Sn and Cu_6Sn_5, the thickness ratio of the Cu layer and Sn layer should be more than 3:1.

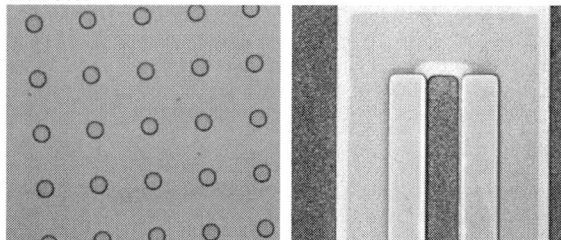

Fig. 3. The enlarged view of RDL for bonding with Cu/Sn.

III. EXPERIMENT OF CU/SN BONDING

Cu-Sn bonding is a method which triggers solid-liquid inter-diffusion into an intermetallic compound (IMC). Bonding can be rapid achieved via migrating to each other at relatively low temperatures converting easily from low melting-point Sn and solid Cu to compounds.

A. Theory about Cu-Sn bonding

Fig.4 shows the equilibrium phase diagram of Cu/Sn binary alloy [8]. It can be seen from the figure that there are several kinds of metal compounds in the Cu-Sn system, and η Phase (Cu_6Sn_5) and ε(Cu_3Sn) phase is the most important mesophase compound. Cu_6Sn_5 will melt to liquid and transform into ε phase at 415 ℃, Cu_3Sn will melt and transform into ɤ phase at 676 ℃. Both compounds can meet the needs of high temperatures environment. The reaction mechanism and reaction process of Cu-Sn have been widely studied [9-17]. K.N.Tu et al [9], studied the diffusion between Cu and Sn in the solid state. According to their research, there are three conclusions about the generation of Cu_6Sn_5 and Cu_3Sn. First, Cu_6Sn_5 begins to grow when Cu film and Sn film contact at 25℃, and its thickness is linearly related to the square root of Cu-Sn diffusion time. Second, when the temperature rises to above 50 ℃, Cu_3Sn begins to form. Third, in Cu_6Sn_5 the diffusion rate of Cu is higher than that of Sn. Experimental verification of Cu-Sn diffusion in different temperature is completed by C. R. Kao et al [18]. The experimental temperature is 240 ℃, 250 ℃, and 275 ℃. They found that the growth of mesophase completely conformed to the parabolic growth rule, and the activation energy of growth was 29kj / mol.

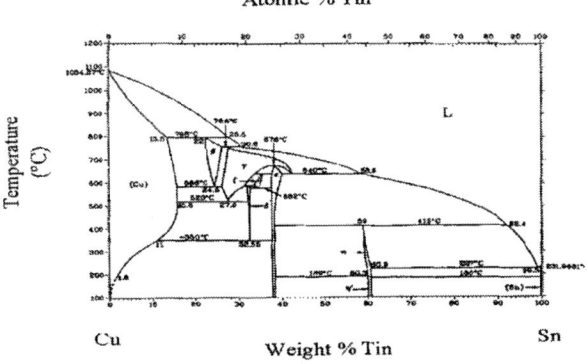

Fig. 4. Cu/Sn equilibrium phase diagram of binary alloy [8].

B. Experimental design and result analysis of Cu/Sn bonding

According to the above theoretical research, 3 sets of bonding experiments are designed, and the bonding parameters are shown in Table I:

TABLE I. BONDING PARAMETER

No.	Pressure (Mpa)	Temperature (℃)	Time (min)	Cu THK (μm)	Sn THK (μm)
1	5.8	350	30	8	2
2	5.8	280	30	8	2
3	5.3	280	30	8	2

The first set of wafer bonding results are shown in Fig.5. It is the SEM inspection image, the bonding layer has no holes, and the bonding surface metal compound is Cu_3Sn without Cu_6Sn_5. The first group of bonding temperature is 350℃, it is too high to keep the performance of MEMS devices. Reduce its bonding temperature to do the second group bonding experiment. The pressure of 5.3 MPa was applied in the first two groups, but the third set of experiments reduces the bonding pressure in order to lower the temperature to reduce the cost.

Fig. 5. The hole for releasing stress, and the bump for bonding.

The second group of wafer bonding results are shown in Fig.6, which shows the condition of whole wafer after bonding. As the picture shows, the two wafer is bonded no creak. Fig.6b is the X-Ray inspection image. As the picture showed, the wafer bonding yield is 100%. Fig.6c is SEM inspection image, the bonding layer has no holes, and the bonding surface metal compounds are Cu_3Sn and Cu_6Sn_5.

Fig. 6. (a) The top view of whole wafer; (b) X-ray inspection;(c) the section checked by SEM.

The third group of wafer bonding results is shown in Fig.7. Fig.7.a shows the condition of whole wafer after bonding, it is checked by X-Ray. As the picture shows, bonded wafers were connected but no creak, in addition, the wafer bonding yield is 100%. Fig.7.b is SEM inspection image, where the metal compounds of bonding interface are Cu_3Sn and Cu_6Sn_5.

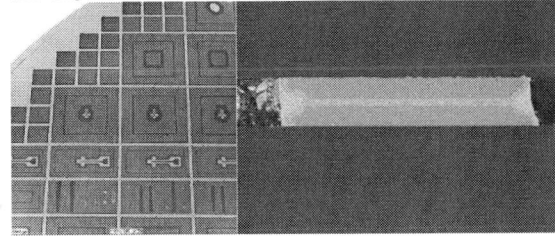

Fig. 7. The inspection shows by X-ray, and the section checked by SEM.

According to these three experiments, in the premise of ensuring the stability of the electrical signal, in this paper the bonding parameter, pressure and time is 280℃, 5.3MPa, and 30min, respectively.

C. Bonding simulation

Take thermal stress for account, the warping of glass wafer had been predicted to warp based on the simulation model in Ansys. On the basis of finite element model analysis, the risk of warpage was expected controllable

978-1-6654-1392-3/21 $31.00 © 2021 IEEE

after bonding. Temporary bonding can reduce the risk, at the meantime maintain whole stability for Cu/Sn bonding.

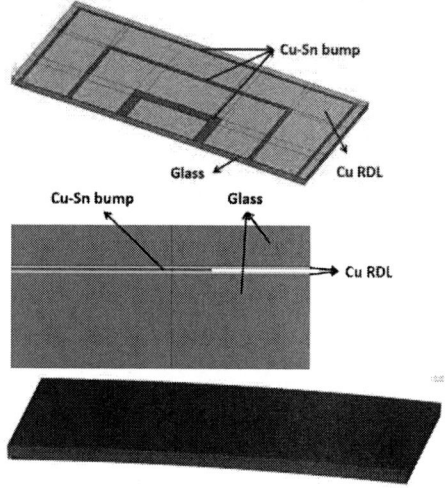

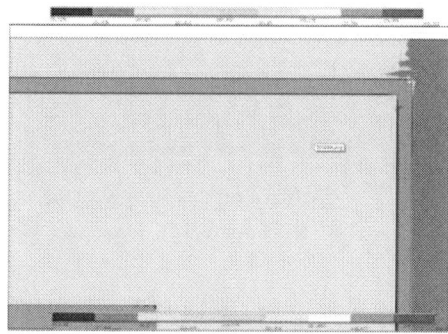

Fig. 8.   Simulation model and result by Ansys.

IV.   CONCLUSIONS

In this paper, the bonding using Cu/Sn for glass wafers with through vias was studied. On the basis of finite element simulation model, temporary bonding was adopted to solve aporia about the warpage of wafer-level bonding. Using glass wafer with TGVs is achieving electrical signal transmission while this method can reduce the loss of electrical signals on vias. Cu/Sn solid-liquid inter-diffusion was employed to realize wafer-level bonding, which can bring benefits such as low cost and high quality. A series of parametric studies are designed to optimize the experimental conditions of wafer-level bonding. The bonding joint are composed of Cu_3Sn and Cu_6Sn_5, which can meet the requirements for post process under high temperatures. Wafer bonding yield is 100%. And the best bonding results are obtained with 280 ℃ bonding temperature, 5.3MPa bonding pressure and 30min bonding time.

V.   REFERENCE

[1]   Rajat Girjashankar Pandey,Upadhyaya Trushit. Substrate integrated waveguide fed dual band quad-elements rectangular dielectric resonator MIMO antenna for millimeter wave 5G wireless communication systems[J]. AEUE - International Journal of Electronics and Communications,2021,137.

[2]   Ma Mingtao,Song Kaixin,Ji Yuping,Hussain Fayaz,Khesro Amir,Mao Minmin,Xue Lingyun,Xu Ping,Liu Bing,Lu Zhilun,Zhou Di,Wang Dawei,Sun Shikuan. 5G microstrip patch antenna and microwave dielectric properties of cold sintered $LiWVO_6$–K_2MoO_4 composite ceramics[J]. Ceramics International, 2021, 47(13).

[3]   Darryl Booth. Building Capacity Through LTE and 5G Wireless[J]. Journal of Environmental Health, 2021, 83(10).

[4]   Joel S. Biyoghe,Vipin Balyan. NOMA Application to Satellite Communication Networks for 5G: A Comprehensive Survey of Existing Studies[J]. Journal of Communications,2021,16(6).

[5]   Donny Jackson. China set to launch world's first in-flight 5G[J]. Urgent Communications,2021.

[6]   Sakuma K , Sueoka K , Kohara S , et al. IMC Bonding for 3D Interconnection[C]// 2010 Proceedings 60th Electronic Components and Technology Conference (ECTC). IEEE, 2010.

[7]   Yaping Lu, Xiaogang Liu, Mingxiang Chen. Multilayer stacked Cu / Sn bonding technology for 3D packaging with TSV structure[J]. semiconductor technology, 2014, 39(1):64-70.

[8]   Maohua Du. Application of Cu / Sn isothermal solidification bonding in MEMS hermetic packaging [D]. Graduate School of Chinese Academy of Sciences.

[9]   Tu K N . Interdiffusion and reaction in bimetallic Cu-Sn thin films[J]. Acta Metallurgica, 1973, 21(4):347-354.

[10] Tu K N , Thompson R D . Kinetics of interfacial reaction in bimetallic Cu-Sn thin films[J]. Acta Metallurgica, 1982, 30(5):947-952.

[11] Wu Jie,Xue Songbai,Yao Zhen,Long Weimin. Study on Microstructure and Properties of 12Ag–Cu–Zn–Sn Cadmium-Free Filler Metals with Trace in Addition[J]. Crystals,2021,11(5).

[12] Yue Wu,Ding Chao,Qin Hongbo,Gong Chenggong,Zhang Junxi. Crystallographic Characteristic Effect of Cu Substrate on Serrated Cathode Dissolution in Cu/Sn–3.0Ag–0.5Cu/Cu Solder Joints during Electromigration[J]. Materials,2021,14(10).

[13] Mechnik V. A.,Bondarenko N. A.,Kolodnitskyi V. M.,Zakiev V. I.,Zakiev I. M.,Gevorkyan E. S.,Kuzin N. O.,Yakushenko O. S.,Semak I. V.. Comparative Study of the Mechanical and Tribological Characteristics of Fe–Cu–Ni–Sn Composites with Different CrB_2 Content under Dry and Wet Friction[J]. Journal of Superhard Materials,2021,43(1).

[14] Sturm Cheryl,Macario Leilane R,Mori Takao,Kleinke Holger. Thermoelectric properties of zinc-doped $Cu_5Sn_2Se_7$ and $Cu_5Sn_2Te_7$.[J]. Dalton transactions (Cambridge, England : 2003),2021,50(19).

[15] Isiyaku Aliyu Kabiru,Ali Ahmad Hadi,Abdu Sadiq G.,Tahan Muliana,Raship Nur Amaliyana,Bakri Anis Suhaili,Nayan Nafarizal. Preparation of Sn doped In_2O_3 multilayer films on n-type Si with optoelectronics properties improved by using thin Al–Cu metals interlayer films[J]. Materials Science in Semiconductor Processing,2021,131.

[16] Canan Aksu Canbay,Oktay Karaduman,Nihan Ünlü,İskender Özkul,Mehmet Ali Çiçek. Energetic Behavior Study in Phase Transformations of High Temperature Cu–Al–X (X: Mn, Te, Sn, Hf) Shape Memory Alloys[J]. Transactions of the Indian Institute of Metals,2021(prepublish).

[17] Ritter Konrad,Gurieva Galina,Eckner Stefanie,Preiß Cora,Ritzer Maurizio,Hages Charles J.,Welter Edmund,Agrawal Rakesh,Schorr Susan,Schnohr Claudia S.. Atomic Scale Structure of $(Ag,Cu)_2ZnSnSe_4$ and $Cu_2Zn(Sn,Ge)Se_4$ Kesterite Thin Films [J]. Frontiers in Energy Research,2021.

[18] S, BADER, W, et al. RAPID FORMATION OF INTERMETALLIC COMPOUNDS BY INTERDIFFUSION IN THE Cu-Sn AND Ni-Sn SYSTEMS[J]. Acta Metallurgica Et Materialia, 1995.

[19] Wang Shuo Ma Kui Yang Fashun.A review on TSV reliability[J].Application of electronic technology, 2021, 47(02):1-6.

[20] LEE J B, AW J L, RHEE M W.3-D TSV six-die stacking and reliability assessment of 20-μ m-pitch bumps on large-scale dies[J]. IEEE Transactions on Components Packaging&Manufacturing Technology, 2017, 7 (1): 1-6.

[21] A. Munding, H. Hubner, A. Kaiser, S. Penka, P. Benkart, E. Kohn. Cu/Sn solid-liquid interdiffusion bonding. Wafer-Level 3D ICs Process Technology, 2008, pp. 136-139.

Research on the Reliability of Board Level Interconnect Solder Joints under Thermal-mechanical Coupling

Xin Liu
Detection Department
China Electronics Technology Group
Corporation NO.58 Research Institute
Wuxi,China
xin_liu305@163.com

Yongjian Yu
Detection Department
China Electronics Technology Group
Corporation NO.58 Research Institute
Wuxi,China
yuyj98@163.com

Weikun Xie
Detection Department
China Electronics Technology Group
Corporation NO.58 Research Institute
Wuxi,China
weikun@163.com

Kaihong Zhang
Detection Department
China Electronics Technology Group
Corporation NO.58 Research Institute
Wuxi,China
kaihong@163.com

Kai Zhu
Detection Department
China Electronics Technology Group
Corporation NO.58 Research Institute
Wuxi,China
1113247886@qq.com

Abstract—In this paper, through the method of finite element simulation, different power density loads are applied to solder joints in the board level interconnection system to make the maximum temperature of the system reach 50 °C, 75 °C and 100 °C respectively, and then the drop simulation is carried out to study the reliability of the board level interconnection solder joints under thermal-mechanical coupling. The simulation results show that compared with single drop impact, the application of thermal load reduces the 1st eigenfrequency of PCB and reduces the solder joint stress. With the increase of system temperature, the 1st eigenfrequency of the PCB and the solder joint stress is decreasing, but the solder joint life is increasing. The stress analysis in the height direction of the solder joint found that the bottom surface of the solder joint connected to the PCB board is the location of the largest stress on the solder joint. It is inferred that when the board level interconnection system is impacted by drop, this position will be damaged first.

Keywords—drop, thermal-mechanical coupling, reliability, simulation

I. INTRODUCTION

In the process of transportation and use of electronic products, board level interconnect solder joints are not only subjected to drop impact load, but also subjected to thermal-mechanical coupling (temperature-drop coupling), which puts forward higher requirements for the reliability of board level interconnect solder joints. In order to improve the reliability of electronic components, it is necessary to conduct research on thermal-mechanical coupling of board level interconnection system and experts and scholars at home and abroad have made great efforts.

Shen et al. [1] used a combination of experiment and simulation to study the effect of electrical load on the temperature field of solder joints and the life of solder joints. The study found that the temperature of solder joints increased with the increase of current density. Under the same vibration load, The solder joint life is significantly reduced. T. T. Mattila et al. [2] studied the reliability of wafer level package under drop impact load, and tested the drop impact life of the package when the maximum temperature of the chip is 25 °C, 75 °C, 100 °C and 125 °C, respectively. The research results show that temperature has a significant effect on the drop impact life of packaged components, from 25 °C to 75 °C, 75 °C to 100 °C, 100 °C to 125 °C solder joints life increased by 300%, 24%, 25%, respectively. Jiao et al. [3] used a combination of experiment and finite element simulation to study the failure mode and failure life of board level interconnect solder joints under thermal-mechanical coupling conditions; The results show that the temperature affects the crack propagation path of the solder joint, the plastic properties of the interconnect micro solder joint increase with the increase of temperature, part of the interfacial stress of the solder joint will be consumed by the deformation of the bulk solder, which affects the failure mode of the solder joint, makes the crack from the pad to the bulk solder, and the solder joint life increases with the increase of temperature.

In this paper, ANSYS finite element software is used to simulate the drop impact of the board level interconnection system at 25 °C (room temperature) and the maximum temperature at 50 °C, 75 °C and 100 °C. The effect of thermal load on the reliability of solder joints of board level interconnection is studied.

II. MODELING AND SIMULATION

A. Geometric Model and Mesh Generation

Fig.1 is a geometric model conforming to JESD22-B111A drop test standard issued by JEDEC in 2016. It can be seen from the figure that four chips are symmetrically distributed on the front of JESD22-B111A standard test board, and there is a chip in the center of the back of the standard test board.

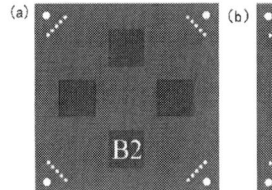

(a) JESD22-B111A front　　(b) JESD22-B111A back
Fig. 1. JESD22-B111A board

The layout of each chip's lower solder joint is 10 x 10, and the mesh of B1 and B2 lower solder joints is refined, as shown in Fig. 2. The geometric dimensions of JESD22-B111A board are shown in Table I.

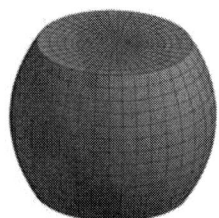

Fig. 2. Solder joint grid refinement

TABLE I.   JESD22-B111A GEOMETRY SIZE

Spec. No.	JESD22-B111A board
PCB size	77 mm×77 mm×1 mm
Solder size	SAC305 solder ball maximum diameter 0.4 mm
Chip size	15 mm×15 mm×1 mm

B. Material parameters

This paper not only established a finite element model for drop impact at room temperature, but also established a finite element model for drop impact of board level systems at different maximum temperatures (50 °C, 75 °C, 100 °C). Since the simulation contains different temperatures, some material properties of PCB components will be affected by temperature, so the PCB component parameters at each temperature are very important. The elastic modulus of PCB board is obtained by DMA test on standard board, which is defined as anisotropic material.

In this paper, the PCB with multilayer structure is selected, which is composed of copper clad layer and epoxy glass cloth layer (FR-4 layer). The solder used is SAC305, and the molding resin is used as the chip material. The specific properties of materials are shown in Table II.

TABLE II.   MATERIAL PROPERTIES

Material	PCB	Solder	Chip
Elastic modulus (GPa)	Ex, Ey:15.9 Ez:3.2(25 °C) Ex, Ey:15.4 Ez:3.2(50 °C) Ex, Ey:14.7 Ez:3.2(75 °C) Ex, Ey:13.8 Ez:3.2(100 °C)	44(25 °C) 41.5(50 °C) 39(75 °C) 36.4(100 °C)	162
Density (Kg/m^3)	2400	7400	2330
Poisson's ratio	0.15	0.3	0.28
Thermal expansion coefficient(K^{-1})	1e^{-5}	2.12e^{-5}	2.3e^{-5}
Thermal conductivity (Wm^{-1}K^{-1})	5	50.6	120

The board level interconnection system will produce a high strain rate during the drop process. Adding the Anand viscoplastic constitutive model to the finite element simulation can improve the accuracy of the simulation. The Anand constitutive model parameters are shown in Table III.

TABLE III.   ANAND CONSTITUTIVE MODEL PARAMETERS

Anand Constant	Units	
A	Sec^{-1}	699.46
Q/R	1/K	4000
m	Dimensionless	0.46
n	Dimensionless	0.0011
ξ	Dimensionless	6
$\hat{s}$	MPa	81
h_0	MPa	223688.16
a	Dimensionless	1.60
S_0	MPa	58.59

C. Boundary conditions and Loads

In this paper, the power density loads of 0, $1.7×10^8$ W/m^3, $3.35×10^8$ W/m^3 and $5.05×10^8$ W/m^3 are applied to the solder joints of the board level interconnection system, and the maximum temperature of the system is 25 °C (room temperature), 50 °C, 75 °C and 100 °C. Then the drop simulation is carried out.

The Input-G method can reduce the error between the drop test and the finite element simulation, make the simulation results more accurate, and greatly shorten the calculation time [4，5]. Therefore, this paper uses the Input-G method to apply the displacement boundary conditions to the four holes of the circuit board, and deduces the displacement boundary conditions through the acceleration.

The peak acceleration of the model is 1500 g, the acceleration pulse width is 0.5 ms, the initial velocity is v=-2.3416 m/s, and the displacement boundary curve is shown in Fig. 3.

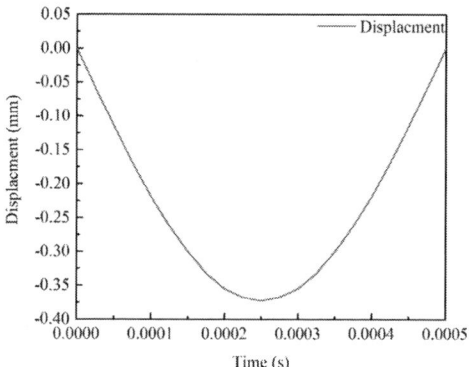

Fig. 3. Displacement boundary curve

III. RESULTS AND DISCUSSION

A. Temperature distribution characteristics

In the finite element simulation, the power density of $1.7×10^8$ W/m^3, $3.35×10^8$ W/m^3 and $5.05×10^8$ W/m^3 is added to the solder joint to make the maximum temperature of the system 50 ℃, 75 ℃ and 100 ℃, and then the drop simulation is carried out.

The temperature distribution cloud chart of the board-level interconnect system with the highest temperature of 50 °C, 75 °C, and 100 °C is shown in Fig. 4. It can be seen from the figure that the temperature of the center position is the highest, the temperature of the corresponding positions of the four chips is lower than that of the center position, and the temperature of the four corners of the PCB is the lowest. This is because when the power density load is applied to the solder joint under the chip, the solder joint generates a higher temperature, and the heat is diffused through the PCB board and air. The solder joints in the center position generate heat, and there are solder joints in four outer directions that generate heat and transfer heat to the center, so the temperature in the middle position is the highest. At the same time, because of the heat generated by the solder joints, the temperature of other parts of the PCB board is higher than the ambient temperature through heat conduction.

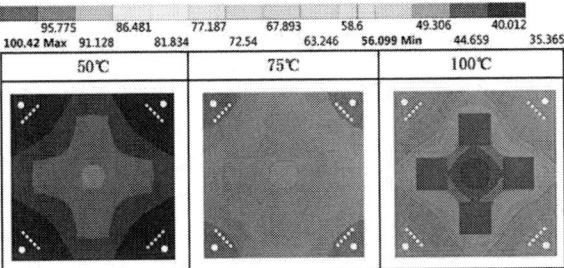

Fig. 4. Temperature distribution cloud chart of thermal-mechanical coupling PCB

B. Analysis of frequency response characteristic

The 1st eigenfrequency of PCB under different temperatures calculated by finite element simulation are shown in Table IV. When the maximum temperature of the system is 25 °C, 50 °C, 75 °C and 100 °C, the 1st eigenfrequency of PCB are 325.46 Hz, 320.65 Hz, 314.81 Hz and 309.81 Hz respectively. The results show that the 1st eigenfrequency of PCB decreases with the increase of the temperature of the board level interconnection system. When the maximum temperature of PCB is 25 °C, the 1st eigenfrequency is 5.05% higher than that when the maximum temperature of PCB is 100 °C.

TABLE IV. THE 1ST EIGENFREQUENCY OF PCB UNDER DIFFERENT CONDITIONS

Temperature	25 °C	50 °C	75 °C	100 °C
The 1st eigenfrequency	325.46Hz	320.65 Hz	314.81 Hz	309.81 Hz

C. PCB strain analysis

In the finite element simulation, the power density of 0, 1.7×10^8 W/m^3, 3.35×10^8 W/m^3 and 5.05×10^8 W/m^3 is added to the solder joints to make the maximum temperature of the system 25 °C, 50 °C, 75 °C and 100 °C, and then the drop simulation is carried out.

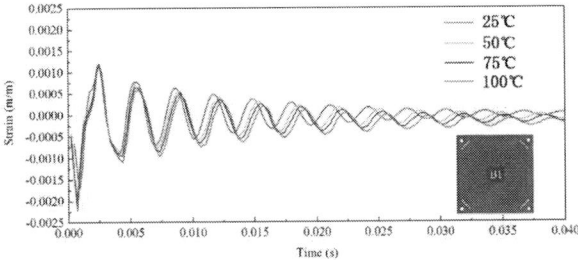

Fig. 5. Comparison of B1 position strain of PCB under different temperatures

The strain of PCB B1 position under different temperatures is shown in Fig. 5. It can be seen from the figure

that at the first peak, the curves of different temperatures (25 °C, 50 °C, 75 °C, 100 °C) reach the maximum strain. When the maximum temperature of the system is 25 °C, 50 °C, 75 °C, 100 °C, the maximum strain of PCB B1 position is -0.001691, -0.001954, -0.002066, -0.002205 respectively. The maximum strain of B1 position in the single drop impact field is less than that of B1 position in the thermal-mechanical coupling field. The maximum strain of B1 position in the thermal-mechanical coupling field increases with the increase of temperature. The maximum strain of B1 position at 100 °C is 12.8% higher than that at 50 °C. It can also be seen from the figure that the strain decay rates of the four curves are similar.

D. Stress analysis of solder joints

The impact of drop will cause the PCB board to bend and vibrate many times, which will lead to the failure of the solder joint. The peeling stress of the key solder joint is the main failure indicator of electronic components. Calculated by the finite element model of the JESD22-B111A board under different temperature at the dangerous solder joint stress at position B1 as shown in Fig. 6. When the temperature is 25 °C, 50 °C, 75 °C and 100 °C, the peak stress of solder joint is 82.64 MPa, 79.77 MPa, 75.62 MPa and 72.01 MPa respectively. The peak stress of solder joint decreases with the increase of the temperature of the board interconnection system. At 25 °C, the peak stress of B1 dangerous solder joint is 14.8% higher than that at 100 °C.

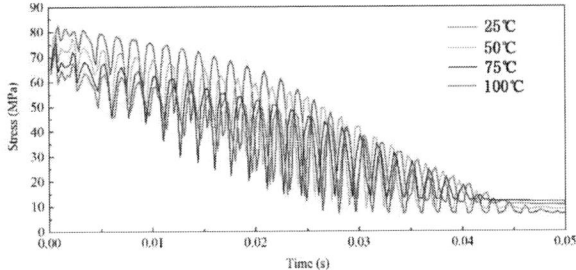

Fig. 6. Comparison of solder joint stress at B1 position under different temperatures

Fig. 7 is a comparison diagram of the solder joint stress at position B2 of the PCB board at different temperatures. When the temperature is 25 °C, 50 °C, 75 °C and 100 °C, the peak stress of solder joint is 80.2 MPa, 77.21 MPa, 72.54 MPa and 69.87 MPa respectively. The order of solder joint stress is the same as that of B1. The peak stress of solder joint decreases with the increase of temperature. In terms of the stress decay rate of solder joints, the stress decay rate of solder joints in the thermal-mechanical coupling field is faster than that in the single drop field, and the stress decay rate of solder joints in the interconnection system with temperature of 50 °C, 75 °C and 100 °C is similar. By comparing with the peak stress of B1, it is found that the stress of B1 is slightly higher than that of B2 in both single drop field and thermal-mechanical coupling field. Due to the thermal load generated by adding power density to the solder joint, the stress of the solder joint decreases obviously, and the solder joint is subject to greater damping rebound at room temperature, which makes the solder joint more prone to failure. The application of power density increases the temperature of solder joint and reduces the stress of solder joints, which further improves the plastic properties of solder joints.

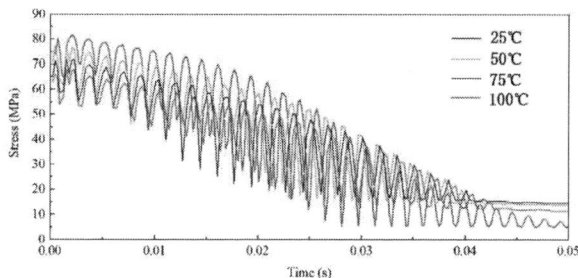

Fig. 7. Comparison of solder joint stress at B2 position under different temperatures

Fig. 8 shows the peak stress cloud chart of dangerous solder joint at B1 position of JESD22-B111A board under four different temperature conditions. The top view is the cloud chart of solder joint and chip connection, and the bottom view is the cloud chart of solder joint and PCB connection. It can be found from the figure that the stress in the middle of the solder joint is much smaller than that in the upper and lower bottom, and the stress of the solder joint decreases with the increase of temperature. The four different conditions of the dangerous solder joint cloud chart have the same situation. The solder joint stress in the bottom view is greater than the solder joint stress in the top view, that is, the position of the largest solder joint stress is at the bottom of the solder joint connected to the PCB. This is because the PCB produces greater flexural deformation during the drop impact. At the same time, due to the different elastic modulus of PCB and chip, the elastic modulus of chip is much larger than that of PCB, which leads to different deflection degree of PCB and chip, resulting in large tensile and compressive stress at the joint of solder joint and PCB. The stress at the joint of solder joint and PCB is the largest, so the crack is mainly initiated and propagated at the joint.

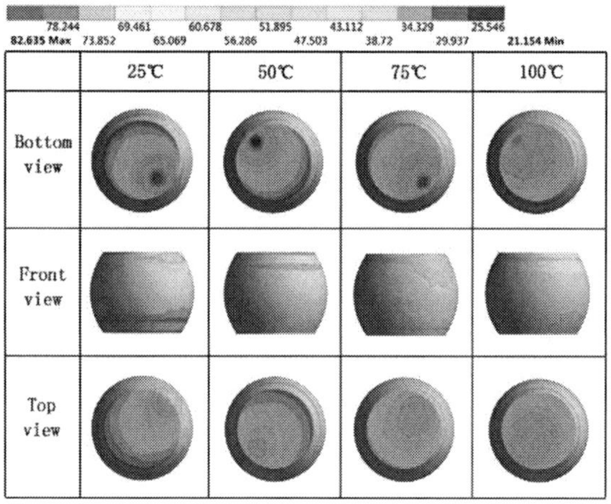

Fig. 8. Stress cloud chart of solder joint at B1 position

E. Analysis of solder joint life

The drop impact performance of the solder joint can be evaluated by the solder joint life. The drop impact performance of the solder joint is the number of times that the solder joint can withstand drop impact. The more drop impacts the solder joint can withstand, the better drop impact performance of the solder joint.

Due to the PCB reciprocating bending under high acceleration, the solder joint is subjected to severe tensile and compressive stress, so the maximum tensile and compressive stress in the drop process is taken as the judgment criterion of the solder joint failure under drop impact. According to the principle of power, a life prediction model is established to link the maximum peel stress with the number of drops during solder joint failure [6]:

$$N = C_1 \sigma_z^{C_2} \tag{1}$$

Where N is the average drop life, σ_z is the maximum tensile stress or the maximum compressive stress (MPa) of the dangerous solder joint. C_1 and C_2 are correlation coefficients [7]. According to the life prediction model, the calculated life of B1 dangerous solder joint is shown in Table V.

TABLE V. SOLDER JOINT LIFE UNDER DIFFERENT TEMPERATURES

Temperature	25 °C	50 °C	75 °C	100 °C
Solder joint stress	82.64 MPa	79.77 MPa	75.62 MPa	72.01 MPa
Average drop life	814	930	1100	1284

The results show that the average drop life of solder joints increases with the increase of temperature. By applying different power densities (0, 1.7×10^8 W/m³, 3.35×10^8 W/m³, 5.05×10^8 W/m³) to solder joints to generate thermal load, the maximum temperature of the system is 25 °C, 50 °C, 75 °C, 100 °C, respectively, the average drop life of the dangerous solder joint at B1 is 814 times, 930 times, 1100 times and 1284 times. The life of solder joint at 100 °C is 57.7% longer than that at 25 °C. The solder joint life at 75 °C is 35.1% longer than that at 25 °C. The life of solder joint at 50 °C is 14.3% longer than that at 25 °C. The solder joint of the thermal-mechanical coupling field have a longer life than the room temperature drop impact solder joint. The reason is that the temperature of solder joints increases, the plastic properties of solder joints are improved, and the drop impact ability of solder joints is enhanced due to the application of thermal load. Therefore, the life of solder joints is improved, and the solder joints in thermal-mechanical coupling field have better ability to withstand drop impact.

IV. CONCLUSION

In this paper, the finite element method is used to simulate the drop of the board level interconnection system under single drop field and thermal-mechanical coupling field, and the reliability of the board level interconnection system under drop impact load is studied. The main conclusions are as follows:

1) Under the thermal-mechanical coupling, the response characteristics of PCB will change accordingly. With the increase of system temperature, the 1st eigenfrequency of PCB decreases. The 1st eigenfrequency of PCB at 25 °C is 5.05% higher than that at 100 °C. Compared with the B1 strain of single drop impact field, the B1 strain of thermal-mechanical coupling is larger than that of single drop impact field, and the B1 strain increases with the increase of temperature.

2) Through the stress analysis of solder joints at different temperatures, it is found that the stress of dangerous solder joints at B1 is slightly higher than that at B2, no matter in the room temperature drop field or thermal-mechanical coupling field. With the increase of temperature, the peak stress of

solder joint decreases. Through the stress analysis of solder joint height direction, it is found that the bottom of solder joint connecting with PCB is the most stress position on solder joint. It is inferred that when the board level interconnection system is impacted by drop, this position will be damaged first.

3）Through the established solder joint life model, it is found that with the increase of temperature, the solder joint life has been improved. The reason for the analysis is that the thermal load generated by the application of power density increases the temperature of the solder joint, the plasticity of the solder joint is improved, and the resistance of the solder joint to drop impact is enhanced.

REFERENCES

[1] Shen J, Zhai D, Cao Z, et al. "Fracture behaviors of Sn-Cu intermetallic compound layer in ball grid array induced by thermal shock," Journal of electronic materials, 2014, 43(2): 567-578.

[2] Mattila T T, James R J, Nguyen L, et al. "Effect of temperature on the drop reliability of wafer-level chip scale packaged electronics assemblies," 2007 Proceedings 57th Electronic Components and Technology Conference, 2007: 940-945.

[3] Jiao H, Liu Y, Sun F,Wu N and Fang H, "Solder interconnects reliability subjected to thermal-vibration coupling loading," Journal of Materials Science: Materials in Electronics, 2019, 30(12): 11482-11492.

[4] Tee T Y, Luan J, Pek E, et al. "Novel numerical and experimental analysis of dynamic responses under board level drop test," 5th International Conference on Thermal and Mechanical Simulation and Experiments in Microelectronics and Microsystems, 2004: 133-140.

[5] Tee T Y, Luan J, Pek E, et al. "Advanced experimental and simulation techniques for analysis of dynamic responses during drop impact," 54th Electronic Components and Technology Conference, 2004, 1: 1088-1094.

[6] Luan J E , Tee T Y , Goh K Y , et al. "Drop impact life prediction model for lead-free BGA packages and modules," International Conference on Thermal, 2005.

[7] Wen L , Fu X , Zhou J , et al. "Dynamic properties testing of solders and modeling of electronic packages subjected to drop impact, " International Conference on Electronic Packaging Technology & High Density Packaging, 2008.

Approach Towards Accurate Modeling of Thermal Resistance in Thermal Management of PCB

Jianghao Wei
School of Mechano-Electronic Engineering, Xidian University
Xi'an, China
2424251661@qq.com

Tao Wan
School of Mechanical Engineering, Tsinghua University
Beijing, China
wantaox@sina.com

Xiangdong Xue
School of Mechano-Electronic Engineering, Xidian University
Xi'an, China
xiangdong_xue@sina.com

Yuming Wang
School of Mechanical Engineering, Tsinghua University
Beijing, China
wangyuming@mail.tsinghua.edu.cn

Abstract—Due to the importance of modeling thermal transmission for microelectronic packages, thermal resistance is considered as a symbolic parameter in thermal management. A great deal of work has been done on developing thermal network models for thermal resistance. Using a network to link junction node and various representative nodes, a network model can take into account of the distribution of heat transfer along paths and also describe the internal uneven situation of components. As a result, thermal network models are flexibly suitable to different situations, binging in reasonably accurate results. This paper reports a modeling study of thermal resistance using a software package, as part of a parallel investigation between experiment and modeling. The focus is put on the two-resistor compact model, though other models are also examined for comparing purpose. Three approaches, computer modeling, theoretical analysis and experimental verification, are involved. The purpose is to gain a realization of the composition and the effect of thermal resistance through an integrative process among the three approaches mentioned above, so that the suitable method can be identified. Different models of thermal resistance are computationally tested and comparably analyzed for a comprehensive understanding of the characteristics of various models. It is found that, after a combinative consideration of effectiveness and efficiency, the two-resistor model can lead to a reasonably good result with adequate cost-efficiency. The parameters of two-resistor model are set up after intensive measures based on theoretical principles and against experimental results. The results agree well with experiments and theoretical analysis. Not only is the current problem solved, but also a realization on how to express thermal resistance in modeling is obtained, paving the way for following on modeling of other types of PCBs. By interpreting the mechanism of thermal resistance and exploring modeling methods, this paper demonstrates an approach for modeling thermal resistance for thermal management of PCBs. Relevant practices and experiences are referable for researchers and engineers in designing heat transfer aisles and improving thermal reliability of PCBs.

Keywords—thermal resistance, modeling and simulation, thermal management of PCB

I. INTRODUCTION

One symbolic phenomenon in information era is the use of microelectronic chips which has become an essential element for smart controlling systems in a number of areas, such as automotive driving, AR, VR, AI, IoT. In the past half a century, chips based microelectronic devices have pushed our world into a new stage. Relevant developments have brought in huge conveniences for human societies. At the same time however, advanced technologies are constantly demanded for satisfying marvelous progresses.

Among various challenges, chip fabrication towards further miniaturization, artificial intelligent engagement for biotechnology approaches and thermal management to safeguard the advent are the three big issues. With regards to the aspect of thermal management, on the one hand every new power module introduced brings new challenge and, on the other hand the trend of dense integrations and 3D packages create new emerging issues. Therefore, thermal management of miniaturized chips is a conventional challenge with modern issues, demanding both an increase of the efficiency of conventional measures and innovative approaches for new developments.

Thermal dissipation efficiency depends on three factors, the power density contained in a microelectronic package, the transmission aisles for thermal energy by means of materials and structural racks, the thermal resistance existing on the paths. Apart from power density determined by system functions, the new development for transmission aisles sees paralleled new routes such as embedded microfluidic channels [1], though increasing the effectiveness of existing aisles is kept as a major target. Existing on the route of thermal dissipation paths in the interior of modules and on boundary interfaces, thermal resistance is a natural existence which can be reduced but cannot be got rid of. It is hence important for researchers, manufacturers and engineers to understand the feature and accurately predict the behavior of various thermal resistances. Only is the concept clearly understood, thermal resistance can be appropriately tested and used [2]. Due to its natural existence in ICs and packages, thermal resistance has become a common basic topic for microelectronic engineering.

Thermal management is the key task of microelectronic packages. Based on the concept of energy conservation, thermal dissipation should be designed from two aspects. One is to reduce the thermal energy of microelectronic systems. The other is to enhance the route of heat transfer by means of conduction, convection and radiation [3]. Among the three methods for thermal dissipation, thermal conduction functions inside packages, where thermal convection and thermal radiation works on system boundary. Comparably thermal radiation is closely related to environments, e.g. aircraft and spacecraft, and often considered as a supplementation factor for package designs. This leads to thermal conduction responsible to internal part of thermal dissipation and thermal convection the external part, though new advances see the application of embedded microchannel cooling system to undertake the task traditionally by conduction.

978-1-6654-1392-3/21 $31.00 © 2021 IEEE

The efficiency of thermal conduction depends on two aspects, the transmission efficiency of materials on thermal transfer routes and the resistant deficiency of thermal resistance on interfaces and in modules. This paper put the focus on the latter aspect. By means of modeling and simulation, a PCB system is extensively studied from modeling and referred from theory and experiments. Through this study, thermal resistance is modelled using different modeling methods and discussed by relevant background basis, so that the right modelling method could be realized to increase the accuracy and saving computational cost at the same time.

II. CONCEPTUAL ANALYSIS

A. Thermal Dissipation of Chips

Both power module/component and electricity circuit produce heat. When temperature is increased beyond permission range, microelectronic systems tend to appear two kinds of damages. On the respect of electronic functions, the moving speed of electrons in semiconductors increases with an increase of temperature, leading to the distortion of signals and the malfunction of devices. On the respect of mechanical reliability, an increase of temperature could cause a number of problems, such as the increase of stresses and deformations, the decrease of mechanical strength and fatigue life, the occurrence of the fatigue damage of solder joints and bonding lines.

As the temperature is an indication of the thermal energy trapped in the system, an improving way for reducing temperature is to enhance the transfer route of the thermal energy by means of conduction, convection and radiation. Working in different fundamentals, i.e. conduction for solid materials, convection for the fluid on solid-fluid boundary and radiation for space, how to combine the three methods for heat transfer is indeed a response to system environments. For the sake of thermal resistance, conduction in materials and among modules is targeted in this study.

The SOP module mounted on a PCB, as shown in Fig.1, is taken as the example to analyze the heat transmission process. The path of heat transfer is marked by arrows in the Fig.1. It can be seen that there are two major heat transfer paths for the thermal dissipation from the cell. One path is to travel upwards from the cell to the top case by conduction and then dissipate thermal energy into ambient by convection. The other path is to travel downwards onto PCB through mounting lines in the sides and materials underneath, before dissipated from the PCB. As the heat transfer paths include either the lamination of different materials or the rack of mounting lines and pads, thermal transmission paths are comprised of the interfaces between materials and boundaries, bringing in different types of thermal resistances.

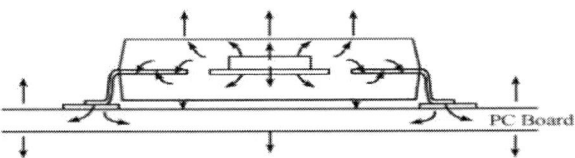

Fig.1 Heat transfer paths of a SOP cell

B. Concept of Thermal Resistance

Thermal resistance corresponds to the transmission ability of thermal energy in materials in a system. Thermal conductivity is obviously the important parameter of solid materials. However, as packaging systems are composed of different types of materials in fabricated ways, thermal resistances among materials and on structural interfaces also play a crucial role for heat transfer. Therefore, thermal management concerns two types of tasks. One is to select materials with high thermal conductivities and the other is to reduce thermal resistances in transfer links. Accurate representation and appropriate understanding of different kinds of thermal resistances are the basis for thermal management of packages.

Thermal resistance refers to the resistant quantity for thermal energy to flow from a high-temperature joint to a low-temperature joint, expressed by the temperature difference between the two joints divided by the thermal energy dissipated [4]. Thermal resistance in a material is closely related to thermal conductivity, as thermal transmission in materials is carried out through the conversion of energy and work, either by means of kinetic energy from electron motions in conductors or potential energy from lattice deflections in isolators. This situation becomes complicated on the interfaces of materials and structures, as the transmission of stress waves is broken among different materials and structures, which usually results in a sudden increase of thermal resistance.

Thermal resistance can be conceptually defined as,

$$Thermal\ resistance = \frac{Temperature\ difference}{Thermal\ energy}$$

The above expression shows that, for a certain of thermal energy transferred, an increase of thermal resistance leads to the increase of temperature difference, resulting in an increase of temperature in the system. Hence, thermal resistance determines the thermal energy blocked which in turn causes an increase of temperature in the transfer path. For one-dimensional transmission, Fourier's law gives

$$Q = -\lambda A \frac{\Delta T}{\Delta X} = -\frac{\Delta T}{\delta/\lambda A}$$

Or

$$\frac{\delta}{\lambda A} = \frac{\Delta T}{-Q}$$

Hence, $R = \delta/\lambda A$ is the thermal resistance, where λ, A, δ are thermal conductivity, cross section and the distance or thickness for heat transfer. It is worth noticing that thermal resistance concerns a general concept in heat transfer and involves all the three models of heat transfer. Therefore, like thermal conduction whose thermal resistance is focused in this study, thermal convection and radiation also have their own thermal resistances.

For the cases among different materials and dissipated in three dimensions, thermal resistance concerns the distribution in different directions, closely related to structural details. Accurate descriptions have to be set based on infinitesimal elements to reflect thermal resistance between any two points. Assumptions are often made in this case to simplify regular cases to one-dimensional transfer if there is a dominative transfer direction. With regards to the situation on boundary interfaces between modules or with PCB, the thermal resistance concerns three materials and affected by both the transmission between materials and surface shape coupling. An obvious increase of thermal contact resistance becomes a certain consequence, requiring specific analysis. In short, thermal resistance concerns the efficiency of heat transfer on transmission paths. An increase of thermal resistance results in the decrease of heat transfer, leading to the increase of temperature.

III. MODEL ESTABLISHMENT

A. Types and Categories of Thermal Resistance

The thermal resistance of a module falls into two categories, the body resistance inside the module and the boundary resistance on the interfaces. The former refers to the thermal resistance from the center of the module to the boundary. This concerns material property of the module and the coupling resistance between materials. The latter refers to the thermal resistance on the boundary interfaces with other modules or PCB. The thermal resistance on boundary interfaces, named as thermal contact resistance, generally higher than in the body due to the representation of three materials on the interface, e.g. two solid boundaries and thermal interface materials (TIM). In a microscope view, the solid contact on the boundary interface is limited, due to surface roughness. TIM thus becomes essential in many cases, though the thermal coupling between TIM and two solid boundaries is often poor due to the different types of materials.

Located in the center of a package, the heat of the chip needs to find ways to transfer to the boundaries of the package. The transmission path may be dominative in one direction but often moves to both the top and the bottom for flat-shaped packages. For the case where heat transfer in side directions is unneglectable, the transfer paths require to be modelled more than two ways. To cope with different situations, various models for thermal resistance have been developed. Figures 2 to 4 show the commonly used three models, single thermal resistor (Fig.2), two thermal resistor (Fig.3) and star thermal resistor (Fig.4). It can be seen that single thermal resistance focuses on the heat transfer from the center to the top surface, whereas two-resistor model takes into account of heat transfer in both directions to the top and the bottom. Comparably two-resistance model is more representative than single-resistor thermal model due to both directions taken into considerations. Different from single and double resistor models, star thermal resistance adds the side direction in heat transfer and thus suitable to block-shaped modules whose side dimension is comparable to vertical dimension.

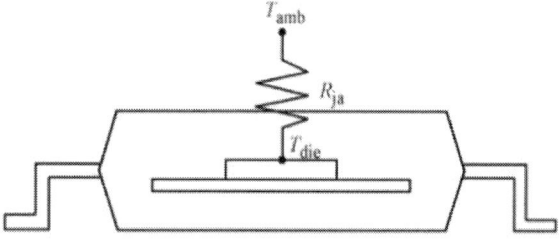

Fig.2 Single resistor model

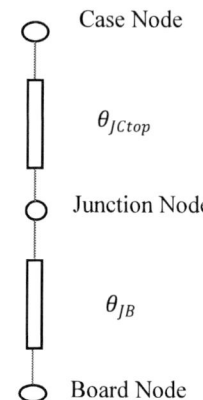

Fig. 3 Physical arrangement of the physical model of two-resistor model

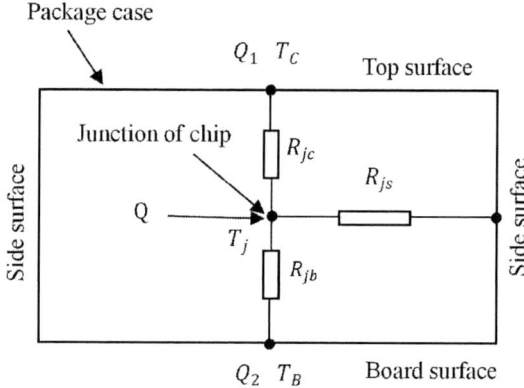

Fig. 4 Star network model -PLCC encapsulation

All the three models in the above figures set a junction node at the center and then determine the directions to link surface nodes with the junction based on thermal flowing directions. In figures 2 to 4, subscripts j, c, b, a and s denote *junction* at the center, *case* at the top, *board* at the bottom, *ambient* for environment and *side* for surrounding boundaries, respectively. Consequently R_{jc}, R_{jb}, R_{ja} and R_{js} represent the thermal resistance of the center junction to the top, bottom, ambient and side, respectively. The rating values specified in the device specifications are generally obtained by the manufacturer based on statistics analysis of standard environmental tests. In the current era, many efforts have be made to increase thermal dissipation efficiency. It therefore becomes a rare case to dissipate thermal energy only along one direction. Single resistor model is thus seldom used nowadays. Comparably, two-resistor model is representative to modules with plate shape and, in principle, becomes widely used.

Giving the dominated pattern of 2D geometry in microelectronic packages, two-resistor model is the widest choice in applications. While the values specified in the device specifications are generally obtained by the manufacturer based on standard environmental tests, the actual application environment is however often very different from the standard test environment. As a result, it is difficult to predict the chip junction temperature based on R_{ja}. Instead, R_{ja} in most cases is used to the horizontal comparison between different package chips and the qualitative comparison of the heat dissipation capacity among different package chips. In applications, shell thermal resistance R_{jc} and junction plate thermal resistance R_{jb} are mainly used to evaluate the heat dissipation ability of devices, which are the parameters of the dual resistor model.

Two/dual resistor model contains two resistances between three nodes, as shown in Fig. 3. This corresponds to a single integrated circuit whose temperature is represented by a single temperature node, the Junction Node. By connecting the Junction Node as the heat resource with the Case Node at the upper surface and the Board Node at the base plate, 1mm away from the component end, a thermal resistance network is formed to represent heat dissipation in two directions. In two-resistor model, the PN junction is connected upwards to the housing and downwards to the substrate by two thermal resistances. The values of the two thermal resistances are set according to the thermal test in JEDEC standard.

Apart from the two thermal resistances to the upper and bottom surfaces in two-resistor model, star network model adds thermal resistor(s) to the side surface to represent thermal dissipation in horizontal plane. Fig.4 shows the simplest star network model [5] where only one side thermal resistance is included. In this star model, heat transfer in the encapsulation is assumed along three directions from the central junction to the top surface, bottom surface and lateral surface, respectively. Star thermal resistance model is largely replaced by Delphi compact network model whose concept can refer literature [6] for details.

Based on the star network model, the thermal resistance between some surface nodes is added in the simplified Delphi model which is then further optimized. At the same time, the concept of "boundary condition independence" is proposed, which means that the model satisfying this condition can be used in any specific application environment, or the component manufacturer can provide a simplified Delphi model in the absence of the actual application environment of the device. As many as 38 groups of boundary conditions are proposed in Delphi project for the purpose of an increase of representativeness of heat flow conditions of simulator in real environment. Each set of boundary conditions is a combination of heat transfer coefficients on the top, bottom, side, pins and other parts of the device package [7].

B. Characteristics of Two-Resistor Thermal Model

Two-resistor thermal model is more accurate than the single resistor model and suitable to the characteristic of packages. On the one hand, two-resistor model can reflect the nature of heat transfer in paired ways by including both thermal resistances of a module to the top and the bottom. On the other hand, most microelectronic packages have two-dimensional shapes, e.g. PCBs, in which thermal dissipation is dominated in the normal direction of the plane shape with little in the lateral sides. Two-resistor model is thus a compact representative model for expressing the main gradients of plane-shaped packages or modules. Two-resistor thermal model is focused in this paper, with a supplementation of other models for specific cases.

In the modelling and simulation of this study, different approaches have been tried and compared to get a fully understanding of the characteristic of different models. It has been found that two-resistor model can gain reasonably good results with a fair cost-efficiency, offering both effectiveness and efficiency. In setting up the model, the three nodes in the model are respectively assigned to shell node on the top surface, module node for die power point and plate node on the board, so that the dominative paths for heat transfer can be represented.

It is worthy noticing that, while two-resistor model can well suit to the type of the modules with high aspect ratio, it has limited ability to express the thermal resistance in lateral directions, i.e. the module with block shapes. For the cases where three dimensions are comparable, other thermal resistance models such as star model containing the expression of heat transfer in side directions are needed to take into account of the feature of the modules.

Two-resistor model is found suitable for modelling power modules after analyzing a number of modelling approaches and comparing with thermal experimental results. In simulation with different cases, two-resistor model produced the results well close to the measuring data by experiments with high efficiency. Some difficulties were encountered in setting up the data for the model. By overcoming the problems encountered, the key factors for setting up the model and relevant background are realized.

Relevant problems are explained as follows.

1) Boundary links: In two-resistor model, single node is set up at module center, top surface and bottom board respectively to represent the temperatures. There is thus an issue for how to make the connection of the single node surface with meshed PCB by transition measures.
2) Resistor value: As a compact model, the accuracy of the values of representative nodes and resistors are the key for the effectiveness of the two-resistor model.
3) Power consumption of modules: The specifications marked on modules represent statistics meaning, leading to inaccuracy for individual uses. The accuracy of the data deducted by tested voltage and current is affected by testing errors and meter accuracy. Both the above situations can prompt inaccuracy of the data.

The solutions for the above problems are listed as follows.

1) Boundary link: The transition problem on the boundary of the module is solved by adopting the Hollow Block setting which can simplify the description of the module. The software package can transfer information between concentrated single node and meshed surfaces.
2) Resistor value: It is found that there actually exists multiple routes in heat transfer. Based on analysis of relevant correlations, we recorded multiple sets of data which are then combined into representative thermal resistors required by two-resistor model. It has been realized that the resistors in the model have representative meanings and the values should be conceptually reasonable and applicably representative. A number of tests have been made to determine the resistors.
3) Power consumption: The rating values of power

modules marked by producers are rectified and rechecked by measuring relevant data, for which the adjustable voltage, instead of the battery, is used.

IV. SIMULATION

A. Simulation Process

Charging bank is taken as the demonstrative case in this paper. The 3D model of the geometry of the PCB system is provided by CAD files which is imputed based on the requirements of mechanical simulation and thermal resistance model. The thermal analysis of the charger is divided into two scenarios, charging state and discharging state. Due to the differences of the involvement of different parts, the thermal resistance of the components needs to be analyzed separately in two states. As the charge was extensively tested, different situations can be verified against the thermal tests.

The model is imported into Icepak from DM with CAD interface, and then certain conditions are set up. The specific operation is shown in the following flow chart.

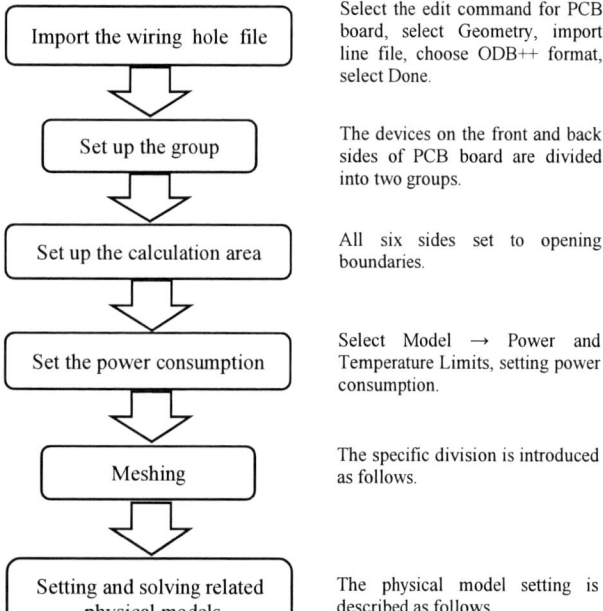

Import the wiring hole file	Select the edit command for PCB board, select Geometry, import line file, choose ODB++ format, select Done.
Set up the group	The devices on the front and back sides of PCB board are divided into two groups.
Set up the calculation area	All six sides set to opening boundaries.
Set the power consumption	Select Model → Power and Temperature Limits, setting power consumption.
Meshing	The specific division is introduced as follows.
Setting and solving related physical models	The physical model setting is described as follows.

B. Meshing and Physical Parameters

The following principles are adopted for ensuring the meshing to be qualified for the requirements of the accuracy.

1) The mesh must be close to the body to ensure that the model is not distorted.
2) All quality standards of the grid meet the requirements of ANSYS Icepak.
3) On the one hand, at least 2-3 grids should be arranged in the thickness direction of PCB and on the other hand, the meshing density is able to express temperature gradient in the thickness direction.
4) In general, unstructured mesh is used for geometrical objects with editable dimensions.

Considering the geometry structure adopted in this simulation is relatively simple, Mesher-HD is used as the basic format with a consideration of some regions on requiring increased dense meshes.

The mesh is set up based on the above principles. The model is meshed with different divisions to check the sensitivity of the model with meshing. It is finally found that the current degree of mesh division can meet the requirements of the simulation.

All the three heat transfer modes, conduction, convection and radiation, are taken into account in this simulation. While thermal conduction mainly concerns material composition and thermal resistance inside the package, natural convection is set up for thermal dissipation on package boundary. Without enforced convection measure, thermal radiation takes an increased role on dissipating heat. The DO (Discrete Ordinates) model is selected to model radiation effect. Environment temperature is set to 25°C.

Two-resistor thermal model is used for power components. To check the effectiveness, single resistor model and, in some cases, star network resistance model are assigned to compare the consequences when different approaches applied. The simulation with two-resistor model is focused and detailed.

Power consumptions with relevant components are assigned respectively based on specifications of rating data and practical measurements using designed circuits. Two scenarios, charging and discharging, are simulated. The effect of different heat consumption periods on thermal dissipation is considered under the two working conditions.

C. Results

Two scenarios, charging and discharging, are simulated. The simulation results are compared with experimental measurements which were carried out beforehand to provide a verification basis. Experimental measurements were obtained with the ambient temperature kept in 25°C. The environment temperature in simulation is also set to 25°C for easy comparisons. As the experiment is to provide a comparable basis to the simulation, many details were designed to be able to approach in parallel ways between the experiment and simulation.

The modelling and simulation work was carried out with a clear feature of case testing. As aforementioned, different meshing strategies have been computationally tested to check the suitable meshing density for stable consequences. Furthermore, in simulation process, different models of thermal resistance and different iteration steps are conducted to investigate the effectiveness of plans and parameters. On the aspect of resistance models, the results of two-resistor model are compared with single thermal resistance model to observe the improvement with the former. On the aspect of iteration steps, a couple of iteration numbers from 100 to 5000 steps are simulated to investigate the sensitivity of iteration step on simulation consequences.

It is found that the results obtained by the two-resistor model are more stable and more accurate than the single resistor model in all the cases, showing a clear advantage of the two-resistor model over single resistor model. In terms of iterative steps, the simulation shows that the differences between iterations of 100 and 5000 is not obvious for a serious consideration, indicating that the results are insensitive to the iteration number beyond 100. Therefore, it is evidenced that after the convergence of 100 steps, the final accuracy is not far away. This shows that the results are relatively convergent. In short, the simulation results agree well with experimental measurements. Figures 5 to 8 show some comparisons between the simulation and the measurement for charging state and discharging state, respectively.

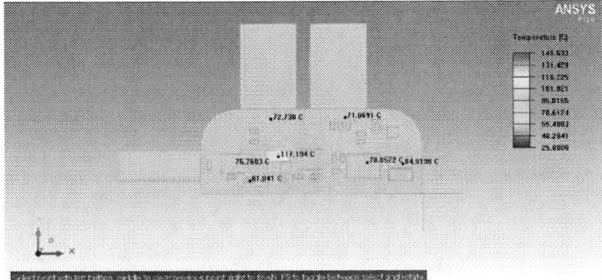

Fig. 5 Simulation results for charging state

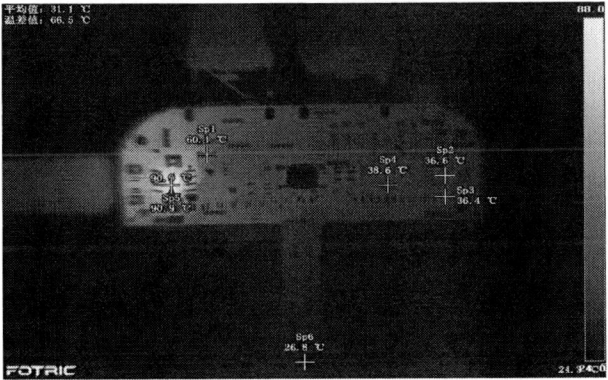

Fig. 6 Measured temperature results for charging state

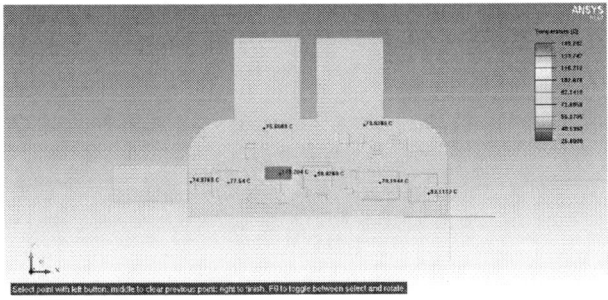

Fig. 7 Simulation results for discharging state

Fig. 8 Measured temperature results for discharging state

After a series of constant refinements, the difference between the simulation temperature and the simulation temperature in near all regions is less than 20℃, achieving the expectation set. Moreover, the results cope well with conceptual understanding, bringing in-deep understanding of the phenomenon and increasing the skills for simulation.

V. CONCLUSIONS

A computational investigation of thermal resistance models is reported in this paper. The simulation results for the PCB of a charging bank agree well with experimental measurements conducted for a verified comparison. Increased understanding and realization are gained through detailed comparisons and virtual tests in modeling. The following conclusions are highlighted.

1) Two-resistor thermal model is found suitable to represent power components. It also shows that the representativeness of the model to the character of the modelled module, e.g. possessing high aspect ratio of plane-shape, is the key for model suitability.

2) This study shows that, on the one hand, modeling study should combine theoretical concept and experimental results for concrete conclusions and, on the other hand, expanding cases may need to be investigated for increased insight.

3) Simplification is a common approach for experienced models, e.g. compact two-resistor model in the current case. The approach for obtaining the simplified parameters with representative meanings however can only be obtained after extensive analyses to set up the basis for simplifying.

Taking the charging bank as an example, a detailed two-resistor thermal model of the heating device was established, leading to a realization for how to express thermal resistance in modeling and, in particular, appropriately make use of two-resistor model for accurate modeling.

REFERENCES

[1] J. Liu, "Definition and measurement techniques for thermal resistance of IC packages[R]." Electronic design resource network, https://www.docin.com/p-853888336.html, 2005-11-16.

[2] X. Qin, "Concept and test method of thermal resistance[J]." Xi 'an Institute of Power Electronics Technology, vol.3, pp. 32-50, 1996.

[3] T. Zhou, X. Lu and Y. Li, "Calculation of thermal resistance of air-cooled radiator for power semiconductor devices[J]." Journal of Bohai University, vol.32(3), pp.228-235, 2011.

[4] EIA/JEDEC STANDARD, "Integrated Circuits Thermal Measurement Method-Electrical Test Method (Single Semi conduct or Device)[J]." JESD51-1:29, 1995.

[5] M. Xie, K. C. Toh and D. Pinjala, "An adaptable compact thermal model for BGA packages[C]." Electronics Packaging Technology Conference, pp.304-311, Singapore, December 2002.

[6] W. Krueger and A. Bar-Cohen, "Thermal Characterization of a PLCC-Expanded Rjc Methodology[J]." IEEE Transactions on Components, Hybrid, and Manufacturing Technology, vol.15(5), pp.691-698, 1992.

[7] H. Pape and G. Noebauer, "Generation and Verification of Boundary Independent Compact Thermal Models for Active Components According to the DELPHI/SEED Methods[C]." Fifteenth IEEE SEMI-THERM Symposium, pp.201-211, San Diego, CA, March 1999.

Study on Warpage and Peeling Mitigation of Wafer Level During Metal Plating Process

Dayang Li
Foundry department
Monolithic Power Systems
Chengdu, China
Dylan.Li@monolithicpower.com

Ming Xiao
Foundry department
Monolithic Power Systems
Chengdu, China
Ming.Xiao@monolithicpower.com

Zhiqi Wang
Foundry department
Monolithic Power Systems
Chengdu, China
Dax.Wang@monolithicpower.com

Abstract—Wafer level packaging is one of the latest trends in microelectronic packaging. Wafer warpage and metal peeling are common problem in wafer level packaging. Large warpage and large area peeling will cause the failure of post packaging and reliability problem. The mechanism of warpage and stress are the mismatch of material's coefficient of thermal expansion (CTE) and the change of temperature in the process. In this paper, after plating a metal layer on passivation (PA) layer, there is a large wafer warpage, which makes the following process unable to continue. Based on this background, plating a metal layer on the back side of wafer is innovatively proposed to overcome warpage. The warpage and peeling of back side metallization (BSM) are also measured. The experiment also compares warpage with and without BSM. The reason of peeling, the location of peeling and the interface of peeling were studied by experiments. Based on the finite element analysis (FEA). The effect of different BSM masks on warpage and peel stress were studied. The effects of local design and BSM thickness on peel stress are also studied. Through the research of this paper. It provides guidance for the metal plating process and design of BSM, and can effectively overcome the problem of wafer warpage and peeling.

Keywords—wafer level packaging; warpage; peeling; backside metallization; simulation

I. INTRODUCTION

With the wider application of wafer level packaging, the warpage of wafer level will become a problem that must be solved. The warpage is mainly caused by the mismatch of CTE between the each layers. In the process of integrated circuit (IC), there are often temperature changes or even temperature cycles (TC). The mismatch of CTE will lead to different thermal deformation of each layer, resulting in different thermal deformation. The thermal deformation of each layer is different, resulting in different thermal stress. Excessive thermal stress will cause failure or damage, such as peeling.

The effect of excessive warpage is that the following process can't continue. Some studies have reported the effect of temperature and structural change on Warpage and stress. J. Lee and A. Mack [1] studied the stress and warpage in the process. The research shows that the stress in the dielectric layer is mainly thermal stress, and thermal stress is the main cause of warpage. Shen Y.-L. [2] also studied the effect of copper and low-k dielectric layer on warpage. The above studies are all aimed at the front end of line (FEOL). In the metal plating process, there is a significant difference from the FEOL, because of the increase of metal film thickness and high temperature. Many scholars have studied the warpage problem at the package level. During the study, the special properties of materials were considered, such as viscoelasticity of adhesive and elastoplasticity of copper. The

results show that accurate characterization of materials can get accurate results [3-5]. In the simulation of wafer level warpage, scholars also have relevant research, such as R. van Silfhout et. of Philips [6] used the method of combining volume element and beam element to simulate the deposition of different films. F. X. Chen [7] proposed to use equivalent material model and sub model to simulate the warpage and stress of wafer. These simulation methods effectively simplify the simulation model, and ensure that the simulation results are consistent with the test results.

There are also a lot of studies on metal peeling. Cher Ming Tan [8] studied aluminum pad peeling after bonding process, and also studied peeling mechanism and solution by simulation method. Vincent Fiori [9] studied the gold wire bonding induced peeling in Cu/low-k interconnect. P Bo[10]obtain the interfacial peeling energy-release rate by using virtual crack-closure technique with dummy nodes.

In this paper, the process of wafer level package is to first grow a seed layer on the surface of PA, and then plating a layer of copper. After copper plating, there is a large warpage, which makes the next process unable to continue. The mitigation method of warpage is studied experimentally and the innovative scheme of plating copper on the backside is proposed. The method can greatly solve the problem of warpage, but it also introduce a large back peeling problem, which seriously affects the reliability. In this paper, the wafer warpage and peel stress are simulated and compared with the experimental results. The causes of warpage and peeling are analyzed, and the influence of BSM design on Warpage and stress is studied.

II. EXPERIMENTAL

Wafer warpage was measured by Shadow Moiré. There are two experimental models in this paper: without BSM and with BSM. Warpage was caused in the process, in which the thin film layer is added and the temperature changes at the same time. Under the combined effect of these two factors, the warpage increases obviously due to the mismatch of materials' CTE. This paper mainly studies the metal plating stage, but before the plating process, i.e. incoming quality control (IQC), there is also a wafer level warpage. In order to better understand the warpage of the plating process, first test the wafer warpage of IQC, as shown in Figure 1. 24 wafers were tested. The warpage average is 107um. That is to say, the wafer has about 107um warpage before the metal plating, which is also the initial warpage.

978-1-6654-1392-3/21 $31.00 © 2021 IEEE

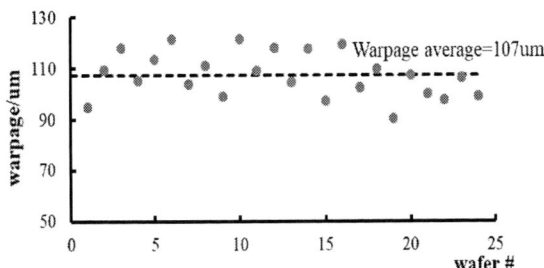

Fig.1 IQC wafer level warpage

Because copper thickness is larger than thin film of Fab process, warpage of copper plating stage will be larger than IQC. IC products need to ensure high reliability performance, so the wafer after copper plating has undergone reflow and TC. The larger temperature difference results in larger warpage in reflow. Therefore, the warpage from 30 to 300 ℃ is measured after copper plating, and the test data and contour are shown in Figure 2. According to the warpage data, the warpage after single-sided metal plating is between 300 and 400um. Because the warpage of this scheme is too large to complete the next process, a layer of metal is plated on the back to reduce the warpage. Comparing the warpage with and without BSM, the experimental results show that the warpage decreases by 50%~80% with add BSM, which indicates that add BSM can effectively overcome the wafer warpage.

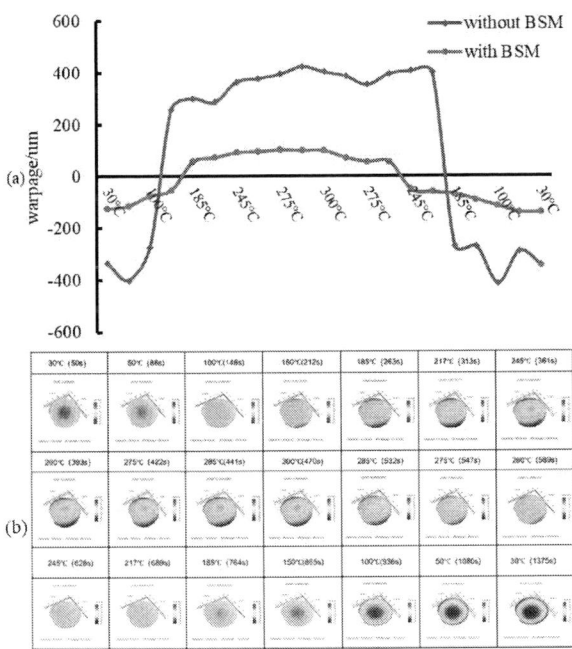

Fig.2 warpage test data and contour: (a)the warpage of with and without BSM (b)warpage with BSM by shadow moiré

From the experimental data of warpage, adding BSM can overcome wafer warpage. As mentioned earlier, in order to ensure the high reliability of the product, it needs to go through the TC. According to the SEM images after TC, there are different degrees of peeling on the backside of wafer after adding BSM, and peeling will lead to the following abnormal process. As peeling mainly occurs after the TC, visual inspection of the peeling wafer after TC shows that peeling is mainly located in the large area copper and occasionally occurs in the stress release hole. In order to make sure that peeling occurs in copper, silicon or interface seed layer, cross section the peeling wafer. As shown in Fig. 3 (a), in the location of large area copper, silicon peeling occurs and silicon is damaged. As shown in Fig. 3 (b), for the peeling interface which is difficult to distinguish, the EDX results show that the proportion of Si in the peeling interface is the highest, followed by the seed layer Ti. From the above two kinds of analysis, the peeling mainly occurs in the silicon on the wafer back surface. The RDL in Figure 3 is the BSM.

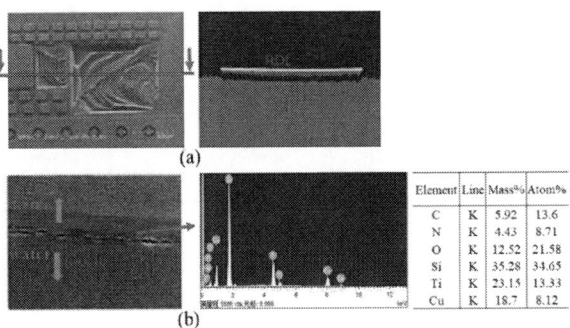

Fig.3 Peeling style: (a)cross section of peeling location (b)EDX of peeling interface

III. FINITE ELEMENT ANALYSIS

Finite element analysis can clearly observe the warpage trend and stress distribution. According to the wafer used in the experiment, the finite element models without BSM and with BSM are established. As shown in Figure 4, the model studied in this paper is 8-inch wafer. Considering that the seed layer Ti is very thin and has little contribution to warpage and stress, and peeling mainly occurs in the silicon layer, it is ignored in the model. The model mainly includes top metal, silicon and BSM. In order to save computing memory, the model is 1/4 wafer, and only three die's detailed BSM layout are built. The sub model technique is used to calculate the stress.

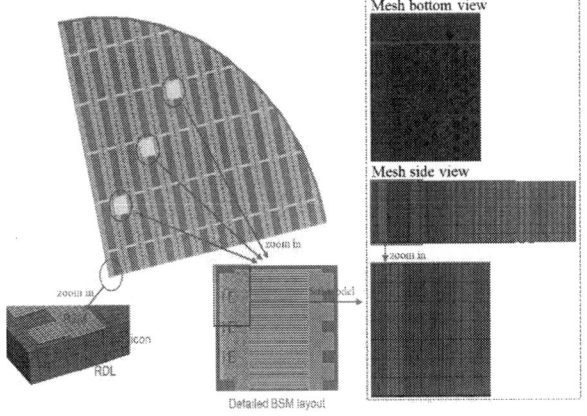

Fig.4 Finite element analysis model

Only copper and silicon are included in the FEA model. The linear elastic material properties is used for silicon and copper when qualitatively calculate the stress, and the linear elastic material parameters are shown in Table 1. In order to accurately characterize the change trend of wafer, the elastoplastic properties of copper was used. The temperature dependent material properties for elastoplastic copper parameters are shown as table 2 [11].

TABLE I. LINEAR ELASTIC MATERIAL PROPERTIES

	Young's modulus MPa	Poisson's ratio	CTE ppm/°C
Silicon	131000	0.278	2.8
Copper	121000	0.34	16.3

TABLE II. ELASTOPLASTIC COPPER MATERIAL PROPERTIES

Temperature	RT	100℃	200℃	300℃
Elastic modulus (Mpa)	121000	107158	95844	73520
Plastic (%)	Plastic stress @RT (MPa)	Plastic stress @100℃ (MPa)	Plastic stress @200℃ (MPa)	Plastic stress @300℃ (MPa)
0	202.45	176.47	161.34	138.14
0.05	233.49	203.54	182.24	156.92
0.1	255.98	222.49	196	170.14
0.15	272.68	235.8	205.48	179.42
0.2	285.66	245.48	212.82	186.11
0.25	296.31	253.12	219.54	191.28
0.3	305.34	259.91	226.46	195.71
0.35	312.79	266.58	233.76	199.92
0.4	318	273.46	240.94	204.12
0.45	319.67	280.43	246.84	208.25

The boundary conditions of simulation are based on the real conditions of warpage and peeling. Warpage mainly occurs in the process and reflow, and the reflow temperature is 30~300℃. Peeling mainly occurs in the TC, and the temperature condition is 150~-65 ℃.

IV. THE RESULTS OF SIMULATION

Warpage is simulated under reflow process, and the warpage results are shown in Figure 5. By comparing warpage's simulation and test data (with and without BSM), the simulation results are consistent with the test results. It is concave at low temperature, convex at high temperature and concave again at low temperature. It can qualitatively evaluate the wafer level warpage through simulation. The simulation data does not consider the initial warpage of IQC. After considering the initial warpage of IQC, the error of warpage is within 20% at the highest and lowest temperature. The simulation can reveal the experimental phenomena well.

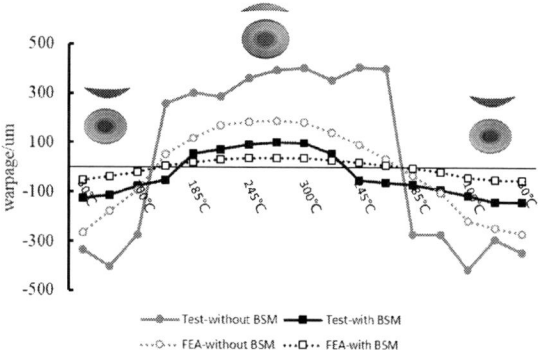

Fig.5 wafer warpage simulation results

The stress is simulated under the experimental condition of temperature cycle. Through the previous EDX analysis, the peeling mainly occurs in the silicon layer, so the maximum principal stress on silicon is extracted by stress simulation. As shown in Figure 6, the simulated stress results are compared with the real peel image of wafer backside. The simulation results show that the large stress is mainly concentrated on the edge of large area copper and the stress release hole. The pictures of peeling after TC are also in the corner of large area copper and the stress release hole. By comparing the results of simulation and experiment, the simulation can qualitatively evaluate the risk of peeling. The BSM mask can also be optimized by simulation to reduce warpage and peeling stress.

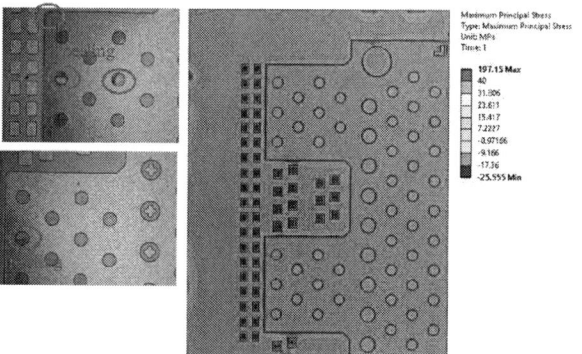

Fig. 6 peeling stress simulation results

A. Effect of different BSM mask on warpage

Three different BSM masks are designed, named as A, B and C mask respectively, and the same warpage simulation method is used for these three masks. As shown in Figure 7, the effect of different BSM masks on the wafer warpage is shown. The results show that without BSM, the warpage is maximum, which is obviously different from the other three curves. After adding BSM, warpage decreased significantly. The influence of different mask on warpage is different. It can be seen that A mask warpage is the smallest. The design of A mask is exactly the same as that of top metal layout. In this design, the warpage is the minimum, because in the process of temperature change, the thermal deformation of the top and back side is basically the same, and the thermal deformation offsets each other. B mask and C mask reduce the area of copper, which leads to a larger area difference between the top and back side, so warpage has a certain increase compared with A mask. Therefore, from the perspective of improving warpage, mask design should ensure that the copper area of the top and back side is as close as possible, and the mask layout of the top and back side is as close as possible.

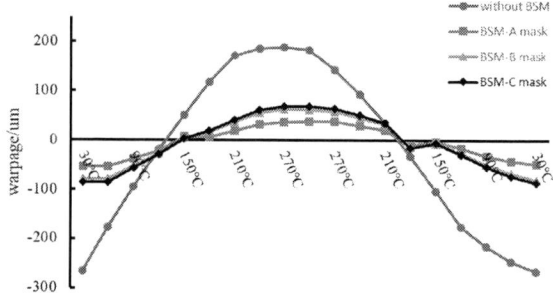

Fig.7 effect of different BSM mask on warpage

B. Effect of different BSM mask on peeling stress

The same stress simulation method is used for A, B and C masks under TC. Figure 8 shows different stress results for different masks. From the stress distribution, the three kinds of mask show the same results, the large stress is concentrated in the edge and corner of the large area copper, and the position of the stress release hole, which are easy to peel from the experiment. From the value of stress, the change of the stress at the edge of large area copper is small, which is kept within 5%. At the position of the stress release hole, the stress changes within 5%, indicating that different masks have an effect on the wafer warpage, but have no effect on the local stress distribution.

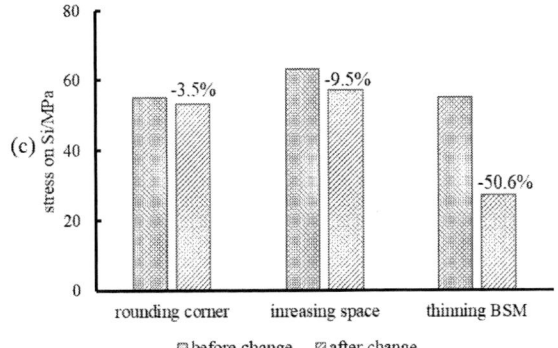

Fig.9 effect of BSM mask design on stress: (a)rounding large area copper corner(b)increasing large area copper space(c)BSM thickness thinning

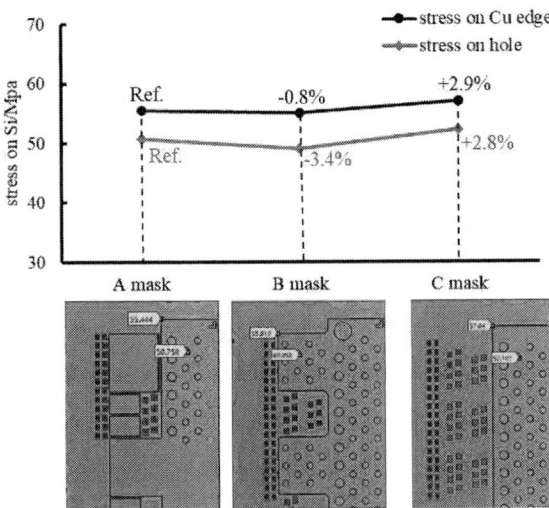

Fig.8 effect of different BSM mask on peeling stress

C. Effect of BSM mask design on stress

According to the previous simulation results, it is found that the location of large stress is very regular, mainly in the edge of large copper and the stress release hole. As shown in Fig. 9 (a), the change rate of stress is about 4% when the corner of large area copper is rounded and the microstructure is added in the stress release hole. Due to the large area copper spacing is too close, there is obvious stress concentration in the spacing, the copper spacing is increased. Results as shown in Figure 9 (b), the stress decreased by about 9.4%. Increasing the copper spacing can reduce the stress. As shown in Fig. 9 (c), when the thickness of BSM is reduced to 1/2 of the original thickness, the stress is greatly improved, and the stress is reduced by 50%. Therefore, the stress can be greatly improved by reducing the thickness of the metal layer.

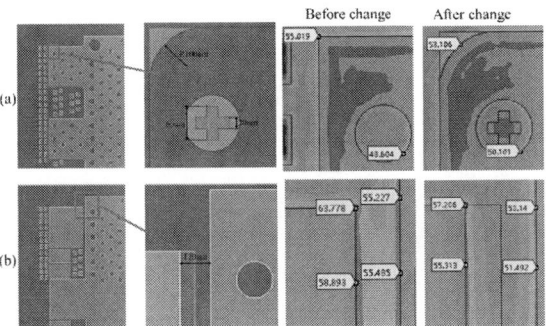

V. CONCLUSION

In this paper, it is found that warpage and peeling in the process of wafer level packaging because of the mismatch of materials' CTE. The experimental results show that the warpage problem can be solved by adding BSM, but at the same time, the peeling was found at back side. Peeling was found to be mainly concentrated in the silicon layer by EDX analysis, and mainly in the large area of copper corners and stress release holes. In this paper, the warpage and peeling are simulated, the results show that the simulation and experiment match well, the trend of warpage is consistent with the experiment, and the position of silicon peeling is also a large stress in the simulation. The simulation also studies the influence of different BSM masks on warpage and peeling. The closer the BSM mask is to the top mask, the smaller the warpage is. Different mask has little effect on the peel stress. When rounding the corner of large area copper, the stress change is not obvious. Increasing the copper spacing can reduce the stress at the edge of copper, and the reduction of BSM thickness has the most obvious effect on stress improvement. The research shows that the silicon peeling problem can be reduced by changing the design of mask.

REFERENCES

[1] LEE J, MACK A S. Finite element simulation of a stress history during the manufacturing process of thin film stacks in VLSI structures[J]. IEEE Transactions on Semiconductor Manufacturing, 2002,11(3): 458-464.

[2] SHEN Y L. Modeling of Thermo-Mechanical Stresses in Copper Interconnect/Low-k Dielectric Systems: ASME 2005 Pacific Rim Technical Conference and Exhibition on Integration and Packaging of MEMS, NEMS, and Electronic Systems collocated with the ASME 2005 Heat Transfer Summer Conference[C], 2005.

[3] ZHANG Z, FAN L, SITARAMAN S K, et al. Four-laser bending beam measurements and FEM modeling of underfill induced wafer warpage: Electronic Components & Technology Conference[C], 2004.

[4] RYU S, JIANG T, LU K H, et al. Characterization of thermal stresses in through-silicon vias for three-dimensional interconnects by bending beam technique[J]. Applied Physics Letters, 2012,100(4): 345-356.

[5] JIANG T, RYU S K, QIU Z, et al. Measurement and analysis of thermal stresses in 3-D integrated structures containing through-silicon-vias[J]. Microelectronics Reliability, 2013,53(1): 53-62.

[6] SILFHOUT R V, DRIEL W V, LI Y, et al. Prediction of back-end process-induced wafer warpage and experimental verification: Electronic Components & Technology Conference[C], 2002.

[7] CHE, F. Development of Wafer-Level Warpage and Stress Modeling Methodology and Its Application in Process Optimization for TSV Wafers[J]. IEEE Transactions on Components, Packaging and Manufacturing Technology, 2012,2(6): 944-955.

[8] TAN C M, GAN Z. Failure mechanisms of aluminum bondpad peeling during thermosonic bonding[J]. IEEE Transactions on Device & Materials Reliability, 2003,3(2): 44-50.

[9] FIORI V, BENG L T, DOWNEY S, et al. Gold Wire Bonding Induced Peeling in Cu/Low-k Interconnects: 3D Simulation and Correlations: Thermal, Mechanical and Multi-Physics Simulation Experiments in Microelectronics and Micro-Systems, 2007. EuroSime 2007. International Conference on[C], 2007.

[10] BO P, HUANG Y A, YIN Z P, et al. Analysis of interfacial peeling in IC chip pick-up process[J]. Journal of Applied Physics, 2011,110(7): 35.

[11] KOCKS U F, MECKING H. Physics and phenomenology of strain hardening: the FCC case[J]. Progress in Materials Science, 2003,48(3): 171-273.

The Study of Effects to the Thermo-Mechanical Performance of the First Level Thermal Interface Materials

Zhenyu Wang[1,2]
1 Shenzhen Institute of Advanced Electronic Materials
Shenzhen Institute of Advanced Technology, Chinese Academy of Sciences Shenzhen , China
2 Shenzhen College of Advanced Technology,
University of Chinese Academy of Sciences
Shenzhen , China
zy.wang4@siat.ac.cn

Linlin Ren[1]
1 Shenzhen Institute of Advanced Electronic Materials
Shenzhen Institute of Advanced Technology, Chinese Academy of Sciences
Shenzhen , China
ll.ren@siat.ac.cn

Xiangliang Zeng[1,3]
1 Shenzhen Institute of Advanced Electronic Materials
Shenzhen Institute of Advanced Technology, Chinese Academy of Sciences
Shenzhen , China
3 College of Chemistry and Chemical Engineering,
Hunan University,
Changsha, China
xl.zeng1@siat.ac.cn

Wenbo Ye[1,4]
1 Shenzhen Institute of Advanced Electronic Materials
Shenzhen Institute of Advanced Technology, Chinese Academy of Sciences
Shenzhen , China
4 Nano Science and Technology Institute
University of science and Technology of China
SuZhou,China
wb.ye@siat.ac

Yonglun Xu[1]
1 Shenzhen Institute of Advanced Electronic Materials
Shenzhen Institute of Advanced Technology, Chinese Academy of Sciences
Shenzhen , China
yl.xu3@siat.ac.cn

Xiaoliang Zeng[1*]
1 Shenzhen Institute of Advanced Electronic Materials
Shenzhen Institute of Advanced Technology, Chinese Academy of Sciences
Shenzhen , China
xl.zeng@siat.ac.cn

YunSong Pang[1]
1 Shenzhen Institute of Advanced Electronic Materials
Shenzhen Institute of Advanced Technology, Chinese Academy of Sciences
Shenzhen , China
ys.pang@siat.ac.cn

Sun Rong[1]
1 Shenzhen Institute of Advanced Electronic Materials
Shenzhen Institute of Advanced Technology, Chinese Academy of Sciences
Shenzhen , China
rong.sun@siat.ac.cn

Abstract- The booming growth of flexible and stretchable electronic devices with increasing power density calls for novel highly efficient thermal interface materials (TIMs) with versatile functions, such as high deformability and excellent resilience. Soft polymer composites such as silicone gap pad are expected to reduce the contact resistance and thermal impedance, but the huge mismatch of the thermal expansion coefficient between the heat sink and the microelectronic chip and stretching behavior make them under stress, especially in the way of tensile stress in stretchable electronic devices. So far, few researchers have reported study about the effect of tensile behavior on the properties of TIMs. In this work, DMA (dynamic mechanical analysis) was used to characterize the resilience, creep and stress relaxation characteristics of silicones gap padsIt was found that as the strain increases, the rebound ratio increases at the first and then decreases, reaching the maximum at 20% of its maximum stretching length. When the term of time is considered, the plasticity of the tensile behavior is not linearly related to time but decelerates as time increases. In five cycles of loading-unloading, the increase in stretch length can enhance the degree of plastic deformation significantly, while the stretch rate has no correspond with it. This research can provide important guidance for the application of TIMs in stretchable electronic devices industry.

Keywords: silicones, gap pads, strain recovery, creep, stress relaxation

I. INTRODUCTION

The development trend of electronic equipment is miniaturization and high integration, so the power density is dramatically increased [1]. A large amount of heat needs to be exported to keep the equipment in good working condition. Efficient thermal management is an effective way to solve the problem of heat dissipation. After decades of hard work, researchers have developed a variety of heat dissipation strategies, including phase change heat dissipation [2], heat pipe heat dissipation [3]and convection heat dissipation [4]. As shown in Fig. 1, thermal interface materials (TIMs) can eliminate the air gap and reduce contact thermal resistance, which is a critical factor for thermal management [5]. TIMs with high thermal conductivity cannot provide optimal thermal management performance, and internal stress and contact conditions of various substrates need to be considered. The huge mismatch of the thermal expansion coefficient between the heat sink and the microelectronic chip or the substrate makes the thermal interface material often under stress, especially in the way of tensile stress. In a word, it is necessary to find an excellent thermal interface material which can adapt to irregular surface, reduce the pressure between two components, eliminate air gap and have good flexibility to improve the overall heat dissipation capacity. [6].

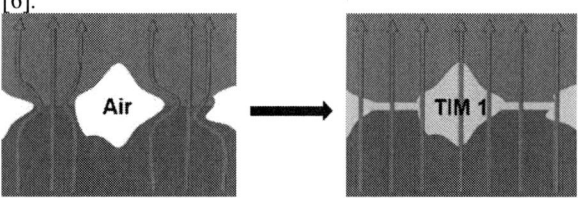

Fig. 1. The above mechanism emphasizes that the thermal interface material can effectively improve the heat transfer between the two uneven interfaces

A lot of efforts have been invested in the compressing thermo-mechanical properties of thermal interface materials, such as modules and compress relaxation, which is significantly affects its thermal contact resistance and long-term reliability. However, in the emerging field of wearable flexible devices [7-9], thermal interface materials are often in a state of large strain tension, and there are few studies on the effects of various factors on the properties of thermal interface materials under large strain conditions. These changes may cause the potential failure of the thermal interface material [10-11]. Therefore, studying the changes of these properties is helpful to realize the long-term stable use of TIMs.

Silicone elastomers are universally used in thermal interface materials because of their low cost and good physical and chemical properties [12]. Not only silicone rubber has good electrical insulation and heat resistance, which makes silicone rubber one of the best choices for preparing thermal interface materials, but also a very low T_g temperature and a low modulus at room temperature. The low modulus helps the thermal interface material to fit well in the gap of the interface under the packaging pressure and overcome the pressure caused by the mismatch of thermal expansion coefficient in service, reducing the contact resistance and thermal impedance. Studying the viscoelasticity of thermal interface materials under different tensile conditions will help to deepen people's understanding of the changes in properties of thermal interface materials when they work, so as to promote the design of thermal interface materials in the future.

The thermo-mechanical properties of silicones thermal interface materials are discussed in this paper. Two thermal interface materials with high elongation were selected, and DMA was used to study the factors affecting their resilience, stress relaxation and creep properties. Finally, the loading-unloading cycle experiment proves that the stretching rate

has little effect on the degree of plastic deformation, but is closely related to the length of the stretching.

II. EXPERIMENTALE

A. Sample preparation

In this study, two silicone rubber thermally conductive gaskets with different elongations were prepared. The main reaction is that the unsaturated vinyl compound undergoes a cross-linking reaction with the hydrogen siloxane cross-linking agent in the presence of the Caster catalyst. The unsaturated vinyl compound is a mixture containing side chains and terminal vinyl groups, and the hydrogen functional group of the crosslinking agent can be in the middle or the end of the main chain. The spherical aluminum powders of different particle sizes are used for mixing to achieve the densest packing and realize the construction of the heat conduction network.

B. Thermal properties measurements

DSC (Differential Scanning Calorimetry, TA Instrument Q20) was used to characterize the thermal properties of sample 1 and sample 2. All tests are performed in a nitrogen atmosphere and the sample is heated from -80°C to 50°C at a rate of 5°C/min.

C. Thermo-mechanical property measurements

DMA (Dynamic Mechanical Analysis, TA Instrument Q800) was used to characterize the rebound ratios, creep and stress relaxation properties of the sample1 and sample2 at different conditions. Tension measurements were performed at a constant stretching rate of 3mm/min. A dumbbell-type sample with 1.5 mm thickness was used. In order to compare the resilience of materials with different elongations, we define (Strain/A_{MAX}) as the ratio of the stretched length to its maximum stretched length, calculated according to Equation (1), where L_e denotes stretching length, while L_0 is initial length (before strained) and A is the maximum tensile strain of the sample.

$$\left(\frac{Strain}{A_{MAX}}\right) = \frac{L_e}{L_0 \times A} \times 100\% \qquad (1)$$

Rebound ratio was expressed in percentage of recovery after strained, calculated according to Equation (2), where L_e denotes stretching length , while L_f is the final length after removing the tension for 5 min.

$$\text{Rebound ratio} = \frac{(L_e - L_f)}{L_e} \times 100\% \qquad (2)$$

III. RESULTS AND DISCUSSIONP

A. Thermally properties silicone gap pads

Silicone has a very low glass transition temperature (T_g), which can realize a low modulus at room temperature and fit well on uneven surfaces. Fig. 2 is a DSC (Differential Scanning Calorimetry) diagram for a typical silicone sample1 and sample 2 in the study. In the temperature range from -80°C to 50°C, neither sample 1 nor sample 2 had obvious glass transition regions. Both of them only have an endothermic peak of crystallization near -46°C [13]. The above results indicate that the T_g points of sample 1 and sample 2 are both lower than -80°C, which means that both materials are soft materials at room temperature，and it is necessary for the application of thermal interface materials.

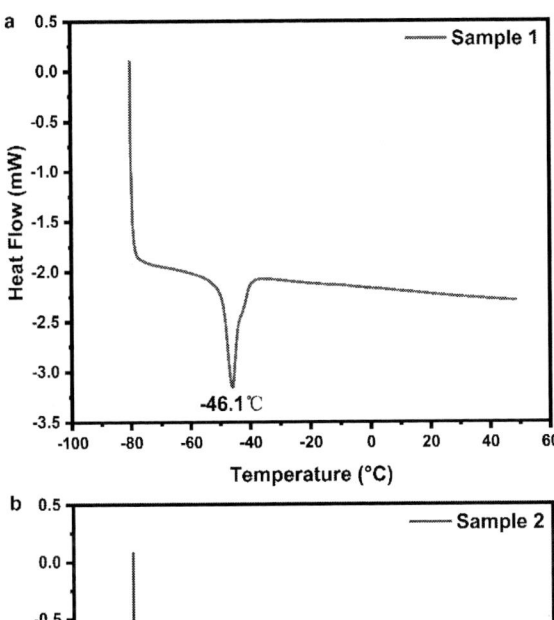

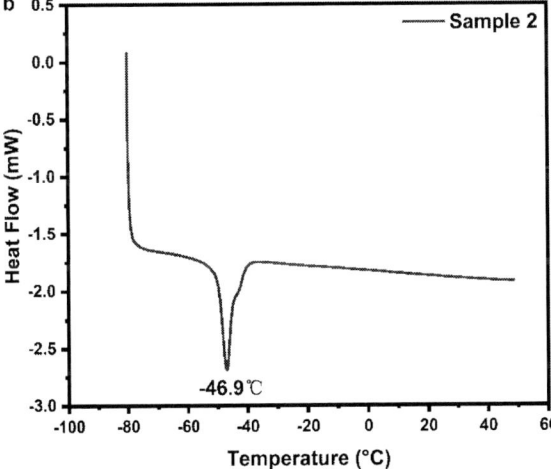

Fig. 2. DSC diagram for a cured silicone gap filler pad a) 1 and b) 2.

B. Strain and creep recovery

Fig. 3a shows that sample 1 has an elongation of 55% and no plastic deformation zone appears. While the sample 2 has an elongation of more than 160%.

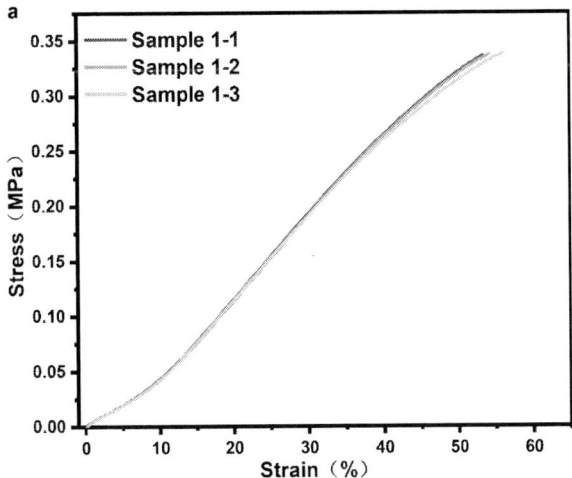

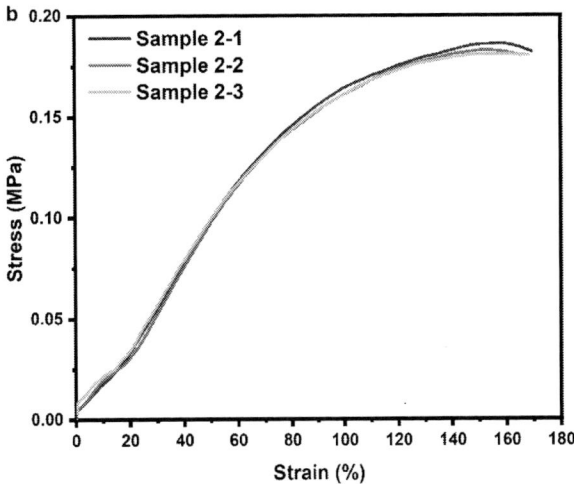

Fig. 3. Typical strain-stress curves of a)sample 1and b)sample 2.

It is worth noting that a clear yield zone appears at the elongation of 150% in sample 2, which means the sample 2 has better flexibility compare to sample1. In others words, sample1 acts like a harder elastomer.

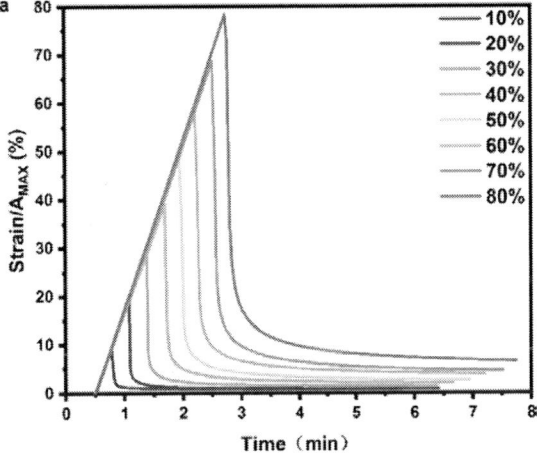

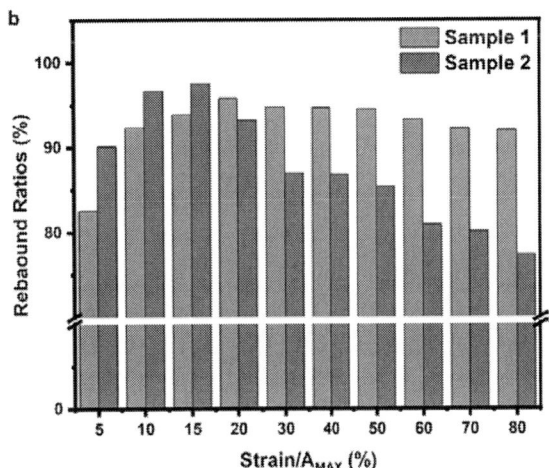

Fig. 4. a) Strain recovery curves of sample1 in a series of elongation. b) Rebound ratios of sample 1 and sample 2 under different deformation .

Fig. 4a shows the resilience curve of sample 1 at

different elongations. It can be seen that even when stretched to 80% of the maximum elongation, sample 1 still has a very excellent resilience rate of over 90%. In addition, it can be seen from Figure 4b that the rebound rate increases at first and then decreases with the increase in elongation, so does the sample 2. The resilience of sample 2 reached a maximum of 95% at 20% of the maximum elongation, while sample 1 reached a maximum of 96% at 15%. This indicated that sample1 and 2 will undergo plastic deformation when stretched in spite of how small the deformation is. This plastic deformation accounts for a relatively large amount under small elongation which manifested as a decrease in rebound ratios.

Fig. 4b shows that with the elongation from 20% to 80%, the resilience rate of sample 1 and sample 2 decreased by 10% and 20%, respectively. A large number of studies have proved that good flexibility is often accompanied by a low degree of cross-linking [14], and sample 2 with good flexibility is more likely to have irreversible deformation such as molecular chain sliding during stretching.

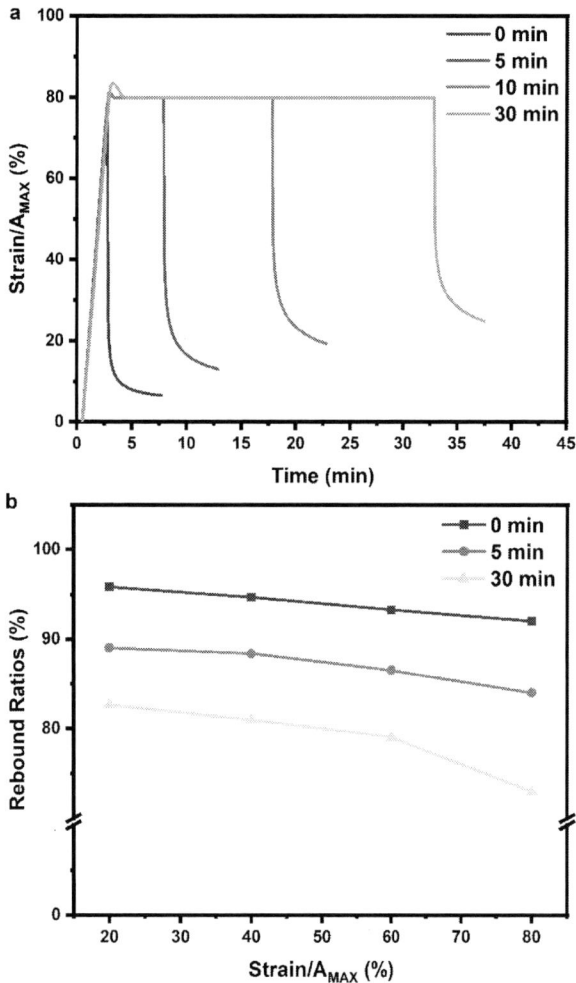

Fig. 5. a) Strain recovery curves of sample1 under different maintainence time at 80% maximum elongation. b) Rebound ratios for sample1 with 20 – 80 % of the maximum elongation and 0,5,30min different maintainence time, respectively.

As shown in Fig. 5a, in this article we explore the rebound ratios after removing the stress for 5 minutes. It can be seen from Fig. 5b that the rebound ratios under different strains decreases as the maintainence time increases. Interestingly, we found that before the elongation reached 60% of the maximum elongation, the resilience rate decreased with a residence time of 5 minutes was almost half of that of 30 minutes. This means that irreversible deformations such as molecular chain movement when the sample is stretched are not linear with time.

C. Creep and Relaxation

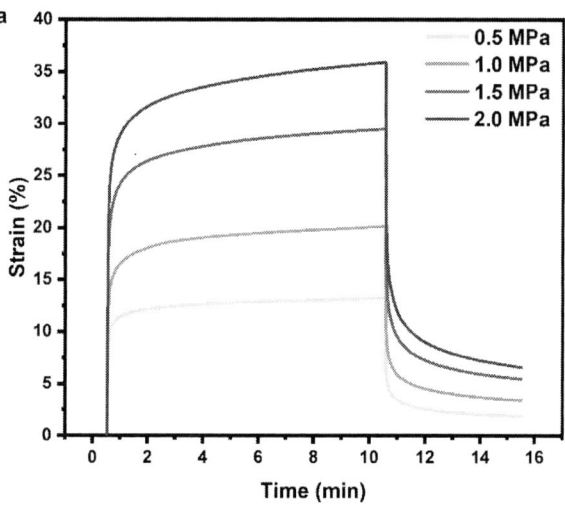

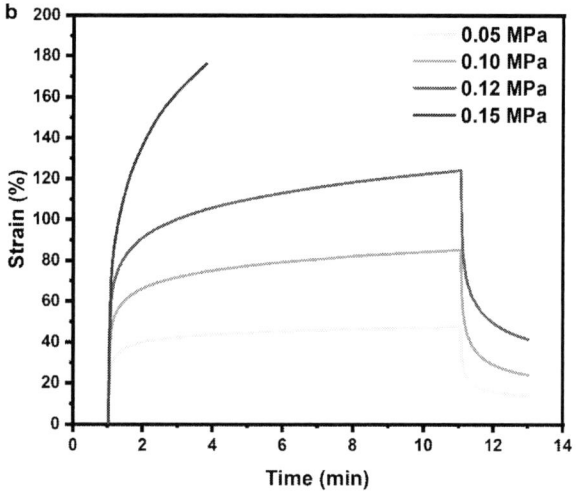

Fig. 6. Creep curves of a)samples 1 and b)sample 2 under a series of stresses.

Fig. 6 shows the creep of the two samples under a series of stresses. Compared with sample 2, sample 1 which behaves as a hard elastomer shows less strain under all stresses. Sample1 shows almost 40% strain compared to 8% of sample2 under a constant stress of 0.15Mpa. For sample 1, it broke instantaneously when a stress exceeding 0.2 MPa was applied in the test, while sample 2 stretched and broke when a stress of 0.15 MPa was applied.

Fig. 7 shows the stress relaxation of the sample1 and 2, which indicated that the sample1 exhibited higher stress relaxation resistance upon 30% compression than sample2. And because sample 2 is a softer thermal interface material, the stress mutation caused by instantaneous stretching is smaller, and it is less likely to cause material failure.

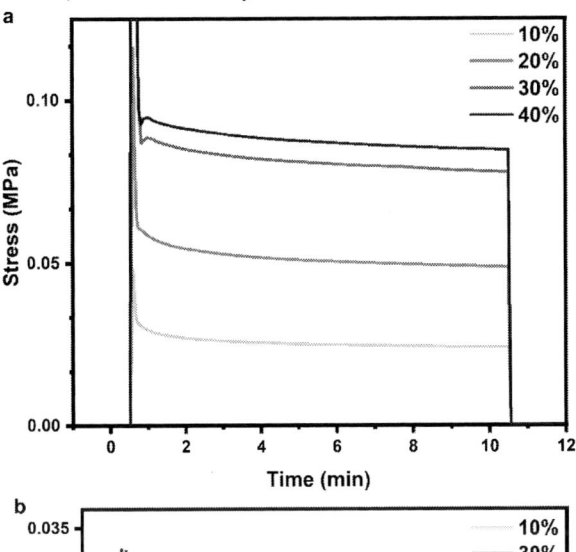

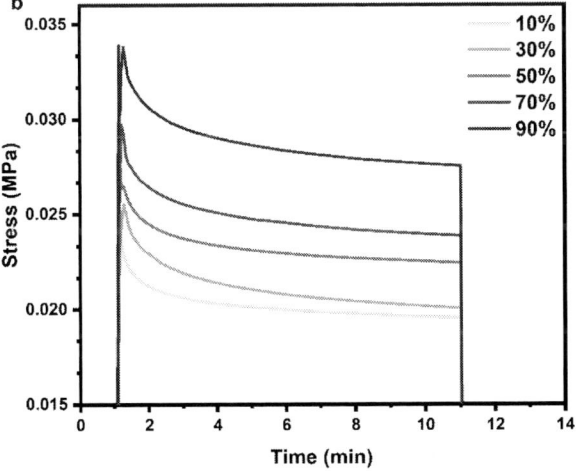

Fig. 7. Stress relaxation of a)sample1 and b)sample2 under a series of constant strain .

D. Loading-Unloading cycle

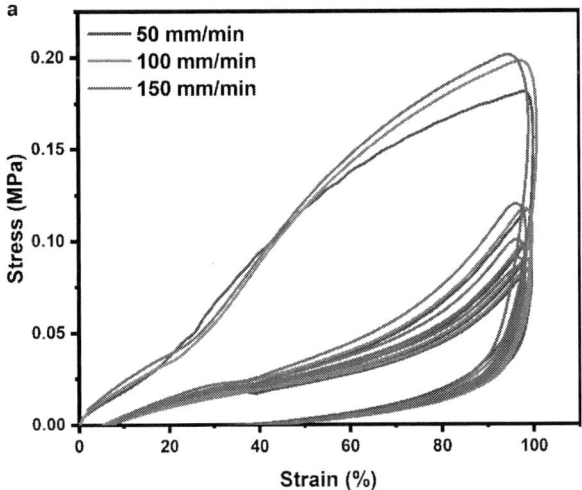

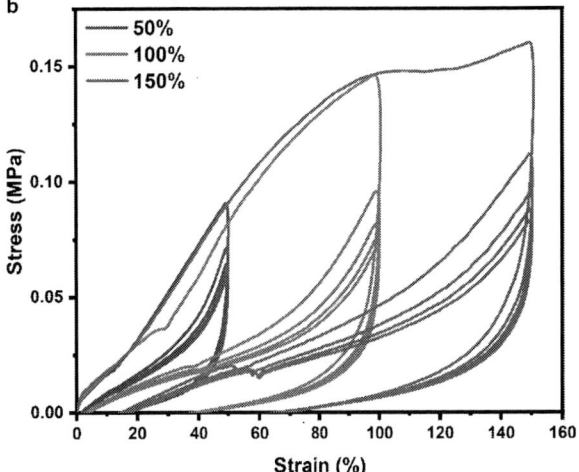

Fig. 8. a) 5 loading-unloading cycles of sample2 at different stretching rates; b) 5 loading-unloading cycles of sample2 at a strain rate of 50%/min in different strain.

Loading and unloading cycle tests are used to investigate the behavior and performance changes of thermal interface materials in practical applications. It can be seen from Figure 8 that when the deformation amount is the same, the stretching rate has a very limited effect on the size of the plastic deformation of the material. This is because the stretched length is in the good elastic length of the material, so it is not prone to greater plastic deformation. When the stretched length changes, as the stretched length increases, the plastic deformation of the sample gradually increases. This is consistent with the previous conclusion that the rebound rate decreases with the increase of the stretched length.

IV. CONCLUSIONS

In the stretchable electronic device industry, thermal interface materials often experience tensile stress due to their tensile behavior. Whether the thermal interface material after stretching can recover its original performance is very important for maintaining good thermal management capabilities. The experiments in this paper showed that 1) silicones gap pad, as the strain increases, the rebound ratios increase at the first and then decreases, reaching the maximum at around 20% of its maximum stretching length. This applies to both the harder elastomer sample 1 and the softer elastomer sample 2. 2) When the term of time is considered, the rebound ratio decreases as the maintenance time increases. And the plasticity of the tensile behavior is not linearly related to time but decelerates as time increases. 3) Cyclic loading-unloading experiments can reflect the reliability of TIMs. In five cycles of loading-unloading, the increase in stretch length can enhance the degree of plastic deformation significantly, while the stretch rate has no correspond with it. This research can provide important guidance for the application of TIMs in stretchable electronic devices industry. Applying the above guidance, it is expected to make the application of silicone gap pad in stretchable electronic devices industry more reasonable.

ACKNOWLEDGMENT

This work was supported by the National Natural Science Foundation of China (Grant No. 52073300), the Guangdong Basic and Applied Basic Research Fund (No. 2019A1515110845), Guangdong Province Key Field R&D Program Project (No. 2020B010190004), Youth Innovation Promotion Association of the Chinese Academy of Sciences (2019354), Shenzhen Science and Technology Research Fund (No.JCYJ20200109114401708 and JCYJ20180507182530279).

REFERENCES

[1] Chang-Ping Feng, Lu-Yao Yang, Jie Yang, Lu Bai, Rui -Ying Bao, Zheng-Ying, et al. "Recent advances in polymer-based thermal interface materials for thermal management: A mini-review." Composites Communications (2020): 100528.

[2] Mudawar, Issam. "Recent advances in high-flux, two-phase thermal management." Journal of Thermal Science and Engineering Applications 5.2 (2013).

[3] Joshua Smitha, Randeep Singhb, Michael Hinterbergera, Masataka Mochizukic. "Battery thermal management system for electric vehicle using heat pipes." International Journal of Thermal Sciences 134 (2018): 517-529.

[4] Puqi Ning, Guangyin Lei, Fred Wang, Khai D. T. Ngo. "Selection of heatsink and fan for high-temperature power modules under weight constraint." 2008 Twenty-Third Annual IEEE Applied Power Electronics Conference and Exposition. IEEE, 2008.

[5] Prasher, Ravi, and Chia-Pin Chiu. "Thermal interface materials." Materials for advanced packaging (2017): 511-535M. Young,

[6] Gowda A, Tonapi S, Reitz B, Gensler G. "Choosing the right thermal interface material." Advanced Packaging 14.3 (2005): 14-18.

[7] Amjadi M, Kyung K U, Park, I, Sitti M. Stretchable, SkinMountable, and Wearable Strain Sensors and Their Potential Applications: A Review. Adv. Funct. Mater. 2016, 26, 1678−1698.

[8] Choi S.; Lee, H.; Ghaffari, R.; Hyeon, T.; Kim, D. H. Recent Advances in Flexible and Stretchable Bio-Electronic Devices Integrated with Nanomaterials. Adv. Mater. 2016, 28, 4203−4218.

[9] Pushparaj, V. L.; Shaijumon, M. M.; Kumar, A.; Murugesan, S.; Ci, L.; Vajtai, R.; Linhardt, R. J.; Nalamasu, O.; Ajayan, P. M. Flexible

[10] Energy Storage Devices Based on Nanocomposite Paper. Proc. Natl. Acad. Sci. U. S. A. 2007, 104, 13574−13577

[11] Wang, S.; Chung, D. Thermal Fatigue in Carbon Fibre Polymer-Matrix Composites, Monitored in Real Time by Electrical Resistance Measurements. Polym. Polym. Compos. 2001, 9, 135−140.

[12] Biron, M. Thermoplastics and Thermoplastic Composites; William Andrew: Waltham, MA, USA, 2012; pp 907−908.

[13] Botterhuis. N E, Van Beek D J M, van Gemert G M L, et al. Self-assembly and morphology of polydimethylsiloxane supramolecular thermoplastic elastomers[J]. Journal of Polymer Science Part A: Polymer Chemistry, 2008, 46(12): 3877-3885

[14] Flory P J, Rabjohn N, Shaffer M C. Dependence of elastic properties of vulcanized rubber on the degree of cross linking[J]. Journal of Polymer Science, 1949, 4(3): 225-245.

Facile Preparation of Cobalt Hydroxide Based Supercapacitor with High Volumetric Energy Density at High Volumetric Power Density

1st Peng Liu
Institute of Materials Research,
Tsinghua Shenzhen International
Graduate School
Tsinghua University
Shenzhen, China
liupeng18@mails.tsinghua.edu.cn

2nd Fangcheng Wang
Institute of Materials Research,
Tsinghua Shenzhen International
Graduate School
Tsinghua University
Shenzhen, China
wfangcheng@sz.tsinghua.edu.cn

3rd Zhuo Zhang
Institute of Materials Research,
Tsinghua Shenzhen International
Graduate School
Tsinghua University
Shenzhen, China
zhangzhuo_199163@163.com

4th Jing Li
Institute of Materials Research,
Tsinghua Shenzhen International
Graduate School
Tsinghua University
Shenzhen, China
1124291742@qq.com

5th Hongjin Fan
School of Physical and Mathematical
Sciences
Nanyang Technological University
Singapore
fanhj@ntu.edu.sg

6th Cheng Yang
Institute of Materials Research,
Tsinghua Shenzhen International
Graduate School
Tsinghua University
Shenzhen, China
yang.cheng@sz.tsinghua.edu.cn

Abstract—**Developing supercapacitor electrodes with an ultra-high specific energy density at high power density and long cycle life is critical to future energy storage devices. However, it is still challenging to fabricate high-performance supercapacitors in a facile and scalable process. In this work, the cobalt nanocone arrays (CNAs) were plated on the copper foil (CF) within 40-second electrodeposition and were transferred into cobalt (Co)/cobalt hydroxide [$Co(OH)_2$] in the alkaline solution after activation of several CV cycles. The flexible $Co/Co(OH)_2$@CF electrode can deliver an ultrahigh specific capacitance of 1043 F cm^{-3} at 1 mA cm^{-2} and excellent cycle stability with a retention of 98% after 20000 cycles in the three-electrode test. $Co/Co(OH)_2$@CF and active carbon were used as the cathode and anode to assemble asymmetric supercapacitors, respectively, in the form of coin and soft package. The soft-package supercapacitor shows an energy density of 28 mWh cm^{-3} at 62 W cm^{-3} with 82% retention after 5000 cycles. And the coin supercapacitor shows a larger energy density of 69 mWh cm^{-3} at 100 W cm^{-3} with retention of 123% after 10000 cycles. Good electrical and mass transport and stable structure contribute to its excellent electrochemical performance. Considering the advantages of the facile and scalable preparation process, low cost, flexibility, and excellent capacitance and stability, this electrode is promising in application to high-performance flexible supercapacitors.**

Keywords—**volumetric specific capacity, stability, self-reconstruction, cobalt hydroxide, cobalt nanocone arrays, supercapacitor, energy density, power density**

I. INTRODUCTION

The fast development of electronic devices puts a high demand on the performance of supercapacitors which is expected to possess higher specific energy density at high power density and longer cycle life [1]. For some applications such as flexible and wearable electronics, the volumetric energy density and power density is especially important [2]. Transition metal oxide such as MnO_2, Co_3O_4, and MoO_3 and hydroxide can deliver a high theoretical specific capacity because of their pseudocapacitive characteristics [3]. Nano-scale oxide and hydroxide such as nanoparticles, nanosheets, and nanowires were prepared to increase their active site density [4]. Traditionally, these materials are usually mixed with conductive additives and binders and stick to the current collector by slurry coating [5]. This process would inevitably introduce non-active materials, and impede the ion/electrical transport because of relatively poor conductivity and less exposed active sites, which would correspondingly deteriorate the capacity, rate performance, and cycle stability [6]. Therefore, in-situ technology has been used to load the active materials on current collectors such as electrodeposition and hydrothermal synthesis [7]. This binder-free strategy has significantly improved the electrochemical performance of these oxides but their stability and capacity still need to be improved [8]. On the one hand, the binding strength between the directly deposited oxide and the smooth metal foil is not strong enough; On the other hand, a closely deposited active material layer leaves no space for cyclical volume expansion and the ion transport of the inner layer during the redox process [9].

In this work, the CNAs were plated on the CF within 40 seconds by one-step electrodeposition and were transferred into $Co/Co(OH)_2$ in the alkaline solution after activation of several CV cycles. Benefiting from the *in situ* self-reconstruction process, the formed thin $Co(OH)_2$ hexagonal platelets were embedded firmly into the Co substrate (good electrical conductivity) and distributed separately with certain space, which helps release the stress from volume expansion and facilitate a thorough mass transfer between the active sites and electrolyte while maintaining good binding with the current collector. Due to the unique strategy, the flexible $Co/Co(OH)_2$@CF electrode shows an ultrahigh specific

Fig. 1. SEM images of cobalt nanocone arrays (CNAs) with different deposition current time: (a) 20 s, (b) 40 s, (c) 80 s, and (d) 160 s; the thickness measurement of Cu foil (e) and Cu foil with CNAs coating (f); SEM image of as-prepared CNAs (i) and corresponding EDS mapping of O (j) and Co (k) and quantitative analysis (l).

capacitance and good cycle stability in the three-electrode test. Exciting performance were also demonstrated in coin and soft package asymmetric supercapacitors, as we would present afterwards. Considering the advantages of the facile and scalable preparation process, low cost, flexibility, and excellent capacitance and stability, this electrode is promising in application to high-performance flexible supercapacitors.

II. EXPERIMENTAL SECTION

A. Fabrication of Cobalt Hydroxide Electrode

The commercial CF was first sonicated in ethanol and light acid for 5 min to clean the surface organic and oxide, respectively. The electrodeposition was carried out in a two-electrode electrochemical cell. The copper foil and platinum platelet were used as the working electrode and countering electrode, respectively. 10 g $CoCl_2 \cdot 6H_2O$, 1.5 g HBO_3 and 10 g NH_4Cl were mixed to form a 50 mL electroplating aqueous electrolyte. Adjusting the pH of electrolyte to 4 at room temperature. To acquire CNAs, the electrodeposition was conducted with a DC Power at 0.4 A cm^{-2} for different times varying from 20 s to 160 s under 60 °C. The CNAs@CF was transferred into $Co(OH)_2$ platelet embedded into the porous Co coating under oxidation potential in the 1 M KOH solution, which was denoted as $Co/Co(OH)_2@CF$.

B. Fabrication of $Co/Co(OH)_2@CF//AC$ Asymmetric Coin and Soft-Package Supercapacitor

The commercial acetylene black, activated carbon, and PVDF (mass ratio 1:8:1) were mixed with NMP (10 wt%) by the magnetic stirrer for 10 h. The mixture paste was dispensed on titanium (Ti) foil and baked for 10 hours. The cellulose membrane and glass fiber film were used as the separator for coin and soft package supercapacitor, respectively. The glass fiber separator (rinsed with 1 M KOH) and electrodes (4 cm^2) were enclosed into the aluminum-plastic film packages by vacuum sealer.

C. Material and Electrochemical Characterization

The FE-SEM was conducted by HITACH S4800, Japan. XRD was conducted by Bruker DS RINT2000/PC.

We use HgO/Hg electrode and graphite rod were as reference and countering electrodes, respectively. The cyclic voltammetry (CV) and galvanostatic charge-discharge (GCD) behavior were investigated using an electrochemical station (VMP3, BioLogic, France).

III. RESULTS AND DISCUSSION

The preparation of CNAs was realized by the electrodeposition. By varying the deposition time from 20 s to 160 s we obtained the CNAs with different morphologies (Fig. 1a-d). With the deposition time increasing, the nanocones become larger and thicker and even form secondary nanocones on the surface of primary nanocones. Through 40 s-deposition we can acquire uniform CNAs on copper foils (CNAs@CF). The CF and CNAs@CF were measured by the section SEM image and micrometer screw gauge to be 6.5 μm and 8.1 μm, respectively, indicating the thickness of the single 40 s-deposited CNAs layer was 0.8 μm (Fig. 1e-h). EDS mapping shows that the CNAs are pure metallic cobalt except slight oxidation on the surface (Fig. 1i-l), which is verified by the XRD pattern (Fig. 2i).

As shown in Fig. 2a-c, the hexagonal platelets, which are about 100-400 nm thick and 200-1000 nm long, appeared after 20 CV cycles between -0.6 and 0.55 V in the 1 M KOH solution and embedded into the substrate. This structure maintained almost unchanged even after 40000 CV cycles, suggesting excellent structure stability. SEM element mapping image shows that oxidation of cobalt occurred on the surface of CF (Fig. 2e-h). Further quantitative element analysis shows that the hexagonal platelet may consist of severely oxidized cobalt while the porous substrate shows that

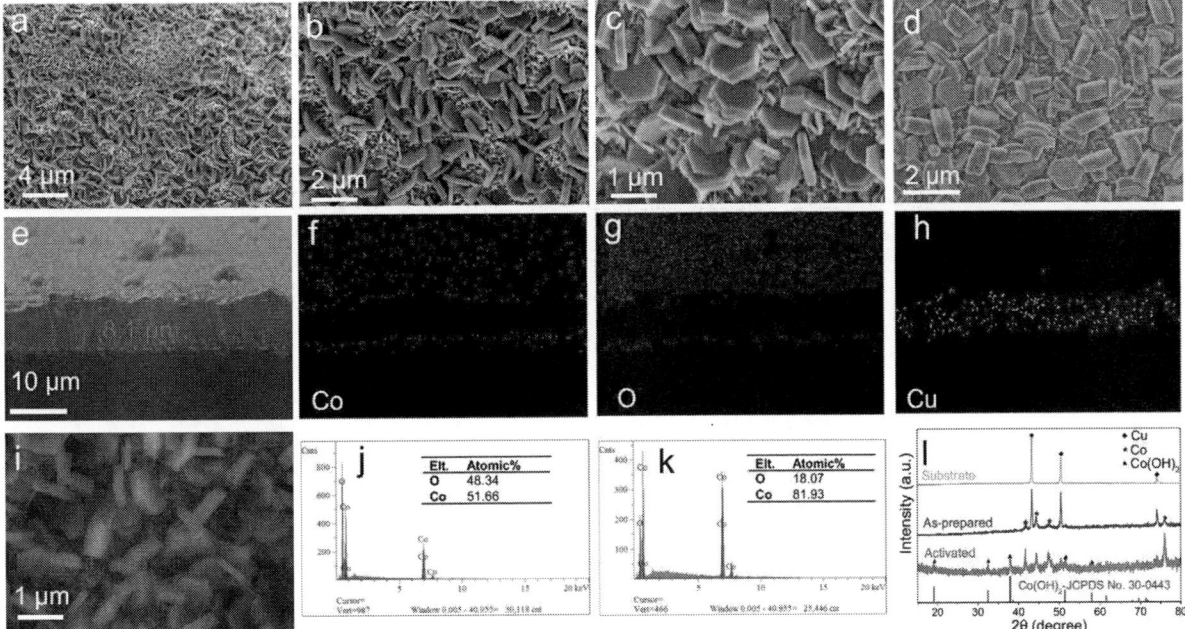

Fig. 2. SEM images of Co/Co(OH)$_2$ before (a-c) and after (d) 40000-cycles testing. SEM image of Co/Co(OH)$_2$@CF (e) and corresponding EDS mapping of Co (f), O (g), and Cu (k) SEM image (i) of Co/Co(OH)$_2$ and corresponding EDS quantitative analysis of 1# (j) and 2# (k) site in (i). (l) XRD analysis.

the cobalt was only slightly oxidized (Fig. i-k). Combined the XRD results, we deduce that the hexagonal platelets are Co(OH)$_2$ (PDF#:30-0443) and the substrate is pure Co (slightly oxidized on the surface, PDF#:05-0727) (Fig. 2i). This Co/Co(OH)$_2$@CF layer still keeps a thickness of around 8.1 μm (Fig. 2e).

As shown in Fig. 3a, the CV curves show a similar shape (three redox peaks) and enclosed area. GCD curves show that the discharge time increases (meaning a larger capacitance) with a longer deposition time (Fig. 3b). However, the specific volumetric capacitance shows a reverse trend because of the increase of thickness, decreasing from the 1108 F cm^{-3} to 223 F cm^{-3} at 2 mA cm^{-2}. Thus we considered the 40-s deposited sample a superior one. 40-s deposited Co/Co(OH)$_2$@CF electrode contains 1.3 mg cm^{-2} active material. Fig. 3c and 3d show that the CV curve still shows excellent reversibility at 1000 mV s^{-1} and keeps a 16% capacitance retention (172 F cm^{-3}), suggesting good electron/ion transport and lower contact resistance characteristics. The three couples of redox peaks may be from the transfer from Co to Co(OH)$_2$ to CoOOH.

These three redox processes provide a wide working potential window ranging from -0.6 to 0.55 V. The GCD curves show three platforms which correspond to three redox process. The capacitance of Co/Co(OH)$_2$@CF electrode based on GCD curves is 1284, 909, 239, and 83 F cm^{-3} (at 1, 2, 5, 10 mA cm^{-2}) (Fig. 3e). This electrode also shows excellent stability as it can retain 97.6 and 91.4 % of the original capacitance even after 20000 and 40000 CV cycles at 100 mV s^{-1}, respectively, outperforming most pseudocapacitive electrodes (Fig. 3g, h) [22-24]. The excellent performance of the Co/Co(OH)$_2$@CF electrode benefits from its unique structure. On the one hand, the *in situ* reconstructions of CNAs into Co/Co(OH)$_2$ nanoplatelets ensure firm binding and fast electron transport. On the other hand, the regularly arrayed cobalt nanocones restrict the layout of the *in situ* formed nanoplatelets which not only guarantee a dense active site but also accomodate enough space for cyclable volume expansion

and the ion transport of the inner layer during the redox process. As shown in Fig. 2d, the nanoplatelets still keep an intact microstructure after 40000 CV cycles. CV curve in Fig. 3f shows a specific area capacitance of 197 mF cm^{-2} for activated carbon.

Asymmetric supercapacitors in the form of coin and soft-package were assembled. The ultrathin Co/Co(OH)$_2$@CF electrode (8.1 μm) is highly flexible and can be tailored to different shapes casually (Fig. 4f). No extra cathode tab is needed for the packaging because the active material was integrated onto the current collector (Fig. 4f). CV curves suggest the working potential of the soft-package supercapacitor can reach 1.6 V (Fig. 4a). The CV curves with three clear redox peaks at different scan rates ranging from 5 mV s^{-1} to 1000 mV s^{-1} demonstrate that the device has good reversibility and can work at a relatively high frequency (Fig. 4b). The GCD curves show the device has capacitances of 281, 240, 221, 163, 110, 79 F cm^{-3} at 1, 2, 4, 8, 20, 40 mA, respectively, demonstrating a good rate performance (Fig. 4c). Further calculation demonstrates that the device has a high energy density of 100, 85, 79, 58, 39, and 28 mWh cm^{-3} at 1.5, 3, 6, 12, 31, and 62 W cm^{-3}, respectively. This soft package supercapacitor also possesses good stability with a retention of 82% after 5000 cycles (Fig. 4d, e), demonstrating a promising application perspective.

The coin supercapacitor shows a similar working potential window and CV curves as compared to that of the soft-package supercapacitor (Fig. 5a, b). However, it presents better performance with a high energy density of 143, 119, 106, 88, 69 mWh cm^{-3} at 5, 10, 20, 50, 100 W cm^{-3} (Fig. 5c), outperforming most pseudocapacitors, which also indicates an excellent rate performance. Its working stability is excellent and the capacitance retention after 10000 CV cycles at 200 mV s^{-1} is as high as 123%.

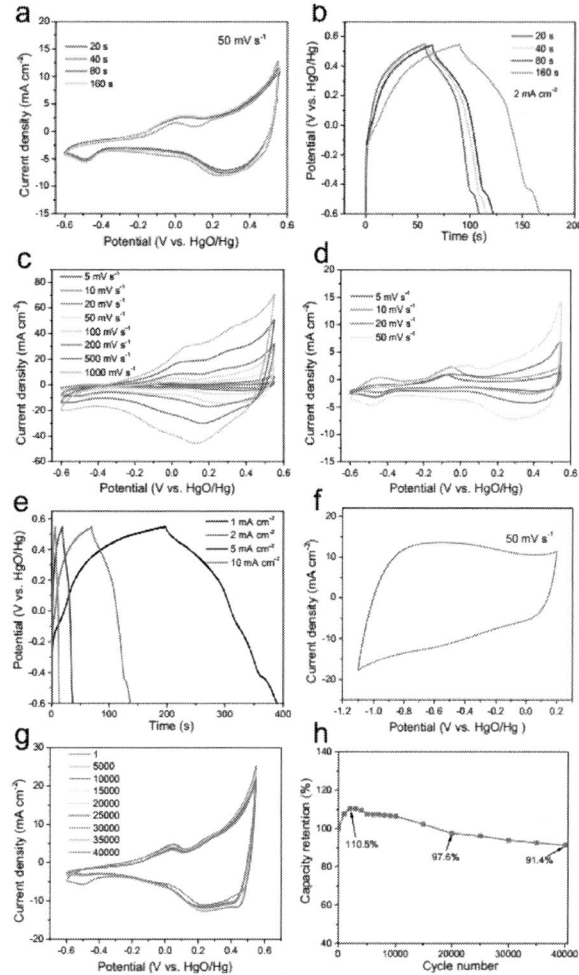

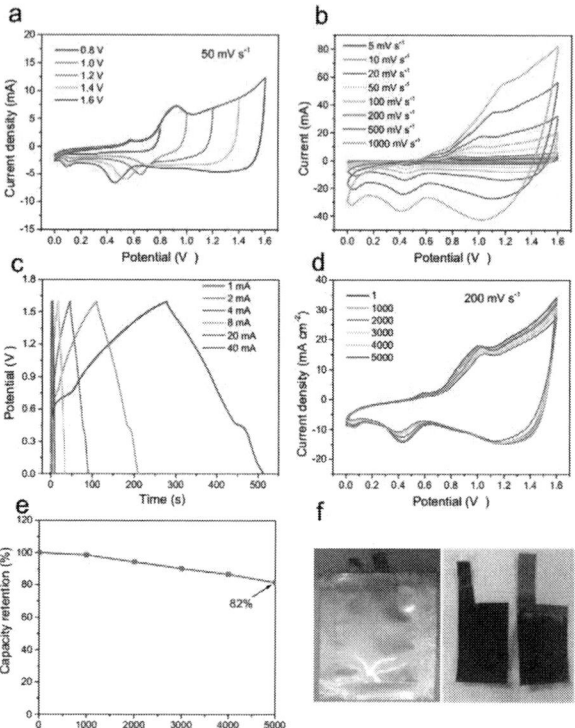

Fig. 3. (a) CV and (b) GCD curves of Co/Co(OH)₂@CF with different deposition times in preparing CNAs. (c-d) CV, (e) GCD, (d) cycling CVs curves, and (e) capacitance retention plot over CV cycling of Co/Co(OH)₂@CF with 40 s deposition time. (f) CV curve of activated carbon.

Fig. 4. (a-b) CV, (c) GCD, (d) cycling CVs curves, (e) capacitance retention plot over CV cycling and (f) optical photographs of Co/Co(OH)₂//AC asymmetric soft-package supercapacitor.

IV. CONCLUSIONS

The Co/Co(OH)₂@CF electrode was successfully prepared by 40-s electrodeposition and electrochemical oxidation of a few minutes in alkaline solution. The in situ reconstructions of CNAs into Co/Co(OH)₂ nanoplatelets ensure their robust connection and fast electron transport with the current collector. The uniform distribution of nanoplatelets also provides enough space to release the stress from cyclical volume expansion and facilitate the ion transport of the inner active material during the redox process meanwhile guarantee a dense active site. Benefiting from the unique structure, the Co/Co(OH)₂@CF electrode shows an ultra-high specific capacitance of 1043 F cm⁻³ at 1 mA cm⁻² and excellent cycle stability with retention of 91.4% after 40000 cycles in the three-electrode test. When applied to asymmetric supercapacitors with active carbon, the soft-package supercapacitor shows an energy density of 28 mWh cm⁻³ at 62 W cm⁻³ with a retention of 82% after 5000 CV cycles, and the coin supercapacitor presents an energy density of 69 mWh cm⁻³ at 100 W cm⁻³ with retention of 123% after 10000 CV cycles. This high-performance, easily-prepared, low-cost electrode is prospective in the application to flexible supercapacitors on a large scale.

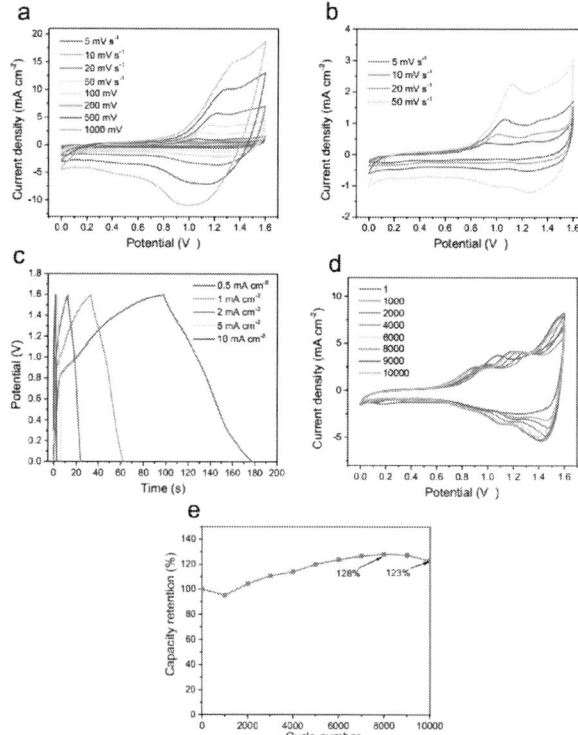

Fig. 5. (a-b) CV curves, (c) GCD curves, and (d) capacitance retention plot over CV cycling of Co/Co(OH)₂//AC asymmetric coin supercapacitor.

ACKNOWLEDGMENT

The authors thank the National Natural Science Foundation of China (52061160482), the Local Innovative and Research Teams Project of Guangdong Pearl River Talents Program (2017BT01N111), Shenzhen Geim Graphene Center, the National Nature Science Foundation of China (Project Nos. 52061160482), Guangdong Province Science and Technology Department (Project No. 2020A0505100014), Shenzhen Government (Project No. JSGG20191129110201725) and Tsinghua Shenzhen International Graduate School Overseas Collaboration Project for financial supports.

REFERENCES

[1] N. A. Kyeremateng, T. Brousse, and D. Pech, "Microsupercapacitors as miniaturized energy-storage components for on-chip electronics," Nat. Nanotechnol., vol. 12, no. 1, pp. 7-15, 2017.

[2] D. P. Dubal, N. R. Chodankar, D.-H. Kim, and P. Gomez-Romero, "Towards flexible solid-state supercapacitors for smart and wearable electronics," Chem. Soc. Rev., vol. 47, no. 6, pp. 2065-2129, 2018.

[3] T. Göhlert, P. F. Siles, T. Päßler, R. Sommer, S. Baunack, S. Oswald, and O. G. Schmidt, "Ultra-thin all-solid-state micro-supercapacitors with exceptional performance and device flexibility," Nano Energy, vol. 33, pp. 387-392, 2017/03/01/, 2017.

[4] A. Ramadoss, B. Saravanakumar, S. W. Lee, Y.-S. Kim, S. J. Kim, and Z. L. Wang, "Piezoelectric-Driven Self-Charging Supercapacitor Power Cell," ACS Nano, vol. 9, no. 4, pp. 4337-4345, 2015.

[5] W. Raza, F. Ali, N. Raza, Y. Luo, K.-H. Kim, J. Yang, S. Kumar, A. Mehmood, and E. E. Kwon, "Recent advancements in supercapacitor technology," Nano Energy, vol. 52, pp. 441-473, 2018.

[6] X. Chang, W. Li, Y. Liu, M. He, X. Zheng, J. Bai, and Z. Ren, "Hierarchical NiCo2S4@NiCoP core-shell nanocolumn arrays on nickel foam as a binder-free supercapacitor electrode with enhanced electrochemical performance," J. Colloid Interface Sci., vol. 538, pp. 34-44, 2019.

[7] Z. Su, C. Yang, C. Xu, H. Wu, Z. Zhang, T. Liu, C. Zhang, Q. Yang, B. Li, and F. Kang, "Co-electro-deposition of the MnO2–PEDOT: PSS nanostructured composite for high areal mass, flexible asymmetric supercapacitor devices," J. Mater. Chem. A, vol. 1, no. 40, pp. 12432-12440, 2013.

[8] L. Li, J. Xu, J. Lei, J. Zhang, F. McLarnon, Z. Wei, N. Li, and F. Pan, "A one-step, cost-effective green method to in situ fabricate Ni(OH)2 hexagonal platelets on Ni foam as binder-free supercapacitor electrode materials," J. Mater. Chem. A, vol. 3, no. 5, pp. 1953-1960, 2015.

[9] X.-h. Xia, J.-p. Tu, Y.-q. Zhang, Y.-j. Mai, X.-l. Wang, C.-d. Gu, and X.-b. Zhao, "Freestanding Co3O4 nanowire array for high performance supercapacitors," RSC Adv., vol. 2, no. 5, pp. 1835-1841, 2012.

Die chipping FDC development at wafer saw process

Dongpeng Xue
SanDisk Semicondictor (Shanghai) Co., Ltd.
Shanghai, China
Dongpeng.Xue@wdc.com

Caiden Zhong
SanDisk Semicondictor (Shanghai) Co., Ltd.
Shanghai, China
Guocheng.Zhong@wdc.com

Elley Zhang
SanDisk Semicondictor (Shanghai) Co., Ltd.
Shanghai, China
Elley.Zhang@wdc.com

Weiting Jiang
SanDisk Semicondictor (Shanghai) Co., Ltd.
Shanghai, China
Weiting.Jiang@wdc.com

Cong Zhang
SanDisk Semicondictor (Shanghai) Co., Ltd.
Shanghai, China
Cong.Zhang@wdc.com

Abstract— **The die chipping which cause function failure always a key issue for BiCSX wafer saw. Traditionally, detecting the defects by die seal ring inspection after it has occurred has been large-scale adoption. However, this method has large limitation due to two factors. First is nuisance defect which may cause by manual classification; second is additional investment from detection machine.**

Fault Detection and Classification (FDC) as one part of Advanced Process Control (APC) has been widely studied in semiconductor manufacturing. FDC can detect the defects like die chipping by record and analysis equipment data during wafer saw. It can not only detect the defects but also help engineer to develop robust process.

In this paper, the mechanism of die chipping and the interaction of equipment data with defects were studied. Successfully create a totally new FDC rule to detect the die chipping.

Keywords—die chipping, FDC

I. INTRODUCTION – PROBLEM DESCRIPTION

Bit Cost Scalable (BiCS) technology which come out in 2007 has been widely used in various memories. It can extremely reduce the bit costs by vertically stacking memory arrays with punch and pug process [1]. With the development of BiCS memory wafer, the metal layer become more and more thick. It is bringing great challenge for the singulation of BiCS wafer. The die chipping (Fig.1) which cause function failure always a key issue during singulation of BiCS wafer.

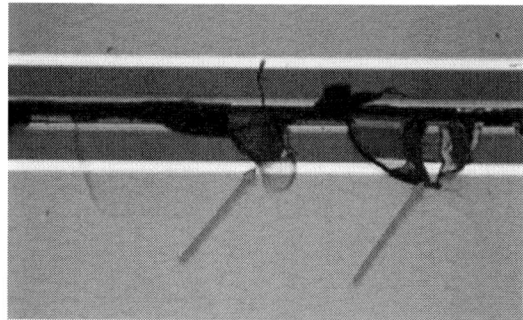

Fig. 1. Die Chipping issue

II. MECHANISM OF DIE CHIPPING

It is important to understand the mechanism of die chipping during singulation. It not only help us to understand the root cause, but also to help to give us clearly improve direction.

Fig. 2 shows that in situ EDS result of blade during singulation. As we all know, the blade contain diamond bond material. Usually, it not contain metal element or Si. But after cutting 10m, the Si, Ni was found. With the increase of cutting length, the amount of Si, Ni become larger. These element are all come from wafer.

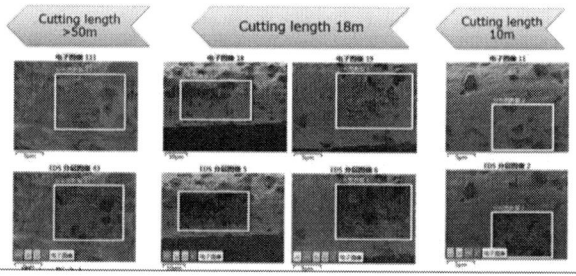

Fig. 2. In situ EDS study of blade during singulation.

It means the blade is blocked during singulation. It lost its self-sharping ability. Fig.3 shows the morphology of blade after blocking. At the yellow circle area, there is some material adhesion on it. its original morphology should like others areas, and diamond can be seen. When the blade under this condition, the yellow circle area has no cutting ability. It can only press the wafer immediately and cause chipping like Fig.1 shows.

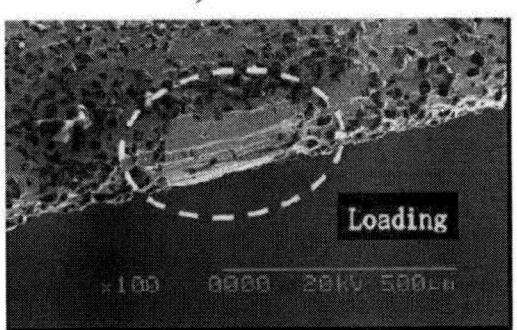

Fig. 3. The morphology of blocked blade.

III. DIE CHIPPING FDC STUDY

Based on mechanism study, it is known that the die chipping is inevitable. But control the yield loss factors is important, especially for 3D integration solution such as systems in package (SIPs) and systems on package (SOPs) [2].

Traditionally, detecting the defects by die seal ring inspection after it has occurred has been large-scale adoption. However, this method has large limitation due to two factors. First is nuisance defect which may cause by manual classification; second is additional investment from detection machine. Fault Detection and Classification (FDC) as one part of Advanced Process Control (APC) has been widely studied in semiconductor manufacturing. FDC can detect the defects like die chipping by record and analysis equipment data during wafer saw [3]. It can not only detect the defects but also help engineer to develop robust process.

During FDC study, first step is benchmark all the equipment data of normal cutting and abnormal cutting. Fig. 4 shows the on signal A, the data of normal cutting and abnormal cutting is obviously different.

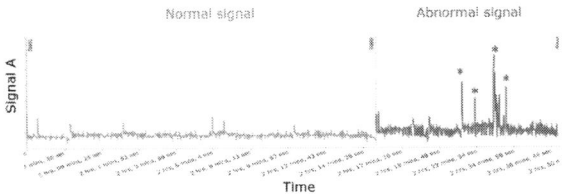

Fig. 4. Comparison of normal and abnormal trace on signal A.

The second step is to analysis how to distinguish the normal and abnormal signal. The table.1 shows a simple method to distinguish the signal. The min, max, mean and standard deviation (STDEV) and others format f(x).

TABLE I.   EQUIPMENT DATA OF NORMAL AND ABNORMAL SIGNAL

Condition	Equipment data-signal A				
	Min	Max	Mean	STDEV	F(x)
Normal signal	650	830	681	14	180
Abnormal signal-1	650	1180	713	49.4	530
Abnormal signal-2	640	1230	700	69.5	590
Abnormal signal-3	650	1200	712	47.2	550

From the Table.1, we can clearly see the signal A STDEV and F(x) are obvious smaller than others run which has die

chipping happen. Similar work will be done on all signal, like signal B, C, D and so on. After all equipment data has been analyses. Then, the control limit can be set on them to distinguish which wafer has die chipping, as a new FDC rule.

In the next step, the new FDC rule will set and use this rule to monitor the rule with different machine and different wafer as Fig. 5 shows. From picture, the control limit of signal A, B, C can effective alarm with defect. However, the control limit of signal D found one alarm on machine II and two alarm on machine III. But actual there is no defect happen. So it is fail. Control limit of Signal E also has same issue. The control limit of signal F can't alarm when defect happen on machine I.

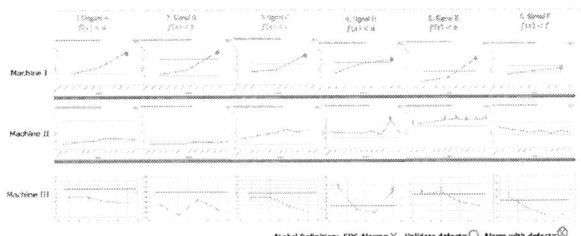

Fig. 5. Analysis of equipment data when defect happen during singulation.

After machine and wafer validate, the accurate FDC rule can be implement in mass production.

IV. RESULT AND CONCLUSION

In this paper, the mechanism of die chipping has been discussed. The major reason cause the chipping is the blocking of blade, leading to loss of self-sharp of blade.

By record and analysis the equipment data, the key signal and control limit can be set. After future test on different machine and different wafer, the accurate FDC rule can be implement in mass production. We provide a detail flow to define die chipping FDC rule. It will help to detect the defects effectively and also help engineer to develop robust process.

ACKNOWLEDGMENT *(Heading 5)*

Thanks for Caiden effort for the die chipping mechanism study and setting new FDC rule on die chipping.

Thanks for Weiting and Elley guide on the FDC study.

Thanks for all team's hard work on the FDC study.

REFERENCES

[1] A. Nitayama and H. Aochi, A. Nitayama and H. Aochi, "Bit Cost Scalable (BiCS) technology for future ultra high density storage memories," 2013 Symposium on VLSI Technology, 2013, pp. T60-T61.

[2] W. H. Teh, D. S. Boning and R. E. Welsch, "Multi-Strata Stealth Dicing Before Grinding for Singulation-Defects Elimination and Die Strength Enhancement: Experiment and Simulation," in IEEE Transactions on Semiconductor Manufacturing, vol. 28, no. 3, pp. 408-423, Aug. 2015.

[3] F. Zhu et al., "Methodology for Important Sensor Screening for Fault Detection and Classification in Semiconductor Manufacturing," in IEEE Transactions on Semiconductor Manufacturing, vol. 34, no. 1, pp. 65-73, Feb. 2021.

Hybrid-Embedded SIP Package Design

Louise Tan
COE of PDL-SIP PTD Design
Western Digital
(Singapore)Pte. Ltd. Taiwan Branch
Taiwan, China
Louis.Tan@wdc.com

Chender Chen
COE of PDL-SIP PTD Design
Western Digital
(Singapore)Pte. Ltd. Taiwan Branch
Taiwan, China
Chender.Chen@wdc.com

CC Liao
COE of PDL-SIP PTD Design
Western Digital
(Singapore)Pte. Ltd. Taiwan Branch
Taiwan, China
Cc.Liao@wdc.com

Abstract— **This paper discusses some of the challenges encountered when moving from a SIP package to Hybrid package technology with combination of flip Chip (controller die) and wire bonding (NAND die) inside one package, where one design has a Hybrid-side by side (Hybrid-SBS) SIP package structure. Due to controller and NAND die size growth with more IOs and capacity, Hybrid-SBS SIP package requires a form-factor change. Thus, Hybrid-Embedded SIP Package structure maybe more suitable for next generation SIP package.**

In this paper, Hybrid-Embedded SIP Package structure is introduced and challenges of this package structure and package routing are discussed.

Keywords— Hybrid; Hybrid-side by side (Hybrid-SBS); Hybrid-Embedded; Flip Chip controller; Substrate

I. INTRODUCTION

When controller dies have been developed to Flip Chip (FC) type, but NAND dies still stay on wire bonding (WB) type, single wire bonding package technology can't be used on this kind of SIP package. We need to use Hybrid process including flip chip and wire bonding for such SIP package with FC controller and WB NAND.

Hybrid-side by side(Hybrid-SBS) SIP Package structure has been applied to real product and run for production. However, controller die size growth has led to development of a next generation SIP package, which is a Hybrid-Embedded SIP Package structure that is embedded flip chip under NAND die stacking and dual side wire bonding for more IOs' fan out.

Compared to Hybrid-SBS SIP, there are 2 main challenges for Hybrid-Embedded SIP design, including package structure and package substrate routing.

II. PHENOMENON

A. Hybrid-side by side(Hybrid-SBS) SIP Package

Controller (FC die) and NAND dies are side by side placement with single side wire bonding.

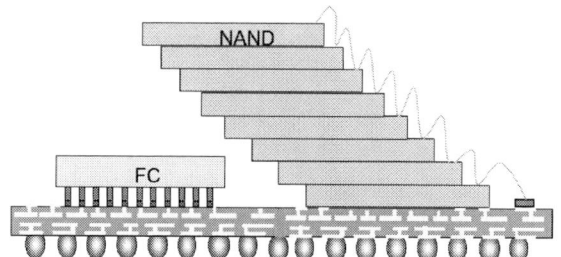

Fig. 1.   Hybrid-SBS SIP Package structure-cross section view

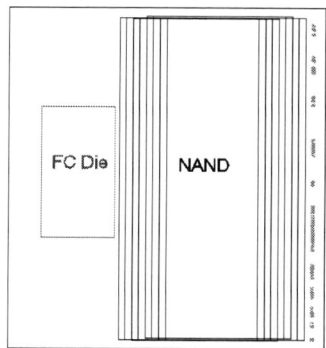

Fig. 2. Hybrid-SBS SIP Package structure-Top view

B. Hybrid-Embedded SIP Package

Controller (FC die) is under NAND dies stacking and dual sides wire bonding with more IOs' inter connection.

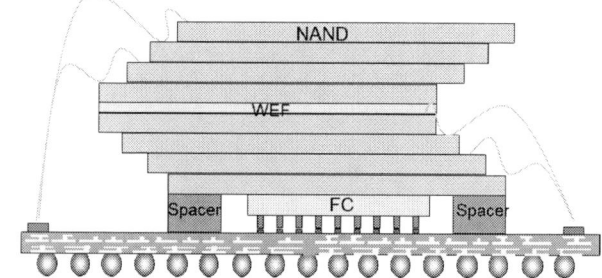

Fig. 3. Hybrid-Embedded SIP Package structure-cross section view

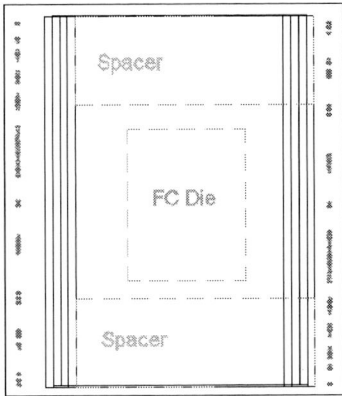

Fig. 4. Hybrid-Embedded SIP Package structure-Top view

III. CHALLENGES

Challenges from package structure and substrate routing need to be considered even Hybrid-Embedded SIP Package structure can provide the solution for more IOs' connection between FC controller and WB NAND so several new technologies have to develop for mitigation.

A. The package structure challenges

Main challenges from package structure.

1) *Embedded structure*: When Controller and NAND placement is not fitted in SBS placement, we need to locate the Controller under NAND, which is called embedded HFC package. This structure face difficult in FCA, UF and MD. The challenges are NAND Die Crack, FC Controller Die Crack, FC Controller Die Bump/ELK crack, delamination between Controller and NAND, No epoxy on FC die, Mold void (Incomplete fill), Die bending, strip warpage during process and finished good unit warpage.

2) *Thin NAND and FC controller die thickness*: To meet package Z height target, controller and NAND die thickness need to be reduced. The challenges are Controller Die Crack and NAND Die Crack.

3) *Wire Bonding*: WB overhang and 7th NAND loop height control. The challenges are Overhang NAND Die Crack and Wire exposed.

B. Substrate routing challenges

There are two challenges of substrate routing to meet SI PI goal.

1) *NAND interface routing:* More IOs routing in the same room to meet SI PI goal, such as impedance control, skew criterial and cross talk. A good placement of controller and NANDs is important for this.

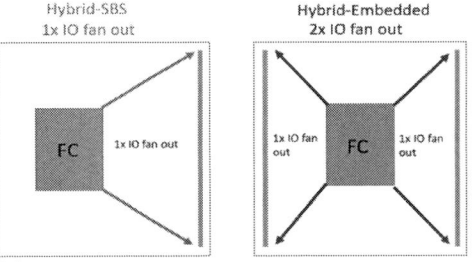

2) *Controller bump deployment*: Need to optimize controller IO bump deployment and controller placement to match with dual sides NAND bonding routing as Fig. 5 and Fig. 6.

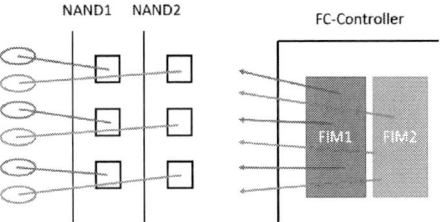

Fig. 5. Interleave IO bump deployment on FC controller

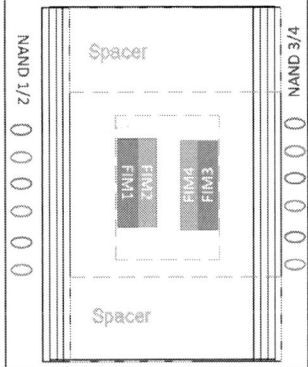

Fig. 6. Grouping IOs at two long sides on FC controller

CONCLUSION

Hybrid-Embedded SIP design has evolved with characterizations like thin FC controller and NAND, HFC-Embedded structure and spacer design optimization for molding tunnel. We work on technology development for all main challenges of Hybrid-Embedded SIP design to eliminate risk.

978-1-6654-1392-3/21 $31.00 © 2021 IEEE

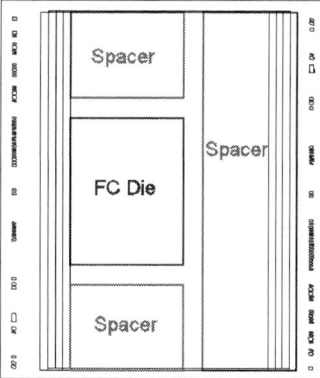

Fig. 7. Hybrid-Embedded SIP with spacer-example-1

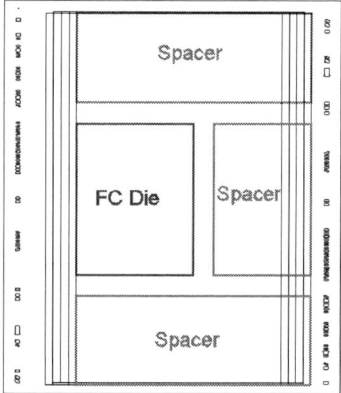

Fig. 8. Hybrid-Embedded SIP with spacer-example-2

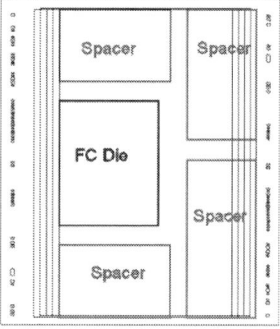

Fig. 9. Hybrid-Embedded SIP with spacer-example-3

In parallel with technology development, FC controller IO bump deployment matching with NAND die pad IO and optimal placement of controller and NANDs play important role in substrate routing to meet SI PI goal.

ACKNOWLEDGMENT

Thanks for Xu Wang/Wei Wang/Cong Zhang/Meiqin Hao to share technology risk and mitigation. Very good team learning and work together.

Thanks for all team's hard work on issue deep dive and recovery.

A novel Cu@Sn@Ag core-shell particles for die attachment in power device packaging

Jiahao Liu
China Electronic Product Reliability
and Environmental Testing Research
Institute
Guangzhou, China
ljh071408@163.com

Hui Xiao *
China Electronic Product Reliability
and Environmental Testing Research
Institute
Guangzhou, China
xiaohui_ceprei@163.com

Xiaotong Guo
China Electronic Product Reliability
and Environmental Testing Research
Institute
Guangzhou, China
guoxiaotong0713@163.com

Xinjie Wang
*Department of Materials Science and
Engineering*
Harbin Institute of Technology
Shenzhen, China
19S155068@stu.hit.edu.cn

Zhijun Yao
*Department of Materials Science and
Engineering*
Harbin Institute of Technology
Shenzhen, China
yaozhijun@stu.hit.edu.cn

Xingchao Mao
*Department of Materials Science and
Engineering*
Harbin Institute of Technology
Shenzhen, China
1723909420@qq.com

Hao liu
Beijing Santel Technology&Trading Corp.
Beijing, China
liuhaohit@163.com

HongTao Chen
*Department of Materials Science and
Engineering*
Harbin Institute of Technology
Shenzhen, China
Chenht@ hit.edu.cn

Abstract—**This paper prepares a novel Cu@Sn@Ag core-shell particles as the die attach material. By the method of Transient Liquid Phase (TLP), the power device can be packaged to realize electrical interconnection and heat conduction. After electroless plating by two-step method, the Cu particles can be coated with Sn layer and Ag layer, successively. The resulting Cu@Sn@Ag particles were compressed into preform. After reflow for 15 min at 250 °C, the interconnection can be completed between devices and Substrates. According to the results of SEM, EDS, and XRD, the resulting bondline consists of three phases, namely, Cu_3Sn、 Ag_3Sn and Cu, which indicates that the sample after reflow can withstand high temperature (>475 °C , the melting point of Ag_3Sn) . Besides, the resulting bondline revealed excellent mechanical properties and thermal properties. After shear testing at high temperature (400 °C), the average shear strengths can be maintained at 25.3 MPa. After thermal shocks for 500 cycles, the interface between bondline and substrate was effectively interconnected, rather than occurred the failure. As a novel material for die attachment, the Cu@Sn@Ag core-shell particles reveal great potential of exploration in the field of power device packaging.**

Keywords—Die attach, Core-shell structure, Sintering, Intermetallic compounds.

I. INTRODUCTION

As the electronic devices develop instantly on the miniaturization, the power density of devices increases sharply. Correspondingly, the demand of high density packaging is urgent. The high power density and high package density will cause a dramatic increase of heat production, which indicates that the interconnection structure needs to withstand higher temperatures. However, the properties of traditional Sn based solders are poor during service at high temperature. Moreover, during the service of power semiconductor represented by SiC And GaN, the local temperature even reaches the melting-point of Sn based solders. Hence, the novel die attachment materials and technologies have been widely studied. s [1-3]. The Transient Liquid Phase (TLP) bonding method is a novel

solutions for operating at high temperature. In the method of TLP, the bondline consisted of at least two phases with different melting-point, and it is sandwiched between substrate and devices. Then the resulting bondline will generate when the temperature of reflow exceeds the melting-point of low-melting-point phase. After the phase with low-melting-point has been completely consumed, the bondline can withstand higher temperature[4,5]. Compared with the specific surface area of flat-plate structure, that of core-shell structure is larger, which can shorten reaction time. Currently, some new systems of core-shell particles, such as Ni@Sn[6], Cu@Sn [7], Ag@Sn [8] has been extensively reported. However, the heat conduction of Ni@Sn particles needs to improve, the oxidizability of Cu@Sn limits the applications, and the high-cost of Ag@Sn cannot match the ues needs of enterprises. Hence, in this work, we prepare a novel Cu@Sn@Ag core-shell particles which are low cost and antioxidant. For simplicty, the Cu@Sn@Ag particles are replaced by CSA particles. The melting-point of the CSA particle is only the melting-pointing of Sn but the remelting-point is much higher than that of Ag-Sn IMCs, which indicates that the bondline can operate during high temperature environment.

II. EXPERIMENTS

A. Electroless plating of the CSA particles

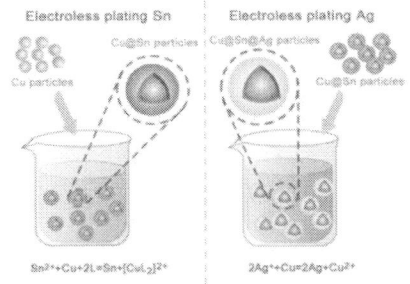

Fig. 1. Procedure diagram of electroless plating CSA particles

The procedure of electroless plating CSA particles was shown in Fig. 1, the coating layer of Sn and Ag was implemented by two-step method. At first, the 5 g rose shaped Cu particles with 10 μm in diameter were cleaned by alcohol solution and 10% sulphuric acid. After removing the oxides and foreign impurities, the Cu particles were pureed into the electroless plating Sn solution in which the Sodium hypophosphite (NaH_2PO_2), Thiourea (CH_4N_2S), Citric acid ($C_6H_8O_7$), Hydrochloric acid (HCl), and Stannous chloride ($SnCl_2$) was 0.51, 1.18, 0.078, 0.205, and 0.079 mol/L, respectively. The solution of electroless plating was 30 °C and the reaction lasted for 60 min with continuous stirring. After cleaning by acetone and de-ioned water, the Cu@Sn particles were obtained. Similarly, the CSA particles were prepared by electroless planting the Cu@Sn particles in the Ag plating solution for 3 min. And in the solution Ammonium carbonate ($CH_8N_2O_3$), Anhydrous salicylic acid ($C_7H_6O_6S$), Silver nitrate ($AgNO_3$), and Alanine ($C_6H_{14}N_2O_2$) was 0.47, 0.18, 0.058, and 0.27 mol/L, respectively.

After cleaning and drying, the CSA particles were transformed into the preform under a pressure of 40 MPa for 2 min. Then, the preform was tailored into the square with the dimensions of 6 mm × 6 mm × 0.2 mm. Finally, the preform was sandwiched between upper Cu substrate and lower Cu substrate, and the dimensions of upper substrate and lower substrate were 6 mm × 6 mm × 3 mm and 12 mm ×12 mm × 3 mm, respectively. After reflow at 250 °C for 15 min, the resulting bondline was prepared which can be used for sheer testing. In order to close the application, the Si precoated 3 μm Ni and 1 μm Ag with dimensions of 5 mm × 3 mm × 0.1 mm as dummy chip was adopted to replace the upper Cu substrate in the test of thermal shock testing.

B. Material characterization

Scanning electron microscopy (SEM, COXEM EM-30AX) was used to characterize the micromorphology, and energy-dispersive X-ray spectroscopy (EDX, GAXD OG4) was uyilized to analysis the element composition. A focused ion beam system (FIB, TESCAN S8000G) was adopted to prepare the cross-sections of the CSA particles. X-ray diffraction (XRD, BRUKER, D2 PHASER) was adopted to analyze the phase composition by using the Cu-Kα radiation (λ = 1.5405 Å) at a scanning speed of 10°/min from 25 ° to 75 °. The vacuum hot-press platform (ETOOL 3030ZK) was used to control the temperature and pressure during reflow process, and a shear testing machine (DAGE 4000PXY) with the shear speed of 30 μm/s was used to measure the shear strength. The rapid temperature change environment test chamber (RIUKAI T260A) was applied for thermal shock testing, and acoustic microscopic detector (SAM, ANALYSIS T156) was used to measure the porosity of the interface before and after thermal shocking.

III. RESULTS AND DISCUSSION

As shown in Fig. 2, the micromorphology and cross-section of the Cu@Sn@Ag particles were revealed, respectively. The Cu@Sn particles shown in Fig. 2(a) were smooth, which the micromorphology of CSA particles in Fig. 2(b) changed obviously, and the surface of CSA particles were rough. The cross-section of the CSA particles was prepared by FIB in Fig. 2(c), and the element composition was determined by EDS. According to the EDS result in TABLE I, it can be distinguished that the outer coating layer and middle coating layer was Ag and Sn, respectively. By

measuring, the thickness of Ag coating and Sn coating were 1.3 μm and 0.3 μm, respectively. Besides, it can be seen in Fig. 2(d), the Ag coating and Sn coating was uncrossed, which indicates that CSA particles were prepared successfully.

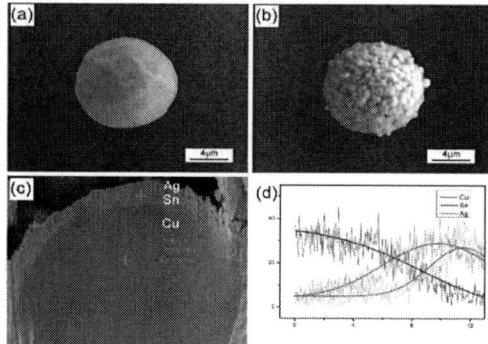

Fig. 2. (a) SEM image of the Cu particle without electroless plating. (b) SEM image of the CSA particle after electroless plating. (c) Cross-sectional image of the CSA particles. (d) line-scan analysis of the CAS particles indicated in Fig. 2 (c).

TABLE I. EDX ANALYSIS RESULTS FOR POSITIONS MARKED IN FIG. 2(C)

Point	Composition（at%）			Corresponding phase
	Cu	Ag	Sn	
1	93.1	0.3	6.6	Cu
2	19.6	3.2	77.2	Sn
3	3.0	16.5	11.3	Ag

As shown in Fig. 3, the phase evolution of CSA particles during the process of reflow was revealed by XRD. Before the reflow process, the peaks in XRD profile of CSA particles correspond to Cu_6Sn_5, Sn, Cu and Ag, respectively. Subsequently, the CSA particles was placed on vacuum hot-press platform at 250 °C for continuous heating. Once heated, the peaks of Cu_3Sn first appeared. Then, the peaks of Ag_3Sn appeared when heated for 1 min. As the time of reflow increased to 10 min, the peak of Sn cannot be observed. In other words, the low-melting-point phase in the bondline was consumed totally, which indicates that the bondlline can withstand higher temperature. When the time of reflow increased to 15 min, only the peaks of Cu, Cu_3Sn, Ag_3Sn were observed in XRD profile. Due to the high-melting-point of intermetallic compounds (IMCs), the bondline can withstand a much higher temperature (> 475 °C, the melting-point of Ag_3Sn).

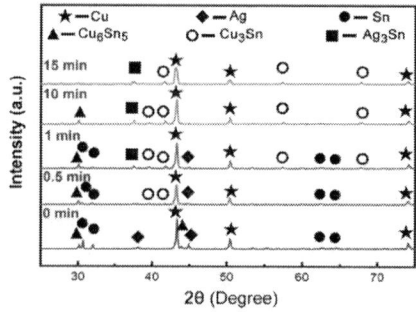

Fig. 3. XRD profile of CSA particles with different reflow time at 250 °C

As the reflow time increased, the molten Sn in the preform began to connect the substrates and filled the voids between particles. After reflow for 15 min, the bondline was grinded and polished, and the cross-sectional of bondline was shown in Fig. 4. As shown in Fig. 4, the CSA particles in the bondline were closely interconnected, and the bondline presented net-work structure. The results of EDS was shown in TABLE II, and only Cu, Cu_3Sn, Ag_3Sn phases existed in the resulting bondline based on CSA preform. Moreover, as shown in Fig.4 (b), the Cu particles were evenly distributed in Cu_3Sn and Ag_3Sn IMCs, and the net-work structure will be conducive to the enhancement of the thermal and mechanical properties of the bondline.

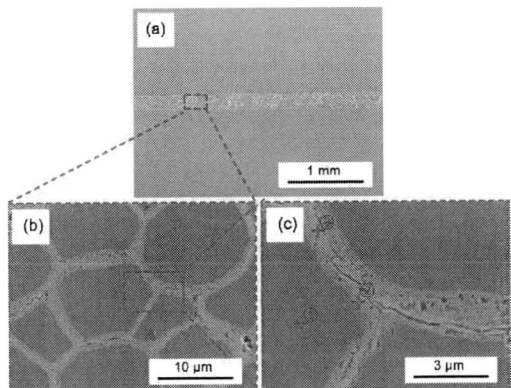

Fig. 4. (a) Cross-section image of bondline. (b) The image zoomed in on the area marked in (a). (c) The image zoomed in on the area marked in (b).

TABLE II.   EDX ANALYSIS RESULTS FOR POSITIONS MARKED IN FIG. 4(C)

Point	Composition（at%）			Corresponding phase
	Cu	Ag	Sn	
1	93.1	0.3	6.6	Cu
2	70.3	4.2	25.5	Sn_3Sn
3	6.3	69.2	24.5	Ag_3Sn

In order to accurately measure the average shear strength of the bondline at ambient temperature and high temperature (400 °C), more than 25 samples were adopted for shear testing, respectively. After shear testing at ambient temperature and high temperature, the average shear strength of the resulting bondline was 30.5 MPa and 25.3 MPa, respectively. Moreover, it can be seen from Fig. 5, the morphology of the bondline after shear testing were revealed. As shown in Fig. 5(a) and (b), the fracture morphology composed of Cu_3Sn, Ag_3Sn, and Cu phases at ambient temperature, and no Sn phase was observed. Unlike the fracture morphology at ambient temperature, as shown in Fig. 5(c) and (d), only Ag_3Sn and Cu_3Sn IMCs phases were found at high temperature (400 °C). Moreover, plastic deformation trace can be observed at high temperature, which might attribute to the plastic deformation of IMCs owing to the high temperature (400 °C = 0.83 T_m of Ag_3Sn). In other words, as the temperature of shear testing continues to increase, the fracture mode will change from brittle fracture to ductile fracture.

In order to further evaluate the performance of the resulting bondline at high temperature, more than 30 samples

were implemented in the thermal shock testing. Meanwhile, the finite element method (FEM) simulation was applied to determine the stress behaviors of the bondline during the thermal shock testing. The creep constants, such as the young's modulus, linear elastic material properties, Poisson's ratio, thermal conductivity, and CTE were obtained from the research of Jeong[9]. The thermal shock temperature range is -25 °C to 150 °C, and the retention time at each extreme was 10 min. The conversion time between two extremes is less than 3 min. The FEM analysis results were shown in Fig. 6. After 3 thermal cycles, the main stress focused on the interface between upper substrate and resulting CSA preform, which indicates that the network structure during resulting CSA preform can effectively relieve the stress concentration, and the interface was the most likely position of failure.

Fig. 5. (a) SEM image of the sample sheared at ambient temperature. (b) The Image zoomed in on the area marked in (a). (c) SEM image of sample sheared at high temperature. (d) The Image zoomed in on the area marked in (c)

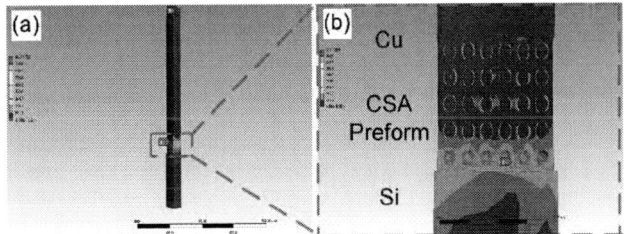

Fig. 6. (a) Total equivalent deformation. (b) The image zoomed in on the area marked in (a).

In order to character the thermal stability of the bondline, the conventional TLP bondline based on Sn3.0Ag0.5Cu (SAC305) solder were tested in the same condition as a control group. According the the results of FEM simulation, the interface between upper substrate and preform was the weakest position. Hence, after 500 cycles of thermal shock testing, the interface of samples was detected by SAM. Compared with the bondline based on CSA preform shown in Fig. 7, the bondline based on SAC305 shown in Fig. 8 seriously cracked, even failure occurred. Moreover, as shown in Fig. 7, the porosity of interface between the Si chip and preform based on CSA particles increased only from 1.0 % to 15.8%, which indicates that the bondline based on CAS preform has better thermal stability.

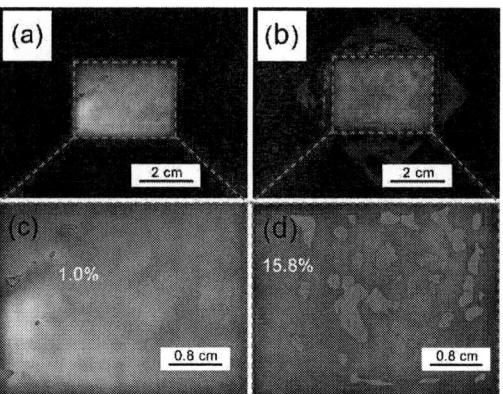

Fig. 7. (a) The SAM image of interface before thermal shock. (b) The SAM image of interface after 500 cycles thermal shock. (c) The image zoomed in on the area marked in (a). (d) The image zoomed in on the area marked in (b).

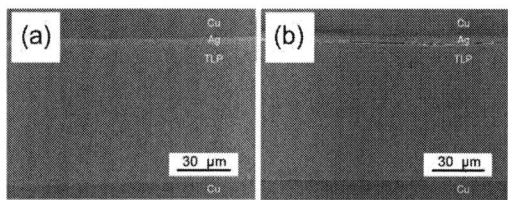

Fig. 8. (a) The SEM image of SAC305 sample before thermal shock. (b) The SEM image of SAC305 sample after 500 cycles thermal shock.

IV. CONCLUSION

(1)After two step method of electroless plating, the CAS core-shell particles were prepared. After reflow at 250 °C for 15 min, the resulting bondline can be obtained, and the re-melting-point of bondline can increased to the melting-point of Ag₃Sn IMC (475 °C).

(2)The average shear strength of the resulting bondline based on CSA preform was measured to be 30.5 MPa and 25.3 MPa at ambient temperature and high temperature (400 °C), respectively. With the temperature of shear testing increasing, the fracture mode changed from the brittle fracture to the ductile fracture.

(3) According to the result of FET analysis, the main stress focused on the interface between Cu@Sn@Ag preform and upper Si substrate. After 500 thermal shock cycles, the porosity of interface between the upper substrate and CSA preform increased from 1.0% to 15.8%, while failure occurred when the samples were interconnected by SAC305 solder.

ACKNOWLEDGMENT

This work is financially supported by by the National Key R&D Program of China under Grant No. 2021YFB1710300, the Science and Technology Program of Guangzhou, China under Grant No. 202002030357，and CEPREI Innovation and Development Fund No. 20Z32, Which were acknowledged.

REFERENCES

[1] N. Jiang, L. Zhang, Z. Q. Liu, el al. Sci. Technol. Adv. Mater. 20(1) (2019) 876.

[2] S. F. Cheng, C. M. Huang, and M. Pecht, el al. Microelectron. Reliab. 75 (2017) 77.

[3] H. W. Zhang, J. Minter, and N. C. Lee. J.Mater. Sci., 48(1) (2019) 201.

[4] H. Kang, A. Sharma, and J. Jung, "Recent progress in transient liquid phase and wire bonding technologies for power electronics". Metals, vol.10, pp.934-952, 2020.

[5] G. Zeng, S. Mcdonald, K. Nogita. Microelectron. Reliab. 52(7) (2012) 1306.

[6] H. Xu, Y. Shen, Y. Hu, el al. J. Microelectron. Electron. Packag. 16(4) (2019) 75.

[7] H. T. Chen, T. Q. Hu, M. Y. Li, el al. IEEE Trans. on Power Electron. 32 (2017) 441.

[8] F. W. Yu, B. Wang, Q. Guo, el al. Adv. Eng. Mater. 20(s) (2018) 1700524.

[9] H. Jeong, D. Min, J. Lee, el al. Microelectron Reliab. 112 (2020) 231.

Effect of dimension of board and micro-bumps on interconnection stress under drop test

Mingtao Lv
State Key Laboratory of High
Performance Complex Manufacturing
College of mechanical and Electrical
Engineering, Central South University
Changsha 410083, China
lvmtao@163.com

Shimei Liu
Changsha Research Institute of Mining
and Metallurgy Co.,Ltd.
Changsha 410083,China
358971237@qq.com

Taotao Chen
Anhui Province Key Laboratory of
Microsystem
East China Research Institute of
Microelectronics
Hefei 230088, China
15256985225@163.com

Yunpeng Liu
Anhui Province Key Laboratory of
Microsystem
East China Research Institute of
Microelectronics
Hefei 230088, China
liuyunpeng1990@126.com

Junfu Liu
Anhui Province Key Laboratory of
Microsystem
East China Research Institute of
Microelectronics
Hefei 230088, China
liujunfu0110@163.com

Hu He
State Key Laboratory of High
Performance Complex Manufacturing
College of mechanical and Electrical
Engineering, Central South University
Changsha 410083, China
hehu.mech@csu.edu.cn

Abstract—The reliability of electronic packaging is one of the most concerned issues, and the board-level drop test has become a key qualification test for assessing the impact reliability of electronic components. In literature, lots of work focus on PCB vibration patterns under different constraints and the influence of PCB dynamic response on solder joint reliability. Several work has compared the package component life with respect to different JEDEC standard boards. They found that the A-type(100x48mm) test board was prone to be failure, and the C-type(77x77mm) test board had longer life. Thus, this work is aiming to further evaluate the influences of variant board on the impact performance of solder interconnects. It was carried out by analyzing the dynamic response of PCBs with different form factors and discussing the effect of the dimension of Cu pillar, IMC and solder layer in orthogonal experiments. Firstly, ANSYS/LS-DYNA software was employed to obtain the X-direction strain curve at the center of PCB bare board with three aspect ratios. The strain curve at the center of PCB revealed more fluctuation as the aspect ratio decreasing, and implied the PCB dynamic response would affect solder joint stress. Through modal analysis and FFT, this phenomenon was mainly attributed to the superposition of mode shapes. Secondly, we conducted an orthogonal test to investigate the influence of size of Cu pillar, IMC and solder layer on the stress of interconnection layer. The Cu pillar on substrate side had the most significant effect on the stress level of solder layer. As the height of Cu pillar increased, the stress of solder layer decreased. Finally, drop test was performed for samples with Cu pillar being 15um and 30um in height and the drop life was calculated. The Weibull probability distribution diagram showed that the experimental results were consistent with our simulation.

Keywords—Reliability; Micro-bumps; Drop impact; Electronic packaging; Simulation

I. INTRODUCTION

Electronic products are developing in the direction of lightweight, multi-functionality, and multi-integration, such as wearable devices, PDAs, aerospace and other equipment. With the increase in portable applications, components may be exposed to destructive environments such as shock and vibration events. In addition, the use of time-delay fuse in aerospace also makes electronic components inevitably subject to the impact of transient shock environments. Board-

level testing is the preferred test tool for evaluating drop reliability. A number of JEDEC standards [1, 2] have been published to provide standardized methods for conducting board-level drop test. However, compared with experiments, numerical simulation is more economical and efficient in structural design and parameter optimization. Tee et al [3, 4] proposed the Input-G method to simulate the transient dynamic responses using ANSYS/LS-DYNA, and the results showed that there is a good correlation between experiment and simulation. Park *et al.* [5] discussed the effect of damping on dynamic responses of PCB under product level drop test and believed that alpha damping is the major parameter. Yeh *et al.* [6] proposed a dynamic modeling approach based on an Input-G loading method using an implicit solver, and its accuracy was proved by dynamic strain. Thukral [7] evaluated the effect of different boards on the reliability of drop impact based on revised version of JESD22-B111A. The result showed that the drop life of solder joints were increased when using the square test board with an aspect ratio of 1:1. In addition, the structure parameters of packaged components have also been studied. For 3D-CoC packages, Cheng *et al.* [8]analyzed the effect of IMC and chip thickness on drop impact reliability. The drop impact reliability can be upgraded with a decreasing IMC and top chip thickness, and increasing bottom chip thickness.

However, the response mechanism of test boards with different form factors has not been studied in depth, and the interconnect layer structure parameters also need to be optimized. In this work, Modal analysis and FFT were used to analyze the strain response of three test boards under drop impact test, aiming to explore the response mechanism. Taguchi method [9, 10]was applied to analyze the effect of geometric parameters on interconnection stress, and the key factor was verified by drop test.

II. MODELLING AND EXPERIMENTAL

ANSYS/LS-DYNA was selected to analyze the dynamic responses of PCB and component. A PWB-Level Drop Tester was used to verify the effect of key factor of the interconnect layer on impact life.

978-1-6654-1392-3/21 $31.00 © 2021 IEEE

A. Finite Element Model

The finite element model of interconnection structure was established as shown in Fig.1. Due to the most serious damage, one package was installed on the center of PCB. Considering the solution time and the symmetry of the model, 1/4 model was adopted. The PCB dimension designed for this study is 132x77mm² with 1.0mm thickness which follow JESE22-B111, The chip size is 6x6mm². A total of 8 layers of structure were built from top to bottom: PCB-substrate-Cu pillar-IMC-solder-IMC-Cu pillar-chip. The PCB is made of FR-4,the solder is made of Sn3.5Ag,while the substrate and die are made of Si. All material parameters are shown in table I. The linear elastic model was simulated for all materials, the whole model was divided into three-dimensional hexahedron elements with mapped mesh characteristics.

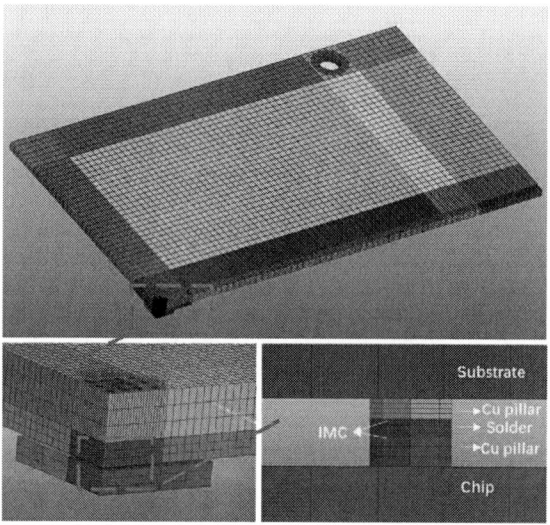

Fig. 1. The finite element model of test board

TABLE I. THE MATERIAL PROPERTY OF MODEL

Component	Material parameters		
	Density (kg/m³)	Elastic modulus(GPa)	Poisson ratio
PCB	1960	E_x/E_y=16.8, E_z=7.4 G_{xz}/G_{yz}=7.59, G_{xy}=3.31	v_{xz}/v_{yz}=0.39 v_{xy}=0.11
Substrate /Die	2300	162.5	0.28
Cu pillar	8940	117	0.36
IMC	8300	92	0.34
Solder	7360	45	0.3

B. Loading and Boundary Conditions

In order to reduce the calculation time and cost, the Input-G method was proposed by Luan et al.[3]. This loading method ignores the overall structure of the drop machine, but directly applies an acceleration loading at the four screw holes of the test board. Considering that acceleration can only be applied to the whole in the WORKBENCH, an Input-V(input velocity) method was developed in this study. The input acceleration was obtained from the experimental data, a half-sine curve, can be described by (1).

$$G(t) = G_m \sin\left(\frac{\pi}{T}t\right) \quad (1)$$

where G_m is the magnitude of acceleration and T is the period of the pulse. The velocity form is integrated from (1). Considering the effects of stress wave and inertia, there are two stages, forced vibration(t<T) and free vibration(t>T) , in the transient response time history of the test board. After eliminating the rigid body movement, the velocity pulse and initial velocity can be described as:

$$V(t) = -\frac{G_m T}{\pi} - \frac{G_m T}{\pi}\cos\left(\frac{\pi}{T}t\right) \quad (2)$$

$$V_{init} = -2\frac{G_m T}{\pi} \quad (3)$$

Where the velocity pulse V(t) was applied to the side of the screw hole as a load. The initial velocity V_{init} was applied to whole test vehicle as initial condition. The 1/4 model was established as shown in Fig.2, and both symmetry planes need to be applied symmetry boundary conditions.

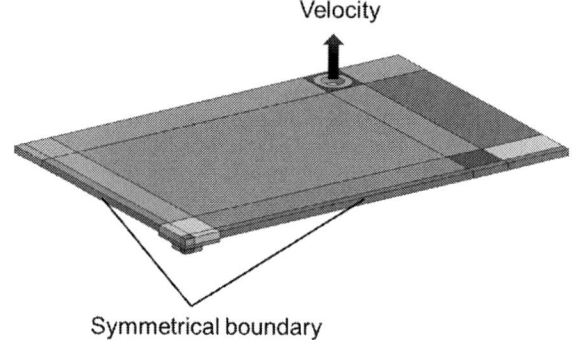

Fig. 2. Load and boundary conditions

C. Sample Preparation and Drop Test Experiment

Following the JEDEC standards, a PWB-Level Drop Tester as shown in Fig.3(a) was used. It can provide the standard pulse with the amplitude of 1500g and the pulse width of 0.5ms. The chip sample, including two daisy chains, was placed in the center of PCB, as shown in Fig.3(b). During the drop test, the daisy chain resistance changes were in real time monitored by a data acquisition system. Failure was defined in accordance with JESD22-B111.

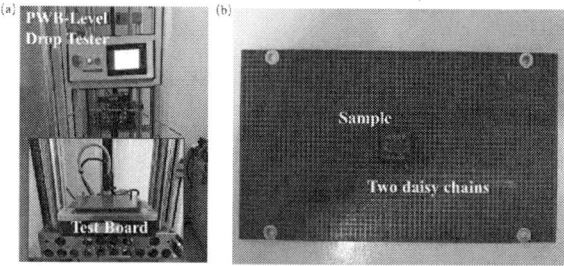

Fig. 3. Experimental equipment: a) PWB-Level Drop Tester, (b) Sample

III. EFFECT OF BOARD ASPECT RATIO

A. Board Characterization

Tee found that PCB bending was the main failure mechanism that caused peeling stress in solder joints. But different test boards may bring different failure mechanisms. In this study, the FEM was carried out by recording the board dynamic response with three different aspect ratios, details are listed in Table II.

TABLE II . TEST BOARDS DESCRIPTION

PCB details	Type-A	Type-B	Type-C
Acceleration	1500G-0.5ms	1500G-0.5ms	1500G-0.5ms
PCB form factor(mm)	100x30	100x48	100x100
PCB thickness(mm)	1.6	1.6	1.6
Number of component	1(center)	1(center)	1(center)

In the drop test, the damping of the PCB plays an important role in the impact response of interconnection structure, and it can be described by the damping ratio. In LS-DYNA, the Rayleigh damping is used in the simulation. The damping matrix in Rayleigh damping is defined as:

$$C = \alpha M + \beta K \qquad (4)$$

where C, M, and K are the damping, mass, and stiffness matrices, respectively. The constants α and β are the mass and stiffness proportional damping constants, which can be defined by keywords DAMPING_GLOBAL and DAMPING_PART _STIFFNESS, respectively.

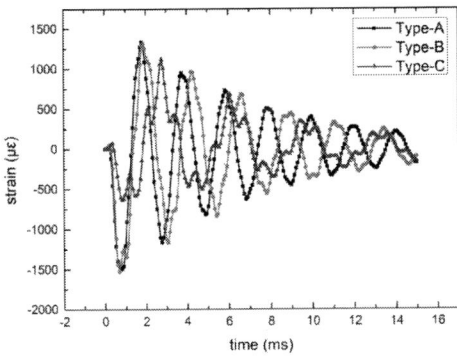

Fig. 4.   Dynamic strains in PCB length direction

TABLE III. PCB DYNAMIC RESPONSE RESULT SUMMARY

Test Methods	Type-A	Type-B	Type-C
Max displacement(mm)	1.59	1.83	2.89
Max strain in length direction (mm)	1410	1530	1130
1st order frequency (Hz)	525	420	315
Frequency ratio	5.3%	10.5%	32%

Figure.4 shows measured time histories of longitudinal strains with different PCB aspect ratios under the JEDEC drop test condition B. The amplitude of PCB vibration decreases gradually with time. In addition, the strain curve of board C fluctuates more than board A and B. Table III concludes PCB vibration features calculated on all test boards. Where the vibration frequencies are obtained from the strain curves by Fast Fourier Transformation(FFT). Fig.5 shows the resonance peaks of vibration response of different test boards. The frequency ratio is defined, which represents the ratio of the contribution of the second-order mode to the first-order mode.

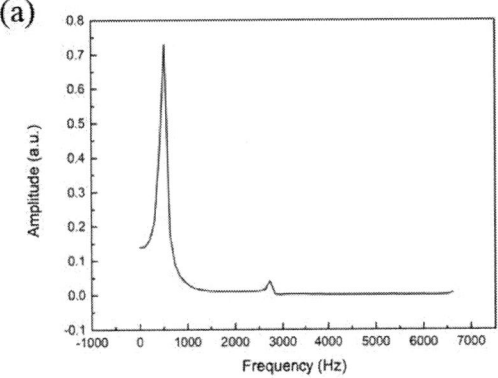

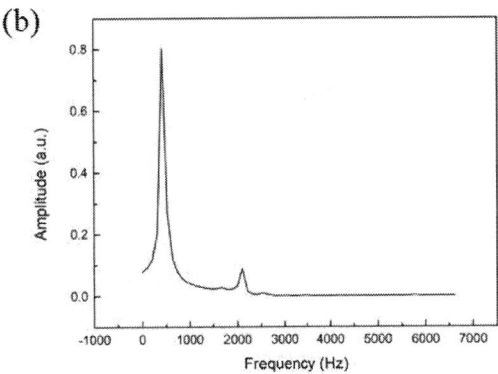

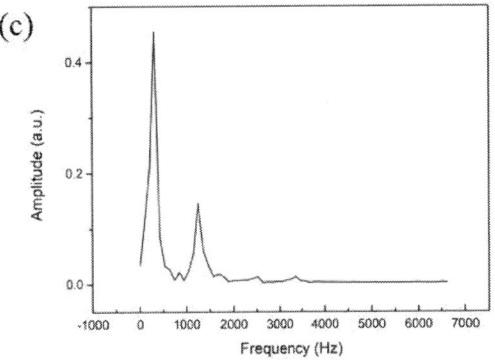

Fig. 5.   Frequency response: (a) Test board A, (b) Test board B, (c) Test board C

The results show that the test board C has the largest out-of-plane displacement, but the max strain in length direction is the smallest. The vibration form of an object is related to its own vibration mode. Under the same load, the lower vibration frequency is more likely to be excited and cause resonance. The first-order frequency of board C is 315 Hz, which is lower than 420 Hz of board B and 525 Hz of board A, and board C is easier to bend under drop impact. At the same time, we should notice that the high-order modes of board C also participate in the vibration. The frequency ratio data shows that the high-order participation rate of board C is significantly higher than that of board A and B. The superposition of multiple modes makes the response curve more fluctuating.

B. Modal Analysis

In order to understand the differences between the three test Boards more clearly, modal analysis was performed. Since the actual research pays more attention to the first few modes, the first 5 modes were acquired, details as Figure.6. Due to the symmetry of the PCB geometry as well as loading, the vibration of board is dominated by symmetrical modes. Namely, board A and B care mode 1 and mode 5 respectively, board C cares mode 1, 4, and 5. Since the frequency ratio of board C is greater, the participation rate of the 4th and 5th modes is also greater, and the modes are superimposed at the PCB center. Although the max displacement is larger, the center tends to be more gentle and max strain is smaller.

Mode	Type-A	Type-B	Type-C
1			
2			
3			
4			
5			

Fig. 6.   Vibration modes of board with Type-A, Type-B, Type-C

IV. EFFECT OF HEIGHT OF MICRO-BUMPS

In addition to the board size, the dimension of packaged components will also affect the reliability of solder joints. Cheng *et al.*[8] have studied the effect of IMC and silicon chip height on impact life. In this work, the effect of height of micro-bumps is more concerned.

A. DOE Matrix

DOE is a low-cost, high-efficiency engineering design method, which can be used to determine key factors that affect the output of results. Two most important tools in DOE are the signal-to-noise ratio based on the quality loss function and the orthogonal experimental design based on key factors. In our study, the orthogonal test for optimizing structural parameters was designed. Substrate, Cu pillar, IMC and solder were selected as the four factors, and each factor has three levels. The level of each factor was reasonably selected based on laboratory samples. Detailed factors and levels are shown in Table IV.

TABLE IV. CONTROL FACTORS AND LEVELS

Structure	Factor	Level 1	Level 2	Level 3
Cu pillar(substrate side)	A	10	20	30
Cu pillar(chip side)	B	20	30	40
solder	C	5	10	15
IMC	D	2	4	6

B. Failure Criteria

Effective stress(Von Mises) and plastic strain are two commonly used metrics in failure criteria. Considering the effect of damping, the amplitude of first cycle is largest, effective stress was selected as the preliminary failure criterion. Figure.6 shows the stress distribution in solder layer at the moment of maximum stress. Solder joints in the width direction are more prone to failure. Figure.8 depicts the magnitude of various stresses at critical solder joint, including normal stress S_x, S_y, S_z, shear stress S_{xy}, S_{xz}, S_{yz}. Among them, the peeling stress(S_z) is close to v-m stress in amplitude, and is much higher than other components. During drop impact, the peeling stress plays an critical role, and it can be used as a failure criterion for structural parameter optimization.

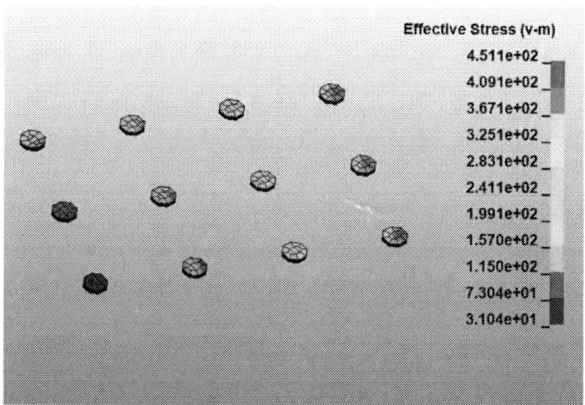

Fig. 7.   Effective stress distribution in solder layer

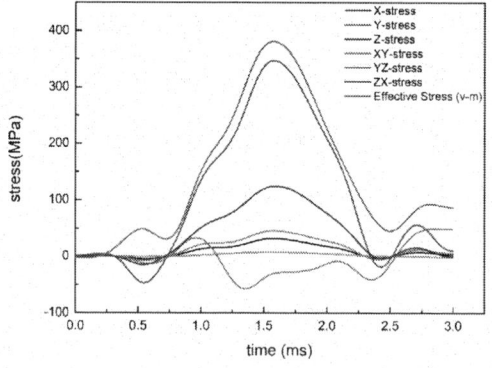

Fig. 8.   Stresses of key solder joint

C. Results and Discussion

According to the control factors and levels in Table IV, the $L_9(3^4)$ orthogonal array was selected to study the structural parameters, it can find out the best parameters through the smallest number of tests. The signal-to-noise ratio is defined as:

$$S/N = -10\,log(\Delta y^2) \qquad (5)$$

where Δy^2 is the average value of multiple experiments. Due to the use of numerical simulation instead of experiment, this paper does not consider the existence of noise factors and Δy can be equivalent to peeling stress. Following the $L_9(3^4)$ orthogonal array, the S/N ratio of each group response is shown in Table V.

TABLE V. TAGUCHI ORTHOGONAL TEST TABLE AND THE EXPERIMENTAL RESULTS

Group	Peeling stress	S/N
A1B1C1D1	429	-52.65
A1B2C2D2	408	-52.21
A1B3C3D3	417	-52.4
A2B1C2D3	263	-48.4
A2B2C3D1	347	-50.81
A2B3C1D2	352	-50.93
A3B1C3D2	230	-47.23
A3B2C1D3	221	-46.89
A3B3C2D1	293	-49.34

In order to understand the effect of four control factors on drop impact reliability, we have studied the effect of S/N ratio of each control factor based on Table V. The influence coefficient is determined by the difference between the maximum S/N ratio and the minimum S/N ratio under different levels. Table VI and Figure 8 show the coefficients of different control factors, and the influence ranks of the four control factors are also listed in Table VI.

TABLE VI. TAGUCHI ORTHOGONAL TEST TABLE AND THE EXPERIMENTAL RESULTS

Control factor	S/N	coefficient	rank
A1	-52.42		
A2	-50.05	4.6	1
A3	-47.82		
B1	-49.43		
B2	-49.97	1.46	3
B3	-50.89		
C1	-50.16		
C2	-49.98	0.18	4
C3	-50.15		
D1	-50.93		
D2	-50.12	1.7	2
D3	-49.23		

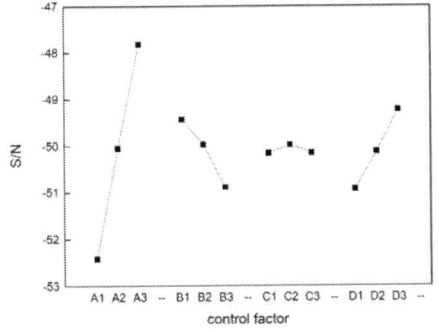

Fig. 9. S/N ratio of each control factor

Fig.9 shows that increasing the height of Cu pillar on substrate side can significantly enhance the impact reliability of interconnect structure within the specified control factor level, while the other three factors have less influence. This means that the height of Cu pillar on substrate side should be paid more attention. In addition, it should be noted that the height of Cu pillar also relates to signal transmission efficiency, and transmission efficiency should also be considered while pursuing high package reliability. All control factors are ranked as the following orders: Cu pillar height on substrate side(Factor A), IMC height(Factor D), Cu pillar height on chip side(Factor B), solder height(Factor C). The peeling stress on solder layer decreases as the height of Cu pillar on substrate side increases. From the energy point of view, PCB transfers energy to package component through stress wave. Increasing the height of Cu pillar on substrate side, more energy is stored in Cu pillar, and the energy transferred to IMC and solder layer are reduced. Therefore, the reliability of drop impact is improved. On the other hand, a negative correlation is reflected in the thickness of IMC and the stress of solder layer. But it should be noted that increasing the height of IMC will make the package component easier to release energy in the form of brittle fracture when impact force is too large. As for the height of Cu pillar on chip side(Factor C), since it is at the end of energy transfer, the main influence on the reliability of interconnect structure comes from inertial force. The larger volume, the greater inertial force, which increases the stress of solder layer and reduces the reliability of drop impact.

Furthermore, an experiment with Cu pillar height 15um and 30um on Factor A has been carried out. 10 samples have been selected for the two sets of experiments. Some samples were damaged due to experimental errors. As shown in Fig.10, the samples with 30um substrate have a drop failure rate of 37%, while the samples with 15um substrate have a drop failure rate of 70%, and the failure rate has increased by 1.89 times. Therefore, the drop life of samples with a height of 30um Cu pillar is greater than 15um,which is consistent with the simulation results. This phenomenon is mainly due to the higher Cu pillar being able to absorb more impact energy.

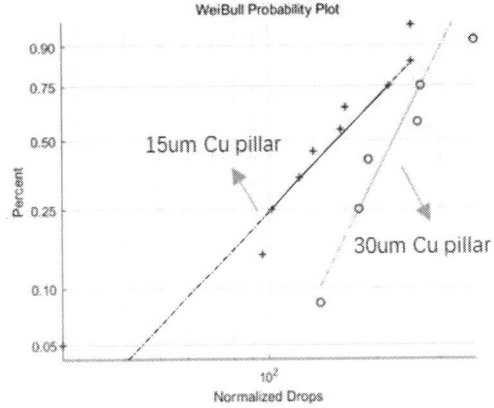

Fig. 10. Weibull probability diagram of drop impact life of components under different Cu pillar height

V. CONCLUSION

This article comprehensively analyzes the effect of dimension of board and micro-bumps on interconnect stress. For the convenience of modeling in WORKBENCH, we proposed the Input-V method. The response mechanism of PCB with three form factors was analyzed more depth. Using Taguchi test method to analyze the key structural parameters of interconnection layer. Some essential concluding remarks can be drawn below:

1) In the case of a fixed length, the reduced aspect ratio causes the vibration frequency to decrease, and the higher-order mode shapes are more easily excited.

2) The superposition of mode shapes could lead to the fluctuation in the strain curve at the center of PCB.

3) The height of the Cu pillar on the substrate side has a significant effect on the stress of solder layer. The optimal size of Cu pillar should be applied based on the balance between drop and cost.

ACKNOWLEDGMENTS

This work was supported by Natural Science Foundation of Hunan Province (2020JJ5728), Innovation-Driven Project of Central South University (2020CX05) and National Natural Science Foundation of China (U20A6004).

REFERENCES

[1] J. S. JESD22, "B104-B," Mechanical shock, 2001.

[2] J. S. JESD22, "B111," Board level drop test method of components for handheld electronic products, 2003.

[3] J.-e. Luan and T. Y. Tee, "Novel board level drop test simulation using implicit transient analysis with input-G method," in Proceedings of 6th Electronics Packaging Technology Conference (EPTC 2004)(IEEE Cat. No. 04EX971), 2004: IEEE, pp. 671-677.

[4] T. Y. Tee, J.-e. Luan, E. Pek, C. T. Lim, and Z. Zhong, "Novel numerical and experimental analysis of dynamic responses under board level drop test," in 5th International Conference on Thermal and Mechanical Simulation and Experiments in Microelectronics and Microsystems, 2004. EuroSimE 2004. Proceedings of the, 2004: IEEE, pp. 133-140.

[5] S. Park, D. Yu, A. Al-Yafawi, J. Kwak, and J. Lee, "Effect of damping and air cushion on dynamic responses of pcb under product level free drop impact," in 2009 59th Electronic Components and Technology Conference, 2009: IEEE, pp. 1256-1262.

[6] S.-S. Yeh et al., "Ultra-Thin Package Board Level Drop Impact Modeling and Validation," in 2019 IEEE 69th Electronic Components and Technology Conference (ECTC), 2019: IEEE, pp. 1550-1555.

[7] V. Thukral, J. Zaal, R. Roucou, J. Jalink, and R. Rongen, "Understanding the impact of PCB changes in the latest published JEDEC board level drop test method," in 2018 IEEE 68th Electronic Components and Technology Conference (ECTC), 2018: IEEE, pp. 756-763.

[8] H.-C. Cheng, H.-K. Cheng, S.-T. Lu, J.-Y. Juang, and W.-H. Chen, "Drop impact reliability analysis of 3-D chip-on-chip packaging: Numerical modeling and experimental validation," IEEE Transactions on Device and Materials Reliability, vol. 14, no. 1, 2013, pp. 499-511.

[9] R.-S. Chen, R.-W. Wu, and W.-C. Liao, "Optimal design of drop impact for TFBGA by concept of average stress within finite grid region," in 2010 12th Electronics Packaging Technology Conference, 2010: IEEE, pp. 841-848.

[10] J. Xia, G. Li, B. Li, and L. Cheng, "Optimal design for vibration reliability of package-on-package assembly using FEA and taguchi method," IEEE Transactions on Components, Packaging and Manufacturing Technology, vol. 6, no. 10, 2016, pp. 1482-1487.

978-1-6654-1392-3/21 $31.00 © 2021 IEEE

Thermomechanical and Electrical Properties of the SiO₂/ZrW₂O₈/Epoxy Composite

Chaofan Li
Shenzhen Institute of Advanced Technology
Chinese Academy of Sciences
Shenzhen, China
Institute of Nano Science and Technology
University of Science and Technology of China
Suzhou, China
cf.li@siat.ac.cn

Suibin Luo
Shenzhen Institute of Advanced Technology
Chinese Academy of Sciences
Shenzhen, China
sb.luo@siat.ac.cn

Shuhui Yu
Shenzhen Institute of Advanced Technology
Chinese Academy of Sciences
Shenzhen, China
sh.yu@siat.ac.cn

Baojin Chu
CAS Key Lab Mat Energy Convers
University of Science and Technology of China
Hefei, China
chubj@ustc.edu.cn

Rong Sun
Shenzhen Institute of Advanced Technology
Chinese Academy of Sciences
Shenzhen, China
rong.sun@siat.ac.cn

Abstract—**Epoxy resin (ER) is commonly used in the insulating dielectric substrates in circuit boards for various devices. However, its coefficient of thermal expansion (CTE) is mostly above 100 ppm/K and seriously mismatched with the connected materials, such as copper (CTE of 17 ppm/K), which usually causes several defects such as cracks, delamination, and warpage in the dielectric layer and then the service life of the devices is affected. Here we report a strategy to reduce CTE of the epoxy based film by using SiO₂/ZrW₂O₈ particles as low/negative CTE nanofillers. For comparison, epoxy composite films separately filled with the SiO₂ and ZrW₂O₈ were also prepared. As the film was filled with both SiO₂ and ZrW₂O₈ at a ratio of 8:2, the CTE achieved a reduction of about 52% compared to the pure epoxy films and was lower than either SiO₂ or ZrW₂O₈ single-phase filled composite film with the same loading (75wt%). At the high frequency of 5143 MHz, the dielectric constant of the SiO₂/ZrW₂O₈/epoxy composite film was 3.73, slightly increased by 29% compared to the pure epoxy film (2.89). The dielectric loss of SiO₂/ZrW₂O₈/epoxy composite film was 0.017, which was decreased by 29.2% of the pure epoxy film (0.024). The performance of low CTE, low permittivity and low dielectric loss is expected to be applied in printed circuit boards for electronic devices.**

Keywords—*Low thermal expansion coefficient; Low dielectric constant; Low dielectric loss; Epoxy composite material*

I. INTRODUCTION

A printed circuit board (PCB), is used to carry various electronic components and provide electrical interconnection between them that make up a circuit in a device. Hence, PCBs are the foundational building block of most modern electronic devices. There are generally three kinds of materials in a PCB: reinforcing material, insulating dielectric material, and copper foil. Glassfiber cloth and glass fiber paper are commonly used as reinforcing materials. Epoxy resin (ER) is widely used as the dielectric base material because of its excellent mechanical and electrical insulation properties[1,2]. With the development of electronic system towards high frequency, high speed and compact integration, it requires the dielectric materials to possess increasingly high heat resistance and dimensional stability. However, the coefficient of thermal

expansion (CTE) of epoxy resin is as high as 100 ppm/K, mismatching with the copper foil and other metal devices[2], which generally leads to warpage, cracking, and other defects of the substrate, and further seriously affects the performance and life of electronic device. Therefore, it is necessary to reduce the thermal expansion of epoxy resin and improve its thermal stability.

Efforts have been made to improve the thermal dimensional stability of epoxy composites. The most immediate and convenient method is to add low or negative CTE filler particles into the substrate. Silicon dioxide (SiO₂) was widely used as the low CTE filler due to its low cost, low density, low thermal expansion coefficient (CTE), low dielectric constant, and non-toxic properties. SiO₂ was employed to improve the dimensional stability of composite materials in many reports[3-5]. For example, it was found that the CTE value of SiO₂/PI composites was reduced by 50% compared to the pure PI[4]. In addition, tantalum tungsten (ZrW₂O₈), as a negative thermal expansion (NTE) filler in a wide temperature range[6], has also attracted much attention. For example, H.Jeon reported that PI composite films containing SiO₂ and ZrW₂O₈ were prepared respectively for comparing the CTE reduction of these two fillers[7]. The CTE value diminished by 24.5% and 17.5% at the loading of 15 wt%, respectively. Some literature reported that the CTE decreased by 64% at the maximum loading of 60 vol% (84 wt%) ZrW₂O₈[8,9]. However, in the above reports, dielectric and mechanical properties of the composite material were not studied. Wu etc. prepared ZrW₂O₈/epoxy composite material with 20 vol% filler content, although CTE declined by 29%, the dielectric constant and dielectric loss increased[10].

Generally, the performance of composites with a single filler can be enhanced in one aspect. By using multiple fillers, more than one property can be improved simultaneously. Jae soon Jang et al.[5] reported that the mechanical damping and thermal dimensional stability were both enhanced by introducing carbon nanofiber (CNF) and micron-sized silica particles as fillers[11]. In this work, we prepared SiO₂/ZrW₂O₈/epoxy composite with low dielectric constant, low dielectric loss, and low CTE, in which nano SiO₂ and micron ZrW_2O_8 are used as multi-scale fillers. ZrW_2O_8 is used as an NTE filler to reduce the thermal expansion of the composite. SiO_2 can further reduce the CTE value, and compensate the negative effect of the increase of dielectric constant and dielectric loss caused by ZrW_2O_8 particles.

II. EXPERIMENTAL

A. Materials and Preparation

ZrW_2O_8 particles (0.4~0.7μm) were provided by Shanghai Dianyang Industry Co. Ltd. SiO_2 particles were purchased from Jiangsu Lianrui New Materials Co. Ltd. MEK (methyl ethyl ketone) was purchased from Shanghai Lingfeng Chemical Reagent Co. Ltd. These materials were used without further treatments.

As the surface-treating agents, film-forming agents have been incorporated into the epoxy resin, the fillers and dispersant were blended with epoxy directly under the appropriate solvent of methyl ethyl ketone (MEK). The mixture was then mixed uniformly by a ball-milling machine for 12h, which has a rotation speed of 500 rpm. The composite slurry was coated on the surface of the PET release film with a thickness of 50 μm by using the scraper coater. The thickness of the composites film could be controlled by adjusting the distance between the scraper and the PET support film. Above all, the coated composite films were dried at 50 °C for 20 min under air atmosphere. The curing process was carried out in the next step, which was divided into two stages. The films were firstly cured at 130 °C for 30 min, then 180 °C for 90 min. Finally, the cured films were peeled from the PET release film and prepared for subsequent performance testing and characterization.

The blending ratio of these two fillers (SiO_2:ZrW_2O_8) is from 8:2 to 2:8 in the prepared SiO_2/ZrW_2O_8/epoxy composite

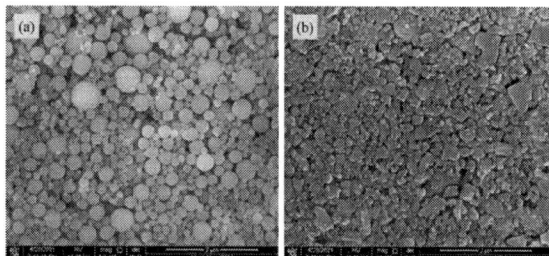

Fig. 1 SEM images of the SiO_2 (a) and ZrW_2O_8 (b).

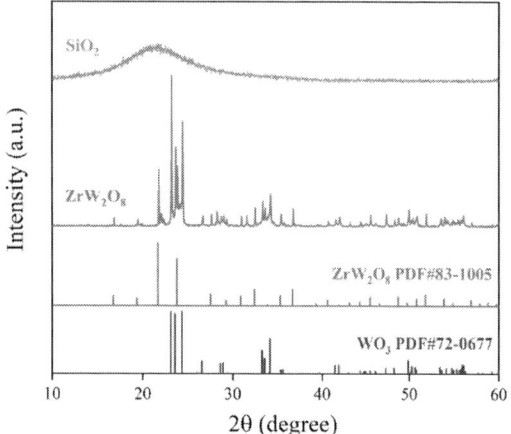

Fig. 2 The XRD patterns of SiO_2 and ZrW_2O_8.

film, the proportion of mixed filler is 75 wt%. As a contrast, the pure epoxy resin film, SiO_2/epoxy (S/ER) and ZrW_2O_8/epoxy (WZ/ER) composite films were also prepared. The amount of single fillers is from 60~75 wt%. It is hereby declared that the SiO_2/ZrW_2O_8/Epoxy composite (S/WZ/ER) mentioned in the following discussion refers to the composites with S:WZ = 8:2, because of its excellent performance.

In the process of slurry preparation, it is necessary to control the amount of MEK for adjusting the viscosity of slurry due to the different densities and content of each kind of filler particle. Otherwise, if the viscosity is low, the wet film will shrink on the PET film. An epoxy membrane will quickly form on the surface of the wet film when the viscosity is high, which will inhibit the volatilization of solvent and then result in forming bubbles on the surface. Therefore, to ensure the integrity of the wet film, it is necessary to continuously adjust the slurry viscosity according to the amount of particles added.

B. Characterization

The CTEs of composite films were determined using a thermal-mechanical analyzer (TMA, TMA-402 F1). The CTEs were measured with an extension probe under a 0.01 N tension force on the films in 25~260 °C at a heating rate of 5 °C/min under nitrogen atmosphere, and the film size is 5 mm wide and 26 mm long. The sectional views of the fillers were analyzed using a scanning electron microscope (SEM, Nova™ NanoSEM 50). Tensile mechanical tests of the composite films were performed by a universal testing machine (UTM, WL2100B) with a uniform strain rate of 5 mm/min at room temperature. The storage modulus and tan delta were acquired using a dynamic thermomechanical analysis (DMA, TA Instruments Q800) with an amplitude of 10 μm in the temperature range of 30~250°C at a heating rate of 5°C/min under air atmosphere. Dielectric constant and dielectric loss were measured by precision impedance analyzer (4294A, 40~110 MHz) at 0.5V. The samples used were 8 mm in diameter, onto which, gold electrodes were sputter coated on each side. The dielectric constant and dielectric loss at 5 GHz were tested by a network analysis tester (E5071C).

III. RESULTS AND DISCUSSION

A. Characterization of the Inorganic Particles

The SiO_2 fillers are spherical particles with sizes of 50 nm~0.8 μm (Fig. 1a), while ZrW_2O_8 particles are irregular particles with size of 0.4~0.7 μm (Fig. 1b). In Fig. 2, the XRD curve of SiO_2 is rather disordered, which indicates its poor crystallinity, "the steamed bun peak" is its characteristic peak besides. In addition to the characteristic peaks of ZrW_2O_8 in its spectrum, it contains the characteristic peaks of WO_3,

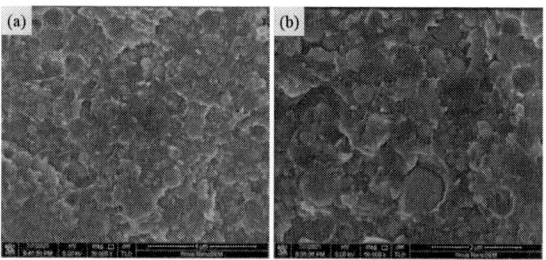

Fig. 3 The cross-section images of the SiO_2/ZrW_2O_8/Epoxy composite film filled with 60 wt% SiO_2 and 15 wt% ZrW_2O_8 at different magnification. (a) The magnification of 30000, (b) The magnification of 50000.

which indicates that the purchased ZrW_2O_8 contains some WO_3 impurities.

B. Cross-section analysis

The cross-section microstructures of the $SiO_2/ZrW_2O_8/$Epoxy composites are shown in Fig. 3. It can be observed that there are some epoxy matrix between fillers, which proves that particles are uniformly dispersed in the matrix. The homogeneously dispersed particles are induced by the following three points: Firstly, the dispersant was used in the slurry preparing process, which can effectively inhibit the aggregation of particles. Secondly, the method of ball milling for 12h makes the dispersion more effective. Finally, a certain amount of solvent is added to the epoxy slurry, therefore, the lower viscosity of the slurry bring about that particles are well-distributed more easily. Inevitably, it is possible that only a very small part of nanoparticles are aggregated, which aspect needs to be studied continuously.

C. Dynamic-mechanical Analysis

Under the external force, DMA can obtain the storage modulus (E'), loss modulus (E"), and mechanical loss factor (tan δ) of composites by analyzing the relationship between strain and stress of the sample with temperature and other conditions. These three physical quantities are important parameters that determine the characteristics of the material. The relationship between the three parameters is:

$$Tan\ \delta = E''/E' \tag{1}$$

Where E' characterizes the ability of a material to resist deformation. The greater the modulus, the greater the stiffness of the material and the less it likely to be deformed; E" reflects the toughness of the material; Tan δ represents the ratio of energy dissipated in the form of heat to stored energy, which represents the ability of a material to lose energy during the deformation process.

The trend of storage modulus and mechanical loss with temperature is shown in Fig. 4. When the temperature is much lower than Tg, the movement of the polymer chain is frozen, the energy loss is small or unchanged, and the storage modulus is high. As the temperature rises, the segments begin to move freely, but the viscosity of the system is still relatively large. Therefore, the molecular chain segments need to consume a lot of energy to overcome the restraint of the filler particles and the viscous movement of the environment, so the peaks of Tan δ and E' begin to appear, that is, glass transition. Viscous flow occurs, the energy dissipated decreases, and storage modulus is also reduced with the temperature rises further. In the single-filler composites, the   increases and the Tan δ decreases as the filler content increases.

The E' of the silica (14680 MPa) and zirconium tungstate (7240 MPa) filled composites are respectively increased by 6 times and 3 times compared to the pure epoxy (2400 MPa), which shows the reinforcing effect of the particles. Moreover, the E' of SiO_2 is much larger than that of ZrW_2O_8. Silica has a large contribution to the storage modulus of composites. Therefore, under the same filler content, the E' and tan δ decreases as the proportion of ZrW_2O_8 increases.

For all the samples, only a single glass transition temperature relaxation can be determined, which reflects the formation of a uniform system. Tg shifts slightly to the high temperature with the increase of the filler, which shows that the addition of the filler is beneficial to the thermal stability of the composites. This may be attributed to the binding effect of

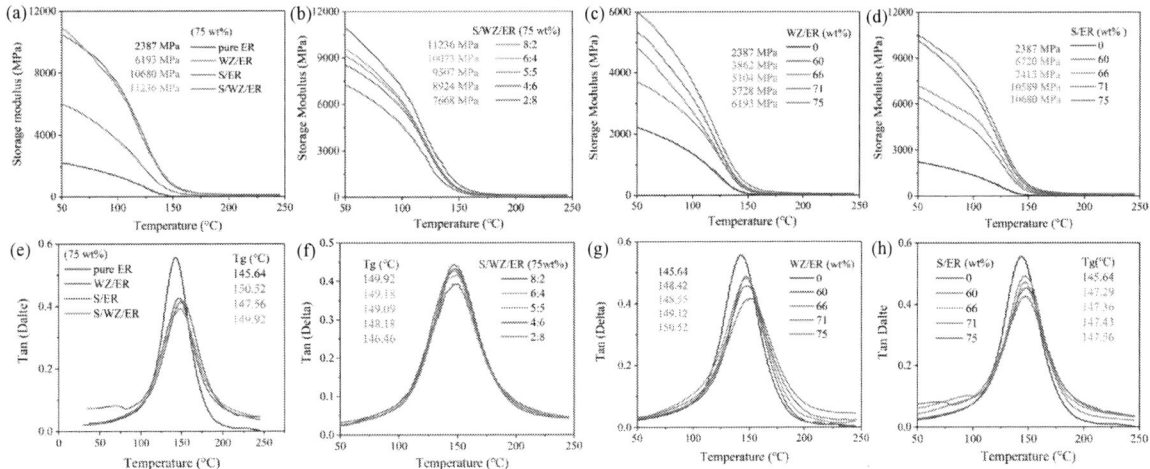

Fig. 4 Dynamic-mechanical study of the filled and unfilled epoxy composite films. (a)-(d) Storage modulus versus temperature and (e)-(f) Tan δ versus temperature.

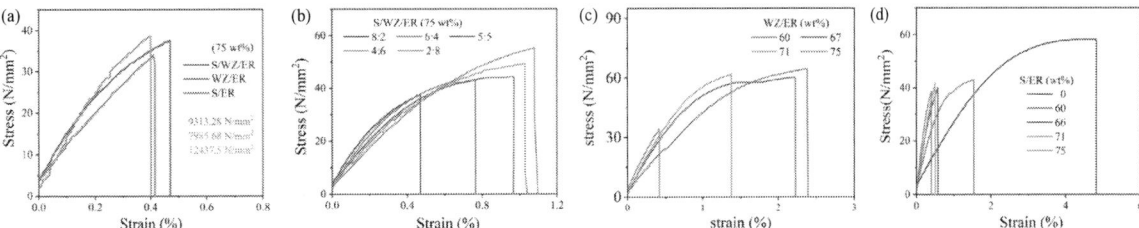

Fig. 5 Mechanical properties of the filled and unfilled epoxy nanocomposites.

the filler on the molecular chain of the epoxy resin. The more of the fillers, the larger of the binding effect, then the more energy needed for the segment movement, and the higher Tg[2]. In addition, the width of the tan δ peak reflects the dispersion of chain segment motion and the compactness of structure.

D. Tensile mechanical Analysis

Slopes of the stress-strain curves represent the Young's modulus and the rigidity of the films. The Young's modulus increases with the adding of fillers[12], and the increasing degree varies with different fillers, which are shown in Fig. 5.

In Fig. 5(c, d), the tensile properties are significantly enhanced with the increase of fillers, which may be due to the fine dispersion of particles. These particles interact with the polymer matrix, which restrict the free movement of the polymer chain, thus increasing the hardness of composites films.

Slopes of the stress-strain curves represent the Young's modulus and the rigidity of the films. The Young's modulus increases with the adding of fillers[12], and the increasing degree varies with different fillers. At 75wt% of fillers, the young's modulus of SiO_2 film, ZrW_2O_8 film and SiO_2/ZrW_2O_8 composite film are 12437, 7986 and 9313 N/mm^2. Compared to the pure epoxy resin (3249 N/mm^2), the young's modulus of these three kinds of films increased 3.8, 2.5 and 2.8 times, respectively. Obviously, SiO_2 contributes greatly to the rigidity of epoxy matrix, and zirconium tungstate increases the toughness of the material. Naturally, too much filler will make the composite membrane brittle and fragile, so the amount of filler should be adjusted according to the application requirements.

E. CTE Values Analysis

The values of CTE (α) can be determined from the slope of the thermal strain curves, as shown below:

$$\alpha = 1/L_0 \cdot dL/dT \quad (2)$$

Where L_0 is the original sample length and dL/dT is the change in sample length with temperature change. The values of α calculated from 25 to 150 °C based on a linear regression of the data. The low-CTE and NTE fillers reduce the CTE of epoxy composites in two main ways. On the one hand, they have the intrinsic low/negative thermal expansion, and their thermal expansion behavior is opposite to that of epoxy resins, which can cancel each other out[13]. On the other hand, the addition of filler particles will restrict the movement of epoxy molecules, so that the deformation of molecules is less affected by the temperature variations. In other words, fillers can restrain the deformation of the epoxy resin matrix. Therefore, reducing the CTE of each component or increasing the volume content of the low/negative thermal expansion material can effectively decline the overall expansion coefficient of the composite material.

As shown in Fig. 6, the CTE value reduction efficiency is various with the filler content increases. The silica composites have a CTE value of 59.5 ppm/K at 75 wt%, which is about 44% lower than that of pure epoxy (107.34 ppm/K). At the same filler content, ZrW_2O_8/epoxy composites (75.73 ppm/K) decreased by 30%. In the blend film, the CTE value decreased the most at the ratio of silica: zirconium tungstate = 8:2, which reduced to 51.82 ppm/K, that is to say, it decreased by 52%. The content of 75 wt% was chosen because too much filler would make the composites brittle.

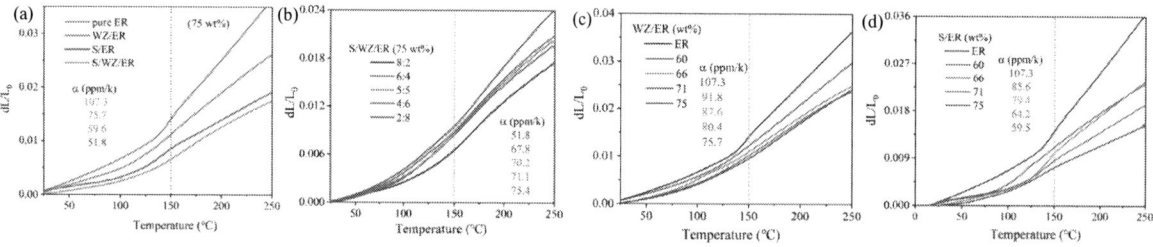

Fig. 6 Temperature dependence of the coefficient of thermal expansion (CTE) for the filled and unfilled epoxy nanocomposites.

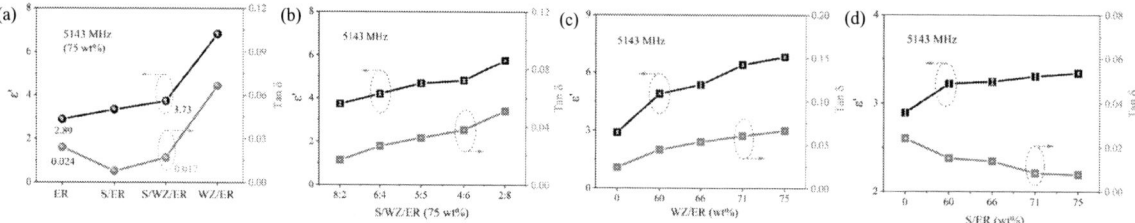

Fig. 7 The dielectric permittivity and dielectric loss for the filled and unfilled epoxy nanocomposites at 5 GHz.

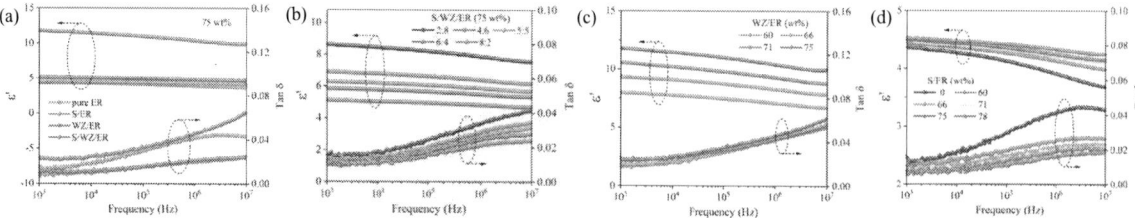

Fig. 8 Frequency dependence of dielectric permittivity and dielectric loss for the filled and unfilled epoxy nanocomposites (10^3-10^7 HZ).

As an NTE filler, the CTE reduction efficiency of ZrW_2O_8 is lower than that with low CTE filler silica. The reason may be that the morphology of zirconium tungstate is irregular, which causes poor bonding between the particles and matrix. In addition, the negative impact of the little WO_3 impurities contained in ZrW_2O_8 may also induce slow reducing of CTE.

F. Dielectric Analysis

In the case of low content of SiO_2, the interfacial zones are formed between the particles and the matrix, which restrict the chain segment movement and the dipole orientation. Moreover, the dielectric constant (ε') of the interfacial regions is small, thus the dielectric constant of the composites decreases. However, there is a limitation for the concentration of fillers[14], if, past this maximum value, the excess particles will lead to a slight increase in ε'. As shown in Fig. 8(d), the filler content is more than 10wt% in our research, so ε' is incremental slightly from 4.4 to 4.5 with the filler increases at 1 kHz, meanwhile, dielectric loss declines successively. The variation trends of ε' and Tan δ in Fig.7 (d) are equal to that in Fig. 8(d), the dielectric constant promotes from 2.9 to 3.3, dielectric loss declines from 0.024 to 0.0077 at 5 GHz.

The increase of the dielectric constant may be the following reasons: The interfacial overlap regions weaken the restriction of the interfacial region on the chain segment, which leads to the increase in ε'. A large number of filler particles introduce more dipole groups into the epoxy compound, which increases the polarization characteristics of the dipole and further increases ε'. Besides, SiO_2 will form a water shell on its surface because of its water absorption, and the polarity of water is high, which will also promote the ε'[14]. In Fig. 8(c), the ε' increases to 13 as the increase of ZrW_2O_8, which maybe on account of its high intrinsic dielectric constant ($\sim$ 10)[5,15] and the rise of interface polarization. In addition, the dielectric loss also increases, which may be thanks to the interface loss caused by excessive interface polarization and the conduction loss caused by the carriers introduced by ZrW_2O_8[16].

Fig. 7(b) and Fig. 8(b) show that the dielectric constant and dielectric loss of the SiO_2/ZrW_2O_8 composite films decrease with the reduction of ZrW_2O_8. Therefore, the SiO_2/ZrW_2O_8 composites possess a relatively low dielectric constant and loss compared to ZrW_2O_8 composite films.

IV. CONCLUSIONS

At the concentration of 75wt%, the CTE of the SiO_2/ZrW_2O_8/Epoxy composites decreased to 51.8 ppm/k, reduced by 52% compared to the pure epoxy, which were lower than either SiO_2 or ZrW_2O_8 epoxy composite film. The reason why the degree of CTE reduction of ZrW_2O_8 composite film is less than that of silica may be due the following two points: (1) The ZrW_2O_8 powder has impurities of WO_3, which affect the ability of CTE reduction; (2) The irregular morphology of ZrW_2O_8. The storage modulus of SiO_2/ZrW_2O_8/Epoxy composite film was up to 11 GPa, which is about 5 times higher than that of the pure epoxy film. At the high frequency of 5143 MHz, the dielectric loss of SiO_2/ZrW_2O_8/Epoxy composite film was 0.017, which decreased by 29.2%; The dielectric constant was 3.73, slightly increased by 29%, while it is still relatively low.

ACKNOWLEDGMENT

This work was financially supported by National Natural Science Foundation of China (51907194 and 51777209), National key R&D Project from Minister of Science and Technology of China (2017YFB0406300).

REFERENCES

[1] Yuanyuan Li, Muqin Tian, Zhipeng Lei, Jianhua Zhang, "Effect of nano-silica on dielectric properties and space charge behavior of epoxy resin under temperature gradient," J. Journal of Physics D: Applied Physics, 2018, 51(12).

[2] Tian Li, Jie Zhang, Huiping Wang, Zhongnan Hu, Yingfeng Yu, "High-Performance Light-Emitting Diodes Encapsulated with Silica-Filled Epoxy Materials," J. ACS applied materials & interfaces, 2013, 5(18).

[3] X. F. Yao, H. Y. Yeh, D. Zhou, Y. H. Zhang, "The structural characterization and properties of SiO_2-epoxy nanocomposites," J. Journal of Composite Materials, 2006, 40(4): 371–381.

[4] Young-Jae Kim, Jong-Heon Kim, Shin-Woo Ha, Dongil Kwon, Jin-Kyu Lee, "Polyimide nanocomposites with functionalized SiO_2 nanoparticles: enhanced processability, thermal and mechanical properties," J. RSC Adv. 2014, 4, 43371–43377.

[5] Jae-Soon Jang, Joshua Varischetti, Gyo-Woo Lee, Jonghwan Suhr, " Experimental and analytical investigation of mechanical damping and CTE of both SiO_2 particle and carbon nanofiber reinforced hybrid epoxy composites," J. Composites Part A, 2011, 42, 98-103.

[6] Juan Yang, Yongsen Yang, Qinqin Liu, Guifang Xu, Xiaonong Cheng, "Preparation of negative thermal expansion ZrW_2O_8 Powders and its application in polyimide/ZrW_2O_8 composites. J. Materials Science Technology, 2010, 26(7), 665-668.

[7] Hyungjoon Jeon, Cheolsang Yoon, Young-Geon Song, Junwon Han, Sujin Kwon, Seungwon Kim, et. al., "Reducing the coefficient of thermal expansion of polyimide films in microelectronics processing using ZnS particles at low concentrations," J. ACS Applied Nano Materials, 2018, 1, 1076−1082.

[8] Lisa M. Sullivan, Charles M. Lukehart, "Zirconium tungstate (ZrW_2O_8)/ polyimide nanocomposites exhibiting reduced coefficient of thermal expansion," J. Chemical Materials, 2005, 17, 2136−2141.

[9] Naoko Yamashina, Toshihiro Isobe, Shinji Ando, "Low thermal expansion composites prepared from polyimide and ZrW_2O_8 particles with negative thermal expansion," J. The Society of Photopolymer Science and Technology, 2012, 25(3), 385−388.

[10] Hongchao Wu, Mark Rogalski, Michael R. Kessler, "Zirconium tungstate/epoxy nanocomposites: Effect of nanoparticle morphology and negative thermal expansivity," J. ACS applied materials & interfaces, 2013, 5(19), 9478-9487.

[11] Yang Hu, Chao Chen, Yingfeng Wen, Zhigang Xue, Xingping Zhou, Dean Shi, et. al., "Novel micro-nano epoxy composites for electronic packaging application: Balance of thermal conductivity and processability," J. Composites Science and Technology, 2021, 209, 108760.

[12] Muhammad Amin, Muhammad Ali, Abraiz Khattak, "Fabrication, mechanical, thermal, and electrical characterization of epoxy/silica composites for high-voltage insulation," J. Science and Engineering of Composite Materials, 2018, 25(4): 753–759.

[13] Hyunaee Chun, Sook-Yeon Park, Su-Jin Park, Yun-Ju Kim, "Preparation of low-CTE composite using new alkoxysilyl-functionalized bisphenol A novolac epoxy and its CTE enhancement mechanism," J. Polymer 2020, 207, 122916.

[14] Jozef Kúdelčík, Emil Jahoda, Juraj Kurimský. "The effect of SiO_2 nano-filler on dielectric properties of epoxy resin," J. The European Physical Journal Applied Physics, 2019, 85(1), 10401.

[15] S. Weiner, H. D.Wagner, "The material bone: Structure-mechanical function relations," J. Annual Review of Materials Science, 1998, 28: 271-298.

[16] Jiongxin Lu, Kyoung-Sik Moon, Jianwen Xua, C. P. Wong, "Synthesis and dielectric properties of novel high-Kpolymer composites containing in-situ formed silver nanoparticles for embedded capacitor applications," J. Journal of Materials Chemistry, 2006, 16: 1543-1548.

978-1-6654-1392-3/21 $31.00 © 2021 IEEE

High efficiency testing system for 5G power amplifier

1st Zongqi Cai
The Testing Center of Electronic Components
China Electronic Product Reliability and Environmental Testing Research Institute
Guangzhou, China
zqcai.uestc@hotmail.com

2nd Sha Tang
The Testing Center of Electronic Components
China Electronic Product Reliability and Environmental Testing Research Institute
Guangzhou, China
tangsha1@163.com

3rd Jun Luo
The Testing Center of Electronic Components
China Electronic Product Reliability and Environmental Testing Research Institute
Guangzhou, China
kyea168@126.com

4th Xing Li
The Testing Center of Electronic Components
China Electronic Product Reliability and Environmental Testing Research Institute
Guangzhou, China
1395464966@qq.com

5th Xiaoqiang Wang
The Testing Center of Electronic Components
China Electronic Product Reliability and Environmental Testing Research Institute
Guangzhou, China
wangxq@ceprei.com

6th Daojun Luo
The Testing Center of Electronic Components
China Electronic Product Reliability and Environmental Testing Research Institute
Guangzhou, China
luodj@ceprei.com

Abstract—In this paper, a high efficiency testing system for 5G high power and high linearity power amplifier (PA) is proposed. Two bidirectional couplers are used in the input/output channels for the return loss, gain and saturated output power measurement at the same time, which can simplify the system construction and calibration process. The date acquirement of S_{11}/S_{22} can be provided by four power meters connected to bidirectional coupler. By using the testing platform, the RF and system-level parameters of sub-6G PAs can be measured at the same time, and the automatic acquisition program generated by the Labview is used for quick access of data. The test period can be reduced dramatically which can save the time and money for the identification & inspection of the third-party testing agency. What is more, the calibration process can be simpler and the consistency between each test process can be better. For validation of the power amplifier measured system, three type of sub-6G PAs operated at 2.595GHz, 3.5GHz and 4.85GHz, respectively were measured by using this platform. The test result shows that the platform can provide reliable and effective test data for different 5G PAs. The test platform including hardware and software can be promoted for practitioners in the designer, user and third-party testing agency which is significant for the development of 5G industry chain.

Keywords—5G, Power amplifier, Testing system, Calibration

I. INTRODUCTION

The vigorously developing of the fifth-generation mobile communication technology (5G) give a solid foundation of modern wireless communication technology for entering the era of industry 4.0[1]. As the key module of the transmitter and the most energy-consuming device in the wireless communication network, the power amplifier (PA) determines the performance of both the transmitter and wireless communication system. In the transmitter front-end circuit, the signal need to be enlarged by the post-stage circuit for the sufficient RF transmit power.

With the advantages of high voltage, high temperature, radiation and corrosion resistance, as well as high thermal conductivity, wide band gad and large electron drift velocity characters, the gallium nitride (GaN) semiconductor has become one of the most suitable materials for the third generation semiconductor which can be used for the design of high frequency and high efficiency PA at high temperature and high voltage environment[2]. As an important part of transmitter, the main technical indicators of RF PA includes output power, output efficiency, gain, harmonic suppression and input/output loss. Besides, the system-level parameters such as adjacent channel power ratio (ACPR) and error vector magnitude (EVM) are critical for the communication of wireless system[3].

Since the birth of radio in 19th century, the generation and testing of radio power have become the main problems and research directions. However, at the beginning of the application stage, the radio frequency signal is only used as a signal carrier for wireless transmission and the measurement of RF signal is focused on whether it is generated, but not signal analysis. With the dramatic development and application of radio technology, the role of radio technology in communication systems, radar systems, aerospace and other fields has become important increasingly. At the same time, the measurement and analysis of the radio signal characteristics would become a necessary Item. Until the 1970s, various latest electronic technologies were applied to test equipment, which greatly promoted the development and upgrading of test equipment[1-3]. Due to the computer's intervention, the test system has entered a rapid development stage, and automated testing has become a development trend. By using computer technology, the test engineer can analyze, process, and store the test data through the test software. With the development of computer technology, test systems with industrial computers as the core have also become an important direction for the development of the current test field. The development of virtual instrument technology has completely subverted the previous concept of test instruments.

According to different test requirements, test engineers design and define personalized test systems to meet changing test requirements. The advantages of virtual instruments are powerful, high-speed, reusable and economical requirements.

Among all kinds of tests, the power test of the radio frequency device is one of the most difficult, the highest technical requirement, and the most error factor test. However, the power test is also the most important application parameter of the radio frequency device.

In this paper, the measured method of RF and system-level parameters is given and used in the 5G PA test. Then, the testing system for 5G high power and high linearity PA is proposed. Three type of sub-6G PAs operated at 2.595GHz, 3.5GHz and 4.85GHz, respectively were measured by using this platform. The test result shows that the platform can provide reliable and effective test data for different 5G PAs.

II. TESTING SYSTEM IMPLEMENTATION OF 5G PA

A. Tehcnology indicators of PA

In order to evaluate the overall performance of PA, the component-level indicators which include DC parameters, RF parameters, and system-level indicators which include ACPR and EVM need to be considered together. The concept of component-level and system-level indicators can be given following:

1) DC voltage and current

The DC bias need to be added when amplifier works, which can be represented as voltage/current or DC power.

2) Frequency

The frequency is the normal frequency range when the amplifier works. in 5G application, the typical center frequency are 2.595GHz, 3.5GHz and 4.85GHz with 160MH，200MHz and 100MHz bandwidth, respectively.

3) Gain

The gain of amplifier represents the ratio of the output signal power to the input signal power, usually expressed in "dB".

4) output power

There are two ways to explain the output power of a power amplifier: the saturation power and 1dB compression point power. Saturation power is the maximum output power by the amplifier, while the 1dB compression point power is the output power when the gain is reduced 1dB. The saturation power of amplifier is generally greater than the 1dB compression point power. There are two types power of pulse amplifiers: peak power and average power. The peak power represents the output power when there is a signal, and the average power is the averaged power over time. The peak power and average power are related to the duty cycle of the input signal.

5) Power added efficiency (PAE)

The power added efficiency is the ratio of the result of the difference between the output power of the power amplifier and the input power to the DC power consumption, which defines the DC conversion efficiency of the power amplifier.

6) Return loss (S_{11}/S_{22})

Return loss is the ratio of the power of the reflected wave to the incident wave, so the input/output return loss S_{11} and S_{22} of power amplifier can be calculated by the input/output bidirectional coupler. The input power and reflected power can be gained by power meter connected to coupled terminal.

7) Harmonics and distortion

The non-linear distortion of the RF power amplifier will generate new frequency components. For example, for the second-order distortion, it will produce the second harmonic and two-tone beat frequency. For the third-order distortion, it will produce the third harmonic and multi-tone beat frequency. If these new frequencies fall within the passband, they will cause direct interference to the transmitted signal, and if they fall outside the passband, they will interfere with other channels' signals. So the RF PA should be linearized, which can solve the problem of signal spectrum regeneration. The principle and method of RF power amplifier linearization technology can use the amplitude and phase of the input signal envelope as a reference, and comparing it with the output signal, and then generating appropriate corrections.

8) Adjacent channel power ratio (ACPR)

ACPR is used to measure the ratio of the power in the main channel to the power in the adjacent channel, which is related to the linearity of the amplifier and the modulation of the signal. ACPR is used Mainly in the research of wide-spectrum signals such as CDMA.

9）Error vector magnitude (EVM)

EVM is generally used to evaluate the modulation quality of the transmitter's transmitted signal, avoiding multiple parameters to characterize the transmitted radio frequency signal. It is a very valuable indicator of the overall signal quality in the development and design process.

B. Measured Method of PA Indicators

When the technology indicators of PA are given, the measured method need to be discussed for the accurate test of output power, gain, PAE, return loss (S_{11}/S_{22}), ACPR and EVM. The schematic diagram of testing system of 5G GaN PA is shown in Fig.1, which includes vector signal generator, spectrum analyzer, DC power supply, Device under test (DUT) and some adapter/accessories. The adapter/accessories include drive amplifier, isolator, bidirectional coupler, power meter, oscilloscope, attenuator and load.

The R&S SMW200A vector signal generator is used to generate 5G NR signal and is amplified by M&T drive amplifier. The 5G NR modulation waveform signal is enlarged to 20-35 dBm, and then is inserted into bidirectional coupler after a isolator (to prevent self-excitation), the output of coupler is connected to the under tested device-power amplifier. The output terminal of PA is connected to another bidirectional coupler, and then absorbed by a load or spectrum analyzer after a large power attenuator.

In this system, the DC voltages is added at the Drain and gate terminal of DUT and can be measured by the DC power supply Keysight N6705C and Chroma 62006P, the current can be test by the current clamp through Tektronix oscilloscope. The input/output power can be measured by the power meter Keysight N1912A from the coupled ports of 40 dB bidirectional coupler. The input/output return loss (S_{11}/S_{22}) also can be calculated by the four power meters connected to the four coupled ports of couplers. The Gain can be gained

978-1-6654-1392-3/21 $31.00 © 2021 IEEE   245

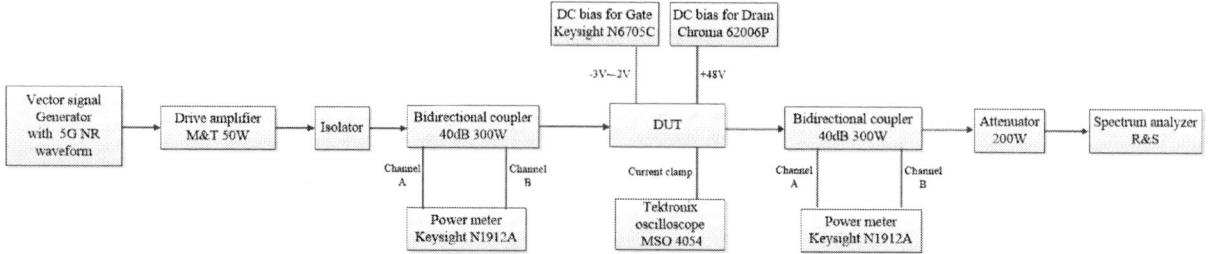

Fig.1 the schematic of 5G PA testing system

Fig.2 The actual testing platform of 5G PAs

through the difference between input and output power. The harmonic suppression and distortion can be seen by spectrum analyzer. For the ACPR and EVM measurement, the 5G NR waveform need to be loaded and the special optional function should be authorized for signal generator and spectrum analyzer. By tuning the voltage of gate, the PA can work at typical situation and the indicators can be measured.

III. MEASUREMENT OF 5G PA TESTING SYSTEM

Table I gives the typical indicators of three PAs following.

TABLE I.   THE TYPICAL INDICATORS OF THREE PAS

Type		A	B	C
RF parameters	f_0(GHz)	2.595	3.5	4.85
	BW(MHz)	160	200	100
	P_{sat}(dBm)	≥50.5	≥50.5	≥47
	η_{sat} (%)	≥68	≥68	≥60
	P(dBm)	P_{out}-8dBm	P_{out}-8dBm	P_{out}-8dBm
	η(%)	≥20	≥20	≥15
	Gain (dB)	≥17	≥17	≥15
	S_{11}(dB)	>10	>10	>10
	S_{22}(dB)	>7	>7	>7
System-level parameters	ACPR (dBc) @10MHz, PAR 7.5dB	≤-30	≤-30	≤-30

Three PAs work at 2.595GHz, 3.5GHz and 4.85 GHz with 160MHz, 200MHz and 100MHz bandwidth, respectively. The typical output power of amplifier A, B and C is larger than 50W (47dBm), so the drive amplifier and high power capacities (isolator, directional coupler, attenuator and load) adaptors need to be used. What is more, in order to reduce the complex of testing system and improve the efficiency, the frequency range of adaptors should involve 2-6GHz. Three PAs can be measured in the same testing system which is helpful the automatic test by computer.

In this paper, 116 PAs of each type PA is measured and the average indicators is listed following in table II. The testing work space can be seen in Fig.2. Each one PA of each type is worked at the typical region which can be seen in Table I. In PA A, the center frequency is 2.515 GHz with 160 MHz bandwidth, so the indicators at 2.515GHz, 2.595 GHz and 2.675 GHz are measured. From Table II, the average saturation power is 51.38 dBm, 51.44 dBm and 51.33 dBm at three frequency points, the gain at linear region are 18.71dB, 18.62 dB and 18.52 dB, respectively. The power added efficiency (PAE) is over 70% at saturation region and over 30% at linear region. The ACPR is more than 30 dB at three frequency points. About PA B and PA C, the measured and average indicators all satisfy the typical indicators.

TABLE II.   THE MEASURED RESULTS OF THREE TYPE PAS

PA type	f_0	P_{sat}	Gain	PAE@ Psat*	PAE@ Psat-8	ACPR@10 MHz
	GHz	dBm	dB	%	%	dBc
A	2.515	51.38	18.71	72.66	33.76	-34.64
	2.595	51.44	18.62	75.06	34.99	-34.70
	2.675	51.33	18.52	72.46	35.49	-34.31
B	3.4	51.05	17.97	74.81	34.87	-31.22
	3.5	51.16	18.3	72.21	33.69	-32.11
	3.6	51.02	17.53	69.63	32.39	-31.69
C	4.8	48.58	15.88	67.46	31.83	-35.34
	4.85	48.61	16.21	69.26	33.08	-35.68
	4.9	48.66	16.16	69.23	33.39	-35.70

P_{sat}-8*: the power of saturated power (in dBm)-8dBm

IV. CONCLUSION

In this paper, three type PAs re measured by a high efficient power amplifier testing system. Two bidirectional couplers are used in the output power, gain S-parameters measurement. The testing period can be reduced dramatically by auto-testing which can save the time and money for the identification & Inspection of the third-party testing agency.

REFERENCES

[1] P. Cerny, "Parameters measurement of power amplifier using vector network analyzer," COMITE, Pardubice, Czech Republic, 2013, pp. 177-181.

[2] M. Marchetti, G. Avolio, M. Squillante, A. K. Doggalli and B. V. Anteverta-Mw, "Wideband Load Pull Measurement Techniques: Architecture, Accuracy, and Applications," ARFTG, Orlando, FL, USA, 2019, pp. 1-7.

[3] W. Fang et al, "A Simple and Universal Measurement Method for the Efficiency of Pulsed RF Power Amplifiers," IEEE Access, vol. 8, pp. 59200-59210, 2020.

Research on High-speed SerDes Interface Testing Technology

Weikun Xie
University of Electronic Science and Technology of China
Chengdu, China
chinagrass@163.com

Guangqiang Cao
University of Electronic Science an Technology of China
Chengdu, China
15295377665@163.com

Weiwei Ji
The 58th Research Institute of China Electronics Technology
Wuxi, China
Groupjiweiwei919@126.com

Abstract—At present, the demand for high-speed data transmission is getting higher and higher. Due to technological progress, the clock frequency is getting higher and higher. Compared with the parallel interface, the serial interface has fewer wires and less interference between wires[1].The transmission rate is improved by continuously increasing the clock frequency. Effective verification and testing of the high-speed serial port SerDes chip is worth studying.

This article introduces a method to test the high-speed SerDes serial interface based on the V93000 test system. A FPGA product with a model of XCKU040 is used as the chip to be tested, and the sending/receiving function test of the SerDes interface chip with a transmission rate of 16Gbps is carried out. A hardware test PCB platform with FPGA as the core was built, and the chip was tested in various working modes, including inner loop mode, outer loop mode, and built-in self-test mode. The hardware test PCB platform based on FPGA is established, and the functions of TX and Rx are tested.

Keywords: SerDes; V93000, automated testing; bit error rate; eye diagram

I. INTRODUCTION

With the advent of the era of big data, data transmission puts forward higher requirements for bus bandwidth. Serial transmission technology, especially serial deserializer (SerDes), can provide higher bandwidth than parallel transmission technology, requires less chip pins, and supports communication, network, data storage, transmission, ultra-high speed and other mainstream standards[1]. High performance CPU, DSP, FPGA and other products at home and abroad are embedded with high-speed serial interface. The development direction of high-speed serial interface technology is to improve the single channel speed and combine multi-channel parallel use, which is widely used in Ethernet, backplane transmission, device bus and other technologies. Therefore, the research of this kind of SerDes interface testing technology is very important.

This paper proposes a high-speed SerDes debugging scheme based on serial port, which is more efficient than the traditional SerDes testing method, convenient for testing and application personnel to debug, and more universal. The scheme is verified on FPGA chip of 13.1 Gbps high speed SerDes.

In this paper, 8B / 10B SerDes is used in the IP core of PCIe and SATA, which can work both transmitting channel and receiving channel at the same time, that is, full duplex mode is adopted. The transmission channel is composed of encoder, parallel serial converter, clock generation circuit and transmitter; The receiving channel is composed of decoder, serial parallel converter, clock recovery circuit and receiver, as shown in Figure 1.

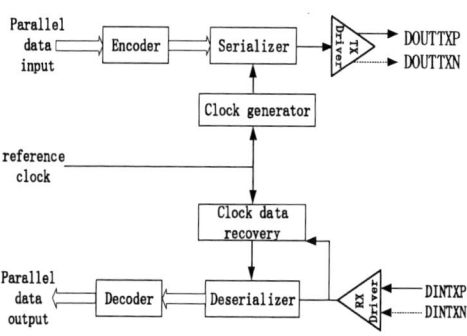

Figure 1. Basic principle and structure of SerDes

II. INTERFACE FUNCTION CONFIGURATION SCHEME DESIGN

Because the chip to be tested is Xilinx xcku040 FPGA chip, many original test methods are limited and cannot be used. Therefore, before testing the high-speed interface of the chip to be tested, it is necessary to configure its function. After that, the high-speed interface of the chip to be tested will be tested.

Specifically, the V93000 automatic test system is used through Slave Select MAP _parallel configurating file to the chip to be tested in parallel mode. The schematic diagram of the connection between V93000 automatic test system and the chip to be tested is shown in Figure 2. PROGRAM_ B is lowered, the program of the chip to be tested will be reconfigured; Chip selection input CSI_ B enables the chip to be tested to slave and configure the bus when CSI_B is pulling up, the chip ignores the slave and configures the interface. At this time, the D[31:00] is in the high resistance state, and the RDWR_B is ignored when CSI_B is in low level, the slave and configuration mode is enabled; RDWR_ B is the control data transmitting or receiving: in order to configure, RDWR_B must be set to write control (RDWR_B = 0）; CCLK is the slave and configures the clock on the data bus; For the slave and parallel configuration mode, 8-bit, 16 bit and 32-bit data bus can be selected. Here, the maximum 32-bit data bus is selected to configure the function for the tester during the test, because it can speed up the configuration time and greatly shorten the test time of the high-speed interface part; The function of CFGBVS is to control the level of configuration pin; The pin of indicator changes from low to high, indicating that the chip under test is configured.

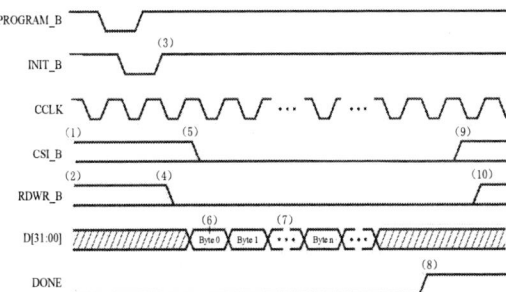

Figure 3. flow of slave and configuration mode

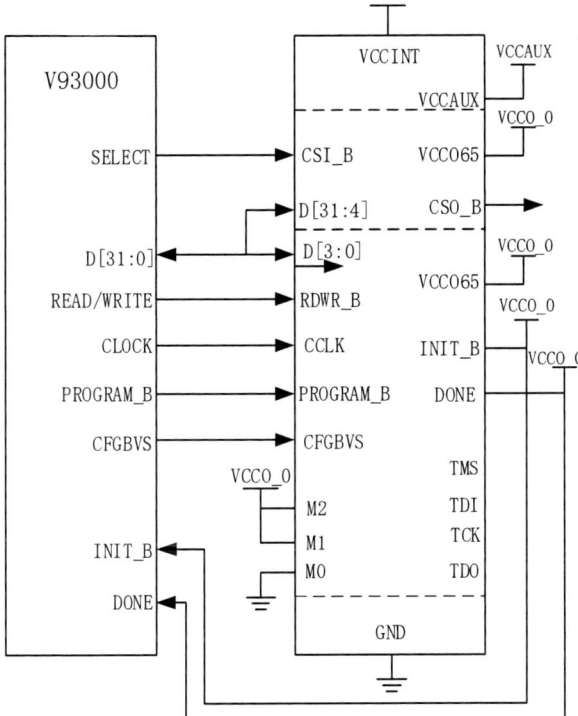

Figure2. Schematic diagram of v93000 and DUT configuration mode

The flow chart of the slave parallel 32-bit continuous configuration mode based on V93000 ATE is shown in Figure 3.

1)Since there is only one chip to be tested on the configuration bus, CSI_ B signal is bound to low level. If V93000 automatic test system is configured, If CSI_B is not low, the configuration operation is invalid.

2) Since the configuration data does not need to be read back, the RDWR_ B is set to low level. And when CSI_ B is low, RDWR_ B should not be switched.

3) When INIT_B becomes high, the chip under test samples the configuration mode pins.

4)Completing the declaration RDWR_ B before CSI_B is low. To avoid configuration failure on the next CCLK.

5) When CSI_B is low, the slave mode data interface is enabled and configured.

6) After CSI_B is low, the first byte of configuration data is loaded on the rising edge of the first CCLK.

7) The configuration code stream is loaded one byte at the rising edge of each CCLK.

8) After the configuration is completed and successful, the done pin will become high.

9) After configuration, CSI_ B signal can be set to invalid.

10) After canceling declaration of CSI_B signal, RDWR_ B can be undeclared。

III. TEST TECHNOLOGY ANALYSIS

A. Function test of transmitter

The transmission function test is the basis of the subsequent sending end parameter test. It mainly verifies the functional logic correctness of the module composed of the transmitter end of the chip to be tested, including: the parallel data input interface of the transmitter, the 8b/10b encoder, the polarity control module, the serial parallel conversion module, the preemphasis / deemphasis control module and the transmission driver.

The internal structure of the chip to be tested is simplified to understand. The dotted line box is the module not used for the transmitter test; After searching the delay time of the sender, the function of the transmitter can be tested[4]. The configuration pin in the figure is the data interface and related control pin of slave configuration mode described in Section 3.1; The clock is the reference clock of FPGA system clock pin and high-speed interface.

The function test process of transmitter is as follows:

1) Under the specified environmental conditions, the device under test is connected to the test system;

2) The specified voltage is applied by DPS power board at the power terminal;

3) The chip is configured to send data mode by the chip configuration port, that is, the 64 bit parallel code stream data is input to the parallel data input interface of the sending end by the V93000 automatic test system. The data is encoded by the 8B / 10B encoder, and then transmitted to the preemphasis / deemphasis control module through the polarity control module of the sending end, and finally to the sending driver;

4) V93000 automatic test system sends 64 bit parallel data to the parallel data input port of transmitter;

5) The PSSL digital board collects the serial data signal from the transmitter and compares it with the expected data.

6) V93000 automatic test system stops sending codes and collecting data, and outputs test results.

1. Jitter Test

Jitter is inevitable when all senders perform the task of sending data. Jitter can reflect the change of time sequence in the transmission data stream. Generally speaking, it is the upper bound jitter data collected under specific test conditions. Before the formal test, it is necessary to determine the test mode and specification[5].

The jitter at the transmitter may be amplified by the channel. After the data mode transformation of 1 and 0, the

978-1-6654-1392-3/21 $31.00 © 2021 IEEE   249

spectrum content of the data changes, and the delay caused by the channel changes into additional jitter at the receiving end. Crosstalk and reflection caused by impedance discontinuity and return loss also cause jitter.

The jitter test schematic diagram of the transmitter is different from the function test of the transmitter. The output data of the transmitter is generated by the internal PRBS generator of the chip to be tested. The reason is that the ate tester does not need to set the expected data code type in advance when testing jitter, and only needs to collect the time when the rising edge and falling edge of the serial signal occur. The test process is as follows:

1) Under the specified environmental conditions, the device under test is connected to the test system;

2) The specified voltage is applied by DPS power board at the power terminal;

3) Through register configuration, the data is generated by the internal PRBS sequence and transmitted to the TX end, that is, the code stream data is generated by the PRBS generator, through the polarity control module of the transmitter, the serial parallel conversion module of the transmitter and the preemphasis / deemphasis control module, and finally sent to the driver for output by the transmitter of the chip under test;

4) Set the judgment level of TX terminal to Vcmt level;

5) Collect all the Vcmt passing time of TX end (the 0-passing time collected in the i-th cycle);

6) According to formula (1), the jitter sequence of the rising / falling edge of each i-th cycle is calculated.

$$t_k = t_i - i \times UI \qquad (1)$$

7) The extremum of the sequence is obtained, which is denoted as and;

8) Calculate according to formula (2)

$$jitter = t_{kmax} - t_{kmin} \qquad (2)$$

Where: is the maximum value in the sequence, is the minimum value in the sequence, and is the jitter amplitude of TX .

2. Rise / Fall time Test

The rise / fall time of the signal sent by the transmitter determines the upper limit of the maximum transmission speed of the interface, so it is necessary to test it.
The test process is as follows:

1) Under the specified environmental conditions, the device under test is connected to the test system;

2) The specified voltage is applied by DPS power board at the power terminal;

3) The chip to be tested is configured to the code stream through register configuration, and the data is input by the parallel data input interface of the transmitter, and then output to the transmitter driver through 8B / 10B encoder, polarity control module of the transmitter, parallel conversion module of the transmitter and preemphasis / de emphasis control module;

4) The 00110011 cyclic sequence is found in the output code of the transmitter driver, and the v93000 test machine runs LLHHLLHHLLHH ;

5) Set the judgment level of TX terminal to 10% or 20% of the original level;

6) Continuously change the comparison time of judgment level, record the time when the test state of test vector changes from no pass to pass, and record as;

7) Set the judgment level of TX terminal to 90% or 80% of the current level;

8) Continuously change the comparison time of judgment level, record the time when the test state of test vector changes from no pass to pass, and record as;

9) According to the formula (3)

$$t_r = t_2 - t_1 \qquad (3)$$

Where: is the start time of voltage rise, unit: s, is the end time of voltage rise, unit: s, is the rise time of differential output, unit: s.

The falling time measurement is similar to the rising time, and the step e) is changed to: set the TX terminal judgment level to 90% or 10% of the rising time; Change step g) to set TX terminal judgment level to 10% or 20%, only steps E) and G) are different from rise time test process.

3. Preemphasis / Deemphasis Test

When high-speed signals transmit data, skin effect and dielectric loss will inevitably occur, especially in high-frequency components. The loss is very serious, and even affects the completion of receiving and decoding tasks at the receiving end. If preemphasis / deemphasis processing is not adopted, eye pattern test is carried out at the receiving end, and the eye pattern can be observed almost or completely closed. For high-speed signal transmission loss caused by signal deviation, and does not affect the final data reception, usually need to use signal compensation technology to ensure that the signal eye diagram of the receiver can meet the relevant requirements[4].
Figure 4 shows the differential signal at the transmitter:

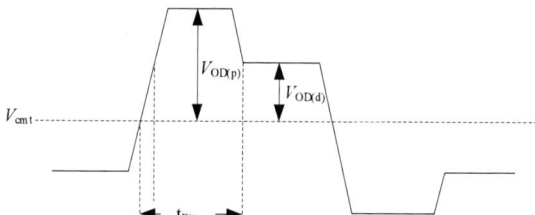

Figure 4. Waveform of preemphasis/deemphasis output amplitude test at transmitter

The preemphasis test process is as follows:

1) Under the specified environmental conditions, the device under test is connected to the test system;

2) The specified voltage is applied by DPS power board at the power terminal;

3) The chip to be tested is configured to the code stream through register configuration, and the data is input by the parallel data input interface of the transmitter, and then output to the transmitter driver through 8B / 10B encoder, polarity control module of the transmitter, parallel conversion module of the transmitter and pre emphasis / de emphasis control module;

4) Find the 00110011 cycle sequence in the output code pattern of the transmitter driver (using this code pattern will be more convenient to test the pre emphasis / de emphasis voltage of the differential signal, the same below), and test the first 1 or the first 0 state in two consecutive high "11" or low "00" in the output code pattern of the transmitter;

5) Change the judging condition of output level from high to low, record the value of differential output amplitude

when the test vector test state changes from pass to fail, and record it as.

The process of de emphasis test is similar to that of pre emphasis test, and step d) is changed to: test the second 1 or the second 0 in two consecutive high "11" or low "00" in the output pattern of the transmitter driver; Change the output level judgment condition from high to low in step e) and record the value of the differential output amplitude when the test vector test state changes from pass to fail.

IV. RECEIVER FUNCTION TEST

Receiving function test mainly verifies the functional logic correctness of the receiver module of the chip to be tested, including RX receiver, equalizer, serial parallel conversion module, polarity control module, 8B/10B decoder and parallel data output interface of the receiver[5].

The function test process of receiver is as follows:

1) Under the specified environmental conditions, the device under test is connected to the test system;

2) The specified voltage is applied by DPS power board at the power terminal;

3) The chip is configured to receive data mode by the chip configuration port, that is, the serial data signal is input to the RX receiver by PSSL digital board of v93000 automatic test system. The serial signal is processed by the equalizer, decoded by the serial parallel conversion module, polarity control module and 8B / 10B decoder, and finally sent out from the parallel data output port of the receiver;

4) The V93000 automatic test system sends serial data to the RX receiver of the chip to be tested;

5) The PS1600 digital board collects the 64 bit parallel data signal from the parallel data output port of the receiver and compares it with the expected data.

6) V93000 automatic test system stops sending codes and collecting data, and outputs test results.

1. Jitter Tolerance Test

The receiver must be able to identify the signal with jitter. Jitter tolerance is the key index to measure the performance of the receiver module [30]. This parameter defines the "worst case" signal that the receiver can receive under normal conditions, and the maximum error does not exceed the acceptable range of BER. There are special specifications for jitter tolerance test, and the choice of test mode and data method for jitter tolerance is very important. Generally speaking, jitter and jitter tolerance are expressed in picoseconds or by unit interval (UI). A unit interval is the cycle time of one bit transmitted on the interface at a single channel line rate. According to the relevant specifications, the jitter of the transmitter is 0.28 UIPP, which means that the peak to peak jitter is less than 28% of the unit time; The jitter tolerance is 0.30uipp, which means that the receiver can accept 30% jitter per unit time at the input.

The schematic diagram of jitter tolerance test at the receiving end is as follows;

1) Under the specified environmental conditions, the device under test is connected to the test system;

2) The specified voltage is applied by DPS power board at the power terminal;

3) The chip is configured to the inner ring PRBS mode by the chip debugging port, that is, S1 and S2 are configured to position 1. The code stream data is input into the chip from the receiving end of the circuit to be tested in serial, then goes through the receiver, equalizer, serial parallel conversion module of the receiving end and polarity control module of the receiving end, and then returns to the polarity control module of the transmitting end The serial to parallel conversion module and the pre emphasis / de emphasis control module are used at the transmitter of the chip to be tested;

4) At the RX end, the PRBS code stream injected into the chip to be tested is sent to the PRBS check end after passing through the inner ring and TX end;

5) Set the judgment level of TX terminal to vcmt level;

6) At the RX end, the data jitter increases continuously, and the PRBS check information is read repeatedly. When the PRBS error occurs, the insertion jitter is recorded as the input jitter tolerance of the circuit.

V. TEST RESULT ANALYSIS

A. Rise / Fall time test results

As shown in Figure 5, the rise / fall time test results show that the rise time TR is 20ps and the fall time TF is 19ps, which meet the design specifications (the maximum rise / fall time is 21ps), and the test items pass.

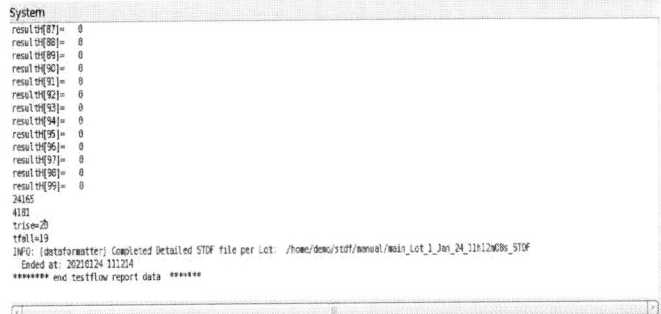

Figure 5. SerDes interface rise / fall time test result report

B. Jitter Test Results

As shown in Figure 6, the effective jitter value is 2.8ps, and the maximum jitter is 15.1ps. Therefore, the maximum jitter of the transmitter is 0.2268UI, which meets the design specification (output jitter 0.28UI), and the test item passes.

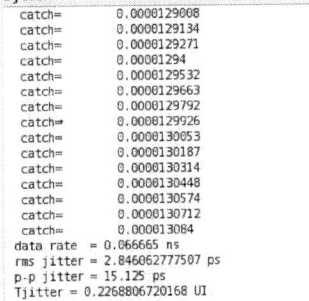

Figure 6. Sender jitter test result report

978-1-6654-1392-3/21 $31.00 © 2021 IEEE

VI. PROBLEMS AND CHALLENGES

A.DUT test board and signal integrity

The signal integrity of DUT test board is mainly due to the interaction of various signal voltage and current waveforms, which affects the stability of the test signal. The main interference will be reflected in the signal reflection, signal timing, switching noise, non monotonicity, attenuation, crosstalk, capacitive load, electromagnetic interference and so on. If the DUT test board which is not designed for signal integrity is applied, the excellent design will be in the state of big test error or stop working completely. When designing the test platform, because the DUT test board includes high-speed signal transmission, signal integrity must be considered in the design. The signal integrity design of the DUT test board mainly includes anti-crosstalk design, impedance matching design, physical isolation design, etc.

When the signal can reach the IC with the required time sequence, duration and voltage amplitude, the circuit has good signal integrity. When the signal can not respond normally, the signal integrity problem appears. Signal integrity is mainly manifested in delay, reflection, crosstalk, timing, oscillation and other aspects. In order to correctly identify and deal with data, IC requires data in a stable state before or after the clock edge, and the duration of this stable state is called setup time and hold time. If the signal changes to an unstable state or changes the state later, the IC may misjudge or lose some data. In the actual high-speed circuit design, the actual signal is often different from the ideal expected waveform, including some unwanted waveforms, which belong to the problem of signal integrity. With the increasing of clock frequency and decreasing of rising edge, the problem of signal integrity becomes more serious.

B. Clock synchronization

Another challenge that cannot be ignored is clock synchronization between external instruments and ATE system. Usually, when measuring low-speed signal from ATE system with external instrument (such as oscilloscope), users only need to provide trigger signal from ATE system to external measuring instrument, which is easily obtained from an unused ATE channel. But for high-speed SerDes test, the main challenge is to ensure the frequency synchronization between the reference clock source and ate system, and to ensure that there is no frequency offset or frequency offset between the two systems. This can be achieved by using the clock of the external device as the master clock of the ATE system, or locking the ate master clock to the external timing reference by using the synchronization signal (usually 10MHz frequency signal). Most instruments or devices have the setting of external reference clock access, which ensures that the clock source is not only synchronized but also has no frequency difference. Because the ate in this paper is more active, the 10 MHz reference clock source of ATE is used as the reference clock, and the external instrument group is connected as the input clock source.

VII. CONCLUSION

1) Compared with the traditional instrument method test, the method adopted in this article provides an effective solution to the automated and high-efficiency test requirements of SerDes high-speed interface chips.
2) This topic is based on the Advantest V93000 test system to study the test methods and test characteristics of the Serdes serial interface, and solve the problems faced by the realization of high-speed signal test methods.

ACKNOWLEDGMENT

This research is supported by The 58th Research Institute of China Electronics Technology.

REFERENCES

[1] An Engineers Guide to Automated Testing of High-Speed Interfaces.2015.

[2] Xu Yi. Research on the test method of 10.3125Gbps high-speed SERDES chip [D]. Sichuan: University of Electronic Science and Technology of China, 2017.

[3] Xie Hanwei. Design and implementation of an integrated ATE platform for high-speed Serdes interface testing[D]. Jiangsu: Nanjing University of Science and Technology, 2018.

[4] H. Werkmann, "PCI Express: ATE Requirements and DFT Features for Functional Test Optimization," IEEE European Test Workshop, May 2003.

[5] H. Werkmann, " Enabling the PCI Express Ramp — ATE Based Testing of PCI Express Architecture," IEC EuroDesignCon, Oct. 2004.

[6] An Engineers Guide to Automated Testing of High-Speed Interfaces.2015.

High-Performance Thermal Grease with the Addition of Silver Particles

1st Xiangliang Zeng
1 *Shenzhen Institute of Advanced Electronic Materials, Shenzhen Institute of Advanced Technology Chinese Academy of Sciences*
Shenzhen, China
2 *College of Chemistry and Chemical Engineering*
Hunan University
Changsha, China
xl.zeng1@siat.ac.cn

4th Linlin Ren*
1 *Shenzhen Institute of Advanced Electronic Materials, Shenzhen Institute of Advanced Technology Chinese Academy of Sciences*
Shenzhen, China
ll.ren@siat.ac.cn

7th Rong Sun
1 *Shenzhen Institute of Advanced Electronic Materials, Shenzhen Institute of Advanced Technology Chinese Academy of Sciences*
Shenzhen, China
rong.sun@siat.ac.cn

2nd Zhenyu Wang
1 *Shenzhen Institute of Advanced Electronic Materials, Shenzhen Institute of Advanced Technology Chinese Academy of Sciences*
Shenzhen, China
zy.wang4@siat.ac.cn

5th Xiaoliang Zeng*
1 *Shenzhen Institute of Advanced Electronic Materials, Shenzhen Institute of Advanced Technology Chinese Academy of Sciences*
Shenzhen, China
xl.zeng@siat.ac.cn

3rd Wenbo Ye
1 *Shenzhen Institute of Advanced Electronic Materials, Shenzhen Institute of Advanced Technology Chinese Academy of Sciences*
Shenzhen, China
yewenbo@mail.ustc.edu.cn

6th Xinnian Xia*
1 *Shenzhen Institute of Advanced Electronic Materials, Shenzhen Institute of Advanced Technology Chinese Academy of Sciences*
Shenzhen, China
xnxia@hnu.edu.cn

Abstract—Thermal grease has been widely used in the thermal management of electronic packaging due to its excellent filling ability between heat resource and heat sink, yet those grease tend to have low thermal conductivity. Increasing the filler content in thermal grease will enhance the thermal transfer capacity but impair its filling performance. Here, we report a thermal grease with optimal thermal conductivity and filling capacity prepared from alkyl silicone oil, hydrogen silicone oil, alumina (Al_2O_3), zinc oxide (ZnO) and silver (Ag). The silver particles effectively connect the fillers in the grease, improving the overall thermal conductivity while ensure proper viscosity. This adjustable interaction between thermal conductivity and viscosity makes it possible to adjust the thermal transfer capacity of the grease without causing a decrease in the filling performance. The prepared thermal grease exhibits a combination of satisfactory thermal conductivity (3.0 W m^{-1} K^{-1}) and lower viscosity (~322.7 Pa·s), those excellent properties indicate their great potential in the application of electronic packaging.

Keywords—*thermal grease; silver; thermal conductivity*

I. INTRODUCTION

Heat dissipation has been a critical technical issue impacting the advancement of electronics [1-2]. In particular, the current high-power-density 5G equipment is posing greater challenges to thermal management. If the heat in the devices cannot be dissipated in time, the internal temperature will rise rapidly, or even exceed the rated temperature, causing seriously affect the reliability and service life of the equipment [3-4]. To address this problem, the development of high-

performance thermal management materials is extremely urgent.

Thermal grease is a kind of thermal management material that can be used to fill the gap between the heat source and the heat sink, thereby improving the heat transfer efficiency. At present, a large number of researchers have devoted themselves to how to improve the thermal conductivity of thermal grease [5]. However, merely improving the thermal conductivity is far from enough, and its filling ability is also significant. Excellent filling ability means that the silicone grease with a low viscosity, which enables the thermal grease to effectively contact the interface of the component, reducing the contact thermal resistance and improving the heat transfer ability. Unfortunately, excellent thermal conductivity and filling capacity are hard to get at the same time, since high thermal conductive filler content leads to high thermal conductivity and high viscosity.

Here, we report a thermal grease with optimal thermal conductivity and filling capacity prepared from alkyl silicone oil, hydrogen silicone oil, alumina (Al_2O_3), zinc oxide (ZnO) and silver (Ag), which shows both high thermal conductivity and low viscosity. The silicone grease was obtained through a facile solvent-free vacuum blending technique. The silver particles work as "bridges" to effectively connect the fillers in the grease to form good thermal conductive pathways, which improves the overall thermal conductivity while ensuring proper viscosity. This adjustable interaction between thermal conductivity and viscosity makes it possible to adjust the thermal transfer capacity of the grease without causing a decrease in the filling performance. The prepared thermal

978-1-6654-1392-3/21 $31.00 © 2021 IEEE

grease exhibits a combination of satisfactory thermal conductivity (3.0 W m^{-1} K^{-1}) and lower viscosity (~322.7 Pa s), those excellent properties indicate their great potential in the application of electronic packaging.

II. EXPERIMENTAL

A. Materials

Spherical alumina (Al$_2$O$_3$, 5 μm) and zinc oxide (ZnO, 400~600 nm) were purchased from Aladdin Reagent Co., Ltd. The silver powder (50 nm and 1 μm) was purchased from Sigma Aldrich. Alkyl silicone oil and hydrogen-containing silicone oil were purchased from Gelest, Inc. All materials were used as received without further processing.

B. Preparation of Ag Content Thermal Grease

Thermal grease is prepared by a facile solvent-free mixing method. Weighted powder (modified Al$_2$O$_3$, modified ZnO and Ag), alkyl silicone oil (10 mPa·s) and hydrogen silicone oil (50 mPa·s) were added to a mixing tank in proportion, then the mixture is mixed uniformly through a planetary vacuum mixer. The prepared paste mixture is the thermal grease.

C. Characterization

Micro morphology of the purchased powders were performed by a scanning electron microscope (FEI Nova NanoSEM 450). Surface condition of the modified powders are tested by a Fourier Transform Infrared (FTIR) spectrometer (Bruker Vector 33, Germany). Thermal conductivity and thermal resistance of the samples was performed by a steady state heat flow thermal conductivity tester (Longwin TIM LW-9389) form Taiwan Long Win Science and Technology Corporation. Rheometer (MCR-302, Anton Paar, Austria) was employed to test the viscosity of prepared thermal grease.

III. RESULTS AND DISCUSSIONS

The morphology of the filler was obtained by SEM, as shown in Fig. 1. The diameter of the Al$_2$O$_3$ sphere is 5 μm (Fig. 1a), ZnO is a polygonal square with a size of 400~600 nm (Fig. 1b), and Fig. 1c and d are Ag particles with a diameter of 50 nm and 1 μm, respectively.

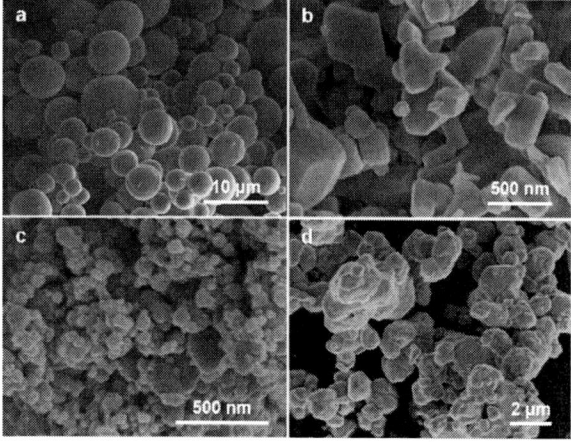

Fig. 1 SEM images of raw material. (a) Spherical Al$_2$O$_3$ with a diameter of 5 μm, (b) Polygonal ZnO with a size of 400~600 nm, (c) Ag powder with a diameter of 50 nm, (d) and 1 μm, respectively.

Fourier infrared spectroscopy measurements were carried out to analysis the modified Al$_2$O$_3$ and ZnO powders (Fig. 2). Peaks near 3600 cm^{-1} and 3300 cm^{-1} are the -OH absorption peaks formed by moisture in the air adsorbed on the surface of Al$_2$O$_3$ (Fig. 2a). Two peaks near 2916 cm^{-1} and 2849 cm^{-1} indicate the existence of methyl and methylene, respectively. Similarly, in the infrared spectrum of ZnO (Fig. 2b), a strong -OH absorption peak was observed at 3697 cm^{-1}, and the absorption peaks at 2922 cm^{-1} and 2850 cm^{-1} are methyl and methylene, respectively. The appearance of methyl and methylene absorption peaks proved the modification of the powder.

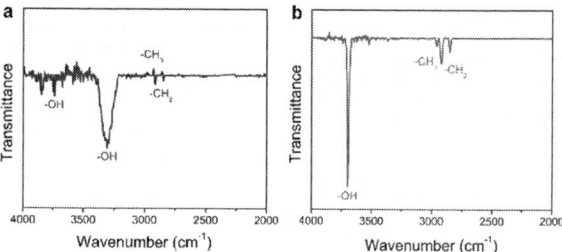

Fig. 2 Fourier infrared spectrum of modified filler. (a) Infrared spectrum of Al$_2$O$_3$; (b) Infrared spectrum of ZnO.

Thermal grease was prepared with the same Ag addition to explore the influence of Ag particle size. Thermal grease with large Ag particles exhibits better thermal conductivity than which with small Ag particles. Although small particles can fill the gap between Al$_2$O$_3$ and ZnO better than large particles to form heat conduction pathways, the filling of large Ag particles will make the thermal conductivity of the thermal grease higher (Fig. 3a). According to the effective medium theory, the size of the filler is related to the interface thermal resistance, which will affect the overall thermal conductivity of the composite material. The relationship between its thermal conductivity and interface thermal resistance is as follows:

$$\frac{\lambda_c}{\lambda_m} = \frac{[\lambda_p(1+2\alpha+2\lambda_m)]+2\Phi[\lambda_p(1-\alpha)-\lambda_m]}{[\lambda_p(1+2\alpha)+2\lambda_m]-\Phi[\lambda_p(1-\alpha)-\lambda_m]} \quad (1)$$

$$\alpha = R_{BD}\frac{\lambda_m}{a} \quad (2)$$

where λ_c, λ_p and λ_m are the thermal conductivity of the composite, filler and matrix respectively, Φ is the volume fraction of the filler, α is the shape factor (used to indicate the degree of influence of the interface thermal resistance on the thermal conductivity), a is the radius of the filler, R_{BD} is the interface thermal resistance. When the filler particles are larger, that is, when α is smaller, the interface thermal resistance has a smaller effect on the thermal conductivity, and vice versa. Therefore, the large Ag particles reduce the influence of the interface thermal resistance on the thermal conductivity, the prepared thermal grease shows a higher thermal conductivity (3.0 W m^{-1} K^{-1}). In addition, small Ag particles will leads to higher surface energy. Therefore, it is difficult for small Ag particles to be uniformly dispersed in the silicone oil and heat conduction pathways cannot be effectively formed in the thermal grease (Fig. 3b). On the contrary, large Ag particles can be uniformly mixed in silicone oil (Fig. 3c).

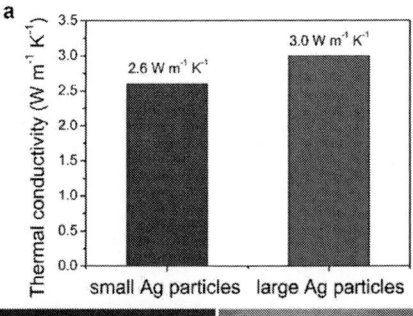

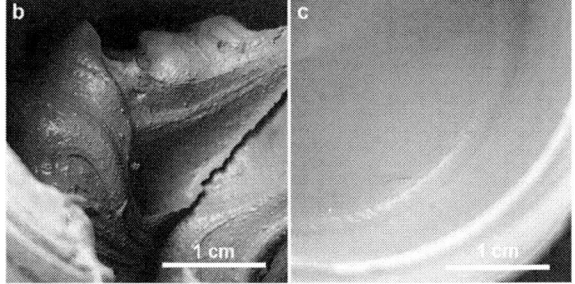

Fig. 3 Effect of Ag particles of different sizes on thermal conductivity and morphology of thermal grease. (a) Thermal conductivity comparison. (b) (c) Morphology comparison.

The large Ag particles are employed to prepare thermal grease. As depicted in Fig. 4a, the thermal conductivity of the prepared grease enhances with the increase of the addition of Ag, since heat conduction pathways are constructed between Al_2O_3 and ZnO with the addition of Ag particles (Fig. 4c and d). Due to the bridge effect of the Ag particles, the thermal conductivity of the prepared grease can be enhanced by more than 3% for every 1.5 wt% increase of Ag. However, add too many Ag particles to thermal grease is not favorable. The main reasons are as follows: (1) Ag is a precious metal, excessive addition will greatly increase the cost of silicone grease, which is not conducive to industrialization; (2) Density of the grease will increase greatly with the addition of a large number of Ag particles, which cannot meet the current miniaturization and light weight demands of electronic equipment. Based on the above analysis, a sample with a thermal conductivity of 3.0 W m^{-1} K^{-1} was selected for further discuss. The thermal resistance of the grease decreases with increasing pressure, since the contact between the filler and filler (test platform) become tightly with the increased pressure which reduces the thermal resistance of the whole system. Besides, the thickness of the thermal grease decreases with the enhanced pressure, thereby reducing the interface that the heat flow need to pass, leading to a reduction in its thermal resistance. As shown in Fig. 4b, the thermal resistance of the prepared grease is about 0.30 cm^2 K W^{-1}, indicating its great application potential in electronic packaging thermal interface materials.

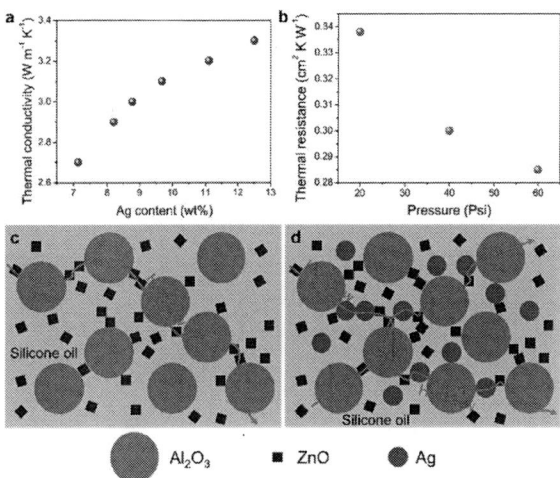

Fig. 4 Thermal properties of prepared thermal grease. (a) Thermal conductivity of prepared thermal grease. (b) Thermal resistance of prepared thermal grease. Schematic diagram of heat flow transmission of silicone grease without (c) and with (d) Ag particles.

The prepared thermal grease was subjected to variable shear rate (the shear rate increased from 0.1 s^{-1} to 10 s^{-1} at a log rate) viscosity tests at room temperature (25°C) (Fig. 5). The grease shows the phenomenon of "shear thinning", that is, the viscosity decreases with the shear rate increases, which can be explained by rubber-like liquid theory. The random clusters of molecular chains are intertwined to form a three-dimensional network, increasing the viscosity of polymer fluids. When the external force applied, the molecular chains in the network are oriented along the direction of the force, which always lead to disentanglement, deform or break of the network structure, thereby reducing the viscosity of the system. In this experiment, the viscosity value (322.7 Pa·s) when the shear rate is 5 s^{-1} is taken as the viscosity of the thermal grease. The satisfactory viscosity indicates that this thermal grease has commercial potential.

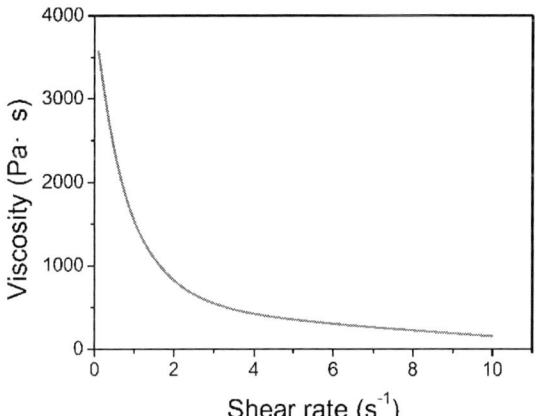

Fig. 5 Viscosity curve of thermal grease (3.0 W m^{-1} K^{-1}) at different shear rates.

IV. CONCLUSIONS

A thermal grease with optimized thermal conductivity (3.0 W m^{-1} K^{-1}) and viscosity (322.7 Pa·s) was prepared. The bridging effect of Ag particles makes the thermal grease have excellent thermal conductivity while ensuring its low viscosity. The satisfactory thermal conductivity and rheological properties of the prepared thermal grease indicate that it has great application potential in the thermal management of electronic packaging.

ACKNOWLEDGMENT

This work was supported by the National Natural Science Foundation of China (Grant No. 52073300), the Guangdong Basic and Applied Basic Research Fund (No. 2019A1515110845), Guangdong Province Key Field R&D Program Project (No. 2020B010190004), Youth Innovation Promotion Association of the Chinese Academy of Sciences (2019354), Shenzhen Science and Technology Research Fund (No. JCYJ20200109114401708, GJHZ20180420180909654 and JCYJ20180507182530279).

REFERENCES

[1] M. M. Waldrop, The chips are down for Moore's law, Nature, 2016, vol. 530, pp. 144-147.

[2] A. L. Moore and L. Shi, Emerging challenges and materials for thermal management of electronics, Materials Today, 2014, vol. 17, pp. 163-174.

[3] W. Dai, T. Ma, Q. Yan, J. Gao, X. Tan, L. Lv, et al, Metal-level thermally conductive yet soft graphene thermal interface materials, ACS Nano, 2019, vol. 13, pp. 11561-11571.

[4] Y. Yao, Z. Ye, F. Huang, X. Zeng, T. Zhang, T. Shang, et al, Achieving significant thermal conductivity enhancement via an ice-templated and sintered BN-SiC skeleton, ACS Appl Mater Interfaces, 2020, vol. 12, pp. 2892-2902.

[5] Cui W, Zhu Y, Yuan XY, H Zhou. Preparation and characterization of silicone grease with high thermal conductivity and high electrical insulation. Rare Metal Materials and Engineering 2011; vol. 40, pp. 443-446.

Sop welding joint bending stress finite element analysis and optimization

1st Gong Jinfeng
School of Electronic Mechanical Engineering
Guilin University of Electronic Technology
guilin, China
a5689735a @qq.com

2nd Huang Chunyue(Corresponding Author)
School of Electronic Mechanical Engineering,
Guilin University of Electronic Technology,
guilin, China
hcymail@163.com

3rd Li Maolin
School of Electronic Mechanical Engineering,
Guilin University of Electronic Technology,
guilin, China
178116415@qq.com

4st Liu Shoufu
School of Electronic Mechanical Engineering
Guilin University of Electronic Technology
guilin, China
1044577910 @qq.com

Abstract—This paper mainly uses ANSYS software to establish the finite element analysis model of SOP welded joints and analyze the stress distribution of SOP welded joints under bending load. A finite element model of the SOP device was developed for different solder joint materials and solder joint morphology parameters, and several sets of simulation data were obtained by model solving, and the results were analyzed based on the data. The results show that: Under bending load, the maximum stress occurs at the solder joints located at the four corners, and the maximum stress at the solder joints is located on the edges of the solder joints close to the chip side; The maximum stress in the solder joints of SOP devices is minimized when the solder joint material is 63Sn37Pb, all other conditions being equal; The optimal combination of optimum solder and solder joint structure parameters is solder 63Sn37Pb, solder joint height 0.085mm, solder joint length 1.74mm, and solder creep height 0.33mm. Theoretical guidance is provided for reducing the maximum bending stress in the solder joints of SOP devices under reduced bending load.

Keywords: SOP package; Solder joint; Orthogonal experiment; Finite element analysis

I. INTRODUCTION

In terms of the current international situation, the level of science and technology determines the international status and strength of a country，And the current level of technology is mainly reflected in the chip. The chip is the product of human intelligence in one, and the reliability of the chip determines whether the product can be stable and operate properly. The reliability of individual components in a product is very important, and the reliability of a component at 0.999 may seem very high. But the product consists of tens of thousands or even hundreds of thousands of components, if the reliability of each component is 0.999,[1] then the overall reliability is not simply 0.999，It will produce a qualitative change. Damage to one component can lead to unstable operation of the entire product, which can have catastrophic consequences. In the process of use, the solder joints are loaded with various stresses, and after cracks are generated, they will expand and finally break, resulting in chip failure and product failure. Solder joints are essential in electronic products, and there are so many of them that a problem with one of them can lead to product unreliability, so it is especially important to study the

reliability of solder joints in chip interconnects. SOP (Small Size Package) is a very common form of component packaging, mostly in plastic packages. The pin is L-type .

After the chip is soldered to the PCB (Printed Circuit Board) board, it will inevitably be subjected to bending loads under various circumstances during transportation and use. Bending stress can lead to a large stress concentration in the connection between the chip and the PCB - the solder joint, Solder joints may fail as a result. Many domestic and foreign experts solder joints for various reliability analyses such as Anli Quan on the SOP solder joints in the climbing height of the solder was studied：It shows that the increase in the climbing height of the solder is conducive to the reduction of the upper limit stress of the solder joint can greatly improve the reliability of the solder joint.[6] Chao Ren conducted a finite element analysis of the effect of the pin structure of SOP devices on device reliability：The best results are achieved with small bending radius and high pin height.[7] The above domestic and foreign scholars have studied the reliability of SOP solder joints and conducted finite element analysis on the morphology of solder joints, SOP components, respectively, so as to conclude that. It also shows that the SOP solder joints of components under bending load are not optimized in terms of solder joint shape and material selection. Therefore, in this paper, the parameters of the HCNW137 SOP chip manufactured by BROADCOM are used as the basis for modeling on ANSYS software. A bending load was also applied to the model to simulate the situation where the product receives a bending load in daily use. After modeling the different cases, it was observed that the maximum stress on the SOP solder joints occurs on the edge of the solder joint near the chip side. A single factor analysis was used to analyze the maximum stress magnitude of different solder joints with the same solder joint shape and the maximum stress of SOP devices to derive the optimal solder selection for the same solder joint shape. The optimal combination of solder joint shape and solder joint material is obtained by orthogonal experimental design analysis. It reduces the maximum bending stress of solder joints, improves the reliability of solder joints, and provides theoretical support for the material morphology selection of SOP solder joints.

1.The Science and Technology Major Project of Guangxi Province
（NO.AA19046004）.

2.This research is supported by the Science Foundation of Guangxi Zh uang AutonomousRegion Government (No.2019JJA160101)

3. Innovation Project of GUET Graduate Education
（2021YCXS009）

978-1-6654-1392-3/21 $31.00 © 2021 IEEE

II. FINITE ELEMENT ANALYSIS OF SOP SOLDER JOINTS

A. Finite element model of SOP device under bending load

SOP devices are deformed when subjected to bending loads, which usually causes deformation of the solder joints while the PCB is deformed, resulting in stress concentration in the solder joints. When the maximum bending stress at the stress concentration point exceeds the tolerable value of the welded joint, or when the process of loading and releasing the bending load repeatedly over a long period of time leads to fatigue of the welded joint, eventually leading to fatigue failure. [2]The SOP solder joint is cracked and the crack expands to form a fracture, resulting in the failure of the solder joint and making the product not work properly，In order to analyze the stress distribution of SOP solder joints when SOP devices are subjected to bending loads, the SOP solder joints were modeled using ANSYS software and their maximum stresses were analyzed by applying bending loads to the model. The SOP solder joint model is created as shown in Figure 1, The model is based on the dimensional parameters of the HCNW137 SOP chip manufactured by BROADCOM. PCB board size is 132mm*77mm*1mm, pcb board material is epoxy resin (FR-4), chip size is 11.23mm*9mm*5.1mm, The material is silicon, the chip has 8 L-shaped pins, and the pin chip pin material is Kova iron-nickel-cobalt alloy.[3] Solder joint material is SAC305 This paper establishes the body of the SOP chip and the type of cell used for the solder joint is SOLID185, and the PCB is SOLID45. As Table Ⅰ shows the material properties of various materials of the model.

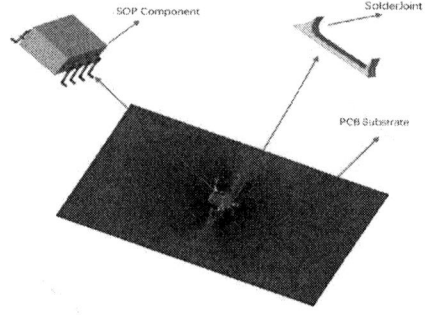

Fig. 1.Finite element model of sop device

TABLE I   MATERIAL PROPERTIES

Materi al	Heat transfer coefficien t W/(m2*k)	Thermal expansio n coefficien t 10- 6/°C	Elastic Modulu s (Gpa)	Poiss o n's ratio μ	Den sity kg/m 3	Specific heat capacity J/ (kg*°C)
si	84	2.6	130	0.28	2320	750
PCB	0.32	13.62	6.50	0.28	1870	1150
sac305	50	25	35.18	0.35	8410	192
kovar	390	18.9	115	0.32	8920	386

B. Bending load application method and bending stress distribution analysis of SOP devices

Fig .2 shows the applied load of the model, where the left and right ends of the PCB are fixed and supported, and a displacement load of 1 mm vertically downward is applied above the chip. After applying the load, the ends of the PCB are bent upward and the chip part is displaced downward. The simulation is that the chip is crushed during transportation or the chip is mistakenly touched during use to subject it to downward pressure, creating a bending load.

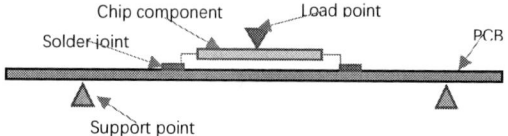

Fig. 2 Bending load applied

Fig. 3 shows the stress distribution of the solder joints of SOP devices under loaded bending stress conditions. The solder joints with maximum bending stress are located at the four corners of the eight pins, The maximum stress in the solder joint occurs at the side of the maximum bending stress solder joint near the chip.

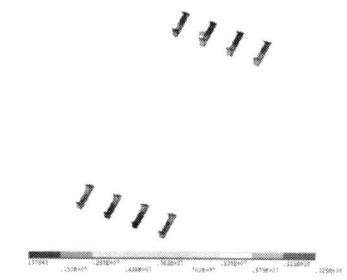

Fig. 3.SOP solder joint stress distribution

Based on the stress distribution diagrams of multiple models, it can be inferred that the stress concentration points of the welded joints do not change when the material of the welded joints and the morphology of the welded joints are changed. [4]This point is the most prone to cracks in the solder joints of SOP devices, where the cracks expand leading to fracture and failure of the solder joints. Therefore, it is important to analyze the bending gravitational force of the solder joint here.

III. BENDING STRESS ANALYSIS OF SOP SOLDER JOINTS BASED ON SINGLE FACTOR ANALYSIS

The selection of solder joint materials is diversified and various materials are available, and the reliability of solder joints can be affected by the solder joint material. Solder joints assume the role of connection, energy transfer and signal transmission. The properties of the material itself determine its impact on the reliability of the solder joint. In order to explore the optimal selection of solder joint materials under this SOP solder joint model, Single-factor analysis of SOP joints was conducted for several different solder joint materials commonly used in the market, and the corresponding models were established.[5] Here, five commonly used solder joint materials were selected for analysis as SAC305,SAC387, 63Sn37pb, 62Sn36Pb2Ag, Sn35Ag0.75Cu. The properties of the five materials are shown in Table Ⅱ.

TABLE II   SOLDER JOINT MATERIAL PROPERTIES

Material	Elastic Modulus (Gpa)	Poisson's ratio μ	Den sity kg/m³
SAC387	47.102	0.35	8400
63Sn37pb	31.395	0.35	8420
sac305	35.18	0.35	8410
62Sn36Pb2Ag	31.306	0.35	8400
Sn35Ag0.75Cu	49.8	0.35	8350

Solve the model separately according to the different material properties, The bending stresses in the welded joints corresponding to different materials are shown in Fig 4，It can be seen that a change in the material of the solder joint does not change the location of the maximum stress in the joint.

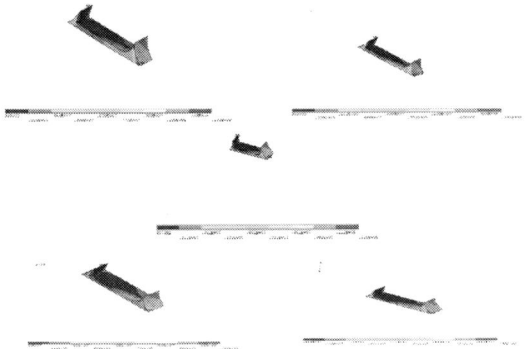

Fig. 4.Bending stress distribution in welded joints

Changing in the material of the solder joint as shown in Table 3 will change the maximum and minimum stress values of the joint, indicating that a change in solder joint material can affect the bending stress of a solder joint under bending load. Chip parameters, solder joint form unchanged, Solder joint material is 63Sn37pb when the maximum bending stress is 12.5Mpa minimum bending stress is 137818pa, It means that the change of material facilitates the reduction of maximum bending stress in the solder joint, The life of the solder joint can be optimized by optimizing the design of the solder joint material and selecting a more reasonable solder joint material.

TABLE III   BENDING STRESS OF SOLDER JOINTS

MATERIAL	SAC305	SAC387	63SN37PB	62SN36PB2AG	SN35AG0.75CU
MAXIMUM BENDING STRESS (MPA)	13.2	14.7	12.5	12.5	15.3
MINIMUM BENDING STRESS (PA)	138304	193799	137818	137843	167261

IV. BENDING STRESS ANALYSIS OF SOP SOLDER JOINTS BASED ON ORTHOGONAL EXPERIMENTS

Changing only the choice of material without considering other factors can reduce the maximum bending stress in the solder joint It is presumed that, while considering the solder joint morphology and solder joint material The maximum bending stress of the welded joint can be optimally combined. This leads to further optimization of the design of the solder joints of SOP devices, The maximum bending stress in the solder joints of SOP devices is further reduced and the reliability when subjected to bending loads is further improved.

A. Orthogonal experimental design

To test the conjecture, by changing the solder joint height, solder joint length, solder climb height, and solder joint material in the solder joint pattern, different groups are combined to model each of them, and perform operations on the model. Finally, the results are compared and the variance is calculated to arrive at the optimal combination.[8] A scheme in which all combinations are modeled would result in 81 sets of experiments The experiment is too cumbersome and not conducive to the experiment, an orthogonal experimental design was used to simplify the experiment, A total of 9 sets of experiments with 3 levels and 4 factors were selected as shown in Table Ⅳ to replace 81 sets of experiments to test the conjecture.

TABLE IV   ORTHOGONAL EXPERIMENT TABLE

GROUP	MATERIALS	SOLDER JOINT HEIGHT	SOLDER JOINT LENGTH	CLIMBING HEIGHT
1	SAC305	0.085	1.7	0.31
2	SAC305	0.087	1.72	0.33
3	SAC305	0.089	1.74	0.35
4	SAC387	0.087	1.74	0.31
5	SAC387	0.089	1.7	0.33
6	SAC387	0.085	1.72	0.35
7	63SN37PB	0.089	1.72	0.31
8	63SN37PB	0.085	1.74	0.33
9	63SN37PB	0.087	1.7	0.35

B. Orthogonal experiments of SOP devices

According to the orthogonal experiment schedule, solder joint length, joint climb height, and joint height as the key factors, and three different joint materials as the horizontal factors. Solder joint length corresponds to 1.70mm, 1.72mm, 1.74mm, the solder joint height corresponds to 0.085mm, 0.087mm, 0.089mm, and the solder creep height uses 0.31mm, 0.33mm, 0.35mm. The three solder joint materials correspond to SAC305, SAC387, 63Sn37Pb, According to the experimental plan of the orthogonal experimental design, 9 groups of models were built using ANSYS respectively, The corresponding maximum bending stress is shown in the experimental results of the visual analysis table in Table Ⅴ.

TABLE V   VISUAL ANALYSIS TABLE

FACTOR	MATERIAL	HEIGHT	HEIGHT	CLIMBING HEIGHT	LRESULTS
1	1	1	1	1	13.2
2	1	2	2	2	12.2
3	1	3	3	3	11
4	2	1	2	3	13.5
5	2	2	3	1	15.4
6	2	3	1	2	12.7
7	3	1	3	2	13.4
8	3	2	1	3	10.7
9	3	3	2	1	13
MEAN1	12.133	13.367	12.200	13.867	
MEAN2	13.867	12.767	12.900	12.767	
MEAN3	12.367	12.233	13.267	11.733	
RANGE	1.734	1.134	1.067	2.134	

According to the extreme difference analysis, the degree of influence on the experimental results are ranked as solder joint length > solder joint material > solder joint climbing height > solder joint height, At the same time can be analyzed from the table to get the minimum bending stress combination for 3213 corresponding to the specific combination of solder 63Sn37Pb, solder joint height 0.085mm, solder joint length 1.74mm, solder creep height 0.33mm solder joint maximum stress of 10.7Mpa. [9]The simultaneous optimization of the weld joint morphology can reduce the maximum bending stress of the weld joint more effectively than the optimization of the weld joint material alone indicates that the maximum bending stress of the weld joint can be reduced by simultaneously changing the weld joint morphology and the weld joint material under the same loading conditions.

V. CONCLUSION

This paper establishes a finite element simulation model of SOP devices on board-level circuits, focusing on the simulation and simulation of solder joints under loaded bending loads, and the maximum bending load of various solder joint materials under bending loads through the single factor analysis method, while for solder joint morphology in the height of the solder joint, the length of the solder joint, the height of the solder climbing with the solder joint material for orthogonal experimental analysis, so as to parameters were optimized and the following conclusions were obtained.

1）Other conditions are the same when the solder joint material is 63Sn37Pb, the maximum bending stress of the SOP device solder joint minimum 12.5Mpa. Solder joint material has a significant impact on the maximum stress of the joint, and attention should be paid to the selection of joint material in practical applications.

2）The optimal combination of the optimal solder and solder joint structure parameters obtained through orthogonal analysis solder 63Sn37Pb, solder joint height 0.085mm, solder joint length 1.74mm, solder creep height 0.33mm . Description of the impact of solder, solder joint height, solder joint climb height solder joint length on the reliability of the solder joint, the parameters of the solder joint need to be

optimized for use, can be very good to improve the reliability of the solder joint.

VI. REFERENCES

[1] Lau John H. State of the Art of Lead-Free Solder Joint Reliability[J]. J. Electron. Packag,2021,143(2).J. Clerk Maxwell, A Treatise on Electricity and Magnetism, 3rd ed., vol. 2. Oxford: Clarendon, 1892, pp.68–73.

[2] Li Xiaoming, Ren Kang, Jiao Chaofeng, Methods and practices for improving the reliability of BGA soldering[J]. Science and Technology Wind,2020(24):140. K. Elissa, "Title of paper if known," unpublished.

[3] Zhao XX. Thermal stress analysis and fatigue life prediction of solder joints for typical packaging devices[D]. Xi'an University of Electronic Science and Technology,2015.

[4] Xiong Gouji Huang Chunyue, Liang Ying, Li Tianming, Tang Wenliang, Huang Wei. Optimal design of silicon through-hole interconnect structure with random vibration based on orthogonal design and gray correlation[J]. Journal of Welding,2016,37(07):22- 26.

[5] Li Chunquan, Research on 3D automatic wiring method and reliability technology for automotive wiring harness. Guangxi Zhuang Autonomous Region, Guilin University of Electronic Science and Technology,2015-07-14.

[6] An Li-Quan, Zheng Jian-Ming. Small size package (SOP) device solder joint reliability research[J]. Aerospace Manufacturing Technology,2009(01):14-17.

[7] Ren Chao, Shao Jiang, Zeng Chenhui, Xue Heping. Analysis of the influence of SOP device pin structure on interconnect reliability[J]. Mechanical Strength,2016,38(03):591-595.

[8] Ren Chao, Zeng Chenhui, Shao Jiang, Wei Lai, Ding Jun. Random vibration reliability test and finite element analysis of SOP devices[J]. Semiconductor Technology,2014,39(09):714-718.

[9] Yiyang Jiang. Crack extension-based fatigue life prediction method and experimental study of solder joints[D]. Xi'an University of Electronic Science and Technology,2014

Comparative Analysis of Temperature-induced Micro-scale Deformation of Package by Experiment and Finite Element Analysis

Cheng Zhong
Shenzhen Institute of Advanced Electronic Materials, Shenzhen Institute of Advanced Technology, Chinese Academy of Science
Shenzhen 518055, China
cheng.zhong1@siat.ac.cn

Chenglong Li
Shenzhen Institute of Advanced Electronic Materials, Shenzhen Institute of Advanced Technology, Chinese Academy of Science School of Mechanical Science and Engineering, Huazhong University of Science and Technology
Shenzhen 518055, China
iclears@163.com

Tao Peng
Shenzhen Institute of Advanced Electronic Materials, Shenzhen Institute of Advanced Technology, Chinese Academy of Science
Shenzhen 518055, China
tao.peng1@siat.ac.cn

Yunxia Wang
Shenzhen Institute of Advanced Electronic Materials, Shenzhen Institute of Advanced Technology, Chinese Academy of Science
Shenzhen 518055, China
yx.wang2@siat.ac.cn

Gang Li
Shenzhen Institute of Advanced Electronic Materials, Shenzhen Institute of Advanced Technology, Chinese Academy of Science
Shenzhen 518055, China
gang.li@siat.ac.cn

Pengli Zhu
Shenzhen Institute of Advanced Electronic Materials, Shenzhen Institute of Advanced Technology, Chinese Academy of Science
Shenzhen 518055, China
pl.zhu@siat.ac.cn

Jibao Lu*
Shenzhen Institute of Advanced Electronic Materials, Shenzhen Institute of Advanced Technology, Chinese Academy of Science
Shenzhen 518055, China
jibao.lu@siat.ac.cn

Rong Sun*
Shenzhen Institute of Advanced Electronic Materials, Shenzhen Institute of Advanced Technology, Chinese Academy of Science
Shenzhen 518055, China
rong.sun@siat.ac.cn

Ching-Ping Wong
School of Materials Science and Engineering, Georgia Institute of Technology
Atlanta, GA 30332, USA
cpwong@cuhk.edu.hk

Abstract—Finite element analysis (FEA) is expected to play an important role in the evaluation of electronic packaging reliability. However, the rationality of the detailed parameter of material properties and boundary conditions in simulation requires sufficient verification to guarantee the accuracy. In-situ micro-scale experimental characterization could effectively check the simulation results and help to deepen the understanding of the root cause of the failure.

In this paper, we investigated the microscopic deformation of the cross section of one flip-chip package during heating process by using the in-situ two-dimensional digital image correlation (2D-DIC) combined with FEA. After a series of verification, the FEA results showed highly consistency with the 2D-DIC experiments. The anisotropy parameters of the substrate, which seldom attracts attention, was demonstrated to have significant effect on the deformation. Our work may help to improve the application of in-situ DIC and FEA in the stress analysis of package structures.

Keywords—two-dimensional digital image correlation, finite element analysis, anisotropy parameter

I. INTRODUCTION

With the rapid development of packaging technology and the continuous reduction of time for product development, it has been very difficult to meet the requirements of product development schedule based on traditional "trial and error" experimental method. In recent years, finite element analysis

(FEA) has been widely used to design or optimize material, structure and process in package, by failure analysis and reliability estimation [1,2]. However, thus "simulation first" approach highly depends on the rationality of the simulation algorithm and the accuracy of the simulation results. As many simplifications and assumptions used in simulation, as well as problems such as nonlinear properties of material, and difficulty in precise setting of boundary conditions, deviations may occur between the simulation results and the real conditions. If simulation results are not fully verified, it will even worsen the reliability of package on the contrary. Therefore, it is extremely important to continuously improve the accuracy of the simulation [1,2], which relies on the accurate characterization of the material properties, as well as deep understanding of the failure mode and mechanism in package [3].

High-precision in-situ observation technology may provide strong support to FEA. First of all, through direct in-situ observation, we have a deep understanding of the actual phenomenon that occurs during the verification test of the package structure. After that, there are higher expectations that the simulation results may accurately reflect phenomena in the real world. Understanding the root cause of failure based on in-situ observation in microscopic scale, has become one of the main challenges in microelectronics technology currently and still in future. Recently, optical non-contact full-field displacement characterization methods have developed

978-1-6654-1392-3/21 $31.00 © 2021 IEEE

rapidly, including digital image correlation (DIC), electronic speckle interferometry, moiré interferometry, holographic interferometry, etc. [4,5]. DIC aims to obtain the displacement and strain at surface of the object, by extracting digital images of object in different states and using regional differentiation and correlated algorithms for analysis. The devices needed for DIC are relatively simple, cheap and easy to implement. [4,5]. In addition to the commercialization of image-related algorithms, DIC has rapidly popularized in the past few years.

In particular, DIC has also shown great prospects in the research of package reliability [6-11], including material characterization [6], warpage evaluation [7,8], and microscopic displacement and strain characterization [9-11] in package. On the macroscopic scale, the warpage characterization by DIC could provide a basis for the accuracy of simulation results, then solder joints reliability were studied in simulation methods [8]. At present, three-dimensional (3D) DIC has been determined as one of the standard methods for measuring warpage of package by JEDEC (Joint Electron Device Engineering Council) [7]. On the microscopic scale, the two-dimensional (2D) DIC is considered to be an effective method for studying the local strain of underfill and solder joint [9,11]. Wang et al. record the full-field displacement of the chip-scale package (CSP) in the corner area of the chip during heating process, and provide effective guidance for the selection of underfill by analyzing the local strain using 2D DIC [9]. Van et al. further use 2D DIC to record the evolution of plastic strain in solder joints during multiple temperature cycles [11]. It is foreseeable that the combining of accurate in-situ experimental observation and numerical simulation could effectively help to enhance the package reliability.

In this study, in-situ 2D-DIC and FEA are combined to study the microscopic deformation at the cross-section of a flip-chip package structure during heating. After a series of verification, the FEA results showed a high degree of consistency with the 2D-DIC experiment, and further proved that the anisotropic parameters of the substrate have an important influence on the deformation, however, which is often ignored in the usual simulation. Our work helps to improve the application of in-situ DIC and FEA in the stress analysis of package structures.

II. EXPERIMENT AND SIMULATION METHODOLOGY

A. Sample Preparation

A typical flip-chip package is studied in this study, as shown in Fig 1. The chip is connected to the substrate by solder joints, while underfill is filled in the gap between the solder joints and fillet is formed on the sidewall of the chip. The dimensions of package obtained from actual measurements are listed in Table 1. The package is cut along the long side of the chip near the center of the first row of solder joints, and a high-quality cross section is obtained after fine grinding and polishing, as shown in Fig 1. After that, black and white speckles are made on the cross section by spraying for 2D DIC analysis.

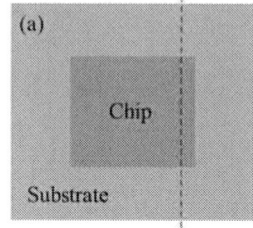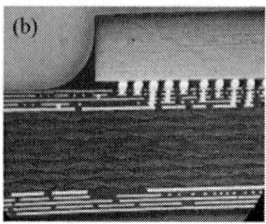

Fig. 1. (a) The package is cut along the long side of the chip, (b) a view of cross section in package.

TABLE 1. THE DIMENSIONS OF PACKAGE IN FEA.

	Length and Width (mm)	Thickness (mm)
Chip	12*1	0.775
Solder	R1=0.11, R2=0.084	0.68
Pad	r1=0.11, r2=0.061, r3=0.08	0.012
Substrate	36*1	1.24

B. Experimental Settings

The cross section of package is placed in laser confocal microscope for observation, where the chip corner areas are moved to the center of the microscope. Through parameter adjustment, clear grayscale images of 2D cross-section of package with speckles are obtained. After verifying the accuracy of the temperature control module, the sample was heated from room temperature (approximately 25 °C) to 125 °C with a rate of nearly 20 °C/min. The 2D cross-sectional grayscale images of the package at different temperatures were continuously recorded during heating.

C. Simulation Settings

The model in simulation is constructed according to the actual structure of the package, as shown in Fig 2. For the convenience of analysis, the materials are set as the linear elastic parameters initially, the values of which are obtained through actual experimental measurement. The specific material parameters are listed in Table 2.

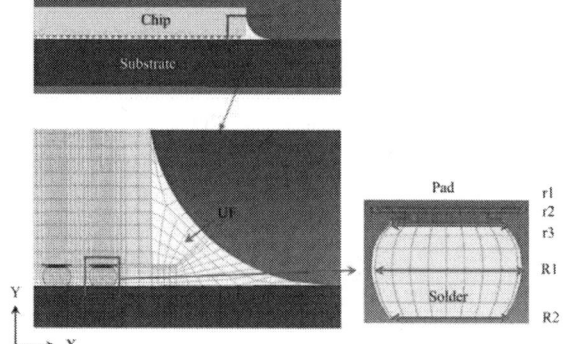

Fig. 2. The model in FEA.

978-1-6654-1392-3/21 $31.00 © 2021 IEEE

TABLE 2. THE SPECIFIC MATERIAL PARAMETERS IN FEA.

Material	CTE (1/°C)	Young's Modulus (MPa)	Poisson's Ratio	Density (Kg/m³)
Silicon	2.8e-6	131000	0.3	2.3e-9
Solder	2.4e-6	T=25°C, 39000 T=200°C, 35000 T=220°C, 100	0.3	1.2e-8
Copper	1.77e-5	117000	0.35	8.93e-9
Underfill	T=25°C, 2.95e-5 T=106°C, 2.95e-5 T=107°C, 9.09e-5 T=260°C, 9.09e-5 (Tg=107°C)	8100	0.3	1.52e-9
Substrate	1.77e-5	20000	0.3	1.9e-9

III. EXPERIMENT AND SIMULATION RESULTS.

After recording the speckled digital images at 25°C and 125°C, the software provided by Correlated Solutions Inc. (CSI) is used to compare and analyze the digital speckles in the image to obtain the relative deformation in the cross section of package [12]. Fig 3(a) and 3(b) show the results of displacement in the X and Y directions by 2D-DIC experiment, respectively. As shown in Fig 2, the horizontal direction is defined as X and the vertical direction is defined as Y. The line in fig 3(a) and 3(b) approximately represents the iso-deformation line, i.e., the value of relative displacement in the position of same line are equal. It shows that the area near the corners of the chip could be clearly divided into different regions by slope of iso-deformation line in X-direction, as shown in Fig 3(a). This phenomenon agrees with the different coefficient of thermal expansion (CTE) and Young's modulus of underfill, chip and substrate. The relative displacement in Y-direction is also similarly divided into different regions, as shown in Fig 3(b).

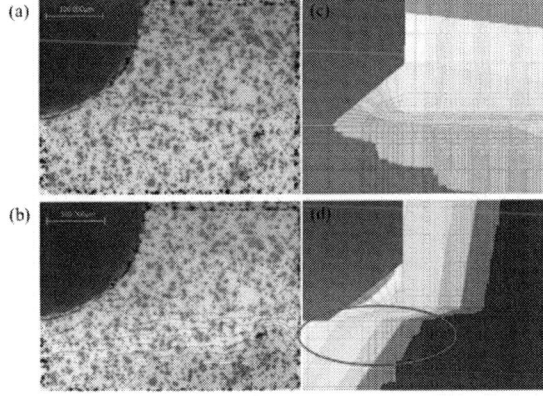

Fig. 3. (a) The displacement (X) of 2D-DIC, (b) the displacement (Y) of 2D-DIC, (c) the displacement (X) of FEA in initial structure, (d) the displacement (Y) of FEA in initial structure.

Similar to the experiment, the temperature in simulation is also set as 25 °C initially, then increased to 125 °C. The simulation results of the displacement in the X and Y directions are shown in Fig 3(c) and 3(d) respectively. Compared with the experimental results, it is found that the distribution of displacement in the X direction are highly similar in simulation, mainly due to the well-characterized input material properties. It implies once again that FEA is a powerful tool for studying the deformation of the package. However, it could also be observed that the slope of iso-

deformation line of the simulation results in the Y direction is quite different from the experimental results, in the position marked by red ellipse in Fig 3(d).

In conjunction with the cross-sectional observation of package structure in Fig 1(b), it is found that the substrate actually has an obvious multi-layer structure, with thinner circuit layers on both sides and a thicker core layer in the middle. However, it is often ignored in most of simulation models. According to the actual multi-layer structure of the substrate, we reconstruct the simulation model with more comprehensive structure of substrate and also increase the temperature from 25 °C to 125 °C. The detailed material parameters of substrate with multi-layer structure are listed in Table 3. The results of relative displacement in new model at 125 °C are shown in Fig 4, and it is clear that the distribution of displacement in simulation corresponds exactly to the experimental results in both X and Y directions. It is demonstrated that the internal fine structure of substrate and the anisotropic properties of materials play a very critical role in package deformation.

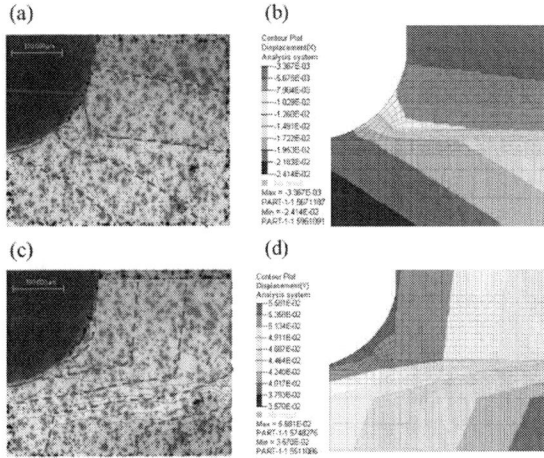

Fig. 4. (a) The displacement (X) of 2D-DIC, (b) the displacement (X) of FEA in reconstructed structure, (c) the displacement (Y) of 2D-DIC, (d) the displacement (Y) of FEA in reconstructed structure.

TABLE 3. DETAILED MATERIAL PARAMETERS OF SUBSTRATE WITH MULTI-LAYER STRUCTURE

Multi-Layer Substrate	CTE (1/°C)	Young's Modulus (MPa)	Poisson's Ratio	Density (Kg/m³)	Thickness (mm)
Upper Circuit Layer	X, 1.77e-5 Y, 9e-5 Z, 1.77e-5	60000	0.3	8.93e-9	0.21
Lower Circuit Layer	X, 1.77e-5 Y, 9e-5 Z, 1.77e-5	60000	0.3	8.93e-9	0.21
Core Layer	1.3e-5	20000	0.3	1.9e-9	0.82

IV. ANALYSIS AND DISCUSSION

The substrate is actually an anisotropic structure, i.e., big differences exist in both Young's modulus and CTE between the value in the thickness direction (ie, the Y direction in Fig

2) and in the plane (ie, the X direction in Fig 2). For example, some studies have mentioned anisotropic material parameters of substrate or printed circuit boards (PCB) [13]. However, though the above content is well known, it is still difficult and complicated to well characterize the internal structure of the substrate without damage the substrate. The accurate thermomechanical parameters of each material in multilayer structure are also very difficult to be obtained by conventional tests of dynamic mechanical analysis (DMA) or thermal mechanical analyzer (TMA). Therefore, in most of simulations, substrate or board-level PCB are usually treated as isotropic materials, which ignore not only the characteristics of the multilayer structure, but also the inconsistency of material properties in different area.

This study provides a solution to accurately characterize the fine material parameters in the thickness direction of substrate. In terms of micro-scale deformation, it might obtain simulated results that are very similar to experiments, as shown in Fig 4. Through the combination of experiment and simulation, as well as iterative verification of the inverse method, it is expected to obtain fine material parameters in the thickness direction of substrate. Moreover, the method that combine experiment and simulation to determine several unknown material properties could also be applicable to other types of scenes which are difficult to analyze using conventional characterization methods, and is worthy of further research.

V. Conclusion

1) In-situ 2D-DIC were used to investigate the microscopic deformation of the cross section of one flip-chip package during the heating process.

2) After a series of verification, FEA results could be highly consistent with the DIC experiments.

3) The anisotropy parameters of the substrate are obtained, which were demonstrated to have significant influence on the deformation.

Acknowledgment

The work was supported by the National Key R&D Program of China (2020YFB0408700), National Natural Science Foundation of China (no. 52003289), the Youth Innovation Promotion Association CAS (no. 2021363), and the SIAT Innovation Program for Excellent Young Researchers (no. 201803).

References

[1] G. Q. Zhang, W. D. Van Driel, and X. J. Fan, "Mechanics of Microelectronics." Springer, 2006.

[2] S. Liu, and Y. Liu, "Modeling and Simulation for Microelectronic Packaging Assembly - Manufacturing, Reliability and Testing ." John Wiley & Sons, 2011.

[3] J. Lau, M. Li, L. Yang, M. Li, Q. X.Yong, Z. Cheng, and R. Lee, "Reliability of fan-out wafer-level packaging with large chips and multiple re-distributed layers." 2018 IEEE 68th Electronic Components and Technology Conference (ECTC). IEEE, 2018, pp. 1574-1582

[4] B. Pan, "Digital image correlation for surface deformation measurement: historical developments, recent advances and future goals." Measurement Science and Technology, 2018, Vol. 29(8): 82001.

[5] M. A. Sutton, J. O. Jean, and S. Hubert, "Image correlation for shape, motion and deformation measurements: basic concepts, theory and applications. " Springer Science & Business Media, 2009.

[6] S. Park, R. Dhakal, L. Lehman, and E. Cotts, "Measurement of deformations in SnAgCu solder interconnects under in situ thermal loading." Acta materialia, 2007, pp. 3253-3260.

[7] J. E. D. E. C. Standard, "Package warpage measurement of surface-mount integrated circuits at elevated temperature." JESD22-B112 , October , 2009.

[8] S. Shao, Y. Niu, J. Wang, R. Liu, S. Park, H. Lee, and L. Yip, "Comprehensive study on 2.5 D package design for board-level reliability in thermal cycling and power cycling." 2018 IEEE 68th Electronic Components and Technology Conference (ECTC), 2018, pp. 1668-1675.

[9] H. Wang, S. Shao, V. Pham, P. Shang, C. Zhong, and S. Park, "Quantification of Underfill Influence to Chip Packaging Interactions of WLCSP." ASME 2018 International Technical Conference and Exhibition on Packaging and Integration of Electronic and Photonic Microsystems, 2018, pp. 1-7.

[10] V. L. Pham, Y. Niu, J. Wang, H. Wang, C. Singh, S. Park, and S. Shao, "Experimentally minimizing the gap distance between extra tall packages and PCB using the digital image correlation (DIC) method." 2018 IEEE 68th Electronic Components and Technology Conference (ECTC), 2018, pp. 1593-1599.

[11] V. L. Pham, J. Xu, K. Pan, J. Wang, S. Park, C. Singh, and H. Wang, "Investigation of underfilling BGAs packages–Thermal fatigue." 2020 IEEE 70th Electronic Components and Technology Conference (ECTC). IEEE, 2020, pp. 2252-2258.

[12] https://www.correlatedsolutions.com

[13] T. T. Nguyen, D. Lee, J. B. Kwak, and S. Park, "Effect of glue on reliability of flip chip BGA packages under thermal cycling." Microelectronics Reliability 50.7 , 2010, pp. 1000-1006.

Active Heat Dissipation By Chip on Thermoelectric Cooler for High-Power LED

Shuang Li
School of Mechanical Science and Engineering
Huazhong University of Science & Technology
Wuhan, China
lishuang@hust.edu.cn

Jinglong Liu
School of Mechanical Science and Engineering
Huazhong University of Science & Technology
Wuhan, China
m201970562@hust.edu.cn

Yang Peng*
School of Aerospace Engineering
Huazhong University of Science & Technology
Wuhan, China
ypeng@hust.edu.cn

Mingxiang Chen*
School of Mechanical Science and Engineering
Huazhong University of Science & Technology
Wuhan, China
chimish@hust.edu.cn

Abstract—**In this paper, we put forward an active heat dissipation by chip on thermoelectric cooler (chip-on-TEC) packaging structure for high-power light-emitting diode (LED). The chip-on-TEC packaging structure was produced by bonding the LED chips on the metal layer on the back side of cold side substrate of the TEC. On account of the Peltier effect of thermoelectric particles and the characteristics of low heat resistance of the packaging device, the heat produced by the chips is able to be absorbed by the cold side of the TEC and distributed into the air availably. The temperature in working of the packaging device when chip currents are 0.2A to 1.0A was researched by the thermal imitation research and package test minutely. The working temperature in TEC-operative of chip-on-TEC packaging structure is indeed decreased in each chip current state in comparison to that in TEC-inoperative. The chip-on-TEC packaging device decreases the temperature of LED chips from 232°C to 114°C when the chip current is 1.0A. Furthermore, the light output power of LED chips is greatly enhanced under various chip currents. The chip-on-TEC packaging structure is able to advance the light intensity effectively, and the light saturated point of the LED chips is increased from 0.8A to 1.0A because of the TEC. The study testify that the chip-on-TEC packaging device is a very feasible active heat dissipation approach for high-power LED.**

Keywords—*chip-on-TEC, heat dissipation, high-power LED, working temperature, light output power*

I. INTRODUCTION

By reason of the merit of energy conservation, environmental protection, long service life, and cramped construction [1], light-emitting diode (LED) have broad application prospects in architectural landscape lighting, display backlighting, aviation lighting, medical services, display screen, and general illumination [2]. The heat generated by LED chips when working is transmitted to a whole device, and plenty of heat passively impacts the service behavior of LED chips in the long term. Moreover, the heat leads LED to the reduction in reliability and lifetime. There is a phenomenon in the working process of LED chips called redshift which is known as that the light intensity would reduces 1% and the light wavelength transforms 0.2nm to 0.3nm when the temperature of LED rises 1°C [3].

therefore, it is necessary to solve the heat dissipation problem for the development of high-power LED.

Nowadays, there are prevailingly two heat dissipation methods applied in the high-power LED. One is passive heat dissipation, such as thermal conductivity substrates, heat pipe radiator, flip chip bonding, and fin heat sink [4], which depend on the thermal conductivity of materials and devices. However, the passive heat dissipation can't meet the working temperature requirements of the high-power LED. The other heat dissipation method is active heat dissipation which is dependent on additional power operation [5]. Ordinarily, active heat dissipation means are united with fins to dissipate the heat generated by high-power LED to the surrounding environment and can decrease the working temperature of high-power LED availably, then improve the performance stability. Forced air cooling [6], piezoelectric fan, liquid cooling [7], vapor chamber [8], and thermoelectric (TEC) cooling are included in common active heat dissipation approaches. Among these methods, TEC stands out due to its advantages of cleanliness, noiselessness, and high reliability. The TEC is based on the Peltier effect that the two joints of different conductors can appear the case of heat absorption and heat release respectively while the direct current flows through the circuit formed of two different semiconductors to realize the aim of cooling [9]. A package structure of silicon-based TEC was put forward as an active heat dissipation method for high-power LED [10]. Li et al. [11] presented a thermal management based on TEC with a fan and it is concluded the effect that the working temperature of the substrate of LED decreases by 17°C due to the TEC. Micro-thermoelectric cooler can enhance the luminous efficiency of the high-power LED by 12.3% and increase the lifetime by 50% [12]. Nevertheless, In the existing TEC-containing heat dissipation method for LED, the LED chips are individually bonded on a substrate, and the substrate is connected to the cold side substrate of the TEC through thermally conductive silicone grease. The silicone grease increase the thermal resistance of package structure on account of its low thermal conductivity. More notably, the thermally conductive silicone grease would influence the lifetime and reliability of LED as silicone grease age thermally. The size of existing

978-1-6654-1392-3/21 $31.00 © 2021 IEEE

TEC-containing heat dissipation devices is also an issue to be considered. These issues are detrimental to the application of TEC in heat dissipation for high-power LED.

In this work, an active heat dissipation by chip on thermoelectric cooler (chip-on-TEC) was put forward for high-power LED. The LED chips and the TEC are on the same substrate since the chips were bonded on metal layer of cold side directly, hence the thermal resistance of the structure decrease. Thanks to the Peltier effect of thermoelectric particles and the low heat resistance of the packaging device, the heat produced from the LED chips would be dispersed efficaciously and actively. The basic performance of the TEC was tested, then the temperature of the packaging structure in working was detailedly researched through emulation and experiment. Furthermore, the optical performance of the LED chips, for instance, light intensity and light output power were tested in TEC-operative and TEC-inoperative.

II. EXPERIMENTS

The Al_2O_3 ceramic substrates with a galvanized copper layer manufactured through direct plated copper (DPC) technique were used for the hot and cold sides of the TEC. The blue LED chips were provided by Shenzhen JingXing Tech. Inc. The P-type thermoelectric particles ($Bi_{0.5}Sb_{1.5}Te_3$) and N-type thermoelectric particles ($Bi_2Te_{2.7}Se_{0.3}$) with the size of $1 \times 1 \times 2$ mm^3 were purchased from Changshan Wangu Electronic Tech. Inc. Fig. 1 shows the preparation procedure of the chip-on-TEC packaging structure. The thermoelectric particles were put in a rubber mould in a staggered order and set them aside for later use. The $Sn_{96.5}Ag_3Cu_{0.5}$ (SAC305) solder paste is brushed on the copper-plated ceramic substrate that was the hot side of the TEC by screen printing. The ceramic substrate brushed with solder paste was aligned with the thermoelectric particles in the rubber mold and then pressure-welded at 350°C for 10s. Two wires are soldered on the remaining copper layer of the hot side substrate of the TEC. The cold side of the TEC was screen-printed SAC305 solder paste and aligned with the soldered particles on hot side substrate, and then pressure-welded at 350°C for 10s. The TEC was finished. The back side of the cold side substrate was electroplated a metal layer for soldering LED chips. $Sn_{48}Bi_{52}$ solder was coated to the four square positions of the metal layer of the cold side and the blue LED chips were placed on the metal layer in order, and then the whole TEC with the four LED chips was placed in a reflow oven for reflow soldering. The input currents of the TEC was adjusted from 0.5A to 2.5A, and the temperature of the cold and hot sides of the TEC was measured with a handheld thermal imager in each input current state, and the temperature difference of the TEC was obtained. Then the temperature of the LED chips was obtained in TEC-operative and in TEC-inoperative when the currents of the LED chips are from 0.2A to 1.0A. The various optical properties of the LED chips, such as light intensity and light output power at currents of 0.2A to 1.0A were tested through an integrating sphere.

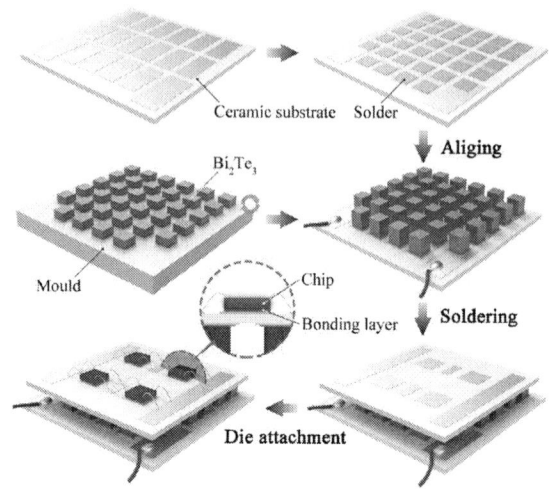

Fig. 1. preparation procedure of the chip-on-TEC packaging structure.

III. RESULTS AND DISCUSSION

Fig. 2(a) displays the hot side of the TEC with the overall dimensions of $11.5 \times 11.5 \times 0.5 mm^3$, 17 pairs of P-type an N-type thermoelectric particles, and two boded wires. The TEC and the cold side substrate with metal layer are shown in Fig. 2(b). The LED chips form a pathway by gold wires and soldered on the metal layer of the cold side, which is the chip-on-TEC as shown in Fig. 2(c). The picture of the blue LED chips in working of the chip-on-TEC packaging structure is displayed in Fig. 2(d).

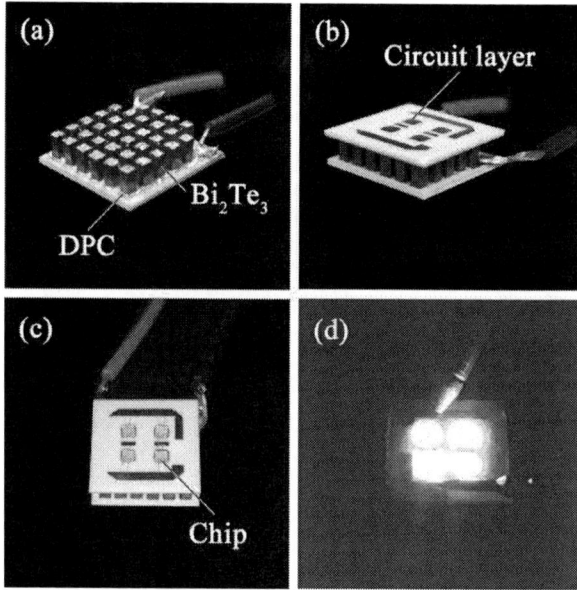

Fig. 2. Pictures of (a) hot side of the TEC, 17 pairs of thermoelectric particles, and two boded wires, (b) TEC and cold side substrate with metal layer, (c) chip-on-TEC, and (d) blue LED chips in working of the chip-on-TEC.

Fig. 3 (a) exhibits the temperature difference between the cold and the hot sides and the voltage values of the TEC under various currents. According to Ohm's law, the voltage is proportional to the current with a certain resistance. It is observed from the Fig. 3(a) that the voltage and current of the TEC are basically linear, which is in line with the theory. The temperature difference of the TEC increases with the increase of the input currents, it can be interpreted as the cooling capacity of the TEC is positively related to the currents of TEC. However, the temperature difference between two sides of the TEC no more increases even slightly decreases at the input current of 2.5A, because the heat generated by the conductor due to the Joule effect is greater than the heat absorbed by the TEC due to the Peltier effect when the current rises to a certain level. Fig. 3 (b)-(e) respectively show the pictures derived from infrared thermal imager of the TEC at the input currents of 0.5A, 1.0A, 1.5A, and 2.0A, in which the lowest temperature and the highest temperature are the temperature of cold side and hot side of the TEC respectively.

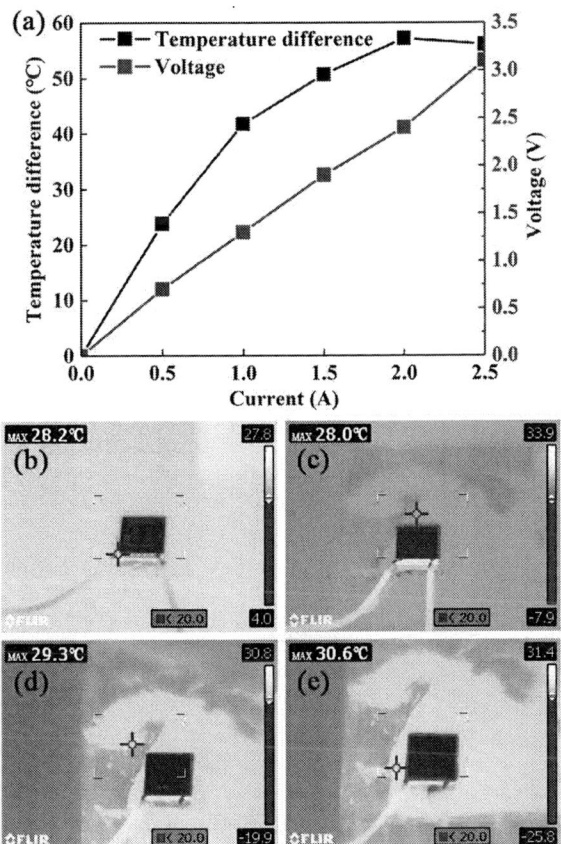

Fig. 3. (a) the temperature difference and the voltage values of the TEC, (b)-(e) pictures derived from infrared thermal image.

The chip-on-TEC packaging structure is emulated by COMSOL Multiphysics software before the experiment of the LED chips, which can predict the experimental results and analyze the influence of TEC on the temperature of the LED chips. Attention of this model is focused on the temperature performances of the LED chips, hence Only the temperature results of the model is discussed. Fig. 4 show the the simulated temperature results of the chip-on-TEC packaging structure in TEC-inoperative, and the highest temperature of the packaging structure is concentrated on the chips. The temperature of the chips is 52.2°C, 91.14°C, 135°C, 184°C, and 244°C when the currents are 0.2A to 1.0A is exhibited in Fig. 4 (a)-(e) respectively. Obviously, the temperature of chips increases with the increase of the

currents, because the increase of input power causes the heat generated by the chips to increase. Fig. 4 (f) is the temperature distribution in cross section of the chip-on-TEC packaging structure, it can be seen that the temperature of the packaging structure from the chip to the TEC is evenly distributed.

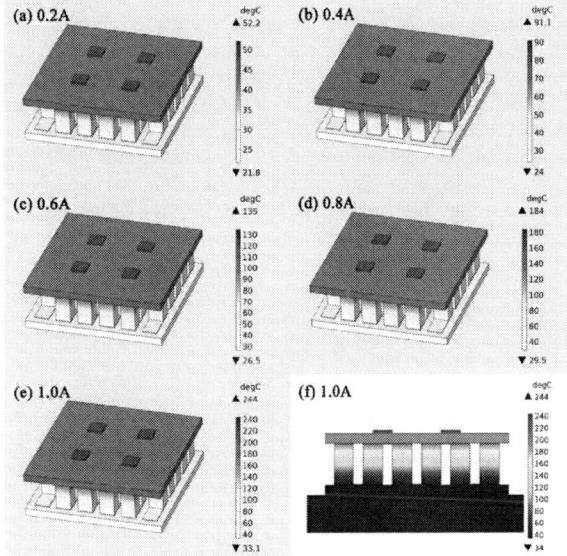

Fig. 4. Simulated temperature results of the packaging structure in TEC-inoperative, (a)-(e) temperature distribution and (f) temperature distribution in cross section.

Fig.5 show the the simulated temperature results of the packaging structure in TEC-operative at the input current of TEC of 1.5A. Due to the cooling of the TEC, the highest temperature of packaging structure doesn't appear on the LED chips at the chip currents of 0.2A and 0.4A, as shown in Fig.5(a) and (b). It can be ascribed by the reason that the heat assimilated by the TEC neutralizes heat generated by chips. The temperature of the chips is the highest temperature in the simulation results at the chip currents of 0.6A to 1.0A, and the chips temperature is 59.8°C, 101°C, and 150°C, respectively. The temperature of chips is reduced by 75.2°C, 83°C, 94°C compared with the above results that in TEC-inoperative. The simulation results show that TEC can play a positive role in the heat dissipation of high-power LED.

The temperature of the chips in TEC-inoperative and TEC-operative (the input current of TEC is 1.5A) at the chip currents of 0.2A to 1.0A is shown in Fig. 6 (a). The experimental results show that the temperature of chips is 48.1°C, 79.4°C, 113.3°C, 164.3°C, and 231.7°C respectively in TEC-inoperative when the currents are 0.2A to 1.0A. And the temperature of chips is 5.27°C, 28.63°C, 47.03°C, 80.37°C, and 114°C, respectively in TEC-operative at the chip currents of 0.2A to 1.0A. Fig. 6(a) exhibits that the temperature of LED chips in TEC-operative is significantly lower than that in TEC-inoperative under various chip currents. The experimental results show that the TEC is able to decrease the temperature of high-power LED effectively. The light output power of the LED chips in TEC-inoperative and TEC-operative (input current of TEC is 1.5A) at the chip currents of 0.2A to 1.0A, as exhibited in Fig. 6(b). The light output power of the LED chips increases as chip currents increase, and the light output power of the LED chips in

TEC-operative is observably bigger compared to that in TEC-inoperative in each current state. The light output power of the LED chips is significantly increased because of the cooling of TEC. Fig. 6(c) exhibits the increase rate of light output power of LED. The increase rate of LED chips is 9.04%, 13.75%, 21.84%, 32.79%, and 35.25% respectively when the chip currents are 0.2A to 1.0A. The increase rate enhances as the chip currents increase, and the increase rate reaches 35.25% when the chip current is 1.0A. The light intensity and wavelength of the chips in TEC-operative and TEC-inoperative when the chip current is 1.0A are shown in Fig. 6(d). The light intensity of the chips in TEC-operative is certainly bigger than that in TEC-inoperative, and the LED chips exist a red shift phenomenon due to the high temperature in TEC-inoperative. It is inferred from Fig. 6, chip-on-TEC packaging structure can afford an very efficient heat dissipation approach for high-power LED.

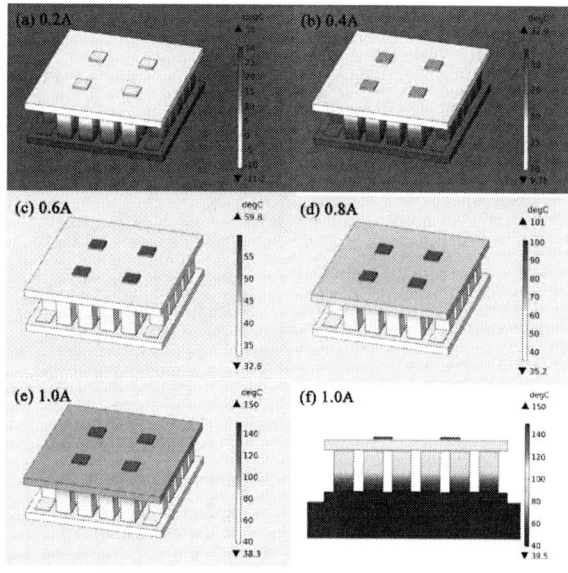

Fig. 5. Simulated temperature results of the packaging structure in TEC-operative, (a)-(e) temperature distribution and (f) temperature distribution in cross section.

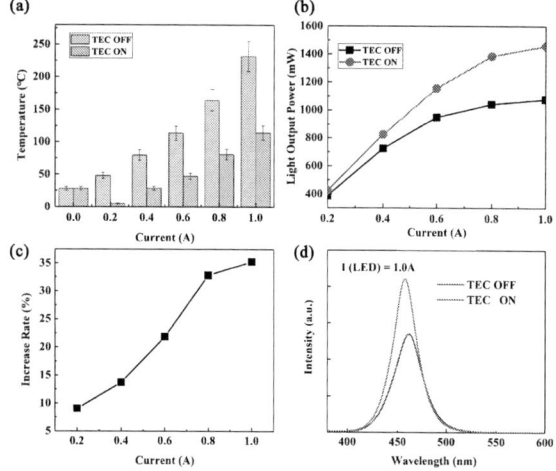

Fig. 6. (a) The temperature of LED chips in TEC-operative and TEC-inoperative, (b) light output power of the chips in TEC-operative and TEC-inoperative, (c) increase rate of LED chips. (d) light intensity and wavelength of the chips.

Fig. 7 displays the light intensity of LED chips in TEC-operative and TEC-inoperative at chip currents of 0.2A to 1.0A. Fig. 7(a) displays that the light intensity in TEC-inoperative improves with currents when the currents are 0.2A to 0.6A, but the light intensity initiates to weaken at currents of 0.8A and 1.0A. Fig. 7(b) exhibits the light intensity of the chips in TEC-operative enhances at the chip currents of 0.2A to 0.8A, but the light intensity of the LED chips initiates to weaken at chip current of 1.0A. It is can be seen from the results that packaging structure certainly advance the light saturated point of the LED chips from 0.8A to 1.0A. The chip-on-TEC packaging structure can indeed advance the optical performance of high-power LED.

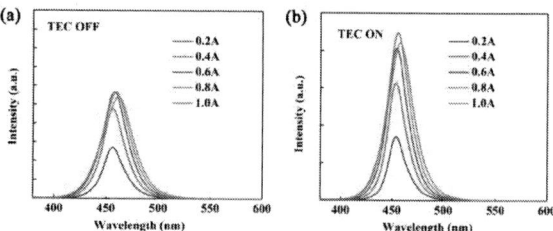

Fig. 7. Light intensity of LED chips in TEC-inoperative (a) and TEC-operative (b).

IV. Conclusions

The chip-on-TEC packaging structure could availably enhance the optical property of LED chips. The packaging structure is expected to become very feasible active heat dissipation approach of high-power LED.

1) The temperature difference of the TEC increases with increase of the input currents at the input currents of 0.5A to 2.0A.

2) The temperature emulation results testify that the temperature of LED chips in working rises from 52.2°C to 244°C when the chip currents are 0.2A to 1.0A in TEC-inoperative. The temperature of LED in working decreases significantly in TEC-operative (the input current of TEC is 1.5A).

3) The experiment demonstrate that the temperature of LED chips in working rises from 48°C to 232°C in TEC-inoperative at chip currents of 0.2A to 1.0A. the temperature of LED chips in working is enhances from 5°C to 114°C in TEC-operative.

4) The light output power of the chips in TEC-operative is indeed bigger than that in TEC-inoperative when the current is 1.0A.

References

[1] Y. Peng, J. Liu, Y. Mou, M. Chen, X Luo, "Heat Dissipation Enhancement of Phosphor-Converted White Laser Diodes by Thermally Self-Managing Phosphor-in-Glass," IEEE Trans Electron Devices. El., vol. 67, no. 10, pp. 4288–4292, Oct 2020.

[2] J. Wang, X. Zhao, Y. Cai, C. Zhang, and W. Bao, "Experimental study on the thermal management of high-power LED headlight cooling device integrated with thermoelectric cooler package," Energy conversion And Management, vol. 101, pp. 532–540, Sep. 2015.

[3] N. Narendran and Y. Gu, "Life of LED-based white light sources," J Disp Technol, vol. 1, no. 1, p. 167, Sep. 2005.

[4] J. Li, L. Han, J. Duan, and J. Zhong, "Microstructural characteristics of Au/Al bonded interfaces," Mater Charact, vol. 58, no. 2, pp. 103–107, Feb. 2007.

[5] V.A.F. Costa and A.M.G. Lopes, "Improved radial heat sink for led lamp cooling," Appl Therm Eng, vol. 70, no. 1, pp. 131–138, Sep. 2014.

[6] T. Açikalin, S.V. Garimella, J. Petroski, and A. Raman, "Optimal design of miniature piezoelectric fans for cooling light emitting diodes," ITherm, vol. 1, pp. 663–671, Jul. 2004.

[7] L. Yan, N. Cordero, F. Barthel, F. Tebbe, J. Kuhn, R. Apfelbeck, and D. Würtenberger, "Liquid Cooling of Bright LEDs for Automotive Applications." Appl Therm Eng, vol. 29, no. 5, pp. 1239–1244, Vol. 2009.

[8] Y. Tang, L. Lin, S. Zhang, J. Zeng, K. Tang, G. Chen, and W. Yuan, "Thermal management of high-power LEDs based on integrated heat sink with vapor chamber," Energ convers Manage, vol. 151, pp. 1–10, Nov. 2017.

[9] B.J. Huang, C.J. Chin, and C.L. Duang, "A design method of thermoelectric cooler," Int J Refrig, vol. 23, no. 3, pp. 208–218, May. 2000.

[10] J.H. Cheng, C.K. Liu, Y. L. Chao, and R.M. Tain, "Cooling performance of silicon-based thermoelectric device on high power LED," I C T, pp. 53–56, Jul. 200.

[11] J. Li, B. Ma, R. Wang, and L. Han, "Study on a cooling system based on thermoelectric cooler for thermal management of high-power LEDs," Microelectron Reliab, vol. 51, no. 12, pp. 2210–2215, Dec. 2011.

[12] D. Sun, G. Liu, L. Shen, H. Chen, Y. Yao, and S. Jin, "Modeling of High Power Light-Emitting Diode Package Integrated with Micro-Thermoelectric Cooler under Various Interfacial and Size Effects," IEEE Trans. Energy Convers. vol. 179, pp. 81–90, Jan. 2019.

978-1-6654-1392-3/21 $31.00 © 2021 IEEE

Tuning the Curing Temperature of Polyimide Precursor: Ploy Amide Ester

1st Kuangyu Wang
Shenzhen Institute of Advanced Electronic Materials, Shenzhen Fundamental Research Institutions Shenzhen Institutes of Advanced Technology, Chinese Academy of Sciences
Shenzhen, China
ky.wang@siat.ac.cn

2nd Liang Shan
Shenzhen Institute of Advanced Electronic Materials, Shenzhen Fundamental Research Institutions Shenzhen Institutes of Advanced Technology, Chinese Academy of Sciences
Shenzhen, China
liang.shan@siat.ac.cn

3rd Guoping Zhang
Shenzhen Institute of Advanced Electronic Materials, Shenzhen Fundamental Research Institutions Shenzhen Institutes of Advanced Technology, Chinese Academy of Sciences
Shenzhen, China
gp.zhang@siat.ac.cn

4th Rong Sun
Shenzhen Institute of Advanced Electronic Materials, Shenzhen Fundamental Research Institutions Shenzhen Institutes of Advanced Technology, Chinese Academy of Sciences
Shenzhen, China
rong.sun@siat.ac.cn

Abstract—Photosensitive polyimide (PSPI) is a kind of dielectric material with dual functions of photosensitivity and heat resistance, which is applied in redistribution layers (RDLs). However, with the development of miniaturization of electronic devices, low curing temperature (<250°C) and low CTE are great in demand, as for the wafer warp in packing process. Herein, we report a kind of PAE, which was synthesized from 4,4'-oxydianiline (ODA) and pyromellitic dianhydride (PMDA) via the poly-iso-imide intermediates, the esterification rate is around 60% which was calculated by ¹H-NMR. With an addition of quinoline at amount of 10%(wt.), the imidization index of the PI cured at 300°C is 98.1%, in comparable with the comparable PAE (no quinoline addition) cured at 350°C. Besides, the mechanical and thermal properties were also invesgated. The PI with quinoline cured at 300°C also exhibit high tensile strength, high Young's modulus and low coefficient of thermal expansion (tensile strength of 120 MPa, modulus of 3.05 GPa and CTE of 38.9 ppm/K) compared to PI cured at 350°C (tensile strength of 140 MPa, modulus of 2.99 GPa and CTE of 33.3 ppm/K). This research extended the range of substrates in quinoline promoted low temperature curing and will have a broad application prospect.

Keywords—*Photosensitive, polyimide, polyamide ester, low curing temperature, low CTE, RDLs*

I. INTRODUCTION

Polyimide (PI), a class of super-engineering plastics, possesses outstanding key properties such as high thermal and chemical stabilities and excellent mechanical and electrical properties, being used as protection and insulation materials[1]. In particular, photosensitive polyimides (PSPI), which contains photo-definable characteristics, has been widely used as passivation stress buffer layer for improved reliability, interlayer dielectrics in Cu redistribution layers (RDL)[2].In order to make the resin precursor photosensitivity, chemists do the esterification to the carboxyl group of the polyamide acid. This means that, higher temperature (>350°C) is required for the imidazation, as for the low leaving ability of alkoxy

anion (Fig. 2)[3]. However, with the development of miniaturization of electronic devices, low curing temperature (<250°C) and low CTE are great in demand, as for the wafer warp in packing process[4].

![Application and mechanism of PSPI in Cu redistribution layers]

Fig.1. Application and mechanism of PSPI in Cu redistribution layers

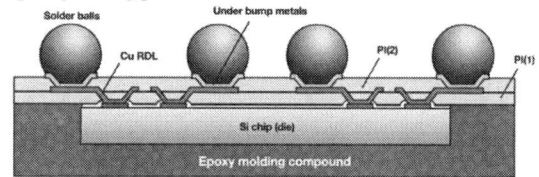

Fig.2. Mechanism of nucleophilic catalysis as curing accelerators

So far, many studies have been examined to reduce the curing temperature. The main methods are as follows: (1) adjusting the monomer structure to increase the flexibility of the resin chain, (2) adding dehydrating agent to remove the water from the reaction and (3) adding curing catalyst. However, increasing the molecular flexibility would lead to the rapid rise of CTE[5], and the excess addition of dehydrating agent would degrade the film performance[6]. By contrast, adding curing catalyst is a better choice to solve this problem with little side effects.

In the previous work, we reported that, quinoline can reduce the curing temperature of polyamide acid obviously, and a reasonable performance of PI film can be obtained[7]. The probable mechanism of this method is that, quinoline can accelerate the nucleophilicity of amide nitrogen atom. From the structure of PSPI, the PAE resin also contains the amide structure, and the quinoline may have the same effect to the amide nitrogen atom, so as to reduce the curing temperature of poly amid ester (PAE). With this idea in mind, we started

978-1-6654-1392-3/21 $31.00 © 2021 IEEE

the following work. In this work, we will analyze effect of quinoline on curing temperature and PI film properties prepared from PAE resin, and this will provide a great inspiration to the PSPI industry.

II. EXPERIMENT SECTIONE

A. Reagents

Pyromellitic Dianhydride (PMDA, >99.0%(T)(HPLC)) was obtained from Aladdin (Shanghai, China), 4,4'-Oxydianiline (ODA, >98.0%(T)(GC)) was purchased from TCI (Tokyo, Japan). Trifluoroacetic anhydride (TFAA, >99.0%), 2-Hydroxyethyl methacrylate (HEMA, >97.0%), Quinoline (QL, >99.0%) and 1-Methyl-2-pyrrolidinone (NMP, >99.0%) were all bought from Energy Chemical (Shanghai, China). All the chemicals and reagents were used without any purification.

B. Experiment procedure

Synthesis of PAE resin: To a solution of ODA (20.02g, 100 mmol) in NMP (400 ml) was added PMDA (22.46g, 103 mmol) in several portions at 0°C. Then, the resultant mixture was gradually warmed up to 80°C, and stirred at 80°C for 12h to get PAA solutions[8]. To this PAA solution was added TFAA (63.09g, 300 mmol) at 0°C. Then, the resultant mixture was gradually warmed up to 50°C, and stirred at this temperature for 2h. Re-colling the solution to 0°C. To this solution, HEMA (260.3g, 2 mol) was added, and then warmed to 50°C, and stirred at this temperature for 12h. The reaction mixture was poured into MeOH (6 L). The residual solid was dissolved in NMP (400 ml). This solution was added to the deionized water dropwise. The residual solid was got through filtration, washed with deionized water twice and then dried under vacuum at 45°C for 48h. The dried PAE resin was got.

Synthesis of resin adhesive solution: the resin was dissolved in NMP to get the resin solution with a solid content of 30%. After the resin dissolved completely, 10%-wt% of quinoline was added to the resin solution, and then stirred at room temperature for 12h.

Synthesis of polyimides: the resin solution was spin-coated on glass wafer, soft-baked at 100°C for 4 min. The samples were then put in a convection oven under air with different temperature program to get polyimide films. The films were peeled from the glass wafer in HF aqueous solution, and then rinsed with ultrapure water. Dried at 100°C under vacuum to get the final polyimide films.

C. Characterrization

Fourier transform infrared (FT-IR;) spectra was recorded with a scan range of 4000-500 cm⁻¹ by a Bruker Vertex 70 spectrometer. The esterification rate is calculated by ¹H-NMR spectra, which were recorded on Bruker AV 400 instruments and calibrated by using residual TMS (δ_H = 0.00 ppm) as internal references. Mechanical properties were measured by DMA Q850 and the test program was set ranged from 1N/min to 18N. Thermogravimetry analysis (TGA) was performed under a nitrogen flow by a TA SDTQ600 thermogravimetric analyzer. The heating process was set from 30°C to 800°C at a rate of 10 K min⁻¹ and the sample mass were about 10 mg. Differential scanning calorimetry (DSC) was tested under a nitrogen condition by a Q20-1173 DSC thermal system. The heating rate was set with a speed of 10°C/min from 30 to 450°C. Coefficients of thermal expansion (CTE) for the films

were carried out using TMA Q800 at a heating rate of 5 °C/min from 20 °C to 400 °C with force 0.001 N in a nitrogen atmosphere. Contact angle measurements were performed on a OCA20 contact angle goniometer.

III. RESULTS AND DISCUSSION

Fig. 3 illustrates the ¹H NMR spectra of the pure PAE resin. From the NMR spectra, we can see that, the chemical shifts from 6.93 to 8.50 ppm are the characteristic peaks of aromatic protons, and the peaks from 5.57 to 6.10 ppm correspond to the HEMA alkene protons. As the ratio of PMDA/ODA was 1.03:1, which was close to 1:1, and each repetitive unit contains ten aromatic protons, so we defined the integral of aromatic protons as 10.00. As each repetitive unit contains two carboxyl group, when the esterification was 100%, the number of HEMA in each unit was two, so the number of alkene protons in each unit was four. From the integral of these two parts peak, esterification rate can be deduced as 60%

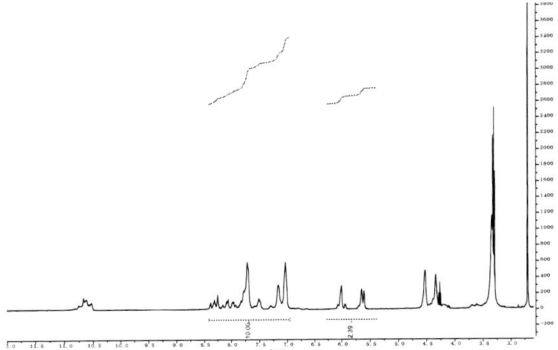

Fig. 3. ¹H-NMR spectra of pure PAE resin dissolved in d6-DMSO

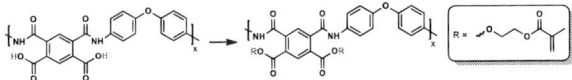

Fig. 4. Mechanism of esterification rate calculation

After we got the PAE resin, we then started to prepared the polyimide films. As shown in Fig. 5, the FTIR spectra of the polyimide films prepared by the two methods. As can be seen from figure 2, these curves share some characteristic peaks, such as the asymmetric stretching vibration peak of C=O bond near 1780 cm⁻¹, the symmetric stretching vibration peak of C=O bond near 1720 cm⁻¹, and the stretching vibration peak of C-N bond near 1378 cm⁻¹. These three peaks represent the existence of imine ring structures. Besides, 1500 cm⁻¹ peak represents the aromatic nucleus absorption. Except for the characteristic peaks, some other bonds absorptions also exist, such as the stretching vibration absorption of C=O bond of carboxylic acid (or ester) near 1633 cm⁻¹. With the temperature raised up to 350°C, the peak of 1633 cm⁻¹ and 1548 cm⁻¹ were disappeared, and the absorption value of peak 1378 cm⁻¹ reach the top, it is noteworthy that imidazation reaction almost completed. From the absorption of these bands, we can calculate the degree of imidization (DOI) of different menbranes. Suppose that the resin was cured completely at 350°C, the degree of imidization (DOI) is defined as[9]:

$$\text{DOI} (\%) = \frac{(peak\ area\ at\ 1378cm^{-1})_T / (peak\ area\ at\ 1500cm^{-1})_T}{(peak\ area\ at\ 1378cm^{-1})_{350℃} / (peak\ area\ at\ 1500cm^{-1})_{350℃}}$$

According to the above, the DOI value with different conditions can be calculated. Compared to the PAA, the PAE requires higher curing temperature (350℃, PI-350). As we can see, the imidization rate of PAE resin with quinoline at 300℃ is 98.1%, and the control group is 79.9%. Other group This indicates that quinoline can significantly accelerate the imidization process.

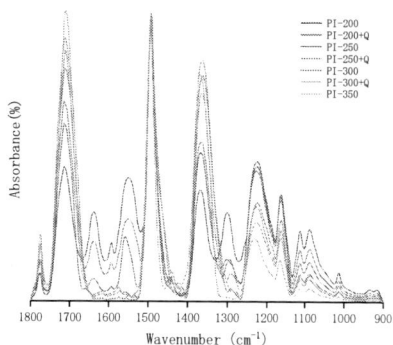

Fig. 5. FT-IR spectrum of PI film synthesized from PAE

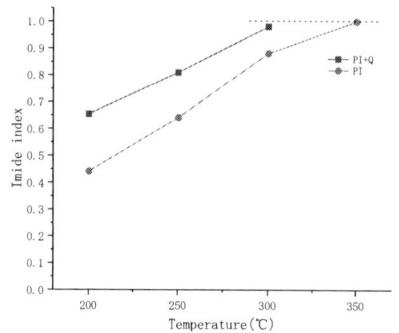

Fig. 6. Degree of imidization in different conditions

Next, the performance of the films was tested to see whether the addition of quinoline had effect to the film performance. First, we did the mechanical property test. The tensile properties of the PI-300+Q was examined. We also measured the tensile properties of pristine PI-350 and PI-300 for comparison. Fig. 7 shows the tensile tests of the PIs in the form of stress-strain curves and the results are summarized in Table 1. From the results, we can see that the PI-350 showed a good result. The tensile modulus was 2.99 GPa, tensile strength was 140.42 MPa, and elongation at break was 22.56%. As a comparison, PI-300+Q also showed good tensile properties: tensile modulus of 3.05 GPa, tensile strength of 119.86 MPa and elongation at break of 22.91%. These results show that, quinoline has a little side effect to the mechanical properties of polyimides, as for the residual quinoline after curing at 300℃. However, the mechanical properties of control group (PI-300) were significantly reduced: tensile modulus, tensile strength and elongation at break were reduced to 2.42 GPa, 95.52 MPa and 15.57% respectively. It was another powerful proof that PAE resin could not cured completely at 300 ℃.

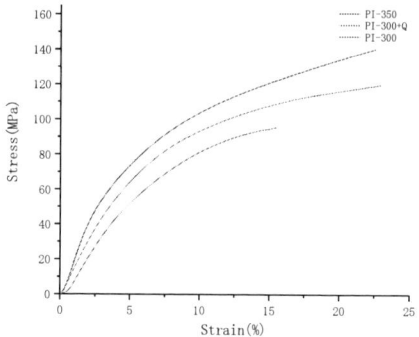

Fig. 7. Stress-strain curves of PI films

Thermal property is also an important indicator. Thermal stability of the PI films was examined by TGA (Fig. 8.) and DSC (Fig. 9.) severally and results are summarized in Table 1. From the results, we can see, the 5% weight loss temperature (T_5) of PI-300+Q was almost the same as the PI-350. By contrast, the T_5 of PI-300 was only 477 ℃, which was 50 ℃ lower than PI-350.

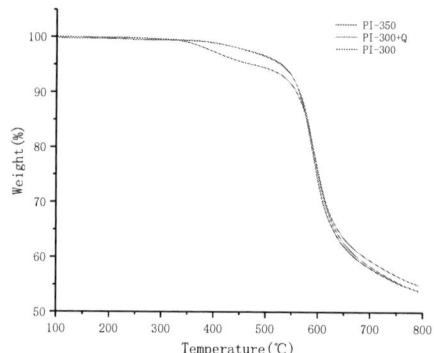

Fig. 8. TGA curves of PI films

The glass transition temperature (T_g) shows the similar tendency, which the T_g of PI-350 and PI-300+Q were 350 ℃ and 338℃ respectively, and the T_g of PI-300 was 300℃, 50℃ lower than the completely cured film. These two curves mean that, quinoline can reduce the curing temperature for 50 ℃ without any side effects.

Table 1. Mechanical, thermal properties and hygroscopicity of PI films

Code	Tensile properties			Thermal properties			Hygroscopicity
	Strength (MPa)	Modulus (GPa)	Elongation (%)	T_5 (℃)	T_g (℃)	CTE (ppm/K)	Contact angles (°)
PI-350	140.42	2.99	22.56	531	350	33.3	81.5
PI-300+Q	119.86	3.05	22.91	529	338	38.9	79.9
PI-300	95.52	2.42	15.57	477	300	29.2	65.1

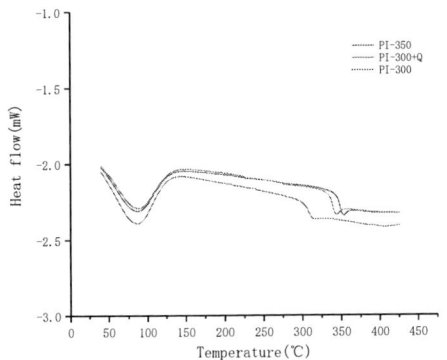

Fig. 9. DSC curves of PI films

The typical CTE value of PI-350 is 33.3 ppm/K (range from 50 to 300°C), which is close to the silicon (25 ppm/K). However, when we added the quinoline to the resin, the CTE was increased to 38.9 ppm/K, which was opposite phenomenon to the previous work[7], but it was still an acceptable value for the polyimides materials.

Contact angle reflects the wettability of a surface. As for the low water absorption requirements in electronic packaging and production, we also investigated the surface properties of the PI film. The results are shown as follows: The contact angle of PI-350 is 81.5°, this shows that the PI has good hydrophobic. As a comparison, the contact angle of PI-300+Q was similar to the PI-350, this explain that the addition of quinoline has no effect to the hydrophily of PI film. However, the contact angle of PI-300 film is only 65.1°, lower than the PI-350. This means that the surface becomes more hydrophilic with a decrease in the degree of imidization, which also reflects the film is not cured completely, as for the unreacted carboxyl group.

Fig. 11. Contact angles of water drops on the surface of PI films

IV. CONCLUSION

In summary, quinoline can also reduce the curing temperature of PAE. By addition of only 10%-wt quinoline, the imidization could be completed at 300°C. Besides, this PI film exhibited the similar thermal and mechanical properties as the PI film cured at 350°C. In a word, the addition of quinoline has a good effect on reducing curing temperature of polyimide materials without any side effects, and will have a broad application prospect.

ACKNOWLEDGMENT

We thank the financially supported of R&D Funds for Basic Research Program of Shenzhen (Grant No. JSGG20160229194437896 and JCYJ20160331191741738), NSFC-Guangdong Jointed Funding (U1601202), NSFC-Shenzhen Robot Jointed Funding (U1613215), Key Laboratory of Guangdong Province (2014B030301014).

REFERENCES

[1] M. Ding, "Polyimide chemistry, relationship between structure, properties and materials," Science Press: Beijing, pp. 1-7, 2005.

[2] D. Lu, C. Wong, "Materials for Advanced Packaging" Springer Press: Switzerland, pp. 5, 2017 doi: DOI 10.1007/978-3-319-45098-8.

[3] Q. Yi, W. Pei, R. Qiu, J. Pei, "Basic organic chemistry," Higher Education Press: Beijing, pp. 603, 2005

[4] D. Lu, C. Wong, "Materials for Advanced Packaging" Springer Press: Switzerland, pp. 403, 2017 doi: DOI 10.1007/978-3-319-45098-8.

[5] W. Jang, M. Seo, J. Seo, S. Park, and H. Han, "Correlation of residual stress and adhesion on copper by the effect of chemical structure of polyimides for copper‐clad laminates," *Polymer International,* vol. 57, no. 2, pp. 350-358, 2008.

[6] M. Oba, "Effect of curing accelerators on thermal imidization of polyamic acids at low temperature," *Journal of Polymer Science: Part A: Polymer Chemistry,* vol. 34, pp 651-658, 1996.

[7] C. Huang, J. Li, G. Zhang, R. Sun, C. Wong, "Development of low temperature curing polyimides with quinoline," *2019 20th ICEPT conference paper,* pp. 1-4, 2019, doi: 10.1109/ICEPT47577.2019.245169.

[8] T. Miwa, Y. Okabe, and M. Ishida, "Effects of precursor structure and imidization process on thermal expansion coefficient of polymide (BPDA/PDA)," *Polymer,* vol. 38, no. 19, pp. 4945-4949, 1997.

[9] T. Sasaki, "Low temperature curable polyimide for advanced package," *Journal of Photopolymer Science and Technology,* vol. 29, no. 3, pp. 379-382, 2016.

Viscoelastic Characterization and Simulation of Thermal Interface Materials

Cheng Zhong
Shenzhen Institute of Advanced Electronic Materials, Shenzhen Institute of Advanced Technology, Chinese Academy of Science
Shenzhen 518055, China
cheng.zhong1@siat.ac.cn

Chenglong Li
Shenzhen Institute of Advanced Electronic Materials, Shenzhen Institute of Advanced Technology, Chinese Academy of Science School of Mechanical Science and Engineering, Huazhong University of Science and Technology
Shenzhen 518055, China
iclears@163.com

Yunxia Wang
Shenzhen Institute of Advanced Electronic Materials, Shenzhen Institute of Advanced Technology, Chinese Academy of Science
Shenzhen 518055, China
yx.wang2@siat.ac.cn

Jibao Lu*
Shenzhen Institute of Advanced Electronic Materials, Shenzhen Institute of Advanced Technology, Chinese Academy of Science
Shenzhen 518055, China
jibao.lu@siat.ac.cn

Linlin Ren*
Shenzhen Institute of Advanced Electronic Materials, Shenzhen Institute of Advanced Technology, Chinese Academy of Science
Shenzhen 518055, China
ll.ren@siat.ac.cn

Rong Sun*
Shenzhen Institute of Advanced Electronic Materials, Shenzhen Institute of Advanced Technology, Chinese Academy of Science
Shenzhen 518055, China
rong.sun@siat.ac.cn

Ching-Ping Wong
School of Materials Science and Engineering, Georgia Institute of Technology
Atlanta, GA 30332, USA
cpwong@cuhk.edu.hk

Abstract—The thermal interface material (TIM) located above the chip in a package is a key factor in determining the heat dissipation of a chip. The deformation or warpage of the package significantly affect the performance of the TIM, which induces stress that causes failures such as cracking and interface delamination. Therefore, a detailed analysis of the stress state of the TIM in the package is essential for the thermal management of the chip.

Most polymer materials used in package are viscoelastic, showing obvious correlation with temperature and time. However, in most finite element analysis (FEA) the polymer materials are usually treated as linear elastic, which may lead to the deviation of simulation from the actual situation.

In this paper, we use inverse analysis method to obtain the viscoelastic constitutive parameters of TIM based on the Prony series derived from the experimental stress relaxation test. The viscoelastic constitutive of the TIM is applied to FEA for studying the cooling process in a flip-chip package. We find that the viscoelastic TIM may absorb more stress than the elastic material does, leading to less deformation from the bottom side of package transferring to the upper lid due to stress relaxation. In addition, the stress and strain distribution in the plane of viscoelastic TIM also shows characteristics different from that derived using linear properties. This work will help to improve the understanding of the actual mechanical behavior of TIM during the deformation of a package.

Keywords—thermal interface material, viscoelastic properties, stress relaxation, finite element analysis

I. INTRODUCTION

The endless demands for higher performance processors lead to a continuous increasing in power density across all kinds of the market products. Increasing power consumption in various chips are main heat sources and cause concerns for the thermal management. A key aspect of electronic package cooling is the thermal interface materials (TIMs) used between the heat generating component and the heat spreader or heat sink, providing a path of low thermal impedance between die and heat sink, to maintain operating temperatures of chips at acceptable level [1].

Various TIMs have been used in industries, such as thermal pastes, gels, adhesives, pads, metal TIMs and so on. The selection of suitable TIMs not only depend on the requirements of heat transfer, but also on manufacturability, cost and reliability [1,2]. Thermal greases are easy to pump out or dry out and metal-based are bad for the price and cracking [1]. As a result, polymeric thermal gel, mainly composed of thermally conductive fillers and polymer matrix, is more popular used in flip chip packages [3-6]. The addition of fillers improves the thermal conductivity, while the polymer matrix retains good flexibility, low cost, and easy processing [1].

However, during packaging process or accelerated testing conditions, there are still many failure modes observed in gel type TIM [2-7]. At high temperature, TIM tends to separate either cohesively or adhesively from the interfaces [4]. During moisture related condition, the hygroscopic expansion along with the thermal expansion of the organic material could also degrade thermal performance of TIMs [7]. Therefore, a detailed analysis of the stress state of the TIM in the package is essential for the thermal management of a chip.

Finite element analysis (FEA) has been applied to predict TIM strains in packages [3-7]. The simulated mechanical strains could be correlated with experimentally observed thermal degradation, and finally a phenomenological model can be set up to predict the thermal performance of an electronic package during thermal cyclic loading [5]. The thermal performance of TIM package is also affected by its stress state and stress history [3,6].

Most polymer materials used in the packaging are viscoelastic, showing obvious correlation with temperature and time. However, in most FEA the polymer materials are usually treated as linear elastic [8], which may lead to the deviation of simulation results from the actual situation. In this paper, inverse analysis method is used to obtain the viscoelastic constitutive parameters of TIM based on the Prony series by experimental stress relaxation test. Then the viscoelastic constitutive of the TIM is applied to FEA for cooling process of a flip-chip package. The difference in package deformation, and the stress and strain distribution of TIM are also investigated between viscoelastic and linear properties.

II. EXPERIMENTAL CHARACTERIZATION

The uniaxial tensile test was used to characterize the stress relaxation behavior of TIMs. In dynamic mechanical analyzer (DMA), the samples with initial effective size of 20*2*1.62mm are elongated by 1mm with a speed of 3mm/min at 30 ℃, i.e., the elongation is about 5%. After that, the stretched state maintains for 2 minutes, then the tensile force is reduced to zero, the deformed samples will gradually recover. The change of samples in length with time is shown in Figure 1. Firstly, the increased length quickly drops to nearly 500um from initial 1000um, then relatively slowly drops to about 350um. After that, it continues to slow down and finally nearly keeps constant at about 300um. Hence, nearly 70% of tensile deformation are recovered.

III. VISCOELASTIC CONSTITUTIVE PARAMETER

One of the main purposes of this study is to determine the viscoelastic constitutive of TIMs through reverse analysis. In FEA, viscoelastic properties are usually presented in form of Prony series. The equation based on generalized Maxwell model is as follows:

$$E(t) = E_\infty + \sum_{i=1}^{N} E_i e^{-\frac{t}{\tau_i}} \qquad (1)$$

Where E_∞ is the final (or equilibrium) modulus, N is the number of items in the equation and a pair of E_i and τ_i is referred to as a Prony pair. Usually, Prony series contains more than 10 items. It has been pointed that the more items of Prony series, the simulated results are more closed to the experimental values [9]. In Prony series, there are many coefficients to be determined, including the E_i and time τ_i of each item. But TIM materials have the following characteristics:

1) TIM is a typical viscoelastic material;

2) The Tg of TIM is generally below -50 °C, i.e., in normal temperature range for the application (-40 to 260 °C, refer to the temperature in temperature cycling and reflow process), the TIM keeps in a rubber state.

Based on the above characteristics, the following assumptions could be made:

1) The modulus of TIM is not sensitive to temperature, so it might be assumed that the modulus of TIM keep constant at different temperatures;

2) The modulus of TIM is very sensitive to time, presenting obvious stress relaxation behavior.

In this way, it is expected to describe viscoelastic constitutive of TIMs by Prony series with less item. Firstly, only one item is considered, but it has been confirmed that viscoelastic properties couldn't be described by Prony series with only one item. Secondly, two items with five coefficients are applied. After several fitting and verifications, the coefficients are obtained, which are in good agreement with the experimental stress relaxation behavior, as shown in Fig 1. The value of E_∞, E_1, E_2, τ_1 and τ_2 are 0.0162, 0.0662, 0.0647, 5 and 200 respectively.

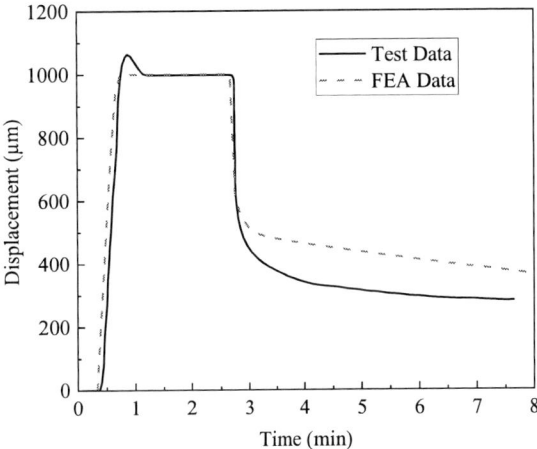

Fig. 1. The result of experimental stress relaxation test and fitting data of TIM.

IV. SIMULATION AND DISCUSSION

The obtained viscoelastic constitutive parameters of TIM are applied to a typical flip-chip structure to evaluate the impact of viscoelastic properties on the mechanical behavior of package.

The package used in FEA is shown in Fig. 2(a). In order to simplify the calculation, the solders and underfill are merged into one part, and equivalent material properties are applied. Except TIM, all of other components are set as linear elastic materials. The detailed material properties are shown in Table 1.

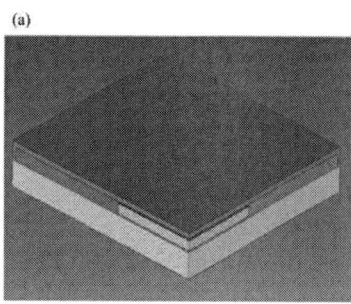

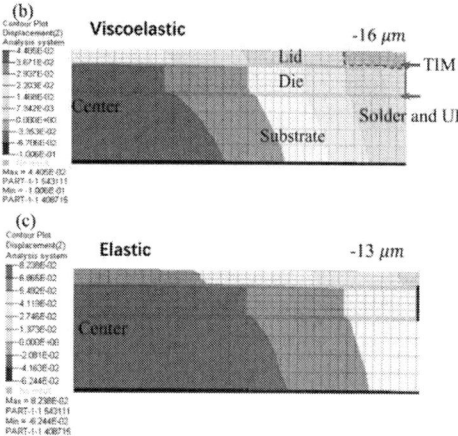

Fig. 2. (a) The package used in FEA, (b) the warpage contour of viscoelastic simulation, (c) The warpage contour of elastic simulation.

TABLE 1. THE SPECIFIC MATERIAL PROPERTIES IN FEA.

Material	CTE (1/°C)	Young's Modulus (MPa)	Poisson's Ratio
Silicon	2.8e-6	131000	0.3
Adhesive	T=25°C, 2.5e-5 T=42°C, 2.5e-5 T=43°C, 1.01e-4 T=165°C, 1.01e-4 (Tg is 42°C)	30	0.35
Copper	1.74e-5	110000	0.3
Solder and Underfill	T=25°C, 2.7e-5 T=106°C, 2.7e-5 T=107°C, 4.0e-5 T=165°C, 4.0e-5 (Tg of underfill is 107°C)	20000	0.3
Substrate	1.3e-5	20000	0.3
TIM	7.1e-5	Viscoelastic	0.3

This study focuses on the deformation of TIMs during cooling process after TIM curing. The initial temperature is set as 165°C then the model is cooled down to 25°C. Deformation of the package and the stress/stain state of the TIM layer are analyzed at 25°C. The sizes of the package and die are 50x50 and 20x20 mm² respectively. When considering the viscoelastic behavior, the cooling time are set as 100s, to evaluate the influence of time effect on the stress of the package. At the same time, TIM with linear elastic property in same model is also carried out for comparison.

The distributions of deformation in thickness direction (marked as "Z") of the package at 25°C with viscoelastic and elastic properties are shown in Fig. 2(b) and 2(c) respectively. Different colors visually represent different degrees of displacement in Z-direction. In cross-sectional view, the warpage in a crying face could be observed, which is mainly caused by the difference of thermal expansion coefficient (CTE) between substrate and lid. It shows that the deformation under TIM layer, including chip and substrate, are almost same in two kinds of materials. However, obvious differences are observed above the TIM layer. With viscoelastic properties, the relative displacement from the upper face to the lower face at the center of TIM layer in Z-direction is -16um, while using elastic properties this value is only -13um, as shown in Fig. 2(b). The results indicate that the TIM layer could absorb more stress with viscoelastic properties, so that less deformation will be transmitted to lid. As viscoelasticity is an intrinsic property of TIM, it needs to be considered when evaluating the deformation and stress of package in FEA.

Next, the stress and strain distributions of TIM layer in plane are also studied, as shown in Fig 3. Compared with elastic, the stress and strain distributions of the TIM layer are a bit different when using viscoelastic materials. As shown in Fig 3, the strain in viscoelastic model is obviously higher than in linear elasticity. However, the stress in viscoelastic model is a little smaller than that in linear elasticity. Furthermore, the maximum stress is a bit shifted in viscoelastic model, which are worth more studying in detail by jointing with the actual failure location in TIMs in experiment.

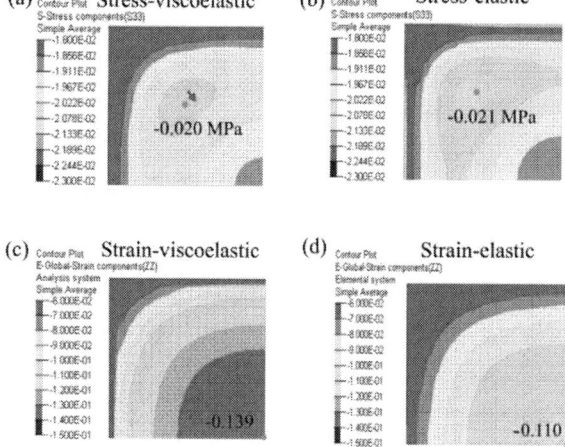

Fig. 3. (a) Stress of TIM layer in plane with viscoelastic parameters; (b) stress of TIM layer in plane with elastic parameters; (c) strain of TIM layer in plane with viscoelastic parameters; (d) strain of TIM layer in plane with elastic parameters.

TIM is a typical viscoelastic material, but it is usually treated as linear elastic in FEA, which may lead to the deviation of simulation results from the actual situation, preventing quantitative evaluation of its reliability in package. The applying of viscoelastic constitutive parameters of TIMs might help to solve this problem. At the same time, the nonlinear relationship between strain and stress could also provide a solution for studying the cumulative damage of the TIMs in temperature cycling, which help to construct a life prediction model about TIM failure. More study on the effect of viscoelastic properties of TIM are still needed.

V. Conclusion

1) Inverse analysis was used to obtain the viscoelastic constitutive parameters of TIM based on the Prony series by stress relaxation experimental test.

2) The viscoelastic constitutive of the TIM was applied to FEA during cooling process of a flip-chip package. It was found that the viscoelastic TIM may absorb more stress so that less deformation will transfer to lid from the bottom, compared to TIM with elastic properties.

3) The stress and strain distribution in the plane of viscoelastic TIM also show different characteristics to the linear properties.

Acknowledgment

The work was supported by the National Natural Science Foundation of China (no. 52003289), the Youth Innovation Promotion Association CAS (no. 2021363), and the SIAT Innovation Program for Excellent Young Researchers (no. 201803).

References

[1] R. Prasher, "Thermal interface materials: historical perspective, status, and future directions." Proceedings of the IEEE, 2006, Vol. 94(8): 1571-1586.

[2] J. Due, and A. J. Robinson, "Reliability of thermal interface materials: A review." Applied Thermal Engineering, 2013, Vol. 50(1): 455-463.

[3] C. Nelson, J. Galloway, C. Henry, and W. Kelley, "Thermal performance of TIMs during compressive and tensile stress states." 2017 33rd Thermal Measurement, Modeling and Management Symposium (SEMI-THERM), 2017, pp. 261-268.

[4] J. Zheng, V. Jadhav, J. Wakil, J. Coffin, S. Iruvanti, R. Langlois, and P. Brofman, "Delamination mechanisms of thermal interface materials in organic packages during reflow and moisture soaking." 2009 59th Electronic Components and Technology Conference, 2009, pp. 469-474.

[5] T. Sinha, J. A. Zitz, R. N. Wagner, and S. Iruvanti, "Predicting thermo-mechanical degradation of first-level thermal interface materials (TIMs) in flip-chip electronic packages." Fourteenth Intersociety Conference on Thermal and Thermomechanical Phenomena in Electronic Systems (ITherm), 2014, pp. 240-250.

[6] Y. Yang, Z. Zhen, and T. Maxat, "Impact of temperature-dependent die warpage on TIM1 thermal resistance in field conditions." 2009 25th Annual IEEE Semiconductor Thermal Measurement and Management Symposium, 2009, pp. 285-292.

[7] X. Liu, J. Zheng, and S. K. Sitaraman, "Hygro-thermo-mechanical reliability assessment of a thermal interface material for a ball grid array package assembly." Journal of Electronic Packaging, 2010, Vol. 132(2):21004-21004.

[8] S. H. Chae, J. H. Zhao, D. R. Edwards, and P. S. Ho, "Characterization of viscoelasticity of molding compounds in time domain." International Electronic Packaging Technical Conference and Exhibition, 2009, Vol. 43598: 435-441.

[9] P. R. Chowdhury, C. S. Jeffrey, and L. Pradeep, "Characterization of Viscoelastic Response of Underfill Materials." 2019 18th IEEE Intersociety Conference on Thermal and Thermomechanical Phenomena in Electronic Systems (ITherm), 2019, pp. 1321-1331

Characterization and Verification of Viscoelastic Constitutive Parameters of Underfill Material

Cheng Zhong[#]
Shenzhen Institute of Advanced Electronic Materials, Shenzhen Institute of Advanced Technology, Chinese Academy of Science
Shenzhen 518055, China
cheng.zhong1@siat.ac.cn

Chenglong Li[#]
Shenzhen Institute of Advanced Electronic Materials, Shenzhen Institute of Advanced Technology, Chinese Academy of Science School of Mechanical Science and Engineering, Huazhong University of Science and Technology
Shenzhen 518055, China
iclears@163.com

Lu Lu
Shenzhen Institute of Advanced Electronic Materials, Shenzhen Institute of Advanced Technology, Chinese Academy of Science
Shenzhen 518055, China
lu.lu@siat.ac.cn

Yunxia Wang
Shenzhen Institute of Advanced Electronic Materials, Shenzhen Institute of Advanced Technology, Chinese Academy of Science
Shenzhen 518055, China
yx.wang2@siat.ac.cn

Gang Li
Shenzhen Institute of Advanced Electronic Materials, Shenzhen Institute of Advanced Technology, Chinese Academy of Science
Shenzhen 518055, China
gang.li@siat.ac.cn

Pengli Zhu
Shenzhen Institute of Advanced Electronic Materials, Shenzhen Institute of Advanced Technology, Chinese Academy of Science
Shenzhen 518055, China
pl.zhu@siat.ac.cn

Jibao Lu*
Shenzhen Institute of Advanced Electronic Materials, Shenzhen Institute of Advanced Technology, Chinese Academy of Science
Shenzhen 518055, China
jibao.lu@siat.ac.cn

Rong Sun*
Shenzhen Institute of Advanced Electronic Materials, Shenzhen Institute of Advanced Technology, Chinese Academy of Science
Shenzhen 518055, China
rong.sun@siat.ac.cn

Ching-Ping Wong
School of Materials Science and Engineering, Georgia Institute of Technology
Atlanta, GA 30332, USA
cpwong@cuhk.edu.hk

Abstract—The continuous miniaturization and intelligentization of electronic products have promoted the development of microelectronic packaging technology. Underfill is one kind of widely used polymer materials in microelectronic packaging, which shows an obvious viscoelastic behavior, i.e., the modulus highly depends on both temperature and time. In many circumstances, the viscoelastic behaviors are often not characterized to reduce the experimental and computational cost. However, accurately characterizing the viscoelastic behavior of the underfill can further obtain a more credible simulation result.

In this paper, frequency dependent dynamic mechanical analysis (DMA) measurements are performed on a self-developed underfill by the double cantilever to characterize its viscoelastic behaviors. Through the assumption of time temperature equivalence (TTE), we can obtain the master curve by shifting the experimental data. Meanwhile, shift factors are carried out by shifting the experimental data using the TTE and Prony pairs parameters based on the generalized Maxwell model are obtained by fitting the master curve. Finally, uniaxial tension is performed to verify the accuracy of viscoelastic material parameters. The simulation results using viscoelastic parameters show good correlation to experimental results, which proves that the developed viscoelastic models could successfully predict the rate-dependent non-linear stress-strain behavior, and might be used to generate more accurate stress behavior in package structure.

Keywords—underfill, viscoelastic properties, frequency dependent, uniaxial tension

I. INTRODUCTION

Nowadays, flip chip is still one of the most important packaging methods in the industry. The interconnections by array of solder joints between chip and substrate may provide high packaging density, short interconnection distance, good electrical performance. Actually, because of the mismatch of coefficients of thermal expansion (CTE) between chip and substrate, the solder joints are often subjected to thermal fatigue loads, which will cause potential failures in package and greatly affects the life of the device [1]. Thence, underfill (UF) is commonly used in the industry to fill into the gaps of solder joints to alleviate the impact of the CTE mismatch on stress in solder joint, so that the reliability of the device will be improved.

The UF is mainly composed of epoxy resin matrix and silica filler. After curing, the UF exhibits high modulus, low CTE to match the solder joints, low moisture absorption and good adhesion to chip and substrate [2]. UF is a viscoelastic material, i.e., the modulus highly depends on both temperature and time [3]. However, UF are usually considered as linear elastic materials in most of numerical simulations [4], which may cause deviations between the simulation results and the experimental results. Ikeda [5] and Sham [6] have pointed out that using viscoelastic parameters of UF materials may obtain more accurate results of stress and strain distributions around solder joints. Therefore, the research on UF materials and correctly constructing their viscoelastic models will further improve the accuracy of numerical simulation, and promote numerical simulation to provide effective guidance for the development and application of UF materials.

This paper mainly focused on the viscoelastic constitutive behaviors of a self-developed UF material. Dynamic mechanical analysis (DMA) instrument is used to characterize the modulus of UF at different temperatures and frequencies, then a viscoelastic model of UF material is successfully constructed based on experimental data. In addition, the accuracy of the viscoelastic model has been carefully verified.

II. EXPERIMENT PROCEDURE

A. Frequency Dependent Dynamic Mechanical Measurements

The DMA instrument provided by TA is used to conduct frequency dependent dynamic mechanical measurements on UF samples by using the double cantilever clamp (Fig. 1b). The length dimension of the fully cured sample remains at 20 mm, the typical width of sample is 5 mm and the thickness is 2 mm. During the measurements, we focus on the sample temperature range of 40 °C to 140 °C, with the temperature interval of 20 °C. Considering that the modulus changes rapidly in the temperature range near glass transition temperature (Tg), reduce the temperature interval to 5 °C near Tg. At each testing temperature, the samples are measured with a frequency increased from 0.05 Hz to 1 Hz.

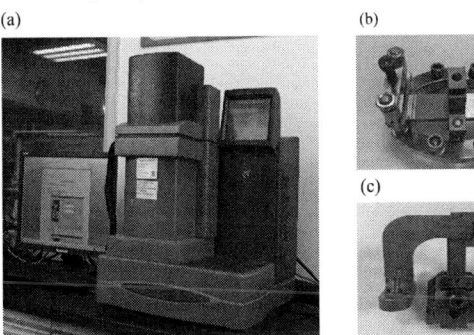

Fig. 1. (a) DMA instrument, (b) Double cantilever clamp,

(c) uniaxial tension clamp.

B. Uniaxial Tension Experiment

In addition, we need to further verify the accuracy of the viscoelastic constitutive parameters. DMA instrument is used to perform a tensile experiment on the UF sample. The results of tensile experiment are compared with the numerical simulation results under same conditions. The size of the sample is 10 x 2.5 x 0.25 mm. Also, the tensile experiments are carried out at different temperatures during 40 °C to 140 °C. The specific loading method as shown below: at each testing temperature, load a tensile force to the sample; the tensile force increases with a speed of 5 N/min and stops when it reaches 10 N. At high temperature, the tensile force may properly be reduced, ensuring that a strain of more than 2 % might be captured at each testing temperature.

III. RESULT AND DISCUSSION

A. Viscoelastic Constitutive Behavior of UF

Through a series of frequency dependent dynamic mechanical measurements at different temperatures, the storage and loss modulus of the material can be obtained respectively. The relationship between the modulus and

frequency of the self-developed UF are shown in Fig 2. The modulus changing with temperature and frequency reflects its viscoelastic characteristics.

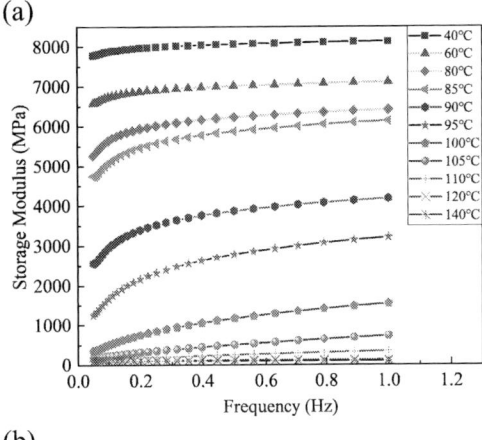

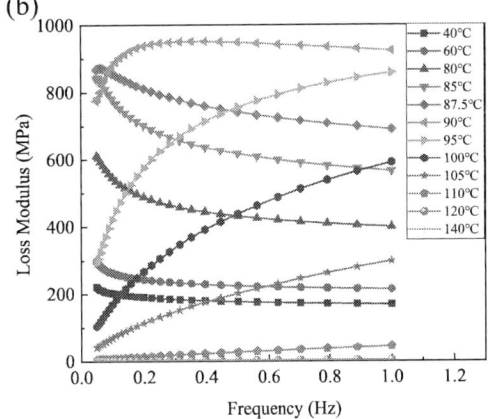

Fig. 2. (a) storage modulus vs frequency, (b) loss modulus vs frequency.

Since the frequency range of these measurements is just from 0.05 Hz to 1 Hz, so that the curve data in Fig 2 are only a part of the master curve at different temperatures. The principle of time-temperature equivalence (TTE) is often used to generate the master curve. By shifting the measurement data, the master curves of the storage modulus and loss modulus at 40 °C can be obtained, as shown in Fig 3.

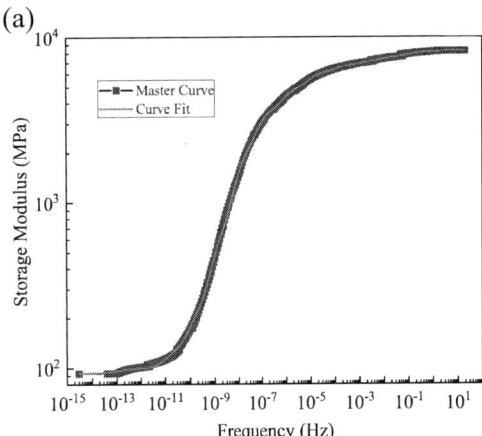

(b)

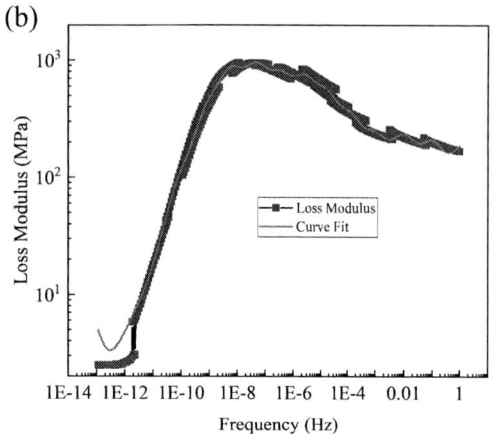

Fig. 3. (a) the master Curve of storage modulus at 40 °C, (b) the master Curve of Loss modulus at 40 °C.

In the logarithmic scales, the temperature shift factor is the horizontal offset distance from each measured data to the master curve, which is often recorded as $\alpha_T(T)$. The shift factor $\alpha_T(T)$ at different temperatures are shown in Fig 4, which could generally be described by the Williams–Landel–Ferry (WLF) equation (Equation 1) [7]:

$$\log \alpha_T(T) = -\frac{C_1(T - T_r)}{C_2 + (T - Tr)} \quad (1)$$

Where $\alpha_T(T)$ is the displacement factor, T is the temperature, T_r is the reference temperature, and C_1 and C_2 are the adaptation parameters. We also use this equation to mathematically fit the temperature shift factor and obtain a better fitting effect, as shown in Fig. 4, where the fitting parameters C_1 and C_2 are 1137.8 and 1000 respectively when Tr is 40 °C.

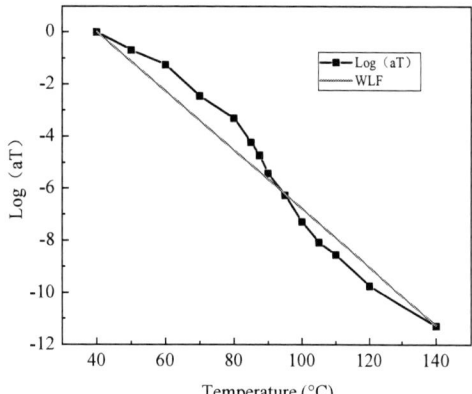

Fig. 4. Log αT(T) Data and the fitting curve of WLF.

In order to apply viscoelastic behaviors in numerical simulation, the master curve of the relaxation modulus of viscoelasticity in the time domain needs to be obtained. The relaxation modulus can be calculated by the storage modulus and loss modulus. The calculation equation is shown in equation 2 and equation 3:

$$E(t) = \frac{2}{\pi} \int_0^\infty \frac{E'(\omega)}{\omega} \cdot \sin \omega t \, d\omega \quad (2)$$

$$E(t) = \frac{2}{\pi} \int_0^\infty \frac{E''(\omega)}{\omega} \cdot \cos \omega t \, d\omega \quad (3)$$

Where E(t) is the relaxation modulus, E'(ω) is the storage modulus, E"(ω) is the loss modulus, ω is the test frequency and t is the relaxation time.

Considering that the solution of equation (2) and equation (3) is very complicated, an empirical formula mentioned by Ninomiya and Ferry at al. [8] [9] is used to approximately solve the above equation. The empirical formula is as follows:

$$E(t) = E'(\omega) - 0.4E''(0.4\omega) + 0.014E''(10\omega) \quad (4)$$

The master curve of relaxation modulus can be obtained by using equation 4, as shown in Fig. 5. The master curve is generally described by the generalized Maxwell constitutive model (Equation 5). The generalized Maxwell equation is as follows:

$$E(t) = E_\infty + \sum_{i=1}^{N} E_i e^{-\frac{t}{\tau_i}} \quad (5)$$

Where E_∞ is the final (or equilibrium) modulus, N is the number of terms in the equation and a pair of E_i and τ_i is referred to as a Prony pair. Through fitting the relaxation master curve in Fig. 5, we may obtain the Prony pairs, as shown in the Fig. 5.

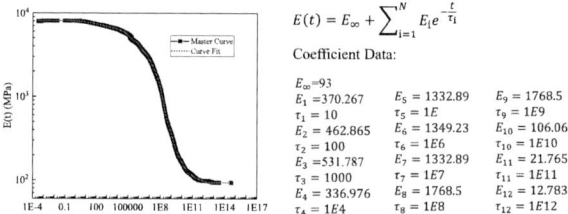

Fig. 5. The master curve of relaxation modulus at 40 °C and the fitting coefficient of Maxwell equation.

If the relaxation modulus is known, the shear modulus and bulk modulus of the UF can be further obtained. The shear modulus is defined as:

$$G(t) = \frac{E(t)}{2(1 + v)} \quad (6)$$

The bulk modulus is defined as:

$$K(t) = \frac{E(t)}{3(1 - 2v)} \quad (7)$$

Where v is Poisson's ratio and v=0.3.

Based on equation 6 and 7, the Prony pairs of shear modulus and bulk modulus can be obtained, as shown in Table 1. So far, through a series of measurements and data processing, a viscoelastic model of UF material has been successfully characterized, which can be used in finite element software ABAQUS.

TABLE 1. THE PRONY PAIRS OF SHEAR MODULUS AND BULK MODULUS.

i	Relaxation Time τ_i (sec)	Shear modules G_i (MPa)	Bulk modulus K_i (MPa)
-	∞	60.45	12.4
1	10	240.6733	49.36887
2	100	300.8624	61.71535
3	1000	345.6613	70.90487
4	10000	219.0342	44.9301
5	100000	857.2255	175.8411
6	1000000	876.9981	179.897
7	1.00E+07	866.3772	177.7184
8	1.00E+08	1149.528	235.8006
9	1.00E+09	261.167	53.57271
10	1.00E+10	68.94492	14.14255
11	1.00E+11	14.14786	2.902125
12	1.00E+12	8.309288	1.704469

B. Verify the Input of Prony Pairs in ABAQUS

Firstly, it is necessary to verify whether the obtained Prony pairs entered into ABAQUS correctly represents the relationship of the relaxation master curve and the WLF equation. We apply instantaneous strain to a bar specimen and maintain a relaxation time of 10^9 s in ABAQUS to simulate the stress relaxation experiment. By calculating the modulus of the sample at different times and temperatures, we can obtain the master curve of the relaxation modulus E(t) in simulation, as shown in Fig 6. By observing the simulation results, the finite element modeling (FEM) results is almost the same as the master curves at every temperature, which means that Prony pairs have been correctly input in ABAQUS.

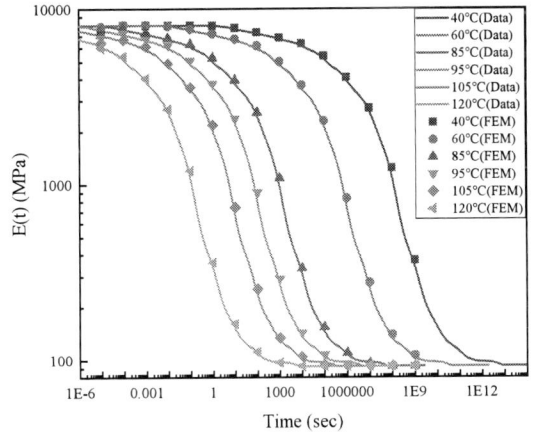

Fig. 6. The master curve of relaxation modulus and the data points obtained by simulation.

C. Comparison of Tensile Experiment and Tensile Simulation

Furthermore, for viscoelastic materials, if the deformation time is long enough, stress relaxation effects may occur during the deformation process, which may appear as a nonlinear stress-strain relationship. However, it is rarely mentioned in previous studies. A combination of experiment and simulation can be used to verify whether the entered viscoelastic parameters can accurately reflect the nonlinear changes in stress-strain.

Tensile experiments of UF samples are applied in DMA at the temperature of 40 °C, 85 °C, 95 °C, 105 °C, 115 °C respectively. The tensile force increases with a speed of 5 N/min and stops when it reaches 10N. Deformation with strain of more than 2 % might be ensured at each testing temperature. Similarly, the tensile simulations are carried out under exactly the same conditions in ABAQUS. The results of tensile experiment and simulation at different temperatures are shown in Fig 7. The straight line is the experimental result and the dashed line is the simulated results. Through comparison, it can be seen that at the low temperature, such as 40 °C, 85 °C, 95 °C, the experimental and simulated data are almost same; while at the high temperature such as 115 °C, there is a slight deviation between the two data; at the temperature near Tg, such as 105 °C, the deviation is relatively larger. The results remind us that the simulation accuracy may still be reduced near Tg even though strict data processing has been performed at each step, which is mainly because the modulus near Tg changes very quickly with temperature. This phenomenon also needs to be considered when using viscoelastic constitutive parameters to analyze problems at temperature near Tg. But in general, the tensile experiments and the simulation results have relatively good consistency, verifying that the viscoelastic model may successfully reflect its rate-dependent nonlinear stress-strain relationship. It can be applied to the package structure to achieve more accurate stress-strain behavior.

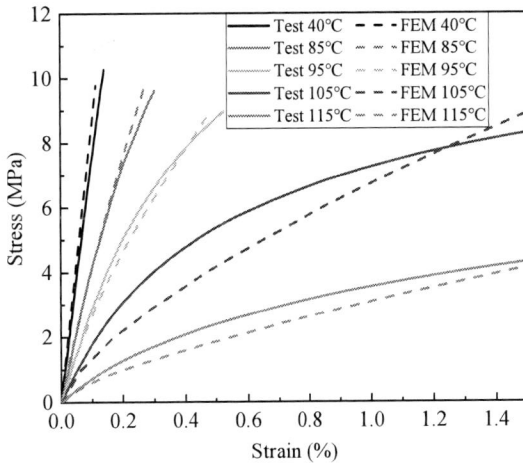

Fig. 7. DMA tensile test curve and tensile simulation data.

IV. CONCLUSION

1) The viscoelastic behavior of one self-developed underfill are characterized using frequency dependent DMA measurements by double cantilever mode.
2) The master curve and shift factors of underfill are carried out using the TTE principle, and Prony pairs can be obtained by fitting the master curve.
3) In uniaxial tension, the simulation results with viscoelastic parameters have a good correlation to experimental results.

ACKNOWLEDGMENT

The work was supported by the National Key R&D Program of China (2020YFB0408700), National Natural Science Foundation of China (no. 52003289), the Youth Innovation Promotion Association CAS (no. 2021363), SIAT Innovation Program for Excellent Young Researchers (no. 201803).

REFERENCES

[1] I. Kim, and S. B. Lee, "Fatigue Life Evaluation of Lead-free Solder under Thermal and Mechanical Loads." Electronic Components & Technology Conference IEEE, 2007, pp. 95-104.

[2] Z. Zhang, and C. P. Wong, "Recent advances in flip-chip underfill: materials, process, and reliability." IEEE transactions on advanced packaging, 2004, Vol. 27(3): 515-524.

[3] Chowdhury, R. Promod, C. S. Jeffrey, and L. Pradeep, "Characterization of Viscoelastic Response of Underfill Materials." 2019 18th IEEE Intersociety Conference on Thermal and Thermomechanical Phenomena in Electronic Systems (ITherm). IEEE, 2019, pp. 1321-1331

[4] W. Wang, and N. Tung, "A Modeling Study of the Effect of Underfill Materials on Solder Joint Thermal Fatigue of Ball Grid Array Package." ASME International Mechanical Engineering Congress and Exposition, 2014, Vol. 46590, p. V010T13A078

[5] T. Ikeda, T. Kanno, N. Shishido, N. Miyazaki, H. Tanaka, and T. Hatao, "Non-linear analyses of strain in flip chip packages improved by the measurement using the digital image correlation method." Microelectronics Reliability, 2013, pp. 145-153.

[6] M. L. Sham, K. Jang-Kyo, and P. Joo-Hyuk, "Numerical analysis of plastic encapsulated electronic package reliability: Viscoelastic properties of underfill resin." Computational materials science, 2007, Vol.40(1): 81-89.

[7] W. N. Findley, J. S. Lai, K. Onaran, and R. M. Christensen, "Creep and relaxation of nonlinear viscoelastic materials with an introduction to linear viscoelasticity." 1977, pp. 364-364.

[8] K. Ninomiya, and J. D. Ferry, "Some approximate equations useful in the phenomenological treatment of linear viscoelastic data." Journal of Colloid Science, 1959, Vol. 14(1): 36-48.

[9] Menard, P. Kevin, and M. Noah, "Dynamic mechanical analysis." Encyclopedia of Analytical Chemistry: Applications, Theory and Instrumentation, 2006, pp. 1-25.

Synthesis and properties study of a thermoplastic polyimide with high glass transition temperature for wafer level package

1st Wen Liu
*Shenzhen Institute of Advanced
Electronic Materials, Shenzhen
Institute of Advanced Technology,
Chinese Academy of Sciences,*
Shenzhen, 518055, China
wen.liu@siat.ac.cn

2nd Jinhui Li*
*Shenzhen Institute of Advanced
Electronic Materials, Shenzhen
Institute of Advanced Technology,
Chinese Academy of Sciences,*
Shenzhen, 518055, China
jh.li@siat.ac.cn

3rd Jinshan Liu
*Shenzhen Institute of Advanced
Electronic Materials, Shenzhen
Institute of Advanced Technology,
Chinese Academy of Sciences,*
Shenzhen, 518055, China
liu.js@siat.ac.cn

4th Tao Wang
*Shenzhen Institute of Advanced
Electronic Materials, Shenzhen
Institute of Advanced Technology,
Chinese Academy of Sciences,*
Shenzhen, 518055, China
tao.wang3@siat.ac.cn

5th Ao Zhong
*Shenzhen Institute of Advanced
Electronic Materials, Shenzhen
Institute of Advanced Technology,
Chinese Academy of Sciences,*
Shenzhen, 518055, China
ao.zhong@siat.ac.cn

6th Guoping Zhang*
*Shenzhen Insititute of Advanced
Electronic Materials, Shenzhen
Institute of Advanced Technology,
Chinese Academy of Sciences,*
Shenzhen, 518055, China
gp.zhang@siat.ac.cn

7th Qiang Liu
*Shenzhen Institute of Advanced
Electronic Materials, Shenzhen
Institute of Advanced Technology,
Chinese Academy of Sciences,*
Shenzhen, 518055, China
qiang.liu@siat.ac.cn

8th Rong Sun
*Shenzhen Institute of Advanced
Electronic Materials, Shenzhen
Institute of Advanced Technology,
Chinese Academy of Sciences,*
Shenzhen, 518055, China
rong.sun@siat.ac.cn

9th Daquan Yu*
*School of Electronic Science and
Engineering,
Xiamen University,*
Xiamen, 361005, China
yudaquan@xmu.edu.cn

Abstract—**Polyimides (PIs) have been used in a wide spectrum of high-tech fields for their excellent thermal property, excellent mechanical property, extremely low dielectric constant, excellent low temperature resistance, low CTE and good chemical resistance. However, the rigid molecular structure combined with strong intermolecular interaction results in insolubility and low meltability. In this work, the non-coplanar and isomeric diamine of 5(6)-amino-1-(4-aminophenyl)-1,3,3-trimethylindane (DAPI) and ether-contained flexible dianhydride of 4,4-oxydiphthalic anhydride (s-ODPA) are selected for the preparation of the soluble thermoplastic polyimide (TPI) with high glass transition temperature. Fourier transform infrared spectrometer indicates that polyimide resin has been successfully synthesized. And thanks to the non-coplanar and flexible ether group, the as-synthesized polyimide exhibits good solubility in high boiler point polar solvents such as m-Cresol, N-methylpyrrolidone (NMP). Then, differential scanning calorimeter shows that glass transition temperature (T_g) is about 300 ºC. In addition, an elongation-at-break of 4-5 % and high tensile modulus above 1500 MPa is also proved. Moreover, the thermal gravity analysis test shows that its 5% thermal weight loss temperature ($T_{5\%}$) is about above 460 ºC. So, the as-prepared TPI shows excellent comprehensive property with superior solubility, thermal, mechanical performance which is suitable for the application in wafer level package.**

Keywords—thermoplastic polyimide, high glass transition temperature, soluble, wafer level package

I. INTRODUCTION

Polyimide (PI) is a kind of high-performance macromolecule which contains imide ring and because of its outstanding integrated performance, it is extensive applied in microelectronics, aeronautics, and many other fields[1]. However, because of its special imide structure, that is the rigid molecular structure combined with strong intermolecular interaction, PI material usually exhibits insolubility and low meltability, which significantly limits the processing method and application rang[2]. In order to extend the processability with low cost, soluble thermoplastic polyimides (TPIs) have been extensively studied[3]. Thermoplastic polyimide refers to the linear molecular chain, which shows the molten state in a specific temperature rang and it is freely soluble in a variety of common or particular organic solvents. TPI shows much better processability, but the heat resistance of glass transition temperature (T_g) is currently relative low, which limits the application in high-temperature process, for example PECVD in advanced electronic package[4]. Therefore, the preparation of soluble TPIs with high T_g is of great application significance.

At present, there are two representative methods for TPI synthesis. The first one is one-step method of which monomers are added to high boiling point solvents such as m-Cresol or NMP for the synthesis of polyamic acid (PAA), and then curing catalyst is added and heated to about 150-180 °C to get the TPI solution. In this process, the water produced during imidization needs to be continuously removed to ensure that high molecular weight polymers are obtained. Another one is a two-step method. The dianhydride and diamine are added into polar solvents to obtain the PAA solution. After spinning coating, high-temperature imidization and redissolution, TPI solution could be obtained.

And in order to obtain the TPIs, the structure of the polyimides have to be carefully designed to realize their solubility. The typical methods includes the introduction of flexible, asymmetric or non-coplanar structures, for example, Cheng-Lin CHUNG[2] et al introduced the ether bonds can increase the flexibility of the chain, reduce the symmetry and

978-1-6654-1392-3/21 $31.00 © 2021 IEEE

regularity of the molecular chain, and increase its solubility, and the T_gs were recorded within the limits of 222 ~ 271 °C by Differential Scanning Calorimeter (DSC). Nilakshi V. Sadavarte[5] et al introduced an asymmetric structure, which broke the order of the original close-packing, and increased the solubility. T_gs were recorded in the rang of 158 ~ 206 °C. Zhiming Qiu[6] et al introduced large side group in the molecular chain which effectively reduce the force between the molecular chains and maintain high temperature resistance. And the T_gs were recorded in the rang of 255 ~ 283 °C. Lihua Wang[7] et al introduced the distorted non-coplanar diamines or dianhydrides and decreased molecular chain regularity which successfully reduced the inter-chain forces and increased their solubility. At the same time, the T_g was recorded of 238 °C. In summary, the existing thermoplastic polyimides generally have poor thermal stability and low glass transition temperature, which limit their application in high-temperature environments and processes. Therefore, the soluble polymer with high glass transition temperature is necessary.

In this paper, a soluble thermoplastic polyimide with excellent comprehensive properties was synthesized by introducing a flexible dianhydride 4,4',-oxydiphthalic anhydride (s-ODPA) with ether linkage and DAPI with a non-coplanar structure containing polymethyl side groups. Its structure, solubility, thermal and mechanical properties were tested by characterization methods. The as-prepared TPI exhibits a high glass transition temperature and good mechanical properties, and is expected to be used in temporary bonding materials for wafer level packaging.

II.   EXPERIMENT

A. Experimental procedure

The s-ODPA was purchased from Chinatech (TIANJIN) Chemical Co., Ltd. The DAPI was purchased from Fuxin Hongji Optoelectronics Material Co., Ltd. The s-ODPA was used after heating to 160 °C for 4 h and the DAPI was used after heating at 60 °C for 6 h in vacuo for drying. N-methyl-2-pyrrolidone (NMP, 98%) was supplied by Energy Co., Ltd.

Polyamic acid (PAA) of s-ODPA-DAPI was synthesized into a three-necked flask, which contains a mechanical stirring device and protected by nitrogen. Firstly, NMP solution and 15 mmol DAPI was added and mixed. And after it was entirely dissolved in the solution, 15 mmol s-ODPA was gradually introduced in two batches. The PAA solution was reacted with ice bath for 5 h, then continue for 20 h at room temperature. At last, the PAA solution with a weight ration of 25 wt% was obtained.

For the preparation of TPI films, the PAA solution was evenly spin-coated on a slippy glass pane at first. And then, stepped heating was put into effect of which the scheduled process is 80 °C stay 1 h, 180 °C stay 1 h, 240 °C stay 1 h and 280 °C stay 0.5 h, 1 h, 2 h respectively, with a ramping rate of 5 °C/min. Then after subsequent time to cool down, the film was detached from the glass plate to obtain the TPI films at last. The TPI samples were named as TPI-280-0.5h, TPI-280-1h and TPI-280-2h, respectively.

B. Characterization

Infrared spectroscopy (FT-IR) was obtained on a Bruker Vertex 7 spectrometer (Bruker, Germany) within the scope of 2000 - 600 cm^{-1}. The molecular weight and distribution of sample were determined by the gel permeation chromatography (GPC, Waters Alliance e2695), using DMF as the eluent at 25 °C. Mechanical properties of TPI films were multiple measured by dynamic mechanical analyzer (DMA Q800, TA Instruments, America). The glass transition temperature (T_g) of TPIs were measured by Differential Scanning Calorimeter (DSC), which was obtained at a heating speed of 10 °C/min from 50 to 400 °C in nitrogen. Data were collected in the second thermal cycle on Mettler Toledo Star System (TA Instruments, New Castle, DE). To explore the heat resistance of TPIs, thermogravimetric analysis (TGA) was detected by using an SDT Q600 (TA Instruments, America) at a heating speed of 10 °C/min from 30 °C to 800 °C in nitrogen purge of 100 mL/min.

The solubility of the prepared TPIs was investigated. The way to test is dissolved 5 mg of the sample in 10 mL of NMP, DMF, N,N-dimethylacetamide (DMAc), m-Cresol, tetrahydrofuran (THF), gamma-butyrolactone (GBL), dimethyl sulfoxide (DMSO) and acetone.

III.   RESULTS AND DISCUSSION

A. Structure Charcaterization

The synthesis process is a condensation polymerization of 4,4',-oxydiphthalicanhydride and DAPI for PAA, and then the water molecules inside the molecule are removed by high-temperature imidization to form TPI. The synthesis process is shown in Fig. 1.

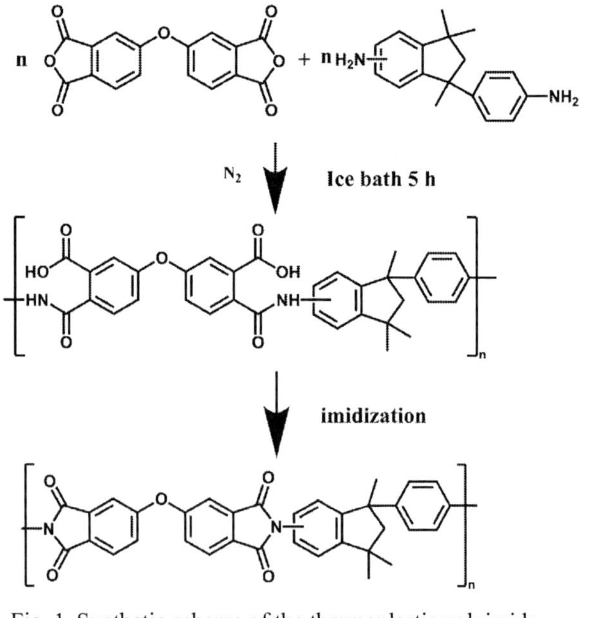

Fig. 1. Synthetic scheme of the thermoplastic polyimide

Then, the FTIR curves of the TPI films are displayed in Fig. 2. Typically, the peaks of 1713cm^{-1} and 1776 cm^{-1} are attribute to of C=O bond in the imine ring. The peak at 1362 cm^{-1} is due to C-N bond stretching vibration of imide ring. The obvious peak at 1230 cm^{-1} is ether bond stretching vibration and the peak at 742 cm^{-1} ought to be the N-H deformation vibration of imine ring. In addition, The number-average molecular weight (M_n) of as-prepared PAA is 56583, the weight-average molecular weight (M_w) is 82985, and the polydispersity is about 1.47. These results imply that the PAA and TPIs were synthesized successfully.

978-1-6654-1392-3/21 $31.00 © 2021 IEEE

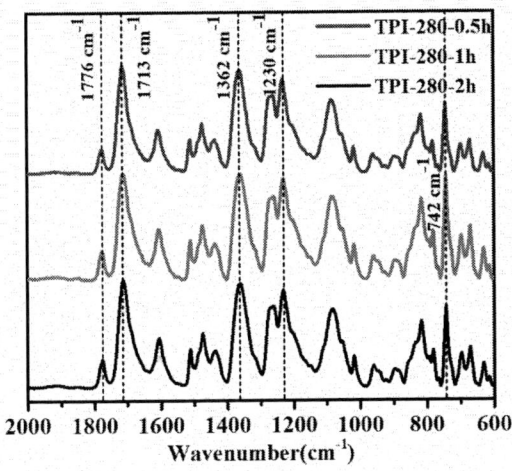

Fig. 2 FT-IR spectra of polyimides

Tab. 1 The GPC data of polyamic acid

Mn	Mw	MP	Mz	Mz+1	PDI
56583	82985	97106	104946	121146	1.4666

B. Solubility of TPIs

After the preparation of the PI films, the solubility was checked at first. The prepared TPI films were put into several kinds of organic solution, and observe their dissolution after 24 hours at room temperature. And some the insoluble samples were raised to 60 °C to further observe the dissolution status. The results were collected and shown in Table 2. It can be seen that all TPIs were dissolved in NMP, DMF, DMAc, m-Cresol and THF at room temperature completely within 24 h. In addition, all TPIs, which can't dissolve at room temperature, became soluble in GBL and DMSO at 60 °C completely within 2 h. However, these TPIs can not dissolve in acetone. So, the as-prepared TPI films proved superior solubility in most of the typical organic solutions.

Tab. 2 The solubility of polyimide in different solvents

	NMP	DMF	DMAc	m-Cresol	THF	GBL	DMSO	acetone
TPI-280-0.5h	++	++	++	++	++	+	+	-
TPI-280-1h	++	++	++	++	++	+	+	-
TPI-280-2h	++	++	++	++	++	+	+	-

++: It can be completely dissolved when placed at room temperature for 24 hours;

+ : Partly dissolved at room temperature, and can be completely dissolved when heated at 60 ℃ for 2 hours;

- : Insoluble at room temperature.

C. Thermal Properties of TPIs

Furthermore, Fig. 3 and Table 3 shows the thermal properties of the TPI films. The temperature of 5% weight loss ($T_{5\%}$) is mirrored in the TGA curves. It can be seen that, the $T_{5\%}$ of all TPIs is above 460 °C, of which $T_{5\%}$ of TPI-280-0.5h is 460.2 °C, $T_{5\%}$ of TPI-280-1h is 465.9 °C and $T_{5\%}$ of TPI-

280-2h is 463.3 °C. The results proved that whole the TPI samples showed prominent thermal stability. And the possible reason is that the non-coplanar structure and multiple side groups of DAPI may benefit for their excellent heat resistance.

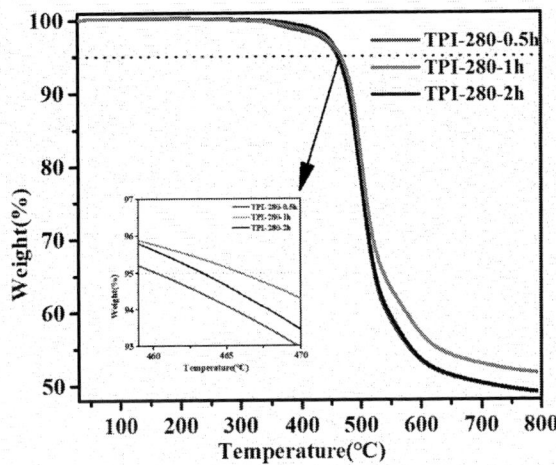

Fig. 3 The TGA curves of TPIs

Then, T_g of TPIs were collected by DSC curves and shown in Fig. 4 and Table 3. The data showed that the T_g of TPI-280-0.5h is 297.6 °C, the T_g of TPI-280-1h is 292.8 °C and the T_g of TPI-280-2h is 298.2 °C, which exhibit superior glass-transition temperature compared to previous studies[4-7].

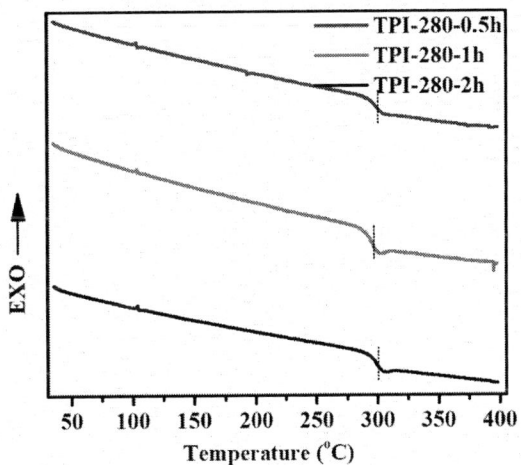

Fig. 4 The DSC curves of TPIs

D. Mechanical Properties of TPIs

At last, Fig. 5 and Table 3 showed the mechanical capability of thermoplasticity polyimides. It can draw a conclusion that for the TPI films of TPI-280-0.5h, TPI-280-1h and TPI-280-2h, the tensile modulus are 59.4 ± 4.0 MPa, 55.8 ± 3.3 MPa, 50.6 ± 5.8 MPa, respectively. And, the elongation at break of these TPI films are 4.77 ± 0.44 %, 4.15 ± 0.30 % and 4.63 ± 0.89 %, respectively. In addition, the tensile modulus of the TPI films are 1.73 ± 0.06 GPa, 1.81 ± 0.10 GPa, 1.47 ± 0.08 GPa, respectively. The results proved similar mechanical properties of all the TPI films.

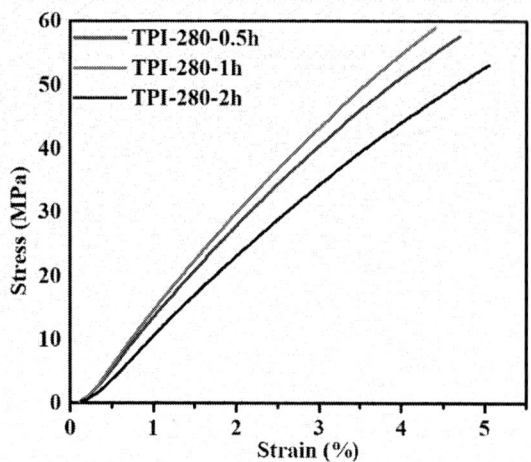

Fig. 5 The DMA curves of TPIs

Tab. 3 Thermal and Mechanical Properties of TPIs

	Tensile strength (MPa)	Elongation-at-break (%)	Tensile modulus (GPa)	T_g (°C)	$T_{5\%}$ (°C)
TPI-280-0.5h	59.4 ± 4.0	4.77 ± 0.44	1.73 ± 0.06	297.6	460.2
TPI-280-1h	55.8 ± 3.3	4.15 ± 0.30	1.81 ± 0.10	292.8	465.9
TPI-280-2h	50.6 ± 5.8	4.63 ± 0.89	1.47 ± 0.08	298.2	463.3

IV. CONCLUSIONS

A series of thermoplastic polyimides with superior solubility in NMP, DMF, DMAc, DMSO, m-Cresol and THF were successfully synthesized. These thermoplastic polyimides possess high T_g of above 290 °C and thermal decomposition temperature ($T_{5\%} > 460$ °C). Moreover, the as-synthesized polyimide also exhibits superior mechanical showing excellent comprehensive property for the application of wafer level package.

ACKNOWLEDGMENT

We acknowledge the financial support of National Natural Science Foundation of China (61904191), Youth Innovation Promotion Association of Chinese Academy of Sciences (2017410), Key R&D Project of Guangdong Province (2020B010180001) and National Key R&D Project from Minister of Science and Technology of China (2017ZX02519).

REFERENCES

[1] F. Li, J. L. Shen, X. F. Liu, Z.H. Cao, X. Cai, J. L. Li, et al, "Flexible QLED and OPV based on transparent polyimide substrate with rigid alicyclic asymmetric isomer," Organic Electronics, vol. 51, pp. 54-61, 2017.

[2] C. L. Chung, W. F. Lee, C. H. Lin, and S. H. Hsiao, "Highly soluble fluorinated polyimides based on an asymmetric bis (ether amine): 1,7-bis (4-amino-2-trifluoromethylphenoxy) naphthalene," Journal of Polymer Science: Part A: Polymer Chemistry, vol. 47, pp. 1756-1770, 2009.

[3] X. Q. Han, P. X. H, J. H. Yao, S. Zhang, X. Y. Cao, J. W. Xiong, et al, "Nitrogen-doped carbonized polyimide microsphere as a novel anode meterial for high performance lithium ion capacitors," Electrochimica Acta, vol. 196, pp. 603–610, 2016.

[4] T. Mori, T. Yamaguchi, Y. Maruyama, K. Hasegawa, S. Kusumoto, "Material Development for 3D Wafer Bond and De-bonding Process," IEEE In 65th Electronic Components and Technology Conference, p. 899-905, 2015.

[5] N. V. Sadavarte, M. R. Halhalli, C. V. Avadhani, P. P. Wadgaonkar, "Synthesis and characterization of new polyimides containing pendent pentadecyl chains," European Polymer Journal, vol. 45, pp. 582-589, 2009.

[6] Z. M. Qiu, S. B. Zhang, "Synthesis and properties of organosoluble polyimides based on 2,2'-diphenoxy-4,4'5,5'-biphenyltetracarboxylic dianhydride," Polymer, vol. 46, pp. 1693-1700, 2005.

[7] L. H. Wang, T. Tian, H. Y. Ding, B. Q. Liu, "Formation of ordered macroporous films from fluorinated polyimide by water droplets templating" European Polymer Journal, vol. 43, pp. 862-869, 2007

The Effect of Thermal-induced Warpage and Degeneration of Thermal Interface Materials on the Thermal Performance of a Flip-chip Package

Ruoyu Jiang[#]
Shenzhen Institute of Advanced Electronic Materials, Shenzhen Institute of Advanced Technology, Chinese Academy of Science University of Science and Technology of China
Shenzhen, China / Suzhou, China
jry@mail.ustc.edu.cn

Cheng Zhong[#]
Shenzhen Institute of Advanced Electronic Materials, Shenzhen Institute of Advanced Technology, Chinese Academy of Science
Shenzhen, China
cheng.zhong1@siat.ac.cn

Haozhe Wang
Shenzhen Institute of Advanced Electronic Materials, Shenzhen Institute of Advanced Technology, Chinese Academy of Science
Shenzhen, China
hz.wang@siat.ac.cn

Chenglong Li
Shenzhen Institute of Advanced Electronic Materials, Shenzhen Institute of Advanced Technology, Chinese Academy of Science School of Mechanical Science and Engineering, Huazhong University of Science and Technology
Shenzhen, China
iclears@163.com

Yi Zheng
Shenzhen Institute of Advanced Electronic Materials, Shenzhen Institute of Advanced Technology, Chinese Academy of Science
Shenzhen, China
yi.zheng@siat.ac.cn

Linlin Ren
Shenzhen Institute of Advanced Electronic Materials, Shenzhen Institute of Advanced Technology, Chinese Academy of Science
Shenzhen, China
rong.sun@siat.ac.cn

Jibao Lu*
Shenzhen Institute of Advanced Electronic Materials, Shenzhen Institute of Advanced Technology, Chinese Academy of Science
Shenzhen, China
jibao.lu@siat.ac.cn

Rong Sun
Shenzhen Institute of Advanced Electronic Materials, Shenzhen Institute of Advanced Technology, Chinese Academy of Science
Shenzhen 518055, China
rong.sun@siat.ac.cn

Ching-Ping Wong
School of Materials Science and Engineering, Georgia Institute of Technology
Atlanta, , USA
cpwong@cuhk.edu.hk

Abstract—With the continuous and rapid development of microelectronics technology, the power consumption of microelectronic device increased with the improvement of performance. The increasing power consumption may cause thermal-induced failures and greatly affect the reliability of the device. To avoid this problem, heat sinks and thermal interface materials (TIMs) are used for effective thermal management. The TIM1 is used inside the device between the chip and the metal lid to eliminate the interstitial air gaps from the interface by conforming to the topography of the surfaces in contact. However, in practical application, TIM1 might be affected by thermal-induced deformation, resulting in the change of bond line thickness (BLT), loss of corner areas, and increase of interface thermal resistance, which may have a significant impact on the heat dissipation in return.

In this paper, the effect of TIM1 on the junction temperature of chip in a flip-chip structure with heat sink and fan are systematically investigated using computational fluid dynamics (CFD) method. The temperature profile is obtained and the heat transfer in the out-of-plane and in-plane directions are analyzed. The effect of reduced in TIM1 coverage, warpage-induced change in bond-line-thickness (BLT) and contact resistance are discussed in detail.

Keywords—thermal interface materials, computational fluid dynamics, junction temperature

I. INTRODUCTION

The continuous demands for higher performance processors lead to increasing power consumption, affecting the operation of the active and passive devices in the integrated circuits. Therefore, it is crucial to effectively control temperature increase in an electronic package [1]. A key method is applying thermal interface materials (TIMs), eliminate the interstitial air gaps from the interface by conforming to the topography of the surfaces in contact. The use of TIMs may provide a path of low thermal impedance between die and heat sink, to maintain component operating temperatures at acceptable level [1].

Higher bulk thermal conductivity, lower Bond Line Thickness (BLT), and lower contact resistance of TIMs are required for lowering the thermal resistance. Thermal pastes, gels, adhesives, pads, metal TIMs have already been used in industries. Thermal gel is the most popular used one in flip chip packages considering the comprehensive performance of heat transfer, manufacturability, cost and reliability [1].

There are various failure modes observed in the gel-type TIM, such as delamination, thermal induced altering in BLT and contact resistance and so on [2-5]. It has been reported that the coverage decreased after reliability test [3]. At high temperature reflow or moisture-related test, delamination or crack may also be observed [4,5], which is similar to coverage decreased in view of heat transfer. Meanwhile, thermal induced altering in BLT and heat resistance has been investigate by experiment and finite element analysis (FEA) [6]. The simulated mechanical strains could be correlated with experimentally thermal degradation and finally setting up a phenomenological model to predict the thermal performance of an electronic package during thermal cyclic loading [7]. The thermal performance of TIM in a package is also dependent on its stress state and stress history [3]. The stress state of TIM impacts the contact resistance, resulting in that the resistance in compression versus tensile stress might be different for the same original BLT [3]. For example, it is found that temperature induced package warpage may dramatically impact the measured Jc values (up to 20%) due primarily to the impact on the spatially BLT [6].

However, in practical application, systematical investigation about the effects of thermal-induced degeneration or warpage of TIMs to thermal performance of package is still lacking. In this paper, the effect of TIM1 on the junction temperature of chip in a flip-chip structure with heat sink and fan are systematically investigated using computational fluid dynamics (CFD) method. The temperature profile is obtained and the heat transfer in the out-of-plane and in-plane directions are analyzed. The effect of reduced in TIM1 coverage, warpage-induced change in Bond Line Thickness (BLT) and contact resistance are discussed in detail.

II. MODELS AND METHODS

A. Model Description

In this study, the CFD commercial software Flotherm is used to simulate the temperature distribution of one flip-chip ball grid array (FCBGA) package under operating condition of the chip. The heat sink and printed circuit board (PCB) with vias are imposed to match the real application scenario closely. The simulation model is shown in Fig 1. TIM1 is located between the chip/die and the metal lid, while TIM2 connects the lid to the heat sink. In this study, a steady-state heat transfer method is used to dissipate heat by forced air cooling, with the direction of wind along the X direction, as shown in Fig 1. The speed of wind is 10m/s along the horizontal direction. The power consumption of chip is 120W and the ambient temperature is 25°C.

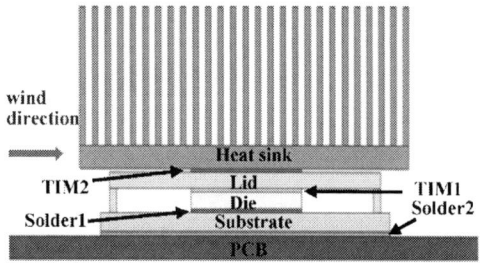

Fig. 1. FCBGA Package Cross Sectional View.

B. Model and Material Properties

The main structure and materials of the model are summarized in Table 1. the PCB and the substrate are applied anisotropic thermal conductivity. A flat structure with rectangular fins is used for the aluminium heat sink. The size of the heat sink is 60*60 mm, the base thickness is 2 mm. The heat sink contains 25 rectangular fins with height of 12 mm,

width of 1 mm and thermal conductivity of 205 W/(m·K). The metal lid consists of two parts, the upper part is 50mm*50mm*0.8mm in size while the lower part consists of four rectangles of 46mm*4mm*0.975mm in size. The lid is made by pure copper with thermal conductivity of 385 W/(m·K).

TABLE I.   THERMAL CONDUCTIVITY SETTING OF PACKAGED COMPONENTS

Description	Dimensions (mm*mm*mm)	Thermal Conductivity W/(m·K)
TIM1	20*20*0.1	2
Die	20*20*0.775	140
Solder1	20*20*0.1	15
Solder2	52.5*52.5*0.4	17
Lid	—	385
Substrate	52.5*52.5*1.2	k_{xy}=30, k_z=0.3
PCB	219.6*113.7*2	k_{xy}=30, k_z=0.3
Heat Sink	—	205
Via	52.5*52.5*2	10

C. Research Programs

Fig 2 shows the design scheme of this study. Firstly, ignoring the interface contact thermal resistance, we focus on the effect of variation in BLT on the chip junction temperature due to deformation induced warpage or decreasing in TIM coverage. After that, we investigate the impact of interface contact thermal resistance on the chip junction temperature.

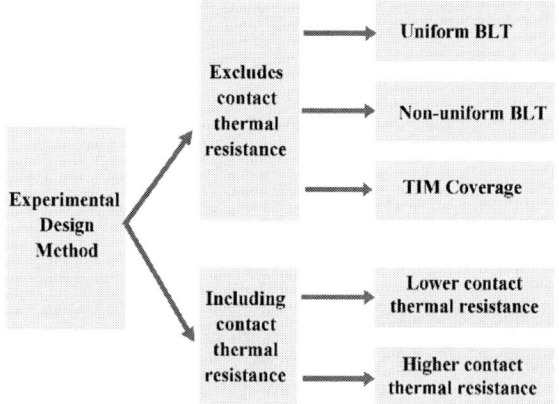

Fig. 2. Experimental protocol design.

1) Perfect contact without considering thermal contact resistance

a) Research benchmark. The thickness of BLT is set to 100um, and the temperature distribution of the flip-chip structure in the vertical direction and the surface plane of the chip are analyzed in detail.

b) The effect of warpage induced Bond Line Thickness variation. When warpage occurs, the BLT thickness is non-uniformly distributed. The layer of TIM might be approximately divided into three regions, as shown in Fig 3. The areas are 100, 125 and 175 um^2 at innermost, second outer and outermost region respectively. While the thicknesses are 135, 100 and 80um at innermost, second outer and outermost region respectively, to simulate the change of warpage on the BLT. Then, the effect of warpage induced BLT thickness variation on the chip junction temperature are studied in this structure.

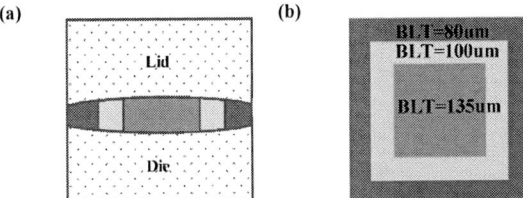

Fig. 3. TIM non-uniform distribution design (a) Side view (b) Top view.

c) TIM coverage impact. Assuming that the reduced TIMs still is covered in the center of the chip with an approximately square shape, as shown in Fig 4, and the TIM coverage ranges alter from 50% to 100% to study the effect to surface temperature of chip at different TIM coverage rates.

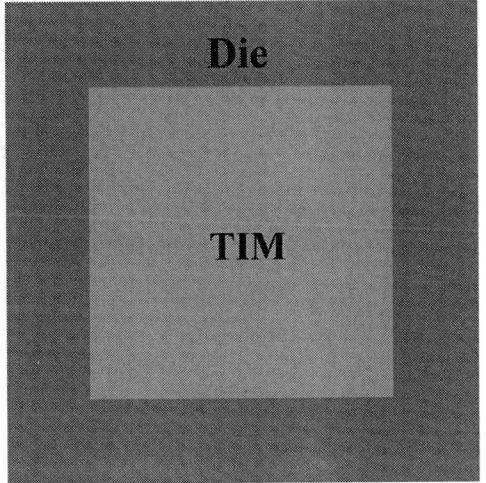

Fig. 4. Distribution of TIM on the chip surface.

2) Consider the effect of contact thermal resistance on-chip heat dissipation

The thermal resistance between TIM and Lid and Die is composed of two parts, containing the body thermal resistance of TIM and the contact thermal resistance, as shown in Equation1

$$R_{TIM} = \frac{BLT}{k_{TIM}} + R_C \qquad (1)$$

where R_{TIM} is the total thermal resistance of the TIM, BLT is the thickness of the TIM layer, and k_{TIM} is the thermal conductivity of TIM, R_C is the contact thermal resistance between TIM and the two contacted solid surfaces. In this study, Rc is the sum of the contact thermal resistance between TIM and Lid and TIM and Die. This study focuses on the effect of the contact thermal resistance of TIM for 0.1 cm2K/W and 0.03 cm2K/W on the chip surface temperature [1].

III. RESULTS AND DISCUSSION

In Flotherm, the corresponding simulation is completed according to the designed research plan.

A. Temperature Distribution in Package

Firstly, the distribution of temperature in the overall structure of the package in the vertical direction and the chip surface plane are analyzed without considering the contact thermal resistance. After inputting the materials and boundary conditions shown in Section II, the distribution of temperature

in package at cross-section and surface of the chip are carried out after the system has been stabilized, as shown in Figs 5a and 5b, respectively. Further, as shown in the red dashed lines marked in 5a and 5b, several points in the thickness direction of the package and the diagonal of the chip are selected and further, as shown in Figs 6 and 7.

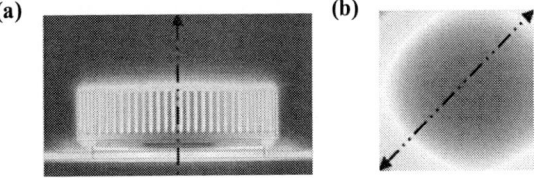

Fig. 5. Simulation results of forced air-cooled FCBGA (a) Temperature cloud of package body with cross-section perpendicular to the wind direction and through the center of the package (b) Temperature distribution cloud of the chip surface.

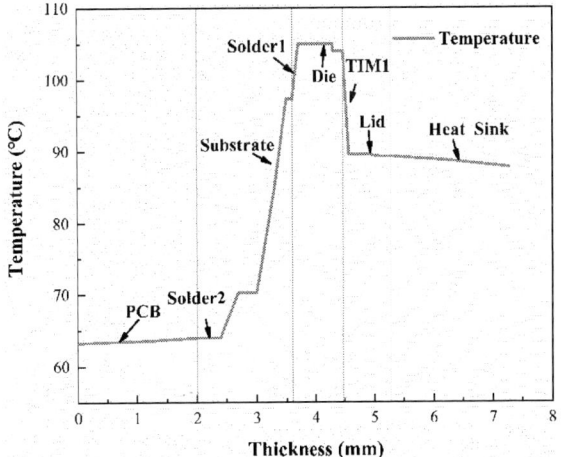

Fig. 6. Heat dissipation curve along the package height direction.

Fig 6 shows the variation of the temperature profile at the center of the package along the package thickness direction (z-direction), encompassing from the lower surface of the PCB at the bottom to the upper surface of the base of heat sink. As shown in fig 6, the chip, as the heat-generating element, is the highest temperature of the entire system, reaching about 105°C. Along both top and bottom sides of the chip, the temperature decreases. TIM1 and substrate are two locations where the temperature decreases more quickly, the value of which are about 15.4°C and 34.7°C respectively. The results clearly indicates that the TIM1 region has a significant impact on the overall heat dissipation of the package, and further illustrates the importance of studying the impact of structural changes in TIM1 on the overall heat dissipation in package.

Next, we focus on the temperature distribution of the chip surface. The temperature distribution along the diagonal of the chip surface is shown in Fig 7, here "0" representing the center of the chip. It might be observed that the temperature distribution is slightly asymmetrical due to the influence of the fan. The right side of the package is away from the fan and in contact with warmer air, resulting in a slight increase in temperature compared to the left side of the package. The temperature at the chip surface is also influenced by hot air. The left and right side of the chip edge temperatures are 95.1 and 98.6°C respectively, while in the center of die the temperature is 104°C. Overall, under the current forced-air cooling condition (wind speed is 10m/s) and package structure, the temperature change on the chip surface is relatively small compared to the package thickness direction. It illustrates again about the importance of TIM1 for effective heat transfer in thickness direction.

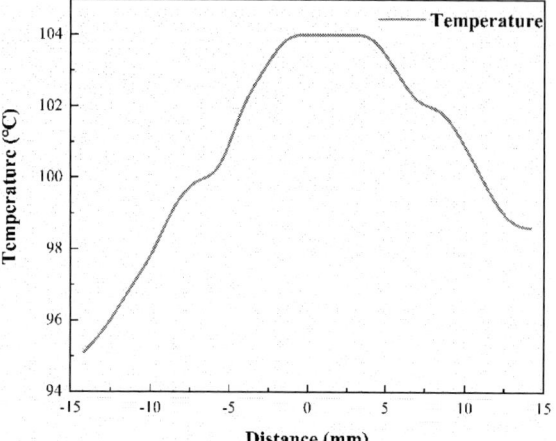

Fig. 7. Chip surface temperature distribution curve.

B. Effect of TIM Coverage on Heat Dissipation

It has been found that the TIM layer will be missing at the edges and corners after reliability tests, such as high-temperature storage, resulting in a decrease in TIM coverage [2]. In addition, if the TIM layer is delaminated locally due to stress, it will lead to a large amount of air entering the internal interface of the package, which will seriously deteriorate the effective heat dissipation of the chip, which can also be approximated as the performance of TIM coverage degradation. This paper focuses on the impact on the chip surface temperature distribution when the TIM coverage varies between 50% and 100%. As mentioned in Fig 4, when the TIM coverage decreases, it still covers the center of the chip in an approximately square shape, and the overall package heat dissipation simulation models are constructed by adjusting the length of the edge to obtain different coverage, the results of which are shown in Fig 8 and Fig 9. When TIM1 fully covers (100%), the temperatures at chip edge are about 96-100°C. While when TIM1 covers only 50%, the chip edge temperature is as high as an impressive 155 °C, while the temperature at chip center will also be increased. The results emphasize again the critical impact of TIM1 structural integrity on the effective heat dissipation of the chip. How to effectively guarantee the structural integrity of TIM1 during the packaging process and reliability testing is worthy of critical attention.

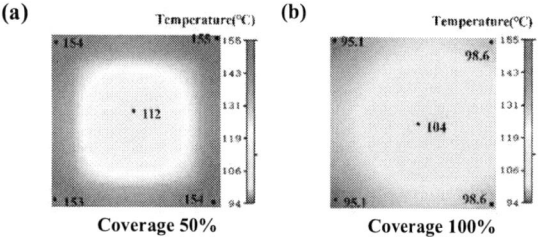

Fig. 8. Temperature cloud on the chip surface for different coverage of TIM (a) Coverage 50% (b) Coverage 100%.

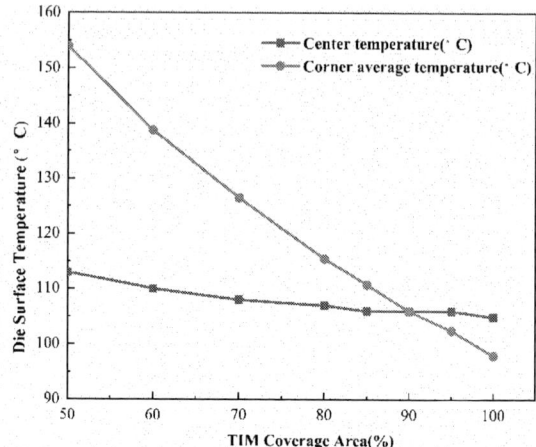

Fig. 9. Chip center and edge temperature profile with TIM coverage change.

Next, we systematically analyze the effect of different TIM coverage on the chip center and edge temperature. The results are shown in Fig 9, where the chip edge temperature is the average value of four points at the corner. It can be seen that both the chip surface center temperature and edge temperature will decrease with the increase of TIM coverage, close to a linear change relationship. However, the center of the chip temperature decreases with a slower rate, while the chip corner temperature decreases much faster. Regarding the chip center temperature at 100% TIM1 coverage as a reference (about 105°C), it may conclude that the TIM coverage needs to be kept above 90% to maintain good heat dissipation in package.

C. Effect of Bond Line Thickness on Heat Dissipation

The effect of B thickness on package heat dissipation has been studied by many cases in the current industry [3, 6]. However, the effect of non-uniform BLT on-chip heat dissipation due to package warpage has not been adequately investigated. The model constructing scheme has been mentioned in Fig 3, which temporarily disregards the contact thermal resistance and focuses only on the differences from non-uniform BLT to a uniform thickness. The TIM material is approximately divided into three regions as mentioned in Fig 3, to make sure the average thickness of such non-uniform thickness distribution is still 100um.

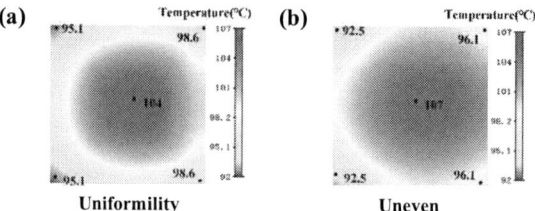

Fig. 10. Heat dissipation cloud on the chip surface with varying Bond Line Thickness (a)Temperature cloud on the chip surface when BLT is uniformly distributed (b) Temperature cloud on the chip surface when BLT is uneven distributed.

The simulation results are shown in Fig 10. Compared with the uniform BLT state, the temperature of the chip center of the sample with non-uniform thickness increases by 3 °C. In contrast, the temperature of the edge position decreases by 2.5°C. It shows that the overall temperature of the package does not change very much only considering the non-uniform effect in BLT. However, warpage may also lead to changes in stress state of TIMs and impact the interface thermal resistance, the effects of which on the overall heat dissipation in package are still needed to evaluate further.

D. Effect of Contact Thermal Resistance on Heat Dissipation

The above simulations have not considered the impact of contact thermal resistance on heat dissipation. Here, the values of 0.03 and 0.1 cm²K/W, generally considered as the range of thermal resistance in reality [1], are applied in simulation respectively, the results of which are shown in Figure 11.

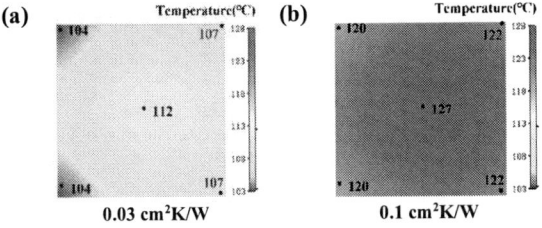

(a) 0.03 cm²K/W

(b) 0.1 cm²K/W

Fig. 11. Effect of changing contact thermal resistance on-chip heat dissipation (a)Temperature cloud on the chip surface when contact thermal resistance is 0.03 cm²K/W (b) Temperature cloud on the chip surface when contact thermal resistance is 0.1 cm²K/W.

As shown in Fig 11, when the TIM contact thermal resistance is set as 0.03 cm²k/W, the temperatures at the chip center and corner temperature will increase by 7 and 7.3 °C, respectively, compared to with the case without considering the contact thermal resistance is not considered. While when the TIM contact thermal resistance is set as 0.1cm²K/W, the chip center and corner temperature will increase further by 15 and 15.5 °C respectively, compared to the case with contact thermal resistance of 0.03 cm²K/W. Referring to Equation 1, the presence of contact thermal resistance can be equivalently regarded as a decrease in TIM thermal conductivity or an increase in BLT, which is more significant for thinner TIM thickness. Similarly, the contact thermal resistance of the interface is also affected by the package stress and reliability conditions [6], which is still worth further evaluation in the future.

IV. CONCLUSION

1) 1) The effects of TIM1 material on the junction temperature of the chip were systematically studied by CFD in a flip-chip structure with heat sink and fan. The temperature profile is was obtained and the heat transfer in the out-of-plane and in-plane directions are were analyzed.

2) 90% or more of the TIM1 coverage area is required to obtain relatively good thermal performance.

3) Warpage-induced change in BLT of TIM1 has a limited effect on the junction temperature, while the decrease in TIM coverage and increase in interface thermal resistance have a great impact.

ACKNOWLEDGMENT

The work was supported by the National Natural Science Foundation of China (no. 52003289), the Youth Innovation Promotion Association CAS (no. 2021363), and the SIAT Innovation Program for Excellent Young Researchers (no. 201803).

REFERENCES

[1] R. Prasher, "Thermal Interface Materials: Historical Perspective, Status, and Future Directions," *Proceedings of the IEEE,* vol. 94, no. 8, pp. 1571-1586, 2006, doi: 10.1109/jproc.2006.879796.

[2] J. Due and A. J. Robinson, "Reliability of thermal interface materials: A review," *Applied Thermal Engineering,* vol. 50, no. 1, pp. 455-463, 2013, doi: 10.1016/j.applthermaleng.2012.06.013.

[3] T. Nordstog, C. Henry, C. Nelson, J. Galloway, P. Fosnot, and Q. Pham, "Junction to case thermal resistance variability due to temperature induced package warpage," in *2017 33rd Thermal Measurement, Modeling & Management Symposium (SEMI-THERM)*, 2017: IEEE, pp. 235-245.

[4] X. Liu, J. Zheng, and S. K. Sitaraman, "Hygro-Thermo-Mechanical Reliability Assessment of a Thermal Interface Material for a Ball Grid Array Package Assembly," *Journal of Electronic Packaging,* vol. 132, no. 2, 2010, doi: 10.1115/1.4001746.

[5] J. Zheng *et al.*, "Delamination mechanisms of thermal interface materials in organic packages during reflow and moisture soaking," in *2009 59th Electronic Components and Technology Conference*, 2009: IEEE, pp. 469-474.

[6] C. Nelson, J. Galloway, C. Henry, and W. Kelley, "Thermal performance of TIMs during compressive and tensile stress states," in *2017 33rd Thermal Measurement, Modeling & Management Symposium (SEMI-THERM)*, 2017: IEEE, pp. 261-268.

[7] T. Sinha, J. A. Zitz, R. N. Wagner, and S. Iruvanti, "Predicting thermo-mechanical degradation of first-level thermal interface materials (TIMs) in flip-chip electronic packages," in *Fourteenth Intersociety Conference on Thermal and Thermomechanical Phenomena in Electronic Systems (ITherm)*, 2014: IEEE, pp. 240-250.

Enhanced Discharged Energy Density in Polyetherimide Composites by Boron Nitride/Aluminum Nitride Hybrid Fillers

Xudong Wu
Department of Materials Science and Engineering
Guangdong Technion Israel Institute of Technology
Shantou, P. R. China.
xudong.wu@gtiit.edu.cn

Shiyi Gao
Department of Materials Science and Engineering
Guangdong Technion Israel Institute of Technology
Shantou, P. R. China.
gao.shiyi@gtiit.edu.cn

Xin Wu
Department of Materials Science and Engineering
Guangdong Technion Israel Institute of Technology
Shantou, P. R. China.
wu.xin@gtiit.edu.cn

Shuo Zhang
Department of Materials Science and Engineering
Guangdong Technion Israel Institute of Technology
Shantou, P. R. China.
zhang.shuo@gtiit.edu.cn

Zhijun Cao
Department of Materials Science and Engineering
Guangdong Technion Israel Institute of Technology
Shantou, P. R. China.
cao.zhijun@gtiit.edu.cn

Daniel Q. Tan *
Department of Materials Science and Engineering
Guangdong Technion Israel Institute of Technology
Shantou, P. R. China.
daniel.tan@gtiit.edu.cn

Abstract—Flexible dielectric polymer composites occupy a significant position in electronic devices. However, many investigations focus on complicated synthetic methods to enhance the energy density of dielectrics. Developing simple and feasible strategy is meaningful and urgent for their acceptance to various applications. Herein, we utilize the high thermal conductivity and electrical insulation characteristic of Boron nitride and Aluminum nitride (BN/AlN) as the hybrid fillers for polyetherimide (PEI) matrix to increase dielectric constant and breakdown strength. This combination resulted in a high discharged energy density of 8.39 J/cm³ at 500 kV/mm with a high efficiency of 90.5% at a low filler fraction. This strategy provides a feasible option for a scalable manufacturing of high performance composite films in the future.

Keywords：*dielectrics, BN/AlN, polyetherimide, discharged energy density*

I. INTRODUCTION

Polymer based dielectrics have played an important role in the fields of electronic packaging, energy storage and conversion, high-voltage insulation, and power electronics [1, 2]. Among passive components, embedded capacitors are a focus due to its versatile characteristics suitable for decoupling, filtering, A/D conversion and energy storage [3]. For instance, microelectronics requires decoupling capacitors with high capacitance density and low dissipation factor.

Nanocomposites are widely used because of their simple processing and outstanding properties. The addition of high percentage of fillers can improve the dielectric constant, but it also introduces the interfacial defects between polymer matrix and fillers [4]. In addition, high-filler content results in the decrease in dielectric breakdown strength due to the formation of multiple conductive path [5]. On the other hand, the low-filler loading strategy developed in recent years has effectively balanced the contradiction between the two competing factors. This approach can improve the breakdown strength and the dielectric constant of the composite films lighting up the possible pathway to industrial application [6-8]. The authors recently reported the unconventional approach using ultra-low filler loading of zinc oxide core shelled with boron nitride with an intention of utilizing their high thermal conductivity for heat dissipation in polyetherimide (PEI). A superior dischargeable energy density of 10.8 J/cm³ and a high efficiency of 92.1% at an electric field of 500 kV/mm were demonstrated in the PEI composite films containing 0.1 vol.% core-shelled ZnO@BN fillers [6]. However, the synthetic procedures for the core-shell structure are complicated giving rise to low yield, which is not conducive to industrial application.

Herein, the authors prepared boron nitride/aluminum nitride hybrid fillers and incorporated in the polyetherimide matrix to further explore the low-filler effect on PEI's energy storage capability. These fillers possess high thermal conductivity and electrical insulation merit. Their combination of platelet and spheres characteristics is expected to further increase the dielectric constant and heat dissipation within the composite films under electrical energization. The addition of the hybrid fillers of 0.6 volume% turns out to enhance the discharge energy density to 8.39 J/cm³ with a high efficiency of 90.5% while presenting a high breakdown strength of at 500 kV/mm simultaneously. This attempt in utilizing the hybrid thermally conductive fillers proves the advantage of improving the breakdown strength, dielectric constant, and flexibility of composite film.

II. EXPERIMENTAL

A. Exfoliation of BN

1g BN (2-3 μm, >99%, Qinhuangdao Eno High-Tech Co. Ltd, China) was dispersed in 300 ml DMF followed by high power sonication for 24h. The mixture was filtered and the sediment was collected and dried at 70°C for 5h in vacuum condition. 100 mg of BN and the 50 mg of AlN (<100 nm, Sigma) were dispersed in 50 ml DMF solvent and subsequently sonicated for 1 h along with a vigorous stirring. After filtration, the solid mixture was collected and dried in a vacuum oven set at 60 °C for 5h.

B. Processing of the composites

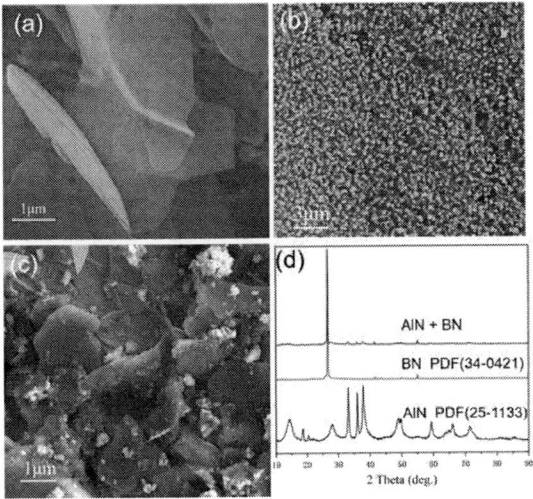

Figure 1. FESEM of exfoliated BN (a) and AlN (b) BN+AlN (c), XRD patterns of BN, AlN, BN+AlN (d).

Composite films start with the dissolution of 1.5 g PEI powders (Sigma Aldrich) into 8 g of 1-methyl-2-pyrrolidinone (NMP) solution followed by a stirring at 50 °C for 12 h. The calculated amount of inorganic filler was dispersed in 2 ml of NMP followed by an ultrasound agitation for 10 minutes to ensure a fully dispersed polymer composite solution. The solution casting method was used to fabricate composite films on a toughened glass sheet followed by an oven drying at 80 °C and then 100 °C for 1 h, respectively. The films were then peeled off from the glass and further dried in a vacuum at 150 °C for 3 h and 200 °C for 2 h, respectively. The final film thickness ranges from 6 to 9μm.

C. Characterization

The surface morphology of the films was inspected using ZEISS Sigma-500 (Germany) field emission scanning electron microscopy (FESEM). The crystallographic structures were analyzed using X-ray diffraction (XRD, Smartlab 9, by Rigaku Corporation in Japan) at 6°/min scan rate, using 150mA current, 40kV voltage, and copper target. Breakdown voltage were measured using a PK-CPE1801 Ferroelectric Polarization Loop and Dielectric Breakdown Test System equipped with a Trek High Voltage Amplifier (Model: 610E), where the sample holder was soaked into a constant-temperature oil bath. Dielectric responses were measured using the Novocontrol Broadband Dielectric Concept 41 in the frequency range of 100 Hz-1MHz.

D. Electrostatic and solid heat transfer simulation.

COMSOL Multiphysics 5.5 was used to simulate the distribution of polarization and heat conduction. The initial voltage and temperature of the electrodes were set at 500 V and 393.15K, respectively. The dielectric constant and thermal conductivities of PEI, BN, and AlN are set as 3.15, 5, and 9, and 0.33 W/(m·K), 60 W/(m·K), and 160 W/(m·K), respectively.

III RESULTS AND DISCUSSION

Figure 1a and 1b display the FESEM images of exfoliated BN nanosheets and AlN nanoparticles. The thickness of the BN ranges from several to dozens of nanometers while the

AlN presents irregular shapes. Figure 1c shows the image of the mixed BN and AlN, where the AlN nanoparticles dispersed randomly among the BN nanosheets. Figure 1d presents the XRD patterns of BN, AlN, and the mixture, where peaks are readily indexed with PDF No.34-0421 and 25-1133 respectively.

Breakdown strength determines the durability and work life of a dielectric material against the voltage stress. How to maintain a high breakdown value while increasing the dielectric constant is one of the goal in this work. Figure 2 shows that the breakdown voltages increase slightly when keeping the filler loading less than 0.6vol.%, but drops with a higher filler content. Generally, high filler contents tend to show negative effect on breakdown voltage resistance whereas the polymer matrix possesses higher values. This work overcomes the disadvantage using low addition of highly thermally conductive fillers. In addition, the hybrid fillers also increase the dielectric constant to 4.0 with a 0.6vol.% loading at 1kHz, as shown in Figure 3. Another noteworthy observation is that the dielectric constant nearly remains independent of frequency of measurements, which is superior over other polymer composites like PVDF. The small dielectric loss is desirable in passive devices for unwanted Joule heating. Interestingly, the BN/AlN filled PEI composites have very dielectric loss of below 0.005 rendering itself an excellent candidate for high performance dielectric application (Figure 3).

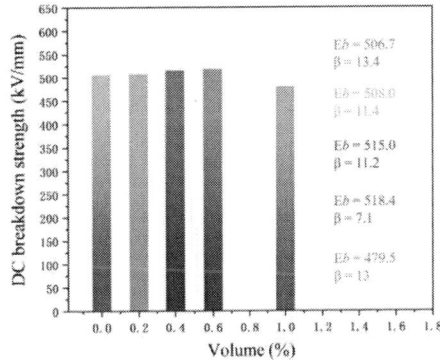

Figure 2. Breakdown strength of PEI composites as a function of the hybrid filler loading fraction.

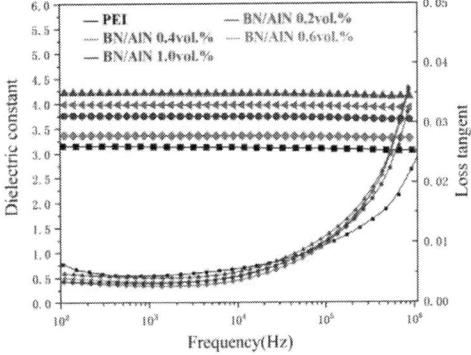

Figure 3. Dielectric constant and loss as a function of frequency for PEI loaded with various fillers.

The polarization behaviors of these composite samples were measured under applied electric fields, as shown in Figure 4a. PEI polymer as a linear dielectric exhibits the linearly increased polarization upon applying electrical fields. The hybrid BN/AlN filler plays a key role for the polarization increment of PEI composites (Figure 4b). At 500 kV/mm, the polarization strength increases from 1.61 to 3.72 $\mu C/cm^2$. This contribution of the hybrid fillers remarkably increases the discharged energy density. Figure 4c shows that the discharge energy density of PEI composites with 0.6vol.% hybrid fillers reaches up to 8.39 J/cm^3 at 500 kV/mm with a high efficiency of 90.5%, much higher than PEI matrix.

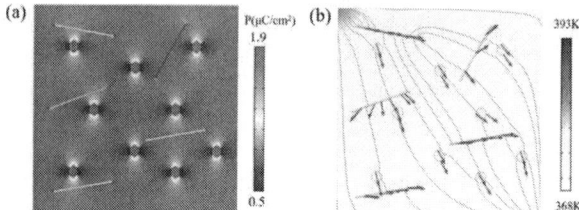

Figure 5. (a) Simulated polarization magnitudes of BN+AlN/PEI; (b) Simulated temperature distribution and of BN+AlN /PEI, the arrow represents the heat flux.

enhancement. The heat production induced under electric field is unavoidable when devices operate continually. The hybrid fillers with high thermal conductivity turns out to be able to transfer the heat out easily, as shown in Figure 5b. The point heat source is set in the upper left corner and the arrow represents the heat flux. The generated heat is mainly dissipated by the hybrid fillers and more rapidly along the nanosheets. The dissipation is seen evenly spreading out to the surroundings of the particles.

IV CONCLUSIONS

The authors prepared the BN/AlN filled PEI composite films elaborately using a solution casting method. Low fraction of hybrid fillers increases the dielectric constant and breakdown strength of PEI matrix simultaneously. This composite approach demonstrates a high discharged energy density of 8.39 J/cm^3 at 500 kV/mm with a high efficiency of 90.5% when adding 0.6vol.% hybrid fillers. The computer simulations also prove the polarization enhancement and heat transfer function attributing to the combinatorial effect of thermally conductive BN platelet and AlN particles. The fabrication of the flexible PEI composite films provides a feasible path for future scale-up manufacturing of composite film rolls.

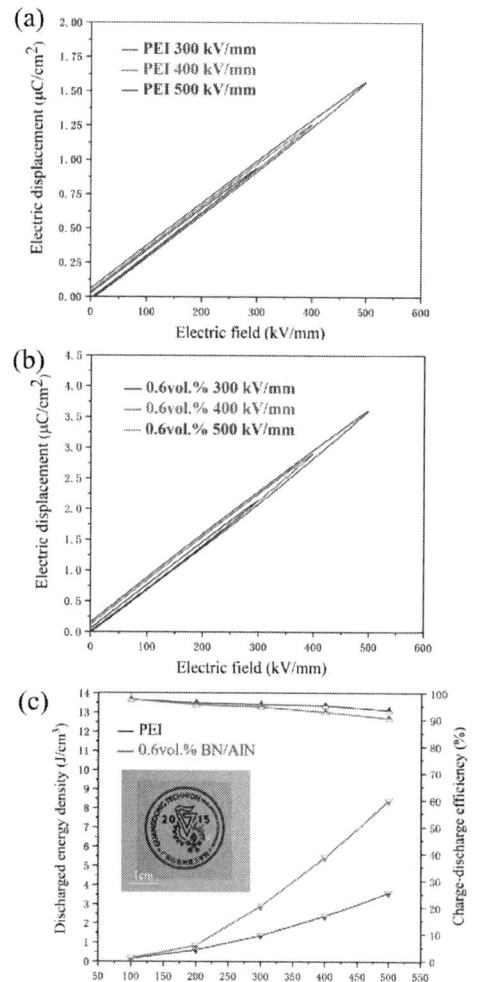

Figure 4. (a-b) P-E loops of PEI and its composites; (c) discharged energy density and charge-discharge efficiency, the inset is composites film sample placed on the GTIIT log.

To clarify the increased polarization effect and the advantage of high thermal conductivity fillers, finite element simulations are carried out as shown in Figure 5. Figure 5a illustrated the polarization strength of PEI composites. The PEI matrix has the lowest and the AlN has the highest polarization. For BN, the strength depends on the orientation with respect to the electrical field. The BN perpendicular to the electric field has smaller polarization, while the BN parallel to the electric field has more significant polarization

ACKNOWLEDGMENT

This work was supported by the Guangdong Basic and Applied Basic Research Foundation – 2019A1515012056.

REFERENCES

[1] D. Q. Tan, "Review of polymer‐based nanodielectric exploration and

film scale‐up for advanced capacitors," Advanced Functional Materials, vol. 30, no. 18, p. 1808567, 2020.

[2] Y.-J. Wan, G. Li, Y.-M. Yao, X.-L. Zeng, P.-L. Zhu, and R. Sun, "Recent advances in polymer-based electronic packaging materials," Composites Communications, vol. 19, pp. 154-167, 2020.

[3] J. Lu and C. Wong, "Recent advances in high-k nanocomposite materials for embedded capacitor applications," IEEE Transactions on Dielectrics and Electrical Insulation, vol. 15, no. 5, pp. 1322-1328, 2008.

[4] Y. Bai, Z.-Y. Cheng, V. Bharti, H. Xu, and Q. Zhang, "High-dielectric-constant ceramic-powder polymer composites," Applied Physics Letters, vol. 76, no. 25, pp. 3804-3806, 2000.

[5] D. Q. Tan, "The search for enhanced dielectric strength of polymer‐based dielectrics: A focused review on polymer nanocomposites," Journal of Applied Polymer Science, vol. 137, no. 33, p. 49379, 2020.

[6] X. Wu, D. Gandla, L. Lei, C. Chen, and D. Q. Tan, "Superior Discharged Energy Density in Polyetherimide Composites Enabled by Ultra-low ZnO@ BN Core-Shell Fillers," Materials Letters, p. 129434, 2021.

[7] T. Zhang et al., "A highly scalable dielectric metamaterial with superior capacitor performance over a broad temperature," Science advances, vol. 6, no. 4, p. eaax6622, 2020.

[8] C. Hou, Z. Bao, H. Sun, Y. Yin, and X. Li, "Improved energy storage performance of nanocomposites with Bi4. 2K0. 8Fe2O9+ δ nanobelts," Journal of Materiomics, vol. 6, no. 2, pp. 371-376, 2020.

The Study on Thermal Aging Mechanism of Silicone Materials for LED Encapsulation

Jiabao Gu
The Fifth Electronics Research Institute
of Ministry of Industry and Information
Technology
Guangzhou, China
121162735@qq.com

Huanxiang Xu *
The Fifth Electronics Research Institute
of Ministry of Industry and Information
Technology
Guangzhou, China
xuhuanxiang@163.com

Bo Peng
The Fifth Electronics Research Institute
of Ministry of Industry and Information
Technology
Guangzhou, China
pengbo@ceprei.com

Zilian Liu
The Fifth Electronics Research Institute
of Ministry of Industry and Information
Technology
Guangzhou, China
lianzi929@163.com

Gang Zhu
The Fifth Electronics Research Institute
of Ministry of Industry and Information
Technology
Guangzhou, China
zhugang05020129@163.com

Abstract—Light-Emitting Diode (LED) has been widely used in various fields owing to their outstanding properties, such as high efficiency, long service life, diverse colors and environmental friendliness. With the rapid development of LED, the requirements of materials for LED encapsulation are getting higher and higher. Currently, silicone resins are widely used for LED encapsulation because of their unique performance. However, in the process of application, silicone resins suffer from aging due to the thermal produced by LED. In this paper, through analyzing the characteristics in aging based on a failure case, the aging mechanism of silicone materials for LED encapsulation is studied. After thermal aging, silicone packaging materials become hard and brittle, leading to the cracking. Furthermore, the proportion of silicon near the crack increases obviously. In the FT-IR spectra, there is an obvious carbonyl absorption peak near $1720cm^{-1}$. In TGA, silicone packaging materials degrade in advance, and the residual mass increase eventually. In DSC, there is an obvious endothermic peak during the first heating process. The fundamental causes of above characteristics can be attributed to the degradation of main chain and the oxidation of side chain.

Keywords—LED; encapsulation; silicone resins; aging; silicone packaging materials

I. Introduction

As the key core of semiconductor lighting technology, Light Emitting Diodes(LEDs) have been developed rapidly all over the world. Under the background of global advocacy of energy conservation and environmental protection, LEDs, as one of the best environmental protection light sources, have a wide application prospect[1]. Compared with the traditional light source such as incandescent lamp and fluorescent lamp, LEDs have the advantages of small size, light weight, flexible application, fast response speed, high luminous efficiency, low energy consumption, long life and so on[2–4]. Therefore, LEDs have been widely used in various fields including lighting, transportation, communications, multimedia electronic products.

Encapsulation is an important process of LED manufacturing, which provides a platform for LED chip[5].

Generally the LED encapsulation materials have a series of excellent physical properties, such as high refractive index, high transmittance, UV radiation resistance, high and low temperature resistance[6]. In addition to sealing and protection, the LED encapsulation materials can also enhance heat dissipation and reduce the temperature of LED chip[7].

At present, the main LED encapsulation materials are silicone materials, epoxy resin, poly(methyl methacrylate) and polycarbonate. Traditional LEDs use epoxy resin as encapsulation material, which possesses excellent adhesion. However, epoxy resin is easy to aging at high temperature and not resistant to UV, making it only suitable for low power LED. With the development of high power LED, silicone materials have been paid more and more attention for their excellent thermal stability and UV stability[8]. Silicone materials have many excellent properties because of its unique molecular structure with Si-O bond as the main chain and organic groups attached to the side chain. Besides thermal stability and UV stability, silicone materials also have the advantages of high light transmittance and high refractive index, making them more widely used in the field of LED encapsulation[9].

The Si-O bond of silicone resin skeleton has high bond energy and stable structure, which is the main reason for the thermal stability of silicone resin. The Si-O bond is partly ionic, and it is easy to crack at high temperature to form oligomer of siloxane. The thermal degradation of silicone resin is mainly composed of Si-O main chain degradation and Si-C side group oxidation[10]. Generally, in the presence of inert gases such as nitrogen, the thermal degradation of silicone resin is mainly the main chain degradation. In the presence of oxygen, main chain degradation and side group oxidation work together to impact the thermal degradation of silicone resin[11]. The main chain degradation leads to the decrease of molecular weight of polysiloxane, which is the main factor affecting the normal use of silicone resin at high temperature. The aging of silicone resin is more complicated with the side group oxidation, resulting in the increase of crosslinking density and decrease of elasticity[12]. In this paper, the aging characteristics of silicone materials for LED encapsulation are analyzed and discussed based on the aging mechanism.

II. The Study on Thermal Aging Mechanism of Silicone Materials for LED Encapsulation

A. Changes of appearance

When the thermal oxidative aging occurs, the silicone materials for LED encapsulation are easy to crack, which is the macro phenomenon of aging (Fig.1). The cracking position is generally located in the encapsulation adhesive above the LED chips where heat is most concentrated. Moreover, the LED encapsulation adhesive became hard and brittle, and the adhesive force decreased significantly.

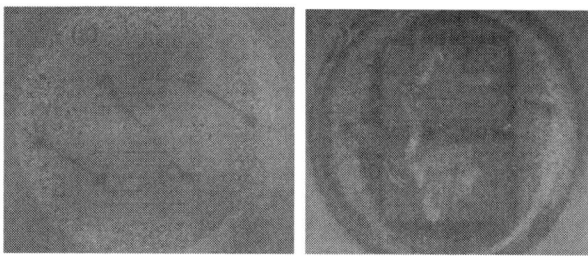

Fig.1. The appearance photos of encapsulation adhesive (a) before and (b) after aging

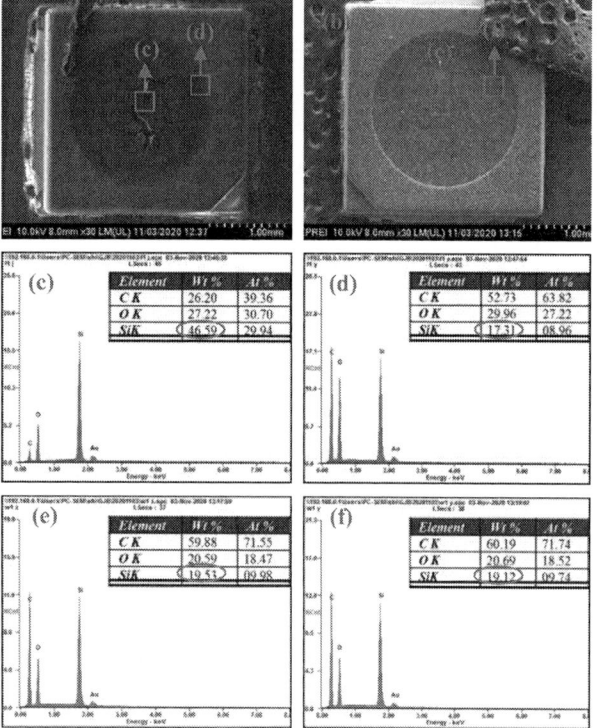

Fig.2. The SEM photos of LED encapsulation adhesive (a) with aging and (b) without aging. The EDS results of aging LED encapsulation adhesive (c) near the crack and (d) far away from the crack. The EDS results of (e) center and (f) edge of unaged LED encapsulation adhesive

B. Changes of Si element

When the LED encapsulation adhesive cracks after aging, the proportion of Si near the crack is obviously higher than that without crack. While in the LED encapsulation adhesive without aging, the proportion of Si in each position is close. As shown in Figure 2, the relative mass percentage of Si near the crack is up to 46.59% which is almost three times that of Si far away from the crack. During the aging of silicone resin, the cyclization depolymerization reaction of Si-O main chain mainly occurs, resulting in cyclic oligomer. Subsequently, the cyclic oligomer diffuses to the surface of silicone resin, leading to the increase of the proportion of Si.

C. Changes of FT-IR spectra

After aging, the side groups of silicone resin oxidize to form carbonyl. Therefore, there is an obvious carbonyl absorption peak near $1720cm^{-1}$ in the FT-IR spectra of aged silicone resin, while no such absorption peak can be detected in unaged silicone resin (Fig.3).

D. Changes of Thermogravimetric Characteristics

Siloxane oligomer with lower decomposition temperature is formed during the aging of silicone resin, leading to the early degradation of silicone resin in Thermogravimetric analysis (TGA). On the other side, small inorganic molecules containing silicon with high decomposition temperature is also formed during the aging, resulting in the increase of final residual mass in TGA. As shown in Figure 4, with the increase of temperature from 250 ℃ to 400 ℃, the residual mass of unaged LED encapsulation adhesive decreases from 99.84% to 98.59%, while the residual mass of aged LED encapsulation adhesive decreases from 97.88% to 95.19% with the early degradation. The final residual mass of unaged LED encapsulation adhesive is 54.30%, while the final residual mass of aged LED encapsulation adhesive increases to 79.81%.

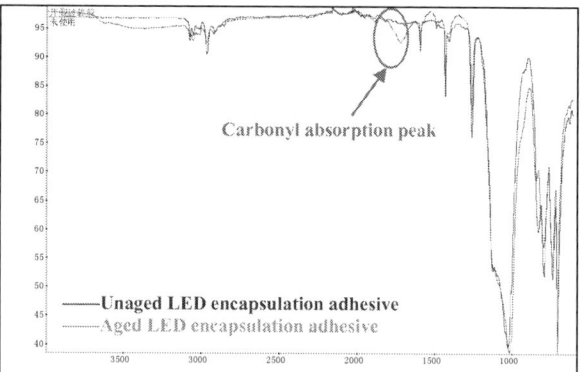

Fig.3. The FT-IR spectra of aged and unaged LED encapsulation adhesive

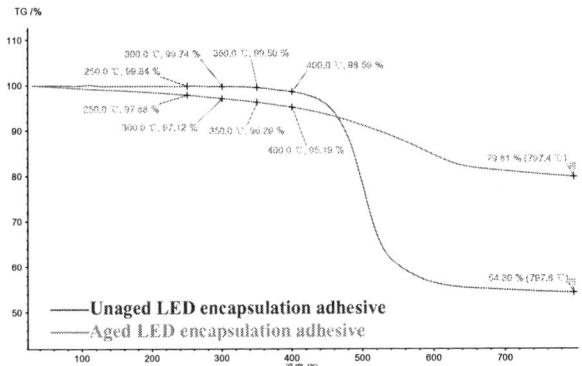

Fig.4. The TGA curves of aged and unaged LED encapsulation adhesive

E. Changes of DSC curves

Compared with unaged LED encapsulation adhesive, a wide endothermic peak can be observed in the differential scanning calorimetry (DSC) curve of aged LED encapsulation adhesive in the first heating process, caused by organic small molecules of aging (Fig.5).

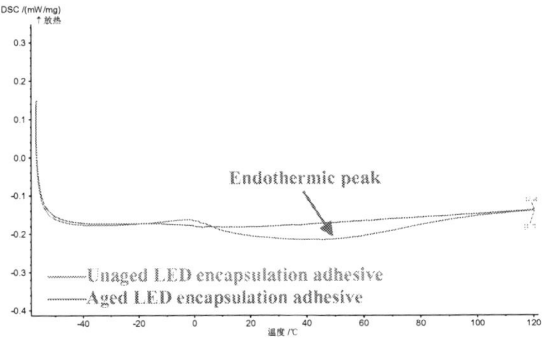

Fig.5. The DSC curves of aged and unaged LED encapsulation adhesive

III.   Conclusion

In summary, the thermal aging mechanism of silicone materials for LED encapsulation is composed of two parts. The first part is the Si-O main chain degradation which mainly leads to the formation of siloxane oligomer and small inorganic molecules containing silicon. The second part is the Si-C side group oxidation which mainly results in the increase of crosslinking density and decrease of elasticity. Affected by aging, some changes can be observed in LED encapsulation adhesive. First of all, the aged LED encapsulation adhesive become hard and brittle, leading to the cracking above the LED chips where heat is most concentrated. Furthermore, the proportion of silicon near the crack increases obviously as the cyclic oligomer diffuses to the surface of silicone resin. Then in the FT-IR spectra, due to the side groups of silicone resin oxidize to form carbonyl, there is an obvious carbonyl absorption peak near $1720cm^{-1}$. In TGA, because of the siloxane oligomer and small inorganic molecules containing

silicon, the aged LED encapsulation adhesive degrade in advance, and the residual mass increase eventually. In DSC, for the organic small molecules of aging, there is an obvious endothermic peak during the first heating process. The above conclusion can provide valuable reference for aging failure analysis of LED encapsulation adhesive.

Acknowledgment

This work was supported in part by grants from the R & D Projects in Key Areas of Guangdong Province (no.2019B010143002).

References

[1] Krames MR, Shchekin OB, Mueller-Mach R, Mueller GO, Zhou L, Harbers G,et al. Status and future of high-power light-emitting diodes for solid-state lighting. J Disp Technol 2007; 3:160–175.

[2] Sun Y P, Gu A J, Liang G Z, et al. Preparation and properties of transparent zinc oxide/silicone nanocomposites for the packaging of high-power light-emitting diodes. Journal of Applied Polymer Science 2011; 121:2018-2028.

[3] Rasoulifard M H, Fazli M, Eskandarian M R. Performance of the light-emitting-diodes in a continuous photoreactor for degradation of direct red 23 using UV-LED/$S_2O_8^{2-}$ process. Journal of Industrial and Engineering Chemistry 2015; 24:121-126.

[4] Pimputkar S, Speck J S, Denbaars S P, et al. Prospects for LED lighting. Naturephotonics 2003; 3:180-182.

[5] Chang M H, Das D, Varde P V, et al. Light emitting diodes reliability review. Microelectronics Reliability 2012; 52: 762-782.

[6] Liu S, Luo X. LED packaging for lighting applications: design, manufacturing, and testing. John Wiley & Sons 2011:23-35.

[7] Tsao J Y. Solid-state lighting lamps, chips and materials for tomorrow. IEEE Circuits&Devices 2004; 20:28-37.

[8] Lei I A, Lai D F, Don T M, et al. Silicone hybrid material useful for the encapsulation of light-emitting diodes. Material Chemistry and Physics 2014; 144:41-48.

[9] Delebecq E, Ganachaud F. Looking over liquid silicone rubbers: (1) network topology vs chemical formulations. ACS Applied Materials and Interfaces 2012; 4:3340-3352.

[10] Hall A D, Patel M. Thermal stability of foamed polysiloxane rubbers: Headspace analysis using solid phase microextraction and analysis of solvent extractablematerial using conventional GC-MS. Polymer Degradation and Stability 2006; 91: 2532-2539.

[11] Camino G, Lomakin S M, Lageard M. Thermal polydimethylsiloxane degradation. Part 2. The degradation mechanisms. Polymer 2002; 43: 2011-2015.

[12] Liptay G, Nagy J, Weis J C . Thermal analysis of silicone polymers and silicone rubbers. Journal of Thermal Analysis 1987; 32:1683-1689.

Defect Localization and Optimization of PIND for Large Size CQFP Devices

Shinan Wang *
China Electronics Technology Group Corporation No.58 Research Institute
Wuxi, China
wsn_123@163.com

Yong Ma
China Electronics Technology Group Corporation No.58 Research Institute
Wuxi, China
mayong_jack@126.com

Kaihong Zhang
China Electronics Technology Group Corporation No.58 Research Institute
Wuxi, China
sally_zhang_cetc@163.com

Yongjian Yu
China Electronics Technology Group Corporation No.58 Research Institute
Wuxi, China
yuyj98@163.com

Yongkang Wan
China Electronics Technology Group Corporation No.58 Research Institute
Wuxi, China
wyk5920@163.com

Weikun Xie
China Electronics Technology Group Corporation No.58 Research Institute
Wuxi, China
chinagrass@163.com

Abstract—CQFP devices have been widely used in aerospace, military and other fields due to the advantages of high reliability[1], but large-size CQFP devices are prone to failure after PING test. This paper uses ANSYS Workbench to locate the defect and verify it by scanning electron microscope (SEM). The verification results are consistent with the simulation results. In addition, a test fixture was used to optimize the PIND test process, and the results show that the use of test fixture can effectively reduce the risk of device failure.

Keywords—Large-Size CQFP Device, PIND, Defect Localization, FEM, Optimization

I. INTRODUCTION

Surface mount technology (SMT) is a widely used electronic assembly technology, which directly attaches components without pins to Printed Circuit Board (PCB), replacing the traditional through-hole mounting method. It has the advantages of light, thin and small, high reliability, easy mass production and so on[2]. Ceramic Quad Flat Pack (CQFP) device is a kind of important surface mount components, using ceramic substrate and gold-plated pin. It has the advantages of small size, light weight, high packaging density, good thermoelectric performance, suitable for surface installation, high reliability, and being suitable for surface mount, which can be widely used in the military, aerospace and aviation fields[3].

During the factory inspection, the device needs to be subjected to a Particle impact noise detection test(PIND)[4,5]. GJB548B-2005 " Test Methods and Procedures for Microelectronic device" Method 2020.1 describes the PIND test that the purpose of this test is to detect the free particles in the device packing cavity, and it is a non-destructive test[6]. A large-size, dual-cavity CQFP device failed after the PIND test. It is preliminarily inferred that during the PIND test, the large-size CQFP device does not remove frame and the frame was suspended out of the PIND test bench. The suspended frame causes stress concentration in the process of shock and vibration, resulting in damages such as cracks and fractures, which eventually leads to failure. This paper uses ANSYS Workbench to simulate the PIND test process, so as to locate the defect.

II. DEFECT LOCALIZATION

A. Model

The package shape of the CQFP204 device is shown in Figure 1. The number of pins is 204, the width of pins is 0.3mm, the thickness of pins is 0.15mm, and the spacing is 0.635mm. The external dimension is 35mm x 35mm x 5.84mm (the size of the shell itself). Due to the variety and a great deal of components, the design of double-sided cavity mount is adopted to optimize the structure layout as much as possible.

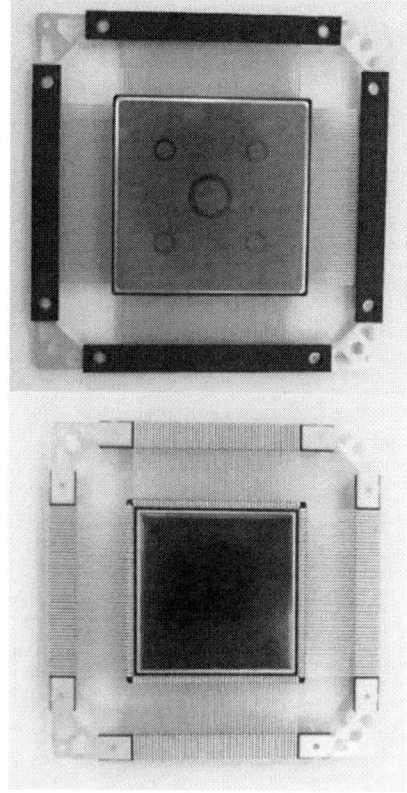

Fig. 1. Picture of CQFP204

According to the dimension information of the device, the corresponding mesh model is established, as shown in Fig. 2. As the pin is the main research object, the mesh of the pin is refined, whereas the mesh of the frame and other non-main structures are simplified appropriately.

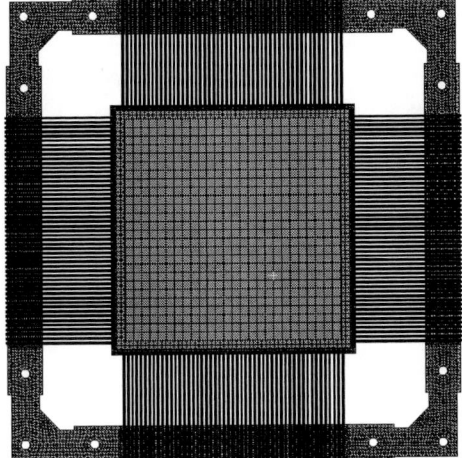

Fig. 2.   Mesh model of CQFP204

B. Material

The substrate material is ceramic.The upper and lower parts are provided with metal heat dissipation plates which made of copper, and the pin is made of 4J29 Kovar alloy. The parameters of each material are shown in Table 1.

TABLE I.   MATERIAL PARAMETER SHEET

Material	Density (kg/m³)	Modulus of elasticity (GPa)	Poisson's ratio
Cu	8942	123	0.35
Ceramic	2190	26.5	0.25
Kovar	7980	138	0.32

C. Test Conditions and Loads

GJB548B-2005 Method 2020.1 specifies the conditions of the PIND test, to provide a basically sinusoidal motion for the device under test:

- Condition 1 : The peak value is 196m/s² at 40Hz ~ 250Hz.

- Condition 2: The peak value is 98m/s² when the frequency is greater than or equal to 60Hz.

According to the detailed requirements of the product, condition 1 is selected in this test. After measurement, we know that the average height of the package cavity is 1.64mm. According to the table of "Relationship between Test Frequency and Effective Height of the Cavity at 196m/s² Acceleration", the test frequency is selected as 100Hz. Thus, the test conditions of this PIND test are: sinusoidal motion with frequency of 100Hz and amplitude of 196m/s²[7].

The displacement-time equation of sinusoidal motion is as follows:

$$S = Asin(2\pi ft) \qquad (1)$$

The acceleration-time equation is obtained by two differentials:

$$a = -4\pi^2 f^2 Asin(2\pi ft) \qquad (2)$$

Through substituting the test conditions into Equation (2), the displacement load equation is obtained:

$$S = 0.5sin(200\pi t) \qquad (3)$$

D. Results and Analysis

The simulation results are shown in Figure 3. Under the above loading conditions, an obvious stress concentration phenomenon appears at the root of the pin, and the maximum Von mises stress is 319.1MPa. The simulation results preliminarily confirm the above speculation and locate the location of the defect at the root of the pin.

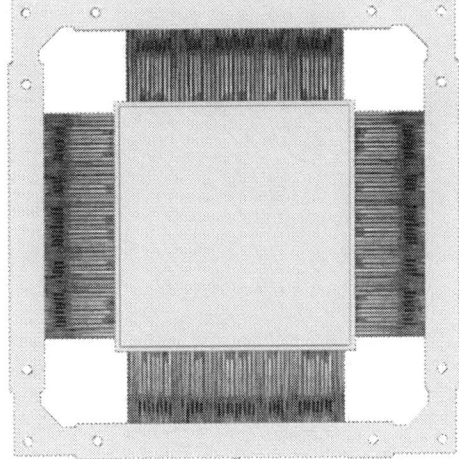

Fig. 3.   Maximum von Mises stress

Scanning electron microscope(SEM) was used to analyze the failed device, and the root of CQFP204 pin did have a crack defect after the PIND test, as shown in Figure 4.

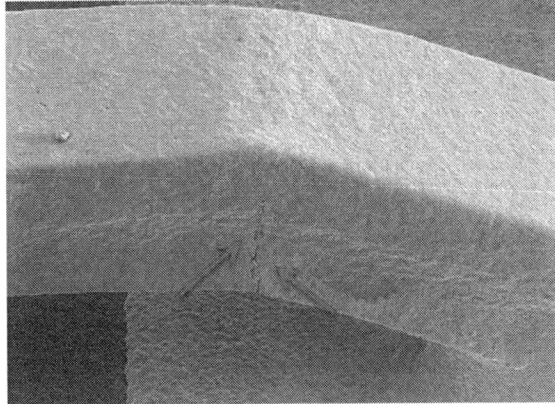

Fig. 4.   Crack defects of pin

III. OPTIMIZATION

A. Fixture Design

In this paper, a fixture is designed to optimize the PIND test process for large CQFP devices, as shown in Fig. 5.

The fixture is made of 6061T6 aluminum magnesium alloy and has the advantages of light weight, good rigidity and high sound conductivity. In the preparation stage of PIND test , the device can be completely confined inside the fixture, and the fixture is bonded to the transducer table through hydrogel. In the process of PIND test, there will be no relative sliding and additional noise, which will not have any impact on the test results, and it can meet the relevant requirements of the fixture in GJB548B-2005 Method 2020.1.

Fig. 5. PIND test fixture

Simulation of PIND test process with fixture was carried out by Ansys Workbench, and the test conditions and loads were consistent with the above conditions.As shown in Fig. 6, the stress concentration at the root of the pin is significantly dropped, and the maximum equivalent stress is 178.4MPa, which is 44.2% lower than that without fixture.The simulation results show that the application of PIND test fixture achieves the expected purpose. And we come to a conclusion that the PIND test process is optimized effectively.

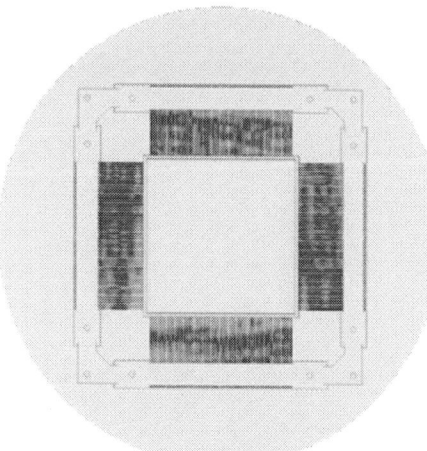

Fig. 6. Von Mises stress using fixture

B. Validation

A PIND test with fixture was carried out again on the CQFP204 device with normal function and no defective pins using the fixture. Then they were tested for electrical properties and examined with SEM. The results show that the electrical performance of the CQFP204 device is normal, and there is no crack defect at the the root of the pin. The scanning electron microscope picture is shown in Figure 7.

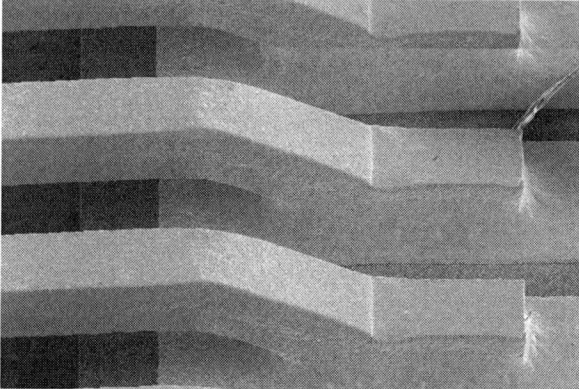

Fig. 7. Pins are normal without cracking

IV. CONCLUSION

In this paper, the failure phenomenon of large CQFP devices after PIND test has been studied. The conclusions are obvious as follows:

(1) In the PIND test of large-size CQFP device, the frame is not removed and the frame was suspended out of the PIND test bench. The suspended frame causes stress concentration in the process of shock and vibration, resulting in damages such as cracks and fractures, which eventually leads to failure.

(2) It can effectively simulate the PIND test process and locate the risk position of stress concentration in the test process by finite element software such as Ansys Workbench.

(3) Proper use of fixtures during PIND testing can alleviate the stress concentration of large CQFP devices, thus reducing the risk of device failure and optimizing the PIND testing process.

REFERENCES

[1] YIN Jiance, LU Wei, CHEN Jiaqiang. High Reliability Assembly of CQFP[J]. Electronics Process Technology, 2014,35(04):230-233.

[2] FEI Jingming, WU Qiong, ZHANG Xiaochao. Study on Assembling Process of Vibration Resistance for Aerospace Power CQFP Packaging [J]. Electronics Process Technology,2018, v.39;No.268(02):84-87.

[3] LIU Shubin, WEN Yuejiao. Stress Analysis of Solder Joint of a CQFP Package Device[J].CHINA SCIENCE AND TECHNOLOGY INFORMATION, 2014,No.499(15):184-185.

[4] Zhang H, Wang S J, Zhai G F. Dynamic model of particle impact noise detection[C].IEEEIECON,2004:2577−2581.

[5] Zhang H, Wang S J, Zhai G F. Test conditions discussion of particle impact noise detection for space relay[C].IEEEIECON,2004: 2566−2572.

[6] Ministry of Machine Building & Electronics Industry of People's Republic of China..GJB548B Test Methods and Procedures for Microelectronic device[S].China Standards Press2009.

[7] CHEN Jinyan, WANG Xiaofei, HUANG Jiaoying, SUN Yue. Simulation analysis of PIND vibration of crystal oscillator based on finite element[J].Modern Electronics Technique, 2012,35(16):156-159.

Exploration of the synthesis method of quaternary copolymerized thermoplastic polyimide

1st Jinshan Liu
Shenzhen Institute of Advanced Electronic Materials, Shenzhen Institute of Advanced Technology, Chinese Academy of Sciences, Shenzhen, 518055, China
Department of Nano Science and Technology Institute, University of Science and Technology of China, Suzhou, 215123, China
liu.js@siat.ac.cn

2nd Jinhui Li*
Shenzhen Institute of Advanced Electronic Materials, Shenzhen Institute of Advanced Technology, Chinese Academy of Sciences, Shenzhen, 518055, China
jh.li@siat.ac.cn

3rd Fangfang Niu*
College of Physics and Optoelectronic Engineering, Shenzhen University, Shenzhen, 518060, China
ffn@szu.edu.cn

4th Tao Wang
Shenzhen Institute of Advanced Electronic Materials, Shenzhen Institute of Advanced Technology, Chinese Academy of Sciences, Shenzhen, 518055, China
tao.wang3@siat.ac.cn

5th Wen Liu
Shenzhen Institute of Advanced Electronic Materials, Shenzhen Institute of Advanced Technology, Chinese Academy of Sciences, Shenzhen, 518055, China
wen.liu@siat.ac.cn

6th Guoping Zhang*
Shenzhen Institute of Advanced Electronic Materials, Shenzhen Institute of Advanced Technology, Chinese Academy of Sciences, Shenzhen, 518055, China
gp.zhang@siat.ac.cn

7th Rong Sun*
Shenzhen Institute of Advanced Electronic Materials, Shenzhen Institute of Advanced Technology, Chinese Academy of Sciences, Shenzhen, 518055, China
rong.sun@siat.ac.cn

Abstract—**Polyimide (PI), especially aromatic polyimide has excellent comprehensive performance, making it widely applied in the field of electronic packaging. However, many aromatic polyimides usually suffer from poor processability, which severely limits their application especially in the flexible copper clad laminate (FCCL) or the low temperature electronic package process of temparory bonding or debonding (TBDB). Herein, a quaternary copolymerized thermoplastic PI (TPI) is studied and the effect of several synthesis methods on the overall properties of TPI films are explored in details. The experimental results show good solubility of TPI films which is ascribed to the synergistic effect of flexible units, low coplanar and distorted dianhydride. Most importantly, our experimental results revealed that the synthesis methods had a limited effect on the thermal properties, compared with the dielectric and mechanical properties. Collectively, this work provides systematic guidelines for synthesis method based on a quaternary copolymerized thermoplastic polyimide system which are promising for high frequency applications of 5G era.**

Keywords—thermoplastic polyimide; synthesis method; mechanical property; dielectric property

I. INTRODUCTION

Polyimide (PI) which contains the imide ring in the polymer chains possess superior thermal properties, excellent mechanical and dielectric properties, adjustable thermal expansion coefficient (CTE), as well as other outstanding characteristics[1]. However, it is difficult for PIs to be used directly as thin films or coatings because they are characterized by insolubility and low meltability in most cases. So, the development of thermoplastic PI (TPI) is highly desirable especially in the flexible copper clad

laminate (FCCL)[2] or the low temperature electronic package process like temparory bonding or debonding[3]. Typically, there are several methods to develop TPIs, such as the introduction of flexible units, asymmetrical structures or of some bulky side groups in the molecular chain. In these methodologies, the introduction of a large number of flexible ether linkages in the PI molecular chains achieved a commercial aviable TPI film named as Kapton®-KJ by DuPont company[4]. Nevertheless, the glass transition temperature (T_g) of 220 °C is relatively low and limits its application in the situation of high temperature environment. Therefore, TPIs with excellent comprehensive performance are highly desired in semiconductor industry.

In addition, it is well acknowledged that polymer chemistry is one of the most effective approaches to achieve overall properties by adjusting their monomers used in synthesis. In terms of PI, the selection of preparation methods for copolymerization of different monomers has already become a hot research topic over the recent years. Accordingly, a great deal of research works have been carried out to obtain the copolymerized TPIs. For instance, Yu et al.[5] successfully prepared a set of organosoluble TPIs based on the combination of 2,3,3′,4′-oxydiphthalic dianhydride (a-ODPA) and 2,3,3′,4′-biphenyltetracarboxylic dianhydride (a-BPDA) with the twisted non-coplanar molecular structures. Meanwhile, by introducing the a-ODPA and 3,4,3′,4′-oxydiphthalic dianhydride (s-ODPA), Cao et al.[6] elaborately prepared a set of novel quaternary copolymerized TPIs with excellent solubility in various industrial polar aprotic solvents and high T_g of 235.3−305.5 °C. In additon, Liu et al.[7] obtained TPIs with

978-1-6654-1392-3/21 $31.00 © 2021 IEEE

rigid isomeric dianhydrides of BPDA as well as containing ether diamines of 4,4'-oxydianiline (ODA) and 1,3-bis(4'-aminophenoxy)benzene (TPE-R) and found that optimal overall performance. However, although the copolymerization has been proved as one important way to prepare TPI which has attracted extensive attention within academia and industry, there are still problems to be discussed and solved, for example, the influence of synthesis methods on the comprehensive properties of TPI.

Herein, the adding order and method of monomers for a quaternary copolymerized TPIs have been probed elaborately. Asymmetric dianhydride of a-BPDA and a-ODPA, as well as two kinds of diamines of 9,9'-bis(4-aminophenyl)fluorene (BAFL) and TPE-R are selected for the preparation of TPIs. Then, the effect of several synthesis methods on the overall properties of TPI films are studied in detail. Our experimental results revealed that synthesis methods of TPIs has more obvious effects on the dielectric and mechanical properties, compared with the thermal properties, which offers a facile way to choice suitable synthesis methods for quaternary copolymerized TPIs.

II. EXPERIMENT SECTION

A. Experimental procedure

a-BPDA was obtained from Energy Chemical. a-ODPA was supplied by TCI, Shang-hai. BAFL and TPE-R were provided by ChinaTech Chemical, Tian-jin. The dianhydrides undergone heat treatment at 160 °C for 7 h and the diamines at 60 °C for 7 h were employed. N, N-dimethylacetamide (DMAc) was obtained from Energy Chemical Co., Ltd.

For all of the TPIs (named as TPI-1 to TPI-4), diamines of BAFL and TPE-R are dissolved in the dimethylacetamide (DMAc) at first. In detils, TPI-1 is prepared by adding dianhydrides solution of a-BPDA and a-ODPA to the DMAc solution of diamines in two batches. TPI-2 is obtained by the addition with the dianhydrides powders of a-BPDA and a-ODPA in two batches. Then, TPI-3 is prepared by adding a-BPDA powder at first and then a-ODPA powder. And TPI-4 is obtained by adding a-ODPA powder at first and then a-BPDA to the DMAc solution of diamines. At last, a 15% solid content PAA solution is obtained, the TPI films are prepared after spin-coating and thermal imidization process with the following heating procedure: 100, 200, 300, 350 °C at a heating rate 5 °C/min and soaking time of 1 h, respectively.

B. Characterization

Fourier transform infrared (FT-IR) spectra was tested by the Bruker Vertex spectrometer, Germany. The molecular weight were obtained from the gel permeation chromatography. The solubility of samples were defined by adding 5 mg of the samples in 10 mL of solvents for 24 h at room temperature. Thermogravimetric analysis (TGA) was characterized via the STD-Q600 machine within the limits of 30 °C to 800 °C. The T_g of the prepared TPI films were tested by a DSC Q20 (TA Instruments, America). The dimensional stability of TPI films was evaluated via coefficient of thermal expansion (CTE) tested by TMA. Mechanical properties were implemented on the TA DMA-

Q800. The dielectric constant at a high frequency of 10 GHz was probed by the vector network analyzer.

III. RESULTS AND DISCUSSION

A. Structure Characterization

The structure of TPI films was characterized by FTIR spectra at first as presented in Fig. 1. Typically, the existence of imide groups was demonstrated by the characteristic absorption peaks at 1777, 1712, 1368, 1218 and 737 cm^{-1}, of which 1777 and 1712 cm^{-1} assigns to the C=O asymmetric and symmetric stretching vibration, respectively. And 1368 cm^{-1} (C–N stretching vibration), 1218 cm^{-1} (C–O–C stretching vibration) and 737 cm^{-1} represents a bending vibration of the imide ring. As a result, the above characteristic absorption peaks appeared proves that TPI-1~TPI-4 had been successfully prepared. Additionally, the molecular weight of PAAs were measured and tabulated in the TABLE I, which shows the prepared PI film of high M_w between 65166 and 67610 g/mol.

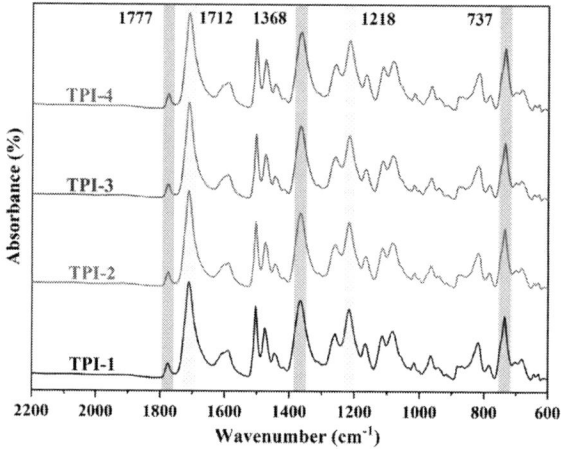

Fig.1. FTIR spectra of the TPI films.

TABLE I. Molecular weights of PAAs.

Groups	TPI-1	TPI-2	TPI-3	TPI-4
M_n [g/mol]	48629	47204	50726	47420
M_w [g/mol]	65574	65166	67610	65420
PDI	1.35	1.38	1.33	1.38

B. Solubility

The solubility of the prepared TPIs are investigated then by adding 5 mg of the samples in 10 mL of solvent. The results show that TPI-1 to TPI-4 are partially soluble at room temperature in DMF, DMSO, DMAc, NMP, and all TPI films become completely soluble under heating at 60 °C for 2 h as shown in TABLE II. All the TPI samples possess good solubility which is ascribed to the mutual effect of flexible ether linkages and relatively low coplanarity of dianhydride distorted.

TABLE II. The solubility of TPI films in common solvents.

Groups	DMF	DMAc	NMP	DMSO	Acetone	THF
TPI-1	Y⁻	Y⁻	Y⁻	Y⁻	N	N
TPI-2	Y⁻	Y⁻	Y⁻	Y⁻	N	N
TPI-3	Y⁻	Y⁻	Y⁻	Y⁻	N	N
TPI-4	Y⁻	Y⁻	Y⁻	Y⁻	N	N

Y⁻: Partially soluble at room temperature within 24 hours and completely soluble under heating at 60 °C for 2 h;
N: insoluble.

In addition, the prepared TPI-1~TPI-4 films have only a broad peak presented in Fig. 2, which shows the amorphous aggregate structure of TPI films. Consequently, the lossed molecular chain stack contributes to their excellent solubility as well.

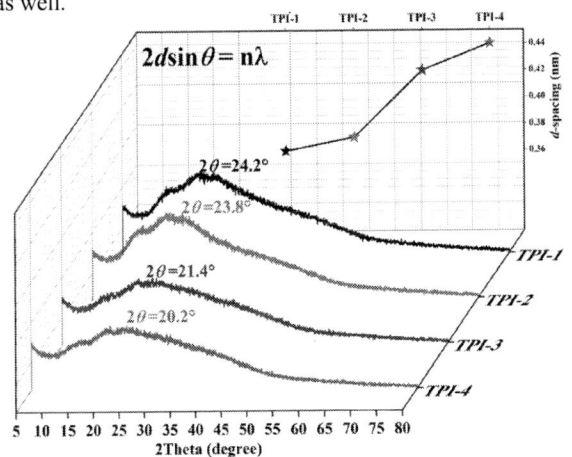

Fig. 2. The WAXD patterns and corresponding d-spacing values.

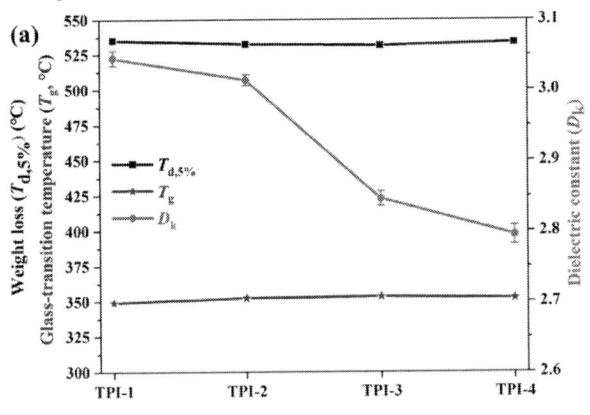

Fig. 3. (a) Thermal and dielectric properties, (b) Mechanical properties of the prepared TPIs with several synthesis methods.

C. Mechanical Properties

Additionally, the rigid biphenyl structure and fluorenyl cardo groups endowed the TPI films with excellent mechanical properties, and the elongation at break (ε_b, %), tensile modulus (E, GPa) and ultimate tensile strength (σ_{max}, MPa) values of TPI-1 to TPI-4 are (8.8, 2.02, 101.4), (7.9, 2.04, 100.7), (10.3, 2.25, 112.6), and (9.2, 2.29, 112.0), respectively. The data summarized in the TABLE III also proves that the seperated adding of dianhydrides (TPI-3 and TPI-4) result in better mechanical properties compared with the mixed addition whether in solution (TPI-1) or powder (TPI-2) as presented in the Fig. 3 (b).

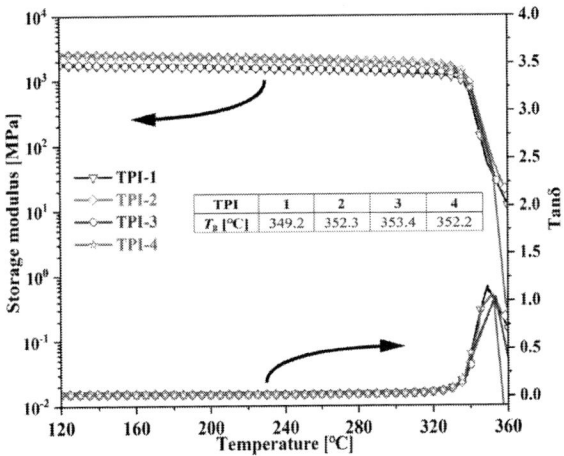

Fig. 4. DMA curves of the TPI films: storage modulus and tanδ.

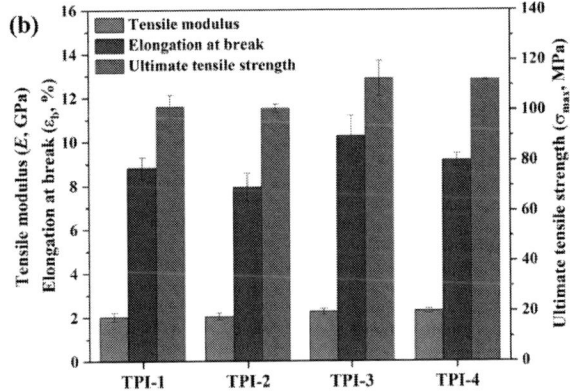

TABLE III. Thermal and mechanical properties of the prepared TPI films with several synthesis methods.

Groups	Thermal properties								Mechanical properties		
	TGA				DSC	DMA	TMA		DMA		
	$T_{d,5\%}$ [°C]	$T_{d,10\%}$ [°C]	T_{max} [°C]	$R_{800°C}$ [%]	T_g' [°C]	T_g'' [°C]	T_g''' [°C]	CTE [ppm/K]	σ_{max} (MPa)	ε_b (%)	E (GPa)
TPI-1	535.2	556.9	585.3	64.81	299.5	349.2	316.6	52.9	101.4±4.7	8.8±0.5	2.02±0.19
TPI-2	532.5	553.7	579.8	65.74	300.8	352.3	313.7	63.7	100.7±1.6	7.9±0.6	2.04±0.16
TPI-3	531.6	552.5	581.7	64.54	300.5	353.4	314.5	52	112.6±7.1	10.3±0.9	2.25±0.13
TPI-4	533.7	553.7	576.3	64.99	301.0	352.2	311.9	68.8	112.0±0.4	9.2±0.3	2.29±0.10

D. Thermal and Dielectric Properties

As shown in the Fig. 4, we can determine that the PIs synthesized are thermoplastic, which ascribes to the sharp decrease above 10^3 MPa in storage modulus near the T_g[7]. Notably, we can see from the Fig. 5 that the thermostability of all the TPI films are excellent of which the temperature of 5% weight loss ($T_{d,5\%}$, °C), retention at 800 °C ($R_{,800°C}$, %) and glass-transition temperature (T_g, °C) values of TPI-1 to TPI-4 are (535.2, 64.81, 349.2), (532.5, 65.74, 352.3), (531.6, 64.54, 353.4), and (533.7, 64.99, 352.2), respectively. Detailed thermal properties of as prepared TPIs are tabulated in TABLE III. Obviously, this results indicate that the synthesis methods show a slight effect on the thermal properties.

Finally, the most important thing to point out is that the TPI-3 and TPI-4 possess lower dielectric constant of 2.79-2.85 at a high frequency of 10 GHz compared to TPI-1 and TPI-2 (3.01-3.04) as shown in Fig. 3 (a). Meanwhile, this can be confirmed by the *d*-spacing values in the XRD pattern. As presented in Fig. 2, the corresponding values are 0.36, 0.37, 0.42 and 0.44 nm from the peak at 24.2°, 23.8°, 21.4° and 20.2°, respectively. The underlying explanation is that the prepared TPI films with a larger *d*-spacing value exhibit lower D_k according to the Clausius-Mosottti equation as follows:

$$D_k = (1 + 2\frac{P}{V})/(1 - \frac{P}{V})$$

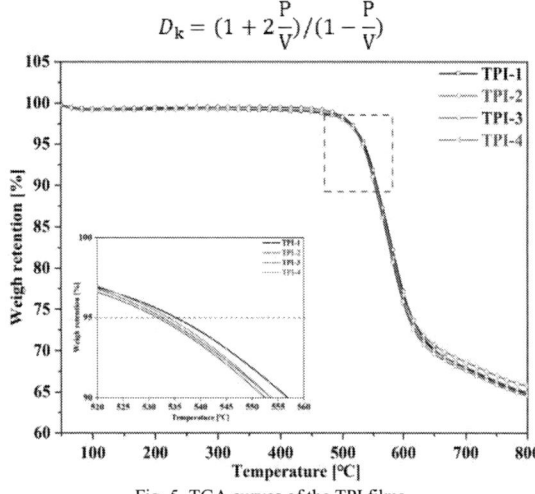

Fig. 5. TGA curves of the TPI films.

IV. CONCLUSIONS

In this study, we have successfully synthesized the quaternary copolymerized TPIs based on different adding order and method of monomers. Through the investigation of overall properties, it is concluded that our experimental results revealed that synthesis methods of TPIs has more obvious effects on the dielectric and mechanical properties, compared with the thermal properties. This work provides systematic guidelines for synthesis method based on a quaternary copolymerized TPI system, which are promising for high frequency applications of 5G era.

ACKNOWLEDGMENT

We acknowledge the financial support of National Natural Science Foundation of China (61904191), Youth Innovation Promotion Association of Chinese Academy of Sciences (2017410), Key R&D Project of Guangdong Province (2020B010180001) and National Key R&D Project from Minister of Science and Technology of China (2017ZX02519).

REFERENCES

[1] Liaw, D. J, et al. Advanced polyimide materials: syntheses, physical properties and applications[J]. Progress in Polymer Science, 2012, 37(7): 907-974.

[2] Cao, X. W, et al. Preparation and properties of adhesive-free double-sided flexible copper clad laminate with outstanding adhesion strength[J]. High Performance Polymers, 2021, doi:10.1177/0954008320988761.

[3] Cheng, C. A, et al. Characterization of temporary bonding and laser release using polyimide and a 300-nm photolysis polymer system for high-throughput 3-D IC applications[J]. IEEE Transactions on Components Packaging and Manufacturing Technology, 2017, 7(3): 456-462.

[4] Kanakarajan K., Kreuz J. A. Flexible multi-layer polyimide film laminates and preparation thereof. U.S. Patent 5298331A, USA (1994).

[5] Yu, P, et al. Influence of different ratios of a-ODPA/a-BPDA on the properties of phenylethynyl terminated polyimide[J]. Journal of Polymer Research, 2018, 25(5).

[6] Cao, X. W, et al. Synthesis and characterization of a novel quaternary copolymerized thermoplastic copolyimide with excellent heat resistance, thermoplasticity, and solubility[J]. Express Polymer Letters, 2020, 14(8): 757-767.

[7] Liu, X, et al. Preparation and properties of a novel quaternary copolymerized thermoplastic polyimide[J]. Express Polymer Letters, 2019, 13(6): 524-532.

Impact Force Control of High-speed Wire Bonding Machine Based on Fuzzy Active Disturbance Rejection Controller

Yachao Liu
State Key Laboratory of Precision Electronic Manufacturing Technolog and Equipment
Guangdong University of Technology
Guangzhou, China
email: 815513904@qq.com

Jian Gao*
State Key Laboratory of Precision Electronic Manufacturing Technolog and Equipment
Guangdong University of Technology
Guangzhou, China
*email: gaojian@gdut.edu.cn

Boyu Zhan
State Key Laboratory of Precision Electronic Manufacturing Technolog and Equipment
Guangdong University of Technology
Guangzhou, China
email: 1442140788 @qq.com

Lanyu Zhang
State Key Laboratory of Precision Electronic Manufacturing Technolog and Equipment
Guangdong University of Technology
Guangzhou, China
email: 571356769 @qq.com

Abstract—**Improving the dynamic performance of impact force control is one of the key factors to ensure the bonding quality in wire bonding processes of IC microelectronics products. A fuzzy active disturbance rejection control (fuzzy ADRC) method is proposed to reduce the impact force and the consequent fluctuation of the bonding head. This control method uses a fuzzy controller to automatically adjust the control parameters of the error feedback controller based on the standard ADRC. With this fuzzy ADRC control method, the overshoot of the bonding head positioning response can be effectively reduced, and the fluctuation of the impact force can be controlled within a small region. The dynamic simulation model of impact force is established, and the design process of fuzzy ADRC control strategy for bonding head servo system is described in detail. Compared with conventional PID and fuzzy PID, simulation and experimental results demonstrate that the fuzzy ADRC control method can achieve better performance in reducing the impact force overshoot and force fluctuation. Therefore, the fuzzy ADRC control method for high dynamic Z-axis motion can improve bonding quality.**

Keywords—ADRC, bonding machine, bonding quality, fuzzy logic, impact force control

I. INTRODUCTION

The development trend in chip miniaturization, high density, high integration, and low cost introduces severe technical challenges to the chip packaging process and equipment, which is the bottleneck restricting the development of semiconductor industry [1]. A wire bonding machine with high-speed and high-acceleration is an important piece of electronic packaging equipment. Usually, a wire bonding machine has a high-acceleration of 10-15g for its XY-table motion and 100g for its Z-axis motion, the velocity of wire bonding can reach 15-20 lines/s, and the positioning accuracy can achieve a level of 1-2μm [2]. This kind of packaging equipment is affected by all kinds of high-frequency mutating signal processing in motion control system, which requires more real-time response to the system than any conventional operational control device. Furthermore, the running process of this packaging machine

includes multiple subsystems such as high dynamic motion between the XY-table and Z-axis, positioning, ultrasonic energy, free air ball (FAB) with electronic flame-off, wire bonding, and position/force section switching control. There is an electrical-magnetic coupling, force heat hybrid, and macro-micro combination between these multiple subsystems.

The influencing factors of the nonlinear disturbance must be considered comprehensively, and the dynamic mathematical model that reflects the characteristic of the mechanical and electrical system of multiple variable, nonlinear and strong coupling must be established [3]. Considering the technical challenge of the wire bonding for the miniature and high-density IC chips, smaller pad sizes and shorter solder line pitches require various new techniques, such as XYZ-table motion control technology, the faster and more stable contact detection control technology, the ultrasonic generator technology, and the precision impact force control technology of Z-axis specially [4].

If the impact force of Z-axis is very strong during wire bonding process, it will produce an oversized squashed bonded ball, as shown in Fig. 1 (a). However, if the impact force is very weak, the bonded ball may cause a virtual welding phenomenon, as shown in Fig. 1 (b). Therefore, the design of an accurate and stable control method for Z-axis impact force is the critical aspect to guaranteeing the wire bonding quality [5].

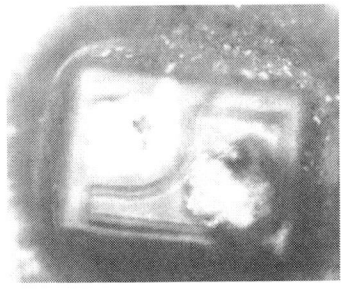

(a)

This work is supported by Guangdong Provincial R&D Key Projects under Grant 2018B090906002, and National Natural Science Foundation under Grant 52075106, and U20A6004.

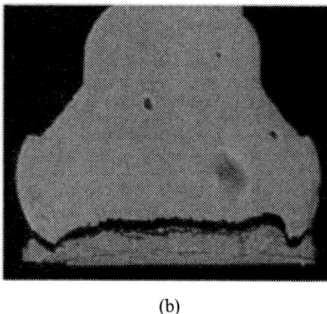

(b)

Fig. 1. The phenomenon of bonded ball (a) oversized bonding, (b) virtual bonding.

Realizing the effective control of the high impact movement of wire bonding machine, and ensuring dynamic response of the bonding servo system performance, has attracted the interest of many researchers. Kim and Yim proposed using the piezo force sensor to compensate impact force, which reduced the peak value [6]. Kim and Park proposed a contact detection algorithm to predict the contact position and adjust the head position/force switching time, and improved the disturbance suppression performance of the wire bonding servo system [7]. To reduce overshoot and settling time of wire bonding control system, Gao and Dai proposed a fuzzy PID control method for force fluctuation control of a LED wire bonding machine [8]. In [9], a Smith-ADRC control algorithm for time-delay of the wire bonding system is proposed, and the superior performance of this method is demonstrated by simulation experiments.

However, the development of micro-spaced microchips has introduced higher requirements for packaging equipment. The high density and three-dimensional packaging processes require a higher precision and stability of the bonding machine [10]. Because the high-speed and high-acceleration motion under the condition of impact force control system is very complicated, the operational process is easily affected by mechanical movement, environmental disturbance, and system vibration, which leads to an inconsistency in bonding quality. The standard ADRC algorithm has many control parameters that requires tuning, which is a challenging task and limits its control performance [11, 12]. To solve these problems, we proposed a fuzzy ADRC based impact force control method for high-speed wire bonding machine. This method realizes online self-tuning of the gain parameters in error feedback controller by fuzzy inference, and thus can achieve precise and stable control for the Z-axis impact force. The experimental results showed that the peak value of impact force can be reduced, and the impact force fluctuation problem can be improved by using this method.

II. FUZZY ADRC CONTROL METHOD

The fuzzy ADRC control method combines standard ADRC algorithm and fuzzy controller, which can realize the online self-tuning of error feedback controller parameters. The proposed control method mainly consists of the following design.

A. Tracking Differentiator (TD)

TD is aimed at changing the completely opposite relationship between the fast response performance and overshoot of system. It can implement the aim of rapid tracking with the input signal and without the overshoot. By using the transient process control method, the controller of the system can be much improved in adaptability and robustness.

In a second-order control system, we can implement the transient process control method through a second-order nonlinear TD. The following sections describe the design of a second-order nonlinear TD. It has one input signal x and two output signals x_1 and x_2, where x_1 is the tracking signal of x, and x_2 is the differential signal of x_1. The discrete version of the second-order nonlinear TD can be expressed as the following equation:

$$\begin{cases} x_1(k+1) = x_1(k) + h \times x_2(k) \\ x_2(k+1) = x_2(k) + h \times fst(x_1(k) - x(k), x_2(k), r, h) \end{cases} \quad (1)$$

where h is the sampling period, r is speed factor. $fst(\cdot)$ is the comprehensive function, which can be expressed as the following equations:

$$fst(x_1 - x, x_2, r, h) = -\begin{cases} r \times \dfrac{a}{d} & |a| \le d \\ r \times sign(a) & |a| > d \end{cases} \quad (2)$$

$$a = \begin{cases} x_2 + \dfrac{x_3}{h} & |x_3| \le d_0 \\ x_2 + \dfrac{(a_0 - d)}{2} \times sign(x_3) & |x_3| > d_0 \end{cases} \quad (3)$$

where $d = rh$, $d_0 = dh$, $x_3 = x_1 + hx_2$ and $a_0 = \sqrt{d^2 + 8r|x_3|}$.

B. Extended State Observer (ESO)

In the design process of the impact force control system, it is inevitable to consider the elimination of various disturbances. The disturbance can be observed using the controlled output, and reflected by the information from the controlled output. Therefore, we can use the idea of state observer to design ESO, expand the disturbance effect that needs controlled into a new state variable, and feedback it into the control system. ESO can estimate the real-time action of various disturbances such as modeling error, uncertainty and external disturbance, and obtain the estimated signals of all state variables. The three-order discrete ESO algorithm is described as follows:

$$\begin{cases} \varepsilon_1 = \tilde{x}_1(k) - y(k) \\ \tilde{x}_1(k+1) = \tilde{x}_1(k) + h(\tilde{x}_2(k) - \beta_{01}\varepsilon_1) \\ \tilde{x}_2(k+1) = \tilde{x}_2(k) + h(\tilde{x}_3(k) - \beta_{02}fal(\varepsilon_1, 0.5, \delta) + bu) \\ \tilde{x}_3(k+1) = \tilde{x}_3(k) + h(-\beta_{03}fal(\varepsilon_1, 0.25, \delta)) \end{cases} \quad (4)$$

where β_{01}, β_{02}, and β_{03} are ESO gain parameters, $\tilde{x}_1$, $\tilde{x}_2$, and $\tilde{x}_3$ are the estimated output of ESO, δ is the error interval length, and u is the final control input. $fal(\cdot)$ is a nonlinear saturation function, which is expressed as:

$$fal(\varepsilon,\alpha,\delta) = \begin{cases} \dfrac{\varepsilon}{\delta^{1-\alpha}}, & |\varepsilon| \le \delta \\[2mm] |\varepsilon|^{\alpha} sign(\varepsilon), & |\varepsilon| > \delta \end{cases} \quad (5)$$

C. Nonlinear State Error Feedback Controller (NSEFC)

The conventional PID control law is a simple linear combination form, which consists of the differentiation of the error, the integration of the error, and the existing state of the error. This structure determines that PID cannot achieve high performance control of complex systems. Therefore, it is necessary to find a more suitable combination form in the nonlinear field to design the error feedback controller. NSEFC uses nonlinear functions to combine the error between the output of TD and the estimated output of ESO, and composes the final control input with the total disturbance compensation. Its algorithm is described as follows:

$$\begin{cases} e_1(k) = x_1(k) - \tilde{x}_1(k) \\ e_2(k) = x_2(t) - \tilde{x}_2(k) \\ u_0 = \beta_1 fal(e_1, \alpha_{01}, \delta) + \beta_2 fal(e_2, \alpha_{02}, \delta) \\ u = u_0 - \dfrac{\tilde{x}_3(k)}{b} \end{cases} \quad (6)$$

where β_1 and β_2 are the gain parameter of NSEFC.

D. Fuzzy ADRC Controller

In the fuzzy ADRC control method, we use the tracking error e_1 and differential error e_2 as input variables to design the two-dimensional fuzzy controller. Based on the variations of e_1 and e_2, the correction quantities $\triangle\beta_1$ and $\triangle\beta_2$ are calculated by fuzzy rules and added to the gain parameters of the NSEFC.

The domain of the input variables e_1 and e_2 are set as [-6, 6] and [-3, 3], and the domain of the output variables $\triangle\beta_1$ and $\triangle\beta_2$ are [-3, 3]. The subset of the fuzzy linguistic variables defined on the domain [-3, 3] is {NB, NM, NS, ZO, PS, PM, PB}, and the subset of the fuzzy linguistic variables defined on the domain [-6, 6] is {NB, NM, NS, NO, PO, PS, PM, PB}. The affiliation function and fuzzy inference are of triangular and Mamdani type, respectively. In the process of defuzzification, the center of gravity method is employed to transform the result. Based on the performance requirements of the Z-axis impact force control system and the parameters settling method of β_1 and β_2, the fuzzy rules are design in Tables I-II.

TABLE I.　　THE FUZZY RULE DESIGN OF $\triangle\beta_1$

e_1	e_2						
	NB	NM	NS	ZO	PS	PM	PB
NB	PB	PB	PM	PM	PS	ZO	ZO
NM	PB	PB	PM	PS	PS	ZO	NS
NS	PM	PM	PM	PS	ZO	NS	NS
NO	PM	PM	PS	ZO	NS	NM	NM
PO	PM	PM	PS	ZO	NS	NM	NM
PS	PS	PS	ZO	NS	NS	NM	NM
PM	PS	ZO	NS	NM	NM	NM	NB
PB	ZO	ZO	NM	NM	NM	NB	NB

TABLE II.　　THE FUZZY RULE DESIGN OF $\triangle\beta_2$

e_1	e_2						
	NB	NM	NS	ZO	PS	PM	PB
NB	NB	NB	NM	NM	NS	ZO	ZO
NM	NB	NB	NM	NS	NS	ZO	ZO
NS	NB	NM	NS	NS	ZO	PS	PS
NO	NM	NM	NS	ZO	PS	PM	PM
PO	NM	NM	NS	ZO	PS	PM	PM
PS	NM	NS	ZO	PS	PS	PM	PB
PM	ZO	ZO	PS	PS	PM	PB	PB
PB	ZO	ZO	PS	PM	PM	PB	PB

After the defuzzification, the correction amounts are obtained. We substitute them in the formula (7) to obtain the ultimate gain coefficient of the NSEFC:

$$\begin{cases} \beta_1 = \beta_1' + \triangle\beta_1 \\ \beta_2 = \beta_2' + \triangle\beta_2 \end{cases} \quad (7)$$

where β_1' and β_2' are initial values of the gain coefficients. Finally, the fuzzy ADRC control method can be achieved by combining the regulating parameter rules and the ADRC algorithm. The control structure is shown in Fig. 2.

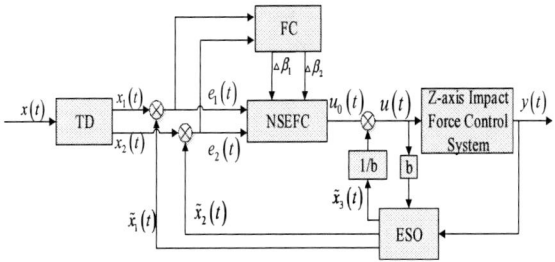

Fig. 2. Structure diagram of the fuzzy ADRC control method.

III. MODELING AND SIMULATION

The main processes of wire bonding are described as follows. Firstly, the reciprocating motion of the Z-axis with a capillary is driven by a voice coil motor. In this process, the capillary moves with high-speed and high-acceleration to the search level in the speed control mode of the voice coil motor, after which it is controlled by a force control mode. Then, the capillary with FAB contacts the chip, and the FAB becomes deformed due to the impact force. Finally, the FAB bonding completes the machining of the chip using ultrasonic energy. The wire bonding process is shown in Fig. 3.

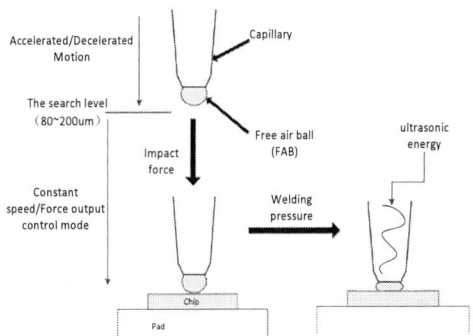

Fig. 3. The wire bonding process.

The impact force caused by the Z-axis high-speed and high-acceleration motion is the key factor influencing the wire bonding quality. According to the existing research, we know that the impact force control subject can be described by a second-order transfer function system, which can be expressed as the following equation:

$$G(s) = \frac{K_e Z_c + K_{fp} Z_d + K_{fi} F_s}{M_z s^2 + (C + K_{fv})s + (K_e + K_{fp})} \quad (8)$$

where K_e, C and M_z are stiffness coefficient, damping and load quality, respectively. K_{fp}, K_{fv}, and K_{fi} are control coefficients, Z_c is the initial height, Z_d is the final height, and F_s is the axial force.

For this kind of second-order system, a step response simulation is carried out to compare the dynamic response performance of the fuzzy ADRC with conventional PID and fuzzy PID. The dynamic response is shown in Fig. 4. Taking setting time (ST), overshoot (OS) and steady-state error (SE) as evaluation indexes, and the specific parameters are shown in Table III.

From the simulation results, we found that the fuzzy ADRC control method can achieve a better performance in terms of ST, OS, and SE. Specifically, compared to the conventional PID and fuzzy PID methods, the settling time of fuzzy ADRC is reduced by 54.5% and 37.5%, the overshoot is reduced by 93.3% and 28.7%, and the steady-state error is reduced by 75% and 88%, respectively.

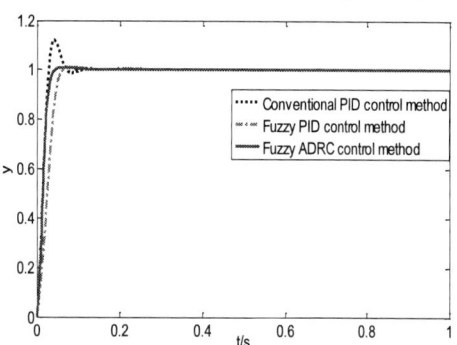

Fig. 4. Curve of step response simulation experiment.

TABLE III. DYNAMIC PERFORMANCE OF STEP RESPONSE

Control method	Indicator of performance		
	ST(s)	OS(mm)	SE(mm)
Conventional PID	0.11	1.218	0.0012
Fuzzy PID	0.08	0.115	0.0025
Fuzzy ADRC	0.05	0.082	0.0003

When the input step signal has a white noise interference with 0.02 signal intensity, keeping the original control parameters unchanged, the response curve is shown in Fig. 5.

From Fig. 5, it can be see that the fuzzy PID control method is sensitive to the noise interference, so it is easy to generate large mutation and cause serious damage to the driving actuator. The fuzzy ADRC and the conventional PID have better disturbance suppression ability for the output disturbances with certain range. Table IV compares the two indexes of IAE and ISE between the conventional PID and the fuzzy ADRC. It can be seen that fuzzy ADRC control method has better input noise suppression ability when the

input signal has noisy interference. The reason is that the TD function in the fuzzy ADRC method has the function of filter, which can filter out the high frequency interference in the original signal and solve the noise amplification phenomenon of the traditional differential process.

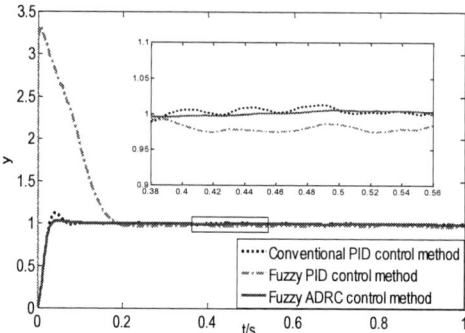

Fig. 5. Response curve of step signal with white noise.

TABLE IV. IAE AND ISE

Indexes	Conventional PID	Fuzzy ADRC
IAE	0.03433	0.01167
ISE	0.00743	0.00113

IV. EXPERIMENTAL ANALYSIS

The experimental work was carried out on a high-speed wire bonding machine. To measure the impact force of the IC pad caused by the bonding head motion, we replaced the IC pad with a PCB-force sensor (type: 209-C11-SN 2559) to measure the contact force of the capillary. The force signal was acquired using the instrument ZonicBook/618E, which transforms the analog voltage signal through the external force sensor. Furthermore, the external force sensor transmits the analog voltage signal to the signal tester. The signal test transmits the analog voltage signal to the industrial PC through the Ethernet. Finally, we can analyze the impact force through the Z-Analyst software in industrial PC. The schematic diagram and experimental setup are shown in Figs. 6-7. The wire bonding machine (type: GD001) was used for LED chip packaging, which can achieve a high bonding speed of 8 wires per second with a Z-axis acceleration up to 120 g.

We designed a series of tests on the impact force of the capillary under different motion speeds. Similar experimental tests were also carried out for the same scenario but with different control methods. In the experiments, the Z-axis was set with a motion of 100g acceleration for the first stage of the motion, and the speed for the search level was set from 15 mm/s to 50 mm/s. At every search speed, we collected 16 groups of experimental data. With the motion settings of the bonding head, the impact force tests were carried out with the bonding head controlled by different control methods. The relevant experimental data are given in Fig. 8.

The results show that the impact force was larger when the speed was higher. However, the impact force can be reduced effectively when the fuzzy ADRC control method is used. To facilitate comparison and analysis, the obtained experimental data are further processed below.

978-1-6654-1392-3/21 $31.00 © 2021 IEEE

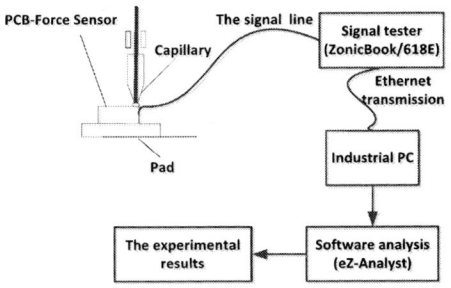

Fig. 6. Schematic diagram of the impact force experiment.

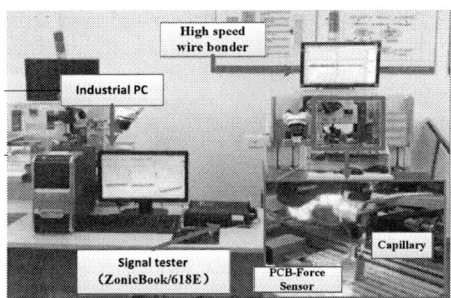

Fig. 7. The experiment setup for the impact force tests.

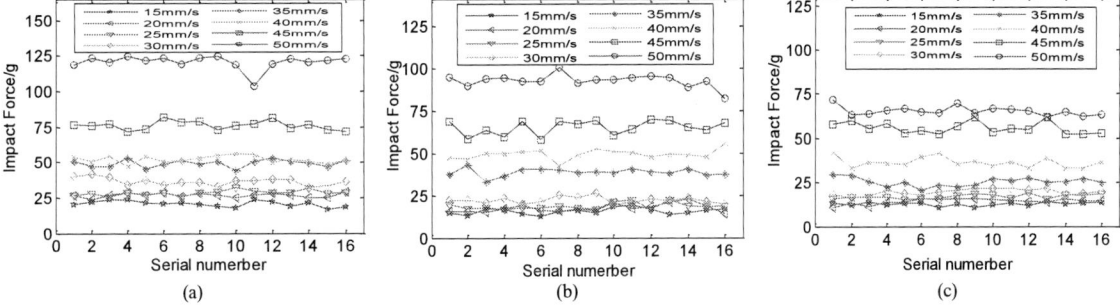

Fig. 8. Line chart of impact experimental data (a) Conventional PID control method, (b) Fuzzy PID control method, (c) Fuzzy ADRC control method.

Table V presents the measured average impact force of the capillary under different control methods. Fig. 9 shows the line chart of average of impact force under various search speeds with different control methods. The results show that the impact force with the fuzzy ADRC control method is much smaller than that of the other two control methods. Compared with the conventional PID control method, the impact force with the fuzzy ADRC control method can be reduced by approximately 26%–48%. Similarly, the impact force can be reduced by approximately 8%–34% in comparison with the fuzzy PID control method.

TABLE V.   AVERAGE DATE OF IMPACT FORCE

Search speed(mm/s)	Control method		
	Conventional PID	Fuzzy PID	Fuzzy ADRC
15	20.60	15.56	12.56
20	26.68	17.43	14.16
25	28.42	18.52	16.95
30	36.24	22.07	19.50
35	49.32	38.82	25.33
40	52.15	49.24	36.26
45	76.11	65.15	55.76
50	120.39	92.62	65.31

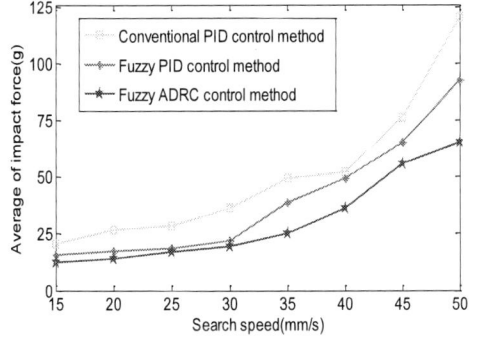

Fig. 9. Line chart of average of impact force.

The fluctuation of impact force is analyzed, and the results are shown in table VI. Fig. 10 contrasts the impact force fluctuation of three control methods. The experimental results show that the fluctuating value of the fuzzy ADRC control method is restricted within the range of 10 grams on each kind of search speed. By comparing with other control methods, the fluctuation of the impact force has been reduced significantly.

TABLE VI.   FLUCTUATING VALUE OF IMPACT FORCE

Search speed(mm/s)	Control method		
	Conventional PID	Fuzzy PID	Fuzzy ADRC
15	7.11	6.39	3.31
20	5.21	8.65	4.89
25	6.50	5.99	4.35
30	9.64	7.08	5.58
35	8.95	9.96	8.76
40	9.16	10.57	9.21
45	10.02	12.32	9.40
50	21.11	18.43	9.46

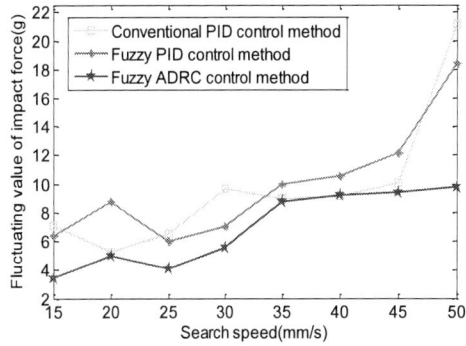

Fig. 10. Line chart of fluctuating value of impact force.

In summary, the experimental results show that the fuzzy ADRC control method effectively guaranteed the stability of

978-1-6654-1392-3/21 $31.00 © 2021 IEEE

the impact force and significantly reduced the peak value of the impact force. Therefore, it is verified that the application of Z-axis impact force control based on fuzzy ADRC control method is advantageous.

V. CONCLUSION

The stability and the peak value of the Z-axis impact force are the key factors to ensure the wire bonding quality. For the problem of impact force control, a fuzzy ADRC control method is proposed in this paper, which has the following advantages:

1) This method can reduce the difficulty of parameter adjustment in traditional ADRC control methods, and obtain good control performance in high-speed and high-acceleration bonding head motion, thus ensuring high bonding quality.

2) The experimental result shows that the fuzzy ADRC control method can effectively reduce the overshoot in the impact force control system. Compared with the fuzzy PID control method, this method can decrease the impact force by 34% and limit the fluctuation range of the impact force to within 10 grams.

REFERENCES

[1] H. K. Kung, and S. H. Ho, "Effect of minute bends and/or kinks of wire bond profile on sag deflection for 3-D and multichip module semiconductor packaging." IEEE T. Semiconduct. M., vol. 29, pp. 168-175, May 2016.

[2] F. Boeren, A. Bareja, T. Kok, O. Tom, "Frequency-domain ILC approach for repeating and varying tasks: with application to semiconductor bonding equipment," IEEE ASME Trans. Mechatron., vol. 21, pp. 2716-2727, Dec 2016.

[3] W. N. Chen, T. L. Wang, T. B. and D. C. Wang, "Research and design for bonding pressure controlling system," Modular Machine Tool & Automatic Manufacturing Technique, vol. 6, pp. 62-64, Jun 2013.

[4] X. Chen, Y. J. Jiang, Y. T. Tan, J. Gao, Z. J. Yang, G. F. Liu, Y. B. He, H. Wang, and Z. X. Li, "Progress and application of key technologies on the electronic packaging equipment development," Journal of Mechanical Engineering, vol. 53, pp. 181-189, Mar 2017.

[5] Y. Ding, J. K. Kim, and P. Tong, "Effects of bonding force on contact pressure and frictional energy in wire bonding," Microelectron. Reliab., vol. 46, pp. 1101-1112, Nov, 2005.

[6] J. H. Kim and C. H. Yim, "An impact force compensation algorithm based on a piezo force sensor for wire bonding processes," Control Eng. Pract., vol. 16, pp. 685–696, Sep 2007.

[7] J. H. Kim and H. J. Park, "A contact detection algorithm of the Z-axis of a wire bonder," Control Eng. Pract., vol. 14, pp. 1035–1043, Jul 2005.

[8] J. Gao, G. H. Dai, Y. J. Jiang, X. C. Wang, Y. Chen, H. Tang, and Y. B. He, "Fuzzy PID control for impact force of high speed wire bonding process," 17th International Conference on Electronic Packaging Technology (ICEPT), 2016, pp. 47-52.

[9] Y. C. Liu, J. Gao, L. Y. Zhang, C. Yun, H. Tang, X. Chen and C. Q. Cui, "Smith-ADRC based z axis impact force control for high speed wire bonding machine," 19th International Conference on Electronic Packaging Technology (ICEPT), 2018, pp. 1003-1008.

[10] F. L. Wang, Y. Chen and L. Han, "Modeling and experimental study of the kink formation process in wire bonding," IEEE Trans. Semicond. Manuf., vol. 27, pp. 51-59, Feb 2014.

[11] J. Li, X. H. Qi, H. Wan, and Y. Q. Xia, "Active disturbance rejection control: theoretical results summary and future researches," Control Theory & Applications, vol. 34, pp. 281-295, Mar 2017.

[12] H. Feng and B. Z. Guo, "Active disturbance rejection control: old and new results," Annu. Rev. in Control, vol. 44, pp. 238-248, May 2017.

Failure Analysis of Anisotropic Conductive Adhesive Packages in Narrow-pitch Flip Chip Packaging

Gui Chen
College of Mechanical and Electrical Engineering
Central South University
Changsha, China
Chengui@csu.edu.cn

Yan Wang*
School of Electronic Science and Engineering
University of Electronic Science and Technology of China
Chengdu,China
wangyancumt@163.com

Xiaoyu Xiao
College of Mechanical and Electrical Engineering
Central South University
Changsha, China
xiao1201@csu.edu.cn

Yamei Yan
College of Mechanical and Electrical Engineering
Central South University
Changsha, China
214742328@qq.com

Wenhui Zhu*
College of Mechanical and Electrical Engineering
Central South University
Changsha, China
zhuwenhui@csu.edu.cn

Abstract—**With the development of microelectronics packaging technology, the interconnection size of chip packaging has gradually decreased to 10μm.ACA (anisotropic conductive adhesive) packaging technology has become one of the key technologies to realize this trend. In this paper, we analyze how to minimize the failure probability of ACA in the case of narrow-pitch chip packaging. The failure mode of ACA can be divided into two kinds: opening failure in the conductivity direction and bridging failure in the insulation direction. Before the bond, it is assumed that the conductive particles obey the Poisson distribution. After the bond, considering the volume curing rate of ACA, the failure probability was calculated by using the probability theory and the improved box model. The prediction model of overall failure for the ACA interconnection of narrow-spacing chips was established. When the pitch is constant, the overall failure probability of pad is evaluated as a function of the optimal diameter and volume fraction of the particles, as well as how the distance between pad and the length of pad should be distributed. This will be useful to optimize the size of the bumps and to choose more suitable ACA material.**

Keywords—Anisotropic conductive adhesive (ACA), Flip chip packaging, Failure probability, Narrow-pitch, Opening, Bridging

I. INTRODUCTION

Microelectronic packaging technology is developing towards miniaturization, high performance, high reliability and low cost. Flip chip bonding technology is the most commonly used and suitable for high-density and narrow-pitch chip packaging due to its advantages of small spacing, simple process and low cost. As the pitch of chip interconnects shrinks to 30 μm and below, the traditional metal interconnection has great problems. The thinning of substrate and chip will lead to the difficulty of ball planting; the thinning of substrate and chip greatly reduces the ability to accept the stress; in the bonding process, too high temperature will lead to the deformation of bumps and pads in the chip, affect the alignment, and affect the reliability of interconnection.

The work was support by National Natural Science Foundation of China (51905554).

Anisotropic conductive adhesive (ACA) for flip chip bonding is considered as one of the most promising packaging materials [1]. Compared with traditional solder, it has the advantages of lead-free, no flux, no reflow, no underfill, low temperature, high reliability, high density and so on. ACA is a composite material consisting of a polymer matrix (epoxy resin) and randomly distributed conductive particles [2]. Under the action of thermal pressure, a stable and reliable interconnection is formed between the chip and the substrate, as shown in Fig. 1. The polymer matrix provides electrical insulation and mechanical connection, and the conductive particles provide electrical connection. Due to its small particle concentration, it is unable to form a resistance network in the plane and presents the transverse insulation property. However, it can make the contact with the bump and the solder pad form a conductive channel in the longitudinal direction by capturing the conductive particles, so as to realize the horizontal insulation and vertical conduction [3].

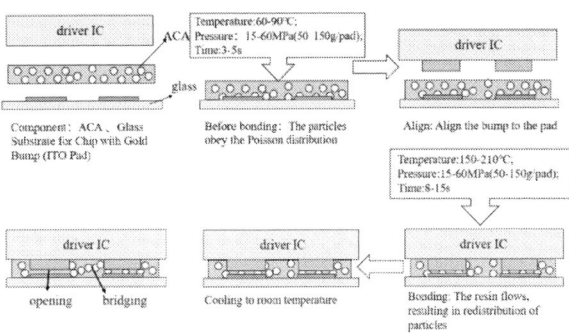

Fig. 1. Bonding process of ACA packaging.

There are two main problems in ACA interconnection: opening failure and bridging failure, as shown in Fig. 1.

1) The number of conductive particles trapped between bump and pad is insufficient, resulting in opening failure.

2) The conductive particles between the bumps form a path, resulting in bridging failure.

In order to realize the reliability interconnection of ACA, it is very important to choose the volume fraction and particle diameter of anisotropic conductive adhesive. Lin et al. [4] proposed box model to calculate the failure probability of ACA interconnection, and analyzed the optimal volume fraction and radius of conductive particles through V-shaped curve model. Ni et al. [5] analyzed the capture rate of conductive particles by Ripley's K function, K-S (Kolmogorov-Smirnov) test and PPP (Poisson Point Process) theory. Radius and volume fraction of conducting particles are mutually affected. Previous studies were based on fixed one factor to get another parameter range, and did not consider the case of ACA curing. In this paper, by improving the box model and considering the curing of ACA, the three-dimensional surface graph of ACA failure probability varying with the volume fraction and radius of conductive particles is obtained. The optimal conductive particle volume fraction and particle diameter when pitch is given, as well as the distribution of pitch between bump size and bump spacing, can reduce the overall failure probability of ACA interconnection, which will contribute to the optimal design of bump size and choose more suitable ACA materials.

II. Opening And Bridging Failure Probability Theory

A. Opening Failure Probability

When the number of particles captured is insufficient between bump and pad, there will be an opening failure. Research shows that when the number of particles captured by bump is less than 5, the on resistance between electrodes is more than 5 Ω, it is judged as opening failure[6]. In ACA, the conductive particles are randomly distributed in the resin matrix. During the bonding process, Poisson distribution is used to predict the probability of particles captured by bump[7].

$$P(n) = \frac{u_1^n e^{-u_1}}{n!} \quad (1)$$

Where $P(n)$ is the probability of the number of n particles captured by each bump; n is the number of particles captured; u_1 is the average number of particles captured. After bonding, the average particle number u_1 captured by bump can be obtained based on the volume fraction of ACA:

$$V_f = \frac{particles\ volume}{gap\ volume} = \frac{u_1 \cdot \frac{4}{3}\pi r^3}{2rl^2} \quad (2)$$

$$u_1 = \frac{3l^2 V_f}{2\pi r^2} \quad (3)$$

Where V_f is the volume fraction of conducting particle in ACA, r is the radius of conducting particle, and l is the edge length of bump, as show in Fig. 2.

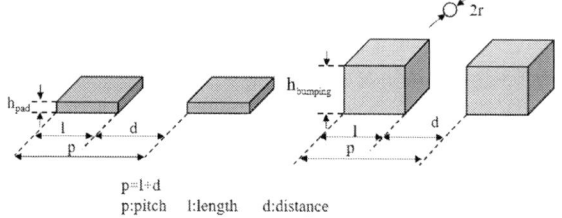

p=l+d
p:pitch   l:length   d:distance

Fig. 2. The size diagram of bump and pad

Thus, the opening failure probability between each bump and pad can be obtained. It can be seen from the formula that the opening failure probability of the ACA interconnection is related to the size of bump, the volume fraction of conducting particle V_f and the radius of conducting particle r.

$$P_O = (n < 5) = \sum_{n=0}^{4} \frac{u_1^n e^{-u1}}{n!} \quad (4)$$

Where P_O is the opening failure probability.

B. Bridging Failure Probability

When the conductive particles between bumps form a path after bonding, the bridging will occur. Studies have shown that almost all resin materials will shrink in volume during curing. For different resins, the volume shrinkage rate will be different, generally between 1%~20%. For different ACA, with the different content and diameter of conducting particles, the curing shrinkage rate will change, and the volume fraction of conducting particles will change before and after curing. According to the composition of commercial ACA and the analysis of laboratory ACA, the solidification shrinkage rate was set as 3%[8]. The volume of ACA required to fill the gap between each bump can be calculated, basing on the the pitch of the chip. According to the curing volume shrinkage of ACA, the actual required volume of ACA, can be calculated by the following formula:

$$V_1 = (l+d)^2 \big(h_{bump} + h_{pad} + 2r\big) - l^2 \big(h_{bump} + h_{pad}\big) \quad (5)$$

$$q = \frac{V_{ACA} - V_1}{V_{ACA}} \quad (6)$$

$$V_{ACA} = \frac{V_1}{1-q} = \frac{(l+d)^2 \cdot h - l^3}{1-q} \quad (7)$$

Where V_1 is the actual volume between bumps; V_{ACA} is the actual ACA volume; d is the distance between bumps; q is the volume curing rate of ACA; h_{pad} is the height of pad; h_{bump} is the height of the bump; Pad is ITO (Indium-Tin Oxide) conductive film with a height of 0.2μm, which can be ignored, $h_{bump} + h_{pad} \approx l$; h is the overall height between the bump and pad after bonding, $h = h_{bump} + h_{pad} + 2r = l + 2r$. So the formula for V_1 can be simplified as:

$$V_1 = (l+d)^2 \cdot h - l^3 \quad (8)$$

According to the actual volume of ACA required, calculate the number of conductive particles between bump and bump, and the probability of conductive particles appearing in each space between bump and bump is obtained.

$$V_{ACA} \cdot V_f = k \cdot \frac{4\pi r^3}{3} \quad (9)$$

$$k = \frac{3V_{ACA} \cdot V_f}{4\pi r^3} = \frac{3V_1 \cdot V_f}{(1-q) \cdot 4\pi r^3} = \frac{3l \cdot d \cdot (l+2r) \cdot V_f}{(1-q) \cdot 4\pi r^3} \quad (10)$$

$$u_2 = \frac{k/2}{N} = \frac{k/2}{d^* \cdot l^* \cdot h^*} \quad (11)$$

Where u_2 is the probability of a conductive particle occurring in each space between bump; N is the number of boxes between bumps; k is the total number of conducting particles around each bump, so the number of conducting particles between bumps is less than $k/2$; d^*、 l^*、 $d^*$is a dimensionless integer，where $d^* = d/2r$; $l^* = l/2r$; $h^* = h/2r$.

Combining with the box model proposed by Lin et al. [9], it can be obtained that the probability of bridging between bump is:

$$P_B = 1 - \prod_{\substack{i_1,i_2:1 \to l^* \\ j_1,j_2:1 \to h^*}} \left(1 - u_2^{\sqrt{(d^*)^2+(i_1-i_2)^2+(j_1-j_2)^2}}\right)$$

$$(12)$$

Where P_B is the bridging failure probability; i_1, i_2, j_1 and j_2 are the number of boxes arranged between adjacent faces in bump interconnect. As shown in Fig. 3.

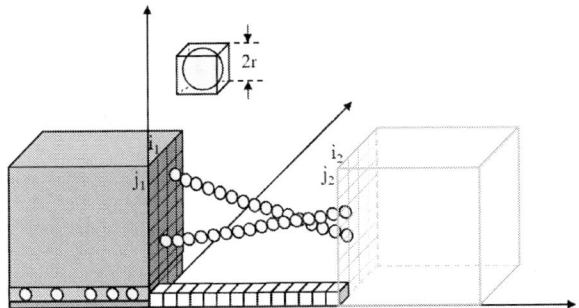

Fig. 3. Box model schematic diagram

C. Total Failure Probality

Considering the failure probability of both opening and bridging, the overall failure probability of ACA interconnection is:

$$P_{OUB} = P_O + P_B - P_{O \cap B} \qquad (13)$$

Where P_{OUB} is the probability of opening or bridging failure； $P_{O \cap B}$ is the probability of both opening and bridging failure.

When opening or bridging failure are independent events, the formula of overall failure probability can be written as follows:

$$P_{O \cap B} = P_B \cdot P_O \qquad (14)$$

$$P_{OUB} = P_O + P_B - P_O \cdot P_B \qquad (15)$$

III. RESULTS AND DISCUSSION

In this paper, the volume curing shrinkage rate of ACA was introduced by improving the box model, and the influence of the volume fraction and radius of conductive particles on the failure probability was considered. It is more accurate than the previously proposed model, which can simultaneously obtain the value norm of the optimal volume fraction and the diameter of the conductive particle, and analyze the different distribution of pitch in the bump size and bump spacing. This will be useful to optimally design the size of the bumps and choose more suitable ACA material.

When the chip pitch is 25μm, according to the current commercial ACA conductive particle concentration and volume fraction, the value range of conductive particle volume fraction V_f and conductive particle radius r is specified as 0%<V_f<30%, 0.5μm<r<3.5μm, respectively.

According to the formula proposed above, the three-dimensional surface diagram of failure probability and two-dimensional contour diagram with the change of volume fraction of electric particle V_f and radius r of conductive particles is obtained. The red plane level for failure probability is 5%. When the three-dimensional diagram of failure probability is on the red horizontal plane, it shows that the failure probability of bump is greater than 5%, otherwise, the failure probability of bump is less than 5%. The two-dimensional contour diagram shows the range of volume fraction and radius of conductive particles under different failure probability. The range of volume fraction and radius of conductive particles can be obtained when the failure probability of opening and bridging failure is less than 5%.

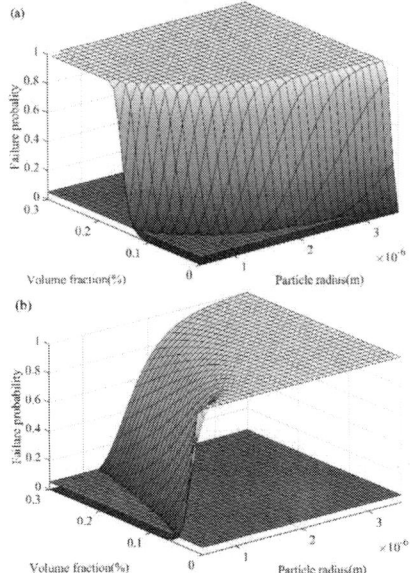

Fig. 4. The failure probability varies with the volume fraction and radius of the conductive particle, when l=10μm, d=15μm. (a) opening failure (b) bridging failure.

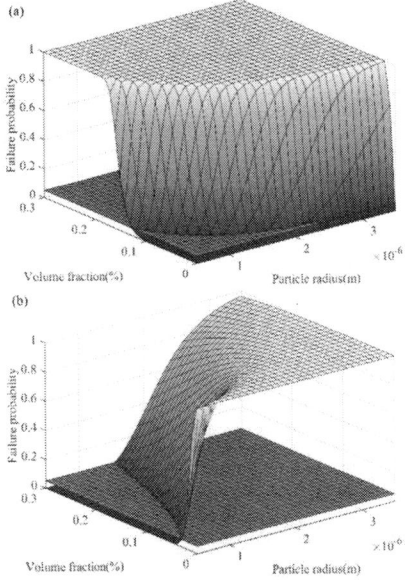

Fig. 5. The failure probability varies with the volume fraction and radius of the conductive particle, when l=12.5μm, d=12.5μm. (a) opening failure (b) bridging failure.

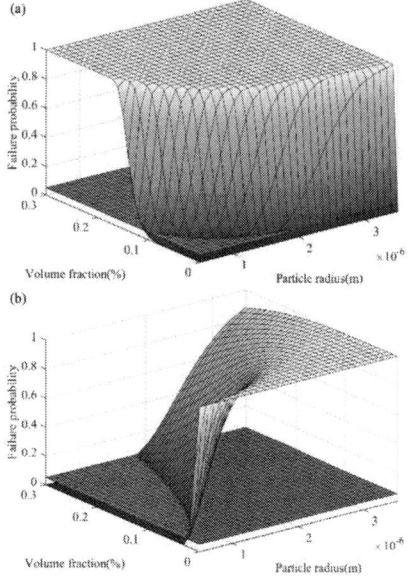

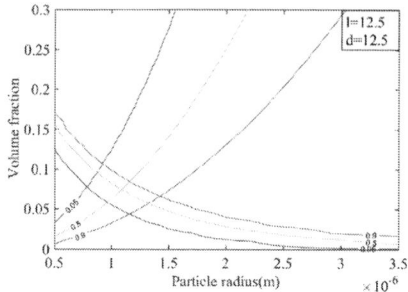

Fig. 9. Contour diagram of opening and bridging failure probability varying with conductive particle volume fraction and radius, when l =12.5μm, d=12.5μm.

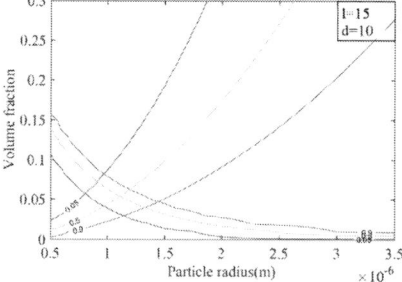

Fig. 10. Contour diagram of opening and bridging failure probability varying with conductive particle volume fraction and radius, when l=15μm, d=10μm.

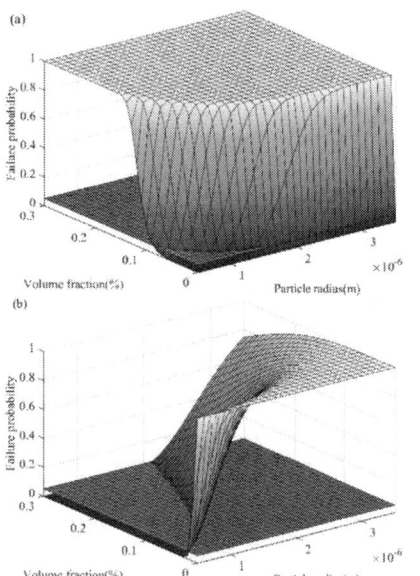

Fig. 6. The failure probability varies with the volume fraction and radius of the conductive particle, when l =15μm, d =10μm. (a) opening failure (b) bridging failure.

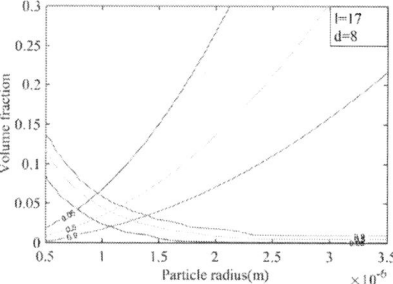

Fig. 11. Contour diagram of opening and bridging failure probability varying with conductive particle volume fraction and radius, when l=17μm, d=8μm.

The distribution of four types of pitch is analyzed Fig. 4-7 shows the failure probability variation of edge length l=10, 12.5, 15 and 17μm, respectively. It can be seen from the figure that both opening and bridging failures increase with the increase of conducting particle radius, and the change of conducting particle radius has a greater impact on the failure probability. For opening failures, the larger the volume fraction of conducting particles, the better, so that more conducting particle numbers can be captured. For bridging failures, the smaller the volume fraction of conducting particles, the better. The smaller the volume fraction, the fewer particles exist between bumps. This limits the volume fraction of conducting particles to a range. Fig. 8-11 is the contour diagram of opening failure and bridge failure for l=10, 12.5, 15 and 17μm respectively. The two red lines are the contour diagram with 5% opening failure and 5% bridging failure respectively. The area where the two lines intersect is the appropriate value range of conductive particle volume fraction and conductive particle radius. As can be seen from the figure, when the failure probability is 5%, the requirements of four pitch for volume fraction and radius of conductive particle are shown in the following table:

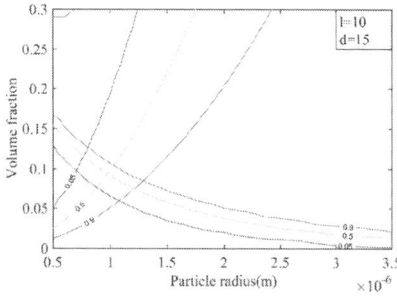

Fig. 7. The failure probability varies with the volume fraction and radius of the conductive particle, when l =17μm, d =8μm. (a) opening failure (b) bridging failure.

Fig. 8. Contour diagram of opening and bridging failure probability varying with conductive particle volume fraction and radius, when l=10μm, d=15μm.

TABLE I.   THE REQUIREMENT OF VOLUME FRACTION AND DIAMETER OF CONDUCTIVE PARTICLES WITH DIFFERENT PITCH DISTRIBUTION

Exp. No.	Size and spacing of bump（μm）	Volume fraction（%）	Particle radius （μm）
1	$l=10$μm, $d=15$μm	$0.05<V_f<0.125$	$r<0.7$
2	$l=12.5$μm, $d=12.5$μm	$0.03<V_f<0.125$	$r<0.79$
3	$l=15$μm, $d=10$μm	$0.025<V_f<0.11$	$r<0.8$
4	$l=17$μm, $d=8$μm	$0.02<V_f<0.085$	$r<0.77$

It can be seen from the data in the above table that with the increase of bump size and the decrease of bump gap, the value range of conductive particle radius r and the conductive particle volume fraction V_f first increases and then decreases. The opening and bridging failure probability of bump interconnects can be less than 5% by selecting any point within the intersection area of 5% contour lines of opening and bridging failure. In the range of value, the volume fraction of conductive particles corresponds to the diameter of conductive particles one by one. In the resin matrix of ACA, the radius of conductive particles is normal distribution. With the decrease of the radius of conductive particles, the manufacturing process of ACA is more difficult. Therefore, when both the volume fraction and the size of conductive particles are limited, the size of conductive particles should be given priority and the larger particle radius should be selected. When pitch is fixed, in order to reduce the failure probability, the size of bump should be increased and the distance between bumps should be reduced. When pitch is equal to 25μm, the optimal pitch distribution should be $l=15$μm, $d=10$μm, the volume fraction of conductive particles should be less than 11%, and the radius of conductive particles should be less than 0.8μm.

IV. CONCLUSION

In this study, the box model for calculating bump failure probability is improved by using Poisson distribution of conductive particles and considering the volume shrinkage of ACA. The three-dimensional surface and two-dimensional contour diagram of the opening and bridging failure probability of bump with the volume fraction and diameter of conductive particles are established. The main conclusions are as follows：

1) The opening failure probability is related to the size of bump, the concentration of conductive particles and the diameter of conductive particles, but has nothing to do with the distance between bumps. In addition, with the increase of the area of bump, the concentration of conducting particle increases, the diameter of conducting particle decreases, and the failure probability decreases.

2) The bridging probability is related to the distance between bumps, the height of bump, the concentration of conductive particles and the diameter of conductive particles.

With the increase of the distance between bumps, the height of bump decreases, the concentration of conductive particles decreases and the diameter of conductive particles decreases.

3) Opening failure and bridging failure are mutually restricted. When the interconnection size is fixed, the concentration and diameter of conductive particles are required. According to the research in this paper, when the pitch is 25μm, the optimal volume fraction of ACA is less than 11%, more than 2.5%, and the radius of conductive particles is less than 0.8μm.

4) When pitch is fixed, in order to reduce the failure probability of flip chip bonding, we should reasonably allocate the relationship between bump and bump, and allocate more sizes on the size of bump. When pitch is 25μm, $l=15$μm, $d=10$μm is the most suitable choice.

ACKNOWLEDGMENT

The work was support by National Natural Science Foundation of China (51905554).

REFERENCES

[1] J. Kim, T. Lee, D. Yoon, T. Kim and K. Paik, "Effects of Anisotropic Conductive Films (ACFs) Gap Heights on the Bending Reliability of Chip-In-Flex (CIF) Packages for Wearable Electronics Applications," 2017 IEEE 67th Electronic Components and Technology Conference (ECTC), Orlando, FL, USA, 2017, pp. 2161-2167.

[2] Y. Pan, L. Song, S. Zhang, X. Cai and K. Paik, " Effects of Polymer Conductive Particle Contents on the Electrical Performance and Reliability of 50-μm Pitch Flex-on-Flex Assemblies Using Anisotropic Conductive Films," in IEEE Transactions on Components, Packaging and Manufacturing Technology, vol. 7, no. 11, pp. 1759-1764, Nov. 2017.

[3] Y. Pan, L. Song, S. Zhang, X. Cai and K. Paik, "Effects of Polymer Conductive Particle Contents on the Electrical Performance and Reliability of 50-μm Pitch Flex-on-Flex Assemblies Using Anisotropic Conductive Films," in IEEE Transactions on Components, Packaging and Manufacturing Technology, vol. 7, no. 11, pp. 1759-1764, Nov. 2017.

[4] Lin CM (2014) Estimation of ACF packaging failure Probability for IC/substrate assemblies with different pad array dimensions. J Mater Sci-Mater Electron 25(2):618–626

[5] Ni Guangming,Liu Lin,Zhang Jing,Liu Juanxiu & Liu Yong.(2017).Analysis of Trapped Conductive Microspheres in LCD FOG Anisotropic Conductive Film Bonding..(eds.)Proceedings of 2017 IEEE 2nd Advanced Information Technology,Electronic and Automation Control Conference (IAEAC 2017)(pp.1470-1476).

[6] ZHAO Yuyuan. Composition and Technical Requirements of Anisotropic Conductive Membrane [J]. Electronic World, 2016(14):131.

[7] S. H. Mannan, D. J. Williams, D. C. Whalley, and A. O. Ogunjimi,"Models to determine guidelines for the anisotropic conducting adhe sives joining process," in Conductive Adhesives for Electronics Pack aging, J. Liu, Ed. Isle of Man, U.K.: Electrochemical Publications Ltd., Jun. 1999, ch. 4.

[8] Chen Xiancai. Research on the formation mechanism of microinterconnect resistance of anisotropic conductive adhesive in flip bonding [D]. Huazhong University of Science and Technology,2012.

[9] Lin CM, Su MH, Chang WJ (2005a) The prediction of failure prob ability in anisotropic conductive adhesive (ACA). IEEE Trans Device Mater Reliab 5:255–261.

Failure Mechanism of Nickel-Chromium thin Film Chip Resistors

Zhiyuan Mao
Reliability Research and Analysis Center
the Fifth Electronics Research Institute of MIIT.
Guangzhou, China
maozhiyuan@ceprei.com

Gaoming Shi*
Reliability Research and Analysis Center
the Fifth Electronics Research Institute of MIIT.
Guangzhou, China
shigm@ceprei.com

Weili Li
Reliability Research and Analysis Center
the Fifth Electronics Research Institute of MIIT.
Guangzhou, China
liweili@ceprei.com

Xianjun Kuang
Reliability Research and Analysis Center
the Fifth Electronics Research Institute of MIIT.
Guangzhou, China
kuangxianjun@ceprei.com

Fuyao Mo
Reliability Research and Analysis Center
the Fifth Electronics Research Institute of MIIT.
Guangzhou, China
mofy@ceprei.com

Abstract—In the current resistor technology, Ni-Cr thin film chip resistors are widely used for their low temperature coefficient, high precision and stability. Unlike thick film chip resistors, which are easy to vulcanize and open circuit, it is found that the common failure modes of thin film resistors are resistance increase out of tolerance and open circuit. Three related failure mechanisms, including over-electricity burnout, electrolytic corrosion, and cracks induced by mechanical stress, are systematically reported.

Keywords—thin film resistor, failure mechanism, over-electricity, electrolytic corrosion, crack

I. Introduction (*Heading 1*)

Thin film chip resistors are made by vacuum deposition to form a resistive material on the surface of the ceramic substrate. Compared with those thick, the thin film resistors have advantages of low temperature coefficient, high stability, low parasitic effects, low noise, and stable high-frequency performance[1]. As the thin film chip resistors with higher stability and lower temperature coefficient develop, they have been widely used in communications, automotive electronics, consumer digital and other fields. Among them, Ni-Cr thin film resistors are the most common thin film chip resistors[2], and their reliability is paid more and more attention. Failure analysis, as an important means to ensure reliability, plays an important role in the quality control and reliability assurance of components. This article analyzes the two most common failure modes of Ni-Cr thin film chip resistors during testing or use: resistance increase out of tolerance and open circuit, and expounds the common failure mechanisms that cause these failures.

II. Structure of Ni-Cr thin film chip resistors

Ni-Cr thin film chip resistors are made by vacuum deposition to form a Nickel-Chromium resistive film on an alumina ceramic substrate, then forming a certain shape through a photolithography process, coating a protective layer (encapsulating material) after trimming the resistance, and printing a logo. The structure of Ni-Cr thin film chip resistor is shown in Fig.1 (in addition to the label in the figure, a logo will be printed on the surface of protective layer).

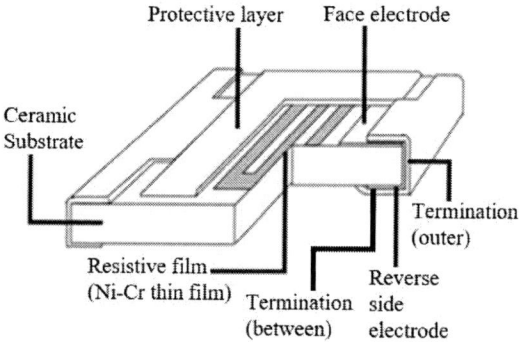

Fig.1 Structure of Ni-Cr thin film chip resistors

III. Failure mechanism of Ni-Cr thin film chip resistors

The most common failure modes of Ni-Cr thin film chip resistors are resistance increase out of tolerance and open circuit. Here we elaborate on three common mechanisms that cause these failures.

A. Over-electricity burnout

When an over-electricity burnout occurs, the failure mode is generally an open circuit.

For Ni-Cr thin film resistors, the over-electricity burnout can be basically divided into two categories. One is long-term over-power burnout. The electric field takes a relatively long time, causing overheat and melt of the resistive film for long-term over-power. The location of over-power burnout is the area where the current density of the resistive film is highest. As shown in Fig.2, the electric field distribution of thin film resistor is distorted after trimming the resistance[3]. The current density in the middle of the resistive film is larger than other parts, so the burnout is mostly concentrated in the middle area of the resistive film. The resistive film in this area is burnt and melted (Fig.3). Generally, the

978-1-6654-1392-3/21 $31.00 © 2021 IEEE

carbonized residue of the encapsulating material can be observed on the surface of the resistive film after unsealing. When the burn is serious, even obvious melting pits can be observed from the appearance. A large area of the resistive film is burned and fused, resulting in failure of the resistive open circuit.

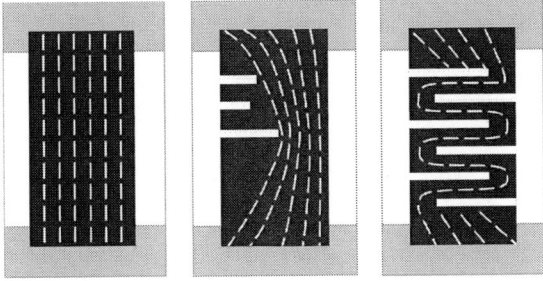

Before Trimming   After Trimming

Fig.2 Schematic diagram of electric field distribution on resistive film before and after resistance trimming

(a)   (b)

Fig.3 (a) Optical photo of long-term over-power resistor

(b) SEM photo of burn-out area

The other one is instantaneous overpower, including ESD, high surge voltage and high pulse current. This over electricity has a short action time, but the instantaneous energy is high, and it will also cause the Ni-Cr thin film resistor to burn. The resistive film structure of the thin film resistor is a conductive film with a thickness of about several hundred nanometers prepared on the surface of the ceramic substrate through thin-film process, which determines that its withstand instantaneous voltage is relatively low[4]. ESD, surge voltage or pulse current brings a large current through the resistive material in a short time, generating a large amount of heat and causing the resistive material to overheat and melt or even vaporize, which results in an open circuit failure of the resistor thereby. Due to the short action time and high energy, although the burned location is also in the area where the current density of the resistive film is the largest, it is generally a small area and mostly linear (Fig.4).

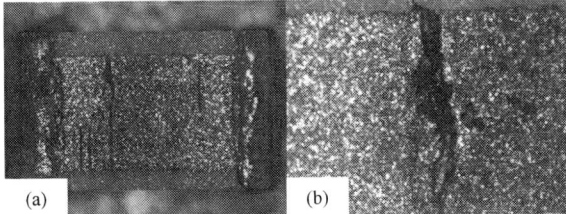

(a)   (b)

Fig.4 (a) Resistive film burn out for pulse current

(b) A small and linear burn-out area

Comparing these two types of burns, the characteristics of long-term over-power burnout include: long current action time, a large burn-out and melting area on the resistive film, and even melting pits on encapsulation layer. The characteristics of instantaneous overpower include: short current action time, a large instantaneous voltage or current passing through the resistive film, and small and mostly linear burn-out area.

In addition to the above-mentioned common over-electricity burnout situations, when the resistance-trim-slot is not etched clean, the remaining resistive film material will bridge two adjacent resistive film lines, forming an additional conductive path. The conductive path is much thinner than the normal resistive film pattern, which causes the larger current density at this position and easier over-current burnout during subsequent use. The burnout will affect the adjacent resistive film line area, leading resistance to increase out of tolerance.

B. Electrolytic corrosion

When electrolytic corrosion occurs, the failure mode of thin film resistors can be manifested as increased resistance or open circuit failure. The general situation is that the resistance increases out of tolerance in the early stage, and then gradually develop into an open circuit failure if it continues to be used. A typical electrolytic corrosion of resistive film is shown in Fig.5. A missing area of the resistive film can be seen under optical microscope, and the charged effect for missing area can be observed under the SEM.

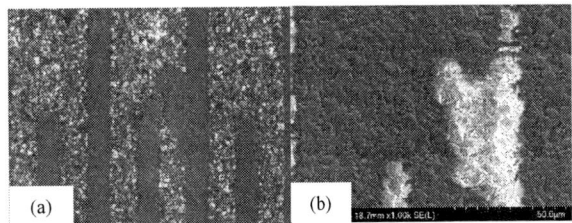

(a)   (b)

Fig.5 (a) Optical photo of electrolytic corrosion

(b) SEM photo of electrolytic corrosion.

For Ni-Cr thin film resistors, the film layer is a nickel-chromium alloy. When the electrolytic corrosion effect occurs, water molecules are adsorbed on the surface of the conductive film through the encapsulation layer. Under the action of electric field, the water is electrolyzed into hydrogen and hydroxide ions, as in (1).

$$H_2O \leftrightarrow H^+ + OH^- \tag{1}$$

Under the action of electric field, the hydroxide ions tend to move towards the positive electrode of the resistor applying voltage, and respectively react with the chromium and nickel in the resistive film to generate chromium trioxide and nickel oxide, as in (2) and (3), which are deposited on the surface of the resistance-trim-slot near the positive terminal of the applied voltage of the resistor.

$$4Cr + 6(OH)^- \rightarrow 2Cr_2O_3 + 2H_2 \uparrow \tag{2}$$

$$2Ni + 2(OH)^- \rightarrow 2NiO + H_2 \uparrow \tag{3}$$

The hydrogen generated by the above chemical reaction escapes through the pores of the encapsulation layer, further enlarging the pores and other defects of the encapsulation layer and accelerating the process of water vapor intrusion into the surface of the conductive film of the resistor, which exacerbates the electrolytic corrosion effect. Moreover, the electrolysis reaction intensifies with the increase of temperature. In the area with the highest current density, a large amount of heat is generated, causing the most serious electrolytic corrosion. Ambient temperature and humidity can also accelerate the occurrence of electrolytic corrosion and induce partial loss of the resistive film, which eventually leads to the resistance increase out of tolerance or even open circuit.

In addition, if the ceramic substrate or encapsulating material contains K^+, Na^+, Cl^- (after dissolving in the water film under an electric field), it will accelerate the formation of OH^- and reduce the resistivity of the water film, thereby intensifying the electrolytic corrosion[5].

Besides, the encapsulation layer of thin film resistors is mostly composed of phenolic resin or epoxy resin. Both of these resins contain hydroxyl ($-OH$), which is relatively fragile and easily breaks to form free hydroxide ions. In addition, the epoxy resin contains a variety of additives, which endows itself with certain moisture absorption and water permeability, and thus water vapor can easily penetrate in a humid environment and corrosion occurs. For electrolytic corrosion, water and electric field are indispensable conditions. Because the water tightness of the encapsulating material cannot completely prevent the intrusion of moisture, the temperature and humidity of the environment require special attention.

When analyzing this type of failure, there should be no obvious burnt or damage points in appearance. After removing the surface encapsulation material, it can be observed that there are some white marks with lighter metallic luster in the resistive film. When the white mark does not completely interrupt the conductive path of the resistive film, the failure mode appears as the resistance increases out of tolerance. When the white mark has completely blocked the conductive path of the resistive film (Fig.6), that is, the resistive film has been completely corroded and disconnected. The failure mode at this time is shown as open circuit failure.

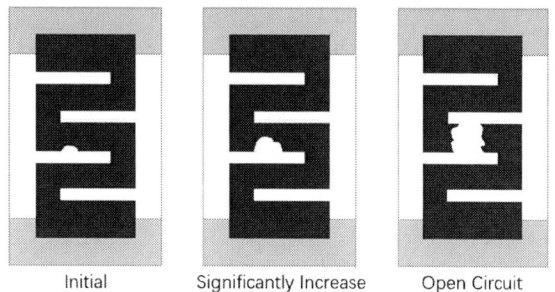

Initial   Significantly Increase   Open Circuit

Fig.6 Schematic diagram of electrolytic corrosion failure process

C. Cracks in ceramic substrate and resistive film

When the ceramic substrate of the thin film resistor cracks, the failure mode can be an increase in resistance or an open circuit, depending on whether the crack has caused the resistive film to be completely disconnected.

For Ni-Cr thin film resistors, the cracking of the ceramic substrate will cause damage, cracking or even loss of the resistive film, resulting in reduction of the effective conductive area or even disconnection of the conductive path. Generally, the cracking and collapse of the ceramic substrate are caused by mechanical stress during production, installation, or use. At first, the micro-cracks will generate in the ceramic substrate and will not affect the resistive film. Therefore, any increase in resistance or open circuit failure cannot be observed during the electrical test at this time. However, when these resistors with micro-cracks on the ceramic substrate are subsequently installed and used, the internal micro-cracks will further expand, causing obvious cracks or breakage of the ceramic substrate and leading to cracking or loss of the resistive film on the surface.

For failures caused by cracking or missing of the resistive film, cracks or damages of the ceramic substrate or protective layer can generally be observed in the appearance during the analysis. The damage position is generally located at the edge of the encapsulating material and the side of the ceramic substrate. After removing the encapsulating material, obvious cracks or missing of the resistive film can be observed near the damage location (Fig.7).

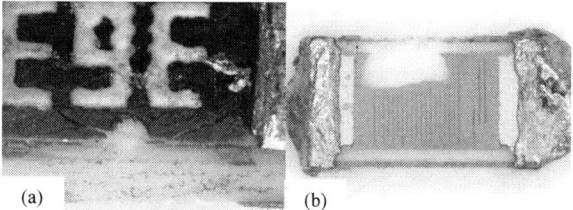

(a)   (b)

Fig.7 (a) Cracks of protective layer and ceramic body

(b) Cracks and missing of resistive film

IV. CONCLUSION

The failure modes of thin film resistors include open circuit, short circuit, and resistance drift. This article takes Ni-Cr thin film chip resistors as the main analysis object and explains the two most common failure modes: resistance increase out of tolerance and open circuit, and three corresponding failure mechanism:

1) The resistive film is burnt out in the area with the highest current density due to long-term overpower, surge voltage, or pulse current.

2) The chromium and nickel in the resistive film undergo an electrolytic reaction under the combined action of water and electric field, and the resistive film is corroded and disconnected.

3) The ceramic substrate is cracked by mechanical stress, causing cracking or collapse of the resistive film.

In addition to the above-mentioned mechanisms, defects in the resistive film itself, foreign objects or defects in the resistance-trim-slot, poor contact or cracking of the termination may all cause the failure of thin film resistors, which were not fully explained in this article.

ACKNOWLEDGMENT

This work is financially supported by the National High Quality Development Program of China under Grant No. 2020-0093-2-1, the Key R&D Program of Guangdong under Grant No. 2019B010143002. Also, we would like to give thanks to Dr. Wu for proofreading.

REFERENCES

[1] X. Wang, Z. Zhang, T. Bai, and Z. Liu, "Thin film chip resistors with high resistance and low temperature coefficient of resistance," Transactions of Tianjin University, 2010, 16(5), pp.348–353.

[2] D. Jiang, M. Tian, Y. Wang, X. Yong-Hong, X. Yang, and W. Da, "Characteristics and research status of nickel-chromium based precision resistance alloy," Electrical Engineering Materials, 2017, 5, pp.23-28.

[3] G. Gong, L. Feng, and S. Chen, "Study on failure mode and mechanism of chip membrane resistors under over-electric stress," Internet of Things Technologies, 2016, 6(2), pp.88–89, 91.

[4] N. Chuang, J. Lin, and H. Chen, "TCR control of Ni–Cr resistive film deposited by DC magnetron sputtering," Vacuum, 2015, 119, pp.200–203.

[5] Q. Ji, "Mechanism analysis of common failure modes of metal film resistors," Electronic Product Reliability and Environmental Testing, 1990, pp.33-37, 15.

Fuzzy Tuning Algorithm for Feedforward Parameter Based on IC Package for Mass Transfer of Micro-LED Equipment XY Motion Platform

Wenbin Fan
State Key Laboratory of Precision Electronic Manufacturing Technology and Equipment
School of Electromechanical Engineering,Guangdong University of Technology
Guangzhou,P.R.China
1391408129@qq.com

Yunbo He
State Key Laboratory of Precision Electronic Manufacturing Technology and Equipment
School of Electromechanical Engineering,Guangdong University of Technology
Guangzhou,P.R.China
heyunbo@gdut.edu.cn

Guofu Qiu
State Key Laboratory of Precision Electronic Manufacturing Technology and Equipment
School of Electromechanical Engineering,Guangdong University of Technology
Guangzhou,P.R.China
1179988928@qq.com

Abstract—With the advent of the next generation display technology, high-speed and high-precision mass transfer has become a core technology to solve high-density Micro-LED transfer packaging. Aiming at the problem of feedforward parameter tuning of XY platform composite controller in Micro-LED mass transfer equipment, the intelligent tuning methods of feedforward parameters were researched, and an intelligent tuning algorithm of feedforward parameters based on fuzzy thought was proposed. The system response of XY motion platform with different speed feedforward and acceleration feedforward parameters was explored through experiments, and the influence law of the two feedforward controllers on the system response indexes was determined. Then, based on the thought of fuzzy control and the dynamic characteristics of the system, the fuzzy intelligent tuning algorithm for feedforward parameters was designed theoretically. Finally, based on the traditional three-loop PID control, the optimal speed feedforward parameters and acceleration feedforward parameters were set by fuzzy tuning algorithm. The feedforward parameters were applied to the X-axis servo control of XY motion platform to verify the effectiveness of the algorithm. The experimental results show that the optimal speed feedforward parameter 24.27 and acceleration feedforward parameter 52.68 can be automatically set in 34s under given conditions. The feedforward parameters obtained by the adjustment act on the X-axis of XY motion platform, which makes the system's adjustment time decrease from 30ms to less than 5ms. All these results show that the proposed algorithm can significantly improve the speed of controller parameter tuning, and the optimization efficiency is much higher than the manual trial and error method.

Keywords—Mass transfer of Micro-LED; XY motion platform; fuzzy tuning algorithm; compound control

I. INTRODUCTION

Micro-LED have become the focus because of its excellent properties in terms of brightness, lifetime, resolution and efficiency[1]. Mass transfer is a powerful solution to realize large-scale Micro-LED displays ,which its main forms include elastomer stamping, electrostatic transfer, electromagnetic transfer, laser-assisted transfer[2]. Thus the higher requirements are put forward for the speed and accuracy of mass transfer equipment to improve the transfer quality. The XY motion platform driven by permanent magnet linear motor (PMLM) has fast response speed, accurate positioning and high reliability, and has been widely used in

the field of giant transfer of Micro-LED. However, owing to the characteristics of nonlinearity and uncertainty, many controller parameters are often difficult to be quantitatively adjusted under the guidance of relevant theories in the actual debugging process of mass transfer equipment.

Usually, the adjustment of parameters is done by manual trial and error method, which severely limits the debugging efficiency of the transfer equipment. In order to resolve the difficulty of parameters turning, scholars have proposed fuzzy self-tuning, iterative learning, neural network self-tuning and other algorithms. Literature [3] designed an adaptive fuzzy end slide film controller, which can adjust the parameters of fuzzy set output of fuzzy mechanism on line without knowing the uncertainty limit in advance. Literature [4] Proposed a model-based adaptive feedforward and PID feedback controller to compensate for the Lorentz force in the voice coil motor automatically. Literature [5] designed a position control algorithm of PMSM based on compound feedforward fuzzy control, and realized the fast position response without overshoot. However, most of these methods are more complex, as well as many still stay in the theoretical level. Furthermore, the on-line tuning algorithm takes a long time, so it is less applied in the actual industrial production process.

To this end, this paper proposes an off-line self-tuning algorithm of feedforward parameters based on fuzzy thought, combining with the traditional PID control and intelligent control theory, to realize the off-line and fast self-tuning of the speed and acceleration parameters of the X-axis feedforward controller of the XY motion platform in Figure 1. The results show that the optimization efficiency is much higher than that of manual trial and error method.

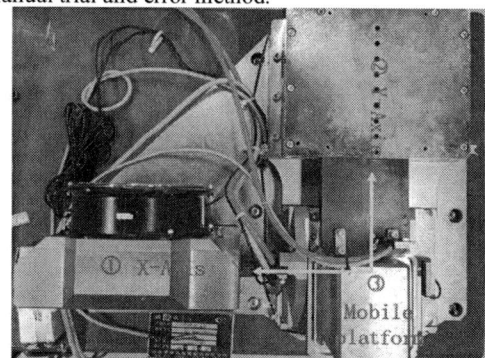

Fig.1. XY motion platform

II. XY MOTION PLATFORM MODEL ANALYSIS

A. Mathematical model of linear motor

The two linear motors composed of XY platform are orthogonal in the horizontal plane, and their movements are independent of each other. Therefore, the coupling effect between the two motors is not considered in the modeling. In this paper, the single-axis (X-axis) linear motor drive system is taken as the object to establish a mathematical model. Affected by factors such as end-effect, temperature variation and electromagnetic coupling, the accurate model of a linear motor is difficult to be established. Therefore, this paper simplifies the linear motor modeling as follows [6]:

(1) Assuming that the stator windings are symmetrically distributed and the resistance and turns of each phase winding are equal.
(2) Assuming that the motor air gap is uniform and constant.
(3) Ignoring magnetic saturation, eddy currents, hysteresis, temperature and end-effects influences.

According to the thought of coordinate transformation, the mathematical model of PMLM is expressed under the d-q shaft-system [6]:

$$\begin{cases} u_d = Ri_d + p\psi_d - \psi_q\omega_r \\ u_q = Ri_q + p\psi_q - \psi_d\omega_r \end{cases} \quad (1)$$

$$\begin{cases} \psi_d = L_d i_d + \psi_f \\ \psi_q = L_q i_q \end{cases} \quad (2)$$

$$\omega_r = \frac{\pi v}{\tau} \quad (3)$$

Where, $u_d, u_q, i_d, i_q, L_d, L_q, \psi_d, \psi_q$ is the stator voltage, current, inductance and stator flux of axis d and axis q, R is the resistance of armature winding, p is the differential operator, ω_s is the electric angular velocity of the actuator, ψ_f is the flux of the permanent magnet, v is the linear velocity of the linear motor, and τ is the polar distance of PMLM. Electromagnetic power of linear motor is:

$$P_e = F_e v = \frac{3}{2}\omega_r(\psi_d i_q - \psi_q i_d) \quad (4)$$

According to formula (1) to (4), the Electromagnetic thrust equation of linear motor is:

$$F_e = \frac{3\pi}{2\tau}[\psi_f i_q + (L_d - L_q)i_d i_q] \quad (5)$$

By using the magnetic field component control strategy, the rotor current vector is orthogonal to the stator magnetic field, that is $i_d = 0$, the thrust equation can be obtained :

$$F_e = \frac{3\pi}{2\tau}\psi_f i_q = K_f i_q \quad (6)$$

Where, K_f is the thrust coefficient.

B. Uniaxial Motion System Model of XY Platform

The motor driving the load of XY motion platform is U-shaped hollow PMLM. Because the actuator of the liner motor has no iron core and no force in the normal direction, the thrust fluctuation of the motor is small. Furthermore, there is a certain gap between the actuator of linear motor and the stator, which means that there is no direct physical contact, so the motor itself has no friction, and most of the system friction is composed of the rolling friction of the linear guide rail.

Therefore, a second order inertial damping system can be used to replace the motion physical model of linear motor. The actuator and its driving load of the X-axis linear motor are simplified into mass block M, and the influence of the friction force between the linear guide rails on the system is simplified into viscous damping force, and all kinds of external interference of the positioning platform are combined into a total interference force. As a result, the physical model of X axis motion system of XY motion platform is established, as shown in Fig.2.

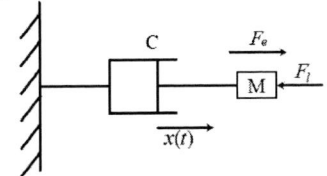

Fig.2. The physical model of X axis motion system

According to Newton's second law, the dynamic equation of linear motor is:

$$F_e = K_f i_q = M\ddot{x} + C\dot{x} + F_l \quad (7)$$

Where, M is the equivalent mass of driving load, C is the equivalent damping of the motion system, F_e is electromagnetic thrust of linear motor, F_l is the resultant interference force and $\ddot{x}, \dot{x}$ is respectively the acceleration and velocity of the electric actuator.

From equations (1) to (7)，the state equation of PMLM can be obtained :

$$\begin{cases} \dot{x}(t) = v(t) \\ \dot{v}(t) = -\frac{C}{M}v(t) + \frac{K_f}{M}i_q + F_l \end{cases} \quad (8)$$

Where，$x(t)$ is the displacement output of PMLM, $v(t)$ is the motion speed of PMLM, $\dot{x}(t)$ is the derivative of $x(t)$, $\dot{v}(t)$ is the derivative of $v(t)$.

According to equation (8), the block diagram of the motion unit of the X-axis linear motor is shown in Fig.3.

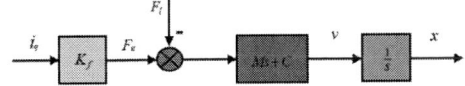

Fig.3. Block diagram of motion unit model of linear motor

Ignoring the disturbance, the transfer function from current to displacement corresponding to the X-axis can be obtained as:

$$G(s) = \frac{K_f}{Ms^2 + Cs} \quad (9)$$

C. Compound Control

The Compound control generally consists of feedforward and feedback control[7]. Feedback control takes the deviation signal as input, and continuously corrects the output through the controller. It has hysteresis and has poor control effect on nonlinear and uncertain systems. The introduction of the feedforward link and the feedback control form a composite control, which can realize the fast response and high-precision position tracking of the servo system. The feedforward link makes up for the shortcomings of the feedback system by directly compensating the input signal, and greatly improves the ability and accuracy of the system to reproduce the input

978-1-6654-1392-3/21 $31.00 © 2021 IEEE

signal. The "feedforward + feedback" compound control principle is shown in Fig.4[9].

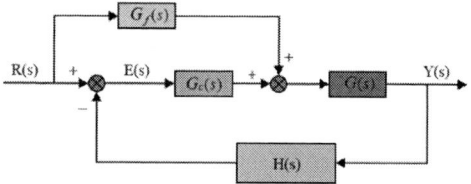

Fig.4. Compound control schematic diagram

Where, $R(s)$ is the system input, $Y(s)$ is the system output, $E(s)$ is the difference between input and output, $G_c(s)$ is the transfer function of the controller, $G(s)$ is the transfer function of the feedforward channel, and $H(s)$ is the transfer function of feedback channel. From Fig.4, the transfer function of the compound control system can be obtained as:

$$\phi(s) = \frac{Y(s)}{R(s)} = \frac{[G_c(s) + G_f(s)]G(s)}{1 + G_c(s)G(s)H(s)} \quad (10)$$

It can be seen from equation (10) that the characteristic equation of the system transfer function does not include the feedforward compensation transfer function, which shows that the introduction of feedforward links not only does not change the stability of the system, but also can improve the dynamic response performance of the system and effectively reduce following error of the system.

In Fig.4, the transfer function between the deviation and the input is:

$$K(s) = \frac{E(s)}{R(s)} = \frac{R(s) - Y(s)}{R(s)} = \frac{1 - G_f(s)G(s)}{1 + G_c(s)G(s)H(s)} \quad (11)$$

When $E(s) = 0$, the transfer function of the feedforward channel can be obtained as:

$$G_f(s) = \frac{1}{G(s)} \quad (12)$$

According to equation (9) and (12), the feedforward channel transfer function of the PMLM is:

$$G_f(s) = \frac{M}{K_f}s^2 + \frac{C}{K_f}s = kaff \cdot s^2 + kvff \cdot s \quad (13)$$

Where, $kaff$ is the acceleration feedforward coefficient, $kvff$ is the velocity feedforward coefficient, $kaff \cdot s^2$ is the acceleration feedforward term and $kvff \cdot s$ is the velocity feedforward term.

Therefore, the fuzzy tuning algorithm in this paper is to find the optimal speed feedforward and acceleration feedforward parameters in the compound controller, so as to compensate the system.

III. THE EFFECT OF FEEDFORWARD ON SYSTEM RESPONSE

Before designing the fuzzy turning algorithm, to analyze the influence law of speed feedforward and acceleration feedforward parameters by experiment on the XY motion platform is essential. In the experiment，the X-axis of the platform in Fig.1 was taken as the control object. The current loop has been designed and packaged in the driver, so this paper only designed the position loop and speed loop at the motion control card level. The control principle block diagram is shown in Fig.5.

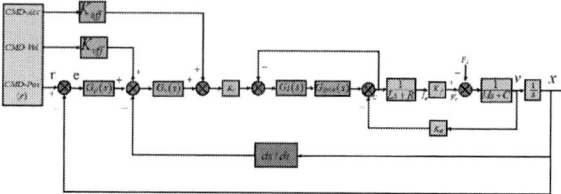

Fig.5. The control principle diagram of X-axis system

Using the control variable method, two parameters of speed feedforward and acceleration feedforward are tested independently. The parameters and motion planning of the controller used in the experiment are shown in Table I and Table II.

TABLE I.   THE PARAMETERS PARAMETERS OF CONTROLLER

Control Loop	Controller parameters		
	P	I	D
Position loop	8	1.5	12
Speed loop	30	0	0

TABLE II.   THE MOTION PLANING

Planning type	Position(m)	Speed(m/s)	Acceleration(m/s²)
third-order S-curve	0.012	0.4	20

A. Experimental Materials

The experimental materials mainly include PC, motion control card（MCC）, terminal board, driver, feedback device and XY motion platform. Among them, the MCC is the googoltech GHN motion control card, which communicates with PC through PCIE agreement and receives instruction signals from the upper computer. The terminal board can transfer multi-axis control signals and feedback signals, which connected with the MCC through Glink-II bus to receive DAC signals sent by the MCC. At the same time, the terminal board and the driver carry out signal transmission through the axis control line. The feedback device is Renishaw grating ruler (resolution is 1μm). The feedback signal realizes the closed-loop control of the system. The driver amplifies the DAC signal from the terminal board, outputs the corresponding current to control the linear motor to drive the XY platform, and receives the feedback signal from the grating ruler and sends it back to the terminal board via the shaft control line. The relationship of each part of the experimental materials is shown in Fig.6.

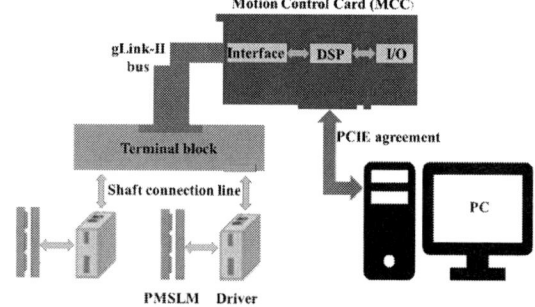

Fig.6. Experimental system block diagram

B. Experimental Process and Result Analysis

(1) The effect of speed feedforward on the system

First, taking the acceleration feedforward parameter as 0, nine groups of parameters are selected as the speed feedforward parameters in the interval (0,50) with variable step size, and then the third-order S-shaped curve movement was carried out on the XY platform respectively. The result is shown in Fig.7.

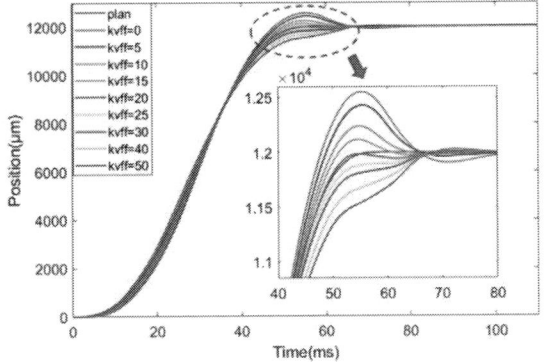

Fig.7. Effects of speed feedforward on the system

From Fig.7, without speed feedforward, there is a large overshoot in the original system response, and the extreme value of position error is large before and after the end of planning. Adding the speed feedforward, the overshoot gradually decreases with the finite increase of $kvff$. When the overshoot is larger than a certain value, the system will even turn into an undershoot system and the degree of undershoot will increase with the increase of $kvff$. It can be judged from the above that, with the increase of $kvff$, the extreme value of position error before and after the end of the planning of the system will decrease first and then increase, and this extreme value determines the length of the adjustment time to a large extent. Therefore, selecting an appropriate speed feedforward parameter can greatly reduce the extreme value of position error before and after the end of the system planning, thus reducing the adjustment time of the system and improving the positioning speed and accuracy.

(2) The effect of acceleration feedforward on the system

Taking the speed feedforward parameter as 22, nine groups of parameters are selected as the acceleration feedforward coefficients in the interval (0,100) with variable step size, and the experiment is carried out following the steps of the speed feedforward experiment. The results are shown in Fig. 8.

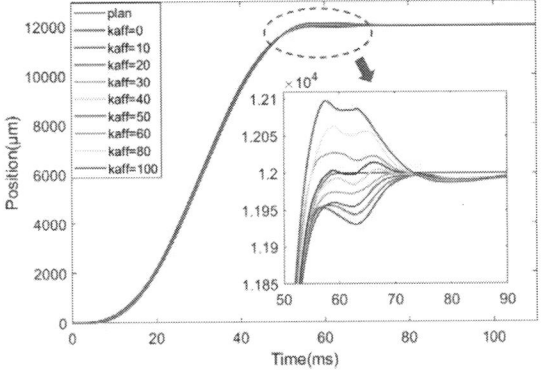

Fig.8. Effects of acceleration feedforward on the system

In Fig.8, before adding acceleration feedforward $kaff$, the position response curve of the system is out of adjustment. And after the end of planning, the position error spends a long time to converges to the allowable error

range. With the increase of $kaff$, the extreme value of position error before and after the end of the planning first decreases and then increases, and the overshoot characteristics of the system become stronger and stronger. In this changing process, when the acceleration feedforward is a certain value, the extreme value of position error before and after the end of the planning will be minimized. At this time, the adjustment time of the system will be reduced, and the positioning speed and accuracy of the system will be further improved.

IV. DESIGN OF FUZZY TUNING ALGORITHM

In this section, a two-dimensional fuzzy feedforward controller is developed based on the fuzzy control thought, combining with the specific law of feedforward parameters' influence on the system in section III.

A. Define variables

Error_max1 is defined as the maximum position error value on the initial first wave peak of the original motion curve, and Error_max2 is defined as the maximum position error value on the initial first wave trough of the original motion curve. The initial first wave crest refers to the area where the actual position curve is convex for the first time after the actual position exceeds the target position for the first time, and its range starting point is artificially defined at the time when the position planning reaches 98%. The first wave valley refers to the area where the actual position curve is first concave after the actual position exceeds the target position for the first time, as shown in Fig.9.

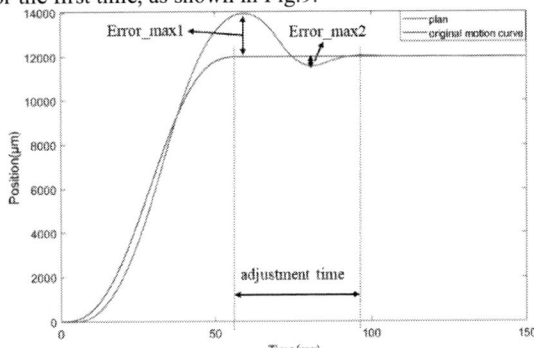

Fig.9. Original motion curve without feedforward

Define the extreme value of position error:

$$Perror = -\max\{|Error\_\max 1|, |Error\_\max 2|\} \quad (14)$$

Theoretically, when Perror before and after the end of planning is 0, the adjustment time Ts is small, and there is no need to change the speed feedforward; When Perror is greater than 0(or less than 0), there is still room for Perror and Ts to decrease, and the speed feedforward can be increased (or decreased), and the amplitude of increase (or decrease) is positively correlated with the absolute value of Perror. Therefore, Perror is regarded as one of the input language variables of fuzzy tuning feedforward parameter algorithm.

Ts is defined as the adjustment time from the end of planning to the first time entering the error band, which determines the strength of the setting action，as shown in Fig.9. There is a positive correlation between adjustment time and setting strength. Therefore, this paper takes the adjustment as the second input language variable of the fuzzy tuning algorithm. After defining the input linguistic variables, the velocity feedforward and acceleration feedforward to be adjusted are taken as the output linguistic variables.

B. Domain division and membership function

The physical domain of Perror\Ts\△$kvff$ \ △$kaff$ are set as table III.

TABLE III.   PHYSICAL DOMAIN DIVISION

Variables	Perror	Ts	△$kvff$	△$kaff$
Physical domain	$[-e_m, e_m]$	$[0, t_m]$	$[-v_m, v_m]$	$[-a_m, a_m]$

Take the fuzzy domain of Perror as [-3,-2,-1,0,1,2,3], its corresponding fuzzy subset is [NB,NM,NS,ZE,PS,PM,PB], at the same time the quantization factor is:

$$k_1 = \frac{3}{e_m} \tag{15}$$

Take the fuzzy domain of Ts as [0,1,2,3,4], its corresponding fuzzy subset is [ZE,PSS,PS,PM,PB], at the same time the quantization factor is:

$$k_2 = \frac{4}{t_m} \tag{16}$$

The values of fuzzy domain and fuzzy subset of △$kvff$ and △$kaff$ are consistent with Perror. The corresponding scale factors are as follows:

$$L_1 = \frac{v_m}{3}$$
$$L_2 = \frac{a_m}{3} \tag{17}$$

The triangle function is selected as the membership function of each variable.

C. Design fuzzy rules

From the analysis of section III, some empirical laws can be concluded as follows:
1) When Perror is positively large, the $kvff$ should be decreased;
2) When Perror is negatively large, the $kvff$ should be increased;
3) When $kvff$ has been set and Perror is still positively large, the $kaff$ should be increased;
4) When Ts is small, the adjusting level should be reduced.

From the variables and empirical laws mentioned previous, the fuzzy rule can be designed as shown in Tables IV.

TABLE IV.   FUZZY RULE OF SPEED/ACCELERATION FEEDFORWARD

Ts	Perror						
	PB	PM	PS	ZE	NS	NM	NB
P B	NB/PB	NM/PM	NS/PS	ZE	PS/PS	PM/ZE	PB/ZE
P M	NB/PB	NM/PM	NS/PS	ZE	PS/PS	PM/ZE	PB/ZE
P S	NM/PM	NS/PS	NS/PS	ZE	PS/PS	PS/ZE	PM/ZE
P S S	NS/PS	NS/PS	NS/PS	ZE	PS/PS	PS/ZE	PS/ZE
Z E	ZE	ZE	ZE	ZE	ZE	ZE	ZE

D. Clarity of fuzzy quantity

In order to obtain the accurate control quantity, the fuzzy method is required to express the calculated results of the output membership function well. In this paper, the barycenter method is used to get the output value. The barycenter method is to take the barycenter bounded by the membership function

curve and the abscissa as the final output value of fuzzy reasoning. Taking $kvff$ as an example, the expression of the method is as follows.

$$\Delta kvff' = \frac{\sum_{k=1}^{35} v_k \mu(v_k)}{\mu(v_k)} \tag{18}$$

$$\Delta kvff = \Delta kvff' * L_1$$

Where, $\mu(v_k)$ is the membership function, μ_i is the clarity value, $\Delta kvff'$ is the final output value of fuzzy inference, L_1 is the scale factor, $\Delta kvff$ is the change in the proportional coefficient in the feedforward controller. The output is accumulated into the controller, and while the final speed feedforward parameter is shown in equation (18):

$$kvff'' = kvff + \Delta kvff \tag{19}$$

Where, $kvff''$ is updated proportional parameter of the fuzzy turning algorithm, and $kvff$ is the initial set value.

V. ALGORITHM VERIFICATION AND RESULT ANALYSIS

Taking the X-axis linear motor of XY motion platform as the feedforward turning object, the turning is carried out according to the principle shown in Fig.10.

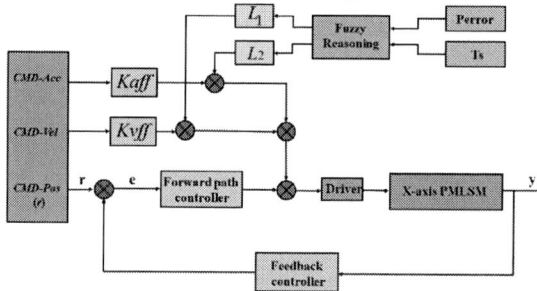

Fig.10. Fuzzy turning principle's block diagram

The fuzzy tuning experiment is divided into two steps. First, the speed feedforward parameter is adjusted. After the speed feedforward parameter is set, the acceleration feedforward is adjusted. The controller parameters and motion planning are the same as the section III.

A. Speed feedforward adjust verification

After the relevant parameter setting, the speed feedforward parameter tuning is carried out at first. The experimental result is shown in Fig.11.

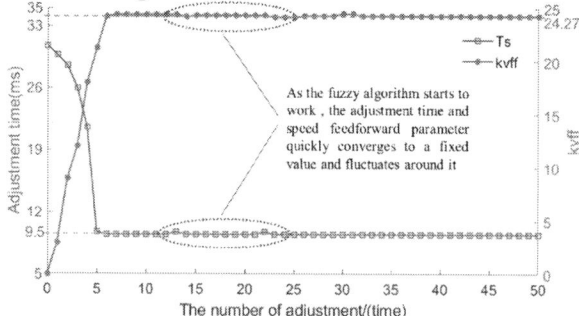

Fig.11. Process of speed feedforward turning

In Fig. 11, $kvff$ is rapidly adjusted in a positive direction after the fuzzy algorithm start to work. As the number of adjustments increases, the range of adjustment becomes smaller and smaller. After six times of adjustment actions, the *kvff* oscillated back and forth around 24.27 and basically fell within the range of 24 ± 0.5. At this time, the adjustment time rapidly decreased from the initial 30ms to 9.5ms. From the perspective of time, it takes less than 8s from the start of adjustment to the first time to find the best value, which is more efficient than traditional manual parameter adjustment.

B. Acceleration feedforward adjust verification

After the speed feedforward fuzzy tuning, the adjusting time is reduced to 81ms, which is still higher than the expected value of 0 set in the experiment. Therefore, on the basis of applying the optimal speed feedforward control, the acceleration feedforward fuzzy tuning is continued. The experimental results are shown in Fig.12.

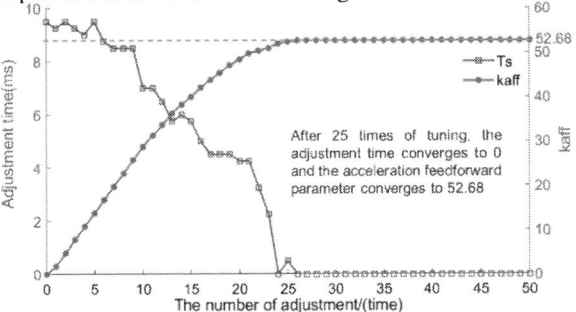

Fig.12. Process of acceleration feedforward turning

From Fig. 12, the result is similar to the speed feedforward fuzzy adjustment. After 24s of turning, the *kaff* oscillated back and forth around 52.68 and basically fell within the range of 53. At this time, the adjustment time rapidly decreased from the initial 8.75ms to 0ms, which can meet the requirement of position error converging to 10μm within 5ms of the end of the planning command. So far the feedforward parameter fuzzy turning is completed.

C. Result analysis

After adjusted by fuzzy algorithm, the optimal speed feedforward parameter is set as 24.27, and the optimal acceleration feedforward parameter is set as 52.68. The motion curves of without feed forward, optimal speed feedforward, optimal speed feedforward and acceleration feedforward are drawn, as shown in Fig. 13.

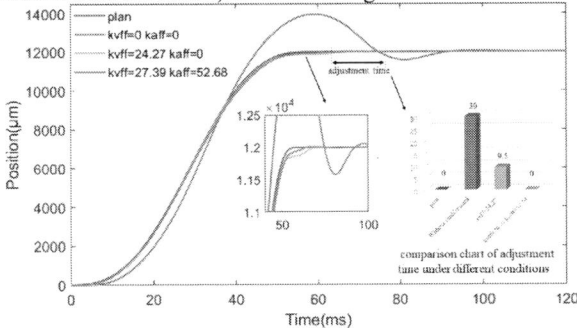

Fig.13. Comparison of position following effect and adjustment time before and after feedforward adjustment

In Fig.13, the position error before and after the end of the planning command is gradually reduced by fuzzy algorithm

setting, and the adjustment time is also reduced from 30ms without feedforward to 0ms. In terms of setting time, it takes 30s for two feedforward parameters to be set one after another, which is much less than that of the traditional manual trial and error method. It proves that the algorithm is effective under the given conditions.

VI. CONCLUSION

(1) Based on the thought of fuzzy control, this paper expresses a fuzzy intelligent tuning algorithm for feedforward parameter. Experimental results show that the fuzzy intelligent tuning algorithm can quickly find the relatively optimal speed feedforward and acceleration feedforward parameters under given conditions, and the efficiency is much higher than the manual trial-and-error method.

(2) The feedforward values obtained from the fuzzy intelligent tuning algorithm for feedforward parameter were applied to the XY motion platform of the micro-LED mass transfer equipment. The positioning stability time and extreme value of position error of the platform system are significantly reduced, which can meet the requirements of the high-speed and high-precision of the massive transfer equipment.

(3) The algorithm presented in this paper is feasible for parameter optimization, and it is of great significance for equipment debugging in the mass production of micro-LED transfer equipment.

ACKNOWLEDGMENT

The work presented in this paper was supported by the Nation Natural Science Foundation of China(No.503200036).

REFERENCES

[1] T. Wu et al., "Mini-LED and Micro-LED: Promising Candidates for the Next Generation Display Technology," Applied Sciences, vol. 8, no. 9, 2018.

[2] Z. J. Liu et al., "Micro-light-emitting diodes with quantum dots in display technology," Light-Science & Applications, vol. 9, no. 1, May 11 2020.

[3] C. W. Tao, J. S. Taur, and M. L. Chan, "Adaptive Fuzzy Terminal Sliding Mode Controller for Linear Systems With Mismatched Time-Varying Uncertainties," IEEE Transactions on Systems, Man and Cybernetics, Part B (Cybernetics), vol. 34, no. 1, pp. 255-262, 2004.

[4] W. Xi, Y. Bintang, and Z. Yu, "Adaptive model-based feedforward to compensate Lorentz force variation of voice coil motor for the fine stage of lithographic equipment," Optik, vol. 135, pp. 27-35, April 2017.

[5] H. A. O. Huan, Q. I. N. Lei, W. U. Shuai, K. Shao-long, and J. I. Ai-ming, "Research on Position Servo System Based on Composite Feedforward Fuzzy PID," (in Chinese), Measurement & Control Technology, vol. 37, no. 12, pp. 128-130,136, 2018.

[6] S. U. N. Chuanqing, "Accurate Position Control Simulation of Fractional-order Iterative Learning Control for Permanent Magnet Linear Synchronous Motor," (in Chinese), Micromotors, vol. 53, no. 9, pp. 59-62, 2020.

[7] Y. U. Xi-da and G. A. O. Jian, "Research on XYθ Positioning Platform Based on Decoupling Control Algorithms," (in Chinese), Modular Machine Tool & Automatic Manufacturing Technique, no. 10, pp. 110-116, 2020.

[8] A. Saleem, M. Mesbah, and M. Shafiq, "Feedback-feedforward control for high-speed trajectory tracking of an amplified piezoelectric actuator," Smart Materials and Structures, vol. 30, no. 2, p. 025033 (12 pp.), February 2021.

[9] H. E. Yun-bo and C. Jia-jun, "Compound control and parameter self-tuning of permanent magnet synchronous motor," (in Chinese), Mechanical & Electrical Engineering Magazine, vol. 36, no. 9, pp. 995-1000, 2019.

Numerical analysis of the microscopic factors influencing the thermal conductivity of Al₂O₃/AlN polymer composites

Nan Cheng[#]
School of Electronic Science & Engineering, Southeast University,
Nanjing, P.R.China
244170663@qq.com

Jibao Lu*
Shenzhen Institutes of Advanced Technology,
Chinese Academy of Sciences
Shenzhen, P.R.China
jibao.lu@siat.ac.cn

Jianbin Xu
Department of Electronics Engineering,
The Chinese University of Hong Kong,
Shatin, New Territories,
Hong Kong 999077, P.R. China
jbxu@ee.cuhk.edu.hk

Xiaoxin Lu[#]
Shenzhen Institute of advanced electronic materials, Shenzhen Institutes of Advanced Technology
Chinese Academy of Sciences
Shenzhen, P.R.China
luxx@siat.ac.cn

Shen Xu*
School of Electronic Science & Engineering, Southeast University,
Nanjing, P.R.China
Xus@seu.edu.cn

Ching-Ping Wong
School of Materials Science and Engineering
Georgia Institute of Technology
Atlanta, USA
cp.wong@mse.gatech.edu

Jiabin Huang
Shenzhen Institute of advanced electronic materials, Shenzhen Institutes of Advanced Technology
Chinese Academy of Sciences
Shenzhen, P.R.China
jb.huang@siat.ac.cn

Sun Rong*
Shenzhen Institute of advanced electronic materials, Shenzhen Institutes of Advanced Technology
Chinese Academy of Sciences
Shenzhen, P.R.China
rong.sun@siat.ac.cn

 # Nan Cheng and Xiaoxin Lu are co-first authors of the article.

Abstract—The thermal interface materials (TIMs) used between a chip and a heat spreader in electronic packaging composes polymeric materials filled with particulate fillers. In this work, we focus on the numerical modeling and design of the particle-laden polymers with high packing density, in which two kinds of particles are mixed. Specifically, Al₂O₃ and AlN, which are commonly used in the electronic packaging industry, are taken as the particulate fillers. Firstly, a series of microstructures of Al₂O₃/AlN filled polymer composites with 75 vol% filler volume fraction are generated in GeoDict software with changing the relative content of Al₂O₃ and AlN. The diameters of the Al₂O₃ and AlN particles obey orthogonal logarithmic distribution and Gaussian distribution, respectively. Then the thermal conductivities of the structures are simulated under various microscopic factors, such as particle-particle and particle-matrix interfacial thermal resistance, etc. The results are analyzed taken advantage of the orthogonal experimental design, showing that the particle-particle interfacial thermal resistance plays dominant role in the thermal properties of the particulate composites. We demonstrate that such technique can be used to optimize the design of particulate TIMs.

Keywords—particulate composites; thermal conductivity; GeoDict software.

I. INTRODUCTION

With the development of microelectronics integration technology, electronic components are becoming more and more miniaturized, leading to more and more heat generated per unit area of the chip. Thus, heat dissipation becomes particularly important to the chip. Polymers have good electrical insulation and excellent mechanical properties, but polymers often have a low thermal conductivity, usually no more than 0.5 $Wm^{-1}K^{-1}$. Particle-laden polymer composites are one of the most widely used thermal interface materials

(TIMs) used for electron packaging [1-3]. Fillings of metals or their oxides/nitride compounds with high thermal conductivity into polymers can significantly improve the thermal conductivity of polymers. Both Al₂O₃ and AlN are excellent and common thermal conductivity fillers [4,5].

The thermal conductivity of polymer composites is mainly affected by the thermal conductivity of fillers, the volume fraction of fillers, the thermal contact resistance between fillers, and the thermal contact resistance between fillers and matrix. Many theories and experiments show that when the volume fraction of the fillers reaches a certain value, a thermal conducting network can be formed between the fillers, leading to a rapid increase in the thermal conductivity of the composites [6]. The high content of the highly-thermally-conductive particles results in significant enhancement of the effective thermal conductivity of the composites. Thus, it is of practical significance to study the thermal conductivity of composites with high filling volume fraction. The up-limit of filling content in the composite with single-particle-size fillers is 64vol% [7]. However, the filling content can reach 80% or even higher by mixed filling with a variety of particle size fillers [8]. Therefore, it is essential to mix various particles with different size distributions to obtain a high filling content for the TIMs. Furthermore, the grading size schemes also play an important role in the effective thermal conductivity of the composites [9]. which requires through exploration.

In this paper, we introduce the particle size distribution of Al₂O₃ particles and AlN particles into the microstructures of the composite systems, which are generated by the GrainGeo Module of GeoDict software [10]. Then, the thermal conductivities of the Al₂O₃/AlN composites are computed based on Fast Fourier transform algorithm by ConductoDict module of GeoDict. During the computation, the contact

978-1-6654-1392-3/21 $31.00 © 2021 IEEE

thermal resistance between particles, and the contact thermal resistance between particles and matrix are taken into account. In order to study the influence of thermal conductivities of Al_2O_3 and AlN, as well as the thermal resistance on the effective thermal conductivity of the composites, we employed the orthogonal experimental design to analyze the data obtained from the simulation, facilitating the validation of main influencing factors.

II. NUMERICAL MODELING

A. Particle size distribution

The size distribution of Al_2O_3 particles is measured by laser particle analyzer, shown by blue points in Fig. 1 (a). The data is then fitted by log-normal distribution denoted by Eq. (1), which is the common distribution in the industrial powder products.

$$f_1 = \frac{1}{\sqrt{2\pi\sigma^2}} e^{[-\frac{1}{2\sigma^2}(lnx-\mu)^2]} \qquad (1)$$

The fitting parameters for the distribution of Al_2O_3 particles obeys $\mu = 1.02$ μm, and $\sigma = 0.3487$ μm ^{2}. The mean of Al_2O_3 particles is 2.95 μm according to the relation:

$$E(x) = e^{(\mu+\sigma^2/2)} \qquad (2)$$

We assume that the distribution of AlN particles obeys Gaussian distribution as Eq. (3), in which $\mu = 30$ μm and $\sigma = 1.6$ μm ^{2}. The probability of AlN distribution is shown in Fig 1 (b). The mean of the AlN particles is 30 μm. Thus, the size ratio between Al_2O_3 and AlN particles is set as 1:10.

$$f_2 = \frac{1}{\sqrt{2\pi\sigma^2}} e^{[-\frac{1}{2\sigma^2}(x-\mu)^2]} \qquad (3)$$

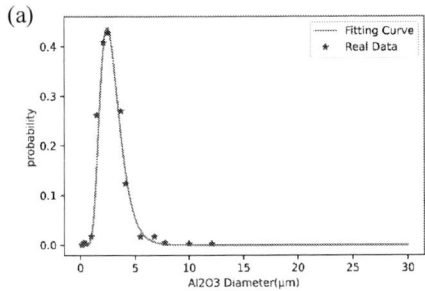

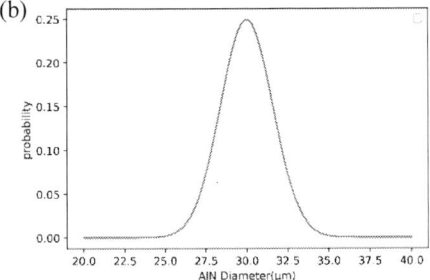

Fig. 1.(a) Al_2O_3 particle size distribution. (b)AlN particle size distribution

B. Generation of microstructures

In this paper, the model of composite material is established by using Geodict software. The representative volume element (RVE) is a cubic box meshed with cubic unit cells (voxels). The size of voxels of RVE is $0.2\times0.2\times0.2$ μm^3, and the number of voxels is $600\times600\times600$. The Al_2O_3 and AlN particles with specific size distribution are dispersed randomly in the RVE, defined as spheres. Four various microstructures of the composites are generated. In all the cases, the total filler volume fraction of Al_2O_3 and AlN particles is set as 75 vol%, while the volume ratio between Al_2O_3 and AlN varies from 0.1:0.9, 0.2:0.8, 0.3:0.7 to 0.4:0.6, respectively. The realizations of the microstructures are shown in Fig 2, in which the red spheres (small) are Al_2O_3 particles and the green spheres (big) are AlN particles.

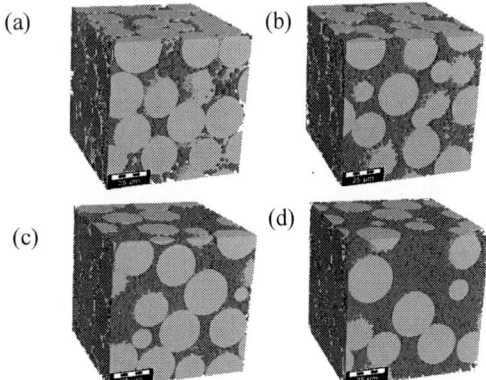

Fig. 2 Microstructures of Al_2O_3/AlN polymer composites, in which the filler volume fraction is 75 vol%. (a) Al_2O_3: AlN = 0.1:0.9; (b) Al_2O_3:AlN = 0.2:0.8; (c) Al_2O_3: AlN = 0.3:0.7; (d) Al_2O_3: AlN = 0.4:0.6.

C. Thermal conductivity simulation

The thermal conductivities of the composites are calculated by ConductoDict module of GeoDict software. With random dispersion of the particles, the effective thermal properties of the composites present isotropic. Therefore, the effective thermal conductivity is obtained by taking the average of the thermal conductivities along three main directions (X, Y and Z). During the computation, the periodic boundary conditions are employed.

There are seven input parameters to compute the thermal conductivity of the composites, i.e., Al_2O_3 and AlN particles, the interfacial thermal resistance between particles and matrix, and the thermal resistance between the same or various particles. In order to explore the effects of all the input parameters on the effective thermal conductivity with reduced computational cost, the orthogonal experimental design is employed in this work. The seven input parameters are taken as main factors with two levels, i.e., high level and low level. The selections of the values for low and high levels are shown in Table I . Here, K_1 and K_2 denote the thermal conductivities of Al_2O_3 and AlN, respectively. R_{ij} corresponds to the interfacial thermal resistance between two components, i and j, where the subscript 0, 1, and 2 denote the polymer matrix, Al_2O_3 particles and AlN particles, respectively. The 1/4 fractional factorial design is employed, leading to 32 trials for the thermal analysis of each microstructure.

TABLE I.   LOW AND HIGH LEVELS OF THE INPUT PARAMETERS

Variables	Low Level	High Level
K_1	20 Wm^{-1}K^{-1}	40 Wm^{-1}K^{-1}
K_2	200 Wm^{-1}K^{-1}	400Wm^{-1}K^{-1}
R_{01}	1×10^{-8}m^{2}KW^{-1}	1×10^{-5} m^{2}KW^{-1}
R_{02}	1×10^{-8} m^{2}KW^{-1}	1×10^{-5} m^{2}KW^{-1}
R_{11}	1×10^{-8} m^{2}KW^{-1}	1×10^{-5} m^{2}KW^{-1}
R_{22}	1×10^{-8} m^{2}KW^{-1}	1×10^{-5} m^{2}KW^{-1}
R_{12}	1×10^{-8} m^{2}KW^{-1}	1×10^{-5} m^{2}KW^{-1}

III. RESULTS AND DISCUSSION

A. Preliminery analysis of the effect of factors on thermal conductivity of the composites

The 4×32 sets of data obtained by Geodict simulation are plotted in Fig 3. It can be observed that the thermal conductivity of composites increases with the increase of the relative content of AlN particles in most cases. This is easy to understand since the thermal conductivity of AlN is an order of magnitude higher than that of Al_2O_3. In addition, the particle size of AlN particles is also an order of magnitude larger than that of Al_2O_3, which can also reduce the thermal barrier resulting from the interfacial thermal resistance.

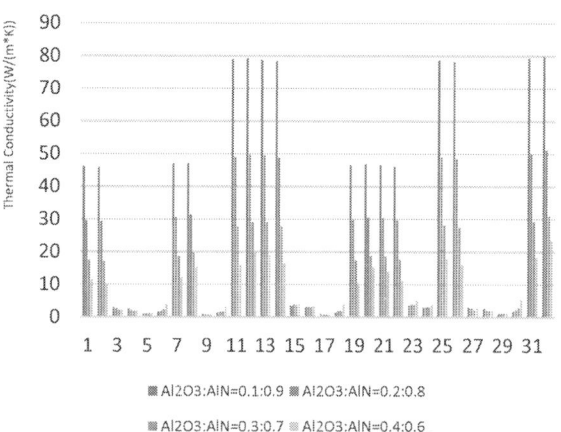

Fig. 3 Numerical simulation results of thermal conductivity. The X-axis denotes the index of 32 trails for 4 various structures.

However, it should be noted that the thermal conductivity of some samples, of which the thermal resistance related to AlN is set high, decreases with the increase of the relative content of AlN, such as trail No. 10, 18, and 30. We take trail No. 30 as an example, $R_{01} = 1 \times 10^{-8}$ m²KW⁻¹, $R_{02} = 1 \times 10^{-5}$ m²KW⁻¹, $R_{11} = 1 \times 10^{-8}$ m²KW⁻¹, and $R_{22} = 1 \times 10^{-5}$ m²KW⁻¹, where the subscript "0" represents matrix, "1" represents Al_2O_3, and "2" represents AlN. Hence, the interfacial thermal resistance associated with Al_2O_3 is three magnitudes lower than that associated with AlN in this trail. Therefore, the effective thermal conductivity of the composites is more affected by the thermal contact resistance than the thermal conductivity and the size of the filler.

B. Orthogonal experiment analysis

The main effect diagrams for the four microstructures obtained from the calculations according to the orthogonal experimental design is shown in Figs 4 (a-d), which presents the influence of the input parameters. Specifically, the positive tangent of the curves means that the high values of the factors can enhance the effective thermal conductivity. In contrast, the negative tangent represents the weaken impact of the parameter. Besides, the sharper the slope of the curve is, the higher impact of its corresponding factor has. It can be seen that the thermal conductivity of AlN has positive effect on the effective thermal conductivity of the composites, while the thermal conductivity of Al_2O_3 has little influence. For all the cases, the interfacial thermal resistance between AlN particles (R_{22}) plays a significant part in the effective thermal conductivities, which means that the decrease R_{22} can dramatically improve the effective thermal conductivity of the composites. When the relative content of Al_2O_3 is less than 30%, the effect of R_{01}, R_{02}, R_{11}, R_{12} can almost be neglected. However, when the relative content of Al_2O_3 reaches 40%, the interfacial thermal resistance related with Al_2O_3 starts to exert an influence.

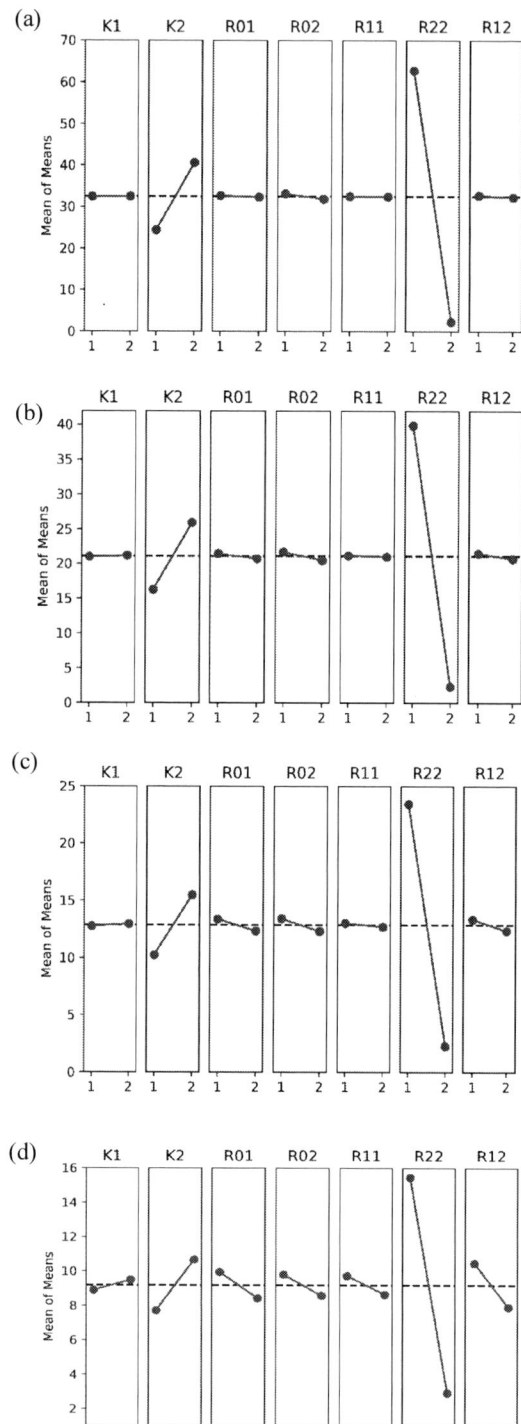

Fig. 4 Main effect diagram of thermal conductivity influencing factors for the four various microstructures (see in Fig. 2).

The corresponding Pareto diagrams are shown in Figs. 5 (a-d), which give the effect of single influencing factor as well as that of the coupled influencing factors. It indicates when the relative content of Al$_2$O$_3$ particles is no more than 30%, the

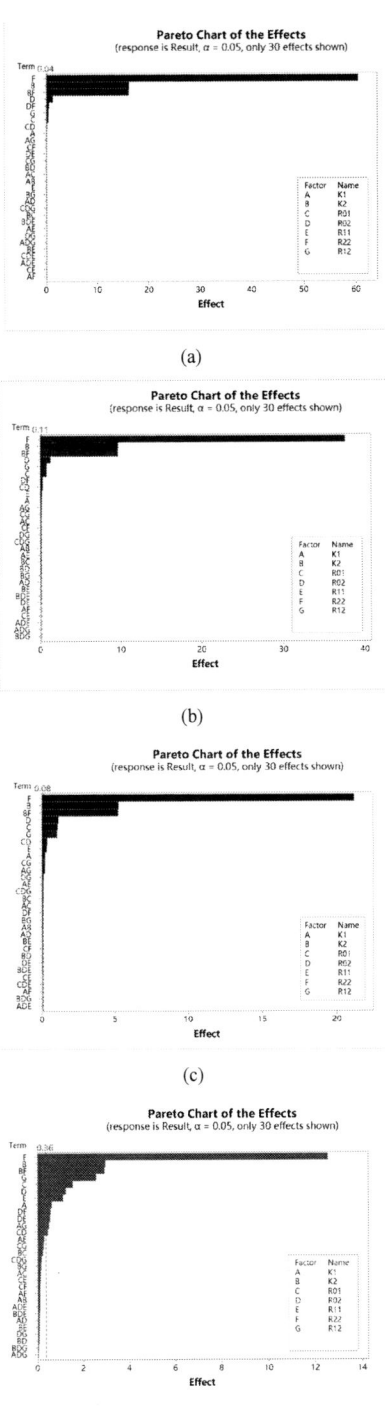

(a)

(b)

(c)

(d)

Fig. 5 Pareto diagrams of influencing factors of thermal conductivity for the four various microstructures.

interfacial thermal resistance between AlN particles has a decisive influence on the effective thermal conductivity of the composites. The effects of the thermal conductivity of AlN comes to the second. Then, the coupled effect of the previous two factors is also obvious. The other input parameters have little impact. When the relative content of Al$_2$O$_3$ reaches 40%, the influencing factors become more, i.e., the thermal conductivity of Al$_2$O$_3$ and AlN, all the interfacial thermal resistance between various components have a significant effect on the thermal conductivity of composites. The results are consistent with the main effects diagrams.

IV. CONCLUSIONS

In this study, the factors affecting the thermal conductivity of Al$_2$O$_3$ /AlN polymer composites were interrogated through numerical simulation and orthogonal design experiments, which has important guiding significance for the design of thermal interface materials. We can draw the following conclusions.

(1) As the relative content of Al$_2$O$_3$ increases, the thermal conductivity of composites generally decreases, which is mainly due to the higher thermal conductivity and larger particle size of AlN. However, when the interfacial thermal resistances related with Al$_2$O$_3$ are small, the thermal conductivity of the composites may increase along with the increase of the relative content of Al$_2$O$_3$.

(2) When the relative content of Al$_2$O$_3$ is less than 30%, the thermal conductivity of composites is mainly affected by the thermal conductivity of AlN and the thermal contact resistance between AlN particles. When the relative content of Al$_2$O$_3$ reaches 40%, apart from the abovementioned two mentioned factors, the other thermal resistances in the composites systems also show effects on the thermal conductivity of composites.

(3) In order to improve the thermal conductivity of polymer composites, large particle fillers with higher thermal conductivity should be selected and the volume content of large particle fillers should be increased. At the same time, appropriate surface modification should be carried out to reduce the thermal contact resistance between particles.

ACKNOWLEDGMENT

The work was supported by the National Natural Science Foundation of China (No. 52003289), Guangdong Basic and Applied Basic Research Foundation (No. 2021A1515010900), the Youth Innovation Promotion Association CAS (No. 2021363), SIAT Innovation Program for Excellent Young Researchers (No. E1G045, 201803).

REFERENCES

[1] K. M. Razeeb, E. Dalton, G. L.W. Cross et al., "Present and future thermal interface materials for electronic devices," International Materials Reviews, vol 63, no. 1, pp. 1-21, 2018.

[2] J. Hansson, T. Nilsson, L. Ye, and J. Liu, "Novel nanostructured thermal interface materials: a review," International Materials Reviews, vol.63, no.1, pp. 22-45, 2018.

[3] W. Dai, L. Le, J. Lu, H. Hou, "A Paper-Like Inorganic Thermal Interface Material Composed of Hierarchically Structured Graphene/Silicon Carbide Nanorods," ACS Nano, vol. 13, no. 2, pp. 1547-1554, 2019.

[4] Y. Hu, G. Du, and N. Chen, "A novel approach for Al2O3/epoxy composites with high strength and thermal conductivity,"Composites Science and Technology, vol. 124, pp. 36-43, 2016.

[5] C R. Yang, C D. Chen, and C. Cheng, "Thermal conductivity enhancement of AlN/PDMS composites using atmospheric plasma

modification techniques," International Journal of Thermal Sciences, vol. 155, p.106431, 2020.

[6] D X. Li, D. Zeng, Q. Chen, M L. Wei, L Song, C G. Xiao, and D H. Pan, "Effect of different size complex fillers on thermal conductivity of PA6 thermal composites," Plastics, Rubber and Composites, vol. 48, no. 8, pp. 347-355, 2019.

[7] G. D. Scott, D. M. Kilgour, "The density of random close packing of spheres," Journal of Physics D: Applied Physics, vol. 2, no. 6, p.863, 1969.

[8] G Roquier " A Theoretical Packing Density Model (TPDM) for ordered and disordered packings," Powder Technology, pp.343-362, 2019.

[9] X. Lu, X. Fu, and J. Lu " Numerical homogenization of thermal conductivity of particle-filled thermal interface material by fast Fourier transform method," Nanotechnology, vol. 32, no. 26, p. 265708, 2021.

[10] GeoDict software, Release 2021 from Math2Market GmbH, Germany, https://www.geodict.com

The Influence Analysis of Geometry on Void in Molded Underfill for Flip Chip

1st Yamei Yan
College of Mechanical and Electrical Engineering
Central South University
Changsha 410083, China
214742328@qq.com

2nd Gui Chen
College of Mechanical and Electrical Engineering
Central South University
Changsha 410083, China
Chengui@csu.edu.cn

3rd Xiaoyu Xiao
College of Mechanical and Electrical Engineering
Central South University
Changsha 410083, China
xiao1201@csu.edu.cn

4th Yan Wang*
School of Electronic Science and Engineering
University of Electronic Science and Technology of China
Chengdu, China
wangyancumt@163.com

5th Wenhui Zhu*
College of Mechanical and Electrical Engineering
Central South University
Changsha 410083, China
zhuwenhui@csu.edu.cn

Abstract—In flip chip packaging，the molded underfill （MUF）integrates the underfilling and molding processes to reduce manufacturing cost and cycle time of process，but is difficlut to obtain high quality filling due to void trapping problem. Virtually, from a technical perspective, it is a microinjection molding process, the MUF process can be thought as a pressure-driven suspension flow. For transfer molding, the main factors affecting the reliability of products include the properties of filling materials, injection process parameters and the geometry of the cavity. In this paper, 3D mold flow modeling of the transfer molding process with MUF using Moldflow is applied. By designing DOE experiments, the MUF process parameters were optimized under different packaging structures of flip chip, and the influence of different methods of chip placement on the molding results was discussed, providing a basis for the design and optimization of process parameters and chip geometry structure.

Keywords—Molded Underfill, Simulation Model, Void, Filling Performance

I. INTRODUCTION

As functional requirements increase and package sizes decrease, it is often necessary to integrate flip-chips into different systems. For traditional flip chip package ,after capillary underfilling(CUF)， the chips then molded. Although material research and dispensing systems are being optimized for underfilling, it is still the most time-consuming step in flip chip encapsulation [1].In order to reduce the time cost and ensure the reliability of the product, the molded underfill package (MUF), which combines the molding and underfilling processes, was proposed. Its advantage is that batch processing operations can be carried out in the production line, reducing the material cost and increasing the output. The MUF material is applied to the flip chip package by transfer molding process, it can not only fills the gap between the chip and the substrate, but also encapsulates the entire chip. It combines underfilling and transfer molding together, which reduces molding time and improves mechanical stability.It is especially suitable for improving the production efficiency of flip chip packaging[2].

In MUF technology, packaging design face the challenge in geometrical structure, for flip chip's structure has higher resistance to the mold flow, such as the smaller bump height , bump pitch or a different way of stacking, etc., make it easier for the air trapped in the chip and form voids, and further lead to encapsulation of popcorn effect and solder extrusion, resulting in failure .Mold temperature, melt temperature, mold closing force, injection pressure and other technological parameters, as well as EMC(Epoxy Molding Compound) composition, filling particle content and size[3], will have an impact on the flow in the filling process.

In the manufacturing production using MUF process, flow and heat transfer are dynamically coupled with the curing reaction [4]. The kinetics of the curing reaction not only affects the EMC conversion rate, but also has a great influence on the flow of the melt due to the change of the curing reaction viscosity. In order to obtain a encapsulation product with complete filling and high reliability, it is usually necessary to conduct DOE on MUF material to obtain a product with voids-free[1].In the past studies, the MUF process was mainly discussed from the material selection, geometric structure parameters and optimization of process parameters. In this paper, the three-dimensional analysis software Moldflow is used to simulate the transfer molding process, mainly to optimize the cavity structure filled by EMC and the arrangement of chip and bump, and to obtain the appropriate process parameters for a commercial MUF material.

II. SIMULATION

A. Governing Equations

The molded underfill process is a three-dimensional, transient and reactive process with moving resin front. The non-isothermal resin flow in mold cavity can be mathematically described by the following equations[5]:

$$\frac{\partial \rho}{\partial t} + \nabla \cdot \rho u = 0 \tag{1}$$

$$\frac{\partial}{\partial t}(\rho u) + \nabla \cdot (\rho u u - \sigma) = \rho g \tag{2}$$

The work was support by National Natural Science Foundation of China (51905554).

978-1-6654-1392-3/21 $31.00 © 2021 IEEE

$$\sigma = -p\boldsymbol{I} + \eta(\nabla\boldsymbol{u} + \nabla\boldsymbol{u}^T) \tag{3}$$

$$\rho C_p\left(\frac{\partial T}{\partial t} + \boldsymbol{u}\cdot\nabla T\right) = \nabla\cdot(k\nabla T) + \boldsymbol{\Phi} \tag{4}$$

where $\boldsymbol{u}$ is the velocity vector, T is the temperature, t is the time, p is the pressure, $\boldsymbol{\sigma}$ is the total stress tensor, $\boldsymbol{\rho}$ is the fluid density, k is the thermal conductivity, C_p is the specific heat, and $\boldsymbol{\Phi}$ is the energy source tem. In this work, the energy source contains two contributions:

$$\boldsymbol{\Phi} = \eta\dot{\gamma} + \dot{\alpha}\Delta H \tag{5}$$

where $\boldsymbol{\eta}$ is the viscosity, $\dot{\boldsymbol{\gamma}}$ is the magnitude of the rate of deformation tensor, $\dot{\boldsymbol{\alpha}}$ is the conversion rate and $\boldsymbol{\Delta H}$ is the exothermic heat of polymerization.

B. Rheokinetic Characterization

When the chip is molded and encapsulated, the flow of the melt is affected by both heat transfer and curing reaction. The curing reaction not only determines the degree of EMC transformation during molding, but also changes the viscosity of the fluid if partial curing reaction occurs during the flow. In addition, due to the non-Newtonian properties of the polymer melt, the velocity and shear rate of the fluid can also change the viscosity. Therefore, to simulate the process of transfer molding, it is necessary to know the properties of melt reaction viscosity and curing reaction kinetics.

An EMC material named Nidem 1 made by Nitto Denko was used in this paper. The EMC was filled with 80% silica. In this study, Cross Castro Macosko's viscosity model [6] was used to describe the correlation between temperature, shear rate and curing of this thermosetting material:

$$\eta(\alpha, T, \dot{\gamma}) = \frac{\eta_0(T)}{1+\left(\frac{\eta_0(T)\dot{\gamma}}{\tau^*}\right)^{1-n}}\left(\frac{\alpha_g}{\alpha_g-\alpha}\right)^{(C_1+C_2\alpha)} \tag{6}$$

Where η is the viscosity, $\dot{\gamma}$ shear rate, T is the temperature, α is the degree of cure, $\eta_0(T) = B\exp(\frac{T_b}{T})$, and the other coefficients are fitted by experimental tests.

TABLE I. NUMERICAL PARAMETER FOR CROSS CASTRO MACOSKO MODEL

Parameter of Kinetics	Value
α_g	0.5007
C_1	96.73
C_2	0.0001793
B (Pa/s)	7.774E-13
T_b (K)	22480
n	0.4842
τ^* (Pa)	6.647E-5

In this study, Kamal's model [7] was used to describe the kinetic properties of curing reaction. The formula and coefficient of the model were as follows:

$$\frac{d\alpha}{dt} = (k_1 + k_2\alpha^m)(1-\alpha)^n \tag{7}$$

$$k_1 = A_1\exp(-E_1/T) \tag{8}$$

$$k_2 = A_2\exp(-E_2/T) \tag{9}$$

Where $\boldsymbol{\alpha}$ is the degree of cure, T is the temperature, t is the time. Other parameters $m, n, \boldsymbol{A_1}, \boldsymbol{A_2}, \boldsymbol{E_1}, \boldsymbol{E_2}$ are constants, which can be obtained by fitting experimental data. During the curing kinetics measurement, the reaction heat is also measured. The heat of the reaction is H J/kg.

TABLE II. NUMERICAL PARAMETER FOR KAMAL'S MODEL

Parameter of Kinetics	Value
m	0.5929
n	1.788
A_1 (s-1)	1.112E11
A_2 (s-1)	1.505E09
E_1 (k)	26660
E_2 (k)	10740
H(J/kg)	25000

C. Geometric parameters and Boundary Conditions

The basic dimensional parameters of the chip used for calculation are given in Table 3. In this paper, two kinds of chip with different bump arrangement are used, one is center array arrangement and another is periphery arrangement, as shown in Fig.1. Considering that there is some requirements on the number of bump in the production, the two different bump arrangements have the same number of 256.

TABLE III. DIMENSIONAL PARAMETERS OF CHIP

Bump Number	256
Chip Size(mm)	7.5×7.5
Chip Thickness(mm)	0.5
Bump Diameter(mm)	0.1
Bump Pitch(mm)	0.2
Bump Height(mm)	0.2

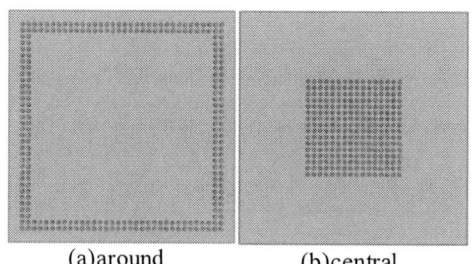

(a)around (b)central

Fig.1. Bump arrangement

In addition, the distribution of the two chips is also discussed, which are the stacking of two chips and the sequential arrangement of two chips. As shown in Fig. 2 , according to the arrangement of the two chips, the mold cavity is changed. When the size is set up, the space around the chips is mainly kept consistent. The design size parameters are shown in the schematic diagram. In this calculation, the boundary condition of the wall surface is no slip, and the pouring port position of the polymer melt is shown in the schematic diagram.

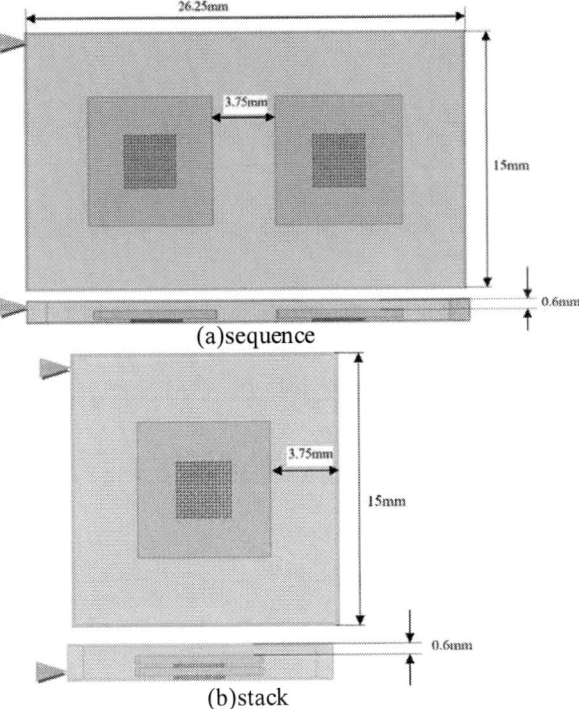

(a)sequence

(b)stack

Fig.2. Chip arrangement

D. Results and Discussion

In the existing MUF process, air is trapped in the cavity due to insufficient flow or small spacing between chip substrate and bumps. With the polymer curing, popcorn effect and solder extrusion are likely to be caused in practical application, which seriously affects the reliability of the product. In this study, flip chip packaging is used to study the impact of different bump structures and chip placement on the void formation. There are four types of structures, named according to table 4. The process parameters are set as the recommended parameters of the selected commercial EMC materials, and the injection rate is calculated according to the micro injection flow rate and gate size. The values of the processing parameters are shown in Table 5.

TABLE IV.   NAME OF DIFFERENT STRUCTURES

Chip arrangement Bump Arrangement	Stack	Sequence
Central	X1	X2
Around	Y1	Y2

TABLE V.   PROCESSING PARAMETERS

Mold Temperature(℃)	170.0
Melt Temperature(℃)	65.0
Injection Rate(mm3/s)	39.3

It is found that the curvature direction of the front edge is different from that of the underfilling driven by capillary force. This is because transfer molding is a process of pressure injection. The injection molding machine provides pressure to move the front edge. At this time, the surface of the cavity, the chip and the bump belong to non-slip boundary, and the shear rate is very small, It can produce viscous force on the flow, and the flow rate in the middle layer of the fluid is higher, thus lead to the shape as shown in Fig.3(a). The correctness of this shape can be verified by referring to the experimental results Fig. 3(b)[8]. For the chip underfilled with capillary force, the wall will also have viscous force, but it can provides capillary force as a power to pull the fluid forward. Coupled with the effect of surface tension, the flow will finally appear as shown in Fig.3(c) [9].

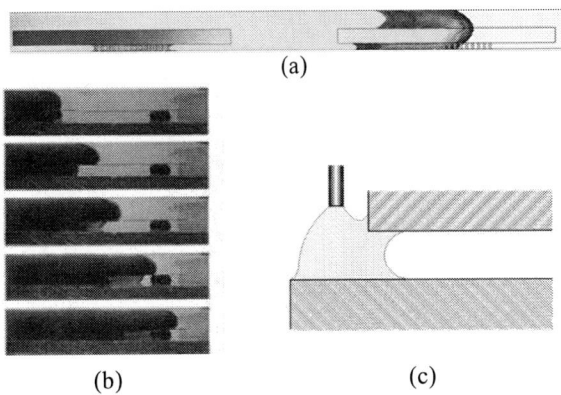

(a)

(b)

(c)

Fig.3. Flow front contrast

Fig. 4 shows the simulation results of four structures under the same process parameters. The color indicates the flowing time of the EMC inside cavity of the flip chip, and the color changes from blue to red, indicating the flowing time changes longer. In the process, the melt flow will be divided into three stream after encountering the chip, which are the flow 1 above the chip, the flow 2 through the bump area and the flow 3 around the chip. As shown in Figure 4(a), the moving velocity of flow 3, i.e the stream passing around the chip is faster than others. The resistance of flow 1 and 2 is larger than flow 3. Especially, flow 2 encounters the bump area under the chip, therefore the resistance is the largest.

It can be seen from the Fig. 4 that void is more likely to occur when the bumps are distributed around. This is because when the bumps are distributed around the chips, flow 2 will encounter the bumps earlier, thus slowing down the flow speed. But at the same time, flow 1 and flow 3 move faster, resulting in the EMC wrapping back from the front, blocking the outlet of the flow through the bumps. The enclosed void area(as shown in Figure1 (a) around) is larger than the central arrangement(as shown in Figure1 (b) central)), such that the reliability is lower.

In the case of stacked chips as shown in the Fig. 4(a) and Fig. 4(c), it is obvious that flow 1 and flow 3 move back after confluence in the front, and the area of voids under the two chips is basically the same. When the chips are placed horizontally in sequence as shown in the Fig. 4(b) and Fig. 4(d), the flow path of flow 3 will be longer, which gives enough time to flow 2 moving through the first chip to fill the bump area. In addition, from the perspective of packaging cost, the stacking method can save packaging area, and can be designed according to product requirements.

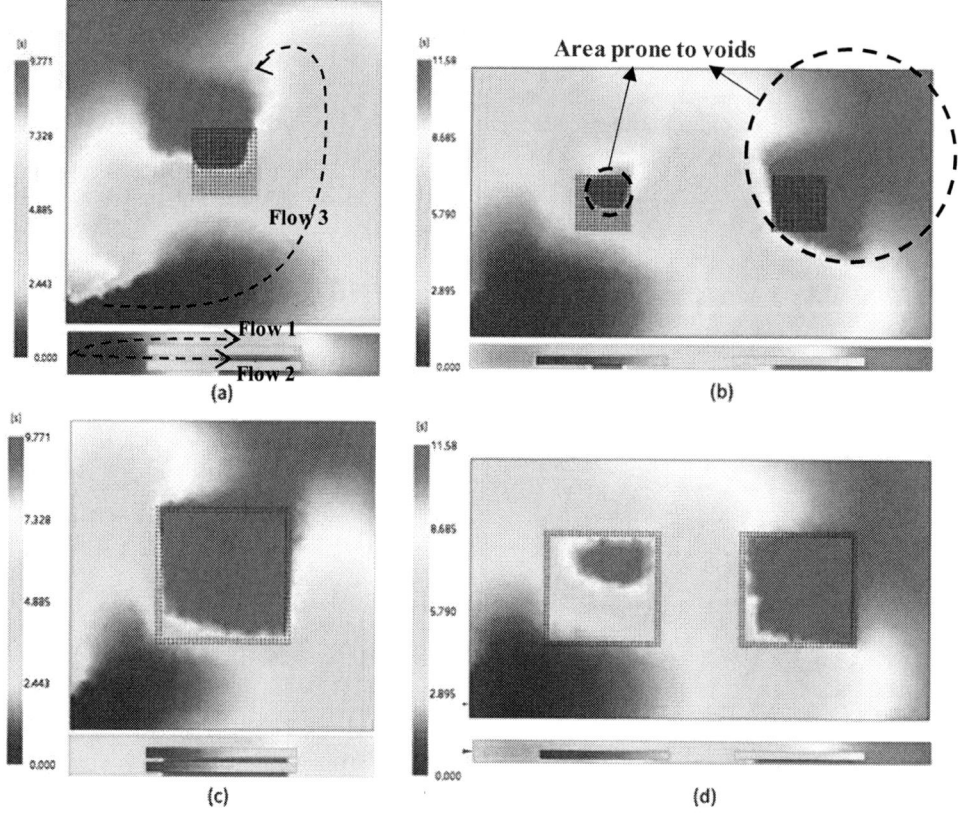

Fig.4. Filling time and flow process of different geometrical structures

III. PROCESS PARAMETER OPTIMIZATION

A. Design of Experiment(DOE)

Through the discussion of the results in the previous section, we can know the flow of EMC in different structures. However, the expected filling is not achieved under the recommended process parameters. So the appropriate geometric structure Type X2 is selected to carry out the experimental design, and the mold temperature, melt temperature and flow rate are selected as variables to optimize the filling process.

TABLE VI.　　　PROCESS PARAMETER SETTING

	Mold Temperature(℃)	Melt Temperature(℃)	Injection Rate(mm3/s)
1	150	45	31.4
2	150	45	47.2
3	150	85	31.4
4	150	85	47.2
5	190	45	31.4
6	190	45	47.2
7	190	85	31.4
8	190	85	47.2
9	150	65	39.3
10	190	65	39.3
11	170	45	39.3
12	170	85	39.3
13	170	65	31.4
14	170	65	47.2
15	170	65	39.3

B. Result and Discussion

To optimize the process parameters, and study the influence of different process parameters on the formation of voids, it is necessary to quantify the results of void formation. As can be seen from Fig. 4, the regions that are all shown in red without color gradient indicates the ends of the flow, and these regions are the most likely to form voids. The bump region is divided into 15×15 square grids by 16×16 bump array. The number of squares occupied by red in the result is the quantization value of the void result, as shown in the Fig. 5.

The effect of melt temperature and flow rate on void formation is not significant, but the void position is impacted by melt temperature and flow rate. The main factor affecting the size of the voids is the temperature of the mold, and the relationship between the two is shown in Fig. 6. It can be seen that there is a large change of voids between 170 ℃ and 185 ℃. This is due to the fact that when the thermosetting plastic reaches the initial curing temperature, the higher the temperature is, the faster the polymer will solidify during the filling time. Therefore, the solidification reaction has occurred in the process of melt flow at high temperature, with higher viscosity and slower flow rate. The flow around the chip and above the chip will slow down, make the melt can also fully fill the height, but may cause the problem of short shot because of early curing. Therefore, the mold cavity should be filled as soon as possible while ensuring the filling quality. Fig. 7 shows the experimental design results of filling time. In addition, it will also lead to the failure of the bumps if the mold temperature is too high.

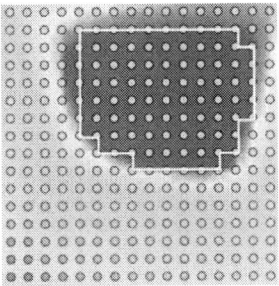

Fig.5. Schematic diagram of void area calculation

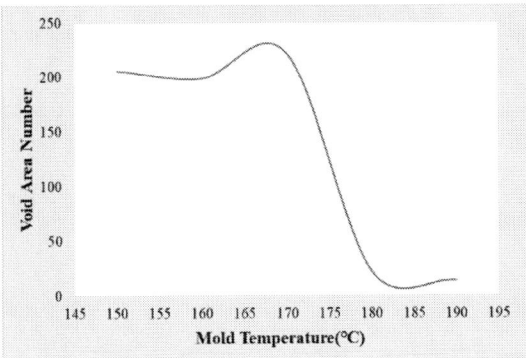

Fig.6. Relationship between mold temperature and void formation

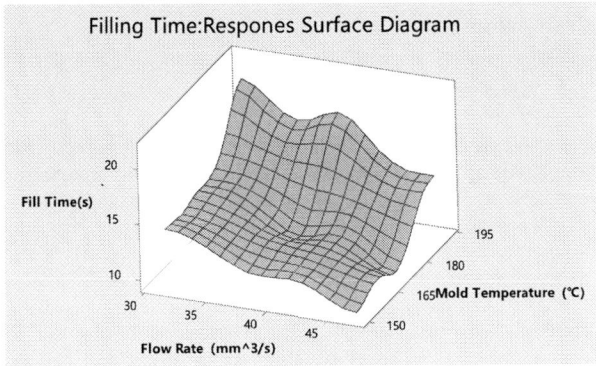

Fig.7. 3D response surface for filling time

IV. CONCLUSION

The conclusions are listed as follows:

The mold flow simulation of the MUF process of the flip chip was carried out to analyze the shape of the flow front，and the influence of the geometric structure on the melt flow process has been discussed. In general, the results show that

Type X2 has the best reliability, but higher area cost compared with X1 and Y1.

Through the experimental design, it is concluded that the main factor affecting the form of void is the mold temperature, and the reason of the variation trend is discussed. It can be used as a reference for process parameter selection.

ACKNOWLEDGMENT

The work was support by National Natural Science Foundation of China (51905554).

REFERENCES

[1] Joshi, M. , Pendse, R. , Pandey, V. , Lee, T. K. , & Lee, H. R. . (2010). Molded Underfill (MUF) Technology for Flip Chip Packages in Mobile Applications. Electronic Components & Technology Conference. IEEE.

[2] Rector LP, Gong S, Miles TR, Gaffney K (2000) Transfer molding encapsulation of flip chip array packages. In: IMAPS proceedings, pp 760–766.

[3] Lee, J. Y. , Oh, K. S. , Hwang, C. H. , Lee, C. H. , & Amand, R. D. S. . (2009). Molded underfill development for flipstack CSP. Electronic Components & Technology Conference. IEEE.

[4] Nguyen, L. T. . (1993). Reactive flow simulation in transfer molding of IC packages. Electronic Components & Technology Conference. IEEE.

[5] Chang, R. Y. , Yang, W. H. , Hwang, S. J. , & Su, F. . (2004). Three-dimensional modeling of mold filling in microelectronics encapsulation process. IEEE Transactions on Components and Packaging Technologies, 27(1), 200-209.

[6] Castro, J. M. , & Macosko, C. W. . (1980). Kinetics and Rheology of Typical Polyurethane Reaction Injection Molding Systems.

[7] Kamal, M. R. , & Ryan, M. E. . (1980). The behavior of thermosetting compounds in injection molding cavities. Polymer Engineering & Science, 20(13), 859–867.

[8] Khor, C. Y. , Abdullah, M. Z. , Ariff, Z. M. , & Leong, W. C. . (2012). Effect of stacking chips and inlet positions on void formation in the encapsulation of 3d stacked flip-chip package. International Communications in Heat and Mass Transfer, 39(5), 670-680.

[9] Wang Hui. Macro mesoscopic multi-scale modeling and calculation of underfill packaging process. (Doctoral dissertation, Huazhong University of science and Technology)

A Cost-saving Thermal Test Chip Design in a Test Vehicle of Large BGA

Jianjun Sun[2*]
Department of Packaging and Testing
ZTE Corporation
Shenzhen, China
sun.jianjun2@sanechips.com.cn

Yuanting Lai[2*]
Department of Packaging and Testing
ZTE Corporation
Shenzhen, China
lai.yuanting1@sanechips.com.cn

Hao Yang[2]
Department of Packaging and Testing
ZTE Corporation
Shenzhen, China
yang.hao11@sanechips.com.cn

Jian Pang[2]
Department of Packaging and Testing
ZTE Corporation
Shenzhen, China
pang.jian@sanechips.com.cn

Tuobei Sun[2]
Department of Packaging and Testing
ZTE Corporation
Shenzhen, China
sun.tuobei@sanechips.com.cn

Keqing Ouyang[1]
State Key Laboratory of Mobile Network and
Mobile Multimedia Technology
Shenzhen, China
ouyangkeqing@sanechips.com.cn

Abstract—Special chip design for thermal test has been developed for a long time for assessment of package thermal characteristics, evaluation of material thermal performance, and validation of thermal models and simulations. Nevertheless, most of the thermal test chip design is independent from the real chip development, leading to thermal test chip hardly be used in the new products with advanced manufacture node and new packaging type. In this paper, we proposed a cost-saving thermal test chip which can be synchronously developed and combined with daisy chain design for package reliability test. This thermal test chip contains metal strip resistors for heating and resistive temperature sensor for temperature monitor. It helps to study the package thermal characteristics and evaluate the performance and reliability of thermal interface material applied in flip-chip ball grid array package.

Keywords—*thermal test chip, thermal resistance test, thermal interface material, packaging*

I. INTRODUCTION

Thermal management of chip is becoming more and more important as the function of chip increases dramatically with higher power consumption, especially in the advanced integrated circuits (ICs) manufacturing process. The power consumption of a chip for communication application can be up to several hundreds Watts, and the power density of whole chip and chip's certain local region can reach up to $0.8W/mm^2$ and $5W/mm^2$, respectively. It is a great challenge for chip heat dissipation. Introducing a high performance and reliable thermal interface material (TIM) to decrease the package thermal resistance is an effective approach to relieve this challenge, especially for flip-chip ball grid array (FCBGA) package with heat spreader. Thermal resistance test is indispensable and essential to evaluate the thermal performance and reliability of a package with a new type of thermal TIM in FCBGA package. For most Metal-Oxide-Semiconductor Field-Effect Transistor (MOSFET) ICs with FCBGA package, we can implement transient dual interface test by T3Ster equipment from Mentor Graphics according JEDEC51-14 [1] to get the thermal resistance of whole package and parts, such as silicon die, TIM and lid. However, it needs a real functional chip manufactured by foundry and packaged by assembly factory, which is relatively backward stage in the full chip development process. So it takes great challenges and high risks to fulfill the reliability requirements within short time-to-market deadlines once the new TIM can't meet the thermal performance required. Therefore, a cost-saving thermal test chip (TTC) is designed

in a test vehicle and it hopes to be helpful to evaluate the FCBGA package thermal resistance and TIM thermal performance synchronously with other package reliability test before the real chip is available.

In this paper, we proposed a cost-saving thermal test chip which can be synchronously developed combined with daisy chain design for package reliability test. This test vehicle die is not only used to validate the package reliability, but also used to implement the thermal test for evaluation of new TIM's thermal performance. Section II summarizes the thermal test chip design from published papers. Section III describes the detail design of this thermal test chip and make an evaluation by simulation comparison. In the end of this part, the printed circuit board also be demonstrated. Finally, the last section summarizes the conclusions of this paper.

II. THERMAL TEST CHIP PUBLISHED

Thermal resistance test has been extensively used for assessment of packaging materials, evaluation of package thermal resistance, and validation of thermal models and simulations. And a thermal test chip can be a substitute for real functional chip for above usages ahead of real chip available. Many thermal test chip has been designed for chip thermal management in literature reported.

Generally, a thermal test chip contains heating elements and temperature sensing elements. Raj Pendse and B.J.Shanker designed a test die, containing power diodes for die heating and a string of emitter-base diodes for temperature sensing, to study the effects of mold compound conductivity, substrate conductivity on junction temperature. They also used 3x3 matric test dies to characterize the temperature distribution of a large die size by activating different heating zone [2]. Bernie Siegal and Jesse Galloway designed a thermal test chip arrayed by unit cell, which contains two metal film resistive heaters and four diode sensors, distributed in chip center, edge center and corner, with 4-wire Kelvin connection to improve the measurement accuracy [3]. Xavier Jorda et al, designed a thermal test chip for power assemblies assessment with poly-silicon strips heater and a Pt resistance temperature detector in the chip center. They used this test chip to evaluate thermal characterization of different kinds substrate and validated the thermal models and simulations [4].

Some versatile and programmable thermal test chips also be designed, in which heater elements can be controlled on/off by software and on-time sensor data can be read out

978-1-6654-1392-3/21 $31.00 © 2021 IEEE

by boundary scan circuitry without additional test apparatus. Zs. Benedek et al, proposed a scale-able multifunctional thermal test chip design. This test chip was built by basic cells in arrays, and each basic cell had four dissipating resistors switched on/off by a MOS transistor and a COMS temperature sensor located in cell center. The main measurement setup contained a personal computer, DC supply and thermal test chip. A special PC software provided full control of the test chip via its boundary scan interface and help data evaluation. And the dissipation patterns of the TTC could programmable by external asynchronous signals set by the software [5]. Suresh Parameswaran et al, designed a versatile thermal test chip arrayed by building-blocks, in which the heating-elements had a mix of resistor heating and ring-oscillator heating, and sensing-elements was ring-oscillator, RTD or diode select-able by user in block center. This thermal test chip was programmable by software to control heat pattern and can do simultaneous temperature measurement during thermal test [6]. This TTC could be used to evaluate temperature profile during the chip floor planning phase, to evaluate package cooling solution ahead of product silicon availability, to identify the critical on-die locations during chip thermal-aware test, as long as the power specifications of a product chip is know.

Previous thermal test chip designs make chip thermal management more flexible and it is convenient for users to buy the existing TTCs from vendors. Considering the expensive cost of TTC fabrication, for most user it is cost effective to buy TTCs ready made by vendors. However, the existing TTCs are hardly be directly used by users. Firstly, as more and more sophisticated and advanced packaging technology emerge in recent years, most of existing TTCs designed by various vendors are not suitable for an advanced packaging, taking consideration of silicon thickness, die size, pad or bumping design rules on die, etc. Secondly, many TTCs are designed and manufactured by 1um process or above. Therefore, the existing TTCs IP design also can't meet all customers' requirements, especially in advanced ICs manufacturing field, due to the difference of process node and design rules between different wafer foundries.

All above facts lead to the existing thermal test die design is independent from a real chip development and is merely applied to specific usage, which limits synchronous TTC development in a new chip to meet customers' requirements.

III. THERMAL TEST CHIP DESIGN AND PACKAGING

A. FCBGA package

This thermal test chip is packaged in a FCBGA with heat spreader. A standard FCBGA with flat type heat spreader package is illustrated in Fig. 1. The flipped silicon die is attached on substrate, electronic connecting by bump soldered on substrate pad after reflow. Epoxy underfill filled in the gap of bumps protects the die after cured. Thermal interface material is necessary to fill the gap between die top surface and heat spreader inner surface, and it will provide the main heat diffusion path for this type package.

This FCBGA package size is up to 65x65mm, which is much larger than the conventional FCBGA size. And the die size is approximately 16x25mm. There is a high risk for packaging reliability due to no similar packaging experience. So a test vehicle is proposed and designed to validate the engineering capability of out sourced assembly and testing factory. Meanwhile, to meet the thermal resistance requirements, some new potential TIMs with high thermal conductivity will be evaluated, including resin type TIM with filler and metal type TIM.

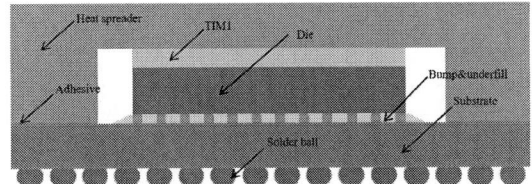

Fig. 1. Construction of FCBGA with heat spreader.

B. Thermal test chip design

This thermal test chip is merely a dummy chip without semiconductor functions, containing several resisters as heaters distributed uniformly for powering the chip and several temperature sensors for monitoring the chip temperature at chip center, edge center and chip corner. This thermal test chip contains 15 heaters and 7 temperature sensors. Actual floorpaln is showed in Fig. 2. Each heater size is about 4.1x5.0mm, and the array of heaters in chip is 5x3. The resistive heater is designed as serpentine strips in a rectangle region and the design resistance value is 200ohms at room temperature. The temperature sensor, generally called resistance temperature detector (RTD), is designed similarly as heater with much small size as 0.2x0.2mm. The design resistance value of temperature sensor is 500ohms. Both the heater and temperature sensor are connected by Kelvin type (4-wire) according to JEDEC JESD51-4 [7] for accurate measurement of voltage and current applied on [8]. The peripheral area of redistribution layer (RDL) in chip is mostly occupied by daisy chain connected to substrate for package reliability test. So resistive conductor of heater and RTD is distributed in top metal layer, which is the upper layer of RDL from the perspective of package. And the vias form top metal layer to RDL for heaters and RTDs keep away from peripheral area to make space for daisy chain.

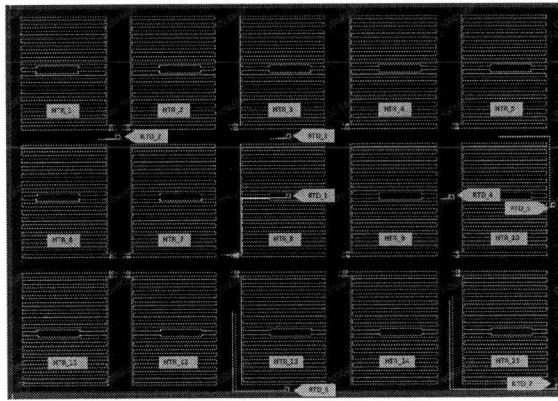

Fig. 2. Heaters and RTDs floorplan.

This test vehicle is intended to not only implement the thermal test for evaluation of new TIM's performance, but also validate the package reliability and reduce the packaging risk of real die. Therefore, from the point of packaging reliability, the test vehicle budget is inevitable in a new project with new packaging without any experience. The thermal test chip design combined with daisy chain will not increase the total cost of test vehicle, except some thermal test expenses. Due to there is no p-n junction in chip, the manufacture process is much easier than real chip. This thermal test chip was manufactured by multi-project wafer considering the manufacture cost. So it is a cost-saving solution for a thermal test chip without additional cost for a new product using advanced manufacturing process.

C. Thermal test chip evaluation by simulation

We compared temperature distribution of the uniform power die and thermal test die by simulation. The simulation is based on ANSYS Icepak platform. The detail chip module is constructed as showed in Fig. 1. To reduce the influence facts, the silicon die thermal conductivity is set as a constant value, rather than a liner fitting with temperature. The power dissipation of two die is both 80W. A constant temperature boundary is set on top surface of package, which is in compliance with junction to case resistance described in JEDEC standard.

Fig. 3 and Fig. 4 show the temperature contours of die source with resin type TIM. Thermal test die maximum temperature at source center is a little higher than that of uniform power die, and the minimum temperature at die corner is a little lower than that of uniform power die. It's due to non-uniform power distribution and heater coverage percentage less than 100%. In the later thermal test, we should take the difference of temperature distribution between two source types into consideration to improve the thermal test accuracy. Fig. 5 shows the the temperature contours of die source with metal type TIM. The maximum temperature is about 32.9°C, which is 2.6°C lower than that of thermal test die with resin type TIM.

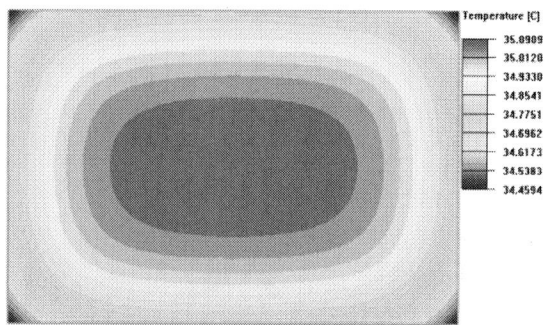

Fig. 3. Temperature contour of uniform power die with resin type TIM.

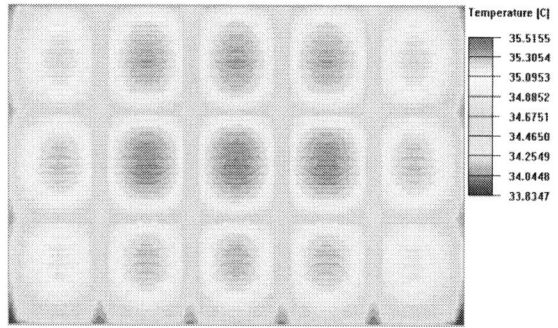

Fig. 4. Temperature contour of thermal die with resin type TIM.

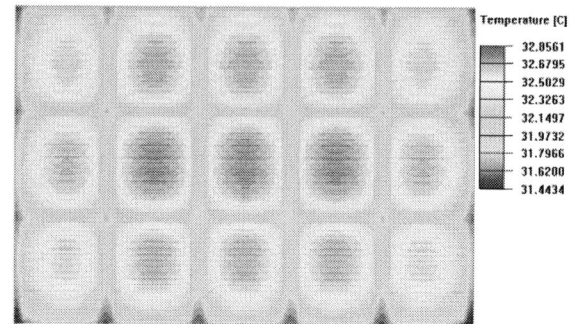

Fig. 5. Temperature contour of thermal die with metal type TIM.

For a given packaging, the temperature difference from source to top surface is proportional to the power dissipation because thermal resistance is a constant value under a known source type. Therefore, during thermal test we can get a suitable temperature difference to decrease the test error by controlling the current applied on heater. 80W heat dissipation will cause approximately 5°C temperature difference form the source to top surface for resin type TIM package, corresponding to about 163mA current applied to each heater. This temperature difference will be enough for thermal resistance test. However, 80W heat dissipation just can generate 2.9°C temperature difference in metal type TIM package due to lower thermal resistance. At this condition, we may need to increase the current applied on heaters for metal type TIM package to get a more significant temperature difference.

D. Printed circuit board design

To implement the thermal test, a printed circuit board designed for corresponding chip is indispensable. Fig. 6 demonstrates the PCB layout for this TTC. There are four gold fingers for each heater. This design can make it possible that we can control the heat dissipation of each heater more accurately.

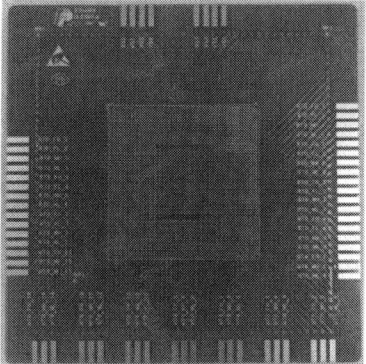

Fig. 6. Thermal test PCB for this test vehicle chip.

IV. CONCLUSION

A thermal test chip design in test vehicle package have been described. This thermal test chip can be designed combining with daisy chain in test vehicle without individual design and manufacture. In a new product development, it's a cost-saving and effective way to evaluate a certain potential TIM's performance and reliability used in FCBGA package ahead of real functional chip available.

REFERENCES

[1] JEDEC Transient Dual Interface Test Methodfor the Measurement of the Thermal Resistance Junction-to-Case of Semiconductor Devices with Heat Flow Through a Single Path Standard JESD51-14, 2010.

[2] R. Pendse and B. J. Shanker, "A study of thermal performance of packages using a new test die," Fourth Annual IEEE Semiconductor Thermal and Temperature Measurement Symposium, 1988, pp. 50-54.

[3] B. Siegal and J. Galloway, "Thermal Test Chip Design and Performance Considerations," 2008 Twenty-fourth Annual IEEE Semiconductor Thermal Measurement and Management Symposium, San Jose, CA, 2008, pp. 59-62.

[4] Jordà, X., Perpinà, X., Vellvehi, M., Madrid, F., Flores, D., Hidalgo, S., & Millán, J., "Low-cost and versatile thermal test chip for power assemblies assessment and thermometric calibration purposes." Applied Thermal Engineering, 31(10), 1664–1672, 2011.

[5] Benedek, Z., Courtois, B., Farkas, G., Kolla´r, E., Mir, S., Poppe, A., Rencz, M., Sze´kely, V., and Torki, K., "A Scalable Multi-Functional Thermal Test Chip Family: Design and Evaluation ." ASME. J. Electron. Packag, 123(4): 323–330, 2001.

[6] S. Parameswaran, S. Balakrishnan and B. Ang, "Versatile chip-level integrated test vehicle for dynamic thermal evaluation," 2018 IEEE International Conference on Microelectronic Test Structures (ICMTS), Austin, TX, 2018, pp. 122-12.

[7] EIA/JEDEC Thermal Test Chip Guideline (Wire Bond Type Chip) Standard EIA/JESD51-4, 1997.

[8] A. Claassen and H. Shaukatullah, "Comparison of diodes and resistors for measuring chip temperature during thermal characterization of electronic packages using thermal test chips," Thirteenth Annual IEEE. Semiconductor Thermal Measurement and Management Symposium, Austin, TX, USA, 1997, pp. 198-209.

Thermodynamic simulation and analysis of metal bumps in flip-chip micro-LED packaging

Xiaoyu Xiao
College of Mechanical and Electrical Engineering
Central South University
Changsha 410083, China
xiao1201@csu.edu.cn

Yamei Yan
College of Mechanical and Electrical Engineering
Central South University
Changsha 410083, China
214742328@qq.com

Gui Chen
College of Mechanical and Electrical Engineering
Central South University
Changsha 410083, China
chengui@csu.edu.cn

Wenhui Zhu*
College of Mechanical and Electrical Engineering
Central South University
Changsha 410083, China
zhuwenhui@csu.edu.cn

Abstract—**Thermal compression bonding provides a good solution for the packaging of micro-LED chips, greatly increases the resolution of micro-led, and can achieve higher density integration and better display effect. In order to study the thermomechanical reliability and display stability of micro-LED devices under thermal cycling load and working power load, this paper first explains the limitation of thermal expansion coefficient (CTE) mismatch on the size of flip chip devices by mathematical method, and proposes a low temperature indium bump flip chip bonding technology which can overcome this defect. Then, the finite element model is established to study the influence of different metal bump materials on the reliability of micro-LED devices under thermal cycling load, and the display stability is analyzed combined with the wavelength drift characteristics of micro-LED devices.**

Keywords—*flip chip; Microdisplay; Fine pitch; wavelength shift; thermal reliability*

I. INTRODUCTION

A. Micro-LED device application situation

As a new generation of mainstream display technology, micro-led inherits the advantages of inorganic led, such as high efficiency, high brightness, high reliability, etc., but also has the advantages of self illumination, small size, high resolution, etc., which are widely used in VR / AR, vehicle display, wearable devices and other fields [1-2]. With the increasing resolution of micro led, the pixel size and pixel spacing of micro LED devices become smaller and smaller, which brings great challenges to the packaging and interconnection of micro LED devices. In traditional flip chip bonding process, solder bumps such as SAC, SnPb and Sn are used to interconnect the micro LED chip with the CMOS driver substrate by thermal compression bonding. However, in flip chip bonding process, the chip often needs to be heated above the metal melting point, which causes the temperature of traditional flip chip bonding process to be about 200-300 ℃, High temperature will cause deformation on the chip and substrate, resulting in pad dislocation, which affects the heterogeneous integration of large-size and high-density micro LED chips [3]. Therefore, the interconnection process of micro led needs to be carried out at low temperature as far as possible. Indium solder can be used to package at 80-120 ℃,

which greatly reduces the process error caused by CTE mismatch.

B. CTE restrictions on flip-chip

For a N*N square matrix with a pixel pitch of a, the theoretical maximum number of pixels Nmax that can guarantee welding reliability at the peak welding temperature "T1" can be directly calculated according to its thermal expansion coefficient. Set the upper chip CTE to c_2 and the lower substrate CTE is c_1.

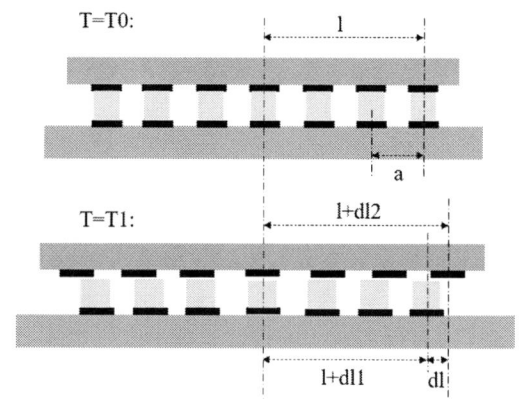

Fig. 1. Misalignment at soldering temperature

As shown in Fig 1, l=distance from the outermost row of pixels to the midpoint of the chip（l=(N-1)/2*a）; Δc =CTE difference between the upper chip and the lower substrate（$\Delta c = c_2 - c_1$）; ΔT =temperature difference during welding（$\Delta T = T_1 - T_0$）.

When the temperature rises to T_1, the deviation of the upper and lower pads at l is:

$$dl = dl2 - dl1 = \left(\frac{N-1}{2}\right) * a * \Delta c * \Delta T \qquad (1)$$

It is generally believed that the connection is reliable when dl<a/4, the interconnection is dangerous when a/4<dl<a/2, and the interconnection fails when dl>a/2.

There are two situations as follows:
1) When dl<a/4,

$$dl_{max} = \left(\frac{N-1}{2}\right) * a * \Delta c * \Delta T = \frac{a}{4} \qquad (2)$$

$$N_{max} = \frac{0.5}{\Delta c * \Delta T} + 1 \qquad (3)$$

2) When a/4<dl<a/2,

$$dl_{max} = \left(\frac{N-1}{2}\right) * a * \Delta c * \Delta T = \frac{a}{2} \qquad (4)$$

$$N_{max} = \frac{1}{\Delta c * \Delta T} + 1 \qquad (5)$$

It can be seen from the above formula that the maximum number of pixels successfully welded is not affected by the pixel pitch, but only depends on $\Delta\alpha$ and ΔT. After the package material is determined, that is, after $\Delta\alpha$ is determined, the smaller the ΔT is,the larger the maximum number of pixels that can be successfully welded, the larger the device size is.

It can be seen that the use of indium bumps can greatly reduce the process temperature in flip-chip bonding, thereby improving the reliability of flip-chip bonding of large-size and high-density Micro-LED chips. Based on this characteristic of indium bumps, in order to further explore the influence of solder on the thermal reliability of Micro-LED devices, the stress and strain energy changes of different solder bumps under the same thermal cycle were compared.

II. MODELING AND SIMULATION

A. Model assumptions

In order to simplify the solution of the problem and avoid the non convergence in the process of solution, the following assumptions are made for the simulation:

1) All packaging materials are homogeneous and isotropic;

2) Except for solder bumps, all other materials are linear elastic materials;

3) All contact surfaces are in complete contact with no defects such as voids and pores;

4) The residual stress in the manufacturing process is not considered;

5) Ignore the influence of heat radiation;

6) The solder ball is a truncated sphere;

7) All solder balls and pads in the array have the same geometric size and are distributed axisymmetrically.

B. Thermal cycle simulation

1) FEM module

A 12×12 Micro-LED array model is established. Fig. 2a is a schematic diagram of the model. The Micro-LED chip and the CMOS drive substrate are connected together by UBM and solder bumps. The finite element analysis uses a quarter A symmetrical 3D model, as shown in Fig. 2b. The size of the micro-LED chip is 480μm × 480μm , the size of the substrate is 500μm × 500μm, the diameter of the solder ball is 20μm and the pitch is 40μm.

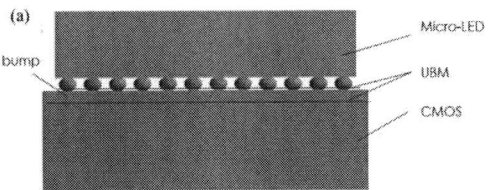

Fig. 2. （a）Simplified model （b）One forth model

2) Material properties

The micro LED module is composed of micro LED chip, pads, solder bumps and CMOS driver substrate. Since the active layer, GaN layer and other components are very thin compared to the thickness of the sapphire substrate, the Micro-LED chip material is regarded as sapphire, the CMOS drive substrate material is silicon, the pad material is Au, and the solder bumps are indium , SAC405, Sn60Pb40 materials, linear elastic material parameters are shown in Table Ⅰ. [4-6].

TABLE I. Linear material properties

Material	E(GPa)	Poisson's ratio	CTE(10^{-6}K^{-1})	ρ(kg/m3)
SAC405	49.0	0.38	21.3	7400
Sn60Pb40	31.7	0.4	29.8	8420
Indium	20.54	0.4326	22.5	7310
Si	170	0.28	2.6	2330
Sapphire	345	0.3	7.5	3980
Au	78.6	0.3	14.1	19320
GaN	290	0.2	5.6	6070

The Anand's constitutive model properties of indium, SAC405 and SN60PB40 solders are shown In TABLE Ⅱ. Since the creep deformation and plastic deformation are caused by dislocation movement, the establishment of unified viscoplastic model can more accurately describe the viscoplastic mechanical behavior of solders.Based on this constitutive model, the thermodynamic properties of different solders under temperature cycling can be well described.

TABLE II. Anand model constants

Description	Sn60Pb40	Indium	SAC405
So（MPa）	53.93	28.3	20
Q/ R（K）	5997	9369.7	10,561
A(SEC^-1)	7341	2.33×10^8	325
ξ	13	49.97	10
m	0.291	0.3	0.32
ho(MPa)	2645	500	800,000
Ŝ(MPa)	134.2	28.3	42
n	0.002	0.005	0.02
a	1.5	1	2.57

3) Loads and boundary conditions

In this paper, the loading method is temperature load, as shown in Fig. 3, using the U.S. government military standard conditions, the specimen is cycled in the temperature range t = - 55 ℃ ~ + 125 ℃. The high and low temperature are kept for 40s, and the heating and cooling rate are both 9 ℃ / s, because the stress and strain of the interconnection solder joints change periodically in the thermal cycle process, and generally tends to be stable in the third and fourth cycle, in order to reduce the amount of calculation, In this paper, four thermal cycles are used in the finite element analysis. The initial room temperature was set at 22 ℃.

According to the above thermal cycle conditions, 12 load steps are adopted, and each step is divided into 20 load sub steps.

In this paper, the center point of the bottom of the model is fixed.

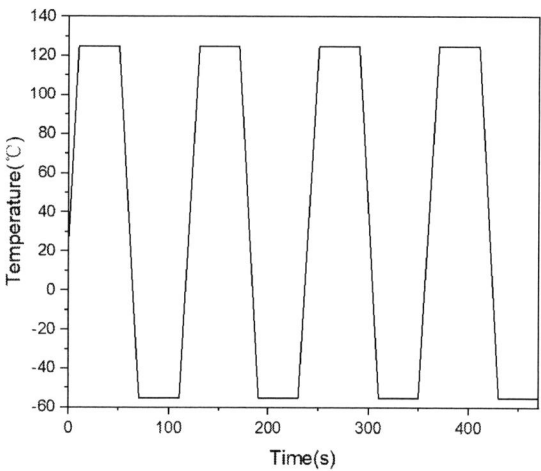

Fig. 3. Temperature profile of thermal cycling

C. Thermal analysis simulation

1) FEM module

As shown in the Fig. 4(a), this model adds a 3μm GaN heat-generating layer between the Micro-LED chip and the upper pad on the basis of the above-mentioned thermal cycle simulation model.

(a)

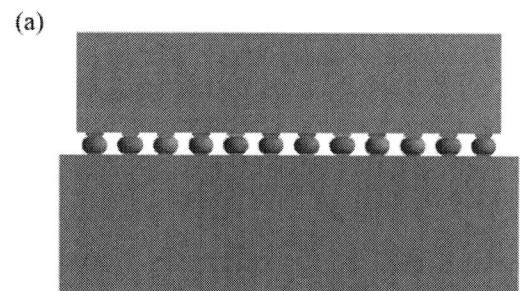

(b)

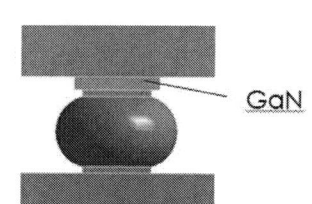

Fig. 4. （a）Simplified model （b）The model details

2) Material properties

The model material is consistent with the model material of the thermal cycle simulation. The supplementary material parameters are shown in the following table:

TABLE III. Thermal analysis material parameters

Material	thermal conductivity(W/m·K)	specific heat capacity(J/Kg·K)
In	81.8	233
Si	150	729
Sapphire	38	753
Au	317	129
GaN	130	490

3) Loading and boundary conditions

When the external ambient temperature is set at 22℃, the heat generation rate of GaN layer is 3.22×10^8 W/m³, and the convection coefficient of the lower surface of the driving substrate is 10 W/m²K.

III. RESULTS AND DISCUSSION

A. Thermal cycle simulation results

Fig. 5 shows the Mises stress distribution of different solder bumps. It can be found that the stress of solder layer increases from the center to the periphery, and the maximum stress appears at the solder joint farthest from the center.The reason is that during the thermal cycle, the thermal expansion coefficient of the upper and lower substrates and the solder bump material does not match [7-8], resulting in a small warpage deformation of the package body, which is proportional to the distance from the symmetric center. The deformation of the center part is small, and the deformation around the edge is the largest, which inevitably causes the outermost solder joint to bear the maximum stress and strain.

(a)

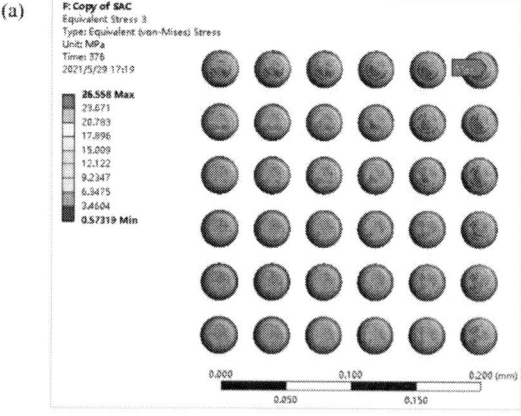

(b)

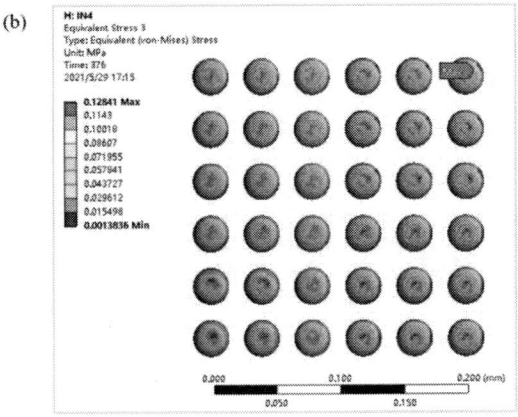

(c)

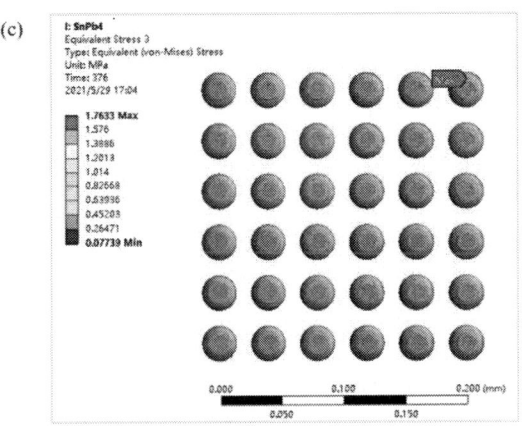

Fig. 5. （a）Equivalent stress distribution of SAC405 solder joint array （b）Equivalent stress distribution of Indium solder joint array （c）Equivalent stress distribution of Sn60Pb40 solder joint array

Fig. 6 is the stress distribution diagram of the central section of the indium dangerous solder joint. It can be seen from the diagram that the stress distribution in the entire model is uneven during the temperature cycle, and the stress distribution on the solder ball has an obvious gradient from the contact surface to the center surface of the solder ball, the stress in the center of the solder ball is the smallest, the stress in the contact surface is large and has some differences, and the joint of the solder joint with the substrate is the most prone to failure.The reason is that compared with Micro-LED chips, the thermal expansion coefficients of solder bumps and CMOS drive substrates are more different.

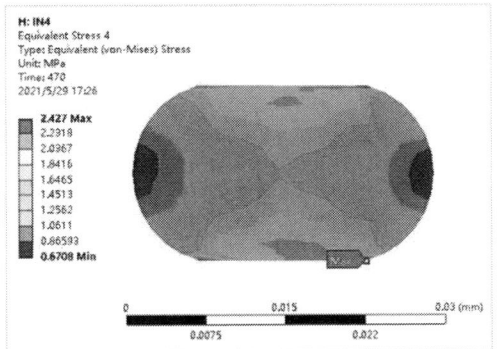

Fig. 6. Equivalent stress distribution diagram of the central section of the dangerous solder joint （Indium）

In order to further study the stress change of solder layer dangerous spot during temperature cycle, the stress data of dangerous spot are extracted for research. Fig. 7 shows the equivalent stress curves of different solders in the process of thermal cycle.The results show that the stress curves of different solders have the same trend, but the amplitude is different. The equivalent stress of the welding ball dangerous point fluctuates the temperature cycle and shows a strong periodicity. In each cycle, the equivalent stress produces a change between the minimum and maximum values. This periodic change causes solder fatigue at this point, and eventually leads to solder layer failure, causing reliability problems for the whole device..

When the environment rises from low temperature to high temperature, the equivalent stress of the dangerous point of the solder layer drops sharply. The reason is that the increase of temperature promotes the viscous flow of the solder layer, which makes the material soften, and the elastic modulus of solder is lower at high temperature.When the temperature is the highest, the solder layer has strong viscoplasticity, the deformation impedance and dynamic recovery are very low, and the equivalent stress at the danger point reaches the minimum value. In the process of environment from high temperature to low temperature, the equivalent stress of the danger point of the solder layer increases sharply, because the elastic modulus of solder is high at low temperature, and the solder reflects the characteristics of cooling back hardening when cooling down, and the increase of brittleness leads to a sharp rise in the equivalent stress.When the temperature is the lowest, the residual plasticity of viscoplastic material will gradually release the accumulated stress in the solder, so that the stress value is slightly reduced, that is, the phenomenon of stress relaxation.It can be seen that the change of the stress at the danger point of the solder layer reflects a strong temperature and time correlation, which conforms to the viscoplastic characteristics of solder. This also indicates that Anand viscoplastic constitutive equation can accurately describe the viscoplastic behavior of solder under temperature cycle.

Comparing the equivalent stress curves of the three kinds of solders, it is found that the equivalent stress of indium solder under the action of thermal cycling is the least, which is 18% of that of Sn60PB40 solder and 4% of that of SAC405 solder. Compared with the traditional tin-lead solder and tin-silver-copper solder, the indium solder reduces the thermal stress caused by heterointegration and has better thermal mechanical reliability.

978-1-6654-1392-3/21 $31.00 © 2021 IEEE   343

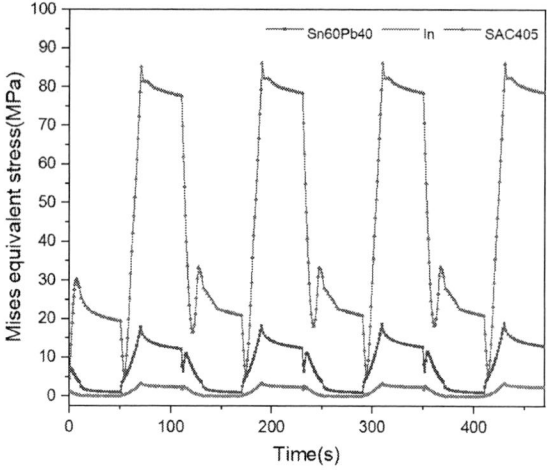

Fig .7. The curve of equivalent stress in different solders under TC simulations

Fig. 8 shows the time history analysis of the plastic work of different solder bumps at the most dangerous unit. The results show that under the same temperature load, the plastic work of different solder bumps has the same trend and different amplitudes over time. As the loading period increases, the plastic work increases continuously. The increase in plastic work during the cooling phase is higher than that during the heating phase, and the plastic work tends to be constant during the heat preservation phase. Among the three solders, indium solder has the smallest plastic work, the smallest plastic energy accumulation, and the highest reliability.

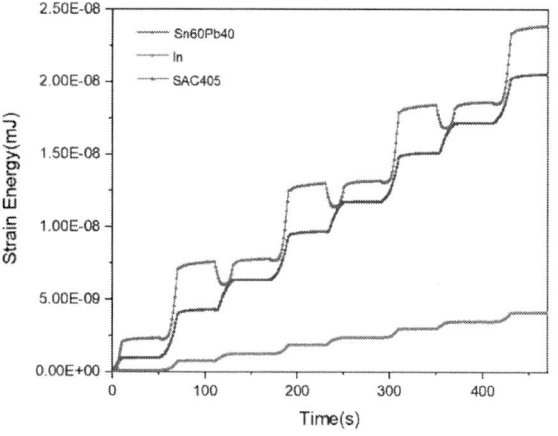

Fig. 8. The curve of strain energy in different solders under TC simulations

In summary, the failure mechanism of the interconnection interface under thermal shock conditions can be inferred and summarized: due to the mismatch between the interconnection interface material and the CTE of the chip and the driving substrate, temperature-related and uneven distribution occurs during the thermal cycle Stress and strain, and continuously accumulate strain energy. After several cycles, cracks appear near the "most dangerous unit" in the area with the largest stress and strain values. The crack grows to a specific failure length, causing the aging failure of the interconnection interface.

B. Thermal analysis results

For GaN based LED devices, with the accumulation of continuous light-emitting time, the ambient temperature increases, and the band gap of the active layer in the PN junction decreases, resulting in the decrease of light energy, red shift of spectral line and the change of light-emitting spectrum.The chromatic aberration value that can be distinguished by human eyes is 0.005.According to the experiment and calculation, when the temperature change of GaN device is within 8℃, the wavelength deviation value is within 5 nm, and the chromatic aberration is lower than 0.005, which can not be perceived by human eyes. Therefore, the maximum chromatic aberration value of backlight should be less than this value, and the temperature rise of the device should be controlled within 8℃ [9].

Fig. 9 shows the overall temperature distribution of micro-LED packaging devices. It can be seen from the figure that under the effect of LED chip power, the device temperature decreases from the center to the edge, and the center temperature is the highest. The heat generated by GaN layer is mainly transmitted to the substrate through solder bumps.The overall temperature rise of the device is 4.116℃, which is less than 8℃. Therefore, the micro-LED device with indium bump has good thermal stability.

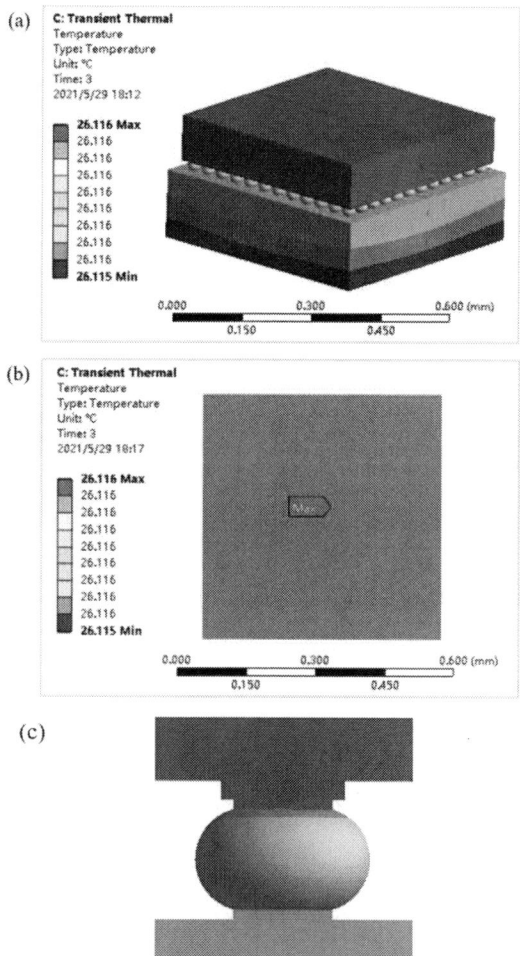

Fig. 9. （a）Overall temperature profile （b）Temperature distribution map of substrate contact surface （c）Soldering point temperature distribution map

IV. CONCLUSION

1) After thermal cycling, the farthest solder joint from the center is the most dangerous solder joint, and the maximum equivalent stress point is located at the joint between the solder joint and the CMOS drive substrate. The equivalent stress of the solder joint changes periodically with the temperature cycle. In one cycle, the equivalent stress decreases rapidly when the temperature increases, and the equivalent stress increases rapidly when the temperature decreases, and there is a stress relaxation effect in the heat preservation stage.

2) Compared with Sn60Pb40 and SAC405 solder joints, Indium solder joints have the least equivalent stress and the least accumulated strain energy in thermal cycle, so it has better thermal reliability.

3) The micro LED device using indium bump has good thermal stability. Under the power load of the device, the temperature rise of the device does not exceed 8°C, and it can maintain good display wavelength stability.The wavelength drift of the micro LED is controlled within 5nm.

REFERENCES

[1] ZHANG S L, GONG Z, MCKENDRY J J, et al. CMOS-controlled color-tunable smart display [J]. IEEE Photonics Journal, 2012,4 (5): 1639-1646.

[2] TIAN P F, MCKENDRY J J, GONG Z, et al. Characteristics and applications of micro-pixelated GaN-based light emitting diodes on Si substrates [J]. Journal of Applied Physics, 2014, 115(3): 033112.

[3] Marion F , Bisotto S , Berger F , et al. A Room Temperature Flip-Chip Technology for High Pixel Count Micro-Displays and Imaging Arrays[C]// IEEE Electronic Components & Technology Conference. IEEE, 2016.K. Elissa, "Title of paper if known," unpublished.

[4] Meng Q , Zhang X , Y Lü, et al. Function reconsideration of indium bump in InSb IRFPAs[J]. Optical and Quantum Electronics, 2019.

[5] Fan X J , Wang H B , Lim T B . Investigation of the underfill delamination and cracking in flip-chip modules under temperature cyclic loading[J]. Components & Packaging Technologies IEEE Transactions on, 2001, 24(1):84-91.

[6] Xudong Chen,Wenbo Zhai,Jingwen Zhang,Renan Bu,Hongxing Wang,Xun Hou.FEM thermal analysis of high power GaN-on-diamond HEMTs[J].Journal of Semiconductors,2018,39(10):50-56.

[7] Marion F , Lasfargues G , Ribot H , et al. Hybrid Heterogeneous Imaging Arrays with Reinforced Functionality and Reliability[C]// Electronics Packaging Technology Conference, 2007. EPTC 2007. 9th. IEEE, 2008.

[8] Davoine C , Fendler M , Marion F , et al. Low temperature fluxless technology for ultra-one pitch and large devices flip-chip bonding[C]// Electronic Packaging Technology Conference, 2005. EPTC 2005. Proceedings of 7th. IEEE Xplore, 2006.

[9] Liu Yuyuan,Luo Yi,Han Yanjun,Qian Keyuan.The influence of LED wavelength uniformity and temperature uniformity on chromatic aberration of backlight[J].Semiconductor Optoelectronics,2010,31(01):104-107.

Stress-strain study of QFN solder joints with different structural parameters under random vibration loading

1st Maolin Li
School of Electronic Mechanical Engineering
Guilin University of Electronic Technology
guilin, China
e-mail : 178116415@qq.com

2nd Chun-yue Huang(Corresponding Author)
School of Electronic Mechanical Engineering,
Guilin University of Electronic Technology,
guilin, China
e-mail :hcymail@163.com

3th Zhuo Wang
School of Electronic Mechanical Engineering
Guilin University of Electronic Technology
guilin, China
e-mail : 1009372429@qq.com

4th Wei Wei
School of Electronic Mechanical Engineering
Guilin University of Electronic Technology
guilin, China
e-mail :1343915534 @qq.com

Abstract—The simulation software ANSYS is used to analyze the stress and strain of the board-level circuit component model under random vibration load. As the selection SAC305 solder material to the solder pad QFN length, width and height as the experimental design variables, in order to stress the solder joints under random vibration load as a target value.The orthogonal experiment design method is used to design 16 groups of different structural parameters and different levels of combinations, and the stress values of the solder joints are obtained by analysis. Using multi-factor range-variance analysis to analyze 16 sets of data, the significant relationship between each factor and the random vibration stress value of QFN solder joints and the optimal structural parameter level combination are obtained: the length was 0.80 mm, the width was 0.30 mm, and the height was 0.06 mm, The simulation verification of this optimal structural parameter level combination shows that the maximum stress of the solder joint is reduced to a certain extent.

Keywords—QFN solder joints; Vibration loading; Finite element analysis; Stress-strain;Orthogonal design ; Variance; Range and variance analysis

I. Introduction

With the vigorous development of microelectronics technology, electronic products have higher requirements for volume control, functions, and performance. The continuous development of new packaging materials, the continuous innovation of technology and process, and the continuous improvement of the performance of electronic products. Electronic packaging and electronic design and manufacturing jointly promote the development of the information society [1]. Continuously improving the reliability of electronic products is an important part of promoting the vigorous development of electronic

technology. Among them, the solder joint reliability interconnection of electronic products adopts a very important part of reliability, so the industry has always attached great importance to the research on the reliability of interconnect solder joints [2]. QFN packaging is a new type of surface mount technology. This packaging method has the characteristics of small packaging area, thin component thickness, extremely low impedance and self-inductance, and excellent thermal performance. Therefore, this packaging method has been rapidly developed. This paper selects QFN solder joints as the research object, establishes the finite element analysis model of the research object, applies random vibration load to it, observes the size and distribution of solder joint stress under different geometric parameter level combinations, and analyzes it. The location most likely to cause the failure. In the daily use of electronic products, they will inevitably be subjected to vibrations, causing internal stress on the substrate and pads, and failures such as joints and cracks may occur, which will lead to failures of electronic products. Many scholars have conducted corresponding studies, such as Dejian. Zhou and Zhaohua Wu on the random vibration response of optoelectronic interconnect PCB; Ziwei conducted a prediction study on the life of the circuit board under thermal cycling and random vibration load [3].

II . Finite Element Analysis of QFN Solder Joints under Random Vibration Load

A. Finite element analysis model of QFN solder joints

This paper uses ANSYS simulation analysis software to build the random vibration model required for this experiment. The analysis model includes printed circuit boards, chips, and QFN solder joints.The PCB circuit board size is 70mm × 40mm × 1mm, the chip size is 10mm × 10mm × 1mm. The component type of PCB board and chip is SOLID185, and SAC305 is used as solder joint material. The length, width, and height of the QFN solder joints are

---

1、 This research is supported by the Science Foundation of Guangxi TheZhuang Autonomous Region Government (No.2019JJA160101) and Science and Technology Major Project of Guangxi Zhuang Autonomou s Region Government （No.AA19046004）

used as experimental variables, and the stress and strain of the QFN solder joints under random vibration loads are used as the experimental results. Figure 1 is a schematic diagram of the geometric model of the simulation model, Table I lists the material parameters of the materials used.

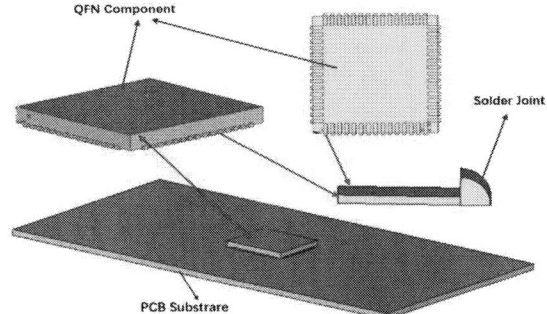

Fig. 1. Finite element model diagram

TABLE I.   MATERIAL PERFORMANCE PARAMETERS

Material name	Density (kg/m3)	Elastic Modulus(Gpa)	Poisson's Ratio(μ)
Solder joint	7300	43	0.35
PCB Substrate	1870	4.2	0.28
Chip	2320	130	0.28

B. Meshing of the finite element analysis model

The analysis model is meshed with the same density and the pad sizes of different sizes, so that the analysis model produces different numbers of elements and nodes. Figure 2 Schematic diagram of this simulation analysis model after meshing.

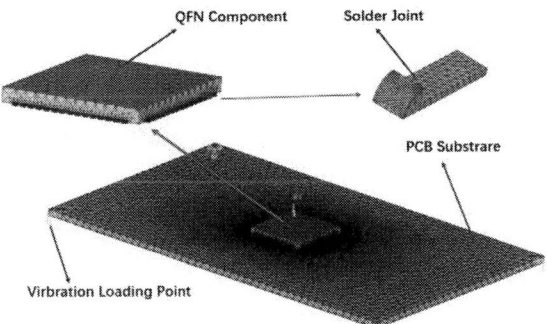

Fig. 2. Schematic diagram of the model after meshing

C. Random vibration load of QFN solder joint

Figure 2 shows the random vibration loading of the trigger components. As shown in Figure 2, apply constraints at the four corners of the PCB, the the rando accelerating power spectral density curve of random vibration load is shown in Figure 3. when m vibration frequency is 20Hz~80Hz, the rising slope of the PSD curve is +3dB/oct, and the corresponding acceleration power spectral density amplitude range is 0.01~0.04g2/Hz, and at 80Hz it is 0.04g2/Hz; When the random vibration frequency is 80Hz~350Hz, the corresponding acceleration power

spectral density amplitude is 0.04g2/Hz, When the random vibration frequency at 350Hz ~ 2000Hz, PSD curve slope -3dB/oct decreases acceleration power spectral density corresponding to the amplitude range of 0.04 ~ 0.01g2 / Hz.

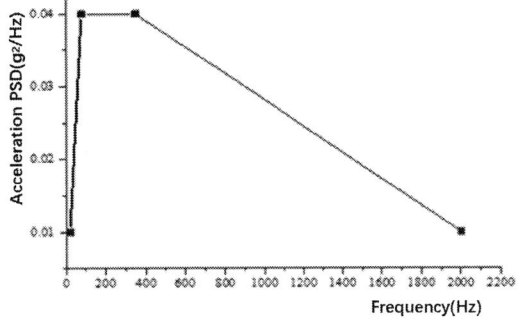

Fig.3. Random vibration acceleration power spectral density curve

D. Stress and strain analysis of QFN solder joints under random vibration loads.

By applying the above random vibration load, it can be seen that the maximum stress and strain of the QFN solder joints appear on the solder joints at the intersection of the two sides, and both appear on the outermost side of the solder joints and are in contact with the PCB circuit board.

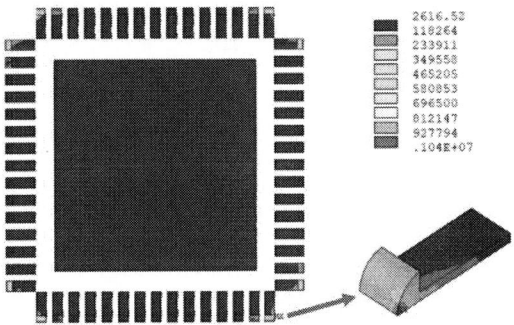

Fig. 4. Schematic diagram of QFN solder joint stress distribution

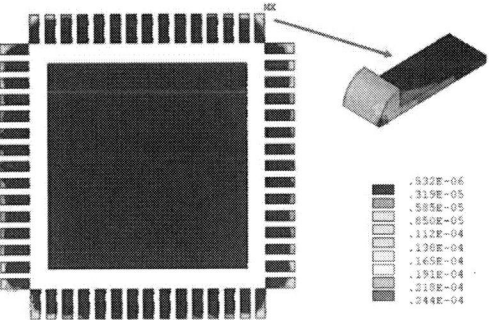

Fig.5 Schematic diagram of strain distribution of QFN solder joints

III. OPTIMIZATION OF QFN SOLDER JOINT STRUCTURE BASED ON VARIANCE-RANGE ANALYSIS METHOD

When solder joints are subjected to random vibration loads, the stress and strain generated will be affected by the structural parameters of the solder joints. Therefore, in order to obtain the structural parameters of QFN solder

978-1-6654-1392-3/21 $31.00 © 2021 IEEE

joints with lower stress, it is necessary to further optimize the structural parameters[4]. Increase its reliability under random vibration loads. The method of combining range analysis and variance analysis used in this paper takes the stress of the solder joint under random vibration load as the optimization target, and obtains ,Through the range analysis optimization of the QFN solder joint structure parameter combination and obtains the butt welding of each factor through the variance analysis Significant relationship between point stress and strain.

A. Orthogonal experimental design

Because this experiment has multiple factors and levels, and there may be interactions between various factors, in order to ensure the scientific and reasonable experiment while reducing the number of experiments, this experiment adopts an orthogonal experimental design [5]. This method is a scientific and efficient experimental design method. It uses the orthogonality of the design method to select some representative level combinations from all experiments for simulation analysis. These selected horizontal combinations are evenly dispersed. In the actual scientific research and production process, if there are too many factors that need to be investigated, such as comprehensive experiments, a large number of experiments are required, which is often impossible to operate [6]. There are 3 factors and 4 levels in this experiment. If you combine all levels to experiment, there will be 64 levels. This kind of experiment has a lot of workload, so in view of these problems, an orthogonal experimental design was chosen. Very reasonable and necessary.

In this experiment, an orthogonal table experimental design method is used, and the same random vibration load is applied to the models constructed with different factor levels, and the stress on the solder joints is taken as the target value. The factors selected in this simulation experiment are the pad length L, the pad width W, and the pad height H of the solder joint. Four level values are selected for these three parameters, and the level combination of solder joint factors is obtained through the orthogonal table experimental design method.

TABLE II. THE TABLE OF LEVELS AND FACTORS

Factor	Level			
	1	2	3	4
Solder joint Length(mm)	0.07	0.80	0.90	1.00
Solder joint width(mm)	0.22	0.26	0.30	0.34
Solder joint height(mm)	0.06	0.07	0.08	0.09

As shown in Table 3, 16 groups of experimental factor level combinations are obtained through orthogonal design, and 16 simulation experimental models were established by

ANSYS. After ANSYS analysis, table III shows the maximum stress value of QFN solder joints corresponding to different combinations.

TABLE III. HORIZONTAL COMBINATION OF SOLDER JOINT FACTORS AND STRESS SIMULATION RESULTS

Experiment number	Solder joint Length (mm)	Solder joint Width (mm)	Solder joint Height (mm)	Maximum stress (MPa)
1	1	1	1	1.36
2	1	2	2	3.04
3	1	3	3	1.48
4	1	4	4	2.10
5	2	1	2	1.93
6	2	2	1	1.88
7	2	3	4	1.88
8	2	4	3	1.51
9	3	1	3	2.52
10	3	2	4	2.48
11	3	3	1	1.21
12	3	4	2	1.86
13	4	1	4	2.75
14	4	2	3	2.06
15	4	3	2	1.37
16	4	4	1	1.09

B. Range analysis and analysis of variance

The range analysis method is one of the commonly used analysis methods in orthogonal design[7]. Through simple calculations and judgments, the optimization results of the experiment can be obtained, such as factors affecting the degree of primary and secondary, and the best combination of levels. Conditions, best technology, best formula and other scientific research and production have been widely used.

From the data in Table III, perform a multivariate range analysis to obtain Table IV. From this table, the maximum mean items of structural parameters L, W, and H are mean 2, mean 3, and mean 1, respectively. The optimal level corresponding to each mean is L=0.80mm, W=0.30mm, H=0.06 mm. Figure 6, Figure 7, Figure 8 respectively show the intuitive analysis diagrams of the structural parameters L, W, H, which can reflect the relationship between the various factors and the average value of the water to a certain extent.

TABLE IV.   RANGE ANALYSIS TABLE

Factor	L	W	H	Blank column	Blank column
Mean 1	1.995	2.140	2.303	1.790	1.680
Mean 2	2.018	1.485	2.050	2.127	2.132
Mean 3	1.800	2.365	1.893	1.745	1.992
Mean 4	1.818	1.640	1.385	1.968	1.825
Range	0.218	0.880	0.918	0.382	0.452

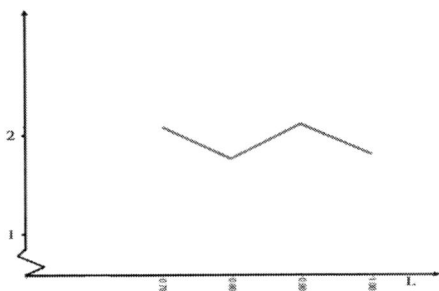

Fig.6 Intuitive analysis diagram of factor L

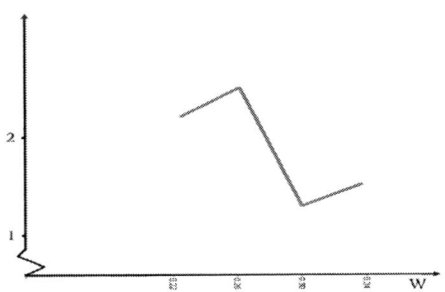

Fig.7 Intuitive analysis diagram of factor W

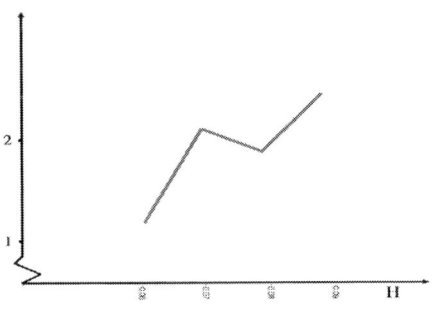

Fig.8 Intuitive analysis diagram of factor H

The analysis of variance method is a mathematical analysis method that distinguishes the difference between the experimental results caused by the changes in factor levels (or interactions) and the fluctuations in the errors caused by the differences between the experimental results. The central point is to divide the total fluctuation of the experimental data into two parts, one part reflects the fluctuation caused by the change of the factor level, and the other part reflects the fluctuation caused by the experimental error. In the orthogonal experiment, it can be used as a supplement to the intuitive range analysis method[8].

Perform a multi-factor analysis of variance on the data in Table III, and choose a=0.05, that is, when 95% confidence is met, Table V is obtained, The data in Table III shows that the order of F ratio of QFN solder joint structure parameters L, W, H is H>W>L, that is, the width of the pad W has the greatest influence on the stress of the QFN solder joint under random vibration load, followed by The width W of the pad has the least influence on the length L of the pad, so the height H of the pad is significant in the analysis of variance.

TABLE V.   ANALYSIS OF VARIANCE TABLE

Factor	Deviation sum of squares	Degree of freedom	F	Signifi-cance
L	0.158	3	0.378	
W	2.054	3	4.307	
H	1.798	3	4.920	Signifi-cant
Error	0.83	6		

C. Use ANSYS simulation software to simulate and analyze the optimal level combination

Based on the multi-factor variance-range analysis method, the optimal level combination is obtained, namely L=0.80mm, W=0.30mm, H=0.06mm. A finite element analysis model is established based on this parameter, and the same load is applied to the model to observe whether the maximum stress of the solder joint under this horizontal combination is reduced. Figure IV shows the stress distribution diagram of the solder joint under this horizontal combination. From the stress distribution data in the figure, the maximum stress value under the same load is 1.04Mp. Compared with the minimum value of 1.09Mp in the horizontal combination data of the 16 groups in Table III, the maximum stress value of the solder joint is reduced by 0.05Mp. The optimal structural parameter level combination obtained by factor range analysis and variance analysis optimization methods minimizes the maximum stress of QFN solder joints under random vibration loads.

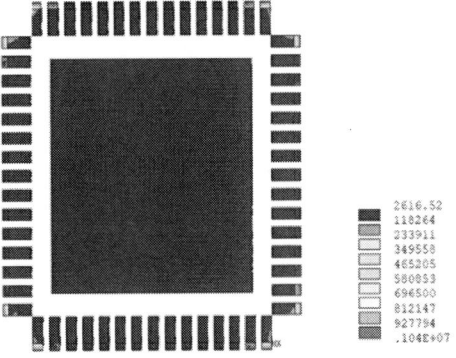

Fig.9. Stress distribution diagram of the optimal horizontal combination of QFN solder joints

IV. CONCLUSION

Based on ANSYS simulation software, set up different experimental factors and level combinations, establish the required analysis model, apply random vibration load to it, and observe its stress value. Design orthogonal test, multi-factor range analysis and variance analysis to optimize QFN solder joint parameters, and draw the following conclusions:

1. Under vibration loading conditions, the maximum stress and strain on a solder joint occurs at its contact with the printed circuit board.So the solder joints and printed circuit board contact is the most vulnerable to the destruction of the solder joints.Figure 2 and Figure 3 show the stress and strain diagrams for a QFN solder joint.

2. Under random vibration loading conditions, the sequence of F ratios of experimental factors L, W, and H is H>W>L, that is, the pad width W has the greatest impact on the QFN solder joint stress under random vibration loading，followed by the width W , The influence of the pad length L is minimal.

3. The range and variance analysis of the experimental data of orthogonal design is carried out, and the degree of influence of each factor on the stress value is obtained, and the best combination of geometric parameters is obtained as a solder joint length of 0.80 mm, width of 0.30 mm, and height of 0.06 mm. It is verified that the maximum stress is reduced by 0.05 Mpa.

V. REFERENCES

[1] QIN Fei,BIE Xiaorui,CHEN Si,AN Tong. Fatigue life model of plasticized ball grid array containing leaded solder joints under random vibration loading[J]. Vibration and Shock,2021,40(02):164-170.

[2] Liu X. Reliability study of board-level interconnect systems under drop shock loading[D]. Harbin Institute of Technology,2020.

[3] LU Liangkun,HUANG Chunyue,LIANG Ying,LI Tianming. Response surface-genetic algorithm-based dual-objective optimization design of random vibration stress and return loss of CSP welded joints[J]. Vibration and Shock,2019,38(21):221-228.

[4] She Chenhui, Liu Yahong, Tan Lipeng, Liu Peisheng. Reliability study of PCBA with different structural parameters under random vibration[J]. Electronic Components and Materials,2019,38(10):58-62.

[5] Chenglei,Zhou Dejian,Wu Zhaohua. Analysis of random vibration response of optoelectronic interconnected PCBs[J]. Journal of Beijing University of Technology,2017,37(06):631-636.

[6] Ma S-P, Zhu X-Q, Chen Z-H, Hu Y-P, Hu E-L, Ren P. Study on random vibration acceleration test of small size PCBA based on finite element simulation[A]. Beijing Society of Mechanics. Proceedings of the Twenty-fourth Annual Meeting of the Beijing Society of Mechanics[C]. Beijing Society of Mechanics:Beijing Society of Mechanics,2018:3.

[7] Y Zwei. Life prediction of circuit boards under thermal cycling and rando vibration loading[J]. Electronic Components and Materias,2019,38(01):89-96.

[8] Liu Chen,Liu Tianxiong,Jiang Wanjie,Fan Yanping. Research on the design method of spacecraft electronic products against Random vibration environment[J]. Spacecraft Engineering,2016,25(03):80-87

Comparison between two numerical methods for the computation of thermal conductivities of particulate composites: FEM and GeoDict

Xiaoxin Lu
Shenzhen Institute of advanced electronic materials, Shenzhen Institutes of Advanced Technology
Chinese Academy of Sciences
Shenzhen, P.R.China
luxx@siat.ac.cn

Jiabin Huang
Shenzhen Institute of advanced electronic materials, Shenzhen Institutes of Advanced Technology
Chinese Academy of Sciences
Shenzhen, P.R.China
jb.huang@siat.ac.cn

Jianbin Xu
Department of Electronics Engineering, The Chinese University of Hong Kong, Shatin, New Territories,
Hong Kong 999077, P.R. China
jbxu@ee.cuhk.edu.hk

Jibao Lu*
Shenzhen Institutes of Advanced Technology,
Chinese Academy of Sciences
Shenzhen, P.R.China
jibao.lu@siat.ac.cn

Sun Rong*
Shenzhen Institute of advanced electronic materials, Shenzhen Institutes of Advanced Technology
Chinese Academy of Sciences
Shenzhen, P.R.China
rong.sun@siat.ac.cn

Ching-Ping Wong
School of Materials Science and Engineering
Georgia Institute of Technology,
Atlanta, USA
cp.wong@mse.gatech.edu

Abstract—The thermal interface material in direct contact with chip, termed as TIM1, is a kind of composite material with thermally conductive particulate fillers. To better understand the thermal conduction mechanism and promote the design of TIM1, numerical studies are very essential. In this work, we compare two numerical methods, finite element method (FEM) and GeoDict, to compute the effective thermal conductivity of the particulate composites, in which the interface thermal resistance is considered. Firstly, we employed the Monte-Carlo algorithm as well as energy minimization method to generate the representative volume elements (RVEs), and the particulate fillers are dispersed randomly in the RVEs. The interface thermal resistance is introduced in the numerical modeling by an imperfect interface between various components. The effective thermal conductivity is evaluated over RVEs by FEM and GeoDict, respectively, with the efficiency and accuracy of the two algorithms compared. The drawback of FEM is that its pre-treatment is quite difficult for particulate composites of high filler content because of the requirements of fine mesh, which also calls for a large demand of memory. In contrast, GeoDict solved the problem using fast Fourier transform algorithm based on pixels of the RVEs, showing high efficiency and can be extended to any kind of additives. Finally, we show that GeoDict could be an efficient tool for high-throughput screening of formulas of TIM1 for advanced electronic packaging.

Keywords—particulate composites, thermal conductivity, finite element method, GeoDict software.

I. INTRODUCTION

The heat dissipation problem has become an important issue due to the shrinking feature size and high power density in the electronic technology [1]. The lifetime and reliability of the device would be reduced dramatically when the junction temperature of the electronic device goes up to the critical value. To reduce the heat accumulation and improve the thermal management in the electronic device, the thermal interfacial material (TIM) used between a chip and a heat spreader plays an important part, termed as TIM1 [2-5]. The design of TIM1 based on experimental trial-and-error

approaches is costly and time-consuming. Thus, analytical and numerical studies have been carried out to accelerate the screening of the formulas of the TIMs, as well as better understand the thermal conduction mechanism in microscopic scale.

The most common analytical models include Maxwell-Garnett effective medium model [6], Bruggeman symmetric model [7], Bruggeman asymmetric model [8], Lewis-Nielsen model [9] et al. However, it should be noted that most of the analytical models are restricted to the low-filler-content composites, in which the fillers are in uniform shape and size and evenly distributed. The full-field numerical simulations, for instance, the finite element method (FEM) and fast Fourier transform algorithm (FFT), are much more advantageous to estimate the effective thermal conductivity and characterize the thermal conduction path of the TIMs. FEM is widely used due to the sophisticated commercial software-ABAQUS, ANSYS and COMSOL. For instance, the effective thermal conductivity of 2D particulate composites is calculated by ABAQUS in the work of Qian et al [10], taking the interfacial thermal resistance into account. Nayak et al. [11] estimated the effective thermal conductivity of 3D particle-filled polymer composites by ANSYS, and further validated the numerical modeling by experimentation. The FEM is accurate in results, but suffer from the high computational cost. Especially for the composites with dense fillers, the large amount of meshes result in a prohibitive demand of memory. The FFT algorithm has been reported to significantly reduce the required memory on the homogenization issues for the effective properties of composites [12,13]. Lu et al. [14] has showed the improved efficiency of FFT method on the computation of thermal conductivity of particulate composites compared with the FEM. The commercial software GeoDict (Release 2021 from Math2Market GmbH, Germany, https://www.geodict.com) employs the FFT method with an

978-1-6654-1392-3/21 $31.00 © 2021 IEEE

optimized parallelization scheme, which is expected to dramatically accelerate the speed of FFT algorithm.

In this work, we systematically evaluate the accuracy and efficiency of the FFT and FEM methods implemented within the commercial COMSOL and GeoDict, respectively, on computing the thermal conductivity of the widely used TIM1 in electronic packaging. Specifically, the TIM1 is a composite material composed of silicone filled with ceramic or metallic particulate particles. We also present an example of GeoDict simulation, showing the potential application in high-throughput screening of formulas of TIM1.

II. NUMERICAL MODELING AND SIMULATION

A. Generation of the microstructures

Focusing on the particle-filled TIM1, the microstructures of representative volume element (RVE) is made up of the spherical particles which are randomly dispersed. The periodic boundary condition is used during the generation of microstructure as well as the following simulation. The microstructures of low filler volume fraction is created by Monte-Carlo algorithm. The procedures are as follows:

1. Randomly disperse the spheritic particles in the RVE without overlapping.

2. Implement large number of iterations. For each step, one particle is randomly chosen whose position is then updated.

3. If the new position of the chosen particle does not overlap with the other particles, it is accept. Otherwise, the displacement is rejected.

The positions are saved during the iterations with regular interval, achieving the random and independent realizations of RVE.

For the RVEs with high filler volume fraction , a method based on energy minimization is employed instead of the Monte-Carlo algorithm. In detail, the interaction between the hard-sphere-like particles is described with a modified standard 12-6 Lennard-Jones potential, which only has steep repulsive interactions (see in [14] for details). The hard-sphere-like particles are firstly dispersed into the RVE cube without considering the overlap problem; then the position of each particle is adjusted through the conjugate gradient (CG) algorithm until the local energy reaches the minimum value.

B. Finite element method

The radius and positions of the random particles are written in COMSOL, with the microstructure as shown in Fig. 1 (a). The size of cubic RVE is $L \times L \times L$. The thermal resistance between the particulate fillers and polymer matrix are considered as an imperfect interface through which the temperature jumps but the heat flux is continuous. We also introduce a cylindrical cushion between the particles if the distance of the two neighboring particles is below ε, whose thermal conductivity and thickness dominant the thermal resistance between the particles. The tetrahedral meshes are used in the simulations.

To evaluate the heat flux and effective thermal conductivity of the composite, a temperature potential ΔT is applied on the corresponding facing surfaces, while the other four faces are adiabatic. The effective heat flux q can be computed, and the effective thermal conductivity is given as

$$k = \frac{qL}{\Delta T} \quad (1)$$

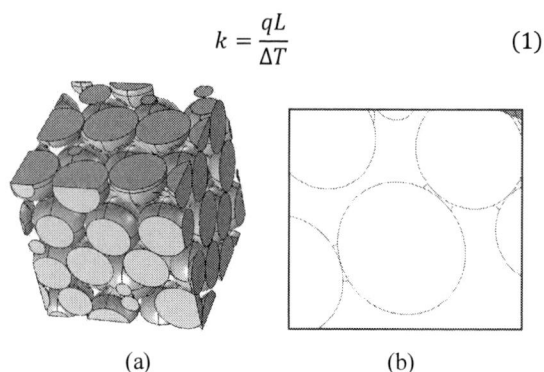

Fig. 1. Scheme of the microstructures in COMSOL. (a) Particles; (b) Cylindrical cushion between particles.

C. Fast Fourier transform-GeoDict software

Using FFT-based iterative algorithm to solve the thermal conduction problem of the particulate composites can extremely reduce the computational time and the required memory compared with FEM. The RVE is meshed by tiny unit cells (pixels) as shown in Fig. 2.

In this work, we employed a commercial software GeoDict®. The ConductoDict module of GeoDict can calculate the conductivity tensors based on the so-called explicit jump immersed interface method, which solves the equations combing FFT and GiGGStab methods. The computations can be done in parallel.

The periodic boundary condition is employed in the computations. The particulate filler/polymer matrix contact resistance (R_{fp}), and the filler/filler contact resistance (R_{ff}), are introduced in the numerical modeling by an imperfect interface. The heat flux at the interface is expressed as:

$$q = h_c \Delta T_c \quad (2)$$

where h_c is the contact conductance, ΔT_c is the temperature difference through the interface. Thus, the contact resistance is defined as

$$R_c = \frac{1}{h_c} \quad (3)$$

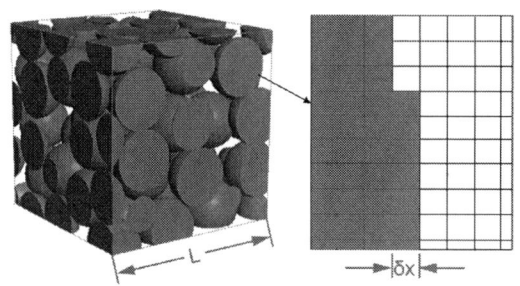

Fig. 2. Scheme of the representative volume element (RVE) and the meshes of unit cells [14].

III. RESULTS AND DISCUSSIONS

A. Comparison of the accuracy between FFT in GeoDict and FEM in COMSOL

In our previous work, the FFT algorithm has been implemented by in-house programming, whose accuracy as well as the efficiency have been validated compared with FEM based on the simulation for close-packed simple cubic model [14]. Here we systematically evaluate the performance of FFT implemented in GeoDict in computing the thermal conductivity of a series of composite with the fillers randomly distributed in the structure.

Firstly, we generate a series of microstructures of the composites. The particles are randomly dispersed in RVEs, whose diameter is 10 μm. The filler volume fraction of the RVEs goes up from 9.7 vol% to 64 vol%, and several examples are shown in Fig. 3. The increase of the filler volume fraction is realized by increasing the particle number in the RVE.

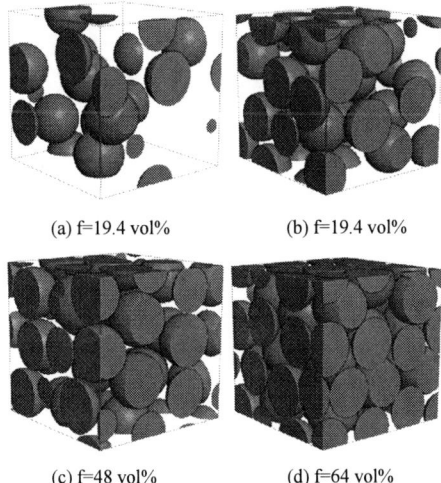

(a) f=19.4 vol%   (b) f=19.4 vol%

(c) f=48 vol%   (d) f=64 vol%

Fig. 3. RVEs of microstructures for the composites with different particle volume fractions. The diameter of particles is 10 μm.

The conductivities of the realizations are computed by both FEM via COMSOL and FFT via GeoDict. In the computations, the conductivities of the polymer matrix and particulate fillers are given as 0.18 W/mK and 30 W/mK, respectively. The particle/matrix interfacial thermal resistance is assumed to be 10^{-7} m^{2}K/W, and the particle/particle interfacial thermal resistance is not considered for similarity.

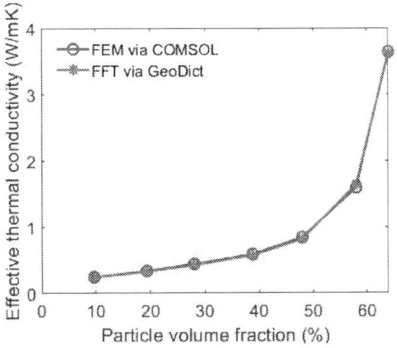

Fig. 4. Comparison of the numerical results between FEM in COMSOL and FFT in GeoDict.

The numerical results of the effective thermal conductivity are plotted and compared in Fig. 4. The results by GeoDict are in good agreement with that of the FEM simulations, indicating the accuracy of the GeoDict software in computing the thermal conductivities of the composites materials.

B. Comparison of the efficiency between FFT in GeoDict and FEM in COMSOL

The computational time and required memory by FEM via COMSOL and FFT via GeoDict for the microstructures with filler volume fractions ranging from low to high are calculated and shown in Fig. 5 and Fig. 6, respectively. In these computations, the size of RVE is 30×30×30 μm^3, and the diameter of the particle is 5 μm. We select a relatively small RVE because we find that the memory requirement of FEM in COMSOL is too large to be handled when the RVE goes too large and complex. As a result, we only compare the efficiency and memory cost of the two methods in serial computing, considering that a parallel run does not make much sense for small systems.

Fig. 5 shows that the computational time required by FFT method implemented in GeoDict is markedly low compared with the FEM in COMSOL at higher volume fractions. For instance, the computational time of FEM reaches 40 times that of FFT in GeoDict when the filler content approaches the up-limit of 64 vol%. The memory requirement of GeoDict keeps relatively constant when the particulate filler content increases due to the fact that the number of the unit cells (pixels) of RVE is not dependent on the filler content (Fig. 6). On the contrary, the computational memory required by FEM in COMSOL grows dramatically when the particle content raises. The reason is that the

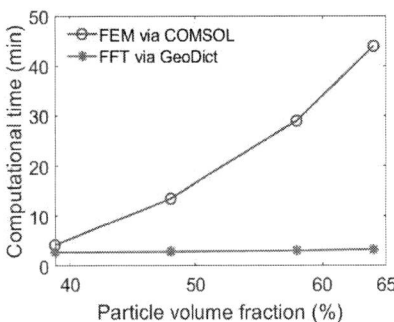

Fig. 5. Comparison of computational time between FEM via COMSOL and FFT via GeoDict.

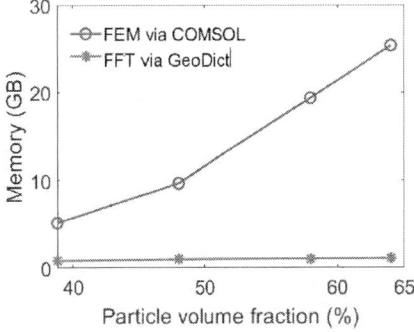

Fig. 6. Comparison of required memory between FEM via COMSOL and FFT via GeoDict.

increasing filler content leads to the increasing number of particles and decreasing distance between them, which requires much more local fine meshes and consequently large amount of tetrahedral meshes in FEM. Thus, the dramatically reduced memory requirement and increased efficiency, as well as the high accuracy, of the FFT in GeoDict makes it a potential numerical method for the high-throughput screening of the formula of thermal interface materials (TIMs).

C. Examples by GeoDict

We now give an example of application of the GeoDict in analyzing the impact factor on the thermal conductivity of TIMs. In our previous work [14] we have investigated the influence of particle volume fraction, particle/particle interface thermal resistance, particle/matrix interface thermal resistance, and particle size on the effective thermal conductivity of single-uniform-particle filled composites. In this work, we extend the discussion to the composites with binary mixed particles. The microstructure of the TIM with binary mixed particles is shown in Fig. 7. The diameter and thermal conductivity of the smaller particles are 1 μm and 30 W/mK, respectively. The diameter and thermal conductivity of the large ones are 30 μm and 300 W/mK, respectively. The volume fractions of the small and large particles are 15 vol% and 60 vol%, respectively. We concentrate on the influence of the thermal resistance at the interface, i.e., R_{sp}, R_{lp}, R_{ss}, R_{ll}, R_{sl}, where the subscripts s, l, p denote small particle, large particles and polymer matrix, respectively. To separately study the effect of each factor, we change the value of the interested factor from 10^{-8} m^{2}K/W to 10^{-5} m^{2}K/W while fixing the values of the rest four ones at 10^{-8} m^{2}K/W. The calculated effective thermal conductivities are presented in Fig. 8. The temperature field and thermal flux field are shown in Fig. 9. We find that the thermal resistance R_{ll} between the larger particles plays the dominant role in hindering the thermal conduction of the composite. The increase of R_{ss} and R_{sl} only causes tiny decrease of the thermal conductivity, while the effect of R_{lp} and R_{sp} is almost negligible.

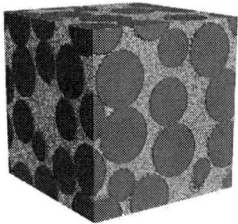

Fig. 7. The microstructure of the composites with double mixed particles. The diameter of small particle (green) and large particles (red) are 1 μm and 30 μm, respectively. The side length of RVE is 100 μm..

IV. CONCLUSION

In this paper, we present the comparison of two numerical methods in computing the thermal conductivity of particulate thermal interface materials, the FFT implemented in GeoDict software and FEM in COMSOL. We find that the FFT algorithm via GeoDict can dramatically cut down the

computational time and memory cost compared with FEM in COMSOL, without undermining the high accuracy.

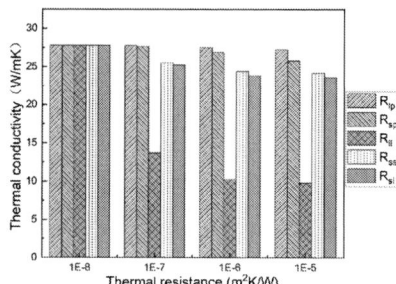

Fig. 8. The influence of R_{sp}, R_{lp}, R_{ss}, R_{ll} and R_{sl} on the effective thermal conductivity of the double mixed composites.

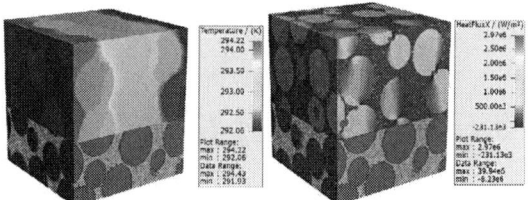

Fig. 9. The temperature field (left) and flux field (right) in the double-mixed composites.

Furthermore, a composite structure corresponding to TIM1 filled with binary mixed particles is generated and analyzed by using the FFT method implemented in GeoDict. The effects of the thermal resistances between different components are discussed in details, showing that the interface thermal resistance between larger particles plays the dominant role in setting up an up-limit boundary of the effective thermal conductivity of the composite. The FFT method could be a potential powerful tool for the high-throughput screening of formulas of thermal interface materials.

ACKNOWLEDGMENT

The work was supported by the National Natural Science Foundation of China (no. 52003289), Guangdong Basic and Applied Basic Research Foundation (no. 2021A1515010900), the Youth Innovation Promotion Association CAS (no. 2021363), SIAT Innovation Program for Excellent Young Researchers (no. E1G045, 201803).

REFERENCES

[1] A.L. Moore, L. Shi, "Emerging challenges and materials for thermal management of electronics", Materials today, vol. 17, no. 4, pp. 163-174, 2014.

[2] R. Prasher, "Thermal interface materials: historical perspective, status, and future directions". Proceedings of the IEEE, vol. 94, no. 8, pp. 1571-1586, 2006

[3] K.M.F. Shahil, A.A. Balandin, "Graphene–multilayer graphene nanocomposites as highly efficient thermal interface materials", Nano letters, vol. 12, no. 2, pp. 861-867, 2012.

[4] K.M. Razeeb, E. Dalton, G.L.W. Cross, et al., "Present and future thermal interface materials for electronic devices", International Materials Reviews, vol. 63, no. 1, pp. 1-21, 2018.

[5] C.P. Feng, L. Bai, R.Y. Bao, et al., "Superior thermal interface materials for thermal management", Composites Communications, vol. 12, pp. 80-85, 2019.

[6] D.P.H. Hasselman, L.F. Johnson, "Effective thermal conductivity of composites with interfacial thermal barrier resistance", Journal of composite materials, vol. 21, no. 6, pp. 508-515, 1987.

[7] R. Landauer, " Electrical conductivity in inhomogeneous media", AIP Conference Proceedings. American Institute of Physics, vol.40, no. 1, pp. 2-45, 1978.

[8] A.G. Every, Y. Tzou, D.P.H. Hasselman, et al., "The effect of particle size on the thermal conductivity of ZnS/diamond composites", Acta Metallurgica et Materialia, vol. 40, no. 1, pp. 123-129, 1992.

[9] T.B. Lewis, L.E. Nielsen, "Dynamic mechanical properties of particulate-filled composites", Journal of applied polymer science, vol. 14, no. 6, pp. 1449-1471, 1970.

[10] L. Qian, X. Pang, J. Zhou, et al., "Theoretical model and finite element simulation on the effective thermal conductivity of particulate composite materials", Composites Part B: Engineering, vol. 116, pp. 291-297, 2017.

[11] R. Nayak, D.P. Tarkes, A. Satapathy, "A computational and experimental investigation on thermal conductivity of particle reinforced epoxy composites", Computational Materials Science, vol. 48, no. 3, pp. 576-581, 2010.

[12] H. Moulinec, P. Suquet, "A numerical method for computing the overall response of nonlinear composites with complex microstructure", Computer methods in applied mechanics and engineering, vol. 157, no. 1-2, pp. 69-94, 1998.

[13] V. Monchiet, G. Bonnet, "A polarization-based FFT iterative scheme for computing the effective properties of elastic composites with arbitrary contrast", International Journal for Numerical Methods in Engineering, vol. 89, no. 11, pp. 1419-1436, 2012.

[14] X. Lu, X.Q. Fu, J.B. Lu, et al., "Numerical homogenization of thermal conductivity of particle-filled thermal interface material by fast Fourier transform method", Nanotechnology, vol. 32, no. 26, p. 265708, 2021.

Research on point-to-point motion control of packaging equipment

Zesheng Li
School of Electromechanical Engineering
Guangdong University of Technology
Guangzhou, China
862696000@qq.com

Yunbo He
School of Electromechanical Engineering
Guangdong University of Technology
Guangzhou, China
heyunbo@gdut.edu.cn

Shujin Liu
School of Electromechanical Engineering
Guangdong University of Technology
Guangzhou, China
2543075618@qq.com

Qihao Qian
School of Electromechanical Engineering
Guangdong University of Technology
Guangzhou, China
1311584566@qq.com

Abstract—In view of the point-to-point motion of the packaging equipment and the requirements for high speed, high acceleration and high precision, the permanent magnet synchronous linear motor of the packaging equipment is mathematically modeled, and the speed planning is carried out. Finally, three closed-loop and feedforward control are adopted. The simulation results show that the three closed-loop and feedforward control methods can meet the high speed, high acceleration and high precision requirements of the packaged equipment.

Keywords—packaging equipment; point-to-point motion; speed planning

I. PREFACE

Manufacturing equipment technology is the strong cornerstone of advanced manufacturing[1], and the semiconductor manufacturing industry cannot do without the support of high-end manufacturing equipment. However, with the development and progress of microelectronic packaging technology, the complexity of integrated circuits and the density of packaging have continuously increased the requirements for chip packaging[2]. Nowadays, packaging equipment has higher and higher requirements for speed, acceleration, and precision, and most of the motion performed by packaging equipment is point-to-point motion,such as bonding wire, wafer taking, dispensing, etc., which requires appropriate speed planning to realize the point-to-point motion.Point-to-point motion control technology has gradually become one of the core technologies of modern high-precision automation systems, such as robots, SMT and other processing equipment, and this type of motion control technology has high requirements for its accuracy and efficiency, and even requires nanometer level accuracy[3]. Acceleration and deceleration control is an important part of the motion control system and one of the key technologies in the development of motion controllers[4]. Zhongqian Zheng of South China University of Technology and others proposed a 5-stage S-curve flexible acceleration and deceleration control method based on acceleration and deceleration time control,and this method can ensure that the acceleration and speed curves are continuous and improve the flexibility of the system,and it is simpler and more intuitive to set parameter[5]. Zhong Qianjin, Wang Ke and others proposed a new S-shaped curve acceleration and deceleration algorithm,and this method uses trigonometric functions to fit a S-shaped speed curve on the trapezoidal speed trajectory,and the algorithm has a small amount of calculation and can obtain continuous speed and acceleration[6]. Ning Peizhi, Bi Qingzhen and others proposed a look-ahead algorithm for preprocessing a specific number of segments based on the speed curve of the quintic polynomial,and simulations and experiments proved that the algorithm can increase the processing speed of continuous small line segments and greatly increase the speed of inter-segment connection,and then improve the processing efficiency and surface quality[7]. If the T-type acceleration and deceleration control is adopted, there will be a large impact and vibration due to the sudden acceleration.If the seven-section S-type acceleration and deceleration control or the higher-order polynomial acceleration and deceleration control is adopted, it will cause huge computational problems.Therefore, this paper adopts four-section S-type acceleration and deceleration control. Peng Rui conducted current, speed and position closed-loop experiments on the permanent magnet synchronous linear motor servo system,and the experimental results show that the DSP/BIOS-based linear motor servo control system has good dynamic and static performance, smooth motor speed control, and high positioning accuracy[8]. Through the combination of DSP and IPM, Zheng Jun used appropriate control strategies to make the overshoot of the AC permanent magnet synchronous linear motor servo system small, fast adjustment speed, and high mover positioning accuracy[9]. The packaging equipment mainly uses permanent magnet synchronous linear motor, which has the advantages of high efficiency, large output torque, small volume, easy control, etc., and has greatly improved the rapid response and motion precision of the motion system[10].

II. THE MATHEMATICAL MODEL OF THE PERMANENT MAGNET SYNCHRONOUS LINEAR MOTOR OF THE PACKAGED EQUIPMENT

Before the mathematical model of the permanent magnet synchronous linear motor of the packaged equipment is modeled, the spatial harmonics were ignored , and the magnetomotive force waves of each phase winding were sinusoidal distributed along the air gap space.Ignoring core saturation and end effect;Vortex and hysteresis damage are ignored;The influence of current frequency and ambient temperature on the resistance and air gap magnetic field is ignored[11].

In view of the fact that the system composed of permanent magnet synchronous linear motor is a multi-variable nonlinear time-varying system, CLARK transformation and PARK transformation are often used in analysis and research, and the model of permanent magnet synchronous linear motor is established in synchronous rotating coordinates in order to adjust individually its torque component and excitation component expediently[12,13].

After CLARK transformation and PARK transformation, the stator voltage balance equation of the permanent magnet synchronous linear motor can be expressed in the synchronous rotating coordinate system as

$$U_d = Ri_d + L_d\frac{di_d}{dt} - \frac{\pi}{\tau}vL_qi_q \qquad (1)$$

$$U_q = Ri_q + L_d\frac{di_q}{dt} - (L_di_d + \varphi_f)\frac{\pi}{\tau}v \qquad (2)$$

In the formula, U_d, U_q, i_d, i_q, L_d, L_q are d-axis voltage, q-axis voltage, d-axis current, q-axis current, d-axis inductance, q-axis inductance, respectively. R is armature resistance. v is mover Linear velocity. φ_f is the flux linkage. And τ is the pole pitch.

The flux linkage equation is

$$\varphi_q = L_qi_q \qquad (3)$$

$$\varphi_d = L_di_d + \varphi_f \qquad (4)$$

In the formula, φ_d, φ_q are d-axis flux linkage and q-axis flux linkage respectively.

The electromagnetic thrust equation of the permanent magnet synchronous linear motor is

$$F_e = \frac{3\pi}{2\tau}(\varphi_di_q - \varphi_qi_d) = \frac{3\pi}{2\tau}[(L_d - L_q)i_d + \varphi_fi_q] \qquad (5)$$

The motion equation of the permanent magnet synchronous linear motor is

$$F_e = F_l + Bv + M\frac{dv}{dt} \qquad (6)$$

In the formula, F_l is the external load, B is the viscous friction coefficient, and M is the total weight of the mover.

III. SPEED PLANNING

Acceleration and deceleration control is one of the key technologies of the open motion control system. It is an important guarantee for high-speed and high-precision machining. On the one hand, it requires accurate and error-free and fast reaching the target position; on the other hand, it requires convenient and simple operation and run. If the T-type acceleration and deceleration control is adopted, there will be a large impact and vibration due to the sudden acceleration.If the seven-section S-type acceleration and deceleration control or the higher-order polynomial acceleration and deceleration control is adopted, it will cause huge computational problems.Therefore, this paper proposes Four-stage S-type acceleration and deceleration control.

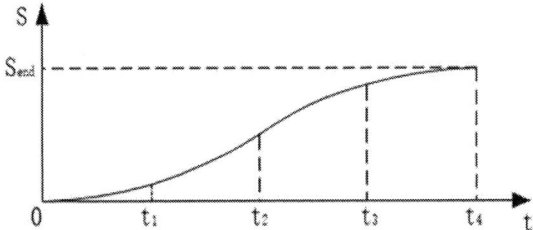

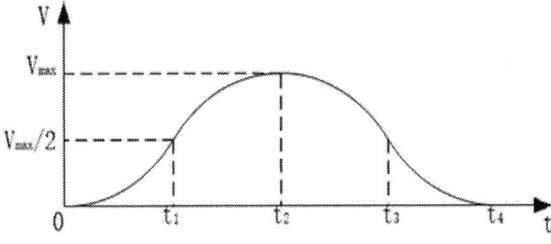

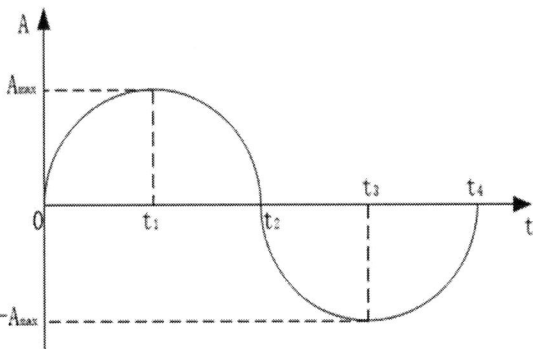

Fig. 1. Four-stage S-type speed planning curve

The four-stage S-type speed planning curve (as shown in Fig. 1) is divided into acceleration and acceleration (0-t_1), deceleration and acceleration (t_1-t_2), acceleration and deceleration (t_2-t_3), and deceleration and deceleration (t_3-t_4). This paper adopts the four-stage S-type speed planning of cubic polynomial. The equation of the velocity curve is

$$V(t) = C_1t^3 + C_2t^2 + C_3t + C_4 \qquad (7)$$

The displacement curve is obtained by integrating of the velocity curve, and the acceleration curve is obtained by differentiation of the velocity curve. According to the boundary conditions shown in Figure 1, the velocity planning curve equation can be obtained. The speed planning curve equation is as follows:

$$S(t) = -\frac{4A_m^3}{27V_m^2}t^4 + \frac{4A_m^2}{9V_m}t^3 \qquad 0 < t \le t_1, t_1 < t \le t_2 \qquad (8)$$

$$S(t) = \frac{3V_m^2}{2A_m} + \frac{4A_m^3}{27V_m^2}t^4 - \frac{4A_m^2}{3V_m}t^3 + 8A_mt^2 - 4V_mt \qquad t_2 < t \le t_3, t_3 < t \le t_4 \qquad (9)$$

$$V(t) = -\frac{16A_m^3}{27V_m^2}t^3 + \frac{4A_m^2}{3V_m}t^2 \qquad 0 < t \le t_1, t_1 < t \le t_2 \qquad (10)$$

$$V(t) = \frac{16A_m^3}{27V_m^2}t^3 - \frac{4A_m^2}{V_m}t^2 + 8A_mt - 4V_m \qquad t_2 < t \le t_3, t_3 < t \le t_4 \qquad (11)$$

$$A(t) = -\frac{16A_m^3}{9V_m^2}t^2 + \frac{8A_m^2}{3V_m}t \qquad 0 < t \le t_1, t_1 < t \le t_2 \qquad (12)$$

$$A(t) = \frac{16A_m{}^3}{9V_m{}^2}t^2 - \frac{8A_m{}^2}{V_m}t + 8A_m \quad t_2 < t \le t_3, t_3 < t \le t_4 \quad (13)$$

In this paper, the speed planning of the wire bonding point-to-point motion of the wire bonding machine in the packaging equipment is carried out, and the 2mm copper wire soldered on the chip of the substrate is selected as the research object, that is, the planned displacement is 0.002m, and the given maximum acceleration is 260m/s² ,and the given maximum speed is 1m/s, and then the system obtains a system maximum speed after calculation. If the system maximum speed is greater than the given maximum speed, then the maximum speed during actual point-to-point motion is the given maximum speed; if the given maximum speed is greater than the system maximum speed, then the maximum speed during actual point-to-point motion is the system maximum speed. This speed planning can quickly obtain the displacement, velocity and acceleration curves according to different given displacement, maximum accelerations and maximum speeds. The four-stage speed planning curve diagram of this article is shown in Fig. 2.

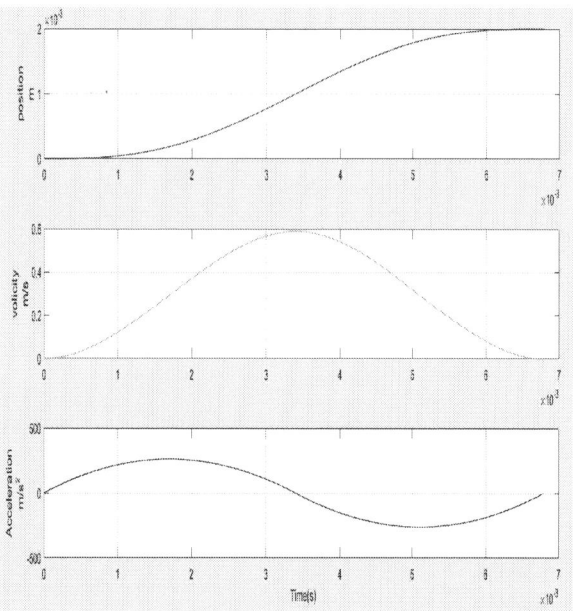

Fig. 2. Four-stage S-type speed planning curve of this article

IV. CONTROLLER DESIGN AND SIMULATION

Depending on the different application of the motor, the vector control mode is also different. The current vector control strategy of synchronous motor mainly has the following four methods: i_d=0 control, $\cos\Phi$=1 control, maximum thrust current ratio control and constant flux linkage control[14]. This article chooses the i_d=0 control scheme. Compared with other control methods, this method is the simplest and easy to implement. The electromagnetic thrust is only proportional to the amplitude of the mover current. By controlling the motor mover current, electromagnetic thrust of the linear proportional relationship can be obtained.So it achieves an effect similar to DC motor control.This is the basic idea of the mover field-oriented vector control that constitutes the permanent magnet synchronous linear motor AC servo system.

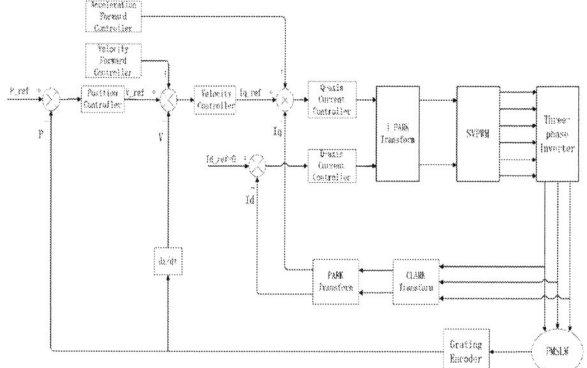

Fig. 3. Three closed-loop control structure of PMLSM

As shown in Fig. 3, this structure is a three closed-loop control structure of PMLSM, from the inside to the outside are current loop, speed loop, and position loop respectively. At the same time, in order to meet the fast tracking of position commands,the control structure also has feedforward control, which are velocity feedforward and acceleration feedforward. The three-closed loop control adopts PID control, which is simple to use and convenient to adjust. The current loop is composed of PI controller, coordinate transformation algorithm, SVPWM algorithm, and current feedback module. It realizes real-time control of current and improves the response performance of the system. The response speed of this loop is the fastest. The speed loop adopts PI controller, so that the system has good followability and anti-disturbance, and steady-state adjustment eliminates the speed error. The position loop adopts PID controller to eliminate the position error in a steady state and make the system follow the command position quickly.

In the Matlab/Simulink simulation environment, a simulation model of the PMLSM servo control system is created. The model includes a feedforward controller, a position loop controller, a speed loop controller, a current loop controller, a coordinate transformation module, a SVPWM module, and a PMLSM module.

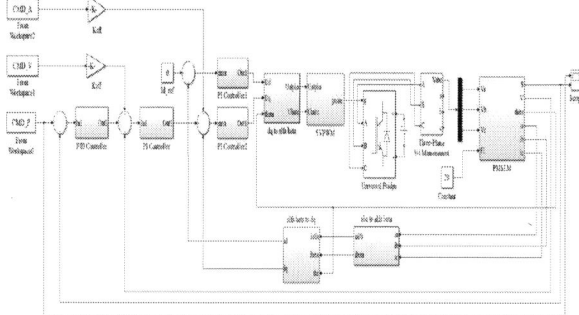

Fig. 4. PMLSM servo control system simulation model

The simulation model of PMLSM servo control system is shown in Fig. 4. PWM carrier frequency is set to 5KHz, bus voltage is 311V. The motor parameters are as follow: mass is 2kg, pole pitch is 48mm, friction coefficient is 0.008, external load disturbance is 20N, flux linkage is 0.2Wb , Resistance is 1.5Ohm, inductance is 4mH.

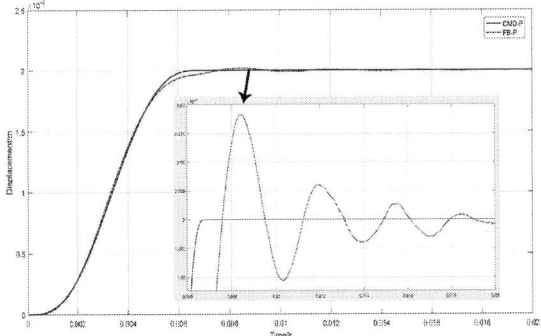

Fig. 5. Displacement response curve

The displacement response curve is shown in Fig. 5, and CMD-P, FB-P are command position and feedback position respectively. The simulation results show that the rise time is 0.0076s, the overshoot is 0.9%, and the time to set to ±0.000004m is 0.0056s. The response speed and steady-state error can meet the requirements of the 2mm copper wire soldered on the chip of the substrate of the wire bonding machine in the packaging equipment. The movement performance of the control system can meet the requirements of rapidity and high precision of the packaging equipment.

V. CONCLUSION

According to the requirements of high-speed and high-precision packaging equipment, this paper designs a four-stage S-type speed planning, and the acceleration of the speed planning is continuous, avoiding shock and vibration during the movement, and improving the movement speed and accuracy of the packaging equipment, and then it could extend the service life of the packaged equipment. The simulation results show that the three-closed loop and feedforward control method can meet the requirements of high-speed, high-acceleration and high-precision point-to-point motion of the packaged equipment.

ACKNOWLEDGMENT

This paper is supported by National Natural Science Foundation of China (No. 61973093).

REFERENCES

[1] Li C W, He X M, "Development trend and industrialization of numerical control technology," Machine Building & Automation, 2010, 39(2), pp. 466–475.

[2] Cai X, "Research on Reliability of High Density Electronic Packaging," Graduate School of Chinese Academy of Sciences (Shanghai Institute of Microsystem and Information Technology), 2002.

[3] Zhu W Q, Chen J H, Liu H L, et al., "Research on a Novel Fourth Order S-Shaped Motion Profile Algorithm and Its Application," Systems Science and Mathematical Sciences, 2017; 37(04), pp. 1034-1048.

[4] Zhang H T, Feng T J, Cao H T, et al, "Key technologies and development trends of high speed machine tools," Machinery vol. 3, pp. 12–14, 2006.

[5] Zheng Z Q, Wang X F, Li S, et al., "Algorithm of S-shape acceleration based on control of anticipation time," School of Mechanical & Automotive Engineering, South China University of Technology, vol. 49, pp. 425–430, April 2014.

[6] Zhong Q J, Wang K, Ding X Z, "Research on a New Type S Acceleration / Deceleration Algorithm," Electric Drive, vol. 49, No. 6, pp. 8–12, June 2019.

[7] Ning P Z, Bi Q Z , Wang Y H, et al., "Research on a New Type S Acceleration / Deceleration Algorithm," Modular Machine Tool & Automatic Manufacturing Technique, No. 4, pp. 15–18, April 2014.

[8] Peng R, "Research on Control Strategy and Drive of the Permanent Magnet Synchronous Liner Moter," Guangdong University of Technoloy, 2015.

[9] He J F, "Research on the DSP-based driving technoloy of the AC permanent magnet synchronous liner moters," Huazhong University of Science and Technoloy, 2007.

[10] Wang W B, Wang D C, "Control Technology of Permanent Magnet Linear Synchronous Motor," Explosion-Proof Electric Machine, vol. 52, No. 5, pp. 13–15, 2017.

[11] Chen Q Y, "Research on Speed Observer of Permanent Magnet Synchronous Linear Motor Based on RBF Neural Network," Huaqiao University, 2020.

[12] BIANCHI N, BOLOGNANI S, CORTE D D, "Tubular Linear Permanent Magnet Motors," IEEE Transactions on Industry Applications, vol. 39, No. 2, pp. 466-475, 2003.

[13] Kung Y S, "Design and Implementation of a High-Performance PMLSM Drives Using DSP Chip," IEEE Transactions on Industrial Electronics, vol. 55, No. 3, pp. 1341-1351, 2008.

[14] Zheng J, "The Disgn and Research of Permanent Magnet Synchronous Linear Motor Servo Control System Based on DSP," Guangdong University of Technoloy, 2012.

978-1-6654-1392-3/21 $31.00 © 2021 IEEE

Numerical analysis on the effect of microstructures on the thermal and mechanical properties of carbon fiber/Al$_2$O$_3$ thermal pad

Shu Liu[#]
School of Electronic Science & Engineering, Southeast University,
Nanjing, P.R.China
shu.liu@siat.ac.cn

Jibao Lu*
Shenzhen Institutes of Advanced Technology,
Chinese Academy of Sciences
Shenzhen, P.R.China
jibao.lu@siat.ac.cn

Jianbin Xu
Department of Electronics Engineering,
The Chinese University of Hong Kong,
Shatin, New Territories,
Hong Kong 999077, P.R. China
jbxu@ee.cuhk.edu.hk

Xiaoxin Lu[#]
Shenzhen Institute of advanced electronic materials, Shenzhen Institutes of Advanced Technology Chinese Academy of Sciences
Shenzhen, P.R.China
luxx@siat.ac.cn

Shen Xu*
School of Electronic Science & Engineering, Southeast University,
Nanjing, P.R.China
Xus@seu.edu.cn

Ching-Ping Wong
School of Materials Science and Engineering,
Georgia Institute of Technology
Atlanta, USA
cp.wong@mse.gatech.edu

Jiabin Huang
Shenzhen Institute of advanced electronic materials, Shenzhen Institutes of Advanced Technology Chinese Academy of Sciences
Shenzhen, P.R.China
jb.huang@siat.ac.cn

Sun Rong*
Shenzhen Institute of advanced electronic materials, Shenzhen Institutes of Advanced Technology Chinese Academy of Sciences
Shenzhen, P.R.China
rong.sun@siat.ac.cn

 # Shu Liu and Xiaoxin Lu are co-first authors of the article.

Abstract—**Thermal interface material (TIM) is pivotal for the heat dissipation in high-density electronic packaging. Recently, carbon-based materials have been used as important TIM located between the integrated heat spreader and heat sink in a flip-chip package, termed as TIM2. Due to the extremely high thermal conductivity of carbon fiber, the thermal pad with array of aligned carbon fibers presents high through-plane thermal conductivity, and the introduction of Al$_2$O$_3$ particles also proves to facilitate the heat transfer and mechanical properties of the composites. In this work, we generate the microstructures of carbon fiber/Al$_2$O$_3$ thermal pad with high filler density (>70 vol%), and simulate their thermal and mechanical properties by GeoDict, an efficient software for modeling of materials, including visualization, property characterization, simulation-based design. To deeply understand the influence of the filler content and formula, intrinsic properties of the additives, and interfacial thermal resistance between various components on the effective properties of the composites, we employ orthogonal experimental design to give a comprehensive perspective with reduced computation cost. The simulation results and numerical analysis provide the potential to develop TIM2 with improved properties.**

Keywords—*Thermal interface material; thermal pad; thermal conductivity; GeoDict software.*

I. Introduction

The requirements for heat dissipation of products are becoming higher and higher by the shrink of the volume of electronic devices. Thermal interface materials (TIM) is the general name of materials used between heated devices and heat spreaders, which has the characteristics of reducing contact thermal resistance and increasing heat conduction

between them.[1-2] High temperatures lead to a damage effect on the reliability of electronic packaging. So the heat dissipation performance becomes the bottleneck of improving the working environment and improving the working state of packaged devices. TIM in flip-chip packages is critical to the mechanical performance and thermal performance of the product. Although the extensive application of TIM has been affirmed, the performance of TIM is restricted to some extent for the uneven heat conduction and poor mechanical properties. Improving the thermal conductivity and mechanical properties of TIM is of great significance in the design and processing of electronic packaging. The carbon-based materials have been used in TIM2 and located between the integrated heat spreader and heat sink in a flip-chip package[3-5], and the carbon fiber has extremely high thermal conductivity[6-7]. Subramani, *et al.* have studied silver, nickel, and their stack prepared at various thicknesses by sputtering process and used as thermal interface material [8]. Munish Sharma, *et al.* compressed the mixture of micrometer-size solder powder and graphite to form a graphite network[9]. Zhao and his team fabricated heat conduction pads by embedding carbon nanotubes into a liquid metal which show distinguish thermal conductivity performance[10].

In this paper, the microstructures of carbon fiber/Al$_2$O$_3$ thermal pad are generated by GrainDict Module of GeoDict software, and the thermal and mechanical properties of the thermal pad are further simulated by ConductoDict and ElastoDict of the software. For the mechanical properties, we study the influence of various filler volume fraction as well as the content of carbon fiber on the elastic modulus of the thermal pads. As for the thermal properties, we focus on the effects of interfacial thermal resistance between various components on the effective thermal conductivity. The simulation results and numerical analysis can provide the potential to develop TIM2 with improved properties.

II. MODELING AND SIMULATION

A. Model And Material

In this study, the microstructures of the carbon fiber/Al$_2$O$_3$ thermal pad are generated by GrainGeo Module of GeoDict software as shown in Fig.1. The size of representative volume element (RVE) is $150 \times 50 \times 50$ μm^3, and the structure is periodic in X and Y directions. The carbon fibers are represented as cylinders and the Al$_2$O$_3$ particles are set as spheres. The positions of carbon fibers are randomly located in the RVE, while the directions of them are aligned. The axial orientations of carbon fibers are uniformly distributed between 0 to 10° around the thickness direction of the thermal pad. The Al$_2$O$_3$ particles are randomly dispersed in the RVE. The sizes and material properties of each component in the thermal pad are shown in TABLE I. It should be noted that the total volume fraction of the fillers (carbon fiber+Al$_2$O$_3$) is denoted by solid volume percentage (SVP), while the percentage of carbon fiber and Al$_2$O$_3$ particles in the fillers are fiber volume percentage (FVP) and particle volume percentage (PVP), respectively.

Fig. 1. Scheme of representative volume element (RVE) of the thermal pad. The greeen cylinders denote carbon fibers and the red spheres denote Al$_2$O$_3$ particles.

TABLE I. MAIN INFORMATION OF THERMAL PADS

Parameter	Value	
Pad Size	Length(μm)	50
	Width(μm)	50
	Thickness(μm)	150
	Solid Volume Percentage	70%-80%
Carbon Fiber	Density(kg/m³)	950
	Young's Modulus E(GPa)	300
	Thermal Conductivity / (W/(mK))	700
	Fiber Length(μm)	150
	Fiber Diameter(μm)	10
	Fiber Volume Percentage	50%-80%
Al$_2$O$_3$	Density(kg/m³)	3790
	Young's Modulus E(GPa)	350
	Thermal Conductivity / (W/(mK))	30
	particle Diameter(μm)	5

Parameter	Value	
Matrix	Particle Volume Percentage	20%-50%
	Density(kg/m³)	1800
	Young's Modulus E(GPa)	4.24
	Thermal Conductivity / (W/(mK))	0.18

B. Calculation of properties

The mechanical and thermal properties of the carbon fiber/Al$_2$O$_3$ thermal pad are simulated by the ElastoDict module and ConductoDict module of GeoDict software, respectively. The partial differential equations are solved by Fast Fourier transform algorithm[11]. In all the following simulations, the periodic boundary condition is employed, and the size of the unit cell (pixel meshes) is set as 0.2 μm. We next present the calculation of several common interested properties from GeoDict by taking a specific composite structure as an example.

From the simulation, we can obtain the local physical characteristics as well as the effective properties of the thermal pad. For instance, we take a realization of the thermal pad whose SVP is 80% , FVP and PVP are 70% and 30% , respectively. Fig.2 shows the Vin Mises stress field in the thermal pad when applying uniform strain of 0.005GPa along Z direction. We can see that the equivalent stress on the particles is larger than that on the carbon fibers because the Young's modulus of the former is larger than that of the latter. In the mechanical simulation of thermal pad, longitudinal Young's modulus (along the thickness direction of the thermal pad), transversal Young's modulus (horizontal to the thickness direction of the thermal pad), transversal plane shear modulus (in X-Y plane) and parallel plane shear modulus (in X-Z or Y-Z plane) are given in TABLE II. It is obvious that the longitudinal Young's modulus is much larger than the transversal Young's modulus according to the contribution of aligned carbon fibers. The shear modulus in X-Y plane and that in X-Z or Y-Z plane does not show big difference.

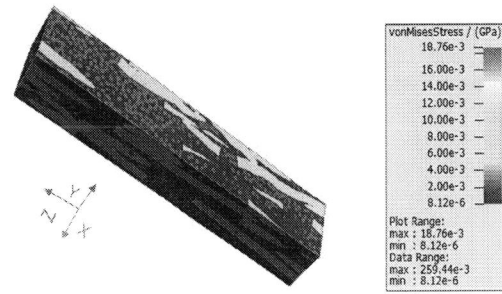

Fig. 2. Von Mises stress field of thermal pad. SVP=80%, FVP=70%, PVP=30%.

Fig.3 provides the temperature field in the thermal pad when we employed a temperature potential of 2K along Z direction (from 294.15 K to 292.15 K). The effective thermal conductivities along three main directions (X, Y, and Z) are presented in TABLE III. The interfacial thermal resistance between various component is not taken into account. It shows that the effective thermal conductivity of carbon fiber/Al$_2$O$_3$ thermal pad is transversely isotropic, and the longitudinal thermal conductivity is dramatically enhanced compared with the transverse thermal conductivity according to the excellent thermal properties of aligned carbon fibers which forms the efficient heat paths in Z direction.

Fig. 3. Temperature field of thermal pad SVP=80%, FVP=70%, PVP=30%.

TABLE II. YOUNG'S MODULUS AND SHEAR MODULUS

(SOILD VOLUME PERCENTAGE IS 80% AND CARBON FIBER VOLUME PERCENTAGE IS 70%)

Modulus	Value (GPa)
Longitudinal Young's Modulus (LYM)	184.36
Transversal Young's Modulus (TYM)	117.06
Transversal Plane Shear Modulus (TPSM)	0.26
Parallel Plane Shear Modulus (PPSM)	0.2634

TABLE III. THERMAL CONDUCTIVITY IN X, Y AND Z DIRECTIONS

(SOILD VOLUME PERCENTAGE IS 80% AND CARBON FIBER VOLUME PERCENTAGE IS 70%)

Directions	Conductivity(W/(mK))
X	2.48
Y	2.48
Z	36.37

III. RESULTS AND DISCUSSION

A. Analysis on thermal properties

The interfacial thermal resistance has been reported to have a significant influence on the effective thermal properties of the composites[12]. In the fiber/Al$_2$O$_3$ thermal pad, there are five kinds of various interfacial resistance between various components. Specifically, R_{f-m} is the interfacial thermal resistance between carbon fiber and matrix, R_{p-m} is the interfacial thermal resistance between Al$_2$O$_3$ particles and matrix, R_{f-f} is the interfacial thermal resistance between carbon fibers, R_{p-p} is the interfacial thermal resistance between Al$_2$O$_3$ particles, R_{f-p} is the interfacial thermal resistance between carbon fiber and Al$_2$O$_3$ particles.

In order to explore the effect of interfacial thermal resistance on the effective thermal conductivity with reduced computational cost, the orthogonal experimental design is employed. In this study, the Taguchi design is generated by Python, and five kinds of interfacial thermal resistance are taken as main factors with two levels, i.e., high level and low level. The choices of the values for high and low levels are shown in TABLE IV. Taking L$_{32}$ orthogonal table, 32 trials for the thermal analysis are generated as shown in TABLE V, in which 1 denotes the low level of the interfacial thermal

resistance corresponding to the value of 10^{-7} (m²K/W) and 2 denotes the high level corresponding to the value of 10^{-5} (m²K/W).

TABLE IV. LOW AND HIGH LEVELS OF INTERFACIAL THERMAL RESISTANCE

Interfacial Thermal Resistance	Values (m²K/W)
R_{p-m}	$10^{-5}, 10^{-7}$
R_{f-m}	$10^{-5}, 10^{-7}$
R_{p-p}	$10^{-5}, 10^{-7}$
R_{f-f}	$10^{-5}, 10^{-7}$
R_{f-p}	$10^{-5}, 10^{-7}$

TABLE V. ORTHOGONAL TABLE FOR THE TAGUCHI DESIGN

Index	R_{p-m}	R_{f-m}	R_{p-p}	R_{f-f}	R_{f-p}
1	1	1	1	1	1
2	1	1	1	1	2
3	1	1	1	2	1
...	...	...	...	...	...
30	2	2	2	1	2
31	2	2	2	2	1
32	2	2	2	2	2

By computing the effective thermal conductivities of the thermal pad (SVP=80%, FVP=70%) according to the parameters in Table V and analyzing the simulation results, we obtain the main effect analysis diagram for the conductivity in Z direction as shown in Fig. 4, which presents the influence of various interfacial thermal resistance. It should be noted that the negative tangent of the curves means that the high values of the factors can weaken the effective thermal conductivity. Moreover, the sharper the slope of the curve is, the higher impact of its corresponding factor has. It indicates that among the five interfacial thermal resistances, interfacial thermal resistance between carbon fibers (R_{f-f}) has a major impact on thermal conductivity and interfacial thermal resistance between carbon fiber and Al$_2$O$_3$ particles (R_{f-p}) comes the second. Therefore, reducing R_{f-f} and R_{f-p} can effectively improve the thermal conductivity in the Z direction.

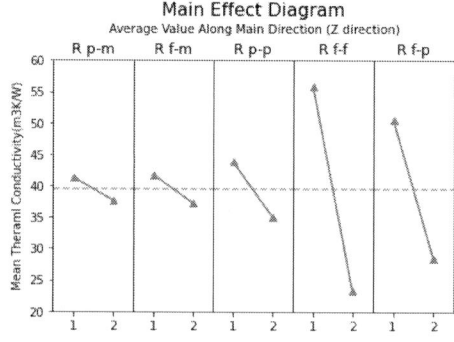

Fig. 4. Main effect diagram of thermal pad for the thermal conductivity along main direction (Z direction).

Figure 5 shows the main effect diagram for the effective transversal thermal conductivity (X or Y direction). It can be seen that the major factors for the transversal thermal conductivities are R_{p-p}, R_{f-p}, and R_{f-f}. Therefore, we can conclude that the thermal conductivity along Z direction is dominated by the interfacial resistance related with fibers, such as fiber-fiber, fiber-particle. That is because the thermal path along Z direction is mainly formed by the aligned fibers. While for the transversal thermal conductivity, the interfacial thermal resistance related with particles plays the important role, e.g., R_{p-p}, R_{f-p}. The reason is that in the transversal plane, the particles makes main contribution to the heat transfer.

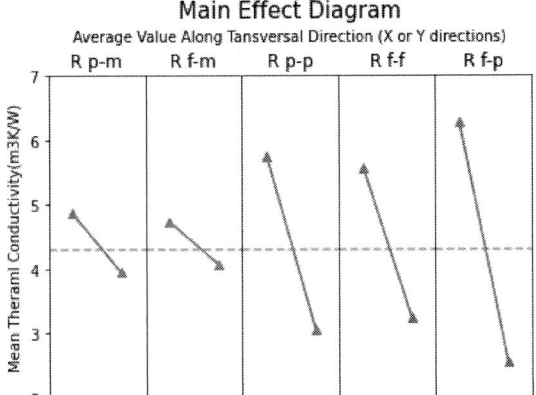

Fig. 5. Main effect diagram of thermal pad for the thermal conductivity along tansversal direction (X or Y directions)

B. Analysis on mechanical properties

The mechanical properties of the thermal pad with various solid volume percentage (SVP) and fiber volume percentage (FVP) are computed. It should be noted that the FVP here denote the fiber volume fraction in the fillers instead of in the composite. The simulation results of four elastic modulus for different cases are shown in TABLE VI. LYM is the longitudinal Young's modulus, TYM is the transversal Young's modulus, TPSM is the transversal plane shear Modulus, PPSM is the parallel plane shear modulus.

TABLE VI. MAIN INFORMATION OF THERMAL PADS

SVP	FVP	LYM (GPa)	TYM (GPa)	TPSM (GPa)	PPSM (GPa)
70%	50%	122.7	69.3	29.4	27.7
	60%	126.1	63.6	25.3	25.3
	70%	135.2	51.3	23.2	20.3
	80%	149.9	53.1	23.1	20.9
75%	50%	148.6	99.8	42.2	39.9
	60%	149.8	91.8	36.3	36.6
	70%	161.6	86.6	36.5	34.5
	80%	168.2	80.2	33.3	31.8
80%	50%	171.7	129.9	53.3	51.8
	60%	175.3	125.1	49.5	49.7
	70%	184.4	117.1	49.2	46.6
	80%	188.6	111.1	45.9	44.1

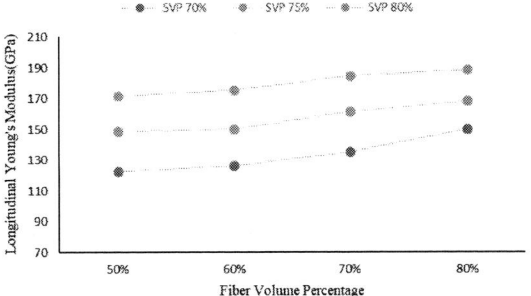

Fig. 6. The effect of SVP and FVP on longitudinal Young's modulus.

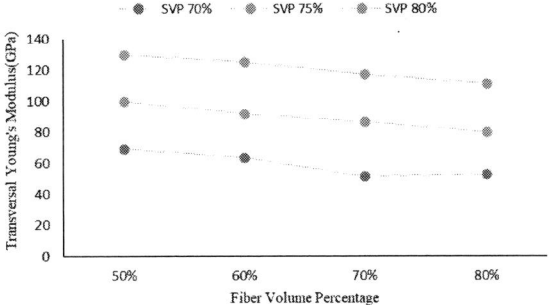

Fig. 7. The effect of SVP and FVP on transversal Young's modulus.

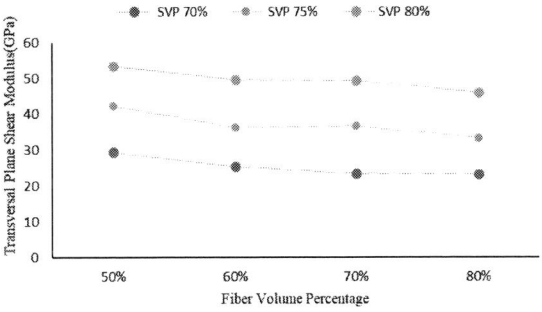

Fig. 8. The effect of SVP and FVP on transversal plane shear modulus.

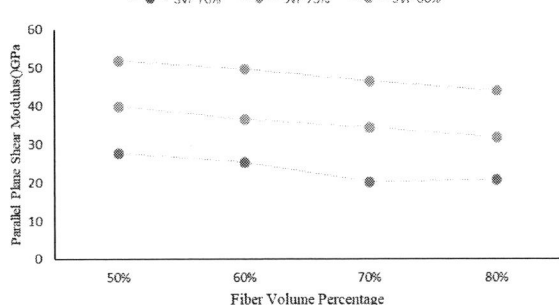

Fig. 9. The effect of SVP and FVP on parallel plane shear modulus.

The data sets are plotted separately in Fig. 6 to Fig. 9, showing the change of individual elastic modulus along versus various filler content. It is obvious that the increase of SVP can enhance all the four elastic modului of the thermal pad. However, with a fixed SVP, the increase of FVP can only

978-1-6654-1392-3/21 $31.00 © 2021 IEEE

improve the longitudinal Young's modulus as shown in Fig. 6, but results in a decrease in the other three elastic modului (see in Fig. 7 to Fig. 9).

IV. CONCLUSION

In this work, we employed the GeoDict software to generate the microstructures of the carbon fiber/ Al_2O_3 thermal pad, as well as simulate their thermal and mechanical properties. The influence of interfacial thermal resistance between various components on the effective thermal properties of the composites are studied by orthogonal experimental design algorithm. Moreover, the effect of various filler volume fraction as well as the content of carbon fiber on the elastic modulus of the thermal pads are discussed in detail. According to the simulation and analysis, the conclusions can be drawn as follows:

(1) The interfacial thermal resistance between the carbon fibers, and that between the carbon fiber and particles, have a major impact on the thermal conductivity in the out-of-plane direction of the thermal pad. For the in-plane thermal conductivity, the dominant factors are the interfacial thermal resistance between the particles, and that between the carbon fiber and particles.

(2) Increasing the total filler volume fraction improves the elastic modulus of the thermal pad. At a fixed total filler content, the increase of the fiber percentage only improves the longitudinal Young's modulus.

ACKNOWLEDGMENT

The work was supported by the National Natural Science Foundation of China (No. 52003289), Guangdong Basic and Applied Basic Research Foundation (No. 2021A1515010900), the Youth Innovation Promotion Association CAS (no. 2021363), SIAT Innovation Program for Excellent Young Researchers (No. E1G045, 201803), the Guangdong Province Key Field R&D Program Project (No.2020B010190004).

REFERENCES

[1] J. Hansson, T. Nilsson, L. Ye, and J. Liu, "Novel nanostructured thermal interface materials: a review," International Materials Reviews, pp. 1-24, 2018

[2] W. Dai, L. Le, J. Lu, H. Hou, "A Paper-Like Inorganic Thermal Interface Material Composed of Hierarchically Structured Graphene/Silicon Carbide Nanorods," ACS Nano, vol. 13, no. 2, pp. 1547-1554, 2019.

[3] J. Xu and T. S. Fisher, "Enhancement of thermal interface materials with carbon nanotube arrays," International Journal of Heat & Mass Transfer, vol. 49, no. 9/10, pp. 1658-1666, 2006.

[4] Q. Liang, X. Yao, W. Wang, Y. Liu, and C. P. Wong, "A three-dimensional vertically aligned functionalized multilayer graphene architecture: an approach for graphene-based thermal interfacial materials," Acs Nano, vol. 5, no. 3, pp. 2392-2401, 2011.

[5] Sun et al., "Investigation on Carbon Nanotubes as Thermal Interface Material Bonded With Liquid Metal Alloy," Journal of heat transfer: Transactions of the ASME, vol. 137, no. 9, 2015.

[6] K. Uetani, S. Ata, S. Tomonoh, T. Yamada, K. Yumura, and K. Hata, "Elastomeric Thermal Interface Materials with High Through‐Plane Thermal Conductivity from Carbon Fiber Fillers Vertically Aligned by Electrostatic Flocking," Advanced Materials, vol. 26, no. 33, pp. 5857-5862, 2015.

[7] J. Hansson, T. Nilsson, L. Ye, and J. Liu, "Novel nanostructured thermal interface materials: a review," International Materials Reviews, pp. 1-24, 2018.

[8] S. Shanmugan, A. N. Jassriatul, and D. Mutharasu, "Structural and thermal performance of Ag, Ni, and Ag/Ni thin films as thermal interface material for light-emitting diode application," Applied Physics A, vol. 123, no. 4, p. 273, 2017.

[9] M. Sharma, "Solder-graphite composites as high-performance thermal interface materials," Dissertations & Theses - Gradworks, 2014.

[10] L. Zhao, S. Chu, X. Chen, and G. Chu, "Efficient heat conducting liquid metal/CNT pads with thermal interface materials," Bulletin of Materials Science, vol. 42, no. 4, pp. 1-5, 2019.

[11] A. Wiegmann and A. Zemitis, "EJ-HEAT: A Fast Explicit Jump Harmonic Averaging Solver for the Effective Heat Conductivity of Composite Materials," 2006.

[12] X. Lu et al., "Numerical homogenization of thermal conductivity of particle-filled thermal interface material by fast Fourier transform method," Nanotechnology, vol. 32, no. 26, p. 265708, 2021.

Quality inspection of optical lens in IC Packaging Equipment based on MTF

Qihao Qian*
State Key Laboratory of Precision Electronic Manufacturing Technology and Equipment
School of Electromechanical Engineering, Guangdong University of Technology
Guangzhou, P.R.China
1311584566@qq.com

Yunbo He
State Key Laboratory of Precision Electronic Manufacturing Technology and Equipment
School of Electromechanical Engineering, Guangdong University of Technology
Guangzhou, P.R.China
heyunbo@gdut.edu.cn

Zesheng Li
State Key Laboratory of Precision Electronic Manufacturing Technology and Equipment
School of Electromechanical Engineering, Guangdong University of Technology
Guangzhou, P.R.China
862696000@qq.com

Shujin Liu
State Key Laboratory of Precision Electronic Manufacturing Technology and Equipment
School of Electromechanical Engineering, Guangdong University of Technology
Guangzhou, P.R.China
2543075618@qq.com

Huilong Liao
State Key Laboratory of Precision Electronic Manufacturing Technology and Equipment
School of Electromechanical Engineering, Guangdong University of Technology
Guangzhou, P.R.China
1471972099@qq.com

Abstract—As IC packaging equipment has higher and higher requirements for vision module, high quality optical lenses play a key role in the image quality. As an important index, the modulation transfer function (MTF) is widely used in qualitative and quantitative expression of an optical imaging system performance. The improved slanted edge method is used to calculate the MTF of the imaging system in the study of the imaging quality. A method of canny operator combined with cubic function fitting is proposed to track the edge, and the accurate sub-pixel edge points are obtained. The five-point linear filter is used to de-noise the sample points, and the improved Hyperbolic function is used to fit the ESF curve. The influences of the angle of edge and noise on the accuracy of MTF measurement are analyzed. Experiments suggested that Edge angle and noise have a great influence on the accuracy of MTF test and the modulation transfer function area (MTFA) method has high efficiency in determining the quality of the lens. Compared with the previous method, the proposed method has higher test accuracy, better test repeatability and stronger noise resistance.

Keywords—IC packaging equipment; Optical lens; MTF; Improved slanted edge method; ESF

I. INTRODUCTION

With the development and progress of microelectronic packaging technology, the complexity of integrated circuits and the density of packaging have continuously increased the requirements for chip packaging.[1] In the work of IC packaging machine, its optical lens is the eyes of the whole system, and its quality will affect the calculation and output results. Therefore, the quality inspection of optical lens of wire bonding machine has been paid more and more attention and research. In the production line, if the imaging quality of the lens is judged manually, there is no unified evaluation standard and can't be digitized. So it is difficult to guarantee the continuity and stability of the results, which will easily cause fatigue and misjudgment of the testers [2].

Modulation transfer function (MTF) is a basic physical

National Natural Science Foundation of China (61973093).

quantity used to describe the image quality of an optical imaging system qualitatively and quantitatively and MTF is the comprehensive performance of imaging system contrast and resolution. It is a more scientific method and evaluation index for analyzing the resolution ability of the lens, and is recognized as the most effective method for evaluating the image quality of the optical system. There are many ways to obtain MTF of an imaging system, such as point light source method, slit method and slant edge method[3]. In the calculation of MTF, a very key step is to accurately obtain the ESF, which even if there is a small error in the calculation of the slope angle, will cause serious error to the calculation of MTF[4]. Therefore, the accurate calculation of ESF is the basis of obtaining the correct MTF, and many researchers have explored this. Tzannes[5] proposed three Fermi function fitting method for ESF, which has strong anti-noise performance and meets the fitting accuracy. Tiecheng Li[6] used the edge method to calculate MTF by adding correction operation, which improved the accuracy of the calculation results. Hang Li[7] used the improved canny operator to obtain the edge position points and then get the edge angle. However, the edge obtained by the canny operator is pixel level and lacked the consideration of sub-pixel edge position points to calculate the angle.

In order to improve the accuracy of MTF calculation, this paper proposes an improved slanted edge algorithm. The Otsu method and the canny operator are used to obtain the pixel level boundary, and the sub-pixel level edge points are obtained by the cubic function fitting after edge tracing. Using an improved hyperbolic function to fit the ESF sample points after noise removal can improve the accuracy of ESF fitting and reduce the amount of calculation. Finally, the method of ISO 12233 and the proposed method are tested and analyzed through simulation experiments.

II. PRINCIPLE AND METHOD OF MTF CACULATION

Classical optical theory usually takes the imaging of a sine wave by an optical imaging system as an example to explain the concept of MTF. Introducing light and dark black and white stripes, the stripes are due to the imaging structure characteristics of the optical system, and the imaging is shown in Figure 1.

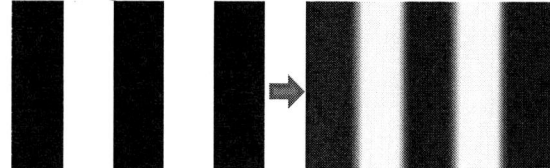

Fig. 1. Before and after black and white fringe imaging

After the fringe passes through the imaging system, the color becomes dim and blurred, and the black and white fringe becomes blurred as a whole, the sharpness decreases, and the resolution ratio decreases. For the ratio of the contrast before and after imaging in the figure, according to the concept of modulation degree, the modulation degree of the target and the image can be obtained, so as to judge the imaging quality of the lens.

A. The calculation principle of MTF

MTF is the response of the system to sinusoidal targets with different spatial frequencies. For example, in the figure 1, the brightness of sinusoidal target changes periodically along a certain direction of space. Let the brightness of the brightest place be I_{max}, the darkest part is I_{min}. To express the contrast degree of light and shade of the image, the system is defined as:

$$M = \frac{I_{max} - I_{min}}{I_{max} + I_{min}} \quad (1)$$

It is known from the formula that the MTF value must be greater than 0 and less than 1. The closer the MTF value is to 1, the better the performance of the lens. Since the brightness of the image will not be less than zero, the value range of M is [0, 1]. The modulation after imaging of the system is lower than that of the object, and the modulation transfer function at frequency f is:

$$M(f) = \frac{M_{Image}(f)}{M_{Object}(f)} \quad (2)$$

Where, $M_{Image}(f)$ represents the modulation of the object and $M_{Object}(f)$ represents the modulation of the image, $M(f)$ is the representation of the transmission performance of the system at different spatial frequencies, including the contour transmission ability corresponding to the object at low frequency, the hierarchical transmission ability corresponding to the object at medium frequency, and the detail transmission ability corresponding to the object at high frequency[8].

For a linearly invariant system, when the system input is $f(x,y)$, the output of the system $g(x,y)$ is the convolution of the input with the impulse response of the $h(x,y)$, like:

$$g(x,y) = f(x,y) * h(x,y) \quad (3)$$

Assuming that the input to the system is a line pulse, the expression is given:

$$f(x,y) = \delta(x) \quad (4)$$

The output response to the bright line is called the line response of the system or the line spread function:

$$L(x) = g(x,y) = \delta(x) * h(x,y) = \int_{-\infty}^{\infty} h(x,\xi)d\xi \quad (5)$$

Where, $h(x,y)$ is called the impulse response or point spread function of the system. The one-dimensional Fourier transform of $L(x)$ is equal to the cross-sectional distribution of the transfer function along the f_x axis:

$$\mathscr{F}[L(x)] = \mathscr{F} \int_{-\infty}^{\infty} [h(x,\xi)d\xi] = H(f_x,0) \quad (6)$$

Where, $H(f_x,f_y)$ is the optical transfer function of the system, f_x,f_y is defined as the spatial frequency of the coordinate axes. When the impulse response is rotationally symmetric, the optical transfer function is also symmetric and can be expressed by an arbitrary cross section $H(f_x,0)$, MTF is the normalized modulus of the optical transfer function, we have:

$$M(f_x,0) = \left| \frac{H(f_x,0)}{H(0,0)} \right| = \left| \frac{\int_{-\infty}^{\infty} L(x)e^{-j2\pi f_x}dx}{\int_{-\infty}^{\infty} L(x)dx} \right| \quad (7)$$

The line response function is the derivative of the edge spread function:

$$L(x) = \frac{dE(x)}{dx} \quad (8)$$

Substituting Equation (1) into Equation (7), the relation between ESF and MTF can be written as[6]:

$$M(f_x,0) = \left| \frac{\int_{-\infty}^{\infty} \frac{dE(x)}{dx} e^{-j2\pi f_x}dx}{\int_{-\infty}^{\infty} L(x)dx} \right| \quad (9)$$

B. ISO 12233 method to calculate MTF

ISO12233 slanted edge method is a standard method for MTF test of electronic still image camera, which main test steps are as follows[9]: First of all, select the Region of Interest (ROI) in the slanted edge image, calculate the derivative on each line ESF, and obtain LSF and its center position as the edge position; Secondly, the angle of edge is obtained by linear fitting of ROI; Thirdly, all pixel points in the ROI are projected to the first row along the direction of the straight line obtained by fitting, and 1/4 of the original sampling interval is taken as the new sampling interval. The projected data points are averaged geometrically to obtain the super-sampled ESF, and derivation of ESF to obtain LSF; Finally, hamming window is used to filter and smooth LSF, then Fourier transform is carried out, and the normalized modulo at zero frequency is obtained to get MTF. Its principle is shown in Figure 2 below:

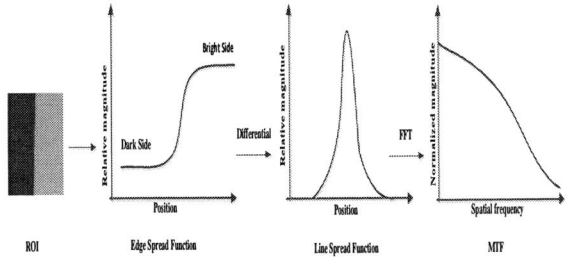

Fig. 2.   ISO 12233 method to get MTF

The ESF obtained by ISO12233 method is greatly affected by the noise in the ROI. When LSF is obtained, the noise of ESF is amplified under derivation, so the calculation accuracy of MTF of the system is reduced.

C. Modified slanted edge method to calculate MTF

Based on the principle of the slanted edge method, the ESF is obtained by using the optimized method. The process is as follows:

(1)After the ROI region is obtained, Gaussian filtering method is used to smooth the ROI image and reduce the influence of image noise.

(2) After using the Otsu method to obtain the best thresholds of the ROI, the canny operator is used to obtain the position of the edge pixel level. Using the idea of edge tracking, a total of seven pixel positions near the border and gray value are collected. Fit the ESF inflection point function of each row with a cubic polynomial[10], its expression is as follows:

$$y = a_1x^3 + a_2x^2 + a_3x + a_4 \qquad (10)$$

Geometrically, when its second derivative is zero, x is an inflection point and also a sub-pixel point with precise edge:

$$y'' = 6a_1x + 2a_2$$
$$x = -a_2 / (3a_1) \qquad (11)$$

(3) Repeat step (2) to obtain the sub-pixel position of the edge in each row. The least squares method is used to fit the edge data points and its effect is shown in Figure 3.

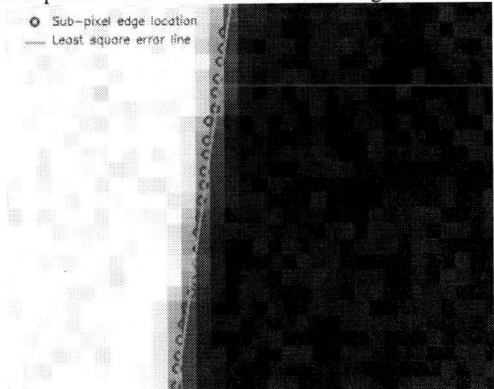

Fig. 3.   Sub-pixel edge location and least square error line

(4) Taking the inflection point as the center, sampling N data near the center of each row, a total of 2N + 1 data. The projection method is adopted to calculate the geometric average of all data points in the same sampling interval in the projected row. A five-point linear filter is performed on the projected ESF data points, which can filter out noise

points without causing curve distortion. As shown in Figure 4, normalize the collected gray-scale data.

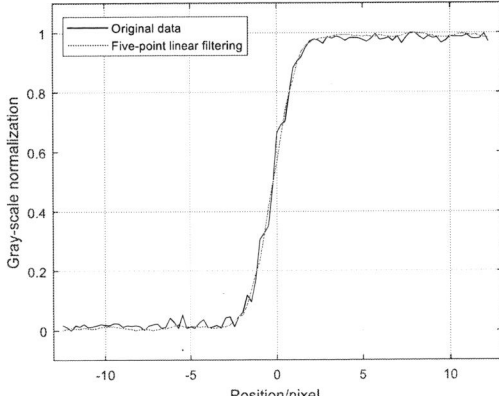

Fig. 4.   Five point linear filter before and after diagram

(5) In order to overcome the ripples of the ESF curve and avoid sharp fluctuations and sharp peaks and valleys[11], etc., this paper adopts the method of function fitting to obtain ESF data. Considering that the ESF curve is generally asymmetric, and it is found that using the improved Hyperbolic function to fit the ESF curve can obtain higher fitting accuracy and less fitting parameter calculation, which expression is as follows:

$$f(x) = a_0 arc\tan(\exp(^{x-b_0}/_c)) + arc\tan(\exp(^{x-a_1}/_{b_1})) + d$$
$$(12)$$

The Levenberg-Marquardt(LM) algorithm is used to solve the nonlinear equation (12), which has the local characteristics of Gauss-Newton method and the global characteristics of gradient method. LM algorithm takes advantage of the approximate second derivative information and is faster than gradient method[12].

D. Evaluation of MTF image quality

The higher the MTF value, the higher the lens' ability to reduce the profile (ie, the contrast); The higher the MTF value corresponding to the nyquist spatial frequency, the higher the capability (ie, the resolution) of detailed features. Modulation Transfer Function Area (MTFA) is a physical quantity used to describe the imaging quality of an imaging system. It can be expressed by the area of the region enclosed by MTF curve and the horizontal and vertical axes. The larger the area enclosed by the curve is, the more information is transmitted in the imaging process of the optical system, and the better the imaging quality is. When evaluating image quality, MTFA is often used as an important evaluation index[13], which can be expressed as:

$$MTFA = \int_{f_0}^{f_1} MTF(x)dx \qquad (13)$$

Where, f_0、f_1 represents the lower limit and upper limit of integration frequency respectively. The evaluation method of MTFA can not only estimate the image quality of the lens through the change degree of the curve, but also quantify it and evaluate the image quality through the precise value.

978-1-6654-1392-3/21 $31.00 © 2021 IEEE

III. EXPERIMENT AND ANALYSIS

A. Simulation experiment

In the simulation experiment, the proposed method, the method of ISO12233 and the method in Reference [14] are compared. A 640x480 image of an ideal slanted edge is generated by MATLAB software. Convolve the ideal slanted edge image with the known Point Spread Function (PSF) to simulate the degradation effect of the real photoelectric imaging system. PSF is simulated by Gaussian function, and its expression is as follows:

$$PSF(x,y) = e^{\frac{-(x^2+y^2)}{2\sigma^2}} \tag{14}$$

Where, let $\sigma=0.8$, change the noise level of the image, and denoted by the Signal to Noise (SNR) of the image. Fig. 5 shows the degraded image with the edge angle of $5°$ inclination angle of 15dB SNR.

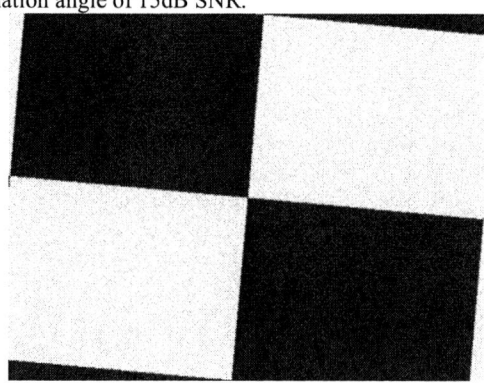

Fig. 5. The fuzzy image of slanted edge with edge angle of $5°$ and SNR of 15dB

Different measurement methods are used to compare image with SNR of 15dB, 20dB, 25dB and 30dB respectively. After averaging values for 50 times for each measurement method and image, the relative error is obtained. The data are shown in Table 1.

TABLE I. RESULTS OF DIFFERENT MEASUREMENT METHODS

Method	15dB	20dB	25dB	30dB
ISO12233	23.2%	9.6%	5.8%	3.6%
Ref[14]	6.5%	5.0%	4.3%	5.4%
Modified	2.4%	1.5%	1.2%	1.2%

From the Table 1, the slanted angle calculated by ISO12233 is greatly affected by image noise and has low accuracy. The gradient method proposed in Reference [14] has better accuracy and anti-noise performance than ISO12233. Compared with the previous two methods, the proposed method has higher accuracy and anti-noise performance and better robustness.

According to equations (5), (6) and (14), the theoretical MTF curve of an image with a slanted edge can be calculated and used as an evaluation criterion. The edge method in ISO12233 and the proposed method are used to calculate the MTF of the image. The abscissa in the figure is the normalized processing of the spatial frequency. MTF calculation errors of the above two methods are shown in Fig. 6 and Fig. 7:

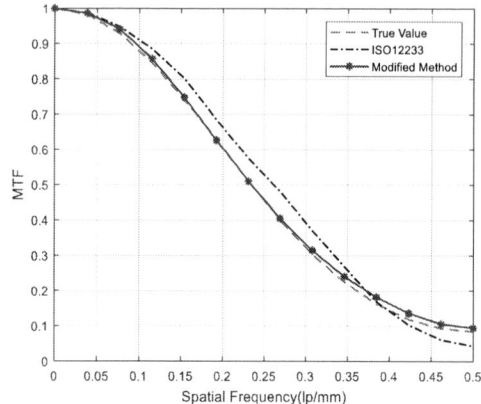

Fig. 6. Comparison of MTF results

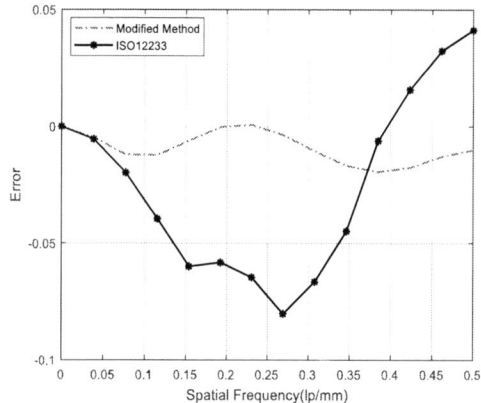

Fig. 7. MTF error curve

The maximum error values, average values and root mean square values calculated by the above two methods are shown in Table 2 below:

TABLE II. ERROR ANALYSIS TABLE

	E_{max}	E_{mean}	E_{rms}
ISO12233	0.080	0.038	0.045
Modified	0.019	0.009	0.011

According to the simulation curve and error table, the improved method is more accurate than ISO12233 in the measurement of inclination angle, which proves the effectiveness of the improved slanted edge method and the accuracy of calculation.

In order to analyze the influence of different tilt angles on the measurement results, graphs with different tilt angles is measured, and the angle range is selected as $3°\sim30°$, the measurement results are shown in Figure 8:

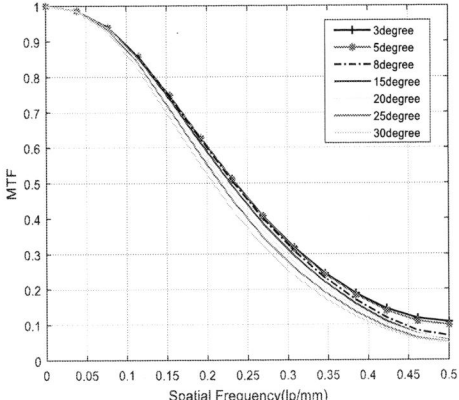

Fig. 8. Influence of different inclination angles on MTF measurement

By analyzing Figure 7, it can be seen that when the tilt angle is small, MTF curves basically coincide. With the increasing angle, the measurement accuracy of MTF decreases. This is because as the angle increases, the oversampling data is insufficient, and some data is repeatedly calculated, resulting in an increase in measurement error. The smaller the angle is, the more data will be obtained and the more accurate the calculation results will be. Therefore, placing the calibration plate at a small angle during testing can improve the testing accuracy.

B. Physical experiment of optical imaging system

In laboratory, a CCD camera with pixel size of p=5.6um is selected. Established an imaging system platform consisting of camera, light source, lens and high-resolution test card, as shown in Figure 9:

Fig. 9. Optical platform

Given the camera pixel size, the nyquist frequency of the spatial frequency is $N_y = \frac{1}{2p} = 89.28 (lp/mm)$ according to the sampling theorem. Under the same test conditions, the MTF value of the qualified optical lens is selected as the evaluation standard. The experiment uses the upper computer software written by VC++ to deal with data. In the new lens test, after the exposure time, light brightness and focus degree are adjusted, Figure 10 is an image collected by the camera, where ROI is the edge region extracted from the image to calculate MTF.

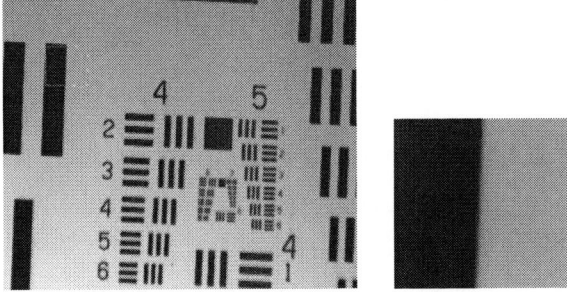

Fig. 10. High resolution test card and ROI

The ROI area around and in the center of the image is obtained to average the MTF data, and Formula (13) is used as the evaluation criterion. The spatial frequency is calculated from zero to the nyquist frequency, and the ideal MTFA is $MTFA_{Ideal}=33.00$. In the actual processing of lenses, assembly error, focusing differences in the testing process and algorithm steps may lead to different MTF calculation value, a 5% tolerance rate is given to judge the quality of the optical lens. MTFA test data are shown in Figure 11 and Table 3.

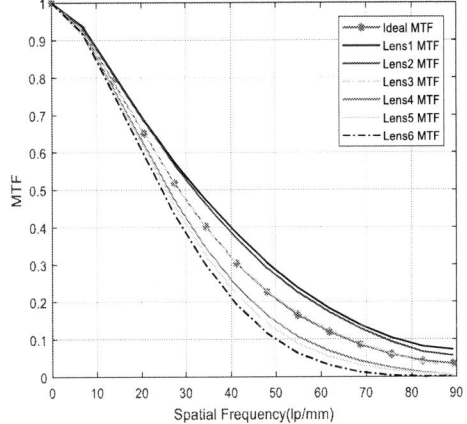

Fig. 11. MTF test data for different lenses

TABLE III. MTFA ANALYSIS TALBE

	Lens1	Lens2	Lens3	Lens4	Lens5	Lens6
MTFA	37.45	36.61	33.08	29.50	28.45	26.94
Judge	113%	110%	100%	89%	86%	81%

From Figure 10 that it is not representative to represent the imaging quality of the whole lens only by the MTF value at nyquist frequency. Analysis from Table 3 shows that the MTFA evaluation value of lens4, lens5 and lens6 is less than 95%, therefore, it is necessary to re-examine the lens components and check various influencing factors, such as wiping, fixing and replacing the main components of the lens, and finally re-assembling the lens, and conducting tests to obtain a qualified lens module. With the help of MTFA data, the trend of MTF value in the whole space frequency can be described and the evaluation standard can be made. MTFA can check the pass rate of the lens on the production line and analyze the defects of the lens components, which can describe the quality of the optical imaging of the lens more realistically.

IV. CONCLUSION

In this paper, the process of calculating MTF by ISO12233 is analyzed, and an improved algorithm is proposed. The accuracy of the proposed method and ISO12233 method are evaluated and analyzed from the angle of edge and noise factors. The results show that the proposed method maintains high stability and accuracy in the measurement of ESF slanted edge angle. Compared with the real value of MTF, the error average, maximum value and root mean square measured by proposed method are all less than ISO12233. Experiments show that the algorithm is affected by the tilt angle, and it is concluded that the tilt angle of $3^{\circ} \sim 8^{\circ}$ can improve the test accuracy. The MTFA calculation method is used to evaluate the imaging quality of the lens of IC packaging equipment and has the characteristics of high accuracy and strong reliability. According to the evaluation criteria, lens module testing can be performed on the production line to record the quality of the lens without causing defective products to flow into the production line and ensuring the production quality of the lens.

ACKNOWLEDGMENT

The work presented in the paper is supported by National Natural Science Foundation of China (61973093).

REFERENCES

[1] Xia Cai. Reliability of high density electronic packaging [D]. Graduate university of Chinese academy of sciences (Shanghai institute of Microsystems and information technology), 2002

[2] Zheyuan Li. Application of improved edge method in measurement of modulation transfer function of security surveillance lens[D]. Zhejiang University of Technology, 2002

[3] Boreman GD. Modulation transfer function in optical and electro-optical systems[J]. Russchemrev. 2001;71(2):159–179.

[4] Greer PB, van Doorn T. "Evaluation of an algorithm for the assessment of the MTF using an edge method." Med Phys. 2000 Sep; 27(9):2048-59.

[5] Tzannes AP,Mooney JM. Measurement of the modulation transfer function of infrared cameras[J]. Optical Engineering. 1995;34(6):1808-1817.

[6] Tiecheng Li, Xiaoping Tao et al.Modulation Transfer Function Calculation and Image Restoration Based on Slanted-Edge Method[J]. Acta Optica Sinica, 2010,30(010):2891-2897.

[7] Hang Li, Changxiang Yan et al.High Accyracy Measurement of the MTF of Electro-optical Imaging System Based ON modified Slanted-edge Method[J]. Acta Optica Sinica, 2016,45(12):82-89.

[8] Yuting Wang,Optimization of Measurement Results of Lens Modulation Transfer Function Based on Artificial Intelligence Algorithm [D]. Northest Normal University, 2018.

[9] Photography-Electronic still-picture cameras-Resolution measurements[J]. Iso. 2000.

[10] Haibin Wen. The Research and Applcation on HJ-1 On-Orbit Satellites Remote Sensing Images MTF Calculation based on the Knife Edge Method [D]. Hunnan University, 2013.

[11] Lihong Yang. Research on Image Restoration of Space Camera with Wide Field of View [D]. Changchun Institute of Optics, Fine Mechanics and Physics, Chinese Academy of Sciences, 2012.

[12] Yuqing Guan, Dongmei Tang et al. Study on LM algorithm in Mueller's ellipsometry calibration method[J]. Infrared and Laser Engineering, 2020,49(08):168-176.

[13] Jiaqi Liu. Digital Image quality assessment based on modulation transfer function[D]. BoHai University, 2013.

[14] Chong Fan, Guanda Li et al. High Accurate Estimation of Point Spread Function Based on Improved Reconstruction of Slant Edge[J]. acta Geodaetica et Cartographica Sinica,2015,44(11):1219 1226.

Stress analysis and parameter optimization of Fine-Pitch BGA solder joints under cantilever plate torsion conditions

1st Zhuo Wang
School of Electronic Mechanical Engineering
Guilin University of Electronic Technology
guilin, China
1009372429@qq.com

line 1: 2nd Chunyue Huang(Corresponding author)
School of Electronic Mechanical Engineering
Guilin University of Electronic Technology
guilin, China
hcymail@163.com

line 1: 3rd Jinfeng Gong
School of Electronic Mechanical Engineering
Guilin University of Electronic Technology
guilin, China
928412627 @qq.com

line 1: 4th Huaiquan Zhang
School of Electronic Mechanical Engineering
Guilin University of Electronic Technology
guilin, China
1599005976 @qq.com

line 1: 5th Shuaidong Liao
School of Electronic Mechanical Engineering
Guilin University of Electronic Technology
guilin, China
1729269081@qq.com

line 1: 6th Shoufu Liu
School of Electronic Mechanical Engineering
Guilin University of Electronic Technology
guilin, China
1044577910@qq.com

Abstract— In this paper, a 3D finite element model of the solder joints of a 165-PIN PLASTIC FBGA (13x15) package was developed using ANSYS software according to the packaging of a static storage device. Then, the model was loaded with cantilever plate torsion model conditions and finite element simulation analysis was performed. In order to analyze the effects of different combinations of parameters with different levels of solder joint material, solder joint height, pad diameter and PCB board thickness on the maximum stress in FBGA solder joints, we designed 16 sets of orthogonal test scenarios using SPSS software and simulated the established finite element models in turn. After multi-factor ANOVA on the experimental data using SPSS software, the results showed that the maximum stresses in the FBGA joints were not uniformly distributed under the torsional loading conditions of the cantilever plate, with the maximum stresses in the joints occurring at the outermost edge of the joints near the side of the applied load that contacts the PCB surface. At 95% confidence level, the PCB board thickness has the most significant effect on the maximum stress in the solder joint among the four factors: solder joint material, solder joint height, pad diameter and PCB board thickness. After optimizing the combination of the given factor levels, the best combination that minimizes the maximum stress in the solder joint is obtained, and the results of the simulation show that the maximum stress in the FBGA solder joint is significantly reduced.

Keywords—FBGA package, Finite element analysis, Solder joint reliability, Cantilever board torsion, Multifactor ANOVA

I. INTRODUCTION

Today, the rapid development of electronic information manufacturing industry, SMT (Surface Mount Technology) surface assembly technology as a new generation of electronic assembly technology, has been widely used in various fields of electronic product assembly. Our requirements for electronic products are also moving in the direction of lighter, cheaper and more environmentally friendly. However, with the miniaturization of the device, the number of pins of the chip increases, and the density of the printed original of the circuit board increases, the packaging method of the device also puts forward higher requirements. The emergence of BGA (Ball Grid Array) packaging technology has solved the problem of high I/O pin count LSI (Large-scale Integration) packaging such as QFP (Quad Flat Package) packaging. BGA package has become a very popular package due to its small package area, low mounting height, good pin coplanarity, high reliability, and low overall cost.[1] With the further development of technology, the fine-pitch ball grid array (FBGA) package, whose package size is close to that of a bare chip, was introduced. The FBGA package is a product of further miniaturization of the BGA package, which becomes smaller, accommodates more pins, has shorter signal transmission delays, and has better thermal performance[2]. As solder joints become more and more numerous and smaller, it becomes difficult to analyze the reliability of solder joints using traditional experimental methods. Using the finite element simulation method can make the study relatively easy, and the accuracy of the finite element simulation method in a certain range has been verified through the test of production practice. As the components are soldered on the printed circuit board through the solder, the reliability of the solder joint is related to the realization of the product function, so the study of the reliability of the solder joint is an important issue of modern electronic packaging technology.[3] Encapsulated devices receive vibration during transportation, and work under thermal and mechanical stresses, which are manifested in the form of bending and twisting on the solder balls. Nowadays, many studies have been conducted by domestic and foreign scholars on the reliability of solder joints in BGA packages, suggesting that material properties, package size, loading conditions, voids, and other factors can have an impact on the reliability of solder joints.[4] Many scholars have studied solder joint reliability, for example, Huang Jiaoying et al. studied fatigue failure of microelectronic packaging solder joints[5], Liang Ying et al. analyzed three-point bending

---

This research is supported by the Science Foundation of Guangxi Zhuang Autonomous Region Government (No.2019JJA160101) , Science and Technology Major Project of Guangxi Zhuang Autonomous Region Government （No.AA19046004） and Innovation Project of GUET 2021 22nd International Conference on Electronic Packaging Technology (ICEPT)

stress-strain of microscale ball grid array solder joints[6], and Wu Liang-Yu et al. studied the thermal stress characteristics of BGA packages under thermal cyclic loading.[7]

When checking the related literature and research, we found that in the actual manufacturing process, due to the error of the fixture or human operation, when clamping, flipping and mounting the IC board, the stress of the cantilever board torsion model will be formed on the IC board. The components and solder joints soldered on the board are affected and stress concentrations can occur on them. This phenomenon has not been studied by scholars, nor has the literature analyzed the differential effects of different levels of solder joint material, solder joint height, pad diameter and PCB board thickness on the maximum stress in FBGA solder joints, and there are no studies on the optimization of parameter combinations of influencing factors under this stress condition. Therefore, in this paper, a three-dimensional finite element model of the solder joint of the 165-PIN PLASTIC FBGA (13x15) package was developed using ANSYS software, and the cantilever plate torsion model conditions were loaded on it. Then the cantilever plate torsion model conditions were loaded, and the effects of different levels of solder joint material, solder joint height, pad diameter and PCB board thickness on solder joint stresses were analyzed and the parameters were optimized. Finally, the results demonstrate the practical significance of the study in reducing the maximum stress in the solder joint and improving the reliability of the solder joint.

II. STRESS ANALYSIS OF FBGA SOLDER JOINTS UNDER CANTILEVER PLATE TORSION CONDITIONS

A. FBGA solder joint model under cantilever plate torsion conditions

In this paper, a finite element model of FBGA package is built using ANSYS software. The model consists of a PCB board, an FBGA chip. The prototype of this chip model is the FBGA package chip UPD44324362F5-E33 of NEC Corporation, its size is 13mm×15mm×1mm, the solder joint array is 11×15 with 165 solder joints, and the PCB board size is 80mm×60mm. The chip is located in the middle of the PCB board. Figure 1 shows the 3D finite element analysis model after meshing.

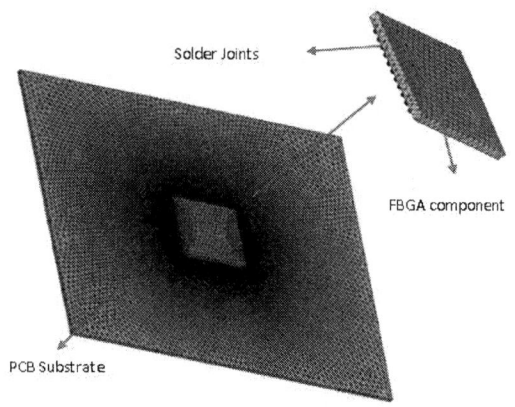

Fig. 1. finite element model diagram after meshing

This paper analyzes the maximum stresses in the solder joints of FBGA packages under torsional loading of the cantilever plate conditions. The PCB in the chip assembly is

fixed and restrained at one end, and a torsional load is applied to the other end. The loading of the chip assembly is shown in Figure 2. This test uses a 0.75mm displacement load simulation analysis. In this loading case, the solder joint between the PCB board and the chip will generate the corresponding model stress, which is the cantilever version of the torsional model stress.

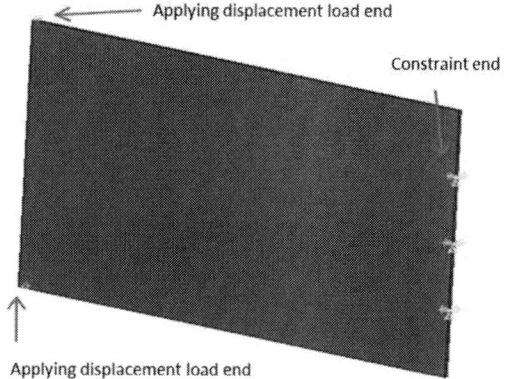

Fig. 2. Load loading diagram of cantilever plate torsion model

B. Factors and levels affecting the maximum stress in FPBGA solder joints

In this paper, the effects of four factors, namely solder joint material, solder joint height, pad diameter and PCB board thickness, on solder joint stress were investigated. As shown in Table 1. The solder joint materials were selected at four levels of SAC387, SAC305, 63Sn37Pb and 62Sn36PbAg, the solder joint heights were selected at three levels of 0.38mm, 0.4mm and 0.42mm, the pad radii were selected at three levels of 0.18mm, 0.20mm and 0.22mm, and the PCB board thicknesses were selected at three levels of 0.8mm, 1.0mm and 1.2mm. The modulus of elasticity and Poisson's ratio used in the analysis are shown in Table 2.

TABLE I.   FACTORS AND LEVERS TABLE

Factor	Level			
Solder joint material	SCA305	SAC387	63Sn37Pb	62Sn36Pb2Ag
Solder joint Height(mm)	0.38	0.40	0.42	/
Pad radius(mm)	0.18	0.20	0.22	/
PCB board thickness(mm)	0.8	1.0	1.2	/

TABLE II.   MATERIAL PERFORMANCE PARAMETERS

Material name	Elastic Modulus(Gpa)	Poisson's ratio
PCB	18.2	0.25
Chip	131	0.28
SCA305	35.18	0.35
SAC387	47.102	0.40
63Sn37Pb	31.395	0.35
62Sn36Pb2Ag	31.306	0.35

C. Finite element simulation analysis

Using ANSYS software to perform finite element simulation analysis of the model, we were able to obtain a cloud of the maximum stress distribution in the FBGA Solder Joints. As shown in Figure 3 and Figure 4.

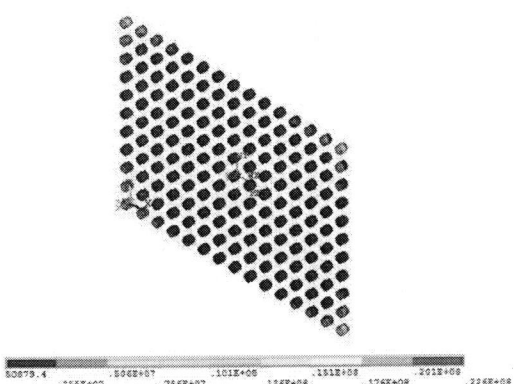

Fig. 3. Stress distribution of all solder joints

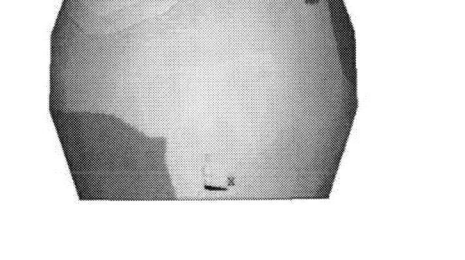

Fig. 4. Solder joint stress distribution

From the figure we can see that under torsional loading conditions of the cantilever beam, the stress distribution in the 165 solder joints is not uniform: the stress in the outer solder joints is greater than the stress in the inner solder joints, and the stress in the solder joints at the four corners is significantly greater than the stress in the solder joints at other locations. By comparison, it was found that under cantilever beam torsional loading conditions, the maximum stress in the FBGA solder joint occurs at the outermost edge near the side of the applied load end that touches the PCB surface.

III. MULTI-FACTOR ANALYSIS OF VARIANCE AND PARAMETER OPTIMIZATION BASED ON SPSS SOFTWARE

A. Mixed-level orthogonal test

It is known from practical experience that a reasonable arrangement of the test can be used to obtain useful information under the premise that the number of tests is small. Orthogonal test is a mathematical method to study multi-factor test problems. This method mainly uses the tool of orthogonal table for the overall design and statistical comparison of the experiment, which can reduce the number of tests, shorten the test cycle and save time. Since this paper studies the effect of four factors on FBGA solder joints: solder joint material, solder joint height, pad diameter and PCB board thickness, and the four factors are taken at four levels, three levels, three levels and three levels, respectively, If a level is selected from the different levels of each of the four factors for the test pairing a total of 4×3×3×3=108 kinds, it is obviously impossible to conduct the test one by one. Therefore, the orthogonal test method was also used for the experimental design of this paper.

Since the number of levels of the four influencing factors studied in this paper are different, it is an experimental design problem of $4^1 \times 3^4$, which belongs to the orthogonal test of mixed levels, and the orthogonal tables are all equal in the number of levels, so we have to apply the orthogonal tables flexibly[8]. We found the orthogonal table L16.4.4 which is close to the one we analyzed, and then we used the "proposed level method", that is, we replaced the fourth level of factor 2, factor 3 and factor 4 with any one of the other levels to generate the mixed level orthogonal table. In this experimental design, we supplemented the second level in Factor 2, the first level in Factor 3, and the first level in Factor 4 in the mixed orthogonal table. As shown in Table 3.

TABLE III.   ORTHOGONAL DESIGN TABLE

Orthogonal design table				
Number	Factor 1	Factor 2	Factor 3	Factor 4
1	1	1	1	3
2	1	2	2	2
3	1	2	3	1
4	1	3	1	1
5	2	1	2	1
6	2	2	1	1
7	2	2	1	3
8	2	3	3	2
9	3	1	1	2
10	3	2	2	1
11	3	2	3	3
12	3	3	1	1
13	4	1	3	1
14	4	2	1	1
15	4	2	1	2
16	4	3	2	3

The simulation analysis of the designed 16 sets of test scenarios was performed using ANSYS software, and the results of the maximum stress in the solder joints were obtained as shown in Table 4.

TABLE IV.   ORTHOGONAL DESIGN TABLE

No.	Solder joint material (mm)	Solder joint Height (mm)	Pad radius (mm)	PCB board thickness	Maximum Stress (Gpa)
1	SAC305	0.038	0.02	0.08	23.6
2	SAC305	0.04	0.018	0.08	27.8
3	SAC305	0.04	0.018	0.12	47.4
4	SAC305	0.042	0.022	0.1	32.2
5	SAC387	0.038	0.018	0.12	36.3
6	SAC387	0.04	0.022	0.08	30.1
7	SAC387	0.04	0.02	0.1	30.5
8	SAC387	0.042	0.018	0.08	27.1
9	63Sn37Pb	0.038	0.018	0.1	29.9
10	63Sn37Pb	0.04	0.02	0.08	27.8
11	63Sn37Pb	0.04	0.022	0.12	39.7
12	63Sn37Pb	0.042	0.018	0.08	24.4
13	62Sn36PbAg	0.038	0.022	0.08	24.0
14	62Sn36PbAg	0.04	0.018	0.1	30.5
15	62Sn36PbAg	0.04	0.018	0.08	26.5
16	62Sn36PbAg	0.042	0.02	0.12	41.5

B. Multi-factor analysis of variance based on SPSS

SPSS software is the world's earliest statistical analysis software, with powerful statistical analysis functions, and is widely used in various fields. ANOVA is a mathematical

978-1-6654-1392-3/21 $31.00 © 2021 IEEE

method that distinguishes differences in experimental results caused by changes in the level of factors from differences between experimental results caused by fluctuations in error. ANOVA can infer whether there is a significant difference between the samples, and if there is a significant difference, then the factor has a significant effect on the overall. It is worth noting that the test used a mixed orthogonal table and performed a multifactorial ANOVA. Multi-factor ANOVA is able to analyze the effects of multiple factors on the dependent variable, find the factors with significant effects, and then find the optimal combination in favor of the dependent variable. The results of the multi-factor ANOVA using SPSS on the data are shown in Table 5.

TABLE V. MULTI-FACTOR ANOVA RESULTS

Multi-factor ANOVA results					
	Sum of squares	df	Mean square	F	p
intercept	12558.270	1	12558.270	1687.916	0.000**
Solder joint material	13.342	3	4.447	0.598	0.639
Solder joint height	0.861	2	0.430	0.058	0.944
Pad radius	586.086	2	293.043	39.387	0.000**
PCB board thickness	44.601	2	22.300	2.997	0.125
Residuals	44.641	6	7.440		
R^2=0.935					
* p<0.05 ** p<0.01					

We used multi-factor ANOVA to investigate the relationship between the four factors of solder joint material, pad radius, solder joint height, and PCB board thickness on the maximum stress As shown in Table 5, the model R-squared value was 0.935, which means that solder joint material, pad radius, solder joint height, and PCB board thickness can explain 93.53% of the variation in the maximum stress in the solder joint. T The analysis shows that the PCB board thickness has a significant effect on the maximum stress value of the solder joint and there is a differential relationship (p<0.05). The material of the solder joint, the radius of the solder pad, and the height of the solder joint do not have a differential relationship on the maximum stress (p>0.05).

C. Optimal combination of influencing factors

After that, we analyzed the effect of four factors, namely, solder joint material, solder joint height, pad radius and PCB board thickness, on the maximum stress in the solder joint using one-way ANOVA, respectively.

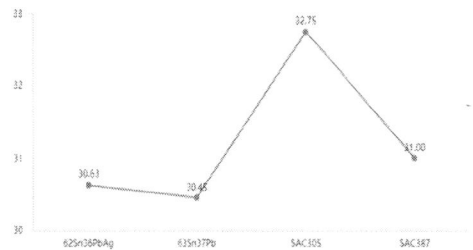

Fig. 5. Comparison of solder joint material and maximum stress variance analysis of solder joints

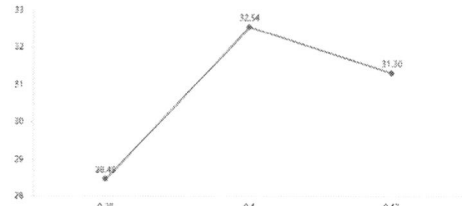

Fig. 6. Comparison of solder joint height and maximum stress variance analysis of solder joints

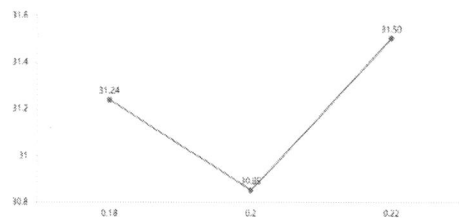

Fig. 7. Comparison of Pad Radius and Solder Joint Maximum Stress Variance Analysis

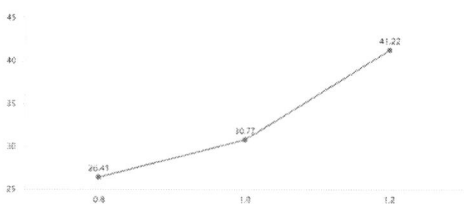

Fig. 8. Comparison of PCB board thickness and maximum stress variance analysis of solder joints

Figure 5, Figure 6, Figure 7, and Figure 8 show the comparison of the factors and the maximum stress variance analysis of the solder joints. As can be seen from the above figure, the maximum stress in the solder joint is minimized when the solder joint material is 63Sn7Pb, followed by 62Sn36PbAg, SAC387 and SAC305. The maximum stress in the solder joint is minimized when the solder joint height is 0.038mm, followed by 0.42mm and 0.4mm. The maximum stress in the solder joint is minimized when the pad radius is 0.02mm, followed by 0.18mm and 0.22mm. The maximum stress in the solder joint is minimized when the PCB board thickness is 0.8 mm, followed by 1.0 mm and 1.2 mm, respectively. The simulation results show that the maximum stress in the solder joint under this combination of parameters is 22.6 MP, which is significantly reduced compared to the maximum stress in the solder joint under other combinations of parameters.

IV. CONCLUSION

The finite element analysis of FBGA solder joints under torsional loading conditions on cantilever plates led to the following conclusions:

1) Conclusion 1: Under the cantilever beam torsional loading conditions, the stresses in the solder joints were unevenly distributed, with the maximum stress occurring at the outermost edge of the solder joint near the side of the applied load that contacts the PCB surface;.

2) Conclusion 2: At the 95% confidence level, the four factors of solder joint material, solder joint height, pad diameter and PCB board thickness, with the PCB board thickness having the most significant effect on the maximum stress in the solder joint After optimization of the given factor levels, the maximum stress at the solder joint is 22.6 MP when the solder joint material is 63Sn7Pb, the solder joint height is 0.038 mm, the pad radius is 0.02 mm, and the PCB board thickness is 0.8 mm.

REFERENCES

[1] Tai Zhou, "Research on the development trend of microelectronic packaging technology", Modern Information Technology, vol. 2(08), pp. 52–53, 2018.

[2] Liu Chen and Jiangang Zhang, "Reliability study of microfine pitch lead-free BGA hybrid solder joints," Electronic Process Technology, vol. 41(01), pp. 32-36 ,2020.

[3] Yudan Zhu, "Thermal reliability study of BGA solder joints," Shanghai Jiaotong University, 2017.

[4] Chen Lili, Li Siyang and Zhao Jinlin. "A review of BGA solder joint reliability research ," Electronic Quality, vol. 09, pp. 22–27, 2012.

[5] Jiaoying Huang, Yang Cao and Cheng Gao, "A review of fatigue failure studies of microelectronic packaging solder joints,". Electronic Components and Materials, vol.39(10), pp. 11-16+24, 2020.

[6] Ying Liang, Chunyue Huang, Rui Yin, Wei Huang, Tianming Li, and Hongwang Zhao, "Three-point bending stress-strain analysis of microscale ball grid array solder joints for microelectronic packaging," Mechanical Strength, vol. 38(04), pp. 744–748, 2016.

[7] Wu Liangyu, Qing Sun, Chenxi Shao, Chengbin Zhang, "Thermal stress characteristics of BGA packages under thermal cyclic loading ,"Thermal Science and Technology, vol. 17(06), pp. 509-516, 2018.

[8] Wenping Wang, "Four-factor mixed-level orthogonal experimental design," Science and Technology Information pp. 653, 2010(23).

The Influence and Optimization of Design Parameters on Integrated Circuits Package Warpage

Qiang Wei
R & D department
HuaTian Technology (Xi'an) Co.,Ltd
Xi'an, China
645088572@qq.com

Ning Sun
R & D department
HuaTian Technology (Xi'an) Co.,Ltd
Xi'an, China
Ning.Sun@ht-tech.com

Huan Yang
R & D department
HuaTian Technology (Xi'an) Co.,Ltd
Xi'an, China
Huan.Yang@ht-tech.com

Cao Ting
R & D department
HuaTian Technology (Xi'an) Co.,Ltd
Xi'an, China
Ting.Cao@ht-tech.com

Qu Fang
R & D department
HuaTian Technology (Xi'an) Co.,Ltd
Xi'an, China
Qu.Fang@ht-tech.com

XiaoJian Ma
R & D department
HuaTian Technology (Xi'an) Co.,Ltd
Xi'an, China
Xiaojian.Ma@ht-tech.com

Weidong Liu
R & D department
HuaTian Technology (Xi'an) Co.,Ltd
Xi'an, China
Wade.liu@ht-tech.com

Jie Liu
R & D department
HuaTian Technology (Xi'an) Co.,Ltd
Xi'an, China
Jie.Liu@ht-tech.com

Abstract—As an important factor affecting the reliability of integrated circuits (IC), warpage has attracted more and more attention. This article uses a combination of theoretical analysis and experimental verification. First, finite element method is used to analyze the impact of IC packaging materials on product warpage, and secondly, the simulation results are verified through experiments. The research results show that for smiling face products, reducing the Coefficient of Thermal Expansion (CTE) of the Epoxy Molding Compound (EMC) and increasing the CTE of the Core can effectively reduce product warpage. At the same time, reducing the thickness of the Core and increasing the thickness of the soldermask (SM) are also conducive to reducing warpage; for crying face products, the opposite is true. The above research results provide a theoretical basis for reducing product warpage in IC packaging.

Keywords—IC, warpage, smiling face, EMC, CTE, SM

I Introduction

Integrated circuits (IC) integration is further improved, die size continues to increase, and the ratio of die size to package size continues to increase, resulting in an increase in package warpage and a decline in product yield, which restricts the further development of IC. Yang Mengke [1] pointed out that most package forms have two causes of warpage: one is the warpage caused by incomplete curing during molding or the uneven shrinkage of plastic curing molding; The other is the warpage caused by the thermal mechanical stress imbalance of the packaging structure. Patel [2-6] and others pointed out in the article that different materials are used in packaging, such as die, substrate (SUB), epoxy molding compound (EMC), etc. These materials have different Coefficients of Thermal Expansion (CTEs). When the entire package undergoes temperature changes, due to the different CTEs of various materials, the shrinkage is inconsistent, resulting in warpage of the packaged product. Zhang Shiyuan [7] studied the influence of material geometry on warpage. The results show that the thinner the SUB thickness is, the greater the warpage of the

whole device is. However, with the increase of SUB thickness, the warpage direction turns downward, and the ratio of SUB to die height is between 1 and 2. There is a critical value, which makes the warpage minimum. However, no experiment has been carried out, The results need to be verified. Wu Xujie [8] studied the influence of printed circuit board (PCB) design parameters on its warpage. The research shows that the warpage value and stress of PCB decrease with the increase of SUB thickness, but the mechanism of warpage reduction is not pointed out. There are various types of IC packaging. In this paper, a typical three-dimensional model of FBGA package products is established. The influence of different design parameters on package warpage is studied through numerical simulation and experimental verification, and the influence law of design parameters on IC package warpage is obtained.

II Mechanical Modeling

A. Finite Element Model

Fine Pitch Ball Grid Array (FBGA) package is widely used in smart phones, storage devices and automotive electronics as a typical IC packaging form [9-11]. This article uses ANSYS to establish a three-dimensional model of four-layer FBGA products, including SUB, die attach film (DAF), die and EMC, the specific structure is shown in Fig. 1, the simulation model geometry is shown in Table I, and the SUB is shown in Table II.

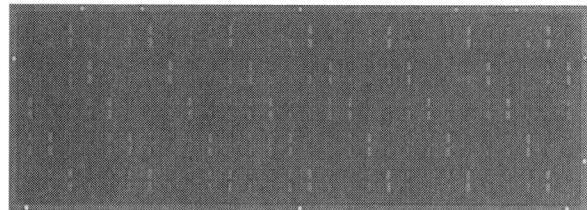

Fig. 1. FBGA package product model diagram

TABLE I. FBGA PACKAGE PRODUCT GEOMETRIC DIMENSIONS

SUB Size (mm)	EMC Size (mm)	Die Size (mm)	DAF Size (mm)
240*76.3*0.4	233*70.25*0.53	1.7*3.5*0.1086 1.7*3.5*0.1086	1.7*3.5*0.02 1.7*3.5*0.02

TABLE II. FOUR-LAYER SUB STRUCTURE PARAMETERS

Layer	Type	Material	Thickness (mm)	Copper content of ratio
SMT	Top soldermask	AUS-410	0.02	/
M1	Metal 1 (top)	Copper	0.02	0.7480
DIELECTRIC	Prepreg1	GHPL-830NS	0.04	/
M2	Metal 2 (Inner)	Copper	0.02	0.7724
DIELECTRIC	Core		0.2	/
M3	Metal 3 (Inner)	Copper	0.02	0.7524
DIELECTRIC	Prepreg2	GHPL-830NS	0.04	/
M4	Metal 4 (Bottom)	Copper	0.02	0.8453
SMB	Bottom soldermask	AUS-410	0.02	/

B. Material Parameters

The material parameters used in the simulation are shown in Table III. The material parameters of each material are determined by consulting the material manual. All materials adopt linear elastic material model.

TABLE III. MATERIAL ATTRIBUTE TABLE

Material type		Young's modulus (MPa)	Poisson's ratio	Tg (°C)	CTE (10⁻⁶/°C)
EMC	G760SWE	27300	0.3	160	10
	G311AC	23000	0.3	155	11.6
	G1250LKDS CL20	23000	0.3	145	13.6
	G760L	25000	0.3	150	15.83
	G1250LKDS U2M CL20	22000	0.3	137	17.6
	A381AAE	19000	0.3	160	18.9
Die	Silicon	131000	0.3	/	2.8
DAF	2100A	1517	0.3	60	48/140
SUB	SM: AUS-410	3200	0.3	105	60/150
	Metal: Copper	117000	0.3	/	17.3
	PP:GHPL-830NS	16000	0.18	255	13
	Core:HL832N SF-LCA	34000	0.18	300	3
	Core:HL832 NSF	34000	0.18	300	5
	Core:HL832 NS-LC	30000	0.18	255	7
	Core:HL832 NS	27000	0.18	255	10
	Core:HL832 NXA	28000	0.19	230	14
	Core:DS-7409HGB	25000	0.18	235	15.5
	Core:DS-7409HGB (GEQ)	23000	0.19	215	18.5

C. Loads and Boundary Conditions

A temperature body load of 175°C→25°C is applied to the FBGA package model. The package model uses three-dimensional solid elements. The meshing results are shown in Fig. 2.

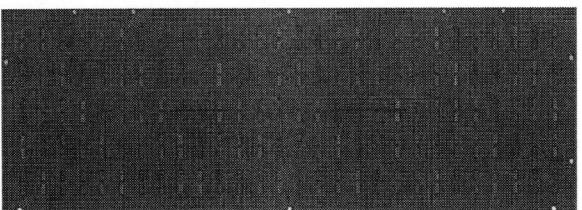

Fig. 2. Mesh division of FBGA package products (EMC not shown)

III FINITE ELEMENT SIMULATION

The CTE of the material has a higher effect on the warpage than the elastic modulus [1], so in this paper, the finite element simulation analysis mainly considers the influence of the material CTE on the warpage of the FBGA package. The warpage shape of the product is shown in Fig. 3.

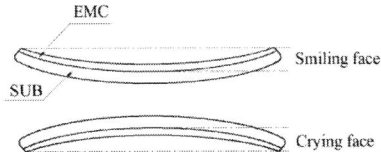

Fig. 3. Schematic diagram of warped shape

A. The Influence of Core's CTE on Warpage under Different EMC

When the SM thickness is 0.02mm and the Core thickness is 0.2mm, the effect of Core's CTE on the warpage of the packaged product under different EMC is studied, and the results are shown in Fig. 4.

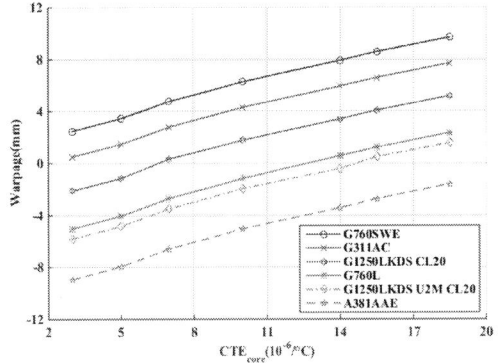

Fig. 4. The influence of Core's CTE on warpage under different EMC conditions

Fig. 4 shows that as the CTE of Core increases, the warpage of the packaged product first decreases and then increases. For smiley products, the Core's CTE should be as large as possible, and the shrinkage of the substrate should be larger, which is conducive to reducing the product Warpage; the smaller the EMC's CTE, the smaller the shrinkage of the EMC, and the smaller the warpage of the product. As the CTE of the EMC gradually decrease, the shrinkage of the EMC continues to decrease, the product will turn into a crying face. Continue to reduce the CTE of the EMC, the warpage will gradually increase.

B. The Influence of SM Thickness on Warpage under Different Core Thickness

G1250LKDS U2M CL20 is used as the EMC, and matched with the Core of HL832NS. The influence of the thickness of the SM on the product warpage was studied under different Core thicknesses. The results are shown in Fig. 5.

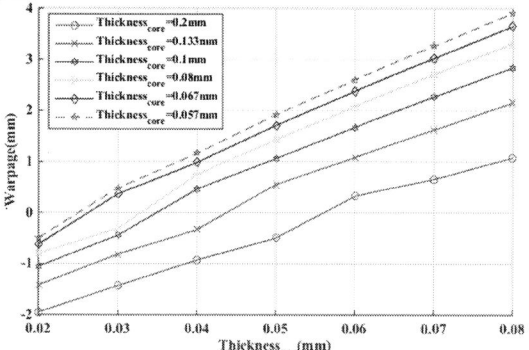

Fig. 5.   The effect of SM thickness on warpage with different Core thicknesses

It can be seen from Fig. 5 that as the thickness of the SM increases, the warpage of the packaged product decreases first and then increases. This is because for smiling face products, the thickness of the SM increases, the shrinkage of the SM increases, the SUB deforms more, and the product warpage decreases, while the crying face products are just the opposite. At the same time, as the thickness of the Core decreases, the warpage of the packaged product gradually decreases. This is because the Core material is reduced in thickness, the Core material plays a smaller role, while the shrinkage of the metal layer and the insulating layer increases, and the product shrinks more downwards, so the product warpage is reduced.

IV EXPERIMENTAL ANALYSIS

In order to verify the accuracy of the simulation results, this article mainly chooses the CTE of the EMC, the CTE of the Core, the Core thickness, and the SM thickness as the control factors for the experiment.

A. Warpage Detection

1) Measuring tool: steel ruler or vernier caliper.

2) Measuring method: place the product plastic package face up on the prescribed measuring warping platform, the vernier caliper depth gauge is perpendicular to the measuring platform, make sure that the depth gauge and the measuring platform are in close contact (see Fig. 6 below), so that the bottom of the main ruler is in contact with the measuring platform. The highest point of the product measuring point is parallel, and read the displayed value.

Fig. 6.   Warpage measurement method

a) Smiling face warped: EMC upwards, measure 4 points (see Fig. 7 below);

Fig. 7.   Smiling face warpage measurement method

b) Crying face warped: EMC upwards, measure 2 points (see Fig. 8 below).

Fig. 8.   Crying face warpage measurement method

B. Analysis of the Influence of the EMC's CTE on Warpage

When the Core material is HL832NS, the thickness is 0.2mm, and the SM thickness is 0.02mm, the warpage of the product with different EMC is studied, and the results are shown in Fig. 9.

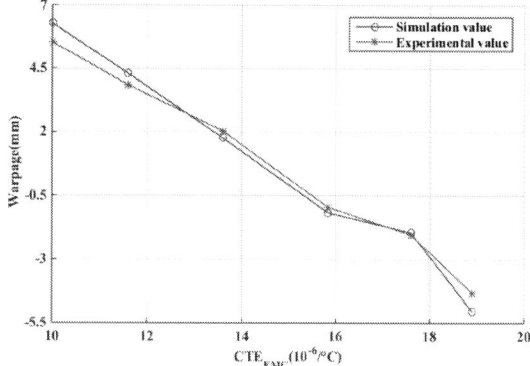

Fig. 9.   The influence of EMC's CTE on warpage

Fig. 9 illustrates the relationship between the CTE of the EMC and the warpage of the FBGA package product. As the CTE of the EMC increases, the product warpage gradually decreases.

C. Analysis of the Impact of the Core's CTE on Warpage

When the EMC material is G1250LKDS U2M CL20, the SM thickness is 0.02mm, and the Core thickness is

0.2mm, the warpage law of the product under different Core is studied, as shown in Fig. 10.

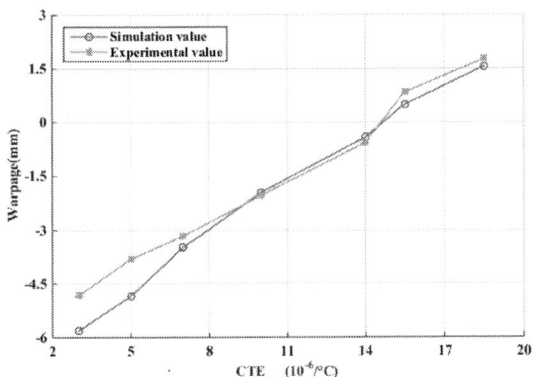

Fig. 10. The influence of Core's CTE on warpage

In the FBGA package, the warpage of the product is related to the CTE of the Core material. As shown in Fig. 10, other conditions being equal, when the CTE of the Core is increased, the crying face warpage of the product also increase.

D. Analysis of the Influence of Core Material Thickness on Warpage

Select the EMC of G1250LKDS U2M CL20, the Core of HL832NS, SM thickness is 0.02mm, and study the influence of Core thickness on warpage, as shown in Fig. 11.

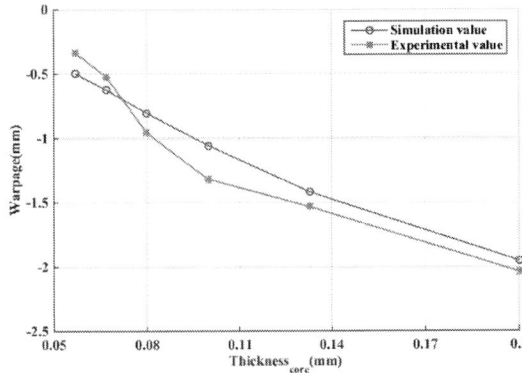

Fig. 11. The influence of Core material thickness on warpage

In Fig. 11, as the thickness of the Core increases, the warpage of the product first decreases and then increases, so the thickness of the Core is an important factor.

E. Analysis of the Influence of SM Thickness on Warpage

Select the EMC of G1250LKDS U2M CL20, the Core of HL832NS, and thickness is 0.2mm, and study the warpage of the product under different SM thicknesses, as shown in Fig. 12.

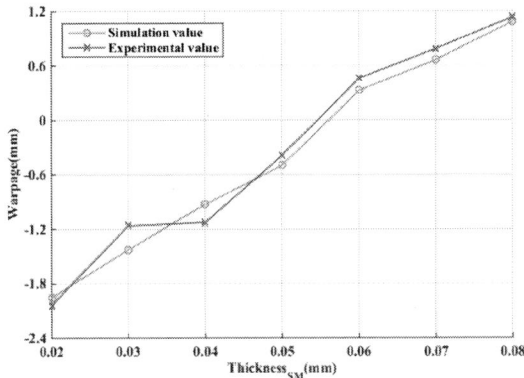

Fig. 12. The influence of SM thickness on warpage

During the experiment, it was found that the thickness of SM has a greater impact on product warpage. As shown in Fig. 12, as the thickness of SM increases, the warpage of the product first decreases and then increases. Therefore, when selecting package design parameters, SM has a greater impact on product warpage. The impact of package reliability is relatively larger.

V CONCLUSION

In this paper, the influence of EMC's CTE, Core thickness, CTE and SM thickness on the warpage of IC package is studied by the method of finite element simulation and experimental verification. The results provide a basis for further packaging parameter design. The conclusions are as follows:

1) For smiling face products: EMC with low CTE and Core with high CTE should be used to reduce the warpage. At the same time, reducing the thickness of Core and increasing the thickness of SM are also conducive to controlling product warpage;

2) For crying face products: the combination of Core with low CTE and EMC with high CTE should be adopted. Increasing the thickness of Core and decreasing the thickness of SM can reduce the warpage after packaging.

REFERENCES

[1] Yang M k, "Research on causes and control methods of WLCSP package wafer warpage," Beijing University of Technology, 2019.

[2] Patel, Kewal, Lall, "Model for inverse determination of process and material parameters for control of package-on-package warpage," IEEE Transactions on Components, Packaging and Manufacturing Technology, 2015, vol. 5, Issue 9, pp. 1358-1375.

[3] Wang M, "Substrate trace modeling for package warpage simulation," Electronic Components & Technology Conference. IEEE, 2016.

[4] Chiu T C, Huang D Y, Lee B S, "Development of a consistent multiaxial viscoelastic model for package warpage simulation," Proceedings Electronic Components & Technology Conference, 2015, pp. 373-379.

[5] Zang M X, Wang T, Liu C, "Optimization design of stack package warpage based on moldflow software," Advanced Materials Research, 2013, pp. 34-37.

[6] Pham V L, Wang H, Xu J, "A study of substrate models and Its effect on package warpage prediction," 2019 IEEE 69th Electronic Components and Technology Conference. IEEE, 2019.

[7] Zhang S Y, Zheng B L, "The influence of material geometry in electronic packaging on warpage," J. Chinese Quarterly of Mechanics, 2005, vol. 26, Issue 3, pp. 506-510.

[8] Wu X J, Niu Z R, Liu K, "Analysis of the influence of printed circuit board design parameters on its warpage," Journal of Hefei University of Technology: Natural Science Edition, 2013, vol. 8, pp. 954-958.

[9] Lin T, Chen C, "Simulation and validation of FBGA package warpage by considering viscoelastic effect," Semiconductor Technology, 2015.

[10] Kang J S, "Parametric study of warpage in PBGA packages," The International Journal of Advanced Manufacturing Technology, 2020, Issue 107, vol. 4.

[11] Liu P, Tong L, Shen H, "Warpage analysis of FBGA packaging during assembling process," Electronic Components and Materials, 2014.

The In-Situ Observation of microstructure, grain orientation evolution and its effect on crack propagation path in SAC305 under extreme temperature changes

KeXin Xu
Harbin Institute of Technology
Shenzhen, China
Science and Technology on Reliability Physics and Application of Electronic Component Laboratory
Guangzhou , China
19s155065@stu.hit.edu.cn

Xing Fu *
Department of Reliability Design Research
China Science and Technology on Reliability Physics and Application of Electronic Component Laboratory
Guangzhou , China
fuxing@ceprei.com

Min Liu
Department of Materials Science and Engineering
Harbin Institute of Technology
Shenzhen, China

ZhiWei Fu
Department of Reliability Design Research
China Science and Technology on Reliability Physics and Application of Electronic Component Laboratory
Guangzhou , China

Si Chen
Department of Reliability Design Research
China Science and Technology on Reliability Physics and Application of Electronic Component Laboratory
Guangzhou , China

YiJun Shi
Department of Reliability Design Research
China Science and Technology on Reliability Physics and Application of Electronic Component Laboratory
Guangzhou, China

Yun Huang
Department of Reliability Design Research
China Science and Technology on Reliability Physics and Application of Electronic Component Laboratory
Guangzhou , China

HongTao Chen
Department of Materials Science and Engineering
Harbin Institute of Technology
Shenzhen, China

Abstract—Microstructures and grain structure evolution of solder joints under extreme temperature changes considerably affects its thermomechanical response and lifetime. Thermal shock tests were conducted to study the thermomechanical responses in Sn-3.0Ag-0.5Cu BGA solder bumps. In this work, the in-situ observation was carried out by EBSD and SEM to examine the microstructure and grain structure evolution during thermal shock tests, which provides the accuracy and continuity of the observation. For the Sn-based solder joint containing limited grains, early failure always occurs in certain solder joints where orientation of grain in contact with substrates exhibited relatively high CTE values parallel to the substrates. Elaborated examination of different stages of recrystallization in the same bump identified localized recrystallization occurred at the place with high CTE mismatch. Moreover, the grain structure had changed from cyclic twin to small random oriented grains, which provides additional degradation mechanism. The grain orientation was critical to reliability and lifetime. Inherent anisotropic properties of β-Sn grain play a vital effect on thermomechanical responses, leading to a change in the initiation propagation path of crack and corresponding reliability problems.

Keywords—Microstructure, Orientation, Extreme temperature, Thermomechanical fatigue

I. INTRODUCTION

As further deep-space exploration requirements of spacecraft continue to increase, the electronic components will work in more harsh and more complex environment in space, such as under the conditions of Mars(133-293K) and Moon(77k-423k), where the fastest temperature change rate can reach 2°C/min[1,2]. Solder joints work as the part to connect the Si die and outside circuit and help heat dissipation, its performance is important to the whole reliability. Sn-rich solder alloys which serve as structure materials of joints contain Sn content over 95 wt.%, resulting in the great dependence of thermomechanical responses on the performance of Sn[3]. β-Sn grain exist in form of body-centered tetragonal (a=b=5.632Å, c=3.182 Å), which displays anisotropic properties in physical, chemical, thermal and electrical performance[4,5]; the linear CTE along c-axis is about twice that along the other axes[6]. Thus, the orientation of Sn grain is critical to the overall solder joint reliability under extreme temperature changes.

Chen et al[7] investigated the change of grain orientation of different solders at thermal cycle test between 0 °C and 100 °C. After 1400 cycles, local recrystallization appeared near the side of PCB. After 3000 cycles, the recrystallization trend increased obviously. IMC particles dispersed in the solder began to aggregate and grow, and tended to distribute at the recrystallization boundary. Li et al[8] studied the microstructure evolution and failure mechanism of PBAG solder bumps under thermal cycling. They estimated that the dispersion distribution of Ag_3Sn is coarsened and the Pb phase is enriched after thermal cycling, its average size has increased by 3.5 times. However, few studies are devoted on the thermomechanical response of Pd-free solder bumps under extreme temperature changes, rather on the temperature of range between -55 ℃ - 125 ℃ . To meet the changing environment of deep space exploration, NASA has established a temperature range of -185 °C to 125 °C to cover all future deep space exploration needs. Under extreme temperature changes, the CTE mismatch in the interface of different materials and between lattices in solder joints is easy to result in thermal mismatch, which leads to the crack initiation in solder joints. For Sn-based solder joints, grain structure has a great effect on the mechanical properties. Thus, a detailed study on the effect of grain structure on the lifetime and failure modes of micro-bumps would be essential and valuable.

Moreover, in-situ observation were seldom utilized in the past studies, which could failure to record microstructure and grain structure continuously and accurately. In this paper, SAC305 solder joints were experimented at thermal shock test between -196 °C and 150 °C. And, the evolution of microstructure, crystal lattice structure of solder bumps were characterized by in-situ SEM and EBSD. Based on these experimental observations, the effect of grain orientation on the thermomechanical response and failure modes of SAC305 solder interconnects was analyzed.

II. EXPERIMENT

A. Sample preparation

As is shown in figure1, all of the solder bumps were linked in a daisy-chain loop with the pre-tested diameter of 300μm. The cross-sections of bumps were obtained by mounted with epoxy, ground, and polished for in-site observations. Ion milling was carried out with the 6 kV and 5 kV beam for 40 min and 20min to removal of surface internal stress for getting better EBSD patterns.

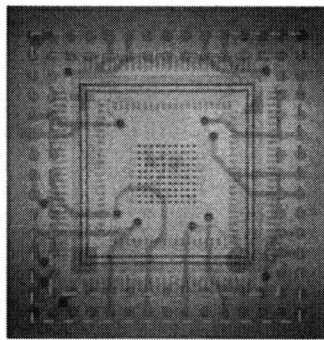

Fig. 1.   X-ray transimission image of component

B. Thermal shock test

The thermal shock tests were carried out in the range of temperature from -196℃ to 120℃ by converting the BGA assemblies between liquid nitrogen container and constant temperature furnace. The temperature ramp rate was 45℃/min and the dwelling time was 20 min at both low temperature and high temperature. After every 100 cycles, SEM and EBSD were used to characterize microstructure and grain structure in thermal-shocked solder joints.

C. Microstructure examination and failure analysis

The SEM and EBSD were used to investigate the microstructure and grain structure, respectively. The scanned data from EBSD which contained grain information was analyzed by Channel 5 software. Due to intrinsic anisotropic properties of Sn grain, it is essential to determine the relationship between different grain orientation. Therefore, what is the most important among the information obtained by Channel 5 software is the angles between c-axes of different Sn grains or the c-axes and substrate.

The orientation of β-Sn grain can be obtained by processing the scanned data from EBSD. θ represents the angle between the c-axes of adjacent grain, which was calculated by the matrix g to transfer Euler angle of the crystal orientation (Φ, φ1, φ2) to the sample stage Cartesian coordinate system.

III. RESULTS AND DISCUSSION

A. Facilitated Cracking by Anisotropic Thermomechanical Responses of Differently Oriented Grains

Before and after thermal shock test, the in-situ SEM and EBSD observation were conducted. Figure 2 showed the initial EBSD of the 1-6 bump and corresponding cross-sectional SEM after 50 cycles of thermal shock tests. It could be found that crack initiation appeared in two polycrystalline solder joints (3 # and 5 #). The corresponding EBSD map showed that the crack growth places were all at high-angle grain boundaries, which was mainly due to the different grain orientations of the two neighboring grains. The anisotropy of β-Sn led to the thermal stress misfit between grains. Besides the internal stress in the high angle boundary, high-angle boundary is considered preferred for grain boundary sliding. Thus, it could be a potential cracking site when internal stress concentrated in the high angle boundary, which provided additional degradation. Table 1-1 shows the CTE of the different materials at the interface. It can be concluded that CTE of c axes of Sn are twice that of a axis (or the b axis) and the CTE of the a axis (or b axis) is almost equal to that of the Cu.

On the other hand, the interfaces between the solder joints and the substrates were also the main location for crack initiation. It was noteworthy that certain solder joints failed earlier than neighboring joints in the interfaces between Sn matrix and substrates. The crack initiation at the interface was mainly related to the CTE between the adjacent materials. For the solder, the CTE was closely related to the grain orientation of Sn. For the bump 1#and2#, the angles between the Cu substrate and Sn grain at the upper interface were 71 ° and 72 ° respectively, and no crack or extrusions were found at the upper interface. It can be seen that two large grains consisted of the 3 # bump, the θ on the left grain was 6 °, and the θ on the right grain was 46 °. Accordingly, crack occurred on the left grain of smaller θ. The same phenomenon happened at the 4 # bump with θ of 14 ° on the left grain and the 6 # bump containing just single grain θ of 13 °. For the 5 # bump, the θ of the Sn grain adjacent to the upper interface was 0.8 °, which mean the c axis was almost parallel to the Cu substrate, and the CTE mismatch was almost maximum. Consequently, cracks at the upper interface of the bump were pronounced than at other bumps, as shown in figure 2-6(i).

The influence of grain orientation on failure modes was analyzed. An analysis of damaged joints through the superimposed crystal lattices in the EBSD maps showed for the Sn-based solder joint containing limited grains, early failure always occurs in certain solder joints where orientation of grain in contact with substrates exhibited relatively high CTE values parallel to the substrates. Maximal mismatch in shear produced by the great differences of CTE between β-Sn along c-axis and substrates at the joint interface was confined. Especially, when the angles between c axes of Sn grain and substrates were below 15°, the solder joints experienced serious thermomechanical fatigue and corresponding early failure. In addition, crack occurred in the Sn bulk earlier than at the interface for certain solder joints. For explaining the effect of grain boundaries on crack initiation, the misorientation angles of the adjoining grains were measured. We noticed that in damaged solder joints, the grain boundaries with adjoining grains exhibiting over 35° misorientation can commonly act as a crack propagation path in the Sn matrix. Furthermore, under thermal shock tests, the grain structure changed owing to thermal stress. Thus, the next part is to focus on the evolution of grain structure.

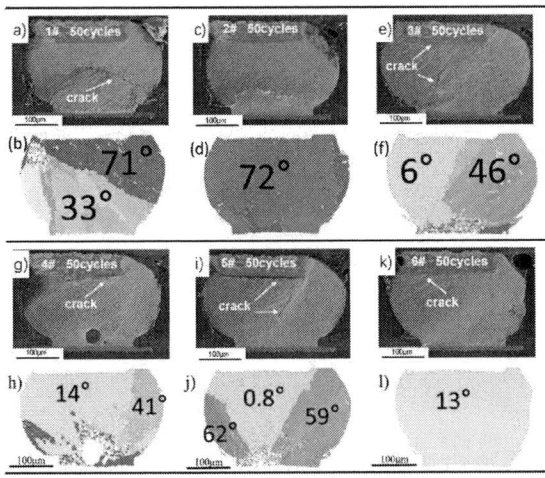

Fig. 2. 1 # ~ 6 # bump after 50 cycles of SEM and corresponding initial EBSD:a-b) : 1 # ; c-d): 2 #; e-f) : 3# g-h) : 4# i-j) :5# k-l) : 6 #

TABLE I. CTE VALUES FOR MATERIALS AT INTERFACE

Axis	β-Sn (°C⁻¹)	Cu (°C⁻¹)
a	14.195	16.5
b	14.195	16.5
c	28.662	16.5

B. Localized Recrystallization Induced and Cyclic Twin structure dispearrance

From the initial EBSD map shown in Fig. 3a and the pole figure of {001}, three predominant orientations of grains were observed in 7#bump. After 100 cycles of thermal shock test, the grain structure of the bump changed obviously, the original pink grain evolved into green grain and blue grain, and a lot of low-angle grain boundaries appeared in the original blue grain. After 200 cycles, the small-angle grain boundary in the bump gradually changed to the large-angle grain boundary (the angle difference is more than 15 °. Combined with SEM and EBSD images, the initiation and propagation of cracks occurred at the localized recrystallization region where original grains evolved from

limited predominant orientations to the more random orientations.

After 100 cycles of thermal shock test, the extrusions occurred obviously, located exactly at the grain boundary of the two original grains according to the EBSD map. It was mainly due to the thermal stress mismatch caused by the different orientation of the two adjacent grains. After 200 cycles, the crack propagation occurred in the recrystallization area, which was connected with the crack in the corner and runs through the whole bump, which showed that the recrystallization provided favorable conditions for the crack initiation and propagation. And a large number of intergranular cracks were produced on the surface of the bump. According to the EBSD figure, these intergranular cracks were generated along the newly formed small-angle grain boundary, the evolution of grain structure was favorable for crack initiation and propagation.

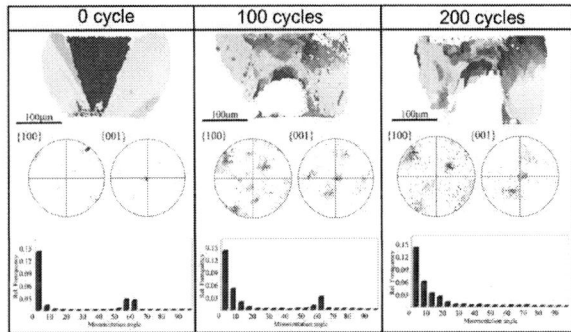

Fig. 3. EBSD orientation map of 7# bump after thermal cycles under extreme temperature changes a)0 cycle; b) after 100 thermal cycles; c) after 200 thermal cycles

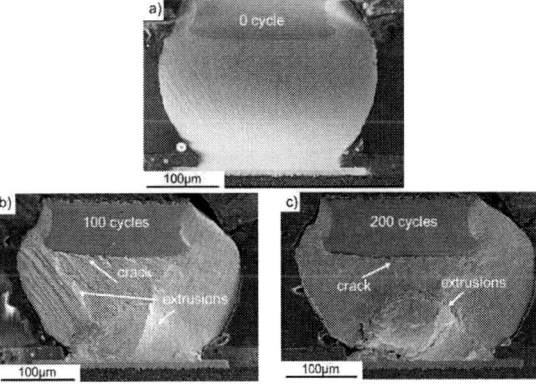

Fig. 4. SEM image of 7# bump after thermal cycles under extreme temperature changes a)0 cycle; b) after 100 thermal cycles; c) after 200 thermal cycles

Cyclic-twin grain structure were also found in as-reflowed bump. As shown in the initial EBSD figures 3 and 5, the two pre-test grain structures had most misorientation at 55°–65° in common. It has been reported that the θ of cyclic-twin grains is 57.2° and 62.8° happening on {101} and {301} crystal plane, respectively [9]. According to the pole figure of {001} plane, only three grains consisted of the 8# bump. After 100 cycles of thermal shock test, the grain structure changed obviously. As shown in Fig. 5(b), the Cyclic-twin grain near PCB side changed into a pale yellow grain, the cyclic twins near the chip and in the center of the bump were replaced by a large number of small grains, most of which are with

978-1-6654-1392-3/21 $31.00 © 2021 IEEE

misorientation between 5 ° and 20 °. From the pole figure of {001} and {100} planes, it can be seen that the three main grains which originally formed the solder joint have been replaced by the random small grains after recrystallization. After 200 cycles, the Cyclic-twin grain structure had basically disappeared. As shown in Fig. 5-11(c), some small angle grain boundaries have been changed into large angle grain boundaries larger than 15 °. The polar graphs of {001} and {100} also became more dispersed, indicating the continuous evolution of lattice structure. In addition, the orientation difference distribution before and after thermal shock shows that the grain boundary proportion of 55-65 ° decreases gradually, and the grain boundary proportion of small angle increases gradually, which also indicates that the twin structure disappears gradually during temperature shock, and with the occurrence of recovery recrystallization. After 100 cycles, the relative frequency of twin-type (57° to 63°) misorientations decreased from 0.081 and 0.089 to 0.046 and 0.061, respectively. Then, after another 100 cycles, the relative frequency of twin-type (57° to 63°) misorientations decreased to 0.037 and 0.039.

Fig. 6(a-c) showed the microstructure of the bump#8 after 0,100 and 200 cycles. After 100 cycles of thermal shock, no obvious deformation and cracks were found, the fatigue property of cyclic-twin grain structure was relatively higher, which was consistent with the research results [10]. However, after 200 cycles of thermal shock, there was obvious crack growth at the lower interface. It was found that the grain structure of this region had changed from cyclic twin to small random grain which had bad fatigue property which provided additional degradation mechanism.

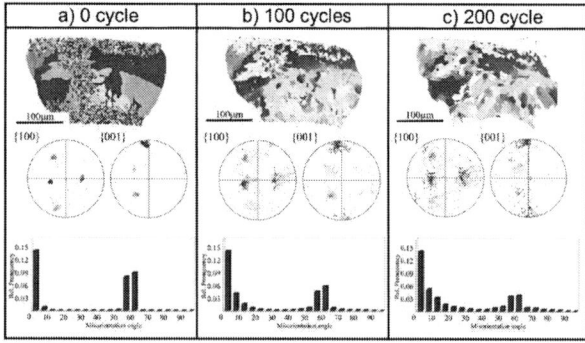

Fig. 5. EBSD orientation map of 8# bump after thermal cycles under extreme temperature changes a)0 cycle; b) after 100 thermal cycles; c) after 200 thermal cycles

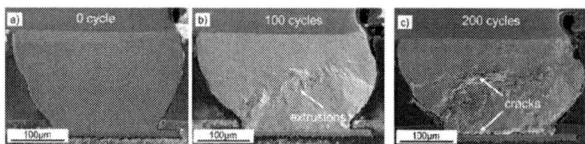

Fig. 6. SEM image of 8# bump after thermal cycles under extreme temperature changes a)0 cycle; b) after 100 thermal shock cycles; c) after 200 thermal cycles conclusion

IV. CONCLUSION

The microstructure, grain orientation play a vital role on initiation and propagation path of cracks. For the Sn-based solder interconnects containing limited grains, early failure always occurs in certain solder joints where orientation of grain in contact with substrates exhibited relatively high CTE values parallel to the substrates. Due to the intrinsic anisotropic thermomechanical responses of the differently oriented grains, the grain boundaries with adjoining grains exhibiting over 35° misorientation can commonly act as a crack propagation path in the Sn matrix. Under extreme temperature changes, the relative frequency of twin-type (57° to 63°) misorientations in Sn-based solder joint decreased approximately by 50%.

ACKNOWLEDGMENT

This work is financially supported by National Natural Science Foundation of China Youth Foundation (No.62004046), the foundation of Science and Technology on Reliability Physics and Application Technology of Electronic Component Laboratory (No. 61428060102-1and No.61428060201) and Development Fund of China Electronic Product Reliability and Environmental Testing Research Institute (No. 19D06).

REFERENCES

[1] Tian, R., Hang, C., Tian, Y., & Xu, J. "Brittle Fracture of Sn-37Pb Solder Joints Induced by Enhanced Intermetallic Compound Growth under Extreme Temperature Changes." Journal of Materials Processing Technology, vol. 268, 2019, pp. 1–9.

[2] Fu, X., En, Y., Zhou, B., Chen, S., Huang, Y., He, X., Yao, R. "Microstructure and Grain Orientation Evolution in SnPb/SnAgCu Interconnects Under Electrical Current Stressing at Cryogenic Temperature." Materials, vol. 12, no. 10, 2019, p. 1593.

[3] Han, J., Guo, F., & Liu, J. "Effects of Anisotropy of Tin on Grain Orientation Evolution in Pb-Free Solder Joints under Thermomechanical Stress." Journal of Materials Science: Materials in Electronics, vol. 28, no. 9, 2017, pp. 6572–6582.

[4] Gu T H, Xu Y L, Gourlay C M, Ben B T. In-Situ Electron Backscatter Diffraction of Thermal Cycling in A Single Grain Cu/Sn-3.0Ag-0.5Cu/Cu Solder Joint[J]. Scripta Materialia, 2020, 175: 55-60.

[5] Xian J W, Zeng G, Belyakov, Gu Q, Nogita K, Gourlay C M. Anisotropic Thermal Expansion of Ni3Sn4, Ag3Sn, Cu6Sn5, and β-Sn[J]. Intermatallics, 2017, 91: 50-64.

[6] Chu Y C, Chen C, Kao N. Effect of Sn Grain Orientation and Strain Distribution in 20-μm-Diameter Microbumps on Crack Formation Under Thermal Cycling Tests[J]. Electronic Materials Letters, 2017, 13(6): 457-462.

[7] Chen, Hongtao, et al. "Localized Recrystallization Induced by Subgrain Rotation in Sn-3.0Ag-0.5Cu Ball Grid Array Solder Interconnects During Thermal Cycling." Journal of Electronic Materials, vol. 40, no. 12, 2011, pp. 2470–2479.

[8] Li, Q., Li, C.-F., Zhang, W., Chen, W., & Liu, Z.-Q. "Microstructural Evolution and Failure Mechanism of 62Sn36Pb2Ag/Cu Solder Joint during Thermal Cycling." Microelectronics Reliability, vol. 99, no. 99, 2019, pp. 12–18.

[9] Lehman, L. P., Xing, Y., Bieler, T. R., & Cotts, E. J. "Cyclic Twin Nucleation in Tin-Based Solder Alloys." Acta Materialia, vol. 58, no. 10, 2010, pp. 3546–3556.

[10] Arfaei, B., Kim, N., & Cotts, E. J. ("Dependence of Sn Grain Morphology of Sn-Ag-Cu Solder on Solidification Temperature." Journal of Electronic Materials, vol. 41, no. 2, 2012, pp. 362–374.

A Vertical Transmission Leadless Surface-mountable Ceramic Package with High Core Proportion

Zhizhuang Qiao
The 13th research institute
CETC
Shijiazhuang, China
qiaozhizhuang@126.com

Linjie Liu
The 13th research institute
CETC
Shijiazhuang, China
chianpackage_rdc@163.com

Yangfan Zhou
The 13th research institute
CETC
Shijiazhuang, China
1017315424@qq.com

Ke Wang
The 13th research institute
CETC
Shijiazhuang, China
Laowu20@126.com

Abstract—This paper presents a 40GHz vertical transmission leadless surface-mountable ceramic package with high core proportion, realized in high temperature co-fired ceramics (HTCC) technologies. The RF transmission terminal of the package adopt vertical hollow via transmission, which can improve the core proportion. Compared with solid transmission, core area can be increased by 25%. The RF transmission structure includes ground coplanar waveguide line-hollow via-ground coplanar waveguide line. By optimizing the discontinuous structures, the application frequency of the package can reach 40GHz, and the board level measurement insertion loss of two ports is less than 1.5dB. At the same time, the package has passed the environmental and mechanical tests. In addition, when the package is applied in the system, it needs to be attached to the PCB. Due to the large difference in thermal expansion coefficient between ceramics and PCB, the reliability risk of board-level installation caused by thermal mismatch between materials is analyzed. And the board level installation was verified without failure. The package shows excellent high frequency transmission characteristics and high reliability.

Keywords—Surface mount; Leadless; Core proportion; Co-fired Ceramic package;

I. INTRODUCTION

With the rapid development of Internet and Mobile Communication, high-frequency and high-speed transmission system requires higher transmission rate. Millimeter wave systems play an important role in new fields such as Wireless LAN, point to point communications, and Wireless Lan. Single Chip Microwave Integrated Circuit (MMIC) is the key active device of this millimeter wave system, and the package is faced with the requirements of high reliability, miniaturization, high core-to-core ratio, low cost, large-scale production and high frequency signal transmission. In recent years, some scholars have studied the vertical transmission structure in millimeter wave field, most of which are based on low temperature co-fired ceramic (LTCC) technology[1-2].

In order to achieve low cost and mass production and meet the high frequency performance requirements of millimeter wave applications, a novel surface mount ceramic package based on high temperature co-fired ceramics (HTCC) is proposed in this paper. Solid vertical transmission hole is used in traditional structure[3]. In order to achieve a larger "core-to-core ratio" , the hollow vertical transmission structure is adopted. In this paper, the input and output ports of the package are grounded coplanar waveguide structure, and the vertical transmission through holes are hollow metal through holes structure, so as to increase the "core ratio" of the package. This paper describes a method to extend the transmission frequency of the case based on HTCC to 40GHz, and verifies the reliability of the free state of the case and the plate level reliability of the secondary installation.

The RF performance of the transition structure is simulated with HFSS software, and the stress distribution is analyzed by the finite element analysis software. The packages were manufactured in a standard HTCC process with screen printed conductor lines. A 90% alumina tape system with tungsten metallization was employed. All the vias were punched with a diameters of 120μm. A relative permittivity value of 9.8 and a loss tangent of 0.003 were used in the simulations. The packages compose of metal base, ceramic transmission structure. Considering the RF performance and reliability, the vias diameter, vias layout, the metal material and so on are optimized and modified. The base material of this package is chosen to Molybdenum copper alloy(Mo-Cu alloy).

To validate the performance of the optimized high speed differential structure, the packages were measured using the Agilent Network Analyzer and probe station with 250 μm pitch GSG air coplanar probes. Furthermore, the reliability tests in free state and secondary installation board level were conducted, which verify the reliability of this package.

II. HIGH FREQUENCY PACKAGE DESIGN RELIABILITY

A. RF Design

When the package is applied, the internal chip is bonded from the chip bonding point to the package bonding finger by a gold wire bonding method to realize the interconnection between the chip and the package level. Then the assembled package is affixed to the printed circuit board, and the interconnect between the packaged device and the board level has realized. Fig. 1 shows the vertical transmission structure of the hollow through-hole.

Fig. 1 Schematic diagram of vertical transmission structure of hollow hole

The simulation model of the RF transmission structure is built using the 3D electromagnetic simulation software HFSS based on the finite-difference time-domain method algorithm. As shown in figure 2, the model includes bonding wire, printed circuit board, ceramic transmission terminal and transmission line. The ground coplanar waveguide structure is used for the signal transmission line, which is convenient for the probe test. The width of coplanar waveguides, the gap between the surface and the ground, the diameter of the holes, and the coplanar waveguides at the welding of the back and the board are optimized by using HFSS software. The wire width of RF signal is widened to compensate the inductance effect of bonding wire, because the inductance effect is introduced at the bonding wire, which leads to impedance mismatch. The transmission structure with excellent performance and minimum transmission loss is obtained.

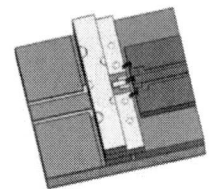

Fig. 2 The diagram of HFSS simulation

The high-frequency transmission structure is applied to the design of the package structure. This paper presents a leadless surface mount package with a 7mm and 7mm square structure, which consists of three layers of alumina ceramic structure, shown in figure 3. There are two RF ports and eight DC ports at the output end of the package. The size of the chip mounting area is 5mm and 5mm, and the "core-to-core ratio" is over 70%. Compared with solid transmission, core area can be increased by 25%. The package samples were prepared by multi-layer high temperature co-firing alumina ceramics and tungsten metallization system. The pattern was printed by screen-printing process with tungsten-based paste, and then nickel-gold was plated on the tungsten-based paste.

Fig. 3 The Photograph of package

III. FABRICATION AND MEASURMENTS

Using Rogers RO4350B with a PCB thickness of 0.254 mm, the lead tin solder paste is printed on the PCB during assembly, then the package is mounted on the printed circuit board and placed in a reflow oven in nitrogen atmosphere to weld the package together with the PCB. The chip mounting area of the package is assembled with 4.9mm long microstrip line, and the interconnection between the two RF ports is realized by gold wire bonding. Vector network analyzer and the probe station are used to test the assembled package, with the 500-micron pitch probes. The S-parameter test curve of the sample is shown in figure 4. The measured dual-port board level results exhibit return and insertion loss values better than 15dB and 1.5dB, respectively, up to 40GHz.

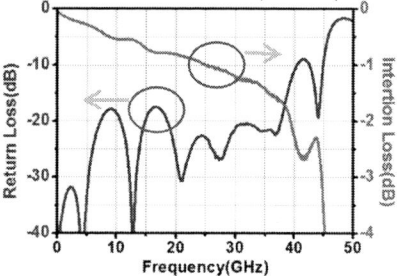

Fig. 4 Measured results of the S-parameter

It can be seen that the package of the hollow vertical transmission structure has excellent transmission characteristics.

IV. RELIABILITY

A. The Reliability Verification of the Package in Free State

The package samples were taken at random, and the samples were melted and sealed with gold-tin Alloy by metal cover plate. After sealing, they were verified by environmental test and mechanical test, including temperature cycling, thermal shock, mechanical vibration and constant acceleration test. The airtightness test and the appearance test are used as the criteria for the failure of the package. The airtightness criterion is less than $1.0 \times 10^{-3}(Pa \cdot cm^3)/s$. The test conditions and results are shown in Table I.

TABLE I   RELIABILITY TEST RESULTS

Test Item	Condition	Result
Temperature Cycle	-65°C to +175°C, 500cycle	0/50
Thermal Shock	-65°C to +150°C, 25cycle	0/50
Mechanical Shock	1500g, Y1 direction	0/50
Constant Acceleration	30000g,Y1 direction	0/50

B. Secondary Installation Board level Reliability Verification

The thermal cycling of the device will produce greater stress, if the materials do not match the coefficient of thermal expansion will cause the thermal fatigue failure of the microelectronic circuits and devices[4]. The case is made of alumina ceramics with a coefficient of thermal expansion value of between 7 and 8×10^{-6} /K, whereas the CTE value of the commonly used industrial printed circuit board material is as high as 15 to 21×10^{-6} /K, on the one hand, a large residual stress is produced in the reflow process, and on the other hand, the thermal cycling stress and reflow residual stress are superimposed during the subsequent temperature loading process, which will lead to accelerated failure of the product, significant reduction in thermal fatigue life of solder joints[5].

The melting point of solder is 183 ℃, which is a low melting point metal. Its deformation behavior is related to temperature and time (or rate), and it is viscoplastic deformation, which usually shows creep and stress relaxation. In the process of temperature cycling, the maximum stress value of dangerous position of solder joint appears at the beginning of low temperature insulation, at which time the solder joint is most likely to crack and fail. The failure mode of tin-lead solder joints is mainly low cycle fatigue failure, and the life mode is characterized by the modified Coffin-Manson equation (C-M equation), that is, the empirical relationship between the low cycle fatigue life (N_f) and the plastic strain range ($\Delta\varepsilon$) is as follows:

$$N_f = \frac{1}{2}\left(\frac{\Delta\gamma}{2\varepsilon_f}\right)^{\frac{1}{c}}$$

In the formula, N_f is the mean life of thermal fatigue failure, $\Delta\gamma$ is the equivalent inelastic shear strain range, $\Delta\gamma= 1.732\Delta\varepsilon$, $\Delta\varepsilon$ is the equivalent inelastic total strain range, ε_f is the fatigue toughness coefficient of 0.325, and coefficient C is generally negative, which needs to be revised according to the experiment. The finite element model of the package was built by using Ansys software, seen figure 4. The corresponding samples were assembled and subjected to thermal cycling at- 65 ℃ ~ 150℃, and the package samples were subjected to thermal fatigue more than 500 times.

Fig. 4 Finite element model of secondary installation

V. CONCLUSION

This paper presents a vertical transfer surface mount ceramic package with high transmission performance, high reliability and high "core-to-core" ratio. The package transmission terminal adopts hollow vertical transmission and coplanar waveguide structure. In the DC~40GHz band, the insertion loss of the whole board-level transmission path is less than 1.5dB. At the same time, the reliability of the package environment and the mechanical reliability were verified. In addition, the reliability of the secondary installation of the package was analyzed and verified. Therefore, the package has excellent transmission characteristics and high reliability, and low cost, suitable for mass.

REFERENCES

[1] Yoshida, K., Shirasaki, T., Matsuzono, S., and Makihara, C, "50 GHz broadband SMT package for microwave applications," Electronic Components and Technology Conference, 2001, pp. 744-749.

[2] R. Valois, D. Baillargeat, S. Verdeyme, M. Lahti and T.Jaakola, "High Performance of Shielded LTCC Vertical Transitions From DC up to 50GHz," IEEE Transactions on Microwave Theory and Techniques, Vol. 53, No.6, June 2005, pp. 2026-2032.

[3] Qiao zhizhuang, Gao ling, Liu linjie, "A Novel High Performance 40GHz Hermetic SMT Ceramic Package for Microwave Applications," 2014 15th International Conference on Electronic Packaging Technology, 2014, pp. 815-819.

[4] Jinan Tang, Design of Multilayer Ceramic Package and ots Some Question in the Appliacation[J]. Electronics and packaging, 2003.

[5] Sung Yi, Guangxing Luo, Kerm Sin Chian, W.T. Chen, "A viscoplastic constitutive model for 63Sn37Pb eutectic solders," International Symposium on Electronic Materials and Packaging (EMAP2000), 2000.

978-1-6654-1392-3/21 $31.00 © 2021 IEEE

Reliability and Thermal Degradation of First-Level Thermal Interface Materials

1st Yunpeng Su[#]
1 *Shenzhen Institute of Advanced Electronic Materials, Shenzhen Institute of Advanced Technology Chinese Academy of Sciences* Shenzhen, China
15541169122@163.com

2nd Junhong Li[#]
1 *Shenzhen Institute of Advanced Electronic Materials, Shenzhen Institute of Advanced Technology Chinese Academy of Sciences* Shenzhen, China
2 *College of Materials Science and Engineering, Shenzhen University,* Shenzhen, China
lijunhong@siat.ac.cn

3rd Qiangqiang Ma
1 *Shenzhen Institute of Advanced Electronic Materials, Shenzhen Institute of Advanced Technology Chinese Academy of Sciences* Shenzhen, China
qq.ma@siat.ac.cn

4th Linlin Ren*
1 *Shenzhen Institute of Advanced Electronic Materials, Shenzhen Institute of Advanced Technology Chinese Academy of Sciences* Shenzhen, China
ll.ren@siat.ac.cn

5th Xiaoliang Zeng*
1 *Shenzhen Institute of Advanced Electronic Materials, Shenzhen Institute of Advanced Technology Chinese Academy of Sciences* Shenzhen, China
xl.zeng@siat.ac.cn

6th Rong Sun
1 *Shenzhen Institute of Advanced Electronic Materials, Shenzhen Institute of Advanced Technology Chinese Academy of Sciences* Shenzhen, China
rong.sun@siat.ac.cn

[#] These authors contributed to the work eqully and should be regarded as co-first authors.

Abstract—**The challenge of the heat dissipation of electronic packaging comes from the continuous increase in power consumption and power density of high-power devices. The demand for the use of thermal interface materials (TIMs) with excellent thermal conductivity between the surface of the heat source and the heat dissipation module is thus increasing. In addition to thermal performance, TIMs are also required to have good reliability and thermal degradation. In this paper, the reliability were evaluated for three types of TIM, including high-temperature storage (HTS), Highly accelerated temperature and humidity stress test (HAST) and high and low temperature cycling (HLTC). The trend of thermal resistance change and thermal degradation under various reliability test conditions were analyzed. The results show that after HAST and high and low temperature cycle evaluation, the sample can maintain the original characteristics. However, the thermal degradation of TIMs is more serious after high-temperature storage test. Therefore TIMs are more sensitive to long-term exposure to high temperature environments.**

Keywords—*thermal interface materials; reliability evaluation; heat dissipation*

I. INTRODUCTION

The high speed and low latency of the 5G bring us a better experience. However, the functions of electronic devices are gradually increasing and becoming more complicated, and the miniaturization of integrated circuits(IC) will increase the power and generate more heat. The performance of heat dissipation directly affects the stability of electronic devices. Therefore, efficient thermal management technology plays an increasingly important role. One of the most important considerations is to provide an effective thermal dissipation between the silicon chip and heat sink [1]. It is well known that when the solid interface is in contact with the solid interface, the actual contact area will be reduced due to the roughness.

Thermal interface materials (TIM) such as silicone or epoxy resin matrix with metal/metal-oxide filler are generally used to fill air gaps between the chip and lid for better contact and thermal path as shown in Fig.1. The ideal TIM should have high thermal conductivity, low interface thermal resistance and possibly thin bond line thickness [2]. Although a lot of work has been done to improve the thermal conductivity of TIM in recent years, the thermal resistance of TIM still dominates in the whole thermal dissipation path, accounting for more than 50% of the total thermal resistance [3-5]. Therefore, the thermal resistance reduction of TIM is still a strong challenging for increasing the overall thermal performance of the system. In addition to the improvement of thermal performance, reliability performance of the TIM is also important as its thermal resistance for better thermal dissipation [6-8].

In the current study, we prepared three samples by bonding silicon chips on copper lids with TIM. Reliability test was carried out through the following three conditions: 150℃ high temperature storage, HAST and thermal cycling. The above three conditions can well evaluated the long term reliability of samples under extreme environment of high temperature, high humidity and thermomechanical loading. The thermal resistance changes and degradation of samples at different times were tested and recorded to analyze the impact of the reliability of samples under the different test conditions.

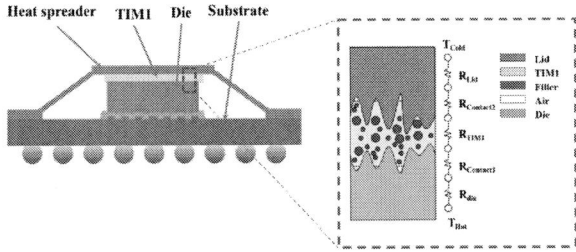

Fig.1. Schematic diagram of Flip-chip BGA and Thermal resistance network of TIM

II. SAMPLE PREPARATION AND EXPERIMENTAL DETAILS

A. Sample Preparation

Three different gel type TIMs, respectively Sample A, Sample B and Sample C, were used in this study. These samples mainly contain different mass fractions of aluminum particles and silicon oil, and other additives including coupling and surfactants. The key materials' properties of three TIMs are listed in TABLE Ⅰ.

TABLE I.   THE KEY MATERIALS' PROPERTIES OF THREE TIMS

Samples	Bulk thermal conductivity (W/m K)	Density (kg/m3)	Type	Filler
A	4.183	2446	Gel	Al+other
B	4.246	2489	Gel	Al+other
C	4.996	2482	Gel	Al+other

In this experiment, 25.4 mm x 25.4 mm silicon wafers with 505 μm thickness were used. The size of the copper lids plated by nickel was 25.4 mm x 25.4mm x 1.01 mm (length x width x thickness). The TIM was placed on lids, and the silicon wafer covered on TIM, then a compressive force was applied on the silicon wafer to obtain Lid/TIM/Si sample with homogenous bond line thickness (BLT). Every TIM was prepared into 16 Lid/TIM/Si samples, in which there are 4 samples with 40 μm, 8 samples with 60 μm and 4 samples with 90 μm.

B. Experimental Details

Thermal resistance of Lid/TIM/Si samples were measured and evaluated using steady state method by ASTM-D5470 in this study. The terms evaluated include the thermal resistance of samples at initial state and at various time intervals of different reliability test conditions, which included 150 ℃ high temperature storage, HAST (135 ℃, 85% RH) and thermal cycling (-45 ℃ to 125 ℃, one cycle per hour). During the thermal resistance test, the silicone grease should be applied to both sides of the Lid/TIM/Si to eliminate the contact thermal resistance with the metal interface of the instrument. The test loading was set at 10 Psi, the temperature of hot source contact interface was set at 80 ℃, and the test took 15 min(Fig. 2). The morphology of Lid/TIM/Si samples under the different reliability test conditions were characterized by D9600™ Acoustic Microscope. Ultrasound is an elastic disturbance that propagates through materials (mainly solids and liquids) at frequencies above 20 KHz. When the ultrasound through solid/liquid interface, some of them is reflected and some of them is transmitted. It will be reflected 100% if ultrasound

through air area. Here we selected the 100MHz transducer and the transmission mode, and the Pixel size was 20 μm, the scan size was 30 mm. The experimental setup is shown in Fig. 3.

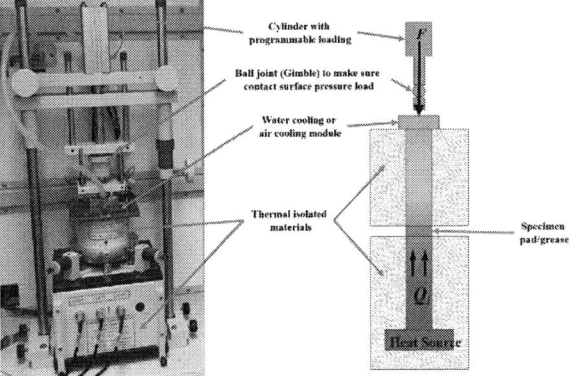

Fig.2. Thermal conductivity measurement setup

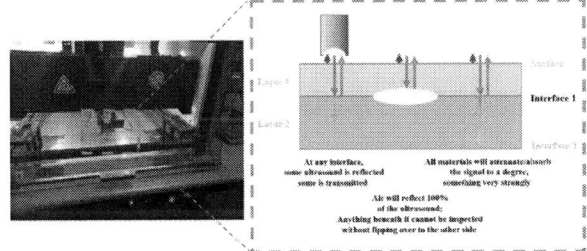

Fig.3. Acoustic microscope setup and schematic diagram

III. RESULTS AND DISCUSSION

A. High Temperature Storage (HTS)

The test of high temperature storage is an important means to verify the effect of continuous high temperature environment on the performance of TIM1. The samples including 4 samples with 40 μm and 4 samples with 90 μm ranging from Sample A to Sample C were put in a constant temperature oven for testing at 150 ℃ for 1000 h. In this process, 250 hours, 500 hours and 1000 hours were used as the points for TIM1 evaluation. The trend of the average thermal resistances of samples with the test time are shown in Fig. 4. Compared with the initial average thermal resistance at 0 h, the average thermal resistance of three samples (Sample A, Sample B and Sample C) with the thickness of 40 μm at HTS 1000 h increased by -4.3%, 2.1% and 4.8%, respectively, with the thickness of 40 μm at HTS 1000 h increased by 5.5%, -0.6% and -1.2%, respectively.

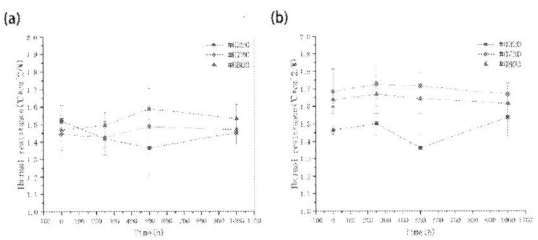

Fig.4. Thermal resistance of thickness of HTS test with different the thickness of (a) 40 μm and (b) 90 μm

As shown in Fig.5, the coverage rates of the three samples A, B, and C decreased significantly over time. The thermal degradation of the Sample A, Sample B and Sample C with a thickness of 40um were 13%, 2.3% and 10% at HTS 1000h, and the samples with a thickness of 90um were 10.4%, 16.2% and 18.7% at HTS 1000h. The acoustic microscope images of the thermal degradation of samples were shown in Fig.6.

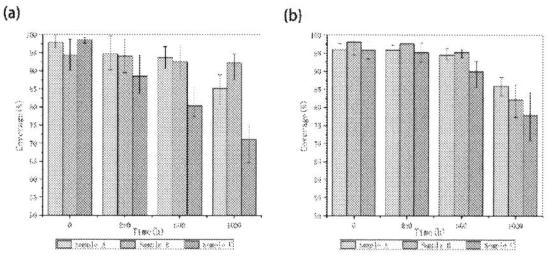

Fig.5. Coverage of HTS test with different the thickness of (a) 40 μm and (b) 90 μm

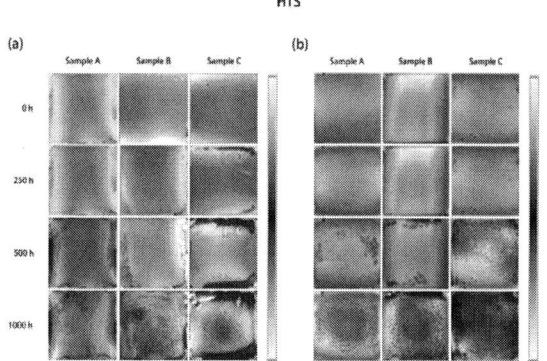

Fig.6. Acoustic microscope images of TIMs of HAST test with different the thickness of (a) 40 μm and (b) 90 μm

B. Highly Accelerated Temperature and Humidity Stress Test (HAST)

The HAST test is to study the effect of moisture on the thermal performance of TIM1 under elevated temperature and high humidity conditions. 4 samples with 60 μm ranging from Sample A to Sample C were subjected to 135 °C/85% RH for up to 256 hours. The read points for this test were 96 h and 128 h. The average thermal resistance of the samples are shown in Fig. 7. The average thermal resistance of three samples (Sample A, Sample B and Sample C) at HAST 256 h increased by 14.8%, 11.5% and 7.5%, respectively, compared with the initial average thermal resistance at 0 h.

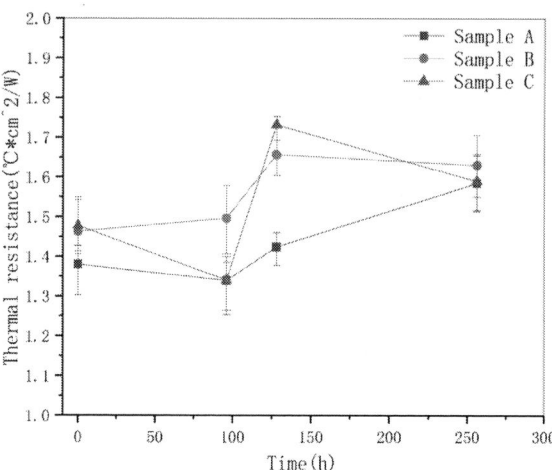

Fig.7. Thermal resistance of HAST test

The thermal degradation of the Sample A, Sample B and Sample C after 256 h were 7.9%, 4.7% and 2.2%, respectively(Fig. 8). The acoustic microscope images of the thermal degradation of samples were shown in Fig. 9, where the local severely degraded areas were marked in red. In a high-temperature and high-humidity environment, the moisture will breaks the Si-O bonds in the silicone material network, and then form hydrophilic Si-OH bonds, which will make the material lose good adhesion to the surface so that leading to local delamination. The thermal degradation of the three samples can further explain that the thermal resistance change of the samples are positively correlated with the thermal degradation, that is, the greater the thermal degradation rate, the greater the thermal resistance increase rate.

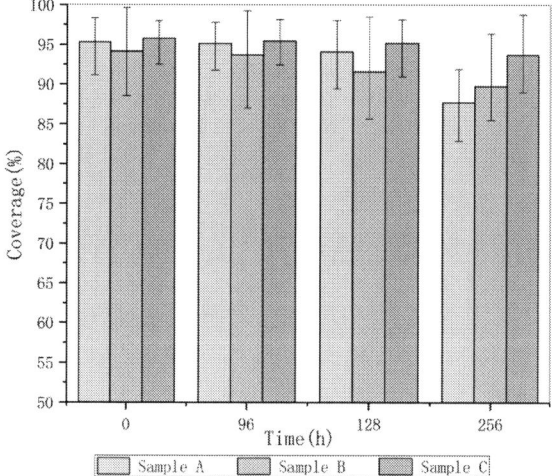

Fig.8. Coverage of HAST test

HAST

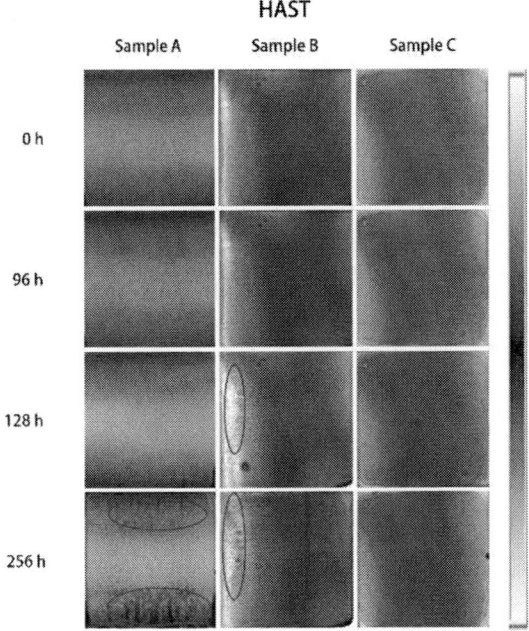

Fig.9. Acoustic microscope images of TIMs of HAST test

C. High and Low Temperature Cycling (HLTC)

The thermal cycling test is used to evaluate the effect of thermomechanical load caused by temperature different. In this study, 4 samples with 60 μm ranging from Sample A to Sample C were subjected to temperature cycles of -45°C / 125°C two cycles per hour. The reading points for this test are 300, 500 and 700 cycles. The average thermal resistances of samples were shown in Fig .10. At the end of thermal cycles of Samples A, B and C, the average thermal resistance increased by 11.5%, 2.5% and -1.2%, respectively, compared to the initial value.

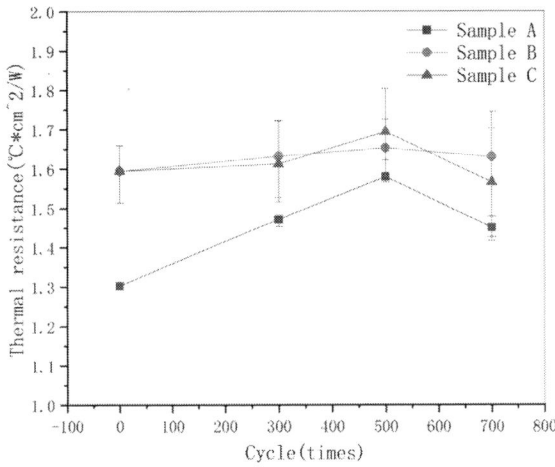

Fig.10. Thermal resistance of thermal cycling test

The thermal degradation of the Sample A after 700 cycles was 3.2%, of the Sample B was 3.1% and of the Sample C was 1.1% (Fig. 11). The acoustic microscope images of the thermal degradation of samples were shown in Fig. 12. According to the data comparison, compared with

HAST test, the thermal degradation of the thermal cycling test is not obvious.

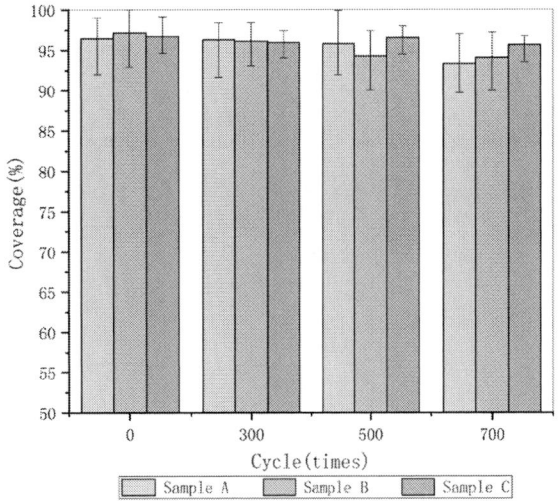

Fig.11. Coverage of thermal cycling test

thermal cycling

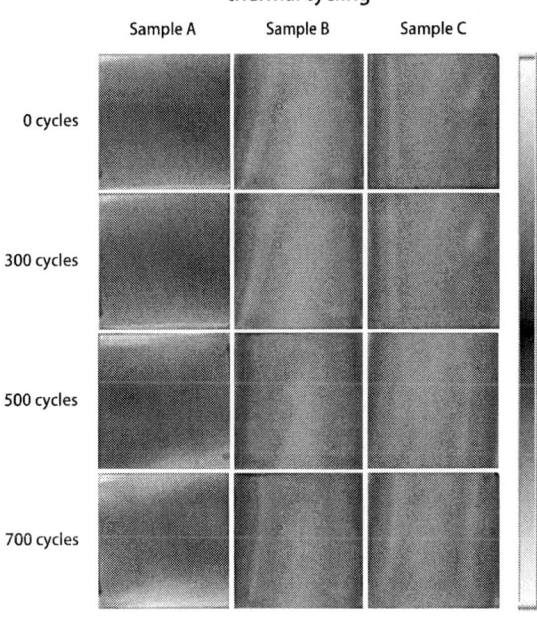

Fig.12. Acoustic microscope images of TIMs of thermal cycling test

IV. CONCLUSION

This work mainly evaluated the change of thermal resistance and thermal degradation of three types of TIMs under different long-term reliability test conditions. The main conclusions are as follows: Ⅰ) The thermal resistance of the three types of TIMs has not changed significantly through the three reliability tests; Ⅱ) Compared with HAST and thermal cycling tests, thermal degrade of TIMs are more obviously under HTS test conditions.

REFERENCES

[1] M.R.S. Shirazy, S. Allard, M. Beaumier, L.G. Frechette, Effect of squeezing conditions on the particle distribution and bond line thickness of particle filled polymeric thermal interface materials, in:

Thermomechanical Phenom. Electron. Syst. -Proceedings Intersoc. Conf. Institute of Electrical and Electronics Engineers Inc. 2014, pp. 251 – 259.

[2] F. Sarvar, D. C. Whalley, and P. P. Conway. Thermal Interface Materials – A Review of the State of the Art. Proc. 1st Electronic Systemintegration Technology Conference, Dresden, Germany, 5 - 7 Sept. 2006, pp. 1292 - 1302.

[3] Jeffery C. C. Lo, Mian Tao, and Yuanjie Cheng. Long-term Stability Evaluation of Thermal Interface Materials (TIMs). 20th International Conference on Electronic Packaging Technology, Hong Kong, China, 12 – 15 Aug. 2019.

[4] K. M. Razeeb, E. Dalton, G. L. W. Cross, and A. J. Robinson, "Present and future thermal interface materials for electronic devices," International Materials Reviews, 2017, DOI: 10.1080/09506608.2017.1296605.

[5] Bajaj, N., Subbarayan, G., and Garimella, S. V., 2012, "Topological Design of Channels for Squeeze Flow Optimization of Thermal Interface Materials," International Journal of Heat and Mass Transfer, 55(13-14) pp. 3560-3575.

[6] J. Hansson, C. Zanden, L. Ye and Johan Liu, "Review of Current Progress of Thermal Interface Materials for Electronics Thermal Management Applications," Proc. 16th International Conference on Nanotechnology, Sendai, Japan, 22-25 Aug 2016, pp. 371-374.

[7] J. Due, and A. J. Robinson, "Reliability of thermal interface materials: A review," Applied Thermal Engineering, vol. 50, 2013, pp. 455-463.

[8] T. H. Wang, HJ. Chen, C. Lee, and Y. Lai, "High-powered thermal gel degradation evaluation on board-level HFCBGA subjected to reliability tests," Microelectronic Engineering, vol. 88, 2011, pp. 3101- 3107.

Coupling Damage Accumulation of Die-Attach Solder Layer with Distributed Void Defects for Power Electronics

Yidian Shi
1 School of Mechanical and Electrical Engineering
Central South University
Changsha, China
sydian@csu.edu.cn

Taotao Chen
2 Anhui Province Key Laboratory of Microsystem
East China Research Institute of Microelectronics
Hefei, China
15256985225@163.com

Hu He*
1 School of Mechanical and Electrical Engineering
Central South University
Changsha, China
hehu.mech@csu.edu.cn

Cheng Peng
1 School of Mechanical and Electrical Engineering
Central South University
Changsha, China
pccheng@csu.edu.cn

Yunpeng Liu
2 Anhui Province Key Laboratory of Microsystem
East China Research Institute of Microelectronics
Hefei, China
liuyunpeng1990@126.com

Wenhui Zhu*
1 School of Mechanical and Electrical Engineering
Central South University
Changsha, China
zhuwenhui@csu.edu.cn

Junfu Liu
2 Anhui Province Key Laboratory of Microsystem
East China Research Institute of Microelectronics
Hefei, China
liujunfu0110@163.com

Abstract—Die-attach solder layer plays key effects in realizing the functions of power devices. The defects induced in soldering process and accumulated damage during service are the main challenges to the lifetime of power modules. Regarding the service conditions of power devices, creep and fatigue mechanisms can be both activated. This paper focuses on the thermomechanical characteristics brought by damage coupling effects of die attach solder layer with different void configurations. The influence of voids on creep strain accumulated during thermal cycles were studied by finite element analysis. A novel way to calculate the creep damage under random loading spectrum is proposed. Meanwhile, the corresponding lifetime to failure are calculated based on the coupled creep-fatigue damage model. The results revealed that the distributed void configurations with the same porosity make no big difference in damage accumulation under real loading spectrum.

Keywords—Voids defects, creep strain, coupling damage, mission profile, solder layer

I. INTRODUCTION

The solder layer in power modules remains one of the critical sites according to the statistical results of failures. The presence of voids induced in the reflow process is inevitable due to the outgassing of flux, which brings reliability problems to the power module. The thermal and mechanical properties of the module will be strongly impacted[1]. In a real service condition, the power module experienced severe conditions from the passive and active thermal swings. The repeated stress induced by the CTE mismatch of different material layers produce alternate strains, which will be accumulated under temperature swings and initiate microcracks when the fracture criterion is reached[2]. Moreover, the solder materials near the void and cracks suffer higher inelastic strain due to the stress concentration[1]. Consequently, the thermo-mechanical effects of voids on the reliability of solder layer is critical. Le et al.[3] have illustrated the size and distribution of voids significantly affect the fatigue life of solder joints. E. Padilla et al.[4] studied the deformation behavior of voids in the solder layer, indicating the influence of pore size on joints failure. K. C. Otiaba et al.[5] have studied the morphology of voids on the fatigue performance by the accumulated plastic work under different thermal cycles. The work shows that the sensitivity of solder joint fatigue life to the voids distribution increases as the porosity increases.

Apart from the fatigue effects, strain caused by the creep effects also existed when the hold period of operating temperature exceeds $0.5T_m$ (T_m is the absolute melting temperature)[6]. Amalu E. H. et al.[7] employed Garofalo–Arrhenius creep model to observe the thermo-mechanical response of solder joints and regarded the accumulate strain energy density in resistor joints as reliability influencing factors. Majid Samavatian et al.[8] have pointed out the voids boundaries acting as the critical regions for damage initiation due to the stress concentration by inducing triaxiality factor into fatigue life prediction. This work considered the creep strain in the solder layer under thermal cycling loadings. P. Wild et al.[9] investigated different void levels on the volume-weighted averagely accumulated creep strain based on the different creep models. The results demonstrated that multiple voids leads to greater reduction compared to a single void the distribution.

Generally, the failure of solder joints is considered to be the combined effects of fatigue and creep. While the fatigue damage and the creep damage of the solder joints are usually focused separately. The influence of the two activation on damage accumulation during the operation or thermal swings is seldom illustrated, which may lead to a large difference in the lifetime prediction. Chen et al.[6] and Samavatian et al.[10] have proposed a coupling damage model including the fatigue and creep effect based on the linear damage rule.

This work was funded by National Natural Science Foundation of China (U20A6004), Natural Science Foundation of Hunan Province (2020JJ5728), Innovation-Driven Project of Central South University (2020CX05) and State Key Laboratory of High Performance Complex Manufacturing (ZZYJKT2019-05).

This method is employed to assess the effect of void distribution on the reliability under a real junction temperature mission profile. In the process of correlating the real random loadings to standard thermal cycling tests by FEM, a clarified method to build the relationship of multiple parameters is necessary. Therefore, this article proposed a detailed way to calculate the accumulated creep damage. And the influence of various voids configurations on coupling damage is obtained.

II. METHODOLOGY

Based on the characteristics of thermal loading, the effect of real mission profile on the damage can be divided into two parts: plastic strains induced during the constant temperature and the temperature ramp up/down, which corresponding to the creep damage and fatigue damage. Namely, the failure mechanism of solder joints is attributed to the coupling effect of fatigue and creep behaviors in a representative loading cycle. The involved models are listed below:

A. Fatigue Damage

The junction temperature swing and the mean junction temperature are the most important factors to the fatigue life prediction of solder layer. Among the previous studies referred to the life prediction model, Conffin-Manson fatigue model is the most representative one and the plastic strain range or the temperature range were regard as the activation. The Coffin–Manson–Arrhenius lifetime model has been employed in this work, which is described as:

$$N_f(T_{mean}, \Delta T_j) = A \times \Delta T_j^{\alpha} \times \exp(Q / RT_m) \quad (1)$$

where, A, α are both constant and device-dependent, R and Q are the gas constant ($8.314 J \cdot Mol^{-1} \cdot K^{-1}$) and internal energy ($7.8 \times 10^2 J \cdot Mol^{-1}$) and T_m is the mean junction temperature of devices in Kelvin. ΔT_j expresses the junction temperature swing of devices. Based on some experiments' data in ref.[10], one can find that $\alpha = -2.505$ and $A = 8.13 \times 10^8$.

Thus, based on the Miner's rule, the fatigue damage accumulated per unit time is presented as:

$$D_F = \sum_{i=1}^{n} N_i(T_m, \Delta T_j) / N_{fi}(T_m, \Delta T_j) \quad (2)$$

where D_F is the accumulated damage of the former n cycles and the N_{fi} is the lifetime cycles under the corresponding T_m and ΔT_j conditions.

B. Lifetime Model

The creep damage is different from the fatigue damage. The creep damage is time-dependent and accumulated in the dwelling time. Monkman-Grant Model is extensively used in describing the creep failure, shown as below:

$$\dot{\varepsilon}_{cr} t_c^{\beta} = C_{MG} \quad (3)$$

where $\dot{\varepsilon}_{cr}$ is the stable creep strain rate, t_c is the creep rupture life, and the exponent β and constant C_{MG} are related to the material properties, the data $\beta = 0.78$, $C_{MG} = 0.5$ are from the experiment in ref.[11].

Thus, the damage accumulated in per unit time is presented as:

$$d_C = 1 / t_c = (\dot{\varepsilon}_{cr} / C_{MG})^{1/\beta} \quad (4)$$

Then, the creep damage accumulated in the dwelling time can be expressed as:

$$D_C = \Delta t / t_c = \Delta t (\dot{\varepsilon}_{cr} / C_{MG})^{1/\beta} \quad (5)$$

where Δt is the dwelling time.

Among the constitutive equations describing the creep strain of SAC solders, Garofalo–Arrhenius creep model is one of the most applicative constitutive models for the evaluation of SAC solder joints. This model refers to a hyperbolic sine creep equation to depict the creep behavior, defined as follows:

$$\dot{\varepsilon}_{cr} = C_1 [\sinh(C_2 \sigma)]^{C_3} \exp(-C_4 / T) \quad (6)$$

where $\dot{\varepsilon}_{cr}$ is the creep strain rate and C_1-C_4 ($C_1 = 2.73 \times 10^5$ (s^{-1}), $C_2 = 0.023$ ($MPa)^{-1}$, $C_3 = 6.3$, $C_4 = 6480.3$) are the constant related to the material properties of SAC solder[10]. σ is the applied stresses induced by temperature swing ΔT and T is the absolute temperature.

C. Coupling Damage Model

Based on the fatigue failure and creep failure model, the global linear damage model will be expressed as follows:

$$D = D_F + D_C \quad (7)$$

D. Rainflow Counting Algorithm

Rainflow Counting method is widely used in counting the life prediction of device under random loading spectrum. This method extracts the loading amplitude and calculates the cycle number of the corresponding loading level. In this work, the four-point cases Rainflow Counting Algorithm is implemented with a MATLAB script. The mission profile of junction temperature is the input parameters, then the junction temperature ΔT_j and mean junction temperature T_m can be obtained.

E. FEM simulation

To obtain the creep strain rates of the solder layer in the temperature swings, a set of thermal cycling tests was performed. Five simplified geometric models containing solder layer with different void configurations were built to study the void distribution on the failure or lifetime of the device. This study implements the static structural FEM simulation by the ANSYS. The critical parameters(C_1-C_4) presenting creep characteristics of solder layer was set in the materials. Different void configurations with a porosity of 10% in solder layer are generated by a MATLAB script. The meshed model was presented in Fig. 1, which contains approximately 42700 elements. To improve the accuracy of the calculation, a finer element size was meshed in the die attach solder region.

Fig. 1 The simplified meshed model containing varied void configurations in die-attach solder layer for FEM analysis

TABLE I.   THE MATERIALS PROPERTIES EMPLOYED IN THE FEM ANALYSIS

Materials	Al$_2$O$_3$	SAC305	Copper	Silicon
Coefficient of Thermal Expansion(K^{-1})	2.2e-5	2.3e5	1.64e5	3e-6
Young's Modulus(GPa)	345	40	117	131
Poisson's Ratio	0.25	0.3	0.34	0.3

A roughly constant dwelling temperature highly impressed the creep degradation. Then the roughly dwelling temperature period is extracted from the junction temperature diagram of power semiconductors in a hybrid vehicle[12]. In this work, the dwelling temperature swing below 2 °C regarded as the constant temperature range and these periods drawn from the load spectrum were been investigated. According to the creep strain and creep strain rate calculated from the thermal cycling results, the creep strain rate varies with the dwelling temperature remaining constant. Consequently, the creep strain rates during the defined constant periods will change with the temperature swings of load spectrum. Namely, the creep strain rate remains changeable corresponding to the dwelling time. As a result, the dwelling temperature and time are key factors affecting the creep strain rate. Then a set of thermal cycling simulations contain of different dwelling temperatures and time are implemented to correlate the desired creep strain rates to the specified simulation results.

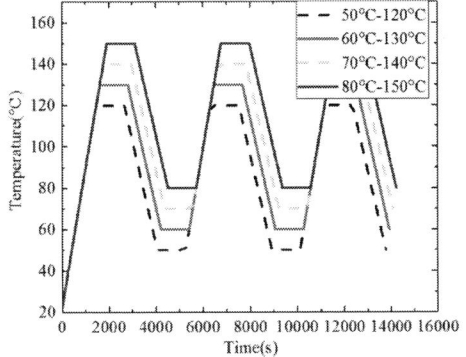

Fig. 2 The applied thermal cycles with different dwelling temperatures.

The accumulated creep strain varies with the thermal cycles. Besides, the creep strain rates at the dwelling temperature T$_{max}$ and T$_{min}$ are time-dependent. As a result, the creep strain rate depends on the dwelling temperature, time and even the history loads (ramping rate). Especially, the roughly constant temperature periods in the real mission profile experienced high temperature ramping rate before the temperature tending to constant. This work ignored the effect of ramping rate on the creep strain rate and all the heating rates were set as 4 °C/min and the cooling rates were 3 °C/min, respectively. The dwelling temperature was set as 20 min. As shown in Fig. 2, the designated simulation conditions are presented. To simply the calculation of creep strain rate of dwelling temperature periods in the real mission profile, a correlation was modeled based on the thermal cycling results with different dwelling temperatures.

III.   RESULTS AND DISCUSSION

Based on the above mechanisms and methods, the lifetime of solder layer containing distributed voids in power modules under real loading spectrum are compared.

A. *Extraction of ΔT_j, T_m and Δt*

As shown in Fig. 3(a), the real junction temperature mission profile is given. The peaks, valleys and diverted points are distinguished from the profile through four-point rainflow algorithm implemented in MATLAB. Then, the amplitude of junction temperature ΔT_j and mean value T$_m$ are calculated. As shown in Fig. 3(b), the histogram of ΔT_j and Tm are displayed, thereby the fatigue damage is obtained according to Eq. (1). Totally, 14 roughly constant temperature periods are extracted at same time according to the assumption in Section II. E. The dwelling temperature T$_{min}$, T$_{max}$ and Δt are detrimental parameters to creep damage based on the Eq. (5)-(6). The representative creep strain results after thermal cycles are presented in Fig. 4 and Fig. 5.

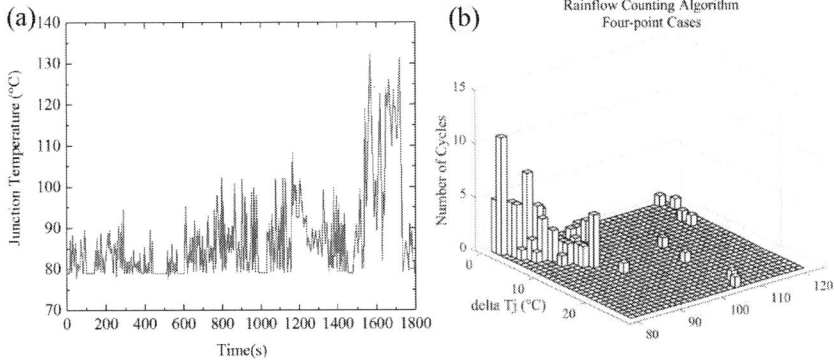

Fig. 3 (a) the junction temperature mission profile in a real hybrid vehicle, and (b) the histogram of ΔT_j and T$_m$ extracted by rainflow counting method.

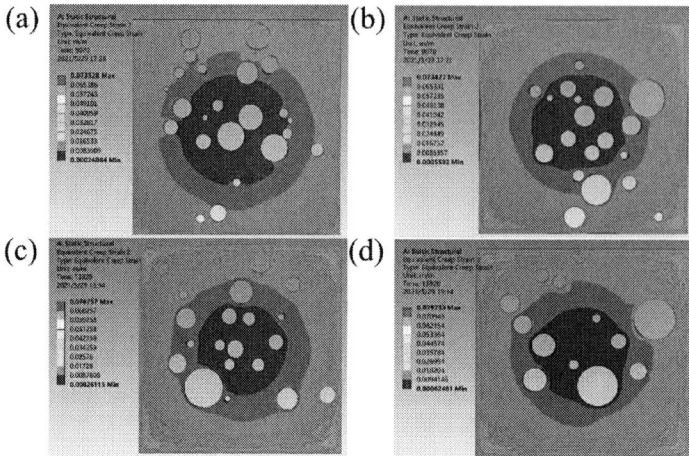

Fig. 4 The creep strain accumulated for (a) Void Configuration 1, (b) Void Configuration 2, (c) Void Configuration 3 and (d) Void Configuration 4 that were generated randomly with constant porosity of 10% after thermal cycling condition 60°C-130°C

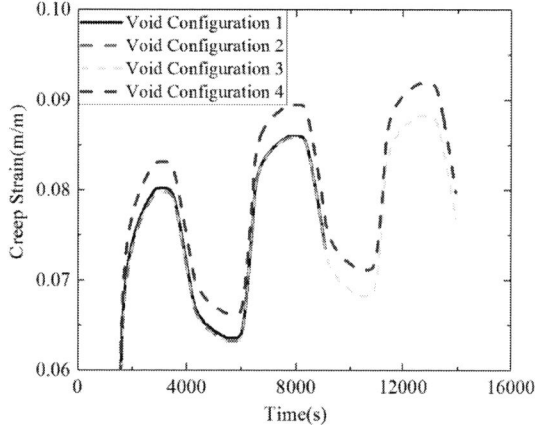

Fig. 5 The creep strain accumulated under thermal cycling condition 60°C - 130°C

B. Damage Evaluation

The fatigue damage depends on temperature swings. Fatigue damage accumulated in the typical loading period of 1800 seconds can be calculated. Then, the lifetime of the solder layer only considering fatigue life can be obtained. The fatigue life is certain to be same of 28737 hours due to the given temperature loading spectrum.

Creep damage accumulation in this observed period is mainly due to the constant temperatures and its dwelling time. As aforementioned, the calculation of creep damage is dependent on the creep strain rates from the results of FEM simulation. With aid of interpolant method, an equation of creep strain rate containing factors of dwelling temperatures and their intervals is built as shown in Fig. 7. To enhance the accuracy, a limited time range of 0-100 seconds is utilized to fit the data. As a result, the creep damage in each temperature dwelling period can be expressed as:

$$D_{Ci} = \int_0^{\Delta t} \left(\frac{\dot{\varepsilon}_{cr}(T_{min}, \tau)}{C_{MG}}\right)^{1/\beta} d\tau \qquad (8)$$

where D_{Ci} is the creep damage of i th dwelling temperature and τ is the dwelling time of the corresponding temperature. The creep damage of the 14 constant temperature intervals in the observed 1800 seconds can be calculated by Eq. (8).

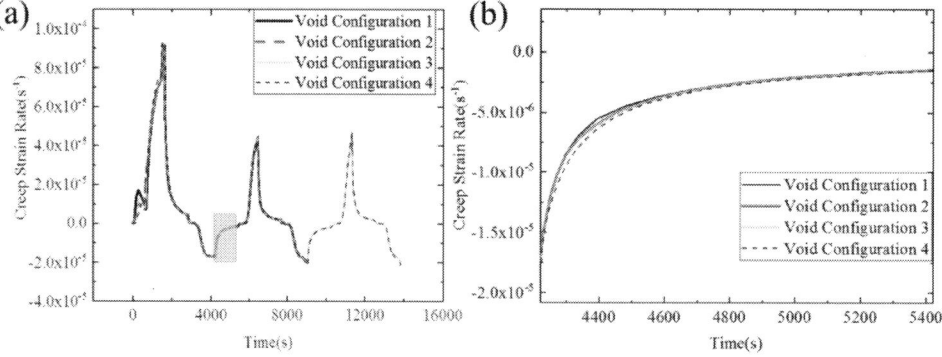

Fig. 6 (a) the creep strain rates for different void configurations under thermal cycles ranging of 60°C-130°C, and (b) the enlarged graph of the red frame shown in (a).

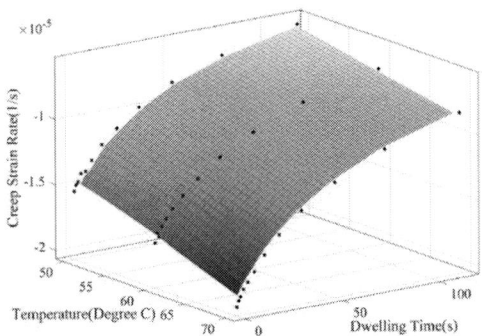

Fig. 7 The creep strain rates of Void Configuration 1 varied with the dwelling temperature and its corresponding retaining time.

C. Effects of Void Configurations on Life Degradation

This study has listed 4 kinds of void configurations and the representative lifetime are calculated. As shown in Fig. 7, the creep damage of Void Configuration 1 in the observed period is 4.39×10^{-4} and the fatigue damage is 1.74×10^{-5}. It indicates the creep damage dominate the failure of the solder layer under the random loading spectrum. The damage accumulated in different void configurations are different, while the Void Configurations 1-4 are of little difference. The Void Configuration 4 presents the biggest damage estimation of 4.44×10^{-4}, which represents the random void configuration with the same porosity has little effect on the reliability of solder joints. A further study will focus on the voids percentage and void concentration in local region.

IV. CONCLUSIONS

To conclude, a fatigue-creep coupling damage model is utilized in evaluating the lifetime of solder layer with different void configurations under real thermal loading spectrum. It takes the temperature swing periods and roughly constant periods as the stages when fatigue damage and creep damage occur respectively. Besides, a method to correlate the creep strain rate calculated from thermal cycling to the uncertain loading spectrum is proposed in this work. The results indicated that the distributed void configurations with a constant void ratio have a small difference on the damage accumulation.

ACKNOWLEDGMENT

This work was supported by National Natural Science Foundation of China (U20A6004), Natural Science Foundation of Hunan Province (2020JJ5728), Innovation-Driven Project of Central South University (2020CX05) and State Key Laboratory of High Performance Complex Manufacturing (ZZYJKT2019-05).

REFERENCES

[1] B. Hu, S. Konaklieva, N. Kourra, M. A. Williams, L. Ran, and W. Lai, "Long-Term Reliability Evaluation of Power Modules With Low Amplitude Thermomechanical Stresses and Initial Defects," IEEE Journal of Emerging and Selected Topics in Power Electronics, vol. 9, no. 1, pp. 602-615, 2021.

[2] H. Xiao, X. Y. Li, Y. Hu, F. Guo, and Y. W. Shi, "Damage behavior of SnAgCu/Cu solder joints subjected to thermomechanical cycling," Journal of Alloys and Compounds, vol. 578, pp. 110-117, 2013.

[3] V. N. Le, L. Benabou, V. Etgens, and Q. B. Tao, "Finite element analysis of the effect of process-induced voids on the fatigue lifetime of a lead-free solder joint under thermal cycling," Microelectronics Reliability, vol. 65, pp. 243-254, 2016.

[4] E. Padilla, V. Jakkali, L. Jiang, and N. Chawla, "Quantifying the effect of porosity on the evolution of deformation and damage in Sn-based solder joints by X-ray microtomography and microstructure-based finite element modeling," Acta Materialia, vol. 60, no. 9, pp. 4017-4026, 2012.

[5] K. C. Otiaba, M. I. Okereke, and R. S. Bhatti, "Numerical assessment of the effect of void morphology on thermo-mechanical performance of solder thermal interface material," Applied Thermal Engineering, vol. 64, no. 1-2, pp. 51-63, 2014.

[6] Y. Chen, Y. Jin, and R. Kang, "Coupling damage and reliability modeling for creep and fatigue of solder joint," Microelectronics Reliability, vol. 75, pp. 233-238, 2017.

[7] E. H. Amalu, N. N. Ekere, M. T. Zarmai, and G. Takyi, "Optimisation of thermo-fatigue reliability of solder joints in surface mount resistor assembly using Taguchi method," Finite Elements in Analysis and Design, vol. 107, pp. 13-27, 2015.

[8] M. Samavatian, V. Samavatian, M. Moayeri, and H. Babaei, "Effect of stress triaxiality on damage evolution of porous solder joints in IGBT Discretes," Journal of Manufacturing Processes, vol. 32, pp. 57-64, 2018.

[9] P. Wild, D. Lorenz, T. Grözinger, and A. Zimmermann, "Effect of voids on thermo-mechanical reliability of chip resistor solder joints: Experiment, modelling and simulation," Microelectronics Reliability, vol. 85, pp. 163-175, 2018.

[10] V. Samavatian, H. Iman-Eini, Y. Avenas, and M. Samavatian, "Effects of Creep Failure Mechanisms on Thermomechanical Reliability of Solder Joints in Power Semiconductors," (in English), Ieee Transactions on Power Electronics, vol. 35, no. 9, pp. 8956-8964, Sep 2020.

[11] Y. Zhu, X. Li, and R. Gao, "Creep failure mechanism and life prediction of lead-free solder joint," Journal of Materials Science: Materials in Electronics, vol. 26, no. 1, pp. 267-272, 2014.

[12] V. Samavatian, H. Iman-Eini, and Y. Avenas, "An efficient online time-temperature-dependent creep-fatigue rainflow counting algorithm," International Journal of Fatigue, vol. 116, pp. 284-292, 2018.

Research on Wire Sweep of Integrated Circuit Packaging Based on Three-dimensional Flow Simulation

1st Fang Qu
R & D department
HuaTian Technology(Xi'an)Co., Ltd
Xi'an, China
18729389793@163.com

2nd Ting.Cao
R & D department
HuaTian Technology(Xi'an)Co., Ltd
Xi'an, China
Ting.Cao@ht-tech.com

3rd Huan.Yang
R & D department
HuaTian Technology (Xi'an)Co., Ltd
Xi'an, China
Huan.Yang@ht-tech.com

4th Ning.Sun
R & D department
HuaTian Technology(Xi'an)Co., Ltd
Xi'an, China
Ning.Sun@ht-tech.com

5th Qiang.Wei
R & D department
HuaTian Technology(Xi'an)Co., Ltd
Xi'an, China
Qiang.Wei2@ht-tech.com

6th Xiaojian.Ma
R & D department
HuaTian Technology(Xi'an)Co., Ltd
Xi'an, China
Xiaojian.Ma@ht-tech.com

7th Weidong Liu
R & D department
HuaTian Technology(Xi'an)Co., Ltd
Xi'an, China
Wade.liu@ht-tech.com

Abstract—Semiconductor manufacturing technology is becoming more and more rapidly. In the process of Integrated Circuit (IC) encapsulation, when wires contact each other, it will cause short circuit. Wire sweep has become the main factor affecting the reliability of the product. Therefore, it is a great challenge to master wire sweep in IC packaging process. This paper takes Low Profile Fine Pitch Ball Grid Array (LFBGA) as the research object, applies moldex3D three-dimensional flow simulation technology to simulate the molding process of electronic products, and predicts the wire sweep problem that may occur in the molding process. The simulation results are compared with the actual experimental results, which shows that there is a good consistency between the two results. The research results show that the wire is close to the packaging gate, and the wire sweep is large; the wire is perpendicular to the melt front flow direction; therefore, when the IC packaging wire layout, long wires are located under the chip (far away from the gate), and short wires are located above the chip (near the gate), so as to reduce wire sweep. Secondly, the flow simulation of packaging products was carried out by changing the epoxy molding compounds, and the results of mold flow analysis of different molding compounds were compared. It was found that the epoxy molding compounds with higher viscosity had greater drag force on the wire, and the molding compound with higher viscosity had greater wire sweep than those with lower viscosity. The problem of wire sweep in IC packaging could be improved by selecting the epoxy molding compound with lower viscosity. This study has a certain reference value and guiding significance for the design and improvement of IC packaging.

Keywords—wire sweep, moldex3D, layout, viscosity, LFBGA

I. INTRODUCTION

With the rapid growth of Integrated Circuit (IC) industry, semiconductor manufacturing technology is facing challenges. In the process of IC encapsulation, wire sweep has become the main factor affecting the reliability of the product. Therefore, it is a great challenge to master the wire sweep in IC packaging process. During the filling process of Epoxy Molding Compounds (EMC) in IC molding process, the high viscosity melt and its rapid flow may cause the wire sweep problem.

After the chip is packaged, the distance between the wires will not be the same as the original design. When the wires deform or contact with other wires, there may be a short circuit problem. Therefore, large deviation of the wires should be avoided in the design and packaging process. Therefore, it is a great challenge to master the wire sweep in IC packaging process. If wire sweep is too large, the adjacent wires will contact each other, the contact point will break or the wires will contact with the chip, resulting in short circuit and damage to the IC products. When IC module is developing towards high density and thin module, it is more important to solve the problem of wire sweep; This phenomenon is especially serious for high pin count and thin package. With the trend of IC lightweight, this problem becomes more and more important, which is an urgent problem to be overcome. Sejin Han analysised wire-sweep based on a three-dimensional model, and comparised with numerical and experimental results [1]. Ramdan, D found that packaging with a lower viscosity shows lower air trap, lower pressure distributions, and lower wire deformation [2]. Wen-Ren Jong; You-Ren Chen found that the high density of bonding wires form a separating layer which will hold back the molding compound flowing through these regions [3]. Huang-Kuang Kung and Bo-Wun Huang found that the effects of bond span and bond height on wire sweep in the elevated-temperature environment can be obtained by ansys analysis [4]. Subramanian N.R. presents an evaluation study on three-dimensional mold filling analysis on a demonstrator package with alternative simulation programs to check the influence of meshing methodologies and on the accuracy of wire-sweep predictions compared with measured wire-sweep data [5]. Ali S. S. Skh found that lower wire loop height and optimum moulding parameters are the essential factors that will improve the wire sweep performance during the mould encapsulation process [6]. Siti Sofiyah Skh Ali found that wire length has the most significant correlation with wire sweep percentage [7]. Skh Ali S.S and Seoh Hian Serene found that the wire sweep characterization carried out on low quad flat package subject to various wire location, mold flow

978-1-6654-1392-3/21 $31.00 © 2021 IEEE

direction, wire length, wire pitch and wire angle [8]. Myung focused failure analysis of a system-in-package component including wire sweep and EMC molding flow during the molding process [9]. Wu Yuelin investigated the electrochemical characterizations in the field of wire metallurgy (Pd concentration) and molding compound chemistry (chloride concentration) to find ways to reduce metallic entities' susceptibility to corrosion at ball-pad interfaces [10]. This paper takes Low Profile Fine Pitch Ball Grid Array (LFBGA) as the research object, applies moldex3D three-dimensional flow simulation technology to simulate the packaging process of electronic products, and explores the possible wire sweep problem in packaging process, so as to optimize the wire layout and improve the wire sweep problem.

II. WIRE SWEEP MECHANISM AND LFBGA MODEL

A. Wire Wweep Mechanism

In the filling stage of IC packaging, the drag force exerts by the viscous fluid of epoxy on the wires will cause the deformation of the wires, which will make the wires contact each other more seriously. There may be a short circuit problem, which will lead to the failure of IC packaging. This phenomenon is the problem of wire sweep. moldex3D can provide a convenient tool to predict wire sweep with stress analysis, and directly display the analysis results in moldex3D project. Wire Sweep Index (WSI) is often used to measure the degree of wire displacement, as shown in Fig. 1. Its definition is as follows:

$$\frac{D_N}{L} = WSI(\%) \tag{1}$$

In which the D_N is the maximum deformation constant for the wire, and the L) is the projection length of the wire [11].

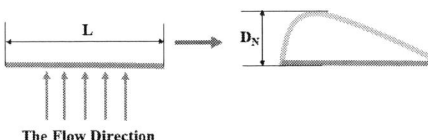

Fig. 1. Definition of WSI

This paper mainly considers the influence of the drag force of EMC the wire in the filling process, and selects the IC packaging analysis module in moldex3D software for simulation analysis. Therefore, the Lamb's model is selected as the mathematical model for wire migration calculation in this paper.

$$C_D = \frac{8\pi}{Re[2.002 - ln(Re)]} \tag{2}$$

In which Re means the Renault coefficient.

The expression of drag force produced by EMC material is:

$$F = \frac{C_D \rho d U^2}{2} \tag{3}$$

In which F is the force on the wire per unit length; ρ is the fluid density; U is the fluid velocity; d is the diameter of wire; CD is the viscosity coefficient.

B. LFBGA Model

The LFBGA packaging model includes: EMC material, chip, substrate and wires. The hybrid grid forming Boundary Layer Mesh (BLM) method is used for grid generation, and the number of grids is 3157836.

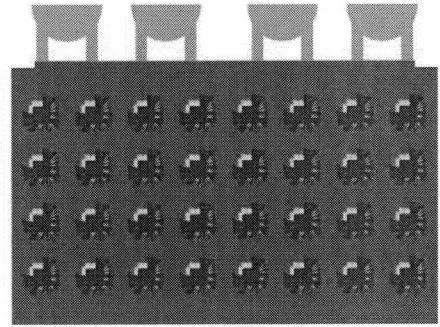

Fig. 2. Structure diagram of LFBGA packaging product

Analysis of LFBGA packaging product with mlodex3D . The length of the product is 120mm, the width is 67mm, and the thickness of the LFBGA package is 0.53mm. The process parameters are shown in Table I. In this paper, the EMCs of KE-G1250M and EME-G770L are selected for molding flow simulation analysis. The viscosity curves of the two kinds of EMCs are shown in Fig. 3 and Fig. 4. The viscosity of EME-G770L is larger than KE-G1250M at 175℃.

TABLE I. LFBGA PACKAGING TECHNOLOGY

Transfer time (sec)	Mold Temperature (℃)	Transfer Pressure(MPa)	EMC
13	175	23.5	KE-G1250M EME-G770L

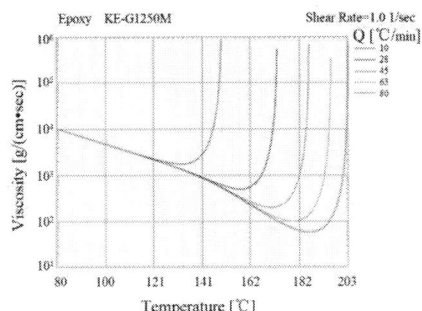

Fig. 3. Viscosity curve of KE-G1250M

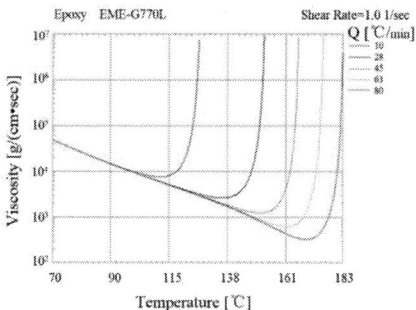

Fig. 4. Viscosity curve of EME-G770L

III. RESULT AND DISCUSSION

The analysis of moldex3D is to simulate the whole process of filling the mold cavity with EMC materials. According to the simulation results, the flow of EMC materials in the model cavity can be obtained, and wire sweep can be observed.

A. Filling Analysis on LFBGA Packaging

The solid mesh of the structure is imported into moldex3D for filling analysis. Fig. 5 and Fig. 6 shows the melt front time diagram of the filling analysis of the two kinds of EMC product structures. The melt front time reflects the flow situation of the EMC material in the mold cavity. It can be seen from the figure that the flow of the EMC in the mold cavity of the two kinds of EMC material is relatively uniform, and there are no defects such as short shot, burning out and high shear. This also shows that the preset process parameters and gate size and location are basically reasonable, and also shows that the designed structure and mold cavity structure are reasonable. KE-G1250M and EME-G770L are used as EMC, the filling time is 12.651s and 12.725s respectively, and there is no significant difference in filling time.

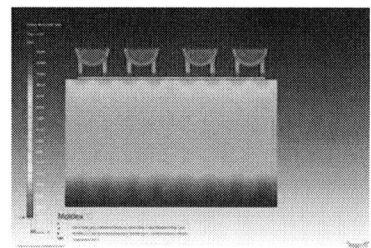

Fig. 5. The melt front time of KE-G1250M packaging product

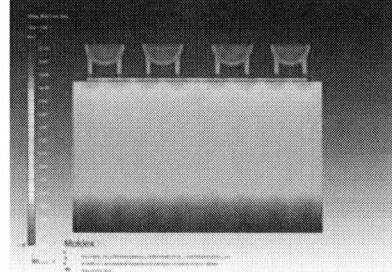

Fig. 6. The melt front time of EME-G770L packaging product

B. Wire Sweep Analysis on LFBGA Packaging

There are many factors that affect wire sweep, one of which is the viscosity of molding compound. In this paper, we choose two kinds of EMC with different viscosities to make the distribution of wire in the mold cavity different. One EMC is KE-G1250M, the other EMC is EME-G770L. Wire sweep results of two packaging products calculated by mlodex3D are as follows:

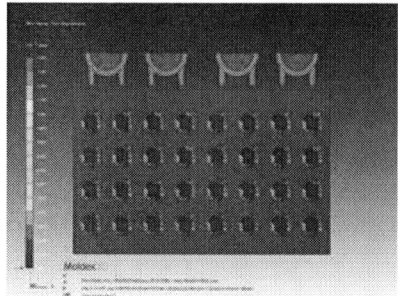

Fig. 7. Total displacement of wire sweep on KE-G1250M packaging product

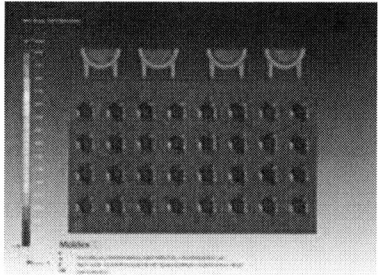

Fig. 8. Total displacement of wire sweep on EME-G770L packaging product

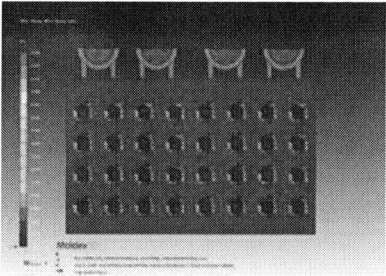

Fig. 9. Wire sweep index on KE-G1250M packaging product

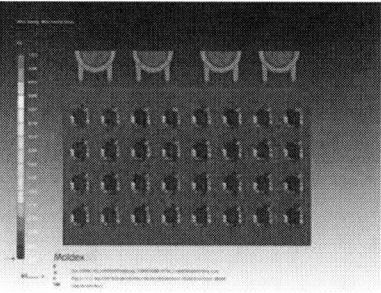

Fig. 10. Wire sweep index on EME-G770L packaging product

Fig. 11. Wire sweep index on near gate of KE-G1250M packaged product

Fig. 12. Wire sweep index on near gate of EME-G770L packaged product

In Fig. 7 and Fig. 8 KE-G1250M packaged product, the total displacement and wire sweep index are 0.01503 mm and 0.498% respectively (because the wire sweep is small, the last four decimal places are reserved for comparison). Fig. 9 and Fig. 10 show the total displacement and wire sweep index on EME-G770L packaging products, with the maximum value of 0.03419mm and 1.123% respectively. The area with large wire sweep is also mainly produced near the gate and the position where the wire is perpendicular to the flow direction of the melt front, which is due to the relatively fast flow speed of the molding material near the gate. Fig. 11 and Fig. 12 are the enlarged schematic diagram of wire sweep near the gate. It can be seen clearly from the figure that the wire sweep in the long welding wire area is larger than that in the short welding wire area. This is because the longer the wire length is, the smaller the stiffness is. When subjected to a certain drag force, the larger the deformation is, and the larger the wire sweep is. It can be seen from the analysis results of wire deviation that EME-G770L has greater influence on wire in filling process than KE-G1250M.

C. Simulation and experiment and Discussion

In order to verify the correctness of the simulation results, X-ray is used to measure wire sweep of LFBGA products. In this measurement, six wires are selected on each side of the chip, a total of twenty-four wires are selected, and all wires are numbered as: ID1, ID2, ID3 ... ID24. The actual measurement figure is shown in Fig.11 and Fig. 12, and the comparison between simulation and experiment of single encapsulation is shown in Fig. 13. The data comparison between simulation results and experimental results of wire sweep index is shown in Fig. 14.

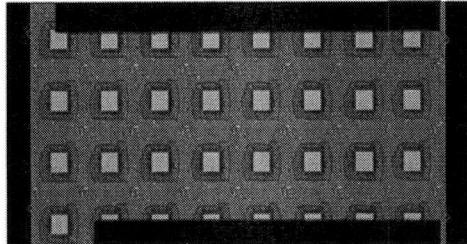

Fig. 13. Actual measurement figure of LFBGA packaged product

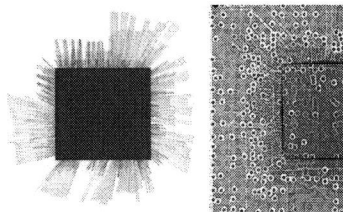

Fig. 14. The comparison figure between simulation and experiment of single LFBGA packaged product

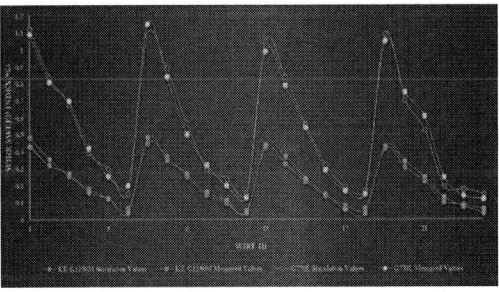

Fig. 15. The data comparison figure between simulation results and experimental results of WSI

It can be seen that the EMC has an important influence on wire sweep, because the resistance generated on the wire comes from the force generated when the polymer melt flows through wire in the cavity. The comparison of simulated and measured wire sweep data shows that the evaluation of wire sweep by two kinds of EMC is in good agreement with the experimental results in Fig. 15. It can be seen that KE-G1250M has a lower wire sweep index than EME-G770L due to its lower viscosity. This trend can also be seen from the simulation data and experimental results. This shows that the current moldex3D three-dimensional flow simulation can solve the actual prediction of wire sweep in the packaging process.

IV. CONCLUSIONS

In this paper, the influence of different EMC on wire sweep is studied by numerical simulation and experimental verification methods. This paper takes LFBGA as the research object, applies moldex3D three-dimensional flow simulation technology to simulate the packaging process of electronic products, and predicts the wire sweep problem that may occur in the molding process. The simulation results are compared with the actual experimental results, which shows that there is a good consistency between the two results. The conclusions are as follows:

978-1-6654-1392-3/21 $31.00 © 2021 IEEE

1) The wire is close to the packaging gate, and the wire sweep is large; the wire is perpendicular to the melt front flow direction; therefore, when the IC packaging wire layout, long wires are located under the chip (far away from the gate), and short wires are located above the chip (near the gate), so as to reduce the wire sweep.

2) The epoxy molding compound with higher viscosity has greater drag force on wires, and the epoxy molding compound with higher viscosity has greater wire sweep than those with lower viscosity.

The problem of wire sweep in IC packaging can be improved by selecting the epoxy molding compound with lower viscosity. This research has a certain reference value and guiding significance for the design and improvement of IC packaging.

ACKNOWLEDGMENT

This work is partially funded by HuaTian Technology (Xi'an)Co., Ltd. The authors specially thank the staff of R & D department in HuaTian Technology(Xi'an)Co., Ltd for their great support.

REFERENCES

[1] Sejin Han, Franco S.etal, "Three-dimensional simulation of wire sweep during semiconductor-chip encapsulation," Society of Plastics Engineers 61st Annual Technical Conference (ANTEC 2003) Materials May 4-8, 2003 Nashville, Tennessee.2003J. Clerk Maxwell, A Treatise on Electricity and Magnetism, 3rd ed., vol. 2. Oxford: Clarendon, 1892, pp. 68–73.

[2] Ramdan, D, "FSI simulation of wire sweep PBGA encapsulation process considering rheology effect," J. Components, Packaging and Manufacturing Technology, IEEE Transactions on, 2012, vol. 2, pp. 593-603.

[3] Wen-Ren Jong,You-Ren Chen, "CAE simulation and verification of wire sweep for BGA 436," ASME international mechanical engineering congress and exposition, 2000.

[4] Huang-Kuang Kung, Bo-Wun Huang, "On the wire sweep experiments and predictions of wire for semiconductor wirebonding technology," International Conference on Advanced Manufacture; 20051128-1202; Taipei(CT). 2006M. Young, The Technical Writer's Handbook. Mill Valley, CA: University Science, 1989.

[5] Subramanian N.R.,Teo Soon Tong, Ian K.L.K., Heinz P,"Improving predictability of wire-sweep in cavity filling analysis," Electronics Packaging Technology Conference (EPTC), 2011 IEEE 13th.2011.

[6] Ali S. S. Skh,Teh S. S. H.,Ang B. C, "Wire sweep improvement study of multi-tier palladium-copper wire bonding on low-quad flat package using low-alpha green mould compound," J. Materials Research Innovations, 2014, 18(Suppla6): pp. 214-219.

[7] Siti Sofiyah Skh Ali, Serene Teh Seoh Hian, Bee Chin Ang, "The effects of wire geometry and wire layout on wire sweep performance using LQFP packages in transfer mold," J. International Journal of Precision Engineering and Manufacturing, 2014, vol. 15.

[8] Skh Ali S.S., Seoh Hian Serene Teh, Ang B. C, "Wire sweep characterization of multi-tier palladium-copper (Pd-Cu) wire bonding on LQFP package using low alpha green mold compound," IEEE Electronics Packaging Technology Conference, 2013.

[9] Myung, Woo-Ram, Bang, Jae-Oh, Ha, Sang-Su, Kim, Dae Up, Jung, Seung-Boo, "Wire Sweep Characterization in System-in-Package (SiP) Component with Au Wire Bonding During Epoxy Molding Compounds Molding Process," J. Nanoscience and Nanotechnology Letters, 2015, vol. 2.

[10] Wu Yuelin, Subramanian K. N., Barton Scott Calabrese, Lee Andre, "Electrochemical studies of Pd-doped Cu and Pd-doped Cu-Al intermetallics for understanding corrosion behavior in wire-bonding packages," J. Microelectronics & Reliability, 2017, 78(nova): pp. 355-361.

[11] Li Chao, "Packaging structure design and molding process analysis of MEMS alcohol sensor. Guilin University of Electronic Science and technology," 2015.

Characteristics of 10-110GHz Transmission Lines on Fused Silica Substrate for Millimeter-wave Modules

Tian Yu
Xiamen university
School of Electronic Science and Engineering
Xiamen Sky Semiconductor
Xiamen, China
36120200155810@stu.xmu.edu.cn

Kai Xue
University of Electronic Science and Technology of China
School of Electronic Science and Engineering
Xiamen Sky Semiconductor
Chengdu, China
uskai@126.com

Ke Li
Xiamen university
School of Electronic Science and Engineering
Xiamen Sky Semiconductor
Xiamen, China
36120200155818@stu.xmu.edu.cn

Yihang Liang
Xiamen university
School of Electronic Science and Engineering
Xiamen Sky Semiconductor
Xiamen, China
36120201150365@stu.xmu.edu.cn

Daquan Yu*
Xiamen university
School of Electronic Science and Engineering
Xiamen Sky Semiconductor
Xiamen, China
yudaquan@xmu.edu.cn

Abstract—**With the development of millimeter-wave (mm-W) wireless communication, the pursuit for high electrical performance package material is crucial. Fused silica is a promising material for ultra-high frequency application due to the excellent electrical properties, higher integration and higher density interconnects. In this paper, the microstrip ring resonators (MRRs) with different radius are fabricated on 6 inch fused silica wafer with 200μm thickness to obtain the dielectric constant from 10GHz to 100GHz. In addition, conductor-backed coplanar waveguide (CBCPW) with a length of 10mm is also designed, fabricated and measured. The measured insertion loss is 0.05dB/mm at frequency of 40GHz and 0.23dB/mm at frequency of 110GHz, which means that fused silica is an ideal substrate for mm-W modules.**

Keywords—millimeter-wave, dielectric constant, microstrip ring resonator, conductor-backed coplanar waveguide, through glass via

I. INTRODUCTION

With the developing diverse high-speed wireless communications applications, such as high-resolution imaging, automotive radars, IoTs and 5G communication, the millimeter-wave (mm-wave) technology has drawn a lot of attention due to the advantages such as wide spectrum bandwidth, high speed, low latency and high integration. Compared with epoxy molding compound (EMC), organic substrate and low-temperature cofired ceramics (LTCC) which are widely used in mm-W modules, glass could provide fine pitch line and good coefficient of thermal expansion (CTE) match with silicon chip [1].

Fused silica is a prime candidate for mm-W application due to its excellent electrical properties. The accurate electrical parameters of fused silica need to be measured for mm-W simulation work. In this paper, the values of dielectric constant at different frequency are extracted by microstrip ring resonator (MRR) method. The MRR is fabricated on 6 inch fused silica wafer with a thickness of 200μm. To get more accurate electrical parameters of fused silica, there are only fused silica and copper in the test structures. From the measured resonant frequencies of MRR test vehicle, the values of dielectric constant and loss tangent are derived. Besides, the conductor-backed coplanar waveguide (CBCPW) structure is realized by through fused silica vias with diameter of 60μm to characterize the electrical performance.

II. PROCESS FLOW OF SINGLE LAYER GLASS SUBSTRATE

As is shown in Fig.1, the fabrication process of the glass substrate with through vias is divided into four steps. Firstly, TGVs with diameter of 60μm are formed in 6-inch fused silica wafer using laser induced deep etching (LIDE) technology [2]. The vias formed in fused silica is relatively vertical. Then the double-sided physical vapor deposition (PVD) is applied to deposit titanium adhesion layer and copper seed layer with 5000A and 10000A thickness respectively. Then the dry films are applied on double side of glass wafer and the photolithography process is applied to realize double-sided patterns. At last, vias are full-filled by double-sided electroplating, and the RDL with a thickness of 5μm is formed meanwhile.

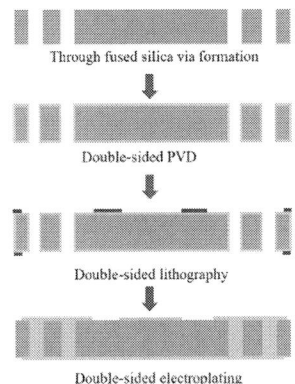

Fig. 1. The process of glass substrate with through vias

The diverse samples on 6 inch wafer is shown in Fig. 2. The 10mm CBCPW line and MRRs with different radius are fabricated after wafer level process.

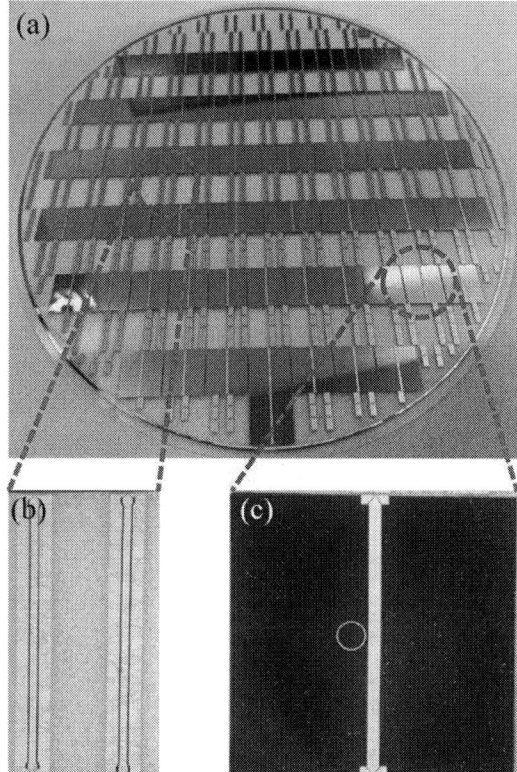

Fig. 2. The fabricated samples on glass substrate: (a) 6 inch fused glass wafer; (b) 10mm CBCPW line and (c) MRR with radius of 0.3mm

III. CHARACTERIZATION OF FUSED SILICA USING MICROSTRIP RING RESONATOR

The ring structure is an ideal solution for dielectric constant measurement in mm-W frequency domain [3]. Ten MRRs with mean radius of 3.015mm, 1.515mm, 1.015mm, 0.765mm, 0.615mm, 0.505mm, 0.43mm, 0.37mm, 0.335mm and 0.3mm and line width of 30μm are fabricated on glass substrate with a thickness of 200μm.

Fig.3 shows that the ring resonator structure consists of CBCPW to microstrip line transition, microstrip line, coupling gap and ring resonator. The line width and spacing of CBCPW are 80μm and 40μm for probe measurement with a pitch of 150μm. The line width of microstrip is 430μm to realize 50Ω impedance and the coupling gap is 120μm.

The effective dielectric constant ε_e can be expressed as equation (1) [4-5]. Where c is the light speed in vacuum, f_1 is the first resonant frequency and r is the mean radius of the ring. Then the relative dielectric constant ε can be derived by equation (2) (3) (4). Where h is the thickness of glass substrate, w is the line width of microstrip line and d is the thickness of copper.

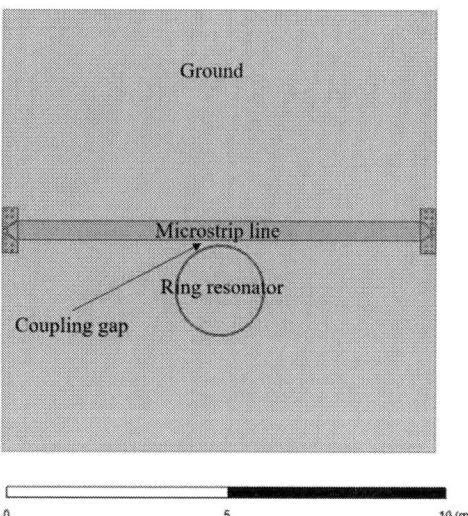

Fig. 3. MRR with radius of 0.49mm

The values of dielectric constant ε are calculated based on the measured resonate frequency and structure parameters, which is shown in Fig.4. The MRRs are aimed to realize the resonance at frequency of 10GHz, 20GHz, 30GHz, 40GHz, 50GHz, 60GHz, 70GHz, 80GHz, 90GHz and 100GHz.

$$\varepsilon_e = \left(\frac{c}{2\pi r f_1}\right)^2 \tag{1}$$

$$\varepsilon_r = \frac{2\varepsilon_e + M - 1}{M + 1} \tag{2}$$

$$M = \left(1 + \frac{12h}{w_e}\right)^{-\frac{1}{2}} \tag{3}$$

$$w_e = w + \frac{1.25d}{\pi}\left[1 + ln\left(\frac{2h}{d}\right)\right] \tag{4}$$

Besides, a fitting approach is applied by finite element analysis, in which the S21 of simulation model is approaching measured results by iterative [6]. Fig.5. shows the fitting results and measured results of 3.015mm MRR and 0.3mm MRR, which are consistent at the resonate frequency. The values of dielectric constant by calculating and fitting do not agree very well. It is considered to be the fabrication error and the lack of test calibration.

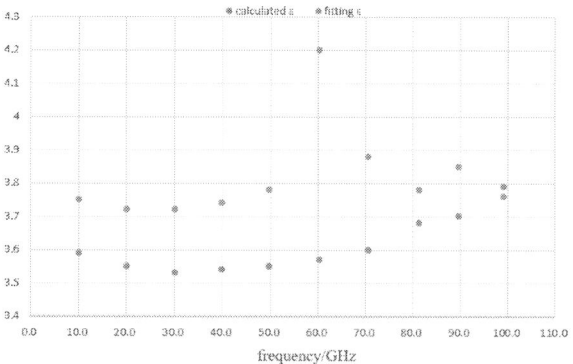

Fig. 4. The fitting ε and calculated ε of ten MRRs with different radius from 10GHz to 100GHz

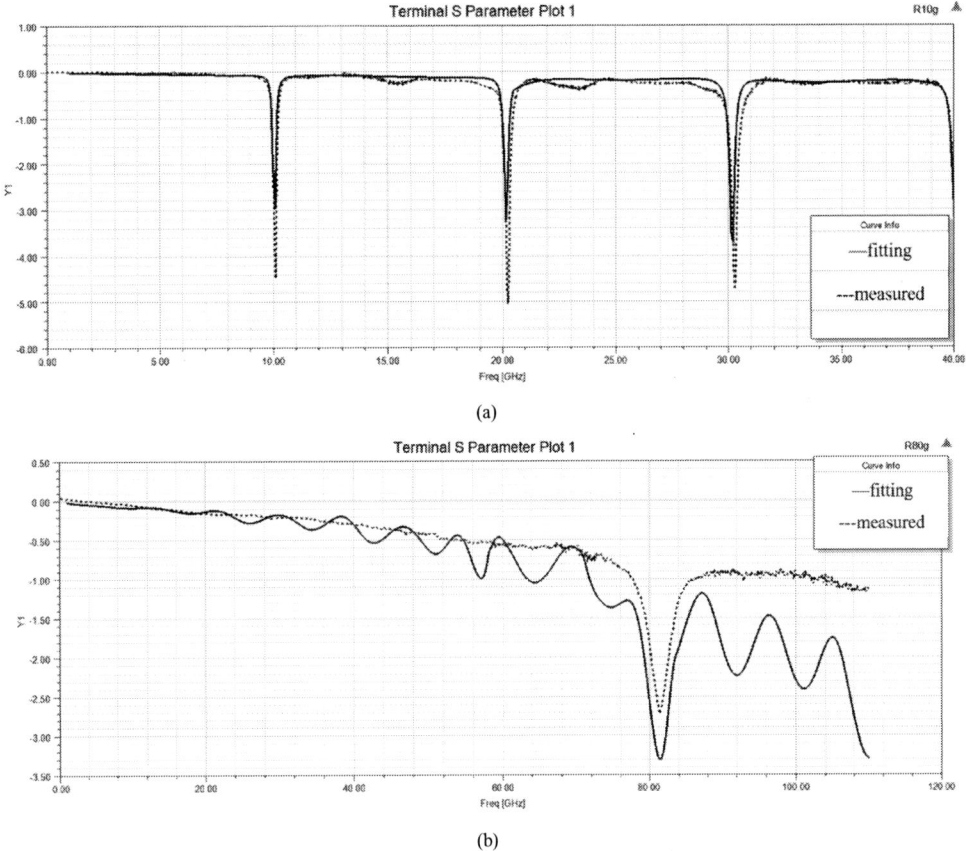

(a)

(b)

Fig. 5. The fitting results and measured results of (a) 3.015mm MRR and (b) 0.37mm MRR

IV. CHARACTERIZATION OF CBCPW

To evaluate electrical performance of transmission line on fused silica, the CBCPW [3] is designed with line width of 230μm and line spacing of 40μm to realize impedance at 50Ω. As is shown in the Fig.6, the schematic view of CBCPW consists of topside lines, back ground and glass substrate. The top-view of CBCPW simulation model with length of 10mm is shown in the Fig.7. TGV with a diameter of 60μm and height of 200μm is used to conduct ground.

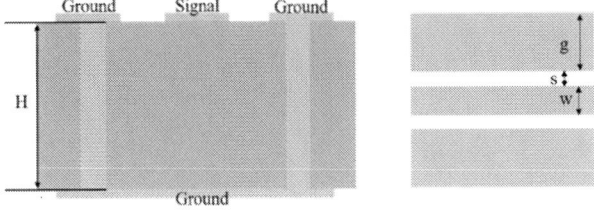

Fig. 6. The schematic sideview and topview of CBCPW

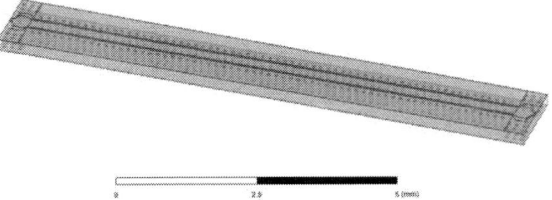

Fig. 7. The CBCPW model with length of 10mm

The measured insertion loss of 10mm CBCPW line is 0.5dB at frequency of 40GHz and 2.3dB at frequency of 110GHz, which is shown in the Fig.8(a). The measured return loss is greater than -20dB at frequency of 40GHz and greater than -10dB at frequency of 110GHz, which is shown in the Fig.8(b).

The comparison of transmission performance on different substrate for high frequency applications is shown in Table.1. Three kinds of glass all have low insertion loss in mm-W frequency domain. Besides, the fine interconnection capability of glass is superior than other substrate material. It shows that fused silica is suitable for mm-W modules. material. The extracted dielectric constant value can be used in design work of mm-W modules.

978-1-6654-1392-3/21 $31.00 © 2021 IEEE

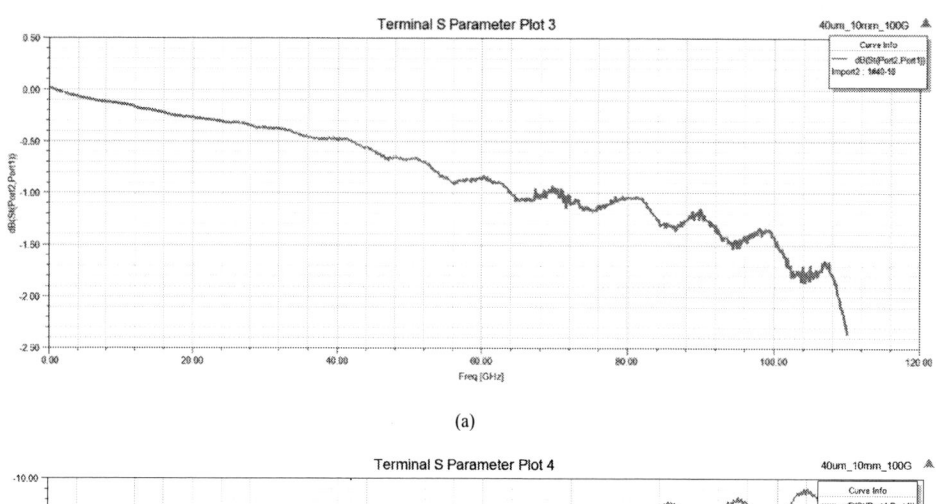

(a)

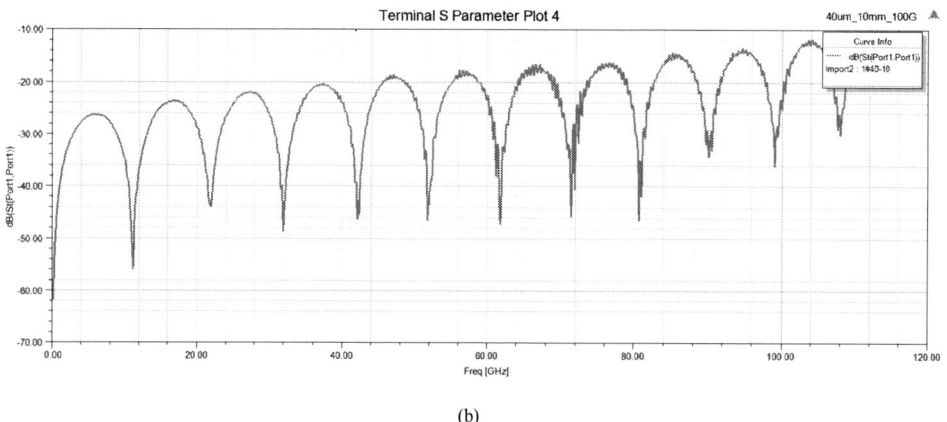

(b)

Fig. 8. The measured S parameter of 10mm CBCPW transmision line from frequency 0 to 110GHz: (a) S21 and (b) S11

TABLE I.   COMPARISON OF TRANSMISSION PERFORMANCE ON DIFFERENT SUBSTRATE FOR MM-W MODULES

Ref.	Substrate material	Structure	Substrate thickness (μm)	f (GHz)	IL (dB/mm)
[8]	EMC	CPW	/	60	0.2
[9]	LTCC	CPW	100	60	0.5
[3]	LCP	CPW	101.6	110	0.1
[1]	Glass (AGC ENA1)	CPW	100	40 110 170	0.085 0.21 0.275
[6]	Glass (SCHOTT D263)	CPW	300	60	0.375
This work	Fused silica	CBCPW	230	40 110	0.05 0.23

V. CONCLUSION

With the development of mm-W applications, the need for high performance material is urgent. In this work, the high frequency characteristic of fused silica are studied. The MRRs with different radius are fabricated and measured to calculate the values of dielectric constant ε and a simulation approach is used to obtain fitting ε. The CBCPW are fabricated and measured to investigate the transmission performance of fused silica. The measured insertion loss of CBCPW is 0.05dB/mm at frequency of 40GHz and 0.23dB/mm at frequency of 110GHz. It shows that fused silica is compatible for future packaging

ACKNOWLEDGMENT

The authors thank Mr. Yujin Zhou for the contribution of design work and measurement work. This research is supported by the National Natural Science Foundation of China (Grant No. 61974121).

REFERENCES

[1] M. u. Rehman, S. Ravichandran, S. Erdogan and M. Swaminathan, "W-band and D-band Transmission Lines on Glass Based Substrates for Sub-THz Modules," 2020 IEEE 70th Electronic Components and Technology Conference (ECTC), Orlando, FL, USA, 2020, pp. 660-665.

[2] Chen, L., Yu, D. Investigation of low-cost through glass vias formation on borosilicate glass by picosecond laser-induced selective etching. J Mater Sci: Mater Electron (2021).

[3] D. C. Thompson, O. Tantot, H. Jallageas, G. E. Ponchak, M. M. Tentzeris and J. Papapolymerou, "Characterization of liquid crystal

polymer (LCP) material and transmission lines on LCP substrates from 30 to 110 GHz," in *IEEE Transactions on Microwave Theory and Techniques*, vol. 52, no. 4, pp. 1343-1352, April 2004

[4] G. Zou, H. Gronqvist, J. P. Starski, and J. Liu, "Characterization of liquid crystal polymer for high frequency system-in-a-package applications," IEEE Trans. Adv. Packag., vol. 25, pp. 503–508, Nov. 2002.

[5] C. Zihao, L. T. Guan, D. H. S. Wee, C. T. Chong and S. Bhattacharya, "Characterization of molding compound material and dielectric layer of RDL," *2017 IEEE 19th Electronics Packaging Technology Conference (EPTC)*, 2017, pp. 1-5.

[6] T. Zhang et al., "Millimeter-Wave Antenna-in-Package Applications Based on D263T Glass Substrate," in *IEEE Access*, vol. 8, pp. 67921-67928, 2020.

[7] D. Thompson, J. Papapolymerou and E. Tentzeris, "W-Band Characterization of Finite Ground Coplanar Transmission Lines on Liquid Crystal Polymer (LCP) Substrates," 2003 ECTC Conference, pp. 1652-1655, New Orleans, LA, May 2003

[8] M. Wojnowski, M. Engl, M. Brunnbauer, K. Pressel, G. Sommer and R. Weigel, "High Frequency Characterization of Thin-Film Redistribution Layers for Embedded Wafer Level BGA," *2007 9th Electronics Packaging Technology Conference*, 2007, pp. 308-314.

[9] M. F. Amiruddin *et al.*, "Low loss LTCC passive transmission line at Millimeter-wave frequency," *2010 IEEE Asia-Pacific Conference on Applied Electromagnetics (APACE)*, 2010, pp. 1-4.

Low temperature bonding by sintering of Ag nanoparticle paste with the assistance of MOD

Xun Liu[1,2]
[1]Shenzhen Institute of Advanced Electronic Materials, Shenzhen Institutes of Advanced Technology, Chinese Academy of Sciences
Shenzhen, China
[2] School of Materials Science and Engineering, Wuhan University of Technology
Wuhan, China

Yulei Yuan[1,2]
[1]Shenzhen Institute of Advanced Electronic Materials, Shenzhen Institutes of Advanced Technology, Chinese Academy of Sciences
Shenzhen, China
[2] Nano Science and Technology Institute, University of Science and Techology of China
Suzhou, China

Junjie Li*
Shenzhen Institute of Advanced Electronic Materials, Shenzhen Institutes of Advanced Technology, Chinese Academy of Sciences
Shenzhen, China
E-mail: lijj@siat.ac.cn

Li Liu
School of Materials Science and Engineering, Wuhan University of Technology
Wuhan, China

Rong Sun
Shenzhen Institute of Advanced Electronic Materials, Shenzhen Institutes of Advanced Technology, Chinese Academy of Sciences
Shenzhen, China

Abstract—To satisfy the performance gains of the third-generation semiconductors represented by SiC and GaN during use, novel high-temperature operation die-attachment material, Ag, have received extensive attention in power devices for its excellent thermal conductivity and reliability with high-temperature (over 250 °C) applications. At present, sintering silver nanoparticles is commonly used to achieve stable die-attach joint in industry. However, the size effect of nanoparticles makes the preparation and storage of very fine nanoparticles (less than 50 nm) extremely difficult, which can increase the cost of paste. In this paper, we introduce a method for preparing sintered solder with MOD (metal organic decomposition) and commercial Ag nanoparticles, which takes the advantage of Ag nucleation of MOD at low temperature to assist commercial Ag nanoparticles in the paste to achieve lower temperature sintering bonding. We used silver oxalate as a precursor and made it complex with 1,2-diaminopropane to prepare the MOD solution ink. The MOD ink was utilized as additive at low temperature to assist the connection of nanoparticles in the paste to form a strong joint. The results suggest that shear strength of sintering joint can reach 32.7 MPa at 175 °C.

Keywords—Silver, paste, nanoparticles, MOD, joint

I. INTRODUCTION

In the past several years, there are increasing appeals for smaller, faster, and more efficient device packages in the emerging electronic filed, such as aerospace, new energy vehicles, optoelectronic devices, which has led the development of high-power density [1]. The devices are required to sustain the extreme conditions including ultra high power and temperature, without compromising lifetime. The WBG have received extensive attention for supporting higher properties [2]. So, compared to the traditional Si-based semiconductor, the WBG semiconductors are able to operate perfectly in high temperatures (over 300 °C). Different mechanical strength, electrical conductivity and thermal expansion coefficient of the various layers can affect the thermoelectric stability and long-term service life of the electronic package under extreme conditions. So, it is important for the bonding materials in the connection of chip-to-substrate and substrate-to-base to meet the demand of effective heat dissipation and long-term reliability [3]. To achieve this, the die-attach materials are required to possess the low processing temperatures (especially, under 300 °C), high mechanical strength, good thermal and power cycling performance and high operating temperature (over 300 °C) [4]. However, conventional Sn-Pb solder paste cannot meet the joint due to its instability in high-temperature conditions above 150 °C (which is the restriction of Si-based) [5].

As a promising material, silver nano-solder play an essential role in the field of high-temperature advanced packaging, while possessing lower processing temperature to meet industrial needs due to a strong driving force provided by a higher surface energy of the nanoparticles. However, the smaller nanoparticles are easier to generate agglomeration due to the size effect, which increases the difficulty of nanoparticles in the process of preparation and storage [6]. What is more, it is unignored for the expensive cost and environmental impact of toxic organic solvent in the synthesis of ultrafine nanoparticles. All of the above factors further limit the research on the direction of synthesizing smaller nanoparticles.

The addition of the burning aid in the silver paste has become the key to solving this problem.. Sn-Ag alloys solders have been employed as a connecting material, due to the Ag-Sn solid solution formed by the reaction between Sn and Ag in low temperature, which can promote the further densification of the sintering structure and increase the shear strength of silver joints. But the presence of two metal elements may form intermetallic compounds to affect the reliability of the joint in long-term service. The research results of Hao Zhang and his workmates suggest that the Ag_2O can be performed as an active additive to bridge the micron-scale silver by in-situ reduced silver nanoparticles in the sintering. The silver joint can achieve the 40 MPa at only 180 °C without any assistance of pressure [7].

So we intend to introduce the metal organic decomposition (MOD) ink of silver to assist sintering of silver particles, because the MOD can decompose at low temperature to form nanoparticles [8]. MOD is often used as a raw material for the preparation of printed circuits due to low temperature decomposition ability. In the study of Junjie Li and his workmates, the MOD ink was prepared with silver oxalate and 1,2-Diaminopropane with PVAc, which can form conductive patterns by sintering at about 150 ℃ under certain stages. It can achieve 5.17 µΩ.cm and high dense microstructure [9].

In this paper, we introduce a method for preparing sintered solder with MOD and commercial silver nanoparticles, which takes the advantage of silver nucleation of MOD at low temperature to assist commercial silver nanoparticles in the paste to achieve lower temperature sintering bonding. We used silver oxalate as a precursor and made it complex with 1,2-diaminopropane to prepare the MOD solution ink. The MOD ink was utilized as additive at low temperature to assist the connection of nanoparticles in the paste to form a stable joint. The results suggest that shear strength of sintering joint can reach 32.7 MPa at 175 ℃.

II. EXPERIMENT

A. Materials

In this paper, we purchased Silver nitrate (AgNO$_3$,>99%), oxalic acid (H$_2$C$_2$O$_4$), ethylenediamine (C$_2$H$_8$N$_2$), ethylene glycol (C$_2$O$_2$H$_6$, EG), isopropyl alcohol (IPA) from Aladdin Reagent Co. Ltd. The commercial silver nanoparticles were obtained from Jiangsu Xian feng nano material technology Co., Ltd. Except for DBC which needs to be ultrasonic cleaned, all other reagents mentioned above do not require further purification.

B. Synthesis of the MOD ink

Before synthesis of the MOD ink, we required to prepare the silver oxalate powder. The method of preparation was as following, firstly we added AgNO$_3$ and 5 g oxalic acid dehydrate to 50 ml of deionized water, then stirring the mixed solution in a magnetic stirrer for 20min. After that, we put the collected silver oxalate precipitate in a vacuum drying oven to dry at room temperature for 24 hours to obtain the silver oxalate powder. Secondly, in order to prepare the MOD ink, we first need to make the organic solvent by using the EG (0.2 g), IPA (0.2 g) with diamino propane (0.1 g). Then silver oxalate is added to the above-mentioned organic solvent to prepare MOD ink. The Synthesis process and MOD ink reaction mechanism diagram is shown in Fig. 1.

C. The process of the preparation of Ag paste and bonding

We mixed MOD ink (0.05 g) and silver paste (0.2 g) into a mixing tank, and rotate in a high-speed mixer at 2000 speeds for 20 minutes. Then, we use the dispensing method to coat the paste on the DBC, and the prepared DBC substrate was sequentially plated with metals such as copper, nickel, and gold. Then we putted the coated DBC substrate in an oven for preheating at 80 °C for 10 minutes, and finally stacked it into a sandwich structure, and performed thermocompression bonding under a certain pressure (10 MPa) and different temperatures (150 °C, 175 °C, 200 °C). The preparation of the paste and the bonding process are shown in Fig. 2.

D. Characterization

The morphological features of commercial Ag particles and the surface of fracture and cross-sectional Ag-Ag bonding joint were detected by the field-emission scanning electron microscope (FE-SEM, FEI Nova Nana SEM 450). The thermal behavior of commercial Ag particles, MOD ink and the silver paste were determined by thermogravimetric analysis (TA-DSC Q600), which was performed with a heating rate of 10 °C /min from 25 °C to 350 °C in an ambient atmosphere. The purities of the sintering ink were investigated by XRD (Rigaku DlMax 2500). The shear strength of Ag-Ag joint was measured by the cutter of the shear tester (DAGE 4000). Also, an ion milling polishing (Gatan 697) was applied to treat the sandwich structure.

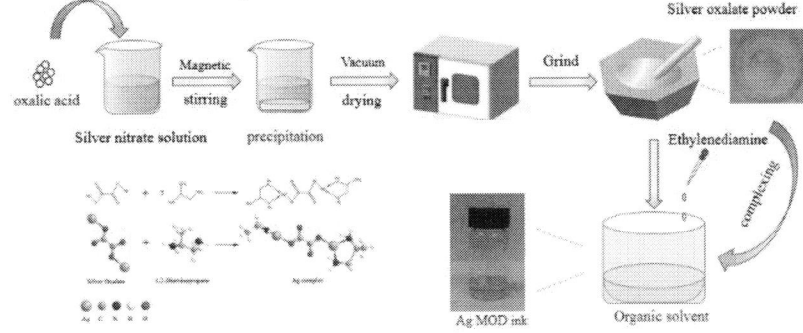

Fig. 1 The diagram of Synthesis process and MOD ink reaction mechanism

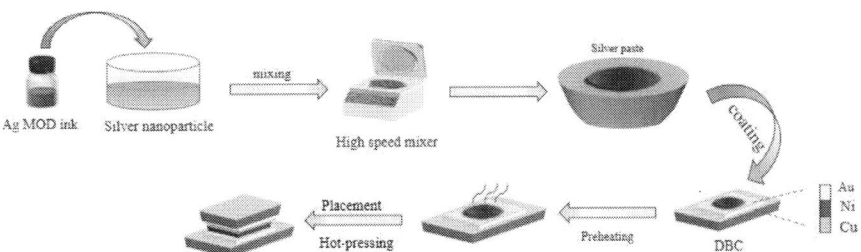

Fig. 2 The diagram of the preparation of the paste and the bonding process

III. RESULTS AND DISSCUSSION

A. Characterization of commercail Ag particles

As shown in Fig. 3 (a) and (b), the average particle size of commercial silver powder is 60 nm. Fig. 3 (c) suggests the commercial silver powder has a high solid content of 98.9 %, and Fig. 3 (d) reflects that the silver powder does not contain other impurities.

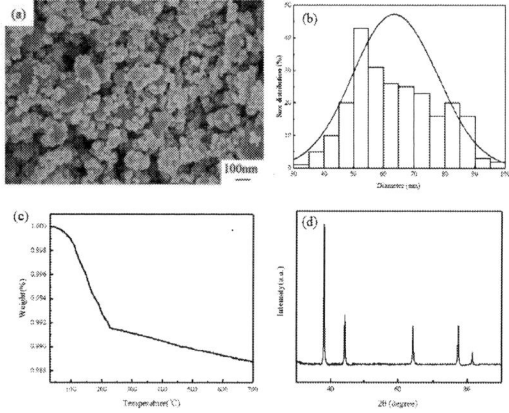

Fig. 3 (a) SEM image (b) relatively volume distribution (c) the TG curves (d) the XRD pattern of commercial Ag particles

B. The charaterization of MOD ink and silver paste

After the silver-ammonia complexation is complete, UV is used to detect the presence of silver particles in MOD ink, as shown in Fig. 4 (a). As expected, there is no absorption peak of silver particles in the wavelength range between 350 and 450 nm. FT-IR spectroscopy is used to observe the reflection of silver-ammonia complexes. From Fig. 4 (b), we can clearly see that the carbon-based peak at 1550 cm^{-1} has a significant shift, which implies the formation of a new bond position between the silver-ammonia complexes. Fig. 4 (c) shows the TG curves of MOD ink, which determines the MOD ink can decompose at 150 °C to generate silver nanoparticles to assist the low-temperature bonding of commercial silver powder in silver paste. At the same time, the XRD pattern is the result of sintering MOD ink at 175 °C for 20 min, which prove the purity of silver nanoparticles produced by MOD ink after sintering, as shown Fig. 4 (d)

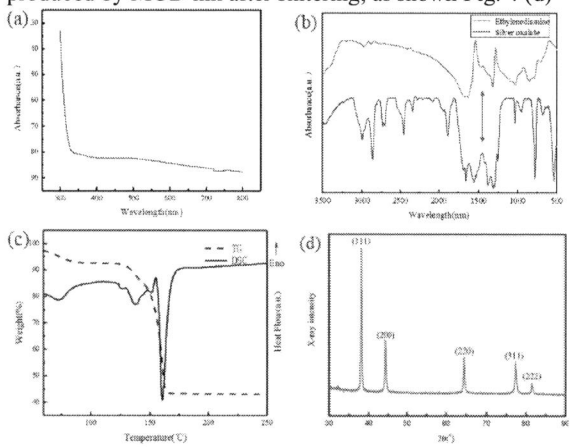

Fig. 4 (a) UV–Vis absorption spectra (b) FT-IR spectra (c) TG-DSC curves of the silver oxalate inks, (d) the XRD pattern of the MOD ink after sintering at 175 °C for 20 min.

C. Shear strength of Ag-Ag joints

The temperature is significant for stability of joints formed by sintering silver paste. The influence of temperature promotes the volatilization of the solvent and the diffusion between the silver nanoparticles, thereby further improving the ability of MOD ink to assist sintering [10]. Fig. 5 shows the temperature rising is able to improve the shear strength of joint. The SEM images were taken to observe the microstructure of fracture and cross-sectional surface.

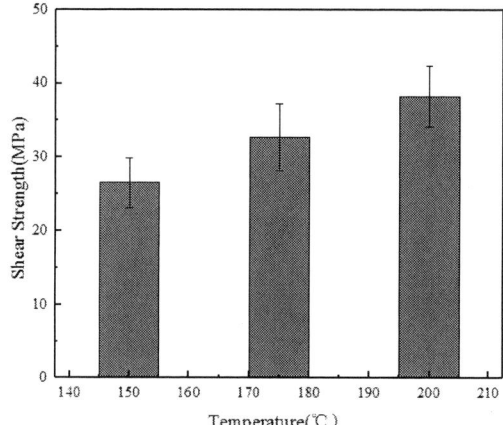

Fig. 5 Die shear strength of bonding joints with different temperature

Fig. 6 presents the SEM images of the fracture and cross-sectional surface at different temperatures. According to Fig. 6 (a-c), plastic deformation continues to increase as the temperature rises. At the same time, the cross-sectional image clearly shows that the sintered structure and density become more and more uniform from 150 °C to 200 °C. The SEM images are consistent with the distribution of shear strength at different temperatures. When the bonding temperature is 175 °C, the shear strength of Ag-Ag joint can achieve 32.7 MPa.

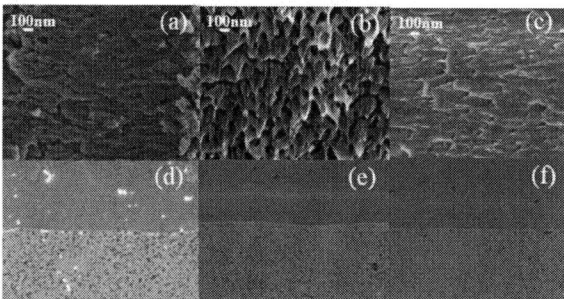

Fig. 6 SEM images of the fracture surface from (a-c) 150 °C to 200 °C respectively; SEM images of the cross-section surface at (d) 150 °C, (e) 175 °C and (f) 200 ° C respectively.

IV. CONCLUSION

In this study, we synthesize MOD ink by complexing silver oxalate and 1,2-propanediamine. Then we use MOD ink's ability to decompose and form silver nanoparticles at low temperatures to assist in sintering commercial silver nanoparticles at low-temperature. The bonded joint could achieve 32.7 MPa at 175 °C. The tested value of shear strength implies the new paste could satisfy the needs of industry. The results of SEM fracture and cross-sectional surface suggest high temperature can generate high stability. This study gives a novel idea for our next research that can further improve the silver in the application of industry.

ACKNOWLEDGMENT

This work is supported by National Natural Science Foundation of China (Grant No. 51805197 and 62004144).

REFERENCES

[1] J. Li et al., "wafer-level packaging," 2017.

[2] J. Liu, "Paste for High Power Device Packaging," pp. 755–761, 2020.

[3] Y. Gao et al., "Novel copper particle paste with self-reduction and self-protection characteristics for die attachment of power semiconductor under a nitrogen atmosphere," Mater. Des., vol. 160, pp. 1265–1272, 2018.

[4] Q. Jia et al., "Sintering Mechanism of a Supersaturated Ag-Cu Nanoalloy Film for Power Electronic Packaging," ACS Appl. Mater. Interfaces, vol. 12, no. 14, pp. 16743–16752, 2020.

[5] C. Chen et al., "Bonding technology based on solid porous Ag for large area chips," Scr. Mater., vol. 146, pp. 123–127, 2018.

[6] J. Yeom et al., "Ag particles for sinter bonding: Flakes or spheres?," Appl. Phys. Lett., vol. 114, no. 25, 2019.

[7] M. Wang, Y. H. Mei, J. Jin, S. Chen, X. Li, and G. Q. Lu, "Pressureless Sintered-silver Die-attach at 180C for Power Electronic Packaging," IEEE Trans. Power Electron., vol. 8993, no. c, 2021.

[8] W. Yang, F. Hermerschmidt, F. Mathies, and E. J. W. List-Kratochvil, "Comparing low-temperature thermal and plasma sintering processes of a tailored silver particle-free ink," J. Mater. Sci. Mater. Electron., vol. 32, no. 5, pp. 6312–6322, 2021.

[9] S. Zhou, X. Qi, Q. Kang, and C. Wang, "Low-temperature direct and indirect bonding using plasma activation for 3D integration," Proc. 2020 IEEE Int. Conf. Integr. Circuits, Technol. Appl. ICTA 2020, no. c, pp. 130–132, 2020.

[10] H.-Y. Kim et al., "Using Ag Sinter Paste to Improve the Luminous Flux and Reliability of InGaN-Based LED Package for Commercial Vehicle Daytime Running Light," ECS J. Solid State Sci. Technol., vol. 10, no. 1, p. 015004, 2021.

Cu-Cu joint formation by sintering of self-reducible Cu nanoparticle paste assisted by MOD under air condition

Yulei Yuan[1,2]
[1]Shenzhen Institute of Advanced Electronic Materials, Shenzhen Institutes of Advanced Technology, Chinese Academy of Sciences
Shenzhen, China
[2] Nano Science and Technology Institute, University of Science and Techology of China
Suzhou, China

Xun Liu[1,2]
[1]Shenzhen Institute of Advanced Electronic Materials, Shenzhen Institutes of Advanced Technology, Chinese Academy of Sciences
Shenzhen, China
[2] School of Materials Science and Engineering, Wuhan University of Technology
Wuhan, China

Junjie Li*
Shenzhen Institute of Advanced Electronic Materials, Shenzhen Institutes of Advanced Technology, Chinese Academy of Sciences
Shenzhen, China
E-mail: lijj@siat.ac.cn

Pengli Zhu
Shenzhen Institute of Advanced Electronic Materials, Shenzhen Institutes of Advanced Technology, Chinese Academy of Sciences
Shenzhen, China

Rong Sun
Shenzhen Institute of Advanced Electronic Materials, Shenzhen Institutes of Advanced Technology, Chinese Academy of Sciences
Shenzhen, China

Abstract—As the electronics industry develops, semiconductor devices increasingly tend to be used in high temperature application. Higher working temperature also puts forward new requirement for die-attach materials, copper nanoparticles (Cu NPs) have shown great promise as a die-attach material due to its excellent thermal and electrical conductivities. However, Cu NPs are easily oxidized. In this article, we introduce a self-reducible Cu NPs paste compose of Cu-based MOD, reducing solvent and commercial Cu NPs (100 nm), and the sintered joints can achieve high shear strength in air. The copper produced by the decomposition of self-reducible MOD helps sintering between copper nanoparticles and has a certain effect of inhibiting oxidation. The influence of MOD content on shear strength and sintered layer structure was explored. Finally, reliable Cu-Cu joints with a shear strength of 20.47 MPa was achieved with 250 ℃ bonding for 15 min in air. This novel Cu NPs paste is promised to be used as a bonding material for electronic packaging and provides a new method for copper sintering in air.

Keywords—Copper nanoparticles, MOD, Cu-Cu bonding joints, sintering

I.   INTRODUCTION

With the rapid development of electric vehicles, aerospace, 5G, high-speed rail and other technologies, high-power semiconductor devices have presented huge market requirements. Compared with Si based semiconductor materials, Wide-bandgap (WBG) semiconductor materials such as SiC and GaN can withstand the service temperature higher than 300 ℃[1], which are suitable for power semiconductor devices. However, traditional die-attach materials such as high-lead solders and tin-based solders can't meet the demand of high temperature serving. Therefore, novel die-attach materials with high electrical

and thermal conductivities which can be stably used above 300 ℃ are highly demanded[2].

In recent years, sintering of metal nanoparticles (NPs), especially Ag NPs has been widely studied, because of the high electrical and thermal conductivities of sintered Ag[3]. Besides, Ag can also be used as a high temperature interconnection material, due to its high melting point. However, the characteristic of high cost and the property of severe electromigration have limited the practical application of Ag NPs. Relatively, Cu NPs has strong advantages of reasonable price and excellent electromigration resistant performance, and the electrical and thermal conductivities of Cu similar to Ag. Therefore, Cu NPs show huge application prospects as a die-attach material[4-6]. Unfortunately, copper are easily oxidized under air condition, the oxide shells would suppress the sintering of copper nanoparticles, which hinder the application of Cu sintering[7]. At present, many methods have been reported, including the use of a protective atmosphere. But these methods tend to increase process cost and complexity. To solve this problem, many approaches have been proposed, including forming fresh copper by in-situ reduction of Cu-based metal organic decomposition (MOD) during sintering[8-10].

Inspired by related researches of MOD, we introduce a self-reducible Cu NPs paste compose of Cu-based MOD, carrier solvent and commercial Cu NPs. The effect of adding MOD in the bonding process was studied, and the effect of MOD content on enhancing the bonding strength and sintered structure was investigated.

II. EXPERIMENTAL

A. Materials

Ethylene glycol (EG), 2-amino-2-methyl-1-propanol (AMP), n-octylamine, formic acid, tert-butyl alcohol, hydrochloric acid and copper nanoparticles (100 nm) were purchased from Aladdin Reagent Co., Ltd, absolute ethyl alcohol and D-glucose monohydrate were purchased from Shanghai Lingfeng Chemical Reagent Co., Ltd. Copper(II) formate tetrahydrate were purchase from Alfa Aesa Reagent Co., Ltd.

B. Preparation of copper nanoparticles paste

Firstly, absolute ethyl alcohol was used to wash Cu NPs for two times. Then Cu NPs were washed with a mixture solution of ethyl alcohol (87.5 vol%) and formic acid (12.5 vol%) for three times to remove the oxide layer on the copper surface. The copper powders prtreated by formic acid were washed with ethyl alcohol for three times. Finally, Cu NPs were dispersed in tert-butyl alcohol, followed by freeze drying for 12 h.

To prepare copper-based MOD, the copper(II) formate tetrahydrate powder was added to 2-amino-2-methyl-1-propanol (AMP) and n-octylamine with a molar ratio of 1:1.5:1.5. In order to complete formation of the MOD, the solution was magnetically stirred for 2 h.

D-glucose monohydrate was dissolving in EG at 70 °C to prepare reducing solvent, the mass ratio of EG to D-glucose monohydrate was 100:35.

To prepare Cu NPs paste, the pre-treated Cu NPs were mixed with Cu-based MOD and reducing solvents with a solid content of 75 %, and the mass ratio of MOD to reducing solvents was 45:55, 55:45 and 65:35, respectively. Meanwhile, Cu NPs paste without MOD was also prepared to investigate the effect of MOD during bonding process. In order to maintain accuracy, the content of reducing solvent was consistent with the 55:45 Cu NPs paste.

C. Fabrication of sintered Cu-Cu joints

First of all, DBC substrates were pretreated with dilute hydrochloric acid, deionized water and absolute ethyl alcohol in sequence. The copper pastes were printed onto DBC substrates to make sandwich structures. The top and bottom DBC substrates are 3 × 3 mm^2 and 5 × 5 mm^2, respectively. The samples were initially pre-heated in air at 110 °C for 5 min. Subsequently, the pre-heated samples were heated to 250 °C for 15 min under air conditional, with a assisting bonding pressure of 9 MPa. Fig. 1 shows all the processes.

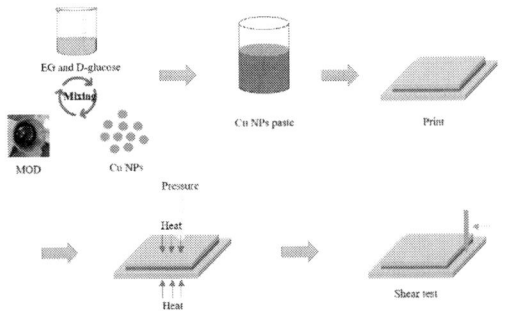

Fig. 1. Diagram of the Cu NPs paste prepare and sintering process

III. RESULT AND DISCUSSION

A. Sintering mechanism of the copper paste

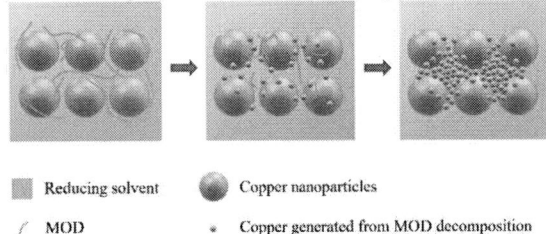

Reducing solvent   Copper nanoparticles

MOD   Copper generated from MOD decomposition

Fig. 2. Schematic illustration of the sintering behaviors of the Cu NPs paste

Generally, the sintering between adjacent copper nanoparticles is carried out by the diffusion of copper atoms, which requires a great driving force. But the introduction of MOD promoted the sintering of copper in another way.

During the heat treatment, the copper-based MOD generates copper through the following equation:

$$Cu(COOH)_2 \rightarrow Cu + 2CO_2 + H_2$$

Based on this equation, Fig. 2 shows the possible sintering mechanism of copper paste with MOD added. In the beginning, the MOD produces fresh metallic copper through thermal decomposition, and the copper nanoparticles in the paste provide heterogeneous nucleation for the process, lead to many tiny copper nuclei are generated on the copper surface. With the complete decomposition of the MOD, the gap between copper nanoparticles will be gradually filled with these copper nuclei, which greatly promote the sintering of copper. In addition, when the gap between copper nanoparticles are filled, the dense structure helps protect the copper nanoparticles from oxidation, which is also beneficial for bonding under air condition. Based on the above mechanism, reliable Cu-Cu joints could be manufactured by adding MOD to assist copper sintering.

B. Characterization of the Cu-Cu bonding joints

The shear strength of fabricated Cu-Cu bonding joints were measured with the speed of 100 µm/s by using a shear force tester (DAGE-series-4000, Nordson Corp.). Fig. 3 show the shear strength of the bonded joints obtained by sintering of copper paste with different MOD contents. When the mass ratio of MOD to reducing carrier solvent was 45:55, the shear strength is 18.95 MPa. Further increase the MOD content to 55:45, the shear strength of the bonded joints is slightly improve to 20.47 MPa. However, continue to increase the MOD content, when mass ratio of MOD to reducing carrier solvent is 65:35, the shear strength is dramatic reduced to 2.25 MPa. Besides, the shear strength of Cu-Cu bonding joints sintered by copper paste without MOD is only 9.15 MPa, which is much lower than 20.47 MPa. Based on the above results, the addition of MOD helps to obtain high shear strength sintered structure, and optimizing the ratio of MOD to reducing solvent was also beneficial to increase the shear strength.

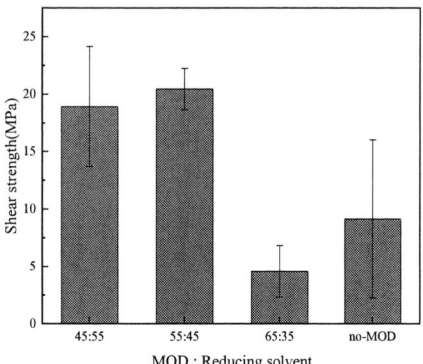

Fig. 3. The shear strength of Cu-Cu bonding joints fabricated by copper paste with different mass ratio of MOD to reducing solvent

In order to further understand the effect of MOD on the bonding, the morphology of the fracture surface was observed. As show in Fig. 4a and b, when the mass ratio of MOD to reducing solvent is 45:55 and 55:45, obvious tensile deformation can be observed, which indicates sufficient large-scale sintering between copper nanoparticles has been achieved. Fig. 4d show the fracture surface of Cu-Cu joints fabricated by no-MOD paste, many copper nanoparticles are separated, which is the main reason for the low shear strength. It can be seen from the result that the addition of MOD does promote the sintering of copper. Fig. 5 shows the corresponding EDS spectra. As show in Fig. 5b and d, the oxygen contents in joints obtained by no-MOD paste are higher, this proves that the addition of MOD has the effect of inhibiting oxidation.

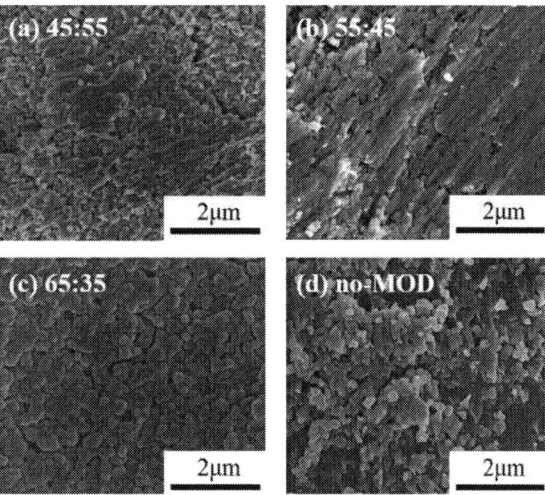

Fig. 4. Fracture surface of sintered Cu-Cu joints obtained by different Cu NPs paste (a)45:55, (b)55:45, (c)65:35 and (d)no-MOD

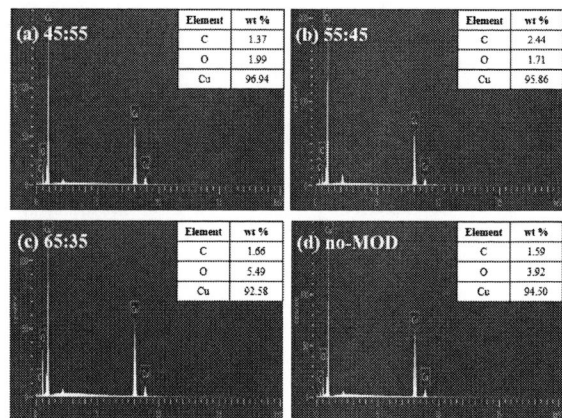

Fig. 5. The EDS spectra of the Cu-Cu bonding joints obtained by different Cu NPs paste (a)45:55, (b)55:45, (c)65:35 and (d)no-MOD

As show in Fig. 4c, in the Cu-Cu bonding joints fabricated from higher MOD content paste, coarse necks between copper nanoparticles could be also observed. However, no tensile deformation has been found. This phenomenon shows that even if there are enough connections between the copper nanoparticles, this connection is still very fragile. This is because of the content of the reducing solvent is decreased, the Cu NPs paste are easily oxidized. As show in Fig. 5c, the oxygen contents in paste (MOD: reducing solvent = 65:35) was higher than others. The small amount of oxidation on the surface of copper and the continuous oxidation during the bonding process have a great negative effect.

These results prove that the introduction of MOD greatly improved the bonding ability of Cu NPs paste, and fresh copper forming by the decomposition of MOD significantly improves the necking between copper nanoparticle. At the same time, the addition of MOD has the effect of inhibiting oxidation. When the mass ratio of MOD to reducing solvent is 55:45, the Cu NPs paste can achieve a highest shear strength of 20.47 MPa after bonding in air.

Fig. 6 show the cross-section images of sintered structure fabricated from copper paste with different MOD content. It can be found that the Cu-Cu joints with MOD added was slightly dense than the Cu-Cu joints without MOD added, this is due to the auxiliary sintering effect of MOD. As show in Fig. 6c, in the Cu-Cu joints prepared from the paste with excessive MOD content (MOD/reducing solvent = 65:35), large crack are observed. The existence of such cracks greatly reduces the strength of the Cu-Cu bonding joints. In the Fig. 6a, b and c, the copper nanoparticles are evenly sintered together, and no obvious gap is found between the sintered structure and DBC substrate. The formation of the crack may be attributed to the oxidation of copper nanoparticles during the sintering process. Even if the copper generate from MOD wraps and connects adjacent copper nanoparticles, the oxide layer reduces the reliability of sintering. Therefore, it is necessary to main the reducibility of the system while introducing MOD into copper paste.

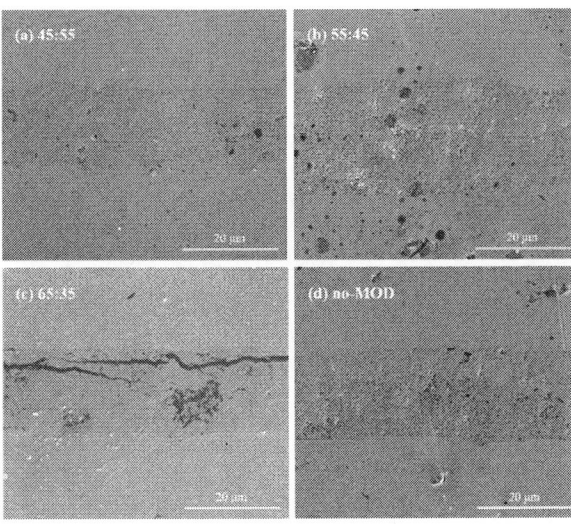

Fig. 6. Cross-section images of Cu-Cu bonding joints fabricated from copper paste with different MOD content

IV. CONCLUSION

Cu-Cu bonding joints with high shear strength were successfully prepared in the air using self-reducible Cu NPs paste with MOD-assisted. Compared to the Cu NPs paste without MOD, Cu NPs paste with MOD added can achieve a higher shear strength after bonding in air. Even under air condition, when the mass ratio of MOD to reducing solvent is 55:45, the shear strength of the bonded joints has reached 20.47 MPa after bonding at 250 °C for 15 min. During the sintering process, the Cu-based MOD generates fresh copper through thermal decomposition. These copper gather on surface of copper nanoparticles, which promotes the sintering of copper. Meanwhile, the dense structure formed helps resist oxidation of the copper nanoparticles. Therefore, the novel Cu NPs paste assisted by MOD is a promising bonding materials for electronic packaging and provides a now method to achieve reliable Cu-Cu joints under air condition .

ACKNOWLEDGMENT

Thanks for the supporting of National Natural Science Foundation of China (Grant No. 51805197).

REFERENCES

[1] H. S. Chin, K. Y. Cheong, and A. B. Ismail, "A Review on Die Attach Materials for SiC-Based High-Temperature Power Devices," Metallurgical and Materials Transactions B-Process Metallurgy and Materials Processing Science, Article vol. 41, no. 4, pp. 824-832, Aug 2010.

[2] H. W. Zhang, J. Minter, and N. C. Lee, "A Brief Review on High-Temperature, Pb-Free Die-Attach Materials," Journal of Electronic Materials, Review vol. 48, no. 1, pp. 201-210, Jan 2019.

[3] S. A. Paknejad and S. H. Mannan, "Review of silver nanoparticle based die attach materials for high power/temperature applications," Microelectronics Reliability, Review vol. 70, pp. 1-11, Mar 2017.

[4] W. Li et al., "The rise of conductive copper inks: challenges and perspectives," Applied Materials Today, vol. 18, p. 100451, 2020.

[5] X. D. Liu and H. Nishikawa, "Low-pressure Cu-Cu bonding using in-situ surface-modified microscale Cu particles for power device packaging," Scripta Materialia, Article vol. 120, pp. 80-84, Jul 2016.

[6] B. Y. Zhang et al., "In-air sintering of copper nanoparticle paste with pressure-assistance for die attachment in high power electronics," Journal of Materials Science-Materials in Electronics, Article vol. 32, no. 4, pp. 4544-4555, Feb 2021.

[7] Y. Gao et al., "Novel copper particle paste with self-reduction and self-protection characteristics for die attachment of power semiconductor under a nitrogen atmosphere," Materials & Design, vol. 160, pp. 1265-1272, 2018.

[8] Y. Choi, K. D. Seong, and Y. Piao, "Metal-Organic Decomposition Ink for Printed Electronics," Advanced Materials Interfaces, Review vol. 6, no. 20, p. 14, Oct 2019, Art. no. 1901002.

[9] Y. Farraj, M. Grouchko, and S. Magdassi, "Self-reduction of a copper complex MOD ink for inkjet printing conductive patterns on plastics," Chemical Communications, Article vol. 51, no. 9, pp. 1587-1590, 2015.

[10] W. L. Li, S. R. Cong, J. T. Jiu, S. Nagao, and K. Suganuma, "Self-reducible copper inks composed of copper-amino complexes and preset submicron copper seeds for thick conductive patterns on a flexible substrate," Journal of Materials Chemistry C, Article vol. 4, no. 37, pp. 8802-8809, Oct 2016.

Research on the board level reliability of CQFJ ceramic package

Zhen-tao Yang
The 13th Research Institute
CETC
Shijiazhuang, China
gattermann@163.com

Fei yu
The 13th Research Institute
CETC
Shijiazhuang, China
15864522419@126.com

Lin-jie Liu
The 13th Research Institute
CETC
Shijiazhuang, China
lj.liu@mail.cetc13.cn

Ling Gao
The 13th Research Institute
CETC
Shijiazhuang, China
Chinapackage_t@163.com

Abstract—In this paper, the finite element method is used to simulate the board level temperature cycling, mechanical impact and random vibration tests of CQFJ ceramic package.The stress and strain distribution of solder joints under different test conditions are analyzed. At the same time, we make special test boards and board level installation. The simulation results of relevant mechanical and temperature tests are verified. After the board level installation, the product meets the requirements of reliability test, forming a guide for the subsequent design of new products. The results show that under the condition of the same device size and total height, the stress relief ability of CQFJ lead out packaging structure is better than CLCC packaging structure. The J-type lead is more conducive to absorb the stress caused by the mismatch of thermal expansion coefficient between the package and PCB. Therefore, the use of ceramic four side J-type lead flat package can improve the reliability at the of premise of compatible pad size, and realize the compatible replacement of the large size CLCC package. Through this research, we have mastered the CQFJ ceramic package board level reliability simulation analysis and verification technology, which can quickly design and manufacture similar ceramic package to meet the needs of different users.

Keywords—CQFJ; Multi-layer ceramic package; Thermal fatigue; Board level reliability; Finite element analysis

I. INTRODUCTION

Compared with other types of ceramic package under the same lead number and overall size, the overall size of ceramic four side J-type lead flat package (CQFJ) is smaller than that of CQFP type ceramic package which lead led out from four sides. On the premise of compatibility of board level pad size, CQFJ ceramic package can realize compatible replacement for large size flat ceramic package without lead (LCC). Compared with CQFP and LCC, CQFJ ceramic packages with the same lead pitch have relatively small welding area, and mainly rely on J-leads to form large solder clad angle. At present, there is little literature and test data reference for the board level installation and reliability research of CQFJ ceramic package. Therefore, this paper carries out board level reliability design technology and verification research on CQFJ ceramic packages through simulation analysis and comparative test[1-3].

II. BOARD LEVEL RELIABILITY DESIGN

In order to verify whether the package can meet the use requirements, it is necessary to carry out the analysis and verification of board level reliability test , include environmental adaptability and mechanical reliability. The temperature cycling is one of the important tests to evaluate the board level environmental adaptability of packaging products. In the process of temperature cycling, periodic stress and strain are generated in the lead wire and board level welding interconnection part. Under the action of long-term alternating thermal stress, the stress and strain increase continuously, which may lead to the cracking of the outer package at the welding position, as shown in Figure 1, This kind of thermal failure often leads to poor circuit performance or even open circuit.

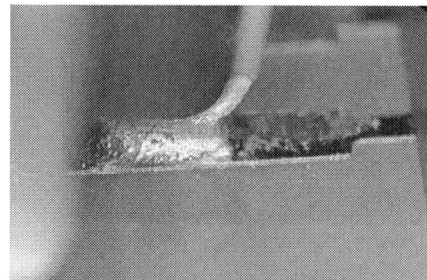

Fig. 1. Thermal fatigue failure mode of CQFJ ceramic package

In this paper, the thermal fatigue life, mechanical impact and random vibration test of CQFJ ceramic package are simulated and analyzed by finite element method, and then the board level installation and related environmental and mechanical tests are carried out by making a special test board to verify the simulation results, which provides a reference for the board level reliability design of CQFJ ceramic package.

III. SIMULATION ANALYSIS

A. Model building

Figure 2 shows the structural model of CQFJ68, with the package area of 23mm×23mm, the lead pitch is 1.27mm. The finite element model consists of four parts: ceramic package, lead wire, welding wrap angle and PCB pad. The ceramic material is alumina, the lead wire is 4J42 alloy, the solder of lead wire and ceramic is $AgCu_{28}$, the material of PCB substrate is FR-4, and the solder of package lead wire and PCB is $Sn_{63}Pb_{37}$. At the same time, according to the different size and feature size of the parts, the local mesh is set, including the division method and mesh size.

978-1-6654-1392-3/21 $31.00 © 2021 IEEE

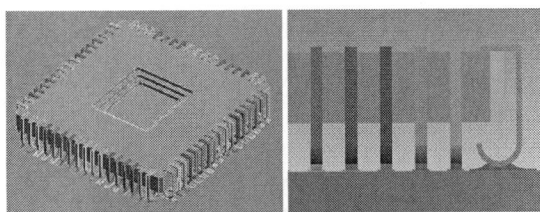

Fig. 2. Schematic diagram of model structure

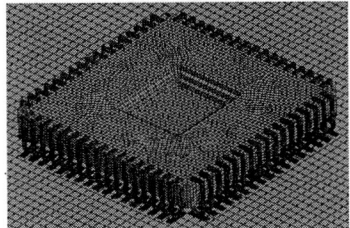

Fig. 3. Finite element model of CQFJ package

B. Simulation of temperature cycle test

In this paper, the simulation models of different wire width, thickness and support height are established, and the temperature cycle test is carried out to verify the influence factors of different wire design parameters on solder joint reliability. The range of temperature load applied in the temperature cycling test is -65℃~175℃, the initial temperature is 25℃, the temperature rise and fall rate is 12℃/min, the high and low temperature are kept for 120 min, the temperature rise and fall time is 15 min, with 5 cycles . The equivalent stress cloud diagram of $Sn_{63}Pb_{37}$ solder joint in cooling stage after simulation is shown in Figure 4.

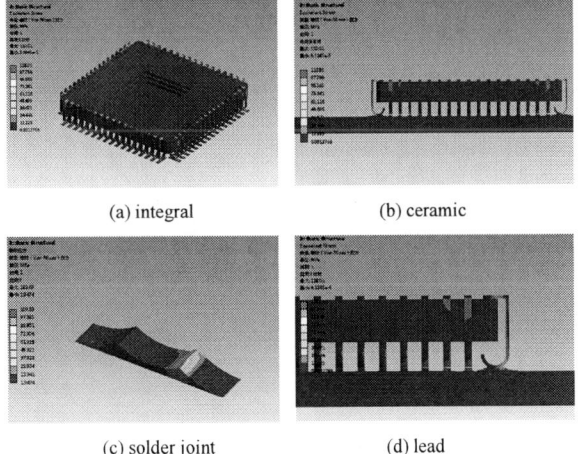

(a) integral   (b) ceramic

(c) solder joint   (d) lead

Fig. 4. Thermal fatigue stress distribution diagram

In the temperature cycling test, the solder joint between the lead wire and PCB board on the edge of the ceramic package is the position with weak stress. In this paper, the position data is selected to obtain the equivalent inelastic strain curve of solder joint with time, and then the fatigue life of solder joint under different models is obtained by calculation.

The calculation of fatigue life of solder joints under temperature cycling is based on the Coffin-Manson equation in the range of inelastic shear strain, N_f is the temperature

cycling fatigue life, the expression of Coffin-Manson equation is as follows:

$$\frac{\Delta\varepsilon_p}{2} = \varepsilon_f \left(2N_f\right)^c$$

N_f is related to the fatigue hysteresis coefficient ε_f, the elastic shear strain range $\Delta\varepsilon_p$ and the fatigue lag index c[4-6].

$$c = -0.442 - 6 \times 10^{-4} T_{sj} + 1.74 \times 10^{-2} \ln(1 + \frac{360}{t_H})$$

Taking the intermediate temperature of the temperature cycle T_{sj} and the high temperature holding time t_H into the above equation, C is -0.451[7]. Then, the simulation value is brought into the Coffin-Manson equation, and the calculated fatigue life results are shown in Table 1.

TABLE I.   COMPARISON OF RESULTS

Model	lead thickness	lead width	support height	Solder joint stress	life
1	0.30mm	0.55mm	3.00mm	185MPa	301
2	0.25mm	0.55mm	3.00mm	166MPa	341
3	0.20mm	0.55mm	3.00mm	151MPa	489
4	0.20mm	0.50mm	3.00mm	144MPa	516
4	0.20mm	0.50mm	3.40mm	132MPa	566
5	0.15mm	0.50mm	3.40mm	109MPa	605

It can be seen from the above table that increasing the support height of the lead wire, reducing the width and thickness of the lead wire can reduce the stress value of the solder joint and improve the fatigue life of the solder joint. In the lead design and processing, we need to consider the lead firmness and mechanical reliability at the same time. Combined with previous experience, the design of 0.20mm lead thickness and 0.50mm lead width is relatively optimal.

C. Mechanical impact simulation

In the temperature cycling test, the optimal model with lead wire thickness of 0.20mm, lead wire width of 0.50mm and support height of 3.40mm is selected to carry out the simulation analysis of mechanical impact test. The test conditions refer to the requirements of GJB548B. The peak impact acceleration is 1500g and the duration of impact pulse is 0.50ms. The stress distribution of the package in the mechanical impact test is shown in Figure 5.

The stress of the ceramic package is mainly distributed in the four corners of the package, and the maximum value is 14.71MPa, which is far less than the allowable stress 187.5MPa of the ceramic package, so the reliability is stable. The maximum stress of the lead in the process of mechanical shock is located in the welding position of the lead and the package pad, and the stress value is 111.5MPa. The maximum stress of solder joint of lead and PCB is at the back end of lead welding wrap angle, and the stress is 17.07MPa.

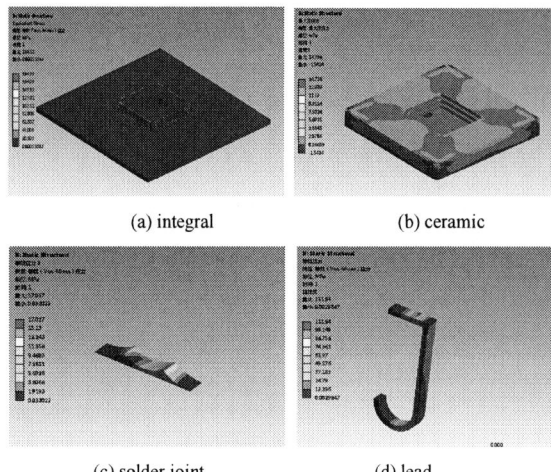

(a) integral   (b) ceramic

(c) solder joint   (d) lead

Fig. 5. Stress distribution cloud diagram in mechanical impact test

D. Random vibration simulation

Before random vibration analysis, modal analysis should be carried out to determine the natural frequency and vibration mode of the structure. Through modal analysis, the first 10 order frequencies of J-type package structure are simulated. The natural frequency data of modal analysis is shown in the figure below. The first order natural frequency is 27676Hz, which is greater than the maximum random vibration frequency of 2000Hz.

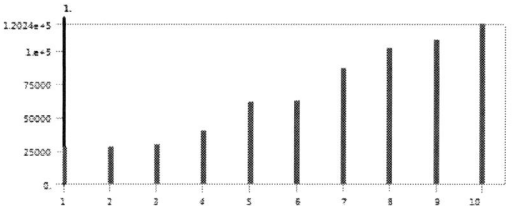

Fig. 6. Natural frequency data of modal analysis

Modal analysis shows that the first natural frequency of J-type package structure is much higher than that of mechanical vibration simulation frequency band, and the corresponding working frequency range is wider, which indicates that the structural model can avoid the resonance caused by the natural frequency close to the external test excitation.

After the modal analysis, the random vibration analysis is carried out, and the random vibration power spectral density curve is shown in Figure 7.

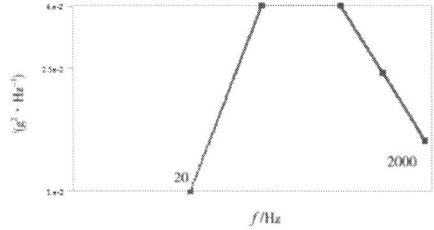

Fig. 7. Power spectral density curve of random vibration

Random vibration simulation analysis results showed that The stress of the lead solder joint is mainly concentrated in the middle and back end of welding position, and the maximum value is 70.91MPa. The stress of the board solder joint is mainly concentrated in the back end of the lead cladding angle, and the maximum value is 84.03MPa. The maximum 3-σstress of random vibration in Y direction is 7.9MPa, the probability that the maximum stress of material in random vibration is greater than 7.9MPa is 0.3%, which is far lower than the allowable stress of ceramic package.

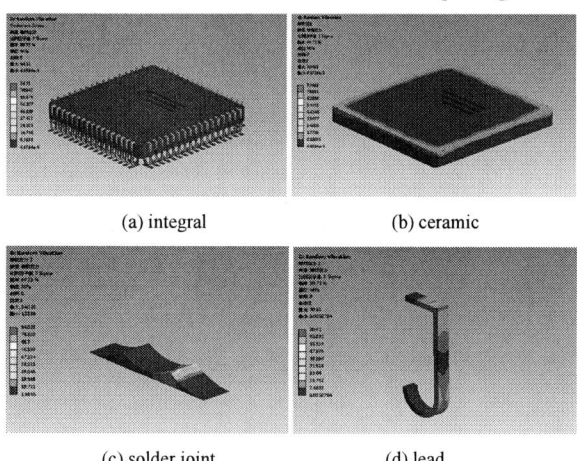

(a) integral   (b) ceramic

(c) solder joint   (d) lead

Fig. 8. Stress distribution cloud diagram in random vibration test

Calculate the deformation of J-lead package in three directions under the combined action of three-dimensional random vibration excitation, as shown in the figure below. Because of the symmetry of the structure, the deformation characteristics in X / Y direction are basically the same, and the deformation in Z direction is significantly higher than that in X / Y direction, about 0.01mm.

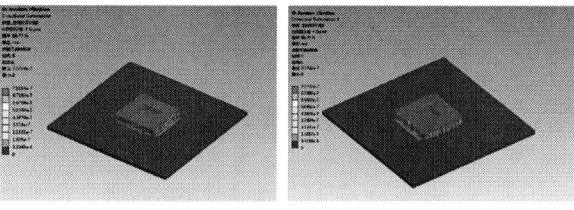

(a) max displacement in X direction   (b) max displacement in Y direction

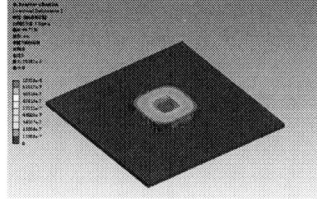

(c) max displacement in Z direction

Fig. 9. Strain distribution cloud diagram in random vibration test

IV. BOARD LEVEL TEST VERIFICATION

In this paper, CQFJ and CLCC package with the same overall dimension which length and width close to 28mm are selected for board level test verification. The appearance of J-type lead package board level welding case is shown in Figure 10.

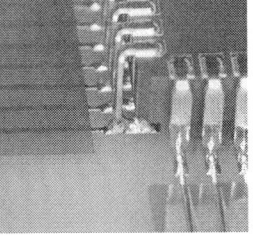

Fig. 10. J-type lead package board level welding sample

Both CQFJ and CLCC are tested by mechanical impact and random vibration, which meet the requirements of national military standard inspection. The environmental test group has done several temperature cycling tests. After the second temperature cycle test of CLCC package, tiny cracks appeared at the welding joint between the leading end and the board level. After the third temperature cycle test, the cracks became larger, and even circuit failure occurred. The appearance is shown in Figure 11.

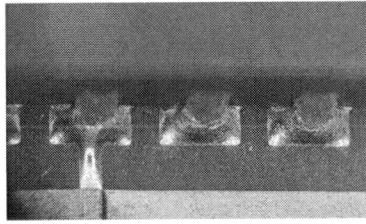

Fig. 11. Crack state of CLCC package after the third temperature cycle test

The experimental results show that the CQFJ lead-out packaging structure has better stress relief ability than CLCC packaging structure under the condition of the same device size and total height. The stronger the stress relief ability is, the higher the board level reliability is. The J-type lead is more conducive to absorb the stress caused by the mismatch of thermal expansion coefficient between the package and PCB, and improve the reliability. Therefore, the use of ceramic four side J-type lead flat package can improve the reliability at the of premise of compatible pad size, and realize the compatible replacement of the large size CLCC package. Through the simulation results of mechanical reliability test, it can be seen that under the vibration condition of CQFJ package, the four edge corner pins bear greater stress and are easy to damage and fracture. Therefore, it is suggested that the process of filling silicone rubber at the bottom of the device, four corner point sealing and integral pouring sealing can be adopted for effective mechanical reinforcement.

V. CONCLUSION

In this paper, the finite element method is used to simulate the board level temperature cycling, mechanical impact and random vibration tests of different models of CQFJ ceramic package.The stress and strain distribution of solder joints under different test conditions are analyzed. At the same time, we make special test boards and board level installation. The simulation results of relevant mechanical and temperature tests are verified. After the board level installation, the product meets the requirements of reliability test, forming a guide for the subsequent design of new products. The test shows that the CQFJ ceramic package can improve the reliability of tube package board level on the premise of compatible pad size, and realize the compatible replacement of CLCC package with larger overall size.

REFERENCES

[1] Li zongya, Tong Liangyu, Li Yao, Jiang Changshun. Design and Simulation of board level temperature cycling reliability of CQFP devices [J]. Electronics and packaging,2014,11 (5) :1681-1070.

[2] Osterman M, Dasgupta A, Han B. A Strain Range Based Model for Life Assessment of Pb-free SAC Solder Interconnects [C]. 2006 Electronic Components and Technology Conference, 2006: 884-890.

[3] Amagi M, Nakao M. Ball Grid Array （BGA） Packages with the Copper Core Solder Balls [C]. IEEE Electronic Components and Technonlgy Conference, 1998: 692-701.

[4] Zhang Liang, Xue Songbai. Effect of different solders on reliability of QFP solder joint [J]. Journal of welding, 2007, 28 (10): 45-49.

[5] GJB 548B-2005. Test methods and procedures of microelectronic devices [s].

[6] Chu Wei Hua, Li Shu Cheng. Fatigue failure and crack propagation of solder joints under vibration environment Analysis [J]. Intensity and environment, 2012, 39 (4): 56-63.

[7] Huang Yanfei. Thermal reliability analysis of key components on PCB Based on finite element method [J]. Micro meter Computer information, 2005, 21 (11-2): 164-165.

978-1-6654-1392-3/21 $31.00 © 2021 IEEE

Thermal And Optical Modeling On Intelligent LED Headlights

Yikang Qin
Guangxi Key Laboratory of Manufacturing System & Advanced Manufacturing Technology School of Mechanical and Electrical Engineering, Guilin University of Electronic Technology
Guilin, China
qykbeicheng97@163.com

Miao Cai*
Guangxi Key Laboratory of Manufacturing System & Advanced Manufacturing Technology School of Mechanical and Electrical Engineering, Guilin University of Electronic Technology
Guilin, China
*caimiao105@163.com

Xindong Chen
Guangxi Key Laboratory of Manufacturing System & Advanced Manufacturing Technology School of Mechanical and Electrical Engineering, Guilin University of Electronic Technology
Guilin, China
874821569@qq.com

Jinyang Li
Guangxi Key Laboratory of Manufacturing System & Advanced Manufacturing Technology School of Mechanical and Electrical Engineering, Guilin University of Electronic Technology
Guilin, China
jinyang_li2021@163.com

Daoguo Yang
Guangxi Key Laboratory of Manufacturing System & Advanced Manufacturing Technology School of Mechanical and Electrical Engineering, Guilin University of Electronic Technology
Guilin, China
daoguo_yang@163.com

Guoqi Zhang
Shenzhen Institute of Wide-Bandgap Semiconductors Delft Institute of Microsystems and Nanoelectronics (Dimes), Delft University of Technology
Shenzhen, China
g.q.zhang@tudelft.nl

Abstract—With the rapid development of lighting technology and auto industry, it has been an important research subject for many motor manufacturers to improve headlight performance. LED light has achieved a dominant position in lighting technology because of the advantages, such as the small size, low energy consumption, and long service life. In a complex working environment, the reliability of light source module is an indispensable parameter for a LED headlight, which can lead to poor energy efficiency, decreased light intensity, and a shorter service life with low reliability. In the working process of headlights, the rising LED temperature will reduce the luminous flux and thermostability, even affect normal operation. Therefore, understanding and evaluating the relevant factors affecting LED reliability is an important measure to ensure the stabilizations of LED headlights.

In this paper, the finite element software and optical simulation software are used to simulate the temperature field, flow field, and optics of the LED headlight module. Firstly, in order to understand the influence of temperature on the stability of the headlight, the temperature field and flow field distribution of the headlight are studied by finite element software. Secondly, we use the optical simulation software to study the optical parameters of headlight at different temperatures. It shows that the highest temperature appears in the LED chip, and the lowest temperature appears in the maximum end face of the headlight shell. For the same power LED, the higher working temperature leads to higher LED junction temperature. Affected by the working environment temperature, the luminous flux of the headlight will decrease with the increase of temperature.

Keywords—finite element, optics, LED headlight, temperature, reliability

I. INTRODUCTION

In recent years, LEDs have been widely used in automotive intelligent headlights. The harsh working environment and strict safety standards make the design and verification of LED automotive headlights face bigger challenges than general LED lighting. The optical power of the LED chip exceeds 1W, even as high as 5W, and the chip area is less than 1mm2, which is equivalent to a heat flux density of 100W/cm2 [1]. With the aging of chip growth technology and manufacturing process, LED reliability problems are more caused by packaging. The factors that cause LED packaging failure include temperature, current, and humidity. Narendran et al. studied the light attenuation of LEDs at different temperatures, and the results showed that as the temperature increases, the light attenuation rate of the LEDs increases, and the LED lifetime decreases exponentially [2]. Mehr et al. studied the influence of the phosphor layer on the life of the LED under high-temperature conditions and found that the life of the LED also decreases exponentially with the temperature increases [3]. Silicone is often used as a lens or packaging material for LED packaging. The accumulated thermal oxidation and hydrolysis reaction of silicone considerably influence the attenuation and color drift of the LED cavity. Therefore, its aging is also a factor that affects the reliability of the package [4].

Thermal management is an important part of the LED lighting system, the major problem of LEDs is that their light output strongly depends on the LED operation temperature [5]. Uncontrolled junction temperature affects the optical performance and reliability of the LED lighting system [6]. In the LED lighting system, there are three main ways of heat transfer from high temperature to low temperature: conduction, convection and radiation [7]. In the heat transfer analysis of headlights, heat convection exists in the lamp cavity, heat conduction exists between the radiator and the LED light source module, the light distribution lens and the reflector have heat radiation to the surrounding, after heat exchange achieves balance, the lights internal temperature field and flows field tend to stable. Aiming at the thermal management problem of LED headlights, there is no way to solve the actual problems with limited functional analysis methods. In the process of flow and heat transfer, finite difference method (FDM), finite element method (FEM), finite analysis method (FAM), finite volume method (FVM) are widely used.

In this paper, the finite element method is used to analyze the internal temperature field and flow field of LED headlight under different ambient temperatures. At the same time, optical analysis software is used to trace the LED modules with environmental factor and different aging times, and the influence of environmental factor on the optical parameters of LEDs is analyzed.

II. INTRODUCTION OF SAMPLES

This research takes OSLON Black Flat, one of OSRAM's high-power LED headlight modules, as the research object, the model is KW H2L531.TE, which meets the AEC-Q102 certification requirements. This type of LED is packaged in SMD epoxy, and there are two LED chips connected in series in one package. When the working current is 1000 mA, the forward voltage required by each LED module is about 6.15V. At 25 °C, under the condition of 1000 mA pulse operation (20-millisecond pulse), the minimum luminous flux can reach 780 lumens. This type of LED headlight module can work in a temperature range of -40 °C to + 125 °C.

III. RESULTS AND DISCUSSION

A. Thermal simulation analysis

Use multi-level meshing models to obtain higher quality meshes, and do temperature distribution and airflow trajectory simulation analysis on LED headlight. The simulation model is shown in Fig. 1.

Fig. 1. simulation model.

The material characteristics of headlight are shown in Table I.

TABLE I.　　MATERIAL CHARACTERISTICS

Material Type	Polycarbonate	GaN	Al	Cu
Density (kg/m3)	1.2	6.1	2.7	8.9
specific heat capacity (J/(g·°C))	1.25	9.75	–	–
Thermal Conductivity (W/(m·k))	0.3	200	240	387.6

The boundary conditions are as follows: 1)Assume that the ambient temperature is 20 °C by default, and a standard atmospheric pressure. 2)The temperature of radiant heat exchange is 20 °C. 3)The chip thermal power is 5.85 W.

Post-processing of results through Icepak, the corresponding component temperature distribution diagram and speed vector diagram can be obtained，Fig. 2 is the

temperature distribution diagram of the entire headlight. it clearly shows the overall temperature distribution of the headlight, and there is a big gap in the overall temperature distribution of LED headlight. The highest temperature is 110.6 °C in the chip part, less than the maximum working temperature of the chip 125 °C. The low-temperature area is located in the lens of the headlight and the surrounding shell, which is close to 21.7 °C of the environment temperature.

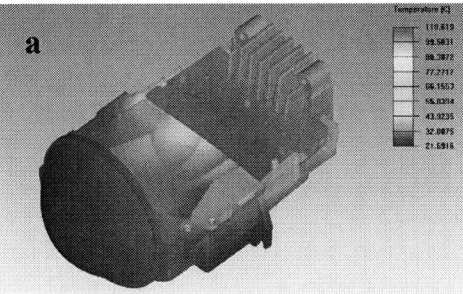

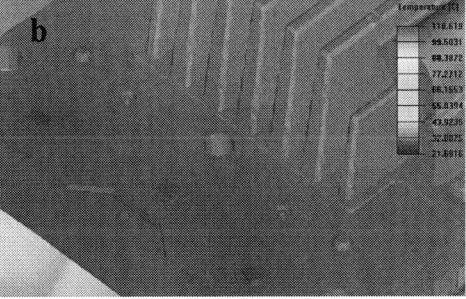

Fig. 2. 20 °C temperature distribution diagram.(a) Model after hiding the shell; (b) chip part

The speed vector diagram inside the headlight is shown in Fig. 3, it can be seen from the figure that there is airflow inside the headlight, the highest flow rate is 0.185 m/s, when lit the LED, the temperature of the LED is heated to a higher temperature, and the surrounding air is also rising rapidly. The hot air rises along with the heatsink to the top of the lamp housing. The flow speed at the bottom of the headlight and away from the light source is slow. The heatsink takes away most of the heat generated by the LED heating through thermal convection.

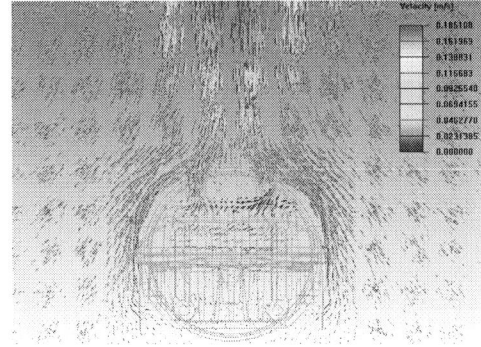

Fig. 3. speed vector diagram.

When the ambient temperature inside the headlight is 95 °C, as shown in Fig. 4, the highest temperature of the headlight is 172.8 °C, and the lowest temperature is 94.6 °C. Compared with the headlight temperature at 20 °C ambient temperature, the working temperature surpasses the chip's

978-1-6654-1392-3/21 $31.00 © 2021 IEEE

bearing temperature, and various indexes of the chip will be greatly reduced. Eventually, reliability will be reduced, even cause permanent damage.

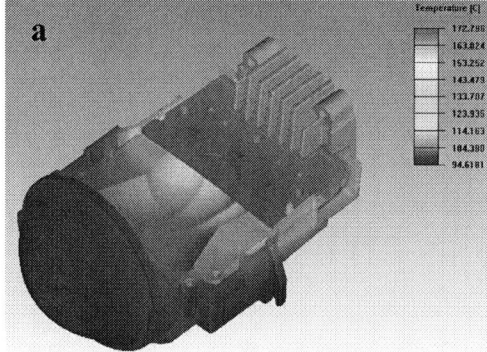

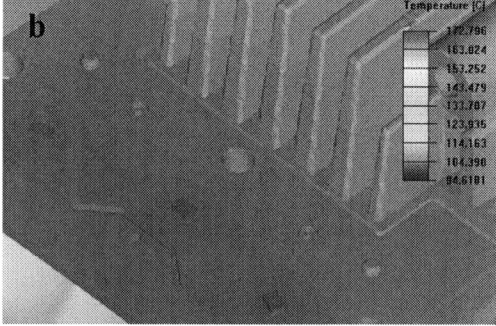

Fig. 4. 95 ℃ temperature distribution diagram. (a) Model after hiding the shell; (b) chip part

B. Integrating sphere data analysis

In the LED industry, luminous flux attenuation is usually used as a standard to measure LED failure. Alliance for Solid-State Illumination Systems and Technologies (ASSIST) standard stipulates that L70 represents the use time when the LED lumen maintenance rate is 70 %, and also represents the LED life. As shown in the Fig. 5, Fig. 5 a and Fig. 5 b are the trajectories of the lumen maintenance degradation of white light LED package devices under different acceleration environments. Fig. 5 a is the trajectory curve under the test conditions of 85 ℃ & 85 RH, and Fig. 5 b is the trajectory curve under the test conditions of 95 ℃.

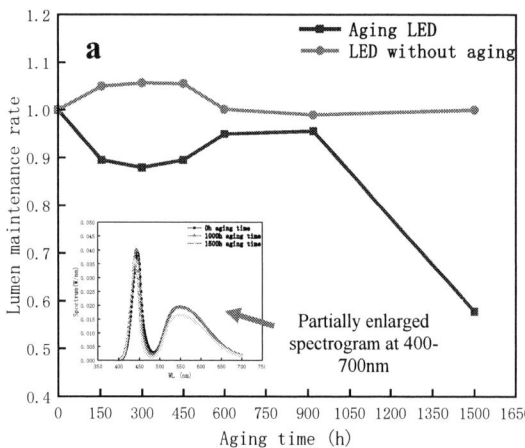

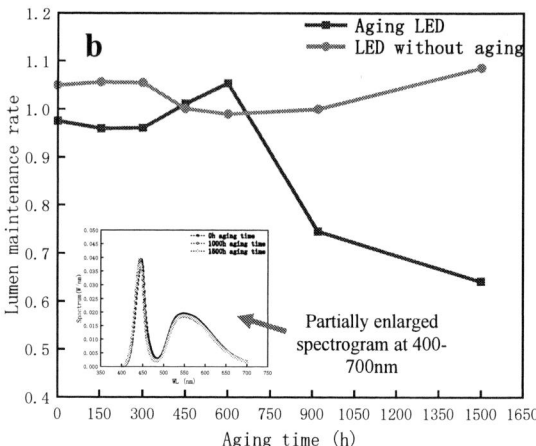

Fig. 5. lumen maintenance degradation trajectory. (a) 85 ℃&85 RH trajectory curve; (b) 95 ℃ trajectory curve

From the two figures, the lumen maintenance rate of the LED gradually decreases with the aging time increases. After 1500 hours of accelerated aging, two test conditions' lumen maintenance rate was reduced to below 70 %. During the test, the lumen maintenance rate will increase in a small range, the reason may be that the phosphor is not stable in the initial stage. During the power-on test, the LED junction temperature rises, and the chip dopants are further activated. With the increase of carriers, the luminous flux also increases, and finally lumen maintenance rate increases. In the same aging time, the higher the temperature is, the faster the lumen maintenance rate will decrease. After the humidity stress is applied, the lumen maintenance rate of the LED degrades more obviously. These results show that LED has strong heat and humidity sensitivity characteristics. Therefore, we need to consider the influence of temperature and humidity factors in the study of LED packaging device's reliability.

C. Optical simulation analysis

There are many types of light simulation software，this article selects Tracepro software to establish the model of the light distribution system. Tracepro has the characteristics of combining real models and powerful optical analysis capabilities, the simulated display light distribution system includes all types. Open the surface source property generator in the software, copy the spectral characteristic curve of the aging test sample obtained from the integrating sphere to the software, print the light distribution curve. In the parameter setting, the light source characteristics of the light source is emission form, add surface light sources with wavelengths of 0.546 μm and 0.4358 μm, set the material refractive index of the lens at 0 K and 300 K to 1.473, the surface absorption rate of the reflector is 0.2, specular reflectance is 0.8. The simulated irradiance analysis diagrams are shown in Figure 6, Figure 7,and Figure 8.

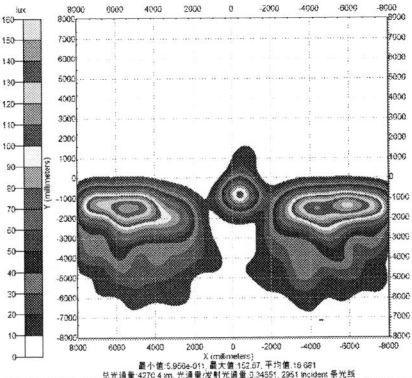

Fig. 6. irradiance analysis diagram without aging.

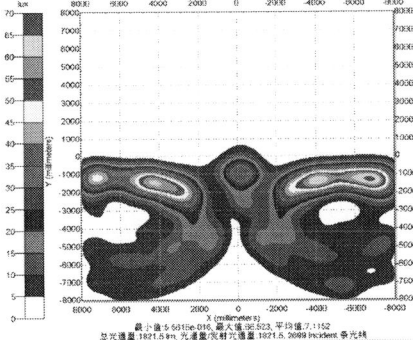

Fig. 7. irradiance analysis diagram after 1500h aging in 85 ℃&85 RH environment.

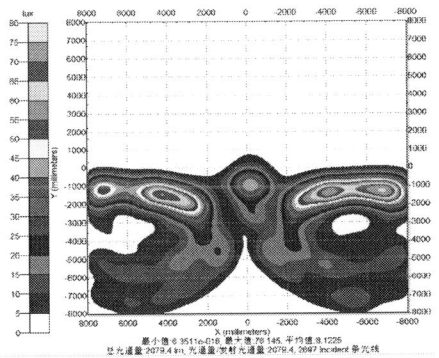

Fig. 8. irradiance analysis diagram after 1500h aging in 95 ℃ environment.

These figures show that the distribution of light on the screen 25 m away is relatively concentrated on the horizontal line，the illuminance starts from the middle and spreads to both sides. Compared with the unaged low beam module, the light distribution of the aging module is more concentrated in the middle, and less on the left and right sides. The luminous flux of the low beam module with added humidity stress decreases more obviously in the same aging time. "LED headlights for automobiles" (GB25990-2010) requires the luminous flux of the low beam should be greater than 1000lm. After 1500h of aging time, the lowest luminous flux of the sample is 1821.5 lm, which meets the requirements of the national standard.

IV. CONCLUSION

In this paper, the finite element method is used to analyze the thermal performance of LED automotive headlamps. The results show that the temperature at the chip is the highest, the temperature of the LED chip working at ambient temperature is less than the maximum allowable temperature of 125 ℃, which meets the design requirements. Due to the influence of the chip temperature, the hot air flows along the direction of the radiator to the top of the lamp housing, avoid internal heat flow accumulation. In addition, the optical performance of LED modules under different acceleration environments is analyzed. The results show that under the pure temperature and added humidity working environment, the luminous flux of the LED chip shows signs of change, explains that LED has strong heat and humidity sensitivity characteristics. The content of this article has great significance to study the reliability failure mechanism of LED headlights and improve their reliability.

ACKNOWLEDGMENT

This research was supported by the National Natural Science Foundation of China (No. 61865004), the Key-Area Research and Development Program of Guangdong Province (No. 2019B010131001), the Innovation-Driven Development Project of Guangxi Province (No. AA182420), and the Guilin Science Research and Technology Development Program (No. 2020010302).

REFERENCES

[1] Yu C, Fan J, Cheng Q, et al. Luminous flux modeling for high power LED automotive headlamp module. 2017 18th International Conference on Electronic Packaging Technology (ICEPT).

[2] Narendran, N, Y. Life of LED-based white light sources[J]. Display Technology, Journal of, 2005.

[3] Mehr M Y , Driel W , Zhang G Q . Accelerated life time testing and optical degradation of remote phosphor plates[J]. Microelectronics Reliability, 2014, 54(8):1544-1548.

[4] Liang Z, Cai M, Yang D, et al. Effect of H 2 O on the Optical Properties of Silicone Materials : a hybrid first principle calculation and molecular dynamic simulation study. 2019 20th International Conference on Electronic Packaging Technology(ICEPT).

[5] Bender V C , Iaronka O , Marchesan T B . Study on the thermal performance of LED luminaire using Finite Element Method[J]. IEEE, 2013.

[6] Shailesh K R, Kurian C P, Kini S G, et al. Review of methods for reliability assessment of LED luminaires using optical and thermal measurements. 2013 International Conference on Green Computing. IEEE, 2013:386-391.

[7] Tsai M Y, Chen C H, Kang C S. Thermal analyses and measurements of low-cost COP package for high-power LED. 2008 56th Electronic Components & Technology Conference. IEEE, 2008:1812-1818.

Research on the design and processing technology of CQFJ ceramic package

Fei yu
The 13th Research Institute
CETC
Shijiazhuang, China
15864522419@126.com

Zhen-tao Yang
The 13th Research Institute
CETC
Shijiazhuang, China
gattermann@163.com

Lin-jie Liu
The 13th Research Institute
CETC
Shijiazhuang, China
lj.liu@mail.cetc13.cn

Ling Gao
The 13th Research Institute
CETC
Shijiazhuang, China
Chinapackage_t@163.com

Abstract— Based on the existing high-density Multi-layer ceramic technology system, this paper successfully broke through the key technology of lead forming of CQFJ ceramic package for the first time in China, formed the lead bending design rules and processing flow of J-type lead ceramic package, and realized the mass production, processing and application of CQFJ ceramic package. The lead structure of J-type lead ceramic package is more complex than the conventional package with lead structure. It needs to be processed and gold-plated before bending the lead. In the process of forming the lead, the direct contact with the mold and surface of the soft gold layer make scratch on the surface of lead coating which leads to the unqualified appearance of the product . Therefore, the lead processing and forming is the key to the design and processing of the package. In this paper, through the research on the key technology and process of lead forming process and bending die design, we have mastered the processing and forming process of J-type lead ceramic package, realized the processing of J-lead ceramic package, and met the requirements of package appearance inspection and lead firmness. By improving the mass production and development capacity of J-type lead ceramic package, the same kind of ceramic package can be designed and manufactured quickly to meet the needs of different users. On the premise of meeting the requirements of Lead board level weld size, the side lead size is widened to enhance the firmness of the lead. At the same time, the welding part of the lead and the package is led out with a flat structure, and a reasonable solder wrap angle is formed through the side hollow hole welding structure design to ensure the welding strength of the lead. At present, there is no clear specification for the lead firmness test method of J-type lead ceramic package in China. This paper introduces it in detail for users' reference.

Keywords—J-type lead; CQFJ; Multi-layer ceramic package; lead firmness; lead molding

I. INTRODUCTION

Four side J-type lead flat ceramic package (CQFJ) is a miniaturized mounting package. The lead at the welding position between the package and PCB board is located at the bottom of the package. The structure is shown in Figure 1. CQFJ package is mainly used in various large-scale integrated circuit packages, such as VLSI, ASIC, ECL and CMOS gate array circuits. It is widely used in amplifiers, drivers, memories, etc. It has the characteristics of small size, light weight, high package density, good Thermoelectric performance, and suitable for surface mounting[1-2].

In terms of structure, the lead of the welding part of the ceramic four side lead flat package (CQFP) is located outside the tube package. Therefore, under the premise of the same number of pins and overall size, the space occupied by CQFJ package after board level installation can be reduced by 0.5mm to 2.0mm compared with CQFP package, which effectively reduces the overall size of the tube package after packaging, and meets the development trend of device miniaturization design.

When the total height of the device is the same, the lead length between the ceramic package and PCB connected by CQFJ package is larger than that of the lead from the bottom of the CQFP structure package, so the ability to buffer and release stress and the board level reliability are higher. Under the premise of the same number of pins and overall dimensions, CQFJ package and four side leadless flat ceramic package (LCC) can be replaced after board level installation[3].

The LCC ceramic package direct welded with PCB board. Because there is no lead buffer, the application of larger LCC ceramic package is limited in the field of high reliability and the field where the ambient temperature changes violently. The CQFJ ceramic package is more conducive to absorb the stress caused by the mismatch of thermal expansion coefficient between the package and PCB, so it has higher reliability.

Fig. 1. Basic structure of CQFJ ceramic package

The lead processing of J-type lead ceramic package is more complex than that of conventional lead, and the lead bending is required after the package is plated with gold. It is difficult to control the size and appearance of the lead in the forming process, so the lead processing and forming is the key to the design and processing of the package. In this paper, through the research on the key technology and process of lead forming process and bending die design, we have mastered the lead processing and forming process of J-type lead ceramic package, and realized the batch production, processing and application of CQFJ ceramic package in China.

II. PACKAGE DESIGN

CQFJ ceramic package is mainly composed of ceramic and leads. The ceramic are processed by multi-layer high-temperature alumina ceramic package processing technology. The front of the pad is distributed on four sides, and the side of the ceramic parts is made of metalized hollow holes. When the leads are welded, the solder wrap angle is formed to strengthen the welding strength. According to the requirements of lead span and bending height, the size of flat lead and the bending size of lead after gilding are designed reasonably, as shown in Figure 2.

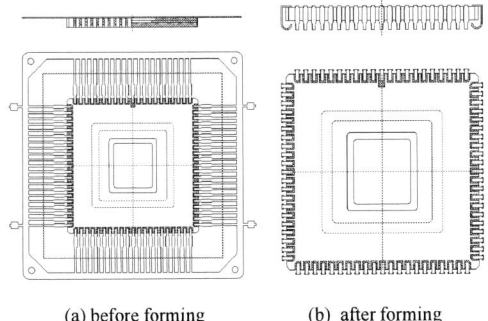

(a) before forming   (b) after forming

Fig. 2.   CQFJ structure diagram of lead before and after forming

The lead thickness c of the package is 0.20 mm, and the lead pitch e is 1.27 mm, 1.016 mm, etc. The welding area of lead and ceramic directly affects the welding strength of lead. Therefore, through design optimization, the minimum welding length is 0.60mm, the width of lead b1 corresponding to 1.27mm pitch is 0.50mm at the welding position of ceramic body, and the width of lead b2 corresponding to PCB board is 0.50mm.

Under the condition of ensuring the insulation of adjacent leads, the width of the middle part of the lead is widened to the maximum extent. The wider the lead, the better the strength of the lead itself. When mechanical impact, random vibration and other mechanical tests are carried out, the stronger the ability to resist the stress caused by the change of external environment, and the higher the reliability. Therefore, the b3 size is designed to be widened to 0.76mm. The values corresponding to the lead pitch of 1.016mm are b1- 0.38mm , b2- 0.30mm ,b3- 0.50mm.

Considering the matching between the dimension tolerance of the ceramic part of the ceramic package and the processing tolerance of the lead, the minimum installation margin L1 is 0.12mm, and the minimum L2 is 0.15mm. The bending angle R1 of lead wire welding directly affects the size and state of cladding angle of package and PCB. Referring to the foreign design value and computer simulation results, the creep process of different R-angle lead wire and PCB welding position is analyzed, and the optimal size is obtained. The range of R1 is 0.64mm minimum and 0.89mm maximum.

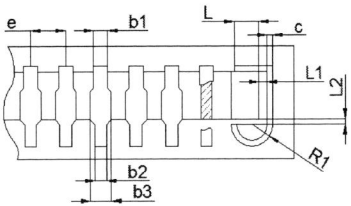

Fig. 3.   Lead forming size of CQFJ ceramic package

III. LEAD FORMING PROCESS

The J-type lead ceramic package in domestic demand is small before and no mass production scale, with the increasing domestic demand, CQFJ ceramic package product category demand is more and more, lead processing and molding are the key to package design and processing, not only to ensure reasonable size design, but also to meet the appearance inspection standard.

The lead of CQFJ ceramic package is formed after welding and gilding the lead and ceramic parts. Firstly, the excess size of the straight lead is cut off after the length for forming is removed, and then the lead end is bent and shaped at a small angle. Then, the lead wire welding is carried out for 60°, and finally using the finishing die to complete 90° bending forming and R-angle forming of lead end bending. The process is shown in Figure 4.

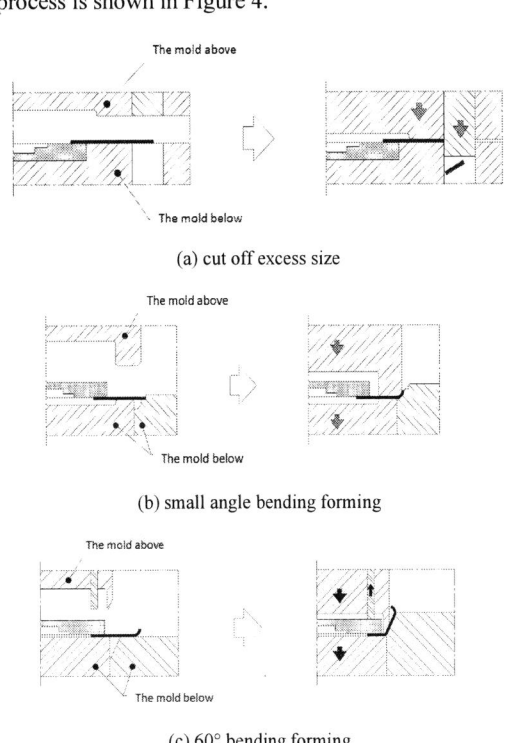

(a) cut off excess size

(b) small angle bending forming

(c) 60° bending forming

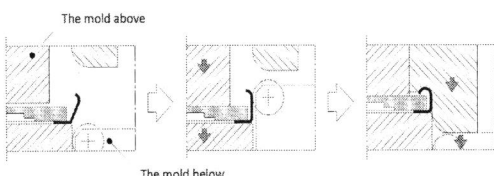

(d) final molding

Fig. 4.   Lead forming processing of CQFJ

978-1-6654-1392-3/21 $31.00 © 2021 IEEE

IV. KEY TECHNOLOGY OF LEAD FORMING

Compared with the lead molding of CQFP package, the J-type lead molding process has more steps, less processing experience, and difficult size control. It is necessary to ensure the dimensional accuracy of lead molding and higher coplanarity requirements (≤ 0.05mm), avoid and eliminate the excessive tension and damage of lead caused by the lead molding process, which affects the lead firmness.

Because the ceramic package of J-type lead needs to be bent and formed after the gold plating on the package, the die directly contacts with the gold plating surface of the package during the forming process of the lead, and the gold layer is relatively soft. The coating on the surface of the lead is easy to scratch after the die extrusion forming, which leads to the unqualified appearance of the product.

In the early forming experiment, it is found that the bending size of the lead wire after processing cannot meet the tolerance requirements of the drawing, the thickness of the bending corner is seriously reduced, there are slight cracks at the bending corner, which affect the strength of the lead wire, and the J-bending part has overall deformation due to die extrusion; There is obvious scratch phenomenon on the surface of the lead by microscope. The nickel layer and even the ceramic substrate are exposed. The deformation of the lead and scratch of the coating are shown in the figure below.

(a) lead deformation

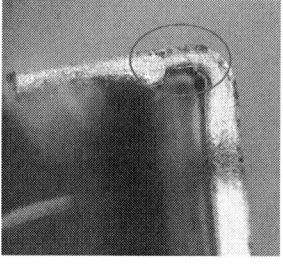

(b) severe thinning at corners

Fig. 5. Deformation of lead wire processing

Fig. 6. Scratch of lead coating

Therefore, in order to control the bending size and lead deformation, the design size of the lead forming die, the gap between the die and the lead, and the chamfer size of the die were analyzed. After the continuous adjustment of the processing die, the problems of lead deformation and corner crack are effectively solved, and the size and coplanarity meet the design requirements.

Considering the particularity of the process sequence of lead forming after gold plating, the finished product after gold plating is protected by spraying binder to avoid the hard contact between the mold and the lead that can eliminate the scratch of the lead. It is need to clean the adhesive of the package after the lead forming, and then carry out relevant test and appearance and performance inspection for the package.

After master the key technology of lead forming for CQFJ ceramic package and realizing the mass production of the package, we prepared a series of CQFJ packages. The number of lead out terminals is mostly 84, the maximum external dimension is 29mm, and the bending height of lead is 3.8mm. Some products are shown in the figure below:

Fig. 7. Product photos of CQFJ ceramic packages

V. TEST METHOD OF LEAD FIRMNESS

At present, there is no clear specification for the lead firmness test method of J-type lead ceramic package in China. This paper introduces it in detail for users reference[4-5].

Lead selection: the tested lead shall be randomly selected from each sample of the test sample and selected from each side of the lead, and only one side of the lead shall be pretreated and tested at one time.

J-type lead straightening: install the package on the fixed fixture, straighten the bending part of the lead with pliers, as shown in Figure 8, and operate carefully to avoid damage to the lead, such as cracks, kinks or deep scratches. Before and after the alignment of the lead, the optical microscope is used to magnify 10 ~ 20 times to check each lead. The damaged lead due to alignment cannot be tested.

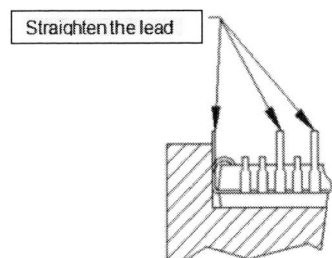

Fig. 8. Straighten the lead

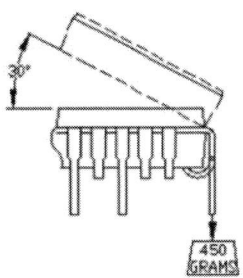

Fig. 9. Rotating package body

Installation of device: use necessary gasket to fix the device on the lead bending test fixture, and the tested lead shall be vertically downward, 90°relative to the bottom of the package.

Test: the lead wire shall be bent outward for three times at an angle sufficient to hold the lead wire for at least 15°(after removing the stress), the bending angle shall be measured from the end of the lead to the installation plane.

Rotating package body: select another untested lead and add a weight of 450g to the lower end of the straightening lead supporting part. Care should be taken to avoid twisting the tested lead or adjacent lead or bending beyond the test requirements. Rotating the package body 30°and return to the original position for three times. One cycle should be completed within 2s ~ 5s, and the rotation of the package must avoid the swing or torsion of the applied force, as shown in Figure 9.

Failure criterion: after the stress is removed, the optical microscope is used to magnify 10 ~ 20 times, and any complete fracture (such as the lead wire is separated from the body) is considered as failure.

VI. CONCLUSION

In this paper, through the research on the key technology and process of lead forming process and bending die design, the key technology of CQFJ ceramic package lead forming was successfully broken through for the first time in China. Through the continuous optimization of mold design and add of lead protection layer, we have mastered the processing and forming process of J-type lead ceramic package lead, met the requirements of package appearance inspection and lead firmness, and realized the mass production, processing and application of CQFJ ceramic package. At the same time, the test method of J-lead ceramic package is introduced for reference.

REFERENCES

[1] Zhenya Li, Yu Zhao. The status and future prospects of System in package technology [J]. Electronics and packaging, 2009,9 (2).5-10.

[2] William E. Lee. Department of Engineering Materials[M].U.K: University of Sheffield,1998:77-89.

[3] Li zongya, Liang yuanyu, et al. Design and simulation of CQFP device panel temperature compliance reliability [J]. Electronics and packaging,2014,14 (44) : 5-8.

[4] GJB 548B-2005. Test methods and procedures of microelectronic devices [s].

[5] GJB1420B-2011. General specification for semiconductor integrated circuits[s].

Research on Double-Layer Networks-on-Chip for Inter-Chiplet Data Switching on Active Interposers

Xiaolong DUAN
*Academy of Smart IC and Networks,
Key Laboratory of the Ministry of
Education for Optoelectronic
Measurement Technology and
Instrument, Beijing Information Science
and Technology University, Beijing,
100101, China
18298482504@163.com*

Min MIAO
*Correspondding author
Academy of Smart IC and Networks,
Key Laboratory of the Ministry of
Education for Optoelectronic
Measurement Technology and
Instrument, Beijing Information Science
and Technology University, Beijing,
100101, China
miaomin@bistu.edu.cn*

Zhuanzhuan ZHANG
*Academy of Smart IC and Networks,
Beijing Information Science &
Technology University, Beijing,
100101, China
1762498233@qq.com*

Liang SUN
*Academy of Smart IC and Networks,
Beijing Information Science &
Technology University, Beijing,
100101, China
3516175325@qq.com*

Abstract—The chiplet-based heterogeneous integration technology implemented on active interposers has been developing rapidly in recent years. However, as the bit widths of interactive data packet is getting larger and larger, for example, the bit widths of HBM (High Bandwidth Memory) has reached 1024 bits, the enormous wiring resources required for parallel data exchange and the complexity of wiring , as well as crosstalk issues inducted by high density interconnection, have necessitated the shift from parallel to serial mode data transmission, so as to effectively mitigate the intimidating challenges from exponential increase of wiring resource demands and interconnection density. However, the advantage of serial transmission is obtained at the cost of several signal integrity and power integrity issues brought by the steep rise/fall edge of the high-speed waveform, in addition to relatively reduced transmission bandwidth under certain occasions. As the bit width continues to increase, this cost becomes more and more unacceptable. NOC (Networks-on-chip) is one of the potential options to solve this problem. This paper proposes a new NOC architecture that is 2D-Mesh structured with a so-called "data station" core, featuring dual-port RAM. Compared with the traditional single-layer NOC, this scheme adopts a double-layer structure, in which the top layer is a network with four dual-port RAMs as the data storage center and circuit-switch as the data exchange network, responsible for data transmission. The bottom layer is a traditional packet-switch NOC used for address packet exchange. The advantage of this design scheme is a largely reduced resource and power consumption of the NOC, making the NOC more attractive for 2.5D integration, especially those based on active interposers.

Keywords—active interposer, chiplet, networks-on-chip, signal integrity, power integrity

I. INTRODUCTION

Since the state of the art IC (integrated circuit) process technology evolves beyond the 20nm/14nm node in around 2016, on-chip integration based on scaling down as demanded by Moore's Law has lost usual attractiveness and popularity, and there come urgent cries for new integration solutions even from top-tier IDMs (integrated device manufacturers). On the other hand, since the year of 2016, high-density and high-performance packaging technology capable of integrating heterogeneous chips designed and manufactured based on different process lines both horizontally and vertically, has

become the most promising option to extend the legend of Moore's Law and are widely accepted as the major platform of More-than-Moore from the perspective of smart microelectronics/microsystem suppliers[1]. In the past 5 years, so called 2.5D and 3D implementation of the heterogeneous integration technology as mentioned, has become the hot spot of research in the semiconductor, among which, those based on the silicon interposer-chiplet combination has been highly valued by the semiconductor industry[2,3], which is to adopt small sized dies called chiplets featuring various standardized functions and passive components, and then assemble them on a silicon interposer that can provide high-speed and short-distance interconnection and even auxiliary functions such as power regulation, to implement a multi-chip module with enhanced performance, manufacturability and yield. At the 2020 ISSCC, CEA-LETI proposed a heterogeneous integration scheme based on an active interposer [4]. The active interposer includes voltage regulator, IO controller, PHY(Physical Layer) and NOC(Networks-On-Chip) for interconnection between chiplets. The chip integrates a total of 96 cores with 6 chiplets. Utilizing the mature 65nm process and low complexity, the active interposer still gains a very high yield. In addition, Intel's embedded interconnect bridges (EMIB) [5] and TSMC's Low-voltage-Package-INterCONnect (LIPINCON) technology [6] are also exploring the benefits of versatile and flexible signaling and heterogeneous integration facilitated by silicon interposer. However, in addition to the high production cost, the current difficulties faced by this technology include the high wiring overhead brought by the increase in the number of integrated chiplets. Moreover, high-density interconnection will also cause problems such as signal integrity and power integrity, as well as hot spots from heat dissipation caused by high power consumption. As one of the potential solutions to this problem, the NOC has received more and more attention in past few years. Widely accepted as an interconnection method for communication within the processors based on the principle of computer network communication, NOC uses routing algorithms and switching technologies to combine multiple processing cores with various network topology, adopts a globally synchronized and locally asynchronous clock architecture, featuring multiple routing nodes distributed over the network structure to communicate in parallel, and replacing the master-slave communication mode of the traditional bus system, so as to overcome the problem of severe global clock and communication congestion in the multi-core system adopting the bus architecture [7~9]. This paper proposes a new NOC architecture that is 2D-Mesh structured with a so-called "data station" core featuring dual-port RAM, that can be integrated on an active interposer. The key concept of this technology is to implement the interchange by a structure called data station. When a node transmits a data packet, it only needs to upload the data to the RAM buffer, and the data interaction between the NOC only demands a transmission of the address of the memory. This technology reduces the interconnection density while increasing the rate of data interchange.

The following is organized as this: 1) Part II introduces the overall architecture of the proposed NOC; 2)Part III introduces the function realization of the NOC; 3)_Part IV shows the experimental results; 4) Part V makes the conclusions and discussion.

II. NETWORKS ON CHIP ARCHITECTURE

A. Overall structure

As shown in Figure 1, this design adopts a double-layer structure, and the bottom layer is a 4×4 mesh structure packet-switch router as shown in Figure 2(a). The router uses the XY dimension sequence algorithm to ensure that no deadlock effect occurs. Its function is to exchange routing information and memory addresses. The top layer is an circuit-switch network composed of four dual-port RAMs as shown in Figure 2(b). It takes the form of four points directly interconnected for data storage and forwarding. The advantage of this is that the NOC not only has the high resource utilization and flexibility of a packet-switch network, but also has the high energy utilization of a circuit-switch network[10]. It greatly reduces the power and die area overhead caused by the FIFO (First In First Out), while simplifying the wiring and reducing the transmission line density per unit area, so as to weaken the coupling effect between the interconnections and improve the signal integrity.

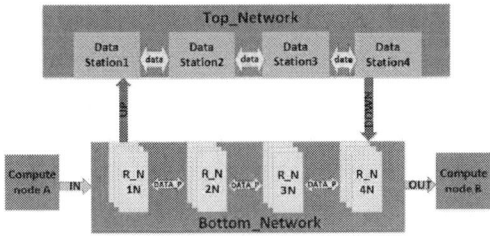

Fig. 1. Structure of Double-Layer NOC

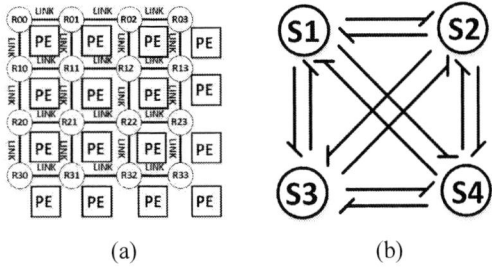

(a)   (b)

Fig. 2. (a)Bottom Router structure, (b) Top Data_Station structure

B. Route Node Structure

Figure 3 shows the router node function diagram of the bottom network, which is mainly composed of the following eight modules:

1) S2P module, whose main function is to convert serial input data into parallel output data;

2) Data_U module is responsible for uploading data to the top network. It is also responsible for packaging and sending the address input from the top network with the destination address of the original data packet;

3) IN_Control module is responsible for managing data input from five directions: local (L), east (E), south (S), west (W), and north (N), and outputs data to the FIFO according to priority.

4) FIFO module is a module for buffering data, which can improve the transmission ability of NOC. In addition, it is also the module that occupies the most resources and consumes the most power in the entire NOC. The destination address will be separated in this module and sent to the R_Compute module;

5) R_Compute module is responsible for using the routing algorithm to assign the data packet to the appropriate data output port according to the destination address from the FIFO, and it can shake hands with the five directions of data output. If the handshake is successful, it will give an output signal to Switch module;

6) Switch module is responsible for outputing data from the FIFO and sending it to the corresponding port according to the instruction sent by the R_Compute module;

7) Data_D module is responsible for downloading the data from the top network according to the memory address separated in the data packet;

8) P2S module, its main function is to convert parallel input data into serial data and output it locally.

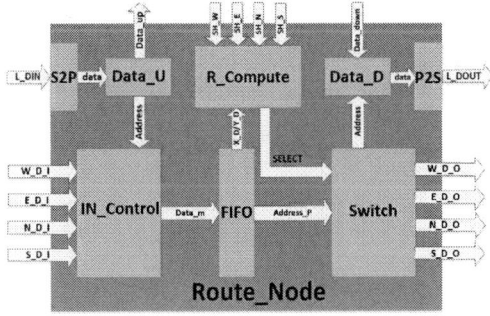

Fig. 3. Structure of Route_Node

C. Data_Station structure

As shown in Figure 4, the structure consists of the following four modules:

1. DATA_WRITE module is responsible for storing the data to the top network and sending the storage address to the bottom network. Since there are multiple data stations, this module also has the function of local data stations, which can send the number of the data station and address to the bottom network;

2. ADDRESS_MANAGEMENT module is responsible for allocating storage addresses for the data and writing the

data in RAM according to the addresses. In addition, the module is also responsible for downloading the data in RAM according to the address provided by the DATA_READ module;

3. Dual-port RAM is a storage module with read and write functions based on addresses;

4. DATA_READ module is responsible for reading the data according to the address sent by the local station or other three data stations. If the address shows that the data is not in the local station, the address is sent to the corresponding station for data reading according to the station number.

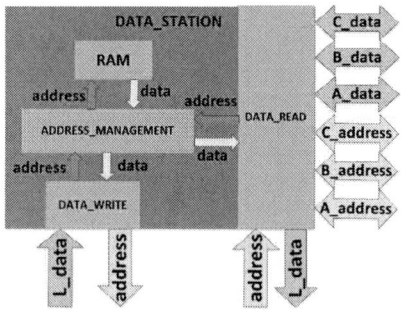

Fig. 4. Structure of Data_Station

III. MECHANISM OF NETWORKS-ON-CHIP

A. Mechanism of bottom network

1. Networks-On-Chip workflow

Figure 5 shows the data communication process of NOC. The data flow of the double-layer NOC has two directions. The first is the exchange of the data body in the data packet between the four data stations; the second is that of small data packets containing data storage address and destination address are transmitted between nodes. The purpose of this arrangement is to prevent the huge data body from flowing between nodes, thereby reducing interconnection density.

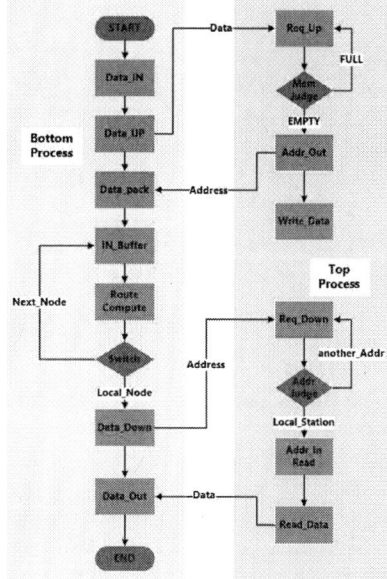

Fig. 5. Workflow of Networks-On-Chip

2. Routing Information Exchange

As shown in Figure 6, there are three stages when the router node works. The first stage is called the input stage. In this stage, the IN_Control module composed of multiple multiplexers makes the handshake signal Valid in the five directions of North, South, West, East and Local to receive the data from the five directions respectively and send data to the buffer module FIFO. The second stage is called the route compute stage. In this stage, at first the FIFO module stores the data in the 16×11 memory array according to the state of the array, and the next cycle 3bit counter scans the entire memory array as the monitoring address. Then, if data is found, the destination address in the data is sent to the R_Compute module, and the data packet is sent to the switch module. Finally, the Route Compute module obtains the sending direction according to the XY dimension sequence algorithm, and at the same time generates the select signal, and sends it to the switch. The XY dimension sequence algorithm code is as follows:

```
if (X_d==X_s_i)
   if(Y_d>Y_s_i)
      i<= 3'b011;          //East direction
   else if(Y_d==Y_s_i)
      i<=3'b111;           //Local direction
      else i<=3'b100;      //West direction
   else if (X_d>X_s_i)
      i<=3'b101;           //South direction
      else i<=3'b110;      //North direction
```

The third stage is called the output stage. In this stage, SWITCH will distribute the data to the corresponding data channel according to the SELECT signal for output.

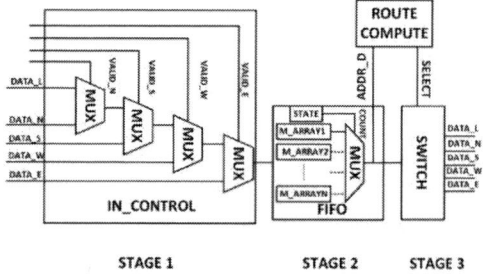

Fig. 6. Mechanism of Router Node

3. Data Upload And Download

After each router node obtains data from the processing core or memory, it can upload the data except routing information to the top network through the Data_U module, and then the storage address of the data in the top network is packaged with the routing information forming a new packet to transmission. When the packet has been transmitted to the corresponding router node, the Data_D module can download the data from the top network according to the storage address in the data packet, and then transmit it to the outside.

In the router node, except for the SWITCH and R_Compute modules that need to wait for the handshake signal, other modules can keep working until the FIFO buffer array is filled. This ability greatly improves the performance of the NOC.

B. Mechanism of top network

1. Data Write Module

This module has only one function, which is to write the data from the router node to the address allocation module for storage, and at the same time send the storage address to the bottom network.

2. Data Read Module

This module have two functions, as shown in Figure 7(a), which is the state machine flow of the first function. The description of each state and the trigger conditions are as follows:

S0: The initial state, this state is triggered when a read request is received, and it goes to the S1 state;

S1: The state for judging, after receiving the request, and monitoring whether there is a read task at this time, that is, whether the read data port is free; if it is busy, continue monitoring; if it is free, the status is triggered, a Valid signal is sent to the bottom network, and the state goes to S2;

S2: Address receiving state, waiting for the bottom network to send the storage address, and then turning to the S3 state;

S3: Slave state machine control state, judging whether the data is stored in the local station according to the storage address, if not, turn on the No. 1 slave state machine, otherwise turn on the No. 2 slave state machine.

S4: Reset state, resetting all handshake signals and registers; then turning to S0, and waiting for the next read operation.

The No. 1 slave state machine is composed of four states:

S0: Judging the state, judging which data station the data is stored in according to the address, and then sending a read request to this station and turning to the corresponding state;

S1, S2, and S3 correspond to the interaction states of the three data stations A, B, and C, they can read data from the corresponding data stations.

After the reading is completed, the No. 1 state machine is closed and returns to S4 of the master state machine;

The function of No. 2 slave state machine is to read data from the local station, and the corresponding state is as follows:

S0: sending request; S1: sending address; S3: receiving data; S4: sending data.

Figure 7(b) is a state machine transition diagram of the other three data stations requesting the data reading function. This function is similar to the No. 2 slave state machine.

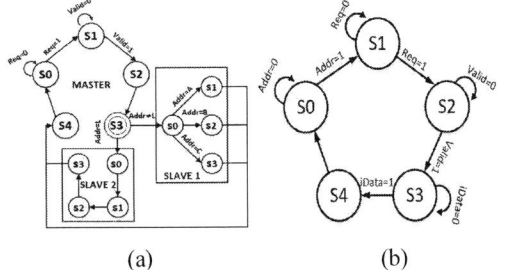

(a)   (b)

Fig. 7. (a) The read function state machine of the local data station access;(b) The read function state machine of other data stations access.

3. Address Management

This module manages the data storage address in the dual-port RAM. When data is written, the module will allocate a free address for storing data according to the memory status. When the data is read, the module will read the data from the RAM according to the input address information, and change the storage state corresponding to the address to free.

IV. EXPERIMENTAL RESULTS AND ANALYSIS

A. Average Number of Leap

As shown in Figure 8, according to the XY dimension sequence algorithm, there are three types of router nodes in the 2D-Mesh structure; (a) is the first type of router node, with the highest leaps count of 6; (b) is the second type of router node with the highest leaps count of 5; (c) is the third type of router node, and the highest number of leaps is 4; Therefore, when calculating the single-leap delay of the NOC, it is necessary to determine which type of router node the injection point is, and calculate the average value of the delay required for different leaps and multiply it by the weight of that type of node. The formula is as follows:

$$t_{leap} = \Sigma(t_i \times q_i) \times Q_i \qquad (1)$$

where t_{leap} represents the time required for a single leap, t_i represents the single leap time required to reach the destination node, q_i represents the proportion of the destination node in the other destination nodes under this type of injection, and Q_i represents the proportion of this type injection node in the proportion of all nodes. The calculation formula of single leap delay t_i is:

$$t_i = \frac{t_{total}}{N_{leap}} \qquad (2)$$

where t_{total} is the total time used in one transmission, and N_{leap} is the total number of leaps in one transmission.

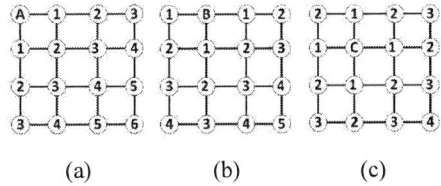

(a)   (b)   (c)

Fig. 8. (a) Number of leaps required for the first type of injection node to the destination node;(b) Number of leaps required for the second type of injection node to the destination node;(c) Number of leaps required for the third type of injection node to the destination node.

B. Performance Of the NOC at Different Injection Rates

In order to test the performance of the NOC, this experiment randomly sends data from local node to others, and then tests the single leap delay used for data transmission under different injection rates. The injection rate is the percentage of sending nodes to all nodes. As shown in Figure 9, as the injection rate increases, the single-leap transmission delay (t) also increases. At 50% injection rate, the performance drops by 50%. At 75% injection, the NOC buffer is full and stops working.

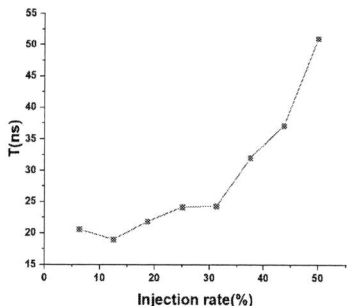

Fig. 9. The performance of the NOC at different injection rates.

C. Performance Of the NOC at Different Clock Frequencies

As shown in Figure 10, test is carried on a Xilinx FPGA AX7020 and the following observations are made : as the operating frequency increases, the performance of the NOC is greatly improved and gradually stabilized. At a clock frequency of 2GHz, the single-leap transmission time reached 20ns. When the transmission bit width is 128bit, the single-leap bandwidth reaches 800MB/s.

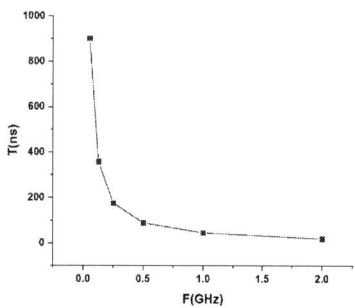

Fig. 10. The performance of the NOC at different clock frequencies.

D. Comparison of Double Layer NOC And Single Layer NOC

The experiment tested the single-leap delay of the two kinds of NOC at different transmission distances under the 2GHz clock. As shown in Figure 11, the longer the transmission distance of the Double Layer NOC, the shorter the single-leap delay; the reasons are as follows :

$$t_{total} = t_{Bottom} + t_{up} + t_{down} + t_{Top} \quad (3)$$

where t_{total} is the total time, t_{Bottom} is the bottom network transmission time, t_{up} is the data upload time, t_{down} is the data download time, and t_{Top} is the top network transmission time.

In the transmission process, t_{up} and t_{down} remain unchanged, and since the top network adopts the form of circuit interconnection, t_{Top} is unrelated with the transmission distance and can be regarded as a fixed value. Therefore, as the transmission distance increases, the single-leap transmission distance of the Double-Layer NOC drops rapidly.

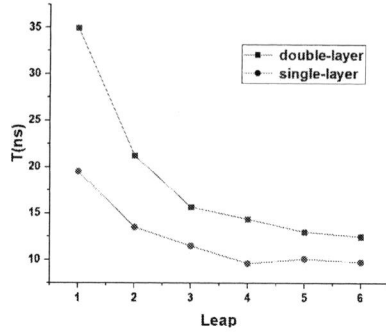

Fig. 11. The performance of the two type NOC under different transmission distances.

TABLE I.   UTILIZATION AND POWER OF TWO TYPE NOC IN DIFFERENT BIT WIDTHS

	Double-layer		Single-layer	
Power-128bit(W)	Signal	Logic	Signal	Logic
	67.253	65.000	119.038	111.117
Register	16bit	128bit	16bit	128bit
	9468	18668	9648	56241

Compared with the Single-Layer NOC, the Double-Layer NOC has the advantage of effectively reducing the utilization and power in FPGA . Registers are the devices that take up the most area and power in digital IC. Therefore, Table I compares the differences in the register number and power of the two type NOC at different bit widths. It can be seen that under the 16 bit width, the number of NOC registers is not very different, but when the bit widths is increased to 128bit, the Single-Layer NOC resource utilization is 3 times that of the Double-Layer NOC, and the power also increases 1 times, which effectively illustrates the advantages of the Double-Layer NOC in reducing wiring complexity and power.

E. Comparison of NOC And other Interconnection Technologies

As shown in Table II , compared to Intel's EMIB technology and TSMC's LIPINCON technology, NOC's most important advantages are high scalability and higher Bandwidth Density, which will not lead Significant wiring resource occupancy with the increase in the number of chiplets. At the same time, because NOC can use active CMOS technology, there will be no signal waveform distortion when data is transmitted in NOC, having a higher signal integrity.

TABLE II.   COMPARISON OF THIS WORK , LIPINCON AND EMIB

	This work	TSMC LIPINCON	Intel EMIB
Technology	Active	Passive	Passive
Scalability	High	Low	Low
Bandwidth Density	High	Low	Low

F. Simulation of Signal Integrity

As the wiring density decreases, the line spacing gradually increases, and the mutual capacitance between lines decreases [11]. In order to verify the relationship between wiring density and signal integrity, we carried out transmission line eye diagram simulation and S21 parameter simulation under different mutual capacitances. As shown in Figure 12, an approximate n-section lumped circuit equivalent model of a pair of tightly coupled transmission lines is established on ADS(Advanced Design System), C12-C21 is the mutual

978-1-6654-1392-3/21 $31.00 © 2021 IEEE

capacitance between the transmission lines; Figure 13 is the simulation of the transmission line eye diagram under different mutual capacitances at 2Gbps bit rate. When the mutual capacitance is reduced by 10 times, it can be seen that the eye width in the eye diagram has been greatly improved; in Figure 14, it can also be seen that as the mutual capacitance is reduced, the crosstalk noise S_{21} is gradually reduced. It is proved that reducing the wiring density can effectively improve the signal integrity.

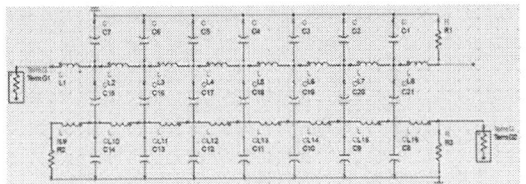

Fig. 12. Approximate n-section lumped circuit equivalent model of a pair of tightly coupled transmission lines.

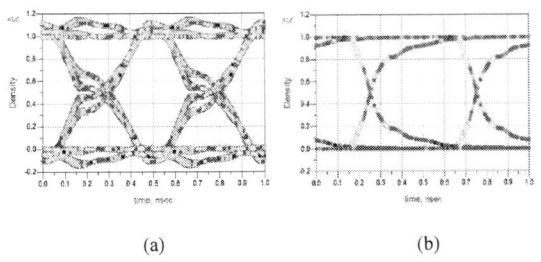

(a)   (b)

Fig. 13. (a) When the mutual capacitance is 100fF, the eye diagram of the transmission line at 2Gbps;(b) When the mutual capacitance is 10fF, the eye diagram of the transmission line at 2Gbps.

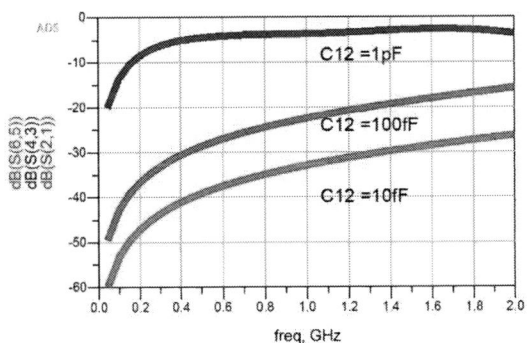

Fig. 14. Crosstalk S_{21} between two tightly coupled transmission lines under different mutual capacitances.

V. CONCLUSION

This paper proposes a networks-on-chip design scheme suitable for active interposer, innovatively using a double-layer structure, in which the top layer is four data stations with electrical interconnection, and the bottom layer is a Mesh network router with a packet forwarding structure. The advantage of this is that the data transmission path can be separated from the routing path, so that the bottom router can improve the performance by increasing the cache or using multiple virtual channels without causing a large resource occupation; The test of FPGA shows that when the transmission bit width is increased from 16bit to 128bit, the resource occupation of the single-layer structure is increased

by 5.83 times, while the resource occupation of the double-layer structure is only increased by 1.97 times, which verifies that the dual-layer NOC reduces the die resource. By modeling the equivalent circuit of a pair of tightly coupled transmission lines on ADS, the simulation of S21 parameters and signal eye diagrams verified that reduced wiring density can improve signal integrity. Compared with the passive interconnection technology proposed by TSMC or Intel, it has higher scalability and a smaller area on the die. The advantage of the double-layer structure comes at the cost of transmission delay. Under 2GHz, 128bit width, the single-leap bandwidth is 800MB/s, and the performance is reduced by half at 50% injection rate. The next work will be done by adding methods such as virtual channels further improve the bandwidth and load capacity in high injection mode, bringing out the advantages of the double-layer structure NOC.

ACKNOWLEDGMENT

The work was supported in part by National Natural Science Foundation of China (Grant No. 62074017, 61674016), in part by the Beijing Nova program Interdisciplinary Studies Cooperative projects (No. Z191100001119013), and the Importation and Development of High-Caliber Talents Project of Beijing Municipal Institutions (Great Wall Scholar, No. CIT&TCD20150320)

REFERENCES

[1] http://eps.ieee.org/hir

[2] J. Lu, "3-D Hyperintegration and Packaging Technologies for Micro-Nano Systems," in Proceedings of the IEEE, vol. 97, no. 1, pp. 18-30, Jan. 2009, doi: 10.1109/JPROC.2008.2007458.

[3] M. Miao et al., "Modeling and Design of a 3D Interconnect Based Circuit Cell Formed with 3D SiP Techniques Mimicking Brain Neurons for Neuromorphic Computing Applications," 2018 IEEE 68th Electronic Components and Technology Conference (ECTC), 2018, pp. 490-497, doi: 10.1109/ECTC.2018.00078.

[4] P. Vivet et al., "2.3 A 220GOPS 96-Core Processor with 6 Chiplets 3D-Stacked on an Active Interposer Offering 0.6ns/mm Latency, 3Tb/s/mm2 Inter-Chiplet Interconnects and 156mW/mm2@ 82%-Peak-Efficiency DC-DC Converters," 2020 IEEE International Solid-State Circuits Conference - (ISSCC), 2020, pp. 46-48, doi: 10.1109/ISSCC19947.2020.9062927.

[5] D. Greenhill et al., "3.3 A 14nm 1GHz FPGA with 2.5D transceiver integration," 2017 IEEE International Solid-State Circuits Conference (ISSCC), 2017, pp. 54-55, doi: 10.1109/ISSCC.2017.7870257.

[6] M. Lin et al., "A 7-nm 4-GHz Arm¹-Core-Based CoWoS¹ Chiplet Design for High-Performance Computing," in IEEE Journal of Solid-State Circuits, vol. 55, no. 4, pp. 956-966, April 2020, doi: 10.1109/JSSC.2019.2960207.

[7] S. Suboh, M. Bakhouya and T. El-Ghazawi, "Simulation and Evaluation of On-Chip Interconnect Architectures: 2D Mesh, Spidergon, and WK-Recursive Network," Second ACM/IEEE International Symposium on Networks-on-Chip (nocs 2008), 2008, pp. 205-206, doi: 10.1109/NOCS.2008.4492739.

[8] H. Matsutani, "Research Challenges on 2-D and 3-D Network-on-Chips," 2013 First International Symposium on Computing and Networking, 2013, pp. 24-25, doi: 10.1109/CANDAR.2013.12.

[9] S. Xiao et al., "NeuronLink: An Efficient Chip-to-Chip Interconnect for Large-Scale Neural Network Accelerators," in IEEE Transactions on Very Large Scale Integration (VLSI) Systems, vol. 28, no. 9, pp. 1966-1978, Sept. 2020, doi: 10.1109/TVLSI.2020.3008185.

[10] Z. Yu and B. M. Baas, "A Low-Area Multi-Link Interconnect Architecture for GALS Chip Multiprocessors," in IEEE Transactions on Very Large Scale Integration (VLSI) Systems, vol. 18, no. 5, pp. 750-762, May 2010, doi: 10.1109/TVLSI.2009.2017912.

[11] Bogatin, Eric. Signal and power integrity--simplified. Pearson Education, 2010.

Investigation of the Influences of Thermal stresses and Joule Heating within a Piezoresistive MEMS Pressure Sensor using the Finite Element Modeling

Chunming Zhou
Dept. of Astronautics Science
and Mechanics
Harbin Institute of Technology
Harbin, Heilongjiang, 150001,
P.R.China
20S118159@stu.hit.edu.cn

Peng Zhou*
Dept. of Astronautics Science
and Mechanics
Harbin Institute of Technology
Harbin, Heilongjiang, 150001, P.R.China
Corresponding author:
zhoup@hit.edu.cn

Yuehua Hu
Dept. of Astronautics Science
and Mechanics
Harbin Institute of Technology
Harbin, Heilongjiang, 150001,
P.R.China
hu983yuehua@126.com

Hao Zhang
Dept. of Astronautics Science
and Mechanics
Harbin Institute of Technology
Harbin, Heilongjiang, 150001, P.R.China
1141195527@qq.com

Abstract—In this paper, the influences of thermal stresses and Joule heating are studied in a piezoresistive MEMS pressure sensor via a 3D finite element model developed in our previous work using FreeFem++. First, a steady state equation of electric currents is solved to determine the electric potential. Second, a quasi-stationary equation of the thermal conduction under the influence of Joule heating is solved to obtain the distribution of the temperature field. Then, the Cauchy-Navier's equations subjected to thermal stresses are solved to find the elastic displacements and thus the elastic strains. As a result, the elastic stresses within the piezoresistors are found; then, using the constitutive equation of the piezoresistive material, the changes in the electrical resistance of the piezoresistors are obtained, which yields the output voltage eventually. During the simulations, the influences on the output voltage from the thermal stresses, Joule heating as well as different size parameters are studied. In the end, a modal analysis of the sensor is performed. The simulation results obtained have the potential to help the design of piezoresistive MEMS pressure sensors and also help to determine the major factors which influence the performance of the sensors.

Keywords—Piezoresistive sensor, Thermal stress, Joule heating, Finite Element Modeling

I. INTRODUCTION

Piezoresistive MEMS pressure sensors have aroused the interests of many researchers due to their smaller sizes and higher accuracies of measurement as compared to traditional sensors[1-3]. Most researches focus on the design and fabrication of sensors, as well as their quality and reliability. The factors which affect the performance of MEMS piezoresistive pressure sensors include size parameters, thermal stresses, Joule heating and so on. However, it's quite

inconvenient to directly investigate the influences of these factors via experiments. Thus, in this paper, a 3D finite element model is developed to simulate a piezoresistive MEMS pressure sensor. Then, the effects of the factors mentioned above are studied. Besides, a modal analysis of the sensor is also performed. This is a follow-up of our previous work[4] which focuses on the influences of size parameters.

This paper is organized in the following way. First, details of the governing equations and the finite element model are presented in section II. Then, simulation results and discussions are provided in section III. Eventually, conclusions are summarized in the last section.

II. GOVERNING EQUATIONS AND THE FINITE ELEMENT MODEL

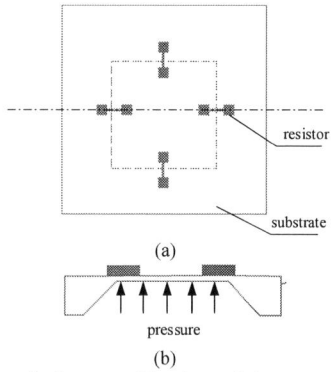

Fig.1 Schematic diagrams of the piezoresistive pressure sensor:
(a) the top view; (b) the cross section.

Shown in Fig.1 are the schematic diagrams of the MEMS piezoresistive pressure sensor simulated in our work. The dimension of the silicon substrate is 1.2 mm×1.2 mm×0.2 mm; the length and width of the silicon diaphragm in the center is 0.45 mm. The thickness of the diaphragm varies from 20 to 40 μm. Four I-shaped piezoresistors are placed at the edge of the silicon diaphragm. The width of resistors varies from 7 to 11 μm and their positions can be relocated. In our model, a Wheatstone bridge is built by connecting the four resistors via Cu interconnects. During the simulations, the displacements at the bottom of the silicon substrate are fixed. The external pressure is applied upward at the lower surface of the silicon diaphragm. The temperatures at the top and bottom surfaces of our model are both fixed, with that at the top being 1 °C higher. Presented below are details of the governing equations.

A. The Electrical and Temperature Field

For the electric conduction, the electric potential ψ can be found by solving

$$\nabla \cdot (\eta \nabla \psi) = 0 , \tag{1}$$

where η is the electric conductivity. Fixed electric potentials are applied as boundary conditions at both ends of the Cu interconnects. Then, a quasi-stationary equation of the thermal conduction under the influences of Joule heating is solved:

$$\nabla \cdot (K \nabla T) + \eta |\nabla \psi|^2 = 0 , \tag{2}$$

where K is the thermal conductivity. Fixed temperatures, T_T and T_B, are applied as boundary conditions at both the top and bottom surfaces of our model.

B. The Elastic Fields

The following quasi-stationary Cauchy-Navier's equations with cubic anisotropy are solved:

$$(C_{11} + C_{44})\partial/\partial x_i (\partial u_i/\partial x_j) + C_{44}(\partial^2 u_i/\partial x_j \partial x_j) + (3C_{12} + 2C_{44})\alpha \partial T/\partial x_i + \tag{3}$$
$$(C_{11} - C_{12} - 2C_{44})\delta_{ijkl}(\partial^2 u_k/\partial x_j \partial x_i - \alpha \delta_{ki}\partial T/\partial x_j) = 0 \,(i = 1,2,3),$$

and the boundary conditions are given by

$$u_{i(x_3 = X_{b_1})} = 0 \quad (i = 1,2,3), \tag{4}$$

$$\sigma_{33(x_3 = X_{b_2})} = t_{33} . \tag{5}$$

Here, u_i (i=1,2,3), which consists of the elastic contribution and the thermal contribution, are the total displacements; C_{11}, C_{12} and C_{44} are elastic constants for the cubic anisotropy; $x_3=x_{b1}$ denotes the bottom surface of our model and the displacements are fixed there; $x_3=X_{b2}$ denotes the lower surface of the silicon diaphragm and t_{33} is the external pressure applied upward there. Usually, the MEMS devices are mounted on certain apparatuses and their vaults are connected to the high-pressure chambers of these apparatuses. Thus, the displacements are fixed at the bottom surface of our model and pressures are applied upward at the lower surface of the silicon diaphragm. Then the elastic strains are calculated by

$$\varepsilon_{kl} = (\partial u_k/\partial x_l + \partial u_l/\partial x_k) - \alpha (T - T_0)\delta_{kl} . \tag{6}$$

Here, α is the coefficient of thermal expansion; T_0 is the room temperature, i.e. 25 °C. This leads to the elastic stresses:

$$\sigma_{ij} = C_{ijkl}\varepsilon_{kl} , \tag{7}$$

$$C_{ijkl} = C_{12}\delta_{ij}\delta_{kl} + C_{44}(\delta_{ij}\delta_{kl} + \delta_{ij}\delta_{kl}) + (C_{11} - C_{12} - 2C_{44})\delta_{ijkl} . \tag{8}$$

Here, C_{ijkl} (i,j,k,l =1,2,3) is the stiffness tensor.

C. The Output Voltage

The output voltage U_{out} is given by[3]

$$U_{out} = |\pi_{44}(\sigma_l - \sigma_t)/2|U_{in} . \tag{9}$$

Here, π_{44} is the piezoresistive coefficient of the P-type silicon, σ_l and σ_t are the longitudinal and transverse stresses calculated on the midplane of the resistors, and U_{in}, set to be 5V, is the input voltage. To determine σ_l and σ_t, sixteen evenly distributed points are set along the direction of the length of the resistor on its midplane. Then, σ_l and σ_t are calculated as the average value of the stresses sampled at these points.

Note that in the above three subsections, all governing equations are steady state equations. First, electrons are highly mobile particles thus the electric field achieves steady states rather quickly. Second, the temperature field of most apparatuses are usually either at equilibriums or varies rather slowly, thus it is reasonable to use a quasi-stationary equation of the thermal conduction. Third, equations of mechanical equilibriums are used for the elastic fields, thus our simulation results are only valid for a constant or slowly varying externally applied pressure.

D. Modal Analysis

To obtain the natural frequency of our simulated device, a modal analysis is performed. Beginning with the dynamic vibration equation in the absence of body forces,

$$\partial/\partial x_j (\sigma_{ij}) = \partial^2 u_i/\partial^2 t , \tag{10}$$

a weak form of the governing equation used to calculate the natural frequency can be obtained,

$$\int_\Gamma C_{ijki}\partial u_i'/\partial x_j \partial v_k/\partial x_i dv - \omega^2 \int_V \rho u_i' v_i dv = 0 , \tag{11}$$

where $u_i = u_i'\sin(\omega t + \varphi)$ with u_i' being the amplitude, ω being the natural frequency and φ being the phase angle; v_i is the weighting function and ρ is the density of mass. The boundary conditions are also given by Eqn (4).

E. Finite Element Modeling

All above governing equations are solved using the free online open-source finite element software, FreeFem++[5]. Values of physical parameters and elastic constants used in our simulations are given in Table I and Table II, respectively. The externally applied pressure t_{33} is set to be 800KPa for all simulation runs.

TABLE I   VALUES OF SOME PHYSICAL PARAMETERS IN OUR MODEL

	$k\left(WK^{-1}m^{-1}\right)$	$\alpha\left(10^{-6}K^{-1}\right)$	$\pi_{44}\left(10^{-11}Pa^{-1}\right)$
Si	141.0	2.33	NA
Cu	397.0	17.0	NA
P-type Si	157.0	2.5	138.1

TABLE II   VALUES OF ELASTIC CONSTANTS IN OUR MODEL

	$C_{11}(GPa)$	$C_{12}(GPa)$	$C_{44}(GPa)$
Si	165.6	63.9	79.5
Cu	202.226	105.576	48.325
P-type Si	165.6	63.9	79.5

TABLE III   THE RATE OF INCREASE dU_{out}/dT WITH DIFFERENT DIAPHGRAM THICKNESSES

Dt (μm)	20	25	30	35	40
dU_{out}/dT (mV/°C)	0.681	0.658	0.635	0.610	0.563

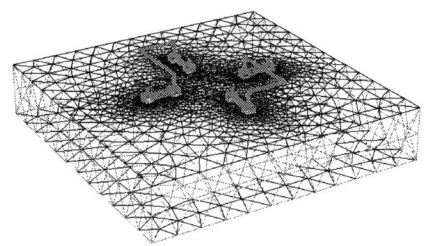

Fig.2 The finite element mesh of our model

Shown in Fig. 2 is a schematic diagram of the finite element mesh of our model. For the actual simulations, the meshes are much finer with over 4×10^5 vertices and 2×10^6 tetrahedral elements. Each simulation run takes about one and half hour on a workstation with Intel Xeon Processors ES-2650 v4.

III. MODELING RESULTS

A. The Influence of Temperatures and Size Parameters on the Output Voltage

Shown in Fig.3 are the variations of the output voltage with different temperatures T and diaphragm thicknesses D_t. Here the width of the P-type silicon resistor W is set to be 10 μm; the distance between the center of the diaphragm and the midpoint of the P-type silicon resistor D_p is set to be 225 μm. As can be seen in plot (a), when the diaphragm thickness remains constant, the output voltage increases linearly as the temperature increases. For an example, when the diaphragm thickness is 20 μm, the output voltage increases from 89.3147 mV to 99.5233 mV. The rate of increase of the ouput voltage w.r.t. the temperature, i.e. dU_{out}/dT, is calculated to be 0.681 mV/°C. The variations of dU_{out}/dT with different diaphragm thicknesses are summarized in TABLE III. It can be seen that dU_{out}/dT decreases evidently with increasing diaphragm thicknesses. In plot (b), it is shown when the temperature is fixed, the output voltage decreases nonlinearly as the diaphragm thickness increases. Furthermore, the output voltages are in the order of 100 mV which also qualitatively agrees with the results from a previous work in the order of magnitude.[6]

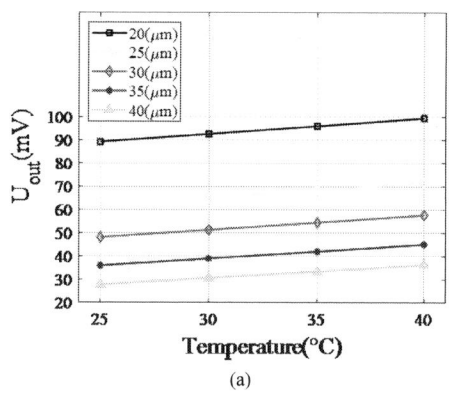

(a)

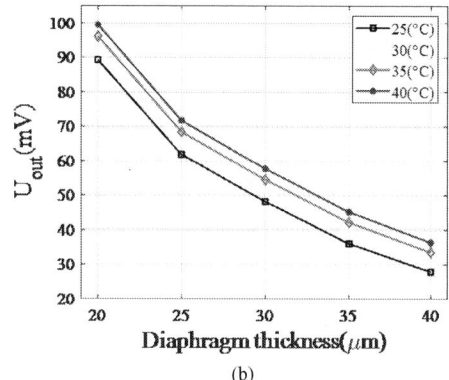

(b)

Fig.3 Variation of the output voltage (a) with increasing temperatures when the diaphragm thicknesses are different; (b)with increasing diaphragm thicknesses when the temperatures are different.

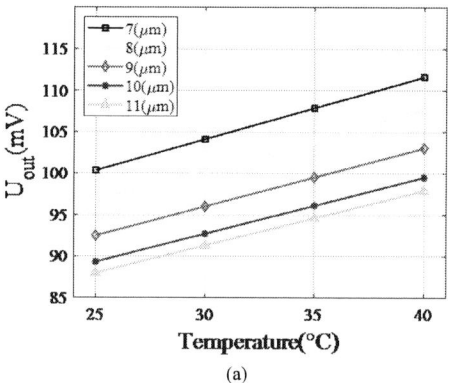

(a)

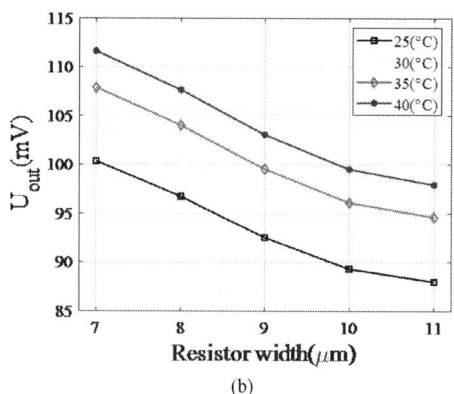

(b)

Fig.4 Variation of the output voltage (a) with increasing temperatures when the widths of the resistors are different; (b)with increasing resistor widths when the temperatures are different.

978-1-6654-1392-3/21 $31.00 © 2021 IEEE   436

TABLE IV THE RATE OF INCREASE dU$_{out}$/dT WITH DIFFERENT WIDTHS OF THE P-TYPE SILICON RESISTOR

W(µm)	7	8	9	10	11
dU$_{out}$/dT(mV/°C)	0.752	0.726	0.702	0.681	0.663

Shown in Fig. 4 are the variations of the output voltage with different temperatures and widths of the resistors. Here the diaphragm thickness is set to be 20 µm; the distance between the center of the diaphragm and the midpoint of the resistor is set to be 225 µm. As can be seen in plot (a), when the resistor width remains constant, the output voltage increases linearly as the temperature increases. For an example, when the resistor width is 7 µm, the output voltage increases from 100.352 mV to 111.632 mV. The rate of increase is calculated to be 0.752 mV/°C. The variations of dU$_{out}$/dT with different widths of the resistors are summarized in TABLE IV. It can be seen that dU$_{out}$/dT decreases evidently with increasing widths of the P-type silicon resistor. In plot(b), it is shown when the temperature remains constant, the output voltage approximately decreases linearly until the resistor width reaches 10 µm and then the rate of decrease slows down.

Shown in Fig.5 are the variations of the output voltage with different temperatures and distances between the center of the diaphragm and the midpoint of the resistor. Here the diaphragm thickness is set to be 20 µm; the width of the resistor is set to be 10 µm. As can be seen in plot (a), when the distance between the center of the diaphragm and the midpoint of the resistor remains constant, the output voltage increases linearly as the temperature increases. For an example, when the distance is 150 µm, the output voltage increases from 14.713 mV to 20.996 mV. The rate of increase is calculated to be 0.419 mV/°C. The variations of dU$_{out}$/dT with different diaphragm thicknesses are summarized in TABLE V. It can be seen that dU$_{out}$/dT grows evidently with increasing distances between the center of the diaphragm and the midpoint of the resistor. In plot(b), it is shown when the temperature remains constant, the output voltage at first increases and then decreases with the increase of the distance. Obviously the maximum of the output voltage occurs when the distance reaches 225 µm, which corresponds to the edge of the diaphragm.

B. The Influence of Joule Heating on the Output Voltage

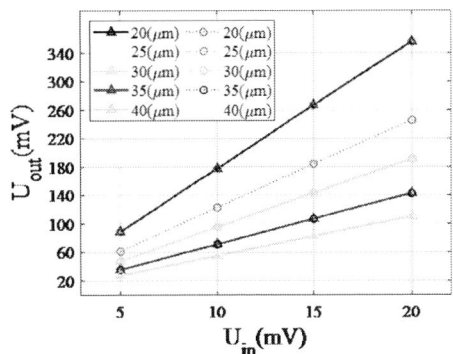

Fig.6 Variation of the output voltage with increasing input voltages when the diaphragm thicknesses are different with (dash lines) and without (solid lines) the effects of Joule heating.

Shown in Fig.6 is the variation of the output voltage with an increasing input voltage when the diaphragm thicknesses are different with (shown by dash lines) and without (shown by solid lines) the effects of Joule heating. Here the width of the resistor is set to be 10 µm; the distance between the center of the diaphragm and the midpoint of the resistor is set to be 225 µm. The effects of Joule Heating to the output voltage are found negligible in our simulations. For example, in one simulation, when the temperature is 50 °C and the input voltage is 15 V, the output voltage only increases by 0.055% when Joule heating is considered.

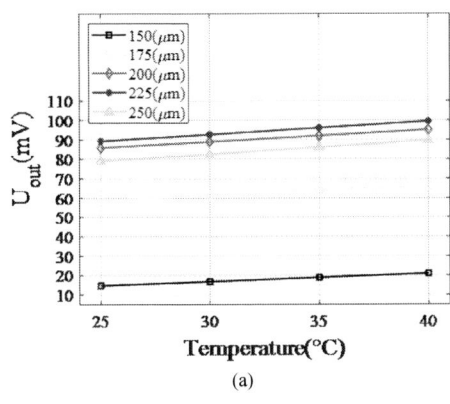

(a)

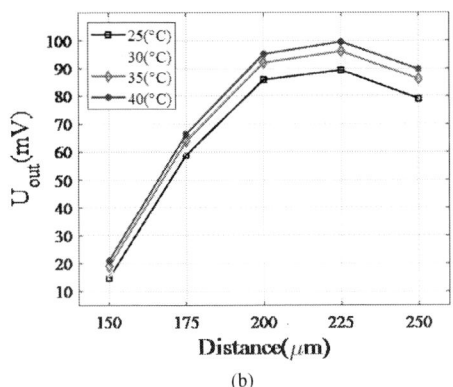

(b)

Fig.5 Variation of the output voltage (a)with increasing temperatures when distances between the center of the diaphragm and the midpoint of the P-type silicon resistor are different; (b)with increasing distances between the center of the diaphragm and the midpoint of the P-type silicon resistor when temperatures are different.

C. Modal Analysis

The natural frequency is an important parameter of the piezoresistive MEMS pressure sensors. Since most sensors work in environments with vibrations, it is an indicator whether the sensor is fit for its working environments or not.

TABLE V THE RATE OF INCREASE dU$_{out}$/dT WITH DIFFERENT DISTANCES BETWEEN THE CENTER OF THE DIAPHRAGM AND THE MIDPOINT OF THE P-TYPE SILICON RESISTOR

D$_p$ (µm)	150	175	200	225	250
dU$_{out}$/dT((mV/°C)	0.419	0.501	0.620	0.681	0.713

Shown in Table VI, VII, VIII are the frequencies of the first vibration mode of the sensors with different diaphragm thicknesses, widths of the resistors and distances between the center of the diaphragm and the midpoint of the resistors. As shown in Table VI, the natural frequency of the piezoresistive MEMS pressure sensors increases evidently with increasing diaphragm thicknesses. However, the influence of the widths of the resistors is found negligible and that of the distances between the center of the diaphragm and the midpoint of the resistor is found minor in our simulations. Shown in Fig.7 are the first vibration modes of the piezoresistive MEMS pressure sensor in one simulation.

TABLE VI   THE NATURAL FREQUENCY OF THE SENSORS WITH DIFFERENT DIAPHGRAM THICKNESSES

D_t (μm)	20	25	30	35	40
F(KHz)	1575.49	1776.56	1906.44	2095.31	2241.42

TABLE VII   THE NATURAL FREQUENCY OF THE SENSORS WITH DIFFERENT WIDTHS OF THE P-TYPE SILICON RESISTOR

W(μm)	7	8	9	10	11
F(KHz)	1545.26	1562.97	1566.81	1575.49	1571.63

TABLE VIII   THE NATURAL FREQUENCY OF THE SENSORS WITH DIFFERENT DISTANCES BETWEEN THE CENTER OF THE DIAPHRAGM AND THE MIDPOINT OF THE P-TYPE SILICON RESISTOR

D_p (μm)	150	175	200	225	250
F(KHz)	1411.15	1507.93	1553.06	1575.49	1571.63

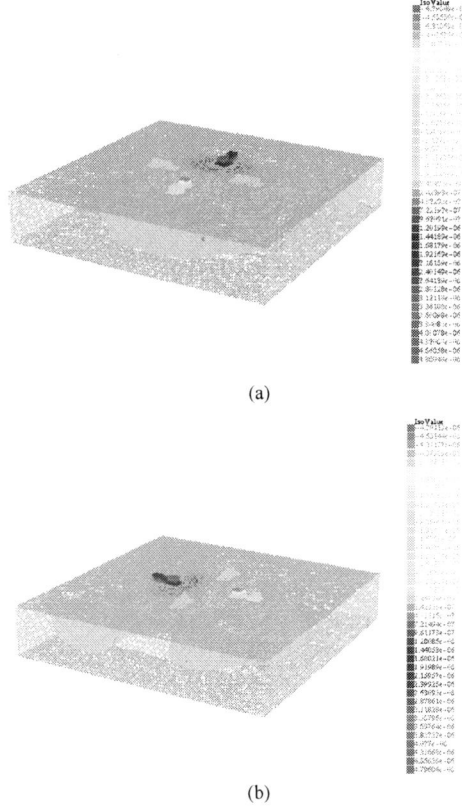

(a)

(b)

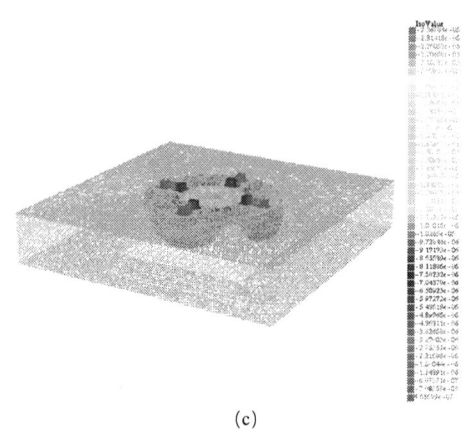

(c)

Fig.7 The first mode of vibration of the piezoresistive MEMS pressure sensors: (a) X direction; (b) Y direction; (c) Z direction.

IV. CONCLUSIONS

In this paper, simulations of a piezoresistive MEMS pressure sensor are performed with FreeFem++ using a 3D finite-element model. First, the electric field and the temperature field are calculated. Then, the elastic stresses under the influence of the thermal stresses are obtained, which leads to the determination of the output voltage eventually. The following conclusions are obtained: (1) the output voltage increases linearly with an increasing temperature. The rate of increase falls with increasing diaphragm thicknesses and widths of the P-type silicon resistor, while it grows with increasing distances between the center of the diaphragm and the midpoint of the P-type silicon resistor; (2) when the temperature remains constant, the output voltage decreases as the diaphragm thickness and the resistor width increases, while it increases to a maximum value when the distance between the center of the diaphragm and the midpoint of the P-type silicon resistor reaches 225 μm , which corresponds to the edge of the diaphragm; (3) the effects of Joule Heating to the output voltage are found negligible in our simulations; (4) the natural frequency of the piezoresistive MEMS pressure sensors decreases with increasing diaphragm thicknesses. Present work does not consider the temperature dependence of the piezoresistive coefficients, which will be investigated in our future work.

ACKNOWLEDGMENT

This work is financially supported by the National Natural Science Foundation of China (Grant No.51201049).

REFERENCES

[1] R. Bogue, Recent developments in MEMS sensors: a review of applications,markets and technologies, Sensor Review 33 (4) (2013) 300–304.

[2] S.S. Kumar, B.D. Pant, Design principles and considerations for the 'ideal' silicon piezoresistive pressure sensor: a focused review, Microsyst. Technol.20 (7) (2014) 1213–1247.

[3] Chang Liu, "Foundations of MEMS", Pearson Education Limited, (2011).

[4] Hao Zhang, Peng Zhou, Jinxin Qian, Guoming Ma, Yubao Zhen and Yuehua Hu, "3D simulations of a Piezoresistive MEMS Pressure Sensor using the Finite Element Modeling", 2019 20th International Conference on Electronic Packaging Technology (ICEPT).

[5] Hecht, F. New development in FreeFem++. J. Numer. Math. 20 (2012), no. 3-4, 251–265. 65Y15.

[6] Sun Xiao-long, "Using Finite Element Simulation Calculate Out-put-voltage of MEMS Piezoresistive Pressure Sensor", Instrument Technique and Sensor, 8, 2010, 12-13.

Accelerated Aging and Lifetime Evaluation of Polyurethane Packaging Material for Optical Fiber Hydrophone

Wenyuan Liao
Science and Technology on Reliability Physics and Application of Electronic Component Laboratory
China Electronic Product Reliability and Environmental Testing Research Institute
Guangzhou, China
wyliaophd@126.com

Canxiong Lai
Science and Technology on Reliability Physics and Application of Electronic Component Laboratory
China Electronic Product Reliability and Environmental Testing Research Institute
Guangzhou, China
lcx305@163.com

Rui Gao
Science and Technology on Reliability Physics and Application of Electronic Component Laboratory
China Electronic Product Reliability and Environmental Testing Research Institute
Guangzhou, China
r.gao90@ceprei.com

Shaohua Yang
Science and Technology on Reliability Physics and Application of Electronic Component Laboratory
China Electronic Product Reliability and Environmental Testing Research Institute
Guangzhou, China
yang01@163.com

Guoguang Lu
Science and Technology on Reliability Physics and Application of Electronic Component Laboratory
China Electronic Product Reliability and Environmental Testing Research Institute
Guangzhou, China
luguog@126.com

Shuwang Li*
Science and Technology on Reliability Physics and Application of Electronic Component Laboratory
China Electronic Product Reliability and Environmental Testing Research Institute
Guangzhou, China
lswang6@mail.ustc.edu.cn

Abstract—Fiber optic hydrophone is an underwater acoustic signal sensor based on modern optical fiber and optoelectronic technology, which is divided into dry end and wet end. Polyurethane is widely used as the coating material for the wet-end encapsulation of fiber-optic hydrophones. The reliability of the polyurethane packaging directly affects the detection sensitivity, connection reliability and the service life of the detection array of the wet end sensor module. Aiming at the reliability problem of polyurethane encapsulation material for optical fiber hydrophone, a dumbbell-shaped polyurethane sample is prepared in this paper. The sample is placed in the aging environment of hot seawater medium with salinity of 3.5%. The accelerated aging experiment is carried out on the samples. The tensile strength of the polyurethane at different aging stages are tested, and the aging kinetics of the tensile strength of the polyurethane in hot seawater medium at 55℃, 65℃, 75℃ and 85℃ are studied. The results show that the elongation at break of polyurethane at different temperatures decreases exponentially with the aging time, and the mechanical properties of polyurethane decrease faster in the thermal seawater aging test at higher temperatures. The lifetime under operating temperature is predicted by using the Arrhenius equation. The activation energy of polyurethane in seawater is calculated to be 0.65eV, and the service life of the polyurethane in seawater at 25℃ is predicted to be 22.81 years.

Keywords—polyurethane, fiber optic hydrophone, accelerated aging, lifetime evaluation

I. INTRODUCTION

Fiber optic hydrophone was developed in the late 1970s[1], which is a new type of underwater acoustic sensor based on fiber optic sensing and optoelectronic technology[2]. Its sensing principle is to obtain the frequency and intensity of sound waves by modulating the intensity, polarization, phase and other parameters of light wave in the optical fiber. Compared with the traditional piezoelectric detector system, the interferometric fiber optic hydrophone has the advantages of high sensitivity, strong resistance to electromagnetic interference and signal crosstalk, large dynamic range, small size, light weight and good fitting [3-5]. More importantly, combined with the existing fiber optic communication technology, fiber optic hydrophone can easily set up a variety of underwater fiber optic sensor networks[6-7], thus providing an ideal technical way to solve the problems in a wide range of applications such as underwater acoustic detection and marine energy exploration.

With the wide application of fiber optic hydrophone system in marine science, the reliability requirement is becoming more and more high. The wet end of fiber optic hydrophone is specifically a waterproof complex optical fiber network, which is composed of a large number of optical fiber passive components in a certain multiplexing mode. Its working environment is harsher and reliability problem is more prominent. There is no doubt that the reliable packaging is key point to ensure the stable operation of fiber optic hydrophone in the decades of service life, and the packaging structure of the fiber optic hydrophone must be carefully designed[8-9]. Polyurethane is widely used as the coating material for the wet-end encapsulation of fiber-optic hydrophones, which is a multi-block copolymer composed of alternating soft and hard segments. The soft segment is composed of oligomer polyol, while the hard segment is formed by the reaction of isocyanate and small molecule chain extender[10]. Because of polyurethane molecular chain which contains carbamate, ester group, biuret group and other groups, under the joint combined action of water, oxygen and heat, the polyurethane material will be hydrolyzed[11]. The molecular chain is hydrolyzed and broken to generate carboxyl groups, which further accelerates the aging of the material and affects the service life of the polyurethane material[12]. The reliability of the polyurethane packaging directly affects the detection sensitivity, connection reliability and the service life of the detection array of the wet end sensor module. Due to the large surface area between polyurethane adhesive and metal, the continuous erosion of water molecules may lead to the failure of bonding between polyurethane and metal matrix in water, especially in harsh seawater environment.

In this paper, a research on accelerated aging and lifetime evaluation of polyurethane packaging material for optical fiber hydrophone is present. First, the variation of elongation at break of polyurethane in seawater with aging time is studied. Then, the activation energy of polyurethane aging in seawater is obtained, and the lifetime of polyurethane is evaluated. The results provide a basis for the effective application of polyurethane in fiber optic hydrophone.

II. EXPERIMENT

A. Sample preparation

According to the design of GB/T 528-2009, the dumbbell-like polyether polyurethane packaging sample is prepared by using the same process conditions of optical fiber hydrophone. The sample thickness is about 4mm, and the length is about 115mm. The structure of the sample is shown in Fig. 1.

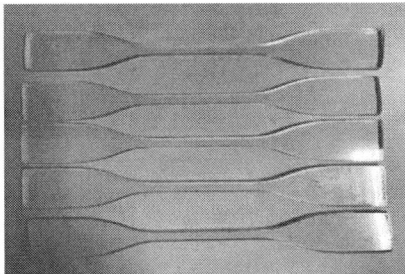

Fig. 1 Dumbbell shaped polyurethane structure.

B. Aging environment and test

The aging environment is hot seawater medium, and the salinity of the seawater medium is 3.5%, as shown in Fig. 2. According to the preliminary situation, the aging temperature is set at 55℃, 65℃, 75℃ and 80℃ respectively. At certain intervals, five samples is taken out and the tensile performance is tested by the universal tensile testing machine. The tensile strength testing equipment is shown in Fig. 3.

Fig. 2 Hot sea water aging equipment.

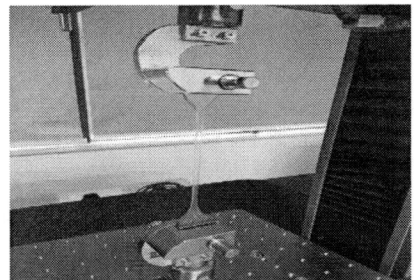

Fig. 3 The tensile strength testing equipment.

III. RESULTS AND DISCUSSIONS

A. Variation of elongation at break of polyurethane aging in seawater

The prepared samples are placed in seawater medium at different temperatures for accelerated aging experiment, and five samples are taken out at certain intervals. The elongation at break of the samples are tested according to GB528-2009 standard. Due to the material inhomogeneity and tensile test error, the test values of the five samples at the same temperature and the same aging time point are different. Take the average of the five sets of data, and the test results are shown in Fig. 4. As can be seen from the figure, When aging in the seawater medium at 50°C, 65°C, 75°C, and 80°C, the elongation at break of the polyurethane decreases exponentially with time. As the aging temperature increases, the aging speed increases. With the increase of aging time, the change of elongation at break tends to be slower.

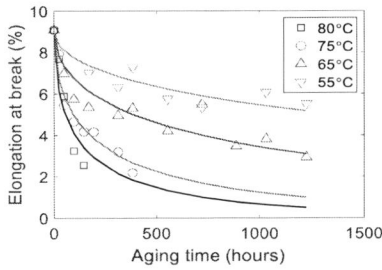

Fig. 4 Variation of polyurethane elongation at break with aging time at different temperatures.

According to the data in the Fig. 4, matlab software is used for fitting to obtain the relationship between the average elongation at break of the sample and aging time, as shown in equation (1) below:

$$E = b \exp(-kt^a) \qquad (1)$$

Where E is the elongation at break, k is the aging rate constant, t is the aging time, b and a are constants independent of temperature.

B. Lifetime evaluation

The change of the aging rate constant with temperature can be described by the Arrhenius equation, as shown in Equation (2).

$$k = A \exp(-\frac{E_a}{RT}) \qquad (2)$$

Where A is the coefficient, E_a is the activation energy, R is the gas constant, and T is the aging temperature.

According to Arrhenius equation, the activation energy of polyurethane aging and the expression of the aging rate constant with temperature are obtained by linear regression of aging rate constant at each temperature point. Then, the aging rate constant at room temperature is obtained and the variation law of polyurethane performance with aging time at room temperature is determined. It is generally believed that when the performance of the rubber material drops to half of the initial performance, the rubber no longer has use value. Therefore, in the process of predicting the service life of polyurethane, the critical value is taken as the value of 50% of the performance retention rate in this article.

The data in the figure 3 are fitted according to equation (1), and the mathematical relationship between elongation at break and aging time of the sample at different seawater temperatures are obtained as shown in Equations (3) to (6). Equations (3) to (6) are the mathematical relationships of elongation at break with aging time at 55°C, 65°C, 75°C, and 80°C seawater temperature, respectively.

$$E = 8.9981 \cdot exp(-0.0144 \cdot t^{0.5140}) \tag{3}$$
$$E = 8.9981 \cdot exp(-0.0276 \cdot t^{0.5140}) \tag{4}$$
$$E = 8.9981 \cdot exp(-0.0571 \cdot t^{0.5140}) \tag{5}$$
$$E = 8.9981 \cdot exp(-0.0751 \cdot t^{0.5140}) \tag{6}$$

Where E is the elongation at break (%), t is the aging time (hour). According to equations (2) to (5), the aging rate constant k at different temperatures can be calculated. Using the aging rate constant k at different temperatures to make linear regression of Arrhenius equation, the change curve of aging rate constant with temperature is obtained, as shown in Fig. 5. Then, the formula (7) of aging rate constant K and temperature is obtained.

$$\ln k = 19.4826 - 7790.5/T \tag{7}$$

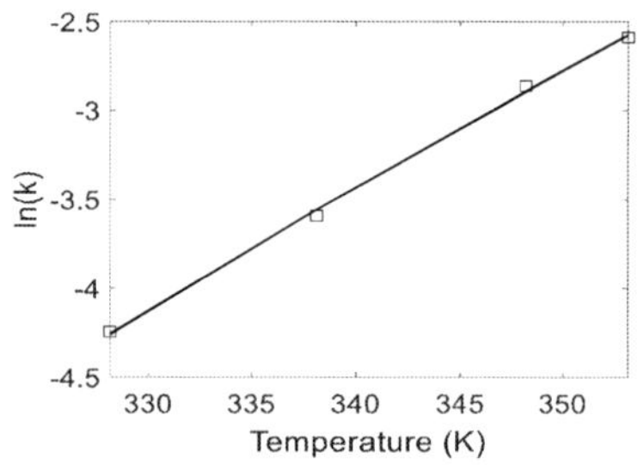

Fig. 5 The logarithm of the aging rate constant changes with temperature.

According to the above formula, when the sample is aged in seawater, the aging activation energy E_a expressed by the change in elongation at break is as follows: E_a=7790.5×8.314/1000=64.77 kJ/mol=0.65 eV. According to formula (7), the aging rate of the sample in 25°C seawater is calculated as $K_{25°C}$=1.28×10^{-3}.

The aging rate constant in seawater at 25°C is substituted into equation (1). Combined with the parameters in equations (3) to (6), the mathematical relationship between the elongation at break and time of the sample aging at room temperature is obtained as shown in equation (8).

$$E = 8.9981exp(-1.28 \times 10^{-3}t^{0.514}) \tag{8}$$

According to the above formula, the time required for the elongation at break to decrease to 50% at 25°C is calculated to be 22.81 years.

IV. CONCLUSIONS

In this work, accelerated aging and lifetime evaluation of polyurethane packaging material for optical fiber hydrophone is demonstrated. We find that the elongation at break of polyurethane at different temperatures decreases exponentially with the aging time, and the mechanical property of polyurethane decreases faster in the thermal seawater aging test at higher temperatures. The activation energy of polyurethane in seawater is calculated to be 0.65eV, and the service life of the polyurethane in seawater at 25°C is predicted to be 22.81 years. In the paper, the service life of polyurethane in practical application condition is calculated, which provides a theoretical basis for the engineering application of polyurethane packaging material for optical fiber hydrophone.

REFERENCES

[1] Bucaro. J. A, Dardy. H. D, Carome. E. Fiber-optic hydrophone [J]. The Journal of the Acoustical Society of America, 1977, 62(5):1302~1304.

[2] Grattan. T. V, Sun. D. T. Fiber optic sensor technology: an overview [J]. Sensors and Actuators, 2000(82):40~61.

[3] T. G. Giallorenzi, J. A. Bucaro and A. Dandridge, et al. Optical Fiber Sensor Technology[J], Quantum Electronics, 1982, 18(4): 626-633.

[4] T. G. Giallorenzi, J. A. Bucaro, A. Dandridge, et al. Optical fiber sensor technology. IEEE Transanction on Microwave Theory and Techniques[J]. 1982, 30(4):472~511.

[5] James. H. Cloes, Joseph A. Bucaro, Clay K. Kirkendall and Anthony Dandridge. "The origin, history and future of fiber-optic interferometric acoustic sensors for US Navy applications," Proceedings of SPIE, 2011, Vol. 7753, pp. 0301-0304.

[6] M. J. F. Digonnet, B. J. Vakoc, C. W. Hodgson, et al. Acoustic Fiber Sensor Arrays[C]. Second European Workshop on Optic Fiber Sensors, SPIE, 2004, vol. 5502: 39~50.

[7] F. Souto. "Fibre optic towed array: the high tech compact solution for naval warfare," Proceedings of Acoustics, Victor Harbor Australia, November 2013: 17-20.

[8] C. Berg, J. Langhammer and P. Nash. "Lifetime stability and reliability of fibre-optic seismic sensors for permanent reservoir monitoring," SEG, 2012, Vol. 1236, pp. 1-5.

[9] P. Nash. "High Reliability Fibre-optic Sensor Architectures for Seismic Permanent Reservoir Monitoring," EAGE, 2010, Vol. 10, pp. 1-5.

[10] K.S. Chian, L.H. Gan. J. Appl. Polym. Sci, 1998, 68: 509–515.

[11] Aglan. H, Calhoun. M, Allie. L. Effect of UV and hygrothermal aging on the mechanical performance of polyurethane elastomers[J]. J Appl Polym Sci, 2008, 108: 558－564.

[12] Boubakri. A, Elleuch. K, Guermazi. N. Investigations on hygrothermal aging of thermoplastic polyurethane material[J]. Mater Desing, 2009, 30: 3958－3965.

Characterizing the die attach layer delamination effect on the heat transferring performance in LED package with entropy generation analysis

Binjie Ai
Guangxi Key Laboratory of
Manfacturing System &Advanced
Manfacturing Technology
*School of Mechanical and Electrical
Engineering ,Guilin University of
Electronic Technology*
Guilin, China
Email:binjieai897@163.com

Miao Cai*
Guangxi Key Laboratory of
Manfacturing System &Advanced
Manfacturing Technology
*School of Mechanical and Electrical
Engineering, Guilin University of
Electronic Technology*
Guilin, China
Email:caimiao105@163.com

Daoguo Yang
Guangxi Key Laboratory of
Manfacturing System &Advanced
Manfacturing Technology
*School of Mechanical and Electrical
Engineering, Guilin University of
Electronic Technology*
Guilin, China
Email:daoguo_yang@163.com

Guangsheng Lu
Guangxi Key Laboratory of
Manfacturing System &Advanced
Manfacturing Technology
*School of Mechanical and Electrical
Engineering ,Guilin University of
Electronic Technology*
Guilin, China
Email:guangshenglu@163.com

Kailin Zhang
Guangxi Key Laboratory of
Manfacturing System &Advanced
Manfacturing Technology
*School of Mechanical and Electrical
Engineering, Guilin University of
Electronic Technology*
Guilin, China
Email:1198583386@qq.com

Guoqi Zhang
Shenzhen Institute of Wide-Bandgap
Semiconductors
*Delft Institute of Microsystems and
Nanoelectronics(Dimes), Delft
University of Technology*
Shenzhen, China
Email:g.q.zhang@tudelft.nl

Abstract—For light-emitting diode (LED) package, interface delamination is a key factor that triggers reliability issues. When the delamination occurs in the thermal path of the die attach (DA) layer, the heat transfer capacity will be blocked and the junction temperature of the device will increase, which will affect various other failure mechanisms and ultimately reduces the life of the LED package. In this paper, the finite element method is used and based on entropy generation analysis. Interfacial delamination is prefabricated to the upper and lower interfaces of a model. The effects of cracks simultaneously appearing at different positions and lengths of the upper and lower interfaces on heat transfer are compared. Based on the simultaneous results, in the same position of the DA layer, the entropy generation value of delamination is larger than that without delamination, and the entropy generation value of delamination in the same position of the upper and lower interfaces is greater than that of not located in the same position. The larger the entropy generation value, the worse the heat transfer performance. And cracks appearing at the edge of the upper and lower interfaces have greater damage on heat conduction than the middle position. In addition, as the delamination length simultaneously increases at the upper and lower interfaces, the heat transfer performance becomes worse.

Keywords—DA layer, Delamination, Thermal effect, Entropy generation analysis

I. INTRODUCTION

LED is an electroluminescent device whose heat cannot be radiated. Narendran et al. [1]found that when the temperature of the LED device rises by 2℃, the reliability drops by 10%. As a first-level package, the die attach (DA) layer is the only way for the temperature to pass from the LED chip to the substrate. In the actual packaging process, the die bonding will inevitably leave voids in the bonding layer. Further, due to the effect of thermal stress, the quality of the solder layer may be degraded. When the voids increase, cracks and even delamination will appear, which reduces the thermal conductivity of the device and hinders heat transfer. In addition, the electro-optical conversion efficiency of LED

chips is low. Only 20% to 30% of the input power is converted into light, and the rest is converted into heat. Interfacial delamination is one of the main failure modes of LED devices during operation. The upper interface is between the chip and DA layer, and the lower interface is between the DA layer and substrate. In recent years, the research on voids, interfacial cracks and delamination in the bonding layer of power devices has received extensive attention from many scholars. Jian Zhang et al. [2] studied the effect of voids in the IGBT solder layer on the thermal reliability of the device. The results show that when the void size is large, the junction temperature of the device will increase. And as the number of voids increases, the junction temperature of the device further rises. Similarly, Pengxin Xie et al. [3] found that the corner void at the upper interface hinders heat dissipation more than the central void, and the maximum stress of the solder layer appears in the corner void. The cohesive zone model was used by Binbin Zhang et al. [4] to simulate the expansion of the LED package interface delamination under thermal stress. The paper by Yuezhu Mo[5] used the volume entropy production rate formula to study the DA layer with delamination of LED. Meanwhile, Peng Cui et al. [6] also studied the delamination prefabricated at the lower interface based on entropy generation. Most of the researches are about the relationship between the expansion of voids or interface delamination and thermal stress. However, when nano-silver solder paste is used as the DA layer and the delamination simultaneously occurs at the upper and lower interfaces, the research on the influence of heat transfer is rare.

In this paper, nano-silver solder paste is used as a DA layer material and is based on entropy generation analysis. The interfacial delamination is prefabricated to the upper and lower interfaces where one crack is prefabricated on the upper interface and the other is simultaneously prefabricated on the lower interface. Research on the influence of delamination with different positions and different lengths on the heat transfer of the DA layer. Firstly, when one crack is prefabricated on the upper interface and the other is simultaneously prefabricated on the lower interface, models

978-1-6654-1392-3/21 $31.00 © 2021 IEEE

of different delamination positions on the upper and lower interfaces are established and the corresponding entropy generation values are compared. Secondly, When one upper interface crack is fixed at the left edge, the other crack is fixed at the right edge, different delamination lengths are prefabricated for the model. And the entropy generation values are compared.

II. MODELING AND METHOD

A. Description of Model

A simplified model of a single-chip LED is established. The basic structure of the model is shown in Fig. 1. The size parameters and material parameters of the model components are shown in Table 1.

The chip has a volume of $0.25 \times 10^{-9} m^3$. The luminous efficiency of the $1w$ LED chip from OSRAM company is 20%. The heat generation rate is defined as the heat flow rate per unit volume. The chip is used as a simulated heat source and is loaded with a heat generation rate load, so the heat generation rate of the LED chip is $3.2 \times 10^9 w / m^3$. The substrate and the heat sink in the model dissipate heat by natural convection with the air, and the convection heat dissipation coefficient is $25w / (m^2 \cdot °C)$.

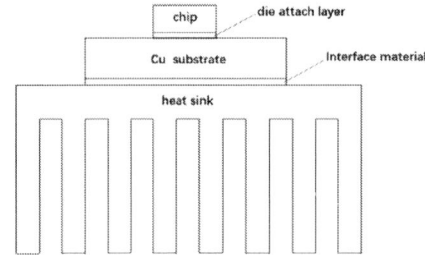

Fig. 1. Model structure

TABLE 1 THE STRUCTURE AND MATERIAL PARAMETERS

Material	Size of material (mm)	Thermal conductivity $(w \cdot m^{-1} \cdot °C^{-1})$	Density $(kg \cdot m^{-3})$
Chip	1× 1 × 0.25	149	2330
Die attach layer	1 × 1 × 0.1	229	7900
Cu substrate	10 × 10 × 1.5	401	8900
Interface material	10 × 10 × 0.1	5	2000
Heat sink	–	217.7	2700

The delamination length in Fig. 2(a) is set to 0.2 mm and the thickness is set to 0.02 mm. Interfacial delamination is prefabricated to the upper and lower interfaces where one crack is prefabricated on the upper interface and the other is simultaneously prefabricated on the lower interface. The heat transfer in the DA layer is shown in Fig. 2(b).

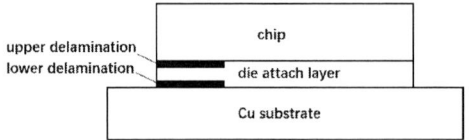

(a)upper delamination and lower delamination

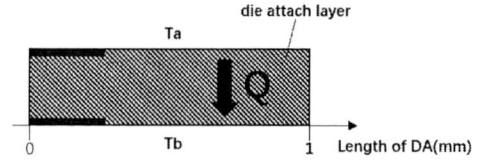

(b) heat transfer process

Fig. 2. Delamination and heat transfer

B. Theory of conductive Entropy Generation

Entropy generation is defined as the size of the system's functional power loss, thus the effect of delamination on the entropy generation of the DA layer can be characterized as the degree of loss of effective heat transfer capacity. The greater the entropy generation, the greater the loss of the effective heat transfer capacity of the DA layer and the lower the heat dissipation capacity. The experiment of Khan et al.[7]has proved that the entropy generation value increases with increasing temperature. According to the existing entropy generation formula for the heat transfer of the DA layer, the heat transfer entropy of the DA layer is as follows[6]:

$$\Delta S = Q(\frac{1}{T_b} - \frac{1}{T_a}) = Q(\frac{T_a - T_b}{T_a T_b}) \tag{1}$$

ΔS is the conduction entropy of the DA layer. Q is the heat generated by the chip. T_a is the temperature of the upper interface of the DA layer, and T_b is the temperature of the lower interface of the DA layer. To facilitate the comparison of the data, the normalized conduction entropy is written as:

$$S = \frac{\Delta S}{\Delta S_0} \tag{2}$$

III. RESULTS AND DISCUSSION

In this paper, When one crack is prefabricated on the upper interface and the other is simultaneously prefabricated on the lower interface, the effect on different positions and lengths of the delamination on the heat transfer of the DA layer is studied. The simulation results evaluate the influence of different delamination positions and lengths on heat transfer performance, which will help engineers recognize the hazards of delamination and improve the reliability of packaged products.

A. Effect of different delamination positions on entropy generation

Different delamination positions will have different degrees of influence on the heat transfer performance of the DA layer and correspond to different entropy generation values. To study the simultaneous occurrence of delamination on the upper and lower interfaces, the effects of different delamination positions on heat transfer were compared. The interfacial delamination is prefabricated to the upper and lower interfaces where one crack is prefabricated on the upper interface and the other is simultaneously prefabricated on the lower interface. As shown in Fig. 3, the location of the upper interface crack is fixed, and the location of the lower interface crack is respectively located at the left edge, middle, and right edge. The simulation results are shown in Fig. 4. Moreover, the entropy generation values corresponding to the models in Fig. 3(a), Fig. 3(e) and Fig. 3(i) are compared in Fig. 5 to study the influence of the edge and the middle delamination on heat transfer. In the simulation result graph, the abscissa represents the position of the DA layer and the ordinate represents the dimensionless entropy generation value.

978-1-6654-1392-3/21 $31.00 © 2021 IEEE

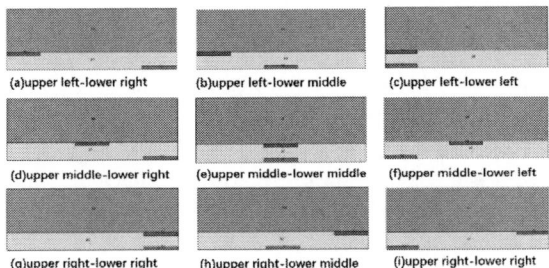

(a)upper left-lower right (b)upper left-lower middle (c)upper left-lower left

(d)upper middle-lower right (e)upper middle-lower middle (f)upper middle-lower left

(g)upper right-lower right (h)upper right-lower middle (i)upper right-lower right

Fig. 3. Different delamination positions

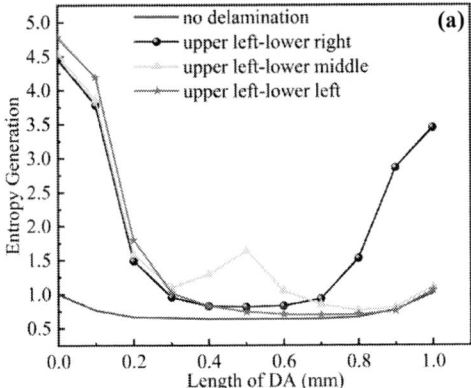

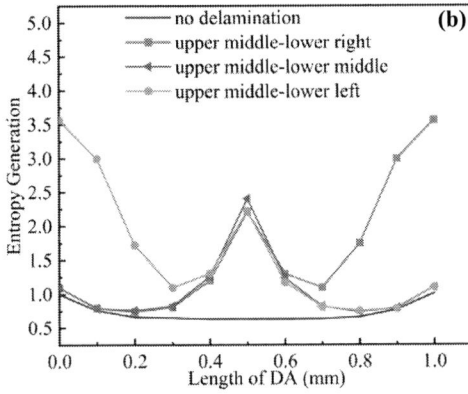

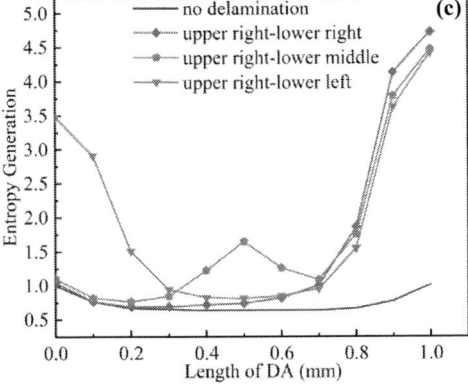

Fig. 4. Comparison of entropy generation values of different delamination positions in (a)、(b)and(c)

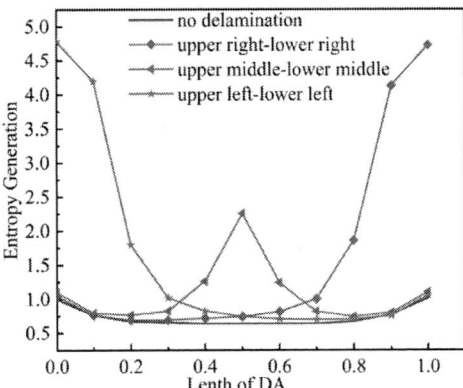

Fig. 5. Comparison of entropy generation values of the edge and middle delamination

Fig. 4 shows that the entropy generation with delamination is significantly higher than the entropy generation without delamination, and the entropy generation will be significantly larger at the location where delamination occurs. In the abscissa axis position where one crack is located on the upper interface and the other crack is located on the lower interface at the same position as the upper interface, the entropy generation of upper and lower interface delamination at the same position is higher than that at a different position. As shown in Fig. 4(a), the entropy generation of the upper left-lower left is the largest and the maximum is 4.78 when the abscissa interval is between 0 and 0.2 mm. In Fig. 4(b), the entropy generation of the upper middle-lower middle is the largest and the maximum is 2.40 when the abscissa interval is between 0.4 and 0.6 mm. In Fig. 4(c), the entropy generation of the upper right-lower right is the largest and the maximum is 4.72 when the abscissa interval is between 0.8 and 1.0 mm. Moreover, the entropy generation values at a location far away from the delamination will be close to or coincide with the entropy generation without delamination. The heat transfer capacity is close to normal. This shows that the delamination will hinder the heat transfer of the DA layer and increase the entropy generation.

The entropy generation values of Fig. 3(a), Fig. 3(e) and Fig. 3(i) are compared. In Fig. 5, the maximum entropy generation of the upper left-lower left is 4.78 in the range of abscissa 0~0.2 mm. Within 0.4~0.6 mm, the maximum entropy generation of the upper middle-lower middle is 2.40. The maximum entropy generation of the upper right-lower right is 4.72 between 0.8 and 1.0 mm. This means that when delamination simultaneously occurs at the same position on the upper and lower interfaces, the entropy generation value of the edge position is greater than the middle position. And the entropy generation value of the right edge is greater than the left edge. That is, when the delamination is simultaneously located at the same position of the upper and lower interfaces, the edge cracks hinder heat transfer more than the middle and the left edge cracks hinder heat transfer more than the right edge.

B. Effect of different delamination lengths on the entropy generation

As the interfacial delamination further expands, the reliability of the device will decline rapidly. When one upper interface crack is fixed at the left edge, the other crack simultaneously is fixed at the right edge, the entropy generation values of different delamination lengths were compared. The model established is shown in Fig. 6. The prefabricated delamination lengths in the model are 0 mm, 0.1 mm, 0.2 mm, 0.3 mm, 0.4 mm, respectively. The delamination length of 0 mm means that there is no prefabricated delamination. Fig. 7 is the simulation result corresponding to entropy generation values.

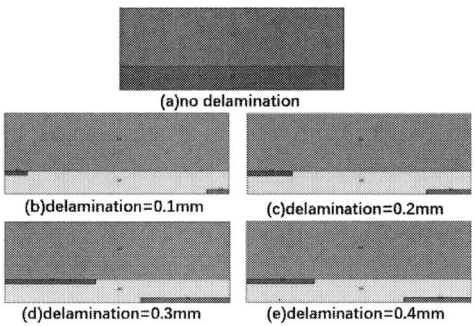

(a)no delamination

(b)delamination=0.1mm   (c)delamination=0.2mm

(d)delamination=0.3mm   (e)delamination=0.4mm

Fig. 6.   Different delamination lengths

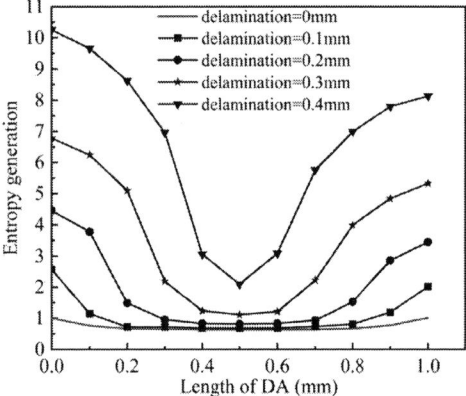

Fig. 7.   Comparison of entropy generation values of different lengths in the same location

Fig. 7 shows that as the delamination length increases, the entropy generation value also increases. And the maximum values are 1, 2.57, 4.46, 6.78, and 10.25, respectively. From the delamination length from 0.3 to 0.4 , the entropy generation has the largest increase and is about 51%. This shows that as the length of the delamination increases, the heat dissipation of the DA layer becomes worse. And this shows that when the delamination length of the upper interface and the lower interface both reach 40% of the DA layer, the hindrance of the delamination to heat transfer has been significantly aggravated. Therefore, it should be avoided that the delamination length of the upper and lower interfaces exceeds 30% of the DA layer. This is consistent with the effect of delamination length on heat transfer when a crack is at the upper interface[6].

IV.   CONCLUSION

After the establishment of the model and analysis of the finite element method, the following important conclusions can be obtained.

Interfacial delamination is simultaneously prefabricated to the upper and lower interfaces, the entropy generation value corresponding to the delamination position will increase significantly. In the abscissa axis position where one crack is located on the upper interface and the other crack is located on the lower interface at the same position as the upper interface, the entropy generation value at the same position is higher than that at different a position. The larger the entropy generation value, the worse the heat dissipation performance.

Delamination simultaneously occurs at the same position on the upper and lower interfaces, the entropy generation value of the edge position is greater than the middle position. And the entropy generation value of the right edge is greater than the left. Therefore, the edge cracks hinder heat transfer more than the middle, and the left edge cracks hinder heat transfer more than the right edge.

One upper interface crack is fixed at the left edge, the other crack simultaneously is fixed at the right edge. As the length of the delamination increases, the entropy generation value becomes larger and the heat transfer performance becomes worse.

ACKNOWLEDGMENT

This research was supported by the National Natural Science Foundation of China (No. 61865004), the Key-Area Research and Development Program of Guangdong Province (No. 2019B010131001), the Guangxi Science and Technology Program (No. AB20159007) and the Guilin Science Research and Technology Development Program (No. 2020010302).

REFERENCES

[1] Narendran N, Gu Y, Freyssiner J P. Deng. Soild-state lighting:failure analysis of white LEDs [J]. Journal of Crystal Growth, 2004,268(3):449-456.

[2] Zhang J , Zhang X , C Lü. Effect of Die Attach Void on IGBT Thermal Reliability[J]. Guti Dianzixue Yanjiu Yu Jinzhan/Research and Progress of Solid State Electronics, 2011, 31(5):517-521.

[3] Xie X, Bi X, and Hu. J. Effects of Voids on Thermal Reliability in Power Chip Die Attachment Solder Layer[J]. Semiconductor Technology, 2009, 34(10):960-959.

[4] Zhang J , Zhang X , C Lü. Effect of Die Attach Void on IGBT Thermal Reliability[J]. Guti Dianzixue Yanjiu Yu Jinzhan/Research and Progress of Solid State Electronics, 2011, 31(5):517-521.

[5] Y.Mo, D.G. Yang, M. Cai, D. Liu and Y. Nie, Thermal transer influence of delamination in the die attach layer of chip-on-board LEDpackage base on entropy generation analysis, IEEE, 2018, pp. 646-651.

[6] P. Cui, M. Cai and D. Yang, Effect of detected behaviour on interfacial heat transferring performance for HP-LED packaging based on entropy generation analysis, IEEE, 2018, pp. 423-427.

[7] A M I K, B T H A, and A S Q. Entropy generation (irreversibility) associated with flow and heat transport mechanism in Sisko nanomaterial-ScienceDirect[J]. Physics Letters A, 2018, 382(34):2343-2353.

Shear Properties and Fracture Behaviors of Cu/Sn-37Pb/Cu Solder Interconnections at Cryogenic Temperatures

Ruyu Tian*
School of Mechanical Engineering
Yangzhou University
Yangzhou, China
tianruyu3@163.com

Chunlei Wang
School of Mechanical Engineering
Yangzhou University
Yangzhou, China
2090879581@qq.com

Yanhong Tian*
State Key Laboratory of Advanced
Welding and Joining
Harbin Institute of Technology
Harbin, China
tianyh@hit.edu.cn

Abstract—**The shear properties and fracture behavior of the Cu/Sn-37Pb/Cu solder interconnections at temperatures ranging from -196 ºC to room temperature were investigated in the present study. With a decrement in the testing temperature, the shear strength of the solder interconnections gradually increased first, but then decreased after it reached a maximum value at -150 ºC. The shear strength of Cu/Sn-37Pb/Cu interconnects at cryogenic temperatures (below -55 ºC) was much higher than that at room temperature. With a decline in the testing temperature, the fracture position of the solder interconnections transformed from in the solder bulk to partially in the solder bulk and partially in the interfacial IMCs.**

Keywords—solder interconnection, shear property, cryogenic temperature, fracture behavior

I. INTRODUCTION

Aerospace electronic products will experience cryogenic temperatures (below -55 ℃) during deep space exploration [1]. For example, the surface temperature of Mars can be as low as -120 ℃, and the night temperature of Moon can drop below -180 ℃. The solder interconnection offers thermal, electrical and mechanical support in electronics assemblies. The mechanical behaviors and reliability of solder interconnections are vital to the normal functioning of electronic assemblies [2, 3]. Hence, it is need to lucubrate the mechanical behaviors of solder interconnection at cryogenic temperatures.

There have been many reports about the mechanical properties and fracture behavior of Cu/Sn-Ag-Cu/Cu solder interconnections at cryogenic temperatures. For instance, Li et al. [4] have investigated the tensile and fracture behaviors of Cu/Sn-3.0Ag-0.5Cu/Cu solder interconnections over the temperature range from -80 ℃ to 25 ℃. In our previous study, the tensile properties and fracture mechanism of Cu/Sn-Ag-Cu/Cu solder interconnections at temperatures ranging from -196 ℃ to 25 ℃ were investigated [5]. However, reports correlating the shear properties and fracture mechanisms of Cu/Sn-37Pb/Cu interconnects under cryogenic temperatures are scarce in the literature.

In this study, shear test at cryogenic temperatures was carried out to evaluate the properties of Cu/Sn-37Pb/Cu interconnections. Additionally, the fracture surface was analyzed to gain an insight into the fracture position of the Cu/Sn-37Pb/Cu solder interconnection at cryogenic temperatures, as well as the fracture mechanism.

II. EXPERIMENTAL DETAILS

A. Sample Preparation

Polycrystalline Cu plate was selected as the substrate, and eutectic Sn-37Pb solder paste (Qualitek) was employed as the solder. The Cu substrate was ground carefully with SiC paper and cleaned with absolute ethyl alcohol before reflow soldering. The Sn-37Pb solder paste has been printed onto the Cu substrate to prepare the specimens for reflow soldering. Fig. 1 presented the structure and dimension of the Cu/Sn-37Pb/Cu solder interconnections. The prepared specimens were clamped and reflowed at a peak temperature of 230 ℃ for 10 min in a reflow furnace (SRO 702, ATV) to form the single-lap solder interconnects.

B. Shear Test and Fracture Surface Analysis

The shear test was performed on Cu/Sn-37Pb/Cu solder interconnections using an Instron 5500 R testing machine which has been specially-modified. The specimens were tested at different temperatures (i. e., -196 ℃, -150 ℃, -100 ℃, -50 ℃ and room temperature). The strain rate of the shear test was $10^{-2}\,\text{s}^{-1}$. For each testing temperature, the average shear strength of the interconnect was obtained from more than 5 specimens. After the shear test, the fracture surface of the solder interconnect was detected by SEM (Quanta 200FEG) equipped with EDS.

III. RESULTS AND DISCUSSION

Fig. 2 showed the shear stress–strain curves of Cu/Sn-37Pb/Cu solder interconnects tested at different temperatures. The shear behavior of the Cu/Sn-37Pb/Cu interconnect varied significantly with the testing temperature. The stress-strain curves of the solder interconnect tested at -100 ℃, -50 ℃ and room temperature contained elastic deformation, plastic deformation, necking and final fracture, which indicated the fracture mode of the Cu/Sn-37Pb/Cu interconnection was ductile fracture. In the case of solder interconnections tested at -196 ℃ and -150 ℃, final fracture occurred abruptly without necking, which suggested that the fracture mode might be different from that of the solder interconnect tested at -100 ℃, -50 ℃ and room temperature.

Fig. 3 displayed the variation in shear strength of the Cu/Sn-37Pb/Cu solder interconnect as a function of the testing temperature. The shear strength of the interconnect at room temperature was 28 MPa. As the testing temperature declined, the shear strength of the Cu/Sn-37Pb/Cu interconnects firstly increased, then decreased. The shear strength reached its

978-1-6654-1392-3/21 $31.00 © 2021 IEEE

maximum at -150 °C, which was 69 MPa. It was noticed that the shear strength of the interconnects at cryogenic temperatures was always higher than that at room temperature.

Fig. 4 displayed the SEM images of facture surfaces of Cu/Sn-37Pb/Cu solder interconnects after shear test. The fracture surface of the solder interconnect tested at room temperature (25 °C) was covered with elongated shear dimples (Fig. 4a,). It was suggested the solder interconnect fractured in the solder bulk and showed a ductile fracture mode [6]. The solder interconnections tested at -100 °C and -50 °C also fractured within the solder bulk. The dimples of the fracture surface became shallower and smaller as the testing temperature declined, revealing that the ductility of the solder interconnect decreased.

In the case of the solder interconnect tested at -150 °C, a few of cracked Cu_6Sn_5 phases were detected in the fracture surface besides elongated shear dimples (Fig. 4d). This indicated the shear fracture of the solder interconnect occurred partially in the solder bulk and partially in the interfacial IMCs, and the fracture mechanism transformed to the ductile-brittle mixed fracture mode. This result was consistent with the inference drawn from the stress-strain curves of Cu/Sn-37Pb/Cu interconnects (Fig. 2). The area proportion of Cu_6Sn_5 IMCs on the fracture surface slightly increased as the testing temperature dropped to -196 °C (Fig. 4e), implying the fracture position changed towards the IMC layer. The transformation of the fracture position can be attributed to the variation of mechanical properties of the Sn-37Pb solder bulk and interfacial Cu_6Sn_5 IMCs with descending temperature.

IV. CONCLUSION

In the present study, the shear test over the temperature range from -196 °C to room temperature was carried out to study the shear properties and fracture mechanism of Cu/Sn-37Pb/Cu interconnects at cryogenic temperatures.

With a decrement in the testing temperature, the shear strength of the Cu/Sn-37Pb/Cu interconnect gradually increased first, but it decreased after it reached a maximum value at -150 °C. The shear strength for the solder interconnect at cryogenic temperatures (below -55 °C) was consistently higher than that at room temperature.

The solder interconnect tested at room temperature fractured within the solder bulk. As the temperature decreased to -150 °C, the fracture position of the Cu/Sn-37Pb/Cu interconnect changed to partially in the solder bulk and partially in the interfacial IMCs.

ACKNOWLEDGMENT

The present work was supported by the Natural Science Foundation of Jiangsu Province (Grant No. BK20200940).

REFERENCES

[1] G. Hu, Y. Zheng, A. Xu, and Z. Tang, IEEE Geosci. Remote. Sens. Lett., vol. 13, pp. 110–114, 2015.

[2] L. Li, B. Jing, and J. Hu, IEEE Access, vol. 8, pp. 204695–204708, 2020.

[3] Y. Chen, B. Jing, J. Li, and X. Jiao, J Hu, IEEE Trans. Compon. Packag. Manuf. Technol., vol. 11, pp. 43–50, 2021.

[4] W. Li, and X. Zhang, 19th International Conference on Electronic Packaging Technology (ICEPT), pp. 324–328, 2018.

[5] R. Tian, Y. Tian, C. Wang, and L. Zhao, Mater. Sci. Eng. A, vol. 684, pp. 697–705, 2017.

[6] S. Kim, K. Son, and S. Hyun, Mater. Sci. Eng. A, vol. 822, pp. 141655, 2021.

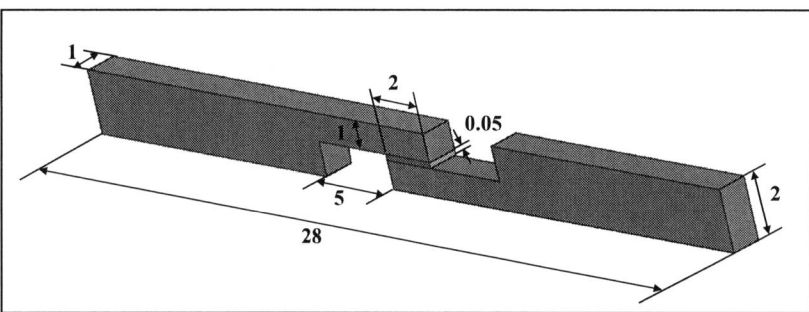

Fig. 1. Schematic illustration of the lap shear specimen of the Cu/Sn-37Pb/Cu solder interconnection.

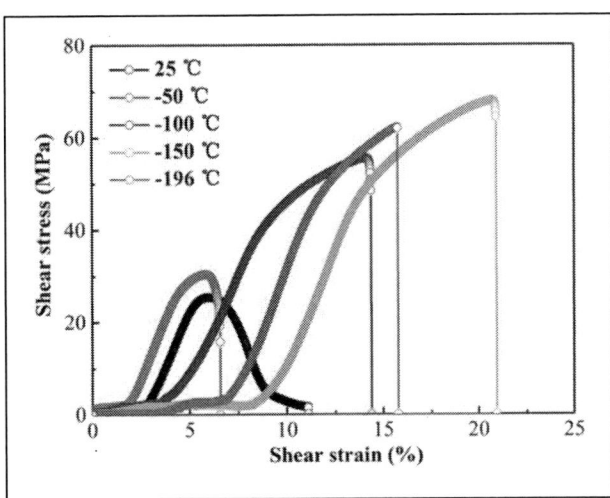

Fig. 2. Shear stress–strain curves of the Cu/Sn-37Pb/Cu solder interconnect tested at different temperatures.

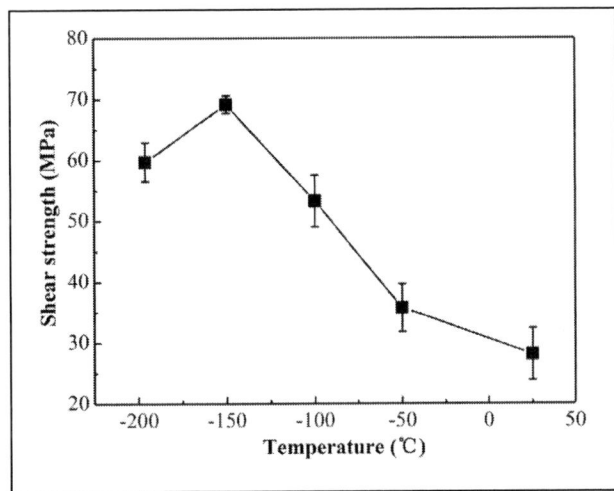

Fig. 3. The variation in shear strength of the Cu/Sn-37Pb/Cu solder interconnect as a function of the testing temperature.

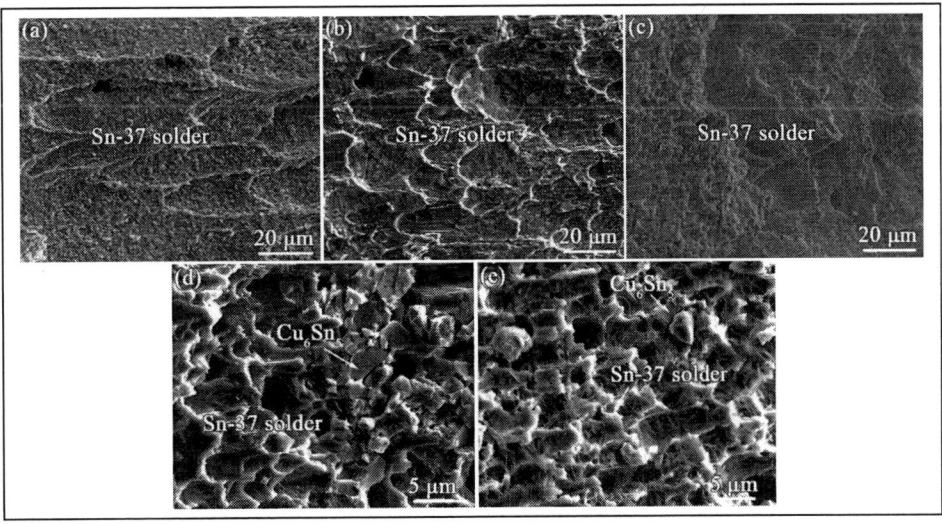

Fig. 4. Top-view SEM images of facture surfaces of Cu/Sn-37Pb/Cu solder interconnects after testing at (a) room temperature, (b) -50 ℃, (c) -100 ℃, (d) -150 ℃, and (e) -196 ℃.

Research on 3D interposer/chip stacking technology and reliability

Ning Zhang
Xi'an Microelectronic Technology
Xi'an,China
13891812096@163.com

Baoxia Li
Xi'an Microelectronic Technology
Xi'an,China
libaoxia163@163.com

Qiucheng Yan
Xi'an Microelectronic Technology
Xi'an,China
405817673@qq.com

Daowei Wu
Xi'an Microelectronic Technology
Xi'an,China
wudaowei1220@163.com

Abstract—**Silicon interposer is an effective way to realize the various kinds of chips integration. This study is based on the characteristics of traditional underfill, by designing a reasonable bump structure, the coplanarity difference in assembly process can be effectively contained. And the key parameters are analyzed, the optimized aseembly scheme is obtained. The product reliability verification results meet the requirement.**

Keywords—chip package/stacking, reliability, traditional underfill

I. INTRODUCTION (*HEADING 1*)

There are many kinds of IC chips involved in military electronic system, which have the characteristics of different sizes, materials, processes and functions. It is urgent and difficult to realize the high-density integration of those chips for a miniaturization of military electronic system. over the years, 3D die/interposer stacking with TSVs has emerged as a solution for a smaller footprint and higher performance. Most 3D chip stacking is applied in memory chips [1-3], but it is difficult to be applied in other chips. It is feasible to integrate many kinds of chips through TSV interposer, some scholars have done some research on silicon interposer stacking. Apple uses POP technology in processor A9 and A19 modules. and in 2011 ECTC stats ChipPAC profiled the study on two-layer TSV interposer. In addition, The Fraunhofer Institute in Europe is also a pioneer in the field of 3D packaging. It claims that the realization of stack of TSV interposer, in its website the schematic diagram of TSV interposer stacking is shown as Fig.1[4], which is three layers, but no more report about practicality interposer stacking. Therefore, it is of great significance to study the stacking technology of TSV interposer.

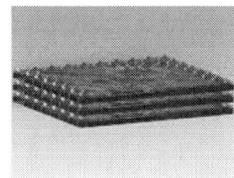

Fig.1 Stacking diagram of TSV interposer of Fraunhofer Institute in Europe

II. EXPERIMENTAL METHODS

A. Layout design

In this study, one or more FC chips are integrated to the TSV interposer firstly, and then the TSV interposer with FC chips are stacked to realize 3D module.

In order to test the electrical performance of 3D module effectively, the series "Daisy Chain" with pad are used, by which can monitor the contact resistance value comprehensively, accurately and effectively. And the resistance value reflects the bonding effect of FC chips and interposer, also reflect interlayer bonding process between two layers interposer. The schematic diagram and design chain of the integrated product are shown in Fig. 2.

Chip A and interposer A are interconnected with the chains A. The interconnection chains between other chips and interposer are similar to chip A. The resistance of chains will be test in part below, so we named all the chains as table1 shows. The interconnection between layers is realized by solder ball, TSVs and test pads. The test pad (of layer A and layer B)is led out to the front of the first layer interposer and the back of the third layer interposer(of layer B and layer C) respectively. Each chain consists of several solder joints, In order to locate the solder joints more quickly, several test pads are inserted in the middle of each chain.

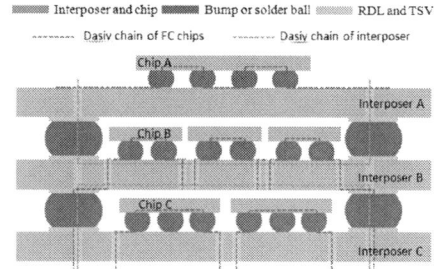

Fig.2 Schematic diagram of TSV interposer stacking

TABLE I DASIY CHAIN BETWEEN CHIPS AND INTERPOSER

Item	Dasiy Chain
Chip A and interposer A	Chain A
Chip B and interposer B	Chain B
Chip C and interposer C	Chain C
Interposer A and interposer B	Chain A-B
Interposer B and interposer C	Chain B-C

B. Structural design

In order to tolerate the coplanarity of balls on the chips, balls on interposer and the thickness difference of chips, the product structure is designed as follows: the normal UBM（Cu5μm/Ni2μm/Au0.05μm）is prepared on the front of the interposer, and the bumps are prepared on the back to adjust the gap between two layers, as shown in Fig.3. Take 100μm chips as an example, solder balls with a diameter of 100μm are implanted on the chip to require the interconnection of signals between chips and interposer, with a diameter of 250μm are implanted on the front of interposer to require the interconnection of signals between different layers. Four kinds of bumps on the back of the interposer are designed, as shown in table II, and the interposer with corresponding bump structure are used to assembly, at last the proper bump height is found with the comparison of results.

TABLE II BUMPS PARAMETERS ON THE BACK OF INTERPOSER

project	Bump 1	Bump 2	Bump 3	Bump 4
Bump specification	Cu5μm Ni2μm Au0.05μm	Cu5μm Ni2μm AgSn20μm	Cu5μm Ni2μm AgSn30μm	Cu5μm Ni2μm AgSn40μm

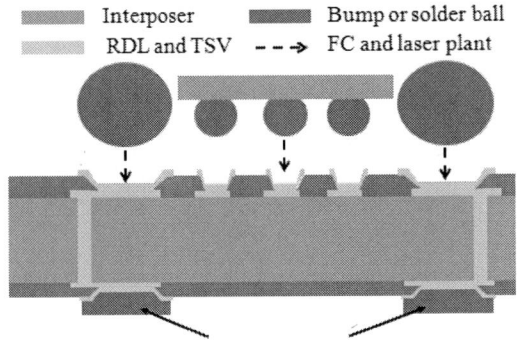

Interposer   Bump or solder ball
RDL and TSV   - - → FC and laser plant

Bump，height can be adjust flexible

Fig.3 Bump specification

III. EXPERIMENT AND RESULTS

A. Optimization of bump height

After the chips and corresponding interposer assembly, the height from the surface of the interposer to the top of the chip of each FC module was measured. And the height of 250μm solder ball after planting was also measured. The datas measured of four corners and the central area of the module are shown in Fig.4, solder ball's height measure location is just as the same as Z in five pots.

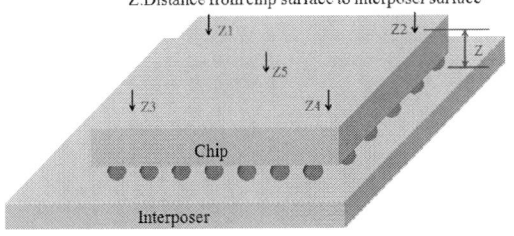

Z:Distance from chip surface to interposer surface

Fig.4 Diagram of measure location

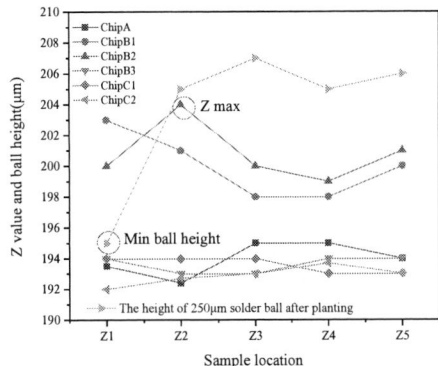

Fig.5 Measure results of Z and 250μm solder ball height after planting

From the data in Fig.4, it can be seen that the difference of Z between the maximum value and the minimum value is about 10μm, and the max value is 204μm. However, the height difference of solder ball after planting is also about 10μm, and the min value is 195μm. The height of the solder ball is little less than that of the FC chip, and the collapse height of the solder ball in welding process has not been calculated. Therefore, it is obviously unreasonable to use the specification bump1, the other three specifications of bumps are used for assembling. But after chip and interposer assembly with specification bump 2 (20 μ m), the solder joint is elongated, as shown in Fig.6(a),the red circle location. When the specification is bump3 (30μ m) and bump 4(40μm), the appearance of solder joint is good, as shown in Fig.6 (b). Among them, the metallographic photos of the section of specification3 (30μm) after welding and IMC layer under SEM are shown in Fig.6(c) and (d).

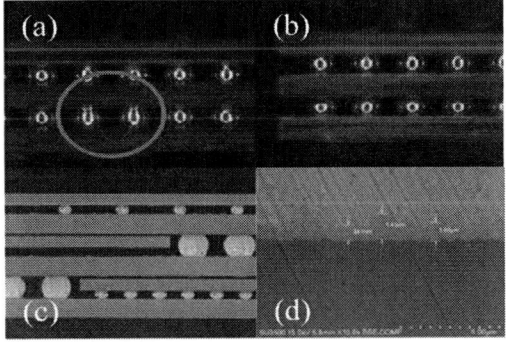

Fig.6 Appearance of solder joint, (a)bumps 2(20μm), (b) bumps 3(30μm),(c) the cross-section shape of solder joint, (b) IMC of solder joint with 30μm bumps

B. Overflow of single layer components

Underfill fluidwidth is an important effect that influences the overflow width. Keep the effects, such as wait time between two dispensing process, dispensing height and temperature in good condition, and using smaller nozzle while the underfill dispensing process, the overflow width of the dispensing edge is about1mm, the width on the opposite side is narrower than that of dispensing edge. However, the distance from the chip edge to the pad/solder ball for layers interconnection is as small as 150μmaccording to the design. Besides, due to the existence of the interlayer interconnection ball in the product, the siphon effect causes the glue flow to the pad/solder ball, the overflow width opposite the dispensing edge increases to about 150μm, Fig.7 shows the overflow appearance of the single layer FC module. and the subsequent assembly cannot be carried out after the interlayer interconnection ball or pad is polluted.

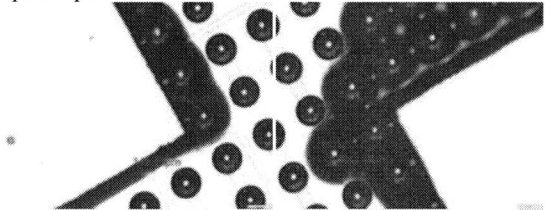

Fig. 7Overflow (a) opposite dispensing edge, (b)dispensing edge overflow

There are ways to solve this problem, multiple pre-applied underfills(PAUF) have been developed by various research teams. But Pre-applied underfill has the common challenges of non-uniform deposition, difficult vision recognition of chip alignment, high investment of the wafer-level facilities, short shelf-life and poor wetting of the under-film[5]. The second way is to increase the distance between chip edge and large solder ball, which requires sacrificing module size. And another is to complete all the gap filling both single layer(chips and interposer) and layers(interposer and interposer) at the same time. The last methods are used in this study, the related experiments and results are shown in the following chapters.

C. Control of glue climbing height between layers

According to the calculation and measurement, the total thickness of the product is about 1.2mm. When three layers （two gaps） of interposer are filled at the same time, the phenomenon of insufficient creepage will occur, resulting in the upper gap not being filled effectively[6], as shown in Fig.8. It is important to find a way to improve the height that the underfill can reach.

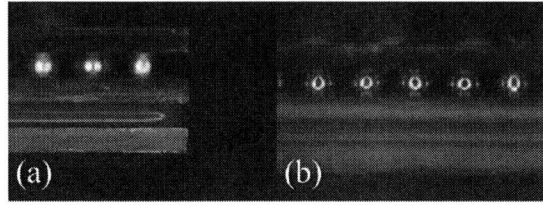

Fig. 8 Insufficient creepage leads to upper layer gap unfilled,(a) edge position of module, (b) middle positionof module

In this paper, we find some film to protect the bottom of product, and use tool to make a cofferdam, so that the underfill can creep higher in a narrow gap, as shown in Fig.9(a). Otherwise, when design the interposer mask, blank area was increased for next layer interposer. So that a pyramid structure is formed after stacked, as shown Fig.9(b), the additional part can be sawed after dispensing process. Fig.10 shows the result of the way proposed above. In this way, the fluid can effectively enter the interlayer.

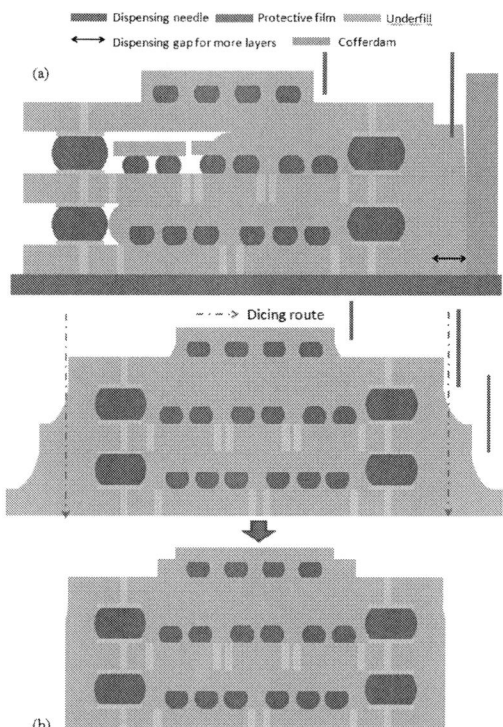

Fig.9 Diagram of dispensing scheme,(a) scheme1, (b)scheme 2

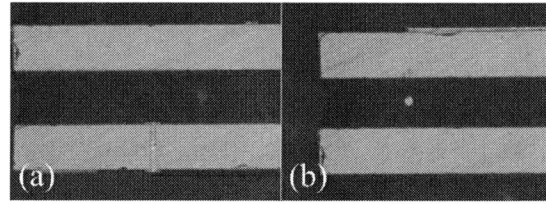

Fig. 10Edge effect of interlayer filling, (a) gap between the second and third layers of adapter plate, (b) gap between the first and second layers of adapter plate

IV. RELIABILITY ASSESSMENT

In order to evaluate the reliability of products, the following reliability tests are carried out to verify the reliability capability of components under certain environmental stress.

A. Welding and filling technics reliability evaluation

According to Appendix B of Q/QJA 416-2016 《Verification specification for flip chip welding process》, X-ray and ultrasonic scanning observation were carried out, all of them met the requirements, as shown in Fig. 11.

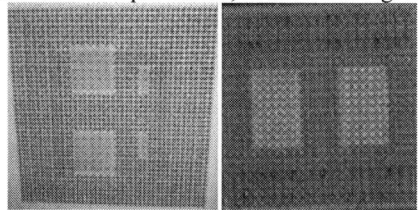

Fig.11 X-ray and ultrasonic scanning result, (b) X-ray detection, (c) scanning result between layer B and C

B. Temperature cycling and stability baking experiments

Three modules were selected for temperature cycling and stability baking experiments. The electrical resistance of each "daisy chain" had little change before and after the experiment, and the maximum resistance change is chain B-C of 23.7Ωand 24.3Ωafter temperature cycle，chainC1 of 9Ωand8.7 Ω after stability baking, the rate was 3.33%and 2.27% respectively, as shown inFig.12 and Fig.13.According to the relevant standards, the failure definition of flip chip solder joint test sample is as follows: (1) the daisy chain resistance of the test sample solder joint increases by more than 20%，（2）The daisy chain resistance of the solder joint of the test sample becomes infinite (that is, the daisy chain of the test sample is interrupted).Therefore, the interconnection is reliable.

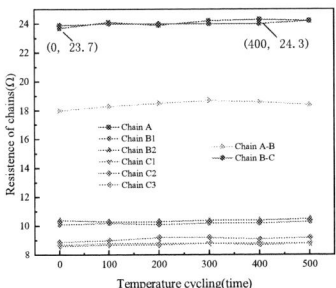

Fig.12 Change trend of resistance value of "Dasiy chain " in temperature cycling test

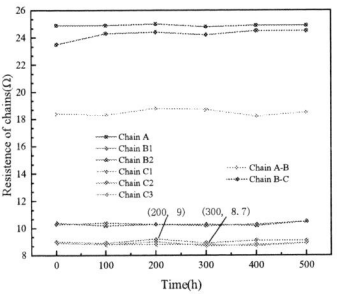

Fig.13Change trend of resistance value of "Dasiy chain" in stability baking test

V. CONCLUSION

In this paper, a structure of interposer was discussed for the integration of kinds of chips. By adopting the structure of solder ball in the front and bump at the back of interposer, the problems of different kinds of chips TTV and coplanarity of solder ball can be tolerated by flexibly adjusting the height of bump. And two ways were adopted for the dispensing process. By using the protect film and cofferdam, the upper gap got a good underfill shape. In addition, dispensing in black area first and sawing it after is also applied. Finally, the reliability analysis of the assembled products meets the requirements, the resistance of "daisy chain" changed little after reliability experiments.

REFERENCES

[1] A 3D-Stacked Logic-in-Memory Accelerator forApplication-Specific Data Intensive Computing.

[2] Application-Specific Data Intensive ComputingPackage stack (POP) technology and its application [J]. Flynn Carson. Integrated circuit applications. 2015 (02).

[3] 3-D Stacked PackageTechnology and Trends[J].FP Carson, YC Kim, IS Yoon.PROCEEDINGS - IEEE. Volume 97, Number 1, 2009.

[4] Fraunhofer Institute for Reliability and Microintegration IZM - 3D Chip Stack.

[5] Daniel Lu and C. P. Wong, Materials for Advanced Packaging, Springer, Atlanta, GA, 2009, pp. 326–332.

[6] Le F , Lee S , Lau K M , et al. Through silicon underfill dispensing for 3D die/interposer stacking[C]// 2014 IEEE 64th Electronic Components and Technology Conference (ECTC). IEEE, 2014.

Highly Conductive Silver Nanowire Transparent Electrodes Hybridized with Laminated Multi-layer MXene

Pengchang Wang
School of Materials Science and Engineering
(Shanghai University)
Key Laboratory of Advanced Display and System Applications
(Shanghai University)
Shanghai, 200072, China
wangpc@shu.edu.cn

Maoliang Jian
Key Laboratory of Advanced Display and System Applications
(Shanghai University)
Shanghai, 200072, China
smujml@163.com

Chi Zhang
Key Laboratory of Advanced Display and System Applications
(Shanghai University)
Shanghai, 200072, China
zhangchi303145@163.com

Majiaqi Wu
Key Laboratory of Advanced Display and System Applications
(Shanghai University)
Shanghai, 200072, China
1633850859@qq.com

Huaying Hu
Key Laboratory of Advanced Display and System Applications
(Shanghai University)
Shanghai, 200072, China
1660987075@qq.com

Lianqiao Yang*
Key Laboratory of Advanced Display and System Applications
(Shanghai University)
Shanghai, 200072, China
yanglianqiao@i.shu.edu.cn

Abstract—Silver nanowires (AgNWs) and new-type two-dimensional material MXene have shown great prospects for applications in new-generation flexible electronic devices due to their brilliant properties. In this work, high-performance silver nanowire/MXene hybrid transparent conductive electrodes (TCEs) were prepared by a solution method. It studied the influence of AgNW concentration on the photoelectric properties of TCEs. Then, several-layer or even single-layer MXene were obtained by a delamination process and were added to the AgNW networks to prepared flexible hybrid electrodes on polyethylene naphtholate (PEN) substrates. The hybrid TCEs showed a low sheet resistance of 24.4 Ω/sq with the transmittance of 90.6% at 550nm. Moreover, the MXene sheets covered on the surface could improve the conductivity of the film and block the oxidation of silver nanowires. The results mean that the novel AgNW/MXene TCEs have great potential for the practical applications of high-performance flexible electronic devices.

Keywords—silver nanowire; Ti₃C₂Tₓ MXene; delamination, hybrid transparent electrode.

I. INTRODUCTION

Transparent conductive electrodes (TCEs) have been widely used in optoelectronic devices such as transparent heaters, organic light-emitting diodes (OLED), transparent heaters, solar cells, liquid crystal displays, and flexible displays[1]. Metal oxide (ITO, FTO) is currently the most common transparent electrode due to its relatively low sheet resistance (10-20 Ω/sq) coupled with high optical transparency (>80% at 550 nm) and favorable work function[2]. However, there are many drawbacks of ITO such as an increased cost due to indium scarcity and fabrication process and sensitivity to both acidic and alkaline environments limit its development in transparent electrodes. Especially, its brittleness makes it unsuitable for the new-generation flexible electronic devices. For these reasons, it has been an essential issue to search for novel Transparent conductive materials to replace ITO in recent years. At present, flexible TCEs have been prepared from carbon nanotubes (CNTs), metal nanowires (AgNW, CuNW), conductive polymers (PEDOT: PSS), graphene, MXene, etc. and hybrids of the above materials[3-4].

The transition-metal carbide and nitride named MXene were new-type two-dimensional materials, which have been studied in many fields due to their special structure and performance since it was discovered in 2011[5]. Several-layer and single-layer $Ti_3C_2T_x$ MXene can be obtained by selective etching and intercalation treatment, where T is the surface functional groups including hydroxyl (OH), oxygen (O), and/or fluorine (F) and so on. Furthermore, MXene has good hydrophilicity, good optical transparency, and metal conductivity, which make it very suitable for the preparation of TCEs by the solution method[6]. However, the conductivity of MXene is not satisfactory compared with ITO and metal materials.

The promising materials silver nanowires (AgNWs) have been attracting more and more attention as flexible, transparent, and stretchable electrodes. It has excellent photoelectric properties and can be easily dispersed in the solvent and coated onto flexible substrates, but the AgNW films possess irregular morphologies and high roughness[7]. The combination of silver nanowires and MXene can complement each other and obtain a transparent electrode with excellent performance. The MXene sheets could connect the nanowires of the network to improve the conductivity and surface flatness.

In this paper, it studied the influence of the concentration of AgNWs on the performance of transparent electrodes. Multi-layer $Ti_3C_2T_x$ MXene was delaminated by DMSO to obtain several-layer or even single-layer MXene. In addition, the uniform and transparent AgNW/MXene hybrid TCEs were prepared via a scalable and simple solution process.

II. EXPERIMENTS

A. Delamination of multilayer MXene

Multilayer MXene was prepared by etching the MAX phase precursors (Ti_3AlC_2) powder with hydrofluoric acid, as described in the previous work[8]. For delamination, 0.12 g of multilayer Ti₃C₂Tx MXene powder was put into 10 mL of dimethyl sulfoxide (DMSO, Aladdin, Shanghai, China), and stirred for 18 h on a magnetic stirrer. After that, the mixture was centrifuged at 8000 rpm for 10 minutes to obtain a precipitate, which was intercalated MXene. This process was repeated two more times with deionized water. Whereafter, the precipitate was added to 30 mL of deionized water and sonicated for 5 h. Then the solution was centrifuged at 3500 rpm for 1 h, to obtain a supernatant, which was a monolayer layered MXene suspension.

B. Preparation of the silver nanowire films

The PEN substrates were ultrasonically cleaned in acetone, ethanol, and deionized water for 10 min, respectively. Then, the substrates were dried by nitrogen and UV treated for 15 min to gain hydrophilic surfaces. Whereafter, the silver nanowire solution (Sigma-Aldrich St. Louis, MO, USA) with a diameter of ∼50 nm and a length of ∼40 μm was diluted with isopropanol to the concentration of 1- 5 mg/ml respectively. Then, the silver nanowires with various concentrations were spin-coated onto the substrates to form silver nanowire conductive networks and annealed at 120 ℃ for 5 minutes.

C. Preparation of the AgNW/MXene hybrid films

the MXene solution was spin-casted onto the silver nanowire network at 2000 rpm for 1 min and dried in the air. The as-prepared silver nanowire film was soaked in the MXene solution for 5-20 min and dried in the air.

III. RESULT AND DISCUSSION

Generally, the transmittance and conductivity of the transparent electrode are mutually restricted. Therefore, we studied the effect of the concentration of silver nanowires on the electrical and optical properties of thin-film electrodes. Moreover, the figure of merit (FoM) expression φ_{TC} determined by Haacke was used to evaluate the the relationship between transmittance and sheet resistance of different TCEs[9].

$$\varphi_{TC} = \frac{T^{10}}{R_{sh}} \quad (1)$$

Where R_{sh} is the sheet resistance and T is the optical transmittance at 550 nm.

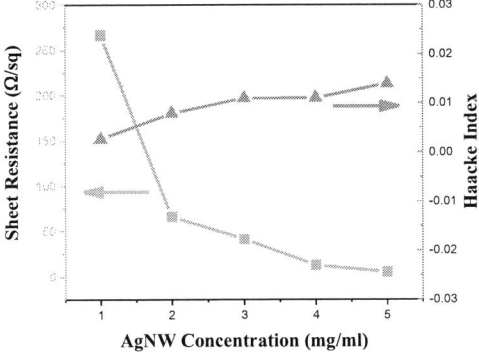

Fig. 1. The sheet resistance and Haacke index of pristine AgNWs transparent conductive electrodes with various concentrations.

Fig. 1 showed the sheet resistance and the Haacke index of AgNW transparent electrodes with different concentrations. It proved that the sheet resistance gradually decreases from

267 to 5.4 Ω/sq with the increase of concentration from 1 to 5 mg/ml. According to Fig, the Haacke index showed a general upward trend. And the FoM is almost equal when the concentration is 3 and 4 mg/ml. As we all know, silver nanowires have the shortcomings of high roughness and high resistance of wire junctions. When the concentration was too high, it would cause nanowire overlapping and low surface evenness which are the key reasons of short circuit and high leakage current of the devices. Therefore, the silver nanowires at a concentration of 3mg/ml were used in the experiments of this paper which could form a uniformly distributed conductive network.

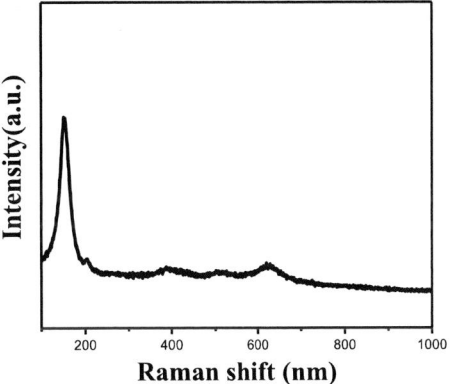

Fig. 2. The Raman spectroscopy of Ti₃C₂Tₓ MXene.

To enhance the photoelectric performance of silver nanowire-based transparent electrodes, MXene was used to prepare silver nanowire composite film electrodes. To characterize the structure of Ti₃C₂Tₓ, Raman spectroscopy was conducted and showed in Fig. 2. It can be seen that the peaks at 156, 384, and 625 cm⁻¹ in the Raman spectra match well that reported in the previous researches [10]. As shown in Fig. 3, multilayer MXene dispersed in deionized water was spin-coated onto the AgNW networks. However, the large size of multilayer MXene just randomly felled on the networks and could not form a complete film resulting in low optical transmittance and higher roughness. Hence, the multilayer MXene could not be used directly to prepare composite film electrodes.

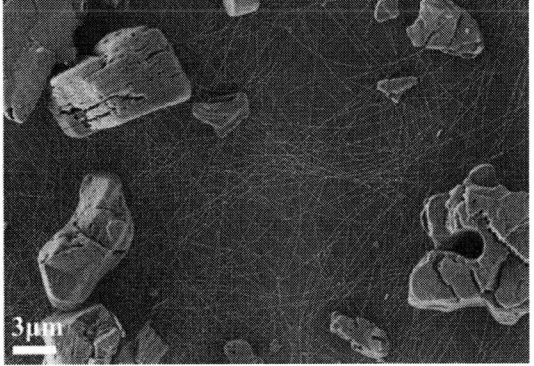

Fig. 4. The SEM image of AgNW hybrid transparent electrodes with MXene before the delamination.

In order to obtain few-layer or even single MXene, the DMSO intercalation experiment was carried out and the

MXene dispersion was shown in Figure. Then, the layered MXene dispersion was coated on the silver nanowire network by spin-coating and dip-coating to prepare AgNW/MXene hybrid transparent electrode. In the paper, it compared the effects of preparation conditions on the performance of composite electrodes. In the dip-coating preparation method, AgNW films were immersed in the MXene solution for 5, 10, 15, and 20 minutes. The tested sheet resistance showed similar values of ~29Ω/sq. In the spin-coating preparation method, MXene was spin-casted onto the silver nanowire network at different speeds (1000rpm, 2000rpm, 3000rpm). By comparing the photoelectric properties of different composite TCEs, the sheet resistance of optimal composite TCEs was 24.4 Ω/sq at 2000 rpm. Furthermore, the transmittance showed a small decrease from 91.6 to 90.6 at 550 nm, as shown in Fig. 6.

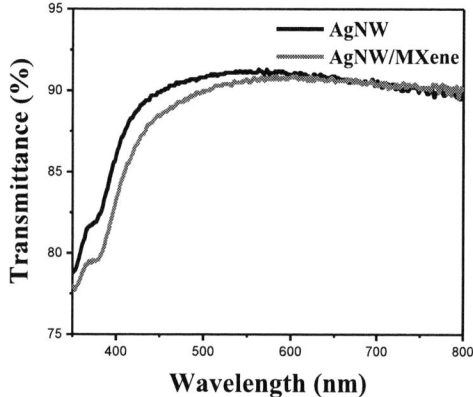

Fig. 4. Transmittance spectra of AgNW and AgNW/MXene transparent conductive electrodes.

SEM was used to characterize the surface morphology of the hybrid TCEs and the image was presented in Fig. 5. It showed that the MXene sheets covering the gaps/voids of the conductive network have connected the nanowires and nanowires, thereby greatly enriching the conductive paths of the electrodes. As a result, the hybrid TCEs showed reduced sheet resistance. Furthermore, the MXene has a uniform distribution on the AgNW network, which could improve the surface roughness of TCE. The water contact angle test was carried to evaluate the wettability of the film surface. These results displayed that after compounding MXene, the water contact angle decreases from 51.5 to 37.2 degrees, as shown in Fig. 6. It proved that the hydrophilicity of the film surface was improved due to the abundant functional groups of MXene.

The easy oxidation of silver nanowires is also a key point to be solved. For AgNW/MXene composite TCEs, MXene sheets could effectively isolate the water vapor and block oxidation. Moreover, the combination of silver nanowires and MXene sheets contribute to the formation of highly hybrid TCEs by a simple and scalable solution method.

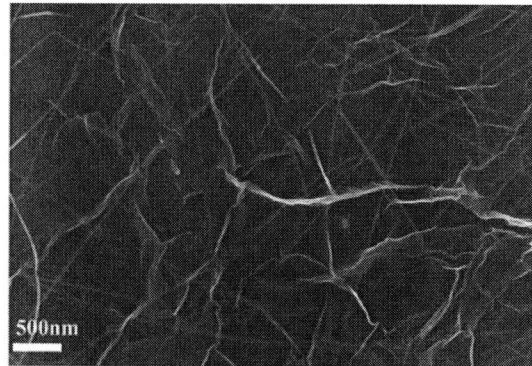

Fig. 5. The SEM image of AgNW hybrid transparent electrodes with MXene after delamination.

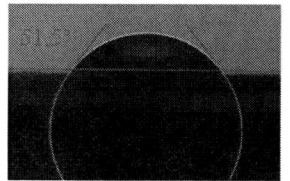 

Fig. 6. The water contact angle results of AgNW and AgNW/MXene.

IV. CONCLUSION

In summary, AgNW/MXene hybrid TCEs were fabricated by a solution process. The favorable concentration of AgNW was selected to prepare dense and flat conductive films. After that, multi-layer MXene was delaminated using DMSO and then coated on the AgNW networks. Therefore, the AgNW/MXene hybrid TCE demonstrated improved performance, showed the sheet resistance of 24.4 Ω/sq and the transmittance of 90.6% (550 nm). It demonstrates a bright application prospects in electronic devices.

ACKNOWLEDGMENT

The work was supported by National Key Research and Development Program of China (2020YFB2008501) and Science and Technology Committee of Shanghai (19142203600).

REFERENCES

[1] McLellan, K.; Yoon, Y.; Leung, S. N.; Ko, S. H., "Recent Progress in Transparent Conductors Based on Nanomaterials: Advancements and Challenges," *Advanced Materials Technologies*, vol. 5 no. 4,pp. 1900939, 2020.

[2] Bai, S.; Guo, X.; Chen, T.; Zhang, Y.; Yang, H., "Solution Process Fabrication of Silver Nanowire Composite Transparent Conductive Films with Tunable Work Function," *Thin Solid Films*, vol. 709 no.,pp. 138096, 2020.

[3] Ricciardulli, A. G., Yang, S., Wetzelaer, G.-J. A. H., Feng, X., Blom, P. W. M., "Hybrid Silver Nanowire and Graphene-Based Solution-Processed Transparent Electrode for Organic Optoelectronics, "*of Alloys and CompoundsAdvanced Functional Materials*, vol. 28 no. 14,pp. 1-6, 2018.

[4] Wang, Y.; Zhang, L.; Wang, D., "Ultrastretchable Hybrid Electrodes of Silver Nanowires and Multiwalled Carbon Nanotubes Realized by Capillary‐Force‐Induced Welding," *Advanced Materials Technologies*, vol. 4 no. 11,pp. 1900721, 2019.

[5] Lei, J. C.; Zhang, X.; Zhou, Z., "Recent advances in MXene: Preparation, properties, and applications, " Frontiers of Physics, vol. 10 no. ,pp. 2015.

[6] Dillon, A. D.; Ghidiu, M. J.; Krick, A. L.; Griggs, J.; May, S. J.; Gogotsi, Y.; Barsoum, M. W.; Fafarman, A. T., "Highly Conductive

Optical Quality Solution-Processed Films of 2D Titanium Carbide," *Advanced Functional Materials*, vol. *26* no. 23,pp. 4162-4168, 2016.

[7] Yang, X.; Du, D. X.; Xie, H.; Wang, Y. H.; Li, J. Z., "Review of Silver Nanowire Based Transparent Conductive Film," *Rare Metal Materials and Engineering,* vol. *48* no. 5,pp. 1707-1716, 2019.

[8] Li, T.; Chen, L.; Yang, X.; Chen, X.; Zhang, Z.; Zhao, T.; Li, X.; Zhang, J., "A flexible pressure sensor based on an MXene–textile network structure," *Journal of Materials Chemistry C*, vol. *7* no. 4,pp. 1022-1027, 2019.

[9] Li, W.; Zhang, H.; Shi, S.; Xu, J.; Qin, X.; He, Q.; Yang, K.; Dai, W.; Liu, G.; Zhou, Q.; Yu, H.; Silva, S. R. P.; Fahlman, M., "Recent progress in silver nanowire networks for flexible organic electronics," *Journal of Materials Chemistry C*, vol. *8* no. 14,pp. 4636-4674, 2020.

[10] Liu, C.; Hao, S.; Chen, X.; Zong, B.; Mao, S., "High Anti-Interference Ti3C2Tx MXene Field-Effect-Transistor-Based Alkali Indicator," *ACS Appl Mater Interfaces,* vol. *12* no. 29,pp. 32970-32978, 2020.

Synthesis of Air-sinterable Copper Nanoparticles for Die-attachment

1st Yue Yao
Shenzhen Institute of Advanced
Electronic Materials, Shenzhen
Institutes of Advanced Technology,
Chinese Academy of Sciences
Shenzhen, China
yue.yao@siat.ac.cn

2nd Liang Xu*
Shenzhen Institute of Advanced
Electronic Materials, Shenzhen
Institutes of Advanced Technology,
Chinese Academy of Sciences
Shenzhen, China
xuliang@siat.ac.cn

3rd Pengli Zhu
Shenzhen Institute of Advanced
Electronic Materials, Shenzhen
Institutes of Advanced Technology,
Chinese Academy of Sciences
Shenzhen, China
pl.zhu@siat.ac.cn

4th Tao Zhao*
Shenzhen Institute of Advanced
Electronic Materials, Shenzhen
Institutes of Advanced Technology,
Chinese Academy of Sciences
Shenzhen, China
tao.zhao@siat.ac.cn

5th Rong Sun
Shenzhen Institute of Advanced
Electronic Materials, Shenzhen
Institutes of Advanced Technology,
Chinese Academy of Sciences
Shenzhen, China
rong.sun@siat.ac.cn

6th Yingchao Huo
Nano Science and Technology
Institute, Suzhou
University of Science and
Technology of China
Hefei, China
yingchhuo@ustc.edu.cn

Abstract—Paste based on Cu nanoparticles (NPs) is a promising die-attachment material for high-power electronic devices due to their high conductivity and inexpensive-cost. However, the rapid oxidation of Cu NPs during sintering in air hinders their application. In this work, we synthesized stable Cu NPs with a diameter of 50~110 nm for die-attachment in air. Unlike most of the previous reports about copper pastes, nano-Cu paste prepared from the synthesized Cu NPs can be sintered at low temperatures without a protective atmosphere. By investigating the effect of different solvents on the conductivity of the sintered copper film, a mixture of glycerin and α-terpineol was selected as a carrier solvent to prepare nano-Cu paste. After sintering, the formed Cu-Cu joints exhibited the high shear strength, which reaches a maximum of 44.5 MPa at 275°C with a pressure of 10 MPa in air. Thus, this work shows a feasibility of the sintering of nano-Cu paste in air and the great potential of nano-Cu paste in the packing of the next-generation power devices.

Keywords—Cu NPs, low-temperature sintering, die-attachment, powder devices, air-sinterable

I. INTRODUCTION

With the rapid development of electronic products, the third-generation semiconductor devices represented by GaN or SiC power devices, which exhibit high switching performance and high temperature sustaining capabilities, have received more and more attentions [1, 2]. The high-power density of these devices makes them generate large amounts of heat during operation, and thus causing the rapid rise in their junction temperature, which can reach more than 200°C [3]. However, existing tin-based lead-free solders only can meet the heat dissipation and the reliability requirements of the low junction temperature devices (<150°C), meaning that they are not suitable for the packaging of high-power devices [4]. Thus, there is an urgent demand to develop new die-attachment materials to serve these high-power devices. Because of their low thermostability derived from small size effect, metal nanoparticles (such as Ag or Cu NPs) can be sintered at low temperatures (200-300°C), making them as the next-generation very promising low-temperature die-

attachment material [5]. Compared with Cu NPs, Ag NPs have attracted much attention of scientific researchers due to their excellent stability and high conductivity [6]. However, the high price and the poor resistance to electromigration of silver heavily limit the application of Ag NPs [7].

In contrast, copper exhibits the higher resistance to electromigration and the much lower price [8, 9]. But it is easily oxidized in air, and the formed oxides will hinder the sintering process of Cu NPs. This makes the sintering of Cu NPs always require an inert atmosphere [10] or a vacuum environment and thus increases the cost and complexity [11]. If Cu NPs can be sintered in air, they can be as a promising alternative to Ag NPs [12]. However, to realize the sintering of Cu NPs in air is still a huge challenge and worthy of our attention.

Herein, we successfully synthesized air-sinterable Cu NPs by a simple wet chemical method and they could form Cu-Cu joints with good electrical conductivity and high strength after a low-temperature sintering process in air. The synthesized Cu NPs were carefully characterized by SEM, XRD, TGA, and XPS. The effect of different solvents on the sintering of the synthesized Cu NPs was systematically investigated, and the mixture of glycerin and α-terpineol was chosen as a carrier solvent to prepare nano-Cu paste. Finally, the sintering process and the mechanical property of the prepared nano-Cu paste were explored in detail.

II. EXPERIMENT

A. Material

Hydrazine monohydrate, isopropanol, ethylene glycol, glycerin, α-terpineol, triethylene glycol monomethyl ether (TGME), triethylene glycol butyl ether (TGBE), triethylene glycol butyl ether acetate(TGBEA), and $CuCl_2$ were acquired from Sigma-Aldrich. All reagents were used without further purification. Direct Bonding Copper (DBC) was supplied by Tong Hsing Electronics Industries.

B. Synthesis of air-sinterable Cu NPs

The synthesis of Cu NPs can be briefly summarized as

978-1-6654-1392-3/21 $31.00 © 2021 IEEE

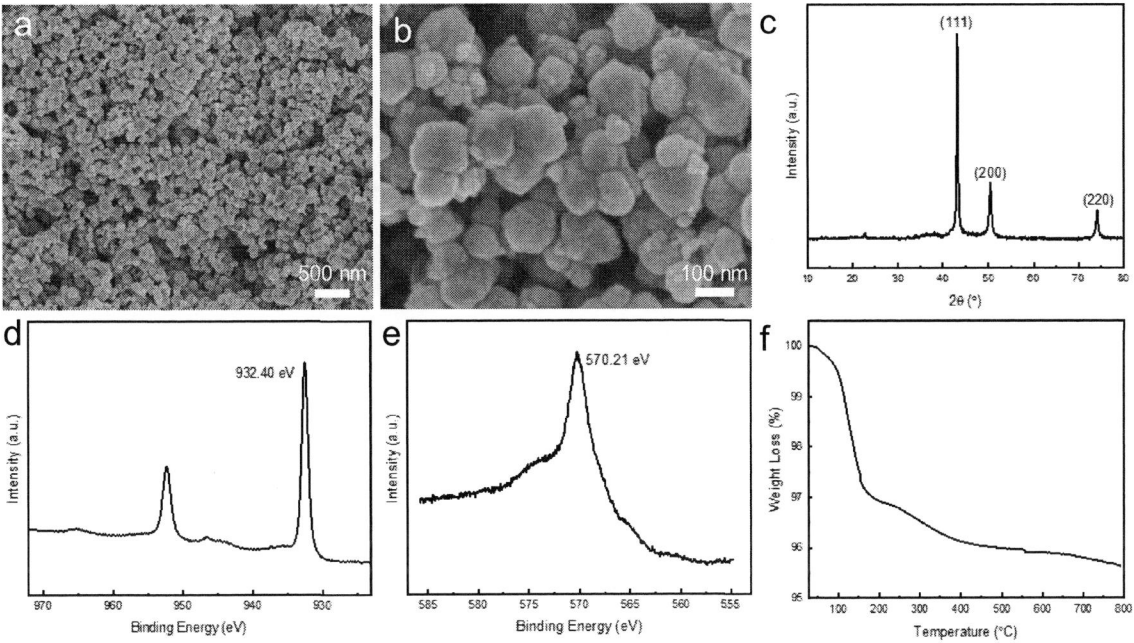

Fig. 1. (a), (b) SEM images of different magnifications, (c) XRD pattern , (d) Cu 2p₃/₂ spectrum, (e) Cu LMM spectrum and (f) TGA curve of the synthesized Cu NPs.

follows: a certain amount of $CuCl_2$ and surfactant was firstly heated to 80°C under vigorous magnetic stirring to form a clear blue solution, then hydrazine hydrate was quickly added into this blue solution. The reaction solution was maintained at 80°C for 15 minutes.

C. Preparation of Conductive paste using Cu NPs

The as-prepared Cu NPs were used as metallic filler to produce nano-Cu paste. Ethanol, ethylene glycol, glycerin, α-terpineol, TGME, TGBE, TGBEA were used as carrier solvents to optimize the formulation of nano-Cu paste. Specifically, nano-Cu pastes consist of the mixture of Cu NPs and different organic solvents with a mass ratio of 75 : 25. The PI films were ultrasonically cleaned in absolute ethanol to remove impurities. The obtained nano-Cu paste was coated on a PI film to form a pattern with a width of 5 mm, and then the pattern was sintered in an annealing furnace at different temperatures to form a conductive Cu film. The resistivity of the sintered Cu film was measured by four-probe method.

For die-attachment bonding, firstly, nano-Cu paste was evenly coated on 5*5 mm² DBC and 3*3 mm² DBC. Then, a sandwich-structure was prepared by placing 3*3 mm² DBC on 5*5 mm² DBC. After that, under 10 MPa applied pressure, the sandwich-structure was heated at different temperatures to prepare Cu-Cu joint.

D. Charracterization

Scanning electron microscope (SEM) was performed to evaluate the morphology of Cu NPs and the microstructure of Cu-Cu joints. The crystalline structure of the Cu NPs was studied by X-ray diffraction (XRD) using Cu K_α radiation. X-ray photoelectron spectroscopy (XPS) was used to analyze the valence state of Cu on the surface of Cu NPs. Thermogravimetric analysis (TGA) is used to study the organic matter content coated on the surface of Cu particles. Shear strength test was conducted via a shear tester (DAGE 4000) at a speed of 100 μm/s.

III. RESULTS AND DISCUSSION

A. Characterization of Cu nanoparticles

SEM images in Fig. 1(a) and (b) reveal that monodispersed Cu NPs with size distribution of 50~110 nm were synthesized via our simple wet chemical method. The XRD pattern of the synthesized Cu NPs is shown in Fig. 1(c): there are three main characteristic diffraction peaks at 43.3 °, 50.3 °, 74.3 °, which represent (111), (200), (220) crystallographic planes of face-centered cubic (FCC) Cu crystal (PDF#04-0836), and no impurities peaks were detected, indicating that the synthesized Cu NPs have high purity. Besides, the sharp and strong peaks signify that Cu NPs have a high crystallinity. Furthermore, in order to identify whether the synthesized Cu NPs were oxidized during the process of washing or storage, XPS analysis of the synthesized Cu NPs was adopted. As shown in Fig. 1(d), the Cu 2p₃/₂ peak of Cu NPs was at 932.40 eV, meaning that the product contains Cu (0) or Cu_2O. However, it difficult to identify Cu^+ (Cu_2O) from Cu (0) in Cu 2p₃/₂ spectrum as the difference between them is only 0.1-0.2 eV. Thus, Cu Auger LMM spectrum was used to identify all three chemical states (Cu, Cu^+ and Cu^{2+}). The synthesized Cu NPs exhibited a peak at 570.09 eV in Cu LMM spectrum (Fig. 1(e)), confirming the presence of Cu_2O on the surface of Cu NPs. However, the characteristic diffraction peaks of Cu_2O were not appeared in the XRD pattern, representing that the oxidation degree of the synthesized Cu NPs was relatively low. TGA was carried out to measure the organic content of the synthesized Cu NPs. In Fig. 1(f), Cu NPs showed a continuous weight loss during 25 ~ 800°C, and the total weight loss was about 4%. A strong weight loss of 3% was observed in the range of 25°C and 200°C, which was due to the decomposition of surfactant coated on the surface of Cu NPs.

B. Effect of different solvents on the resistivity of the sintered Cu film

The solvent is another very important component for metal paste except metal filler, which determines the rheological property and the sintering process of metal paste to a certain extent. Some organic solvents, such as α-terpineol, can effectively promote the sintering of metal NPs at low temperatures [5]. A variety of organic solvents were selected here to explore their influence on the resistivity of nano-Cu paste after sintering. As shown in Fig. 2, the resistivity of the sintered Cu paste decreased as the heating temperature increased. Among them, Cu paste used glycerin as a carrier solvent (paste-G) always exhibited the lowest resistivity after sintering at different temperatures. In addition, when α-terpineol was used as a carrier solvent, the resistivity of sintered Cu was relatively lower at 250°C. Therefore, glycerin and a mixture of glycerin and α-terpineol were finally selected as carrier solvents to prepare nano-Cu paste.

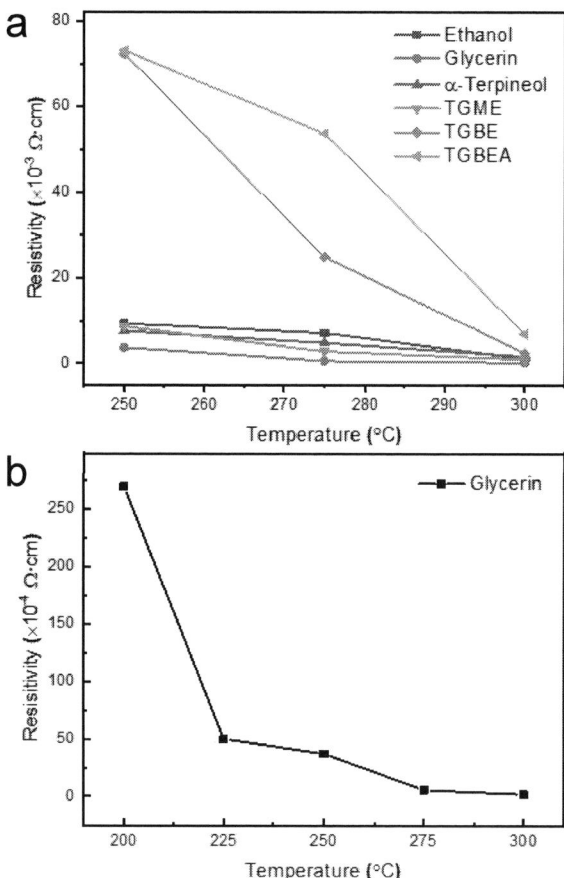

Fig. 2. Resistivities of Cu films formed by sintering different nano-Cu pastes at different temperatures for 10 min in N₂: (a)using various carrier solvents; (b) using glycerin as a carrier solvent.

C. Strength of Sintered nano-Cu Paste

The shear strength of Cu-Cu joints formed at various sintering temperatures was given in Fig. 3. Obviously, the increase of the sintering temperature was beneficial to the enhancement of the mechanical property of the joints. The nano-Cu paste-G showed a shear strength of 14.76 MPa at 225°C. As the temperature increased to 250°C, the shear strength noticeably increased to 19 MPa, similar to that of conventional Pb-5Sn die-attachment joints[6]. The shear strength of the sintered Cu-Cu joints reached the highest value (about 34.05 MPa) at 300°C.

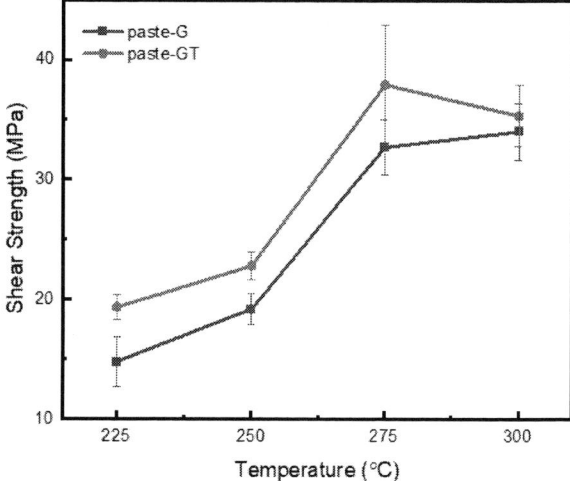

Fig. 3. Shear strength of Cu-Cu joints formed at different sintering temperatures for 10 min in air.

In addition, the composition of carrier solvent of nano-Cu pastes also heavily affected the shear strength of Cu-Cu joints derived from them. To investigate the influences of the composition of solvent on the shear strength of the sintered joints, a mixture of glycerin and α-terpineol (5:5) instead of glycerin was used as a carrier solvent. The obtained nano-Cu paste (paste-GT) was sintered under the same condition. As shown in the Fig. 3, the maximum shear strength of Cu-Cu joints for this formulation can reach 44.55 MPa at 275°C, which is comparable to the sintering performance of nano-Ag paste [6].

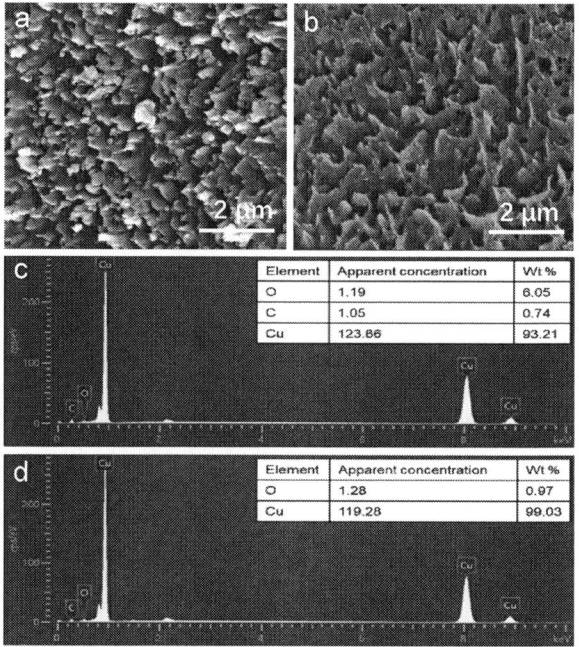

Fig. 4. SEM images and EDS spectrograms of the fracture surface of the sintered Cu−Cu joints (300°C for 20 min in air) with different solvent: (a), (c) glycerin and (b), (d) the mixture of glycerin and α-terpineol.

The fracture surfaces of Cu-Cu joints sintered at 300ºC were shown in Fig. 4, and the ductile deformation was clearly observed in these fracture surfaces. However, some nanoparticles still existed in the fracture surface of Cu-Cu joints, meaning that the sintering of Cu NPs was insufficiency. The corresponding EDS spectrograms show that the content of carbon and oxygen of these sintered Cu-Cu joints is relatively low. Interestingly, when the carrier solvent was the mixture of glycerin and α-terpineol, no carbon element was detected in the fracture surface of Cu-Cu joint, indicating that organics in this paste evaporated completely during the sintering process. The above results prove that the presence of α-terpineol can promote the sintering of nano-Cu paste at low temperatures. To investigate the microstructure of the sintered joints, Cu-Cu joints were milled and polished. As shown in Fig. 5, the cross-section of Cu-Cu joint sintered at 275ºC exhibited a low porosity of 16%. While the porosity of Cu-Cu joint formed at 300ºC increased 20%, and some irregular voids were found at this cross-section, resulting in its weaker mechanical property. This difference in the microstructure revealed that the suitable temperature can promote the sintering of nano-Cu slurry.

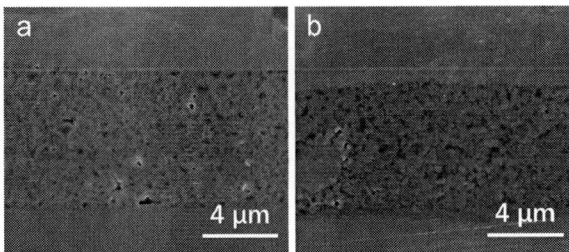

Fig. 5. SEM images of the cross-section of Cu-Cu joints formed at different sintering temperatures: (a) 275°C and (b) 300°C.

IV. CONCLUSION

In summary, the sintering behavior of nano-Cu paste made from the synthesized Cu NPs in air under various temperature was investigated in this study. Air-sinterable Cu NPs were synthesized by a simple liquid phase reduction method. The influences of different solvents on the sintering of the synthesized Cu NPs were systematically investigated, and the mixed solution of glycerin and α-terpineol was chosen as the carrier solvent to prepare nano-Cu paste. After sintering at 275ºC in air, Cu-Cu joints derived from the prepared nano-Cu paste displayed a high shear strength of 44.5 MPa. The microstructure analysis showed that the extensive plastic deformation area was found at the fracture surface of Cu-Cu joints, which explained their high shear strength. Therefore, we think that this work shows a feasibility of the sintering of nano-Cu paste in the absence of inert gas or reducing atmosphere and a great potential of nano-Cu paste for application in the packing of the next-generation power devices.

ACKNOWLEDGMENT

This work was financially supported by the National Natural Science Foundation of China (21805189, 51805197, and 61704182), the Natural Science Foundation of Guangdong (2021A1515011284 and 2018A030310617), Shenzhen basic research plan (JCYJ20190807154409372), and SIAT Innovation Program for Excellent Young Researchers (E1G014).

REFERENCES

[1] J. Kahler, N. Heuck, A. Wagner, A. Stranz, E. Peiner, and A. Waag, "Sintering of Copper Particles for Die-Attachment," *IEEE Transactions on Components, Packaging and Manufacturing Technology,* vol. 2, no. 10, pp. 1587-1591, 2012.

[2] Y. Gao, S. Takata, C. Chen, S. Nagao, K. Suganuma, A. S. Bahman et al., "Reliability analysis of sintered Cu joints for SiC power devices under thermal shock condition," Microelectronics Reliability, vol. 100-101, 2019. pp. 113456-113461, 2019.

[3] H. Tatsumi, H. Yamaguchi, T. Matsuda, T. Sano, Y. Kashiba, and A. Hirose, "Deformation Behavior of Transient Liquid-Phase Sintered Cu-Solder-Resin Microstructure for Die-Attachment," *Applied Sciences,* vol. 9, no. 17, pp. 2476-2490, 2019.

[4] S. K. Bhogaraju, O. Mokhtari, F. Conti, and G. Elger, "Die-attachment bonding for high temperature applications using thermal decomposition of copper(II) formate with polyethylene glycol," Scripta Materialia, vol. 182, pp. 74-80, 2020.

[5] S. Koga, H. Nishikawa, M. Saito, and J. Mizuno, "Fabrication of Nanoporous Cu Sheet and Application to Bonding for High-Temperature Applications," *Journal of Electronic Materials,* vol. 49, no. 3, pp. 2151-2158, 2020.

[6] Y. Kobayashi, T. Maeda, Y. Yasuda, and T. Morita, "Metal–metal bonding using silver/copper nanoparticles," Applied Nanoscience, vol. 6, no. 6, pp. 883-893, 2015.

[7] Q. Fu, M. Stein, W. Li, J. Zheng, and F. E. Kruis, "Conductive films prepared from inks based on copper nanoparticles synthesized by transferred arc discharge," *Nanotechnology,* vol. 31, no. 2, pp. 025302-025314, 2020.

[8] D. Deng, Y. Cheng, Y. Jin, T. Qi, and F. Xiao, "Antioxidative effect of lactic acid-stabilized copper nanoparticles prepared in aqueous solution," Journal of Materials Chemistry, vol. 22, no. 45, pp. 23989-23995, 2012.

[9] Y. Kamikoriyama, H. Imamura, A. Muramatsu, and K. Kanie, "Ambient Aqueous-Phase Synthesis of Copper Nanoparticles and Nanopastes with Low-Temperature Sintering and Ultra-High Bonding Abilities," *Sci Rep,* vol. 9, no. 1, pp. 899-909, 2019.

[10] S. K. Bhogaraju, F. Conti, H. R. Kotadia, S. Keim, U. Tetzlaff, and G. Elger, "Novel approach to copper sintering using surface enhanced brass micro flakes for microelectronics packaging," Journal of Alloys and Compounds, vol. 844, pp. 156043-156049, 2020.

[11] S. Kwon, T.-I. Lee, H.-J. Lee, and S. Yoo, "Improved sinterability of micro-scale copper paste with a reducing agent," Materials Letters, vol. 269, pp. 127656-127658,2020.

[12] B. Zhang, Damian. A, Zijl. J, van Zeijl. H, Zhang. Y, Fan. J et al., "In-air sintering of copper nanoparticle paste with pressure-assistance for die-attachment in high-power electronics," *Journal of Materials Science: Materials in Electronics,* vol. 32, no. 4, pp. 4544-4555, 2021.

The Particle Interaction Analysis for Nanoparticles in Underfill for Flip-Chip Packaging*

Mingyong Du
Shenzhen Institute of Advanced
Electronic Materials
Shenzhen Institute of Advanced
Technology, Chinese Academy of
Sciences
ShenZhen, China
my.du@siat.ac.cn

Ning Wang
Shenzhen Institute of Advanced
Electronic Materials
Shenzhen Institute of Advanced
Technology, Chinese Academy of
Sciences
ShenZhen, China
wangning@siat.ac.cn

Xiaomeng Du
Shenzhen Institute of Advanced
Electronic Materials
Shenzhen Institute of Advanced
Technology, Chinese Academy of
Sciences
ShenZhen, China
xm.du1@siat.ac.cn

Tao Zhao
Shenzhen Institute of Advanced
Electronic Materials
Shenzhen Institute of Advanced
Technology, Chinese Academy of
Sciences
ShenZhen, China
tao.zhao@siat.ac.cn

Pengli Zhu
Shenzhen Institute of Advanced
Electronic Materials
Shenzhen Institute of Advanced
Technology, Chinese Academy of
Sciences
ShenZhen, China
pl.zhu@siat.ac.cn

Rong Sun
Shenzhen Institute of Advanced
Electronic Materials
Shenzhen Institute of Advanced
Technology
Chinese Academy of Sciences
ShenZhen, China
rong.sun@siat.ac.cn

Abstract—The introduction of inorganic nanoparticles into polymer matrix improves the performance of the underfill in flip-chip packaging, for example it improves the mechanical properties of the polymer and reduces the coefficient of thermal expansion (CTE). However, the nanoparticles are not always compatible with the polymer resin and hard to be well dispersed in the colloidal suspension. The phenomenon is caused by the very small diameter, the large specific surface area, and the outstanding surface energy of nanoparticles. As a result, nanoparticles are easy to interact with each other and form agglomerates in the colloid, and lead to mobility problem. To resolve this problem, this paper investigated the particle interaction for raw nanoparticles and surface modified nanoparticles in epoxy resin, and distinguished the functional group impact on the particle interactions. The Fourier transform infrared spectroscopy (FT-IR) is used to characterize the specific chemical groups of modified nanoparticles. The FI-IR result verified that silane coupling agents (SCA) were grafted onto the nanoparticles. Raw nanoparticles and surface modified nanoparticles were suspended into epoxy resin to investigate the rheological behaviour. The rheological flow curves were well fitted by the Herschel-Bulkley model. The dynamic rheology of these colloidal suspensions was examined as well. According to the frequency spectrum, the elastic modulus (G') dominants over the whole angular frequency range for raw nanoparticle epoxy resin colloidal suspension. Nanoparticles interact with each other via hydrogen bonding and van der Waals attraction strongly in the suspension. However, surface modification for nanoparticles reduced the fluidity according to the rheological flow curves. The dynamic rheological property demonstrates that surface modification improves the elasticity of the suspension, which means van der Waals attraction between nano-silica particles becomes stronger while the particle-resin interaction is weaker for the SCA modified nano-silica suspension.

Keywords—nanoparticles, particle interaction, rheology, van der Waals attraction, colloidal suspension

I. INTRODUCTION

Nano-silica with large specific surface area and large surface energy are easy to agglomerate [1]. The van der Waals (vdW) attractive forces between nano-silica particles drive nanoparticles to aggregate and form unstable structure in the underfill formulation. The interfacial interaction between the nanoparticle and epoxy resin dominates the equilibrium and dynamic behaviour of the underfill[2]. According to previous report, the particle-particle interaction and particle-resin interaction can be modified and controlled by either changing the surface property of the nano-silica, or modifying the organic resin properties. A lot of efforts have been paid to disperse the nano-silica into the epoxy resin to formulate the desired underfill with low viscosity, high fluidity and improve the material properties.

Chemical modification with silane coupling agent (SCA) is a widely used particle surface treatment technique to change the surface tension of particles and improve the wettability of epoxy [3]. The ideal treated nano-silica should disperse well in the epoxy resin and form a nanocomposite underfill with good rheological properties to assure the material dispensing and chip collapsing. Rheology reflects the relation between the deformation and mechanical response of a material. Rheological behaviour of a suspension reveals the colloidal particle interactions[4, 5]. The repelling particles in a suspension result in a well-dispersed colloid, whereas attractive forces among particles is the driving force for agglomeration. Rheology investigation provides the information analyzing particle interaction in the suspension[6]. This paper aims to systematically study the rheological properties of as-received nano-silica and surface modified nano-silica with SCA.

II. EXPERIMENTAL

A. Scanning Electron Microscope (SEM)

SEM images of nano-silica were taken with the Thermo Scientific Apreo 2. Nano-silica sample was coated with gold to get rid of the charging effect. The coated sample was scanned at 5 kV.

B. *Fourier transform infrared spectroscopy (FT-IR)*

Bruker OPUS-VERTEX70 was utilized to scan FT-IR spectra of nano-silica. Infrared diffuse reflection mode was used to scan the sample from wavelength from 4000 to 400 cm^{-1} for 320 times.

C. *Rheology*

Aton Paar MCR 302 cone-and-plate rheometer was used to measure the rheology property of the suspensions. The suspension was sheared at 1000 s^{-1} to reach a stable stage before data recording. The flow properties were measured as a function of shear rate at 25 °C. The linear viscoelastic region (LVER) was obtained from the dynamic strain sweep measurement. Dynamic frequency sweep tests of the suspensions were measured in this region. Angular frequency decreases from 100 rad/s to 0.1 rad/s, the elastic modulus (G') and viscous modulus (G") were recorded.

III. RESULTS AND DISCUSSION

A. *SEM*

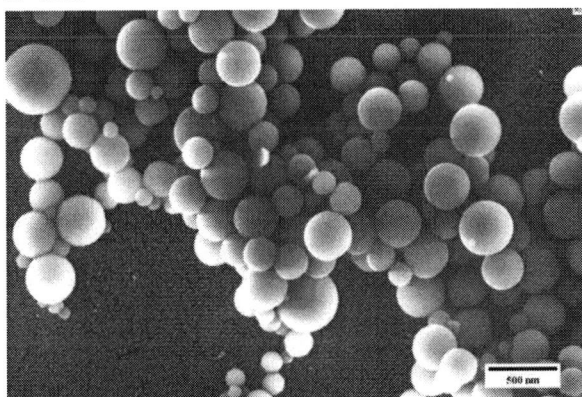

Fig. 1. SEM images of nano-silica

The SEM image shows that nano-silica has excellent degree of sphericility. The diameter ranges from 76 to 315 nm in the visible area. Nano-silica particles interact with each other and form inhomogeneous particle packing, which confirms the argument that nanoparticles with large specific surface areas tend to agglomerate in order to reduce the surface free energy[7].

B. *FT-IR*

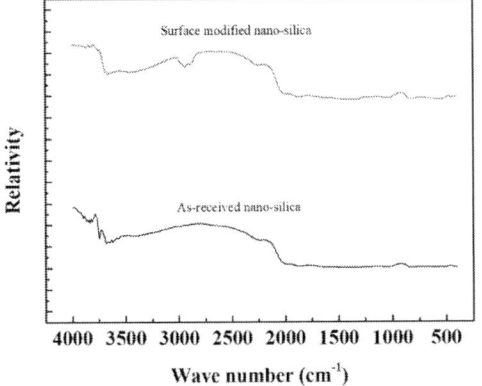

Fig. 2. FT-IR spectra of as-received nano-silica and surface modified nano-silica

The FT-IR spectra of the as-received nano-silica exhibits signals of Si-O from 1020 to 840 cm^{-1}of surface modified nano-silica demonstrates strong peak signals for -CH$_3$ and -CH$_2$- between 3020 and 2825 cm^{-1}.

C. *Rheology*

The flow property of nano-silica-epoxy resin suspensions were studied. Fig. 3a shows the shear stress curves of as-received silica and surface modified nano-silica suspensions. The initial shear stress at 1000 s^{-1}for as received nano-silica is 1748 Pa. Shear stress decreases rapidly to ~100 Pa as shear rate decreases from 1000 to 10 s^{-1}. However, surface modified nano-silica suspension demonstrates higher shear stress while the suspension is sheared at a high speed. Herschel-Bulkley model [8, 9] is widely applied in fitting the rheological flow curves. The model is described as follows:

$$\tau = \tau_{HB} + \eta_{HB}(\dot{\gamma})^n \quad \text{if } \tau \geq \tau_{HB}$$

$$\dot{\gamma} = 0 \qquad\qquad \text{if } \tau \leq \tau_{HB} \qquad (1)$$

Where τ_{HB} is the Herschel-Bulkley yield stress, and the η_{HB} is viscosity. If $n=1$, this equation equals to Bingham model; if $n>1$, the fluid behaves shear-thickening (yield dilatant) property, if $n<1$, the fluid is shear-thinning (yield pseudoplastic).

TABLE I. HERSCHEL-BULKLEY MODEL PARAMETERS

	τ_{HB} (Pa)	η_{HB} (Pa·s)	n
As-received nano-silica suspension	15.91	26.26	0.60
Surface modified nano-silica suspension	30.55	19.10	0.76

The parameters obtained from the Herschel-Bulkley model in Table I confirmed that the prepared suspensions are pseudoplastic (shear-thinning) suspensions with yield stress. Both suspensions cannot flow until the critical yield stress τ_{HB} is exceeded, as the continuous structured network cannot be broken down. The critical yield stress is higher for surface modified nano-silica suspension, which means the gel strength is higher for this suspension.

Both suspensions demonstrate strong shear-thinning property, as the viscosity decreases while shear rate increases, revealed by the viscosity curves in Fig. 3b. According to the previous report, this type of viscosity property could be fitted by a power law model[10], described as follows:

$$\mu = K\dot{\gamma}^{n-1} \qquad (2)$$

where, K represents the consistency index; $\dot{\gamma}$ represents shear rate; n represents power law index. For shear thinning behaviour n value is smaller than 1. The extent shear-thinning property increases as n decreases. For Newtonian behaviour, $n = 1$. The power law indexes are 0.537 and 0.649 for as-received and surface modified nano-silica suspensions. The modified nano-silica suspension displays a less shear-thinning property, which means the gel strength is higher. The conclusion is in correspondence with the one drawn from shear stress data.

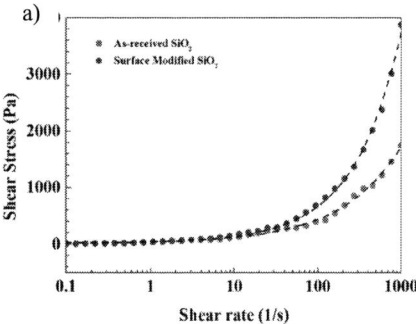

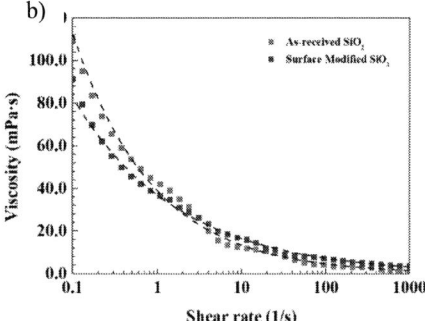

Fig. 3. Flow properties of nano-silica-epoxy resin suspensions

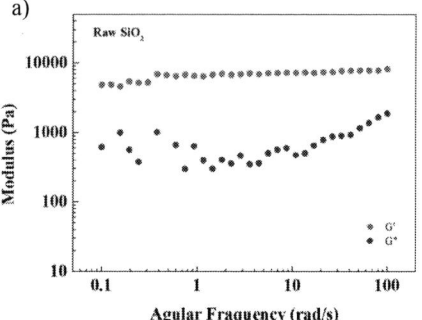

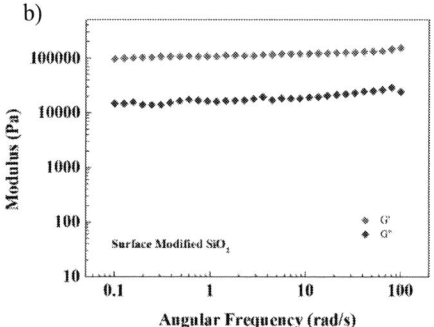

Fig. 4. Rheological viscoelastic property of as-received nano-silica suspension (a) and surface modified nano-silica suspension (b).

Frequency sweep test was used to characterize viscoelastic properties of the suspensions in the LVER. The red plot and blue plot represent elastic modulus (G') and viscous modulus (G") respectively as a function of angular frequency. Both suspensions are more elastic than viscous, with no crossover. The modulus is weakly relied on angular frequency, especially the G'. The property indicates that the suspension behaviour is un-disturbed over a range of time scales. We draw the conclusion that samples are weak physical gels composed of a 3-dimentional network of physical bonds by the feature that G' and G" were constant or increased slightly, together with no crossover point for G' and G"[11]. Nano-size silica with high surface energy has strong interfacial interaction between each other and forms a flocculated solid-like suspension. Hydrogen bonding between nanoparticle surfaces dominants rather than the nanoparticle and epoxy molecule interaction. Compared with as-received nano-silica suspension, the surface modified nano-silica suspension demonstrates even higher G' and G", and more solid-like elasticity. we conclude that the interaction between nano-silica and epoxy molecules is weaker in this suspension. Nanoparticles interact with each other via hydrogen bonding and van der Waals attraction strongly in the suspension. However, surface modification for nanoparticles reduced the fluidity according to the rheological flow curves. The dynamic rheological property demonstrates that surface modification improves the elasticity of the suspension, which means van der Waals attraction between nano-silica particles becomes stronger while the particle-resin interaction is weaker for the SCA modified nano-silica suspension.

IV. CONCLUSION

The suspension composed of nano-silica and epoxy resin demonstrates shear-thinning property and elasticity because of high gel strength. Nano-silica particles interact with each other mainly by van der Waals attraction. The particles tend to contact with each other to reduce the surface energy instead of contact with epoxy resin molecules. The surface of nano-silica was modified by SCA. However, the surface modified nano-silica displays stronger gel strength and stronger elasticity, which means that more particle-particle interactions present. It becomes more difficult to overcome the attraction between particles to interact with resin. It is high recommended to study the physical chemistry property of SCA and epoxy resin so as to choose the proper SCA to modify the nano-silica. Moreover, more efforts should be made to develop a reasonable design of experiment to improve the grafting efficiency.

ACKNOWLEDGMENT

This work was supported by the National Key R & D Project from Minister of Science and Technology of China (2020YFB0311800) , Shenzhen basic research plan (JCYJ20190807154409372), the Guangdong Basic and Applied Basic Research Foundation (No. 2019A1515010743) and the SIAT Innovation Program for Excellent Young Researchers (No. E1G012).

REFERENCES

[1]   Q. Zeng, T. Mao, H. Li, Y. Peng, Thermally insulating lightweight cement-based composites incorporating glass beads and nano-silica

aerogels for sustainably energy-saving buildings, Energy and Buildings 174 (2018) 97-110.

[2] F.C. Ng, A. Abas, Z. Gan, M.Z. Abdullah, F.C. Ani, M.Y.T. Ali, Discrete phase method study of ball grid array underfill process using nano-silica filler-reinforced composite-encapsulant with varying filler loadings, Microelectronics Reliability 72 (2017) 45-64.

[3] B. Pang, Y. Zhang, G. Liu, W. She, Interface properties of nanosilica-modified waterborne epoxy cement repairing system, ACS applied materials & interfaces 10(25) (2018) 21696-21711.

[4] M. Du, P. Liu, J. Wong, P. Clode, J. Liu, Y. Leong, Colloidal forces, microstructure and thixotropy of sodium montmorillonite (SWy-2) gels: roles of electrostatic and van der Waals forces, Applied Clay Science 195 (2020) 105710.

[5] M. Du, J. Liu, P. Clode, Y.-K. Leong, Microstructure and rheology of bentonite slurries containing multiple-charge phosphate-based additives, Applied Clay Science 169 (2019) 120-128.

[6] M. Du, J. Liu, P.L. Clode, Y.-K. Leong, Surface chemistry, rheology and microstructure of purified natural and synthetic hectorite suspensions, Physical Chemistry Chemical Physics 20(28) (2018) 19221-19233.

[7] Z. Shahedi, M.R. Jafari, A.A. Zolanvari, Synthesis of ZnQ 2, CaQ 2, and CdQ 2 for application in OLED: optical, thermal, and electrical characterizations, Journal of Materials Science: Materials in Electronics 28(10) (2017) 7313-7319.

[8] A. Saasen, J.D. Ytrehus, Rheological properties of drilling fluids: use of dimensionless shear rates in herschel-bulkley and power-law models, Applied Rheology 28(5) (2018).

[9] H. Vaidya, M. Gudekote, R. Choudhari, K. Prasad, Role of slip and heat transfer on peristaltic transport of Herschel-Bulkley fluid through an elastic tube, Multidiscipline Modeling in Materials and Structures (2018).

[10] M. Du, P. Liu, P.L. Clode, J. Liu, B. Haq, Y.-K. Leong, Impact of additives with opposing effects on the rheological properties of bentonite drilling mud: Flow, ageing, microstructure and preparation method, Journal of Petroleum Science and Engineering 192 (2020) 107282.

[11] M. Du, C. Dai, A. Chen, X. Wu, Y. Li, Y. Liu, W. Li, M. Zhao, Investigation on the aggregation behavior of photo-responsive system composed of 1-hexadecyl-3-methylimidazolium bromide and 2-methoxycinnamic acid, RSC advances 5(84) (2015) 68369-68377.

A 3D TSV-MEMS Based Heterogeneous Integration Technology for RF Application

Min Huang
*National Key Laboratory of Science
and Technology on Monolithic
Integrated Circuits and Modules*
Nanjing, China
179278182@qq.com

Tinglei Wang
*Shanghai Radio Equipment Research
Institute*
Shanghai, China

Fan Hou
*State Key Laboratory of Millimeter
Waves, School of Information Science
and Engineering
Southeast University*
Nanjing, China

Ping Su
*Shanghai Radio Equipment Research
Institute*
Shanghai, China

Chao Sun
Nanjing Electronic Devices Institute
Nanjing, China

Huakai Luan
Nanjing Electronic Devices Institute
Nanjing, China

Abstract—Advanced radio frequency systems demand better performance in more compact volume, especially in high frequency and broad bandwidth applications. In this paper, we present a 3D TSV-MEMS based heterogeneous integration technology for radio frequency microsystem. Up to four layers of silicon interposers fabricated by MEMS technology can be stacked vertically in this configuration. MEMS based filters can be integrated within silicon interposers, providing ultimate radio frequency performance. A 6-channel 6-18 GHz switchable filter bank is presented in this paper to demonstrate the 3D TSV-MEMS based architecture. The size of the switchable filter bank is 20 mm × 11 mm × 1 mm, only 1.5% in volume comparing to its conventional counterpart. This architecture is not limited to switchable filter banks. It is suitable for more complex radio frequency systems up to 60 GHz.

Keywords—heterogeneous integration, MEMS, RF system, switchable filter bank

I. INTRODUCTION

Future advanced radio frequency (RF) systems demand more and greater integration of higher frequency and broader bandwidth microwave functions into smaller volumes [1]. Recent significant advances in heterogeneous integration along with three dimension (3D) stacking have led to novel integrated system architectures with increasingly better performance, greater miniaturization and enhanced reliability [2]. However, the biggest issue on RF system miniaturization remains in the RF receiver module, especially when switchable filter banks are applied. Filters are normally treated as independent components and integrated using wire-bond or flip-chip assemblies [3-5]. Some interposers enable integrated passive devices (IPD), but fail to offer high-quality performance in high frequency and wide band applications [6].

In this paper, we present a 3D TSV-MEMS based heterogeneous integration technology for RF micro-system. Up to four layers of silicon interposers can be stacked vertically using wafer-level bonding in this architecture. Each layer is fabricated using conventional MEMS process. The interconnecting precision is within 2μm by delicate wafer-scale alignment, making it possible to achieve lateral and vertical compactness required of high-density interconnects for high frequency applications. MEMS filters can be designed and fabricated using two stacked interposers. MEMS based IPDs provide ultimate performance in microwave applications. Moreover, two layers of MEMS filters can be realized in four stacked interposers, further pushing the compactness of RF microsystem. In order to demonstrate the 3D TSV-MEMS based architecture, a compact switchable filter bank is presented in this paper. The filter bank contains 6 channels covering 6-18 GHz in frequency. 6 band-pass filters are fabricated within four layers of MEMS interposers and electrically isolated using metallic TSV fences. 2 PIN switches are heterogeneously integrated into these interposers. Parasitic effects and impedance mismatch are minimized in this design. The size of the switchable filter bank is 20mm × 11mm × 1mm, only 1.5% in volume comparing to its conventional counterpart.

The proposed architecture is not limited to switchable filter banks. It is suitable for RF systems/modules up to 60GHz in frequency. More complex circuit functionality can be achieved by integrating amplifier, mixer, ICs and antenna, further extending the scope of its application. The 3D TSV-MEMS based technology offers two unique benefits. On one hand, high-quality MEMS filters are fabricated and stacked within the interposers. The precise wafer-scale alignment and scaled-down interspacing is highly suitable for high frequency applications. On the other hand, this architecture is built on relatively mature and straightforward MEMS process, making it more feasible and cost effective for mass production.

II. 3D TSV-MEMS ARCHITECTURE AND PROCESS

A. Proposed 3D TSV-MEMS Architecture

The 3D TSV-MEMS based heterogeneous integration architecture is shown in Fig. 1. It consists of four silicon interposers. In order to reduce insertion loss, interposer is made of high-resistivity silicon (HRS). Silicon wafer resistivity is chosen to be ~10000 Ω·cm. Thickness of the interposer is 250 μm. Metalized through-silicon vias (TSVs) with 120μm in diameter are formed in the interposer. TSVs interconnect signals vertically, and shield electromagnetic field in some cases. Two layers of redistribution layer (RDL) can be patterned on both side of the interposer, connecting signals horizontally. RDLs combined with TSVs build a 3D transmission network for both RF and digital signals.

Functional chips such as compound semiconductor (CS) chip and ICs can be integrated by all means of chip to wafer (C2W) methods. Conventional multichip module (MCM) techniques (for example, fusion bonding, flip chip, wire bond, etc.) can be applied. Heterogeneous integration methods such

978-1-6654-1392-3/21 $31.00 © 2021 IEEE

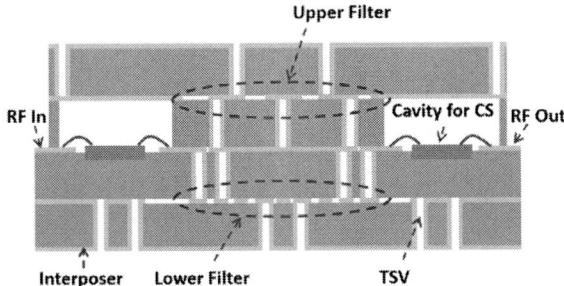

Fig. 1. Cross-sectional view of the 3D TSV-MEMS based heterogeneous integration architecture.

as high density μ-bump interconnect are also compatible with this architecture. Cavities can be formed around these chips to preserve the RF performance.

MEMS RF filters can be formed by using two interposers. Two sets of filters can be realized by four stacked interposers. Other IPDs such as capacitors and resistors can also be integrated in the interposer. The input/output of the module can be grounded coplanar waveguide (GCPW), ball grid array (BGA), or quad flat no-lead package (QFN), depending on different applications. GCPW structure is shown in Fig. 1.

Interposers are stacked vertically using wafer level bonding technique. Specifically, every two interposers are stacked by Au/Au fusion bonding, and two stacked modules are bonded by Au/Sn based eutectic bonding. Four interposers support two layers of MEMS filters and functional chips at the same time, providing enough design variables for most RF applications. However, more layers of interposers can be stacked if needed, since Au/Sn Compound metal is formed at around 300 ℃ and melt at much higher temperature.

B. 3D TSV-MEMS Fabrication Process

The 3D TSV-MEMS based heterogeneous integrated module is implemented using MEMS process and C2W stacking, as shown in Fig. 2. The whole process can be divided into four parts: interposer fabrication, wafer to wafer (W2W) bonding, C2W integration and low temperature W2W technique.

Interposer fabrication flow is shown in Fig. 2 (a). Every interposer shares the same fabrication process. We start with a standard 6-inch double side polished silicon wafer. Resistivity of the wafer is ~10000 Ω·cm. Wafer is cleaned using RCA first. TSVs are formed by photo lithography and deep reactive ion etching (DRIE). Chip cavities can be patterned and etched using similar process before TSV formation. Chip cavities are needed in two situations. One is for chip fan-out assembly and the other is to create more space when 250 μm thick cavity on top is not enough. Then the wafer is thinned to 250 μm. 2000 Å silicon oxide layer is deposited on both side of the wafer. Followed by 1000 Å titanium (Ti) and 2000 Å gold severing as seed layer. After that, 3 μm gold layer is electroplated on both sides of the wafer, and inside the TSVs as well. Finally, RDLs are patterned by lithography and dry etch. If more than one layer of RDLs are involved, RDLs can be patterned at the beginning. The rest process is the same as described above.

Every two interposers are bonded using wafer level bonding process. Two different bonding methods are used in our process: the Au/Au fusion bonding and the Au/Sn eutectic

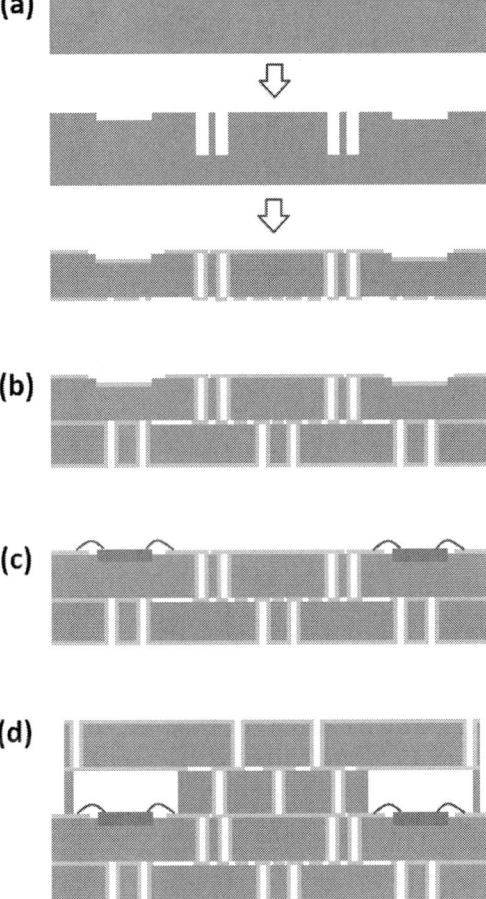

Fig. 2. 3D TSV-MEMS Fabrication Process. (a) MEMS interposer process. (b) Wafer to wafer fusion bonding process. (c) Die to wafer process. (d) Wafer to wafer eutectic bonding process.

bonding. Au/Au bonding offers better RF transmission performance, while Au/Sn is carried out at lower temperature. Two interposers are stacked first using Au/Au bonding, as shown in Fig. 2 (b). Fusion compression at 300°C for 2 hours is applied and form the lower and upper parts of the module. Bonding accuracy is within 3 μm.

Then C2W integration process is used to assemble the CS or IC chips onto the interposer. Different integration techniques can be employed. Fig. 2 (c) shows a typical method for MMIC chip integration. The chip can be solder bonded into the cavity. The interconnection between the chip and the interposer is realized by wire bonding. After wafer level electrical testing, a low temperature bonding process is performed to finish the 3D TSV-MEMS module. As shown in Fig. 2 (d), the wafers are bonded by Au/Sn eutectic bonding at 300 °C for 20 minutes.

III. 6-18GHZ SWITCHABLE FILTER BANK MODULE

A. 6-18 GHz Switchable Filter Bank Design

A 6-18 GHz broadband switchable filter bank is designed based on the 4-layer TSV-MEMS architecture. Schematic view of the filter bank is shown in Fig. 3 (a). For the proposed switchable filter, six bandpass filters (BPFs) are implemented.

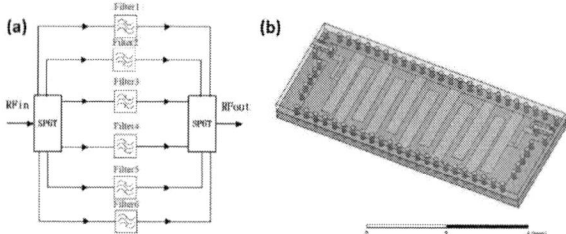

Fig. 3. (a) Schematic view of 6-18 GHz Switchable Filter Bank. (b) Proposed 3D filter with SIW cavity.

Three filters are realized in the stacked 1st and 2nd interposer, while the other three are realized in 3rd and 4th layer. Cavities are etched on the 2nd and 3rd layers respectively to integrate two single-pole six-through (SP6T) switches. The input and output of the switchable filter bank are fed by GCPW. Introduction of the cavities make switches and GCPW at the same height level, minimizing the bonding path of the gold wires. RF parasitic effect is alleviated in such configuration. Strip-line based MEMS filters are employed in this filter bank, as shown in Fig. 3 (b). TSV fences are fabricated around the bandpass filter, serving as a metal cavity. The diameter of TSV is 120μm and the pitch is 300μm.

B. Measurement Results and Discussion

The proposed switchable filter bank is fabricated based on the 3D TSV-MEMS process, as shown in Fig. 4 (a). The overall size is 20mm×11mm×1mm. Fig. 4 (b) shows the cross-sectional view of the 4-layer stacked module. SP6T switch is integrated on the 2nd layer. Cavity in the 3rd layer provide around 200 μm air space, ensuring the MMICs working properly. Switches are connected to the 2nd layer via wire bond. Insets in Fig. 4 (b) shows the wire bonded SP6T switch. Filters are connected to the switches via TSVs.

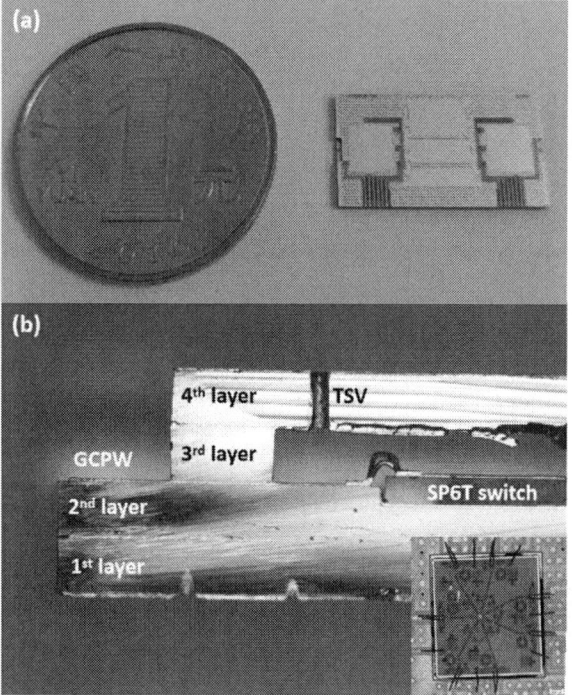

Fig. 4. 6-18 GHz switchable filter bank: (a) Top view of the module, (b) Cross-sectional view of the module.

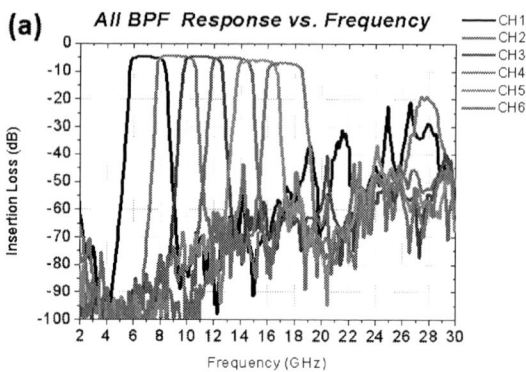

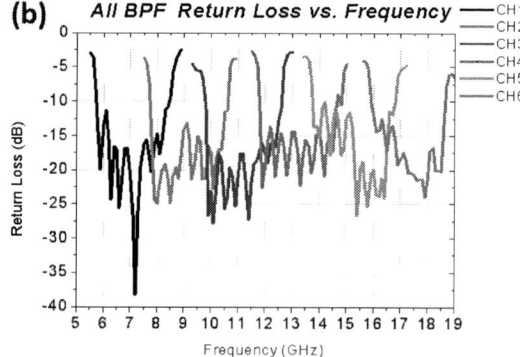

Fig. 5. Measurement results of 6-18 GHz switchable filter bank: (a) Insertion loss, (b) Return loss.

Notches are etched on the 3rd and 4th layer, revealing the GCPW.

The 6-18GHz switchable filter is measured using vector network analyzer with ground-signal-ground RF probes. As shown in Fig. 5 (a), insertion losses of six channels are 6 dB, 6 dB, 6 dB, 6 dB, 7 dB and 10 dB, respectively. The measured insertion loss include the loss introduced by two switches, which is around 2 dB each. The 16-18 GHz filter can be further optimized. The filter bank shows a wide tuning range, and the measured in-band flatness is within 1.5 dB for the first five channels and 2.5dB for the last one. Due to the TSV based SIW shielding, the measured isolation is higher than 40 dB in most cases. This performance meets the industrial specifications for many applications. Fig. 5 (b) shows the return losses of six channels. All channels are measured to be higher than 15 dB at their operational passbands. The performance of the filter bank can be further improved by minimizing fabrication tolerance and reducing parasitic effect of components.

More detailed measurement results are listed in Table I. Table II summarizes the above measured results and compares them with conventional switchable filter banks. The switchable filter bank built by 3D TSV-MEMS process achieves competitive performances in terms of insertion loss, isolation and return loss comparing to the conventional filter bank made by MCM. Moreover, the volume and weight of the proposed filter bank are reduced by almost two orders of magnitude.

TABLE I.   6-18 GHz FILTER BANK MEASUREMENT RESULTS

Channel	CH1	CH2	CH3	CH4	CH5	CH6
Frequency (GHz)	6.0-8.0	8.0-10.0	10.0-12.0	12.0-14.0	14.0-16.0	16.0-18.0
In-band flatness (dB)	≤1.5	≤1.5	≤1.5	≤1.5	≤1.5	≤2.5
Insertion loss (dB)	≤6.0	≤6.0	≤6.0	≤6.0	≤7.0	≤10.0
Isolation (dBc)	≥40@5.2 GHz ≥40@8.9 GHz ≥50@5.05 GHz ≥50@9.0 GHz	≥40@7.35 GHz ≥40@11.05 GHz ≥50@7.2GHz ≥50@11.3 GHz	≥40@9.15 GHz ≥40@13.1 GHz ≥50@9.0GHz ≥50@13.2 5GHz	≥40@11.2 GHz ≥40@15.3 GHz ≥50@11.0 GHz ≥50@15.4 5GHz	≥40@13.0 5GHz ≥40@17.2 GHz ≥50@12.9 GHz ≥50@17.4 GHz	≥40@15.3 5GHz ≥40@19.3 GHz ≥50@15.1 5GHz ≥50@19.4 5GHz
VSWR	≤2.0					

TABLE II.   COMPARISION BETWEEN THE TSV-MEMS BASED FILTER BANK AND CONVENTIONAL FILTER BANK

	TSV-MEMS based filter bank	Conventional filter bank
Insertion loss (dB)	6-10	6-8
Isolation between filters (dB)	40	NA
Return loss (dB)	>17	>10
Filter configurations	Four-layer stacked silicon substrate	Single-layer planar filters on Rogers 3210
Self-package	YES	NO
Size	20mm×11mm×1mm	40mm×36mm×10mm

straightforward MEMS process. It is highly feasible and cost effective for mass production.

IV.   CONCLUSION

We present a 3D TSV-MEMS based heterogeneous integration architecture for compact RF micro-system. This architecture provides an excellent way to realize embedded, high-quality passives to highly integrated, compact and packaged RF module and system. This technology is suitable for a broad range of RF applications up to 60 GHz. A 6-18 GHz switchable filter bank is demonstrated based on the 3D TSV-MEMS architecture. The size of the module is 20mm × 11mm × 1mm, only 1.5% in volume comparing to its counterpart made by conventional method. The architecture presented in this paper is built on relatively mature and

REFERENCES

[1] S.Rangan, T. S Rappaport and E. Erkip, "Millimeter-wave cellular wireless networks: potentials and challenges," Proceedings of the IEEE 2014, vol. 102, pp. 366-385, 2014.

[2] D. S. Green, C. L. Dohrman, D. J emmin, Y. Zheng and T. H. Chang "A Revolution on the Horizon from DARPA: Heterogeneous Integration for Revolutionary Microwave\/Millimeter-Wave Circuits at DARPA: Progress and Future Directions," IEEE Microwave Magazine, vol. 18, pp. 44-59, 2017.

[3] J. Xu, X. Zhang, H. Li, et al, "Ultracompact multichannel bandpass filter based on trimode dielectric-loaded cavities," IEEE Trans. Microw. Theory Tech., vol. 5, pp. 1668-1677, 2020.

[4] D. Shojaei-Asanjan and R. R. Mansour, "The sky's the limit: switchable RF MEMS filter," IEEE Microwave Magazine, vol. 18, pp. 100–106, 2017.

[5] J. Xu and Y. Zhu, "Tunable bandpass filter using a switched tunable diplexer technique," IEEE Trans. Ind. Electron., vol. 64, pp. 3118-3126, 2017.

[6] E. Crespin, et al. "Fully integrated switchable filter banks," Proc. IEEE/MTT-S Int. Microwave Symp., pp. 17-22, 2012.

The effect of annealing time on the mechanical properties of TSV-Cu

Yadong LI
Institute of Electronics Packaging
Technology & Reliability
Beijing University of Technology
Beijing, China
yadong_li@emails.bjut.edu.cn

Pei CHEN*
Institute of Electronics Packaging
Technology & Reliability
Beijing Key Laboratory of Advanced
Manufacturing Technology
Beijing University of Technology
Beijing, China
peichen@bjut.edu.cn

Fei QIN
Institute of Electronics Packaging
Technology & Reliability
Beijing Key Laboratory of Advanced
Manufacturing Technology
Beijing University of Technology
Beijing, China
qfei@bjut.edu.cn

Abstract—Relying on the advantages like high density integration and stacking, through silicon via (TSV) as an advanced technology has been widely used for three-dimensional integrated circuits (3D ICs). Copper filled in through silicon via (TSV-Cu) should be annealed to eliminate defects and residual stress after electroplating, which is beneficial to improve the performance of transistors. The subsequent deformation and thermo-mechanical reliability of TSV devices would be affected due to the change of the mechanical properties of TSV-Cu via annealing. With the help of nanoindentation characterization and finite element method (FEM) inversion, this paper studied the stress-strain relationship of TSV-Cu on five different annealing time at 400°C. The results showed that TSV-Cu had been softened by annealing and this phenomenon would be intensified with annealing time extended. However, up to 200 min annealing time resulted in indentation load and yield strength increasing instead of decreasing. This work provided theory guide for calculating the deformation and stress of TSV structure and reducing thermo-mechanical reliability issues.

Keywords—TSV-Cu, annealing time, nanoindentation, mechanical properties, constitutive relationship

I. INTRODUCTION

TSV is a critical technique used in 3D micro-electronical stacking and packaging due to its advantages of higher electrical performance, lower form and multifunction. After forming holes in silicon (Si) substrates by laser or dry etching, it is necessary to electroplate and fill conductive material to accomplish electrical interconnection and mechanical support of devices. Copper (Cu) is one of the most common filling materials due to its excellent electrical and mechanical performances [1]. There are many defects such as holes or pores, and high residual stress in TSV-Cu after electroplating, which is harmful to devices and transistors [2-4]. Defects and residual stress can be eliminated by annealing. Besides, TSV-Cu has also suffered thermal loading in the subsequent process and normal service. The coefficient of thermal expansion (CTE) of Cu and Si are 17×10^{-6} /°C and 2.8×10^{-6} /°C, respectively, the huge thermal mismatch results in that thermal stress is inevitable and emerges some risk issues such as TSV-Cu protrusion, multiply interface cracking and so on [5].

The stress and deformation states of TSV-Cu is subjected to its service conditions, like temperature and mechanical loads, and mechanical parameters, like Young's modulus (E) and yield strength (σ_y). Different preparation processes, such as electroplating solution, current waveform and density, etc., and post-processing, like annealing temperature and time, have influenced on the mechanical properties of TSV-Cu. Okoro et al. [6] researched the evolution of mechanical properties and microstructure of TSV-Cu on three electroplating solutions and impurity particles, and they found the introduction of impurity particles was beneficial free-void filling but would cause a higher residual stress. Chen et al. [7] found that annealing resulted in Cu protrusion and reduced the yield strength of TSV-Cu by increasing its grain size. Budiman et al. [8] studied the distribution of mechanical stresses in Cu and Si before and after annealing. The results suggested that the stresses in TSV structure had reduced after post-annealing, which was beneficial to the performance of devices. Therefore, annealing can be regarded as a feasible method to improve performance on TSV devices by controlling the microstructure and mechanical properties TSV-Cu, and annealing time will also affect them, but related research is not sufficient.

Nanoindentation can be used to characterize mechanical properties of TSV-Cu due to its submicron structure results in traditional testing would not be suitable. The continuous stiffness measurement (CSM) on nanoindentation proposed by Oliver and Pharr [9] has been a mature characterization method for film and bulk materials [10], load-displacement curve (*P-h* curve), hardness (*H*), contact stiffness (*S*) and *E* can be obtained by this method. However, stress-strain curve cannot be directly described by nanoindentation test. Antunes et al. [11] presented a new approach of inversion analysis by using numerical simulation in nanoindentation processes to obtain the elastoplastic behaviors of materials. The validity of the inversion analysis method also had been verified by experiments results. Using the similar method, constitutive relationship of TSV-Cu also can be determined in our research.

In this paper, we focused on the constitutive behavior of TSV-Cu before and after annealing. The *P-h* curves, *H*, *E* and *S* of TSV-Cu on five annealing time at 400°C were successfully measured by nanoindentation. An axisymmetric two-dimension FEM model had been used for inverse analysis. The power law stress-strain relationships of TSV-Cu on five annealing time were determined finally, which was convenient to calculate the deformation and stress of TSV devices in subsequent processes.

II. EXPERIMENTS AND INVERSION

A. Samples Preparation

Fabrication of TSVs on a Si wafer requires 6 key steps, as shown in Fig. 1.

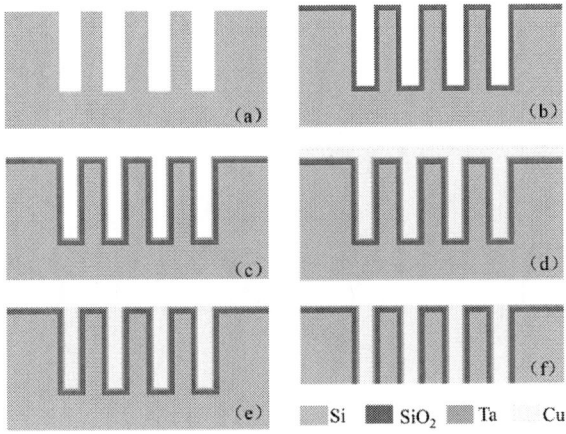

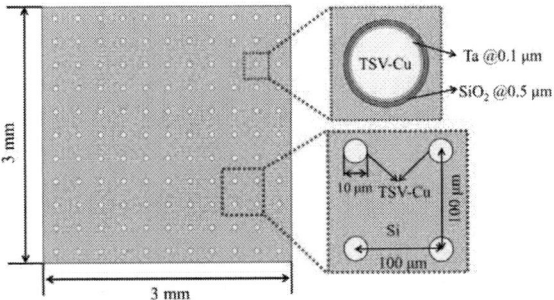

Fig. 1. The flow chart of the TSV fabrication process (a) forming vias, (b) deposition of insulating layer, (c) deposition of barrier and seed layers, (d) filling of conductive material, (e) frontside CMP, (f) backside thinning.

a) Forming vias: TSV blind vias can be fabricated on a side of the Si wafer by deep reactive ion etching (DRIE) of laser ablation, shapes of vias including straight and tapered.

b) Deposition of insulating layer: Silicon dioxide is a commonly insulating material that can be deposited on via walls using thermal oxidation or plasma enhanced chemical vapor deposition (PECVD).

c) Deposition of barrier and seed layers: Titanium (Ti), tantalum (Ta) and their nitrides are common barrier materials. Both barrier and Cu seed layers can be deposited by physical vapor deposition (PVD).

d) Filling of conductive material: As an excellent conductive material, Cu should be filled in TSV by electroplating to achieve electrical interconnection.

e) Frontside CMP: Overburden Cu due to overfilling should be removed by chemical mechanical polishing (CMP).

f) Backside thinning: TSV on wafer backside should be exposed by grinding and thinning.

The entire Si wafer is divided into individual chip with 3 $\times$ 3 mm² after TSV fabrication completed. The diameter and depth of TSV are 10 μm and 100 μm, respectively. Tantalum (Ta) with 0.1 μm thickness and silicon dioxide (SiO₂) with 0.5 μm thickness are used as barrier and insulting layers, as shown in Fig. 2.

B. Annealing

The annealing experiments were completed by the TL1200 tube furnace, which was made in Nanjing Boyuntong Instrument Technology Co., Ltd. Annealing should to be carried out in a vacuum environment to prevent TSV-Cu being oxidation and reduce the errors of experiments. With ramp rate of 5°C/min, TSV samples were heated to 400°C from room temperature. After keeping at 400°C for a while, and they should be cooled naturally to room temperature. Five annealing time (refers to holding time at 400°C in this paper) was used for testing, including 30, 60, 90, 120, and 200 min, respectively. It is necessary to remove the residual protrusion of TSV-Cu after annealing. No less than #3000 mesh sandpaper was used for rough grinding and fine grinding successively, and 0.5 μm diamond polishing liquid is used to provide a smooth surface for nanoindentation tests.

Fig. 2. The schematic diagram of TSV arrays and structure.

C. Nanoindentation

Nanoindentation is one of the promising methods to characterize mechanical properties of materials, which also be named depth-sensing indentation. In this paper, a Berkvich indenter made of diamond assembled in Agilent Nano Indenter G200 was used for testing. *P-h* curves, *H*, *E*, *S* can be obtained directly based on nanoindentation test. The maximum indention depth (h_m), loading rate and strain rate were set as 800 nm, 0.5 mN/s and 0.05 s⁻¹, respectively. No less than five repeated tests in each condition were conducted to ensure the accuracy of experiments, and the average value of *H* and *E* can be calculated in the range of 600-700 nm according to the measurement principle of nanoindentation CSM technique.

D. FEM Model and Inversion Process

Based on nanoindentation response and FEM model built by ABAQUS commercial software, the plastic parameters of the sample can be obtained by an inverse analysis. In nanoindentation process, due to the barrier and insulation layers have less effect on the mechanical response of TSV-Cu, and whose thickness is much smaller than the radius of TSV-Cu, so both of them can be ignored in FEM model. A two-dimension axisymmetric equivalent FEM model as shown in Fig. 3(a) was used to simulate the indentation and inversion process.

The meshes near indenter should be refined to save storage space and improve calculation speed and accuracy. The radius of TSV-Cu, thickness of Si and model height are set at 5 μm, 30 μm and 35 μm, respectively. The berkvich indenter is equivalent to a cone with an inclination of 70.3°. The power law constitutive relationship as shown in (1) was used to depict stress-strain relationship of TSV-Cu and build FEM model. In the simulation process, indentation depth, mechanical and displacement boundaries condition should be as same as experiments.

$$\begin{cases} \sigma = E\varepsilon & \sigma < \sigma_y \\ \sigma = \sigma_y \left(1 + \frac{E}{\sigma_y}\varepsilon_p\right)^n & \sigma \geq \sigma_y \end{cases} \quad (1)$$

where, the deformation process divided in elastic and plastic strengthen stages, stress σ and strain ε follow a liner positive correlation and the proportionality factor is E when σ is less than the yield strength σ_y. The plastic strain ε_p is defined to describe the plastic strength process, and n is the hardening exponent, when σ is more than σ_y.

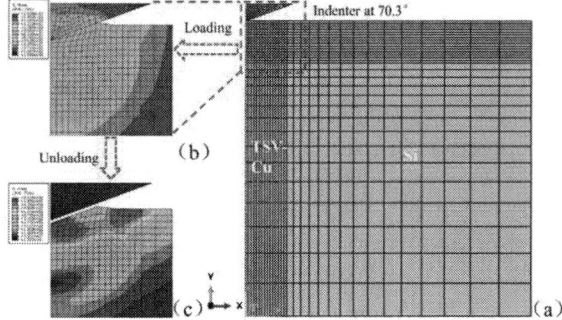

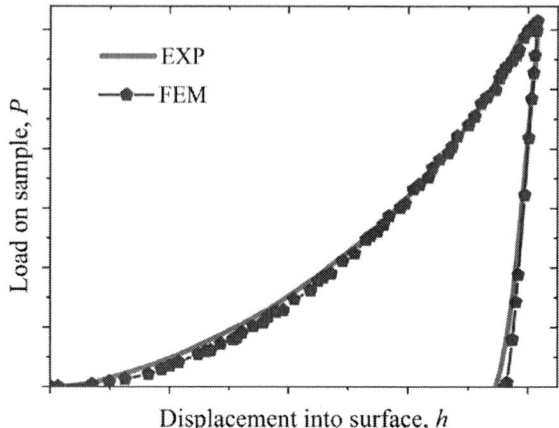

Fig. 3. (a) the mesh of FEM model, (b) partial contour plot of von Mises stress distribution for TSV-Cu at maximum indentation depth, (c) partial contour plot of von Mises stress distribution for TSV-Cu after unloading.

Fig. 5. Comparison to the *P-h* curves of FEM and experiment.

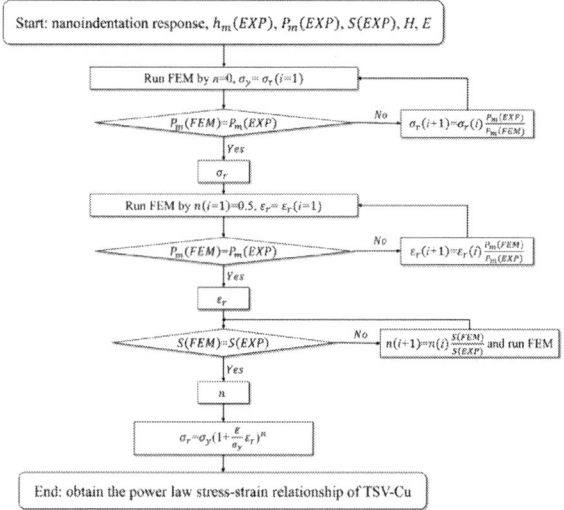

Fig. 4. FEM inversion flowchart to determine stress-strain relationship of TSV-Cu

Fig. 4 illustrates the algorithm flowchart to obtain stress-strain parameters of TSV-Cu based on FEM inversion. The principle of inversion analysis is that the power law relationship can be determined by obtaining representative point (σ_r, ε_r) using FEM and solving (1). The specific process is as following.

1) According nanoindentation response (*P-h* curve, $h_m(EXP)$, $P_m(EXP)$, $S(EXP)$, H, E), build FEM model.

2) Assume material as ideal elastoplastic (n=0), using (2) to calculate initial representative stress σ_r and regard it as initial yield stress, run FEM model to obtain *P-h* curve and $P_m(FEM)$.

$$\frac{E_r}{H} = 0.231\left(\frac{E_r}{\sigma_r}\right) + 4.910 \quad (2)$$

$$\frac{1}{E_r} = \frac{1-v^2}{E} + \frac{1-v_i^2}{E_i} \quad (3)$$

where, E_r can be calculated by (3) and named reduced modulus, H represents the hardness of sample, Poisson's ratios v is 0.3 for Cu , E_i and v_i are 1141 GPa and 0.07, and represent Young's modulus and Poisson's ratios of indenter.

3) Using iteration to adjust σ_r repeatedly, compare with maximum load P_{max} in maximum indentation depth of FEM and experiment until $P_m(EXP)=P_m(FEM)$ (relative error of two less than 0.5%, same below), Fig. 3(b) and (c) show the distribution of von Mises stress at maximum indentation depth and after unloading. The maximum von Mises stress is located directly below the indenter and the indentation edge of TSV-Cu is slightly raised.

4) Assume material as power law constitutive model (*n* considered as 0.5 initially), using (4) to calculate initial representative strain ε_r, run FEM model to obtain *P-h* curve, $P_m(FEM)$ and $S(FEM)$.

$$\varepsilon_r = exp(-3.19 + 166.7/(E_r/\sigma_r + 177.3)) \quad (4)$$

5) Using iteration to adjust ε_r repeatedly, compare with maximum load P_{max} in maximum indentation depth of FEM and experiment until $P_m(EXP)=P_m(FEM)$.

6) Using iteration to adjust *n* repeatedly, compare S in maximum indentation depth (slope of unloading curve in initial stage) of FEM and experiment until $S(EXP)=S(FEM)$. In Fig. 5, it is obvious that two *P-h* curves of FEM and experiment coincide better, which represents materials parameter in FEM can represent the true value.

7) Obtain power law stress-strain equation of TSV-Cu by inputting σ_r, ε_r and *n* into plastic stage in (1).

III. RESULTS AND DISCUSSION

Fig. 6 describes the nanoindentation response of TSV-Cu before and after annealing, annealing time are 30, 60, 90, 120, and 200 min, respectively. Fig. 6(a) illustrates that a higher indentation depth results in a higher load, and the load keeps declining as annealing time extended from 30min to 120 min, except it at 200 min. Fig. 6(b) and (c) describe the variation of H and E with indentation depth, both of two have violent fluctuations in start 100 nm due to the influence of surface roughness, and then become steady when indentation depth is over 100 nm. Both H and E in the range of 600-700 nm keep decreasing with annealing time extended from 30 min to 200 min. Fig. 6(d) quantitatively represents the values of H and E on different annealing time. It is obvious that H and E decrease after annealing, which means TSV-Cu has softened.

978-1-6654-1392-3/21 $31.00 © 2021 IEEE

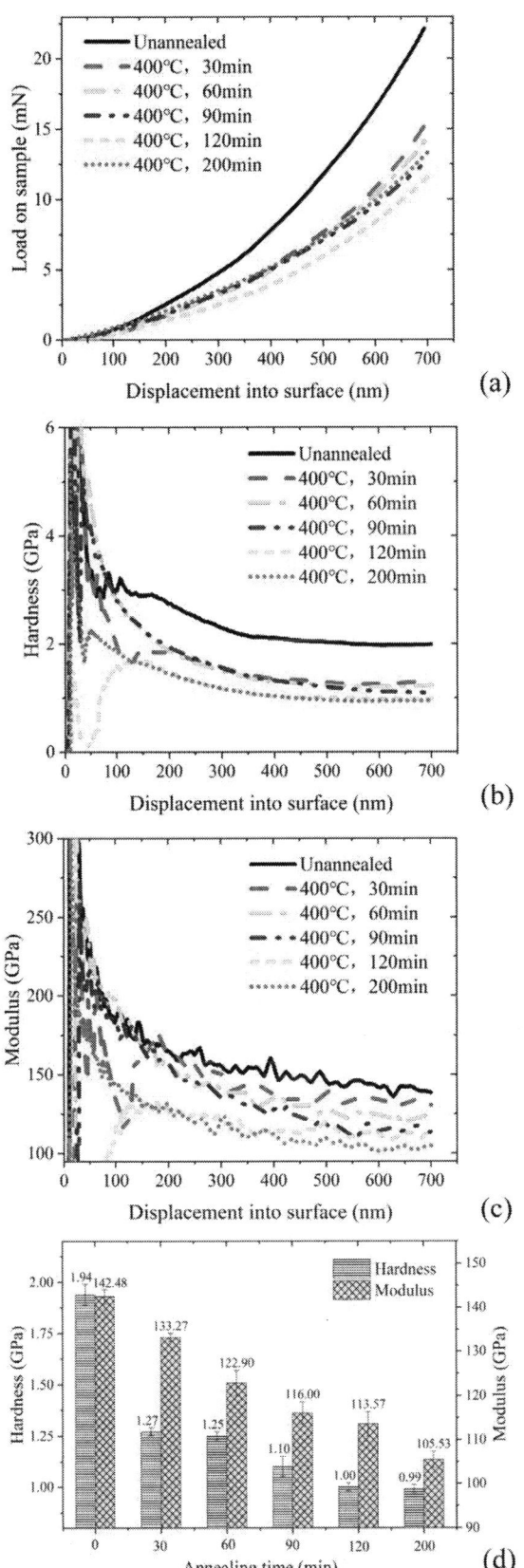

Compared with unannealed sample, the value of H decreased from 1.94 GPa to 1.27 GPa with 34.5% reduction, and the value of E decreased from 142.48 GPa to 133.27 GPa with 6.4% reduction after annealing at 400°C, 30 min. As the annealing time is extended from 30 min to 200 min, H decreased from 1.27 GPa to 0.99 GPa with 22% reduction, and E decreased from 133.27 GPa to 105.53 GPa with 20.8% reduction, which means that the softening phenomenon became more and more obvious. It should not be ignored that the value of H has less variation from 120 min to 200 min, but E keeps decreasing.

According to reverse analysis method shown in Fig.4, Fig. 7 and table I show the power law stress-strain curves and mechanical parameters of TSV-Cu on five annealing time, respectively. It can be seen that the value of σ_y keeps decreasing from 30 to 120 min but rebound in 200 min, which have the same variation on load. A higher annealing time results in a higher n. Fig. 7 illustrates the stress-strain curves of TSV-Cu on five annealing time and the inner picture in Fig. 7 shows the variation of σ_y.

The phenomenon of grain growth after annealing had reported in previous research [12]. Okoro et al. found the average grain size of copper filled in TSV increases from 1.7 μm to 2.3 μm after 420°C and 300°C annealing, which resulted in the hardness reduced from 2.2 GPa to 1.9 GPa. The mechanical properties of material strongly depended on its microstructure, reduction in grain boundaries of TSV-Cu is a main reason for reduction in resistance of dislocation movement, which expressed as reduction in H, E and σ_y macroscopically.

With the extension of annealing time, TSV-Cu will keep softening drove by total free energy, whose principle is the same as before. When annealing time arrived to 200 min, however, the value of σ_y increased, even exceeding the value at 90 min. This phenomenon is probably because the excessively long annealing time results in the accumulation of residual stress in TSV-Cu and changing the nanoindentation response. Thus, the TSV-Cu becomes easier to deform suffered annealing, which is advantageous to reduce the risk of 3D packaging structure. Therefore, the annealing time as long as 120 min is preferable, but higher annealing time exceeding 120 min is not recommended, because it probably causes new risks for structure.

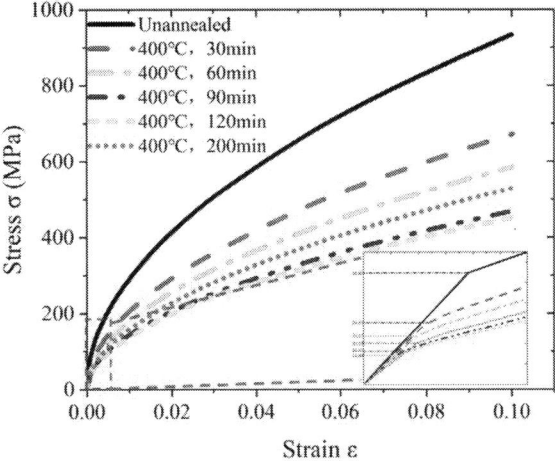

Fig. 7. The power law stress-strain curves of TSV-Cu on five annealing time.

Fig. 6. Nanoindentation results on five annealing time at 400°C (a) displacement-load curves, (b) displacement-hardness curves, (c) displacement-modulus curves, (d) the value of hardness and modulus.

978-1-6654-1392-3/21 $31.00 © 2021 IEEE

TABLE I.   THE MECHANICAL PARAMETERS OF TSV-CU ON FIVE ANNEALING TIME

Annealing Time /min	Young's Modulus E /GPa	Yield Strength σ_y /MPa	Hardening Exponent n
Unannealed	142.48	54.82	0.5091
30	133.27	29.43	0.5119
60	122.90	23.87	0.5124
90	116.00	16.02	0.5131
120	113.57	14.49	0.5168
200	105.53	20.19	0.5237

IV. CONCLUSIONS

This paper concentrated on the evolution of mechanical properties of TSV-Cu before and after 400°C annealing, and five annealing time was used to study. H and E of TSV-Cu were characterized by nanoindentation, and its power law constitutive behavior was determined by FEM inversion. The results showed that TSV-Cu had obvious softening phenomenon after annealing, and H, E and σ_y were all reduced, which is because the Cu grains re-crystallized and the grain boundaries reduced after annealing resulted in the resistance to dislocation movement reduced. The softening phenomenon continued to occur as the annealing time extended from 30 min to 200 min. When the annealing temperature reached 200 min, the P-h curve and σ_y rebounded, which was probably due to the changes of residual stress and nanoindentation response on a longer annealing time.

ACKNOWLEDGEMENT

The authors are thankful to the National Natural Science Foundation of China (No. 11672009) and HiSilicon Technologies CO., LIMITED for their supports.

REFERENCES

[1] Z. Wang, "Microsystems using three-dimensional integration and TSV technologies: Fundamentals and applications," Microelectron. Eng. vol. 210, pp. 35-64, March 2019.

[2] N. Lin, J. Miao and P. Dixit, "Void formation over limiting current density and impurity analysis of TSV fabricated by constant-current pulse-reverse modulation," Microelectron. Reliab. vol. 53, pp. 1943-1953, June 2013.

[3] F. Li, C. Xiao, H. He, J. Li and W. Zhu, "Investigation on the defect induced thermal mechanical stress for TSV," 17th International Conference on Electronic Packaging Technology (ICEPT), pp. 713-715, 2016.

[4] C. Rao, T. Wang, Y. Peng, J. Cheng, Y. Liu, S.K. Lim and X. Lu, "Residual Stress and Pop-Out Simulation for TSVs and Contacts in Via-Middle Process," IEEE Trans. Semicond. Manuf. vol. 30, pp. 143-154, May 2017.

[5] Z. Cheng, Y. Ding, L. Xiao, X. Wang and Z. Chen, "Comparative evaluations on scallop-induced electric-thermo-mechanical reliability of through-silicon-vias," Microelectron. Reliab. vol. 103, 113512, November 2019.

[6] C. Okoro, R. Labie, K. Vanstreels, A. Franquet, M. Gonzalez, B. Vandevelde, E. Beyne, D. Vandepitte and B. Verlinden, "Impact of the electrodeposition chemistry used for TSV filling on the microstructural and thermo-mechanical response of Cu," J. Mater. Sci. vol. 46, pp. 3868-3882, February 2011.

[7] S. Chen, F. Qin, T. An, P. Chen, B. Xie and X.Q. Shi, "Protrusion of electroplated copper filled in through silicon vias during annealing process," Microelectron. Reliab. vol. 63, pp. 183-193, May 2016.

[8] A.S. Budiman, H.A.S. Shin, B.J. Kim, S.H. Hwang, H.Y. Son, M.S. Suh, Q.H. Chung, K.Y. Byun, N. Tamura, M. Kunz and Y.C. Joo, "Measurement of stresses in Cu and Si around through-silicon via by synchrotron X-ray microdiffraction for 3-dimensional integrated circuits," Microelectron. Reliab. vol. 52, pp. 530-533, November 2012.

[9] W.C. Oliver and G.M. Pharr, "An improved technique for determining hardness and elastic modulus using load and displacement sensing indentation experiments," J. Mater. Res. vol. 7, pp. 1564-1583, January 1992.

[10] M. Dao, N. Chollacoop, K.J.V. Vliet, T.A. Venkatesh and S. Suresh, "Computational modeling of the forward and reverse problems in instrumented sharp indentation," Acta Mater. vol. 49, pp. 3899–3918, August 2001.

[11] J. Antunes, J. Fernandes, L. Menezes and B. Chaparro, "A new approach for reverse analyses in depth-sensing indentation using numerical simulation," Acta Mater. vol. 55, pp. 69-81, October 2007.

[12] C. Okoro, K. Vanstreels, R. Labie, O. Lühn, B. Vandevelde, B. Verlinden and D. Vandepitte, "Influence of annealing conditions on the mechanical and microstructural behavior of electroplated Cu-TSV," J. Micromech. Microeng. vol. 20, 045032, February 2010.

Study on Current Carrying Capacity of a Novel Interconnect Material ZrTe$_3$

Xiaokun Wen[1,2]
[1] *Center for Joining and Electronic Packaging, School of Materials Science and Engineering, State Key Laboratory of Materials Processing and Die&Mould Technology, Huazhong University of Science and Technology*
Wuhan, China
2 *Shenzhen R&D Center of Huazhong University of Science and Technology*
Shenzhen, China
351936812@qq.com

Liangyi Ni[1,2]
[1] *Center for Joining and Electronic Packaging, School of Materials Science and Engineering, State Key Laboratory of Materials Processing and Die&Mould Technology, Huazhong University of Science and Technology*
Wuhan, China
2 *Shenzhen R&D Center of Huazhong University of Science and Technology*
Shenzhen, China
735497822@qq.com

Wenyu Lei[1,2]
[1] *Center for Joining and Electronic Packaging, School of Materials Science and Engineering, State Key Laboratory of Materials Processing and Die&Mould Technology, Huazhong University of Science and Technology*
Wuhan, China
2 *Shenzhen R&D Center of Huazhong University of Science and Technology*
Shenzhen, China
308563494@qq.com

Li Yang[1,2]
[1] *Center for Joining and Electronic Packaging, School of Materials Science and Engineering, State Key Laboratory of Materials Processing and Die&Mould Technology, Huazhong University of Science and Technology*
Wuhan, China
2 *Shenzhen R&D Center of Huazhong University of Science and Technology*
Shenzhen, China
1332159920@qq.com

Yuan Liu[1,2]
[1] *Center for Joining and Electronic Packaging, School of Materials Science and Engineering, State Key Laboratory of Materials Processing and Die&Mould Technology, Huazhong University of Science and Technology*
Wuhan, China
2 *Shenzhen R&D Center of Huazhong University of Science and Technology*
Shenzhen, China
1934445155@qq.com

Pengzhen Zhang[1,2]
[1] *Center for Joining and Electronic Packaging, School of Materials Science and Engineering, State Key Laboratory of Materials Processing and Die&Mould Technology, Huazhong University of Science and Technology*
Wuhan, China
2 *Shenzhen R&D Center of Huazhong University of Science and Technology*
Shenzhen, China
1581566719@qq.com

Haixin Chang[1,2]
[1] *Center for Joining and Electronic Packaging, School of Materials Science and Engineering, State Key Laboratory of Materials Processing and Die&Mould Technology, Huazhong University of Science and Technology*
Wuhan, China
2 *Shenzhen R&D Center of Huazhong University of Science and Technology*
Shenzhen, China
hxchang@hust.edu.cn

Wenfeng Zhang*[1,2]
[1] *Center for Joining and Electronic Packaging, School of Materials Science and Engineering, State Key Laboratory of Materials Processing and Die&Mould Technology, Huazhong University of Science and Technology*
Wuhan, China
2 *Shenzhen R&D Center of Huazhong University of Science and Technology*
Shenzhen, China
wfzhang@hust.edu.cn

Abstract—Recently, the discussion about new candidate materials that can replace traditional Cu for future interconnects in integrated circuit (IC) fabrication attracts growing interests. In this paper, we have systematically investigated the breakdown current characteristic of a novel transition metal trichalcogenide (TMTCs) ZrTe$_3$ by the finite element simulation. First, an electro-thermal coupling model was established, which the Joule heat of the ZrTe$_3$ nanoribbon acts as the heat source, and both the solid heat transfer and the natural convection heat transfer are considered for the finite element simulation. Then, using the established model and experimental data, the breakdown temperature of the ZrTe$_3$ nanoribbons in the air environment has been determined. Furthermore, the distribution of both temperature and current density of ZrTe$_3$ nanoribbons during the breakdown process was investigated for failure analysis. Finally, according to the size-independent resistivity of TMTCs, the breakdown characteristic of the ZrTe$_3$ with 10nm width was investigated by further finite element simulation. Its breakdown current density can reach 83.8 MA/cm^2, which demonstrates the high potential of ZrTe$_3$ nanoribbon as a new type of interconnects material of replacing copper interconnects.

Keywords—Cu interconnect, TMTCs, ZrTe$_3$, finite element simulation

I. INTRODUCTION

The nonlinear increase in the resistivity of copper (Cu) with scaling below 40 nm would cause high power loss and instability, which becomes a serious problem for the back-end of line (BEOL) interconnection in integrated circuit (IC) fabrication[1]. Although industry has made a lot of efforts to address this challenge, continuous scaling of Cu is impossible. Thus, seeking for new materials to replace Cu attracts growing interests. Recently, transition metal trichalcogenides (TMTCs) has attracted broad attention as candidate material of replacing Cu benefitting from its metallic nature and unique quasi-one-dimensional (1D) structure, which implies less grain boundary scattering and dangling bonds relative to traditional interconnect materials. For example, TaSe$_3$ has been reported to maintain the same resistivity as the bulk even after scaling to 7 nm width[2], indicating its high competitivity with Cu at the nanometer scale. Among TMTCs family, ZrTe$_3$ is also appealing for its low resistivity, high breakdown current density and good low-frequency noise characteristics[3, 4]. However, up to now, no detailed investigation of ZrTe$_3$ for potential interconnect has been reported. In this paper, we have systematically investigated the breakdown current characteristic of ZrTe₃ nanoribbons below 10 nm by finite element simulation with Comsol software based on the size-independent resistivity of TMTCs.

In order to simulate the breakdown process of ZrTe₃ nanoribbons at the scale with experimental data by Comsol software, it is necessary to determine specific failure process during the simulation. Although it is difficult to directly simulate the transient behavior of ZrTe₃ nanoribbons during the breakdown process, it can be simplified as that the breakdown occurs when the highest temperature of ZrTe₃ nanoribbons satisfies the critical requirement, which can be defined as the failure temperature. To access that, first of all, we need to establish a thermoelectric coupling model in simulation, which should be well consistent with the real characterization environment of ZrTe₃ nanoribbons. Secondly, on the basis of the experimental data of ZrTe₃ nanoribbons characterization, the complete breakdown process can be simulated with the established model, thus to further obtain the temperature and current density distribution. Taking the highest temperature during the breakdown process as the failure temperature, the breakdown process of ZrTe₃ nanoribbons with 10 nm diameter is further simulated under the assumption that ZrTe₃ has the same size-independent resistivity similar with TaSe₃. The detailed process is as follows.

II. RESULTS AND DISCUSSION

A. Synthesis of ZrTe₃ nanoribbons and device fabrication

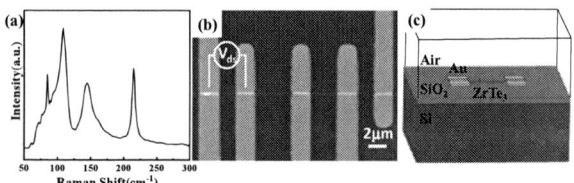

Fig. 1. (a) Raman spectra of as-synthesized ZrTe₃ single crystal; (b) The SEM image of a typical ZrTe₃ nanoribbon device. (c)Geometric model based on the dimension of ZrTe₃ nanoribbon.

We has synthesized the ZrTe₃ single crystal by traditional chemical vapor transport (CVT) method using iodine as transport medium [5]. Then, the as-synthesized ZrTe₃ single crystal were exfoliated onto Si/SiO₂ substrate with tape to obtain ZrTe₃ nanoribbons. Raman characterizations with 532nm was used to verify the quality of the as-synthesized ZrTe₃ nanoribbons. As shown in **Fig. 1.(a)**, four main Raman peaks located at 84 cm⁻¹(A_g), 108cm⁻¹(E_{2g}), 1144cm⁻¹(E_{2g}) and 217cm⁻¹(A_g) were observed, which is consistent with the reference [5]. To fabricate the ZrTe₃ nanoribbon device, standard photolithography process was adopted to make pattern, which was followed by the electrode deposition of 10/100nm Cr/Au via electron beam evaporation. The scanning electron microscope (SEM) image of a typical ZrTe₃ nanoribbon device is shown in the **Fig. 1. (b)**, which the length and width were determined as 2 um and 242 nm, respectively. The height of the nanoribbon was further measured by atomic force microscope (AFM) to be 37.6nm. Based on the experimental characterization above, The geometrical schematic diagram of the established model for Comsol simulation is shown in the **Fig. 1. (c)**.

B. The establishment of three-dimensional electro-thermal coupling model for simulation

In order to investigate the complete breakdown process of ZrTe₃ nanoribbons, an electro-thermal coupling model was established by Comsol software, in which the Joule heat as a heat source and both the heat transfer by natural convection and conduction are considered. The basic governing Equations for this model is as follows:

The continuity equation for the electrical potential:

$$\nabla \cdot \mathbf{J} = 0 \tag{1}$$

$$\mathbf{E} = -\nabla V \tag{2}$$

$$\mathbf{J} = \sigma(\mathbf{T})\mathbf{E} \tag{3}$$

Where V is the electric potential, $\mathbf{E}$ is the electric field, $\mathbf{J}$ is the current density vector and $\sigma(\mathbf{T})$ is the specific electric conductivity.

Generally, the conductivity of metal decreases with the increase of temperature. Thus, the electrical conductivity (σ) of metallic ZrTe₃ nanoribbon for the simulation can be defined as:

$$\sigma(\mathbf{T}) = 1/(\rho_0(1+\alpha(\mathbf{T}-\mathbf{T_0}))) \tag{4}$$

where ρ_0 is the resistivity at T_0, and α is the temperature coefficient of resistivity.

Joule heat generated by the current described with equation (3) can be used as the heat source ($\mathbf{Q_e}$) for the heat transfer model. The thermal field equation ignoring radiation can be described as:

$$\mathbf{Q_e} = \mathbf{J} \cdot \mathbf{E} \tag{5}$$

$$\rho C_P \frac{\partial T}{\partial t} = \nabla \cdot (k\nabla \mathbf{T}) + \mathbf{Q_e} \tag{6}$$

Where ρ is the density of ZrTe₃ nanoribbon, C_p is the heat capacity at a constant pressure, k is the thermal conductivity, $\mathbf{Q_e}$ is the heat sources, which is the Joule heat in our study.

Regardless of viscous dissipation, the governing equation of the fluid can be described as:

$$\rho_f \nabla \cdot \mathbf{u} = 0 \tag{7}$$

$$\rho_f \frac{\partial \mathbf{u}}{\partial t} + \rho(\mathbf{u} \cdot \nabla)\mathbf{u} = -\nabla p + \nabla(\mu(\nabla \mathbf{u} + (\nabla \mathbf{u})^T)) + \rho_f g \tag{8}$$

Where ρ_f is the fluid density, p is the fluid pressure, $\mathbf{u}$ is the fluid velocity field, μ is the dynamic viscosity, and g is the gravity.

To solve the system of the equations above, the following boundary conditions and initial values have been set:

i. The initial temperature of the entire system is 293K;

ii. The Si/SiO₂ substrate boundary temperature is 293K;

iii. The initial air pressure is 1 atmospheric pressure;

iv. The initial air velocity field is 0 m/s;

v. Set all the boundaries of the air as open boundaries, and the gas can flow in or out freely.

Meanwhile, various physical parameters of the ZrTe₃ nanoribbons are also required.

First, as shown in **Fig. 2**, by the temperature-dependent resistivity of the ZrTe₃ nanoribbon measured with the Physical Property Measurement System (PPMS), ρ_0= 234.69 uΩ·cm and α=0.72 uΩ·cm/K have been calculated. Since the

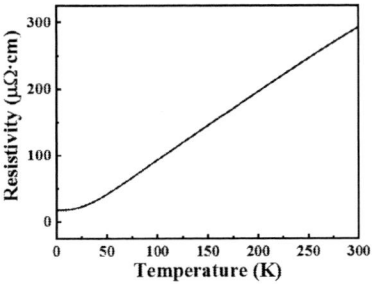

Fig. 2. The resistivity of ZrTe₃ nanoribbon as a function of temperature.

as-synthesized ZrTe₃ nanoribbons exhibit metallic property, its thermal conductivity can be obtained by the following Wiedemann–Franz law:

$$\frac{\kappa}{\sigma}=LT \tag{9}$$

Where κ is the thermal conductivity of ZrTe₃ nanoribbon, σ is the conductivity of ZrTe₃ nanoribbon, L is the Lorentz number.

Considering the van der Waals gap of two-dimensional materials, there will be contact thermal resistance (R_{eq}) at the interface between ZrTe₃ nanoribbons and SiO₂/Si substrate. A summary of the physical properties related to ZrTe₃ nanoribbon are shown in Table I. The physical parameters of other materials are set by Comsol software built-in database.

TABLE I. PHYSICAL PROPERTIES OF ZRTE₃

PHYSICAL PROPERTIES	PHYSICAL PROPERTIES OF ZrTe₃		
	Symbol	vale	units
Density[6]	ρ	6970	kg/m3
Specific heat[7]	Cp	0.116	J/kg·K
the resistivity at T_0	$\rho 0$	234.69	uΩ·cm
resistivity temperature coefficient	α	0.72	uΩ·cm/K
Electric conductivity	$\sigma(T)$	$\dfrac{1}{\rho 0(1+\alpha(T-T_0))}$	uΩ·cm
Lorentz number	L	2.44×10^{-8}	W·Ω·K^{-2}
Thermal conductivity	κ	$\sigma(T)$ LT	W/m·K
Contact thermal resistance	R_{eq}	10^{-8}	K·m^2/W

C. Failure temperature determination of ZrTe₃ nanoribbon

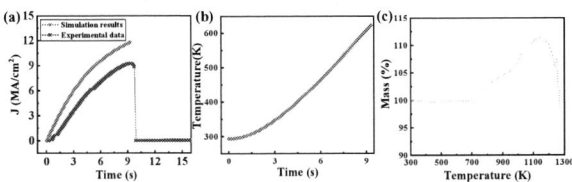

Fig. 3. (a) Comparison of the simulation result and experimental measurement on current density as a function of time of voltage. (b) The corresponding relationship of temperature as a function of time. (c) The thermogravimetric test of ZrTe₃ in air environment.

To simulate the real breakdown process, a bias voltage has been applied to the electrode for 9.3 seconds set at the same rate of 0.01V/s with the practical experimental test. **Fig. 3. (a)** presents the comparison of the current density (J) of ZrTe₃ nanoribbon as a function of time between the simulation and experimental measurement. Experimentally, the ZrTe₃ nanoribbon has been observed to reach the maximum current density 9.25 MA/cm² at 9.27 second, and then quickly broken down. The Comsol simulation exhibits that the current density is 11.7 MA/cm² at 9.27 second, which is a bit higher than that of experimental result. Such small discrepancy can be attributed to the variation of the voltage drop caused by the gold electrode and contact resistance for the experiment measurement. The temperature curve in **Fig. 3. (b)** shows that the maximum temperature during the breakdown process of the ZrTe₃ nanoribbons is 622K, which is slightly less than the thermogravimetric characterization (**Fig. 3. (c)**) with an oxidation temperature of 711 K, which may be due to the exacerbated failure at high temperature caused by the high current density. Although there are a bit differences between the simulation result and the experimental measurement, the established Comsol electro-thermal coupling model is still effective and reliable considering that the physical properties of ZrTe₃ at high temperature cannot be accurately calculated. Thus, we set 622K as the failure temperature of the ZrTe₃ nanoribbon in the air environment, implying that the ZrTe₃ nanoribbon can be considered failed if the maximum temperature reaches such value for the subsequent simulation.

D. The failure analysis of ZrTe₃ nanoribbon

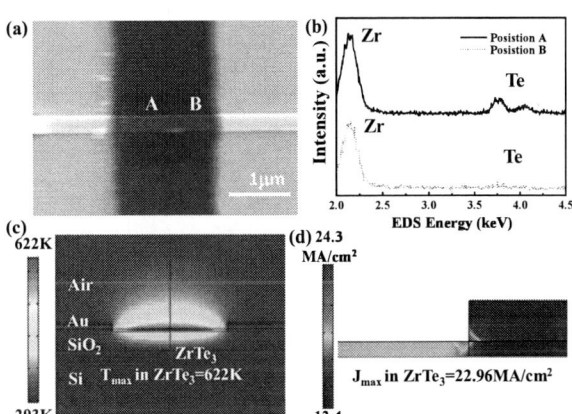

Fig. 4. (a) The SEM image of a typical ZrTe₃ nanoribbon device which has been broken down. (b) EDS spectra at different locations. (c) The temperature and (d) current density distribution of ZrTe₃ nanoribbon during the breakdown process.

To intuitively evaluate the failure of ZrTe₃ nanoribbon, SEM and corresponding energy sispersive spectrometer (EDS) analysis have been performed on the failed device. First, as shown in **Fig. 4. (a)**, the obvious contrast (conductivity) difference located at the breakdown position from other positions can be clearly observed. Further EDS spectra for A and B position were shown in **Fig. 4. (b)**, which clearly exhibites the reduction of Te atoms, thus indicating that the escape of Te atoms caused by high temperature is mainly responsible for the device failure. The distribution of temperature (**Fig. 4. (c)**) and current density (**Fig. 4. (d)**) for the ZrTe₃ nanoribbon with the dimension as shownd in Fig. 1.(b) during the breakdown process have been simulated with the established model above, and the results indicate that the temperature at both ends is significantly lower than the middle for the investigated ZrTe₃ nanoribbon due to the severe heat accumulation at the electrodes. However, the current density distribution shows an opposite trend with the temperature distribution (**Fig. 4. (d)**), which implying that the failure of ZrTe$_3$ nanoribbon should be the synergic effect of both temperature and current density. Such result is also consistent with the previous SEM observation that the failure occurs neither in the middle of the ZrTe$_3$ nanoribbons with the highest temperature of 622 K, nor the junction with the highest current density of 22.96 MA/cm^2. In this regard, improved current-carrying capacity can be expected with encapsulation or suitable substrate selection with good heat dissipation capability.

E. Further extention to 10 nm ZrTe$_3$ nanoribbons

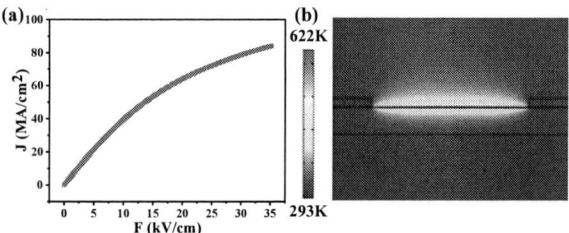

Fig. 5. (a) Current density as function of electrical field of 10 nm ZrTe$_3$ nanoribbon. (b) Temperature distribution of 10 nm ZrTe$_3$ nanoribbon during the breakdown process.

Previous report has indicated that TMTCs such as TaSe$_3$ exhibit the characteristic of the size-independent resistivity [2]. Under the assumption that the ZrTe$_3$ possesses the same characteristic, the breakdown characteristic of 10 nm ZrTe$_3$ nanoribbon was further investigated with the thermoelectric coupling model established above. As shown in **Fig. 5. (a)**, the current density (J) as a function of electric field (F) for 10 nm ZrTe$_3$ nanoribbon exhibits a nonlinear increase with an maximum breakdown current density ~84.1MA/cm^2 under the electric field of 35.2 kV/cm. Such result is quite inspiring since the performance is competitive with nanoscale Cu. The temperature distribution of 10 nm ZrTe$_3$ nanoribbon during the breakdown process is further shown in **Fig. 5. (b)**. Interestingly, a more uniform distribution compared with ZrTe$_3$ nanoribbon with large diameter (**Fig. 4. (c)**) was observed, which can be attributed to the effective alleviation of the heat accumulation. The current impressive performance suggest the great potential of ZrTe$_3$ nanoribbon in higher-density interconnection application of replacing Cu for the strategy of more than Moore's law.

III. CONCLUSION

In this work, an effective electro-thermal coupling model was established for finite element simulation with Comsol software. With that, firstly, the failure temperature of ZrTe$_3$ nanoribbon in air atmosphere has been determined to be 622K. Since TMTCs generally exhibit size-independent resistivity, the breakdown characteristic of 10 nm ZrTe$_3$ nanoribbon was further investigated. The results show that the breakdown current density reach 84.1 MA/cm^2, implying the great potential of ZrTe$_3$ as a novel interconnect material to replace copper during IC fabrication.

ACKNOWLEDGMENT

The authors thank support from the National Natural Science Foundation of China (grant no. 62074061, 61674063), the Foundation of Shenzhen Science and Technology Innovation Committee (Grant Nos.JCYJ20180504170444967).

REFERENCES

[1] K. Croes, Ch. Adelmann, C. J. Wilson, H. Zahedmanesh, O. Varela Pedreira, C. Wu, et al, "Interconnect metals beyond copper: reliability challenges and opportunities," IEEE International Electron Devices Meeting, San Francisco, CA, USA5.3.1-5.3.4, 2018

[2] T. A. Empante, A. Martinez, M. Wurch, Y. Zhu, A. K. Geremew,; K. Yamaguchi, et al, "Low resistivity and high breakdown current density of 10 nm diameter van der Waals TaSe$_3$ nanowires by chemical vapor deposition," Nano letters, vol. 19, no. 7, pp. 4355–4361, 2019.

[3] A. Geremew, A. M. Bloodgood, E. Aytan, K. W. B.Woo, R. S. Corber, G. Liu, et al, "Current Carrying Capacity of Quasi-1D ZrTe 3 Van Der Waals Nanoribbons," IEEE Electron Device Lett., vol. 39, no. 5, pp. 735–738, 2018

[4] A. K. Geremew, S. Rumyantsev, M. A. Bloodgood, T. T. Salguero, , A. A. Balandin, "Unique features of the generation-recombination noise in quasi-one-dimensional van der Waals nanoribbons," Nanoscale, vol. 10, no. 42, pp. 19749–19756, 2018

[5] Y. Hu, F. Zheng, X. Ren, J. Feng, Y. Li, "Charge density waves and phonon-electron coupling in ZrTe3," Phys. Rev. B, vol. 91, no. 14, 2015

[6] K. Stöwe, F. R. Wagner, "Crystal Structure and Calculated Electronic Band Structure of ZrTe3," Journal of Solid State Chemistry, vol. 138, no. 1, pp. 160–168, 1998

[7] K. Yamaya, S. Takayanagi, S. Tanda, "Mixed bulk-filament nature in superconductivity of the charge-density-wave conductor ZrTe3," Phys. Rev. B, vol. 85, no. 18, 2012

Interaction of silane coupling agents with nano-silica probed by nano-IR*

Pengli Zhu
Shenzhen Institute of Advanced
Electronic Materials
Shenzhen Institute of Advanced
Technology, Chinese Academy of
Sciences
ShenZhen, China
pl.zhu@siat.ac.cn

Jianjun Ruan
NOVORAY Corporation
Lianyungang, China
tony@novoray.com

Mingyong Du*
Shenzhen Institute of Advanced
Electronic Materials
Shenzhen Institute of Advanced
Technology, Chinese Academy of
Sciences
ShenZhen, China
my.du@siat.ac.cn

Ning Wang
Shenzhen Institute of Advanced
Electronic Materials
Shenzhen Institute of Advanced
Technology, Chinese Academy of
Sciences
ShenZhen, China
wangning@siat.ac.cn

Xiaomeng Du
Shenzhen Institute of Advanced
Electronic Materials
Shenzhen Institute of Advanced
Technology, Chinese Academy of
Sciences
ShenZhen, China
xm.du1@siat.ac.cn

Tao Zhao
Shenzhen Institute of Advanced
Electronic Materials
Shenzhen Institute of Advanced
Technology, Chinese Academy of
Sciences
ShenZhen, China
tao.zhao@siat.ac.cn

Xiaodong Li
NOVORAY Corporation
Lianyungang, China
lixd@novoray.com

Jiakai Cao
NOVORAY Corporation
Lianyungang, China
caojk@novoray.com

Jianping Zhang
NOVORAY Corporation
Lianyungang, China
zhangjp@novoray.com

Xiaoyao Sun
NOVORAY Corporation
Lianyungang, China
sunxy@novoray.com

Abstract—The nanoparticle-polymer composite is a widely used electronic packaging material. To satisfy the miscibility of nanoparticles with polymer, the surface of nanoparticles needs to be modified. The particle surface modification has been widely studied. However, the interaction between the silane coupling agent (SCA) and the nanoparticle surface is not clear yet. This article aims to investigate the surface modification impact on the nano-SiO_2 particle surface and to clarify the interaction between chemical reagent and the particle surface. In this investigation nano-SiO_2 was characterized by thermal gravity analysis (TGA), Fourier transform infrared spectroscopy (FT-IR), and nano-IR. The FT-IR results showed that the specific peak of SCA appears in the synthesized nano-silica. The TGA experiments were carried out to test the weight loss of SCA modified nano-SiO_2 as well as the grafting efficiency of the surface modification. The FT-IR and TGA results confirmed the SCA chemical grafting onto the nanoparticle surfaces. The nano-IR test provided detailed information for nano-silica surface. It could not only show the topographical AFM images but also the corresponding nanoscale IR adsorption spectra of the nano-silica surface. The topographic image distinguished the nanoparticle from the substrate. The discrete domains spectra shows the existence of SCA on the silica surface, which indicated that the SCA was grafted onto the silica surface with chemical bonding.

Keywords—Electronic packaging; nano-SiO_2; surface modification; nano-IR; silane coupling agent

I. INTRODUCTION

Flip-Chip packaging is widely used for integrated circuit fabrication, as it has many advantages including high input/output (I/O) density, smaller footprint, high throughout, self-alignment, low profile, etc. compared with traditional packaging technology [1]. However, the major concerns including the thermal stress occurrences while flip-chip assembly, generated on the solder joints during thermal cycling. This is caused by the great coefficient of thermal expansion (CTE) difference between the widely used organic substrate (18-24 ppm/°C) and the silicon die (2.5 ppm/°C)[2]. The development of underfill guarantees the application of low-cost organic substrate in flip-chip packages. Underfill, generally composed of epoxy resin, fillers, curling agents, etc. provides thermo-mechanical protection as well as environmental protection. Nanoparticles especially nano-SiO_2 are generally used as a filler to improve mechanical and thermal properties, especially to reduce overall CTE of the underfill to improve the device reliability[3]. However, as a filler, nano-SiO_2 particles are easy to adhere and form agglomerations, and exhibit poor compatibility with the epoxy resin. As a result, high filler loading rate always has poor filler-matrix interface.

Surface modificationmethod is widely used to improve the filler-epoxy interface interaction. Silane coupling agent (SCA) is a generally used surface modifier to improve the surface of silica fillers. The ideal SCA could reduce particle-particle interaction but enhance the particle-epoxy resin interaction. One end of SCA can react with the silanol group of silica particles, whereas the other hydrophobic functional group is compatible with the organic resin. The well grafted SCA could react and remove the silanol groups, as well as obtain a hydrophobic surface. The surface modification efficiency has been widely studied[4], such as SCA type, SCA concentration, process time, temperature. However, there is few research studying the interaction between SCA and nanoparticles.

Fourier transform infrared spectroscopy (FT-IR) is used for chemical bonding analysis. It has been used to trace the interactions between protein modifiers and biomaterials and inorganic matters. However, it is still impossible to be used to probe the interaction between SCA and nanoparticles, as the limited diffraction resolution. Atomic force microscope (AFM) combined with nanoscale infrared spectroscopy technique[5] is developed to overcome the diffraction-limited resolution of FT-IR. The infrared light is scattered at a scanning probe tip, typically a metalized atomic force microscope (AFM), to capture nanoscale spatial resolution infrared images. Scattering-type scanning near-field optical microscopy (s-SNOM) is the fundamental basis for this technique. This technique, called nano-IR, has the advantage of the chemical specificity by IR analysis as well as high spatial resolution of AFM.

The objective of this study is to further investigate the application of nano-IR for probing the interaction between SCA and nano-SiO$_2$ in conjunction with FT-IR and Thermal Gravimetric Analysis (TGA). TGA is a widely used analytical technique detecting the information about the physical phenomena and chemical phenomena. FT-IR and TGA are used to preliminary judge the surface modification. The interaction between SCA and nano-SiO$_2$ was investigated by nano-IR. Our studies demonstrate the potential analyzing the interaction between SCA and nanoparticles.

II. Experimental

A. FT-IR

The FT-IR Spectroscope was used to study the surface modification of nano-silica by SCA. The nano-silica were scanned at an infrared diffuse reflection mode in FT-IR for 320 times from wavelength 4000 to 400 cm^{-1}.

B. TGA

The surface adsorption of nano-SiO$_2$ after surface modification was measured with TGA. The nano-SiO$_2$ was dried in a vacuum drier on 150°C for 5 hours before experiment. Particles were heated from 30 °C to 800 °C at the ramping rate of 10 °C/min.

C. Nano-IR

Particles were mounted onto the golden base. Bruker Anasys nanoIR3s was used to carry out the experiment. The spectra scanning range is 780-1900 cm^{-1}, with the resolution of 2 cm^{-1}. The AFM mapping area is 5 μm × 5 μm, with the resolution of 256 pts × 256 pts.

III. Results And Discussion

A. FT-IR

The nano-SiO$_2$ was surface modified by SCA. The FT-IR spectra of raw and surface modified nano-SiO$_2$ were

shown in Figure 1. The spectra show that raw SiO$_2$ has broad peaks of Si-O from 1320 to 950 cm^{-1} and -OH from 3700 to 3400 cm^{-1}. The vibration peaks of -CH$_3$, -CH$_2$- and -CH$_3$ are detected at 2980, 2940 and 2878 cm^{-1} in the SCA treated nano-SiO$_2$. This ascertained that the SCA is successfully grafted onto the surface.

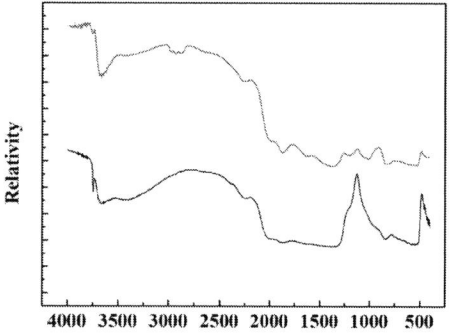

Fig. 1. FI-IR image of raw SiO$_2$ (black line) and surface modified SiO$_2$ (red line)

B. TGA

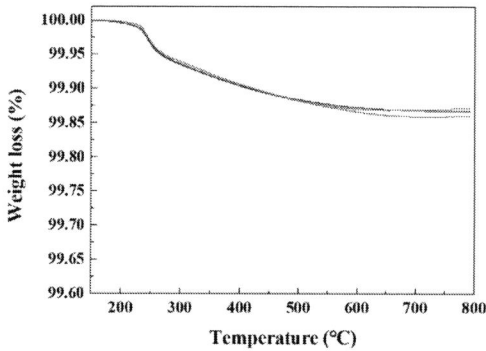

Fig. 2. Weight loss of SCA grafted nano-SiO$_2$

The weight loss test of SCA grafted nano-SiO$_2$ was carried out three times with TGA. The nano-SiO$_2$ was dried under 150°C for 5 hours in vacuum drying oven to remove the physical adsorbed water. The nano-SiO$_2$ did not start to lose weight before 230 °C. Then it started to lose weight continuously with temperature increasing. The overall lost weight for each sample is 0.13%, 0.13% and 0.14% respectively. This was resulted from the debonding and degradation of grafted SCA on the nano-SiO$_2$ surface.

C. AFM and Nano-IR

Fig. 3 shows the topographical AFM image as well as the corresponding bulk IR adsorption spectra. The inset image in Fig. 3b shows that the height of the nano-SiO$_2$ particle is ~700 nm. The peaks from 1103 to 1029 cm^{-1} is characterized as the stretching vibration arising from Si-O group.

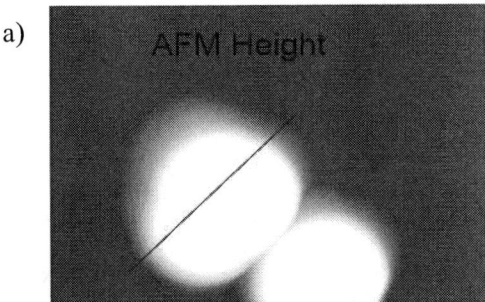

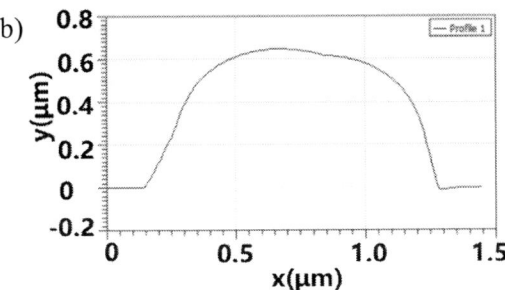

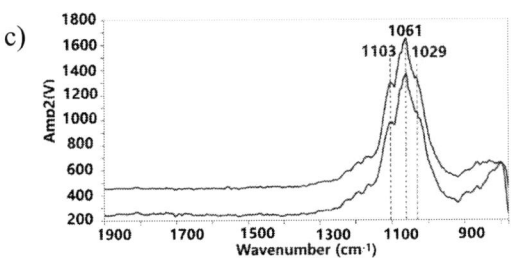

Fig. 3. AFM and nano-IR spectra of raw nano-SiO₂ (a. topographical AFM image; b. height information of the nano-SiO₂ particle; c. the corresponding bulk IR adsorption spectra)

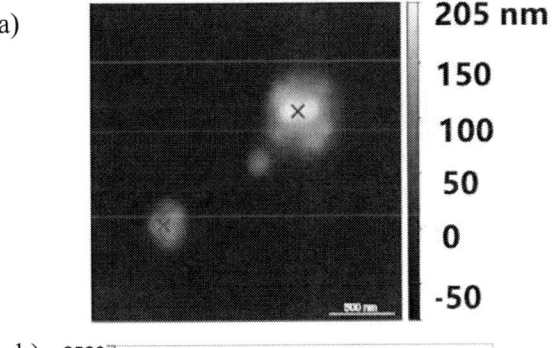

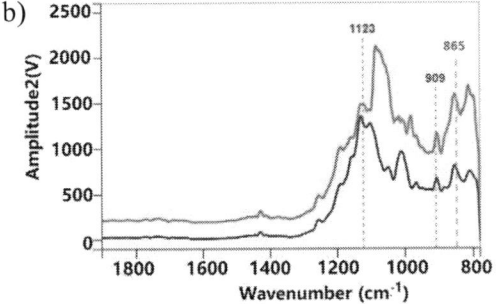

Fig. 4. AFM and nano-IR spectra of surface modified SiO₂ (a. topographical AFM image; b. the corresponding bulk IR adsorption spectra)

Fig. 4 demonstrates the tomographic AFM image and nano-IR spectra of surface modified nano-SiO₂. The spectra taken from different surface locations on the surface modified SiO₂ show specular peaks at 1123, 909 and 865 cm-¹, representing Si-O group, epoxy group, and -CH₃ group. The AFM and nano-IR results show that the chemical composition of the SCA modified nano-SiO₂ can be confirmed. In order to check if SCA is grafted onto the surface of the particle and the SCA distribution, the IR adsorption phase images of SCA modified nano-SiO₂ at two different frequencies were taken. Fig. 5 shows the IR adsorption phase images taken at 1120 and 910 cm⁻¹. In Fig. 5a, the phase image shows strong contrast on the nano-SiO₂ particle surfaces owing to the Si-O-Si adsorption at 1120 cm⁻¹. Similarly, the phase image shows strong adsorption around the particles at 910 cm⁻¹ in Fig. 5b. There are also some strong adsorption spots at 910 cm⁻¹ for particles with smaller size. The overlay AFM height and infrared spectra image in Fig. 5c shows clearly that SCA was successfully grafted onto the surface of nano-SiO₂. Meanwhile, the red spots demonstrate that SCA was grafted onto smaller particles as well.

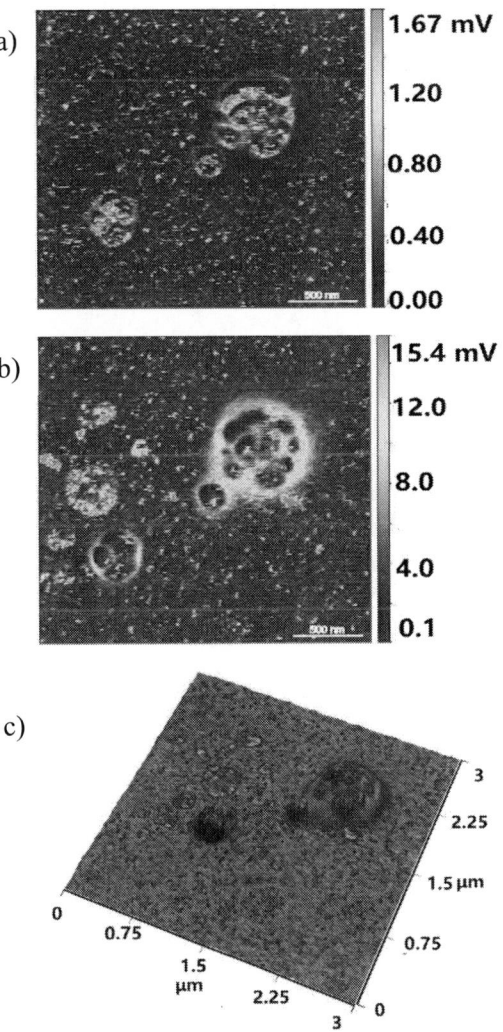

Fig. 5. IR adsorption phase images of HY SCA modified nano-SiO₂ at two different frequencies, 1120 (a) and 910 cm-1 (b), and AFM-nano-IR image (c) of SCA modified nano-SiO₂ at 910 cm⁻¹.

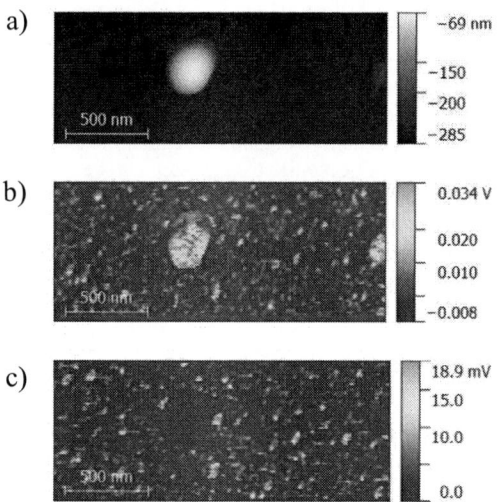

Fig. 6. AFM tomography (a) and IR adsorption phase images of raw nano-SiO$_2$ at two different frequencies, 1120 (b) and 910 cm^{-1} (c)

The tomographic height image and IR adsorption spectra for raw nano-SiO$_2$ was scanned in Fig. 6. The particle shows strong contrast at 1120 cm^{-1}, which confirmed the existence of Si-O, whereas no adsorption signal was detected at the frequency of 910 cm^{-1}.

IV. CONCLUSION

FT-IR and TGA test verified the existence of SCA in the surface modified nano-SiO$_2$. Nano-IR technique is an reliable technique probing the interaction between SCA and nano-SiO$_2$. Nano-IR confirms the assumption that SCA was grafted onto the surface of nano-SiO$_2$.

ACKNOWLEDGMENT

This work was supported by the National Key R & D Project from Minister of Science and Technology of China (2020YFB0311800) , Shenzhen basic research plan (JCYJ20190807154409372), the Guangdong Basic and Applied Basic Research Foundation (No. 2019A1515010743) and the SIAT Innovation Program for Excellent Young Researchers (No. E1G012).

REFERENCES

[1] Y. Sun, Z. Zhang, C. Wong, Study on mono-dispersed nano-size silica by surface modification for underfill applications, Journal of Colloid and Interface Science 292(2) (2005) 436-444.

[2] C. Wong, S. Luo, Z. Zhang, Flip the chip, Science 290(5500) (2000) 2269-2270.

[3] P. Dittanet, R.A. Pearson, P. Kongkachuichay, Thermo-mechanical behaviors and moisture absorption of silica nanoparticle reinforcement in epoxy resins, International Journal of Adhesion and Adhesives 78 (2017) 74-82.

[4] L. Huang, S. Yang, J. Chen, J. Tian, Q. Huang, H. Huang, Y. Wen, F. Deng, X. Zhang, Y. Wei, A facile surface modification strategy for fabrication of fluorescent silica nanoparticles with the aggregation-induced emission dye through surface-initiated cationic ring opening polymerization, Materials Science and Engineering: C 94 (2019) 270-278.

[5] M. Roman, T.P. Wrobel, C. Paluszkiewicz, W.M. Kwiatek, Comparison between high definition FT‐IR, Raman and AFM‐IR for subcellular chemical imaging of cholesteryl esters in prostate cancer cells, Journal of biophotonics 13(5) (2020) e201960094.

Lightweight and Compressible Expandable Polymer Microspheres/Silver Flakes Composites for High-efficiency Electromagnetic Interference Shielding

Jianhong Wei
Shenzhen Institute of Advanced Electronic Materials,
Shenzhen Institute of Advanced Technology, Chinese Academy of Sciences
Shenzhen 518055, China
weijh@siat.ac.cn

Yadong Xu
Shenzhen Institute of Advanced Electronic Materials,
Shenzhen Institute of Advanced Technology, Chinese Academy of Sciences
Shenzhen 518055, China
yd.xu@siat.ac.cn

Zhiqiang Lin
Shenzhen Institute of Advanced Electronic Materials,
Shenzhen Institute of Advanced Technology, Chinese Academy of Sciences
Shenzhen 518055, China
zq.lin@siat.ac.cn

Zuomin Lei
Shenzhen Institute of Advanced Electronic Materials,
Shenzhen Institute of Advanced Technology, Chinese Academy of Sciences
Shenzhen 518055, China
zm.lei@siat.ac.cn

Xianzhu Fu
College of Materials Science and Engineering,
Shenzhen University
Shenzhen 518055, China
xz.fu@szu.edu.cn

Yougen Hu*
Shenzhen Institute of Advanced Electronic Materials,
Shenzhen Institute of Advanced Technology, Chinese Academy of Sciences
Shenzhen 518055, China
yg.hu@siat.ac.cn

Rong Sun*
Shenzhen Institute of Advanced Electronic Materials,
Shenzhen Institute of Advanced Technology, Chinese Academy of Sciences
Shenzhen 518055, China
rong.sun@siat.ac.cn

Abstract— **It is important to exploit high-efficiency electromagnetic shielding materials to eliminate electromagnetic interference (EMI) pollution in electronic equipment. Here, a lightweight and compressible EMI shielding composite was successfully fabricated via mixing expandable polymer microspheres (EPM) with silver flakes (AgF), followed by heating expansion in a close mold. The composites with only loading 2.57 vol% AgF exhibits a superb electrical conductivity (973.56 S/m), and its EMI SE is up to 100 dB when the thickness is 5 mm in the ultra-waveband of 8 - 40 GHz. The EPM were squeezed and bonded with each other, and the AgF powders were uniformly distributed among the interfaces of the EPM, which offers a segregated conductive network, low density (0.328 g/cm³ at 2.57 vol% of AgF), and high mechanical strength. In particular, it can be concluded that the EPM/AgF composites have promising potential applications in the EMI shielding materials of advanced electronic equipment via near-field testing. The average NF-EMI of EPM/AgF-2.75vol% is up to 62.7 dB, which is better than that of copper foil in the 1-9 GHz frequency range. Owing to the high elastic resilience and compressibility, the composites are a promising multifunctional material for lightweight electromagnetic shielding material.**

Keywords— *expandable polymer microspheres, lightweight, electromagnetic interference shielding, heating-foam, near-field*

I. INTRODUCTION

With the advent of the 5G era, electromagnetic interference (EMI) has become one of the most serious environmental problems worldwide. EMI will not only interfere with the normal functions of electronic devices but also endanger human health[1]. To adapt to the deteriorating EMI environment, it is urgent to improve the electromagnetic shielding ability of various electronic equipment. EMI shielding materials play a significant role in eliminating EMI. Therefore, electromagnetic shielding materials are widely used in the field of electronic packaging. Compared with traditional metal materials, conductive polymer composites (CPCs) composed of polymer matrix and conductive fillers have many advantages, such as lightweight, corrosion resistance, and adjustable conductivity, CPCs become a promising alternative EMI shielding material[2].

Due to the high conductivity of metals, metals have become common conductive fillers for CPCs. Lee *et al.* fabricated 3D PS/Cu/Ag composites which shown EMI SE of 110 dB at a thickness of 0.5 mm and a thermal conductivity of 16.1 W·m⁻¹K⁻¹ with 13 vol % of filler[3]. Xu *et al.* flexible T-ZnO/Ag/WPU film that exhibited an EMI SE of 42 dB with a thickness of 0.02 mm[4].

Polymers such as WPU, PDMS, Epoxy, PMMA, PVA, PS, etc. are a kind of matrix widely used in the research of

978-1-6654-1392-3/21 $31.00 © 2021 IEEE

lightweight CPCs[5] [6]. However, traditional methods for preparing lightweight CPCs (such as directional freeze-drying, physical foaming, chemical foaming, templates, etc.) have problems such as complicated processes and environmental pollution[7]. Based on the thermal foaming properties of expandable polymer microspheres, it is suggested that this is a new method for preparing lightweight CPCs. Expandable polymer microspheres (EPM) are composed of acrylate polymer shells containing liquid hydrocarbon[8]. The hydrocarbon gasifies and the shells soften at heat exposure, resulting in the size of EPM increases to several times[9]. EPM were widely used as physical foaming agents to adjust the mechanical property of composites[10]. For instance, Bao *et al.* used a small amount of EPM as foaming agents to reduce the density of epoxy/multi-walled carbon nanotubes/nickel-plated carbon fibers composites[11]. EPM can be also used for the construction of a segregated structure to improve the conductive and EMI shielding performance[12]. But, there are few reports on EMI shielding materials using EPM as the matrix of composites.

In this study, a lightweight EMI shielding composite was successfully fabricated via mixing expandable polymer microspheres with silver flakes (AgF), followed by heating expansion in a close mold. There is a porous structure in the EPM/AgF sample. The EPM were squeezed and bonded with each other, and the AgF powders were uniformly distributed among the interfaces of the EPM, which offers a segregated conductive network, low density, and high mechanical strength. The compressive strength of the EPM/AgF composite increases with the addition of AgF. The results suggest that the as-prepared EPM/AgF composites possess lightweight, high compressibility and recovery rate, high mechanical strength, and outstanding EMI shielding performance at a low AgF loading, indicating that the composites have promising applications in EMI shielding fields. This work also exhibits a method for preparing lightweight EMI shielding materials, which has a wide application prospect.

II. MATERIALS AND EXPERIMENTS

2.1 Materials

Silver flakes were provided by CNMC Ningxia New Materials Co., Ltd. Expandable polymer microspheres (EPM, 031DU40) with initial particle size of 10-16 μm were purchased from AkzoNobel Inc.

2.2 Preparation of EPM/AgF

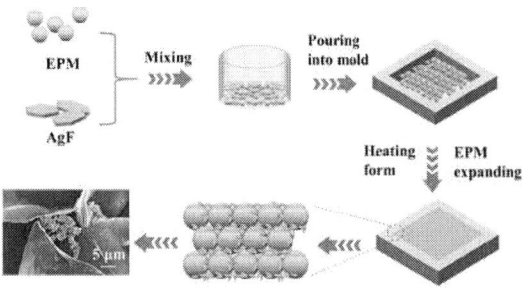

Fig. 1. Schematic diagram of the preparation process of the EPM/AgF

The fabrication procedure of expandable polymer microspheres/silver flakes (EPM/AgF) composites with

segregated structure via heat-foaming method is described in Fig. 1. Firstly, the AgF and expandable polymer microspheres were mixed together. Herein, a homogeneous mixture of EPM/AgF was gained by mechanical stirring. Secondly, the uniform mixture (the mass of the EPM powder was 0.2 g) of EPM/AgF was directly poured into the mold. Finally, as the volume of the microspheres expands at high temperatures, the EPM/AgF composites with a thickness of 5.0 mm were gained by the heat foaming method in a closed mold at 95 °C for 1.5 h. The EPM/AgF composites with different AgF loadings (1.69, 2.25, 2.40, 2.57, 2.75 vol%) were finally obtained. AgF-2.40 vol% represented the EPM/AgF composite loaded with 2.40 vol% of AgF.

2.3 Characterization

The morphologies were characterized via a scanning electron microscope (SEM, Apreo 2) at 5.0 kV. The electrical conductivity and compression of samples were measured using an electronic universal testing machine (MAX-1kN-P-2) combined with a digital multimeter at 2 mm/min. Near-field shielding properties of composites were characterized via EMI near-field scan technology using Amber Precision Instruments (EMI Scanner SmartScan-350), which picked up fields at reactive near field (NF) region.

The shielding effectiveness (SE) can be calculated using the electromagnetic wave scattering parameters (S_{11}, S_{21}, S_{12}, and S_{22}) that were tested operating a vector network analyzer (VNA, Keysight PNA-E5227B). SE_R, SE_A, and the total value of EMI SE (SE_T) can be calculated with the following formulas:

$$R = \left| S_{11} \right|^2$$

$$T = \left| S_{21} \right|^2$$

$$A = 1 - T - R$$

$$SE_T = SE_R + SE_A + SE_M$$

$$SE_T = -10 \log T$$

$$SE_R = -10 \log\left(1 - R\right)$$

$$SE_A = -10 \log\left(\frac{T}{1-R}\right)$$

where R, T, and A are the power coefficients of reflection, transmission, and absorption. EMI SE is generally expressed in decibel (dB), the multiple reflections (SE_M) are customarily ignored when $SE_T > 10$ dB[13].

III. RESULTS AND DISCUSSION

3.1. Morphology

The SEM image of the unexpanded EPM is shown in Fig.2a, indicating that the unfoamed EPM is spherical, the unexpanded powder with a particle size of 10-16 μm. The morphology of the flake silver powder is shown in Fig. 2b, and its size is less than 5 μm. From the cross-sectional SEM image of Fig. 2c, it can be observed that there is a porous structure of the EPM/AgF samples. The EPM were squeezed and bonded with each other, and the AgF powders were uniformly distributed among the interfaces of the EPM, which offers a segregated conductive network, low density (0.29 g/cm³ at 2.25 vol% of AgF), and high mechanical strength. Fig. 2d visually demonstrates the lightweight and low density of the EPM/AgF composite, which can be easily supported by a leaf without obvious bent. Fig. 2e further clearly proves the uniform distribution of AgF in the interfaces of EPM. In summary, lightweight EPM/AgF composites with segregated structures were successfully fabricated.

3.2 Mechanical properties of the EPM/AgF composite.

The EPM/AgF composites exhibit high compression strength and excellent recovery rate. The compression strength of EPM/AgF with 0, 1.69, 2.40 and 2.75 vol% of AgF is about 0.46, 0.47, 0.51 and 0.58 MPa at compression strain of 30%, respectively, and their recovery rate is about 90.3%, 69.7%, 75% and 67.3% after the pressure released (Fig. 3a). Fig. 3b visually demonstrates the high compressibility of the EPM/AgF composite, the compression ratio of EPM/AgF-2.40 vol% is up to 85%. The stress-strain curves were shown in Fig. 3c, and large hysteresis loops appeared, illustrating that the EPM/AgF had effective energy dissipation. The conductivity of EPM/AgF-2.40 vol% at the first ten cycles of the loading-unloading test were much smaller changes (Fig. 3d), it shows that the internal conductive structure of the material does not change significantly under external force. Because of the elasticity of polymer shells and the compressibility of the gas, the EPM/AgF shows excellent compressibility and resilience. The EPM/AgF had effective energy dissipation verified from the loading-unloading cycle[14]. Finally, this may be due to the higher hardness of the metal, which makes the compressive strength of the EPM/AgF composite increase under the loading of AgF.

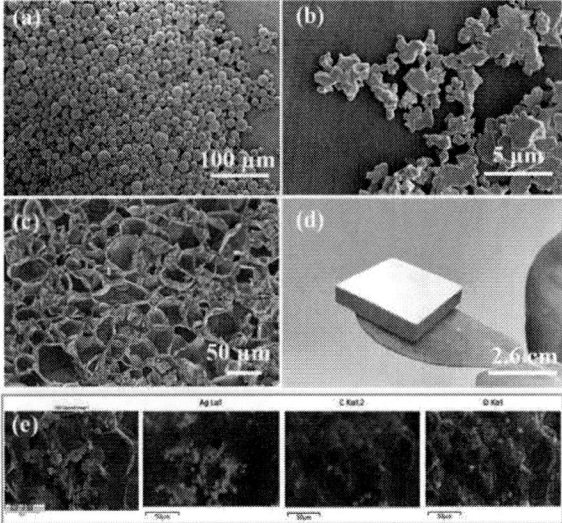

Fig. 2. SEM images of (a) original expandable polymer microspheres, (b) Silver flakes, (c) EPM/AgF composites. (d) Digital photos of EPM/AgF composites. (e) Elemental mapping of the EPM/AgF composites

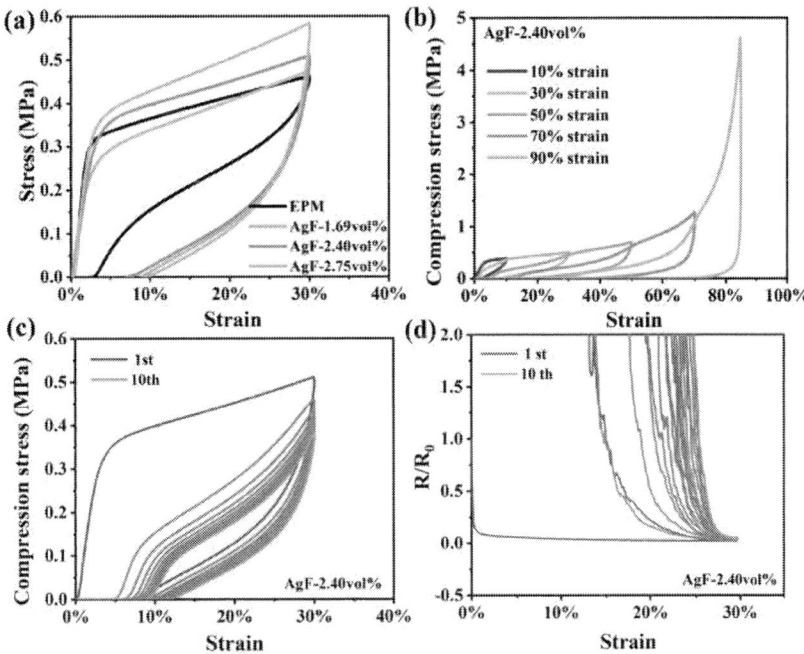

Fig. 3. (a) Compression stress-strain curves of the EPM/AgF with different AgF vol% at 30% strain. Compression stress-strain curves of EPM/AgF-2.40vol% composite under (b) different strain and (c) 30% cyclic strain. (d) The conductivity changes of EPM/AgF-2.40vol% under loading-unloading test.

3.3 EMI SE and electrical conductivity of the EPM/AgF

In the future, electronic devices of various electromagnetic wavebands will be widely used, and more

EMI will be generated in a wider frequency range. Therefore, the EMI SE of EPM/AgF composites were tested in an ultra-wideband of 8 - 40 GHz. In Fig. 4a, the average value of EMI SE of EPM/AgF with 1.69, 2.25, 2.40, and 2.57 vol% of AgF is about 5.74, 26.3, 51.5, and 97.7 dB at 8-12 GHz, respectively. The EMI SE value of EPM/AgF-2.25 vol% is greater than 20 dB, which can meet commercial application requirements. In high-frequency electromagnetic fields (Fig. 4b-4d), the conduction and polarization of carriers are enhanced, so the EMI SE of EPM/AgF composite is higher in the higher frequency band[15].

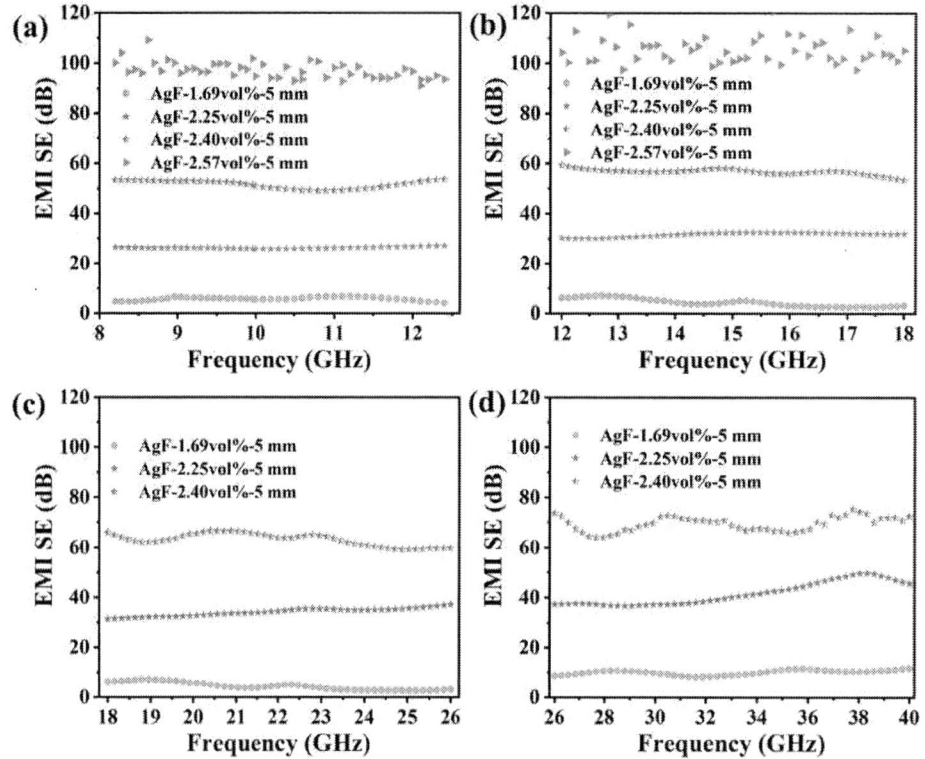

Fig. 4. EMI SE of the EPM/AgF composite with 5mm thickness in the (a) X-band, (b) Ku-band, (c) k-band, and (d) Ka-band.

When exploring the shielding effect of materials, the thickness of the sample is an important parameter. Therefore, the EMI SE of the EPM/AgF-2.57 vol% composite with a thickness of 2 mm, 3 mm, and 5 mm in the X-band were studied (Fig. 5a). The average EMI SE of the EPM/AgF-2.57 vol% with 2 mm thickness was up to 35 dB. Excellent electrical conductivity often gives the composite material excellent electromagnetic shielding performance. In Fig. 5d, it is shown that the composite material has excellent electrical conductivity, which is due to the high conductivity of AgF. EPM/AgF-2.57 vol% with high electrical conductivity (973.56 S/m) show the excellent EMI SE. To further study the shielding mechanism of the EPM/AgF composites, SE_R, SE_A, SE_T, R, A, and T of composites are shown in Fig. 5b and 5c, respectively. The EPM/AgF-2.25 vol%, EPM/AgF-2.40 vol% have large absorption coefficient of 0.67 and 0.32, respectively. This indicates that the composite has a large SE_A loss. The reason why the composite material has high absorption loss is that the Ag powder is uniformly distributed at the interface of the microspheres inside the composite material, thus forming a conductive network with a segregated structure. As shown in Figure 5e, the segregated structure can effectively cause multiple reflection effects of electromagnetic waves, resulting in a large absorption loss for the composite material. However, the R-value of EPM/AgF-2.57 vol% reached 0.92, which may be due to the high Ag loading, which enriched a large amount of AgF powders on the sample surface, resulting in a high reflection coefficient. SSE/d is often used to express the effectiveness of lightweight shielding materials. SSE/d is obtained by dividing EMI SE by density and thickness[5]. The high SSE/d and low density are the research focus for electromagnetic shielding materials. The SSE/d of EPM/AgF-2.57 vol% with low density (0.328 g/cm^3) was up to 608 dB cm^2 g^{-1}, indicating that the EPM/AgF has higher shielding effectiveness and lower density. The EPM were squeezed and bonded with each other, and the AgF powders were uniformly distributed among the interfaces of the EPM, which offers a segregated conductive network and low density. The segregated conductive network can effectively increase the multiple reflections of electromagnetic waves, resulting in the material having a larger absorption coefficient[12]. High conductivity and low density make the EPM/AgF composite have high SSE/d, which shows that EPM/AgF is a lightweight composite.

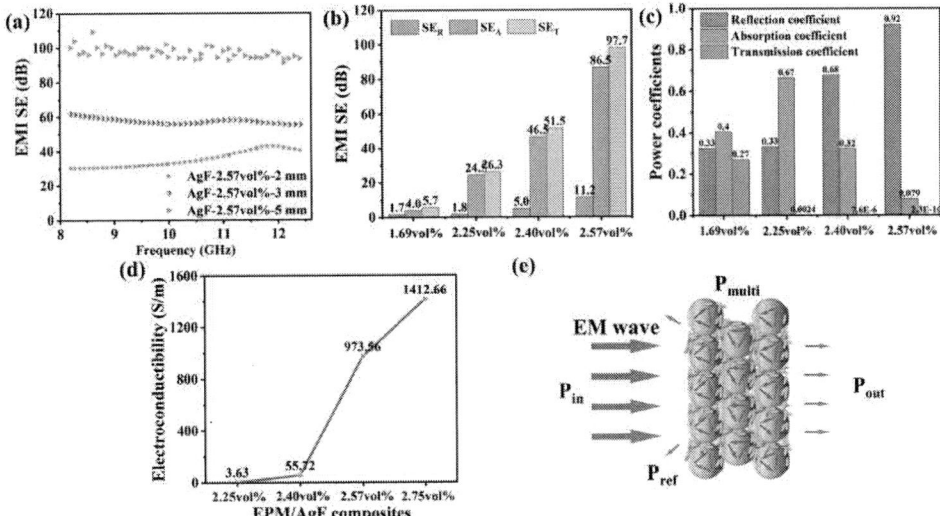

Fig. 5. (a) EMI SE of the EPM/AgF-2.57vol% composite with 2mm, 3mm, and 5mm thickness at the X-band. (b) SE_R, SE_A, and SE_T of EPM/AgF at the X-band. (c) R, A, and T of the EPM/AgF in the X-band. (d) Electrical conductivity of the EPM/AgF with various contents of AgF. (e) Schematic diagram of the EMI shielding mechanism of the samples.

3.4 Near field-EMI shielding performance of the EPM/AgF composite

The application of EPM/AgF in the fields of EMI shielding was discussed based on a near-field test, which was often used by engineers to test the actual radiation leakage of various antennas and devices[16]. Near-field testing can locate the true source of radiation, which is different from far-field testing. The near-field test system consists of the network analyzer, scanner probe, and antenna board (Fig. 6a). The near-field EMI SE (NF-EMI SE) testing principle was shown in Fig. 6b. The near-field antenna emits electromagnetic waves signal, which is weakened by the sample, and the scanning probe scans the signal at each position of the sample. After software calculation, the NF-EMI SE of each position can be obtained. The NF-EMI SE data of all testing points can form a 2D shielding effectiveness mapping, as shown in Fig. 6c. Figure 6c shows the 2D average shielding effectiveness mapping of EMI SE in a frequency range of 1-9 GHz. The 2D mapping shows that the EPM/AgF composites materials have excellent near-field shielding ability, and the average shielding effectiveness of EPM/AgF-2.75 vol% is much better than that of copper foil with 0.06 mm thickness. The shielding data of the center point of the 2D shielding effectiveness mapping is plotted in Figure 6d, which shows that EPM/AgF-2.40 vol% can meet the daily electromagnetic shielding requirements. In figure 6d, the average NF-EMI of EPM/AgF-2.75 vol% is up to 62.7 dB, which is better than that of copper foil in a frequency range of 1-9 GHz. For compressible EPM/AgF composite materials, under external pressure, it can effectively interconnect with the packaging frame of the antenna, leading to a lower electromagnetic wave leakage.

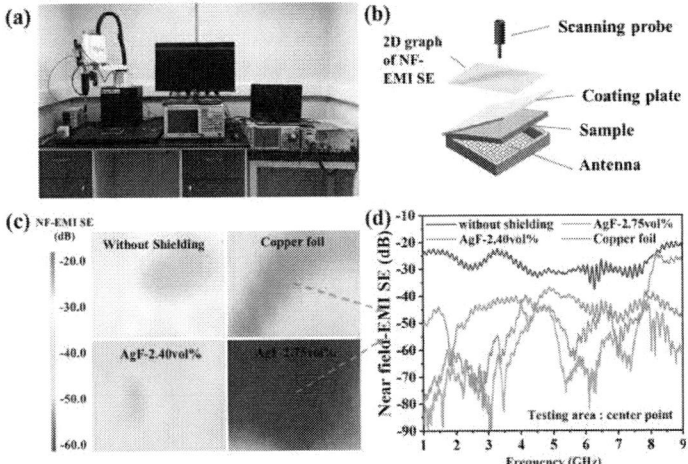

Fig. 6. (a) The digital picture of near field-EMI SE testing system. (b) Schematic depicting the near-field EMI SE testing principle. (c) The 2D mapping of NF EMI SE of the samples with the thickness of 5 mm. (d) The curves of NF EMI SE at the center point of the samples with the thickness of 5 mm.

IV. CONCLUSION

Electromagnetic shielding materials play an important role in the field of electronic packaging. A lightweight and compressible EMI shielding composite was fabricated via mixing expandable polymer microspheres (EPM) with silver flakes (AgF). The composite with low AgF loading of 2.57 vol% and the thickness of 5 mm exhibit a superb electrical conductivity (973.56 S/m), and its EMI SE is up to 100 dB in the ultra-wide frequency range of 8-40 GHz. The SSE/d of EPM/AgF-2.57vol% with low density (0.328 g/cm^3) is up to 608 dB·cm^{2}g^{-1}. The AgF powders were uniformly distributed among the interfaces of the EPM, which offers a segregated conductive network, low density, and high mechanical strength. Near-field tests prove the feasibility of EPM/AgF in actual electromagnetic shielding. These results can show that this highly compressible and lightweight EMI shielding composite is expected to apply in the field of high-frequency electromagnetic shielding. The method for preparing lightweight EMI shielding materials based on thermal foaming shows the characteristics of a simple process and has bright application prospects.

ACKNOWLEDGMENT

The authors are grateful for the financial support from the National Natural Science Foundation of China (62074154), China Postdoctoral Science Foundation (2020M682983), Guangdong Basic and Applied Basic Research Fund (2020A1515110962, 2020A1515110154), and SIAT Innovation Program for Excellent Young Researchers (E1G035).

REFERENCES

[1] THOMASSIN J-M, JéRôME C, PARDOEN T, et al. Polymer/carbon based composites as electromagnetic interference (EMI) shielding materials [J]. Materials Science and Engineering: R: Reports, 2013, 74 (7): 211-232.

[2] WANG M, TANG X-H, CAI J-H, et al. Construction, mechanism and prospective of conductive polymer composites with multiple interfaces for electromagnetic interference shielding: A review [J]. Carbon, 2021, 177 377-402.

[3] LEE S H, YU S, SHAHZAD F, et al. Low percolation 3D Cu and Ag shell network composites for EMI shielding and thermal conduction [J]. Composites Science and Technology, 2019, 182 107778.

[4] XU Y, YANG Y, YAN D-X, et al. Flexible and conductive polyurethane composites for electromagnetic shielding and printable circuit [J]. Chemical Engineering Journal, 2019, 360 1427-1436.

[5] ZHAO B, HAMIDINEJAD M, WANG S, et al. Advances in electromagnetic shielding properties of

composite foams [J]. Journal of Materials Chemistry A, 2021, 9 (14): 8896-8949.

[6] ZHU S, CHENG Q, YU C, et al. Flexible Fe3O4/graphene foam/poly dimethylsiloxane composite for high-performance electromagnetic interference shielding [J]. Composites Science and Technology, 2020, 189 108012.

[7] ZHANG L-Q, YANG S-G, LI L, et al. Ultralight Cellulose Porous Composites with Manipulated Porous Structure and Carbon Nanotube Distribution for Promising Electromagnetic Interference Shielding [J]. ACS Applied Materials & Interfaces, 2018, 10 (46): 40156-40167.

[8] CHEN S-Y, SUN Z-C, LI L-H, et al. Preparation and characterization of conducting polymer-coated thermally expandable microspheres [J]. Chinese Chemical Letters, 2017, 28 (3): 658-662.

[9] AGLAN H, SHEBL S, MORSY M, et al. Strength and toughness improvement of cement binders using expandable thermoplastic microspheres [J]. Construction and Building Materials, 2009, 23 (8): 2856-2861.

[10] CAI J-H, LI J, CHEN X-D, et al. Multifunctional polydimethylsiloxane foam with multi-walled carbon nanotube and thermo-expandable microsphere for temperature sensing, microwave shielding and piezoresistive sensor [J]. Chemical Engineering Journal, 2020, 393 124805.

[11] PAN J, XU Y, BAO J. Epoxy composite foams with excellent electromagnetic interference shielding and heat-resistance performance [J]. Journal of Applied Polymer Science, 2017, 135 46013.

[12] GE C, WANG G, ZHAO G, et al. Lightweight and flexible poly(ether-block-amide)/multiwalled carbon nanotube composites with porous structure and segregated conductive networks for electromagnetic shielding applications [J]. Composites Part A: Applied Science and Manufacturing, 2021, 144 106356.

[13] KONG D, LI J, GUO A, et al. High temperature electromagnetic shielding shape memory polymer composite [J]. Chemical Engineering Journal, 2021, 408 127365.

[14] GAO H-L, ZHU Y-B, MAO L-B, et al. Super-elastic and fatigue resistant carbon material with lamellar multi-arch microstructure [J]. Nature Communications, 2016, 7 (1): 12920.

[15] SHEN B, LI Y, YI D, et al. Microcellular graphene foam for improved broadband electromagnetic interference shielding [J]. Carbon, 2016, 102 154-160.

[16] GAO L, WEI X-C, HUANG Y-T, et al. Analysis of Near-Field Shielding Effectiveness for the SiP Module [J]. 2018, 60 288-291.

Study of Efficient Automatic Detection of Coplanarity and Position of CCGA Devices

Qi Zhang*
Xi'an Microelectronic Technology Institute
Xi'an, China
374354293@qq.com

Xiaoyan Liu
Xi'an Microelectronic Technology Institute
Xi'an, China

Leida Chen
Xi'an Microelectronic Technology Institute
Xi'an, China

Xujing Nan
Xi'an Microelectronic Technology Institute
Xi'an, China.

Abstract—An efficient and fast method to detect the position and coplanarity of pins is obtained by combining optics and laser in this study. By using the optical camera to position, enlarge and auto focus the ceramic column grid array (CCGA) device, adjusting the light source of lens to realize the transformation of light and dark field, the pin can be accurately identified, and the pin image within the field of view can be collected through the rapid automatic contour capture technology, the pin coordinate can be efficiently extracted, and the device position data can be calculated quickly. Then, the coplanarity calculation would be figured out by scanning the pin height data with a line laser. Through a variety of means such as measuring microscope, the parameters of the detection method and equipment settings are optimized. The coplanarity and position of CCGA devices are detected conveniently, efficiently and quickly, and the detection efficiency and accuracy are significantly improved.

Keywords—*CCGA device, coplanarity and position, automatic detection*

I. INTRODUCTION

With the development of electronic packaging in the direction of miniaturization, multi-function and high reliability, CCGA devices have been widely used in the fields of aerospace and other harsh working environment with its unique advantages of high temperature resistance, high pressure resistance, multiple I/O quantity and high signal transmission efficiency[1]. The board level interconnection of CCGA devices is a crucial step to realize the function of the devices, and the position and coplanarity of CCGA devices are essential factors affecting the assembly. When the position and coplanarity of the device are poor, there will be bad contact between the pins and printed circuit board (PCB) pad in the assembly process, which is easy to cause problems such as lack of weld and false welding[2], leading to product failure directly in serious cases. Therefore, it is necessary to detect the position and coplanarity of CCGA devices effectively.

At present, position and coplanarity mostly can only be checked offline manually from experience. As we all know, CCGA devices usually possess a large number of pins, the efficiency of manual visual inspection is extremely low, and the accuracy of data obtained varies from person to person, resulting in poor repeatability and low traceability of the data, which is difficult to meet the quality control requirements in the aerospace field[3]. Consequently, it is very urgent to establish a high efficiency and high precision measurement method for CCGA device position and coplanarity.

In this study, an efficient and fast method to detect the position and coplanarity of CCGA devices is realized by combining optics and laser. The high precision and fast detection of the position and coplanarity accomplished the data of various CCGA devices. The measurement results can meet the actual requirements and improve detecting efficiency and accuracy of CCGA devices position and coplanarity inspection.

II. PRINCIPLE OF POSITION MEASUREMENT WITH OPTICS

Firstly, the optical imaging system collects the center coordinates of the mutually symmetrical pads on the four corners of the CCGA device, connects the center of the diagonal pads to form diagonal lines, and takes the intersection of the two diagonal lines as the origin to establish a coordinate system. According to the origin, the theoretical position of the welding columns can be imported to obtain the coordinates (X, Y) of the pins theoretical center point. The CAD measurement model is shown in Fig. 1. Then, with the help of adjustable light source, the optical imaging system could realize the distribution of light and dark fields on the top surface of the welding columns and the device itself, as shown in Fig. 2, so as to quickly identify the actual position of the welding columns in the field of view and obtain the coordinates of the center point (X1, Y1). According to formula (1), the position of the device can be calculated:

$$f = max[|X1 - X|], |Y1 - Y| \qquad (1)$$

From GB/T 36479-2018: Integrated circuits-Test methods for column grid array, the inspection standards for pin position of CCGA devices can be obtained, as shown in Table I[4].

TABLE I. INSPECTION STANDARD OF POSITION

Pad Pitch/mm	Maximum Position/μm
⩾1.27	150
1.00	125
0.80	100
⩽0.65	80

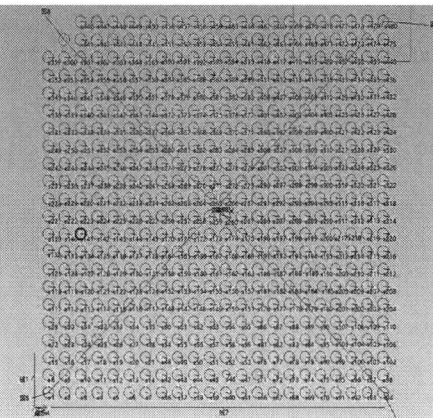

Fig.1. The CAD measurement model of CCGA device

Fig.2. The image for rapid automatic contour capture of pins

III. PRINCIPLE OF COPLANARITY DETECTION BY LASER

The coplanarity of CCGA device is measured by scanning the height of each column with line laser, and then the three highest columns vertices are selected to form a plane. If the plane formed by these three points contains the geometric center of the device, then the plane is selected as the base plane. If the constructed plane does not contain the geometric center of the device, the next highest column and two of the previous highest columns are used to reconstruct the available base plane. Taking the established base plane as the origin of Z axis, the Z value of each column vertex is obtained, and the coplanarity of the device is calculated according to formula (2) (Fig. 3):

$$Z = Zmax - Zmin \qquad (2)$$

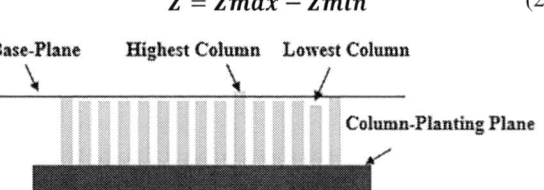

Fig.3. Coplanarity of CCGA devices

According to GB/T 36479-2018: Integrated circuits-Test methods for column grid array, if the coplanarity of CCGA device is more than 150 μm, this device is regarded as defective product.

IV. AUTOMATIC DETECTION METHOD OF POSITION AND COPLANARITY

The optics system is used to locate, enlarge and auto focus the device to be tested, adjusting the light source to realize the transformation of light and dark field, which can achieve the accurate recognition of pads and pins. The pins image within the field of view is collected by the rapid automatic contour capture technology, which can attend the efficient extraction of the pin center point coordinates, and the device position data can be obtained after calculation. Then, by scanning the pins height data with line laser system, base plane could be established to complete the coplanarity calculation. The specific steps of detection are as follows:

1) The optical system scans the symmetrical pads on the four corners of the device, and gets four circular elements in the program;

2) Invoking the center of four circles and connect them to form diagonals, and invoking the intersection of the diagonal as the coordinate origin;

3) Selecting array to create the theoretical pin circle of the device and set it as the elements to be measured;

4) Invoking all pins circles to create a plane, defined as the line laser scanning plane;

5) According to the formulas (1) and (2) to edit the output formula of calculation logic about the position and coplanarity;

6) Run the program for automatic detection, and output the detection report automatically after measurement completed. The report format can be word, Excel, PDF, TXT, etc.

After first detection, the measurement program can be saved, and the subsequent measurements can be automatically detected by using the measurement program directly. At the same time, in order to further improve the detection efficiency, the special detection tooling is made. Through a variety of measurement methods to modify the parameters and optimize the methods, the accurate automatic cycle detection of multiple devices can be realized, and the detection speed of the same batch of devices is greatly accelerated.

V. TEST ANALYSIS AND VERIFICATION

Selecting 10 CCGA devices with 484 pins and pitch of 1.27 mm in the same batch to detect the position and coplanarity with optical laser measuring system and a measuring microscope appraised by a third party simultaneously. The positional degree and coplanarity data obtained are shown in Table II.

TABLE II. POSITION AND COPLANARITY DATA OF 10 CCGA484 (PITCH 1.27)

Serial Number	Optical Laser		Measuring Microscope	
	Position /μm	Coplanarity /μm	Position /μm	Coplanarity /μm
1	56.8	53.7	57.2	55.1
2	41.7	58.9	40.5	54.6
3	26.2	29.2	25.4	30.3
4	32.7	31.4	35.2	33.7
5	57.4	39.8	55.8	36.5
6	39.4	52.8	36.6	51.4

Serial Number	Optical Laser		Measuring Microscope	
	Position /μm	Coplanarity /μm	Position /μm	Coplanarity /μm
7	94.8	49.9	95.7	52.2
8	65.3	51.2	63.5	50.3
9	77.6	68.3	75.1	69.6
10	81.3	60.7	80.5	64.3

It can be seen from the comparison that although there are some differences between the position and coplanarity data obtained by the optical laser measuring system and the measuring microscope, the measurement deviation is within 5 μm. Therefore, the measurement results of the optical laser measuring system can achieve satisfactory results. Moreover, the rapid automatic contour capture technology of the optical part can recognize and capture 35 pins at the same time, and complete the coordinates data acquisition of 484 pins in 45s. The height data acquisition of 484 pins by the line laser part can finish in 15s, and then the position and coplanarity data of the device can be obtained after automatic calculation.

In order to further verify the repeatability of the detection data of the optical laser measuring system, the CCGA484 device of No. 1 in Table II was repeatedly detected for 10 times, the position and coplanarity data arranged in Table III.

TABLE III.　10 TIMES POSITION AND COPLANARITY DATA OF SAME CCGA484 DEVICE

Detecting Time	Position/μm	Coplanarity/μm
1	56.8	53.7
2	55.9	53.1
3	58.3	52.2
4	57.6	50.7
5	54.1	54.0
6	56.4	53.9
7	54.7	52.6
8	53.4	51.3
9	57.2	51.8
10	55.6	54.5
Repeatability/μm	4.9	3.8

It can be seen from the measurement results that the repeatability of position and coplanarity of 10 times are 4.9 μm and 3.8 μm, respectively. The results show that the repeatability of the optical laser measuring system is very superior, and it can achieve high precision and high efficiency automatic detection of CCGA devices

Meanwhile, the results of CCGA devices in Table II and Table III all fulfill the requirements of GB/T 36479-2018: Integrated circuits-Test methods for column grid array, and the position < 150 μm, coplanarity < 150 μm.

In order to ensure the performance of a CCGA device, the position and coplanarity are essential inspection items. At the same time, the quality inspection of solder joints is also indispensable. Solder joint voids have a great influence on the welding reliability of CCGA devices, which can be inspected and identified by X-ray and strength tests.

Through the X-ray inspection of the solder joint, the solder joint has no voids, and the column and pad are well welded. The X-ray pictures of some solder joints are shown in Fig. 4.

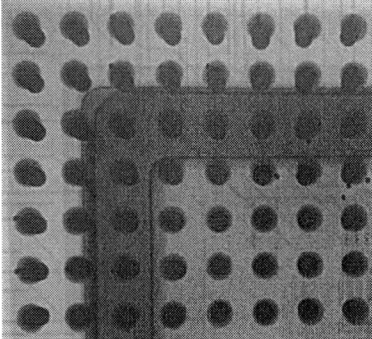

Fig.4. X-ray pictures of some solder joints

Furthermore, for checking the strength of solder joints, the shear and pull-off tests of CCGA484 are carried out, and the data obtained in Table IV.

TABLE IV.　SHEAR STRENGTH AND PULL-OFF STRENGTH OF CCGA484 WELDING COLUMN

Shear Strength/N	Pull-off Strength/N
10.51	6.48
10.39	6.06
10.34	6.41
10.42	6.13
10.71	7.05
10.33	6.21
10.38	6.91
10.53	6.13
10.41	6.17
10.46	6.33

The diameter of the columns used in this CCGA484 device is 0.51 mm, and its shear and pull-off strength should not be less than 5.6 N. It can be seen from the data in Table IV that the shear and pull-off strength far exceed the standard. The results of X-ray inspection and strength tests both showed that the reliability of solder joint is excellent.

In summary, the optical laser measuring system can achieve high efficiency and high precision automatic detection of CCGA devices, and the performances of CCGA484 devices used in the test process can meet the quality requirements of national standards.

VI. CONCLUSION

Through the establishment of CAD model, the CCGA device can be positioned quickly and the automatic detection of the device can be realized. The rapid automatic contour capture technology of the optics system can capture 35 pins coordinates data at the same time, complete the extraction of 484 pins coordinates within 45 seconds, and get the position data by operation, with repeatability less than 5 μm. The line laser system can collect the height of 484 pins and acquire coplanarity in 15 seconds, and the repeatability is less than 4 μm. The optical laser technology is applied to the automatic detection of the position and coplanarity of CCGA devices, which solves the problems of high difficulty and poor repeatability, significantly improved the detection accuracy and efficiency, enriched the detection means of CCGA devices, and could be conducive to the further application of CCGA devices.

REFERENCES

[1] Weiping Zhang, Liming Pang, Peng Huang, et al. Disassembly and welding process of electrical components of military photoelectric instruments[J]. Electronic Technology, 2015, 36 (5): 275-290.

[2] Cheng Yang, Yingying He, Chen Tan, et al. Research on coplanarity detection of airfoil lead surface mount integrated circuits[J]. Electronics and Packaging, 2017, 17 (5): 12-14.

[3] Tpipathi S., Patil H.M., Patil S.A., et al. Ceramic column grid array assembly qualification and reliability analysis for space missions [J]. IEEE Transactions on Components, Packaging and Manufacturing Technology, 2015, 5(2): 279-286.

[4] State Administration of market supervision and administration, China National Standardization Administration. Integratedcircuits—Test methods for column grid array: GB/T 36479-2018[S]. 2018.

Modeling and Simulation of Interconnection Structure Compensation Design in High-speed Modules

Qi Zheng
Microsystem Packaging
Research Center
Institute of
Microelectronics of
Chinese Academy of
Sciences
Beijing，China
zhengqi@ime.ac.cn

Huimin He
Microsystem Packaging
Research Center
Institute of
Microelectronics of
Chinese Academy of
Sciences
Beijing，China
hehuimin@ime.ac.cn

Lijuan Bai
Microsystem Packaging
Research Center
Institute of
Microelectronics of
Chinese Academy of
Sciences
Beijing，China
bailijuan@ime.ac.cn

Yubo Wang
Beijing smartchip
microelectronics
technology company
limited
Beijing，China
wangyubo@sgitg.sgcc.com
.cn

Dejian Li
Beijing smartchip
microelectronics
technology company
limited
Beijing，China
lidejian@sgitg.sgcc.com.c
n

Shunfeng Han
Beijing smartchip
microelectronics
technology company
limited
Beijing，China
hanshunfeng@sgitg.sgcc.c
om.cn

Bofu Li
Beijing smartchip
microelectronics
technology company
limited
Beijing，China
libofu@sgitg.sgcc.com.cn

Dameng Li
Beijing smartchip
microelectronics
technology company
limited
Beijing，China
lidameng@sgitg.sgcc.com.
cn

Fengman Liu
Microsystem Packaging
Research Center
Institute of
Microelectronics of
Chinese Academy of
Sciences
Beijing，China
liufengman@ime.ac.cn

Liqiang Cao
Microsystem Packaging
Research Center
Institute of
Microelectronics of
Chinese Academy of
Sciences
Beijing，China
caoliqiang@ime.ac.cn

Abstract—**As a basic component of optical communication network, the speed of optical modules is evolving towards 56G bit/s, which brings a great challenge to the signal integrity of interconnect structures. At present, optical chip and PCB board or substrate are interconnected mainly through gold wire bonding. Gold wire bonding has the property of low pass. The compensation method is studied in this paper. It is confirmed that the inductive compensation after the capacitive compensation using low pass filter theory can improve the performance of the transmission line.**

Keywords—Signal integrity, Wire bonding, Compensation

I. INTRODUCTION

With the rapid development of modern information industry, as well as the arrival of the era of Internet of Things, cloud computing and big data, the demand for information data capacity and processing speed is growing rapidly. As a medium of information transmission, light has achieved remarkable success in long-distance optical fiber communication systems because of its large bandwidth, small delay, low power consumption and free from electromagnetic interference.

Generally speaking, the interconnection between optical chip and PCB or substrate is realized by gold wire in optical module. Recently, with the application of advanced packaging technology, FC-BGA and other technologies have been gradually applied to the packaging of optical modules.

However, due to the difficulty in solving optical coupling problems, gold wire bonding still plays a dominant role in the packaging of optical modules.

With the further increase of the communication rate of optical modules, the gold wires bond faces great challenges in signal integrity. Gold wire bonding has the property of low pass. With the increase of the frequency, the effect of parasitic inductance will be obviously intensified[1].

In order to improve the high frequency transmission characteristics of the gold wire bond line, it is common practice to increase the capacitive component of the bond wire pad, such as increasing the width of the gold wire bonding pad, reducing the distance between the reference ground and the pad, etc[2]. However, since the impedance of the RF transmission line is basically matched by 50 ohm or differential 100 ohm impedance, such compensation structure tends to change the impedance at this point, forming a new discontinuity point, thus increasing the return loss[3].

The compensation method of single-ended signal line and differential signal line at the gold wire bonding is studied in this paper. The insertion loss and reflection of the uncompensated signal line are compared with that of the compensated signal line. The simulation results show a good consistency, which has a positive guiding significance for the package interconnection scheme of optical modules using gold wire bonding.

978-1-6654-1392-3/21 $31.00 © 2021 IEEE

The following is the organization of the paper:

In Section I, the challenge of signal integrity due to the further increase of the optical module rate is analyzed. The theoretical model and compensation model of gold wire bond are introduced, and the change of parameters of compensation method is analyzed in Section II. In Section III, 3D full wave electromagnetic field simulation is performed for different compensation methods and the thesis is summarized in Section IV.

II. THE THEORETICAL ANALYSIS

A. Theoretical model of gold wire bonding

The interconnection structure of the bond lines is shown in Figure 1. with optical chip/electric chip in the middle and PCB board or substrate on both sides, which form the traditional interconnection architecture of most optical modules

The equivalent circuit model of the gold wire bond line is shown in the Figure2. The pads on both sides can be equivalent to capacitance. Due to the small diameter of the gold wire, there is a large inductance effect, and at the same time, there is an equivalent resistance on the gold wire.

As can be seen from Fig. 2, the gold wire bond line has the characteristics of low pass. The inductance is related to the length of the gold wire, so we want the chip to be as close to the PCB or substrate as possible to reduce the parasitic inductance L1[4].

B. Theoretical model of compensation structure

In order to improve the high frequency transmission characteristics of the gold wire bond line, the capacitive component of the bond can be increased appropriately, as we have done before[5]. However, in order to further improve the high frequency performance, we introduce a low-pass filter model to further compensate the lead bond structure.

According to the filter principle, the cutoff frequency can be increased by increasing the order of the low-pass filter. Therefore, a third-order filter model is introduced, as shown in Figure 3. In an impedance matched interconnect, a thickened line corresponds to a capacitive transmission line and a narrowed line corresponds to an inductive transmission line. Therefore, the inherent inductance L1 of the bond line in Figure 2 is added to a third-order filter model, with a thick interconnect for capacitance and a thin interconnect for inductance to increase the bandwidth of the structure[6]-[7], as shown in Figure 4.

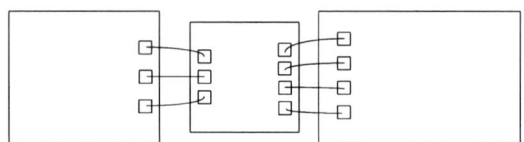

Fig.1. Schematic diagram of bonding line interconnection structure

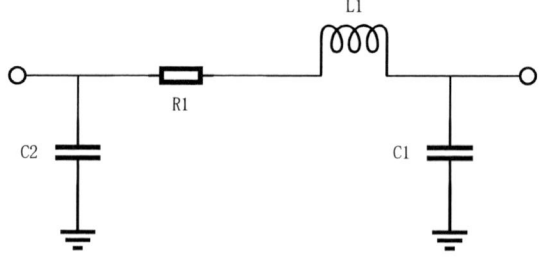

Fig.2. Equivalent circuit model of gold wire bond line

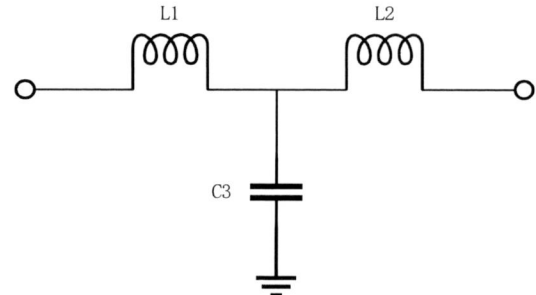

Fig.3. Third order filter model circuit diagram

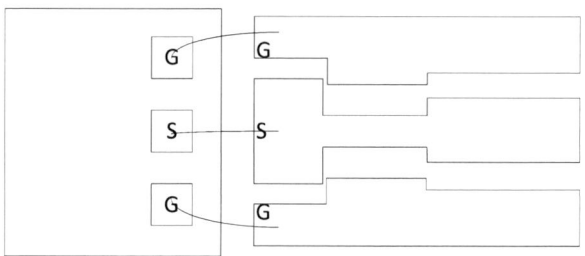

Fig.4. Schematic diagram of interconnection compensation structure

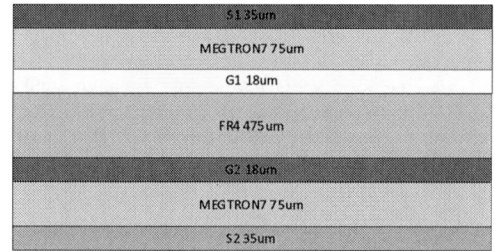

Fig.5. Schematic diagram of PCB lamination

Fig. 6. 3D model of GSG compensation structure

III. COMPENSATION STRUCTURE DESIGN AND SIMULATION

According to the theoretical analysis in the second part, a three-dimensional full-wave electromagnetic field simulation of the bond line interconnection structure based on this compensation structure is carried out.

In order to reduce the length of the bonding line, the distance between the chip pad and the chip edge is set to 100um. The maximum simulation frequency is 40GHz. The optical module adopts four layers of PCB board, and the laminated structure is shown in the Figure 5.

A. GSG compensation simulation

The compensation model of single-ended signal line (GSG) bonding is shown in the Figure6. After impedance matching, the normal line width is set to 120um and line space is set to 109um. The diameter of the ground hole is 150um.

When there is no compensation optimization, the high frequency performance of the structure is poor, with S11 close to -5dB and S21 close to -2.2dB. The inductive compensation method after inductive compensation is adopted at the bonding pad.

After optimization, the effect is the best when w1(the width of the first section) is 0.14mm, l1(the length of the first section) is 0.45mm and w2(the width of the second section) is 0.08mm, l2(the length of the second section) is 1.25mm. S11 is basically below -8dB and S21 is below -1.7dB. The comparison results for S11 are shown in Figure 7. It can be seen from this that the high frequency compensation has obvious effect, but based on the fabrication process of PCB board, the second narrow line width cannot be reduced to a smaller value.

To solve this problem, the reference ground plane on both sides of the second part is hollowed out, which can significantly increase the distance between the signal line and the reference ground in the coplanar waveguide, thus reducing the capacitance. The Figure 8 shows better result. In this way, by increasing the line distance, the difficult problem of manufacturing technology is solved.

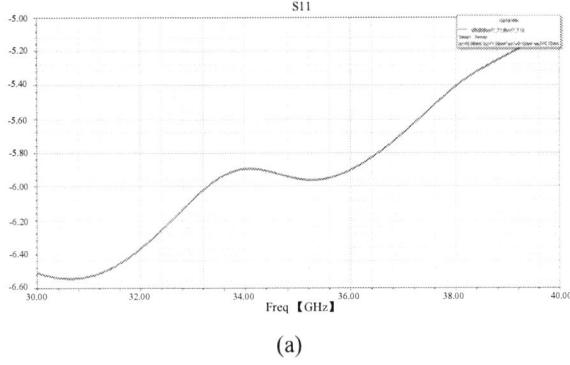

(a)

B. GSSG compensation simulation

Similarly, the gold wire bond line simulation of the differential structure is carried out. Figure 9 is the schematic diagram of the model structure. After impedance matching, the normal line width is set to 105um, the differential line space is set to 120um and line to ground space is set to 109um . The diameter of the ground hole is 150um.

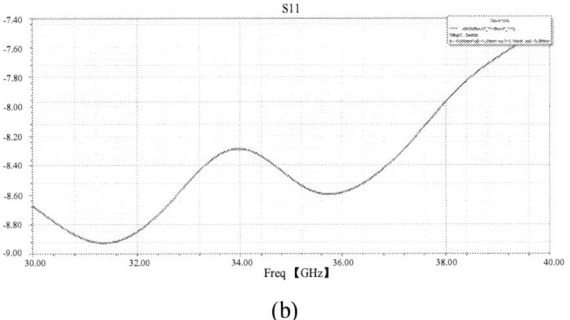

(b)

Fig.7. S11 (a)without compensation optimization (b)with compensation optimization

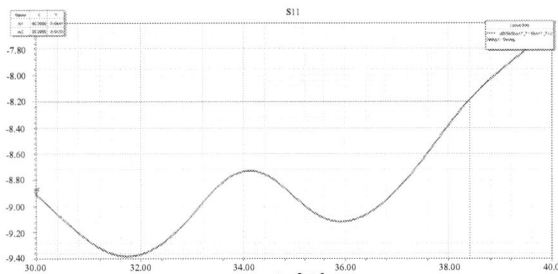

Fig.8. The reflection by hollowing out the reference ground on both sides

The simulation results have similar characteristics with GSG structure. The high frequency characteristics of the interconnect structure can be improved by adopting the compensation method of fore-inductive and post-inductive. Fig. 10 shows the difference of reflection coefficients in the case of non-optimized, and two-stage compensation structure （w1:0.255um, l1:0.6um, w2: 0.025um, l2: 0.4um）. It can be seen that the high frequency characteristic is the best through a multi-stage compensation structure.

At the same time, the simulation results show that the smaller the line width of the second segment is, the lower the reflection is. However, due to the technical limitations, the effect of inductance can only be realized by widening the line distance.

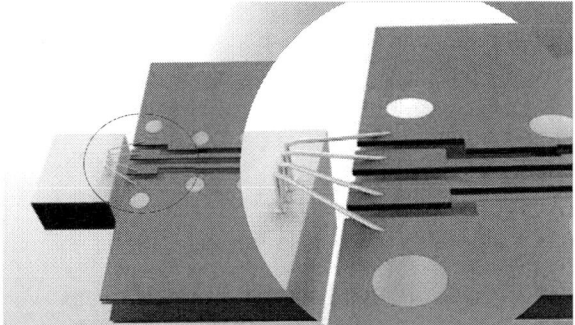

Fig.9. 3D model of GSSG compensation structure

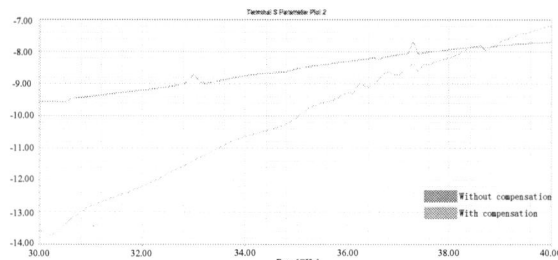

Fig.10. S11 of GSSG structure

IV. CONCLUSION

Gold wire bonding has a bottleneck in the increasing bandwidth of optical modules due to the larger inductive parasitic effect at high frequencies. In this paper, the circuit model of gold wire bonding is analyzed, and the method of interconnection compensation structure at high frequency is given according to the theory of filter.

According to the guidance of the circuit model, the three-dimensional full-wave electromagnetic field simulation of the gold wire bond line and its interconnection compensation structure is carried out. The results show that the compensation structure of the third-order low-pass filter has a good compensation effect on the performance of gold wire bonding at high frequencies. In order to meet the design requirements of PCB or substrate technology, the method of indenting the reference ground plane is used to realize the inductive compensation of the second section, and the simulation results show that it has a good effect. The results have a positive guiding significance for the further evolution of optical modules to high bandwidth

REFERENCES

[1] W. Jiang, M. Tang, J. Mao, H. Yue and Y. Tang, "A Novel Compensation Technique for Bonding-Wire Interconnection Based on Single-Objective Optimization in Millimeter-Wave Band," 2018 International Conference on Microwave and Millimeter Wave Technology (ICMMT), Chengdu, China, 2018, pp. 1-3

[2] H. Zhu, Y. Sun and X. Wu, "Investigation of the capacitance compensation structure for wire-bonding interconnection in multi-chips module," 2017 IEEE Electrical Design of Advanced Packaging and Systems Symposium (EDAPS), Haining, China, 2017, pp. 1-3.

[3] Y. P. Zhang and D. Liu, "Antenna-on-Chip and Antenna-in-Package Solutions to Highly Integrated Millimeter-Wave Devices for Wireless Communications," in IEEE Transactions on Antennas and Propagation, vol. 57, no. 10, pp. 2830-2841

[4] Hai-Young Lee, "Wideband characterization of a typical bonding wire for microwave and millimeter-wave integrated circuits," in IEEE Transactions on Microwave Theory and Techniques, vol. 43, no. 1, pp. 63-68, Jan. 1995, doi: 10.1109/22.363006.

[5] Q. Zheng et al., "Research on Optical interconnect transmitter based on Mach-Zehnder silicon optical chip," 2020 21st International Conference on Electronic Packaging Technology (ICEPT), 2020, pp. 1-4, doi: 10.1109/ICEPT50128.2020.9202969.

[6] T. P. Budka, "Wide-bandwidth millimeter-wave bond-wire interconnects," in IEEE Transactions on Microwave Theory and Techniques, vol. 49, no. 4, pp. 715-718, April 2001, doi: 10.1109/22.915447.

[7] F. Alimenti, P. Mezzanotte, L. Roselli and R. Sorrentino, "Modeling and characterization of the bonding-wire interconnection," in IEEE Transactions on Microwave Theory and Techniques, vol. 49, no. 1, pp. 142-150, Jan. 2001, doi: 10.1109/22.899975.

ACKNOWLEDGMENT

This paper is funded by the peoject of the state grid corporation of china in 2019 "The key technologies research of the special mulit-core SoC architecture for relay protection equipments". The number of the project is 5700-201941501A-0-0-00.

Study on Gold Wire Sweep in Cantilever Chip-Stacked Package during Molding Process

Sicheng Cao
School of Mechanical and Electrical
Engineering , Guilin University of
Electronic Technology
Guilin, China
cao_sicheng@163.com

Daoguo Yang
School of Mechanical and Electrical
Engineering , Guilin University of
Electronic Technology
Guilin, China
daoguo_yang@163.com

Wangyun Li*
School of Mechanical and Electrical
Engineering , Guilin University of
Electronic Technology
Guilin, China
li.wangyun@guet.edu.cn

Xiyou Wang
School of Mechanical and Electrical
Engineering , Guilin University of
Electronic Technology
Guilin, China
wxy_07@126.com

Shirui Xue
School of Mechanical and Electrical
Engineering , Guilin University of
Electronic Technology
Guilin, China
shiruixue4268@163.com

Zhanfei Yun
School of Mechanical and Electrical
Engineering , Guilin University of
Electronic Technology
Guilin, China
zf_yun@163.com

Abstract—With the continuous improvement on the function of electronics, the IC integration density is gradually increased. Meanwhile, the stacked chip packaging technology appeared to meet the requirements of chip integration. Among them, the cantilever stacked chip structure not only increases the package density, but also satisfies the need of vertically integrated chips with the same size. However, the gold wire sweep induced during the transfer molding process is still one of the critical issues for the reliability of the cantilever stack structure. Therefore, it is of great significance to explore the wire offset in the process of transfer molding to enhance the reliability of stacked chip packaging.

In this paper, a mold-flow analysis model of cantilevered chip-stacked structure is setup. The viscoelastic properties were measured by using a DMA analyzer. For studying the deviation of wire loop in transfer molding, the process parameters including the molding temperature, filling time and melt temperature are analyzed by finite element simulation. Firstly, the effect caused by different positions of the gate on the wire offset also is studied. Secondly, the influence of the above process parameters on the position deviation of gold wire is studied following an orthogonal experiment route. The results show that the inhomogeneous temperature field caused by transfer molding has an affect the quality of the chip wire bonding, while the melt temperature has a great influence on the wire offset. Due to the influence of the flow rate of transfer molding, the closer the wire to the gate position is, the larger the offset will be, especially for the wire perpendicular to the gate position.

Keywords—stacked chip, cantilever, wire sweep, transfer molding

I. INTRODUCTION

Chip-Stacked packaging technology is a widely used three-dimensional packaging technology. It can double the capacity of storage devices and diversify the capabilities of chips by stacking multiple chips vertically [1]. Since the cantilever stacked chip structure can integrate chips of the same size vertically, it is an effective way to encapsulate storage, logic and other chips, which has gradually become a research hotspot of stack chip package technology [2]. However, due to the structure of the package form, the defects such as the wire sweep and the warpage deformation of the package appearing during the process of package are the main factors affecting the reliability of the package structure [3]. Taking wire sweep as an example, when the degree of wire sweep is too high, it is easy to cause short circuit or circuit break, declination of the chip reliability, package failure and other problems.

Among the existing packaging schemes of stacked chips, wire bonding is still an economical and reliable way of interconnection [4]. As the height of chip stacking gradually increases, the requirement for the reliability of the lead is raised higher and higher in the transfer molding. Ali et al. [5] researched the effect of wire sweep by describing the flow when the epoxy molding compound was poured into single LQFP. It caused a bigger wire sweep index because of the faster flow at the gate. Besides, they also found that the 90° flow turning effect occurs when the melt flows along the edge of the mold to the corners of the mold. Melt affected by the flow turning effect joined melt undeflected at the edge, which led to that the wire received more impact force. Ramdan et al. [6] studied the deformation degree of a single die and stacked die in PBGA package with different wire spans through experiments and FSI simulation. The result showed that the wire sweep index of stacked die increased 57.4% compared to a single die. Meanwhile, it proved that the biggest wire sweep index has occurred at the place where wire bonding was perpendicular to the direction of melt flow.

To sum up, this article adopts Moldflow to simulate the process of melt flow inside the package, to explore how process parameters affect wire sweep in the process of transfer molding. First of all, analyzing how gates at different locations influence the direction of melt flow around the chip. Next, according to filling time, molding temperature, melt temperature, and other process parameters, researching the effect of wire sweep by simulating the melt flow. Finally, the best group of process parameters will be set up through orthogonal experiment. It is beneficial to provide guidance for the production.

978-1-6654-1392-3/21 $31.00 © 2021 IEEE

II. MODEL SETUP

EMC (Epoxy Molding Compound) is a composite material which is made up of epoxy resin, fillers, curing agents, etc. It is formed by cross-linking reaction under high temperature. Its performance depends on the epoxy matrix and the amount of additives added. To fit the deviation of the wire sweep in the actual production, it is necessary to measure the material properties of EMC and calculate the resistance of the wire sweep. The model for mold flow analysis should be set up.

A. Performance Test of EMC

The packaging materials used in this study are produced by Showa Denko, which belong to the series named CEL-9240HF10. In the preparation, it was first in-mold cured 0.5 hour at 175 °C. Afterward, it was put in the high-temperature oven and cured for 3 hours at 200 °C to obtain fully cured material. After that, the fully cured sample was processed into cuboid around at 45 mm × 8.5 mm × 1.5 mm. The viscoelastic properties of the sample were measured by the DMA Q800 analyzer. The examination conditions of the epoxy sealing material sample are shown in Table 1. Fig. 1 demonstrates the curves of storage modulus and loss modulus at different temperatures.

TABLE I.　　DMA PARAMETER SETTING

Geometry size	45 mm×8.5 mm×1.5 mm
Module	DMA Multi-Frequency-Strain
Method	Temperature ramp
Frequency	1 hz, 3 hz, 5 hz, 10 hz, 20 hz
Amplitude	8 μm
Ramp	3.00 °C/min to 180 °C

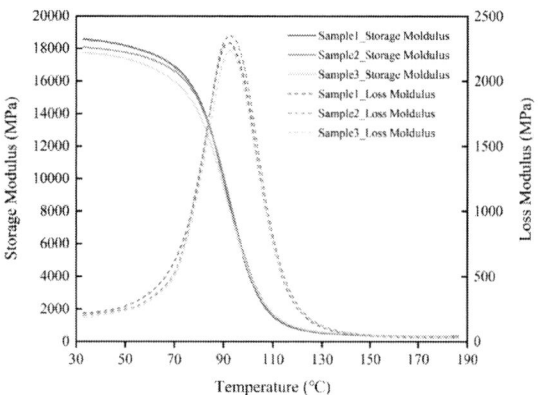

Fig. 1. Material performance curve of EMC

B. Modeling of Stacked Chip

The paper mainly investigates how different process parameters affect the wire sweep in the transfer molding. For the simplicity reason, the circuit distribution on the substrate surface is omitted and the number of wires is reduced to simplify model without affecting the result. The thickness of chip is 200 μm and the diameter of gold wire is 25 μm. To avoid short circuit of wire contact caused by insufficient height between chips, the height of wire is designed as 150 μm. A 3D model is established by SolidWorks, as shown in Fig. 2.

Fig. 2. Cantilevered stacked chip structure

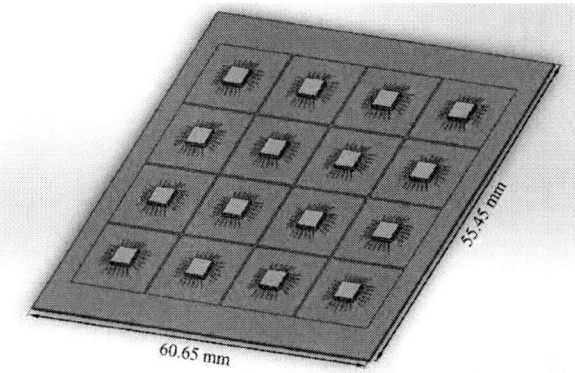

Fig. 3. Chips on the substrate

C. Wire Sweep Theory

The wire sweep is one of the inevitable issues in transfer molding, and excessive wire sweep index (WSI) can lead to ineffectiveness of the device. Therefore, the section will elaborate briefly about theories relevant with wire sweep and judgment criteria to provide the theoretical basis for following simulated analysis. The melt flow will generate a drag force on the gold wire during the transfer molding, the wire sweep index is an indicator to evaluate the degree of the wire deformation. To ensure the reliability of package, WSI should not exceed 5 %.

$$F = \frac{C_P \rho U^2 d}{2} \qquad (1)$$

$$WSI(\%) = D_N/L \qquad (2)$$

Where F is the force on the gold wire per unit length; P is the density of molding compound; U is the fluid velocity; d is the diameter of gold wire; C_P is the viscosity coefficient; L represents the projected length of wire; D_N is the biggest offset perpendicular to the wire.

III. RESULTS AND ANALYSIS

A. Gate Location Analysis

The gate connects channels and mold cavity. The quality of the transfer molding is affected by the gate parameters, such as location, shape and size. The gate is made of top and bottom molds as shown in Fig. 4. The gate position determines the balance of melt flow. However, the unreasonable gate position could cause the defect of excessive wire deformation in the transfer molding. In this article, two gate locations are considered, as shown in Fig. 5.

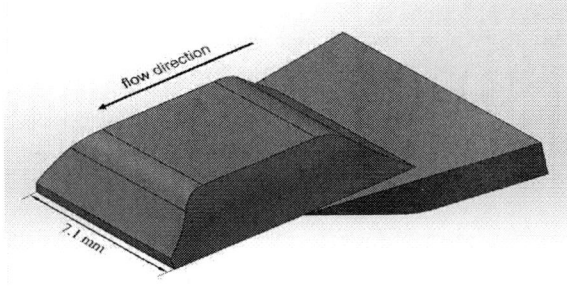

Fig. 4. Gate model

Under the default process parameters (filling time 10.5 s, molding temperature 175 °C, melt temperature 80 °C, Cure time 120 s), when keeping the filling time consistent, the melt flow rate is ensured to be stable in a certain range. And then simulating the transfer molding process of the two gate locations. Fig. 5 demonstrates the process of the filling area with time.

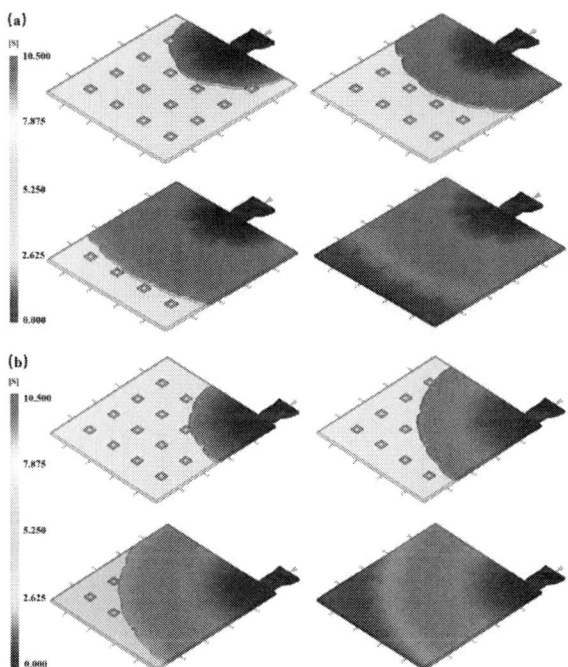

Fig. 5. Filling area with time (a) middle location; (b) side location

Fig. 6 shows the melt temperature change near the gate. As the color of the filling area becomes brighter, the better the fluidity of the melt, the impact ability of the melt flow on the wire sweep gradually decreases. Therefore, it is important to focus on the chip near gates.

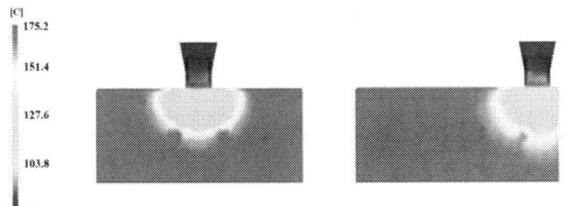

Fig. 6. Melt temperature with flling area

Fig. 7 shows the distribution of wire sweep and shear stress with the gold wire, which is about the location of gates. Fig. 7 (a) and Fig. 7 (b) are about the former location. Fig. 7 (c) and Fig. 7 (d) are about the later location. The maximum deformation occurs in the middle of gold wire. In either case, the maximum wire deformation occurred on the top chip, reaching 16.7 μm and 30.5 μm respectively, which are corresponding to WSI 0.92% and 1.69% both in the acceptable range of 5%. Wire spans in the top chip are larger than others in the bottom chip. Therefore, the area of the gold wire impacted by the melt is larger, and the gold wire is easier to deform with the melt flow. Moreover, the maximum shear stress concentrates on the bonding point of gold wires. That's due to bonding points were fixed and the impact force of melt is relatively large. The middle of the gold wire can shift with melt flow, so it receives less shear stress. In addition, the shear stress of gold wire reaches 92.52 MPa in former, while it reaches 174.4 MPa in latter. Compared with the former, there is a possibility of fillet lifting in latter.

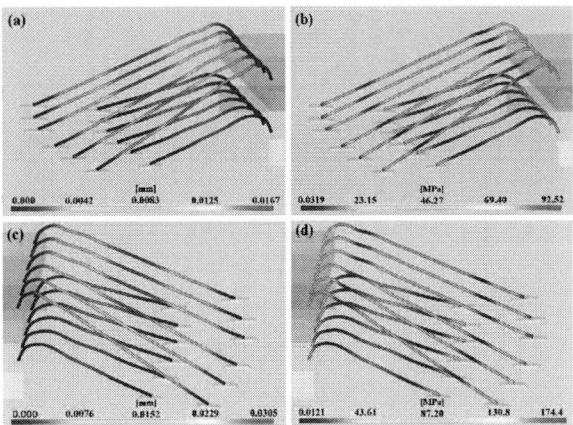

Fig. 7. The distribution of wire sweep and shear stress

When gate is lied in the middle of feed side, the melt flow is more average, even the wire sweep deformation decreases 13.8 μm. So, to improve the packaging reliability of the structure, the gate may be close to the middle of feed side.

B. Optimization of the Process Parameters

There is an indispensable relationship between the packaging quality and various process parameters in the transfer molding, and the degree of influence varies. For different locations of gates, three factors of filling time (A), molding temperature (B), and melt temperature (C) are analyzed by the orthogonal experiment. The molding temperature is controlled at 170-180 °C, filling time 9.5-11.5 s and melt temperature 70-90 °C. These parameters are shown in Table 2. The analysis will refer intermediate gate on the feed side as Design A, and the corner gate on the feed side as Design B for the convenience of expression.

TABLE II. WIRE SWEEP PERFORMANCE ORTHOGONAL TEST FACTOR DESIGN

Level	Experimental Factor		
	Filling time (s) (A)	Molding temperature (°C) (B)	Melt temperature (°C) (C)
1	9.5	170	70
2	10.5	175	80
3	11.5	180	90

TABLE III. WIRE SWEEP PERFORMANCE ORTHOGONAL TEST DATA

Experiment Number	Experimental Factor			Experimental Result	
	A	B	C	Maximum wire sweep in Design A (μm)	Maximum wire sweep in Design B (μm)
1	1	1	1	21.43	49.7
2	1	2	2	17.61	36.4
3	1	3	3	18.95	26.8
4	2	1	2	17.50	35.0
5	2	2	3	16.73	28.9
6	2	3	1	21.32	29.4
7	3	1	3	15.73	27.4
8	3	2	1	18.93	28.0
9	3	3	2	20.51	27.0

TABLE IV. WIRE SWEEP PERFORMANCE ORTHOGONAL EXPERIMENTAL RANG ANALYSIS DATA

Experiment Number		Experimental Factor		
		A	B	C
Design A	K1	19.330	18.220	20.560
	K2	18.517	17.757	18.540
	K3	18.390	20.206	18.460
	R1	0.940	2.503	3.423
Design B	K1	37.633	37.367	35.700
	K2	31.100	31.100	32.800
	K3	27.467	27.733	27.700
	R1	10.166	9.634	8.000

Range analysis indicates that the optimal group of process parameters for transfer molding corresponding to the wire sweep at different gate positions is different. As for Design A, the degree of influence with various factors on the wire sweep shows: melt temperature (C) > molding temperature (B) > filling time (A). Because the gate is located in the middle of feed side and the melt flow does not directly affect the gold wire. However, there is still a gap between melt temperature and molding temperature. The viscoelastic properties of EMC are not stable and the fluidity is poor. Therefore, the wire near the gate on the chip will bear a large drag force, resulting in a larger deformation of the gold wire. As for Design B, the degree of influence of various factors on wire sweep shows: filling time (A) > molding temperature (B) > melt temperature (C). The melt flow directly will impact the gold wire when the gate is closed to the chip near the corner. With the decrease of filling time and the melt flow rate increases, the effect on the wire sweep is greater than melt viscosity.

Based on the variation trend of the maximum wire sweep, it can be concluded that Design A is better than Design B. In Design A, the influence of the melt temperature and mold temperature is much larger than the filling time. But, the effect of filling time is not ignored，due to that filling time

directly affects the melt flow velocity, thereby impacting the melt temperature when it contacts the gold wire. Therefore, when designing and applying this kind of cantilever stacked chip considering the influence of process parameters on packaging reliability. Therefore, we should follow this order which is melt temperature (C) > molding temperature (B) > filling time (A) to analyze the packaging reliability.

IV. CONCLUSION

This paper uses Moldflow to simulate the gold wire sweep in cantilever stacked chip package during the molding process, with the purpose to predict the actual gold wire sweep and reduce unnecessary mold tests. Under the same process parameters, different gate positions can significantly affect the shear stress and the degree of gold wire sweep. In the gate location analysis, due to the low melt temperature, the melt fluidity is low. The maximum wire deformation and the maximum shear stress occur at the wire of top chip near the gate. The shear stress on bonding points and the deformation in the middle of gold wires should be paid attention to avoid defects such as fillet lifting and short circuit. Through the orthogonal experiment, the conclusion can be drawn that the gate in the middle of feed side is better than the gate in the corner of feed side. For the Above results, the degree of influence of various factors on the wire sweep in the middle of feed side show in the following order: melt temperature > molding temperature > filling time. From the perspective of actual production, the rationality of gates should be considered firstly. And then, some process parameters should be optimized.

ACKNOWLEDGMENT

The research was supported by the Innovation-Driven Development Project of Guangxi Province (grant No. GuiKeAA21077015), and the Natural Science Foundation of Guangxi Province under Grant No. 2018GXNSFBA281065.

REFERENCES

[1] C.Y. Khor, M.Z. Abdullah, Z.M. Ariff, and W.C. Leong, "Effect of stacking chips and inlet positions on void formation in the encapsulation of 3D stacked flip-chip package," Int. Commun. Heat Mass, vol. 39, pp. 670-680, 2012.

[2] Y. Tang, S.M. Luo, G.Y. Li, Z. Yang, R. Chen, Y. Han et al., "Optimization of the thermal reliability of a four-tier die-stacked SiP structure using finite element analysis and the Taguchi method," Microelectron. J., vol. 73, pp. 18-23, 2018.

[3] H. K. Huang and C. L. Hsieh, "Effects of stacked layers and stacked configurations on wire sweep and wire sag of advanced overhang/ pyramid stacked packages," J. Electron. Packing, vol. 139, 2017.

[4] W. K. Mak, Y. C. Lin, C. Chu and T. C. Wang, "Pad Assignment for Die-Stacking System-in-Package Design," IEEE T. Comput. Aid. D., vol. 31, pp. 1711-1722, 2012.

[5] S. S. S. Ali, S. T. S. Hian and B. C. Ang, "The effects of wire geometry and wire layout on wire sweep performance using LQFP packages in transfer mold," Int. J. Precis. Eng. Man., vol. 15, pp. 1793-1799, 2014.

[6] D. Ramdan, M. Z. Abdullah, C. Y. Khor, W. C. Leong, W. K. Loh and C. K. Ooi et al., "Fluid/Structure Interaction Investigation in PBGA Packaging," IEEE T. Comp. Pack. Man., vol. 2, pp. 1786-1795, 2012.

X-shaped Through Glass Via and Its Transmission Performance in Ka Band

1st Qiangwen Wang
The 38th Research Institute
China Electronics Technology Group
Corporation
Hefei, China
wangqiangwen0722@163.com

2nd Yuhua Guo
The 38th Research Institute
Key Lab of Aperture Array and Space
Application (KLAASA)
China Electronics Technology Group
Corporation
Hefei, China
13866778611@163.com

3rd Rui Wang
The 38th Research Institute
China Electronics Technology Group
Corporation
Hefei, China
7534727@qq.com

Abstract—Based on unique material characteristics e.g., low dielectric loss, high insulation property and low cost, glass substrate has incomparable advantages over silicon and organic substrates, which has been paid more attention in high frequency microwave applications, such as antennas on quartz wafer and T/R modules on glass interposers. Through glass via (TGV) is the key technology of glass substrate and the fabrication methods have been further studied in recent years. In order to reduce the difficulties of filling metal in the through hole, laser induced deep etching (LIDE) technology is applied for the formation and produces a kind of X-shaped TGV. For microwave applications, it is important to investigate the vertical transmission performance of this kind of TGV. In this paper, transmission structures based on X-shaped TGVs have been designed, simulated, fabricated, and measured by using SCOTT AF32 glass wafer as the substrate and Ka band (32~38GHz) as test frequency band. From the photo of the cross-section of the fabricated X-shaped TGV has a good copper filling without void. And the SEM image shows that the top side width is 42~ 45μm and the middle waist width is about 20μm. The transmission performance such as return loss and insert loss are tested by prober and vector network analyzer. The result shows that the insert loss is within 0.2dB per via in the range of 32~38GHz, which can meet the requirements of high frequency microwave applications.

Keywords—interconnection, TGV, transmission performance

I. INTRODUCTION

As the increasing demands of high integration, low cost and miniaturized RF front-end modules for phased array radar or 5G applications, next generation of RF integration architecture by using new materials and 3D integration methods has been widely investigated. To choose appropriate substrate material is key to the success. Based on unique material characteristics, glass substrate has incomparable advantages over silicon and organic substrates[1]. Firstly, the dielectric loss of glass substrate can be as low as 1/10000(e.g. quartz), which is one order of magnitude lower than that of silicon substrate. It is very suitable for high frequency microwave applications such as microstrip antenna on quartz wafer. Secondly, the manufacturing cost of a glass substrate is much less than a silicon one. Due to the high insulation property of glass (10^{13}~10^{16}Ω•cm), there is no need of insulation layer in the processing. At the same time, glass can process a large area (>10m^2) at one time, which greatly improves the yield compared with silicon wafer with a maximum of 12 inches. Thus, in the 3D integration of T/R modules, using glass substrate becomes a more promising method[2,3].

Through glass via (TGV) is one of the key technologies of glass substrate and the fabrication methods have been

further studied in recent years[4-6]. Because the metal filling of TGV is by electroplating in the through hole, it is more difficult to get void-free filling than the plating in the blind hole[7-9]. In order to improve via quality and reduce the difficulties of via filling, laser induced deep etching (LIDE) technology is applied for the via formation[10,11]. And the Bessel beam of picosecond laser is usually used following with wet etching and corrosion reaming, which produces a kind of X-shaped TGV. From the cross-section of X-shaped TGV, there are wide openings both on the top side and the bottom side, and a narrow waist in the middle. Compared to the through hole, it is easier to be covered by the seed layers in physical vapor deposition (PVD) process and to fill the metal in the following electroplating. With more and more use of LIDE technology, it is important to investigate the vertical transmission performance of this kind of TGV especially for high frequency microwave applications[12].

In this paper, the design and transmission performance simulation analysis are completed by modeling firstly. The 3D model of coplanar waveguide (CPW) transmission structure including two X-shaped TGVs and three transmission lines is built. At the same time, the model of a straight transmission line with equal length is also built to calculate the loss of a single X-shaped TGV. The model is optimized by ANSYS High Frequency Structure Simulator (HFSS) to fix the dimensions of TGV structures. Secondly, the glass substrate with the designed transmission structures are produced by LIDE technology using SCOTT AF32 glass wafer. Last, the transmission performances of the fabricated transmission structures are tested in Ka band (32~38GHz). The return loss and insert loss of the transmission structure with two X-shaped TGVs and the straight transmission line are tested by the prober and vector network analyzer. Thus the insert loss of a single X-shaped TGV can be calculated.

II. DESIGN OF TGV TRANSMISSION STRUCTURE

Compared with other types of glass, AF32 glass has the possibility of ultra-thin processing and good electrode adhesion. The dielectric constant is 5.1 and it is stable with the increase of frequency. The dielectric loss is relatively small (~10^{-3}@5GHz) and the coefficient of thermal expansion (CTE) is 3.2×10^{-6} /K and well match that of silicon. Based on the above excellent characteristics, it is very suitable for glass substrate.

The 3D model of CPW transmission structure is designed on SCOTT AF32 glass substrate, as shown in Fig. 1. The signal part in the middle is consisted of two X-shaped TGVs, two transmission lines on the surface and a transmission line on the bottom. The other part is the large-area electrical ground plane. The ground planes on the surface and the bottom are interconnected by a large number of TGVs, also in the form of X-shape by using the same fabrication method. At the same time, the model of a straight transmission line with equal length is also built to calculate the insert loss of a single X-shaped TGV.

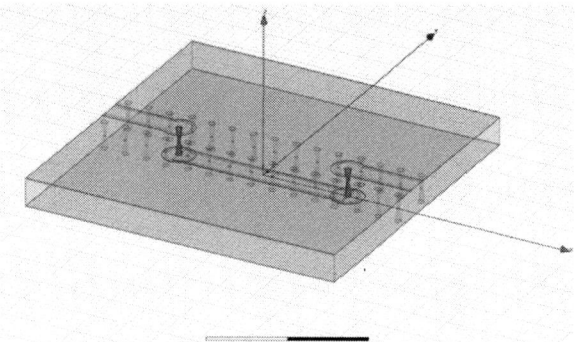

Fig. 1. 3D model of transmission structure with X-shaped TGVs.

The initial thickness of AF32 glass is 300μm. In the fabrication process especially in the step of chemical mechanical polishing (CMP), the thickness of glass substrate will reduce to ~180μm. Thus, the straight transmission line with CPW transmission form is designed on a AF32 glass wafer in the thickness of 180μm. The specific structure and dimensions optimized by HFSS simulation of the straight transmission line is shown in Fig. 2. The total length is 2.2mm and the width is 120μm. The space between the transmission line and the large-area ground plane on the surface side is 25μm. The grounding on the surface and bottom is interconnected by a number of TGVs. The diameter of grounding TGVs is 50μm. The center space between the adjacent TGVs is 150μm.

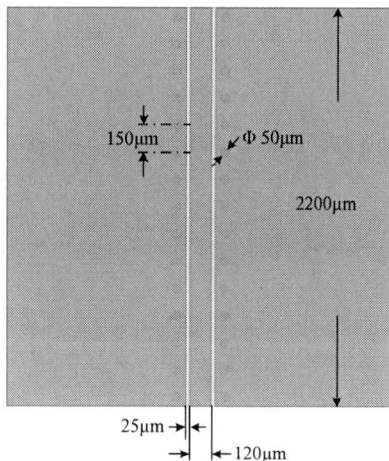

Fig. 2. The dimension of a straight transmission line.

The CPW transmission structure is also designed on a AF32 glass wafer in the thickness of 180μm. The 3D model of CPW transmission structure is shown in Fig. 1 and the specific dimensions optimized by HFSS simulation are shown in Fig. 3. The total length of three transmission lines is 2.2mm which is equal to the straight transmission line. The width of transmission line is 120μm. The space between the transmission line and the large-area ground plane on the surface side is 25μm. The diameter of grounding TGVs is

50μm. The center space between the adjacent TGVs is 150μm. The diameter of two signal X-shaped TGVs is also 50μm. The center space between the two signal X-shaped TGVs is 1200μm. The ground plane surrounding the signal TGVs is designed as a circle with diameter of 220μm.

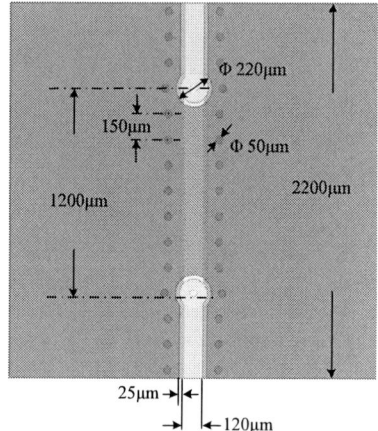

Fig. 3. The dimension of TGV Transmission structure.

The dimensions of the transmission structure and the transmission line given above have been optimized by means of ANSYS HFSS simulation. The simulated return loss and insert loss of transmission structure with two X-shaped TGVs are shown in Fig. 4 and Fig. 5. The simulation result shows that S_{11} value of transmission structure is about -56 ~ -38dB and the S_{21} value of transmission structure is 0.203 ~ 0.229 dB in the frequency range of 32 ~ 38 GHz, which indicates good RF performance.

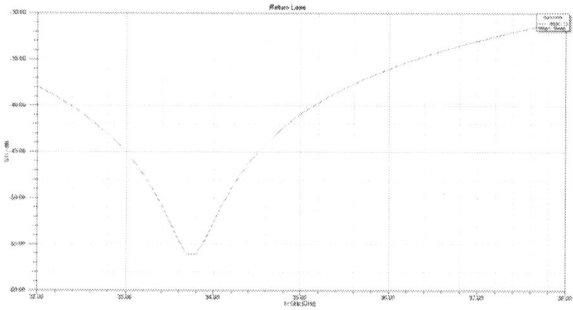

Fig. 4. The simulated return loss of X-shaped TGV transmission structure.

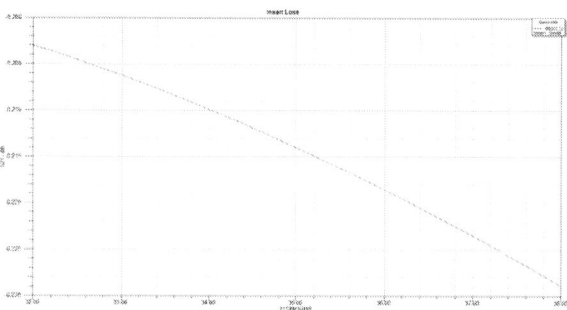

Fig. 5. The simulated insert loss of X-shaped TGV transmission structure.

III. FABRICATION OF TRANSMISSION STRUCTURE

The glass substrate with the designed transmission structure and transmission line is fabricated by LIDE technology using 4-inch SCOTT AF32 glass wafer with the thickness of 300μm. The diagram of the fabrication process is shown in Fig. 6.

Cleaning the glass wafer is the first step. Acetone and IPA solution are used to rinse the wafer assisted with ultrasonic. After deionized water cleaning, the glass is baked in oven at 180℃.

Step 2 is laser induced modification on AF32 glass wafer. As shown in Fig.7, a picosecond laser Bessel beam focused by an axicon lens has a larger focal depth (L) compared with Gaussian beam, which is very suitable for micro via formation in the full thickness range of glass wafer. The micro- activated via by Bessel beam is a biconical shape with a diameter of 2~3μm on the top and bottom and a diameter of 1~2μm in the waist. The laser modified micro vias has a higher selectivity (≥100:1) to strong alkali corrosion solution compared with other regions without modification.

In step 3, micro vias is formed by wet etching. Then through-holes with bigger diameter (e.g.50μm) is formed by corrosion reaming in step 4. The etched micro vias inherit the waist shape of the Bessel beam and thus form X-shaped through vias. The basis of glass is SiO_2, so the main composition of corrosion solution is HF. The corrosion rate of not diluted HF (49 wt.%) can be 1.8μm/min, so buffered HF is used usually. The constituent of buffered oxide enchant (BOE) is NH_4F(40 wt.%):HF(49 wt.%)=7:1 mix solution. The higher the ratio of BOE, the lower the corrosion rate of glass (SiO_2). Mixing a small quantity of diluted HCL (9 wt.%) in BOE corrosion solution can make the corrosion surface smoother and the corrosion rate can be more than 1μm/min. The designed thickness of the glass substrate is 200μm. Since corrosion is isotropic, each surface of glass substrate will be corroded about 25μm. In consideration of following chemical mechanic polishing (CMP) process, the glass wafer with a thickness 300μm is chosen.

And then the X-shaped vias is deposited with metal seed-layers by PVD of Ti/Cu (50nm/100nm) from both sides and filled by copper electroplating in step 5 and 6. To get a void-free TGV, a small current (0.5ASD) is chosen to fill in the vias. Different from through hole electroplating, at the first stage of plating the electroplated copper will connect quite quickly in the middle waist of the X-shaped via and forms a support base for both sides' copper growth, which greatly reduces the risk of voids in the subsequent plating process. The over-plated copper will be polished by chemical mechanic polishing (CMP) which reduce the total thickness down to 180μm.

Circuits of the transmission lines on the surface and ground plane on the bottom are fabricated by thin-film processing at last, which includes PVD, lithography, electroplating process and etching metal layers. Ti/Au seed layer (50nm/100nm) is sputtered on both sides of substrate first. Then AZ4620 photoresist is spinned and photoresist pattern is patterned by double side aligned lithography. After Au (thickness 5μm) is electroplated, photoresist is cleaned and redundant Ti/Au seed layer is removed by dry etching.

The surface photo of the TGVs in the transmission structure fabricated by LIDE process is shown in Fig. 8. There are two signal TGVs and grounding TGVs array on both sides.

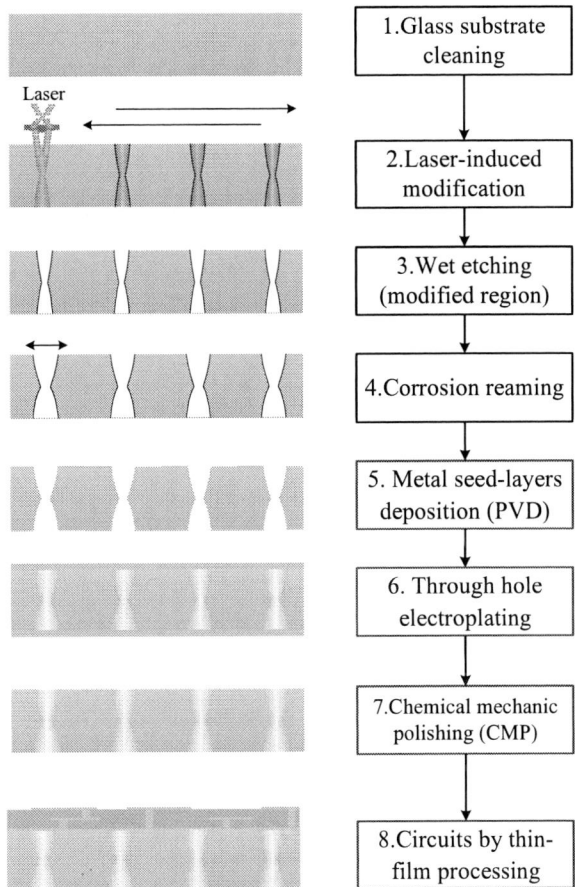

Fig.6. The diagram of LIDE fabrication process.

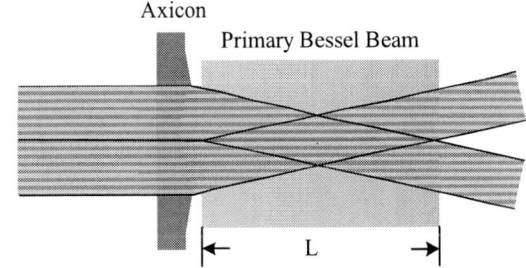

Fig.7. The diagram of laser induced modification in the full thickness range by a picosecond laser Bessel beam.

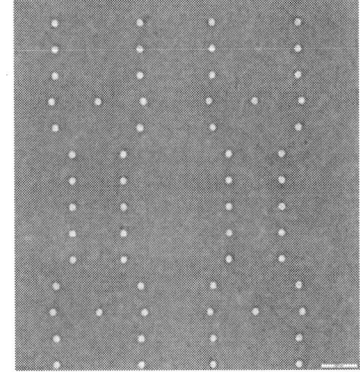

Fig.8. The surface photo of the TGVs in the transmission structure.

The cross-section of fabricated X-shaped TGVs is shown in Fig. 9. There is a good copper filling without void inside the fabricated X-shaped TGV. And the SEM image shows that the top side width is 42~ 45μm and the middle waist width is about 20μm.

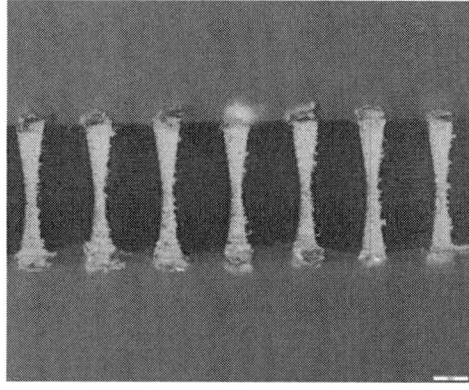

a)

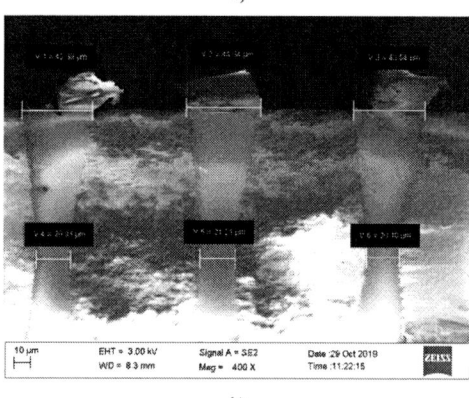

b)

Fig. 9. The cross-section of fabricated X-shaped TGVs: a) the optical photo b) SEM image.

After thin-film processing, the photos of straight transmission line and transmission structure with X-shaped TGVs fabricated are shown in Fig. 10 (a) and (b).

a)

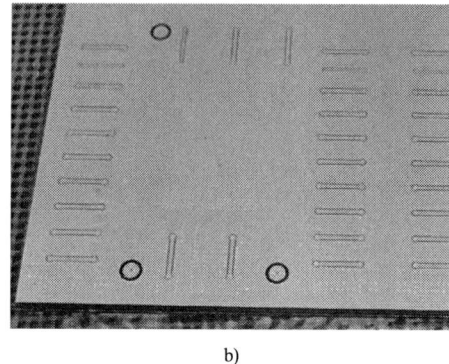

b)

Fig.10. The photos of glass substrate a) circuits on the surface and b) circuits on the bottom.

IV. TESTING OF TGV TRANSMISSION STRUCTURE

The transmission performance of straight transmission line and transmission structure with X-shaped TGVs is tested by GSG prober and vector network analyzer.

Firstly, the return loss of straight transmission line and transmission structure with X-shaped TGVs is tested. The measured curves are shown in Fig. 11 and Fig. 12. In the range of Ka band (32~38GHz), the return loss of straight transmission line is -21~-28dB, which shows good return loss performance. It can be noticed that there is a peak value on return loss curve at 27 GHz in Fig. 10. It is because when tested by vector network analyzer, prober contacts with straight transmission line or transmission structure directly, a RLC oscillating filter is formed partly, a number of resonance points will appear. In Fig. 10 resonance point appear at 27 GHz which forms an obvious peak value, while resonance point appears at 33 GHz in Fig. 11 which forms a peak value. Moreover, impedance matching is a little bit worse influenced by two X-shaped TGVs. The return loss of transmission structure is 4~6 dB lower than straight transmission line in the range of -17~-22 dB which indicates further optimization of more details in the TGV processing.

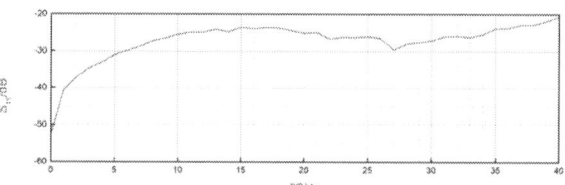

Fig.11. The measured return loss of the straight transmission line.

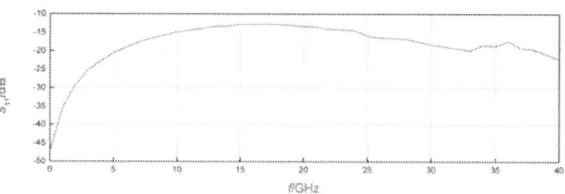

Fig.12. The measured return loss of the transmission structure.

Secondly, the insert loss of straight transmission line and transmission structure with X-shaped TGVs is tested. The measured curves are shown in Fig. 13 and Fig. 14. In the range of Ka band (32~38GHz), s parameters are obtained by vector network analyzer. The insert loss of straight transmission line (IL_1) and transmission structure with X-shaped TGVs (IL_2) are analyzed by ADS software.

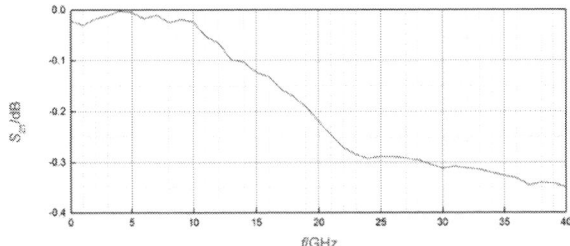

Fig.13. The measured insert loss of the straight transmission line.

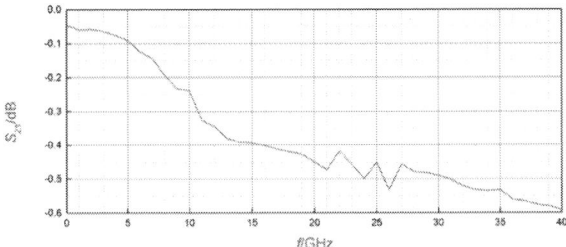

Fig.14. The measured insert loss of the transmission structure.

On the basis of Fig. 13 and Fig. 14, the insert loss measurement results of the straight transmission lines and the transmission structures in Ka band (32~38GHz) are shown in Table I and Table II.

TABLE I.   INSERT LOSS MEASUREMENT RESULTS OF THE STRAIGHT TRANSMISSION LINES (IL₁)

Frequency(GHz)	Max.(dB)	Mean.(dB)	Min.(dB)
32	0.182	0.143	0.112
33	0.304	0.256	0.226
34	0.352	0.298	0.222
35	0.438	0.396	0.326
36	0.295	0.285	0.273
37	0.171	0.148	0.126
38	0.174	0.156	0.124

TABLE II.   INSERT LOSS MEASUREMENT RESULTS OF THE TRANSMISSION STRUCTURES WITH X-SHAPED TGVs (IL₂)

Frequency(GHz)	Max.(dB)	Mean.(dB)	Min.(dB)
32	0.459	0.392	0.347
33	0.624	0.524	0.459
34	0.607	0.523	0.436
35	0.6	0.561	0.476
36	0.614	0.531	0.483
37	0.522	0.467	0.423
38	0.427	0.356	0.306

The insert loss of single X-shaped TGV is calculated as $(IL_2 - IL_1)/2$. The calculated insert loss results of single X-shaped TGV in working frequency band (32~38GHz) are shown in Table III. The insert loss of single X-shaped TGV is 0.042~0.199 dB.

TABLE III.   THE CALCULATED INSERT LOSS RESULTS OF THE SINGLE X-SHAPED TGV

Frequency(GHz)	Max.(dB)	Mean.(dB)	Min.(dB)
32	0.174	0.125	0.083
33	0.199	0.134	0.078
34	0.193	0.113	0.042
35	0.137	0.083	0.019
36	0.171	0.123	0.094
37	0.198	0.16	0.126
38	0.152	0.1	0.066

V. CONCLUSION

In this paper, the transmission structures based on X-shaped TGVs have been designed, simulated, fabricated, and measured in Ka band (32~38GHz). The cross-section of the fabricated TGVs shows that there are wide openings both on the top side and bottom side, and a narrow waist in the middle which is a typical X-shaped TGV has a good copper filling without void. The tested result of transmission performance shows that the insert loss is within 0.2dB per X-shaped TGV in the range of 32~38GHz which can meet the demand of low insert loss requirement in microwave and millimeter wave system integration

REFERENCES

[1] M. Töpper, I. Ndip, R. Erxleben, L. Brusberg, N. Nissen, and H. Schröder, 3-D Thin film interposer based on TGV (through glass vias): An alternative to Si-interposer, Electronic Components & Technology Conference 2010; 66–73.

[2] A. E. Amrani, K. Demir, M. Bouya, M. Faqir, A. Hadjoudja, and M. Ghogho, Fabrication, assembly and testing of a glass interposer-based 3D systems in package. Microelectronic Engineering 2016; 165:6-10.

[3] Y. Sun, D. Q. Yu, R. He, F. W. Dai, X. F. Sun, and L. X. Wan, The development of low cost through glass via (TGV) interposer using additive method for via filling. 13th International Conference on Electronic Packaging Technology and High-Density Packaging (ICEPT-HDP) 2012; 49–51.

[4] R. Delmdahl and R. Paetzel, Laser drilling of high-density through glass vias (TGVs) for 2.5D and 3D packaging. Journal of Microelectronic Package Society 2014; 21:53–57.

[5] L. A. Hof and J. A. Ziki, Micro-hole drilling on glass substrates—A review. Micromachines 2017; 8,53:1-23.

[6] L. Brusberg, M. Queisser, C. Gentsch, H. Schröder, and K. D. Lang, Advances in CO_2-laser drilling of glass substrates. Physics Procedia 2012; 39:548-555.

[7] Z. Chen, Y. Peng, H. Cheng, Z. Z. Yang, and M. X. Chen, Void-free and high-speed filling of through ceramic holes by copper electroplating. Microelectronics Reliability 2017; 75:171-177.

[8] S. S. Wu, H. Q. Ling, Y. T. Xie, M.Li, and D. Q. Yu, Synergistic effect of additives on filling of tapered TGV vias by copper electroplating. 21st International Conference on Electronic Packaging Technology (ICEPT) 2020.

[9] S. Shi, X. F. Wang, C. L. Xu, J. J. Yuan, J. Fang, and S. Liu Simulation and fabrication of two Cu TSV electroplating methods for wafer-level 3D integrated circuits packaging. Sensors and Actuators A: Physical 2013; 203:52–61.

[10] A. K. Dubey, and V. Yadava, Laser beam machining—A review. International Journal of Machine Tools and Manufacture 2008; 48:609–628.

[11] Y. Sato, N. Imajyo, K. Ishikawa, R. Tummala, and M. Hori, Laser-drilling formation of through-glass-via (TGV) on polymer-laminated glass. Journal of Materials Science: Materials in Electronics 2019; 30: 10183-10190.

[12] H. Cai, J. Yan, S. L. Ma, R. F. Luo, Y. M. Xia, and J. W. Li, Design, fabrication, and RF property evaluation of a TGV interposer for 2.5D RF integration. Journal of Micromechanics and Microengineering 2019; 29, 075002:1-10.

978-1-6654-1392-3/21 $31.00 © 2021 IEEE

Process Development and Failure Analysis of Super-Size Embedded Silicon Fan-out Package

Dongzhi Fu
Huatian Technology (Kunshan)
Electronics Co., Ltd.
Kunshan, China
dongzhi.fu_ks@ht-tech.com

Shuying Ma
Huatian Technology (Kunshan)
Electronics Co., Ltd.
Kunshan, China
shuying.ma_ks@ht-tech.com

Jiao Wang
Huatian Technology (Kunshan)
Electronics Co., Ltd.
Kunshan, China
jiao.wang_ks@ht-tech.com

Abstract—With the rapid development of mobile consumer electronics, Internet of Things, 5G, artificial intelligence, new energy vehicles and other technologies, the demand for high integration, high performance, fast heat dissipation and miniaturization of electronic packaging is becoming more and more urgent. In this paper, based on the traditional embedded silicon fan-out (eSiFO) package with a package size of less than 5mm*5mm, we have developed a super-large eSiFO packaging technology, in which the package size is up to 30mm*40mm, the embedded chip size is up to 14mm*14mm, and the number of I/Os is as high as 1808. After completing the successful development of critical process such as cavity etching, die attaching, and vacuum lamination, the super-size eSiFO package we manufactured has an electrical yield rate of 96%, and has passed the electrical test in uHast-B 264 hours and TC-B 1000 cycles. Regarding the key technical problems in the super-size eSiFO package, a systematic study was launched and some practical optimization suggestions were provided.

Keywords—*super-size eSiFO, failure analysis, wafer level package*

I. INTRODUCTION (*HEADING 1*)

More than half a century has passed since Moore's Law was proposed by Moore Gordon in 1965. It has been guiding computer processor chips to continue technological iteration in accordance with a certain time rhythm. The proposal of Moore's Law has greatly promoted the development of information technology. However, as the process nodes of the chip continue to approach the physical limit, the cost of research and development has risen sharply, and its difficulty continues to increase. In the post-Moore era, while people continue to strive to advance the chip process nodes, many people turn their attention to the improvement of chip system functions through the innovation of packaging technology. Since the beginning of this century, a variety of packaging technologies with system integration capabilities such as 2.5D IC, 3DIC, and Fan-out have continued to emerge. TSMC, Samsung, Intel and other major global foundries have begun to expand their business scope to the development of the above-mentioned new advanced packaging technologies. TSMC began to introduce 2.5D-based Chip on Wafer on Substrate (CoWoS) packaging technology in 2011, and formally entered the advanced packaging field, followed by the introduction of fan-out-based Integrated Fan-out (InFO) packaging technology in 2012. Meanwhile the mainland OSAT represented by Huatian Technology introduced embedded silicon fan-out (eSiFO) packaging technology in 2016, and began to continuously explore in the field of Fan-out.

In this paper, based on the small-size eSiFO technology developed by Huatian, we developed a super-size eSiFO packaging technology. The package size of the device is 30mm*40mm, and two chips with a thickness of 70um or 90um and a size of 14mm*14mm are embedded. In our packaging process, 15um/15um line width and line spacing are used, with the package thickness being 425um or 625um, and the solder ball pitch being 500um. In order to test the performance and process stability of the package, we designed four test units, including daisy chain unit, Kelvin test unit, leakage test unit, on off test unit across the gap, and performed ongoing reliability tests on the test vehicle. The circuit probe (CP) electrical yield rate of the super-size eSiFO package we developed has reached 96%, and the package passed the electrical tests both in uhast-B-264 hours and TC-B-1000cycle. A systematic analysis on the appearance failure in process flow and reliability test was conducted. We believe that our research will lay a solid foundation for the future application of this type of technology in mass production.

II. PACKAGE AND PROCESS

A. Package Information

The layout of the test unit of the super-size silicon-based fan-out model we designed is shown in Fig.1(a). The package size is 30mm×40mm, and two test chips with a size of 14mm×14mm are embedded. Table I lists the specific packaging information. In order to study the impact of different package thickness and embedded chip thickness on product yield and reliability, we designed two package thicknesses of 450um and 650um (excluding the height of solder balls), and designed two types of embedded chips, 70um and 90um thickness. The package structure we designed includes two passivation layers and one RDL layer. The minimum line width and line spacing of the RDL is 15um/15um, and different test units are also designed on the RDL layer. The I/O quantity of the whole package structure is 1808, and the pitch between the solder balls is 500um.

Fig. 1(b) shows all the test units used in this package, including the most commonly used daisy chain unit for circuit continuity testing, a leakage test unit with staggered dense wiring, on off test unit across the gap, and the Kelvin test unit for measuring the resistance. All these test units are evenly distributed in different positions of the package according to their functions.

978-1-6654-1392-3/21 $31.00 © 2021 IEEE

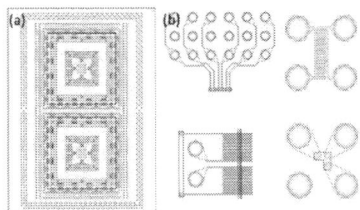

Fig. 1. (a) The layout of the test unit, (b) various test units

TABLE I.   PACKAGE INFORMATION

Items	Specification
Package body size(mm)	30mm×40mm
Package thickness(μm) （Excludeing solder ball）	450um；650um
Embedded die size(mm)	14mm×14mm
Embedded chip thickness(μm)	70um；90um
Embedded chip amount	2
Min RDL width/ Spacing	15um/ 15um
Bump I/O	1808
Bump pitch(μm)	500um

B. Process Flow

Figure 2 depicts the process flow of the eSiFO package used in this study. Firstly, the silicon-based carrier is ground to obtain a clean surface. The photolithography process is used to define the cavity for embedding the chip, and then the cavity is formed through the Bosch etching process. After stripper and cleaning, the preparation of silicon-based carrier is completed. At the same time, all the preparations are completed after the incoming wafer is ground and cut into individual chips to be embedded.

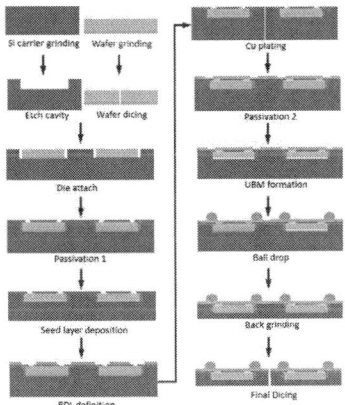

Fig. 2. Process flow of eSiFO

The test vehicle are embedded in the cavity by die attaching (DA), and both are connected by die attaching film (DAF), with a 30um width gap being formed between the chip and the sidewall of the cavity. The surface of the embedded chip is basically flat with the surface of the silicon carrier, and the height difference is within ±10um. After DA, an insulating layer needs to be covered on the front side of the package to isolate the RDL layer from the test chip and silicon wafer. In this study, a photosensitive dry film is laminated to fill the gap and flatten the surface of the reconstructed wafer. At the pad position of the test chip, an opening is formed on the dry film through a photolithography process to facilitate subsequent RDL and pad connection. The subsequent process is to form the RDL layer. First, sputtering is used to form a Ti 0.1um/Cu 0.3um metal layer as a seed layer for subsequent electroplating. Next, define the pattern of the RDL layer on the seed layer by photolithography. Afterwards, an electroplating process is performed to electroplate a layer of metal copper with a thickness of 5um±1um at the opening of the photoresist to form an RDL pattern. Finally, after stripper and seed layer etching, the RDL layer is formed. Then, a second passivation layer is formed on the surface by similar coating, exposure and development processes. The passivation layer plays a role of blocking water and oxygen and protecting the RDL layer, usually using PI photoresist materials. The second passivation layer forms openings at the positions where the solder balls are to be planted. Then, using a method similar to that of manufacturing RDL layer, sputtering the seed layer, defining the under ball metal (UBM) pattern, and electroplating the UBM to form the UBM structure. It is generally believed that UBM can improve the reliability of package.

After the UBM is manufactured, the ball is dropped on the UBM, and the solder ball is closely combined with the UBM after applying flux and reflow. After that, the whole 12 inch wafer is grinded to about 450um and 650um respectively. The purpose of back grinding is to reduce the package volume, which is conducive to the miniaturization and thinning of the end product. Finally, a diamond cutter wheel is used to cut the 12-inch wafer into hundreds or thousands separate small dies.

III. RESULTS

A. Super-Size eSiFO Package Manufacturing

1) Cavity Formation and Die Attaching

Compared with the molding fan-out package which embeds the chip in the molding material, the eSiFO package is characterized by embedding the chip in the silicon wafer. First, a cavity with a three-dimensional size similar to that of the embedded chip and slightly larger than the size of the latter is formed on the silicon-based carrier. Then the test chip is embedded in the cavity with the front side of the chip facing up, and its surface height is basically the same as that of the silicon wafer. The formation of cavity is a critical process. There are several parameters that must be controlled during the process, including depth, bottom footing, and sidewall angle. Fig. 3(a) and (b) show the profile of the cavity formed by Bosch etching, and Fig. 3(c) is the cross-section of the cavity after all the processes are completed. The difference between the cavity depth and the target value cannot exceed 10um. If the cavity is too deep or too shallow, the first passivation layer cannot guarantee the effective leveling of the front of the packaging structure, which is not conducive to the RDL wiring. In Table II, we have listed values of the cavity depths from reconstructed wafer #1-#6 measured with a 3D microscope, and all measured depths are within specifications. The bottom of the cavity is not completely horizontal. There is a small slope extending inward from the bottom of the side wall of cavity, which is called footing. There are several different ways to define the footing height. Here we define it as the height difference of 15um and 65um respectively from the bottom of the sidewall to the center of the groove. The footing height specification that we control is ≤7um. As

978-1-6654-1392-3/21 $31.00 ©2021 IEEE

shown in Table III, the footing height values we measured are all within the specification. The sidewall angle of the cavity will affect the silica coverage rate in the subsequent PECVD process. The usual control specification is 86-90°. As shown in Table III, our measured values are all within specifications.

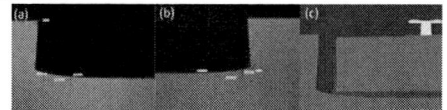

Fig. 3. (a) (b) Profile of the cavity, (c) SEM image of cross-section near the gap

TABLE II. THE DEPTH OF SI CARRIER CAVITY

Wafer	Spec (um)	1	2	3	4	5
#1	80±5um	79.5	82.3	82.1	83.0	82.8
#2	80±5um	80.5	82.4	79.7	81.6	82.4
#3	80±5um	76.8	78.7	80.1	78.8	77.4
#4	100±5um	98.8	103.7	105.8	100.5	100.1
#5	100±5um	99.0	100.0	102.4	101.3	98.1
#6	100±5um	100.3	103.1	103.3	104.8	103.4

TABLE III. THE ANGLE AND FOOTING OF SI CARRIER CAVITY

Items	Spec	1	2	3	4	5
Angle	86-90°	87.1	87.2	86.1	87.5	86.4
footing	≤7um	3.7	3.9	5.2	5.0	5.8

The incoming wafer is divided into separate chips through grinding and cutting processes, and a layer of DAF with a thickness of 10um is attached to the bottom of the test chip. Using ASM's DA machine, two test chip in a size of 14mm*14mm are embedded in the cavity, and the DAF was tightly connected to the bottom of the cavity. Fig. 4 shows the appearance of the reconstructed wafer after DA. In the DA process, the control of the quality will be achieved by debugging the platform temperature, attaching pressure and time. The key items that need to be inspected in the DA process are shift, height difference and tilt, which are the important indicators of manufacturing accuracy of eSiFO package. We mark the four corners of the cavity as A, B, C and D respectively, and then define the height difference (HD) between the embedded chip surface and the silicon carrier surface near each corner as the height difference. Therefore, there will be four height differences for each inspected test chip, corresponding to four corners A, B, C and D, respectively. Meanwhile we defined the maximum minus the minimum of the four height differences as the tilt of the embedded chip. Generally, all height differences are required to be within ±10um, and the tilt of all embedded chips is required to be ≤5um. These two points can ensure that the thickness of the passivation layer on the embedded chip after the passivation 1 process is uniform and controllable. Table IV show the shift, height difference and tilt measured from silicon wafer #1 and #5. Since the size of the embedded chip is much larger than that of the mass production chip (<5mm), some of the values appear to be out of specification, especially tilt. Compared with the small-sized eSiFO package, the super-size eSiFO package shows more obvious undulations at the bottom of the cavity, and the probability of bumps at the bottom of cavity is significantly increased. These factors will cause the height

difference and tilt to easily exceed the original specifications. When dealing with dummy embedded chip in DA process, the mark captured by die attaching machine is dummy pad made by wet etching, and its size has a fluctuation of ±3um. What`s more, the optical contrast of the dummy pad is not as high as that of the real customer chip, which leads to reduction of the precision of DA process. Therefore, compared with the real customer chip, the shift of dummy chip in DA process is easier to be out of the specification. Finally, the void under the DAF after placement is the top problem for super-size eSiFO.

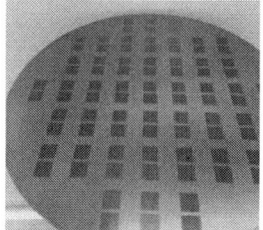

Fig. 4. Outlook of the reconstructed wafer

TABLE IV. THE SHIFT, HEIGHT DIFFERENCE AND TILT OF EMBEDDED CHIP

Wafer	Items	1	2	3	4	5
#1	Shift_X	5.9	2.0	2.3	2.1	4.9
	Shift_Y	1.0	2.7	0.1	0.2	8.0
	HD_A	-4	-3.5	-1.5	-5.5	-3
	HD_B	-3.5	-4.5	-1	-5.5	-4.5
	HD_C	2.5	0.5	1	-1	-1
	HD_D	1	-2	0	-3	0
	Tilt	6.5	5	2.5	4.5	4.5
#5	Shift_X	2.6	0.3	1.3	1.8	0.9
	Shift_Y	0.3	4.5	0.8	4.3	1.3
	HD_A	4.5	3	1.5	3.5	3.5
	HD_B	2.5	-1	2.5	-0.5	0
	HD_C	11	6	7.5	2.5	6.5
	HD_D	7	-0.5	3	3	1
	Tilt	8.5	7	6	4	6.5

2) Front-Side Process Flow of eSiFO

After the DA process, a dry film is first laminated on the reconstructed wafer surface to fill the gap between the test chip and the cavity, and to act as a passivation layer. Through the photography process, a window is opened in the pad position of the embedded chip to form a connection with the RDL to be made later. According to our design, the opening of the first passivation layer is 40um. Fig. 5(a) shows the optical picture of the dry film opening, and Fig. 5(b) is the cross section. The resolution of this dry film is very high, and the opening of 20um can be achieved. The next step RDL layer will be manufactured. The purpose is to connect the pad on the test chip with the solder ball, and transform the centralized distribution of pad on the chip into the matrix distribution of solder ball. First a metal layer of Ti 0.1um/ Cu0.3um is sputtered on dry film surface, as shown in Fig. 5(c). The thin Ti layer can enhance the adhesion and prevent the diffusion of copper. Meanwhile copper layer can be used as the seed layer of electroplating process. Next, we need to define the RDL pattern on the top

978-1-6654-1392-3/21 $31.00 © 2021 IEEE

of the seed layer with photoresist, as shown in Fig. 5(d). Finally, the RDL patter is formed by copper plating process shown in Fig. 5(e). Table V shows the line width, line spacing and thickness of RDL line, and the above data are within the specification.

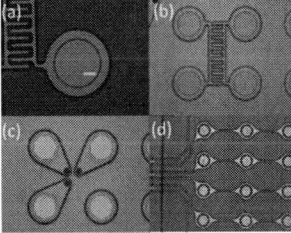

Fig. 5. (a-b) The image and SEM cross section of Passivation 1 opening, (c) the seed layer, (d) litho patterns of RDL, (e) plating of RDL

TABLE V. LINE WIDTH, LINE SPACING AND THICKNESS OF RDL AFTER PLATING PROCESS

Wafer	Items	SPEC	1	2	3	4	5
#1	Line width	15±3um	18.0	17.2	17.5	16.5	17.3
	Line spacing	15±3um	12.8	12.5	12.8	13.1	13.1
#2	Line width	15±3um	13.8	14.9	14.9	14.3	13.6
	Line spacing	15±3um	16.2	15.1	15.7	15.9	15.9
#3	Line width	15±3um	13.3	13.6	13.3	13.3	13.6
	Line spacing	15±3um	16.7	17.5	16.7	17.0	15.9
#4	Thickness	5±1um	5.5	5.4	5.4	5.4	5.5
#5	Thickness	5±1um	5.7	5.6	5.6	5.8	5.8
#6	Thickness	5±1um	5.8	5.6	5.6	5.8	5.8

After the preparation of RDL layer, the second passivation layer should be covered on it to prevent water and oxygen. As shown in Fig. 6(a), the second insulating layer will form an opening with a diameter of 200um at UBM position. In order to improve the reliability of the package, as shown in Fig. 6(b-d), a process similar to RDL preparation is used to form a layer of UBM with a thickness of about 8um on the opening of second passivation.

Fig. 6. (a) Opening of passivation 2, (b) the seed layer, (c) litho patterns of UBM, (d), (e) plating of UMB

After the preparation of UBM, the ball planting process is carried out on it. Fig. 7(a) and Fig. 7(b) show the appearance of the ball after reflow and the cross section of the ball drop position, respectively. The ball diameter and height after reflow are 258um / 198um, respectively. After reflow, X-ray will be used for void detection, as shown in Fig. 7(c). Then, after laser marking, grinding and dicing, a super-size package in a size of 30mm*40mm is obtained.

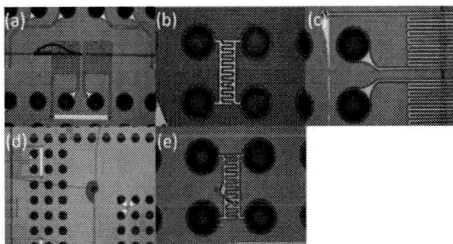

Fig. 7. (a) Image of solder ball, (b) SEM cross section of solder ball, (c) X-ray image of solder ball

B. Package Level Failure Analysis

1) Failure Analysis in CP Test

After the completion of the whole process, we carried out CP test on the test vehicle, and the electrical yield reached 96%. The main reasons for the loss of electrical yield are summarized in Table VI: open circuit induced by bubbles at gap produced in dry film laminating process (Fig.8(a)), open or short circuit caused by large particles (Fig. 8(b)), open circuit induced by dummy pad damage (Fig. 8(c)), open circuit caused by die crack (Fig.8(d)) happened in DA process, and short circuit led to by RDL deformation (Fig. 8(e)). For the above defects, we have carried out some improvement verification. For example, by optimizing the temperature and time of vacuum laminating, the bubble at gap has been solved. There are many sources of large particles of foreign matter. We will add the automatic optical inspection (AOI) after each process to lock the node where the foreign object appears, and then give effective improvement measures. The open circuit caused by dummy pad damage and die crack can be solved by adding AOI test before DA process.

TABLE VI. SUMMARY OF ELECTRICAL TEST FAILURE

Electrical test failure	Yield Loss
Open Circuit Caused By Air Bubbles In Dry Film	1.62%
Open Circuit Or Short Circuit Caused By Large Size Particles	0.97%
Dummy Pad Damage	0.65%
Die Crack	0.32%
RDL Deformation	0.32%

Fig. 8. (a) Air bubbles in dry film, (b) particles, (c) dummy pad damage, (d) chip crack, (e) RDL deformation

In addition to the electrical defects mentioned above, there are also several kind of appearance defects. As shown in Table VII, the yield loss of dry film deformation near the gap (Fig. 9(a)) reached 18.4%, and the defective rate of dry film bubbles (Fig. 9(b)) reached 5%. At the same time, the defective rate of RDL deformation (Fig. 9(c)) reached 3.0%. Among the above appearance defects, dry film bubbles and RDL deformation are also critical causes of part of electrical test failure, and the former has been improved, which will not be repeated here. The RDL deformation (Fig9-c) is presumed to be mainly induced by the poor topography of the photoresist before electroplating, and the specific reason cannot be determined temporarily. Finally, we will focus on the dry film deformation near the gap, which is the most serious appearance defect appearing in the package process of large size eSiFO. Fig. 10(a) and (e) show the optical picture of the deformation of the dry film at the gap. This kind of defect only occurs at the gap, and only appears during the reflow process. In addition, the dry film deformation was proved to be related to the void under DAF by scanning acoustic tomography (SAT) and cross section. Fig. 10(b) and (f) is an SAT image of bad die with dry film deformation, where the dark black part indicates that there is a void between the DAF and the bottom of the cavity. It can be found that there are many voids around the embedded chip, especially at the four corners. Fig. 10(c) /(d) and Fig. 10(g)/ (h) show the cross section corresponding to the defect location. From the cross section, it can be seen that there is a large amount of foreign matter filled into the void under the DAF. It is inferred that the debris was stuffed into the void during the cross section sample preparation process. A large number of SAT and cross section analysis were conducted, and the conclusions are consistent: the dark black part of the SAT image indicates the existence of void under DAF, and whose sizes obtained by SAT or cross section are approximating to each other. The existence of void indicates that there is residual gas between the DAF and the bottom of the cavity during the DA process. The voids are always distributed in the four corners of the embedded chip. We speculate that it is related to the suction nozzles we used for super-size embedded chip DA. The size of the embedded chip is 14mm*14mm, while that of the suction nozzle is 9mm*9mm, which indicates insufficient force in the 2.5mm range around the chip when the chip is being attached. For a square embedded chip, the four corners are less stressed than the center of the four sides during DA process, so the voids are more prone to appear at four corners. The correlation between this dry film deformation and the reflow process is mainly determined by the physical characteristics of the dry film materials. Specifically, the Tg of this type of dry film after curing is about 230°C. Among the entire process flow, only the reflow temperature exceeds 230°C, which reaches about 260°C. When reflowing, because the temperature exceeds the Tg of the dry film, the dry film will gradually become soft after entering the glass state. When the gas in the void under the DAF is heated and expands, the chip above the void will be pushed upward, thereby forming a pulling force on the dry film at the Gap. At this moment, the dry film in the glass state is easily deformed. In response to this kind appearance defect, we will first start with the design of the suction nozzle and carry out detailed research work in the follow-up.

TABLE VII.   SUMMARY OF APPEARANCE FAILURE

Appearance Failures	Yield Loss
Dry Film Deformation	18.4%
Air Bubbles In Dry Film	5.0%
RDL Deformation	3.0%

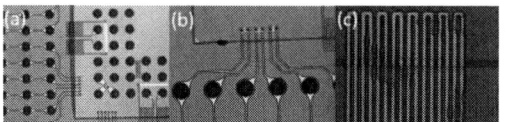

Fig. 9. (a) Dry film deformation, (b) air bubbles in dry film, (c) RDL deformation

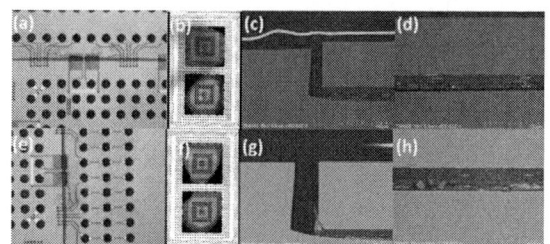

Fig. 10. (a/e) Image of dry film deformation, (b/f) SAT image of dry film deformation area, (c/d/g/h) SEM cross section of dry film deformation area

2) Failure Analysis in Ongoing Reliability Test

After the CP test, 115 good Dies were selected for reliability test. The specific test conditions and test results are shown in Table VIII. The super-size eSiFO package has passed the electrical tests in all reading points of TC-B 250/ 500/ 1000 cycles and uHast-B 96/168/264 hours. Although there were no failure die in the electrical test after reliability, two types of obvious appearance abnormalities were found, and both of them occurred in the TC-B test: one is the rainbow pattern, the other is the solder ball deformation. As shown in Fig. 11(a) and (c), the rainbow pattern all appear at the edge of the package. The rainbow pattern is usually an optical phenomenon caused by the separation of adjacent layer. The CTE of silicon is 2.5 ppm/°C, while that of the dry film material and PI material we used are 65 ppm/°C and 60 ppm/°C, respectively. Since the CTE of the first passivation layer and the second one are close to each other, there is little peeling stress between the two passivation layers during temperature cycling. However the difference in CTE between dry film and silicon is very serious, and there will be a greater peeling stress between the two kinds of materials in the temperature cycling test. Obviously, due to the contact with air, the probability of peeling at the edge is higher. Fig. 11(b) and (d) show the cross section of the corresponding position of the rainbow pattern. It can be seen that there is a gap of 1-2um between the dry film and the surface of the carrier wafer at the defect position, which is consistent with our aforementioned speculation. Due to the significant increase in package size, area of risk area at the edge is significantly enlarged. Compared with the small size eSiFO, the large size eSiFO is more likely to have rainbow patterns on the edges. To improve the rainbow pattern defect, it is necessary to strengthen the bonding force between the silicon wafer surface and the dry film at the edge of the package. We have envisaged a possible solution, that is, while etching the cavity on the carrier wafer etching several rows of straight holes with small diameter along the edge of the package, and optionally etching a circle of straight grooves around the edge of the package. When the dry film is being laminated, it will be filled into the straight hole or the straight groove, enhancing the bonding force between the silicon wafer and the dry film. The above scheme needs to be further verified by us.

TABLE VIII.   RELIABILITY TEST RESULTS

Test Items	Read Points	Samples	Electrical test Failure	Appearance Failure
Bake (125°C)	24h	115	None	None
Soak (60°C /60% RH)	40h	115	None	None
Reflow	×3	115	None	None

	(245°C)			
TC-B (-55°C to 125°C) Carbon fiber tray	250cycles	57	None	Rainbow pattern
	500cycles	57	None	Rainbow pattern
	1000 cycles	57	None	Rainbow pattern&Solder ball deformation
uHAST-B (110°C/85% RH)	96H	58	None	None
	168H	58	None	None
	264H	58	None	None

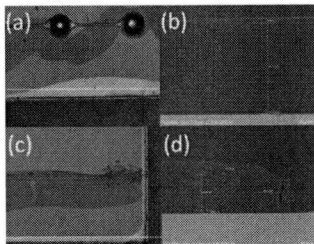

Fig. 11. (a/c) Image of rainbow pattern, (b/d) SEM cross section of rainbow pattern area

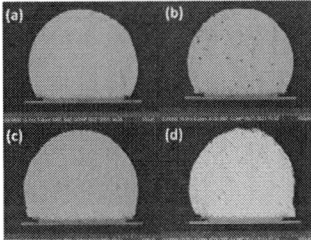

Fig. 12. (a-d) SEM cross section of the solder ball for pre-con, TC-B 250 cycles, TC-B 500 cycles, TC-B 1000 cycles, respectively

Another common appearance defect that appears in the TC-B test is the solder ball deformation. As shown in Fig. 12(a-d), after pre-con test, the cross section of the solder ball is normal, and no deformation is found; after TC-B 250 cycles test, the position where the edge of the solder ball contacts the UBM begins to deform, forming a tiny bump; after TC-B 500 cycles test, a little bigger bump was formed at the same position; after TC-B 1000 cycles test, a larger bump was formed at same position, and the bump is large enough to touched the PI layer. Meanwhile there are obvious signs of oxidation and corrosion on the surface of the solder ball. As the solder balls are exposed to the air, corrosion and oxidation on the surface is difficult to avoid. The bump on the edge of the solder ball becomes bigger with the increase in the number of temperature cycles. It is speculated that the temperature cycle accelerates the release of the residual stress inside the solder ball. Since the stress at the contact position of the solder ball and UMB is relatively concentrated, this position is more prone to deformation, which needs to be further confirmed by simulation.

IV. CONCLUSION

In this article, we have developed a super-size eSiFO packaging technology in a package size of 30mm*40mm, in which two test chips with a size of 14mm*14mm are embedded. The CP electrical yield rate of the super-size eSiFO package reaches 96%, and it has passed the electrical test in the uHast-B 264 hours and TC-B 1000 cycles reliability experiment, which shows that the super-size eSiFO package technology we developed is functionally feasible. In response to the appearance defects that occurred during the manufacturing process and the reliability experiment, we conducted a systematic analysis and provided improvement countermeasures and suggestions as much as possible. The super-size eSiFO packaging technology provides an excellent platform for multi-chip system integration. Due to its many advantages such as high integration, high precision, fast heat dissipation, and strong process compatibility, it is believed that it will play a significant role in applications in artificial intelligence, 5G, Internet of Things and other fields.

REFERENCES

[1] G. E. Moore, "Cramming more components onto integrated circuits," in Proceedings of the IEEE, vol. 86, no. 1, pp. 82-85, Jan. 1998, doi: 10.1109/JPROC.1998.658762.

[2] B. Banijamali, S. Ramalingam, K. Nagarajan and R. Chaware, "Advanced reliability study of TSV interposers and interconnects for the 28nm technology FPGA," 2011 IEEE 61st Electronic Components and Technology Conference (ECTC), 2011, pp. 285-290, doi: 10.1109/ECTC.2011.5898527.

[3] C. C. Liu et al., "High-performance integrated fan-out wafer level packaging (InFO-WLP): technology and system integration," 2012 International Electron Devices Meeting, 2012, pp. 14.1.1-14.1.4, doi: 10.1109/IEDM.2012.6479039.

[4] D. Yu, Z. Huang, Z. Xiao, L. Yang and M. Xiang, "Embedded Si fan out: a low cost wafer level packaging technology without molding and de-bonding processes," 2017 IEEE 67th Electronic Components and Technology Conference (ECTC), 2017, pp. 28-34, doi: 10.1109/ECTC.2017.166.

[5] C. Chen, D. Yu, T. Wang, Z. Xiao and L. Wan, "Warpage prediction and optimization for embedded silicon fan-out wafer-level packaging based on an extended theoretical model," in IEEE Transactions on Components, Packaging and Manufacturing Technology, vol. 9, no. 5, pp. 845-853, May 2019, doi: 10.1109/TCPMT.2019.2907295.

[6] S. Ma et al., "Progress and applications of embedded system in chip (eSinC®) technology," 2020 IEEE 70th Electronic Components and Technology Conference (ECTC), 2020, pp. 1671-1676, doi: 10.1109/ECTC32862.2020.00262.

[7] X. Gu, S. Ma, J. Wang and Y. Hao, "Process development of large size embedded silicon fan-out (eSiFO) package," 2020 21st International Conference on Electronic Packaging Technology (ICEPT), 2020, pp. 1-5, doi: 10.1109/ICEPT50128.2020.9202937.

Surface modification of graphite and its effect on thermal and mechanical properties of graphite-based thermal interface materials

1st Yuexing Zhang[1,2]
1 Shenzhen Institute of Advanced
Electronic Materials
Shenzhen Institutes of Advanced
Technology, Chinese Academy of
Sciences
Shenzhen, China
2 School of Mechanical and Electrical
Engineering
Guilin University of Electronic
Technology,
Guilin, Guangxi, China
zhang.yx@siat.ac.cn

2nd Hong He[2]
School of Mechanical and Electrical
Engineering
Guilin University of Electronic
Technology
Guilin,Guangxi, China
he.h@siat.ac.cn

3rd Junwei Li[1]
Shenzhen Institute of Advanced
Electronic Materials
Shenzhen Institutes of Advanced
Technology, Chinese Academy of
Sciences
Shenzhen, China
li.jw@siat.ac.cn

4th Chenxu Zhang[1*]
Shenzhen Institute of Advanced
Electronic Materials
Shenzhen Institutes of Advanced
Technology, Chinese Academy of
Sciences
Shenzhen, China
zhang.cx@siat.ac.cn

5th Rong Sun[1]
Shenzhen Institute of Advanced
Electronic Materials
Shenzhen Institutes of Advanced
Technology, Chinese Academy of
Sciences
Shenzhen, China
rong.sun@siat.ac.cn

6th Meng Han[1*]
Shenzhen Institute of Advanced
Electronic Materials
Shenzhen Institutes of Advanced
Technology, Chinese Academy of
Sciences
Shenzhen, China
meng.han@siat.ac.cn

7th Ping Zhang[2*]
School of Mechanical and Electrical
Engineering
Guilin University of Electronic
Technology,
Guilin, Guangxi, China
pingzhang@guet.edu.cn

Abstract—Graphite has been widely used in the preparation of polymer-based thermal interface materials with excellent performance due to its high thermal conductivity. However, there is a huge difference in the surface energy between graphite and polymer matrix, which can lead to the aggregation of graphite powder in the composite system. Because of the agglomeration, the thermal transport paths in the thermal interface materials are blocked. The main strategy to solve such a problem is to perform surface modification on graphite, which can improve the dispersion of graphite and build more transfer paths. In this work, the graphite was hydroxylated by oxidation firstly, then the dodecyl trimethoxy silane, hexadecyl trimethoxy, and titanate coupling agent were grafted onto the graphite surface by wet modification. Then, the thermal interface material was prepared by adding original and modified graphite in the polymer, and their thermal and mechanical properties were studied and compared. Experimental results showed that the thermal properties of the two groups had no significant difference. With the graphite content increasing from 45 wt.% to 60 wt.%, the thermal conductivity increased from 11 W/(m·K) to 21.5 W/(m·K). After three cycles of compressions, the stress of modified samples was smaller under the same strain. This means that the modified sample has better compression resilience and can work more stably under certain cyclic compression. This work provides important information for the surface modification of

graphite to improve its thermal and mechanical behaviors in thermal interface materials.

Keywords—Thermal interface materials, Graphite, Surface modification, Compressive properties

I. INTRODUCTION

Over the last few decades, the frequent replacement of integrated circuits has aroused great attention in thermal interface materials, which can reduce the temperature of electronic devices in operation, thereby ensuring their long-term stability[1]. Therefore, the development of advanced thermal interface materials (TIMs) has been strengthened. In an electronically packaged unit, these units dissipate heat by attaching the TIMs to the device as usual. However, The thermal resistance of the TIM itself and the contact thermal resistance between the TIM and the packaging system, form hot spots in the device and reduce the efficiency of the device[2,3]. The reason for this phenomenon is that some materials have a poor affinity with devices, and the thermal expansion coefficients of the two are quite different. Polymer-based TIM are the most expectant to resolve this different due to its flexibility[4,5]. As we all know, bulk polymer materials have a certain degree of oxidation resistance[6,7]. Carbon-based filters, alumina, magnesium Oxide, and nanosilver are widely used to reduce the thermal resistance of polymer matrix composites. Among them, metal fillers usually have low intrinsic thermal resistance, but they do not meet the economic principle; metal oxide does not have these defects, but they can induce high thermal resistance. Carbon-based filler has good thermal conductivity and is economical [8].

Because of the excellent thermal performance of graphite, graphene, expanded graphite (EG), graphite nanosheets (GNPs), and other carbon-based materials, these materials are widely used as fillers in composites. In the process of preparing composites, it is necessary to adjust the groups on the surface of carbon materials for different matrix materials to increase their dispersion performance in the matrix materials. Among many surface modification methods, the simplest and most commonly used method is a chemical modification way, including substitution-doped and covalent-doped ones. No-covalently modified graphene is a star in the application of carbon materials. The physical and chemical activities of graphene produced by chemical vapor deposition (CVD) can be adjusted in various ways. A cycloaddition reaction occurs to the anchor group of the modified CVD graphene and the polyethylene glycol with a carboxyl group at the end. Such a modification by changing the functional group of the graphene surface is a covalent modification [9]. Teng et.al [10] adopt a non-covalent functionalization method to realize the functionalization of graphene nanosheets (GNSs) by superimposing pyrene molecules with functionalized segmented polymer chains π-π. GNSs absorb poly glycidyl methacrylate (Py-PGMA) through the π-π bond, and the functional group on Py-PGMA-GNS can form a covalent bond with the epoxy matrix, and then in the Py -PGMA-GNS/epoxy composites form a cross-linked structure. The Py-PGMA on the surface of GNS not only promotes the uniform dispersion of GNS in the polymer matrix but also improves the interaction between GNS and the polymer, thereby obtaining a high contact area. Gu et.al [11] introduced the use of methane sulfonic acid/y-glycidoxy propyl trimethoxy silane (MSA/KH-560) to functionalize the surface of GNPs (fGNPs) with a two-step method. With the increase of the amount of fGNPs added, the thermal conductivity of fGNPs/E-51 nanocomposites increase, indicating that the surface functionalization of GNPs by MSA/KH-560 has a positive effect on the thermal conductivity and mechanical properties of the composites.

Therefore, surface modification of fillers can greatly improve their dispersion stability in the matrix and improve the performance of the composites. However, large-scale production of graphene and graphene oxide is difficult and costly, and it is not economic to use them as fillers to produce thermal interface materials. Graphite, as studied in this paper, has the advantages of high thermal conductivity of more than 300 W/(m·K) and is relatively economic. In this paper, an effective wet modification method was used to hydrophobically modify graphite to achieve the purpose of good dispersion in the polymer matrix. FTIR (Fourier transform infrared spectroscopy) and XPS (X-ray photoelectron spectroscopy) were used to characterize the changes of functional groups and element types before and after graphite modification. At the same time, the thermal conductivity and compression properties of the graphite/polymer composites were discussed in detail.

II. EXPERIMENTAL

A. Materials

Graphite film was purchased from Hefei Aoqi Electronic Technology Co., Ltd., with a thickness of 25-40μm, hydrogen peroxide (H_2O_2, analytical purity) was purchased from Shanghai Wokai Biotechnology Co., Ltd., concentrated sulfuric acid (H_2SO_4, analytical purity) was purchased from Dongguan Dongjiang Chemical Reagent Co., Ltd., isopropyl titanate triisostearate (TTS, 96%) was purchased from Yangzhou Lida Resin Co., Ltd., dodecyl trimethoxy silane (DTS, 96%) was purchased from Qufu Chenguang Co., Ltd., Hexadecyl trimethoxy silane (HTS, 96%) was purchased from Shandong Sike New Materials Co., Ltd. Silicone was purchased from Foshan Yingzhi Organic silicone Materials Co., Ltd.

B. Modification of graphite

Graphite microchips are made from smash of graphite film, with a particle size between 300-600 μm. Then surface modification is performed on the obtained graphite microchips. Firstly, the graphite microchips were hydroxylated, that is, the mass ratio of graphite: concentrated sulfuric acid: hydrogen peroxide = 1:22.5:1 is uniformly mixed at 300 rpm for 4 hours to wash off the surface impurities of the graphite microchips. And it forms a series of hydrophilic groups on the surface of the graphite microplates. Then deionized water was used to clean the hydroxylated graphite flakes to neutrality, which was further dried in an electric blast oven at 150 °C for 5 h. Finally, add graphite to a fixed mass ratio of coupling agent, deionized water, anhydrous ethanol mixture under the conditions of 450 rpm and 20 h to change its group. Then, the modified-graphite microchips are washed 5-7 times with anhydrous ethanol to remove excess coupling agent on the surface and finally dried to obtain modified graphite microchips.

C. Preparation of graphite microchips /silicone composites

Graphite microchips /silicone composites are prepared by adding graphite microchips to a mixture of silicone, antioxidant, and coupling agent with fixed ratios, with adjusted the graphite microchips mass fraction of 45%, 55% and 60%, and then an appropriate amount of acetone is added. Then the mixture is mixed uniformly at a speed of 1800r/min, and vacuumed to 30 Pa at to remove acetone. The mixture is evenly spread on the centrifugal paper and the orientation of graphite in the polymer is obtained by the mechanical shearing force of double roller calendaring. Both the mixtures with modified graphite microchips samples and unmodified graphite microchips samples are produced with the same procedures, only the fillers are different. The two sets of samples were put into the same molds. The molds were prepressed and put into a vacuum drying oven, vacuum for 8 h, at 30 Pa, 70 °C, and then cured at 140 °C and 4 h to obtain the final thermal interface materials.

III. RESULTS AND DISCUSSION

A. Analyzation of functional group

Fig. 1 shows the FTIR curve of the modified graphite microchips. The FT-IR test can accurately reveal the types of functional groups on the surface of the graphite microchips with showing corresponding peaks. The 3460cm^{-1} peak corresponds to the stretching vibration peak of the -OH bond, which is the peak shape of graphite powder during the hydroxylation process; the stretching vibration peak of -CH₃ appears at the position of 2745 cm⁻¹, which is grafted with the coupling agent introduced; the stretching vibration peak at 1722 cm⁻¹ is attributed to -C=O, which may be introduced during the hydroxylation process, or maybe due to the three carbonyl groups contained in the molecular structure of titanate. At 1100 cm⁻¹ the corresponding peaks on the left and right sides are the stretching vibration peaks of other elements with carbon, and other elements with oxygen. Among them, the two peaks of 1142 cm⁻¹ and 1124 cm⁻¹ correspond to the stretching vibration peaks of Si-O-C and Si-O-Si respectively [4]. The appearance of these two bond types indicates that through chemical modification, silicon and titanium elements appear on the surface of the corresponding modified graphite microchips.

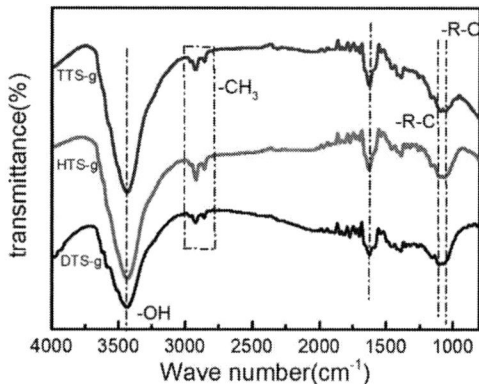

Fig. 1. FTIR curve of modified graphite.

B. XPS diagram of modified graphite

X-ray photoelectron spectroscopy (XPS) is used to further characterize the element content on the surface of graphite powder and to determine the chemical bond corresponding to each element by fitting the peaks. Fig. 2 shows the XPS map of modified graphite microchips. The corresponding curves of the two silane coupling agents in the modified graphite powder XPS full spectrum showed silicon element, and titanium element appeared on the surface of the titanate modified graphite microchips, indicating that the coupling agent was grafted on the graphite surface. Among them, the content of silicon in dodecane-modified graphite powder is 1.69%, the content of silicon in hexadecane-modified graphite microchips is 1.75%, and the content of titanium in titanate-modified graphite powder is 1.14%, as shown in Table 1. Figure 2 (b-d) shows the peak fitting results. After peak fitting, $C1_S$ is divided into three characteristic peaks. The C1 peaks in the three figures all appear at 284.6 eV, which is the Sp^2 hybridization of carbon atoms, and represents the original graphite structure; the C2 peaks all appear at 285.4 eV, which are formed by the hydroxylation of graphite through concentrated sulfuric acid and hydrogen peroxide. The C3 peak in figure b appears at 287.9 eV, which is derived from the C=O bond in the molecular structure of the titanate coupling agent. The C3 peak appearing at 285.4 eV in figures c and d is the C-OH bond formed by the grafting of the silane coupling agent on the surface of the graphite powder. The proportion of each element is shown in Table 1. Carbon accounts for a relatively large proportion, which also

indicates that the introduction of other elements does not change the original structure of graphite microchips too much.

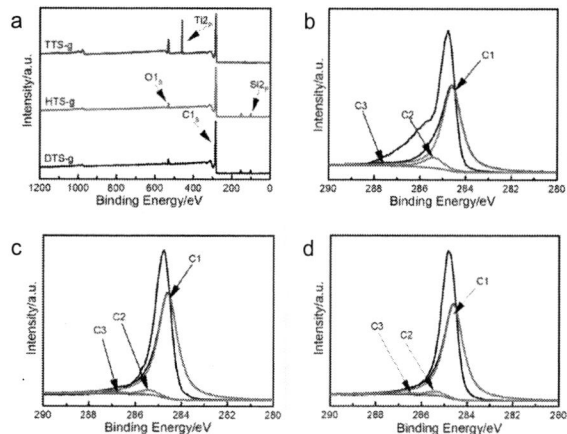

Fig. 2. XPS diagram of modified graphite: XPS spectrum of modified graphite (a); C1S peak diagram of titanate modified graphite (a); C1S peak diagram of hexadecane modified graphite (b); dodecane modified Graphite C1S peaks (d)

Table1. Analysis of XPS element content of three kinds of coupling agent

Sample	Element	Peak (eV)	Content (%)
DTS-g	C1S	284.6	94.31%
	O1S	285.4	3.99%
	Si2P	286.4	1.69%
HTS-g	C1S	284.6	94.15%
	O1S	285.4	4.1%
	Si2P	286.9	1.75%
TTS-g	C1S	284.6	87.72%
	O1S	285.4	11.13%
	Ti2P	288.0	1.14%

C. Thermal conductivity

The thermal conductivity is measured with the LW-9389 TIM Thermal Resistance &Thermal Conductivity Measurement Apparatus, according to the ASTM D5470 standard. A complete thermal conductivity measurement will need at least three continuous measurements of total thermal resistance on the same sample, but with different thicknesses. The total thermal resistances are then plot versus samples thickness, followed by a linear fitting. The reciprocal of the slope is the thermal conductivity of the sample. The intercept of the linear fitting is composed of two contact thermal resistances between the sample and the instrument surfaces. The thermal conductivity obtained by this experimental method is the intrinsic thermal conductivity of the sample. The thermal conductivity of the two sets of samples is shown in Fig.3. When the mass fraction of graphite is 45%, 55%, and 60%, the thermal conductivity of the graphite/polymer composite with unmodified graphite is 10.93W/(m·K) and 14.63 W/(m·K), 21.77 W/(m·K) respectively, while the thermal conductivity with modified graphite is 11.86 W/(m·K), 15.01 W/(m·K), 21.51 W/(m·K). Unfortunately, the difference in thermal conductivity between composites with modified and unmodified graphite is relatively small, indicating that the effect of surface modification on thermal transport improvement is not significant. Moreover, Fig. 3 shows that the thermal conductivity increases as the number of graphite increases, for both sets of samples. The difference is that when the amount of addition is less than 55%, the rate of increase is relatively small, while after 55%, the additional 5% causes an increase of thermal conductivity at a rate of 145%.

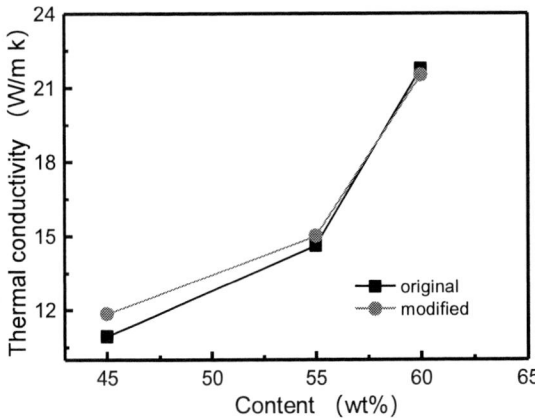

Fig. 3 Thermal conductivity of graphite/polymer composites

D. mechanical property

A dynamic mechanical analysis (DMA) instrument was used to test the mechanical properties of the two sets of samples. Since the graphite/polymer composite materials in this article are mainly used as TIM2 in electronics packaging, they were subjected to compression cycle testing. In the testing process, the strains under different stresses are obtained by controlling the deformations of the sample. Three compression cycles are performed, the first compression to 5% and then slowly removing the stress, the second compression to 10% and then release, and the third compression to 15%. The compression rebound curve is shown in Fig.4. It can be seen that the starting points of the three-cycle compressions on the six samples are almost the same, and the maximum deformation occurs on the same curve, indicating that they all have good compression and rebound properties. When the two groups of graphite are added at 55wt% and 60wt%, the compression cycle curve shows a step. This is because the maximum pressure exceeds the test range of the DMA instrument is 18N. Comparing the two sets of samples, the modified samples have lower stress and lower work loss under the same deformation. That means the modified samples have the better mechanical property to adapt to the packaging.

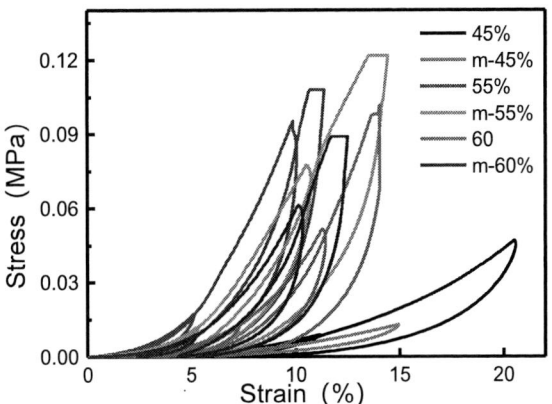

Fig. 4. Compression cycle curve

CONCLUSION

In summary, we performed surface modification on graphite microchips and prepared polymer-based thermal interface materials with modified and unmodified graphite as fillers. The results showed that there is no significant difference in thermal conductivity between the two groups of samples. This may be because the coupling agent added to the composites with unmodified graphite also help build links between graphite and polymer matrix, inducing similar functions to those grafted to the graphite. Moreover, we found that the thermal conductivity of composites increases rapidly as filler content increases, especially at higher filler content. The mechanical properties were then tested by cyclic compression and rebound. The results show that the samples with modified graphite microchips have better compression performance as they have lower stress and lower under the same deformation. This phenomenon is particularly obvious in the two samples with fillers content of 45wt%. For the samples with a filler content of 55wt% and 60wt%, the maximum deformation of modified graphite is larger. This indicates that surface modification of graphite fillers can affect the mechanical properties of composite materials, significantly.

ACKNOWLEDGMENT

This work was supported by the Guangdong Province Key Field R&D Program Project (No. 2020B010179002), Youth Innovation Promotion Association of the Chinese Academy of Sciences (2019354), Shenzhen Science and Technology Research Funding (No. JCYJ20200109114401708 and JCYJ20180507182530279).

REFERENCES

[1] Zhang, Y., et al., Recent advanced thermal interfacial materials: A review of conducting mechanisms and parameters of carbon materials. Carbon, 2019. 142: p. 445-460.

[2] Kim, K., H. Ju and J. Kim, Surface modification of BN/Fe3O4 hybrid particle to enhance interfacial affinity for high thermally conductive material. Polymer, 2016. 91: p. 74-80.

[3] Li, H., et al., Enhanced thermal conductivity of graphene/polyimide hybrid film via a novel "molecular welding" strategy. Carbon, 2018. 126: p. 319-327.

[4] Xu, X., et al., Thermal Conductivity of Polymers and Their Nanocomposites. Advanced Materials, 2018. 30(17): p. 1705544.

[5] Huang, X., et al., Polyhedral Oligosilsesquioxane-Modified Boron Nitride Nanotube Based Epoxy Nanocomposites: An Ideal Dielectric Material with High Thermal Conductivity. Advanced Functional Materials, 2013. 23(14): p. 1824-1831.

[6] Hill, R.F. and P.H. Supancic, Thermal Conductivity of Platelet-Filled Polymer Composites. Journal of the American Ceramic Society, 2002. 85(4): p. 851-857.

[7] Chen, Y. and J. Ting, Ultra high thermal conductivity polymer composites. Carbon (New York), 2002. 40(3): p. 359-362.

[8] Garnier, B. and A. Boudenne, Use of hollow metallic particles for the thermal conductivity enhancement and lightening of filled polymer. Polymer Degradation and Stability, 2016. 127: p. 113-118.

[9] Komissarov, I.V., et al., Nitrogen-doped twisted graphene grown on copper by atmospheric pressure CVD from a decane precursor. Beilstein journal of nanotechnology, 2017. 8(1): p. 145-158.

[10] Teng, C., et al., Thermal conductivity and structure of non-covalent functionalized graphene/epoxy composites. Carbon, 2011. 49(15): p. 5107-5116.

[11] Gu, J., et al., Functionalized graphite nanoplatelets/epoxy resin nanocomposites with high thermal conductivity. International Journal of Heat and Mass Transfer, 2016. 92: p. 15-22.

Design, Fabrication, and Test of an Embedded Si-Glass Microchannel Heat Sink for High-power RF Application

Jianyu Du
School of Engineering and Technology,
China University of Geosciences (Beijing)
Institute of Microelectronics, Peking
University
Beijing, China

Weihao Li
Institute of Microelectronics, Peking
University
Beijing, China

Xu Gao
Institute of Microelectronics, Peking
University
Beijing, China

Deyin Zheng
Institute of Microelectronics, Peking
University
Beijing, China

Yuchi Yang
Institute of Microelectronics, Peking
University
Beijing, China

Zetian Wang
Institute of Microelectronics, Peking
University
Beijing, China

Haoran Zhao
School of Engineering and Technology,
China University of Geosciences (Beijing)
Institute of Microelectronics, Peking
University
Beijing, China

Jiajie Kang*
School of Engineering and Technology,
China University of Geosciences (Beijing)
Beijing, China
kangjiajie@cugb.edu.cn

Wei Wang*
Institute of Microelectronics, Peking
University
Beijing, China
w.wang@pku.edu.cn

Abstract—The heat dissipation of power amplifier (PA) chips is one of the biggest challenges in the development of miniaturized state of art glass-based high-power RF modules. Glass has excellent electrical properties, but the extremely poor thermal conductivity of it also brings many barriers in the application. Its (quartz glass) thermal conductivity is only 1/93 of that of silicon, so there will be a problem of poor heat dissipation. In recent years, there has been an increasing amount of literature on microfluidic cooling technology and this method was demonstrate as an efficient way in cooling application. In this article, we designed, fabricated, and tested a Si-Glass microchannel heat sink, which took advantage of the high thermal conductivity of silicon to deal with the insufficient thermal conductivity of the glass interposer. Finite element simulation was used to study the thermal property of the Si-Glass heat sink and a multi-parameter optimal method was used to design the geometrical parameters, including the number of flow channels and other geometrical parameters of the heat sink. Then the aforementioned microchannel heat sinks were fabricated using cleanroom fabrication on 4-inch silicon and glass wafers. To complete the thermal test, the Thermal Demonstration Vehicles (TDVs) were fabricated by bonding the sample onto a customized PDMS holder for fluid connections with the flow loop. A programmable power source was used to heat the TDV in a stepwise manner, and a syringe pump was used to supply the liquid to cool the heat sink. Results show that the heat sink can dissipate heat flux greater than 150W/cm² with substrate temperature lower than 100°C.

Keywords—Microfluidic Cooling Channel, Si-Glass Bonding, Embedded Heat Sink

I. Introduction

With the development of communication technology, the demand for high-frequency, high-performance, and high-power RF micro-system packaging is increasing. However, the bad electrical performance of traditional silicon interposer in the high-frequency region has gradually emerged. To get

better electrical performance, the scientists began to look for alternative materials for interposer fabrication, such as organic materials, ceramics and glass. Among these materials, glass has attracted attention because of its excellent electrical performance and low cost. But the poor thermal conductivity of the glass makes the thermal management of the glass-based high-power RF modules become a serious problem. To overcome this problem, many published studies have investigated it and made huge progress. In previous research, the cooling technology for glass-based high-power radio frequency modules can be roughly divided into two categories: passive cooling technology and active cooling technology [1~3]. The passive cooling technology of the glass adapter plate mainly uses thermal vias of metal, that is, through glass vias (TGV) are processed on the glass with lasers or other ways, and then the glass vias are filled with metals with good thermal conductivity, such as copper, to transfer heat to the substrate. In the active cooling technology, in addition to the fan cooling method that is generally used to enhance the heat dissipation capacity of the heat sink, the microfluidic cooling technology has attracted attention because it has a stronger heat dissipation capacity than conventional heat sinks. Microfluidic cooling takes away heat by using a cooling liquid such as water to flow in the channels embedded in the interposer. The heat dissipation capacity of microfluidic cooling depends on the physical properties of the fluid itself, the size of the flow channel, the flow rate of the fluid, the structure of the interposer and other factors. The results in the previous report, the heat dissipation performance of the microfluidic cooling can usually reach hundreds or thousands of watts per square centimeter [4-6].

In this article, we designed and fabricated a microchannel heat sink with embedded silicon microchannel and glass inlet/outlet and test the thermal property of the Si-Glass heat sink.

978-1-6654-1392-3/21 $31.00 © 2021 IEEE

II. METHOD

The third part describes the fabrication processes of the aforementioned Si-glass heat sink.

Starting with a 500μm 4-inch glass wafer, 300μm diameter through glass vias (TGVs) were fabricated on the glass wafer using an etching process, serving as the liquid inlet and outlet of the heat sink. Then a 4-inch 525μm silicon wafer was thinned by 340μm by a polishing process. To ensure the accuracy and the quality of the silicon wafer, the polishing process was implemented in three steps. 300μm, 30μm, 10μm height silicon were thinned during three steps respectively. Fig.1 is the photograph of the thinned silicon wafer.

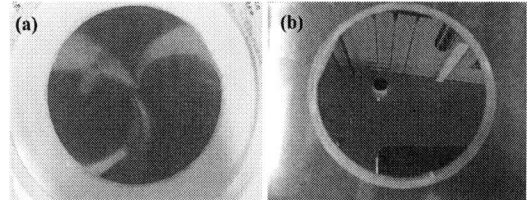

Fig. 1. (a) The silicon wafer after thinned. (b) The silicon wafer after polish.

The thinned silicon wafer then patterned by photolithography, and the 85μm embedded microchannels were deep reactive-ion etched on one side of the silicon wafer using the Bosch process and use photoresist as the mask. The silicon-glass anodic bonding process was used to bond the silicon wafer and glass wafer together.

On the opposite side of the silicon wafer, the Pt heater was patterned using a lift-off process. The heater consists of a 200-nm layer of Pt deposited on top of a 20-nm adhesion layer of Ti. And the other area of the silicon wafer was etched by the Bosch process. After the above processes, the wafer was diced into 3.2mm*1.8mm small chips. The Si-Glass heat sink process flow chart is shown in Fig.2 and Fig.3 shows the photographs of the Si-Glass heat sink.

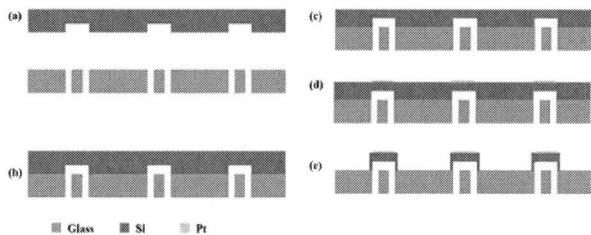

Fig. 2. The Si-Glass heat sink processing flow chart.

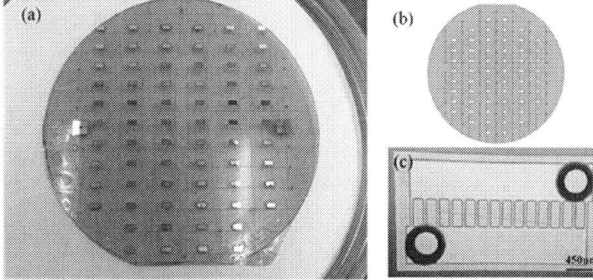

Fig. 3. (a) The photographs of the Si-Glass bonding wafer after cut into small chips. (b)The layout of the chips. (c) The photographs of the Si-Glass heat sink under the microscope.

III. RESULT

To complete the thermal test, the Thermal Demonstration Vehicles (TDVs) were fabricated by bonding the sample onto a customized Polydimethylsiloxane (PDMS) holder for fluid connections with the flow loop as shown in Fig4 (a). Two PDMS layers were bonding together by oxygen plasma bombardment and the Si-Glass heat sink was bond to the top of the PDMS layer by PDMS-Glass bonding technology.

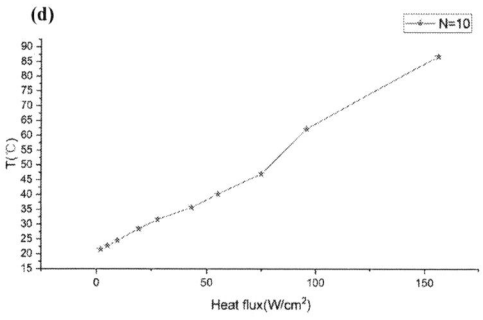

Fig. 4. (a) The photographs of the TDV fabricate process. (b) The photographs of the TDV during the thermal test. (c) The photograph of the test system. (d) The result of the thermal experiment.

A programmable power source was used to heat the TDV in a stepwise manner, and a syringe pump was used to supply the liquid to cool the heat sink. In this experiment, the flow rate is 5ml/min. The room temperature and the inlet coolant temperature were 20°C. The probes were used to power up the sample. The temperature rise was calculated by the change in Pt resistance value.

Results show that when the power is 4W, the temperature of the heat source is 47.11°C. The heat sink can dissipate heat flux greater than 150W/cm² with substrate temperature lower than 100°C.

IV. DISCUSSION

The aim of this chapter is to introduce the optimization of the geometrical parameters of the microchannel heat sink by the finite element method.

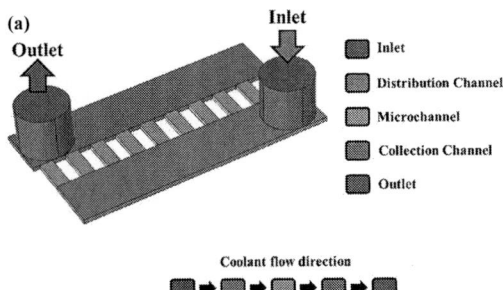

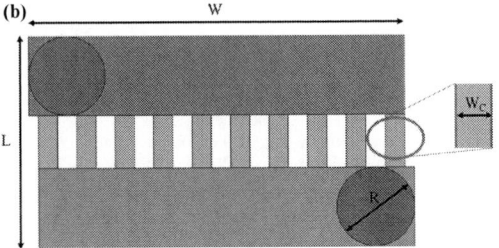

Fig. 5. The microchannel structure diagram.

Fig.5 shows an overview of the microchannel heat sink, the heat sink was consists of two layers, the inlet and the outlet were fabricated in the glass layer and the distribution channel, collection channel and microchannel were fabricated in the silicon layer. The coolant enters from the inlet and then flow through the distribution channel, collection channel and microchannel. Finally, the heated coolant flows out from the outlet. L and W is the length and the width of the heat sink respectively. In this case, L=1.8mm and W=3.2mm. W_C is the width of the microchannel, R is 300μm and the height of the glass layer and silicon are 500μm and 100μm respectively.

The basic equations for heat transfer and flow in the laminar flow of this simulation are given as follows.

Laminar heat transfer module total equation:

$$\rho C_P u \cdot \nabla \cdot q = Q \tag{1}$$

$$q = -k\nabla T \tag{2}$$

Where C_P is the specific heat capacity and ρ is the coolant density.

$$\rho C_P u \cdot \nabla T + \nabla \cdot q = Q + Q_{ted} \tag{3}$$

Equation(3) is the solid heat transfer equation, where Q_{ted} is heat flux heat.

$$\rho C_P u \cdot \nabla T + \nabla \cdot q = Q + Q_P + Q_{vd} \tag{4}$$

Equation(4) is the fluid heat transfer equation, where Q_P is the heat carried in the liquid.

$$\rho(u \cdot \nabla)u = \nabla \cdot \left[-pl + \mu(\nabla u + (\nabla u)^T) - \frac{2}{3}\mu(\nabla \cdot u) \cdot l\right] + F \tag{5}$$

Equation(5) is laminar flow general equation, where u is the fluid flow rate, μ is the hydrodynamic viscosity, ρ is fluid density, and F is the inlet pressure.

In this case, a multi-parameter optimal method was used to investigate how the width and the number of the microchannels influence the cooling performance of the heat sink.

TABLE I.   BOUNDARY CONDITION FOR SIMULATION

Condition	Parameter	Value
Heat source	Heat power	4W/10W
Initial	Temperature	20°C
Inlet	Flow rate	5ml/min
	Temperature	20°C
	Coolant	DI water
Outlet	Pressure	0kPa

Fig.6(a) shows the temperature distribution of the heat sink when the total heat power is 10W and the corresponding heat flux is 173W/cm². Water was selected as the coolant and the temperature at the inlet is 20°C. Fig.6(b) describes the relationship between the temperature and the width of the microchannel when the heat sink has 5 microchannels and the flow rate is 5ml/min. The results of the relationship between the number of the microchannel and the temperature are shown in Fig.6(c). From the simulation, when the power of the heat source is 4W, the temperature of the heat sink is 47°C which consists well with the thermal experiment result. Together these results provide that the number and the width of the microchannel have little effect on the heat performance of this heat sink.

The thermal resistance of this heat sink can be calculated by equation(6).

$$R_{total} = \frac{\Delta T \cdot S}{P_{total}} \tag{6}$$

Where R_{total} is the total thermal resistance of this heat sink, ΔT is the temperature rise and the P_{total} is the total power of the heat source. When the power is 4W, and the size of the chip is 3.2mm*1.8mm and the temperature rise is 27°C, so the thermal resistance is only 0.388 cm²·°C/W.

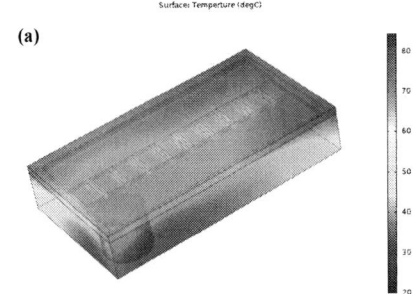

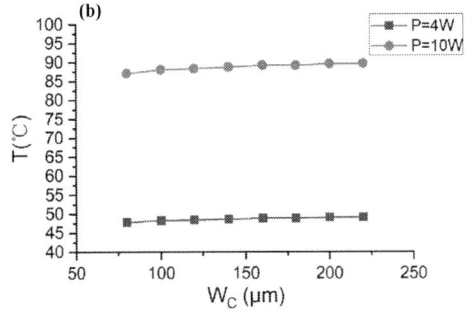

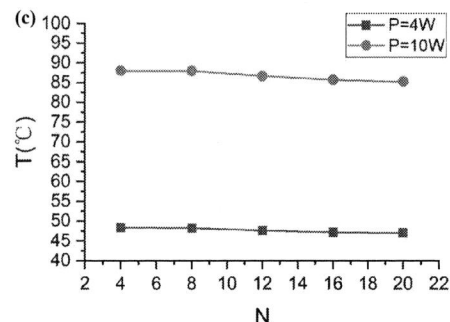

Fig. 6. (a) Simulation of the temperature distribution of the Si-Glass heat sink. (b) The width of microchannel and heat source temperature (c)The number of microchannel and heat source temperature

V. CONCLUSIONS

The serious thermal management problem of the glass-based high-power RF modules has brought many barriers in the application. In this paper, a Si-Glass microchannel heat sink was fabricated, designed and tested. From the date of the thermal experiment, the heat dissipation performance of the Si-Glass heat sink is much better than the traditional passive cooling technology. When the flow rate is 5ml/min

and the heat flux great than 150W/cm², it also can keep the temperature of the devices at a proper range. This work demonstrates the effectiveness of the embedded Si-Glass microchannel heat sink for handling the thermal management issue of glass-based high-power RF modules.

REFERENCES

[1] Qian L , Xia Y , Shi G , et al. Electrical-thermal characterization of through packaging vias in glass interposer[J]. IEEE Transactions on Nanotechnology, 2017:1-1.

[2] Cho S, Joshi Y, Sundaram V, et al. Comparison of thermal performance between glass and silicon interposers[C]//2013 IEEE 63rd Electronic Components and Technology Conference. IEEE, 2013: 1480-1487.

[3] Cho S , Sundaram V , Tummala R R , et al. Impact of Copper Through-Package Vias on Thermal Performance of Glass Interposers[J]. IEEE Transactions on Components, Packaging and Manufacturing Technology, 2015, 5(8):1075-1084.

[4] M. Yang, M. T. Li, et al. "Experimental study on single-phase hybrid microchannel cooling using HFE-7100 for liquid-cooled chips" International Journal of Heat and Mass Transfer 160 (2020) 120230

[5] Tuckerman D B, Pease R F W. High-performance heat sinking for VLSI[J]. IEEE Electron device letters,1981, 2(5): 126-129.

[6] Muszynski T, Andrzejczyk R. Heat transfer characteristics of hybrid microjet–Microchannel cooling module[J]. Applied Thermal Engineering, 2016, 93: 1360-1366.

A High-Q Inductor based on Fan-Out Panel Level Package Technology

Xulei Niu
Fujian Provincial Engineering Technology Research Center of Photoelectric Sensing Application,College of Photo and Electronic Engineering, Fujian Normal University
Fuzhou City,Fujian Province, 350007, China
17839192858@163.com

Zixu Wang
Fujian Provincial Engineering Technology Research Center of Photoelectric Sensing Application,College of Photo and Electronic Engineering, Fujian Normal University
Fuzhou City,Fujian Province, 350007, China
2302473442@qq.com

Shaopan Lin
Fujian Provincial Engineering Technology Research Center of Photoelectric Sensing Application,College of Photo and Electronic Engineering, Fujian Normal University
Fuzhou City,Fujian Province, 350007, China
694674342@qq.com

Guochi Huang
Fujian Provincial Engineering Technology Research Center of Photoelectric Sensing Application,College of Photo and Electronic Engineering, Fujian Normal University
Fuzhou City,Fujian Province, 350007, China
guochi.huang@fjnu.edu.cn

Tingyu Lin
Guangdong Xinhua Microelectronics Co., Ltd. 2/F Block A Buddha High-tech Think Tank Center Nanhai High-tech Zone
Foshan City, Guangdong Province, 528000, China
tingyulin@ncap-cn.com

Abstract—In this paper, a planar spiral inductor is presented based on the fan-out panel-level packaging (FOPLP) technology. The inductor is constructed by two redistribution layers (RDL) of the FOPLP process. The size of the designed planar spiral inductor is 250um x 250um. Compared to the on-chip inductors with the same size, the proposed inductor using fan-out panel-level packaging technology has much higher quality factor and nearly same inductance. Due to the thicker metal used for the proposed inductor, the self-resonating frequency (SRF) is slightly lower than that of the on-chip counterpart. The maximum value of the inductor quality factor designed using fan-out panel-level packaging technology is 16.61, while the on-chip inductor achieve 13.27 Q-factor, and both of the two inductors present 1.3nH inductance. Obviously, if the on-chip inductor is replaced by the FOPLP inductor, not only higher Q-factor and larger inductance can be obtained, but also a large chip size can be saved.

Keywords—Planar spiral inductor; quality factor; fan-out panel-level packaging

I. INTRODUCTION

Since the commercialization of 5G, with the development of high-capacity, high-speed, and low-latency services, the communication frequency band has continued to extend to the mm-Wave direction, which has stimulated the upsurge in the design of radio frequency transceivers. On-chip spiral inductors are important components of integrated circuit modules such as Voltage Controlled Oscillators (VCO), Low Noise Amplifiers (LNA) and passive filters. High selectivity and signal-to-noise ratio require a higher quality factor (Q - factor). However, conventional on-chip spiral inductors have lower Q and lower self-resonant frequency (SRF) due to the capacitance and electromagnetic coupling with the substrate at high frequencies, and the on-chip inductors occupy a larger chip area.

FOPLP technology is one of the next generation fan-out packaging technologies, and is a breakthrough technology that

extends Fan-out Wafer-Level Packaging (FOWLP)[1-2]. In terms of cost, FOPLP is cheaper than FOWLP. Moreover, when the size of the package becomes larger and larger, the area utilization rate of the carrier can be more effectively used when using FOPLP compared to FOWLP. FOPLP technology realizes fan-out wiring on large-size panels by integrating LCD, advanced packaging technology and IC carrier technology. Because the RDL of FOPLP technology has a thicker metal layer and better dielectric material than the inside of the chip, a spiral inductor design with higher Q and inductance can be realized.

Inductor in the package is the mainstream direction of research to replace traditional on-chip inductance. Circular spiral inductors designed with FOWLP technology can effectively reduce the chip area while achieving high Q-factor[3]. This paper proposes to use lower-cost FOPLP technology to design inductors. FOPLP technology can provide thicker metal layers and better dielectric materials than chip technology.

In this paper, a single ended spiral inductor with 1.3nH inductance using FOPLP technology is present. The proposed design of the inductor is composed of two layers of RDL and achieves a peak Q-factor of 16.61 and SRF of 23GHz at 8GH. The paper is organized as follows. The Section II proposes the design of square spiral inductors based on FOPLP technology. Simulation and analysis of the spiral inductors in package and on-chip spiral inductors are presented in section-III. Conclusions are made in Section-IV.

II. SPIRAL INDUCTOR DESIGN

A. Spiral inductor using FOPLP technology

The spiral inductor using FOPLP technology is composed of two layers of RDL, as shown in Fig. 1. RDL1 is used for the main spiral structure wiring of the inductor, and RDL2 is used

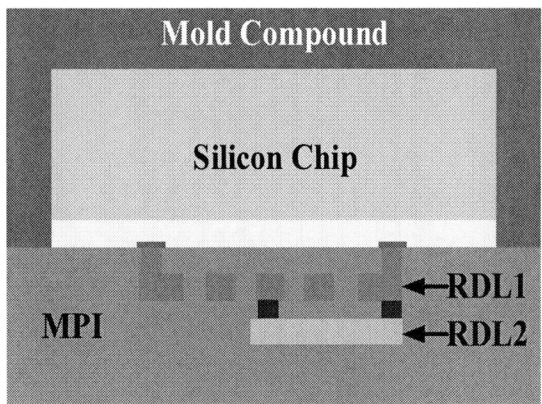

Fig. 1. illustration of Fan-Out Panel Level Package solution of inductor

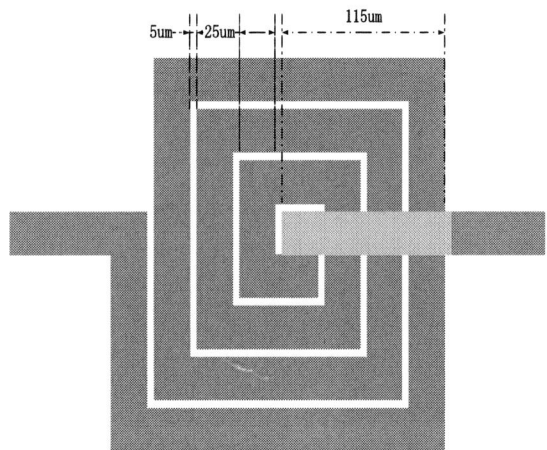

Fig. 2. The structure of the inductor.

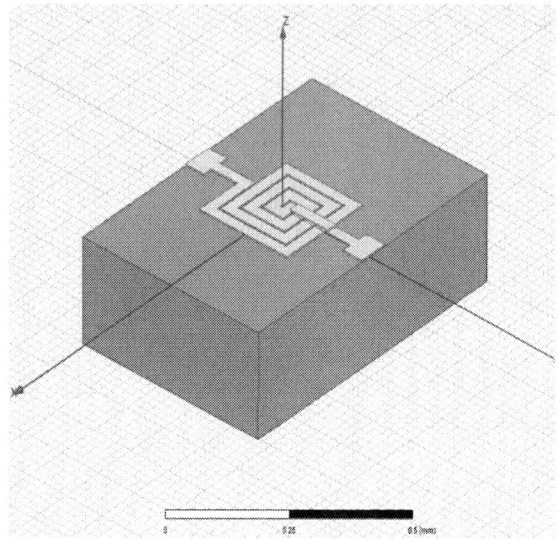

Fig. 3. Inductor model in HFSS

TABLE I.   THE MATERIAL AND THICKNESS

The material and thickness	
Material	*Thickness*
Mold Compound	600um
MPI	100um
RDL1	15um
RDL2	15um

B. Spiral inductor on-chip

In order to accurately reflect the performance of the spiral inductor designed using Fan-Out Pan Level Package technology, the parameter configuration of the on-chip spiral inductor is the same as that of the package spiral inductor, as shown in Fig. 2. Based on SMIC 130nm CMOS RF technology, the on-chip spiral inductor is designed to have three-turn square spiral windings (M5) on the top layer of metal, and the second layer (M4) is used for the lower end channel and is connected to the first layer through VIA conversion. In particular, due to process limitations, the thickness of the M4 and M5 metal layers are not the same (RDL2 for M4, RDL1 for M5), which limits the Q-factor of the inductor.

III. SIMULATION RESULTS

In order to verify the high Q-factor and inductance of the proposed spiral inductor based on the FOPLP technology, this section will illustrate through simulation and comparison experiments with on-chip spiral inductors of the same size.

A. Simulation

The materials and thickness of the proposed spiral inductor based on FOPLP technology are shown in Table I. and the structure is shown in Fig. 3. In particular, the dielectric constant (ε) of the dielectric (MIP) is 3, and tgδ is 0.005, and low tgδ can achieve a higher Q-factor. Simulate and extract S parameter files through HFSS. Figure 3 shows the 3D model

for the lead out of the other end of the inductor. Both ends of the spiral inductor are connected to the PAD through RDL1.

The modified polyimide (MPI) material is used as the dielectric in the horizontal package of the fan-out panel, as shown in Figure 1. The thickness of MPI is 30um, and both RDL1 and RDL2 are 15um, as shown in Table 1. The thickness of RDL layers can be precisely controlled and changed which gives out one more design parameter during optimizing the performance of the inductor., Fig. 2 shows the parameter configuration of the spiral inductor proposed based on the FOPLP technology. The inductor is design into a square spiral structure, which is the most common type in real design and provides more convenience in manufacture. The designed inductor includes three-turn square spiral windings on the RDL1, and RDL2 is used for the lower port channel to connect to the RDL1 layer through a via. The wiring width of the winding group is 25um, the spacing is 5um, and the inductance area is 250umX250um. These parameters are chose in order to achieve the same inductor size which is normal and available on CMOS process. The inductor type can be any structure other than square-shape, such as octagonal inductor or, which depends on the requirement of the inductor quality-factor. The closer to circular type, the higher the inductor quality –factor is.

978-1-6654-1392-3/21 $31.00 © 2021 IEEE   522

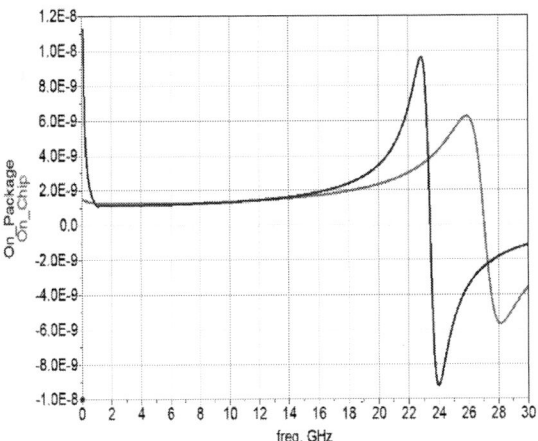

Fig. 4. Comparison of inductance values between on-chip inductors and inductors in package.

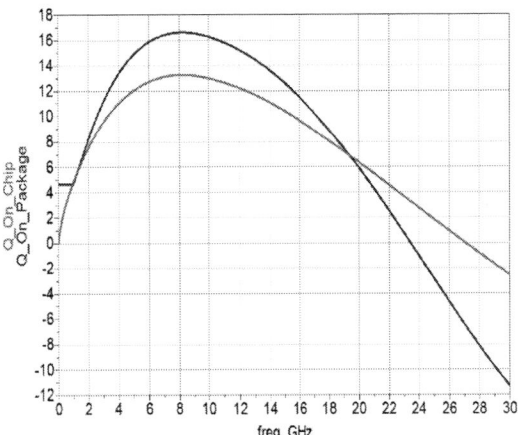

Fig. 5. Comparison of Q-factor between on-chip inductors and inductors in package.

diagram of the spiral inductor in HFSS. The winding of the spiral inductor is set on the reference ground to obtain a better Q-factor. The HFSS simulation frequency ranges from 0-30GHz. It should be noted that a higher inductance value can be obtained by reducing the wiring spacing or reducing the winding line width to reduce the parasitic capacitance between the windings. The trace width of the lower channel of the RDL2 layer should be reduced as much as possible under the condition of meeting the limit current to reduce the parasitic capacitance between it and the winding.

TABLE 2. The material and thickness

The material and thickness			
Material	*dielectric constant(ε)*	*Sheet resistance(Ω/sq)*	*Thickness*
M5(Cu)	—	0.007	3um
M4(Cu)	—	0.0202	0.9um
USG	4.2	—	3.63um
SiN	7	—	0.07um

According to the structure size of the proposed spiral inductor in package, Cadence Virtuoso is used to design the on-chip spiral inductor of the same size. EM simulation of the inductor is done using EMX and the S-parameter is extracted for calculating the inductance, quality-factor and the self-resonant frequency. The on-chip spiral inductor is designed using SMIC's 0.13um process. Table 2 shows the main material parameters of the on-chip spiral inductor. In particular, compared with packaged inductors, the materials and parameters of on-chip inductors are quite different, and it is these differences that enable packaged spiral inductors to obtain better indicators.

B. Result and Analysiss

In this section, import the extracted S-parameter files of the spiral inductor in package and the on-chip spiral inductor into ADS for simulation and comparison. Fig. 4. and Fig. 5. show the inductance and Q value of the inductor in package and the on-chip inductor, respectively. Use the following method to extract the inductance and quality factor (Q factor)

of a single-ended spiral inductor, where A is the input admittance of the inductor:

$$L = \frac{Im\left(\frac{1}{y11}\right)}{\omega} \tag{1}$$

$$Q = \frac{Im[y11]}{Re[y11]} \tag{2}$$

Fig. 4. shows the change curve of the inductance value with frequency in the range of 0~30GHz. In 0~18GHz, the on-chip spiral inductor and the spiral inductor in package have the same inductance. But the SRF of the on-chip inductor is higher than the spiral inductor in package 4GHz. This is mainly caused by the excessively large winding wire width of the spiral inductor in package. In the design process, the SRF is increased by reducing the winding trace width, which will also reduce the area of the spiral inductor on package.

Fig. 5. Comparison of Q-factor between on-chip inductors and inductors in package. At 8GHz, the Q-factor of the spiral inductor in package is higher than that of the on-chip spiral inductor 4. Therefore, the comparison shows that the spiral inductor in package using FOPLP technology can achieve higher Q- factor.

IV. CONCLUSIONS

This paper proposes a single-ended multilayer square spiral inductor based on low-cost FOPLP technology. Compared to traditional CMOS on-chip spiral inductors, it can achieve a higher Q-factor. The mental layer of traditional on-chip inductors are replaced by FOPLP technology using RDL layer during packaging process, which can effectively reduce the chip area and then the chip cost. MPI material is used as the dielectric in the FOPLP technology, which has lower dielectric loss and guarantees a higher inductor Q-factor. In particular, the thickness of MPI and RDL can be precisely controlled. Therefore, the design of large inductance can be realized, which provides greater convenience for the design of high-performance millimeter-wave circuits, such as voltage controlled oscillator, power amplifier, low noise amplifier, and et al..

ACKNOWLEDGMENT

Thanks to Professor Guochi Huang for his systematic explanation of FOPLP and revision of the paper. The most important thing is to point out the research direction for the author. At the same time, I would like to thank Fujian Normal University for providing scientific research sites and Guangdong Fozhixin Microelectronics Technology Research Co., Ltd. for technical support.

REFERENCES

[1] L. Chen, X. Sun and F. Chen, "Development of 500mm×500mm Fan-out panel level packaging for heterogeneous chip integration," 2020 21st International Conference on Electronic Packaging Technology (ICEPT), Guangzhou, China, 2020, pp. 1-6.

[2] G. Huang, Y. Dai, N. Liu, Z. Duan, Y. Wu and T. Lin, "A low cost 60GHz antenna in Fan-Out Panel Level Package for millimeter-wave radar application," 2020 21st International Conference on Electronic Packaging Technology (ICEPT), Guangzhou, China, 2020, pp. 1-4.

[3] K. S. Murugesan et al., "High Quality Integrated Inductor in Fan-out Wafer-Level Packaging Technology for mm-Wave Applications," 2020 50th European Microwave Conference (EuMC), 2021, pp. 89-92,

[4] E. Ashenafi and M. H. Chowdhury, "Noise voltage analysis of spiral inductor for on-chip buck converter design," 2017 IEEE International Symposium on Circuits and Systems (ISCAS), Baltimore, MD, USA, 2017, pp. 1-4.

[5] S. C. Chong, V. S. Rao, K. Yamamoto and S. L. Seow Huang, "Development of RDL-1stFan-Out Panel-Level Packaging (FO-PLP) on 550mm x 650mm size panels," 2020 IEEE 22nd Electronics Packaging Technology Conference (EPTC), Singapore, Singapore, 2020, pp. 425-429.

[6] M. Khodapanahandeh, M. Pazhooh, S. Afrang and H. B. Ghavifekr, "Novel High Q Single Layer Bridgeless RF MEMS Inductors," 2020 28th Iranian Conference on Electrical Engineering (ICEE), 2020, pp. 1-5.

Effects of cetyltrimethylammonium bromide (CTAB) on electroplating twin-structured copper interconnects

Yi Dong
Shenzhen Institute of Advanced
Electronic Materials,
Shenzhen Institute of Advanced
Technology, Chinese Academy of
Sciences (CAS),
Shenzhen 518055, China
College of Engineering,
Southern University of Science and
Technology,
Shenzhen 518055, China

Zhe Li*
Shenzhen Institute of Advanced
Electronic Materials,
Shenzhen Institute of Advanced
Technology, CAS,
Shenzhen 518055, China
zhe.li@siat.ac.cn

Li-Ying Gao
Shenzhen Institute of Advanced
Electronic Materials,
Shenzhen Institute of Advanced
Technology, CAS,
Shenzhen 518055, China

Xiao Li
Shenzhen Institute of Advanced
Electronic Materials,
Shenzhen Institute of Advanced
Technology, CAS,
Shenzhen 518055, China

Zhi-Quan Liu*
Shenzhen Institute of Advanced
Electronic Materials,
Shenzhen Institute of Advanced
Technology, CAS,
Shenzhen 518055, China
zqliu@siat.ac.cn

Rong Sun
Shenzhen Institute of Advanced
Electronic Materials,
Shenzhen Institute of Advanced
Technology, CAS,
Shenzhen 518055, China

Abstract—Twin-structured copper possesses great potential in advanced interconnect electroplating because of high performance and high reliability. Whereas, the issue of surface nonuniformity restricts its practices in integrated circuits and demands a prompt solution. A novel additive strategy based on coexisting cetyltrimethylammonium bromide (CTAB) and gelatin is proposed and effects of CTAB to bath electrochemistry and material properties are investigated. With an optimum concentration of CTAB, the electroplating formular enables improved surface leveling and filling height coplanarity for twin-structured copper filling in 15/15μm line/space redistribution layer (RDL) patterns.

Keywords—copper electroplating, twin structure, CTAB, RDL

I. INTRODUCTION

Copper nowadays plays an irreplaceable role as electric interconnect material in integrated circuits manufacturing and packaging, relying on advantages of superior electric conductivity and electromigration resistance compared to aluminum [1], and above all, compatibility to electroplating techniques for flawless bottom-up filling in micro- and nano-scaled patterns (e.g. trenches, vias, holes, etc.).[2] The covering power and throwing power of copper interconnects depends on the uses and interplays of functional organic additives. Particularly, levelers, including nitrogenous heterocyclic compounds and quaternary amides, can adsorb to high current density protrusions to neutralize excess charge and inhibit local copper deposition to flatten plating profile. Innovation of copper electroplating formular centers on syntheses and screenings of novel levelers for controllable surface profile. [3-5]

Twin-structured copper refers to those microstructures of a high-proportioned twin boundaries. As the density of twin boundaries increases and the thickness of twin lamellar reduces to nanoscale, these so-called nano-twins endow copper with ultrahigh strength and thermal stability but uncompromised ductility and electric conductivity.[6, 7] With the presence of promotor, for example gelatin, growth of twin boundaries is achievable through direct-current electroplating and moreover facilitated under specific process parameters such as low bath temperature [8], severe convection [9], and high current density [10]. Efforts have been made to further tune the twin-structure to achieve improve mechanical properties, from aspects such as grain size and orientation or twin lamellar orientation and distribution.[11, 12]

Twin-structured copper is therefore expected to be the next-generation interconnect material for high performance and high reliability.[13] However, the spiral growth of twin boundaries within grains inevitably leaves pyramid endings at the top [14], which leads to zigzagged surface topography and restricts its applications in scenarios of demanding surface uniformity and pattern coplanarity. Limited work has been reported to address the problem of surface roughness. Despite co-addition of commercial leveler may relieve this problem, there is also a chance that leveler competes with gelatin and take dominant place in cathodic adsorption, which impacts or even eliminates formation of twin boundaries.[15]

Cetyltrimethylammonium bromide (CTAB) is a cationic surfactant. It can modify the chain-like gelatin molecules to become curled and grafted with cationic heads through electrostatic interaction.[16] In this study, CTAB is formulated with gelatin as in a twin-structured copper electroplating process reported previously.[17] The formation

978-1-6654-1392-3/21 $31.00 © 2021 IEEE

and properties of twin boundaries are found unaffected while the surface uniformity and pattern coplanarity of electroplated redistribution layer (RDL) interconnects are significantly improved accompanied with reduced columnar grain size. This novel additive formulation points out a feasible direction for morphology control of twin-structured electroplated copper with quaternary ammonium.

II. EXPERIMENTAL DETAILS

A. Electroplating

The copper electroplating bath was based on a proprietary formular consisting of 160g/L $CuSO_4 5H_2O$, 50g/L H_2SO_4, 40ppm chloride, 30ppm gelatin, and various concentrations of CTAB (Aladdin) ranging from 0 to 10 ppm. The plating baths were electrochemically characterized by steady-state cathodic polarization on a electrochemical workstation (CH Instruments 760E). Linear sweep voltammetry (LSV) was performed based on a three-electrode electrolytic cell, with a platinum disk ($\emptyset$ = 3mm) as the working electrode, a platinum plate (2×2 cm^2) as the counter electrode, and a Ag/AgCl reference (+0.2223V vs SHE).

The twin-structured copper blanket films and RDL patterned dies were electroplated with a direct-current power supply (Tektronix PWS4323) and above-mentioned plating baths in a 500ml Teflon cell. A pure titanium plate or a proprietary RDL patterned die of 15/15 µm line/space was used as the cathode, and a phosphor copper plate (0.035~0.065 wt.% of phosphorous) as the anode. Electroplating was carried out at room temperature at a current density of 30 mA/cm^2, with 300 rpm of magnetic stirring applied. The plating duration was set to 20 mins for blanket films to obtain 15 µm film thickness and 6 mins for RDL patterned dies to fill up 6 µm trench depth, respectively.

B. Materials Characterization and Testing

The crystal structures of the film areas on the RDL patterned dies were examined on a X-ray diffractometer (XRD, Bruker D8Advance), equipped with Cu target (λ = 0.15406 nm). The theta-2theta scan mode was applied at a step size of 0.2 deg. and rate of 10 deg./min. The cross-sectional microstructure of the blanket films were observed on a focused ion beam (FIB, FEI SCIOS).

Blanket films were trimmed into dog-bone shape (Min. width = 2.5 mm) for tensile tests, carried out on a dynamic thermomechanical analyzer (DMA, TA Q800) at room temperature. A preload of 0.1N was applied and the elongation rate was set to 1N/min.

RDL patterned dies after electroplating were immersed into acetone solution (50 wt.%) and ultrasonicated for 5 mins to remove photoresists. The morphology of patterned copper interconnects was observed on a scanning electron microscope (SEM, FEI Nova Nano). The surface profile was measured on a laser scanning confocal microscope (LSCM, Keyence VK-X1100) for calculation of plating coplanarity.

III. RESULTS AND DISCUSSION

A. Electrochemistry

The electrochemical behaviors of additive CATB and co-existing gelatin are analyzed by steady-state cathodic polarization (Fig. 1). Indicated by reduced copper reduction current and negatively-shifted onset potential of deposition, the cathodic polarization is enhanced after addition of gelatin.

By introduction of increased amount of CTAB from 2 to 10 ppm, the onset potential of deposition negatively to shift from -0.1 to -0.25 V, and the electrode potential at the plating current density of 30 mA/cm^2 also negatively shits from -0.18 to -0.23 V, showing further strengthened polarization degree. The greater overpotential due to raised concentration of CTAB stimulates nucleation relative to growth during copper electrocrystallization, and a finer grain size and improved surface leveling can be expected.[18]

The enhanced cathodic polarization and inhibiting effects on copper deposition by additive CTAB and gelatin can be attributed to molecular adsorption at electrochemical double-layer. The polar amide nitrogen from gelatin can form chemical bond with surface copper atoms [19], while the positively-charged ammonium nitrogen from CTAB electrostatically adsorb to the electron-rich cathode [19], and these two processes occur simultaneously on cathode and compete. Moreover, free CTAB molecules can complex with co-existing gelatin through electrostatic interaction and modify the chain-like gelatin molecules to become curled, which alters the adsorption behaviors of gelatin. Since CTAB molecules possess a significantly smaller spatial size and faster directed migration, the presence of CTAB should displace gelatin during dynamic adsorption/desorption to a certain extent and occupy more active sites of adsorption.

Additionally, the reduction peak appearing on the overall reduction wave is possibly due to hydrogen evolution. CTAB aggressively suppresses copper reduction, and as its content rises, the excessive cathodic polarization drives the overpotential of copper deposition close to or beyond that of hydrogen evolution potential.[20] The concomitant proton reduction is therefore more obvious compared to organic-additive-free electrolyte (VMS) or gelatin-added-only counterpart.

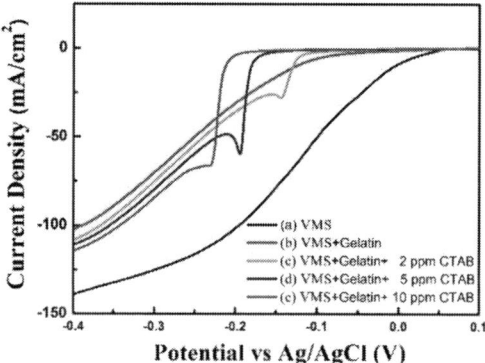

Fig. 1. Linear sweep voltammograms of plating baths containing: (a) VMS; (b) VMS and 30 ppm gelatin; (c)-(e) VMS, 30 ppm gelatin and 2~10 ppm CTAB. Scan rate: 1mV/s

B. Microstructure

The cross-sectional microstructures of blanket films electroplated in baths of various CTAB concentrations are observed by FIB (Fig. 2). The typical twin-structure obtained from the proprietary formular (CTAB-excluded) consists of a primary columnar-grain region (electrolyte-side) and a transitional equiaxed-grain region (substrate-side) along the growth direction.[21] Columnar grains are about 12µm in length and 1-4 µm in diameter, showing high density of twin lamellar oriented perpendicular to the growth direction. The 1 μm or submicron sized equiaxed grains are generated due to accelerated nucleation at initial stage of deposition and twin lamellar are not formed there.

When the concentration of CTAB is raised to 10 ppm, the diameter of the columnar grains gradually reduce to 1-3 μm and becomes more uniform along the axial direction. This is closely related to above-discussed electrocrystallization that the nucleation rate is manipulated by CTAB to become mildly higher, which restricts the lateral growth of columnar grains. No distinct differences can be observed for the volume of twin-structured region or the orientation of twin-lamellar, indicating a negligible impact of CTAB to twin boundary formation.

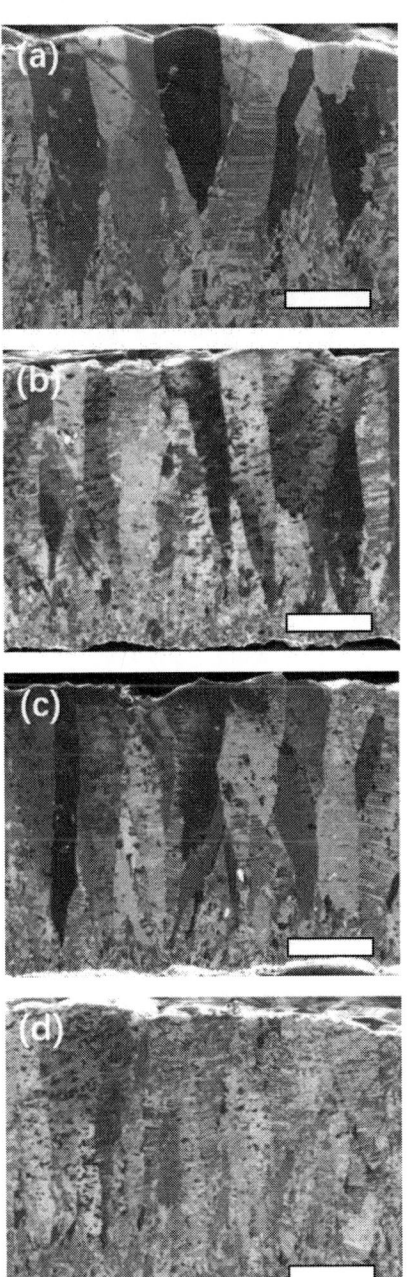

Fig. 2. Cross-sectional FIB images of blanket films electroplated in baths containing VMS, 30 ppm gelatin, and: (a) no CTAB; (b) 2 ppm CTAB; (c) 5 ppm CTAB; (d) 10 ppm CTAB. Scale bar: 5μm.

C. Crystal Orientation

The X-ray diffraction patterns of twin-structured blanket films electroplated with the presence of various CTAB concentrations are compared (Fig. 3). Diffraction peaks of Cu (111) plane (2theta = 43.297 deg., JCPDS#04-0836) and Si (400) plane (2theta = 69.130 deg., JCPDS#27-1402) are detected. The strong intensity of (111) peak shows all these films are (111)-textured. It is because preferred adsorption of CTAB on higher surface energy lattice planes except (111) plane blocks the copper deposition there and allows a primary deposition along (111) plane, which also facilitates the growth of coherent twin boundaries at (111) plane. Increasing CTAB content from 0, 2, 5 to 10 ppm, the intensity ratio of Cu (111)/ Si (400) accordingly changes from 5.7, 22.1, 10.1 and 7.9, respectively. The decline of (111) texture is likely ascribed that some CTAB complex with gelatin and desorb at raised concentrations.

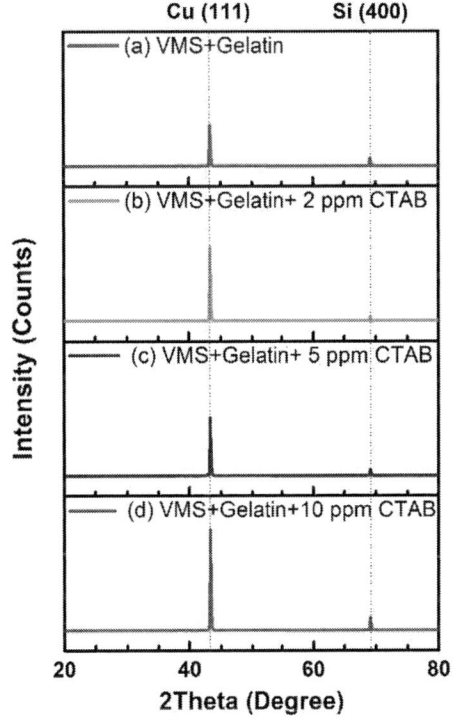

Fig. 3. XRD patterns of blanket films electroplated in baths containing VMS, 30 ppm gelatin, and: (a) no CTAB; (b) 2ppm CTAB; (c) 5ppm CTAB; (d) 10 ppm CTAB.

D. Mechanical Properties

Tensile tests are performed on DMA using twin-structured thin film specimens electroplated with varying CTAB contents (Fig. 4). Three out of five measured tensile curves are selected to avoid random error, and tensile strength and elongation at break are plotted against the amounts of CTAB added. Bulk films of copper has a tensile strength about 150-250 MPa and an elongation at break about 20-30%. However, because of the size effect, thin films generally exhibit higher strength and lower ductility compared the bulk counterparts. Besides, the existence of twin-structure also greatly contributes to material strengthening.

Twin structured copper exhibit brittle facture behaviors and no yield plateau is seen. The presence of CTAB causes dual drop of tensile strength from 419 to 299 MPa and elongation at break from 2.42 to 0.73%. Then, as more CTAB is added, the tensile strength and elongation at break slightly recover to 332 MPa and 0.87%. These are closely related to addition of CTAB and its concentrations: first, when the diameter of columnar grains drops, the length of stronger twin boundaries reduces while the weaker grain boundaries grows more，which could lower the strength; second, the inverse of Cu (111)/ Si (400) intensity ratio is in agreement with changes of strength, in contrast to the common understanding that highly (111)-textured copper displays better strength due to close-packed lattice structures and resistance to deformation [22], but the reason behind is not clear yet; last, the hydrogen evolution becomes nonnegligible at excessive deposition overpotential, which gives rise to hydrogen embrittlement and harms both strength and ductility. The decline of mechanical properties is probably a result of joint effect of above mechanisms.

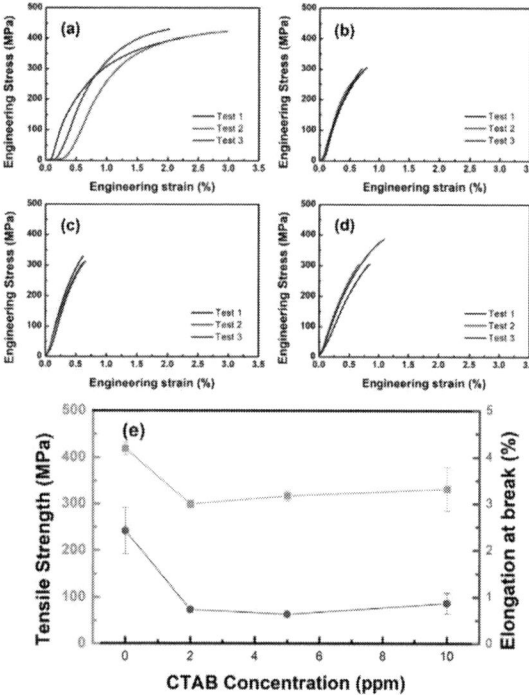

Fig. 4.   DMA tensile curves of specimens electroplated in baths containing VMS, 30 ppm gelatin, and: (a) no CTAB; (b) 2ppm CTAB; (c) 5ppm CTAB; (d) 10 ppm CTAB. And (e) summarized tensile strength and elongation at break plotted against CTAB concentrations.

E. RDL Pattern Filling

Based on mechanical performances, the additive formular of 30 ppm gelatin and 10 ppm CTAB is selected for twin-structured copper electroplating in 15/15 μm line/space RDL patterns. Evidenced by top-view SEM images (Fig. 5), CTAB-excluded formular leads to a rough topography of broad grain morphology distribution. Uneven-sized copper nodules appear, originated from abnormal nucleation and follow-up growth. In contrast, CTAB-containing formular brings about a smoother profile and no nodule is observed in the field of examination. This CTAB-stimulated surface leveling is in agreement with the prediction by electrochemical analyses. The coplanarity (C) of RDL filling height among the 10 measured lines within die is described through equation:

$$C = (H_{max} - H_{min})/ 2H_{avg} \times 100\% \qquad (1)$$

Where, H_{max}, H_{min} and H_{avg} are the maximum, minimum and average filling height. Coplanarities are calculated to be 11.9% and 5.2% for CTAB- excluded and containing formulars, respectively, indicating a prominent upgrade of coplanarity. This is also ascribed to effective leveling of CTAB on RDL patterned copper interconnects throughout the die.

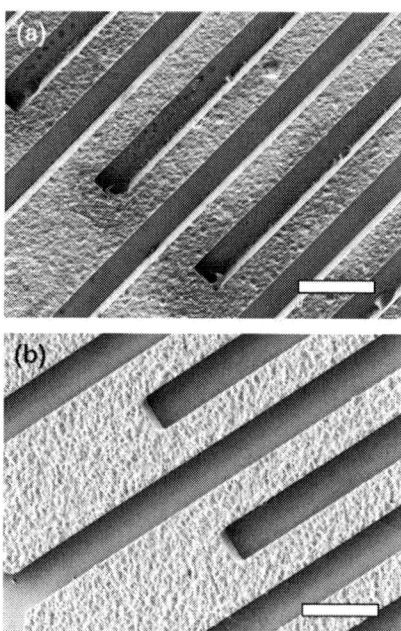

Fig. 5.   Top-view SEM images of 15/15μm line/space RDL patterns electroplated in baths containing VMS, 30 ppm gelatin, and: (a) no CTAB; (b) 10 ppm CTAB. Scale bar: 20 μm.

IV. Conclusion

In summary, a novel additive strategy, consisting of a quaternary ammonium CTAB and gelatin, is demonstrated for electroplating twin-structured copper. Effects of CTAB to electroplating bath and material properties are revealed: when a raised CTAB concentration from 0 to 10 ppm is present with existing 30 ppm gelatin, the cathodic polarization and deposition overpotential are prominently enhanced; the origin, density and orientation of twin boundaries are not impacted while the diameter of columnar grains is reduced; the tensile strength and elongation at break, along with inverse of (111) texture, first decline and then recover. For 15/15μm line/space RDL pattern electroplating, this additive formular enables improved surface leveling and achieves a small filling height coplanarity index at an optimum 10 ppm CTAB.

Acknowledgment

This work was supported by Center for Materials Service Reliability, Shenzhen Institute of Advanced Electronic Materials. The authors gratefully acknowledge Ms. Jing Huang in performing mechanical tests, Mr. Zhi-Gang Lei in assisting electrochemical analysis, and Mr. Zhen-Jia Peng in sample grinding and polishing.

References

[1] Y.-S. Wang, W.-H. Lee, S.-C. Chang, J.-N. Nian, and Y.-L. Wang, "An electroplating method for copper plane twin boundary manufacturing," *Thin Solid Films,* vol. 544, pp. 157-161, 2013.

[2] U. G. Stöckgen, S. Wehner, J. Heinrich, A. Kiesel, and R. Liske, "(Invited) Integration Challenges for Copper Damascene Electroplating," *ECS Transactions,* vol. 25, no. 38, pp. 3-17, 2019/12/17 2019.

[3] X. Wang *et al.*, "Quinacridone skeleton as a promising efficient leveler for smooth and conformal copper electrodeposition," *Dyes and Pigments,* vol. 181, 2020.

[4] K. Wang *et al.*, "Engineering aromatic heterocycle strategy: Improving copper electrodeposition performance via tuning the bandgap of diketopyrrolopyrrole-based leveler," *Tetrahedron,* vol. 76, no. 5, 2020.

[5] J. Li *et al.*, "Copolymer of Pyrrole and 1,4-Butanediol Diglycidyl as an Efficient Additive Leveler for Through-Hole Copper Electroplating," *ACS Omega,* vol. 5, no. 10, pp. 4868-4874, Mar 17 2020.

[6] R. Niu and K. Han, "Strain hardening and softening in nanotwinned Cu," *Scripta Materialia,* vol. 68, no. 12, pp. 960-963, 2013.

[7] J. Ye, Y. Wang, T. Barbee, and A. Hamza, "Orientation-dependent hardness and strain rate sensitivity in nanotwin copper," *Applied Physics Letters,* vol. 100, 06/29 2012.

[8] G. Cheng, H. Li, W. Zhang, G. Xu, and L. Luo, "Controllable large scaled nanotwin formation in Cu film at lower temperatures," *Journal of Physics D: Applied Physics,* vol. 49, no. 40, p. 40LT01, 2016.

[9] C.-H. Tseng and C. Chen, "Growth of Highly (111)-Oriented Nanotwinned Cu with the Addition of Sulfuric Acid in CuSO4 Based Electrolyte," *Crystal Growth & Design,* vol. 19, no. 1, pp. 81-89, 2018.

[10] C.-C. Lin and C.-C. Hu, "The Ultrahigh-Rate Growth of Nanotwinned Copper Induced by Thiol Organic Additives," *Journal of The Electrochemical Society,* vol. 167, no. 8, 2020.

[11] T.-L. Lu, Y.-A. Shen, J. A. Wu, and C. Chen, "Anisotropic Grain Growth in (111) Nanotwinned Cu Films by DC Electrodeposition," *Materials,* vol. 13, no. 1, p. 134, 2020.

[12] Y. J. Li, K. N. Tu, and C. Chen, "Tensile Properties and Thermal Stability of Unidirectionally <111>-Oriented Nanotwinned and <110>-Oriented Microtwinned Copper," *Materials (Basel),* vol. 13, no. 5, Mar 8 2020.

[13] K.-N. Tu and T. Tian, "Metallurgical challenges in microelectronic 3D IC packaging technology for future consumer electronic products," *Science China Technological Sciences,* vol. 56, no. 7, pp. 1740-1748, 2013.

[14] X. Zhan *et al.*, "Preparation of highly (111) textured nanotwinned copper by medium-frequency pulsed electrodeposition in an additive-free electrolyte," *Electrochimica Acta,* vol. 365, 2021.

[15] J. Huang, L.-Y. Gao, and Z.-Q. Liu, "The electrochemical behavior of leveler JGB during electroplating of nanotwinned copper," in *2020 21st International Conference on Electronic Packaging Technology (ICEPT),* 2020, pp. 1-4: IEEE.

[16] Ji. Yun, X.-H. Zhang, and R. Guo, "Interaction between gelatin and cationic surfactant CTAB," *Acta Chimica Sinica-Chinese Edition,* vol. 62, no. 4, pp. 345-350, 2004.

[17] F.-L. Sun *et al.*, "Electrodeposition and growth mechanism of preferentially orientated nanotwinned Cu on silicon wafer substrate," *Journal of materials science & technology,* vol. 34, no. 10, pp. 1885-1890, 2018.

[18] Y.-J. Shih, Z.-L. Wu, C.-Y. Lin, Y.-H. Huang, and C.-P. Huang, "Manipulating the crystalline morphology and facet orientation of copper and copper-palladium nanocatalysts supported on stainless steel mesh with the aid of cationic surfactant to improve the electrochemical reduction of nitrate and N2 selectivity," *Applied Catalysis B: Environmental,* vol. 273, 2020.

[19] C. Meudre, L. Ricq, J.-Y. Hihn, V. Moutarlier, A. Monnin, and O. Heintz, "Adsorption of gelatin during electrodeposition of copper and tin–copper alloys from acid sulfate electrolyte," *Surface and Coatings Technology,* vol. 252, pp. 93-101, 2014.

[20] B. Bozzini, L. D'Urzo, M. Re, and F. De Riccardis, "Electrodeposition of Cu from acidic sulphate solutions containing cetyltrimethylammonium bromide (CTAB)," *Journal of Applied Electrochemistry,* vol. 38, no. 11, pp. 1561-1569, 2008.

[21] Z.-G. Li, L.-Y. Gao, and Z.-Q. Liu, "The effect of transition layer on the strength of nanotwinned copper film by DC electrodeposition," in *2020 21st International Conference on Electronic Packaging Technology (ICEPT),* 2020, pp. 1-3: IEEE.

[22] M. Mieszala *et al.*, "Orientation-dependent mechanical behaviour of electrodeposited copper with nanoscale twins," *Nanoscale,* vol. 8, no. 35, pp. 15999-16004, 2016.

978-1-6654-1392-3/21 $31.00 © 2021 IEEE

Analysis for thermal contact resistance of Press-Pack IGBTs

1st Yakun Zhang
Institute of Electronics Packaging Technology and Reliability
Beijing Key Laboratory of Advanced Manufacturing Technology
Faculty of Materials and Manufacturing
Beijing University of Technology
Beijing, China
zhangyakun@emails.bjut.edu.cn

2nd Tong An*
Institute of Electronics Packaging Technology and Reliability
Beijing Key Laboratory of Advanced Manufacturing Technology
Faculty of Materials and Manufacturing
Beijing University of Technology
Beijing, China
antong@bjut.edu.cn

3rd Fei Qin
Institute of Electronics Packaging Technology and Reliability
Beijing Key Laboratory of Advanced Manufacturing Technology
Faculty of Materials and Manufacturing
Beijing University of Technology
Beijing, China
qfei@bjut.edu.cn

4th Yanpeng Gong
Institute of Electronics Packaging Technology and Reliability
Beijing Key Laboratory of Advanced Manufacturing Technology
Faculty of Materials and Manufacturing
Beijing University of Technology
Beijing, China
yanpeng.gong@bjut.edu.cn

5th Chen Liang
Beijing New Energy Vehicle Technology Innovation Center Co., Ltd
Beijing, China
Beijing Automotive Research Center Co., Ltd,
Beijing, China
liangchen@nevc.com.cn

Abstract—**Press-pack insulated gate bipolar transistors (IGBT) have a multilayer structure, which is clamped by pressure. The IGBT chip generates heat and forms the temperature gradient in the multilayer structure during work. Due to different coefficient of thermal expansion of multilayer structure, temperature and clamping force will lead constant changes in the internal stress of the device, resulting in interface wear and roughness. The thermal contact resistance and the heat dissipation will be affected. And then reduce the reliability of the device. In this paper, the finite element Analysis (FEA) is used to analyze the heat transfer in a three-dimension model of PP IGBT. Based on the contact surface of the chip and the emitter molybdenum plate, the submodel technology is used to investigate the effects of several factors on the thermal contact resistance, including real contact areas, the height of contact joint, and the position of the third-level submodel. And then, the method of single factor analysis is used to analyze the effect on thermal contact resistance. The results show that the thermal contact resistance decrease with the increase of real contact areas, the thermal contact resistance increase with the increase of the height of the contact joint, and the position of the third-level submodel have little effect on thermal contact resistance.**

Keywords—PP IGBT, thermal contact resistance, FEA, joint

I. INTRODUCTION

Insulated gate bipolar transistor (IGBT) modules are widely used in strategic industries such as high-speed railway, new energy vehicles, wind power generation and so on. Press-pack IGBTs can be divided into spring contact and pedestal style direct contact from the point of internal layout [1]. In this paper, the former PP IGBTs are studied and called PP IGBTs for short. PP IGBTs are composed of multilayer structure by pressure clamping. Compared with the traditional wire-bonded IGBT, PP IGBTs have the advantages of high thermal cycling capability, double-side cooling, high power density, high reliability and short circuit failure mode. And the section view of single chip PP IGBT is shown in Fig. 1.

Thermal resistance is a scale of heat dissipation. In the multilayer structure, thermal contact resistance is a significant parameter. It can be affected by the material properties, surface roughness, contact pressure, temperature and so on. It can also be affected by external factors, clamping force, media between interfaces and so on. When two rough surfaces are in contact, real contact occurs at the top of the joint. The real contact area account for approximately 0.01%-0.1% of the apparent contact area, and the proportion only increase to 1%-2% under a contact pressure of 10MPa [2,3].

In PP IGBTs, the total thermal resistance R_{th} is composed of thermal bulk resistance $R_{th,b}$ and thermal contact resistance $R_{th,c}$. There is thermal contact resistance between each layer. Thermal contact resistance accounts for approximately 50% of the total thermal resistance [3]. Erping Deng studied the effect of temperature and clamping force on the thermal contact resistance. The microhardness of a joint will reduce with the temperature variation, leading the roughness reduced. The clamping force increases in the appropriate range, the thermal contact resistance will be affected and decrease. And the bulk thermal resistance stays same [3].

In this paper, the third-level contact model between aluminum and molybdenum emitter plate in PP IGBT is established by submodel technology. And single factor analysis is used to study several factors affecting the thermal contact resistance.

II. METHODOLOGY

A. Finite Element Analysis

In the paper, the coupled thermal-electrical-structural model of the single chip PP IGBT is simulated through ABAQUS. The interaction between each two layer is ignored at present. The interface between the chip and the emitter molybdenum plate is analyzed. There is an aluminum metallization layer on the emitter side of the chip, so the two kinds of materials in contact at the interface are aluminum and molybdenum. The submodel technology is used. It is a finite element technique to obtain more accurate solutions in some regions of the model. The global model is as shown in Fig. 2(a). The first-level submodel is aluminum metallization layer and molybdenum plate as shown in Fig. 2(b). The second-level submodel in Fig. 2(c) is to select two contact models from the center and the corner of the first-order submodel. And the size of the secondary submodel is $400\mu m \times 400\mu m \times 100\mu m$. The third-level submodel is the second-level submodel which changes the interface contact as shown in Fig. 2(d). The three levels of submodels are shown in the Fig. 2. The material properties of three submodels are shown in Tab. II. Load and boundary conditions are used to the model as shown in Fig. 3 to simulate the process of power cycle. The load DC current is 50A and flows in from the collector and out from the emitter. The cycle period is 4 s, and duty ratio is 0.5. The clamping force is applied to the surface of collector. Apply 45°C of temperature boundary to simulate the radiator and ambient temperature is set to 23°C according to the actual situation. The total number of nodes is 81459. The total number of elements is 72206, and the elements type is Q3D8.

Supposed that the height of rough surface profile obeys Gaussian distribution, and the contact between two rough surfaces is equivalent to the contact between a rough surface and a smooth rigid surface [5]. The influencing factors of thermal contact resistance A-C are real contact areas, the height of contact joint, and the position of the third-level submodel respectively. The shape of joint is cylinder. Real

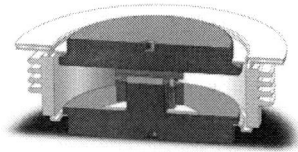

Fig. 1.   The section view of single chip PP IGBT

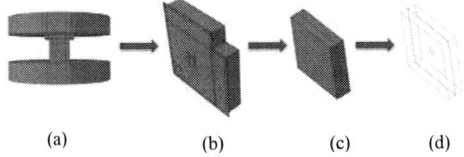

Fig. 2.   The three levels of submodels

TABLE I.   THE LEVELS OF FACTORS

No.	Notation	Unit	Levels		
			1	2	3
1	A	%	1	3	5
2	B	μm	0.5	0.75	1
3	C	-	center	corner	-

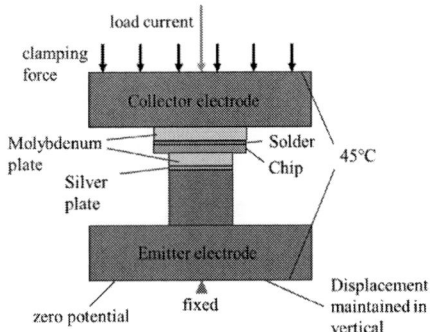

Fig. 3.   The general view of boundary conditions and load

contact areas is account for 1-2% of apparent area when the interface apparent pressure reaches 10 MPa [3]. And after calculation, the interface we studied is 14.225 MPa. Therefore, the selecting factor levels of real contact areas are 1%, 3% and 5% of the apparent area.

The height of contact joint is determined by the surface roughness of the two surfaces measured by the experiment. The selecting factor levels of the height of contact joint are 0.5 μm, 0.75 μm and 1 μm. The equivalent roughness, σ, can be calculated by [4]:

$$\sigma = \sqrt{\sigma_1{}^2 + \sigma_2{}^2} \tag{1}$$

The stress field in the IGBT chip is uneven as shown in Fig. 6. The position of the third-level submodel is selected to verify whether the stress field has an effect on the heat transfer. The center and the corner position are selected, and the stress fields of two positions are different, as shown in the red line in Fig. 2. The different levels of all factors are shown in the Tab. I.

B. Calculation

There are three modes in heat transfer. They are thermal conduction, thermal convection and thermal radiation. In this paper, thermal conduction is studied. Thermal convection can be ignored, because the thermal convection requires a medium between the two interfaces. However, the untouched part of the interface is in a vacuum. And the thermal radiation can also be ignored, because the temperature is lower than 700K during working, and the capacity of thermal radiation in vacuum is poor in this time. These two modes of heat transfer can be ignored in this paper.

The thermal contact resistance R_{th} between interfaces can be calculated using:

$$R_{th,c} = (T_A - T_B)/P \tag{2}$$

Where T_A, T_B, P are respectively the temperature of higher surface, the temperature of the lower surface and the heat flow. Heat flow can be calculated using:

$$P = q \times A \tag{3}$$

Where q, A are respectively the heat flux and the apparent contact area.

TABLE II. MATERIAL PROPERTIES

Material	Thermal conductively (W/(m*k))	Conductivity (S/mm)	Thermal expansion coefficient (1/°C)	Specific heat (10^{-6}/(kg*K))	Young's modulus (MPa)	Poisson's ratio	Density (10^{12}kg/mm3)
Si	150	0.0001299	2.60E-06	7.00E+08	1.70E+05	0.28	2.33E-09
Cu	400	59.52	1.75E-05	3.85E+08	1.10E+05	0.35	8.93E-09
Al	238	37.73	2.30E-05	9.00E+08	7.00E+04	0.33	2.70E-09
Sn3.0Ag0.5Cu	57	9.62	2.24E-05	2.30E+08	5.40E+04	0.35	7.30E-09
Mo	138	19.23	5.10E-06	2.50E+08	3.12E+05	0.30	1.02E-08
Ag	429	63.29	1.89E-05	2.35E+08	8.30E+04	0.37	1.05E-08

III. RESULT AND DISCUSSION

Fig. 4(a) shows the temperature profiles of the global model of PP IGBT at 10.01s. There is the temperature gradient in PP IGBT that can be concluded in Fig. 4. Due to the various thermal conductivity of material of each layer, the temperature gradient distribution is very obvious. The highest temperature occurs at the center of the active region of chip in heating phase. The temperature of the center of the active region of chip is changing with time periodically in transient analysis is shown in Fig. 5. And junction temperature is 53.52°C, and the difference of temperature is about 8°C. Fig. 6 shows the stress profiles of the chip. Fig. 6(a) and (b) are emitter and collector side of the chip respectively.

Fig. 7 shows the temperature profiles and heat flux in vertical direction of the second-level submodel in corner. Fig. 8 shows the temperature profiles and heat flux in vertical direction in first group of the third-level submodel position in corner. The difference between the two level submodels is the different interface, the second-level submodel is full contact. And the third-level submodel is point contact. The different contact interface has different thermal conductivity, it also affects the thermal contact resistance and heat conduction. In Fig. 7(a) and Fig. 8(a), the two interface condition results two temperature distribution. It is clear that the heat conduction has been influenced by the changed interface. Compare the Fig. 7(b) and Fig 8(b), the heat flux is uniform in second-level submodel, the heat flux is concentrated on the surface of joint.

The stress field of the third-level submodel is applied to a model that geometric is same as the third-level submodel as shown in Fig. 8 without any other field. Then, the new temperature boundary conditions are applied to the model. The new temperature boundary condition is that 70°C in the upper surface and 60°C in the lower surface. The $R_{th,c}$ can be calculated by the formula (2) in section II.

In order to research the effect of real contact areas, three groups of simulation are carried out. The height of the joint is 1 μm, and the model is in the corner of the chip. Keep the other factor unchanged, the level of real contact areas is 1%, 3%, 5% respectively. And the $R_{th,c}$ is 0.03645 K/W, 0.02161 K/W, 0.01629 K/W respectively. In Fig. 9, the results show that with the increase of real contact areas, the thermal contact resistance decrease. And within this range, their mathematical

relationship is not linear. The rate of decrease gets slow with the increase of the area.

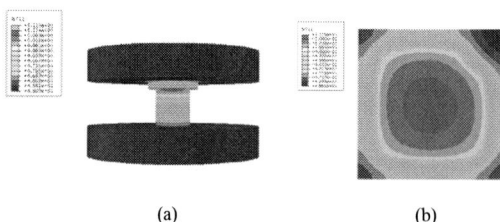

(a)   (b)

Fig. 4. The temperature distribution of PP IGBT

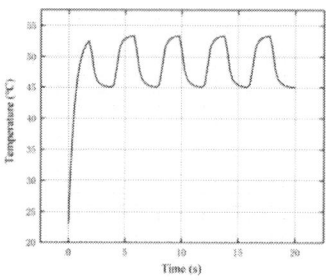

Fig. 5. The temperature of the chip center

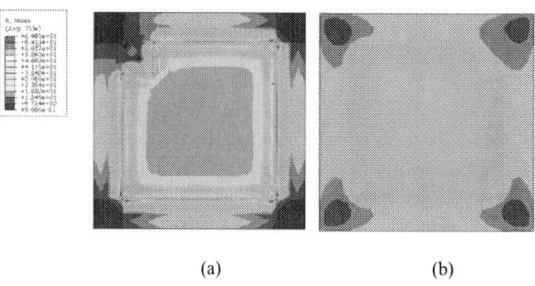

(a)   (b)

Fig. 6. The stress distribution of the chip

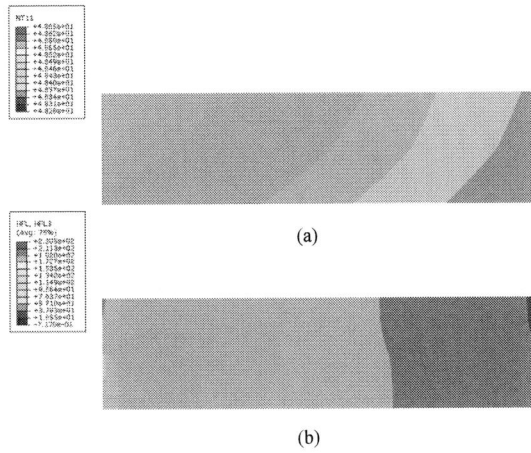

(a)

(b)

Fig. 7. The temperature distribution and heat flux of the second-level submodel

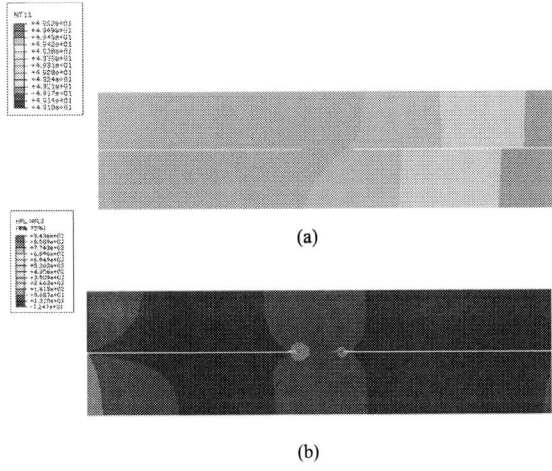

(a)

(b)

Fig. 8. The temperature distribution and heat flux of the third-level submodel

In order to research the effect of the height of contact joint, three groups of simulation are carried out. The real contact area is 1%, and the model is in the corner of the chip. Keep the other factor unchanged, the level of the height of contact joint is 0.5 μm, 0.75 μm, 1 μm respectively. And the $R_{th,c}$ is 0.03485 K/W, 0.03545 K/W, 0.03645 K/W respectively. In Fig. 10, the results show that with the increase of the height of contact joint, the thermal contact resistance increase slightly.

In order to research the effect of the position of the third-level model, two groups of simulation are carried out. The real contact area is 1%, and the height of the joint is 1 μm. Keep the other factor unchanged, the level of the position of third-level submodel is corner and center respectively. And the $R_{th,c}$ is 0.03645 K/W, 0.03628 K/W respectively. And it can be seen from Fig. 11, the position of the third-level submodel has little influence on thermal contact resistance. And the difference of position in center and in corner can be ignored.

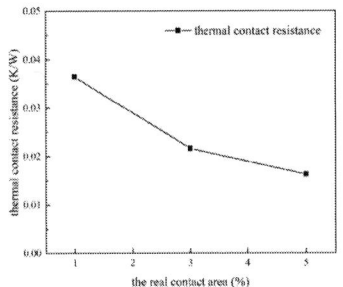

Fig. 9. The influence of real contact area on R_{th}

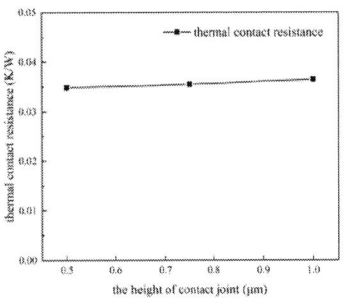

Fig. 10. The influence the height of contact joint of on R_{th}

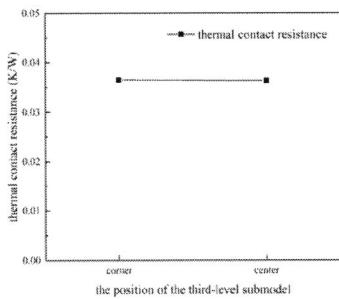

Fig. 11. The influence of the position of the third-level submodel on R_{th}

IV. CONCLUSION

This paper establishes the coupled thermal-electrical-structural model to simulate the power cycling and gain the temperature profiles of PP IGBT with time. Temperature gradient is formed in the structure of PP IGBT. The temperature profiles on the chip is not uniform. The temperature gets lower from the center to the edge. The junction temperature appears at the center of the active region of IGBT chip. And the temperature of the center of the active region of chip changing periodically. Then establish the third-level submodel of the interface between the chip and the emitter molybdenum plate according to the three groups. The interface change caused the thermal contact resistance to change. The temperature distribution is influenced. The heat flux concentrate on the surface of joint. And the thermal contact resistance is calculated. Through comparison and analysis, results show that the thermal contact resistance decrease with the increase of real contact areas, the thermal

978-1-6654-1392-3/21 $31.00 © 2021 IEEE   533

contact resistance increase with the increase of the height of the contact joint, and the position of the third-level submodel have little effect on thermal contact resistance.

ACKNOWLEDGMENT

This research was supported by the National Natural Science Foundation of China (NSFC) No. 11872078.

REFERENCES

[1] E. P. Deng, Z.B. Zhao, Q.M. Xin, J.W. Zhang, Y.Z. Huang, "Analysis on the difference of the characteristic between high power IGBT modules and press pack IGBTs." Microelectron. Reliab. vol.78, pp.25–37, Nov 2017.

[2] Bahrami, M., J. R. Culham, and M. M. Yovanovich, "AIAA 2004-0822 Thermal Resistances of Gaseous Gap for Non-Conforming Rough Contacts.".

[3] E.P. Deng, Z.B. Zhao, P. Zhang, Y.Z. Huang, J.Y. Li, "Optimization of the thermal contact resistance within press pack IGBTs." Microelectron. Reliab. vol.69, pp.17-28, Feb 2017.

[4] M. M. Yovanovich, "Four decades of research on thermal contact, gap, and joint resistance in microelectronics." IEEE Trans. Compon. Packag. Manuf. Technol. vol.28.2, pp. 182-206, 2005.

[5] M. Bahrami, J. R. Culham, M. M. Yovanovich, G. E. Schneider, "Thermal Contact Resistance of Nonconforming Rough Surfaces, Part 1: Contact Mechanics Model." J. Thermophys Heat Transfe vol.18.2, pp.209-217, 2004.

Analysis on the Thermal Stress of Al-Si Thin Film using DIC Method

Huiming Pan
The Institute of Technological Sciences
Wuhan University
Wuhan, China
panhuiming@whu.edu.cn

Guoliang Xu
The Institute of Technological Sciences
Wuhan University
Wuhan, China
2019286520033@whu.edu.cn

Zhiwen Chen
The Institute of Technological Sciences
Wuhan University
Wuhan, China
zwchen_lu@163.com

Chongming Zhang
The Institute of Technological Sciences
Wuhan University
Wuhan, China
2757715014@qq.com

Chao Sun
The Institute of Technological Sciences
Wuhan University
Wuhan, China
20183000003075@whu.edu.cn

Sheng Liu
The Institute of Technological Sciences
Wuhan University
Wuhan, China
victor_liu63@126.com

Li Liu
School of Materials Science and
Engineering
Wuhan University of Technology
Wuhan, China
l.liu@whut.edu.cn

Abstract—In electronics, silicon-based metal thin films are a sort of common material structure. In practical applications, electronic packaging materials should show stable mechanical properties under temperature changes, as well as predicted and preventing the fatigue failure. In this paper, Digital Image Correlation method was employed to measure the thermal strain and thermal stress of silicon-based metal thin film. It has been investigated that the feasibility of DIC Method to determine the thermal stress of thin film and the impact of the temperature range on measurement. The heating process of thin film was simulated by a quarter thin film model. The results shows that the DIC method was well suitable to determine the thermal stress of thin film with low error. Besides, the longer the temperature range is, the more accurate the measuring is. This provides a method of undamaged test on the thermal stress of thin film.

Keywords—Digital Image Correlation; Thermal stress; Al-Si thin film; Finite element analysis; Ansys

I. INTRODUCTION

Thin film materials are widely applied in many fields like sensors, battery, electronics, medical equipment, automotive, building, as well as in household and consumer products[1-5]. Thin films adhering to the substrate, such as glass or polymers, for example, enable the substrate to gain beneficial functions which can improve some of its properties like energy saving, optical transparency, operational life span, ductility and so on[1,4-5]. The diversity of application environments makes the study into the thermal failure meaningful. This work employed Digital Image Correlation (DIC) method as a nondestructive method of determining the thermal stress of Al-Si thin film.

DIC is a non-contacting image-based optical method for full-field shape, displacement and deformation measurements[6]. DIC performs full-field deformation measurement by first acquiring high-quality digital images with an optical imaging system, and then processing these images using image correlation algorithms[7]. Broadly, DIC techniques can be divided into three main categories 2D DIC method, Stereo-DIC method and DVC method. A two-dimensional (2D) DIC method using a single fixed camera is limited to in-plane deformation measurement of nominal planar objects[6].

In this paper, the thermal strain of Al-Si thin film was readily determined by 2D DIC method. The thermal strain was converted into thermal stress through computational formula[9]. The calculation results were in good agreement with the simulating results which were obtained via a finite element model.

II. EXPERIMENT PROCEDURES

A. Material Size and Speckle Fabrication

Al-Si thin film was adopted as our experimental material, whose shape and size were illustrated in Fig. 1. The thickness of Al and Si layer are 100nm and 500um respectively. The planar size of it is 10mm times 10mm.

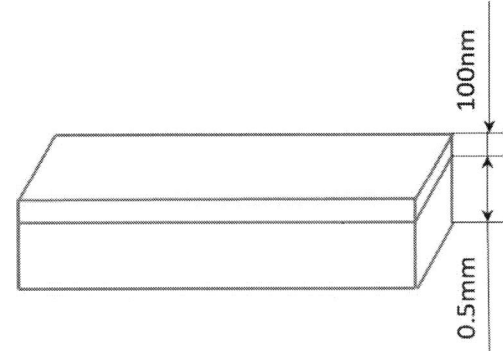

Fig. 1. Schematic diagram of Al-Si thin film

In the speckle fabrication process, the white ceramic powder of ZrO_2 (with the radius of around 1μm) was added to the absolute alcohol liquid with the mass-volume concentration of 0.4mg/mL. After thoroughly mixing, the liquid mixture was dropped on the specimen surface[8]. For an even distribution of speckle pattern, the specimen was processed by an ultrasonic oscillator before the absolute ethyl alcohol volatilized. Therefore, the ceramic powder adhered to the specimen surface, which generated a random speckle pattern. The image of a specimen with speckles is shown in Fig. 2(b). In contrast, original specimen without speckles is shown in Fig. 2(a). The grey part in Fig. 2(a) was the specimen, while the black part is the background.

B. Experimental Procedure

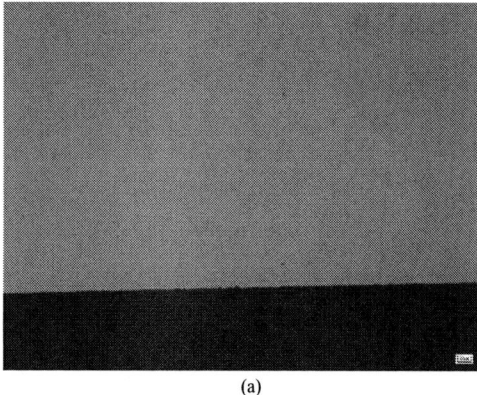

(a)

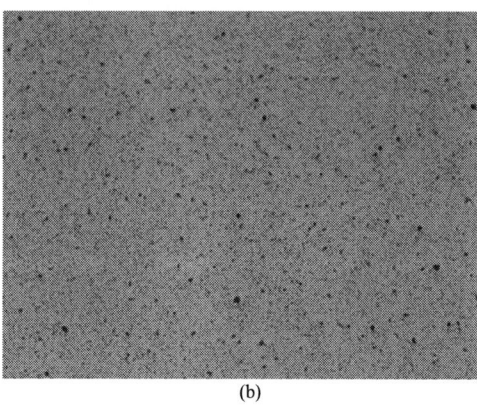

(b)

Fig. 2. Specimen image (a)without speckles, (b)with speckles

In the experimental process, experimental samples were observed with an optical microscope at 20x lens while they were being heated by a cooling and heating stage. The specimen was first heated to a temperature of 25℃. After being kept warm at 25℃ for one minute, it was imaged in a stable condition. Then, the specimen resumed being heated at a rate of ±10℃/min. The maximum temperature was 125℃. The temperature of the specimen started to go down when it reached the peak. The temperature-time curve was shown in Fig. 3.

An image was captured at a change of 10℃ during the process. Therefore, a total of 21 images were collected for subsequent analysis. These imaged were fed to our DIC software named VIC-2D.

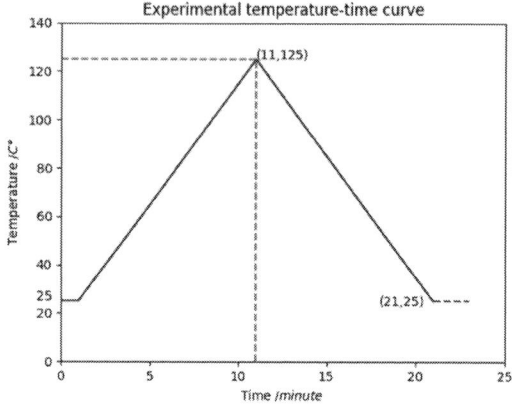

Fig. 3. The temperature change in heating and cooling stage

The Area of Interest, which feature a first-rank speckle pattern, was selected manually in the software. VIC-2D was able to automatically compute the deformation and strain of the specimen. The relevant data was exported into a CSV file and then imported to a script to compute the stress. The computation formula was as follows:

$$\sigma = E/(1-\mu)(\alpha\Delta T - \varepsilon) \qquad (1)$$

Here, σ was the thermal stress which was located in the Al layer and what we needed; E was the Young modulus of Al; μ was the Poisson's ratio of Al; α was the linear expansion coefficient of Al; ΔT was the temperature change for current image; ε was the strain which was obtained by VIC-2D and located in the Al layer as well. It was notable that both σ and ε are temperature-dependent. By applying this formula and data we got, 21 thermal stress values corresponding to 21 images was calculated. The values of E, μ and α were consistent with Table I.

III. SIMULATION DETAILS

ANSYS was used to simulate the thermal deformation process of an Al-Si thin film. Because solid film and heat were involved in the heating process, the model involved static structural module and steady-state module. A quarter thin film model was built to simulate the heating process(Fig. 4(a)). Two symmetry planes pass through the center of the entire model and are perpendicular to the plane of the film. The thickness of Al and SI were 100nm and 500μm respectively. Fig. 4(b) illustrates the boundary conditions of our model where we set the bottom face (face 1) and symmetry planes (face 2 and 3) with zero displacement in the Y-axis, Z-axis and X-axis respectively. The other three planes are free.

The Major material properties of Al and Si are summarized in Table 1. Relevant properties are in accord with formula (1).

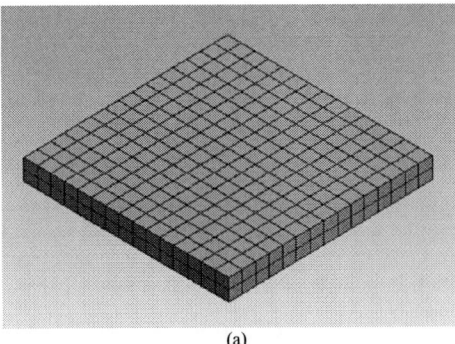

(a)

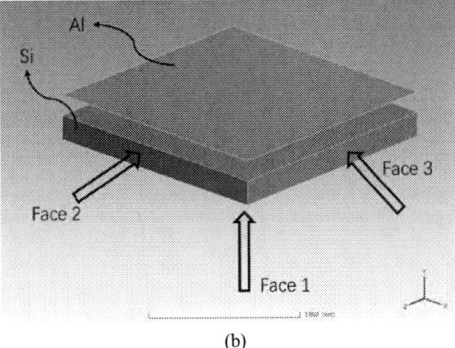

(b)

Fig. 4 Schematic of (a) quarter symmetric model and (b) boundary conditions

In the Steady-State Thermal module, we set the entire model with temperate changing linearly from 25°C to 125°C during 600s. The two modules are coupled before computing.

TABLE I.   Properties of Al and Si at 23°C

Materials (23°C)	Density (kg /m³)	Coefficient of thermal expansion (10⁻⁶)	Young modulus (10⁴MPa)	Poisson's ratio
Aluminum	2770	23.6	7	0.32
Silicon	2329	2.578	16.27	0.27

To simplify the model, several assumptions were also proposed in building the model:

1. The linear expansion coefficient, Young's modulus, Poisson's ratio of Al and Si are isotropy.

2. The linear expansion coefficient of Al and Si don't change with temperature.

IV. RESLUTS AND DISCUSSION

A. Feasibility of DIC Method to Determine the Thermal Stress

In the simulation model, the minimum normal stress values of all time are minus and in Al layer. The range is 0~-229.32 MPa. In contrast, the maximum values are positive and always in Si layer whose range is 0~0.3187MPa. These two points suggested that the normal stress in Al film and Si film are compressive stress and tensile stress respectively. The normal stress in Al film is what needed and regarded as the result of simulation. Fig 5. Showed the normal stress in Al film in 125°C.

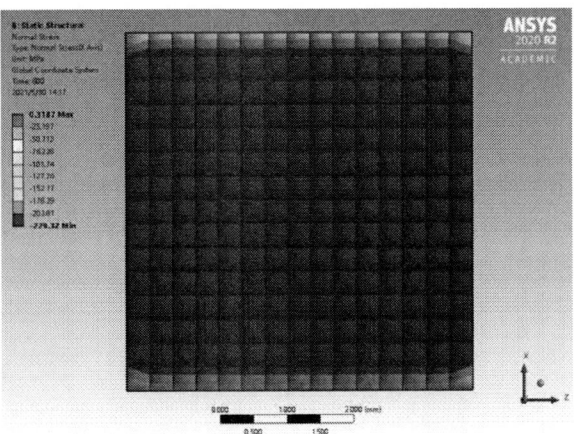

Fig. 5 The normal stress in Al film in 125°C

With temperature as X-axis and stress as the Y-axis, the experimental curve has a slanting downward trend. Through least-square method, a fitting straight line is obtained which clearly reflect the characteristic of experimental data. To compare the experimental fitting curve and simulation result, both two lines were put together as shown in Fig 6.

The mean and standard deviation of the bias between DIC result and DIC fitting curve are -2.199×10^{-15} MPa and 9.988MPa. Likewise, the mean and standard deviation of the bias between Simulation result and DIC fitting curve are -19.679MPa and 2.505MPa.

The slope of fitting line and simulation result was −2.1975 and −2.2932 respectively. The error between the slopes is only 4.17%.

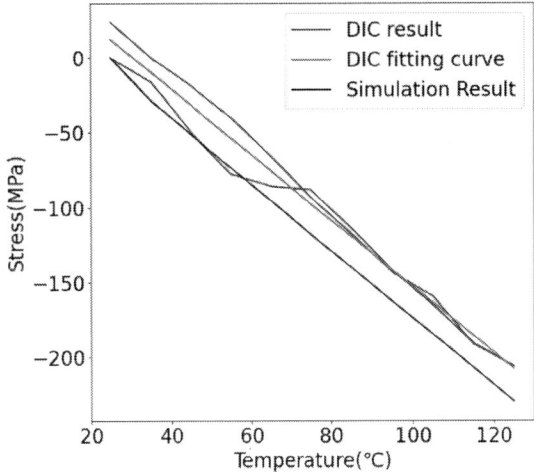

Fig. 6 Experimental data curve and simulation result

B. The Impact of the Temperature Range on Measurement

To study the stability the DIC method to determine the thermal stress of Al-Si thin film, experiments were conducted with the different temperature range. Two different heating temperature ranges used were: 25~125°C and 25~325°C. For a clear comparison, all lines were lengthened to 325°C with slopes and Y-intercepts constant. As shown in Fig. 7, the wider temperature range could decrease the error between the simulation result and experimental result. Fitting line 1 corresponds to temperature range of 25~125°C, while fitting line 2 corresponds to 25~325°C. The slopes of the fitting line 1, fitting line 2 and simulation result are -2.1975, -2.2138, -2.2932 respectively. The error between the slopes of fitting line 2 and simulation result is 3.46%, whereas the counterpart of fitting line 1 is 4.17%. Moreover, the mean and standard deviation of the bias between Simulation result and DIC fitting line 1 are 26.5128MPa and 8.5565MPa. With comparison, the counterpart of fitting line 2 are 8.9723MPa and 7.0984MPa. These results indicate that DIC method could be more precise with wider temperature range.

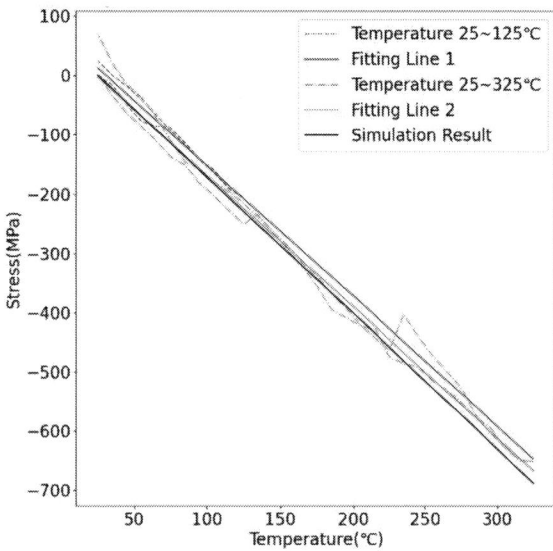

Fig. 7 Results with different temperature ranges

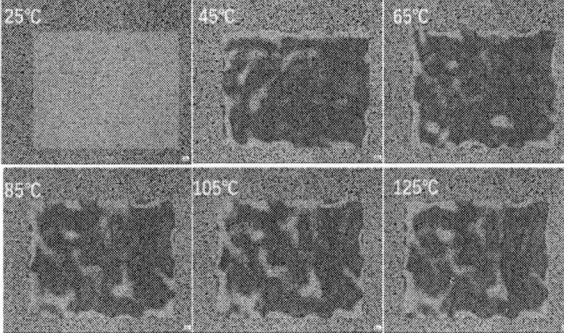

Fig. 8 The Distribution of Principle Strain in Different Temperatures

C. The Measurement Process of Thermal Strain

The thermal strain of Al-Si thin film here is also the principle strain in either direction. In the heating process, it was obtained directly by using DIC method through software

VIC-2D. Fig 8 consists of the cloud figures in different temperatures in which each figure showed the distribution of principle strain in different temperatures. . In the process of heating, the distribution of the strain showed stability in different subfigures. However, it's easily found that the distribution of thermal strain is kind of random when looing at a particular image. The stability is a natural feature in the heating process. Whereas the randomness of distribution may result from the speckle pattern. The speckle particle size and distribution, viz. the speckle quality, have a great influence on the regularities of the thermal strain distribution.

V. CONCULSIONS

DIC method was applied to measure the thermal stress of Al-Si thin film. The thermal strain was directly obtained from software VIC-2D, and the thermal stress was gained indirectly by mechanical formula. Simulation based on quarter axisymmetric model showed the thermal stress of Al film. Experiments indicated similar results. It indicated that the DIC method is suitable for the measurement of thermal stress of Al-Si thin film which could be applied for other kinds of multiple-layers thin films.

In future, the temperature will be enlarged to a higher level to verify the feasibility of DIC method further. The kinds of materials will also be varied as well as making the material properties more accurate.

REFERENCES

[1] Piegari, Angela, and François Flory. "Optical Thin Films and Coatings."

[2] Xie, Lilia S., Grigorii Skorupskii, and Mircea Dincă. "Electrically conductive metal–organic frameworks." Chemical reviews 120.16 (2020): 8536-8580.

[3] Li, Hai-Yang, et al. "Functional metal–organic frameworks as effective sensors of gases and volatile compounds." Chemical Society Reviews 49.17 (2020): 6364-6401.

[4] Rabaia, Malek Kamal Hussien, et al. "Environmental impacts of solar energy systems: A review." Science of The Total Environment 754 (2021): 141989.

[5] Kim, Chan, et al. "Damage-free transfer mechanics of 2-dimensional materials: competition between adhesion instability and tensile strain." NPG Asia Materials 13.1 (2021): 1-11.

[6] Pan, Bing. "Digital image correlation for surface deformation measurement: historical developments, recent advances and future goals." Measurement Science and Technology 29.8 (2018): 082001.

[7] Yu, L., and B. Pan. "Overview of High-temperature Deformation Measurement Using Digital Image Correlation." Experimental Mechanics (2021): 1-22.

[8] Yin, YuanJie, HuiMin Xie, and Wei He. "In situ SEM-DIC technique and its application to characterize the high-temperature fatigue crack closure effect." Science China Technological Sciences 63.2 (2020): 265-276.

[9] Peters, W. H., and W. F. Ranson. "Digital imaging techniques in experimental stress analysis." Optical engineering 21.3 (1982): 213427.

Flexible Thermal Interface Materials Through Controlling the Ratio of Silicone Oil Functional Groups

1st Wendian Tu
1 Shenzhen Institute of Advanced Electronic Materials
Shenzhen Institutes of Advanced Technology, Chinese Academy of Sciences
Shenzhen, China
2 School of Materials Science and Engineering
University of Nanchang
Nanchang, China
wd.tu@siat.ac.cn

2nd Linlin Ren*
Shenzhen Institute of Advanced Electronic Materials
Shenzhen Institutes of Advanced Technology, Chinese Academy of Sciences
Shenzhen, China
ll.ren@siat.ac.cn

3rd Guoping Du*
School of Materials Science and Engineering
University of Nanchang
Nanchang, China
guopingdu@ncu.edu.cn

4th Rong Sun
Shenzhen Institute of Advanced Electronic Materials
Shenzhen Institutes of Advanced Technology, Chinese Academy of Sciences
Shenzhen, China
rong.sun@siat.ac.cn

5th Xiaoliang Zeng
Shenzhen Institute of Advanced Electronic Materials
Shenzhen Institutes of Advanced Technology, Chinese Academy of Sciences
Shenzhen, China
xl.zeng@siat.ac.cn

6th Junwei Li
1 Shenzhen Institute of Advanced Electronic Materials
Shenzhen Institutes of Advanced Technology, Chinese Academy of Sciences
Shenzhen, China
2 School of Materials Science and Engineering
University of Nanchang
Nanchang, China
ljw@siat.ac.cn

Abstract—**Thermal interface materials (TIMs) have become important due to the development of electronic devices towards high integration. The materials face dilemmas of thermal conductivity and flexibility. Herein, we demonstrate a kind of TIMs via a simple method , which exhibits excellent elongation at break of 114.0%, high thermal conductivity of 4.3 $W \cdot m^{-1} \cdot K^{-1}$, and low viscosity of 254 Pa·s. In addition, it exhibits great potential in the future and emerging applications such as wearable electronics since it is cost-effective and economical.**

Keywords—thermal interface materials, flexibility, silicone rubber composites

I. INTRODUCTION

Owing to the rapid development of science and technology, heat is the bottleneck of high-frequency and high-power density electronic products [1-5]. As we all know, 50% of the failure of electronic equipment is caused by the temperature exceeding the limit value. It brings higher requirements for heat loss so that thermal management becomes more important. Nowadays, there are kinds of thermal interface materials (TIMs) such as thermal gel which fill the gap between a heat source's uneven surface and heat sink. People hope that it could not only provide a reliable heat dissipation path for electronic components but also have a certain degree of flexibility [6-10].

Currently, some methods have been used to increase the flexibility of thermal gel. Lin et al. [11] found that the mechanical properties of silicone rubber composite were improved owing to the modification of Al_2O_3 powder and the addition of carbon nanotubes. However, few people have studied the silicone rubber composites from the angle of adjusting the polymer matrix.

Thus, in this article, the silicone rubber matrix has been changed to adjust the molar ratio of vinyl to silicone-hydrogen bond and the silicone-hydrogen bond molar ratio between chain extender and crosslinker. We demonstrate that changes in the ratio could be of merit of flexibility.

TABLE I. VARIOUS PROPERTIES OF THE MATERIAL WITH THE CHANGES OF THE RATIO OF VINYL TO SILICONE-HYDROGEN BOND.

Number	The ratio of vinyl to silicone-hydrogen bond while the silicone-hydrogen bond molar ratio between chain extender and crosslinker is 0.19:1	Viscosity Pa·s	Storage modulus KPa	Loss modulus KPa	Tensile strength MPa	Elongation at break %	Thermal conductivity W·m⁻¹·K⁻¹	Cross-linking density E-5mol·ml⁻¹
1	0.6:1	198	3220.0	186.0	0.28	6.5	3.8	1.6
2	0.8:1	266	3250.0	190.0	0.36	7.3	3.9	5.9
3	1:1	277	2580.0	175.0	1.35	8.2	3.7	9.7
4	1:0.8	270	1780.0	152.0	1.16	13.1	3.8	5.7
5	1:0.6	188	442.0	42.3	0.76	20.1	4.0	4.2
6	1:0.5	260	421.0	40.2	0.54	24.3	3.7	2.1

II. EXPERIMENTAL

A. Materials

Divinyl terminated polydimethylsiloxane, side-chain polysiloxane polydimethylsiloxane, and disilyl hydrogen-terminated polydimethylsiloxane were purchased from Hubei Chengfeng Chemical Co., Ltd, China. Al powder was provided from Luoyang Discoverer Aluminum Industry Co., Ltd, China. ZnO powder was provided from Shandong Ruiqi Chemical Co., Ltd, China. Pt catalyst was purchased from Shin-Etsu Chemical Co., Ltd, Japan.

B. Preparation of silicone rubber composites

The content of the silicone oil based on the vinyl-to-silicone-hydrogen-bond molar ratios of 0.6:1, 0.8:1, 1:1, 1:0.8, 1:0.6, 1:0.55, 1:0.5 and the condition that the silicone-hydrogen bond molar ratio between chain extender and crosslinker was 0.19:1 was calculated Meanwhile, the filler consisted of Al powder and ZnO powder was in a certain proportion. The filler and the silicone oil were mixed in the planetary mixer (XJB-2.5) stirring at 363 K for 1 h. When it was cooling to room temperature, the Pt catalyst was added to it. The stirring was continued for 1 h. After mixing, the silicone rubber compound was coated to the release film, compressed to a fixed thickness, and curing at 423 K for 1 h. The sample which was for rheological measurement was with no compression and curing. After finding the most flexible ratio in this set of experiments, we fixed the molar ratio of vinyl to silicone-hydrogen bond and adjusted the silicone-hydrogen bond molar ratio between chain extender and crosslinker in polymer composite. Similarly, the content of silicone oil which met the condition that the silicone-hydrogen bond molar ratio between chain extender and crosslinker was 0.19:1, 0.38:1, 1:1, 1.5:1, 2:1 to find the ratio which showed the best flexibility was calculated. Subsequent experiments were consistent with the previous ones.

C. Characterization

Emission scanning electron microscopy (Apreo2, Thermo Scientific) was used to observe the cross section. The rheological behavior of composites was measured on a rheometer (MCR-302, Anton Paar) at 298 K. The viscosity, the storage modulus, and the loss modulus were measured by MCR302. The storage modulus and the loss modulus were measured at 393 K. Heat conduction instrument (LW9389, Longwin) was used to measure the thermal conductivity at 10 PSI. The cross-linking density was measured through the swelling behaviors of the silicone rubber in toluene without the filler. The mechanical properties were measured on the folding resistance testing machine (UTM2102, Shenzhen suns) when the samples met standard GB/T 9865.1 4 with the extensometer.

III. RESULT AND DISCUSSION

A. Morphology

From Fig. 1, it could be seen that the fillers are distributed in the matrix uniformly. Because of the high filler content (90 wt%), it is very dense but the bonding of the fillers and the matrix is bad which shows that the outcomes of interface modification are generally bad. From Fig. 2, it could be seen that the No. 9 sample has a flat surface while the No. 10 sample has an uneven surface (Fig. 3). Through Fig.3, we suspect that the No. 10 sample is not an elastomer.

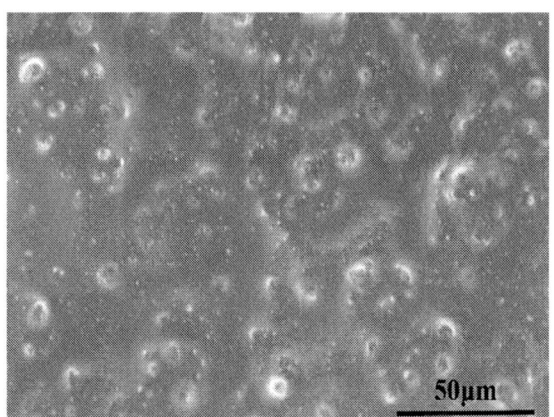

Fig.1. SEM photograph of No. 9 sample.

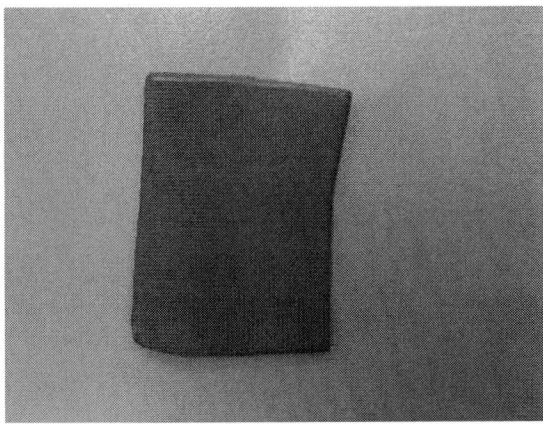

Fig.2. Photograph of No. 9 sample.

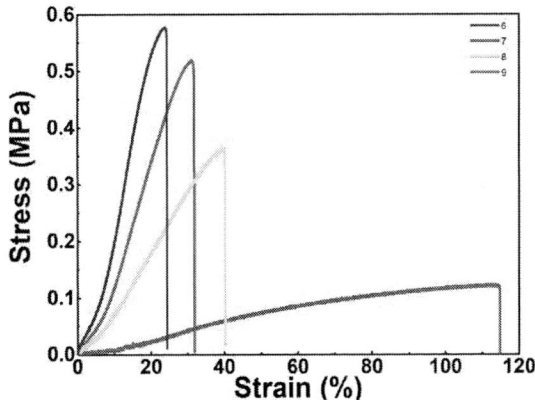

Fig.5. Tensile stress–strain curves of No. 6 to No. 9.

B. Viscosity and modulus

The properties of two series of silicone rubber composites are shown in TABLE I and TABLE II. The viscosity is linked to the content of the filler and the silicone oil of different viscosity. When the content of silicone oil is varied, the viscosity changes with it. TABLE I and TABLE II show that the each viscosity of No.1 to No.10 samples is less than 400 Pa·s, and therefore it is convenient to use. Meanwhile, the storage modulus and the loss modulus are related to the cross-linking density. TABLE I shows that with the increment of the density, the modulus gets bigger generally which demonstrates the material is more rigid.

C. Thermal conductivity

When the polymer matrix changes, the thermal conductivity does not change noticeably (TABLE I and TABLE II). The dissipation of heat is majorly based on the filler which could transport heat, and the thermal conductivity is less linked to the cross-linking density of the silicone rubber.

D. Mechanical properties

The relation between the mechanical properties and the cross-linking density is shown in TABLE I and TABLE II. As shown in that, the tensile strength reveals a positive relationship with the cross-linking density. When the density is decreasing, the entanglement between the molecular chains is reduced, making mutual movement easier. Meanwhile, when the density firstly gets bigger, the elongation at break firstly gets bigger because it starts to form a 3D macromolecular network during this period until the density exceeds a certain value. Then, elongation increases with the reduction of the density because the 3D macromolecular network prevents motions of the molecular chain. Similarly, the changes in cross-linking density are obvious when the polymer matrix is changed. Because of the varied molar ratio of vinyl to silicone-hydrogen bond (TABLE I), the cross-linking density firstly increases and then decreases. Firstly, with the increment of the ratio, more and more vinyl groups are involved in the reaction probably until the molar ratio of vinyl to silicone-hydrogen bond is 1:1. When the molar ratio is 1:1, the density is the biggest among those. Then, when the ratio continues to increase, the density gets lower owing to the complete reaction of the silicone-hydrogen bond. When the ratio of silicone-hydrogen bond

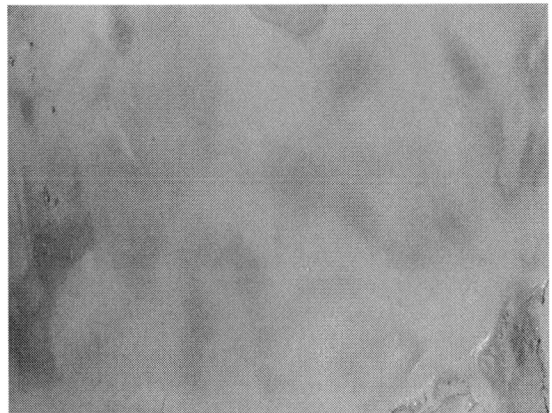

Fig.3. Photograph of No. 10 sample.

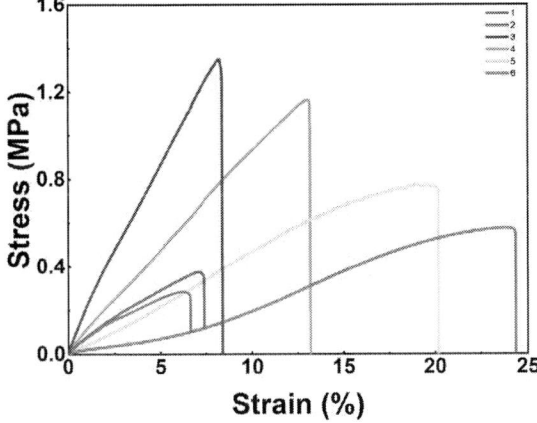

Fig.4. Tensile stress–strain curves of No. 1 to No. 6.

TABLE II. VARIOUS PROPERTIES OF THE MATERIAL WITH THE CHANGE OF THE RATIO OF THE SILICONE-HYDROGEN BOND BETWEEN CHAIN EXTENDER AND CROSSLINKER WHILE THE RATIO OF VINYL TO SILICONE-HYDROGEN BOND IS 1:0.5.

Number	The ratio of silicone-hydrogen bond between chain extender and crosslinker when the ratio of vinyl to silicone-hydrogen bond is 1:0.5	Viscosity Pa·s	Storage modulus KPa	Loss modulus KPa	Tensile strength MPa	Elongation at break %	Thermal conductivity $W \cdot m^{-1} \cdot K^{-1}$	Cross-linking density $E\text{-}5mol \cdot ml^{-1}$
6	0.19:1	260	421.0	40.2	0.54	24.3	3.7	2.1
7	0.38:1	253	73.0	54.0	0.51	31.9	3.8	1.6
8	1:1	282	47.7	13.9	0.36	40.2	3.9	0.7
9	1.5:1	254	13.8	5.3	0.12	114.0	4.3	0.1
10	2:1	253	1.1	5.7	—	—	—	—

between chain extender and crosslinker gets bigger and the ratio of vinyl to silicone-hydrogen bond is fixed, the cross-linking density gets smaller. The mechanical properties are varied as cross-linking density varied. .

With the ratio of the silicone-hydrogen bond between chain extender and crosslinker varied (TABLE II), the cross-linking density gets lower which leads to better elongation at break. When ratio of the silicone-hydrogen bond between chain extender and crosslinker is 1.5:1 and ratio of vinyl to silicone-hydrogen bond is 1:0.5 (No. 9), the elongation is 114.0% which is more than 17 times than No. 1 sample. The viscosity of it is 254 Pa·s which could be advantageous for good use. Its thermal conductivity is 4.3 $W \cdot m^{-1} \cdot K^{-1}$. When the ratio of silicone-hydrogen bond between chain extender and crosslinker increases to 2:1 and the ratio of vinyl to silicone-hydrogen bond is 1:0.5 (No. 10), the storage modulus is less than the loss modulus which indicates that it isn't an elastomer so it is difficult to measure its mechanical properties. Therefore, No. 9 exhibits the best flexibility. The flexibility favors filling the gap between the chip and sink material which could decrease the thermal resistance. In addition, it is economical. Its future is promising which has low cost and excellent properties so that it could be applied in many things such as wearable electronics.

IV. CONCLUSIONS

The present experiment demonstrates a kind of TIMs with high thermal conductivity (4.3 $W \cdot m^{-1} \cdot K^{-1}$), large elongation at break (114.0%), and low viscosity (254 Pa·s) which is obtained by adjusting the molar ratio of vinyl to silicone-hydrogen bond and the silicone-hydrogen bond molar ratio between chain extender and crosslinker in the polymer matrix. Because of its simple production method and low cost, it could be put into industrialization. In summary, it has good overall performance which could be applied to emerging applications such as wearable electronics.

ACKNOWLEDGMENT

This work was supported by the National Natural Science Foundation of China (Grant No. 52073300), the Guangdong Basic and Applied Basic Research Fund (No.

2019A1515110845), Guangdong Province Key Field R&D Program Project (No. 2020B010190004), Youth Innovation Promotion Association of the Chinese Academy of Sciences (2019354), Shenzhen Science and Technology Research Fund (No. JCYJ20200109114401708 and JCYJ20180507182530279).

REFERENCES

[1] J. Song, Z. Peng, and Y. Zhang, "Enhancement of thermal conductivity and mechanical properties of silicone rubber composites by using acrylate grafted siloxane copolymers," Chemical Engineering Journal, vol. 391, July 2020.

[2] K. C. Otiaba, N. N. Ekere, R. S. Bhatti, S. Mallik, M. O. Alam, and E. H. Amalu, "Thermal interface materials for automotive electronic control unit: Trends, technology and R&D challenges," Microelectronics Reliability, vol. 51, pp. 2031–2043, December 2011.

[3] Z. G. Wang, et al. "Highly thermal conductive, anisotropically heat-transferred, mechanically flexible composite film by assembly of boron nitride nanosheets for thermal management," Composites, vol. 180, January 2010.

[4] X. F. Xu, J. Chen, J. Zhou, and B. W. Li, "Thermal Conductivity of Polymers and Their Nanocomposites," Advanced Materials, vol. 30, March 2018.

[5] S. Lee, et al. "Carbon nanotube covalent bonding mediates extraordinary electron and phonon transports in soft epoxy matrix interface materials," Carbon, vol. 157, pp. 12–21, February 2020.

[6] B. Fan, Y. Liu, D. He, and J. B. Bai, "Enhanced thermal conductivity for mesophase pitch-based carbon fiber/modified boron nitride/epoxy composites," Polymer, vol. 122, pp. 71–76, July 2017.

[7] C. C. Teng, C. C. M. Ma, K. C. Chiou, and T. M. Lee, "Synergetic effect of thermal conductive properties of epoxy composites containing functionalized multi-walled carbon nanotubes and aluminum nitride," Composites Part B Engineering, vol. 43, pp. 265–271, March 2012.

[8] H. Wu, and L. T. Drzal, "High thermally conductive graphite nanoplatelet/polyetherimide composite by precoating: Effect of percolation and particle size," Polymer Composites, vol. 34, pp. 2148–2153, August 2013.

[9] L. Mao, J. Han, D. Zhao, N. Song, L. Shi, and J. Wang, "Particle Packing Theory Guided Thermal Conductive Polymer Preparation and Related Properties," ACS Applied Materials & Interfaces, vol. 10, September 2018.

[10] R. M. Minas'yan, "One-component silicon adhesives-sealants," Polym. Sci. Ser. D, vol. 1, pp. 286–288, November 2008.

[11] J. L. Lin, S. M. Su, Y. B. He, and F. Y. Kang, "Improving the thermal and mechanical properties of an alumina filled silicone rubber composite by incorporating carbon nanotubes," New Carbon Materials, vol. 35, pp. 66–72, February 2020.

978-1-6654-1392-3/21 $31.00 © 2021 IEEE

Simulation and Optimization of 3D Heterogeneous Integration of Inertial Micro-System

Nanxin Wang
Shenzhen Graduate School of Peking University
Shenzhen, China
1901213022@pku.edu.cn

Shenglin Ma*
Department of Mechanical & Electrical Engineering, Xiamen University
Xiamen, China
mashenglin@xmu.edu.cn

Yufeng Jin*
Shenzhen Graduate School of Peking University
Shenzhen, China
yfjin@pku.edu.cn

Chaoyang Xing
Beijing Aerospace Control Device Institute
Beijing, China
mems13@163.com

Nannan Li
Beijing Aerospace Control Device Institute
Beijing, China
melinan@pku.edu.cn

Peng Sun
Beijing Aerospace Control Device Institute
Beijing, China
pengsun2008@sina.com

Abstract— This paper presents a new 3D heterogeneous integration scheme for inertial Micro-system, where a TSV interposer with a cavity is used to separate two aligned TSV interposers that are mounted with inertial MEMS (Micro Electro-Mechanical System) devices, ASIC (Application Specific Integrated Circuit) chips, and passive components receptively. A finite element simulation method that uses equivalent elements and sub-model methods is developed to do design for reliability in Thermal Cycling Test (TCT). Comparing with the normal simulation method, an agreement is obtained in the stress distribution in the MEMS chips, TSVs and micro bumps. In the meantime, the simplified simulation method provides a competitive accuracy, the difference is 0.88% in the maximum deformation of the whole model between the two methods, the difference is 0.41% in the maximum deformation of the TSV and 0.65% in the micro bump at the most dangerous position. Base on this simplified simulation method, a 3D heterogeneous inertial micro-system is analyzed and optimized. The design of ladder TSV with the usage of underfill and gold balls is selected. The simulation results also showed that RDL (Redistribution Layer) can make significant influence on the stress distribution in the systems. If considering RDL, the maximum stress inside the micro bump at the most dangerous location increases by 62.79% while the max stress in TSV decreases by 14.03%, which means that stress optimization should also be carried out after designing the RDL layout.

Keywords—Inertial Micro-System; MEMS; 3D heterogeneous integration; Finite element analysis; Equivalent element method; Sub-model; RDL; TCT.

I. INTRODUCTION

An inertial micro-system, which is composed of inertial MEMS devices, ASIC chips and passive components and is capable to sense acceleration and angular rate for predicting the position or the moving state of an object, has potential applications in the navigation system. The 3D heterogeneous integration scheme is an effective way to realize a miniaturized, intelligent inertial system. The heterogeneous integration schemes can be divided into three main categories: 1) The MEMS chip is stacked with the responding ASIC chip in form of chip-to-chip, where wire bonding is utilized to realize electrical connections between them. 2) The inertial MEMS devices and their responding ASIC chips are assembled on a TSV interposer with flip-chip bonding with micro bumps or micro solder in the form of chip-to-wafer. 3) The inertial MEMS and its responding ASIC are realized in a stacked form in the wafer-level process, where vertical electrical connections are implemented with TSVs in the substrate of MEMS or ASIC. Among them, the first is a method enabled by package technology with a relatively low density, while the third has the highest integration density but confronts constraints due to the co-design of inertial MEMS devices and ASIC, which means it's more challenging to implement. The second provides a moderate integration density and the most flexibility to implement, which is gaining increasing attention across the industry and academic circle [1-5].

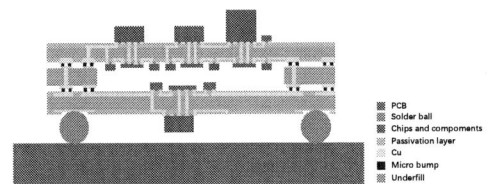

Fig. 1. Schematic diagram of the integration scheme.

This paper proposes a 3D heterogeneous integration scheme for inertial micro-system shown in Fig. 1. To reduce the parametric effect and ensure accuracy, the MEMS chips and ASIC chips are placed on the different sides of the TSV interposer and electrically connected with Cu TSVs as a basic unit. One layer or two layers of TSV interposer with a cavity is used to separate and support a pair of aligned TSV interposers mounted with chips on double sides to provide enough space for the thick MEMS chips. The stacking of TSV interposers can be realized by flip-chip bonding with Cu-Sn micro bump, the whole system is mounted on a PCB by solder.

The design for reliability (DFR) in thermal cycling experiment is necessary and difficult because the 3D integrated inertial micro-system consists of kinds of micro-structures made of varying materials, such as Cu TSVs, micro-bumps, RDLs, ranging from a few sub-microns to a few thousand microns, which make the modeling and simulation very challenging. There are many publications about the method to do DFR in thermal cycling experiments [6-9]. They use the equivalent element method to analyze the deformation and stress of the quarter model. Besides, in order to reduce the difficulty and computation of simulation, many simulation works treated all materials within the structure as linear elastic materials instead of elastoplastic material. They ignored the RDL or treated it as a layer of equivalent material and calculate its equivalent material parameters by simply

978-1-6654-1392-3/21 $31.00 © 2021 IEEE

multiplying the material parameters of each component by the material proportion. These will lead to inaccurate results.

In this article, a simplified simulation method based on the equivalent element method and sub-model method is developed and verified. Analysis and optimization are done with a 3D integrated inertial micro-system.

II. THE EQUIVALENT ELEMENT METHOD

In the 3D integrated inertial micro-system, there are plenty of micro bumps and TSVs which have a significant influence on thermo-mechanical behavior. In this paper, we used the equivalent element method to simplify them.

TABLE I.   DETAILED MATERIAL PARAMETERS

	Young's modulus (GPa)		CTE (ppm/°C)	Poisson's ratio
Silicon [10]	169.5		3	0.28
Copper [11]	*Elastic*	*Plastic*	17	0.35
	120@25°C 110@260 °C	Yield strength/Tangent modulus 0.12/3.33@25°C 0.11/3.46@260°C		
Under-fill	8@25°C 0.1@190°C		5	0.33
IMC [12]	142.7		14.6 @25°C 16 @160°C	0.28
Nickel [13]	207		13.4 @25°C 16.8 @525°C	0.31
Alumi-num [14]	64.7		24.2	0.36
Tin[10]	*Elastic*	*Plastic*	22.5	0.4
	17.7 @-55°C 13.26 @-25°C 11.646 @25°C 8.09@75°C 5.158 @125°C	Yield strength/Tangent modulus 0.036/0.8208 @-55°C 0.031/0.6686 @-25°C 0.024/0.2951@25°C 0.016/0.2051@75°C 0.01/0.1281@125°C		
Gold	78		0.142	0.42
PCB	22		18	0.28
Passi-vation (Si₃N₄) [15]	155		5	0.28
Solder ball	47		21.7	0.3

A. Parameters Extraction

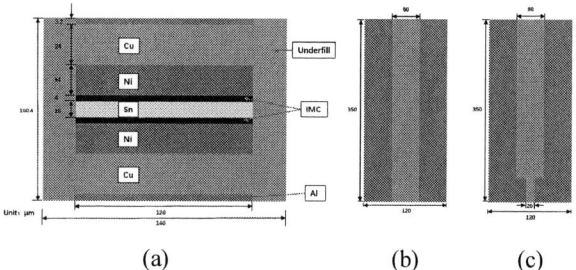

Fig. 2. Cross-sectional view of the FEA model: (a) micro bump unit. (b) straight-through TSV. (c) ladder TSV.

The micro bump and the TSV units shown in Fig. 2 are utilized to extract the equivalent material parameters. Table I summarizes the material properties. It is worth noting that, copper and tin are treated as bilinear elastoplastic materials here to increase the accuracy of the simulation. Table II has summarized the extracted equivalent parameters.

TABLE II.   EXTRACTED PARAMETERS OF EQUIVALENT ELEMENTS

Parameters	Equivalent elements		
	Micro bump	*Straight-through TSV*	*Ladder TSV*
E_x(GPa)	7.22	157.32	158.97
E_y(GPa)	43.82	152.92	153.65
E_z(GPa)	7.22	157.32	158.94
ν_{xy}	0.289	0.299	0.297
ν_{yz}	0.283	0.299	0.298
ν_{xz}	0.343	0.305	0.303
G_{xy}(GPa)	1.36	435.48	437.74
G_{xz}(GPa)	4.25	37.4	38.13
G_{yz}(GPa)	3.94	13.81	13.90
α_x(ppm/°C)	21.8	5.91	5.81
α_y(ppm/°C)	20.7	5.51	5.35
α_z(ppm/°C)	21.9	5.90	5.81

B. Verification of Equivalent Element Method

An accelerator is flip-chip bonded on a TSV interposer is used as a test vehicle. The four corners of the bottom surface of the TSV interposer are fully constrained to prevent rigid body displacement as shown in Fig.3. According to GB/T 2423.27-2020 (Environmental testing—Part2: Test methods—Test method and guidance: Combined temperature or temperature and humidity with low air pressure tests), a thermal cycle from -55°C to 125°C at a rate of 0.2°C/s is applied.

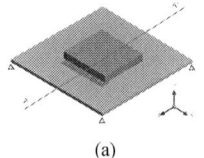

(a)

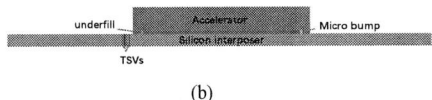

(b)

Fig. 3. (a) The model used to verify the simplified method and the boundary conditions. (b) Cross-section of the verification model.

The simulation results show that the stress and strain of the test vehicle reach the maximum value at 125°C as shown in Fig. 4. A similar distribution is observed in the overall deformation and the deformation on the boundaries of TSVs and micro bumps. The differences in the maximum value of these parameters are 0.26%, 0.97%, 0.20%, respectively.

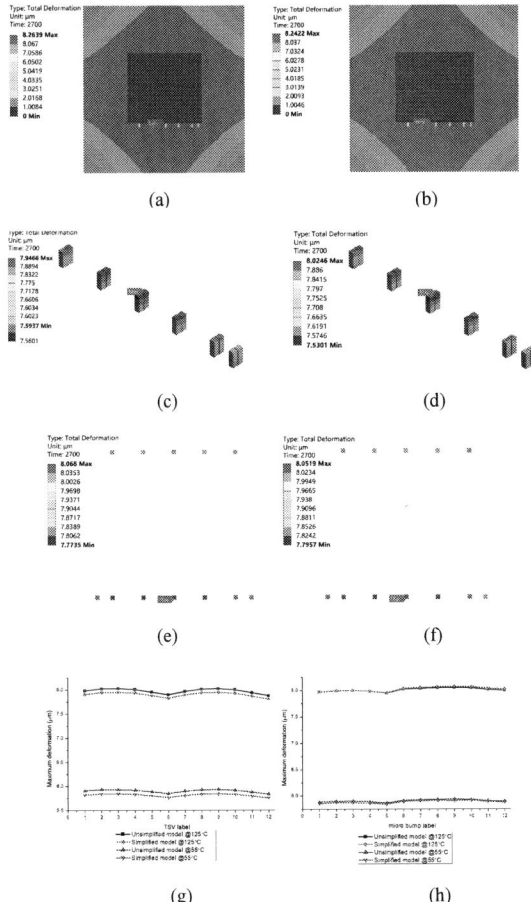

(a)   (b)

(c)   (d)

(e)   (f)

(g)   (h)

Fig. 4. Comparison between results of simplified overall model and unsimplified model@ 125°C: (a), (b) overall deformation. (c), (d) deformation on the boundaries of TSVs. (e), (f) deformation on the boundaries of micro bumps. (g), (h) the maximum deformation distribution on the boundary of each TSV and micro bump.

Then, the extracted deformation data from the global model based on the equivalent material method is input into the refined local sub-model as the boundary condition, and the simulation results are shown in Fig. 5. The simulation results that the maximum stress points in TSV are located at the interface of Cu and Si on the lower surface of the TSV interposer. This is because the CTE of Cu and Si is greatly different and the deformation of the lower surface is larger than that of the upper surface. The maximum stress points in micro bump are located at the bottom surface of the lower Al layer. This is because the differences between Al and Si are much larger than those between other components. The results

all show that the order of stress is: Al>Ni>IMC>Cu>Sn>Underfill in the most dangerous micro bump.

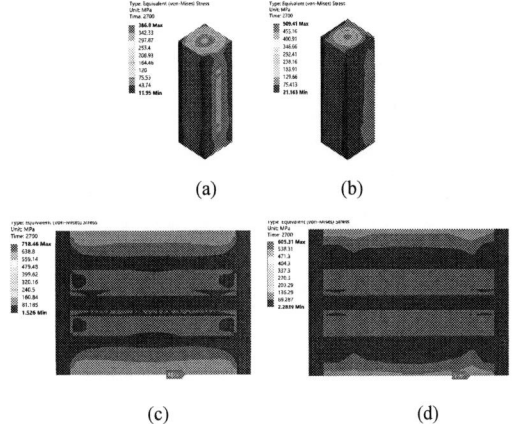

(a)   (b)

(c)   (d)

Fig. 5. Comparison between results of unsimplified model and simplified model: (a), (b) stress distribution in the most dangerous TSV. (c), (d) stress distribution in the most dangerous micro bump.

III. SIMULATION AND OPTIMIZATION OF 3D HETEROGENEOUS INERTIAL MICRO-SYSTEM

Fig. 6 shows the isometric view and the cross-section view of the 3D heterogeneous integrated inertial micro-system. This model consists of two accelerometer chips bonded to two aligned TSV interposers which are connected by two TSV interposers with a cavity. The signal is transmitted vertically through the micro bumps and TSVs and finally is transmitted to the PCB board through the solder on the bottom. Simulation for DFR in a thermal cycling experiment is carried out as before with that the bottom surface of the PCB is fixed.

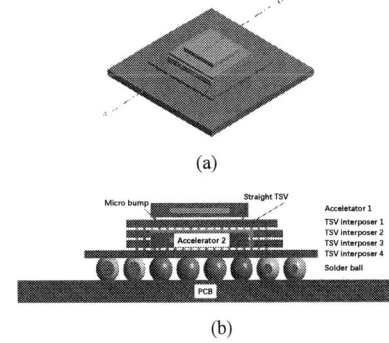

(a)

(b)

Fig. 6. (a) The isometric view and (b) the cross-section view of the stacking integration model.

The simulation results show that the maximum stress value occurs at 125°C as shown in Fig. 7. Due to the CTE mismatch, the stress values at the positions where the two accelerometers contact the micro bumps are greater than the ones at other positions. The maximum stress in accelerator 2 is 6.402Mpa, increases by 16.89% compared to the maximum stress in accelerator 1 (5.6009MPa). The result of the sub-model of the most dangerous TSV shows that the maximum stress appears at the interface of Cu and Si on the lower end surface (237.23MPa), and the stress decreases gradually from the outside to the inside. In the sub-model of the most dangerous micro bump, the interface of Al and Si has the maximum stress

978-1-6654-1392-3/21 $31.00 © 2021 IEEE   545

(530.54MPa) and the stress decreases from the upper and lower boundary to the middle.

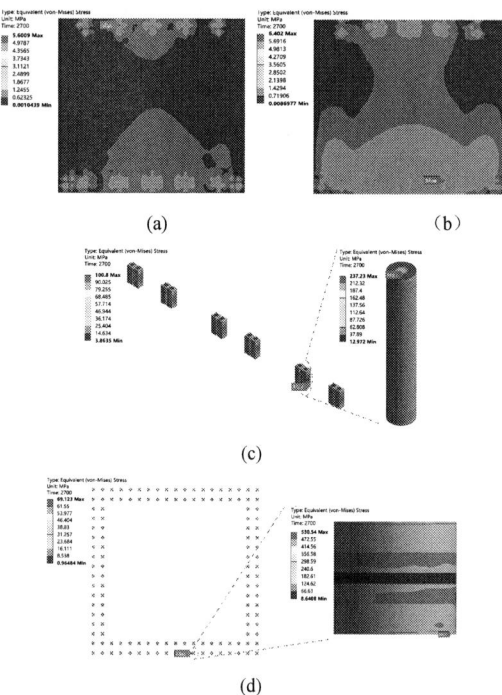

(a)

(b)

(c)

(d)

Fig. 7. Simulation results of the test vehicle @125°C: (a), (b) Stress distribution in accelerator 1 and 2. (c)Stress distribution on the boundaries of TSVs in the lower silicon interposer and inside the most dangerous TSV. (d) Stress distribution on the boundaries of micro bumps under the lower silicon framework and inside the most dangerous micro bump.

A. The influence of ladder TSVs

The Design of Ladder TSV is favorable for fabrication than straight TSVs because it can be achieved by etching on both sides with a big tolerance on misalignment. Therefore, a comparison between the normal straight TSV design and ladder TSV design is finished as shown in Fig. 2 (c).

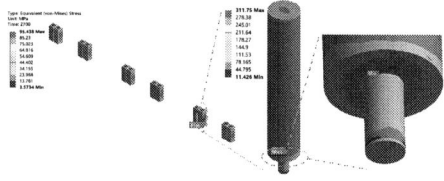

Fig. 8. Stress distribution in the most dangerous TSV @125°C

On the boundary of TSVs in the structure, although the distribution of stress is basically unchanged, the maximum stress values are reduced from 100.8MPa to 95.438MPa, a decrease of 5.32%. This is mainly because Si accounts for a larger proportion in the ladder TSV equivalent element than that in the straight TSV, so the stiffness of the whole equivalent element is larger, and the stress is smaller. Comparing the result shown in Fig. 8 and Fig. 7 (c), it is obvious that the stress value at most locations of the ladder TSV is smaller than that of the straight-through TSV. However, due to the obvious geometric change in the ladder TSV, the stress concentration will occur at the junction of the large hole and the small hole, so the maximum stress in the

ladder TSV (311.75MPa) increases by 24% compared with that in the straight TSV (237.23MPa).

B. The influence of gold balls

In this paper, the effects of the gold ball and the micro bump on the thermomechanical performance are also compared. The diameter of the gold ball is 100μm, and its interface with the pad is a circle with a diameter of 60μm.

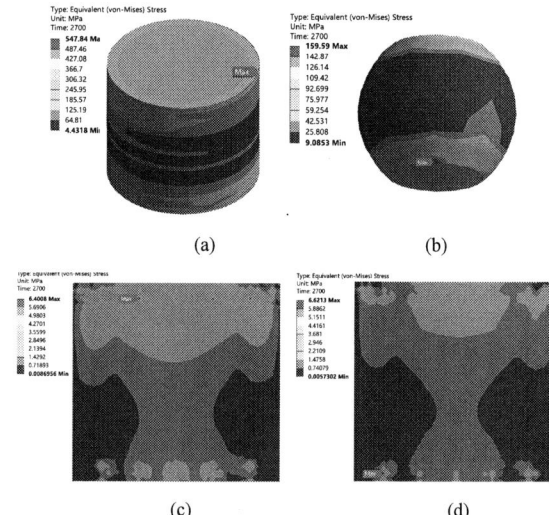

(a)

(b)

(c)

(d)

Fig. 9. The simulation results of model without and with gold balls@125°C: (a), (b) the stress distribution in the most dangerous micro bump and gold ball; (c), (d) the stress distribution in the accelerator 2.

According to the simulation results shown in Fig. 9, the maximum stress inside the gold ball (159.59MPa) is 70.87% lower than that inside the micro bump at the same position (547.84MPa). This is because although there is a large CTE mismatch between the gold ball and silicon, the inside of the gold ball is homogeneous, and the elastic modulus of gold is far greater than that of tin and aluminum. However, the gold ball will introduce more stress on the accelerator2. Compared with the chips bonded by micro bumps, the maximum stress value of the chip bonded with gold ball is 3.33% higher. In conclusion, although chips bonded by gold balls may suffer from larger stress, which may cause a decline in the performance of the chip, the performance of the micro-system with gold balls will still be better because gold balls have better thermal stability compared to micro bumps.

C. The influence of underfill

The influence of underfill is also studied. The model has 5 layers of underfill between each TSV interposer and between TSV interposers and chips.

According to the simulation results shown in Fig. 10, With the addition of the underfill, the maximum stress value on the boundary of the micro bump increases from 67.505MPa to 80.076MPa. However, the addition of the underfill can effectively make the stress distribution on the micro bumps in different positions more uniform, so as to increase the overall reliability of the micro bumps in the thermal cycle. In addition, the maximum stress on the gold ball is greatly reduced after adding underfill, thus the reliability can be improved. Moreover, underfill can reduce the overall deformation of the

978-1-6654-1392-3/21 $31.00 © 2021 IEEE

test vehicle, which can improve the overall reliability and stability.

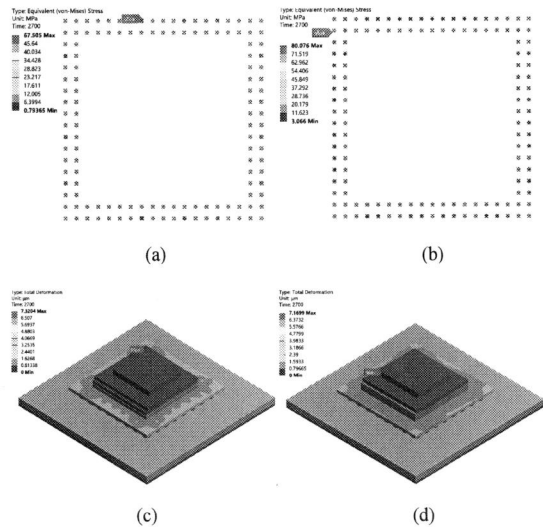

Fig. 10. The simulation results of model without and with underfill@125℃: (a), (b) stress distribution on the boundary of bumps under the lower silicon framework. (c), (d) overall deformation.

D. The influence of RDL

This paper has compared the simulation results of a test vehicle that ignores the RDL and includes the RDL. In the model, there is a layer of RDL on both sides of each TSV interposer and silicon frame, and there are 8 layers of RDL in total. Each layer of RDL is composed of Cu wiring and Si_3N_4 passivation, and the distribution of RDL in the FEA model is basically the same as that in the actual manufactured sample. At the same time, in order to reduce the number of grids in the simulation model and reduce the amount of computation, the shell element is used to mesh RDL in the simulation.

Fig. 11 shows the FEA model with RDL and the simulation results. According to the simulation results, after considering the RDL, the maximum stress value on the boundary of TSVs and micro bumps decreases from 171.22MPa to 83.11MPa and from 57.314MPa to 44.703MPa, respectively. However, these stresses are transferred to the interface of RDL and silicon, resulting in a maximum stress value of 218.21MPa in the RDL under the TSV interposer 1. In addition, the stress distribution in RDL is very uneven, the stress of Cu is far greater than that of the passivation layer, and the stress of the part over TSVs is far greater than that of the part over micro bumps. For the TSV at the most dangerous position, considering the RDL makes the maximum stress value decreases by 14.03%. This may because that when RDL is ignored, copper in TSV directly contacts the underfill and there is a large CTE mismatch between the underfill and copper. When considering RDL, the upper and lower end faces of the copper in TSV are in contact with the copper in RDL, so the stress in TSV at this time is smaller. For the micro bump at the most dangerous position, considering the RDL makes the maximum stress increases by 62.79%. These results all showed that ignoring RDL will greatly underestimate the stress in micro bumps, resulting in a false assessment of reliability. The significant influence of RDL also means that the RDL layout should be optimized in the stress without affecting the electrical performance.

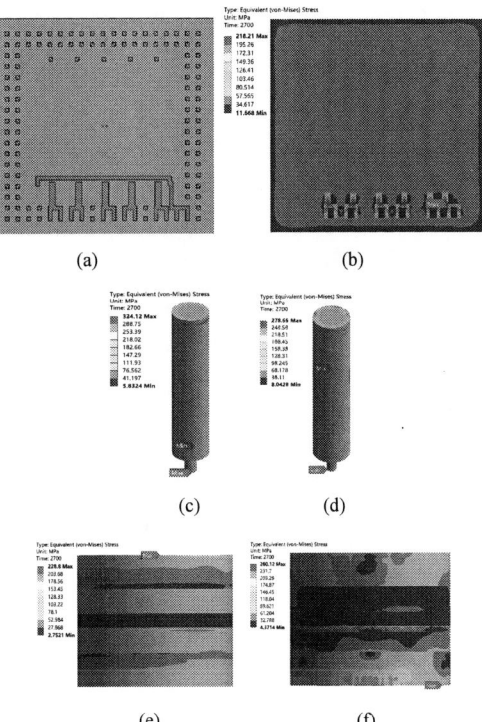

Fig. 11. Simulation model and results of test vehicle without and with RDL @125℃: (a) the model of the RDL over the interposer 1. (b)stress distribution on the RDL under the interposer 1. (c), (d)stress distribution in the most dangerous TSV in the model.(e), (f) stress distribution in the most dangerous micro bump in the model.

IV. CONCLUSION

This paper proposed a 3D integration scheme for inertial micro-systems, a simplified simulation method based on the equivalent element is developed to perform the design for reliability in TCT. The difference between the simplified simulation and the unsimplified model is less than 1% in the maximum deformation. Based on the simplified simulation method, a 3D heterogeneous integrated inertial micro-system is analyzed and optimized. The results show that the ladder TSV design can reduce the stress, but the maximum stress in the ladder TSV increases by 24% compared with that in the straight TSV design because of the stress concentration problem. Using flip-chip bonding with gold ball increases the stress of the accelerator by 3.33% compared with that using micro bumps, but the stress inside the micro bump is 243.27% greater than that inside the gold ball. The use of underfill can effectively reduce the stress changes experienced by the micro bumps during the thermal cycling experiment and makes the stress distribution more uniform. In addition, the use of underfill can also reduce the maximum stress of the golden ball and the deformation of the whole micro-system, so the introduction of underfill is greatly helpful for improving the reliability of the system. Therefore, the integrated scheme with the design of ladder TSV, flip-chip bond with gold balls, usage of underfill are utilized to build a 3D heterogeneous inertial micro-system. In addition, the simulation shows that the influence of RDL on the stress in the micro-system is significant, especially inside the most dangerous micro bump and TSV where the stress increases and decreases by 62.79% and 14.03% if considering RDL. This means that RDL should be carefully dealt with in DFR.

978-1-6654-1392-3/21 $31.00 © 2021 IEEE

In the following work, our team will focus on comparing the simulation results with the manufactured inertial microsystem to optimize the simulation method and 3D integrated inertial microsystem.

ACKNOWLEDGMENT

This research was funded by the Shenzhen Science and Technology Innovation Committee (grant number: ZDSYS201802061805105 and JCYJ20190808155007550), and project No. 2019-JCJQ-JJ-433 of National Basic Strengthening Program Technology Field Foundation of China. Among them grant ZDSYS201802061805105 is the TSV 3D Integrated Micro/Nanosystem Laboratory program.

REFERENCES

[1] T. Waber, W. Pahl, M. Schmidt, G. Feiertag, S. Stufler, R. Dudek, A. Leidl, "Flip-chip packaging of piezoresistive barometric pressure sensors," Proc. SPIE 8763, Smart Sensors, Actuators, and MEMS VI, 87632D, 17 May 2013.

[2] G. Xu, P. Yan, X. Chen, W. Ning, L. Luo and J. Jiao, "Wafer-level chip-to-Wafer (C2W) integration of high-sensitivity MEMS and ICs," 2011 12th International Conference on Electronic Packaging Technology and High Density Packaging, 2011, pp. 1-5.

[3] W. K. Choi et al., "A novel die to wafer (D2W) collective bonding method for MEMS and electronics heterogeneous 3D integration," 2010 Proceedings 60th Electronic Components and Technology Conference (ECTC), 2010, pp. 829-833.

[4] Y. Hsu et al, "New capacitive low-g triaxial accelerometer with low cross-axis sensitivity," Journal of Micromechanics and Microengineering, vol. 20, (5), 2010, pp. 055019.

[5] B. Vigna, E. Lasalandra and T. Ungaretti, "Motion MEMS and sensors, today and tomorrow," in Anonymous New York, NY: Springer New York, 2012, pp. 117-127.

[6] C. Lee, "Effect of wafer level underfill on the microbump reliability of ultrathin-chip stacking type 3D-IC assembly during thermal cycling tests," Materials, vol. 10, (10), 2017, pp. 1220.

[7] J. W. Baek et al, "Representative volume element analysis for wafer-level warpage using Finite Element methods," Materials Science in Semiconductor Processing, vol. 91, 2019, pp. 392-398.

[8] C. Lee et al, "Demonstration of an Equivalent Material Approach for the Strain-Induced Reliability Estimation of Stacked-Chip Packaging," IEEE Transactions on Device and Materials Reliability, vol. 20, (2), 2020, pp. 475-482.

[9] Q. Zeng et al, "Influence of Viscoelastic Underfill on Thermal Mechanical Reliability of a 3-D-TSV Stack by Simulation," 2017.

[10] C. Lee, T. L. Tzeng, and P. C. Huang, "Development of equivalent material properties of microbump for simulating chip stacking packaging," Materials, vol. 8, no. 8, pp. 5121–5137, August 2015.

[11] R. Iannuzzelli, "Predicting plated-through-hole reliability in high temperature manufacturing process," in Proc. IEEE Electron. Compon. Technol. Conf. (ECTC), pp. 410–421, May 1991.

[12] G. Y. Jang, J. W. Lee, and J. G. Duh, "The nanoindentation characteristics of Cu_6Sn_5, Cu_3Sn, and Ni_3Sn_4 intermetallic compounds in the solder bump," J. Electron. Mater., vol. 33, no. 10, pp. 1103–1110, October 2004.

[13] A. Lis, S. Kicin, F. Brem, and C. Leinenbach, "Thermal stress assessment for transient liquid-phase bonded Si chips in high-power modules using experimental and numerical methods," J. Electron. Mater., vol. 46, no. 2, pp. 729–741, February 2017.

[14] A. Lis, S. Kicin, F. Brem, and C. Leinenbach, "Thermal stress assessment for transient liquid-phase bonded Si chips in high-power modules using experimental and numerical methods," J. Electron. Mater., vol. 46, no. 2, pp. 729–741, February 2017.

[15] A. Lis, S. Kicin, F. Brem, and C. Leinenbach, "Thermal stress assessment for transient liquid-phase bonded Si chips in high-power modules using experimental and numerical methods," J. Electron. Mater., vol. 46, no. 2, pp. 729–741, February 2017.

Study on image alignment technology based on CCD thermal reflection method

He Yang
Micro-nano Fabrication Technology Department, Institute of Electrical Engineering Chinese Academy of Sciences
University of Chinese Academy of Science
Beijing, China
yh@mail.iee.ac.cn

Weikang Si
Optical Navigation Department Shanghai Aerospace Control Technology Institute，
Shanghai, China
dyxiaokang@126.com

Dazheng Wang
Micro-nano Fabrication Technology Department, Institute of Electrical Engineering Chinese Academy of Sciences
Xi'an Jiaotong University
Beijing, China
dzwang@mail.iee.ac.cn

Yingying Gao
Micro-nano Fabrication Technology Department, Institute of Electrical Engineering Chinese Academy of Sciences
University of Chinese Academy of Science
Beijing, China
gaoyy@mail.iee.ac.cn

Libing Zheng*
Micro-nano Fabrication Technology Department, Institute of Electrical Engineering Chinese Academy of Sciences
University of Chinese Academy of Science
Beijing, China
ieezlb@mail.iee.ac.cn

Abstract—Thermoreflectance imaging technique is a non-contact measurement technology that has attracted much attention in the field of thermal characteristics measurement of semiconductor devices. This technology is often used for steady-state temperature measurement and transient temperature measurement of semiconductor devices, with the advantages of high spatial resolution, high temperature resolution, and high time resolution. However, during the measurement of the device, the thermal expansion of the temperature control table and the sample caused by the temperature rise makes the optical image appear defocused. In order to ensure that the optical image collected by the CCD is not distorted, it is necessary to design an image alignment system for focusing so that the semiconductor device can be measured normally. In this paper, an image alignment system consisting of an imaging module, a high-precision motion control component and an image alignment algorithm is proposed to solve the problem of image defocusing caused by thermal expansion, which can provide references for the optimal design and precise measurement of the thermal reflection imaging system based on CCD.

Keywords—thermoreflectance imaging, image alignment, thermal characterization

I. INTRODUCTION

With the improvement of living standards, people's demand for electronic products increases, and the performance requirements for electronic products are also increasing. In order to meet the needs of the public, the electronic equipment industry continues to develop in the direction of miniaturization, integration, and multi-function. At the same time, integrated circuit manufacturing technology is also constantly developing. With the reduction of feature size, the degree of integration is improved, that is, the number of integrated devices per unit area is increasing and the power density is also increasing [1].The heat generated by the semiconductor devices increases, but the heat distribution in different parts of the device is not uniform, and the failure of the device may occur because the local temperature of the device is too high.

In response to this problem, designers usually increase the size of the heat sink when designing the circuit of the electronic device, so that the temperature of each position of the device is lower than the specified threshold, to ensure the normal use of the semiconductor device. However, the increase of the size of the heat sink will affect the further reduction of the characteristic size of the semiconductor device. There may also be a situation where the temperature of the local point of the device itself is less than the threshold value and no heat dissipation is needed, so that the radiator does not effectively play the maximum role. In order to achieve a good heat dissipation for the device without affecting the further reduction of the device size, the measurement and characterization of thermal characteristics of micro-scale devices is particularly important. The hot spots of microscale devices are usually very small, making it difficult to measure the temperature of each specific region or location of the device. Therefore, the accuracy of thermal characteristics measurement technology of semiconductor devices is highly required [2].

At present, the existing technologies used to measure the operating temperature and thermal resistance of micron, submicron and nanoscale semiconductor devices can be divided into three categories: electrical measurement method, physical contact method and optical method [3].The optical method can be further divided into thermal imaging technology, internal thermal detection technology and spectral analysis temperature measurement method. Each method has both advantages and limitations. The spatial resolution, time resolution, temperature resolution, working conditions, applicable samples and other factors are considered. Compared with other methods, Thermoreflectance imaging technique is currently a kind of non-contact measurement technology which is widely used in the field of thermal characterization and measurement of semiconductor devices [4].This technology has the advantages of high spatial resolution, high temperature resolution and high time resolution [5].

978-1-6654-1392-3/21 $31.00 © 2021 IEEE

When using a thermal reflection imaging system for measurement, the temperature of the semiconductor device will rise during operation, causing the expansion of the semiconductor device. The thermal expansion of the device will cause the horizontal and vertical displacement of the collected optical image, which in turn causes the reference image collected by the CCD to change the field of view of the current image, resulting in the defocus of the optical image. Defocusing problem is an important factor restricting the measurement accuracy of thermal reflection imaging system. If the optical image defocusing problem image is not solved, the thermal measurement error of semiconductor devices will be large, which will affect the thermal resolution of the imaging system. In this paper, image alignment technology is introduced on the basis of thermal reflection imaging system. The technology is mainly composed of imaging module, high-precision motion control module and image alignment algorithm. It is of great significance to solve the defocus problem caused by thermal expansion of devices and ensure the normal measurement of semiconductor devices.

II. IMAGE ALIGNMENT SYSTEM

A. Imaging module

The imaging module of image alignment system is composed of CCD and optical microscopic element.

CCD cameras are divided into color cameras and black and white cameras. The color camera decomposes the white light into red light, green light and blue light, which is absorbed by the sensor, and the color image is obtained after calculation according to the corresponding interpolation algorithm. The black-and-white camera directly absorbs the white light, converts it into the corresponding electrical signal and then quantizes it into the corresponding gray-scale image. The color CCD camera will cause the loss of image information due to the effect of the filter, so the black and white CCD camera has a relatively high sensitivity. This system uses a 14-bit 1360*1024 resolution black-and-white CCD camera, and its related performance indicators are shown in TABLE I.

TABLE I.   CCD CAMERA RELATED PERFORMANCE INDICATORS

The camera parameters	Index	Remark
Model	Daheng GT1380	N/A
Resolution	$1360(H) \times 1024(V)$ $1.4MP$	N/A
Pixel size	$6.45\mu m \times 6.45\mu m$	N/A
Sensor size	$8.8mm \times 6.6mm$	N/A
Maximum pixel depth	12/14bit	N/A
Image buffer	128MB	N/A
Image color values	black and white	Mono8, Mono12, Mono14
data interface	Cable/network port	Camera control, data transmission and power supply

The imaging module is composed of a Nikon MUE21200 objective lens and a CCD camera. The numerical aperture of the objective lens is 0.4N.A and the magnification is 20X.

A detection light source with a wavelength of 530nm is adopted, and the theoretical spatial resolution calculated according to the Abbe diffraction limit formula is as follows:

$$d = \frac{0.5\lambda}{N.A} = \frac{0.5 \times 530}{0.4} = 662.5nm\,/\,pixel \qquad (1)$$
$$= 0.6625\mu m\,/\,pixel$$

The spatial resolution of the imaging module can be expressed by the following formula:

$$N = \frac{S_{sample}}{N_{camera}} = \frac{FOV\,/\,P}{N_{camera}} = \frac{N \times N_{camera}\,/\,P}{N_{camera}} \qquad (2)$$
$$= \frac{6.45 \times 1360\,/\,20}{1360} = 0.3225\mu m\,/\,pixel$$

Where, N_{camera} represents the number of pixels of the camera in the vertical direction, FOV represents the visual field of the camera, and S_{sample} represents the visual field range of the sample to be tested.

Because $0.3225\mu m\,/\,pixel < 0.6625\mu m\,/\,pixel$, the spatial resolution of the imaging system fully meets the resolution requirements of the thermal reflection imaging system, which confirms the feasibility of the imaging module.

B. High precision motion control module

The high-precision motion control component of the image alignment system is composed of a sample stage and a three-dimensional displacement platform. The sample platform is fixed on a three-dimensional displacement platform, and the sample platform can be moved in X, Y and Z directions with a moving accuracy of 1 μm by means of a cooperative drive between the stepping motor and the lead screw. The main function of the motion control component is to control the 3D displacement platform to move in the corresponding direction by using the minimum step distance, so the field of view of the current image and the reference image are overlapped. and the problem caused by the movement of the field of view area caused by thermal expansion is solved.

C. Image alignment algorithm

The image alignment algorithms used in this study include the evaluation algorithm to determine the best focus (mean square error method), the corner position detection algorithm to determine the corner position detection algorithm (optimized Harris corner detection algorithm), and the corner position dynamic tracking algorithm (pyramid Lucas-Kanade optical flow algorithm).

Firstly, the mean square error method is introduced. The purpose of this algorithm is to detect whether the camera is in focus state, that is, to judge whether the current position is the best position of focus. This method has many advantages such as unimodality, certainty, repeatability, universal applicability, insensitivity to other parameters, simplicity and easy implementation [6].The basic principle of the algorithm is as follows:

$$AF = \frac{1}{M*N} \sum_{M} \sum_{N} (I(x,y) - \bar{I})^2 \qquad (3)$$

Where, $I(x,y)$ represents the grayscale value of pixel points in specific positions, $\bar{I}$ represents the average grayscale value of all pixel points in the whole image, and M, N represents the height and width of the image. Through this evaluation function, we can judge whether the focusing requirements are met. The larger the change of pixel gray value, the larger the variance, the larger the AF value, and the more obvious the image definition.

In order to verify the feasibility of the focusing evaluation function, 180 pictures are collected to draw the focusing curve of the evaluation function, as shown in Fig.1.The horizontal axis of the figure is the image sequence collected by CCD. By changing the z position of the vertical axis of the microscope, the focus curve of the mean square error function is drawn. The sharpest image corresponds to the position where the sharpest value of the focusing function is highest, and this position is the best focusing position. In addition, the monotony of the image before and after this position is good, the overall performance is superior, and the calculation amount is small, which proves that the algorithm is suitable for the focus of thermal reflection imaging system.

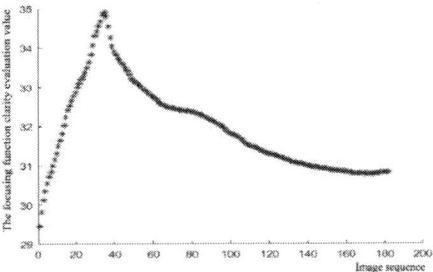

Fig. 1.   Image mean square method focusing curve

The next step is to carry out corner detection, with the purpose of identifying the position of each corner point as a reference position, so as to compare with the position of the corner points [7] after the thermal expansion of the image offset, so as to determine the direction and distance of the image movement caused by thermal expansion.

Generally, the intersection point of two boundaries or the boundary with different principal directions in the neighborhood is defined as a corner point. Harris corner detection algorithm is adopted in this paper for corner detection [8]. The basic principle is to introduce a small window and determine the position of corner points according to the change of grayscale when moving in each direction in the small window. When the small window moves in all directions of the optical image, there is no obvious change in the gray scale, then this area is called a smooth area (as shown in Fig.2a). If the gray level changes significantly when moving in one direction, it is called the edge area (as shown in Fig.2b). If the gray level changes significantly in multiple directions, it is called the corner area (as shown in Fig.2c). Harris corner detection is to judge whether the position is a corner by using the gray change degree of the window in multiple directions.

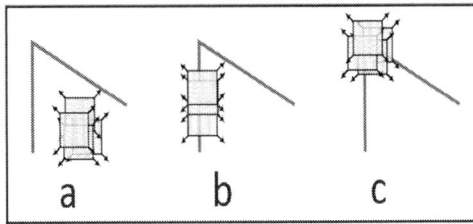

Fig. 2.   Harris corner detection schematic

When this theory is applied to image processing, each pixel in the image can be used as a small window, so the position of corner can be determined by calculating the gradient change of the pixel. If the absolute gradient value of the pixel in both directions is higher than the set threshold, this particular pixel value is called a corner point. Harris corner detection algorithm is applied to the field of image processing to establish a mathematical model, and the calculation formula can be expressed as:

$$E(u,v) = \sum_{x,y} w(x,y)[I(x+u,y+v) - I(x,y)]^2 \quad (4)$$

Where, $I(x,y)$ is the gray value of the original pixel, $w(x,y)$ is the window function, is generally the Gaussian function, the moving distance in the direction x is u, the moving distance in the direction y is v, then $E(u,v)$ is the gray difference value of the target pixel. Therefore, if the value of $E(u,v)$ is almost unchanged, the pixel point is non-corner point; If the value of $E(u,v)$ changes greatly, then the pixel point is a corner point.

Take the first-order Taylor expansion of $E(u,v)$ and write it in matrix form:

$$E(x,y) = [u,v]M\begin{bmatrix} u \\ v \end{bmatrix} \quad (5)$$

$$M = \sum_{x,y} w(x,y)\begin{bmatrix} I_x^2 & I_xI_y \\ I_xI_y & I_y^2 \end{bmatrix} = w(x,y) \otimes \begin{bmatrix} I_x^2 & I_xI_y \\ I_xI_y & I_y^2 \end{bmatrix} \quad (6)$$

Among them, matrix M is called autocorrelation matrix, because the process of finding eigenvalue (λ_1, λ_2) of matrix M is particularly complex. Therefore, this paper uses the response function of the target pixel, which is a relatively simple method in the calculation process, to judge the corners.

Define the response function as:

$$R(x,y) = Det(M) - kTr(M)^2 \quad (7)$$

Where, $Det(M) = \lambda_1 \cdot \lambda_2$ is the determinant of matrix M and $Tr(M) = \lambda_1 + \lambda_2$ is the trace of the matrix. The value of k is determined according to experience, generally ranging from 0.04 to 0.06.Usually we also need to set an artificial threshold. When the value of $R(x,y)$ is greater than the threshold, the pixel at position (x,y) is called a corner.

Therefore, the basic flow of corner detection using Harris corner detection algorithm in the field of image processing can be summarized as follows: First, determine the initial position of feature points on the collected grayscale image, and calculate the pixel gradient values I_x and I_y in directions x and y .Then the autocorrelation matrix M is obtained after filtering by Gaussian function of $w(x,y)$ .After comparing the response function $R(x,y)$ of each pixel with the self-set threshold, the candidate corner points are screened out. Finally, the non-maximum suppression method is used to remove the sticky corners and determine the final corner. Its flow chart is shown in Fig.3.

978-1-6654-1392-3/21 $31.00 © 2021 IEEE

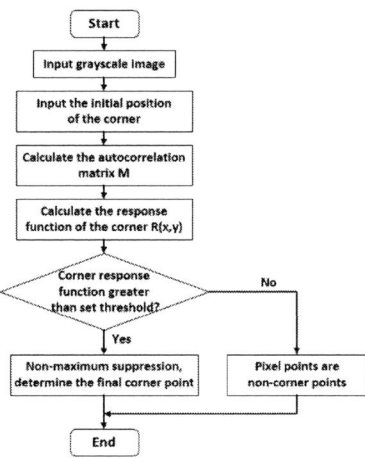

Fig. 3. Harris corner detection flow chart

Because the image alignment needs real-time, the movement of the optical image is detected at any time, so the detection accuracy is very high. Harris corner detection algorithm is based on the pixel level. In order to improve the accuracy, this paper introduces the minimization error function iterative algorithm to detect corner points at the sub-micron level [9].

The basic theory of submicron level detection is orthogonal vector theory. When the selected point is a corner point, the vector from this point to the pixel point in its neighborhood has a vertical relationship with the direction of the gray gradient of the neighborhood. However, in the actual operation, due to the influence of noise, the result of orthogonal vector dot product is often not zero, but there is a certain error. The above principles can be expressed by mathematical formula as follows:

$$\nabla I_{p_i}^T \cdot (q - p_i) = \varepsilon_i \qquad (8)$$

By multiplying both sides of the above formula by ∇I_{p_i}, the equation for solving the sub-pixel coordinate position of the corner point in the entire neighborhood can be obtained:

$$\sum_i \nabla p_i \nabla p_i^T q - \sum_i \nabla p_i \nabla p_i^T p_i = \varepsilon \qquad (9)$$

The sub-pixel coordinates of the corner can be calculated by determining the number of iterations and the value of the final error accuracy.

Because the thermal expansion of the device is a dynamic process, it is necessary to track the position of corner points. In this paper, the pyramid Lucas-Kanade optical flow algorithm is used to track the position of corner points. This algorithm is an improvement on the Lucas-Kanade optical flow algorithm used for tracking small dynamic changes [10].

The implementation process of the pyramid Lucas-Kanade optical flow algorithm [11] can be expressed as follows: Suppose L is the level of the pyramid image, where $L = 0 \ldots L_m$. Starting from the highest level L_m of the pyramid, the optical flow and affine transformation matrix are calculated according to the corresponding corner features. Then the final calculation results of this level are transferred to the next level to continue the calculation. It iterates one by one until it reaches the lowest level of the pyramid 0. The algorithm needs to track the dynamic changes of corner points in as many directions as possible.

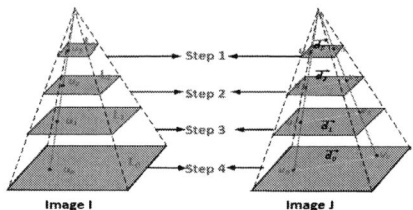

Fig. 4. Pyramid Lucas-Kanade optical flow model diagram

As shown in Fig.4, a three-layer pyramid with two frames of images is taken as an example to illustrate the pyramid Lucas-Kanade optical flow algorithm . For the given corner point u in Figure I, find the corresponding position $v = u + d$ in Figure J, where $d = [dx, dy]$ is the displacement of u to v. Local affine transformation matrices of image I and image J are created near corner points u and v, respectively. Where, L is the level of pyramid image, for $L = 0, L_1, L_2, \cdots, L_m$, the corresponding coordinate of corner point u in pyramid image I_L can be expressed as $u_L = \dfrac{u}{2^L}$. The specific implementation process is to calculate the optical flow and affine transformation matrix [12] from pyramid image L_3, calculate the pixel displacement, and pass the above calculated value as the initial value to L_2. The same calculation process is passed in turn until the iteration reaches the bottom layer 0.

III. Software Platform Design

The software control platform of the image alignment system uses Visual C++ programming to establish the MFC program framework, and through calling the VmbAPI library of the CCD camera and the OpenCV library of the computer vision application program to establish the human-computer interaction interface. The software design of the image alignment system is shown in Fig.5, and the design of the human-computer interaction interface is shown in Fig.6. The human-computer interaction interface mainly includes modules such as image acquisition, image alignment, image focusing, thermal expansion detection and image saving.

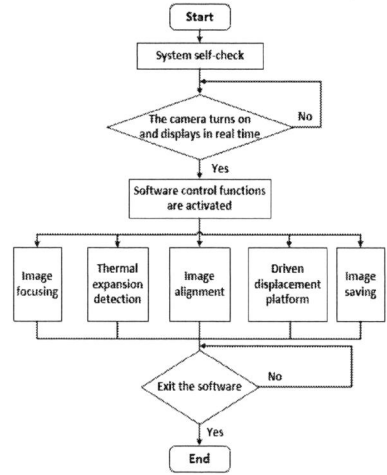

Fig. 5. Design of image alignment system software

978-1-6654-1392-3/21 $31.00 © 2021 IEEE

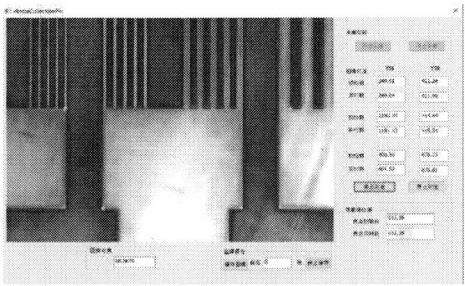

Fig. 6. Design of the human-computer interaction interface
of the image alignment system

Fig.6 shows the design of the human-computer interaction interface of the image alignment system software. Click the "Start Collection" button in the image acquisition module on the right to obtain the optical image of the sample. Next, the corner points are selected. Generally, the position with a large gray gradient is selected. The three green points in the figure above are the corner points. The specific position of each corner can be obtained through the image alignment module on the right. There is also a thermal expansion detection module on the right side of the main page to detect the dynamic changes of the device in real time. At the bottom of the main page are image focusing modules used to evaluate the sharpness of focusing and image saving modules.

IV. THE SAMPLE MEASUREMENT

Fig.7 shows the basic structure of the CCD thermal reflection imaging system with high resolution. Based on this system, this research group built an experimental setup for temperature measurement, as shown in Fig.8. The basic components of the system include CCD camera, LED detection light source, micro objective lens assembly, temperature control table, signal generator, displacement platform, etc.

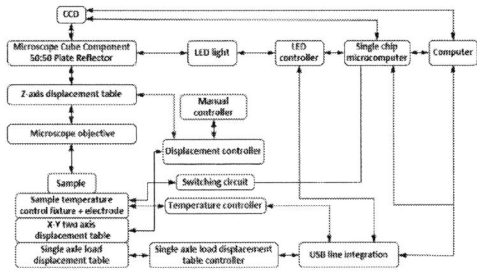

Fig. 7. Fundamental structure diagram of a high resolution thermal
reflection imaging system based on CCD imaging

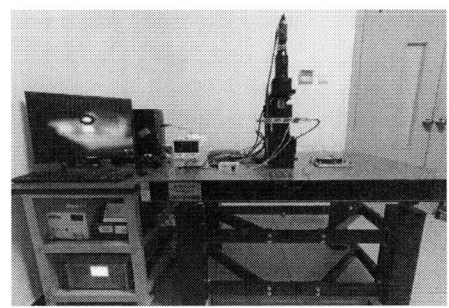

Fig. 8. Experimental setup of high resolution thermal reflection
imaging system based on CCD

The device under test in this study is a micro-resistance wire made of gold as the main material. Its structure is shown in Fig.9. The device size is $1 \times 1 cm^2$, it consists of a $500\ \mu m$ thick substrate covered with a layer of $200\ nm$ oxide passivation layer. The resistance wires have widths of $3\ \mu m$, $5\ \mu m$, $10\ \mu m$, $15\ \mu m$, $20\ \mu m$, $30\ \mu m$, $35\ \mu m$ and thickness of $150\ \mu m$ respectively. They are attached to the silicon oxide substrate with a substrate of $200\ nm$ thickness. There is an adhesive layer of $5\ nm$ thick (Ti) between the micro resistance wires and the silicon oxide. Because resistance lines have different widths, their resistance also varies, so each resistance line produces different joule heat.

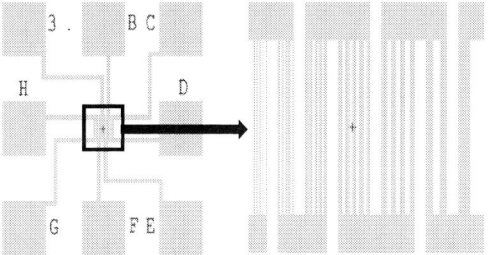

Fig. 9. The geometry of micro resistance wires

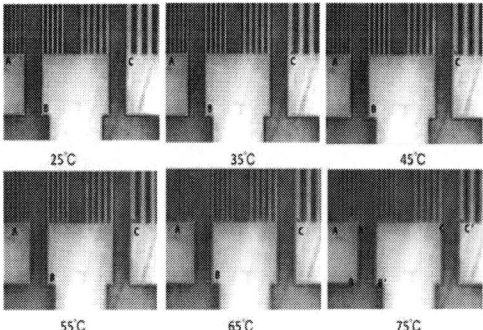

Fig. 10. Defocusing images of microresistor samples
at different temperatures

Fig.10 shows the defocusing images of the micro-resistance wire samples at different temperatures. It can be clearly seen that as the temperature increase, the device thermally expands and the optical image becomes blurred. Points A, B and C marked at 25℃ on the original drawing are the initial positions of the three selected three corners. At the same time, it founds that as the temperature rises, the image moves, resulting in the change of corner position. Corner tracking adopts pyramid Lucas-Kanade optical flow algorithm, which is represented by the red line in the image to determine the moving distance of the corner. Taking the image at 75 ° C as an example, we can clearly see that the position of the corner points moves from A, B, and C to A ', B ', and C '. The moving distance of the corner is determined by tracking the position of the corner after it moves. These data can be used as the basis for image alignment of moving 3D displacement platform.

In order to verify the accuracy of the corner detection of the optimized Harris corner detection algorithm, the test is carried out on the detection of the line width of the micro-resistance line.

978-1-6654-1392-3/21 $31.00 © 2021 IEEE

TABLE II. MICRO RESISTANCE LINE WIDTH DETECTION

Line width	Theoretical pixel value	Pixel value between two points	Measured value of pixel accuracy	Error
15 μm	46.51	46.09	14.86 μm	0.93%
20 μm	62.02	62.11	20.03 μm	0.15%
30 μm	93.02	92.74	29.91 μm	0.30%
50 μm	155.03	152.74	49.28 μm	1.40%
325 μm	1007.75	993.11	320.28 μm	0.31%

It can be concluded from the experimental results that the error of corner detection of the device under test is within a reasonable range without considering the influence of the preparation error of the micro-resistor wire, and the algorithm fully meets the requirement of corner detection.

In order to verify the tracking ability of corner points of pyramid Lucas-Kanade optical flow algorithm, corner points after thermal expansion at different temperatures were tracked. The tracking results are as follows.

TABLE III. THE TRACKING RESULT OF THE IMAGE CORNER

T(℃)	Clarity evaluation	Z-axis travel distance	Coordinates	Distance
25	73.532	0 μm	（975.91，375.68）	936.78
35	70.401	5.1 μm	（988.20，381.57）	935.86
45	67.611	11.4 μm	（1003.18，386.00）	935.16
55	65.740	14.6 μm	（1020.13，390.22）	934.81
65	64.031	18.4 μm	（1037.21，393.70）	936.15

According to the experimental data, as the temperature rises, the position coordinates of the corners change and the distance between the two corners will narrow. This indicates that the measured sample moves upward during thermal expansion, resulting in a shorter distance to the focal plane, thus resulting in defocus.

According to the data, corner tracking failed when the temperature is 65 ℃ .Therefore, the degree of temperature change should not be greater than 40℃. Rapid temperature change will lead to serious image deviation, and the algorithm cannot track the corner points, resulting in wrong experimental data. The above experiment proves that the image tracking algorithm is very reliable in a certain temperature range.

Finally, by heating the device under test to a certain extent and moving the 3D displacement platform to make the current corner point coincide with the target corner point, the accuracy of matching and calibration between the current image and the reference image is verified. The experimental results are as follows.

TABLE IV. THE IMAGE CALIBRATION RESULTS OF SYSTEM

T(℃)	Clarity evaluation	Coordinates	Distance
25	73.532	（975.91，375.68）	936.78
35	71.941	（975.38，375.61）	937.16
45	70.780	（975.72，375.43）	937.26
55	69.901	（975.45，375.21）	937.32
65	69.850	（975.27，395.46）	937.34

According to the experimental data, it can be concluded that the sharpness evaluation value of the image becomes smaller and smaller when the current corner is superposed with the target corner by the 3D displacement platform. It shows that the accuracy of the displacement platform fully meets the requirements, and it is also confirmed that the thermal expansion phenomenon does occur when the device is working, and the present invention can solve this problem well.

V. CONCLUSION AND DISCUSSION

In this paper, the image alignment technology has solved the device optical defocused image problems caused by the thermal expansion, through the corner points location monitoring and further track and exactly get optical image relative to the reference images, moving direction and distance through the three dimensional displacement platform for precise manual focus, on the minimum error is 0.15%,Thus, the influence of thermal expansion on the resolution of thermal imaging is basically eliminated. It provides reference for the optimization design and accurate measurement of thermal reflection imaging system based on CCD, and is of great significance for the high precision measurement of device temperature. Through the camera focusing evaluation function, the sharpness evaluation value tends to be stable, and the sharpness of the image does not change greatly, which proves the reliability of the image tracking algorithm.

REFERENCES

[1] Christofferson J, Maize K, Ezzahri Y, Shabani J, Wang X and Shakouri A 2008 Microscale and nanoscale thermal characterization techniques J. Electron. Packag. 130 041101

[2] A.R. Hefner, D.L. Blackburn. Simulating the dynamic electro-thermal behavior of power electronic circuits and systems[P]. Computers in Power Electronics, 1992., IEEE Workshop on,1992.

[3] Wang Linping,Poque Sylvain,Valkonen Jari P T. Phenotyping viral infection in sweetpotato using a high-throughput chlorophyll fluorescence and thermal imaging platform.[J]. Plant methods,2019,15.

[4] Z. Y. Wang, J. Sheu, J. Y. Qian et al.. Measurement of internal junction temperature of AlGaAs/GaAs heterojunctionbipolar transistors with hyperspectral imaging.

[5] Ju S, Kading O W, Leung Y K, et al. Short-timescale thermal mapping of semiconductor devices[J]. IEEE Electron Device Letters, 2002, 18(5): 169-171.

[6] Groen F C A. A comparison of different focus functions for use in autofocus algorithms[J]. Cytometry, 1987, 6.

[7] Loncomilla, Patricio. Object Recognition using Local Invariant Features for Robotic Applications: A Survey[J]. Pattern Recognition, 2016: S0031320316301054.

[8] Ying L, Zhang J D, Zhou W H. Optimization of Harris Corner Detection Algorithm[J]. Journal of Yunnan University of Nationalities, 2011.

[9] Yu F J, Luan X, Song D L, et al. A Novel High Accuracy Sub-Pixel Corner Detection Algorithm for Camera Calibration[J]. Applied Mechanics & Materials, 2013, 239-240: 713-716.

[10] Ayvaci A, Raptis M, Soatto S. Sparse Occlusion Detection with Optical Flow[J]. International Journal of Computer Vision, 2012, 97(3): 322-338.

[11] Ohta N, Kanatani K, Kimura K. Moving Object Detection from Optical Flow without Empirical Thresholds[J]. Ieice Trans.inf. & Syst.d, 1998, 81(2): 243-245.

[12] Lucas B, Kanade T. An Iterative Image Registration Technique with an Application toStereo Vision[C]. Proceedings of the 7th International Joint Conference on ArtificialIntelligence, 1997.

A thermal network model for thermal analysis in automotive IGBT modules

1st Yanzhong Tian
Institute of Electronics Packaging Technology and Reliability
Beijing Key Laboratory of Advanced Manufacturing Technology
Faculty of Materials and Manufacturing
Beijing University of Technology
Beijing, China
tianyz@emails.bjut.edu.cn

2nd Tong An
Institute of Electronics Packaging Technology and Reliability
Beijing Key Laboratory of Advanced Manufacturing Technology
Faculty of Materials and Manufacturing
Beijing University of Technology
Beijing, China
antong@bjut.edu.cn

3rd Fei Qin
Institute of Electronics Packaging Technology and Reliability
Beijing Key Laboratory of Advanced Manufacturing Technology
Faculty of Materials and Manufacturing
Beijing University of Technology
Beijing, China
qfei@bjut.edu.cn

4th Yanpeng Gong
Institute of Electronics Packaging Technology and Reliability
Beijing Key Laboratory of Advanced Manufacturing Technology
Faculty of Materials and Manufacturing
Beijing University of Technology
Beijing, China
yanpeng.gong@bjut.edu.cn

5th Chen Liang
Beijing New Energy Vehicle Technology Innovation Center Co., Ltd
Beijing Automotive Research Center Co., Ltd
Beijing, China
liangchen@nevc.com.cn

Abstract—With the developing of insulated gate bipolar transistor (IGBT), a method to accurately solve the detailed problems caused by thermal behaviors in different locations and layers of automotive IGBT modules is necessary. The paper proposes an *RC* thermal network for automotive IGBT modules. The thermal effects are modeled among chips and key layers. And particularly boundary conditions are considered, including the heat dissipation conditions. It is demonstrated that the model makes it possible to estimate temperature quickly and accurately of automotive IGBT modules in the real normal running conditions. Compared with the results of finite-element-based simulation and infrared temperature measurement, the proposed thermal model is verified.

Keywords—RC thermal network, insulated gate bipolar transistor (IGBT), reliability, thermal analysis, finite-element method (FEM)

I. INTRODUCTION

With the deepening of global energy crisis, the new energy vehicle industry is developing rapidly. As the core component of new energy vehicles, automotive insulated gate bipolar transistor (IGBT) is facing new opportunities and challenges. Compared with the general industrial IGBT modules, the working conditions of automotive IGBT modules are more complex. Therefore, the reliability requirements of latter are higher than that of former [1]. Generally speaking, the working environment of automotive IGBT modules is very harsh, and they often have to withstand severe temperature (-40° C ~ 150° C). Moreover, the power loss of the modules during operation will cause the junction temperature (T_j) to rise, which will affect the performance of the modules, and even cause the permanent failure of the module chips [2]. Consequently, how to obtain the junction temperature of IGBT modules quickly and accurately is of great significance for optimizing the heat dissipation and improving reliability of automotive IGBT modules.

Currently, there are two main methods to obtain the junction temperature of IGBT modules. One is experimental measurement method, and the other is simulation calculation method [3, 4]. Although the former method can get more accurate junction temperature, it is usually necessary to change the module packaging structure and build a special measuring circuit, which is difficult to realize for the actual working condition. Therefore, the latter method is generally used in actual working conditions, and *RC* thermal network model is a choice in the method. *RC* thermal network model obtain parameters through theoretical calculation, experimental measurement or numerical simulation, and it can quickly calculate the junction temperature, so it has a wide range of applications.

This paper puts forward an accurate and quick *RC* thermal network model for automotive IGBT modules. It contains some key material and critical geometry information of the devices. And the structure of the paper is as follows. The finite element modeling and parameter extraction are introduced first. Subsequently, the suggested method to model the *RC* thermal network is described. Finally, the whole model is confirmed by both the finite element simulation and the experiments.

II. PREPARATION BEFORE MODELING

A. Modeling in ABAQUS

This paper mainly concentrates on an automotive IGBT module that is rated 450 A and 1200 V. The IGBT module consists of two identical sections, upper and lower half bridge, and each section has three IGBT chips and three diodes. In each half bridge, the chips are connected in parallel. The IGBT module utilizes the most current assembly technology. It uses a direct copper bonded (DCB) substrate to function as the electrical insulator, current channels and thermal conductor. And as shown in Fig. 1, it uses bond wires to achieve interconnections between chips, DCB substrate and terminals. The IGBT module is simulated using a finite-element-based

978-1-6654-1392-3/21 $31.00 © 2021 IEEE

software named ABAQUS. The material properties in FE analysis are shown in Table I.

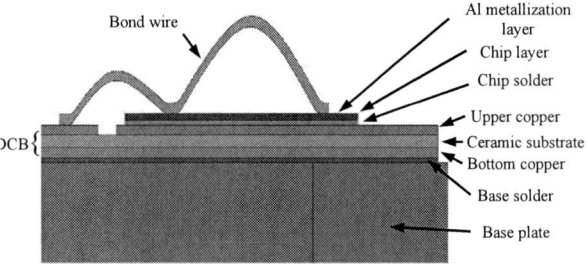

Fig. 1. Cross-section schematic of the IGBT module.

TABLE I   MATERIAL PROPERTIES IN FE ANALYSIS

Materials	Thermal conduction (mW/mm·K)	Electrical Resistivity (mΩ·mm)	Density (kg/m³)	Specific heat (J/kg·K)
Aluminum	237	2.65×10^{-2}	2.70×10^{3}	900
Silicon (IGBT)	148	7.7×10^{3}	2.33×10^{3}	700
Solder (SAC305)	57	1.04×10^{-1}	7.30×10^{3}	230
Ceramic (Al$_2$O$_3$)	20	1×10^{18}	3.96×10^{3}	753
Copper	400	1.68×10^{-2}	8.92×10^{3}	380

Without losing of generality, the finite element simulation is simplified in this paper. With the purpose of decreasing the step time in simulation, it has been supposed that the base copper pad is set to a fixed temperature (45° C). In this way, the influence of baseplate and air convection heat transfer is ignored. Moreover, the ambient temperature of the whole model is fixed to 23° C in simulations. Then, an electrothermal FE analysis is performed to obtain the junction temperature. Here, eight-node linear coupled thermal-electrical elements named DC3D8E-type elements are used in the simulation. And a fine mesh is defined using elements, and the model included 578642 elements and 665535 nodes.

B. Obtaining the Junction Temperature

In the electrothermal FE analysis, the electrical current is injected at the one side of upper copper. The current flows through the upper copper layer, the chip solder, the IGBT chip, the Al metallization layer, and the bond wires and finally into the side of bottom copper, as shown in Fig. 2. The main heat is produced at the IGBT chips and then dissipates through the entire IGBT module. In this way, we can acquire the junction temperature of chip surface and each critical layer.

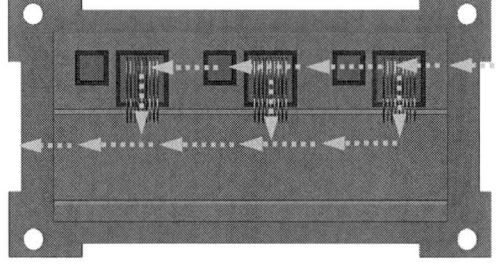

Fig. 2. Flow direction of current in FE analysis.

After all the conditions are ready, the finite element simulation is started to get the T_j of IGBT chip and each layer. The module has three chips, in order to simplify the calculation, we choose one of them as the research object. In order to make the results more accurate, we choose to obtain the temperature curves of the middle point of the chip and the corresponding layers below this point. The results of each layer are shown in Fig. 3.

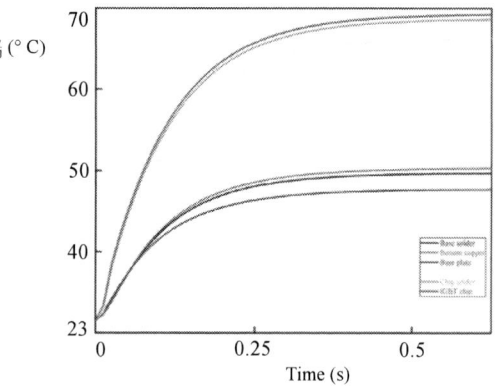

Fig. 3. The junction temperatures of each layer

III. MODELING OF THERMAL NETWORK

The common RC thermal network models mainly include Cauer model and Foster model. Among them, Foster model cannot reflect the thermal resistance and heat capacity characteristics of the real physical structure, and cannot predict the temperature of each layer structure because it ignores the internal heat transfer structure of the device. But it is widely used because it is easy to acquire the model parameters. However, the Cauer thermal network model reflects the thermal resistance and heat capacity characteristics of the real physical structure of the device, and can predict the temperature of each layer structure. The model parameters can be obtained through theoretical calculation combined with the structure size and material parameters of the device. But it is difficult to know the accurate heat transfer process inside the device in practice, so it is very difficult to validate it through experiments.

Considering that the Cauer model can directly reflect the physical nature of internal heat transfer of the device, and the influence of the structure size and material parameters of the device on its heat transfer characteristics, the paper adopts the Cauer thermal network to quickly predict the T_j of automotive IGBT modules, as shown in Fig 4. With the purpose of establishing the model, the thermal impedance curve $Z_{th}(t)$ should be obtained according to the junction temperature of each layer calculated by finite element simulation, and the thermal resistance R_{th} and heat capacity parameters C_{th} needed by the thermal network model should be obtained by fitting the curve results. In addition, the thermal coupling effects are considered, and particularly boundary conditions are also considered, including the heat dissipation conditions.

The distance and power magnitude of heat sources affect the coupling effect from other chips [5]. It can be researched that the coupling effects between cells are negligible in each half-bridge, because there is no obvious different temperature [6]. So, the study can be concentrated on a half-bridge part. In addition, thanks to symmetry, the effects of coupling can be concentrated on a pair of IGBT/diode for simplicity.

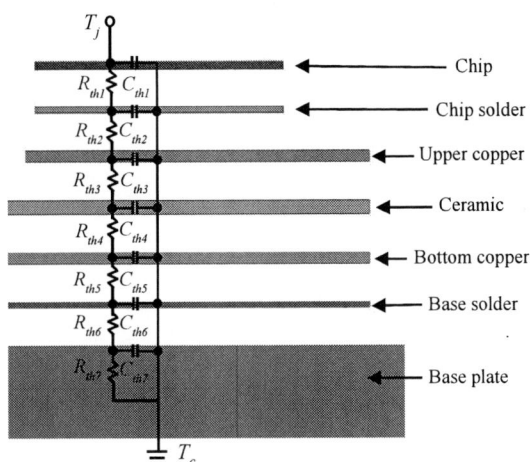

Fig. 4.   Thermal network module of automotive IGBT modules.

A. Fitting Thermal Impedance Curve

In general, the temperature rise of the IGBT modules is proportional to the power dissipation in both transient and steady state [7]. The temperature rise in steady state can be calculated using thermal resistance R_{th} between the reference point and the target point. Similarly, the temperature in dynamic state is calculated applying the transient thermal impedance curve $Z_{th}(t)$ between the target point and the reference point. The R_{th} and the $Z_{th}(t)$ equations are written as

$$R_{\text{th}(a\text{-}b)} = \frac{T_a - T_b}{P_{\text{self}}} \tag{1}$$

$$Z_{\text{th}(a\text{-}b)}^{\text{self}}(t) = \frac{T_a(t) - T_b(t)}{P_{\text{self}}} \tag{2}$$

where T_a and T_b are the temperatures in the target point and the reference point, and P_{self} is the power losses generated in the same chip [6].

Similarly, the coupling thermal impedance equations can be given by

$$Z_{\text{th}(a\text{-}b)}^{\text{couple}}(t) = \frac{T_a(t) - T_b(t)}{P_{\text{couple}}} \tag{3}$$

where P_{couple} is the couple power losses generated in the neighbor chip [6].

According to the junction temperature of each layer obtained in the FEM analysis and (2) and (3), the $Z_{th}(t)$ of each critical layer can be calculated, so as to prepare for obtaining the parameters of the RC thermal network model.

B. Acquisition of Parameters in the Thermal Network

The RC networks can obtain temperatures efficiently. This will decrease the simulation time, but the accuracy is acceptable compared with the finite-element-method. The method applied in this paper to establish the thermal network model is found on the extraction of transient $Z_{th}(t)$ from the finite-element-method and transformation of these thermal impedance curves into the equivalent RC thermal networks.

With the purpose of using Z_{th} in the temperature calculation, a sum of exponential functions is used to curve-fitted the transient thermal impedance curves mathematically, as shown in (4), so as to gain an equivalent thermal network of the thermal impedance curve.

$$Z_{\text{th}}(t) = R_{\text{th}}\left(1 - e^{\frac{-t}{R_{\text{th}}C_{\text{th}}}}\right) \tag{4}$$

where $Z_{th}(t)$ is the transient thermal impedance, R_{th} is the equivalent thermal resistance and C_{th} is the equivalent heat capacitance.

In this paper, the R_{th} and C_{th} magnitude in the thermal network are calculated mathematically in MATLAB [8]. By fitting the thermal impedance curve with (4), the parameters of thermal resistance R_{th} and heat capacity C_{th} can be obtained, as shown in table II. Similarly, the coupled R_{th} and the coupled C_{th} can also be obtained by this method, as shown in table III.

TABLE II   PARAMETERS OF CAUER NETWORK

Structure of layer	R_{th} (° C/W)	C_{th} (J/° C)
IGBT chip	7.83×10^{-3}	6.03×10^{-2}
Chip solder	1.52×10^{-2}	4.60×10^{-2}
Upper copper	4.11×10^{-3}	0.21
Ceramic substrate	8.93×10^{-2}	0.19
Bottom copper	3.99×10^{-3}	0.28
Base solder	2.20×10^{-2}	0.11
Base plate	2.12×10^{-2}	3.58

TABLE III   COUPLING THERMAL RESISTANCE AND CAPACITANCE

Coupling structure	R_{thc} (°C/W)	C_{thc} (J/°C)
Diode	8.55×10^{-2}	0.48

C. Modeling of Thermal Network

Combined with the above-mentioned Cauer thermal network model of IGBT module and the coupled thermal impedance between chips, the RC thermal network model of IGBT module considering the thermal coupling effect between chips is established as shown in Fig. 5.

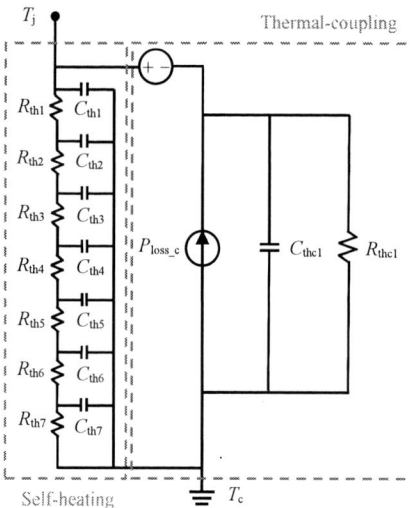

Fig. 5.   *RC* thermal network model .

Finally, A thermal network model for thermal analysis in automotive IGBT modules is established in MATLAB/Simulink. The RC thermal network model includes three parts: self-heating part, thermal-coupling part and cooling condition part. In the self-heating part, the Cauer thermal network is applied to calculate the junction temperature T_j of the chip surface. In the thermal-coupling part, the simplified coupled thermal network model is used to estimate the influence of the coupled chip on the T_j of IGBT chip, and then the actual T_j of IGBT chip surface is calculated. In addition, the cooling conditions are considered in the thermal network model, which makes the calculation of T_j accord with the real loading situation and the results are more accurate.

As shown in the Fig. 6, firstly, the power losses P_{loss} of IGBT and diode are calculated by input current I_c, as shown in (5).

$$P_{loss} = V_{ce} \cdot I_c\left(t\right) \qquad (5)$$

where P_{loss} is the power losses, V_{ce} is the collector-emitter saturation voltage drop, and I_c is the input current.

Then the junction temperature changes of self-heating part and thermal-coupling part are calculated by thermal network model.

Finally, the junction temperature T_j of IGBT chip is obtained by summation. Last but not least, because the boundary conditions are taken into account in the networks, the calculation results are more accurate.

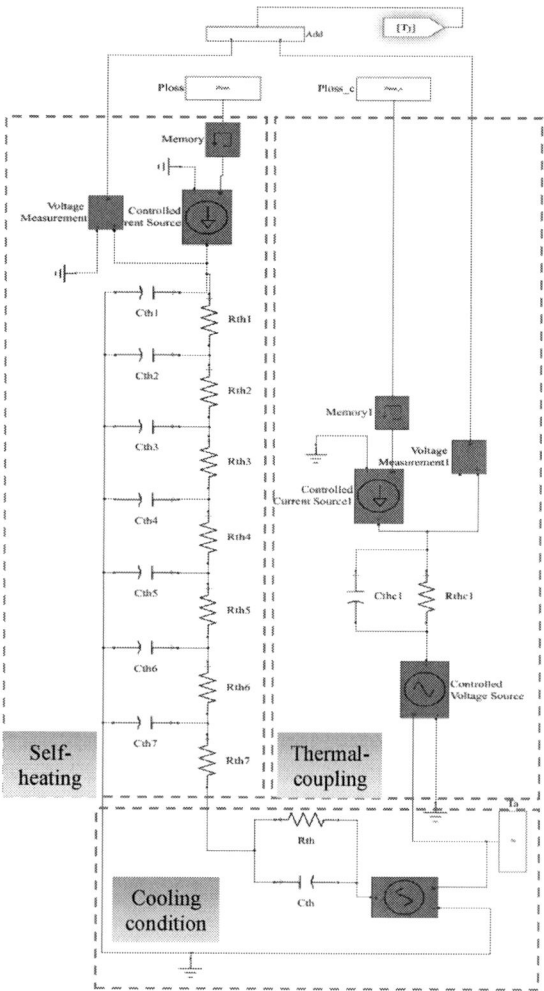

Fig. 6.   *RC* thermal network model in MATLAB/Simulink

IV.   MODEL VERIFICATION

A.   FEM Verification

As shown in Fig. 7, the calculation results of the mentioned thermal network are compared with FEM results (ABAQUS) in this section. It can be observed that there is little difference between the finite element calculation results and the results of the thermal network model, which indicates the accuracy of the presented thermal network model.

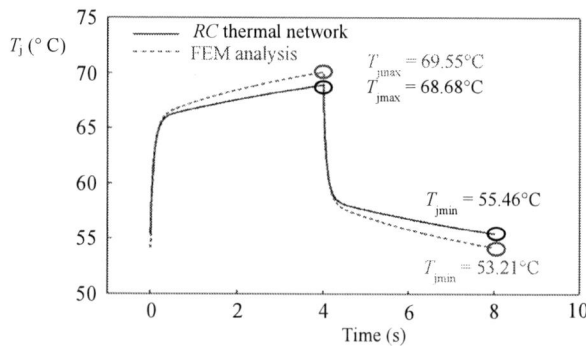

Fig. 7. T_j of IGBT chip surface calculated by thermal network and FEM

978-1-6654-1392-3/21 $31.00 © 2021 IEEE

Moreover, so as to prove the superiority of the thermal network model, the calculation time of the two methods is counted. The computational time of thermal model and FEM model is 10 seconds and 40 minutes respectively. It can be discovered that the calculation time for the *RC* thermal network is more quickly than the finite-element analysis.

B. Experimental Verification

In this section, results of the presented thermal network model are compared with the measurements from the IR camera, which is shown in Fig. 8. It can be observed that there is little difference between the experimental measurement results and the results of the thermal network model, which indicates the accuracy of the presented thermal network model.

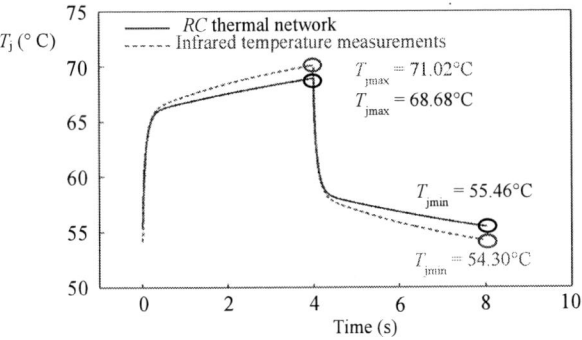

Fig. 8. T_j of IGBT chip surface calculated by thermal network and experiments

It can be discovered that the maximum junction temperature T_{jmax} of *RC* thermal network model is slightly lower than that of infrared temperature measurement, which is because *RC* thermal network model obtains the average junction temperature of chip surface, while infrared temperature measurement obtains the T_{jmax} of chip surface.

In conclusion, the presented *RC* thermal network model is validated by finite-element-method and experimental measurements. Results calculated by the *RC* thermal network model is highly consistent with the finite-element-method and shows low errors compared with the experimental measurements.

V. CONCLUSION

In this paper, a thermal network model for thermal analysis in automotive IGBT modules has been proposed. The presented thermal network includes the self-heating part and thermal coupling part among chips and critical layers. And the cooling conditions are also considered. So, the proposed model can obtain the surface temperature of the chips quickly and accurately. The *RC* parameters of thermal network are extracted by FEM simulation. By accurate curve fitting, the extracted temperature are modeled as self-heating and thermal coupling Cauer networks. Finally, the proposed thermal model is validated by finite-element-method simulation and experimental measurements. The model shows little difference compared with the experimental results and is highly consistent with the FEM simulation. Moreover, the thermal network can obtain the temperature results of the chips quickly and accurately. The fast temperature calculation can help to better estimate the reliability of the IGBT module .

ACKNOWLEDGMENT

This research was supported by the National Natural Science Foundation of China (NSFC) No. 11872078.

REFERENCES

[1] Majumdar G, and Minato T. "Recent and future IGBT evolution". IEEE Power Conversion Conference-Nagoya, vol. 07, pp. 355-359, 2007.

[2] Lutz J, Schlangenotto H, Scheuermann U, and Doncker R D. "Semiconductor Power Devices: Physics, Characteristics, Reliability". Berlin: Springer-Verlag Berlin Heidelberg, pp. 380-409, 2011.

[3] SHENG K, FINNEY S J, and WILLIAMS B W. "A new analytical IGBT model with improved electrical characteristics". IEEE Transactions on Power Electronics, vol. 14(1), pp. 98-107, 1999.

[4] BLAABJERG F, JAEGER U, and MUNK-NIELSEN S. "Power losses in PWM-VSI inverter using NPT or PT IGBT devices". IEEE Transactions on Power Electronics, vol. 10(3), pp. 358-367, 1995.

[5] A. S. Bahman, K. Ma, and F. Blaabjerg, "Thermal impedance model of high power IGBT modules considering heat coupling effects," in Proc. Int. Electron. Appl. Conf. Expo., nov. 2014, pp. 1382–1387.

[6] Bahman, A. S. , Ma, K. , Ghimire, P. , Iannuzzo, F. ,and Blaabjerg, F. "A 3D lumped thermal network model for long-term load profiles analysis in high power igbt modules". IEEE Journal of Emerging & Selected Topics in Power Electronics, pp. 1050-1063, 2016.

[7] I. Swan, A. Bryant, P. A. Mawby, T. Ueta, T. Nishijima, and K. Hamada,"A fast loss and temperature simulation method for power converters, part II: 3-D thermal model of power module," IEEE Trans. PowerElectron., vol. 27, no. 1, pp. 258–268, Jan. 2012.

[8] MATLAB Version 8.1.0.604, MathWorks Inc., Natick, MA, USA, 2013.

Novel water-soluble protective adhesive for wafer's laser dicing

Deliang Sun
Shenzhen Institute of Advanced Electronic Materials, Shenzhen Institute of Advanced Technology, Chinese Academy of Sciences
Shenzhen, Chin
dl.sun@siat.ac.cn

Jinhui Li
Shenzhen Institute of Advanced Electronic Materials, Shenzhen Institute of Advanced Technology, Chinese Academy of Sciences
Shenzhen, China
jh.li@siat.ac.cn

Yuxi Yi
Research and Development Department Shenzhen Samcien Semicoductor Materials Co., Ltd
Shenzhen, China
yyx@samcien.com

Guoping Zhang
Shenzhen Institute of Advanced Electronic Materials, Shenzhen Institute of Advanced Technology, Chinese Academy of Sciences
Shenzhen, China
gp.zhang@siat.ac.cn

Rong Sun
Shenzhen Institute of Advanced Electronic Materials, Shenzhen Institute of Advanced Technology, Chinese Academy of Sciences
Shenzhen, China
rong.sun@siat.ac.cn

Mingqi Huang
Research and Development Department Shenzhen Samcien Semicoductor Materials Co., Ltd
Shenzhen, China
hmq@samcien.com

Abstract—In the field of high brightness LED, the requirements of miniaturization, output and yield are urgent. At the same time, low k dielectric layers are more and more widely applied in integrated circuit filed, in order to solve the problem of crosstalk and stray capacitance. Laser grooving technology is developed and be seen as an ideal solution for high brightness LED wafer and low-k wafer dicing. Because the chips dicing by non-contact laser grooving technology always have fewer defects than traditional blade sawing technology, such as surface chipping, backside chipping, metal peeling, inside cracks, and delamination. In this paper, two water-soluble adhesives containing different light absorbents are used in laser dicing process. Both adhesives have excellent absorption of special wave length light, and perfect wetting ability with substrates. In the LED wafers UV laser dicing, adhesive named LGP1020 can achieve a good dicing protective efficacy and no crack formed. The low-k wafer protected by LGP2020 after green laser grooving, chips are all in perfect condition. And the protective adhesive would not to be serious carbonized which difficult to clean, even if the power of the green light laser up to 7W. Experiment results show that both adhesives have a perfect protective effect and a widely usage range of laser power.

Keywords—laser grooving, laser protective adhesive, low-k wafer, no residual

I. INTRODUCTION

Blade dicing is a traditional technology in semiconductor packaging area[1]. In the dicing process, the blade is rotated at a high speed to grind the wafer by diamond abrasive grits embedded on the edge of a rotary annular blade, until a full cut or partial cut through the wafer formation[2, 3]. However, this technology shows its shortcomings in 3D stacked IC packaging fields over the past years. Most advance packaging methods require chips with thickness of less than 100μm. As the ultra thin device wafers becoming too flexibly and fragile which tend to warp and fold[4], blade dicing could cause severe wafer damage and hence reducing the yield[5, 6].

A similar situation to ultra-thin wafers, blade dicing also has been unable to meet the requirements of low-k wafer process. In order to reduce RC delay and power loss of ICs, more and more low dielectric constant inter-layer dielectric film (low-k layer) materials have been used in wafer manufacture[7, 8]. The low-k materials are typically more brittle and have lower adhesion than silicone dioxide. Poor adhesion and thermo-mechanical properties of low-k wafer have been identified as a major cause of delamination and cracking during blade dicing[9, 10]. In addition, the mechanical strength of low-k wafer is low, which caused the topside layer of the wafer become more and more crisp[11]. Thus the peeling and topside chipping issue are easier occurred during the dicing process[12]. These failures from blade dicing may be reduced by two-step blade dicing technology but can't be completely eliminated by process optimization[13].

In addition, blade dicing requires wide cutting paths, which reduces the utilization of wafer area and increases manufacturing costs. As the size of LED chip becomes smaller, the requirement for the utilization of wafer area becomes higher. Especially for LED chips, blade cutting does not have any advantages.

Laser grooving has been regarded as the solution of ultra-thin wafer and low-k wafer dicing[14, 15]. Laser dicing has huge advantages in providing higher cutting speed, lower damage, and smaller kerf width[16]. Multi-step cutting method has been applied to low-k wafer's laser grooving. At first, dual narrow beams are applied to locate the cut position of kerf. Then dual wide beams will scan kerf to full cut the low-k layer between two narrow beams in the saw street. At last a blade is used to cut the whole street inside the grooving kerf position which will not touch the grooving edge. Then get the good performance on the topside chipping is received[17].

Although laser grooving has many advantages, various technical challenges still remain to be solved. One of the challenges is laser cutting protective adhesive which must have a good protective effect on wafers even if under high laser power ablation. This paper presents two types novel water-soluble protective adhesive using in laser dicing. These protective adhesives with good wetting ability, process operation, protection and easy cleaning properties, could improve the yield of low-k wafer grooving and LED chips laser dicing.

II. PROCESS FLOW AND EXPERIMENT

Polyvinyl alcohol, polyvinyl pyrrolidone and several functional additives were used to prepare adhesives. Those laser protective adhesives with trade name LGP1020 or LGP2020 could be purchased from Shenzhen Samcien Semicoductor Materials Co., Ltd. The difference between the two type adhesives is only that the different tiny light absorption auxiliaries using for different wave light absorption.

Relatively cheaper UV laser equipment (DSI-LC608, Han's Laser Technology Industry Group Co., Ltd, China) was used to dicing LED wafer to control the chips total cost. The laser grooving equipment (Dsi-Gv5232, Han's Laser Technology Industry Group Co., Ltd, China) with 532nm wave length green ray was used to groove low-k wafer.

A. Process flow of LED wafer dicing
The process flow of LED wafer dicing is more easier than low-k wafer. And the whole flow is shown in Fig.1.
1) Protective adhesive LGP1020 coated on low-k wafer by spinning
2) Laser dicing
3) Protective adhesive washed by water with spinning process

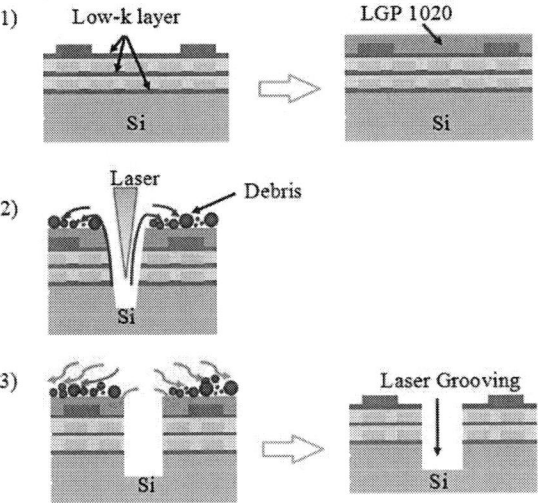

Fig.1 Process flow of LED wafer dicing

Process flow of low-k wafer dicing

The whole process is shown in Fig.2, in which:
1) Protective adhesive LGP2020 coated on low-k wafer by spinning
2) Located the dicing position using two narrow laser beams
3) Low-k material between two narrow beams traces was grooved by dual wide beams
4) Protective adhesive washed by water with spinning process
5) Silicon which under low-k layer was diced by blade method

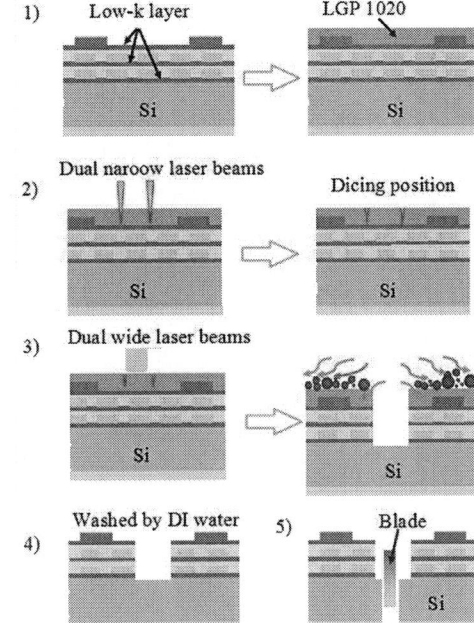

Fig.2 Process flow of low-k wafer dicing

III. RESULT AND DISCUSSION

A. Contact angle and thickness of protective adhesive

The contact Angle is an important performance index to evaluate the wetting ability of adhesives and substrates. Good wetting ability can avoids cavity informing at the interface and achieves sufficient protective effect. Further more, perfect wetting ability ensures that the adhesive forms a uniformly coating film on the substrate. As shown in Fig. 3, LGP1020 adhesive has a small contact angle with sapphire or silicon wafer. It is practicability to be used as laser dicing protective material. The test result of LGP2020 is identical with LGP1020.

The range of coating film thickness is from 160nm to 260nm at different spinning speed. The thickness of the adhesive layer can be slightly adjusted according to the crystal shape of the device, according to data shown in Fig.4.

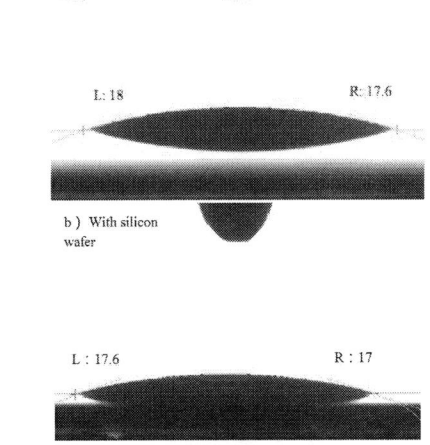

Fig.3 Contact Angle between laser protective adhesive and sapphire and silicon wafer

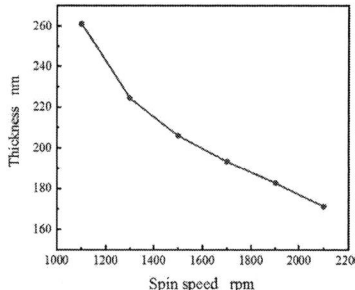

Fig.4 Relationship of spinning speed and adhesive thickness (LGP1020)

B. The protective effect of LGP1020 for UV laser dicing

As shown in Fig.5, adhesive named LGP1020 has a significant absorption peaks at 352nm. And that adhesive can absorb excess energy of UV laser. The parameter Settings of the laser equipment are wave length at 355nm, power at 3W, and dicing width at 8.3μm. The dicing speed of dual laser is 200mm/s forth moving and 350mm/s backing.

When laser energy is high enough to break down the lattice in ablation process, silicon will absorb photons of laser and start to melt and vaporize. At the same time, a plasma plume vaporized material are formed. The shock waves caused by plasma expansion and vaporized material is the driving force of molten material expulsion. Then ejected molten material accumulates at the edge of the kerf. The amount of molten substance can be reduced by optimizing laser parameters. And appropriately increasing the thickness of the protective adhesive layer is conducive to cleaning the ejected molten material.

The surface of LED after washed as shown in fig.6, is clean. No crack, peeling and other failures are found in micro-graph. The width of dicing path coincide with set value. All the evidence suggests that LGP1020 has achieved good protective effect in UV laser dicing.

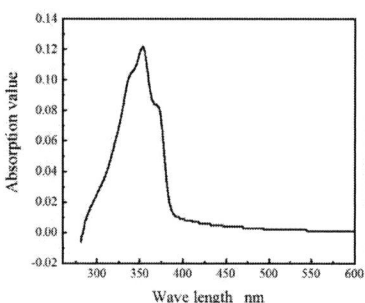

Fig.5 UV-spectra of LGP1020

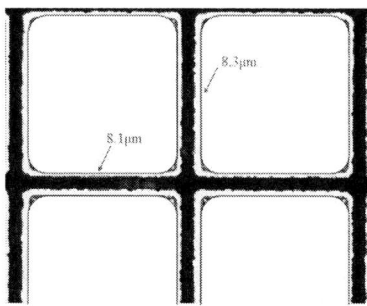

Fig.6 Micro-graph of UV laser dicing LED wafer after DI water washed

C. The protective effect of LGP2020 for green laser grooving

The visible absorption spectra of LGP2020 is shown in Fig.7. Adhesive has a wide absorption peaks at from 400 to 600nm. Low-k wafer are always very expensive, the green laser used to groove low-k material can avoid the damage caused by excessive laser energy. The parameter settings of the green laser is shown in Table.1, which using in the grooving experiment.

Table.1 Parameter settings of the green laser grooving

Parameter		Dual narrow laser beams	Dual wide laser beam
Frequency	KHz	1000	800
Power	W	5	5
Speed	mm/s	220	200

After laser grooving, lots of debris exist on the surface of low-k wafer. But after water washed, the surface is clean as shown in Fig.7 b. laser kerfs are not only no crack failure, but also orderliness. LGP2020 has a good protection for low-k material in green laser grooving. And even if laser power up to 7W, protective adhesive is not to be carbonized which could contaminates wafer surface.

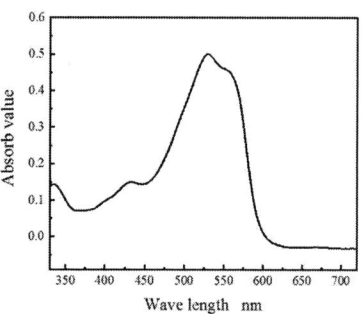

Fig.7 Visible absorption spectra of LGP2020

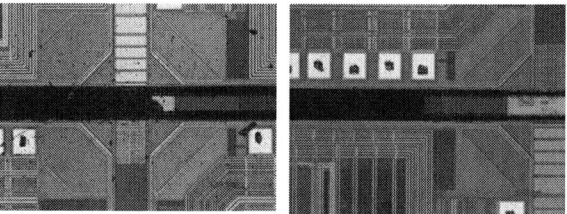

Fig.7 Micro-graph of green laser grooving low-k wafer, a: before washed; b: after washed

IV. CONCLUTION

The protective adhesives for laser dicing have been studied and investigated successfully. The following results were obtained:

1) Both adhesives have good performance in spin coating process and wetting ability.

2) The laser protective adhesive named LGP1020 can be used for LED wafer dicing, to increase the output and yield. This adhesive have a good protective effect in UV laser dicing.

3) The adhesive named LGP2020 used for green laser dicing, is excellent when it be used in low-k wafer's

grooving. At a widely range of laser power, no residual remains after cleaning.

ACKNOWLEDGMENT

Thanks are due to National Natural Science Foundation of China (61904191), Youth Innovation Promotion Association of Chinese Academy of Sciences (2017410), Key R&D Project of Guangdong Province (2020B010180001) and National Key R&D Project from Minister of Science and Technology of China (2017ZX02519).

REFERENCES

[1] Efrat U . Optimizing the wafer dicing process. International Electronic Manufacturing Technology Symposium. IEEE, 1993..

[2] Kroninger, W. and F. Mariani, Thinning and singulation of silicon: Root causes of the damage in thin chips. Vol. 2006. 2006. 6 pp.

[3] Lei, W., A. Kumar, and R. Yalamanchili, Die singulation technologies for advanced packaging: A critical review. Journal of Vacuum Science & Technology B Microelectronics and Nanometer Structures, 2012. 30: p. 040801-1~040801.

[4] Jouve, A., et al., Facilitating Ultrathin Wafer Handling for TSV Processing. 2009. 45-50.

[5] Ryningen, B., et al., Capillary forces as a limiting factor for sawing of ultrathin silicon wafers by diamond multi-wire saw. Engineering Science and Technology, an International Journal, 2020. 23.

[6] Sudani, N., V. Krishnan, and B. Tan, Laser singulation of thin wafer: Die strength and surface roughness analysis of 80μm silicon dice. Optics and Lasers in Engineering, 2009. 47: p. 850-854.

[7] Shamiryan, D., et al., Low-K dielectric materials. Materials Today, 2004. 7: p. 34-39.

[8] Liu, F., et al., Advances in High Performance RDL Technologies for Enabling IO Density of 500 IOs/mm/layer and 8-μm IO Pitch Using Low-k Dielectrics. 2020. 1132-1139.

[9] Hussein, M.A. and H. Jun, Materials' impact on interconnect process technology and reliability. IEEE Transactions on Semiconductor Manufacturing, 2005. 18(1): p. 69-85.

[10] Mercado, L.L., et al., Impact of flip-chip packaging on copper/low-k structures. IEEE Transactions on Advanced Packaging, 2003. 26(4): p. 433-440.

[11] Luo, S. and Z. Wang, Studies of chipping mechanisms for dicing silicon wafers. International Journal of Advanced Manufacturing Technology, 2008. 35: p. 1206-1218.

[12] Liu, H., et al. Investigation of single cut process in mechanical dicing for thick metal wafer. in 2016 17th International Conference on Electronic Packaging Technology (ICEPT). 2016.

[13] Dong, Z. and Y. Lin, Ultra-thin wafer technology and applications: A review. Materials Science in Semiconductor Processing, 2020. 105: p. 104681.

[14] Lau Teck, B., et al. Laser grooving process development for low-k / ultra low-k devices. in 2008 33rd IEEE/CPMT International Electronics Manufacturing Technology Conference (IEMT). 2008.

[15] Borkulo, J., R. Hendriks, and P. Dijkstra, Comparison between Single & Multi Beam Laser Grooving of Low-K layers. International Symposium on Microelectronics, 2012. 2012: p. 000433-000439.

[16] Bovatsek, J. and R. Patel, Highest-speed dicing of thin silicon wafers with nanosecond-pulse 355nm q-switched laser source using line-focus fluence optimization technique. Proceedings of SPIE - The International Society for Optical Engineering, 2010. 7585: p. 19.

[17] Koh, W.S., et al., Multi beam laser grooving process parameter development and die strength characterization for 40nm node low-K/ULK wafer. 2014. 1-8.

A TGV-based Antenna in Package for 5G mm-Wave Application

Sha Xu

School of Information Engineering
Guangdong University of Technology
Guangzhou, China

sally.xu@gdut.edu.cn

Dianyang Shi

China Electronic Product Reliability and
Environmental Testing Research Institute
Guangzhou, China

18520662706@163.com

Chunbing Guo*

School of Information Engineering
Guangdong University of Technology
Guangzhou, China

cbguo@gdut.edu.cn

Abstract —Antenna in Package (AiP) relies on 3D packaging technology, which greatly shortens the feed line length, thereby reducing interconnection loss and improving system power efficiency. This paper briefly evaluated the performance of glass substrate as the 5G mm-wave packaging and then discusses the losses of 3 methods of the feed line. A TGV-based patch antenna is simulated at 28 GHz. The CPW feeding line between package and antenna is TGV. The TGV structure can reduce the interconnection loss and provide low dielectric loss. The proposed TGV-based patch antenna has the advantages of simple structure, bandwidth, and low loss, which can be applied for 5G communication.

Keywords—through-glass-via(TGV), Antenna-in-Pacakge (AiP), millimeter-wave

I. INTRODUCTION

In recent years, new technologies such as 5G communications, cloud computing, artificial intelligence, and the Internet of Things have pushed up the demand for computing power, transmission rate and transmission bandwidth. Due to the advantages of high speed, high capacity, high precision, and rich spectrum resources, millimeter-wave communication is the basis for the development of the above technologies. As the frequency increases to the millimeter-wave frequency band, a series of new problems are rising due to the limitations of the material properties and physical rules. For example, the delay of chip interconnection has exceeded the gate delay. The loss of chip and package interconnection has become more and more prominent and has even become a bottleneck restricting the performance of the chip[1]. With the increase of frequency, the separated packaging of antenna and RF chip is facing challenges of excessive interconnection loss and low integration density, which result in degraded system performance and difficulty in achieving large-scale system integration for future communications. Therefore, the antenna and RF front-end co-design and co-packaging, in form of AoC (Antenna-on-Chip), AiP (Antenna-in-Package), or a hybrid manner, is widely regarded as a viable solution for communication systems at millimeter-wave and beyond. Compared with AoC, the AiP solution exhibits excellent trade-offs among process difficulty, antenna performance, and cost, and it can be flexibly integrated into the wireless system via heterogeneous integration. AiP relies on 3D packaging technology, which greatly shortens the feed line length, thereby reducing interconnection loss and improving system power efficiency. AiP is being widely used in millimeter-wave applications such as 5G communications, radar, imaging, and detection, etc[2] – [4].

There are several approaches to the millimeter-wave AiP packaging. One is to stack and integrate antenna and package vertically. The antennas are vertically stacked on the chip package, which greatly shortens the feed length, thereby reducing transmission loss and size. The other is to use packaging materials with lower dielectric loss to replace existing ceramics, organic materials, and silicon. For example, glass is a promising candidate due to its advantages of insulation, ultra-thinness, high rigidity, high stability, and adjustable thermal expansion coefficient. In particular, the millimeter-wave electromagnetic properties of glass are much better than silicon and can achieve ultra-low dielectric loss. Compared with organic materials, glass can use finer design rules to realize precision circuits, with lower loss, ultra-thinness, and stability. In addition, the glass package can be produced with a large panel and does not require the deposition of an insulating layer, so the cost is lower. When the thickness is below 100 µm, the glass substrate can also be made into a flexible material, which is very suitable for making wearable devices.

The choice of packaging method and feeding structure is crucial. Through-glass-via (TGV) is more suitable for the application of glass AiP in the millimeter-wave band[5] – [9]. In this work, millimeter-wave packaging technology based on TGV is studied to prevent excessive transmission loss. In terms of TGV-based transmission line modeling, a three-dimensional electromagnetic model of TGV is constructed, using the finite element simulation method to analyze the loss caused by the incorporated TGV structure. In terms of design and application, a TGV-based patch antenna is designed and investigated. Simulations are performed using HFSS software.

II. TGV-BASED TRANSMISSION LINE

As mentioned in the introduction, glass material has a superior microwave and RF performance, which is suitable for the RF chip packaging, especially in higher frequency, such as millimeter-wave range. The conventional materials to form the RF module includes LTCC, LCP, and FR4. Glass-based RF packaging has been investigated in recent years, because of its advantages such as low loss, dimensional stability for precision geometries, and the ability to form low-loss

978-1-6654-1392-3/21 $31.00 © 2021 IEEE

through-vias. When TGV is used as the interconnection of a three-dimensional package or package antenna, it usually appears in the form of "ground-signal-ground"(GSG) three ports or the form of TGV "signal-shielding array". This type of structure has great high-frequency advantages over a single TGV structure. But this further led to an increase in the number of TGVs. Therefore, the electromagnetic simulation results in the accuracy of TGV-based GSG transmission line package design in practical applications. For feeding structures, a Coplanar waveguide (CPW)is an effective approach for high-frequency applications. CPW transmission line has superior performance and convenient processing, which is playing an increasingly important role in MMIC circuits. Especially in the millimeter-wave frequency band, CPW has better performance than microstrip lines. Compared with the conventional microstrip transmission line, the CPW can realize the serial and parallel connection of passive and active devices in the MMIC and can increase the circuit density, which has greater flexibility when connected to various devices. We simulated the CPW transmission line, and compare the mm-wave property of glass and FR4 as packaging substrates.

For our work, a straight-through transmission line and a transmission line with a TGV structure are designed respectively. The simulated S11 and S21 are compared, to investigate the loss of the incorporated TGV structure. For the straight-through transmission line, the CPW transmission line is chosen, since it has better electromagnetic performance under the mm-wave frequency range. The designed working frequency is 28GHz.

A straight-through CPW transmission line is studied, the thickness of the glass substrate is 200 μm, with a tan δ is 0.0037, and a dielectric constant of 4.6. Figure 1 shows the CPW transmission line composed of a metal strip on the top surface of the glass substrate as the signal channel and two metal sheets as the ground. There is a symmetrical gap between the signal line and the ground. Two wave ports, which is a suitable excitation method, are set to the ports of the CPW.

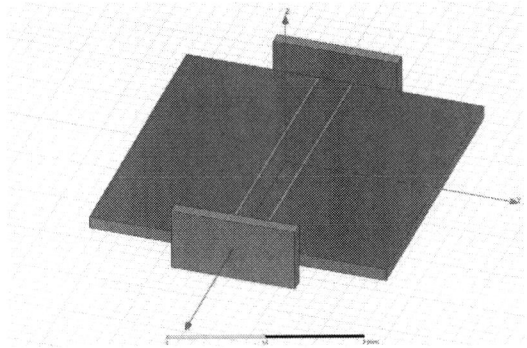

Figure 1. The simulated CPW transmission line

Firstly, the S11 of the straight-through transmission line is obtained and evaluated. The high-frequency transmission performance of the glass substrate-based

CPW transmission line, the S11 curves are shown in Figure 2. It can be seen from Figure 2 that the performance of the return loss of the millimeter-wave (10 ~ 40GHz) is smaller than -18.56 dB. At the same time, the S11 curve in Figure 2 has a peak value of 37 GHz, which is a resonance point is generated near 37 GHz and form a relatively obvious peak.

Secondly, the S21 of the straight-through transmission line is obtained and evaluated. The S21 of the straight-through CPW transmission line is shown in Figure 2. In the 10 ~ 40GHz working frequency band, it can be seen from the simulation results that the insertion loss S21 of straight-through CPW transmission line in the test frequency band of 10 ~ 40GHz.

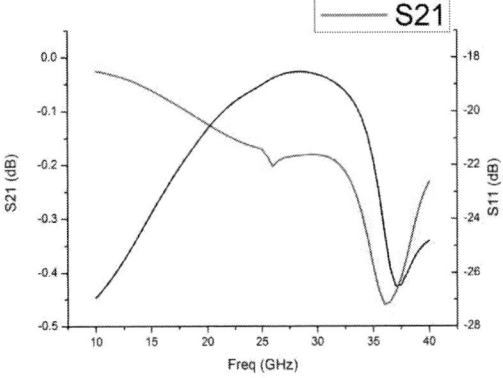

Figure 2. Simulated insertion loss and return loss of glass CPW transmission line

To compare with the straight-through CPW transmission line, the CPW with TGV structure is simulated. The schematic structure is shown in Figure 3. The CPW transmission line is deposited on the top and bottom surface of the glass substrate, and the transmission line is connected by TGV as the vertical interconnects.

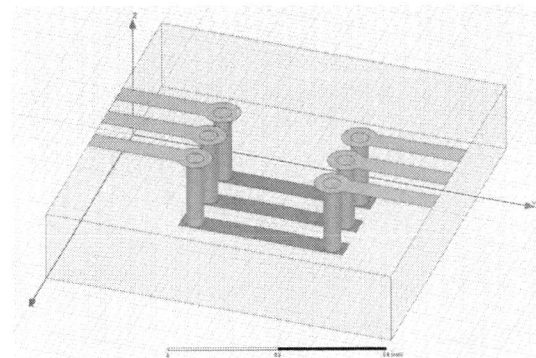

Figure 3. The simulated CPW with TGV as vertical interconnects

Firstly, the S11 of the straight-through transmission line is obtained and evaluated. The high-frequency transmission performance of the glass substrate-based CPW transmission line, the S11 curves are shown in figure 4. It can be seen from Figure 4 that the performance of the return loss of the millimeter-wave (10 ~ 40GHz) is smaller than -17.10dB. At the same time, the S11 curve has better performance in the lower working frequency. Secondly, the insertion loss of the CPW transmission line with TGV is obtained and evaluated. The S21 of the straight-through CPW transmission line is shown in Figure 4. In the 10 ~ 40GHz working frequency band, it can be seen from the simulation results that the insertion loss S21 of straight-through CPW transmission line in the test frequency band of 10 ~ 40GHz.

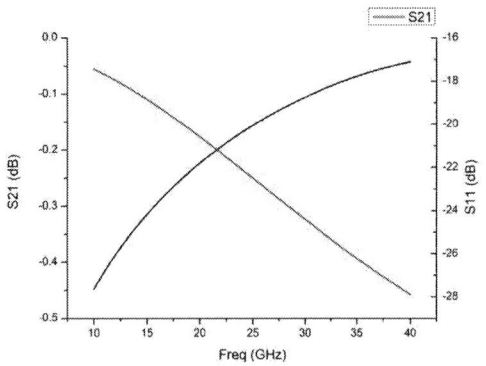

Figure 4. Simulated insertion loss and return loss of glass CPW transmission line with TGV

To compare with the straight-through transmission line, the Grounded CPW (GCPW) with TGV structure is also simulated. The design is shown in Figure 5. The CPW transmission line is deposited on the top and bottom surface of the glass substrate, and the transmission line is connected by TGV as the vertical interconnects. The two ground lines are connected as a Co-grounded structure.

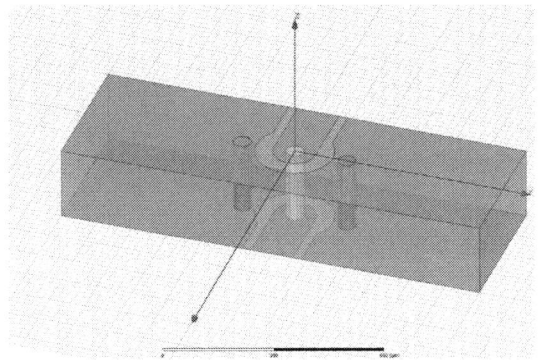

Figure 5. The simulated GCPW transmission line with TGV as vertical interconnects

Firstly, the S11 of the straight-through transmission line is obtained and evaluated. The high-frequency transmission performance of the glass substrate-based CPW transmission line, the S11 curves are shown in Figure 6. It can be seen from Figure 6 that the performance of the return loss of the millimeter-wave (10 ~ 40GHz) is smaller than -28.54dB. At the same time, the

return loss curve has better performance in the lower working frequency. Secondly, the S21 of the CPW transmission line with TGV is obtained and evaluated. The S21 curves of the straight-through CPW transmission line is shown in Figure 6. In the 10 ~ 40GHz working frequency band, it can be seen from the simulation results that the insertion loss S21 of straight-through CPW transmission line in the test frequency band of 10 ~ 40GHz.

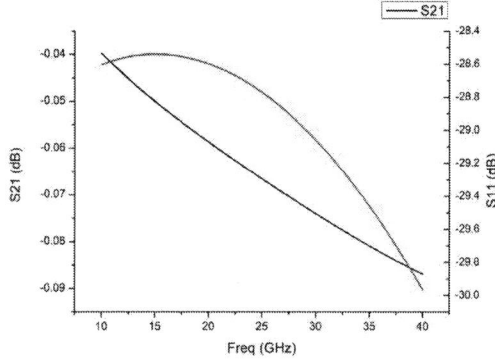

Figure 6. Simulated insertion loss and return loss of glass GCPW transmission line with TGV

In addition, the dielectric loss of the glass substrate can be further analyzed according to the insertion loss value through the transmission line, and the loss of the which is caused by the dielectric substrate generate a dielectric loss, including the loss tangent of the glass substrate, skin effect of conductors, conductivity, etc.

III. TGV-BASED PATCH ANTENNA

Figure 7 demonstrates a package on package integration solution of antenna and package. The basic structure of the AiP proposed in this paper consists of three parts. The upper surface of the glass substrate is the antenna, the patch antenna is distributed on the upper glass substrate. The middle part is another glass substrate for fan-out packaging. One or more cavities can be realized in the substrate by etching, which can be used for chip placing. The bottom of the glass substrate is RDL, and the chip is connected to the next-level carrier board through RDL traces. The lower surface is the grounded metal surface. The CPW transmission line with TGVs is connected between the patch antenna and the package to excite the antenna. The feed part of the antenna is interconnected with the chip through TGVs.

Figure 7. package on package AiP solution

978-1-6654-1392-3/21 $31.00 © 2021 IEEE

Figure 8 shows the schematic of the TGV-based patch antenna for AiP application, including the antenna on the top surface of the glass substrate and the TGV vertically go through the glass substrate. By drilling TGVs in the glass substrate, a millimeter-wave patch antenna can be realized, which features a simple structure and low transmission loss. We use a glass-based packaging process to achieve the antenna module, and introduce TGVs into the transmission line to improve the packaging density. The excitation amplitude distribution of the antenna will be determined, according to the antenna radiation requirements. Then calculate and determine the offset of the TGV feed point. It is necessary to fine-tune the key parameters of the package, such as the TGV spacing, cavity size, glass substrate height, etc. The designed TGV-based patch antenna will be used in the AiP design.

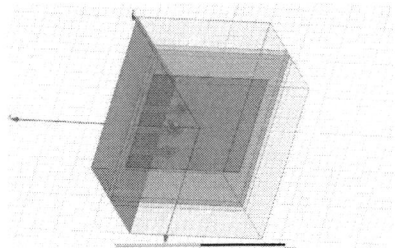

Figure 8. proposed TGV-based patch antenna for AiP application

Fig 9 is the simulated S11 of the TGV-based patch antenna, and peak return losses smaller than -29dB within the scanning frequencies. The resonant radiation frequency is 25.12 GHz, which is attributed to the reflection of the vertical structure. In addition, affected by the vertical transition of the two TGVs, the impedance matching is slightly poor, and there are further optimizations.

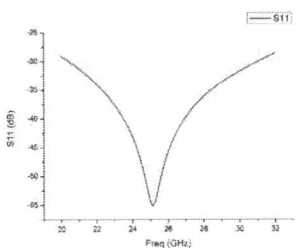

Figure 9 The simulated return loss of TGV-based patch antenna

IV. CONCLUSION

This paper briefly evaluated the performance of glass substrate as the 5G mm-wave packaging and then discusses the losses of 3 methods of the feed line. A TGV-based patch antenna is simulated at 28 GHz. The CPW feeding line between package and antenna is TGV. The TGV structure can reduce the interconnection loss and provide low dielectric loss.

ACKNOWLEDGMENT

This work is funded by Key-Area Research and Development Program of Guangdong Province under Grant 2018B010115001 and by the Guangdong Provincial Key Laboratory of Millimeter-Wave and Terahertz opening fund, No. 2019B030301002KF2003.

REFERENCES

[1] V. Sukumaran "Through-Package-Via Hole Formation, Metallization, and Characterization for Ultra-Thin 3D Glass Interposer Packages," no. August 2014.

[2] Y. K. Yoon, S. Hwangbo, A. Rahimi, S. P. Fang, and T. Schumann, " Glass interposer integrated millimeter-wave antennas for inter-/intra chip communications, " 2015 Int. Work. Antenna Technol. iWAT 2015, pp. 62-64, 2015, doi: 10.1109/IWAT.2015.7365312.

[3] S. Hwangbo, A. Rahimi, C. Kim, H. Y. Yang, and Y. K. Yoon, " Through Glass Via (TGV) disc loaded monopole antennas for millimeter-wave wireless interposer communication, " Proc. - Electron. Components Technol. Conf., vol. 2015-July, pp. 999-1004, 2015, DOI: 10.1109/ECTC.2015.7159717.

[4] S. Hwangbo, Y. K. Yoon, and A. B. Shorey, " Millimeter-wave wireless chip-to-chip (C2C) communications in 3D system-in-packaging (SiP) using compact through glass via (TGV) - Integrated Antennas," Proc. - Electron. Components Technol. Conf., vol. 2018-May, pp. 2074 - 2079, 2018, DOI: 10.1109/ECTC.2018.00311

[5] T. Yu, et al., " Development of Embedded Glass Wafer Fan-Out Package with 2D Antenna Arrays for 77GHz Millimeter-wave Chip," Proc. - Electron. Components Technol. Conf., vol. 2020-June, pp. 31-36, 2020.

[6] H. Cai et al., " Design, fabrication, and radiofrequency property evaluation of a through-glass-via interposer for 2.5D radio frequency integration, " J. Micromechanics Microengineering, vol. 29, no. 7, 2019.

[7] A. O. Watanabe et al., " Glass-Based IC-Embedded Antenna-Integrated Packages for 28-GHz High-Speed Data Communications, " in Proceedings - Electronic Components and Technology Conference, 2020, vol. 2020-June, pp. 89-94.

[8] X. Zhao, et al., " High Aspect Ratio TGV Fabrication Using Photosensitive Glass Substrate," 2018 14th IEEE Int. Conf. Solid-State Integr. Circuit Technol. ICSICT 2018 - Proc., vol. 4, pp. 1 - 3, 2018.

Loading rate on mode II fracture toughness of sintered silver

Yanning LI
Institute of Electronics Packaging Technology and Reliability
Faculty of Materials and Manufacturing
Beijing University of Technology
Beijing,China
liyanning@emails.bjut.edu.cn

Yanwei DAI*
Institute of Electronics Packaging Technology and Reliability
Faculty of Materials and Manufacturing
Beijing University of Technology
Beijing,China
ywdai@bjut.edu.cn

Fei QIN
Institute of Electronics Packaging Technology and Reliability
Faculty of Materials and Manufacturing
Beijing University of Technology
Beijing,China
qfei@bjut.edu.cn

Shuai ZHAO
Institute of Electronics Packaging Technology and Reliability
Faculty of Materials and Manufacturing
Beijing University of Technology
Beijing,China
zhaoshuai@emails.bjut.edu.cn

Abstract—The failure mechanism of sintered Ag/Cu interface was investigated based on ENF test, so as to study the change law of mode II fracture toughness of sintered Ag/Cu interface at different low strain rates. The sandwich structure ENF specimens were prepared with 100 μm-thick nano-silver paste and two pieces of bare copper. The sandwich structure was sintered at a sintering temperature of 280°C, held for 60 minutes, and at a heating rate of 10°C per minute with a pressure of 10 KPa. The ENF specimens were tested on the Instron 5948 micro tester with a three point bending load applied four different strain rates at room temperature. The Hitachi S4000 SEM and optical microscopy (OM) were used to analyze the fracture morphology and microstructure of sintered silver layer after ENF test. Cracks all crack from the sintered Ag/Cu interface at different strain rates. The compliance based beam method (CBBM) was used by researchers for fracture analysis of ENF specimens. Under the certain range of quasi-static loading conditions, as the strain rate increases, the fracture toughness of sintered Ag/Cu interface decreases.

Keywords—nano-silver paste, strain rate, mode II fracture toughness, ENF test

I. Introduction

Wideband gap semiconductor materials, such as SiC GaN, have excellent electronic, physical, chemical and mechanical properties and can be used in high temperature and high power environments. The development trend of wideband gap semiconductor devices is the continuous miniaturization and integration of the volume, and its performance has been rapidly improved. While realizing the progress of science and technology, it facilitates the daily life of human beings and promotes the growth of the world economy. The performance and reliability of electronic systems are closely related to the field of electronic packaging. Chip connection material, as a packaging material that has a primary interconnection with the chip, is the main channel for the conduction and heat dissipation of the chip. Therefore, the development and research of the new type of connection material that can be applied to high-power and high-temperature devices are urgent. Nano-silver paste has the characteristics of high melting point (961°C), excellent electrical and thermal conductivity,which could be operated at temperatures over 500°C, enough to bond most high-power semiconductor equipment. It has become a new type of chip interconnection material for high-density packaging instead of traditional solder alloy and conductive adhesive. As solder joints in electronic devices, the bonding layer materials usually bear shear loads during service. The mechanical properties of sintered silver layer as solder joint play a very important role in determining the reliability of the whole integrated circuit.

Xin Li [1] designed a fixture to test the mechanical properties of Ag nanopaste joint, the mechanical properties of the joints at room temperature and high temperature have carried on the comprehensive test and theoretical study. Isothermal cyclic shear tests were carried out on lap joints by means of stress or strain control, and the effects of average stress, stress amplitude and ambient temperature on the reliability of the joints were investigated. In addition, many scholars have studied the sintering process and properties of nano-sized silver paste.Sintered silver is a kind of porous material. They believe that changing the sintering temperature, heating rate and sintering holding time can change the microstructure and density of sintered silver, thus affecting the porosity.The porosity has a significant effect on the thermal, electrical and mechanical properties of silver.

Since the thermal expansion coefficient of each module of electronic devices is different, the main force that the equipment receives in the process of operation is the shear force, so the bonding strength and sintering quality of the joint interface are usually judged by shear experiment. Tensile tests are are often used to characterize the resistance to elongation and fracture properties of materials. But the mode II fracture toughness has not been studied much. Under the operation condition, the power device is usually subjected to different strain rate loading, which will cause the fracture failure between sintered silver layer and copper layer on DBC substrate [2]. To solve this problem, the reliability of sintered Ag/Cu interface in the sintered Ag /DBC structure of high-temperature power SiC device is studied in this work. The fracture properties of sintered Ag/Cu interface under different strain rates under the same sintering process parameters are measured by theoretical and experimental means. The fracture morphology and microstructure of the sintered silver joint were observed and analyzed to find out the failure mechanism, and the influence law of different strain rates on the fracture toughness of sintered silver/copper interface was found.

II. EXPERIMENTAL PROCEDURE

The ENF test was used in this work to perform the mode II fracture toughness [3]. The ENF sample was composed of two pieces of bare copper and a thin layer of Ag nanopaste, which shows a sandwich structure, as shown in Fig. 1. The model of Ag nanopaste is LOCTITE ABLESTIK SSP 2020. The copper sheet dimensions are 100 mm long, 8 mm wide and 1.2 mm thick. First of all, the surface of the copper sheet was polished with 2000-mesh sandpaper to remove the oxide layer, wipe clean the surface impurities with dust-free cloth, and then the surface of the copper sheet is polished with 1μm single-crystal diamond water-based suspended substance. Then the copper substrate was cleaned by ultrasonic vibration in acetone and ethanol in turn for 15 minutesand. Finally, the cleaned copper sheet is placed on a dust-free cloth for natural air drying. The SUS304 stainless steel mask with a thickness of 100 μm is prepared to cover a piece of copper substrate, and the silver paste is evenly printed with a blade. The opening size of the metal mask is 60mm×8mm. Finally put another piece of copper over it.

After bonding, all the specimens were sintered at 280°C for 60 min with a pressure of 10 KPa, the heating rate was 10°C/min. The heating equipment is a nitrogen lead-free reflow over T200N heating stage. Its temperature setting system is controlled by computer. The temperature can be adjusted from room temperature to 360°C, and the temperature accuracy is ±2°C.

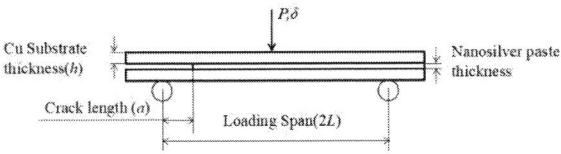

Fig. 1. Schematic diagram of ENF specimen structure and three-point bending experimental loading

The Instron 5948 micro tester is used to test the mode II fracture toughness. It can realize high precision and ultra-high precision micro tensile, compression, deflection, low period fatigue test of materials under quasi-static and dynamic conditions. For example, tensile and shear tests can be performed on solder joints on chips and IC boards. The displacement range of the fixture head is 0.01 mm ~ 100 mm, the load range is 2 mN ~ 2 KN, and the loading rate range is 0.00001 mm/min to 1500 mm/min. Select an appropriate initial height and displacement rate for the cross head at the beginning of use at room temperature. Three-point bending test was performed on the sample, as shown in Fig. 2. The rate of downward movement of the cross head is 5, 2, 0.2 and 0.05 mm/min. Those corresponding values of strain rates are $1.875 \times 10^{-4} s^{-1}$, $0.75 \times 10^{-4} s^{-1}$, $0.75 \times 10^{-5} s^{-1}$, and $1.875 \times 10^{-6} s^{-1}$, respectively, which can be calculated as below [4].

$$\dot{\varepsilon} = \frac{d\varepsilon_f}{d_t} = \frac{6 \times V_T \times 2h}{(2L)^2} \quad (1)$$

In this equation, the loading span (2L=80mm) can be ensured by adjusting the position of the three-point bending test fixture, ε_f is the strain at the center of the lower surface of the copper sheet, t is the loading time, V_T is the velocity of the cross-head, $2h$ is the thickness of the whole specimen. and the initial crack length of ENF specimen can be determined as 20mm by screw micrometer.

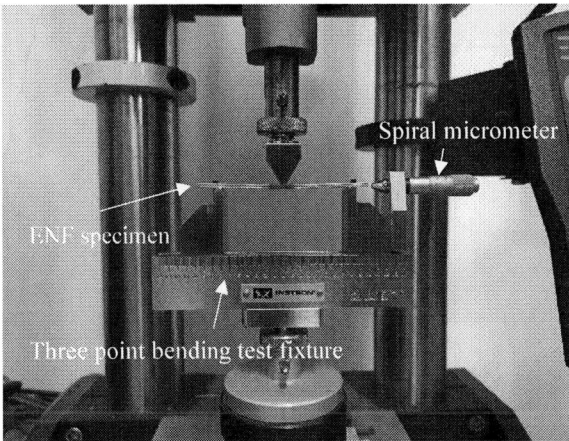

Fig. 2. Set-up for end-notched flexure (ENF) fracture tests by Instron 5948.

Combined with the matching Bluehill software, the loading displacement and loading level of the three-point bending movement were recorded. The software can accurately obtain the maximum load, displacement and other relevant data of the specimen and the corresponding curve, and draw the load-displacement diagram of each specimen. The compliance-based beam method (CBBM) was used in this paper for fracture analysis of ENF specimens. The mode II fracture toughness can be calculated as following [5].

$$G_{II} = \frac{9P^2 C_{0corr}}{2B(3a_0^3 + 2L^3)} \left[\frac{C_{corr}}{C_{0corr}} a_0^3 + \frac{2}{3} \left(\frac{C_{corr}}{C_{0corr}} - 1 \right) L^3 \right]^{\frac{2}{3}} \quad (2)$$

In this equation P is the load applied to the ENF sample, B represents sample width, a_0 represents the initial crack length, C_{corr} is related to the load and displacement values at the vertex with the highest load in the load-displacement diagram. C_{0corr} is related to the slope fitted by the first ascending curve in the load-displacement diagram, L is the half-span length of ENF specimen.

After the sintering and three-point bending test, the sample is inlaid with Goral's cold mounting resin (epoxy binder). The curing time of the resin is 8 hour. First, the epoxy resin layer was ground off with 240 mesh coarse sandpaper to expose the section of ENF sample. Then, the section was finely ground with 400 mesh, 800 mesh, 1500 mesh and 2000 mesh sandpaper successively. Finally, the section was polished with 1μm single-crystal diamond water-based suspended substance to eliminate the scratches on the sample surface, so that the microstructure of sintered silver layer could be observed more accurately. The fracture morphology and microstructure of sintered silver layer were observed by a scanning electron microscope (SEM) and an optical microscopy (OM).

III. RESULTS AND DISCUSSION

A. Effect of strain rate on mode II fracture toughness of sintered Ag/Cu interface

According to the experimental data of three-point bending recorded by Bluehill software, the load-displacement curve as shown in Fig. 3 is drawn. Fig. 3(a) shows the load-displacement diagrams of four samples at the horizontal strain rate of $0.75 \times 10^{-5} s^{-1}$, and Fig. 3(b) shows the load-displacement curve comparison of samples under four different strain rates. The critical loads for the ENF specimens with different strain rate are quite different where the maximum load is obtained under the minimum strain rate loading, up to 78.15N, and the minimum load corresponds to the maximum strain rate, which is 54.03N.

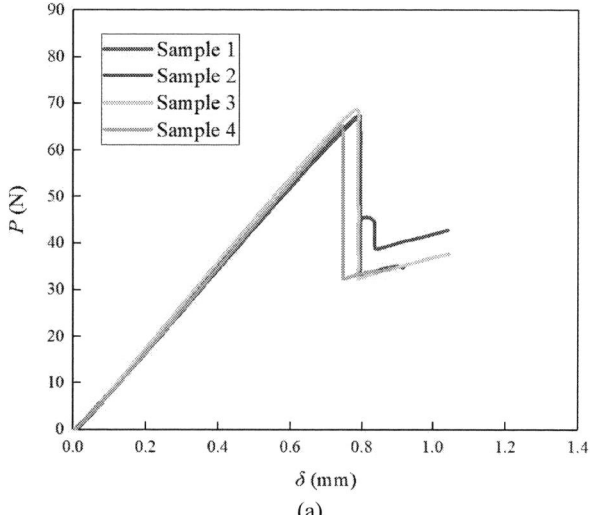

(a)

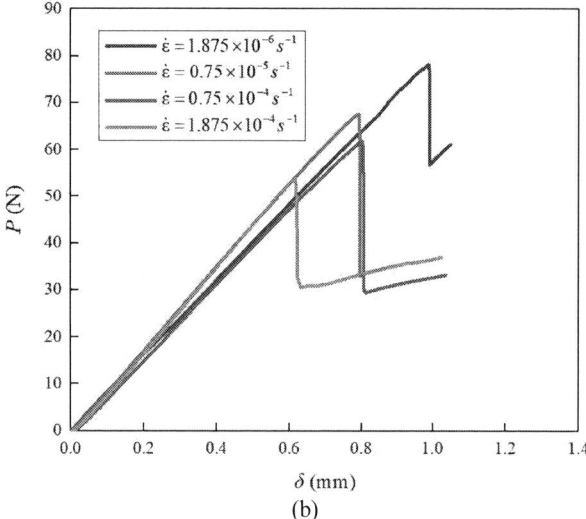

(b)

Fig. 3. (a) P-δ curves for the ENF specimens at a strain rate of $0.75 \times 10^{-5} s^{-1}$, (b) P-δ curves comparison under four different strain rates

In the initial stage of all P-δ curves, a uniform linear increase trend was observed, indicating that elastic deformation occurred at the beginning of the bond between the copper sheet and the Ag nanopaste of ENF specimen. The initial slope of the load-displacement curve is roughly the same. When the maximum loading was reached, the curve dropped sharply, and the cracks in the adhesive layer began to expand. Then the load of the ENF specimen began

to increase when the load dropped to a certain value. The deformation of the substrate recovered completely after the ENF specimen was completely unloaded, indicating that no plastic deformation occurred in the substrate during the three-point bending test.

In the three-point bending test, the crack propagation process is very small, and it is difficult to measure under high pressure, and it is easy to produce large test error, resulting in low fracture toughness value. Therefore, when analyzing the original P-δ curve, the CBBM method is chosen. Meanwhile, it does not need to consider the process of crack growth, and the experimental error has little influence on the data results. Fig. 4 shows the influence of four strain rates on the mode II fracture toughness. It represents the mean and variance of fracture toughness of three test samples for each strain rate. It can be seen that the specimens with minimum strain rate show a highest average G_{IIC} of 0.1174 ± 0.027 N/mm, whereas the average G_{IIC} of specimens with maximum strain rate only as low as 0.0472 ± 0.00185 N/mm.

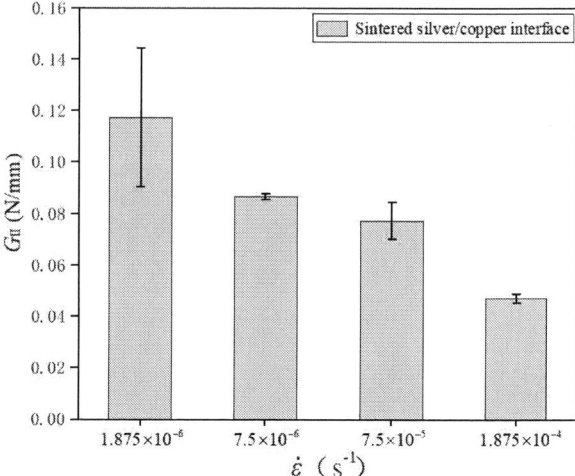

Fig. 4. Influence of four strain rates on the mode II fracture toughness

B. Effect of strain rate on fracture mode

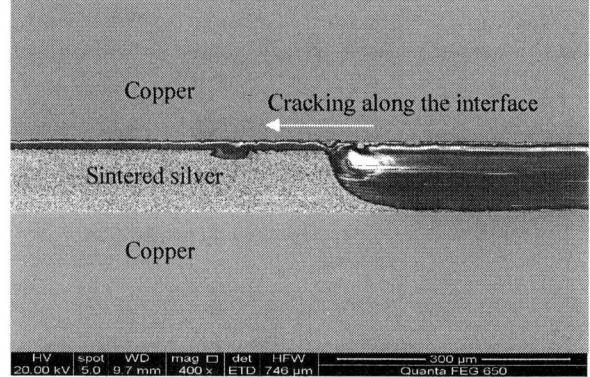

Fig. 5. Fracture morphology and microstructure of sintered silver layer.

The fracture morphology of the specimens with strain rate of $0.75 \times 10^{-5} s^{-1}$ was taken by a scanning electron microscope (SEM) as shown in Fig. 5. The densification degree of the silver layer is high, the pores are small and the particle size is large, which proves that the selection of sintering process is very appropriate. The porosity of sintered silver is as low as 27.12%. The crack can be clearly seen to crack along the interface. The fracture modes of the samples at other strain rates are the same, all cracking along the interface, present "interface delamination".

IV. Conclusions

In this paper, the action law of fracture toughness of sintered silver joint under four kinds of low strain rate conditions is studied, and the failure mode of sintered silver/copper interface is observed. The LOCTITE ABLESTIK SSP 2020, was used to make ENF samples. All the specimens were sintered at 280°C for 60 min with a pressure of 10 KPa. Three-point bending test was performed on the ENF sample, at strain rates of $1.875 \times 10^{-4} s^{-1}$, $0.75 \times 10^{-4} s^{-1}$, $0.75 \times 10^{-5} s^{-1}$ and $1.875 \times 10^{-6} s^{-1}$. The fracture mode of sintered silver/copper structure at the low strain rate level set in this paper is cracking along the interface. The CBBM method is used to calculate the fracture toughness. It can be found that the fracture toughness of sintered Ag/Cu interface decreases with the increase of strain rate within the strain rate range presented in this paper. This work enables us to understand the influence of low strain rate on the fracture toughness of sintered Ag/Cu interface, which is of great significance for improving the reliability of joints and electronic devices.

Acknowledgment

The authors acknowledge the supports from the Beijing Natural Science Foundation (2204074).

References

[1] Xin Li, Gang Chen, Xu Chen, Guo-Quan Lu, Lei Wang, Yun-Hui Mei, High temperature ratcheting behavior of nano-silver paste sintered lap shear joint under cyclic shear force. Microelectronics Reliability 2013; 53(1): 174-181.

[2] H. Zheng, D. Berry, K. D. T. Ngo and G. Lu, Chip-Bonding on Copper by Pressureless Sintering of Nanosilver Paste Under Controlled Atmosphere. IEEE Transactions on Components, Packaging and Manufacturing Technology 2014; 4(3): 377-384.

[3] J.A. Sousa, A.B. Pereira, A.P. Martins, A.B. de Morais.Mode II fatigue delamination of carbon/epoxy laminates using the end-notched flexure test,Composite Structures 2015; 134: 506-512.

[4] P.N.B. Reis, L. Gorbatikh, J. Ivens, S.V. Lomov, Strain-rate sensitivity and stress relaxation of hybrid self-reinforced polypropylene composites under bending loads, Composite Structures 2019,209:802-810.

[5] J.C.P. Figueiredo, R.D.S.G. Campilho, E.A.S. Marques, J.J.M. Machado, L.F.M. da Silva, Adhesive thickness influence on the shear fracture toughness measurements of adhesive joints, International Journal of Adhesion and Adhesives 2018; 83: 15-23.

Comparative Research of Infrared Thermography and Electrical Measurement Method for the Thermal Characteristics Test of GaN HEMT Devices

Zhiwei Fu
Science and Technology on Reliability Physics and Application of Electronic Component Laboratory
China Electronic Product Reliability and Environmental Testing Research Institute
Guangzhou, China
fzw19940124@163.com

Bingjie Zheng
Science and Technology on Reliability Physics and Application of Electronic Component Laboratory
China Electronic Product Reliability and Environmental Testing Research Institute
Guangzhou, China

Xu Huang
School of electrical engineering
Chongqing University
Chongqing, China

Bin Zhou
Science and Technology on Reliability Physics and Application of Electronic Component Laboratory
China Electronic Product Reliability and Environmental Testing Research Institute
Guangzhou, China

Xiaofeng Yang*
Science and Technology on Reliability Physics and Application of Electronic Component Laboratory
China Electronic Product Reliability and Environmental Testing Research Institute
Guangzhou, China
yxf004@hotmail.com

Huaixin Guo*
Science and Technology on Monolithic Integrated Circuits and Modules Laboratory
Nanjing Electronic Devices Institute
Nanjing, China
guohuaixin@163.comave

Abstract—In this study, the infrared method and the electrical test method are compared through the thermal resistance measurement of GaN HEMT device which is flip-mounted to the PCB board. In order to ensure the identical heat dissipation, measuring system of electrical method is integrated with heating platform of thermal microscope. The comparison results show that in the infrared test calibration process, the inconsistency in temperature of the device surface and the heating platform would cause the deviation from the infrared test results. This paper proposes a compensation method to correct the infrared thermal test results. After the temperature compensation, junction temperature was slightly lower than the electrical test results (162℃) instead of higher than that, which was consistent with the trend of theoretical analysis. In addition, the test shows that the bottom temperature of the device was not equal to the temperature of heating platform under both dry and wet condition. Compared with the electrical method, the surface temperature of the device is suggested to being measured under the dry contact surfaces when using the infrared method.

Keywords—GaN HEMT, Infrared Thermography, Electrical Measurement Method, Junction Temperature, Thermal Resistance

I. INTRODUCTION

With the rapid development and application of the wide bandgap(WBG) semiconductor power devices, the integration and power density of GaN HEMT devices are continuously increasing [1-2]. In order to meet the performance development demand of the WBG semiconductor devices, GaN devices with flip package have gradually become a research focus. Flip-chip GaN devices without using wire bonding for interconnection have less the capacitive reactance/impedance interference between interconnection lines for better interconnection characteristics [3-5]. In addition, flip-chip interconnection greatly reduces the packaging volume of GaN devices, which is more conducive to miniaturization and high-density integration of GaN devices.

The operating junction temperature of GaN HEMT devices is higher due to the improvement of power density. The junction temperature, an important parameter to characterize the thermal performance of GaN HEMT, is mainly measured by infrared thermography and electrical method. Infrared thermal imaging test requires unsealing the device under test, which changes the heat dissipation path of the device, and only the surface temperature distribution of the device can be analyzed [6-7]. When performing thermal resistance test, a thermocouple is also required to test the temperature of the bottom side. The electrical method test is a non-destructive test method without the ability to obtain the temperature distribution of the device. It's not easy to compare and evaluate the test results of the above two methods because of the different bottom heat dissipation boundary conditions. On the other hand, during the infrared thermal test, the flip-chip structure makes the bigger temperature difference between device surface and temperature control platform, which in turn affects the calibration process.

Xuan Li [8] measured the temperature of GaN HEMT device by using electrical measurement method. A temperature-sensitive electrical parameter suitable for electrical testing was proposed, the validity of the parameter has been verified by using the infrared thermography. Lei Chi [9] compared the results of infrared thermography and electrical measurement method under the same condition of temperature in the device and at the same power density. The results showed that the temperature of the device under different microscopes were different. However, the above comparative research results from different methods are all under the different test platforms and boundaries of heat dissipation, which leads to the incomparability of the test results. Otherwise, calibration processes of infrared thermography with flip-chip structure, for the large temperature differences between the surface of the device and the temperature control platform are also not taken into consideration.

In this study, flip-chip structure GaN HEMT device were tested to investigate infrared thermography and electrical measurement method. In order to ensure the identical heat dissipation, measuring system of electrical method was integrated with temperature control platform of thermal microscope. In the electrical method test process, the infrared thermal imaging test was carried out simultaneously to ensure the heat dissipation conditions and the consistency of the device status. For the large temperature difference of the flip-chip structure device between the surface of the device and the temperature control platform during the calibration process of the infrared test, a thermal compensation method is used to compensate the infrared thermal test results. After the temperature compensation, junction temperature and thermal resistance could be slightly lower than the electrical test results. It was more consistent with the trend of theoretical analysis.

II. EXPERIMENT

In this paper, GaN devices were soldered on adapter PCBs due to their package. On the front side of the chip, the active region is connected to the circuit on the PCB through micro solder joints, the details are shown in the fig.1. The V_{ds} voltage of the GaN device is selected as the temperature sensitive parameter in the electrical method test. In the process of tests, the V_{gs} voltage has been set to 5V to ensure the device to be in a fully-on state. Test current is loaded between the Drain and Source, and the change in the temperature of the device can be characterized by the change in the drain-source voltage V_{ds}. The measurement current is 200mA, and the voltage drop V_{ds} generated by the device is less than 20mV, and the mW-level power generated by the test current is not enough to make the GaN device produce self-heating effect.

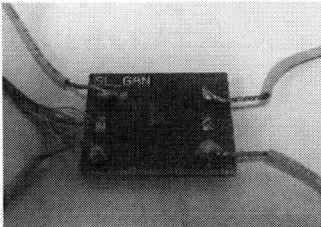

Fig. 1. GaN HEMT devices flipped on PCB

According to the JESD51-14 standard [10], which has possesses the electrical measurement method usually requires the device to be placed on a cold plate during test which has a good heat dissipation performance. In the infrared thermography test process, it is necessary to heat the temperature control platform below the device to about 70°C for emissivity calibration. It can be seen that in the two tests, the heat dissipation boundary conditions of the device are completely different. Therefore, the corresponding results of the two tests are difficult to compare. In the experiment, we connected the infrared test system and the electrical

measurement system to the same device above temperature control platform. The schematic diagram is shown in Fig.2.

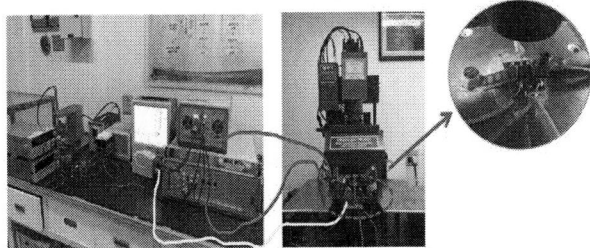

Fig. 2. Testing equipments of infrared thermography and electrical method

The DUT is fixed on the infrared temperature control platform, and a thermocouple is placed on the bottom of the PCB to monitor the case temperature of the device. The thermal resistance of the GaN device is carried out with the dual interface method and the infrared thermal imaging simultaneously records the stable surface temperature of the DUT. In the infrared thermal imaging test, the emissivity of the GaN device surface needs to be calibrated. Taking into account the temperature difference between the device surface temperature and the bottom temperature of the temperature control platform during calibrating process, the thermocouple is used to test the above difference before calibration for the purpose of compensating the infrared test results later.

III. RESULTS AND DISCUSSION

A. Results of electrical method

As shown in the Fig. 3, the relationship between the V_{ds} voltage and temperature of the GaN device were obtained with the start temperature of 25°C, step size of 25°C and the measurement current of 200mA. The results show that it has a good linear relationship and high coincidence between the V_{ds} voltage and temperature in all three tests. At 100°C, the power generated by the measurement current was about 3mW, which is the biggest power of the measurement during the test. Therefore, it can be considered that the test current hardly produces self-heating effect and has a negligible impact on the test results.

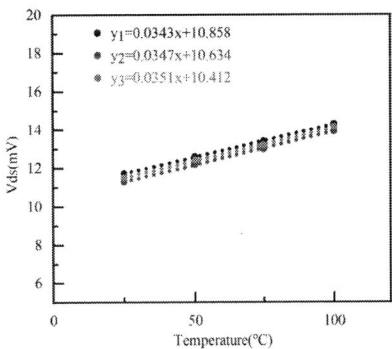

Fig. 3. Forward voltage V_{ds} versus temperature for the devices

Refer to the JESD51-14 standard, we use the dual interface method to test the thermal resistance of the device with the T3Ster equipment. The device was used with wet and dry methods to contact the surface of the infrared temperature control platform (with and without thermally conductive glue). The heating current I_{ds} of the device was 5A. In order to ensure that the device was close to the temperature equilibrium state after heating and cooling, the heating and test time respectively were 180s and 120s. During the experiment, the temperature of the temperature control platform above the device was set at 70°C, and the temperature drop curve was recorded as shown in the fig.4.

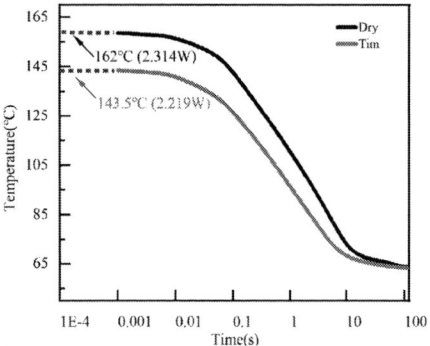

Fig. 4.   Temperature drop curves of the device under dry and wet condition

A measurement delay time t between the load currents was switched off and the sense current was applied which was inevitable. The t induced a maximum junction temperature offset ΔT. In order to reduce this temperature offset ΔT and obtain the highest temperature of the device in a steady state, the method of root t was usually used to correct the temperature of the device. The temperature fell linearly with the square root of the cooling time t as shown below.

$$\Delta T(t) = \frac{P_H}{A} k_{therm} \sqrt{t} \qquad (1)$$

$$k_{therm} = \frac{2}{\sqrt{\pi c \rho \lambda}} \qquad (2)$$

Where P_H is power loss in W, A the chip active area in mm^2, c is the specific heat in J/(kg*K), ρ the density in kg/m^3, λ the thermal conductivity in W/(m*K), and t is measurement delay time in s. Therefore, the maximum junction temperature of the device can be obtained by linear derivation.

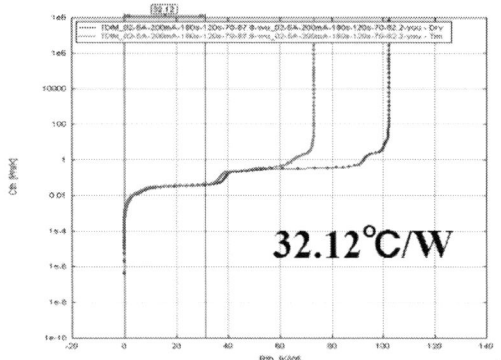

Fig. 5.   Integral structure function curves of GaN HEMT devices

The Fig.5 showed that when the heating current was set to be 5A, the power of the GaN device recorded by the system in the dry test (without thermally conductive glue) was 2.314W, which was slightly larger than the wet condition (2.219W). This was mainly because the heat dissipation conditions

changed during the two tests, resulting in a significant change in the junction temperature and resistance R_{dson} between the Drain and Source of the device at almost the same power. Based on the structure function algorithm and the separation point of the integral function curve, the thermal resistance from the junction area of the GaN device to the bottom of the PCB board was about 32.12°C/W.

B. Results of infrared thermography method

During the test with the electrical method, the infrared method simultaneously tested the temperature when the device reached steady-state thermal equilibrium under the dry and wet test conditions respected, as shown in the Fig.6.

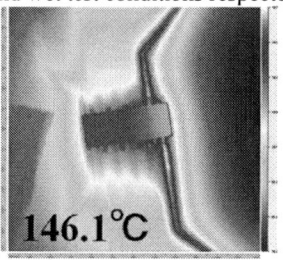 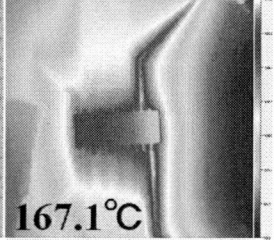

(a) with heat-conducting glue   (b) without heat-conducting glue

Fig. 6.   Surface temperature distribution of GaN HEMT devices

Under the dry and wet test conditions, the average surface temperature of the device was 167.1°C and 146.1°C, respectively in the infrared test, which were slightly higher than the 161.98°C and 143.5°C measured by the electrical test method. In the infrared test, the active area of the flip-chip GaN device was inverted in the silicon wafer that makes the temperature of infrared test closer to the surface average temperature of the DUT. The electrical method, however, corresponded to the average temperature of the active area. Therefore, theoretically, the junction temperature tested by the electrical method should be higher than the infrared method. The bottom temperature of the PCB monitored by the thermocouples was 87.8°C and 82.2°C in the dry and wet conditions. From the following thermal resistance calculation formula, the thermal resistance of the device using the infrared test in the dry and wet conditions were 34.27°C/W and 28.80°C/W, respectively. The difference between the two thermal resistance calculation results was considered to be caused by the difference in the dry and wet contact between the bottom of the PCB and the thermocouple.

$$R_{th-jx} = \frac{T_j - T_x}{P} \qquad (3)$$

Where T_j is the junction temperature of the device, T_x is the temperature of special location in the device, in this paper the T_x is the temperature of PCB. P is power loss of the device and R_{th-jx} is the thermal resistance of the device between the junction and the special location.

C. Comparison of test results

Under the dry and wet contact between GaN device and the infrared temperature control platform, the thermocouple was used to measure the surface temperature of the DUT. The emissivity correction of the infrared method had been processed when the temperature control platform was 70°C, the temperature difference between the surface of the device and the temperature control platform was about 8.1°C in dry

978-1-6654-1392-3/21 $31.00 © 2021 IEEE

contact and 6.1°C in wet contact respectively. Therefore, it was necessary to perform surface temperature compensation for the actual test results. The specific results were compared as shown in the following table.

TABLE I.   COMPARISON OF TEST RESULTS

Test items		Test condition	
		Dry test	Wet test
Power Dissipation(W)		2.314 W	2.219 W
T_j	Infrared Thermography	167.1 °C	146.1 °C
	Infrared Thermography (corrected value)	159.0 °C	140.0 °C
	Electrical Measurement Method	162.0 °C	143.5 °C
T_x		87.8 °C	82.2 °C
R_{th}	Infrared Thermography	34.27 °C/W	28.80 °C/W
	Infrared Thermography (corrected value)	30.71 °C/W	26.05 °C/W
	Electrical Measurement Method	32.12 °C/W	

By comparing the data in the table, it can be found that the junction temperature results of the infrared method after temperature compensation are 159.0°C and 140.0°C under dry and wet conditions, which were slightly smaller than the electrical junction temperature test results. It was more consistent with the theoretical analysis, therefore, when the infrared method was used for junction temperature testing, the temperature compensation was necessary to improve the accuracy of the test results. On the other hand, we can find that when the infrared method was used for the thermal resistance test, the thermal resistance result of the wet test was significantly smaller than the dry test. Because of the difference in contact between the bottom of the PCB board and the thermocouple during the two conditions, the temperature of the PCB measured by the thermocouple was different, which also led to different thermal resistance results. Using the junction temperature of the electrical method and the thermocouple test results of the PCB, using the formula (3) to solve the thermal resistance, the thermal resistance of dry and wet condition were 32.07°C/W and 27.62°C/W, respectively. The test result of dry method was closer to the electrical method test result. Therefore, in order to avoid the influence of heating platform dissipation, the temperature at the bottom of devices was suggested being measured in the dry condition.

IV. CONCLUSIONS

In this article, a comparative test study of electrical and infrared methods with the flip-chip structure GaN is carried out. The main conclusions are as follows:

1) Through the integration of temperature control platforms of the thermal microscope into the electrical method system, identical test conditions including the heat dissipation and heating power were achieved. It significantly

increased the comparability of results for different measuring methods.

2) A new method of temperature compensation in infrared thermal test was proposed to improve measure precision. After the temperature compensation, junction temperature was reduced from 167.1°C to 159.0°C, slightly lower than the electrical test results (162°C). It was consistent with the theoretical analysis.

3) When the infrared method was used for thermal resistance test, temperature at the bottom of device should be measured by thermocouple because it was not equivalent to the temperature of the heating platform. In order to avoid the influence of heating platform dissipation, the temperature at the bottom of device was suggested being measured in the dry contact surfaces.

ACKNOWLEDGMENT

This research was supported by the National Key R&D Program of China (No.2020YFB2008900), the Opening Project of Science and Technology on Monolithic Integrated Circuits and Modules Laboratory(No.614280303022005), and Key-Area Research and Development Program of Guangdong Province(No.2020B010173001).

REFERENCES

[1] LI Ruguan, LIAO Xueyang, YAO Bin, et al. Progress of technologies and applications of temperature measurements for GaN-based HEMTs [J]. Electronic components and materials, 2017 (9): 1-8.

[2] Heinz Pape, Dirk Schweitzer, Liu Chen, et al. Development of a Standard for Transient Measurement of Junction-to-Case Thermal Resistance[J]. Microelectron Rel, 2012, 52: 1272−1278.

[3] Chen K J, Haberlen O, Lidow A, et al. GaN-on-Si power technology: devices and applications[J]. IEEE Transactions on Electron Devices, 2017, 64 (3): 779-795.

[4] Zhang M, Ma X H, Mi M H, et al. Improved on-state performance of AlGaN/GaN Fin-HEMTs by reducing the length of the nanochannel[J]. Applied Physics Letters, 2017, 110(19):193502.

[5] Jo Y W, Son D H, Won C H, et al. AlGaN/GaN FinFET with Extremely Broad Transconductance by Side-wall Wet Etch[J]. IEEE Electron Device Letters, 2015, 36(10): 1-1.

[6] Farzaneh M, Maize K, L"uerßen D, et al. CCD-based thermoreflectance microscopy: principles and applications [J]. Journal of Physics D:Applied physics, 2009, 42 (14): 1-20.

[7] Sarkany Z, Farkas G, Rencz M. Thermal transient characterization of pHEMT devices [J]. International Workshop on Thermal Investigations, 2012, 11 (4): 1-4.

[8] Li X, Feng S, Liu C, et al. A Drain-Source Connection Technique: Thermal Resistance Measurement Method for GaN HEMTs Using TSEP at High Voltage[J]. IEEE Transactions on Electron Devices, 2020, 67 (12): 5454-5459.

[9] Lei C, Ru Z, Liang T, et al. Electrical Measurement Method for Thermal Characteristics of GaN HEMT Devices[J]. Semiconductor Technology, 2017.

[10] Electronic Industries Association. JESD51-14 Transient dual interface test method for the measurement of thermal resistance junction-to-cas e of semiconductor devices with heat flow through a single path [S/O L] (2010-10-01). http://standards.globalspec.com/std/1288922/jedec-j esd-51-14.

978-1-6654-1392-3/21 $31.00 © 2021 IEEE

Research Progress of Extreme Low Temperature Reliability of Typical Electronic Interconnection Structures

Zhaoning Sun
Reliability Analysis and Research Centre
China Electronic Product Reliability and Environmental Testing Research Institute
Guangzhou, China
sunzhaoning@ceprei.com

Xiaotong Guo*
Reliability Analysis and Research Centre
China Electronic Product Reliability and Environmental Testing Research Institute
Guangzhou, China
guoxiaotong0713@163.com

Zhenbo Zhao
Reliability Analysis and Research Centre
China Electronic Product Reliability and Environmental Testing Research Institute
Guangzhou, China
Zhenbozhao83@163.com

Yiqing Ni
Reliability Analysis and Research Centre
China Electronic Product Reliability and Environmental Testing Research Institute
Guangzhou, China
leon.ni@qq.com

Guanghui He
Reliability Analysis and Research Centre
China Electronic Product Reliability and Environmental Testing Research Institute
Guangzhou, China
heguanghui@ceprei.com

Abstract—The temperature range of general military and civil electronic reliability research is - 55 ℃ ~ 400 ℃. But in deep space exploration, the extreme low temperature and large changing condition without thermal guarantee will give a great challenge to the service reliability of electronic interconnection structures and normal operation of the equipment. The low temperature brittleness is the serious problem of the Sn-Pb solder or Sn-Ag, Sn-Cu and Sn-Ag-Cu lead-free solders. In-Pb solder which has good toughness at extremely low temperature is a promising solder for deep space exploration. In this paper, the evolution mechanism of microstructure and mechanical properties of solder alloys in extreme conditions, and the reliability of solder joints and welded structures both domestic and abroad are reviewed.

Keywords—Extreme low temperature condition; Solder, Ductile-Brittle Transition, Reliability

I. INTRODUCTION

With the rapid development of space technology, human being's exploration of the unknown field of the universe has expanded from the earth orbit to the deeper and more complex universe. The development of deep space exploration missions such as lunar exploration project, manned lunar landing project, Mars exploration and space station is of great strategic significance to the development of technology, politics and economy of the countries. Since 2004, China has officially launched the lunar exploration project, named the Chang'e Project, which is the first step of deep space exploration. Since then, a series of breakthroughs have been made in the field of lunar exploration and space station construction [1-2]. Compared with near-earth orbit spacecraft, the Important extravehicular electronic equipment of deep space probe suffers from extreme temperature, ultra-high vacuum and radiation environment. Table 1 shows the surface temperature of the near-earth planet [3], which indicates that the deep space detector should not only face

extreme high and low temperature environment, but also face the changes of extreme temperature. The reliability of electronic equipment directly affects the smooth progress of the detection task.

II. SOLDER JOINT AND RELIABILITY

Solder joint plays a key role in mechanical support and connection of electronic equipment, mainly including pad, solder, lead wire and intermetallic compound formed by welding, as shown in Fig.1. Solder is the key material in electronic assembly process, and Sn-Pb has been used widely. In recent years, due to the influence of lead-containing solder on human body and environment, green lead-free solder and its welding technology have become inevitable choice. However, in the field of deep space exploration, lead-containing solder can be exempted because of its superior performance. The main study objects include Sn-Pb, Sn-Pb-Ag, Sn-Pb-Bi solder alloys. The research of lead-free solder mainly focuses on Sn-Ag, Sn-Ag-Cu and Sn-Bi solder alloys. The addition of Ag and Cu generally improves the mechanical properties, wettability and reliability.

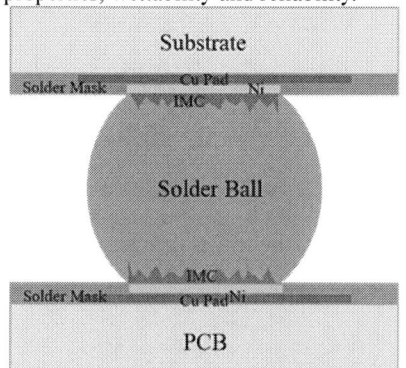

Fig. 1. The structure of solder joint

0.5Cu (wt%) (SAC305)solder alloy has become the research hot spots. In addition to the solder materials, the application environment must also be considered when studying the reliability of interconnection structure.

Generally, the failure of electronic equipment' solder joint is caused by the different thermal expansion of materials used in the solder joint structure under various thermal or mechanical stresses, or the growth brittleness increase of intermetallic compounds during the high-temperature aging process. The failure mode and service reliability of electronic equipment solder joint will be more complex, so design and selection of materials become the key point.

Table 1 The surface temperature of near-earth planet

Planet	Temperature(°C)	Planet	Temperature(°C)
Moon	-180~150	Uranus	-212
Mercury	-196	Saturn	-185
Mars	-140~20	Neptune	-225

The reliability service temperature range of general military and civil electronic products is -55℃~400℃.Since the early 20th century, the National Aeronautics and Space Administration, Belgium Microelectronics Research Center and other major aerospace research institutions have started to study the reliability of electronic assembly solder joints for deep space exploration. In China, the research about solder materials and solder joint structure in extremely low temperature and large temperature variation conditions is still in initial stage. In this paper, the mechanical behavior, microstructure evolution and failure mechanism of solder joints or solders materials used in electronic interconnection in deep space exploration extreme environment at home and abroad are discussed, and some suggestions are put forward.

III. STUDY ON MICROSTRUCTURE AND PROPERTIES OF SOLDER ALLOY AT EXTREMELY LOW TEMPERATURE

The research on the microstructure and properties of Sn-based solder alloy mainly includes the following aspects：Firstly, the mechanical research of different Sn-based solder alloys in extremely low temperature environment shows brittle behavior and brittle failures. It is difficult to meet the stability requirements of deep space exploration service. Secondly, the second phase particles, such as Ag_3Sn and Cu_6Sn_5, produced by lead-free solder will influence the microstructure of the solder and further ductile brittle transition temperature of the solder. However, the mechanism of the above problems is not unified. Thirdly, the fracture morphology, mechanical behavior of IMC layer, or the mechanism of interfacial stress and ductile brittle transition of solder joints in extremely low temperature environment are still under exploration. Fourthly, the packaging forms and packaging materials have a direct impact on the service life of electronic equipment in extreme environment.

Ratchev[4] who is from Interuniversity Microelectronics Center (IMEC) studied the fracture toughness of pure Sn, Sn-Pb and lead-free solder alloys with Ag and Cu contents in the range of - 195 ℃ ~ 100 ℃ by the modified Charpy energy test. Fig.2 shows that the fracture toughness of solder alloys, except Sn-37%Pb alloy, increases obviously with the decrease of temperature, and decreases rapidly after the maximum value. The results showed that typical ductile-to-

brittle transition (DBT) phenomenon occurs in solder alloy, and the different ductile-brittle transition temperatures are obtained. The transition temperature of Sn-based solder without Ag is greatly reduced by adding Cu element, and the safe use temperature reaches - 130 ℃ . In contrast, the addition of Ag can increase the ductile-brittle transition temperature of lead-free solder, which is between - 78 ℃ and 45 ℃. This is mainly due to the formation of Ag_3Sn particles in solder, which hinders the movement of dislocation and fracture in alloy.

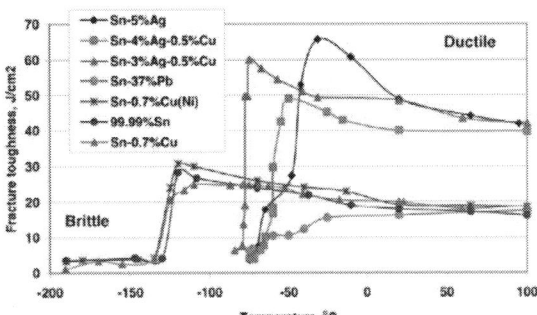

Fig.2 Fracture toughness of the studied solders as the function of temperature

Lupinacci [5]who's from NASA Jet Laboratory studied the fracture toughness of Sn-Pb solder alloys with different lead content in the range of - 185 ℃ ~ room temperature through Charpy test. During the decrease process of temperature , the ductile brittle transition occurs in most Sn-Pb solder alloys. The ductile brittle transition temperature of Sn-Pb eutectic alloy（63Sn37Pb） is - 80 ℃, and with the increase of Pb content, the ductile brittle transition phenomenon is gradually blurred. When the Pb content reaches 90wt%, the ductile brittle transition phenomenon disappears. Comparing the above two results, it can be seen that ductile-brittle transition on Sn-Pb eutectic solder is no unified conclusion.

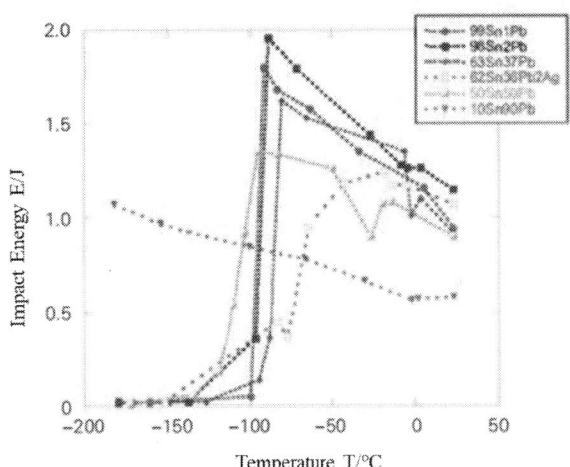

Fig.3 The ductile brittle transition temperature of Sn-Pb solder alloy

Table 2 The ductile-brittle transition temperature of different solders

Solder	Ductile-Brittle Transition temperature (℃)
Pure Sn（99.99%）	-130
Sn-0.7%Cu	-130
Sn-0.7%Cu（Ni）	-130
Sn-3%Ag-0.5%Cu	-78~-45
Sn-4%Ag-0.5%Cu	-70~-60
Sn-37%Pb	~-80

The micro morphology can directly show the process of solder alloy ductile-brittle fracture. When the temperature of Sn-37% Pb eutectic solder decreases from 20 ℃ to - 120 ℃, the fracture morphology shows dimple like pattern, and the size of the dimple decreases. At - 90 ℃ and - 120 ℃, there are side-by-side stripes and directional steps between the stripes. The similar "river pattern" and "step" morphology indicates that local brittle fracture and ductile-brittle fracture transformation occurs in the solder alloy.

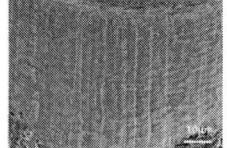

a) fracture morphology at -90℃   b) fracture morphology at -90℃

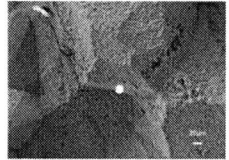 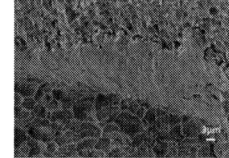

c) veins morphology at -120℃   d) river morphology at -120℃

Fig.4 The micro brittle fracture morphology of Sn37%Pb eutectic solder alloy at different temperatures

 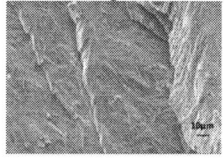

a) fracture morphology at -55℃   b) fracture morphology at -60℃

c) veins morphology at -60℃   d) river morphology at -60℃

 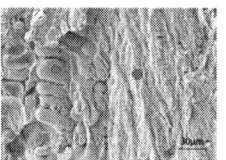

e) fracture morphology at -85℃   f) fracture morphology at -55℃

Fig.5 The micro brittle fracture morphology of SAC305 eutectic solder alloy at different temperatures

The granular or rod-shaped Ag_3Sn brittle particles will be introduced into SAC305 solder as internal defects, which will affect the crack propagation path along the cleavage plane or may act as new crack source, resulting in the distortion of lattice shape and the formation of quasi-cleavage rupture. Compared with the pure Sn solder, the resistence to the dislocation will lead to the decrease of toughness and the increase of ductile-brittle transition temperature. Toughness is the plastic deformation ability of the material, which depends on the type of chemical bond and the orientation or symmetry of crystal[6-7]. Therefore, the more symmetrical the metal lattice structure is, the more easily ductile fracture occurs. It has been proved that 50In50Pb and 50In30Pb solders still have good toughness at - 196 ℃ or even - 269 ℃, as shown in Fig.6[8]. β - Sn is a body centered cube with asymmetric crystal structure and poor plastic hinge. At very low temperature, The p-n force and yield strength in the crystal lattice of β-Sn solder alloy increase obviously, and the resistance strength of dislocation movement increases. Meanwhile, the tensile stress overcomes the bonding force between atoms. Finally, brittle cleavage fracture occurs in the alloy.

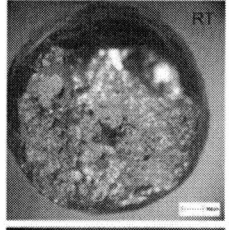

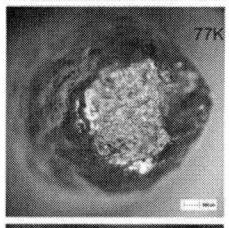

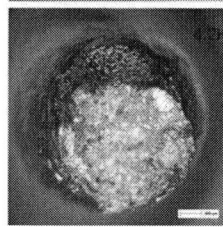

Fig.6 Fracture surfaces of ductile solder (50In50Pb) performed with optical microscope

The temperature change is the main reason for the periodic creep and accelerated thermal fatigue aging of materials. In deep space exploration, some electronic equipments and components in spacecraft have to experience the extreme low and high temperature environment. The solder joints which are sensitive to the thermal stress and mechanical stress will change greatly in the large temperature change environment, suffer damage and lead to failure. Rajeshuni rameshamf of NASA Jet Propulsion Laboratory chose high density chip CCGA with Pb37 and SAC305 as filler metals to study the intermittent failure under the environment of - 185 ℃ ~ 125 ℃ (temperature rise and fall rate of 5 ℃ / min, holding time of 15 min). The failure occurs at low temperature stage and recovers at high temperature stage. The failure position is mainly between the cylindrical pin and the printed circuit board The crack caused by the displacement between the cylindrical pin and the ceramic matrix.

IV. STUDY ON RELIABILITY OF SOLDER JOINTS AT EXTREMELY LOW TEMPERATURE

The results show that low temperature will lead to the enhancement of atoms bonding tightness, the decrease of alloy elasticity and weaker ability to resist the impact of stress. The reliability of solder joint is determined by solder material and intermetallic compounds formed in welding process. The reliability of solder joint determines the reliability of the whole welding structure. In the temperature range of $25\,℃ \sim -196\,℃$, Tian [9] has studied the tensile mechanical properties and microstructure of SAC305/Cu solder joint. At 25 ℃ and -50 ℃, the fracture is composed of a large number of microporous dimples, and the second phase particles of Ag_3Sn (Fig. 7 (a) and (b)) are distributed at the bottom of the dimples. It is found that the fracture mode is ductile fracture. When the temperature drops to - 100 ℃, the morphology of the quasi-cleavage fracture shows "crystal sugar flower" like pattern, pits and scallop like Sn_6Cu_5 particles (Fig.7 (c) ~ (e)). The fracture of the solder joint is transformed into mixed ductile-brittle fracture happening inside solder and solder/IMC interface. At -150℃, the dimples disappear and cleavage surface appears which indicates fracture of solder joint is brittle fracture happening between IMC interface and IMC layer (Fig.7 (f)). At the extreme low temperature of -196 ℃, the fracture surface is composed of Cu_5Sn_6 and cluster Cu_3Sn, and brittle fracture occurs in IMC layer completely (Fig.7 (g) and (h)). Fig. 8 shows the tensile strength and fracture location of SAC305/Cu solder joints at different temperatures. Firstly, the difference of thermal expansion coefficient and elastic modulus between solder and IMC layer leads to the decrease of bonding force. Secondly, IMC shows sintrinsic brittleness and β- Sn is difficult to lead to slip or fracture.

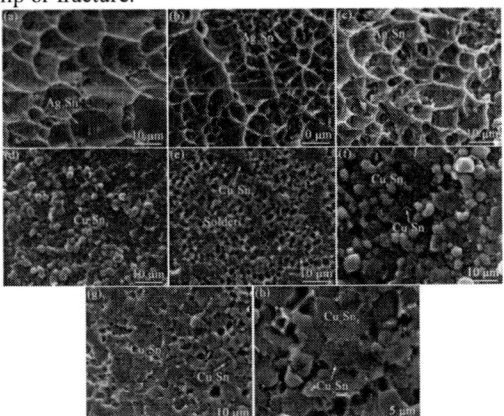

Fig.7 SEM images of fracture surface of SAC305/Cu solder joints at different temperatures
(a) 25°C, (b) -50 °C, (c) -100 °C, (d) -100 °C, (e) -100 °C, (f) -150 °C, (g) -196 °C, (h) -196 °C

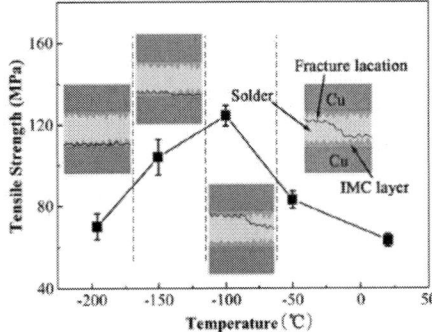

Fig.8 Schematic illustration of tensile strength and fracture location of SAC solder joints at different temperatures.

Compared with Sn-based solder, In-based solder with face-centered cubic structure and having no isomer transformation is more applicable in extreme low temperature environment. Chang [10] has firstly studied the influence of ultra-low temperature on the fatigue properties of In-based solder and its joints. Based on different large-scale chip welding structures, the three-point cyclic bending tests were carried out at - 150 ℃ and - 55 ℃ under isothermal condition. Different with creep and stress relaxation under high temperature thermal fatigue, mixed ductile brittle fracture or brittle fracture mainly occurs at the interface of intermetallic compound under extremely low temperature mechanical loading. According to the micro morphology in Fig. 9, the crack starts from the outermost edge and propagates to the main body of in solder through the IMC interface. The brittleness and porosity of IMC layer may lead to crack initiation and provide a way for crack propagation along the interface. At the same time, due to the different thickness of IMC, the fracture mode can be divided into ductile brittle mixed fracture and brittle fracture. When the IMC thickness is less than 15um, the IMC layer will not affect the fatigue life of solder joints.

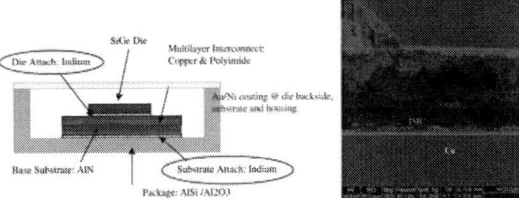

Fig. 9. The drawing of packaged structure tensile fracture morphology of In-based solder joints.

V. CONCLUSION

In conclusion, only a few researches have deal with the service reliability of solder alloy and joint in the field of deep space exploration. Low temperature brittleness is the main problem. Reducing the hardness and improving the ductility of solder are the main solutions. However, most of Sn-based solder alloys show low-temperature brittleness and it's difficult to meet the stability requirements in deep space. In-Pb solder has good toughness and it's a promising solder for deep space exploration. In the future, we need to continue to develop new solders suitable for deep space exploration.

The research on extreme low temperature reliability of typical electronic interconnection structures is mainly focused on the life test, and lack of its mechanical behavior or fracture mechanism. Difference of test equipment system and measurable methods increases the inaccuracy of the research results.

Solder joint reliability directly affects the stability of electronic equipment, as well as the packaging form, size, surface coating materials under extreme conditions. Therefore, it is necessary to deeply study the service reliability performance of the system.

ACKNOWLEDGMENT

This study was supported by the Science and Technology Program of Ministry of Industry and Information Technology, China under Grant No. 2020-0093-2-1, CEPREI Innovation and Development Fund No. 20Z32, which was acknowledged.

REFERENCES

[1] JIA Y, Li Y, JI L. "Demands of Mars exploration missions on environment simulation technologies[J]. " Spacecraft Environment Engineering, 2015, 32(5): 464-468.

[2] TAN, C L, HU T B, WANG D P. "Analysis on foreign sapcecraft in3-oribit failures[J]. Spacecraft Engineering, "2011, 20(4): 130-136.

[3] ZHANG Q X, WANG L. "Extreme space environment and its effects on interplanetary exploration mossions[J]." Spacecraft Engineering, 2007, 16(6): 61-66

[4] Peter Ratchev, Tony Loccufier, Bart Vandevelde. Bert Verlinden, Steven Teliszewski. "A Study of Brittle to Ductile Fracture Transition Temperatures in Bulk Pb-Free Solders," EMPC 2005, June 12-15. Brugge. Belgium.

[5] A. Lupinacd, A.A. Shapiro, J.O. Suh, A.M. Minor . "A study of solder alloy ductility for cryogenic applications", 2013 IEEE International Symposium on Advanced Packaging Materials, Session 2-5 (Paper No. 35)

[6] Qi An, Chunqing Wang, Hong Wang and Xiangxi Zhao. The Mechanism Study of Low-Temperature Brittle Fracture of Bulk Sn-Based Solder. 18 th International Conference on Electronic Packaging Technology.

[7] Qi An .Sthdy on brittle fracture mechaism of Sn-based solder alloys at low temperature. Harbin, Harbin Institute of Technology,2017.

[8] M. Fink, Th. Fabing, M. Scheerer, E. Semerad, B. Dunn, Eeasurement of mechanical properties of electronic materials at temperatures down to 4.2 K. Cryogenics, 2008, 48:497 − 510.

[9] Tian RY, Tian YH, Wang CX, Zhao LY. Mechanical properties and fracture mechanisms of Sn-3.0Ag-0.5Cu solders and joints at cryogenic temperatures[J]. Materials Science and Engineering: A, 2017, 684: 697-705.

[10] CHANG R W, MCCLUSKEY P F. Reliability assessment of indium solder for low temperatur electronic packaging[J]. Cryogenics, 2009, 49(11): 630-634

Properties of room temperature bonded and UV cured temporary bonding adhesive for ultra-thin wafer's handling

Xujun Li
Shenzhen Institute of Advanced
Electronic Materials
Shenzhen Institute of Advanced
Technology, Chinese Academy of
Sciences
Shenzhen 518055, China
xj.li1@siat.ac.cn

Qiang Liu
Shenzhen Institute of Advanced
Electronic Materials
Shenzhen Institute of Advanced
Technology, Chinese Academy of
Sciences
Shenzhen 518055, China
qiang.liu@siat.ac.cn

Deliang Sun
Shenzhen Institute of Advanced
Electronic Materials
Shenzhen Institute of Advanced
Technology, Chinese Academy of
Sciences
Shenzhen 518055, China
dl.sun@siat.ac.cn

Zhipeng Li
Shenzhen Institute of Advanced
Electronic Materials
Shenzhen Institute of Advanced
Technology, Chinese Academy of
Sciences
Shenzhen 518055, China
zp.li@siat.ac.cn

Guoping Zhang*
Shenzhen Institute of Advanced
Electronic Materials
Shenzhen Institute of Advanced
Technology, Chinese Academy of
Sciences
Shenzhen 518055, China
gp.zhang@siat.ac.cn

Rong Sun
Shenzhen Institute of Advanced
Electronic Materials
Shenzhen Institute of Advanced
Technology, Chinese Academy of
Sciences
Shenzhen 518055, China
rong.sun@siat.ac.cn

Abstract—With the development of advanced packaging towards more complex heterogeneous integration, larger package carrier, thinner chip and smaller package size, the new temporary bonding material and technology is required, such as lower bonding temperature, higher heat resistance, lower warpage and less/no cleaning solvent. Herein, a novel temporary bonding and de-bonding (TBDB) material has been developed and the properties of room temperature bonded and ultraviolet (UV) cured temporary bonding adhesive for ultra-thin wafer's handling is introduced. The curing process with UV light could be completed within 30 seconds which successfully improves the production efficiency and reduces energy consumption. Besides, as the bonding material is liquid, it is easy to flow without any residual stress during the bonding process. Then, low TTV and warpage of the bonding pair has also been observed, the results of ultrasonic scanning microscope shows there is no failure can be found between the interfaces of the bonding pair. The UV curing characteristic was investigated by photo-DSC and the heat resistance was investigated by TGA. It shows the 5% weight loss temperature of the material is more than 320 °C. At last, after de-bonded, the cured bonding material on the device wafer can be easily removed by adhesive tape without any other cleaning process proved by energy dispersive X-ray spectroscopy (EDS).

Keywords—Temporary bonding adhesive, UV cured, Room temperature bonded, No-clean

I. INTRODUCTION

Today there are two important characteristics of the integrated circuits, on the one hand, the development of Moore's law slows down due to the proximity of the front circuit manufacturing to the physical limit. On the other hand, the diversification of micro-electronic products blossom, such as intelligent mobile terminal, internet of things, artificial intelligence, 5G communication. To meet the multi-function and high performance of this products, 3D integrated circuits (3DIC), system-level packaging, heterogeneous integration and other new technologies have emerged [1, 2]. Among the above technologies, the thinning and handling technology of chips or packages is becoming more and more important, and all of them are faced with the handling problem of ultra-thin wafers. As a key process of advanced manufacturing and packaging, temporary bonding technology has attracted more and more attentions. After more than 20 years development and application, There are many types of adhesives available today. In terms of their de-bonding method, there are four mainstream classifications: thermo-sliding, mechanical, Infrared laser and UV laser de-bonding [3~5].

Herein, a novel temporary bonding and de-bonding (TBDB) material has been developed, it's a room temperature bonded and UV cured temporary bonding adhesive (TB 5130) for ultra-thin wafer's handling, it can be used with the UV lase release layer (WLP LB 210 from Shenzhen Samcien Semiconductor Materials Co., Ltd.) as shown in Fig. 1, which can realize the bonding and de-bonding processes at room temperature and with high throughput. The UV laser release layer is spin-coated on the glass carrier wafer and baked to be a dry film, while the reported temporary bonding material TB 5130 is spin-coated on the device wafer without baking. The device wafer and glass carrier wafer are bonded together in vacuum at room temperature, and then cured with UV light just for 30 seconds, After a series of backside processes, such as grinding, lithography, etching, passivation, electroplating, the manufactured ultra-thin device wafer can be separated at room temperature without stress after radiated by UV laser, which induced photochemical reaction and breakdown the

978-1-6654-1392-3/21 $31.00 © 2021 IEEE

chemical bond of the UV laser release layer to decrease the adhesion strength and realize the high efficiency, stress-free separation. Using adhesive tape to remove UV-cured temporary bonding material on the ultra-thin device wafer and auxiliary cleaning agent to remove organic matter residue on the glass carrier wafer, which successfully improves the production efficiency and reduces energy consumption.

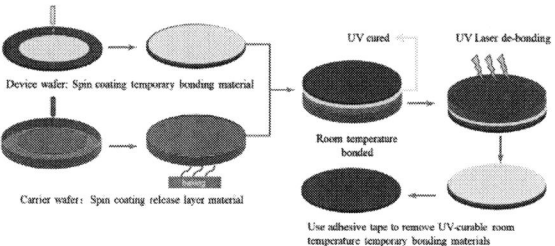

Fig. 1. Schematic diagram of UV cured and room temperature bonded temporary bonding process

II. EXPERIMENT AND RESULT

A. curing property of the temporary bonding material

At first, the UV curing characteristic of bonding material is investigated by photo differential scanning calorimetry (photo-DSC)[6]. As shown in Fig. 2, the UV curing characteristic of TB 5130 with or without UV laser release layer have been studied. The results show that these material are fully cured within 30 seconds after directly radiated by UV light with energy density at 50 mW/cm². But the UV light act on the TB 5130 temporary bonding material after through the UV laser release layer on the carrier wafer in practice, and part of UV light is absorbed by the laser release layer, so the energy of the UV light acted on the TB 5130 will decrease. That is why the photo-DSC curve of TB 5130 obstructed by UV laser release layer slows down at the same energy density in Fig.2. The exothermic area curves which are calculated by integration of heat flow are showed in Fig.3, the TB 5130 covered with the UV laser release layer shows lower exothermic area after UV radiated, which indicates the UV laser release layer slows the curing speed and reduces the degree of the curing. In the practice application, it's better to ensure the enough UV energy to realize the completed cure process.

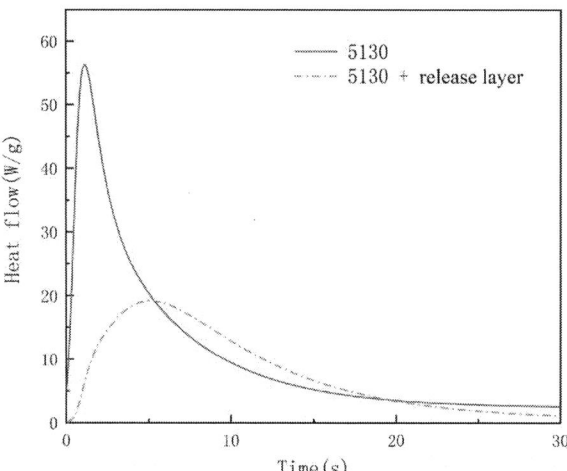

Fig. 2. The heat flow(w/g) of room temperature bonded temporary bonding material

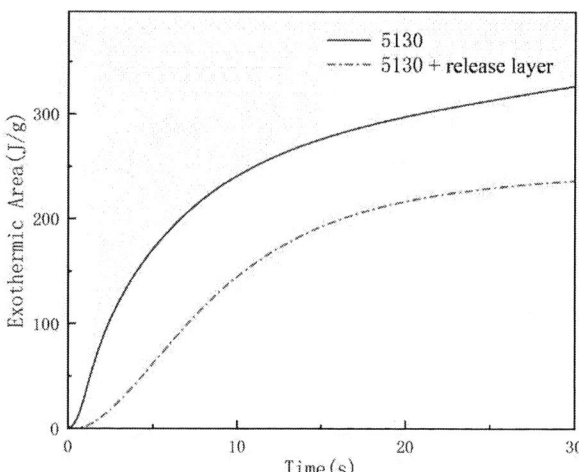

Fig. 3. The exothermic area curve of room temperature bonded temporary bonding material

B. Thermal property of the temporary bonding material

The heat resistance is a key performance for the temporary bonding material which shows great influence on the high temperature and high vacuum backside process capacity of bonding pair [7]. The thermal property of the TB 5130 has been investigated with the thermogravimetric analysis (TGA). The thermal stability and decomposition profiles of the UV cured TB 5130 was measured. The sample was cured with 365 nm UV light and loaded into an aluminum ceramic pan, and heated from 30 °C to 600 °C at a constant heating rate of 10 °C/min in an inert air atmosphere. As shown in Fig. 4, the UV cured TB 5130 keeps stable below 300 °C and the 5% weigh loss temperature is 320 °C. In order to further simulate the actual application condition, the TGA at constant temperature of the UV-cured TB5130 is collected and shown in Fig. 5. The sample was heated from 30 °C to 150 °C at a constant heating rate of 10 °C/min in an inert air atmosphere , and then kept for 1 hour. And there are 0.75% weigh loss during this treatment. Then it is heated to 200 °C and kept for another1 hour, which shows only 0.62% weigh loss during this isothermal treatment.

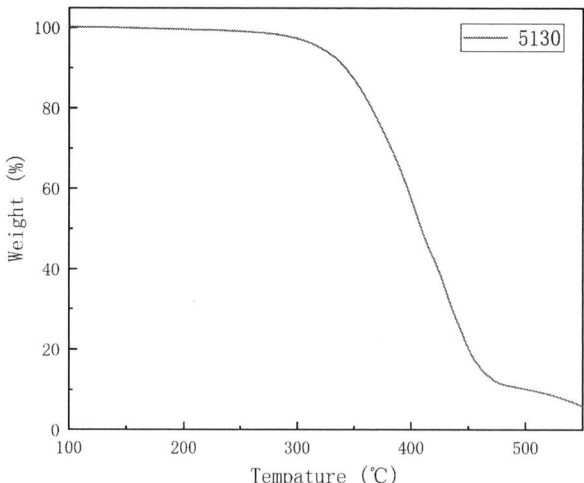

Fig. 4. TGA of the UV-cured TB5130 in air with a ramp rate of 10 °C/min

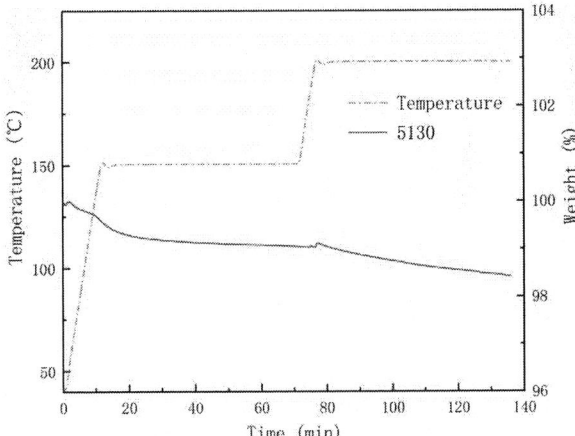

Fig. 5. TGA at constant temperature of the UV-cured TB5130 (in air, ramp rate of 10 °C/min, 150 °C for 1 h and 200 °C for 1 h)

C. The property of bonding pair

The TB 5130 was spin-coated on the silicon wafer, and a smooth film with low TTV can be achieved. The thickness distribution for the adhesive on the wafer was measured with the Filmetrics F50, as shown in Fig. 6, the average film thickness was 73.59 um with the spin-coating speed of 1200 rpm for 30 seconds, and the accelerated speed was 5000 rpm/s. The UV laser release layer was spin-coated on the glass carrier wafer, the wafer pair was bonded in vacuum at 500 N for 1 min, and then cured with 365 nm UV light with 600 mW/cm² for 30 seconds.

Scanning acoustic microscopy (SAM) was used to investigate the quality of the bonding pair before and after the treatment at 150 °C in vacuum oven (20 Pa) for 1 hour. As shown in Fig. 7, there is no obvious different between image A and image B, and there is no void cannot be found in both two images. It means that there is no defect in the interface of the bonding pair before and after the treatment in high temperature and vacuum. So, TB 5130 can withstand the backside processes with high temperature and high vacuum.

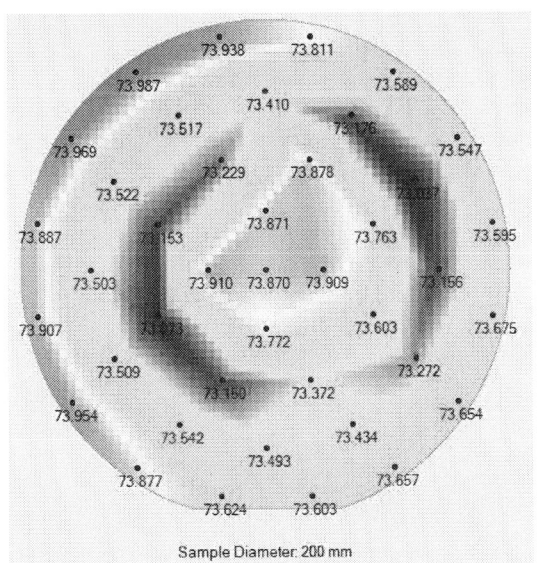

Fig. 6. Thickness distribution of spin-coated TB 5130 (1200 rpm for 30 seconds with accelerated speed 5000 rpm/s)

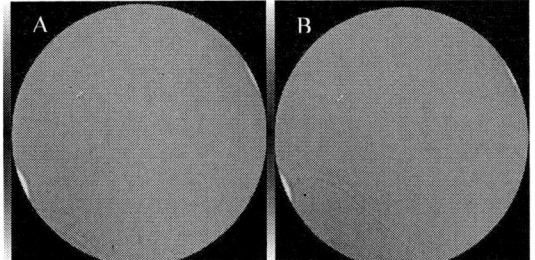

Fig. 7. SAM images of the boned wafers: (A)bonded wafer; (B)bonded wafer treated for 1hr at 150 °C in vacuum

D. The residual of the device wafer after debonding

The above-mentioned bonding pair was de-bonded after radiated by 355nm UV laser, which induced photochemical reaction and breakdown the chemical bond of the material in UV laser release layer to decrease the adhesion strength and realize the high efficiency, stress-free separation. Then, adhesive tape is adopted to remove UV-cured temporary bonding material on the silicon wafer without any more cleaning process. And the adhesive residual is investigated with the energy dispersive X-ray spectroscopy (EDS). As shown in Table 1 and Fig. 8, there is no difference in the EDS spectrum between de-bonded wafer and the new clean wafer. It indicates that there is no residual can be found on the de-bonded wafer after removed TB 5130.

TABLE I. EDS DISTRIBUTION OF MAIN ELEMENTS

Wafer	Elements and Content (%)			
	Carbon	*oxygen*	*silicon*	*Others*
A[a]	1.31	0.13	97.68	0.88
B[b]	1.19	0.12	96.40	2.29

[a] Clean wafer.

[b] De-bonded wafer after removed UV-cured temporary bonding material

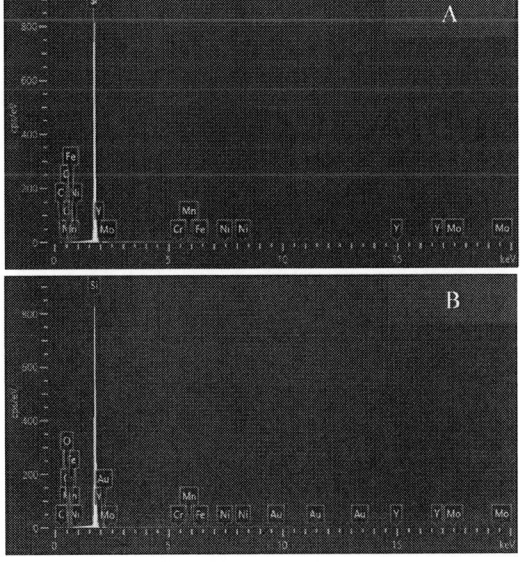

Fig. 8. EDS images of the wafers: (A) Clean wafer; (B) De-bonded wafer after removed UV-cured temporary bonding material

III. CONCLUSION

The temporary bonding material TB 5130 can be bonded at room temperature, and then cured with UV light just for 30 seconds, which improves the production efficiency and reduces energy consumption effectively. The temporary bonding material proves good comprehensive properties to cover the most backside process without failure. The energy dispersive X-ray spectroscopy (EDS) analysis shows that the cured bonding material on the device wafer has been removed clearly without cleaning solvent which will be more environment friendly.

ACKNOWLEDGMENT

This work was financially supported by National Natural Science Foundation of China (61904191), Youth Innovation Promotion Association of Chinese Academy of Sciences (2017410), Key R&D Project of Guangdong Province (2020B010180001) and National Key R&D Project from Minister of Science and Technology of China (2017ZX02519).

REFERENCES

[1] X. Shuai, R. Sun, G. Zhang, and L. Deng, "A novel temporary adhesive for thin wafer handling," in *International Conference on Electronic Packaging Technology (ICEPT), 2014 IEEE 15th*, 2014, pp. 256-261.

[2] Q. Liu, J. Xia, X. Li, D. Sun, M. Huang, W. Chen, et al. "Temporary bonding materials solution for ultra-thin device processing," in *Journal of Integration Technology*, vol. 10, pp. 23-34, 2021.

[3] J. Li, Q. Liu, G. Zhang, B. Zhao, R. Sun and C. Wang, "Thermally reversible and crosslinked polyurethane based on diels-alder chemistry for ultrathin wafer temporary bonding at low-temperature," in *Electronic Components and Technology Conference (ECTC), 2017 IEEE 67th*, 2017, pp.746-751.

[4] M. Fowler, J. Massey, T. Braun, S. Voges, R. Gemhardt, "Investigation and methods using various release and thermoplastic bonding materials to reduce die shift and wafer warpage for eWLB chip-first processes," in *Electronic Components and Technology Conference (ECTC), 2019 IEEE 69th*, 2019, pp.363-369.

[5] Y. Yang, K. Hwang, R. Gorrell, "Laser releasable temporary bonding film with high thermal stability," in *Electronic Components and Technology Conference (ECTC), 2019 IEEE 69th*, 2019, pp.330-333.

[6] S. Lee, J. Park, C. Park, D. Lim, H. Kim, J. Song, et al. "UV-curing and thermal stability of dual curable urethane epoxy adhesives for temporary bonding in 3D multi-chip package process," in *International Journal of Adhesion & Adhesives*, vol.44, pp.138–143, 2013.

[7] A. Lee, J. Su, H. Chang, C. Chien, B. Wang, L. Tsai, et al. " Optimization for temporary bonding process in PECVD passivated micro-bumping technology," in *Electronics packaging Technology Conference (EPTC), 2013 IEEE 15th*, 2013, pp.673-676.

Effect of thermomigration on evolution of interfacial intermetallic compounds in Cu/Ni/Sn-3.5Ag microsolder joints for 3D interconnection

Chu Tang
School of Mechanical and Electrical
Engineering, Central South University
Changsha, China
tangchu@csu.edu.cn

Wenhui Zhu
State Key Laboratory of High
Performance Complex Manufacturing,
Central South University
School of Mechanical and Electrical
Engineering, Central South University
Changsha, China
zhuwenhui@csu.edu.cn

Zhuo Chen
State Key Laboratory of High
Performance Complex Manufacturing,
Central South University
School of Mechanical and Electrical
Engineering, Central South University
Changsha, China
zhuochen@csu.edu.cn

Abstract—In recent years, the increasing emphasis on higher package density, richer functionality and smaller size in commercial electronics has spurred the advancement of three-dimensional integrated circuits (3DICs). However, the high power density and severe heat generation in 3DIC packages often require liquid cooling to dissipate heat, which induces higher temperature gradients than in 2DIC packages. To study the growth of Ni_3Sn_4 intermetallic compound (IMC) at the interface of 100 µm diameter microsolder joints, we used a heat sink and heat source device to create a temperature gradient of 16667 °C/cm and an average temperature of 150 °C in the solder joint. Due to the temperature gradient, Ni atoms migrated from the hot side to the cold side and the Ni_3Sn_4 IMC at the interface grew asymmetrically, with the Ni_3Sn_4 layer growing faster at the cold side than at the hot side. After applying a temperature gradient to the solder joint for 100 hours, significant depletion of the Ni layer was noticed at the hot end in a serrated shape. The kinetic equations of IMC growth are used to obtain the diffusion fluxes of Ni atoms in microsolder joints under temperature gradient and isothermal aging conditions, respectively. The consumption rate of the Ni barrier layer is explored to provide guidance for the thermal reliability of microsolder joints in 3D packages.

Keywords—*Intermetallics; thermomigration; solid state reactions; diffusion; temperature gradient; 3D interconnection*

I. Introduction

The trend toward miniaturization and multifunctionality of electronics has led the microelectronics industry to move toward three-dimensional integrated circuits (3DICs) for electronic packaging, which has also led to a dramatic reduction in solder joints (ball grid array balls or solder joints) from the millimeter to the micron level. Solder joints are often considered to be the weakest part of packaged systems and electronic assemblies, and chip service process thermal migration (TM) is critical to the integrity and reliability of solder joints [1]. As power density continues to increase, the chip's thermal management will face a huge challenge. At this time, heat dissipation measures must be taken to remove the heat that endangers the normal working of the chip, so a temperature gradient is established on the chip. The atoms will undergo relative diffusion under the effect of the temperature gradient, called thermal migration. In addition, under a certain temperature gradient, the under bump metallization (UBM) of Cu or Ni dissolves rapidly on the hot side, causing exceptional growth of intermetallic compounds (IMCs) on the cold side of the solder [2,3]. Cu and Ni films, which are extensive applicated as microbump and diffusion barrier layers in electronic packaging, react with brazing Sn alloys to produce Cu-Sn and Ni-Sn intermetallic compounds (IMCs). Interfacial reactions under isothermal aging and electromigration have been widely investigated [4, 5]. Coupled with the increasingly smaller solder joint sizes nowadays, the temperature gradient in three-dimensional integrated circuit (3DIC) packages is expected to be even larger. 3DICs use thermocompression bonded micro-bumps, which are typically 10 µm in height, and the temperature gradient (1000 °C/cm) is enough to cause thermal migration of the solder when the temperature difference is 1 °C [6], which further affects the interfacial reaction of IMCs in microbumps. Therefore, it is crucial to investigate the influence of temperature gradient on IMC growth in microbumps.

In this paper, the IMC growth patterns of Cu/Ni/Sn-3.5Ag microbumps were investigated at high power densities and temperature gradients greater than 10000 °C/cm considering both heat migration driving force and chemical concentration gradient driving force dual action conditions. The Ni layer consumption rate at 150°C was calculated to provide guidance for the lifetime prediction of Cu/Ni/Sn-3.5Ag microbumps. Meanwhile, the kinetic equations of IMC growth are used to obtain the diffusion fluxes of Ni atoms in microsolder joints during thermal migration and isothermal aging, respectively, to provide guidance for the thermal reliability of microsolder joints for 3D packaging.

II. Material and Methods

A. Test Vehicle

The sample chip includes an upper chip and a lower chip with Cu/Ni/Sn-3.5Ag microsolder joints between the two chips. The two chips are bonding by thermal compression process with an Athlete CB-600 flip-chip bonding machine. Figure 1 shows the design diagrams of the test chip with (a) top and (b) bottom chips having 100 µm diameter bumps. The copper pillar height of the upper and lower chip is 25 µm and 15 µm. The Sn cap height of the upper and lower chip is both 15 µm. There is a 2 µm Ni barrier between the Cu/Sn-3.5Ag layer. The Sn-3.5Ag layer is reflowed into a sphere after plating. Both chips have the same thickness of 0.5 mm.

978-1-6654-1392-3/21 $31.00 © 2021 IEEE

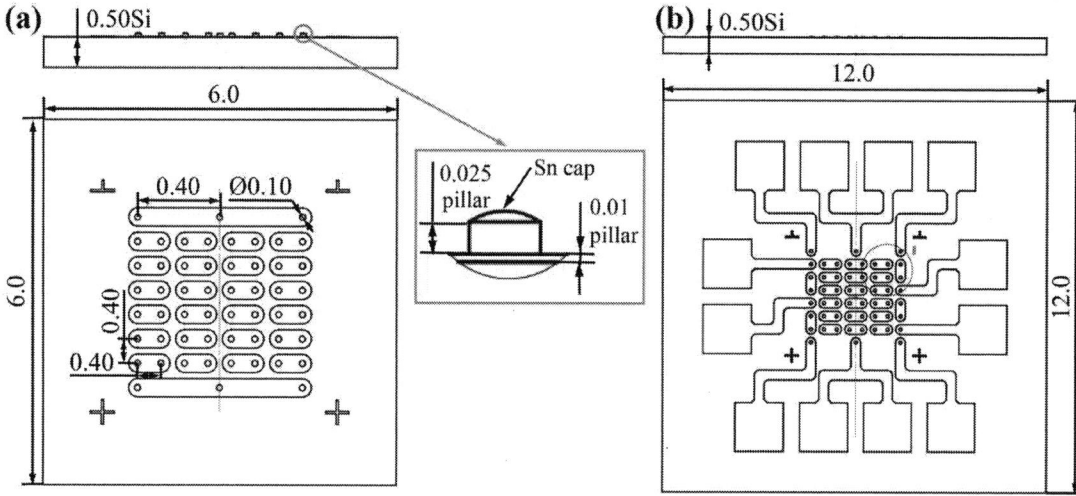

Fig. 1. Design diagrams of chip: (a) the top chip; (b) the bottom chip.

B. Execution of thermomigration test

Three experimental conditions were set up in this study. As shown in Figure 2, in conditions 1 and 2, the bottom and top ends of the samples were equipped with a heat generator and a heat spreader, respectively, thus creating a temperature gradient. Thermocouples are used at the hot and cold sides of the sample to determine the temperature. The lower chip in condition 1 is the hot end. The temperatures of the heat generator and heat spreader were held at 200°C ± 2 °C and 100 ± 2 °C. Whereas in condition 2, the hot end is located in the upper part of the chip and the temperature gradient of the specimen is inverted. As a comparison, the other samples were isothermally aged at 150 °C. Because of the small size and complex structure of the microbumps, it is harder to measure the temperature gradient between the microbumps. In this paper, the finite element method is applied to solve the temperature distribution of the micro-solder joint. The 3D model and the numerical simulation mesh of the microbump are shown in Fig. 3(a). The thermal boundary conditions in the numerical simulation of the micro-solder joints are set based on the thermocouple temperatures recorded by the two devises.

The specimens were fixed with epoxy after bonding and then polished by metallographic grinding and diamond spray to reveal the cross-section. Beside, using SEM observe the microscopic morphology of the micro-solder joint interface.

III. RESULT AND DISCUSSION

A. Growth of Ni₃Sn₄ IMC in microbump under isothermal aging and a temperature gradient

For the heat migration test, the sample was first isothermally aged for 100 h in order to observe more clearly the changes in IMC at both hot and cold sides of the joint, so that a layer of IMC of relatively similar thickness was generated at both ends of the microbump.

Figure 3 shows the simulated temperature distribution of the microbump when a temperature gradient (condition 1) is loaded. As shown in Figure 3(c), the upper and lower interface temperatures of the microbump near the Ni/Sn-3.5Ag interface are 125 °C and 175 °C, with a temperature

difference of 50 °C (ΔT). This results in a temperature gradient of 16667 °C/cm in the microbump, and the average temperature of the microbump is 150 °C, which we can assume to be the operating temperature.

Figure 4 shows the morphology of the interfacial IMC at the cold side and hot side in the Cu/Ni/Sn-3.5Ag/Ni/Cu solder joint by applying a temperature gradient of

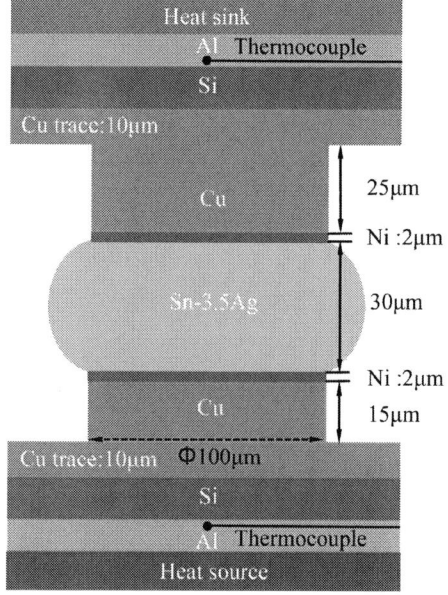

Fig. 2 Illustration of sample and test setup for condition 1.

16667 °C/cm to the sample for 50 h and 100 h. At the cold end (shown in Fig. 4(a), (c)), the interfacial IMC at the cold end gradually thickens with increasing TM test time. Meanwhile, the Ni substrate remains intact at the cold end. However, the IMC at the hot end interface gradually thins and the Ni matrix layer is consumed. The results indicate that the hot side interface IMC continuously dissolves into the

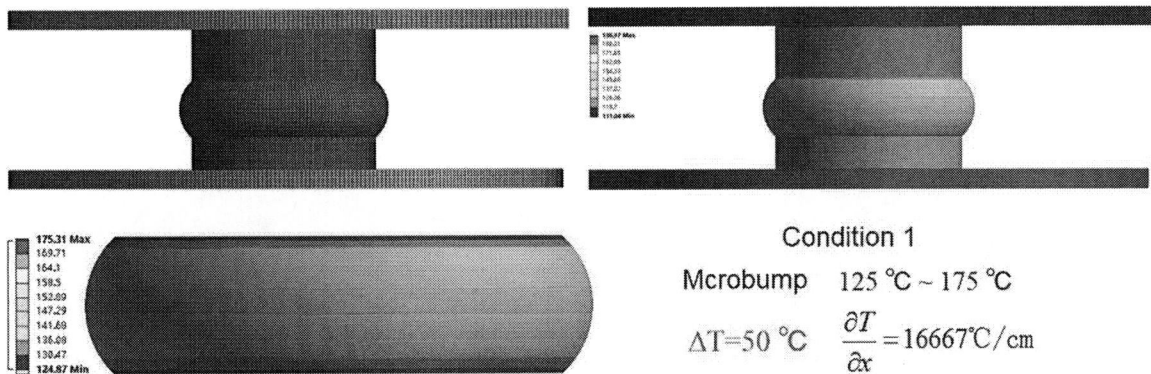

Condition 1

Microbump   125 °C ~ 175 °C

$\Delta T = 50\ °C$ $\dfrac{\partial T}{\partial x} = 16667\ °C/cm$

Fig. 3 (a) model and numerical simulation element mesh of a microsolder joint, and (b), (c) simulated temperature distribution of the microsolder joint for condition 1.

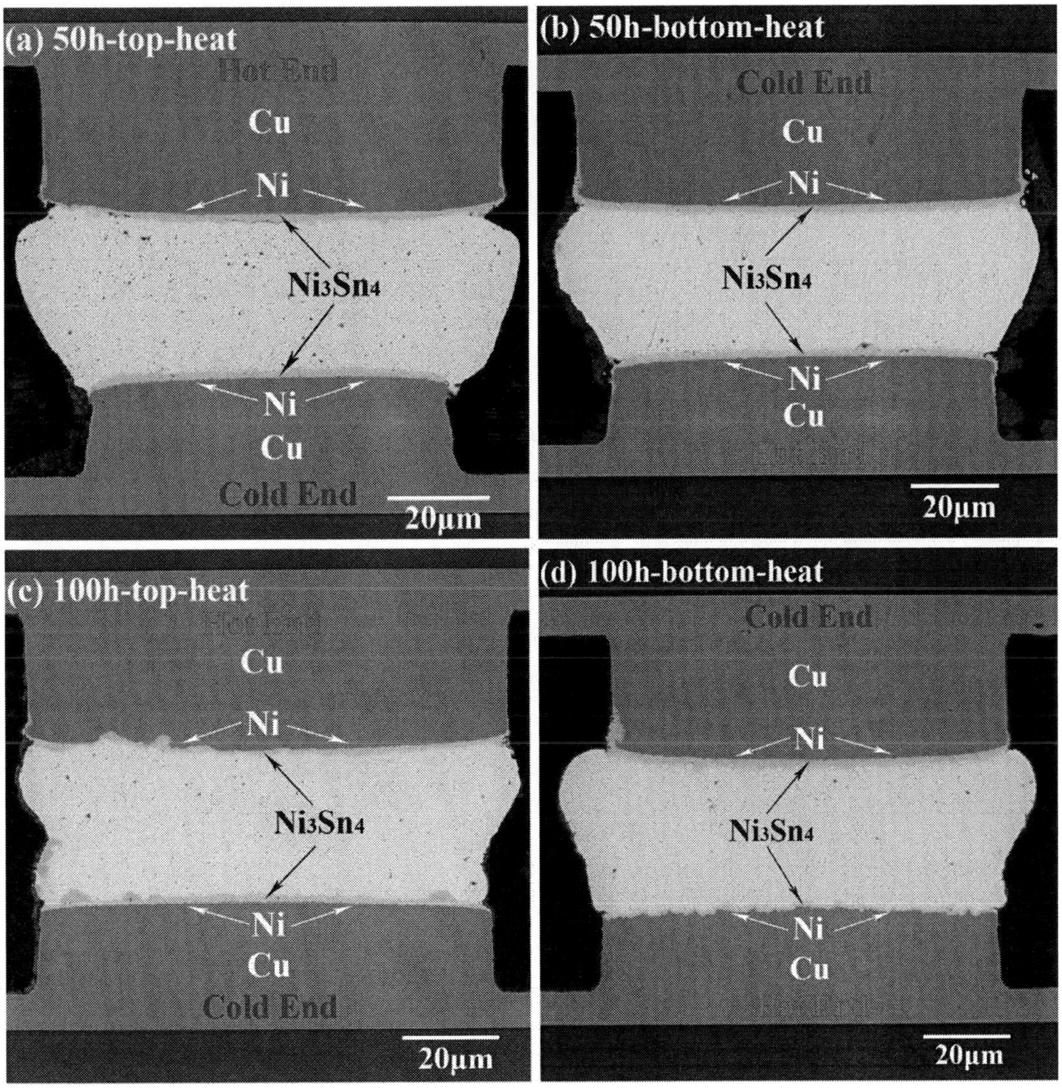

Fig. 4 Cross-sectional SEM images of a microbump under thermomigration at 145 °C for (a), (b), (c), (d) 50 h, (e), (f), (g), (h) 100 h.

solder, leading to an increase of Ni content in the solder. Besides, the interface between Ni UBM and Ni_3Sn_4 intermetallic compound is serrated. the serrated shape of the Ni/Ni_3Sn_4 interface may be caused by the inhomogeneous

consumption of Ni at the intermetallic compound grain boundaries [7].

Similar behavior was found on the samples when the hot and cold ends of the chip were reversed (condition 2). Figure

4(b), (d) shows that the growth of the interfacial IMC is asymmetric and the cold side IMC is thicker than the hot side IMC.

previous observations. Figure 6 shows the graph of IMC thickness at the interface of microbump at the hot end and cold end with TM time.

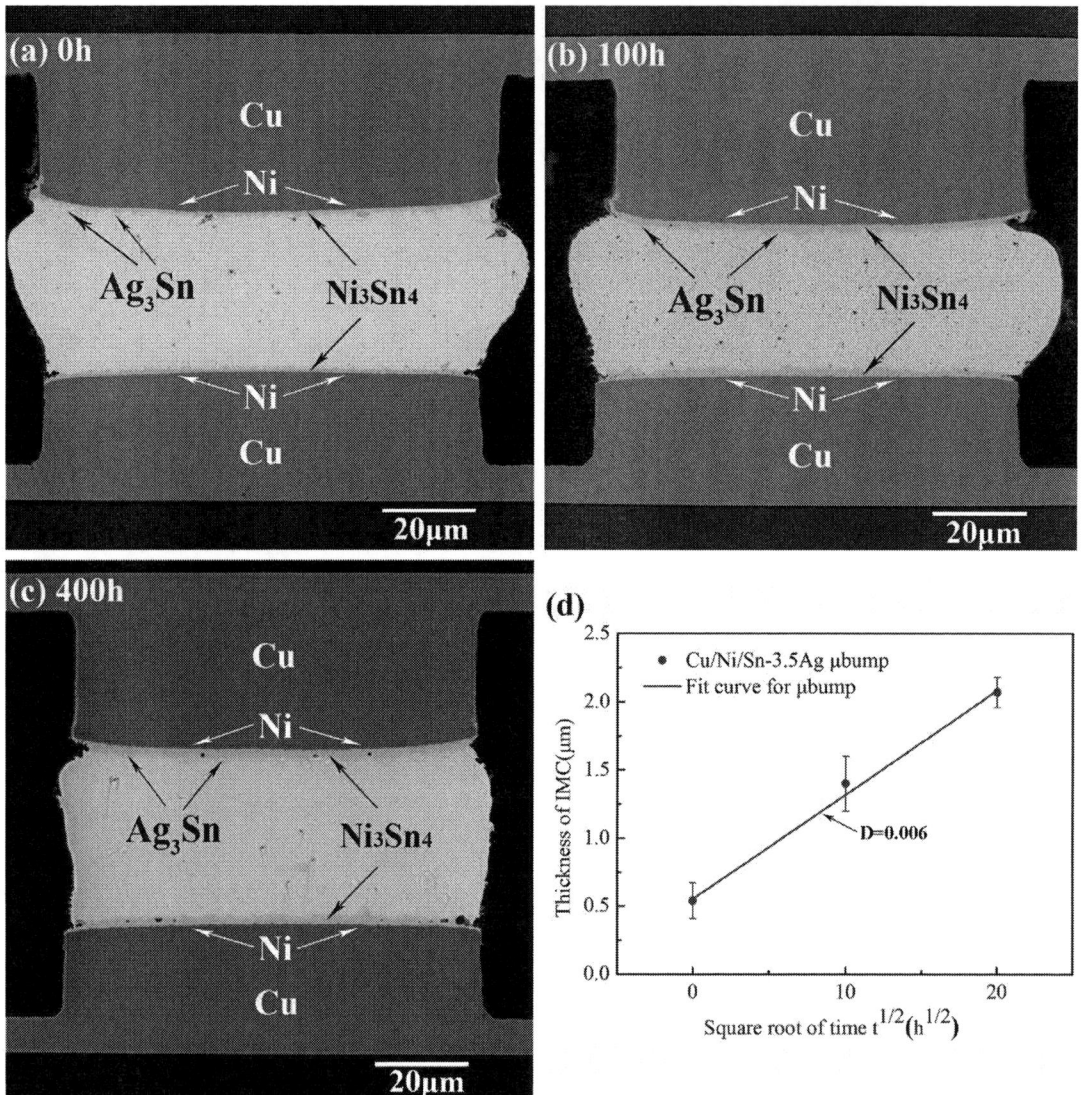

Fig. 5 Cross-sectional microstructure of Cu/Ni/Sn-3.5Ag sample after aging at 150 ℃ for (a) 0 h, (b) 100 h, (c) 400 h and (d) variation of IMC thickness with aging time square root.

The growth behavior of IMC at the cold and hot ends of the interface of the microsolder joints was different under the temperature gradient. As can be seen from Figure 4 (condition 1), the thickness of the Ni_3Sn_4 at the cold end was 1.87 μm and 1.60 μm after TM 50 and 100 hours, respectively, which were 0.18 μm and 0.45 μm thicker than the original thickness. While the thickness of the Ni_3Sn_4 intermetallic compound on the hot side decreases by 0.11 μm and 0.34 μm, indicating that the cold-side intermetallic compound growth was mainly derived from the hot-side IMC as well as the dissolution of the Ni matrix. The results of condition 2 show that the thickness of the Ni_3Sn_4 IMC at the lower interface significantly increased by 0.23 μm and 0.51 μm after 50 h and 100 h, indicating that the IMC growth is faster at the colder interface, which is consistent with the

As shown in Figure 5, the interfacial IMC of Cu/Ni/Sn-3.5Ag microbump is steadily increasing after isothermal aging at 150 °C for 0 h, 100 h, and 400 h. The average values were 0.54 μm, 1.40 μm and 2.07 μm, respectively. The aging temperature were adopted from the mean temperature of the solder joints in the heat migration test. The interfacial IMC of the solder joint after bonding is less. With isothermal aging, the Ni/Sn-3.5Ag interface reacts to form Ni_3Sn_4 IMC and grows symmetrically within the micro solder joint, as shown in Fig. 5(b), (c). the Ni_3Sn_4 thickness grows linearly with the square root of time, and the fitting results are shown in Fig. 5(d).

Figures 6 shows that the growth rate of IMCs is faster than isothermal aging under thermal migration. As is well known, it is more difficult to generate IMC at the Ni/Sn interface than at the Cu/Sn interface during isothermal aging. [8]. However, as shown by our experimental results, the presence of temperature gradients is accelerating the formation of all-IMC solder joints, which significantly affects the thermo-mechanical reliability and lifetime of the micro-interconnected solder joints.

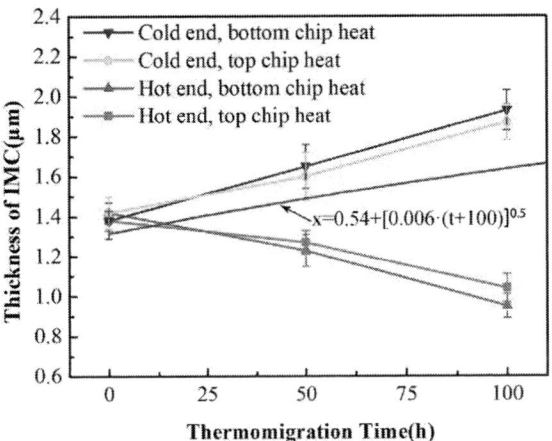

Fig. 6 The relationship between the Ni_3Sn_4 layer thickness and thermomigration time.

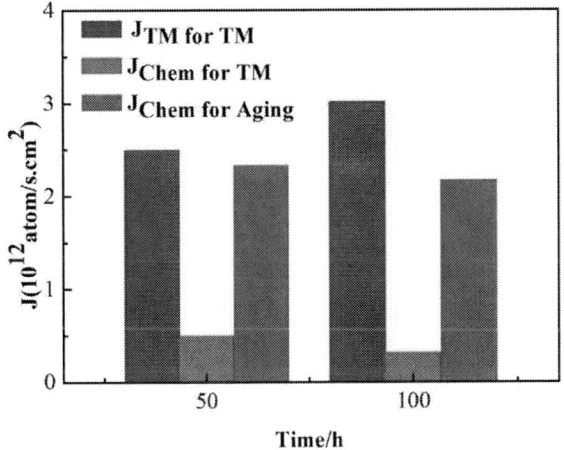

Fig. 7 The fluxes caused by chemical potential gradient and temperature gradient as a function of time.

B. The growth kinetic of interfacial Ni_3Sn_4 during temperature gradient

From Fig. 5(d), it can be seen that the linear slope is the square root of the growth rate constant D. The growth rate value is D = 0.006. The equation for the growth of Ni_3Sn_4 thickness in the micro-bump is given in (1).

$$x = 0.54 + \sqrt{0.006t} \qquad (1)$$

Figure 6 concludes the mean thickness with heat migration time of Ni_3Sn_4 IMC at the hot and cold sides of the micro-bump. The fitted curves of the unilateral IMC growth thickness for isothermal aging samples are also shown in Figure 6. Since the sample has been isothermally aged for 100 h before the thermal migration experiment, the time parameter in the unilateral IMC growth curve of the isothermally aged sample in the figure is t=t+100. The

specimens tested by thermal migration exhibit a trend of Ni_3Sn_4 growth at the cold side and IMC dissolution at the hot side. And the cold end IMC thickness is close to linear growth, which is greater than the growth rate of IMC of isothermal aging samples. And the thickness of Ni_3Sn_4 at the hot side is similarly close to a linear decrease. When the heat sourcing and heat spreaders are inverted, the sample IMC growth also shows the same trend.

The brittle Ni_3Sn_4 thickness at the cold side of the specimen after 100h isothermal aging and 100h thermomigration test is thicker than the unilateral IMC after 400h isothermal aging. It indicates that temperature gradients in the solder joint have a far more significant damaging effect on the joint than constant temperature work and is well worthy of concern.

In this paper, the effect of temperature gradients on the growing of IMC layers is investigated by studying the growth of Ni_3Sn_4 layers at the Ni/Sn-3.5Ag interface. When temperature gradient is applied in brazing materials, the interfacial reaction actuated by the chemical potential gradient and temperature gradient must be regarded. During the interfacial reaction, the Ni atomic flux (J_{TM}) motivated by the temperature gradient is from the hot side to the cold side. In contrast, the Ni atomic flux (J_{Chem}) driven by the chemical concentration gradient flows from the Ni substrate to the solder. the thickness variation of Ni_3Sn_4 is governed by the net flux into the IMC layer, as shown in equations (2) and (3).

$$J_{Ni,cold} = J_{Chem} + J_{TM} \qquad (2)$$

$$J_{Ni,hot} = J_{Chem} - J_{TM} \qquad (3)$$

The net atomic fluxes at both sides are substituted into equations (2) and (3) to obtain J_{Chem} and J_{TM}. It is assumed that all solubilized Ni atoms at the hot side diffuse to the cold side and that J_{Chem} is the same at the hot and cold side interfaces. The calculations of J_{Chem} and J_{TM} for different times are plotted in Fig. 7. After TM 50 h and 100 h, the J_{TM} values for microsolder joints are 2.495 atom/cm^2·s and 3.015 atom/cm^2·s, respectively. The J_{Chem} values for microsolder joints are 0.495 atom/cm^2·s and 0.315 atom/cm^2·s, respectively. Besides, After Aging 50 h and 100 h, the J_{Chem} values for microsolder joints are 2.33 atom/cm^2·s and 2.17 atom/cm^2·s, respectively.

In thermomigration test, there is flux competition between the chemical potential gradient and the temperature gradient at the hot side of the sample. When the flux of atoms driven by the temperature gradient is greater than that by the chemical potential ($J_{TM} > J_{Chem}$), the IMC decreases. However, the atomic flux driven by the chemical potential and motivated by the temperature gradient is in the same direction at the cold side of the sample. The IMC always increases. Compared to isothermal aging, thermal migration has an additional temperature gradient as a driving force, so IMC growth at the cold side of the microsolder joint will be faster than isothermal aging. In addition, the Ni barrier layer at the hot side will be consumed rapidly and is more easily exhausted than isothermal aging, affecting the reliability and lifetime of the solder joint and even the whole device.

IV. CONCLUSION

Based on the experimental results of thermal migration and isothermal aging of microsolder joints with a diameter of

100 μm and a solder height of 30 μm, the following two important conclusions were obtained:

(1) The IMC at the hot and cold sides of the joint grows abnormally under the effect of temperature gradient. The cold end IMC growth is faster than isothermal aging, while the hot end IMC and Ni barrier layer dissolve rapidly.

(2) In this paper, the diffusion flux of Ni atoms in the Sn matrix under the temperature gradient is obtained by calculating the Ni_3Sn_4 IMC growth rate J. The diffusion flux of Ni atoms will be larger than in isothermal aging due to the presence of two driving forces, the temperature gradient and the chemical concentration gradient, in the thermal migration experiment. Therefore, the phenomenon of temperature gradient accelerating cold-end IMC growth can be explained by IMC growth kinetics. The temperature gradient also accelerates the dissolution of the Ni barrier layer, which seriously affects the reliability and lifetime of the microsolder joints and even the whole device.

ACKNOWLEDGMENT

This work is supported by the Fundamental Research Funds for the Central Universities of Central South University, China, under Grant 1053320200169 and Grant 2021zzts0144.

REFERENCES

[1] Chen C , Hsiao H Y , Chang Y W , et al. Thermomigration in solder joints[J]. Materials Science & Engineering R, 2012, 73(9-10):85-100.

[2] Ouyang F Y , Tu K N , Lai Y S , et al. Effect of entropy production on microstructure change in eutectic SnPb flip chip solder joints by thermomigration[J]. Applied Physics Letters, 2006, 89(22):1045.

[3] Ouyang F Y , Kao C L . In situ observation of thermomigration of Sn atoms to the hot end of 96.5Sn-3Ag-0.5Cu flip chip solder joints[J]. Journal of Applied Physics, 2011, 110(12):022110.

[4] Yeh E , Choi W J , Tu K N , et al. Current-crowding-induced electromigration failure in flip chip solder joints[J]. Appl.phys.lett, 2002, 80(4):580-582.

[5] Gan H , Tu K N . Polarity effect of electromigration on kinetics of intermetallic compound formation in Pb-free solder V-groove samples[J]. Journal of Applied Physics, 2005, 97(6):337-344.

[6] Huang A T , Gusak A M , Tu K N , et al. Thermomigration in SnPb composite flip chip solder joints[J]. Applied Physics Letters, 2006.

[7] Ke J H , Chuang H Y , Shih W L , et al. Mechanism for serrated cathode dissolution in Cu/Sn/Cu interconnect under electron current stressing[J]. Acta Materialia, 2012, 51(5):2082-2090.

[8] M.S. Park, S.L. Gibbons, R. Arroyave, Phase-field simulations of intermetallic compound growth in Cu/Sn/Cu sandwich structure under transient liquid phase bonding conditions, Acta Materialia 60(18) (2012) 6278-6287.

978-1-6654-1392-3/21 $31.00 © 2021 IEEE

Improvement of Au electrode by glass optimization for LTCC application

line 1: 1st Tingnan Yan
line 2: *Shenzhen Institute of Advanced Electronic Materials, Shenzhen Institute of Advanced Technology*
line 3: *Chinese Academy of Sciences*
line 4: Shenzhen 518055, China
line 5: yantingnan2021@163.com

line 1: 2nd Dawei Wang
line 2: *Shenzhen Institute of Advanced Electronic Materials, Shenzhen Institute of Advanced Technology*
line 3: *Chinese Academy of Sciences*
line 4: Shenzhen 518055, China
line 5: wangdawei102@gmail.com

line 1: 3rd Yuanyuan Wang
line 2: *School of Materials Science*
line 3: *Shanghai University of Engineering Science*
line 4: Shanghai 201620, China
line 5: wangyy6140@163.com

line 1: 4th Jinhao Jia
line 2: *College of Aerospace Science and Engineering*
line 3: *National University of Defense Technology*
line 4: Changsha, 410073, China
line 5: jiajinhao12@foxmail.com

Abstract—The quality of low temperature co-fired ceramics (LTCC) modules can be optimized by increasing the conductivity of electrode and the bonding strength of electrode/ceramic. In this work, the properties of Au electrode in LTCC technology were optimized by different glasses. Two kinds of glasses with similar high-temperature viscosity were synthesized by using the traditional conventional melt quenching technique. High-temperature wettability of substrate by glass melt were studied by scanning electron microscopy (SEM). The optimized performance of Au electrode sheet resistivity of 2.49 mΩ/sq and wire bonding strength of 10.22 g was achieved, when the high-temperature wettability between substrate and glass was increased.

Keywords—high-temperature wettability, sheet resistivity, wire bonding, Au electrode, LTCC

I. INTRODUCTION

The rapid growth of high-frequency communications such as the fifth generation of mobile communication (5G) requires advanced high-performance ceramic materials with high signal-transmission speed, low loss, high thermal conductivity and excellent chemical stability substrate [1, 2]. LTCC meets the above demands ascribe to the modularization packaging, excellent dielectric properties, design flexibility, and well compatibility with Cu, Au and Ag electrodes [2]. Au electrode own many advantages over Ag and Cu in some environments, such as space, humid and heat environment, which is mainly due to its high electromigration resistance and oxidation–reduction stability [2, 3].

The sheet resistivity and wire bonding strength are the most important performance for Au electrode in LTCC practical applications [4]. The loss of signal in Au electrode will be significantly decreased with increasing the Au electrode sheet resistivity. Increasing the interfacial bonding strength will improve the reliability of Au/ceramic structure. Recent studies showed that reducing the content of glass could increase the conductivity of Au electrode [4], but decreasing the content of glass would reduce the wire bonding strength of Au electrode remarkably and lead to the mismatch of cofiring between Au electrode and substrates It is well-known that the glasses used in Au electrodes can change the sintering shrinkage behavior and increase the strength of Au/ceramic interfacial bonding, which cannot be improved by only reducing the content of glass in practical applications.

It was reported that the residual glass in Au electrode damaged the conductivity of samples [5]. When the Au electrode is cofiring with substrates, a part of glass in Au electrode will flow downward to the substrates and form a frit bonding in the Au/ceramic structure interface, which can increase the conductivity and wire bonding strength of electrode [2, 5]. So far, to our knowledge, there is no research reported that the effect of glass's high-temperature wettability on the flow behavior of Au electrode, which is important for guiding the design of Au electrode for LTCC technology.

In this work, to study the relationship between the high-temperature wettability of substrates and the properties of Au electrode, two kinds of glasses with similar thermodynamic properties and high temperature viscosity were designed and synthesized for the preparation of Au electrode paste. The high-temperature wettability of substrate by glass melt and the cofiring behavior of Au/ceramic were studied by scanning electron microscope (SEM). The results show that decrease of the high-temperature wettability angle between glass and substrate will promotes the glass flow to the substrate, which can increase conductive of Au electrode, and enhance the bonding strength of Au/ceramic.

II. EXPERIMENTAL

The ingredients of substrate and glasses in this paper were shown in Table 1. To prepare the glasses, the chemical reagent were melted at 1550 °C for 150 min. The molten glasses were quenched in the distilled water. The glassy slag crushed to powder by ball mill. The substrates were produced by casting process. The dielectric loss (tanδ), dielectric constant (ε_r), mechanical strength (σ) and coefficient of thermal expansion (CTE) of the laminated substrates sintered in 870 °C for 15 min was summarized in the Table 2.

The Au powder was manufactured according to the article of F. UCHIKOBA [6]. The SEM picture of Au powder was illustrated in Fig. 2(c). The preparation process of Au electrode paste is as follows: Firstly, the resins, organic solvent, dispersant and thixotropic agent was mixed at 90 °C for 120 min to produce resin solution. Secondly,

978-1-6654-1392-3/21 $31.00 © 2021 IEEE

the resin solution was fully mixed with the Au powder and glass by a speed Mixer (FlackTek, Germany). Mass ratio of resin solution: glass: Au powders were 27: 3: 70. Finally, the mixed compound was mixed by a roll mill to uniformly disperse all compositions in the Au paste. The Au electrode containing Glass-1 and Glass-2 was referred to as Au-Glass-1 and Au-Glass-2, respectively. To manufactured the Au/ceramic multilayer, the Au paste was silk-screen printed on the substrate to form the electrode. The size of Au electrode for sheet resistivity test was 100 mm^2. The substrates were then laminated at 65 °C/25 MPa for 300 seconds. The substrates were sintering in 500 °C with a heat rate of 3.0 °C/min for 2 h for debinder.

The sintering shrinkage behaviour of glasses and substrate were tested by a dilatometry (NETZSCH 402C, Germany). High-temperature viscosity of glasses was tested by a rheometer (Anton Paar MCR302, Germany). The amplitude, frequency and heating rate of complex dynamic shear viscosity ($|\eta^*|$) test was 0.2 %, 2 Hz and 5 °C/min, respectively. The SEM images of the samples and the high-temperature wettability of substrate by glass melt were tested by a scanning electron microscope (JEOL JSM–7900F, Japan). The wire bonding testing using an Au wire of 25 μm was welded to the Au electrode by a Au wire bonder (WESTBOND-7700D, America).

TABLE 1 THE INGREDIENTS AND PERFORMANCE OF GLASSES AND SUBSTRATE

Properties	CaO (mol. %)	B$_2$O$_3$ (mol. %)	SiO$_2$ (mol. %)	CaF$_2$ (mol. %)
Glass-1	40	20	36	4
Glass-2	42	24	34	0
Substrate	42.6	12.4	45	0

TABLE 2 THE PERFORMANCE OF SUBSTRATE SINTERED AT 870 °C

Properties	ε_r	tanδ (10^{-3})	CTE (ppm/°C)	σ (MPa)
Substrate	6.1	1.2	5.7	≥150

III. RESULTS AND DISCUSS

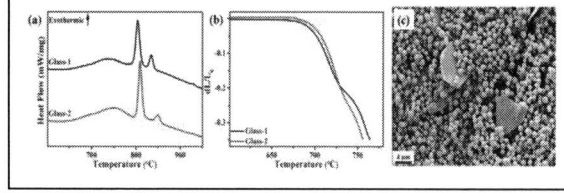

Fig. 1. The (a) DSC curve, (b) sintering shrinkage of glasses and (c) the SEM of Au powder.

The thermodynamic properties and high temperature viscosity of glasses have significant effect on the cofiring behavior of the Au/ceramic structure [4, 6]. The DSC curve and sintering shrinkage of glasses are shown in Fig. 1 (a, b). The initial crystallization temperature (T$_{c, onset}$) and softening point (T$_s$) of Glass-1 and Glass-2 are 795 °C and 696 °C, 800 °C and 700 °C, respectively. It is obvious that the difference in the T$_{c, onset}$ and T$_s$ between Glass-1 and Glass-2 is unremarkable. The microstructure of Au powder is shown in Fig. 1 (c), the size of particles are about 1.6 μm, and the

flaky Au powders can enhance yield stress and improve the printing properties of Au paste.

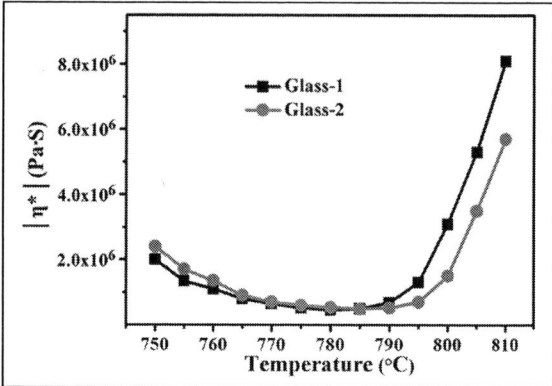

Fig. 2. The shear viscosity curves of glasses.

The viscosity curves of glass are shown in Fig. 2. With the increase of temperature, the viscosity of glass decline first and then enhance. It is reported that the precipitated crystal in glass will lead to the increase of viscosity with the rise of temperature [5]. Fig. 2 shows that when the temperature is lower than the crystallization temperature of samples, no remarkable difference in the high-temperature viscosity is observed between Glass-1 and Glass-2, which indicates that the similar thermodynamic properties and high-temperature viscosity of Glass-1 and Glass-2 will have unobvious influence on the flow behavior of both glasses in Au electrode.

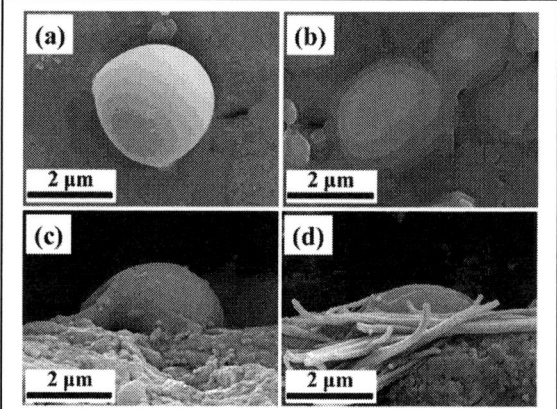

Fig. 3. SEM images of high-temperature wettability of substrate by glass melt at 770 °C, surface: (a) Glass-1 and (b) Glass-2; cross section: (c) Glass-1, (d) Glass-2.

The flow behavior of glasses in the Au electrode during cofiring process could be influenced by the high-temperature wetting behavior between glass and substrates. Fig. 3 shows the SEM images of high-temperature wettability of substrate by glass melt. The high-temperature wettability of substrate by Glass-2 is better than Glass-1, and the wetting angle of Glass-1 and Glass-2 are about 85 ° and 30 °, respectively. The viscosity and interfacial tension of glass are the main factors affecting the high temperature wettability of substrate by glass melt [4, 5]. The difference of high-temperature viscosity between Glass-1 and Glass-2 is small, as discussed in Fig. 2. As a result, the lower interfacial tension between Glass-2 and substrate will increase the high temperature wettability of substrate by Glass-2 melt.

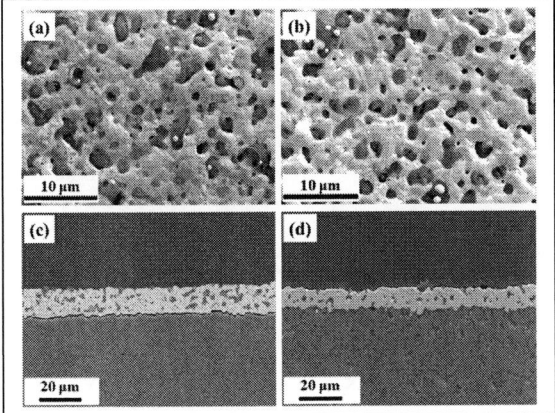

Fig. 4. SEM images of surface: (a) Au-Glass-1 and (b) Au-Glass-2; cross section: (c) Au-Glass-1, (d) Au-Glass-2 cofiring at 850 °C for 10 min.

To reveal the flow behavior of glass in Au electrode, the distribution of glass in Au electrode is observed by SEM. Fig. 4 shows the SEM surface image of Au-Glass-1 after cofiring with substrate in 850 °C. It is obvious that the content of glass in the surface and cross section of Au-Glass-2 is obviously decreased compared with Au-Glass-1 after co-firing with substrate, which indicates that the increased wettability of substrate by glass melt promotes the glass flow to the substrate in the co-firing process. It is proposed that the improvement of high-temperature wettability of substrate by glass melt is beneficial to the glass penetrating into the interior of substrate and spreading on the surface of substrate, promoting the flowing of glass in Au electrode down to the substrate.

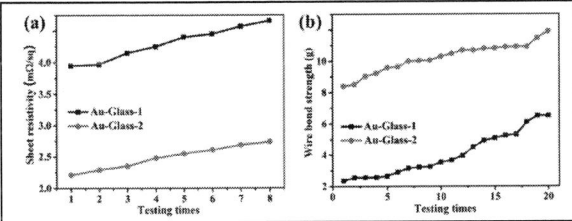

Fig. 5. The (a) sheet resistivity and (b) wire bonding test of Au electrode.

Fig. 5 (a) shows the Au electrode sheet resistivity after co-firing with substrate, and the thickness of Au electrode was 25 μm. The average sheet resistivity of Au-Glass-1 and Au-Glass-2 is 4.30 mΩ/sq and 2.49 mΩ/sq, respectively. It is obvious that the conductivity is increased with reduce glass content in the Au electrode, along with the results in Fig. 4. The wire bonding strength of Au electrode after co-firing with substrate is shown in Fig. 5 (b). The average strength of Au-Glass-1 and Au-Glass-2 in wire bonding test are 4.03 g and 12.22 g, respectively. The average strength of Au-Glass-2 is much higher than that of Au-Glass-1, which is due to that the Glass-2 flows to substrate and forms more frit bonding compared to Glass-1, increasing the bonding strength of Au-Glass-2/ceramic cofired structure.

IV. CONCLUSIONS

In this paper, the high-temperature wettability of substrate influence by two different glass melts on the quality of Au electrode for LTCC technology was studied. Increase in the high-temperature wettability between substrate and glass will promote the glass flow to substrate in the cofiring process. The glass flowing out of Au electrode forms much more frit bonding in Au/ceramic structure interface, not only enhancing the conductive of Au electrode, but also increasing the interface strength of Au/ceramic cofired structure. This study of the high-temperature wettability of substrate on the performance of Au electrode will open an opportunity for choosing the glass of Au electrode for LTCC practical applications.

REFERENCES

[1] D. Nowa, A. Dziedzic, "LTCC package for high temperature applications," Microelectron Reliab, vol. 51, pp. 1241–1244, 2011.

[2] Y. Imanaka, "Multilayered Low Temperature Cofired Ceramics (LTCC) Technology," Springer Science: Boston, 2005, pp. 59–79.

[3] V.R. Manikam, K.Y. Cheong, "Die attach materials for high temperature applications: a review," IEEE T COMP PACK MAN, vol. 1, pp. 457–478, 2011.

[4] T. N Yan, W. J Zhang, X. Y Chen, Z. F Liu, H. J Mao, F. L Wang, "Improvement of Au electrode conductivity after cofiring with CaO–B_2O_3–SiO_2 green tapes for LTCC application," Ceram. Int, vol. 46, pp. 493–499, 2020.

[5] T. N Yan, X. Y Chen, W. J Zhang, F. L Wang, Z. F Liu, K. L Zhang, "Decrease in the camber degree of Au/ceramic co-fired structure for LTCC technology," J. Mater. Sci. Mater, vol. 31, pp. 17225–17232, 2020.

[6] K. Lutz, M. Golla, Process for the Production of Au Powder, (1975) US3885955.

Study on The Heat Dissipation Performance of Symmetrical Broken-line Microchannel Radiator

Pengfei Wang
School of Material Science and
Engineering,Xiamen University of
Technology
Xiamen, China
China Electronic Product Reliability
and Environmental Testing Research
Institute
Guangzhou, China
wpf1995a@163.com

Hongyue Wang
China Electronic Product Reliability
and Environmental Testing Research
Institute
Guangzhou, China
wanghongyue@pku.edu.cn

Yijun Shi*
China Electronic Product Reliability
and Environmental Testing Research
Institute
Guangzhou, China
syj20094870@sina.com

Xiangjun Lu*
School of Material Science and
Engineering,Xiamen University of
Technology
Xiamen, China
luxiangjun0531@163.com

Jile Xu
School of Material Science and
Engineering,Xiamen University of
Technology
Xiamen, China
Guangzhou, China
xujile1996@163.com

Abstract—In this paper, the heat dissipation of symmetrical broken-line microchannel radiator is analyzed. Firstly, the symmetrical broken-line microchannel model is established based on the ordinary broken-line microchannel model, and two models are simulated and compared under the same conditions to compare their heat dissipation capacity. Then, the symmetrical broken-line microchannel model is orthogonally optimized based on factors of the outer corner radius, pipe spacing and pipe width-height ratio. The heat dissipation capacity is analyzed with the average temperature of heat source as the characterization. This study provides theoretical guidance for subsequent microchannel cooling technology.

Keywords—Cooling capacity, simulation, heat dissipation uniformity, symmetrical broken-line microchannel

I. INTRODUCTION

At present, with the continuous development of the IC technology, the IC chips are on the increased level of integration, which will greatly increase the power dissipation density, subsequently increasing the temperature of the IC chips and threatening the characteristic and reliability. Increasing the heat propagation velocity can effectively reduce the temperature of the IC chips. The traditional air cooling uses fans to blow air to achieve the purpose of cooling, So it has disadvantages such as vibration and high noise, and the heat dissipation capacity is not very ideal due to the low thermal conductivity of the air itself, it is only suitable for the heat dissipation of ordinary chips with low integration and computing speed[1].Because there are no air fans in the liquid cooling radiator with microchannel, it will not produce vibration and the noise will be relatively smaller. In addition, there are several waterways inside the microchannel, which can give full play to the advantages of water cooling and take away more heat[2].

Since the concept of microchannel cooling was first proposed by Tuckerman and Pease in the 1980s[3], researchers have carried out related work on microchannel radiators. Lan-Ying Zhang et al[4]. investigated three microchannel structures integrated on low-temperature co-fired ceramic substrates, including straight, helical and I-

shaped. It is found that the helical microchannel with 1mm thermal through hole has the best heat transfer performance, and the copper thermal through hole significantly reduces the temperature gradient of the I-shaped microchannel. Xu Guoqiang et al[5]. studied the Y-shaped microchannel structure and optimized its structural parameters. It was found that, under the same hydraulic diameter and same convective heat transfer area, the Y-shaped microchannel structure had higher heat transfer efficiency and lower inlet and outlet pressure drop compared with the straight-row microchannel. Finally, it is concluded that the best heat dissipation performance is obtained when the channel series is 3, the bifurcation Angle is 60° and the number of branches is 2. M. Marzougui et al[6]. made a comparative study of the ordinary broken-line microchannel and the straight-row microchannel, and compared the heat dissipation performance of the two by taking the thermal resistance and heat transfer coefficient as the main parameters. The study found that the ordinary broken-line microchannel was more effective in cooling than the straight-row microchannel.

The researchers have studied the different groove structures and the corresponding conclusions are obtained. However, little attention is paid to the temperature distribution of microchannel radiator. This paper will focus on the temperature distribution of microchannel radiator and take the structure parameters of the symmetrical broken-line microchannel as the research object to analyze the influence of structural parameters on heat dissipation capacity. This work put forward a better heat dissipation scheme according to the optimization combination, providing theoretical guidance for the follow-up research on microchannel heat dissipation.

II. MODEL BUILDING AND SIMULATION ANALYSIS

A. Physical model

In this paper, a symmetrical broken-line microchannel model based on ordinary broken-line microchannel is established. The model includes a heat source, a substrate and a channel. The size of the heat source is 5000 um×5000 um×100 um, the size of the substrate is 6000 um×6000 um×600 um, the width and height of the microchannel are 100 um and 300 um, and the distance between each channel is 120 um. The model is shown in Fig. 1. In the model, the heat source is made of platinum, the substrate is made of silicon, and the coolant is made of water and the contact between heat source and radiator is air.

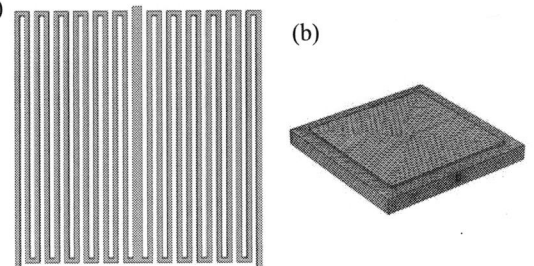

Fig.1. Physical model:(a) Microchannel model;(b) Mesh generation

In the simulation, the coolant flow mode is turbulence, and the applied load and boundary conditions are shown in Table I.

TABLE I. LOADING AND BOUNDARY CONDITIONS

Environment temperature	25°C
Inlet temperature	25°C
Inlet mass flow	13m/s
Coefficient of natural convection heat transfer in air	10W/(m²·k)
Thickness of contact gap between heat source and radiator	5nm
Dissipation power	75W

The following assumptions are made:

- Ignoring radiation heat transfer

- fluid incompressible, single-phase flow

- Ignoring gravity and other forces

- The channel wall is the non-slip boundary condition

- The physical properties of liquids and solids are

  independent of temperature

- Fluid flow and heat transfer are three dimensional

  steady states

In order to test the influence of grid number on simulation accuracy, we take half of the symmetrical broken-line microchannel model as an example to test the mesh sensitivity. In the examination, three kinds of quality grids with rough, conventional and refined are employed for symmetrical broken-line microchannel model, the test results are shown in Table II. This shows that the relative errors of conventional and refined grids on the substrate average temperature are less than 1%. Therefore, the conventional grid can reach the required accuracy .

TABLE II. GRID INDEPENDENCE TEST RESULTS

Mesh quality	Computing time	Substrate average temperature(°C)	Relative error
Rough	2min12s	74.833	/
Conventional	7min2s	74.884	0.06%
Refined	11min18s	74.650	0.3%

B. Theoretical model

Because the radiation heat transfer is ignored, the main heat transfer methods in this paper are conduction heat transfer and convection heat transfer. The total thermal resistance of the model consists of three parts: contact thermal resistance(R_{cont})[7], conduction thermal resistance(R_{cond}) and convective thermal resistance(R_{conv}), It can be expressed by the following formula:

$$R_{th} = R_{cont} + R_{cond} + R_{conv} \qquad (1)$$

When the heat transfer is stable, the contact resistance and conduction resistance tend to be stable, so only the convection resistance is considered. The calculation formula can be express as:

$$R_{conv} = \frac{1}{hA} \qquad (2)$$

Where h is the convective heat transfer coefficient (W/(m²·k)); A is convective heat transfer area(m²).

According to Newton's cooling formula, the convective heat transfer coefficient can be calculated as follows:

$$h = \frac{q}{T_w - T_f} \qquad (3)$$

Where q is the conduction heat flux (W /m²); T_w is the average wall temperature (K);T_f is the average water temperature (K).

According to the empirical formula[8], the convective heat transfer coefficient can be expressed as:

$$h = \frac{\lambda}{d} Nu \qquad (4)$$

$$Nu = 0.012(Re^{0.87} - 280) Pr^{0.4}[1 + (\tfrac{d}{L})^{2/3}] (\tfrac{Pr}{Pr_w})^{0.11} \qquad (5)$$

Where λ is the thermal conductivity of water(W/(m·k)); d is the hydraulic diameter(m); Re is the Reynolds number; Nu is the Nusselt number; Pr is the Prandtl number; L is the microchannel length(m).

C. Simulation analysis

A comparative simulation analysis was made between the ordinary broken-line microchannel and the symmetrical broken-line microchannel, and the simulation results were shown in Fig. 2 and Fig. 3. From the temperature cloud map of the two, it can be seen that the symmetrical broken-line microchannel has better heat dissipation capacity, and the maximum temperature of the heat source is 65.96°C, while the maximum temperature of the heat source of the ordinary broken-line microchannel is 89.60°C, and the heat dissipation is obvious. In addition, it can be seen from the temperature cloud map that the temperature distribution of the symmetrical broken-line microchannel is more excellent.

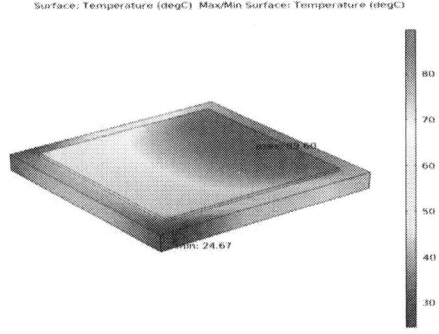

Fig.2. Ordinary broken-line temperature nephogram

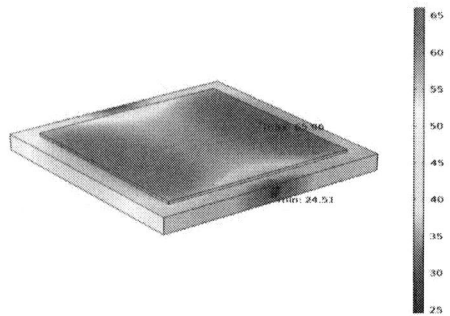

Fig.3. Symmetrical broken-line temperature nephogram

According to literature [9], the uniformity of heat source temperature distribution can be calculated by the following formula:

$$\omega = \frac{\sqrt{(T_{b,max}-T_{b,avg})(T_{b,avg}-T_{b,min})}}{qA} \quad (6)$$

Where $T_{b,max}$, $T_{b,avg}$, $T_{b,min}$ are the maximum, average and minimum temperatures of the heat source (℃); q is the heat flux of the heat source (W /m²); A is the area of the heat source (m²).

The calculation results are shown in Table III.

TABLE III.   UNIFORMITY RESULTS

Indicators	Ordinary	Symmetrical
Temperature uniformity	0.334℃/W	0.146℃/W

The results also confirm that the symmetrical broken-line microchannel has a better temperature distribution, which is also a factor of its excellent cooling effect.

Fig. 4 is a comparison chart of convective heat transfer coefficient calculated by empirical formula and simulation formula. The results show that the error between theoretical calculation and simulation calculation is within 20%, which is an acceptable error range, indicating the accuracy of the numerical mode.

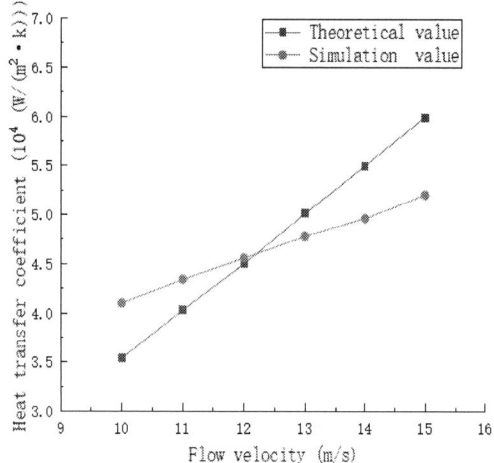

Fig.4. Validation results

III. ORTHOGONAL OPTIMIZATION

Orthogonal experimental design is a kind of experimental design method which studies many factors and many levels. According to the orthogonality, some representative points

were selected from the comprehensive test to carry out the test. These representative points have the characteristics of uniform dispersion and neat comparison. Orthogonal experimental design is the main method of fractional factorial design. It is an efficient, rapid and economical experimental design method.

In this paper, orthogonal experiment is used to optimize the structure of the symmetrical broken-line microchannel. The structural parameters include the outer corner radius of the channel, the pipe width-height ratio and the pipe spacing. These three parameters are taken as three factors, and three levels are selected for each factor. The horizontal relationship between factors and level pairs is shown in the table below.

TABLE IV.   FACTORS AND LEVELS CORRESPOND TO TABLES

Factors	Levels
Outer corner radius(um)	0、50、100
Pipe width-height ratio	1:1、1:2、1:3
Pipe spacing(um)	100、120、140

According to the steps of orthogonal experiment design, the orthogonal test table is designed. The experiment consisted of three factors and three levels, and the orthogonal table obtained was shown in Table V, including a total of nine parameter combinations. Each combination corresponds to a finite element simulation model, and the average temperature of the heat source in each model is obtained through simulation. The simulation results obtained are shown in the last column of Table V.

TABLE V.   THREE FACTORS AND THREE LEVELS ORTHOGONAL TEST TABLE

Serial number	Outer corner radiu（um）	Pipe width-height ratio	Pipe spacing（um）	Average temperature（℃）
1	0	1:1	100	87.893
2	0	1:2	120	63.553
3	0	1:3	140	60.091
4	50	1:1	120	95.247
5	50	1:2	140	70.493
6	50	1:3	100	57.658
7	100	1:1	140	99.764
8	100	1:2	100	67.429
9	100	1:3	120	57.540

A. range analysis

The range analysis of the average temperature of the heat source obtained under the conditions of different parameter combinations in Table V is carried out, and the range analysis results obtained are shown in Table VI.

TABLE VI.   RANGE ANALYSIS TABLE

Index average	Outer corner radius (um)	Pipe width-height ratio	Pipe spacing (um)
Mean1	70.512	94.301	70.993
Mean2	74.466	67.158	72.113
Mean3	74.911	58.430	76.783
Range	4.399	35.871	5.790

As can be seen from the analysis table, the order of range from large to small is pipe width-height ratio, pipe spacing and outer corner radius. It can be seen that among the three structural parameters, the pipe width-height ratio has the greatest influence on the average temperature of the heat source, followed by the pipe spacing, and then the outer corner radius.

Fig.5-7 shows the effect curves of each structural parameter, from which we can intuitively see the influence intensity of each structural parameter, It is clear that the pipe width-height ratio has the greatest effect.

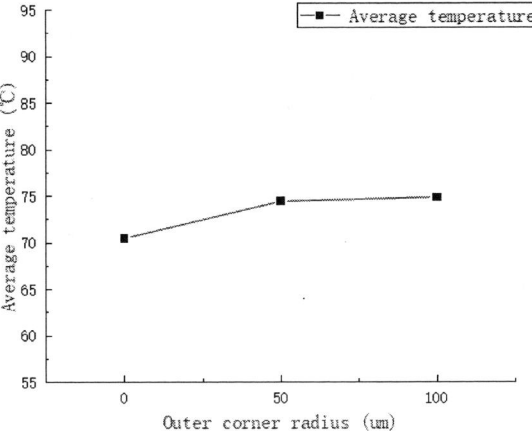

Fig.5. Outer corner radius effect curve

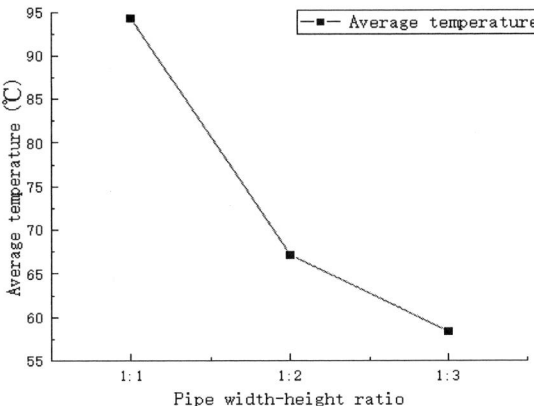

Fig.6. Pipe width-height ratio effect curve

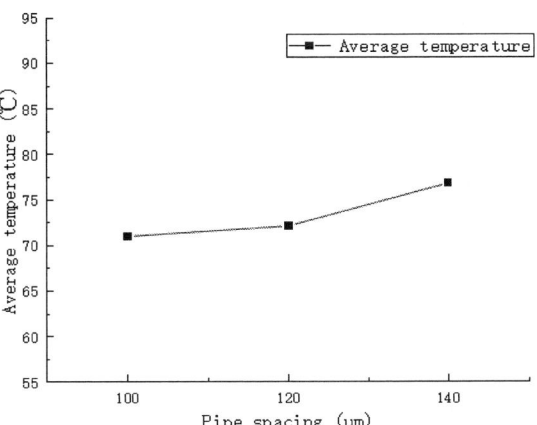

Fig.7. Pipe spacing effect curve

In addition, it can also be concluded from the range analysis table that the optimal combination to achieve the lowest average temperature is the radius of the outer corner of 0 um, the pipe width-height ratio of 1:3, and the pipe spacing of 100 um.

IV. CONCLUSION

1) Under the same boundary conditions and the same structure size, the heat dissipation performance of the symmetrical broken-line microchannel is better than that of the ordinary broken-line microchannel. The max temperature of the symmetrical broken-line microchannel is lower 20°C than that of the ordinary broken-line microchannel, and the surface temperature is more uniform.

2) The orthogonal optimization analysis results show that the average temperature of the microchannel radiator is most affected by the pipe width-height ratio of the channel, followed by the distance between the channels and the outer corner radius. According to the range analysis, the optimal structure combination with the minimum average temperature can be obtained: the radius of the outer corner is 0 um, the pipe width-height ratio of the channel is 1:3, and the distance between the channel is 100 um.

ACKNOWLEDGMENT

This work was supported by the National Key Laboratory stability support funding project under Grant:JBSY212800040.

REFERENCES

[1] Liu Yu, Gao Hongyan. Discussion on electronic chip cooling technology and its application [J]. Heilongjiang Science and Technology Information, 2010. (In Chinese)

[2] Zhang Xiaojing, Yi Zhihua. New Technology and New Process, 2008(01):10-11+51-53. (In Chinese)

[3] Tuckerman D B , Pease R . High-performance heat sinking for VLSI[J]. IEEE Electron Device Letters, 1981, 2(5):126-129.

[4] Zhang L Y , Zhang Y F . Simulation on heat transfer of microchannels and thermal vias for high power electronic packages[C]// International Conference on Electronic Packaging Technology. IEEE, 2014.

[5] Xu Guoqiang, Wang Meng, Wu Hong, et al. Numerical analysis of heat transfer characteristics in Y-shaped microchannel flow [J]. Journal of Beijing University of Aeronautics and Astronsutics, 2009, 35(3):313-317. (In Chinese)

[6] Marzougui M , Hammami M , Maad R B . Experimental study on thermal performance of heat sinks: the effect of hydraulic diameter and geometric shape[J]. Heat and Mass Transfer, 2016, 52(10):1-10.

[7] Wu Shengtao, Zeng Kejie, Liu Heng. Calculation of Thermal Contact Resistance and Numerical Simulation in ICEPAK Environment [J]. Communications Technology, 2013(01):105-108. (In Chinese)

[8] Huang min. Fundamentals of thermal and fluid mechanics [M]. China Machine Press, 2003

[9] Vinodhan V L , Rajan K S . Fine-tuning width and aspect ratio of an improved microchannel heat sink for energy-efficient thermal management[J]. Energy Conversion and Management, 2015, 105(NOV.):986-994.

FCCSP(MUF) Mold-flow Void Risk Prediction with Different Substrate Surface and Bump Height Design

Freedman Yen
Cooperate R & D
SPIL
Taichung, Taiwan, China
Freedman@spil.com.tw

Nicholas Kao
Cooperate R & D
SPIL
Taichung, Taiwan, China
Nicholas kao@spil.com.tw

David Lai
Cooperate R & D
SPIL
Taichung, Taiwan, China
Davidl@spil.com.tw

Yu Po Wang
Cooperate R & D
SPIL
Taichung, Taiwan, China
Ywang@spil.com.tw

Abstract Today's flip chip scale packaging (FCCSP) microelectronics products are becoming more and more complex, especially in the packaging process using molded underfill manufacturing, which reduces the cost of flip chip technology such as molded underfill (MUF) High, provides molding ability in molding. The problem is to leave effective space under the chip. Generally speaking, this requires a lot of experimental DOE (chip size, substrate design, and epoxy resin type) to solve this problem. For the reasons mentioned above, mold-flow simulation software can be applied to use different flow condition to obtain the best solution as the MUF FCCSP of the substrate structure design. Such a simulation method can help shorten the product development cycle.

This article uses the commonly used 3D molding process simulation software for transferring manufacturing process parameters. This article will introduce a comparison of two different structures. A MUF is FCCSP, it is matched with different bump height (SOH) structure comparison (control the flow of different chip bottom space), and observe the difference of the flow melt-front. The second is that there are two different designs on the surface of the substrate (welding masks with completely open or partially open patterns) through the table to enter the previously separated welding mask samples (welding masks with 10um depth structure) to maintain different modes. As this research, we can get a lot of conclusion that improve the molding capability of molded underfill FCCSP in the transfer molding process. If the bottom of the molded underfill FCCSP chip is matched with a structure with a height of 55 μm, then the molding epoxy can easily flow under the area with a large flow space, so the risk of entrapment can be reduced. In addition, the molding compound also has good melt front fluidity, and the fully open substrate surface design of the substrate solder resist layer can achieve more 10μm space underneath the die area. Then, the CAE output are consistent as the experiment and it can to predict whether the product has a risk assessment of void.

Keywords—FCCSP; MUF; Solder mask; Partial Open

INTRODUCTION

Use the MUF (molding mold) process for the mold transfer process of the flip chip device. As shown in Pic. 1, the use of traditional transfer molding technology can provide better assembly solutions. Compared with the general capillary underfill package, it can reduce the process time, material and equipment cost of these several advantages. Because this application uses a lot of mold property technology and traditional transfer molding machine type molding experience. However, the gap between the die and the substrate is very narrow, and it is still hardly to obtain steady MUF property characteristics and optimal molding conditions. Due to this serious problems of void capture and void capture [1], the challenge is the small gap in this impact region underneath mold. This leads to an imbalance between the melt front and the flow friction. As a lot of previous case studies, scanning sound microscopy was also used to observe the capture phenomenon [2].

Pic.1 FC Die w/ MUF structure package

The combination of DOE matrix designed in many experiments is usually used to solve the problem of entrapment. However, optimizing mold design is just a "trial and error method." However, due to the complex interaction between the fluid flow, which is very difficult. As described above, the epoxy compound (EMC) and the heat of polymerization. Advances in hardware and software of the computer-aided engineering (CAE) has become a powerful software to do complex physical phenomena microchip contained in plastic packaging.

The main objective of this paper is the use of CAE technology to develop a useful tool for linking pre-analysis. Packaging and structuring and post-processing provide integrated solutions for flip-chip packaging.

Herein, using Moldex3D R12 MUF transfer molding process using mold 3D flow simulation apply for molding development, in order to reduce defects and increase production tool. -Macosko MUF for measuring the viscosity of the epoxy resin acts, using a parallel plate rheometer and DSC (differential scanning calorimetry) to obtain rheological parameters.

Three-dimensional flow analysis

A. Governing Equations:

In principle, the chip package is a three-dimensional, and the

978-1-6654-1392-3/21 $31.00 © 2021 IEEE

reaction of the process transient problem, affecting the flow of the resin melt front moving. Flows into the cavity of the non-isothermal resin by the following formula [3] described mathematically:

$$\frac{\partial \rho}{\partial t} + \nabla \cdot \rho \mathbf{u} = 0 \tag{1}$$

$$\frac{\partial}{\partial t}(\rho \mathbf{u}) + \nabla \cdot (\rho \mathbf{u}\mathbf{u} - \boldsymbol{\sigma}) = \rho \mathbf{g} \tag{2}$$

$$\boldsymbol{\sigma} = -p\mathbf{I} + \eta\left(\nabla \mathbf{u} + \nabla \mathbf{u}^T\right) \tag{3}$$

$$\rho C_P\left(\frac{\partial T}{\partial t} + \mathbf{u} \cdot \nabla T\right) = \nabla(k\nabla T) + \Phi \tag{4}$$

u: velocity vector,   T: temperature
T: time,   p: pressure
σ: total stress tensor   ρ: fluid density,
k: thermal conductivity   Cp: specific heat,
Φ: energy ,

Energy contains two contributions:

$$\Phi = \eta\dot{\gamma} + \dot{\alpha}\Delta H \tag{5}$$

η:viscosity,   Ẏ: magnitude of the rate
α: conversion rate   ΔH:exothermic heat of polymerization.

B. Rheokinetic Characterization of Molded Underfill

In this transfer molded process, the dynamic flow and heat transfer occurred curing reaction [4]. Kinetics of the curing reaction not only impacts the level of conversion of the epoxy compound, but it is also flow mode with a great influence, because the curing reaction increases the viscosity. Viscosity is also mainly affected by temperature and shear rate. Thus, molded plastic rheological behavior modeling is very important for the molding process.

To characterize the rheological behavior of commercial EMC, the use of the test sample and the Panasonic industrial molding process. For analysis, measurement of the physical characteristics, like as cure kinetics, viscosity, heat capacity and thermal conductivity. At three different heating rates, the use of a non-isothermal DSC scan mode measurement curing kinetics. Curing conversion (0 <α <1.0) using experimental data by numerical parameter fit relationship Lotus [5,6]. Fitting parameters as shown in Table 1. The numerical fitting experimental data line and the heating rate of 5,10,20°C / min good agreement, as shown in Pic. 2 & Pic. 3.

$$\frac{d\alpha}{dt} = \left(k_1 + k_2\alpha^m\right)(1-\alpha)^n$$

$$k_1 = A_1 \exp\left(-\frac{E_1}{RT}\right)$$

$$k_2 = A_2 \exp\left(-\frac{E_2}{RT}\right)$$

Where α is the conversion of reaction, *A1 A2 E1 E2 m n* are model parameters.

Parameter of Kinetics	Unit	Value
m	-	4.65E-01
n	-	1.30E+00
A	1/s	3.57E+03
B	1/s	3.47E+05
Ta	k	8.33E+03
Tb	k	6.33E+03
H	J/kg	3.10E+08

Table1 data of the Curing fitting parameters

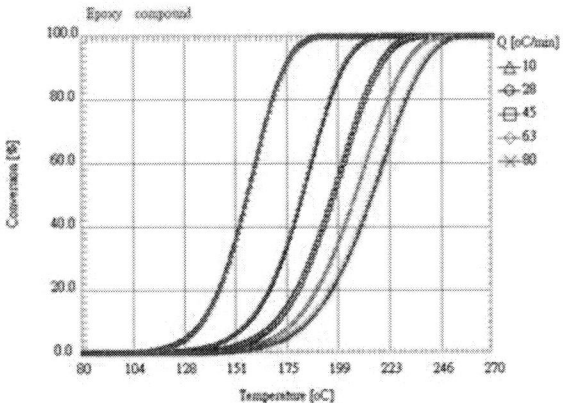

Pic.2 Curing Kinetics Curves_1

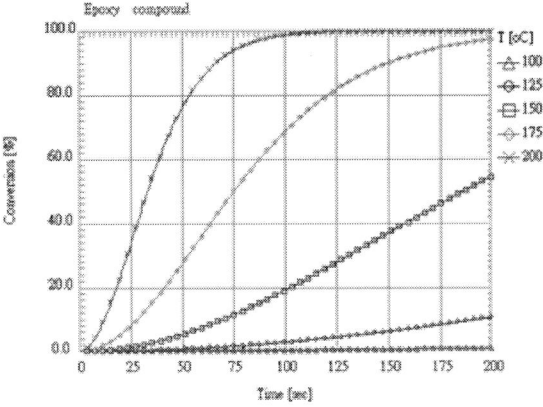

Pic.3 Curing Kinetics Curves_2

Related processing conditions the viscosity of a parallel plate rheometer at different temperatures, shear rate and degree of cure viscosity to a viscosity change over time. The following viscosity measurement Castro Macosko Cross model [7] is compatible. The fitting parameters as Table 2, data sets and curve fitting results shown in Pic. 4 and 5.

Parameter of Kinetics	Unit	Value
α_g	-	5.15E-01
C 1	-	4.60E+00
C 2	1/s	-4.83E+00
B	1/s	1.18E-16
T b	k	1.73E+04
n	k	4.91E-01
τ^*	J/kg	0.90E+01

Table 2 Cross Castro Macosko model

$$\eta = \frac{\eta_0}{1 + \left(\frac{\eta_0 \gamma}{\tau^*}\right)^{1-n}} \left(\frac{\alpha_g}{\alpha_g - \alpha}\right)^{c_1 + c_2 \alpha}$$

$$\eta_0 = B e^{\frac{T_b}{T}}$$

Where γ is the shear rate, α is a transformation, n being the power law exponent, η_0 is the zero shear viscosity, τ^* is a parameter representative of the transition region between the zero shear rate viscosity curve and the power law region.

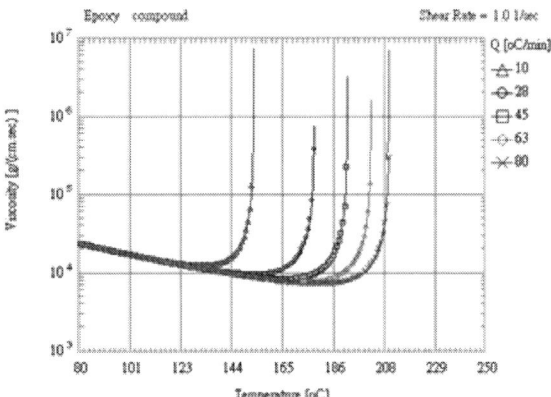

Pic.4 Viscosity Curves: Viscosity & Temperature

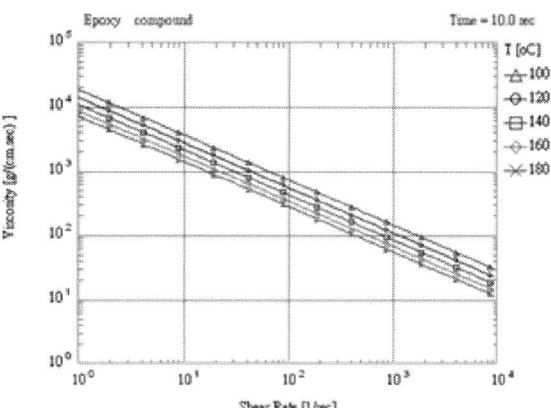

Pic.5 Viscosity & Shear rate

Results and discussions

Typically, manufacturing a PKG assembly can result in delamination, solder extrusion, so FCCSP package void problem is a major problem. When the package improves reliability of the temperature cycle test.

In this paper, we use the MUF (molded underfill) molding process FCCSP selected package type, and the last location area and flow distribution void was observed. Accordingly, the present paper describes two structures. One is the surface of a circuit substrate surface pattern design, wherein the solder resist layer layout is completely open and the partially open design.

Another is the bump dimension. FCCSP carrier having a large bump height dimension structure can increase a space formed between the substrate and the mold, thus having a large bump height support significant improvement melt front balance and reduce flow resistance.

As shown, the two types of the solder resist table. 3 (open and all open portions) for different bump height dimension of 25, 35, 45 and 55 um comparison.

Test Device	SM Open design	Bump Standoff height (um)
Vehicle1	Partial Open	25
		35
		45
		55
Vehicle2	All Open	25
		35
		45
		55

Table3 DOE for SM surface and SOH size

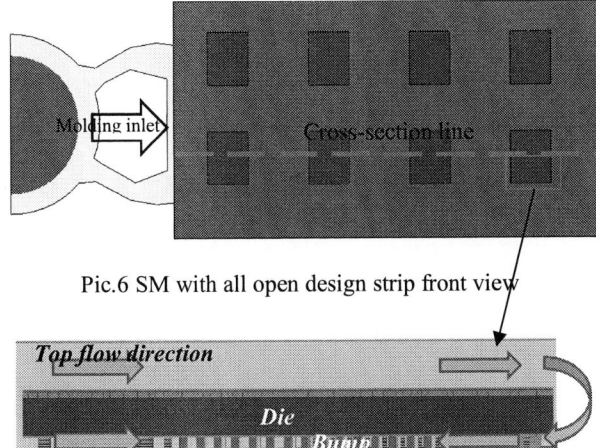

Pic.6 SM with all open design strip front view

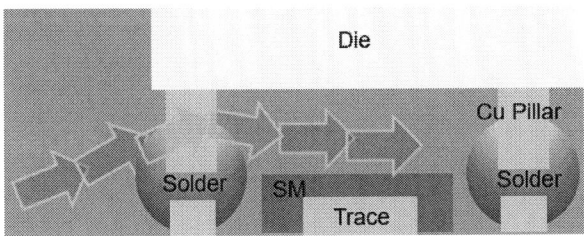

Pic.7 SM with all open design strip cross-section

To compare the all open and partially open pattern design, the design uses a partially open design of 5um solder resist layer (SM) as shown in Pic.8. SM partially open design results in a greater resistance to flow molding process forming.

Pic.8 SM w/ partial open design

Pic.9 SM w/ Partial Open & SOH as 55um

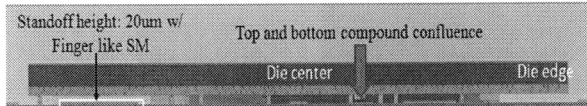

Pic.10 SM w/ all open design & SOH as 55um

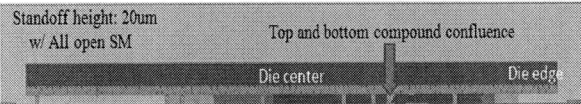

Pic.11 SM w/ partial open design & SOH as 25um

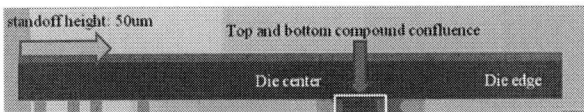

Pic.12 SM w/ all open design & SOH as 25um

Molded test results following vehicle 2, when the bump height of the support is reduced from 55um to 25um, top and bottom of the flow merging position below the melt front forming die becomes close to the central region of the mold, since the smaller 25um SOH exhibit significant flow resistance and melt front imbalance, leading to the merging close to the mold center. The flow pattern is shown in Pic. 13 to Pic. 16.

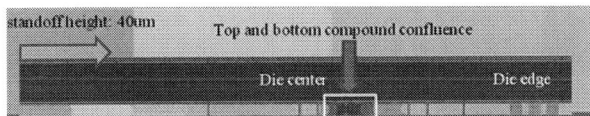

Pic.13 SM w/ all open design & SOH as 55um

Pic.14 SM w/ all open design & SOH as 45um

Similar to the molding results of test vehicle 1 above, on the surface of the SM substrate with a partially open design, the epoxy melt front under the die at the confluence of the upper and lower flows is away from the edge of the die. Increases the risk of void problems.

Pic.15 SM w/ all open design & SOH as 35um

Pic.16 SM w/ all open design & SOH as 25um

As shown in Pic 17, the abutment FCCSP of 55um bump height, the last row of the two packages appeared empty, but when the seat height of the bump is 25um, the void size will grow and appear in more packages. Therefore, the molded melt front under the die in the upper and lower flow confluence positions near the center of the die raises the FCCSP with the risk of air voids during the MUF process.

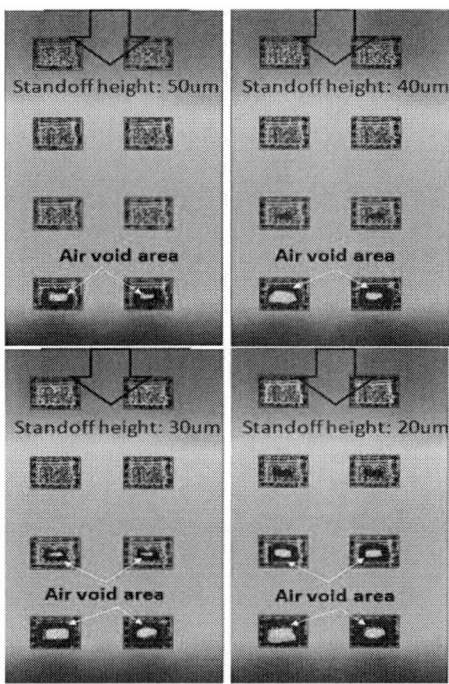

Pic.17 SM with all open design & SOH 55~25um
(Bottom substrate view)

Conclusions

In this paper, we have shown that 3D simulation software can be applied to predict the molding process of flip-chip package using MUF. Proper design of the bump standoff height is an important factor in improving the forming capacity of the upper and lower flow confluences for FCCSP, which has a large bump standoff height dimension in the package structure.

Close the die edge to reduce the risk of void problems. Conversely, all open-surface design boards perform a more balanced melt at the front of the upper and lower flow confluence under the die area, which increases the flow space between the dies.

References

[1] M. Joshi, "Molded Underfill (MUF) Technology for Flip Chip Packages in Mobile Applications," IEEE ECTC proceedings 2010, pp. 1250-1257, 2010.Components.

[2] Cheng Xu, et al, "Void risk prediction for molded underfill technology on flip chip packages" Proceedings of the 2017 18th International Conference on Electronic Packaging Technology.

[3] George J. Scot, "Heterogeneous Integration Using Organic Interposer Technology," 2020 IEEE 70th Electronic Components and Technology Conference (ECTC), pp. 885-892.

[4] M.R. Kamal and M.R. Ryan, "Injection and Compression Molding Fundamentals." chap. 4, A.I.Isayev (Ed.), Marcel Dekker, 1987.

[5] T. Hasegawa, "Wafer Level Compression Molding Compounds," Proceedings of 2012 Electronic Components & Technology Conference, pp. 1400-1405.

[6] J. M. Castro, and C. W. Macosko, "Kinetics and Rheology of Typical Polyurethane Reaction Injection Molding Systems," SPE Technical Paper, Vol. 26, pp. 434-438, 1980.

[7] Thomas Schreier-Alt, Frank Rehme, Frank Ansorge and Herbert Reichl, "Simulation and experimental analysis of large area substrate overmolding with epoxy molding compounds," Microelectronics Reliability, Vol. 51, Issue 3, pp. 668-675, 2011.

Research on Electromigration Behavior of Cu Pillar Bumps under Pulse Current Stress

JiLe Xu[1,2]
[1]School of Materials Science and Engineering, Xiamen University of Technology
[2]The Science and Technology on Reliability Physics and Application of Electronic Component Laboratory
Xiamen, China
xujile1996@163.com

XiangJun Lu*
School of Materials Science and Engineering, Xiamen University of Technology
Xiamen, China
luxiangjun0531@163.com

ZhiWei Fu*
The Science and Technology on Reliability Physics and Application of Electronic Component Laboratory
Guangzhou, China
fzw19940124@163.com

ChenBing Qu
The Science and Technology on Reliability Physics and Application of Electronic Component Laboratory
Guangzhou, China
quchenbing@126.com

Xiao Luo
The Science and Technology on Reliability Physics and Application of Electronic Component Laboratory
Guangzhou, China
957748731@qq.com

PengFei Wang
School of Materials Science and Engineering, Xiamen University of Technology
Xiamen, China
wpf1995a@163.com

Abstract—The electromigration (EM) of Cu/Sn-Ag1.8/Ni/Cu flip chip Cu pillar bumps structure under DC stress and bidirectional pulse current stress is studied. The temperature, thermal gradient and stress distribution of Cu pillar bumps under different current stress are obtained by ANSYS finite element simulation. Two groups of EM experiments under different current stress were carried out. At 125℃ and $3.5 \times 10^4 A/cm^2$, the resistance change rate of Cu pillar bumps under bidirectional pulse current stress was about 1/5 of the DC stress. After 220h of EM experiment, the bumps were completely alloyed into Cu_6Sn_5 and Cu_3Sn under DC stress, and the polarity phenomenon appeared on the Ni layer and Cu pad at the cathode side. However, under the stress of bidirectional pulse current, the growth rate of intermetallic compound (IMC) is obviously slow, there is a small amount of solder remaining in the bump and the solder is biased to the side of the Ni layer, and the Ni layer does not show obvious polarity phenomenon. At the same time, combined with the simulation, it is found that the Cu trace may also become a weak link of the reliability of the Cu pillar bump.

Keywords—Cu pillar bump, Pulse current stress, ANSYS, Microstructure

I. INTRODUCTION

As the development of advanced packaging technology, micro-bumps have become an important carrier and technical guarantee to support the miniaturization of high-density integrated devices. And the micro-interconnected Cu pillar bump has been widely used in the field of high-density flip chip packaging due to its advantages of good electrical conductivity, strong heat dissipation and high density[1-2]. Compared with the traditional solder joint, the diameter of the Cu pillar bump is smaller, and the current density to be carried by the bump is larger. Due to the difference between the Cu pillar bump structure and the traditional solder joint, the rapid consumption of under bump metal (UBM) and the formation of interface voids caused by current crowding effect are no longer the decisive failure mechanism of Cu pillar bumps

interconnection structure[3-5]. Less solder in Cu pillar bumps are easy to be completely alloyed, resulting in volume shrinkage and defects such as voids and cracks, leading to a series of serious electromigration (EM) reliability problems[6-8].

In recent years, there are many researches on thermoelectric reliability of Cu pillar bumps. Hui-Cai Ma et al. studied the EM life of Cu pillar bumps under DC stress and observed the microstructure evolution of interconnect structure under current stress[9]. Zhiwei Fu et al. carried out the experimental research on EM of micro-interconnected Cu pillar bumps structure under DC stress, and established the reliability model of micro-bump interconnection[10-11]. At present, the main method of Cu pillar bumps EM research is still using constant DC stress to evaluate its EM reliability. However, the load current stress of interconnect structure is not constant under practical conditions. In the case of CMOS devices, interconnect solder joints in the structure work mainly under pulse stress with varying size and direction. Compared with the constant DC stress, the migration rule of metal atoms which in the Cu pillar bumps structure under pulse current stress variation, so the corresponding EM failure mechanism is different. The failure mechanism under pulse current stress is not clear, so it is of great significance to study its failure mechanism.

In this paper, the EM behavior of Cu pillar bumps under pulsed current stress is studied. Firstly, based on ANSYS finite element simulation, the physical field distribution of Cu pillar bumps under constant DC and pulse current stress is compared and analyzed. Secondly, a multi-channel EM experimental system was established. The EM experiments of Cu pillar bumps under DC stress and bidirectional pulse stress were carried out. The microstructure changes and differences of bumps under DC stress and pulse stress were compared. Combined with the finite element simulation and microstructure analysis, the rule and mechanism of the microstructure change of the Cu pillar bump were discussed.

978-1-6654-1392-3/21 $31.00 © 2021 IEEE

II. FINITE ELEMENT SIMULATION

A. Finite element model

ANSYS simulation software was used to carry out thermally-electric-mechanical coupling simulation on the Cu pillar bump structure. Fig. 1(a) shows the simplified model of the sample structure. The Cu pillar bump structure includes plastic package, silicon chip, Cu trace, Cu pillar, Ni layer, Sn solder, intermetallic compound (IMC), BT substrate. Table I shows some material parameters of the model[12-13]. The three-dimensional finite element model of Cu pillar bumps is shown in Fig. 1(b). The current loads applied to the model are DC, PDC and bidirectional pulse current.

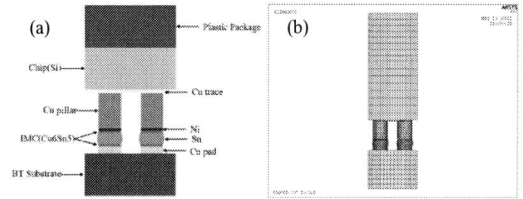

Fig. 1. The structure of the Cu Pillar Bump (a) Schematic diagram, (b) 3D finite element model.

TABLE I. MATERIAL PARAMETERS OF MODEL

Material	Young's Module (GPa)	Poisson's ratio	Thermal Conductivity (W/m·K)	CTE (10^{-6}/K)
Plastic Package	5	0.3	0.7	21
Si	190	0.27	150	2.8
Cu	124	0.31	398	16.5
Ni	210	0.312	91	13
Sn	41	0.36	67	23
Cu_6Sn_5	140	0.3	34.1	16.3
BT Substrate	17.9(xy) 7.85(z)	0.11(xy) 0.39(z)	0.6	12.4(xy) 57.0(z)

B. Simulation result

The results of the finite element simulation show that the distribution of temperature, thermal gradient and stress of Cu pillar bump simulated by finite element simulation under three different stress conditions are not significantly different in the overall structure of Cu pillar bumps.

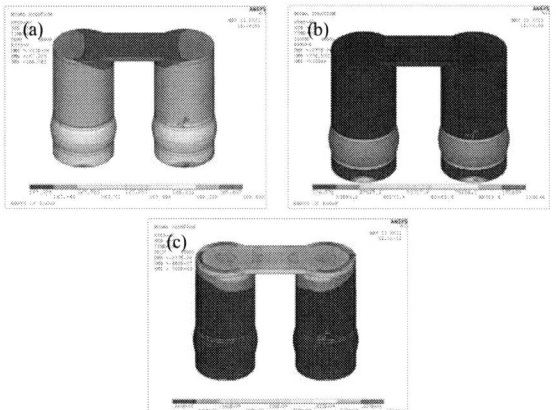

Fig. 2. Simulation results of the Cu pillar bump under bidirectional pulse stress (a) Temperature distribution contour plot, (b) Thermal gradient distribution contour plot, (c) Stress distribution contour plot.

As shown in Fig. 2, under the current size of 0.63A bidirectional pulse stress, the finite element simulation results show that the maximum temperature of the Cu pillar bump structure is located in the Cu trace, and the temperature reaches about 168℃. The solder with the Cu pillar bump structure and the IMC layer at both ends of the solder have a high thermal gradient, the thermal gradient value is about 270-530K/cm. The high thermal stress distribution area is mainly at the interface between the Cu trace and the chip. Due to the relatively high temperature in this area and the difference of thermal expansion coefficient (CTE) of different materials, large thermal stress is generated at the interface. The maximum thermal stress is 0.7×10^9Pa.

III. ELECTROMIGRATION EXPERIMENT

In this research, a lead-free flip-chip Cu pillar bump package test chip was adopted. The test sample is shown in Fig. 3(b). The samples were divided into group A and group B according to the experimental conditions. The experimental current type of group A was DC current, and that of group B was bidirectional pulse current. The positive and negative duty ratios of the bidirectional pulse current were both 50%, and the frequency was 1Hz. The experimental environment temperature of A and B was 125 °C, and the current density was 3.5×10^4A/cm^2. The applied actual current was 0.63 A. The current density was calculated by dividing the applied current by the cross-sectional area of the Cu pillar.

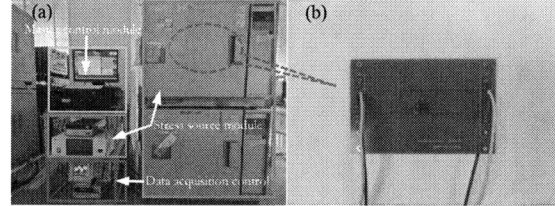

Fig. 3. EM experiment (a) System device, (b) The test sample

The multi-channel EM experimental system designed and built by ourselves was adopted in the experiment. As shown in Fig. 3(a), the EM experiment system is divided into three parts, the main control module, the stress source module and the data acquisition module. The main control module includes computer hardware equipment and software control program. The stress source module includes DC power, bidirectional pulse power and high temperature experiment chamber. Data acquisition module includes 2700 multimeter, oscilloscope, etc.

IV. RESULTS AND DISCUSSIONS

A. The resistance data

Fig. 4 shows the variation curve of sample resistance with time under DC stress and bidirectional pulse stress. Under the condition of DC stress, the resistance value increases slowly and linearly with time from 0 to 180h, with a growth rate of 21.9% compared with the initial value. Then, the resistance value increases rapidly until the open-circuit failure in a short period of time. However, under the bidirectional pulse current stress, it can be observed that the resistance growth rate is lower than that of the direct current stress, and the resistance increases by about 20% compared with the initial value in the time from 0 to 180h. The resistance change rate of Cu pillar bumps under bidirectional pulse current stress is about 1/5 of that of DC.

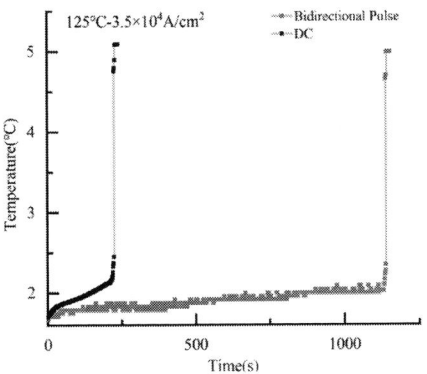

Fig. 4. The resistance curve of the Cu pillar bump

B. Microstructure analysis

The samples of group A and group B after 220 h of EM experiment were treated, and the interface microstructure of Cu pillar bumps was observed by scanning electron microscope (SEM).

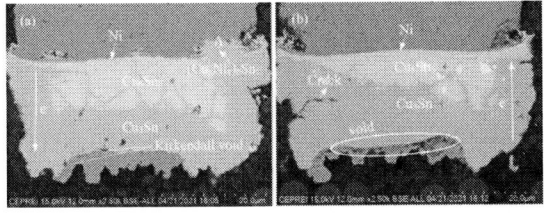

Fig.5. Microstructure of Cu pillar bumps of group A sample

Fig. 5 shows the microscopic morphology of Cu pillar bumps of samples in group A after 220h. (a) and (b) are solder joints at both ends of adjacent interconnected Cu pillar bumps. The electron flow direction of (a) is from Cu pillar end to Cu pad end, and that of (b) is from Cu pad end to Cu pillar end.

The Cu pillar bump showed obvious EM phenomenon under DC stress, and the bump solder had been completely alloyed into metal compounds Cu_6Sn_5 and Cu_3Sn. As shown in (a), a large amount of Cu_3Sn is generated at the solder/Cu pad interface as the anode, and accompanied by a small amount of voids. The Ni layer at the cathode side shows obvious dissolution and consumption. When the local area of the Ni layer is completely consumed to form A notch, A large number of IMCs are generated in the notch area. The composition of IMC at the position A is Cu-Ni-Sn ternary phase metal compound. Since the Ni layer hinders the diffusion of Cu atoms at the Cu pillar end to a certain extent, the Ni layer as the cathode only generates obvious Cu_3Sn compounds in the notch region. As shown in (b), when the electrons flow from the end of the Cu pad to the end of the Cu pillar, the Ni layer at the anode can effectively isolate the direct contact between the Cu pillar and the solder. The alloying reaction between the Cu pillar and the solder is blocked, resulting in the Sn solder mainly reacting with Cu atoms in the Cu pad, the Cu pad as the cathode is consumed

greatly. Therefore, the corrosion of the cathode Cu pad is more serious and accompanied by a large number of voids. The Ni atom on the anode side inhibits its diffusion to the cathode due to the effect of electron wind, showing that the layered morphology is intact. There is no obvious erosion of the Ni layer, but only the Cu pillar in the corner is eroded.

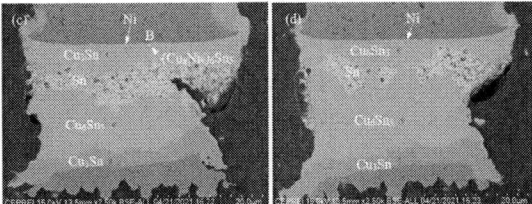

Fig. 6. Microstructure of Cu pillar bumps of group B sample

Fig. 6 shows the microscopic morphology of Cu pillar bumps of samples in group B after 220h. (c) and (d) are solder joints at both ends of adjacent interconnected Cu pillar bumps.

It is observed that the growth rate of IMCs at the Cu pillar bumps is obviously inhibited, and there is no obvious Ni layer erosion phenomenon. During one cycle of the bidirectional pulse current, the Cu pillar bump is subjected to continuous current action in both directions. When the electrons flow from the Cu pillar to the Cu pad, that is, the Cu pad side is the anode, the electron wind promotes the diffusion of Ni atoms and Cu atoms on the Cu pillar side. With the consumption of Ni atoms and the continuous decomposition and migration of IMCs at the cathode side, the limiting force of Ni layer and Ni-containing compounds on Cu atoms at the end of the Cu pillar is weakening. The growth rate of IMCs at both ends is accelerating until the Ni layer is completely consumed. With the continuous accumulation of Cu atoms in the anode, the growth of IMCs on the anode side was further accelerated. When the current is reversed, the electronic wind will inhibit the consumption of Ni atoms and the continuous decomposition of IMCs at the cathode side, so there is no obvious erosion of the Ni layer as the cathode compared with that under DC stress. However, the reverse current cannot completely inhibit the diffusion of Ni atoms. There are also Cu-Ni-Sn ternary metal compounds at the position B near the Ni layer, so the reverse current prolongs the time of Ni layer as a barrier layer to a certain extent.

When electrons flow from the Cu pad end to the Cu pillar end, the Cu pad side is the cathode. The Ni layer in the anode can effectively isolate the direct contact between the Cu pillar and the solder, and block the Cu pillar from participating in the alloy reaction of the solder. The Sn solder mainly reacts with the Cu atoms in the Cu pad, the Cu pad as the cathode is consumed in large quantities. However, when the current is reversed, the electronic wind force inhibit the migration of Cu atoms in the Cu pad to the end of the Cu pillar, resulting in the weakening of the reaction between Sn solder and Cu atoms in the Cu pad.

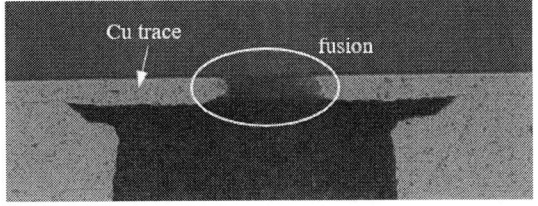

Fig.7. SEM images of Cu trace

During the experiment, it is found that the Cu trace may also be a hidden danger of the reliability of the Cu pillar bump, except that the continuous expansion of the void during the IMCs reaction leads to the cracking of the solder part of the Cu pillar bump. As shown in Fig. 6, when the current density of the chip sample is $4 \times 10^4 A/cm^2$ at 125°C, the circuit would be failed due to the Cu trace fusing. Combined with the ANSYS temperature simulation distribution, the Cu trace is the region with the highest temperature in the Cu pillar bump interconnection. When the current density is $3.5 \times 10^4 A/cm^2$, the highest temperature reaches about 168 °C, which confirms that the copper trace may also become a weak link in the reliability of the Cu pillar bump.

CONCLUSIONS

(1) ANSYS finite element simulation shows that the distribution of temperature, thermal gradient and stress of Cu pillar bumps under bidirectional pulse current stress and direct current stress are not significantly different in the whole structure. The maximum temperature is located in the Cu trace area, and the thermal gradient of the IMC layer at both ends of the weld ball and the welding ball is relatively high. The high thermal stress distribution area is mainly located at the interface between the Cu trace and the chip.

(2) At 125℃ and $3.5 \times 10^4 A/cm^2$, the resistance of the sample under the conditions of DC stress and bidirectional pulse stress with time curve shows that the resistance growth rate under the pulse current stress is lower than that under the DC stress, and the resistance change rate of the sample under the bidirectional pulse current stress is about 1/5 of that under the DC stress.

(3) The polarity of Ni layer and Cu pad is obvious due to the EM behavior under DC current stress. The Ni layer at the cathode side is partially eroded, and the cathode Cu pad is consumed and a large number of voids are produced. Under the stress of bidirectional pulse current, the continuous bidirectional current changes the original migration mechanism of Cu and Ni atoms, which leads to the growth rate of IMCs slow down. At the same time, due to the blocking property of Ni layer, the growth of IMCs on the Cu pad side is faster than that on the Cu pillar side, so that the remaining solder tends to the Ni layer side.

(4) It is found that the Cu trace may also be a hidden danger of the reliability of the Cu pillar bump. When the current density exceeds $4 \times 10^4 A/cm^2$, the circuit would be failed due to Cu trace fusing. Combined with ANSYS simulation, the Cu trace is the region with the highest temperature distribution. When the current exceeds a certain range, fuse failure is easy to occur.

ACKNOWLEDGMENT

This research was supported by the Development Fund of China Electronic Product Reliability and Environmental Testing Research Institute (No. 19D06), the Opening Project of Science and Technology on Reliability Physics and Application Technology of Electronic Component Laboratory (No.61428060201) and the Natural Science Foundation of Guangdong Province, China (No. 2021A1515011996).

REFERENCES

[1] M. Gerber, et al., "Next generation fine pitch Cu Pillar technology– Enabling next generation silicon nodes," 2011 IEEE 61st Electronic Components and Technology Conference (ECTC), Lake Buena Vista, FL, 2011, pp. 612-618.

[2] Ma H C, Guo J D, Chen J Q, et al. Reliability and failure mechanism of copper pillar joints under current stressing[J]. Journal of Materials Science: Materials in Electronics, 2015, 26(10):7690-7697.

[3] J. W. Nah, K. W. Paik, J. O. Suh, and K. N. Tu, "Mechanism of electromigration–induced failure in the 97Pb-3Sn and 37Pb-63Sn composite solder joints," Journal of Applied Physics, vol. 94, no. 12, pp. 7560-7566, 2003.

[4] D. Kim, J. -h. Chang, J. Park, and J. J. Pak, "Formation and behavior of Kirkendall voids within intermetallic layers of solder joints," Journal of Materials Science: Materials in Electronics, vol. 22, no. 7, pp. 703-716, 2011/07/01 2011.

[5] T. Laurila, J. Karppinen, J. Li, V. Vuorinen, and M. Paulasto-Kröckel, "Effect of isothermal annealing and electromigration pre-treatments on the reliability of solder interconnections under vibration loading," Journal of Materials Science: Materials in Electronics, vol. 24, no. 2, pp. 644-653, 2013/02/01 2013.

[6] K. N. Tu, "Recent advances on electromigration in very-large-scale-integration of interconnects," Journal of Applied Physics, vol. 94, no. 9, pp. 5451-5473, 2003.

[7] B. Kwak, S. Kim and Y. Park, "Current stressing effect on interfacial reaction characteristics of Cu pillar/Sn-3.5Ag micro-bumps for 3D integration," 2011 IEEE International 3D Systems Integration Conference (3DIC), 2011 IEEE International, 2012, pp. 1-2.

[8] B.-J. Kim et al., "Intermetallic Compound Growth and Reliability of Cu Pillar Bumps Under Current Stressing," Journal of Electronic Materials, vol. 39, no. 10, pp. 2281-2285, 2010/10/01 2010.

[9] H.-C. Ma et al., "Reliability and failure mechanism of copper pillar joints under current stressing," Journal of Materials Science: Materials in Electronics, vol. 26, no. 10, pp. 7690-7697, 2015/10/01 2015.

[10] Fu Z W, Zhou B, Yao R, et al. Research on thermal-electric coupling effect of the copper pillar bump in the flip chip packaging[C]. International Conference on Electronic Packaging Technology. 2016:1377-1380.

[11] B. Zhou, Z. Fu, Y. Huang, R. Yao and J. Zhang, "Thermo-electric coupling reliability model of copper pillar bump based on Black equation," 2017 18th International Conference on Electronic Packaging Technology (ICEPT), 2017, pp. 769-773.

[12] Z. Leng, M. Xiao, M. He, W. Xia and B. Wang, "Electromigration simulation of Cu pillar interconnect microstructure of 3D packaging," 2016 17th International Conference on Electronic Packaging Technology (ICEPT), Wuhan, 2016, pp. 234-239.

[13] L. Meinshausen, K. Weide-Zaage, B. Goldbeck, A. Moujbani, J. Kludt and H. Frémont, "Electromigration reliability of cylindrical Cu pillar SnAg3.0Cu0.5 bumps," 2014 15th International Conference on Thermal, Mechanical and Mulit-Physics Simulation and Experiments in Microelectronics and Microsystems (EuroSimE), Ghent, 2014, pp. 1-6.

Simulation on TSV Protrusion from Atomic to Micron Scales

Xiaoting Luo
The Key Laboratory of Low-carbon Chemistry & Energy Conservation of Guangdong Province, and School of Materials Science and Engineering
Sun Yat-sen University
Guangzhou, China
luoxt29@mail2.sysu.edu.cn

Zhiheng Huang*
The Key Laboratory of Low-carbon Chemistry & Energy Conservation of Guangdong Province, and School of Materials Science and Engineering
Sun Yat-sen University
Guangzhou, China
hzh29@mail.sysu.edu.cn

Yuezhong Meng
The Key Laboratory of Low-carbon Chemistry & Energy Conservation of Guangdong Province, and School of Materials Science and Engineering
Sun Yat-sen University
Guangzhou, China
mengyzh@mail.sysu.edu.cn

Shan Ren
The Key Laboratory of Low-carbon Chemistry & Energy Conservation of Guangdong Province, and School of Materials Science and Engineering
Sun Yat-sen University
Guangzhou, China
stsrs@mail.sysu.edu.cn

Hui Yan
School of Computer Science and Engineering
Sun Yat-sen University
Guangzhou, China
yanh26@mail.sysu.edu.cn

Qizhuo Li
School of Computer Science and Engineering
Sun Yat-sen University
Guangzhou, China
liqzh7@mail.sysu.edu.cn

Abstract—**Data-centric computing including data analytics, machine learning and AI is the main driving force for high-end performance 3D microelectronic packaging. TSV-based technology has enabled the Foveros 3D packaging from Intel, the 3D system integration SoIC from TSMC, and the Xtacking from YMTC. However, thermally induced protrusion of TSV still remains a reliability challenge and the routine finite-element based thermo-mechanical modeling and analysis can only find its limitations as variation in materials and packaging is becoming a much bigger and complex problem as process nodes pushed towards 3/2 nm nodes and beyond. While the simulation results from the previous PFC studies have shed some light on the atomic scale mechanisms underlying protrusion, further information regarding the distribution of internal stress and strain accompanying the atomic scale microstructural evolution and plastic deformation in deep-trench TSVs is missing. In addition, the sizes of the simulated TSVs are in the order of a few nanometers in diameter. It is necessary, therefore, to further extend the capability of the PFC model and make it capable of modeling TSVs and interconnects with dimensions covering the whole spectrum from atomic scale to nanoscale, and from nanoscale to micron scale. This paper aims to simulate TSV protrusion from atomic to micron scales using the recently developed APFC models and derive the strain and stress distributions in the TSVs accompanying the microstructural evolution. It is significantly important to obtain such data to assist the further understanding on the cross-scale mechanisms of TSV protrusion and the consequences of its variations on the functionality and reliability of individual devices, and even the integrated systems at the current and future more advanced process nodes.**

Keywords—*Advanced process node, 3D-TSV, protrusion, atomic scale microstructure, APFC model*

I. INTRODUCTION

Driven by the era of 5G, Internet of Things (IoTs), cloud computing and big data, the demand for chips is increasing and the requirements for chip performance are getting dramatically high. In recent years, the semiconductor process node has advanced from 7 nm to 5 nm and now is progressing towards the 3/2 nm node and beyond [1]. On May 9th, IBM just announced the world's first chip using 2 nm transistors. Interconnects, as part of the full 2-nm technology, however, remains yet to be correctly scaled as well [2]. Interconnect scaling is now starting to play a more important role in achieving improvements in product performance [3]. Last few years have seen leading research institutes and major companies actively developing cutting-edge technologies for 3D microelectronic packaging, and EMIB [4], COWOS [5] and X-Stacking [6] are typical examples.

Along with the rapid advancement of semiconductor process node, the feature size of interconnects are also forced to shrink. And with this dimension reduction, the material issues, in particular, microstructure and its evolution, in interconnects become particularly important. Among many interconnect technologies, through silicon via (TSV) is one of the enabling technologies to achieve 3D integration. Thermal management and reliability control of 3D-TSVs with diameters of a few microns or even nano TSVs have been conceived in advanced 3D integration, behavior of the filler materials at the nano and even down to the atomic and electron scales and interactions with surrounding materials have to be taken into consideration. Under the service condition and owing to the mismatch of the coefficients of thermal expansion between the filler material and the surrounding silicon, thermal induced stresses cause TSVs to deform and accumulation of plastic deformation could lead to protrusion, which is a serious reliability concern because it can destroy the redistribution layers (RDLs) of the interconnected devices, resulting in short or open circuits and thus the failure of chips [7].

To address the reliability problem of TSV protrusion, experimental technologies for accurate strain and stress characterization have been developing. Okoro *et al.* utilized synchrotron-based X-ray microdiffraction technique for stress of deep-trench Cu-TSV [8]. Raman Spectra was also used by Li *et al.* to nondestructively probe how the residual stress varies with depth of deep trenches [9]. While Su *et al.* adopted a micro-infrared photoelasticity system to evaluate stress filed in TSV [10]. However, those techniques are still limited in one way or another for probing the evolution of microstructure and the coupled temporal evolution of stress distribution in TSV, thus numerical simulation becomes an important and necessary remedy to overcome such difficulties.

The information regarding the evolution of dislocations, e.g. motion, proliferation and annihilation occurring over a short period of time, cannot easily be observed, which necessitates new simulation techniques and tools with resolution at the atomic scale in the study of TSV protrusion. Simulation using the phase field crystal (PFC) models has been reported by Liu *et al.* to reproduce the whole process of protrusion, in which the FCC crystal structure of copper and the microstructural evolution coupled with the external mechanical loading was fully considered [11-13]. However, the maximum size of the TSV system simulated in Liu's work only reached approximately 10 nm. Although the strain distribution was derived and plotted in Liu's work, the information on stress distribution is missing.

Goldenfeld *et al.* developed so-called amplitude expansion of PFC model, i.e. APFC, and further developments had made it capable of modeling materials with different crystalline symmetries [14,15]. Skaugen *et al.* derived and studied the crystal plasticity behavior in the PFC model from a perspective of mechanics [16] and proposed to impose an elastic equilibrium condition when solving the PFC model for plastic motion [17]. Salvalaglio *et al.* presented a coarse-grained description of deformation in crystals and plasticity based on the APFC model and introduced the implementation of the governing equations under the framework of the adaptive FEM toolbox AMDiS [18-21].

Based on the above-mentioned progresses, this paper simulates TSV protrusion from atomic to micron scales using the APFC models aiming to firstly substantially increase the size of the modelled TSVs and secondly derive the stress and strain distributions in the TSVs accompanying the microstructural evolution from the atomic to micron scales. It is significantly important to obtain such data to assist the further understanding on the cross-scale mechanisms of TSV protrusion and the consequences of its variations on the functionality and reliability of individual devices, and even the integrated systems at the current and future more advanced process nodes.

II. MODELLING METHOD

The PFC model describes a crystal by using a periodic field, i.e. the atomic density filed ψ, and the free energy functional of the system as a function of ψ can be written as [19]:

$$F_\psi = \int_\Omega \left[\frac{\Delta B_0}{2}\psi^2 + \frac{B_0^x}{2}\psi\left(1+\nabla^2\right)^2\psi - \frac{t}{3}\psi^3 + \frac{v}{4}\psi^4 \right]d\vec{r} \quad (1)$$

The evolution of the order parameter with time can be obtained by solving the following governing equation [15]:

$$\frac{\partial \psi}{\partial t} = M\nabla^2 \frac{\delta F}{\delta \psi} \quad (2)$$

where M is the mobility of boundaries, and $\delta F/\delta\psi$ denotes the first variation of the free energy functional versus the order parameter. In the crystallization state, ψ can be:

$$\psi = \psi_0 + \sum_q A_q e^{iq\cdot r} = \psi_0 + \sum_j^N A_j e^{iq_r\cdot r} + c.c. \quad (3)$$

where ψ_0 indicates the average atomic density of the crystalline state, which is set to be 0 in this study.

In the APFC model, the free energy functional can be formulated directly in terms of A_j and A_j^*, by the amplitude approximation for the atomic density, and F_ψ in Eq.(2) is replaced by F_A as following [19]:

$$F_A = \int_\Omega \left[\begin{array}{l} \frac{\Delta B_0}{2}\Phi + \frac{3v}{4}\Phi^2 + \sum_{j=1}^N \left(B_0^x \left|g_j A_j\right|^2 - \frac{3v}{2}\left|A_j\right|^4 \right) \\ + f^s\left(\{A_j\},\{A_j^*\}\right) \end{array} \right]d\vec{r}$$
$$(4)$$

where $g_j = \nabla^2 + 2i\vec{q_j}\cdot\nabla$ and $\Phi = 2\sum_{j=1}^N \left|A_j\right|^2$. The term $f^s\left(\{A_j\},\{A_j^*\}\right)$ is determined by the crystal structure and a functional with respect to A_j and A_j^*.

Then the governing equation Eq. (2) is transformed to [22]:

$$\frac{\partial A_j}{\partial t} = -\left|q_j\right|^2 \frac{\delta F}{\delta A_j^*} \quad (5)$$

In this study, a general-purpose platform integrating a PDE solver based on the Finite Element Method (FEM) – COMSOL Multiphysics, and a programming environment for scientific computing MATLAB was used for solving the governing equations of APFC model and the post-processing of simulation results.

We consider the configuration in both the geometries model named geometry 1 and 2, and place grains with different orientations in the disordered liquid phase. The complex-valued amplitudes of those nuclei which rotates by an angle θ ranging from 15° to 85° are given by [22]:

$$A_j = \phi_0 e^{i\delta q_j(\theta)\cdot r} \quad (6)$$

where

$$\delta q_j(\theta) = \left[q_j^x(\cos\theta-1)-q_j^y\sin\theta\right]+\left[q_j^x\sin\theta-q_j^y(\cos\theta-1)\right] \quad (7)$$

The strain and rotation field can be connected to the complex amplitude functions A_j [23]

$$A_j = \phi_j e^{ik_j\cdot u} \quad (8)$$

Equation (8) can be rewritten as [23]:

$$\varphi_j = q_j \cdot u \quad (9)$$

with $\varphi_j = \arg(A_j) = \arctan[\mathrm{Im}(A_j)/\mathrm{Re}(A_j)]$. The 2D deformation field u_i (with $i = x, y$) can be determined by inverting Eq. (9):

$$u_i^{2D} = \frac{e_{ij}}{\left|q_l\times q_m\right|}\left[q_m^j\varphi_l - q_l^j\varphi_m\right] \quad (10)$$

where e_{ij} is the 2D Levi-Civita symbol.

For a 2D system, the components of strain tensor ε can be calculated according to the following equations [23]:

$$\varepsilon_{xx} = \frac{1}{\left|q_l\times q_m\right|}\left[q_m^y\frac{\partial\varphi_l}{\partial x} - q_l^y\frac{\partial\varphi_m}{\partial x}\right] \quad (11)$$

$$\varepsilon_{yy} = \frac{1}{|q_l \times q_m|} \left[q_l^x \frac{\partial \varphi_m}{\partial y} - q_m^x \frac{\partial \varphi_l}{\partial y} \right] \quad (12)$$

$$\varepsilon_{xy} = \frac{1}{2|q_l \times q_m|} \left[q_m^y \frac{\partial \varphi_l}{\partial y} - q_l^y \frac{\partial \varphi_m}{\partial y} + q_l^x \frac{\partial \varphi_m}{\partial x} - q_m^x \frac{\partial \varphi_l}{\partial x} \right] \quad (13)$$

Analogously, ω can be given by [17]:

$$\omega = \frac{1}{2|q_l \times q_m|} \left[q_m^y \frac{\partial \varphi_l}{\partial y} - q_l^y \frac{\partial \varphi_m}{\partial y} - q_l^x \frac{\partial \varphi_m}{\partial x} + q_m^x \frac{\partial \varphi_l}{\partial x} \right] (14)$$

The variables in this study are dimensionless and no-flux boundary conditions are used consistently throughout this paper.

III. RESULTS AND DISCUSSION

A. Numerical computation details

The solidification model shown in Fig. 1 has been tested on different workstations and the numerical details are shown in Table I. It is noted that the tests are conducted in a share memory parallel computing mode. The performance tests suggest that the CPU clock frequency and the number of cores in the CPU are key parameters that determine the computing speed. The physical memory in the workstation should be large enough to support the calculation, i.e., the calculation of a model with more degrees of freedom (DOFs) requires a larger memory. As for the efficiency of the linear solvers, MUMPS is superior to PARDISO. Then in our next-mentioned study, the workstation with a faster speed and MUMPS solver is used.

TABLE I. PERFORMANCE BENCHMARK OF THE SOLIDIFICATION MODEL IN FIGURE 1 WITH 1,600K DOFS ON DIFFERENT WORKSTATIONS

Workstation number		1	2	3
CPU Information		2 × Intel Xeon E5-2687W @3.10 GHz	Intel Xeon E5-1620 @3.50 GHz	2 × Intel Xeon E5649 @2.53 GHz
Number of cores		20	4	12
Physical memory (GB)		64	32	24
Computing time (s)	**MUMPS**	22756	29726	40656
	PARDISO	30014	35944	71478

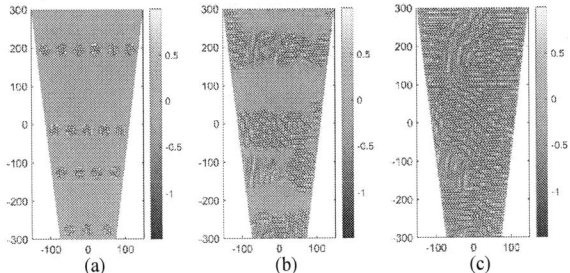

Fig. 1. The APFC performance benchmark model showing the snapshots of the solidified microstructure at (a) $t = 0$, (b) $t = 50$, and (c) $t = 250$.

B. Solidification process

During the solidification process, grains with different orientations behave differently depending on the orientations of adjacent grains. Fig. 2 illustrates two detailed examples. At the very beginning, 18 nuclei, whose radii are set to 15 and orientations ranging from 15° to 85°, are placed in the TSV, as shown in Fig. 2(a). The reason why an initial radius of 15

was chosen is because a smaller nucleus cannot initiate the solidification process. The explanation is that the APFC model focuses mainly on the amplitude of the atomic density field and ignores other details, so that the critical radius for nucleation should be much larger compared with the PFC model. In order to observe the grain boundaries (GBs) more clearly, we set the misorientation between two neighboring nuclei to be greater than 15°. And with time advances, the microstructural evolution and the formation of defects over time in the TSV can be obtained. Initially, the nuclei are surrounded by a large amount of supercooled liquid that provides material supply for the subsequent growth of the nuclei. But the growth rates amongst those nuclei are different, and some grains even disappear during this stage, which is ascribed to the anisotropy of the simulated crystal with a triangular lattice. As solidification proceeds, the crystal grains grow and come into contact with other grains or reach the edges of the TSV.

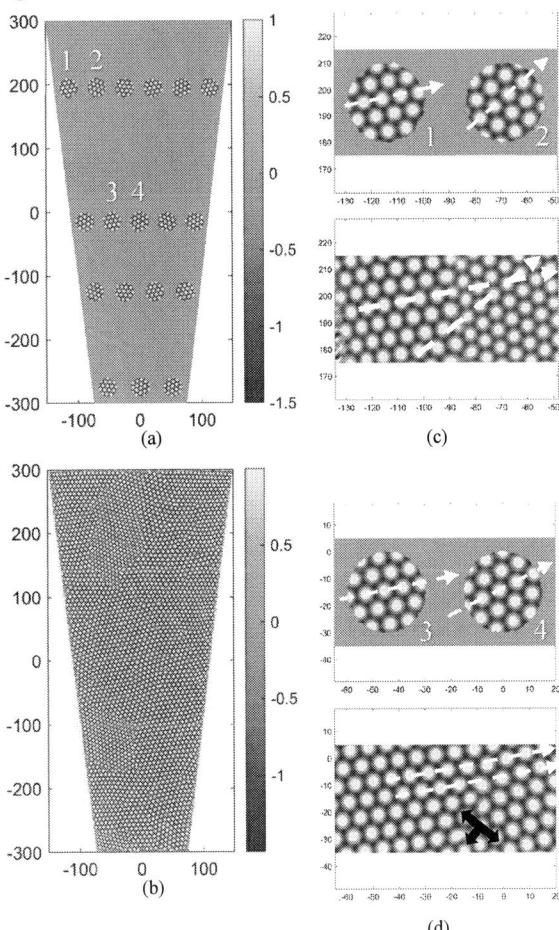

Fig. 2. Solidification microstructure in a TSV. (a) $t = 0$, (b) $t = 250$, (c) zoom in of grains 1 and 2 at $t = 0$ (top) and $t = 250$ (bottom), (d) zoom in of grains 3 and 4 at $t = 0$ (top) and $t = 250$ (bottom).

The formation of GBs and dislocations can be observed in the APFC simulation, as shown Figs. 2(b) and 2(c). Grains 1 and 2 grow, with a misorientation of 30° and the white arrows marking out the grain orientations, grow with time and gradually contact with each other. Since the orientation within each grain changes little, a high-angle GB is eventually formed between grains 1 and 2. For another two neighboring grains 3 and 4, however, the orientations in each grain evolves with time during solidification. Since the neighboring grains begin to adjust their orientation after their contact, the misorientation between those two grains gradually becomes smaller as solidification proceeds, which finally leads to the phenomenon of grain coalescence and only some dislocations are found to form along the place where the two grains join together. In summary, the nuclei initially placed in the supercooled liquid phase gradually grow and finally all liquid in the TSV solidifies to form a polycrystalline solid.

C. Different initial conditions and geometries

For the purpose of studying the diversity of microstructure that possibly exists in TSV, we consider to change the initial condition of the APFC model, i.e. the number and position of the seed grains. Fig. 3 shows two different initial configurations of the seed grains in the TSV, where we randomly select and take away or move several crystal nuclei. Owing to the disappearance and movement of those nuclei, the size of the formed solid crystals are larger than that in Figs. 1 and 2, therefore the number of solidified grains is reduced in the same TSV. This result indicates that the microstructure in TSVs are susceptible to change and can exhibit a diverse nature. In summary, TSVs with highly heterogeneous structures can be obtained from the APFC simulations under different initial conditions, e.g. different grain structure, amount of defects, defect distributions as well in the TSV. With such a diversity of microstructure existing in the TSVs, significant differences in mechanical behavior can be anticipated.

In fact, TSVs are structures with high aspect ratios. To take the advantages of the APFC model over the PFC model, we can further consider a new geometry with an aspect ratio of 4 by increasing the height of TSV. The simulation result using this new geometry is plotted in Fig. 4. Noted that the height of the simulated TSV is doubled compared to Liu's PFC work.

Fig. 5 shows the simulation result using a geometry with an aspect ratio of 5 by referring to the SEM image of a real TSV reported by [24]. It is worth pointing out that the same number of initial nuclei but with a more uniform distribution is considered in this case. It is found that the formed grains come into contact with adjacent grains or the TSV edges more quickly, which reduces the time for grain growth and thus resulting finer grains. Then we enlarge the size of the model by a factor of 1.5, keep the aspect ratio and initial condition the same, and obtain the simulation results as presented in Figs. 5(d) to 5(f).

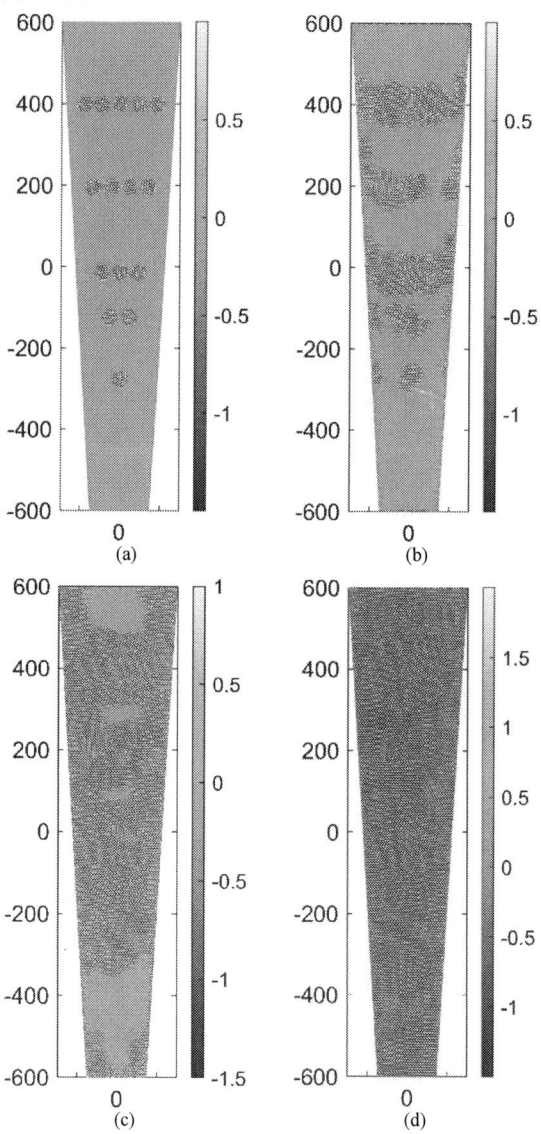

Fig. 4. Solidification process in a TSV with an aspect ratio of 4 at (a) $t = 0$, (b) $t = 50$, (c) $t = 100$, and (d) $t = 250$.

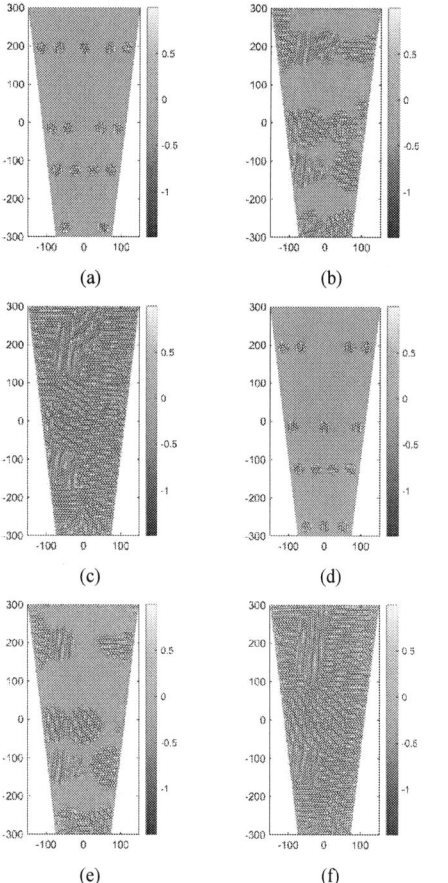

Fig. 3. Solidification process in TSVs with different configurations of the initial nuclei. (a) and (d) $t = 0$, (b) and (e) $t = 50$, (c) and (f) $t = 250$.

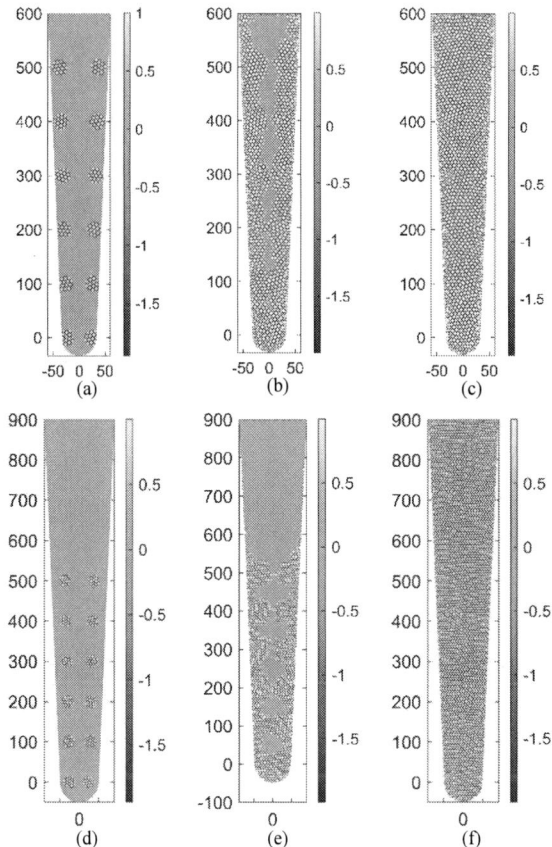

Fig. 5. Solidification process in TSVs with geometry referring to a real TSV geometry. The size of the TSV in (d)-(e) is 1.5 of that in (a)-(c). (a) and (d) $t = 0$, (b) and (e) $t = 20$ and (c) and (f) $t = 250$.

D. Strain and rotation field

The elastic strain ε and rotation field ω in the TSV as a result of the displacements of the atoms during solidification can be calculated from the complex amplitudes A_js according to Salvalaglio *et al.* in [23], based on the cross-scale linkage between continuum elasticity (CE) theory and lattice deformation. Fig. 6 plots the distribution of strain and rotation field. It is obvious that strain concentrates in places where defects accumulate regardless of the TSV geometry. During the stage of grain growth, any two neighboring grains compete to grow with each other after the fronts of solidification meet. As a consequence, the atoms inside the grains shift slightly relative to their perfect lattice position, resulting in small grain deformation. Figs. 6(b), 6(d), and 6(f) shows that the rotation ω keeps nearly constant inside the grains but changes abruptly in defects, which clearly outlines the grain structure in the TSV.

IV. SUMMARY AND OUTLOOK

This paper adopts the APFC model to simulate a diversity of polycrystalline structures that possibly exist in TSVs. In contrast to the previous studies using the PFC model, the contributions of this work mainly lie in following two aspects. Firstly, we successfully extend the dimension and the aspect ratio of the modelled TSV systems to around 20 nm and 5, respectively. Secondly, the strain and rotation field in the TSV are derived directly from the complex amplitudes A_js. For future work, mechanical loadings will be applied to the

TSVs with coarse-grained atomic scale structures. The process of TSV protrusion as well as the accompanying evolution of microstructure, strain and stress fields will be produced and studied to further advance our understanding on this complex and challenging reliability problem.

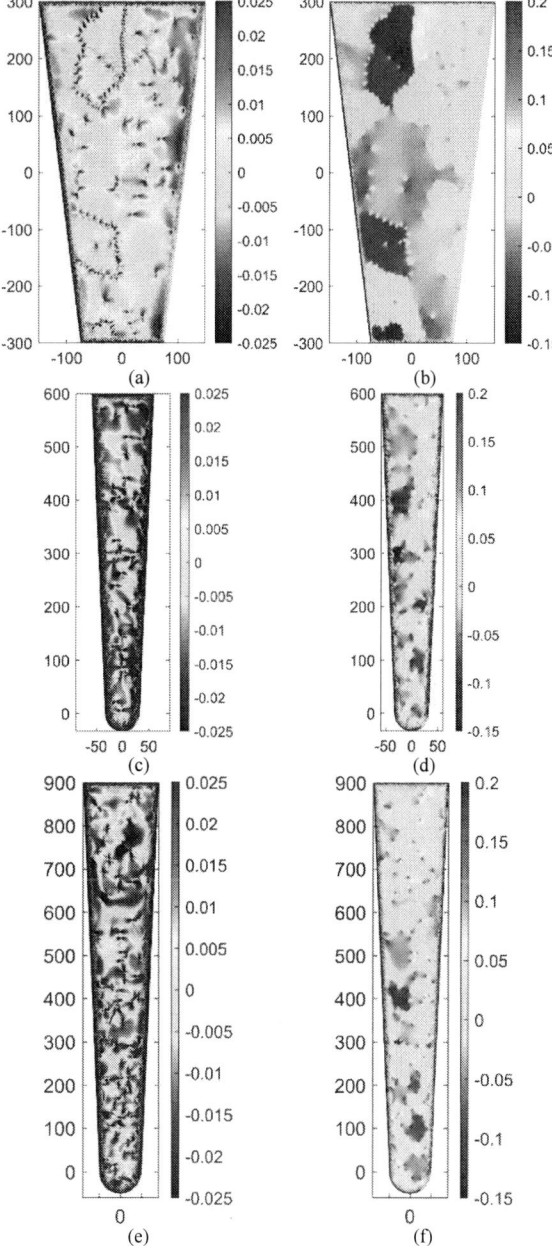

Fig. 6. The pseudo-color plots of the strain field ε_{xy} in (a), (c) and (e), and the rotation field ω in (b), (d) and (f) in TSVs with different sizes and geometries at $t = 250$.

ACKNOWLEDGMENT

The authors acknowledge financial support from the National Natural Science Foundation of China (NSFC) under grant 51832002 and the Guangdong Natural Science Foundation under grant 2015A030312011. Last but not least, we thank Dr. G. Wang from COMSOL China for discussion and support on the implementation of the APFC model in COMSOL Multiphysics.

REFERENCES

[1] E. Sperling, "Predicting reliability at 3/2nm and beyond," Semiconductor Engineering, December 2020.

[2] P. Shen, C. Su, Y. Lin, et al. "Ultralow contact resistance between semimetal and monolayer semiconductors," Nature 593, 211–217 2021.

[3] S. Ward-Foxton, "IBM unveils world's first 2 nm chip," EETimes, May 2021.

[4] R. Mahajan, R. Sankman, N. Patal, et al. "Embedded multi-die interconnect bridge (EMIB)---A high density, high bandwidth packaging interconnect," 2016 IEEE 66th Electronic Components and Technology Conference (ECTC), 31 May-3 June 2016, Las Vegas, Nevada, USA.

[5] M.-S. Lin et al., "A 7nm 4GHz Arm-core-based CoWoS chiplet design for high performance computing," 2019 Symposium on VLSI Circuits, 2019, pp. C28-C29.

[6] Xtacking: Modular approach to accelerate process development and shorten manufacturing cycle time. http://www.ymtc.com/index.php?s=/cms/cate/69.html. Accessed 30 May 2021.

[7] J.H. Lau, Heterogeneous Integrations, Springer Nature Singapore, 2019.

[8] C. Okoro, L. E. Levine, R. Xu, K. Hummler and Y. S. Obeng, "Nondestructive measurement of the residual stresses in copper through-silicon vias using synchrotron-based microbeam X-ray diffraction," IEEE Transactions on Electron Devices, vol. 61, no. 7, pp. 2473-2479, July 2014.

[9] C. Li, X. Si, L. Chen, et al, "Non-destructive measurement of residual stress distribution as a function of depth in sapphire/Ti6Al4V brazing joint via Raman spectra," Ceramics International, vol. 45, issue 3, pp. 3284-3289, 2019.

[10] F. Su, T. Lan, X. Pan and Z. Zhang, "Development and application of a micro-infrared photoelasticity system for stress evaluation of through-silicon Vias (TSV)," 2015 IEEE 65th Electronic Components and Technology Conference (ECTC), 26 May-29 May 2015, San Diego, CA, USA, pp. 1789-1794.

[11] J. Liu, Z. Huang, P.P. Conway, and Y. Liu, "Microstructural evolution and protrusion simulations of Cu-TSVs under different loading conditions," ASME. J. Electron. Packag., vol. 142, 011009, March 2020.

[12] J. Liu, Z. Huang, Y. Zhang, and P.P. Conway, "Mechasims of copper protrusion in through-silicon-via structures at the nanoscale," Jpn. J. Appl. Phys., vol. 58, 016502, November 2018.

[13] J. Liu, Z. Huang, P.P. Conway, Y. Liu, "Processing-structure-protrusion relationship of 3-D Cu TSVs: control at the atomic scale," IEEE J. Electron Devices Soc., vol. 7, pp. 1270-1276, October 2019.

[14] N. Goldenfeld, B.P. Athreya, J.A. Dantzig, "Renormalization group approach to multiscale simulation of polycrystalline materials using the phase field crystal model," Phys. Rev. E, vol. 72, 020601, August 2005.

[15] N. Goldenfeld, B.P. Athreya, J.A. Dantzig, "Renormalization group approach to multiscale modelling in materials science" J. Stat. Phys,vol. 125, pp. 1015-1023, March 2006.

[16] A. Skaugen, L. Angheluta, J. Vinals, "Dislocation dynamics and crystal plasticity in the phase-field crystal model," Phys. Rev. B, vol. 97, 054113, Feburary 2018.

[17] A. Skaugen, L. Angheluta, J. Vinals, "Separation of elastic and plastic timescales in a phase field crystal model," Phys. Rev. Lett., vol. 121, 255501, December 2018.

[18] S. Praetorius, M. Salvalaglio, and A. Voigt, "An efficient numerical framework for the amplitude expansion of the phase-field crystal model," Modelling Simul. Mater. Sci. Eng., vol. 27, no. 4, 044004, April, 2019.

[19] M. Salvalaglio, L. Angheluta, Z. Huang, et al. "A coarse-grained phase-field crystal model of plastic motion," Journal of the Mechanics and Physics of Solids, vol. 137, 103856, 2020.

[20] S. Vey and A. Voigt, "AMDiS: adaptive multidimensional simulations," Comput. Visual. Sci., vol. 10, pp. 57-67, December 2006.

[21] T. Witkowski, S. Ling, S. Praetorius, And A. Voigt, "Software concepts and numeical algorithms for a scalable adaptive parrallel finite element method," Adv. Comput. Math., vol. 41, pp. 1145-1177, January 2015.

[22] M. Salvalaglio, A. Voigt, K.R. Elder, "Controlling the energy of defects and interfaces in the amplitude expansion of the phase-field crystal model," Phys. Rev. E, vol. 96, 023301, 2017.

[23] M. Salvalaglio, A. Voigt, K.R. Elder, "Closing the gap between atomic-scale lattice deformations and continuum elasticity," NPJ Computational Materials, 2019.

[24] W.W. Shen, K.N. Chen, "Three-dimensional integrated circuit (3D IC) key technology: through-silicon via (TSV)," Nanoscale Research Letters, vol. 12, 56, 2017.

978-1-6654-1392-3/21 $31.00 © 2021 IEEE

BGA chip torsion finite meta analysis at high temperature.

line 1: 1st YUE Yu-Qing
line 2: *School of Electro-Mechanical Engineering，Guilin University of Electronic Technology*
line 3: *School of Electro-Mechanical Engineering，Guilin University of Electronic Technology*
line 4: GuiLin,GuangXi, China
line 5: 1332621832@qq.com

line 1: 2nd JIANG Chao
line 2: *School of Electro-Mechanical Engineering，Guilin University of Electronic Technology*
line 3: *School of Electro-Mechanical Engineering，Guilin University of Electronic Technology*
line 4: GuiLin,GuangXi, China

line 1: 2rd CHEN Yan-ting
line 2: *School of Electro-Mechanical Engineering，Guilin University of Electronic Technology*
line 3: *School of Electro-Mechanical Engineering，Guilin University of Electronic Technology*
line 4: GuiLin,GuangXi, China

line 1: 2rd WU Lv
line 2: *School of Electro-Mechanical Engineering，Guilin University of Electronic Technology*
line 3: *School of Electro-Mechanical Engineering，Guilin University of Electronic Technology*
line 4: GuiLin,GuangXi, China

line 1: 2rd WEI Jing
line 2: *School of Electro-Mechanical Engineering，Guilin University of Electronic Technology*
line 3: *School of Electro-Mechanical Engineering，Guilin University of Electronic Technology*
line 4: GuiLin,GuangXi, China

Abstract—

With the rapid development of information industrialization, the rapid development of electronic technology industry, chip component packaging form in continuous improvement , electronic chip solder joint packaging is widely used in electronic products, these electronic chips have the characteristics of thin, efficient and high performance, users in the process of making great savings in human and material resources and other resources.

Electronic products are used in high temperature environment for a long time, and due to the weight of the chip and other factors will cause its twisting deformation, but also make the welding joint produce different degrees of stress strain, if the stress strain is too large will lead to the damage of the welding joint, further leading to the failure of components.

Liang Yongfeng obtained Sn3Ag0 through the pure torque fatigue test. The fatigue characteristic data of 5Cu and Sn0. 7Cu were compared with the fatigue characteristics of brazing pure torque.

Liang Ying et al. studied the effect of solder spot material, solder joint diameter, solder joint height and pad diameter on the torsional stress strain of welding spot by establishing a microscale chip size package solder point finite analysis model.

Maia Filho et al. forecasted the torsion fatigue life of BGA solder joints by combining torsion experiments with finite meta-analysis;

John et al. studied the effect of PCB distortion deformation on BGA solder joints through a four-point distortion test.

Seung et al. studied the failure mode of many different CSP solder joints under the effect of distorted load using the cycle distortion test.

Quayle Chen et al. used a distortion method to study the distortion performance of electronic products such as flexible printed circuit boards and Super Twisted Nematic: STN LCD displays to arrive at the main factors affecting their reliability.

The chip size package solder joint finite factor analysis model is established and its high temperature torsional stress strain simulation and orthosectance design analysis are made. Through simulation analysis, it is found that the stress is mainly distributed at the edge of the region and on the upper side of the solder joint, the middle part of the welding spot stress is small and the edge stress, the stress gradually increases from the center part to the edge part, and the maximum stress is greater in contact with the PCB board position.

The orthosective analysis method is used to design 9 sets of solder joint models with different horizontal combinations, and the torsional stress is obtained in the corresponding weld point high temperature environment. Through the very poor analysis, it is found that the effect of the stress size of the weld point is from large to small: the diameter of the pad, the diameter of the solder joint, the height of the pad. The optimal weld point structure parameters are combined horizontally as: 0.55mm solder diameter, 0.42mm pad diameter, and 0.50mm solder point height, and the optimal solder point parameter combination is established, which is verified by simulation;

Keywords—Twist the heat; Stress strain; Finite meta-analysis; Very poor analysis; A VARlance analysis.

I. INTRODUCTION (*HEADING 1*)

Because electronic products are used in high temperature environment for a long time, chip weight and other factors will lead to its distortion and deformation. At the same time, it will also cause different degrees of stress and strain in the welded joint. If the stress and strain is too large, the welded joint will be damaged, which will further lead to component failure. Zhang Haomin [6] in the solder joint failure analysis of BGA package, by means of X-ray scanning, dye penetration, metallographic analysis, scanning electron microscope and thermal analysis, analyzed the existing problems of BGA package, such as the specific analysis of the failed solder joint and the gravity concentration at the stress discordance, which led to crack initiation and propagation. In order to solve the influence of over-voltage on the performance of electronic products, it is necessary to carry out finite element analysis and optimization design of solder joints. The torsional stress of high temperature welded joint with chip package size was simulated by orthogonal design method. Liang Yongfeng [1] obtained sn3ag0 by pure torque fatigue test. Fatigue characteristic data of 5Cu and sn0. The fatigue characteristics of 7cu and pure brazing torque were compared. By establishing the finite element analysis model of micro scale package solder joint, the effects of pad diameter, solder joint

978-1-6654-1392-3/21 $31.00 © 2021 IEEE

height, solder joint diameter and other factors on the torsional stress and strain of solder joint were studied. For the establishment of finite element analysis of micro scale package solder joints, scholars at home and abroad have carried out relevant research work. Xia zhuojie [7] explored the shortcomings and solutions of finite element method in lead-free solder joint reliability research, and explained the application of finite element numerical simulation in BGA device solder joint reliability research. Seon [8] compares BGA components with embedded chips with traditional BGA packages by finite element method. It is found that the maximum creep of lead-free solder joints of BGA with embedded chips is different from that of traditional BGA packages. The maximum creep of lead-free solder joints of BGA with embedded chips is transferred to the central area of the chip, but the maximum creep does not change. The results show that the finite element method and orthogonal design method are used to analyze the stress and strain of BGA weld under high temperature torsion. However, there are still some shortcomings in the research on the stress-strain analysis of BGA Solder Joints under high temperature torsion by domestic and foreign scholars, such as the existing research only indicates that the temperature is summarized, The influence of different shear rate and underfill binder on the reliability of BGA under a single change condition has not been further studied. Based on the orthogonal experimental design method, this paper selects the relevant structural parameters which affect the stability of BGA Solder Joints, designs 9 sets of horizontal combined solder joints model, and obtains the torsional stress under the corresponding high temperature environment of the welding points, and conducts the extreme difference analysis and variance analysis of the obtained data, the main and secondary factors affecting the torsional stress and strain of welded joints are analyzed, which provides a theoretical basis for further improving the torsional stability of welded joints.

II. EXPERIMENT

With the multi-directional and high-density development of electronic products, the key chip devices often use large-size BGA devices. In the SMT welding process, the large-size BGA chips appear too large warpage deformation, which leads to poor welding and other problems. At present, moire interference test is generally used to measure BGA warpage [9]. The BGA Solder Ball spacing is 0.5mm and the package size is 12mm × 12mm, the diameter of solder ball is 0.3mm, and the number of ball array is 22 × 22. In the high temperature torsion test, the experimental temperature is set at 125. C. The experimental time was 4, 9 and 16 days respectively; use

Quanta 200feg field emission environment scanning electron microscopy (SEM) was used to observe the microstructure of the mixed solder joint before and after heating and before and after different heating test time, including the morphology distribution of pores, voids, IMC and crack propagation in the solder joint. Energy dispersive X-ray spectrometer (EDS) was used to qualitatively analyze the distribution of Pb in the solder joint. The experimental results are shown in Fig. 1, 2 3.

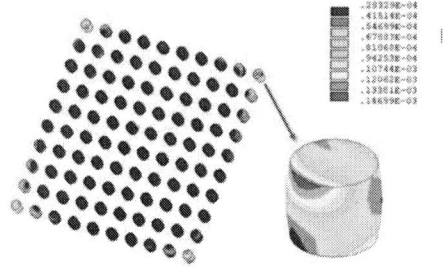

Figure 1: elastic strain of BGA Solder Joint

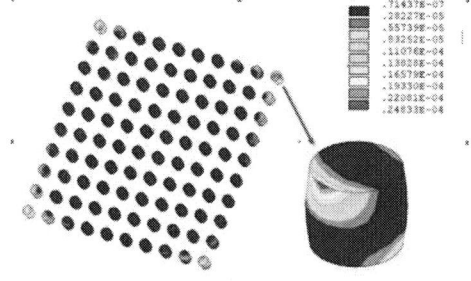

Figure 2: plastic strain of BGA Solder Joint

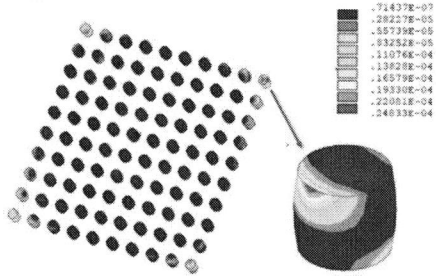

Fig. 3. The equivalent force of the BGA solder joint.

When using the finite element method to calculate the thermal deformation, it is generally necessary to solve the temperature field first, and then get the distribution of thermal deformation through thermal mechanical coupling. In reflow soldering, the heat source of BGA device mainly comes from the finite element modeling of convection conduction:

The complete model structure is shown in Figure 4 and figure 5:

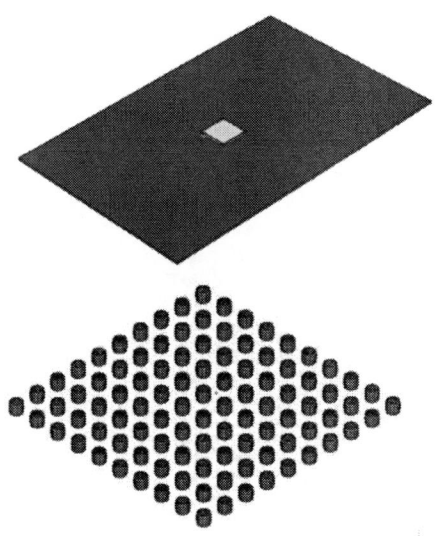

Fig. 4.5. Complete model structure diagram

III. FINITE ELEMENT ANALYSIS OF STRESS IN HIGH TEMPERATURE TWIST SOLDER JOINT OF BGA CHIP

A. Finite element analysis model

(finite element model diagram after meshing) the finite element model diagram after meshing is established based on ANSYS software and actual chip, as shown in Figure 6.

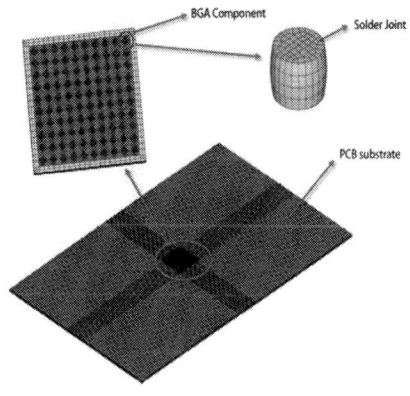

Fig. 6. finite element model diagram after meshing.

For the convenience of problem analysis, the copper pad between solder joint and PCB board is not considered here, and the solder joint is assumed to have no defects such as air hole and air hole [11]. The finite element model consists of three parts: BGA chip, micro scale BGA Solder Joint and PCB substrate. There are 100 BGA chips (10 × 10), as shown in Figure 2. In the finite element stress-strain simulation analysis, the boundary conditions imposed on the model are: in the case of natural air convection, the temperature of the surrounding air is 25 ℃. C。 Because of the transient calculation, we can see the distribution of temperature field and deformation field at any time during the whole reflux process, and the deformation results are shown in Figure 123 above. Combined with the cloud figure, it can be seen that through the simulation analysis, it is found that the stress is mainly distributed in the edge and the upper end of the solder joint, the stress in the middle of the solder joint is small, the edge stress increases

gradually, the stress increases gradually from the center to the edge, and the maximum stress is larger when it contacts with the PCB. The influence of solder joint diameter on the stress and strain of BGA was analyzed. The residual stress of BGA Solder Joint after high temperature torsion was analyzed by simulation analysis and orthogonal design analysis. Nine sets of horizontal combined solder joint models were designed by using orthogonal analysis method, and the torsional stress was obtained under the corresponding high temperature environment.

B. Based on orthogonal experimental design method, the influence of high temperature torsion solder joint diameter, solder joint height and pad diameter on stress and strain of BGA Solder Joint

Design of orthogonal design test table

According to the actual application parameters of solder joint diameter, solder joint height and pad diameter, the relevant factors affecting the high temperature torsion stress and strain of BGA Solder joint are selected as: solder joint diameter, solder joint height and pad diameter. The three factors are all at three levels, and the setting of each factor and its corresponding level is shown in Table 5. L9 (34) orthogonal table is used to arrange the orthogonal test of three factors and three levels, The horizontal orthogonal table is shown in Table 7.

Range analysis

Table solder joint size solder joint height pad diameter stress The range analysis of high temperature torsion stress and strain under different parameter level combinations in Table 7 is carried out, and the range analysis results based on orthogonal design are shown in Table 8. It can be seen from the table that the order of the factors according to the range from large to small is: pad diameter, solder joint diameter and pad height. Therefore, it can be seen that the pad diameter has the greatest influence on the stress of the butt joint, followed by the diameter of the solder joint, and finally the height of the solder joint.

In the column	1	2	3	
Table Head	Spot size	Solder joint height	Bonding pad diameter	Stress
Experiment 1	0.45	0.38	0.42	5.2100
Experiment 2	0.45	0.42	0.46	4.8352
Experiment 3	0.45	0.46	0.50	4.4409
Experiment 4	0.5	0.38	0.46	4.4349
Experiment 5	0.5	0.42	0.50	3.7418
Experiment 6	0.5	0.46	0.42	5.2972
Experiment 7	0.55	0.38	0.50	3.4903
Experiment 8	0.55	0.42	0.42	4.4062
Experiment 9	0.55	0.46	0.46	4.5764

variance analysis

According to the theory of variance analysis of orthogonal test [12] and the data of temperature simulation results in Table 7, the degree of freedom, variance estimation value and variance ratio (i.e. F value) of each factor can be calculated, as shown in Table 8. It can be seen from table 8 that the corresponding F value of pad diameter is 86.571, which is larger than the corresponding critical value F0.1 (2,4) (i.e. 19.000), so it is at the significant level α In the case of 0.10 (i.e. 90% confidence), pad diameter has the most significant effect on the stress of solder joint; The corresponding F value of solder joint diameter factor is 32.190, which is also greater than the corresponding critical value F0.1 (2,4). Therefore, in the case of 90% confidence, this factor also has a significant effect on the stress of solder joint; The corresponding F value of solder joint height is 16.857, which is less than the corresponding critical value F0.1 (2,4). Therefore, the influence of solder joint height on solder joint is not significant. According to the F value of each influencing factor in Table 7, it can also be concluded that the order of factor significance is pad diameter > solder joint diameter > solder joint height, which is consistent with range analysis. Through the above variance analysis, it can be seen that the pad diameter has the most significant effect on the stress of the solder joint among the factors investigated. Therefore, when designing BGA Solder joint parameters, we need to pay attention to the design of pad diameter, so as to design the pad solder joint with the best stress-strain performance under high temperature torsion more effectively.

Table Head	Table Column Head				
	Sum of squares of deviation	Degree of freedom	F 比	Fthe critical value	Signficance
Spot size	0.676	2	32.190	19.000	*
Solder joint height	0.354	2	16.857	19.000	
Bonding pad diameter	1.818	2	86.571	19.000	*
error	0.02	2			

IV. CONCLUSION

Through the finite element model analysis of the stress and strain of the solder joint under high temperature torsion, the following conclusions can be obtained：

1) Under certain conditions, increasing the pad diameter and solder joint diameter is beneficial to reduce the stress and strain of BGA Solder joint.

2) Range analysis shows that the pad diameter has the greatest influence on the stress of BGA high temperature torsion solder joint, followed by the diameter of solder joint, and finally the height of solder joint.

3) Variance analysis shows that: in the case of 90% confidence, pad diameter and solder joint diameter have

significant effect on the stress of BGA high-temperature torsion welding joint, and solder joint height has no significant effect on the stress of BGA high-temperature torsion welding joint.

REFERENCES

[1] [1] Liang Yong-feng. Torsion Fatigue and Multiaxial Ratcheting Fatigue of Lead-Free Solder ［D］. Tianjin: Tianjin University, 2010.

[2] Maia F W C, Brizoux M, Frémont H, etal. Lifetime prediction of BGA assemblies with experimental torsion test and finite element analysis ［A］. International Conference on Thermal, Mechanical and Multi-Physics Simulation and Experiments in Microelectronics and Micro-Systems ［C］. New York: IEEE, 2008. 1 — 7.

[3] Lau J H. Solder joint reliability of flip chip and plastic ball grid array assemblies under thermal, mechanical, and vibration conditions ［J］. IEEE Transactions on Components, Packaging, and Manufacturing Technology, 1996, 19 (4): 723 — 735.

[4] Seung W Y, Jun K H, Hwa J K, et al. Board-level reliability of Pb-free solder joints of TSOP and various CSPs ［J］. IEEE Transactions on Electronics Packaging Manufacturing, 2005, 28(2): 168 — 175.

[5] Chen Q, Xu L, Salo A. Reliability of flexible display by simulation and strain gauge test ［A］. 10th Electronics Packaging Technology Conference ［C］. New York: IEEE, 2008. 322 — 327.

[6] Zhang Hao Ming, Li Xiao Qian ,et al. Solder joint failure analysis of BGA package ［A］. 1th ELECTRONIC PRODUCT RELIABILITY AND ENVIRONMENTAL TESTING ［C］. 2021. 02 — 39.

[7] Xia Zhuo Jie, Zhang Liang, et al. Application of finite element numerical simulation in solder joint reliability of BGA / QFP / CCGA devices. The application in the research ［A］. 2th ELECTRONICS & PACKAGING ［C］. 2020. 02 — 20

[8] YU S Y, KWON Y M, KIM J S, et al. Studies on the thermal cycling reliability of BGA system- in-package (SiP) with an embedded die [J].IEEE Trans. Components, Packaging and Manufacturing Technology,2012,2(4):625-633.

[9] WangBinLin,LiuZhe,Fu HongZhi. Warpage simulation of large size BGA on PCB ［A］.

[10] Liang Yin, Huang Chun Yue, et al. Micro scale CSP based on torsional load Stress strain analysis and optimization of solder joint. Acta Electronical Sinica,2020,10(48):

[11] Hang Chun Jin, Tian Yian Hong, et al. Study on Microstructure of solder joint for high temperature aging test of mixed device. Acta metallurgica Sinica,2013,7(49):

[12] LIANG Ying, HUANG ChunYue, YIN R ui. Three — point bending stress strain analysis of micro - scale ball grid array solder joint of microelectronic package ［J］. Journal of Mechanical Strength, 2016, 38(4): 744-748

Thermal Analysis of High-Power Light-Emitting Diode Using Thermoreflectance Thermography

Dazheng Wang
*Micro-nano Fabrication Technology
Department, Institute of Electrical
Engineering Chinese Academy of
Sciences
Xi'an Jiaotong University*
Beijing, China
dzwang@mail.iee.ac.cn

Libing Zheng
*Micro-nano Fabrication Technology
Department, Institute of Electrical
Engineering Chinese Academy of
Sciences
University of Chinese Academy of
Science*
Beijing, China
ieezlb@mail.iee.ac.cn

Weikang Si
*Micro-nano Fabrication Technology
Department, Institute of Electrical
Engineering Chinese Academy of
Sciences*
Beijing, China
siweikang@mail.iee.ac.cn

He Yang
*Micro-nano Fabrication Technology
Department, Institute of Electrical
Engineering Chinese Academy of
Sciences
University of Chinese Academy of
Science*
Beijing, China
yh@mail.iee.ac.cn

Yingying Gao
*Micro-nano Fabrication Technology
Department, Institute of Electrical
Engineering Chinese Academy of
Sciences*
Beijing, China
gaoyy@mail.iee.ac.cn

Abstract—**High temperature has a serious impact on the reliability and optical properties of high-power light-emitting diodes. It is of great significance to obtain the transient thermal distribution and peak temperature of high-power LED devices during operation for component testing, design, and structure optimization. Thermal microscopic imaging based on visible light and thermoreflectance has not only high temporal-spatial resolution, but also high temperature resolution. Therefore, this technology can be used in thermal measurement of precision semiconductor devices. In this paper, we proposed a timing control scheme for separating electroluminescence and thermal signals, measured the transient thermal distribution and peak temperature of high-power light-emitting diode devices by using a thermal microscopy imaging device based on visible light and thermoreflectance, verified the feasibility of the proposed method in thermal distribution characterization and material defect diagnosis of high-power LED devices.**

Keywords—transient thermal distribution, high-power LED, microscopic imaging, thermal measurement, thermoreflectance

I. INTRODUCTION

Compared with traditional light sources such as incandescent lamps, halogen tungsten lamps and fluorescent lamps, high-power light-emitting diode (LED) has the advantages of small size, less power consumption, long life, high safety, and high electro-optical conversion efficiency, and its rated power can be up to tens of watts[1]. The packaging of high-power LED mainly involves optics, heat, electricity, structure, and technology, etc. These factors are independent of each other and influence each other. In LED's packaging work, light is the purpose, heat is the key, and the others are the means, and performance is the embodiment of packaging level[2]. High-power LED is a thermal sensitive component, so the reliability of heat dissipation is the main factor affecting its application, and has a great influence on the brightness, color stability and life. Without heat dissipation measures, the core temperature of high-power LED will rise rapidly and even catastrophic failure, when the temperature of junction (TJ) rises beyond the maximum allowable value, which is generally 150 °C[3]. In addition, limited by the development of semiconductor

technology, more than 60% of the electric energy will be released as heat energy. Therefore, it is important to obtain the thermal distribution for the optimal design of high-power LED[4].

Infrared thermal imaging is a common method in the semiconductor field, but it is affected by several aspects when measuring high-power LED[6]. First, infrared cannot penetrate the package of the LED, so that the infrared camera cannot measure directly. And then, the low spatial and temporal resolution of infrared thermal imaging, cannot meet the needs of high-power LED thermal measurement. Thermoreflectance thermography technology overcomes the above shortcomings. It is a non-contact, non-destructive method, with microsecond/ nanometer resolution, which can be used for hot spot detection and fault prediction of high-power LED[7].

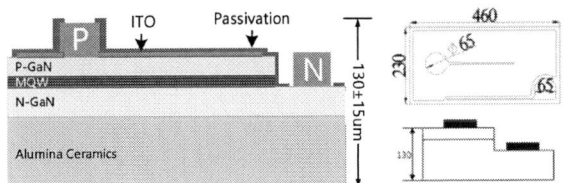

Fig. 1. Structure diagram of the sample under test, a commercial ceramic substrate LED, MODEL 5015, designed and produced by INTELED Company

The sample under test (SUT) is a commercial ceramic substrate LED (MODEL 5015) designed and produced by INTELED Company. The basic structure of SUT is shown in Fig. 1, where P-type and N-type GaN semiconductors are stacked vertically on the aluminum ceramics substrate, which surface is covered with indium tin oxide protective layer, and two electrodes are led out. When current flows through the PN junction, electrons in the N-type semiconductor will be pushed to the P area, collide with the holes in the P-type semiconductor, and then release energy in the form of photons. At room temperature, the LED can emit 220lm blue light (spectral range from 445 to 500nm) under the action of 20mA current. The size of the device is shown in Fig. 1, the

978-1-6654-1392-3/21 $31.00 © 2021 IEEE

external size is 460μm×230μm, and the overall height is 130μm.

II. EXPERIMENT METHOD

A. Thermoreflectance Thermography

The schematic diagram of the thermal microscopy imaging device based on thermoreflectance is shown in Fig. 2. It mainly consists of illumination, CCD camera, filter, optical microscopy components, temperature control sample table, excitation source, timing controller and other parts. The light of illumination is vertically incident to the surface of SUT through optical assembly, which is vertically reflected back to the microscopic objective lens of the sample surface, and finally is received and imprinted by the CCD sensor. The device can obtain transient thermal distribution image with high spatial and temporal resolution through proper timing control of CCD exposure, illumination trigger and excitation period[8].

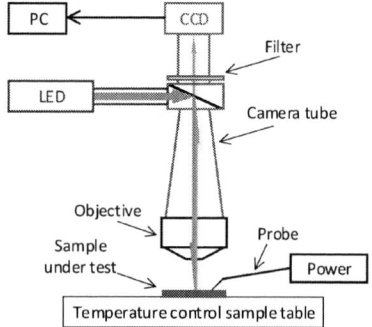

Fig. 2. Schematic diagram of the thermal microscopy imaging device based on thermoreflectance tehermography

In a certain temperature range, the change of the material's reflectivity to visible light has a linear relationship with the change of temperature, the thermoreflectance coefficient, C_{TR}, of the material according to:

$$\frac{\Delta R}{R_0} = C_{TR}\Delta T \qquad (1)$$

where, ΔR is the relative change in reflectivity of the SUT surface, which is proportional to the change in the gray value of the image (the mean of repeated experiments), $\overline{\Delta c}$, in this device. R_0 is the reflectivity of SUT surface at a given temperature, T_0, which is proportional to the gray value of the SUT surface image, $\overline{c_0}$, under T_0. The unit of C_{TR} is K^{-1}, and the value ranges from 10^{-5} to $10^{-4}K^{-1}$. Accurate thermoreflectance coefficient allows accurate temperature variations, depending on the wavelength of the illumination, the material type and roughness of the SUT surface. According to the above description, the formula can be summarized as:

$$\frac{\overline{\Delta c}}{\overline{c_0}} = C_{TR}(T - T_0) \qquad (2)$$

B. Thermoreflectance Coefficient Calibration

For any material with a definite composition, the thermoreflectance coefficient is not significantly affected by the surface treatment or deposition process, so it is usually not necessary to calibrate for each material under test. It can be seen from Fig. 1 that there is a passivation layer on the surface of our SUT. Although this layer is thin and

transparent to visible light, it will also introduce optical interference to the measurement of real temperature and cause the oscillation of thermal resistance signal.

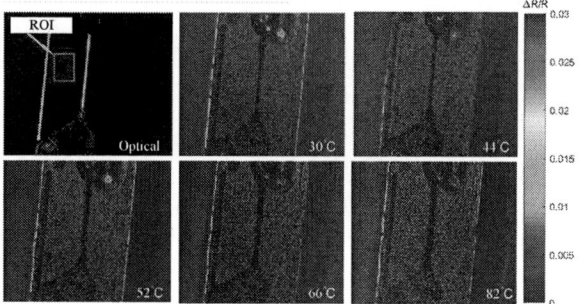

Fig. 3. Variation of SUT's relative thermal reflectance with temperature

The SUT is uniformly heated by controlling the temperature controller and the temperature is adjusted with a gradient of 10K. After the temperature stabilized, CCD started to collect thousands of images. According to (1), the images collected are used to calculate the relative thermal reflection images at different temperatures on the surface of SUT, as shown in Fig. 3, as the temperature gradually increases, the relative thermal reflectance, $\Delta R/R$, of the SUT becomes larger and larger. The region of interest (ROI) is selected from the optical image at 25°C to calculate the thermoreflectance coefficient C_{TR}. As shown in Fig. 4, under the 530nm illumination of SUT, the temperature change in the ROI has a linear change with $\Delta R/R$. The slope obtained by linear fitting of the results is the thermoreflectance coefficient of the surface material of SUT, which can be obtained as $9.9315 \times 10^{-5}K^{-1}$.

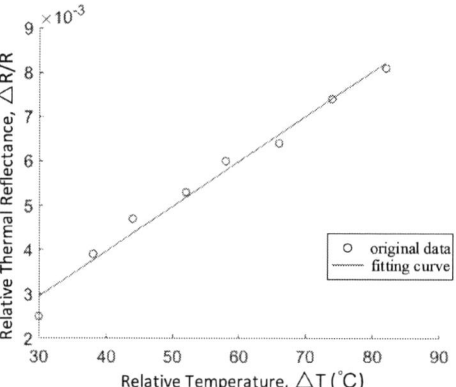

Fig. 4. Relative thermal reflectance with temperature curve

C. Separation of EL and Thermal Signal

As the SUT is an LED, it will emit light when it is stimulated. To distinguish the illumination in the device, we call the light emitted by SUT electroluminescence (EL). The existence of EL will cover the emission signal of the thermal signal in the process of testing, and when its intensity reaches a certain magnitude, it will even lead to the CCD overexposure. To minimize the influence of the EL of SUT, the band-pass filter can be used to filter the EL under the premise of ensuring the high transmittance of the illumination. However, in practical application, it is found that optical filter is not enough to improve the signal-to-noise ratio of the system signal. The adoption of filter can solve the problem of CCD overexposure caused by EL. However, due to the wide wavelength range of EL and partial overlap with the illumination, weak EL will also affect the measurement results[9]. Therefore, it is necessary to filter EL through algorithms.

By controlling the timing sequence of SUT excitation, illumination and CCD exposure, the accumulation and superposition of illumination and EL were imaged, respectively. The timing diagram shown in Fig. 5, where I_0 is the numerical accumulation of light intensity of Illumination and EL in the same image, I_1 is the sum of the light intensity of illumination and EL, which includes light intensity change due to heat generation, I_2 is the image of illumination. During the test, the SUT was placed on the radiator fin to prevent changes in the SUT luminous flux due to heat accumulation.

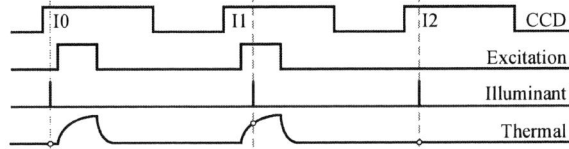

Fig. 5. Timing diagram of SUT transient image sequence

By controlling the delay of illumination relative to SUT excitation pulse, the entire thermal process of SUT heating and cooling after excitation can be traversed. The temperature change can be calculated by:

$$\Delta T = \frac{I_1 - I_0}{I_2 C_{TR}} \qquad (3)$$

The intensity of EL can be obtained by:

$$I_{EL} = I_0 - I_2 \qquad (4)$$

III. RESULTS AND DISCUSSION

Using the thermoreflectance thermography setup, we measured the transient thermal distribution in the heating process of the SUT and separated the EL and thermal signal successfully under the timing control method proposed above, which excitation period is 1000us and duty cycle is

10%. The pulse width of illumination was 10us, implying the minimum temporal resolution in the test. Using a 20x objective, the setup has a FOV of 438um×330um. As shown in Fig. 6, there are optical image collected by the system, EL image calculated and thermal image at 100us of excitation applied. EL is mainly concentrated around the electrode P, and it has great attenuation due to the installation of band-pass filter.

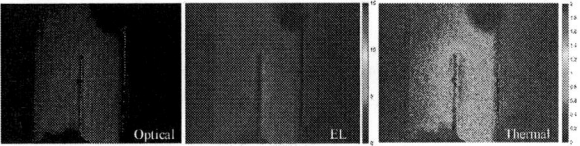

Fig. 6. Timing diagram of SUT transient image sequence

A. Image Separation Test

By applying different excitation voltages, the EL intensity is changed to verify the effect of the separation algorithm of EL and thermal signal. Fig. 7 shows the EL images and thermal signal of the SUT when the excitation voltage is 10V, 12V and 14V. The higher the excitation voltage is, the stronger EL is, and the thermal signal intensity also increases. However, the overall intensity of thermal signal is weaker than that of EL, which is about 40%. It is proved that the timing control method can effectively remove the EL from the measurement images.

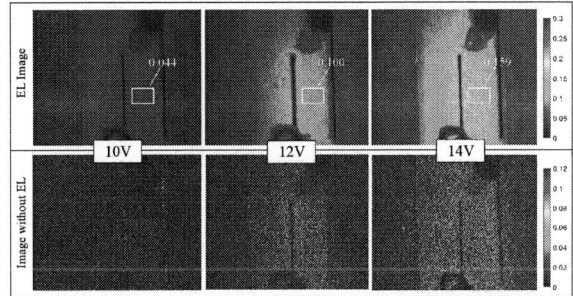

Fig. 7. EL images and thermal signal images without EL under different excitation voltage conditions

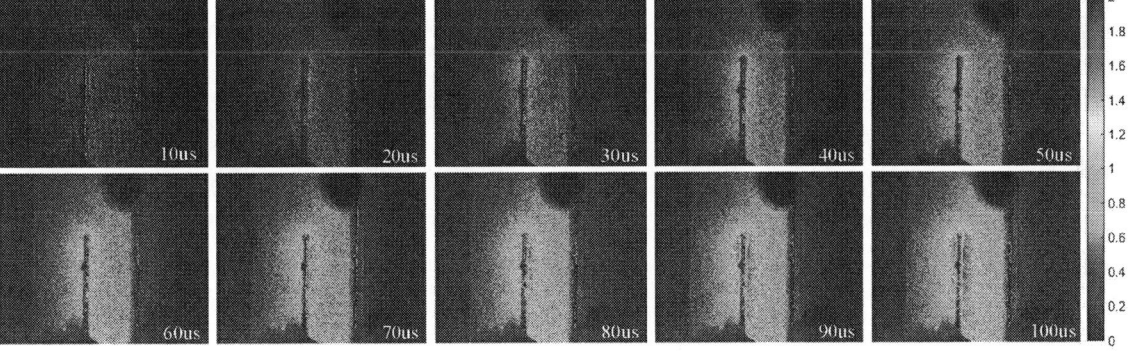

Fig. 8. Transient thermal images of 10-100us excitation applied to the sample

B. Transient Thermal Images

The heating area is not uniform as shown in the thermal image (Fig. 6), where is higher near the P electrode than that near the N electrode. Since the thermoreflectance coefficient of the electrode material has not been calibrated, the thermal image of the electrode position has no reference and is only

used for area limitation. Fig. 8 shows the transient thermal images of 10-100us excitation applied to the sample. With the increase of excitation time, the temperature on the surface of SUT gradually increases, and the heating rate is fast near the P electrode, and then expands to the N electrode. This is mainly due to the electro-optical conversion that is

978-1-6654-1392-3/21 $31.00 © 2021 IEEE

not complete and leads to heat, and where the EL intensity is stronger, the heat generation is more. The internal quantum efficiency is not high, and when the electron binds to the hole, the photon generation efficiency cannot reach 100%. Due to the current leakage, the carrier recombination rate in the PN junction region is reduced, and this part of energy is converted into heat energy[10]. In addition, part of the photons generated inside the chip cannot be emitted to the outside, also converted to heat. This part of the conversion rate is called external quantum efficiency, which is the main source of heat. We selected the marked position by white rectangle in the image to calculate the average temperature and draw the change trend chart of 10-100us excitation applied (Fig.9).

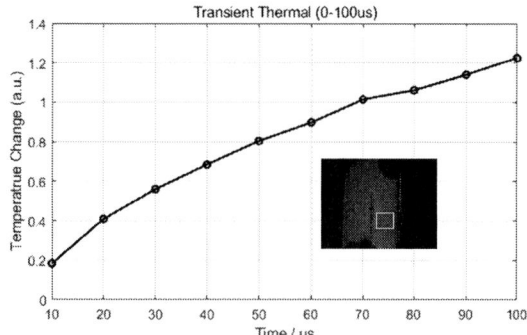

Fig. 9. The thermal transient of a single element in the SUT

C. Thermal Profile

Fig. 10 scatter diagram shows the temperature profile across the diode (from electrode P to electrode N) at 100us in the thermal transient. The maximum temperature appears at 14.8μm from the electrode P, and decreases gradually with the increase of the distance from the electrode P. It reaches the lowest and tends to be stable at 64.6μm from the electrode P, and its value is 50% of the maximum temperature. The right plot of Fig. 10 shows the fitting data for transient thermal profiles at different times. With the increase in excitation time, the overall temperature rises rapidly. The temperature near the two electrodes rises to another region, and the highest point appears at 14.8μm from electrode P and 23.6μm from electrode N, respectively.

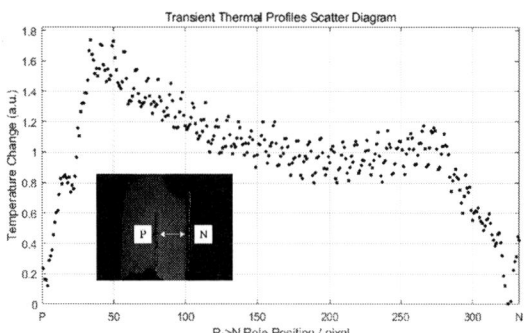

Fig. 10. Scatter diagram of the thermal profiles across the SUT from electrode P to N

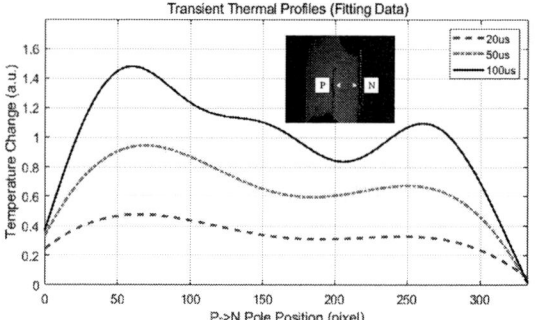

Fig. 11. Fitting data curve of the thermal profiles across the SUT from electrode P to N

IV. CONCLUSION

In this paper, we proposed a kind of method to separate the EL and thermal signal based on thermoreflectance thermography system. An experiment using this method was designed to measure the transient thermal distribution of high-power LED, which separated the EL and the thermal signal successfully in the acquired images. High spatial resolution images of thermal distribution and EL are obtained, and the transient thermal profile of the device is analyzed in detail. It has been proved that the thermoreflectance thermography method is feasible in the transient thermal measurement of high-power LED, which is of great significance for the subsequent packaging design, heat dissipation design and thermal testing of high-power LED devices.

REFERENCES

[1] Lin X, Mo S, Mo B, et al. Thermal management of high-power LED based on thermoelectric cooler and nanofluid-cooled microchannel heat sink[J]. Applied Thermal Engineering, 2020, 172: 115165.

[2] Arik M, Petroski J, Weaver S. Thermal challenges in the future generation solid state lighting applications: light emitting diodes[C]// Conference on Thermal & Thermomechanical Phenomena in Electronic Systems. IEEE, 2002.

[3] Tsai M Y, Chen C H, Kang C S. Thermal measurements and analyses of low-cost high-power LED packages and their modules[J]. Microelectronics Reliability, 2012, 52(5):845–854.

[4] Wang H, Jing D, Liu Z, et al. The analysis of measurement methods for high power LED thermal resistance[C]// International Forum on Strategic Technology. 2011.

[5] Y Sümer, Karaman O, Karaman C. Design and Thermal Analysis of High Power LED Light. 2021.

[6] Kendig, Yazawa, Shakouri. Thermal imaging of encapsulated LEDs[J]. Annual IEEE Semiconductor Thermal Measurement & Management Symposium, 2011, 15(4):310-313.

[7] Farzaneh M, Maize K, D Lüer?En, et al. CCD-based thermoreflectance microscopy: principles and applications[J]. Journal of Physics D Applied Physics, 2009, 42(14):143001.

[8] Wang D, Liu Z, Zheng L, et al. A High-resolution Thermoreflectance Imaging Technique based on Visible light[C]// 2019 20th International Conference on Electronic Packaging Technology(ICEPT). 2019.

[9] Ling J, Tay A. Measurement of LED junction temperature using thermoreflectance thermography[C]// 2014 15th International Conference on Electronic Packaging Technology (ICEPT). IEEE, 2014.

[10] Poppe A, Molnar G, Temesvolgyi T. Temperature dependent thermal resistance in power LED assemblies and a way to cope with it[C]// Semiconductor Thermal Measurement & Management Symposium. IEEE, 2010.

Low-temperature Cu/SiO$_2$ hybrid bonding using a novel two-step cooperative surface activation

Qiushi Kang
State Key Laboratory of Advanced Welding and Joining
Harbin Institute of Technology
Harbin, China
kangqiushi@stu.hit.edu.cn

Chenxi Wang*
State Key Laboratory of Advanced Welding and Joining
Harbin Institute of Technology
Harbin, China
wangchenxi@hit.edu.cn

Ge Li
State Key Laboratory of Advanced Welding and Joining
Harbin Institute of Technology
Harbin, China
1172910108@stu.hit.edu.cn

Shicheng Zhou
State Key Laboratory of Advanced Welding and Joining
Harbin Institute of Technology
Harbin, China
zhoushichengwow@163.com

Yanhong Tian
State Key Laboratory of Advanced Welding and Joining
Harbin Institute of Technology
Harbin, China
tianyh@hit.edu.cn

Abstract—Compared with continuous node scaling in a two-dimensional (2D) plane, advanced three-dimensional (3D) integration finds a new pathway to expand Moore's Law in the vertical direction. The essence of advanced 3D integration technology relies on the dense vertical interconnection, and state-of-the-art metal/oxide hybrid bonding provides such an ideal fine-pitch structure (≤ 1 μm) by eliminating microbump and underfill. Of the various hybrid bonding platform, Cu/SiO$_2$ hybrid bonding structure is the most promising candidate due to the excellent electrical and mechanical properties of Cu and SiO$_2$, respectively. However, the feasible Cu/SiO$_2$ hybrid bonding technology often requires high temperature (~400 °C) currently, which is not desirable for temperature-sensitive chips. Here, we develop a novel two-step cooperative surface activation method to overcome this bottleneck. Based on the combination of plasma activation and acid treatment of this cooperative surface activation, the atomic smooth Cu and SiO$_2$ surface with hydrophilic layers were obtained, and the strong homogeneous bonding of Cu-Cu and SiO$_2$-SiO$_2$ was realized at 200 °C. Eventually, the Cu/SiO$_2$ hybrid bonding device was successfully achieved, which void-free and atomically interconnected Cu-Cu, SiO$_2$-SiO$_2$, and even Cu-SiO$_2$ interfaces were obtained simultaneously. This hybrid bonding structure realized by cooperative surface activation brings unprecedented 3D integration feasibility and flexibility.

Keywords—Cu/SiO$_2$ hybrid bonding, two-step cooperative, low temperature, interface

I. INTRODUCTION

Nowadays, three-dimensional (3D) integration has opened up new pathways for expanding Moore's Law in the vertical direction. In the future, the density of interconnection is bound to increase further for next-generation advanced chips. Compared with the conventional 3D integration method assisted by microbump and underfill, state-of-the-art metal/oxide hybrid bonding can effectively scale down interconnect pitch to 1 μm via eliminating microbump and underfill at the interface [1-2]. As one of the most promising platforms, the Cu/SiO$_2$ hybrid structure can be fabricated on Si-based or other non-Si-based substrates, thus providing unprecedented feasibility and flexibility for multifunctional integration. Ziptronix Inc. took the lead in developing Cu/SiO$_2$ hybrid bonding technology, which realized interconnection at 125 °C~350 °C [3]. CEA-Leti has further reduced the interconnect pitch, achieving hybrid bonding with a pitch of

National Natural Science Foundation of China under grant 51975151.
Heilongjiang Provincial Natural Science Foundation of China under grant LH2019E041

1μm at 400 °C [4]. Currently, It is known that the robust Cu/SiO$_2$ hybrid bonding was usually achieved at ~400 °C. However, the high bonding temperature may degrade the performance of temperature-sensitive chips [5-6]. The essence of lowering the bonding temperature is to control the chemical structures on Cu and SiO$_2$ surfaces, respectively. Nevertheless, the desirable chemical states for Cu-Cu and SiO$_2$-SiO$_2$ low-temperature bonding are different (for example, non-oxidized Cu surface and SiO$_2$ surface with sufficient dangling bonds), which cannot be directly achieved by a single activation method. Therefore, a novel surface activation method is urgently needed to reconcile this discrepancy.

In this paper, we develop a two-step cooperative surface activation method to circumvent the dilemma, realizing Cu/SiO$_2$ hybrid bonding at 200 °C successfully. By combining the process of Ar/O$_2$ plasma activation and formic acid (FA) immersion, a hydrophilic layer was established on the Cu and SiO$_2$ surfaces simultaneously. With the aid of low-temperature annealing, the dehydration polymerization reaction occurred between the −OH groups. The bonding strengths of Cu-Cu and SiO$_2$-SiO$_2$ homogeneous structures were high enough. Observing the microstructure of the Cu-Cu bonding interface, the original interface diminished, the O element at the interface diffused into the matrix, and the substantial Cu grain growth was confirmed. Also, the pristine SiO$_2$-SiO$_2$ interface cannot be distinguished and the atomic-level interconnection was obtained. This method provides feasibility for the future ultra-fine pitch 3D integration.

II. EXPERIMENTS

A. Materials

Regarding Cu, 500 nm-thick Cu layers were deposited on Si wafers using magnetron sputtering, and Ti layers with a thickness of 30 nm were also sputtered as barrier metal. Regarding SiO$_2$, 500 nm-thick thermal oxide layers were fabricated on Si substrates by dry oxidation. The above-mentioned wafers were cut into 10×10 mm^2 in size for homogeneous bonding. Additionally, Cu/SiO$_2$ hybrid chips (20×20 mm^2 in size) with planarized SiO$_2$ dielectric layer and Cu connections were provided by Shanghai Anji Co., Ltd. to realize the final hybrid bonding.

B. Methods

The process flow of the two-step cooperative surface activation method is shown in Fig. 1. For the substrates to be bonded, formic acid solution immersion was conducted for 20 min. After that, the treated samples were cleaned with deionized water immediately and dried with flowing N_2. Subsequently, these samples were activated by Ar/O_2 mixed plasma for 30 s to 150 s. So far, the samples have undergone the complete cooperative activation procedure. In order to obtain the robust homogeneous and hybrid bonding interface, the thermal compression bonding (TCB) process and following annealing treatment were carried out at 200 °C in the atmospheric environment.

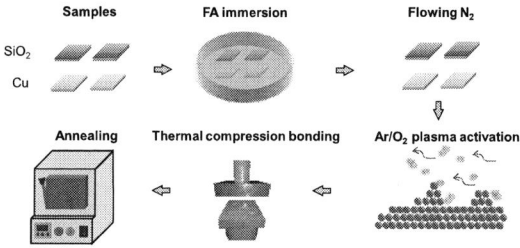

Fig. 1. Process flow of the two-step cooperative surface activation.

III. RESULTS AND DISCUSSION

A. Surface morphology characterization

After $FA \rightarrow Ar/O_2$ activation, the physical and chemical structures of Cu and SiO_2 surfaces were investigated to find out the root cause of the realization of low-temperature bonding. The surface roughness of surfaces varying with Ar/O_2 plasma activation time contained in the cooperative activation process is measured by atomic force microscopy (AFM) and shown in Fig. 2. It turns out that the roughness average (Ra) value of bare Cu surface exceeded 3 nm, and rougher Cu surfaces were obtained with prolonging plasma activation time. However, the opposite results were observed for the SiO_2 surface. When the SiO_2 substrates were treated by $FA \rightarrow Ar/O_2$, the Ra values rapidly decreased from 0.72 nm to less than 0.5 nm, and the relatively smooth surface was obtained when the Ar/O_2 activation time was extended to 120s.

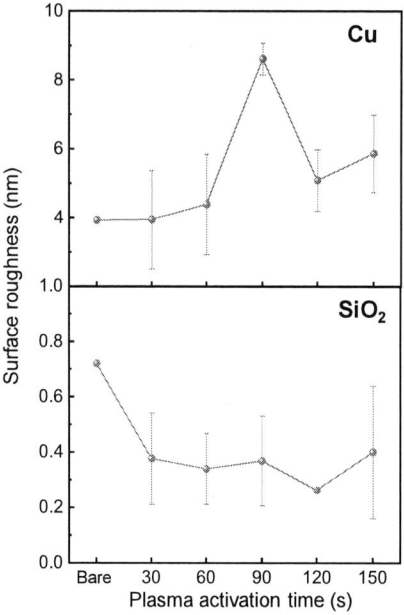

Fig. 2. Surface roughness of Cu and SiO_2 before and after $FA \rightarrow Ar/O_2$ activation.

Also, the three-dimensional surface morphology was also recorded by AFM to further evaluate the surface structure. As displayed in Fig. 3 (a) and (b), the increased peak-to-valley value (15.1 nm to 22.8 nm) concordances with the rougher surface morphology, which indicated that oxides were introduced on the surface [7]. In addition, many nano-ripple structures generated on the Cu surface activated by $FA \rightarrow Ar/O_2$, which is attributed to the high-energetic bombardment of Ar plasma [8]. Regarding SiO_2, Fig. 3 (c) and (d) present that the surfaces were flattened effectively after cooperative activation, and this smooth surface is beneficial to increase the real contact area in the subsequent hydrophilic bonding.

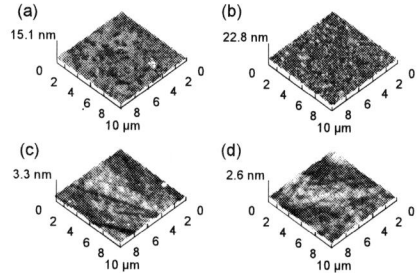

Fig. 3. Three-dimensional surface morphology of (a-b) Cu and (c-d) SiO_2 before and after $FA \rightarrow Ar/O_2$ (120 s) activation.

B. Chemical structure investigation

To investigate the chemical composition of activated surfaces, a water contact angle (CA) test was carried out for Cu and SiO_2 surfaces before and after $FA \rightarrow Ar/O_2$ activation. The images of water droplets spreading on the surfaces are shown in Fig. 4. Regardless of Cu and SiO_2, the higher CA values of bare surfaces indicated that hydrophobic structures were adsorbed on the surface, such as organic contaminants and Cu_2O (CA > 45°) [9]. Surprisingly, the CA values dropped immediately after $FA \rightarrow Ar/O_2$ activation, even if the contained Ar/O_2 plasma activation time was just 30 s. It indicates that the hydrophobic components were peeled from the surface and a hydrophilic layer (e.g., Si-OH, CuO, $Cu(OH)_2$) was established.

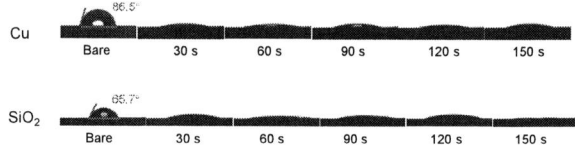

Fig. 4. Images of droplet spreading on the surface of Cu and SiO_2 after $FA \rightarrow Ar/O_2$ activation with different plasma treatment times

Subsequently, the chemical structures were further verified by timely Fourier transform infrared spectroscopy (FT-IR). Fig. 5 (a) shows that a strong vibration peak was detected at the wavelength of 666 cm^{-1} on the bare Cu surface, which represented the Cu-O bonds of copper oxide [7]. In addition, some peaks assigned to C-O and C-H were also found, demonstrating the adhering of organics [10]. After $FA \rightarrow Ar/O_2$ activation, the peaks of Cu-O, C-O, and C-H disappeared, but signals related to hydrophilic functional groups were not detected due to the low content. Regarding SiO_2, the intensity of the Si-O peak at 1050 cm^{-1} [11] was improved after activation, as shown in Fig. 5 (b), which provides more sites for the formation of -OH groups and is in accordance with CA results.

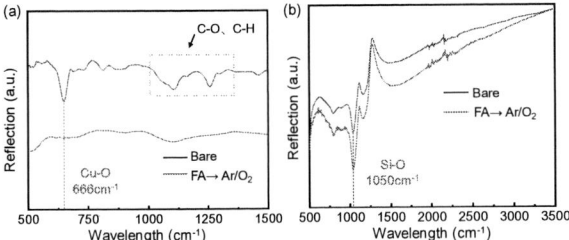

Fig. 5. FT-IR spectrum of the (a) Cu and (b) SiO₂ surfaces before and after FA→Ar/O₂ (120 s) activation.

To obtain direct evidence of the generation of $Cu(OH)_2$ on the activated Cu surface, an X-ray photoelectron spectroscopy (XPS) measurement was also conducted. Fig. 6 (a) and (b) show that the oxides on the bare Cu surface consisted of Cu_2O, CuO, and $Cu(OH)_2$. However, the relative content of Cu_2O and CuO (Fig. 6 (b) and (d)) increased from 14.76% to 39.65% after FA→Ar/O₂ activation. It seems like activation caused the strong oxidation on the Cu surface, which is not desirable for the subsequent low-temperature bonding and is inconsistent with the result of FTIR. It can be attributed to the activated surface was in a thermodynamically metastable state owing to the high-energetic bombardment of Ar plasma, which could easily adsorb O_2 in the air and result in secondary oxidation [8]. Therefore, a large amount of oxide on the activated surface could be introduced due to the inevitable storage in ambient air before XPS testing.

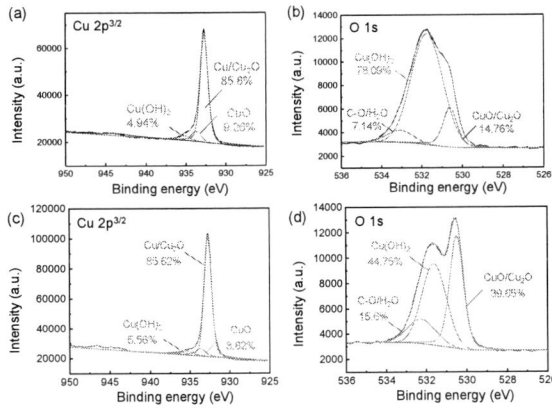

Fig. 6. XPS spectra of Cu surface (a)(b)before and (c)(d)after FA→Ar/O₂ (120 s) activation.

C. Bonding interfacial performance

Based on AFM, CA, and FT-IR results, the relatively clean and smooth Cu and SiO₂ surfaces with a hydrophilic layer were formed using FA→Ar/O₂ activation. Subsequently, TCB and low-temperature annealing were applied to realize the homogeneous bonding of Cu-Cu and SiO₂-SiO₂. Fig. 7 shows that the bonding strength of the Cu-Cu interface reached the maximum (12.35 MPa) when the plasma activation time extended to 120 s. The optical image of the fracture interface was also inserted. It can be seen that the Cu

layer on one side surface was adhered to the other side surface, indicating that the bonding strength between the Cu-Cu interface is much higher than that of the Cu-Ti interface. Hence, hydrophilic layers on the Cu surface play a positive role in strengthening the interface.

Similarly, the mechanical performance of the FA→Ar/O₂-activated SiO₂-SiO₂ interface was measured and recorded in Fig. 8. The optimal bonding strength (~ 4 MPa) was also achieved when plasma activation time contained in cooperative surface activation was 120 s. Although the strength of the SiO₂-SiO₂ interface was lower than that of the Cu-Cu interface, it is strong enough to provide robust mechanical support for the subsequent hybrid bonding. Combining the interfacial mechanical property of two kinds of homogeneous bonding interfaces, this cooperative activation method is feasible to achieve hybrid bonding at low temperatures.

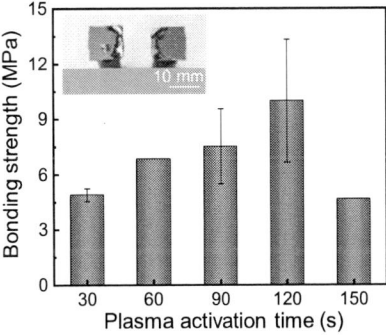

Fig. 7. Bonding strength of Cu-Cu samples obtained by FA→Ar/O₂ activation with different plasma treatment times.

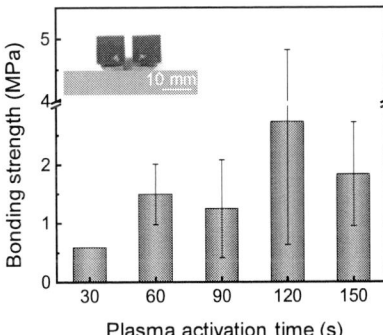

Fig. 8. Bonding strength of SiO₂-SiO₂ samples obtained by FA→Ar/O₂ activation with different plasma treatment times.

Until now, we explored the optimal cooperative activation parameters to form Cu/SiO₂ hybrid bonding. The bonding interface was evaluated by scanning electron microscopy (SEM). Fig. 9 (a) displays the hybrid bonding interface realized by the cooperative activation route, which possesses intimate and seamless Cu-Cu, and SiO₂-SiO₂ interfaces simultaneously. In addition to the homogeneous interface, the formation of heterogeneous Cu-SiO₂ interconnection is inevitable due to the mismatch between the top and bottom chips. Actually, some cracks can be observed at the relatively weak Cu-SiO₂ interface. Because this heterogeneous interface did not play a dominating role, the entire hybrid bonding interface was sufficiently strong.

978-1-6654-1392-3/21 $31.00 © 2021 IEEE

To further characterize the interfacial structure at the atomic level, transmission electron microscopy (TEM) observation was performed. As presented in Fig. 10 (a) and (b), there were no microvoids observed, and the original bonding interface disappeared. It demonstrated that the substantial atomic diffusion occurred across the interface. Moreover, the elemental analysis across the Cu-Cu interface (Fig. 10 (c)) shows that no O element was detected. According to the previous analysis, an ultrathin hydrophilic layer was formed on the FA→Ar/O₂-activated Cu surface, which may comprise CuO and Cu(OH)₂. However, it has been reported that CuO would inhibit atomic diffusion, leading to the demand for high temperatures for Cu-Cu bonding. Since the oxygen-free Cu-Cu interface was obtained through a lower energy (200 °C) input, we believe that hydrophilic layers on the Cu surface mainly composed of Cu(OH)₂, which not only did not inhibit but also facilitated the atomic diffusion across the interface. Moreover, Fig. 10 (d) shows that the atomically interconnected SiO₂-SiO₂ interface was also achieved on the hybrid bonding interface at the same time. However, there was a discontinuous intermediate layer with a thickness of several nanometers. The formation of the interlayer could be attributed to the use of organic acid in the activation process, resulting in the physical or chemical adsorption of HCOOH.

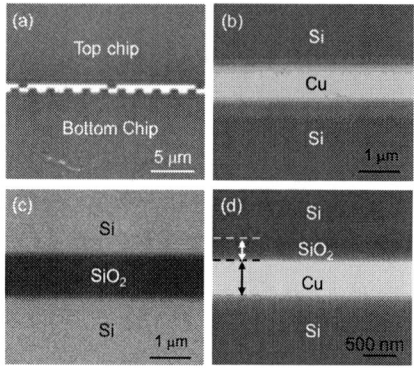

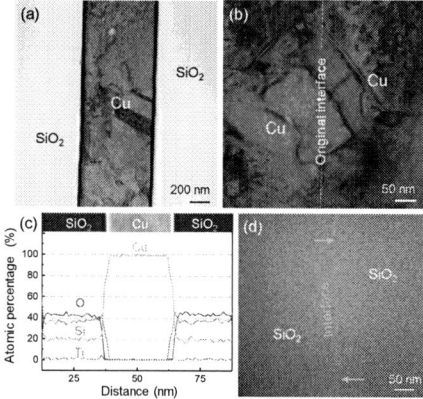

Fig. 9. The SEM cross-sectional image of (a) Cu/SiO₂ hybrid bonding interface with the included homogeneous (b) Cu-Cu and (c) SiO₂-SiO₂ and heterogeneous (d) Cu-SiO₂ interface.

Fig. 10. The TEM characterization of (a-b) Cu-Cu interface and (d) SiO₂-SiO₂ interface from Cu/SiO₂ hybrid bonding device. The elemental analysis across the bonding interface of (a) is shown in (c).

D. Bonding Mechanism

Here, a model of low-temperature bonding was proposed in Fig. 11. As discussed above, a hydrophilic -OH layer was established on both Cu and SiO₂ via the two-step cooperative surface activation. When the activated surfaces were contacted to each other by the imposing of thermal and pressure during TCB, the bonding between SiO₂ surfaces occurred owing to the Si-OH polymerization reaction and smooth surface morphology. Simultaneously, the Cu surface was also covered by hydrophilic -OH layers and a similar polymerization between Cu-OH groups also happened at the Cu-Cu interface and even the Cu-SiO₂ interface [12]. Hence, the hybrid interface composed of Cu-O-Cu, Si-O-Si, and Cu-O-Si bonds was obtained. To stabilize the microstructure and improve the interfacial performance, the low-temperature annealing process was necessary to promote the O element contained in Cu-O-Cu diffusing into the matrix [13]. Eventually, the robust hybrid bonding interface was realized at 200 °C.

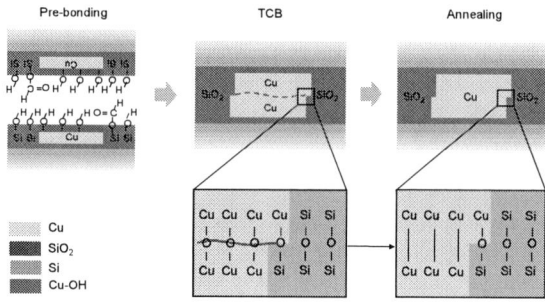

Fig. 11. Low-temperature bonding mechanism of Cu/SiO₂ hybrid bonding.

IV. CONCLUSION

In conclusion, we develop a novel FA→Ar/O₂ cooperative surface activation method to achieve Cu/SiO₂ hybrid bonding at 200 °C successfully. After activation, a hydrophilic layer was established on both Cu and SiO₂ surfaces. With the assistant of this layer, the robust homogeneous bonding of Cu-Cu and SiO₂-SiO₂ was realized, and their maximum bonding strengths reached 12.35 MPa and 4.27 MPa, respectively. From the hybrid bonding interface, it is possible to obtain the intimate and void-free Cu-Cu and SiO₂-SiO₂ interfaces simultaneously, and even a Cu-SiO₂ interconnection. This hybrid bonding structure realized by cooperative surface activation brings unprecedented 3D integration feasibility and flexibility.

REFERENCES

[1] J. H. Lau, "Overview and outlook of three-dimensional integrated circuit packaging, three-dimensional Si integration, and three-dimensional integrated circuit integration," J. Microelectron. Electron. Packag., vol. 136.4, 2014.

[2] J. H. Lau, "The future of interposer for semiconductor IC packaging," Chip Scale Rev, vol. 18, 2014, pp. 32-36.

[3] P. Enquist, G. Fountain, C. Petteway, A. Hollingsworth, H. Grady, "Low cost of ownership scalable copper direct bond interconnect 3D IC technology for three dimensional integrated circuit applications," IEEE International Conference on 3D System Integration. IEEE, 2009.

[4] A. Jouve, V. Balan, N. Bresson, et al. "1μm Pitch direct hybrid bonding with< 300nm wafer-to-wafer overlay accuracy," IEEE SOI-3D-Subthreshold Microelectronics Technology Unified Conference (S3S). IEEE, 2017.

[5] Y. Kagawa, N. Fujii, K. Aoyagi, et al. "Novel stacked CMOS image sensor with advanced Cu2Cu hybrid bonding," IEEE International Electron Devices Meeting (IEDM). IEEE, 2016.

[6] C. T. Ko and K. N. Chen, "Low Temperature Bonding Technology for 3D Integration," Microelectron. Reliab, vol. 52 (2), pp. 302–311, 2012.

[7] C. C. Chiang, M. C. Chen, L. J. Li, Z. C. Wu, S. M. Jang, and M. S. Liang, "Effects of O2 and N2 Plasma Treatments on Copper Surface," Jpn. J. Appl. Phys., Vol, 43, pp. 7415, 2004.

[8] M. Park, S. Baek, S. Kim and S. E. Kim, "Argon Plasma Treatment on Cu Surface for Cu Bonding in 3D Integration and Their Characteristics," Appl. Surf. Sci., vol. 324, pp. 168–173, 2015.

[9] S. H. Tu, C. C. Wu, H. C. Wu, S. L. Cheng, Y. J. Sheng and H. K. Tsao, "Time-Varying Wetting Behavior on Copper Wafer Treated by Wet-Etching," Appl. Surf. Sci. vol. 341, pp. 37–42, 2015.

[10] A. León, P. Reuquen and C. Garín, et al. "FTIR and Raman Characterization of TiO2 Nanoparticles Coated with Polyethylene Glycol as Carrier for 2-Methoxyestradiol," Appl. Sci. vol. 7 (1), pp. 49, 2017.

[11] S. P. Zhdanov, L. S. Kosheleva and T. I. Titova, "IR Study of Hydroxylated Silica," Langmuir, vol. 3 (6), pp. 960–967, 1987.

[12] R. He, M. Fujino, A. Yamauchi, T. Suga, "Combined Surface-Activated Bonding Technique for Low-Temperature Hydrophilic Direct Wafer Bonding," Jpn. J. Appl. Phys. Vol. 55 (4S), pp. 04EC02, 2016.

[13] R. He, M. Fujino, A.Yamauchi, Y. Wang and T. Suga, "Combined Surface Activated Bonding Technique for Low-Temperature Cu/Dielectric Hybrid Bonding," ECS J. Solid State Sci. Technol. Vol. 5 (7), P419, 2016.

Key factor analysis of nano silica on the dispersion in underfill

Xiaomeng Du
Shenzhen Institute of Advanced
Electronic Materials
Shenzhen Institute of Advanced
Technology, Chinese Academy of
Sciences
Shenzhen, China
xm.du1@siat.ac.cn

Ning Wang
Shenzhen Institute of Advanced
Electronic Materials
Shenzhen Institute of Advanced
Technology, Chinese Academy of
Sciences
Shenzhen, China
wangning@siat.ac.cn

Mingyong Du
Shenzhen Institute of Advanced
Electronic Materials
Shenzhen Institute of Advanced
Technology, Chinese Academy of
Sciences
Shenzhen, China
my.du@siat.ac.cn

Leicong Zhang
Shenzhen Institute of Advanced
Electronic Materials
Shenzhen Institute of Advanced
Technology, Chinese Academy of
Sciences
Shenzhen, China
lc.zhang@siat.ac.cn

Tao Zhao
Shenzhen Institute of Advanced
Electronic Materials
Shenzhen Institute of Advanced
Technology, Chinese Academy of
Sciences
Shenzhen, China
tao.zhao@siat.ac.cn

Pengli Zhu
Shenzhen Institute of Advanced
Electronic Materials
Shenzhen Institute of Advanced
Technology, Chinese Academy of
Sciences
Shenzhen, China
pl.zhu@siat.ac.cn

Rong Sun
Shenzhen Institute of Advanced
Electronic Materials
Shenzhen Institute of Advanced
Technology, Chinese Academy of
Sciences
ShenZhen, China
rong.sun@siat.ac.cn

Jiakai Cao
NOVORAY Corporation
Lianyungang, China
caojk@novoray.com

Jianjun Ruan
NOVORAY Corporation
Lianyungang, China
tony@novoray.com

Abstract—The miniaturization of the electronic devices requires smaller packaged chips, and the distance between the chip and the substrate becomes smaller, which make it necessary to introduce the nano-silica in the filler for advanced packaging. However, the nano-silica is easy to settle in the underfill and block the packing, because it tends to agglomerate due to its large specific area. Unfortunately, the addition of nano-silica in the underfill could also largely improve the viscosity, giving rise to the unacceptable low flowability. Therefore, it is necessary to investigate the key factors that could influence the dispersion of nano-silica in the epoxy resin.

In this paper, the time domain nuclear magnetic resonance spectrometer (NMR) technique was used to study the dispersion of silica fillers. This research studied the nano-silica particle size impact and the surface modification impact on the dispersion of nano-silica in epoxy resin. The shape and size distribution of three types of particles were characterized before studying the relaxation property. The three types of particles were then modified with silane coupling agents. The results of time domain NMR relaxation rate showed that the surface modified nano-silica particles have lower relaxation time, meaning the better dispersity in the epoxy. Meanwhile, after surface modification, larger fillers obtained better dispersity compared with smaller fillers, which result in the uniform dispersion in the resin.

Keywords—underfill; filler; surface modification; dispersion

I. INTRODUCTION

Underfill is a key material composed of epoxy resin and the silica filler, which is used between the chip and the substrate for flip-chip packaging. After curing, the underfill forms a low-CTE (coefficient of thermal expansion) thermosetting composite material that could reduce the CTE mismatch between the silicon chip and the organic substrate, and therefore release the thermal-mechanical stress on the solder balls, and thus improve the reliability of the devices [1]. In general, micron-size silica occupies the better flowability due to their relatively smaller specific surface area compared with the nano-size counterparts. However, the micron-size fillers are easy to cause the delamination, due to the weak interaction with the resin matrix. More importantly, with the rapid development of the electronic industry, electronic devices are becoming more and more functional and miniaturized. The conventional micron sized fillers have a high tendency to be sandwiched between the solder bumps on the chip and the contact plate of the substrate[2], which may reduce the devices' life. Therefore, nano-silica particles are necessary to be used as the filler in the underfill for the advanced electronic packaging.

However, compared with the micro-size silica, nano sized silica prepared by sol-gel method has more silanol surface silanol groups and larger specific surface area, leading to easier particle agglomeration, and the significantly increased viscosity. In order to reduce the viscosity of the underfill and increase the degree of filler loading, it is necessary to perform the surface treatment of the nano silica fillers, which could give rise to the better compatibility and dispersion of the filler in the epoxy resin. Silane coupling agent has a unique bifunctional structure, one end can react with the silanol group on the silica surface, and the other end is compatible with the resin matrix, which should be the good choice for the surface treatment of silica fillers[3]. The grafted silane coupling agent could form a "bridge" between the filler and the epoxy, and the compatibility between particles and epoxy resin could be improved. For the characterization after modification, the quality is generally determined directly according to the CTE of the final sample, which is time-consuming and a waste of raw materials. Liang N [4] introduced the use of NMR technology to study the adsorption, desorption and diffusion behavior of polymer on the surface of SiO_2 particles in liquid phase environment, and the liquid NMR technology could be used to study the absorption process of water in contaminated soil and soil wettability. It is emphasized that Acorn Ratio Surface Analyzer based on NMR technology can directly measure the relaxation time of materials in liquid phase environment and reflect the surface properties and structure characteristics of materials.

In order to analyze the influence of particle size and surface modification on the properties of silica fillers, mono-disperse spherical SiO_2 particles with diameters 300 nm/500nm/800nm were firstly prepared by the classical stöber method. Using trimethoxy [3-(phenylamino) propyl] silane as coupling agent, aniline group was grafted onto the surface of SiO_2. Unmodified SiO_2 (S-300, S-500 and S-800, respectively) and modified SiO_2 (SBA-300, SBA-500 and SBA-800, respectively) were added into the epoxy resin matrix as fillers. Diffuse reflectance infrared spectroscopy (DR-IR) and thermogravimetric analysis (TGA) were used to confirm the surface modificatio, and low - field NMR analysis was used to investigate the key factors for the silica dispersibility.

II. Experimental

A. Materials

Tetrathyl orthosilicate (TEOS, 98%), absolute ethanol (EtOH, 99%) , de-ionized (DI) water and ammonia (NH_4OH) were used for the synthesis of SiO_2. Trimethoxy [3-(phenylamino) propyl] silane (≥98%, Shanghai Aladdin Biochemical Technology Co., Ltd.) were used as coupling agents for silica surface modification. Sodium hydroxide (≥98%, pellets, Aladdin) , hydrochloric acid, sodium chloride (AR, 99.5%, Aladdin), Bisphenol-F type epoxy (Hexion Specialty Chemicals) were used for preparing the underfill .

B. Surface modification of nano silica

In the design of experiment, the effects of different silica particle size(300nm, 500nm, 800nm) and modification on the dispersion were investigated. The surface of the synthesized silica was modified by industrial dry modification method. Firstly, 100 g of silica was placed in a heated container at 130°C and stirred. Secondly, 0.5 g of the coupling agent was dispersed in 300 mL of absolute ethanol. Finally, the mixed solution was sprayed into the container at a certain rate. After cooling down, the attained silica powder was collected for further characterizations.

C. Preparation of underfill with original SiO_2 nanoparticles and modified SiO_2 nanoparticles

Firstly, the solvent-free epoxy resin matrix was prepared to prevent the adverse effects of organic solvents. Then, the epoxy resin nanocomposites filled with different types of silica were prepared according to the following steps. The composite material can be obtained by mixing epoxy resin and silica filler according to a certain mass ratio. The high speed mixer was used to mix the composite material fully under vacuum, and the rotating speed was 2000 rpm. In the vacuum

condition, the bubble generated in the mixing process could be fully removed.

D. Characterization

Diffuse reflection infrared spectroscopy (Bruker, Vertex70) was used to characterize the surface modification of SiO_2 and verify whether the coupling agent could be successfully grafted onto the spherical surface of SiO_2. A thermogravimetric analyzer (TGA, TA, SDT-Q600) was used to analyze the amount of coupling agent on SiO_2 spheres by sealing the sample to be tested in a crucible made of high temperature Al_2O_3 and obtaining TGA curves from 200 °C to 800 °C under air atmosphere. Scanning electron microscopy (SEM, Nova Nano SEM 450) was used to characterize the surface morphology of silica before and after the surface modification. The number of surface silanol groups before and after modification was studied by automatic titrator (Metrohm, 905 Titrando).The relaxation time of the sample was measured on a low-field NMR (Xigo, Acorn area).

III. Results And Discussion

A. Detection of Surface modification of SiO_2 filler

The surface of SiO_2 spheres contains silanol groups, and most of the particles are connected by hydrogen bonds to form secondary particles, leading to agglomeration. In this study, we can increase the interface compatibility and affinity between SiO_2 spheres and the resin matrix through dry modification.

The morphology of SiO_2 before and after surface treatment was characterized by SEM. It can be depicted from the SEM figure (Fig. 1 a-f) that the synthesized SiO_2 is a relatively uniform spherical particle. As shown in Fig. 1, the modified SiO_2 has a good dispersion, and it can be seen that there exists a thin film (coupling agents) on the silica sphere surface. Therefore, it can be proved that the surface modification of SiO_2 is successful.

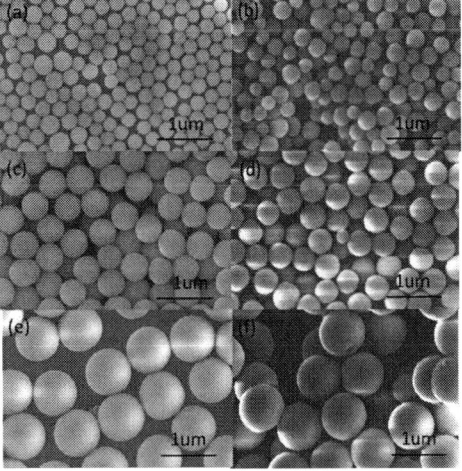

Fig. 1. The SEM images of S-300 (a), BAS-300 (b), S-500 (c), BAS-500 (d), S-800 (e), BAS-800 (f).

The surface chemical properties of silica before and after silane treatment were studied by diffuse reflectance infrared spectroscopy(DR-IR). Fig. 2 shows the DR-IR spectra of silane treated and untreated SiO_2. It can be seen from Fig. 2 that the DR-IR spectrum of unmodified SiO_2 only shows the stretching vibration absorption peak of -OH at 3600 cm^{-1}. However, for the spectrum of modified SiO_2, there is not only a stretching vibration absorption peak of 3600 cm^{-1}, but also shows the asymmetric and symmetric stretching vibration absorption peaks of methylene (-CH$_2$) at 2940 cm^{-1} and 2850 cm^{-1}, and the C-H stretching vibration absorption peaks on phenyl at 3050 cm^{-1}. Moreover, The stretching vibration absorption peak of the secondary amine group (-NH-) at 3440 cm^{-1} is also observed. Therefore, it is further indicated that the surface graft of SiO$_2$ is successful.

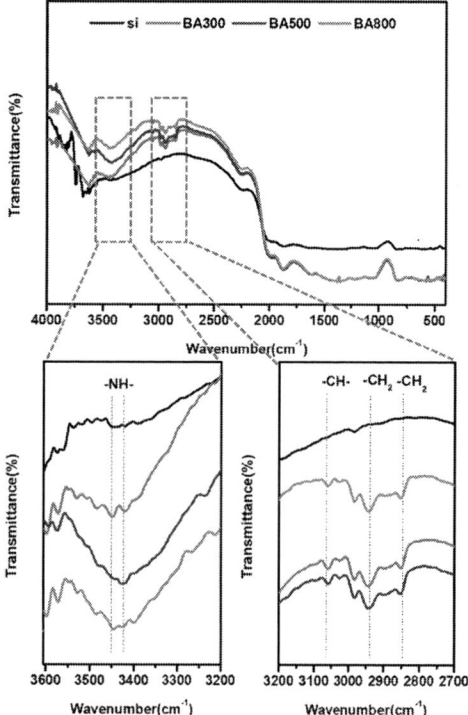

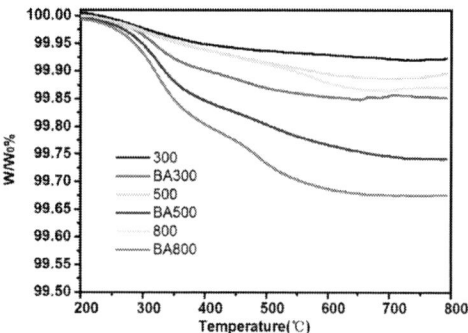

Fig. 2. The DR-IR spectra of untreated and silane treated SiO$_2$.

Fig. 3. TGA image before and after modification of silica with different particle sizes.

The surface coupling agent content on the surface of SiO$_2$ was evaluated by measuring TGA. Fig. 3 shows the TGA curves of SiO$_2$ and BASiO$_2$ from 200 to 800 °C. It can be seen that the TGA curves of SiO$_2$ can be divided into two stages, from 200 to 300 °C and after 300 °C. The weight loss between 200 °C and 300 °C is caused by the dehydration reaction of hydroxyl groups on the surface of SiO$_2$. The weight loss after 300 °C is due to the dehydration of the internal hydroxyl groups of SiO$_2$. In contrast, BASiO$_2$ has different TGA curves. It can be seen that the weight loss of BASiO$_2$ after 300 °C is much greater than that of SiO$_2$ in the same range. Therefore, in this temperature range, the larger weight loss of BASiO$_2$

after 300 °C can be attributed to the decomposition of the coupling agent. Table I summarizes the weight loss of SiO$_2$ and BASiO$_2$ in different temperature stages. The surface coupling agent content (W$_{BA}$ %) can be roughly calculated according to the following formula:

$$W_{BA} \% = W_{BASiO2} \% - W_{SiO2}\% \qquad (1)$$

Among them, W$_{BASiO2}$% and W$_{SiO2}$ % are the weight loss percentages of BASiO$_2$ and SiO$_2$ after 300 °C, respectively. According to Equation (1), W$_{BA}$% of BA300, BA500 and BA800 are 0.059wt%, 0.134wt% and 0.153wt%, respectively.

Table. I TGA results of the SiO$_2$ and BASiO$_2$.

	300	BA300	500	BA500	800	BA800
Weight Loss 200-300 °C, wt%	0.023	0.035	0.03	0.051	0.027	0.069
Weight Loss > 300 °C, wt%	0.054	0.113	0.073	0.207	0.102	0.255

B. The content of silanol group on the surface of dioxide

The presence of silanol group on the surface of silica makes the surface of silica powder hydrophilic and it is difficult to disperse evenly in epoxy. The silanol group distributed on the surface exists in three forms, namely, isolated free, geminal free, and vicinal [5]. In some applications, it is necessary to find out the number of silanol groups on the surface of silica. For example, the modification of silica can be judged by analyzing the change of the number of silanol groups on the surface before and after the modification[3].

The method is similar to that in the article of Ouyang Zhaohui. Specifically, it is described as follows: add silica into the system of anhydrous ethanol and sodium chloride solution with stirring evenly, and adjust the pH value of the system to 3.0 with hydrochloric acid solution. The sodium hydroxide solution was slowly added to the above solution system using potentiometric titrator to raise the pH value to 9.0, and the volume of the sodium hydroxide solution used to record the pH from 4 to 9 was recorded. Finally, formula (2) is used to calculate the number of silanol groups on the surface of silica. The Number of silanol groups:

$$N=(C \times (V_{pH4-9}-V_B) \times Na \times 0.001)/(S \times m) \qquad (2)$$

Where, C is the concentration of standard solution of NaOH (mol/L), V$_{pH4-9}$ is the volume (mL)of NaOH solution consumed to increase pH from 4 to 9 , V$_B$ is the volume of NaOH solution consumed from pH 4 to 9 in blank sample (mL), Na is the Avogadro constant, S is the Specific Surface Area (m^2/g), m is the silica mass (g).

The graft density can be calculated by formula (2), as shown in Table II. It can be seen that the graft density increases with the increase of particle size. It is possible that due to the large particle size, the steric hindrances between the hydroxyl groups are reduced and grafted more easily.

Table. II Results of titration of silicon surface hydroxyl.

	300	BA300	500	BA500	800	BA800
Number of surface silanol, OH/nm2	1.22	1.01	1.36	1.08	1.75	1.40
Grafting number of surface silanol, OH/nm2	--	0.21	--	0.28	--	0.35

C. Relaxation time

Table. III NMR results of the SiO_2 and $BASiO_2$.

	EP	300	BA300	500	BA500	800	BA800
relaxation time, ms	16.2	13.5	12	13	11.8	11.7	11.2
relaxation rate	--	0.2	0.35	0.25	0.37	0.38	0.45

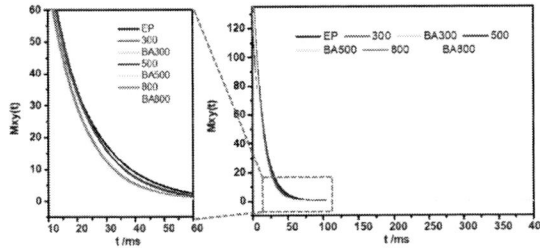

Fig. 4. Mxy fit for T_2 measurement.

In the Low-field NMR, the relaxation time of particle surface molecular motion can be directly used to detect the adsorption between particles and medium molecules, which is of great significance for the study of homogeneous and heterogeneous slurry systems[6]. Liquid can move freely in space, and the relaxation time of hydrogen atoms in liquid should be the longest. With the addition of particles, the movement of the adsorbed liquid on the surface of particles is limited, so the relaxation time of hydrogen atoms should become shorter. Through this principle, the dispersibility of particles in the liquid epoxy could be measured with respect to the relaxation time. In a well dispersed system, the moving of the liquid molecules absorbed on the silica surface is restricted uniformly, giving rise to the shorter relaxation time, whereas in a poor dispersed system, the particles will quickly moved to the bottom of the solution, and the liquid molecules'

movement could not be effectively restricted, resulting in a longer relaxation time.

By testing the relaxation time before and after modification, as shown in table III, it can be seen that the relaxation time becomes shorter after modification, which proves that the dispersion is better after modification than before modification. Unfortunately, due to the relatively close particle sizes, the relaxation time difference is small, so it is impossible to compare the effect of particle sizes on dispersion.

IV. Conclusion

In this work, we have successfully modified the surface of three types of silica with different particle size. The surface modification effect of nano silica was characterized systematically. The diffuse reflectance infrared spectra showed that the silane functional groups were successfully grafted onto the surface of silica. The results of silanol titration showed that the graft density increased with the increase of particle size. The low field NMR analysis showed that the surface modification improves the dispersion of the filler in the underfill.

ACKNOWLEDGMENT

This work was supported by the National Key R & D Project from Minister of Science and Technology of China (2020YFB0311800) , Shenzhen basic research plan (JCYJ20190807154409372), the Guangdong Basic and Applied Basic Research Foundation (No. 2019A1515010743）and the SIAT Innovation Program for Excellent Young Researchers (No. E1G012).

REFERENCES

[1] C.-C. Tuan, N.P. James, Z. Lin, Y. Chen, Y. Liu, K.-S. Moon, Z. Li, C.P. Wong, Self-Patterning of Silica/Epoxy Nanocomposite Underfill by Tailored Hydrophilic-Superhydrophobic Surfaces for 3D Integrated Circuit (IC) Stacking, ACS Applied Materials & Interfaces 9(10) (2017) 8437-8442.

[2] S.H. Shi, C.P. Wong, Effects of the complexed moisture in metal acetylacetonate on the properties of the no-flow underfill materials, Journal of Applied Polymer Science 73(1) (1999) 103-111.

[3] E.F. Vansant, P. Van Der Voort, K.C. Vrancken, Preface, in: E.F. Vansant, P. Van Der Voort, K.C. Vrancken (Eds.), Studies in Surface Science and Catalysis, Elsevier1995, pp. v-vii.

[4] Liang N, Zhang D. The new applications of NMR technology in the field of characterization of surface properties of the material[J]. Spectroscopy and Spectral Analysis, 2015, 35(2): 497-501.

[5] L.T. Zhuravlev, The surface chemistry of amorphous silica. Zhuravlev model, Colloids and Surfaces A: Physicochemical and Engineering Aspects 173(1) (2000) 1-38.

[6] Karpovich A, Vlasova M, Sapronova N, et al. Exfoliation dynamics of laponite clay in aqueous suspensions studied by NMR relaxometry[J]. Oriental Journal of Chemistry, 2016, 32(3): 1679

Effect of filler, toughening agent and coupling agent on the curing shrinkage of epoxy-based underfills.

Xiaohui Peng
Shenzhen Institute of Advanced Technology
Chinese Academy of Sciences
Shenzhen university
ShenZhen, China
xh.peng1@siat.ac.cn

Jinbao Yang
Shenzhen Institute of Advanced Electronic Materials
Shenzhen Institute of Advanced Technology
Chinese Academy of Sciences
ShenZhen, China
jb.yang@siat.ac.cn

Tao Peng
Shenzhen Institute of Advanced Electronic Materials
Shenzhen Institute of Advanced Technology
Chinese Academy of Sciences
ShenZhen, China
tao.peng1@siat.ac.cn

Pengli Zhu
Shenzhen Institute of Advanced Electronic Materials
Shenzhen Institute of Advanced Technology
Chinese Academy of Sciences
ShenZhen, China
pl.zhu@siat.ac.cn

Xing Ouyang
Shenzhen university
Shenzhen university
ShenZhen, China
oyx@szu.edu.cn

Yan Pan
Shenzhen Institute of Advanced Electronic Materials
Shenzhen Institute of Advanced Technology
Chinese Academy of Sciences
ShenZhen, China
yan.pan@siat.ac.cn

Gang Li
Shenzhen Institute of Advanced Electronic Materials
Shenzhen Institute of Advanced Technology
Chinese Academy of Sciences
ShenZhen, China
gang.li@siat.ac.cn

Rong Sun
Shenzhen Institute of Advanced Electronic Materials
Shenzhen Institute of Advanced Technology
Chinese Academy of Sciences
ShenZhen, China
rong.sun@siat.ac.cn

Yajing Yang
Huawei Corporation
Huawei Corporation
ShenZhen, China
yyj_0902@163.com

Liang Peng
Shenzhen Institute of Advanced Electronic Materials
Shenzhen Institute of Advanced Technology
Chinese Academy of Sciences
ShenZhen, China
liang.peng@siat.ac.cn

Abstract—Epoxy-based underfills were commonly used in microelectronic packages to re-distribute stress in solder bumps induced by the thermal mismatch. In this case, the volume shrinkage caused by the epoxy has an important effect on the internal stress of encapsulation, which is often related to the reliability problems, such as deformation warpage and fracture. In our work, an improved real-time monitoring method was conducted to measure the curing shrinkage of different underfills. The influence of the filler content, particle size, types of toughening agent and coupling agent on the curing shrinkage of composites were systematically investigated. The crosslinking density of composites was determined by swelling degree, and the relationships between crosslinking density and shrinkage were also evaluated. Our experimental results explicates that the shrinkage of epoxy-based underfills has positive correlation with the cross-linking density of polymer matrix.

Keywords—Underfill, shrinkage, epoxy, cross-linking density

I. Introduction

Epoxy-based underfills are commonly used in microelectronic packages to re-distribute stress in solder bumps. During the polymer curing process, thermal expansion (or shrinkage) and fixed shrinkage will occur simultaneously, which will inevitably lead to warpage and internal stress[1]. Thermal expansion and curling shrinkage result in polymer composites volume change. The macroscopic volume change determines the surface morphology, warpage and size of the final product[2]. To optimize the process, the volume variation of the material need to be considerably studied, as it is helpful to predict and control the warpage, and the internal stress of the material accurately[3]. In the past, Hiroshi Yamguchi[4] used the volume shrinkage to predict the curing stress variation during the curing process, and found a good coherence between the shrinkage and warpage induced by contraction stress. Although some experiments have been performed on the curing shrinkage variation of resin composites by changing single component[5,6]. The effect of multiple components and their coupling on the shrinkage of polymer composites have never been fully understood.

So far, viscoelastic properties of the resin while curing have been extensively studied[7], but there are few public report utilizing these data to predict the relationship between crosslinking density and curing shrinkage. In order to investigate the relationships between crosslinking density and shrinkage, we conducted in-situ measurements of volume variation in the process of resin curing. The crosslinking density was obtained by measuring swelling degree by cured samples.

II. EXPERIMENTS

A. Preparation of Epoxy/SiO2 Nanocomposites

The epoxy resin, and bisphenol-F (Hexion Specialty Chemicals) were used as the matrix material. The aromatic amine was used as a curing agent for the epoxy. The bisphenol-F epoxy resin was added into aromatic amine curing agent with ratio of 10: 4 and thoroughly mixed using a high-speed magnetic stirrer for about four minutes; Then 0, 4, 6, 8, 10 and 12 g SiO_2 were added into 20 g epoxy-hardener mixture respectively. The samples were named as neat epoxy, 20wt% SiO_2, 30wt% SiO_2, 40wt% SiO_2, 50wt% SiO_2 and 60wt% SiO_2 with respect to different SiO_2 contents.

B. Preparation of underfill materials filled with differents silane toughening agents

The toughening agents, accounting for 0.5~1% of the total content, were added into the mixed epoxy and curing agent system, and evenly mixed up. Afterwards, 50% of the total content of silica filler is added to the system, and the underfill sample can be obtained after uniformly mixing. We selected five types of toughening agents from our products, which have different characteristics as indicated in Table I.

TABLE I

	Epoxy	Curing agents	Filler	Toughening agents(1)	Additive
A	34.4	13.6	50	--	1
B	34.4	13.6	50	POSS	1
C	34.4	13.6	50	M52N	1
D	34.4	13.6	50	KMP605	1
E	34.4	13.6	50	W35	1
F	34.4	13.6	50	CTBN	1

C. Preparation of underfill materials filled with differents silane coupling agents

We added 0.5% of the total content of coupling agent into the mixed system composed of epoxy resin and curing agent, and mixed well; Then, 50% of the total content of silica was added into the system to prepare the underfill materials with different coupling agents.

TABLE II

	Epoxy	Curing agents	Filler	Coupling agents (0.5)
P0	35.4	14.1	50	--
P1	35.4	14.1	50	KH560
P2	35.4	14.1	50	KH550
P3	35.4	14.1	50	Triethoxyphenylsilane
P4	35.4	14.1	50	Anilino-methyl-triethoxysilane

D. Resin shrinkage

In this article, the curing shrinkage of the underfill was measured by the densitometric method. Float viscometer was use the to measure the density of the resin after deaeration. A certain amount of resin was take into the volumetric flask. After a period of standing time, we added the silicone oil into the bottle along the bottle wall until the graduation line. Finally, immerse the entire volumetric flask into an oil bath with a inserted pipette, and record the initial graduation.

As the oil bath is heated up to the curing temperature of the resin (165°C), the liquid level of the pipette will rise due to thermal expansion. After a while, the resin starts to cure, at the same time, shrinkage will be generated. When the pipette interface no longer changes, record the liquid level scale to determine the curing shrinkage rate of the resin.

E. Preparation of crosslinking density test spline

T The crosslinking density of composites were determined by swelling degree test. The freshly prepared underfill was injected into the mold (length x width x thickness: 10 mm x 10 mm x 3 mm); Then, the samples were placed into the oven, and were heated from 25 °C to 165 °C at a heating rate of 5 °C/min and then post-cured at 200°C for a further 2h. The resultant composites were then allowed to cool slowly to room temperature. The cured samples were weighed and put into a glass bottle containing 50 mL acetone. They were completely immersed in the solvent. The bottle was corked tightly and placed for 24h, 48h and 72h, respectively. The mass of the sample was weighed with an analytical balance.

III. RESULTS AND DISCUSSION

Shrinkage occurs while epoxy resins are curing. The shrinkage is caused by the formation of polymeric and crosslinked structures.

A. The effect of filler loading on shrinkage and crosslinking density

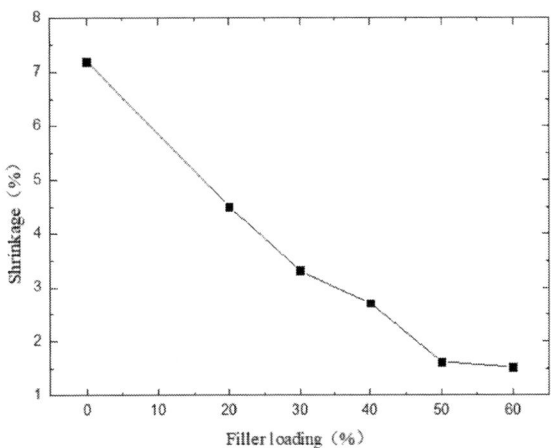

Fig.1 Curing shrinkage as a function of filler content

Fig.1 shows that the addition of SiO_2 filler reduced the overall shrinkage of the composites. SiO_2 dilutes the reaction volume of the composites, which in turn decreases the volume shrinkage of epoxy/SiO_2 composites. In addition, the presence of dense nano-filler may increase the viscosity of the system, thus gradually reduces the relative volume shrinkage. Furthermore, high surface volume ratio of nano-SiO_2 guarantees the epoxy polymer/oligomer to adhere to the surface of the filler, which generates a confining polymer on the filler surface. Overall, rigid SiO_2 particles prohibit the confining layers of these polymer chains from shrinking or limit the shrinkage extent.

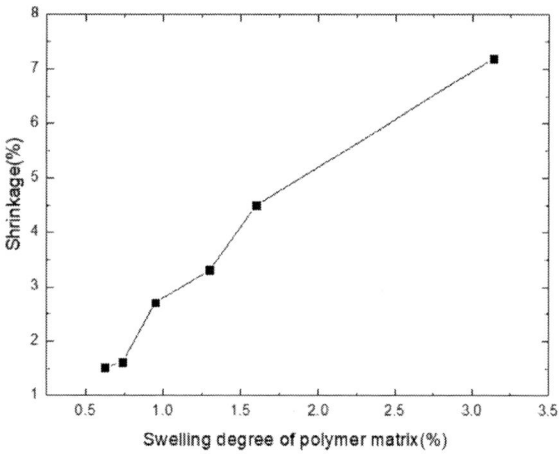

Fig. 2. Variation of curing shrinkage with the swelling degree of polymer matrix

As we can see in Fig. 2, the shrinkage of epoxy-based underfills was positively correlated with the cross-linking density of polymer matrix. This is because the SiO_2-epoxy interaction reduces the crosslinking degree attained by the polymer matrix.

B. The shrinkage and crosslinking density of underfills filled with different toughening agents

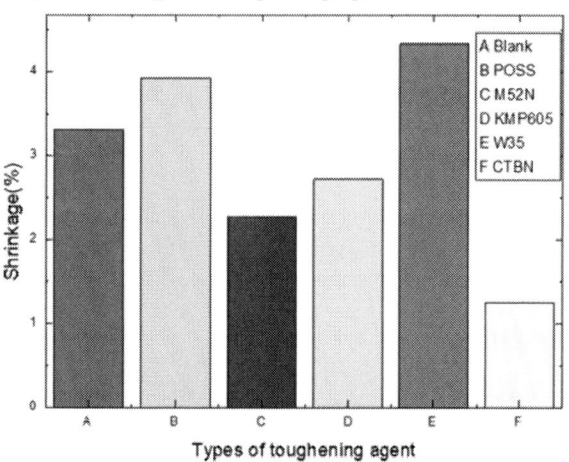

Fig.3. Effect of the toughening agents on the shrinkage.

Generally speaking, toughening agents affect the curing shrinkage. It can be seen from Fig. 3 that most of toughening agents reduce the shrinkage due to the steric hindrance. The addition of reactive toughening agents affects the curing mechanism and reduces the crosslinking density, which could reflect by reduced shrinkage. The reactive toughening agents partially incorporated in the crosslinked epoxy phase and vice versa. In this way it is possible to modify the morphology-related properties as well as to control the degree of the unwanted shrinkage.

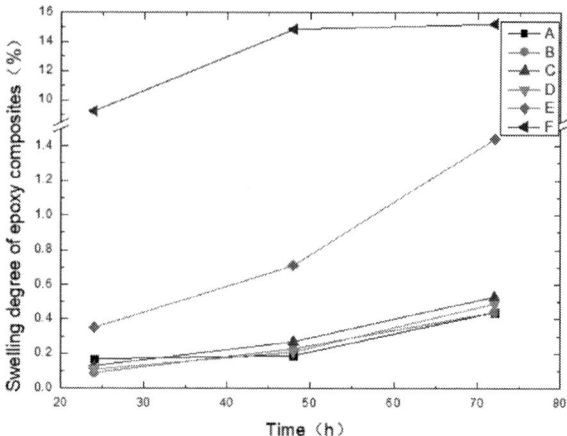

Fig. 4. Swelling degree of epoxy composites as function of time.

Fig. 4 shows that the swelling degree increased with the increase of acetone soaking time. The swelling of curve F is about to reach saturation after 48 h acetone immersion. It is noticeable that the shrinkage has a positive correlation with cross-linking density.

C. The shrinkage and crosslinking density of underfills filled with different coupling agents

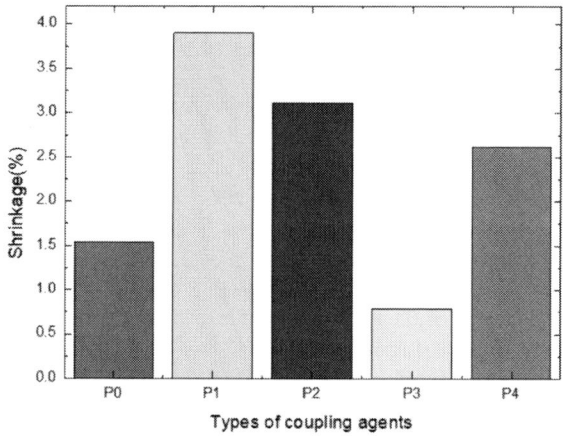

Fig. 5. Effect of the coupling agents on the shrinkage

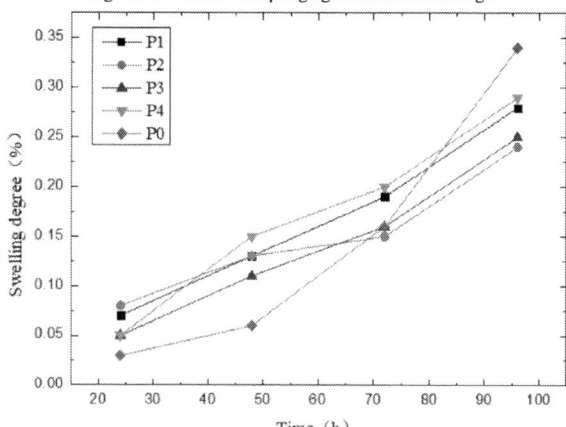

Fig. 6. Swelling degree of epoxy composites as function of time.

We analyzed this shrinkage of the underfill consisting of bisphenol-F (DGEBF), aromatic amine, and an active or a non-active coupling agents. The addition of coupling agents increased the shrinkage except for the inert one as shown in Fig. 5. The interaction between the reactive groups and hydroxide radical on SiO_2 surface enriches the shrinkage. Furthermore, the active coupling agent could help to increase the dispersion of SiO_2 and reduce shrinkage.

IV. CONCLUSION

In this work, the addition of SiO_2 filler reduced the overall shrinkage of the composites, and the curing shrinkage decreased with the increase of filler content. However, this phenomenon cannot be simply explained by the substitution of the filler volume to the polymer volume. On the other hand, dense nano-SiO_2 may increase the viscosity of the system and hence progressively decrease the relative volume shrinkage. Toughening agents had a great effect on the curing shrinkage, and most of them reduced the shrinkage due to the steric hindrance. The shrinkage can be decreased by adding coupling agents except for the inert one, due to the interaction between the reactive groups and hydroxide radical on SiO_2 surfaces. Furthermore, the active coupling agent could help to increase the dispersion of SiO_2 and reduce shrinkage. This study found that the shrinkage of epoxy-based underfills was positively correlated with the cross-linking density of polymer matrix.

ACKNOWLEDGMENT

This work was supported by the National Key R & D Project from Minister of Science and Technology of China (2020YFB0311800) ， Shenzhen basic research plan (JCYJ20190807154409372), the National Key R & D Project from Minister of Science and Technology of China 2020YFB0408703 ， Shenzhen Science and Technology Program (Grant No. RCBS20200714114859112)

REFERENCES

[1] X. Wang, C. Wang, Y. Jia, L. Luo, and P. Li, "Cure-volume-temperature relationships of epoxy resin and graphite/epoxy composites," *Polymer,* vol. 53, no. 19, pp. 4152-4156, 2012.

[2] Boyard N, Vayer M, Sinturel C, "Modeling PVTX diagrams: Application to various blends based on unsaturated polyester—Influence of thermoplastic additive, fillers, and reinforcements," Journal of applied polymer science, 2004, 92(5): 2976-2988.

[3] C. Li, K. Potter, M. R. Wisnom, and G. Stringer, "In-situ measurement of chemical shrinkage of MY750 epoxy resin by a novel gravimetric method," Composites Science and Technology, vol. 64, no. 1, pp. 55-64, 2004.

[4] Yamaguchi, H., Enomoto, T., & Sato, T. Stress variation analysis during curing process of epoxy underfill. In 2014 International Conference on Electronics Packaging (ICEP) (pp. 507-510). IEEE.

[5] Parameswaranpillai J, George A, Pionteck J, et al. Investigation of cure reaction, rheology, volume shrinkage and thermomechanical properties of nano-TiO2 filled epoxy/DDS composites[J]. Journal of Polymers, 2013, 2013.

[6] Pionteck J, Müller Y, Häußler L. Reactive Epoxy‐CTBN Rubber Blends: Reflection of Changed Curing Mechanism in Cure Shrinkage and Phase Separation Behaviour[C]//Macromolecular Symposia. Weinheim: WILEY‐VCH Verlag, 2011, 306(1): 126-140.

[7] Masashi OHORI, Chiaki SATO and Kozo IKEGAMI, "Viscoelastic Properties and internal Stress of Thermosetting Resin in Curing Process," J. Soc. Mat. Sci., Japan, vol. 43, No.484, pp. 18-22, Jan 1994.

Solid-Liquid Mixing-state Organic Lenses for Deep-Ultraviolet Light-Emitting Diodes to Enhance the Light-Extraction Efficiency

Zihao Deng
National & local joint engineering research center of semiconductor display and optical communication devices, South China University of Technology,
Guangzhou, China
scutzihao@foxmail.com

Jiexin Li
National & local joint engineering research center of semiconductor display and optical communication devices, South China University of Technology,
Guangzhou, China
jiexinli.jeson@foxmail.com

Jiayong Liang
National & local joint engineering research center of semiconductor display and optical communication devices, South China University of Technology,
Guangzhou, China
scutjiayong@foxmail.com

Jiayi Li
National & local joint engineering research center of semiconductor display and optical communication devices, South China University of Technology,
Guangzhou, China
13413451620@163.com

Jiasheng Li
National & local joint engineering research center of semiconductor display and optical communication devices, South China University of Technology,
Guangzhou, China
jiasli@foxmail.com

Xinrui Ding
National & local joint engineering research center of semiconductor display and optical communication devices, South China University of Technology,
Guangzhou, China
dingxr@scut.edu.cn

Zongtao Li*
National & local joint engineering research center of semiconductor display and optical communication devices, South China University of Technology,
Guangzhou, China
meztli@scut.edu.cn

Abstract—Deep ultraviolet light-emitting diodes (DUV-LEDs) have been considered as the most potential ultraviolet light sources, however, the traditional packaging with flat quartz glass has severe total internal reflection (TIR) loss, largely reducing the efficiency of DUV-LED. It is an efficient method to address this problem using organic lens, but traditional organic materials cannot be applied in DUV-LEDs because they are easily cracked by short-wave ultraviolet light. In this report, solid-liquid mixing state (SLMS) organic lens is proposed to enhance the light extraction efficiency and reliability of DUV-LEDs. Results show that the radiant power of DUV-LEDs using SLMS organic lens is enhanced by 53.86 % under the injection current of 150 mA compared to DUV-LEDs (275nm)encapsulated with conventional flat quartz glass. Moreover, after 279 hours continuous operation, DUV-LEDs encapsulated with SLMS organice lens can still radiate 8.289 mW, which is only 3.94% lower than the initial power. Consequently, an efficient, low-cost and universal strategy has been proposed to enhance the light-extraction efficiency of DUV-LEDs, which has a great application prospect in packaging industry of DUV-LEDs.

Keywords—Deep ultraviolet light-emitting diodes; light extraction efficiency; solid-liquid mixing-state

I. INTRODUCTION

DUV-LEDs, with a wavelength range of 200-300nm, have been used in various applications such as curing engineering, communication security and air purification. Nowadays, the UV light is mainly produced by a low-pressure mercury lamp, which is made from the mercury that is pernicious to human health [1]. Moreover, according to the Minamata Convention on Mercury signed in 2013, the manufacture, import, and export of large quantities of mercury containing products have been prohibited since 2020. Owing to the non-mercury and high-reliability of DUV-LEDs, they have been considered as the most potential ultraviolet light sources. Generally, the traditional packaging of DUV-LEDs with flat quartz glass reduces the light extraction from chips due to the TIR loss, severely limiting the optical performance [2-8]. The organic lens can efficiently reduce the TIR loss and enhance the light extraction efficiency of devices. However, traditional organic materials which is common in the encapsulation of white LEDs, such as silica gel and epoxy resin, cannot be applied in DUV-LEDs [9,10]. Because the methyl functional groups in common organic materials will be cracked by short-wave ultraviolet light, sharply reducing the output efficiency of DUV-LEDs [11]. For deep UV-LEDs packaging, it remains challenging to improve the light extraction efficiency and reliability of DUV-LEDs.

In this paper, we demonstrated an efficient, low-cost and universal strategy to ameliorate these drawbacks, enhancing the optical performance of DUV-LEDs. Solid-liquid mixing state (SLMS) organic lens can be obtained through mixing method and applied to the devices. We experimentally and theoretically investigated the deep ultraviolet transparent property of SLMS structure and finally investigated the optical performance and reliability of DUV-LEDs using SLMS organic lens.

978-1-6654-1392-3/21 $31.00 © 2021 IEEE

II. EXPERIMENT

To fabricate a series of cured films, first, various amount of silicone oil was mixed with PDMS to obtain a slurry. The mass ratio of silicone oil to total mass ranges from 0% to 85wt.%. Then the slurry was stirred in the planetary stirring machine in a vacuum for 3 min to homogenize the slurry by removing the air bubbles. Due to stripping the film is easy to lead to the fragmentation of the film, we coated the slurry in flat quartz glass which barely absorb ultraviolet waves. The thickness of the films is 1 mm, which is controlled by the apparatus. These flat quartz glass coated with slurry was kept in an oven at 90°C for 90 min to cure the slurry. The baseline of the transmittance spectrum was calibrated by using clean flat quartz glass. Similarly, the flat quartz glass with those slurry after curing in the oven, the radiant power and efficiency of these devices was tested. And the DUV-LEDs that coated with a certain mass of slurry were heated in an oven at 90°C for 90 min. Those devices were weld to the plum plate for stability testing.

III. RESULTS AND DISCUSSIONS

In order to investigate the ultraviolet light transparent property of SLMS structure, the films gained with various silicone concentrations were fabricated and the corresponding transmittance spectra were measured using a fluorescence spectrometer. The transmittance spectrum of PDMS film, OE6650 film and PDMS film with silicone oil concentrations of 20, 40, 60, 85wt.% are given in Fig. 1(a). The transmittance of traditional organic encapsulation materials, such as OE6650 is similar to the PDMS film with various silicone oil concentrations in UVA and UVB range. The transmittance almost declines to zero in the UVC range, which is due to the strong deep UV absorption ability of traditional organic encapsulation materials. However, the SLMS films show different trend of the transmittance. The higher the ratio of silicone oil in the film, the gentler the decline in UVC range. The SLMS films with silicone oil concentrations of 85 wt. % still has a transmittance up to 90 % at the wavelength of 245 nm. For subsequent investigation, the concentration of silicone oil was set as 85 wt.% to holding excellent ultraviolet light transparent property.

The viscosity of the silicone oil was also investigated and the corresponding transmittance spectrum are given in Fig. 1(b). It can be observed that the viscosity of silicone oil has little effect on the SLMS film's transmittance. Due to the little effect of silicone oil's viscosity on SLMS film, the formability of SLMS structure can be easily controlled by adjusting the viscosity of silicone oil while keeping high DUV transparent property.

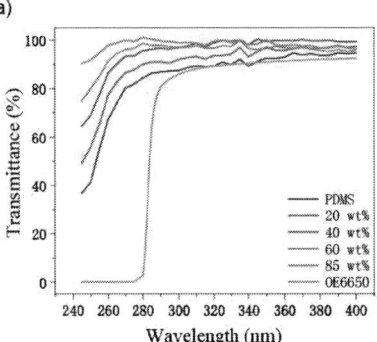

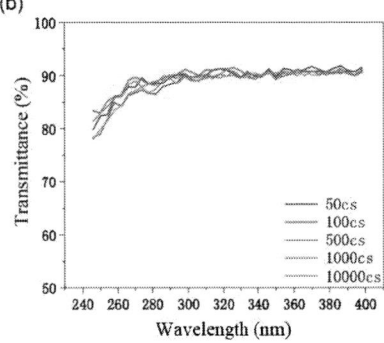

Fig. 1. (a) Transmittance spectrum of PDMS film, OE6650 film, and PDMS film with silicone oil concentrations of 20 wt. %, 40wt. %, 60 wt. %, 85wt. %, respectively. (b) Transmittance spectrum of SLMS film (with 85 wt. % silicone oil) with the viscosity of 50 cs, 100 cs, 500 cs, 1000 cs, 10000 cs, respectively.

To verify the cross-linking reaction between silicone oil and PDMS and silicone oil can remain liquid state in the cross-linking network, we experimentally and theoretically investigated the SLMs films. Generally, silicone oil is composited of Si-O and Si-CH₃ functional groups. The main chain of silicone oil is bonded by Si-O whose bond energy is 108 Kcal/mol and is higher than UVC and UVB energy which is 102 Kcal/mol and 91 Kcal/mol respectively. Thus, it is hard to be cracked by the short-wave ultraviolet light. But the main chain of cured PDMS is bonded by C-C whose bond energy is only 85 Kcal/mol [12]. Furthermore, the infrared transmittance spectrum of PDMS, Ethylene-PDMS, silicone oil, cured PDMS and cured Silicone oil-PDMS mixture are given in Fig. 2. It is observed that there is an absorption peak at 3080 cm⁻¹ in the infrared transmittance spectrum of PDMS, which is due to the exist of C=C bonds in PDMS. After curing the PDMS slurry, the absorption peak doesn't appear in the infrared transmittance spectrum of cured PDMS. Because the C=C bond in PDMS will react with the Si-H bond in curing agent, then PDMS will be cross-linked due to the hydrosilylation reaction between C=C bonds and Si-H bonds. The absorption peak of C=C bonds doesn't appear in the infrared transmittance spectrum of SLMS tructure. As discussed above, the hydrosilylation reaction of PDMS is completely complete even with the addition of silicone oil. Thus, the SLMS structure contains a large number of the original liquid silicone oil in its networks because the liquid silicone oil did not participate in the reaction and thus maintain the high DUV transparent property.

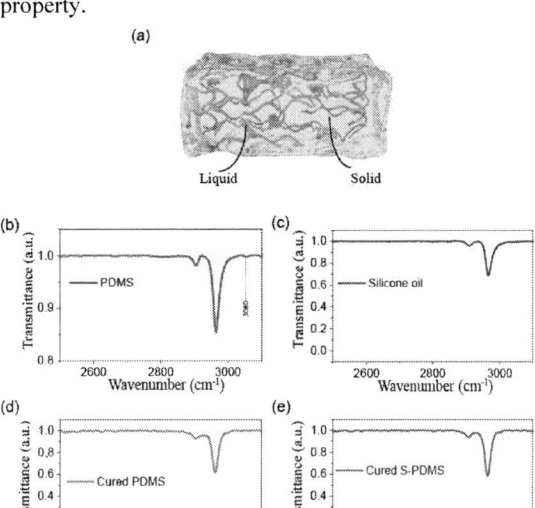

Fig. 2. (a) Structure diagram of SLMS composite. (b)-(e) Infrared Transmittance spectrum of PDMS, Silicone oil, Cured PDMS and Cured S-PDMS.

Generally, the quartz encapsulation is widely utilized in the encapsulation of DUV-LEDs and it can be processed by either directly sealing the chip with a thin quartz plate or forming a single sintered monolithic silica from glass frits. Although quartz glasses can provide excellent light transmittance and gas barrier performance, the quartz glass encapsulation has the disadvantages of limited light extraction, which is due to the existence of chip/void interface and chip damage during the high temperature sintering process [5]. Additionally, the liquid encapsulation with high DUV transparent property can enhance the light extraction, but cannot be easily implemented in the application of DUV-LEDs owing to the leakage of liquid. As discussed above, the SLMS structure can combined the advantages of the quartz and liquid encapsulation, exhibiting excellent DUV transparency and forming ability. It is suitable for the encapsulation of DUV LEDs and can be conveniently prepared as transparent lens to enhance the light extraction.

The SLMS organic lens with various curvatures can be fabricated by controlling the mass of slurry. Firstly, the optical performance of SLMS organic lens are optimized by ray-trace simulation. Fig. 3(b) shows the radiant efficiency of DUV-LEDs using SLMS organic lens with various curvature. The radiant efficiency of DUV-LEDs with organic lens made of PDMS/silicone oil composite is significantly higher than that of those just made of PDMS, and the difference between them becomes lager and lager with the increase of curvature of the organic lens. The maximum difference reaches 22.5 %. This indicates the feasibility of using PDMS/ silicone oil composite to fabricate the SLMS organic lens. In order to confirm the feasibility of SLMS organic lens that can improve the optical performance of DUV-LEDs, the absorption of organic lens and substrate are simulated respectively, as shown in Fig. 3(c) and Fig. 3(d). The absorption of organic lens that made of PDMS/silicone oil composite is at most 27.49 % lower than that made of PDMS, and the difference between them increases with the increase

of the curvature of organic lens. When the curvature of the organic lens is greater than 0.37 mm^{-1}, the absorption of the organic lens to the chip light increases slightly with the increase of the curvature of the organic lens. The absorption of organic lens with the curvature of 0.79 mm^{-1} is only 1.91% higher than that of the organic lens with the curvature of 0.37 mm^{-1}. The absorption of substrate with the organic lens made of PDMS/silicone oil composite is higher than that made of PDMS, but when the curvature of the organic lens is greater than or equal to 0.51 mm^{-1} which is almost maintained at 8.8 %, and the difference between them is maintained at 4.0 %. Then the DUV-LEDs with SLMS organic lens packaged were photographed by the apparatus as the inset diagram shown in Fig. 3(a), the curvature of the organic lens was calculated, and a curve of the relationship between the curvature of the organic lens and the mass of the slurry was fitted as shown in Fig. 3(a). Therefore, it is demonstrated that the organic lens with a curvature of 0.7 mm^{-1} is more beneficial to improve the optical performance of DUV-LEDs.

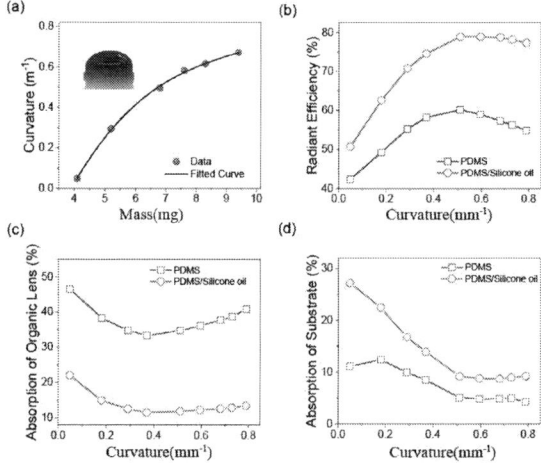

Fig. 3. (a) Fitting relationship between SLMS and the mass of slurry. The inset shows the photograph of SLMS organic lens. (b) Simulation of Radiant Efficiency spectra of PDMS and PDMS/Silicone oil composite (with 85 wt. % silicone oil). (c)(d) Simulation of Absorption spectra of Organic Lens and Substrate of PDMS and PDMS/Silicone oil composite (with 85 wt. % silicone oil).

To further verify the feasibility of SLMS organic lens that can improve the optical performance of DUV-LEDs. The SLMS organic lenses with the curvature of 0.7 mm^{-1} are applied to DUV-LEDs with the central emission wavelength of the chip light is 275 nm. Similarly, the devices that encapsulated without lens(chip) and encapsulated with flat quartz glass, PDMS organic lens and OE6650 organic lens were tested. As shown in Fig. 4(a), the peak width at half height of the chip light is 25 nm, and the inset diagram shows the DUV-LEDs with SLMS organic lens for subsequent testing. The radiant power of DUV-LEDs encapsulated without lens and encapsulated with flat quartz glass, PDMS organic lens and OE6650 organic lens are given in Fig. 4(b). It can be observed that the radiant power of DUV-LEDs encapsulated with SLMS organic lens is higher than that encapsulated with PDMS organic lens. The radiant power of DUV-LEDs encapsulated without lens is higher than that encapsulated with flat quartz glass is due to TIR which reduces the light extraction from chips. Under the injection current of 150 mA, the DUV-LEDs with SLMS organic lens has a 25.763 mW radiant power while the devices with conventional flat quartz glass only reach 16.744 mW, largely increased by 53.86 %. Therefore, SLMS organic lens can enhance the efficiency of light extraction. As Fig. 5 shows, the radiant power of DUV-LEDs that encapsulated with SLMS organic lens is fluctuate slightly. After 279 hours continuous operation, the radiant power can still reach 8.289 mW which is only 3.94 % lower than the initial power. Compared to the devices using flat quartz glass, the average radiant power of DUV-LEDs using SLMS organic lens is 2.799 mW higher. Consequently, these devices can operate steadily for long periods of time while achieving higher radiant power.

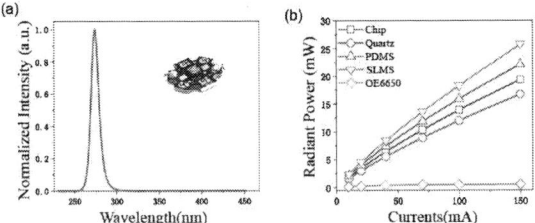

Fig. 4. (a) Spectra of UV device with emission wavelength of 275 nm. The inset shows the photograph of DUV-LEDs with SLMS organic lens. (b) Radiant power of chip, quartz, PDMS, silicone oil and OE6650 under various injected currents.

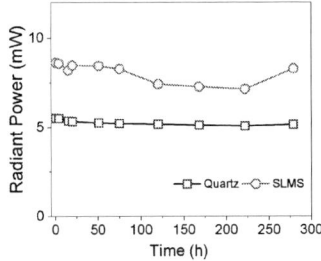

Fig. 5. The stability test of DUV-LEDs that encapsulated with glat quartz glass and SLMS organic lens, respectively.

IV. CONCLUSIONS

We proposed a high light-extraction efficiency DUV-LEDs device with solid-liquid mixing state organic lens. The high transmittance of SLMS composite in deep UV band was proved by a series of SLMS films. Also, the infrared transmittance spectrum prove that PDMS and curing agent form a cross-linked network structure after curing, which can store liquid silicone oil. The results indicate that the radiant power of DUV-LEDs encapsulated with SLMS organic lens has a 25.763 mW radiant power under the injection current of 150 mA which is 53.86 % more than that encapsulated with conventional flat quartz glass. The stability test shows that DUV-LEDs using SLMS organic lens can operate for long periods of time while the radiant power can still reach 8.299 mW which is 3.133 mW higher than that using conventional flat quartz glass. Consequently, this study

provides an efficient method to improve the efficiency of DUV-LEDs with high stability and it has a great application prospect in packaging industry of DUV-LEDs.

ACKNOWLEDGMENT

This work was supported by the Science & Technology Program of Guangdong Province (No. 2019B010130001), the Project of the National and Local Joint Engineering Research Center of Semiconductor Display and Optical Communication, Zhongshan Branch (No. 190919172214566), the Postdoctoral Science Foundation of China (No. 2020M680122) and the Natural Science Foundation of Guangdong Province (No. 2018B030306008).

REFERENCES

[1] J.S. Petruci, P.R. Fortes, V. Kokoric, A. Wilk, I.M. Raimundo, A.A. Cardoso, B. Mizaiko Ff Real-time monitoring of ozone in air using substrate-integrated hollow waveguide mid-infrared sensors, Scientific Reports.

[2] Y. Muramoto, M. Kimura, S. Nouda, Development and Future of Ultraviolet Light-Emitting Diodes: UV-LED Will Replace the UV Lamp, Semiconductor Science and Technology 29(8) (2014) 084004.

[3] M. Shatalov, W. Sun, R. Jain, A. Lunev, X. Hu, A. Dobrinsky, Y. Bilenko, J. Yang, G.A. Garrett, L.E. Rodak, High power AlGaN ultraviolet light emitters, Semiconductor Science & Technology 29(8) (2014) 1779-1781.

[4] M. Shur, M. Shatalov, A. Dobrinsky, R. Gaska, Deep Ultraviolet Light-Emitting Diodes, GaN and ZnO-based Materials and Devices2012.

[5] W. Sun, V. Adivarahan, M. Shatalov, Y. Lee, W. Shuai, J. Yang, J. Zhang, M.A. Khan, Continuous Wave Milliwatt Power AlGaN Light Emitting Diodes at 280 nm, Japanese Journal of Applied Physics 43(11A) (2004) L1419-L1421.

[6] Y. Taniyasu, M. Kasu, T. Makimoto, An aluminium nitride light-emitting diode with a wavelength of 210 nanometres, Nature 441(7091) (2006) 325-8.

[7] I.C. Wu, S. Hao-Yi, J. Chun-Ping, M.Y. Lu, Y.T. Chen, M.T. Wu, K. Chie-Tong, T. Yu-Yuan, H.C. Wang, Early identification of esophageal squamous neoplasm by hyperspectral endoscopic imaging, Scientific Reports 8(1) (2018) 13797-.

[8] H.C. Wang, N. Ngoc-Viet, R.Y. Lin, J. Chun-Ping, Characterizing Esophageal Cancerous Cells at Different Stages Using the Dielectrophoretic Impedance Measurement Method in a Microchip, Sensors 17(5) (2017) 1053-.

[9] W. Huang, Y. Zhang, Y. Yu, Y. Yuan, Studies on UV-stable silicone–epoxy resins, Journal of Applied Polymer Science (2007).

[10] J.C. Huang, Y.o. Chu, W. Ming, R.D. Deanin, Comparison of epoxy resins for applications in light-emitting diodes, Advances in Polymer Technology 23(4) (2010) 298-306.

[11] C.C. Lu, C.P. Wang, C.Y. Liu, C.P. Hsu, The Efficiency and Reliability Improvement by Utilizing Quartz Airtight Packaging of UVC LEDs, IEEE Transactions on Electron Devices 63(8) (2016) 1-4.

[12] Kang, Chieh-Yu, Lin, Chih-Hao, Wu, Tingzhu, Lee, Po-Tsung, Chen, Zhong, A Novel Liquid Packaging Structure of Deep-Ultraviolet Light-Emitting Diodes to Enhance the Light-Extraction Efficiency, Crystals (2019).

Numerical Simulation Analysis of Flexible Printed Circuits Under Bending Conditions

Yongchao Liu[1,2#], Haozhe Wang[1,2#], Xianqin Hu[3], Chao Peng[3], Xu Long[4], Jibao Lu[1,2*], Rong Sun[1,2], Ching-Ping Wong[5]

1 Shenzhen Institute of Advanced Technology, Chinese Academy of Science, Shenzhen, 518055, China
2 Shenzhen Institute of Advanced Electronic Materials, Shenzhen, 518103, China
3 Avary Holding (Shenzhen) Co., Ltd, Shenzhen, 518105, China
4 School of Mechanics, Civil Engineering and Architecture, Northwestern Polytechnical University, Xi'an, 710072, China
5 School of Materials Science and Engineering, Georgia Institute of Technology, Atlanta, GA 30332, USA

Abstract—**Flexible Printed Circuits (FPC) have been extensively used in daily life. It is inevitable to perform high-frequency reciprocating bending when using FPC, which induces deviation of the circuit board structure, separation of the adhesive layer and finally damage of the device. To solve the problem, it is suggested to change the FPC laminate design and adjust the internal structure and materials of the FPC so that the stress of the Cu layer is concentrated in the neutral layer during the bending process of the FPC and reducing the strain of the adhesive layer.**

In this paper, we systematically investigate the impact factors to the bending process of the FPC by using the finite element analysis (FEA). The characterized factors include the stacked structure and elastic modulus of each layer, the thickness of the glue/Cu layer, and the bending radius of the FPC. We find that different stack structures drastically influence the position of the stress neutral layer; the stress increases as the elastic modulus of material increases, but decreases with reducing the adhesive layer. Whereas, the elastic modulus of the material and the thickness of the adhesive layer have little effect on the position of the stress neutral layer of the FPC. Furthermore, the increasement in the thickness of the Cu layer increases the stress of the FPC and moving the position of the stress neutral layer downward. As the bending radius increases, the stress on FPC decreases, while the position of the neutral layer hardly changes. We also find that the smaller elastic modulus and thinner glue/Cu layer reduce the stress of the Cu layer. These simulation results provide a guidance for the design of structures and material selection of FPC, which may help solving the mechanical problems during the bending process.

Keywords—Flexible Printed Circuits; stacked structure; neutral layer; finite element analysis

I. INTRODUCTION

The flexible circuit board is mainly made of polyimide or other dielectric materials as the substrate, and is made of conductive materials such as Cu, which has good flexibility and reliability[1]. Its main function is to connect various electronic components and make them form a functional whole. With the growth of portable electronic communication equipment and high-density IC packaging semiconductor applications, electronic products continue to develop in the direction of multi-function, lightness and thinness. For example, smart phones, notebook computers, digital photography cameras, etc. are becoming thinner and lighter, so that FPC applications have a broader development space[2].

As electronic products tend to be lighter, thinner, and smaller, flexible circuit boards are used more and more widely, and flexible circuit boards are used in almost every electronic product. The failure of FPC brings unimaginable consequences to electronic products in various application fields. Therefore, studying the failure of flexible circuit boards is of important practical engineering significance.

II. EXPERIMENT AND SIMULATION

A. Analytical Model

At present, the theoretical formulas for FPC bending analysis are divided into two categories: single-layer boards and multi-layer boards. The thickness and elastic modulus of the material in the analytical solution are the main factors affecting the bending strain. The analytical solution of the single-layer board is shown in formula (1).

$$\varepsilon = \frac{b}{R} \quad (1)$$

The strain in the convex surface is equal to the distance b from the neutral plane divided by R[3].

In a complicated multilayer structure, the component layers can be modeled as a composite beam subject to a bending curvature. The distance b is given by[4-6]:

$$b = \frac{\sum_{i=1}^{n} \overline{E_i} h_i \left[\left(\sum_{j=1}^{i} h_j \right) - \frac{h_i}{2} \right]}{\sum_{i=1}^{n} \overline{E_i} h_i} \quad (2)$$

where n is the total number of layers; hi is the thickness of the ith layer; when the devices are narrow, E_i is equal to the Young's modulus of each layer; when the biaxial deformation should be considered, E_i is equal to the plane strain modulus $\overline{E_i} = E_i(1 - v_i^2)$, where E_i and v_i denote the Young's modulus and the Poisson's ratio of the ith layer, respectively.

B. Simulation Experiment Verification

According to the theory of single-layer board, the bending experiment of single-layer Cu layer is carried out in combination with the film tensile testing machine in the early stage, as shown in Fig.1, and the corresponding force and displacement curves are obtained[7]. At the same time, the finite element analysis of the entire bending process is carried out using the finite element analysis method, and the theoretical solution, the bending strain calculated by the experiment and the simulation are compared. It is found that the strain calculated according to the theoretical formula is similar to the strain result obtained by simulation, and the error is smaller. It shows that the means of simulation can well

978-1-6654-1392-3/21 $31.00 © 2021 IEEE

characterize the bending process of the Cu layer. Therefore, this paper uses finite element analysis to characterize the dynamic process of FPC bending[8-9].

In this experiment, the total length of the Cu layer is 77mm, and the length of the Cu layer after the fixture is fixed is 63mm. The two ends of the clamp move to the middle at the same time, the moving speed is 200mm/min, and the bending radius obtained by calculation is 8.25mm, so both sides of the clamp move 23.25mm to the middle. The maximum strain value of the experiment output is 1.587×10^{-3}. The simulation is set up completely according to the relevant experimental parameters. In order to make the simulation convergence better, the simulation experiment did not use the two ends of the experiment to be fixed. During the simulation, only one end was fixed and the other end moved. The moving speed was 400mm/min, which was consistent with the experimental theory. The simulated strain value is 1.716×10^{-3}, which is slightly larger than the experimental value, but the error is within the allowable range.

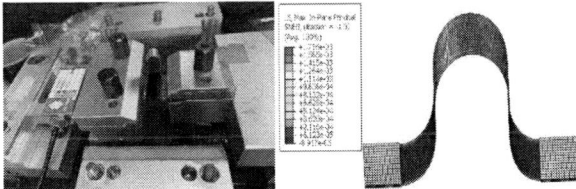

Fig.1. Cu layer bending experiment and simulation

III. RESULT AND DISCUSSION

In the bending process, there are many factors that affect the bending stress, such as the stacked structure of the material, material stiffness, thickness, and bending radius. In this article, a detailed finite element analysis is carried out for each of the different factors. However, multi-factor analysis means that a lot of model calculations and comparisons are required. If the modeling analysis is performed manually, most of the time will be spent in the pre-processing stage, in order to speed up the calculation rate of the finite element simulation. In this paper, a customized secondary development of finite element software is carried out, and calculation files are generated in batches for modeling, material assignment, contact settings, boundary conditions, and mesh settings using Python language. As shown in Fig.2, the secondary development window of FPC bending analysis studied in this paper has been independently developed. The parameters of interest can be set freely, and the FPC model can be quickly parameterized, which greatly improves the calculation efficiency.

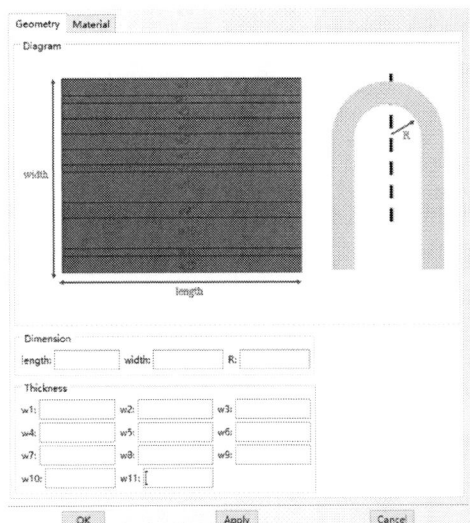

Fig.2. Parameterized window of secondary development based on finite element software

A. Effect of Material Stack Structure (Airgap)

In flexible electronic technology, in addition to the development of anti-bending laminated materials, mechanical structure design based on the principle of reducing bending stress also plays a vital role. For example, whether the Airgap structure is added to the FPC, as shown in Fig.3, which will make the stress and strain of the FPC bending part change greatly. Therefore, this section explores the influence of Airgap structure on FPC bending performance.

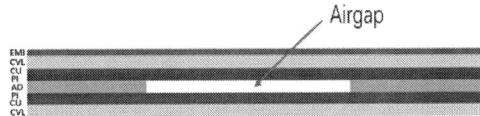

Fig.3. FPC finite element model with Airgap

As shown in Fig.4, when there is no Airgap structure in the FPC structure, the maximum stress generated by bending reaches 202.1MPa. After the Airgap structure is installed in the FPC structure, the maximum stress generated by bending is reduced to 180.0MPa, a decrease of 10.9%. It is worth noting that the maximum stress of the FPC structure without Airgap is concentrated on the interface between the Cu layer and the PI layer, which is also the main reason why the Cu layer and the PI layer are prone to interface peeling and fracture failure. Therefore, a reasonable setting of Airgap is beneficial to reduce the stress concentration effect caused by bending and reduce the probability of FPC failure.

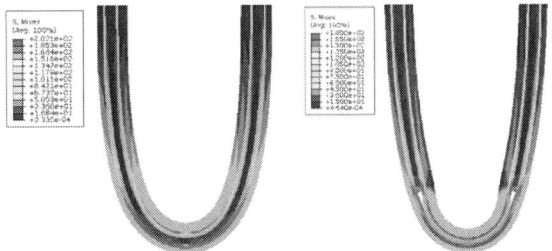

Fig.4. Bending stress cloud diagram of FPC with or without airgap

B. Effect of PI Material Stiffness

The PI layer is located in the part adjacent to the Cu layer in the FPC structure. During the bending process, stress concentration in this area often leads to excessive bending deformation. It is necessary to adjust the material parameters of the PI layer and explore the effect of its stiffness on the stress and strain of the Cu layer. In this paper, the elastic moduli of 4400MPa, 5200MPa, 6000MPa, 6800Mpa and 7600MPa are selected for comparative analysis.

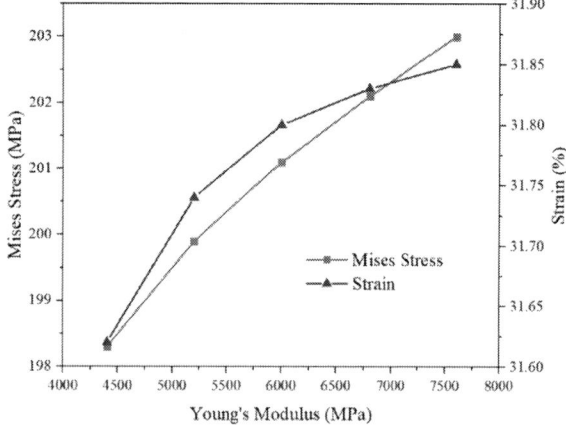

Fig.5. The influence of PI with different elastic modulus on the bending stress and strain of FPC

It can be seen from Fig.5 that with the continuous increase of the elastic modulus of the PI layer material, the stress and strain generated by the bending will gradually increase slightly. The change of PI elastic modulus has relatively little effect on the stress and strain generated during the bending process. Therefore, a PI material with a larger elastic modulus will cause the stress and strain of the FPC to increase during the bending process, which is not conducive to the optimization of the bending process. It is recommended to choose a material with a relatively small elastic modulus as the PI layer structure of the FPC.

C. Effect of Cu Layer Thickness

The Cu layer is the most important structure in the FPC, and the FPC transmits signals mainly through the Cu layer. Defects and fractures in the Cu layer will affect the performance of the entire FPC and cause FPC failure or even more serious consequences. For the analysis of the thickness of the copper layer, this paper selects 8μm, 10μm, 12μm, 14μm, etc., and different thicknesses are used to analyze the influence on the bending of the FPC.

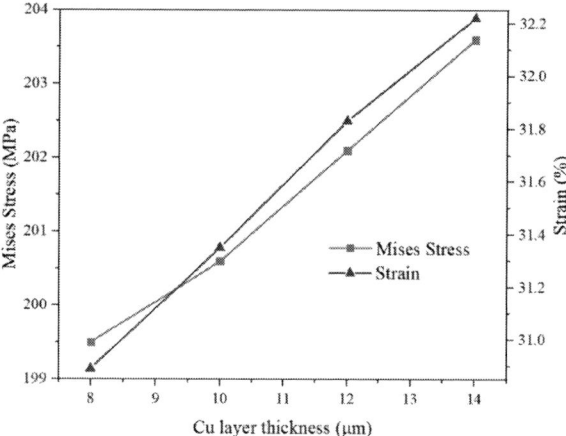

Fig.6. The influence of different thickness of Cu layer on the bending stress and strain of FPC

It can be seen from Fig.6 that during the bending process, the stress and strain both increase with the increase of the thickness of the Cu layer, mainly because the position of the neutral layer changes. As the thickness of the Cu layer increases, the position of the neutral layer shifts, and Cu gradually moves away from the neutral layer, causing its bending strain and stress to gradually increase.

D. Effect of Bending Radius

The bending radius is an important factor affecting the bending of FPC. In this paper, two different bending radii are selected to explore their influence, and 0.2mm and 0.4mm bending radii are respectively selected for finite element analysis.

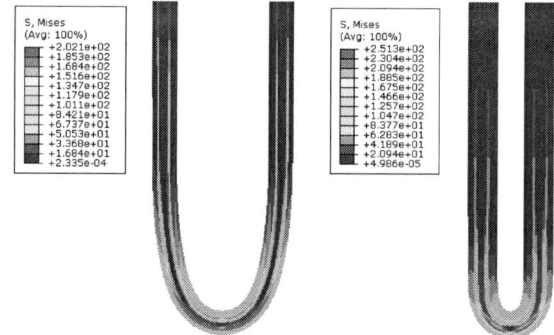

Fig.7. FPC bending stress cloud diagram when the bending radius is 0.2mm and 0.4mm

As shown in Fig.7, the stress is mainly concentrated at the bending part during the bending process. When the bending radius is 0.2mm, the maximum stress reaches 251.3MPa. When the bending radius is 0.4mm, the maximum stress is reduced by 19.6% to only 202.1MPa. It can be seen that the bending radius has a more obvious influence on the stress of FPC bending, and it is the main factor that affects the bending performance of FPC.

IV. CONCLUSION

This paper uses finite element software ABAQUS for a customized secondary development for characterization of bending process of FPC, and explores the impact factors such as material stacking structure, PI material stiffness, Cu layer thickness, bending radius, etc. The following conclusions can be obtained through the comparative analysis:

(1) When the Airgap structure is set in the FPC structure, the stress and strain generated by the FPC during bending can be significantly reduced, as well as the probability of FPC failure.

(2) The selection of PI material affects the performance of the FPC. As the elastic modulus of PI material increases from 4400MPa to 7600MPa, the stress and strain of FPC during the bending increases, but with only a small magnitude of change.

(3) When the thickness of the Cu layer increases from 8μm to 14μm, the FPC bending stress and strain become larger. The main reason is that the position of the neutral layer shifts as the thickness of the Cu layer changes. The Cu layer gradually moves away from the neutral layer, leading to the increase of stress and strain.

(4) The bending radius has significant influence on the bending of the FPC structure. When the bending radius changes from 0.4mm to 0.2mm, both the stress and strain increase, and the stress-increase is nearly by 20%. Therefore, the bending radius is one of the most important factors to be considered when designing the performance of the FPC structure for bending.

ACKNOWLEDGMENT

The work was supported by the National Natural Science Foundation of China (no. 52003289), the Youth Innovation Promotion Association CAS (no. 2021363), and the SIAT Innovation Program for Excellent Young Researchers (no. 201803).

REFERENCES

[1] Arruda, L. , et al. "Experimental and numerical analyses of flexible PCBs under various loading conditions." International Conference on Thermal, Mechanical and Multi-Physics simulation and Experiments in Microelectronics and Microsystems;EuroSimE 2009.

[2] Lee, S. , et al. "Enhanced Light Extraction from Mechanically Flexible, Nanostructured Organic Light‐Emitting Diodes with Plasmonic Nanomesh Electrodes." Advanced Optical Materials 3.9(2015).

[3] Suo Z , Ma E Y , Gleskova H , et al. "Mechanics of rollable and foldable film-on-foil electronics". Applied Physics Letters, 1999, 74(8):1177-1179.

[4] Kim R H , Bae M H , Kim D G , et al. "Stretchable, Transparent Graphene Interconnects for Arrays of Microscale Inorganic Light Emitting Diodes on Rubber Substrates." Nano Letters, 2015, 11(9):3881-3886.

[5] Kim D H , Ahn J H , Choi W M , et al. "Stretchable and Foldable Silicon Integrated Circuits. " Science, 2008, 320(5875):507-511.

[6] Park S L , Le A P , Jian W , et al. "Light emission characteristics and mechanics of foldable inorganic light-emitting diodes." Advanced Materials, 2010, 22(28):3062-3066.

[7] Lee, S. M. , et al. "A Review of Flexible OLEDs Toward Highly Durable Unusual Displays." IEEE Transactions on Electron Devices (2017):1922-1931.

[8] Wang, W. , W. Shangguan , and X. Duan . "A study on the effects of hyperelastic constitutive models on the static characteristic prediction of rubber isolator." Qiche Gongcheng/Automotive Engineering 34.6(2012):544-550+539.

[9] Lee, C. C. , et al. "Development of robust flexible OLED encapsulations using simulated estimations and experimental validations." Journal of Physics D Applied Physics 45.27(2012):275102-275109(8).

Tensile deformation mechanism of Sn-37Pb solder alloy at cryogenic temperatures

Xiaotong Guo[1, 2, *]
[1]Chongqing CEPREI Industrial
Technology Research Institute Co., Ltd
Chongqing 401332, China
[2]China Electronic Product Reliability
and Environmental Testing Research
Institute
Guangzhou 510610, China
guoxiaotong0713@163.com

Kun Zhang[2, *]
China Electronic Product Reliability
and Environmental Testing Research
Institute
Guangzhou 510610, China
cliviazk@163.com

Jiahao Liu[2]
China Electronic Product Reliability
and Environmental Testing Research
Institute
Guangzhou 510610, China
ljh071408@163.com

Yong Li[1, 2]
[1]Chongqing CEPREI Industrial
Technology Research Institute Co., Ltd
Chongqing, China
[2]China Electronic Product Reliability
and Environmental Testing Research
Institute
Guangzhou 510610, China
403ly@163.com

Xinlang Zuo[2, **]
China Electronic Product Reliability
and Environmental Testing Research
Institute
Guangzhou 510610, China
Sinazuo@163.com

Hui Xiao[2]
China Electronic Product Reliability
and Environmental Testing Research
Institute
Guangzhou 510610, China
xiaohui_ceprei@163.com

Guanghui He[2, **]
China Electronic Product Reliability
and Environmental Testing Research
Institute
Guangzhou 510610, China
shirleyhe@ceprei.com

*: Co-first author

**: Corresponding author

Abstract—Traditional solder alloy Sn-37Pb is still widely used in some special industries, such as aerospace and military fields, although the use of Pb containing solder is prohibited in the civil market. However, some of the applications, such as external wires of earth satellites and military application equipment of North and South poles, will make the electronic devices face the cryogenic temperatures (up to -150 ℃) service conditions, and thus the reliability of solder alloys is seriously threatened. In the field of cryogenic temperature, attentions are always paid to the ductile brittle transition of metal materials, which is also the focus of this study. Thus, we measured the tensile properties of Sn-37Pb at temperatures of -150~0 ℃ and a tensile rate of 6.67×10^{-4} s^{-1}. The yield strength and ultimate strength of Sn-37Pb alloy increased gradually with decreasing the testing temperature. However, the elongation of the alloy was above 8.9% in the temperature range of -100~0 ℃, while decreased to 4.8% at -150 ℃. Sn-37 Pb alloy rapidly suffered brittle fracture after crossing the ultimate point at -150 ℃. In the temperature range of -100~0 ℃, there were a lot of dimples in the tensile fracture, which is in good agreement with the ductile fracture of the alloy. At -150 ℃, there were a lot of plate-like phases in the fracture besides some dimples, this implies that the formation of plate-like phases was the cause of the ultimate brittle fracture. This work provides guidance for the reliability of solder alloy at cryogenic temperatures

Keywords—Sn-37Pb alloy, Tensile property, Cryogenic temperature, Microstructure, Deformation mechanism.

I. INTRODUCTION

Sn-Pb solder alloy has low price, excellent weldability and conductivity, good mechanical and mechanical properties.

Therefore, it has been still widely used as brazing materials in electronic packaging [1-3]. In the process of application, people gradually realize the harm of lead and lead-containing substances to human body, and lead-free electronic packaging materials. It has become an inevitable trend in the development of electronic industry [4-6]. However, in the field of aerospace and military fields, Sn-Pb solder alloys are widely used. It is still the most commonly used welding material [7, 8]. The composition of Sn-Pb eutectic solder alloy is Sn-37Pb (wt.%), and its melting point is 183 ° C. Therefore, the welding temperature is relatively low; the molten Sn has a high surface area, and the Sn-Pb eutectic solder alloy has a lower surface tension with the addition of Pb [9, 10].

Solder joints are often withstand mechanical loads during the service process, and the environmental resistance of solder alloys directly affect the reliability of electronic products. The Sn-Pb alloy generally has mechanical properties in comparison with those at room temperature, and brings new challenges to the reliability of spacecraft electronic products. Although the application of Sn-37Pb alloy is very mature, the research on its microstructure and properties at cryogenic temperatures is still very limited. In this work, we studied the tensile behavior at cryogenic temperatures, and studied its reliability at low temperature.

II. EXPERIMENTS

Sn-37Pb solder alloy ingot from Qianzhu Metal Industry Co., Ltd. was used as raw material in the tensile test of solder alloy. According to ASTM standard, the solder alloy ingot was processed into dog bone tensile sample by wire cutting. Figure 1 shows the tensile specimen size, and the gauge length was 60 mm. The tensile temperatures are 0 °C, -50 °C, -100 °C and -150 °C, respectively, and tensile rate was set as 6.67×10^{-4} s^{-1}.

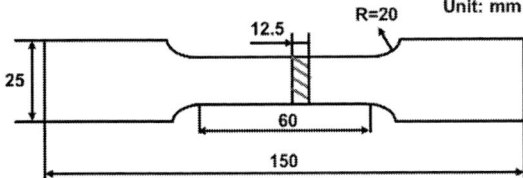

Fig. 1 Sketch of standard plate-type specimen for tensile test

Samples were sectioned, grinded, and polished for microstructural characterization. The lower and higher magnification microstructures are observed using LV 150 optical microscope (OM) and a Inspect F50 scanning electron microscope (FE-SEM).

III. RESULTS AND DISCUSSION

Fig. 2 shows the as-received microstructure of Sn-37Pb alloy. The microstructure mainly contains Sn matrix and Pb rich phase.

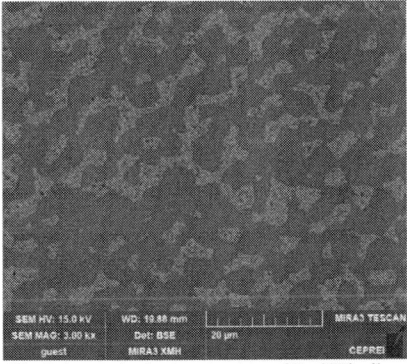

Fig. 2 As-received microstructure of Sn-37Pb alloy

Tensile stress-strain curves of Sn-37Pb alloy in the temperature range of -150 °C∼0 °C is depicted in Fig. 2. It is shown that deformation stages including elastic and plastic deformation can be observed on the tensile stress-strain curve of Sn-37Pb solder alloy at -100 °C, -50 °C and 0 °C, as well as the local necking deformation, indicating the fracture mode is ductile fracture in the corresponding temperature range. However, the stress-strain curve undergoes a plastic deformation stage after the elastic deformation, and decreases rapidly after reaching the maximum stress without necking at -150 °C. The fracture mode of Sn-37Pb solder alloy is likely to change between -100 °C and -150 °C. The results show that Sn-37Pb solder alloy exhibited a very little plastic deformation before fracture at low temperature, indicating that the change of fracture mode of Sn-37Pb solder alloy at cryogenic temperatures.

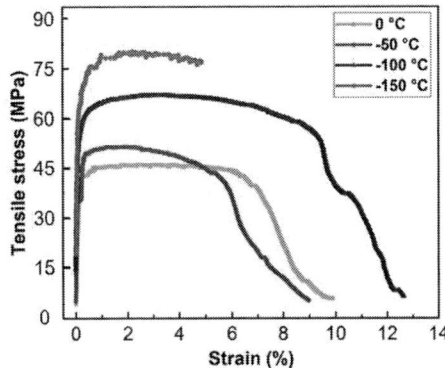

Fig. 3 Tensile stress-strain curves of Sn-37Pb alloy at cryogenic temperatures

Table 1 lists the tensile properties of Sn-37Pb alloy at cryogenic temperatures. With the decrease of temperature, the tensile strength of Sn-37Pb solder alloy increases gradually. The yield strength and ultimate strength was 17.1 MPa and 18.7 MPa at 0 °C, respectively, and increased to 29.5 MPa and 33.8 MPa at -150 °C. It should be noted that the tensile elongation of the alloy is more than 8% at temperature range between - 100 °C and 0 °C, but it decreased to 4.8% at - 150 °C. This is consistent with the law of tensile stress-strain curve.

TABLE I: YIELD STRENGTN, ULTIMATE STRENGTH AND ELONGATION OF SN-37 PB ALLOY AT CRYOGENIC TEMPERATURES

Temperature (°C)	Yield strength (MPa)	Ultimate strength (MPa)	Elongation (%)
0 °C	37.6±1.6	41.4±2.2	10.0±1.8
-50 °C	40.5±1.5	47.0±1.8	8.9±1.3
-100 °C	59.0±4.2	64.0±2.8	12.0±2.3
-150 °C	65.5±2.1	80.5±5.0	4.8±1.4

Fig. 4 are SEM-SE images showing the tensile fracture surfaces of Sn-37 Pb alloy at cryogenic temperatures. A large number of dimples are distributed on the fracture surface of Sn-37Pb solder alloy at 0 °C, and the dimples are dense and deep (Fig. 4a), showing the characteristics of ductile fracture, which belongs to micropore aggregation fracture. With the decrease of temperature, the dimples become smaller and shallower, indicating the gradually decreased toughness of solder alloy, as shown in Fig. 4a-d. At -150 °C, the tensile fracture surface of Sn-37Pb solder alloy is characterized by quasi cleavage brittle fracture, the tearing edge and cleavage facet coexist on the fracture surface, and a certain number of dimples can be seen at the same time. It is obvious that the tensile fracture of Sn-37Pb solder alloy is a ductile brittle mixed fracture composed of ductile fracture and quasi cleavage fracture. Additionally, There were some plate-like phases on the fracture surface, but these precipitates did not appear at other tensile temperatures (Fig. 4d).

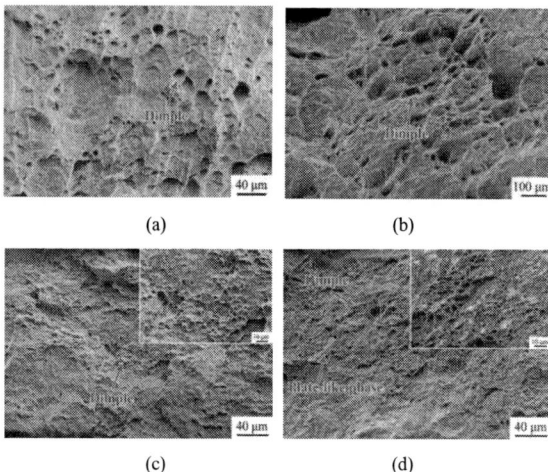

Fig. 4 SEM-SE images of fracture surfaces of Sn-37 Pb alloy of tensile tests at cryogenic temperatures (a) 0 °C; (b) -50 °C; (c) -100 °C (a higher magnification inset); (d) -150 °C (a higher magnification inset)

Fig. 5 are SEM-SE images of microstructures at the longitudinal sections of tensile specimens of Sn-37Pb alloy. The cracks were distributed at the interface between eutectic precipitates and matrix (Fig. 5a and 5d). In the matrix far away from the fracture, a large number of tensile holes were also found, which were still located at the interface between precipitates and matrix. This means that the interfaces between eutectic phase and matrix are weak regions at cryogenic temperatures, which are easy to initiate tensile holes and cracks. At the same time, the size of eutectic precipitates at - 150 °C is obviously lower than 0 °C, which means that the size of eutectic precipitates still increases during the tensile process. The lower the tensile temperature is, the slower the growth rate is.

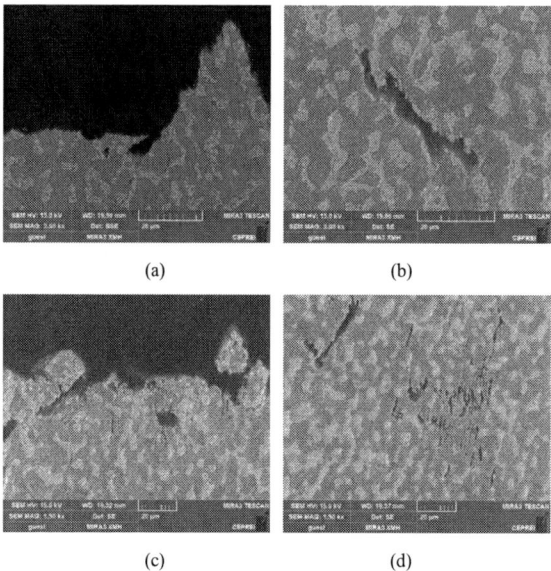

Fig. 5 SEM-SE images of microstructures at the longitudinal sections of tensile specimens of Sn-37Pb alloy
(a) (b) 0 °C; (c) (d) -150 °C

The above results indicates that Sn-37Pb alloy exhibits a decreased toughness with decreasing tensile temperature when testing under -150 °C~0 °C. The alloy suffered ductile fracture at -100 °C, -50 °C and 0 °C, while transferred to brittle fracture at -150 °C. Toughness largely depends on the type and orientation of chemical bonds (ionic bond, covalent bond, metal bond) and the symmetry characterization of crystal structure. For metals, the material tends to ductile fracture due to the symmetrical of the crystal structure. Highly symmetrical face centered cubic (FCC) metals are mainly ductile fracture, but not ductile fracture β-Sn [11]. The crystal structure of Sn is asymmetric body centered tetragonal (BCT) structure. Under certain conditions (such as low enough temperature), it is easy to cleave, resulting in brittle fracture of the material. However, the brittleness of Sn-37Pb alloy brings great difficulties to the preparation of TEM samples, and the related work is still in progress

IV. CONCLUSION

(1) In the range of -150~0 °C, the tensile strength of the Sn-37Pb alloy gradually increased with the decrease of temperature. However, the elongation of the alloy was above 8.9% in the temperature range of -100~0 °C, while decreased to 4.8% at -150 °C.

(2) The Sn-37Pb suffered ductile fracture in the temperature range of -100~0 °C, while rapidly suffered brittle fracture after crossing the ultimate point at -150 °C. Lots of brittle plate-like phases were found on the fracture surface when tested at -150 °C, and cause the ultimate brittle fracture

(3) With the temperature of shear testing increasing, the fracture mode of the Sn-37Pb alloy transferred from the brittle fracture to the ductile fracture. The cracks were mainly located at the interface of matrix/eutectic precipitates, and the size of eutectic phase decreases with the decrease of tensile temperature.

ACKNOWLEDGMENT

This work is financially supported by the National Key R&D Program of China under Grant No. 2020YFB1710300, the Science and Technology Program of Guangzhou, China under Grant No. 202002030357, CEPREI Innovation and Development Fund No. 20Z32, Which were acknowledged.

REFERENCES

[1] Haseeb A S M A, Arafat M M, Tay S L, et al. Effects of metallic nanoparticles on interfacial intermetallic compounds in Tin-based solders for microelectronic packaging. Journal of Electronic Materials, 2017, 46(10): 5503-5518.

[2] Wang F J, O'Keefe M, Brinkmeyer B. Microstructural evolution and tensile properties of Sn-Ag-Cu mixed with Sn–P solders. Journal of Alloys and Compounds, 2009, 477(1): 267-273.

[3] Wang F, Li D, Tian S, et al. Interfacial behaviors of Sn-Pb, Sn-Ag-Cu Pb-free and mixed Sn-Ag-Cu/Sn-Pb solder joints during electromigration[J]. Microelectronics Reliability, 2017, 73: 106-115.

[4] Wood E P, Nimmo K L. In search of new lead-free electronic solders. Journal of Electronic Materials, 1994, 23(8): 709-713..

[5] Wang M, Wang J, Ke W. Corrosion behavior of Sn-3.0Ag-0.5Cu lead-free solder joints. Microelectronics Reliability, 2017, 73: 69-75.

[6] Tan A T, Tan A W, Yusof F. Influence of high-power-low-frequency ultrasonic vibration time on the microstructure and mechanical properties of lead -free solder joints. Journal of Materials Processing Technology, 2016, 238: 8 -14.

[7] Ramesham R. Reliability of Sn/Pb and lead-free (SnAgCu) solders of surface mounted miniaturized passive components for extreme

emperature -185 ℃ to +125 ℃) space missions. roceedings of Society of Photo-Optical Instrumentation Engineers (SPIE), 2011: 79280F.

[8] Ghaffarian R, Kim N P. Ball grid array reliability assessment for aerospace applications. Microelectronics Reliability, 1999, 39(1): 107-112.

[9] Wang H, Ma X, Gao F, et al. Sn concentration on the reactive wetting of high-Pb solder on Cu substrate. Materials Chemistry and Physics, 2006, 99(2): 202-205.

[10] Cho M G, Seo S-K, Lee H M. Wettability and interfacial reactions of Sn-based Pb-free solders with Cu–xZn alloy under bump metallurgies. Journal of Alloys and Compounds, 2009, 474(1): 510-516.

[11] Lambrinou K, Maurissen W, Limaye P, et al. A novel mechanism of embrittlement affecting the impact reliability of Tin-Based lead-free solder joints. Journal of Electronic Materials, 2009, 38(9): 1881-1895.

The Reliability Assessment of Pulse-Driven Light Emitting Diodes

1st *Shen Yaoyang*
School of Information Engineering
Guangdong University of Technology
Guangzhou, China
2478976713@qq.com

2nd *Sun Bo*
School of Information Engineering
Guangdong University of Technology
Guangzhou, China

Abstract - **In this paper, effect of the input current frequency on the reliability of LEDs is investigated experimentally. A type of common-used white light LED is selected as test samples. LED aging tests were conducted by using inputs with different frequencies. Meanwhile, to avoid measurement errors in the long-term test, a group of LEDs without aging are tested as well. During the aging process, LEDs' I-V curves and luminous output at different time points were measured. Then, via parameter fitting, degradations of key parameters along aging duration, such as the saturation current, Ideal Factor, equivalent series resistance, luminous efficacy, can be extracted. By analyzing the degradations of key parameters in different conditions, impacts of the input current frequency on LEDs' reliability can be determined. It has found that the AC input with half-wave rectification may have significant impact on reliability of LEDs' interconnections inside packages. The results of the conventional method will only match the reality when the frequency exceeds a specific value. The AC inputs with half-wave rectification have limited impacts on LED chips. It is difficult to simulate all actual usage conditions with conventional DC aging methods. Therefore, it is necessary to use AC inputs with similar frequency as practical applications in reliability assessments of LED packages.**

Keywords—LED Degradation; Reliability; Current Ripple;

I. INTRODUCTION

The LED is gradually replacing conventional mainstream light sources (such as incandescent and fluorescent lamps) [1-2] for its their low power consumption, long life and high energy efficiency. Compared to traditional light sources, LEDs are not only limited in lighting, but also widely used in many emerging applications, such as visible light communication, sterilization and automotive lights [3-5]. In the past year, research on UV LEDs for coronavirus killing has attracted a lot of attention [6]. The LED-based Visible Light Communication (VLC) technology become one of the most popular technologies of LED lighting [8]. The LED light source brings more functions to automotive head-lights [5].

As a widely used lighting source, the failure mechanism of LED light sources is complex. Numerous studies have been carried out on LED failures lead by temperature, humidity, vibration, UV light and other factors. For instance, different DC currents and temperatures have been used to analyze the failure modes of PC-LEDs [9]. Constant DC input were usually used to investigate the effect of reverse bias currents on LED reliability [10]. It has found that the ambient humidity has a significant impact on the lifetime of optical components of LED package [11]. Blue light and UV light have been used to age optical components as well [12-13]. Most studies on LED failure and reliability use DC inputs to drive LEDs. In many practical applications, an ever-changing input are used to drive LEDs. For example, visible light communication uses Pulse Width Modulation (PWM) signals to strobe the light emission and thus transmit data [14]. Many lamps control the intensity and color of light output by adjusting the duty cycle of the PWM signal [15-16]. It was found that the duty cycle and frequency of PWM current affect the junction temperature of the LED [17]. Experiment results have shown that low frequency and high duty cycle of input current have significant effect on LED's junction temperature [18]. Current ripple accelerates LED degradations and color shift, reducing LED lifetime [19]. Therefore, a systematic study is needed for the impact of high frequency drive current on LED's reliability.

In this paper, effect of the input current frequency on the reliability of LEDs is investigated experimentally. A type of common-used white light LED is selected as test samples. LED aging tests were conducted by using inputs with different frequencies. Meanwhile, to avoid measurement errors in the long-term test, a group of LEDs without aging are tested as well. During the aging process, LEDs' I-V curves and luminous output at different time points were measured. Then, via parameter fitting, degradations of key parameters along aging duration, such as the saturation current, Ideal Factor, equivalent series resistance, luminous efficacy, can be extracted. By analyzing the degradations of key parameters in different conditions, impacts of the input current frequency on LEDs' reliability can be determined.

II. METHNOLOGY

The I-V curve describes the electronic performance an LED chip, As shown in Eq. 1:

$$I_D = I_S \cdot EXP\left(V_D/nv_t - 1\right) \qquad (1)$$

where, I_D is the applied current, I_s is the reverse saturation current. The hot spot voltage v_t, which equals to $\kappa T/q$, is about 25mV at room temperature. n is the Ideal Factor of the PN junction. Since there is an equivalent series resistance R, the voltage drop on the equivalent resistance should be taken into consideration. Therefore, relationship between the measured forward voltage V_F and diode voltage V_D can be describes in the following function:

$$V_F = V_D + I_D \cdot R \qquad (2)$$

where R is the equivalent resistance of the LED package. In this work, the saturation current I_s, the Ideal Factor n and the luminous flux of the LED will be measured experimentally.

978-1-6654-1392-3/21 $31.00 © 2021 IEEE

In this work, a type of common-used white light LED is selected as test samples as show in Fig.1.

Fig.1. The selected white light LED

TABLE I Sample Conditions and Quantities

Input	Sample Numbles	Temp. & Humidity
No Current	12	25 C & 65% RH
DC/0 Hz	12	85 C 65% RH
100 Hz	12	
1000 Hz	12	

The ageing test applies four different current conditions: no current, rated direct current (DC), 100Hz and 1000Hz Alternating Current (AC) with half-wave rectification. The sample size and aging condition of each sample are shown in the Table I.

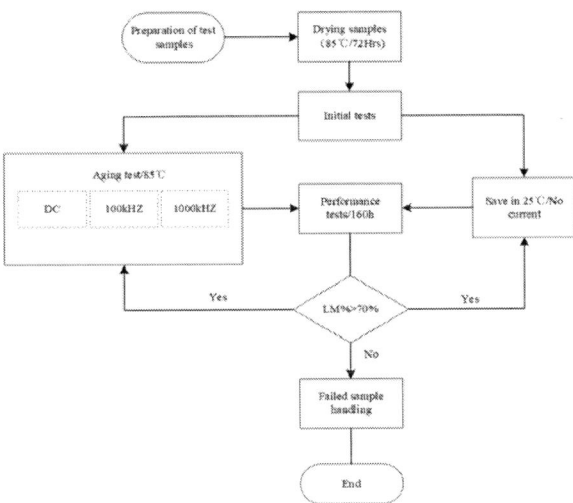

Fig.2. The flowchart of aging tests

Fig.2 shows the flowchart of the ageing experiment. Before aging, test samples were baked for 72 hours in 85°C to remove moisture. Then, the I-V curve and luminous flux of each sample were measured. LEDs were aged in the four different current conditions. The RMS values of all driving current were kept at the rated value of the LED. During the aging process, the I-V curve and luminous flux of each sample were measured every 160 hours by a 50cm integrating sphere system. When the lumen maintenance drops below 70%, the LED is considered failed. Particularly, the lumen maintenance drops to 0%, it is considered that a catastrophic failure occurs.

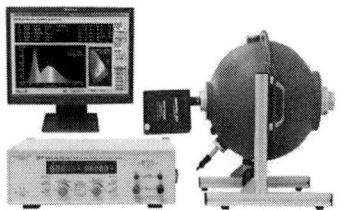

Fig.3. The test platform

III.   EXPERIMENT RESULTS

A. Catastrophic failure

During the ageing process, some catastrophic failures of LEDs were found. Table II shows number of catastrophic failures in 336 hours.

TABLE II Number of Catastrophic Failures

Input	Number of Catastrophic Failure	%
No Current	0	0.00%
DC/0 Hz	1	0.83%
100 Hz	5	41.67%
1000 Hz	5	41.67%

According to Table II, unbiased LEDs do not show any catastrophic failure in 336 hours. The number of catastrophic failures with DC, 100Hz and 1000Hz AC pulses are 1, 5 and 5 respectively. Failure rate of DC input is about 0.83%. When the AC pulses are applied, failure rates of both groups rise to about 41.67%. Such a result indicates that AC input with half-wave rectification cause more damage to LEDs than direct current.

To determine the failure mode, the ultrasonic scanning (C-Scan) is used to inspect the interconnections of failed samples. Fig.4 displays the C-Scan image. As shown in Fig.3, an opened wire bonding of a failed LED can be found. By pressing from top of package, the LED is briefly restored to operation. The major reason of the catastrophic failure is interconnection opening. Thus, the AC input with half-wave rectification may have significant impact on reliability of LEDs' interconnections inside packages. Conventional aging methods, which use DC input, may underestimate interconnection failures of LEDs.

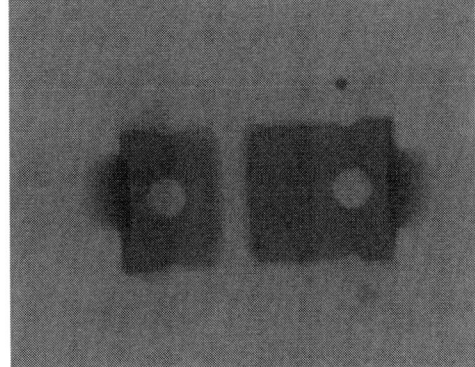

Fig.4. Cracked interconnect wires inside the LED

B. Error analysis

In order to eliminate the measurement errors caused by long-term experiments, this paper analyses measured forward voltage and lumen flux of unbiased LEDs. Theoretically, if the input current is fixed, forward voltage and lumen flux should not change over time.

Mean values of forward voltage of unbiased LEDs in different time are shown in Fig.5. The initial voltage is about 2.61V. Due to the errors of equipment and operators, measured forward voltage increases with time and rises to about 2.68V in 2000 hours. These errors in this experiment may lead to inaccuracies in extractions of saturation current, equivalent resistance and ideal factor. Therefore, this work corrects the errors of measured forward voltage by using the initial voltage as the reference.

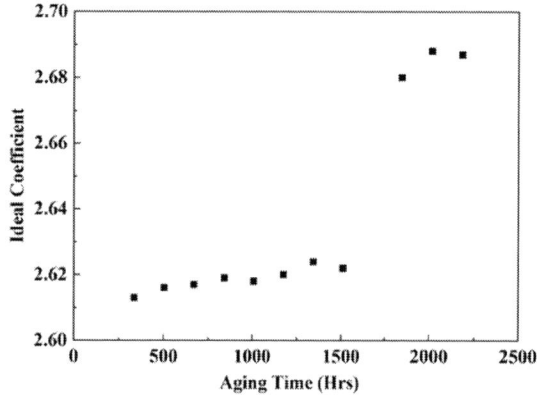

Fig.5. Mean values of forward voltage of unbiased LEDs

Theoretically, the mean value of luminous flux over time should keep unchanged, since unaged samples are measured in this test. However, Fig.6 shows the change in luminous flux over time for the unaged LEDs. Even without aging, it also shows that the luminous flux of the LEDs drops with time. Therefore, such these errors should be introduced into test results by the measurement system, including the 50cm integrating sphere system, operators and ambient conditions. This work uses the initial luminous flux as the reference, and corrects the errors of following time points linearly.

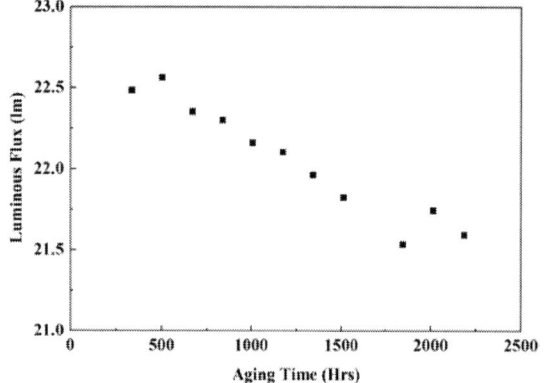

Fig.6. Mean value of luminous flux of unbiased LEDs

C. I-V Characteristics

In order to analyze impact of current pulse on LEDs electronic characteristics, LED's I-V characteristics in different aging conditions are measured during the aging tests. Then, via fitting measured I-V curves by Eq.(1) and (2), the key factors, such as the ideal factor and equivalent resistance, are extracted. In an LED package, the ideal factor and equivalent resistance are indicators of reliability of LED chip and interconnections in a package respectively.

Firstly, LEDs' ideal factor as a function of current frequency and aging time are fitted by the least square method. Fig.6 shows the mean value of ideal factors of 12 samples in each aging conditions. The ideal factor curve of unbiased LEDs shows a similar concave trend with LEDs applied in 100Hz current pulse. Their mean ideal factor values decrease at the beginning, touch the minimum value around 1000 hours, and then slightly increase. In contrast, the ideal factor curves of LEDs in DC and 1000Hz input show a similar convex trend. Their mean ideal factor values increase at the beginning, reach the maximum value around 1000 hours, and then significantly decrease. These results indicate that different frequencies have different impact on LED chips' ideal factor. The effect of low frequency current (100Hz AC) is quite slight and close to zero current aging. Since the bandwidth of the selected LED is limited, the impact of high frequency current (1000Hz AC) is close to DC input. If applications apply low frequency current to LED, reliability assessment results with DC input may have significant errors with user's experiences. The results of the conventional method will only match the reality when the frequency exceeds a specific value. Thus, it is difficult to simulate all actual usage conditions with conventional DC aging methods

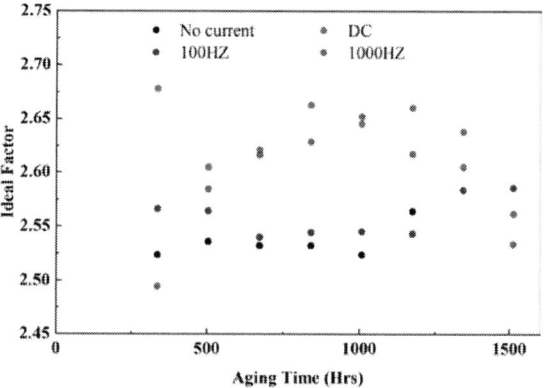

Fig.7. Mean value of ideal factors in different aging conditions

The equivalent resistance is a common-used indicator of reliability of interactions in an LED packages. In order to assess impact of the AC inputs on interconnections of the selected LED's package, the LED equivalent resistance as a function of aging conditions and time is analyzed. As shown in the Fig.8, the mean values of equivalent resistance in different conditions rise with time. The equivalent resistance curves of LEDs in zero current and 1000Hz input show a similar trend. The increment of equivalent resistance curves of LEDs in DC and 1000Hz input are about 1.28×10^{-4} ohm/hrs, 1.29×10^{-4} ohm/hrs and 1.36×10^{-4} ohm/hrs.

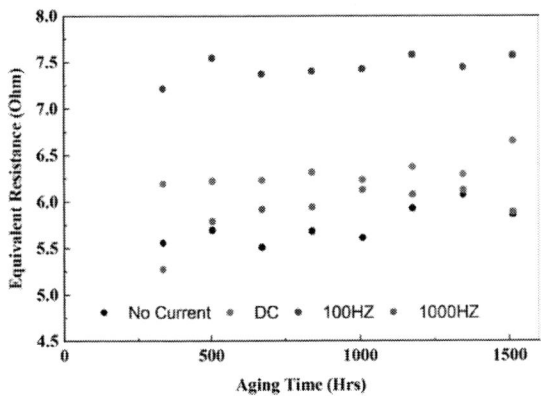

Fig.8. Mean value of equivalent resistance in different aging conditions

D. Luminous flux

Mean values of luminous flux in different aging conditions is shown in Fig.9. The lumen degradations of LEDs in 2500 hours are quite slight. In identical RMS values of input current, impacts of different aging conditions on the lumen degradations is too small to be distinguished. Since the luminous flux determined by the reliability of LED chips, the AC input has limited impact on the LED chips.

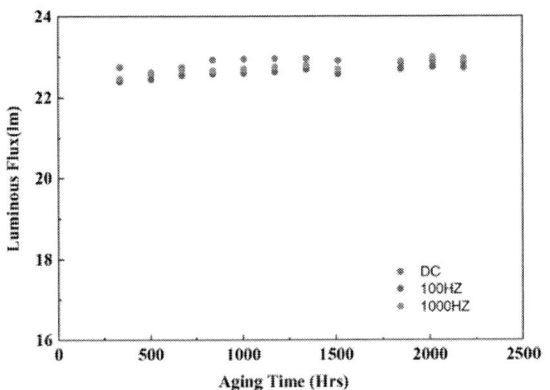

Fig.9. Mean value of luminous flux in different aging conditions

IV. CONCLUSION

In this paper, effect of the input current frequency on the reliability of LEDs is investigated experimentally. A type of common-used white light LED is selected as test samples. LED aging tests were conducted by using inputs with different frequencies. Meanwhile, to avoid measurement errors in the long-term test, a group of LEDs without aging are tested as well. During the aging process, LEDs' I-V curves and luminous output at different time points were measured. Then, via parameter fitting, degradations of key parameters along aging duration, such as the saturation current, Ideal Factor, equivalent series resistance, luminous efficacy, can be extracted. By analyzing the degradations of key parameters in different conditions, impacts of the input current frequency on LEDs' reliability can be determined. According to the testing results, the following conclusions have been found:

The AC inputs with half-wave rectification have limited impacts on LED chips. In identical RMS values of input

current, impacts of different aging conditions on the lumen degradations is too small to be distinguished.

The AC input with half-wave rectification may have significant impact on reliability of LEDs' interconnections inside packages. When the AC pulses are applied, interconnection failure rates of both groups rise from 0.83% to about 41.67%. Conventional aging methods, which use DC input, may underestimate interconnection failures of LEDs.

If applications apply low frequency current to LED, reliability assessment results with DC input may have significant errors with user's experiences. The results of the conventional method will only match the reality when the frequency exceeds a specific value. Thus, it is difficult to simulate all actual usage conditions with conventional DC aging methods.

Acknowledgment

The authors would like to acknowledge the support of the Youth Program of National Natural Science Foundation of China (Grant No. 61904041). We also would like to thank the editors and reviewers of the conference for their time and efforts.

References

[1] W. Driel, X. J. Fan, and G. Q. Zhang, *Solid State Lighting Reliability Part 2*. Solid State Lighting Reliability Part 2, 2018.

[2] W. V. Driel, "Solid State Lighting Reliability Components to Systems."

[3] M. Z. Afgani, H. Haas, H. Elgala, and D. Knipp, "Visible light communication using OFDM," in *International Conference on Testbeds & Research Infrastructures for the Development of Networks & Communities*, 2006.

[4] M. Z. Afgani, H. Haas, H. Elgala, and D. Knipp, "Visible light communication using OFDM," in *International Conference on Testbeds & Research Infrastructures for the Development of Networks & Communities*, 2006.

[5] W. Wojtkowski, "Automotive LED Lighting with Software PWM Generators," in *Lighting Conference of the Visegrad Countries*.

[6] Y. Gerchman, H. Mamane, N. Friedman, M. J. J. o. P. Mandelboim, and P. B. Biology, "UV-LED disinfection of Coronavirus: Wavelength effect," vol. 212, p. 112044, 2020.

[7] Q. Lin, C. Wang, and Y. Tian, "Thermal design of a LED multi-chip module for automotive headlights," in *Electronic Packaging Technology and High Density Packaging (ICEPT-HDP), 2012 13th International Conference on*, 2012.

[8] T. Komine and M. J. C. E. I. T. o. Nakagawa, "Fundamental analysis for visible-light communication system using LED lights," vol. 50, no. 1, pp. 100-107, 2004.

[9] L. Trevisanello, F. D. Zuani, M. Meneghini, N. Trivellin, and G. Meneghesso, "Thermally activated degradation and package instabilities of low flux LEDS," in *IEEE International Reliability Physics Symposium*, 2009.

[10] Meneghini and M. J. I. T. o. E. Devices, "A Review on the Physical Mechanisms That Limit the Reliability of GaN-Based LEDs," 2010.

[11] M. Cai *et al.*, "Determining the thermal stress limit of LED lamps using highly accelerated decay testing," vol. 102, pp. 1451-1461, 2016.

[12] K. Gandhi, C. L. Hein, R. V. Heerbeek, J. E. J. P. D. Pickett, and Stability, "Acceleration parameters for polycarbonate under blue LED photo-thermal aging conditions," vol. 164, no. JUN., pp. 69-74, 2019.

[13] S. K. Yang, P. M. Tu, S. C. Huang, Y. W. Lin, and C. Y. Chang, "Investigation of efficiency droop in GaN-based UV LEDs with N-type AlGaN underlayer," in *Semiconductor Electronics (ICSE), 2012 10th IEEE International Conference on*, 2012.

[14] L. Matheu19.s, A. B. Vieira, L. Vieira, M. Vieira, O. J. I. C. S. Gnawali, and Tutorials, "Visible Light Communication: Concepts, Applications and Challenges," pp. 1-1, 20

[15] Y. T. Hsieh, J. F. Wu, C. L. Fang, H. H. Tsai, and Y. Z. Juang, "Design of fast stabilized LED driver IC with low overcurrent,"

978-1-6654-1392-3/21 $31.00 © 2021 IEEE

in *Power Electronics and Drive Systems (PEDS), 2013 IEEE 10th International Conference on*, 2013.

[16] Y. T. Ling, C. W. Guo, and S. T. Mei, "A chromatic aberration correction method for RGB LED based on PWM," in *2017 Progress in Electromagnetics Research Symposium - Fall (PIERS - FALL)*, 2017.

[17] N. Yang, B. Shieh, T. Zeng, and S. Lee, "Analysis of Pulse-Driven LED Junction Temperature and its Reliability," in *2018 15th China International Forum on Solid State Lighting: International Forum on Wide Bandgap Semiconductors China (SSLChina: IFWS)*, 2018.

[18] S. Buso, G. Spiazzi, M. Meneghini, G. J. I. T. o. D. Meneghesso, and M. Reliability, "Performance Degradation of High-Brightness Light Emitting Diodes Under DC and Pulsed Bias," vol. 8, no. 2, pp. 312-322, 2008.

[19] P. H. Pathak, X. Feng, P. Hu, and P. Mohapatra, "Visible Light Communication, Networking, and Sensing: A Survey, Potential and Challenges," *IEEE Communications Surveys & Tutorials,* vol. 17, no. 4, pp. 2047-2077, 2015.

Sequential Analysis of Drop Impact and Thermal Cycling of Electronic Packaging Structures

Yongchao Liu[1,2], Xu Long[3*], Haozhe Wang[1,2], Jibao Lu[1,2*], Rong Sun[1,2], Ching-Ping Wong[4]

1 Shenzhen Institute of Advanced Technology, Chinese Academy of Science, Shenzhen, 518055, China
2 Shenzhen Institute of Advanced Electronic Materials, Shenzhen, 518103, China
3 School of Mechanics, Civil Engineering and Architecture, Northwestern Polytechnical University, Xi'an, 710072, China
4 School of Materials Science and Engineering, Georgia Institute of Technology, Atlanta, GA 30332, USA

Abstract—Electronic packaging structures go through complicated loading history such as power on-off, vibration and drop impact during the storage and serving conditions processes. All these loading applications result in the deformation and possible damage accumulation in the packaging structure of the electronic device. Considering that the thermal cycling is a standard procedure for evaluating the mechanical reliability of a packaging, we systematically investigate the combination effect of the drop and thermal cycling on the reliability of the packaging structures by using the finite element analysis. Compared with the transient analysis of drop impact, the analysis duration for thermal cycling is significantly prolonged.

A representative Ball Grid Array (BGA) packaging structure is adopted to perform the sequential analysis. The strain rate and temperature dependent visco-plastic properties are considered for the Sn-3.0Ag-0.5Cu (SAC305) solder material at the solder joints. Those material properties have been validated against published experimental data. Firstly, the numerical analysis of the packaging structures subject to drop impact is conducted to obtain the distribution of plastic deformation and strain. As the plastic deformation and strain are not recoverable after the loading application, those distributions can be utilized as initial strain conditions to be imported into the subsequent analysis under thermal cycling. In the second analysis step, a temperature profile between -50°C and 125°C is applied to the packaging structure to predict the deformation and strain during the thermal cycling. The thermal fatigue life is evaluated using the Coffin-Manson model based on the predicted increment of equivalent plastic strain in the critical solder joints. More importantly, the effect of drop-impact damage on the thermal deformation and thus on the fatigue life is discussed by correlating the drop impact, thermal cycling and fatigue life as a closed loop.

Keywords—*finite element analysis drop impact; thermal cycling; Coffin-Manson model; fatigue life*

I. INTRODUCTION

With the rapid development of modern technology, electronic devices are gradually developing in the direction of high ductility, high performance, and small size[1]. More and more electronic devices will fail during use, such as reciprocating bending, thermal cycling, drop impact and other special conditions. In the design of miniature electronic packaging, it is very necessary to test and inspect the structural performance of electronic packaging. In order to explore the performance of the BGA package structure under the combined action of temperature cycling and drop impact, this paper uses the ABAQUS[2] finite element analysis software to conduct a finite element simulation of the BGA package

structure under the combined action of the two working conditions. The stress and equivalent plastic strain distribution under three working conditions are mainly compared and analyzed: thermal cycle; drop shock; the combined effect of thermal cycle and drop shock. The Coffin-Manson model is used by more scholars and researchers to evaluate the thermal fatigue life under various working conditions. In this paper, corresponding modified models are proposed for temperature cycling and drop impact conditions to provide guidance for actual projects.

II. FINITE ELEMENT ANALYSIS

The finite element modeling of the BGA package structure is first performed in this paper, as shown in Fig.1. Including the main BGA simulation structure: Mold Compound; Solder Joint; Die; Substrate; Underfill. Based on the model in Fig.1, the thermal cycle and drop simulation are performed. During the entire thermal cycle, the selected thermal cycle curve is shown in Fig.2. In order to make the simulation results more credible, the initial test temperature is 25°C at room temperature, the rate of heating (cooling) is 15°C/min, the residence time at high temperature (low temperature) is 5min, and 3 thermal cycles are selected for simulation.

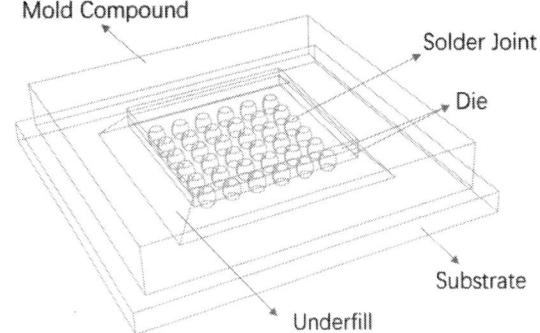

Fig.1. Schematic diagram of BGA package structure model

This section simulates the changes in the stress and equivalent plastic strain of the BGA package structure when the thermal cycle alone, the drop impact alone, and the thermal cycle and low drop impact work together. And carried out a comparative analysis, as shown in Fig.2.

978-1-6654-1392-3/21 $31.00 © 2021 IEEE

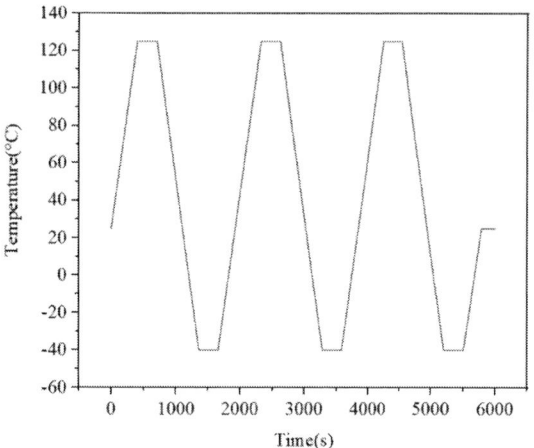

Fig.2. Thermal cycle curve

During thermal cycle simulation, the BGA package structure will be affected by the thermal stress. Due to the different thermal expansion coefficients of the materials of different components of the package structure, the amount of warpage of each part is different, which will cause the failure of the entire BGA package structure. For solder joints, the maximum stress appears on the solder joints at the corners. Similarly, the maximum equivalent plastic strain also appears in the same position. In Fig.3, no matter from the stress or equivalent plastic strain, after the thermal cycle of the BGA package structure, the solder joints located at the four peripheral corners of the package structure are relatively most prone to failure.

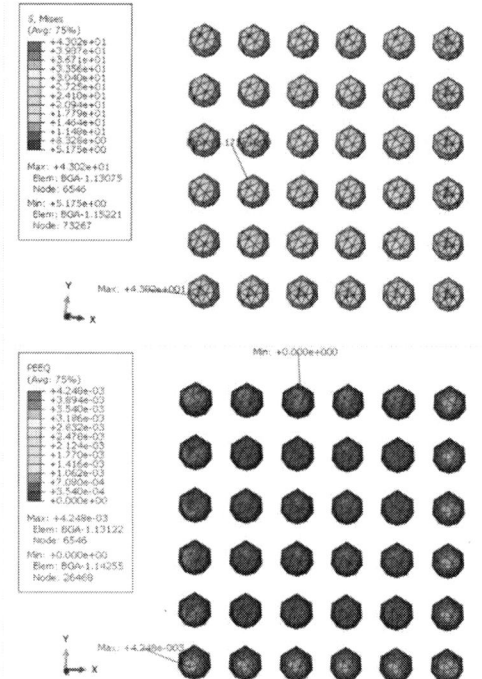

Fig.3. Distribution of solder joint stress (left) and equivalent plastic strain (right) during temperature cycling

In the process of the drop simulation test, the selected drop height is 1.50m, the drop impact direction is the negative direction of the Y axis, and the speed close to the rigid board is 5.42m/s. As shown in Fig.4, the stress and equivalent plastic strain distribution of the entire solder joint of the BGA structure after the drop impact. It can be seen intuitively that the stress on the solder joints that first contact the rigid board is greater than the solder joints above, and the position with the greatest stress is concentrated on the outermost solder joints. It shows that during the drop impact process, the solder joint that first contacts the outermost side of the rigid board is the most prone to failure. In addition, from the distribution of equivalent plastic strain, the fatigue life of the solder joint that first contacts the rigid plate during the entire drop impact process is the lowest.

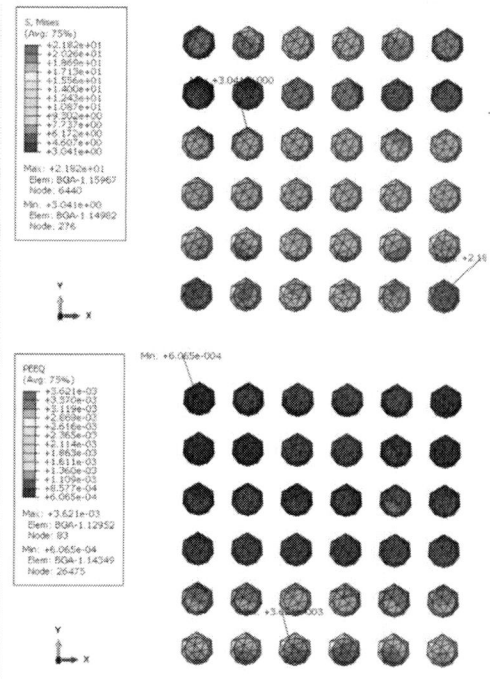

Fig.4. Distribution of solder joint stress (left) and equivalent plastic strain (right) under drop impact

Under the combined action of temperature cycling and drop shock, the convergence of the simulation test is relatively more complicated. In this paper, the sequential coupling method is adopted, and the simulation result of thermal cycle is added on the basis of the drop impact simulation test. In Fig.5, after the combined action of thermal cycle and drop impact, the distribution form of thermal stress and accumulated equivalent plastic strain on the solder joint of the BGA package structure is roughly the same, which presents a similar cloud picture to that of a single drop impact. And it is worth noting that the corresponding values have increased significantly. So, the fatigue life of solder joints under the combined action of temperature cycling and drop impact is shorter, and failure is more likely to occur.

978-1-6654-1392-3/21 $31.00 © 2021 IEEE

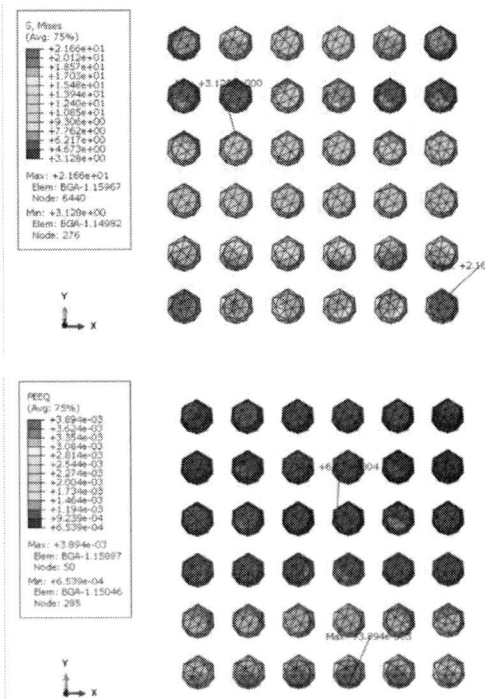

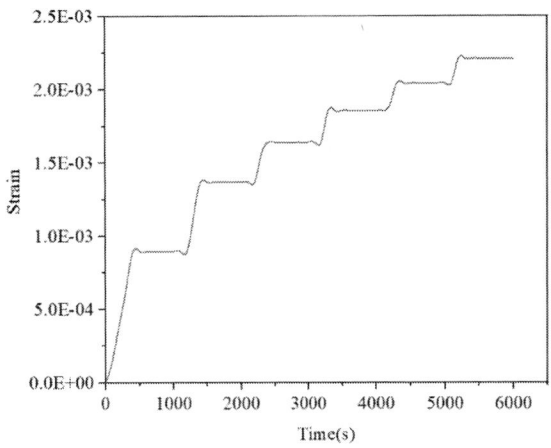

Fig.6. Stress and equivalent plastic strain during temperature cycling

III. FATIGUE LIFE PREDICTION

There are many fatigue models related to solder joint fatigue life and thermal-mechanical stress analysis: Coffin-Manson model[3], Goodman model[4], Smith-Waston-Topper model[5] and so on. Among them, the fatigue model based on plastic strain for low cycle fatigue analysis is mainly the Coffin-Manson model[6]. Based on the results of the previous nonlinear finite element analysis, the maximum equivalent plastic strain value accumulated in the solder joint can be determined. According to the Coffin-Manson model, the position with the largest equivalent strain is the position where the solder joint is most prone to failure, and it is also the position that we should focus on analysis. In this paper, the Coffin-Manson model modified by Engelmaier is used, and the equation is shown in Eq. (1).

$$N_f = \frac{1}{2}\left(\frac{\Delta\gamma}{2\varepsilon_f}\right)^{\frac{1}{c}} \qquad (1)$$

Among them, N_f is the predicted life of the solder joint after thermal cycle fatigue failure; $\Delta\gamma$ is the inelastic strain range of the solder joint, and also the shear plastic strain range of the solder joint. The analytical formula is shown in Eq. (2); ε_f is the fatigue toughness coefficient of the solder joint, and the value is constant in the revised model of this paper, which is 0.325; c is the fatigue toughness index of the solder joint, which is jointly defined by Eqs. (2)-(4), and the calculated result is -0.3979.

$$\Delta\gamma = \sqrt{3}\Delta\varepsilon \qquad (2)$$

$$c = -0.442 - 6 \times 10^{-4} T_m + 1.74 \times 10^{-2} \ln(1 + f) \quad (3)$$

$$T_m = (T_{max} + T_{min})/2 \qquad (4)$$

In Eq. (2), $\Delta\varepsilon$ refers to the equivalent value of the shear plastic strain of the solder joint, which is generally obtained from numerical simulation results. In Eq. (3), f is the frequency of a complete thermal cycle, and the frequency unit selected here is cycles/day. In Eq. (4), T_m refers to the average temperature of the thermal cycle in the experiment. Correspondingly, T_{max} and T_{min} refer to the highest and lowest temperature in the experimental thermal cycle, respectively.

Fig.5. Distribution of solder joint stress (left) and equivalent plastic strain (right) under the combined action of temperature cycling and drop impact

Fig.6 shows the change trend of the solder joint stress and equivalent plastic strain over time during the thermal cycle. It can be seen that the stress of the solder joint shows a similar change trend with the temperature curve, and the high and low peaks all represent the temperature residence stage in the thermal cycle. The accumulation of equivalent plastic strain is also getting bigger and bigger with time, and presents a form of step-like increase.

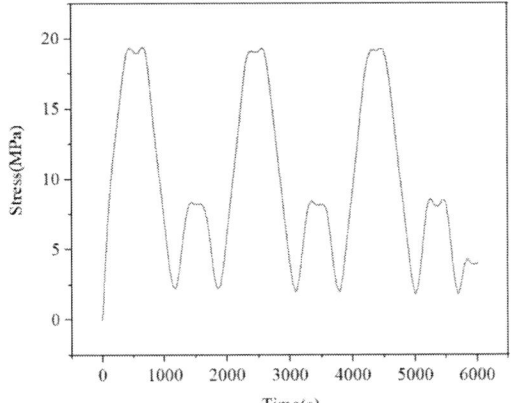

Through the Coffin-Manson model introduced above, we can calculate that the fatigue life of the solder joint after thermal cycling is 47146 cycles, the fatigue life of the drop impact is 58126 cycles, and the fatigue life of the two conditions is 38911 cycles. It can be seen that the thermal stress generated by the thermal cycle has the greatest effect on the fatigue life of the BGA package structure. In addition, the thermal stress generated by the thermal cycle also has a great influence on the drop. When the two working conditions work together, the fatigue life is significantly reduced.

IV. CONCLUSION

This paper explores the fatigue life of solder joints when the BGA package structure is dropped and impacted in the working state. The effects of drop impact damage and thermal cycling on the fatigue life of solder joints are discussed. The thermal fatigue life is evaluated using the Coffin-Manson model based on the predicted increment of equivalent plastic strain in the critical solder joints. It was found that both drop impact and temperature cycling have an important influence on the reliability of the solder joints of the BGA package. According to the analysis, the fatigue life of solder joints reached 47146 cycles under thermal cycling conditions. Under the drop impact condition, the fatigue life of the solder joint reached 58126 cycles. Under the combined effect of the two working conditions, the fatigue life of the solder joints is drastically reduced to 38911 cycles. It can be found that thermal cycling and drop impact are necessary factors to consider the reliability of the BGA package structure, and the combination of the two has a destructive effect on the package structure. Relative thermal cycling and drop impact work alone, the fatigue life when the two work together is reduced by 17.5% and 33.1%. This is of great significance for the design and prediction of the fatigue life of electronic package interconnections in high-power electronic devices.

ACKNOWLEDGMENT

The work was supported by the National Key R&D Program of China (2020YFB0408700), the National Natural Science Foundation of China (no. 52003289), the Youth Innovation Promotion Association CAS (no. 2021363), and the SIAT Innovation Program for Excellent Young Researchers (no. 201803).

REFERENCES

[1] X. Long, Y. Liu, Y. Yao, F. Jia, C. Zhou, Y. Fu, and Y. Wu, Constitutive behaviour and life evaluation of solder joint under the multi-field loadings, AIP Adv. 8, 085001 (2018).

[2] Dassault Systemes Simulia Corp., ABAQUS User's Manual 2017 (Hibbitt, Karlsson & Sorensen, Rhode Island, 2017)

[3] W. Engelmaier, "Fatigue life of leadless chip carrier solder joints during power cycling," IEEE Transactions on Components Hybrids & Manufacturing Technology 6, 232–237 (2003).

[4] E., and Altus. "A cohesive micromechanic fatigue model. Part II: Fatigue-creep interaction and Goodman diagram." Mechanics of Materials (1991).

[5] A. Ince, and G. Glinka . "A modification of Morrow and Smith–Watson–Topper mean stress correction models." Fatigue & Fracture of Engineering Materials & Structures 34.11:854-867 (2011).

[6] T.J. Kilinski, J.R. Lesniak, B.I. Sandor, Solder Joint Reliability: Theory and Applications (1991), pp.384–405 (1991)

Enhanced Optical Performance and Thermal Stability of Quantum Dot Converters for Laser Source

Jiayong Liang
National and local joint engineering research center for semiconductor display and optical communication devices, South China University of Technology,
Guangzhou, China
scutjiayong@foxmail.com

Jiexin Li
National and local joint engineering research center for semiconductor display and optical communication devices, South China University of Technology,
Guangzhou, China
jiexinli.jeson@foxmail.com

Zihao Deng
National and local joint engineering research center for semiconductor display and optical communication devices, South China University of Technology,
Guangzhou, China
scutzihao@foxmail.com

Yihua Qiu
National and local joint engineering research center for semiconductor display and optical communication devices, South China University of Technology,
Guangzhou, China
1920937102@qq.com

Zongtao Li
National and local joint engineering research center for semiconductor display and optical communication devices, South China University of Technology,
Guangzhou, China
meztli@scut.edu.cn

Jiasheng Li*
National and local joint engineering research center for semiconductor display and optical communication devices, South China University of Technology,
Guangzhou, China
jiasli@foxmail.com

Abstract—Quantum dots (QDs) are a revolution in the field of photoelectric and have a crucial potential for future applications as a color converter in micro/mini-LEDs. However, the reported QD converter efficiency (~20-40%) has not reached the level on par with that of the conventional phosphor color converters (~40-60%). In this study, we fabricate the QD nanowires (NWs) templated in the nanoporous anodic aluminum oxide (AAO) substrate using a combination of inkjet printing and vacuum-deposition. Owing to the light couple and waveguide effect of the nanopores cavity structure, we demonstrate a 1.25-fold enhancement in the luminous flux and a 2.04-fold increase in the photoluminescence (PL) intensity compared with the traditional QD planar converter. Furthermore, the stability test showed a 2.34 times improvement in lifetime resulting from the heat dissipation and moisture protection of the nanopores. The inkjet printing fabrication strategy combined with the nanostructure provides new insights for mass-produce high-quality QD NWs, which is promising for modern lighting applications in the micro/mini-lighting-emitting diodes (LEDs).

Keywords—Quantum dots, packaging structure, color converter, thermal and optical performance

I. INTRODUCTION

Quantum Dots (QDs) have emerged as promising candidates for highly efficient color conversion layer of micro/mini-LEDs due to high intense photoluminescence quantum yield (PLQY), narrow full-width at half maximum (FWHM), and tunable emission spectra[1-3]. Green and red QDs packaged with micro/mini-LED chips have become the most advanced technique for producing ultra-high-resolution displays[4].

Generally, QDs are dispersed into a transparent polymer matrix to form QD-polymer composites and utilized as color converters[5]. When operated in the heat and flux of the on-chip environment, it is difficult for the QD converter to withstand the continuous high-density power as a result of the poor thermal conductivity of the polymer, leading to the thermal quenching of QDs[6]. The poor thermal stability is one of the crucial obstacles for its commercial applications.

Another one is the low conversion efficiency of the QD converters[7]. Great efforts have been made to increase it by applying various packaging structures and designs. Recently, researchers have found that the utilization of optical microcavity can assure both high photoluminescence quantum yield and outstanding thermal stability of lead halide perovskite quantum dots through the electromagnetic field enhancement from the resonance effect[8]. Besides, some reports reveal that the light extraction with microlens or hollow fiber structures successfully achieves a few times enhancement for OLED performance[9], but this approach has not been adopted for QD converters. Herein, we introduce the nanoporous AAO substrate as a packaging structure of the QD color converter.

In our work, we introduce a facile and scalable method to fabricate QD nanowires (NWs) array within the nanoporous AAO substrate (the Meso QD@AAO). The optical performance tests were conducted to verify the improved light extraction of the Meso QD @AAO compared with the traditional QD film (the QD @PDMS). Finally, the stability comparisons were carried out via the infrared thermal method. The strategy represented here provides a new way to mass-produce QD nanowires of good quality for large-scale lighting applications and highly integrated optoelectronics in the future.

II. EXPERIMENT

Figure 1 shows the QD NWs fabrication process. First, a nanoporous AAO substrate (2 cm(length)×2 cm(width), purchased from Jiangyin Youkun Experimental Equipment Co., Ltd.) was consecutively cleaned with deionized water with 500 s of sonication. Then, certain amounts of QDs were dissolved in chlorobenzene and ultrasonicated for 300s at room temperature(22℃) to form a uniform and transparent ink, where the QD was fixed at 1 mol/L in solutions.

978-1-6654-1392-3/21 $31.00 © 2021 IEEE

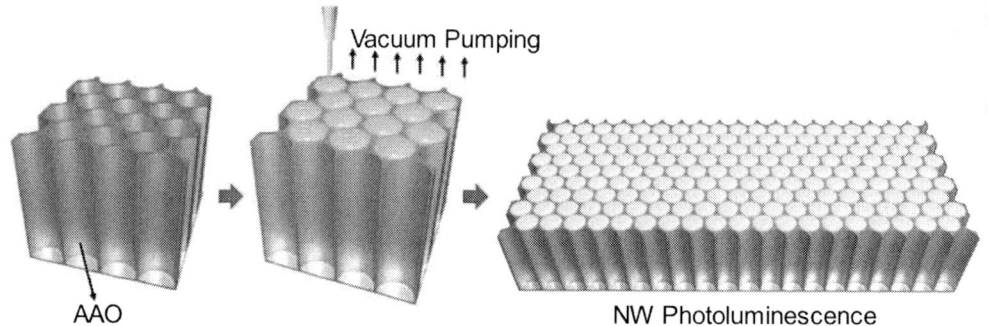

Fig.1. Schematic diagram of the preparation process for Quantum Dots Nanowire in nanoporous AAO for lighting.

Then 0.2 ml QD solutions were slowly inkjet-printed on the AAO substrate until the surface of the AAO substrate was fully covered. After that, vacuum suction was intended on the top of the AAO substrate to accelerate the evaporation of chlorobenzene solution. As the ink dried associated with the evaporation of the chlorobenzene, it reached a saturation state and then induced the crystallization and deposition of QD, thus forming a QD nanowire array embedded in the AAO. The green CdSe/ZnS Quantum Dots used in this paper were purchased from Beijing Beidajubang Science Technology Co., Ltd., and their peak emission wavelength and PLQY were 525 nm and 85%, respectively. The chlorobenzene was supplied from Aladdin.

In previous studies, the QDs were dispersed in organ silicate matrix by the solvent evaporation method[10], therefore, we produced a traditional QD film with the same QD quantity as the QD NWs within the nanoporous AAO to compare the color-converted efficiency between them. First, 0.1 g QDs were dissolved in 2 ml hexane solution and mixed with 1 g polydimethylsiloxane (PDMS) for 5 minutes ultrasonic treatment. After that, the QD slurry was coated on a cleaned sheet of glass(2 cm(length)×2 cm(width)) and cured at 90℃ for 40 min. PDMS, which serves as a QDs dispersion matrix, was purchased from Dow Corning Co., Ltd. All processes were carried out under laboratory conditions at a constant ambient temperature of 22℃.

III. RESULTS AND DISCUSSIONS

The presence of QD NWs in the AAO nanopores was confirmed by cross-sectional scanning electron microscopy (SEM). Fig. 2(a) is a high-resolution top-view SEM image that showing good filling rate of QDs in the AAO pores, while the inset shows empty nanopores with unprinted AAO substrate. The nanoporous AAO substrate here has a depth of 60 μm and a diameter of 400 nm.

A. Optical performance of the QD NWs

Our previous report has revealed that the strong cluster-induced scattering (AIS) effect of QDs in the silicone matrix results in severe backscattering and reabsorption loss, thus leading to the low efficiency of the LED devices with high QD concentration[6]. Fig. 3(a)(b) schematically demonstrates the QD NWs templated in the nanoporous AAO (Meso QD @AAO) and the traditional QD films (QD @PDMS). With the same amount of QDs, they uniformly dispersed in the nanoporous AAO substrate, while aggregate and form large particles in the PDMS. Due to the large effective size of QDs and the difference in refractive index between the QDs and the silicone matrix (PDMS), the diffraction effect leads to a strong

scattering along the edges of the QDs, which causes severe optical loss confirmed by the PL spectra given in Fig. 3(c). The PL intensity of the Meso QD @AAO is much higher than that of the QD @PDMS under the same quantity of QD. It achieves a maximum PL enhancement of 2.04-fold compared with the QD @PDMS.

One reasonable explanation is that the AAO nanopores fundamentally prevent the aggregation of QDs, which is served as the dispensed substrate instead of the PDMS. Another one is that the nanopores in the AAO substrate act as the light coupler. The inner wall of the nanopores of AAO has a high metal reflection index, which helps to focus the light on the QD NWs and to cause more conversion events, thus leading to enhanced light out-extraction. Besides, some QD light can bypass the QDs in the nanopore of the AAO, which

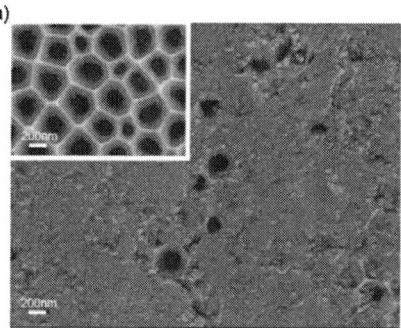

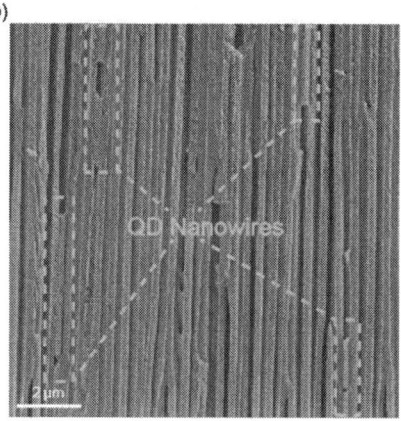

Fig.2. SEM image of the QD NWs in the AAO nanopores. (a) Top view and (b) Cross-sectional SEM image of filled QD NWs in AAO.

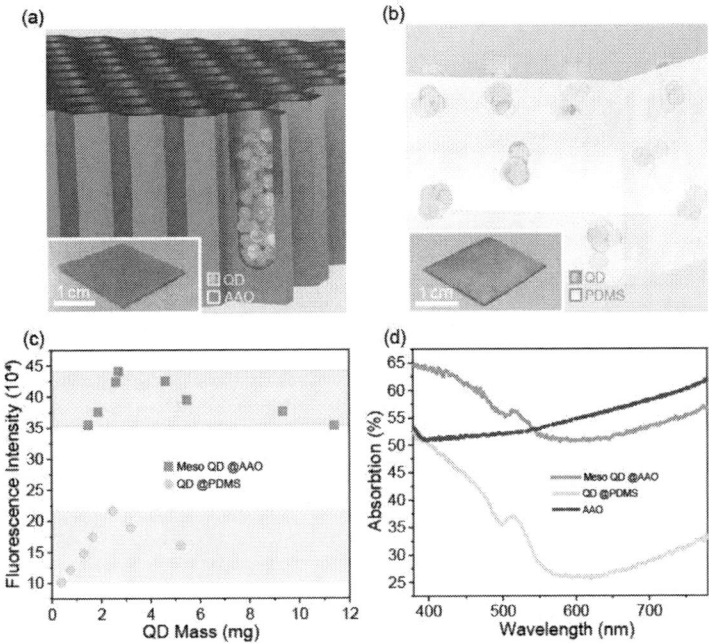

Fig.3. Characterization of the QD NWs templated in nanoporous AAO and the traditional QD films. Cross-sectional views schematic of (a) the Meso QD @AAO and (b) the QD @PDMS. The insets are the photographs of them. (c) PL spectra and (d) Reflection spectra of the Meso QD @AAO and the QD @PDMS, respectively.

successfully reduces the self-reabsorption loss. Another reasonable explanation is the waveguide effect of the nanopores of the AAO substrate. Compared with the QD @PDMS, light can directly enter the QDs in the nanopores of the AAO without passing through an organic matrix interface with a refractive index of 1.41, thus improving the light absorption and utilization. As can be seen from Fig. 3(d), the absorption of the Meso QD @AAO is much higher than that of the QD @PDMS, even the bare AAO substrate does.

However, the PL intensity of the Meso QD @AAO also decreases with the increasing QD mass seen from Fig. 3(c). One explanation is that the propagation direction of QD light is disturbed by the internal scattering effect when the light incident at the boundary of the nanopore. When the

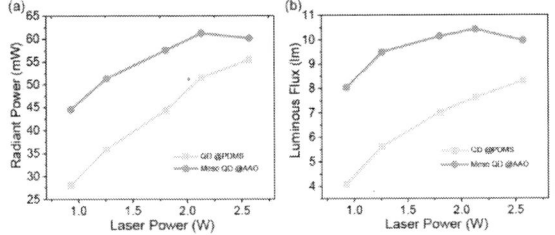

Fig.4. Optical efficiency of the QD NWs templated in AAO nanopores and the traditional QD planar TF CCL under different power excitation. (a) Radiant power. (b) Luminous flux.

characteristic size of the nanopores' boundary is equivalent to the wavelength of the QD light, the strong scattering effect results in greater interference of the QD light and more serious total internal reflection. Even though, the lowest PL intensity of the Meso QD @AAO is still 63.6% higher than the highest one of the QD @PDMS.

Their radiant power and luminous flux under different laser power are compared, as shown in Fig. 4. Intriguingly, the Meso QD @AAO always has higher radiant power than the QD @PDMS. It can be simply attributed to the fact that the waveguide effect of the nanopore of the AAO. Moreover, the luminous flux can be greatly improved by replacing the silicone matrix (PDMS) with the nanoporous AAO, which achieves a maximum luminous flux enhancement of 25.5%.

B. Stability of the QD NWs

Stability is a particularly significant characteristic of the QD converter since the severe thermal quenching phenomenon of QDs easily happens under a high temperature. Therefore, we performed stability tests on two kinds of samples: (i) the QD @PDMS, (ii) the Meso QD @AAO. Fig. 5(a) shows the time-varing maximum temperatures of the QD @PDMS and the Meso QD @AAO excited by the external laser diode at 0.93 W of injected power, respectively. Both are inspired from 0 seconds, and then cease to excite when the maximum temperatures stabilize. As seen from Fig. 5(a), there is a dramatic increase at the beginning, while a dramatic decrease at the end. The maximum temperature of the Meso QD @AAO is 36.974℃, just 0.5-fold that of the QD @PDMS. The nanoporous AAO cavity structure for heat dissipation offers a great amount of heat transfer channels. Fig. 5(b) is the stability under the action of a high-density laser and a harsh closed environment without any heat dissipation. The T_{half} values for the Meso QD @AAO are 11.6664 min, while 4.9998 min for the QD @PDMS, which shows 2.34 times of lifetimes. The enhancement under the continuously high-density laser power stability can be attributed to the cavity on the AAO that helps to against the water and oxygen diffusion.

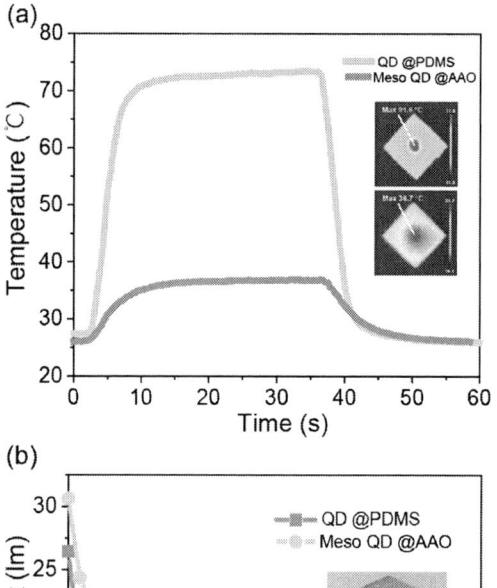

Fig.5. Stability tests of the QD NWs templated in AAO nanopores and the traditional QD planar TF CCL. (a) Time-dependent Surface temperature distributions, the insets are the stable infrared image of the QD @ PDMS and the Meso QD @AAO. (b) Time-dependent Luminous flux, the insets are thee photograph of them. All devices were measured at an injection power of 0.93 W.

IV. CONCLUSION

In summary, we proposed a facile strategy to fabricate QD NWs array by inkjet-printing and vacuum-depositing within a nanoporous AAO substrate. The morphology of the AAO nanopores is applied to prevent QDs from aggregation, thus reducing the AIS effect under high concentration. The cavity structure of the AAO nanopores acts as the light coupler, which causes more conversion events and reduces self-reabsorption loss. Furthermore, the waveguide effect of the AAO nanopores improves light utilization and extraction. It achieves a maximum PL enhancement of 2.04-fold compared with the QD @PDMS. Additionally, the AAO nanopores

provide heat transfer channels and protection from moisture and oxygen diffusions for the QDs, which contributes to the Meso QD @AAO 2.3334 times the lifetime of the traditional QD film. The study described above provides a strategy for mass-producing QD NWs and simultaneously solves critical issues on optical performance and thermal stability, which suggests a general guide for future designs on the QD converters integrated with other intelligent equipment, such as micro/mini-LED.

ACKNOWLEDGMENT

This work was supported by the Postdoctoral Science Foundation of China (No. 2020M680122), by the Natural Science Foundation of Guangdong Province (No. 2018B030306008), and the Project of the National and Local Joint Engineering Research Center of Semiconductor Display and Optical Communication, Zhongshan Branch (No. 190919172214566).

REFERENCES

[1] P. Pust, P.J. Schmidt, W. Schnick, A revolution in lighting, Nature Materials 14(5) (2015) 454.

[2] X. Dai, Z. Zhang, Y. Jin, N. Yuan, H. Cao, X. Liang, L. Chen, J. Wang, X. Peng, Solution-processed, high-performance light-emitting diodes based on quantum dots, Nature.

[3] H.V. Han, H.Y. Lin, C.C. Lin, W.C. Chong, J.R. Li, K.J. Chen, P. Yu, T.M. Chen, H.M. Chen, K.M. Lau, Resonant-enhanced full-color emission of quantum-dot-based micro LED display technology, Optics Express (2015).

[4] Optical cross-talk reduction in a quantumdot-based full-color micro-light-emitting-diode display by a lithographic-fabricated photoresist mold, Photonics Research 05(v.34;No.191) (2017) 32-37.

[5] S. Yu, B. Fritz, S. Johnsen, D. Busko, B.S. Richards, M. Hippler, G. Wiegand, Y. Tang, Z. Li, U. Lemmer, Enhanced Photoluminescence in Quantum Dots–Porous Polymer Hybrid Films Fabricated by Microcellular Foaming, Advanced Optical Materials 7(12) (2019) 1900223.

[6] J.-X. Li, Z.-H. Deng, J.-Y. Liang, J.-S. Li, Unraveling the Origin of Low Optical Efficiency for Quantum Dot White Light-Emitting Diodes From the Perspective of Aggregation-Induced Scattering Effect, IEEE Transactions on Electron Devices PP (2021) 1-8.

[7] B. Xie, Quantum Dots-Converted Light-Emitting Diodes Packaging for Lighting and Display: Status and Perspectives.

[8] J. Hao, Z. Dai, M. Guan, P. Dang, G. Li, Simultaneous enhancement of luminescence and stability of lead halide perovskites by a diatomite microcavity for light-emitting diodes, Chemical Engineering Journal (5) (2020) 128056.

[9] Z. Chen, C. Zhang, X.F. Jiang, M. Liu, R. Xia, T. Shi, D. Chen, Q. Xue, Y.J. Zhao, S. Su, High‐Performance Color‐Tunable Perovskite Light Emitting Devices through Structural Modulation from Bulk to Layered Film, Advanced Materials 29(8) (2016) 1603157.1-1603157.8.

[10] J.S. Li, Y. Tang, Z.T. Li, L.S. Rao, X.R. Ding, B.H. Yu, High efficiency solid–liquid hybrid-state quantum dot light-emitting diodes, Photonics Research 6(12) (2018).

Micron-Scale Silver Flake Paste Sintering Without Pressure for Power Electronic Die Attachment

Shijun Huang
School of Materials, and state Key Laboratory of Optoelectronic Materials and Technologies
Sun Yat-sen University
Shenzhen, China
huangshj28@mail2.sysu.edu.cn

Ruidong Luo
School of Materials, and state Key Laboratory of Optoelectronic Materials and Technologies
Sun Yat-sen University
Shenzhen, China
luord5@mail2.sysu.edu.cn

Zhen Wu
School of Materials, and state Key Laboratory of Optoelectronic Materials and Technologies
Sun Yat-sen University
Shenzhen, China
wuz73@mail2.sysu.edu.cn

Cai-Fu Li
School of Materials, and state Key Laboratory of Optoelectronic Materials and Technologies
Sun Yat-sen University
Shenzhen, China
licaifu@mail.sysu.edu.cn

Abstract—**With the development of power electronics, metal particle sintering, especially Ag particle sintering technology has gained much attentions. In this work, different sintering parameters such as sintering time, temperature and assisted pressure were taken into consideration to evaluate the performances of micron-scale Ag flake paste. The shear strength of the sintered joints prepared with micron-scale silver flake paste reached as high as 26.5 MPa after sintering at 250°C for 30 min without pressure under atmosphere condition.**

Keywords—Micron-scale silver flake paste, Shear strength, Microstructure

I. INTRODUCTION

Power electronic devices are widely used in smart power supply system, new energy vehicles and rail transits, et al. The power density of the devices becoming greater, and the operation temperature becoming much higher than 250 °C. The most used traditional soldering material cannot meet the increasing demand mentioned above[1, 2], or there are varying degrees of restrictions in pratical applications. For instance, it is known that high-leaded solder joints exhibit excellent electrical, mechanical and thermal properties, nevertheless, the utilization of leaded solders is entirely restricted, due to the toxicity of Pb element to environment and human body. Other high temperature solders such as gold-based, zinc-based, and bismuth-based solders, show higher cost and reliability problems such as oxidation and brittleness. Although transient liquid phase method can realize low temperature bonding but high temperature application, it requires very complicated procedures and long sintering time. Apart from that, the bonding interface of joints is easily affected by intermetallic compounds[3, 4]. Compared with the aforementioned materials and methods, silver(Ag) paste sintering technology has become a promising alternative for interconnection materials of power electronics due to its excellent thermal conductivity, fabulous electrical conductivity, high melting point and good mechanical reliability[5]. Moreover, it allows for a low sintering temperature and high operating temperature.

Nanoscale silver paste has been extensively studied since the appearance of low-temperature sintering technology in 1980s. Due to the large surface energy of nanoparticles, the nanoparticles need less additional driving force for sintering, which means that the nanoparticle paste can be sintered under a relatively low temperatures below 250°C[1, 6, 7]. No doubt that nano-silver paste performs well in this field, however, the cost of large-scale production of Ag nanoparticles is high. Furthermore, the surface energy of nanoparticles is so high that they are easy to agglomerate. Accordingly it is necessary to add dispersants, thinners and binders while preparing silver nanopaste, which extends the cost. It was reported that the evaporation of organic solvents is more difficult owing to the small gap between nanoparticles[8]. Therefore, research on micron-scale Ag paste draws much attention. In recent years, there has been progress in pressure assisted sintering of micron-scale Ag paste as well as pastes sintered without pressure[8, 9]. Sintering temperature, time, pressure, heating rate, will affect the properties of the sintered joints. In this work, we investigated the shear strength of sinter Ag joining under different sintering conditions with the aim of exploring new paste and searching optimal sintering parameters.

II. EXPERIMENTAL PROCEDURE

A. Preparation of micron-scale Ag paste

In this study, only micron-scale Ag flake particles, shown in Fig.1(a), were selected to make the paste. The tap density of the Ag particles was 3.0 g·cm⁻³ and the average size was 7 μm. Ethylene glycol (EG, 3A Chemistry) was used as a solvent to adjust the viscosity and fluidity of the silver paste. The weight ratio of Ag particles and solvent was set as 8:1, i.e. the Ag loading of the paste was 88.89 wt%. The paste was fabricated by evenly mixing the micron silver particles and EG solvent via stirring in vacuum mixer(MZ-8, Thinky).

B. Fabrication of die-attached structures

Ag-coated Cu plates were adopted as substrates, which was first immersed in ethanol and cleaned with an ultrasonic cleaning machine. Then the prepared Ag pastes were uniformly stencil printed onto substrates with a thickness of about 120 μm. After the printing template was removed, Ag-plated Si chips with the dimension of 3 mm × 3 mm were orderly mounted on the Ag paste to form sandwich structures. The specimens were sintered on a hotplate (NDK-1K, As one) at different temperatures (T = 200 °C, 225 °C, 250 °C, 275 °C, 300 °C) in air under pressure-assisted or pressureless condition. The sintering time ranged from 10 min to 60 min.

Talented program of Sun Yat-Sen University.

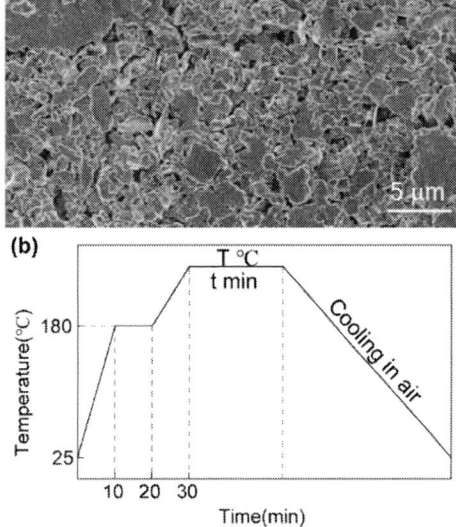

Fig. 1. (a) The SEM image of Ag particles, (b) the sintering process of the Ag paste.

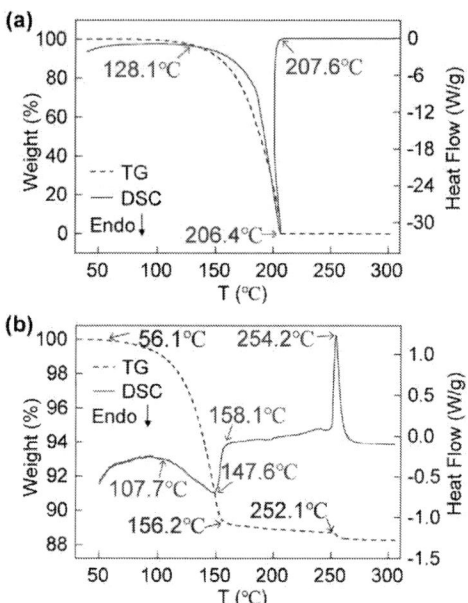

Fig. 2. TG-DSC results for (a) EG solvent and (b) the Ag paste in air.

Finally, the joints were cooled naturally. The detailed heating program is depicted in Fig.1(b).

C. Characteristic analysis of the paste and the joints

Thermal analysis of the paste was done with thermal Gravimetric and differential scanning calorimetry instrument (TGA/DSC3+, Mettler) from room temperature to 300 °C with the heating rate of 10 °C/min in air, in order to determine the weight loss and energy variations of the silver paste due to the volatilization of the organic system and sintering process.

Shear strength of sintered joints was evaluated by a multifunction bond tester (MFM1200, Try precision) to investigate the adhesion behavior. The shear speed was set as 10 μm/s. The microstructures of the silver particles and joints were recorded by Scanning Electronic Microscope (SEM, Nova nano430, FEI).

III. RESULTS AND DISCUSSION

A. Thermal analysis of the paste

TG-DSC curves of the EG solvent, and the silver paste in air are illustrated in Fig.2, individually. As shown in Fig.2(a), EG solvent started to lose weight at around 128.1 °C and end at around 206.7 °C. There was a rapid weight loss and a large endothermic peak since about 180 °C, which mean that the EG solvent evaporated quite fast near this temperature. Fig.2(b) suggested that the weight loss of the silver paste was able to isolate into 2 stages. The primary and main stage, mainly caused by the evaporation of the EG solvent, approximately started at 56.1 °C and finished at 156.2 °C. After the first loss, the weight of the Ag paste remained 89.04 %, which was somewhat bigger than the Ag loading of the paste, indicating that the EG solvent didn't evaporate totally. As the solid line shown in Fig.2(b), an endothermic peak at about 147.6 °C (ranged from 107.7 to 158.1 °C) also corresponded to the evaporation of the EG solvent. As the temperature increased, the small amount of residual EG solvent gradually evaporated completely without obvious endothermic peak. The second weight loss was very little at around 252.1 °C, corresponding

the decomposition and evaporation of the organic materials on the Ag surface, as it were with the sum of 0.72 %. And a large exothermic peak at about 254.2 °C, mainly resulted from densification of the sintered joints.

B. Influence of holding time on sintered joints

It is shown that the shear strength increased with the increase of holding time in Fig. 3(a). The average shear strength of the silver joints reached 23.5 MPa and 28.5 MPa after holding at 250 °C for 10 min and 30 min, and was as high as 32.3 MPa at the holding time of 60 min. Sinter is a process controlled by diffusion, and the degree of diffusion is time dependent. The longer the time, the more sufficiently the diffusion can proceed, and the greater the sintering degree between adjacent Ag flakes. Moreover, as the holding time increases, grain boundaries move, voids aggregate and shrink, and the densification of the silver paste is achieved. The shear strength of the solder joint corresponds to its density. With the holding time extended from 10 min to 30 min, the shear strength increased by 21.16 %, but only increased by 13.32 % when the time was extended from 30 min to 60 min, indicating that the densification rate was not consistent in each time period. To be specific, the densification was faster in earlier stage, whereas it slowed down later because the vacancy concentration difference of the necks decreased with the growth of necks.

C. Influence of sintering pressure on sintered joints

In this work, with the sintering temperature of 250 °C and the time of 30 min, we have searched the sintering condition without pressure or under small pressure of 0.28 MPa and 0.37 MPa, respectively. It can be seen that increasing pressure will result in higher shear strength in Fig.3(b). The shear strength increased by 6.36 % with the sintering pressure extended from 0 MPa to 0.28 MPa, and 27.28 % from 0 MPa to 0.37 MPa, indicating that pressure has a more obvious contribution to the densification of the sintered joints. According to the sintering therapy, the free energy tends to reduce in sintering process,

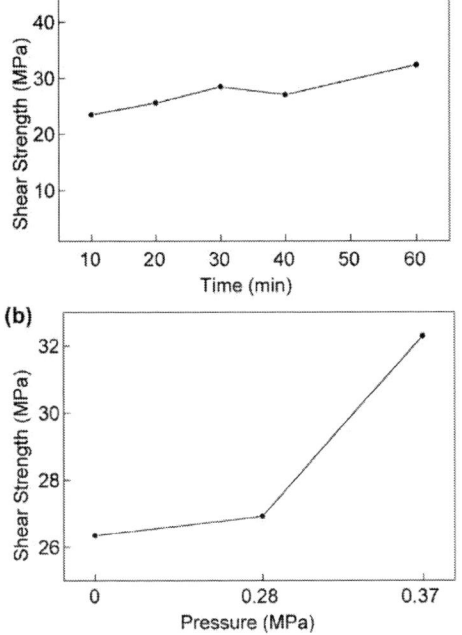

Fig. 4. The shear strength of Ag joints sintered at 250 °C, (a) different sintering time under pressure of 0.37 MPa in air, (b) different pressure in air.

and applying pressure can provide additional driving force for sintering but not for the decomposition of the organic shell of silver particles. By applying pressure, the contact between the silver particles, the Ag paste and the chip/substrate can be improved. The Ag atoms can diffuse faster between the two interfaces to form a stronger bond. But applying pressure will squeeze part of the silver solder paste out of the sandwich structure, and too much pressure is possible to crush the chips and also makes the sintering process more complicated. Moreover, now we have achieved a shear strength as high as 26.4 MPa even without pressure during the sintering process. So, the following exploration of sintering temperature was also carried out under pressureless conditions.

D. Influence of temperature on sintered joints

The sintering time was set as 30 min. The Ag pastes were sintered without pressure at 200 °C, 225 °C, 250 °C, 275 °C 300 °C, respectively. As shown in Fig.4(a), the shear strength of micron-scale silver joints rose with the increase of temperature, reaching a maximum of 34.03 MPa at 300 °C. With the sintering temperature extended from 200 °C to 250 °C, the shear strength increased by nearly 305 %, indicating that the sintering temperature had an obvious and important influence on the mechanical property of silver joints. But it only increased by around 34 % when the temperature was extended from 250 °C to 300 °C, demonstrating that much higher temperatures increase the degree of solder joint shear strength is limited, as it may cause grain coarsening.

At 200 °C, the average shear strength of sinter Ag joints has been just 5.61 MPa, demonstrating that the paste failed to bond the Ag coated Si die and the Ag coated Cu substrate. According to the sintering theory, surface diffusion dominates at low temperatures, which costs the driving force but does not contribute to densification. Conversely, the Ag joints sintered above 250 °C had an excellent shear strength over 26.4 MPa.

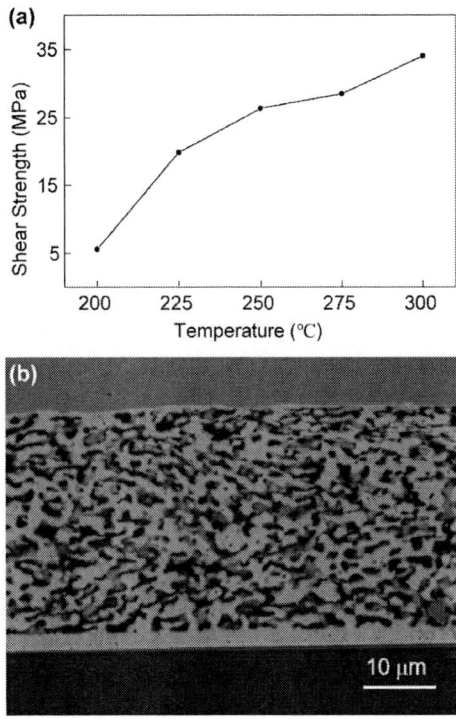

Fig. 3. (a) The average shear strength of joints at different sintering temperature in air, (b) the cross-sectional microstructure of a joint (no pressure, 250 °C, 30 min).

Fig.4(b) shows a porous cross-sectional microstructure of micron-scale Ag joint at the sintering temperature of 250 °C. The image reveals that all the Ag particles had sintered together and the Ag paste was tightly bonded to the Ag metallization layer of the die and the substrate without interface delamination. As during the relatively high temperature, the organic shell breaks down and evaporates completely, and the diffusion types are mainly grain boundary diffusion and lattice diffusion, these diffusion types are the process of densification. On the other hand, the growth of voids is realized by the diffusion of vacancies, which is essentially the movement of vacancies from high-concentration to low-concentration areas, that is, the vacancies of small voids will diffuse to large voids. At higher temperature, the growth rate of voids increase, resulting that the number of voids decrease and the scale of voids increase. However, compared with the coarsening of voids caused by elevated temperature, the voids refinement brought by the increase in the density of the structure dominates. Therefore, the number of voids in the sintered structure at high temperature is nearly the same as low temperature, but the average size is reduced.

IV. CONCLUSION

An interesting die-attach material, i.e. micron Ag particle paste was successfully fabricated to make power electronic joints. The influence of the sintering parameters, including sintering temperature, sintering time and pressure, to the mechanical property of the silver joints were researched in this work. The shear strength of micron-scale Ag joints increased with the increase of sintering time. Even with a small pressure of 0.37 MPa, the paste realized a shear strength over 30 MPa sintering at 250 °C for 30 min in air, significantly improved compared to unpressurized. And the shear strength of micron-scale silver joint was positively correlated with sintering temperature. It reached as high as 26.4 MPa after sintering at 250 °C for 30 min without pressure under atmosphere condition. This silver paste is helpful for the development of power electronic packaging.

ACKNOWLEDGMENT

The authors gratefully acknowledge the support of talented program from Sun Yat-Sen University.

REFERENCES

[1] P. Zhang, X. Jiang, P. Yuan, H. Yan and D. Yang. "Silver nanopaste: Synthesis, reinforcements and application," Int. J. Heat Mass Transf., vol. 127, pp. 1048-1069, Dec. 2018.

[2] V. R. Manikam and C. Kuan Yew. "Die attach materials for high temperature applications: A review," IEEE Trans Compon Packaging Manuf Technol, vol. 1(4), pp. 457-478, Apr. 2011.

[3] L. Sun, M. Chen, L. Zhang, P. He and L. Xie. "Recent progress in slid bonding in novel 3d-ic technologies," J. Alloys Compd., vol. 818, pp. 152825, Mar. 2020.

[4] H. L. J. Pang, K. H. Tan, X. Q. Shi and Z. P. Wang. "Microstructure and intermetallic growth effects on shear and fatigue strength of solder joints subjected to thermal cycling aging," Mater. Sci. Eng. A, vol. 307(1), pp. 42-50, Jun. 2001.

[5] S. A. Paknejad and S. H. Mannan. "Review of silver nanoparticle based die attach materials for high power/temperature applications," Microelectron Reliab, vol. 70, pp. 1-11, Mar. 2017.

[6] K. Sasaki, N. Mizumura, A. Tsuno, S. Yagci and G. Kopp. "Development of low-temperature sintering nano-silver die attach materials for bare cu application," 21st European Microelectronics and Packaging Conference, Warsaw, pp. 5, Sept. 2017.

[7] D. Wang, Y. Mei, H. Xie, K. Zhang, K. S. Siow, X. Li and G. Lu. "Roles of palladium particles in enhancing the electrochemical migration resistance of sintered nano-silver paste as a bonding material," Mater. Lett., vol. 206, pp. 1-4, Nov. 2017.

[8] S.-C. Fu, M. Zhao, H. Shan and Y. Li. "Fabrication of large-area interconnects by sintering of micron ag paste," Mater. Lett., vol. 226, pp. 26-29, Sept. 2018.

[9] C. Choe, S. Noh, C. Chen, D. Kim and K. Suganuma. "Influence of thermal exposure upon mechanical/electrical properties and microstructure of sintered micro-porous silver," Microelectron Reliab, vol. 88-90, pp. 695-700, Sept. 2018.

Warpage measurement of substrates and printed circuit boards with Shadow Moiré

Xingjia Huang
School of Automotive Engineering
Geely University of China
Chengdu, China
huangxingjia@bgu.edu.cn

Changping Ou
EFA Lab
CNSBG, Foxconn Technology Group
Shenzhen, China
efa-nsd-03@mail.foxconn.com

Shengcong Zhu
EFA Lab
CNSBG, Foxconn Technology Group
Shenzhen, China
sheng-cong.zhu@foxconn.com

Yixiu Huang
NSDI
CNSBG, Foxconn Technology Group
Shenzhen, China
david.huang@foxconn.com

Abstract—**Warpage of ball grid array substrate and printed circuit board is a common issue during reflow process due to the mismatch of coefficients of thermal expansion. With the development of the substrate becoming larger and larger, there is a higher risk of reflow defects caused by warpage. In this paper, an open soldering failure, head-in-pillow, caused by warpage is reported when ball grid array package was reflowed onto printed circuit board. Shadow moiré technique was then used for dynamic warpage analysis on both sides of ball grid array substrate and printed circuit board. The dynamic warpage measurement results indicate that A1 corner is the highest risk location for head-in-pillow failure, which matches that of actual head-in-pillow defect location very well. Design of experiment on reflow profile and stencil aperture design were conducted. Based on design of experiment results and considering the reflow process control window, medium temperature profile (U1 peak temperature of 240°C) with aperture design of the stencil 2 was chosen for manufacturing.**

Keywords—*warpage, printed circuit board, ball grid array substrate, Shadow Moiré, head-in-pillow*

I. INTRODUCTION

Ball grid array (BGA) substrates and printed circuit boards (PCBs) are getting more and more complex due to the requirement of more powerful function, higher component density, smaller line width and pitch, and marginal process window. Quality of BGA substrates and printed circuit boards (PCBs) directly impacts the yield of printed circuit board assembly (PCBA).

BGA substrate and PCB consist of organic composites and metallic circuit layers laminated together. Organic composites are composed of resin (for substrate usually bismaleimide-triazine (BT), for PCB commonly FR4) and glass fibers. While metallic circuit layers are usually copper. Coefficients of thermal expansion (CTEs) of organic composites and metal layers are different. CTE mismatch of organic composite and metal layer will lead to substrate and PCB warpage during the reflow process even though the bare PCB and BGA substrate are flat before assembly [1-6]. Substrate and PCB warpage is a serious issue, which will result in, in worst case, reflow defects such as open soldering and/or solder bridging, as shown in Fig. 1 [1]. For surface mount manufacturing supplier, when reflow defects occur due to the warpage during reflow process, it is imperative to know the location and extent

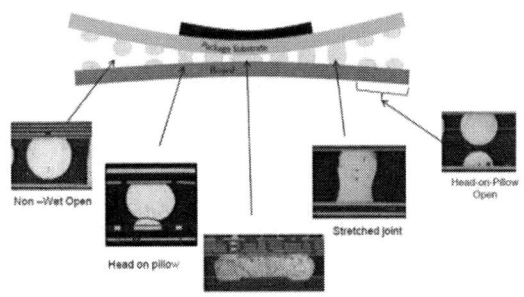

Fig. 1 BGA solder joint defects due to BGA and/or PCB warpage

of warpage so as to find out a way to mitigate and/or eliminate the reflow defects caused by warpage.

Shadow Moiré is a non-contact, full-field optical technique that uses geometric interference between a reference grating and its shadow on a sample to measure relative vertical displacement (warpage) at each pixel position in the resulting image. It requires a Ronchi-ruled grating, a white line light source at approximately 45 degrees to the grating and a camera perpendicular to the grating [7]. Its optical configuration is shown in Fig. 2 [8]. A technique, known as phase stepping, is applied to Shadow Moiré to increase measurement resolution and provide automatic ordering of the interference fringes. This technique is implemented by vertically translating the sample relative to the grating. Shadow Moiré has the capability to simulate the

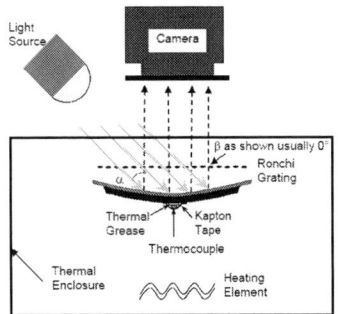

Fig. 2 Schematic of the optical configuration of Shadow Moiré

978-1-6654-1392-3/21 $31.00 © 2021 IEEE

reflow process, and then the dynamic warpage change of BGA substrate or PCB during the reflow process can be measured [7-8]. Shadow Moiré technique is getting more and more widely used to study dynamic warpage of IC package and/or PCB, solder joints defects during reflow process [3-6, 9-12].

In this paper, an open soldering failure, head-in-pillow (HIP), caused by warpage is reported when BGA package was reflowed onto PCB. X-ray inspection, and Dye and pry test were performed. Moreover, to find out potentially high risk locations of HIP defect, dynamic warpage of both BGA substrate and PCB was measured using Akrometrix TherMoiré system. Dynamic warpage measurement results indicate that BGA A1 corner is the highest risk area for HIP failure which matches that of HIP defect location well. Design of experiment (DOE) for reflow profile and stencil aperture design were then conducted. Based on DOE results and considering the reflow process control window, medium temperature profile (BGA U1 peak temperature of 240°C) with aperture design of the stencil 2 was chosen for manufacturing. With the optimized reflow profile and stencil aperture design, HIP issue was solved.

II. FAILURE ANALYSIS ON HIP ISSUE

A. Open soldering failure - HIP issue

During process validation build of a new PCBA product, reflow defects occurred. Defect locations were narrowed down to BGA solder joints through preliminary failure analysis. Dage 7500 X-ray machine was used to do X-ray inspection on BGA solder joints. Fig. 3 is an X-ray graph. HIP defects were observed, as indicated by red arrows in Fig. 3. Majority of HIP defects was located in BGA A1 corner area and there were a few of HIP defects in BGA AW1 corner area.

To validate X-ray inspection results and find out if there were HIP defects at other locations, dye and pry test was then conducted. Dye and pry results confirmed that HIP defects occurred mainly in BGA A1 corner area and a few of HIP defects were found in AW1 corner as well. Fig. 4 is a representative graph showing HIP defect at B1 solder joint of A1 corner area.

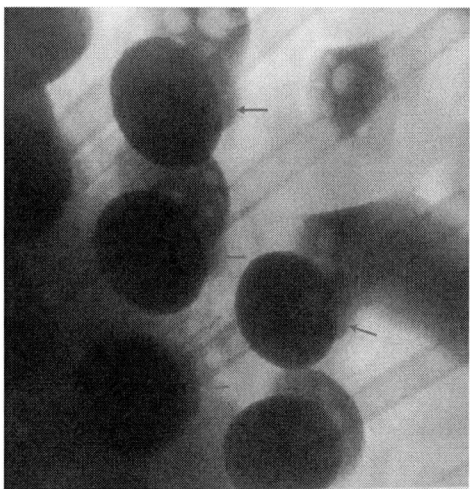

Fig. 3 HIP defects indicated by red arrows in A1 corner area

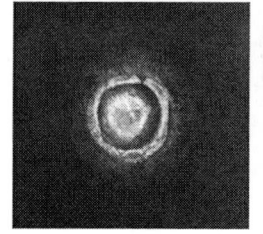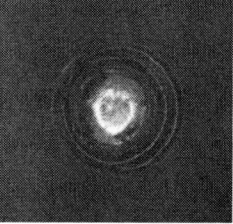

BGA side   PCB side

Fig. 4 Dye and pry results showing HIP at B1 solder joint

B. Dynamic warpage measurement with Shadow Moiré

In the present study, the dimension of BGA package is 40x40 mm. Fig. 5 is the BGA package U1 used for dynamic warpage measurement. BGA solder balls were removed mechanically before the measurement with Shadow Moiré. Fig. 6 shows the corresponding BGA U1 area on PCB side.

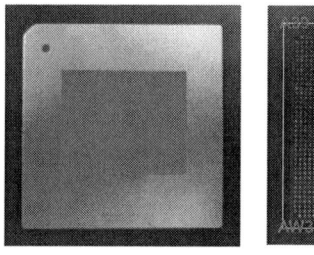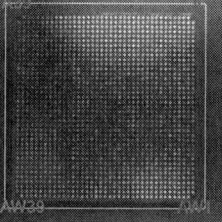

Top view   Bottom view

Fig. 5 BGA U1 for dynamic warpage measurement

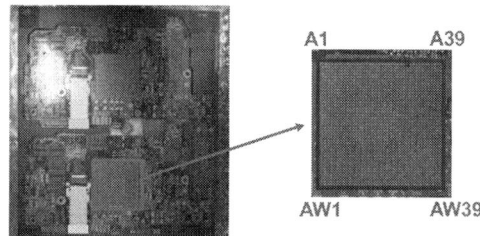

Fig. 6 Corresponding BGA U1 area on PCB side

Dynamic warpage of both BGA U1 package and PCB was measured using Akrometrix TherMoiré PS600S system in a temperature range of 30°C to 245°C to simulate the actual reflow temperature profile. The objective of dynamic warpage measurement is to try to find out potentially high risk locations of HIP defect for BGA U1 on PCB so as to mitigate and/or eliminate the HIP defect.

Fig. 7 shows 3D warpage contour plots of BGA substrate at various temperatures from Shadow Moiré measurement. It can be seen that above the melting point of solder alloy, all of four corners of BGA U1 substrate warp inside the paper. Furthermore, BGA A1 corner shows the most serious warpage and the second most serious warpage is AW1 corner.

Fig. 8 shows 3D warpage contour plots of BGA U1 area on PCB surface at various temperatures from Shadow Moiré measurement. It can be seen that above 200°C, among four corners of BGA area on PCB side, A1 corner shows the most serious warpage inside the paper, implying that A1 corner is the highest risk area of potentially having HIP defect.

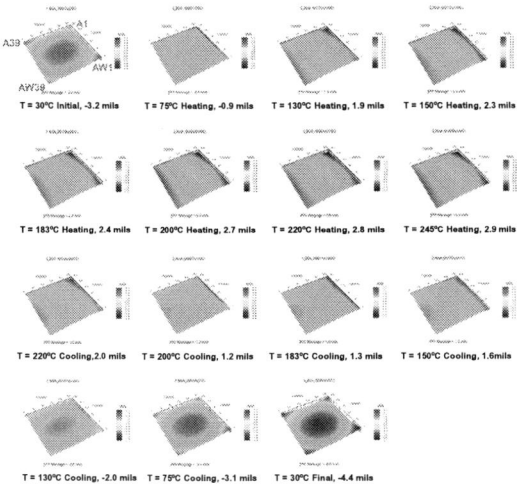

Fig. 7 3D warpage contour plots of BGA U1 substrate from Shadow Moiré measurement

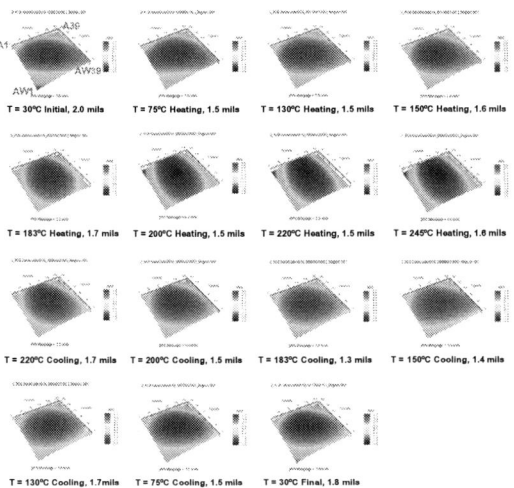

Fig. 8 3D warpage contour plots of BGA U1 area on PCB surface

Furthermore, interface analysis on BGA and PCB was performed to have a better view of dynamic warpage between dynamic surfaces of BGA and PCB at temperatures of 200°C, 220°C and 245°C, respectively. As can be seen in Fig. 9, interface analysis results show clearly that all of four BGA corners are potentially high risk locations of HIP defect, and A1 corner is the most probable location for having HIP defect during the reflow process and AW1 corner is the second one. Shadow Moiré results match that of actual HIP failure locations very well.

III. Design Of Experiment For Hip Issue

For HIP issue, common ways to mitigate and/or eliminate HIP defect are (1) to change solder paste, (2) to dip BGA solder balls into flux, (3) to change stencil design, and (4) to change the reflow profile. For a contract manufacturing supplier of surface mount technology, the solder paste is usually already specified by a customer. Then the easier way is to change the stencil aperture design and the reflow profile.

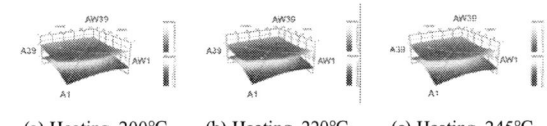

| (a) Heating, 200°C | (b) Heating, 220°C | (c) Heating, 245°C |

Fig. 9 Interface analysis on BGA and PCB. The red brown area is BGA substrate, while the blue area is PCB.

For the present study, the reflow profile is of soaping type, as shown in Fig. 10. For design of experiment (DOE), three reflow peak temperatures of 235°C, 240°C and 245°C for BGA U1 were selected. Two stencil aperture designs, stencil 1 and stencil 2, are shown in Table I.

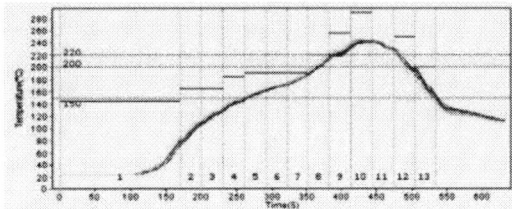

Fig. 10 Reflow profile is of soaking type

TABLE I DETAILS OF BGA U1 STENCIL APERTURE DESIGN

	Pad area	Diameter (mm)		
		PCB pad	Stencil 1	Stencil 2
BGA U1 opening	Red	0.53	0.635	0.64
	Orange	0.51	0.635	0.62
	Green	0.48	0.635	0.60
	White	0.48	0.41	0.58

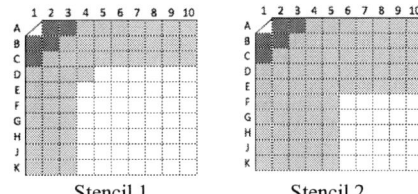

Stencil 1   Stencil 2

DOE description and results are listed in Table II. For each DOE, sample size was 20. One HIP defect was found for DOE 1, while the other three DOEs all passed. Based on DOE results and after considering the reflow process control window, medium temperature profile (BGA U1 peak temperature of 240°C) with the aperture design of the stencil 2 was chosen for manufacturing.

TABLE II DOE RESULTS

	Description	Result
DOE 1	1. Low temperature profile (U1 peak temperature ~ 235°C) 2. Stencil 1	1 HIP
DOE 2	1. Medium temperature profile (U1 peak temperature ~ 240°C) 2. Stencil 1	Pass
DOE 3	1. High temperature profile (U1 peak temperature ~ 245°C) 2. Stencil 1	Pass
DOE 4	1. Medium temperature profile (U1 peak temperature ~ 240°C) 2. Stencil 2	Pass

IV. CONCLUSIONS

In this paper, an open soldering failure, HIP defect, caused by warpage is reported when BGA package of 40x40 mm dimension was reflowed onto PCB. X-ray inspection and dye and pry test were conducted. Shadow moiré technique was then used for dynamic warpage analysis on both sides of BGA substrate and PCB. DOE for reflow profile and stencil aperture design were performed for HIP issue. Below is a summary of analytical results.

1. X-ray inspection revealed that HIP defects were located mainly in BGA A1 corner area and a few of HIP defects were in AW1 corner.

2. Dye and pry test confirmed X-ray results.

3. To find out the potentially high risk locations of HIP defect, dynamic warpage of both BGA substrate and PCB was measured using Akrometrix TherMoiré PS600S system. Dynamic warpage measurement results reveals that for BGA area on PCB side, A1 corner is the highest risk location for HIP issue. While for BGA package, all four corners show higher risk. When BGA package is reflowed onto PCB, the combination of dynamic warpage between BGA and PCB makes A1 corner area the highest risk location for HIP defect and AW1 corner is the second one, which matches that of actual HIP defect locations very well.

4. DOE for HIP issue was performed. Three reflow peak temperatures of 235°C, 240°C and 245°C for BGA U1, and two stencil aperture designs, stencil 1 and stencil 2, were selected for DOE. Based on DOE results and considering the reflow process control window, medium temperature profile (BGA U1 peak temperature of 240°C) with the aperture design of stencil 2 was chosen for manufacturing.

ACKNOWLEDGMENT

The authors would like to acknowledge the contribution and collaboration of colleagues in EFA Lab and other departments of CNSBG, Foxconn Technology Group.

REFERENCES

[1] R. Aspandiar, "FCBGA package warpage," HDP User Group Meeting, Feb. 2013.

[2] Y. Yang, "Discussion and failure analysis of PCB warpage," 2019 20th International Conference on Electronic Packaging Technology, August 2019.

[3] Yao Bin, Wang Xiaofeng, and Zou Yabing, "The study of thermally induced warpage of BGA package during reflow soldering," Proceedings of 2018 19th International Conference on Electronic Packaging Technology, pp.1411-1414, October 2018.

[4] O. Albrecht, K. Meier, and H. Wohlrabe, "Study on the effect of the warpage of electronic assemblies on their reliability," Proceedings of 2020 IEEE 8th Electronics System-Integration Technology Conference, September 2020.

[5] S. J. Oon, K. S. Tan, T. Y. Tou, S. S. Yap, C. S. Lau, and Y. T. Chin, "Warpage studies of printed circuit boards with Shadow Moiré and simulations," 2018 IEEE 38th International Electronics Manufacturing Technology Conference, October 2018.

[6] P. Geng, T. Bandorawalla, S. Cho, H. Hsiao, J. Kuchy, G. Long, R. Martinson, A. McAllister, M. Mello, K. Meyyappan, R. Williams, and L. Zhu, "Application of Shadow Moiré technique to board level manufacturing technologies," Proceedings of Electronic Components and Technology Conference, v 2006, pp.1816-1820, 2006.

[7] "Vision measurement technologies," https://akrometrix.com/products/measurement-techniques/shadow-moire/

[8] JESD22-B112A, "Package warpage measurement of surface-mount integrated circuits at elevated temperature," October 2009.

[9] M. Ying, Y. C. Chia, A. Mohtar, T. F. Yin, and S. P. Chuah, "Thermal induced warpage characterization for printed circuit boards with Shadow Moiré system," Proceedings of the Electronic Packaging Technology Conference, pp.265-270, 2006.

[10] Y. Um and J. Khim, "Review on the high temperature warpage measurement using Shadow Moiré," 2010 IEEE CPMT Symposium Japan, ICSJ10, 2010.

[11] R. W. Kulterman, W. K. Loh, H. Fu, and M. Tsuriya, "Comparison of advanced package warpage measurement metrologies," Proceedings of the IEEE/CPMT International Electronics Manufacturing Technology (IEMT) Symposium, November 2016.

[12] D. W. Wang, H.-S. Huang, S.-C. Ho, A.-H. Liu, and D.-S. Liu, "Study of warpage characteristics of molded stacked-die MCP using Shadow Moiré and micro Moiré techniques," Proceedings of Electronic Components and Technology Conference, pp.1968-1973, 2010.

Reliability Analysis of Thermal Interface Materials (TIMs) in large size FCBGA package

Yi Zheng
Shenzhen Institute of advanced
electronic materials
Shenzhen Institutes of Advanced
Technology
Shenzhen, China
yi.zheng@siat.ac.cn

Cheng Zhong
Shenzhen Institutes of Advanced
Technology
Chinese Academy of Sciences
Shenzhen, China
cheng.zhong1@siat.ac.cn

Haozhe Wang
Shenzhen Institute of advanced
electronic materials
Shenzhen Institutes of Advanced
Technology
Shenzhen, China
hz.wang@siat.ac.cn

Jibao Lu*
Shenzhen Institutes of Advanced
Technology
Chinese Academy of Sciences
Shenzhen, China
jibao.lu@siat.ac.cn

Rong Sun
Shenzhen Institutes of Advanced
Technology
Chinese Academy of Sciences
Shenzhen, China
rong.sun@siat.ac.cn

Ching-Ping Wong
School of Materials Science and
Engineering
Georgia Institute of Technology
Atlanta, USA
cp.wong@mse.gatech.edu

Abstract—In large size package, high temperature affects the stability, reliability and service life of electronic components. The thermal interface material (TIM) plays an important role of effectively enhancing the heat transfer in Flip Chip Ball Grid Array (FCBGA) package; therefore reliability of TIM is indispensable especially when the package device needs to withstand wide high and low temperature periods. In this paper, we conduct a detailed analysis of TIM's reliability in large-size FCBGA package by simulating the stress and strain of TIM layer with the finite element software ABAQUS. The stress of TIM in the package is complex and can be affected by multiple factors such as device structure, material parameters, etc. It is necessary to check the model structure and material parameters to correctly understand the stress of TIM in FCBGA. The statistics method is adopted for analyzing the influence of various parameters.

Keywords—large size package, FCBGA, TIM, stress

I. INTRODUCTION

Nowadays die size and package size of FCBGA for networking are increasing which exceeds Moore's Law [1]. With the development of electronic devices, the power density in electronic system is increasing, and the heating problem becomes more and more serious. High temperature has an important impact on stable and reliable performance of electronic components, resulting in decline of service life. For some special electronic devices, service life decreases exponentially with the increase of working temperature.

The contact between rough solid surfaces results in the existence of interface thermal resistance, which seriously affects the interface heat transfer efficiency and the heat dissipation of the electronic device. The thermal interface material increases the contact area by filling the voids of the solid contact surface, which can effectively reduce the interface thermal resistance and enhance the interface heat transfer efficiency.

The thermal interface material (TIM) is widely used in the package for heat dissipation purpose and can be defined as a material inserted between heat-producing and heat-dissipating components [2]. A wide variety of TIMs have emerged with years of development, from traditional polymer-based composite materials to phase change materials, metals, and different types of carbon structures (carbon nanotube,

graphene, graphite, etc.). Numerous researchers have studied the thermal transport through the interfaces between the TIM and the adjacent layers in the package [3][4], however, the thermo-mechanical reliability of TIM is much less discussed.

Orthogonal experimental design is a method for studying multi-factors and multi-levels. It selects the representative points based on the orthogonality to conduct experiments, which is highly efficient and fast. Orthogonal experimental design generally includes the following steps: determine the research factors and levels; make an orthogonal test table; implement the test; analyze the test results.

In this paper we give a detailed analysis of thermo-mechanical reliability of TIM in terms of the orthogonal experiment based on finite element analysis. First, we list the factors that may affect the reliability of the TIM, which are mainly divided into two categories, the structural parameters in the package and the physical parameters of different materials. It can be summarized as follows: the structural factors that may affect the reliability of TIM include lid thickness, TIM thickness, lid attach thickness, substrate thickness, package size; physical parameters that may affect the reliability of TIM include lid Young's modulus, lid CTE, substrate Young's modulus, substrate CTE, lid attach Young's modulus. In our orthogonal experiments, we comprehensively analyze these factors and expect to find the factors that have the greatest impact on the reliability of TIM.

II. MODEL DESCRIPTION

A numerical model based on finite element method is established by using ABAQUS software. Figure 1 shows the quarter symmetry finite element model of FCBGA package, and main components include substrate, die, TIM, lid, underfill, and lid glue. The original package size is 50mmx50mm, chip size is 20mmx20mm, TIM thickness is 0.12mm and lid thickness is 0.4mm. Due to the computational efficiency reason, solder bumps in underfill are simplified. As warpage and deformation of package will be observed due to the mismatch of the CTE between different components, in this study, the boundary condition of temperature varies from 165℃ to 25℃. The final number of elements used in the model is 91586 and number of nodes is 113565.

978-1-6654-1392-3/21 $31.00 © 2021 IEEE

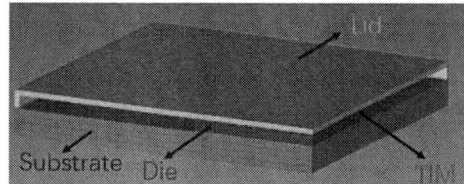

Fig. 1. Finite element model of large size FCBGA package.

III. VALIDATION OF THE FINITE ELEMENT MODELING

From finite element results as Fig. 2 shows, displacement of the corner is small, and the middle chip area is arched upward. The TIM layer stress contour during temperature change is shown in Fig. 3. Point A represents the corner point of TIM, point D represents the center point of TIM in the packaging system, obviously the maximum stress position locates near the corner of the TIM. For our finite element model, the result of TIM stress contour is corresponding to the research of C. Nelson as Fig. 4 shows [5]. In [5], the void of TIM is near the edges and corners of the die. In our model, the corners are the locations where the TIM bears the greatest stress and it is the vunerable part and prone to failure.

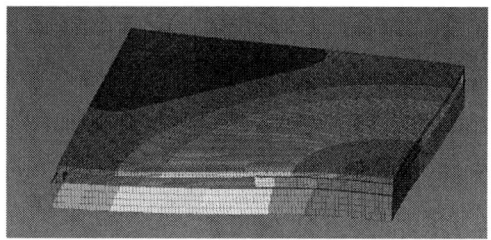

Fig. 2. Displacement of FCBGA package during temperature decreasing period.

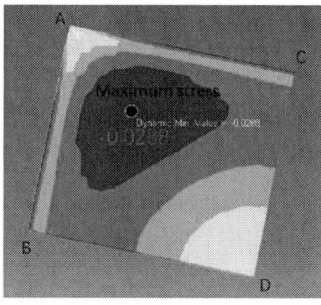

Fig. 3. Stress contour of TIM in FCBGA package.

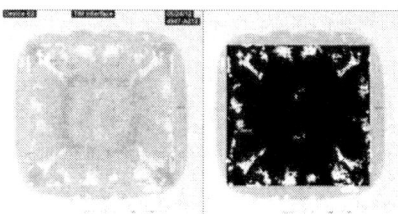

Fig. 4. TIM layer images from a Confocal ScanningAcoustic Microscope (CSAM) [5].

IV. SENSITIVITY ANALYSIS AND ORTHOGONAL EXPERIMENT FOR THE TIM RELIABILITY

A. Single factor analysis for the TIM reliability

Single factor analysis for the TIM reliability in the large-size package is conducted. The thickness of lid is increased from 0.4mm to 2mm, four points of the quarter model are analyzed.

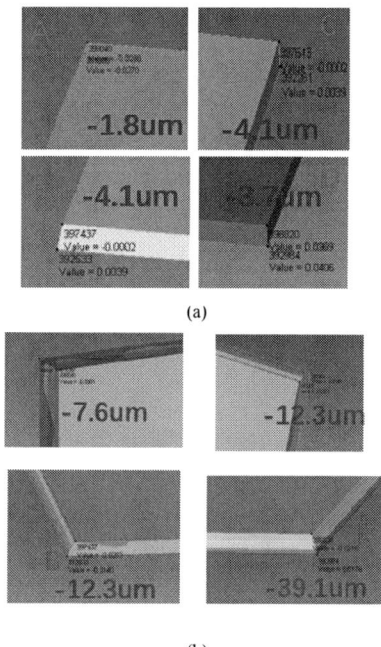

Fig. 5. (a) BLT change in 0.4mm lid thickness package, (b) BLT change in 2mm lid thickness package.

Fig. 5 (a) and (b) show variation of bond line thickness (BLT) in 0.4 mm lid thickness package and 2mm lid thickness package in temperature decreasing process. For 0.4 mm lid thickness package, the BLT change is -1.8 μm at corner point A, the negative sign means that the TIM layer is compressed and the thickness becomes smaller. For 2 mm lid thickness package, the BLT change is -7.6 μm at corner point A. For center point D, the BLT change is -3.7 μm for 0.4 mm lid thickness package, -39.1μm for 2mm lid thickness package. Except corner point A and center point D, the two points B and C also be analyzed which located at the center of the TIM side edge. For point B and C, the BLT change is -4.1 μm for 0.4 mm lid thickness package, -12.3μm for 2mm lid thickness package.

Fig. 6 (a) and (b) show the stress contour in 0.4mm lid thickness package and 2mm lid thickness package in temperature decreasing process. For 0.4 mm lid thickness package, the TIM stress is -0.0197MPa at corner point A, the negative sign also means that the TIM layer stress direction. For 2 mm lid thickness package, the TIM stress is 0.047MPa at corner point A. For center point D, the TIM stress is -0.0109MPa for 0.4mm lid thickness package, -0.3455MPa for 2 mm lid thickness package. The other two points B and C, the TIM stress is -0.0249MPa for 0.4 mm lid thickness package, -0.0953MPa for 2 mm lid thickness package.

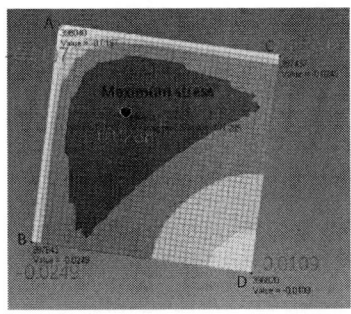

(a)

(b)

Fig. 6. (a) stress contour in 0.4mm lid thickness package, (b) stress contour in 2mm lid thickness package.

From the result of Fig. 5 and Fig. 6, both BLT and TIM stress change as the lid thickens from 0.4 to 2mm. Lid thickness has observable variation in both stress and BLT. When lid thickens, the TIM layer is subjected to greater compressive stress, especially in the middle part. The reason can be analyzed as follows: during the process of temperature change, package arches upward, however, the upper side of the package is covered by the thickened lid and it is difficult to deform, therefore the middle part of TIM bears greater compression stress.

B. Orthogonal experiment for the TIM reliability

From single factor analysis, the lid thickness affect both BLT and TIM stress, the reliability of TIM is reduced as lid thickens. In actual situations, TIM reliability is affected by more than one factor, therefore it is meaningful to find the TIM reliability when multi factors change together.

Orthogonal experimental design is a method which uses orthogonal table to arrange and analyze multi-factors. Through a few representative tests, we can find out the influence of each factor on TIM reliability, determine the primary and secondary factors, and find out the best parameter combination for TIM reliability.

Based on FCBGA structure and material of different components, the factors may affect TIM reliability are summarized as lid thickness, TIM thickness, lid attach thickness, substrate thickness, package size, lid Young's Modulus, lid CTE, substrate Young's Modulus, substrate CTE, lid attach Young's Modulus. Table I shows the structural factors for TIM reliability analysis and Table II shows the physical parameter factors for TIM reliability analysis.

TABLE I.   STRUCTURAL FACTORS FOR TIM RELIABILITY ANALYSIS [6][7][8][9]

Model Configuration	Baseline (mm)	Range (mm)
Lid thickness	2	1-3
TIM thickness	0.1	0.03-0.15
Lid attach thickness	0.05	0.03-0.07
Substrate thickness	2	1-3
Package size	50*50	25*25-75*75

TABLE II.   PHYCICAL PARAMETER FACTORS FOR TIM RELIABILITY ANALYSIS [6][7][8][9]

Model Configuration	Baseline	Range
Lid Young's modulus	188000MPa	20000-188000MPa
Lid CTE	8 ppm/°C	8-17 ppm/°C
Substrate Young's Modulus	24500MPa	24500-270000MPa
Substrate CTE	16 ppm/°C	5.5-16 ppm/°C
Lid attach Young's Modulus	600MPa	600-2000MPa

We have designed an orthogonal experiment with ten factors and two levels. Finite element models which based on orthogonal table are established, results are compared and analyzed to find out the parameter that has the greatest impact on TIM performance.

Figure 7 shows TIM stress contour from part of the orthogonal experiment. From Fig. 7, it can be deduced that TIM stress is considerable complex if multiple factors change together. TIM stress can be influenced by both structural factors and physical parameter factors.

Figure 8 shows the main effect diagram for maximum stress. The main effect diagram shows how each factor affects the response characteristics (maximum stress). It is possible to find which factor has most significant influence on TIM maximum stress. As Fig. 8 shows, TIM thickness, package size and lid Young's modulus have non-negligible effect for maximum stress of TIM layer. Lid attach thickness, substrate thickness, substrate CTE, lid attach Young's modulus have less impact comparing with other factors.

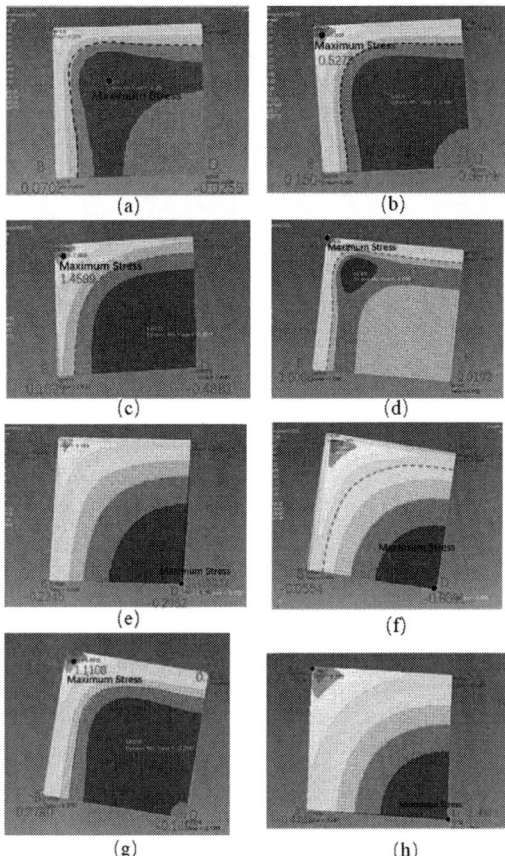

Fig. 7. TIM stress contour for multiple factor DOE experiments.

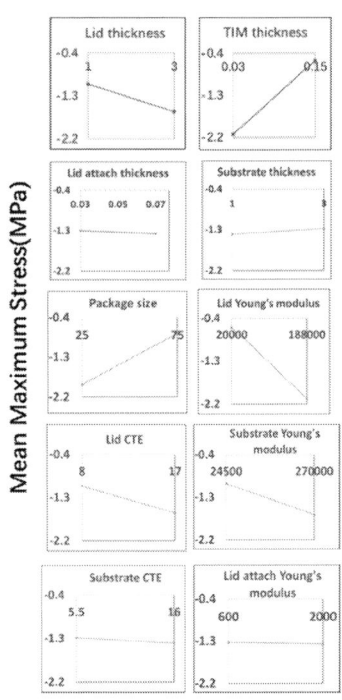

Fig. 8. Main effect figure of multiple factor DOE experiments.

V. CONCLUSION

In this study, we conduct sensitivity analysis and orthogonal experiment for the TIM reliability. From the single factor analysis, the stress and TIM BLT variation have an obvious proportional relationship. If the thickness of the lid is increased from 0.4mm to 2mm, both stress and TIM BLT variation increases at four point selected in finite element model. In the middle point of TIM, the stress changes from -0.0109MPa to -0.3455MPa and BLT change from -3.7μm to -39.1μm. In the corner point of TIM, the stress changes from -0.0197MPa to 0.047MPa and the BLT change from -1.8μm to -7.6μm. The TIM layer is subjected to greater stress and BLT variation in the middle part. Based on the orthogonal experiment, we identify that TIM thickness, package size, and the Young's modulus of lid are main factors affecting TIM maximum stress which particular attention should be paid.

ACKNOWLEDGMENT

The work was supported by the National Natural Science Foundation of China (No. 52003289), the National Key R&D Project from Minister of Science and Technology of China (No. 2017YFB0406000), the Youth Innovation Promotion Association CAS (No. 2021363), the SIAT Innovation Program for Excellent Young Researchers (No. 201803).

REFERENCES

[1] F. Tung, M. Lu, A. Lan, S. Pan, "Assembly Challenges for 75x75mm Large Body FCBGA with Emerging High Thermal Interface Material (TIM)," IEEE 67th Electronic Components and Technology Conference, 2017, pp 130-135.

[2] R. Prasher, "Thermal interface materials: Historical perspective, status, and future directions," Proceedings of the Ieee, 2006, pp. 1571-1589.

[3] R. Kenney; V. Oruganti; A. Ortega; D. Nguyen, M. Brooks, "Experiments on the thermal resistance of deformable thermal interface materials under mechanical loading," 2017 33rd Thermal Measurement, Modeling & Management Symposium (SEMI-THERM), 2017.

[4] K. Matsumoto, S. Ibaraki, K. Sueoka, K. Sakuma, H. Kikuchi, Y. Orii, F. Yamada, "Experimental thermal resistance evaluation of a three-dimensional (3D) chip stack," 27th Annual IEEE Semiconductor Thermal Measurement and Management Symposium, 2011.

[5] C. Nelson, J. Galloway, C. Henry, W. Kelley, "Thermal Performance of TIMs during Compressive and Tensile Stress States," 33rd SEMI-THERM Symposium, 2017.

[6] G. Xu, B. Guenin, M. Vogel, "Extension of air cooling for high power processors", The Ninth Intersociety Conference on Thermal and Thermomechanical Phenomena In Electronic Systems, 2004.

[7] X. Dai, N. Pan;A. Castro, J. Culler, M. Hussain, R. Lewis, T. Michalka "High I/O Glass Ceramic Package Pb-free BGA Interconnect Reliability", Proceedings Electronic Components and Technology, 2005.

[8] M. Hsieh, C. Lee, L. Hung, V. Wang, H. Perng, "Thermo-mechanical stress analysis and optimization for 28nm extreme low-k large die fcBGA," 6th International Microsystems, Packaging, Assembly and Circuits Technology Conference (IMPACT), 2011.

[9] W. Dauksher, J. Lau "A Finite-Element-Based Solder-Joint Fatigue-Life Prediction Methodology for Sn–Ag–Cu Ball-Grid-Array Packages," IEEE Transactions on Device and Materials Reliability, 2009.

Study on Warpage after Post Solidifying of Ultra-thin Fingerprint Package Products

Ning Sun
Huatian Technology (Xi'an) Co., LTD.,
Xi'an, China
Ning.Sun@ht-tech.com

Qiang Wei
Huatian Technology (Xi'an) Co., LTD.,
Xi'an, China
Qiang.Wei2@ht-tech.com

Weidong Liu
Huatian Technology (Xi'an) Co., LTD.,
Xi'an, China
Wade.liu@ht-tech.com

Xiaojian Ma
Huatian Technology (Xi'an) Co., LTD.,
Xi'an, China
Xiaojian.Ma@ht-tech.com

Huan Yang
Huatian Technology (Xi'an) Co., LTD.,
Xi'an, China
Huan.Yang@ht-tech.com

Ting Cao
Huatian Technology (Xi'an) Co., LTD.,
Xi'an, China
Ting.Cao@ht-tech.com

Fang Qu
Huatian Technology (Xi'an) Co., LTD.,
Xi'an, China
Fang.Qu@ht-tech.com

Abstract—With the decrease of mobile electronic product volume and weight, fingerprint packaging products used for unlocking need thinner structure. Warpage may occur in ultra-thin plastic seal products which seriously affects the producibility of the product. Because of the difference of thermal expansion coefficient of each material in fingerprint packaging products, warping deformation is easy to occur after curing. This paper studies the key factors (material characteristics, product structure, etc.) which affect the warpage of fingerprint products. For a fingerprint packaging product, ANSYS is used Simulation analysis software, the material parameters and product structure are simulated and optimized, and the experimental results show that the combination of simulation and experiment can effectively control and optimize the warpage of the package, which is of great significance to ensure the production of fingerprint products.

Keywords—Fingerprint Packaging Products, ANSYS, Warpage

I. INTRODUCTION

In recent years, electronic products tend to be thinner and miniaturized, and the packaging technology of electronic devices is the key link restricting the development of integrated circuits[1]. However, in the process of manufacturing and using electronic devices, they are often subjected to multiple large thermal loads (such as changes in thermal loads caused by welding, ambient temperature and power heat consumption, etc.). The changes in temperature will cause thermal expansion and cold contraction between different materials that constitute the devices corresponding to their thermal expansion coefficients. But if this heat bilges cold shrink phenomenon due to external constraints, or internal deformation coordination and not free to happen, would be the additional stress in the device [2], it will cause additional stress in the device, which will cause producibility problems, such as packaging warpage caused by mismatch of thermal expansion coefficient of packaging materials, unreasonable packaging structure parameters and curing shrinkage of packaging materials.

In this paper, taking an ultra-thin fingerprint plastic packaging device as an example, using ANSYS simulation analysis software, the warpage phenomenon of ultra-thin fingerprint product in the packaging process is studied from the aspects of material characteristics, chip thickness and substrate thickness.

II. PRODUCT FAILURE DESCRIPTION

The warpage of an ultra-thin fingerprint packaging product after post curing exceeds the warpage upper limit of the subsequent process, as shown in Fig.1. Warpage exceeding the standard is mainly caused by CTE mismatch of packaging materials, unreasonable packaging structure parameters, curing shrinkage of plastic packaging materials and other factors. After measurement, the warpage of failure fingerprint circuit reaches 25 mm (greater than the allowable limit of conventional warpage of 5 mm). Therefore, the optimization of this fingerprint packaging product mainly focuses on warpage control.

Fig. 1. Measurement warpage of fingerprint products

III. FINITE ELEMENT SIMULATION OF WARPAGE DEFORMATION

A. Geometric Model

According to the structure of fingerprint circuit in the assembly process, the simulation model is established. The structure diagram of the simulation model is shown in Fig.2, which mainly includes chip, adhesive layer, plastic sealant and substrate.

978-1-6654-1392-3/21 $31.00 © 2021 IEEE

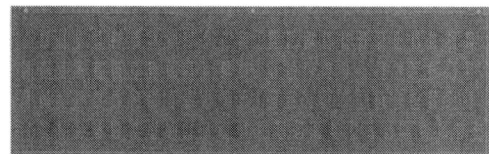

Fig. 2. Geometry model of fingerprint product

B. Material Properties

The material properties of the model are shown in Table I.

TABLE I.　　MATERIAL PROPERTIES

	Material Properties			
	Young's modulus (MPa)	Poisson's ratio	Tg (°C)	CTE (ppm/°C)
Die	131000	0.28	/	3.6
DAF	2299	0.3	50	53.8/206
Solder mask1	2400	0.29	100	60/130
Solder mask2	9000	0.29	155	17.5/75
Copper	119000	0.33	/	18
Core 1	27000	0.2	255	10
Core 2	34000	0.2	300	3
EMC	24000	0.26	140	9/35

Among them, the thickness of Core is 100μm, the thickness of copper is 15-20μm, and the thickness of the solder mask is about 30μm. The thickness of the wiring layer remains basically unchanged, the thickness of the Core layer can be adjusted in the range of 60~150μm, and the thickness of the solder mask can be adjusted in the range of 20-30μm.

C. Finite Element Simulation

The equivalent model is used to simulate and analyze the failure circuit[3]. The results show that the warpage of the substrate after curing is 19.17mm, which is in good agreement with the warpage of the failure circuit (25mm), thus verifying the effectiveness of the simulation model. The warpage distribution of the substrate obtained by simulation is shown in Fig.3.

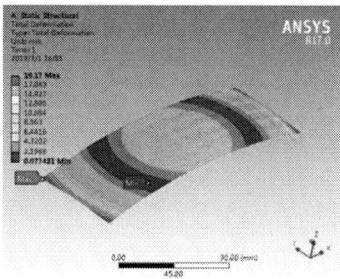

Fig. 3. Simulation results of fingerprint product warpage

According to the existing research results, the change of substrate to chip thickness ratio has a great influence on the overall warpage [4]. Therefore, the subsequent analysis mainly considers the influence of substrate material and substrate thickness.

IV. ANALYSIS OF INFLUENCING FACTORS

In order to improve the warpage of the substrate, the finite element simulation of different substrate cores and solder resist layers was carried out, and the warpage values of different substrate materials after curing were obtained. Reducing the CTE of the substrate core and solder mask can improve the warpage, but it can not meet the requirement of less than 5mm warpage. The warpage values of different substrate materials after curing are shown in Table II.

TABLE II.　　WARPAGE SIMULATION RESULTS OF DIFFERENT SUBSTRATE MATERIALS

	Simulation Results			
	Core type	Solder mask type	Strip Warpage /mm	conclusion
Core thickness 100μm	Core 1	SM 1	19.17	unqualified
	Core 1	SM 2	5.18	qualified

The results show that, in this case, the smaller the CTE of the material, the smaller the warpage value of the substrate after post-curing. According to the above analysis results, considering the shortage of materials in the board factory, it is further analyzed that when Core 2 is selected for the core layer of the substrate, SM 1 is selected for the solder mask, and the core thickness is 60μm, 100μm and 150μm, the core thickness of the substrate will affect the circuit warpage. The results are shown in Table III.

TABLE III.　　WARPAGE SIMULATION RESULTS OF DIFFERENT SUBSTRATE MATERIALS

	Simulation Results		
	Core thichness/μm	Strip Warpage /mm	conclusion
Core 2, SM 1	60	14.64	unqualified
	100	9.12	unqualified
	150	3.80	qualified

It can be seen from table 2 that the smaller the CTE of the core layer and solder mask layer of the substrate, the smaller the warpage of the circuit, that is, the smaller the CTE mismatch between the substrate and the chip and the plastic packaging material, which helps to reduce the packaging warpage; At the same time, it can be seen from table 3 that when the CTE of the core layer remains unchanged, the thickness of the core layer increases from 60 μm to 150 μm. It helps to reduce package warpage. Based on the above analysis results, for the ultra-thin fingerprint packaging products, core 2 is selected for the core layer of the substrate and SM1 is selected for the solder mask layer. The simulation results of the final optimization scheme are shown in Fig.4, and the experimental verification results of the final optimization scheme are shown in Fig.5. The warpage problem of the ultra-thin fingerprint packaging products in the packaging process is successfully solved.

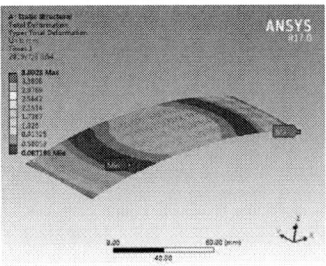

Fig. 4. Simulation results of the final optimization plan

Fig. 5. Experimental results of the final optimization plan(Warpage 4mm)

V. CONCLUSION

In this paper, the key factors affecting the warpage of ultra-thin fingerprint packaging products in the packaging process are analyzed. The results show that the warpage of ultra-thin fingerprint products is mainly affected by the thickness of the core layer and solder mask layer of the substrate and CTE. The optimization design method combining simulation with experiment can effectively optimize and control the warpage of packaging.

REFERENCES

[1] GAO Nayan, WANG Jianfeng, CHEN Bo, et at. The effect of the material and size of the flip -chip plastic package on warping [EB/OL]. [2019-07-20].

[2] Cho K, Jeon I. Numerical analysis of the warpage problem in TSOP[J] . Microelectronics Reliability, 2004, 44:621 -626J. Clerk Maxwell, A Treatise on Electricity and Magnetism, 3rd ed., vol. 2. Oxford: Clarendon, 1892, pp.68–73.

[3] Cho K, Jeon I. Numerical analysis of the warpage problem in TSOP[J]. Microelectronics Reliability, 2004,pp.621-626.

[4] Sawada Y, Harda K, Fujioka H. Study of package warp behavior for high-performance flip-chip. [J]. Microelectronics Reliability, 2003,pp. 465-471.

Influence of IMC morphology on fatigue stress, strain and life of solder layer between SiC chip and DBC substrate in IGBT under thermal cycling

Guang Yang
School of Materials Science and Engineering
Huazhong University of Science and Technology
Wuhan, China
M202070919@hust.edu.cn

*Fengshun Wu
School of Materials Science and Engineering
Huazhong University of Science and Technology
Wuhan, China
fengshunwu@hust.edu.cn

Longzao Zhou
School of Materials Science and Engineering
Huazhong University of Science and Technology
Wuhan, China
lzzhou@hust.edu.cn

Xinghe Luan
School of Materials Science and Engineering
Huazhong University of Science and Technology
Wuhan, China
D201980285@hust.edu.cn

Xinrui Zou
School of Materials Science and Engineering
Huazhong University of Science and Technology
Wuhan, China
M202070914@hust.edu.cn

Hui liu
School of Materials Science and Engineering
Huazhong University of Science and Technology
Wuhan, China
zxhuiliu@qq.com

Yang Wan
Hubei Key Laboratory of Micro/Nanocrystals Processing Technology
Suizhou, China
wy@sztkd.com

Xiaowei Zhang
Hubei Key Laboratory of Micro/Nanocrystals Processing Technology
Suizhou, China
zxw@sztkd.com

Bin Wang
Hubei Key Laboratory of Micro/Nanocrystals Processing Technology
Suizhou, China
wb@sztkd.com

Abstract—The influence of IMC morphology between solder layer and DBC substrate on fatigue stress, strain, and life of the solder in IGBT module was studied by ANSYS workbench. The IGBT model (SiC chip–SAC305–DBC (Cu/Al_2O_3/Cu)) for FEM was built. Under thermal cycling, both the equivalent stress and plastic strain of the solder are concentrated near the corner of the interface of solder/substrate. Then the sub-models with scallop-like IMCs (Cu_6Sn_5) built with sine function as profile were located at the corner. In the sub-model, the stress and plastic strain concentration of solder occur in the recessed solder/IMC interface caused by scallop-like IMCs. The model based on plastic strain energy for the prediction of fatigue life was used to analyze the influence of different IMC roughness and thickness on the fatigue life of the solder layer. It is found that both the decrease in IMC roughness and the increase in IMC thickness will lead to the shortening of the fatigue life of the solder layer.

Keywords—SiC based IGBT, IMC morphology, thermal fatigue simulation

I. INTRODUCTION

The power device-insulated gate bipolar transistor module (IGBT) has been widely used in many fields, such as rail transit, smart grid, electric vehicles, new energy, high-power radar, lasers and so on. And SiC based IGBT (SiC-IGBT) is a SiC chip-solder-substrate sandwich structure that has the advantages of low conduction loss, high switching frequency, and high-temperature resistance [1], so its development is very rapid. But the mismatches of the coefficient of thermal expansion (CTE) and Young's modulus among SiC, solder

The National Natural Science Foundation of China (NSFC No. 62074062) and Hubei Province's key research and development program "High-performance, ultra-high frequency quartz crystal device (VCXO) manufacturing process research and development (2020BAA4).

layer, and substrate lead to the reliability problems of the SiC-IGBT interconnection in service. The intermetallic compounds (IMCs) growing between the solder and substrate are the key to a reliable connection between substrate and solder. However, they may concentrate the thermal stress and strain, which would accelerate the solder layer's fatigue failure [2].

The IMCs grow and their morphologies change during soldering and service. At the SAC305/Cu interface, the hemispherical Cu_6Sn_5 grows at first. Then as the hemispherical Cu_6Sn_5 grows, it will gradually merge and become scallop-like [3]. The morphology of Cu_6Sn_5 has a certain influence on the plastic stress and strain of the SAC305. Wolfgang H. Müller built a two-dimensional idealized scallop-like IMC model and the simulation results by sub-model method showed that the increase in the curvature of the tip of IMC scallops would lead to an increase in stress concentration of the solder [4]. On its basis, Wang Liquan explored the change of Cu_6Sn_5 morphology during the aging process, found that the morphology of the IMCs changes with aging time increases, leading to the stress and strain of the solder layer increasing [5].

There are articles that mainly study the influence of IMCs on stress and strain of the solder layer. And few deeply study the influence of IMC morphology on the fatigue life of IGBT modules through the plastic strain energy method. Therefore, in this paper, the IGBT model (SiC chip-SAC305-DBC (Cu/Al_2O_3/Cu)) for the Finite Element Method (FEM) is built. Between the solder layer and DBC substrate, sub-model method is used by ANSYS Workbench to study the effect of scallop-like IMC (Cu_6Sn_5) morphology on the fatigue stress and strain of the solder (SAC305) in the IGBT module under thermal cycling. And the IMCs between SiC and solder are ignored because the metals (Ni, Au) coated on SiC are pretty thin. The change of scallop-like IMC morphology is simplified as the change of surface roughness and thickness. The plastic strain energy, which is believed to determine the occurrence and propagation of cracks [6], is used to predict the fatigue life of solder layer.

II. MODELING

A. Coarse model (Overall model)

The overall model (1/4) of SiC-IGBT is shown in Fig. 1. The dimensions of the model are as follows: SiC chip is a cube with a size of 10 mm×10 mm×0.1 mm; the thickness of SAC305 is 0.08 mm; Length, width, and height of DBC are 30 mm, 30 mm, and 1.235 mm, in which the heights of Cu are 0.3mm and Al_2O_3 is 0.635 mm. Thermal cycling load is applied on all bodies. The temperature changes from -55 °C to 150 °C, the rate of change is 13.66 °C/min, and the holding time of the highest/lowest temperature is 15min (As shown in Fig. 2.). Fatigue failure is most likely to occur at the most severely stressed position in the solder layer of the coarse model and a sub-model will be built at this position for further local analysis. The result of the coarse model is used as the load on the sub-model. Previous studies by scholars have shown that the stress and strain concentration generally occur at the corners of the solder layer where is most prone to fatigue failure [7].

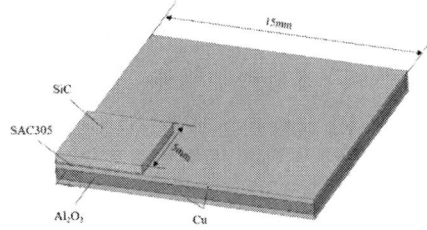

Fig. 1. The plot of 1/4 IGBT model-coarse model

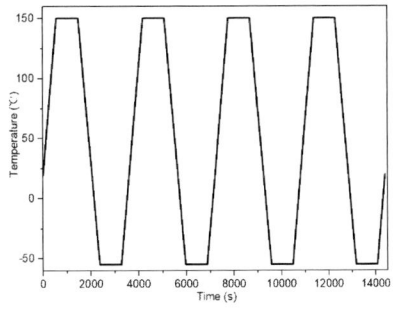

Fig. 2. The curve of temperature versus time on IGBT model

B. Sub-models

The sub-model method is based on the Saint-Venant principle, also known as the cutting boundary displacement method [5]. As shown in Fig. 3, the sub-model can be understood as cut out from the coarse model. In the sub-model, IMCs with scallop-like morphology are built for studying the influence of morphology on the stress and strain of solder. The scallop-like IMCs in the sub-model are shown in Fig. 4. According to the shape of the actual IMCs [3], a sine function is used to imitate it. The amplitude of the sine function is the roughness of the IMC layer. And the thickness of the IMC layer is the overall average thickness: the offset of the sin

function from the bottom surface. Displacement boundary conditions obtained from FEM analysis of the coarse model is implemented on sub-model for detailed local analysis.

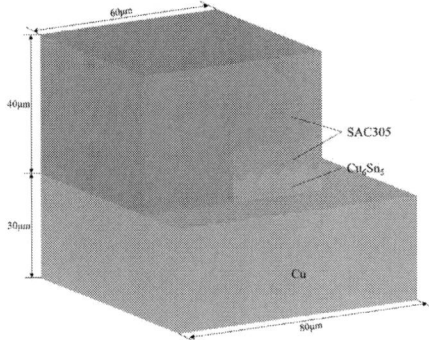

Fig. 3. The diagram of the geometry and size of the sub-model

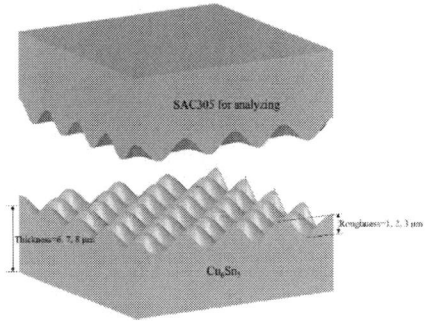

Fig. 4. The diagram of IMC mophorogy represented by sine function

C. Material models

In this paper, five types of materials are included in the SiC-IGBT module for FEM. They are SiC, SAC305, Cu, Al_2O_3, and Cu_6Sn_5 respectively. Among these materials, SiC, Cu, Al_2O_3, Cu_6Sn_5 exhibit elastic performance, their material properties are shown in Table I. As a visco-plastic material, SAC305 could be described by Anand constitutive equations [8]:

$$\dot{\varepsilon}_p = A \exp\left(-\frac{Q}{RT}\right)\left[\sinh\left(\xi \frac{\sigma}{s}\right)\right]^{\frac{1}{m}} \quad (1)$$

$$\dot{s} = \left[h_0\left(|B|\right)^a \frac{B}{|B|}\right]\dot{\varepsilon}_p \quad (2)$$

$$B = 1 - \frac{s}{s^*} \quad (3)$$

$$s^* = \hat{s}\left[\frac{\dot{\varepsilon}_p}{A}\exp\left(-\frac{Q}{RT}\right)\right]^n \quad (4)$$

Where $\dot{\varepsilon}_p$ is the inelastic strain rate, σ is equivalent stress for a steady plastic flow, s is the deformation resistance, s^* is the saturation value of the deformation resistance, and the meanings and specific values of other related parameters are shown in Table II. Equation (1) is the flow equation and (2) is the evolution equation of s.

978-1-6654-1392-3/21 $31.00 © 2021 IEEE   675

TABLE I.   MATERIAL PROPERTIES USED FOR SIMULATION

Material	Material Properties			
	Density (kg/m³)	CTE (×10⁻⁶k⁻¹)	Young's modulus (GPa)	Poisson's Ratio
Al_2O_3	3900	6.8	380	0.27
Cu	8300	18.0	110	0.34
SiC	3210	4.0	420	0.14
Cu_6Sn_5	7890	19.8	160	0.30
SAC305	7404	21.8	51	0.36

TABLE II.   THE PARAMETERS AND MEANINGS OF ANAND VISCO-PLASTIC MODEL OF SAC305 [8]

Model Parameters	Content
Viscoplasticity coefficient A (s⁻¹)	5.87×10^6
Activation energy Q (J/mol)	62022
Multiplier of stress ξ	2
Strain rate sensitivity of stress m	0.0942
Hardening constant h_0 (MPa)	9350
Strain rate sensitivity of hardening a	1.5
Initial deformation resistance s_0 (MPa)	45.9
Coefficient for deformation resistance saturation $\hat{s}$ (MPa)	58.3
Strain rate sensitivity of saturation (deformation resistance) n	0.015

D. Fatigue life prediction model

Manson-Coffin model which is based on plastic strain amplitude and Darveaux model which is based on strain energy are widely used to evaluate the solder's fatigue life. It is mainly caused by the growth of cracks, and strain energy is the driving force for the initiation and grow of cracks. Qin Hongbo improved the Darveaux model and believed that the growth rate of fatigue cracks should be a surface rate rather than a line rate. The model could be described as bellow [9]:

$$N_f = N_0 + N_p \quad (5)$$

Where N_f is the solder's fatigue life, N_0 is the number of cycles of cracks initiation, N_p is the number of cycles of fatigue cracks growth. Initial cracks generally grow at the location where the strain energy is most concentrated. So N_0 is calculated from the maximum strain energy density ΔW_{max}:

$$N_0 = C_1 \left(\Delta W_{max} \right)^{C_2} \quad (6)$$

Where ΔW_{max} is the maximum plastic strain energy density accumulated in a cycle. And C_1, C_2 are constants related to material properties such as strength and strain hardening.

When cracks grow, it leads to the fracture of the solder layer, a fracture surface forms. And cracks are prone to grow at the region where strain energy is concentrated. Therefore, N_p is related to the cross-sectional area of the solder layer A_0 and the average strain energy density per cycle (ΔW_{ave}) of the strain energy concentrated region:

$$N_p = \frac{A_0}{C_3} \left(\Delta W_{ave} \right)^{-C_4} \quad (7)$$

Where C_3, C_4 are constants related to material properties. The value of C_1, C_2, C_3, and C_4 of SAC305 are shown in table III. To calculate ΔW_{ave} in ANSYS, the volume weighted average (VWA) method is used:

$$\Delta W_{ave} = \frac{\sum_i \Delta W_e^i V_e^i}{\sum_i V_e^i} \quad (8)$$

Where ΔW_e^i is the accumulated strain energy density in one cycle in element i, and V_e^i is the volume of element i. Through (5)-(8), the fatigue life could be calculated.

TABLE III.   CRACK GROWTH CONSTANTS OF C_1 TO C_4 [9]

Constants	C_1	C_2	C_3	C_4
Value	8.850×10^5	-0.615	1.959×10^{-13}	0.667

III. RESULTS ANALYSIS AND DISCUSSION

According to the magnitude of the stress and strain of the solder layer in the coarse model, the location of the sub-model is determined. And the deformation of the coarse model is implemented as the displacement condition of the sub-model. By establishing sub-models with different IMC morphologies, the influence of IMC morphology on the fatigue failure of the solder layer was studied.

A. Results of coarse model

The curves of the maximum equivalent stress and plastic strain in solder changing versus time is shown in Fig. 5. It shows that the equivalent stress of the solder hardly changes in the four cycles, and the equivalent plastic strain of the solder gradually accumulates. The variation range of equivalent plastic strain in one cycle begins to stabilize after the third thermal cycling.

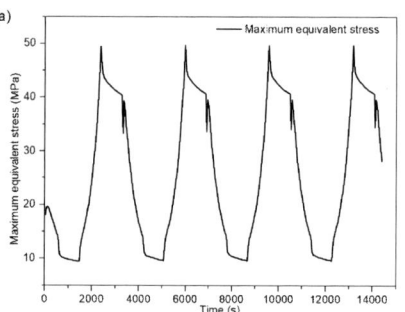

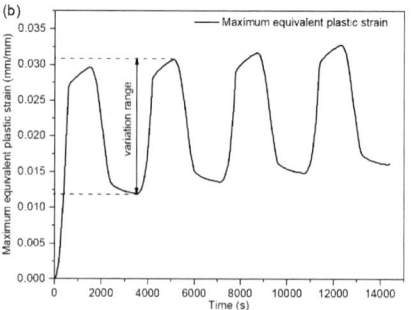

Fig. 5.   The curves of the maximum equivalent stress (a) and maximum equivalent plastic strain (b) of solder versus time

In Fig. 6, this is the equivalent stress and plastic strain of the solder layer in coarse model after 4 cycles. Both the solder's equivalent stress and plastic strain are concentrated at corner of the substrate side which makes the strain energy the biggest, where the cracks are the most prone to initiate. And this is the location the sub-model would be built.

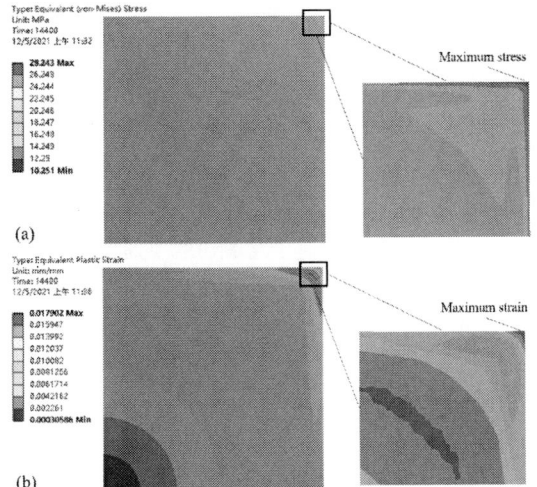

(a)

(b)

Fig. 6. Plots of equivalent stress (a) and equivalent plastic strain (b) of solder layer after 4 cycles

B. Stress and plastic strain analysis of solder in sub-model

Since the morphology of IMC is scallop-like, which would influence the stress and strain distribution on the solder surface, the distribution of the stress and strain is discussed. Because the stress and plastic strain of the solder mainly concentrate near Cu_6Sn_5, this region is specially cut out for analysis (Fig. 4), which is also the strain energy concentration region for calculating ΔW_{ave}. The stress and plastic strain of the analyzed region is shown in Fig. 7, the results show that the stress and plastic strain of solder in the sub-model have a tendency to concentrate at the corner that is similar to the coarse model. Due to the scallop-like shape of Cu_6Sn_5, there are many depressions in solder, and the stress and plastic strain concentration also occur in these depressions. The the solder's maximum equivalent plastic strain at corner may be caused by the irregular shape of the Cu_6Sn_5 morphology at the boundary.

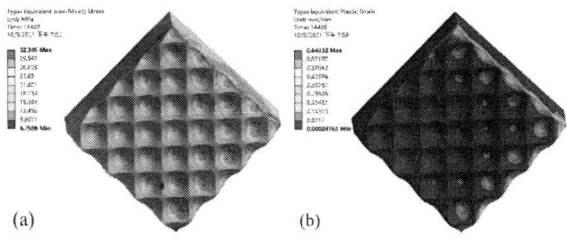

(a)   (b)

Fig. 7. Plots of equivalent stress (a) and equivalent plastic strain (b) of solder for analysis in sub-modle after 4 cycles

C. Influence of IMC roughness on fatigue life

The scallop-like Cu_6Sn_5 would ripen in service, They gradually grow up and merge, resulting in the decrease in the roughness of the Cu_6Sn_5. The influence of Cu_6Sn_5 roughness is discussed below.

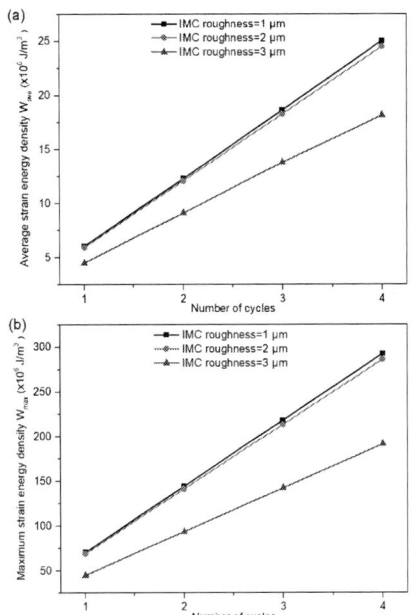

Fig. 8. Curves of W_{ave} (a) and W_{max} (b) of the analyzed area versus number of cycles in sub-modes with different IMC roughnesses during 4 cycles

TABLE IV.   CALCULATION RESULTS OF SUB-MODELS WITH DIFFERENT IMC ROUGHNESSES

IMC roughness (μm)	1	2	3
ΔW_{max} (×10⁶ J/m³)	6.32	6.20	4.57
ΔW_{ave} (×10⁶ J/m³)	73.85	72.29	48.76
N_0	13	13	17
N_p	3713	3714	4609
N_f	3726	3777	4626

The average strain energy density W_{ave} calculated from the analyze area and maximum strain energy density W_{max} during 4 cycles are shown in Fig. 8. It shows that there is a linear relationship between strain energy density and the number of cycles, which have also been confirmed in other articles [2]. Therefore, the strain energy density that accumulated in a cycle can be obtained from the slope of each curve. The results are shown in table IV. Because of the large ΔW_{max}, cracks are easily initiated, leading to the small N_0. And the full propagation time of the crack in the solder layer N_p is decisive for N_f. With the roughness of Cu_6Sn_5 decreasing, the N_f of solder get shorter. It gets stable after the roughness is less than 2 μm.

D. Influence of average IMC thickness on fatigue life

The offset of the sin function from the bottom surface is the average of the Cu_6Sn_5 thickness. With time going by, Cu_6Sn_5 would grow, and the thickness of it would increase. So the influence of the thickness of Cu_6Sn_5 on solder's fatigue life is worth being discussed.

The W_{ave} and W_{max} of sub-model with different Cu_6Sn_5 thicknesses are shown in Fig. 9. There is also a linear relationship between the strain energy density and the number of cycles. The corresponding calculation results are shown in table V. With the increase of Cu_6Sn_5 thickness, ΔW_{ave} increases, cracks are easier to grow, N_p gradually decreases.

And the ΔW_{max} increases first and then decreases, so does N_0. The sum of N_p and N_0 i.e. N_f is reduced. The thickness increases by 1μm, N_f is reduced by approximately 228 cycles.

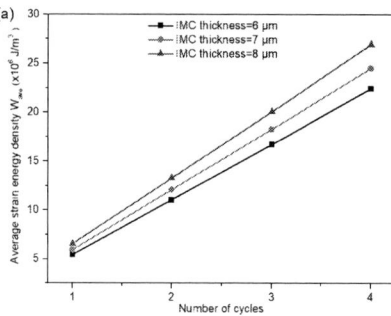

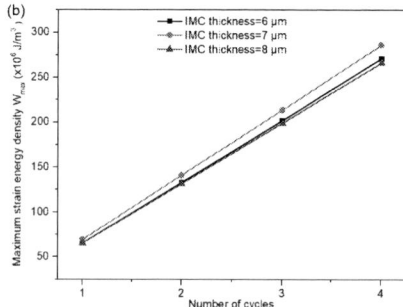

Fig. 9. Curves of W_{ave} (a) and W_{max} (b) of the analyzed area vesus number of cycles in sub-modes with different average IMC thicknesses during 4 cycles

TABLE V.   CALCULATION RESULTS OF MODELS WITH DIFFERENT IMC THICKNESSES

IMC thickness (μm)	6	7	8
ΔW_{max} ($\times 10^6$ J/m^3)	5.67	6.20	6.80
ΔW_{ave} ($\times 10^6$ J/m^3)	68.63	72.29	67.10
N_0	13	13	14
N_p	3992	3714	3536
N_f	4005	3777	3550

IV. CONCLUSION

In this paper, the sub-model method has been used to study the influence of the morphology of IMC (Cu$_6$Sn$_5$) on the solder layer's fatigue life in SiC-IGBT module. The main conclusions obtained are as follows:

Under thermal cycling, the equivalent stress and equivalent plastic strain in the overall SAC305 concentrate at the corner. As a result, it is most prone to grow cracks. The maximum equivalent stress is 28.24 MPa and the equivalent plastic strain gets stable after the third cycle.

The scallop-like shape of Cu$_6$Sn$_5$ would cause many depressions in the solder. The equivalent stress and plastic strain concentrations would occur in these depressions. At the corner, the plastic strain of the solder in contact with Cu$_6$Sn$_5$ gets pretty large: from 0.017 to 0.64, and the equivalent stress gets bigger: from 28.24 MPa to 32.40 MPa.

The reduction in Cu$_6$Sn$_5$ roughness (3μm to 1μm) leads to a shorter fatigue life (4626 to 3726 cycles) of the solder. And

the N_f decreases few (3777 to 3726 cycles) with the roughness decreasing from 2μm to 1μm. As the thickness of the Cu$_6$Sn$_5$ increases (6μm, 7μm, 8μm), N_f decreases (4005, 3777, 3550 cycles) at a rate of approximately 228 cycles per micron.

ACKNOWLEDGMENT

We sincerely acknowledge the financial support provided by the National Natural Science Foundation of China (NSFC No. 62074062) and the funding of Hubei Province's key research and development program "High-performance, ultra-high frequency quartz crystal device (VCXO) manufacturing process research and development (2020BAA4)".

REFERENCES

[1] G. Liu, Y. Wu, K. Li, Y. Wang, and C. Li, "Development of high power SiC devices for rail traction power systems," Journal of Crystal Growth, vol. 507, pp. 442-452, 2019/02/01/ 2019.

[2] F. X. Che and J. H. L. Pang, "Characterization of IMC layer and its effect on thermomechanical fatigue life of Sn–3.8Ag–0.7Cu solder joints," Journal of Alloys and Compounds, vol. 541, pp. 6-13, 2012/11/15/ 2012.

[3] K.-N. Tu, Solder Joint Technology: Materials, Properties, and Reliability (Springer Series in Materials Science). Springer, New York, NY, 2007, pp. 62-77.

[4] W. H. Mullera, T. Hannach H, and Albrechtb, "FE-investigation of the stress/strain and fracture mechanics properties of intermetallic phase regions in leadfree solder interconnects," in 2006 8th Electronics Packaging Technology Conference, 2006, pp. 114-120.

[5] Wang Liquan, "The Study of Interface Morphology Change of Micro Solder Joints during Aging and the Simulation Study of its Impact on Performance," in Huazhong University of Science and Technology, Wuhan, China, 2013.

[6] D. E. Martin, "An Energy Criterion for Low-Cycle Fatigue," Transaction of the American Society of Mechanical Engineers Journal of Basic Engineering, vol. 83, no. 4, p. 565, 1961.

[7] V.-N. Le, L. Benabou, Q.-B. Tao, and V. Etgens, "Modeling of intergranular thermal fatigue cracking of a lead-free solder joint in a power electronic module," International Journal of Solids and Structures, vol. 106-107, pp. 1-12, 2017/02/01/ 2017.

[8] H. Yongle, L. Yifei, X. Fei, L. Binli, and T. Xin, "Physics of failure of die-attach joints in IGBTs under accelerated aging: Evolution of micro-defects in lead-free solder alloys," Microelectronics Reliability, vol. 109, p. 113637, 2020/06/01/ 2020.

[9] Qin Hongbo, "Study of Size Effects on Mechanical Property, Fatigue and Electromigration Behavior of Microscale Lead-free Solder Interconnects," in South China University of Technology, Guangzhou, China, 2014.

Research on Reliability of Solder Layer in IGBT Module Packaging

Panwang Chi
School of Material Science and Engineering
Shanghai Jiao Tong University
Shanghai, China
CHIPANWANG@sjtu.edu.cn

Shengru Lin
School of Material Science and Engineering
Shanghai Jiao Tong University
Shanghai, China
andrew5600@sjtu.edu.cn

Yesu Li
School of Material Science and Engineering
Shanghai Jiao Tong University
Shanghai, China
liyesu@sjtu.edu.cn

Yiping Liu
School of Material Science and Engineering
Shanghai Jiao Tong University
Shanghai, China
liuyp@sjtu.edu.cn

Jicun Lu
Fudan University
Shanghai, China
jerrylu2008@yahoo.com

Ming Li
School of Material Science and Engineering
Shanghai Jiao Tong University
Shanghai, China
mingli90@sjtu.edu.cn

Liming Gao*
School of Material Science and Engineering
Shanghai Jiao Tong University
Shanghai, China
liming.gao@sjtu.edu.cn

Abstract—Sn-3.0Ag-0.5Cu (SAC305) and Pb-8Sn-2Ag are two common solders in IGBT module packaging. In this paper, the tensile tests of these two kinds of solders at strain rates of 2E-4/s, 4E-4/s, 1.2E-3/s, 2E-3/s and 4E-3/s were carried out at -40℃, 25℃ and 150℃. The physical properties and fracture characteristics of the two solders were compared. The Young's modulus and UTS (Ultimate Tensile Strength) of SAC305 solder were higher than those of Pb-8Sn-2Ag. In addition, the reliability of the IGBT module under thermal cycling conditions is simulated and analyzed based on the finite element software Abaqus. The results show that the stress distribution of the SAC305 solder layer is always higher than that of the Pb-8Sn-2Ag solder layer, but the equivalent creep strain is smaller than that of the Pb-8Sn-2Ag solder layer, which characterizes the solder layer in the process of long-term service, the SAC305 solder layer has a harsher stress environment than the Pb-8Sn-2Ag solder layer, but the lead-tin solder has poorer creep resistance. The IMC (Intermetallic Compound) layer between each solder layer and copper mainly plays a role in slowing down the creep, but has little effect on the stress distribution of solder layer. Finally, suggestions are made for the selection of IGBT solder.

Keywords—IGBT module, solder layer, intermetallic compound layer, reliability, failure mechanism

I. INTRODUCTION

In recent years, IGBT (Insulated Gate Bipolar Transistor) module, as a key component of electronic power devices, has the characteristics of high efficiency and energy saving, green environmental protection, intelligent control and so on, and is widely used in new energy vehicles, new energy power generation, power transmission and other fields[1,2]. At the same time, more and more researchers pay attention to the reliability of IGBT module[3]. The connection layer between IGBT chip and centrotherm DBC (Direct Bonding Copper) plays a supporting and heat dissipation role in the packaging structure, which is the weak part of the packaging

structure[4]. Under large dynamic load and high working temperature, thermal fatigue damage of solder layer often leads to module failure[5].

Pb-Sn alloy has high ductility, large electrical and thermal conductivity, good fatigue resistance, low price and easy processing, which is a common solder layer in IGBT packaging[6]. But lead is a kind of heavy metal element harmful to human body[7]. In view of the concern for human safety and environment, the development of lead-free solder has carried out important research work. Sn-Ag-Cu (SAC) solder is considered to be the most promising alternative to traditional Sn Pb solder[8]. Sn-Ag-Cu solder is developed from Sn-Ag binary eutectic alloy. The addition of Cu can reduce the corrosion of Cu matrix, reduce the melting point of the alloy, improve the wettability and fatigue resistance of solder, increase the Ag content can increase the wettability, and the dispersed Ag3Sn in the matrix can improve the mechanical properties, which can slow down the thermal stress in the thermal cycle[9]. At present, Sn-Ag-Cu solder has been widely used in power device packaging.

In the actual working environment of power device IGBT, due to the different thermal expansion coefficient of each layer material in IGBT, the deformation degree of each layer structure is different under the constant heat flow impact. Limited by the fixed constraints between each other, the periodic shear stress occurs between the layers, and the irrecoverable plastic deformation will occur in the solder layer. With the increase of the number of cycles, irreversible plastic deformation will continue to accumulate, resulting in damage deformation and even failure of solder layer[10]. Solder layer failure is mainly divided into two failure forms: crack and cavity. IGBT module failure first appears in the form of crack in solder layer[11]. Generally, the cracks of solder layer initiate from the edge and extend to the center with fatigue aging, and voids will be produced in severe cases[12].

Creep constitutive model and viscoplastic constitutive model are often used in the finite element analysis of solder layer of IGBT module. Dian Hao Zhang et al. Established a thermo mechanical finite element model at cyclic temperature. Based on the creep constitutive model, the cyclic stress and cumulative creep strain of Sn-Ag-Cu solder in SiC-IGBT power module were estimated. The thermal fatigue of SiC-IGBT power module is predicted from creep strain and strain energy, and the main failure modes are discussed[13]. Xiping Wang et al. Quantitatively analyzed the effect of different power cycle conditions on the viscoplastic strain of chip solder. The average value of viscoplastic strain increment under large thermal load is calculated by integral method. Combined with the Coffin-Manson model, the real fatigue life of IGBT chip solder can be obtained[14].

In order to improve the soldering reliability between the solder layer and the chip and other packaging materials, this paper uses tensile tests to characterize the thermal performance of lead-tin solder and SAC305 solder under temperature changes and uses Abaqus finite element analysis to construct an IGBT model. Combined with the Anand viscoplastic model of solder, the performance of the two solders under temperature cycling conditions was compared, and solder selection recommendations were given.

II. EXPERIMENT AND SIMULATION

A. Solder Sample Preparation

In order to ensure the maturity of the process and the consistency of the actual solder, the experimental solders (SAC305 and Pb-8Sn-2Ag) were purchased from PFARR Solder Preform (Changzhou) Co., Ltd., and their purity was above 99.9%. After the solder has cooled, the solder is rolled to a thickness of 4 mm. Finally, the solder alloy was wire cut to form dog bone tensile specimen, as shown in Fig. 1. The main part of the solder sample used for the tensile test is 50mm×10mm, and the thickness is 4mm. The size of the clamping part at both ends of the sample is 32×32mm.

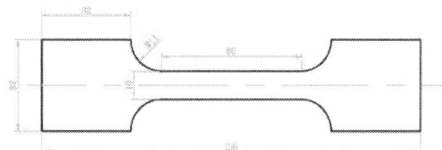

Fig.1. Schematic of tensile test sample in 4mm thickness (unit: mm).

B. Tensile Test

Zwick universal material testing machine was used for tensile test. The variables of the tensile test are temperature and strain rate. The temperature of the tensile test is -40°C, 25°C, and 150°C, respectively. The strain rates are 2E-4/s, 4E-4/s, 1.2E-3/s, 2E-3/s and 4E-3/s. The relationship curve between tensile stress and strain of solder is recorded and drawn.

C. Microstructure Analysis

A field emission scanning electron microscope (SEM) MIRA3 was used to detect the solder fracture of the tensile specimen under five different strain rates.

D. Description of Simulation Model

In this study, a typical IGBT structure is selected. The packaging structure of the model is composed of chip, solder layer, DBC, substrate and lead frame. The whole structure is encapsulated by EMC epoxy molding compound. The physical model of IGBT module is constructed by using finite element analysis software, as shown in Fig. 2.

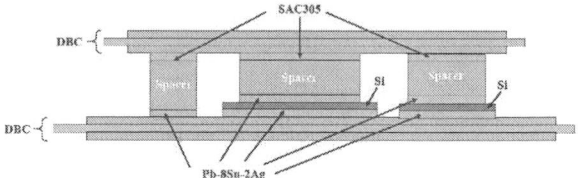

Fig.2. Schematic diagram of IGBT module.

III. RESULTS AND DISCUSSION

A. Tensile Test Results

Fig. 3 and Fig. 4 show the tensile test results of SAC305 solder and Pb-8Sn-2Ag solder at five different strain rates at room temperature (25°C). Pb-8Sn-2Ag does not show all the curves at the lowest strain rate due to its good ductility.

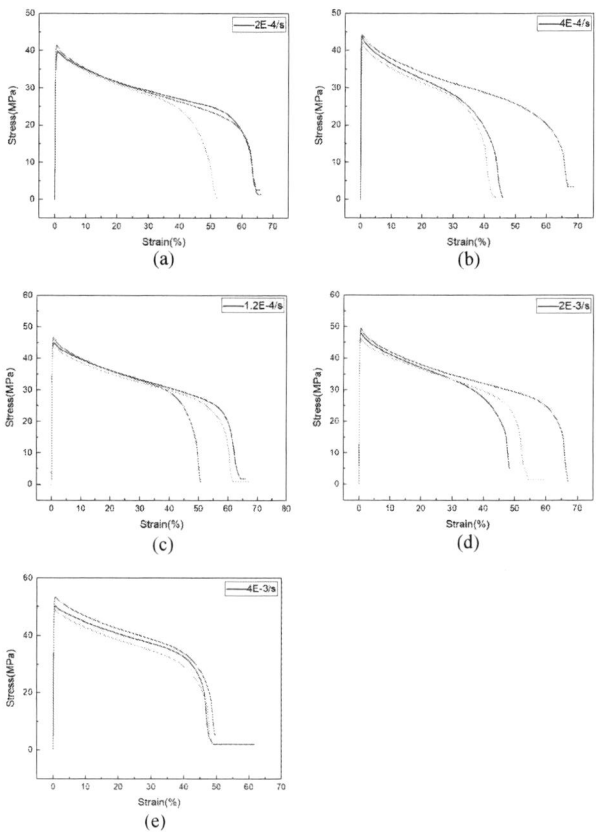

Fig. 3. Tensile curves of SAC305 solder.

(a) 2E-4/s; (b) 4E-4/s; (c) 1.2E-3/s; (d) 2E-3/s; (e) 4E-3/s

Based on the results of the tensile test, the stress of both solders obviously increased rapidly at 0.01-0.02 (1%-2%) strain, reaching the UTS (Ultimate Tensile Strength). Subsequently, the stress gradually decreases with the increase of strain and enters the plastic state. It can be considered that plastic deformation occurs under all non-zero stress states, and the material has no obvious yield stage. At the same time, solder has significant rate-dependent characteristics. The higher the strain rate, the higher the yield strength and tensile strength, and the greater the range of stable plastic flow.

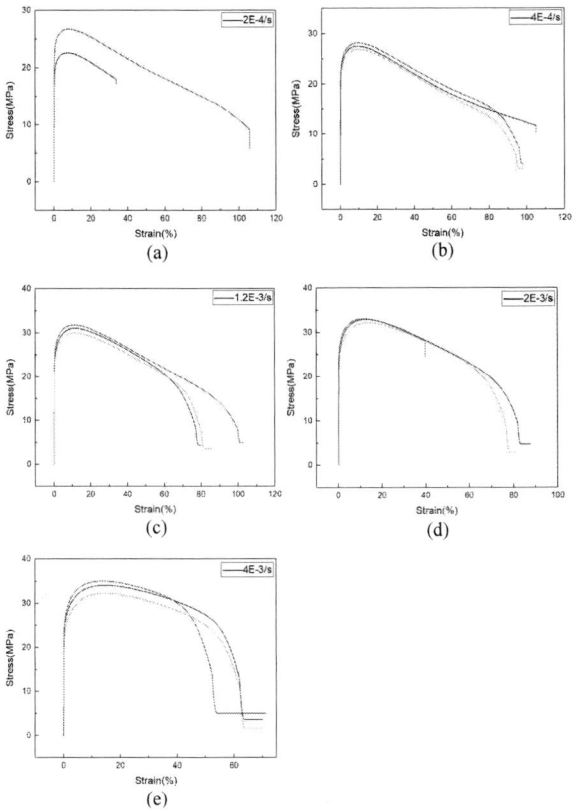

Fig. 4. Tensile curves of Pb-8Sn-2Ag solder.

(a) 2E-4/s; (b) 4E-4/s; (c) 1.2E-3/s; (d) 2E-3/s; (e) 4E-3/s

Fig. 5 and Fig.6 summarize the Young's modulus and UTS (ultimate tensile stress) of the two solders at different temperatures and strain rates. The ultimate tensile stress is the maximum peak tensile stress at each temperature and strain rate. Since the tensile test under each condition is repeated three times, the final value is the average value after excluding some abnormal values.

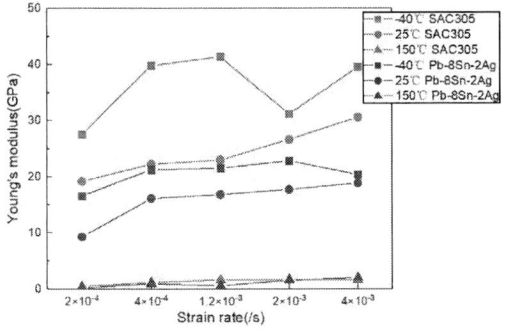

Fig. 5. Young's modulus of SAC305 and Pb-8Sn-2Ag.

From the Young's modulus and UTS under various experimental conditions, it can be seen that the strength and stiffness of SAC305 solder are higher than Pb-8Sn-2Ag under the same conditions. At the same time, when the temperature increases, the yield strength, tensile strength and stable plasticity The flow range is reduced accordingly.

Therefore, it can be seen from the above qualitative analysis that the stress-strain curve conforms to the Anand model law[15].

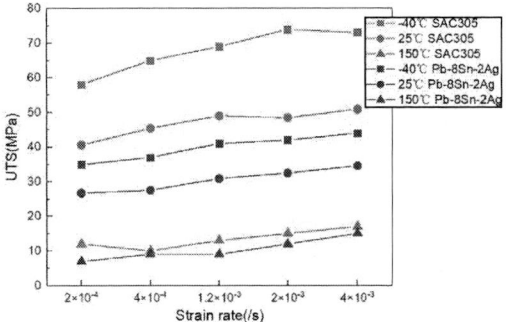

Fig. 6. Ultimate tensile strength of SAC305 and Pb-8Sn-2Ag.

B. SEM Analysis

The tensile fracture of solder sample was detected by scanning electron microscope (SEM). The surface morphology and fracture characteristics were studied, and the fracture modes of the two solders were compared and evaluated.

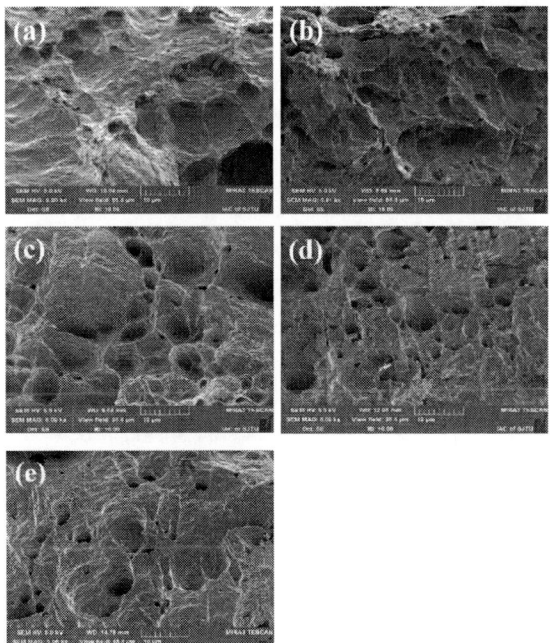

Fig. 7. Morphology of SAC305 fracture surface after 25℃ tensile testing.

(a) 2E-4/s; (b) 4E-4/s; (c) 1.2E-3/s; (d) 2E-3/s; (e) 4E-3/s

Fig. 7 shows the fracture morphology of SAC305 after stretching at 25℃. It can be seen from the figure that at low strain rate (2E-4/s-2E-3/s), the fracture surface of SAC305 solder shows obvious ductile fracture characteristics, with a large number of dimples, which is the result of nucleation, growth and coalescence of micropores.

The SEM images of the fracture show that there are debonding Ag3Sn and Cu6Sn5 grains at the root of some initial dimples, which indicates that the debonding between ag3n and Cu6Sn5 grains and the matrix is the fundamental cause of the fracture. At a strain rate of 4×E-3/s, a river-shaped pattern appears on the fracture. This is due to work hardening. As the strain rate increases, ductile fracture gradually transfers to brittle fracture.

Fig. 8 shows the fracture morphology of Pb-8Sn-2Ag solder after tensile test at 25°C. It shows typical toughness at five strain rates at room temperature.

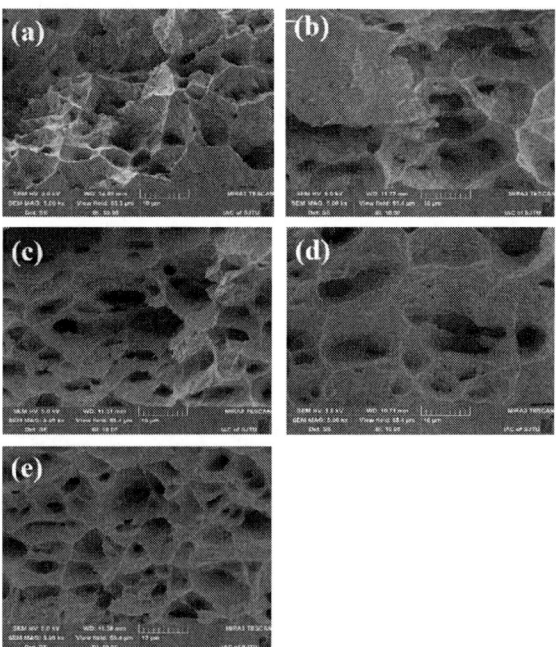

Fig. 8. Morphology of Pb-8Sn-2Ag fracture surface after 25°C tensile testing.

(a) 2E-4/s; (b) 4E-4/s; (c) 1.2E-3/s; (d) 2E-3/s; (e) 4E-3/s

C. Simulation of IGBT Temperature Cycle

The temperature distribution of the module under the condition of temperature cycle is simulated, and the load applied to the whole IGBT module is shown in Fig. 9, in which the high temperature is 120°C, the low temperature is -40°C, the high and low temperature are maintained for 10 minutes respectively, and the high and low temperature conversion time is also 10 minutes. The whole model is placed in a uniform temperature environment, and the high and low temperature conversion time is long enough to ensure the temperature uniformity of the whole module.

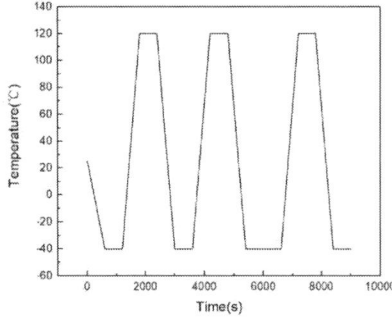

Fig. 9. Temperature load applied to the IGBT module.

The tensile test shows that the solder conforms to the Anand viscoplastic model, and this constitutive model is brought into the calculation.

Fig. 10 shows the stress distribution of the solder layer at 2100 seconds, and Fig. 11 shows the equivalent creep strain distribution of the solder layer at 2100 seconds. Through further observation of the stress and creep strain of the solder layer with temperature, it can be found that the stress distribution of the upper solder layer (SAC305 layer) is always higher than that of the lower solder layer (Pb-8Sn-2Ag layer), so during the long-term service of the solder layer, The stress environment of the upper solder layer (SAC305 layer) is more severe than that of the lower solder layer (Pb-8Sn-2Ag). However, based on the previous basic physical properties of solder, it can be found that the overall strength of SAC305 solder is slightly higher than that of Pb-8Sn-2Ag solder; At the same time, the equivalent creep of SAC305 solder layer during the whole temperature cycle is less than that of Pb-8Sn-2Ag solder, which shows the excellent creep resistance of higher lead solder.

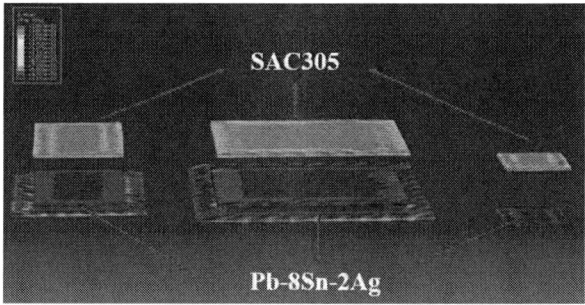

Fig.10. Stress distribution of solder layer at 2100s.

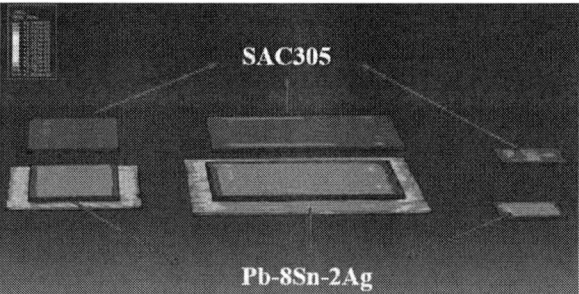

Fig.11. Equivalent creep strain of solder layer at 2100s.

In order to conform to the actual situation of the solder layer, a thin layer of Cu6Sn5 intermetallic compound (IMC) is placed above and below the SAC305 solder layer, and a thin layer of Cu3Sn intermetallic compound is placed above and below the Pb-8Sn-2Ag solder layer.

Fig. 12 shows the variation of maximum stress of solder layer with and without IMC layer with temperature cycle time, and Fig. 13 shows the variation of equivalent creep strain of solder layer with and without IMC layer with temperature cycle time. It can be found that IMC, as a brittle and hard phase, does not slow down the stress distribution of the whole solder layer, but based on the comparison of equivalent creep strain, it can be found that IMC can slow down the internal creep of solder layer, With the further development of temperature cycling aging, brittle and hard IMC can slow down the occurrence of creep, but also increase the risk of cracks.

978-1-6654-1392-3/21 $31.00 © 2021 IEEE

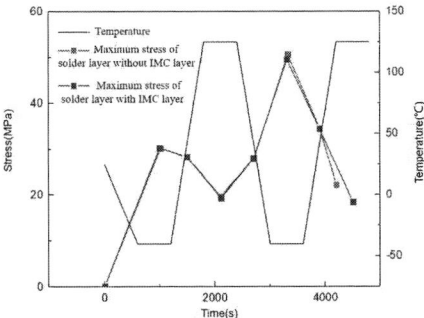

Fig. 12. Maximum stress of solder layer with time.

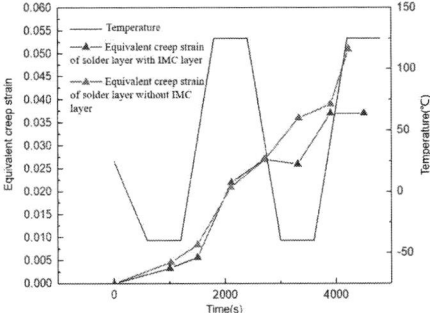

Fig. 13. Equivalent creep strain of solder layer with time.

IV. CONCLUSION

In this paper, tensile tests of SAC305 and Pb-8Sn-2Ag solders were carried out at different temperatures and strain rates. The fracture surface of solder was observed by SEM. The results show that the strength and stiffness of SAC305 solder are higher, but the ductility is not as good as Pb-8Sn-2Ag solder.

Based on Anand viscoplastic model, the finite element analysis of the above two kinds of solder in IGBT module temperature cycle was carried out. It was found that the stress distribution of SAC305 solder was always higher than that of Pb-8Sn-2Ag solder layer, but the equivalent creep of SAC305 solder layer was less than that of high lead solder layer during the whole temperature cycle. Therefore, for the selection of high lead solder (Pb-8Sn-2Ag) and lead-free solder (SAC305), it is necessary to comprehensively consider their physical parameters, and combined with the specific use scenarios, such as high temperature stress, high stress scenario and creep scenario which need to consider the long-term service of solder, the selection focus of solder will be different. Finally, based on the simulation of solder IMC layer, it is found that the main function of IMC layer is to slow down the creep strain of solder layer.

ACKNOWLEDGMENT

Thanks for the equipment support from the Analysis and Test Center of Shanghai Jiao Tong University.

REFERENCES

[1] J.G. Kassakian and T. M. Jahns, "Evolving and emerging applicationsof power electronics in systems," IEEE Journal of Emerging and Selected Topics in Power Electronics, vol. 1, no. 2, pp. 47-58, Jun 2013.

[2] A. Stippich, "Key components of modular propulsion systemsfor next generation electric vehicles," CPSS Transactions on Power Electronics and Applications, vol. 2, no.4, pp. 249-258, 2017.

[3] Choi. UM, Blaabjerg. F and Lee. KB, "Study and Handling Methods of Power IGBT Module Failures in Power Electronic Converter Systems," IEEE Transactions on Power Electronics, vol. 30, no. 5, pp. 2517-2533, May 2015.

[4] H. Li, Y. Hu, S. Liu, Y. Li, X. Liao, Z. Liu, "An improved thermal network model of the IGBT module for wind power converters considering the effects of base-plate solder fatigue," IEEE Transactions on Device & Materials Reliability, vol. 16, no. 4, pp570-575, 2016.

[5] Huang XG, "Simulation on the interfacial singular stress-strain induced cracking of microelectronic chip unde rpower on-off cycles," Journal of Microelectronics, Electronic Components and Materials, vol. 49, no. 2, pp.69-77.

[6] Chi. Pu Lin, Chih. Ming Chen, Yee. Wen Yen, Hsin. Jay Wu, Sinn-Wen Chen, "Interfacial reactions between high-Pb solders and Ag," Journal of Alloys and Compounds, vol 509, pp. 3509-3514, February 2011.

[7] Mulugeta Abtew, Guna Selvaduravb, "Lead-free Solders in Microelectronics," Materials Science and Engineering, vol. 27, February 2000.

[8] Keming Liu, Erxian Yao, Jinlong Yang, Yiwen Wang, Xiaogang Hu, "The research and development of soldering materials applied in IGBT modules packaging," Electrical and Material Application, 2019.

[9] Shnawah. D. A, SabriMF. M and Badruddin. I. A, "A review on thermal cycling and drop impact reliability of SACsolder joint in portable electronic products," Microelectronics Reliability, vol. 52, no. 1, pp. 90-99, 2012.

[10] Smet. V, Forest. F, "Ageing and failure modes of IGBT modules in high temperature Power Cycling," IEEE Transactions on Industrial Electronics, vol. 58, no. 10, pp. 4931-4941, 2011.

[11] Liu. C, Brem. F, Riedel. G, Eichelberger. E, Hofmann. N, "The influence of thermal cycling methods on theinterconnection reliabilityevalu ation within IGBT modules," ElectronicSystem-Integration Technology Conference (ESTC), 2012.

[12] Vermeersch. B, Mey. G. D, "Influence of substrate thickness on thermal impedanceof microelectronic structures," Microelectronics Reliability, vol. 47, no. 2-3, pp. 437-443, 2007.

[13] D. H. Zhang, X.G. Huang, B. L. Cheng, N. Zhang, "Numerical analysis and thermal fatigue life prediction of solder layer in a SiC-IGBT power module," Fracture and Structural Integrity, vol. 55, pp. 316-326, 2021.

[14] X. P. Wang, Z. G. Li, F. Yao, S. X. Tang, "Prediction of chip solder fatigue in IGBTs," IEEJ Trans, vol. 16, pp. 188-198, 2021.

[15] Anand L, "Constitutive equations for the rate-dependent deformation of metals at elevated temperatures," Journal of Engineering Materials and Technology, vol. 104, no. 1, pp. 12-17, 1982.

The Study on Electromigration of Solder Joints under Thermal Cycling Load

Leyi Niu
School of Mechano-Electronic Engineering
Xidian University
Xi'an , China
2540796531@qq.com

Xiaodi Tian
School of Mechano-Electronic Engineering
Xidian University
Xi'an , China
3312357650@qq.com

Fei Jia*
School of Mechano-Electronic Engineering
Xidian University
Xi'an , China
fjia@xidian.edu.cn

Abstract—As the main components of aerospace, home appliances and smart equipment, electronic packaging devices are gradually developing towards miniaturization, lightweight and multi-function, which leads to more and more reliability problems. BGA solder joints play the role of mechanical support and electrical connection between the chip and the PCB board. It is prone to electromigration failure under the condition of high current density, which is one of the most important problems restricting the development of chips. In this study, the electromigration failure of solder joint in WLCSP package with BGA interconnects is investigated by establishing the multi-physical fields. Based on the finite element simulation software COMSOL, the current density, temperature, stress, and concentration on the electromigration behavior of solder joints are taken into account by fully-coupled transient analysis. And the birth and death element method is applied to simulation of voids. Meanwhile, the effects of thermal cycling load and thermal conductivity of voids are considered. According to mass diffusion theory and voids formation and expansion criterion, the thermal characteristics and failure behavior of solder joint under the thermoelectric condition were studied. At the last, the failure time of solder joint is determined according to the failure criterion of electromigration. The result shows that the temperature, stress, and atomic concentration of solder joint change periodically with the thermal cycling load. During the ramp rate stage, the temperature and stress of the solder joints change accordingly, while there is no remarkable effect on atomic concentration and voids generation rate. At the high-temperature dwell time, the maximum temperature and stress of the solder joint occur on the cathode side. Its atomic concentration drops rapidly, which is the main stage of voids generation. Voids mainly formed at the cathode corner of the solder joint, which will cause the current density continue to increase and stress concentration. Thus, electromigration is accelerated and the failure life of solder joint became shorter. When the current is 1.5A, the failure time of the solder joint is about 516 cycles.

Keywords—electromigration, thermal cycling load, voids, mass diffusion theory, failure life

I. INTRODUCTION

With increasing demands for electronic devices operating at harsh environment, electronic packaging reliability has become a major technical challenge in electronics industry [1-2]. The current advanced packaging technology platform, including Flip-Chip, WLCSP, Fan-Out, Embedded IC, TSV technology, etc [3-4]. Wafer Level Chip Scale Packaging (WLCSP) is a low cost and small packaging structure. It is widely used in the end products such as laptop, mobile phones, disk drives and other portable products [5]. Because of the small size and low I/O count of WLCSP, it is used for products that require a high current to be transferred through solder balls [6]. Meanwhile,

there will be a serious problem of electromigration failure in balls. There are also many scholars research on EM failure of solder joints, and made a great contribution in this field. Chang et al. [7] investigated voids propagation process in flip chip solder balls and found that voids growth rapidly with the increase of current crowding region. Huang et al. [8] used synchrotron radiation real-time imaging technology to experiment and found that Cu_6Sn_5 grains dissolved into the liquid solder during the heating stage, and the diffusion of atomic concentration is the main reason of grain growth.

Power devices will inevitably experience harsh environment in application, they will be affected by environmental stress and current changes. Long-term temperature cycling causes thermomechanical damage to the product. It is of great importance to investigate the device performance under harsh environments. Van [9] discussed voids effect on the fatigue reliability of lead-free solder joint under thermal cycling. Result shows that the critical sites for damage is located at the corners of the joint and voids facilitate initiation of damage significantly. Jiao et al. [10] investigated the fatigue life of solder joints under temperature cycling and current density, revealed the thermo-mechanical coupling effect of thermal cycling and current, and mechanical properties of different structure of solder balls. In conclusion, most of the study on EM failure are at a steady temperature, without considering the effect of environmental stress. Because the research of thermal cycling has complex factors, long time period and large transient analysis calculation.

In the present work, with the assistance of super computing servers in the laboratory, FEM is used to establish a meaningful relation between EM, current density and thermal cycling. The EM failure of solder joints in WLCSP package with BGA interconnects is studied by establishing the multi-physical fields.

II. NUMERICAL AND FINITE MODEL

A. Basic electromigration formulation

Solder joints in WLCSP packaging are used as electrical connection structures. Under the high current density, metal atoms not only migrate in the electric field, but also are affected by temperature, stress, and concentration. The directional diffusion of atoms will cause voids and cracks in the structure, and the change of microstructure may induce non-uniform current density distribution, increased difference of temperature and stress. Thermal, electric, stress and atomic concentration interact through the physical field sources provided by each other. Therefore, a numerical model is established for electromigration of solder joints, considering the effects of current density, temperature, stress

978-1-6654-1392-3/21 $31.00 © 2021 IEEE

and concentration. Based on Black's equation [11], the atomic flux can be expressed as following:

$$\overline{q_{Tol}} = \overline{q_{Ew}} + \overline{q_{Th}} + \overline{q_s} + \overline{q_c} \tag{1}$$

where $\overline{q_{Ew}}, \overline{q_{Th}}, \overline{q_s}, \overline{q_c}$ are the atomic flux due to electronic wind force, thermal, stress and concentration gradient, which are calculated as:

$$\overline{q_{Ew}} = \frac{cD}{k_B T} e \rho Z^* \overline{j} \tag{2}$$

$$\overline{q_{Th}} = -\frac{cD}{k_B T} Q^* \frac{\nabla T}{T} \tag{3}$$

$$\overline{q_s} = -\frac{cD}{k_B T} \Omega \nabla \sigma_H \tag{4}$$

$$\overline{q_c} = -D\nabla C \tag{5}$$

where T is absolute temperature, k_B is Boltzmann constant, D is diffusion coefficient, $D = D_0 \exp[-E_a/(k_B T)]$, Z^* is effective charge number, and e is electron charge number. Q^* is the heat of transmission, Ω is atomic volume, σ_H is hydrostatic stress in the material and C is atomic concentration.

The variation of atomic in electromigration process follows mass diffusion theory [12]. Thus, the partial differential equation can be established as follows:

$$\frac{\partial c}{\partial t} + \nabla q = 0 \tag{6}$$

where $c = C/C_0$ is the normalized atomic concentration, the initial atomic concentration in the absence of a stress field can be expressed as $c_0 = 1$, q is total atomic flux. For blocking boundary condition, $q_0 = 0$.

Taking Equation (1) into (6), and transform it to obtain partial differential equations that can be solved by COMSOL.

$$\frac{\partial c}{\partial t} + \nabla \bullet \left\{ \left(\frac{D}{k_B T} (e\rho Z^* \overline{j} - Q^* \frac{\nabla T}{T} - \Omega \nabla \sigma_H) \right) c - D\nabla c \right\} = 0 \tag{7}$$

B. The finite element model of WLCSP

WLCSP packaging is the mainstream of packaging technology in the future, which integrates the technology of thin-film passive devices and large-area manufacturing. The package effectively reduces the size of package and the inductance between IC and PCB. It has a lighter and thinner appearance, and greatly improves thermal conductivity. The structure of WLCSP including die, solder balls, Cu RDL, PCB and so on. In this study, the final WLCSP package with BGA interconnects is illustrated in Fig.1.

This WLCSP package has a die size of 1.5 mm ×1 mm × 0.15mm, 2×3 BGA ball array, ball diameter is 260μm and ball pitch of 0.5mm with built-in electromigration structures. The Cu trace width and pad diameter are 300μm. Current is transmitted through ball-to-trace.

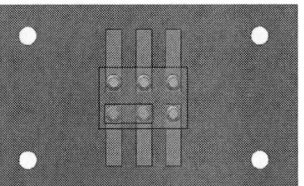

Fig.1 Schematic diagram of WLCSP

The material of solder joint is SAC305, which has viscoplastic properties. The ANAND model parameters and electromigration parameters are used to describe thermo-electric behavior of bumps[13-14].

The materials parameters of the sample are listed in Table I.

TABLE I. MATERIAL PROPERTIES FOR FINITE ANALYSIS

Material	Density /(kg/m³)	CTE (K⁻¹)	Resistivity (Ω • m)
Silicon	2329	2.8×10^{-6}	1×10^8
Copper	8960	17.1×10^{-6}	$1.58 \times 10^{-8} \times (1+4.3e\text{-}3\,\Delta T)$
PCB	1900	16×10^{-6}	1×10^{10}
SnAgCu	7390	23×10^{-6}	$13.3 \times 10^{-8} \times (1+2.8e\text{-}3\,\Delta T)$
Air	1.29	-	1×10^{17}

The sample is subjected to environmental stress. According to thermal cycling simulation standard of ML-STD-883 [15], thermal cycling load is shown in Fig.2. The temperature range is - 55 ℃ ~ 125 ℃. The dwell time is 5min, and ramp rate is 20 ℃ /min. The reference temperature is 20 ℃. The surface of circuit board is natural convection boundary, and PCB board is fixed through bolt holes.

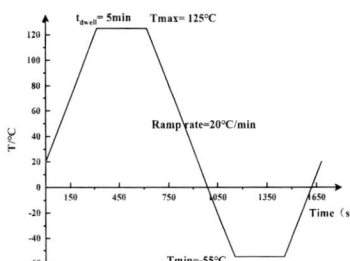

Fig.2. Thermal loading profiles

III. RESULTS AND DISCUSSIONS

A. Transient response analysis

The atomic concentration of solder joint will change under the effect of current density. When concentration is less than critical atom concentration, the void will be formed. Therefore, the formation of voids need two stages. The first is incubation period of voids, and solder joint structure does not change. Followed by voids propagation period and solder joint structure changes. The growth and expansion of voids will have a certain impact on current, stress, and atomic concentration distribution. In order to simulate the formation and propagation of voids, and to consider the

978-1-6654-1392-3/21 $31.00 © 2021 IEEE

effect of voids on the structure. This study established a multi-physics coupling analysis, including electric-thermal-structural and concentration. In the electromigration test, ball-on-trace style construction is selected. The EM test die contains two solder joints, and its layout is on one edge of the die. Selecting two-bumps pair structure because it is simple, low resistance and Joule heating, and opposite electron flow in neighboring balls. A current of 1.5A is applied to two solder joints of chip through Cu trace for electrical connection, and its current density distribution is represented in Fig.3.

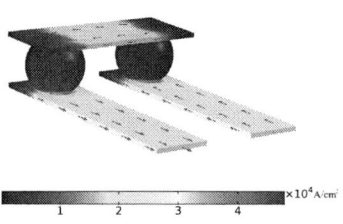

(a) the direction of current flow

(b) corner solder joint

Fig. 3. Current density distribution at t=0s

The arrows in Fig.3(a) shows the direction of current flow. It can be seen that current crowding region occurs at the electrons flow in and out between solder joint and Cu trace. The maximum current density is located at the cathode of corner solder joint, where is easily cause electromigration damage. During the void incubation stage, the current density of solder joint remains 2.59×10^4 A/cm^2.

During the EM test, the sample is in a thermal cycling load. As a result, temperature, stress and atomic concentration of solder joint show periodic variation. Select the first four cycles for analysis, and the transient response of maximum temperature with time is shown in Fig.4. The temperature change of bump is consistent with thermal cycling load, which increases with ambient temperature and reaches the maximum at high-temperature holding time. Due to current Joule heating effect, the temperature of solder joint is higher than environment temperature.

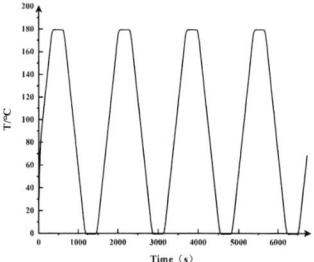

Fig. 4. Maximum temperature transient response of solder joint

The stress profile with time of solder joint is shown in Fig.5. Under the current and thermal cycling loads, the thermal expansion coefficient of various materials are mismatched and stress changes periodically. The stress of bump increases during ramp up, and the maximum stress appears in the high-temperature dwell time. With the decrease of temperature, stress is reduced continuously and keeps stable in the low-temperature dwell time. During cycling before the void is generated, the maximum stress of solder joint remains a constant of 124.88MPa.

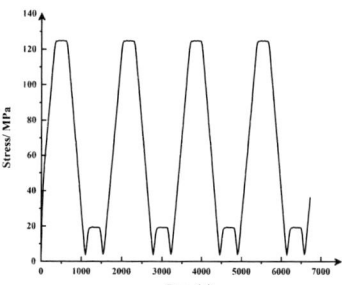

Fig. 5. Maximum stress transient response of solder joint

Through the principle of electromigration, the atomic concentration of solder joint is affected by electron wind force, thermal gradient, stress gradient and itself. Select a certain position of the cathode of bump for analysis, the atomic concentration of solder joint shows a stepped downward trend from Fig.6. There is no obvious change in the low temperature and ramp rate stage, but it decreases obviously in the high-temperature dwell time. After each cycling, the atomic concentration decreases about 3×10^{-4} mol/m^3. It shows that atomic concentration is significantly affected by ambient temperature, and atoms are more active at high temperature, which is beneficial to promote electromigration and leading to the continuous consumption of atoms in the cathode. The failure time of bump also decreases with the increase of high-temperature holding time.

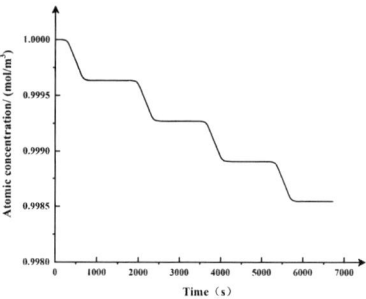

Fig. 6. Variation of atomic concentration at the cathode of solder joint.

B. Electromigration failure analysis

Through the simulation of EM test, voids are observed on the bump at the location with maximum current crowding, where is electron flowed into cathode. Typically, the current-carrying area of solder joint will decrease during voids propagation. According to the law of resistance ($R = \rho l / S$), the decrease in cross-sectional area leads to an increase in resistance, which will impede electrical signals transmission. The percent increase (%ΔR/R) or absolute increase (ΔR) in resistance is used as the failure criteria in the industry. Through the analysis of relevant research, the failure criteria is used an onset of resistance increase with 10% [5]. Figure 7 shows the temperature distribution when t=867300s and current-carrying area decreases by 10%. The EM failure time is just in the high-temperature dwell stage. The maximum temperature of bump is 178.69°C, and thermal gradient promotes atom migration to the anode. Electric current results into Joule heating phenomenon which may be also affected by ambient temperature. And compared with the maximum temperature at the initial stage, it is found that the influence of voids on the temperature is not obvious.

Fig. 7. The distribution of temperature

Fig.8 shows the distribution of current density, it can be seen that current at the cathode appears obvious crowding effect. The current density continues to increase with voids growth. At the end of 516 cycles, current density reaches a maximum of 4.02×10^4 A/cm^2, which is 55.21% higher than the initial time. This is due to the growth of voids resulting an overall reduction in the current-carrying area. As current passes through this area, current density increases rapidly in the remaining area. The increase in current density will promote the electron wind force of atoms, thus further accelerating voids growth.

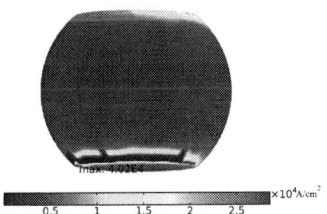

Fig. 8. The distribution of current density at t=867300s

The differences in the thermal expansion coefficients of solder joints, Cu, PCB and die, materials deformation will be constrained to produce larger stress. Figure 9 shows the stress distribution at EM failure time of solder joint. It can be seen that stress is mainly concentrated on the surface that contacts with Cu trace. Voids propagation further caused stress accumulation. The maximum stress of bump is 178.75 MPa, which is 43.14% higher than void incubation stage. The stress causes atom to migrate towards stress gradient.

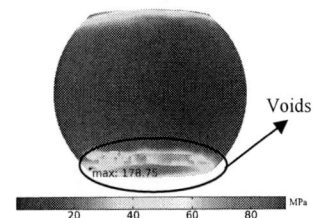

Fig. 9. The distribution of stress at t=867300s

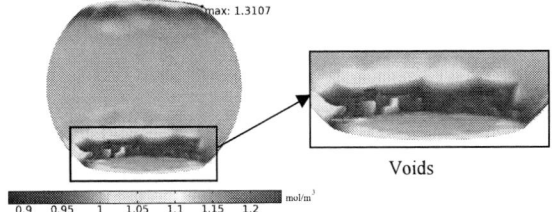

Fig. 10. Atomic concentration at t=867300s

As shown in Figure 10, atom migration reduces cathode concentration and voids are easy to form. Conversely, the anode tends to form whiskers as the atomic concentration increases [16]. At the same time, the non-uniform of atomic concentration is aggravated under the current and thermal cycling. According to Le Chatelier's principle, concentration balance is destroyed, it will move in the direction of weakening this change. Therefore, atomic concentration itself has an inhibitory effect on electromigration, and its effect will be more and more obvious with the increase of atomic concentration imbalance. Moreover, the decrease of atom concentration will promote voids formation, and the EM failure time of solder joint decreases as the volume fraction of voids increases.

IV. CONCLUSION

The thermal characteristics of the WLCSP package is studied under the current and thermal cycling load. The electromigration effects contributed to the microstructure variations and greatly determined the reliability performance of the solder joint. Based on the simulation, the following conclusions are obtained:

1) The temperature and stress of the solder joint varies with the thermal cycling load, the maximum temperature is 178.69 ℃ and the stress could reach 124.88MPa in the high temperature dwell time before the void formation. Meanwhile, the atomic concentration drops significantly on the cathode side of the solder joint and its failure life decreases with the increase of high temperature dwell time. The value of high temperature plays a key role in the diffusion of atomic concentration in the temperature cycle parameters. And the low temperature and ramp rate have no significant effect on EM.

2) The electronic wind force is the main reason for the formation of voids. High temperature and stress gradient will promote electromigration, but atomic concentration inhibit electromigration and its effect gradually increases with time.

3) The current density has a crowding effect where electrons flow in and out, and voids appear in the cathode of the solder joints. Voids induced by electromigration would lead to the change of solder joint structure, causing current crowding, stress concentration, and even accelerating electromigration.

ACKNOWLEDGMENT

This work was supported by the Natural Science Basic Research Program of Shaanxi (No. 2021JM-123), the China Postdoctoral Science Foundation (No. 2016M600766) and the National Natural Science Foundation of China (No. 51606137).

REFERENCES

[1] Madanipour H, Kim Y R, Kim C-U, et al. Effect of Intermetallic Compound Growth on Electromigration Failure Mechanism in Low-Profile Solder Joints [M]. 2019 IEEE 69th Electronic Components and Technology Conference (ECTC). 2019: 1316-23.

[2] Hyodong Ryu, Kirak Son, Jeong Sam Han, et al. The role of a nonconductive film (NCF) on Cu/Ni/Sn-Ag microbump interconnect reliability[J]. Journal of Materials Science: Materials in Electronics,2020,31(18).

[3] Zuo Y, Bieler T R, Zhou Q, et al. Electromigration and Thermomechanical Fatigue Behavior of Sn0.3Ag0.7Cu Solder Joints [J]. Journal of Electronic Materials, 2017, 47(3): 1881-95.

[4] Baolei Liu, Yanhong Tian, Jingkai Qin, et al. Degradation behaviors of micro ball grid array (μBGA) solder joints under the coupled effects of electromigration and thermal stress[J]. Journal of Materials Science: Materials in Electronics,2016,27(11).

[5] A. Syed, K. Dhandapani, C. Berry, et al. "Electromigration reliability and current carrying capacity of various WLCSP interconnect structures," 2013 IEEE 63rd Electronic Components and Technology Conference, 2013, pp. 714-724.

[6] R.L.J.M. Ubachs. Electromigration in WLCSP solder bumps[J]. Microelectronics Reliability,2010,50(9).

[7] Yuan-Wei Chang, Yin Cheng, Feng Xu, et al. Study of electromigration-induced formation of discrete voids in flip-chip solder joints by in-situ 3D laminography observation and finite-element modeling[J]. Pergamon,2016,117.

[8] M.L. Huang, F. Yang, N. Zhao, et al. In situ study on dissolution and growth mechanism of interfacial Cu6Sn5 in wetting reaction[J]. Materials Letters,2015,139.

[9] Van Nhat Le, Lahouari Benabou, Victor Etgens, et al. Finite element analysis of the effect of process-induced voids on the fatigue lifetime of a lead-free solder joint under thermal cycling[J]. Microelectronics Reliability,2016,65.

[10] Yufeng Jiao, Yufeng Jiao, Kittisak Jermsittiparsert, et al. Interaction of thermal cycling and electric current on reliability of solder joints in different solder balls[J]. Materials Research Express, 2019, 6(10): 106302 (10pp).

[11] Liang Lihua, Zhang Yuanxiang, Liu Yong. Prediction of Electromigration Failure of Solder Joints and Its Sensitivity Analysis[J]. Journal of Electronic Packaging,2011,133(3).

[12] Yahong Du, Li-Yin Gao, Daquan Yu, et al. Comparison and mechanism of electromigration reliability between Cu wire and Au wire bonding in molding state[J]. Journal of Materials Science: Materials in Electronics,2020,31(4).

[13] Fu Xing, En Yunfei, Zhou Bin, et al. Microstructure and Grain Orientation Evolution in SnPb/SnAgCu Interconnects Under Electrical Current Stressing at Cryogenic Temperature.[J]. Materials (Basel, Switzerland),2019,12(10).

[14] Ni, Jiamin, Ring, et al. Modeling Effect of Grain Orientation on Degradation in Tin-Based Solder—Part II: Electromigration Experiments[J].IEEE Transactions on Components, Packaging and Manufacturing Technology,2019,9(10):1993-1999.

[15] WANG Y F. On Reformation of American Military Standard and Civil‐military Integration Development [J]. Ship& Boat, 2016, 27(02)：93‐96.

[16] Dániel Straubinger, Attila Géczy, András Sipos, et al. Advances on high current load effects on lead-free solder joints of SMD chip-size components and BGAs[J]. Circuit World,2019,45(1).

Evaluation of Fatigue Crack Growth in Solder Layer of IGBT Module under Power Cycle by Using J-integral Method

Kai Yang
School of Materials Science and Engineering
Huazhong University of Science and Technology
WuHan,China
M202070911@hust.edu.cn

Longzao Zhou*
School of Materials Science and Engineering
Huazhong University of Science and Technology
WuHan,China
lzzhou@hust.edu.cn

Fengshun Wu
School of Materials Science and Engineering
Huazhong University of Science and Technology
WuHan,China
fengshunwu@hust.edu.cn

Yi Zhang
China Construction Third Bureau First Engineering Co., LTD
WuHan,China
826794449@qq.com

Yang Han
China Construction Third Bureau First Engineering Co., LTD
WuHan,China
593554600@qq.com

Zhou Zhang
China Construction Third Bureau First Engineering Co., LTD
WuHan,China
824883007@qq.com

Yang Wan
Hubei Key Laboratory of Micro/Nanocrystals Processing Technology
SuiZhou,China
wy@sztkd.com

Xiangmiao Huang
Hubei Key Laboratory of Micro/Nanocrystals Processing Technology
SuiZhou,China
hxm@sztkd.com

Dayong Huang
Hubei Key Laboratory of Micro/Nanocrystals Processing Technology
SuiZhou,China
hdy@sztkd.com

Abstract—**In this paper, finite element simulation and J-integral method were used to evaluate the fatigue crack growth in solder layer of IGBT module under power cycle. The variations of J-integral value with different crack lengths in SAC305 solder were calculated. The maximum J-integral values of crack for solder layer in Si/SAC305/Cu, SiC/SAC305/Cu，and Si/Pb92.5Sn5Ag2.5 /Cu were calculated, respectively. The results show that when the crack tip is located at the edge of the solder layer, the crack growth rate is larger than that at the center of the solder layer. The change value of J-integral tends to be stable when the crack length is greater than 1 mm. The maximum J-integral value of crack for SAC305 is larger than that for Pb92.5Sn5Ag2.5. The maximum J-integral value of crack for solder layer in SiC/SAC305/Cu is smaller than that in Si/SAC305/Cu. The results of this paper can help to screen out the IGBT module with crack defects but still meet the use requirements, and improve its reliability.**

Keywords—igbt module, j-integral, fatigue crack growth, power cycle

I. INTRODUCTION

Insulated gate bipolar transistor (IGBT) has been widely used as a core electrical device in new energy vehicles, rail transit, inverters, wind power generation, and other fields due to its advantages of high voltage resistance, high switching frequency, and stable working state. However, the IGBT module is a laminated package structure, which has high power consumption and serious heat generation in service. Different materials have different coefficients of thermal expansion. Great cyclic thermal stress will occur between layers. Cyclic thermal stress will cause fatigue defects such as cracks and delamination in the solder layer of the IGBT module, especially in the under-chip solder layer with a higher temperature than the under-copper solder layer.

The defects in the solder layer of the IGBT module are very harmful, which will increase the thermal resistance of the module, reduce the bonding strength of the chip-solder interface and substrate-solder interface, some scholars have also carried out research on this. Literature [1] uses the finite element simulation method to preset cracks with different lengths in the solder layer, and studies the variation of thermal resistance of modules under different crack lengths, but does not explore the growth rate with different crack lengths. K Fakpan calculated the relationship between the change value of J-integral and C* integral and fatigue crack growth rate of SAC305 and Sn37Pb solders, but the J-integral is not calculated using a more realistic three-dimensional model [2]. The existence of cracks will reduce the service life of the module, so some scholars have studied the fatigue life of the module when cracks exist in the solder layer. Sasaki estimated the fatigue life of IGBT module solder joints by using the change value of stress intensity factor in fracture mechanics, but whether the stress intensity factor is suitable for viscoplastic materials is still unknown [3].

The J-integral method is a path-independent integral, Dowling believes that under cyclic conditions, the variation value of J-integral can be used to calculate the crack growth rate [4]. At present, there are little researches on measuring crack propagation of solder layer by using three-dimensional J-integral. In this paper, J-integral and its variation value were used to measure the crack growth rate in different lengths, and the maximum J-integral value of cracks for solder in Si/SAC305/Cu, SiC/SAC305/Cu, and Si/Pb92.5 Sn5Ag2.5/Cu were calculated.

The National Natural Science Foundation of China (62074062) and Hubei Province's key research and development program "High-performance, ultra-high frequency quartz crystal device (VCXO) manufacturing process research and development (2020BAA4)".

II. MODELING

The thickness of the solder layer is very small, it is very difficult to measure the stress and strain field of the solder layer. Therefore, the finite element simulation method is used to simulate the power cycle process of the IGBT module.

A. J-integral Theory

To measure the crack propagation in elastic-plastic materials, Rice proposed a path-independent integral called J-integral [5]. The variation amplitude of J-integral in each cycle can be used to describe the yield of the plastic zone at the crack tip. In the two-dimensional case, the calculation formula of J-integral is as follows:

$$J = \lim_{\Gamma \to 0} \int_\Gamma n \cdot (WI - \sigma \cdot \frac{\partial u}{\partial x}) \cdot q d\Gamma \tag{1}$$

Where Γ is the counterclockwise integral path from the lower surface of the crack to the upper surface of the crack, n is the outer normal direction of Γ, W is the elastic strain energy density plus plastic dissipation, I is the identity matrix, σ is the stress acting on $d\Gamma$, u is the strain acting on $d\Gamma$, q is the virtual crack propagation direction. The formula for calculating W is as follows:

$$W = \int \sigma_{ij} d\varepsilon_{ij} \tag{2}$$

Where σ_{ij} is stress and ε_{ij} is strain.

In the case of three dimensions, the calculation of J-integral is more complicated, and the calculation formula is as follows:

$$J = -\int_V \frac{\partial}{\partial x}[m \cdot (WI - \sigma \cdot \frac{\partial u}{\partial x}) \cdot \overline{q}]dV - \int_{A_{ends}+A_{cracks}} t \cdot \frac{\partial u}{\partial x}\overline{q}dA \tag{3}$$

Where V is the volume enclosed by the closed surface, m is the normal vector outside the surface, $t = m \cdot \sigma$, σ is the normal stress, A_{ends} and A_{cracks} are shown in Fig.2, $\overline{q}$ is the weight function (the modulus of $\overline{q}$ is 0 on A_t, $\lambda(s)$ on A_0, and changes smoothly between these two values in the surface A, A_t and A_0 are shown in Fig. 2).

Paris formula is used to calculate the crack growth rate and the formula is as follows:

$$\frac{da}{dN} = C_1 (\Delta J)^{C_2} \tag{4}$$

Where a is the crack length, N is the number of cycles, C_1 and C_2 are the material constants, ΔJ is calculated as follows:

$$\Delta J = J_{max} - J_{min} \tag{5}$$

Where J_{max} is the maximum value of J integral in a period and J_{min} is the minimum value of J integral in a period.

B. Finite Element Model

Due to the complexity of the complete IGBT module package structure, and the focus of this article is on the solder layer under the chip. If a model exactly the same as the actual one is established, it will undoubtedly increase a lot of useless calculations. Therefore, this paper simplifies the geometric model, only IGBT chip, upper solder layer, DBC substrate, lower solder layer, and copper substrate are established. To reduce the amount of computation, the symmetric model is adopted. The established quarter model and mesh are shown in Fig.1, and the size parameters of each layer structure are shown in Table I.

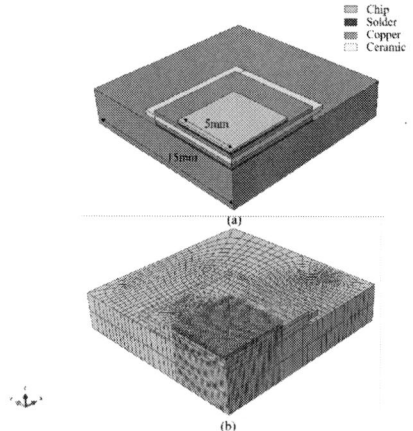

Fig. 1. Quarter geometric model and mesh diagram.

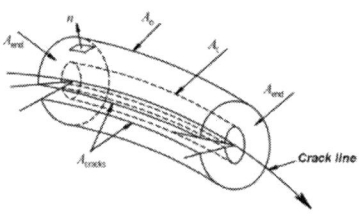

Fig. 2. J-Integral calculation area diagram [6].

TABLE I. GEOMETRIC DIMENSION PARAMETERS

Name	Dimensions		
	Length(mm)	Width(mm)	Height(mm)
Chip	10	10	0.2
Upper solder	10	10	0.15
DBC copper	15	15	0.3
DBC ceramic	17	17	0.6
Lower solder	15	15	0.15
Copper substrate	30	30	3

Owing to the low melting point of the solder layer and the high working temperature of IGBT chips, the conventional elastic-plastic model cannot accurately describe the deformation behavior of the solder layer. In this paper, a unified Anand constitutive model is adopted. The deformation impedance s has a stress unit, which is related to dislocation density, solid solution strengthening and grain size, and s is proportional to stress. The specific formula is as follows:

$$\sigma = c \cdot s \quad (5)$$

$$c = \frac{1}{\xi} \sinh^{-1} \left[\left\{ \frac{\dot{\varepsilon}_{in}}{A} \exp(\frac{Q}{RT}) \right\}^m \right] \quad (6)$$

Where c is the material parameter (under constant strain rate, c is a constant), ξ is the stress multiplier, ε_{in} is the inelastic strain, A is a constant, m is the strain sensitivity index, Q is the activation energy, R is the gas constant, and T is the absolute temperature. From this, the expression of inelastic strain can be obtained:

$$\dot{\varepsilon}_{in} = A[\sinh(\xi \frac{\bar{\sigma}}{s})]^{\frac{1}{m}} \exp(-\frac{Q}{RT}) \quad (7)$$

The evolution equation of internal variable impedance can be expressed as

$$\dot{s} = \left\{ h_0 \left| 1 - \frac{s}{s^*} \right|^\alpha \cdot sign(1 - \frac{s}{s^*}) \right\} \cdot \dot{\varepsilon}_{in} \quad (8)$$

$$s^* = \hat{s}[\frac{\dot{\varepsilon}_{in}}{A} \exp(\frac{Q}{RT})]^n \quad (9)$$

Where h_0 and α are deformation hardening sensitivity coefficients, s^* is the saturation value of internal variables at a given temperature and strain rate, $\hat{s}$ is a coefficient, and n is an exponent.

The viscoplastic parameters of SAC305 and Pb92.5 Sn5Ag2.5 solder are shown in Table II, and other material parameters are shown in Table III and Table IV.

In the power cycle process of the IGBT module, the chip will generate heat due to power consumption, and this paper focuses on thermal fatigue. Therefore, the electric-thermal coupling process is simplified, and the electric power of the IGBT chip is converted into heat source density. The conversion formula is as follows:

$$H = \frac{P}{V} \quad (10)$$

Where H is the heat source density, V is the chip volume, and P is the power consumption. The period of the power cycle is 60 s, the duty cycle is 0.5, and the power loading curve is shown in Fig.3.

TABLE II. VISCOPLASTIC PARAMETERS [7,8]

Parameters	Material	
	SAC305	Pb92.5Sn5Ag2.5
Q/R (℃)	7460	11010
A ($1/\sec$)	5.87×10^6	1.03×10^5
s_0 (MPa)	45.9	23.07
$\hat{s}$ (MPa)	58.3	33.07
h_0 (MPa)	9350	1432
m	0.0942	0.241
ξ	2	7
α	1.5	1.3

Parameters	Material	
	SAC305	Pb92.5Sn5Ag2.5
n	0.015	0.002

The heat of the IGBT chip is conducted downward, the heat dissipation of the module is mainly through the heat sink connected with the copper substrate. Therefore, a convective heat transfer coefficient of 3000 $W \cdot m^{-2}℃^{-1}$ is applied to the bottom of the copper substrate, and a convective heat transfer coefficient of 20 $W \cdot m^{-2}℃^{-1}$ is applied to the side of the copper substrate. While the module is in a closed environment, heat is difficult to be conducted through air, so heat dissipation at other positions is not considered.

TABLE III. MECHANICAL PARAMETERS OF MATERIALS [7,8]

Material	Mechanical parameters		
	Density (kg/m^3)	Young's modulus (GPa)	Poisson's ratio
SiC	3210	420	0.14
Si	2330	170	0.28
Ceramics	3900	380	0.27
Cu	8700	120	0.35
SAC305	7404	51	0.36
Pb92.5Sn5Ag2.5	11020	24.7	0.35

In addition, the IGBT module is fixed on the base, thus restricting all degrees of freedom of the bottom surface of the copper substrate so that it cannot move.

TABLE IV. THERMAL PARAMETERS OF MATERIALS [7,8]

Material	Thermal parameters		
	Thermal conductivity ($W/(m \cdot K)$)	Specific heat capacity ($J/(K \cdot kg)$)	Thermal expansion coefficient (ppm/℃)
SiC	490	690	4
Si	124	713	2.6
Ceramics	25	880	6.4
Cu	385	397	16.4
SAC305	57	217	21.6
Pb92.5Sn5Ag2.5	26	180	24

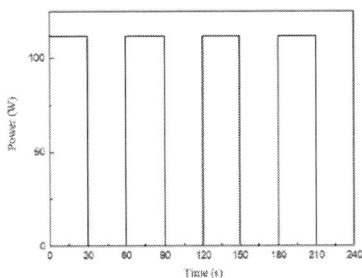

Fig. 3. Power loading diagram.

C. Location of Crack Tip

The stress and strain distribution in different areas of the solder layer is different, so the difficulty of crack propagation is also different. Reference [9] and [10] introduces the initial position of cracks in the solder layer by experiments and believes that cracks will propagate from the edge of the solder layer to the center. In order to explore the crack growth rate at different positions in the solder layer, cracks with different lengths are preset at the edge of the solder layer, and the crack width is 0.02 mm. the distribution is shown in Fig.4. In this paper, we do not pay attention to the stress at the crack tip, but only pay attention to J integral, so the mesh at the crack tip is dense square. the distribution and crack mesh are shown in Fig.4.

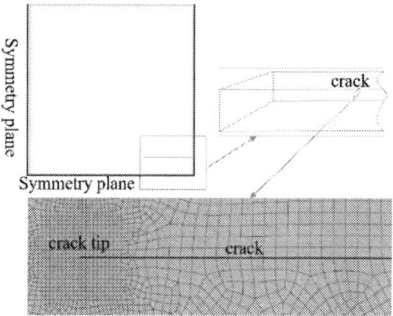

Fig. 4.   Distribution and mesh diagram of cracks.

III.   RESULTS AND DISCUSSION

The finite element simulation method is used to simulate the power cycle process of the IGBT module after presetting cracks in the solder layer under the chip, so as to obtain the stress and strain of the solder layer and the J-integral at the crack tip.

A.   Characteristics of Stress and Strain

The temperature change curve of the solder layer on the IGBT module is shown in Fig. 5. A crack with a length of 1 mm is present at the edge of the solder layer. The stress nephogram of the solder layer after four cycles is shown in Fig. 6. As can be seen from the Figure, the overall residual stress of the solder layer is not high, and the stress at the corners and four sides of the solder layer are higher than that at other positions. The stress curve of a node in the center of the solder layer with time is shown in Fig. 7. As can be seen from Fig.8, there is a stress relaxation phenomenon in the node, and the stress gradually decreases with the increase of deformation, which is consistent with the description of the Anand model, and the Mises stress value at high temperature is less than the Mises stress value at low temperature, while the stress mutation phenomenon in Fig.8 is due to sudden expansion and sudden contraction of solder layer at the beginning of each cycle.

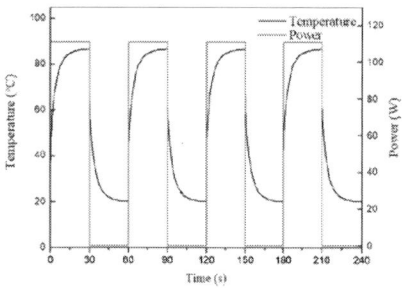

Fig. 5.   Temperature and power cycle diagram.

After Anand viscoplastic constitutive model is selected, the deformation in the solder layer mainly creeps strain, and the change of creep with time at a certain point is shown in Fig.8. Therefore, the strain varies with the periodic loading of the heat source, and at the end of each cycle, the strain value is larger than that of the previous cycle.

And at the beginning of the power cycle, due to thermal expansion, the solder layer will change from a free state without stress to a state with large internal stress. However, in the previous cycles, the stress change of the solder layer is very severe, so the calculated J-integral value will have a large error. The stress-strain hysteretic curve in the solder layer tends to stabilize after the fourth cycle, as shown in Fig. 9. However, in the second cycle, the change value of the area enclosed by the stress-strain ring is not large, so the second cycle can already be used to calculate the J-integral.

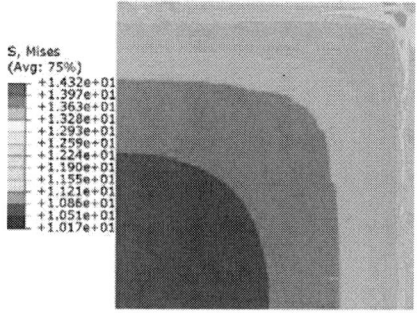

Fig. 6.   Stress nephogram at 240 seconds.

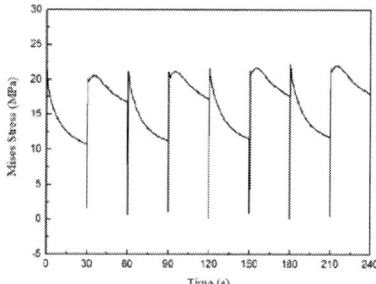

Fig. 7.   Diagram of stress change with time.

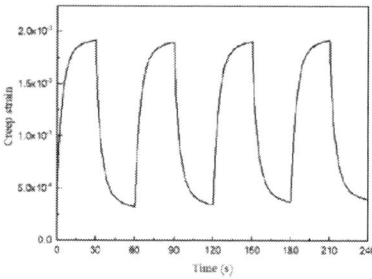

Fig. 8.   The diagram of creep strain change with time.

B.   J-integral and Cracks Growth Rate at The Edge of Solder Layer

The value of J-integral changes periodically with the periodic loading of power. When a crack with a length of 1mm is located at the edge for the solder layer of Si/SAC305/Cu, the variation of J-integral value is shown in Fig.10. At high temperatures, the solder layer cannot expand normally due to the constraint of chip and copper, and there is compressive stress in the solder layer, and the crack is in a closed state. At low temperature, the solder layer shrinks more than the chip and copper, but the shrinkage is also hindered by restraint, so the solder layer has tensile stress and cracks open, so the J-integral value is larger than at high temperature.

However, the unrecoverable plastic deformation of the solder layer in each cycle. Therefore, the value of J-integral is increasing all the time and cannot be used to measure the crack growth rate, but it can be calculated by using its periodic change value.

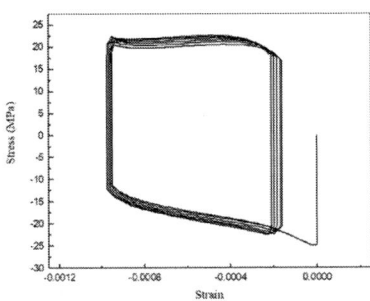

Fig. 9. Stress-strain cycle diagram.

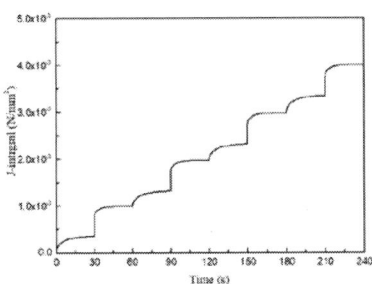

Fig. 10. The diagram of J-Integral with period.

In order to explore the growth rate of cracks with different lengths, the J-integral is calculated by presetting cracks at the edge of the solder layer and changing the crack length. In ABAQUS, different paths can be defined to calculate the J-integral of different lengths from the crack tip, and the J-integral value tends to be stable at a position far enough from the crack tip, which can be used to evaluate the crack growth rate.

The variation value of J-integral for cracks with different lengths is shown in Fig. 11. It can be seen that when the crack length is short, the change value of J-integral is large, and the critical crack length is 0.6 mm, while when the crack length is greater than 2.5 mm, the change value of J-integral tends to be stable. This is because the stress at the edge of the solder layer is greater, so the J-integral changes more when the crack tip is located at the edge. However, the stress and strain values in the middle area of the solder layer are stable, so the calculated J-integral change value will also be stable.

For three-dimensional J-integral, it is necessary to divide the calculated ΔJ by crack thickness to obtain the growth rate of crack length. According to reference [8], the constant $C_1 = 0.0027$ and the constant $C_2 = 1.8$ in Equation 4. The growth rates of cracks with different lengths are shown in Table V.

C. J-Integral Value under Different Material Combinations

The same crack is preset in the solder of Si/SAC305/Cu, Si/Pb92.5Sn5Ag2.5/Cu, and SiC/SAC305/Cu, and the J-integral value is calculated, as shown in Fig.12. As can be seen from Fig.12, the variation trend of the maximum J-integral value of cracks with different lengths in the three combinations is the same, while the maximum J-integral value of Si/Pb92.5Sn5Ag2.5/Cu is smaller than that of Si/SAC305/Cu.

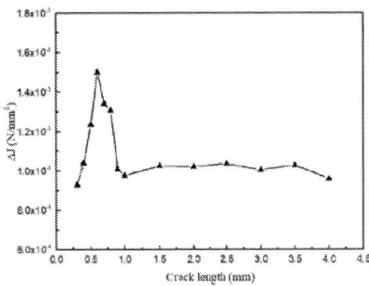

Fig. 11. Variation diagram of J-integral with different crack length in SAC305 solder.

TABLE V. GROWTH RATE OF CRACKS WITH DIFFERENT LENGTHS FOR SOLDER LAYER OF SI/SAC305/CU

Crack length(mm)	Growth rate(mm/cycle)
0.5	1.80×10^{-5}
0.6	2.55×10^{-5}
1.0	1.17×10^{-5}
1.5	1.28×10^{-5}
2.0	1.27×10^{-5}
2.5	1.31×10^{-5}
3.0	1.23×10^{-5}

This is because the stress value in the Pb92.5Sn5Ag2.5 solder layer is small. However, the maximum J-integral of cracks in the solder of SiC/SAC305/Cu is slightly smaller than that in Si/SAC305/Cu combined solder layer. This shows that the cracks in the Si/SAC305/Cu composite solder layer are easier to propagate.

In addition, as can be seen from Fig.12, the maximum J-integral value of the three combinations is the smallest compared to the other lengths when the crack length is 1mm, and the maximum J-integral values of cracks for solder layer in Si/SAC305/Cu, and SiC/SAC305/Cu Si/Pb92.5Sn5Ag 2.5/Cu are 1.78×10^{-4}, 1.84×10^{-4}, and 1.12×10^{-4} N/mm², respectively.

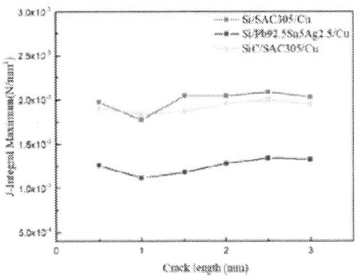

Fig. 12. Variation diagram of maximum value of J-integral in different combinations.

IV. CONCLUSION

In this paper, the power cycle process of the IGBT chip was simulated, and the J-integral value of cracks with different lengths of crack in different solder and chip interconnection were studied. The conclusions were as follow:

(1) The J-integral value is larger when the crack tip is located in the region with higher stress.

(2) When the length of cracks in the solder layer of Si/SAC305/Cu is 0.6 mm, the crack growth rate is up to 2.55×10^{-5} mm/cycle, which is the fastest among all conditions.

(3) When the crack length is 1mm, the maximum J-integral values of cracks for solder layer in Si/SAC305/Cu, Si/Pb92.5Sn5Ag2.5/Cu, and SiC/SAC305/Cu are 1.78×10^{-4}, 1.12×10^{-4} and 1.84×10^{-4} N/mm^2, respectively.

ACKNOWLEDGMENT

The authors acknowledge the financial support provided by the National Natural Science Foundation of China (NSFC No.62074062) and the funding of Hubei Province's key research and development program "High-performance, ultra-high frequency quartz crystal device (VCXO) manufacturing process research and development (2020BAA4)".

REFERENCES

[1] J. B. Zhou, B. Li, Y. G. He, W. B. Yuan, J. C. Liu, and H. D. Ni, "Electro-Thermal-Mechanical Multiphysics Coupling Failure Analysis Based on Improved IGBT Dynamic Model," IEEE Access, vol. 7, pp. 174155-174166, 2019.

[2] K. Fakpan, Y. Otsuka, Y. Mutoh, S. Inoue, K. Nagata, and K. Kodani, "Creep-Fatigue Crack Growth Behavior of Pb-Containing and Pb-Free Solders at Room and Elevated Temperatures," J Electron Mater, vol. 41, no. 9, pp. 2463-2469, 2012.

[3] K. Sasaki et al., "Thermal and structural simulation techniques for estimating fatigue life of an IGBT module," Int Sym Pow Semicond, pp. 181-184, 2008.

[4] J.R. Rice, "A path independent integral and the approximate analysis of strain concentration by notches and cracks," J Applied Mechan, vol.35, pp.379-388, 1964.

[5] N. E. Dowling, J. A. Begley, "Fatigue Crack Growth During Gross Plasticity and the J-Integral," Mechan Crack Growth, ASTM STP 590, pp.82-103,1976.

[6] SIMULIA User Assistance 2021, Dassault Systemes Simulia Corp., 2021.

[7] Y. L. Huang, Y. F. Luo, F. Xiao, B. L. Liu, and X. Tang, "Physics of failure of die-attach joints in IGBTs under accelerated aging: Evolution of micro-defects in lead-free solder alloys," Microelectron Reliab, vol. 109, Jun 2020.

[8] S.H. Zhang et al., "Reliability of PbSnAg solder layer of power modules under thermal cycling in electronic packaging," The Chinese Journal of Nonferrous Metals, Vol. 11, No. 1, pp. 120-124,2001.

[9] K. Stinson-Bagby, D. Huff, and D. Katsis, "Thermal performance and microstructure of lead versus lead-free solder die attach interface in power device packages," IEEE Int Symp Electr, pp. 27-32, 2004.

[10] Y. L. Huang, Y. F. Luo, F. Xiao, and B. L. Liu. "Failure Mechanism of Die-Attach Solder Joints in IGBT Modules Under Pulse High-Current Power Cycling," IEEE Journal of Emerging and Selected Topics in Power Electronics, vol. 7, no. 1, pp. 99-107, Mar, 2019.

Study on leadframe overflow prevention of soldering paste using fluid-structure coupling analysis

Guangsheng Lu
School of Mechanical and Electrical Engineering , Guilin University of Electronic Technology
Guilin, China
guangshenglu@163. com

Daoguo Yang*
School of Mechanical and Electrical Engineering , Guilin University of Electronic Technology
Guilin, China
daoguo_yang@163. com

Wangyun Li
School of Mechanical and Electrical Engineering , Guilin University of Electronic Technology
Guilin, China
leemwy@163. com

Xiyou Wang
School of Mechanical and Electrical Engineering , Guilin University of Electronic Technology
Guilin, China
wxy_07@126. com

Xiangli Wei
School of Mechanical and Electrical Engineering , Guilin University of Electronic Technology
Guilin, China
weixiangli_guet @163. com

Binjie Ai
School of Mechanical and Electrical Engineering , Guilin University of Electronic Technology
Guilin, China
binjieai897@163. com

Abstract—With the development of social demand for power semiconductor packaging, the leadframe's design developed with high density manufacturing quality and low cost. The market demands packaging technology with high performance/ price, handiness size, excellent thermal and electrical performance, optimizing packaging space, and reducing the lead spacing. On the other hand, we can also reduce the packaging size by reducing the die-attach area of the leadframe placement as much as possible. However, the overflow of solder paste in the baking process has always been the main challenge faced by QFN packaging. Based on the principle of fluid-structure coupling analysis of solder paste on the leadframeleadframe, We had analyzed the flow state of solder paste on the groove structure and selected the structure that can effectively prevent solder paste from overflowing. According to the manufacturing process of the leadframe, different models such as V-shaped grooves, rectangular grooves, and trapezoidal groove had established on the position of the frame where the chip had been placed, and different parameters are setting for these structures, e. i. The volume, spacing and depth of the grooves analyze the influence of the above parameters on solder flow. The results of simulation and experiment show that the rectangular groove is the most effective structure to prevent the overflow of solder paste. The wider the groove spacing at the edge of the leadframe, the shorter the solder flow distance. And when the depth and width of the groove are the same, the solder paste overflow is more effectively prevented. The leadframeleadframe's anti-overflow design prevents the short circuit of power dies to reduce the package device reliability and is used as a reference for the packaging process of power devices to solve electronic products' high-density packaging.

Keywords—Leadframe, Soldering paste, QFN, Fluid-structure coupling analysis

I. INTRODUCTION

With the development of IC manufacturing technology, the traditional packaging is not applied to the high performance, high integration, and high reliability of integrated circuits[1]. The chip integration and the packaging technology continuously improve due to the reduction of circuit frame structure size, which means higher quality requirements for the chip[2]. However, large package size has been bothering researchers during the packaging process[3].

To improve the integration density of the device, on the one hand, reducing the size of the bare silicon chip; on the other hand, shortening the pin spacing and even the device size is close to miniaturization.

We will take the quad flat no-lead package (QFN) device as the research object in this work. By reducing the size of the leadframe, we can reduce the packaging volume of the QFN device. In order to achieve high-density packaging, it is an effective method that was reducing the size of the leadframe. However, when the leadframe pad chip (LFP) size had been fixed, excessive reduction of LFP area will lead to solder paste overflow, which will affect the packaging process and cause chip failure. In this work, based on the principle of fluid-structure coupling analysis of solder paste on the leadframe, the overflow of solder paste had been simulated and analyzed by designing different structures on the edge of the leadframe and the adequate structure to prevent the overflow of solder paste had been selected. Based on the leadframe's fabrication process[4], Set the edge parameters (such as the width and aspect ratio of the groove) of different leader frames by different models (such as rectangular groove and V-shaped groove) frame position where the chip has been placed. At last, simulate the flow state of solder paste in the actual situation and combined it with the die attach experiment to test its accuracy of the simulation.

II. MODEL PRINCIPLE

A. Setup of the simulation model

This work designs a packaging model of QFN style, a 4x4mm package model with 16 pins. The size of the chip is 0. 78mmX0. 64mmX0. 22mm. The leadframe had used JEDEC standard silver-plated copper, which overall thickness is about 200 μm, the simulating model as shown in Fig. 1. The leadframe is 98% copper, 2% silver layer had been plated on the surface of the pin and the outer ring of the LFP to wire bonding easily. Fluid-structure coupling analysis of solder paste by COMSOL[5]finite element software. The environment refers to the standard baking temperature of JEDEC. The temperature in the furnace was heating from room temperature ($25°$ C) to $175°$ C within 30 minutes and then holding to $175°$ C for 60 minutes and at last, cooling to room temperature within 30 minutes. In order to improve the speed of simulation analysis calculation, radiation heat dissipation had been ignored, and the temperature is heating to 175° C at 1 minute. Thermal radiation and flow were used to simulate the heat transfer.

Fig. 1. Schematic diagram of leadframe structure

B. Model principle

The analysis of solder paste fluid-structure coupling is based on The Double Domain ALE Model, as schematically shown in Fig. 2. Based on the transient heat law (1), mass (2) and momentum (3) conservation in both the metal and the gas domains. combined with the model of samokhin[6], an analytical model for estimating the magnitude of steady-state metal vapor velocity was derived. Assuming the solder paste is in a closed space, melt and steam flow coupling is feasible. The influence of solid heat transfer of the frame has been ignored to solve the energy and mass balance. In order to speed up the calculation of the simulation, we had simplified the 3D model of solder paste fluid-structure coupling is to a 2D axisymmetric shape.

$$\rho c_p^{eq} \frac{\partial T}{\partial t} + \rho c_p (\vec{u} \cdot \vec{\nabla} T) = \vec{\nabla} \cdot (k\vec{\nabla} T) \qquad (1)$$

$$\frac{\partial p}{\partial t} + \vec{\nabla} \cdot (\rho \vec{u}) = 0 \qquad (2)$$

$$\rho \frac{\partial \vec{u}}{\partial t} + \rho (\vec{u} \cdot \vec{\nabla}) \vec{u} = \vec{\nabla}$$

$$\cdot \left\{ -pI + \mu \left[\vec{\nabla} \vec{u} + (\vec{\nabla} \vec{u})^T \right] - \frac{2}{3} (\vec{u} \cdot \vec{\nabla}) I \right\} + k\vec{u} \\ + \rho \vec{g} \qquad (3)$$

ρ[kg/m3]: density; cp [J/kg/K]: specific heat; k [W/m/K]: thermal conductivity; μ [Pa. s]: dynamic viscosity

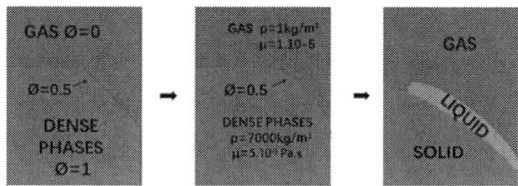

Fig. 2. Schematics of Double Domain ALE Model

III. SIMULATION DETAILS

The solder paste on the leadframe flows occurs during the heating process. The flowing speed and distance have been affected by the weight of the chip, the ambient temperature, the material of the leadframe, and the viscosity of the solder paste. The above parameters are uncontrollable factors, which is difficult to improve the flow state of solder paste by changing the above factors. Different structures are etched on

the edge of the leadframe using the etching process to prevent the solder paste from overflowing out of the frame area, which provides a reference for improving the integration design of device packaging. The experimental variables are the shape (such as rectangle, semicircle, V-shape, and trapezoid), distance, and the ratio of the depth to width of the groove. The other conditions remain unchanged, and the parameters are shown in TABLE I. The viscosity of Solder paste is 11000cps, and the thixotropic index is 4. 0.

TABLE I. MAIN OVEN CURE PROCESS PARAMETERS

Curing time (min)	Vacuum nitrogen charging (L/min)	Heating time (min)	Insulation value (℃)	Cooling time （min）
120	100	30	175	30

A. Influence of leadframe structure on solder paste flow

As shown in Fig. 4, Different groove structures had been designed, such as rectangular structure (Fig. 4 (a)), V-shaped structure(Fig. 4 (b)), semicircular structure (Fig. 4 (c)), and trapezoidal structure (Fig. 4 (d)). In order to ensure a single variable, the spacing between each groove and the volume of grooves of different structures are identical, respectively. Fig. 5 and Fig. 6shows the comparison results of the simulation. Under the same external environment, the triangle groove structure has the best fluidity, and the rectangular groove structure has the shortest flow distance. The height of the solder paste on the side of the chip is around to1/4 to 3/4 of the chip when the solder paste has superior wet ability, which accord with the standard of JEDEC solder climbing, and which means that the etching structure of the frame does not affect the wettability of the solder paste. The simulation results show that the rectangular groove structure has the best effect of preventing the flow of solder paste, and it is the structure with the lowest cost to machining in among the four structures.

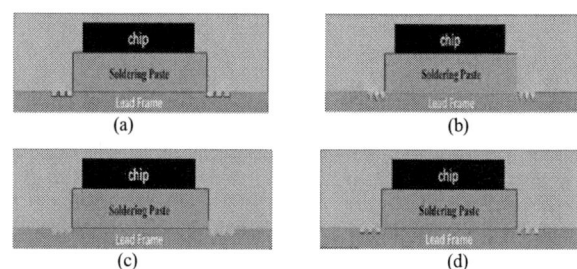

Fig. 4. Schematics of (a) rectangle (b) V shape (c) circular(d)trapezoid

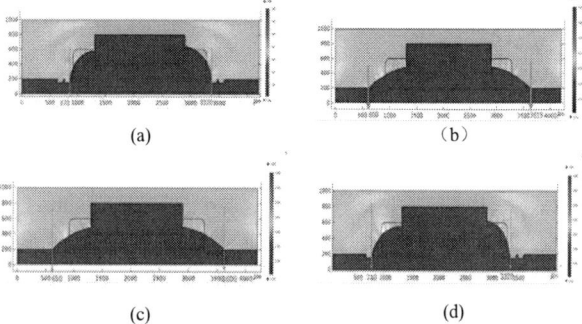

Fig. 5. Results of mode flow analysis for
(a) rectangle (b) V shape (c) circular(d) trapezoid

TABLE II. EFFECT OF STRUCTURE ON SOLDER PASTE OFFSET

Groove structure type	Cross sectional area (μm) 2	Left lateral coordinate (μm)	Right lateral coordinate (μm)	Average offset (μm) $\overline{X}$
rectangle	5000	870	3320	2450
V shape	5000	600	3615	3015
Round	5000	650	3600	2950
trapezoid	5000	740	3320	2580

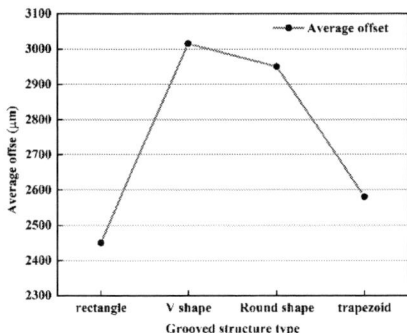

Fig. 6. Relationship between solder offset and groove structure

B. Influence of groove spacing on solder paste flow

Fig. 7 shows that change the spacing of rectangular grooves when the length of the leadframe remains unchanged. Fig. 8 and TABLE IIIshowed that the flow distance of solder paste becomes shorter when the distance between grooves increases, and the solder climbing is in accord with the standard of JEDEC. Under the condition of specified distance, the effect of preventing the solder paste from flowing out of the frame with the longer the distance between grooves is more prominent.

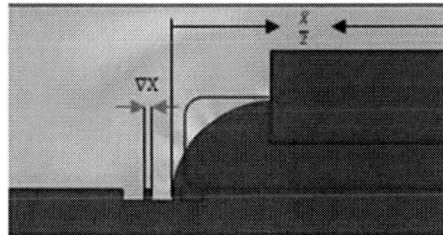

Fig. 7. Schematic diagram of groove parameters

TABLEIII. EFFECT OF GROOVE SPACING ON SOLDER PASTE OFFSET

Groove structure type	Groove spacing (μm) ∇X	Left lateral coordinate (μm)	Right lateral coordinate (μm)	Average offset (μm) $\overline{X}$
rectangle	20	870	3320	2450
V shape	30	600	3615	3015
Round	40	650	3600	2950
trapezoid	50	740	3320	2580

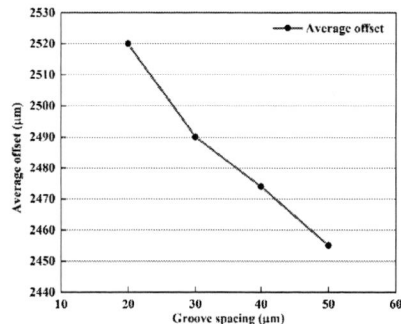

Fig. 8. Relationship between solder paste offset and spacing

C. Influence of groove depth width ratio on solder paste flow

TABLE IV shows that change the aspect ratio of rectangular grooves under the same condition that the space between grooves and the volume of grooves, respectively. The flow distance of solder paste is the shortest when the aspect ratio is 1:1, Fig. 9 shows the effect of solder paste flow distance and groove depth width ratio. The ratio of the depth and width of the rectangular groove is 1:1 when the LFP of the frame is fixed. This structure is the best one to prevent the overflow of solder paste and reduce the overall size of the frame more effectively.

TABLE IV. Effect of spacing on solder paste offset

Groove structure type	Ratio of depth width	Left lateral coordinate (μm)	Right lateral coordinate (μm)	Average offset (μm) $\overline{X}$
rectangle	0. 25	810	3402	2592
rectangle	0. 5	830	3321	2491
rectangle	0. 75	842	3312	2470
rectangle	1	850	3305	2455
rectangle	1. 2	846	3318	2472
rectangle	1. 4	840	3327	2487
rectangle	1. 5	812	3351	2539
rectangle	1. 8	780	3380	2600
rectangle	2	750	3410	2660

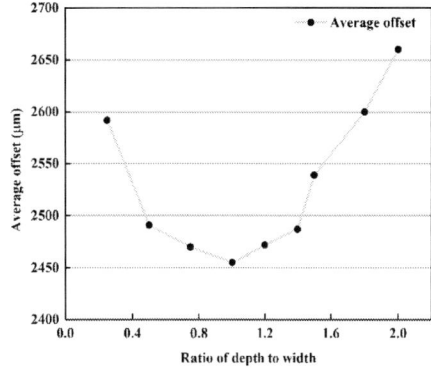

Fig. 9. Relationship between solder paste offset and proportion

IV. SIMULATION EXPERIMENT VERIFICATION

Compared with the fluid-structure coupling analysis of the first three simulation experiments, it is helpful to the flow by etching the position around the LFP under the condition of ensuring the stress intensity of the frame and not affecting the wettability and climbing height of the solder paste. In addition, a group of experiments was designed to change the depth width ratio parameters of the groove to verify the experiment's accuracy further. The laser marking machine (precision 10 μm) had been used to etch the groove with three parameters in the middle of the frame (Fig. 11a), and the same pressure was used for dispensing through the crystal fixing machine (Fig. 10), and then the chip (Fig. 10) (0. 78mmx0. 64mmx0. 22mm) was attached on. Finally, put it into the oven for curing, and then using the SEM to Observation distance. According to the uncertain fluidity of the solder paste, the overflow distance of the four sides of the solder paste had been averaged. Therefore, design the grooves' depth, and width ratio is 1:1 and 1:2 for each 200. The Fig. 10 shows that the average overflow distance of rectangular groove with an aspect ratio of 1:1 is the shortest, which coincide with the simulation analysis of mode flow. Therefore, the experiment and simulation are correct.

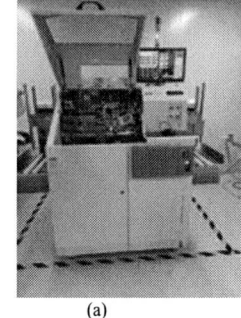 

(a)   (b)

Fig. 11. (a) Die Bonder(b)Curing Oven

V. CONCLUSION

This paper studied the effect of solder paste overflow on the leadframe of 3 factors as Groove shape, groove spacing, and groove depth width ratio. The experimental results show that:

- The grooves with different structures around the LFP can effectively prevent the overflow of solder paste, and the rectangular groove has the most significant effect on the overflow of solder paste;

- the larger the spacing of the same number of grooves, the better the effect of preventing the solder paste overflow Under a specific frame size.

- When the volume of the rectangular groove is fixed, and the ratio of depth to width of the rectangular groove is 1:1, prevent the flow of solder paste is the best.

ACKNOWLEDGMENT

This research was supported by the Key R & D Plan Project of Guangxi Province (grant No. GuiKe AB20159038) , and the Natural Science Foundation of Guangxi Province under Grant No. 2018GXNSFBA281065, Science and Technology Planning Project of Guangxi Province under Grant No. GuiKeAD18281021.

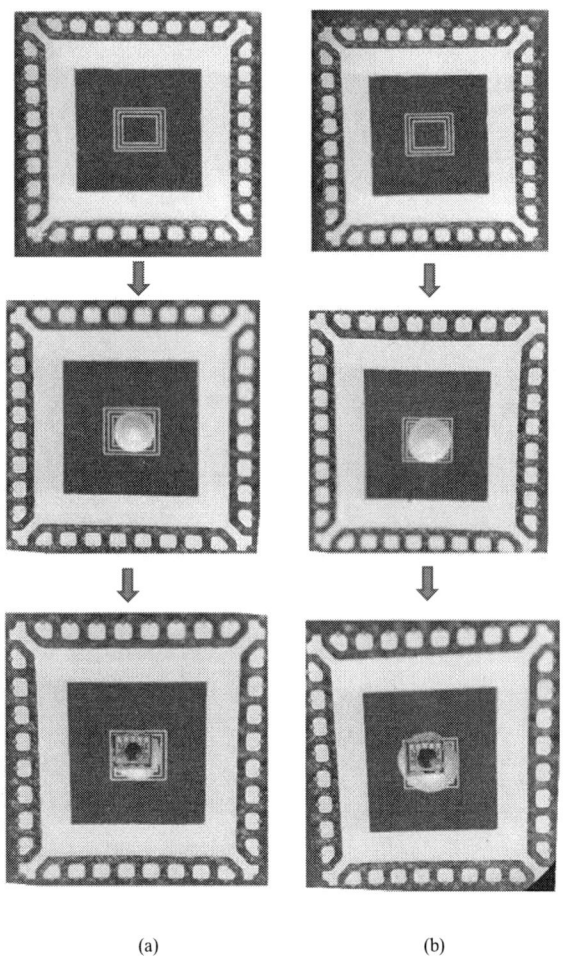

(a)   (b)

Fig. 10. (a)The shape of solder paste with the ratio of depth to width of 1 (b)The shape of solder paste with the ratio of depth to width of 1. 2

REFERENCES

[1] KESER B. Advances in Embedded and Fan-Out Wafer Level Packaging Technologies[M]. USA: WILEY, 2019:1-53.

[2] GHAFFARIAN R. Microelectronics Packaging Technology Roadmaps, Assembly Reliability, and Prognostics[M]. 2016:543-611.

[3] S W Y. Challenges of advanced wafer level packaging technology: Cost-effectiveness, integration and scalability. 2012 IEEE 14th Electronics Packaging Technology Conference(EPTC)[C]. 2012:451-455.

[4] Materials Research; Researchers at Henan University of Science and Technology Publish New Study Findings on Materials Research (Optimal Hot-Dipped Tinning Process Routine for the Fabrication of Solderable Sn Coatings on Circuit Lead Frames)[J]. Journal of Technology, 2020.

[5] COMSOL Multiphysics®. User's Guide.

[6] Samokhin. Effect of Laser Radiation on Absorbing Condensed Matter. Proceedings of the Institute of General Physics (1990).

Simulation Study on Thermomechanical Reliability in Embedded Die Package Fabrication Process

Zhou Zhou
Nexperia Hong Kong Limited
Hong Kong SAR, China
zhou.zhou@nexperia.com

Haibo Fan
Nexperia Hong Kong Limited
Hong Kong SAR, China
haibo.fan@nexperia.com

Yuning Shi
Nexperia Hong Kong Limited
Hong Kong SAR, China
yuning.y.shi@nexperia.com

Abstract—EDP (Embedded Die Package) is promising in future years to become the key technology for power electronics because of its compact layer structure fulfilling integration requirements, its minimized parasitic effect with enhanced power conversion efficiency, as well as its benefits in shielding and thermal dissipation. Understanding the new elements presented by EDP and assessing their individual and collective impact on reliability are keys to successfully launching this new packaging technology into applications. Simulation works in this research present a detailed layer stacking fabrication process with risk causes analysis at each assembly step, with status tracking for each component, and with timing highlighted for risk presence. ANSYS Workbench has been utilized as the simulation tool, birth & death elements technique and thermal loadings have been properly applied on corresponding assembly procedures.

Different with typical study solely focusing on component local behavior, a complete fabrication process has been considered with corresponding model established for each assembly step in this study, which enables us to investigate each single component on global package stand of view, and to discover mutual impacts between spatially neighboring components, as well as to explore the interactions between related components regarding assembly procedures. Based on simulation data, assessment on final products reliability is achievable. Models and results in this study could be references for further structure or layout optimization, material selection, and process plan modification. The analysis approach regarding multi-step multi-elements fabrication procedures could also be expanded to explore units and cases with similar 3D integration process flow.

Keywords—Embedded Die Package; Package Reliability; Thermomechanical Stress; 3D Integration

I. INTRODUCTION

In recent years, thinner, smaller package outline and higher power converting efficiency become the essential requirements for portable electronic devices represented by mobile phone, tablet, and ultrathin notebook. Especially in new generation product design, both manufacture and consumer expect more available area for higher level of functionality and design flexibility. New elements highlighted on 3D integration structures of EDP, such as thin die, copper filled micro-via, compact layer structure with shorter tracks are exactly fulfilling above integration requirements.

On the other side, to satisfy enormous demands on high-voltage, high-current power modules, which will be massively applied on electric or hybrid electric automotive, renewable energy generation, the emerging next generation wide-bandgap devices are placed in the spotlight with tens or even hundreds faster switch rate than conventional silicon devices [1].

Although there is obvious advantage for GaN and SiC power semiconductor on low dynamic losses, a minimized parasitic effect of neighboring layers becomes to the primary consideration to fully utilize wide-bandgap benefits on dynamic performance, so that entirely transfer their switching efficiency to device advantage. EDP appropriately fits the requirement on low parasitic. Multiple Cu micro-via matrixes are applied in EDP as the connection in package, to form electric paths from chip I/O pad to outside. Compare with conventional wire bond or clip bond package as depicted in Fig. 1, Cu micro-via matrixes ensure shorter loop with faster electrical response. Wide width Cu micro-via and thick Cu layer can achieve more than one order of magnitude improvement on parasitic characteristics [2], including lower resistance and lower inductance.

By enhancing power conversion efficiency within a narrower space, as well as further benefits in aspect of shielding and thermal dissipation [2], EDP is promising in future years to become the key packaging technology for fast switching power device, metal-oxide semiconductor field-effect transistor (MOSFET) and insulated gate bipolar transistor (IGBT) [3], so that decreases our carbon dioxide emission and achieve carbon neutrality as early as possible.

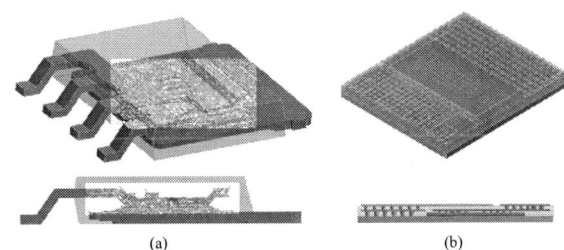

Fig. 1. (a) Conventional package technology with a clip; (b) EDP (Embedded Die Package).

II. FABRICATION PROCESS OF EDP

EDP in this study is formed base on fundamental layer contains die pad, die attach and die. After 1st lamination is created and drilled as required patterns, Cu plating is followed to fill those patterns, where 1st Cu micro-via matrix and 1st Cu layer are produced. In same way 2nd lamination, 2nd Cu micro-via and 2nd Cu layer are fabricated. In PCB loading process, EDP is placed to the Cu trace on PCB board with PCB solder connection. Cu trace position is assigned to match the source, drain and gate zone of 2nd Cu layer. Corresponding fabrication sequence has been illustrated in Fig. 2. The EDP samples are fabricated accordingly and presented in two views by Fig. 3.

978-1-6654-1392-3/21 $31.00 © 2021 IEEE

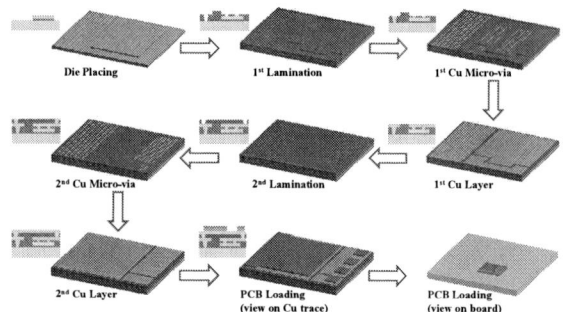

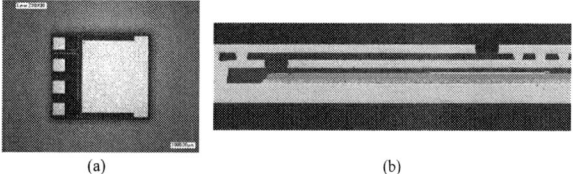

Fig. 2. EDP fabrication process (components assembly and PCB loading)

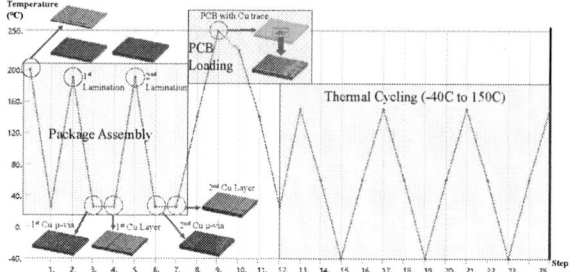

Fig. 4. Thermal condition of EDP fabrication process

Fig. 3. Fabricated EDP sample (a) top view; (b) cross-section view

The EDP fabrication procedures contain repeated high-low temperature process including laminate layer formation at high temperature, copper plating at room temperature, and PCB loading at high temperature for solder reflow. The thermal process finally causes thermomechanical stress on various packaging materials with different coefficients of thermal expansions (CTE), and induces package warpage. Excessive package warpage always comes with several failures including delamination [4]-[6], solder fatigue [7]-[11], Cu micro-via failure [5], [12]-[14], laminate cracking [14], [15], die cracking [5], [16], and misalignment issue [4], [5], [8]. Therefore, package warpage during fabrication process is a critical property to be recognized [17]-[19] and properly controlled for an acceptable package lifetime.

A basic design idea for such Cu micro-via interconnected power electronics is carefully optimizing entire EDP structure to maintain the construction symmetry, and deploying compatible materials to balance and mitigate thermomechanical stress under dynamic thermal loading. Thus, further than overall package warpage estimation, understanding each components, especially emerging new elements presented by EDP, and assessing their individual and collective impacts on reliability are keys to successfully launching such new packaging technology into applications.

III. NUMERICAL MODEL DISCRIPTION

In this study, simulations are performed by ANSYS workbench. Thermal loadings and element activations are properly applied corresponding to three thermal stages including package assembly stage, PCB loading stage and thermal cycling stage, so that examining each element mechanical features and interactions among components introduced by EDP. Detailed layer stacking fabrication procedures are completely presented with the utilizing of birth & death elements technique.

Fig. 4. demonstrates the thermal loading conditions regarding EDP fabrication process, which has been divided into three stages. First stage (step 0-step 7) is package assembly, which start with a die bonded to die pad by Ag sintering die attach at 200°C (step 0), and cool down to room

temperature 25°C (step 1), then increase temperature to 190°C and activate 1st lamination (step 2), then cool down to 25°C and sequentially activate 1st Cu micro-via (step 3) and 1st Cu layer (step 4), then increase to 190°C again and activate 2nd lamination (step 5), and cool down to 25°C again to activate 2nd Cu micro-via (step 6) and 2nd Cu layer (step 7) in sequence. All components inside package have been activated in the first stage. The following step with the completion of package assembly is PCB loading (step 8-12), which is the second stage: temperature is increased to 250°C, simultaneously activate PCB, Cu trace and PCB solder (step 9), and followed by cooling down until 25°C (step 12). At this moment, all elements in model have been involved and prepared to enter the third stage, thermal cycling (step 13-25). Thermal cycling process is set ranging from -40°C to 150°C. Totally three thermal cycles (step 13-17, step 17-21, and step 21-25) are applied in this study.

Above Model has been applied with a homogeneous thermal condition which refers a uniformly distributed temperature at each time step. All interfaces are treated as perfect adhesion without delamination in the model. Since delamination area, location, pattern, expansion rate and path, etc. all can change local stress greatly, delamination scenario becomes a topic require specific investigate in future. Die, die attach, lamination and PCB are set as linear elastic materials; non-linear behavior is considered for die pad, Cu micro-via, Cu layer, Cu trace and PCB solder. Stress-free temperature for die, die attach and die pad is 200°C, for lamination is 190°C, for Cu micro-via and Cu layer are 50°C, for PCB, Cu trace and PCB solder are 250°C. The relevant material property details are listed in Table I.

TABLE I.   MATERIAL PROPERTIES

Components	Physical Properties			
	Young's Modulus (GPa)	Poisson's Ratio	CTE (ppm)	Yield Stress (MPa)
Die	169	0.23	2.6	/
Die Pad	120.7	0.34	17.1	470
Die Attach	7.5 (25°C) 0.9(250°C)	0.35	α1=40, T<140°C α2=85, T≥140°C	/
Lamination	26.5(25°C) 25.5(100°C) 3.12(180°C) 0.76(200°C) 0.74(250°C)	0.35	In-plane: α=13 Out-plane: α1=25, T<180°C α2=145, T≥180°C	/
Cu Micro-via / Cu Layer / Cu Trace	105	0.34	17.7	85
PCB	In-plane: 22 Out-plane: 10	0.28	In-plane: 18 Out-plane: 70	/
PCB Solder	50	0.36	21.6	42[a]

a. Tangent modulus: 120 MPa

IV. RESULT AND DISCUSSION

Different with typical thermomechanical packaging reliability study solely focusing on component local behavior, model in this study covering above complete fabrication process, with components activation at each corresponding assembly step. This model enables us to investigate single component behavior on global package stand of view, to discover mutual impacts between spatially neighboring components, as well as to explore the interactions between related components regarding assembly procedures. In this way, risk cause analysis at each assembly step, status tracking for each component, and timing highlighted for risk presence are achievable, with all together information we could build up a panoramic thermomechanical stress portrait of EDP fabrication process.

A. Package Overview

The package overall warpage direction and warpage extent of each step within three stages can be observed from Fig. 5 on package sideview. For better observation on Cu micro-via and die, parts of lamination are hidden. PCB board is hidden as well. Contour color represents von-Mise stress. When temperature decreasing or rising, the shrinking and expanding effect of 1st laminate layer is more apparent than die, since silicon die has the lowest CTE in whole system. The warpage direction for lamination around die is concave up at high temperature (step 2, 5) and concave down at low temperature (step 3-4, step 6-7). As further Cu micro-via, Cu layer and lamination are assembled at different temperature, the local package warpage behavior becomes complex. Especially at 2nd Cu micro-via and 2nd Cu layer loading step (step 6-7), the warpage direction of lamination away from die is concave up, which opposite to the scenario around die. Base on this we can have a glance that how later introduced elements impact on package local warpage.

After 1st stage, besides the obvious package extension (step 9, 13, 17, 21) at high temperature and contraction (step 12, 15, 19, 23) at low temperature, the primary package warpage shape become stable. It is worth to understand that the shape and pre-load stress at the beginning of 3rd stage will accompany with the package all along, thus relevant deformation and stress distribution prediction is meaningful.

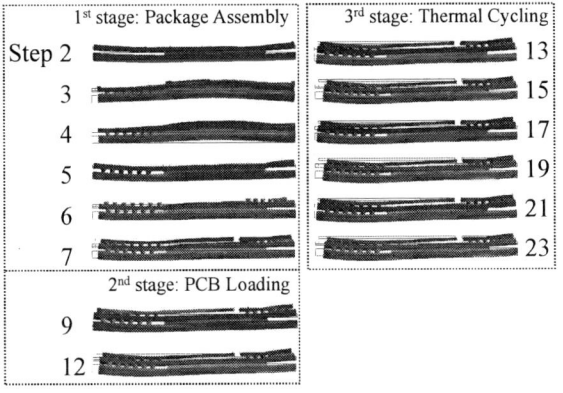

Fig. 5. Package warpage (enlarged by 10 times) at assembly, PCB loading and thermal cycling stages

B. Thin Die

A crucial novo-technology for EDP is thin die, which potentially subject to significant mechanical stress introduced during packaging process [5]. This stress effect in

Fig. 6. Die von-Mise stress distribution at step 15 turn has an impact on components performance and reliability.

The highest von-Mise stress of thin die occurs at low temperature steps of 3rd stage (-40°C) as shown in Fig. 6. High stress areas locate at 4 upper corners of die. A further force analysis illustrated by Fig. 7 indicates the die corner high stress point are collectively contributed by Y-axis normal stress toward downside, and shear stress regarding XY plane and YZ plane towards die center for both top and bottom surface. The shear stress directions demonstrate the consistent contraction status of materials surrounding die at low temperature. In view1 of Fig. 7, an asymmetric die bending with downside at left following the package middle part warpage trend during 3rd stage. The die warpage shape is a combined effect of low temperature environment at step 15, as well as pre-loadings during former fabrication steps.

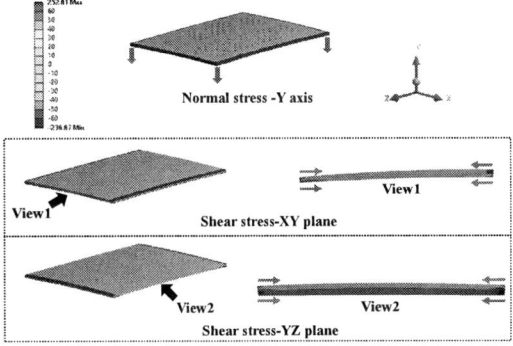

Fig. 7. Die normal/shear stress analysis at step 15 (asymmetric bending in view1, symmetric bending in view2, warpage enlarged by 10 times)

As the brittle material, die failure risk is primarily attributed to 1st principal stress, especially the positive peeling stress. Fig. 8 (a) shows the highest stress happens at step 9 (250°C) of 2nd stage. The detail stress distribution in Fig. 8 (b) reveals that high peel stress (>200MPa) locations are situated within contact area between die and Cu micro-via, indicate a strong interaction at this moment under perfect adhesion assumption. Minimum 1st principal stress has been reached at low temperature steps (step 15, 19, 23) of 3rd stage, which is negative compressive stress with no risk on die crack, as the example at step 23 shown by Fig. 8 (c).

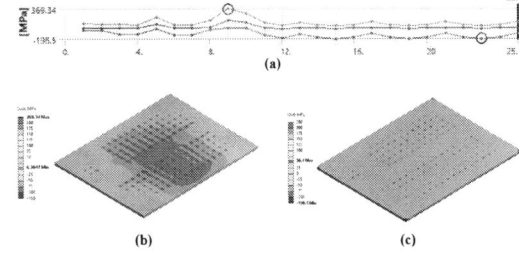

Fig. 8. (a) Die 1st principal stress variation with steps (Max. value on green line; Ave. value on blue line; Min. value on red line); (b)~(c) Die 1st principal stress distribution at b-step 9, c-step 23

C. Lamination

Lamination acts as the buffer layer in package, which stress mostly comes from the low temperature moments of three stages when package temperature largely deviates from lamination stress free temperature (190°C). Fig. 9 reveals that high stress areas at 1st laminate layer are die edges nearby area and upper edge of large via matrix of 1st Cu-micro via. For 2nd laminate layer, the upper edge of 2nd Cu micro-via and nearby zones are high stress area. Possible lamination failures include cracking and delamination, which are frequently reported in former 3D integration studies [4]-[6], [14], [15]. Those failures disable lamination function as buffer layer and create local stress accumulation. A compatible lamination material with low Tg, less CTE mismatch with enclosed Cu via is preferred to mitigate them.

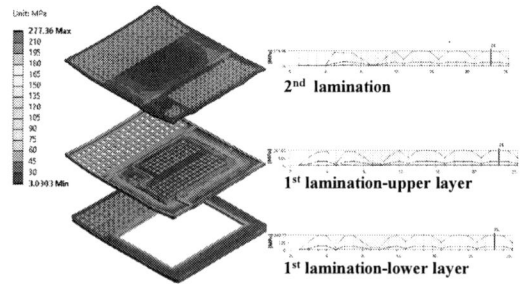

Fig. 9. Laminations von-Mise stress distribution at step 23, and von-Mise stress variation with steps (Max. value on green line; Ave. value on blue line; Min. value on red line; warpage enlarged by 10 times)

The warpage format of 2nd laminate layer, as portrayed by Fig. 10, is representative to reflect whole package extreme bending shape at low temperature moment of 3rd stage. Apparently, lamination at this moment is subjected to a complex warpage condition, thus three profile lines are depicted for bi-angle observation. View1 in Fig. 10 describes a stronger shrinkage that take place at left side and right end of profile line-1, causing concave-up bending curvatures. While at middle of profile line-1 where close to lamination follows die concave-down warpage format, implies the die impact on package warpage. In Fig. 10 view2, the difference between profile line-2 and profile line-3 provide more viewing angle on package warpage. Four-bumper shape curve at profile line-2 with small radius of curvature follows

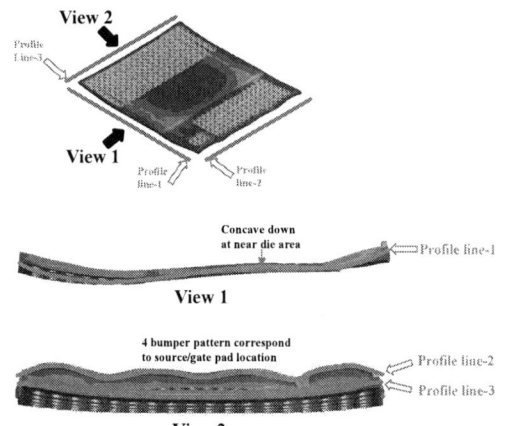

Fig. 10. 2nd Lamination warpage status analysis from two views (at step 23, warpage enlarged by 20 times)

the location of three source pads and gate pad, which further denotes the PCB and Cu trace impact on package warpage.

EDP package warpage is a comprehensive outcome induced by all elements' CTE, modulus, thickness, loading moment, and their contact format. In this study, two dominate warpage format factors are die and PCB relevant components, with influence scope on die surrounding zone and on 2nd lamination/2nd Cu micro-via layer, respectively.

D. Cu Micro-via

Efforts on Cu micro-via reliability research are greatly developed because of its prominent advantage on low parasitic effect. However, relevant Cu micro-via failures, including crack, disconnection, or void, will directly lead whole package breakdown. Besides unstable processing conditions, those failures are commonly induced by thermomechanical stress variation during either micro-via assembly process or following post-procedures.

Fig. 11 (a) introduces the location of 1st Cu micro-via inside package. The small via matrix of 1st Cu micro-via directly locate above die, connecting die and 1st Cu layer; the large via matrix linking die pad and 1st Cu layer. Both small and large matrix of 1st Cu micro-via deform along with the package warpage status from 1st stage to 3rd stage. Through Fig. 11 (b) we can observe the maximum bending scenario, warpage direction of left side large via matrix away from die is different with right side small via matrix above die, indicates that 1st Cu micro-via warpage behavior is affected by their relative position to die.

The impact induced by relative position to die can be further noticed from the 1st Cu micro-via plastic strain accumulation process that depicted by Fig. 12 (b)-(d). Large CTE difference between Si/Cu, and high modulus of die bring the largest plastic strain to those die-contacted small via matrix, of which strain accumulate slower at die center and rapidly at die edges. This is the outcome of severe package warpage around die edge and moderate warpage at center. Therefore, a stronger connection design at small

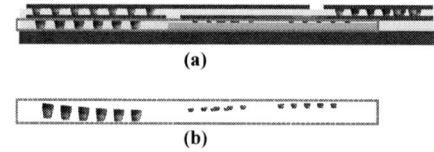

Fig. 11. (a) 1st Cu micro-via location; (b) 1st Cu micro-via deformation status at step 23 (warpage enlarged by 10 times)

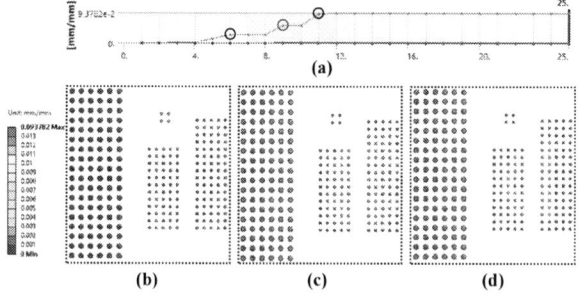

Fig. 12. (a) 1st Cu micro-via accumulated plastic strain variation with steps (Max. value on green line; Ave. value on blue line; Min. value on red line); (b)~(d) Accumulated plastic strain of 1st Cu micro-via from bottom view at b-step 6, c-step 9, d-step 11

matrix corner area is recommended, such as applying large via or plan dense via array there. Avoid to place via array outline closing to die edge. By contrast, because of similar CTE between die pad and Cu micro-via, strain accumulation on larger via matrix is lower, among which the relative higher values are noted on matrix right side that near to die, and the highest two points are remarked at two corners.

2nd Cu micro-via location is exhibited in Fig. 13 (a). Via distance from die center is not the dominant factor on 2nd Cu micro-via strain anymore. Instead, 2nd Cu micro-via strain pattern is affected by neighboring PCB relevant components. Fig. 13 (b) reveals its warpage status with concave up direction. As displayed by Fig. 14 (b)-(d), higher strain areas are exhibited at step 9 when PCB loaded, which locate right underneath the gate pad, drain pad edge, and source pad corner. Those areas indicate the major interaction locations between PCB and package, where strains further develop and expand in later steps. After PCB loaded, a new stress balance is achieved with updated warpage format.

According to Fig. 12 (a), two key moments for 1st Cu micro-via plastic strain accumulation involve assembly stage regarding 2nd lamination (step 5) and 2nd Cu micro-via (step 6), during which nearly 30% of total strain accumulated, as well as PCB loading stage (step 8-12) with more than 70% strain accumulated. Particularly consider opposite bending direction between 2nd lamination and die which been discussed in section "*C. Lamination*", as the connecting elements, small via matrix of 1st Cu micro-via were consequently subjected to maximum plastic strain in whole package, thus are identified as the high failure risk area. Fig. 14 (a) displays that primary strain accumulation process for later assembled 2nd Cu micro-via is PCB loading stage (step 8-12), during which strain accumulation occupying 90% of total strain. The reason that 1st and 2nd Cu micro-via strain substantially accumulated at PCB loading stage attributes to the enormous increasing of laminate layer out-plane expansion when temperature go over lamination Tg (180°C). The huge gap between lamination and copper expansion in

vertical direction generates a large tension force that deforms copper plastically. To relieve the overall internal stress in package fabrication process, a stress release step is suggested after high strain accumulation steps.

E. PCB Solder

Due to the thermal expansion mismatch between the package and PCB, the excessive plastic deformation on PCB solder might be produced and consequently result in the crack initiation and propagation. In this reason PCB solder joint crack has been summarized as one of major package failure mode [20].

A similarity relationship can be detected by comparing the strain variation trend of PCB solder and neighboring 2nd micro-via at step 9 (Fig. 14 (c) and Fig. 15 (b)), at which components strain accumulate apparently faster on those areas close to gate pad and area close to left two corners of drain pad. The high strain area correspondence also applicable on step 11 (Fig. 14 (d) and Fig. 15 (c)), during which the high strain area for 2nd micro-via and PCB solder extends from drain pad two corners to a more extensive edge area and occurred at source pad corner. Fig. 15 (d)~(e) further display the PCB solder high strain area expanding status in thermal cycling stage. Although all source/gate pads strains growing to high level in the end, PCB solder at gate zone is identified with highest failure risk because of its highest plastic strain accumulation. Since 2nd micro-via is surrounded by 2nd lamination, its plastic deformation can also be analyzed by referring 2nd lamination warpage status which been discussed in section "*C. Lamination*".

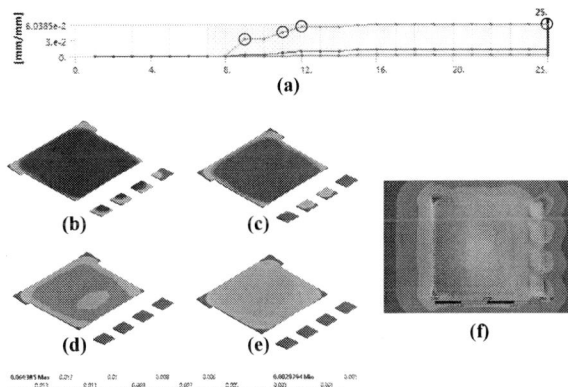

Fig. 15. (b)~(e) PCB solder accumulated plastic strain of at b-step 9, c-step 11, d-step 12, e-step 25 (warpage enlarged by 10 times); (f) PCB board elastic strain distribution as the imprint of package warpage (warpage enlarged by 30 times)

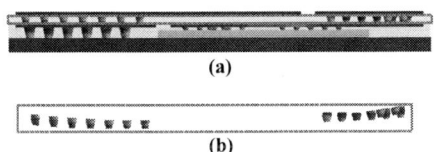

(a)

(b)

Fig. 13. (a) 2nd Cu micro-via location; (b) 2nd Cu micro-via deformation status at step 23 (warpage enlarged by 10 times)

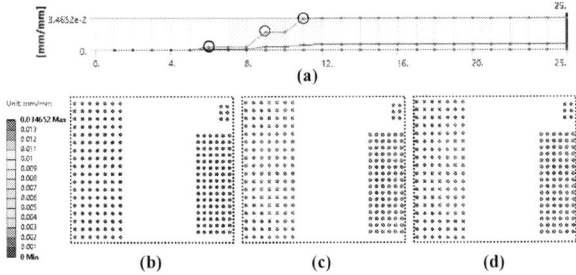

Fig. 14. (a) 2nd Cu micro-via accumulated plastic strain variation with steps (Max. value on green line; Ave. value on blue line; Min. value on red line); (b)~(d) Accumulated plastic strain of 2nd Cu micro-via from bottom view at b-step 6, c-step 9, d-step 11

The imprint of package warpage is revealed by PCB board strain distribution as displayed by Fig. 15 (f). The concave zone of PCB board demonstrates a consequential compressive strain at drain pad edge and source/gate pad. A tensile strain on PCB is reflected by the convex at drain pad center area. By linking PCB board and EDP package, PCB solder accumulated more than 90% strain at PCB loading stage. Fig. 15 (a) verifies that over 50% PCB solder strain already existed initially (step 9) as the result of counteracting package deformation at 250°C.

V. CONCLUSIONS

In this paper, warpage and stress variations of EDP package and components have been recorded and analyzed during three stages: package assembly, PCB board loading, and thermal cycling. A panoramic thermomechanical stress portrait of EDP fabrication process has been built up. Package warpage, thin die stress and bending, lamination stress and warpage have been discussed in detail. Plastic strain accumulation features of Cu micro-via layers and PCB solder are depicted and compared, those features relationship with adjacent components (e.g., lamination, die, PCB board) bending behavior have been investigated. Results predict overall existing package warpage and internal stress after all components assembled and PCB loaded. Major conclusions and design guidelines are summarized as below:

- High failure risks have been identified for corner via of 1st Cu micro-via small matrix, and been identified for PCB solder at gate zone. Stronger connection (thicker via, dense via array layout) at those areas, and layout design avoid die edge are recommended.

- Plastic element strain accumulation features are affected by adjacent components. The primary impact factor for 1st Cu micro-via strain is via's distance from die center, the impact factor for 2nd Cu micro-via strain is nearby PCB components position, the impact factors for PCB solder are neighboring 2nd Cu micro-via as well as 2nd lamination layout. Design optimization on adjacent impact factor component is a feasible way to diminish plastic elements failure.

- The critical strain accumulation steps for 1st Cu micro-via involve the 2nd lamination assembly step and following steps in PCB loading stage; the critical strain accumulation steps for 2nd Cu micro-via and PCB solder only involve PCB loading stage. Stress release step is suggested after each high strain accumulation steps to mitigate the overall internal stress in package.

- After assembly completed and PCB loaded, the existing warpage and internal stress will accompany with the package all along. 2nd lamination warpage format is a representative example reflecting whole package bending shape during thermal cycling stage.

Base on simulation data, reliability assessment on final products is achievable. Models and results in this study could be references for further structure or layout optimization, material selection, and process plan modification. The analysis approach regarding multi-step multi-elements fabrication procedures could also be expanded to explore units and cases with similar 3D integration process flow.

ACKNOWLEDGMENT

The author acknowledges the contributions of colleagues including Charles Tang from packaging research team for providing the fabricated EDP samples, and King Man Tai from R&D lab for imaging.

REFERENCES

[1] A. Alderman, L. Burgyan, B. Narveson, and E. Parker, "3D Embedded Technology - Analyzing its needs and challenges," *IEEE Power Electr. Mag.*, vol. 2, no. 4, pp. 30-39, Dec. 2015, doi: 10.1109/MPEL.2015. 2485359.

[2] K. S. Essig, C. Chiu, J. Kuo, P. Chen and J. Yannou, "High efficiency power solutions by chip embedding," in *2016 International Symposium on 3D Power Electronics Integration and Manufacturing (3D-PEIM)*, 2016, pp. 1-16, doi: 10.1109/3DPEIM.2016.7570575.

[3] Y. Chiu et al., "New Developments of Copper Plating Technology for Embedded Power Chip Packages Challenges," in *2019 IEEE 69th Electr. Comp. Tech. Conf. (ECTC)*, 2019, pp. 1426-1431, doi: 10.1109/ECTC.2019.00219.

[4] W. Ki et al., "Chip Stackable, Ultra-thin, High-Flexibility 3D FOWLP (3D SWIFT® Technology) for Hetero-Integrated Advanced 3D WL-SiP," in *2018 IEEE 68th Electr. Comp. Tech. Conf. (ECTC)*, 2018, pp. 580-586, doi: 10.1109/ECTC.2018.00092.

[5] L. Li, P. Ton, M. Nagar and P. Chia, "Reliability Challenges in 2.5D and 3D IC Integration," in *2017 IEEE 67th Electr. Comp. Tech. Conf. (ECTC)*, 2017, pp. 1504-1509, doi: 10.1109/ECTC.2017.208.

[6] K. Knadle, "Reliability and Failure Mechanisms of Laminate Substrates in a Pb-free World," in *Proc. IPC Apex Expo*, Las Vegas, NV, USA, March. 31-April. 2, 2009

[7] P. Fruehauf, A. Munding, K. Pressel, M. Vogt and P. Schwarz, "Chip-package-board reliability of System-in-Package using laminate chip embedding technology based on Cu leadframe," in *2018 7th Electr. Sys. Integ. Tech. Conf. (ESTC)*, 2018, pp. 1-7, doi: 10.1109/ESTC.2018.8546426.

[8] J. K. Lee et al., "Three-Dimensional Integrated Circuit (3D-IC) Package Using Fan-Out Technology," in *2019 IEEE 69th Electr. Comp. Tech. Conf. (ECTC)*, 2019, pp. 35-40, doi: 10.1109/ECTC.2019.00013.

[9] M. Shih, Y. Lee, R. Chen, D. Tarng and C. P. Hung, "Parameters study of thermomechanical reliability of board-level fan-out package," in *2017 Int. Conf. Electr. Pack. (ICEP)*, 2017, pp. 66-70, doi: 10.23919/ICEP.2017.7939326.

[10] K. Kao, S. Wu, Y. Hung, T. Chang, R. Cheng and T. Chen, "Application of numerical analysis to the reliability assessment of a novel package on package (PoP) structure for memory stacking," in *2010 5th Int. Microsys. Pack. Assem. Circ. Tech. Conf. (IMPACT)*, 2010, pp. 1-4, doi: 10.1109/IMPACT.2010.5699573.

[11] F. Hou et al., "Thermo-mechanical reliability study for 3D package module based on flexible substrate," in *2013 14th International Conference on Electronic Packaging Technology (ICEPT)*, 2013, pp. 1296-1300, doi: 10.1109/ICEPT.2013.6756695.

[12] J. Kim, J. Cho, J. S. Pak, J. Kim, J. Yook and J. C. Kim, "High-frequency through-silicon Via (TSV) failure analysis," in *2011 IEEE 20th Conf. Electr. Perf. Electr. Pack. Sys. (EPEPS)*, 2011, pp. 243-246, doi: 10.1109/EPEPS.2011.6100237.

[13] Y. Ning, M. H. Azarian and M. Pecht, "Effects of Voiding on Thermomechanical Reliability of Copper-Filled Microvias: Modeling and Simulation," in *IEEE Trans. Dev. Mat. Relia.*, vol. 15, no. 4, pp. 500-510, Dec. 2015, doi: 10.1109/TDMR.2015.2476823.

[14] A. Munding, A. Kessler, T. Scharf, B. Plikat and K. Pressel, "Laminate Chip Embedding Technology - Impact of Material Choice and Processing for Very Thin Die Packaging," in *2017 IEEE 67th Electr. Comp. Tech. Conf. (ECTC)*, 2017, pp. 711-718, doi: 10.1109/ECTC.2017.261.

[15] U. Rahangdale, P. Rajmane, A. Doiphode and A. Misrak, "Structural integrity optimization of 3D TSV package by analyzing crack behavior at TSV and BEOL," in *2017 28th Adv. Semic. Manuf. Conf. (ASMC)*, 2017, pp. 201-208, doi: 10.1109/ASMC.2017.7969230.

[16] X. Zhang, Jason Chan, L. Cao, Fengze Hou, Hongwen He and L. Wan, "Thermal-mechanical simulation of embedded module based on organic substrate," in *2013 IEEE Int. Symp. Adv. Pack. Mat.*, 2013, pp. 126-136, doi: 10.1109/ISAPM.2013.6510396.

[17] M. Frewein et al., "Package Level Warpage Simulation of a Fan Out System in Board Module," in *2019 20th Int. Conf. Therm. Mech. Phys. Sim. Exp. Microelectr. Microsys. (EuroSimE)*, 2019, pp. 1-7, doi: 10.1109/EuroSimE.2019.8724518.

[18] V. Pham, H. Wang, J. Xu, J. Wang, S. Park and C. Singh, "A Study of Substrate Models and Its Effect On Package Warpage Prediction," in *2019 IEEE 69th Electr. Comp. Tech. Conf. (ECTC)*, 2019, pp. 1130-1139, doi: 10.1109/ECTC.2019.00175.

[19] K. J. Lee, M. Damani, R. V. Pucha, S. K. Bhattacharya, R. R. Tummala and S. K. Sitaraman, "Reliability Modeling and Assessment of Embedded Capacitors in Organic Substrates," in *IEEE Trans. Comp. Pack. Tech.*, vol. 30, no. 1, pp. 152-162, March 2007, doi: 10.1109/TCAPT.2007.892059.

[20] Y. Liu, B. Yang, S. Huang, N. Ye, S. Bhagath and R. Shukla, "Board Level Reliability Enhancement with Considerations of Solder Ball, Substrate and PCB," in *2020 IEEE 70th Electr. Comp. Tech. Conf. (ECTC)*, 2020, pp. 249-256, doi: 10.1109/ECTC32862.2020.00050.

978-1-6654-1392-3/21 $31.00 © 2021 IEEE

Wettability Improvement of Solder in Fluxless Soldering under Formic Acid Atmosphere

Yuhao Bi
School of Mechanical and Electrical Engineering, Guilin University of Electronic Technology
Guangxi Key Lab of Manufacturing System and Advanced Manufacturing Technology
Guilin, China
1017829917@qq.com

Siliang He*
School of Mechanical and Electrical Engineering, Guilin University of Electronic Technology
Guangxi Key Lab of Manufacturing System and Advanced Manufacturing Technology
Guilin, China
siliang_he@guet.edu.cn

Wangyun Li*
School of Mechanical and Electrical Engineering, Guilin University of Electronic Technology
Guangxi Key Lab of Manufacturing System and Advanced Manufacturing Technology
Guilin, China
li.wangyun@guet.edu.cn

Daoguo Yang
School of Mechanical and Electrical Engineering, Guilin University of Eletronic Technology
Guangxi Key Lab of Manufacturing System and Advanced Manufacturing Technology
Gulin, China
daoguo_yang@163.com

Hiroshi Nishikawa
Joining and Welding Institute
Osaka University
Department of Manufacturing Process
Osaka, Japan
nisikawa@jwri.osaka-u.ac.jp

Abstract—**Soldering is applied to the electronic packaging industry. During soldering, flux is the chemical cleaning agent frequently to remove the oxide layers at the surfaces of the solder and the soldered metallic pad for their sufficient interfacial reactions. However, flux residue is a corrosive issue lowing the reliability of solder joint. Additionally, the flux residue is hardly removed in the micro-bump of three-dimensional integrated circuit packaging. Although some studies have demonstrated the fluxless soldering technique with formic acid (FA) atmosphere, in-situ observation on chemical reduction during the soldering has not been studied well, especially for various metals of the soldered pad. In this study, an Sn-Ag-Cu solder was soldered on various substrates under FA atmosphere. Their wettability of Sn-Ag-Cu (SAC) solder was estimated by the measurements of their contact angle and spreading area. The substrates are Cu, Ni, and electroless nickel immersion gold (ENIG), which was frequently used for electronic packaging. Additionally, a soldering process with commercial rosin mildly activated (RMA) flux with Cu substrate was set as the reference. From the results, it demonstrate that the contact angle of solder on Cu substrate under FA atmosphere was no substantial difference to that of solder using RMA flux. The Pre-heat time affects the wettability, but not obvious, and the wettability was increased with increasing the peak temperature. The effect of the FA atmosphere on the wettability of ENIG was much more significant than those of Cu and Ni because the soldering spreading area of ENIG was ten times larger than those of Cu and Ni. Additionally, the pressure of the FA atmosphere was considered as the effect on the soldering wettabilities. Thus, the soldering was tested at different atmospheric pressures of the FA (1kPa, 10kPa, 80kPa). The findings show the contact angle of each soldering system was slightly affected by the FA pressures.**

Keywords—*fluxless soldering，wettability，formic acid*

I. INTRODUCTION

Flux has been widely used in soldering process, due to its strong ability to remove the oxide film on the metal surface to make the pure atom of metals in contact and improve the

wettability of the liquid molten solder on substrate. After soldering, a cleaning process is necessary to remove the flux residues because the residues cause some electrical and mechanical connection problems in solder joints then lead long-term reliability of device decreasing. And some researchers mentioned the solvents using in cleaning process are not Eco-friendly and the volatile organic compounds(VOC) in the solvents are harmful to human beings[1]. To solve these problems, no-clean flux and low VOC flux were developed in recent years. However, some new problems were found, such as flux spattering on in-circuit testing pads, poor reduction ability comparing with traditional fluxes and harsh Pre-heat temperature control in reflow process[1, 2]. Meanwhile, in modern electronics, according to the rapidly advancement of electronic, with the increasing of the interconnection density and reductions in the gap between the chip and substrate gap, the spaces between various components can become roadblocks for the flux cleaning process[3]. Furthermore, fine-pitch interconnection is the way to improve chip performance, it is difficult to spread flux in advance sub-micron fine-pitch structure because the solder was fabricated by electroplating[4]. Therefore, fluxless soldering/bonding is becoming an active area of research, especially in micro-electromechanical system (MEMS), sensors, optical devices packaging area[5].

The key to achieve a sufficient wettability fluxless soldering in electronic packaging is to remove oxides of metals and create a low oxygen atmosphere during soldering process. Hence, in-situ reduction during soldering using reduction atmosphere has a high potential in future due to the simple procedures. The mixed atmosphere of hydrogen and nitrogen can reduce the metal oxide film very well at high temperature. However, the instability of hydrogen at high temperatures, formation of toxic carbon monoxide in the reduction reaction with metals, and high cost limited the mix atmosphere application to industrialization[6]. An Formic acid (FA) atmosphere as an effective reducing agent is a valid alternative to liquid flux in soldering process[7]. The carboxyl in FA replaces the oxygen atom to form formates at about

150°C, and then formates decomposed at 200°C. After soldering, formic acid will decompose spontaneously and will not corrode the solder joints and cause long-term reliability problems[8].

In this study, the wettability of Sn-Ag-Cu(SAC) solder on different substrates under FA atmosphere was evaluated via contact angle and spreading area measurement. The effects of Pre-heat time, peak temperature, substrate metallization and atmosphere pressure on wettability were considered.

II. EXPERIMENTAL PROCEDURE

A. Wettability

As shown in Fig.1,a Sn-mass3.0Ag-mass0.5Cu(SAC305) solder ball (diameter:0.76mm) and a substrate(size:15mm × 15mm × 0.5mm) which indicates three different common substrate(Cu, Ni and ENIG) were prepared for wettability test. For this process, FA vapour was used as the reduction atmosphere(5vol% formic acid + 95 vol% N_2). The contact angle was measured using a 3D laser scanning confocal microscope (VK- 9710, KEYENCE). The spreading area of solder on substrates were measured by an optimal microscope (OM, KEYENCE VHX-900) after heating.

B. Heating processes

To obtain the optimal condition for the substrate and solder ball, the process of pre-treatment before soldering is inevitable. The substrates were infiltrated in 4% HCl solution for 120s. Then, the solder and substrates were ultrasonically cleaned in an ethanol solution for 300 s. After that, the SAC balls were heated on the substrates under a FA atmosphere. Five heating processed with different Pre-heat time and peak temperature were used in wettability test, as shown in Table I. SAC solder was heated using RMA flux as reference. In this study, reflow process curve of solder use RMA flux shown in Fig. 2.

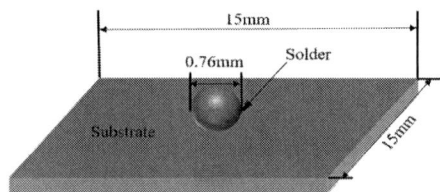

Fig. 1. Schematic diagram of substrate using for experiment

TABLE I. SUMMARY OF SOLDER HEATING PROCESSES IN FA ATMOSPHERE.

Heating process	1	2	3	4	5
Pre-heat temperature (°C)	185	185	185	185	185
Pre-heat time (min)	1	5	10	5	5
Peak heating temperature(°C)	275	275	275	255	295
Peak heating time(min)	2	2	2	2	2
Heating rate（°C/s）	2.5	2.5	2.5	2.5	2.5

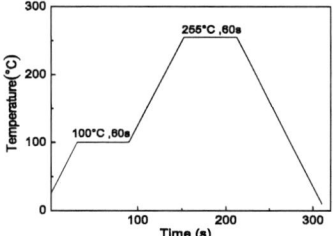

Fig. 2. Reflow process of RMA flux

III. RESULT AND CONCLUSION

A. Contact angle analysis

The equilibrium contact angle was decided by counterpoising the surface tension base on Young's equation[9]:

$$\gamma_{SG} - \gamma_{SL} = \gamma_{LG}\cos\theta \qquad (1)$$

Young's equation is valid upon the suppose of ideal surface(smooth, homogeneous and non-deformable) in the case of thermodynamic equilibrium. The process of wetting and spreading is affected by various factors. For example, the viscosity of the liquid and IMC generated between the solder and different substrate.

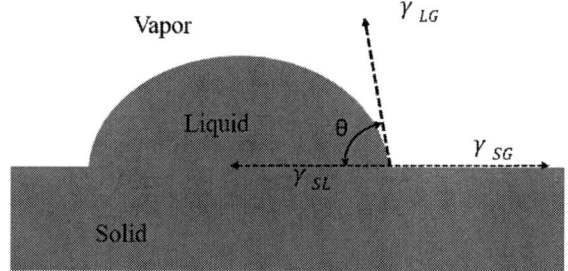

Fig. 3. Schematic of a liquid-drop contact angle system

The wetting of molten SAC solder on the Cu substrate is a classic reactive wetting process, in which IMC is regularly formed. However, the Cu substrate is easily oxidized in the air process, forming a dense Cu oxide film, which hinders the wetting process. Therefore, it's cruical to remove the oxide film for wetting.

In previous studies, Conti. et al[10] purposed a two-step reaction of common metal oxides in the FA reduction atmosphere. First, the reaction of MO with FA to form metal formate (Eq(1)), then the metal formate decompose at higher temperature (Eq.(2))

$$\text{HCOOH}_{(g)} + \text{MO}_{(s)} \rightarrow M(COOH)_{(s)} + H_2O_{(g)} \qquad (2)$$

$$M(COOH)_{(s)} \rightarrow M_{(s)} + CO_{2(g)} + H_{2(g)} \qquad (3)$$

In this study, the formed contact angles between of SAC solder balls and Cu substrate under different processes were measured. It was found the contact angle between solder and Cu substrate under an FA atmosphere with a peak temperature of 295°C was the best(34°), and the result under process 4(255°C) is worst. The results show that with the increase of peak temperature, the contact angle between solder and Cu substrate becomes better, as shown in Fig. 4.

The reactive wetting system is particularly sensitive to temperature as temperature affects many properties of liquid and solid such as viscosity, surface tension and oxidation behavior. The viscosity and surface tension of the liquid solder reduced when the wetting system temperature increased.. Hence, the value of γ_{LG} in Young's Equation becomes smaller when peak temperature became higher. The value of $\cos\theta$ have to be larger to keep the equal of Young's Equation, leading contact angle becomes smaller. On the other hand, Conti et al claimed the temperature should reach 256 °C to remove the strong oxidation of Cu effectively using FA atmosphere[10]. The peak temperature of Process 4 was below 256 °C may also be one reason to the worst contact angle result.

The effect of the Pre-heat time of FA reflow soldering on contact angle was also studied. The contact angle results acquired by process 1, process 2 and process 3 indicate that contact angle increases with the Pre-heat time increasing under the FA atmosphere, but it's not obvious. This result is indicates one minute at 185°C is enough for Pre-heat in a fluxless soldering process under an FA atmosphere. Because FA not only react with oxides of solder, but also react with solder, the content of FA decreases due to the continuous reaction between solder and FA in the sealed chamber. A longer Pre-heat time may lead lack of FA to remove the oxides on Cu substrate because FA effectively react with Cu oxides need higher temperature[10]. The Cu oxides prevent the wetting between SAC305 and Cu substrate.

The result of the contact angle between solder and Cu substrate using RMA flux is also shown in Fig. 4. There was no significant difference comparing with that under an FA atmosphere.

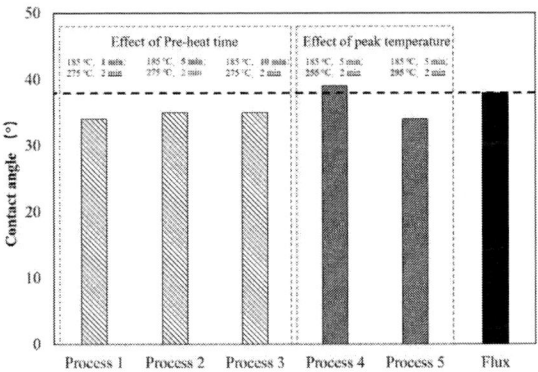

Fig. 4. Contact angle of SAC solder on Cu substrates with different Process in FA atmosphere.

B. Spreading area analysis

In addition to the above mentioned contact angle, the spreading area of the solder on the substrate is also an important evaluation method to demonstrate the wetting behavior of liquid solder on substrate.

Fig. 5 shows the average spreading area of solder joints on Cu substrates by changing the Pre-heat time (same as previous section)under FA atmosphere. The results on Cu substrate acquired by process 1, process 2 and process 3 are 2.93, 2.91 and 2.86mm^2, indicated that spreading area increases as the Pre-heat time decrease under the FA atmosphere which match contact angle results very well. The difference of wetting results under different Pre-heat time is not very obvious.

Base on the above verification, the change in the Pre-heat time of the process has a limited effect on the wettability. However, the increase of the peak temperature lead increasing of the contact angle between the solder and the Cu substrate under FA atmosphere. To confirm this phenomenon, the spreading area of solder on substrate was also measured. As shown in Fig. 6, the average spreading area of Cu substrates using different peak temperature(255, 275 and 295°C), are 2.75, 2.91 and 3.01mm^2, respectively. The same conclusion

mentioned in the previous section of the wetting angle experiment was obtained: increasing the peak temperature of heating process will improve the wettability of SAC solder on the Cu substrate. To explore the metallization of substrate effect on wettability, Ni and ENIG substrates were prepared. The same trendency could be observed on Ni substrates and ENIG substrates. Rozhitsina et al.[11] found that as the temperature increases, the viscous flow activation energy of SAC305 eutectic melt decrease. Accordingly, the viscosity of solder SAC305 decrease with rising of peak temperature, making it easier to spread the molten solder on the substrate.

Otherwise, as shown in Fig. 6, the average spreading area of SAC solder on Cu, Ni and ENIG substrates at the peak temperature of 275° are 2.91, 2.68 and 42.05 mm^2, respectively. The data of spreading area shows a gap of more than ten times between ENIG substrate and Cu, Ni substrate. It's well known that the metal oxide film is an important factor affecting the spreading effect of the interface between solder and substrate. Either the oxide film on the solder surface or the oxide film on the substrate surface will have a great impact on its wetting behavior. The oxide film of solder and base metal would be a barrier to prevent wetting. The oxide film on the Cu substrate and Ni substrate will hinder the spreading process, and the spreading area of the solder on the substrate will be greatly reduced. In contrast, the ENIG finish on the substrate without any oxide film, which improved wettability greatly. It has been verified by a large number of studies that using ENIG substrates for micro-connection is a common method to improve wettability[12].

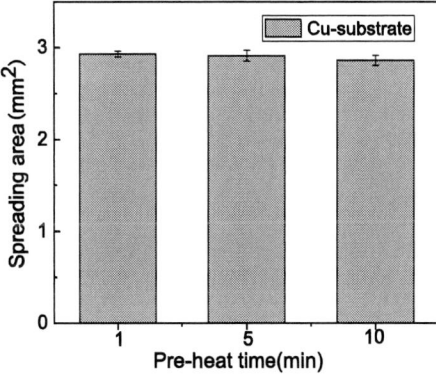

Fig. 5. Spreading area of SAC solder on Cu substrates with different Pre-heat time(1,5 and 10 min) in FA atmosphere.

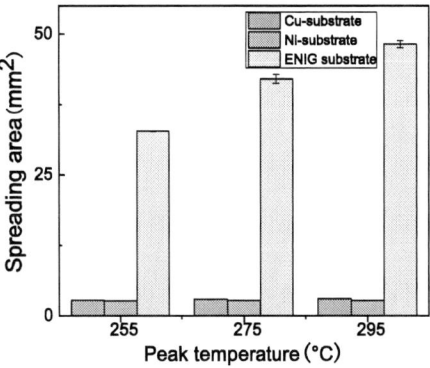

Fig. 6. The spreading area of SAC solder on Cu, Ni and ENIG substrates at different peak temperatures(255, 275 and 295°C) under FA atmosphere.

C. The pressure of formic acid atmosphere analysis

Pressure of atmosphere is also a factor that affects liquid wetting. Three approximately proportional pressures of FA atmosphere were set in this study. Considering about the safety of experimenter and hermeticity of chamber, we used 80kPa instead of 100kPa. Wettability of solder on Cu substrate under same pressures of nitrogen atmosphere was also observed. As shown in Table II, the results under different pressures (1kPa, 10kPa and 80kPa) of FA atmosphere are almost the same. However, solder was not capable to wet on Cu substrate under any pressure of nitrogen atmosphere set in this study. This result indicate the FA atmosphere could effectively improve the wettability of solder and the effect of FA atmosphere pressure on wettability was unobvious. This result is important to apply the fluxless soldering process under a FA atmosphere in electronic packaging because decreasing atmosphere pressure is a common method to reduce the void in solder joint and improve the reliability of solder bumps.

TABLE II. SUMMARY OF CONTACT ANGLE RESULTS OF DIFFERENT PRESSURES IN DIFFERENT ATMOSPHERE

	1kPa FA	10kPa FA	80kPa FA	1kPa N_2	10kPa N_2	80kPa N_2
Contact Angle (°)	42	41	41	Not Wetting	Not Wetting	Not Wetting

IV. CONCLUSION

In this study, the effects of an FA atmosphere on the wettability of SAC with Cu, Ni, and ENIG were examined. The contact angle of the SAC on the Cu under different pressures of the FA atmosphere was also investigated.

According to the result, FA can remove oxides on solder and substrate to improve wettability, the contact angle of solder on Cu substrates under FA atmosphere can approach that of solder using RMA flux. Samples with one minute Pre-heat time acquired better contact angle results. The increase in the peak temperature can significantly improve the wettability of SAC. In this study, the wettability of molten solder on the substrate is the best when the peak temperature reached 295 °C.

Moreover, the spreading area of the ENIG finish substrates are ten times larger than those of the Cu and Ni substrates, showing that ENIG finish substrates can improve wettability obviously. Besides, the FA atmosphere under different pressures has no significant effect on the wettability of SAC/Cu.

ACKNOWLEDGMENT

This work was supported by Guangxi Natural Science Foundation(No.2021GXNSFBA075027,2019GXNSFAA245 059 and Guike AD20297022), Guangxi Key Laboratory of Manufacturing System & Advanced Manufacturing Technology (Grant No.20-065-40-003Z).

Wangyun Li thanks the support by National Natural Science Foundation of China under Grant No. 51805103.

REFERENCES

[1] Songbo, X.et al., *Development of Green Welding Technology in China During the Past* Decade, Materials Reports, 2019. 33(9): p.2813-2830. in Chinese.

[2. Dusek, K. and D. Busek, *Problem with No-clean Flux Spattering on In-circuit Testing Pads Diagnosed by EDS Analysis.* Microelectronics Reliability, 2016. 56: p. 162-169.

[3] Samson, M., et al., *Fluxless Chip Join Process Using Formic Acid Atmosphere in a Continuous Mass Reflow Furnace.* 2016 IEEE 66th Electronic Components and Technology Conference (ECTC), 2016: p. 574-579.

[4] Morita, M., et al., *Formation of Solder Cap on Cu* Pillar Bump *using Formic Acid Reduction.* Proceedings of the 2012 Ieee 14th Electronics Packaging Technology Conference(EPTC), 2012: p. 602-607.

[5] Mu, F.W., et al., *Nano-Cu Paste Sintering in Pt-catalyzed Formic Acid Vapor for Cu Bonding at a Low Temperature.* 2019 International Conference on Electronics Packaging (ICEP), 2019: p. 365-366.

[6] Zhixian,M.et al.,*Studies on Flux-free Soldering Technology.*Eletronic Process Technology,2014.35(1):p. 19-25.in Chinese.

[7] He, S., et al., *In-situ Observation of Fluxless Soldering of Sn-3.0Ag-0.5Cu/Cu under a Formic Acid Atmosphere.* Materials Chemistry and Physics, 2020. 239.

[8] H.Siliang, N.Hiroshi, *Effect of thermal aging on the impact* strength *of soldered bumps under formic acid atmosphere.*Quarterly Journal of The Japan Welding Society, 35(2017) : p. 127s-131s.

[9] Noor, E.E.M., N.F.M. Nasir, and S.R.A. Idris, *A Review: Lead Free Solder And Its Wettability Properties.* Soldering & Surface Mount Technology, 2016. 28(3): p. 125-132.

[10] Conti, F., et al., *Thermogravimetric Investigation on the Interaction of Formic Acid With Solder Joint Materials.* New Journal of Chemistry, 2016. 40(12): p. 10482-10487.

[11] Rozhitsina, E.V., et al., *Dynamic Viscosities of Pure Tin and Sn-Ag, Sn-Cu, and Sn-Ag-Cu Eutectic Metals.* Russian Metallurgy (Metally), 2011. 2011(2): p. 118-121.

[12] Siliang He. and Hiroshi Nishikawa, *Effect of Substrate Metallization on the Impact Strength of Sn-Ag-Cu* Solder *Bumps Fabricated in a Formic Acid Atmosphere.* 2017 International Conference on Electronics Packaging (ICEP), 2017: p. 381-385.

The Effect of Flux on Si-Al Wire Bonding Reliability

Yao Zhang*
Hybrid integrated circuit department
Xi'an Microelectronics Technology Institute
Xian, China
523814939@qq.com

Wuxing Cao
Hybrid integrated circuit department
Xi'an Microelectronics Technology Institute
Xian, China
cao_wuxing@163.com

Jun Zhang
Hybrid integrated circuit department
Xi'an Microelectronics Technology Institute
Xian, China
zjdamon@sina.com

Liu Yang
Hybrid integrated circuit department
Xi'an Microelectronics Technology Institute
Xian, China
14790809@163.com

Pei Zhang
Hybrid integrated circuit department
Xi'an Microelectronics Technology Institute
Xian, China
343629522@qq.com

Jiao Yang
Hybrid integrated circuit department
Xi'an Microelectronics Technology Institute
Xian, China
418811455@qq.com

Abstract—**The rosin flux is commonly used in the welding process of the thick-film components for DC/DC power supply circuits. Under certain conditions, the organic acids in the flux react with Si-Al wire and hence diminish the wire bonding strength. Based on the reaction mechanism between Si-Al wire and flux, this paper investigates the impact of rosin flux on the bonding strength of Si-Al wire at the interface of Au. In particular, the micro-structure change of the bonding interface versus standing time of flux is researched with microscopic methods. The experimental results show that the rosin flux has a great influence on the bonding strength of the thick-film gold interface when the standing time is more than 6 hours. The wire bonding strength declines due to the dendritic structure formed in Si-Al wire. It is found that, by accurately controlling standing time of the flux, the reliability of wire bonding strength is significantly improved.**

Keywords—flux, wire bonding strength, reliability

I. INTRODUCTION

Thick film power supply is composed of discrete semiconductor chips, magnetic components, resistance and capacitance components on thick film substrate. It has the characteristics of high assembly density, good electrical performance and flexible design. In the process of assembly, power chip, resistance capacitance component and magnetic component are mainly assembled by soldering. Flux is used as auxiliary material in soldering, the main function of which is to remove the oxide on the surface of solder and base metal to make the metal surface clean, reduce the surface tension of solder and improve the welding performance, then the residual flux is removed by a cleaning process after welding[1]. Due to the mechanism of corrosion between the activator in the flux and the aluminum metal, if the residual flux is not cleaned timely and effectively after welding, it may cause corrosion to the aluminum wire in the circuit, leading to the decline of the reliability of bonding[2].

This paper starts from the reaction mechanism of flux and aluminum metal, mainly studies the influence of rosin flux on the bonding strength of Si-Al wire under the gold interface, and studies the change of morphology of Si-Al wire and the bonding interface under different residence time of flux from the microscopic level. Through the test, the cleaning effect of different placement time of flux and bonding reliability is verified, which is of great significance to the optimization of post-welding process conditions.

978-1-6654-1392-3/21 $31.00 © 2021 IEEE

II. CHARACTERISTICS OF FLUX

A. Selection of flux

In the process of soldering, rosin flux is usually used to assist welding. Rosin flux refers to the rosin or modified rosin as the main composition of flux, generally composed by the activator, solvent and surfactant, there are some special ingredients, including corrosion inhibitor, antioxidant and film forming agent. Flux activators are generally composed by hydrogen, inorganic salts, acids, amines, and their complex compositions. Adding halogen-containing (F, Cl, Br) active agents into the flux can significantly improve its solderability and welding effect, but too much content will cause a series of corrosion problems and lead to cleaning difficulties[3]. Therefore, according to the national military standard, thick film power supply products usually choose RMA moderately active rosin flux, halogen content is generally not more than 0.15%, which can effectively remove the oxides on the welding surface and is also conducive to the cleaning of surface residues.

B. Reaction mechanism between flux and aluminum metal

Thick film power supply products mainly use Si-Al wire ultrasonic bonding process to realize the electrical interconnection between the component pin and the substrate. The Si-Al wire is mainly composed of Al and doped with 1% Si. Under the action of ultrasonic energy and pressure, the Si-Al wire rubs rapidly at the bonding interface and plastic deformation occurs, completing the bonding in a relatively short time. The bonding point morphology of Si-Al wire at the bonding interface is shown in Figure 1:

Fig. 1. Morphology of Si-Al wire bonding point

As can be seen from the morphology of the bonding point, the root of the Si-Al wire bond point has the largest deformation, which is the most vulnerable position of the whole bonding wire. Moreover, there will be microcrevice between the bonding point and the bonding interface, which is most likely to cause the flux residue.

Although the halogen content in the RMA type rosin flux is strictly controlled according to the requirements of the national military standard, the halogen composition still exists. When the flux and the Si-Al wire contact long enough, the halogen composition in the flux will corrode the Si-Al wire, the specific chemical reaction equations as in (1) and (2).

$$Al+4Cl \longrightarrow Al(Cl)_4^-+3e^- \qquad (1)$$
$$2Al(Cl)_4^-+6H2O \rightarrow 2Al(OH)_3+6H^++8Cl^- \qquad (2)$$

Chemical corrosion will lead to the formation of dendritic structure on the Si-Al wire, resulting in the decline of bonding strength, thus reducing the product reliability.

III. EXPERIMENT CONTENTS AND RESULTS

In order to further verify the corrosion effect of the residual time of flux on the Si-Al wires and the cleaning effect after welding, the paper studies the influence of flux residue on bonding reliability by analyzing the micro-structure of bonding interface under different placement time of flux residue and testing the bonding strength data of the bonding wire.

A. experimental material

The test substrate is Al2O3 thick film ceramic substrate; the bonding interface is gold; the solder pad is palladium silver; the flux used in the test is liquid rosin flux; the bonding wire use Φ40μm Si-Al wire; the vapor phase cleaning process is used after welding.

B. experimental content

The test sample was prepared and bonded with Φ40μm Si-Al wire on the gold interface. Welding is performed on the pads around the bonding wire, so that the flux can soak to the bonding point. The samples were cleaned by vacuum vapor phase cleaning after placement with flux for 2hours, 4hours and 6hours. Specific test groups are shown in Table 1.

TABLE I. SAMPLE GROUPS FOR DIFFERENT PLACEMENT TIMES

Sample Number	Test group
1#	Placed for 2h after welding
2#	Placed for 4h after welding
3#	Placed for 6h after welding

C. experimental results

1) Comparison of cleaning effect at different placement time of flux

The morphology of 1#-3# test samples before and after vapor phase cleaning was compared, as shown in Figure 2. It can be found that after the vapor phase cleaning of sample 1#, there was no flux residue mark on the gold surface and the bonding wire, indicating that the flux residue could be effectively removed by cleaning within 2 hours after welding. After the vapor phase cleaning of sample 2#, there was no flux residue mark on the bonding wire, but the color of gold surface slightly changed, indicating that the flux residue could not be completely cleaned after 4 hours placement after welding. On sample 3#, there were obvious white flocculent residue remained at the root of many bonding wires, indicating that the flux had chemically reacted with the Si-Al wire after 6 hours after welding, and white dendritic structure was formed around the bonding point.

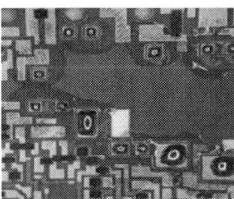

a. 1# flux residue

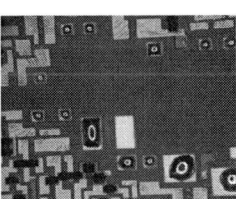

b. 1# after cleaning

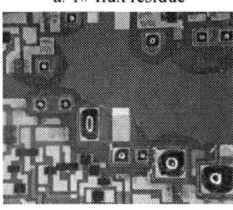

c. 2# flux residue

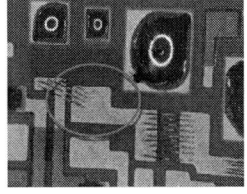

d.2# after cleaning

e.3# flux residue

f.3# after cleaning

Fig. 2.   Interface morphology of test sample before and after vapor phase cleaning

SEM and energy spectrum analysis were performed on the bonding interface of 1#-3# test samples after cleaning, and the analysis results were shown in Figure 3. The appearance of the bonding point and bonding interface of sample 1# is normal, the main component of the bonding interface is gold, indicating that the flux residue can be effectively removed by vapor phase cleaning within 2 hours after welding. Bonding point morphology of sample 2# is normal, but the bonding interface has a slight discoloration. In addition to gold, the bonding interface also contains obvious carbon and oxygen elements, that further verified the flux residue cannot be cleaned up thoroughly after placing for 4 hours after welding, the longer the flux residue stays, the more difficult it is to clean. In sample 3#, white flocculent residue was found at the root of many bonding points, and the bonding interface was discolored. The spectrum analysis of the white flocculent shows that the main components are aluminum, oxygen and chlorine. According to the reaction mechanism between the flux and the Si-Al wire, it can be concluded that when placed 6 hours after welding, the chloride ion in the flux will react with the Si-Al wire to generate $Al(OH)_3$ white flocculent. The longer soaking time between the flux and the Si-Al wire is, the more sufficient reaction is, the $Al(OH)_3$ white flocculent dendritic structure is generated around the bonding point, which cannot be cleaned up by the vapor phase cleaning.

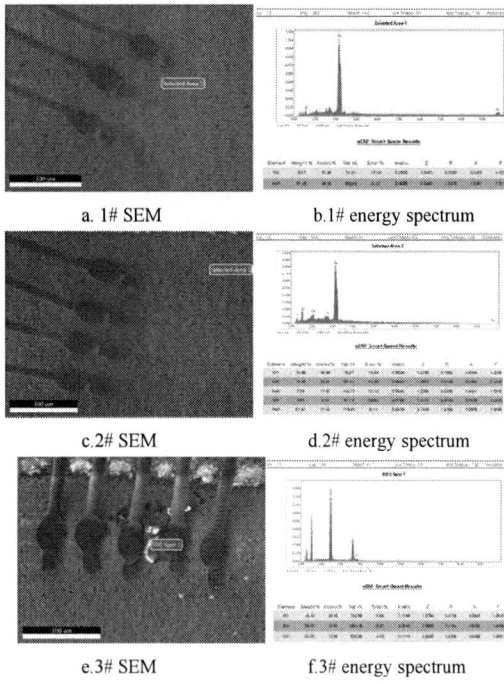

a. 1# SEM   b.1# energy spectrum

c.2# SEM   d.2# energy spectrum

e.3# SEM   f.3# energy spectrum

Fig. 3. SEM and energy spectrum analysis of the bonding interface

2) Comparison of bonding strength under different flux placement time

The bonding strength of 1#-3# samples were tested at room temperature and high temperature of 300 ℃. The test results are shown in Figure 4 and Figure 5.

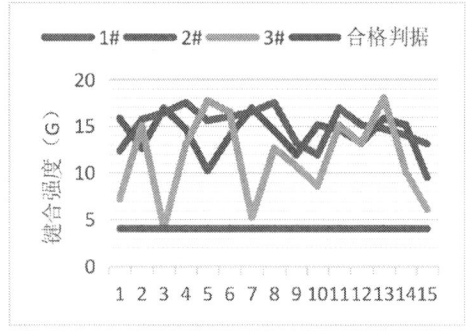

Fig. 4. Bond strength data at room temperature

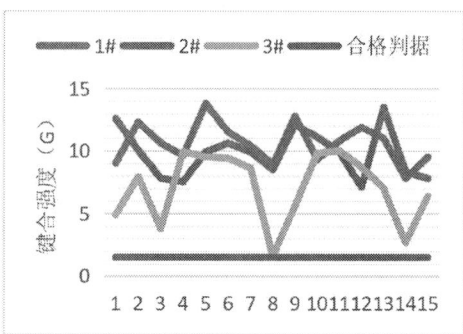

Fig. 5. Bond strength data at high temperature

It can be seen from the bonding strength test results, the bonding strength of 1 # samples under the condition of room temperature and high temperature meet the requirement of standard without bonding-off failure mode, and the consistency of bonding strength is good. The data show that when the test sample clean within 2 hours after welding, flux residues can be removed effectively. The bonding strength of 2 # samples under the condition of room temperature and high temperature can satisfy the requirements of the standard, there is no bonding-off failure mode in room temperature, but the bonding-off failure rate reach to 13.3% under the condition of high temperature, the results indicate that the residual flux exists after 4 hours placement, which accelerates the formation of Au/Al MIC at high temperature and forms Kokendall void, which leads to the failure of bonding. Although the bonding strength of 3# sample at room temperature and high temperature are within the range of qualified criteria, there are several critical low values, and the bond strength fluctuates greatly, moreover, the bonding-off failure rate at high temperature is as high as 33.3%. The results show that most of the root of the bonding point has been corroded by flux when the sample is placed for 6 hours after welding. The Al(OH)3 white dendritic structure was formed in the reaction, which resulted in the root damage of the bonding point and decreased the bonding strength. Under high temperature condition, the reaction between the bonding wire and the flux is intensified and the bonding strength is further reduced.

IV. CONCLUSION

In this paper, the influence of RMA type rosin flux on Si-Al wire is taken as the research object, the cleaning effect of different placement time of the flux and the influence on bonding reliability are studied, and the mechanism of the corrosion of Si-Al wire is confirmed. According to the test results, in order to further improve the reliability of power supply products, three control requirements for welding process were proposed: 1) Avoid designing the soldering pads near the bonding area from the product layout design source; 2) Control the amount of flux in the welding process to reduce the overflow range of flux; 3) Strictly control the placement time after welding to ensure that the vapor phase cleaning is carried out within 2 hours after welding, which can effectively guarantee the cleaning effect of the circuit and effectively reduce the flux residue at the bonding interface, thus improving the reliability of bonding.

ACKNOWLEDGMENT

I would like to express my gratitude to all those who helped me during the writing of this thesis.

My deepest gratitude goes first to the chief technologist Yali Xi, who provided professional modification suggestions to the thesis.

Second, I would like to extend my sincere gratitude to my colleagues. Without their helpful advisement and discussion, this thesis could not have reached its present form.

Last my thanks would go to my beloved family for their support and trust. I also owe my sincere gratitude to my friend Youzhe Fan who gave me the help in my English writing.

REFFERENCE

[1] Guanghui Sun, "The effects of flux residues on the PCB", J. Inspection and Reliability, 2011 -0209-05, pp. 209-213.

[2] Morten S. Jellesen, etc. "Corrosion failure due to flux residues in an electronic add-on device", J. Engineering Failure Analysis, 2010(17), pp. 1263-1272.

[3] Si Gao, Shengrong Liu, "Brief Introduction to the Roles of Components in Fluxes",J.Printed Circuit Information, 2010 No.1, pp. 52-55.

Finite element analysis of thermal contact resistance in Press-Pack IGBT module

1st Tong An*
Institute of Electronics Packaging Technology and Reliability
Beijing Key Laboratory of Advanced Manufacturing Technology
Faculty of Materials and Manufacturing
Beijing University of Technology
Beijing, China
antong@bjut.edu.cn

2nd Rui Zhou
Institute of Electronics Packaging Technology and Reliability
Beijing Key Laboratory of Advanced Manufacturing Technology
Faculty of Materials and Manufacturing
Beijing University of Technology
Beijing, China
1193393036@qq.com

3rd Fei Qin
Institute of Electronics Packaging Technology and Reliability
Beijing Key Laboratory of Advanced Manufacturing Technology
Faculty of Materials and Manufacturing
Beijing University of Technology
Beijing, China
qfei@bjut.edu.cn

4th Yanpeng Gong
Institute of Electronics Packaging Technology and Reliability
Beijing Key Laboratory of Advanced Manufacturing Technology
Faculty of Materials and Manufacturing
Beijing University of Technology
Beijing, China
yanpeng.gong@bjut.edu.cn

5th Chen Liang
Beijing New Energy Vehicle Technology Innovation Center Co., Ltd
Beijing Automotive Research Center Co., Ltd
Beijing, China
liangchen@nevc.com.cn

Abstract—**PP IGBT is a new type of electronic device, which has different characteristics, such as compact structure, double-sided heat dissipation, and large scale parallel chip can be realized in a single device. It is widely used in power system engineering applications such as flexible direct current transmission, flexible alternating current transmission, and photovoltaic access. PP IGBT are generally used in series applications requiring external pressure to maintain electrical and mechanical connections between the internal components. For PP IGBT, thermal resistance is an important indicator related to its thermal characteristics. The thermal resistance of PP IGBT includes not only the thermal resistance of the material of each component, but also the thermal contact resistance. A plurality of contact surfaces is formed between each material layer inside the device to form a thermal contact resistance. Insufficient contact surface will increase the thermal contact resistance, resulting in high temperature or uneven temperature distribution. Therefore, the thermal contact resistance will have an important effect on its performance. In this paper, through the establishment of single die PP-IGBT module three-dimensional finite element model, thermal-electrical-structural coupling analysis was carried out on PP IGBT. Calculate and compare the values of the thermal contact resistance of the devices before and after 250000 power cycles experiment. The effect of the contact surface morphology on the thermal contact resistance is mainly studied. Furthermore, the effect of the thermal contact resistance of all material layers on the overall thermal resistance of the PP-IGBT is obtained.**

Keywords—press pack Insulated gate bipolar transistor (PP IGBT), thermal-electrical-structural coupling, thermal contact resistance, finite element simulation

I. INTRODUCTION

The new type of PP IGBT has the different characteristics from the wire-bonded high power IGBT, and the unique advantage of the PP IGBT devices is that the internal leads are contacted by crimping which means without connection of bonding wires. Therefore, it reduces the risk of solder joint cracking caused by large amount of heat generated by the power cycle and the thermal cycle. In addition, because of the crimp package has a more compact structure, a larger-scale chip parallel can be realized. And the PP IGBT module series technology will make the structure of the main circuit simpler, make the device more compact and lighter, and have unique advantages in the application of VSC-HVDC, photovoltaic power generation is connected to the grid and other power system engineering applications [1,2]. The internal components (Copper block, molybdenum plate on both sides of collector and emitter, as well as IGBT chip, emitter silver gasket) of the PP IGBT device are stacked in sequence through a certain external pressure and kept mechanical connect with electricity. For PP IGBT devices, thermal resistance is a critical thermal characteristic parameter. The thermal resistance of PP IGBT includes not only the thermal resistance of the material of each component, but also the thermal resistance between the contact surfaces. Multiple contact surfaces are formed between the various material layers inside the device, thereby forming thermal contact resistance. Insufficient contact interface will increase thermal contact resistance, resulting in excessive temperature or uneven temperature distribution. Therefore, the thermal contact resistance will have an important impact on the performance of PP IGBT. The research of the influence of contact surface morphology changes on thermal contact resistance is of great significance for solving the reliability problems of PP IGBT devices.

Poller et al. proposed a method of measuring temperature, and carried out experimental measurement and finite element simulation to determine the electrical resistance and thermal resistance of the contact interfaces of the PP IGBT [3]. Deng Erping established a finite element model to analyze the thermal resistance of the internal contact interfaces of the PP IGBT, and verified the influence of temperature on the thermal resistance of the contact interfaces of the PP IGBT through experiments[4]. Bahrami M et al. established an

978-1-6654-1392-3/21 $31.00 © 2021 IEEE

approximate analysis model to predict the amount of heat that the gas existing in the contact interface gap can transfer. The main input parameters that affect the thermal resistance of the contact interfaces and the thermal resistance of the gap gas were studied[5].

This paper mainly establishes a three-dimensional finite element model of single die PP IGBT module, conducts an electro-thermal-mechanical coupling analysis of the device, calculates the value of the thermal resistance of the device before and after 250000 power cycles, and compares the values. The focus is on the influence of the change of the contact surface morphology on the thermal resistance of the contact interfaces. Finally, we can get the effect of the change of the thermal resistance between the contact surfaces of all components on the overall thermal resistance of the PP IGBT device is obtained.

II. BASIC THEORETICAL PREPARATION

A. PP IGBT device structure

The PP IGBT module shell is packaged in a ceramic case. There are several sub-module units inside the module, which are all independent sub-module structures. And it is distributed in a specific way and in a specific ratio. In the independent sub-modules, the IGBT chip and the FRD chip have independent plastic bases. The single die PP IGBT module includes the following components: chip, molybdenum sheets, silver sheet, plastic frame, gate thimble (Figure 1) [6]. Among them, the silver sheet is located between the pedestal of the device and the molybdenum sheet underneath, which serves as a connection. The plastic frame has a certain degree of corrosion resistance and can fix various components in the sub-module. The gate thimble serves as a connection between the gate of the chip and the gate PCB of the device. Each component in the sub-module maintains electrical and mechanical connections through mechanical pressure.

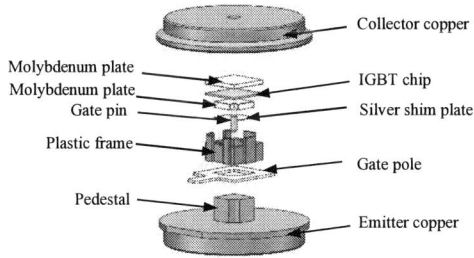

Fig. 1. Structure of PP IGBT

B. Thermal contact resistance

The microstructure of contact interface between any parts inside the PP IGBT is shown in Figure 2. The upper and lower interfaces are not completely in contact, and the other uncontacted places will be filled with gas. The thermal conductivity of air is about 0.023W·m⁻¹·K⁻¹, which is very small and negligible compared to the thermal conductivity of the contact place. Therefore, the main conduction method of heat flow between the interfaces is contact conduction. When two interfaces are in contact, and the interface is rough, their tiny contacts will be randomly distributed in the interface, other voids are filled with air, and the thickness of such voids is relatively small. With such a small gap thickness, the Grashof number is generally less than 2000, so the heat convection can be ignored. When the device is in service, the

surface temperature of its components is less than 700K, and the heat radiation phenomenon is not obvious and can be ignored. In summary, the thermal conductivity of the contact interface only needs to consider the case of heat conduction [4].

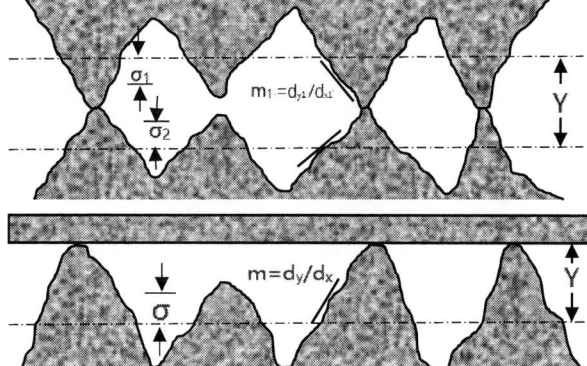

Fig. 2. Microstructure of contact surface.

The area of the contact surface of the internal components of the PP IGBT affects the thermal conductivity of the device, so the contact area affects the thermal resistance between the contact surfaces. The larger contact area can better transfer heat between the two contact surfaces, and the thermal resistance between the contact surfaces is also reduced. It is difficult to directly measure the area of the contact surface in the device, and the degree of the contact area can be characterized by the surface roughness.

In the equivalent contact surface, the effective surface roughness σ and mean absolute slope m as given by

$$\sigma = \sqrt{\sigma_1^2 + \sigma_2^2} \tag{1}$$

$$m = \sqrt{m_1^2 + m_2^2} \tag{2}$$

Among them, 1 and 2 respectively represent two contact surfaces, σ is a characteristic parameter root mean square height in surface roughness and m is related to machining error and surface roughness. The approximate relationship between m and σ is as follows.

$$m = 0.125 \ (\sigma \times 10^6)^{0.402} \tag{3}$$

In a single die PP-IGBT module , assuming that the rough surface has only plastic deformation and the distance between the interface contact points is sufficiently large, the contact thermal conductivity can be obtained.

$$TCC = 0.125 k_s \frac{m}{\sigma} \ \left(\frac{p}{H_c}\right)^{0.95} \tag{4}$$

$$k_s = \frac{2k_1 k_2}{k_1 + k_2} \tag{5}$$

In the above formula, k_s is the effective thermal conductivity of the contact surface, p is the contact pressure of the joint, k_1 and k_2 are the thermal conductivity of the upper and lower interface materials respectively, and H_c is the hardness of the soft material of the two contact surface materials[7,8].

III. FINITE ELEMENT SIMULATION

A. Modeling in ABAQUS

Using ABAQUS finite element simulation software, in order to save calculation time, the device model is simplified under the premise of ensuring the calculation accuracy. And establish a three-dimensional finite element model of single die PP IGBT module, and conduct an electro-thermal-mechanical coupling analysis. The actual shape of each component is considered when building the model. The simplified model is shown in the figure 3.

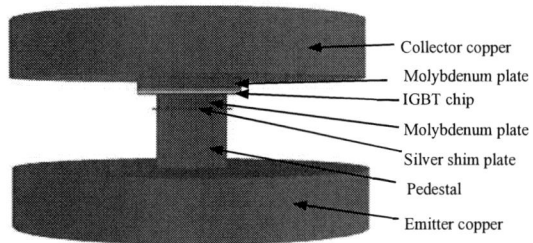

Fig. 3.   Model of PP IGBT.

Establish the material properties of all components of the device, as shown in the table 1.

TABLE I   MATERIAL PROPERTIES

Materials	Thermal conduction (W/m·K)	Eelectrical Conductivity(S/mm)	Density (10^{12}kg/ mm^3)	Specific heat (J/kg·K)
Aluminum	237	37.73	2.70×10^{-9}	900
Silicon (IGBT)	148	0.0001299	2.33×10^{-9}	700
Solder (SAC305)	57	9.615	7.30×10^{-9}	230
Silver	429	63.19	1.05×10^{-8}	235
Molybdenum	138	19.23	1.02×10^{-8}	250
Copper	400	59.52	8.93×10^{-9}	380

B. Contact conditions

The internal chip of the PP IGBT device is connected with the upper molybdenum sheet through a solder layer, and the copper boss is connected with the emitter copper block. The internal contact interfaces of the device are the contact surface between the collector copper block and the collector molybdenum sheet, the contact surface between the IGBT chip and the emitter molybdenum sheet, the contact surface between the emitter molybdenum sheet and the silver shim plate, and the contact surface between the silver shim plate and the pedestal. There are four contact interfaces in total. The surface roughness of the internal components of the device can be measured in experiments firstly. Then the thermal contact resistance of the mutual contact interface can be obtained through calculation. After 250000 power cycles of the device, the surface roughness of each component in the device is different from the initial state, and the thermal resistance between the contact surfaces of the device also changes, as shown in the following table 2. In the simulation, the contact thermal conductivity and contact conductivity are set for each component contact.

TABLE II   THERMAL CONTACT RESISTANCE

Contact surface	Initial state （K/W）	250000 power cycles(K/W）
Collector Cu-Mo	0.0460	0.0522
IGBT chip-Emitter Mo	0.0179	0.0197
Emitter Mo – Ag	0.0116	0.0131
Ag- Emitter Cu	0.0057	0.0067

C. Boundary condition and loads

The PP IGBT device maintains the mechanical and electrical connections of the components in the device through externally loaded pressure. In the power cycle experiment, the pressure of the device is applied by an external fixture. According to the actual situation of the device in the experiment, certain the boundary conditions are set for the finite element model established in this paper. During the simulation, the emitter copper block side is fixed, and a displacement load is applied to the collector copper block side. This displacement load causes the pressure on the device to be about 1.2KN. Under the service condition of the device, the chip temperature will rise and cause thermal expansion, but the displacement change caused by this expansion is much smaller than the displacement caused by external pressure loading and can be ignored. In the power cycle experiment, the current flows in on the collector and flows out on the emitter, so the surface current is set on the collector side, the magnitude is about $0.0273\mathrm{A}/mm^2$, and the zero potential is set on the emitter side. In the actual experiment, the heat generated by the device is transferred to both sides, and the temperature of the heat sink is 45°C. In the simulation, the temperature on both sides of the collector and emitter is set to 45°C. The device does power cycling at room temperature, and the initial ambient temperature is set to 20°C.

D. Solution results

Use ABAQUS software to conduct electro-thermal-mechanical coupling analysis, and obtain the junction temperature curve of the PP IGBT in the initial state and after 250000 power cycles(Figures 4 and 5).

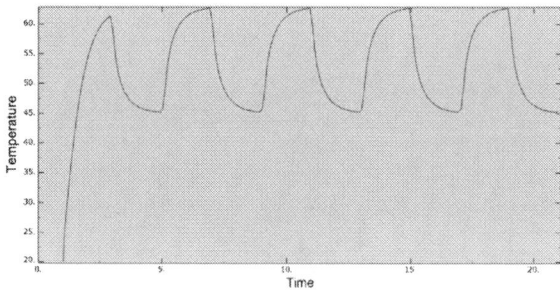

Fig. 4.   Initial junction temperature curve

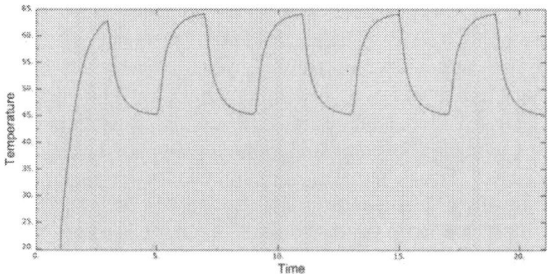

Fig. 5.   250000 power cycles junction temperature curve

According to the junction temperature curve obtained by simulation results, the maximum temperature of the chip inside the device before and after the power cycle is 62.6°C and 64.2°C respectively. The reason for the temperature change is that the surface roughness of each contact interface inside the device changes before and after the power cycle, and the thermal contact resistance has changed accordingly. In the initial state, the surface roughness of the contact surfaces of all components is small, so the thermal resistance of the contact surface is also relatively small. After 250000 power cycles, the surface roughness of the contact surfaces of all components is relatively large, so the thermal resistance of the contact interfaces is also large.

After the maximum temperature of the PP IGBT chip is obtained, the heat transfer model of the PP IGBT device can be established to obtain the heat flow on both sides, and get overall the thermal resistance of the device by calculation. R is the overall thermal resistance (unit, K/W), T, T' are the junction temperature of the internal chip of the device and the case temperature of the device (unit, °C), Q_1 and Q_2 are the heat flow between the device chip and the case (unit, W/m^2).

$$R = \frac{T - T'}{\frac{Q_1 + Q_2}{2}} \tag{6}$$

Finally, it is calculated that the overall thermal resistance of the device in the initial state is about 0.0942 K/W and the overall thermal resistance after 250000 power cycles is about 0.1024 K/W. The thermal contact resistance of the contact surfaces of all components of the device has been changed after the power cycle experiment, resulting in a change in the overall thermal resistance of the device.

IV. CONCLUSION

In the introduction above, a three-dimensional finite element model of single die PP-IGBT module is established to conduct an electro-thermal-mechanical coupling analysis. The change of the surface morphology of the parts that are in contact with each other inside the device changes the thermal contact resistance between the components, and the overall thermal resistance is also affected. The following conclusions were initially obtained:

(1) The surface roughness of the contact surfaces of the internal components of the PP IGBT has increased after the power cycle experiment, and the corresponding thermal contact resistance has also increased. As a result, the maximum temperature of the PP IGBT chip will increase under service conditions.

(2) The thermal resistance between the internal contact surfaces of the PP IGBT increases, and the thermal resistance of the entire device also increases.

ACKNOWLEDGMENT

This research was supported by the National Natural Science Foundation of China (NSFC) No. 11872078.

REFERENCES

[1] Dou Z C , Stevens R , Xin L Y , et al. Design and Characteristic Analyss of Novel Press-Contact IGBT Module[J]. Electric Drive for Locomotives, 2013.

[2] Tinschert L , Ardal A R , Poller T , et al. Possible failure modes in Press-Pack IGBTs[J]. Microelectronics Reliability, 2015, 55(6):903-911.

[3] Poller, T., Lutz, J., D'Arco, S., & Hernes, M. (2013). Determination of the thermal and electrical contact resistance in press-pack IGBTs. *2013 15th European Conference on Power Electronics and Applications, EPE 2013*, 2013 15th European Conference on Power Electronics and Applications, EPE 2013, 2013.

[4] [1] Deng E , Zhao Z , Peng Z , et al. Influence of the Temperature on the Thermal Contact Resistance Within Press Pack IGBTs[J]. Semiconductor Technology, 2016.

[5] Bahrami M , Culham R , Yovanovich M . Thermal Resistances of Gaseous Gap for Conforming Rough Contacts[C]// Aiaa Aerospace Sciences Meeting & Exhibit. 2013.

[6] [1] Deng E , Zhao Z , Zhang P , et al. Clamping Force Distribution within Press Pack IGBTs[J]. Transactions of China Electrotechnical Society, 2017, 32(6):201-208.

[7] Poller T , Lutz J , D 'Arco S , et al. Determination of the thermal and electrical contact resistance in press-pack IGBTs[C]// European Conference on Power Electronics & Applications. IEEE, 2013.

[8] Rajaguru P , Lu H , Bailey C , et al. Electro-thermo-mechanical modelling and analysis of the press pack diode in power electronics[C]// International Workshop on Thermal Investigations of Ics & Systems. IEEE, 2015.

978-1-6654-1392-3/21 $31.00 © 2021 IEEE

Orthogonal Experiment for Analyzing the Impact of Thermal Stress on the Reliability of an EMC Package

Yulong Li[#]
*Shenzhen Institute of Advanced
Electronic Materials, Shenzhen
Institute of Advanced Technology,
Chinese Academy of Science*
Shenzhen, China
yl.li@siat.ac.cn

Yi Zheng[#]
*Shenzhen Institute of Advanced
Electronic Materials, Shenzhen
Institute of Advanced Technology,
Chinese Academy of Science*
Shenzhen, China
yi.zheng@siat.ac.cn

Haozhe Wang
*Shenzhen Institute of Advanced
Electronic Materials, Shenzhen
Institute of Advanced Technology,
Chinese Academy of Science*
Shenzhen, China
hz.wang@siat.ac.cn

Jibao Lu*
*Shenzhen Institute of Advanced
Electronic Materials, Shenzhen
Institute of Advanced Technology,
Chinese Academy of Science*
Shenzhen, China
jibao.lu@siat.ac.cn

Rong Sun*
*Shenzhen Institute of Advanced
Electronic Materials, Shenzhen
Institute of Advanced Technology,
Chinese Academy of Science*
Shenzhen, China
rong.sun@siat.ac.cn

Wenhui Zhu*
*Central South University
The College of Mechanlcal and
Electrical Engineering*
Changsha, China
zhuwenhui@csu.edu.cn

Ching-Ping Wong
*School of Materials Science and
Engineering, Georgia Institute of
Technology*
Atlanta, GA, USA
cp.wong@cuhk.edu.hk

Abstract—In recent years, plastic packaging has become the main form of electronic packaging technology due to its low cost, miniaturization, and simple producing process. Epoxy molding compound (EMC) are widely used in electronic packaging as encapsulants of electronic devices. However, in the actual service environment, due to the material properties of epoxy molding compounds, there are many factors cause the package structure failure.

In this paper, we design the orthogonal experiment to systematically investigate the impact of thermal stress on the reliability of a molded underfill (MUF) FCBGA (flip-chip ball grid array) structure by using the finite element analysis. In the orthogonal experiment, the F test is used to estimate whether the experiment is reliable, and the T test is used to select the most important factors from the samples. The statistical analysis indicates that the coefficient of thermal expansion (CTE) and modulus of EMC have the significant influence on the thermal stress of the EMC package, and static max thermal stress is at the interface between the chip and EMC. In addition, we find that the node of the static max stress and the one of the max displacement are different.

Keywords—*EMC, MUF, DOE, statistical analysis, FEA, thermal influence*

I. INTRODUCTION

With the development of the integrated circuit (IC) industry, the electronic information technology has greatly changed people's way of life and work, which is one of the important symbols of a country's international influence. At present, the process of chip design, manufacturing and packaging gets a lot of attention. Gordon Moore, one of the founders of Intel, proposed Moore's Law [1] : Chip integrated circuit generation every four years, each generation of integration quadrupled, chip line width about 30% smaller, working speed increased 1.5 times.

However, currently it is very difficult to improve the integration of the chip, enforcing people to consider the innovation in electronic packaging. Packaging technology can be divided into three categories according to material: ceramic packaging, metal packaging and plastic packaging. In recent years, plastic packaging technology develops rapidly due to its lower cost, miniaturization, and simple producing process. And EMC (epoxy molding compound) takes very large proportion in plastic package and are widely used in electronic packaging as encapsulants of electronic devices. In order to improve the stability of the EMC package, a lot of scholars have done research on it. Fan, X et al. provided a physical model to calculate the vapor pressure [2]. Yong, T. P. et al. pointed that the mismatch CTE (coefficient of thermal expansion) of EMC, substrate, chip and adhesive materials cause the warpage of the package structure [3]. D Yang et al. concluded that thermal aging of EMC has significant influence on the rubbery modulus and the glass transition temperature [4]. Li R et al. drew a conclusion that different mechanical properties of the materials, such as their elasticity module, thermal expansion coefficients, Poisson's ratios, etc., caused a large internal stress formed at the interface, which eventually led to cracking and failure of the EMC package structure [5].

From what have been mentioned above, it comes to a conclusion that there are still many problems of EMC package to be solved. And this article provide a method which combines statistical analysis and experiment in order to reveal which factors have the greatest thermal impact on the EMC packaging structure by using the finite element analysis. The most significant impact factors are determined from the orthogonal tests.

II. STATISTICAL ANALYSIS

In addition to doing experiments, the design of experiment and the processing of experimental results are also very important. In order to obtain the results with theoretical support, we need to select several factors that may have influence on the experimental results and set up an orthogonal experimental table which consists of different level of the chosen factors. And the experimental data will be processed by statistical analysis, including F test and T test to screen out the factors that have significant influence on the thermal result of experiment.

A. F test

F test is used to determine whether there are factors that have significant influence on the experiment. Calculate the F value of the experiment group and compare it with the critical F value which can be ensured by df_R and df_E, if the F value is greater than the critical F value, it indicates that there are significant influence factors in this experimental group. And we can get the F value from the following equations:

$$SS_R = \sum_{i=1}^{n}(\hat{y}_i - \overline{y})^2 \qquad (1)$$

$$df_R = k \qquad (2)$$

$$SS_E = \sum_{i=1}^{n}(y_i - \hat{y}_i)^2 \qquad (3)$$

$$df_E = n - k - 1 \qquad (4)$$

$$F = {MS_R}/{MS_E} = \frac{SS_R/df_R}{SS_E/df_E} = \frac{MS_R}{MS_E} \qquad (5)$$

In the formula, $\overline{y}$ is the mean value of the experiment result, $\hat{y}_i$ is the fitted value of the experiment result, y_i is the result of each experiment, n is the number of the experiment, k is the number of selected factors. dependent document.

Furthermore, if F test is failed, it indicates that none of the chosen factors is significant to the experiment result, so it will be required to select again, and repeat F test until the conditions are met.

B. T test

T test is used to find out which chosen factor has the significant influence on the result of experiment. Calculate the T value of the each chosen factor and compare it with the critical T value which can be ensured by df_E, if the T value of one factor is greater than the critical T value, it indicates that it is the significant influence factor in this experimental group. And we can get the T value from the following equations:

$$coef = {(\overline{y}_h - \overline{y}_l)}/{2} \qquad (6)$$

$$s = \sqrt{{MS_E}/{n}} \qquad (7)$$

$$T = {coef}/{s} \qquad (8)$$

In the formula, $\overline{y}_h$ is the average of the experimental results when the factor is taken to a high level, $\overline{y}_l$ is the average of the experimental results when the factor is taken to a low level.

III. MODEL AND SIMULATION

Mechanical simulation can be used effectively as a tool to estimate the change of packaging structure in the reflow process, and the simulation results can give guidance to design packaging structure or select material on a certain extent.

A. Geometry and Material Parameters

The model of molded underfill (MUF) flip-chip ball grid array (FCBGA) is established. The geometry and material parameters are shown in TABLE I. and TABLE II.

TABLE I. GEOMETRY PARAMETERS

Items	MUF FCBGA
Die size (mm)	8×8×0.775
Substrate size (mm)	20×20×1.16
EMC size (mm)	20×20×1.5
Bump size (mm)	r=0.035, R=0.06, h=0.08
Bump pitch (mm)	0.5

TABLE II. MATERIAL PARAMETERS

Material	Young's modulus (MPa)		CTE (ppm/°C)		v
	High	Low	High	Low	
EMC	29000 (T=25) 10490 (T=125) 900 (T=220)	11700 (T=25) 3300 (T=125) 343 (T=220)	13.6 (T=25) 34 (T=125)	6 (T=25) 24.7 (T=125)	0.25
Substrate	18600	2450	16	1.76	0.2
Die	148000		2.62		0.28
Bump	50000		24		0.4

B. Model and Methodology

CAE tool is a very effective method to rapidly compare various geometries, components and materials without the cost of purchasing and building physical parts. And the FEA simulation can estimate the internal stress between the interface of different parts in the packaging structure.

The model geometry was built in Hypermesh and translated into Abaqus to simulate the reflow process. Perfect adhesion between materials was assumed, several different parts were created and united with bonded contact elements. Quarter-symmetry models approximately three hundred thousand elements were used for these simulations (see Fig. 1 and Fig. 2).

Linear-elastic and temperature dependent material properties were used for the simulation model. Object mass and weight were ignored, since the forces due to gravity at the package level are much smaller than the stiffness of the package.

The models used a stress-free starting temperature of 25°C, with stress and displacement results measured at 260°C. The result of thermal stress on EMC and displacement of the packaging structure were measured through the whole reflow process simluation.

Fig. 1. The quarter-symmetry FEA model of MUF FCBGA.

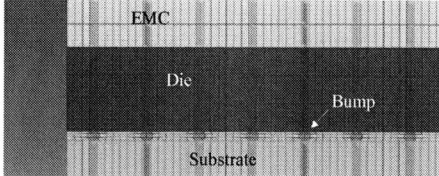

Fig. 2. The section view of the FEA model.

IV. RESULTS AND ANALYSIS

According to the selecting factors, the simulation experiment can be designed as an orthogonal test shows in TABLE III. 1 represent high level, -1 represent low level. The number of the experiment depends on the chosen factors and its level. For the two level of each chosen factors, the number of experiments is equal to two to the power of n.

TABLE III. ORTHOGONAL TEST

No.	EMC Young's modulus	EMC CTE	Substrate Young's modulus	Substrate CTE
1	-1	1	-1	-1
2	-1	-1	-1	1
3	-1	-1	1	-1
4	-1	1	-1	1
5	-1	-1	1	1
6	-1	1	1	-1
7	-1	1	1	1
8	1	-1	-1	-1
9	1	1	-1	1
10	1	-1	-1	1
11	1	1	1	-1
12	1	1	1	1
13	1	1	-1	-1
14	1	-1	1	-1
15	1	-1	1	1
16	-1	-1	-1	-1

A. Simluation Results

During the process of reflow, the packaging structure is easy to fail, especially plastic packaging. Due to the different materials property, the CTE mismatch often occurs in the packaging structure. With the rise of temperature, the interface between different material will generate internal stress, which causes the warpage of the packaging structure. And if the stress reaches to the critical value, the whole structure will collapse.

The thermal stress of EMC results are reported in Fig. 3 (a). and Fig. 4 (a). The node of static max stress number is 48562, and it locates on the interface of die and EMC due to the different material properties. Fig. 3 (b). and Fig. 4(b). show the results of static max displacement of the packaging structure, the node of static max displacement is on the corner of the structure, and its node number is 340983, quite different from the node number of static max stress, so there is no connection between them.

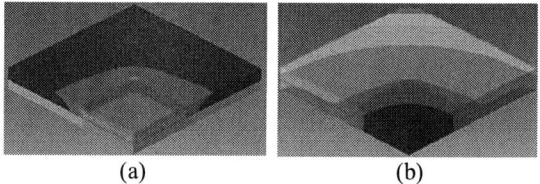

(a)   (b)

Fig. 3. (a) Stress contour of EMC; (b) Displacement contour of packaging structure

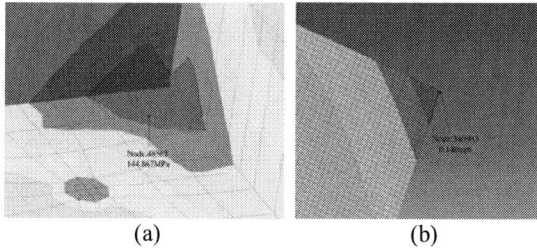

(a)   (b)

Fig. 4. (a) The node of static max stress; (b) The node of static max displacement

B. Statistical Analysis

Fig. 5 shows each experiment Dynamic stress during the simulation process. The twelfth group has the max stress among the experiments, and its factor level is high. Because of the CTE mismatch in this packaging structure, the internal stress between different parts is very complicated, the static max stress doesn't match the terminal of dynamic stress.

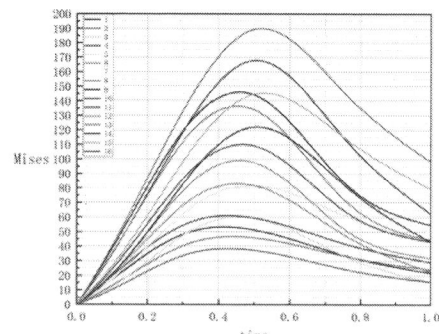

Fig. 5. Dynamic stress of the experiment

Statistical analysis provide theoretical support, and it can draw an accurate conclusion rapidly by using the equations that have been mentioned above. For this experiment, k is 4, n is 16, the F value of this experiment can be calculated to be equal to 112.4. According to df_E and df_R, the critical F value is 3.357 which is less than 112.4, so there are significant impact factors in this experiment. Then the T value of each factor can be calculated to be equal to 17.81, 7.04, 8.24, 3.83. According to df_E, the critical T value is 2.2 which is less than 3.83, so all the factors have significant influence on the thermal stress of EMC package. Fig. 6. Shows the connection of T value of each factors and critical T value. A represents EMC young's modulus, B represents EMC CTE, C represents substrate young's modulus, D represents substrate CTE.

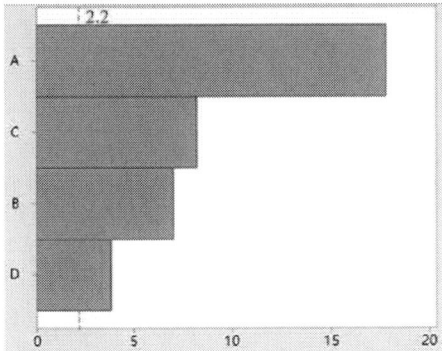

Fig. 6. Dynamic stress of the experiment

V. CONCLUSIONS

We provide a systematic orthogonal experiment to investigate the impact of thermal stress on the reliability of a molded underfill (MUF) FCBGA (flip-chip ball grid array) structure by using the finite element analysis. The F test is used to confirm whether there are significant factors in the experiment and T test is used to find out which factor has the significant influence on the experimental results.

We choose four factors that may have an influence on the EMC package and simulate the reflow process. The results show that the Young's modulus of the EMC plays the most important role in affecting the reliability, though all the factors has an influence. Besides, there is no connection between the static max stress and static max displacement. Due to the CTE mismatch of different parts, the static max stress doesn't match the terminal of dynamic stress.

ACKNOWLEDGMENT

The work was supported by the National Key R&D Program of China (2020YFB0408700), the National Natural Science Foundation of China (no. 52003289), the Youth Innovation Promotion Association CAS (no. 2021363), and the SIAT Innovation Program for Excellent Young Researchers (no. 201803).

REFERENCES

[1] Moore GE. Cramming more components onto integrated circuits. Electronics, 1965, pp.114.

[2] Fan X , Zhao J H . Moisture diffusion and integrated stress analysis in encapsulated microelectronics devices. International Conference on Thermal. IEEE Xplore, 2011.

[3] Yong T P, Kang T M. Numerical Analysis of Residual Warpage of FBGA Package during EMC Curing Process. Key Engineering Materials, 2007, 334/335(Pt1):385-388.

[4] D Yang, Cui Z. The effect of thermal aging on the mechanical properties of molding compounds and the reliability of electronic packages. International Conference on Electronic Packaging Technology & High Density Packaging. IEEE, 2011.

[5] Li R , Yang D , Zhang P , et al. Effects of High-Temperature Storage on the Elasticity Modulus of an Epoxy Molding Compound. Materials, 2019, 12(4).

Optical Performance Analysis of UVC-LED Package Structure Based on Ray-tracing Simulation

Jinyang Li
School of Mechanical and Electrical Engineering , Guilin University of Electronic Technology
Guilin, China
jinyang_li2021@163.com

Wangyun Li
School of Mechanical and Electrical Engineering , Guilin University of Electronic Technology
Guilin, China
li.wangyun@guet.edu.cn

Daoguo Yang*
School of Mechanical and Electrical Engineering , Guilin University of Electronic Technology
Guilin, China
daoguo_yang@163.com

Yikang Qin
School of Mechanical and Electrical Engineering , Guilin University of Electronic Technology
Guilin, China
qykbeicheng97@163.com

Sicheng Cao
School of Mechanical and Electrical Engineering , Guilin University of Electronic Technology
Guilin, China
cao_sicheng@163.com

Feixiang Liu
School of Mechanical and Electrical Engineering , Guilin University of Electronic Technology
Guilin, China
674173923@qq.com

Abstract—**UVC-LED medical sterilization technology is widely used nowadays. Due to the refractive index of the UVC-LED light-emitting layer is too different from the refractive index of air, the light difficult for light to escape from the inside of the LED. The light extraction efficiency of the UVC-LED chip is very low, and the light penetration of this wavelength band is extremely weak, so the luminous flux of the UVC-LED chip is very low. This article optimizes the existing package structure to increase the luminous flux. First, a traditional package structure model was built and analyzed using the method of ray-tracing simulation. Secondly, a new model including brackets with different angles and different reflective layers was built. Different reflective coatings were used and evaluated in the simulation. Finally, the simulation data were sorted out and the luminous flux of different reflective coatings under the same bracket tilt angle and the luminous flux of the same reflective coating with different bracket tilt angles were compared. Through the comparison of simulation results, it can be concluded that with the same reflective coating, the luminous flux effect is better when the tilt angle of the bracket is between 40°and 70°. T he luminous flux will be increased by about 30% after using the reflective coating under the same structure. For a circular package, the best radiation intensity occurs near the tilt angle of 45°. For the square pack age, the best radiation intensity occurs near the tilt angle of 60°. Finally, from the perspective of overall optical performance, the packaging structure of the circular cavity is more suitable for UVC-LED packaging than the square cavity. This research guides the development of UVC-LED packaging routes.**

Keywords—*UVC-LED, ray tracing simulation, reflective layer, luminous flux*

I. INTRODUCTION

Compared with mercury lamps, ultraviolet light-emitting diodes (UVC-LEDs) based on AlGaN materials have the advantages of small size, high efficiency, low energy consumption, safety, environmental protection, and adjustable wavelength [1-2]. Therefore, UVC-LED is one of the best choices to replace the traditional mercury ultraviolet lamp [3]. According to research, a dose of 40 mj/cm² is su fficient to inactivate most bacteria, spores, and viruses [4]. However, the UVC-LED cannot directly use the traditional LED packaging structure due to the limitation of the packaging process.

Because the wavelength emitted by the UVC-LED is an idiographic short wave, this poses a new challenge to the packaging of the UVC-LED [5]. The search for high-efficiency and reliable UVC-LED packaging technology has become an essential link in promoting the large-scale industrial application of UVC-LEDs [6]. Researchers have proposed many packaging methods to improve the light extraction efficiency of deep-ultraviolet light-emitting diodes, such as micro-machining the quartz surface of the light outlet or doping the quartz. Although these methods can promote light extraction by reducing the light loss inside the package structure, the processing process is relatively complicated. The light energy radiated to the sidewalls is hard to form a specular reflection and easily absorbed by the packaging material, resulting in a decrease in the overall efficiency of the chip. Therefore, the optimization of the sidewall structure is considered in this research to improve the light extraction rate of the deep ultraviolet light-emitting diode. We compared the two models of UVC-LED packaging from the angle of the sidewall and the internal coating of the package: the circular cavity package and the square cavity package. Through optical simulation, the influence of the reflectivity and inclination of the inner surface of the LED holder on the light output efficiency of the UVC-LED is discussed. Finally, the optical performance of the UVC-LED with the optimized package structure is summarized.

II. SIMULATION MODEL

In this chapter, we present the three-dimensional models of the square lumen product and the circular lumen product. And the main dimension parameters of the square support model and the circular support model are introduced.

Fig. 1 shows two models of UVC-LED packaged for experiments. This research mainly used the optical simulation carried out by the optical software Tracepro and investigated the optical performance of the above two UVC-LED package models.

978-1-6654-1392-3/21 $31.00 © 2021 IEEE

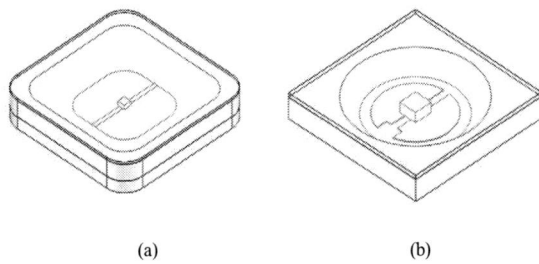

(a)   (b)

Fig. 1. (a) Circular cavity UVC-LED model; (b) Square cavity UVC-LED model.

Fig. 2 (a) shows the geometric model of the round package structure. The bottom of the cavity of the circular cavity package is a circle with a radius R of 2.5mm, and the cavity depth Is 0.4mm. The wall inclination angle θ is set at every 15° from 30° to 90°.

Fig. 2 (b) shows the geometric model of the square package structure. The bottom of the cavity of the square cavity package is a square with a side length of 3.5mm, the cavity depth is 0.7mm, and the sidewall inclination angle β is from 30° to 90°. Research nodes are set every 15°.

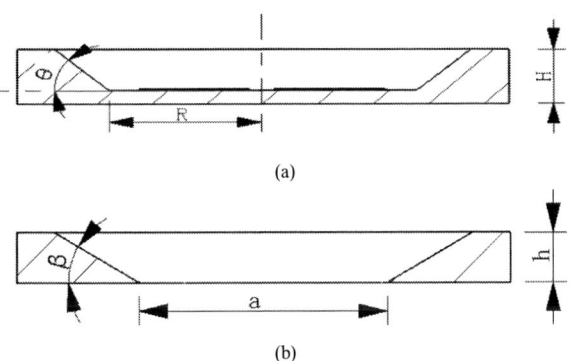

(a)

(b)

Fig. 2. (a) Round package geometry;(b) Square package geometry.

At all research nodes, we set up three comparative experiments. The experiment included no processed, only internal bottom surface processed, and all internal surface processed. We use special aluminum paint for surface processed to improve the reflectivity of the surface. Through the simulation results of ray tracing, we summarized the influence of the inclination angle of the sidewall and the surface processed position on the LED light efficiency. We optimized and adjusted the UVC-LED model structure based on the simulation results. We select the node with the best effect from all the research nodes and refine the inclination angle of the nearby sidewall to find the best position.

III. SIMULATION ANALYSIS

A. Simulation analysis of total luminous flux

Fig. 3 shows the luminous flux of a circular cavity with the same bottom area and different sidewall angles. Under the same conditions, the luminous flux simulation results of three different experiments. Only processing the internal bottom surface can increase the luminous flux by about 10%. After processing all the internal surfaces, an improvement of more than 16% can be obtained between 30° and 75°. Especially in a cavity close to 45°, the increase in luminous flux after all-

aluminum processing reaches its peak, which is 28% higher than the luminous flux of the original model.

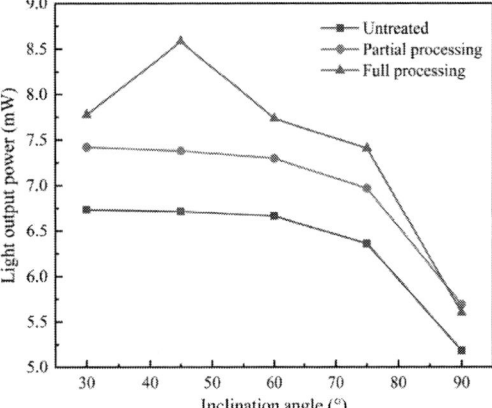

Fig. 3. The total luminous flux of different surface treatment areas of circular cavity models with the same bottom area and different sidewall angles.

Fig. 4 shows the luminous flux of the aluminum-clad model of a circular cavity with the same bottom area at an inclination angle of 40° to 46°. We can get that when the sidewall angle is about 42°, the luminous flux reaches its peak.

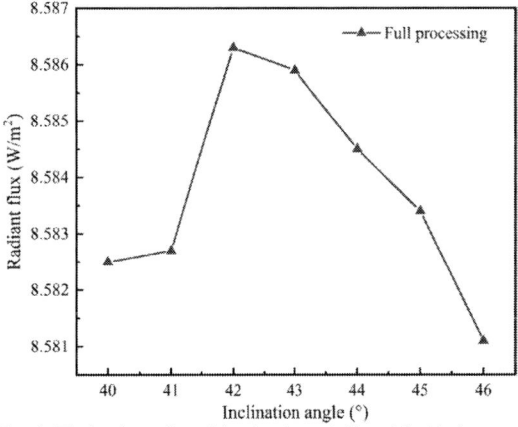

Fig. 4. The luminous flux of the aluminum shell model with the same bottom area and circular cavity after the angle is refined.

Fig. 5 shows the irradiance and luminous flux of the 42° model with all aluminum-clad inside the circular cavity.

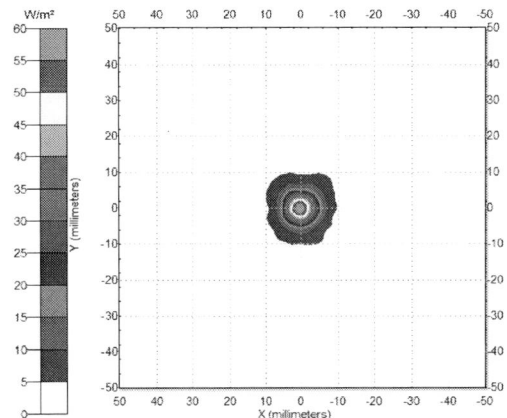

Fig. 5. Irradiance and luminous flux of UVC-LED model.

Fig. 6 shows the luminous flux of a circular cavity with the same light exit area and different sidewall angles. Under the same conditions, the luminous flux simulation results of three different experiments. Only processing the internal bottom surface can increase the luminous flux by about 5％. After processing all the internal surfaces, an improvement of more than 26％can be obtained between 30° and 75°.

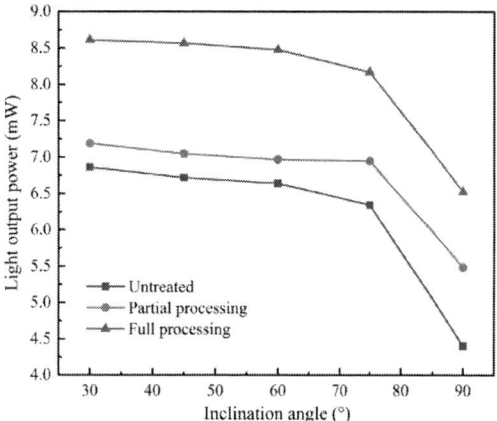

Fig. 6. The total luminous flux of different surface treatment areas of a circular cavity model with the same light exit area and different sidewall angles

Fig. 7 shows the luminous flux of a square cavity with the same bottom area and different sidewall angles. Under the same conditions, the luminous flux simulation results of three different experiments. Treating only the inner bottom surface can increase the luminous flux by about 14%. After processing all the inner surfaces, an improvement of more than 30% can be obtained between 30° and 75°.

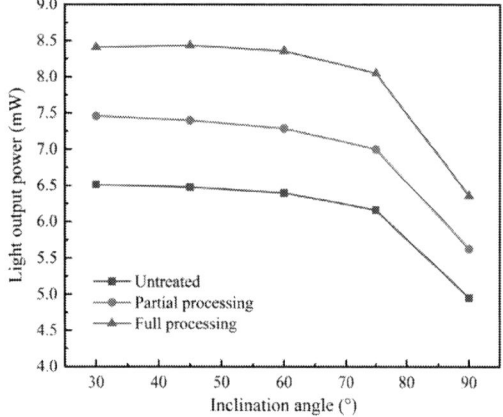

Fig. 7. The total luminous flux of different surface treatment areas of a square cavity model with the same bottom area and different sidewall angles.

B. Simulation analysis of radiation intensity

When UVC LED is used for disinfection, there is another important influencing factor, that is, radiation intensity. Fig. 8 shows a circular cavity with the same bottom area and different sidewall angles. Under the same conditions, the radiation intensity simulation results of three different experiments. The radiation intensity of the all-copper cavity, aluminum-clad bottom, and all aluminum-clad in the cavity under the same conditions. It can conclude that the radiation intensity of the three experiments all reached the peak at the

inclination angle of 45°.In the cavity close to 45°, the radiation intensity after all-aluminum processing increases by 50%.

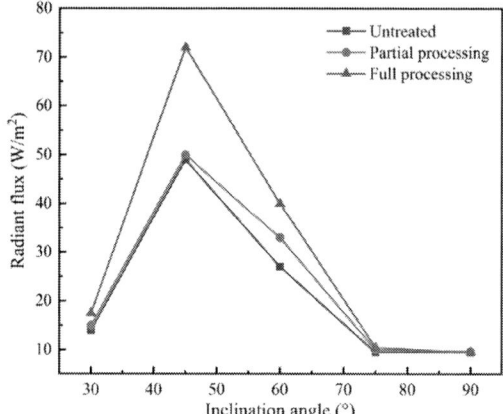

Fig. 8. Radiation intensities of different surface treatment areas of circular cavity models with the same bottom area and different sidewall angles.

Fig. 9 shows a circular cavity with the same light exit area and different sidewall angles. Under the same conditions, the radiation intensity simulation results of three different experiments. It can conclude that the radiation intensities of the three experiments all reached the peak at the 45° tilt angle. For the cavity with an inclination angle between 45° and 60°, the radiation intensity after the all-aluminum treatment increases by about 40%.

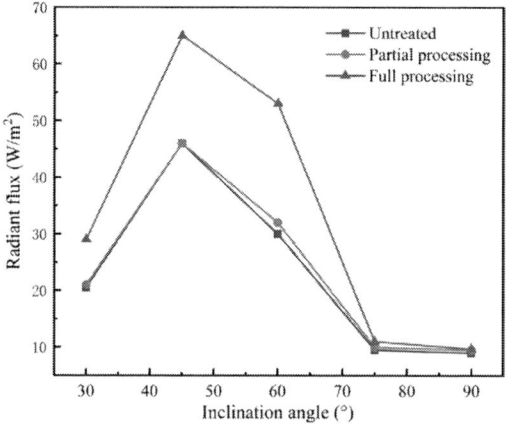

Fig. 9. Radiation intensities of different surface treatment areas of circular cavity models with the same light exit area and different sidewall angles.

Fig. 10 shows a square cavity model with the same bottom area and different sidewall angles. Under the same conditions, the radiation intensity simulation results of three different experiments. It can conclude that the radiation intensities of the three experiments all reach the peak value at the tilt angle of 60°. Treating only the inner bottom surface can increase the radiation intensity by about 10%. After processing all the inner surfaces, the radiation intensity can improve by more than 56%.

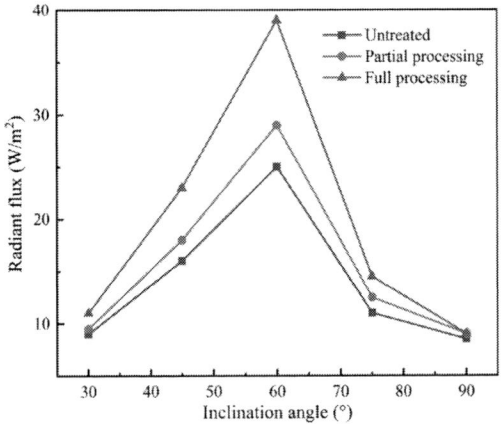

Fig. 10. Radiation intensities of different surface treatment areas of square cavity models with the same bottom area and different sidewall angles.

Fig. 11 shows that the highest radiation intensity is $71\,W/m^2$ when the sidewall inclination angle is $45°$, the highest radiation intensity is $17.5\,W/m^2$ when the sidewall inclination angle is $30°$. The highest radiation intensity when the sidewall tilt angle is $45°$ is 4.1 times that the tilt angle is $30°$. When the sidewall tilt angle is $45°$, the central radiation intensity is the strongest. It can conclude that the inclination angle processed of the sidewall can effectively increase the central radiation intensity, make the ultraviolet light more concentrated, and improve the light extraction capability of the deep ultraviolet light-emitting diode.

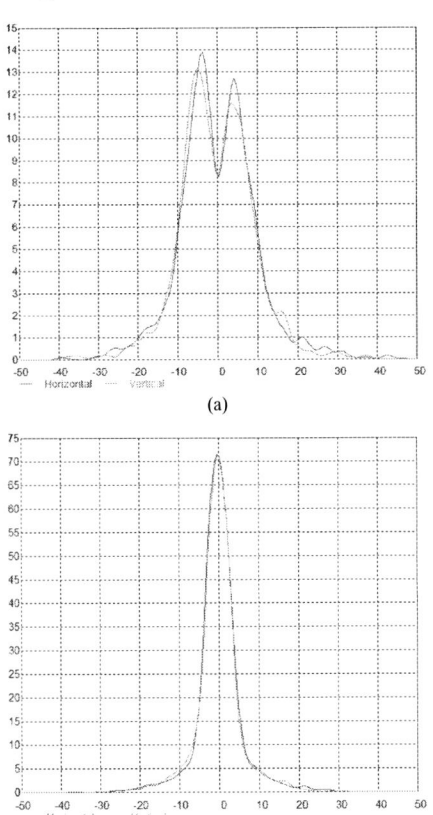

Fig. 11. Profile of irradiance and luminous flux of the same base area model with the inner surface covered by all aluminum and the tilt angles of (a) $30°$ and (b) $45°$

IV. CONCLUSION

This study uses optical simulation to compare two types of UVC-LED device packaging structures: circular cavity packaging structure and square cavity packaging structure. The simulation found that when the reflection coefficient of the reflective paint on the inner surface of the LED sidewall increased from 0.6 to 0.9, the total luminous flux of the circular package structure and the square package structure increased to 1.28 times and 1.3 times, respectively. This phenomenon can be attributed to the internal surface treatment to help the package structure reduce the loss of light. In addition to improving the reflectivity of the bracket, the tilt angle of the LED sidewall will also affect the optical performance of the model. When the inclination angle of the LED sidewall changed from $90°$ to $30°$, the luminous flux of the circular cavity package structure increases by 16%, and the luminous flux of the square cavity package structure increases by 30%. We also discussed the impact of model package tilt angle on UVC-LED radiation intensity. For a circular package, the best radiation intensity occurs near the tilt angle of $45°$. For the square package, the best radiation intensity occurs near the tilt angle of $60°$. Finally, from the perspective of overall optical performance, the packaging structure of the circular cavity is more suitable for UVC-LED packaging than the square cavity. This research guides the development of UVC-LED packaging routes.

ACKNOWLEDGMENT

This research was supported by the Natural Science Foundation of Guangxi Province (No. 2017GXNSFDA198006).

REFERENCES

[1] Y. N. Guo, Y. Zhang, J. C. Yan, H. Z. Xie, L. Liu, X. Chen, M. J. Hou and et al. "Light extraction enhancement of AlGaN-based ultraviolet light-emittingdiodes by substrate sidewall roughening," Appl. Phys. Lett, vol. 111, pp. 011102, 2017.

[2] Z. H. Ma, H. C. Cao, S. Lin, X. D. Li, and L. X. Zhao, "Degradation and failure mechanism of AlGaN-based UVC-LEDs," Solid. State. Electron, vol. 156, pp. 92-96, 2019.

[3] A. Khan, K. Balakrishnan, and T. Katona, "Ultraviolet light-emitting diodes based on group three nitrides," Nat. Photonics, vol. 2, pp. 77-84, 2008.

[4] M. Kneissl, T. -Y. Seong, J. Han, and H. Amano, "The emergence and prospects of deep-ultraviolet light-emitting diode technologies," Nat. Photonics, vol. 13, pp. 233-244, 2019.

[5] C. Y. Kang, C. H. Lin, T. Wu, P. -T. Lee, Z. Chen, and H. -C. Kuo, "A Novel Liquid Packaging Structure of Deep-Ultraviolet LightEmitting Diodes to Enhance the Light-Extraction Efficiency," Crystals, vol. 9, pp. 203, 2019.

[6] R. Liang, J. Dai, L. Xu, J. He, S. Wang, Y. Peng, and et al. "High light extraction efficiency of deep ultraviolet LEDs enhanced using nanolens arrays," IEEE. T. Electron. Dev, vol. 65, pp. 2498-2503, 2018.

978-1-6654-1392-3/21 $31.00 © 2021 IEEE

Determination of Parameters in Mixed-Mode Cohesive Zone Models for Modified Button Shear Tests by Particle Swarm Optimization

Wenyu Wu
School of System Design and Intelligent Manufacturing
Southern University of Science and Technology
Shenzhen, China
11930200@mail.sustech.edu.cn

Ke Xue*
School of System Design and Intelligent Manufacturing
Southern University of Science and Technology
Shenzhen, China
xuek@sustech.edu.cn

Weijing Dai
School of System Design and Intelligent Manufacturing
Southern University of Science and Technology
Shenzhen, China
daiwj@sustech.edu.cn

Dashun Liu
Center for Engineering Materials and Reliability
HKUST Fok Ying Tung Research Institute
Guangzhou, China
dsliu@ust.hk

Dali Yang
FENGHUA RESEARCH INSTITUTE(GUANG ZHOU) CO.,LTD.
Guangzhou, China
dali.yang@china-fenghua.com

Abstract—Delamination is a common failure mode and widely concerned in microelectronic package. In this paper, we focus on the underfill/Si interface in flip-chip package. To predict the risk of delamination, a modified button shear test with pre-delamination varying shearing height is taken. In finite element analysis, virtual crack closure technique is applied to calculate the critical strain energy release rate and phase angle. In addition, BK law is fitted to specify the mode-mix criterion of the interface. Finally, a corresponding cohesive zone model is built to fully define the initiation and propagation of interfacial delamination. The model parameters are determined by particle swarm optimization, a heuristic algorithm, that can automatically find the optimal solution and reduce repeated manual attempts.

Keywords—button shear test, finite element analysis, delamination, mixed-mode cohesive zone model, particle swarm optimization

I. INTRODUCTION

Delamination is a common failure mode and widely concerned in microelectronic package [1-3], especially at interface sites. It is usually triggered by the mismatch of the coefficient of thermal expansion (CTE) between adjacent layers in package. As the mismatch increases, interfacial thermo-mechanical stress at the respective sites strengthens and may exceed delamination threshold, causing package to fail. Since the devices for 5G application will generate excessive heat, leading to large thermal expansion. Thus, the interfacial thermo-mechanical stress becomes an critical issue influencing the 5G devices. To ensure the stable service of those devices, throughout investigation of the interfacial thermo-mechanical characteristics is on demand.

In order to predict the risk of delamination, adhesion strength and fracture toughness of interfaces need to be evaluated. However, most of the tests for adhesive quality are complex and time-consuming, such as double cantilever beam (DCB), four point bending (FPB), and end notch flexure (ENF). In addition, they are generally unable to get the adhesive parameters directly. Hence, it is necessary to set up an efficient and reliable method to evaluate the interfacial characteristics based on experimental and numerical consideration.

Button shear test is an effective method and widely used in industry due to its convenient specimen preparation and short test duration [4-5]. Whereas the standard button shear test can only qualitatively compare the adhesive strength of different interfaces. For quantitative comparison purpose, fracture toughness of interfaces is generally used and obtained by the application of finite element analysis. In this way, a modified button shear test is implemented, in which quantified pre-delamination is introduced along the investigated interface. As a result, the critical strain energy release rate (SERR) can be extracted numerically by virtual crack closure technique (VCCT) to represent the fracture toughness of the interface. Theoretically, both shearing and tensile stress contribute to interfacial delamination, and the proportion between these two types of stresses determines the modes of delamination and the respective critical SERR. To characterize this relationship, a series of tests and simulation with varying shearing height will be conducted, and eventually, a fitted function will be derived explicitly.

Based on the derived function, mixed-mode cohesive zone model (CZM) can be further built up for predicting delamination initiation and propagation in finite element analysis. However, the best suitable parameters in the model are hard to obtain directly from experiments. The common practice is to try different combination of parameters according to engineering experience, but this is obviously time consuming. Therefore, it is reasonable to employ some optimization algorithms which can find the optimal parameters iteratively according to well-defined computational strategies instead of repeated manual trials. Heuristic algorithms, including Ant Colony Optimization (ACO), Simulated Annealing (SA) and Genetic Algorithm (GA), are mainly based on the natural phenomena implying optimization strategies, they have advantage in solving the implicit black-box problem with complex process. Particle

978-1-6654-1392-3/21 $31.00 © 2021 IEEE

Swarm Optimization (PSO) is implemented here to determine the parameters of CZM automatically and efficiently.

In this paper, adhesive quality between underfill and silicon is studied quantitatively through the methodology proposed above. Underfill has been widely applied in flip-chip (FC) package to alleviate thermal mismatch between silicon chip and organic laminated substrate as well as improve solder joint reliability [6-7], therefore, a well-established method to characterize underfill interfaces is critical to evaluate the stability of package.

II. BUTTON SHEAR TEST COMBINED WITH FEA FOR SERR

A. Experimental Setup

The commercial underfill material investigated in this paper is named UF-A, whose curing condition is 165 ℃ for 2 hours. Since only the delamination at room temperature is studied, underfill is considered as linear elastic , and its elastic modulus is measured by three point bending test at room temperature. The stress-strain curve measured by three point bending is shown in Fig. 1, and its elastic modulus is about 11542 MPa.

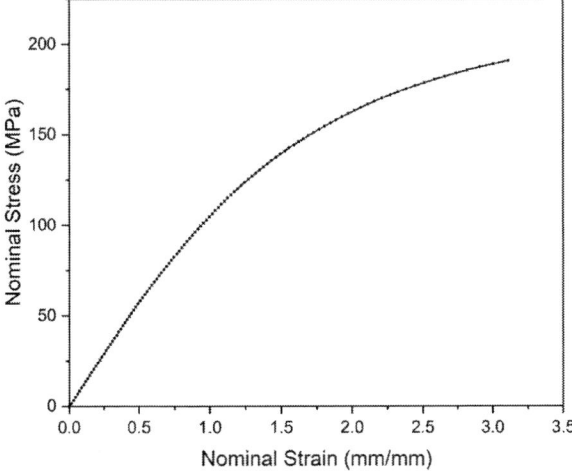

Fig. 1. Stress-strain curve of UF-A measured by three point bending at room temperature.

A cylindrical button of investigated underfill with radius of 1mm and height of 4mm is made onto a silicon substrate. The pre-delamination of 0.5mm in length and 0.05 mm in opening height is created by Teflon stripping method between the button and substrate. The modified button shear tests are carried out using DAGE series 4000, in which the test speed is set up as 85 $\mu m/s$ and the maximum range of the force cell is 1000 N. For determining the critical SERR as function of the mode mixture of delamination, the adhesion of similar interface is tested with different shearing heights: 0.1, 0.2, 0.3, 0.4 and 0.5 mm, and 6 samples are measured for each height. Fig. 2 exhibits the equipment and the diagram of the test.

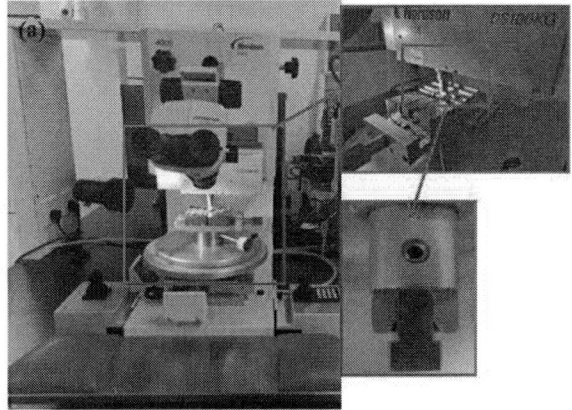

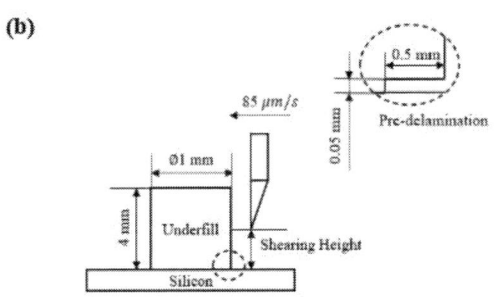

Fig. 2. Equipment and test conditions. (a) DAGE series 4000; (b) Diagram of the modified button shear test.

B. Experimental Results

The typical experimental result is shown in Fig. 3(a). The maximum force is recorded, and the corresponding point is taken as the initial delamination at the interface. Stiffness begins to degrade after the critical point, which represents the process of delamination propagation. As illustrated in Fig. 3(b), the force required to initiate the delamination decreases when increasing the shearing height. It indicates that the mode mixture gradually change from mode II to mode I as the shearing height increases. It is worth mentioning that the deviation of the results of each shearing height mainly comes from the quality of the specimen. The length of the pre-delamination and the parallelism between the crack tip and silicon edge can also influence the maximum force in tests. In the follow-up work, the shape of the button and the design of the mold will be thoughtfully considered to reduce the deviation of the specimen size.

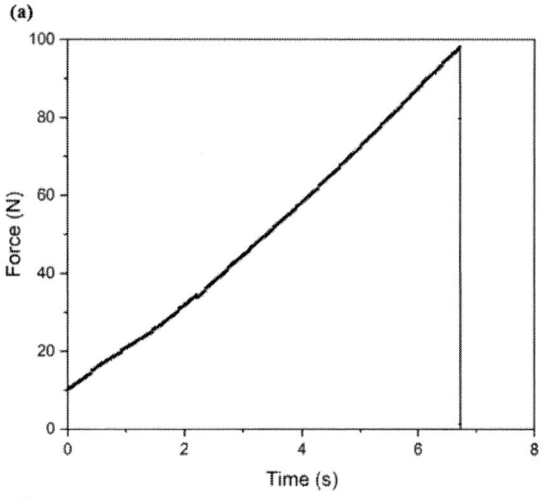

(a)

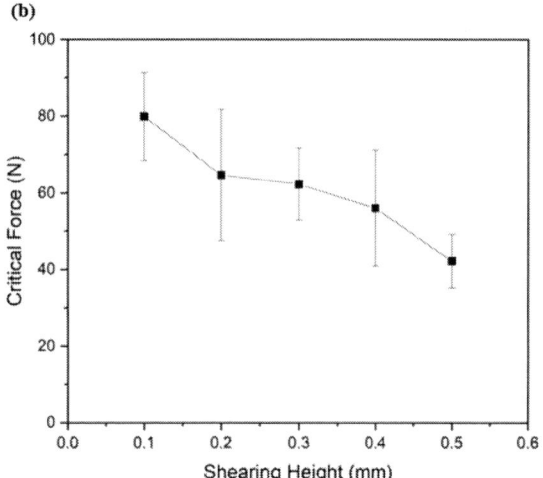

(b)

Fig. 3. Experimental results. (a) Typical force-displacement curve of the button shear test; (b) Critical forces corresponding to different shearing heights.

C. Numerical Modeling

Different energy-based simulation methods can be used to assess the delamination risk. Virtual crack closure technique (VCCT) is adopted in this paper. It was first proposed by Rybicki and Kanninen in 1977 for linear elastic, homogeneous and isotropic materials, assuming that the energy required to extend a crack by a finite distance is equal to the energy required to close the crack to its initial state [8].

The energy release rate is given by G = G1 + G2, where the normal and tangential components G1 and G2 for four-node elements are given by Eq. (1) and (2):

$$G_1 = -\frac{1}{2\Delta a}\left[Z_i(w_l - w_{l^*})\right] \tag{1}$$

$$G_2 = -\frac{1}{2\Delta a}\left[X_i(u_l - u_{l^*})\right] \tag{2}$$

in which Δa represents the length of the element ahead of the crack tip, X_i and Z_i are the forces at the crack tip, difference in brackets are the relative displacements behind the crack tip calculated by nodal displacements at the upper and lower crack face, as shown in Fig. 4.

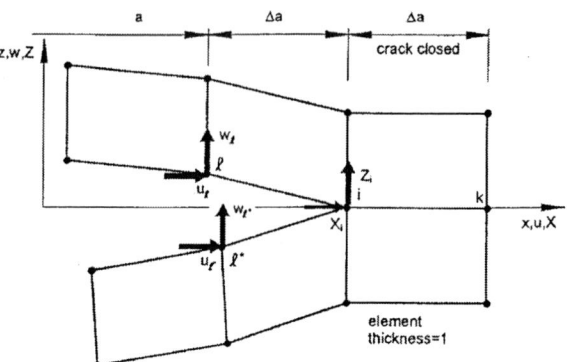

Fig. 4. Virtual crack closure technique for four-node element (excerpted from Ref. 7).

An approximation of the mixed mode angle called ERR angle is given by Eq. (3):

$$\varphi = arctan\left(\sqrt{\frac{G_2}{G_1}}\right) \tag{3}$$

This modified button shear test is modeled using commercial software ABAQUS 2019. Silicon and underfill are both set to linear elastic with elastic modulus of 121000 MPa and 11542 MPa respectively. Nodes on both sides of the substrate are fixed to imitate the clamping condition in tests. Moreover, shearing tool is modeled as rigid body and added force loading, the magnitude of which is equal to the maximum force recorded before. VCCT is defined in the area where button and substrate bond initially. The values in VCCT property are set significantly large to assume that the interface is firmly bonded so that the calculated values of SERR are critical strain energy release rate. Element type is chosen as C3D8 in the model. The finite element model in ABAQUS is shown in Fig. 5.

978-1-6654-1392-3/21 $31.00 © 2021 IEEE

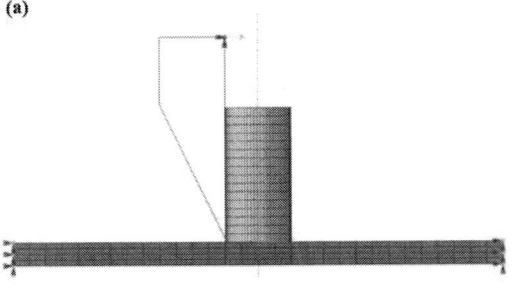

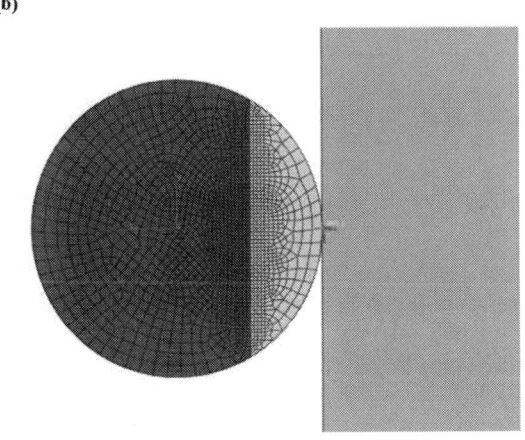

Fig. 5. The finite element model of button shear test. (a) Front view; (b) View of the interface, that the reddish part represents the area set VCCT property.

D. Numerical Results

Fig. 6(a)(c) exhibits the contour plot of interfacial tensile and shear stress respectively.

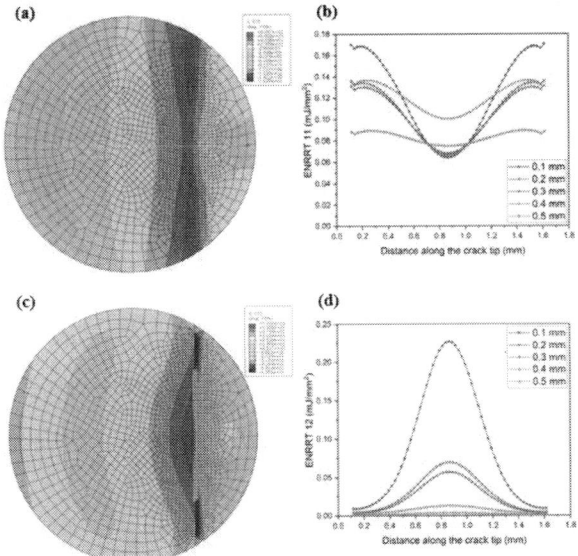

Fig. 6. Numerical results. (a) Contour plot of interfacial tensile stress; (b) Calculated G_1 along the crack tip at different shearing height; (c) Contour plot of interfacial shear stress; (d) Calculated G_2 along the crack tip at different shearing height.

It can be seen that interfacial stress is symmetrically distributed. Near the crack tip, the edge region is mainly tensioned while the central region is mainly sheared. At each shearing height, the values of G1 and G2 along the crack tip are calculated using VCCT, as shown in Fig. 6(b)(d). Similarly, G1 gradually decreases towards the center while G2 has the opposite trend. They all reach the extreme value at the midpoint at which is chosen as the critical SERR for subsequent calculation.

As demonstrated in Fig. 7, the value of critical SERR and mixed mode angle decreases while the shearing height is increased. It is corresponding to the anticipation that the mode mixity transform from mode II to mode I with the increasing shearing height in button shear test. Moreover, Fig. 8 shows the plot of the critical SERR versus mixed mode angle. BK law [9] is chosen to specify the mode-mix criterion of the interface in ABAQUS 2019.

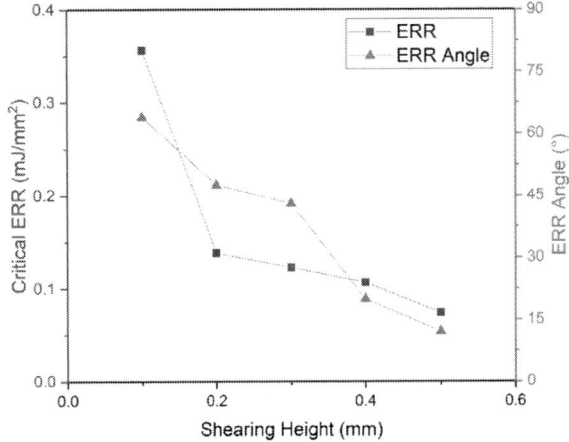

Fig. 7. Trend of critical strain energy release rate and mixed mode angle as shearing height increases.

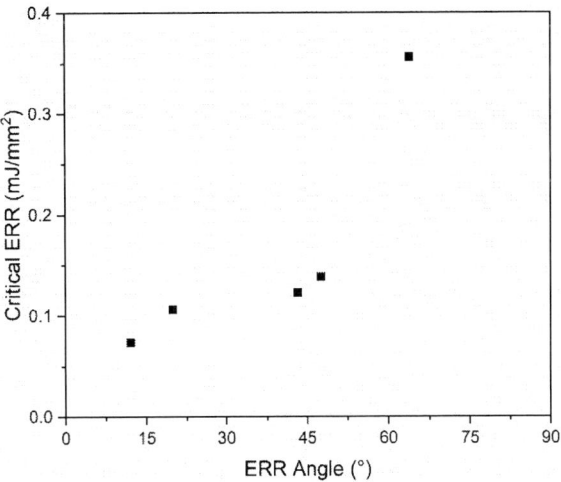

Fig. 8. Scatter plot of critical strain energy release rate versus mixed mode angle.

Ignoring SERR components along the out-of-plane direction (mode III), the BK law model can be described by the following Eq. (4):

$$G_{Tc} = G_{1c} + (G_{2c} - G_{1c})\left(\frac{G_2}{G_1 + G_2}\right)^{\eta} \qquad (4)$$

where G_{1c}, G_{2c}, G_{Tc} are the mode Ⅰ, mode Ⅱ and total critical strain energy release rate respectively.

To introduce the mixed mode angle in the formula, $\frac{G_2}{G_1 + G_2}$ in Eq. (4) can be reformulated as:

$$\frac{G_2}{G_1 + G_2} = \frac{1}{\frac{G_1}{G_2} + 1} = \frac{1}{\frac{1}{\tan^2 \varphi} + 1} = \sin^2 \varphi$$

And then, the BK law in Eq. (4) can be expressed as:

$$G_{Tc} = G_{1c} + (G_{2c} - G_{1c})(\sin^2 \varphi)^\eta$$

In this case, three required parameters in BK law were obtained using the least squares method to fit the curve, as exhibits in Fig. 9, the calculated results are given in Table 1 for subsequent simulation.

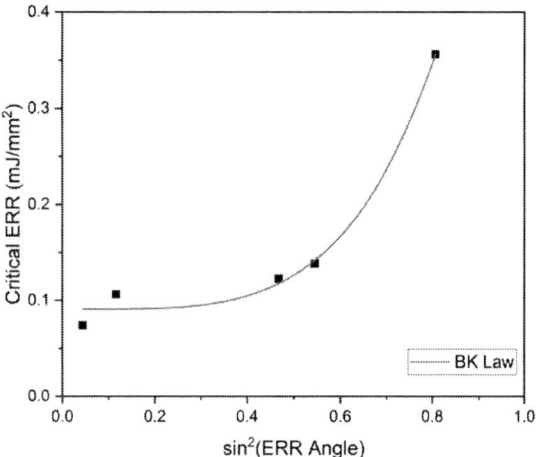

Fig. 9. BK law obtained by least squares fitting.

TABLE I.   CALCULATED VALUE OF PARAMETERS IN BK LAW

G_{1c} (mJ/mm²)	G_{2c} (mJ/mm²)	η
0.09089	0.74552	4.21155

III. DETERMINATION OF PARAMETERS IN MIXED-MODE CZM USING PSO

A. Cohesive Zone Model (CZM)

CZM (Cohesive Zone Model) is a common finite element method for studying interface delamination numerically [10-11]. It determines the initiation and propagation of cracks by defining the Traction-Separation criterion in the tiny area of the crack tip. Fig. 10 is a commonly used bilinear model. The area enclosed by the triangle is equal to the fracture toughness required for interface delamination.

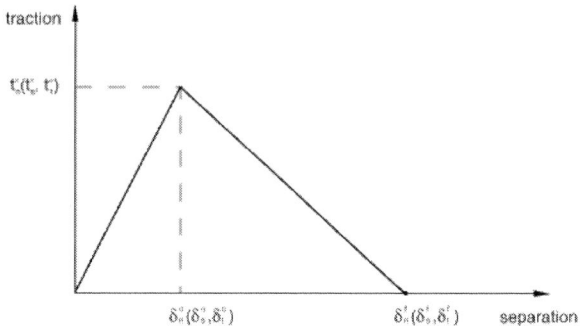

Fig. 10. Typical traction-separation response (excerpted from ABAQUS Documentation).

The model requires three parameters to be determined, namely the initial stiffness K, the maximum traction σ_{max}, and the fracture toughness G_c. Without considering the out-of-plane direction separately, delamination is generally a mixed mode of Mode Ⅰ and Mode Ⅱ, so a complete definition of CZM requires at least 7 parameters, namely K_{nn}, K_{ss}, σ_{normal}, σ_{shear}, G_{1c}, G_{2c} and η.

B. Particle swarm optimization (PSO)

Particle Swarm Optimization (PSO) was first proposed by Eberhart and Kennedy in 1995 [12], it is inspired from the behavior of bird flock foraging.

Assuming that there exists n particles in the N-dimensional search space, the current position of each particle is a candidate solution of the corresponding optimization problem. Particles have only two attributes: velocity and position, which lead to the next movement. Specifically, the optimal solution searched by each particle named individual extreme value will be compared and the best individual extreme value in the particle swarm is regarded as the current global optimal solution. According to the individual and global optimal solution, the velocity and position are updated for the next search, in this way, the final optimal solution can be obtained iteratively.

C. Implementation and Results

Since G_{1c}, G_{2c} and η have been obtained in section Ⅱ, unknown parameters in CZM are reduced to four. In addition, K_{nn} and K_{ss} defined in cohesive behavior in ABAQUS have negligible effect on the result if they were not set large enough (approximately larger than 100 times the modulus). In this paper we set them 1E6 and 3E6, respectively. Consequently, the parameters to be determined are further reduced to σ_{normal} and σ_{shear}.

Python scripts are adopted to employ PSO algorithm in Abaqus for automatically determining CZM parameters. The search space is a 2-dimensional positive real number space of σ_{normal} and σ_{shear}. The particles are the ABAQUS jobs created during each iteration. The fitness function is defined as the relative error between the maximum force value obtained by the test and by the simulation of button shear tests. PSO updates the parameter values according to the relative error of each particle, i.e., each simulation, calculated from last iteration, and uses the updated parameter in the simulation model to recalculate the maximum force. When the relative error is less than the preset expected value or the number of iterations exceeds the preset maximum value, the iteration exits and the final result is output. Fig. 11 demonstrates the flow chart of inversing parameters in ABAQUS using PSO.

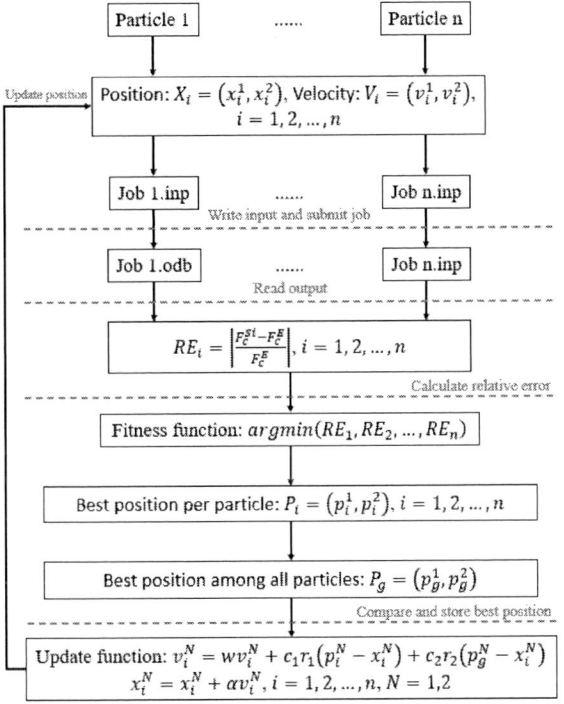

Fig. 11. Flow chart of inversing parameters in ABAQUS using PSO.

Button shear test with no-delamination with shearing height 0.05 mm is modeled in ABAQUS. Maximum force tested is 101.55 N. Considering the computational consumption, the number of particles is set to 6, the preset expected value is 5%, and the maximum number of iterations is 10. The optimal solution is obtained after 4 iterations, and final results of CZM parameters are given in Table 2.

Overall, the relative error is gradually reduced during the iteration, as demonstrated in Fig. 12, which is in line with the algorithm's expectations. To obtain a more accurate solution, the preset expected value can be narrowed down and the number of particles can be increased, if so, the number of iterations and computational cost will inevitably increase correspondingly.

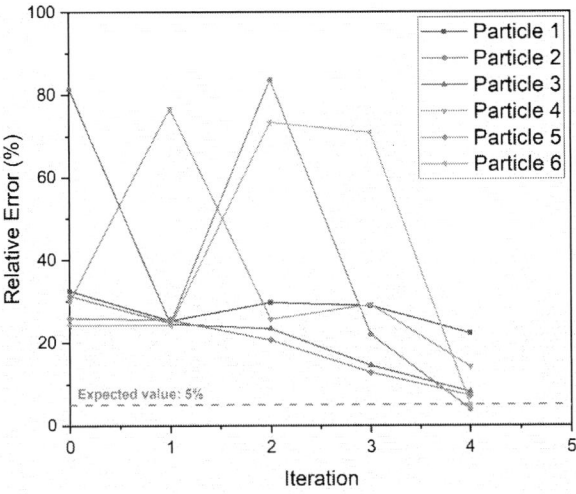

Fig. 12. Trends of relative error per particle during the interation.

TABLE II. DETERMINED VALUE OF COHESIVE ZONE MODEL

K_{nn}	K_{ss}	σ_{normal} (MPa)	σ_{shear} (MPa)
1E6	3E6	100.378322421	728.200265152

IV. CONCLUSIONS

A systematic methodology for determine parameters in mixed mode cohesive zone model is proposed. Maximum forces at different shearing heights are recorded in button shear test with pre-delamination, and then used to calculate critical strain energy release rate and phase angle, respectively, by using virtual crack closure technique with finite element analysis. In addition, BK law is specified and fitted for mixed mode behaviors. Parameters for damage initiation (quadratic traction) in CZM are treated as an optimization problem and inversed using particle swarm optimization, which can iteratively search for optimal solution automatically rather than by repeated manual operation.

ACKNOWLEDGMENT

This work was financially supported by the Science and Technological Bureau of Guangzhou Huangpu District (Project No. 2019GH03).

REFERENCES

[1] W.D. van Driel, M.A.J. van Gils, R.B.R. van Silfhout, G.Q. Zhang, "Prediction of delamination related IC & packaging reliability problems," Microelectronics Reliability, vol. 45, Issues 9–11, 2005, pp. 1633-1638.

[2] T. Y. Lin et al., "Failure analysis of full delamination on the stacked die leaded packages," 53rd Electronic Components and Technology Conference, 2003. Proceedings., 2003, pp. 1170-1175.

[3] C. A. Le Gall, Jianmin Qu and D. L. McDowell, "Delamination cracking in encapsulated flip chips," 1996 Proceedings 46th Electronic Components and Technology Conference, 1996, pp. 430-434.

[4] W. K. Szeto, M. Y. Xie, J. K. Kim, M. M. F. Yuen, P. Tong and S. Yi, "Interface failure criterion of button shear test as a means of interface adhesion measurement in plastic packages," International Symposium on Electronic Materials and Packaging (EMAP2000) (Cat. No.00EX458), 2000, pp. 263-268.

[5] L. W. Lee and L. W. Cheat, "Adhesion Study on Different Surface Treatment by Button Shear Test," 2020 4th IEEE Electron Devices Technology & Manufacturing Conference (EDTM), 2020, pp. 1-4.

[6] S. Jakschik, F. Feustel and E. Meusel, "Mechanism and growth rate of underfill delaminations in flip chips," 2001 Proceedings. 51st Electronic Components and Technology Conference (Cat. No.01CH37220), 2001, pp. 98-103.

[7] M. Hsieh, C. C. Lee and L. C. Hung, "Comprehensive Thermo-Mechanical Stress Analyses and Underfill Selection of Large Die Flip Chip Ball Grid Array," in IEEE Transactions on Components, Packaging and Manufacturing Technology, vol. 3, no. 7, pp. 1155-1162, July 2013.

[8] R. Krueger , "Virtual crack closure technique: History, approach, and applications ." ASME. Appl. Mech. Rev. March 2004; 57(2): 109–143.

[9] M.L. Benzeggagh, M. Kenane, "Measurement of mixed-mode delamination fracture toughness of unidirectional glass/epoxy composites with mixed-mode bending apparatus," Composites Science and Technology, vol. 56, Issue 4, 1996, pp. 439-449.

[10] Blackman, B.R.K., Hadavinia, H., Kinloch, A.J. et al., "The use of a cohesive zone model to study the fracture of fibre composites and adhesively-bonded joints," Int J Fract 119, 25–46 (2003).

[11] Min Jung Lee, Tae Min Cho, Won Seock Kim, Byung Chai Lee, Jung Ju Lee, "Determination of cohesive parameters for a mixed-mode cohesive zone model," International Journal of Adhesion and Adhesives, vol. 30, Issue 5, 2010, pp. 322-328.

[12] J. Kennedy and R. Eberhart, "Particle swarm optimization," Proceedings of ICNN'95 - International Conference on Neural Networks, 1995, pp. 1942-1948.

Design and Simulation of 3D Antenna Based on Conical Via Structure

Ziyu Liu*
School of Microelectronics,
Fudan University
Shanghai, China
liuziyu@fudan.edu.cn

Junhao Wang
School of Electronic Information
Engineering, Southwest University,
Chongqing, China

Zhiyuan Zhu
School of Electronic Information
Engineering, Southwest University,
Chongqing, China

Lin Chen
School of Microelectronics,
Fudan University
Shanghai, China

Qingqing Sun
School of Microelectronics,
Fudan University
Shanghai, China

Abstract—In this paper, a 3D monopole antenna based on TSV structure applied for network on chip (NoC) is proposed to increase the bandwidth and reduce the resonant frequency of short-distance antenna. The 3D antenna has the signal vias and the grounding vias. To optimize the performance, the cylindrical and conical signal vias are compared. Then the upper and lower radius size of the conical via is analyzed to reveal their influence on the bandwidth and resonant frequency. The return loss/ insert loss were used to characterize the performance of 3D antenna simulated by HFSS. The transmission distance between the transmitting and receiving antenna was 10 mm. The antenna with conical via of 1 mil upper radius and 7 mil lower radius had an minimum insertion loss of -7.46 dB at 26.4 GHz, which had a return loss of -24.08 dB. Besides, the conical antenna above also had two bandwidths with one bandwidth of 2.22 GHz between 25.45 GHz and 27.67 GHz，and the other bandwidth of 2.54 GHz between 30.15 GHz and 32.69 GHz. The 3D antenna also had working frequencies less than 32 GHz.

Keywords—*3D integration, 3D antenna, conical via, monopole*

I. INTRODUCTION

With the increasing system complexity and integration density, improving the performance and lowering the power are the demand of integrated circuit (IC). Networks-on-chip (NoC) is proposed to reduce the interconnection latency and power [1]. However, short-distance wireless communication between different chips in the NoC system is still needed to be improved. Three-dimensional (3D) antenna was reported to meet the demand and greatly increase the performance of wireless interconnection [2].

In the current research, some attempt in 3D antenna applied for wireless communication in NoC has been reported [2-6]. Kim K. proposed a 3D antenna and integrated the antenna in the silicon substrate [2]. The performance of monopole antenna applied for wireless NoCs (WNoC) application was also evaluated [3-4]. Lin demonstrated the fabricated zigzag dipole antenna used in WNoCs, which can realize 100kb/s radio communication [5]. Vasil demonstrated the prototype of 3D antenna based on TSV structures in the printed circuit board (PCB) to improve the insertion loss [6].

However, 3D antenna always has the narrow bandwidth and high resonance frequency. In this study, a 3D monopole antenna based on conical via structure applied for wireless communication NoC system is proposed to increase the bandwidth and lower resonance frequency. The 3D antenna had the signal and ground via. Thus, the geometry of signal via and the radius of signal via in the 3D antenna were separately simulated by High Frequency Structure Simulator (HFSS) to reveal the effect of these parameters on the 3D antenna characteristics.

II. DESIGN

The 3D antenna in the PCB with signal vias and grounding vias mimicking through sillicon via(TSV) is designed in this study. The structure and size design in the HFSS models are divided into four parts in the following.

A. Substrate Material Design

Tangent loss is a very important factor to determine the substrate material. It can be calculated as follows:

$$\tan \delta = \frac{\sigma}{\omega * \varepsilon} \tag{1}$$

The σ is the conductivity of the material, the ω is the angular frequency, and the ε is the dielectric constant of the meterial. Assuming that the tangent loss of silicon is 2.6×10^{-3} at 20 GHz. The PCB is made of Rogers RO4003 material with a dielectric constant of 3.55 so the dielectric tangent loss is 2.7×10^{-3}. PCB has the same tangent loss as high resistance (HR) silicon. PCB has low-cost and mature fabraication technology. Therefore, it is first to fabricate antenna on PCB in this study and the substrate is made of Rogers RO4003.

B. Overall Structure Design of 3D Antenna

Figure 1 (a) shows the cross section of the overall design. Two conical signal vias are used as the transmitting and receiving antennas, in which one is treated as the transmitting end and the other as the receiving end in the WNoC. To improve the signal performance and signal integrity, the top and bottom ground planes are designed as the shielding structure. In addition to the top and bottom ground planes (both with the thickness of 1 mil), the middle ground plane (1 mil thickness) is added in the shielding structure. The coplanar waveguide (CPW) is used for feeding.

Figure 1 (b) is the top view of the design and shows the detail parameters of antenna structure.

The detailed structure of conical signal via is shown in Fig. 2. The conical via has an upper radius of 4 mils, a lower radius of 7 mils and a height of 34 mils. The air gap size is 8 mils and the disc size is 13.5 mils as showed in Fig. 2. The height of cylindrical grounding via is 35 mils and the radius is 10 mils, as shown in Fig. 1 (a). The upper and lower radius of conical signal via will be varied in the part D, which is used to optimize the transmitting/receiving performance of signal via.

The signal via is placed in the inner layer of Rogers RO4003 high frequency PCB. The top and bottom grounding layers can increase the signal transmission gain and improve the antenna performance. The top grounding layer is used as the reflection surface of the signal via and the propagation path of electromagnetic wave. It also reduces the interference between the metal layer and Rogers RO4003 material.

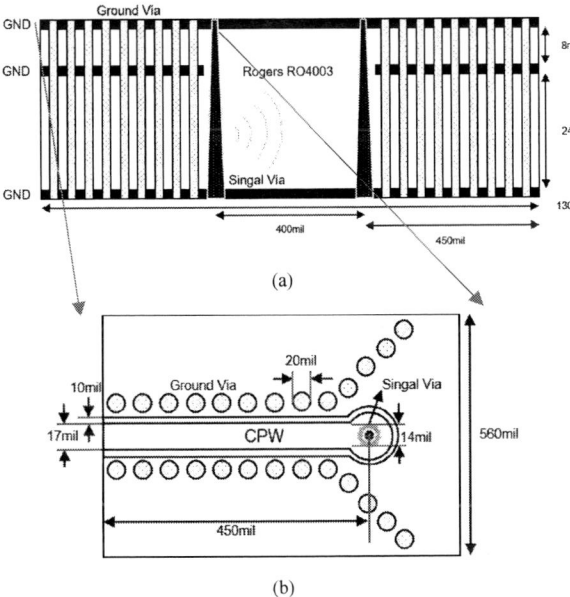

(a)

(b)

Fig. 1. Overall structure diagram of via antenna in PCB
(a) Section (b) Top view (symmetrical structure)

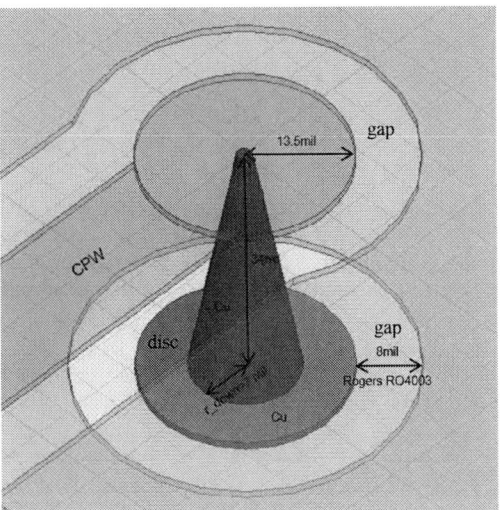

Fig. 2. Signal via structure diagram

C. CPW Test Structure Design

The CPW is shown in Fig. 1 and transmitting and receiving antennas both have a CPW for feeding. CPW is designed to transmit signals to conical signal vias. CPW has the linear and circular part. The length of linear CPW is 450 mils and the width is 17 mils. The circular part of CPW has a radius of 13.5 mil as shown in Fig. 2. The space between the grounding layer and the CPW trace on the top layer is 10 mils. The CPW for transmitting or receiving antenna is shown in Fig. 1 (b).

D. Structure Design of Three Dimensional Via Antenna

The structure of 3D antenna is shown in Fig. 1, which includes two design:

Geometry design as shown in Fig. 3: the signal via with the cylindrical shape showed in Fig. 3(a) and conical shape showed in Fig. 3(b) is designed, and the S parameters is used to characterize the performance. The cylindrical structure has a radius of 7 mils. The conical via has an upper radius of 4 mils and a lower radius of 7 mils.

Radius design: based on the conical structure, the aspect ratio of the conical via is investigated and the R_ Up / R_ Down representing upper radius and lower radius of conical via are separately analyzed. The R_ Up will set the range of 1-7 mils with the R_ Down is 7mil. The R_ Down will set the range of 3-10 mils with the R_ Up is 4mil.

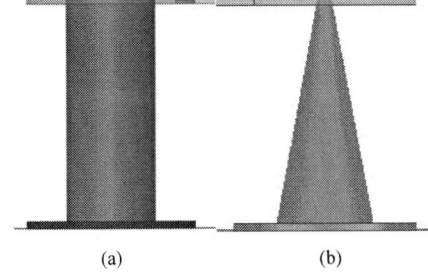

(a)   (b)

Fig. 3. Via structures with different geometric shapes
(a) cylindrical structure (b) conical structure

III. SIMULATION RESULT OF SIGNAL VIA

The influence factors of signal via were studied in detail, including the geometry and radius in order to determine the best physical conditions of conical via antenna. In section A, the data of S-parameters and resonance frequency points of cylindrical and conical structure are compared at the geometric level. In Section B, the influence of the upper and lower radius for conical via on signal performance was studied.

A. Simulation of Signal Via Geometry

As shown in Fig. 3 (a), the general signal via is cylindrical structure. However, considering the discharge principle of the antenna tip, conical via as a monopole antenna is also designed, which can more easily transmit signal along the PCB and has better directional transmitting performance. To verify this assumption, S11 and S21 of these two antennas are simulated. The results of the two geometries are shown in Fig. 4.

- Distribution of resonance frequency points. The results in Fig. 4 blue line shown that the resonant

978-1-6654-1392-3/21 $31.00 © 2021 IEEE

frequencies of the cylindrical via are about 29.5 GHz, 33.51 GHz, 39.51 GHz, and 43.99 GHz. The minimum return loss was about -18.97 dB near 39.51 GHz, and the maximum insertion loss was -2.48 dB near 43.99 GHz; The results with red line shown that the main measured resonant frequencies of conical via were about 22.9 GHz, 26.4 GHz, and 29.8 GHz. The minimum return loss was about -24.08 dB near 26.4 GHz, and the maximum insertion loss was -8.72 dB near 22.9 GHz.

- Bandwidth. According to the S11 and S21 in Fig. 4, the minimum bandwidth of cylindrical structure was 0.41 GHz, and the maximum was 1.17 GHz. The minimum bandwidth of conical via structure was 1.24 GHz, and the maximum bandwidth was 2.45 GHz about twice that of cylindrical structure.

- Advantages and disadvantages. Cylindrical signal via had advantages in the insertion loss and the number of resonance frequency points. The maximum insertion loss was -2.48 dB. There were at least five frequency points in the sweep frequency range, and even more in the high frequency range. Conical via had a larger bandwidth. Also it controlled the working frequency below 32 GHz, which greatly reduced the working frequency. To obtain the higher bandwidth and lower resonance frequency, conical via antenna was a better select.

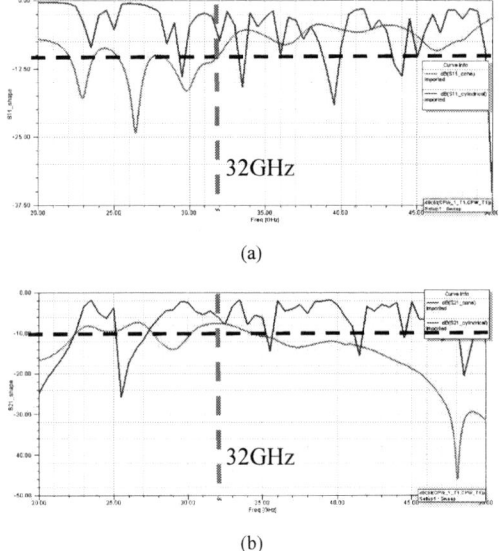

(a)

(b)

Fig. 4. S-parameter simulation diagram of cylindrical structure and conical structure (a) return loss (b) insertion loss

B. Simulation of Signal Via Radius

The structure parameters of conical via studied in the previous section was R_ Up=4 mil and R_ Down=7 mil. In this section, we would explore the influence of R_ Up and R_ Down in the conical via structure on signal performance.

1) The Impact of Upper Radius

Figure 5 showed the simulation results of R_ Up.

- The resonance frequencies exhibited directional shifts as the upper radius of conical via increased from 1 mil to 7 mil with lower radius of 7 mil. Different

radius have different colors in Fig. 5. Below the central frequency of 32 GHz as the star in the Fig. 5 (a), the resonance frequencies shifted to the left as the left two arrows in Fig. 5 (a). Above the central frequency of 32 GHz, the resonance frequencies shifted to the right as the right three arrows in Fig. 5 (a). The resonance frequency point also reduced. However, the S parameter at higher frequencies (> 32 GHz) is not good.

- Conical via had low return loss at low resonance frequency. The minimum return loss was about -30.89 dB and the maximum insertion loss was -7.28 dB at 26.70 GHz as shown in Fig. 5 (a) and (b), which was better than the planar log-periodic antenna in the same communication distance (return loss of -15 dB and insertion loss of -23 dB at 44 GHz)[7]. So the conical via can work at lower frequency with the minimum frequency of 23 GHz.

- Insertion loss did not change much with the upper radius increase as shown in Fig. 5 (b).

- The bandwidth decreased with the upper radius R_ Up increasing from 1 mil to 7 mil. The conical via with 1 mil upper radius and 7 mil lower radius had quite larger bandwidth than the cylindrical via. The bandwidth of conical via antenna was 2.22 GHz at 25.45 ~ 27.67 GHz and 2.54 GHz at 30.15 ~ 32.69 GHz.

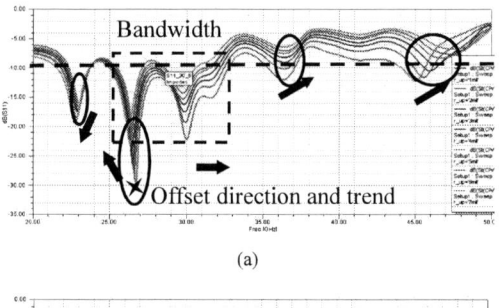

(a)

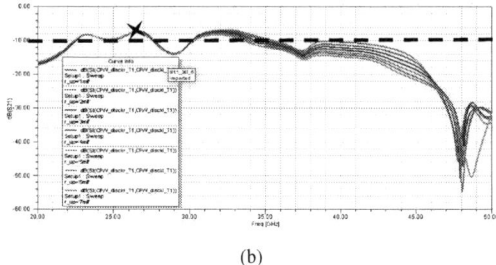

(b)

Fig. 5. R_ Up change simulation diagram (a) return loss (b) insertion loss

2) The Impact of Lower Radius Change

When the lower radius changed from 3 mil to 10 mil (upper radius was 4 mil), the effect of the lower radius on the S11 and S21 was similar with that of the upper radius as shown in Fig. 6. The overall situation was that the return loss decreased with the lower radius near 23 GHz, and increased with the lower radius at the other frequencies. In the shifting direction aspect, the return loss and insertion loss kept the same trend. Below 32 GHz, the resonant frequency shifted to the left with the lower radius, and the minimum return loss was -40 dB near 27 GHz. Above 32GHz, the resonant frequency shifted to the right and the resonant frequency point decreased with the lower radius. However, at higher frequencies (> 32 GHz), the S parameter was not good. And the return loss decreases with the lower radius as seen in Fig. 6 (a) and moves down in the vertical direction, which increase the effective bandwidth range. The insertion loss was consistent with the upper radius, and did not change much.

In Fig. 6 (b), with the increase of the lower radius, the insertion loss shifts to both sides, and the effective interval was slightly reduced. Therefore, according to S11 and S21, the bandwidth was also slightly reduced and the decrease magnitude was less than 0.3 GHz. Whether it was the upper radius or the lower radius, the insertion loss in the higher frequency were not very good.

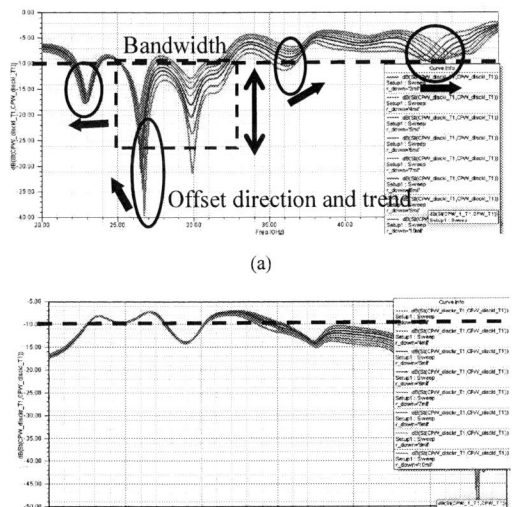

(a)

(b)

Fig. 6. R_Down change simulation diagram
(a) return loss (b) insertion loss

IV. CONCLUSIONS

This paper studied and evaluated a 3D monopole antenna mimicking TSV structure applied for wireless communication on NoC system. The signal source transmitted and received in the PCB with short distance communication, which was different from the general antenna on-chip. It was not on the surface, but inside the system or chip. The Rogers RO4003 high-frequency PCB was used as the substrate for the antenna because it had the same tangent loss with silicon and was easily to be demoed. The 3D antenna could increase the wireless communication distance, increase the bandwidth, and decrease the resonance frequency. In the future, it will be an ideal antenna choice for high integration and complex system in the cloud computing and AI.

In order to verify the function and performance, the 3D antenna of PCB prototype was simulated by HFSS. The results showed that the conical via had obvious superiority in improving the bandwidth and lowering the frequency. The bandwidth of conical via antenna was quite large, which was 2.22 GHz (between 25.45 GHz and 27.67 GHz) and 2.54 GHz (between 30.15 GHz and 32.69 GHz). Compared with other 3D antenna, the working frequency of conical via antenna (mainly below 32 GHz) was reduced by about 30 GHz. Besides, the conical via antenna had the minimum return loss was about - 30.89 dB and the maximum insertion loss was -7.28 dB at 26.70 GHz. The maximum insertion loss was also much better than other 3D antenna.

The conical structure also had the advantage of saving the area on the top side of the substrate with smaller via size. The saved area can be used for the other components integration and greatly improve the integration density.

ACKNOWLEDGMENTS

This work was supported by National Natural Science Foundation of China under Grant No. 61904037, 61804132, 62074132, and Shanghai Science and Technology Innovation Action Plan under Grant No. 19511131900.

REFERENCES

[1] V. F. Pavlidis and E. G. Friedman, "Interconnect-Based Design Methodologies for Three-Dimensional Integrated Circuits," in Proceedings of the IEEE, vol. 97, no. 1, pp. 123-140, Jan. 2009.

[2] K. Kim and K. Ko, "Integrated dipole antennas on silicon substrates for intra-chip communication," IEEE Antennas and Propagation Society International Symposium. 1999 Digest. Held in conjunction with: USNC/URSI National Radio Science Meeting (Cat. No.99CH37010), Orlando, FL, USA, 1999, pp. 1582-1585 vol.3.

[3] Wu J , Kodi A K , Kaya S , "Monopoles Loaded with 3-D-Printed Dielectrics for Future Wireless Intra-Chip Communications," IEEE Transactions on Antennas & Propagation, 2017, 65(12):6838-6846.

[4] Tasolamprou A C , Mirmoosa M S , Tsilipakos O , " Intercell Wireless Communication in Software-defined Metasurfaces," 2018 IEEE International Symposium on Circuits and Systems (ISCAS). IEEE, 2018 , pp. 1−5.

[5] Lin J J , Wu H T , Su Y , "Communication Using Antennas Fabricated in Silicon Integrated Circuits," IEEE Journal of Solid-State Circuits, 2007, 42(8):1678-1687.

[6] Pano V, Tekin I, Liu Y, "TSV-Based Antenna for On-Chip Wireless Communication," IET Microwaves Antennas & Propagation, 2019, 14(4).

[7] Samaiyar A , Deb S , Ram S S . "Millimeter-wave planar log periodic antenna for on-chip wireless interconnects," The 8th European Conference on Antennas and Propagation (EuCAP 2014). IEEE, 2014, pp. 1007−1009.

Comparative Study on the Effects of Fe and Ni Additions on the Electromigration Properties of Sn58Bi Solder Joints

Zhuangzhuang Hou
School of Materials
Science&Engineering
Beijing Institute of Technology
Beijing, China
hzzpaper@163.com

Yaru Dong
School of Materials
Science&Engineering
Beijing Institute of Technology
Beijing, China
dongyaru52@163.com

Lingyao Sun
School of Materials
Science&Engineering
Beijing Institute of Technology
Beijing, China
sunlingyao_513@163.com

Ying Liu
School of Materials
Science&Engineering
Beijing Institute of Technology
Beijing, China
yingliu@bit.edu.cn

Yongjun Huo
School of Materials
Science&Engineering
Beijing Institute of Technology
Beijing, China
huoyongjun@bit.edu.cn

Yingxia Liu
School of Materials
Science&Engineering
Beijing Institute of Technology
Beijing, China
yingxia.liu@bit.edu.cn

Xiuchen Zhao *
School of Materials
Science&Engineering
Beijing Institute of Technology
Beijing, China
zhaoxiuchen@bit.edu.cn

Abstract—**Sn58Bi eutectic solder is the preferred material for low temperature solder in multistage packaging. However, the brittleness and reliability problems of Sn58Bi eutectic solder, especially the phase separation of electromigration, bring great challenges to its application for multi-stage package interconnection. The modification of Sn58Bi eutectic solder by alloying has many significant technical advantages, such as effective control of interfacial wetting, intermetallic compounds phase transformation, growth kinetics, and improvement of resistance to electromigration. In this paper, Sn58Bi, Sn58Bi-1Fe/1Ni solder were prepared by melting method and fabricated to the solder joint for electromigration test. The electromigration test of solder joint were carried out with $0.5\times10^4 A/cm^2$ current density at 50 ℃, and the influence of Fe, Ni alloy elements addition on interfacial IMC growth and microstructure evolution of solder joint interface were studied. The results show that Fe and Ni alloy elements react with Sn to form $FeSn_2$ and Ni_3Sn_4 IMC, respectively. $FeSn_2$ are triangle or square, and distributed near the interface and embedded in bismuth rich phase and tin rich phase after electromigration test. Compared these two kinds solder joints, it can be seen that the Fe alloy elements can promote the growth of interface IMC and the aggregation of bismuth rich phase, while the Ni alloy elements can greatly inhibit the growth of interface IMC and the aggregation of bismuth rich phase. It may be due to the existence of $FeSn_2$ near the interface, which changes the growth behavior of bismuth rich phase and bin rich phase near the interface of Sn58Bi-1Fe/Cu solder joints and promotes their growth. As for the inhibition of Ni alloy elements on the growth of interfacial IMC and bismuth rich phase in the anode under the condition of EM, the reason is that Ni_3Sn_4 forms a barrier in the solder matrix, which hinders the diffusion of atoms.**

Keywords—alloying element addition, IMC growth, electromigration

I. INTRODUCTION

The multi-level interconnection of planar array needs Sn-based solder with different melting points, which puts forward an urgent demand for Sn-based solder with low melting point to realize low-temperature interconnection.[1-2] In this case, Sn58Bi eutectic solder is widely concerned because of its low melting temperature, low cost and outstanding mechanical properties.[3] Sn58Bi has the better thermal fatigue resistance, yield strength and shear strength than conventional Sn37Pb eutectic solder.[4] However, owing to the low melting temperature, under the combined action of high homologue temperature and high density current, atoms are easy to diffuse in the lattice, so atoms migration in Sn58Bi eutectic solder will be considerable.[5] According to Gu et. al [6], bismuth rich layer is formed at the anode and tin rich phase is formed at the cathode. and the thickness of IMC at the cathode interface is greater than that at the anode interface. The thickness of bismuth rich layer increases linearly with the stress time. Compared to the configuration or geometry of Sn-Bi BGA interconnects, the effect of non-uniform eutectics on current density distribution is more pronounced. The current density in Sn phase is much higher than that in bismuth phase. The Bi atoms in the tin rich phase will not migrate directly along the interface between the bismuth rich and tin rich phases, but tend to migrate to the anode first. Therefore, the destruction of tin rich phases rather than phase interfaces or bismuth rich phases can usually be observed in experiments[7]. Accordingly, researchers applied themselves to solve the problems of Bi-phase aggregation and interfacial IMC overgrowth caused by electromigration(EM). Alloying and heterogeneous phase addition were regarded as two critical modification method to alleviating EM in Sn-Bi eutectic solder joints. J. Kim's[8] work indicate that the Ag MWCNT can acts as a diffusion barrier in the Sn58Bi eutectic solder to inhibit the growth of Bi-rich layer, Cu_6Sn_5 and Cu_3Sn IMC. So, Ag MWCNT is expected to improve the performance of

EM in the Sn58Bi solder joint. L. Hua et.al[9] think the doping of Sb can inhibit the growth of Cu6Sn5 IMC, but the formation of SnSb compound reduces the reliability of EM and solder joint life. Moreover, Ge element has a skin effect, that is, it is highly enriched and separated at the edge of SnBiAg solder, so as to protect Sn in SnBiAg solder from electrochemical corrosion. T. Laurila [10] thinks that in Sn/Cu system, Ni can greatly reduce the thickness of Cu3Sn, thus reducing the total thickness of IMC layer. The effect of Au is similar, but obviously weaker. However, Fe and Ag had little effect.

Fe, Ni alloy elements are face centered cubic structure, the structure and properties are very similar, and they can act with Sn elements to generate FeSn2, Ni3Sn4 respectively. Currently, the research of the Fe, Ni elements' effect on EM focus on Sn/Cu system, but, what roles they played in EM problem in Sn58Bi eutectic solder joints are unknown. Thus, it is of great interest to compare the effects of Fe, Ni elements additions on the EM properties of Sn58Bi eutectic solder joints. In this paper, we add the Fe, Ni elements into Sn58Bi eutectic solder to comparative study the effects of Fe and Ni additions on the EM properties of Sn58Bi eutectic solder joints..

II. EXPERIMENTAL

As shown in Table 1, solder composition were Sn58Bi, Sn58Bi-1Fe and Sn58Bi-1Ni, respectively. That is to say, the alloy elements are added into the SnBi solder with eutectic composition ratio. The purity of all these metal materials are above 99.9%. These Sn58Bi-based solder was prepared by fusible salt protection melting method. First, put the weighed metal materials according to table 1 into alumina crucible. To protect the metal materials from oxidation, we covered the metal materials in alumina crucible with molten salt which consist of KCl and LiCl (1.3:1 wt. %) before melting process. And then, put the alumina crucible into muffle furnace, and keep the temperature at 600 ℃ for 2 hours with stirring every 10 minutes used alumina ceramic rod. At the end of heat preservation process, alumina crucible was cooling in air. After solidified salt was washed away by water, Sn58Bi, Sn58Bi-1Fe and Sn58Bi-1Ni solder alloy were obtained.

Table 1 Composition of Sn58Bi, Sn58Bi-1Fe and Sn58Bi-1Ni solder

	Sn/g	Bi/g	Fe/g	Ni/g	Total/g
SB	42	58	0	0	100
SBF	41.58	57.42	1	0	100
Ni	41.58	57.42	0	1	100

In this work, we fabricated the interconnection circuit by FR-4 copper clad laminate and the diameter of copper pad is 600 μm. Prepared solder was divided into 3 mg pieces and reflow in 170℃ with 5min to form the solder balls. Then placed the solder balls on the copper pad in interconnection circuit. Use FlipChip equipment (Fin-placer145, Fintech, Germany) to align the pads of the upper and lower circuits and then run a reflux process with the peak temperature of 170℃ for 120 s under nitrogen atmosphere. The spacing between the upper and lower circuits is controlled at 400 μm with gaskets during reflow process and then we obtained the test circuit with two solder joints as shown in Fig. 1. In EM test, the current density at solder pad is $0.5 \times 10^4 A/cm^2$, time of power on are 0, 24, 48, 120, 240 h and the ambient temperature is 50℃. The EM test current direction are shown in Fig. 1. After EM test, the cross section of the solders joints was observed by scanning electron microscopy (SEM, S4800) and the element composition of each part is obtained by energy dispersive spectrometry (EDS).

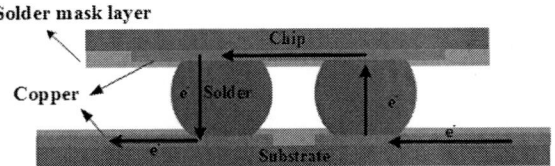

Fig. 1 Schematic diagram of EM test circuit and current direction

III. RESULTS AND DISCUSSION

Fig. 2 shows the microstructure morphology of Sn58Bi, Sn58Bi-1Fe and Sn58Bi-1Ni solder joints and the distribution of interface microstructure after interconnection. It can be seen from the Fig. 2 (a-c) that FeSn2 and Ni3Sn4 IMC are formed by the reaction of Fe and Ni with tin respectively, and the microstructure of solder is refined to a certain extent. After interconnection, it can be observed that the thickness of interface IMC increases with the addition of Fe, but no obvious effect with the addition of Ni. It is worth noting that only FeSn2 IMC is found near interfacial IMC of Sn58Bi-1Fe/Cu joint, while Ni3Sn4 IMC is not found near interfacial IMC of Sn58Bi-1Ni/Cu joint.

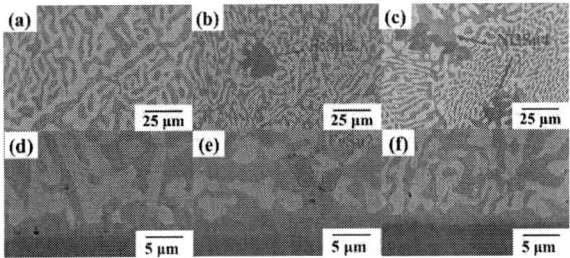

Fig. 2 (a-c) The microstructure and morphology of the prepared solder matrix: (a) Sn58Bi, (b) Sn58Bi-1Fe, (c) Sn58Bi-1Ni; (d-f) Cross section of solder joint interface after interconnection: (d) Sn58Bi/Cu, (e) Sn58Bi-1Fe/Cu, (f) Sn58Bi-1Ni/Cu

To investigate the growth behavior of IMC between Sn58Bi, Sn58Bi-1Fe and Sn58Bi-1Ni solder with Cu substrate in the process of EM. Cross section images of interconnection interface of solder joints in EM test circuit were obtained, and interfacial IMC thickness was further measured. Fig. 3 shows the average thickness of interface IMC in Sn58Bi/Cu, Sn58Bi-1Fe/Cu, Sn58Bi-1Ni/Cu solder joints, and the growth rules of interfacial IMC along with charge time. Form the Fig. 3(a), all interfacial IMC became thicker with the power on time. Anode and cathode interfacial IMC have the similar growth law, basically keeps a linear growth with the power on time. However, it can be seen from Fig. 3 (b) and (c) that under the condition of EM, the growth behavior of interface IMC does not change after the addition of Ni and Fe alloy elements, and the thickness of interfacial IMC at anode and cathode almost equal, showing a similar growth law. This is obviously different from other tin based solders in that the IMC of anode grows rapidly and the IMC of cathode interface thins gradually under the condition of EM. In contrast, Bi atoms migrate in large scale and form a bismuth rich layer near the anode interface during the EM process in Sn58Bi eutectic solder joints. Therefore, we infer that the formation of bismuth rich phase may be one of the factors that cause the difference of IMC growth behavior between Sn58Bi eutectic solder bumps and other Sn based solders under the condition of EM.

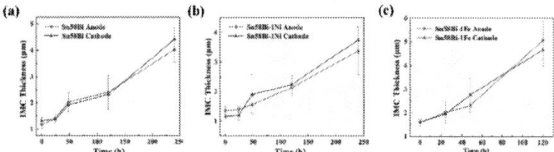

Fig. 3 Comparison of IMC thickness of anode cathode interface with EM time: (a) Sn58Bi/Cu, (b) Sn58Bi-1Ni/Cu, (c) Sn58Bi-1Fe/Cu

In order to further understand the influence of Fe and Ni alloy elements on the growth of interfacial IMC, we compared the growth rules of IMC at the anode and cathode interface of three kinds of solder joints respectively, as shown in Fig. 4. As shown in Fig. 4(a), the growth rate of IMC at the anode interface increased significantly with the addition of Fe alloy elements, increased by 48.6% from 1.191×10^{-2} μm/h to 1.771×10^{-2} μm/h, and decreased by 38.8% from 1.191×10^{-2} μm/h to 0.729×10^{-2} μm/h after the addition of Ni. In addition, it can be seen from Fig. 4(b) that the growth rate of IMC at the cathode interface is increased to 2.253×10^{-2} μm/h by about 83.4% with the addition of Fe alloy elements, while the growth rate of IMC at the cathode interface is reduced to 1.004×10^{-2} μm/h by about 18.2% with the addition of Ni alloy elements, which are consistent with the effect of Fe and Ni alloy elements on the growth of IMC at the anode interface. Therefore, we can conclude that the addition of Fe alloy elements can promote the growth of IMC at the interface of Sn58Bi/Cu solder joint, while the addition of Ni alloy elements has the opposite effect.

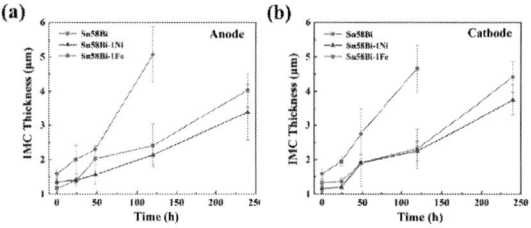

Fig.4 Comparison of IMC thickness of three kinds of solder joints with EM time: (a) anode, (b) cathode

The existence of bismuth rich phase will make there is almost no Sn phase near the anode interfacial IMC, which greatly inhibit the growth of anode interfacial IMC. Due to the blocking of the material transport channel near the anode, the Sn atoms near the cathode will not be exported. On the contrary, the migration of Bi atoms to the anode leads to the formation of Sn rich phase near the cathode interfacial IMC, which enhances the driving force of IMC growth at the cathode interface. Therefore, the IMC thickness of cathode interface does not decrease, but is basically consistent with that of anode interface, and increases with the EM time.

Fig. 5(a) shows the change of bismuth rich phase concentration thickness at the anode interface of in Sn58Bi/Cu, Sn58Bi-1Fe/Cu, Sn58Bi-1Ni/Cu solder joints with the power on time. It can be seen from Fig. 5(a) that with the extension of power on time, the bismuth rich phase aggregation thickness at the anode of Sn58Bi/Cu joint increases gradually, and the bismuth rich phase aggregation speed increases suddenly after power on for 120h, showing some signs of failure. With the addition of Fe alloy elements,

the bismuth rich phase in the anode of Sn58Bi-1Fe/Cu interconnector increases significantly, and the bismuth rich phase begins to thicken rapidly after 48 h being electrified. This indicates that the addition of Fe greatly accelerates the aggregation rate of bismuth rich phase at the anode interface, makes the phase separation of Sn58Bi eutectic solder joints more serious during the process of EM, and the resistance to EM is damaged to a certain extent. In contrast, the addition of Ni alloy elements can significantly inhibit the bismuth rich phase aggregation at the anode interface of solder joints. During the whole period of EM, the bismuth rich phase aggregation rate maintains at a low level, and the bismuth phase aggregation rate is going down after 120 h. Fig. 5(b) shows the thickness of bismuth rich phase at the anode interface of the three solder joints 48 h after EM. At this time point, the aggregation rate of bismuth rich phase has not changed suddenly. The thickness of bismuth rich phase in Sn58Bi-1Fe/Cu solder joint is the largest, followed by that in Sn58Bi/Cu solder joint, and the thickness of bismuth rich phase in Sn58Bi-1Ni/Cu solder joint is the smallest. This shows that compared with Fe alloy elements, the addition of Ni alloy elements can effectively alleviate the bismuth rich phase aggregation in the process of EM, and improve the anti-EM performance of Sn58Bi eutectic solder joint.

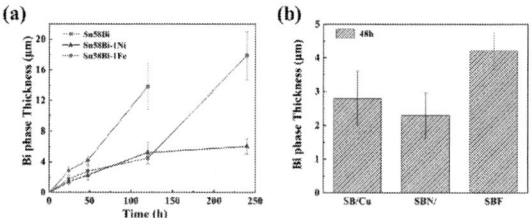

Fig. 5 Comparison of Bi rich phase thickness at anode interface of three kinds of solder joints after 120 h EM test

Fig. 6 is cross section of solder joint interface after 120 h EM (a) Sn58Bi-1Fe anode interface, (b) Sn58Bi-1Fe cathode interface, (c) Sn58Bi-1Ni anode interface, (d) Sn58Bi-1Ni cathode interface. It can be seen from the Fig. 6 that bismuth rich phase is formed at the anode interface of Sn58Bi-1Ni/Cu and Sn58Bi-1Fe/Cu solder joint after 120h EM test, and the thickness of bismuth rich phase in Sn58Bi-1Fe/Cu solder joint is much thicker than that in Sn58Bi-1Ni/Cu interconnector. Concurrently, tin rich phase appears obviously at Sn58Bi-1Fe/Cu joint cathode interface. It is worth mentioning that FeSn2 is discovered in both anode bismuth rich phase and cathode tin rich phase of Sn58Bi-1Fe/Cu solder joint, which is triangular or square in shape. However, there is no obvious Ni3Sn4 near the interface IMC and bismuth rich phase. It can be seen from Fig. 6 that bismuth rich phase is formed at the anode interface of Sn58Bi-1Ni/Cu and Sn58Bi-1Fe/Cu solder joint after 120h of EM, and the thickness of bismuth rich phase in Sn58Bi-1Fe/Cu solder joint is much thicker than that in Sn58Bi-1Ni/Cu interconnector. Simultaneously, obvious tin rich phase appears at Sn58Bi-1Fe/Cu joint cathode interface, while no tin rich phase was found in the Sn58Bi-1Ni/Cu solder joint. So, it may be due to the existence of FeSn2 near the interface, which changes the growth behavior of bismuth rich phase and tin rich phase near the interface of Sn58Bi-1Fe/Cu solder joints and promotes their growth. As for the inhibition of Ni alloy elements on the growth of interfacial IMC and bismuth rich phase in the anode under the condition of EM, the reason is that Ni3Sn4 forms a barrier in the solder matrix, which hinders the diffusion of atoms.

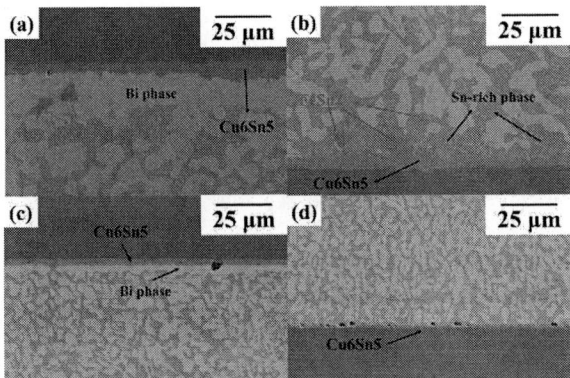

Fig. 6 Cross section of solder joint interface after 120 h EM (a) Sn58Bi-1Fe anode interface, (b) Sn58Bi-1Fe cathode interface, (c) Sn58Bi-1Ni anode interface, (d) Sn58Bi-1Ni cathode interface.

IV. CONCLUSION

(1) Fe and Ni in Sn58Bi eutectic solder react with Sn to form FeSn2 and Ni3Sn4 IMC respectively, and refine the solder structure to a certain extent.

(2) After interconnection, FeSn2 mainly exists near the interface IMC in Sn58Bi-1Fe/Cu solder joint, which is triangular and square, while Ni3Sn4 is not found near the interface of Sn58Bi-1Ni/Cu joint. After the EM test, FeSn2 is located near the interface, embedded in the anode bismuth rich phase and cathode tin rich phase, while Ni3Sn4 has not been found near the interface.

(3) The addition of Fe alloy elements can promote the growth of IMC and bismuth rich phase at the anode interface of Sn58Bi/Cu solder joint, while the addition of Ni alloy elements has the opposite effect.

REFERENCES

[1] S. Annuar, R. Mahmoodian, Hamdi M, KN. Tu, "Intermetallic compounds in 3D integrated circuits technology: a brief review," Sci. Tech. Adv. Mater, vol. 18(1), pp. 693-703, 2017.

[2] KN. Tu, Y. Liu, "Recent advances on kinetic analysis of solder joint reactions in 3D IC packaging technology," Mater. Sci. Eng, vol. 136(APR.), pp. 1-12, 2019.

[3] L. Chen, C. Chen, "Electromigration study in the eutectic SnBi solder joint on the Ni/Au metallization," J. Mater. Res, vol. 21(4), pp. 962-969, 2006.

[4] D. Ye, C. Du, M. Wu, Z. Lai, "Microstructure and mechanical properties of Sn-xBi solder alloy," J. Mater. Sci. Mater. Electron, vol. 26, pp. 3629–3637, 2015 .

[5] C. Chen, S.W. Liang, "Electromigration issues in lead-free solder joints," J. Mater. Sci. Mater. Electron, vol. 18(1-3), pp. 259-268, 2006.

[6] X. Gu, Y.C. Chan, "Electromigration in Line-Type Cu/Sn-Bi/Cu Solder Joints," J. Electron. Mater, vol 37(11), pp. 1721-1726, 2008.

[7] H. Qin, X. Luan, W. Yue, "Influence of Phase Inhomogeneity on Electromigration Behavior in Cu/Sn-58Bi/Cu Solder Joint," J. Electron. Mater. vol. 48(6), pp. 3410-3414, 2019

[8] J. Kim, K. H. Jung, J. H. Kim, C. J. Lee, S. B. Jung, "Electromigration behaviors of Sn58%Bi solder containing Ag-coated MWCNTs with OSP surface finished PCB," J. Alloys. Compd, vol. 775, pp. 581-588, 2019

[9] L. Hua, H. N. Hou, "Electrochemical corrosion and electrochemical migration of 64Sn-35Bi-1Ag solder doping with xGe on printed circuit boards," Microelectron. Reliab, vol. 75, pp. 27-36, 2019.

[10] T. Laurila, J. Hurtig, V. Vuorinen, J.K. Kivilahti, "Effect of Ag, Fe, Au and Ni on the growth kinetics of Sn–Cu intermetallic compound layers," Microelectron. Reliab, vol. 49(3), pp. 242-247, 2009.

Low Temperature and Short Time Au/Sn Solid-liquid Diffusion Bonding for 3D Integration

Ziyu Liu *
School of Microelectronics,
Fudan University
Shanghai, China
liuziyu@fudan.edu.cn

Wang Wenchao
School of Electronic Information
Engineering, Southwest University
Chongqing,China
985621465@qq.com

Zhu Zhiyuan
School of Electronic Information
Engineering, Southwest University
Chongqing,China
zyuanzhu@swu.edu.cn

Chen Lin
School of Microelectronics,
Fudan University
Shanghai, China
linchen@fudan.edu.cn

Sun Qingqing
School of Microelectronics,
Fudan University
Shanghai, China
qqsun@fudan.edu.cn

Abstract—High temperature and long-time solid-liquid diffusion bonding (SLID) will barrier the scaling down of multi-chip interconnection and increase the total time of 3D integration. A wafer-level temperature gradient bonding technology (TGB) was used in the Au/Sn SLID bonding to lower the temperature and shorten the bonding time. Au/Sn bonding with different pad sizes were designed. The results showed that the Au/Sn bump with different pad size effectively reduce the Sn overflow than that with same pad size. In addition, suitable electroplating current density could lower the Au bump surface roughness and evaporation process could also make Sn flat. With all these optimized design, the bonding was successfully achieved in a vacuum bonding chamber. The hot end temperature was set to 250 °C and the cold end temperature was set to 150 °C with the bonding pressure of 5000 N. The bonding condition was maintained for 20 minutes. Then X-ray detection, shearing test and scanning electron microscope (SEM) were used to study the quality of the bonding interface and the overflow of Sn. Misalignment was not detected based on the wafer-level and chip-level X-ray. The chips in the middle of the bonded wafer had a higher shear force compared to the surrounding part of the wafer. SEM images indicated the fracture all occurred on the grain boundary and no dimple were founded. Daisy chain structure were observed when the cross sectional image of the bonded chip was detected.

Keywords—TGB, SLID, Au/Sn bump, overflow, brittle fracture

I. INTRODUCTION

With the continuous miniaturization of microelectronic systems, 3D integration technology has attracted much attention. Chip interconnection is the key to 3D integration, which mainly includes intra-chip and inter-chip interconnection. Solid-liquid diffusion bonding (SLID) is one of the widely-used inter-chip interconnection technologies [1]. It has been successfully used to connect two chips for many years. Compared with other bonding technologies, it does not require large bonding pressure and can be applied in high density integration. This process combines two metal layers with a high and a low melting-point. By heating to the temperature exceeding the low melting-point, the low-melting-point metal with the liquid state will quickly diffuse and react with the high low-melting-point metal to form a high melting-point intermetallic compound (IMC). But the low-melting-point metal such as Sn is easily squeezed out

with the bonding pressure. This overflow phenomenon is likely to make the adjacent micro-bumps connected and causes the interconnection short [2]. The bonding pitch then can not shorten and integration density can not increase.

Au/Sn bonding can generate IMC with good mechanical performance, excellent electrical and thermal conductivity at a lower temperature. Lower bonding temperature can reduce the overflow during bonding but always needs longer bonding time, which hinders the widespread application in the multi-chip interconnection. Thermal gradient bonding (TGB) was reported to speed up the Au/Sn bonding and lower the temperature [3-4]. Thus, TGB can be applied in the Au/Sn bonding and could further lower the bonding temperature and bonding pressure.

In traditional packaging technology, stencil printing and ball placement are always used to form the Au/Sn solder in the bonding [5]. The disadvantage of these two technologies is that the bonding pitch is quite large, which is always larger than 100 μm. Since Au/Sn solder is pre-formed, it usually has a high oxide content, which is harmful to bonding and a barrier to reduce bonding pitch to tens of micrometers. Electroplating has the advantages of low cost and evaporation in the vacuum can get high surface roughness. Thus, the Au layer was formed by the electroplating and the Sn layer was deposited by the evaporation in this study, so that the Au/Sn bump did not have lower oxide content but also had lower roughness surface, which was beneficial to improve the subsequent bonding quality. This paper designed and studied a wafer-level low-temperature bonding process with optimized thermal gradient process [6].

II. EXPERIMENTAL PROCEDURES

A. Preparation of the bonded wafer

Figure 1 shows the bonding structure and two kinds of Au/Sn-Sn/Au are designed. One design contains same size Au/Sn pad and the other has smaller Sn pad than Au. Then the fabrication process are designed as followed [7]. First, a dicing groove is etched on the back-side of the double-polished oxide silicon wafer to facilitate subsequent bonding alignment and dicing. An Al layer with 1 μm thickness is sputtered on the front-side of the silicon wafer and a wiring layer pattern is formed after photolithography. Then the patterned wafer is placed in the high concentration H_3PO_4 at

978-1-6654-1392-3/21 $31.00 © 2021 IEEE

65℃ for 6 minutes and the Al wiring is etched. Daisy chain and Kelvin testing structure are designed to test the electrical resistance. Each daisy chain contains 3463 micro-interconnection structures and each Kelvin structure includes three micro bonding points. There are many daisy chains and Kelvin structures on a bonded chip. Before the next metal deposition process, the silicon wafer is subjected to plasma pretreatment for 3 minutes, to remove the organic compound on the surface. Later, Ti and Au layers with thicknesses of 50 nm and 100 nm are sputtered, respectively, in which Ti is as the adhesion layer and Au as the seed layer. The roughness has a great influence on the bonding result. In the third part of the article, we did an optimization experiment to determine the electroplating parameters. Therefore, the

current density of 1 mA/cm^2 was selected during the electroplating process to obtain a 4 μm thick Au layer at 60 ℃. Finally, the Sn layer with the thickness of 2 μm is evaporated. After the bump seed layer is removed by ion plasma etching (IBE), all the fabrication processes are completed. The reason to use evaporation is that the Sn surface roughness by evaporation is smaller than electroplating, and a better bonding surface roughness can be obtained.

B. Thermal gradient bonding (TGB)

Figure 2 shows the thermal gradient bonding structure. Before the bonding, the sample is treated by Ar plasma with a gas flow of 200 sccm and a power of 200 W for 2 minutes to activate Au/Sn bumps and remove oxides.

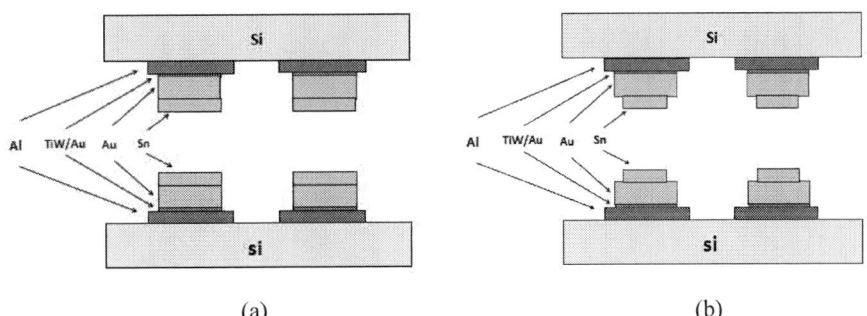

(a)   (b)

Fig. 1. Au/Sn-Sn/Au bump structure with Au bump size of 12 μm ×12 μm and different Sn bump size of (a) 12×12μm, (b) 10 μm×10 μm.

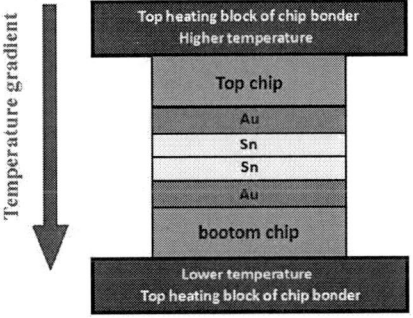

Fig. 2. The thermal gradient bonding (TGB) process with different temperatures on the top and bottom heating blocks.

The two wafers are first aligned in the SUSS MA6 alignment machine, and then bonded in the SUSS CB6L with a vacuum environment. The bonded area is 12.3 cm^2 and the bonding force of 5000 N is applied (bonding pressure of 4.065 MPa). The temperature of the hot and cold end is separately set to 250 ℃ and 150 ℃. Bonding temperature maintained under this condition for 20 minutes. Then it takes 20 minutes to cool to room temperature after bonding.

C. Dicing Line

The bonded wafer pair will experience two dicing process to separate the top chip and bottom chip. Two sets of scribing slots are designed as shown in Fig. 3, in which the first dicing line for the bottom chip and the second dicing line for the top chips. The space between the two dicing lines is used to expose the test pad on the bottom chip. The total thickness of the bonded wafer is about 1124 μm (the thickness of the single Si wafer is 550 μm, the thickness of

the bonded bump is about 12 μm), The cutting depths of the first group and the second group of scribe grooves are set to 1150 μm and 530 μm, respectively. After cutting the wafer with the DAD3350 cutting machine, the bonded chips with a size of 12 X 12 mm are obtained, and the bonding quality are evaluated accordingly.

D. Bonding quality evaluation

Firstly, the quality of the bonding interface is studied with an X-ray detector, and then a shear testing is performed to obtain bonding strength and optimize bonding process parameters. When the bonding strength is larger than the adhesion strength of the seed/adhesive layer under the bumps, it indicates the bonding strength the fracture mode and location, and explore the bonding strength of different areas of the entire wafer. By optimizing the bonding process, using a scanning electron microscope (SEM) to analyze the bonding interface and observe the overflow of Sn.

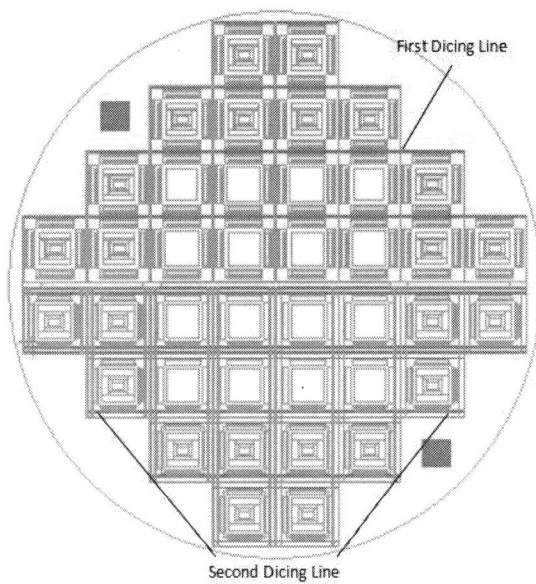

Fig. 3. Dicing line location.

III. RESULTS AND DISCUSSION

A. Evaluation of electroplating Au and evaporating Sn

Low surface roughness and uniform bump height were a premise to achieve high quality Au/Sn bonding. Therefore, the electroplating process deposited by the Au bumps was first optimized. The effect of different electroplating densities on Au surface morphology were shown in Figs. 4 (a) and (b). It was seen that the bumps on the wafer were uneven with the current density of 0.7 mA/cm^2 for 75 minutes at the temperature of 60 ℃ and the highness was as poor as 0.5 μm. When the current density increased to 1 mA/cm^2 and Au electroplating for 75 minutes at the temperature of 60 ℃ was carried out, the Au surface was uniform and the roughness was less than 100 nm. The above results indicated that the current density was an important factor affecting Au morphology and uniformity. When the current density was lower, the electroplating efficiency was low, and the low bump surface roughness would barrier the subsequent bonding process.

In the conventional SLID process, the Sn layer is made by electroplating and the surface roughness of Sn layer was quite large. If the electroplating area was large, the bump roughness across the wafer would be uneven. To solve this problem, the authors proposed a method to evaporate an Sn layer on Au and increased the surface roughness. Figure 5 showed SEM images of Sn layer surface by electroplating and evaporation. Sn deposited by electroplating had many nanoparticles in the grain boundaries, which made the surface roughness was large. The Sn surface roughness by the evaporation deposition was small because the grain boundary was very clean and the grain size showed more uniform. The surface roughness of micro bump after the whole process flow was shown in Fig. 6. Figure 6 (a) showed the micro-bump in the daisy chain around the chips and Fig.

6 (b) presented the large Sn bump in the center of the chips.

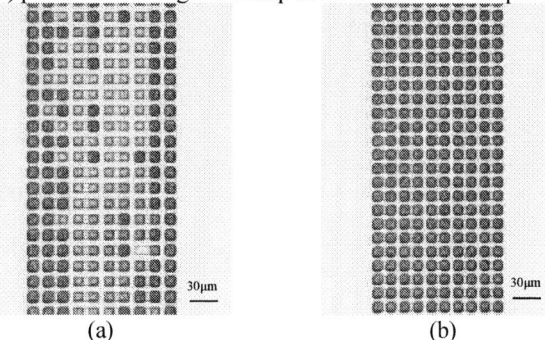

Fig. 4. Au surface morphology with the current density of (a) 0.7 mA/cm^2, and (b) 1 mA/cm^2.

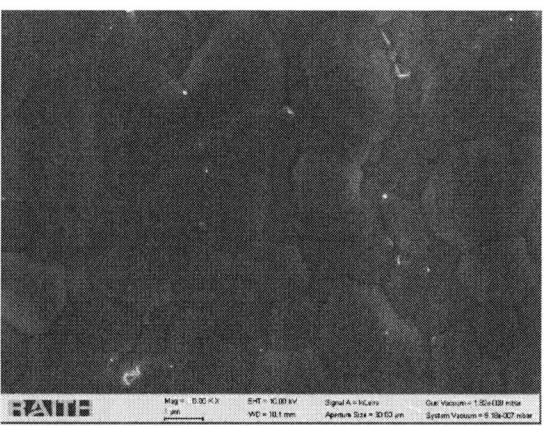

(a)

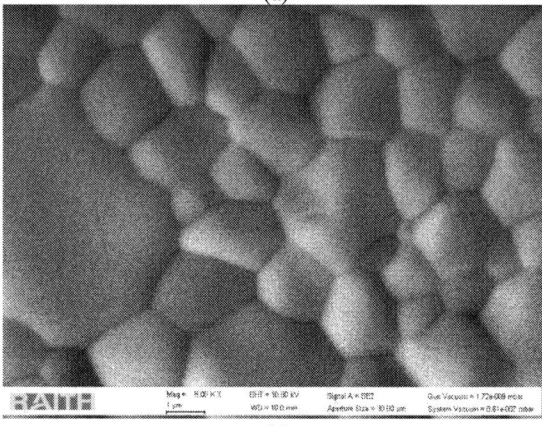

(b)

Fig. 5. Sn surface roughness by SEM (a) electroplating (b) evaporation.

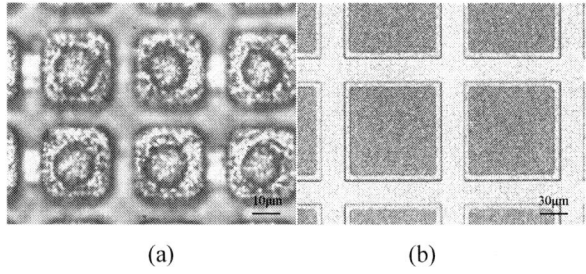

Fig. 6. (a) Sn bumps in the daisy chain around the chips, and

(b) large square Sn bumps in the center of the chips.

B. X-ray detection after the bonding process

The X-ray detection is one of the best qualitative methods to evaluate the wafer-level alignment. In addition, it was an effective mean to access the micro-voids in the bonding interface. Thus, X-ray detection was used to study the quality of the bonding interface first at wafer level. Figures 7 (a) and (b) showed the X-ray inspection images of the wafers after the bonding was completed. On the whole, misalignment of the micro bumps were not detected, and the bonding alignment accuracy was high. There were no micro voids in the interface of the bonded bumps, which indicated there were no defects with micro-meter.

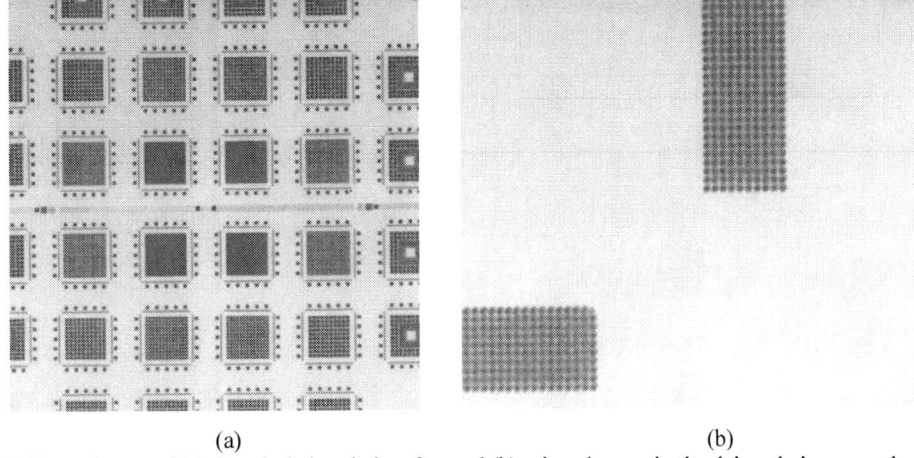

(a)   (b)

Fig. 7. X-ray image of (a) the whole bonded wafer, and (b) micro-bumps in the daisy chains around one chip.

C. Shear testing

The bonding strength of Au/Sn bonding was evaluated by measuring the shearing force of the bonded chip. Many sample groups across the wafer was were statistically analyzed as shown in Fig. 8.

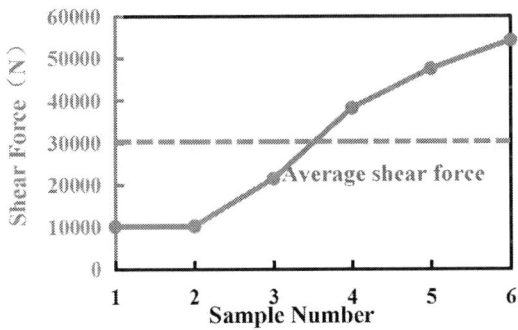

Fig. 8. Shear force range for 5 pairs of bonded chips.

Here, the shear force was used to evaluate the bonding quality of the chip as other literatures reported. The total thickness of the bonded chip was 1124 μm, so the shear height was set to 520 μm, which was higher than the bottom chip and the bonding bump. The results of the shear experiment as shown in Fig. 8 presented the shear force varied from 10.24 KN to 54.22 KN at different location of the wafer. Due to the difference in the bonding area of the samples, the difference of shear force was relatively large. However, it could be seen from the shear force that the bonding quality of the bonded chip was very high. Figure 9 showed the fracture surface of the bonded chips after shearing test by SEM. It was found that the fracture surfaces were all fractured along the grain boundary and no dimple was detected. Thus, the fracture mode was brittle fracture.

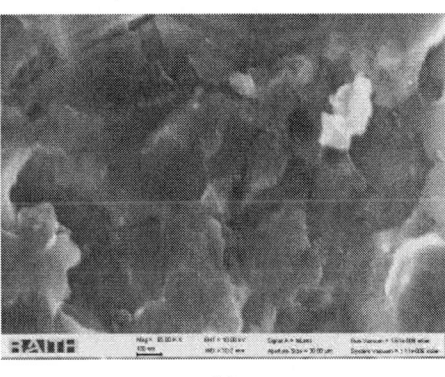

(a)

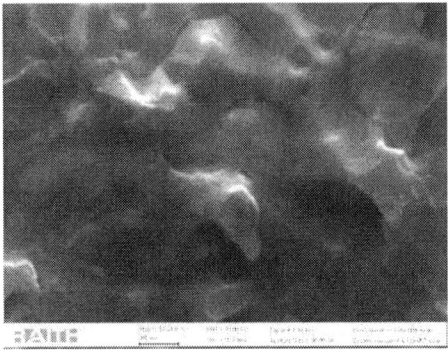

(b)

Fig. 9. SEM images of the fracture surface for (a) sample 1 and (b) sample 2.

After the measurement, the bonding force distribution diagram of the entire wafer was also analyzed. Figure 10 showe the approximate bonding strength of some areas in the wafer, and a~e indicated the bonding strength from low to high. For example, "a" area had the lowest shearing force and "e" area had the highest shearing force. We believed that the reasons might be as followed: (1) The uneven pressure of the wafer bonding equipment caused this phenomenon. (2) Due to the bump height difference between the large chip area and micro-bumps in the unstable voltage source of the electroplating equipment, which caused the uneven electroplating. We would continue to optimize this phenomenon in future experiments.

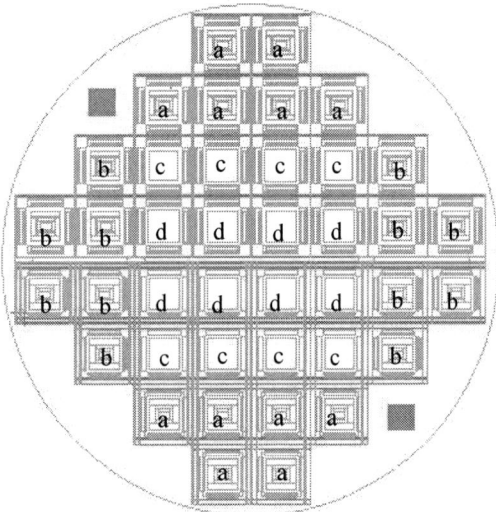

Fig. 10. Schematic diagram of bonding strength distribution.

D. Sn overflow situation analysis

The minimum pitch between the bumps was 20 μm. The higher temperature in SLID would cause the liquid Sn to be squeezed out, which was called Sn overflow. This phenomenon might cause short circuits in the connection, increased the risk of interconnection failure, and hindered the pitch shrinking of the interconnection. When designing the bump structure, the Sn bump size was designed to be smaller than the Au bump size to provide a solution for Sn overflow. Figure 11 showed the Au/Sn bump fracture morphology in the two bump structures after the shear experiment. The results showed that the design of Au bump size larger than Sn bump size could effectively reduce the Sn overflow during the bonding process, which will be benefit to the research of ultra-fine pitch micro bumps in the future. Figure 11 (e) also showed the daisy chain of the bonded structure.

(a)

(b)

(c)

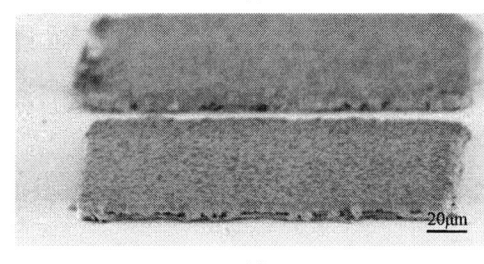

(d)

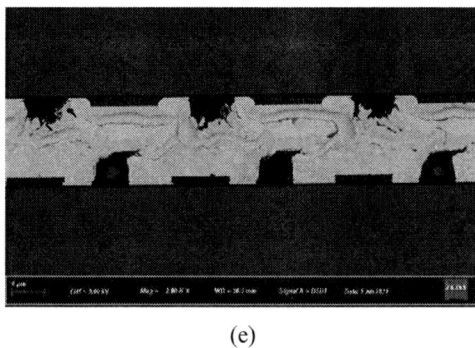

(e)

Fig. 11. SEM images of (a) different Au/Sn bump size in a daisy chain, (b) large square interconnects with different Au/Sn bump sizes; and (c) same Au/Sn bump size in the daisy chain, (d) large square interconnects with the same Au/Sn bump size, (e) cross section of the bonded Au/Sn bumps with different size.

IV. CONCLUSION

In this study, based on the traditional SLID process, a temperature gradient bonding process (TGB) was used to overcome the shortcomings of the Au/Sn SLID bonding. Au bump was deposited by electroplating and Sn bump was evaporated to improve the surface roughness. During the bonding process, the hot end temperature was set to 250℃, and the cold end temperature was set to 150℃. The wafer-level bonding was completed for 20 minutes under a pressure of 5000 N with the designed bonding area of 12.3 cm². Two the bump structure with different Au/Sn sizes were designed to barrier Sn overflow. The experimental results showed that the design structure with the same Au-Sn bump size has Sn overflow and the design structure with different Au/Sn bump sizes was effective to reduce Sn overflow.

Besides, the effect of electroplating current density on Au surface morphology was explored. Under the condition of electroplating at a temperature of 60 ℃ and a current density of 1mA/cm² for 75 minutes, Au bumps had lower surface roughness and more uniform height in the wafer level. Sn bumps by evaporation deposition had a flat surface and small roughness, which was important to achieve large shear force in the work.

X-ray images proved the bonding alignment was good. Large shearing force proved the bonding had strong shear strength. Fracture images of the bonded chips after shearing test revealed the fracture mode of the fracture surface. Cross section of the bonded chips directly showed the daisy chain was connected during the bonding. Thus, the fast low temperature Au-Sn wafer level bonding was achieved with the hot and cold end of 250 and 150 ℃ for 20 minutes.

ACKNOWLEDGMENT

This work was supported by National Natural Science Foundation of China (Grant No. 61804132, 62074132) and Shanghai Science and Technology Innovation Action Plan under Grant 19511131900.

REFERENCES

[1]   Xu H, Suni T, Vuorinen V, et al. Wafer-level SLID bonding for MEMS encapsulation[J]. Advances in Manufacturing, 2013, 03:32-41.

[2]   Cai J , Wang J , Wang Q . Experimental and computational investigation of low temperature Cu Sn solid-state-diffusion bonding for 3D integration[J]. Microelectronic Engineering, 2021, 236:111479.

[3]   T. L, Yang, T, et al. Full intermetallic joints for chip stacking by using thermal gradient bonding[J]. Acta Materialia, 2016, 113:90-97.

[4]   Sun L, Chen M H, Zhang L, et al. Recent progress in SLID bonding in novel 3D-IC technologies[J]. Journal of Alloys and Compounds, 2019, 818:152825.

[5]   Cyh A, Cwl B, Yyl B, et al. Interfacial evolution and mechanical properties of Au-Sn solder jointed Cu heat sink during high temperature storage test - ScienceDirect[J]. Materials Letters, 2020, 275:128103.

[6]   Xu H, Vuorinen V, Dong H, et al. Solid-state reaction of electroplated thin film Au/Sn couple at low temperatures[J]. Journal of Alloys & Compounds, 2015, 619:325-331.

[7]   Munding A , H Hübner, Kaiser A , et al. Cu/Sn Solid‑Liquid Interdiffusion Bonding. Springer US, 2008.

Copper adhesion promoters for polyimide: Heterocyclic compounds additives containing amino and hydroxyl groups

Yingying Li
Shenzhen Institute of Advanced Electrical Materials
Shenzhen Institute of Advanced Technology
Chinese Acedamy of Sciences
Shenzhen, China
yy.li2@siat.ac.cn

Guoping Zhang*
Shenzhen Institute of Advanced Electrical Materials
Shenzhen Institute of Advanced Technology
Chinese Acedamy of Sciences
Shenzhen, China
gp.zhang@siat.ac.cn

Jinhui Li*
Shenzhen Institute of Advanced Electrical Materials
Shenzhen Institute of Advanced Technology
Chinese Acedamy of Sciences
Shenzhen, China
jh.li@siat.ac.cn

.Changqing Li
Shenzhen Institute of Advanced Electrical Materials
Shenzhen Institute of Advanced Technology
Chinese Acedamy of Sciences
Shenzhen, China
cq.li@siat.ac.cn

Ao Zhong
Shenzhen Institute of Advanced Electrical Materials
Shenzhen Institute of Advanced Technology
Chinese Acedamy of Sciences
Shenzhen, China
ao.zhong@siat.ac.cn

Deliang Sun
Shenzhen Institute of Advanced Electrical Materials
Shenzhen Institute of Advanced Technology
Chinese Acedamy of Sciences
Shenzhen, China
dl.sun@siat.ac.cn

Abstract—Fan Out Wafer Level Packaging (FOWLP) has been applied for various types of semiconductor devices, such as power controller, RF device, and application processor of smart phone. However, issues of adhesion on the interface between polyimide (PI) and Cu in redistribution layers (RDL) of FOWLP remain unsolved. In this study, we conducted a comparative study to improve the adhesion strength of PI to Cu and realize higher reliability by utilizing two heterocyclic compounds additives containing amino and hydroxyl groups as adhesion promoters. First, the effect of adhesion promoters on interfacial adhesion strength between Cu and PI were quantitatively measured by peel test. Second, the mechanical, thermal and dielectric properties of the as-synthesized PI films were measured. The results proved that the heterocyclic compounds additives of DHP and DAHP could be excellent candidates for Cu adhesion promoters of PI, and the as-synthesized PI film also proved excellent mechanical, thermal and dielectric properties.

Keywords—polyimide, Cu adhesion strength, redistribution layers, reliability

I. INTRODUCTION

In 5G era, huge data are generated from high-performance computing (HPC) applications such as server, machine and learning data center, and transported to applications such as mobile devices and autonomous driving. HPC relies on high density package integration, and the electrical performance of the packages are determined by the interconnection density. The interconnection density is provided by the key components and structure of redistribution layers (RDLs). As shown in Fig. 1, it forms connections between logic and high-bandwidth memories or chiplets utilizing the polyimide (PI) dielectric material layers and fine pitch Cu vias and traces, thus completing the interconnections among electrical interfaces[1], increasing the I/O density, and enabling the existing FOWLP application in 5G era.

The RDL interconnections requires reduced insertion loss, low impedance, and high reliability in horizontal and vertical

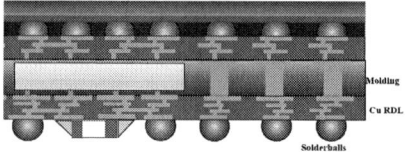

Fig. 1. Typical structure of Fan out wafer level packaging (FOWLP)

directions. However, because of the electrical skin effect, large surface roughness of RDL causes high transmission loss[2], and finer pitch RDL leads to worse transmission loss. The electrical skin effect can be eliminated when the surface roughness of Cu is less than 0.1μm[3], but following is the issue of poor adhesion between Cu and PI. The poor adhesion of PI on Cu arises from the coefficient of thermal expansion (CTE) mismatch between Cu and PI, which results in cracking and delamination when going through high temperature storage test, temperature & humidity test, high accelerated temperature & humidity stress test, high impact test, and thermal cycle test[4].

In this paper, we consider two heterocyclic compounds containing amino and hydroxyl groups, 2,4-diamino-6-hydroxypyrimidine (DAHP) and 4,6-dihydroxypyrimidine (DHP), as adhesion promoters for the comparative study. 1,2,4,5-benzenetetracarboxylic anhydride (PMDA) and 4,4'-diaminodiphenyl ether (ODA) are selected for the synthesis of PI precursors. We made a comprehensive study of the as-synthesized PI films, including interfacial adhesion strength, thermal, mechanical, dielectric properties.

II. EXPERIMENTAL

A. Polyamic acid (PAA) preparation

0.01 mol PMDA and 0.01 mol ODA were weighted and dissolved in N-methyl pyrrolidone (NMP) solvent. The polymerization was performed at 0 °C for 24 h. Then, 5 wt.%

DHP or DAHP were added thereupon respectively, and stirred at 0 °C for another 6 h to prepare comparative PAA samples.

B. PI Films preparation

The as-synthesized PAA solutions were spin-coated on Cu foil and pre-baked at 100 °C for 10 min to evaporate most of the solvent. The films were cured at 350 °C under N_2 atmosphere which were prepared for adhesion strength measurement.

The PAA solutions were also cured at 350 °C on glass under air atmosphere for the mechanical, dielectric, and thermal properties tests which were named as Blank PI, PI-DHP and PI-DAHP, respectively.

C. Measurement

The adhesion strength was measured by peel test with the IPC-TM-650 test method. The mechanical properties were measured by DMA. Values of the in-plane CTE were measured by TMA. The storage modulus curves were tested by DMA and glass transition temperatures (T_g)were obtained from the corresponding peaks of tan delta curves. The temperatures at 5% weight loss ($T_{d, 5\%}$) in N2 were tested by TGA. The dielectric properties were measured by e5071c keysight ENA vector network analyzer.

III. RESULTS AND DISCUSSION

A. Effect of adhesion promoters on interfacial adhesion strength

The interfacial adhesion strength between the PI and Cu was quantitatively measured by 90 ° peeling test as shown in Fig. 2. Box-plot analysis of displacement from 20 mm to 70 mm were shown in Fig. 3. The adhesion strength of PI-DHP could achieve around 0.50 N/mm, and PI-DAHP could achieve around 0.95 N/mm, whereas blank PI without any adhesion promoters could only achieve 0.2 N/mm. The improved adhesion strength came from the π electrons on heterocycle of pyrimidine, lone pair electrons on amino and hydroxyl groups, which could form conjugated bonds and coordination bonds with Cu surfaces. Besides, these groups could form hydrogen bonds with the imide or the residual carbonyl groups of PI to improve the interface strength. Moreover, the amino groups of DAHP could also form chemical bonds with PAA and thereby "joined in" the polymer chains. Thus, DHP and DAHP acted as junctions between PI and Cu.

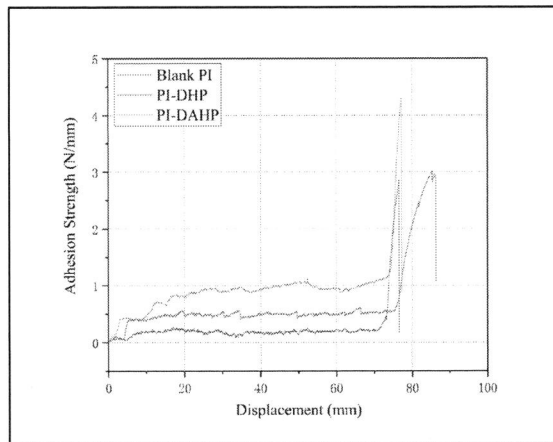

Fig. 2 90 ° peel test for PI-DHP, PI-DAHP and blank PI

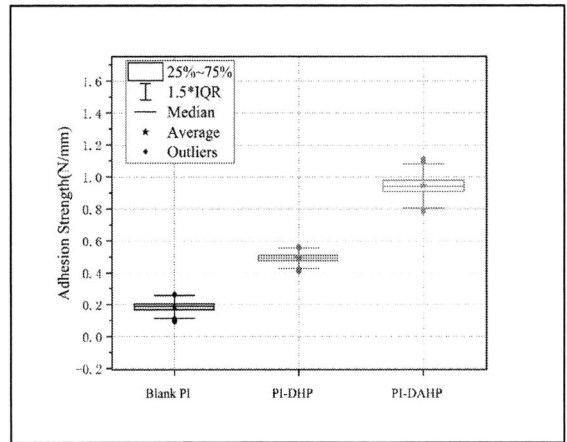

Fig. 3 Box-plot analysis of PI-DHP, PI-DAHP and blank PI

As a result, the interfacial reliability could be thus improved in the usage. The results proved that the heterocyclic compounds additives of DHP and DAHP containing amino and hydroxyl groups could be excellent candidates for Cu adhesion promoters of PI.

B. Effect of adhesion promoters on mechanical properties

Table 1 and Fig. 4 show the mechanical properties of PI-DHP, PI-DAHP and blank PI, including stress, elongation, and modulus. PI-DHP shows approximate tensile strength (104.80 MPa) and modulus (2.12 GPa) which is similar to those of blank PI (tensile strength of 102.80 MPa, modulus of 1.88 GPa). Surprisingly, PI-DAHP showed much higher tensile strength (144.72 MPa) and elongation (89.28%) than those of PI-DHP (14.12%) and blank PI (30.54%). This might because of the bonding of DAHP in the PAA molecules, and the following crosslinking of the polymer chains by the diamino groups on each DAHP molecule.

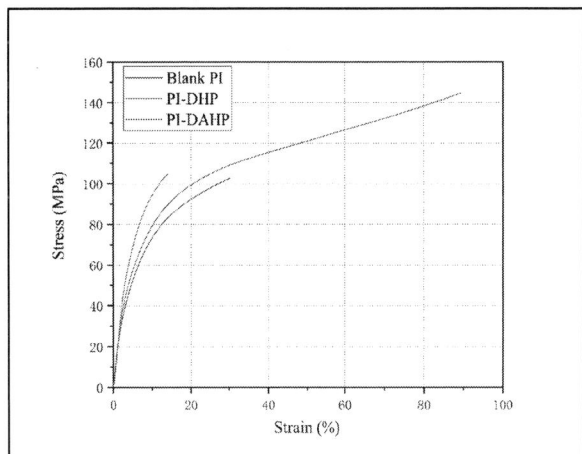

Fig. 4 Stress-strain analysis for PI-DHP, PI-DAHP and blank PI

TABLE I Mechanical properties of PI-DHP, PI-DAHP and blank PI

	Sample		
	PI-DHP	**PI-DAHP**	**Blank PI**
Tensile strength (MPa)	104.80	144.72	102.80
elongation(%)	14.12	89.28	30.54
Modulus （GPa）	2.12	1.97	1.88

C. Effect of adhesion promoters on thermal properties

The T_g, $T_{d, 5\%}$ and coefficient of thermal expansion (CTE) are shown in Fig. 5-7 and Table II. PI-DAHP exhibits highest T_g (421 °C), and lowest in-plane CTE (27.12 ppm/K), while blank PI has lowest T_g (370±4 °C[5]), and highest CTE (36.19 ppm/K). Generally, the thermal properties are attributed to the chemical structural rigidity of the molecular backbones[6]. These results indirectly proved that DAHP and DHP may crosslinked into the PI chains, disturbed the backbone structure and made the linkages less flexible and decreased the possibility of bending and rotation. Thus, T_g values were increased while CTE values were decreased of PI-DAHP and PI-DHP. Meanwhile, the thermal stabilities of blank PI, PI-DHP and PI-DAHP were similar. Adding more volatile DHP and DAHP would not cause the decrease of thermal stability of PI-DHP and PI-DAHP, which could also prove the crosslinking of polymer chains in them.

The adhesion strength will decrease because of the thermal stress due to the mismatch of CTEs of dielectric material and Cu, which would get worse with the increase of curing temperature. Thus, the lower CTE of PI-DAHP and PI-DHP also contributed to the higher adhesion strengths of Cu and PI interface.

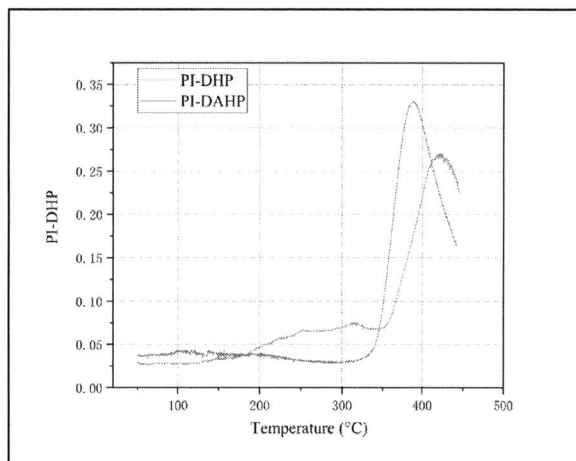

Fig. 5 T_g of PI-DAHP and PI-DHP

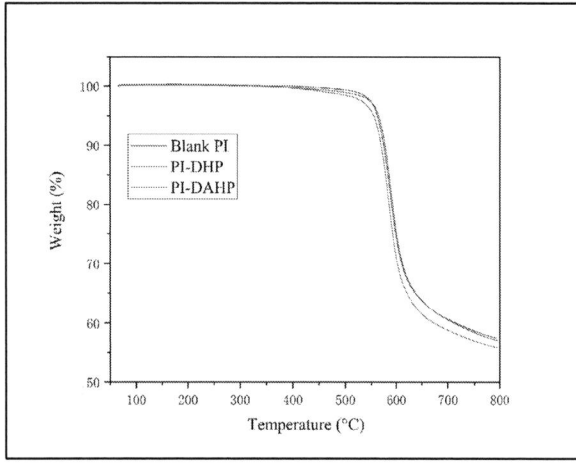

Fig. 6 T_d of PI-DAHP, PI-DHP and blank PI

TABLE II Thermal properties of PI-DHP, PI-DAHP and blank PI

	Sample		
	PI-DHP	**PI-DAHP**	**Blank PI**
T_g (°C)	390	421	370±4[5]
T_d,5% (°C)	554.3	563.8	561.2
In-plane CTE （ppm/K）	31.66	27.12	36.19

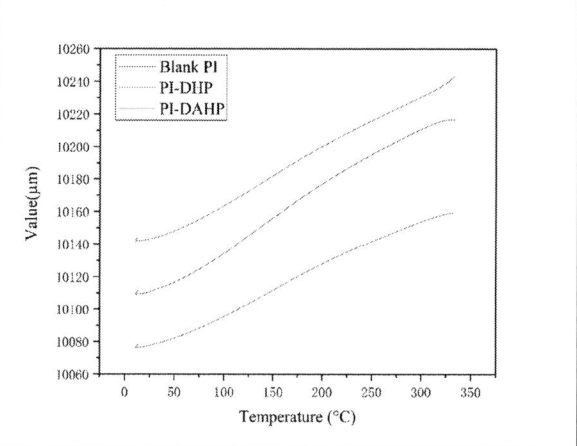

Fig. 7 CTE of PI-DAHP, PI-DHP and blank PI

D. Effect of adhesion promoters on dielectric properties

Materials with chemical bonds of higher polarizability will lead to higher dielectric properties, and the polar molecules in DHP and DAHP may increase the polarizability, and hence increase the permittivity and dielectric loss[7]. The dielectric properties of PI-DHP, PI-DAHP and blank PI at 5GHz are tested and shown in Table III.

PI-DHP and PI-DAHP contain polar molecules of DHP and DAHP, which increased the dipole strength and polarizability in the PI-DHP (D_k=3.51, D_f=0.0179) and PI-DAHP （D_k=3.59, D_f=0.0201), while blank PI shows lower D_k (3.31) and D_f (0.0165). Furthermore, the possible crosslinking of chains by DAHP and DHP made the chains more co-planar and strengthened the charge transfer (CT) interactions in the structure, thus, the dielectric constant and loss were relative higher.

TABLE III Dielectric properties of PI-DHP, PI-DAHP and blank PI at 5 GHz

	Sample		
	PI-DHP	**PI-DAHP**	**Blank PI**
Permittivity （D_k)	3.51	3.59	3.31
Dielectric loss (D_f)	0.0179	0.0201	0.0165

IV. CONCLUSIONS

Heterocyclic compounds additives of DHP and DAHP containing amino and hydroxyl groups have been proved to be excellent adhesion promoters. The interfacial reliability of Cu and PI has been improved significantly. The results show the adhesion strength of PI-DHP could achieve around 0.50 N/mm, and PI-DAHP could achieve around 0.95 N/mm, whereas blank PI without any adhesion promoters could only achieve 0.2 N/mm. Furthermore, PI films with the adhesion promoters also proved excellent comprehensive properties with superior mechanical, thermal and dielectric properties for the application in RDL of FOWLP.

ACKNOWLEDGMENT

This work was financially supported by National Natural Science Foundation of China (61904191), Youth Innovation Promotion Association of Chinese Academy of Sciences (2017410), Key R&D Project of Guangdong Province (2020B010180001) and National Key R&D Project from Minister of Science and Technology of China (2017ZX02519).

REFERENCES

[1] Y. Lin *et al.*, "Multilayer RDL Interposer for Heterogeneous Device and Module Integration," in *2019 IEEE 69th Electronic Components and Technology Conference (ECTC)*, 2019, pp. 931-936.

[2] F. Liu *et al.*, "Advances in High Performance RDL Technologies for Enabling IO Density of 500 IOs/mm/layer and 8-μm IO Pitch Using Low-k Dielectrics," in *2020 IEEE 70th Electronic Components and Technology Conference (ECTC)*, 2020, pp. 1132-1139.

[3] C. H. Yu *et al.*, "High Performance, High Density RDL for Advanced Packaging," in *2018 IEEE 68th Electronic Components and Technology Conference (ECTC)*, 2018, pp. 587-593.

[4] C. Nair, H. Lu, K. Panayappan, F. Liu, V. Sundaram, and R. Tummala, "Effect of Ultra-Fine Pitch RDL Process Variations on the Electrical Performance of 2.5D Glass Interposers up to 110 GHz," in *2016 IEEE 66th Electronic Components and Technology Conference (ECTC)*, 2016, pp. 2408-2413.

[5] Y. Jung, Y. Yang, S. Kim, H.-S. Kim, T.-g. Park, and B. W. Yoo, "Structural and compositional effects on thermal expansion behavior in polyimide substrates of varying thicknesses," *European Polymer Journal,* vol. 49, no. 11, pp. 3642-3650, 2013/11/01/ 2013.

[6] H. Inoue, Y. Sasaki, and T. Ogawa, "Properties of copolyimides prepared from different tetracarboxylic dianhydrides and diamines," vol. 62, no. 13, pp. 2303-2310, 1996.

[7] Y. Zhang and W. Huang, "Chapter 8 - Soluble and Low-κ Polyimide Materials," in *Advanced Polyimide Materials*, S.-Y. Yang, Ed.: Elsevier, 2018, pp. 385-463.

Underfill Filler Settling Effect on the Adhesive Force of Flip Chip Packages

Guolin Zhao
Shenzhen Institute of Advanced Electronic Materials
Shenzhen Institute of Advanced Technology, Chinese Academy of Sciences;
School of Mechanical and Electrical Engineering
Central South University
Shenzhen, China / Changsha, China
gl.zhao@siat.ac.cn

Houya Wu*
Shenzhen Institute of Advanced Electronic Materials
Shenzhen Institute of Advanced Technology, Chinese Academy of Sciences
Shenzhen, China
hy.wu1@siat.ac.cn

Yuanyuan Yang
Shenzhen Institute of Advanced Electronic Materials
Shenzhen Institute of Advanced Technology, Chinese Academy of Sciences
Shenzhen, China
yy.yang@siat.ac.cn

Gang Li
Shenzhen Institute of Advanced Electronic Materials
Shenzhen Institute of Advanced Technology, Chinese Academy of Sciences
Shenzhen, China
gang.li@siat.ac.cn

Pengli Zhu*
Shenzhen Institute of Advanced Electronic Materials
Shenzhen Institute of Advanced Technology, Chinese Academy of Sciences
Shenzhen, China
pl.zhu@siat.ac.cn

Rong Sun
Shenzhen Institute of Advanced Electronic Materials
Shenzhen Institute of Advanced Technology, Chinese Academy of Sciences
Shenzhen, China
rong.sun@siat.ac.cn

Wenhui Zhu*
School of Mechanical and Electrical Engineering
Central South University
Changsha, China
zhuwenhui@csu.edu.cn

Abstract—As microelectronic packaging technology is developed toward miniaturization and high density, flip chip is one of the most accepted packaging forms at present. In order to solve the reliability problem caused by the mismatch of thermal expansion coefficient between the silicon substrate and other structures in the flip chip, the underfill is used to solve the stress-strain problem caused by the thermal mismatch. However, due to the gravity, the filler of underfill is prone to settle down during the filling and the curing process. This study focuses on the effect of the filler settling on the adhesive strength of underfill. A simplified sample of sandwich structure was used to mimic the microstructure around the corner of the copper pad and the tin bump in the flip chip. The underfill was filled and cured in the gap (100 μm) between the copper and tin blocks. Three patterns of filler settlement were produced to fill the copper/tin gap by a special experimental method. The adhesive strength of the underfill with three filler settling patterns is measured by the shearing test (Dage 4000) at room temperature and 160 ℃, respectively. The results indicate that the filler settling has a significant effect on the adhesive strength of underfill. The adhesive strength decreases dramatically when segregation is formed. Therefore, it is important to ensure the uniform filler distribution of the underfill for the reliability improvement of the chip.

Keywords—underfill, filler settlement, adhesive force, advanced packaging materials

I. INTRODUCTION

With the development of microelectronic packaging technology to miniaturization, high density, high heat dissipation and high reliability, flip chip packaging technology is born to meet this change[1]. Flip chip has high packaging density, short signal path and good electrical and mechanical properties, which has a wide application prospect in microelectronic packaging. Due to its high packaging density, flip chip will produce high heat in the working process. With the increase of temperature, the mismatch of thermal expansion coefficient between chip and substrate material will cause cracks and even deformation of flip chip devices due to thermal stress, which becomes a hidden danger of product safety. During the thermal cycle, this kind of deformation will gradually accumulate, resulting in the relative displacement between the chip and the substrate, mechanical fatigue, poor welding connection, and eventually reliability problems. Underfill can effectively reduce the deformation of flip chip[2]. After the underfill cured, a protective layer is formed between the chip and the substrate, which can not only relieve the thermal stress between the chip and the substrate but also enhance the adhesion between the chip and the substrate. It would improve the stability of flip chip packaging and the service life of the product. Underfill is an epoxy resin material composed of thermosetting polymer and silica filler. It is a special colloidal dispersion system of silica particles in organic liquid with relatively low CTE. However, due to the gravity, the filler of underfill is prone to

978-1-6654-1392-3/21 $31.00 © 2021 IEEE

settle down during the filling and the curing process. This leads to a worse adhesive performance of the underfill, which may cause severe reliability problems. There are mainly three categories of filler distribution in the underfill: Pattern (I), uniform dispersion, where the filler particles are uniformly dispersed throughout the filling layer, and the filler particles dispersion density is consistent; Pattern (II), bilayered filler settling, where clear resin rich and filler rich zones can be found; Pattern (III), gradual filler settling, where the density of the fillers decreases gradually from top to bottom. The schematic diagram of filler settlement patterns is shown in Fig. 1[3]. Many characteristics of underfill are related to the content of filler. Filler settlement also affects the thermal expansion coefficient, thermal stability, toughness, and adhesion of underfill after curing. In the resin rich region, the thermal expansion coefficient is larger due to its less solid fillers; while in the silicon dioxide rich region at the bottom, the thermal expansion coefficient is smaller. The reduction of the matrix material will also reduce the bonding strength between the sealing layer and the substrate, and it is easy to cause the peeling phenomenon of the filling layer during the use of the chip.

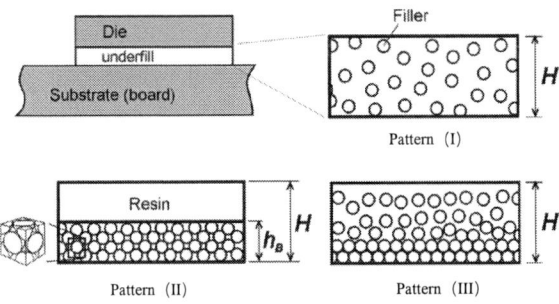

Fig. 1. Schematic diagram of filler settlement patterns

As reported recently, epoxy resin with silica filler and epoxy resin without silica filler were used as underfill materials, and thermal shock reliability tests were carried out. It is found that the overall failure rate of the test chip with SiO_2 filler settlement increases slightly. High viscosity underfill has higher reliability than low viscosity underfill[4]. In terms of filler particle size distribution, the reliability of using 1 um filler is improved compared with 5 um filler[5]. Serious filler settlement will lead to cracks in the thermal cycle, which will affect the reliability of flip chip packaging. At present, reliability researches of underfill mainly focus on the fluidity and thermal matching of underfill. It is rarely studied on the settlement of fillers. Therefore, this study focuses on the influence of filler settlement on underfill adhesive force. A simplified model is used to mimics the metal connection structure in flip chip. The underfill material is used to fill the gap between copper and tin blocks. Under the same experimental conditions, three kinds of filler settlement patterns (I), (II), (III) are produced by special experimental approach. In order to ensure the accuracy of the experiment, the samples were sliced and polished, and the settlement pattern was confirmed by a scanning electron microscope. The adhesion of the three filler settlement patterns at room temperature and high temperature was tested by welding strength tester.

II. EXPERIMENT AND METHOD

We conducted a series of experiments to explore the influence of settlement on the bonding strength between metal interconnection structures of flip chips.

A. Experimental Procedure

The experimental procedure is shown in the Fig. 2. The underfill are firstly prepared, and the properties of which are measured. Then, sandwich structures consisted of metal blocks and underfill are made. The cured sandwich samples are grinded out of the cross section on the grinder, the filler distribution pattern is observed using an electron scanning microscope, and then the samples with different settlement patterns are placed on the shear stress tester (Dage 4000).

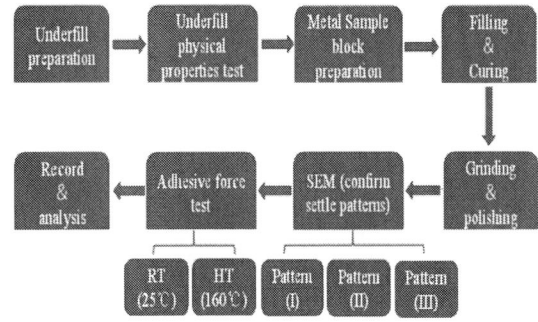

Fig. 2. The schematic flow of experimental procedure

B. Properties of Underfill Material

The underfill selected in this study is made of spherical silica filler particles and acid anhydride epoxy resin. The content of silica particles is 60 wt%, where the average particle size is 1 μm (cut at 5μm). The multiscale distributed fillers are beneficial for better fluidity of the underfill. The material properties of the underfill are shown in Table I. The CTE of the underfill is 29.5 ppm/K below the glass conversion temperature which is 122.5 °C. The viscosity at room temperature (25 °C) is 7.9 Pa.s, the viscosity decreases to 0.16 Pa.s during filling (90 °C), and the viscosity is lower than 0.06 Pa.s at 150 °C. Fig.3 shows the relationship between the modulus and temperature.

TABLE I. UNDERFILL MATERIAL PROPERTIES

Gel Time (150℃)	CTE(α1) < T_g	CTE(α2) > T_g	T_g	Flex Moldulus
s	ppm/K	ppm/K	°C	Gpa
620	29.5	90.9	122.5	8.1

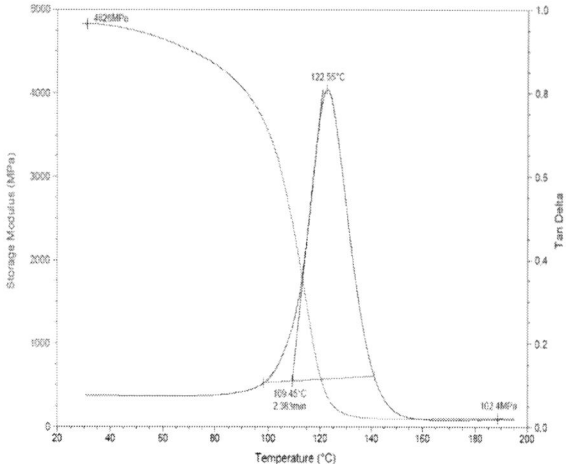

Fig. 3. The modulus versus temperature

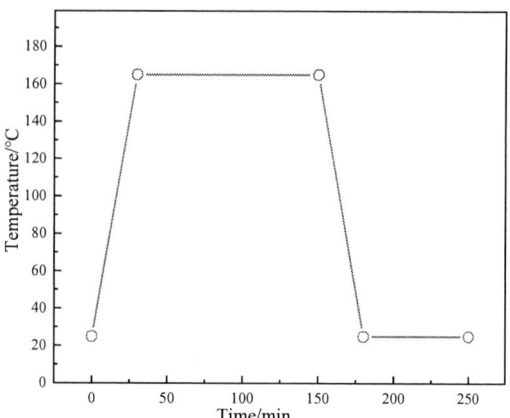

Fig. 5. Temperature change during curing

C. Sample preparation

The experimental sample consists of a copper block on top and a tin block underneath, with a gap of 100 μm between the two blocks, as shown in Fig. 4. Both of the blocks are with the dimension of 5 mm × 5 mm × 1.5 mm , The copper and tin blocks are both high purity metal blocks with a content of more than 99 %. Before making the sample, the metal blocks are polished and cleaned to improve the surface cleanliness, and hence reduce other factors possibly affecting the experiment results.

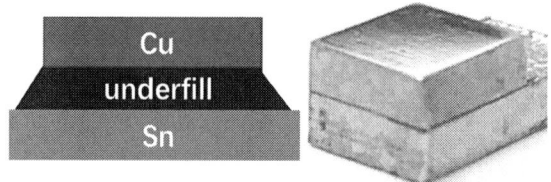

Fig. 4. Experimental sample and its schematic diagram

The prepared metal blocks are subsequently used to make sandwich structure containing underfill in the gap. The process is as follow:

1. The metal blocks are first baked for 30 minutes in a 65 °C thermostatic oven to remove residual moisture.

2. The underfill is taken out from a refrigerator and placed at room temperature for one hour to improve its fluidity before dispensing.

3. The metal blocks are preheated at 90 °C for 10 minutes before the underfilling process, aiming increase the fluidity of the underfill.

4. The underfill is dispensed to fill the gap between the metal blocks with a dispensing path of "I" Pattern. The filled sample is then post heated for 10 minutes to ensure the gap fully filled.

5. Putting the filled sample into the oven, the curing program of the oven as shown in Fig. 3, heating up 30 minutes to 165 °C, and at 165 °C for 2 hours, then cooling down to room temperature within 30 minutes.

D. Die shear test

Die shear test is a useful method to test the adhesive or bonding strength of sandwich structures. As shown in Fig. 6, the sample is placed on a heated clamping table at 25 °C and 160 °C to simulate room temperature and high temperature, respectively. During the test process, it should be noted that the sample should be placed horizontally, and the push knife should be parallel to the edge of the sandwich. It is important to ensure there is no obstacle on the path of the push knife. Or it will cause unreasonable errors if the push knife touch other parts outside the sample body.

Fig. 6. The test of adhesive force

III. RESULT AND DISCUSSION

A. Filler distribution patterns

After the curing process, the underfilled samples are polished. An SEM is used to observe the cross section of the sample to confirm settlement pattern. Three patterns of filler settlement have been found, as seen in Fig.7. which corresponds to Fig. 1. It can be seen that Pattern (I) has no settlement occurred, where the fillers are evenly distributed among the copper-tin blocks. From the local enlargement picture, it can be seen that the filler particles are also randomly and evenly distributed. Pattern (II) and Pattern (III) show different degrees of settlement, which present clear dividing lines between the epoxy and the fillers. The local magnification of Pattern (II) shows that there is no filler distributed in the upper region. Pattern (III) shows that the filler density gradually decrease from the top region, where only small filler particles are distributed. These filler

978-1-6654-1392-3/21 $31.00 © 2021 IEEE

distribution results meet the experimental requirements of this research. Next, die shear test will be performed to investigate the effect of filler settlement on the adhesive strength of the underfill.

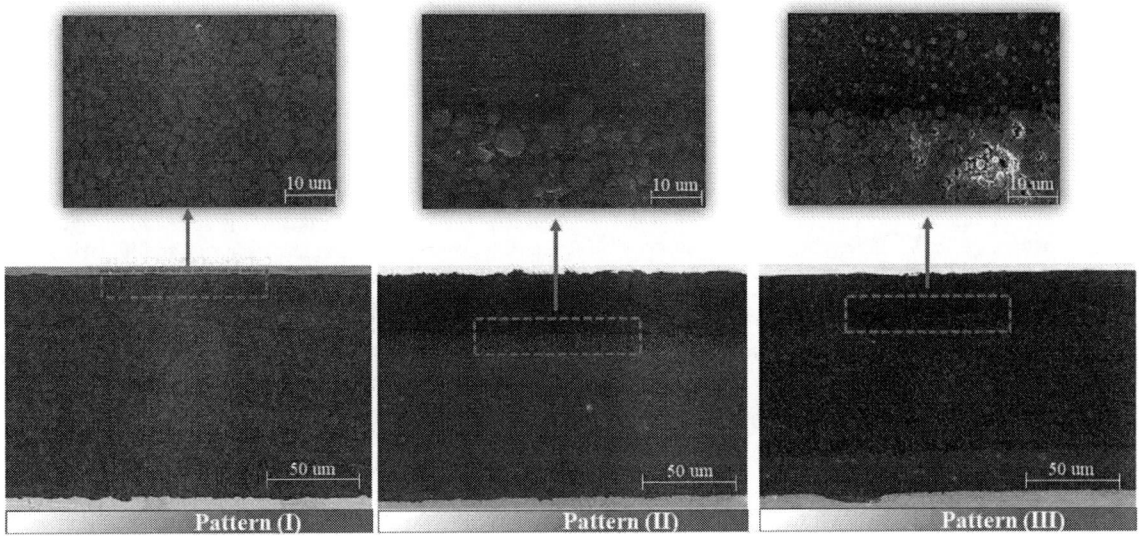

Fig. 7. Settlement results by SEM

B. Shear test at room temperature

The above samples are classified according to the settlement pattern, and the adhesive force is tested using the weld strength tester. The adhesive force results are shown in Fig. 8. At room temperature, the adhesive force of the sample without filler settlement is 375 N, Pattern (II) with bilayered filler settling is 247 N, and Pattern (III) with gradual filler settling is 318 N. Among them, the adhesive force of Pattern (I) is the highest, corresponding to it uniformly filler distribution; in contrast, the adhesive force of Pattern (II) is the lowest, corresponding to its complete settlement. Pattern (III) is a state in between. In the upper region of Pattern (III), there is a lower filler particle density, leading to a slight decrease in the adhesive force. Therefore, a possible conclucion is that uniform distribution of the fillers is beneficial to improve the adhesive strength of the underfill; otherwise, the adhesion will be weak when clear settlement are formed.

C. Shear test at high temperature

As shown in Fig. 9, at 160 °C, the adhesive force of Pattern (I) is 60 N, Pattern (II) is 37 N, and Pattern (III) is 59 N. The adhesive force at high temperature is consistent with that at room temperature, Pattern (I) and Pattern (II) are the highest and the lowest adhesive force, respectively. It is worth noting that the adhesive force of Patter (III) decreases slightly, which is basically the same as Pattern (I). As the temperature increases, the adhesive force at high temperature decreases significantly compared with that at room temperature. The adhesive force becomes less than 100 N at high temperatures. It is possible that the chemical bonds between the underfill and the metals are broken by the high thermal kinetic energy of the molecules at high temperatures. At the same time, copper, tin, and other element oxides can promote thermal and oxidative decomposition of underfill under high temperature aerobic conditions, which also reduces the mechanical properties of cured underfill.

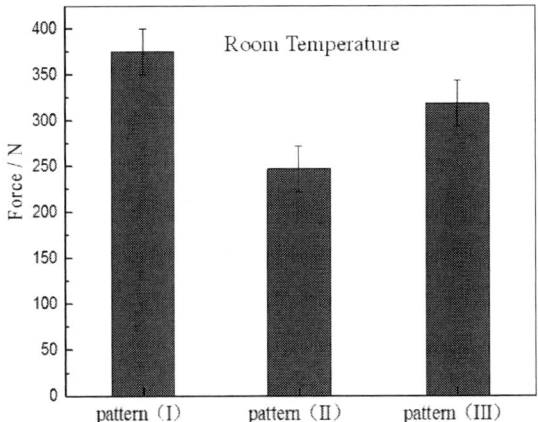

Fig. 8. The shearing force of three settlement patterns at room temperature

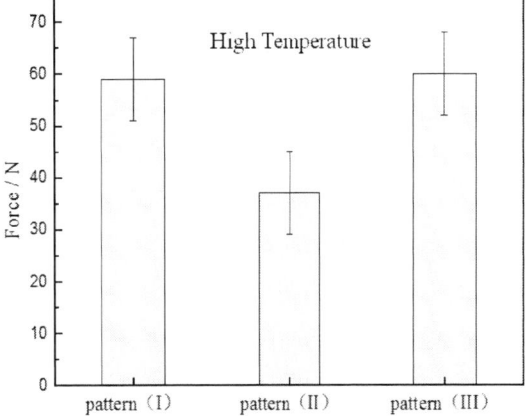

Fig. 9. The shearing force of three settlement patterns at high temperature

In summary, from the point of view by settlement pattern, the occurrence of settling will result in a decrease in adhesive force. Especially in the Pattern of completely settling Pattern (II), which decreases by more than 34% compared with Pattern (I) both at room and high temperature. Pattern (III) with a decrease in the density of the upper filler, which leads to a slight decrease in the adhesive force.

D. Discussion on elastic potential energy

During the adhesive force test, the deformation of underfill test sample always accompanied by energy input, transfer and response. For the same energy input, different filler distribution patterns show different energy transfer characteristics and response modes due to their differences in microstructures and fillers contact modes. Finally, it is reflected in the adhesive force of shear tests.

The total differential equation of thermodynamics for the system containing filler particles is[6]:

$$dw = \pi_{ij} d\varepsilon_{ij}^e + T_g dS_g + TdS + \sum_{a=S,L} (u_{ca} d\rho^a + v_i^a d\, m_i^a)$$

$$\pi_{ij} = \partial w / \partial \varepsilon_{ij}^e$$

Where w is the total energy density, m_i^a is momentum, ε_{ij}^e is the elastic strain, π_{ij} is the effective stress, u_{ca} is the chemical potential energy, S_g is thermodynamic entropy of filler, T is temperature, T_g is the dissipative force causing entropy increase, ρ^a is density, S is thermodynamic entropy.

As silica filler particles are solid, the elastic potential energy density function of granular materials is:

$$w_e = B\, (\varepsilon_v^e)^{0.5} \left[\frac{2}{5} (\varepsilon_v^e)^2 + \frac{1}{\zeta} (\varepsilon_s^e)^2 \right]$$

In the formula, ε_s^e is the deviatoric strain, B describes the degree of particle filling in the material, which is in the same dimension as the stress, and ζ is the material parameter, similar to Poisson's ratio.

In the adhensive force test, the filler of pattern (I) is evenly distributed in the whole space, which can be regarded as having the maximum value of B, the corresponding elastic potential energy is also larger; The filler of pattern (II) is concentrated in the bottom layer, which indicates that value of B is smaller and elastic potential energy is the smallest. The larger the elastic potential energy is, the more energy can be absorbed in the shear test, it also means that the underfill sample block has the larger adhesive force.

IV. CONCLUSION

This study investigates the influence of filler settlement on the adhesive force of underfill. A simplified sample of sandwich structure was used to mimic the microstructure around the corner of the copper pad and the tin bump in the flip chip. The adhesive strength of the underfill with three filler settling patterns is measured by shearing test. The most important finding of this study shows below:

1) Three patterns of the filler distribution were found.The filler distribution should to be three patterns: no filler settling, bilayered filler settling, and gradual filler settling.

2)The results were showed that high temperature can significantly reduce the adhesive force, the filler settlement pattern also has a significant effect on the adhesive force of underfill.

3)The filler settling weakened the adhesive strength of the underfill, especially when clear resin rich and filler rich zones are formed. The adhesive force of completely settling pattern decreases by more than 34% compared to the pattern with no filler settling.

4)The assumption of elastic potential energy is put forward that underfill with uniform distribution of filler particles has higher elastic potential energy, can absorb more energy when resisting external deformation, and shows greater adhesive force in shear stress test.

ACKNOWLEDGMENT

This work is supported by the National Key R & D Project from Minister of Science and Technology of China (2020YFB0311800), Shenzhen basic research plan (JCYJ20190807154409372), the National Natural Science Foundation of China (61704182), and the GuangDong Basic and Applied Basic Research Foundation (2020A1515111003).

REFERENCES

[1] Kim Gunrae, Ha Sangsu, Pae Sangwoo. "Reliability Impacts on Flip Chip Packages: Moisture Resistance, Mechanical Integrity and Photo-Sensitive Polyimide (PSPI) Passivation". Science of advance materials 2020; 12 (4): 577-582.

[2] Siyi Zhou, Ying Sun. "Multiscale, Multiphysics Model of Underfill Flow for Flip-Chip Packages". IEEE Transactions on components packaging and manufacturing technology 2012; 2(6): 893-902.

[3] Chengfu Chen, Karulkar Pramod C. "Underfill Filler Settling Effect on the Die Backside Interfacial Stresses of Flip Chip Packages". Journal of Electronic Packaging 2008; 130(3), 031005.

[4] Lin, Po Yao, Lee. "Modeling and Characterization of Cure-Dependent Viscoelasticity of Molded Underfill in Ultrathin Packages". IEEE Transactions on components packaging and manufacturing technology 2020; 10(9): 1491-1498.

[5] Srikanth, Narasimalu. "Warpage analysis of epoxy molded packages using viscoelastic based model". Journal of materials science 2006; 41(12): 3773-3780.

[6] Yang G C, Bai B. "A thermodynamic model to simulate the thermo-mechanical behavior of fine-grained gassy soil". Bulletin of Engineering Geology and the Environment 2020; 79(5):2325.

The Influence of Different Phosphor Coating Methods on the Temperature of LED

Kun Chen
School of Materials Science and Engineering
Xiamen University of Technology
Xiamen, China
chenkun19980917@163.com

Deming Hu
School of Materials Science and Engineering
Xiamen University of Technology
Xiamen, China
hdm9229@163.com

Liang Yang
School of Materials Science and Engineering
Xiamen University of Technology
Xiamen, China
yangliang86@xmut.edu.cn

Abstract—High-power light-emitting diodes (LEDs) with its energy saving, environmental protection, long life and other excellent performance gradually penetrated into modern lighting. In order to study the influence of different phosphor coating methods on the temperature of LED, high-power LED devices with different phosphor coating methods were designed, and the thermodynamic models are established by ANSYS software for simulation, which compares the simulation with the measured results. The main method to realize the white LED is to apply yellow phosphor to the surface of the blue-ray chip, and to get the blue-yellow mixed white light by exciting the yellow phosphor on the surface of the chip. However, in the process of blue light excitation of yellow phosphor, there will be a part of the light convert into thermal energy, due to the low thermal conductivity of the phosphor layer, phosphor is often in a high temperature state, resulting in high-power white LED light decay faster. The experimental prediction result is that the analog temperature is lower than the phosphor thermal load and the measurement temperature, and the amount of phosphors are applied affects the chip junction temperature, aluminum substrate temperature and lens top temperature. It also measures the temperature at the bottom of the aluminum substrate and the top of the lens, the calculated chip temperature is closer to the simulated phosphorus heat load temperature. In this paper, comparing the chip temperature measured by different phosphor coating methods with the simulation data, it is concluded that the temperature of the chip is the highest when the phosphor is directly in contact with the chip.

Keywords—light emitting diodes, packaging,phosphor

I. INTRODUCTION

High-power and high-brightness light-emitting diodes (LED) are the most promising new type of cold light source in the 21st century. It has attracted people's attention because it has significant advantages in the field of solid-state lighting and is recognized as the fourth generation of green environmental protection lighting products. Its luminescence mechanism is that the electrons in the PN junction transition between energy bands to generate light energy. When it is under the action of an external electric field, the radiation of electrons and holes recombine and produce electro-induced effects to convert part of the energy into light energy without radiation.The crystal lattice oscillation produced by the recombination converts the remaining energy into heat energy[1]. The existing high-power LED packaging method uses phosphor and silica gel to be mixed and uniformly coated on the LED chip. This packaging method is less efficient because the phosphor distributed in the silica gel is only about 10%, when excited by the chip just 60% conversion efficiency. The existing main method to realize white light LED is to coat yellow phosphor powder on the surface of blue light chip, to obtain blue-yellow mixed white light by exciting the yellow phosphor powder on the chip surface. The production process is simple and the cost is low. During the process, part of the light energy is converted into heat energy. Due to the low thermal conductivity of the phosphor layer, the phosphor is often in a high temperature state, resulting in a faster light decay of the high-power white LED. Moreover, in the existing packaging process, the phosphor is above the chip, and when excited by blue light, it generates heat by itself, which is superimposed with the heat generated by the chip, resulting in a more serious decline in the luminous performance of the phosphor. This paper mainly uses simulation and experiment to study the influence of phosphors on the thermal characteristics of high-power LED devices, using ANSYS software to establish a thermodynamic model to simulate the thermal load of phosphors and the thermal load of no phosphors. Compare the influence of different coating methods of phosphor heat generation on the temperature distribution of LED devices. Compare the actual result with the simulation result, and analyze the influence of the heat generated by the phosphor in the light energy conversion process on the temperature distribution of the LED device.

II. PRINCIPLE

Compared with thermal radiation, the luminescence principle of phosphors is a process of generating light with very little heat. Appropriate materials absorb high-energy radiation and then emit light. The energy of the emitted light is lower than the energy of the excitation radiation. When the luminescent material is solid, the material is usually called phosphor. The high-energy radiation that excites the phosphor can be electrons or high-speed ions, or photons ranging from gamma rays to visible light. At present, most of the phosphor actually used for lighting purposes are powder photoluminescent. It uses the electronic transitions of isolated ions in oxide crystals to emit light. When luminescent material is irradiated by LED blue light, part of blue light is reflected, scattered, and transmitted, the rest is absorbed. Among the absorbed light, some act as luminescence transitions, emitting photons, and some act as lattice vibrations, resulting in quenching.

978-1-6654-1392-3/21 $31.00 © 2021 IEEE

Luminescence and quenching are two independent and mutually competitive processes in luminescent materials. When quenching is dominant, the luminescence is weak and the efficiency is low. The fluorescence quantum efficiency is the ratio of the number of photons emitted from the secondary radiation to the number of photons from the primary radiation absorbing the excitation light per unit time. The reason why the phosphor generates heat is firstly due to the selected blue light chip. When the blue light is converted from the phosphor into yellow light, part of the energy is converted into heat, which is also called Stokes displacement; the second is the error of energy conversion loss[2].

III. SIMULATION

Assuming that the LED chip is a constant heat source, the heat generation rate is 80%, and the heat power is 0.8W. The external environment temperature is set to 10℃, and the LED device adopts natural air convection to exchange heat with the environment. This simulation experiment uses LED lights from National Star Optoelectronics[3].

TABLE Ⅰ. THERMAL CONDUCTIVITY OF LED PACKAGING MATERIALS

Material	Material						
	Ch ip	copp er	alumin um	Ther mal Greas e	le ns	Phosp hor	Sili ca gel
λ / [W· (m· K) $^{-1}$]	120	380	237	2	0.1 8	13	0.2

（a） Boundary conditions: environmental conditions 25℃. Set the direction of gravity. LED chip is 0.8W heat source

（b） On the basis of (a), remove the silicon filler.

The simplify 3D model created in SOLIDWORKS and import it into ANSYS. When establishing the finite element model, considering the meshing, the fillets, small holes and some small features that have little effect on the result were deleted. To mesh the model, the number of HD mesh grids is 454717, and the grid nodes are 521943[6].

Fig 1. Three-dimensional model of LED lights

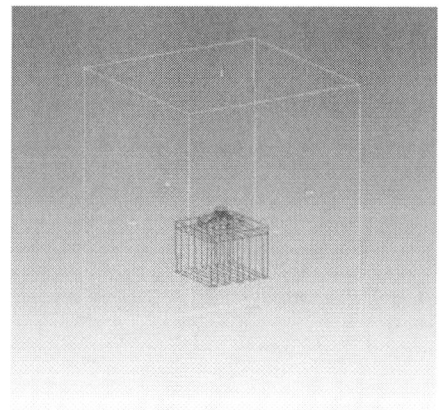

Fig2. The imported 3D model

After all initialization conditions and boundary conditions are set, iterative calculations can be performed. The residual curve diagram obtained after the calculation completed is shown in the figure. The residual is the sum of the fluxes of each surface unit. After convergence, theoretically speaking, at this time, there is no source in the unit and the flux inflow from each surface, that is, the sum of the transport of physical quantities should be zero. The RSM residual or maximum residual reflects the gap between the flow field and the flow field to be simulated. Theoretically, the smaller the residual, the better. However, due to accuracy problems, it is impossible to obtain a zero residual. For single precision, it is generally low. It is better to be less than $1e^{-2}$ of the initial residual. Among them, continuity represents the residual of the continuity equation (mass conservation equation), x-velocity, y-velocity, and z-velocity represent the velocity in the three directions of the Cartesian coordinate system, and energy represents the residual of the energy equation[4][5].

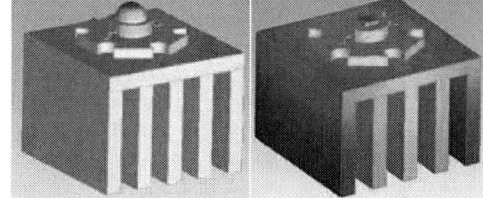

Fig 3. The temperature distribution cloud diagram of the model(remove the shell) （a）

Fig 4. The temperature distribution cloud diagram of the model(remove the shell) （b）

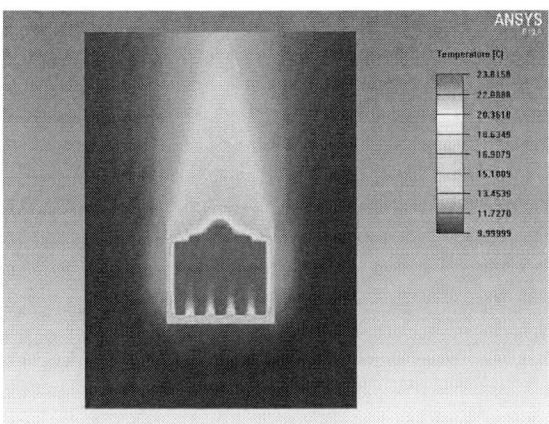

Fig 5.Cloud map of the air field around the lamp(a)

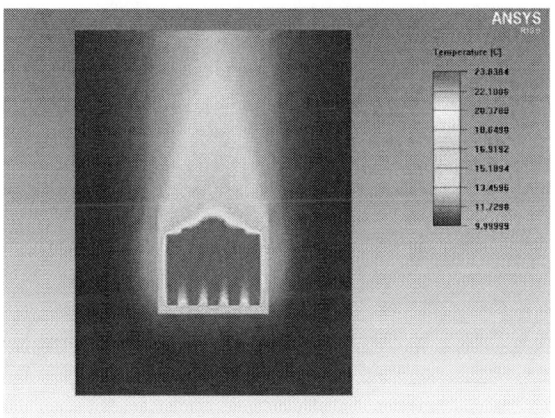

Fig 6.Cloud map of the air field around the lamp(b)

From the comparison of (a) and (b), it can be concluded that different phosphor coating methods have a certain impact on the thermal efficiency of the LED. It can be seen that the maximum junction temperature of a and b is 23.8350 and 23.8575, the specific performance is when there is filler inside, the heat energy of the phosphor is transferred from the filler to the chip. When there is no filler, the phosphor directly transfers heat energy to the chip, which makes the thermal efficiency of the lamp bead larger, but the temperature field distribution of the two in this simulation is not much different, and the analysis may be due to the lower power setting[8].

IV. EXPERIMENT

Choose the model of OE6550 AB glue, and mix A glue and B glue in a ratio of 1:1. Choose BM304D YAG fluorescent powder, mix the evenly mixed AB glue and fluorescent powder, stir evenly, and degas in vacuum.

1# No glue is filled in the lamp beads;

2#The filling glue is not in contact with the chip;

3# The filling glue is in contact with the chip.

According to the above three different coating methods, the concentration and coating amount of phosphors are also controlled in time to prepare luminous flux and color temperature. Table 2 shows the parameters of high-power

LEDs with three different phosphor coating methods when the input power is 0.8W.

TABLE II.PARAMETERS OF FIVE DIFFERENT PHOSPHOR COATING METHODS

LED	LED		
	NO.1	*NO.2*	*NO.3*
Phosphor concentration /wt%	10.4	10.5	12.2
Coating amount/g	0.0341	0.0312	0.0323
Chip temperature/℃	55	53	58
Lens temperature/℃	68	66	57

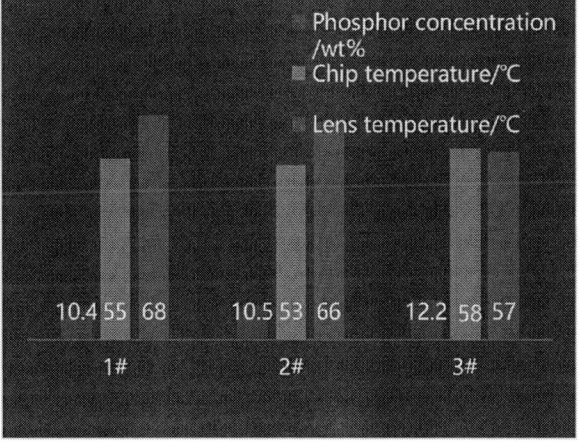

Fig 7.Temperature histogram

It can be seen from Figure 7 that the coating method where the filler does not contact the chip has a lower temperature than the chip where the filler contacts the chip; in addition, the temperature of the top of the lens 2# is greater than 3#. It can be concluded that the influence of the heat generated by the phosphor on the heat distribution of the aluminum substrate cannot be ignored[9][10].

V. CONCLUSION

The experimental results show that different phosphor coating methods have a certain impact on the thermal efficiency of the LED. When there is a filler between the phosphor and the chip, the temperature of the chip is lower than the temperature of the phosphor and the chip without the filler. On the contrary, the temperature of the top of the lens will be higher than the temperature of the filler between the phosphor and the lens. The simulation data also confirmed this conclusion.

ACKNOWLEDGEMENT

Nature Science Foundation Project of China (NSFC619041 56); Natural Science Foundation of Fujian Key Project (202 0J02049); Xiamen Major Science and Technology Projet (3 502Z20201003, 3502ZCQ20201001).

REFERENCES

[1] Tian Xi, Zhang Chi, Li Jiangbo, Fan Shuxiang,Yang Yi,Huang Wenqian, "Detection of early decay on citrus using LW-NIR hyperspectral reflectance imaging coupled with two-band ratio and improved watershed segmentation algorithm,"[J] Food Chemistry, 2021.

[2] Li Wen Yu, Zhu Qiong Bin, Jin Lu Ya,Yang Yi, Xu Xiao Yan,Hu Xing Yue, "Exosomes derived from human induced pluripotent stem cell-derived neural progenitor cells protect neuronal function under ischemic conditions,"[J]Neural Regeneration Research, 2021.

[3] Fu Mengjing, Liang Yijing, Lv Xue, Li Chengnan,Yang Yi Yan,Yuan Peiyan, Ding Xin, "Recent advances in hydrogel-based anti-infective coatings,"[J]Journal of Materials Science&Technology, 2021.

[4] Zhang Jing,Yang Yi, Huang Xiaohui, Shan Qian, Wu Wei, "Novel preparation of high-yield graphene and graphene/ZnO composite,"[J]Journal of Alloys and Compounds,2021.

[5] TodorovT K, Reuter K B, Mitzi D B,"High-efficiency solar cell with earth-abundant liquid-processed absorber,"[J]Adv.Mater, 2010, 22(220):156-159.

[6] WooK,KimY,MoonJ,"A non toxic,solution- processed, earth abundant absorbing layer for thin- film solar cells,"[J]Energy & Environmental Science , 2012.5(1): 5340-5345.

[7] Ki W, HillhouseH W,"Earth-abundant elementphotovoltaics directly from soluble precursors with high yield using a non-toxic solvent,"[J].Adv. Energy Mater, 2011, 1(5): 732-735.

[8] Steele R V,"LED automotive headlamps move closer to market L,"[J] Laser Focus World, 2005, 41(11):91-95.

[9] Xiaobing Luo, Bulong Wu, Sheng Liu, "Effects of Moist Environments on LED Module Reliability,"[J]IEEE Transactions on Device and Materials Reliability (S1530-4388), 2010,10(2): 182-186.

[10] Richard K. Ulrich, William D. Brown,"Advanced electronic packaging,"[M]2nd Edition. New York: John Wiley & Sons, 2010.

An Infrared Laser Temporary Bonding Material Used for Device Wafer Thinning and Completion of Backside Processing Technology

Zhenwen.Ye
R&D Department
Shenzhen Samcien Semiconductor
Materials Co., Ltd
ShenZhen,China
zw.ye@samcien.com

Deliang,Sun
Wafer Level Packaging Materials
Center
Shenzhen Institute of Advanced
Electronic Materials, Shenzhen
Institute of Advanced Technology,
Chinese Academy of Sciences,
Shenzhen 518055, China
ShenZhen,China
dl.sun@siat.ac.cn

Mingqi.Huang
R&D Department
Shenzhen Samcien Semicoductor
Materials Co., Ltd
ShenZhen,China
mq.huang@samcien.com

Guoping.Zhang
Wafer Level Packaging Materials
Center
Shenzhen Institute of Advanced
Electronic Materials, Shenzhen
Institute of Advanced Technology,
Chinese Academy of Sciences,
Shenzhen 518055, China
ShenZhen,China
gp.zhang@siat.ac.cn

Jianwen.Xia*
R&D Department
Shenzhen Samcien Semicoductor
Materials Co., Ltd
ShenZhen,China
xjw@samcien.com

Abstract—Temporary bonding/de-bonding cooperation is the core technology of thin wafer holding. Through the innovative application of temporary bonding materials, the device wafer is fixed on the carrier wafer to provide sufficient mechanical support for the ultra-thin device wafer to ensure The device wafer can smoothly and safely complete subsequent processes, such as photolithography, etching, passivation, sputtering, electroplating, and reflow soldering. With the rapid development of advanced packaging processes, temporary bonding/de-bonding technology has been vigorously developed and widely used in the field of wafer-level packaging (WLP), such as PoP stack packaging, fan-out packaging, eWLB, silicon Through hole (TSV), 2.5D/3D package, etc. Therefore, temporary bonding materials will play a vital role in the development of advanced packaging. In this paper, WLP LB310 is used as an infrared laser thermal decomposition material and temporary bonding glue for device wafer thinning and completing the backside processing technology (lithography, etching, passivation, electroplating, reflow soldering, etc.). After all the manufacturing processes are completed, the infrared laser irradiates the laser thermal decomposition material to undergo photothermal conversion, which causes the WLP LB310 to thermally decompose, and finally realizes the efficient and low-stress separation of thin wafers, and uses supporting cleaning agents to remove residual organic matter on the wafers. The results show that the temperature of the infrared laser pyrolysis material when the thermal mass loss of 1% in nitrogen is greater than 406 ℃; the bonding pair formed with the temporary bonding material has good chemical resistance; the film thickness is 1.4 nm infrared The laser pyrolysis material has no residual peeling after de-bonding under 1064nm Gaussian light, power 10w, and frequency 5000Hz.

Keywords: temporary bonding, pyrolysis ,infrared laser de-bonding, no residual peeling

I. INTRODUCTION

With the continuous emergence of new technologies such as 2.5D/3D, SIP, and heterogeneous integration, the thinning technology of chips or packages is becoming more and more important[1,2], but these new technologies are facing the problem of holding ultra-thin wafers, and temporary bonding technology as a key process of advanced manufacturing and packaging has received more and more attention, mainly because of the following aspects: 1. the processing accuracy of flexible ultra-thin devices or chips, in order to hold the flexible ultra-thin device manufacturing process temporary bonding technology must be used for thin wafers; 2. in Fan-out wafer-level packaging, the wafer warps after molding, which requires temporary bonding technology to improve packaging accuracy and flatness[3,4]; 3. the "additive manufacturing" of ultra-thin devices also requires the support of temporary bonding technology; 4. the current compound semiconductor material itself is relatively brittle. Ultra-thin wafer holding technology attracts more and more attention of researchers. And the temporary bonding and de-bonding technology is seen as a excellent solution with good operability. According to the different bonding methods, the temporary bonding process can be divided into thermal slip de-bonding method, chemical de-bonding method, mechanical de-bonding method and laser de-bonding method. These methods have their own advantages[5]. For example, the cost of the thermal slip de-bonding method is relatively low, but its limitation is that the temperature of the device needs to be above 200 ℃ during separation, such as image sensors, MEMS sensors, etc,living in such a high temperature, can't pass the manufacturing process. In contrast, the material used in laser release method has a

978-1-6654-1392-3/21 $31.00 © 2021 IEEE

expert thermal stability even above 400℃ and can be de-bonded at low temperature by UV laser[6]. Therefore, laser releasable temporary bonding system has draw more and more attention, because it's great advantage to treat large area and ultra-thin wafer by a non-contact, stress-less, and high efficiency way.

The development of photo-polymer is a very important part of laser de-bonding technology. The in-depth study of material properties is the basis to achieve the reliability of multiple processes. For example, the curing conditions and curing conversion rate of the polymer have a significant impact on the reliability of the material in the manufacturing process.

In this paper, material and temporary bonding technology for 1064nm wavelength were shown. A novel release material with intense absorption at 1064nm wavelength was prepared. The curing process and thermal stability were investigated. And then, the reliability, process capability, laser debonding, and cleanability of the material were all evaluated. The results showed that this adhesive has excellent performance and widely application prospects in laser releasable temporary bonding .

II. EXPERIMENTAL

A. Temporary bonding adhesive system

The temporary bonding system contains two adhesives. One is a novel photo-material (WLP LB310 provided by Samcien Semiconductor Materials Co., Ltd) which coated on the glass wafer for laser de-bonding. And another is a adhesive (WLP TB5130 provided by Shenzhen Samcien Semiconductor Materials Co., Ltd) on the device wafer to boding wafer stack.

B. Temporary bonding and laser de-bonding processes

The flow chart of temporary bonding is shown in Figure 1. Adhesive named TB5130 and laser release polymer named LB310 were spined on the device wafer and glass wafer, respectively. And then both wafers were bonded together formed a wafer stack. After backside processing, the bonded wafer stack was irradiated using 1064nm wavelength laser. Glass and device wafer were separated successfully. The glass wafer could be recycle after cleaning.

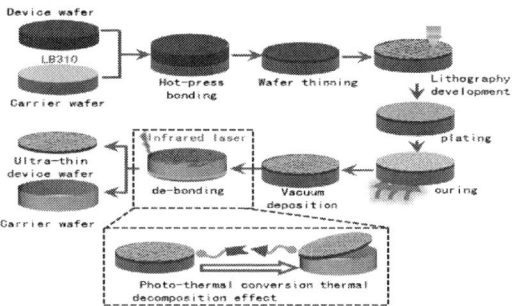

Fig. 1. Temporary bonding and laser de-bonding processes

C. Methods and Characterization

The coating speed of adhesives were set at 1500 rpm and 20 second by spin coater (WS-650-8B, Laurell). Then the device wafer with TB5130 layer was baked at 115°C and 150°C for 5 minutes respectively, to volatilize solvent.

While the glass wafer with LB310 layer was baked at 180°C for 5min. All of baking processes were heated on blast oven. After wafers bonded together (EVG510,EVGroups, 185°C, 5min), some backside process for device wafer was carried out. Then, the wafer stack was de-bonded by laser with wavelength of 1064 nm. Thermal gravity analysis (TGA, TA, SDT Q600) was used to evaluate the thermal stability of LB310. And UV-visible transmission spectra (Shimadzu, UV-3600) was used to recorded absorption value from 200nm to 1200nm.

III. RESULTS AND DISCUSSION

D. Thermal stability

In order to improve the ability of the temporary bonding material to be stable without decomposition or denaturation at high temperatures for a long time, we need to pay attention to its thermal stability, and the result of TGA curve is shown in Fig.2. In order to show the superiority of LB310 in thermal stability, we have compared with similar temporary bonding adhesives on the market. The specific performance is listed in Table1. The 1% heat weight loss temperature ($T_{1\%}$) of LB310 = 406°C, 5% heat weight loss temperature ($T_{5\%}$) = 523°C. In contrast, 1% heat weight loss temperature ($T_{1\%}$) = 244°C for similar products, 5% heat weight loss temperature ($T_{5\%}$) = 336°C. It is found that LB310 is in thermal stability significant improvement. TGA equipment allows the byproducts of decomposition escape and reduce the sample mass[7]. But the laser release polymer is in the middle of wafer stack and not exposed in the atmosphere. The data of TGA can not directly represent the actual use of the material. The actual usable temperature of LB310 would be higher than the test value.

TABLE.1 Thermal weight loss of LB310 and the similar

Item	weigh loss temperature of release materials	
	LB310	Similar temporary bonding adhesive
Thermal weight loss under nitrogen	$T_{1\%}$, $T_{5\%}$	$T_{1\%}$, $T_{5\%}$
Temperature (℃)	406，523	244，336
Curing condition	180℃，5-10min	180℃，30min

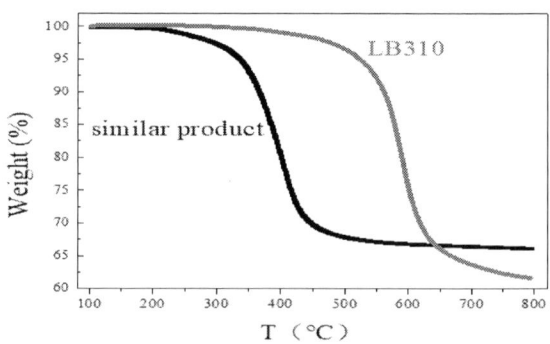

Fig. 2. TGA curve of release material with LB310 and the similar

E. Infrared transmission

The principle of infrared release material de-bonding is that the material absorbs infrared laser light, absorbs light and converts it into heat, thereby generating high temperature in the bonding interface, and then decomposing to produce gas, increasing the pressure of the bonding interface to help device wafers and glass carrier wafers peel off, the infrared transmittance of the infrared emitting material is as shown in the figure3.

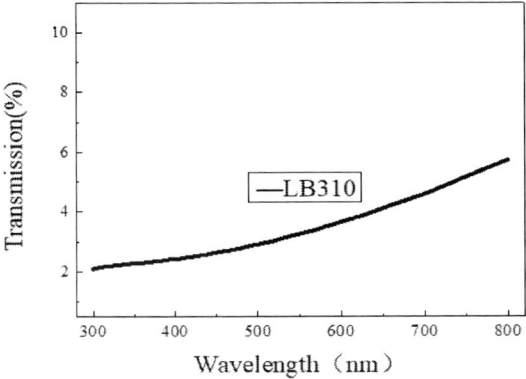

Fig. 3. Infrared transmission of LB310

F. Cohesive force

Since the carrier glass and the device wafer need to go through thinning, photolithography, etching, passivation, electroplating, ball planting and other processes after bonding, it is necessary that the bonding force between the two is sufficiently high, and the maximum bonding force is measured. The common method is the Hundred Grid method[8]. The test standard is "GBT9286-1998 Paint and Varnish Film Scratch Test". The test results are shown in the figure4. After the initial Hundred Grid method, the edges of the film are smooth and there is no peeling off. . After three vertical pulling tests of the 3M tape, there was a slight burr on the edge of the film, and there was no obvious shedding.

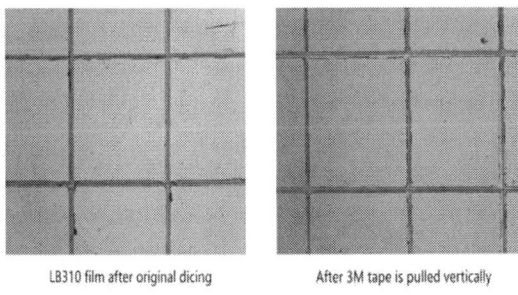

Fig. 4. LB310 uses 3M tape to tear vertically before and after

G. Application

In order to verify the feasibility of the infrared laser de-bonding material LB310 in bonding and de-bonding, our temporary bonding material baking temperature is selected at 180℃. Take 8-inch glass as the carrier(as shown in Figure 5) after baking, and thinning after bonding with the wafer carrier.

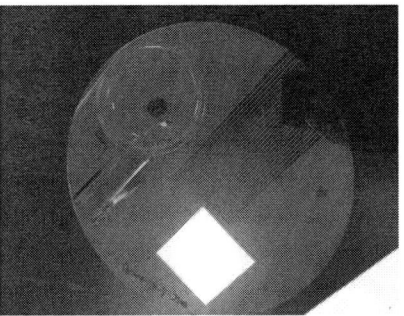

Fig. 5. LB310 coated and baked on glass

Finally, through the temporary bonding and laser bonding wafer stripping process, it can be easily separated by 1064nm laser irradiation, and there is no glue and carbon dust on the surface of the glass carrier and the wafer, as shown in Figure 6.

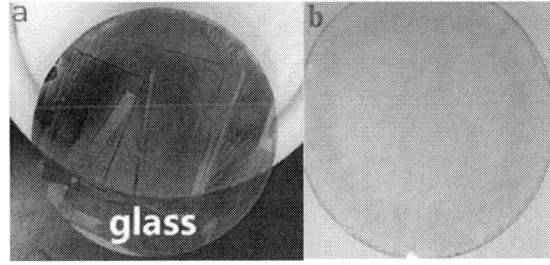

Fig. 6. Temporary bonding wafers after thinning process and a) de-bonding by laser could separate easily; b) glass wafer could be recycled after wiping with a lint-free cloth.

IV. CONCLUSION

As a full-spectrum absorbing material, LB310 solves the problem of weak absorption of traditional ultraviolet laser debonding materials above 400nm wavelength by adding nano-materials, and also solves the problem of poor film formation and low adhesion of organic thin films after adding nano-materials.

ACKNOWLEDGMENT

This product was supported by R&D Department of Shenzhen Samcien Semicoductor Materials Co., Ltd and Shenzhen Institute of Advanced Electronic Materials, Shenzhen Institute of Advanced Technology, Chinese Academy of Sciences.

REFERENCES

[1] Tsai, W.-L., et al. How to select adhesive materials for temporary bonding and de-bonding of 200mm and 300mm thin-wafer handling for 3D IC integration .IEEE..

[2] Xu, D., et al. A novel design of temporary bond debond adhesive technology for wafer-level assembly. IEEE.

[3] Stefan Pargfrieder; Paul Kettner; Mark Privett; Jack Ting. Temporary Bonding and DeBonding Enabling TSV Formation and 3D Integration for Ultra-thin Wafers. IEEE.

[4] Thomas Uhrmann; Matthias Pichler; Julian Bravin; Daniel Burgstaller; Boris Povazay. Laser Debonding Enabling Ultra-Thin Fan-Out WLP Devices. IEEE.

[5] Hsiao Hsiang-Yao; David Ho Soon Wee. Effect of laser on passivation photo-dielectric during laser de-bonding process for Fan-Out Wafer Level Package. IEEE.

[6] Jianwen Xia; Guoping Zhang. The effect of curing process on laser releasable debonding temporary material for 3D packages.IEEE.

[7] Yang, Y.-s., K.-s. Hwang, and R. Gorrell. Laser Releasable Temporary Bonding Film with High Thermal Stability. IEEE.

[8] Julian Schirmer;Jewgeni Roudenko; et al. Adhesion Measurements for Printed Electronics: A Novel Approach to Cross Cut Testing. IEEE.

The Studys of Adhesion and Contact Thermal Resistance of TIM1

Yunsong Pang*
Shenzhen Institute of Advance Electronic Materials
Chinese Academy of Sciences
Shenzhen, China
ys.pang@siat.ac.cn

Meng Han
Shenzhen Institute of Advanced Technology
Chinese Academy of Sciences
Shenzhen, China
meng.han@siat.ac.cn

Ting Liang
Shenzhen Institute of Advanced Electronic Materials
Chinese Academy of Sciences
Shenzhen, China
ting.liang@siat.ac.cn

Xue Bai
Shenzhen Institute of Advance Electronic Materials
Chinese Academy of Sciences
Shenzhen, China
xue.bai@siat.ac.cn

Liang Li
Key Laboratory of Materials Processing and Mold Ministry of Education
Zhengzhou University
Zhengzhou, China
liliangabc1234@163.com

Yonglun Xu
Shenzhen Institute of Advanced Electronic Materials
Chinese Academy of Sciences
Shenzhen, China
yl.xu@siat.ac.cn

Bin He
Shenzhen Institute of Advance Electronic Materials
Chinese Academy of Sciences
Shenzhen, China
bin.he@siat.ac.cn

Daifeng Ai
Shenzhen Institute of Advanced Electronic Materials
Chinese Academy of Sciences
Shenzhen, China
df.ai@siat.ac.cn

Liuxin Wang
Shenzhen Institute of Advanced Technology
Chinese Academy of Sciences
Shenzhen, China
lx.wang@siat.ac.cn

Linlin Ren*
Shenzhen Institute of Advance Electronic Materials
Chinese Academy of Sciences
Shenzhen, China
ll.ren@siat.ac.cn

Xiaoliang Zeng
Shenzhen Institute of Advanced Technology
Chinese Academy of Sciences
Shenzhen, China
xl.zeng@siat.ac.cn

Rong Sun
Shenzhen Institute of Advanced Technology
Chinese Academy of Sciences
Shenzhen, China
rong.sun@siat.ac.cn

Abstract—As the filler between the flip chip and lid, thermal interface material (TIM1) plays a role in heat dissipation to maintain the good performance of chips, lower power during operation. However, as the performance of the device itself continues to improve, in order to maintain reliability, the design demands for packaging materials such as TIM1 are increasing accordingly. Through the observation and analysis of TIM1 after working for a long time, the failure of the material is often due to the separation between the TIM1 and the chip or lid, which indicates that the interface interaction of the designed TIM1, that is, the adhesion is weak. Such problems can lead to a huge thermal resistance so that the thermal performance of TIM1 will be affected. In addition to the thermal resistance resulted from material failure, the contact thermal resistance of TIM1 also can defeat its ability of thermal transport. Herein, the adhesion and contact thermal resistance of TIM1 are studied. In order to learn the term of contact thermal resistance, two types thermal testing method, laser flash and photo-thermal radiation are proposed. The research result shows that the adhesion has negative correlation with contact thermal resistance, two proposed method can work for the contact thermal resistance measurement. A method is probably to be used to improve the quality of TIM1 that is introducing adhesion promoter into it to enhance the adhesion at interface and defeat contact thermal resistance simultaneously.

Keywords—thermal interface material, adhesion, contact thermal resistance, laser flash, photo-thermal radiation

I. INTRODUCTION

Thermal interface material especially the one of first level (TIM1) acts as a critical role in the heat dissipation of semiconductor packages that guarantee the chip can work under an acceptable temperature environment to assure its superior calculation and reliable performance over the long term [1]. Because of the relatively high intrinsic thermal conductivity, TIM1 can help transferring the heat generated by the chip to the heat sink, lid in real time so that achieving the purpose of cooling down.

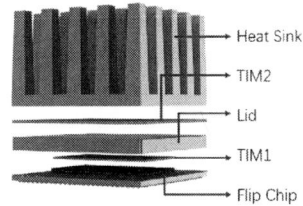

Fig. 1. Schematic illustration for package construction

However, TIM1 cannot work in an ideal environment and any uncertain condition may defeat its performance. For instance, since the coefficient of thermal expansion of TIM1 and die are different. The warpage always can occur at the interface [2], even leads to the failure. Such problem can

978-1-6654-1392-3/21 $31.00 © 2021 IEEE

bring a huge thermal resistance effect the TIM1's thermal performance. In order to avoid such incident, the adhesion of TIM1 should be improved.

Beside such incident can affect the TIM1's thermal performance occasionally, the thermal performance is mainly dominated by thermal resistance that comes from TIM1. Such thermal resistance concludes the material inherent thermal resistance and the contact thermal resistance. Herein, contact thermal resistance need to be learnt since it is dominant.

In this research, the adhesion and contact thermal resistance of TIM1 that at interface is studied. According to the mechanism of laser flash and photo-thermal radiation techniques, two specific testing method are developed to be applied to detect the contact thermal resistance. The experimental result shows that the adhesion has negative correlation with contact thermal resistance, and a method used to improve the overall ability of heat transfer of TIM1 is proposed, that is defeat the contact thermal resistance by enhancing the adhesion at interface.

It should be pointed out that our work exhibits the pioneering effort to measure the contact thermal resistance of TIM1. The methods that applying photo-thermal to measure contact thermal resistance and processing the data tested by laser flash are both proposed for the first time in this field.

II. EXPERIMENTAL

Materials

Our TIM1 is prepared by mixing metallic particle fillers and silicones materials. The primary physical properties of our TIM1 are summarized in Table 1. In order to obtain the TIM1 samples with various adhesion capacity, different contents of adhesion promoter are introduced into the precursor mixture of TIM1 during the step of synthesis. Once TIM1 is obtained, it will be prepared into suitable form for the mechanical and thermal tests that will be discussed below.

TABLE I.   PHSICAL PROPERTIES OF TIM1

Viscosity，μ	$405\ Pa \cdot s\ @\ 5/s$
Coefficient of Thixotropy	6.31
Density, ρ	$2.5\ g/cm3$
Thermal Conductivity, k	$3.8\ W/mK$
Storage Modulus, G'	$60\ KPa$
Loss Modulus, G''	$15\ KPa$
Tensile Strength, σ	$0.2\ MPa$
Elongation, ε	$\geq 150\%$

Adhesion Performance Test

A method named lid pull adhesion test is employed for evaluating the adhesive performance of TIM1. For this testing method, the TIM1 is firstly mounted into a sandwich that contacting lid and silicon wafer to simulate the real environment of TIM1 during the work. The thickness of the mounted TIM1 is about 60 micrometers. After the high temperature curation, the mounted sandwich is then installed to the self-designed holding piece which this piece can help the instruments used for tensile testing to meet the requirement of the specialized fixtures to the sandwich specimen. The final lid pull adhesion tests are conducted under ambient conditions

using an electromechanical load frame, Shimadzu. The mounted specimens are loaded at a crosshead speed of 3 mm/min until failure. The ultimate tensile strength is calculated from the measured force and cross-sectional of specimen. Measurements are reported as the mean (± standard deviation) of replicates (n = 5-10 specimens/group).

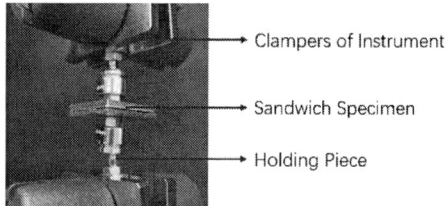

Fig. 2. Lid pull adhesion test configuration

Laser Flash Test

Laser flash tests are conducted by the instrument of Netzsch with LFA 467 series. In order to meet the requirement of the 3 layers-model of the built-in testing program in the analysis software, the TIM1 is prepared into 6 sandwich specimens with different thickness and wafers as both sides. Graphite is sprayed onto one side of cured sandwich specimen to maintain the absorption of laser during the testing. The testing results are proceed via "Three-layered heat loss & Pulse correction" model and then provide the heat diffusivity, α of TIM1.

Photo-thermal Radiation Test

The testing sample for photo-thermal method is prepared as followed, the TIM1 is bladed onto the surface of wafer with solid thickness ($\sim 50\ \mu m$) and then to be cured with high temperature. Graphite is sprayed onto the top surface of TIM1 to form a 3-layer (graphite + TIM1 + wafer) system, even though the thickness of graphite layer is tiny, $\sim 4\ \mu m$. For the measurement, a modulated laser is applied to heat the graphite sprayed side. The absorbed laser energy penetrates the sample through cross-plane heat conduction and emits radiation into ambient environment. The amplitude and phase changes of the radiation signal are detected to monitor the temperature profile along with the time of the sample surface. The conduct thermal resistance of the TIM1 sample can be extracted by fitting the recorded data to a theoretical thermal model. In this work, the range of modulated laser beam frequencies is set from 17 Hz to 20K Hz.

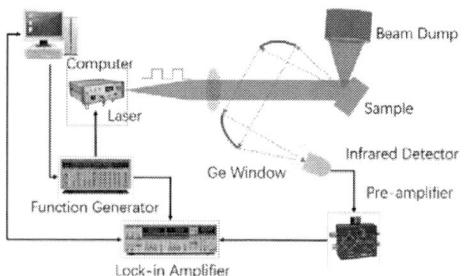

Fig. 3. Schematic of the photo-thermal radiation experimental setup [3]

III. RESULTS & DISCUSSION

Fig. 4 displays the adhesion summary of the tested samples. The tested adhesion range of our self-made TIM1 is about from 0.23 to 0.35 MPa that can match to a type of classic commercial product (0.30 ± 0.02 MPa), or even better than it, the detailed information is available from authors. For Sample A, since without any modification by promoter, the failure occurs at the interface of lid and TIM1 (Fig. 5) which exhibit the weakest adhesion among such tested samples. As the promotor is introduced into TIM1 gradually, the adhesion increases, and the separation area shifts from the interface of lid/TIM1 to the inherent area of TIM1 simultaneously. Therefore, the promoter does have significant influence on the adhesion promotion for TIM1. It is a feasible way that adding different contents of promoters to obtain a sufficient amount of TIM1 samples with different adhesion for us to conduct research on contact thermal resistance.

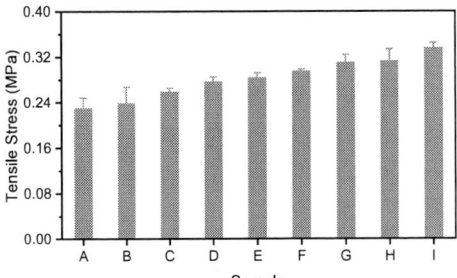

Fig. 4. TIM1 sample with various adhesion

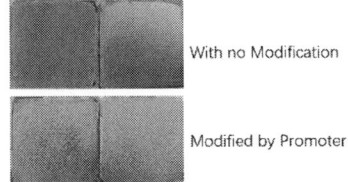

Fig. 5. The adhesion faluire of tested TIM1 samples

For the term of laser flash test, the thermal resistance R_{total} that the sum of intrinsic thermal resistance, R_{TIM1} and interfacial contact thermal resistance of TIM1 to wafer, $R_{contact}$ can be calculated according to the following equations:

$$k_{total} = \alpha \rho C_p \qquad (1)$$

$$R_{total} = d/k_{total} \qquad (2)$$

$$R_{total} = R_{TIM1} + R_{contact} \qquad (3)$$

where ρ and C_p are the density and specific heat capacity of TIM1 which can be measured through Archimedean method and differential scanning calorimetry accordingly, d is the thickness of TIM1.

Fourier's Law describing 1-D heat flow defines the thermal resistance of a material as:

$$R_{total} = d/k_{TIM} + R_{contact} \qquad (4)$$

Thus, the laser flash can help testing the R_{total} as a function of thickness of the TIM1. The tested plot is linear, and the slop of the fitting line represents the proportional to $1/k_{TIM}$, and the intercept refers to the measure of $R_{contact}$ [4], which needed (Fig. 6).

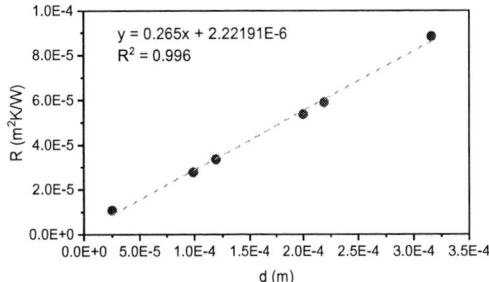

Fig. 6. The Tested Data and Fitting Linear Model

Fig. 7 shows the contact thermal resistance of partial tested samples. The magnitude of tested results has agreement with previous studies [5][6], which means the laser flash method can be employed to test the contact thermal resistance of TIM1. Through making the comparison between the tested results of adhesion and contact thermal resistance, it is noticed that as the adhesion is promoted, the contact thermal resistance defeat along. The similar results are obtained from the portion of photo-thermal radiation measurements.

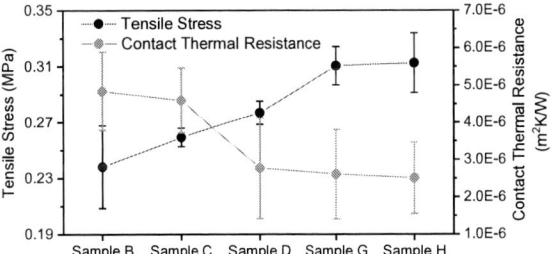

Fig. 7. The contact thermal resistance measured by laser flash technique and adhesion of samples

In terms of the measurement of photo-thermal radiation, Fig.8 depicts the fitting between the experimental data and the theoretical model values is satisfied which means the prepared specimen with the form of 3-layers can stand the test to absorb/emit signal with sufficient intensity so that can meet the minimum requirements for the tested data processing. When de signal is obtained through the experimental measurement, the least square method is applied to fit the experimental phase shift and the theoretical phase shift calculated over the specified modulation frequency range. The trail value that gives the least square deviation between the theoretical phase shifts and the experimental ones is taken as the real property of materials.

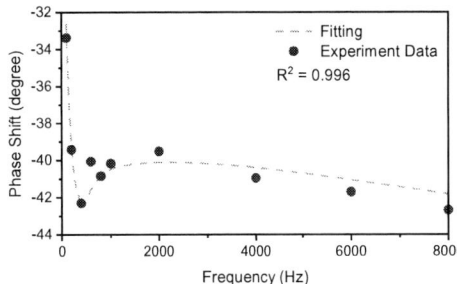

Fig. 8. Data fitting of phase shift for the thermal radiation from the graphite film surface.

The photo-thermal radiation testing data has the same tendency with laser flash measurements and the magnitude of the tested contact thermal resistance can match to the previous study as well (Fig. 9). Moreover, such testing results present more precision comparing with laser flash one. The reason why laser flash testing results possess the relevant large error may come from the preparation of tested specimens. Although there exists error, it is still clear to notice that the adhesion has the negative correlation with contact thermal resistance. The testing results by either measurement can reveal such conclusion.

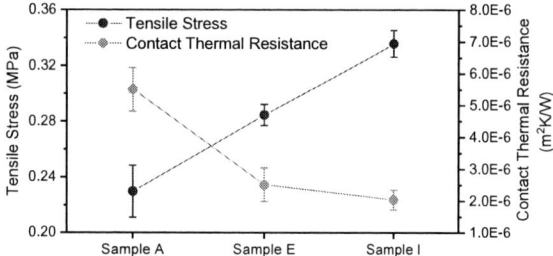

Fig. 9. The contact thermal resistance measured by photo-thermal radiation technique and adhesion of samples

The contribution of the adhesion promotion mainly comes from bonding formation at interface. When the promoter is introduced into TIM1, since the reaction among the promoter, TIM1 and substrate (lid/wafer), covalent bonds are formed so that the heat can cross the interface efficiently leading to the higher interface thermal conductance and lower contact thermal resistance. Beside help transporting heat, the formed covalent bonds also can help holding TIM1 and substrate tightly result in the van der Waals' force and hydrogen bonding are dominant at interface area and either can improve the heat transportation capability once again cross the interface.

It is a wise way to help improving the quality of our TIM1, which promoting the interface adhesion to prolong its service life and simultaneously decreasing the contact thermal resistance to enhance the ability of TIM1's heat transportation.

IV. Conclusion

Some sub-conclusions are drawn as followed:

- TIM1 with various adhesion for this research can be obtained by tunning the adding content of adhesion promoter.

- Two types of testing method (based on the techniques of laser flash and photo-thermal radiation) are developed to be employed for the contact thermal resistance detection. Both testing results can meet the agreement of others' work and photo-thermal radiation one is more accurate.

- The testing results reveal that the adhesion has negative correlation with contact thermal resistance.

- Adhesion promoter has dual function that not only can enhance adhesion but also can be applied to improve the ability of TIM1's heat transport. Such work can provide a guideline for researchers to do the relevant research in the same filed.

Acknowledgment

The authors would like to acknowledge the financial support from Youth Innovation Promotion and Association of the Chinese Academy of Sciences (2019354), Guangdong Province Key Field R & D Program Project (No. 2020B010179002 and No. 2020B010190004), Guangdong Basic and Applied Basic Research Fund (No. 2019A1515110845), Shenzhen Science and Technology Research Fund ((No. JCYJ20200109114401708 and GJHZ20180420180909654).

References

[1] M. Mahalingam, "Thermal management in semiconductor device packaging," in Proceedings of the IEEE, vol. 73, no. 9, pp. 1396-1404, Sept. 1985.

[2] Y. Yang, Z. Zhang and M. Touzelbaev, "Impact of temperature-dependent die warpage on TIM1 thermal resistance in field conditions," 2009 25th Annual IEEE Semiconductor Thermal Measurement and Management Symposium, 2009, pp. 285-292.

[3] Han, Meng, "Cross-plane thermal transport in graphene-based structures" (2018). Graduate Theses and Dissertations. 16370.

[4] C. J. M. Lasance, C. T. Murray, D. L. Saums and M. Rencz, "Challenges in thermal interface material testing," Twenty-Second Annual IEEE Semiconductor Thermal Measurement And Management Symposium, 2006, pp. 42-49.

[5] A. Gowda, D. Esler, S. Tonapi, K. Nagarkar and K. Srihari, "Voids in thermal interface material layers and their effect on thermal performance," Proceedings of 6th Electronics Packaging Technology Conference (EPTC 2004) (IEEE Cat. No.04EX971), 2004, pp. 41-46.

[6] A. Gowda et al., "Design of a high reliability and low thermal resistance interface material for microelectronics," Proceedings of the 5th Electronics Packaging Technology Conference (EPTC 2003), 2003, pp. 557-562.

AUTHOR INDEX

Ai, Binjie 443, 695
Ai, Daifeng 763
An, Tong 530, 555, 714
Bai, Jie 1131
Bai, Lijuan 493
Bai, Xue 1, 763, 832
Ban, Yu 986
Bao, Shuchao 999
Baojun, Qiu 148
Bedner, Dave 1435
Bi, Yuhao 705
Bo, Sun 646
Boowei, Tan 963
Braun, Robin 1169
Cai, Miao 420, 443, 854, 990
Cai, Zhikuang 1064
Cai, Zongqi 244, 1002
Cao, Chunyan 928
Cao, Guangqiang 248
Cao, Jiakai 479, 626
Cao, Liqiang 493, 1196
Cao, Rui 1153
Cao, Sicheng 497, 722
Cao, Ting 398, 671
Cao, William 1111
Cao, Wuxing 709
Cao, Xiuhua 1006
Cao, Zhijun 292
Cen, Kai 1097
Chang, Chao 823
Chang, Haixin 475
Chang, Xiaotong 809, 823, 836
Chang, Xufeng 827
Chao, Shuanshe 1494
Chen, Chender 226
Chen, Feng 1288
Chen, Gui 311, 331, 340
Chen, Haibin 1407, 1473
Chen, Hao 21
Chen, Hongtao 229, 381
Chen, Jiajia 1422
Chen, Jian 1521
Chen, Jing 121
Chen, Jin 1422
Chen, Kai 1069, 1134, 1139
Chen, Kun 755, 928
Chen, Leida 489
Chen, Liang-Pin 1504

Chen, Lijun 1288
Chen, Lin 732
Chen, Longfei 29
Chen, Mengyu 1346
Chen, Ming-Qiang 880, 1054
Chen, Mingxiang 34, 265
Chen, Pei 470
Chen, Phoebe 185
Chen, Qing 1025
Chen, Shiqi 776, 859
Chen, Shuai 67, 107, 1258
Chen, Si 48, 381, 767, 1040, 1323, 1489, 1521
Chen, Taotao 233, 393, 1173
Chen, Tao 958, 978, 1025
Chen, Tiezhu 961
Chen, Weijian 1273
Chen, Wen 864
Chen, Xiangxu 169, 1518
Chen, Xindong 420
Chen, Xinpeng 159
Chen, Xin 940, 1206
Chen, Xuanlong 1269
Chen, Yan-Ting 613
Chen, Yarong 1153
Chen, Yiming 1241
Chen, Yi 1148
Chen, Yong 1489
Chen, Yun 940, 1206
Chen, Yuqian 958, 978
Chen, Zhifeng 932
Chen, Zhiwen 58, 535
Chen, Zhuo 88, 585, 1241
Chen, Zuohuan 1525, 1529
Cheng, Chen 1478
Cheng, Hao 71
Cheng, Nan 326
Cheng, Xingwang 827
Cheng, Yuanjie 1121
Chenjun, Zhu 950
Chi, Panwang 679
Chin, Stella Wong Wun 1297
Chow-Khong, Tan 963
Chu, Baojin 239
Chu, Liu 1169
Chuantao, Hou 1382
Chunyue, Huang 257
Chunyue 371
Cu, Chengqiang 1459

Cui, Chengqiang	75, 1187, 1507, 1511
Cui, Hao	44, 1489
Cui, Zirui	1230
Dai, Fengwei	1288
Dai, Weijing	726, 1148
Dai, Xuanjun	1224
Dai, Yanwei	568, 1305
Dai, Zongbei	1246, 1315
Daojun, Luo	148
Deng, Chuanjin	972, 1002, 1323
Deng, Fei	1
Deng, Jinrong	928
Deng, Ronghua	1302
Deng, Rui	972, 1002, 1323
Deng, Yun-Kai	847
Deng, Zihao	634, 655
Di, Zhao	1088
Ding, Chao	1246, 1315, 1327
Ding, Kunpeng	1332
Ding, Lei	7, 121
Ding, Shuquan	1206
Ding, Wanchun	1449
Ding, Xinrui	634
Ding, Yuhan	967
Dong, Changlong	1518
Dong, Chong	169, 1518
Dong, Dong	950
Dong, Guoshuai	1302
Dong, Yaru	736, 1445
Dong, Yi	525
Du, Guoping	539
Du, Jianyu	517
Du, Mingyong	462, 479, 626
Du, Xiaomeng	462, 479, 626
Duan, Xiaolong	428
Fan, Chenrui	928
Fan, Haibo	699
Fan, Hongjin	99, 219
Fan, Jiajie	1044
Fan, Jingyu	180
Fan, Jinhu	1050
Fan, Tao	958, 978
Fan, Wenbin	320
Fan, Yuqing	103
Fan, Zhekun	1173
Fang, Chao	936
Fang, Jianming	1269
Fang, Mingang	88
Fang, Qu	376
Fang, Yi	1366, 1370, 1374
Fei, Jiu-Bin	870
Feng, Changqi	990
Feng, Chenzefang	12
Feng, Colin	185
Feng, Guan-Lin	1040
Feng, Jiayun	1012
Feng, Junbo	25
Feng, Xuegui	44
Fu, Dongzhi	506
Fu, Huaqiang	1378
Fu, Shuai	1210
Fu, Xianzhu	483
Fu, Xing	381
Fu, Zhenxiao	1006
Fu, Zhiwei	48, 381, 572, 603, 1040, 1521
Gai, Wei	1262
Gan, Guisheng	776, 859
Gao, Chenshan	1144
Gao, Guohua	1449
Gao, Haitao	1352
Gao, Jian	115, 305, 771, 940, 1206
Gao, Li-Ying	525
Gao, Li-Yin	1453
Gao, Liming	679
Gao, Ling	416, 424
Gao, Meng	177, 864
Gao, Qiu	7
Gao, Rui	440
Gao, Shiyi	292
Gao, Xu	517
Gao, Yingke	1361
Gao, Yingying	549, 617
Gao, Ziyang	1387
Ge, Bangtong	25
Geng, Fei	55
Gommers, Pieter	1297
Gong, Jinfeng	371
Gong, Weixi	1002
Gong, Xinjian	1422
Gong, Yanpeng	530, 555, 714
Gu, Erdan	1131
Gu, Jiabao	295
Gu, Jionajiong	1478
Gu, Ling	1343
Gu, Zhenyu	189
Guan, Yuan	1292
Guan, Zunyu	1473
Guangjie, Liu	912
Guo, Chunbing	564
Guo, Huaixin	572
Guo, Jingdong	1075
Guo, Kaiyu	1219
Guo, Sihua	1422
Guo, Xiaotong	229, 576, 642, 1441

Guo, Ying 809, 1518
Guo, Yufeng 1064
Guo, Yuhua 501
Guo, Yuxin 1117
Guodan, Zhou 1311
Han, Meng 512, 763
Han, Shouyu 1191
Han, Shunfeng 493, 1292
Han, Yang 689
Han, Yinhui 885
Han, Zhehao 1366, 1370, 1374
Hang, Yuan 185
Hao, Lichao 902
He, Aaron 185
He, Bin 763
He, Daping 1378
He, Di 1250
He, Guanghui 576, 642
He, Hengjian 854
He, Hongwen 1029
He, Hong 512
He, Huimin 493, 1292
He, Hu 233, 393, 1235
He, Jinming 17
He, Laisheng 25
He, Sifeng 1412
He, Siliang 705, 1336
He, Yunbo 144, 320, 356, 365, 1206, 1401
He, Zhiyuan 902
Hou, Bin 1059
Hou, Fan 466
Hou, Maoxiang 940, 1206
Hou, Shuhan 136
Hou, Xuewei 44
Hou, Zhuangzhuang 736, 1445
Hu, Deming 755, 928
Hu, Huaying 454
Hu, Jin 103
Hu, Qinghua 832
Hu, Shanwen 1343
Hu, Shaowei 944, 947
Hu, Wei-Lin 847, 870
Hu, Xianqin 638
Hu, Yougen 17, 483, 891, 1481
Hu, Yuehua 434
Hua, Hu Qing 898
Hua, Xueming 967
Huang, Chun-Yue 346
Huang, Chunyue 126, 131
Huang, Dayong 689
Huang, Feifei 936
Huang, Guochi 521

Huang, Hai-Jun 907, 1054
Huang, Haojie 94
Huang, Jiabin 326, 351, 360
Huang, Jianhong 791
Huang, Jiaqiang 823
Huang, Mingliang 936
Huang, Mingqi 560, 759
Huang, Min 466, 1158
Huang, Peng 932
Huang, Shijun 659
Huang, Tao 809, 823, 836
Huang, Tian 859
Huang, Wei 63
Huang, Xiangmiao 689
Huang, Xingjia 663
Huang, Xueyin 1153
Huang, Xu 572
Huang, Yixiu 663
Huang, Yiyong 193, 791
Huang, Yuhua 12
Huang, Yun 381, 1224
Huang, Zhiheng 607
Huang, Zhongwei 75, 1187, 1507, 1511
Huddar, Vinod Arjun 85
Hui, Li 950
Huo, Jia Ren 1034
Huo, Ruixia 159
Huo, Yinachao 458
Huo, Yongiun 736
Huo, Yongjun 1357, 1445
Ibrahim, Mesfin S. 1044
Ji, Liangzheng 796
Ji, Weiwei 248
Jia, Fei 684
Jia, Jinhao 591
Jia, Zhaowei 1101
Jian, Maoliang 454
Jian, Pang 1311
Jiang, Chao 613
Jiang, Feng 1525
Jiang, Jing 1034
Jiang, Junbo 1250
Jiang, Kulun 1006
Jiang, Kun 972, 1002
Jiang, Liheng 1336
Jiang, Miaomiao 1126
Jiang, Ruoyu 287
Jiang, Weiting 224
Jiang, Wenyu 80
Jiang, Zhaoqi 776, 859
Jiaxin, Liu 918
Jie, Bai 1088

Jin, Nong 1255
Jin, Xing 1258
Jin, Yinuo 1165
Jin, Yufeng 543
Jinfeng, Gong 257
Jing, Chen 898
Jing, Zhou 1044
Jo, Eunsol 852
Joosten, Annelies 1297
Jun, Tong 1382
Jung, Cheong-Ha 852
Kaixue, Ma 148
Kang, Jiajie 517
Kang, Qiushi 621
Kao, Nicholas 598
Ke, Chang-Bo 870
Kim, Gu-Sung 852
Kuang, Xianjun 316
Lai, Aaron 185
Lai, Canxiong 440
Lai, David 598
Lai, Haiqi 1459
Lai, Yuanting 336
Lan, Xin 38
Le, Fred Fuliang 1473
Lee, Ning-Cheng 1435
Lee, S. W. Ricky 1121, 1302
Lei, Chuyi 1059
Lei, Wenyu 475
Lei, Zuomin 17, 483
Lezhi, Ye 912
Li, Baoxia 450, 1305
Li, Bin 902, 972
Li, Bofu 493, 1292
Li, Bo 1006
Li, Cai-Fu 659
Li, Ce 827
Li, Changqing 746, 954
Li, Chaofan 239
Li, Cheng-Bo 907
Li, Chenglong 261, 274, 278, 287
Li, Cheng 1346
Li, Dameng 493, 1292
Li, Dayang 208, 1139
Li, Dejian 493, 1292
Li, Gang 261, 278, 630, 750, 781, 813, 923
Li, Ge 621
Li, Guoyuan 767, 995
Li, Haolin 1401
Li, Hao 1201
Li, Huacong 75
Li, Jiasheng 634, 655

Li, Jiayi 634
Li, Jiedong 1412
Li, Jiexin 634, 655
Li, Jing 219, 1340
Li, Jinhui 283, 301, 560, 746, 954, 1469
Li, Jinming 12
Li, Jinyang 420, 722
Li, Jin 885
Li, Junhong 388
Li, Junhui 1173
Li, Junjie 408, 412
Li, Junlong 875
Li, Junwei 512, 539
Li, Jun 1144, 1161, 1426
Li, Kai 767
Li, Ke 403, 999, 1273
Li, Kui 1305
Li, Kun 1262, 1265
Li, Liang 763, 864
Li, Lingyun 1265
Li, Lu 1277
Li, Maolin 346
Li, Menglin 29
Li, Mingyu 944, 947, 1366, 1370, 1374
Li, Ming 679
Li, Nannan 543
Li, Qizhuo 607
Li, Qi 1352
Li, Quanbing 1478
Li, Ruining 1429
Li, Rui 1081
Li, Shanshan 1514
Li, Shenglong 1361
Li, Shuang 265
Li, Shuwang 440
Li, Tao 103
Li, Tina 185
Li, Wangyun 497, 695, 705, 722, 843, 1059, 1327, 1336
Li, Weihao 517
Li, Weili 316
Li, Weiming 1441
Li, Wei 791
Li, Wenfeng 1346
Li, Wenqi 1393, 1397
Li, Xiaodong 479
Li, Xiao 525, 1453
Li, Xing 244
Li, Xujun 581
Li, Yadong 470
Li, Yangyang 1126
Li, Yanning 568
Li, Yan 1025

Li, Yesu 679
Li, Yingying 746, 954
Li, Yong 642
Li, Yulong 718
Li, Yun-Wei 880
Li, Zesheng 144, 356, 365
Li, Zewei 1064
Li, Zhankun 1173
Li, Zhe 525, 1453
Li, Zhipeng 581
Li, Zijian 940
Li, Zongtao 634, 655
Liang, Bunv 1340
Liang, Chen 530, 555, 714
Liang, Jiayong 634, 655
Liang, Jingyang 1246
Liang, Peilin 75, 1507, 1511
Liang, Ting 763
Liang, Yihang 403
Liang, Yi 1179
Liang, Zeng Xiao 898
Liang, Zhentang 995
Liang, Zhi 1332
Liao, Ao 1094
Liao, Cc 226
Liao, Chengyu 1029
Liao, Huilong 144, 365
Liao, Shuaidong 126, 131, 371
Liao, Wenyuan 440
Lin, Chen 740
Lin, Haoliang 781, 813, 923
Lin, Junshu 177
Lin, Qingping 1183
Lin, Shaopan 521
Lin, Shengru 679
Lin, Tao 1426
Lin, Tingyu 136, 521
Lin, Xiaohui 1401
Lin, Xinyi 1494
Lin, Yuanwei 1019
Lin, Zhiqiang 483, 1481
Liu, Cong 859
Liu, Dashun 726, 1069, 1134
Liu, Debo 1034, 1144
Liu, Dongcheng 38
Liu, Feixiang 722
Liu, Fengman 493, 1292
Liu, Fengmei 1352
Liu, Gai 895
Liu, Guanghui 800
Liu, Haiyan 1277
Liu, Hao 229

Liu, Hefeng 21
Liu, Huan 55
Liu, Huicong 805
Liu, Hui 674
Liu, Jiahao 229, 642, 1441
Liu, Jianhui 1117, 1161
Liu, Jiaxin 34
Liu, Jie 376, 986
Liu, Jinglong 265
Liu, Jing 1429
Liu, Jinshan 283, 301
Liu, Johan 180, 1422
Liu, Junfu 233, 393, 1173
Liu, Kai 7, 121
Liu, Lihong 1478
Liu, Lin-Jie 416, 424
Liu, Linjie 385, 895, 1255
Liu, Li 408, 535
Liu, Lu 1029, 1064
Liu, Mingjie 99
Liu, Min 381
Liu, Pan 786, 817
Liu, Pengfei 1158
Liu, Peng 219
Liu, Qiang 75, 283, 581, 1187
Liu, Richeng 1069, 1134
Liu, Sheng 58, 153, 535, 1081
Liu, Shimei 233, 1235
Liu, Shnjin 144
Liu, Shoufu 126, 131, 371
Liu, Shujin 356, 365
Liu, Shu 360
Liu, Tianhan 1246, 1315, 1327, 1336
Liu, Wansheng 159
Liu, Weidong 376, 398, 671
Liu, Wei 1183
Liu, Wen 283, 301
Liu, Xiaoyan 489
Liu, Xiaoying 51
Liu, Xin 197
Liu, Xuan 1498
Liu, Xuebin 1481
Liu, Xun 408, 412
Liu, Yachao 305
Liu, Yangzhi 1094
Liu, Yingxia 736, 1357, 1445
Liu, Ying 736, 1357, 1445
Liu, Yiping 679
Liu, Yongchao 638, 651
Liu, Yuan 475
Liu, Yunpeng 233, 393, 1173
Liu, Yu 1187

Liu, Zhi-Quan	525, 1453
Liu, Zhidan	67, 107
Liu, Zhigao	1059
Liu, Zilian	295
Liu, Ziyu	732, 740
Liu, Zuyao	1029
Lo, Jeffery C. C.	1121
Long, Junyu	1206
Long, Wang	1382
Long, Xu	638, 651, 809, 823, 836
Lou, Liang	805
Lu, Dong	1069, 1134, 1139, 1148
Lu, Guangsheng	443, 695, 843
Lu, Guoguang	440
Lu, Jibao	261, 274, 278, 287, 326, 351, 360, 638, 651, 667, 718
Lu, Jicun	679
Lu, Lu	278
Lu, Pei	180
Lu, Qian	1126
Lu, Xiangjun	594, 603
Lu, Xiaoxin	326, 351, 360
Luan, Huakai	466
Luan, Xinghe	674, 1059
Luh, Ding-Bang	1418
Luo, Bin	1191, 1393, 1397
Luo, Daojun	244, 1015
Luo, Dongxue	1215
Luo, Jun	244, 1015
Luo, Le	1265
Luo, Ruidong	659
Luo, Suibin	239
Luo, Xiaoting	607
Luo, Xiao	603
Luo, Yan	7, 121
Luo, Yuheng	771
Lv, Hongfeng	1015
Lv, Meijuan	1075
Lv, Mingtao	233
Lv, Xiaomeng	1094
Ma, Haitao	169, 1518
Ma, Haoran	169, 1518
Ma, Jusha	967
Ma, Peng	776
Ma, Qiangquiang	388
Ma, Shenglin	543
Ma, Shuying	506
Ma, Xiaojian	376, 398, 671
Ma, Xiao	907
Ma, Xu-Liang	1453
Ma, Yong	298
Ma, Yupa	94
Ma, Zhaolong	827
Mao, Mao	1323
Mao, Xingchao	229
Mao, Zhiyuan	316
Maolin, Li	257
Mei, Na	1494
Meng, Dominic Koey Poh	1297
Meng, Meng	1153
Meng, Yuezhong	607
Miao, Min	428, 885
Min, Zhixian	58
Mingxiang, Chen	918
Mo, Fuyao	316
Mou, Yun	34
Nan, Xujing	489
Neng, Liqiang	982
Ng, Wai Leong	1387
Ni, Liangyi	475
Ni, Yiqing	576
Ning, Minjie	1246
Nishikawa, Hiroshi	705
Niu, Fangfang	301, 1469
Niu, Leyi	684
Niu, Pingjuan	1131
Niu, Xulei	521
Oian, Qihao	144
Ou, Changping	663
Ou, Zhengping	1206
Ouyang, Keqing	111, 336, 982, 1311, 1494
Ouyang, Xing	630
Pan, Fei	1107
Pan, Huiming	58, 535
Pan, Kai-Lin	63
Pan, Kuang	1179
Pan, Lei	1064
Pan, Liu	898
Pan, Qingyu	1277
Pan, Yan	630
Pang, Jian	111, 336, 982
Pang, Yunsong	1, 213, 763
Park, Jung-Rae	852
Peng, Bo	295
Peng, Chao	638
Peng, Cheng	393
Peng, Fei	1366, 1370, 1374
Peng, Liang	630
Peng, Tao	261, 630, 923
Peng, Ting	1126
Peng, Xiaohui	630
Peng, Yang	265
Peng, Zhang	1097
Philipsen, Jos	1297

Pingfan, Ning 1088
Pingjuan, Niu 1088
Pingsheng, Zhang 950
Pu, Jie 1097
Qi, Zhangguo 898
Qian, Bin 967
Qian, Jiyu 94
Qian, Qihao 356, 365
Qiang, Jiang 1311
Qiao, Lan 1094
Qiao, Zhizhuang 385, 895, 1255
Qin, Fei 470, 530, 555, 568, 714, 1305, 1529
Qin, Haotong 958, 978
Qin, Hongbo 1059, 1246, 1315, 1327, 1336
Qin, W. L. 1085
Qin, Yijing 1069, 1134, 1139, 1148
Qin, Yikang 420, 722
Qing, Wang 918
Qingqing, Sun 740
Qiu, Guofu 320
Qiu, Xing 1121
Qiu, Yihua 655
Qu, Chenbing 603, 1040
Qu, Fang 398, 671
Rai, Pradeep 1111
Ran, Honglei 1219, 1230
Ran, Teng 958
Ren, Jie 1262, 1265
Ren, Kuili 193
Ren, Linlin 1, 173, 213, 253, 274, 287, 388, 539, 763
Ren, Shan 607
Ren, Xiaolei 51
Ren, Yan 902, 1323
Ren, Yulong 55
Ren, Zan 1255
Rong, Sun 213, 326, 351, 360
Ruan, Jianjun 479, 626
Sa, Zicheng 1012
Shan, Liang 270, 1469
Shang, Jintang 1191, 1393, 1397
Shang, Min 169, 1518
Shang, Panju 875
Shao, Dongdong 1332
Shao, Guangping 982
Shen, Chen 967
Shen, Jun 1201
Shen, Minghao 1250
Shen, Tianhua 1478
Shen, Ziyi 836
Sheng, Can 1081
Shi, Dachuang 940
Shi, Dianyang 564

Shi, Gaoming 316
Shi, Haitao 1429
Shi, Hongbin 809, 823, 836
Shi, Jiajia 1169
Shi, Ru-Zeng 1054
Shi, Weiguang 932
Shi, Xinrong 1481
Shi, Yidian 393
Shi, Yijun 48, 381, 594, 1521
Shi, Yuning 699
Shieh, Brian 1302
Shoufu, Liu 257
Shu, Xiayun 928
Shuai, Zhou 148
Si, Weikang 549, 617
Si, Yu 189
Song, Guan Qiang 1034
Song, Lei 854
Song, Tingting 982
Su, Dezhi 71, 80
Su, Hao-Hang 1210
Su, Ping 466
Su, Tianxiong 823
Su, Yangquan 193
Su, Yunpeng 388
Su, Yutai 809, 823, 836
Su, Zhaoxi 1191, 1393, 1397
Suga, Tadatomo 875
Sun, Chao 466, 535
Sun, Chen 1040
Sun, Chuanchuan 1361
Sun, Daoheng 1429
Sun, Deliang 560, 581, 746, 759
Sun, Fei 1418
Sun, Guoli 1305
Sun, Jianjun 336
Sun, Junfeng 1215
Sun, Liang 428
Sun, Lingyao 736, 1445
Sun, Li 870
Sun, Ning 376, 398, 671
Sun, Peng 55, 543
Sun, Qingqing 732
Sun, Rong 1, 17, 173, 239, 253, 261, 270, 274,
.... 278, 283, 287, 301, 388, 408, 412, 458, 462, 483, 512, 525,
539, 560, 581, 626, 630, 638, 651, 667, 718, 750, 763, 832,
891, 923, 1006, 1320, 1453, 1469, 1481
Sun, Tuo Bei 1494
Sun, Tuobei 111, 336
Sun, Xiaofeng 140
Sun, Xiaoyao 479
Sun, Yameng 1081

Sun, Yue 1418
Sun, Zhaoning 576
Sun, Zhefei 839
Tan, Daniel Q. 292
Tan, Liangchen 1514
Tan, Louise 226
Tan, Shwu Miin 1282
Tan, Xiaopeng 1258
Tang, Chu 88, 585
Tang, Hui 1412
Tang, Jiaqi 1327
Tang, Jiuyang 786
Tang, Linjiang 140
Tang, Qinglin 38
Tang, Qi 1139
Tang, Sha 244, 1323
Tang, Ying 1111
Tang, Yufeng 776
Tang, Zirong 1050
Tian, Chuang 1332
Tian, Dingkun 891
Tian, Jun 1429
Tian, Kun 1025
Tian, Ruyu 447
Tian, Wenchao 44, 1489
Tian, Xiaodi 684
Tian, Yanhong 447, 621, 1012
Tian, Yanzhong 555
Ting, Cao 376
Tong, Jin 75, 1507, 1511
Tong, Jun 21
Tong, Zhihao 805
Tu, Bingyi 809, 836
Tu, Qixuan 180
Tu, Wendian 539
Tuobei, Sun 982, 1311
Van Der Meulen, Rinse 1473
Wakeel, Saif 1297
Wan, Chengan 140
Wan, Dayuan 1148
Wan, Tao 202
Wan, Weikang 1196
Wan, Yang 674, 689
Wan, Yongkang 298, 1463
Wan, Zhaorong 1069, 1134
Wang, Bei 1029
Wang, Bingguang 827
Wang, Bin 674, 781, 813, 923
Wang, Cen 71, 80
Wang, Chengqian 1097
Wang, Chenxi 621
Wang, Chuanwei 1179

Wang, Chunlei 447
Wang, Congsi 1429
Wang, Daochang 1179
Wang, David H. 1165
Wang, David 1101
Wang, Dawei 591
Wang, Dazheng 549, 617
Wang, Dun 1463
Wang, Fangcheng 99, 219
Wang, Fuliang 88, 1241
Wang, Fuxin 80
Wang, Haozhe 287, 638, 651, 667, 718
Wang, Hong-Guang 870
Wang, Hongjie 1081
Wang, Hongkun 71
Wang, Hongyue 48, 594, 995
Wang, Huanhuan 854
Wang, Huihui 180
Wang, Jianhong 1277
Wang, Jian 1101, 1165, 1315, 1327
Wang, Jiao 506
Wang, Jie 1050
Wang, Junhao 732
Wang, Juntao 1034, 1161
Wang, Jun 1219, 1230, 1485
Wang, Ke 385, 895, 1255
Wang, Kuangyu 270
Wang, Langkun 1235
Wang, Lei 1485
Wang, Lichun 7, 121
Wang, Liuxin 763
Wang, Liwei 1040
Wang, Mei 189, 902
Wang, Ming-Sheng 839
Wang, Min 967, 1269
Wang, Nanxin 543
Wang, Ning 462, 479, 626
Wang, Pengchang 454
Wang, Pengfei 594, 603
Wang, Ping'An 189
Wang, Qiangwen 501
Wang, Qidong 1196, 1418
Wang, Qing 34
Wang, Rui 501
Wang, Ruolei 972, 1002
Wang, Shang 1012
Wang, Shinan 298
Wang, Shizhao 1081
Wang, Tao 283, 301, 1469
Wang, Tinglei 466
Wang, Weiyin 1426
Wang, Wei 55, 517

Wang, Wendong 1075
Wang, Wenlong 67, 1258
Wang, Xiaoqiang 244, 972, 1015
Wang, Xinjie 229
Wang, Xiyou 497, 695, 843
Wang, Xu 1111
Wang, Yan 311, 331, 1429
Wang, Yinghui 875
Wang, Yong 891
Wang, Yu Po 598
Wang, Yuanyuan 591, 1422
Wang, Yubo 493, 1292
Wang, Yuming 163, 202
Wang, Yunpeng 51
Wang, Yunxia 261, 274, 278
Wang, Zetian 517
Wang, Zhenyu 173, 213, 253
Wang, Zhen 1262, 1265
Wang, Zhibin 1153
Wang, Zhichao 967
Wang, Zhihai 1429
Wang, Zhiqin 58
Wang, Zhiqi 208, 1148
Wang, Zhizhe 1323
Wang, Zhuo 346, 371
Wang, Ziji 1393, 1397
Wang, Zixu 521
Wei, Jiahui 1305
Wei, Jianghao 163, 202
Wei, Jianhong 483
Wei, Jing 613
Wei, Qiang 376, 398, 671
Wei, Song 990
Wei, Tao 94
Wei, Wei 346
Wei, Xiangli 695, 843
Wei, Yiqiao 1241
Wei, Yuanyang 115
Wen, Jian 75, 1187, 1459
Wen, Minzhen 1044
Wen, Xiaokun 475
Wen, Zhibin 173
Wenchao, Wang 740
Weng, Zhangzhao 1015
Wong, Ching-Ping .... 261, 274, 278, 287, 326, 351, 360, ... 638, 651, 667, 718, 1469
Wu, Cheng-Tar 1250
Wu, Daowei 159, 450
Wu, Fengshun 674, 689
Wu, Haomiao 1401
Wu, Hao 1302
Wu, Heng 1273, 1525

Wu, Houya 750, 781, 813, 923
Wu, Hua 1449
Wu, Jingshen 1069, 1134, 1407, 1473
Wu, Lv 613
Wu, M. H. 1085
Wu, Majiaqi 454
Wu, Qi 1346
Wu, Shaocheng 169
Wu, Wenyu 726
Wu, Xin 292
Wu, Xudong 292
Wu, Xueting 1277
Wu, Yanchen 854
Wu, Yanhong 1262
Wu, Yanpei 809, 823, 836
Wu, Yan 1158, 1215
Wu, Yilong 1094
Wu, Yukun 1179
Wu, Zhaohui 902
Wu, Zhen 659
Wu, Zhipeng 805
Xia, Dongmei 1029
Xia, Hongbin 1282
Xia, Jianwen 759
Xia, Minglu 1387
Xia, Pengcheng 1097
Xia, Wei 153
Xia, Xinnian 253
Xiang, Chen 1250
Xiang, Weiwei 1126
Xiao, Fei 1485
Xiao, Hui 229, 642, 1441
Xiao, Jinbo 791
Xiao, Jinqing 1173
Xiao, Ming 208
Xiao, Xiaoyu 311, 331, 340
Xiao, Yong 1378
Xiao, Zeping 29
Xiaoqiang, Wang 148
Xie, An 928
Xie, Jiacheng 1201
Xie, Weikun 197, 248, 298, 1463
Xie, Xiaoming 1265
Xing, Chaoyang 543, 1393, 1397
Xiong, Chunshui 153
Xu, Binbin 1064
Xu, Gaowei 1262, 1265
Xu, Guanzhe 177
Xu, Guoliang 535
Xu, Hongyan 800, 1498
Xu, Huanxiang 295
Xu, Jian-Bin 1

Xu, Jianbin	326, 351, 360, 832
Xu, Jile	594, 603
Xu, Ju	800, 1498
Xu, Kexin	381
Xu, Liang	458
Xu, Lu	1191
Xu, Qianzhu	776, 859
Xu, Sean	1277
Xu, Sha	564
Xu, Shen	326, 360
Xu, Weihua	1094
Xu, Xiangtao	776, 859
Xu, Xiaowei	1315
Xu, Yadong	483, 891
Xu, Yang	875
Xu, Yonglun	1, 213, 763
Xuan, Hui	1449
Xuanjie, Song	912
Xue, Cheng	1340
Xue, Dongpeng	224
Xue, Kai	403
Xue, Ke	726, 1069, 1134, 1139, 1148
Xue, Lianghao	1081
Xue, Shirui	497, 843
Xue, Song	1429
Xue, Xiangdong	163, 202
Yan, Chenkan	1463
Yan, Hui	607
Yan, Qiucheng	450
Yan, Shuxia	932
Yan, Tingnan	591
Yan, Yamei	311, 331, 340
Yan, Yan	936
Yan, Z. P.	1085
Yang, Bing-Xian	847, 870
Yang, Cheng	99, 219
Yang, Dali	726
Yang, Dan	1494
Yang, Daoguo	420, 443, 497, 695, 705, 722, 843, 854, 990, 1336
Yang, Donghua	958, 978
Yang, Guang	674
Yang, Guannan	75, 1187, 1459
Yang, Guoming	1302
Yang, Hao	336
Yang, He	549, 617
Yang, Huan	376, 398, 671
Yang, Huihui	80
Yang, Jianxin	163
Yang, Jiao	709
Yang, Jinbao	630, 923
Yang, Jinglei	1407, 1473

Yang, Junli	1094
Yang, Kai	689
Yang, Liang	755, 928
Yang, Lianqiao	454
Yang, Liu	709
Yang, Li	475
Yang, Nana	932
Yang, Qian	864
Yang, Shaohua	440
Yang, Wanchun	944, 947
Yang, Xiaofeng	48, 572, 767, 1521
Yang, Yajing	630
Yang, Yiren	990
Yang, Yong	1224
Yang, Yuanyuan	750, 781, 813, 923
Yang, Yuchi	517
Yang, Zhen-Tao	416, 424
Yang, Zheng	1361
Yao, Peilun	1407
Yao, Yimin	1
Yao, Yue	458
Yao, Yu	1230
Yao, Zhijun	229
Yaoyang, Shen	646
Ye, Huaiyu	1034, 1144
Ye, Huijie	1126
Ye, Le	177, 864
Ye, Wenbo	173, 213, 253
Ye, Yu	1153
Ye, Zhenwen	759
Yen, Freedman	598
Yi, Yaoyong	1352
Yi, Yuxi	560
Yilin, Wu	918
Yilong, Wu	950
Yin, Zhihao	800
Yu, Chengyu	1097
Yu, Daquan	193, 283, 403, 791, 999, 1273, 1525, 1529
Yu, Dianru	800
Yu, Fei	416, 424
Yu, Guangliang	1012
Yu, Jincheng	1387
Yu, Kunpeng	1429
Yu, Miao	1215
Yu, Qixing	1165
Yu, Rongying	177
Yu, Shuhui	239, 1006
Yu, Tian	403
Yu, Xuecheng	1320
Yu, Ye Huai	898
Yu, Yongjian	197, 298, 1463
Yu, Zheng	1449

Yuan, Yulei 408, 412
Yue, Wu 1340
Yue, Yu-Qing 613
Yueping, Zhang 1382
Yuhong, Li 1088
Yumin, Zhao..912
Yun, Minghui..990
Yun, Zhanfei..497
Yunhui, Mei 1088
Zehai, Wen..950
Zeng, Baoshan 12
Zeng, Xiangliang 173, 213, 253
Zeng, Xiaoliang 1, 173, 213, 253, 388, 539, 763, 832
Zeng, Yanping..38
Zeng, Zejun..817
Zhai, Xiang..958
Zhan, Bihong 153
Zhan, Boyu 305
Zhang, Baotan 891
Zhang, Binbin 1025
Zhang, Chenwei 1034
Zhang, Chenxu 512
Zhang, Chi 454
Zhang, Chongming 535
Zhang, Chunhong 978
Zhang, Cong 224
Zhang, Donglin 1357
Zhang, Elley 224
Zhang, Fei 67, 107
Zhang, Guoping 270, 283, 560, 581, 746, 759, 954, 1469
Zhang, Guopinz 301
Zhang, Guoqi 420, 443, 786, 990, 1044
Zhang, Hao 434
Zhang, Hongze 1158
Zhang, Houdun 791
Zhang, Huaiquan.............................. 126, 131, 371
Zhang, Huibin 1463
Zhang, Jiabo 1179
Zhang, Jianfeng...1397
Zhang, Jiangtao...111
Zhang, Jianping...479
Zhang, Jian..63, 1126
Zhang, Jindi 771
Zhang, Jing 786, 796, 817
Zhang, Jin 1393
Zhang, Jun 709, 1518
Zhang, Kaihong 197, 298, 1463
Zhang, Kailin...443, 990
Zhang, Kai 55
Zhang, Kang 1250
Zhang, Kun 642
Zhang, Lanyu.................................... 115, 305, 771

Zhang, Leicong 626, 1320
Zhang, Lei 1006
Zhang, Lejun 71, 80
Zhang, Li 111
Zhang, Luhui 17
Zhang, Mingchuan 193, 999, 1525
Zhang, Minghua...140, 1025
Zhang, Ning 450, 1012
Zhang, Pei 709
Zhang, Pengzhen 475
Zhang, Ping 512
Zhang, Qiuchen 1282
Zhang, Qi 489
Zhang, Ruolin 1111
Zhang, Shuo 292
Zhang, Songsong 805
Zhang, Weijie...48, 1144
Zhang, Wei 1498
Zhang, Wenfeng 475
Zhang, Wenjing 1241
Zhang, Xianshun 29
Zhang, Xiaowei 674
Zhang, Xin-Ping 847, 870, 880, 907, 1054
Zhang, Xinlei 1343
Zhang, Xueying 1352
Zhang, Yakun 530
Zhang, Yan 180
Zhang, Yao 709
Zhang, Ye 963
Zhang, Yiming 1489
Zhang, Yinghui 1191
Zhang, Yi 689
Zhang, Yong..1422
Zhang, Yu-Bo 1453
Zhang, Yuehua 1343
Zhang, Yuexing 512
Zhang, Yuting 1258
Zhang, Yu 75, 1187, 1459, 1507, 1511
Zhang, Zebo 177
Zhang, Zhenyu 1340
Zhang, Zhihao 839
Zhang, Zhitao 1343
Zhang, Zhou 689
Zhang, Zhuanzhuan 428
Zhang, Zhuo 99, 219
Zhangzhao, Weng 148
Zhao, Chen 912
Zhao, Dinglei 58
Zhao, Fanny 1302
Zhao, Guangyao 99
Zhao, Guolin 750
Zhao, Haoran...517

Zhao, Heng ... 25
Zhao, Jin ... 1525, 1529
Zhao, Junxiang ... 71
Zhao, Ming ... 1126
Zhao, Ning ... 51
Zhao, Shuai ... 568, 1529
Zhao, Tao ... 458, 462, 479, 626, 891
Zhao, Wenzhong ... 1258
Zhao, Xiaowei ... 1117
Zhao, Xiuchen ... 736, 1357, 1445
Zhao, Yong ... 1094
Zhao, Yue ... 121
Zhao, Yulin ... 1418
Zhao, Yunfu ... 1361
Zhao, Zhenbo ... 576
Zhao, Zhiping ... 107
Zheng, Bingjie ... 572
Zheng, Bo ... 1165
Zheng, Deyin ... 517
Zheng, Dongfei ... 29
Zheng, Libing ... 549, 617
Zheng, Lihua ... 38
Zheng, Qi ... 493
Zheng, Wei ... 947
Zheng, Yi ... 287, 667, 718
Zheng, Yuxiang ... 1196
Zhiyuan, Zhu ... 740
Zhiyue, Wang ... 912
Zhong, Ao ... 283, 746
Zhong, Caiden ... 224
Zhong, Cheng ... 261, 274, 278, 667
Zhong, Jianfeng ... 1514
Zhong, Keju ... 1507, 1511
Zhong, Sung-Hua ... 1504
Zhong, Yi ... 999
Zhong, Yongbin ... 115
Zhong, Yong ... 1069, 1134
Zhonz, Cheng ... 287
Zhou, Bin ... 48, 572, 767, 995, 1224, 1521
Zhou, Cheng ... 1429
Zhou, Chunming ... 434
Zhou, Jie ... 1107
Zhou, Jinzhu ... 189
Zhou, Liang ... 961
Zhou, Longzao ... 674, 689
Zhou, Min-Bo ... 870, 880, 907, 1054
Zhou, Peng ... 434
Zhou, Qing ... 1107
Zhou, Qin ... 1320
Zhou, Quan ... 1144, 1262, 1265
Zhou, Rui ... 714
Zhou, Shicheng ... 621

Zhou, Shuai ... 1015
Zhou, Yangfan ... 385, 895, 1255
Zhou, Yi ... 7
Zhou, Zhiwei ... 771
Zhou, Zhou ... 699
Zhu, Deliang ... 17
Zhu, Fulong ... 12
Zhu, Gang ... 295
Zhu, Jian ... 1158
Zhu, Kai ... 197, 1463
Zhu, Pengli ... 261, 278, 412, 458, 462, 479, 626, ... 630, 750, 781, 813, 923, 1320
Zhu, Shengcong ... 663
Zhu, Wenbo ... 1366, 1370, 1374
Zhu, Wenhui ... 88, 311, 331, 340, 393, 585, 718, 750, 1241
Zhu, Zhiyuan ... 732
Zhu, Zhongyuan ... 1412
Zi, Chunfang ... 1183
Zito, Elaina ... 1435
Zongwei, Wang ... 1311
Zou, Longjiang ... 51
Zou, Xinrui ... 674
Zou, Yabing ... 1441
Zuo, Xinlang ... 642